SOLID STATE PHYSICS LITERATURE GUIDES
Volume 6

FERROELECTRICS LITERATURE INDEX

Solid State Physics Literature Guides

Prepared under the auspices of the Research Materials Information Center
Oak Ridge National Laboratory

General Editor: T. F. Connolly

Solid State Division
Oak Ridge National Laboratory*
Oak Ridge, Tennessee

Volume 1: Ferroelectric Materials and Ferroelectricity—1970

Volume 2: Semiconductors—Preparation, Crystal Growth, and Selected Properties—1972

Volume 3: Groups IV, V, and VI Transition Metals and Compounds— Preparation and Properties—1972

Volume 4: Electrical Properties of Solids—Surface Preparation and Methods of Measurement—1972

Volume 5: Bibliography of Magnetic Materials and Tabulation of Magnetic Transition Temperatures—1972

Volume 6: Ferroelectrics Literature Index—1974

---

*Oak Ridge National Laboratory is operated by Union Carbide Corporation for the U.S. Atomic Energy Commission.

SOLID STATE PHYSICS LITERATURE GUIDES
Volume 6

FERROELECTRICS LITERATURE INDEX

Compiled by
T. F. Connolly
Research Materials Information Center
Solid State Division
Oak Ridge National Laboratory
Oak Ridge, Tennessee

and

Donald T. Hawkins
Libraries and Information Systems Center
Bell Telephone Laboratories
Murray Hill, New Jersey

With a Foreword by
S. C. Abrahams
Bell Telephone Laboratories
Murray Hill, New Jersey

Springer Science+Business Media, LLC

Library of Congress Cataloging in Publication Data

Connolly, T   F
    Ferroelectrics literature index.

    (Solid state physics literature guides, v. 6)
    1. Ferroelectricity — Bibliography. 2. Ferroelectricity — Indexes. I. Hawkins,
Donald T., joint author. II. Title. III. Series.
Z7144.S58S65 vol. 6 [Z7144.F4]   016.5304'1'08s
ISBN 978-1-4684-6212-8   ISBN 978-1-4684-6210-4 (eBook)
DOI 10.1007/978-1-4684-6210-4

FOREWORD

Research on ferroelectricity and ferroelectric materials started in 1920 with the discovery by Valasek that the variation of spontaneous polarization in Rochelle salt with sign and magnitude of an applied electric field traced a complete and reproducible hysteresis loop. Activity in the field was sporadic until 1935, when Busch and co-workers announced the observation of similar behavior in potassium dihydrogen phosphate and related compounds. Progress thereafter continued at a modest level with the undertaking of some theoretical as well as further experimental studies. In 1944, von Hippel and co-workers discovered ferroelectricity in barium titanate. The technological importance of ceramic barium titanate and other perovskites led to an upsurge of interest, with many new ferroelectrics being identified in the following decade. By 1967, about 2000 papers on various aspects of ferroelectricity had been published.

The bulk of this widely dispersed literature was concerned with the experimental measurement of dielectric, crystallographic, thermal, electromechanical, elastic, optical, and magnetic properties. A critical and excellently organized compilation based on these data appeared in 1969 with the publication of *Landolt–Börnstein,* Volume III/3. This superb tabulation gave instant access to the results in the literature on nearly 450 pure substances and solid solutions of ferroelectric and antiferroelectric materials. Continuing interest in ferroelectrics, spurred by the growing importance of electrooptic crystals, resulted in the publication of almost as many additional papers by the end of 1969 as had been surveyed in *Landolt–Börnstein.* Fortunately for those working in this field, a bibliography of nearly all these papers was compiled by T. F. Connolly and the late E. S. Turner and published as Volume 1 of the present series.

Activity in ferroelectrics and related properties has remained at a high level, as evidenced by the appearance of at least a further 3000 papers since 1970. A guide to this most recent literature is a necessity if advances in understanding and the planning of new experiments are to be optimized. The continued collaboration between the Research Materials Information Center of Oak Ridge National Laboratory and the Libraries and Information Systems Center of Bell Laboratories has provided a most efficient approach in organizing this literature, and the present volume will undoubtedly be put into immediate use by the large number of researchers currently engaged in investigating the properties of ferroelectric crystals.

February, 1974   S. C. Abrahams

FOREWORD

Research on ferroelectricity and ferroelectric materials started in 1920 with the discovery by Valasek that the variation of spontaneous polarization in Rochelle salt with sign and magnitude of an applied electric field traced a complete and reproducible hysteresis loop. Activity in the field was sporadic until 1935, when Busch and co-workers announced the observation of similar behavior in potassium dihydrogen phosphate and related compounds. Progress thereafter continued at a modest level with the understanding of some theoretical as well as further experimental studies. In 1944, von Hippel and co-workers discovered ferroelectricity in barium titanate. The technological importance of ceramic barium titanate and other perovskites led to an upsurge of interest with many new ferroprojects being identified. In the following decade, by 1955, about 2000 papers on various aspects of ferroelectricity had been published.

The bulk of this widely dispersed literature was concerned with the experimental measurement of dielectric, crystallographic, thermal, electromechanical, elastic, optical, and magnetic properties. A critical and excellently organized compilation based on these data appeared in 1969 with the publication of Landolt-Börnstein, Volume III/3. This superb tabulation gave instant access to the results in the literature on nearly 750 pure substances and solid solutions of ferroelectric and antiferroelectric materials. Continuing interest in ferroelectrics, spurred by the growing importance of electrooptic crystals, resulted in the publication of almost as many additional papers by the end of 1969 as had been surveyed in Landolt-Börnstein. Fortunately for those working in this field, a bibliography of nearly all these papers was compiled by T. F. Connolly and the late E. S. Turner and published as Volume 1 of the present series.

Activity in ferroelectrics and related properties has remained at a high level, as evidenced by the appearance of at least a further 3000 papers since 1970. A guide to this most recent literature is a necessity if advances in understanding and the planning of new experiments are to be optimized. The continued collaboration between the Research Materials Information Center at Oak Ridge National Laboratory and the Literature and Information Systems Center at Bell Laboratories has provided a most efficient approach in organizing this literature, and the present volume will undoubtedly be put into immediate use by the large number of researchers currently engaged in investigating the properties of ferroelectric crystals.

February, 1974   S. C. Abrahams

PREFACE

Volume 1 of the *ORNL Solid State Physics Literature Guides,* entitled "Ferroelectricity and Ferroelectric Materials," by T. F. Connolly and E. S. Turner, was published in 1970. The present volume, containing 3681 references, is a continuation of the joint effort of the Research Materials Information Center (RMIC) of Oak Ridge National Laboratory and the Libraries and Information Systems Center (LISC) of Bell Telephone Laboratories (BTL) to index the rapidly growing ferroelectrics literature.

Since 1963, the RMIC has been answering inquiries on the availability, preparation, and properties of inorganic solid-state research materials. An important function of RMIC is the preparation of bibliographies, and the continuing interest in ferroelectrics on the part of RMIC and BTL has led to a compilation of the literature on that subject. The 1962 book *Ferroelectric Crystals,* by Jona and Shirane, was taken as a cutoff point for Volume 1, and all papers received through mid-1969 by RMIC were included in that volume.

Besides a literature searching function, the LISC has developed a proprietary computer system called BELDEX,* which formats and indexes bibliographic and other types of material. This system, which has evolved over more than a decade, was used to process RMIC's ferroelectric references.

Coverage of this volume is mainly from mid-1969 through December 1972, but it is not limited to the cutoff date for Volume 1. About 750 references from the period 1960–1969 missing from Volume 1 have been included here. In addition, the few 1970 references in Volume 1 have been repeated for completeness. (Most of these are from the 1969 Kyoto conference on ferroelectrics, and page numbers of the Proceedings are now given instead of paper numbers.)

February, 1974   T. F. Connolly
                                                                                                         D. T. Hawkins

---

*The BELDEX program has been copyrighted by Bell Telephone Laboratories. For further information, see W. K. Lowry, Use of computers in information systems, *Science* **175**: 841–846 (1972)

PREFACE

Volume 1 of the ORNL Solid State Physics Literature Guides, entitled "Ferroelectricity and Ferroelectric Materials," by T. F. Connolly and E. S. Turner, was published in 1970. The present volume, containing 3661 references, is a continuation of the joint effort of the Research Materials Information Center (RMIC) of Oak Ridge National Laboratory and the Libraries and Information Systems Center (LISC) of Bell Telephone Laboratories (BTL) to index the rapidly growing ferroelectrics literature.

Since 1963, the RMIC has been answering inquiries on the availability, preparation, and properties of inorganic solid-state research materials. An important function of RMIC is the preparation of bibliographies, and the continuing interest in ferroelectrics on the part of RMIC and BTL has led to a compilation of the literature on that subject. The 1962 book Ferroelectricity, by Jona and Shirane, was taken as a cutoff point for Volume 1, and all papers received through mid-1969 by RMIC were included in that volume.

Besides a literature searching function, the LISC has developed a proprietary computer system called BELDEX,* which formats and indexes bibliographic and other types of material. This system, which has evolved over more than a decade, was used to process RMIC's ferroelectrics references.

Coverage of this volume is mainly from mid-1969 through December 1972, but it is not limited to the cutoff date for Volume 1. About 750 references from the period 1960-1969 missing from Volume 1 have been included here. In addition, the few 1973 references in Volume 1 have been repeated for completeness. (Most of these are from the 1969 Kyoto conference on ferroelectrics, and page numbers in the Proceedings are now given instead of paper numbers.)

February, 1974.

T. F. Connolly
D. T. Hawkins

*The BELDEX program has been commented by Bell Telephone Laboratories. For further information, see W. K. Lowry, Use of a computer in information systems, Science, Ser. P. 3, Vol. 81, 1967.

CONTENTS

Introduction xi
Acknowledgments xiii
1 Bibliographies, Data Compilations, Books, and Conferences 1
2 Reviews 4
3 Theory 13
4 Barium Titanate 44
5 Lead Titanate 84
6 Strontium Titanate 89
7 Lead Zirconate 106
8 Other Perovskites 108
9 Mixed Titanates–Zirconates (including PZT, PLZT) 127
10 Tungsten Trioxide and Tungstates 139
11 Lithium Niobate 142
12 Lithium Tantalate 157
13 Tungsten Bronze Type and Layer Structure Oxides 161
14 Boracites 175
15 Antimony Sulfoiodide and Related Compounds 178
16 Nitrates and Nitrites 190
17 Potassium Dihydrogen Phosphate (KDP) and Related Phosphates 201
18 Potassium Dihydrogen Arsenate (KDA) and Related Arsenates 227
19 Ammonium Sulfate, Ammonium Fluoberyllate, and Related Compounds .. 230
20 Alums 235
21 Guanidinium Aluminum Sulfate Hexahydrate (GASH) and Related Compounds 236
22 Selenites and Selenates 238
23 Colemanite 243
24 Potassium Ferrocyanide 244
25 Thiourea 246
26 Formates and Propionates 248
27 Triglycine Sulfate (TGS) and Related Compounds 251
28 Rochelle Salt and Related Compounds 271
29 Tartrates 278
30 Molybdates 280
31 Hydrogen Halides 287
32 Lithium Hydrazinium Sulfate and Fluoberyllate 289
33 Cesium Lead Trichloride, Potassium Manganese Trifluoride, and Related Halides 291
34 Ice 296
35 Lead Germanate and Bismuth Germanate 303
36 Other Materials 306
37 Device Applications of Ferroelectrics 315
Permuted Title Index 329
Author—Title Index 615

INTRODUCTION

Coverage

This is primarily a materials bibliography which includes nonferroelectric properties of ferroelectric materials. It includes theoretical and experimental papers, as well as reviews and data compilations. A separate chapter on device applications of ferroelectrics has been added. Patents are not covered. Most materials included in *Landolt–Börnstein* (Reference 2) are covered, plus materials discovered to be ferroelectric since the appearance of those tables. Materials stated to be possibly ferroelectric (e.g., V_3Si and some tellurides) are not included.

The excellent ongoing ferroelectrics bibliography by K. Toyoda (Reference 32) has been a source of many of the references in this bibliography. Toyoda's bibliography appears at regular intervals in the journal *Ferroelectrics.* It is unusually comprehensive, although not indexed.

Although papers published after 1972 have not been included, three significant compilations appearing in 1973 deserve mention here. They are:

Compilation of the Static Dielectric Constants of Inorganic Solids
Young, K. F., and Frederikse, H.P.R.
J. Phys. Chem. Ref. Data **2**(2): 313–409 (1973)
(Comprehensive; tabular and graphic data on more than 300 inorganic solids, with added annotations for many of the graphs. Introduced by a discussion of definitions, measurement methods, and criteria used in data evaluation.)

Ferroelectric and Antiferroelectric Materials
Subbaro, E. C.
Ferroelectrics **5**: 267–280 (1973)
(A concise compilation of ferroelectric and antiferroelectric materials known up to 1971 is presented, together with the Curie and other transition temperatures. Spontaneous polarization values are given in the case of ferroelectrics.)

Literature Guide to Pyroelectricity 1970–1972
Lang, S. B.
Ferroelectrics **5**(1): 125–151 (1973)

Layout

The bibliography section is arranged in thirty-three chapters organized on a materials basis, plus four additional chapters on books, conferences, bibliographies, and data compilations; reviews; theory; and device applications of ferroelectrics. The thirty-three materials chapters follow the classification

order of *Landolt–Börnstein*. Within each chapter, entries are arranged by author. (Only the first three characters of the first author's surname were used as a sorting key, so minor anomalies exist.)

Each reference is assigned a sequential number, preceded by the number of the chapter in which it first appears. When a reference is pertinent to more than one chapter, duplicate entries have been made in the other pertinent chapters. In these cases, no number has been assigned to the entry, and no duplicate entries appear in the indexes. To prevent undue proliferation of such cross-references, entries pertinent to more than three chapters have been arbitrarily classed as reviews and placed in Chapter 2. The three-digit numbers appearing after many of the references are RMIC reference numbers.

Indexes

A permuted title index serves as a detailed subject index and provides additional entry points into the bibliography. Parenthetical notes supplement many titles in order to enhance this index. Indexing under the term FERROELECTRIC... has been suppressed. Terms beginning with the common prefixes BI . . . , DI . . . , DODECA . . . , HEXA . . . , META . . . , MONO . . . , ORTHO . . . , OXY . . . , PENTA . . . , SESQUI . . . , SUB . . . , TETRA . . . , and TRI . . . are indexed under the prefixed word. (An exception to this practice has been made in the case of the frequently appearing term TRIGLYCINE.)

ACKNOWLEDGMENTS

In addition to the compilations mentioned at the beginning of the Introduction, and Toyoda's bibliography, the reader should be aware of the journal *Ferroelectrics,* edited by G. W. Taylor and I. Lefkowitz, which is an important source of information on these materials.

S. C. Abrahams was a source of encouragement and advice and provided many helpful suggestions, as well as the foreword to this bibliography. Most of the pre-1969 references, forming a nucleus of this bibliography, were compiled by the late E. S. Turner.

1. BIBLIOGRAPHIES, DATA COMPILATIONS, BOOKS AND CONFERENCES

01-0001 ALL-UNION CONFERENCE ON FERROELECTRICITY, 7TH, 1970, VORONEZH, USSR, PROC, PARTS 1-3.
BULL ACAD SCI USSR, PHYS SER 34(12): 2191-353 (1970),
35(9): 1611-813 (1971), 35(12): 2353-72 (1971)

01-0002 LANDOLT-BÖRNSTEIN, NEUE SERIE. VOLUME III/3.   FERRO- AND ANTIFERROELECTRIC SUBSTANCES.
SPRINGER, 1969, 584P.

01-0003 CERAMIC ACOUSTIC DETECTORS. (BARIUM TITANATE)
ANANEVA AA
CONSULTANTS BUREAU, NY, 1965, 122P.

01-0004 FERROELECTRICITY AND FERROELECTRIC MATERIALS: A COMPREHENSIVE BIBLIOGRAPHY.
CONNOLLY TF + TURNER ES
PLENUM, NY, 1970.

01-0005 SYMPOSIUM ON APPLICATIONS OF FERROELECTRICS.
CROSS LE + KURTZ SK (EDS)
FERROELECTRICS 3: 67-341 (1972)
IEEE TRANS SON ULTRASON 19: 67-341 (1972)

01-0006 FERROELECTRIC CERAMICS.
DERI M
MACLAREN, LONDON, 1966, 95P.
(TRANSLATION OF HUNGARIAN EDITION)

01-0007 INTERNATIONAL MEETING ON FERROELECTRICITY, PRAGUE, PROC.
DVORAK V + FOUSKOVA A + GLOGAR P (EDS)
INST PHYS CZECH ACAD SCI, PRAGUE, 1966. (2 VOLS)

01-0008 REPORT ON THE SECOND INTERNATIONAL MEETING ON FERROELECTRICITY. A EUROPEAN VIEW. AN
AMERICAN VIEW.
DVORAK V + FOUSEK J + LAWLESS WN
FERROELECTRICS 1: 57-61 (1970)

01-0009 CHEMICAL PHYSICS OF ICE.
FLETCHER NH
CAMBRIDGE UNIV PRESS, 1970, 271P.

01-0010 FIFTY YEARS OF FERROELECTRICITY: A REVIEW OF THE PROCEEDINGS OF THE SECOND
INTERNATIONAL MEETING ON FERROELECTRICITY.
FLEURY PA
SCIENCE 171: 564 (1971)

01-0011 PHASE TRANSITIONS IN FERROELECTRICS.
FRITSBERG VYA + ROLOV BN + KRUCHAN YAYA (EDS)
IZDATELSTVO ZINATNE RIGA (1971) (IN RUSSIAN) 206P

01-0012 EUROPEAN MEETING ON FERROELECTRICITY, 2ND PROC, DIJON, 1971.
GODEFROY L + GODEFROY G (EDS)
J PHYS (PARIS), 33, SUPPL 4, COLLOQ 2 (1972)
294

01-0013 AN INTRODUCTION TO THE PHENOMENOLOGICAL THEORY OF FERROELECTRICITY.
GRINDLAY J
PERGAMON, 1970, 256P.

01-0014 ELECTRIC CONDUCTIVITY OF FERROELECTRICS. (ELECTRICAL CONDUCTIVITY AND RELATED
PROPERTIES OF 60 COMPOUNDS, 58 SOLID SOLUTIONS, AND 10 TECHNICAL FERROCERAMICS, CURIE
TEMPERATURES OF 290 COMPOUNDS TABULATED)
GUREVICH VM
ISRAEL PROGRAM FOR SCIENTIFIC TRANSLATIONS, JERUSALEM, 1971
(TRANSLATED FROM RUSSIAN, COMM OF STANDARDS, MEASURES AND MEASURING
INSTRUM OF THE USSR COUNCIL OF MINISTERS, MOSCOW, 1969)
CHEM ABSTR 73: 71260C
276

1. BIBLIOGRAPHIES, DATA COMPILATIONS, BOOKS, CONFERENCES

01-0015 PIEZOELECTRIC CERAMICS. (NONMETALLIC SOLIDS, VOL-3, REVIEW)
JAFFE B + COOK WR + JAFFE H (EDS)
ACADEMIC PRESS, 1971, 317P.
273

01-0016 LINEAR ELECTROOPTICAL MATERIALS. (REVIEW, 138 REFS)
KAMINOW IP + TURNER EH
P447-59 OF CHEMICAL RUBBER CO, CLEVELAND, HANDBOOK OF LASERS,
PRESSLEY RJ(ED), 1971, 631P.

01-0017 FERROELECTRICS.
KAWABE K
KYORITSU SHUPPAN CO, LTD, TOKYO, 1971, 214P. (IN JAPANESE)
248

01-0018 ANTIFERROELECTRICITY IN COMPOUNDS WITH PEROVSKITE STRUCTURE. (EXAMINATION OF PUBLISHED
DATA, CLASSIFICATION, LISTING AND DISCUSSION OF STRUCTURES)
KRAINIK NN
BULL ACAD SCI USSR, PHYS SER 28(4): 550-5 (1964)

01-0019 MODERN ASPECTS OF PHYSICS OF SOLIDS. (FERROELECTRICITY)
MERCOUROFF W
MASSON, PARIS, 1969, 144P. (IN FRENCH)
CHEM ABSTR 75: 40710M (1971)

01-0020 ANTIFERROELECTRICITY AND ANTIFERROELECTRIC MATERIALS. (TABLE OF NEEL TEMPERATURES)
MILEK JT
ELECTRONIC PROPERTIES INFORMATION CENTER, CULVER CITY, CALIF, 1970,
9P. AD735629

01-0021 LINEAR ELECTROOPTIC MODULATOR MATERIALS. (POTASSIUM OR AMMONIUM DIHYDROGEN PHOSPHATE OR
ARSENATE, POTASSIUM DIDEUTERIUM PHOSPHATE, POTASSIUM TANTALATE OR NIOBATE, LITHIUM
NIOBATE OR TANTALATE, BARIUM SODIUM NIOBATE, CUPROUS CHLORIDE, PROUSTITE, CALCIUM
PYRONIOBATE, AND BISMUTH(12) GERMANIUM OXIDE(20). (REVIEW)
MILEK JT + WELLES SJ
ELECTRONIC PROPERTIES INFORMATION CENTER, LINEAR ELECTROOPTIC
MODULATOR MATERIALS, CULVER CITY, CALIF, 1970, 260P.

US GOVT RES DEVELOP REP 70(12): 204-5 (1970)
187

01-0022 SEMICONDUCTING AND NONSTOICHIOMETRIC BARIUM TITANATE. A REFERENCE GUIDE.
MILEK JT
ELECTRONIC PROPERTIES INFORMATION CENTER, CULVER CITY, CALIF,
FEB 1971.
231

01-0023 EUROPEAN MEETING ON FERROELECTRICITY, PROC.
MUSER HE + PETERSSON J (EDS)
WISS VERLAG GMBH, STUTTGART, 1970, 396P.

01-0024 A NOTE ON THE CLASSIFICATION OF FERROELECTRICS. (BY CURIE-WEISS CONSTANT)
NAKAMURA E + MITSUI T + FURUICHI J
J PHYS SOC JAP 18: 1477-81 (1963)

01-0025 II-VI SEMICONDUCTING COMPOUNDS - DATA TABLES. (MECHANICAL, CRYSTALLOGRAPHIC, PHYSICAL,
THERMAL MAGNETIC, ELECTRONIC AND OPTICAL PROPERTIES OF EACH OF THE II-VI BINARY
SEMICONDUCTING COMPOUNDS)
NEUBERGER M
ELECTRONIC PROPERTIES INFORMATION CENTER, CULVER CITY, CALIF,
1969, 196P.
CHEM ABSTR 73: 39950S (1970) US GOVT RES DEVELOP REP 70(4): 175 (1970)
174

01-0026 PHYSICS OF ICE. (PROC OF INT SYMP, 9-14 SEP 1968 MUNICH)
RIEHL N + BULLEMER B + ENGELHARDT H (EDS)
PLENUM PRESS, 1969, 642P.

01-0027 FERROELECTRICS AND OXIDE SEMICONDUCTORS.
SINYAKOV EV (ED)
REDAKTSIONNO-IZDATELSKII OTDEL DNEPROPETROVSKII GOSUDARSTVENNYI
UNNIVERSITET, DNEPROPETROVSK, 1971, 100P. (IN RUSSIAN)

1. BIBLIOGRAPHIES, DATA COMPILATIONS, BOOKS, CONFERENCES

01-0028   MICROWAVE ACOUSTICS HANDBOOK. VOL 1: SURFACE WAVE VELOCITIES. (BARIUM SODIUM NIOBATE,
          BISMUTH(12) GERMANIUM OXIDE(20), LITHIUM NIOBATE, LITHIUM TANTALATE, AND OTHER
          NONFERROELECTRIC SUBSTANCES)
          SLOBODNIK AJ + CONWAY ED (EDS)
          AIR FORCE CAMBRIDGE RES LABS, MASS, MAR 1970, 354P. AFCRL-70-0164.
               238

01-0029   FERROELECTRICS AND ANTIFERROELECTRICS.
          SMOLENSKII GA + BOKOV VA + ISUPOV VA + KRAINIK NN + PASYNKOV RE + SHUR MS
          NAUKA, LENINGRAD, 1971, 476P (IN RUSSIAN)
               CHEM ABSTR 76: 8086X (1972)
               233

01-0030   FERROELECTRICS AND ANTIFERROELECTRICS.
          SMOLENSKII GA + KRANIK NN
          NAUKA, MOSCOW, 1968, 183P. (IN RUSSIAN)

01-0031   INTERNATIONAL MEETING ON FERROELECTRICITY, SECOND PROC.
          TAKAGI Y (ED)
          PHYS SOC JAP, TOKYO, 1970, 460P. (J PHYS SOC JAP 28, SUPPL)
          (548.85/I61) 141982

01-0032   BIBLIOGRAPHY OF FERROELECTRICS.
          TOYODA K
          FERROELECTRICS 1: 43-55, 103-10, 179-88, 269-80 (1970), 2: 63-74,
          163-74, 227-34, 307-16 (1971), 3: 59-66 (1971), 4: 57-63, 121-30,
          299-315 (1972)

01-0033   BIBLIOGRAPHY ON FERROELECTRICS. (1949-1964, 3048 REFERENCES, SUBJECT AND AUTHOR
          INDEXES)
          TOYODA K
          RESEARCH GROUP OF FERROELECTRIC PHYSICS IN JAPAN, WASEDA UNIV, 1965.
               210

01-0034   SYMPOSIUM ON FERROELECTRICITY, 1966, GENERAL MOTORS RES LABS, WARREN, MICHIGAN, PROC,
          FERROELECTRICITY.
          WELLER EF (ED)
          ELSEVIER, 1967, 318P.

2. REVIEWS

02-0035   MEASUREMENT OF DIELECTRIC CONSTANTS FOR A FEW FERROELECTRIC MATERIALS AT LOW
           TEMPERATURES.
           ABE R + TOKUMARU Y + MASAKI H
           REP TOYODA PHYS CHEM RES INST 23: 6-8 (1970) (IN JAPANESE)

02-0036   COUPLED FERROELASTIC CRYSTALS. (GADOLINIUM MOLYBDATE, KDP, ORTHOFERRITES, DYSPROSIUM
           ANTIMONIDE, RUBIDIUM IRON TRIFLUORIDE, BORACITES, BISMUTH(4) TITANIUM(3) OXIDE(12),
           HAFNIUM DIVANADIDE)
           ABRAHAMS SC
           ACTA CRYSTALLOGR A 28: S230 (1972)

02-0037   FERROELASTICITY.
           ABRAHAMS SC
           MATER RES BULL 6: 881-90 (OCT 1971)

02-0038   FERROELECTRIC CERAMIC OXIDES.
           AINGER FW
           P147-75 OF COCKAYNE B AND JONES DW(EDS), MODERN OXIDE MATERIALS;
           PREPARATION, PROPERTIES AND DEVICE APPLICATIONS, ACADEMIC PRESS,
           1972, 315P.
               269

02-0039   FERROELECTRICS IN OPTICAL MEMORIES AND DISPLAYS: A CRITICAL APPRAISAL.
           ANDERSON LK
           IEEE TRANS SON ULTRASON 19: 69-79 (1972)

02-0040   OPTICAL PROPERTIES OF FERROELECTRICS. DIGRESSIONS FROM PHENOMENOLOGICAL THEORY.
           (AMMONIUM SULFATE, AMMONIUM FLUOBERYLLATE, GUANIDINIUM ALUMINUM SULFATE HEXAHYDRATE)
           ANISTRATOV AT
           REV PHYS APPL, SUPPL J PHYS 7: 77-9 (JUN 1972) (IN FRENCH)
               293

02-0041   POINT DEFECTS IN PEROVSKITE TYPE FERROELECTRICS. (REVIEW, 77 REFS)
           AREND H
           P231-54, VOL 2 OF DVORAK V + FOUSKOVA A + GLOGAR P (EDS),
           INT MEETING ON FERROELECTRICITY, PRAGUE, 1966, PROC.
           INST PHYS CZECH ACAD SCI, PRAGUE, 1966.

02-0042   SYNTHESIS OF OXIDE BRONZES. (REVIEW, 88 REFS)
           BANKS E + WOLD A
           P237-68 OF JOLLY WL, PREPARATIVE INORGANIC REACTIONS, VOL 4,
           INTERSCIENCE, NY, 1968.

02-0043   ACOUSTIC PROPERTIES OF MATERIALS OF PEROVSKITE STRUCTURE. (RELATION OF FERROELECTRICITY
           TO LATTICE DYNAMICS, REVIEW, 68 REFS)
           BARRETT HH
           P65-108 OF MASON WP + THURSTON RN(EDS), PHYSICAL ACOUSTICS,
           PRINCIPLES AND METHODS, VOL 6, ACADEMIC PRESS, 1970, 386P.

02-0044   PHASES AND POLARIZATION OF LUMINOUS WAVES IN CUBIC ELECTROOPTIC CRYSTALS. (NONLINEAR
           OPTICAL PROPERTIES, ELECTROOPTIC COEFFICIENTS)
           BARKOVSKII LM + FEDOROV FI
           J APPL SPECTROSC 10(1): 83-6 (1969)
               162

2. REVIEWS

02-0045 SINGLE CRYSTAL TITANATES AND ZIRCONATES. (METHODS OF GROWTH)
BEALS MD
P99-116, PART 2, CHAPTER 3 OF ALPER AM(ED), HIGH TEMPERATURE OXIDES.
(REFRACT MATER 5: 1970), ACADEMIC PRESS, 1970, 276P.

234

02-0046 IMPROVEMENT IN THE PYROELECTRIC INFRARED RADIATION DETECTOR. (TRIGLYCINE SULFATE,
TRIGLYCINE FLUOROBERYLLATE, LITHIUM SULFATE, BARIUM STRONTIUM NIOBATE)
BEERMAN HP
FERROELECTRICS 2: 123-8 (1971)

02-0047 SYNTHESIS AND CERTAIN PROPERTIES OF NEW COMPOUNDS WITH A PYROCHLORE STRUCTURE.
(NIOBATES, TANTALATES)
BELYAEV IN + AVERYANOVA IN + SOLOVEV LA + EZHOV VM + GOLTSOV YUI
SOV PHYS CRYSTALLOGR 17: 98-101 (1972)

02-0048 CHEMISTRY AND PHYSICS OF (ALKALI) METAL TANTALATES. (PREPARATION, REVIEW)
BERGMANN H
CHEM ZTG, CHEM APP 95(4): 177-9 (1971) (IN GERMAN)
CHEM ABSTR 74: 150434M (1971)
242

02-0049 NONLINEAR OPTICAL MATERIALS (REVIEW, DATA COLLECTION, MATERIALS NEEDS, 118 REFS)
BERGMAN JG + KURTZ SK
MATER SCI ENG 5(5): 235-50 (1970)
208

02-0050 FERROELECTRICS. (THEORY, DYNAMICAL ORDER DISORDER PROPERTIES, KDP, TGS, ROCHELLE SALT,
SILVER PERIODATE, SODIUM TRIHYDROGEN SELENITE)
BLINC R
ACTA CRYSTALLOGR A 28: S238 (1972)

02-0051 LATTICE DYNAMICS BY NEUTRON SCATTERING. PARTS 1 AND 2. (REVIEW)
BOFFI S + CAGLIOTI G
ENERG NUCL (ITALY) 16(6): 369-77 (1969); 16(7): 425-36 (1969)
282

02-0052 NUCLEAR QUADRUPOLE RESONANCE OF LITHIUM IN FERROELECTRIC COMPOUNDS. (LITHIUM AMMONIUM
TARTRATE, LITHIUM THALLIUM TARTRATE, LITHIUM TANTALATE, LITHIUM NIOBATE, LITHIUM
HYDRAZINIUM SULFATE)
BURNS G
PHYS REV 127: 1193-96 (1962)

02-0053 PROPERTIES OF TUNGSTEN BRONZE FERROELECTRICS. (REVIEW, 33 REFS)
BURNS G + GRESS EA + O'KANE DF + SCOTT BA + SMITH AW
P153-8 OF INT MEETING ON FERROELECTRICITY, 2ND, 1969, KYOTO, PROC,
(J PHYS SOC JAP, VOL 28 SUPPL). PHYS SOC JAP, 1970, 459P.

02-0054 TUNGSTEN BRONZE TYPE FERROELECTRICS FOR OPTICAL USE (REVIEW, 22 REFS).
BURNS G
OYO BUTSURI 39(1): 3-10 (1970) (IN JAPANESE)
223

02-0055 NONLINEAR OPTICAL MATERIALS.
CARRUTHERS JR
P623-31 OF ENCYCL OF CHEM TECHNOL, SUPPL VOLUME, KIRK RE +
OTHMER DF(EDS), INTERSCIENCE, 1970.

02-0056 (HIGH SPEED, SMALL APERTURE) MODULATORS FOR OPTICAL COMMUNICATIONS (REVIEW EMPHASIZING
ELECTROOPTIC MODULATION, LITHIUM TANTALATE, LITHIUM NIOBATE, KDP, 113 REFS).
CHEN FS
PROC IEEE 58(10): 1440-57 (1970)
220

2. REVIEWS

02-0057 SYNTHESIS AND CHARACTERIZATION OF THIN FERROELECTRIC AND SEMICONDUCTING FILMS. (THE MAJOR PORTION OF THIS REPORT IS CONCERNED WITH ION IMPLANTATION OF SINGLE CRYSTAL CADMIUM SULFIDE)
CHERNOW F
UNIV COLORADO, PHOTOCOND SEMICONDUCTORS AND DEV LAB, 1970, 143P.
AD706097, N70-38389
SCI TECH AEROSP REP 8(21): 4006 (1970)
201

02-0058 FERROELECTRIC, PIEZOELECTRIC, AND ELECTROOPTIC MATERIALS. (REVIEW, 991 REFS)
COOK WR
P306-408 OF NAT RES COUNC, DIV ENG INDUST RES, DIGEST OF LITERATURE
ON DIELECTRICS, VOL 32, 1968, WASHINGTON, DC.

02-0059 NEUTRON INELASTIC SCATTERING AT STRUCTURAL PHASE TRANSITIONS. (STRONTIUM TITANATE, POTASSIUM TANTALATE, BARIUM TITANATE, POTASSIUM NIOBATE, SODIUM NITRITE, POTASSIUM DIDEUTERIUM PHOSPHATE)
COWLEY RA
J PHYS (PARIS) 33, SUPPL 4, COLLOQ 2: 7-10 (1972)
294

02-0060 FERROELECTRICITY AND CONDUCTION IN FERROELECTRIC CRYSTALS. (LITHIUM THALLIUM TARTRATE MONOHYDRATE, ELECTRIC CONTROL OF ELASTIC COMPLIANCE, NICKEL IODINE BORACITE OPTICAL ANISOTROPY, BISMUTH TITANATE, POSSIBLE FERROELECTRIC SILICATE)
CROSS LE
MATER RES LAB, PENNSYLVANIA STATE UNIV, 1970, 219P. AD721479,
(SEE ALSO FINAL REPORT DATED 31 MAR 1972, AD744834)
GOVT REP ANNOUNCE 71(10): 174 (1971)
222

02-0061 FERROELECTRICITY AND CONDUCTION IN FERROELECTRIC CRYSTALS. (BISMUTH TITANATE, LITHIUM THALLIUM TARTRATE MONOHYDRATE, LEAD(5) GERMANIUM(3) OXIDE(11))
CROSS LE
MATER RES LAB, PENNSYLVANIA STATE UNIV, 1972, 128P. AD744834
(SEE ALSO INTERIM REPORT DATED 30 NOV 1970, AD721479)
GOVT REP ANNOUNCE 72(16): 220 (1972)
267

02-0062 AGING OF BARIUM TITANATE AND LEAD ZIRCONATE TITANATE FERROELECTRIC CERAMICS. (REVIEW, 31 REFERENCES)
DE LAUNAY J + SMITH PL
NAVAL RES LAB, 1970, 31P. N71-18809, AD715624
SCI TECH AEROSP REP 9(8): 1279 (23 APR 1971)
238

02-0063 GENERATION OF THE SECOND HARMONIC AND FERROELECTRICITY. (THEORY)
DOLINO G + LAJZEROWICZ J + VALLADE M
J PHYS (PARIS) 33, SUPPL 4, COLLOQ 2: 21-3 (1972) (IN FRENCH)
294

02-0064 FERROELECTRIC RARE EARTH MOLYBDATES. (REVIEW, FORMATION, CRYSTAL GROWTH, PHASE TRANSITIONS, STRUCTURE, 49 REFS)
DROBYSHEV LA + RABINOVICH AZ + VENEVTSEV YUN
BULL ACAD SCI USSR, PHYS SER 34(12): 2250-61 (1970)
238

02-0065 ANISOTROPIC, SOLID DIELECTRICS AND FERROELECTRICITY. (106 REFS)
EYRAUD L
GAUTHIER-VILLARS, PARIS, 1967, 184P. (IN FRENCH)
CHEM ABSTR 70: 52353G (1969)
141

02-0066 SWITCHING PROPERTIES IN FERROELECTRICS OF THE FAMILY BISMUTH(4) BARIUM(M-2) TITANIUM(M+1) OXIDE(3) (M+2) (M=2,3,4). (DATA FOR THE TITANIUM(3) OXIDE(12), BARIUM TITANIUM(4) OXIDE(15), AND BARIUM(2) TITANIUM(5) OXYGEN(18) COMPOUNDS)
FANG PH + FATUZZO E
J PHYS SOC JAP 17: 238 (1962)

02-0067 OPTICAL HARMONICS AND NONLINEAR PHENOMENA. (THEORY, REVIEW)
FRANKEN PA + WARD JF
REV MOD PHYS 35: 23-39 (1963)

02-0068 DIELECTRIC INSTABILITIES. (THEORETICAL RELATIONSHIP TO FERROELECTRICITY)
FROHLICH H
P9-15 OF SYMP ON FERROELECTRICITY, 1966, GENERAL MOTORS
RES LABS, WARREN, MICHIGAN, PROC, FERROELECTRICITY.
WELLER EF(ED), ELSEVIER, 1967, 318P.

02-0069 FERROELECTRIC SOLID SOLUTIONS.
FUKUDA T
J CRYSTALLOGR SOC JAP 12: 212-20 (1970)

02-0070 ULTRASONIC INVESTIGATION OF PHASE TRANSITIONS AND CRITICAL POINTS. SECTION-5:
FERROELECTRIC AND ANTIFERROELECTRIC TRANSITIONS. (REVIEW)
GARLAND CW
P92-110, 140-8 OF MASON WP AND THURSTON RN(EDS), PHYSICAL ACOUSTICS,
PRINCIPLES AND METHODS, VOL 7, ACADEMIC, 1970, 380P.

02-0071 STRUCTURE AND POINT DEFECTS OF ICE: THEIR EFFECT ON THE ELECTRICAL AND MECHANICAL
PROPERTIES. (REVIEW, 56 REFS)
GLEN JW
SCI PROGR 57: 1-21 (1969)
        160

02-0072 REVIEW ON THE PROBLEMS OF THE PHYSICS OF ICE.
GRANICHER H
P1-18 OF INT SYMP ON PHYSICS OF ICE, 3RD PROC, 1968,
RIEHL N + BULLEMER B + ENGELHARDT H(EDS), PLENUM, NY,
1969, 642P.

02-0073 SOME RECENT ADVANCES IN FERROELECTRICS. (ELECTROOPTIC MATERIALS, SINGLE CRYSTALS OF
POTASSIUM TANTALATE NIOBATE, LITHIUM NIOBATE, AND BISMUTH TITANATE)
HERBERT JM
P157-99 OF BRITISH CERAMIC SOC, ELECTRICAL MAGNETIC CERAMICS CONF,
2ND, SEP 1969, WARWICK, ENGLAND, PROC. BRITISH CERAMIC SOC, 1970
        SCI ABSTR A 74: 2555 (1971)
        238

02-0074 DOPING PROBLEMS IN (TITANATE) FERROELECTRIC SEMICONDUCTORS. (ORIGIN AND EFFECTS OF
GRAIN BOUNDARY BLOCKING LAYERS, ANTIMONY DOPANT, REVIEW)
HEYWANG W
OYO BUTSURI 39(4): 294-300 (1970) (IN JAPANESE)
        CHEM ABSTR 73: 60132M (1970)
        238

02-0075 ENERGY STORAGE IN FERROELECTRIC CERAMICS: SURVEY OF MATERIALS BASED UPON BARIUM
TITANATE.
HILL GJ
ELEC RES ASSOC, LEATHERHEAD, SURREY, ENGLAND, 1969, 18P. 69-35
        SCI ABSTR A 74: 28575 (1971)
        238

02-0076 NEW PIEZOELECTRIC MATERIALS.
ICHINOSE N + EGAMI H + TAKAHASHI T
TOSHIBA REV 27: 482-6 (1972) (IN JAPANESE)

02-0077 ACCURATE STRUCTURE ANALYSIS OF FERROELECTRICS IN CONNECTION WITH THEIR PHASE
TRANSITIONS. PART-1: TGS. PART-2: SILVER SODIUM DINITRITE. PART-3: ROCHELLE SALT.
PART-4: LITHIUM AMMONIUM TARTRATE HYDRATE.
ISHIDA K + MITSUI T + HINAZUMI H + SHIOZAKI Y
ACTA CRYSTALLOGR A 28: S184-5 (1972)

02-0078 NONLINEAR OPTICAL PROPERTIES OF MOLECULAR CRYSTALS.
JERPHAGNON J
BULL SOC FRANC MINERAL CRISTALLOGR 95(2): 262-7 (1972) (IN FRENCH)
        286

02-0079 EPR STUDY OF X-RAY DAMAGED FERROELECTRICS.
JIMENEZ B
ELECTRON FIS APLIC 14: 265-70 (1971) (IN SPANISH)

02-0080 MODERN APPLICATIONS OF FERROELECTRICITY.
KELL RC
BRIT J APPL PHYS 14: 249-55 (1963)

2. REVIEWS

02-0081   NONLINEAR OPTICAL AND ELECTROOPTICAL PROPERTIES OF DIELECTRICS AND FERROELECTRICS.
(REVIEW, 195 REFERENCES)
KIELICH S
FERROELECTRICS 4: 257-82 (1972)

02-0082   OPTICAL ANOMALIES IN FERROELECTRIC CRYSTALS.
KOBAYASHI J
NIPPON KESSHO GAKKAISHI (J JAP CRYSTALLOGR SOC) 4(3): 47-   (1962)
(IN JAPANESE)

02-0083   PYROELECTRIC INVESTIGATIONS OF THE DOMAIN STRUCTURE STABILITY IN FERROELECTRICS.
(ROCHELLE SALT, TRIGLYCINE SULFATE, SELENATE, FLUOBERYLLATE AND BARIUM TITANATE,
SUMMARY ONLY)
KOPSIK VA + GAVRILOVA ND + NOVIK VK
P382 OF INT MEETING ON FERROELECTRICITY, 2ND, 1969,, KYOTO, PROC,
(J PHYS SOC JAP, VOL 28 SUPPL). PHYS SOC JAP, 1970, 459P.

02-0084   FERROELECTRIC SEMICONDUCTORS. (ANTIMONY SULFOIODIDE, REVIEW, 57 REFS)
KREHER K
WISS Z KARL MARX UNIV LEIPZIG, MATH NATURWISS REIH 20(2): 287-301
(1971)
       CHEM ABSTR 75: 113587P (1971)
       263

02-0085   INTERBAND MECHANISMS FOR FERROELECTRIC PHASE TRANSITIONS.
KRISTOFFEL NN + KONSIN PI
BULL ACAD SCI USSR, PHYS SER 35(9): 1611-15 (1971)

02-0086   RECENT DEVELOPMENTS IN ELECTROOPTIC CERAMICS (HOT PRESSED LEAD ZIRCONATE TITANATE,
EFFECT OF GRAIN SIZE, DOPANTS, REVIEW).
LAND CE
INT J NONDESTRUCT TEST 1: 315-36 (1970)
       188

02-0087   CRITICAL SLOWING DOWN PROCESS OF (ANOMALOUS) DIELECTRIC RELAXATION IN SEVERAL CRYSTALS
OF ORDER DISORDER TYPE. (COPPER FORMATE TETRAHYDRATE, AMMONIUM IRON AND CHROMIUM ALUMS,
METHYL AMMONIUM ALUMINUM SULFATE DODECAHYDRATE, COLEMANITE, SODIUM TRIDEUTERIUM
DISELENITE)
MAKITA Y + SEO I + SUMITA M
P268-70 OF INT MEETING ON FERROELECTRICITY, 2ND, 1969, KYOTO, PROC,
(J PHYS SOC JAP, VOL 28 SUPPL). PHYS SOC JAP, 1970, 459P.

02-0088   INFLUENCE OF THE QUALITY OF TRIGLYCINE SULFATE SINGLE CRYSTALS ON THEIR DIELECTRIC
PROPERTIES. (REVIEW)
MALEK Z + MORAVEC F + STRAJBLOVA J + NOVOTNY J + JANTA J + MASTNER J
P181-9 OF EUROPEAN MEETING ON FERROELECTRICITY, 1969, SAARBRUCKEN,
MUSER HE AND PETERSSON J(EDS), WISSENSCH VERLAGSGES, STUTTGART,
1970, 396P.

02-0089   EFFECT OF CRYSTALLIZATION CONDITIONS ON THE SHAPE AND COMPOSITION OF FERROELECTRIC
CRYSTALS (SOLUTION GROWTH, TGS, TRIGLYCINE SELENATE OR FLUOBERYLLATE, POTASSIUM OR
RUBIDIUM DIHYDROGEN PHOSPHATE, SODIUM, POTASSIUM(2), OR LITHIUM THALLIUM TARTRATE,
LITHIUM AMMONIUM TARTRATE, AND MIXED CRYSTALS, BARIUM TITANATE, AND BARIUM STRONTIUM
TITANATE)
MARECEK V + NOVAK J
KRISTALL TECH 5(1): 109-19 (1970)
       227

02-0090   FIFTY YEARS OF FERROELECTRICITY (REVIEW).
MASON WP
J ACOUST SOC AMER 50: 1281-98 (NOV 1971)

02-0091   ENERGY BANDS FOR POTASSIUM NICKEL TRIFLUORIDE, STRONTIUM TITANATE, POTASSIUM MOLYBDATE,
AND POTASSIUM TANTALATE.
MATTHEISS LF
PHYS REV B 6: 4718-40 (1972)

2. REVIEWS

02-0092 FROM BARIUM TITANATE TO DNA. (FERROELECTRICITY, BIOLOGICAL MECHANISMS, MEMORY, ORGANICS)
MATTHIAS BT
MATER RES BULL 6: 1005-6 (OCT 1971)
247

02-0093 QUANTITATIVE ANALYSIS OF MICROSTRUCTURE AND ELECTRICAL PROPERTIES OF FERROELECTRIC MATERIALS. (FINE POWDERS PREPARED BY DECOMPOSITION OF ORGANIC COMPOUNDS)
MAZDIYASNI KS + BROWN LM
BULL AMER CERAM SOC 50: 798 (1971)

02-0094 ROLE OF THE REGULAR OCTAHEDRON AND LARGER POLYHEDRA IN FERROELECTRIC MATERIALS. (HETEROTYPES, REVIEW)
MEGAW HD + SAKOWSKI-COWLEY AC + LUKASZEWICZ K + LEFKOWITZ I + KATZ L + BRANDON JK + DARLINGTON AN
P27-35 OF EUROPEAN MEETING ON FERROELECTRICITY, 1969, SAARBRUCKEN,
MUSER HE AND PETERSSON J (EDS), WISSENSCH VERLAGSGES, STUTTGART,
1970, 396P.

02-0095 ABSOLUTE SIGNS OF NONLINEAR OPTICAL (SECOND HARMONIC GENERATION) COEFFICIENTS OF POLAR (PYROELECTRIC) CRYSTALS (LITHIUM NIOBATE OR TANTALATE OR GALLATE, BARIUM TITANATE, ZINC OXIDE, CADMIUM SULFIDE).
MILLER RC + NORDLAND WA
OPT COMMUN 1(8): 400-2 (1970)
207

02-0096 ABSOLUTE SIGNS OF SECOND HARMONIC GENERATION COEFFICIENTS OF PIEZOELECTRIC CRYSTALS.
MILLER RC + NORDLAND WA
PHYS REV B 2(12): 4896-902 (15 DEC 1970)

02-0097 NONLINEAR OPTICAL DEVICES. (BASED IN SECOND HARMONIC GENERATION AND OPTICAL PARAMETRIC OSCILLATION IN FERROELECTRIC CRYSTALS, REVIEW, 41 REFS)
MILLER RC
OYO BUTSURI 39(1): 11-17 (1970) (IN JAPANESE)
CHEM ABSTR 72: 84356T (1970)

02-0098 RELATIVE SIGNS OF NONLINEAR OPTICAL COEFFICIENTS OF POLAR CRYSTALS (LITHIUM NIOBATE OR TANTALATE, BARIUM TITANATE, ZINC OXIDE, AND LITHIUM GALLIUM OXIDE).
MILLER RC + NORDLAND WA
APPL PHYS LETT 16: 174-6, 372 (1970)
181

02-0099 DIELECTRIC AND X-RAY STUDIES OF THE PHASE TRANSITIONS IN ROCHELLE SALT, BARIUM TITANATE, ANTIMONY SULFOIODIDE, THIOUREA, TRIGLYCINE SULFATE, HYDROGEN AMMONIUM MONOCHLORO ACETATE, DICALCIUM STRONTIUM HEXAACETATE.
MITSUI T + NAKAMURA E + SHIOZAKI Y + MOTEGI H + SEKIDO T + ICHIKAWA M + TAKAMA T + HOSOYA M + SHIBUKAWA K + WAKAKI N + FURUICHI J
P220-4, VOL 1 OF DVORAK V + FOUSKOVA A + GLOGAR P (EDS),
INT MEETING ON FERROELECTRICITY, PRAGUE, 1966, PROC.
INST PHYS CZECH ACAD SCI, PRAGUE, 1966.

02-0100 FUTURE PROBLEMS ON FERROELECTRICS.
MITSUI T
J CRYSTALLOGR SOC JAP 12: 64-75 (1970) (IN JAPANESE)

02-0101 INFORMATION ON HIGH DIELECTRIC CONSTANT MATERIALS GAINED BY ELECTRON PARAMAGNETIC RESONANCE. (ABSTRACT ONLY, 11 REFS)
MUELLER KA
P293-4 OF EUROPEAN MEETING ON FERROELECTRICITY, 1969, SAARBRUCKEN,
MUSER HE AND PETERSSON J (EDS), WISSENSCH VERLAGSGES, STUTTGART,
1970, 396P.

02-0102 FERROELECTRICITY AND THE CHEMICAL BOND IN PEROVSKITE TYPE OXIDES. (43 REFS)
NELSON CW
MIT LAB INSULATION RES, MAY 1963, 34P. N63-17564
SCI TECH AEROSP REP 1(16): 1147 (1963)

02-0103 PREPARATION OF TRANSPARENT FERROELECTRIC CERAMICS. (HOT PRESSING)
O'BRYAN HM + THOMSON J
BULL AMER CERAM SOC 49: 833 (1970)

2. REVIEWS

02-0104   STUDY OF OPTICAL PROPERTIES AND COLLECTIVE OSCILLATIONS IN NEW SOLID STATE MATERIALS
          (INCLUDING FERROELECTRICS) AS A FUNCTION OF TEMPERATURE USING INFRARED AND RAMAN
          TECHNIQUES.
          PERRY CH
          NORTHEASTERN UNIV, BOSTON, 1970, 119P. N70-29596
               205

02-0105   PIEZOELECTRIC CERAMIC SUBSTANCES. (REVIEW, 23 REFS)
          PETZSCH K + TOPF KJ
          BER DEUT KERAM GESELL 47: 681-6 (1970) (IN GERMAN)

02-0106   ELECTROOPTICAL MATERIALS.
          PINNOW DA
          P478-88 OF CHEMICAL RUBBER CO, CLEVELAND, HANDBOOK OF LASERS WITH
          SELECTED DATA ON OPTICAL TECHNOLOGY, PRESSLEY RJ(ED), 1971, 631P.

02-0107   DETERMINATION OF ABSOLUTE SIGNS OF MICROWAVE NONLINEAR SUSCEPTIBILITIES. (BARIUM
          TITANATE, LITHIUM TANTALATE, LITHIUM NIOBATE, LITHIUM IODATE, KDP, ADP, AND OTHER
          NONFERROELECTRIC MATERIALS)
          POLLACK MA + TURNER EH
          PHYS REV B 4: 4578-82 (1971)

02-0108   STUDY OF MICROWAVE DIELECTRIC DISPERSION IN FERROELECTRICS OF VARIOUS TYPES. (BARIUM
          TITANATE AND TITANATE ZIRCONATE, TRIGLYCINE SULFATE AND ROCHELLE SALT)
          POPLAVKO YUM
          P171-9, VOL 2 OF DVORAK V + FOUSKOVA A + GLOGAR P (EDS),
          INT MEETING ON FERROELECTRICITY, PRAGUE, 1966, PROC.
          INST PHYS CZECH ACAD SCI, PRAGUE, 1966.

02-0109   ELECTRICAL CONDUCTION IN METAL OXIDES. (REVIEW, 319 REFS)
          RAO CNR + SUBBARAO GV
          PHYS STATUS SOLIDI A 1: 597-652 (1970)

02-0110   ADVANCES IN PRODUCTION OF (COMMERCIAL) PIEZOELECTRIC CERAMICS AND (TABLES OF) THEIR
          PROPERTIES.
          REDDIG H
          SPRECHSAAL KERAM GLAS EMAIL SISIKATE 101(24): 1125-8, 1131-2 (1968)
          (IN GERMAN)
               CHEM ABSTR 70: 99216E (1969)

02-0111   SOME ASPECTS OF THE CURRENT STUDY AND USE OF SEVERAL TECHNICALLY IMPORTANT PROPERTIES
          OF FERROELECTRICS.
          REZ IS
          BULL ACAD SCI USSR, PHYS SER 34(12): 2274-80 (1970)
               238

02-0112   FLASH EVAPORATION. (REVIEW OF TECHNIQUE AND APPLICATION TO THIN FILMS OF PEROVSKITES
          AND III-V COMPOUNDS, 8 REFS).
          RICHARDS JL
          P71-86 OF NATO ADVANCED STUDY RES INST, 1965, LONDON, USE OF THIN
          FILMS IN PHYSICAL INVESTIGATIONS, LECTURES AND DISCUSSIONS,
          ANDERSON JC(ED), ACADEMIC PRESS, 1966, 462P.

02-0113   INFRARED ABSORPTION SPECTROPHOTOMETRIC STUDY OF MONOVALENT METAL NIOBATES.
          ROCCHICCIOLI-DELTCHEFF C + DEPUIS T
          COMPT REND B 270: 268-71 (1970) (IN FRENCH)

02-0114   THE GRUNEISEN PARAMETER OF THE SOFT FERROELECTRIC MODE IN THE CUBIC PEROVSKITES.
          (STRONTIUM TITANATE, POTASSIUM TANTALATE, BARIUM TITANATE, LEAD TITANATE, LEAD
          ZIRCONATE, LEAD HAFNATE)
          SAMARA GA
          FERROELECTRICS 2: 177-82 (1971)

02-0115   NEUTRON INELASTIC SCATTERING STUDY ON SOFT MODES. (REVIEW OF BROOKHAVEN WORK ON
          POTASSIUM TANTALATE, BARIUM AND STRONTIUM TITANATES, KDP, 22 REFS)
          SHIRANE G
          P20-5 OF INT MEETING ON FERROELECTRICITY, 2ND, 1969, KYOTO, PROC,
          (J PHYS SOC JAP, VOL 28 SUPPL). PHYS SOC JAP, 1970, 459P.

2. REVIEWS

02-0116 PHONONS, ELECTRONS, AND PROTONS IN FERROELECTRIC MATERIALS.
SILVERMAN BD
P168-86 OF COLLOQUE AMPIRE, 15TH, 1968, GRENOBL, PROC. MAGNETIC
RESONANCE AND RADIOFREQUENCY SPECTROSCOPY, NORTH-HOLLAND, 1969

   238

02-0117 MAGNETICALLY ORDERED FERROELECTRIC MATERIALS (REVIEW, 251 REFS).
SKINNER SM
IEEE TRANS PARTS MATER PACKAG 6(2): 68-90 (1970)
   191

02-0118 NEW CLASSES OF FERROELECTRICS OF DISPLACEMENT TYPE. (BORACITES, MOLYBDATES, TUNGSTEN
BRONZES, NIOBATES)
SMOLENSKII GA + ISUPOV VA
J PHYS (PARIS) 33, SUPPL 4, COLLOQ 2: 25-31 (1972)
   294

02-0119 PHYSICAL PHENOMENA IN FERROELECTRICS WITH DIFFUSED PHASE TRANSITION. (POLARIZATION
MECHANISMS, ELECTROOPTICAL PROPERTIES, REVIEW OF RUSSIAN WORK, 44 REFS)
SMOLENSKII GA
P26-37 OF INT MEETING ON FERROELECTRICITY, 2ND, 1969, KYOTO, PROC,
(J PHYS SOC JAP, VOL 28 SUPPL). PHYS SOC JAP, 1970, 459P.

02-0120 RAMAN SCATTERING IN FERROELECTRICS. (REVIEW)
STEIGMEIER EF
J PHYS (PARIS) 33, SUPPL 4, COLLOQ 2: 15-16 (1972)
   294

02-0121 X-RAY DIFFRACTION TOPOGRAPHY OF FERROELECTRIC CRYSTALS.
TAKAGI M + SUZUKI S
J CRYSTALLOGR SOC JAP 13: 311-19 (1971) (IN JAPANESE)

02-0122 APPLICATIONS OF FERROELECTRICS AND RELATED MATERIALS: REVIEW OF DEVELOPMENTS IN EUROPE.
THOMAS LA
FERROELECTRICS 3: 231-8 (1972);
IEEE TRANS SON ULTRASON 19: 231-8 (1972)
   263

02-0123 FERROELECTRIC MATERIALS. (REVIEW, 31 REFS)
BRIEBWASSER S
P343-400 OF MODERN MATERIALS; ADVANCES IN DEVELOPMENT AND
APPLICATIONS, VOL3, 1962. HAUSNER HH(ED), ACADEMIC PRESS.

   CHEM ABSTR 58: 10697 (1963)

02-0124 MILLIMETER WAVELENGTH DISPERSION IN THE DIELECTRIC CONSTANT ABOVE THE CURIE POINT IN
VARIOUS FERROELECTRICS AND ANTIFERROELECTRICS. (ZIRCONATES, NIOBATES, TUNGSTATES)
TSYKALOV VG + POPLAVKO YUM
BULL ACAD SCI USSR, PHYS SER 34(12): 2302-5 (1970)

02-0125 BEHAVIOR OF THE COMPLEX PERMITTIVITY (AUDIO FREQUENCIES) OF SOME FERROELECTRIC CRYSTALS
(TRIGLYCINE SULFATE, POTASSIUM DIHYDROGEN PHOSPHATE, AMMONIUM SULFATE, ROCHELLE SALT)
AT THE TRANSITION TEMPERATURES.
UNRUH HG
P214-19, VOL 1 OF DVORAK V + FOUSKOVA A + GLOGAR P (EDS),
INT MEETING ON FERROELECTRICITY, PRAGUE, 1966, PROC.
INST PHYS CZECH ACAD SCI, PRAGUE, 1966.

02-0126 EARLY HISTORY OF FERROELECTRICITY (1920-1943).
VALASEK J
FERROELECTRICS 2(4): 239-44 (1971)

02-0127 FUNDAMENTAL PROPERTIES OF METAL- FERROELECTRIC- METAL JUNCTION SYSTEMS. (INFLUENCE OF
THE ELECTRODE MATERIAL ON THE PYROELECTRIC EFFECT)
VEKHTER BG + GIFEISMAN SHN + KREMENCHUGSKII LS + PERLIN YUE + SAMOILOV VB
SOV PHYS SOLID STATE 13(1): 74-8 (JUL 1971)
   237

2. REVIEWS

02-0128   NEW FERROELECTRICS AND SEIGNETTO MAGNETS OF THE PEROVSKITE AND PYROCHLORE TYPE.
          SYNTHESIS AND STUDY OF THE CRYSTAL STRUCTURE AND PROPERTIES. (REVIEW OF KARPOV INST
          WORK, 44 REFS)
          VENEVTSEV YUN + KAPYSHEV AG + SHVORNEVA LI + TOMASHPOLSKII YUYA + VISKOV AS
          + ZUBOVA YEV + IVANOVA VV + DEMUROV DG + DZHMUKHADZE DF + SALNIKOV VD + ZHDANOV GS
          P139-44 OF INT MEETING ON FERROELECTRICITY, 2ND, 1969, KYOTO, PROC,
          (J PHYS SOC JAP, VOL 28 SUPPL). PHYS SOC JAP, 1970, 459P.

02-0129   FLUX GROWTH OF SOME COMPLEX OXIDE MATERIALS. (NIOBATES, TANTALATES, STANNATES,
          TITANATES, CHROMATES, COBALTATES)
          WANKLYN BM
          J MATER SCI 7: 813-21 (1972)

02-0130   CERAMIC FERROELECTRICS.
          WAYE BE
          CERAMICS 22(266): 18-24 (1971)
                254

02-0131   PREPARATION AND PROPERTIES OF RARE EARTH COMPOUNDS. (WITH MAGNETIC AND FERROELECTRIC
          ORDERING)
          WOOD VE + AUSTIN AE + BROG KC
          BATTELLE COLUMBUS LABS, 1972, 42P. AD754775
                297

02-0132   FERRO- AND ANTIFERROELECTRIC PHASE TRANSITIONS AND RELATED SCATTERING EXPERIMENTS.
          (REVIEW)
          YAMADA Y
          PROC PHYS SOC JAP 25: 354-65 (1970) (IN JAPANESE)

02-0133   NEUTRON DIFFRACTION INVESTIGATIONS OF THE ATOMIC STRUCTURES OF INORGANIC MATERIALS.
          (REVIEW, 244 REFS)
          YAMZIN II + LOSHMANOV AA
          INORG MATER 8(1): 1-20 (JAN 1972)
                278

02-0134   DIPOLE PATTERNS IN THE STRUCTURES OF SOME FERROELECTRICS AND ANTIFERROELECTRICS.
          (BARIUM TITANATE, KDP, POTASSIUM SODIUM TARTRATE, TGS, AMMONIUM FLUOBERYLLATE, SODIUM
          NITRITE, ALUMS)
          ZHELUDEV IS
          PROC INDIAN ACAD SCI A 57: 361-78 (1963)

3. THEORY

03-0135 STRUCTURAL BASIS OF FERROELECTRICITY AND FERROELASTICITY. (SYMMETRY RELATIONS)
ABRAHAMS SC + KEVE ET
FERROELECTRICS 2: 129-54 (1971)
235

03-0136 CONSIDERATIONS OF PARTIALLY FERROELASTIC AND PARTIALLY ANTIFERROELASTIC CRYSTALS AND
PARTIALLY FERROELECTRIC AND PARTIALLY ANTIFERROELECTRIC CRYSTALS (THEORY).
AIZU K
J PHYS SOC JAP 28(3): 717-22 (1970)
190

03-0137 DETERMINATION OF THE STATE PARAMETERS AND FORMULATION OF SPONTANEOUS STRAIN FOR
FERROELASTICS (THEORY).
AIZU K
J PHYS SOC JAP 28(3): 706-16 (1970)
190

03-0138 DETERMINATION OF KIND (TYPE, CLASS) FOR ACTUAL FERROELECTRICS TOGETHER WITH EXAMINATION
OF CHARACTERS OF THEIR PARAELECTRIC PHASE TRANSFORMATIONS. (AND SUGGESTIONS FOR
DISCOVERING MEMBERS OF VACANT CLASSES)
AIZU K
PHYS REV 136(3): 753-8 (1964)

03-0139 FERROELECTRIC TRANSFORMATIONS OF TENSORIAL PROPERTIES IN REGULAR FERROELECTRICS.
AIZU K
PHYS REV 133: 1350-9 (1964)

03-0140 GENERAL CONSIDERATION OF FERROELECTRICS AND FERROELASTICS SUCH THAT THE ELECTRIC
SUSCEPTIBILITY OR ELASTIC COMPLIANCE IS TEMPERATURE INDEPENDENT IN THE PROTOTYPIC
PHASE.
AIZU K
J PHYS SOC JAP 33(3): 629-40 (1972)
278

03-0141 MECHANISM OF THE FERROELECTRIC PHASE TRANSFORMATION OF DIAMMONIUM DICADMIUM TRISULFATE,
A MODEL IN WHICH THE POLARIZATION VECTOR, P(S), ARISES PERPENDICULAR TO THE WAVE NORMAL
OF THE SOFT WAVES. (THEORY)
AIZU K
J PHYS SOC JAP 32(4): 1033-6 (1972)
278

03-0142 POSSIBLE SPECIES OF FERROMAGNETIC, AND FERROELASTIC CRYSTALS (PREVIOUS THEORY EXTENDED
TO MAGNETIC CRYSTALS AND TO INCLUDE PARTIAL CASES, "FERROIC CRYSTALS").
AIZU K
PHYS REV B 2(3): 754-72 (1970)
225

03-0143 POLARIZATION, PYROELECTRICTY, AND FERROELECTRICITY OF IONIC CRYSTALS. (THEORY, NO REFS)
AIZU K
REV MOD PHYS 34: 550-76 (1962)

03-0144 PHENOMENOLOGICAL LATTICE DYNAMICAL THEORY OF FERROELASTICITY.
AIZU K
J PHYS CHEM SOLIDS 32: 1959-69 (1971)
240

03-0145 PRESENTATION AND DISCUSSION OF EXAMPLES OF FERROELECTRICS AND FERROELASTICS HAVING THE
INDEX OF FAINTNESS UNEQUAL TO THE CELL MULTIPLICITY.
AIZU K
J PHYS SOC JAP 33(5): 1390-5 (NOV 1972)
280

3. THEORY

03-0146 SUPPLEMENT TO A PREVIOUS THEORY OF WEAK FERROELASTICS AND WEAK FERROELECTRICS.
AIZU K
J PHYS SOC JAP 29: 1401 (1970)
218

03-0147 BEHAVIOR OF MAGNETICALLY ORDERED FERROELECTRICS IN A RAPIDLY OSCILLATING ELECTRIC
FIELD.
AKHIEZER IA + DAVYDOV LN
SOV PHYS SOLID STATE 13(6): 1499-502 (DEC 1971)
253

03-0148 COUPLED ELECTROMAGNETIC SPIN WAVES IN MAGNETICALLY ORDERED FERROELECTRICS.
AKHIEZER IA + DAVYDOV LN
SOV PHYS SOLID STATE 12(11): 2563-5 (1971)
234

03-0149 INSTABILITY OF SPIN WAVES IN FERROELECTRIC ANTIFERROMAGNETS IN EXTERNAL FIELDS.
(THEORY)
AKHIEZER IA + DAVYDOV LN
JETP LETT 13: 271-3 (1971)

03-0150 POSSIBILITY OF EXCITATION OF SPIN WAVES IN MAGNETICALLY ORDERED FERROELECTRICS.
(THEORY)
AKHIEZER IA
SOV PHYS JETP 32: 549-51 (1971)
UKR FIZ ZH 15: 1747-9 (1970) (IN RUSSIAN)

03-0151 A SIMPLE MECHANICAL MODEL WITH FERROELECTRIC BEHAVIOR.
ALBERS J
P381-6 OF EUROPEAN MEETING ON FERROELECTRICITY, 1969, SAARBRUCKEN,
MUSER HE AND PETERSSON J(EDS), WISSENSCH VERLAGSGES, STUTTGART,
1970, 396P.

03-0152 THERMODYNAMIC THEORY OF IRRADIATED TRIGLYCINE SULFATE.
ALEMANY C + MENDIOLA J + KIMENEZ B + MAURER E
ACTA CRYSTALLOGR A 28: S233 (1972)

03-0153 AN APPARATUS FOR INVESTIGATING THE ELECTRICAL PROPERTIES OF FERROELECTRICS.
ANDRUSHKO AI + SAIFULLIN LI + USACHEV EP
IZV VUZ FIZ 1972(10): 152-3 (IN RUSSIAN)

OPTICAL PROPERTIES OF FERROELECTRICS. DIGRESSIONS FROM PHENOMENOLOGICAL THEORY.
(AMMONIUM SULFATE, AMMONIUM FLUOBERYLLATE, GUANIDINIUM ALUMINUM SULFATE HEXAHYDRATE)
ANISTRATOV AT
REV PHYS APPL, SUPPL J PHYS 7: 77-9 (JUN 1972) (IN FRENCH)

03-0154 DYNAMICAL EFFECTS IN SOLID STATE PHASE TRANSFORMATIONS. (FERROELECTRICS)
AXE JD
P749-76, PART B OF HENCH LL + DOVE DB(EDS), PHYS OF ELECTRON
CERAMICS, DEKKER, NY, 1972.
262

03-0155 NEUTRON STUDIES OF DISPLACIVE STRUCTURAL PHASE TRANSFORMATIONS. (INELASTIC DIFFUSION IN
FERROELECTRICS)
AXE JD
TRANS AMER CRYSTALLOGR ASSOC 7: 89-106 (1971)
299

03-0156 NEUTRON STUDIES OF DISPLACIVE STRUCTURAL PHASE TRANSFORMATIONS. (IN FERROELECTRICS)
AXE JD
TRANS AMER CRYSTALLOGR ASSOC 7: 89-106 (1971)
299

03-0157 INTERACTION OF LIGHT WITH MAGNETOACOUSTIC WAVES IN A FERRODIELECTRIC (THEORY).
BAKAI AS + YANKELEVICH RP
SOV PHYS SOLID STATE 11(11): 2749-50 (1970)
194

03-0158 INTERACTION OF LIGHT WITH SPIN WAVES IN FERROELECTRICS. (THEORY)
BAKAI AS + YANKELEVICH RP
SOV PHYS JETP 29(3): 464-6 (1969)

03-0159   ATTENUATION OF CRITICAL VIBRATIONS AND DIELECTRIC LOSSES IN DISPLACIVE TYPE
          FERROELECTRICS (THEORY).
          BALAGUROV BYA + VAKS VG + SHKLOVSKII BI
          SOV PHYS SOLID STATE 12(1): 70-8 (1970)
               201

03-0160   HIGH FREQUENCY SOUND ATTENUATION AND THERMAL CONDUCTIVITY IN DISPLACIVE TYPE
          FERROELECTRICS. (THEORY, BARIUM TITANATE)
          BALAGUROV BYA
          SOV PHYS JETP 34(4): 868-72 (1972)

03-0161   LOCAL ELECTRON STATES IN DOMAIN WALLS OF FERROELECTRICS.
          BALKAREI YUI
          SOV PHYS SOLID STATE 14(6): 1583-4 (DEC 1972)
               279

03-0162   POSSIBILITY OF GENERATING INFRARED RADIATION BY POLARIZATION REVERSAL OF A
          FERROELECTRIC.
          BALKAREI YUI + CHENSKII EV
          JETP LETT 13: 190-2 (1971)

03-0163   RAMAN SPECTROSCOPY OF PHONON INSTABILITY IN FIRST ORDER FERROELECTRIC PHASE
          TRANSITIONS.
          BALKANSKI M + TENG MK
          INDIAN J PURE APPL PHYS 9: 863-9 (1971)

03-0164   SPIN WAVE PHENOMENOLOGICAL THEORY IN FINITE SAMPLES OF FERRODIELECTRICS WITH AN
          ARBITRARY MAGNETIC SYMMETRY.
          BALAKHONOV NF + KURBATOV LV + CYRLIN VP
          UKR FIZ ZH 16(2): 244-52 (1971) (IN RUSSIAN)
               242

03-0165   FAR INFRARED DISPERSION AND RAMAN SPECTRA OF FERROELECTRIC CRYSTALS. (THEORY)
          BARKER AS
          P247-96 OF NATO ADVAN STUDY INST, 1968, DELFT, NETHERLANDS, PROC,
          FAR INFRARED PROPERTIES OF SOLIDS, MITRA SS + NUDELMAN S(EDS), PLENUM
          1970, 605P.
               208

03-0166   QUANTUM THEORY OF VIBRATIONS IN FERROELECTRIC FERROMAGNETICS WITHOUT A SYMMETRY CENTER.
          BARIYAKHTAR VG + CHUPIS IE
          UKR FIZ ZH 17: 652-9 (APR 1972) (IN RUSSIAN, ENGLISH ABSTRACT)

03-0167   POLARIZATION DISTRIBUTION IN THE VICINITY OF A FERROELECTRIC SEMICONDUCTOR INTERFACE.
          BATRA IP + SILVERMAN BD
          BULL AMER PHYS SOC 17: 245 (1972)

03-0168   GENERALIZED FERROELECTRIC MODEL ON A SQUARE LATTICE.
          BAXTER RJ
          STUD APPL MATH 50(1): 51-69 (1971)

03-0169   EFFECT OF ELECTRIC FIELD ON FERROELECTRIC AND ANTIFERROELECTRIC CERAMICS. PART-1:
          DEFORMATION AND PHASE TRANSFORMATION. PART-2: ELECTRIC CONDUCTION IN THE INDUCTION
          REGION.
          BENGUIGUI L
          ONDE ELEC 49(8): 889-93; 49(10): 1087-91 (IN FRENCH)
               SCI ABSTR B 73: 3868; A 73: 25903
               198

03-0170   MODEL OF THE FERROELECTRIC MODE.
          BENDIK OG
          SOV PHYS SOLID STATE 14(4): 849-56 (OCT 1972)
               276

03-0171   SOME REMARKS ABOUT ANTIFERROELECTRIC PHASE TRANSITIONS (TO PARAELECTRIC OR
          FERROELECTRIC, THEORY).
          BENGUIGUI L
          PHYS LETT A 33(2): 79-80 (1970)
               224

03-0172 THERMODYNAMIC THEORY OF THE MORPHOTROPIC PHASE TRANSITION TETRAGONAL- RHOMBOHEDRAL IN
THE PEROVSKITE FERROELECTRICS.
BENGUIGUI I
SOLID STATE COMMUN 11: 825-8 (1972)
278

03-0173 COLLINEAR AND NONCOLLINEAR FERRO- AND ANTIFERROELECTRICS. (REPRESENTATION THEORY
APPLIED, 18 REFS)
BERTAUT EF
P7-21 OF EUROPEAN MEETING ON FERROELECTRICITY, 1969, SAARBRUCKEN,
MUSER HE AND PETERSSON J(EDS), WISSENSCH VERLAGSGES, STUTTGART,
1970, 396P.

03-0174 ELECTROOPTICAL EFFECT IN FERROELECTRIC CRYSTALS HAVING A SMEARED PHASE TRANSITION.
BEREZHNOI AA
SOV PHYS SOLID STATE 14(7): 1756-60 (JAN 1973)
282

03-0175 MICROTHEORY OF SPONTANEOUS POLARIZATION AND PHASE TRANSITIONS IN CRYSTALS OF PEROVSKITE
TYPE AND OF ONES WITH TETRAHEDRAL STRUCTURAL UNITS.
BERSUKER IB + VEKHTER BG + MUZALEVSKII AA
J PHYS (PARIS) 33, SUPPL 4, COLLOQ 2: 139-40 (1972)
294

03-0176 STRUCTURAL PHASE TRANSITIONS AND SOFT MODES.
BERRE B + FOSSHEIM K
P255-69 OF NATO ADVAN STUDY INST, GEILO, NORWAY, STRUCTURAL
PHASE TRANSITIONS AND SOFT MODES, SAMUELSEN EJ + ANDERSEN E +
FEDER J(EDS), 1971, UNIVERSITETSFORLAGEN, BOSTON, 1972.

03-0177 VOLTAGES AND CURRENTS DURING DEPOLARIZATION OF A FERROELECTRIC MATERIAL. (THEORY)
BERNARD M + FETIVEAU Y + GRANGE G + PERRIGOT J
J PHYS (PARIS) 33, SUPPL 4, COLLOQ 2: 260-2 (1972)
294

03-0178 DETERMINATION OF THE GROWTH RATE OF FERROELECTRIC CRYSTALS FROM BINARY SYSTEMS.
(THEORY)
BIRMAN BI
P285-94, VOL 1 OF DVORAK V + FOUSKOVA A + GLOGAR P (EDS),
INT MEETING ON FERROELECTRICITY, PRAGUE, 1966, PROC.
INST PHYS CZECH ACAD SCI, PRAGUE, 1966.

03-0179 DYNAMICS OF ORDER DISORDER TYPE FERROELECTRICS AND ANTIFERROELECTRICS
BLINC R + ZEKS B
ADVAN PHYS 21: 693-757 (1972)

03-0180 MAGNETIC RESONANCE IN FERROELECTRICS. (MODEL FOR POTASSIUM DIHYDROGEN PHOSPHATE
PRESENTED, TRIGLYCINE SULFATE AND BARIUM AND STRONTIUM TITANATE DISCUSSED)
BLINC R
P333-57, VOL 2 OF DVORAK V + FOUSKOVA A + GLOGAR P (EDS),
INT MEETING ON FERROELECTRICITY, PRAGUE, 1966, PROC.
INST PHYS CZECH ACAD SCI, PRAGUE, 1966.

03-0181 PROTON LATTICE INTERACTIONS IN HYDROGEN BONDED FERROELECTRIC CRYSTALS. (THEORY, KDP
TYPE)
BLINC R + RIBARIC M
PHYS REV 130: 1816-21 (1963)

03-0182 ELECTRIC FIELD DISTRIBUTIONS IN DIELECTRICS, WITH SPECIAL EMPHASIS ON NEAR SURFACE
REGIONS IN FERROELECTRICS.
BLOOMFIELD PE + LEFKOWITZ I + ARONOFF AD
PHYS REV B 4: 974-87 (1971)

03-0183 PHYSICS OF TRANSITIONS. (THERMODYNAMICS, THEORY, MAGNETISM, FERROELECTRICITY, ORDERED
ALLOYS, SUPERFLUIDITY, SUPERCONDUCTIVITY)
BACCARA N
PRESSES UNIV FRANCE, PARIS, 1970, 127P. (IN FRENCH)
241

03-0184 EXPERIMENTAL CRITERION OF FERROELECTRICITY. (SHARP OR DIFFUSE TRANSITIONS)
BOGDANOV SV
SOV PHYS SOLID STATE 5(3): 591-6 (1963)

3. THEORY

03-0185 NONLINEARITY OF DIELECTRIC POLARIZATION AND FERROELECTRIC PROPERTIES OF MATERIALS.
BOGDANOV SV
SOV PHYS SOLID STATE 5(3): 588-90 (1963)

03-0186 NONLINEAR SADOVSKII EFFECT (THEORY, PROPAGATION OF RADIATION AND FIRST HARMONIC,
CALCULATIONS FOR LITHIUM NIOBATE).
BOKUT BV + SERDYUKOV AN
J APPL SPECTROSC 11: 1059-61 (1969)
198

03-0187 NUCLEAR QUADRUPOLE SPIN LATTICE RELAXATION AND DYNAMICAL PROPERTIES OF FERROELECTRIC
CRYSTALS. (THEORY)
BORSA F + BONERA G + RIGAMONTI A
P295-303 OF EUROPEAN MEETING ON FERROELECTRICITY, 1969, SAARBRUCKEN,
MUSER HE AND PETERSSON J(EDS), WISSENSCH VERLAGSGES, STUTTGART,
1970, 396P.

03-0188 TEMPERATURE DEPENDENCE OF DIELECTRIC CONSTANTS OF CUBIC IONIC COMPOUNDS. (HALIDES AND
OXIDES, RELATION BETWEEN CURIE CONSTANT AND CURIE TEMPERATURE FOR FERROELECTRICS AND
ANTIFERROELECTRICS BARIUM TITANATE, COMPLEX LAYER OXIDES)
BOSMAN AJ + HAVINGA EE
PHYS REV 129: 1593-1600 (1963)

03-0189 THEORETICAL STATIC MODEL FOR BARIUM AND LEAD TITANATES - APPLICATION TO CALCULATING THE
WALL ENERGY AND THE DIELECTRIC CONSTANTS.
BOUILLOT J + MACHET R
P111-17 OF EUROPEAN MEETING ON FERROELECTRICITY, 1969, SAARBRUCKEN,
MUSER HE AND PETERSSON J(EDS), WISSENSCH VERLAGSGES, STUTTGART,
1970, 396P.

03-0190 THERMODYNAMIC THEORY OF DOMAIN WALLS IN FERROELECTRIC MATERIALS WITH THE PEROVSKITE
STRUCTURE.
BULAEVSKII LN
SOV PHYS SOLID STATE 5: 2329-32 (1964)

03-0191 TEMPERATURE DEPENDENCE OF THE SHAPE OF THE DOMAIN WALL IN FERROMAGNETICS AND
FERROELECTRICS.
BULAEVSKII LN + GINZBURG VL
SOV PHYS JETP 18: 530-5 (1964)

03-0192 ENTROPY OF AN ORDER DISORDER TRANSITION. (IN FERROELECTRICS, MECHANISM)
BURFOOT JC
J PHYS (PARIS) 33, SUPPL 4, COLLOQ 2: 79-81 (1972)
294

03-0193 (MACROSCOPIC AND MICROSCOPIC) SYMMETRY OF FERROELECTRIC ENERGY FUNCTIONS. (SOME
COMMENTS)
BURFOOT JC
P140-43, VOL 2 OF DVORAK V + FOUSKOVA A + GLOGAR P (EDS),
INT MEETING ON FERROELECTRICITY, PRAGUE, 1966, PROC.
INST PHYS CZECH ACAD SCI, PRAGUE, 1966.

03-0194 POWDER RAMAN SPECTRA: APPLICATION TO DISPLACIVE FERROELECTRICS. (LATTICE VIBRATIONS)
BURNS G
MATER RES BULL 6: 923-30 (OCT 1971)
247

03-0195 POLARIZATION OF FERROELECTRIC FILM BY BENDING AND BENDING DEFORMATION IN THE ELECTRIC
FIELD. (OBSERVATION REPORTED)
BURSIAN EV + ZAIKOVSKII OI + MAKAROV KV
P416 OF INT MEETING ON FERROELECTRICITY, 2ND, 1969, KYOTO, PROC,
(J PHYS SOC JAP, VOL 28 SUPPL). PHYS SOC JAP, 1970, 459P.

03-0196 PHONON SPECTRUM OF FERROELECTRICS STUDIED BY LIMITATION OF THE CRYSTAL SIZE.
BURSIAN EV + GIRSHBERG YAG
J PHYS (PARIS) 33, SUPPL 4, COLLOQ 2: 69-71 (1972)
294

03-0197 OSCILLATIONS OF FERROELECTRIC BODIES. (WITH FREQUENCY OF APPLIED FIELD CAUSED BY
SHEARING DUE TO POLARIZATION INDUCED STRAIN)
CALLABY DR + FATUZZO E
J APPL PHYS 35(8): 2443-51 (1964)

3. THEORY

03-0198 STATISTICAL MECHANICS OF FERROELECTRIC PHASE TRANSITIONS.
         CAMP WJ
         BULL AMER PHYS SOC 17: 673 (1972)

03-0199 X-RAY DIFFRACTION STUDIES OF ELECTRON POLARIZATION IN FERROELECTRICS.
         CANUT ML
         SOUTHERN ILLINOIS UNIV, 1970, 149P. AD722067

03-0200 CIRCLE THEOREM FOR ICE RULE FERROELECTRIC MODELS.
         CHANG KS + WANG SY + WU FY
         PHYS REV A, GEN PHYS 4: 2324-7 (DEC 1971)

03-0201 INSTABILITY PHENOMENA IN FERROELECTRIC SEMICONDUCTORS. (THEORY)
         CHENSKII EV
         SOV PHYS SOLID STATE 11(3): 534-9 (1969)

03-0202 SINGLE DOMAIN POLARIZATION OF A FERROELECTRIC HAVING A SECOND ORDER PHASE TRANSITION
         (THEORY).
         CHENSKII EV
         SOV PHYS SOLID STATE 12(2): 446-50 (1970)
              206

03-0203 AN ATTEMPT TO CORRELATE THE FERROELECTRIC PROPERTIES OF ABO(3) AND A(0.5)BO(3) TYPE
         PHASES AND THE IONIC MASS OF THE B IONS.
         CHINCHOLKAR VS
         P135-40 OF EUROPEAN MEETING ON FERROELECTRICITY, 1969, SAARBRUCKEN,
         MUSER HE AND PETERSSON J (EDS), WISSENSCH VERLAGSGES, STUTTGART,
         1970, 396P.

03-0204 DEUTERON MAGNETIC RESONANCE STUDY OF SOME CRYSTALS CONTAINING O-D...O BOND. (INCLUDING
         FULLY DEUTERATED AMMONIUM DIHYDROGEN PHOSPHATE)
         CHIBA T
         J CHEM PHYS 41: 1352-8 (1964)

03-0205 PROTON DYNAMICS IN HYDROGEN BONDED FERROELECTRICS. PART-2: APPROXIMATED SOLUTIONS TO
         THE KINETIC EQUATIONS.
         CHOCK DP + DAGONNIER R
         PHYSICA 53(3): 393-411 (1971)

03-0206 PROTON DYNAMICS IN HYDROGEN BONDED FERROELECTRICS ABOVE THE CURIE TEMPERATURE.
         CHOCK DP
         PHYS LETT A 34(1): 17-18 (1971)
              238

03-0207 FERROMAGNETIC RESONANCE LINE WIDTH IN FERROELECTRIC FERROMAGNETS.
         CHUPIS IE + PLYUSHKO NYA
         SOV PHYS SOLID STATE 13: 1890-3 (1972)

03-0208 DYNAMICAL (DIELECTRIC) AND SCATTERING PROPERTIES OF FERROELECTRICS. (THEORY)
         COCHRAN W
         P1-5 OF EUROPEAN MEETING ON FERROELECTRICITY, 1969, SAARBRUCKEN,
         MUSER HE AND PETERSSON J (EDS), WISSENSCH VERLAGSGES, STUTTGART,
         1970, 396P.

03-0209 GENERALIZED SUSCEPTIBILITY AND THE SCATTERING PROPERTIES OF FERROELECTRICS. (ABSTRACT
         ONLY, DISCUSSION)
         COCHRAN W
         P237-8 OF INT MEETING ON FERROELECTRICITY, 2ND, 1969, KYOTO, PROC,
         (J PHYS SOC JAP, VOL 28 SUPPL). PHYS SOC JAP, 1970, 459P.

03-0210 DIELECTRIC RESPONSE OF CRYSTALS. (FERROELECTRIC FLUCTUATIONS)
         COOMBS GJ + COWLEY RA
         J PHYS (PARIS) 33, SUPPL 4, COLLOQ 2: 57-8 (1972)
              294

03-0211 SOLUTION OF AN ELASTIC FERROELECTRIC MODEL.
         COPLAN LA
         PHYS LETT A 35: 309-10 (1971)

3. THEORY

03-0212 DIELECTRIC PROPERTIES OF AN ANHARMONIC CRYSTAL. (EXTENSION OF "SOFT MODE" THEORY)
COWLEY RA
P239-41 OF INT MEETING ON FERROELECTRICITY, 2ND, 1969, KYOTO, PROC,
(J PHYS SOC JAP, VOL 28 SUPPL). PHYS SOC JAP, 1970, 459P.

03-0213 (MICROSCOPIC AND THERMODYNAMIC) THEORY OF FERROELECTRICITY AND ANHARMONIC EFFECTS IN
CRYSTALS. (AND CALCULATIONS BASED ON STRONTIUM TITANATE MODEL)
COWLEY RA
PHIL MAG 11: 673-706 (1965)

03-0214 THERMODYNAMIC PHENOMENOLOGY OF FERROELECTRICITY IN SINGLE CRYSTAL AND CERAMIC SYSTEMS.
CROSS LE
P707-48, PART B OF HENCH LL + DOVE DB(EDS), PHYS OF ELECTRON
CERAMICS, DEKKER, NY, 1972.
262

03-0215 HARMONICS GENERATION IN FERROELECTRIC MEDIA.
DA COSTA RCT + LOBO R
FERROELECTRICS 4: 1-4 (1972)
267

03-0216 DOMAIN STATISTICS AND FERROELECTRIC TRANSIENTS.
DALTON NW + JACOBS JT + SILVERMAN BD
FERROELECTRICS 2: 21-9 (JAN 1971)

03-0217 RELATION BETWEEN THE MAIN ELECTROPHYSICAL AND STRUCTURAL PARAMETERS OF FERROELECTRIC
PIEZOELECTRIC CERAMICS AND METHODS OF THEIR CHANGING. (SPONTANEOUS DEFORMATION RELATED
TO THEIR CHARACTERISTICS)
DANTSIGER AYA + FESENKO EG
P325-7 OF INT MEETING ON FERROELECTRICITY, 2ND, 1969, KYOTO, PROC,
(J PHYS SOC JAP, VOL 28 SUPPL). PHYS SOC JAP, 1970, 459P.

03-0218 PHENOMENOLOGICAL THEORY OF ANTIFERROELECTRICITY AND FERROELECTRICITY APPLIED TO SODIUM
NIOBATE AND THE SYSTEM POTASSIUM NIOBATE- SODIUM NIOBATE.
DARLINGTON CNW
J APPL PHYS 43(12): 4951-6 (1972)
282

03-0219 SPONTANEOUS POLARIZATION NOISE IN POLAR DIELECTRICS. (THEORY, EINSTEIN THERMODYNAMIC
METHOD)
DAVIS L
J APPL PHYS 35: 2004-10 (1964)

03-0220 INTERNAL FIELD IN GENERAL DIPOLE LATTICES. (THEORY APPLIED TO POTASSIUM FERROCYANIDE
TRIHYDRATE)
DE WETLE FW + SCHACHER GE
PHYS REV 137(1): 78-91 (1965)

03-0221 VIBRATIONAL APPROACH TO THE ANHARMONIC LATTICE PROBLEM. (WITH AN APPLICATION TO THE
THEORY OF FERROELECTRIC TRANSITIONS)
DONIACH S
SOLID STATE COMMUN 1: 176-7 (1963)

03-0222 ENERGY GENERATION FROM IMPACT ON PIEZOELECTRIC (OR FERROELECTRIC) MATERIALS.
DOVE RC + BAKER WE + VALATHUR M
KIRTLAND AFB, NEW MEXICO, DEC 1971, 201P. N72-23755
SCI TECH AEROSP REP 10(14): 1924 (1972)
271

03-0223 DIELECTRIC SUSCEPTIBILITY AND THE ORDER OF FERROELECTRIC PHASE TRANSITIONS.
DRAEGERT DA + SINGH S
SOLID STATE COMMUN 9: 595-7 (1971)

03-0224 DYNAMIC SHELL THEORY FOR FERROELECTRIC CERAMICS.
DRUMHELLER DS + KALNINS A
J ACOUST SOC AMER 47(5, PART 2): 1343-53 (MAY 1970)
213

3. THEORY

03-0225   DYNAMIC PIEZOELECTRIC THEORY FOR FERROELECTRIC CERAMIC SHELLS.
          DRUMHELLER DS
          LEHIGH UNIV PHD THESIS, 1969, 168P. N71-25564
              DISS ABSTR INT B 30: 5532 (1970)
              238

03-0226   A NEW TYPE OF SECOND ORDER PHASE TRANSITION DERIVED FROM DEVONSHIRE"S THEORY OF
          FERROELECTRICS.
          DUKEK G + FALK G
          Z PHYS 240(2): 93-9 (1970)
              225

03-0227   LIFETIME OF THE "FERROELECTRIC" PHONON IN SOME PEROVSKITE CRYSTALS (STRONTIUM TITANATE,
          POTASSIUM TANTALATE) AT LOW TEMPERATURES. (THEORY)
          DVORAK V
          P61-8, VOL 1 OF DVORAK V + FOUSKOVA A + GLOGAR P (EDS),
          INT MEETING ON FERROELECTRICITY, PRAGUE, 1966, PROC.
          INST PHYS CZECH ACAD SCI, PRAGUE, 1966.

03-0228   PHENOMENOLOGICAL THEORY OF ULTRASONIC ATTENUATION IN FERROELECTRICS.
          DVORAK V
          CZECH J PHYS B 21(8): 836-45 (1971)
              262

03-0229   PHASE TRANSITION IN KDP AND THE ROLE OF THE DEPOLARIZING ENERGY IN IT. (THEORY)
          DVORAK V
          P252-4 OF INT MEETING ON FERROELECTRICITY, 2ND, 1969, KYOTO, PROC,
          (J PHYS SOC JAP, VOL 28 SUPPL). PHYS SOC JAP, 1970, 459P.

03-0230   SOFT MODE BEHAVIOR IN IMPROPER FERROELECTRICS. (THEORY, KDP)
          DVORAK V + PETZELT J
          PHYS LETT A 35: 209-10 (1971)

03-0231   THERMODYNAMIC THEORY OF THE CUBIC- ORTHORHOMBIC PHASE TRANSITION IN BORACITES.
          DVORAK V
          CZECH J PHYS B 21(12): 1250-61 (1971)
              278

03-0232   THERMODYNAMIC THEORY OF GADOLINIUM MOLYBDATE.
          DVORAK V
          PHYS STATUS SOLIDI B 46: 763-72 (1971)
              244

03-0233   TUNNELING RESULTS ON INDIUM STRONTIUM TITANATE BARRIERS INTERPRETED AS CONFIRMATION OF
          THE THEORY OF TRANSITIONS FROM LARGE TO NEARLY SMALL POLARONS.
          EAGLES DM
          PHYS STATUS SOLIDI B, BASIC RES 48(1): 407-17 (1971)

03-0234   ANHARMONIC MODE MIXING IN IONIC FERROELECTRICS.   (LATTICE DYNAMICS)
          ENGLMAN R + YATOM H
          P41-50, VOL 1 OF DVORAK V + FOUSKOVA A + GLOGAR P (EDS),
          INT MEETING ON FERROELECTRICITY, PRAGUE, 1966, PROC.
          INST PHYS CZECH ACAD SCI, PRAGUE, 1966.

03-0235   MAGNETIC PROPERTIES OF FERRODIELECTRIC CATALYSTS.
          EVDOKIMOV VB + KUZNECOVA MN
          RUSS J PHYS CHEM 44(12): 1742 (1970)
              238

03-0236   THEORY OF X-RAY DIFFRACTION LINE SHAPE EFFECTS DUE TO DOMAIN FORMATION IN
          FERROELECTRICS.
          EVENSON WE + BARNETT JD
          BULL AMER PHYS SOC 16: 415 (1971)

03-0237   ELASTIC ENERGY AND THE FERROELECTRIC PARAELECTRIC TRANSITION.
          EYRAUD L + MALECOT G + RICHARD M + GOUTLE R
          COMPT REND 255: 1581-2 (1962) (IN FRENCH)

03-0238   CRITICAL STATES OF FERROELECTRICS. (THEORY)
           FALK G
           P135-8 OF BATTELLE INST MATER SCI, COLLOQ, 5TH PROC, GENEVA AND
           GSTAAD, SWITZERLAND, CRITICAL PHENOMENA IN ALLOYS, MAGNETS, AND
           SUPERCONDUCTORS, MILLS RE + ASCHER E + JAFFEE RI(EDS),
           MCGRAW-HILL, 1971, 661P.

03-0239   FERROELECTRIC SWITCHING AND THE SIEVERT INTEGRAL.
           FANG PH + STEGUN IA
           J APPL PHYS 34: 284-6 (1963)

03-0240   THERMODYNAMICS OF AN ELASTIC FERROELECTRIC MODEL.
           FAURE P + HINTERMANN A + OHANESSIAN SP
           HELV PHYS ACTA 45: 911 (1972)

03-0241   THEORY OF A STRUCTURAL PHASE TRANSITION IN PEROVSKITE- TYPE CRYSTALS, PART-2,
           INTERACTION WITH ELASTIC STRAIN (STRONTIUM TITANATE AND LANTHANUM ALUMINATE).
           FEDER J + PYTTE E
           PHYS REV B 1: 4803-10 (15 JUN 1970)
                216

03-0242   THEORY OF STRUCTURAL PHASE TRANSITION IN PEROVSKITES.
           FEDER J
           P171-88 OF NATO ADVAN STUDY INST, GEILO, NORWAY, STRUCTURAL
           PHASE TRANSITIONS AND SOFT MODES, SAMUELSEN EJ + ANDERSEN E +
            FEDER J(EDS), 1971, UNIVERSITETSFORLAGEN, BOSTON, 1972.

03-0243   PEROVSKITE FAMILY AND CHEMICAL BOND. (COVALENCY, DIRECTIVITY CONDITIONS FOR ELECTRIC
           ORDERING AND STABILITY)
           FESENKO EG + SHUVAYEV AT
           P64-6 OF INT MEETING ON FERROELECTRICITY, 2ND, 1969,, KYOTO, PROC,
           (J PHYS SOC JAP, VOL 28 SUPPL). PHYS SOC JAP, 1970, 459P.

03-0244   THEORY OF THE MOBILITY OF STRUCTURAL DEFECTS IN ICE.
           FISCHER SF + HOFACKER GL + RATNER MA
           P385-400 OF INT SYMP ON PHYSICS OF ICE, 3RD PROC, 1968,
           RIEHL N + BULLEMER B + ENGELHARDT H(EDS), PLENUM, NY,
           1969, 642P.

03-0245   INTERPRETATION OF TEMPERATURE AUTOSTABILIZATION OF FERROELECTRIC CRYSTAL. (DIELECTRIC
           HEATING AND POLARIZATION REVERSAL, COMPLEX PERMITTIVITY CALCULATION)
           FOUSEK J
           J APPL PHYS 36(2): 588-98 (1965)

03-0246   (FACTORS INFIUENCING) TYPE AND ORIENTATION OF THE EQUILIBRIUM DOMAIN STRUCTURES IN
           FERROELECTRICS.
           FOUSEK J + SAFRANKOVA M
           P59-68, VOL 2 OF DVORAK V + FOUSKOVA A + GLOGAR P (EDS),
           INT MEETING ON FERROELECTRICITY, PRAGUE, 1966, PROC.
           INST PHYS CZECH ACAD SCI, PRAGUE, 1966.

03-0247   PERMISSIBLE DOMAIN WALLS IN FERROELECTRIC SPECIES (THEORY AND TABULATION)
           FOUSEK J
           CZECH J PHYS B 21: 955-68 (1971)
                283

03-0248   RECENT PROBLEMS IN DOMAIN STATICS AND DYNAMICS. (56 REFS)
           FOUSEK J
           P147-65 OF EUROPEAN MEETING ON FERROELECTRICITY, 1969, SAARBRUCKEN,
           MUSER HE AND PETERSSON J(EDS), WISSENSCH VERLAGSGES, STUTTGART,
           1970, 396P.

           OPTICAL HARMONICS AND NONLINEAR PHENOMENA. (THEORY, REVIEW)
           FRANKEN PA + WARD JF
           REV MOD PHYS 35: 23-39 (1963)

03-0249   INVESTIGATION OF POLARIZATION NONLINEARITY NEAR THE PHASE TRANSITION IN PEROVSKITE LIKE
           FERROELECTRICS, USING AN ANHARMONIC OSCILLATOR MODEL.
           FRITSBERG VYA
           P7-21 OF FAZOVYE PEREKHODY V SEGNETOELEKTRIKAKH (PHASE
           TRANSFORMATIONS IN FERROELECTRICS), FRITSBERG VYA + ROLOV BN +
           KRUCHAN YY(EDS), IZDATELSTVO ZINATNE RIGA, 1971, 206P. (IN RUSSIAN)

3. THEORY

03-0250   PARTICULAR PROPERTIES OF PHASE TRANSFORMATION IN FERROELECTRIC SOLID SOLUTIONS. (THEORY
            OF COMPOSITION DEPENDENCE OF PHASE TRANSITION)
            FRITZBERG WJ
            P163-71, VOL 1 OF DVORAK V + FOUSKOVA A + GLOGAR P (EDS),
            INT MEETING ON FERROELECTRICITY, PRAGUE, 1966, PROC.
            INST PHYS CZECH ACAD SCI, PRAGUE, 1966.

03-0251   THEORY OF SPACE CHARGE LIMITED CURRENTS IN FERROELECTRICS.
            FRIDKIN VM + KREHER K
            PHYS STATUS SOLIDI A 2: 281-5 (1970)

03-0252   THEORY OF SUPEREXCHANGE REACTION. PART-1: POTASSIUM MANGANESE TRIFLUORIDE.
            FUCHIKAMI N
            J PHYS SOC JAP 28: 871-90 (1970)

03-0253   INFLUENCE OF LONG RANGE DIPOLE DIPOLE FORCES ON ATTENUATION AND VELOCITY OF ULTRASOUND
            IN FERROELECTRICS NEAR SECOND ORDER TRANSITION POINT.
            GEGUZINA SYA + KRIVOGLAZ MA
            SOV PHYS SOLID STATE 9(11): 2441-7 (1968)
               219

03-0254   PHENOMENOLOGICAL THEORY OF PHASE TRANSITIONS IN SODIUM NITRITE.
            GESI K
            J PHYS SOC JAP 20(10): 1764-72 (1965)

03-0255   ELECTRON MICROSCOPE TRANSMISSION IMAGES OF COHERENT DOMAIN BOUNDARIES. PART-1:
            DYNAMICAL THEORY. PART-2: OBSERVATIONS. (FERROELECTRIC DOMAINS IN BARIUM TITANATE)
            GEVERS R + DELAVIGNETTE P + BLANK H + AMELINCKX S
            PHYS STATUS SOLIDI 4: 383-410; 5: 595-633 (1964)

03-0256   PHASE TRANSITIONS IN A MODEL FERROELECTRIC.
            GILLIS NS
            PHYS REV B SOLID STATE 5(5): 1925-32 (1 MAR 1972)
               260

03-0257   PHASE TRANSITIONS IN A MODEL FERROELECTRIC.
            GILLIS NS + KOEHLER TR
            PHYS REV B SOLID STATE 5: 1925-32 (MAR 1, 1972)

03-0258   ANALYTICAL PROPERTIES OF A CLASS OF PLANAR FERROELECTRIC MODELS. (THEORY, KDP TYPES)
            GLASSER ML
            P165-75 OF BATTELLE INST MATER SCI, COLLOQ, 5TH PROC, GENEVA AND
            GSTAAD, SWITZERLAND, CRITICAL PHENOMENA IN ALLOYS, MAGNETS, AND
            SUPERCONDUCTORS, MILLS RE + ASCHER E + JAFFEE RI(EDS),
            MCGRAW-HILL, 1971, 661P.

03-0259   CHARGE MECHANISM IN FERROELECTRICS. (LEAD ZIRCONATE TITANATE, STOICHIOMETRY EFFECTS)
            GLOWER DD + BURT JG
            OHIO STATE UNIV, 1970, 6P. AD709694
               SCI ABSTR A 74: 21777
               238

03-0260   NEUTRON SCATTERING BY A FERROELECTRIC SOLID SOLUTION. (THEORY, APPLIED TO BARIUM
            STRONTIUM TITANATE AND POTASSIUM NIOBATE TANTALATE)
            GLOGAR P
            P92-9, VOL 1 OF DVORAK V + FOUSKOVA A + GLOGAR P (EDS),
            INT MEETING ON FERROELECTRICITY, PRAGUE, 1966, PROC.
            INST PHYS CZECH ACAD SCI, PRAGUE, 1966.

03-0261   THEORETICAL DYNAMIC STUDY OF CUBIC PHASE BARIUM TITANATE.
            GNININVI L + BOUILLOT J
            J PHYS (PARIS) 33(11-12): 1049-58 (1972) (IN FRENCH)
               SCI ABSTR A 76: 20826 (1973)
               285

03-0262   FERROELECTRIC FREE ENERGY EXPANSION COEFFICIENTS FROM DOUBLE HYSTERESIS LOOPS.
            GONZALO JA + RIVERA JM
            FERROELECTRICS 2: 31-5 (JAN 1971)

03-0263   STATISTICAL THEORY FOR FERROELECTRICITY IN TRIGLYCINE SULFATE.
            GONZALO JA + LOPEZ-ALONSO JR
            J PHYS CHEM SOLIDS 25(3): 303-10 (1964)

3. THEORY

03-0264   CRITICAL POINT INDICES OF FERROELECTRIC AND FERROMAGNETIC PHASE TRANSITIONS.
          GOPAL ESR + GOVINDARAJAN K
          CZECH J PHYS B 21(1): 4-6 (1971)
              238

03-0265   EFFECT OF ELASTIC STRESSES ON PARAMETERS OF SOLID SOLUTIONS WITH FERROELECTRIC
          PROPERTIES.
          GRANOVSKII VG
          IZV VUZ FIZ 1964(5): 131-4 (IN RUSSIAN)
              CHEM ABSTR 62: 4742C (1965)

03-0266   MAGNETIC RESONANCE AND THE PROBLEM OF THE DEFINITION OF ANTIFERROELECTRICITY.
          GRANICHER H
          J PHYS (PARIS) 33, SUPPL 4, COLLOQ 2: 187-8 (1972)
              294

03-0267   THEORY OF ELECTROSTRICTION IN CUBIC CRYSTALS WITH APPLICATION TO BARIUM TITANATE.
          GRINDLAY J
          P433-41, VOL 1 OF DVORAK V + FOUSKOVA A + GLOGAR P (EDS),
          INT MEETING ON FERROELECTRICITY, PRAGUE, 1966, PROC.
          INST PHYS CZECH ACAD SCI, PRAGUE, 1966.

03-0268   ANOMALIES IN THE ELECTRICAL CONDUCTIVITY OF A FERROELECTRIC SEMICONDUCTOR NEAR THE
          CUBIC POINT.
          GUBANOV AI + SHUR MS
          SOV PHYS SOLID STATE 12(2): 517-18 (1970)
              206

03-0269   THEORY OF PHASE TRANSITIONS IN BORACITES.
          GUFAN YUM + SAKHNENKO VP
          SOV PHYS SOLID STATE 14(7): 1660-5 (JAN 1973)
              282

03-0270   PHENOMENOLOGICAL THEORY OF FERROELECTRIC AND ANTIFERROELECTRIC CRYSTALS.
          GUSEINOV NG
          VEST AKAD NAVUK BELARUSS SSR, SER FIZ TEKH MAT NAUK 3-4: 135-40)
          (1971) (IN RUSSIAN)

03-0271   EFFECT OF THE SURFACE ON POLARIZATION OF A DIELECTRIC (THEORY).
          HARBRINK B
          Z PHYS 232(2): 108-25 (1970)
              216

03-0272   DIELECTRIC RELAXATION PROCESSES NEAR THE CRITICAL POINT IN FERROELECTRICS.
          HATTA I
          KOTAI BUTSURI 6: 14-26 (1971)

03-0273   DOMAIN WALL MOTION IN FERROELECTRIC SWITCHING.
          HAYASHI M
          MEM FAC ENG, NAGOYA UNIV 24: 216-69 (1972)

03-0274   KINETICS OF DOMAIN WALL MOTION IN FERROELECTRIC SWITCHING. PART-1: GENERAL FORMULATION.
          HAYASHI M
          J PHYS SOC JAP 33(3): 616-28 (SEP 1972)
              278

03-0275   ACOUSTIC ATTENUATION AND AMPLIFICATION IN PIEZOELECTRIC AND FERROELECTRIC
          SEMICONDUCTORS.
          HERVOUET C
          PHYS LETT A 28(5): 311-12 (1968)
              146

03-0276   ABRUPT (STEPWISE) CHANGE IN PROBABILITY OF THE MOSSBAUER EFFECT AT THE PHASE TRANSITION
          IN FERROELECTRICS.
          HIEN PZ + VISKOV AS + SHPINEL VS + VENEVTSEV YUN
          SOV PHYS JETP 17: 1465-6 (1963)

03-0277   COMPILATION OF CONVERSION FORMULAS INTERRELATING THE LINEAR ELASTIC AND ELECTROELASTIC
          EQUATIONS OF STATE: ISOTROPIC AND ORTHOTROPIC (FERROELECTRIC CERAMIC) CASES.
          HOLLAND R
          SANDIA LABS, ALBUQUERQUE, 1967, 30P. SC-RR-67-157
              213

3. THEORY

03-0278 PIEZOELECTRIC EFFECTS IN FERROELECTRIC CERAMICS (AND THE THEORY OF PIEZOELECTRIC ENERGY
CONVERSION).
HOLLAND R
IEEE SPECTRUM 7(4): 67-77 (1970)
213

03-0279 QUANTUM FIELD THEORETICAL TREATMENT OF A DYNAMICAL MODEL OF THE FERROELECTRIC KDP.
PART-1. PART-2.
HOUSTON GD + BOLTON HC
J PHYS C SOLID STATE PHYS 4: 2097-108, 2894-902 (1971)

03-0280 EFFECT OF THE CHANGE OF ELECTROSTATIC CONSTRAINTS ON THE PHASE TRANSITIONS OF SOME
PHENOMENOLOGICAL MODELS OF FERROELECTRICS.
HUANG CC + GRINDLAY J
CAN J PHYS 48: 847-51 (1 APR 1970)
195

03-0281 CRITICAL DYNAMICS OF THE ORDER DISORDER TRANSFORMATION IN FERROELECTRICS. (THEORY)
HUBER DL
PHYS REV B SOLID STATE 6(9): 3379-84 (1 NOV 1972)
278

03-0282 PHENOMENOLOGICAL THEORY OF FERROELECTRIC POLARIZATION REVERSAL.
HUSIMI K
P337-9 OF INT MEETING ON FERROELECTRICITY, 2ND, 1969, KYOTO, PROC.
(J PHYS SOC JAP, VOL 28 SUPPL). PHYS SOC JAP, 1970, 459P.

03-0283 THE BEHAVIORS OF THE MICRODOMAINS ON THE DIFFUSE PHASE TRANSITION FERROELECTRICS.
ICHINOSE N + YOKOMIZO Y + TAKAHASHI T
ACTA CRYSTALLOGR A 28: S187 (1972)

03-0284 RESONANCE OF A LOW FREQUENCY SPIN WAVE WITH AN ACOUSTIC OR ELECTROMAGNETIC WAVE IN A
FERROELECTRIC ANTIFERROMAGNETIC MATERIAL.
IOFFE IV + KAZAKOV AL
SOV PHYS SOLID STATE 13(7): 1806 (JAN 1972)
254

03-0285 DYNAMIC MODEL OF THE SLATER KDP FERROELECTRIC.
IRVINE RD
AUSTRAL J PHYS 23: 833-45 (1970)
236

03-0286 A THEORY OF THE PHASE TRANSITION IN AMMONIUM DIHYDROGEN PHOSPHATE.
ISHIBASHI Y + OHYA S + TAKAGI Y
J PHYS SOC JAP 33(6): 1545-50 (DEC 1972)
282

03-0287 COMPUTER EXPERIMENT ON LINEAR CHAIN OF ATOMS LYING IN DOUBLE MINIMUM POTENTIAL. (BARIUM
TITANATE)
ISHIBASHI Y + TAKAGI Y
J PHYS SOC JAP 33(1): 1-5 (JUL 1972)
278

03-0288 DISTRIBUTION OF RELAXATION TIMES IN SOME FERROELECTRICS.
ISHIBASHI Y + TAKAGI Y
J PHYS SOC JAP 31: 54-5 (JUL 1971)

03-0289 THE ROLE OF LORENTZ FACTORS IN ROCHELLE SALT. (THEORY)
ISHIBASHI Y + TAKAGI Y
J PHYS SOC JAP 32(3): 723-8 (1972)
278

03-0290 X-RAY DIFFRACTOMETRIC DETERMINATION OF THE CURIE TEMPERATURE AND TEMPERATURE DEPENDENCE
OF SPONTANEOUS POLARIZATION OF HEXAGONAL (RHOMBOHEDRAL) FERROELECTRICS.
ISMAILZADE IH
PHYS STATUS SOLIDI B 46: K39-41 (1971)

03-0291 CAUSES OF PHASE TRANSITION BROADENING (DIFFUSNESS) AND THE NATURE OF DIELECTRIC
POLARIZATION RELAXATION IN SOME FERROELECTRICS (SOLID SOLUTIONS AND MIXED CATION
COMPOUNDS, KANZIG REGIONS, PEROVSKITES)
ISUPOV VA
SOV PHYS SOLID STATE 5: 136-40 (1963)

3. THEORY

03-0292   SCATTERING OF CONDUCTION ELECTRONS BY TRANSVERSE AND LONGITUDINAL OPTICAL PHONONS IN
          IONIC FERROELECTRIC SEMICONDUCTORS.
          ITSKOVSKII MA
          BULL ACAD SCI USSR, PHYS SER 35(9): 1622-8 (1971)

03-0293   CONCERNING THE CRITERION OF FERROELECTRICITY IN THE MICROSCOPIC THEORY.
          IVANCHIK II
          SOV PHYS SOLID STATE 4: 2369-71 (1963)

03-0294   EFFECT OF AUTOMATIC TEMPERATURE STABILIZATION IN FERROELECTRIC OSCILLATORY SYSTEMS.
          IVANOV IV
          FERROELECTRICS 4: 29-37 (1972)
              267

03-0295   MACROSO MACROSCOPIC THEORY OF FERROELECTRICS.
          IVANCHIK II
          SOV PHYS SOLID STATE 3: 2705-12 (1962)

03-0296   ASYMMETRIC MINOR LOOP AND TRANSIENT OPEN LOOP COMPUTATIONS FOR FERROELECTRICS.
          JACOBS JT + SILVERMAN BD
          FERROELECTRICS 1(4): 265-7 (OCT 1970)
              234

03-0297   INELASTIC SCATTERING CROSS SECTION OF NEUTRONS IN FERROELECTRICS.
          JAISWAL VK + SHARMA PK
          PHYSICA 44: 69-80 (1969)
              178

03-0298   ANOMALIES NEAR FERROELECTRIC TRANSITIONS FROM THE POINT OF VIEW OF THE THEORY OF
          CRITICAL POINTS.
          JANOVEC V
          P103-10 OF EUROPEAN MEETING ON FERROELECTRICITY, 1969, SAARBRUCKEN,
          MUSER HE AND PETERSSON J(EDS), WISSENSCH VERLAGSGES, STUTTGART,
          1970, 396P.

03-0299   GENERALIZATION OF THE PIPPARD-GARLAND RELATIONS AND THEIR APPLICATION TO FERROELECTRIC
          (LAMBDA) PHASE TRANSITIONS.
          JANOVEC V
          P144-50, VOL 1 OF DVORAK V + FOUSKOVA A + GLOGAR P (EDS),
          INT MEETING ON FERROELECTRICITY, PRAGUE, 1966, PROC.
          INST PHYS CZECH ACAD SCI, PRAGUE, 1966.

03-0300   INFLUENCE OF THE SHAPE OF DOMAINS ON THE FERROELECTRIC HYSTERESIS LOOP.
          JANTA J
          FERROELECTRICS 2(4): 299-302 (1971)

03-0301   SIDEWAYS DOMAIN WALL MOTION AND (MECHANISM OF THE) HYSTERESIS LOOP OF FERROELECTRICS.
          JANTA J
          P340-2 OF INT MEETING ON FERROELECTRICITY, 2ND, 1969,, KYOTO, PROC,
          (J PHYS SOC JAP, VOL 28 SUPPL). PHYS SOC JAP, 1970, 459P.

03-0302   THERMODYNAMIC THEORY OF SOLID SOLUTIONS OF ISOMORPHOUS FERROELECTRICS.
          JANOVEC V
          P172-5, VOL 1 OF DVORAK V + FOUSKOVA A + GLOGAR P (EDS),
          INT MEETING ON FERROELECTRICITY, PRAGUE, 1966, PROC.
          INST PHYS CZECH ACAD SCI, PRAGUE, 1966.

03-0303   THERMODYNAMIC CLASSIFICATION OF SECOND ORDER FERROELECTRIC PHASE TRANSITIONS.
          JANOVEC V
          P129-33 OF EUROPEAN MEETING ON FERROELECTRICITY, 1969, SAARBRUCKEN,
          MUSER HE AND PETERSSON J(EDS), WISSENSCH VERLAGSGES, STUTTGART,
          1970, 396P.

03-0304   OPTICAL MODES OF THE PEROVSKITE STRUCTURE. (THEORY, STRONTIUM TITANATE DATA)
          JOSEPH RI + SILVERMAN BD
          J PHYS CHEM SOLIDS 24: 1349-55 (1963)

03-0305   POLARONS IN ANISOTROPIC ENERGY BANDS (THEORY, CALCULATIONS FOR STRONTIUM TITANATE).
          KAHN AH
          PHYS REV 172(3): 813-15 (1968)
              197

3. THEORY

03-0306 POLARIZATION WAVES IN SOLIDS. (INCLUDING DEVELOPMENTS IN THEORY OF FERROELECTRIC
        PROPERTIES OF CRYSTALS, 58 REFS)
        KAHANE A
        J PHYS (PARIS) 24: 394-404 (1963) (IN FRENCH)
            CHEM ABSTR 59: 12265A (1963)

03-0307 ELASTOOPTICAL EFFECT IN FERROELECTRIC WITH DIFFUSE PHASE TRANSITION.
        KAMZINA LS + KRAINIK NN + GENE VV + MYLNIKOVA IE
        BULL ACAD SCI USSR, PHYS SER 35(9): 1693-5 (1971)

03-0308 DISTRIBUTION OF ZEROES OF THE PARTITION FUNCTION FOR THE SLATER MODEL OF
        FERROELECTRICITY (KDP MODEL, ANTIFERROELECTRICITY F MODEL, WU KDP MODEL, CALCULATIONS).
        KATSURA S + ABE Y + OHKOHCHI K
        J PHYS SOC JAP 29(4): 845-50 (1970)
            216

03-0309 KINETIC EQUATION FOR QUADRUPOLE RESONANCE IN FERROELECTRICS NEAR THE PHASE TRANSITION
        POINT.
        KESSEL AR + KORCHEMKIN MA
        SOV PHYS SOLID STATE 13(10): 2522-4 (APR 1972)
            263

03-0310 MACROSCOPIC ANALOG OF THE SPIN- ECHO EFFECT IN POLYCRYSTALLINE FERROELECTRICS. (THEORY)
        KESSEL AR + SAFIN IA + FOLDMAN AM
        SOV PHYS SOLID STATE 12: 2488-9 (1971)

03-0311 NATURE OF PHASE TRANSITIONS IN FERROELECTRIC AND ANTIFERROELECTRIC MATERIALS.
        KESSEL AR + KORCHEMKIN MA
        PHYS LETT A 37: 95-6 (1971)
        SOV PHYS SOLID STATE 13(11): 2858-9 (MAY 1972)
            265

03-0312 PIEZOELECTRIC EFFECT ON THE REFLECTION OF A TRANSVERSE ACOUSTIC WAVE FROM DOMAIN
        BOUNDARIES IN FERROELECTRICS.
        KESSENIKH GG + SANNIKOV DG + SHUVALOV LA
        SOV PHYS CRYSTALLOGR 17(2): 292-6 (SEP-OCT 1972)
            275

03-0313 REFLECTION AND REFRACTION OF TRANSVERSE SOUND WAVES AT DOMAIN BOUNDARIES IN
        FERROELECTRICS. (THEORY)
        KESSENIKH GG + SANNIKOV DG + SHUVALOV LA
        SOV PHYS CRYSTALLOGR 15(5): 888-92 (1971)
            232

03-0314 REFLECTION AND REFRACTION OF QUASI LONGITUDINAL AND QUASI TRANSVERSE SOUND WAVES BY 180
        DOMAIN BOUNDARIES IN FERROELECTRICS.
        KESSENIKH GG + SANNIKOV DG + SHUVALOV LA
        SOV PHYS CRYSTALLOGR 16: 287-91 (SEP-OCT 1971)

03-0315 SURFACE ELASTOPOLARIZATION WAVES AT DOMAIN BOUNDARIES IN A FERROELECTRIC.
        KESSENIKH GG + LYUBIMOV VN + SANNIKOV DG
        SOV PHYS CRYSTALLOGR 17: 512-14 (1972)

03-0316 DIELECTRIC, ELASTIC AND MAGNETIC PROPERTY, SYMMETRY, AND PHASE TRANSITION
        REPRESENTATION. (18 REFS)
        KEVE ET + ABRAHAMS SC
        FERROELECTRICS 1(4): 243-6 (OCT 1970)
            234

03-0317 ADMIXTURE OF THE 4T1G STATE TO THE 6A1G(6S) STATE OF THE MANGANESE(2+) ION IN A CUBIC
        ANTIFERRODIELECTRIC AN EXTERNAL MAGNETIC FIELD.
        KHARKYANEN VN + PETROV EG
        PHYS STATUS SOLIDI B, BASIC RES 52(2): K61-4 (1972)

03-0318 CONNECTION BETWEEN THE STARTING FIELD FOR PARKHAUSEN STEPS AND THE CRITICAL
        POLARIZATION AND REPOLARIZATION FIELDS IN FERROELECTRICS
        KHARITONOV YUN + KOMLYAKOVA NS
        IZV VUZ FIZ 1970(3): 130-1 (IN RUSSIAN)

03-0319 COMPLEX PHASE TRANSITIONS IN CRYSTALS (APPLICATION TO SINGLE DOMAIN FERROELECTRICS
        BELOW CURIE POINT).
        KHMELNITSKII DE
        SOV PHYS JETP 34(5): 1125-31 (1972)

03-0320   LOW TEMPERATURE DISPLACEMENT TYPE PHASE TRANSITION IN CRYSTALS. (THEORY,
          FERROELECTRICS, PIEZOELECTRICS)
          KHMELNITSKII DE + SHNEERSON VL
          SOV PHYS SOLID STATE 13: 687-94 (1971)

03-0321   THEORY OF DOMAINLESS, NONHOMOGENEOUS, SPONTANEOUS POLARIZATION OF FERROELECTRIC PLATES
          NEAR THE CURIE POINT.
          KHOLODENKO LP
          SOV PHYS SOLID STATE 5: 660-7 (1963)

03-0322   THERMODYNAMICAL THEORY OF BARIUM TITANATE TYPE FERROELECTRICS.
          KHOLODENKO LP
          IZDATELSTVO, RIGA, USSR, 1971, 227P. (IN RUSSIAN)

03-0323   ON SURFACE STATE AND POLARIZATION REVERSAL OF FERROELECTRIC CRYSTALS.
          KINASE W + SUZUKI K + KAZAMI A
          PROGR THEOR PHYS 30: 266-8 (1963)

03-0324   90 DEGREE TYPE DOMAIN WALL OF BARIUM TITANATE CRYSTAL. (THEORY, CALCULATIONS)
          KINASE W + OKAYAMA H + YOSHIKAWA A + TAGUCHI N
          P383-5 OF INT MEETING ON FERROELECTRICITY, 2ND, 1969,, KYOTO, PROC,
          (J PHYS SOC JAP, VOL 28 SUPPL). PHYS SOC JAP, 1970, 459P.

03-0325   STATISTICAL AND ELECTRONIC PROBLEMS IN FERROELECTRICITY. (THEORY OF ENERGY STATES
          INVOLVED)
          KINASE W
          P29-40, VOL 1 OF DVORAK V + FOUSKOVA A + GLOGAR P (EDS),
          INT MEETING ON FERROELECTRICITY, PRAGUE, 1966, PROC.
          INST PHYS CZECH ACAD SCI, PRAGUE, 1966. (548.85/I61) 133246-7

03-0326   THEORY OF POLARIZATION REVERSAL OF BARIUM TITANATE CRYSTAL.
          KINASE W + KAZAMI A + YOKOYAMA S
          P343-5 OF INT MEETING ON FERROELECTRICITY, 2ND, 1969,, KYOTO, PROC,
          (J PHYS SOC JAP, VOL 28 SUPPL). PHYS SOC JAP, 1970, 459P.

03-0327   CURIE POINT SHIFT WITH APPLIED FIELD IN FERROELECTRICS POSSESSING FIRST ORDER
          TRANSITION. (THEORY, CALCULATION)
          KISZKOWSKI P + SZWEYCER H
          P151-3, VOL 1 OF DVORAK V + FOUSKOVA A + GLOGAR P (EDS),
          INT MEETING ON FERROELECTRICITY, PRAGUE, 1966, PROC.
          INST PHYS CZECH ACAD SCI, PRAGUE, 1966.

03-0328   ELASTIC GIBBS FUNCTION FOR A FERROELECTRIC CRYSTAL.
          KISZKOWSKI P + SZWEYCER H
          BULL SOC AMIS SCI LETT POZNAN, SER B 21: 57-9 (1970)
              234

03-0329   INDUCED POLARIZATION AND A ONE POLARIZATION MODEL OF FERROELECTRICITY.
          KISZKOWSKI P + SZYWYCER H
          BULL SOC AMIS SCI LETT POZNAN, SER B 21: 71-5 (1970)
              252

03-0330   (EFFECT OF TEMPERATURE ON) QUASI SYMMETIRICAL HARD IONS (PHOSPHATE, TITANATE, SULFATE,
          FLUOBERYLLATE, ETC) OF SOME MONOCLINIC CRYSTALS IN THE FERROELECTRIC TRANSITION
          MECHANISM.
          KISLOVSKII LD + SHUVALOV LA + IVANOV NR + GALLANOV EK
          P205-13, VOL 1 OF DVORAK V + FOUSKOVA A + GLOGAR P (EDS),
          INT MEETING ON FERROELECTRICITY, PRAGUE, 1966, PROC.
          INST PHYS CZECH ACAD SCI, PRAGUE, 1966.

03-0331   QUASISTATIC GENERATION OF HARMONICS IN A FERROELECTRIC CRYSTAL WITH A SECOND ORDER
          TRANSITION ABOVE THE CURIE POINT.
          KISZKOWSKI P + DANIELEWICZ-FERCHMIN I
          ELECTRON TECHNOL 5: 39-47 (1972)

03-0332   TWO POLARIZATION MODEL THEORY OF FERROELECTRICITY.
          KISZKOWSKI P + SZWEYCER H
          BULL SOC AMIS SCI LETT POZNAN, SER B 21: 65-70 (1970)
              234

3. THEORY

03-0333   SIMULTANEOUS MEASUREMENT OF PYROELECTRIC COEFFICIENT AND DIELECTRIC CONSTANT OF FERROELECTRICS.
        KLADKEVICH MD + KREMENCHUGSKII LS + MALNEV AF
        UKR FIZ ZH 13(4): 629-32 (1968) (IN RUSSIAN)
           141

03-0334   THEORY OF RAYLEIGH AND BRILLOUIN SCATTERING NEAR THE PHASE TRANSITION.
        KLEIN R
        J PHYS (PARIS) 33, SUPPL 4, COLLOQ 2: 11-14 (1972)
           294

03-0335   PHASE TRANSFORMATIONS OF IMPROPER FERROELECTRICS.
        KOBAYASHI J
        J CRYSTALLOGR SOC JAP 14: 385-97 (1972) (IN JAPANESE)

03-0336   POLAR- POLAR AND POLAR- NEUTRAL CRYSTAL STRUCTURES. (CLASSIFICATION OF ANISOTROPIC FERROELECTRIC AND ANTIFERROELECTRIC STRUCTURES)
        KOPTSIK VA
        SOV PHYS CRYSTALLOGR 8: 249-54 (1963)

03-0337   BARKHAUSEN EFFECT IN FERROELECTRICS UNDER CONTINUOUS X-RAY DIFFRACTION.
        KOMLYAKOVA NS + KHARITONOV YUN
        IZV VUZ FIZ 1970(2): 158-60 (IN RUSSIAN)

03-0338   RANDOM PHASE APPROXIMATION FOR THE BLINC-DE GENNES MODEL FOR HYDROGEN BONDED FERROELECTRICS.
        KONWENT H
        ACTA PHYS POLON A 41(6): 717-25 (1972)
           288

03-0339   SOME ADDITIONAL FEATURES OF FERROELECTRIC PHASE TRANSITIONS IN VIBRONIC MODEL.
        KONSIN PI + KRISTOFFEL NN
        TOIM EESTI NSV TEAD AKAD 20(1): 37-47 (1971) (IN RUSSIAN)
           SCI ABSTR A 75: 16278 (1972)
           264

03-0340   MICROSCOPIC THEORY OF STATIC DOMAIN STRUCTURES OF FERROELECTRIC CRYSTALS. (CRYSTALLOGRAPHIC VIEWPOINT)
        KOPTSIK VA
        P20-30, VOL 2 OF DVORAK V + FOUSKOVA A + GLOGAR P (EDS),
        INT MEETING ON FERROELECTRICITY, PRAGUE, 1966, PROC.
        INST PHYS CZECH ACAD SCI, PRAGUE, 1966.

03-0341   OBSERVATION OF DOMAIN STRUCTURE IN LOW TEMPERATURE FERROELECTRICS BY SOLID DEW TECHNIQUE.
        KOPTSIK VA + TOSHEV SD
        BULL ACAD SCI USSR, PHYS SER 29(6): 960-4 (1965)

03-0342   MECHANICAL MODEL OF LONG WAVELENGTH VIBRATIONS IN A (SINGLE DOMAIN) FERROELECTRIC TYPE CRYSTAL.
        KOSEVICH AM + BOGACHEK EN
        UKR FIZ ZH 15(3): 477-86 (1970)
           226

03-0343   CRITICAL PHENOMENA IN ANTIFERROELECTRICS. (EQUATION OF STATE)
        KOZLOVSKII VKH
        IZV VUZ FIZ 1972(9): 27-31 (IN RUSSIAN)
           292

03-0344   DYNAMICAL THEORY OF RIGID LATTICES OF ANTIFERROELECTRICS.
        KOZLOVSKII VKH
        SOV PHYS CRYSTALLOGR 8: 659-64 (1964)

03-0345   NONSTATIONARY POLARIZATION IN THE ANHARMONIC VIBRATOR MODEL. (FERROELECTRICS)
        KOZLOVSKII VKH
        BULL ACAD SCI USSR, PHYS SER 35(9): 1619-22 (1971)

03-0346   PHASE TRANSFORMATIONS IN FERROELECTRIC CRYSTAL IN PRESENCE OF DOMAIN WALLS. (THEORY)
        KOZLOVSKII VKH
        BULL ACAD SCI USSR, PHYS SER 29(6): 888-92 (1965)
           219

3. THEORY

03-0347 QUANTUM EFFECTS IN FERROELECTRIC MATERIALS WITH HYDROGEN BONDS.
         KOZLOVSKII VKH
         SOV PHYS SOLID STATE 5: 2411-15 (1964)

03-0348 SPACE CHARGE LIMITED CURRENTS IN A METAL FERROELECTRIC METAL SYSTEM (THEORY,.
         KRAPIVIN VF + CHENSKII EV
         SOV PHYS SOLID STATE 12(2): 454-8 (1970)
              206

03-0349 THEORY OF VIBRONIC PHASE TRANSFORMATIONS IN WIDE GAP FERROELECTRICS.
         KRISTOFFEL NN + KONSIN PI
         SOV PHYS SOLID STATE 13(9): 2113-18 (MAR 1972)
              261

03-0350 THERMAL EXPANSION ANOMALIES AND THE DIRECTION OF SPONTANEOUS POLARIZATION IN
         FERROELECTRICS. (ABSTRACT ONLY)
         KRISHNAN RS + NARAYANAN PS + DEVANARAYANAN S
         P166 OF INT MEETING ON FERROELECTRICITY, 2ND, 1969,, KYOTO, PROC,
         (J PHYS SOC JAP, VOL 28 SUPPL). PHYS SOC JAP, 1970, 459P.

03-0351 THEORY OF FERROELECTRIC PHASE TRANSITIONS WITH VIBRONIC MECHANISM.
         KRISTOFFEL NN + KONSIN PI
         ANN UNIV TURKU SER A1 146: 3-14 (1971)
              SCI ABSTR 74: 59095 (1971)
              254

03-0352 DIFFUSION POTENTIALS IN BARIUM TITANATE AND THE THEORY OF PTC MATERIALS.
         KULWICKI BM + PURDES AJ
         FERROELECTRICS 1(4): 253-63; 2: 176 (1970-1)

03-0353 ACOUSTOELECTRIC INTERACTION IN FERROELECTRIC SUBSTANCES.
         KUNIGELIS VF + SAMULIONIS VI + GARSHKA EP
         BULL ACAD SCI USSR, PHYS SER 35(5): 843-5 (1971)
              253

03-0354 FERRO- AND ANTIFERROELECTRIC COUPLING AND DIELECTRIC RELAXATION.
         LAJZEROWICZ J
         DIELECTRICS 1(3): 150-8 (1963)
              SCI ABSTR A 67: 10104 (1964)

03-0355 FIRST AND SECOND ORDER PHASE TRANSITIONS IN FERROELECTRICS: A SELF CONSISTENT
         POLARIZATION SCHEME FOR ANHARMONIC POTENTIALS. (THEORY)
         LAO BY
         BULL AMER PHYS SOC 16: 416 (1971)

03-0356 FOURIER TRANSFORM FAR INFRARED REFLECTION SPECTRA OF PARAELECTRIC POTASSIUM DIHYDROGEN
         PHOSPHATE AND POTASSIUM DIDEUTERIUM PHOSPHATE. (SOFT MADE THEORY)
         LAVRENCIC B
         FIZIKA 3(1): 67-9 (1971)
              SCI ABSTR A 74: 37536 (1971)
              241

03-0357 ON THE SYMMETRY OF THE SOFT MODES IN THE FERROELECTRIC PHASE TRANSITIONS.
         LAVRENCIC B + BLINC R
         PHYS STATUS SOLIDI B 49: K119-24 (1972)
              262

03-0358 SMALL SIGNAL PERMITTIVITY OF THE STATIONARY (100)-180 DEG DOMAIN WALL IN BARIUM
         TITANATE (THEORY, CALCULATIONS).
         LAWLESS WN + FOUSEK J
         J PHYS SOC JAP 28(2): 419-24 (1970)
              189

03-0359 INPUT ADMITTANCE AND FIELDS RADIATED BY AN OPEN WAVEGUIDE FILLED WITH HIGH PERMITTIVITY
         DIELECTRIC. (THEORY)
         LEFEUVRE S + JONGEJANS A
         J PHYS (PARIS) 33, SUPPL 4, COLLOQ 2: 99-101 (1972)
              294

PAGE   29

03-0360   NEW TWO- AND THREE PHASE FERROELECTRIC FERROMAGNETIC MATERIALS. SOME PROBLEMS AND
           RESULTS.
           LEIBLER K + ISUPOV VA + BIELSKA-LEWANDOWSKA H
           ACTA PHYS POLON A 40(6): 815-27 (DEC 1971)
               264

03-0361   FREQUENCY AND TEMPERATURE DEPENDENCES OF THE COMPLEX DIELECTRIC CONSTANT OF
           FERROELECTRICS NEAR THE CURIE POINT. (THEORY)
           LEVANYUK AP + SHCHEDRINA NV
           SOV PHYS SOLID STATE 14(4): 1027-32 (OCT 1972)
               276

03-0362   PHENOMENOLOGICAL THEORY OF ANOMALIES OF THERMODYNAMIC PARAMETERS NEAR SECOND ORDER
           TRANSITION POINTS IN FERROELECTRICS.
           LEVANYUK AP
           BULL ACAD SCI USSR, PHYS SER 29(6): 885-7 (1965)
               219

03-0363   PHENOMENOLOGICAL THEORY OF DIELECTRIC ANOMALIES IN FERROELECTRIC MATERIAL WITH SEVERAL
           PHASE TRANSITIONS AT TEMPERATURES CLOSE TOGETHER. (ROCHELLE SALT)
           LEVANYUK AP + SANNIKOV DG
           SOV PHYS JETP 33: 600-4 (1971)
               262

03-0364   THEORY OF SECOND ORDER PHASE TRANSITIONS.
           LEVANYUK AP
           SOV PHYS SOLID STATE 5: 1294-8 (1964)

03-0365   TWO DIMENSIONAL FERROELECTRIC MODELS.
           LIEB EH + WU FY
           PHASE TRANSITIONS AND CRITICAL PHENOMENA 1: 331-490 (1972)

03-0366   COMPARISON OF FERROELECTRICITY IN ISOMORPHIC LITHIUM NIOBATE AND LITHIUM TANTALATE
           USING AUTHOR'S STATISTICAL THEORY. (SPONTANEOUS POLARIZATION, BONDING)
           LINES ME
           PHYS REV B 2(3): 698-705 (1970)
               217

03-0367   INTERCELL CORRECTIONS FOR IONIC MOTION IN DISPLACEMENT FERROELECTRICS (LITHIUM
           TANTALATE EXAMPLE).
           LINES ME
           PHYS REV B 2(3): 690-7 (1970)
               219

03-0368   POLARIZATION FLUCTUATIONS NEAR A FERROELECTRIC PHASE TRANSITION.
           LINES ME
           PHYS REV B 5(9): 3690-702 (1 MAY 1972)
               264

03-0369   STATISTICAL THEORY FOR DISPLACEMENT FERROELECTRICS.
           LINES ME
           P175-7 OF INT MEETING ON FERROELECTRICITY, 2ND, 1969,, KYOTO, PROC,
           (J PHYS SOC JAP, VOL 28 SUPPL). PHYS SOC JAP, 1970, 459P.

03-0370   (EFFECT OF TEMPERATURE GRADIENT WITHIN THE SAMPLE ON THE) MICROWAVE DISPERSION OF
           FERROELECTRICS NEAR THE TRANSITION POINT (THEORY, MEASUREMENTS ON TGS).
           LUTHER G + MUESER HE
           Z ANGEW PHYS 29(3): 237-40 (1970)
               204

03-0371   ELASTIC AND ELECTROMAGNETIC WAVES IN PIEZOELECTRIC CRYSTALS.
           LYUBIMOV VN
           BULL ACAD SCI USSR, PHYS SER 34(12): 2222-9 (1970)

03-0372   FERROELECTRICITY AND ANTIFERROELECTRICITY IN POLAR CRYSTALS. (THEORY, LEAD ZIRCONATE)
           LYUBIMOV VN + VENEVTSEV YUN + ZHDANOV GS
           SOV PHYS SOLID STATE 4: 1554-57 (1963)

03-0373   DIELECTRIC DISPERSION IN ORDER DISORDER FERROELECTRICS. (THEORY)
           MANSINGH A + LIM KO
           J PHYS SOC JAP 33(3): 747-9 (SEP 1972)
               278

3. THEORY

03-0374 QUASI PHENOMENOLOGICAL THEORY OF ELASTIC ANOMALY IN FERROELECTRIC CRYSTALS.
MARUTAKE M
P214-16 OF INT MEETING ON FERROELECTRICITY, 2ND, 1969, KYOTO, PROC,
(J PHYS SOC JAP, VOL 28 SUPPL). PHYS SOC JAP, 1970, 459P.

03-0375 VIBRATIONS OF LONGITUDINALLY POLARIZED FERROELECTRIC CYLINDRICAL TUBES.
MARTIN GE
J ACOUST SOC AMER 35: 510-20 (1963)

03-0376 VIBRATIONS OF COAXIALLY SEGMENTED, LONGITUDINALLY POLARIZED FERROELECTRIC TUBES.
MARTIN GE
J ACOUST SOC AMER 36: 1496-1506 (1964)

FROM BARIUM TITANATE TO DNA. (FERROELECTRICITY, BIOLOGICAL MECHANISMS, MEMORY,
ORGANICS)
MATTHIAS BT
MATER RES BULL 6: 1005-6 (OCT 1971)

03-0377 SUPERCONDUCTIVITY VERSUS FERROELECTRICITY (SEEM MUTUALLY EXCLUSIVE, THEREFORE
FERROELECTRIC, BUT NOT SUPERCONDUCTIVE ORGANICS EXIST).
MATTHIAS BT
MATER RES BULL 5: 665-8 (1970)
        201

03-0378 SOME THEORETICAL CONSIDERATIONS RELATING TO SWITCHING AND REMANENCE IN FERROELECTRIC
PHOTOCONDUCTOR MEMORY DEVICES.
MEHTA RR
FERROELECTRICS 4: 5-18 (1972)
        267

03-0379 NONLINEAR OPTICAL PROPERTIES OF CRYSTALS.
MEISNER LB + REZ IS
BULL ACAD SCI USSR, PHYS SER 35(9): 1639-43 (1971)

03-0380 EXTENSION OF THE POULET- LOUDON APPROXIMATION IN THE THEORY OF OPTICAL LATTICE
VIBRATIONS. (QUARTZ, BARIUM TITANATE)
MERTEN L + LAMPRECHT G
Z NATURFORSCH A 26(2): 215-19 (1971) (IN GERMAN)
        241

03-0381 A MORE EXACT DYNAMIC METHOD OF DETERMINING POLARIZATION IN FERROELECTRICS.
MIHAILOV MP
COMPT REND ACAD BULG SCI 24(2): 153-4 (1971)
        CHEM ABSTR 75: 11975W (1971)
        269

03-0382 DEPENDENCE OF EFFECTIVE POLARIZATION IN FERROELECTRICS ON THE INTENSITY OF THE ELECTRIC
FIELD.
MIHAILOV MP
COMPT REND ACAD BULG SCI 24(1): 23-5 (1971)
        263

03-0383 NONLINEAR OPTICAL PROPERTIES OF FERROELECTRIC CRYSTALS. (SECOND HARMONIC GENERATION,
OPTICAL PARAMETRIC OSCILLATION)
MILLER RC
P15-19 OF INT MEETING ON FERROELECTRICITY, 2ND, 1969, KYOTO, PROC,
(J PHYS SOC JAP, VOL 28 SUPPL). PHYS SOC JAP, 1970, 459P.

03-0384 THEORY OF HYDROGEN BONDED FERROELECTRICS. PART-1: (FORMALISM OF LOCAL MODE MODEL.)
PART-2: (DYNAMICAL PROPERTIES, TRANSVERSE ISING MODEL, SOFT MODE COUPLING TO TRANSVERSE
OPTICAL PHONON). PART-3: (POTASSIUM DIHYDROGEN ARSENATE, CESIUM DIHYDROGEN ARSENATE)
MOORE MA + WILLIAMS HCWL
J PHYS C 5: 3168-84, 3185-221, 3222-44 (1972)
        279

03-0385 ON RELAXATION PROCESSES NEAR THE SECOND ORDER PHASE TRANSITION POINT.
MORI H
PROGR THEOR PHYS 30: 576-8 (1963)

03-0386 SOME RESULTS ABOUT PHOTOEMISSION ON FERROELECTRIC CRYSTALS. (DOMAIN IMAGING, ABSTRACT
         ONLY)
         MORLON B + COQUET E
         P367 OF EUROPEAN MEETING ON FERROELECTRICITY, 1969, SAARBRUCKEN,
         MUSER HE AND PETERSSON J (EDS), WISSENSCH VERLAGSGES, STUTTGART,
         1970, 396P.

03-0387 THERMODYNAMIC POTENTIALS AND SECONDARY EFFECTS IN FERROELECTRIC CRYSTALS. (THEORY)
         MUSER HE
         ANN PHYS 14(3-4): 176-200 (1964) (IN GERMAN)
             CHEM ABSTR 62: 11261E (1965)

03-0388 LATTICE STATISTICS OF HYDROGEN BONDED CRYSTALS, PART-1, THE RESIDUAL ENTROPY OF ICE,
         PART-2, HYDROGEN BONDED FERROELECTRIC AND ANTIFERROELECTRIC MODELS.
         NAGLE JF
         YALE UNIV PHD THESIS, 1965, 113P.
             DISS ABSTR 26: 5490-1
             219

03-0389 ONE DIMENSIONAL KDP MODEL IN STATISTICAL MECHANICS.
         NAGLE JF
         AMER J PHYS 36(12): 1114-7 (1968)
             181

03-0390 FERROELECTRIC DOMAIN STRUCTURE AND ITS INTERACTION WITH CRYSTAL DISLOCATIONS. (DOMAIN
         ORIGIN, SHAPES, MOVEMENT, 34 REFS)
         NAKAMURA T
         P3-19, VOL 2 OF DVORAK V + FOUSKOVA A + GLOGAR P (EDS),
         INT MEETING ON FERROELECTRICITY, PRAGUE, 1966, PROC.
         INST PHYS CZECH ACAD SCI, PRAGUE, 1966.

03-0391 NOTE ON THE DIELECTRIC RELAXATION IN FERROELECTRICS (THEORETICAL DISCREPANCY).
         NAKAMURA E + ISHIDA K
         J PHYS SOC JAP 29(3): 695-7 (1970)
             213

03-0392 INFLUENCE OF SCREENING ON THE FERROELECTRIC CURIE POINT (UNIAXIAL FERROELECTRIC
         PHOTOCONDUCTOR).
         NATTERMANN T
         PHYS STATUS SOLIDI B, BASIC RES 51(1): 395-405 (1972)

03-0393 ULTRASONIC ATTENUATION ANOMALY IN DISPLACIVE TYPE FERROELECTRICS.
         NATTERMANN T
         PHYS STATUS SOLIDI B, BASIC RES 54(2): K123-6 (1972)

03-0394 PHASE TRANSITION FROM NORMAL FERROELECTRIC TO MAGNETICALLY ORDERED FERROELECTRIC STATE.
         (THEORY)
         NEDLIN GM
         BULL ACAD SCI USSR, PHYS SER 29(6): 896-9 (1965)
             CHEM ABSTR 63: 9179E (1965)

03-0395 THEORY OF SECOND ORDER PHASE TRANSITION FROM THE FERROMAGNETIC TO THE FERROMAGNETIC AND
         FERROELECTRIC STATE.
         NEDLIN GM
         SOV PHYS SOLID STATE 4: 2612-5 (1963)

03-0396 ANALYSIS OF PHONON TUNNELING MODES IN THE RAMAN SPECTRUM OF TETRAGONAL POTASSIUM
         DIHYDROGEN PHOSPHATE. (THEORY)
         NETTLETON RE
         J PHYS SOC JAP 33(3): 641-6 (SEP 1972)
             278

3. THEORY

03-0397 FERROELECTRIC PHASE TRANSITIONS: A REVIEW OF THEORY AND EXPERIMENT. PART-1:
INTRODUCTION AND MAGNETIC RESONANCE. PART-2: THERMAL CONDUCTIVITY. PART-3: NEUTRON
SCATTERING. PART-4: INFRARED AND RAMAN SCATTERING. PART-5: MOSSBAUER EFFECT. PART-6:
ELECTRICAL CONDUCTIVE, SURFACE, AND ELECTROOPTIC PROPERTIES. PART-7: DIELECTRIC AND
SWITCHING MEASUREMENTS. PART-8: ULTRASOUND PROPAGATION. PART-9: SPECIFIC HEAT. PART-10:
PHOTON RENORMALIZATION AND THEORY OF SPONTANEOUS POLARIZATION. PART-11: CONCLUSIONS.
NETTLETON RE
FERROELECTRICS 1: 87-91, 93-101, 111-19, 121-5, 127-35, 207-20, 221-6
(1970), 2: 5-9, 77-91, 93-6 (1971)
     191
     234
     235

03-0398 LATTICE DYNAMICAL THEORY OF SWITCHING IN BARIUM TITANATE SINGLE CRYSTALS.
NETTLETON RE
J APPL PHYS 38(7): 2775-86 (1967)

03-0399 STATISTICAL PERTURBATION THEORY OF ORDER DISORDER FERROELECTRICS: FIRST PERTURBATIVE
CORRECTION.
NETTLETON RE
PHYS STATUS SOLIDI B 48: 689-98 (1971)

03-0400 PROPOSED MECHANISM FOR THERMOPHOTOTROPIC BEHAVIOR IN PEROVSKITE STRUCTURED TITANATES.
NEUDER SM
NASA GODDARD SPACE FLIGHT CENTER, GREENBELT, MD, 1964, 17P. N64-17296
    CHEM ABSTR 61: 14014 (1964) SCI TECH AEROSP REP 2(9): 1177 (1964)

03-0401 FREQUENCY SPECTRUM OF HYDROGEN BONDED FERROELECTRICS. (INFRARED ABSORPTION, NEUTRON
INELASTIC SCATTERING, POTASSIUM DIHYDROGEN PHOSPHATE, DEUTERATED POTASSIUM DIHYDROGEN
PHOSPHATE, THEORY)
NOVAKOVIC L
P127-39, VOL 1 OF DVORAK V + FOUSKOVA A + GLOGAR P (EDS),
INT MEETING ON FERROELECTRICITY, PRAGUE, 1966, PROC.
INST PHYS CZECH ACAD SCI, PRAGUE, 1966.

03-0402 QUANTUM THEORY OF FERROELECTRICITY, PART-1: SPECTRAL ANALYSIS (PROTON DYNAMICS IN
HYDROGEN BONDED FERROELECTRICS).
NOVAKOVIC L
J PHYS CHEM SOLIDS 31: 431-46 (1970)
     190

03-0403 QUANTUM THEORY OF FERROELECTRICITY. PART-2: THERMODYNAMIC BEHAVIOR OF FERROELECTRIC
SPINS.
NOVAKOVIC I + STAMENKOVIC S + VLAHOV A
J PHYS CHEM SOLIDS 32: 487-97 (FEB 1971)

03-0404 PHENOMENOLOGICAL THEORY OF ANTIFERROELECTRIC TRANSITION. PART-2: FIRST ORDER
TRANSITION.
OKADA K
P58-60 OF INT MEETING ON FERROELECTRICITY, 2ND, 1969, KYOTO, PROC,
(J PHYS SOC JAP, VOL 28 SUPPL). PHYS SOC JAP, 1970, 459P.

03-0405 DYNAMIC SUSCEPTIBILITY OF CLASSICAL ANHARMONIC OSCILLATOR. A UNIFIED OSCILLATOR MODEL
FOR ORDER DISORDER AND DISPLACIVE FERROELECTRICS.
ONODERA Y
PROGR THEOR PHYS 44(6): 1477-99 (1970)
     254

03-0406 FREE ENERGY AND SPECIFIC HEAT IN FERROELECTRIC PHASE TRANSITION IN TERMS OF A SINGLE
MODE ANHARMONIC OSCILLATOR MODEL.
ONODERA Y
PROGR THEOR PHYS 45(3): 986-8 (1971)
    CHEM ABSTR 75: 55101N (1971)
     263

03-0407 RAMAN SCATTERING IN PIEZOELECTRIC CRYSTALS WITH TETRAGONAL SYMMETRY.
OVANDER LN
SOV PHYS SOLID STATE 4: 1078-80 (1962)

3. THEORY

03-0408 EFFECT OF FREE CARRIERS ON THE INFRARED SPECTRUM OF A FERROELECTRIC. (THEORY, ABSTRACT ONLY)
PASYNKOV RE
P236 OF INT MEETING ON FERROELECTRICITY, 2ND, 1969, KYOTO, PROC.
(J PHYS SOC JAP, VOL 28 SUPPL). PHYS SOC JAP, 1970, 459P.

03-0409 THERMODYNAMIC THEORY OF FERROELECTRIC SEMICONDUCTORS.
PASYNKOV RE
BULL ACAD SCI USSR, PHYS SER 34(12): 2191-204 (1970)

03-0410 THERMODYNAMIC THEORY OF FERROELECTRIC SEMICONDUCTORS.
PASYNKOV RE
BULL ACAD SCI USSR, PHYS SER 34(12): 2191-203 (1970)
238

03-0411 SOME STRUCTURAL MECHANISMS IN FERROELECTRICITY.
PEPINSKY R
P567-706, PART B OF HENCH LL + DOVE DB(EDS), PHYS OF ELECTRON
CERAMICS, DEKKER, NY, 1972.
262

03-0412 DIELECTRIC PERMITTIVITY DISPERSION PECULIARITIES IN FERROELECTRICS WITH ORDER DISORDER TYPE PHASE TRANSITION.
PEREVERZEVA LP
BULL ACAD SCI USSR, PHYS SER 35(12): 2369-72 (1971)
264

03-0413 DIELECTRIC PROPERTIES AND OPTICAL PHONONS IN PARA- AND FERROELECTRIC PEROVSKITES. (THEORY, ABO(3) TYPE)
PERRY CH
P557-91 OF MCLLER KD + ROTHSCHILD WG(EDS), FAR INFRARED
SPECTROSCOPY, INTERSCIENCE, 1971, 797P.

03-0414 EXTENDED FINE STRUCTURE IN X-RAY ABSORPTION SPECTRA OF CERTAIN PEROVSKITES (STRONTIUM OR CALCIUM TITANATE OR ZIRCONATE, TEST OF SHORT OR LONG RANGE ORDER THEORIES).
PEREL J + DESLATTES RD
PHYS REV B 2(5): 1317-23 (1970)
217

03-0415 DIELECTRIC DISPERSION AND THERMODYNAMICS OF IRREVERSIBLE PROCESSES. (APPLIED TO FERROELECTRIC TRANSITIONS, 31 REFS)
PETERSSON J
P69-89 OF EUROPEAN MEETING ON FERROELECTRICITY, 1969, SAARBRUCKEN,
MUSER HE AND PETERSSON J(EDS), WISSENSCH VERLAGSGES, STUTTGART,
1970, 396P.

03-0416 EXCITON MODEL OF SPIN AND OPTICAL EXCITATIONS OF AN ANTIFERRODIELECTRIC.
PETROV EG
PHYS STATUS SOLIDI B 48: 367-79 (1971)
259

03-0417 EFFECT OF COLLECTIVE INTERACTION OF DIPOLES ON THE NATURE OF DIELECTRIC DISPERSION IN THE FERROELECTRIC PHASE TRANSITION REGION.
PETROV VM
BULL ACAD SCI USSR, PHYS SER 35(12): 2365-70 (1971)
264

03-0418 ELECTRONIC AND SPIN EXCITATIONS IN ANTIFERRODIELECTRIC ELECTRIC AND MAGNETIC MOMENTS OF OPTICAL TRANSITIONS.
PETROV EG
INST TEOR FIZ, AKAD NAUK UKR SSR, KIEV, 1969, 34P. ITF-69-81

03-0419 ON THE THEORY OF LIGHT ABSORPTION BY ANTIFERRODIELECTRICS IN THE FREQUENCY RANGE OF DOUBLE ELECTRONIC EXCITATIONS OF MOLECULES (IONS).
PETROV EG + LOKTEV VM + GAIDIDEI YUB
PHYS STATUS SOLIDI 41: 117-27 (1970)
225

03-0420 THERMODYNAMIC THEORY OF DISPERSION PHENOMENA. (FERROELECTRICITY)
PETERSSON J
FERROELECTRICS 4: 221-34 (1972)

3. THEORY

03-0421　PIEZOELECTRIC SURFACE WAVES ON A C-DOMAIN WALL OF A FERROELECTRIC (CLASS 6MM).
PEUZIN JC
SOLID STATE COMMUN 9: 1639-41 (1971) (IN FRENCH)

03-0422　THEORY OF POLARIZATION REVERSAL IN FERROELECTRICS. (CALCULATIONS OF SWITCHING CURRENT)
PEUZIN JC + LE FAOU C
SOLID STATE COMMUN 8: 505-12 (1970)
　　　192

03-0423　CRITICAL BEHAVIOR OF THE PSEUDOSPIN MODEL FOR ORDER DISORDER TYPE FERROELECTRICS WITH A
SHORT RANGE INTERACTION. (THEORY)
PFEUTY P + ELLIOTT RJ
P53-62 OF EUROPEAN MEETING ON FERROELECTRICITY, 1969, SAARBRUCKEN,
MUSER HE AND PETERSSON J(EDS), WISSENSCH VERLAGSGES, STUTTGART,
1970, 396P.

03-0424　SELF CONSISTENT LATTICE DYNAMICAL THEORY OF STRUCTURAL PHASE TRANSITIONS IN PEROVSKITE
TYPE CRYSTALS. (STRONTIUM TITANATE, LANTHANUM ALUMINATE)
PIETRASS B
PHYS STATUS SOLIDI B 53: 279-86 (1972)
　　　276

03-0425　GUIDE LINES FOR THE SELECTION OF ACOUSTOOPTIC MATERIALS.
PINNOW DA
IEEE J QUANTUM ELECTRON 6(4): 223-38 (1970)
　　　286

03-0426　IMPURITY MODES IN AN INDUCED MOMENT SYSTEM AND HYDROGEN BONDED FERROELECTRICS (KDP
TYPE, THEORY).
PINK DA
CAN J PHYS 49: 243-56 (1971)
　　　225

03-0427　EFFECTS OF ELECTRON PHONON INTERACTION IN FERROELECTRIC CRYSTALS (CYCLOTRON RESONANCE,
ABSORPTION AND LUMINESCENCE SPECTRA, INFRARED, OPTICAL, AND ACOUSTIC SCATTERING).
POKATILOV EP
UKR FIZ ZH 15(3): 519-21 (1970)
　　　226

03-0428　DIELECTRIC DISPERSION IN FERROELECTRICS.
POPLAVKO YUM
BULL ACAD SCI USSR, PHYS SER 34(12): 2289-97 (1970)

03-0429　PROBLEM OF THE ACCURATE MEASUREMENT OF THE DIELECTRIC PROPERTIES OF FERROELECTRICS AT
MICROWAVE FREQUENCIES.
POPLAVKO YUM
SOV PHYS JETP 16: 566-8 (1963)

03-0430　CONTRIBUTION TO THE THEORY OF DOMAIN STRUCTURES IN MAGNETS AND FERROELECTRICS.
PRIVOROTSKII IA
SOV PHYS JETP 33: 825-30 (1971)

03-0431　STIMULATED COMBINATION SCATTERING OF SOUND IN PIEZOELECTRICS. (THEORY, LITHIUM NIOBATE
EXAMPLE)
PUSHKINA NI + KHOKHLOV RV
SOV PHYS DOKL 17: 235-6 (1972)

03-0432　THEORY OF PEROVSKITE FERROELECTRICS (HAMILTONIAN IN TERMS OF LOCALIZED D-STRAIN AND
SOFT- NORMAL MODE COORDINATES AND TEMPERATURE INDEPENDENT PARAMETERS).
PYTTE E
PHYS REV B SOLID STATE 5: 3758-69 (MAY 1, 1972)
　　　264

03-0433　THEORY OF SECOND ORDER PHASE TRANSITIONS AT LOW TEMPERATURES. (TIN TELLURIDE, POTASSIUM
TANTALATE)
RECHESTER AB
SOV PHYS JETP 33: 423-30 (1971)

03-0434　SYMMETRY LIMITATIONS TO POLARIZATION OF POLYCRYSTALLINE FERROELECTRICS.
REDIN RD + MARKS GW + ANTONIAK CE
J APPL PHYS 34: 600-10 (1963)

3. THEORY

03-0435   CALORIMETRIC INVESTIGATIONS OF SECOND ORDER FERROELECTRIC TRANSITIONS.
REESE RL + MAY LF
FERROELECTRICS 4: 65-9 (1972)

03-0436   ELECTRICAL SUSCEPTIBILITY OF ADP (THEORY).
RICHMOND P
PHYS LETT A 32(1): 52-3 (1970)
      209

03-0437   AC POLARIZATION AND DIELECTRIC HYSTERESIS NEAR THE TRANSITION POINT OF DIFFUSE
FERROELECTRIC PHASE. (KAENZIG MODEL)
ROLOV BN + ROMANOVSKIS T
LATV PSR ZINAT AKAD VESTIS, FIZ TEH ZINAT SER NO 2: 14-19 (1970)
(IN RUSSIAN)
     CHEM ABSTR 73: 50108U (1970)
      231

03-0438   ACCOUNTING FOR INTERACTION IN THE KANZIG REGIONS MODEL FOR DIFFUSE FERROELECTRIC PHASE
TRANSITIONS.
ROLOV BN + YURKEVICH VE
P23-5 OF FAZOVYE PEREKHODY V SEGNETOELEKTRIKAKH (PHASE
TRANSFORMATIONS IN FERROELECTRICS), FRITSBERG VYA + ROLOV BN +
KRUCHAN YY(EDS), IZDATELSTVO ZINATNE RIGA, 1971, 206P. (IN RUSSIAN)

03-0439   A MODEL OF KANZIG REGIONS TAKING INTO ACCOUNT DIFFERENT ORIENTATIONS OF THE SPONTANEOUS
POLARIZATION.
ROLOV BN
P37-45 OF FAZOVYE PEREKHODY V SEGNETOELEKTRIKAKH (PHASE
TRANSFORMATIONS IN FERROELECTRICS), FRITSBERG VYA + ROLOV BN +
KRUCHAN YY(EDS), IZDATELSTVO ZINATNE RIGA, 1971, 206P. (IN RUSSIAN)

03-0440   DIELECTRIC CONSTANT OF FERROELECTRIC SOLID SOLUTIONS WITH ALLOWANCE FOR CONCENTRATION
FLUCTUATIONS.
ROLOV BN
BULL ACAD SCI USSR, PHYS SER 34(12): 2218-22 (1970)

03-0441   EFFECT OF AN EXTERNAL ELECTRIC FIELD ON A DIFFUSE FERROELECTRIC PHASE TRANSITION,
TALKING INTO ACCOUNT THE INTERACTING KANZIG REGIONS.
ROLOV BN + YURKEVICH VE
P47-54 OF FAZOVYE PEREKHODY V SEGNETOELEKTRIKAKH (PHASE
TRANSFORMATIONS IN FERROELECTRICS), FRITSBERG VYA + ROLOV BN +
KRUCHAN YY(EDS), IZDATELSTVO ZINATNE RIGA, 1971, 206P. (IN RUSSIAN)

03-0442   EFFECT OF HYDROSTATIC PRESSURE ON THE NATURE OF DIFFUSE FERROELECTRIC PHASE
TRANSITIONS.
ROLOV BN
P55-65 OF FAZOVYE PEREKHODY V SEGNETOELEKTRIKAKH (PHASE
TRANSFORMATIONS IN FERROELECTRICS), FRITSBERG VYA + ROLOV BN +
KRUCHAN YY(EDS), IZDATELSTVO ZINATNE RIGA, 1971, 206P. (IN RUSSIAN)

03-0443   HEAT CAPACITY IN THE FERROELECTRIC TRANSITION REGION (THEORY, RELATION TO KAENZIG
DOMAIN MODEL).
ROLOV BN + ROMANOVSKII TB
LATV PSR ZINDT AKAD VESTIS FIZ TEH ZINAT 1969(6): 53-6 (IN RUSSIAN)
      216

03-0444   SOME POSSIBILITIES OF DETERMINING THE VOLUME OF KANZIG REGIONS IN FERROELECTRICS.
ROLOV BN
J PHYS (PARIS) 33, SUPPL 4, COLLOQ 2: 257-8 (1972)
      294

03-0445   THERMODYNAMICS AND STATISTICAL MECHANICS OF DIFFUSE FERROELECTRIC PHASE TRANSITIONS.
ROLOV BN + ROMANOVSKII TB
BULL ACAD SCI USSR, PHYS SER 34(12): 2215-18 (1970)

03-0446   ESTIMATION OF THE DIMENSIONS OF KANZIG REGIONS ON THE BASIS OF THE DYNAMIC THEORY OF
CRYSTAL LATTICES.
ROMANOVSKII TB + ROLOV BN
P67-74 OF FAZOVYE PEREKHODY V SEGNETOELEKTRIKAKH (PHASE
TRANSFORMATIONS IN FERROELECTRICS), FRITSBERG VYA + ROLOV BN +
KRUCHAN YY(EDS), IZDATELSTVO ZINATNE RIGA, 1971, 206P. (IN RUSSIAN)

3. THEORY

03-0447   NATURE OF DISSIPATIVE EFFECTS WHICH ACCOMPANY DISACCOMODATION TYPE PHENOMENA IN
           FERRITES AND FERROELECTRICS.
           ROZENBAUM LB
           SOV PHYS SOLID STATE 14(1): 196-8 (JUL 1972)
               270

03-0448   MECHANISMS FOR BARKHAUSEN JUMPS AND BEHAVIOR OF THE BARKHAUSEN EFFECT IN FERROELECTRIC
           CRYSTALS.
           RUDYAK VM
           BULL ACAD SCI USSR, PHYS SER 34(12): 2313-16 (1970)

03-0449   EFFECT OF VIOLATION OF ELECTRIC NEUTRALITY ON THE PHASE TRANSITION IN A FERROELECTRIC
           WITH HYDROGEN BONDS (THEORY, WU MODEL).
           RYAZANOV GV
           SOV PHYS JETP 32: 544-6 (1970)
               225

03-0450   SECOND ORDER PHASE TRANSITION IN THE THREE DIMENSIONAL MODEL OF A FERROELECTRIC.
           RYAZANOV GV
           SOV PHYS JETP 34: 1367-70 (JUN 1972)

03-0451   TYPE-II PHASE CHANGE IN THREE DIMENSIONAL FERROELECTRIC MODEL. (THEORY)
           RYAZANOV GV
           SOV PHYS JETP 34: 1367-70 (1972)
               278

03-0452   ON THE QUANTUM THEORY OF THE FERROELECTRIC STATE IN ELECTRIC DIPOLE SYSTEMS (SPIN WAVES
           IN HYDROXYL DIPOLE SYSTEM IN POTASSIUM CHLORIDE LATTICE, FERROELECTRIC STATE AT LOW
           TEMPERATURES).
           SABUROVA RV
           PHYS STATUS SOLIDI 41(1): 129-34 (1970)
               225

03-0453   RELATIONSHIP BETWEEN IONIC DISPLACEMENT AND THE CURIE TEMPERATURE AND SPONTANEOUS
           POLARIZATION IN DISPLACIVE FERROELECTRICS.
           SAMARA GA
           FERROELECTRICS 2(4): 271-6 (1971)

03-0454   DISPERSION IN FERROELECTRICS. (THEORY, BARIUM TITANATE)
           SANNIKOV DG
           SOV PHYS JETP 14: 98-101 (1962)

03-0455   ELECTROMAGNETIC AND ACOUSTIC WAVES IN FERROELECTRICS. (THEORY, BARIUM TITANATE,
           ROCHELLE SALT)
           SANNIKOV DG
           SOV PHYS SOLID STATE 4: 1187-92 (1962)

03-0456   DOUBLE HYSTERESIS LOOPS AND FIRST ORDER PHASE TRANSITIONS IN FERROELECTRIC CRYSTALS.
           SAWAGUCHI E + CROSS LE
           PHYS STATUS SOLIDI B 50: K89-92 (1972)
               280

03-0457   THERMODYNAMICS AND THE PHENOMENOLOGY OF FERROELECTRICITY.
           SCHMIDT G
           PHYS STATUS SOLIDI 3: 1281-317 (1963) (IN GERMAN)
               CHEM ABSTR 59: 12284F (1963)

03-0458   INFLUENCE OF COMPENSATING CHARGES ON THE C-DOMAIN STRUCTURE OF FERROELECTRICS.
           SELYUK BV
           SOV PHYS CRYSTALLOGR 16: 292-6 (SEP-OCT 1971)

03-0459   WEIGHT OF SHOTTKY BARRIERS IN FERROELECTRICS.
           SEPLYUK BV
           BULL ACAD SCI USSR, PHYS SER 35(9): 1635-8 (1971)

03-0460   NATURE OF THE TEMPERATURE DEPENDENCE OF NMR SPECTRA SECOND MOMENT IN ORDER DISORDER
           TYPE FERROELECTRICS.
           SERGEEV NA + KUBAREV YUG
           PHYS STATUS SOLIDI B 53: K99-102 (1972)
               280

3. THEORY

03-0461   ELECTROOPTIC EFFECT IN OPTICALLY HETEROGENEOUS CRYSTALS (THEORY AND METHOD OF
           DETERMINING HALF WAVE VOLTAGE).
           SHAMBUROV VA
           SOV PHYS CRYSTALLOGR 14(6): 875-79 (1970)
               194

03-0462   THEORETICAL AND PRACTICAL STUDIES OF DIELECTRIC APPLICATIONS OF FERROELECTRICS.
           SHAH A
           EIDGENOSSISCHE TECH HOCHSCHULE PHD THESIS, ZURICH, 1969, 108P.
           N70-24641
               SCI TECH AEROSP REP 8(11): 2104 (1970)
               216

03-0463   CRYSTALLOPHYSICAL CLASSIFICATION OF FERROELECTRICS. (BASED ON DOMAIN STRUCTURE AND ITS
           APPLICATIONS)
           SHUVALOV LA
           BULL ACAD SCI USSR, PHYS SER 28(4): 567-72 (1964)

03-0464   (MACROSCOPIC) SYMMETRY ASPECTS OF FERROELECTRICITY. (DESCRIPTIVE AND PREDICTIVE
           POSSIBILITIES)
           SHUVALOV LA
           P38-51 OF INT MEETING ON FERROELECTRICITY, 2ND, 1969, KYOTO, PROC,
           (J PHYS SOC JAP, VOL 28 SUPPL). PHYS SOC JAP, 1970, 459P.

03-0465   SYMMETRY OF CRYSTALS IN WHICH FERROMAGNETIC AND FERROELECTRIC PROPERTIES APPEAR
           SIMULTANEOUSLY.
           SHUVALOV LA + BELOV NV
           SOV PHYS CRYSTALLOGR 7: 150-1 (1962)

03-0466   THEORY OF ELECTRIC INDUCTION WAVES IN FERROELECTRIC SEMICONDUCTOR CRYSTALS.
           SHUR MS
           SOV PHYS SOLID STATE 10(12): 2827-31 (1969)
               146

03-0467   DIELECTRIC RESONANCE AND RELAXATION IN ELECTRICALLY BIASED KDP AND DEUTERATED
           ISOMORPHS. (THEORY)
           SILVERMAN BD
           P255-7 OF INT MEETING ON FERROELECTRICITY, 2ND, 1969, KYOTO, PROC,
           (J PHYS SOC JAP, VOL 28 SUPPL). PHYS SOC JAP, 1970, 459P.

03-0468   TEMPERATURE DEPENDENT SPECTRUM OF AN ANTIFERROELECTRIC LINEAR CHAIN MODEL.
           (PEROVSKITES)
           SILVERMAN BD
           PHYS REV 128: 638-45 (1962)

03-0469   MEASUREMENT OF THE PYROELECTRIC COEFFICIENT AND PERMITTIVITY FROM THE PYROELECTRIC
           RESPONSE TO STEP RADIATION SIGNALS IN FERROELECTRICS.
           SIMHONY M + SHAULOV A
           APPL PHYS LETT 21: 375-7 (1972)

03-0470   INTRODUCTION TO THE THEORY OF FERROELECTRICITY.
           SLATER JC
           P1-8 OF SYMP ON FERROELECTRICITY, 1966, GENERAL MOTORS
           RES LABS, WARREN, MICHIGAN, PROC, FERROELECTRICITY.
           WELLER EF(ED), ELSEVIER, 1967, 318P.

03-0471   LIGHT SCATTERING BY ACOUSTIC PHONONS IN FERROELECTRIC AND PIEZOELECTRIC CRYSTALS.
           SMOLENSKII GA + HASHKOZEV ZM
           MATER RES BULL 6: 1065-74 (OCT 1971)

03-0472   THERMODYNAMIC THEORY OF CRYSTALS POSSESSING FERROELECTRIC AND FERROMAGNETIC PROPERTIES.
           SMOLENSKII GA
           SOV PHYS SOLID STATE 4: 807-9 (1962)

03-0473   ELECTRON MIRROR MICROSCOPIC OBSERVATION OF FERROELECTRIC DOMAINS.
           SOMEYA T
           KOTAI BUTSURI 7: 101-6 (1972) (IN JAPANESE)

3. THEORY

03-0474   OBSERVATION OF FERROELECTRIC DOMAIN STRUCTURES, MODULATED BY A PHASE TRANSITION, IN A
          SCANNING ELECTRON MICROSCOPE.
          SPIVAK GV + ANTOSHIN MK
          RADIO ENG ELECTRON PHYS 16: 1580-2 (SEP 1971)

03-0475   DYNAMICAL THEORY OF HYDROGEN BONDED FERROELECTRICS. PART-1: DAMPINGS AND (FREQUENCY)
          SHIFTS (OF PROTONIC COLLECTIVE MOTION, TUNNELING, CALCULATIONS, INTERPRETATION).
          STAMENKOVIC S + NOVAKOVIC L
          PHYS STATUS SOLIDI 41: 135-50 (1970)
                221

03-0476   DYNAMICAL THEORY OF AMMONIUM DIHYDROGEN PHOSPHATE TYPE ANTIFERROELECTRICS WITH HYDROGEN
          BONDS.
          STASYUK IV + LEVITSKII RR
          BULL ACAD SCI USSR, PHYS SER 35(9): 1616-19 (1971)

03-0477   EPR STUDY OF PARAMAGNETIC ION COMPLEXES IN FERROELECTRIC CRYSTALS.
          STANKOWSKI J
          FIZ DIELEK RADIOSPEKTROSK 2: 295-316 (1972) (IN POLISH)

03-0478   THEORY OF COHERENT NEUTRON SCATTERING BY HYDROGEN BONDED FERROELECTRICS AT LOW
          TEMPERATURES. PART-1: GENERAL EXPRESSION FOR INELASTIC COHERENT SCATTERING OF SLOW
          NEUTRONS AND EFFECTIVE THERMAL FACTORS. (KDP TYPE) PART-2: SCATTERING CHARACTERISTICS
          AND VARIOUS CONCEPTIONS OF THE TUNNELING QUASI SPIN MODEL.
          STAMENKOVIC S
          J LOW TEMP PHYS 9: 475-83, 485-94 (1972)
                283

03-0479   ON THE FREQUENCY DEPENDENCE OF THE TEMPERATURE AUTOSTABILIZATION EFFECT IN
          FERROELECTRICS.
          STRAJBLOVA J + FIALA J
          PHYS LETT 8: 306-7 (1964)

03-0480   CRYSTALLOGRAPHICAL CLASSIFICATION OF FERROELECTRICS, FERROELECTRIC PHASE TRANSITIONS
          AND SPECIAL FEATURES OF DOMAIN STRUCTURE AND SOME PHYSICAL PROPERTIES OF FERROELECTRICS
          OF DIFFERENT CLASSIFICATION CLASSES.
          SUVAOV LA
          ACTA CRYSTALLOGR 16(SUPPL): A186 (1963) (ABSTRACT ONLY)

03-0481   ZEROES OF THE PARTITION FUNCTION FOR THE HEISENBERG, FERROELECTRIC, AND GENERAL ISING
          MODELS.
          SUZUKI M + FISHER ME
          J MATH PHYS 12: 235-46 (1971)

03-0482   MICROSCOPIC THEORY OF THE PHASE TRANSITIONS IN ROCHELLE SALT.
          TAKAGI Y + ISHIBASHI Y
          J PHYS SOC JAP 33(5): 1381-9 (NOV 1972)
                280

03-0483   THEORY OF BARIUM TITANATE.
          TAKAHASHI H
          J PHYS SOC JAP 16(9): 1685-9 (1961)

03-0484   THEORY OF NEUTRON SCATTERING FROM LATTICE VIBRATIONS (REDEVELOPED FROM FIRST
          PRINCIPLES, APPLIED TO DISPLACIVE TYPE FERROELECTRICS).
          TANI K
          J PHYS SOC JAP 29(3): 594-605 (1970)
                214

03-0485   ENERGY SPECTRUM OF AN ORDER DISORDER FERROELECTRIC CONTAINING DISLOCATIONS.
          TARASENKO VV + KHARITONOV VD
          SOV PHYS SOLID STATE 12(11): 2653-5 (1971)
                234

03-0486   EFFECT OF POINT DEFECTS ON THE ENERGY SPECTRUM OF HYDROGEN BONDED FERROELECTRICS.
          TARASENKO VV + KHARITONOV VD
          SOV PHYS SOLID STATE 12: 270-6 (1970)

03-0487   EFFECTS OF THERMALLY DIFFUSED IMPURITIES ON PROPAGATION OF ELASTIC SURFACE WAVES ON
          FERROELECTRIC CERAMICS. (IMPROVEMENT)
          TODA K + TAKAHASHI K + OKAZAKI K
          APPL PHYS LETT 18(9): 396-8 (1 MAY 1971)
                231

3. THEORY

03-0488   ANOMALY OF THE SPIN LATTICE RELAXATION TIME NEAR THE TRANSITION TEMPERATURE IN THE
          ORDER DISORDER TYPE FERROELECTRICS. (THEORY)
          TOKUNAGA M + YOSHIMITSU K
          P117-19 OF INT MEETING ON FERROELECTRICITY, 2ND, 1969, KYOTO, PROC,
          (J PHYS SOC JAP, VOL 28 SUPPL). PHYS SOC JAP, 1970, 459P.

03-0489   MODELS OF THE PHASE TRANSITION IN POTASSIUM DIHYDROGEN PHOSPHATE TYPE FERROELECTRICS.
          (24 REFS)
          TOKUNAGA M
          FERROELECTRICS 1(4): 195-206 (OCT 1970)
              234

03-0490   (METHODS OF) STRUCTURAL STUDIES OF FERROELECTRICS AND FERROELECTRIC MAGNETS. (ELECTRON
          MICRODIFFRACTION, SUMMARY ONLY)
          TOMASHPOLSKII YUYA + VENEVTSEV YUN + ZHDANOV GS
          P301 OF INT MEETING ON FERROELECTRICITY, 2ND, 1969, KYOTO, PROC,
          (J PHYS SOC JAP, VOL 28 SUPPL). PHYS SOC JAP, 1970, 459P.

03-0491   SHOCK WAVE INDUCED ELECTROMECHANICAL CONVERSION IN FERROELECTRIC MATERIALS.
          TROCCAZ M + FETIVEAU Y + EYRAUD L + BERNARD J + DAVID J + LANDURE Y
          COMPT REND B 275(14): 481-4 (1972) (IN FRENCH)
              CHEM ABSTR 78: 9171Z (1973)
              288

03-0492   RELAXATION ABSORPTION OF SOUND NEAR THE CURIE POINT OF FERROELECTRIC SEMICONDUCTORS.
          TSEKVAVA BE
          SOV PHYS SEMICOND 5(4): 702-3 (OCT 1971)
              245

03-0493   STATISTICAL METHOD FOR INVESTIGATING (DYNAMICS OF) POLARIZATION REVERSAL IN
          FERROELECTRIC CERAMICS.
          TURIK AV
          SOV PHYS SOLID STATE 5(9): 1751-3 (1964)

03-0494   THEORY OF POLARIZATION AND HYSTERESIS OF FERROELECTRICS.
          TURIK AV
          SOV PHYS SOLID STATE 5: 885-6 (1963)

03-0495   INFLUENCE OF POINT DEFECTS ON DIELECTRIC PROPERTIES, DOMAIN DYNAMICS, AND PHASE
          TRANSITIONS OF FERROELECTRICS. (GENERAL DISCUSSION)
          UNRUH HG + SAILER E
          P437-8 OF INT MEETING ON FERROELECTRICITY, 2ND, 1969, KYOTO, PROC,
          (J PHYS SOC JAP, VOL 28 SUPPL). PHYS SOC JAP, 1970, 459P.

03-0496   ON THE INFLUENCE OF CRYSTAL DEFECTS UPON FERROELECTRIC PROPERTIES. (DISPERSION,
          RELAXATION TIME AND STRENGTH, HYSTERESIS, 26 REFS)
          UNRUH HG
          P167-80 OF EUROPEAN MEETING ON FERROELECTRICITY, 1969, SAARBRUCKEN,
          MUSER HE AND PETERSSON J (EDS), WISSENSCH VERLAGSGES, STUTTGART,
          1970, 396P.

03-0497   CORRELATION EFFECTS IN DISPLACEMENT TYPE PHASE TRANSITIONS IN FERROELECTRICS. (THEORY
          APPLIED TO BARIUM TITANATE)
          VAKS VG
          SOV PHYS JETP 31(1): 161-7 (1970)
              231

03-0498   NONLINEAR EFFECTS IN THE PROPAGATION OF FINITE AMPLITUDE HYPERSONIC WAVES IN
          FERROELECTRIC CRYSTALS.
          VALITOVA NR + GONCHAROV KV
          SOV PHYS SOLID STATE 14(3): 768-70 (SEP 1972)
              276

03-0499   DEBYE SCREENING OF A NONUNIFORM ELECTRIC FIELD IN A FERROELECTRIC.
          VENDIK OG + MIKHALEVSKII AN
          SOV PHYS SOLID STATE 13(8): 1849-53 (FEB 1972)
              258

3. THEORY

03-0500   ELECTRICAL NONLINEARITY OF A FERROELECTRIC SEMICONDUCTOR IN A NONUNIFORM ELECTRIC
          FIELD.
          VENDIK OG + MIKHALEVSKII AN
          SOV PHYS SOLID STATE 12(2): 466-7 (1970)
              206

03-0501   MICROSCOPIC THEORY OF SURFACE PHENOMENA IN FERROELECTRIC CRYSTALS. (CALCULATION OF
          FIELD NEAR SURFACE)
          VENDIK OG + ROSENBERG LA
          P413-15 OF INT MEETING ON FERROELECTRICITY, 2ND, 1969, KYOTO, PROC,
          (J PHYS SOC JAP, VOL 28 SUPPL). PHYS SOC JAP, 1970, 459P.

03-0502   PARAELECTRIC SEMICONDUCTOR IN A NONUNIFORM ELECTRIC FIELD.
          VENDIK OG + MIKHALEVSKII AN
          BULL ACAD SCI USSR, PHYS SER 35(12): 2356-9 (1971)
              264

03-0503   PHENOMENOLOGICAL THEORY OF MICROWAVE LOSSES IN FERROELECTRICS. (AT TEMPERATURES ABOVE
          THE CURIE POINT)
          VENDIK OG + PLATONOVA LM
          P61-3 OF INT MEETING ON FERROELECTRICITY, 2ND, 1969, KYOTO, PROC,
          (J PHYS SOC JAP, VOL 28 SUPPL). PHYS SOC JAP, 1970, 459P.

03-0504   SOME ASPECTS OF THE PROBLEM OF FERROELECTRICITY AND ANTIFERROELECTRICITY IN
          PEROVSKITES.
          VENEVTSEV YUN + LEBEDEV VM
          BULL ACAD SCI USSR, PHYS SER 34(12): 2243-9 (1970)

03-0505   DOMAIN STRUCTURE OF POLYCRYSTALLINE FERROELECTRICS WITH DIFFUSE PHASE TRANSITIONS.
          VERBITSKAYA TN + LAVERKO EN + POLYAKOV SM + ROZORENOVA LA + RAEVSKAYA EB
          BULL ACAD SCI USSR, PHYS SER 35(9): 1792-4 (1971)

03-0506   ENERGY PARAMETERS AND PROTON TUNNELING IN POTASSIUM DIHYDROGEN PHOSPHATE TYPE
          FERROELECTRICS. (THEORY)
          VILLAIN J + AUBRY S
          P41-5 OF EUROPEAN MEETING ON FERROELECTRICITY, 1969, SAARBRUCKEN,
          MUSER HE AND PETERSSON J (EDS), WISSENSCH VERLAGSGES, STUTTGART,
          1970, 396P.

03-0507   SELF CONSISTENCY OF LANDAU'S MODEL IN THE TRANSITION FROM PIEZO TO FERROELECTRICITY
          (THEORY).
          VILLAIN J
          SOLID STATE COMMUN 8: 295-7 (1970)
              190

03-0508   NATURE OF THE ANOMALOUS ELECTRIC CONDUCTIVITY OF FERROELECTRICS IN THE PHASE TRANSITION
          REGION.
          VINETSKII VL + ITSKOVSKII MA + KUKUSHKIN LS
          PHYS STATUS SOLIDI 39: K23-7 (1970)
              208

03-0509   RESONANCE EFFECTS IN MAGNETICALLY EQUIAXIAL FERRODIELECTRIC SINGLE CRYSTALS HAVING
          DOMAIN STRUCTURE.
          VLASOV KB + ONOPRIYENKO LG
          PHYS METALS METALLOGR 15(1): 41-51 (1963)

03-0510   DO WE REALLY UNDERSTAND (THE CAUSE OF) FERROELECTRICITY. (ITS NATURE AND THE PREDICTION
          OF ITS OCCURANCE)
          VON HIPPEL AR
          P1-6 OF INT MEETING ON FERROELECTRICITY, 2ND, 1969, KYOTO, PROC,
          (J PHYS SOC JAP, VOL 28 SUPPL). PHYS SOC JAP, 1970, 459P.

03-0511   FIELD EFFECT AT THE CONTACT BETWEEN A SEMICONDUCTOR AND A C-DOMAIN FERROELECTRIC
          (THEORY, TITANATES DISCUSSED).
          VUL BM + GURO GM + IVANCHIK II
          SOV PHYS SEMICOND 4(1): 128-31 (1970)
              200

3. THEORY

03-0512 ELECTRICAL CONTACT TO N- AND P-TYPE FERROELECTRIC OXIDES.
WEMPLE SH
P128-37 OF SCHWARTZ B(ED), OHMIC CONTACTS TO SEMICONDUCTORS,
ELECTROCHEM SOC, NY, 1969.

03-0513 THEORY OF THE ELASTOOPTIC EFFECT IN NONMETALLIC CRYSTALS. (STRAIN DEPENDENT CURIE
TEMPERATURE, POLARIZATION FLUCTUATIONS, LITHIUM NIOBATE, ALKALI HALIDES)
WEMPLE SH + DIDOMENICO M
PHYS REV B 1: 193-202 (1970)

03-0514 MEASUREMENT OF THE DISSIPATION ASSOCIATED WITH THE ELECTROMECHANICAL COUPLING IN
PIEZOELECTRIC CERAMICS.
WOOLLETT RS
UNIV CONNECTICUT PHD THESIS, 1967, 142P. N69-10026
    DISS ABSTR B 28: 3291 (1968)
    141

03-0515 APPARENT VIOLATION OF THE UNIVERSALITY HYPOTHESIS IN A LATTICE MODEL OF A PHASE
TRANSITION.
WU FY
NORTHEASTERN UNIV DEPT PHYSICS, 1972, 12P. N72-18684
    SCI TECH AEROSP REP 10(9): 1237 (1972)
    264

03-0516 CRITICAL BEHAVIOR OF HYDROGEN BONDED FERROELECTRICS.
WU FY
PHYS REV LETT 24: 1476-8 (29 JUN 1970)
    208

03-0517 PHASE TRANSITION IN A SIXTEEN VERTEX LATTICE MODEL (FERROELECTRIC SQUARE LATTICE).
WU FY
PHYS REV B SOLID STATE 6: 1810-3 (SEP 1, 1972)

03-0518 PHENOMENOLOGICAL THEORY OF REFRACTIVE INDEX AND ELECTROOPTICAL PHENOMENA IN CRYSTALS OF
PEROVSKITE STRUCTURE (WITH APPLICATION TO BARIUM TITANATE).
WUNSCH DC
NEW MEXICO STATE UNIV SCD THESIS, 1969, 184P. N70-39085
    DISS ABSTR INT B 30(5): 2379 (1969)
    219

03-0519 PHOTOEFFECTS AT THE PHOTOCONDUCTOR- FERROELECTRIC BOUNDARY.
WURFEL P + RUPPEL W
P379-83 OF INT CONF ON PHOTOCONDUCTIVITY, 3RD PROC, PELL EM(ED),
1969, PERGAMON, 1971, 410P.

03-0520 EXACT SOLUTION OF MODEL OF TWO DIMENSIONAL FERROELECTRICS IN ARBITRARY EXTERNAL
ELECTRIC FIELD.
YANG CP
PHYS REV LETT 19(10): 586-8 (1967)

03-0521 DISTRIBUTION OF RELAXATION TIME IN FERROELECTRICS NEAR THE CURIE TEMPERATURE (POLY
VERSUS MONODISPERSIVE BEHAVIOR).
YOSHIMITSU K + HATTA I + MATSUBARA T
PROGR THEOR PHYS 43(5): 1411-12 (1970)
    218

03-0522 DIELECTRIC CONSTANT BEHAVIOR IN THE NEIGHBORHOOD OF THE PHASE TRANSITION IN
CHARACTERISTICALLY FERROELECTRIC SOLIDS.
YURKEVICH VE + ROLOV BN
LATV PSR ZINAT AKAD VESTIS FIZ TEH ZINAT SER NO 6: 38-43 (1971)
(IN RUSSIAN)
    CHEM ABSTR 76: 91602F (1972)
    278

03-0523 THERMODYNAMICS OF FERROELECTRIC SOLID SOLUTIONS. (THEORY)
YURKEVICH VE + ROLOV BN
LATV PSR ZINAT AKAD VESTIS FIZ TEH ZINAT SER NO 6: 29-37 (1971)
(IN RUSSIAN)
    CHEM ABSTR 76: 77498V (1972)
    278

03-0524   THERMODYNAMICS OF FERROELECTRIC SOLID SOLUTIONS. PART-1. PART-2.
           YURKEVICH VE + ROLOV BN
           PHYS STATUS SOLIDI B 52: 335-43, 683-91 (1972)
               275

03-0525   MEASUREMENT OF THE TEMPERATURE DEPENDENCE OF THE ELECTRICAL CONDUCTIVITY OF
           FERROELECTRICS.
           ZHDAN AG + ARTOBOLEVSKAYA ES
           INSTRUM EXPER TECH 2: 562-4 (1972)

03-0526   CHARACTERISTICS OF THE STRUCTURAL CHANGES INVOLVED IN PHASE TRANSFORMATIONS IN
           ANTIFERROELECTRICS.
           ZHELUDEV IS + LYUBIMOV VN
           SOV PHYS CRYSTALLOGR 8: 255-8 (1963)

03-0527   CHANGE OF COMPLETE SYMMETRY OF THE CRYSTALS DURING THE PHASE TRANSITIONS IN
           FERROELECTRICS AND FERROMAGNETICS.
           ZHELUDEV IS
           PROC INDIAN ACAD SCI A 59: 168-84 (1964)

03-0528   FERROELECTRICITY AND SYMMETRY. (41 REFS)
           ZHELUDEV IS
           P429-64 OF SOLID STATE PHYS, VOL 26, EHRENREICH H + SEITZ F +
           TURNBULL D(EDS), ACADEMIC PRESS, NY, 1971.
               264

4. BARIUM TITANATE

04-0529　LEED STUDY OF SURFACE STRUCTURES ON THE (001) FACE OF BARIUM TITANATE.
ABERDAM D + BOUCHET G + DUCROS P
SURFACE SCI 27(3): 559-70 (1971)

04-0530　LEED STUDIES ON THE (001) FACE OF BARIUM TITANATE CRYSTALS.
ABERDAM D + GAUBERT C + DUCROS P
P288-9 OF INT MEETING ON FERROELECTRICITY, 2ND, 1969, KYOTO, PROC.
(J PHYS SOC JAP, VOL 28 SUPPL). PHYS SOC JAP, 1970, 459P.

04-0531　STUDY OF THE TEMPERATURE DEPENDENCE OF LEED INTENSITIES FROM THE (001) SURFACE OF
BARIUM TITANATE BETWEEN 20 DEGREES C AND 650 DEGREES C.
ABERDAM D + GAUBERT C
SURFACE SCI 27(3): 571-85 (1971)

04-0532　TEMPERATURE FACTORS OF TETRAGONAL AND CUBIC BARIUM TITANATE.
ABERDAM D
PHYS STATUS SOLIDI B 47: 159-69 (1971)
　　　249

04-0533　(FAST ION SCATTERING) SHADOW EFFECT IN (DISPLACIVE) PHASE TRANSITIONS. (BARIUM
TITANATE)
AGRANOVICH VM + RYABOV VA
SOV PHYS SOLID STATE 12(1): 207-9, (5): 1258 (1970)
　　　215

04-0534　STUDYING SMALL DEFORMATIONS OF A CRYSTAL LATTICE BASED ON THE SHADOW EFFECT. (NEUTRON
DIFFRACTION, BARIUM TITANATE)
AKHMETOVA BG + PLETS YUM + TULINOV AF
SOV PHYS JETP 29(3): 442-4 (SEP 1969)
　　　156

04-0535　CRITICAL X-RAY INCOHERENT SCATTERING BY BARIUM TITANATE NEAR THE CURIE TEMPERATURE.
ALEXANDROPOULOS NG
PHYS LETT A 34: 83-4 (1971)

04-0536　PIEZOELECTRIC EFFECT IN ALUMINUM DOPED BARIUM TITANATE CERAMIC.
AMIN M + TAWFEIK A
INDIAN J PHYS 46: 227-32 (1972)

CERAMIC ACOUSTIC DETECTORS. (BARIUM TITANATE)
ANANEVA AA
CONSULTANTS BUREAU, NY, 1965, 122P.

04-0537　DIELECTRIC PROPERTIES AND DOMAIN TEXTURE OF BARIUM TITANATE CRYSTALS.
ANDONOV P + LAMBERT M + QUITTET AM
P69-75, VOL 2 OF DVORAK V + FOUSKOVA A + GLOGAR P (EDS),
INT MEETING ON FERROELECTRICITY, PRAGUE, 1966, PROC.
INST PHYS CZECH ACAD SCI, PRAGUE, 1966.

04-0538　GENERATION OF SECOND OPTICAL HARMONIC IN THE CUBIC PHASE OF BARIUM TITANATE.
ARABIDZE AA + CHANTURIYA GF + KOKEVA VP
OPT SPECTROSC 31: 346 (1971)

04-0539　COMPENSATION OF HETEROVALENT FERRIC IONS IN THE LATTICE OF BARIUM TITANATE SINGLE
CRYSTALS.
ARÉND H + COUFCVA P
CZECH J PHYS B 13: 55-61 (1963)

04-0540 CHEMICAL PROCESSES DURING HYDROGEN REDUCTION OF BARIUM TITANATE CRYSTALS.
         AREND H + COUFOVA P + NOVAK J
         P255-61, VOL 2 OF DVORAK V + FOUSKOVA A + GLOGAR P (EDS),
         INT MEETING ON FERROELECTRICITY, PRAGUE, 1966, PROC.
         INST PHYS CZECH ACAD SCI, PRAGUE, 1966.

04-0541 ON THE PHASE COMPOSITION CF REDUCED AND REOXIDIZED BARIUM TITANATE.
         AREND H + KIHLBORG L
         P339-41 OF EUROPEAN MEETING ON FERROELECTRICITY, 1969, SAARBRUCKEN,
         MUSER HE AND PETERSSON J (EDS), WISSENSCH VERLAGSGES, STUTTGART,
         1970, 396P.

04-0542 DETERMINATION OF CHROMIUM VALENCY FROM STUDIES OF EPR SPECTRA AND STATIC MAGNETIC
         SUSCEPTIBILITY OF CHROMIUM DOPED BARIUM METATITANATE.
         ARIYA SM + VERBITSKAYA TN + ENDEN NM + WINTRUFF W
         P131-3 OF INT MEETING ON FERROELECTRICITY, 2ND, 1969, KYOTO, PROC.
         (J PHYS SOC JAP, VOL 28 SUPPL). PHYS SOC JAP, 1970, 459P.

04-0543 STATE OF CHROMIUM ADDED TO BARIUM METATITANATE.
         ARIYA SM + VERBITSKAYA TN + VINTRUFF V AND OTHERS
         INORG MATER 6: 1130-2 (JUL 1970)

04-0544 EFFECT OF FERRIC OXIDE ON THE 120 DEGREE PHASE TRANSITION IN BARIUM TITANATE SINGLE
         CRYSTALS.
         ARNOLD RT + STANFORD AL
         J CHEM PHYS 41: 1520-1 (1964)

04-0545 STUDY OF THE AGING PHENOMENON IN BARIUM TITANATE SINGLE CRYSTALS.
         BABA PD
         RUTGERS UNIV PHD THESIS, 1961, 129P.

         HIGH FREQUENCY SOUND ATTENUATION AND THERMAL CONDUCTIVITY IN DISPLACIVE TYPE
         FERROELECTRICS. (THEORY, BARIUM TITANATE)
         BALAGUROV BYA
         SOV PHYS JETP 34(4): 868-72 (1972)

04-0546 FAR INFRARED DIELECTRIC DISPERSION OF PEROVSKITES. (BARIUM TITANATE, POTASSIUM
         TANTALATE)
         BALLANTYNE JM
         P55-60, VOL 1 OF DVORAK V + FOUSKOVA A + GLOGAR P (EDS),
         INT MEETING ON FERROELECTRICITY, PRAGUE, 1966, PROC.
         INST PHYS CZECH ACAD SCI, PRAGUE, 1966.

04-0547 COUPLED OPTICAL PHONON MODE THEORY OF INFRARED DISPERSION IN BARIUM AND STRONTIUM
         TITANATES AND POTASSIUM TANTALATE.
         BARKER AS + HOPFIELD JJ
         PHYS REV 135(6): 1732-7 (1964)

04-0548 TEMPERATURE DEPENDENCE OF THE RAMAN CROSS SECTIONS IN BARIUM TITANATE AND STRONTIUM
         TITANATE.
         BARBOSA GA + CHAVES A + PORTO SPS
         SOLID STATE COMMUN 11: 1053-5 (1972)
             279

04-0549 ULTRASONIC STUDIES OF FIELD INDUCED PHASE TRANSITION IN BARIUM TITANATE.
         BARTZBEDNARCZYK D + BEDHARCZYK J + JANSKIEWICZ A
         ACTA UNIV BRATISL 58: 117-27 (1967)
             257

04-0550 TIME DEPENDENCE OF THE DIELECTRIC PERMITTIVITY OF HYDROXYL DOPED BARIUM TITANATE.
         BEIGE H + SCHMIDT G
         PHYS STATUS SOLIDI A 6: K151-3 (1971)
             244

04-0551 METHOD FOR ACHIEVING HIGH DIELECTRIC CONSTANT TEMPERATURE STABLE FERROELECTRICS.
         (BARIUM TITANATE, STRONTIUM STANNATE)
         BELDING WA + SMOKE EJ
         BULL AMER CERAM SOC 47(4): 390 (APR 1968)
             143

04-0552 PHOTOFERROELECTRIC EFFECTS IN A(V)B(VI)C(VII) AND BARIUM TITANATE TYPE FERROELECTRICS.
         BELYAEV LM + FRIDKIN VM + GREKOV AA + KOSONOGOV NA + RODIN AI
         J PHYS (PARIS) 33, SUPPL 4, COLLOQ 2: 123-5 (1972)
              294

04-0553 TOP SEEDED SOLUTION GROWTH OF OXIDE CRYSTALS FROM NONSTOICHIOMETRIC MELTS. (BARIUM
         TITANATE, STRONTIUM TITANATE, GERMANATES)
         BELRUSS V + KALNAJS J + LINZ A + FOLWEILER RC
         MATER RES BULL 6: 899-906 (1971)
              247

04-0554 DOUBLE INJECTION IN BARIUM TITANATE CERAMIC.
         BENGUIGUI L
         P326-30, VOL 2 OF DVORAK V + FOUSKOVA A + GLOGAR P (EDS),
         INT MEETING ON FERROELECTRICITY, PRAGUE, 1966, PROC.
         INST PHYS CZECH ACAD SCI, PRAGUE, 1966.

04-0555 FERROELECTRIC LOSSES IN (PURE) BARIUM TITANATE CERAMIC.
         BENGUIGUI L + MORICE M
         PHYS STATUS SOLIDI 37(1): K19-21 (1970)

04-0556 POLARITONS IN TETRAGONAL BARIUM TITANATE: THEORY OF THE RAMAN LINE SHAPE.
         BENSON HJ + MILLS DL
         SOLID STATE COMMUN 8: 1387-93 (1970)
              214

04-0557 SPACE CHARGE LIMITED CURRENT IN POLYCRYSTALLINE BARIUM TITANATE. (CURRENT VOLTAGE
         CHARACTERISTICS, 150-375 DEG C)
         BENGUIGUI L
         COMPT REND B 262(9): 642-5 (1966) (IN FRENCH)
              CHEM ABSTR 64: 16753F (1966)

04-0558 PERTURBED DIRECTIONAL CORRELATION OF SCANDIUM-44 IN THE FERROELECTRIC PHASES OF BARIUM
         TITANATE.
         BERTAIN LF
         UNIV NEVADA PHD THESIS, 1970, 96P
              DISS ABSTR INT B 31(9): 5559 (1971)
              234

04-0559 MICROWAVE DIELECTRIC LOSSES IN OXYGEN OCTAHEDRON FERROELECTRICS ABOVE THE CURIE
         TEMPERATURE. (BARIUM(0-1.0) STRONTIUM(1.0-0) TITANATE)
         BETHE K
         P259-71 OF EUROPEAN MEETING ON FERROELECTRICITY, 1969, SAARBRUCKEN,
         MUSER HE AND PETERSSON J (EDS), WISSENSCH VERLAGSGES, STUTTGART,
         1970, 396P.

04-0560 PREPARATION AND PROPERTIES OF (BARIUM, STRONTIUM) TITANATE SINGLE CRYSTALS.
         BETHE K + WELZ F
         MATER RES BULL 6: 209-17 (1971)

04-0561 SINGLE CRYSTALS OF THE SYSTEM (BARIUM, STRONTIUM) TITANATE. (SYNTHESIS AND LOW
         FREQUENCY DIELECTRIC PROPERTIES)
         BETHE K
         P343-50 OF EUROPEAN MEETING ON FERROELECTRICITY, 1969, SAARBRUCKEN,
         MUSER HE AND PETERSSON J (EDS), WISSENSCH VERLAGSGES, STUTTGART,
         1970, 396P.

04-0562 EFFECT OF DOPING BARIUM TITANATE WITH MANGANESE DIOXIDE.
         BHIDE VG + MULTANI MS
         PHYSICA 29: 23-32 (1963)

04-0563 IONIC CHARACTERS FROM MOSSBAUER ISOMER SHIFTS. (DEGREE OF COVALENCY IN IRON DOPED
         BARIUM TITANATE)
         BHIDE VG + SHENOY GK + MULTANI MS
         SOLID STATE COMMUN 2: 221-3 (1964)

04-0564 MOSSBAUER STUDIES OF LOWER TRANSITIONS IN BARIUM TITANATE. (LATTICE INSTABILITIES)
         BHIDE VG + DURGE VV
         SOLID STATE COMMUN 10: 401-3 (1972)
              263

04-0565　ELECTRIC AND MAGNETIC MEASUREMENTS OF CERTAIN TWO PHASE FERROELECTRIC FERROMAGNETIC
　　　　　MIXTURES. (BARIUM TITANATE)
　　　　　BIELSKA-LEWANDOWSKA H
　　　　　ACTA PHYS POLON A 39(6): 687-93 (1971)
　　　　　　　263

04-0566　DC BIAS EFFECTS IN BARIUM TITANATE CERAMICS.
　　　　　BIGGERS JV + SCHULZE WA
　　　　　BULL AMER CERAM SOC 51: 620-4 (1972)

04-0567　SURFACE STUDIES OF BARIUM TITANATE CERAMICS USING THE SCANNING ELECTRON MICROSCOPE.
　　　　　BIGGERS JV
　　　　　BULL AMER CERAM SOC 49: 418 (1970)

04-0568　ELECTRICAL PROPERTIES OF CRYSTALLITES AND INTERCRYSTALLITE LAYERS IN THE SEMICONDUCTOR
　　　　　CERAMIC BARIUM TITANATE. (CERIUM DOPED)
　　　　　BOGATIN AS + PROKOPALO OI
　　　　　IZV VUZ FIZ 1970(3): 96-100 (IN RUSSIAN)
　　　　　(TRANSLATION TO APPEAR IN SOV PHYS J)
　　　　　　　CHEM ABSTR 73: 39938
　　　　　　　213

04-0569　SURFACE CONDUCTIVITY IN POLARIZED BARIUM TITANATE CRYSTALS.
　　　　　BOGATKO VV + KOVTONYUK NF
　　　　　SOV PHYS SOLID STATE 12(2): 459-60 (1970)
　　　　　　　206

04-0570　THERMOSTABILIZING PROPERTIES OF SEMICONDUCTING BARIUM TITANATE.
　　　　　BOGATIN AS + PROKOPALO OI
　　　　　BULL ACAD SCI USSR, PHYS SER 34(12): 2332-7 (1970)

04-0571　IONIC CONTRIBUTION TO INJECTION AND CONDUCTION IN BARIUM TITANATE (CONDENSER CHARGE-
　　　　　DISCHARGE METHOD).
　　　　　BOISSIERE C + CHANUSSOT G
　　　　　COMPT REND B 271(14): 729-32 (1970) (IN FRENCH)
　　　　　　　227

04-0572　INVERSION OF ELECTRICAL PROPERTIES OF SEMICONDUCTING BARIUM TITANATE.
　　　　　BOIS GV + ZHUKOVSKII VI + MIKHAILOVA NA
　　　　　BULL ACAD SCI USSR, PHYS SER 35(9): 1785-8 (1971)

04-0573　CHANGE OF THE MOSSBAUER EFFECT PROBABILITY ON TIN-119 IN BARIUM TITANATE STANNATE SOLID
　　　　　SOLUTIONS IN THE VICINITY OF THE FERROELECTRIC PHASE TRANSITION.
　　　　　BOKOV VA + ROMANOV VP + VERKIN BI + KAZAKEVICH LI + CHEKIN VV
　　　　　P80-6, VOL 1 OF DVORAK V + FOUSKOVA A + GLOGAR P (EDS),
　　　　　INT MEETING ON FERROELECTRICITY, PRAGUE, 1966, PROC.
　　　　　INST PHYS CZECH ACAD SCI, PRAGUE, 1966.

04-0574　PHOTOCONDUCTIVITY AND PHOTOFERROELECTRIC PHENOMENA IN BARIUM TITANATE.
　　　　　BOLK TR + GREKOV AA + KOSONOGOV NA AND OTHERS
　　　　　SOV PHYS CRYSTALLOGR 16: 198-200 (JUL-AUG 1971)

04-0575　DIELECTRIC PROPERTIES OF BARIUM TITANATE CONTAINING NIOBIUM AND THE EFFECT OF
　　　　　ADDITIVES.
　　　　　BONSACK JP
　　　　　BULL AMER CERAM SOC 50: 488-92 (MAY 1971)

04-0576　DIELECTRIC AND OPTICAL PROPERTIES OF TRANSPARENT FERROELECTRIC GLASS CERAMIC SYSTEMS.
　　　　　BORRELLI NF + LAYTON MM
　　　　　J NONCRYST SOLIDS 6: 197-212 (1971)

04-0577　PHASE BOUNDARIES IN BARIUM TITANATE SINGLE CRYSTALS NEAR CURIE POINT.
　　　　　BORODIN VZ + GORELOV MI + BORODINA VA
　　　　　BULL ACAD SCI USSR, PHYS SER 35(9): 1698-702 (1971)

04-0578　ELECTROSTATIC FIELD IN A SLIGHTLY ORTHORHOMBIC IONIC CRYSTAL. APPLICATION TO THE
　　　　　CALCULATION OF TETRAGONAL BIREFRINGENCE IN BARIUM AND LEAD TITANATES.
　　　　　BOUILLOT J + MACHET R + TSALLIS C
　　　　　PHYS STATUS SOLIDI 38: 313-16 (1970)
　　　　　　　190

4. BARIUM TITANATE

THEORETICAL STATIC MODEL FOR BARIUM AND LEAD TITANATES - APPLICATION TO CALCULATING THE
WALL ENERGY AND THE DIELECTRIC CONSTANTS.
BOUILLOT J + MACHET R
P111-17 OF EUROPEAN MEETING CN FERROELECTRICITY, 1969, SAARBRUCKEN,
MUSER HE AND PETERSSON J(EDS), WISSENSCH VERLAGSGES, STUTTGART,
1970, 396P.

04-0579   AGING IN TETRAGONAL FERROELECTRIC BARIUM TITANATE.
           BRADT RC + ANSELL GS
           BULL AMER CERAM SOC 47(4): 355 (APR 1968)
               143

04-0580   DIELECTRIC BREAKDOWN IN BARIUM TITANATE.
           BRANWOOD A + HURD JD + TREDGOLD RH
           BRIT J APPL PHYS 13: 528 (1962)

04-0581   GRAIN BOUNDARY BARRIER LAYERS IN (N TYPE, ANTIMONY DOPED) BARIUM TITANATE CERAMIC WITH
           A HIGH EFFECTIVE DIELECTRIC CONSTANT (DUE TO ADDING CUPROUS OR FERRIC OXIDE).
           BRAUER H
           Z ANGEW PHYS 29(5): 282-7 (1970) (IN GERMAN)
               216

04-0582   GRAIN SIZE DEPENDENCE OF DIELECTRIC AND STRUCTURAL CHARACTERISTICS OF BARIUM TITANATE.
           BRADLEY FN
           BULL AMER CERAM SOC 50: 396 (1971)

04-0583   PRODUCTION AND STUDY OF THE INNER BARRIERS IN SEMICONDUCTING BARIUM TITANATE CERAMICS.
           BRAUER H + KUSCHKE R
           KERAM Z 22: 735-41 (1971) (IN GERMAN)

04-0584   RESISTIVITY ANOMALY IN SEMICONDUCTING BARIUM TITANATE. (POSITIVE TEMPERATURE
           COEFFICIENT OF RESISTIVITY)
           BRAHMECHA BG + SINHA KP
           JAP J APPL PHYS 10(4): 496-504 (1971)
               267

04-0585   INTERACTION BETWEEN THE 90 DEG AND 180 DEG DOMAINS IN BARIUM TITANATE.
           BREZINA B + FOUSEK J
           SOV PHYS SOLID STATE 4: 1030-7 (1962)

04-0586   CONCENTRATION DEPENDENCE OF THE CURIE-WEISS CONSTANT IN POLYCRYSTALLINE SOLID SOLUTIONS
           IN THE (BARIUM, STRONTIUM) TITANATE SYSTEM.
           BROK AYA + VETSSILE ZA
           P109-15 OF FAZOVYE PEREKHODY V SEGNETOELEKTRIKAKH (PHASE
           TRANSFORMATICNS IN FERROELECTRICS), FRITSBERG VYA + ROLOV BN +
           KRUCHAN YY(EDS), IZDATELSTVO ZINATNE RIGA, 1971, 206P. (IN RUSSIAN)

04-0587   EFFECT OF OXIDE IMPURITIES ON THE ELECTRICAL RESISTIVITY OF LANTHANUM DOPED BARIUM
           TITANATE.
           BROWN DJ + SLY FAW + ARTHUR G
           PROC BRIT CERAM SOC NO 10: 195-203 (MAR 1968)
               PHYS ABSTR 71: 49229 (1968)
               141

04-0588   SHOCK WAVE STUDIES OF SURFACE LAYERS IN BARIUM TITANATE.
           BRODY PS + LEAVITT RP
           BULL AMER PHYS SOC 16: 1427 (1971)

04-0589   SHOCK INDUCED TRANSITION IN BARIUM TITANATE.
           BRODY PS
           HARRY DIAMOND LABS, WASHINGTON, DC, 1970, 27P. AD717551

04-0590   BINARY AND PSEUDOBINARY PHASES BETWEEN NIOBIUM PENTOXIDE AND LEAD, BARIUM, AND
           STRONTIUM OXIDES.
           BRUSSET H + GILLIER-PANDRAUD H + MAHE R + VOLIOTIS SD
           MATER RES BULL 6: 413-24 (1971) (IN FRENCH)

04-0591   TEMPERATURE DEPENDENCE OF THE DIELECTRIC CONSTANT OF FERROELECTRIC SUBSTANCES. (BARIUM
           TITANATE CERAMICS)
           BRUECKMANN H + ENGELS W
           PHYS LETT 2: 282-4 (1962) (IN GERMAN)

4. BARIUM TITANATE

04-0592   SATURATION OF VOLT- AMPERE CHARACTERISTICS OF CADMIUM SELENIDE LAYERS ON BARIUM
           TITANATE- STRONTIUM TITANATE SUBSTRATES AT SUPERSONIC DRIFT VELOCITY OF CURRENT
           CARRIERS.
           BRYUZGIN AR + MISELYUK EG + NEKRASOV MM AND OTHERS
           SOV PHYS DOKL 16(6): 466-7 (1971)

04-0593   ACOUSTIC EMISSION FROM FERROELECTRIC CRYSTALS. (BARIUM TITANATE, TGS)
           BUCHMAN P
           SOLID STATE ELECTRON 15: 142-4 (JAN 1972)

04-0594   EFFECT OF FERROELECTRIC (BARIUM TITANATE) POLARIZATION FIELDS ON (GERMANIUM OR
           TELLURIUM) SEMICONDUCTOR FILMS. (AND USE OF THESE FILMS TO STUDY SURFACE STRAINS IN THE
           FERROELECTRIC CRYSTAL)
           BUCHMAN P + DIAMOND H
           P313-25, VOL 2 OF DVORAK V + FOUSKOVA A + GLOGAR P (EDS),
           INT MEETING ON FERROELECTRICITY, PRAGUE, 1966, PROC.
           INST PHYS CZECH ACAD SCI, PRAGUE, 1966.

04-0595   EFFECTS OF GRAIN GROWTH ON THE DISTRIBUTION OF NIOBIUM IN BARIUM TITANATE CERAMICS.
           BUESSEM WB + KAHN M
           J AMER CERAM SOC 54: 458-61 (SEP 1971)

04-0596   SURFACE BREAKDOWN IN VACUUM ON BARIUM TITANATE.
           BUGAEV SP + KREMNEV VV + TERENTEV YUI + SHPAK VG + YURIKE YAYA
           SOV PHYS TECH PHYS 17: 1547-51 (1972)

04-0597   DIELECTRIC AND ELASTIC PROPERTIES OF SINGLE CRYSTALS OF BARIUM LEAD TITANATE SOLID
           SOLUTIONS.
           BUNINA LK + KUDZIN AYU
           SOV PHYS SOLID STATE 10(10): 2496-7 (1969)
               141

04-0598   EFFECT OF IMPURITIES AND GROWTH CONDITIONS ON THE STRUCTURE OF BARIUM TITANATE CRYSTALS
           GROWN FROM SOLUTION IN POTASSIUM FLUORIDE MELT.
           BUNINA LK + GUENOK EP + KUDZIN AYU
           P3-11 OF SEGNETOELEKTRIKAKH I OKISNYE POLUPROVODNIKI (FERROELECTRICS
           AND OXIDE SEMICONDUCTORS), SINYAKOV EV (ED), REDAKTSIONNO-IZDATELSKII
           OTDEL DNEPROPETROVSKII GOSUDARSTVENNYI UNNIVERSITET, DNEPROPETROVSK,
           1971, 100P. (IN RUSSIAN)

04-0599   INFLUENCE OF PHASE TRANSITIONS ON ELECTRICAL RESISTANCE OF BARIUM TITANATE.
           BUNGET I + LEONOVICI MR + NICOLAU P
           REV ROUM PHYS 17: 423-32 (1972)

04-0600   DIELECTRIC CONSTANT AND LATTICE DYNAMICS OF A THIN FILM OF A FERROELECTRIC (BARIUM
           TITANATE DOWN TO 3 MICRONS).
           BURSIAN EV + GIRSHBERG YAG + MAKAROV KV + ZAIKOVSKII OI
           SOV PHYS SOLID STATE 12(6): 1472-3 (1970)
               221

04-0601   DISPERSION AT MILLIMETER WAVELENGTHS IN BARIUM TITANATE ABOVE THE TRANSITION POINT.
           BURSIAN EV + RYCHGORSKII VV + GIRSHBERG YAG
           SOV PHYS SOLID STATE 13(2): 433-5 (AUG 1971)
               240

04-0602   ENERGY STORAGE IN CERAMIC DIELECTRICS. (STRONTIUM TITANATE, BARIUM TITANATE)
           BURN I + SMYTH DM
           J MATER SCI 7: 339-43 (1972)
               261

04-0603   ELECTRICAL AND OPTICAL PROPERTIES OF (BARIUM, STRONTIUM) MIXED TITANATE THIN SECTIONS.
           BURFOOT JC + SANVORDENKER V
           P424-32, VOL 1 OF DVORAK V + FOUSKOVA A + GLOGAR P (EDS),
           INT MEETING ON FERROELECTRICITY, PRAGUE, 1966, PROC.
           INST PHYS CZECH ACAD SCI, PRAGUE, 1966.

04-0604   GROWTH AND STRUCTURE OF BARIUM TITANATE EVAPORATED FILMS.
           BURFOOT JC + SLACK JR
           P363-6 OF EUROPEAN MEETING ON FERROELECTRICITY SAARBRUCKEN
           MARCH 1969. ED BY H.E. MUESER, J. PETERSSON. WISSENSCH VERLAGSGES
           STUTTGART, 1970
               205

4. BARIUM TITANATE

04-0605   GROWTH STRUCTURE (MICROSTRUCTURE) AND ELECTRICAL PROPERTIES OF FLASH EVAPORATED BARIUM
           TITANATE THIN FILMS (SEPARATED FROM SUBSTRATE).
           BURFOOT JC + SLACK JR
           P417-20 OF INT MEETING ON FERROELECTRICITY, 2ND, SEP 1969, KYOTO,
           PROC (J PHYS SOC JAPAN VOL 28 SUPPL 1970)
                207

04-0606   INTERPRETATION OF THE DYNAMICS OF THE BARIUM TITANATE TRANSITION IN TERMS OF
           NUCLEATION. (CONCEPTS)
           BURFOOT JC + PARKER TJ
           P123-7 OF EUROPEAN MEETING ON FERROELECTRICITY, 1969, SAARBRUCKEN,
           MUSER HE AND PETERSSON J (EDS), WISSENSCH VERLAGSGES, STUTTGART,
           1970, 396P.

04-0607   MAGNITUDE OF SPONTANEOUS POLARIZATION IN THIN CRYSTALS AND FINE PARTICLES OF BARIUM
           TITANATE.
           BURSIAN EV + ZAIKOVSKII OI + SMIRNOVA NP + SHAPKIN VV
           P387-93, VOL 1 OF DVORAK V + FOUSKOVA A + GLOGAR P (EDS),
           INT MEETING ON FERROELECTRICITY, PRAGUE, 1966, PROC.
           INST PHYS CZECH ACAD SCI, PRAGUE, 1966.

04-0608   NUCLEATION MODELS FOR A FIRST ORDER TRANSITION FRONT, AND OBSERVATIONS IN BARIUM
           TITANATE.
           BURFOOT JC + PARKER TJ
           P364-5 OF INT MEETING ON FERROELECTRICITY, 2ND, 1969, KYOTO, PROC,
           (J PHYS SOC JAP, VOL 28 SUPPL). PHYS SOC JAP, 1970, 459P.

04-0609   POLARON MECHANISM OF CONDUCTIVITY IN BARIUM TITANATE.
           BURSIAN EV + GIRSHBERG YAG + STAROV EN
           BULL ACAD SCI USSR, PHYS SER 35(9): 1782-5 (1971)

04-0610   POWDER RAMAN STUDY FOR BARIUM TITANATE.
           BURNS G + SCOTT BA
           BULL AMER PHYS SOC 16: 415 (1971)

04-0611   RAMAN SCATTERING IN THE FERROELECTRIC SYSTEM LEAD BARIUM TITANATE.
           BURNS G + SCOTT BA
           SOLID STATE COMMUN 9: 813-17 (1971)
                239

04-0612   SINGLE CRYSTAL BARIUM TITANATE FILMS GROWN FROM THE MELT IN AN OXYGEN ATMOSPHERE.
           BURSIAN EV + SMIRNOVA NP
           SOV PHYS SOLID STATE 4: 1231-2 (1962)

04-0613   SMALL POLARONS IN CONDUCTING BARIUM TITANATE CRYSTALS.
           BURSIAN EV + GIRSHBERG YAG + STAROV EN
           PHYS STATUS SOLIDI B 46(2): 529-33 (1971)

04-0614   TEMPERATURE DEPENDENCE OF CARRIER MOBILITY IN BARIUM TITANATE.
           BURSIAN EV + GIRSHBERG YAG + STAROV EN
           SOV PHYS SOLID STATE 14(4): 872-5 (OCT 1972)
                276

04-0615   EFFECT OF "A" DOMAINS UPON "C" DOMAIN SWITCHING IN BARIUM TITANATE.
           CALLABY DR
           P105-11, VOL 2 OF DVORAK V + FOUSKOVA A + GLOGAR P (EDS),
           INT MEETING ON FERROELECTRICITY, PRAGUE, 1966, PROC.
           INST PHYS CZECH ACAD SCI, PRAGUE, 1966.

04-0616   DIELECTRIC PROPERTIES OF SINGLE DOMAIN MELT GROWN BARIUM TITANATE.
           CAMLIBEL I + DIDOMENICO M + WEMPLE SH
           J PHYS CHEM SOLIDS 31: 1417-19 (1970)

04-0617   THE ORTHORHOMBIC PHASE IN MODIFIED BARIUM TITANATE. (EFFECT OF SUBSTITUTIONS)
           CASTELLIZ LM
           BULL AMER CERAM SOC 47(4): 390 (APR 1968)
                143

04-0618   SOME CRYSTAL CHEMICAL AND PHYSICAL PROPERTIES OF SINGLE CRYSTALS OF BARIUM TITANATE
           WITH COBALT ADDED.
           CAUFOVA P + AREND H + NOVAK J
           SOV PHYS CRYSTALLOGR 9: 92 (1964)

04-0619   COUPLED POLARITIONS OF A1 SYMMETRY IN BARIUM TITANATE.
          CHAVES A + ANDRADE PR + KATIYAR RS + PORTO SPS
          BULL AMER PHYS SOC 17: 1190 (1972)

04-0620   EXPERIMENTAL AND THEORETICAL STUDY OF DEFECTS IN TRIGLYCINE SULFATE AND BARIUM
          TITANATE.
          CHANUSSOT G
          PHD THESIS, DIJON, 1970, 197P. ARCH ORIG CENTRE DOCUMENT CNRS, 4706,
          23 JUN 1970, (IN FRENCH)
               238

04-0621   PHASE CHANGES IN TRIGLYCINE SULFATE AND BARIUM TITANATE FERROELECTRICS. (26 REFS)
          CHANUSSOT G
          P125-41 OF SUCHET JP(ED), INFLUENCE DES CHANGEMENTS DE PHASE SUR LES
          PROPERIETES PHYSIQUES DES CORPS SOLIDS. MASSON, PARIS, 1970.
          (IN FRENCH)
               CHEM ABSTR 72: 137545N (1970)
               238

04-0622   SOME PROPERTIES OF THE THERMOCURRENTS IN BARIUM TITANATE AND TITANIUM DIOXIDE SINGLE
          CRYSTALS.
          CHANUSSOT G + PIA G + MICHERON F + GODEFROY L
          P313-24 OF EUROPEAN MEETING ON FERROELECTRICITY, 1969, SAARBRUCKEN,
          MUSER HE AND PETERSSON J(EDS), WISSENSCH VERLAGSGES, STUTTGART,
          1970, 396P.

04-0623   STRUCTURAL DI- AND FERROELECTRIC PROPERTIES OF X POTASSIUM DINIOBATE AND (1-X) BARIUM
          TITANATE.
          CHINCHOLKAR VS
          Z ANGEW PHYS 28: 288-90 (1970)

04-0624   THE PRODUCTION OF BARIUM TITANATE FROM BARIUM SULFATE AND TITANIUM DIOXIDE IN THE
          PRESENCE OF SODIUM CARBONATE AT HIGH TEMPERATURE.
          CHILVERS JO + BARHAM D
          J CAN CERAM SOC 40: 55-7 (1971)

04-0625   HYDROTHERMAL PREPARATION OF BARIUM TITANATE BY TRANSPORT REACTIONS (INVOLVING TITANIUM
          ESTERS TO YIELD SMALL CRYSTAL PRISMS).
          CHRISTENSEN AN
          ACTA CHEM SCAND 24: 2447-52 (1970)
               227

04-0626   HYDROTHERMAL PREPARATION OF BARIUM TITANATE BY TRANSPORT REACTIONS.
          CHRISTENSEN AN
          ACTA CRYSTALLOGR A 25: S6 (1969)

04-0627   VARIATION IN CERTAIN ELECTRICAL PROPERTIES OF SINTERED BARIUM TITANATE AS A FUNCTION OF
          GRAIN SIZE.
          CIARRAPICO JO + LEVIALDI A + SHOIJET M
          COMPT REND 254: 1219-21 (1962) (IN FRENCH)

04-0628   INFLUENCE OF CRITICAL POLARIZATION FLUCTUATIONS ON THE PHOTOELASTIC BEHAVIOR OF BARIUM
          TITANATE.
          COHEN MG + DIDOMENICO M + WEMPLE SH
          PHYS REV B 1(11): 4334-7 (1970)
               205

04-0629   LINEAR DISORDER IN CRYSTALS (CASE OF SILICON, QUARTZ AND FERROELECTRIC PEROVSKITES).
          (BARIUM TITANATE AND POTASSIUM NIOBATE)
          COMES R + LAMBERT M + GUINIER A
          ACTA CRYSTALLOGR A 26(2): 244-54 (1970) (IN FRENCH)
               199

04-0630   STRUCTURE DISORDER OF BARIUM TITANATE TYPE FERROELECTRICS. BASED ON NEW TREATMENT AND
          INTERPRETATION OF X-RAY DATA)
          COMES R + LAMBERT M + GUINIER A
          P195-8 OF INT MEETING ON FERROELECTRICITY, 2ND, 1969, KYOTO, PROC.
          (J PHYS SOC JAP, VOL 28 SUPPL). PHYS SOC JAP, 1970, 459P.

04-0631   NONSTOICHIOMETRY IN BARIUM TITANATE.
          CONGER GJ + ANDERSON HU
          J AMER CERAM SOC 55: 539 (OCT 1972)

04-0632   MINIMUM DETECTABLE POWER OF A PYROELECTRIC THERMAL RECEIVER. (USING BARIUM TITANATE
          CRYSTAL)
          COOPER J
          REV SCI INSTRUM 33: 92-5 (1962)

04-0633   CRYSTAL CHEMICAL PROBLEMS OF (BARIUM, STRONTIUM) TITANATE SOLID SOLUTIONS.
          COUFOVA P + NOVAK J
          KRISTALL TECH 6: K1-K6 (1971)

04-0634   DIELECTRIC PROPERTIES OF BARIUM TITANATE SINGLE CRYSTALS DOPED BY HYDROXIDE GROUPS.
          COUFOVA P + JANOUSEK V
          P357-62 OF EUROPEAN MEETING ON FERROELECTRICITY, 1969, SAARBRUCKEN,
          MUSER HE AND PETERSSON J(EDS), WISSENSCH VERLAGSGES, STUTTGART,
          1970, 396P.

04-0635   DIELECTRIC PROPERTIES OF BARIUM(1-X) STRONTIUM(X) TITANATE SINGLE CRYSTALS GROWN FROM A
          POTASSIUM FLUORIDE FLUX.
          COUFOVA P + JANOUSEK V + NOVAK J
          CZECH J PHYS B 22: 485-9 (1972)
              274

04-0636   HYDROXYL AS DEFECT IN BARIUM TITANATE.
          COUFOVA P + NOVAK J
          P262-6, VOL 2 OF DVORAK V + FOUSKOVA A + GLOGAR P (EDS),
          INT MEETING ON FERROELECTRICITY, PRAGUE, 1966, PROC.
          INST PHYS CZECH ACAD SCI, PRAGUE, 1966.

04-0637   THICKNESS DEPENDENCE OF THE PERMITTIVITY OF BARIUM TITANATE SINGLE CRYSTALS.
          COUFOVA P + AREND H
          CZECH J PHYS B 12: 308-12 (1962)

04-0638   DOUBLE SPACE CHARGE INJECTION IN SOLIDS. (USING BARIUM TITANATE ABOVE CURIE POINT)
          COX GA + TREDGOLD RH
          PHYS LETT 4: 199-200 (1963)

04-0639   SURFACE DOMAIN REORIENTATION PRODUCED BY ABRASION AND ANNEALING OF POLYCRYSTALLINE
          BARIUM TITANATE.
          CUTTER IA + MCPHERSON R
          J AMER CERAM SOC 55: 334-6 (JUL 1972)

04-0640   ELECTRON PARAMAGNETIC RESONANCE IN NONSTOICHIOMETRIC BARIUM TITANATE.
          DANILYUK YUL + KHARITONOV EV
          SOV PHYS SOLID STATE 6: 260-1 (1964)

04-0641   STRUCTURE OF 90 DEGREE DOMAIN WALLS IN BARIUM TITANATE.
          DARINSKII BM + FEDOSOV VN
          SOV PHYS SOLID STATE 13(1): 17-20 (JUL 1971)
              237

04-0642   CALCULATION OF THE ELECTROSTATIC INTERACTIONS IN FINE GRAINED PEROVSKITE FERROELECTRIC
          CERAMICS. (BARIUM TITANATE)
          DE LAUNAY J
          NAVAL RES LAB, 1971, 143P. AD728995

04-0643   MOVING STRIPE PATTERN IN BARIUM TITANATE.
          DEGUCHI K + NAKAMURA E
          J PHYS SOC JAP 30: 301 (1971)

04-0644   ABSORPTION AND REFLECTION SPECTRA OF BARIUM TITANATE IN THE LONG WAVELENGTH INFRARED
          REGION.
          DEMESHINA AI + MURZIN VN
          SOV PHYS SOLID STATE 4: 2185-6 (1963)

04-0645   DIELECTRIC PERMEABILITY OF BARIUM TITANATE SINGLE CRYSTALS IN THE PHASE TRANSITION
          REGION.
          DEMYANOV VV + SOLOVEV SP
          SOV PHYS CRYSTALLOGR 17(3): 494-503 (NOV-DEC 1972)
              280

04-0646   DIELECTRIC DISPERSION IN BARIUM TITANATE.
          DEMYANOV VV + SOLOVEV SP
          BULL ACAD SCI USSR, PHYS SER 34(12): 2297-301 (1970)
              241

04-0647   DIELECTRIC SPECTRUM OF BARIUM TITANATE. (AND MICROSCOPIC MODEL OF LOCALLY ORDERED
           "NUCLEI", PHASE STRUCTURE ABOVE CURIE POINT, RAMAN SPECTRUM OF STRONTIUM TITANATE)
           DEMYANOV VV + SOLOVEV SP
           P181-4 OF INT MEETING ON FERROELECTRICITY, 2ND, 1969, KYOTO, PROC,
           (J PHYS SOC JAP, VOL 28 SUPPL). PHYS SOC JAP, 1970, 459P.

04-0648   DIELECTRIC DISPERSION OF IRRADIATED BARIUM TITANATE NEAR THE PHASE TRASNITION.
           DEMYANOV VV + SOLOVEV SP
           REV PHYS APPL, SUPPL J PHYS 7: 81-2 (JUN 1972)
                 286

04-0649   EXPERIMENTAL CHECK OF THE THEORY OF MICROWAVE DISPERSION OF THE PERMITTIVITY OF BARIUM
           TITANATE TYPE FERROELECTRICS.
           DEMYANOV VV
           SOV PHYS SOLID STATE 12(8): 1921-5 (1971)
                 227

04-0650   KINETICS OF THE PHASE TRANSITIONS OF BARIUM TITANATE.
           DEMYANOV VV
           INORG MATER 7: 875-80 (JUN 1971)

04-0651   DOMAIN INTERACTIONS DURING PHASE TRANSITIONS IN BARIUM TITANATE.
           DENNIS MD
           BULL AMER CERAM SOC 51: 355 (1972)

04-0652   DIELECTRIC DISPERSION OF IRRADIATED BARIUM TITANATE IN THE PHASE TRANSITION REGION.
           DENYANOV VV + SOLOVEV SP
           INORG MATER 8(11): 1802-4 (1972)

04-0653   BINARY LIGHT BEAM DEFLECTION IN SINGLE CRYSTAL BARIUM TITANATE.
           DRAKE MD + JOSEPH RI + PALMER CH
           J APPL PHYS 42: 1950-7 (1971)
           IEEE TRANS SON ULTRASON 19: 340 (1972)

04-0654   BINARY LIGHT BEAM DETECTION IN SINGLE CRYSTAL BARIUM TITANATE.
           DRAKE MD + JOSEPH RI + PALMER CH
           FERROELECTRICS 3: 340 (1971)

04-0655   ANTIPARALLEL SWITCHING IN ALPHA DOMAIN BARIUM TITANATE CRYSTALS.
           DUDA VM + DUDNIK EF + SINYAKOV EV
           SOV PHYS SOLID STATE 14(7): 1844-5 (JAN 1973)
                 282

04-0656   SIZE EFFECTS IN FERROELECTRICS. (BARIUM TITANATE)
           DUDKEVICH VP + FESENKO EG + MARGOLIN AM + BONDARENKO VS + GARBUZ NG + NOVOSILTSEV VN
           BULL ACAD SCI USSR, PHYS SER 35(9): 1773-6 (1971)

04-0657   PREPARATION OF SODIUM AND BARIUM NIOBATES AND LEAD MOLYBDATE.
           DUMAS J + VILLELA G + BONIORT JY
           DGRST, 1971, 13P. FINAL REPT CONTRACT 7002170.
                 268

04-0658   ANGULAR ENERGY DEPENDENCE OF 180 DEG DOMAIN WALL IN BARIUM TITANATE. (THERMODYNAMIC
           CALCULATIONS)
           DVORAK V + JANOVEC V
           JAP J APPL PHYS 4(6): 400-2 (1965)

04-0659   ON THE INTERPRETATION OF INFRARED SPECTRA AND THE LATTICE DYNAMICS OF CUBIC BARIUM
           TITANATE.
           DVORAK V + JANOVEC V
           CZECH J PHYS B 12: 461-70 (1962)

04-0660   EFFECT OF SILICA ADDITIONS ON THE SEMICONDUCTING PROPERTIES OF DOPED BARIUM TITANATE.
           EASTMAN C + ELYARD CA + WARREN D
           PROC BRIT CERAM SOC NO 18: 77-85 (1970)

04-0661   INVESTIGATION OF ANISOTROPY IN THE DIELECTRIC PROPERTIES OF FERROELECTRICS, SUBJECTED
           BEFOREHAND TO AN ELECTRIC FIELD, IN STRONG ELECTRIC FIELDS. (ANISOTROPY DEVELOPED,
           BARIUM TITANATE AND BARIUM(0.95) CALCIUM(0.05) TITANATE USED)
           ELGARD AM
           SOV PHYS SOLID STATE 4: 967-70 (1962)

4. BARIUM TITANATE

04-0662   VIBRONIC EFFECT IN THE TEMPERATURE DEPENDENCE OF THE SOFT FREQUENCY OF BARIUM TITANATE.
ENGLMAN R
P51-4, VOL 1 OF DVORAK V + FOUSKOVA A + GLOGAR P (EDS),
INT MEETING ON FERROELECTRICITY, PRAGUE, 1966, PROC.
INST PHYS CZECH ACAD SCI, PRAGUE, 1966.

04-0663   ELECTRICAL CONDUCTIVITY CF SINGLE CRYSTALLINE BARIUM TITANATE.
EROR NG + SMYTH DM
BULL AMER CERAM SOC 47(4): 354 (APR 1968)
     143

04-0664   OXYGEN STOICHIOMETRY OF DONOR DOPED BARIUM TITANATE AND TITANIUM DIOXIDE.
(STOICHIOMETRIC COMPENSATION)
EROR NG + SMYTH DM
P62-74 OF INST FOR ADVANCED STUDY ON THE CHEMISTRY OF
EXTENDED DEFECTS IN NON-METALLIC SOLIDS, 1969,
SCOTTSDALE, ARIZ. EYRING I AND O"KEEFFE M(EDS),
NORTH-HOLLAND, 1970

04-0665   FIELD DEPENDENCE OF ABSORPTION BANDS OF FERROELECTRICS IN FAR INFRARED. (APPLIED TO
BARIUM TITANATE)
FATUZZO E
PROC PHYS SOC 84(541, PT5): 709-17 (1964)

04-0666   POLARIZED STATES OF FERROELECTRIC CERAMICS. (BARIUM TITANATE)
FEDCHENKO ED + IBRAIMOV NS + KUZMIN EN
INORG MATER 6: 1496-8 (SEP 1970)

04-0667   PRODUCTION AND STUDY OF PURE AND DOPED BARIUM TITANATE SINGLE CRYSTAL
FELTZ A + LANGBEIN H
KRISTALL TECH 6: 359-66 (1971) (IN GERMAN)

04-0668   PREPARATION AND PROPERTIES OF PURE AND DOPED BARIUM TITANATE SINGLE CRYSTALS.
FELTZ A + LANGBEIN H
KRISTALL TECH 6: 359-66 (1971) (IN GERMAN)

04-0669   PREPARATION AND PROPERTIES OF THIN BARIUM TITANATE FILMS.
FEUERSANGER AE + HAGENLOCKER AH + SOLOMON AL
J ELECTROCHEM SOC 111: 1387-91 (1964)

04-0670   PREPARATION AND PROPERTIES OF HIGH PERMITTIVITY THIN FILM DIELECTRICS. (BARIUM AND LEAD
TITANATES, EFFECT OF FERROELECTRIC CHARACTERISTICS ON ELECTRIC PROPERTIES)
FEUERSANGER AE
P209-36 OF VRATNY F(ED), THIN FILM DIELECTRICS, ELECTROCHEM SOC,
NY, 1969, 669P.

04-0671   ACOUSTIC SOFT OPTIC MODE INTERACTIONS IN FERROELECTRIC BARIUM TITANATE.
FLEURY PA + LAZAY PD
PHYS REV LETT 26(21): 1331-4 (24 MAY 1971)
     235

04-0672   LIGHT EMISSION FROM BARIUM TITANATE CRYSTALS SUBJECTED TO UNIAXIAL STRESS PULSES.
FLEROVA SA + SAMCHENKO YUI
SOV PHYS SOLID STATE 14(2): 498-9 (AUG 1972)
     274

04-0673   CALCULATION OF (DOMAIN) WALL ENERGY IN BARIUM TITANATE.
FONTAINE G
J PHYS CHEM SOLIDS 28(11): 2199-206 (1967)

04-0674   EFFECTS OF ANISOTROPY OF ELASTIC CONSTANTS ON DISLOCATIONS IN FERROELECTRIC BARIUM
TITANATE.
FONTAINE G
PHYS STATUS SOLIDI 5(1): 203-6 (1964) (IN FRENCH)
     CHEM ABSTR 61: 3760D (1964)

04-0675   LINEAR DISORDER AND TEMPERATURE DEPENDENCE OF RAMAN SCATTERING IN BARIUM TITANATE.
FONTANA MP + LAMBERT M
SOLID STATE COMMUN 10: 1-4 (1972)
     260

4. BARIUM TITANATE

04-0676 STRUCTURE DISORDER AND RAMAN SCATTERING IN BARIUM TITANATE AND POTASSIUM NIOBATE.
FONTANA MP + LAMBERT M + QUITTET AM
J PHYS (PARIS) 33, SUPPL 4, COLLOQ 2: 54 (1972)
294

04-0677 (STEP OR SMOOTH) DOMAIN WALL STRUCTURE IN BARIUM TITANATE DURING SWITCHING
(CALCULATIONS COMPARED WITH DATA).
FOUSKOVA A
CZECH J PHYS B 20: 790-6 (1970)
215

04-0678 STRUCTURE OF A MOVING DOMAIN WALL. (MECHANISM OF A WALL MOTION IN BARIUM TITANATE)
FOUSKOVA A
P235-8 OF EUROPEAN MEETING ON FERROELECTRICITY, 1969, SAARBRUCKEN,
MUSER HE AND PETERSSON J(EDS), WISSENSCH VERLAGSGES, STUTTGART,
1970, 396P.

04-0679 NEUTRON DIFFRACTION STUDIES OF FERROELECTRICS. (BARIUM TITANATE AND ROCHELLE SALT AS
EXAMPLES)
FRAZER BC
J PHYS SOC JAP 17(SUPPL B-II): 376-82 (1962)
IN PART II P376-82 OF INT CONF ON MAGNETISM AND CRYSTALLOGRA, PHY,
1961, KYOTO, PROC. PHYS SOC JAPAN, 1962

04-0680 DIELECTRIC PROPERTIES OF FERROELECTRIC (BARIUM, STRONTIUM) TITANATE SOLID SOLUTIONS
NEAR THE PHASE TRANSITION, UNDER HIGH PRESSURES.
FRITSBERG VYA
P117-22 OF FAZOVYE PEREKHODY V SEGNETOELEKTRIKAKH (PHASE
TRANSFORMATIONS IN FERROELECTRICS), FRITSBERG VYA + ROLOV BN +
KRUCHAN YY(EDS), IZDATELSTVO ZINATNE RIGA, 1971, 206P. (IN RUSSIAN)

04-0681 EXTRINSIC PHOTOCONDUCTIVITY IN FERROELECTRICS DUE TO SURFACE LAYERS. (BARIUM TITANATE,
ANTIMONY SULFOIODIDE)
FRIDKIN VM + GREKOV AA + SAVTCHENKO EA + VOLK TR
J PHYS (PARIS) 33, SUPPL 4, COLLOQ 2: 127-9 (1972);
PHYS STATUS SOLIDI A 8: K55-9 (1971)
294

04-0682 PHOTODOMAIN EFFECT IN BARIUM TITANATE.
FRIDKIN VM + GREKOV AA + KOSONOGOV NA
FERROELECTRICS 4(3): 169-75 (1972)

04-0683 CERAMIC DIELECTRICS. (BARIUM TITANATE, STRONTIUM TITANATE)
FUJIWARA S + SHIRAIWA T
TDK ELECTRONICS CO LTD, US PAT 3,345,189, JAP APPL 30 OCT 1963,
PUBL 3 OCT 1967, 3P.; US PAT 3,427,173, APPL 8 JUN 1964, PUBL
11 FEB 1969, 4P.; US PAT 3,440,067, APPL 8 JUN 1964, PUBL
22 APR 1969, 4P.
CHEM ABSTR 69: 47381Q (1968); 70: 80582Z (1969); 71: 6293K (1969)

04-0684 FORMATION OF BARIUM TITANATE CERAMIC SEMICONDUCTING LAYER BY DIFFUSION.
FUKAMI T
TRANS INST ELECTRON COMMUN ENG JAP C 55: 186-8 (1972) (IN JAPANESE)

04-0685 CHANGE IN THE DECAY CONSTANT OF ZIRCONIUM-89 IN BARIUM TITANATE (BARIUM CHLORIDE FLUX
GROWTH IN PRESENCE OF ZIRCONIUM-89).
GAGNEUX S + HUBER P + LEUENBERGER H + NYIKOS P
HELV PHYS ACTA 43(1): 39-58 (1970)
229

04-0686 STUDY OF THE PYROELECTRIC EFFECT IN THIN FILMS OF TRIGLYCINE SULFATE AND BARIUM
TITANATE.
GAVRILOVA ND + ZVIRGZD YUA + NOVIK VK + POSHIN VG
SOV PHYS SOLID STATE 13(6): 1506-7 (DEC 1971)
254

04-0687 CHANNELING OF PROTONS IN THIN BARIUM TITANATE CRYSTALS AT TEMPERATURES ABOVE AND BELOW
THE FERROELECTRIC CURIE POINT.
GEMMELL DS + MIKKELSON RC
PHYS REV B SOLID STATE 6(5): 1613-35 (1 SEP 1972)
273

04-0688 CHANNELING OF LIGHT IONS IN THE CUBIC AND TETRAGONAL (FERROELECTRICALLY POLARIZED PHASE
         OF BARIUM TITANATE.
         GEMMELL DS + MIKKELSON RC
         RADIAT EFF 12: 21-33 (1972)

04-0689 MOBILITY DETERMINATIONS FROM WEIGHT MEASUREMENTS IN SOLID SOLUTIONS OF (BARIUM,
         STRONTIUM) TITANATE.
         GERTHSEN P + HARDTL KH + CSILLAG A
         PHYS STATUS SOLIDI A 13: 127-33 (1972)
              280

04-0690 ELECTROOPTIC PROPERTIES OF SOME ABO(3) PEROVSKITES IN THE PARAELECTRIC PHASE.
         (POTASSIUM TANTALATE AND TANTALATE NIOBATE, BARIUM AND STRONTIUM TITANATES)
         GEUSIC JE + KURTZ SK + VAN UITERT LG + WEMPLE SH
         APPL PHYS LETT 4: 141-3 (1964)

         ELECTRON MICROSCOPE TRANSMISSION IMAGES OF COHERENT DOMAIN BOUNDARIES. PART-1:
         DYNAMICAL THEORY. PART-2: OBSERVATIONS. (FERROELECTRIC DOMAINS IN BARIUM TITANATE)
         GEVERS R + DELAVIGNETTE P + BLANK H + AMELINCKX S
         PHYS STATUS SOLIDI 4: 383-410; 5: 595-633 (1964)

04-0691 PERTURBED DIRECTIONAL CORRELATION OF SCANDIUM-44 WITHIN THE LATTICE OF FERROELECTRIC
         BARIUM TITANATE.
         GLASS JC
         UNIV NEVADA PHD THESIS, 1968, 93P.
              DISS ABSTR B 29: 4309
              198

04-0692 STATIC QUADRUPOLE INTERACTION OF SCANDIUM-44 IN BARIUM TITANATE.
         GLASS JC + KLIWER JK
         NUCL PHYS A 115(1): 234-40 (1968)
              CHEM ABSTR 69: 40257V (1968)

04-0693 DEPENDENCE OF COERCIVE FIELD OF BARIUM TITANATE SINGLE CRYSTALS ON THEIR THICKNESS.
         GLOGAR P + JANOVEC V
         CZECH J PHYS B 13: 261-5 (1963)

04-0694 THERMAL STUDIES OF CALCIUM(X) BARIUM(1-X) TITANATE.
         GLOWER DD + WALLACE DC
         J PHYS SOC JAP 18: 679-84 (1963)

         THEORETICAL DYNAMIC STUDY OF CUBIC PHASE BARIUM TITANATE.
         GNININVI L + BOUILLOT J
         J PHYS (PARIS) 33(11-12): 1049-58 (1972) (IN FRENCH)
              SCI ABSTR A 76: 20826 (1973)

04-0695 CONDUCTIVITY OF BARIUM TITANATE, PURE SINGLE CRYSTALS AND DOPED (IRON, HYDROXYL) SINGLE
         CRYSTALS.
         GODEFROY L + COUSON B
         J PHYS (PARIS) 33, SUPPL 4, COLLOQ 2: 120-2 (1972)
              294

04-0696 DIELECTRIC PROPERTIES OF BARIUM TITANATE DOPED WITH TUNGSTEN OXIDE.
         GODYCKI E + HODGKINS CE
         BULL AMER CERAM SOC 50: 797 (1971)

04-0697 (PHASE TRANSITIONS AND) THERMOCURRENTS IN TITANIUM DIOXIDE AND BARIUM TITANATE SINGLE
         CRYSTALS.
         GODEFROY G + PIA G + CHANUSSOT G
         P172-3 OF INT MEETING ON FERROELECTRICITY, 2ND, 1969, KYOTO, PROC,
         (J PHYS SOC JAP, VOL 28 SUPPL). PHYS SOC JAP, 1970, 459P.

04-0698 NOISE MEASUREMENTS IN FERROELECTRICS. (TGS, ROCHELLE SALT, BARIUM TITANATE)
         GODEFROY L
         J PHYS (PARIS) 33, SUPPL 4, COLLOQ 2: 44-8 (1972)
              294

4. BARIUM TITANATE

04-0699   "VIBRATING WIRE" STUDY OF POTENTIAL DISTRIBUTION (STABILITY) IN BARIUM(X)
          STRONTIUM(1-X) TITANATE CERAMICS.
          GODEFROY G + ORMANCEY G
          P174 OF INT MEETING ON FERROELECTRICITY, 2ND, 1969, KYOTO, PROC.
          (J PHYS SOC JAP, VOL 28 SUPPL). PHYS SOC JAP, 1970, 459P.

04-0700   ELECTROSTRICTIVELY GENERATED AND MODULATED FRACTURE IN BARIUM TITANATE CERAMICS.
          GOODMAN G
          J APPL PHYS 35: 2725-6 (1964)

04-0701   CHARACTERIZATION OF SURFACE AND GRAIN BOUNDARY LAYER OF BARIUM TITANATE.
          GOSWAMI AK
          NORTHROP LABS, HAWTHORNE, CALIF, 1970, 78P. AD715097

04-0702   PROPERTIES OF BLACK SEMICONDUCTING BARIUM TITANATE.
          GOSWAMI AK
          BULL AMER CERAM SOC 50: 390 (1971)

04-0703   KERR EFFECT STUDIES OF ELECTRIC CONDUCTION IN IRON DOPED BARIUM TITANATE SINGLE
          CRYSTALS.
          GOTO Y + KACHI S
          J PHYS CHEM SOLIDS 32: 889-95 (APR 1971)

04-0704   ELECTRICAL PROPERTIES OF HIGH PURITY POLYCRYSTALLINE BARIUM TITANATE. (DIELECTRIC
          CONSTANT, CONDUCTIVITY)
          GRAHAM HC + TALLAN NM + MAZDIYASNI KS
          J AMER CERAM SOC 54(11): 548-53 (1971)

04-0705   CRYSTAL STUDY OF BARIUM TITANATE WITH PARALLEL INTERFERENCE FRINGES.
          GRENOUILLEAU P + MINN SS
          COMPT REND B 274: 819-22 (1972) (IN FRENCH)

04-0706   CRYSTAL STRUCTURE OF BARIUM TITANATE PRODUCING SQUARE INTERFERENCE FRINGES.
          GRENOUILLEAU P + MINN SS
          COMPT REND B 275: 461-4 (1972) (IN FRENCH)

          THEORY OF ELECTROSTRICTION IN CUBIC CRYSTALS WITH APPLICATION TO BARIUM TITANATE.
          GRINDLAY J
          P433-41, VOL 1 OF DVORAK V + FOUSKOVA A + GLOGAR P (EDS),
          INT MEETING ON FERROELECTRICITY, PRAGUE, 1966, PROC.
          INST PHYS CZECH ACAD SCI, PRAGUE, 1966.

04-0707   RELAXATION AND FERROELECTRIC PROPERTIES OF SOLID SOLUTIONS IN THE SYSTEM BARIUM
          TITANATE- STRONTIUM NIOBIUM(0.8) OXIDE(3).
          GROZNOV IN + VISKOV AS + PETROV VM
          INORG MATER 7: 1078-81 (JUL 1971)

04-0708   INVESTIGATION OF THE FINE STRUCTURE OF BARIUM TITANATE CRYSTALS FLUX GROWN FROM
          POTASSIUM FLUORIDE).
          GUENOK EP + ZABARA YUV + KUDZIN AYU + FOMICHEV OI
          SOV PHYS SOLID STATE 12(3): 751-2 (1970)
              206

04-0709   PECULIARITIES OF POLARIZATION OF BARIUM TITANATE SINGLE CRYSTALS WITH DOUBLE HYSTERESIS
          LOOP.
          GUENOK EP + KUDZIN AYU + LEVKINA AP
          BULL ACAD SCI USSR, PHYS SER 29(11): 1851-5 (1965)
              219

04-0710   PHASE EQUILIBRIA IN THE SYSTEM BARIUM TITANATE- BARIUM GERMANATE.
          GUHA JP + KOLAR D
          J MATER SCI 7: 1192-6 (1972)

04-0711   LOWERING THE CURIE TEMPERATURE IN REDUCED BARIUM TITANATE.
          HARDTL KH + WERNICKE R
          SOLID STATE COMMUN 10: 153-7 (1972)
              260

04-0712   NEUTRON SCATTERING STUDY CF SOFT MODES IN CUBIC BARIUM TITANATE.
          HARADA J + AXE JD + SHIRAND G
          PHYS REV B SOLID STATE 4: 155-62 (JUL 1, 1971)

4. BARIUM TITANATE

04-0713   X-RAY AND NEUTRON DIFFRACTION STUDY OF TETRAGONAL BARIUM TITANATE.
HARADA J + PEDERSEN T + BARNEA Z
ACTA CRYSTALLOGR A 26: 336-44 (1970)

04-0714   POLARIZATION REFLECTION TYPE LIGHT MODULATOR USING FERROELECTRICS. (BARIUM TITANATE)
HASEGAWA T + SATO H
FERROELECTRICS 3: 183-90 (1972;
IEEE TRANS SON ULTRASON 19: 183-90 (1972)
      263

04-0715   SPECIFIC HEAT OF BARIUM TITANATE.
HATTA I + IKUSHIMA A
PHYS LETT A 40(3): 235-6 (17 JUL 1972)
      278

04-0716   SHELL MODEL OF INTERIONIC INTERACTIONS FOR BARIUM TITANATE.
HAVINGA EE
J PHYS CHEM SOLIDS 28(1): 55-64 (1967)

04-0717   ATTEMPT TO PREPARE HIGH DOPED BARIUM STRONTIUM TITANATE SEMICONDUCTOR WITH PZT.
HAYASHI J + YANASE M ++ CONO K
MITSUBISHI DENKI LAB REPT 12(4): 133-52 (1971) (IN JAPANESE)
      297

04-0718   EFFECT OF PRESSURE (TO ABOUT 2KBAR) ON POLARIZATION REVERSAL IN BARIUM TITANATE.
HAYASHI M
P348-50 OF INT MEETING ON FERROELECTRICITY, 2ND, 1969,
KYOTO, PROC (J PHYS SOC JAP, VOL 28 SUPPL). PHYS SOC JAP,
1970.
      197

04-0719   POLARIZATION OF SOLID BARIUM(X) STRONTIUM(1-X) TITANATE COMPOUNDS BETWEEN 4 AND 100
DEGREES K.
HEBENGARTH E + MARTIN G
WISS Z TECH UNIV DRESDEN 20: 495-9 (1971) (IN GERMAN)

04-0720   DIELECTRIC AND THERMAL (SPECIFIC HEAT, ELECTROCALORIC EFFECT) STUDIES OF FERROELECTRIC
CERAMICS AT LOW TEMPERATURES. (PCLYCRYSTALLINE STRONTIUM TITANATE AND BARIUM(X)
STRONTIUM(1-X) TITANATE SOLID SOLUTIONS)
HEGENBARTH E
PHYS STATUS SOLIDI 2: 1544-51 (1962) (IN GERMAN)
    CHEM ABSTR 58: 9710G (1963)

04-0721   INVESTIGATION OF THE THERMAL CONDUCTIVITY IN CONNECTION WITH OTHER PHYSICAL PROPERTIES
OF SOME FERROELECTRICS BELOW 80 DEG K. (SINGLE CRYSTAL STRONTIUM TITANATE,
POLYCRYSTALLINE, STRONTIUM TITANATE, AND SOLID SOLUTIONS BARIUM STRONTIUM TITANATE)
HEGENBARTH E
P104-9, VOL 1 OF DVORAK V + FOUSKOVA A + GLOGAR P (EDS),
INT MEETING ON FERROELECTRICITY, PRAGUE, 1966, PROC.
INST PHYS CZECH ACAD SCI, PRAGUE, 1966.

04-0722   SHIFTING OF THE TRANSITION TEMPERATURE OF BARIUM(0-0.1) STRONTIUM(0.9-1) TITANATE SOLID
SOLUTIONS BY HYDROSTATIC PRESSURE. (TO 1500 ATM)
HEGENBARTH E
P407-9 OF INT MEETING ON FERROELECTRICITY, 2ND, 1969, KYOTO, PROC,
(J PHYS SOC JAP, VOL 28 SUPPL). PHYS SOC JAP, 1970, 459P.

04-0723   MOSSBAUER EFFECT OF COBALT-57 IN BARIUM TITANATE.
HEIMAN N
BULL AMER PHYS SOC 17: 546 (1972)

04-0724   AGING IN COMPLEX CERAMICS BASED ON BARIUM TITANATE. (CAPACITANCE, PERMITTIVITY)
HERBERT JM
P303-6, VOL 2 OF DVORAK V + FOUSKOVA A + GLOGAR P (EDS),
INT MEETING ON FERROELECTRICITY, PRAGUE, 1966, PROC.
INST PHYS CZECH ACAD SCI, PRAGUE, 1966.

04-0725   ELECTRICAL PROPERTIES OF MICROCRYSTALLINE FERROELECTRIC MATERIALS CRYSTALLIZED FROM
GLASS. (60-90 VOL PERCENT, FERROELECTRIC, COMPARED WITH CERAMIC BARIUM TITANATE)
HERCZOG A + LAYTON MM + ALLEN RE
P4-7 OF ELEC INSULATION CONF (IEEE), MATER APPLIC 5TH, TECH PAP,
IEEE, NY, 1963.

4. BARIUM TITANATE

04-0726 VIBRATIONAL AMPLITUDES IN FERROELECTRIC TETRAGONAL BARIUM TITANATE.
HEWAT AW
J PHYS SOC JAP 32: 1156 (1972)
278

04-0727 CURIE SHIFT IN BARIUM TITANATE BY HETEROSUBSTITUTION.
HEYWANG W
FERROELECTRICS 1: 177-8 (1970)

04-0728 CORRELATION BETWEEN STRUCTURE AND ELECTRIC PROPERTIES OF FERROELECTRIC SUBSTANCES.
(BARIUM TITANATE)
HEYWANG W
BER DEUT KERAM GESELL 47: 674-9 (1970) (IN GERMAN)

04-0729 PIEZORESISTIVITY OF DOPED BARIUM TITANATE. (OR BARIUM STRONTIUM TITANATE, DOPANT NOT
NAMED)
HEYWANG W + GUNTERSDORFER M
P307-12, VOL 2 OF DVORAK V + FOUSKOVA A + GLOGAR P (EDS),
INT MEETING ON FERROELECTRICITY, PRAGUE, 1966, PROC.
INST PHYS CZECH ACAD SCI, PRAGUE, 1966.

04-0730 SEMICONDUCTING BARIUM TITANATE. (DIELECTRIC CONSTANT, CONTACT RESISTANCE)
HEYWANG W
J MATER SCI 6: 1214-24 (1971)
246

04-0731 CHANGES IN THE FERROELECTRIC PROPERTIES OF BARIUM TITANATE CERAMICS CAUSED BY
IRRADIATION WITH CADMIUM TRANSMITTED NEUTRONS.
HILCZER B
PHYS STATUS SOLIDI 5(3): K113-15 (1964)

04-0732 EFFECT OF NEUTRON IRRADIATION ON THE FERROELECTRIC PROPERTIES OF BARIUM TITANATE
CERAMICS.
HILCZER B
PHYS STATUS SOLIDI 2: 447-55 (1962)

04-0733 ENERGY STORAGE IN HIGH PERMITTIVITY CERAMICS: SURVEY OF MATERIALS BASED ON BARIUM
TITANATE.
HILL GJ
PROC BRIT CERAM SOC NO 18: 201-20 (1970)

04-0734 A SEMICONDUCTING BARIUM TITANATE CERAMIC DOPED WITH NIOBIUM AND MANGANESE IONS AND ITS
APPLICATIONS.
HIROSE N + MATSUOKA T + MATSUO Y + HAYAKAWA S + ISHIKAWA K + NAGANO K + MIYAMOTO M
+ MIFUNE H
NAT TECH REP 18: 426-37 (1972) (IN JAPANESE)

04-0735 EFFECT OF GRAIN SIZE ON THE RESISTIVITY ANOMALY IN SEMICONDUCTIVE BARIUM TITANATE
CERAMICS.
HIROSE N + SASAKI H
J AMER CERAM SOC 54: 320 (JUN 1971)

04-0736 DETERMINATION OF THE DEVIATION FROM OXYGEN STOICHIOMETRY OF BARIUM TITANATE SINGLE
CRYSTALS.
HLASIVCOVA N + COUFOVA P + NOVAK J
KRISTALL TECH 6: K11-15 (1971)

04-0737 OPTICAL EVIDENCE FOR POLARIZATION FLUCTUATIONS IN PARAELECTRIC BARIUM TITANATE.
HOFMANN R + WEMPLE SH + GRANICHER H
P265-7 OF INT MEETING ON FERROELECTRICITY, 2ND, 1969, KYOTO, PROC,
(J PHYS SOC JAP, VOL 28 SUPPL). PHYS SOC JAP, 1970, 459P.

04-0738 TEMPERATURE DEPENDENCE OF ELECTRONIC POLARIZABILITY OF STRONTIUM AND BARIUM TITANATE.
HOFMANN R
ABHANDLUNG DOKT. NATURWISSENSCH. ZURICH, 79P. JURIS DRUCK AND VERLAG,
ZURICH, 1968
187

04-0739 EFFECT OF CLAY AND DOLOMITE ADDITIONS ON BARIUM TITANATE CERAMICS. CRYSTAL STRUCTURE
AND SINTERABILITY.
HOSHINO M
NEC RES DEVELOP 27: 76-88 (1972)

4. BARIUM TITANATE

04-0740   X-RAY AND NEUTRON DIFFRACTION STUDY OF TETRAGONAL BARIUM TITANATE.
          HARADA J + PEDERSEN T + BARNEA Z
          ACTA CRYSTALLOGR A 26(3): 336-44 (MAY 1970)

04-0741   BARIUM TITANATE FILMS PREPARED BY RF SPUTTERING ONTO INDIUM ANTIMONIDE OR GALLIUM
          ARSENIDE.
          IIDA S + KATAOKA S
          APPL PHYS LETT 18: 391-2 (MAY 1, 1971)

04-0742   ELECTRON PROBE X-RAY MICROANALYSIS ON IMPURITIES IN EVAPORATED BARIUM TITANATE FILMS.
          IIJIMA Y
          JAP J APPL PHYS 9: 852-3 (1970)
              202

04-0743   PREPARATION OF BARIUM TITANATE FILM BY EVAPORATION.
          IIJIMA Y
          ZAIRYO 18(192): 819-23 (1969) (IN JAPANESE)
              CHEM ABSTR 71: 116119R (1969)
              201

04-0744   FAR INFRARED REFLECTIVITY OF BARIUM TITANATE.
          IKEGAMI S + UEDA I + KISAKA S + MITSUICHI A + YOSHINAGA H
          J PHYS SOC JAP 17: 1210-11 (1962)

04-0745   MECHANISM OF MICROWAVE DIELECTRIC DISPERSION IN POLYCRYSTALLINE BARIUM TITANATE.
          IKEGAMI S
          J PHYS SOC JAP 18: 1203-13 (1963)

04-0746   RAMAN SPECTRUM OF BARIUM TITANATE.
          IKEGAMI S
          J PHYS SOC JAP 19: 46-51 (1964)

04-0747   ELECTRON SPIN RESONANCE OF MANGANESE(2+) IN BARIUM TITANATE.
          IKUSHIMA H + HAYAKAWA S
          J PHYS SOC JAP 19: 1986 (1964)

          COMPUTER EXPERIMENT ON LINEAR CHAIN OF ATOMS LYING IN DOUBLE MINIMUM POTENTIAL. (BARIUM
          TITANATE)
          ISHIBASHI Y + TAKAGI Y
          J PHYS SOC JAP 33(1): 1-5 (JUL 1972)

04-0748   FORCE CONSTANTS IN BARIUM TITANATE AND STRONTIUM TITANATE.
          ISHIBASHI Y + TAKAGI Y
          J PHYS SOC JAP 31(6): 1712-18 (1971)
              268

04-0749   MICROWAVE DYNAMIC NONLINEARITY OF BARIUM TITANATE SINGLE CRYSTALS.
          IVANOV IV + MOROZOV NA
          P53-5 OF INT MEETING ON FERROELECTRICITY, 2ND, 1969,, KYOTO, PROC,
          (J PHYS SOC JAP, VOL 28 SUPPL). PHYS SOC JAP, 1970, 459P.

04-0750   X-RAY AND THERMOLUMINESCENCE OF BARIUM TITANATE.
          IVANII GM
          P58-63 OF SEGNETOELEKTRIKAKH I OKISNYE POLUPROVODNIKI (FERROELECTRICS
          AND OXIDE SEMICONDUCTORS), SINYAKOV EV(ED), REDAKTSIONNO-IZDATELSKII
          OTDEL DNEPROPETROVSKII GOSUDARSTVENNYI UNNIVERSITET, DNEPROPETROVSK,
          1971, 100P. (IN RUSSIAN)

04-0751   CHARACTERISTICS OF BARIUM TITANATE PYROELECTRIC DETECTOR FOR OPTICAL AND INFRARED
          RADIATION (0.5 TO 2 MICRONS).
          IWASAKI H
          J RADIO RES LAB 17(90): 147-51 (1970)
              216

04-0752   ANOMALOUS DECAY EFFECT IN POLYCRYSTALLINE BARIUM TITANATE.
          JASKIEWICZ A + TERPILOWSKI J
          ACTA PHYS POLON 23: 407-9 (1963)

04-0753   FIELD INDUCED NUCLEATION AT THE (CUBIC TETRAGONAL) PHASE TRANSITION IN BARIUM TITANATE.
          JASKIEWICZ A
          ACTA PHYS POLON 22(SUPPL): 165-72 (1962)

04-0754   NUCLEATION PROCESS IN BARIUM TITANATE AT CUBIC- TETRAGONAL PHASE TRANSITIONS.
          JASKIEWICZ A
          ACTA PHYS POLON 22: 489-98 (1962)

04-0755   DISPERSION OF ELECTROOPTIC EFFECT IN BARIUM TITANATE. (BIREFRINGENCE, PRINCIPAL
          INDICES)
          JOHNSTON AR
          J APPL PHYS 42(9): 3501-7 (AUG 1971)
                239

04-0756   DISPERSION OF THE ELECTROOPTIC EFFECT IN BARIUM TITANATE.
          JOHNSTON AR
          J OPT SOC AMER 61: 679 (1971)

04-0757   NATURE OF AGING IN FERROELECTRIC CERAMICS. (BARIUM TITANATE, LEAD ZIRCONATE TITANATE)
          JONKER GH
          J AMER CERAM SOC 55(1): 57-8 (JAN 1972)
                258

04-0758   SOME ASPECTS OF SEMICONDUCTING BARIUM TITANATE.
          JONKER GH
          SOLID STATE ELECTRON 7: 895-903 (1964)

04-0759   MEASUREMENT OF THE PERMITTIVITY OF BARIUM TITANATE SINGLE CRYSTALS AT MICROWAVE
          FREQUENCIES.
          KAATZE U
          PHYS STATUS SOLIDI B 50: 537-45 (1972) (IN GERMAN)
                CHEM ABSTR 76: 146433J (1972)
                280

04-0760   X-RAY INVESTIGATION OF THE EFFECT OF HYDROSTATIC PRESSURE ON THE STRUCTURE OF BARIUM
          TITANATE.
          KABALKINA SA + VERESHCHAGIN LF + SHULENIN BM
          SOV PHYS DOKL 7: 527-9 (1962)

04-0761   INFLUENCE OF MADELUNG ENERGY AND COVALENCY ON THE STRUCTURE OF AB(5) OXIDE COMPOUNDS.
          (BARIUM TITANATE, RUBIDIUM NICBATE)
          KAFALAS JA
          P287-93 OF MATER RES SYMP, 5TH PROC, 1971, SOLID STATE CHEMISTRY,
          ROTH RS + SCHNEIDER SJ (EDS), NBS SPEC PUB 364, 1972, 783P.

                271

04-0762   EFFECT OF HEAT TREATMENT ON THE PTCR ANOMALY IN SEMICONDUCTING BARIUM TITANATE.
          KAHN M
          BULL AMER CERAM SOC 50: 676-80 (1971)

04-0763   INFLUENCE OF GRAIN GROWTH ON DIELECTRIC PROPERTIES OF NIOBIUM DOPED BARIUM TITANATE.
          KAHN M
          J AMER CERAM SOC 54: 455-7 (SEP 1971)

04-0764   OHMIC CONTACTS TO SEMICONDUCTING BARIUM TITANATE.
          KAHN M
          BULL AMER CERAM SOC 50: 396 (1971)

04-0765   PREPARATION OF SMALL GRAINED AND LARGE GRAINED CERAMICS FROM NIOBIUM DOPED BARIUM
          TITANATE.
          KAHN M
          J AMER CERAM SOC 54: 452-4 (SEP 1971)

04-0766   ELECTROCALORIC EFFECT IN POLYCRYSTALLINE BARIUM TITANATE.
          KARCHEVSKII AI
          SOV PHYS SOLID STATE 3: 2249-54 (1962)

04-0767   ULTRAHIGH FREQUENCY DIELECTRIC ANISOTROPY OF A POLARIZED FERROELECTRIC CERAMIC. (BARIUM
          TITANATE, BARIUM TIN TITANATE, LEAD BARIUM DINIOBATE)
          KARGOPOLOVA NP + POPLAVKO YUM + ISUPOV VA
          SOV PHYS SOLID STATE 12: 477-9 (1970)

4. BARIUM TITANATE

04-0768   INFLUENCE OF THE AMPLITUDE OF FREE VIBRATIONS ON THE PERMITTIVITY OF BARIUM TITANATE
           SINGLE CRYSTALS.
           KASHCHENKO EP + KUDZIN AYU
           SOV PHYS SOLID STATE 14(4): 1085-6 (OCT 1972)
              276

04-0769   FERROELECTRIC PHOTOCONDUCTOR OPTICAL STORAGE MEDIUM UTILIZING BISMUTH TITANATE.
           KENEMAN SA + TAYLOR GW + MILLER A
           FERROELECTRICS 1: 227-41 (1970)

           THERMODYNAMICAL THEORY OF BARIUM TITANATE TYPE FERROELECTRICS.
           KHOLODENKO LP
           IZDATELSTVO, RIGA, USSR, 1971, 227P. (IN RUSSIAN)

04-0770   MACROSCOPIC THEORY OF BARIUM TITANATE TYPE FERROELECTRIC CRYSTALS IN THE FORM OF THICK
           PLATES.
           KHOLODENKO LP
           SOV PHYS SOLID STATE 5(8): 1524-31 (1964)

04-0771   PIEZOELECTRIC ACTIVITY OF BARIUM TITANATE SINGLE CRYSTALS IN THE PARAELECTRIC PHASE.
           KHOTCHENKOV AG
           IZV VUZ FIZ 1970(12): 152-4 (IN RUSSIAN)
           (TRANSLATION TO APPEAR IN SOV PHYS J)

04-0772   ELECTRICAL PROPERTIES OF THICK FILM BARIUM TITANATE DIELECTRICS PRODUCED BY FLAME
           SPRAYING (CAPACITORS).
           KIMURA S
           ELEC COMMUN JAP 54: 104-12 (JUN 1971)

04-0773   FLAME SPRAYED BARIUM TITANATE AS A CAPACITOR DIELECTRIC.
           KIMURA S
           IEEE TRANS PARTS MATER PACKAG 6: 3-11 (1970)

04-0774   STUDIES OF FLAME SPRAYED BARIUM TITANATE CAPACITORS. PART-10: STRUCTURE OF
           SEMICONDUCTIVE TITANIUM DIOXIDE AND TITANIUM DIOXIDE- BARIUM TITANATE COATINGS PRODUCED
           BY FLAME SPRAYING. PART-11: ELECTRICAL CHARACTERISTICS OF SEMICONDUCTIVE BARIUM
           TITANATE COATINGS PRODUCED BY FLAME SPRAYING. PART-12: STRUCTURE OF SEMICONDUCTIVE
           BARIUM TITANATE COATINGS PRODUCED BY FLAME SPRAYING.
           KIMURA S
           KINZOKU HYOMEN GIJUTSU 21: 14-19, 327-34, 363-7 (1970) (IN JAPANESE)

           THEORY OF POLARIZATION REVERSAL OF BARIUM TITANATE CRYSTAL.
           KINASE W + KAZAMI A + YOKOYAMA S
           P343-5 OF INT MEETING ON FERROELECTRICITY, 2ND, 1969,, KYOTO, PROC,
           (J PHYS SOC JAP, VOL 28 SUPPL). PHYS SOC JAP, 1970, 459P.

           90 DEGREE TYPE DOMAIN WALL OF BARIUM TITANATE CRYSTAL. (THEORY, CALCULATIONS)
           KINASE W + OKAYAMA H + YOSHIKAWA A + TAGUCHI N
           P383-5 OF INT MEETING ON FERROELECTRICITY, 2ND, 1969,, KYOTO, PROC,
           (J PHYS SOC JAP, VOL 28 SUPPL). PHYS SOC JAP, 1970, 459P.

04-0775   ON THE LATTICE VIBRATION OF CUBIC BARIUM TITANATE.
           KINASE W + ISHIBASHI Y + MATSURA K
           J PHYS SOC JAP 19: 264-8 (1964)

04-0776   THEORY OF THE DIELECTRIC CONSTANTS OF BARIUM TITANATE.
           KINASE W
           J PHYS SOC JAP 17: 70-84 (1962)

04-0777   THICKNESS OF DOMAIN WALLS IN FERROELECTRIC AND FERROELASTIC CRYSTALS. (BARIUM TITANATE,
           GADOLINIUM MOLYBDATE)
           KITTEL C
           SOLID STATE COMMUN 10: 119-21 (1972)
              260

04-0778   CHARACTERISTICS OF THIN, MULTILAYERED FERROELECTRIC CERAMICS OBTAINED BY PROJECTION
           WITH A PLASMA TORCH OR FRITTING. (BARIUM TITANATE)
           KLEIMANN H + PALETTO J + RICHARD M
           J PHYS (PARIS) 33, SUPPL 4, COLLOQ 2: 270-1 (1972)
              294

04-0779 LOW LOSS (FERROELECTRIC CERAMICS) WITH GREAT DIELECTRIC PERMITTIVITY. (BARIUM TITANATE
TYPE MATERIALS)
KLEIMANN H + PALETTO J + FETIVEAU Y AND OTHERS
COMPT REND B 272: 1037-40 (1971) (IN FRENCH)

04-0780 EFFECT OF LITHIUM OXIDE ON THE PROPERTIES OF A SERIES OF FERROELECTRIC MATERIALS.
(BARIUM TITANATE, LEAD FERRATE(0.5) NIOBATE(0.5))
KLIMOV VV + DIDKOVSKAYA OS + ZVONIK VA
INORG MATER 6: 161-2 (1970)

04-0781 EFFECT OF THERMAL NEUTRON IRRADIATION ON THE RELATIVE PERMITTIVITY OF BARIUM TITANATE
SINGLE CRYSTAL.
KOBAYASHI S + TSUNODA H
TRANS INST ELECTRON COMMUN ENG JAP C 55: 71-2 (1972) (IN JAPANESE)

04-0782 X-RAY IRRADIATION EFFECT ON DIELECTRIC PROPERTIES OF SOLID SOLUTIONS OF BARIUM TITANATE
TYPE MATERIALS.
KOBAYASHI S + YATABE K
RADIOISOTOPES 20: 603-6 (1971) (IN JAPANESE)

04-0783 PREPARATION AND PROPERTIES OF GLASS- CERAMICS CONTAINING FERROELECTRIC CRYSTALS (BARIUM
OR LEAD TITANATE, FROM SYSTEMS BARIA OR LITHIA, TITANIA, ALUMINA, SILICA).
KCKUBO T
BULL INST CHEM RES KYOTO UNIV 47(6): 553-71 (1969)
216

04-0784 ORIGIN OF LATTICE DEFECTS IN (BARIUM) TITANATE MATERIALS AND THEIR INFLUENCE ON
BEHAVIOR.
KOLLER A
P477-81 OF STEWART GH(ED), SCIENCE OF CERAMICS, VOL 4, BRIT CERAM SOC
1968.
222

04-0785 EFFECT OF CRYSTALLINE GRAIN SIZE ON THE PHYSICAL PROPERTIES OF CERAMIC BARIUM TITANATE.
KOMAROV VD + TURILE AV
IZV VUZ FIZ 1970(9): 138-40   (IN RUSSIAN)
261

04-0786 SOLID SOLUTIONS BASED ON BARIUM TITANATE.
KOMAROV VD + MOLCHANOVA RA
INORG MATER 6: 43-6 (JAN 1970)

04-0787 SPONTANEOUS DESTRUCTION OF FERROCERAMICS. (BARIUM TITANATE)
KOMAROV VD + MOLCHANOVA RA
INORG MATER 7: 1145-6 (1971)

04-0788 DISTRIBUTION OF POTENTIAL ALONG TITANATE CERAMICS PREVIOUSLY POLARIZED.
KOUKAL V + GODEFROY G
COMPT REND B 272: 131-4 (1971) (IN FRENCH)

04-0789 DEPENDENCE OF DIELECTRIC PROPERTIES OF BARIUM TITANATE CERAMICS ON THE PHYSICO-
CHEMICAL NATURE OF THE SOLID SOLUTION FORMED FROM TWO TITANATES.
KOZLOVSKII LV + TOMORA EL + IVANOVA TSV + IORDANOVA M
11P OF CONF ON CERAMICS FOR ELECTRONICS, 4TH, 1971, SPINDLERUV MLYN,
CZECHOSLOVAKIA VYZKUMNY USTAV ELEKTROTECHNICKE KERAMIKY, 1971
(IN RUSSIAN)
SCI ABSTR A 75: 16288 (1972)
264

04-0790 WET SYNTHETIC METHOD OF MAKING PURE, EXACTLY STOICHIOMETRIC BARIUM TITANATE.
KUBO T + KATO M + FUJITA T
J CHEM SOC JAP 71(1): 114-18 (1968)

04-0791 AMPLITUDE DEPENDENCE OF THE INTERNAL FRICTION IN BARIUM TITANATE SINGLE CRYSTALS.
(PIEZOELECTRIC MODULUS)
KUDZIN AYU + KASHCHENKO EP
SOV PHYS SOLID STATE 13(7): 1551-3 (JAN 1972)
254

4. BARIUM TITANATE

04-0792   DYNAMIC FATIGUE OF SINGLE CRYSTALS OF BARIUM TITANATE.
          KUDZIN AYU + PANCHENKO TV
          IZV VUZ FIZ 1972(6): 92-6 (IN RUSSIAN)
          (TRANSLATION TO APPEAR IN SOV PHYS J)
            284

04-0793   EFFECT OF ADDING NICKEL ON THE PROPERTIES OF BARIUM TITANATE SINGLE CRYSTALS.
          KUDZIN AYU
          SOV PHYS SOLID STATE 4: 1006-7 (1962)

04-0794   INTERNAL FRICTION IN BARIUM TITANATE SINGLE CRYSTALS.
          KUDZIN AYU + KASHCHENKO EP
          BULL ACAD SCI USSR, PHYS SER 35(9): 1743-6 (1971)

04-0795   STABILIZATION OF SPONTANEOUS POLARIZATION OF BARIUM TITANATE SINGLE CRYSTALS.
          KUDZIN AYU + PANCHENKO TV
          SOV PHYS SOLID STATE 14(6): 1599-600 (DEC 1972)
            279

04-0796   TEMPERATURE DEPENDENCE OF INTERNAL FRICTION OF BARIUM TITANATE SINGLE CRYSTALS.
          KUDZIN AYU + KASHCHENKO EP + KOSTRUB VI
          SOV PHYS SOLID STATE 13: 3162-3 (1972)

          DIFFUSION POTENTIALS IN BARIUM TITANATE AND THE THEORY OF PTC MATERIALS.
          KULWICKI BM + PURDES AJ
          FERROELECTRICS 1(4): 253-63; 2: 176 (1970-1)

04-0797   DEVELOPMENT OF BARIUM TITANATE WITH CONSISTENT DIELECTRIC PROPERTIES.
          KULKARNI AK + THIAGARAJAN K + UPADHYAYA DD + MOORTHY VK
          TRANS INDIAN CERAM SOC 30: 1-8 (1971)

04-0798   INFLUENCE OF MICROSTRUCTURES (PREPARATION CONDITIONS) ON DIELECTRIC PROPERTIES OF
          BARIUM TITANATE.
          KULKARNI AK + PRASAD R + VYAS BR + MOORTHY VK
          TRANS INDIAN CERAM SOC 27(4): 130-9 (1968)
            CHEM ABSTR 71: 16816X (1969)

04-0799   LANTHANUM DOPED BARIUM TITANATE CERAMICS WITH BISMUTH OXIDE AS A GRAIN BOUNDARY
          ELEMENT.
          KUWABARA M + YANAGIDA H
          J CERAM SOC JAP 80: 269-76 (1972) (IN JAPANESE)

04-0800   TWO-STEP ANOMALOUS INCREASE OF RESISTIVITY ON LANTHANUM DOPED BARIUM TITANATE- BISMUTH
          TRIOXIDE COMPOSITE CERAMICS WITH SURFACE BARRIER LAYER.
          KUWABARA M + YANAGIDA H
          JAP J APPL PHYS 11(8): 1130-8 (1972)
            278

04-0801   TWO STEP ANOMALOUS INCREASE OF RESISTANCE ON LANTHANUM(0.002) BARIUM(0.998) TITANATE-
          BISMUTH OXIDE COMPOSITE CERAMICS.
          KUWABARA M + YANAGIDA H
          JAP J APPL PHYS 10: 805 (1971)

04-0802   BARIUM TITANATE CERAMIC: A NEW ELECTRON EMITTER.
          LAUSCH W + ELNAGGAR S
          RADIOCHEM RADIOANAL LETT 12(2-3): 177-84 (1972)
            283

04-0803   RAMAN SCATTERING LINE SHAPE OF THE SOFT E POLARITON MODE IN BARIUM TITANATE.
          LAUGHMAN L + DAVIS LW + NAKAMURA T
          PHYS REV B SOLID STATE 6(9): 3322-6 (1 NOV 1972)
            276

04-0804   180 DEG DOMAIN WALL CONTRIBUTION TO THE DIELECTRIC CONSTANT OF BARIUM TITANATE.
          (CALCULATIONS, SUMMARY ONLY, DISCUSSION)
          LAWLESS WN + FOUSEK J
          P346-7 OF INT MEETING ON FERROELECTRICITY, 2ND, 1969,, KYOTO, PROC,
          (J PHYS SOC JAP, VOL 28 SUPPL). PHYS SOC JAP, 1970, 459P.

SMALL SIGNAL PERMITTIVITY OF THE STATIONARY (100)-180 DEG DOMAIN WALL IN BARIUM
TITANATE (THEORY, CALCULATIONS).
LAWLESS WN + FOUSEK J
J PHYS SOC JAP 28(2): 419-24 (1970)

04-0805   TEMPERATURE DEPENDENT POLARIZABILITIES AND COCHRAN MODE ASSIGNMENTS IN BARIUM AND
          STRONTIUM TITANATES.
          LAWLESS WN + GRANICHER H
          P69-79, VOL 1 OF DVORAK V + FOUSKOVA A + GLOGAR P (EDS),
          INT MEETING ON FERROELECTRICITY, PRAGUE, 1966, PROC.
          INST PHYS CZECH ACAD SCI, PRAGUE, 1966.

04-0806   ACOUSTIC PHONON SOFT OPTIC PHONON INTERACTION IN BARIUM TITANATE.
          LAZAY PD + FLEURY PA + RUVALDS J
          BULL AMER PHYS SOC 16: 29 (1971)

04-0807   TEMPERATURE DEPENDENCE OF THE BRILLOUIN- RAMAN SPECTRUM OF BARIUM TITANATE.
          LAZAY PD + FLEURY PA
          P406-10 OF INT CONF ON LIGHT SCATTERING IN SOLIDS, 2ND, 1971,
          BALKANSKI M(ED), FLAMMARION, PARIS, 1971.
              SCI ABSTR A 75: 16525 (1972)
              264

04-0808   DIRECT OBSERVATION OF BARIUM TITANATE CRYSTALS BY SCANNING ELECTRON MICROSCOPE.
          LE BIHAN R + MAUSSION M
          COMPT REND B 273: 1075-8 (1972) (IN FRENCH)
              269

04-0809   DIELECTRIC ABSORPTION IN THE RHOMOHEDRAL PHASE OF BARIUM TITANATE.
          LE TRAON F + LE TRAON A + LE MONTAGNER S
          COMPT REND 255: 1078-80 (1962) (IN FRENCH)
              CHEM ABSTR 59: 1362H

04-0810   PHOTOEMISSION MICROSCOPIC OBSERVATION OF THE SURFACE OF BARIUM TITANATE CRYSTALS.
          LE BIHAN R
          COMPT REND B 275(1): 29-32 (1972) (IN FRENCH)
              CHEM ABSTR 77: 157693R (1972)
              288

04-0811   STUDY OF DOMAINS IN FERROELECTRIC CRYSTALS WITH A PHOTOEMISSION MICROSCOPE (TUNGSTEN
          TRIOXIDE, BARIUM TITANATE).
          LE BIHAN R
          COMPT REND B 270: 741-4 (1970) (IN FRENCH)
              197

04-0812   SEM OBSERVATION OF THE FERROELECTRIC SURFACE OF BARIUM TITANATE CRYSTALS.
          LE BIHAN R + MAUSSION M
          COMPT REND B 274: 1075-8 (1972) (IN FRENCH)

04-0813   COMMENTS ON THE NEUTRON SPECTRUM OF FERROELECTRIC AND PARAELECTRIC BARIUM TITANATE.
          LEFKOWITZ I
          PROC PHYS SOC 80: 868-72 (1962)

04-0814   CONDUCTIVITY INJECTION AND EXTRACTION IN POLYCRYSTALLINE BARIUM TITANATE.
          LEHOVEC K + SHIRN GA
          J APPL PHYS 33: 2036-44 (1962)

04-0815   TWO-PHASE FERROELECTRIC SYSTEM. PART-1: BARIUM TITANATE- METAL.
          LEIBLER K + BRANSKI W
          ACTA PHYS POLON 20: 447-53 (1961)

04-0816   TWO PHASE FERROELECTRIC SYSTEMS. PART-2: FERROELECTRICS- SILVER (BARIUM TITANATE AND
          BARIUM TITANATE- MAGNESIUM STANNATE).
          LEILBLER K + BRANSKI W
          ACTA PHYS POLON 23: 279-85 (1963)

04-0817   REJUVENATION OF A BARIUM(0.72) STRONTIUM(0.28) TITANATE CERAMIC.
          LEVIALDI A + SCHOIJET M
          PHYS STATUS SOLIDI 3: 322-8 (1963)

4. BARIUM TITANATE

04-0818   UHF DIELECTRIC PROPERTIES OF SOME POLYCRYSTALLINE FERROELECTRICS (BARIUM TITANATE AND
SAME WITH 1 PERCENT SILICON REPLACING TITANIUM).
LIBERMAN ZA + OGURTSOV SI + ZHUKOV OK + KLEMENT'EV FM
BULL ACAD SCI USSR, PHYS SER 33(7): 1101-2 (1969)
     225

04-0819   RESISTIVITY AND SINTERING OF BARIUM TITANATE.
LINGSCHEIT JN + MCGEE TD
BULL AMER CERAM SOC 47(4): 390 (APR 1968)
     143

04-0820   TEMPERATURE DEPENDENCE OF THE COMPLEX PERMITTIVITY OF SOME TITANATES BETWEEN 20 AND
1000 DEG C. (BARIUM TITANATE, STRONTIUM TITANATE AND OTHER MATERIALS)
LIPAEVA GA
SOV PHYS SOLID STATE 4: 1183-86 (1962)

04-0821   CALCULATIONS OF ELECTRONIC CONDUCTIVITY IN NONSTOICHIOMETRIC BARIUM TITANATE.
LONG SA
BULL AMER CERAM SOC 49: 830 (1970)

04-0822   TITANIUM RICH NONSTOICHIOMETRIC BARIUM TITANATE. PART-1: HIGH TEMPERATURE ELECTRICAL
CONDUCTIVITY MEASUREMENTS. (INTRINSIC BAND GAP, PTCR (POSITIVE TEMPERATURE COEFFICIENT
OF RESISTIVITY) DEVICES) PART-2: ANALYSIS OF DEFECT STRUCTURE.
LONG SA + BLUMENTHAL RN
J AMER CERAM SOC 54: 515-19, 577-83 (1971)
     246

04-0823   EFFECT OF PARTICLE SIZE AND SHAPE ON THE INFRARED ABSORPTION SPECTRA OF BARIUM TITANATE
AND STRONTIUM TITANATE POWDERS.
LUXON JT + MONTGOMERY DJ + SUMMITT R
J APPL PHYS 41(6): 2303-7 (1970)
     196

04-0824   CALCULATION OF ELECTRIC FIELDS IN IONIC CRYSTALS. (BARIUM TITANATE TYPE)
LYUBIMOV VN + VENEVTSEV YUN + KOIRANSKAYA EYU
SOV PHYS CRYSTALLOGR 7: 768-70 (1963)

04-0825   STABILIZED BARIUM TITANATE CERAMICS FOR CAPACITOR DIELECTRICS.
MACCHESNEY JB + GALLAGHER PK + DI MARCELLO FV
J AMER CERAM SOC 46: 197-202 (1963)

04-0826   STATIC SHELL MODEL FOR BARIUM AND LEAD TITANATES.
MACHET R + BOUILLOT J + GODEFROY L
P366-8 OF INT MEETING ON FERROELECTRICITY, 2ND, 1969,, KYOTO, PROC,
(J PHYS SOC JAP, VOL 28 SUPPL). PHYS SOC JAP, 1970, 459P.

04-0827   INVESTIGATION OF THE FERROELECTRIC PHASE CHANGE IN BARIUM TITANATE USING ELECTRON SPIN
RESONANCE AND THE MOSSBAUER EFFECT.
MAGUIRE HG + REES LVC
J PHYS (PARIS) 33, SUPPL 4, COLLOQ 2: 173-6 (1972)
     294

04-0828   PHYSICAL AND ELECTRICAL PROPERTIES OF THIN FILM BARIUM TITANATE PREPARED BY RF
SPUTTERING ON SILICON SUBSTRATES.
MAHER GH + DIEFENDORF RJ
IEEE TRANS PARTS HYBRIDS PACKAG 8(3): 11-15 (1972)

04-0829   BARIUM TITANATE CERAMICS WITH HIGH DIELECTRIC CONSTANT. PART-1: EFFECTS OF ADDITION OF
GLASSES CONTAINING BARIUM OXIDE AND TITANIUM DIOXIDE (WITH SILICA AND ALUMINA) ON
PROPERTIES OF BARIUM TITANATE CERAMICS. PART-2: EFFECTS OF THE ADDITION OF GLASS
POWDERS CONTAINING BARIUM OXIDE AND TITANIUM DIOXIDE ON THE TEMPERATURE CHARACTERISTICS
OF THE DIELECTRIC PROPERTIES OF BARIUM TITANATE CERAMICS.
MAKI T + TASHIRO M
YOGYO KYOKAI SHI (J CERAM ASSOC JAP) 75(865): 278-83 (1967);
76(877): 320-4 (1968) (IN JAPANESE)
     CHEM ABSTR 69: 79811W, 70: 80538Q

04-0830   PHONON TRANSPORT IN BARIUM TITANATE.
MANTE AJH + VOLGER J
PHYSICA 52(4): 577-604 (1971)
     253

04-0831   ENERGY LOSS NONPOLAR STATE POLYCRYSTALLINE BARIUM TITANATE. (150-500 DEG C, 20-50
           KILOHERTZ, LISSAJOUS FIGURE)
           MARKS GW + SKAAR DL
           IEEE TRANS PARTS MATER PACKAG 6: 110-18 (SEP 1970)
                218

04-0832   STRUCTURE AND PROPERTIES OF BARIUM TITANATE FILMS PRODUCED BY CATHODE SPUTTERING.
           MARGOLIN AM + BARABANOVA LA + BONDARENKO VS + FESENKO EG + DUDKEVICH VP
           BULL ACAD SCI USSR, PHYS SER 35(9): 1768-72 (1971)

04-0833   DIELECTRIC CERAMICS WITH BOUNDARY LAYER STRUCTURE FOR HIGH FREQUENCY APPLICATION.
           (BARIUM, STRONTIUM TITANATE)
           MASUNO K + MURAKAMI T + WAKU S
           FERROELECTRICS 3: 315-9 (FEB 1972)

04-0834   THIN FILMS: PREPARATION OF FERROELECTRIC THIN FILMS OF BARIUM TITANATE BY VACUUM
           EVAPORATION.
           MASSON S + MINN SS
           P421-3 OF INT MEETING ON FERROELECTRICITY, 2ND, 1969, KYOTO, PROC,
           (J PHYS SOC JAP, VOL 28 SUPPL). PHYS SOC JAP, 1970, 459P.

04-0835   EXAGGERATED GRAIN GROWTH IN LIQUID PHASE SINTERING OF BARIUM TITANATE.
           MATSUO Y + SASAKI H
           J AMER CERAM SOC 54: 471 (1971)

04-0836   FERROELECTRIC TRANSFORMATION IN BARIUM TITANATE. (MODEL GIVES NEARLY COMPLETE
           DESCRIPTION)
           MATTES BL + NIX WD
           P387-91 OF EUROPEAN MEETING ON FERROELECTRICITY, 1969, SAARBRUCKEN,
           MUSER HE AND PETERSSON J (EDS), WISSENSCH VERLAGSGES, STUTTGART,
           1970, 396P.

04-0837   PTCR (POSITIVE TEMPERATURE COEFFICIENT OF RESISTANCE) BEHAVIOR OF BARIUM TITANATE WITH
           NIOBIUM PENTOXIDE AND MANGANESE DIOXIDE ADDITIVES.
           MATSUOKA T + MATSUO U + SASAKI H AND OTHERS
           J AMER CERAM SOC 55: 108 (FEB 1972)

04-0838   ELECTRET EFFECT IN A MIXTURE OF POLY METHACRYLATE AND BARIUM TITANATE
           MAZUR K + HANDEREK J + PIECH T
           ACTA PHYS POLON A 37: 31-40 (1970)

04-0839   MICROSTRUCTURE AND ELECTRICAL PROPERTIES OF SCANDIUM TRIOXIDE DOPED, RARE EARTH OXIDE
           DOPED, AND UNDOPED BARIUM TITANATE.
           MAZDIYASNI KS + BROWN LM
           J AMER CERAM SOC 54: 539-43 (1971)

04-0840   SYNTHESIS OF NIOBIUM PENTOXIDE DOPED BARIUM TITANATE WITH IMPROVED ELECTRICAL
           PROPERTIES. (DIELECTRIC CONSTANT, DISSIPATION FACTOR)
           MAZDIYASNI KS + BROWN LM
           J AMER CERAM SOC 55(12): 633-4 (1972)
                285

04-0841   FRACTURE TOUGHNESS OF COMMERCIAL BARIUM TITANATE TRANSDUCER MATERIAL.
           MCKINNEY KR + SMITH HL
           BULL AMER CERAM SOC 51: 650 (1972)

           EXTENSION OF THE POULET- LOUDON APPROXIMATION IN THE THEORY OF OPTICAL LATTICE
           VIBRATIONS. (QUARTZ, BARIUM TITANATE)
           MERTEN I + LAMPRECHT G
           Z NATURFORSCH A 26(2): 215-19 (1971) (IN GERMAN)

04-0842   POLARITON DISPERSION OF TETRAGONAL FERROELECTRIC BARIUM TITANATE. (CALCULATIONS)
           MERTEN L
           P273-80 OF EUROPEAN MEETING ON FERROELECTRICITY, 1969, SAARBRUCKEN,
           MUSER HE AND PETERSSON J (EDS), WISSENSCH VERLAGSGES, STUTTGART,
           1970, 396P.

04-0843   PREPARATION AND PROPERTIES OF THIN FILMS OF BARIUM TITANATE.
           MESNARD MG + TREILLEUX M + METRAT G
           THIN SOLID FILMS 10: 21-9 (1972) (IN FRENCH)
                265

04-0844   AUTOMATIC IMPEDANCE MEASUREMENTS USING THERMAL NOISE ANALYSIS (APPLICATION TO BARIUM
          TITANATE).
          MICHERON F + GODEFROY L
          REV SCI INSTRUM 43: 1460-5 (OCT 1972)

04-0845   BAND STRUCTURE OF THE CUBIC PHASE OF BARIUM TITANATE. (IMAGINARY DIELECTRIC CONSTANT)
          MICHEL-CALENDINI F + MESNARD MG
          PHYS STATUS SOLIDI B 44: K117-22 (1971) (IN FRENCH)
               SCI ABSTR A 74: 41178 (1971)
               269

04-0846   BAND STRUCTURE OF CUBIC AND TETRAGONAL BARIUM TITANATE.
          MICHEL-CALENDINI F
          UNIV CLAUDE BERNARD LYON SCD THESIS, 1971, 156P. (IN FRENCH)
               269

04-0847   ELECTRICAL CONTROL OF FIXATION AND ERASURE OF HOLOGRAPHIC PATTERNS IN FERROELECTRIC
          MATERIALS. (BARIUM TITANATE, POTASSIUM NIOBATE)
          MICHERON F + BISMUTH G
          APPL PHYS LETT 20: 79-81 (1972)

04-0848   HOLOGRAPHIC STORAGE, ELECTRICAL FIXING AND ERASING IN DOPED BARIUM TITANATE CRYSTALS.
          MICHERON F + BISMUTH G
          J PHYS (PARIS) 33, SUPPL 4, COLLOQ 2: 149-50 (1972)
               294

04-0849   INTERNAL STRESSES IN BARIUM TITANATE CERAMICS. (AND DIFFERENCE IN TEMPERATURES FOR
          MAXIMUM DIELECTRIC CONSTANT AND LOSS)
          MICHERON F + GODEFROY L
          P333-8 OF EUROPEAN MEETING ON FERROELECTRICITY, 1969, SAARBRUCKEN,
          MUSER HE AND PETERSSON J (EDS), WISSENSCH VERLAGSGES, STUTTGART,
          1970, 396P.

04-0850   RELAXATION ABSORPTION OF ULTRASOUND IN SINGLE CRYSTALS OF BARIUM TITANATE IN THE
          PRESENCE OF A STEADY ELECTRIC FIELD.
          MIKHELSON LM
          P75-81 OF FAZOVYE PEREKHODY V SEGNETOELEKTRIKAKH (PHASE
          TRANSFORMATIONS IN FERROELECTRICS), FRITSBERG VYA + ROLOV BN +
          KRUCHAN YY (EDS), IZDATELSTVO ZINATNE RIGA, 1971, 206P. (IN RUSSIAN)

          SEMICONDUCTING AND NONSTOICHIOMETRIC BARIUM TITANATE. A REFERENCE GUIDE.
          MILEK JT
          ELECTRONIC PROPERTIES INFORMATION CENTER, CULVER CITY, CALIF,
          FEB 1971.

04-0851   POTENTIAL BARRIERS ON SEMICONDUCTING BARIUM TITANATE. (PTCR DEVICES)
          MILLER CA
          J PHYS D APPL PHYS 4: 690-6 (MAY 1971)
               250

04-0852   QUANTITATIVE STUDIES OF OPTICAL HARMONIC GENERATION IN CADMIUM SULFIDE, BARIUM
          TITANATE, AND POTASSIUM DIHYDROGEN PHOSPHATE TYPE CRYSTALS.
          MILLER RC + KLEINMAN AD + SAVAGE A
          PHYS REV LETT 11: 146-9 (1963)

04-0853   PRESSURE DEPENDENCE OF TRANSITION TEMPERATURES AND ELECTROSTRICTIONS IN PEROVSKITE
          BARIUM TITANATE.
          MINOMURA S + TANAKA M + OKAI B + NAGASAKI H
          P404-6 OF INT MEETING ON FERROELECTRICITY, 2ND, 1969, KYOTO, PROC,
          (J PHYS SOC JAP, VOL 28 SUPPL). PHYS SOC JAP, 1970, 459P.

04-0854   TRANSFORMATION OF ELECTROMAGNETIC RADIATION INTO MOTION. (PYROELECTRIC EFFECT IN TGS,
          ETHYL DIAMINE TARTRATE, BARIUM TITANATE, SALICYLIDENE- 4-BROMO ANILINE.
          MOCKEL P
          NATURWISSENSCHAFTEN 57(5): 240 (1970) (IN GERMAN)
               216

4. BARIUM TITANATE

04-0855 MEASUREMENT OF DIELECTRIC PROPERTIES OF FERROELECTRIC MATERIALS AT CENTIMETER AND
         DECIMETER WAVELENGTHS EMPLOYING A SYMMETRIC STRIPLINE. (BARIUM TITANATE AND BARIUM
         CALCIUM TITANATE)
         MOLCHANOV VI + POPLAVKO YUM
         IZV VUZ RADIOELEKTRON 1971(2): 158-62 (IN RUSSIAN)
         (TRANSLATION TO APPEAR IN RADIO ELECTRON COMMUN)
              238

04-0856 PRESSURE DEPENDENCE OF THE DIELECTRIC CONSTANTS OF PARAELECTRIC MATERIALS. (BARIUM
         TITANATE, STRONTIUM TITANATE)
         MORENO M + GRANICHER H
         HELV PHYS ACTA 37(7-8): 625 (1964) (IN GERMAN)
              SCI ABSTR A 68: 18002 (1965)

04-0857 SURFACE LAYER IN BARIUM TITANATE (X-RAY DIFFRACTION, DC FIELD EFFECT, PARAELECTRIC AND
         FERROELECTRIC PHASES).
         MOTEGI H + HOSHINO S
         J PHYS SOC JAP 29: 524 (1970)
              208

04-0858 X-RAY STUDY OF THE SURFACE LAYER ON BARIUM TITANATE SINGLE CRYSTAL.
         MOTEGI H
         J PHYS SOC JAP 32(1): 202-9 (1972)
              276

04-0859 EFFECT OF SURFACE MOISTURE ON THE DIELECTRIC BEHAVIOR OF ULTRAFINE BARIUM TITANATE
         PARTICULATES.
         MOUNTVALA AJ
         J AMER CERAM SOC 54: 544-8 (NOV 1971)

04-0860 ELECTRICAL PROPERTIES OF RUBIDIUM AND CESIUM DOPED BARIUM TITANATE CERAMICS.
         MOUNTVALA AJ
         J AMER CERAM SOC 53: 53-4 (1970)

04-0861 PREPARATION OF BARIUM TITANATE AND OTHER CERAMIC POWDERS BY COPRECIPITATION OF CITRATES
         IN AN ALCOHOL.
         MULDER BJ
         BULL AMER CERAM SOC 49: 990-3 (1970)

04-0862 FORMATION PROCESS OF BARIUM TITANATE THIN FILM EVAPORATED IN VACUO.
         MURAYAMA Y + UEDA R + SUZUKI T + TOKUGAWA F
         OYO BUTSURI 33(1): 456-60 (1964) (IN JAPANESE)
              CHEM ABSTR 63: 73F (1965)

04-0863 TEMPERATURE STUDIES OF INFRARED REFLECTION SPECTRA OF BARIUM AND STRONTIUM TITANATES IN
         THE 2-1000 MICRON REGION.
         MURZIN VN + DEMESHINA AI
         SOV PHYS SOLID STATE 5: 1716-17 (1964)

04-0864 VIBRATIONAL SPECTRA OF STRONTIUM, BARIUM AND CALCIUM TITANATES.
         MURZIN VN + DEMESHINA AI + BOGDANOV SV
         BULL ACAD SCI USSR, PHYS SER 29(6): 926-30 (1965)
              219

04-0865 VIBRATIONAL SPECTRA OF THE TITANATES OF STRONTIUM, BARIUM, AND CALCIUM. (75 REFS)
         MURZIN VN
         TR FIZ INST PN LEBEDEVA, AKAD NAUK SSSR 48: 145-203 (1969)
         (IN RUSSIAN)
              CHEM ABSTR 72: 94858X (1970)
              198

04-0866 HIGH L/F NOISE ANOMALY IN SEMICONDUCTING BARIUM STRONTIUM TITANATE.
         MYTTON RJ + BENTON RK
         PHYS LETT A 39: 329-30 (1972)

04-0867 MECHANISM FOR FORMING INSULATION LAYERS ON THE GRAIN BOUNDARIES OF BARIUM TITANATE
         SEMICONDUCTIVE CERAMICS.
         NAKAHARA M
         ELEC COMMUN LAB TECH J 20: 2681-4 (1971) (IN JAPANESE)
              256

4. BARIUM TITANATE

04-0868   PERMITTIVITY CHANGE OF BARIUM TITANATE UNDER VERY SLOWLY CHANGING FIELD.
           NAKAMURA T + ISHIBASHI Y + KUROKI H
           P131-7, VOL 2 OF DVORAK V + FOUSKOVA A + GLOGAR P (EDS),
           INT MEETING ON FERROELECTRICITY, PRAGUE, 1966, PROC.
           INST PHYS CZECH ACAD SCI, PRAGUE, 1966.

04-0869   FLUORINE CONTAINING BARIUM TITANATE. (ELECTRICAL PROPERTIES)
           NEKRASOV MM + SAVOSHCHENKO VS + SYCH AM
           INORG MATER 6: 1907-9 (1970)

04-0870   NONLINEAR PROPERTIES OF FERROELECTRICS AT MICROWAVE FREQUENCIES. (TRIGLYCINE SULFATE,
           POLYCRYSTALLINE BARIUM TITANATE)
           NEKRASOV MM
           P193-9, VOL 2 OF DVORAK V + FOUSKOVA A + GLOGAR P (EDS),
           INT MEETING ON FERROELECTRICITY, PRAGUE, 1966, PROC.
           INST PHYS CZECH ACAD SCI, PRAGUE, 1966.

           LATTICE DYNAMICAL THEORY OF SWITCHING IN BARIUM TITANATE SINGLE CRYSTALS.
           NETTLETON RE
           J APPL PHYS 38(7): 2775-86 (1967)

04-0871   CHANGE IN THE DECAY CONSTANT OF TECHNETIUM-99 IN BARIUM TITANATE BY THE FERROELECTRIC
           TRANSITION.
           NISHI M + SHIMIZU S
           PHYS REV B 5(8): 3218-21 (15 APR 1972)
                264

04-0872   CHEMICAL CHANGES IN BARIUM TITANATE FIRED UNDER NITROGEN IN THE PRESENCE OF CARBON.
           NITTA T + NAGASE K + SASAKI H + HAYAKAWA S
           J AMER CERAM SOC 54: 220 (APR 1971)

04-0873   DEFECT STRUCTURE OF (BARIUM(1-X) STRONTIUM(X)) TITANATE- MIXED CRYSTALS FROM A
           POTASSIUM FLUORIDE MELT.
           NOVAK J + COUFOVA P + HLASIVCOVA N + JANOUSEK V
           COLLECT CZECH CHEM COMMUN 35: 782-7 (MAR 1970)
           (IN GERMAN)
                CHEM ABSTR 72: 94170S (1970)
                204

04-0874   HETEROVALENT SUBSTITUTION AND OXYGEN NONSTOICHIOMETRY OF BARIUM TITANATE AND HEMATITE
           SINGLE CRYSTALS.
           NOVAK J
           KRISTALL TECH 6: K5-9 (1971)

04-0875   OPTICAL AND DIELECTRIC PROPERTIES OF SINGLE CRYSTAL BARIUM TITANATE WITH NIOBIUM
           ADDITIONS.
           NOVAK J + COUFOVA P + DOBIASOVA L + RYSAVA N
           2P OF CONF ON CERAMICS FOR ELECTRONICS, 4TH, 1971, SPINDLERUV MLYN,
           CZECHOSLOVAKIA VYZKUMNY USTAV ELEKTROTECHNICKE KERAMIKY, 1971
           (IN CZECH)
                SCI ABSTR A 75: 16287 (1972)
                264

04-0876   VARIATION OF THE DECAY CONSTANT OF STRONTIUM-85 IN BARIUM TITANATE.
           NYIKOS P + GAGNEUX S + HUBER P + KOBEL HR + LEUENBERGER H
           HELV PHYS ACTA 43: 412 (1970) (IN GERMAN)

04-0877   FORBIDDEN LINES IN THE HYPERFINE SPECTRUM OF THE ELECTRON PARAMAGNETIC RESONANCE OF
           MANGANESE(+2) IN BARIUM TITANATE.
           ODEHNAL M
           CZECH J PHYS 13: 566-72 (1963)

04-0878   POLARITON DISPERSION RELATION IN CUBIC BARIUM TITANATE.
           OHAMA N + OKAMOTO Y + MATUMURA O
           J PHYS SOC JAP 29: 1648 (1970)
                222

04-0879   ELECTRICAL PROPERTIES OF HOT PRESSED FERROELECTRIC (BARIUM TITANATE TYPE) CERAMICS. (IN
           RELATION TO DENSIFICATION AND GRAIN GROWTH)
           OKAZAKI K + TAKAHASHI K
           P328-30 OF INT MEETING ON FERROELECTRICITY, 2ND, 1969, KYOTO, PROC,
           (J PHYS SOC JAP, VOL 28 SUPPL). PHYS SOC JAP, 1970, 459P.

4. BARIUM TITANATE

04-0880   SPACE CHARGE POLARIZATION AND AGING OF BARIUM TITANATE CERAMICS.
          OKAZAKI K + SAKATA K
          ELECTROTECH J JAP 7: 13-18 (1962)

04-0881   PIEZOELECTRIC CHARACTERISTIC OF BARIUM TITANATE CERAMIC EXHIBITED WHEN IT IS GIVEN AN
          IMPACT.
          ORIKASA K + KUBOTA Y + TAKASU M
          MEM MURORAN INST TECHNOL SCI ENG 7: 287-96 (1970) (IN JAPANESE)

04-0882   INFLUENCE OF WATER VAPOR ON BARIUM TITANATE CERAMICS.
          ORMANCEY G
          5P OF CONF ON CERAMICS FOR ELECTRONICS, 4TH, 1971, SPINDLERUV MLYN,
          CZECHOSLOVAKIA VYZKUMNY USTAV ELEKTROTECHNICKE KERAMIKY, 1971
                SCI ABSTR A 75: 16285 (1972)
                264

04-0883   INFLUENCE OF HUMIDITY ON THE POTENTIAL DISTRIBUTION IN BARIUM TITANATE CERAMICS AND
          CRYSTALS.
          ORMANCEY G
          J PHYS (PARIS) 33, SUPPL 4, COLLOQ 2: 114-15 (1972)
                294

04-0884   CURRENT EXCITATION IN GAMMA-IRRADIATED BARIUM TITANATE.
          PANCHENKOV GM + PLUZHNIKOV VM + KAUSHANSKII DA AND OTHERS
          BULL ACAD SCI USSR, PHYS SER 35(9): 1725-6 (1971)

04-0885   HIGH TEMPERATURE THERMODYNAMIC AND X-RAY DIFFRACTION STUDY OF NONSTOICHIOMETRIC BARIUM
          TITANATE.
          PANLENER RJ + BLUMENTHAL RN
          BULL AMER CERAM SOC 49: 830 (1970)

04-0886   TITANIUM RICH NONSTOICHIOMETRIC BARIUM TITANATE. PART-3: HIGH TEMPERATURE THERMODYNAMIC
          AND X-RAY DIFFRACTION MEASUREMENTS.
          PANLENER RJ + BLUMENTHAL RN
          J AMER CERAM SOC 54: 610-3 (DEC 1971)

04-0887   CUBIC TETRAGONAL (MARTENSITIC TYPE) PHASE TRANSITIONS IN HIGH RESISTIVITY BARIUM
          TITANATE SINGLE CRYSTALS. (AND TWINNING TENDENCY)
          PARKER TJ + BURFOOT JC
          P362-3 OF INT MEETING ON FERROELECTRICITY, 2ND, 1969, KYOTO, PROC.
          (J PHYS SOC JAP, VOL 28 SUPPL). PHYS SOC JAP, 1970, 459P.

04-0888   DIFFUSIONLESS PHASE TRANSITIONS AND THE STABLE DOMAIN STRUCTURE OF (HIGH SENSITIVITY)
          BARIUM TITANATE.
          PARKER TJ + BURFOOT JC
          P119-22 OF EUROPEAN MEETING ON FERROELECTRICITY, 1969, SAARBRUCKEN,
          MUSER HE AND PETERSSON J (EDS), WISSENSCH VERLAGSGES, STUTTGART,
          1970, 396P.

04-0889   SOLUBILITY LIMIT OF LANTHANUM(0.5) LITHIUM(0.5) TITANATE IN BARIUM TITANATE.
          PATIL PV + CHINCHOLKAR VS
          CURR SCI 39(15): 348 (1970)
                209

04-0890   INFLUENCE OF VIBRATIONS ON THE DIELECTRIC PROPERTIES OF BARIUM TITANATE CERAMICS.
          PAUDERT R + STELLENBERGER K
          HERMSDORFER TECH MITT 32: 1001-5 (1971) (IN GERMAN)

04-0891   EFFECT OF INTERNAL BOUNDARIES ON THE DIELECTRIC PROPERTIES OF BARIUM TITANATE CERAMICS.
          PAYNE DA + CROSS LE
          BULL AMER CERAM SOC 49: 414 (1970)

04-0892   TEMPERATURE DEPENDENCE OF THE DEBYE- WALLER FACTORS OF BARIUM TITANATE (BRAGG
          REFLECTIONS C-AXIS, 20-180 DEG C, CALCULATIONS).
          PEDERSEN T
          J PHYS SOC JAP 29(5): 1314-21 (1970)
                218

4. BARIUM TITANATE

04-0893   PRESSURE DEPENDENCE OF ACOUSTIC MODE- SOFT OPTIC MODE INTERACTIONS IN FERROELECTRIC
          BARIUM TITANATE.
          PEERCY PS + SAMARA GA
          PHYS REV B SOLID STATE 6(7): 2748-51 (1 OCT 1972)
               275

04-0894   SPECIFIC FEATURES OF THE TEMPERATURE DEPENDENCE OF THE VELOCITY OF ULTRASOUND IN
          POLYCRYSTALLINE SOLID SOLUTIONS OF BARIUM AND STRONTIUM TITANATES.
          PERRO IT + GRINVALD GZH + FRITSBERG VYA
          P97-107 OF FAZOVYE PEREKHCDY V SEGNETOELEKTRIKAKH (PHASE
          TRANSFORMATIONS IN FERROELECTRICS), FRITSBERG VYA + ROLOV BN +
          KRUCHAN YY(EDS), IZDATELSTVO ZINATNE RIGA, 1971, 206P. (IN RUSSIAN)

04-0895   DEPENDENCE OF DIELECTRIC DISPERSION IN BARIUM TITANATE SINGLE CRYSTALS ON THEIR DOMAIN
          STRUCTURE.
          PETROV VM + MARTYNOVA SV
          P91-9 OF SEGNETOELEKTRIKAKH I OKISNYE POLUPROVODNIKI (FERROELECTRICS
          AND OXIDE SEMICONDUCTORS), SINYAKOV EV(ED), REDAKTSIONNO-IZDATELSKII
          OTDEL DNEPROPETROVSKII GOSUDARSTVENNYI UNNIVERSITET, DNEPROPETROVSK,
          1971, 100P. (IN RUSSIAN)

04-0896   EFFECT OF LINEAR COMPRESSION ON THE UHF DIELECTRIC PROPERTIES OF BARIUM TITANATE SINGLE
          CRYSTALS.
          PETROV VM + MARTYNOVA SV
          BULL ACAD SCI USSR, PHYS SER 34(12): 2310-12 (1970)

04-0897   STUDY OF CRYSTALS OF BARIUM TITANATE HAVING PARALLEL INTERFERENCE FRINGES.
          PHILIPPE G + SENSIK M
          COMPT REND B 274: 819-22 (1972) (IN FRENCH)
               SCI ABSTR A 75: 35980 (1972)
               264

04-0898   INFLUENCE OF PRESSURE ON PHASE TRANSITIONS AT LOW TEMPERATURES: STRONTIUM TITANATE AND
          ((BARIUM(X) STRONTIUM(1-X) TITANATE (RECIPROCAL DIELECTRIC CONSTANT FOLLOWS QUADRATIC
          TEMPERATURE LAW).
          PIETRASS B + HEGENBARTH E
          J LOW TEMP PHYS 7(3-4): 201-9 (1972)

04-0899   DIELECTRIC ABSORPTION IN THE ORTHORHOMBIC AND QUADRATIC PHASES OF BARIUM TITANATE.
          PILET JC
          UNIV RENNES PHD THESIS, 1972, 93P.
               299

04-0900   DIELECTRIC ABSORPTION IN THE ORTHORHOMBIC AND QUADRATIC PHASES OF BARIUM TITANATE.
          PILET JC
          UNIV RENNES PHD THESIS, 1972, 93P. (IN FRENCH)
               299

04-0901   FERROELECTRIC PROPERTIES OF BARIUM TITANATE IN THE CENTIMETER RANGE OF SUPERHIGH
          FREQUENCIES (MICROWAVES).
          POPLAVKO YUM
          SOV PHYS SOLID STATE 4: 787-9 (1962)

04-0902   MICROWAVE DISPERSION MECHANISM IN BARIUM TITANATE TYPE FERROELECTRICS.
          POPLAVKO YUM
          BULL ACAD SCI USSR, PHYS SER 29(11): 1856-61 (1965)
               219

04-0903   PERMITTIVITY DISPERSION OF BARIUM TITANATE TYPE FERROELECTRICS. (SINGLE CRYSTALS AND
          CERAMICS)
          POPLAVKO YUM
          SOV PHYS SOLID STATE 6(1): 45-9 (1964)

04-0904   THERMAL CHARACTERISTICS OF A FERROELECTRIC (BARIUM TITANATE TYPE) CERAMIC WITH
          CONTROLLED MICROWAVE PERMITTIVITY.
          POPLAVKO YUM
          SOV PHYS SOLID STATE 4: 1911-12 (1963)

04-0905   UHF (WAVEGUIDE RESONATOR) METHOD OF STUDYING TEMPERATURE CHARACTERISTICS OF ROCHELLE
          TYPE CERAMICS. (BARIUM CALCIUM TITANATE CERAMIC)
          POPLAVKO YUM + YAZYTSKII BYA
          INSTRUM EXPER TECH 1967(4): 841-3
               219

4. BARIUM TITANATE

04-0906   CHARACTERISTICS OF RF SPUTTERED BARIUM TITANATE THIN FILMS.
          PRATT IH
          PROC IEEE 59: 1440-7 (1971)

04-0907   FABRICATION OF RF SPUTTERED BARIUM TITANATE THIN FILMS.
          PRATT IH + FIRESTONE S
          J VACUUM SCI TECHNOL 8(1): 256-60 (JAN-FEB 1971)
              238

04-0908   MECHANISM OF ELECTROCONDUCTIVITY IN BARIUM TITANATE.
          PROKOPALO OI
          BULL ACAD SCI USSR, PHYS SER 35(9): 1777-81 (1971)

04-0909   EFFECT OF SOME ADMIXED NICKEL ON SOME PHYSICAL PROPERTIES OF BARIUM TITANATE SINGLE
          CRYSTALS. (LATTICE PARAMETERS)
          PSHEDMOISKII YT
          SOV PHYS CRYSTALLOGR 17(1): 187-8 (JUL-AUG 1972)
              274

04-0910   DIELECTRIC MEASUREMENTS ON PILE-IRRADIATED BARIUM TITANATE SINGLE CRYSTALS.
          QUITTET AM + LAMBERT M
          J PHYS (PARIS) 33, SUPPL 4, COLLOQ 2: 142 (1972)
              294

04-0911   AGING PROCESSES IN BARIUM TITANATE (DECREASE IN NUMBER OF 180 DEG DOMAINS).
          RAPOPORT SL + DONTSOVA LI
          SOV PHYS CRYSTALLOGR 15(2): 327-9 (1970)
              208

04-0912   COMPARATIVE STUDY OF THE CRYSTALLOGRAPHIC, DIELECTRIC, AND NONLINEAR OPTICAL PROPERTIES
          OF THE ABC NIOBATES (WHERE A AND B ARE CALCIUM, STRONTIUM, OR BARIUM, AND C IS SODIUM
          OR POTASSIUM) PHASES OF THE TETRAGONAL TUNGSTEN BRONZE OXIDE TYPE.
          RAVEZ J + BUDIN JP + HAGENMULLER P
          J SOLID STATE CHEM 5: 239-46 (1972) (IN FRENCH)
              CHEM ABSTR 77: 119223E (1972)
              276

04-0913   IMAGING ELECTRIC MICRO FIELDS (OF SOLID SURFACES) IN THE EMISSION ELECTRON MICROSCOPE
          (BARRIER LAYERS IN DOPED BARIUM TITANATE CERAMICS).
          REHME H
          Z ANGEW PHYS 29(3): 173-80 (1970) (IN GERMAN)
              207

04-0914   PREPARATION AND PROPERTIES OF THIN FILM MATERIALS WITH PEROVSKITE TYPE STRUCTURE.
          (MIXED OXIDES ABO(3)) (BARIUM TITANATE AND LITHIUM NIOBATE FILMS ON PLATINUM OR
          SILICON)
          REIBER LM
          J PHYS (PARIS) 33, SUPPL 4, COLLOQ 2: 265-7 (1972)
              294

04-0915   REMARKS ABOUT SWITCHING IN BARIUM TITANATE AND TRIGLYCINE SULFATE SINGLE CRYSTALS.
          REIBER M + GODEFROY L
          P117-21, VOL 2 OF DVORAK V + FOUSKOVA A + GLOGAR P (EDS),
          INT MEETING ON FERROELECTRICITY, PRAGUE, 1966, PROC.
          INST PHYS CZECH ACAD SCI, PRAGUE, 1966.

04-0916   TWO WAVE SHOCK STRUCTURES IN THE FERROELECTRIC CERAMICS BARIUM TITANATE AND LEAD
          ZIRCONATE TITANATE.
          REYNOLDS CE + SEAY GE
          J APPL PHYS 33: 2234-41 (1962)

04-0917   EFFECT OF GRAIN SIZE ON THE FERROELECTRIC PARAELECTRIC TRANSITION OF BARIUM TITANATE.
          RICHARD M + EYRAUD L + FETIVEAU M + RIVIERE R
          COMPT REND 255: 2917-19 (1962) (IN FRENCH)

04-0918   ELECTRICAL CONDUCTIVITY (SEEBECK COEFFICIENT, CURIE POINT) OF REDUCED BARIUM TITANATE
          CRYSTALS (POLARON HOPPING INTERPRETATION).
          RIDPATH DL + WRIGHT DA
          J MATER SCI 5: 487-91 (1970)
              197

4. BARIUM TITANATE

04-0919 SOME SEMICONDUCTING PROPERTIES OF BARIUM TITANATE. (RESISTIVITY, SEEBECK COEFFICIENT, OPTICAL ABSORPTION)
RIDPATH DL
UNIV DURHAM, ENGLAND, PHD THESIS, 1969.

04-0920 ELECTRON PARAMAGNETIC RESONANCE OF TRIVALENT GADOLINIUM IONS IN STRONTIUM AND BARIUM TITANATES.
RIMAI L + DEMARS GA
PHYS REV 127: 702-10 (1962)

04-0921 THERMAL HYSTERESIS IN THE TETRAGONAL CUBIC TRANSITION OF BARIUM TITANATE.
RIVERA JM + GONZALO JA
REV MEX FIS 19(3): 297-302 (1970) (IN SPANISH)
      CHEM ABSTR 74: 103855Q (1971)
      254

04-0922 ETCH PITS IN BARIUM TITANATE SINGLE CRYSTALS.
ROTHWELL WS
J AMER CERAM SOC 47: 409-11 (1964)

04-0923 AUGER-LIKE RESONANT INTERFERENCE IN RAMAN SCATTERING FROM ONE AND TWO PHONON STATES OF BARIUM TITANATE.
ROUSSEAU DL + PORTO SPS
PHYS REV LETT 20(24): 1354-7 (1968)

04-0924 EFFECT OF ULTRASOUND ON THE POLARIZATION AND REPOLARIZATION PROCESSES IN TRIGLYCINE SULFATE CRYSTALS AND BARIUM TITANATE CERAMIC.
RUDYAK VM + BARANOV AI
BULL ACAD SCI USSR, PHYS SER 29(6): 955-9 (1965)
      CHEM ABSTR 63: 10801H (1965)

04-0925 PHOTON PHONON INTERACTION IN THE NEAR INFRARED. (BARIUM TITANATE)
RUPPRECHT G
PHYS REV LETT 12: 580-3 (1964)

04-0926 DIFFUSION COEFFICIENT OF NIOBIUM IN BARIUM TITANATE BY A DIELECTRIC MEASUREMENT TECHNIQUE.
RUTT TC
BULL AMER CERAM SOC 50: 396 (1971)

04-0927 ELECTRON PARAMAGNETIC RESONANCE OF IRON(3+) IN BARIUM TITANATE AT LOW TEMPERATURES.
SAKUDO T
J PHYS SOC JAP 18: 1626 (1963)

04-0928 ELECTRON PARAMAGNETIC RESONANCE IN BARIUM TITANATE NEAR THE CURIE TEMPERATURE.
SAKUDO T + UNOKI H + MAEKAWA S
J PHYS SOC JAP 18: 913 (1963)

04-0929 ELECTRON PARAMAGNETIC RESONANCE OF IRON(3+) IN BARIUM TITANATE IN THE RHOMBOHEDRAL AND THE CUBIC PHASES.
SAKUDO T + UNOKI H
J PHYS SOC JAP 19: 2109-18 (1964)

04-0930 CHARACTERISTICS OF RF SPUTTERED BARIUM TITANATE FILMS ON SILICON.
SALAMA CAT + SICIUNAS E
J VACUUM SCI TECHNOL 9(1): 91-6 (1972)
      265

DISPERSION IN FERROELECTRICS. (THEORY, BARIUM TITANATE)
SANNIKOV DG
SOV PHYS JETP 14: 98-101 (1962)

ELECTROMAGNETIC AND ACOUSTIC WAVES IN FERROELECTRICS. (THEORY, BARIUM TITANATE, ROCHELLE SALT)
SANNIKOV DG
SOV PHYS SOLID STATE 4: 1187-92 (1962)

04-0931 A NEW PTC THERMISTOR MATERIAL. (BARIUM TITANATE)
SASAKI H + MATSUO Y + HAYAKAWA S
BULL AMER CERAM SOC 47(4): 385 (APR 1968)
      143

04-0932   FORMATION OF FLUORINE CONTAINING SOLID SOLUTIONS BASED ON BARIUM TITANATE.
           SAVOSHCHENKO VS + GOLUB AM + NEKRASOV MI + VOITKO II + GORBATYUK VA
           INORG MATER 8: 643-6 (1972)

04-0933   FERROELECTRIC AND ANTIFERROELECTRIC SYMMETRY GROUPS. (AND THEIR COMPARISON, APPLICATION
           TO BARIUM TITANATE)
           SCHELKENS R
           PHYS STATUS SOLIDI 37: 739-43 (1970)
                189

04-0934   INCORPORATION OF ANTIMONY INTO THE BARIUM TITANATE LATTICE. (STRUCTURE, AND PCT
           RESISTORS)
           SCHMELZ H
           P351-6 OF EUROPEAN MEETING ON FERROELECTRICITY, 1969, SAARBRUCKEN,
           MUSER HE AND PETERSSON J (EDS), WISSENSCH VERLAGSGES, STUTTGART,
           1970, 396P.

04-0935   ON THE ANOMALOUS CRYSTALLOGRAPHIC PROPERTIES OF SMALL BARIUM TITANATE PARTICLES.
           SCHOIJET M
           BRIT J APPL PHYS 15: 719-23 (1964)

04-0936   SPACE CHARGE LIMITED CURRENTS IN HEXAGONAL BARIUM TITANATE(0.75) PLATINATE(0.25)
           CRYSTALS.
           SCHOIJET M
           BRIT J APPL PHYS 15: 1124-26 (1964)

04-0937   STATIC ELECTRIC QUADRUPOLE INTERACTION OF TANTALUM AND HAFNIUM IONS IN BARIUM AND LEAD
           TITANATE.
           SCHAFER G + HERZOG P + WOLBECK B
           Z PHYS 257(4): 336-52 (1972)

04-0938   BARIUM TITANATE PIEZOELECTRIC MICROPROBE FOR STUDYING ULTRASONIC FIELDS.
           SEGARD N + POULIQUEN J
           COMPT REND 255(7): 1187-92 (1962) (IN FRENCH)
                SCI ABSTR A 66: 1777 (1963)

04-0939   EXPERIMENTAL STUDY OF EVAPORATED BARIUM TITANATE FILMS.
           SEKINE E + TOYODA H
           REV ELEC COMMUN LAB 10: 457-73 (1962)

04-0940   MOTION OF DOMAIN BOUNDARIES DURING SWITCHING IN SINGLE CRYSTAL FILMS OF BARIUM TITANATE
           (ELECTRON MICROSCOPY).
           SHAKMANOV VV + SPIVAK GV + YAKUNIN SI
           SOV PHYS SOLID STATE 12(8): 1827-30 (1971)
                228

04-0941   NONLINEAR OPTICAL PROPERTIES OF LEAD TITANATE AND BISMUTH TITANATE CRYSTALS.
           SHALDIN YUV + BARSUKOVA ML + KUZNETSOV VA + LOBACHEV AN
           SOV PHYS CRYSTALLOGR 16(1): 114-17 (JUL-AUG 1971)
                240

04-0942   OBSERVATION OF THE CHANGE IN THE POLARIZATION OF SINGLE CRYSTAL FILMS OF BARIUM
           TITANATE BY MEANS OF A STROBOSCOPIC TRANSMISSION ELECTRON MICROSCOPE.
           SHAKMANOV VV + YAKUNIN SI + SPIVAK GV + VASILEVA NV + MELAMED VSH
           SOV PHYS CRYSTALLOGR 17(2): 297-300 (SEP-OCT 1972)
                275

04-0943   ELECTROCONDUCTIVITY AND HALL EFFECT IN SEMICONDUCTING SOLID SOLUTIONS OF ISO- AND
           HETEROVALENT CATIONIC SUBSTITUTIONS IN BARIUM TITANATE.
           SHEFTEL IT + TEKSTER-PROSKURYAKOVA GN + LEIKINA BB + STOGOVA VA
           BULL ACAD SCI USSR, PHYS SER 35(9): 1789-92 (1971)

04-0944   RELATIONSHIP BETWEEN THE BREAKDOWN MECHANISM IN BARIUM TITANATE AND COLORATION.
           (INJECTED COLOR CENTERS)
           SHEFER NI + BURSIAN EV
           FOREIGN TECHNOL DIV, WRIGHT-PATTERSON AFB, OHIO, 1970, 12P.
           AD710251, N70-42534 (TRANSLATED FROM GOS PED INST, UCH ZAP PED INST,
           UCH ZAP ORENBURG GOS PEDAGOG INST NO 21: 180-7 (1967))
                US GOVT RES DEVELOP REP 70(19): 202 (1970)
                225

4. BARIUM TITANATE

04-0945   CATALYTIC OXIDATION OF CARBON MONOXIDE ON SEMICONDUCTIVE BARIUM TITANATE.
           SHIMIZU T + HARA H + SHIMADA K + YAMAMOTO O + YANAI H
           J CHEM SOC JAP, INDUST CHEM SECT 74: 825-8 (1971) (IN JAPANESE)

04-0946   INELASTIC NEUTRON SCATTERING FROM SINGLE DOMAIN BARIUM TITANATE. (LATTICE DYNAMICS)
           SHIRANE G + AXE JD + HARADA J + LINZ A
           PHYS REV B 2(9): 3651-7 (1970)
                233

04-0947   OXIDATION OF CARBON MONOXIDE ON LANTHANUM OXIDE DOPED BARIUM TITANATE.
           SHIMIZU T + HARA H
           J AMER CERAM SOC 55: 533 (OCT 1972)

04-0948   PREPARATION OF THIN BARIUM TITANATE FILMS BY DC DIODE SPUTTERING (ONTO PLATINUM OR
           FUSED QUARTZ).
           SHINTANI Y + TADA O
           J APPL PHYS 41(6): 2376-80 (1970)
                196

04-0949   THICKNESS DEPENDENCE OF POLARIZATION REVERSAL IN BARIUM TITANATE SINGLE CRYSTALS.
           SHIBATA H + TOYODA H
           J PHYS SOC JAP 17: 404-5 (1962)

04-0950   FERROELECTRICITY (IN BARIUM TITANATE TYPE STRUCTURES) AND LATTICE DYNAMICS.
           SILVERMAN BD
           P3-28, VOL 1 OF DVORAK V + FOUSKOVA A + GLOGAR P (EDS),
           INT MEETING ON FERROELECTRICITY, PRAGUE, 1966, PROC.
           INST PHYS CZECH ACAD SCI, PRAGUE, 1966.

04-0951   CRITICAL FIELD IN CERAMIC BARIUM TITANATE.
           SINYAKOV EV + FLEROVA SA
           SOV PHYS SOLID STATE 12(11): 2680-1 (1971)
                234

04-0952   DIRECT OBSERVATION OF 180 DEG DOMAINS IN BARIUM TITANATE BY MEANS OF A POLARIZING
           MICROSCOPE.
           SINYAKOV EV + DUDA VM + DUDNIK EF + YAKUNIN SI
           SOV PHYS SOLID STATE 13(9): 2149-51 (MAR 1972)
                261

04-0953   ELECTROOPTICAL PROPERTIES OF SINGLE CRYSTALS OF SOLID SOLUTIONS OF BARIUM TITANATE AND
           ZINC OXIDE.
           SINYAKOV EV + GOLOVYANKO AA + FERIN VD
           SOV PHYS CRYSTALLOGR 15(5): 875-8 (1971)
                232

04-0954   EFFECT OF EXTERNAL ACTIONS ON THE DOMAIN STRUCTURE OF SINGLE CRYSTALS FORMED BY BARIUM
           TITANATE BASE SOLID SOLUTION.
           SINYAKOV EV + DUDNIK EF + FLEROVA SA
           SOV PHYS CRYSTALLOGR 16(4): 681-4 (JAN-FEB 1972)
                254

04-0955   ELECTROOPTICAL PROPERTIES OF BARIUM TITANATE- TANTALUM PENTOXIDE SOLID SOLUTION SINGLE
           CRYSTALS.
           SINYAKOV EV + GOLOVYANKO AA + SAVCHENKO VG
           SOV PHYS CRYSTALLOGR 16(3): 468-70 (NOV-DEC 1971)
                251

04-0956   EFFECT OF TRANSVERSE COMPRESSION OF DIELECTRIC PROPERTIES OF BARIUM TITANATE SINGLE
           CRYSTALS.
           SINYAKOV EV + FLEROVA SA + GAVRILISHINA AI
           BULL ACAD SCI USSR, PHYS SER 35(9): 1731-5 (1971)

04-0957   EFFECT OF EXTERNAL ACTIONS ON THE REPOLARIZATION OF BARIUM TITANATE SINGLE CRYSTALS.
           SINYAKOV EV + DUDNIK EF + FLEROVA SA
           P34-40 OF SEGNETOELEKTRIKAKH I OKISNYE POLUPROVODNIKI (FERROELECTRICS
           AND OXIDE SEMICONDUCTORS), SINYAKOV EV(ED), REDAKTSIONNO-IZDATELSKII
           OTDEL DNEPROPETROVSKII GOSUDARSTVENNYI UNNIVERSITET, DNEPROPETROVSK,
           1971, 100P. (IN RUSSIAN)

4. BARIUM TITANATE

04-0958   ELECTROOPTICAL EFFECT IN SINGLE CRYSTALS OF BARIUM TITANATE- ZINC TITANATE SOLID
           SOLUTIONS.
           SINYAKOV EV + GOLOVYANKO AA + SAVCHENKO VG
           UKR FIZ ZH 16: 1371-4 (1971) (IN RUSSIAN)

04-0959   EFFECT OF NONISOVALENT SUBSTITUTION OF TITANIUM(4+) IONS ON THE ELECTROOPTICAL
           PROPERTIES OF BARIUM TITANATE SINGLE CRYSTALS.
           SINYAKOV EV + GOLOVYANKO AA + SAVCHENKO VG
           BULL ACAD SCI USSR, PHYS SER 35(12): 2360-4 (1971)

04-0960   GROWING SINGLE CRYSTALS OF SOLID SOLUTION BASED ON BARIUM TITANATE, WHERE BARIUM(2+)
           AND TITANIUM(4+) IONS ARE REPLACED BY ISOVALENT AND NONISOVALENT CATIONS.
           SINYAKOV EV + DUDNIK EF + FRASLAVSKAYA IP
           P12-21 OF SEGNETOELEKTRIKAKH I OKISNYE POLUPROVODNIKI (FERROELECTRICS
           AND OXIDE SEMICONDUCTORS), SINYAKOV EV(ED), REDAKTSIONNO-IZDATELSKII
           OTDEL DNEPROPETROVSKII GOSUDARSTVENNYI UNNIVERSITET, DNEPROPETROVSK,
           1971, 100P. (IN RUSSIAN)

04-0961   INFLUENCE OF UNIAXIAL TENSILE STRESSES ON THE ELECTRICAL PROPERTIES AND DOMAIN
           STRUCTURE OF BARIUM TITANATE SINGLE CRYSTALS.
           SINYAKOV EV + FLEROVA SA
           SOV PHYS SOLID STATE 12(9): 2196-7 (1971)
                227

04-0962   PIEZOELECTRIC PROPERTIES OF POLARIZED BARIUM TITANATE CERAMICS SUBJECTED TO UNIAXIAL
           TRANSVERSE COMPRESSION.
           SINYAKOV EV + GAVRILISHINA AI + FLEROVA SA
           SOV PHYS SOLID STATE 14(5): 1302-3 (NOV 1972)
                278

04-0963   REPOLARIZATION PROCESSES OF BARIUM TITANATE SINGLE CRYSTALS IN ORTHORHOMBIC AND
           RHOMBOHEDRAL PHASES.
           SINYAKOV EV + DUDNIK EF + DUDA VM
           SOV PHYS CRYSTALLOGR 16: 480-3 (NOV-DEC 1971)

04-0964   ROLE OF JAHN TELLER EFFECTS IN ORIGIN OF FERROELECTRICITY, OCCURRENCE OF ORDERED PHASE
           IN PEROVSKITE TYPE FERROELECTRICS. (BARIUM TITANATE)
           SINHA KP + SINHA AP
           INDIAN J PURE APPL PHYS 2(3): 91-4 (1964)
                ENG INDEX 1964: 412

04-0965   SEMICONDUCTOR PROPERTIES OF BARIUM (TITANATE, STANNATE) SOLID SOLUTIONS DOPED BY
           TUNGSTEN TRIOXIDE.
           SINYAKOV EV + KOLESNICHENKO KA
           P69-74 OF SEGNETOELEKTRIKAKH I OKISNYE POLUPROVODNIKI (FERROELECTRICS
           AND OXIDE SEMICONDUCTORS), SINYAKOV EV(ED), REDAKTSIONNO-IZDATELSKII
           OTDEL DNEPROPETROVSKII GOSUDARSTVENNYI UNNIVERSITET, DNEPROPETROVSK,
           1971, 100P. (IN RUSSIAN)

04-0966   EFFECT OF EXCESS TITANIA ON THE ELECTRIC RESISTANCE OF SEMICONDUCTIVE BARIUM TITANATE
           CERAMICS.
           SIVOVA R + SIMEONOV K
           COMPT REND ACAD BULG SCI 22(9): 1015-8 (1969)
                208

04-0967   ELECTRICAL PROPERTIES CF FLASH EVAPORATED FERROELECTRIC BARIUM TITANATE THIN FILMS.
           (THICKNESS DEPENDENCE)
           SLACK JR + BURFOOT JC
           J PHYS C 4: 898-909 (1971)
                247

04-0968   FLASH EVAPORATION OF FERROELECTRIC THIN FILMS OF POLYCRYSTALLINE BARIUM TITANATE.
           SLACK JR + BURFOOT JC
           THIN SOLID FILMS 6: 233-7 (1970)
                231

04-0969   CATION NONSTOICHIOMETRY IN TERNARY OXIDES. (BARIUM TITANATE)
           SMYTH DM
           BULL AMER CERAM SOC 51: 653 (1972)

4. BARIUM TITANATE

04-0970   EFFECTS OF (MONOENERGETIC) ELECTRON AND (COBALT-60) GAMMA-IRRADIATION ON BARIUM
          TITANATE, (COMPARED TO STUDY DAMAGE MECHANISM)
          SOLOVEV SP + KUZMIN II + KHARCHENKO VA
          P277-87, VOL 2 OF DVORAK V + FOUSKOVA A + GLOGAR P (EDS),
          INT MEETING ON FERROELECTRICITY, PRAGUE, 1966, PROC.
          INST PHYS CZECH ACAD SCI, PRAGUE, 1966.

04-0971   (IN PILE, FAST NEUTRON) RADIATION INDUCED PHASE TRANSFORMATIONS IN (CERAMIC) BARIUM AND
          LEAD TITANATES. (SUMMARY ONLY)
          SOLOVEV SP + KUZMIN II + ZAKURKIN VV
          P441 OF INT MEETING ON FERROELECTRICITY, 2ND, 1969, KYOTO, PROC,
          (J PHYS SOC JAP, VOL 28 SUPPL). PHYS SOC JAP, 1970, 459P.

          FERROELECTRICS 1: 19-22 (1970)
              191

04-0972   RADIATION PHYSICS OF FERROELECTRICS OF BARIUM TITANATE TYPE.
          SOLOVEV SP + KUZMIN II
          BULL ACAD SCI USSR, PHYS SER 34(12): 2320-6 (1970)
              252

04-0973   ELECTRON MIRROR MICROSCOPIC ASPECTS OF FERROELECTRIC DOMAINS OF BARIUM TITANATE AND
          DICALCIUM STRONTIUM PROPIONATE.
          SOMEYA T + KOBAYASHI J
          P478-9 OF ANNU MEET ELECTRON MICROSC SOC AMER, 28TH PROC, 1970,
          HOUSTON, CONDENSED PAP, CLAITORS PUB DIV, BATON ROUGE, 1970.
              SCI ABSTR A 75: 10120 (1972)
              264

04-0974   ELECTRON MIRROR MICROSCOPIC OBSERVATION OF FERROELECTRIC DOMAINS. (GOLD COATED BARIUM
          TITANATE)
          SOMEYA T + AZUMI T + KOBAYASHI J
          P374-6 OF INT MEETING CN FERROELECTRICITY, 2ND, 1969, KYOTO, PROC,
          (J PHYS SOC JAP, VOL 28 SUPPL). PHYS SOC JAP, 1970, 459P.

04-0975   FAR INFRARED DIELECTRIC DISPERSION IN BARIUM AND STRONTIUM TITANATES AND TITANIUM
          DIOXIDE.
          SPITZER WG + MILLER RC + KLEINMAN DA + HOWARTH LE
          PHYS REV 126: 1710-21 (1962)

04-0976   PARAMAGNETIC RESONANCE OF AN F CENTER IN NONSTOICHIOMETRIC BARIUM TITANATE.
          SROUBEK Z + ZKDANSKY K
          CZECH J PHYS B 13: 309-10 (1963)

04-0977   PARAMAGNETIC RESONANCE OF PLATINUM IONS IN NIOBIUM DOPED BARIUM TITANATE SINGLE
          CRYSTAL.
          SROUBEK Z + ZDANSKY K + SIMANEK E
          PHYS STATUS SOLIDI 3(1): K1-4 (1963)

04-0978   DETECTION OF ELECTROMAGNETIC RADIATION USING PYROELECTRIC EFFECT. (BARIUM TITANATE,
          TRIGLYCINE SULFATE, AT 5.6 AND 10.5 GHZ)
          STANFORD AL
          SOLID STATE ELECTRON 8(9): 747-55 (1965)

04-0979   NUCLEATION AND GROWTH OF FERROELECTRIC DOMAINS IN BARIUM TITANATE AT FIELDS FROM 2 TO
          450 KV/CM.
          STADLER HL + ZACHMANIDIS PJ
          J APPL PHYS 34: 3255-60 (1963)

04-0980   RELATION BETWEEN ELECTRICAL PROPERTIES AND MICROSTRUCTURE OF BARIUM TITANATE.
          STADELMAIER HH + DERBYSHIRE SW
          P57-65 OF RES CONF ON STRUCTURE AND PROPERTIES OF ENGINEERING
          MATERIALS, STADELMAIER HH + AUSTIN WW (EDS), PLENUM, NY, 1963.

              CHEM ABSTR 60: 9996

04-0981   SUSCEPTIBILITY INCREASE DURING POLARIZATION REVERSAL. (IN BARIUM TITANATE AND
          TRIGLYCINE SULFATE)
          STADLER HL & NAKAMURA T & ISHIBASHI Y & KUROKI H
          P138-43, VOL 2 OF DVORAK V + FOUSKOVA A + GLOGAR P (EDS),
          INT MEETING ON FERROELECTRICITY, PRAGUE, 1966, PROC.
          INST PHYS CZECH ACAD SCI, PRAGUE, 1966.

4. BARIUM TITANATE

04-0982 EMISSION OF LIGHT BY SURFACE MICROPLASMA OF CERAMIC (BARIUM, LEAD) TITANATE NEAR THE
         CURIE TEMPERATURE.
         SUJAK B + BIEDRZYCKI K
         ACTA PHYS POLON A 42: 49-54 (JUL 1972)
              289

04-0983 INFRARED SPECTRA OF FLUORINE CONTAINING BARIUM TITANATE.
         SYCH AM + SAVOSHCHENKO VS + DEMYANENKO VP + NEKRASOV MI
         UKR FIZ ZH 16: 922-6 (1971) (IN RUSSIAN)

04-0984 PREPARATION OF FLUORINE CONTAINING BARIUM TITANATE.
         SYCH AM + BILYK DI + BARCHUK LF
         INORG MATER 8: 1130-3 (1972)

04-0985 INFLUENCE OF DOMAIN STRUCTURE OF CERAMIC FERROELECTRICS ON THEIR MECHANICAL PROPERTIES.
         (BARIUM TITANATE, LEAD ZIRCONATE TITANATE, LEAD METANIOBATE, 20-150 DEG C, PRESSURE TO
         1200 KG/SQ CM)
         SYRKIN LN + ELGARD AM
         SOV PHYS SOLID STATE 7(4): 967-71 (1965)

04-0986 NONLINEARITY OF ELECTROMECHANICAL AND MECHANICAL PROPERTIES OF FERROELECTRICS. (BARIUM
         TITANATE CERAMICS)
         SYRKIN LN + POPOV SN
         P316-17 OF INT MEETING ON FERROELECTRICITY, 2ND, 1969, KYOTO, PROC.
         (J PHYS SOC JAP, VOL 28 SUPPL). PHYS SOC JAP, 1970, 459P.

04-0987 ON THE INFLUENCE OF DEFECTS ON THE TEMPERATURE HYSTERESIS OF BARIUM TITANATE.
         SZYMANSKA J
         POZNAN TOW PRZYJ NAUK, WYDZ MAT PRZYR 11: 197-201 (1962) (IN POLISH)
         CHEM ABSTR 61: 13966A

04-0988 ANOMALY OF RESISTANCE IN REDUCED BARIUM TITANATE CRYSTAL.
         TADA O + SHINTANI Y + YOSHIHARA T
         JAP J APPL PHYS 10: 1639 (NOV 1971)

         THEORY OF BARIUM TITANATE.
         TAKAHASHI H
         J PHYS SOC JAP 16(9): 1685-9 (1961)

04-0989 THERMAL CONDUCTIVITY OF STRONTIUM TITANATE, POTASSIUM TANTALATE AND BARIUM TITANATE
         NEAR THE CURIE POINTS.
         TAKEMURA M + TANI K
         J PHYS SOC JAP 31: 151-4 (JUL 1971)

04-0990 TWO SETS OF ELECTRON SPIN RESONANCE OF GADOLINIUM(3+) IN BARIUM TITANATE CERAMIC
         SEMICONDUCTOR.
         TAKADA T + WATANABE A
         J PHYS SOC JAP 19: 1742 (1964)

04-0991 BARIUM TITANATE BY CONVERGENT BEAM ELECTRON DIFFRACTION.
         TANAKA M + LEHMPFUHL G
         JAP J APPL PHYS 11: 1755-6 (NOV 1972)

04-0992 DYNAMICAL BEHAVIOR OF DISPLACIVE TYPE FERROELECTRICS NEAR THE TRANSITION POINT. (BARIUM
         TITANATE, STRONTIUM TITANATE)
         TANI K + TAKEMURA M
         J PHYS SOC JAP 30: 328-37 (FEB 1971)

04-0993 ELECTRON MICROSCOPIC STUDIES ON FERROELECTRIC DOMAINS OF PEROVSKITE TYPE OXIDES. (LEAD
         TITANATE, POTASSIUM NIOBATE, BARIUM TITANATE)
         TANAKA M + YATSUHASHI T + HONJO G
         P386-8 OF INT MEETING ON FERROELECTRICITY, 2ND, 1969, KYOTO, PROC.
         (J PHYS SOC JAP, VOL 28 SUPPL). PHYS SOC JAP, 1970, 459P.

04-0994 HIGH FIELD (1.5-6 KV/CM) POLARIZATION REVERSALS IN LIQUID ELECTRODED BARIUM TITANATE
         CRYSTALS (CALCULATIONS).
         TAYLOR GW
         AUSTRAL J PHYS 15: 549-67 (1962)
              216

04-0995 SEMICONDUCTING BARIUM AND STRONTIUM TITANATES WITH POSITIVE TEMPERATURE COEFFICIENTS OF RESISTIVITY.
TEKSTER-PROSKURYAKOVA GN + SHEFTEL IT
SOV PHYS SOLID STATE 5: 2548-8 (1964)

04-0996 EFFECT OF OXIDE ADDITIONS ON THE SINTERING OF BARIUM TITANATE COMPACT
THOME JR
BULL AMER CERAM SOC 49: 420 (1970)

04-0997 TIME DEPENDENCE OF THE ELECTRICAL CONDUCTIVITY OF BARIUM TITANATE SINGLE CRYSTALS HEATED IN OXYGEN (EFFECT TIED TO GOLD ANODES).
THOMAS RT
J PHYS D APPL PHYS 3: 1434-7 (1970)
218

04-0998 A FLOATING ZONE SINGLE CRYSTAL GROWING APPARATUS. (BARIUM TITANATE)
TIEN TY + GARRETSON RM
J CRYST GROW 16: 177-80 (1972)
281

04-0999 INFLUENCE OF OXYGEN PARTIAL PRESSURE ON PROPERTIES OF SEMICONDUCTING BARIUM TITANATE.
TIEN TY + CARLSON WG
J AMER CERAM SOC 46: 297-300 (1963)

04-1000 PREPARATION AND STUDY OF THIN FILMS OF (BARIUM, STRONTIUM) TITANATE.
TOMASHPOLSKII YUYA + SOROKINA LA + VENEVTSEV YUN
INORG MATER 6: 1203-4 (JUL 1970)

04-1001 VACUUM EVAPORATION OF BARIUM TITANATE.
TOMASHPOLSKII YUYA
INORG MATER 8: 127-36 (1972)

04-1002 INFLUENCE OF FIRING OF BARIUM TITANATE SINTERS ON ELECTRICAL AND TECHNOLOGICAL PROPERTIES OF FERROELECTRICS.
TOROPOV AN + SEMILETOVA DV + KRASOTKIN IS AND OTHERS
J APPL CHEM USSR 43: 2348-50 (OCT 1970)

04-1003 REACTION SEQUENCE IN FORMING BARIUM METATITANATE IN AN INDUSTRIAL ROTARY FURNACE.
TOROPOV AN + BORISENKO AI + SEMILETOVA DV + YANKOVSKAYA LI
J APPL CHEM USSR 41(11): 2247-52 (1968)

04-1004 OPTICALLY INDUCED REFRACTIVE INDEX CHANGES IN BARIUM TITANATE (EXPERIMENTS TO CHARACTERIZE THE OPTICAL DAMAGE).
TOWNSEND RL + LAMACCHIA JT
J APPL PHYS 41(13): 5188-92 (1970)
220

04-1005 OPTICALLY INDUCED REFRACTIVE INDEX CHANGES IN BARIUM TITANATE.
TOWNSEND RL + LAMACCHIA JT
P27 OF OPTICAL SOC AMER, OCT 1969, CHICAGO, ANNU MEET, PROC. OPT SOC AMER, 1969.
SCI ABSTR A 73: 62600 (1970)
222

04-1006 BARIUM TITANATE SINGLE CRYSTALS SUBJECTED TO MECHANICAL IMPULSES.
TOYODA H + SHIBATA H
JAP J APPL PHYS 2: 556-60 (1963)

04-1007 DISLOCATIONS IN BARIUM TITANATE SINGLE CRYSTALS.
TOYODA H + WAKU S + NIIZEKI N
ACTA CRYSTALLOGR 16(SUPPL): A110 (1963) (ABSTRACT ONLY)

04-1008 ELECTROSTATIC FIELD IN A PSEUDOCUBIC IONIC STRUCTURE. APPLICATION TO ALL BARIUM TITANATE STRUCTURE TYPES.
TSALLIS C + MACHET R + BOUILLOT J
J PHYS (PARIS) 32: 171-5 (1971)

04-1009 DIELECTRIC DISPERSION IN BARIUM TITANATE TYPE FERROELECTRICS IN THE METER BAND.
TURIK AV + CHERNYSHEV KR + KOMAROV VD
SOV PHYS SOLID STATE 13(10): 2619-20 (APR 1972)
263

4. BARIUM TITANATE

04-1010 DIELECTRIC, ELASTIC AND PIEZOELECTRIC PROPERTIES OF BARIUM TITANATE SINGLE CRYSTALS
WITH LAMINAR DOMAIN STRUCTURE.
TURIK AV
P318-19 OF INT MEETING ON FERROELECTRICITY, 2ND, 1969, KYOTO, PROC,
(J PHYS SOC JAP, VOL 28 SUPPL). PHYS SOC JAP, 1970, 459P.

SOV PHYS SOLID STATE 12(3): 688-93 (1970)
    206

04-1011 EFFECT OF GRAIN SIZE ON THE PIEZOELECTRIC AND ELASTIC PROPERTIES OF BARIUM TITANATE.
TURIK AV + KOMAROV VD
BULL ACAD SCI USSR, PHYS SER 34(12): 2337-42 (1970)

04-1012 INFLUENCE OF A CONSTANT ELECTRIC FIELD ON THE DIELECTRIC PROPERTIES OF POLYCRYSTALLINE
BARIUM TITANATE.
TURIK AV + SIDORENKO EN + ZHESTKOV VE + KOMAROV VD + TAISEVA VA
IZV VUZ FIZ 1972(10): 122-4 (IN RUSSIAN)

04-1013 PROBLEM OF THE SURFACE LAYER IN BARIUM TITANATE SINGLE CRYSTALS.
TURIK AV
SOV PHYS SOLID STATE 5: 1748-50 (1964)

04-1014 UHF REVERSAL PROPERTIES OF BARIUM TITANATE TYPE FERROELECTRIC CERAMICS.
TURIK AV + SIDORENKO EN + ZHESTKOV VG + KOMAROV VD
BULL ACAD SCI USSR, PHYS SER 34(12): 2306-9 (1970)

04-1015 DIELECTRIC BREAKDOWN OF POLYCRYSTALLINE BARIUM TITANATE.
UEDA I + TAKIUCHI M + IKEGAMI S + SATO H
J PHYS SOC JAP 19: 1267-73 (1964)

04-1016 PHASE TRANSITIONS IN THE MIXED CRYSTAL SYSTEM STRONTIUM(1-X) BARIUM(X) TITANATE.
UWE H + FUJII Y + UNOKI H + SAKUDO T
ACTA CRYSTALLOGR A 28: S178 (1972)

CORRELATION EFFECTS IN DISPLACEMENT TYPE PHASE TRANSITIONS IN FERROELECTRICS. (THEORY
APPLIED TO BARIUM TITANATE)
VAKS VG
SOV PHYS JETP 31(1): 161-7 (1970)

04-1017 EFFECT OF CHARGED LATTICE IMPERFECTIONS ON THE DIELECTRIC PROPERTIES OF MATERIALS.
(BARIUM TITANATE TYPE)
VENDIK OG + PLATONOVA LM
SOV PHYS SOLID STATE 13(6): 1353-9 (DEC 1971)
    254

04-1018 FERROELECTRIC FAMILY OF BARIUM TITANATE.
VENEVTSEV YUN
MATER RES BULL 6: 1085-95 (OCT 1971)

04-1019 DIELECTRIC AND NONLINEAR MICROWAVE PROPERTIES OF POLYCRYSTALLINE SOLID SOLUTIONS BASED
ON BARIUM TITANATE. (BARIUM STRONTIUM TITANATE, BARIUM TITANATE STANNATE, BARIUM
TITANATE ZIRCONATE)
VERBITSKAYA TN + IVANOV IV + MOROZOV NA
SOV PHYS SOLID STATE 12(5): 1250-1 (1970)
    215

04-1020 PECULIARITIES OF THE ELECTRON PARAMAGNETIC RESONANCE SPECTRA IN A FERROELECTRIC CERAMIC
CONTAINING TITANIUM. (BARIUM TITANATE, STRONTIUM TITANATE)
VOTINOV MP + DEMIDENKO NI
SOV PHYS SOLID STATE 4: 2351-3 (1963)

04-1021 DEPENDENCE OF THE PIEZOEFFECT IN BARIUM TITANATE ON THE MEANS OF SCREENING THE
SPONTANEOUS INDUCTION.
VUL BM + GURO GM + IVANCHIK II
SOV PHYS SOLID STATE 14(3): 602-7 (SEP 1972)
    276

04-1022 DISCUSSION ON THE CRYSTAL GROWTH OF BARIUM TITANATE SINGLE CRYSTALS IN POTASSIUM
FLUORIDE FLUX.
WAKU S
REV ELEC COMMUN LAB 10: 1-19 (1962)

4. BARIUM TITANATE

04-1023 ETCH PITS CORRESPONDING TO DISLOCATIONS IN BARIUM TITANATE SINGLE CRYSTALS.
WAKU S
J PHYS SOC JAP 17: 1068-9 (1962)

04-1024 FERROELECTRIC PROPERTIES OF (INSULATING, BOUNDARY LAYER) BARIUM TITANATE CERAMICS AT
GRAIN BOUNDARY.
WAKU S + MURAKAMI T + YAMAJI A
P457-9 OF INT MEETING ON FERROELECTRICITY, 2ND, 1969, KYOTO, PROC,
(J PHYS SOC JAP, VOL 28 SUPPL). PHYS SOC JAP, 1970, 459P.

04-1025 STUDIES ON THE BARIUM STRONTIUM TITANATE BOUNDARY LAYER CERAMIC DIELECTRICS.
WAKU S + UCHIDATE M + KIKUCHI K
REV ELECTRON COMMUN LAB 18: 681-93 (1970)

04-1026 INFLUENCE OF ADDITIVES ON DENSIFICATION AND PROPERTIES OF SINTERED AN PRESSURE SINTERED
BARIUM TITANATE.
WALKER BE + RICE RW + SPANN JR
BULL AMER CERAM SOC 49: 420 (1970)

04-1027 CORRELATION BETWEEN RHEOLOGICAL AND ELECTRICAL BEHAVIOR IN SINTERED BARIUM TITANATE.
WEINTRITT DJ + STOCKLER HA
BULL AMER CERAM SOC 50: 463 (1971)

04-1028 POLARIZATION FLUCTUATIONS AND THE OPTICAL ABSORPTION EDGE IN BARIUM TITANATE.
WEMPLE SH
PHYS REV B 2(7): 2679-89 (1970)
        228

04-1029 SINGLE FIRE PROCESS FOR BARIUM TITANATE BARRIER LAYER DIELECTRICS.
WICHANSKY H
BULL AMER CERAM SOC 49: 998-9 (1970)

04-1030 FERROELECTRIC PHASE TRANSITION OF FIRST ORDER IN BARIUM TITANATE.
WISSEL C
ANN PHYS 27(2): 207-21 (1971)

        PHENOMENOLOGICAL THEORY OF REFRACTIVE INDEX AND ELECTROOPTICAL PHENOMENA IN CRYSTALS OF
        PEROVSKITE STRUCTURE (WITH APPLICATION TO BARIUM TITANATE).
        WUNSCH DC
        NEW MEXICO STATE UNIV SCD THESIS, 1969, 184P. N70-39085
             DISS ABSTR INT B 30(5): 2379 (1969)

04-1031 MICROSTRUCTURE OF DOMAINS AND DOMAIN WALLS IN SINGLE CRYSTAL FILMS OF BARIUM TITANATE.
YAKUNIN SI + SHAKMANOV VV + SPIVAK GV + VASILEVA NV
SOV PHYS SOLID STATE 14(2): 310-13 (AUG 1972)
        274

04-1032 CRITICAL (POLARIZATION) FLUCTUATIONS IN BARIUM TITANATE (ABOVE THE FERROELECTRIC CURIE
TEMPERATURE, AS SHOWN) BY NEUTRON SCATTERING.
YAMADA Y + SHIRANE G + LINZ A
PHYS REV 177(2): 848-57 (1969)
        133

04-1033 MEASUREMENT OF OXYGEN DIFFUSION CONSTANT IN SOME TITANATES. (STRONTIUM TITANATE, BARIUM
TITANATE)
YAMAJI A
ACTA CRYSTALLOGR A 28: S167 (1972)

04-1034 ELECTRON MICROSCOPIC STUDY OF FERROELECTRIC DOMAINS OF BARIUM TITANATE IN LOW
TEMPERATURE PHASES.
YATSUHASHI T + HONJO G
ACTA CRYSTALLOGR A 28: S236 (1972)

04-1035 ANISOTROPY OF THE ELECTRICAL CONDUCTIVITY OF BARIUM TITANATE SINGLE CRYSTALS GROWN FROM
SOLUTION IN A POTASSIUM FLUORIDE MELT.
ZABARA YUV + KUDZIN AYU
SOV PHYS CRYSTALLOGR 16: 140-2 (JUL-AUG 1971)

04-1036 POSITIVE TEMPERATURE COEFFICIENT OF RESISTANCE OF BARIUM TITANATE SINGLE CRYSTALS IN
THE REGION OF THE CURIE TEMPERATURE.
ZABARA YUV + KUDZIN AYU + BARABAN VS
IZV VUZ FIZ 1972(11): 134-7 (IN RUSSIAN)

4. BARIUM TITANATE

04-1037 ABSORPTION AND AMPLIFICATION OF ULTRASOUND IN A TWO-LAYER SYSTEM CONSISTING OF A
PIEZOELECTRIC CERAMIC (BARIUM TITANATE) WITH LARGE DIELECTRIC CONSTANT AND A
SEMICONDUCTOR.
ZHABITENKO NK + KUCHEROV IYA + MISELYUK EG AND OTHERS
JETP LETT 14: 312-14 (1971)

04-1038 STRUCTURAL ELECTRON MICRODIFFRACTION STUDIES OF FERROELECTRICS AND FERROELECTRIC
MAGNETS. (LEAD AND BARIUM TITANATES, BISMUTH FERRATE, POTASSIUM NIOBATE)
ZHDANOV GS + TOMASHPOLSKII YUYA + PLATONOV GL + VENEVTSEV YUN
P322-31, VOL 1 OF DVORAK V + FOUSKOVA A + GLOGAR P (EDS),
INT MEETING ON FERROELECTRICITY, PRAGUE, 1966, PROC.
INST PHYS CZECH ACAD SCI, PRAGUE, 1966.

04-1039 INVESTIGATION OF THE EFFECT OF SAMPLE THICKNESS ON THE DOMAIN STRUCTURE AND THE
PYROELECTRIC PROPERTIES OF SINGLE CRYSTALS OF BARIUM TITANATE.
ZVIRGZD YUA
P139-47 OF FAZOVYE PEREKHODY V SEGNETOELEKTRIKAKH (PHASE
TRANSFORMATIONS IN FERROELECTRICS), FRITSBERG VYA + ROLOV BN +
KRUCHAN YY (EDS), IZDATELSTVO ZINATNE RIGA, 1971, 206P. (IN RUSSIAN)

5. LEAD TITANATE

05-1040   MOSSBAUER EFFECT FOR IRON-57 IN FERROELECTRIC LEAD TITANATE.
           BHIDE VG + HEGDE MS
           PHYS REV B 5(9): 3488-99 (1 MAY 1972)
               264

           ELECTROSTATIC FIELD IN A SLIGHTLY ORTHORHOMBIC IONIC CRYSTAL. APPLICATION TO THE
           CALCULATION OF TETRAGONAL BIREFRINGENCE IN BARIUM AND LEAD TITANATES.
           BOUILLOT J + MACHET R + TSALLIS C
           PHYS STATUS SOLIDI 38: 313-16 (1970)

           THEORETICAL STATIC MODEL FOR BARIUM AND LEAD TITANATES - APPLICATION TO CALCULATING THE
           WALL ENERGY AND THE DIELECTRIC CONSTANTS.
           BOUILLOT J + MACHET R
           P111-17 OF EUROPEAN MEETING ON FERROELECTRICITY, 1969, SAARBRUCKEN,
           MUSER HE AND PETERSSON J(EDS), WISSENSCH VERLAGSGES, STUTTGART,
           1970, 396P.

           DIELECTRIC AND ELASTIC PROPERTIES OF SINGLE CRYSTALS OF BARIUM LEAD TITANATE SOLID
           SOLUTIONS.
           BUNINA IK + KUDZIN AYU
           SOV PHYS SOLID STATE 10(10): 2496-7 (1969)

           RAMAN SCATTERING IN THE FERROELECTRIC SYSTEM LEAD BARIUM TITANATE.
           BURNS G + SCOTT BA
           SOLID STATE COMMUN 9: 813-17 (1971)

05-1041   RAMAN STUDIES (TEMPERATURE DEPENDENCE) OF UNDER DAMPED SOFT MODES IN LEAD TITANATE.
           BURNS G + SCOTT BA
           PHYS REV LETT 25(3): 167-70 (1970)
               222

05-1042   SOLVENT ZONE GROWTH OF FERROELECTRIC CRYSTALS. (STRONTIUM TITANATE, LEAD(X)
           STRONTIUM(1-X) TITANATE, LEAD(X) TITANIUM(1-X) TITANATE)
           DI BENEDETTO B + MLAVSKY AI + WOLFF GA
           ACTA CRYSTALLOGR A 25: S10 (1969)

05-1043   EFFECT OF GERMANIUM ON THE STRUCTURE AND PROPERTIES OF FERROELECTRIC LEAD TITANATE
           (GERMANATE SOLID SOLUTIONS TO 7 OR 8 MO1).
           DIDKOVSKAYA OS + KLIMOV VV + VENEVTSEV YUN
           INORG MATER 4(1): 331-2(1968)
               216

05-1044   THE SYSTEM LEAD TITANATE- STRONTIUM COPPER NIOBATE. (CERAMIC SYNTHESIS)
           DZHMUKHADZE DF + VENEVTSEV YUN + ZHDANOV GS
           SOV PHYS CRYSTALLOGR 16(1): 136-9 (JUL-AUG 1971)
               240

05-1045   THE SYSTEM LEAD TITANATE- STRONTIUM CUPRATE(0.33) NIOBATE(0.67).
           DZHMUKHADZE DF + VENEVTSEV YUN + ZHDANOV GS
           SOV PHYS CRYSTALLOGR 16: 136-9 (1971)

05-1046   FERROELECTRIC PROPERTIES OF LEAD TITANATE SINGLE CRYSTALS.
           FESENKO EG + GAVRILYACHENKO VG + ZAROCHENTSEV EV
           BULL ACAD SCI USSR, PHYS SER 34(12): 2262-9 (1970)

05-1047   GROWTH OF LEAD TITANATE CRYSTALS AND EXAMINATION OF THEIR DOMAIN STRUCTURE.
           FESENKO EG + GAVRILYACHENKO VG + SPINKO RI + MARTYNENKO MA + GRIGOREVA EA + FERONOV AD
           SOV PHYS CRYSTALLOGR 17(1): 122-5 (JUL-AUG 1972)
               274

5. LEAD TITANATE

05-1048   RECIPROCAL DOMAINS IN LEAD TITANATE CRYSTALS.
          FESENKO EG + GAVRILYACHENKO VG
          J PHYS (PARIS) 33, SUPPL 4, COLLOQ 2: 169-71 (1972)
               294

          PREPARATION AND PROPERTIES OF HIGH PERMITTIVITY THIN FILM DIELECTRICS. (BARIUM AND LEAD
          TITANATES, EFFECT OF FERROELECTRIC CHARACTERISTICS ON ELECTRIC PROPERTIES)
          FEUERSANGER AE
          P209-36 OF VRATNY F(ED), THIN FILM DIELECTRICS, ELECTROCHEM SOC,
          NY, 1969, 669P.

05-1049   PIEZOELECTRIC EFFECT IN LEAD TITANATE SINGLE CRYSTALS.
          GAVRILYACHENKO VG + FESENKO EG
          SOV PHYS CRYSTALLOGR 16: 549-50 (NOV-DEC 1971)

05-1050   SPONTANEOUS POLARIZATION AND COERCIVE FIELD OF LEAD TITANATE (FLUX GROWN SINGLE
          CRYSTALS).
          GAVRILYACHENKO VG + SPINKO RI + MARTYNENKO MA + FESENKO EG
          SOV PHYS SOLID STATE 12(5): 1203-4 (1970)
               215

05-1051   POSITIVE TEMPERATURE COEFFICIENT OF RESISTANCE IN DOPED LEAD TITANATE.
          GOLTSOV YUI + PROKOPALO OI + BELOVA LA
          SOV PHYS SOLID STATE 14(3): 805-6 (SEP 1972)
               276

05-1052   TIME EFFECTS IN THE HYSTERESIS LOOP OF (LEAD, STRONTIUM) TITANATE.
          GRIFFITHS CR + RUSSELL R
          J AMER CERAM SOC 55(2): 110-11 (1972)
               260

05-1053   TIME EFFECTS IN THE HYSTERESIS LOOP OF LEAD STRONTIUM TITANATE.
          GRIFFITHS CR
          BULL AMER CERAM SOC 49: 833 (1970)

05-1054   APPLICATION OF DIELECTRIC MIXTURE FORMULAS TO (LEAD TITANATE) GLASS- CERAMIC SYSTEMS.
          GROSSMAN DG + ISARD JO
          J PHYS D 3: 1058-67 (1970)

05-1055   PERTURBED DIRECTIONAL CORRELATION OF SCANDIUM-44 IN LEAD TITANATE.
          HAAS MR + GLASS JC
          PHYS REV B SOLID STATE 4: 147-50 (JUL 1, 1971)

05-1056   SEEBECK EFFECT IN METAL LEAD TITANATE METAL CONTACT SYSTEMS.
          HANDEREK J + DUDEK J + ESOP B
          ROZPR ELEKTROTECH 18: 67-86 (1972) (IN POLISH)

05-1057   THERMOELECTRIC EFFECTS IN THE SYSTEM METAL- LEAD TITANATE- METAL. (METHOD OF STUDYING
          THE TYPE OF CONDUCTIVITY, PLATINUM ELECTRODES)
          HANDEREK J + PIECH TJ + DUDEK J
          PHYS STATUS SOLIDI A 5: 237-45 (1971)
               237

05-1058   DISTRIBUTION OF VACANCIES IN LANTHANA DOPED LEAD TITANATE.
          HENNINGS D + HARDTL KH
          PHYS STATUS SOLIDI A 3: 465-74 (1970)

05-1059   X-RAY STRUCTURE INVESTIGATION OF LANTHANUM MODIFIED LEAD TITANATE WITH A-SITE AND
          B-SITE VACANCIES.
          HENNINGS D + ROSENSTEIN G
          MATER RES BULL 7: 1505-13 (1972)

05-1060   ELECTROMECHANICAL PROPERTIES OF LEAD TITANATE CERAMICS CONTAINING LANTANUM AND
          MANGANESE.
          IKEGAMI S + UEDA I + NAGATA T
          J ACOUST SOC AMER 50: 1060-6 (1971)
               255

5. LEAD TITANATE

05-1061   POLARIZATION AND PIEZOELECTRIC PROPERTIES OF LEAD TITANATE CERAMICS. (CONTAINING 1 MOLE
          PERCENT MANGANESE DIOXIDE)
          IKEGAMI S + UEDA J
          P331-3 OF INT MEETING ON FERROELECTRICITY, 2ND, 1969,, KYOTO, PROC,
          (J PHYS SOC JAP, VOL 28 SUPPL). PHYS SOC JAP, 1970, 459P.

05-1062   PIEZOELECTRIC PROPERTIES OF LEAD TITANATE CERAMICS.
          IKEGAMI S + UEDA I + NAGATA T
          TRANS INST ELECTRON COMMUN ENG JAP C 55: 165-71 (1972) (IN JAPANESE)

05-1063   PHYSICAL PHENOMENA IN PIEZOCERAMICS BASED ON LEAD TITANATE AND ZIRCONATE. (SOLID
          SOLUTIONS LEAD MAGNESATE(0.5) TUNGSTATE(0.5)), EFFECTS OF POSITIVE OR NEGATIVE HIGH DC
          FIELDS SUMMARY ONLY)
          ISUPOV VA + STOLYPIN YUE
          P312 OF INT MEETING ON FERROELECTRICITY, 2ND, 1969,, KYOTO, PROC,
          (J PHYS SOC JAP, VOL 28 SUPPL). PHYS SOC JAP, 1970, 459P.

05-1064   AN X-RAY STUDY OF THE EFFECT OF HYDROSTATIC PRESSURE ON THE STRUCTURE OF LEAD TITANATE.
          KABALKINA SA + VERESHCHAGIN LF
          SOV PHYS DOKL 7: 310-12 (1962)

          PREPARATION AND PROPERTIES OF GLASS- CERAMICS CONTAINING FERROELECTRIC CRYSTALS (BARIUM
          OR LEAD TITANATE, FROM SYSTEMS BARIA OR LITHIA, TITANIA, ALUMINA, SILICA).
          KOKUBO T
          BULL INST CHEM RES KYOTO UNIV 47(6): 553-71 (1969)

05-1065   SEIGNETTO MAGNETIC SOLID SOLUTIONS OF THE LEAD FERRATE(0.67) TUNGSTATE(0.33)- LEAD
          TITANATE SYSTEM.
          LEVINA SS + PARASHCHUKOV MP
          IZV VUZ FIZ 1972(8): 131-3 (IN RUSSIAN)

          STATIC SHELL MODEL FOR BARIUM AND LEAD TITANATES.
          MACHET R + BOUILLOT J + GODEFROY L
          P366-8 OF INT MEETING ON FERROELECTRICITY, 2ND, 1969,, KYOTO, PROC,
          (J PHYS SOC JAP, VOL 28 SUPPL). PHYS SOC JAP, 1970, 459P.

05-1066   FURTHER MEASUREMENTS OF ABSOLUTE SIGNS OF SECOND HARMONIC GENERATION COEFFICIENTS OF
          PIEZOELECTRIC CRYSTALS. (LEAD TITANATE, BARIUM STRONTIUM NIOBATE, LEAD NIOBIUM(4)
          OXIDE(11), SILVER GALLIUM DISULFIDE, SILICON CARBIDE)
          MILLER RC + NORDLAND WA
          PHYS REV B 5: 4931-4 (1972)

05-1067   ANOMALOUS FERROELECTRIC HYSTERESIS LOOPS. (LEAD STRONTIUM TITANATE)
          MURDOCH FJ
          ARMY ELECTRON COMMAND, FT MONMOUTH, NJ, 1971, 58P. AD732336
              SCI TECH AEROSP REP 10(7): 948 (1972)
              268

05-1068   DIRECT CHELATOMETRIC DETERMINATION OF LEAD AND TITANIUM IN LEAD TITANATE AND SIMILAR
          COMPOUNDS. STUDIES ON ANALYSIS OF ELECTROCERAMICS AND ITS RAW MATERIALS. PART-1.
          PART-2.
          MURAHTA M + KITAO A
          JAP ANALYST 21: 907-11, 1231-3 (1972) (IN JAPANESE)

05-1069   LEAD TITANATE CERAMIC RESONATORS OPERATING IN VHF BAND.
          NAGATA T + NAKAJIMA Y + SASAKI R
          TRANS INST ELECTRON COMMUN ENG JAP C 55: 345-50 (1972) (IN JAPANESE)

05-1070   DIELECTRIC AND PIEZOELECTRIC PROPERTIES IN THE TERNARY SYSTEM LEAD (ZINCATE NIOBATE)-
          BARIUM (ZINCATE NIOBATE)- LEAD TITANATE. (PHASE DIAGRAM, LATTICE CONSTANTS)
          NOMURA S + ARIMA H
          JAP J APPL PHYS 11(3): 358-64 (1972)
              279

05-1071   DIELECTRIC PROPERTIES OF LEAD TITANATE- LEAD (FERRATE NIOBATE) SOLID SOLUTIONS AT HIGH
          PRESSURES.
          POLANDOV IN + LEZHEPEKOV AP + DIDKOVSKAYA OS + KLIMOV VV
          SOV PHYS SOLID STATE 12(12): 2938-9 (JUN 1971)
              240

5. LEAD TITANATE

05-1072 (FLUX) GROWTH AND FERROELECTRIC PROPERTIES OF HIGH RESISTIVITY SINGLE CRYSTALS OF LEAD
TITANATE.
REMEIKA JP + GLASS AM
MATER RES BULL 5(1): 37-46 (1970)

05-1073 TESTING A POINT DIPOLE ELECTROOPTIC MODEL BY AN APPLICATION TO LEAD TITANATE.
SANCHEZ AS
AIR FORCE INST TECHNOL, MS THESIS, 1972, 71P. AD753467

STATIC ELECTRIC QUADRUPOLE INTERACTION OF TANTALUM AND HAFNIUM IONS IN BARIUM AND LEAD
TITANATE.
SCHAFER G + HERZOG P + WOLBECK B
Z PHYS 257(4): 336-52 (1972)

05-1074 CRYSTAL GROWTH AND TEMPERATURE DEPENDENT RAMAN SPECTRA OF LEAD TITANATE.
SCOTT BA + BURNS G
J ELECTROCHEM SOC 117: 195C (1970)

NONLINEAR OPTICAL PROPERTIES OF LEAD TITANATE AND BISMUTH TITANATE CRYSTALS.
SHALDIN YUV + BARSUKOVA ML + KUZNETSOV VA + LOBACHEV AN
SOV PHYS CRYSTALLOGR 16(1): 114-17 (JUL-AUG 1971)

05-1075 CHARACTERISTICS OF DEFECT LEAD TITANATE.
SHIRASAKI S + TAKAHASHI K + MANABE K
BULL CHEM SOC JAP 44(11): 3189-90 (1971)
278

05-1076 DEFECT LEAD TITANATES WITH DIVERSE CURIE TEMPERATURES.
SHIRASAKI S
SOLID STATE COMMUN 9: 1217-20 (1971)
241

05-1077 SOFT FERROELECTRIC MODES IN LEAD TITANATE (NEUTRON INELASTIC SCATTERING).
SHIRANE G + AXE JD + HARADA J
PHYS REV B 2(1): 155-9 (JUL 1, 1970)
225

05-1078 APPARENT SOFT MODE LINE WIDTH DIVERGENCE IN LEAD TITANATE.
SILVERMAN BD
SOLID STATE COMMUN 10: 311-14 (1972)
262

05-1079 NONLINEAR OPTICAL PROPERTIES OF FERROELECTRIC LEAD TITANATE.
SINGH S + REMEIKA JP + POTOPOWICZ JR
APPL PHYS LETT 20(3): 135-7 (1 FEB 1972)
257

05-1080 NONLINEAR OPTICAL PROPERTIES OF FERROELECTRIC LEAD TITANATE.
SINGH S + REMEIKA JP + POTOPOWICZ JR
J OPT SOC AMER 61: 669 (1971)

(IN PILE, FAST NEUTRON) RADIATION INDUCED PHASE TRANSFORMATIONS IN (CERAMIC) BARIUM AND
LEAD TITANATES. (SUMMARY ONLY)
SOLOVEV SP + KUZMIN II + ZAKURKIN VV
P441 OF INT MEETING ON FERROELECTRICITY, 2ND, 1969, KYOTO, PROC,
(J PHYS SOC JAP, VOL 28 SUPPL). PHYS SOC JAP, 1970, 459P.

FERROELECTRICS 1: 19-22 (1970)

05-1081 X-RAY INVESTIGATION OF IRRADIATED LEAD TITANATE.
SOLOVEV SP + DUDAREV VYA + ZAKURKIN VV + KUZMIN II
BULL ACAD SCI USSR, PHYS SER 35(9): 1752-6 (1971)

EMISSION OF LIGHT BY SURFACE MICROPLASMA OF CERAMIC (BARIUM, LEAD) TITANATE NEAR THE
CURIE TEMPERATURE.
SUJAK B + BIEDRZYCKI K
ACTA PHYS POLON A 42: 49-54 (JUL 1972)

5. LEAD TITANATE

ELECTRON MICROSCOPIC STUDIES ON FERROELECTRIC DOMAINS OF PEROVSKITE TYPE OXIDES. (LEAD TITANATE, POTASSIUM NIOBATE, BARIUM TITANATE)
TANAKA M + YATSUHASHI T + HONJO G
P386-8 OF INT MEETING ON FERROELECTRICITY, 2ND, 1969, KYOTO, PROC,
(J PHYS SOC JAP, VOL 28 SUPPL). PHYS SOC JAP, 1970, 459P.

05-1082   STRUCTURAL MICROELECTRON DIFFRACTION STUDY OF FERROELECTRIC AND FERROMAGNETIC
          SUBSTANCES. (LEAD TITANATE)
          TOMASHPOLSKII YUYA + VENEVTSEV YUN + ZHDANOV GS
          SOV PHYS CRYSTALLOGR 13(3): 425-7 (1968)

05-1083   TEMPERATURE DEPENDENT OPTICAL PHONONS IN LEAD TITANATE.
          TORNBERG NE + PERRY CH
          J CHEM PHYS 53: 2946-55 (1970)

05-1084   EFFECTS OF ADDITIVES ON PIEZOELECTRIC AND RELATED PROPERTIES OF LEAD TITANATE CERAMICS.
          (DIELECTRIC CONSTANT, RESISTIVITY)
          UEDA I
          JAP J APPL PHYS 11(4): 450-62 (1972)
              279

05-1085   ELECTROMECHANICAL PROPERTIES AND APPLICATION OF LEAD TITANATE CERAMICS.
          UEDA I + KOBAYASHI S + IKEGAMI S
          NAT TECH REP 18: 423-25 (1972) (IN JAPANESE)

05-1086   ELECTRON SPIN RESOURCE STUDIES AT DEFECTS IN FLUORITES. (LEAD TITANATE)
          WESSEL GK
          DEP OF PHYS, SYRACUSE UNIV, 1971, 92P. AD737314
              GOVT REP ANNOUNCE 72(7): 206 (1972)
              255

05-1087   LEAD TITANATE PYROELECTRIC INFRARED DETECTOR.
          YAMAKA E + HAYASHI T + MATSUMOTO M
          INFRARED PHYS 11: 247-8 (1971)

          STRUCTURAL ELECTRON MICRODIFFRACTION STUDIES OF FERROELECTRICS AND FERROELECTRIC
          MAGNETS. (LEAD AND BARIUM TITANATES, BISMUTH FERRATE, POTASSIUM NIOBATE)
          ZHDANOV GS + TOMASHPOLSKII YUYA + PLATONOV GL + VENEVTSEV YUN
          P322-31, VOL 1 OF DVORAK V + FOUSKOVA A + GLOGAR P (EDS),
          INT MEETING ON FERROELECTRICITY, PRAGUE, 1966, PROC.
          INST PHYS CZECH ACAD SCI, PRAGUE, 1966.

6. STRONTIUM TITANATE

06-1088   ELECTRIC PROPERTIES OF HYDROGEN DOPED STRONTIUM TITANATE.
           AEGERTER S + LIBBY WF
           HELV PHYS ACTA 43: 499 (1970)

06-1089   MEASUREMENT OF DIELECTRIC PROPERTIES OF FERROELECTRICS AT SUBMILLIMETER WAVELENGTHS.
           (STRONTIUM TITANATE)
           ALESHECHKIN VN + MERIAKRI VV + USHATKIN EF AND OTHERS
           BULL ACAD SCI USSR, PHYS SER 35(9): 1746-8 (1971)

06-1090   ELECTRON THERMAL DIFFUSION EFFECTS (INTERNAL GENERATION OF STRONG FIELDS) DURING
           HOLOGRAM RECORDING IN CRYSTALS (IRON DOPED STRONTIUM TITANATE).
           AMODEI JJ
           APPL PHYS LETT 18(1): 22-4 (1971)
                222

06-1091   PHOTOCONDUCTIVITY AND TRAPPING PHENOMENA IN STRONTIUM TITANATE.
           AMODEI JJ + ROACH WR
           P93-7 OF INT CONF ON PHOTOCONDUCTIVITY, 3RD, 1969, PROC, PELL EM(ED),
           PERGAMON, 1971, 410P.

06-1092   STUDY OF THE DYNAMICS OF PHOTOCHROMIC SWITCHING IN STRONTIUM TITANATE MATERIALS.
           AMODEI JJ
           UNIV PENN PHD THESIS, 1968, 143P.
                DISS ABSTR B 29(10): 3880
                185

06-1093   SOFT MODE SUPERCONDUCTIVITY IN STRONTIUM TITANATE(3-X) AND CALCIUM(Y) STRONTIUM(1-Y)
           TITANATE(3-X).
           APPEL J
           PHYSICA 55: 577-84 (1971)
                CHEM ABSTR 76: 78183A

           TEMPERATURE DEPENDENCE OF THE RAMAN CROSS SECTIONS IN BARIUM TITANATE AND STRONTIUM
           TITANATE.
           BARBOSA GA + CHAVES A + PORTO SPS
           SOLID STATE COMMUN 11: 1053-5 (1972)

           COUPLED OPTICAL PHONON MODE THEORY OF INFRARED DISPERSION IN BARIUM AND STRONTIUM
           TITANATES AND POTASSIUM TANTALATE.
           BARKER AS + HOPFIELD JJ
           PHYS REV 135(6): 1732-7 (1964)

06-1094   FAR INFRARED FERROELECTRIC VIBRATION MODE IN STRONTIUM TITANATE.
           BARKER AS + TINKHAM M
           PHYS REV 125: 1527-30 (1962)

06-1095   THERMAL CONDUCTIVITY IN PEROVSKITES (3 PHONON INTERACTION PROPOSED FOR STRONTIUM
           TITANATE AND POTASSIUM TANTALATE).
           BARRETT HH + HOLLAND MG
           PHYS REV B 2(8): 3441-3 (1970)
                226

06-1096   PRESSURE DEPENDENCE OF THE ELASTIC CONSTANTS OF STRONTIUM TITANATE.
           BEATTIE AG + SAMARA GA
           J APPL PHYS 42: 2376-81 (MAY 1971)

06-1097   PRESSURE DEPENDENCE OF THE ELASTIC CONSTANT OF STRONTIUM TITANATE.
           BEATTIE AG + SAMARA GA
           BULL AMER PHYS SOC 15: 381 (1970)

6. STRONTIUM TITANATE

TOP SEEDED SOLUTION GROWTH OF OXIDE CRYSTALS FROM NONSTOICHIOMETRIC MELTS. (BARIUM TITANATE, STRONTIUM TITANATE, GERMANATES)
BELRUSS V + KALNAJS J + LINZ A + FOLWEILER RC
MATER RES BULL 6: 899-906 (1971)

MICROWAVE DIELECTRIC LOSSES IN OXYGEN OCTAHEDRON FERROELECTRICS ABOVE THE CURIE TEMPERATURE. (BARIUM(0-1.0) STRONTIUM(1.0-0) TITANATE)
BETHE K
P259-71 OF EUROPEAN MEETING ON FERROELECTRICITY, 1969, SAARBRUCKEN,
MUSER HE AND PETERSSON J(EDS), WISSENSCH VERLAGSGES, STUTTGART,
1970, 396P.

SINGLE CRYSTALS OF THE SYSTEM (BARIUM, STRONTIUM) TITANATE. (SYNTHESIS AND LOW FREQUENCY DIELECTRIC PROPERTIES)
BETHE K
P343-50 OF EUROPEAN MEETING ON FERROELECTRICITY, 1969, SAARBRUCKEN,
MUSER HE AND PETERSSON J(EDS), WISSENSCH VERLAGSGES, STUTTGART,
1970, 396P.

PREPARATION AND PROPERTIES OF (BARIUM, STRONTIUM) TITANATE SINGLE CRYSTALS.
BETHE K + WELZ F
MATER RES BULL 6: 209-17 (1971)

06-1098   ELECTROCOLORATION IN STRONTIUM TITANATE: VACANCY DRIFT AND OXIDATION REDUCTION OF TRANSITION METALS.
BLANC J + STAEBLER DL
PHYS REV B 4: 3548-57 (1971)

06-1099   OPTICAL ABSORPTION EDGE OF STRONTIUM TITANATE AROUND THE 105 DEGREES K PHASE TRANSITION.
BLAZEY KW
PHYS REV LETT 27: 146-8 (1971)

06-1100   BAND STRUCTURE OF STRONTIUM TITANATE AS MEASURED BY X-RAY PHOTOELECTRON SPECTROSCOPY (ESCA).
BOARD R + WEAVER H + HONIG JM
P595-601 OF SHIRLEY DA(ED), ELECTRON SPECTROSCOPY, AMER ELSEVIER,
1972, 916P.

06-1101   ENERGY BAND CHANGES IN PEROVSKITES DUE TO LATTICE POLARIZATION. (LCAO ANALYSIS OF BAND SCHEME, STRONTIUM TITANATE)
BREWS JR
PHYS REV LETT 18(16): 662-4 (1967)

CONCENTRATION DEPENDENCE OF THE CURIE-WEISS CONSTANT IN POLYCRYSTALLINE SOLID SOLUTIONS IN THE (BARIUM, STRONTIUM) TITANATE SYSTEM.
BROK AYA + VETSSILE ZA
P109-15 OF FAZOVYE PEREKHODY V SEGNETOELEKTRIKAKH (PHASE TRANSFORMATIONS IN FERROELECTRICS), FRITSBERG VYA + ROLOV BN + KRUCHAN YY(EDS), IZDATELSTVO ZINATNE RIGA, 1971, 206P. (IN RUSSIAN)

BINARY AND PSEUDOBINARY PHASES BETWEEN NIOBIUM PENTOXIDE AND LEAD, BARIUM, AND STRONTIUM OXIDES.
BRUSSET H + GILLIER-PANDRAUD H + MAHE R + VOLIOTIS SD
MATER RES BULL 6: 413-24 (1971) (IN FRENCH)

SATURATION OF VOLT- AMPERE CHARACTERISTICS OF CADMIUM SELENIDE LAYERS ON BARIUM TITANATE- STRONTIUM TITANATE SUBSTRATES AT SUPERSONIC DRIFT VELOCITY OF CURRENT CARRIERS.
BRYUZGIN AR + MISELYUK EG + NEKRASOV MM AND OTHERS
SOV PHYS DOKL 16(6): 466-7 (1971)

ENERGY STORAGE IN CERAMIC DIELECTRICS. (STRONTIUM TITANATE, BARIUM TITANATE)
BURN I + SMYTH DM
J MATER SCI 7: 339-43 (1972)

ELECTRICAL AND OPTICAL PROPERTIES OF (BARIUM, STRONTIUM) MIXED TITANATE THIN SECTIONS.
BURFOOT JC + SANVORDENKER V
P424-32, VOL 1 OF DVORAK V + FOUSKOVA A + GLOGAR P (EDS),
INT MEETING ON FERROELECTRICITY, PRAGUE, 1966, PROC.
INST PHYS CZECH ACAD SCI, PRAGUE, 1966.

06-1102   EFFECT OF UNIAXIAL STRESS ON RAMAN SCATTERING IN STRONTIUM TITANATE.
           BURKE WJ + PRESSLEY RJ
           BULL AMER PHYS SOC 15: 327 (1970)

06-1103   RAMAN SCATTERING AND PHASE TRANSITIONS IN STRESSED STRONTIUM TITANATE.
           BURKE WJ + PRESSLEY RJ + SLONCZEWSKI JC
           SOLID STATE COMMUN 9: 121-4 (1971)

06-1104   STRESS INDUCED FERROELECTRICITY IN STRONTIUM TITANATE.
           BURKE WJ + PRESSLEY RJ
           SOLID STATE COMMUN 9: 191-5 (1971)

06-1105   ANOMALOUS BEHAVIOR OF THE DIELECTRIC NONLINEARITY COEFFICIENT IN STRONTIUM TITANATE
           NEAR THE PHASE TRANSITION AT 110 DEG K.
           BUZIN IM + IVANOV IV + RUKIN EI + CHUPRAKOV VF
           SOV PHYS SOLID STATE 14(7): 1770-3 (JAN 1973)
               282

06-1106   DIELECTRIC NONLINEARITY OF STRONTIUM TITANATE SINGLE CRYSTALS NEAR PHASE TRANSITION AT
           110 DEGREES K.
           BUZIN IM + IVANOV IV + KOROBOV AI AND OTHERS
           BULL ACAD SCI USSR, PHYS SER 35(9): 1696-7 (1971)

06-1107   ABSORPTION EDGE AND STRUCTURAL PHASE TRANSITION IN STRONTIUM TITANATE.
           CAPIZZI M + FROVA A
           SOLID STATE COMMUN 10: 979-82 (1972)
               267

06-1108   DETERMINATION OF THE NATURE OF THE OPTICAL GAP OF STRONTIUM TITANATE.
           CAPIZZI M + FROVA A
           NUOVO COMENTO B 5(2): 181-203 (11 OCT 1971)
               275

06-1109   DISTORTION ENHANCED OPTICAL ABSORPTION IN STRONTIUM TITANATE AT THE CUBIC TO TETRAGONAL
           TRANSITION.
           CAPIZZI M + FROVA A
           PHYS REV LETT 29(26): 1741-4 (25 DEC 1972)
               283

06-1110   IS THERE A UNIFIED MECHANISM UNDERLYING URBACH'S RULE. (STRONTIUM TITANATE, POTASSIUM
           TITANATE)
           CAPIZZI M + FROVA A + DUNN D
           SOLID STATE COMMUN 10: 1165-9 (1972)
               270

06-1111   OPTICAL GAP OF STRONTIUM TITANATE. (DEVIATION FROM URBACH TAIL BEHAVIOR)
           CAPIZZI M + FROVA A
           PHYS REV LETT 25(18): 1298-302 (1970)
               225

06-1112   DIRECT OBSERVATION OF SINGLE DOMAIN STRONTIUM TITANATE (OPTICAL STUDY UNDER UNIAXIAL
           STRESS).
           CHANG TS + HOLZRICHTER JF + IMBUSCH GF + SCHAWLOW AL
           APPL PHYS LETT 17(6): 254-7 (1970)
               209

06-1113   DOMAIN STRUCTURE OF STRONTIUM TITANATE UNDER UNIAXIAL STRESSES.
           CHANG TS
           J APPL PHYS 43(8): 3591-5 (1972)
               270

06-1114   EFFECT OF DOMAIN STRUCTURE ON THE FLUORESCENCE OF CHROMIUM(3+) DOPED STRONTIUM
           TITANATE.
           CHANG TS + IMBUSCH GF
           J APPL PHYS 42: 4704-7 (1971)

06-1115   POLARIZED FLUORESCENCE STUDY OF CHROMIUM(3+) THROUGH A STRESS INDUCED PHASE TRANSITION
           IN STRONTIUM TITANATE.
           CHANG TS + HOLZRICHTER JF + IMBUSCH GF + SCHAWLOW AL
           SOLID STATE COMMUN 8: 1179-81 (1970)

6. STRONTIUM TITANATE

DIELECTRIC PROPERTIES OF BARIUM (1-X) STRONTIUM (X) TITANATE SINGLE CRYSTALS GROWN FROM A
POTASSIUM FLUORIDE FLUX.
COUFOVA P + JANOUSEK V + NOVAK J
CZECH J PHYS B 22: 485-9 (1972)

CRYSTAL CHEMICAL PROBLEMS OF (BARIUM, STRONTIUM) TITANATE SOLID SOLUTIONS.
COUFOVA P + NOVAK J
KRISTALL TECH 6: K1-K6 (1971)

06-1116   BIREFRINGENCE OF STRONTIUM TITANATE PRODUCED BY THE 105 DEGREE K STRUCTURAL PHASE
TRANSITION.
COURTENS E
PHYS REV LETT 29: 1380-3 (1972)
     292

(MICROSCOPIC AND THERMODYNAMIC) THEORY OF FERROELECTRICITY AND ANHARMONIC EFFECTS IN
CRYSTALS. (AND CALCULATIONS BASED ON STRONTIUM TITANATE MODEL)
COWLEY RA
PHIL MAG 11: 673-706 (1965)

06-1117   DIELECTRIC AND ELECTROOPTIC STUDIES OF THE POLARIZATION PROCESS IN STRONTIUM TITANATE
AT LOW TEMPERATURE. (BIREFRINGENCE, 20-275 DEG K)
CROSS LE + CHAKHAVORTY D
P394-404, VOL 1 OF DVORAK V + FOUSKOVA A + GLOGAR P (EDS),
INT MEETING ON FERROELECTRICITY, PRAGUE, 1966, PROC.
INST PHYS CZECH ACAD SCI, PRAGUE, 1966.

06-1118   ULTRASONIC STUDIES OF CRITICAL PHENOMENA IN INSULATING AND SEMICONDUCTING STRONTIUM
TITANATE. (ABSTRACT)
DEIS DW + ASHKIN M + HULM JK + JONES CK
BULL AMER PHYS SOC 15(1): 102-3 (1970)
     174

SOLVENT ZONE GROWTH OF FERROELECTRIC CRYSTALS. (STRONTIUM TITANATE, LEAD(X)
STRONTIUM(1-X) TITANATE, LEAD(X) TITANIUM(1-X) TITANATE)
DI BENEDETTO B + MLAVSKY AI + WOLFF GA
ACTA CRYSTALLOGR A 25: S10 (1969)

06-1119   ERASE MODE RECORDING CHARACTERISTICS OF PHOTOCHROMIC CALCIUM FLUORIDE, STRONTIUM
TITANATE, AND CALCIUM TITANATE CRYSTALS.
DUNCAN RC
RCA REV 33: 248-72 (1972)

LIFETIME OF THE "FERROELECTRIC" PHONON IN SOME PEROVSKITE CRYSTALS (STRONTIUM TITANATE,
POTASSIUM TANTALATE) AT LOW TEMPERATURES. (THEORY)
DVORAK V
P61-8, VOL 1 OF DVORAK V + FOUSKOVA A + GLOGAR P (EDS),
INT MEETING ON FERROELECTRICITY, PRAGUE, 1966, PROC.
INST PHYS CZECH ACAD SCI, PRAGUE, 1966.

TUNNELING RESULTS ON INDIUM STRONTIUM TITANATE BARRIERS INTERPRETED AS CONFIRMATION OF
THE THEORY OF TRANSITIONS FROM LARGE TO NEARLY SMALL POLARONS.
EAGLES DM
PHYS STATUS SOLIDI B, BASIC RES 48(1): 407-17 (1971)

06-1120   DISPERSION IN STRONTIUM TITANATE- LANTHANUM FERRATE DIELECTRIC CERAMICS.
EDAHIRO T + NAKAHARA M
ELEC COMMUN LAB TECH J 20: 2686-90 (1971) (IN JAPANESE)

06-1121   HIGH PRESSURE COMPRESSIBILITY AND GRUNEISEN PARAMETER OF STRONTIUM TITANATE.
EDWARDS LR + LYNCH RW
J PHYS CHEM SOLIDS 31: 573-4 (1970)
     190

06-1122   NONCOLLINEAR PARAMETRIC 4 PHOTON RECIPROCAL EFFECT IN CADMIUM SULFIDE AND STRONTIUM
TITANATE.
EICHLER H + FERY H + HERMANN F
OPT COMMUN 6: 152-5 (1972) (IN GERMAN)

06-1123   SHARED HOLES TRAPPED BY CHARGE DEFECTS IN STRONTIUM TITANATE.
ENSIGN TC + STOKOWSKI SE
PHYS REV B 1: 2799-810 (15 MAR 1970)
     201

6. STRONTIUM TITANATE

06-1124　LATTICE DYNAMICS ABOVE STRUCTURAL PHASE TRANSITIONS: STRONTIUM TITANATE.
ENZ CP
PHYS REV B 6(12): 4695-702 (15 DEC 1972)
281

06-1125　ELECTRON PARAMAGNETIC RESONANCE SPECTRUM OF MOLYBDENUM(5+) IN STRONTIUM TITANATE: AN
EXAMPLE OF THE DYNAMIC JAHN TELLER EFFECT.
FAUGHMAN BW
PHYS REV B 5: 4925-31 (1972)

06-1126　PHOTOCHROISM IN TRANSITION METAL DOPED STRONTIUM TITANATE.
FAUGHMAN BW
PHYS REV B SOLID STATE 4: 3623-36 (1971)

THEORY OF A STRUCTURAL PHASE TRANSITION IN PEROVSKITE- TYPE CRYSTALS, PART-2,
INTERACTION WITH ELASTIC STRAIN (STRONTIUM TITANATE AND LANTHANUM ALUMINATE).
FEDER J + PYTTE E
PHYS REV B 1: 4803-10 (15 JUN 1970)

06-1127　ULTRASONIC PROPAGATION, STRESS EFFECTS, AND INTERACTION PARAMETERS AT THE DISPLACIVE
TRANSITION IN STRONTIUM TITANATE.
FOSSHEIM K + BERRE B
PHYS REV B 5(8): 3292-308 (15 APR 1972)
264

06-1128　NORMAL VIBRATIONS OF STRONTIUM TITANATE LATTICE.
FRAITOVA D + ZENTKOVA A
CZECH J PHYS B 13: 670-9 (1963)

06-1129　ELECTRONIC TRANSPORT (CONDUCTIVITY AND HALL AND SEEBECK COEFFICIENTS) IN STRONTIUM
TITANATE (4.2-300 DEG K)
FREDERIKSE HPR + THURBER WR + HOSLER WR
PHYS REV 134(2): 442-5 (1964)

DIELECTRIC PROPERTIES OF FERROELECTRIC (BARIUM, STRONTIUM) TITANATE SOLID SOLUTIONS
NEAR THE PHASE TRANSITION, UNDER HIGH PRESSURES.
FRITSBERG VYA
P117-22 OF FAZOVYE PEREKHODY V SEGNETOELEKTRIKAKH (PHASE
TRANSFORMATIONS IN FERROELECTRICS), FRITSBERG VYA + ROLOV BN +
KRUCHAN YY(EDS), IZDATELSTVO ZINATNE RIGA, 1971, 206P. (IN RUSSIAN)

CERAMIC DIELECTRICS. (BARIUM TITANATE, STRONTIUM TITANATE)
FUJIWARA S + SHIRAIWA T
TDK ELECTRONICS CO LTD, US PAT 3,345,189, JAP APPL 30 OCT 1963,
PUBL 3 OCT 1967, 3P.; US PAT 3,427,173, APPL 8 JUN 1964, PUBL
11 FEB 1969, 4P.; US PAT 3,440,067, APPL 8 JUN 1964, PUBL
22 APR 1969, 4P.
CHEM ABSTR 69: 47381Q (1968); 70: 80582Z (1969); 71: 6293K (1969)

06-1130　INTERFEROMETRIC DETERMINATION OF THE QUADRATIC ELECTROOPTIC COEFFICIENTS IN STRONTIUM
TITANATE (SINGLE) CRYSTAL.
FUJII Y + SAKUDO T
J APPL PHYS 41(10): 4118-20 (1970)
209

06-1131　SECOND HARMONIC GENERATION IN STRESS INDUCED FERROELECTRIC STRONTIUM TITANATE.
FUJII Y + UWE H + UNOKI H + SAKUDO T
ACTA CRYSTALLOGR A 28: S230 (1972)

06-1132　SPECIFIC HEAT OF STRONTIUM TITANATE NEAR THE STRUCTURAL TRANSITION.
GARNIER PR
PHYS LETT A 35(6): 413-14 (12 JUL 1971)
244

MOBILITY DETERMINATIONS FROM WEIGHT MEASUREMENTS IN SOLID SOLUTIONS OF (BARIUM,
STRONTIUM) TITANATE.
GERTHSEN P + HARDTL KH + CSILLAG A
PHYS STATUS SOLIDI A 13: 127-33 (1972)

ELECTROOPTIC PROPERTIES OF SOME ABO(3) PEROVSKITES IN THE PARAELECTRIC PHASE.
(POTASSIUM TANTALATE AND TANTALATE NIOBATE, BARIUM AND STRONTIUM TITANATES)
GEUSIC JE + KURTZ SK + VAN UITERT LG + WEMPLE SH
APPL PHYS LETT 4: 141-3 (1964)

6. STRONTIUM TITANATE

06-1133 DIELECTRIC SUSCEPTIBILITY OF QUASI FERROELECTRIC STRONTIUM TITANATE.
GINTEL J
BULL SOC AMIS SCI LETT POZNAN, SER B 21: 29-34 (1968-1969)
CHEM ABSTR 74: 117246D (1971)
234

06-1134 TEMPERATURE DEPENDENCE OF THE DIELECTRIC CONSTANT OF STRONTIUM TITANATE.
GINTEL J
POZNAN TOW PRZYJ NAUK PR KOM MAT-PRZYR, FIZ DIELEK RADIOSPEKTROSK
2: 293-4 (1972) (IN POLISH)

"VIBRATING WIRE" STUDY OF POTENTIAL DISTRIBUTION (STABILITY) IN BARIUM(X)
STRONTIUM(1-X) TITANATE CERAMICS.
GODEFROY G + ORMANCEY G
P174 OF INT MEETING ON FERROELECTRICITY, 2ND, 1969, KYOTO, PROC,
(J PHYS SOC JAP, VOL 28 SUPPL). PHYS SOC JAP, 1970, 459P.

06-1135 THERMAL EXPANSIVITY AND ULTRASONIC PROPAGATION NEAR THE STRUCTURAL TRANSITION OF
STRONTIUM TITANATE.
GOLDING B
PHYS REV LETT 25(20): 1439-42 (1970)
225

06-1136 THERMAL EXPANSIVITY OF STRONTIUM TITANATE NEAR THE STRUCTURAL TRANSITION.
GOLDING B
BULL AMER PHYS SOC 17: 673 (1972)

06-1137 ULTRASONIC VELOCITY ABOVE TANTALUM IN STRONTIUM TITANATE.
GOLDING B
BULL AMER PHYS SOC 15: 362-3 (1970)

TIME EFFECTS IN THE HYSTERESIS LOOP OF (LEAD, STRONTIUM) TITANATE.
GRIFFITHS CR + RUSSELL R
J AMER CERAM SOC 55(2): 110-11 (1972)

TIME EFFECTS IN THE HYSTERESIS LOOP OF LEAD STRONTIUM TITANATE.
GRIFFITHS CR
BULL AMER CERAM SOC 49: 833 (1970)

06-1138 MEASUREMENT OF ELECTRON ENERGY LOSS FUNCTION IN BULK STRONTIUM TITANATE.
HAMMER JM
PHYS REV B 2: 1261-3 (1970)

06-1139 DETERMINATION OF THE NORMAL VIBRATIONAL DISPLACEMENTS OF SEVERAL PEROVSKITES (POTASSIUM
TANTALATE, STRONTIUM TITANATE, RUBIDIUM MANGANESE TRIFLUORIDE) BY INELASTIC NEUTRON
SCATTERING.
HARADA J + AXE JD + SHIRANE G
BULL AMER PHYS SOC 15: 102 (1970)

ATTEMPT TO PREPARE HIGH DOPED BARIUM STRONTIUM TITANATE SEMICONDUCTOR WITH PZT.
HAYASHI J + YANASE M ++ OONO K
MITSUBISHI DENKI LAB REPT 12(4): 133-52 (1971) (IN JAPANESE)

POLARIZATION OF SOLID BARIUM(X) STRONTIUM(1-X) TITANATE COMPOUNDS BETWEEN 4 AND 100
DEGREES K.
HEBENGARTH E + MARTIN G
WISS Z TECH UNIV DRESDEN 20: 495-9 (1971) (IN GERMAN)

DIELECTRIC AND THERMAL (SPECIFIC HEAT, ELECTROCALORIC EFFECT) STUDIES OF FERROELECTRIC
CERAMICS AT LOW TEMPERATURES. (POLYCRYSTALLINE STRONTIUM TITANATE AND BARIUM(X)
STRONTIUM(1-X) TITANATE SOLID SOLUTIONS)
HEGENBARTH E
PHYS STATUS SOLIDI 2: 1544-51 (1962) (IN GERMAN)
CHEM ABSTR 58: 9710G (1963)

INVESTIGATION OF THE THERMAL CONDUCTIVITY IN CONNECTION WITH OTHER PHYSICAL PROPERTIES
OF SOME FERROELECTRICS BELOW 80 DEG K. (SINGLE CRYSTAL STRONTIUM TITANATE,
POLYCRYSTALLINE, STRONTIUM TITANATE, AND SOLID SOLUTIONS BARIUM STRONTIUM TITANATE)
HEGENBARTH E
P104-9, VOL 1 OF DVORAK V + FOUSKOVA A + GLOGAR P (EDS),
INT MEETING ON FERROELECTRICITY, PRAGUE, 1966, PROC.
INST PHYS CZECH ACAD SCI, PRAGUE, 1966.

PECULIAR FERROELECTRIC BEHAVIOR OF POTASSIUM IODATE.
HELG U + GRANICHER H
P169-71 OF INT MEETING ON FERROELECTRICITY, 2ND, 1969, KYOTO, PROC,
(J PHYS SOC JAP, VOL 28 SUPPL). PHYS SOC JAP, 1970, 459P.

06-1140   PHOTOCHROMIC MATERIALS RESEARCH. (STRONTIUM TITANATE PHOTOCONDUCTIVITY, CALCIUM
          FLUORIDE PHOTOCONDUCTIVITY AND OPTICAL PROPERTIES)
          HEYMAN PM + STAEBLER DL
          RCA LABS, PRINCETON, NJ, DEC 1969, 136P. AD705195
              CHEM ABSTR 73: 104297A (1970) US GOVT RES DEVELOP REP 70(12): 219 (1970)
              200

          TEMPERATURE DEPENDENCE OF ELECTRONIC POLARIZABILITY OF STRONTIUM AND BARIUM TITANATE.
          HOFMANN R
          ABHANDLUNG DOKT, NATURWISS, ZURICH, 79P. JURIS DRUCK VERLAG,
          ZURICH, 1968

          FORCE CONSTANTS IN BARIUM TITANATE AND STRONTIUM TITANATE.
          ISHIBASHI Y + TAKAGI Y
          J PHYS SOC JAP 31(6): 1712-18 (1971)

06-1141   IONIC CONDUCTION IN SINTERED OXIDES BASED ON CALCIUM TITANATE OR STRONTIUM TITANATE.
          IWAHARA H + TAKAHASHI T
          DENKI KAGAKU 39: 400-5 (1971) (IN JAPANESE)

06-1142   ATOMIC DISPLACEMENTS IN PEROVSKITE STRONTIUM TITANATE (AT OPTICAL FREQUENCIES,
          CALCULATIONS).
          JAISWAL VK + SHARMA PK
          ANN PHYS 24(7-8): 321-6 (1970)
              213

06-1143   DIELECTRIC RELAXATION IN (CERAMIC) STRONTIUM TITANATES CONTAINING (TRIVALENT) RARE
          EARTH IONS.
          JOHNSON DW + CROSS LE + HUMMEL FA
          J APPL PHYS 41(7): 2828-33 (1970)
              202

          OPTICAL MODES OF THE PEROVSKITE STRUCTURE. (THEORY, STRONTIUM TITANATE DATA)
          JOSEPH RI + SILVERMAN BD
          J PHYS CHEM SOLIDS 24: 1349-55 (1963)

          POLARONS IN ANISOTROPIC ENERGY BANDS (THEORY, CALCULATIONS FOR STRONTIUM TITANATE).
          KAHN AH
          PHYS REV 172(3): 813-15 (1968)

06-1144   STRONG AXIAL ELECTRON PARAMAGNETIC RESONANCE SPECTRUM OF IRON(3+) IN STRONTIUM TITANATE
          DUE TO NEAREST NEIGHBOR CHARGE COMPENSATION.
          KIRKPATRICK ES + MUELLER KA + RUBINS RS
          PHYS REV 135: 86-90 (1964)

06-1145   STRUCTURE OF POLYCRYSTALLINE SOLID SOLUTIONS OF STRONTIUM AND BISMUTH TITANATES.
          KISELEVA KV + BOGDANOV SV
          SOV PHYS SOLID STATE 5(11): 2294-6 (MAY 1964)

06-1146   (TEMPERATURE AND) ELECTRIC FIELD DEPENDENCE OF A HYDROGEN IMPURITY MODE IN STRONTIUM
          TITANATE.
          KLUKUHN AFW + BRUINING J + KLOOTWIJK B + VAN DER ELSKEN J
          PHYS REV LETT 25(6): 380-3 (1970)
              217

06-1147   ESR OF CHROMIUM(3+) IN CHROMIUM DOPED STRONTIUM TITANATE SINGLE CRYSTALS.
          LAGENDIJK A + MOREL RJ + GLASBEEK M + VAN VOORST JDW
          CHEM PHYS LETT 12: 518-21 (1972)

06-1148   DIELECTRIC (CONSTANT AND LOSS) BEHAVIOR OF REDUCED STRONTIUM TITANATE SINGLE CRYSTALS
          (40 HZ-10 MHZ, EFFECTS OF REDUCTION METHODS).
          LAL HB
          INDIAN J PURE APPL PHYS 8: 81-5 (1970)
              207

6. STRONTIUM TITANATE

06-1149   BRILLOUIN SCATTERING IN STRONTIUM TITANATE SINGLE CRYSTALS IN THE TEMPERATURE RANGE 5
          DEG K-300 DEG K.
          LAUBEREAU A + ZUREK R
          Z NATURFORSCH A 25(3): 391-401 (1970) (IN GERMAN)
              226

          TEMPERATURE DEPENDENT POLARIZABILITIES AND COCHRAN MODE ASSIGNMENTS IN BARIUM AND
          STRONTIUM TITANATES.
          LAWLESS WN + GRANICHER H
          P69-79, VOL 1 OF DVORAK V + FOUSKOVA A + GLOGAR P (EDS),
          INT MEETING ON FERROELECTRICITY, PRAGUE, 1966, PROC.
          INST PHYS CZECH ACAD SCI, PRAGUE, 1966.

06-1150   THREE APPLICATION AREAS FOR STRONTIUM TITANATE GLASS CERAMICS.
          LAWLESS WN
          IEEE TRANS SON ULTRASON 19: 287-93 (1972)

06-1151   ELECTRONIC CONDUCTION IN SLIGHTLY REDUCED STRONTIUM TITANATE AT LOW TEMPERATURES.
          LEE C + YAHIA J + BREBNER JL
          PHYS REV B 3: 2525-33 (1971)

06-1152   TRANSPORT PROPERTIES IN REDUCED STRONTIUM TITANATE. (ABSTRACT)
          LEE C + YAHIA J
          BULL AMER PHYS SOC 15(1): 30-1 (1970)
              174

          REJUVENATION OF A BARIUM(0.72) STRONTIUM(0.28) TITANATE CERAMIC.
          LEVIALDI A + SCHOIJET M
          PHYS STATUS SOLIDI 3: 322-8 (1963)

          TEMPERATURE DEPENDENCE OF THE COMPLEX PERMITTIVITY OF SOME TITANATES BETWEEN 20 AND
          1000 DEG C. (BARIUM TITANATE, STRONTIUM TITANATE AND OTHER MATERIALS)
          LIPAEVA GA
          SOV PHYS SOLID STATE 4: 1183-86 (1962)

06-1153   MOSSBAUER STUDY OF COLOR CENTERS IN STRONTIUM TITANATE.
          LUISKUTTY CT + OUSEPH PJ
          BULL AMER PHYS SOC 17: 261 (1972)

06-1154   FAR INFRARED REFLECTIVITY AND TRANSMISSION MEASUREMENTS ON STRONTIUM TITANATE FILMS.
          LURIO A + PENNEBAKER WB
          BULL AMER PHYS SOC 16: 312 (1971)

          EFFECT OF PARTICLE SIZE AND SHAPE ON THE INFRARED ABSORPTION SPECTRA OF BARIUM TITANATE
          AND STRONTIUM TITANATE POWDERS.
          LUXON JT + MONTGOMERY DJ + SUMMITT R
          J APPL PHYS 41(6): 2303-7 (1970)

06-1155   SOUND PROPAGATION NEAR THE STRUCTURAL PHASE TRANSITION IN STRONTIUM TITANATE.
          LUTHI B + MORAN TJ
          PHYS REV B 2: 1211-14 (1970)

06-1156   X-RAY DIFFRACTOMETRY OF LOW TEMPERATURE PHASE TRANSFORMATIONS IN STRONTIUM TITANATE.
          LYTLE FW
          J APPL PHYS 35(7): 2212-15 (1964)

06-1157   COUPLED MODES (OPTICAL, FERROELECTRIC- ACOUSTIC) IN THE PHONON SPECTRA OF STRONTIUM
          TITANATE AND POTASSIUM TANTALATE (LATTICE DYNAMICS).
          MANTE AJH
          SOLID STATE COMMUN 8: 1415-17 (1970)
              214

          DIELECTRIC CERAMICS WITH BOUNDARY LAYER STRUCTURE FOR HIGH FREQUENCY APPLICATION.
          (BARIUM, STRONTIUM TITANATE)
          MASUNO K + MURAKAMI T + WAKU S
          FERROELECTRICS 3: 315-9 (FEB 1972)

06-1158   EFFECT OF THE 110 DEGREE K PHASE TRANSITION ON THE STRONTIUM TITANATE CONDUCTION BANDS.
          MATTHEISS LF
          PHYS REV B 6: 4740-53 (1972)

06-1159   ANOMALOUS SPECIFIC HEAT OF STRONTIUM TITANATE.
          MCCORMIC WD + TRAPPE KI
          BULL AMER PHYS SOC 16: 850 (1971)

06-1160   TEMPERATURE DEPENDENCE OF THE FIRST ORDER ELASTIC CONSTANTS OF STRONTIUM TITANATE.
          MEEKS EL + ARNOLD RT
          PHYS REV B 1(3): 982-8 (1970)
              197

06-1161   CHARGE COMPENSATION BY OXYGEN(2-) VACANCIES IN CHROMIUM(3+) DOPED STRONTIUM TITANATE
          (ESR MEASUREMENTS, MODEL PROPOSED).
          MEIERLING HD
          PHYS STATUS SOLIDI 43: 191-7 (1971)
              226

06-1162   EPR INVESTIGATION OF POLYCRYSTALLINE STRONTIUM TITANATE CONTAINING MANGANESE DIOXIDE.
          MESHCHERYAKOV NA + BERDOV GI + GINDULINA VZ
          INORG MATER 8: 254-6 (1972)

06-1163   DEFECTS IN OXIDE CERAMICS. (STRONTIUM TITANATE)
          MILLER GR
          UNIV UTAH, 1971, 9P. AD731071

          PRESSURE DEPENDENCE OF THE DIELECTRIC CONSTANTS OF PARAELECTRIC MATERIALS. (BARIUM
          TITANATE, STRONTIUM TITANATE)
          MORENO M + GRANICHER H
          HELV PHYS ACTA 37(7-8): 625 (1964) (IN GERMAN)
              SCI ABSTR A 68: 18002 (1965)

06-1164   MONODOMAIN STRONTIUM TITANATE (OCCURANCE IN THIN (110) PLATES BELOW CUBIC- TETRAGONAL
          TRANSITION, AND PRODUCTION BY SHAPING AND STRESSING).
          MUELLER KA + BERLINGER W + CAPIZZI M + GRANICHER H
          SOLID STATE COMMUN 8: 549-53 (1970)
              192

06-1165   PARAMAGNETIC RESONANCE AND OPTICAL ABSORPTION OF TRANSITION ELEMENT IONS IN STRONTIUM
          TITANATE AND LANTHANUM ALUMINATE.
          MUELLER KA
          P17-43 OF INT CONF ON PARAMAGNETIC RESONANCE, 1ST PROC, HEBREW
          UNIV JERUSALEM, 1962, 128P.

06-1166   CRITICAL ASYMMETRY IN LOCAL FLUCTUATIONS IN STRONTIUM TITANATE ABOVE THE CRITICAL
          TEMPERATURE.
          MULLER KA + BERLINGER W
          PHYS REV LETT 29: 715-18 (1972)

06-1167   EFFECTIVE CHARGE OF TITANIUM(4+) FROM THE PARAMAGNETIC RESONANCE OF MANGANESE(4+) IN
          STRONTIUM TITANATE.
          MULLER KA
          P369-72, VOL 2 OF DVORAK V + FOUSKOVA A + GLOGAR P (EDS),
          INT MEETING ON FERROELECTRICITY, PRAGUE, 1966, PROC.
          INST PHYS CZECH ACAD SCI, PRAGUE, 1966.

06-1168   INTERPRETATION OF ELECTRIC FIELD EFFECTS IN EPR OF GADOLINIUM(3+) DOPED STRONTIUM
          TITANATE.
          MULLER KA
          SOLID STATE COMMUN 9: 373-7 (1971)
              232

06-1169   MONODOMAIN STRONTIUM TITANATE.
          MULLER KA + BERLINGER W + CAPIZZI M + GRANICHER H
          SOLID STATE COMMUN 8: 549-53 (1970)

06-1170   ORDER PARAMETER AND PHASE TRANSITIONS OF STRESSED STRONTIUM TITANATE. (EPR OF IRON-
          VANADIUM PAIRS)
          MULLER KA + BERLINGER W + SHUNCZEWSKI JC
          PHYS REV LETT 25: 734-7 (14 SEP 1970)
              218

06-1171   PHOTOCHROMIC IRON IN STRONTIUM TITANATE. EVIDENCE FROM PARAMAGNETIC RESONANCE.
          MULLER KA + VON WALDKIRCH TH + BERLINGER W AND OTHERS
          SOLID STATE COMMUN 9: 1097-101 (1971)

6. STRONTIUM TITANATE

VIBRATIONAL SPECTRA OF STRONTIUM, BARIUM AND CALCIUM TITANATES.
MURZIN VN + DEMESHINA AI + BOGDANOV SV
BULL ACAD SCI USSR, PHYS SER 29(6): 926-30 (1965)

TEMPERATURE STUDIES OF INFRARED REFLECTION SPECTRA OF BARIUM AND STRONTIUM TITANATES IN
THE 2-1000 MICRON REGION.
MURZIN VN + DEMESHINA AI
SOV PHYS SOLID STATE 5: 1716-17 (1964)

ANOMALOUS FERROELECTRIC HYSTERESIS LOOPS. (LEAD STRONTIUM TITANATE)
MURDOCH FJ
ARMY ELECTRON COMMAND, FT MONMOUTH, NJ, 1971, 58P. AD732336
    SCI TECH AEROSP REP 10(7): 948 (1972)

VIBRATIONAL SPECTRA OF THE TITANATES OF STRONTIUM, BARIUM, AND CALCIUM. (75 REFS)
MURZIN VN
TR FIZ INST PN LEBEDEVA, AKAD NAUK SSSR 48: 145-203 (1969)
(IN RUSSIAN)
    CHEM ABSTR 72: 94858X (1970)

HIGH L/F NOISE ANOMALY IN SEMICONDUCTING BARIUM STRONTIUM TITANATE.
MYTTON RJ + BENTON RK
PHYS LETT A 39: 329-30 (1972)

06-1172 EFFECTS OF SILICA ON MANGANESE DOPED STRONTIUM TITANATE CERAMICS.
NAKAHARA M + EDAHIRO T
ELEC COMMUN LAB TACH J 20: 2684-6 (1971) (IN JAPANESE)
    256

06-1173 ANOMALOUS RESONANCE OF STRONTIUM TITANATE.
NEVILLE RC + HOENEISEN B + MEAD CA
J APPL PHYS 43(10): 3903-5 (1972)
    276

06-1174 PERMITTIVITY OF STRONTIUM TITANATE.
NEVILLE RC
J APPL PHYS 43(5): 2124-31 (MAY 1972)
    265

06-1175 SURFACE BARRIER ENERGIES ON STRONTIUM TITANATE.
NEVILLE RC + MEAD CA
J APPL PHYS 43: 4657-63 (NOV 1972)

06-1176 SOME ELECTRONIC PROPERTIES OF ZINC OXIDE AND STRONTIUM TITANATE. (SURFACE BARRIER
SYSTEMS, METAL CONTACTS)
NEVILLE RC
CALIF INST TECHNOL PHD THESIS, 1971, 105P.
    255

DEFECT STRUCTURE OF (BARIUM(1-X) STRONTIUM(X)) TITANATE- MIXED CRYSTALS FROM A
POTASSIUM FLUORIDE MELT.
NOVAK J + COUFOVA P + HLASIVCOVA N + JANOUSEK V
COLLECT CZECH CHEM COMMUN 35: 782-7 (MAR 1970)
(IN GERMAN)
    CHEM ABSTR 72: 94170S (1970)

06-1177 ELASTIC CONSTANT ANOMALIES IN STRONTIUM TITANATE.
O'SHEA DC
BULL AMER PHYS SOC 15: 383 (1970)

06-1178 TEMPERATURE DEPENDENCE OF THE SOFT MODE IN STRONTIUM TITANATE ABOVE THE 105 DEG K
TRANSITION.
OTNES K + RISTE T
SOLID STATE COMMUN 9: 1103-6 (1971)
    241

EXTENDED FINE STRUCTURE IN X-RAY ABSORPTION SPECTRA OF CERTAIN PEROVSKITES (STRONTIUM
OR CALCIUM TITANATE OR ZIRCONATE, TEST OF SHORT OR LONG RANGE ORDER THEORIES).
PEREL J + DESLATTES RD
PHYS REV B 2(5): 1317-23 (1970)

6. STRONTIUM TITANATE

SPECIFIC FEATURES OF THE TEMPERATURE DEPENDENCE OF THE VELOCITY OF ULTRASOUND IN POLYCRYSTALLINE SOLID SOLUTIONS OF BARIUM AND STRONTIUM TITANATES.
PERRO IT + GRINVALD GZH + FRITSBERG VYA
P97-107 OF FAZOVYE PEREKHODY V SEGNETOELEKTRIKAKH (PHASE TRANSFORMATIONS IN FERROELECTRICS), FRITSBERG VYA + ROLOV BN + KRUCHAN YY(EDS), IZDATELSTVO ZINATNE RIGA, 1971, 206P. (IN RUSSIAN)

06-1179 DETERMINATION OF X-RAY ABSORPTION COEFFICIENTS AND THE THICKNESS OF NONUNIFORM THIN SAMPLES (STRONTIUM TITANATE AS EXAMPLE).
PERFL J
J PHYS E 3(4): 268-70 (1970)
216

06-1180 ELASTIC PROPERTIES OF YTTRIUM DOPED STRONTIUM TITANATE SINGLE CRYSTALS.
PERSON JS + WANG J + PETERS RD + ARNOLD RT
BULL AMER CERAM SOC 15: 1382 (1970)

06-1181 ULTRASONIC THIRD HARMONIC GENERATION IN STRONTIUM TITANATE SINGLE CRYSTALS (NONLINEAR ELASTIC PROPERTIES).
PETERS RD + ARNOLD RT
J APPL PHYS 42(1): 980-2 (1971)
230

06-1182 EFFECT OF STRESS ON THE SUPERCONDUCTING TRANSITION TEMPERATURE OF STRONTIUM TITANATE.
PFEIFFER ER + SCHOOLEY JF
J LOW TEMP PHYS 2: 333-52 (1970)

INFLUENCE OF PRESSURE ON PHASE TRANSITIONS AT LOW TEMPERATURES: STRONTIUM TITANATE AND ((BARIUM(X) STRONTIUM(1-X) TITANATE (RECIPROCAL DIELECTRIC CONSTANT FOLLOWS QUADRATIC TEMPERATURE LAW).
PIETRASS B + HEGENBARTH E
J LOW TEMP PHYS 7(3-4): 201-9 (1972)

SELF CONSISTENT LATTICE DYNAMICAL THEORY OF STRUCTURAL PHASE TRANSITIONS IN PEROVSKITE TYPE CRYSTALS. (STRONTIUM TITANATE, LANTHANUM ALUMINATE)
PIETRASS B
PHYS STATUS SOLIDI B 53: 279-86 (1972)

06-1183 ANALYSIS OF THE ELASTIC ANOMALIES AT THE STRUCTURAL PHASE TRANSITION OF STRONTIUM TITANATE NEAR 105 DEG K.
PIETRASS B
PHYS STATUS SOLIDI B 47: 495-500 (1971)
253

06-1184 FIELD INDUCED SWITCHING OF TETRAGONAL DOMAINS IN STRONTIUM TITANATE AT LOW TEMPERATURES.
PIETRASS B
PHYS STATUS SOLIDI A, APPL RES 9(1): K39-42 (1972)

06-1185 LATTICE DYNAMICS OF CUBIC PEROVSKITE STRUCTURES, IN PARTICULAR STRONTIUM TITANATE.
RAJAGOPAL AK + SRINIVASAN R
J PHYS CHEM SOLIDS 23: 633-8 (1962)

COMPARATIVE STUDY OF THE CRYSTALLOGRAPHIC, DIELECTRIC, AND NONLINEAR OPTICAL PROPERTIES OF THE ABC NIOBATES (WHERE A AND B ARE CALCIUM, STRONTIUM, OR BARIUM, AND C IS SODIUM OR POTASSIUM) PHASES OF THE TETRAGONAL TUNGSTEN BRONZE OXIDE TYPE.
RAVEZ J + BUDIN JP + HAGENMULLER P
J SOLID STATE CHEM 5: 239-46 (1972) (IN FRENCH)
CHEM ABSTR 77: 119223E (1972)

06-1186 FUNDAMENTAL ABSORPTION EDGE OF STRONTIUM TITANATE.
REDFIELD D + BURKE WJ
PHYS REV B SOLID STATE 6(8): 3104-9 (15 OCT 1972)
277

06-1187 OPTICAL ABSORPTION EDGE OF STRONTIUM TITANATE.
REDFIELD D + BURKE WJ
BULL AMER PHYS SOC 15: 394 (1970)

06-1188 REINTERPRETATION OF WAVELENGTH MODULATED ABSORPTION IN STRONTIUM TITANATE WITHOUT COEXISTING PHASES.
REDFIELD D + BURKE WJ
PHYS REV LETT 28: 435-7 (1972)

6. STRONTIUM TITANATE

06-1189   ANOMALOUS ULTRASONIC ATTENUATION AT THE 105K TRANSITION IN STRONTIUM TITANATE.
           REHWALD W
           SOLID STATE COMMUN 8: 607-11 (1970)
               192

06-1190   LOW TEMPERATURE ELASTIC MODULI OF STRONTIUM TITANATE (10 TO 130 DEG K).
           REHWALD W
           SOLID STATE COMMUN 8: 1483-85 (1970)
               214

06-1191   ULTRASONIC PROPERTIES OF STRONTIUM TITANATE AT THE 105 DEGREE K TRANSITION.
           REHWALD W
           PHYS KONDENS MATER, PHYS MATER CONDENSEE 14(1): 21-36 (1971)

           ELECTRON PARAMAGNETIC RESONANCE OF TRIVALENT GADOLINIUM IONS IN STRONTIUM AND BARIUM
           TITANATES.
           RIMAI L + DEMARS GA
           PHYS REV 127: 702-10 (1962)

06-1192   ELECTRON PARAMAGNETIC RESONANCE OF SOME RARE EARTH IMPURITIES IN STRONTIUM TITANATE.
           RIMAI L + DEMARS GA
           P51-8 OF INT CONF ON PARAMAGNETIC RESONANCE. 1ST PROC, HEBREW
           UNIV JERUSALEM, 1962, 128P.

06-1193   EFFECT OF TEMPERATURE AND PRESSURE ON THE ELECTRON PARAMAGNETIC RESONANCE SPECTRA OF
           SUBSTITUTIONAL IMPURITIES IN CUBIC STRONTIUM TITANATE.
           RIMAI L + DEUTSCH T + SILVERMAN BD
           PHYS REV 133: 1123-34 (1964)

06-1194   CRITICAL BEHAVIOR OF STRONTIUM TITANATE NEAR THE 105 K PHASE TRANSITION.
           RISTE T + SAMUELSEN EJ + OTNES K AND OTHERS
           SOLID STATE COMMUN 9: 1455-8 (SEP 1971)

06-1195   CRITICAL NEUTRON SCATTERING FROM STRONTIUM TITANATE.
           RISTE T + SAMUELSEN EJ + OTNES K
           P395-407 OF NATO ADVAN STUDY INST, GEILO, NORWAY, STRUCTURAL
           PHASE TRANSITIONS AND SOFT MODES, 1971 PROC, SAMUELSEN EJ +
           ANDERSEN E + FEDER J(EDS), UNIVERSITETSFORLAGEN, BOSTON, 1972.

06-1196   SOFT MODE STUDY OF THE STRESS INDUCED PHASE TRANSITIONS NEAR THE TRANSITION TEMPERATURE
           IN STRONTIUM TITANATE.
           ROKNI M + WALL LS
           J CHEM PHYS 55: 435-8 (1971)

06-1197   INERTIAL ELECTRICAL POLARIZATION OF CRYSTALS (ALKALI HALIDES, STRONTIUM TITANATE).
           ROST A + SCHMIDT G
           ANN PHYS 28(2): 147-53 (1972) (IN GERMAN)

06-1198   ELECTRICAL AND OPTICAL PROPERTIES (CONDUCTIVITY AND MOBILITY VERSUS TEMPERATURE,
           ABSORPTION SPECTRA) OF SEMICONDUCTING STRONTIUM TITANATE SINGLE CRYSTALS (HYDROGEN
           REDUCED, CERIUM OR NIOBIUM DOPED).
           ROZHDESTVENSKAYA MV + SHEFTEL IT + STOGOVA VA + KOZYREVA MS + KRAYUKHINA EK
           SOV PHYS SOLID STATE 12(3): 674-78 (1970)
               206

06-1199   MICROWAVE LOSSES IN STRONTIUM TITANATE ABOVE THE PHASE TRANSITION.
           RUPPRECHT G + BELL RO
           PHYS REV 125: 1915-20 (1962)

06-1200   DIELECTRIC PROPERTIES OF STRONTIUM TITANATE AT LOW TEMPERATURES.
           SAIFI MA
           PENN STATE UNIV PHD THESIS, 1968, 169P.
               DISS ABSTR B 29: 4658
               189

06-1201   DIELECTRIC PROPERTIES OF STRONTIUM TITANATE (ANNEALED SINGLE CRYSTALS) AT LOW
           TEMPERATURE (5-300 DEG K).
           SAIFI MA + CROSS LE
           PHYS REV B 2(3): 677-84 (1970)
               221

6. STRONTIUM TITANATE

06-1202   DIELECTRIC PROPERTIES OF STRONTIUM TITANATE AT LOW TEMPERATURES. (TEMPERATURE
           DEPENDENCE, ANISOTROPY)
           SAKUDO T + UNOKI H
           PHYS REV LETT 26(14): 851-3 (1971)
                235

06-1203   ELECTRIC FIELD INDUCED OPTICAL HARMONIC GENERATION IN STRONTIUM TITANATE AND POTASSIUM
           TANTALATE.
           SAKUDO T + FUJII Y
           P87-9 OF INT MEETING ON FERROELECTRICITY, 2ND, 1969, KYOTO, PROC,
           (J PHYS SOC JAP, VOL 28 SUPPL). PHYS SOC JAP, 1970, 459P.

06-1204   CRITICAL DYNAMICS OF STRONTIUM TITANATE ABOVE THE CRITICAL TEMPERATURE.
           SCHWABL F
           PHYS REV LETT 28: 500-3 (1972)

06-1205   DYNAMICS AT STRUCTURAL TRANSITIONS ABOVE FERROELECTRIC CRITICAL TEMPERATURE (ROTATIONS
           OF BO_6 OCTAHEDRA IN $ABO(3)$ PEROVSKITES, STRONTIUM TITANATE).
           SCHWABL F
           Z PHYS 254(1): 57-70 (1972)

06-1206   ELECTROMECHANICAL PROPERTIES OF STRONTIUM TITANATE SINGLE CRYSTALS.
           SCHMIDT G + HEGENBARTH E
           PHYS STATUS SOLIDI 3: 329-38 (1963) (IN GERMAN)
                CHEM ABSTR 59: 4614C (1963)

06-1207   SOFT MODE RAMAN SPECTROSCOPY: COUPLED MODES (FERROELECTRICS, CESIUM DIHYDROGEN
           ARSENATE, STRONTIUM TITANATE)
           SCOTT JF
           P387-92 OF INT CONF ON LIGHT SCATTERING IN SOLIDS, 2ND, 1971,
           BALKANSKI M(ED), FLAMMARION, PARIS, 1971.
                SCI ABSTR A 75: 15995 (1972)
                264

06-1208   CRITICAL NEUTRON SCATTERING IN STRONTIUM TITANATE AND POTASSIUM MANGANESE TRIFLUORIDE.
           SHAPIRO SM + AXE JD + SHIRANE G
           PHYS REV B 6(12): 4332-41 (1 DEC 1972)
                279

06-1209   EFFECTS OF TEMPERATURE AND PHASE TRANSITIONS ON THE HYDROXYL STRETCHING VIBRATION IN
           STRONTIUM TITANATE AND TITANIUM DIOXIDE.
           SHANER JW + SEWARD WD
           BULL AMER PHYS SOC 17: 102 (1972)

06-1210   NEW OBSERVATIONS FROM NEUTRON SCATTERING STUDIES OF THE STRUCTURAL PHASE TRANSITION IN
           STRONTIUM TITANATE.
           SHAPIRO SM + AXE JD + SHIRANE G
           P155 OF NUSIMOVICI M(ED), PHONONS, INT CONF, RENNES, 1971 PROC,
           FLAMMARION, PARIS, 1971, 502P.

06-1211   EFFECT OF THE QUARTIC ANHARMONICITY ON THE SOFT PHONON INSTABILITY OF STRONTIUM
           TITANATE.
           SILBERGLITT R
           BULL AMER PHYS SOC 17: 102 (1972)

06-1212   MICROWAVE ABSORPTION IN CUBIC STRONTIUM TITANATE.
           SILVERMAN BD
           PHYS REV 125: 1921-30 (1962)

06-1213   MICROSCOPIC CALCULATION OF THE RESPONSE FUNCTION OF THE SOFT ZONE BOUNDARY PHONON IN
           STRONTIUM TITANATE.
           SILBERGLITT R
           SOLID STATE COMMUN 11: 247-51 (1972)
                274

06-1214   FREE CARRIER OPTICAL ABSORPTION IN NIOBIUM DOPED STRONTIUM TITANATE.
           SIMANEK E + LIU NLH + WILD RL
           J PHYS CHEM SOLIDS 33: 951-4 (APR 1972)

6. STRONTIUM TITANATE

06-1215 ANALYSIS OF STRESS AND TEMPERATURE DEPENDENCE OF FLUORESCENCE IN CHROMIUM (3+) DOPED
         STRONTIUM TITANATE.
         SLONCZEWSKI JC
         PHYS REV B 2: 4646-55 (1970)

06-1216 INTERACTION OF ELASTIC STRAIN WITH THE STRUCTURAL TRANSITION OF STRONTIUM TITANATE.
         SLONCZEWSKI JC + THOMAS H
         PHYS REV B 1(9): 3599-608 (1970)
              234

06-1217 PREPARATION AND CHARACTERIZATION OF ALKOXY-DERIVED STRONTIUM ZIRCONATE AND STRONTIUM
         TITANATE (OF HIGH PURITY AND SURFACE ACTIVITY VIA HYDROLYTIC DECOMPOSITION).
         SMITH JS + DOLLOFF RT + MAZDIYASNI KS
         J AMER CERAM SOC 53: 91-5 (1970)
              183

06-1218 MECHANICAL RELAXATION AND NONLINEARITY IN STRONTIUM TITANATE SINGLE CRYSTALS.
         SORGE G + HEGENBARTH E + SCHMIDT G
         PHYS STATUS SOLIDI 37: 599-603 (1970)

06-1219 PRESSURE DEPENDENCE OF THE ELASTIC COMPLIANCE OF STRONTIUM TITANATE NEAR THE TRANSITION
         TEMPERATURE.
         SORGE G + SCHMIDT G + HEGENBARTH E + FRENZEL C
         PHYS STATUS SOLIDI 37(1): K17-18 (1970)

06-1220 TEMPERATURE DEPENDENCE OF THE ELASTIC COMPLIANCE OF STRONTIUM TITANATE SINGLE CRYSTALS
         IN THE TEMPERATURE RANGE 20 TO 45 DEG K.
         SORGE G + HEGENBARTH E
         PHYS STATUS SOLIDI 33: K79-82 (1969)
              160

06-1221 ENERGY BAND STRUCTURE OF STRONTIUM TITANATE FROM A SELF CONSISTENT FIELD TIGHT BINDING
         CALCULATION.
         SOULES TF + KELLY EJ + VAUGHT DM AND OTHERS
         PHYS REV B SOLID STATE 6: 1519-32 (AUG 15, 1972)

         FAR INFRARED DIELECTRIC DISPERSION IN BARIUM AND STRONTIUM TITANATES AND TITANIUM
         DIOXIDE.
         SPITZER WG + MILLER RC + KLEINMAN DA + HOWARTH LE
         PHYS REV 126: 1710-21 (1962)

06-1222 ELECTRON TUNNELING AND BAND STRUCTURE OF (NIOBIUM DOPED) STRONTIUM TITANATE AND
         (CALCIUM DOPED) POTASSIUM TANTALATE (SCHOTTKY BARRIERS WITH INDIUM).
         SROUBEK Z
         PHYS REV B 2(8): 3170-5 (1970)
              228

06-1223 STATIC SCALING HYPOTHESIS AND ITS IMPLICATIONS FOR THE PLANAR SPIN MODEL OF STRONTIUM
         TITANATE.
         STANLEY HE
         P271-8 OF NATO ADVAN STUDY INST, GEILO, NORWAY, STRUCTURAL
         PHASE TRANSITIONS AND SOFT MODES, 1971 PROC, SAMUELSEN EJ +
         ANDERSEN E + FEDER J(EDS), UNIVERSITETSFORLAGEN, BOSTON, 1972.

06-1224 LATTICE DYNAMICS OF STRONTIUM TITANATE.
         STIRLING WG + COWLEY RA
         J PHYS (PARIS) 33, SUPPL 4, COLLOQ 2: 135-7 (1972) (IN FRENCH)
              284

06-1225 NEUTRON INELASTIC SCATTERING STUDY OF THE LATTICE DYNAMICS OF STRONTIUM TITANATE:
         HARMONIC MODELS.
         STIRLING WG
         J PHYS C 5: 2711-30 (1972)
         J PHYS (PARIS) 33, SUPPL 2: 135-7 (1972)
              279

         THERMAL CONDUCTIVITY OF STRONTIUM TITANATE, POTASSIUM TANTALATE AND BARIUM TITANATE
         NEAR THE CURIE POINTS.
         TAKEMURA M + TANI K
         J PHYS SOC JAP 31: 151-4 (JUL 1971)

6. STRONTIUM TITANATE

DYNAMICAL BEHAVIOR OF DISPLACIVE TYPE FERROELECTRICS NEAR THE TRANSITION POINT. (BARIUM TITANATE, STRONTIUM TITANATE)
TANI K + TAKEMURA M
J PHYS SOC JAP 30: 328-37 (FEB 1971)

06-1226 SPECIFIC HEAT ANOMALY ASSOCIATED WITH THE 110 DEG K PHASE TRANSITION IN STRONTIUM TITANATE. (ABSTRACT)
TAYLOR DJ + SEWARD WD
BULL AMER PHYS SOC 15(12): 1624 (1970)
218

SEMICONDUCTING BARIUM AND STRONTIUM TITANATES WITH POSITIVE TEMPERATURE COEFFICIENTS OF RESISTIVITY.
TEKSTER-PROSKURYAKOVA GN + SHEFTEL IT
SOV PHYS SOLID STATE 5: 2548-8 (1964)

PREPARATION AND STUDY OF THIN FILMS OF (BARIUM, STRONTIUM) TITANATE.
TOMASHPOLSKII YUYA + SOROKINA LA + VENEVTSEV YUN
INORG MATER 6: 1203-4 (JUL 1970)

06-1227 THERMALLY STIMULATED CONDUCTIVITY IN STRONTIUM TITANATE SINGLE CRYSTALS.
TSIKIN AN + SHTURBINA NA
SOV PHYS SOLID STATE 12(11): 2723-5 (1971)
234

06-1228 ELECTRON SPIN RESONANCE STUDY IN STRONTIUM TITANATE.
UNOKI H
ELECTROTECHNICAL LAB, TOKYO, JAPAN, NOV 1970, 61P. N72-13708
(IN JAPANESE, ENGLISH SUMMARY)
SCI TECH AEROSP REP 10(4): 528 (1972)
261

06-1229 ELECTRIC FIELD EFFECT OF ELECTRON SPIN RESONANCE IN STRONTIUM TITANATE. (DOPED WITH GADOLINIUM(3+) OR IRON(3+))
UNOKI H + SAKUDO T
P125-7 OF INT MEETING ON FERROELECTRICITY, 2ND, 1969, KYOTO, PROC,
(J PHYS SOC JAP, VOL 28 SUPPL). PHYS SOC JAP, 1970, 459P.

06-1230 NEW ESR EVIDENCE FOR THE 65 DEGREES K ANOMALY IN STRONTIUM TITANATE.
UNOKI H + SAKUDO T
PHYS LETT A 32: 368-9 (1970)

PHASE TRANSITIONS IN THE MIXED CRYSTAL SYSTEM STRONTIUM(1-X) BARIUM(X) TITANATE.
UWE H + FUJII Y + UNOKI H + SAKUDO T
ACTA CRYSTALLOGR A 28: S178 (1972)

06-1231 ANALYSIS OF THE IRON OXYGEN VACANCY CENTER IN THE TETRAGONAL PHASE OF STRONTIUM TITANATE.
VON WALDKIRCH TH + MULLER KA + BERLINGER W
PHYS REV B 5(11): 4324-34 (1 JUN 1972)
265

06-1232 EPR STUDIES AND DYNAMICS IN STRONTIUM TITANATE IN THE NEIGHBORHOOD OF THE PHASE TRANSFORMATION.
VON WALDKIRCH TH + MULLER KA + BERLINGER W
HELV PHYS ACTA 45: 866 (1972) (IN GERMAN)

06-1233 FLUCTUATIONS AND CORRELATIONS IN STRONTIUM TITANATE ABOVE THE CRITICAL TEMPERATURE.
VON WALDKIRCH TH + MULLER KA + BERLINGER W + THOMAS H
PHYS REV LETT 28: 503-6 (1972)

06-1234 THE IRON(3+) - VACANCY CENTER IN THE TETRAGONAL PHASE OF STRONTIUM TITANATE.
VON WALDKIRCH TH + MULLER KA + BERLINGER W
HELV PHYS ACTA 45: 866 (1972) (IN GERMAN)

PECULIARITIES OF THE ELECTRON PARAMAGNETIC RESONANCE SPECTRA IN A FERROELECTRIC CERAMIC CONTAINING TITANIUM. (BARIUM TITANATE, STRONTIUM TITANATE)
VOTINOV MP + DEMIDENKO NI
SOV PHYS SOLID STATE 4: 2351-3 (1963)

6. STRONTIUM TITANATE

STUDIES ON THE BARIUM STRONTIUM TITANATE BOUNDARY LAYER CERAMIC DIELECTRICS.
WAKU S + UCHIDATE M + KIKUCHI K
REV ELECTRON COMMUN LAB 18: 681-93 (1970)

06-1235 CUBIC TO TRIGONAL STRESS INDUCED PHASE TRANSITION IN STRONTIUM TITANATE.
WALL LS + ROKNI M + SCHAWLOW AL
SOLID STATE COMMUN 9: 573-7 (1971)

06-1236 RAMAN STUDIES OF STRESS INDUCED PHASE TRANSITION NEAR CURIE TEMPERATURE IN STRONTIUM
TITANATE. (ABSTRACT)
WALL LS + ROKNI M + SCHAWLOW AL
BULL AMER PHYS SOC 15(12): 1622 (1970)
218

06-1237 NUCLEAR MAGNETIC RESONANCE STUDY OF THE PHASE TRANSITION IN STRONTIUM TITANATE.
WEBER MJ + ALLEN RR
J CHEM PHYS 38: 726-9 (1963)

06-1238 LATTICE DISTORTIONS IN A STRONTIUM TITANATE SINGLE CRYSTAL ELECTRET.
WEIK H + LAMBERT VL
PHYS REV LETT 10: 51-3 (1963)

06-1239 ULTRASONIC DISPERSION IN STRONTIUM TITANATE.
WESSON RA
NY UNIV PHD THESIS, 131P, 1970.
    DISS ABSTR INT B 31: 7538 (1971)
238

06-1240 PHOTOCONDUCTIVITY (TRANSIENTS) IN PHOTOCHROMIC STRONTIUM TITANATE (ON COLORING AND
BLEACHING).
WILLIAMS R
J APPL PHYS 42(3): 1131-5 (1971)
230

06-1241 ELECTRIC FIELD DEPENDENCE OF OPTICAL PHONON FREQUENCIES (RAMAN SCATTERING IN CUBIC
PEROVSKITE STRONTIUM TITANATE, 8-250K, 0.2-12 KV/CM).
WORLOCK JM + FLEURY PA
PHYS REV LETT 19(20): 1176-9 (1967)
219

06-1242 LIGHT SCATTERING STUDIES OF THE SOFT MODES NEAR THE CUBIC TO TETRAGONAL PHASE
TRANSITION IN STRONTIUM TITANATE.
WORLOCK JM + OLSON DH
P410-14 OF INT CONF ON LIGHT SCATTERING IN SOLIDS, 2ND PROC,
BALKANSKI M(ED), PARIS, 1971, FLAMMARION, PARIS, 1971, 518P.

06-1243 EFFECTS OF LONGITUDINAL MAGNETIC FIELD ON ULTRASOUND IN STRONTIUM TITANATE.
WU CC + TSAI J
J PHYS C SOLID STATE PHYS 5: 2419-26 (1972)
277

MEASUREMENT OF OXYGEN DIFFUSION CONSTANT IN SOME TITANATES. (STRONTIUM TITANATE, BARIUM
TITANATE)
YAMAJI A
ACTA CRYSTALLOGR A 28: S167 (1972)

06-1244 DEFECT STUDIES IN STRONTIUM TITANATE.
YAMADA H + MILLER GR
BULL AMER CERAM SOC 51: 320 (1972)

06-1245 OPTICAL ABSORPTION SPECTRA OF REDUCED STRONTIUM TITANATE.
YAMADA H + MILLER GR
BULL AMER PHYS SOC 17: 618 (1972)

06-1246 STUDIES ON THE (STRONTIUM, CALCIUM) TITANATE SOLID SOLUTION CERAMIC WITH BOUNDARY LAYER
STRUCTURE.
YAMAJI A
TRANS INST ELECTRON COMMUN ENG JAP C 55: 69-71 (1972) (IN JAPANESE)

06-1247 STUDIES ON THE (STRONTIUM, CALCIUM) TITANATE BOUNDARY LAYER CERAMIC DIELECTRICS.
YAMAJI A + KOGA K
ELEC COMMUN LAB TECH J 21: 769-71 (1972) (IN JAPANESE)

6. STRONTIUM TITANATE

06-1248   STUDIES ON STRONTIUM TITANATE BOUNDARY LAYER DIELECTRICS.
           YAMAJI A + WAKU S
           REV ELECTRON COMMUN LAB 20: 747-63 (1972)

06-1249   PHOTOCONDUCTIVITY ANOMALY AS AN EVIDENCE OF THE LOCAL FERROELECTRIC PHASE TRANSITION
           AROUND 47 DEG K IN STRONTIUM TITANATE.
           YASUNAGA H
           P454-6 OF INT MEETING ON FERROELECTRICITY, 2ND, 1969, KYOTO, PROC,
           (J PHYS SOC JAP, VOL 28 SUPPL). PHYS SOC JAP, 1970, 459P.

06-1250   INFRARED SPECTRA OF SINGLE AND POLYCRYSTALLINE STRONTIUM TITANATE.
           ZAKHAROV VP + KNYAZEV AS + POPLAVKO YUM AND OTHERS
           BULL ACAD SCI USSR, PHYS SER 35(9): 1652-5 (1971)

7. LEAD ZIRCONATE

07-1251   FERROELECTRICITY AND ANTIFERROELECTRICITY IN PURE AND NIOBIUM DOPED LEAD ZIRCONATE.
BENGUIGUI L
J SOLID STATE CHEM 3: 381-6 (AUG 1971)

07-1252   FIELD ENFORCED FERROELECTRICITY IN GLASS BONDED LEAD ZIRCONATE. (PHASE TRANSITION 200
DEG C BELOW THE CURIE POINT)
BURN I
BULL AMER CERAM SOC 50 (5): 501-5 (1971)
      235

07-1253   MOSSBAUER EFFECT STUDIES OF THE PHASE TRANSITIONS IN ANTIFERROELECTRIC LEAD ZIRCONATE
AND FERROELECTRIC LEAD TITANATE ZIRCONATE.
CANNER JP + YAGNIK CM + GERSON R + JAMES WJ
J APPL PHYS 42 (12): 4708-11 (NOV 1971)
      248

07-1254   EVALUATION OF LEAD LANTHANATE ZIRCONATE CERAMICS FOR APPLICATIONS IN OPTICAL
COMMUNICATIONS.
CHEN FS
OPT COMMUN 6 (3): 297-300 (NOV 1972)
      281

07-1255   THE INTERNAL ELECTRIC FIELD GRADIENT IN ANTIFERROELECTRIC LEAD ZIRCONATE STUDIED BY
PERTURBED ANGULAR CORRELATIONS.
FORKER M + HAMMESFAHR A
Z PHYS 255: 196-205 (1972)

07-1256   COMPLEX DIELECTRIC CONSTANT OF LEAD ZIRCONATE UNDER POLARIZING FIELD, 370-520 DEG K.
GOULPEAU L + LEMONTAGNER S
COMPT REND 255: 849-51 (1962) (IN FRENCH)

07-1257   RELAXATION OF ANTIFERROELECTRIC DOMAIN WALLS, EFFECTIVE FIELD, AND ACTIVATION ENERGY.
APPLICATION TO LEAD ZIRCONATE.
GOULPEAU L
UNIV RENNES SCD THESIS, 1970, 108P.
      243

07-1258   ATHERMAL PHASE TRANSITION IN LEAD ZIRCONATE.
HANG KW + COOK RL
BULL AMER CERAM SOC 50: 396 (1971)

07-1259   EFFECT OF PILE NEUTRON IRRADIATION ON THE DIELECTRIC PROPERTIES OF LEAD ZIRCONATE.
HILCZER B + KULEK J
ACTA PHYS POLON A 40: 229-37 (AUG 1971)

          PHYSICAL PHENOMENA IN PIEZOCERAMICS BASED ON LEAD TITANATE AND ZIRCONATE. (SOLID
SOLUTIONS LEAD MAGNESATE(0.5) TUNGSTATE(0.5)), EFFECTS OF POSITIVE OR NEGATIVE HIGH DC
FIELDS SUMMARY ONLY)
ISUPOV VA + STOLYPIN YUE
P312 OF INT MEETING ON FERROELECTRICITY, 2ND, 1969,, KYOTO, PROC,
(J PHYS SOC JAP, VOL 28 SUPPL). PHYS SOC JAP, 1970, 459P.

07-1260   TEMPERATURE DEPENDENT OPTICAL MODE IN ANTIFERROELECTRIC LEAD ZIRCONATE BY THE MOSSBAUER
EFFECT.
JAIN AP + SHRINGI SN + SHARMA ML
PHYS REV B 2: 2756-9 (1970)

          FERROELECTRICITY AND ANTIFERROELECTRICITY IN POLAR CRYSTALS. (THEORY, LEAD ZIRCONATE)
LYUBIMOV VN + VENEVTSEV YUN + ZHDANOV GS
SOV PHYS SOLID STATE 4: 1554-57 (1963)

7. LEAD ZIRCONATE

07-1261 DIPOLE STRUCTURE AND INTERNAL ELECTRIC FIELDS IN LEAD ZIRCONATE.
         LYUBIMOV VN + VENEVTSEV YUN + SOLOVEV SP + ZHDANOV GS + BAKUSHINSKII AB
         SOV PHYS SOLID STATE 4: 2594-9 (1963)

07-1262 ANTIFERROELECTRIC MODE IN LEAD ZIRCONATE. (LATTICE VIBRATIONS)
         MANI KK + SHRINGI SN
         PHYS STATUS SOLIDI B 44: K49-52 (1971)
              235

07-1263 EFFECTS OF HYDROSTATIC PRESSURE ON FERROELECTRIC PROPERTIES. (SUMMARY WITH NEW RESULTS
         ON LEAD ZIRCONATE AND OTHER ANTIFERROELECTRICS)
         SAMARA GA
         P399-403 OF INT MEETING ON FERROELECTRICITY, 2ND, 1969, KYOTO, PROC,
         (J PHYS SOC JAP, VOL 28 SUPPL). PHYS SOC JAP, 1970, 459P.

07-1264 CRYSTAL GROWTH AND OBSERVATION OF THE FERROELECTRIC PHASE OF LEAD ZIRCONATE.
         SCOTT BA + BURNS G
         J AMER CERAM SOC 55: 331-3 (JUL 1972)

07-1265 GROWTH AND DIELECTRIC PROPERTIES OF SINGLE CRYSTAL LEAD ZIRCONATE
         SCOTT BA + BURNS G
         BULL AMER CERAM SOC 50: 755 (1971)

07-1266 HIGH TEMPERATURE (TO 236 DEG C) PHASE TRANSITIONS IN (HIGH PURITY) LEAD ZIRCONATE.
         TENNERY VJ
         J AMER CERAM SOC 49(9): 483-6 (1966)

07-1267 THERMOELECTRIC EFFECT IN SEMICONDUCTIVE LEAD ZIRCONATE.
         UJMA Z + OLECH J + WROBEL Z
         ACTA PHYS POLON A 41: 179-85 (FEB 1972)

8. OTHER PEROVSKITES

08-1268   DIELECTRIC CONSTANTS OF THE SYSTEM SAMARIUM NIOBATE- LANTHANUM NIOBATE.
          CURR SCI 40(15): 400-1 (5 AUG 1971)
              SCI ABSTR A 75: 1559 (1972)
          268

08-1269   EFFECT OF PRESSURE ON THE STATIC DIELECTRIC CONSTANT OF POTASSIUM TANTALATE.
          ABEL WR
          PHYS REV B SOLID STATE 4: 2696-701 (1971)

08-1270   FERROELASTIC EFFECT IN LANTHANUM ORTHOFERRITE.
          ABRAHAMS SC + BARNS RL + BERNSTEIN JL
          SOLID STATE COMMUN 10: 379-81 (1972)
          262

08-1271   PIEZOELECTRIC PROPERTIES OF POTASSIUM TANTALATE NIOBATE SINGLE CRYSTAL.
          ADACHI M + KAWABATA A
          JAP J APPL PHYS 11: 1855 (1972)

08-1272   DIELECTRIC PROPERTIES AND OPTICAL ABSORPTION OF POTASSIUM TANTALATE SINGLE CRYSTALS AND
          (POTASSIUM SODIUM TANTALATE MIXED CRYSTALS (100 HZ-10 MHZ, 30-350 C).
          AGRAWAL MD + RAO KV
          J PHYS C SOLID STATE PHYS 3: 1120-6 (1970)
          208

08-1273   STRUCTURES OF SODIUM NIOBATE BETWEEN 480 AND 575 DEG C, AND THEIR RELEVANCE TO SOFT
          PHONON MODES. (X-RAY DIFFRACTION)
          AHTEE M + GLAZER AM + MEGAW HD
          PHIL MAG 26(4): 995-1014 (1972)
          279

08-1274   DISPLACEMENT PHASE TRANSITIONS IN PEROVSKITES.
          ALEKSANDROV KS + ZINENKO VI + RESHCHIKOVA LM AND OTHERS
          BULL ACAD SCI USSR, PHYS SER 35(9): 1655-60 (1971)

08-1275   CRYSTALLIZATION OF LEAD METANIOBATE FROM GLASS.
          ANDERSON RC + RIEDBERG AL
          P29-34 OF SYMP ON NUCLEATION AND CRYSTALLIZATION IN GLASSES AND MELTS
          1961, TORONTO, PAPERS. RESER MK(ED), AMER CERAM SOC, 1962

              CHEM ABSTR 60: 2612C (1964)

08-1276   NUCLEAR QUADRUPOLE INTERACTIONS IN PEROVSKITE TYPE COMPOUNDS OF HAFNIUM-181 STUDIED BY
          THE PERTURBED ANGULAR CORRELATION TECHNIQUE. (LEAD HAFNATE, TIN HAFNATE, CALCIUM
          HAFNATE, STRONTIUM HAFNATE)
          ANDRADE PR + FORKER M + ROGERS JD + KUNZLER JV
          PHYS REV B 6: 2560-5 (1972)

08-1277   BISMUTH TITANATE SOLID SOLUTIONS.
          ARMSTRONG RA + NEWNHAM RE
          MATER RES BULL 7: 1025-34 (1972)
          276

08-1278   POTASSIUM IODATE CRYSTAL STRUCTURE STUDIED BY INFRARED SPECTROSCOPY AND NUCLEAR
          QUADRUPOLE RESONANCE.
          BAISA DF + BARABASH AI + DEMYANENKO VP + PUCHKOVSKAYA GA + REZ IS
          UKR FIZ ZH 17: 1346-52 (1972) (IN RUSSIAN)

8. OTHER PEROVSKITES

FAR INFRARED DIELECTRIC DISPERSION OF PEROVSKITES. (BARIUM TITANATE, POTASSIUM
TANTALATE)
BALLANTYNE JM
P55-60, VOL 1 OF DVORAK V + FOUSKOVA A + GLOGAR P (EDS),
INT MEETING ON FERROELECTRICITY, PRAGUE, 1966, PROC.
INST PHYS CZECH ACAD SCI, PRAGUE, 1966.

THERMAL CONDUCTIVITY IN PEROVSKITES (3 PHONON INTERACTION PROPOSED FOR STRONTIUM
TITANATE AND POTASSIUM TANTALATE).
BARRETT HH + HOLLAND MG
PHYS REV B 2(8): 3441-3 (1970)

08-1279   HYDROTHERMAL CRYSTALLIZATION AND SOME PROPERTIES OF BISMUTH TITANATES
BARSUKOVA ML + KUZNETSOV VA + LOBACHEV AN + SHALDIN YUV
J CRYST GROW 13-14: 530-4 (1972)

METHOD FOR ACHIEVING HIGH DIELECTRIC CONSTANT TEMPERATURE STABLE FERROELECTRICS.
(BARIUM TITANATE, STRONTIUM STANNATE)
BELDING WA + SMOKE EJ
BULL AMER CERAM SOC 47(4): 390 (APR 1968)

THERMODYNAMIC THEORY OF THE MORPHOTROPIC PHASE TRANSITION TETRAGONAL- RHOMBOHEDRAL IN
THE PEROVSKITE FERROELECTRICS.
BENGUIGUI L
SOLID STATE COMMUN 11: 825-8 (1972)

MICROTHEORY OF SPONTANEOUS POLARIZATION AND PHASE TRANSITIONS IN CRYSTALS OF PEROVSKITE
TYPE AND OF ONES WITH TETRAHEDRAL STRUCTURAL UNITS.
BERSUKER IB + VEKHTER BG + MUZALEVSKII AA
J PHYS (PARIS) 33, SUPPL 4, COLLOQ 2: 139-40 (1972)

08-1280   LIGHT BEAM DEFLECTION CONTROLLED WITH A PRISM OF LEAD MAGNESIO NIOBATE CRYSTALS.
BEREZHNOI AA
OPT SPECTROSC 31: 432-4 (1971)

08-1281   MICROWAVE BEHAVIOR OF NONLINEAR DIELECTRICS (EXPERIMENT AND DISCUSSION OF BARIUM (OR
LEAD OR CALCIUM) STRONTIUM TITANATE, BARIUM TITANATE ZIRCONATE (OR STANNATE), POTASSIUM
TANTALATE NIOBATE, 228 REFS).
BETHE K
PHILLIPS RES REP SUPPL 2, 1970 145P. (IN GERMAN)
(TECH UNIV AACHEN PHD THESIS, 1969)
          202

08-1282   MOSSBAUER MEASUREMENTS OF BISMUTH FERRATE- BISMUTH FERRATE- LEAD ZIRCONATE SYSTEMS.
BIRAN A + MONTANO PA + SHIMONY U
J PHYS CHEM SOLIDS 32: 327-34 (1971)

08-1283   NMR STUDY OF PHASE TRANSITIONS IN PEROVSKITES CRYSTALS. (NIOBIUM-93 AND SODIUM-23
RESONANCE IN POTASSIUM NIOBATE AND SODIUM NIOBATE)
BONERA G + BORSA F + RIGAMONTI A
J PHYS (PARIS) 33, SUPPL 4, COLLOQ 2: 195-7 (1972)
          294

08-1284   EIGENFREQUENCIES AND EIGENVECTORS OF POLARITONS WITH APPLICATION TO LEAD NIOBATE.
PART-1: LONG OPTICAL PHONONS.
BORSTEL G + MERTEN L
Z NATURFORSCH A 26: 653-60 (1971)

BINARY AND PSEUDOBINARY PHASES BETWEEN NIOBIUM PENTOXIDE AND LEAD, BARIUM, AND
STRONTIUM OXIDES.
BRUSSET H + GILLIER-PANDRAUD H + MAHE R + VOLIOTIS SD
MATER RES BULL 6: 413-24 (1971) (IN FRENCH)

08-1285   STRUCTURAL STUDY OF THE TRANSITION OF THE (ORTHORHOMBIC) FERROELECTRIC (LEAD DINIOBATE
INTO THE NONFERROELECTRIC (RHOMBOHEDRIC) FORM.
BRUSSET H + GILLIER-PANDRAUD H + MAHE R
COMPT REND C 270(3): 302-5 (1970) (IN FRENCH)
          CHEM ABSTR 72: 104903Z (1970)
          198

08-1286   STUDY OF DIVALENT BINARY AND TERNARY NIOBATES IN THE SOLID STATE.
BRUSSET H + MAHE R + KYI UA
MATER RES BULL 7: 1061-73 (1972) (IN FRENCH)

8. OTHER PEROVSKITES

08-1287 PRECISION DETERMINATION OF THE LATTICE PARAMETERS AND THE COEFFICIENTS OF THERMAL
         EXPANSION OF BISMUTH FERRATE.
         BUCCI JD + ROBERTSON BK + JAMES WJ
         J APPL CRYSTALLOGR 5: 187-91 (1972)
             269

08-1288 FORBIDDEN ZONE WIDTHS IN POLYCRYSTALLINE DIELECTRICS WITH PEROVSKITE STRUCTURE
         (TITANATES AND ZIRCONATES).
         BUDIM NI + KOZYREVA MS
         INORG MATER 7: 1290-1 (AUG 1971)

08-1289 (CRYSTALLOGRAPHIC AND DIELECTRIC) PROPERTIES OF BISMUTH PEROVSKITES. (AND THE
         FERROELECTRIC STATE, BISMUTH(0.5) (POTASSIUM OR SODIUM(0.5) TITANATES AND PHASE
         DIAGRAMS FOR BISMUTH(X) POTASSIUM(X) BARIUM(1-2X) TITANATE, BISMUTH(0.5) POTASSIUM(X)
         SODIUM(0.5-X) TITANATE, BISMUTH(X) (POTASSIUM OR SODIUM)(X) LEAD (1-2X) ZIRCONATES).
         BUHRER CF
         J CHEM PHYS 36: 798-803 (1962)

         THERMODYNAMIC THEORY OF DOMAIN WALLS IN FERROELECTRIC MATERIALS WITH THE PEROVSKITE
         STRUCTURE.
         BULAEVSKII LN
         SOV PHYS SOLID STATE 5: 2329-32 (1964)

08-1290 ABSOLUTE STRUCTURE OF LITHIUM IODATE CRYSTALS.
         CAMPBELL ID + MATHIESON AMCL
         ACTA CRYSTALLOGR B 25: 1214-15 (1969)
             172

08-1291 (CONTINUOUS WAVE) SPONTANEOUS PARAMETRIC SCATTERING OF LIGHT IN LITHIUM IODATE (INDEX
         OF REFRACTION 1.2-7 MICRONS).
         CAMPILLO AJ + TANG CL
         APPL PHYS LETT 16(6): 242-4 (1970)
             187

         IS THERE A UNIFIED MECHANISM UNDERLYING URBACH'S RULE. (STRONTIUM TITANATE, POTASSIUM
         TITANATE)
         CAPIZZI M + FROVA A + DUNN D
         SOLID STATE COMMUN 10: 1165-9 (1972)

08-1292 STRUCTURE AND DIELECTRIC PROPERTIES OF LITHIUM CONTAINING PEROVSKITES. (STRONTIUM
         LITHIUM NIOBATE, STRONTIUM LITHIUM TANTALATE, STRONTIUM(1-X) LEAD(X) LITHIUM NIOBATE)
         CHAN VT + BADER VI + KRAINIK NN + MYLNIKOVA IE + TUTOV AG
         INORG MATER 8: 1431-3 (1972)
             292

         STRUCTURAL DI- AND FERROELECTRIC PROPERTIES OF X POTASSIUM DINIOBATE AND (1-X) BARIUM
         TITANATE.
         CHINCHOLKAR VS
         Z ANGEW PHYS 28: 288-90 (1970)

08-1293 FERROELECTRIC PROPERTIES OF TETRAGONAL (BARIUM, LEAD) (NIOBATE, ZIRCONATE). (CURIE
         TEMPERATURES)
         CHINCHOLKAR VS + BISWAS AB + MENEZES CA
         INDIAN J PURE APPL PHYS 8(11): 707-9 (1970)
             238

08-1294 POLARITONS IN LITHIUM IODATE.
         CLAUS R
         Z NATURFORSCH A 25(2): 306-7 (1970)
             197

08-1295 DIELECTRIC AGING IN TETRAGONAL SOLID SOLUTIONS OF CALCIUM TITANATE IN BARIUM TITANATE.
         COHEN A + BRADT RC + ANSELL GS
         J AMER CERAM SOC 53: 396-8 (1970)
             199

         LINEAR DISORDER IN CRYSTALS (CASE OF SILICON, QUARTZ AND FERROELECTRIC PEROVSKITES).
         (BARIUM TITANATE AND POTASSIUM NIOBATE)
         COMES R + LAMBERT M + GUINIER A
         ACTA CRYSTALLOGR A 26(2): 244-54 (1970) (IN FRENCH)

08-1296 RELATION OF PHYSICAL PROPERTIES TO THE SYMMETRY OF POTASSIUM IODATE. (NONLINEAR OPTIC,
         REFRACTIVE INDICES)
         CRANE GR
         J APPL CRYSTALLOGR 5: 360-5 (1972)
              279

08-1297 THERMODYNAMICS OF FERROELECTRICITY IN BISMUTH TITANATE.
         CROSS LE
         P56-7 OF INT MEETING ON FERROELECTRICITY, 2ND, 1969, KYOTO, PROC,
         (J PHYS SOC JAP, VOL 28 SUPPL). PHYS SOC JAP, 1970, 459P.

08-1298 A NEW METHOD OF OPTICALLY READING DOMAINS IN BISMUTH TITANATE AND MEMORY APPLICATIONS.
         CUMMINS SE + LUKE TE
         IEEE TRANS ELECTRON DEV 18: 761-8 (1971)

08-1299 EFFICIENT WHITE LIGHT READING OF DOMAIN PATTERNS IN BISMUTH TITANATE.
         CUMMINS SE + LUKE TE
         FERROELECTRICS 3: 125-30 (1972);
         IEEE TRANS SON ULTRASON 19: 125-30 (1972)
              263

08-1300 ELECTRON BEAM WRITING OF FERROELECTRIC DOMAINS IN BISMUTH TITANIUM OXIDE SINGLE
         CRYSTALS.
         CUMMINS SE + HILL BH
         PROC IEEE 58: 938-9 (JUN 1970)
              216

08-1301 OPTICAL STUDY OF DOMAIN WALLS IN BISMUTH TITANATE.
         CUMMINS SE
         P396-8 OF INT MEETING ON FERROELECTRICITY, 2ND, 1969, KYOTO, PROC,
         (J PHYS SOC JAP, VOL 28 SUPPL). PHYS SOC JAP, 1970, 459P.

08-1302 MATERIAL AND ELECTROOPTIC PROPERTIES OF THE (LEAD, LANTHANUM) (HAFNATE, TITANATE)
         SYSTEM.
         CUTCHEN JT + HAERTLING GH
         BULL AMER CERAM SOC 51: 354 (1972)

08-1303 INFRARED AND RAMAN SPECTROSCOPY OF POTASSIUM METANIOBATE.
         DAO NQ + HUSSON E + REPELIN Y
         COMPT REND C 275: 609-12 (1972) (IN FRENCH)

         PHENOMENOLOGICAL THEORY OF ANTIFERROELECTRICITY AND FERROELECTRICITY APPLIED TO SODIUM
         NIOBATE AND THE SYSTEM POTASSIUM NIOBATE- SODIUM NIOBATE.
         DARLINGTON CNW
         J APPL PHYS 43(12): 4951-6 (1972)

08-1304 DIFFUSE SCATTERING IN SODIUM NIOBATE AT ROOM TEMPERATURE.
         DARLINGTON CNW
         FERROELECTRICS 3: 9-15 (1971)
              262

08-1305 MOVEMENT OF 180 DEGREE DOMAIN BOUNDARIES IN PEROVSKITE FERROELECTRICS.
         DARINSKII BM + FEDOSOV VN
         BULL ACAD SCI USSR, PHYS SER 35(9): 1633-5 (1971)

08-1306 DIELECTRIC PROPERTIES AND SOFT MODES IN THE FERROELECTRIC MIXED CRYSTALS POTASSIUM(1-X)
         SODIUM(X) TANTALATE.
         DAVIS TG
         PHYS REV B SOLID STATE 5(7): 2530-7 (1 APR 1972)
              261

08-1307 CHARACTER OF PHASE TRANSITION OF FERROELECTRIC POTASSIUM TANTALATE. (SECOND ORDER)
         DEMUROV DG + VENEVTSEV YUN
         SOV PHYS SOLID STATE 13(3): 553-6 (SEP 1971)
              244

08-1308 POTASSIUM TANTALATE LEAD TITANATE SYSTEM IN THE TEMPERATURE RANGE 4.2-300 DEG K
         (LATTICE PARAMETERS VERSUS CURIE POINTS).
         DEMUROV DG + VENEVTSEV YUN + ZHDANOV GS
         SOV PHYS CRYSTALLOGR 15(3): 498-500 (1970)
              217

08-1309   SYNTHESIS, X-RAY ANALYSIS, AND STUDY OF THE DIELECTRIC PROPERTIES OF NEW
          CADMIUM-CONTAINING PEROVSKITES AND PYROCHLORES.
          DEMUROV DG + VENEVTSEV YUN
          SOV PHYS CRYSTALLOGR 16: 133-5 (JUL-AUG 1971)

08-1310   X-RAY ANALYSIS AND DIELECTRIC PROPERTIES OF SOLID SOLUTIONS BASED ON THE FERROELECTRIC
          POTASSIUM TANTALATE.
          DEMUROV DG + VENEVTSEV YUN + ZHDANOV GS
          SOV PHYS CRYSTALLOGR 16(2): 297-300 (SEP-OCT 1971)
                245

08-1311   ANISOTROPIC CRITICAL X-RAY DIFFUSE SCATTERING FROM SODIUM NIOBATE CRYSTALS (NEAR THE
          640 DEG C PHASE TRANSITICN).
          DENOYER F + COMES R + LAMBERT M
          SOLID STATE COMMUN 8: 1979-1981 (1970)
                220

08-1312   ANISOTROPIC CRITICAL X-RAY DIFFUSE SCATTERING FROM POTASSIUM MANGANESE TRIFLUORIDE AND
          SODIUM NIOBATE SINGLE CRYSTALS.
          DENOYER F + LAMBERT M
          J PHYS (PARIS) 33, SUPPL 4, COLLOQ 2: 132 (1972)
                294

08-1313   X-RAY DIFFUSE SCATTERING FROM SODIUM NIOBATE AS A FUNCTION OF TEMPERATURE.
          DENOYER F + COMES R + LAMBERT M
          ACTA CRYSTALLOGR A 27: 414-20 (1971)
                220

08-1314   ELECTRICAL PROPERTIES AND DEFECT STRUCTURE OF POTASSIUM TANTALATE.
          DEPUTY GO + VEST RW
          BULL AMER CERAM SOC 51: 320 (1972)

08-1315   DOMAINS IN FERROELECTRIC POTASSIUM NIOBATE SINGLE CRYSTALS.
          DESHMUKH KG + INGLE SG
          CURR SCI 39(16): 368-9 (1970)
                225

08-1316   DOMAIN STRUCTURES IN POTASSIUM NIOBATE SINGLE CRYSTALS ASSOCIATED WITH STEP LADDERS ON
          PSEUDOCUBIC (001) PLANES.
          DESHMUKH KG + INGLE SG
          J PHYS D 4: 1633-6 (1971)

08-1317   DOMAINS IN POTASSIUM NIOBATE SINGLE CRYSTALS FROM PSEUDOCUBIC (001) CLEAVAGE PLANES.
          DESHMUKH KG + INGLE SG
          INDIAN J PURE APPL PHYS 10: 29-33 (1972)

08-1318   INTERFEROMETRIC STUDIES OF DOMAIN STRUCTURES IN POTASSIUM NIOBATE SINGLE CRYSTALS.
          DESHMUKH KG + INGLE SG
          J PHYS D APPL PHYS 4: 124-32 (1971)
                237

08-1319   INTERFEROMETRIC STUDIES CF DOMAIN STRUCTURE IN POTASSIUM NIOBATE SINGLE CRYSTALS.
          DESHMUKH KG + INGLE SG
          J PHYS D 4: 124-32 (1971)

08-1320   OPTICAL STUDIES OF DOMAIN STRUCTURE OF POTASSIUM NIOBATE SINGLE CRYSTALS.
          DESHMUKH KG + INGLE SG
          INDIAN J PURE APPL PHYS 10: 353-5 (1972)

          LIFETIME OF THE "FERROELECTRIC" PHONON IN SOME PEROVSKITE CRYSTALS (STRONTIUM TITANATE,
          POTASSIUM TANTALATE) AT LOW TEMPERATURES. (THEORY)
          DVORAK V
          P61-8, VOL 1 OF DVORAK V + FOUSKOVA A + GLOGAR P (EDS),
          INT MEETING ON FERROELECTRICITY, PRAGUE, 1966, PROC.
          INST PHYS CZECH ACAD SCI, PRAGUE, 1966.

08-1321   GROUP ANALYSIS OF LATTICE VIBRATIONS OF CUBIC PEROVSKITE ABO(3).
          DVORAK V
          PHYS STATUS SOLIDI 3: 2235-40 (1963)

08-1322   CIELECTRIC PROPERTIES OF NIOBATES AND TUNGSTATES.
          EMMENEGGER FP + ROETSCHI H
          J PHYS CHEM SOLIDS 32: 787-90 (APR 1971)

8. OTHER PEROVSKITES

08-1323 GROWTH AND EVALUATION OF SINGLE CRYSTAL BISMUTH TITANATE AND RELATED COMPOUNDS.
(GADOLINIUM MOLYBDATE, GROWTH BY RF CZOCHRALSKI AND FROM TOP SEEDED SOLUTION)
EPSTEIN DJ
MIT, AUG 1970, 96P. AD715312
    CHEM ABSTR 75: 54390U (1971) US GOVT RES DEVELOP REP 71(2): 169 (1971)
    215

THEORY OF STRUCTURAL PHASE TRANSITION IN PEROVSKITES.
FEDER J
P171-88 OF NATO ADVAN STUDY INST, GEILO, NORWAY, STRUCTURAL
PHASE TRANSITIONS AND SOFT MODES, SAMUELSEN EJ + ANDERSEN E +
FEDER J(EDS), 1971, UNIVERSITETSFORLAGEN, BOSTON, 1972.

PEROVSKITE FAMILY AND CHEMICAL BOND. (COVALENCY, DIRECTIVITY CONDITIONS FOR ELECTRIC
ORDERING AND STABILITY)
FESENKO EG + SHUVAYEV AT
P64-6 OF INT MEETING ON FERROELECTRICITY, 2ND, 1969,, KYOTO, PROC,
(J PHYS SOC JAP, VOL 28 SUPPL). PHYS SOC JAP, 1970, 459P.

08-1324 CRYSTALLOCHEMISTRY OF PEROVSKITES. (EXISTENCE CONDITIONS AND CLASSIFICATION)
FESENKO EG + FILIPEV VS + KUPRIYANOV MF
P306-13, VOL 1 OF DVORAK V + FOUSKOVA A + GLOGAR P (EDS),
INT MEETING ON FERROELECTRICITY, PRAGUE, 1966, PROC.
INST PHYS CZECH ACAD SCI, PRAGUE, 1966.

08-1325 SYNTHESIS AND STUDY OF LEAD SCANDIUM NIOBATE SINGLE CRYSTALS.
FESENKO EG + GRIGOREVA EA + DANTSIGER AYA + GOLOVKO YUI + DUDKINA SI
BULL ACAD SCI USSR, PHYS SER 34(12): 2287-9 (1970)

08-1326 ELECTRIC FIELD INDUCED RAMAN EFFECT IN PARAELECTRIC CRYSTALS. (POTASSIUM TANTALATE)
FLEURY PA + WORLOCK JM
PHYS REV LETT 18(16): 665-7 (1967)

STRUCTURE DISORDER AND RAMAN SCATTERING IN BARIUM TITANATE AND POTASSIUM NIOBATE.
FONTANA MP + LAMBERT M + QUITTET AM
J PHYS (PARIS) 33, SUPPL 4, COLLOQ 2: 54 (1972)

08-1327 ORIENTATION OF COMPOSITION PLANES IN ANTIFERROELECTRICS. (APPLIED TO ADP, TUNGSTEN
TRIOXIDE, SODIUM NIOBATE)
FOUSEK J + JANOVEC V
P380-1 OF INT MEETING ON FERROELECTRICITY, 2ND, 1969, KYOTO, PROC,
(J PHYS SOC JAP, VOL 28 SUPPL). PHYS SOC JAP, 1970, 459P.

08-1328 OPTICALLY INDUCED PHYSICAL DAMAGE TO LITHIUM NIOBATE, PROUSTITE, AND LITHIUM IODATE.
FOUNTAIN WD + OSTERINK LM + MASSEY GA
P91-7 OF DAMAGE IN LASER MATERIALS, 1971 PROC, GLASS AJ +
GUENTHER AH(EDS), NBS SPEC PUB 356, 1971.

08-1329 GROWTH AND PROPERTIES OF EPITAXIAL FILMS OF FERROELECTRIC BISMUTH TITANATE.
(ELECTROOPTICS)
FRANCOMBE MH + TAKEI WJ + WU SY
RES DEVELOP CENTER, WESTINGHOUSE ELEC CORP, PITTSBURGH, PA, 1970, 56P
AFAL-TR-71-59
    235

INVESTIGATION OF POLARIZATION NONLINEARITY NEAR THE PHASE TRANSITION IN PEROVSKITE LIKE
FERROELECTRICS, USING AN ANHARMONIC OSCILLATOR MODEL.
FRITSBERG VYA
P7-21 OF FAZOVYE PEREKHODY V SEGNETOELEKTRIKAKH (PHASE
TRANSFORMATIONS IN FERROELECTRICS), FRITSBERG VYA + ROLOV BN +
KRUCHAN YY(EDS), IZDATELSTVO ZINATNE RIGA, 1971, 206P. (IN RUSSIAN)

08-1330 INVESTIGATING THE RELATIONS BETWEEN COMPOSITION AND PROPERTIES IN FERROELECTRIC SOLID
SOLUTIONS OF THE PEROVSKITE TYPE.
FRITSBERG VYA
BULL ACAD SCI USSR, PHYS SER 34(12): 2342-9 (1970)

08-1331 HIGH PRESSURE SYNTHESIS OF LEAD METAL- OXYGEN(3) TYPE PEROVSKITE.
FUJITA T + FUKUNAGA O + NAKAGAWA T + NOMURA S
MATER RES BULL 5: 759-64 (1970)
    214

08-1332   PREPARATION OF POTASSIUM NIOBATE SINGLE CRYSTAL FOR OPTICAL APPLICATION.
          FUKUDA T + UEMATSU Y
          JAP J APPL PHYS 11(2): 163-9 (1972)
              278

08-1333   FERROELECTRIC PROPERTIES OF BISMUTH FERRATE (PREPARATION AND STRUCTURE) AND RELATED
          MATERIALS (SYSTEM WITH LEAD ZIRCONATE AND WITH LEAD ZIRCONATE TITANATE MOSSBAUER
          SPECTROSCOPY.
          GERSON R + JAMES WJ
          UNIV MISSOURI, PB192909, JUN 1970, 65P
              212

08-1334   TIGHT BONDING BAND CALCULATIONS ON RHENIUM TRIOXIDE, SODIUM TUNGSTATE, AND POTASSIUM
          TANTALATE.
          GERSTEIN BC + KARIAN HG
          BULL AMER PHYS SOC 15: 311 (1970)

          ELECTROOPTIC PROPERTIES OF SOME ABO(3) PEROVSKITES IN THE PARAELECTRIC PHASE.
          (POTASSIUM TANTALATE AND TANTALATE NIOBATE, BARIUM AND STRONTIUM TITANATES)
          GEUSIC JE + KURTZ SK + VAN UITERT LG + WEMPLE SH
          APPL PHYS LETT 4: 141-3 (1964)

08-1335   EFFECT OF PRESSURE ON THE DIELECTRIC PROPERTIES OF A PARAELECTRIC MATERIAL. (POTASSIUM
          TANTALATE)
          GILLIS NS
          SOLID STATE COMMUN 10: 887-9 (1972)
              268

08-1336   HIGH TEMPERATURE PHASE TRANSITIONS IN SODIUM NIOBATE AND THE USE OF TILTING SCHEMES IN
          THE SOLUTIONS OF PEROVSKITE STRUCTURES.
          GLAZER AM + AHTEE M + MEGAW HD
          ACTA CRYSTALLOGR A 28: S179 (1972)

08-1337   THE STRUCTURE OF SODIUM NIOBATE T(2) AT 600 DEG C, AND THE CUBIC- TETRAGONAL TRANSITION
          IN RELATION TO SOFT PHONCN MODES.
          GLAZER AM + MEGAW HD
          PHIL MAG 25(5): 1119-35 (1972)
              267

08-1338   OPTICAL PARAMETRIC OSCILLATION IN LITHIUM IODATE.
          GOLDBERG LS
          APPL PHYS LETT 17(11): 489-91 (1970)
              225

08-1339   COMPLEX DIELECTRIC CONSTANT OF LEAD HAFNATE, 320-520 DEG K.
          GOULPEAU L
          J PHYS (PARIS) 24: 406-7 (1963) (IN FRENCH)
              CHEM ABSTR 59: 12268D (1963)

08-1340   DIELECTRIC STUDY OF THE LEAD ZIRCONATE LEAD HAFNATE (PHASE) DIAGRAM.
          GOULPEAU L + LE MONTAGNER S + LIMOU P
          COMPT REND 259(5): 1095-7 (1964) (IN FRENCH)
              CHEM ABSTR 61: 15795 (1964)

08-1341   CHARACTER OF THE CHEMICAL BONDS IN ABO(3) FERROELECTRIC CRYSTALS WITH A PEROVSKITE TYPE
          STRUCTURE.
          GRANOVSKII VG
          SOV PHYS CRYSTALLOGR 7: 484-7 (1963)

          RELAXATION AND FERROELECTRIC PROPERTIES OF SOLID SOLUTIONS IN THE SYSTEM BARIUM
          TITANATE- STRONTIUM NIOBIUM(0.8) OXIDE(3).
          GROZNOV IN + VISKOV AS + PETROV VM
          INORG MATER 7: 1078-81 (JUL 1971)

08-1342   EPR OF COBALT(2+) IN POTASSIUM TANTALATE.
          HANNON DM
          PHYS STATUS SOLIDI 43: K21-3 (1971)
              226

08-1343   ELECTRON PARAMAGNETIC RESONANCE OF MANGANESE(2+) IN POTASSIUM TANTALATE.
          HANNON DM
          PHYS REV B 3(7): 2153-8 (1 APR 1971)
              234

08-1344 RAMAN SCATTERING FROM FERROELECTRIC SODIUM NITRITE.
HARDWIG CM + WIENAR EA + PORTO SPS
BULL AMER PHYS SOC 15: 327 (1970)

DETERMINATION OF THE NORMAL VIBRATIONAL DISPLACEMENTS OF SEVERAL PEROVSKITES (POTASSIUM
TANTALATE, STRONTIUM TITANATE, RUBIDIUM MANGANESE TRIFLUORIDE) BY INELASTIC NEUTRON
SCATTERING.
HARADA J + AXE JD + SHIRANE G
BULL AMER PHYS SOC 15: 102 (1970)

08-1345 FERROELECTRICITY OF POTASSIUM IODATE. (UNUSUAL FERROELECTRIC BEHAVIOR)
HELG U
P300-5, VOL 1 OF DVORAK V + FOUSKOVA A + GLOGAR P (EDS),
INT MEETING ON FERROELECTRICITY, PRAGUE, 1966, PROC.
INST PHYS CZECH ACAD SCI, PRAGUE, 1966.

08-1346 POTASSIUM IODATE, A FERROELECTRIC WITH NONREVERSIBLE, BUT TILTABLE (TO 3 STABLE
POSITIONS, SPONTANEOUS) POLARIZATION (ON APPLYING AN ELECTRIC FIELD).
HELG U
Z KRISTALLOGR 131: 241-77 (1970) (IN GERMAN)
        221

08-1347 PECULIAR FERROELECTRIC BEHAVIOR OF POTASSIUM IODATE.
HELG U + GRANICHER H
P169-71 OF INT MEETING ON FERROELECTRICITY, 2ND, 1969, KYOTO, PROC,
(J PHYS SOC JAP, VOL 28 SUPPL). PHYS SOC JAP, 1970, 459P.

08-1348 POTASSIUM IODATE - A DISORDERED FERROELECTRIC. (DEPENDENCE OF DIELECTRIC CONSTANT ON
THERMAL AND FIELD HISTORY)
HELG U
P325-31 OF EUROPEAN MEETING ON FERROELECTRICITY, 1969, SAARBRUCKEN,
MUSER HE AND PETERSSON J (EDS), WISSENSCH VERLAGSGES, STUTTGART,
1970, 396P.

08-1349 RANGE OF EXISTENCE OF PEROVSKITE PHASES IN THE SYSTEM LEAD OXIDE- TITANIUM DIOXIDE-
LANTHANUM TRIOXIDE.
HENNINGS D
MATER RES BULL 6: 329-40 (1971)
        233

08-1350 NUCLEAR QUADRUPOLE RESONANCE, PHASE TRANSFORMATIONS, AND FERROELECTRICITY OF ALKALI
IODATES.
HERLACH F
HELV PHYS ACTA 34: 305-30 (1961) (IN GERMAN)

08-1351 LOW FREQUENCY ZONE BOUNDARY MODES IN PEROVSKITE FERROELECTRICS INDICATED BY ANISOTROPIC
DEBYE- WALLER FACTORS. (POTASSIUM TANTALATE NIOBATE)
HEWAT AW
PHYS STATUS SOLIDI B 53: K33-5 (1972)
        276

08-1352 PREPARATION OF POLED, TWIN-FREE CRYSTALS OF FERROELECTRIC BISMUTH TITANATE, (BY
ANNEALING).
HOPKINS MM + MILLER A
FERROELECTRICS 1: 37-42 (1970)
        191

08-1353 (TOP SEEDED FLUX) CRYSTAL GROWTH AND NEUTRON CHARACTERIZATION OF POTASSIUM NIOBATE.
HURST JJ + LINZ A
MATER RES BULL 6: 163-8 (1971)
        226

08-1354 FORMATION PROCESSES OF LEAD MAGNESIO NIOBATE IN THE SOLID STATE REACTION.
IMOTO F + IIDA H
J CERAM SOC JAP 80: 197-203 (1972) (IN JAPANESE)

08-1355 NATURE OF THE HIGH PERMITTIVITY OF CERIUM ALUMINATES WITH PEROVSKITE TYPE STRUCTURE.
IOFFE VA + LEONOV AI + YANCHEVSKAYA IS
SOV PHYS SOLID STATE 4: 1313-18 (1963)

8. OTHER PEROVSKITES

08-1356 PREPARATION AND PHYSICAL PROPERTIES OF PEROVSKITE TYPE COMPOUNDS, LANTHANUM(1-X)
CALCIUM(X) MANGANATE (PHASE TRANSFORMATIONS CONFIGURATIONAL ENERGY).
ISERENTANT CM + ROBBRECHT GG
P465-76 OF STEWART GH(ED), SCIENCE OF CERAMICS, VOL 4,
BRIT CERAM SOC, 1968.
220

08-1357 ANISOTROPIC DIFFUSE X-RAY SCATTERING AND PHASE TRANSITIONS IN SODIUM NIOBATE.
ISHIDA K + HONJO G
J PHYS SOC JAP 32: 1441 (1972)

08-1358 DIFFUSE X-RAY STUDY OF 640 DEGREES C PHASE TRANSITION IN SODIUM NIOBATE.
ISHIDA K + HONJO G
J PHYS SOC JAP 30: 899-900 (1971)

08-1359 RH DOPED LITHIUM NIOBATE AS AN IMPROVED NEW MATERIAL FOR REVERSIBLE HOLOGRAPHIC
STORAGE.
ISHIDA A + MIKAMI O + MIYAZAWA S + SUMI M
APPL PHYS LETT 21: 192-3 (1972)

08-1360 MAGNETOELECTRIC EFFECT IN FERROELECTRIC- ANTIFERROMAGNETIC BISMUTH TITANIUM FERRATE.
ISMAILZADE IH + YAKUPOV RG + MELIK-SHANAZAROVA TA
PHYS STATUS SOLIDI A 6: K85-7 (1971)
244

08-1361 NEW DATA FROM AN X-RAY STUDY OF PHASE TRANSFORMATIONS IN SODIUM NIOBATE.
ISMAILZADE IG
SOV PHYS CRYSTALLOGR 8: 284-7 (1963)

CAUSES OF PHASE TRANSITION BROADENING (DIFFUSNESS) AND THE NATURE OF DIELECTRIC
POLARIZATION RELAXATION IN SOME FERROELECTRICS (SOLID SOLUTIONS AND MIXED CATION
COMPOUNDS, KANZIG REGIONS, PEROVSKITES)
ISUPOV VA
SOV PHYS SOLID STATE 5: 136-40 (1963)

08-1362 X-RAY DETERMINATION OF THE SYMMETRY OF THE FERROELECTRIC COMPOUNDS (POTASSIUM OR
SODIUM)(0.5) BISMUTH(0.5) TITANATE AND THE HIGH TEMPERATURE PHASE TRANSITIONS IN
POTASSIUM(0.5) BISMUTH(0.5) TITANATE.
IVANOVA VV + KAPYSHEV AG + VENEVTSEV YUN + ZHDANOV GS
BULL ACAD SCI USSR, PHYS SER 26(3): 358-60 (1962)

08-1363 DIPOLE PATTERNS IN ORTHORHOMBIC AND TRIGONAL PHASES OF ABO(3) SUBSTANCE. (WHEN THE B
ION IS FERROELECTRICALLY ACTIVE)
JASKIEWICZ A + KONWENT H
ACTA PHYS POLON 25: 543-50 (1964)

08-1364 FERROELECTRICALLY ACTIVE A-IONS IN ABO(3) STRUCTURES. (ANTIFERROELECTRICITY ORIGINS)
JASKIEWICZ A + KONWENT H
ACTA PHYS POLON 21: 509-21 (1962)

08-1365 ELECTRON TUNNELING INTO POTASSIUM TANTALATE SCHOTTKY BARRIER JUNCTIONS.
JOHNSON KW + OLSON DH
PHYS REV B SOLID STATE 3: 1244-8 (1971)

08-1366 INFRARED ACTIVE MODES IN POTASSIUM TANTALATE.
JOSEPH RI + SILVERMAN BD
J PHYS CHEM SOLIDS 25: 1125-8 (1964)

INFLUENCE OF MADELUNG ENERGY AND COVALENCY ON THE STRUCTURE OF AB(5) OXIDE COMPOUNDS.
(BARIUM TITANATE, RUBIDIUM NIOBATE)
KAFALAS JA
P287-93 OF MATER RES SYMP, 5TH PROC, 1971, SOLID STATE CHEMISTRY,
ROTH RS + SCHNEIDER SJ(EDS), NBS SPEC PUB 364, 1972, 783P.

08-1367 CRYSTALLIZATION OF POTASSIUM IODATE.
KASATKIN AP + PETROV TG + TREIVUS EB
SOV PHYS CRYSTALLOGR 7: 770-1 (1963)

8. OTHER PEROVSKITES

08-1368 ELECTRON SPIN RESONANCE (SPECTRA, TEMPERATURE DEPENDENCE) IN SOME FERROELECTRICS AND
         FERROELECTRIC FERROMAGNETS. (LEAD FERRATE(0.67) TUNGSTATE(0.33) LEAD FERRATE(0.5)
         NIOBATE(0.5), BISMUTH FERRATE AND ITS SOLID SOLUTIONS WITH BARIUM TITANATE, AND
         LANTHANUM FERRATE)
         KASHLINSKII AI + VENEVTSEV YUN + CHECHERNIKOV VI
         P373-7, VOL 2 OF DVORAK V + FOUSKOVA A + GLOGAR P (EDS),
         INT MEETING ON FERROELECTRICITY, PRAGUE, 1966, PROC.
         INST PHYS CZECH ACAD SCI, PRAGUE, 1966.

08-1369 SURFACE CHARGE LAYERS ON FERROELECTRIC POTASSIUM NIOBATE CRYSTALS.
         KATPATAL AG + DESHMUKH KG
         J PHYS D APPL PHYS 5: 1937-44 (1972)
             276

08-1370 NOTE ON THE SPACE GROUP OF POTASSIUM HYDROGEN IODATE.
         KEMPER G + VOS A
         ACTA CRYSTALLOGR B 26(3): 302 (1970)
             216

08-1371 BISMUTH TITANATE FERROELECTRIC- PHOTOCONDUCTOR OPTICAL STORAGE MEDIUM FURTHER
         DEVELOPMENTS.
         KENEMAN SA + MILLER A + TAYLOR GW
         FERROELECTRICS 3: 131-7 (1972);
         IEEE TRANS SON ULTRASON 19: 131-7 (1972)
             263

08-1372 ENERGY BANDS OF POTASSIUM NIOBATE.
         KENNEY JF
         BULL AMER PHYS SOC 15: 1378 (1970)

08-1373 FERROELECTRIC SINGLE CRYSTALS WITH LARGE NONLINEARITY. (BARIUM TITANATE HAFNATE GROWN
         FROM POTASSIUM FLUORIDE FLUX)
         KHODAKOV AL + SHOLOKHOVICH ML
         SOV PHYS DOKL 6: 964-6 (1962)

08-1374 THE INVERSE PIEZOELECTRIC EFFECT IN THE REPOLARIZATION PROCESS OF FERROELECTRICS WITH
         THE PEROVSKITE STRUCTURE.
         KHOTCHENKOV AG
         IZV VUZ FIZ 1971(6): 136-8 (IN RUSSIAN)

08-1375 MICROWAVE INVESTIGATION OF SOME FERROELECTRICS AND ANTIFERROELECTRICS OF PEROVSKITE
         TYPE. (POLYCRYSTALLINE AND SINGLE CRYSTAL LEAD MAGNESATE(0.33) NIOBATE(0.67), AND LEAD
         FERRATE(0.5) NIOBATE(0.5))
         KHUCHUA NP
         P161-70, VOL 2 OF DVORAK V + FOUSKOVA A + GLOGAR P (EDS),
         INT MEETING ON FERROELECTRICITY, PRAGUE, 1966, PROC.
         INST PHYS CZECH ACAD SCI, PRAGUE, 1966.

         STRUCTURE OF POLYCRYSTALLINE SOLID SOLUTIONS OF STRONTIUM AND BISMUTH TITANATES.
         KISELEVA KV + BOGDANOV SV
         SOV PHYS SOLID STATE 5(11): 2294-6 (MAY 1964)

         EFFECT OF LITHIUM OXIDE ON THE PROPERTIES OF A SERIES OF FERROELECTRIC MATERIALS.
         (BARIUM TITANATE, LEAD FERRATE(0.5) NIOBATE(0.5))
         KLIMOV VV + DIDKOVSKAYA CS + ZVONIK VA
         INORG MATER 6: 161-2 (1970)

08-1376 X-RAY STUDY OF THE SYSTEM BISMUTH OXIDE FERRIC OXIDE.
         KOIZUMI H + NIIZEKI N + IKEDA T
         JAP J APPL PHYS 3: 495-6 (1964)

08-1377 RELATIONSHIP BETWEEN THE ATOMIC DISPLACEMENTS AND SPONTANEOUS DEFORMATION IN THE
         FERROELECTRIC PHASES OF OXIDES WITH THE PEROVSKITE STRUCTURE.
         KOLESOVA RV + FESENKO EG + SAKHNENKO VP
         SOV PHYS CRYSTALLOGR 16(2): 301-4 (SEP-OCT 1971)
             246

8. OTHER PEROVSKITES

08-1378 DIELECTRIC PROPERTIES AND THERMAL EXPANSION OF SOME FERROELECTRICS AND
ANTIFERROELECTRICS WITH HIGH CURIE TEMPERATURES. (BISMUTH FERRATE AND SYSTEMS BASED ON
IT, AND SINGLE CRYSTAL LITHIUM NIOBATE)
KRAINIK NN + KHUCHUA NP + ZHDANOVA VV + MYLNIKOVA IE + PARFENOVA NN
P377-86, VOL 1 OF DVORAK V + FOUSKOVA A + GLOGAR P (EDS),
INT MEETING ON FERROELECTRICITY, PRAGUE, 1966, PROC.
INST PHYS CZECH ACAD SCI, PRAGUE, 1966.

08-1379 ORDERING OF MAGNESIUM AND NIOBIUM IN THE OCTAHEDRAL POSITIONS OF THE "CUBIC" PEROVSKITE
STRUCTURE OF LEAD MAGNESIUM NIOBATE.
KRAUSE HB + GIBBON DL
Z KRISTALLOGR 134: 44-53 (1971)

08-1380 INVESTIGATION OF THE DYNAMICS OF POLARIZATION OF FERROELECTRIC SOLID SOLUTIONS OF THE
PEROVSKITE TYPE WITHIN A WIDE RANGE OF TEMPERATURES AND ELECTRIC FIELD INTENSITIES AND
TIMES OF APPLICATION.
KRUMIN AE
P155-206 OF FAZOVYE PEREKHODY V SEGNETOELEKTRIKAKH (PHASE
TRANSFORMATIONS IN FERROELECTRICS), FRITSBERG VYA + ROLOV BN +
KRUCHAN YY (EDS), IZDATELSTVO ZINATNE RIGA, 1971, 206P. (IN RUSSIAN)

08-1381 DIELECTRIC CONSTANT OF POTASSIUM NIOBATE SINGLE CRYSTALS UNDER BIASING CONDITIONS.
KULKARNI RH + INGLE SG
J PHYS D APPL PHYS 5: 1474-7 (1972)
271

08-1382 AN X-RAY STUDY OF THE PHASE TRANSITION IN STRONTIUM FERRATE TANTALATE
KUPRIYANOV MF + FESENKO EG
SOV PHYS CRYSTALLOGR 7: 246-7 (1962)

08-1383 VISIBLE AND ULTRAVIOLET OPTICAL PROPERTIES (REFLECTANCE SPECTRA AND BAND STRUCTURE) OF
SOME ABO(3) FERROELECTRICS. (PHENOMENOLOGICAL MODEL OF ELECTROOPTICAL EFFECT)
KURTZ SK
P413-23, VOL 1 OF DVORAK V + FOUSKOVA A + GLOGAR P (EDS),
INT MEETING ON FERROELECTRICITY, PRAGUE, 1966, PROC.
INST PHYS CZECH ACAD SCI, PRAGUE, 1966.

08-1384 DIELECTRIC, PIEZOELECTRIC AND PYROELECTRIC PROPERTIES OF THE LEAD DINIOBATE- BARIUM
DINIOBATE SYSTEM.
LANE R + MACK DL + BROWN KR
TRANS J BRIT CERAM SOC 71: 11-22 (JAN 1972)

08-1385 DIELECTRIC, PIEZOELECTRIC AND PYROELECTRIC PROPERTIES OF THE LEAD METANIOBATE- BARIUM
NIOBATE SYSTEM.
LANE R + MACK D + BROWN KR
ADMIRALTY MATER LAB, 1971. AD892373

08-1386 CRYSTALLIZATION OF SOLID SOLUTIONS OF SODIUM NIOBATE FROM GLASS.
LAYTON MM
BULL AMER CERAM SOC 50: 756 (1971)

08-1387 HIGH TEMPERATURE X-RAY INVESTIGATION OF THE PEROVSKITE MODIFICATION OF CADMIUM TITANATE
(AND THE ANTIFERROELECTRIC- FERROELECTRIC PHASE CHANGE).
LEBEDEV VM + VENEVTSEV YUN + ZHDANOV GS
SOV PHYS CRYSTALLOGR 15(2): 318-20 (1970)
207

08-1388 RADIATION AND ACOUSTIC EXCITATION ON ANOMALOUS LAYERS OF PEROVSKITE FERROELECTRICS.
LEFKOWITZ I
NATURE 198: 657-9 (1963)

SEIGNETTO MAGNETIC SOLID SOLUTIONS OF THE LEAD FERRATE(0.67) TUNGSTATE(0.33)- LEAD
TITANATE SYSTEM.
LEVINA SS + PARASHCHUKOV MP
IZV VUZ FIZ 1972(8): 131-3 (IN RUSSIAN)

08-1389 INTERNAL ELECTRIC FIELDS IN CRYSTALS OF SODIUM TANTALATE AND CADMIUM TITANATE.
LYUBIMOV VN + VENEVTSEV YUN + ZHDANOV GS
SOV PHYS CRYSTALLOGR 7: 9-14 (1962)

8. OTHER PEROVSKITES

08-1390 ORIGIN OF DIPOLE CONFIGURATIONS IN SOME (PEROVSKITE) STRUCTURES WITH SPECIAL DIELECTRIC
         PROPERTIES.
         LYUBIMOV VN + VENEVTSEV YUN
         SOV PHYS CRYSTALLOGR 7: 641-3 (1963)

         COUPLED MODES (OPTICAL, FERROELECTRIC- ACOUSTIC) IN THE PHONON SPECTRA OF STRONTIUM
         TITANATE AND POTASSIUM TANTALATE (LATTICE DYNAMICS).
         MANTE AJH
         SOLID STATE COMMUN 8: 1415-17 (1970)

08-1391 RAMAN SPECTRUM OF POTASSIUM TANTALATE NIOBATE. (POTASSIUM TANTALATE(0.65)
         NIOBATE(0.35), CURIE TEMPERATURE EQUALS 10 DEG C, ABSTRACT)
         MANLIEF SK + FAN HY
         BULL AMER PHYS SOC 15(12): 1623 (1970)
             218

08-1392 GROWTH OF IODIC ACID AND IODATE CRYSTALS BY A SOLUTION GROWTH METHOD.
         MATSUMURA S
         OYO BUTSURI 39: 604-9 (1970) (IN JAPANESE)

08-1393 HIGH PRESSURE SYNTHESIS (SOLID STATE REACTION, SINTERING) AND PROPERTIES OF PEROVSKITE
         TYPE LEAD ZINCATE(0.33) NIOBATE(0.67).
         MATSUO Y + SASAKI H + HAYAKAWA S + KANAMARA F + KOIZUMI M
         P410-12 OF INT MEETING ON FERROELECTRICITY, 2ND, 1969, KYOTO, PROC,
         (J PHYS SOC JAP, VOL 28 SUPPL). PHYS SOC JAP, 1970, 459P.

08-1394 VIBRATION FREQUENCIES OF TITANATES.
         MATOSSI F
         ANN PHYS 11(1-6): 22-8 (1963) (IN GERMAN)
             CHEM ABSTR 60: 116F (1964)

08-1395 STRUCTURE OF SODIUM NIOBATE HETEROTYPES.
         MEGAW HD
         P314-21, VOL 1 OF DVORAK V + FOUSKOVA A + GLOGAR P (EDS),
         INT MEETING ON FERROELECTRICITY, PRAGUE, 1966, PROC.
         INST PHYS CZECH ACAD SCI, PRAGUE, 1966.

08-1396 STRUCTURE AND TRANSITIONS IN PEROVSKITES. (ORTHOFERRITES, SODIUM NIOBATE)
         MEGAW HD
         J PHYS (PARIS) 33, SUPPL 4, COLLOQ 2: 1-5 (1972)
             294

08-1397 TRANSFORMATIONS IN SODIUM NIOBATE.
         MEGAW HD
         ACTA CRYSTALLOGR A 25: S227 (1969)

08-1398 ATOMIC STRUCTURE IN THE PEROVSKITIC FERRATE.
         MICHEL C + MOREAU JM + GERSON R + JAMES WJ
         INT CONGR CRYSTALLOGR, 8TH, 1969, NY, ABSTR. AMER INST PHYS, 1969.
             SCI ABSTR B 73: 61842 (1970)
             219

08-1399 DETERMINATION OF THE ELASTIC CONSTANTS OF POTASSIUM NITRATE.
         MICHARD F + PLICQUE F
         COMPT REND B 272: 848-50, 1159 (1971) (IN FRENCH)

08-1400 CHANGE OF RESONANCE ABSORPTION SPECTRA OF 23.8 KEV GAMMA-RAYS OF TIN-119 DURING PHASE
         TRANSITIONS OF THE BISMUTH FERRATE STRONTIUM STANNATE(0.33) MANGANATE(0.67) SYSTEM.
         MITROFANOV KP + VISKOV AS + DRIKER GYA + POLINKOVA MV + HIEN PZ + VENEVTSEV YUN
         + SHPINEL VS
         SOV PHYS JETP 19: 260-2 (1964)

08-1401 MOSSBAUER STUDY AND THEORETICAL ESTIMATION OF INTERNAL ELECTRIC FIELD GRADIENTS IN
         FERROELECTRIC FERROMAGNETS. (PEROVSKITES)
         MITROFANOV KP + PLOTNIKOVA MV + VISKOV AS + TOMASHPOLSKII YUYA + VENEVTSEV YUN
         + SHPINEL VS
         P87-91, VOL 1 OF DVORAK V + FOUSKOVA A + GLOGAR P (EDS),
         INT MEETING ON FERROELECTRICITY, PRAGUE, 1966, PROC.
         INST PHYS CZECH ACAD SCI, PRAGUE, 1966.

8. OTHER PEROVSKITES

08-1402   FERROELECTRIC BISMUTH FERRATE X-RAY AND NEUTRON DIFFRACTION STUDY. (ANTIFERROMAGNETIC)
MOREAU JM + MICHEL C + GERSON R + JAMES WF
J PHYS CHEM SOLIDS 32: 1315-20 (1971)
241

08-1403   CRYSTAL CHEMISTRY OF THE ABX(3) FAMILY WITH PARTICULAR EMPHASIS ON PEROVSKITES.
MULLER O + ROY R
BULL AMER CERAM SOC 50: 372 (1971)

08-1404   FERROELECTRIC PROPERTIES OF STRONTIUM NIOBATE SINGLE CRYSTAL.
NANAMATSU S + KIMURA M + DOI K AND OTHERS
J PHYS SOC JAP 30: 300-1 (JAN 1971)

08-1405   LARGE NONLINEAR OPTICAL COEFFICIENT AND PHASE MATCHED SECOND HARMONIC GENERATION IN
LITHIUM IODATE.
NATH G + HAUSSUEHL S
APPL PHYS LETT 14(5): 154-6 (1969)
154

PROPOSED MECHANISM FOR THERMOPHOTOTROPIC BEHAVIOR IN PEROVSKITE STRUCTURED TITANATES.
NEUDER SM
NASA GODDARD SPACE FLIGHT CENTER, GREENBELT, MD, 1964, 17P. N64-17296
CHEM ABSTR 61: 14014 (1964) SCI TECH AEROSP REP 2(9): 1177 (1964)

08-1406   STRUCTURAL BASIS OF FERROELECTRICITY IN THE BISMUTH TITANATE FAMILY.
NEWNHAM RE + WOLFE RW + DORRIAN JF
MATER RES BULL 6: 1029-39 (OCT 1971)

08-1407   NEUTRON STUDY OF DIFFUSE SCATTERING IN CUBIC POTASSIUM NIOBATE.
NUNES AC + AXE JD + SHIRANE G
FERROELECTRICS 2(4): 291-7 (1971)

08-1408   PHOTOCONDUCTION OF POTASSIUM TANTALATE.
OHI K + TAKAHASHI S
J PHYS SOC JAP 31: 614 (AUG 1971)

08-1409   CRYSTAL GROWTH OF SODIUM NIOBATE.
OKA K + UNOKI H + SAKUDO T
BULL ELECTROTECH LAB 36: 851-9 (1972) (IN JAPANESE)

08-1410   EFFECT OF INTENSE LIGHT ON THE REFRACTIVE INDEX OF A CRYSTALLINE POTASSIUM NIOBATE.
OSTROWSKY D + ROYER O + SPITZ E
COMPT REND B 270(6): 415-16 (1970) (IN FRENCH)
CHEM ABSTR 72: 126767K (1970)
198

08-1411   INFRARED AND RAMAN SPECTRA OF SEVERAL PEROVSKITE ZIRCONATES.
PASTO AE + CONDRATE RA
BULL AMER CERAM SOC 50: 365 (1971)

08-1412   SPACE CHARGE LIMITED CURRENTS IN SEMICONDUCTING POTASSIUM NIOBATE CRYSTALS.
PAUL R
KRISTALL TECH 6: 405-16 (1971) (IN GERMAN)

08-1413   MEASUREMENT OF THE RELATIVE NONLINEAR COEFFICIENTS OF POTASSIUM DIHYDROGEN PHOSPHATE,
RUBIDIUM DIHYDROGEN PHOSPHATE, RUBIDIUM DIHYDROGEN ARSENATE, AND LITHIUM IODATE.
PEARSON JE + EVANS GA + YARIV A
OPT COMMUN 4(5): 366-7 (JAN 1972)
262

DIELECTRIC PROPERTIES AND OPTICAL PHONONS IN PARA- AND FERROELECTRIC PEROVSKITES.
(THEORY, ABO(3) TYPE)
PERRY CH
P557-91 OF MOLLER KD + ROTHSCHILD WG(EDS), FAR INFRARED
SPECTROSCOPY, INTERSCIENCE, 1971, 797P.

08-1414   DIELECTRIC PROPERTIES OF TITANATE- ALUMINA CERAMICS.
PERRY GS
P63-6 OF CONF ON DIELEC MATER. MEAS AND APPLIC, UNIV OF LANCASTER,
1970. IEE, LONDON, 1970, 339P.

PAGE   120

8. OTHER PEROVSKITES

08-1415 PHOTOEFFECTS IN POTASSIUM TANTALATE.
PETERSEN PE
BULL AMER PHYS SOC 17: 673 (1972)

08-1416 ELASTIC CONSTANT OF ORTHORHOMBIC POTASSIUM NIOBATE BY X-RAY DIFFUSE SCATTERING.
PHATAK SD + SRIVASTAVA RC + SUBBARAO EC
ACTA CRYSTALLOGR A 28: 227-31 (1972)

08-1417 ELECTRICAL CONDUCTIVITY OF STRONTIUM ZIRCONATE AND HAFNATE AT HIGH TEMPERATURES.
PIVOVAR TL + TOLSTAYA-BELIK VYA
HIGH TEMP 8(6): 1227-8 (1970)

08-1418 HIGH FIELD POLARIZATION REVERSAL IN POLYCRYSTALLINE FERROELECTRICS. (8 SUBSTITUTED LEAD
TITANATE ZIRCONATES, 2 SUBSTITUTED LEAD ZIRCONATE STANNATES, COMPLEX PEROVSKITES)
PLUMLEE RH
P358-9 OF INT MEETING ON FERROELECTRICITY, 2ND, 1969, KYOTO, PROC.
(J PHYS SOC JAP, VOL 28 SUPPL). PHYS SOC JAP, 1970, 459P.

08-1419 THERMODYNAMICS OF FERROELECTRICITY IN BISMUTH TITANATE.
POHANKA RC + CROSS LE
BULL AMER CERAM SOC 49: 417 (1970)

08-1420 DEFORMATION AND RELAXATION BEHAVIOR IN FERROELECTRIC POTASSIUM NIOBATE SINGLE CRYSTALS.
PRASAD VCS + SUBBARAO EC
MATER SCI ENG 10: 297-9 (1972)

08-1421 MECHANICAL SWITCHING OF 60 DEGREE DOMAIN WALLS IN FERROELECTRIC POTASSIUM NIOBATE
SINGLE CRYSTALS.
PRASAD VCS + SUBBARAO EC
SOLID STATE COMMUN 10: 811-14 (1972)
           268

08-1422 COEXISTENCE OF THE ANTIPOLARIZED AND POLARIZED STATE IN BISMUTH TITANATE.
PULVARI CF
P392-5 OF INT MEETING ON FERROELECTRICITY, 2ND, 1969, KYOTO, PROC.
(J PHYS SOC JAP, VOL 28 SUPPL). PHYS SOC JAP, 1970, 459P.

THEORY OF PEROVSKITE FERROELECTRICS (HAMILTONIAN IN TERMS OF LOCALIZED D-STRAIN AND
SOFT- NORMAL MODE COORDINATES AND TEMPERATURE INDEPENDENT PARAMETERS).
PYTTE E
PHYS REV B SOLID STATE 5: 3758-69 (MAY 1, 1972)

THEORY OF SECOND ORDER PHASE TRANSITIONS AT LOW TEMPERATURES. (TIN TELLURIDE, POTASSIUM
TANTALATE)
RECHESTER AB
SOV PHYS JETP 33: 423-30 (1971)

08-1423 ELECTROOPTICAL PROPERTIES OF LITHIUM IODATE SINGLE CRYSTALS.
REZ IS + MEISNER LB + NEFOMNYASHCHAYA VN + MARCHENKO EI + FILIMONOV AA + KALITINA MP
BULL ACAD SCI USSR, PHYS SER 34(12): 2285-7 (1970)

08-1424 ABOUT THE CURIE POINT AND NATURE OF DIELECTRIC PROPERTIES IN PEROVSKITE MODIFICATION OF
CADMIUM TITANATE. (SOLID SOLUTIONS WITH STRONTIUM TITANATE LITHIUM TANTALATE, LITHIUM
NIOBATE)
RODICHEVA EN + LEBEDEV VM + SHAPIRO ZI + FEDULOV SA + VENEVTSEV YUN
P370-6, VOL 1 OF DVORAK V + FOUSKOVA A + GLOGAR P (EDS),
INT MEETING ON FERROELECTRICITY, PRAGUE, 1966, PROC.
INST PHYS CZECH ACAD SCI, PRAGUE, 1966.

08-1425 COEXISTENCE OF ANTIFERROMAGNETIC AND SPECIAL DIELECTRIC PROPERTIES IN THE BISMUTH-
LANTHANUM FERRATE SYSTEM.
ROGINSKAYA YUE + VENEVTSEV YUN + ZHDANOV GS
SOV PHYS JETP 17: 954-5 (1963)

08-1426 NUCLEAR QUADRUPOLE RESONANCE OF IODINE-127 IN SILVER PERIODATE.
ROOS J + KIND R
HELV PHYS ACTA 45: 867 (1972) (IN GERMAN)

8. OTHER PEROVSKITES

ELECTRIC FIELD INDUCED OPTICAL HARMONIC GENERATION IN STRONTIUM TITANATE AND POTASSIUM TANTALATE.
SAKUDO T + FUJII Y
P87-9 OF INT MEETING ON FERROELECTRICITY, 2ND, 1969, KYOTO, PROC,
(J PHYS SOC JAP, VOL 28 SUPPL). PHYS SOC JAP, 1970, 459P.

08-1427 ELECTRICAL AND OPTICAL CONSTANTS OF FERROELECTRIC POTASSIUM IODATE.
SALJE E
Z KRISTALLOGR 134(1-2): 107-15 (1971) (IN GERMAN)

08-1428 GROWTH OF SINGLE CRYSTALS OF VARIOUS PEROVSKITES WITH COMPLEX COMPOSITIONS, AND THE STUDY OF THEIR STRUCTURES AND DIELECTRIC PROPERTIES.
SALNIKOV VD + TITOVA SA + TOMASHPOLSKII YUYA + VENEVTSEV YUN + BOZHEVOLNOV EA
SOV PHYS CRYSTALLOGR 16(5): 916-17 (MAR-APR 1972)
    261

08-1429 GROWTH AND STUDY OF NEW SINGLE CRYSTAL CADMIUM COMPOUNDS WITH PEROVSKITE STRUCTURE.
SALNIKOV VD + VENEVTSEV YUN
BULL ACAD SCI USSR, PHYS SER 35(9): 1672-4 (1971)

08-1430 POLAR AND LATTICE DYNAMIC CHARACTERISTICS OF FERROELECTRIC POTASSIUM IODATE CRYSTALS.
SALJE E
TECH HOCHSCHULE HANNOVER PHD THESIS, 1971, 57P. N72-28769
(IN GERMAN)
    SCI TECH AEROSP REP 10(19): 2605 (1972)
    278

08-1431 PIEZOELECTRIC AND ELECTROOPTIC MODULI OF FERROELECTRIC POTASSIUM IODATE.
SALJE E
Z KRISTALLOGR 136: 135-43 (1972) (IN GERMAN)
    286

08-1432 SOME FATIGUING EFFECTS IN 8/65/35 PLZT FINE GRAINED FERROELECTRIC CERAMIC.
SALANECK WR
FERROELECTRICS 4: 97-101 (1972)

08-1433 PRESSURE AND TEMPERATURE DEPENDENCE OF THE DIELECTRIC PROPERTIES AND PHASE TRANSITIONS OF THE ANTIFERROELECTRIC PEROVSKITES LEAD ZIRCONATE AND LEAD HAFNATE.
SAMARA GA
PHYS REV B 1(9): 3777-86 (1970)
    207

08-1434 PRESSURE AND TEMPERATURE DEPENDENCE OF THE DIELECTRIC PROPERTIES AND PHASE TRANSITIONS OF THE FERROELECTRIC PEROVSKITES: LEAD TITANATE AND BISMUTH TITANATE.
SAMARA GA
FERROELECTRICS 2(4): 277-89 (1971)

08-1435 PRESSURE STUDIES OF FERROELECTRIC PROPERTIES. (PEROVSKITES, KDP TYPE)
SAMARA GA + DRICKAMER HG
COMMENT SOLID STATE PHYS 3: 1-7 (1970)

08-1436 DIELECTRIC BEHAVIOR (PERMITTIVITY, SPONTANEOUS POLARIZATION, COERCIVE FIELD) OF BISMUTH TITANATE AT LOW TEMPERATURE (5-350K).
SAWAGUCHI E + CROSS LE
MATER RES BULL 5: 147-52 (1970)
    180

08-1437 CALCIUM CONCENTRATION VS NET IONIZED DONOR CONCENTRATION IN SINGLE CRYSTAL POTASSIUM TANTALATE. (CONDUCTIVE, KYROPOULOS GROWN)
SENHOUSE LS + DE PAOLIS MV + LOOMIS TC
APPL PHYS LETT 8(7): 173-4 (1966)

08-1438 PREPARATION AND INVESTIGATION OF FERROELECTRICS OF ILMENITE STRUCTURE. (ABO(3) NIOBATES, TANTALATES, AND MIXTURES WITH THEM)
SHAPIRO ZI + FEDULOV SA + RIGERMAN LG + VENEVTSEV YUN
P277-84, VOL 1 OF DVORAK V + FOUSKOVA A + GLOGAR P (EDS),
INT MEETING ON FERROELECTRICITY, PRAGUE, 1966, PROC.
INST PHYS CZECH ACAD SCI, PRAGUE, 1966.

8. OTHER PEROVSKITES

TEMPERATURE DEPENDENT SPECTRUM OF AN ANTIFERROELECTRIC LINEAR CHAIN MODEL.
(PEROVSKITES)
SILVERMAN BD
PHYS REV 128: 638-45 (1962)

SEMICONDUCTOR PROPERTIES OF BARIUM (TITANATE, STANNATE) SOLID SOLUTIONS DOPED BY
TUNGSTEN TRIOXIDE.
SINYAKOV EV + KOLESNICHENKO KA
P69-74 OF SEGNETOELEKTRIKAKH I OKISNYE POLUPROVODNIKI (FERROELECTRICS
AND OXIDE SEMICONDUCTORS), SINYAKOV EV(ED), REDAKTSIONNO-IZDATELSKII
OTDEL DNEPROPETROVSKII GOSUDARSTVENNYI UNNIVERSITET, DNEPROPETROVSK,
1971, 100P. (IN RUSSIAN)

08-1439 INTERATOMIC DISTANCES IN OXIDES WITH THE PEROVSKITE STRUCTURE.
SKHNENKO VP + FESENKO EG + SHUVAEV AT + SHUVAEVA ET + GEGUZINA GA
SOV PHYS CRYSTALLOGR 17: 268-73 (1972)

PREPARATION AND CHARACTERIZATION OF ALKOXY-DERIVED STRONTIUM ZIRCONATE AND STRONTIUM
TITANATE (OF HIGH PURITY AND SURFACE ACTIVITY VIA HYDROLYTIC DECOMPOSITION).
SMITH JS + DOLLCFF RT + MAZDIYASNI KS
J AMER CERAM SOC 53: 91-5 (1970)

08-1440 A MICROSCOPIC THEORY FOR THE DIELECTRIC PROPERTIES OF LEAD MAGNESIUM NIOBATE.
SMITH JW + CROSS LE
FERROELECTRICS 1: 137-40 (1970)

08-1441 ANTIFERROMAGNETIC PROPERTIES OF SOME PEROVSKITES. (BISMUTH FERRATE)
SMOLENSKII GA + YUDIN VM + SHER ES + STOLYPIN YUE
SOV PHYS JETP 16: 622-4 (1963)

08-1442 CALCULATIONS OF THE INTERNAL ELECTRIC FIELDS AND ELECTRIC FIELD GRADIENTS IN THE
PEROVSKITE TYPE COMPOUNDS WITH SPECIAL DIELECTRIC PROPERTIES.
SOLOVEV SP + LUBIMOV VN + VENEVTSEV YUN + ZDANOV GS
ACTA CRYSTALLOGR 16(SUPPL): A187-8 (1963) (ABSTRACT ONLY)

08-1443 X-RAY AND DIELECTRIC STUDIES OF IRRADIATED PEROVSKITE TYPE COMPOUNDS.
SOLOVEV SP + KUZMIN II + ZAKURKIN VV + DUDAREV VYA
J PHYS (PARIS) 33: 443-5 (APR 1972)

ELECTRON TUNNELING AND BAND STRUCTURE OF (NIOBIUM DOPED) STRONTIUM TITANATE AND
(CALCIUM DOPED) POTASSIUM TANTALATE (SCHOTTKY BARRIERS WITH INDIUM).
SROUBEK Z
PHYS REV B 2(8): 3170-5 (1970)

THERMAL CONDUCTIVITY OF STRONTIUM TITANATE, POTASSIUM TANTALATE AND BARIUM TITANATE
NEAR THE CURIE POINTS.
TAKEMURA M + TANI K
J PHYS SOC JAP 31: 151-4 (JUL 1971)

08-1444 IONIC CONDUCTION IN PEROVSKITE TYPE OXIDE SOLID SOLUTION AND ITS APPLICATION TO THE
SOLID ELECTROLYTE FUEL CELL. ((CALCIUM, STRONTIUM) LANTHANUM TITANATE- ALUMINATES)
TAKAHASHI T + IWAHARA H
ENERG CONVER 11: 105-11 (1971)

ELECTRON MICROSCOPIC STUDIES ON FERROELECTRIC DOMAINS OF PEROVSKITE TYPE OXIDES. (LEAD
TITANATE, POTASSIUM NIOBATE, BARIUM TITANATE)
TANAKA M + YATSUHASHI T + HONJO G
P386-8 OF INT MEETING ON FERROELECTRICITY, 2ND, 1969, KYOTO, PROC,
(J PHYS SOC JAP, VOL 28 SUPPL). PHYS SOC JAP, 1970, 459P.

08-1445 BISMUTH TITANATE PHOTOCONDUCTOR OPTICAL STORAGE MEDIUM. (HIGH RESOLUTION READ WRITE)
TAYLOR GW + MILLER A + KENEMAN SA
RCA CORP, 1970, 180P. AFAL-TR-70-61
          191

08-1446 DEPOLING OF SINGLE DOMAIN BISMUTH TITANATE.
TAYLOR GW + KENNEMAN SA + MILLER A
FERROELECTRICS 2: 11-20 (JAN 1971)

08-1447 FEASIBILITY OF ELECTROOPTIC DEVICES UTILIZING FERROELECTRIC BISMUTH TITANATE.
TAYLOR GW + MILLER A
PROC IEEE 58: 1220-9 (1970)

8. OTHER PEROVSKITES

08-1448   SELF REVERSAL AND WAITING TIME EFFECTS IN SINGLE CRYSTAL BISMUTH TITANATE.
TAYLOR GW
P389-91 OF INT MEETING ON FERROELECTRICITY, 2ND, 1969, KYOTO, PROC,
(J PHYS SOC JAP, VOL 28 SUPPL). PHYS SOC JAP, 1970, 459P.

     217

08-1449   DIELECTRIC HYSTERESIS IN (FLUX GROWN) SINGLE CRYSTAL BISMUTH FERRATE (CONFIRMS FERROELECTRIC NATURE).
TEAGUE JR + GERSON R + JAMES WJ
SOLID STATE COMMUN 8: 1073-74 (1970)
     205

08-1450   STRUCTURAL PHASE TRANSITIONS IN PEROVSKITE TYPE CRYSTALS. (EXTENDED ABSTRACT)
THOMAS H + FEDER J + PYTTE E + MUELLER KA
P23-5 OF EUROPEAN MEETING ON FERROELECTRICITY, 1969, SAARBRUCKEN,
MUSER HE AND PETERSSON J(EDS), WISSENSCH VERLAGSGES, STUTTGART,
1970, 396P.

08-1451   SUPERSTRUCTURE OF BISMUTH METAL OXYGEN(3) PEROVSKITES. (WHERE METAL IS SCANDIUM, CHROMIUM, MANGANESE, IRON, COBALT, NICKEL, YTTRIUM)
TOMASHPOLSKII YUYA + VENEVTSEV YUN
SOV PHYS CRYSTALLOGR 16(5): 905-8 (MAR-APR 1972)
     261

08-1452   STRUCTURE AND DIELECTRIC LOSSES OF FERROELECTRIC PEROVSKITES SYNTHESIZED AT HIGH PRESSURES.
TOMASHPOLSKII YUYA + VENEVTSEV YUN + ZUBOVA YEV
INORG MATER 6(8): 1362-4 (AUG 1970)
     234

08-1453   SYNTHESIS OF PEROVSKITE COMPLEX COMPOSITIONS: LEAD MANGANATE(0.5) NIOBATE(0.5) AND LEAD MANGANATE(0.33) NIOBATE(0.67).
TSUBOUCHI N + YONEZAWA M
BULL AMER CERAM SOC 51: 353 (1972)

08-1454   NONLINEAR OPTICAL PROPERTIES OF POTASSIUM NIOBATE SINGLE CRYSTALS. (REFRACTIVE INDICES)
UEMATSU Y + FUKUDA T
JAP J APPL PHYS 10(4): 507 (1971)
     267

     SOME ASPECTS OF THE PROBLEM OF FERROELECTRICITY AND ANTIFERROELECTRICITY IN PEROVSKITES.
VENEVTSEV YUN + LEBEDEV VM
BULL ACAD SCI USSR, PHYS SER 34(12): 2243-9 (1970)

08-1455   CALCULATION OF INTERNAL ELECTRIC FIELDS AND THEIR GRADIENTS IN ABO(3) PEROVSKITE COMPOUNDS WITH DISTINCTIVE DIELECTRIC PROPERTIES.
VENEVTSEV YUN + LYNBIMOV VN + SOLOVEV SP + ZHDANOV GS
BULL ACAD SCI USSR, PHYS SER 28(4): 537-42 (1964)

08-1456   CERTAIN ASPECTS OF THE PROBLEM OF FERROELECTRICITY AND ANTIFERROELECTRICITY IN THE PEROVSKITES.
VENEVTSEV YUN + LEBEDEV VM
BULL ACAD SCI USSR, PHYS SER 34(12): 2243-9 (1970)

08-1457   CONDITIONS FAVORING THE ORIGINATION OF SPONTANEOUS POLARIZATION IN PEROVSKITES. (ABSTRACT ONLY)
VENEVTSEV YUN
P149 OF INT MEETING ON FERROELECTRICITY, 2ND, 1969, KYOTO, PROC,
(J PHYS SOC JAP, VOL 28 SUPPL). PHYS SOC JAP, 1970, 459P.

08-1458   CRYSTAL STRUCTURE AND PROPERTIES OF NEW CADMIUM AND THALLIUM CONTAINING PEROVSKITES. (TITANATES, HAFNATES, MIXED NIOBATES AND TANTALATES)
VENEVTSEV YUN + KAPYSHEV AG + LEBEDEV VM + SALNIKOV VD + ZHDANOV GS
J PHYS (PARIS) 33, SUPPL 4, COLLOQ 2: 241-2 (1972)
     294

08-1459   PEROVSKITE TYPE SEIGNETTO MAGNETS.
VENEVTSEV YUN + LYUBIMOV VN + IVANOVA VV + ZHDANOV GS
J PHYS (PARIS) 33, SUPPL 4, COLLOQ 2: 255-6 (1972)
     294

08-1460 STRUCTURE AND PROPERTIES CF SOME NEW FERROELECTRIC MATERIALS WITH PEROVSKITE STRUCTURE
         (PEROVSKITE COMPOUNDS CONTAINING EITHER BISMUTH, COPPER, OR LEAD).
         VENEVTSEV YUN + KAPYSHEV AG + VISKOV AS + SHVORNEVA LI + LEVEDEV VM + PETROV VM
         + ZHDANOV GS
         BULL ACAD SCI USSR, PHYS SER 31(7): 1086-91 (1967)
             219

08-1461 ELECTRICAL PROPERTIES OF NONLINEAR FERROELECTRIC CERAMICS WITH THE PEROVSKITE
         STRUCTURE.
         VERBITSKAYA TN + ALEXANDROVA LM + SOKOLOVA LS
         J PHYS (PARIS) 33, SUPPL 4, COLLOQ 2: 273-5 (1972)
             294

08-1462 NEW FERROELECTRICS WITH PEROVSKITE AND PYROCHLORE STRUCTURES. (BARIUM AND BISMUTH(0.5))
         (NIOBIUM, TANTALUM, OR VANADIUM)(0.5) OXIDE, (BARIUM OR LEAD) BISMUTH(0.67) (TUNGSTEN
         OR MOLYBDENUM)(0.33) OXIDE)
         VISKOV AS + VENEVTSEV YUN + ZHDANOV GS
         SOV PHYS DOKL 10(5): 391-3 (NOV 1965)
             CHEM ABSTR 63: 3733F (1965)

08-1463 SYNTHESIS OF THALLIUM CONTAINING PEROVSKITES AT HIGH PRESSURES AND THEIR X-RAY
         ANALYSIS.
         VISKOV AS + ZUBOVA YEV + BURDINA KP + VENEVTSEV YUN
         SOV PHYS CRYSTALLOGR 15(5): 932-4 (1971)
             232

08-1464 PIEZOELECTRIC AND PHOTOELASTIC PROPERTIES OF LITHIUM IODATE.
         WARNER AW + PINNOW DA
         J ACOUST SOC AMER 47(3): 791-4 (1970)
             198

08-1465 PIEZOELECTRIC AND PHOTOELASTIC PROPERTIES OF LITHIUM IODATE.
         WARNER AW + BERGMAN JG + PINNOW DA + CRANE GR
         J ACOUST SOC AMER 47: 791-4 (1970)

08-1466 ISOTHERMAL PHASE TRANSITIONS IN CERAMIC LEAD ZIRCONATE.
         WEIRAUCH DF + TENNERY VJ
         J AMER CERAM SOC 53(5): 229-32 (1970)
             191

08-1467 SPLIT INTERSTITIAL CONFIGURATION INDUCED IN PEROVSKITE TYPE SINGLE CRYSTALS BY NEUTRON
         BOMBARDMENT.
         WEIK H
         PHYS LETT 9: 92-3 (1964)

08-1468 OPTICAL PROPERTIES (BIREFRINGENCE AND REFRACTIVE INDICES) OF POTASSIUM NIOBATE (LINEAR
         ELECTROOPTICS).
         WIESENDANGER E
         FERROELECTRICS 1: 141-8 (1970)
             226

08-1469 NIOBIUM-93 AND SODIUM-23 IN POLYCRYSTALLINE SODIUM NIOBATE.
         WOLF FJ + KLINE D + STORY HS
         J CHEM PHYS 53: 3538-43 (1970)

08-1470 DOMAIN STRUCTURE AND POLARIZATION REVERSAL IN FILMS OF FERROELECTRIC BISMUTH TITANATE.
         WU SY + TAKEI WJ + FRANCOMBE MH
         FERROELECTRICS 3: 217-24 (1972);
         IEEE TRANS SON ULTRASON 19: 217-24 (1972)
             263

08-1471 MOSSBAUER STUDIES OF BISMUTH FERRATE LEAD TITANATE PEROVSKITE TYPE SOLID SOLUTIONS
         (ROOM TEMPERATURE TO 600 DEG C).
         YAGNIK CM + GERSON R + JAMES WJ
         J APPL PHYS 42(1): 395-400 (1971)
             224

08-1472 LONGITUDINAL ELECTROOPTIC EFFECT IN OBLIQUE CUT STRONTIUM BARIUM NIOBATE PLATES.
         YAZAKI T + KANATANI K
         JAP J APPL PHYS 10: 522-3 (1971)
             267

8. OTHER PEROVSKITES

08-1473   THE EFFECT OF PRESSURE ON THE TRANSITION OF LEAD ZINCATE(0.33) NIOBATE(0.67).
           YOSHIMOTO J + OKAI B + NOMURA S
           J PHYS SOC JAP 31: 307 (JUL 1971)

08-1474   EFFECT OF IRRADIATION OF DIELECTRIC PROPERTIES OF SOME FERRO-, ANTIFERRO- AND
           PARAELECTRIC PEROVSKITES.
           ZAKURKIN VV + SOLOVEV SP + KUZMIN II
           BULI ACAD SCI USSR, PHYS SER 35(9): 1810-13 (1971)

08-1475   LATTICE DYNAMICS OF PHASE TRANSITIONS IN PEROVSKITE TYPE ANTIFERROELECTRICS.
           ZEIN NE + ZINENKO VI + SCHNEIDER VE
           SOV PHYS SOLID STATE 12: 1858-63 (1971)

           STRUCTURAL ELECTRON MICRODIFFRACTION STUDIES OF FERROELECTRICS AND FERROELECTRIC
           MAGNETS. (LEAD AND BARIUM TITANATES, BISMUTH FERRATE, POTASSIUM NIOBATE)
           ZHDANOV GS + TOMASHPOLSKII YUYA + PLATONOV GL + VENEVTSEV YUN
           P322-31, VOL 1 OF DVORAK V + FOUSKOVA A + GLOGAR P (EDS),
           INT MEETING ON FERROELECTRICITY, PRAGUE, 1966, PROC.
           INST PHYS CZECH ACAD SCI, PRAGUE, 1966.

9. MIXED TITANATES–ZIRCONATES (INCLUDING PZT, PLZT)

09-1476   TRANSPARENT CERAMIC DEVELOPED BY SANDIA LABORATORIES. (LEAD LANTHANUM ZIRCONATE
          TITANATE)
          CERAM BULL 49(6): 618-19 (1970)
               194

09-1477   PHOTOMIXING AT 10.6 MICRONS WITH STRONTIUM BARIUM NIOBATE PYROELECTRIC DETECTORS.
          ABRAMS RL + GLASS AM
          APPL PHYS LETT 15(8): 251-3 (1969)
               163

09-1478   POINT DEFECTS AND SINTERING OF LEAD ZIRCONATE TITANATE.
          ATKIN RB + FULRATH RM
          J AMER CERAM SOC 54: 265-70 (MAY 1971)

09-1479   SOLUBILITY OF ALUMINUM IN LEAD ZIRCONATE TITANATE.
          ATKIN RB + FULRATH RM
          J AMER CERAM SOC 53: 51-2 (1970)
               181

09-1480   SUBSTITUTION OF BISMUTH AND NIOBIUM IONS IN LEAD ZIRCONATE(0.53) TITANATE(0.47)
          (EXPERIMENT AND STRUCTURAL DISCUSSION).
          AITKIN RB + HOLMAN RL + FULRATH RM
          J AMER CERAM SOC 54(2): 113-15 (1971)
               225

09-1481   SINTERING AND FERROELECTRIC PROPERTIES OF LEAD ZIRCONATE TITANATE CERAMICS.
          ATKIN RB
          LAWRENCE RAD LAB, UNIV CALIF, BERKELEY, SEP 1970, 49P. N71-17603
               SCI TECH AEROSP REP 9(7): 1048 (1971)
               243

09-1482   HYDROTHERMAL CRYSTALLIZATION OF LEAD TITANATE ZIRCONATE SOLID SOLUTIONS.
          BARSUKOVA ML + KUZNETSOV VA + LOBACHEV AN + LIDER VV
          SOV PHYS CRYSTALLOGR 16(1): 178-80 (JUL-AUG 1971)
               240

09-1483   FERROELECTRIC CERAMICS AND CONVERSION OF MECHANOELECTRIC ENERGY. (LEAD ZIRCONATE
          TITANATE)
          EAUER F + FETIVEAU Y
          COMPT REND B 272: 193-6 (1971) (IN FRENCH)

09-1484   BEHAVIOR OF MODIFIED LEAD ZIRCONATE LEAD TITANATE PIEZOELECTRIC CERAMICS UNDER HIGH
          ELECTRIC FIELDS. (SUBSTITUTION OF STRONTIUM OR CALCIUM FOR LEAD, IRON FOR ZIRCONIUM AND
          TITANIUM, EFFECT ON ACOUSTIC RADIATION POWER)
          BELDING JH + MCLAREN MG
          BULL AMER CERAM SOC 49(12): 1025-9 (1970)
               221

09-1485   EFFECTS OF HIGH ELECTRIC FIELDS ON MODIFIED LEAD ZIRCONATE- LEAD TITANATE CERAMICS FOR
          PIEZOELECTRIC APPLICATIONS.
          BELDING JH
          RUTGERS UNIV PHD THESIS, 1969, 147P. N71-24124
               DISS ABSTR INT B 30: 5447 (1970)
               238

09-1486   ANTIFERROELECTRIC- FERROELECTRIC TRANSITION IN THE LEAD (ZIRCONATE, STANNATE, TITANATE)
          SOLID SOLUTIONS. (ELECTROSTRICTION CONSTANTS)
          BENGUIGUI L
          MEMOR ARTILL FRANC 44(173): 547-72 (1970) (IN FRENCH)
               241

9. MIXED TITANATES-ZIRCONATES (INCLUDING PZT, PLZT)

09-1487 ELECTROSTRICTIVE CONSTANTS IN FERROELECTRIC AND ANTIFERROELECTRIC PHASES. (LANTHANUM OR NIOBIUM DOPED LEAD ZIRCONATE STANNATE TITANATE)
BENGUIGUI L
P141-5 OF EUROPEAN MEETING ON FERROELECTRICITY, 1969, SAARBRUCKEN,
MUSER HE AND PETERSSON J (EDS), WISSENSCH VERLAGSGES, STUTTGART,
1970, 396P.

09-1488 RELEASE OF ELECTRIC ENERGY IN LEAD NIOBIUM ZIRCONATE TITANATE STANNATE BY TEMPERATURE AND PRESSURE ENFORCED PHASE TRANSITIONS.
BERLINCOURT D + JAFFE H + KRUEGER HHA + JAFFE B
APPL PHYS LETT 3: 90-2 (1963)

09-1489 VIBRATION OF ELECTROELASTIC CONSTANTS OF POLYCRYSTALLINE LEAD TITANATE ZIRCONATE WITH THOROUGHNESS OF POLING.
BERLINCOURT D
J ACOUST SOC AMER 36: 515-20 (1964)

09-1490 DC BIAS EFFECTS IN LEAD ZIRCONATE TITANATE CERAMICS.
BIGGERS JV + SCHULZE WA
BULL AMER CERAM SOC 50: 396 (1971)

09-1491 TEMPERATURE AND FIELD DEPENDENCE OF THE ANTIFERROELECTRIC FERROELECTRIC PHASE BOUNDARY IN PLZT CERAMICS.
BIGGERS JV + SCHULZE WA
BULL AMER CERAM SOC 51: 650 (1972)

09-1492 BARIUM ZIRCONATE LEAD TITANATE FERROELECTRIC CERAMICS.
BRATSCHUN WR
J AMER CERAM SOC 46: 141-4 (1963)

09-1493 COLD PRESSING AND LOW TEMPERATURE SINTERING OF ALKOXY-DERIVED PLZT AND PLZT.
BROWN LM + MAZDIYASNI KS
J AMER CERAM SOC 55: 541-4 (NOV 1972)

09-1494 EFFECTS OF CALCINING ON SINTERING OF LEAD ZIRCONATE TITANATE CERAMICS.
BUCKNER DA + WILCOX PD
BULL AMER CERAM SOC 51: 218-22 (MAR 1972)

09-1495 OXYGEN CONCENTRATION CELL AND ELECTRICAL CONDUCTIVITY MEASUREMENTS ON PZT FERROELECTRICS.
BURT JG + KRAKOWSKI RA
J AMER CERAM SOC 54: 415-9 (SEP 1971)

09-1496 PZT-5 UNDER PRESSURE, DIELECTRIC AND PIEZOELECTRIC PROPERTIES.
BURFOOT JC + MARTIRENA HT
J PHYS (PARIS) 33, SUPPL 4, COLLOQ 2: 249-50 (1972)
294

09-1497 RAMAN SPECTRA OF POLYCRYSTALLINE SOLIDS: APPLICATION TO THE LEAD TITANATE ZIRCONATE SYSTEM (MEASUREMENT OF SOFT OPTICAL PHONON MODES IN TETRAGONAL FERROELECTRIC REGION)
BURNS G + SCOTT BA
PHYS REV LETT 25(17): 1191-4 (1970)
222

MOSSBAUER EFFECT STUDIES OF THE PHASE TRANSITIONS IN ANTIFERROELECTRIC LEAD ZIRCONATE AND FERROELECTRIC LEAD TITANATE ZIRCONATE.
CANNER JP + YAGNIK CM + GERSON R + JAMES WJ
J APPL PHYS 42(12): 4708-11 (NOV 1971)

09-1498 MOSSBAUER EFFECT STUDIES OF FERROELECTRIC PHASE TRANSITIONS IN THE LEAD ZIRCONATE- LEAD TITANATE- BISMUTH FERRATE.
CANNER JP
UNIV MISSOURI PHD THESIS, 1969, 180P. N71-27314
DISS ABSTR INT B 31: 349 (1970)
238

09-1499 ORIGIN OF THE MAXIMUM IN THE ELECTROMECHANICAL ACTIVITY IN LEAD ZIRCONATE(X) TITANATE(1-X) CERAMICS NEAR THE MORPHOTROPIC PHASE BOUNDARY.
CARL K + HARDTL KH
PHYS STATUS SOLIDI A, APPL RES 8(1): 87-98 (1972)

9. MIXED TITANATES-ZIRCONATES (INCLUDING PZT, PLZT)

09-1500 ORIGIN OF THE MAXIMUM IN THE ELECTROMECHANICAL ACTIVITY IN LEAD ZIRCONATE(X)
TITANATE(1-X) CERAMICS NEAR THE MORPHIC PHASE BOUNDARY.
CARL K + HARDTL KH
PHYS STATUS SOLIDI A 8: 87-98 (1971)

09-1501 POLARIZATION, ELECTROSTRICTION, AND PIEZOELECTRIC ACTIVITY OF LEAD ZIRCONATE(X)
TITANATE(1-X) CERAMICS.
CARL K + HARDTL KH
J PHYS (PARIS) 33, SUPPL 4, COLLOQ 2: 251-3 (1972)
        294

09-1502 STRUCTURE AND ELECTROMECHANICAL PROPERTIES OF LANTHANUM DOPED LEAD TITANATE ZIRCONATE
CERAMICS.
CARL K + HARDTL KH
BER DEUT KERAM GESELL 47: 687-9 (1970) (IN GERMAN)

09-1503 THIN FILM FERROELECTRIC PHOTOCONDUCTOR MEMORY DEVICE. (PLZT)
CHAPMAN DW
J VACUUM SCI TECHNOL 9: 425-31 (1972)

09-1504 A FERROELECTRIC PIEZOELECTRIC RANDOM ACCESS MEMORY. (LEAD ZIRCONATE- LEAD TITANATE
CERAMIC MEMORY PLATE)
CRAWFORD JC
IEEE TRANS ELECTRON DEV 18: 951-8 (1971)

09-1505 FERROELECTRIC FIELD EFFECT STUDIES AT LOW TEMPERATURES (WITH TIN OXIDE FILMS ON LEAD
ZIRCONATE(0.65) TITANATE(0.35) CERAMIC).
CRAWFORD JC
FERROELECTRICS 1: 23-30 (1970)
        191

09-1506 FERROELECTRIC FIELD EFFECT STUDIES. (ON TIN OXIDE FILMS ON LEAD(0.97) BISMUTH(0.02)
TITANATE (0.35) ZIRCONATE(0.65), SUMMARY ONLY)
CRAWFORD JC
P320 OF INT MEETING ON FERROELECTRICITY, 2ND, 1969, KYOTO, PROC,
(J PHYS SOC JAP, VOL 28 SUPPL). PHYS SOC JAP, 1970, 459P.

09-1507 TERNARY SYSTEM LEAD TITANATE- LEAD ZIRCONATE- LEAD NIOBATE(0.67) ZINCATE(0.33).
DANTSIGER AYA + DEVLIKANEVA RO + DUDKINA SI + MORDANOV BP + ROGACH TV + KUPRIYANOV MF
+ FESENKO EG
BULL ACAD SCI USSR, PHYS SER 35(9): 1801-5 (1971)

09-1508 FERROELECTRIC CERAMICS AND ENERGY CONVERSION. (LEAD ZIRCONATE TITANATE)
DAVID J + FETIVEAU Y + LANDURE Y + TROCCAZ M
COMPT REND B 274: 445-8 (1972) (IN FRENCH)

09-1509 REALIZATION OF A 40 MHZ COLOR TELEVISION INFRARED RESPONSE USING SURFACE WAVE
TRANSDUCERS ON LEAD ZIRCONATE TITANATE.
DEVRIES AJ + ADLER R + DIAS JF + WOJCIK TJ
IEEE TRANS SON ULTRASON 17: 61 (1970)

09-1510 PIEZOCERAMICS BASED ON LEAD ZIRCONATE TITANATE WITH COMPLEX ADDITIVES CONTAINING
GERMANIUM.
DIDKOVSKAYA OS + KLIMOV VV + VENEVTSEV YUN
INORG MATER 6: 541 (1970)

09-1511 VARIATION OF SOUND VELOCITY AT FERROELECTRIC-1, FERROELECTRIC-2 TRANSITION IN (NIOBIUM
DOPED) ZIRCONATE(0.95) TITANATE(0.05).
DORAN DG + GOETTELMAN R
APPL PHYS LETT 2: 22 (1963)

09-1512 OXYGEN CONCENTRATION CELL MEASUREMENTS OF IONIC TRANSPORT NUMBERS IN PZT (LEAD
ZIRCONATE TITANATE) FERROELECTRICS.
EZIS A + BURT JG + KRAKOWSKI RA
J AMER CERAM SOC 53(9): 521-4 (1970)
        211

09-1513 ELASTIC CONSTANTS OF FERROCERAMICS.
FEDOTOV II
SOV PHYS CRYSTALLOGR 15(6): 1096 (MAY-JUN 1971)
        234

9. MIXED TITANATES-ZIRCONATES (INCLUDING PZT, PLZT)

09-1514   TIME AND TEMPERATURE DEPENDENCES OF THE PARAMETERS OF A FERROELECTRIC CERAMIC.
FELDMAN NB + KONGAROVA NI + SMAZHEVSKAYA EG + SEGALLA AG + DENISOVA VM + SOLOVEVA ES
+ TYULYAEVA NG
BULL ACAD SCI USSR, PHYS SER 34(12): 2349-53 (1970)

09-1515   FERROELECTRIC DISPLAYS. (BISMUTH TITANATE, LEAD ZIRCONATE TITANATE)
FRANCOMBE MH + WU SY + TAKEI WJ
WESTINGHOUSE RES LAB, 1972, 89P. AD755746

09-1516   IMPROVED AGING AND SWITCHING OF LEAD ZIRCONATE- LEAD TITANATE CERAMICS WITH INDIUM
ELECTRODES.
FRASER DB + MALDONADO JR
BULL AMER CERAM SOC 49: 412 (1970)

09-1517   IMPROVED AGING AND SWITCHING OF LEAD ZIRCONATE- LEAD TITANATE CERAMICS WITH INDIUM
ELECTRODES.
FRASER DB + MALDONADO JR
J APPL PHYS 41: 2172-6 (1970)

09-1518   PROPERTIES OF NEW TRANSPARENT FERROELECTRIC CERAMICS. (LEAD ZIRCONATE- LEAD TITANATE
TYPE)
FRASER DB + O'BRYAN HM
BULL AMER CERAM SOC 49: 833 (1970)

09-1519   PIEZOELECTRIC AND DIELECTRIC PROPERTIES OF LEAD TITANATE ZIRCONATE CERAMICS AT LOW
TEMPERATURES.
GERSON R
J APPL PHYS 33: 830-2 (1962)

CHARGE MECHANISM IN FERROELECTRICS. (LEAD ZIRCONATE TITANATE, STOICHIOMETRY EFFECTS)
GLOWER DD + BURT JG
OHIO STATE UNIV, 1970, 6P. AD709694
    SCI ABSTR A 74: 21777

09-1520   EFFECTS OF RADIATION INDUCED DAMAGE CENTERS IN LEAD ZIRCONATE TITANATE CERAMICS.
GLOWER DD + HESTER DL + WARNKE DF
J AMER CERAM SOC 48(8): 417-21 (1965)

09-1521   PHASE CHANGES INDUCED BY HYDROSTATIC PRESSURE IN A FERROELECTRIC MATERIAL. (PZT)
GONNARD P + FETIVEAU Y + BAUER F + EYRAUD L
COMPT REND B 275: 633-6 (1972) (IN FRENCH)

09-1522   DIELECTRIC NONLINEARITY AND LOSS IN THE BARIUM (OR LEAD, 4 COMPOSITIONS EACH) STRONTIUM
TITANATE SYSTEMS. (FOR VOLTAGE DEPENDENT CAPACITORS)
GRIFFITHS CR
OHIO STATE UNIV PHD THESIS, 1969, 126P.
    DISS ABSTR INT B 30(4): 1656 (1969)
    200

09-1523   A NEW LONGITUDINAL DISPLAY MODE FOR CERAMIC ELECTROOPTIC DEVICES. (PLZT)
HAERTLING GH + MCCAMPBELL CB
PROC IEEE 60: 450-1 (1972)

09-1524   HOT PRESSED FERROELECTRIC LEAD ZIRCONATE TITANATE CERAMICS FOR ELECTROOPTICAL
APPLICATIONS.
HAERTLING GH
BULL AMER CERAM SOC 49(6): 564-7 (1970)
    201

09-1525   HOT PRESSED LEAD ZIRCONATE TITANATE CERAMICS CONTAINING BISMUTH.
HAERTLING GH
BULL AMER CERAM SOC 43(12): 875-9 (1964)

09-1526   HOT PRESSED (LEAD, LANTHANUM) (ZIRCONATE, TITANATE) FERROELECTRIC CERAMICS FOR
ELECTROOPTIC APPLICATIONS.
HAERTLING GH + LAND CE
J AMER CERAM SOC 54(1): 1-11 (1971)
    204

9. MIXED TITANATES-ZIRCONATES (INCLUDING PZT, PLZT)

09-1527 HOT PRESSED FERROELECTRIC CERAMICS FOR ELECTROOPTICAL APPLICATIONS. (LEAD ZIRCONATE
TITANATE AND SODIUM POTASSIUM NIOBATE)
HAERTLING GH
BULL AMER CERAM SOC 47(4): 389 (APR 1968)
143

09-1528 IMPROVED HOT PRESSED ELECTROOPTIC CERAMICS IN THE LEAD LANTHANUM ZIRCONATE TITANATE
SYSTEM.
HAERTLING GH
J AMER CERAM SOC 54: 303-9 (JUN 1971)
236

09-1529 RECENT IMPROVEMENTS IN THE OPTICAL AND ELECTROOPTIC PROPERTIES OF PLZT CERAMICS.
HAERTLING GH + LAND CE
IEEE TRANS SON ULTRASON 19: 269-80 (1972)

09-1530 DISTRIBUTION OF A-SITE AND B-SITE VACANCIES IN (LEAD, LANTHANUM) (TITANATE, ZIRCONATE)
CERAMICS.
HARDTL KH + HENNINGS D
J AMER CERAM SOC 55: 230-1 (MAY 1972)

09-1531 STRUCTURE AND ELECTROMECHANICAL PROPERTIES OF LANTHANUM DOPED LEAD TITANATE ZIRCONATE
CERAMICS.
HARDTL KH
BER DEUT KERAM GESELL 47: 389 (1970) (IN GERMAN)

09-1532 USE OF FERROELECTRIC CERAMICS IN COMPUTERS AND OTHER DEVICES. (LEAD ZIRCONATE- LEAD
TITANATE AND PLZT)
HARRISON WB
BULL AMER CERAM SOC 50: 805 (1971)

09-1533 FLUX GROWN KTN SINGLE CRYSTALS AND THEIR ELECTRICAL PROPERTIES.
HIGASHI N + TANIGUCHI I
SCI ENG REV DOSHISHA UNIV 13: 152-63 (1972) (IN JAPANESE)

09-1534 INTRINSIC NONSTOICHIOMETRY IN SINGLE PHASE LEAD ZIRCONATE TITANATE.
HOLMAN RL + FULRATH RM
J AMER CERAM SOC 55(4): 192-5 (APR 1972)
262

09-1535 INTRINSIC NONSTOICHIOMETRY IN THE LEAD ZIRCONATE TITANATE SYSTEM.
HOLMAN RL + FULRATH RM
BULL AMER CERAM SOC 50: 797 (1971)

09-1536 INTRINSIC AND EXTRINSIC NONSTOICHIOMETRY IN THE LEAD ZIRCONATE TITANATE SYSTEM. (DEFECT
STRUCTURE)
HOLMAN RL
UNIV CALIF, BERKELEY, PHD THESIS, 1972, 97P. LBL-880
291

09-1537 MEASUREMENT OF PIEZOELECTRIC PHASE ANGLES IN A FERROELECTRIC CERAMIC (THEORY, APPLIED
TO LEAD(0.99) NIOBIUM(0.02) ZIRCONATE(0.65) TITANATE(0.35)).
HOLLAND R
IEEE TRANS SON ULTRASON 17(2): 123-4 (1970)
199

09-1538 SOLUTION KINETICS OF LUTETIUM AND SCANDIUM IONS IN LEAD TITANATE(0.5) ZIRCONATE(0.5).
HOLMAN RL + FULRATH RM
BULL AMER CERAM SOC 51: 727 (1972)

09-1539 PIEZOELECTRIC CERAMICS OF LEAD ZIRCONATE TITANATE MODIFIED WITH SODIUM OR POTASSIUM
ANTIMONATE, NIOBATE, OR BISMUTHATE, OR BY BISMUTH OR LANTHANUM FERRATE, ALUMINATE OR
CHROMATE.
IKEDA T + OKANO T
JAP J APPL PHYS 3: 63-71 (1964)

09-1540 PRECIPITATION OF ZIRCONIA PHASE IN NIOBIUM MODIFIED CERAMICS OF LEAD ZIRCONATE
TITANATE.
IKEDA T + TANAKA Y + AYAKAWA T + NOAKE H
JAP J APPL PHYS 3: 581-7 (1964)

9. MIXED TITANATES-ZIRCONATES (INCLUDING PZT, PLZT)

09-1541   DIELECTRIC POLARIZATION OF LEAD TITANATE ZIRCONATE SOLID SOLUTIONS (THERMODYNAMIC
           EXISTENCE CONDITIONS FOR MORPHOTROPIC PHASE BOUNDARY, STOICHIOMETRY DEPENDENCE).
           ISUPOV VA
           SOV PHYS SOLID STATE 12(5): 1084-8 (1970)
               215

09-1542   HOT PRESSING OF POTASSIUM- SODIUM NIOBATE.
           JAEGER RD + EGERTON L
           J AMER CERAM SOC 45: 209-13 (1962)

           NATURE OF AGING IN FERROELECTRIC CERAMICS. (BARIUM TITANATE, LEAD ZIRCONATE TITANATE)
           JONKER GH
           J AMER CERAM SOC 55(1): 57-8 (JAN 1972)

09-1543   DEMONSTRATION OF AGING EFFECTS IN BARIUM TITANATE- BARIUM ZIRCONATE.
           JONKER GH
           BULL AMER CERAM SOC 50: 397 (1971)

09-1544   NEW PIEZOELECTRIC CERAMICS. (LEAD TITANATE- LEAD FERRATE (0.5) NIOBATE(0.5), LEAD
           TITANATE- LEAD ZIRCONATE)
           KLIMOV VV + DIDKOVSKAYA OS + SAVENKOVA GE + VENEVTSEV YUN
           J PHYS (PARIS) 33, SUPPL 4, COLLOQ 2: 243-5 (1972)
               294

09-1545   ACOUSTIC DETECTION OF FERROELECTRIC PHASE TRANSITIONS IN PLZT CERAMICS.
           KRAUSE JT + O'BRYAN HM
           J AMER CERAM SOC 55: 497-9 (OCT 1972)

09-1546   NONLINEAR EFFECTS IN FERROCERAMICS. (PZT-8, LITHIUM NIOBATE)
           KRAUT EA + LIM TC
           J ELECTROCHEM SOC 119: 98C (1972)

09-1547   DISPERSION STUDY OF DOMAIN WALL MOBILITY AND MECHANICAL DAMPING IN LEAD TITANATE
           ZIRCONATE TRANSDUCER CERAMICS.
           KRUEGER HHA
           CLEVITE CORP, CLEVELAND, 1970, 23P, AD706793; 1971, 59P, AD717964

09-1548   FERROELECTRIC PROPERTIES OF LEAD (STANNATE(0.5) NIOBATE (0.5) (1-X) ZIRCONATE(X)
           PEROVSKITE TYPE SOLID SOLUTIONS.
           KUCHAR F + VALENTA MW
           PHYS STATUS SOLIDI A, APPL RES 6(2): 525-33 (1971)
               244

09-1549   DIELECTRIC AND PIEZOELECTRIC PROPERTIES OF LEAD COBALTATE (0.33) NIOBATE(0.67)-
           TITANATE- ZIRCONATE SOLID SOLUTION CERAMICS.
           KUDO T + YAZAKI T + NAITO F + SUGAYA S
           J AMER CERAM SOC 53(6): 326-8 (1970)
               194

           RECENT DEVELOPMENTS IN ELECTROOPTIC CERAMICS (HOT PRESSED LEAD ZIRCONATE TITANATE,
           EFFECT OF GRAIN SIZE, DOPANTS, REVIEW).
           LAND CE
           INT J NONDESTRUCT TEST 1: 315-36 (1970)

09-1550   ELECTROOPTIC CERAMICS- NEW MATERIALS FOR INFORMATION STORAGE AND DISPLAY (HOT PRESSED
           LEAD ZIRCONATE TITANATE, TUTORIAL, 17 REFS).
           LAND CE + THACHER PD
           W ELEC ENG 14(1): 13-24 (JAN 1970)
               188

09-1551   ELECTROOPTIC EFFECTS IN FERROELCTRIC CERAMICS. (LEAD ZIRCONATE TITANATE)
           LAND CE + HOLLAND R
           IEEE SPRCTRUM 7(2): 71-8 (1970)
               272

09-1552   FERROELECTRIC CERAMICS FOR INFORMATION STORAGE AND DISPLAY. (PZT)
           LAND CE
           MET TRANS 2: 781-8 (1971)

09-1553   ISOTHERMAL GRAIN GROWTH AND ELECTRICAL BEHAVIOR OF FULLY DENSE PLZT CERAMICS.
           LANGMAN RA + RUNK RB + BUTLER SR
           BULL AMER CERAM SOC 51: 646 (1972)

09-1554 OPTICAL PROPERTIES (ELECTRICALLY VARIABLE BIREFRINGENCE) OF (HOT PRESSED) FERROELECTRIC
         CERAMICS. (PZT)
         LAND CE + HAERTLING GH
         P96-9 OF INT MEETING ON FERROELECTRICITY, 2ND, 1969, KYOTO, PROC,
         (J PHYS SOC JAP, VOL 28 SUPPL). PHYS SOC JAP, 1970, 459P.

09-1555 SINTERING SCANDIUM AND NIOBIUM MODIFIED LEAD ZIRCONATE TITANATE.
         LEE DC
         UNIV CALIF, BERKELEY, MS THESIS, 1970, 35P. UCRL-19188

09-1556 NEW COMPOSITE CERAMIC PIEZOELECTRIC TRANSDUCER MATERIAL. (ALUMINA WHISKERS IN LEAD
         ZIRCONATE TITANATE)
         LESTER WW
         OFF NAV RES 1970, 11P. AD714494
             GOV REP ANNOUNCE 71(1): 97 (1971)
             239

09-1557 INFLUENCE OF DOMAIN STRUCTURE ON THE PROPERTIES OF POLYCRYSTALLINE LEAD TITANATE
         ZIRCONATE COMPOSITIONS NEAR THE MORPHOTROPIC PHASE TRANSITION.
         LEZGINTSEVA TN + FELDMAN NB
         P144-8, VOL 2 OF DVORAK V + FOUSKOVA A + GLOGAR P (EDS),
         INT MEETING ON FERROELECTRICITY, PRAGUE, 1966, PROC.
         INST PHYS CZECH ACAD SCI, PRAGUE, 1966.

09-1558 PYROELECTRIC PROPERTIES OF THE LANTHANUM DOPED FERROELECTRIC PLZT CERAMICS.
         LIU ST + HEAPS JD + TUFTE ON
         FERROELECTRICS 3: 281-5 (1972;
         IEEE TRANS SON ULTRASON 19: 281-5 (1972)
             263

09-1559 FERROELECTRICITY IN SPUTTERED LEAD TITANATE ZIRCONATE FILMS.
         MACBETH JW + LUPFER DA
         BULL AMER CERAM SOC 47(4): 389 (APR 1968)
             143

09-1560 HIGH D*, FAST, LEAD ZIRCONATE TITANATE PYROELECTRIC DETECTORS.
         MAHLER RJ + PHELAN RJ + COOK AR
         INFRARED PHYS 12: 57-9 (1972)

09-1561 DISPLAY APPLICATIONS OF PLZT FERROELECTRIC CERAMICS.
         MALDONADO JR
         IEEE INT CONV DIGEST 1972: 190-1

09-1562 ELECTROOPTIC DEVICES USING STRAIN BIASED PLZT FERROELECTRIC CERAMICS.
         MALDONADO JR
         WESCON TECH PAP 15 (SESSION 31, PAPER 3): 1-8 (1971)

09-1563 FERROELECTRIC DOMAIN SWITCHING IN RHOMBOHEDRAL PHASE PLZT CERAMICS.
         MALDONADO JR + MEITZLER AH
         FERROELCTRICS 3: 169-75 (1972);
         IEEE TRANS SON ULTRASON 19: 169-75 (1972)
             263

09-1564 STRAIN BIASED FERROELECTRIC- PHOTOCONDUCTOR IMAGE STORAGE AND DISPLAY DEVICES OPERATED
         IN A REFLECTION MODE. (PLZT)
         MALDONADO JR + ANDERSON LK
         IEEE TRANS ELECTRON DEV 18: 774-7 (1971)

09-1565 STRAIN BIASED FERROELECTRIC PHOTOCONDUCTOR IMAGE STORAGE AND DISPLAY DEVICES. (PLZT)
         MALDONADO JR + MEITZLER AH
         PROC IEEE 59: 368-82 (1971)

09-1566 AN ADAPTIVE FERROELECTRIC TRANSFORMER - A SOLID STATE ANALOG MEMORY DEVICE. (USING, FOR
         EXAMPLE, PZT)
         MCCUSKER JH + PERLMAN SS
         IEEE TRANS ELECTRON DEV 17: 534-40 (1970)

09-1567 MEASUREMENT OF THE PYROELECTRIC COEFFICIENT OF 65-35 PZT. (LEAD ZIRCONATE TITANATE)
         MCGEVNA VG + SACCENTI JC + MCNEILLY JH
         BALLISTIC RES LAB, ABERDEEN PROVING GROUND, 1972, 43P. AD755520

09-1568   FERROELCTRIC- PHOTOCONDUCTOR MEMORY DEVICE. (PLZT)
           MEHTA RR
           J APPL PHYS 42: 1842-5 (1971)

09-1569   DEVICE APPLICATIONS OF PLZT CERAMICS.
           MEITZLER AH
           IEEE TRANS SON ULTRSON 19: 339 (1972)

09-1570   ENHANCED ORDERING IN PLZT CERAMICS.
           MEITZLER AH + O'BRYAN HM
           BULL AMER CERAM SOC 50: 397 (1971)

09-1571   FERROELASTIC BEHAVIOR OF LEAD LANTHANUM ZIRCONATE TITANATE CERAMICS WHEN SUBJECTED TO
           LARGE TENSILE STRAINS.
           MEITZLER AH + O'BRYAN HM
           APPL PHYS LETT 19(4): 106-8 (15 AUG 1971)
              241

09-1572   IMAGE STORAGE AND DISPLAY DEVICES USING FINE GRAIN FERROELECTRIC CERAMICS. (LEAD
           ZIRCONATE- LEAD TITANATE)
           MEITZLER AH + MALDONADO JR + FRASER DB
           BELL SYST TECH J 49: 953-67 (1970)

09-1573   LANTHANUM DEPENDENCE OF ELASTIC AND PIEZOELECTRIC PROPERTIES OF PLZT CERAMICS WITH A
           ZIRCONIUM TO TITANIUM RATIO OF 65 TO 35.
           MEITZLER AH + O'BRYAN HM
           J AMER CERAM SOC 55: 504-6 (OCT 1972)

09-1574   ATOMIC STRUCTURES OF THE FERROELECTRIC COMPOUNDS LEAD ZIRCONATE(X) TITANATE(1-X) FOR
           TWO RHOMBOHEDRAL PHASES. (X=0.9 AND 0.58) (CORRELATION OF CATION SHIFTS WITH BISMUTH
           FERRATE, LITHIUM NIOBATE AND LITHIUM TANTALATE)
           MICHEL C + GERSON R + JAMES WJ
           INT CONGR CRYSTALLOGR, 8TH, 1969, NY, ABSTR. AMER INST PHYS, 1969.
              SCI ABSTR B 73: 65271 (1970)
              219

09-1575   INSCRIPTION OF HOLOGRAPHIC RESULTS IN TRANSPARENT CERAMIC FERROELECTRICS. (LEAD
           LANTHANUM ZIRCONIUM TITANATE)
           MICHERON F + HERMOSIN A + BISMUTH G + NICHOLAS J
           COMPT REND B 274: 361-4 (31 JAN 1972) (IN FRENCH)
              261

09-1576   PROPERTIES OF OXYGEN DEPLETED CERAMIC LEAD ZIRCONATE TITANATE (PLUS 1 PERCENT NIOBIUM
           PENTOXIDE) SURFACE LAYERS.
           MILLER DW + GLOWER DD
           FERROELECRTICS 3: 295-303 (1972);
           IEEE TRANS SON ULTRASON 19: 295-303 (1972)
              263

09-1577   HIGH TEMPERATURE PHASE EQUILIBRIA IN THE LEAD ZIRCONATE TITANATE SYSTEM.
           MOON RL + FULRATH RM
           J AMER CERAM SOC 54: 124-5 (1971)
              225

09-1578   VAPORIZATION OF LEAD ZIRCONATE- LEAD TITANATE MATERIALS. PART-2: HOT PRESSED
           COMPOSITIONS AT NEAR THEORETICAL DENSITY.
           NORTHROP DA
           J AMER CERAM SOC 51(7): 357-61 (1968)
              143

09-1579   ENHANCED ORDERING OF FERROELECTRIC DOMAINS IN PLZT CERAMICS.
           O'BRYAN HM + MEITZLER AH
           BULL AMER CERAM SOC 51(5): 479-85 (1972)
              265

09-1580   ENHANCED ORDERING OF FERROELECTRIC DOMAINS IN PLZT CERAMICS.
           O'BRYAN HM + MEITZLER AH
           BULL AMER CERAM SOC 51: 479-85 (1972)

09-1581   PHASE DIAGRAM OF LEAD ZIRCONATE(1-X) TITANATE(X) PLUS 0.8 PERCENT TUNGSTEN TRIOXIDE.
           PALETTO J + TROCCAX M + GONNARD P + GRANGE G + EYRAUND L
           COMPT REND B 275: 657-60 (1972) (IN FRENCH)
              282

9. MIXED TITANATES-ZIRCONATES (INCLUDING PZT, PLZT)

09-1582   INFLUENCE OF THE CONCENTRATION OF POINT DEFECTS ON THE INTERNAL FRICTION IN
           POLYCRYSTALLINE LEAD ZIRCONATE TITANATE.
           PAVLOV VS + TURKOV SK + BESSONOVA EN + POSTNIKOV VS
           SOV PHYS SOLID STATE 13(3): 745-6 (SEP 1971)
                244

09-1583   ADAPTIVE FERROELECTRIC TRANSFORMERS WITH IMPROVED TEMPERATURE CHARACTERISTICS. (PZT-5H)
           PERLMAN SS + MCCUSKER JH
           IEEE TRANS ELECTRON DEV 19: 383-4 (1972)

09-1584   INTERNAL FRICTION (AND CURIE TEMPERATURES) IN (LEAD ZIRCONATE TITANATE) FERROELECTRICS
           (WITH AND WITHOUT NIOBIA ADDITION) DUE TO INTERACTION OF DOMAIN BOUNDARIES AND POINT
           DEFECTS (TORSION PENDULUM AND RESONANCE TECHNIQUES, 20-400C).
           POSTNIKOV VS + PAVLOV VS + TURKOV SK
           J PHYS CHEM SOLIDS 31: 1785-91 (1970)
                212

09-1585   ULTRASONIC INTERFEROMETER WITH LEAD ZIRCONATE TITANATE AS TRANSDUCER.
           RAMACHANDRA RS
           INDIAN J PURE APPL PHYS 8: 848-9 (1970)

           TWO WAVE SHOCK STRUCTURES IN THE FERROELECTRIC CERAMICS BARIUM TITANATE AND LEAD
           ZIRCONATE TITANATE.
           REYNOLDS CE + SEAY GE
           J APPL PHYS 33: 2234-41 (1962)

09-1586   TEMPERATURE DEPENDENT FERROELECTRIC DOMAIN ALIGNMENT IN LEAD ZIRCONATE- LEAD TITANATE-
           LANTHANUM OXIDE (PLZT) CERAMIC.
           SALANECK WR
           J APPL PHYS 43(11): 4468-73 (1972)
                278

09-1587   OXIDATION REDUCTION PHENCMENA IN THE LEAD TITANATE- LANTHANUM TITANATE SYSTEM.
           SASAKI H + MATSUO Y
           BULL AMER CERAM SOC 51: 164-6 (FEB 1972)

09-1588   EFFECT OF BISMUTH AND MANGANESE CONTAINING ADDITIVES ON THE PROPERTIES OF SOLID
           SOLUTIONS OF LEAD ZIRCONATE TITANATE.
           SAVENKOVA GE + DIDKOVSKAYA OS + KLIMOV VV AND OTHERS
           INORG MATER 7: 881-4 (JUN 1971)

09-1589   A SEMICONDUCTOR- FERROELECTRIC MEMORY DEVICE. (PLZT)
           SAWYER DE + SANDSTROM DB
           PROC IEEE 59: 87-8 (1971)

09-1590   PHASE BOUNDARY AND DEFECT STRUCTURE STUDIES IN PLZT CERAMICS.
           SCHULZE WA + DARLINGTON CNW + BIGGERS JV
           BULL AMER CERAM SOC 51: 355 (1972)

09-1591   EFFECT OF FERROELECTRIC PHASE TRANSITION CONDITIONS ON DIELECTRIC AND PIEZOELECTRIC
           PROPERTIES OF A CERAMIC (LEAD STRONTIUM TITANATE ZIRCONATE DOPED WITH CHROMIUM OXIDE).
           SEGALLA AG + SMAZHEVSKAYA EG + FELDMAN NB + SOLOVEVA ES + ALEKSANDROVA IA + KHARASH EV
           BULL ACAD SCI USSR, PHYS SER 35(9): 1806-9 (1971)

09-1592   THE ROLE OF IRON OXIDE ADULTRANT IN A LEAD ZIRCONATE LEAD TITANATE CERAMICS.
           SEUFERT SW + HARSELL WB
           BULL AMER CERAM SOC 51: 354 (1972)

09-1593   RETENTION IN THE THIN FERROELECTRIC FILMS. (PLZT)
           SHARMA BS
           BULL AMER PHYS SOC 17: 102 (1972)

09-1594   FERROELECTRIC MEMORY DEVICE PARAMETER STUDY. (LEAD ZIRCONATE TITANATE)
           SHEFELBINE HC
           SANDIA LABS, ALBUQUERQUE, 1970, 81P. SC-RR-69-384

09-1595   CURIE TEMPERATURES OF SOLID SOLUTIONS IN A LEAD ZIRCONATE- LEAD TITANATE- LEAD
           METANIOBATE TERNARY SYSTEM.
           SIROTA NN + CHOBOT MA
           DOKL AKAD NAUK BELORUS SSR 15: 113-15 (1971) (IN RUSSIAN)

09-1596   CRYSTAL STRUCTURE AND PHASE RELATIONS IN THE LEAD ZIRCONATE LEAD TITANATE- LEAD
           METANIOBATE TERNARY SYSTEM.
           SIROTA NN + CHOBOT MA
           DOKL AKAD NAUK BELORUS SSR 14: 997-9 (1970) (IN RUSSIAN)

09-1597   THERMAL VARIATION OF SPONTANEOUS POLARIZATION AND COERCIVE FORCE OF THE PEROVSKITES IN
           THE SYSTEM LEAD ZIRCONATE TITANATE METANIOBATE.
           SIROTA NN + CHOBOT MA
           DOKL AKAD NAUK BELORUS SSR 15(7): 582-4 (1971) (IN RUSSIAN)
               CHEM ABSTR 75: 134548S (1971)
               268

09-1598   ELECTROELASTIC CONSTANT MEASUREMENTS FOR PLZT CERAMICS.
           SMITH WD
           IEEE TRANS SON ULTRASON 18: 49 (1971)

09-1599   SCATTERING MODE FERROELECTRIC PHOTOCONDUCTOR IMAGE STORAGE AND DISPLAY DEVICES. (PLZT
           CERAMICS)
           SMITH WD + LAND CE
           APPL PHYS LETT 20: 169-71 (1972)

09-1600   FABRICATION OF ELECTROOPTIC PLZT CERAMICS BY ATMOSPHERE SINTERING.
           SNOW GS
           BULL AMER CERAM SOC 51: 355 (1972)

09-1601   SOME PHYSICAL AND ELECTRICAL SWITCHING CHARACTERISTICS OF A LEAD ZIRCONATE TITANATE
           FERROELECTRIC CERAMIC.
           STEWART WC + COSENTINO LS
           FERROELECTRICS 1: 149-67 (1970)

09-1602   PREPARATION AND ELECTRICAL PROPERTIES OF SOME CERAMIC MATERIALS AT MICROWAVE
           FREQUENCIES. (TITANATE ZIRCONATE GARNET MIXTURES)
           STIGLITZ ME
           AIR FORCE CAMBRIDGE RES LABS, 1970, 17P. AD717686
               CHEM ABSTR 75: 52350G (1971); GOVT REP ANNOUNCE 71(6): 87 (1971)
               225

09-1603   DIELECTRIC SYSTEM POLARIZATION AND THERMAL EXPANSION OF POLARIZED CERAMICS IN THE LEAD
           TITANATE ZIRCONATE MAGNESATE(0.5) TUNGSTATE(0.5).
           STOLYPIN YUE + ISUPOV VA
           SOV PHYS SOLID STATE 9(9): 2059-63 (1968)

09-1604   MOSSBAUER STUDIES OF LEAD ZIRCONATE TITANATE STANNATE. (CERAMIC DOPED WITH NIOBIUM AND
           TITANIUM)
           SUNDARAM VA + SUBBARAO EC
           P134-6 OF INT MEETING ON FERROELECTRICITY, 2ND, 1969, KYOTO, PROC,
           (J PHYS SOC JAP, VOL 28 SUPPL). PHYS SOC JAP, 1970, 459P.

           INFLUENCE OF DOMAIN STRUCTURE OF CERAMIC FERROELECTRICS ON THEIR MECHANICAL PROPERTIES.
           (BARIUM TITANATE, LEAD ZIRCONATE TITANATE, LEAD METANIOBATE, 20-150 DEG C, PRESSURE TO
           1200 KG/SQ CM)
           SYRKIN LN + ELGARD AM
           SOV PHYS SOLID STATE 7(4): 967-71 (1965)

09-1605   ELECTROMECHANICAL PROPERTIES OF CERAMIC FERROELECTRICS (BARIUM TITANATE ZIRCONATES) IN
           STRONG ELECTRIC FIELDS AND AT HIGH PRESSURE.
           SYRKIN LN + ELGARD AM
           SOV PHYS SOLID STATE 6(11): 2586-90 (MAY 1965)

09-1606   DETERIORATION OF THE ELECTROMECHANICAL PROPERTIES OF LEAD ZIRCONATE(0.52)
           TITANATE(0.48) CERAMICS CAUSED BY THE ADDITION OF NIOBIUM PENTOXIDE (BELOW 0.2
           PERCENT).
           TAKAHASHI M + TAKAHASHI S
           JAP J APPL PHYS 9: 1009 (1970)
               207

09-1607   ELECTRICAL RESISTIVITY OF LEAD ZIRCONATE TITANATE CERAMICS CONTAINING IMPURITIES.
           TAKAHASHI M
           JAP J APPL PHYS 10: 643-51 (MAY 1971)

9. MIXED TITANATES-ZIRCONATES (INCLUDING PZT, PLZT)

09-1608 EFFECTS OF IMPURITIES ON THE MECHANICAL QUALITY FACTOR OF LEAD ZIRCONATE TITANATE
         CERAMICS.
         TAKAHASHI S + TAKAHASHI M
         JAP J APPL PHYS 11(1): 31-5 (1972)
              278

09-1609 SPACE CHARGE EFFECT IN LEAD ZIRCONATE TITANATE CERAMICS CAUSED BY THE ADDITION OF
         IMPURITIES.
         TAKAHASHI M
         JAP J APPL PHYS 9(10): 1236-46 (1970)
              217

09-1610 LINEAR ELECTROOPTIC EFFECT IN FERROELECTRIC CERAMICS: PLZT 12-40-60. (USED FOR SENSING
         LOW VOLTAGES)
         THACHER PD
         FERROELECTRICS 3: 147-56 (1972);
         IEEE TRANS SON ULTRASON 19: 147-56 (1972)
              263

09-1611 STABILIZATION EFFECTS AT THE PHASE TRANSITION TETRAGONAL- RHOMBOHEDRAL IN LEAD
         ZIRCONATE TITANATE CERAMICS.
         THOMANN H
         J PHYS (PARIS) 33, SUPPL 4, COLLOQ 2: 281-3 (1972)
         FERROELECTRICS 4(3): 141-6 (1972)
              294

09-1612 ELASTIC SURFACE WAVE ON THE PZT CERAMIC PLATE.
         TODA K + SAKURAI S + KAWABATA A + TANAKA T
         TRANS INST ELECTRON COMMUN ENG JAP A 53: 89-96 (1970) (IN JAPANESE)

09-1613 PROPAGATION CHARACTERISTICS OF SURFACE ELASTIC WAVES ON BARIUM TITANATE(0.95)
         ZIRCONATE(0.05) CERAMIC PLATES.
         TODA K
         J APPL PHYS 43(2): 261-5 (FEB 1972)
              257

09-1614 SURFACE WAVE DELAY LINES WITH INTERDIGITAL TRANSDUCERS ON UNPOLARIZED PZT CERAMIC
         PLATES.
         TODA K + KAWABATA A + TANAKA T
         JAP J APPL PHYS 10: 671-7 (1971)

09-1615 METHOD OF MEASURING POLARIZATION VIBRATIONS OF A PREPOLARIZED CERAMIC FERROELECTRIC.
         (PZT)
         TROCCAZ M + PERRIGOT J + GONNARD P + FETIVEAU Y + EYRAUD L
         COMPT REND B 275: 597-600 (1972) (IN FRENCH)

09-1616 LEAD TITANATE PIEZOELECTRIC CERAMICS. (MATSUSHITA LEAD TITANATE)
         UEDA I + IKEGAMI S
         NAT TECH REP 15(6): 643-55 (1969) (IN JAPANESE, ENGLISH ABSTRACT)
              ENG INDEX 1970: 492

09-1617 FERROELECTRIC AND PIEZOELECTRIC PROPERTIES IN A LEAD TITANATE- LEAD ZIRCONATE- LEAD
         (ZINCATE(0.33) NIOBATE(0.67)) SYSTEM.
         UGRYUMOVA MA + ANANEVA AA
         SOV PHYS DOKL 16(9): 767-9 (1972)

09-1618 ELECTROOPTIC CERAMICS AS WAVELENGTH SELECTION DEVICES IN DYE LASERS (PLZT).
         VARNADO SG + SMITH WD
         IEEE J QUANTUM ELECTRON 8: 88-9 (1972)

09-1619 X-RAY ANALYSIS OF MODIFIED LEAD ZIRCONATE- LEAD TITANATE CERAMICS.
         VINCENT SM
         BULL AMER CERAM SOC 50: 755 (1971)

09-1620 STUDY OF PHASE TRANSITIONS IN THE LEAD ZIRCONATE TITANATE SYSTEM.
         WEIRAUCH DF
         UNIV ILLINOIS PHD THESIS, 1968.
              DISS ABSTR INT B 30(1): 185 (1969-70)
              200

9. MIXED TITANATES-ZIRCONATES (INCLUDING PZT, PLZT)

09-1621 STUDY OF PHASE TRANSITIONS IN THE LEAD ZIRCONATE- LEAD TITANATE SYSTEM.
WEIRAUCH DF
UNIV ILLINOIS PHD THESIS, 1968, 109P.
    DISS ABSTR INT B 30(1): 185 (1969)
    200

09-1622 EFFECTS OF CALCINING ON THE FIRING OF LEAD ZIRCONATE TITANATE CERAMIC
WILCOX PD + BUCKNER DA
BULL AMER CERAM SOC 49: 833 (1970)

09-1623 INTERNAL FRICTION OF MODIFIED LEAD ZIRCONATE- LEAD TITANATE CERAMICS. (PROPRIETARY
COMPOSITIONS)
YAMAUCHI F + TAKAHASHI M
P313-15 OF INT MEETING ON FERROELECTRICITY, 2ND, 1969, KYOTO, PROC,
(J PHYS SOC JAP, VOL 28 SUPPL). PHYS SOC JAP, 1970, 459P.

NEC RES DEVELOP NO 17: 52-5 (APR 1970)
    209

09-1624 STUDIES ON THE SINTERING BETWEEN LEAD MONOXIDE AND PZT. PART-1. PART-2: EFFECTS OF
AVERAGE PARTICLE SIZE OF RAW POWDER ON SINTERING. PART-3: SINTERING PROCESS AND
MECHANISM OF THE REACTION BETWEEN LEAD MONOXIDE AND PZT. PART-4: EFFECT OF BORON OXIDE
ADDITION ON THE PRODUCT OF PZT. PART-5: EFFECT OF THE FORMING PRESSURE ON SINTERING.
PART-6: FORMATION OF PZT AND SINTERING UNDER OXYGEN ATMOSPHERE. PART-7: EFFECT OF
CALCINATION ON SINTERING. PART-8: MECHANOCHEMISTRY OF THE SINTERING BODY. PART-9:
EFFECT OF MOTHER SALT OF LEAD MONOXIDE ON THE SINTERING. PART-10: PROPERTIES OF LEAD
OXY HYDROXIDE POWDER.
YAMAGUCHI O + HARUMI F + SHIMOMURA S + YASUMOTO M + TAKEOKA K + SHIMIZU K + TOKITO Y
+ OHASHI Y + MIWA K + SUZUKI H
FUNTAI OYOBI FUNMATSU YAKIN (J JAP SOC POWDER MET) 17: 116-21, 251-5,
(1971); 18: 68-72, 73-4, 95-9, 100-4, 145-9, 181-6, 266-74 (1972);
19: 148-56 (1972) (IN JAPANESE)
    256

09-1625 CHROMATOGRAPHIC METHOD FOR ANALYZING LEAD ZIRCONATE TITANATES CONTAINING BISMUTH AND
STRONTIUM ADDITIVES.
ZHECHKOVA LA + XENNIKOVA TI
GLASS CERAM 28(6): 380-1 (1971)
    256

10. TUNGSTEN TRIOXIDE AND TUNGSTATES

10-1626   THERMAL EXPANSION AND THE ORTHORHOMBIC- TETRAGONAL TRANSITION OF THE TUNGSTEN TRIOXIDE
PHASE.
ACKERMANN RJ + SORRELL CA
HIGH TEMP SCI 2: 119-30 (1970)

10-1627   EXAMINATION OF SUBSTOICHIOMETRIC TUNGSTEN OXIDE(3-X) CRYSTALS BY ELECTRON MICROSCOPY.
ALLPRESS JG + TILLEY RJD + SIENKO MJ
J SOLID STATE CHEM 3: 440-51 (1971)

10-1628   EPR STUDY OF THE CATALYTIC ACTIVITY OF NONSTOICHIOMETRIC TUNGSTEN TRIOXIDE.
ALQUIE AM + LAMY C
COMPT REND C 275: 1207-10 (1972) (IN FRENCH)

10-1629   EFFECT OF OXYGEN DEFICIENCY ON ELECTRICAL TRANSPORT PROPERTIES OF TUNGSTEN TRIOXIDE
CRYSTALS.
BERAK JM + SIENKO MJ
J SOLID STATE CHEM 2: 109-33 (1970)

10-1630   CS (CRYSTALLOGRAPHIC SHEAR) FAMILIES DERIVED FROM THE RHENIUM TRIOXIDE STRUCTURE TYPE:
AN ELECTRON MICRCSCOPE STUDY OF REDUCED TUNGSTEN TRIOXIDE AND RELATED PSEUDOBINARY
SYSTEMS.
BURSILL LA + HYDE BG
J SOLID STATE CHEM 4: 430-46 (1972)

10-1631   STUDY OF THE X-RAY PHOTOELECTRON SPECTRUM OF TUNGSTEN- TUNGSTEN OXIDE AS A FUNCTION OF
THICKNESS OF THE SURFACE OXIDE LAYER.
CARLSON TA + MCGUIRE GE
J ELECTRON SPECTROSC 1: 161-8 (1972)

DIELECTRIC PROPERTIES OF NIOBATES AND TUNGSTATES.
EMMENEGGER FP + ROETSCHI H
J PHYS CHEM SOLIDS 32: 787-90 (APR 1971)

10-1632   THE COLOR PROBLEM OF TUNGSTEN TRIOXIDE. X-RAY STUDIES.
FARAG MS + HANAFI Z + KHILLA MA
Z PHYS CHEM 76: 265-72 (1971)

10-1633   INVESTIGATION OF PHASE TRANSITION IN LEAD CADMIUM TUNGSTATE.
FILIPEV VS + FESENKO EG
BULL ACAD SCI USSR, PHYS SER 29(6): 900-1 (1965)
     219

ORIENTATION OF COMPOSITION PLANES IN ANTIFERROELECTRICS. (APPLIED TO ADP, TUNGSTEN
TRIOXIDE, SODIUM NIOBATE)
FOUSEK J + JANOVEC V
P380-1 OF INT MEETING ON FERROELECTRICITY, 2ND, 1969, KYOTO, PROC,
(J PHYS SOC JAP, VOL 28 SUPPL). PHYS SOC JAP, 1970, 459P.

TIGHT BONDING BAND CALCULATIONS ON RHENIUM TRIOXIDE, SODIUM TUNGSTATE, AND POTASSIUM
TANTALATE.
GERSTEIN BC + KARIAN HG
BULL AMER PHYS SOC 15: 311 (1970)

10-1634   OPTICAL TRANSITION TO THE FERMI LEVEL IN SODIUM TUNGSTATE.
GIULIANI A + GUSTINETTI A + STELLA A
PHYS LETT A 38: 515-16 (1972)

10-1635   RAMAN SPECTROSCOPY OF TUNGSTEN TRIOXIDE.
HANNON DM
BULL AMER PHYS SOC 15: 297 (1970)

10-1636   THE COLOR PROBLEM OF TUNGSTEN TRIOXIDE ELECTRICAL CONDUCTIVITY.
          HANAFI Z + KHILLA MA AND OTHERS
          Z PHYS CHEM (FRANKFURT) 82(5-6): 209-16 (DEC 1972)
              299

10-1637   ELECTRICAL CONDUCTIVITY OF TUNGSTEN TRIOXIDE.
          HIROSE T + KAWANO I + NIINO M
          J PHYS SOC JAP 33: 272 (1972)
              278

10-1638   ELECTROMECHANICAL PROPERTIES IN THE BINARY SYSTEM LEAD TITANATE- LEAD CADMATE(0.5)
          TUNGSTATE(0.5).
          ICHINOSE N + TAKAHASHI T
          JAP J APPL PHYS 11: 1224-5 (AUG 1972)

10-1639   AN EFFECT OF HEAT TREATMENT ON TUNGSTEN TRIOXIDE CRYSTAL. PART-2: GREEN STRIPES.
          IWAI T + HORIE T
          J PHYS SOC JAP 17: 1142-7 (1962)

10-1640   THALLIUM PEROVSKITES. THALLIUM A(1-X) TUNGSTATE OR MOLYBDATE. (WHERE A IS ZIRCONIUM,
          HAFNIUM, CERIUM, THORIUM, TITANIUM, SODIUM, MAGNESIUM, CADMIUM, YTTRIUM, GADOLINIUM,
          DYSPROSIUM, IRON, SELENIUM, NIOBIUM, COBALT, SCANDIUM, OR COPPER)
          KAPYSHEV AG + VENEVTSEV YUN
          BULL ACAD SCI USSR, PHYS SER 35(9): 1675-7 (1971)
              269

10-1641   ON FERROELECTRICITY AND ANTIFERROELECTRICITY OF THE AO3 TYPE CRYSTAL. (TUNGSTEN
          TRIOXIDE)
          KINASE W + ISHIBASHI Y + KURASAWA Y
          J PHYS SOC JAP 19: 273-81 (1964)

10-1642   BOMBARDMENT INDUCED AMORPHIZATION IN TUNGSTEN TRIOXIDE AND ITS USE IN DEDUCING MEAN
          DAMAGE RANGES. (STUDIES ON BOMBARDMENT INDUCED DISORDER. PART-6.)
          LAM NQ + KELLY R
          CAN J PHYS 50: 1887-95 (1972)

          STUDY OF DOMAINS IN FERROELECTRIC CRYSTALS WITH A PHOTOEMISSION MICROSCOPE (TUNGSTEN
          TRIOXIDE, BARIUM TITANATE).
          LE BIHAN R
          COMPT REND B 270: 741-4 (1970) (IN FRENCH)

10-1643   (PREPARATION AND ELECTRICAL AND OPTICAL) STUDY OF TUNGSTEN TRIOXIDE AT LOW TEMPERATURE.
          LE BIHAN R + VACHERAND C
          P159-61 OF INT MEETING ON FERROELECTRICITY, 2ND, 1969,, KYOTO, PROC,
          (J PHYS SOC JAP, VOL 28 SUPPL). PHYS SOC JAP, 1970, 459P.

10-1644   PREPARATION OF TUNGSTIC OXIDE SINGLE CRYSTALS (SUBLIMATION IN AIR OF TUNGSTEN
          TRIOXIDE).
          LE BIHAN R + VACHERAND C
          P147-57 OF SUCHET JP(ED), CROISSANCE DE COMPOSES MINERAUX
          CRYSTALLINS: GROWTH OF SINGLE CRYSTAL MINERAL COMPOUNDS, MASSON,
          1969, 169P. (IN FRENCH)
                CHEM ABSTR 72: 71459
                    178

10-1645   PHOTOEMISSION MICROSCOPIC OBSERVATION OF THE SURFACE OF TUNGSTEN TRIOXIDE CRYSTALS.
          LE BIHAN R
          COMPT REND B 274(26): 1428-31 (1972) (IN FRENCG)
                CHEM ABSTR 77: 145243F (1972)
                    288

10-1646   STUDY OF DOMAINS OF FERROELECTRIC CRYSTALS WITH SCANNING ELECTRON MICROSCOPE. (TUNGSTEN
          TRIOXIDE SINGLE CRYSTAL PLATES)
          LE BIHAN R + SELLA C
          P377-9 OF INT MEETING ON FERROELECTRICITY, 2ND, 1969,, KYOTO, PROC,
          (J PHYS SOC JAP, VOL 28 SUPPL). PHYS SOC JAP, 1970, 459P.

10-1647   EVAPORATED FILMS OF TUNGSTEN OXIDE.
          MIURA S + HARADOME M + BAN I + TAKETA Y
          OYO BUTSURI 32: 817-   (1963) (IN JAPANESE)

10. TUNGSTEN TRIOXIDE AND TUNGSTATES

10-1648 STRUCTURAL AND PHASE RELATIONSHIPS AMONG TRIVALENT TUNGSTATES AND MOLYBDATES.
NASSAU K + SHIEVER JW
P445-56 OF MATER RES SYMP, 5TH PROC, 1971, SOLID STATE CHEMISTRY,
ROTH RS + SCHNEIDER SJ(EDS), NBS SPEC PUB 364, 1972, 783P.

271

10-1649 STRUCTURAL AND PHASE RELATIONSHIPS AMONG TRIVALENT TUNGSTATES AND MOLYBDATES.
NASSAU K + SHIEVER JW + KEVE ET
J SOLID STATE CHEM 3: 411-19 (1971)

10-1650 GROWTH, CRYSTALLOGRAPHY AND DIELECTRIC PROPERTIES OF BISMUTH TUNGSTATE.
NEWKIRK HW + QUADFLIEG P + LIEBERTZ J + KOCKEL A
FERROELECTRICS 4: 51-5 (1972)
267

10-1651 STRUCTURAL AND DIELECTRIC MODIFICATIONS OF TUNGSTEN TRIOXIDE DUE TO NEUTRON
IRRADIATION.
SAVIN W + FINK T + GAYDOS W + LEFKOWITZ I + SHIELDS M
BULL AMER PHYS SOC 16: 499 (1971)

10-1652 A NEW FORM OF TUNGSTEN TRIOXIDE.
SCHRODER FA + FELSER H
Z KRISTALLOGR 135: 391-8 (1972) (IN GERMAN)

10-1653 SOME ASPECTS OF THE STRUCTURE OF TUNGSTEN TRIOXIDE AND A CONTRIBUTION TO THE
UNDERSTANDING OF THE SO-CALLED "SHEAR STRUCTURES".
SCHRODER FA + HARTMAN P
Z NATURFORSCH B 27: 902-8 (1972)

10-1654 RAMAN STUDY OF METALLIC TUNGSTEN BRONZES. (SODIUM AND RUBIDIUM TUNGSTATES)
SCOTT JF + LEHENY RF + SWEEDLER AR
PHYS REV B 2: 3883-7 (1970)

10-1655 NEUTRON DIFFRACTION STUDY OF (TETRAGONAL- CUBIC PEROVSKITE) STRONTIUM IRON TUNGSTATE .
(MAGNETIC G STRUCTURE ANALOGOUS TO ANTIFERROELECTRIC STRUCTURE)
SOLOVEV SP + SMIRNOV VP + FADEEVA NV + IVANOVA VV + KAPISHEV AG
P286-7 OF INT MEETING ON FERROELECTRICITY, 2ND, 1969, KYOTO, PROC,
(J PHYS SOC JAP, VOL 28 SUPPL). PHYS SOC JAP, 1970, 459P.

10-1656 THE BISMUTH OXIDE- TUNGSTEN TRIOXIDE SYSTEM.
SPERANSKAYA EI
INORG MATER 6: 127-9 (1970)

10-1657 DIELECTRIC (AND TEMPERATURE) CHARACTERISTICS OF SOLID SOLUTIONS OF TUNGSTEN OXIDE.
TANAKA M
OYO BUTSURI 33(4): 260-3 (1964) (IN JAPANESE)
CHEM ABSTR 62: 8475 (1965)

10-1658 PHASE TRANSITION IN TUNGSTEN TRIOXIDE UNDER HIGH PRESSURE.
TANAKA M + MINOMURA S + SAWADA S
P238-43, VOL 1 OF DVORAK V + FOUSKOVA A + GLOGAR P (EDS),
INT MEETING ON FERROELECTRICITY, PRAGUE, 1966, PROC.
INST PHYS CZECH ACAD SCI, PRAGUE, 1966.

10-1659 FORMATION OF SHEAR STRUCTURES IN SUBSTOICHIOMETRIC TUNGSTEN TRIOXIDE.
TILLEY RJD
MATER RES BULL 5: 813-23 (1970)

11. LITHIUM NIOBATE

11-1660 FERROELECTRICS IN THE LITHIUM POTASSIUM NIOBATE SYSTEM.
AINGER FW + BESWICK JA + BICKLEY WP AND OTHERS
FERROELECTRICS 2: 183-99 (JUL 1971)

11-1661 HOLOGRAPHIC PATTERN FIXING IN ELECTROOPTIC CRYSTALS. (LITHIUM NIOBATE, BARIUM SODIUM
NIOBATE)
AMODEI JJ + STAEBLER DL
APPL PHYS LETT 18: 540-2 (1971)

11-1662 HOLOGRAPHIC RECORDING IN LITHIUM NIOBATE.
AMODEI JJ + STAEBLER DL
RCA REV 33: 71-93 (1972)

11-1663 FRACTURE OF NONLINEAR CRYSTALS (KDP AND LITHIUM NIOBATE) BY RADIATION FROM A RUBY
LASER.
ANANIN OB + BYKOVSKII YUA + PETROVSKII AN + REZ IS
SOV PHYS TECH PHYS 17: 659-61 (1972)

11-1664 INHOMOGENEITY OF THE OPTICALLY INDUCED INDEX OF REFRACTION IN LITHIUM NIOBATE AND
LITHIUM TANTALATE.
ANGERT NB + PASHKOV VA + SOLOVEVA NM
SOV PHYS JETP 35(5): 867-9 (1972)
                    290

11-1665 PHASE MATCHING ANGLES AND TEMPERATURE OF LITHIUM METANIOBATE CRYSTALS OF DIFFERENT
STOICHIOMETRIES.
ANGERT NB + BUTYAGIN OF + ZORENKO VP + KUDRYAVTSEVA AP + KUSHNIR VR + RUSTAMOV SR
SOV J QUANTUM ELECTRON 1: 542-3 (1972)

11-1666 ACOUSTIC DIPOLE AND QUADRUPOLE SATURATION OF THE LITHIUM-7 NMR SIGNALS IN LITHIUM
NIOBATE.
ANTOKOLSKII GL + SARNATSKII VM + SHUTILOV VA
SOV PHYS SOLID STATE 12: 1831-3 (1971)

11-1667 LAYER WAVE AMPLIFICATION IN A CADMIUM SULFIDE FLUID- LITHIUM NIOBATE STRUCTURE.
ARAGHI MN + DAS P
APPL PHYS LETT 18: 133-5 (1971)

11-1668 HEAT TREATMENT INDUCED POINT DEFECTS IN LITHIUM METANIOBATE SINGLE CRYSTALS.
ARSENEV PA + PARSHTENDIKER VL
KRISTALL TECH 7: K73-6 (1972)

11-1669 PROPERTIES OF THE IRON TRANSITION GROUP IN THE LATTICE OF SINGLE CRYSTALLINE LITHIUM
NIOBATE.
ARSENEV PA + BARANOV BA
PHYS STATUS SOLIDI A 9: 673-7 (1971)

11-1670 LOW TEMPERATURE COEFFICIENT OF LITHIUM NIOBATE RESONATORS BY USE OF TEMPERATURE
SENSITIVE CAPACITANCES.
ASHIDA T
PROC IEEE 58: 146-7 (1970)

11-1671 X-RAY TOPOGRAPHICAL STUDY OF THE CONTRAST AT FERROELECTRIC DOMAIN BOUNDARIES (DOPED AND
UNDOPED TGS, LITHIUM NIOBATE, KDP, ABSTRACT ONLY).
AUTHIER A + PETROFF JF
P373 OF INT MEETING ON FERROELECTRICITY, 2ND, 1969, KYOTO, PROC,
(J PHYS SOC JAP, VOL 28 SUPPL). PHYS SOC JAP, 1970, 459P.

11. LITHIUM NIOBATE

11-1672 GROWTH TWINNING (UNDER DIRECT CURRENT) IN C-AXIS LITHIUM NIOBATE CRYSTALS (CZOCHRALSKI METHOD).
AZARBAYEJANI GH
J CRYST GROW 7: 327-8 (1970)
217

11-1673 X-RAY DETERMINATION OF POLARITY SENSE BY ANOMALOUS SCATTERING AT AN ABSORPTION EDGE.
(LITHIUM NIOBATE, BARIUM MANGANESE TETRAFLUORIDE)
BARNS RL + KEVE ET + ABRAHAMS SC
J APPL CRYSTALLOGR 3: 27-32 (1970)

11-1674 NEODYMIUM YAG LASER IRRADIATION INDUCED DAMAGE TO LITHIUM NIOBATE AND KDP.
BASS M
IEEE J QUANTUM ELECTRON 7: 350-9 (1971)

11-1675 NONCOLLINEAR PHASE MATCHING EFFECTS IN LITHIUM NIOBATE.
BATES HE
J OPT SOC AMER 61: 904-9 (1971)

11-1676 THREE LAYER BROAD BAND ANTIREFLECTION COATINGS FOR LITHIUM NIOBATE.
BERTHOLD JW
APPL OPT 9: 1490-1 (1970)

11-1677 RELATIVE MEASUREMENT OF THE OPTICAL NONLINEARITIES OF POTASSIUM DIHYDROGEN PHOSPHATE,
AMMONIUM DIHYDROGEN PHOSPHATE, LITHIUM NIOBATE, AND ALPHA- IODIC ACID.
BJORKHOLM JE
IEEE J QUANTUM ELECTRON 5(5): 260 (1969)
171

11-1678 COLLINEAR INTERACTION OF LONGITUDINAL ELASTIC WAVES IN LITHIUM NIOBATE CRYSTALS.
BOGDANOV VL + LEMANOV VV + YUSHIN NK
SOV PHYS SOLID STATE 13: 1208-11 (1971)

NONLINEAR SADOVSKII EFFECT (THEORY, PROPAGATION OF RADIATION AND FIRST HARMONIC,
CALCULATIONS FOR LITHIUM NIOBATE).
BOKUT BV + SERDYUKOV AN
J APPL SPECTROSC 11: 1059-61 (1969)

11-1679 DISORDER OF LITHIUM NIOBATE CRYSTALS.
BOLLMANN W + GERNAND M
PHYS STATUS SOLIDI A 9: 301-8 (1971)

11-1680 COLLINEAR INTERACTION BETWEEN A LONGITUDINAL ACOUSTIC WAVE AND TWO LUMINOUS WAVES ALONG
THE (100) DIRECTION OF LITHIUM NIOBATE.
BRIDOUX E + DELANNOY M + MORIAMEZ M + ROUVAEN JM
COMPT REND B 272: 717-19 (1971) (IN FRENCH)

11-1681 DETERMINATION OF A NONLINEAR PARAMETER FOR ACOUSTIC SURFACE WAVE CONVOLUTION IN LITHIUM
NIOBATE.
BRIDOUX E + MORIAMEZ M + ROUVAEN JM + DIEULESAINT E
ELECTRON LETT 8: 257-8 (1972)

11-1682 GENERATION OF SECOND AND THIRD HARMONICS BY A TRANSVERSE ACOUSTIC WAVE PROPAGATING
ITSELF IN THE (001) DIRECTION OF LITHIUM NIOBATE.
BRIDOUX E + THERY P + MORIAMEZ C + DEROYON JP
COMPT REND B 270: 933-6 (1970) (IN FRENCH)

11-1683 PHASE EQUILIBRIA AND CRYSTAL GROWTH IN THE LITHIUM OXIDE- NIOBIUM PENTOXIDE SYSTEM.
BRIDENBAUGH PM
J ELECTROCHEM SOC 117: 196C (1970)

11-1684 SPATIALLY UNIFORM AND ALTERABLE SHG PHASE MATCHING TEMPERATURES IN LITHIUM NIOBATE
(CRYSTAL GROWTH BY ADDING MAGNESIUM OXIDE TO A CONGRUENT MELT).
BRIDENBAUGH PM + CARRUTHERS JR + DZIEDZIC JM + NASH FR
APPL PHYS LETT 17(3): 104-6 (1970)
202

11-1685 THERMAL EXPANSION OF LITHIUM NIOBATE IN THE TEMPERATURE RANGE OF 70-300 K.
BROWDER JS + BALLARD SS
J OPT SOC AMER 62: 1405 (1972)

11. LITHIUM NIOBATE

11-1686   WIDE RANGE TEMPERATURE COMPENSATION BY ADDITION OF TWO CRYSTAL RESONATOR FREQUENCIES: APPLICATION TO QUARTZ AND LITHIUM TANTALATE.
BRUNNER J
ELECTRON LETT 8: 639-40 (1972)

11-1687   INFLUENCE OF A TRANSVERSE INHOMOGENEITY OF THE REFRACTIVE INDEX OF A NONLINEAR CRYSTAL ON SECOND HARMONIC GENERATION. (LITHIUM NIOBATE)
BUTYAGIN OF + ZORENKO VP + ILINSKII YUA
SOV J QUANTUM ELECTRON 1(4): 393-6 (1972)
    289

11-1688   GROWTH OF HIGH QUALITY LITHIUM NIOBATE CRYSTALS (FOR NONLINEAR OPTICS WITH LOW BIREFRINGENCE VARIATIONS) FROM THE CONGRUENT MELT.
BYER RL + YOUNG JF + FEIGELSON RS
J APPL PHYS 41(6): 2320-5 (1970)
    196

11-1689   ROTATING WAVEPLATE OPTICAL FREQUENCY SHIFTING IN LITHIUM NIOBATE.
CAMPBELL JP + STEIER WH
IEEE J QUANTUM ELECTRON 7: 450-7 (1971)

11-1690   MIXING OF NONCOLLINEAR ELASTIC SURFACE WAVES ON LITHIUM NIOBATE.
CARR PH
J APPL PHYS 42: 5330-2 (DEC 1971)

11-1691   NONSTOICHIOMETRY AND CRYSTAL GROWTH OF LITHIUM NIOBATE.
CARRUTHERS JR + PETERSON GE + GRASSO M + BRIDENBAUGH PM
J APPL PHYS 42: 1846-51 (1971)

11-1692   ACOUSTIC SURFACE WAVE CONVOLUTION ON CRYSTALS OF CADMIUM SULFIDE, LITHIUM NIOBATE AND BISMUTH(12) GERMANIUM OXIDE(20).
CHAMBERS J + MASON IM + TURNER CW
ELECTRON LETT 8: 314-16 (1972)

11-1693   THERMAL DEPENDENCE OF DIELECTRIC, PIEZOELECTRIC, AND ELASTIC PROPERTIES OF LITHIUM NIOBATE SINGLE CRYSTALS.
CHKALOVA VV + BONDARENKO VS + FOKINA GO AND OTHERS
BULL ACAD SCI USSR, PHYS SER 35(9): 1712-15 (1971)

11-1694   ASSIGNMENTS OF OPTICAL PHONON MODES IN LITHIUM NIOBATE.
CLAUS R + BORSTEL G + WIESENDANGER E + STEFFAN L
PHYS REV B 6(12): 4878-9 (15 DEC 1972)
    281

11-1695   DIRECTIONAL DISPERSION AND ASSIGNMENT OF OPTICAL PHONONS IN LITHIUM NIOBATE.
CLAUS R + BORSTEL G + WIESENDANGER E + STEFFAN L
Z NATURFORSCH A 27: 1187-92 (1972)

11-1696   DIRECTIONAL DISPERSION OF PHONON FREQUENCIES AND POLARITON SCATTERING IN LITHIUM NIOBATE.
CLAUS R + SCHROTTER HW
P244-7 OF INT CONF ON LIGHT SCATTERING IN SOLIDS, 2ND PROC,
BALKANSKI M(ED), PARIS, 1971, FLAMMARION, PARIS, 1971, 518P.

11-1697   LIGHT SCATTERING BY POLARITONS ASSOCIATED WITH ORDINARY PHOTONS IN LITHIUM NIOBATE.
CLAUS R + BORSTEL G + MERTEN L
OPT COMMUN 3: 17-18 (1971)

11-1698   SURFACE WAVE DIFFRACTION IN LITHIUM NIOBATE.
CRABB JC + MAINES JD + OGG NR
ELECTRON LETT 7: 253-5 (1971)

11-1699   ACOUSTIC RADIATION BY INTERDIGITAL GRIDS ON LITHIUM NIOBATE.
DANIEL MR + EMTAGE PR
J APPL PHYS 43: 4872-5 (1972)

11-1700   SURFACE WAVE CONVOLUTION USING A CADMIUM SULFIDE FLUID LITHIUM NIOBATE STRUCTURE.
DAS P + ARAGHI MN
APPL PHYS LETT 21: 373-4 (1972)

11. LITHIUM NIOBATE

11-1701 MEASUREMENTS OF PHOTON CORRELATIONS OF SECOND HARMONIC GENERATED LIGHT (FROM LITHIUM
NIOBATE CRYSTAL, AGREEMENT WITH THEORY).
DAVIDSON F + KLEBBA J + LAURENCE C + TITTEL FK
APPL PHYS LETT 17(3): 117-20 (1970)
202

11-1702 MEASURED ELECTRICAL CHARACTERISTICS OF INTERDIGITAL SURFACE WAVE TRANSDUCERS ON LITHIUM
NIOBATE.
DE KLERK J + DANIEL MR
APPL PHYS LETT 16: 219-21 (1970)

11-1703 DOMAIN STRUCTURE IN LITHIUM NIOBATE SINGLE CRYSTALS.
DESHMUKH KG + SINGH K
J PHYS D APPL PHYS 5: 1680-5 (1972)
275

11-1704 GROUP THEORETICAL SELECTION RULES FOR INELASTIC NEUTRON SCATTERING WITH APPLICATIONS TO
LITHIUM NIOBATE.
DEVINE SD + PECKHAM G
J PHYS C 4: 1091-1104 (1971)

11-1705 RADIATION DAMAGE IN LITHIUM NIOBATE. (CAPACITANCE, OPTICAL SPECTRA, ESR, CELL
PARAMETERS)
DOWELL MB + LEFKOWITZ I + TAYLOR GW
P442-4 OF INT MEETING ON FERROELECTRICITY, 2ND, 1969, KYOTO, PROC,
(J PHYS SOC JAP, VOL 28 SUPPL). PHYS SOC JAP, 1970, 459P.

11-1706 DISPERSION OF ELASTIC WAVES IN FERROELECTRICS. (LITHIUM NIOBATE)
ERMILIN KK + KRASILNIKOV VA + PROKHOROV VM
SOV PHYS SOLID STATE 14(1): 251-2 (JUL 1972)
270

11-1707 GENERATION OF SUPERHIGH FREQUENCY ACOUSTIC HARMONICS IN A (SINGLE DOMAIN) LITHIUM
NIOBATE CRYSTAL (NONLINEAR ELASTIC PROPERTIES).
ERMILIN KK + ZAREMBO LK + KRASILNIKOV VA
SOV PHYS SOLID STATE 12(5): 1045-8 (1970)
215

11-1708 DOMAIN STRUCTURE OF LITHIUM METANIOBATE CRYSTALS.
EVLANOVA NF + RASHKOVICH LN
SOV PHYS SOLID STATE 13(1): 223-4 (JUL 1971)
237

11-1709 ELECTRICAL PROPERTIES OF LITHIUM NIOBATE. (ANOMALY AT 263 DEG K IS STRUCTURAL
TRANSITION)
FERMOR JH + KJEKSHUS A
ACTA CHEM SCAND 23: 1581-7 (1969)
236

OPTICALLY INDUCED PHYSICAL DAMAGE TO LITHIUM NIOBATE, PROUSTITE, AND LITHIUM IODATE.
FOUNTAIN WD + OSTERINK LM + MASSEY GA
P91-7 OF DAMAGE IN LASER MATERIALS, 1971 PROC, GLASS AJ +
GUENTHER AH(EDS), NBS SPEC PUB 356, 1971.

11-1710 TRANSIENT ELECTROOPTIC EFFECTS AND Q SWITCHING PERFORMANCE IN LITHIUM NIOBATE AND
POTASSIUM DIDEUTERIUM PHOSPHATE POCKELS CELLS. (COMMENTS)
FOUNTAIN WD
APPL OPT 10: 972-3 (1971)

11-1711 GROWTH AND PROPERTIES OF FERROELECTRIC POTASSIUM LITHIUM TANTALATE(X) NIOBATE(1-X).
FUKUDA T + HIRANO H + KOIDE S
J CRYST GROW 6: 293-6 (1970)

11-1712 GROWTH AND PROPERTIES OF LITHIUM NIOBATE AND LITHIUM TANTALATE FILMS.
FUKUNISHI S + KAWANA A + UCHIDA N
ACTA CRYSTALLOGR A 28: S143 (1972)

11-1713 ABSORPTION AND LUMINESCENCE SPECTRA AND ENERGY LEVELS OF NEODYMIUM(3+) AND ERBIUM(3+)
IONS IN LITHIUM NIOBATE CRYSTALS.
GABRIELYAN VT + KAMINSKII AA + LI L
PHYS STATUS SOLIDI A 3: K37-42 (1970)
234

11-1714   MEASUREMENT OF SPUTTER ETCHING RATES FOR MONOCRYSTALLINE LITHIUM NIOBATE, SILICON
          DIOXIDE, AND SILICON.
          GARVIN HL
          BULL AMER PHYS SOC 16: 836 (1971)

11-1715   CUTS OF LITHIUM NIOBATE CRYSTALS SUITABLE FOR LIGHT MODULATORS WITH A LOW HALF WAVE
          VOLTAGE.
          GISIN BV
          SOV J QUANTUM ELECTRON 1: 686-7 (1972)

11-1716   CONTROL OF LASER DAMAGE IN LITHIUM NIOBATE.
          GLASS AM + PETERSON GE + BRIDENBAUGH PM
          IEEE TRANS SON ULTRASON 19: 341 (1972)

11-1717   TEMPERATURE AND FREQUENCY DEPENDENCES OF THE ELECTRICAL PROPERTIES OF LITHIUM
          METANIOBATE.
          GLUSHKOVA TM + KISELEV DF + FIRSOVA MM
          SOV PHYS SOLID STATE 13(9): 2299-2301 (MAR 1972)
               261

11-1718   TEMPERATURE DEPENDENCE OF HYPERSONIC WAVE DAMPING IN RUBY AND LITHIUM NIOBATE IN THE
          THREE CENTIMETER RADIO FREQUENCY RANGE.
          GRIGOREV MA + ZYURYUKIN YUA + NAYANOV VI AND OTHERS
          BULL ACAD SCI USSR, PHYS SER 35(5): 876-9 (1971)

11-1719   TEMPERATURE DEPENDENCES OF THE ABSORPTION OF 9.4 GHZ HYPERSOUND IN ALUMINUM OXIDE AND
          LITHIUM NIOBATE SINGLE CRYSTALS.
          GRIGOREV MA + ZYURYUKIN YUA + NAYANOV VI + POLOGNYAGIN VA + SHEVCHIK VN
          SOV PHYS SOLID STATE 12: 2449-50 (1971)

11-1720   BISMUTH(12) GERMANIUM OXIDE(20) VERSUS LITHIUM NIOBATE IN THE SURFACE WAVE AMPLIFIER. A
          THEORETICAL AND EXPERIMENTAL COMPARISON.
          HAGON PJ + SALLEE CF + LAKIN KM
          IEEE TRANS SON ULTRASON 18: 54 (1971)

11-1721   RADIATION EXPOSURE OF A LITHIUM NIOBATE CRYSTAL AT HIGH TEMPERATURES.
          HALVERSON SL + ANDERSON TT + GAVIN AP + GRATE T
          IEEE TRANS NUCL SCI 17: 335-40 (1970)

11-1722   TEMPERATURE DEPENDENCE OF THE LITHIUM NMR SPECTRUM AND ATOMIC MOTION IN LITHIUM
          NIOBATE.
          HALSTEAD TK
          J CHEM PHYS 53: 3427-35 (1970)

11-1723   ACOUSTOELECTRIC AMPLIFICATION OF SURFACE WAVE STRUCTURE OF A CADMIUM SELENIDE FILM ON
          LITHIUM NIOBATE.
          HANEBREKKE H + INGEBRIGTSEN KA
          ELECTRON LETT 6: 520-1 (1970)

11-1724   LITHIUM TANTALATE AND LITHIUM NIOBATE PIEZOELECTRIC RESONATORS IN THE MEDIUM FREQUENCY
          RANGE WITH LOW RATIOS OF CAPACITANCE AND LOW TEMPERATURE COEFFICIENTS OF FREQUENCY
          (DEVICE DESIGN AND CHARACTERIZATION).
          HANNON JJ + LLOYD P + SMITH RT
          IEEE TRANS SON ULTRASON 17(4): 239-46 (1970)
               216

11-1725   AN EPR INVESTIGATION OF IRON(3+) AND MANGANESE(2+) IN LITHIUM NIOBATE.
          HERRINGTON JB + DISCHLER B + SCHNEIDER J
          SOLID STATE COMMUN 10: 509-11 (1972)

11-1726   APPARATUS FOR GROWTH OF SINGLE CRYSTAL, SINGLE DOMAIN LITHIUM NIOBATE.
          HILTON RM
          AIR FORCE CAMBRIDGE RES LABS, 1969, 27P. AD699551
               CHEM ABSTR 72: 115438C (1970) US GOVT RES DEVELOP REP 70(6): 168 (1970)
               178

11-1727   TRANSIENT ELASTOOPTIC EFFECTS AND Q SWITCHING PERFORMANCE IN LITHIUM NIOBATE AND KDP
          POCKELS CELLS.
          HILBERG RP + HOOK WR
          APPL OPT 9: 1939-40 (1970)

11-1728   MELT COMPOSITION (STOICHIOMETRY) DEPENDENCE OF POCKELS EFFECT IN LITHIUM NIOBATE
           CRYSTALS.
           HIRANO H
           P90-2 OF INT MEETING ON FERROELECTRICITY, 2ND, 1969, KYOTO, PROC,
           (J PHYS SOC JAP, VOL 28 SUPPL). PHYS SOC JAP, 1970, 459P.

11-1729   LUMINESCENCE FROM LITHIUM NIOBATE.
           HORDVIK A + SCHLOSSBERG H
           IEEE J QUANTUM ELECTRON 8: 593 (1972)
           APPL PHYS LETT 20: 197-9 (1972)

11-1730   BIREFRINGENCE OF CERTAIN CRYSTALS IN THE MILLIMETER WAVELENGTH RANGE. (LITHIUM NIOBATE)
           IRISOVA NA + KOZLOV GV
           SOV PHYS CRYSTALLOGR 15(5): 941-2 (1971)
                232

11-1731   ACOUSTOOPTICAL MEASUREMENTS OF ACOUSTIC WAVE GENERATION IN PIEZOELECTRICS COUPLED TO A
           GUNN OSCILLATOR. (LITHIUM NIOBATE)
           ISHIDA A + SUMI M
           APPL PHYS LETT 18: 252-3 (1971)

11-1732   AN ELECTROOPTIC SWITCH USING ROTATED Y PLATE OF LITHIUM NIOBATE.
           ITAKURA M
           JAP J APPL PHYS 10: 957-8 (1971)

11-1733   ELECTRICAL CONDUCTIVITY OF LITHIUM NIOBATE SINGLE CRYSTALS.
           IVLEVA LI + KUZMINOV YUS + OSIKO VV
           INORG MATER 7: 1224-7 (AUG 1971)

11-1734   NONLINEAR OPTICAL POLARIZABILITY OF THE NIOBIUM- OXYGEN BOND. (LITHIUM NIOBATE, LITHIUM
           TANTALATE, POTASSIUM LITHIUM NIOBATE, BARIUM SODIUM NIOBATE, BARIUM STRONTIUM NIOBATE)
           JEGGO CR + BOYD GD
           J APPL PHYS 41: 2741-3 (1970)

11-1735   (MECHANISM OF) OPTICAL INDEX DAMAGE IN LITHIUM NIOBATE AND OTHER PYROELECTRIC
           INSULATORS.
           JOHNSTON WD
           J APPL PHYS 41(8): 3279-85 (1970)
                207

11-1736   POWER AND LINE WIDTH OF TUNABLE STIMULATED FAR INFRARED EMISSION IN LITHIUM NIOBATE.
           JOHNSON BC + PUTHOFF HE + SOOHOO J + SUSSMAN SS
           APPL PHYS LETT 18(5): 181-3 (1971)
                227

11-1737   POINT DEFECTS IN LITHIUM NIOBATE SINGLE CRYSTALS USED IN LASER MODULATION.
           JORGENSEN PJ + BARTLETT RW
           STANFORD RES INST, 1969, 38P. AD686721
                US GOVT RES DEVELOP REP 69(13): 187 (1969)
                157

11-1738   ELECTRON BEAM SENSING OF SURFACE ELASTIC WAVES FOR SIGNAL PROCESSING PURPOSES.
           (TECHNIQUE APPLIED TO LITHIUM NIOBATE AND QUARTZ)
           JOSHI SG + EPSTEIN M + SERAFIN RJ + VAN DEN HEUVEL AP
           P72-5 OF EUROPEAN MICROWAVE CONF, 1969 PROC. IEE, LONDON, 1970, 570P.

11-1739   SIMPLIFIED CHARACTERISTIC EQUATION FOR PLANE PIEZOELECTRIC VIBRATIONS OF LITHIUM
           NIOBATE AND LITHIUM TANTALATE CRYSTALS.
           KALISKI S
           BULL ACAD POLON SCI, SER SCI TECH 19: 127-32 (1971)

11-1740   SOME PROPERTIES OF THE EQUATIONS OF PIEZOELECTRIC VIBRATION OF LITHIUM NIOBATE AND
           TANTALATE CRYSTALS.
           KALISKI S
           PROC VIBRATION PROBLEMS 12: 63-70 (1971)

11-1741   LASER AND SPECTROSCOPIC PROPERTIES OF ACTIVATED FERROELECTRICS. (LITHIUM NIOBATE AND
           GADOLINIUM MOLYBDATE)
           KAMINSKII AA
           SOV PHYS CRYSTALLOGR 17(1): 194-207 (JUL-AUG 1972)
                274

11. LITHIUM NIOBATE

11-1742   PRECISION MACHINING OF LITHIUM NIOBATE CRYSTALS FOR ELECTROOPTIC ELEMENTS.
KASAI T + NODA J + IDA I
ELEC COMMUN LAB TECH J 20: 915-26 (1971) (IN JAPANESE)

11-1743   MANDELSTAM BRILLOUIN SCATTERING OF LIGHT IN LITHIUM NIOBATE (FROM TRANSVERSE (MORE
INTENSE) AND LONGITUDINAL PHONONS).
KHASHKHOZHEV ZM + LEMANOV VV + PISAREV RV
SOV PHYS SOLID STATE 12(1): 101-3 (1970)
    201

11-1744   SCATTERING OF LIGHT ON ACOUSTIC PHONONS IN LITHIUM NIOBATE AND LITHIUM TANTALATE.
KHASHKHOZHEV ZM + LEMANOV VV + PISAREV RV
BULL ACAD SCI USSR, PHYS SER 35: 911-15 (1971)

11-1745   PHOTOELASTIC CONSTANTS OF LITHIUM NIOBATE CRYSTALS.
KLUDZIN VV
SOV PHYS SOLID STATE 13(2): 540-1 (AUG 1971)
    245

11-1746   PARAMETRIC LUMINESCENCE AND LIGHT SCATTERING BY POLARITONS. (LITHIUM NIOBATE)
KLYSHKO DN + PENIN AN + POLKOVNIKOV BF
JETP LETT 11: 5-8 (5 JAN 1970)
    204

11-1747   AMPLIFICATION OF RAYLEIGH ULTRASONIC WAVES IN A LAYERED STRUCTURE CONSISTING OF LITHIUM
NIOBATE AND P-TYPE SILICON.
KMITA AM + KOTELYANSKII IM + MEDVED AV
SOV PHYS SEMICOND 5: 873-4 (1971)

11-1748   TRANSVERSE ACOUSTOELECTRIC EFFECT IN A LAYERED LITHIUM NIOBATE- SILICON STRUCTURE.
KMITA AM + MEDVED AV
JETP LETT 14: 310-12 (1971)

11-1749   DETERMINATION OF ELASTIC AND PIEZOELECTRIC CONSTANTS OF LITHIUM NIOBATE SINGLE
CRYSTALS.
KOROLYUK AP + MATSAKOV LYA + VASILCHENKO VV
SOV PHYS CRYSTALLOGR 15(5): 893-6 (1971)
    232

    DIELECTRIC PROPERTIES AND THERMAL EXPANSION OF SOME FERROELECTRICS AND
ANTIFERROELECTRICS WITH HIGH CURIE TEMPERATURES. (BISMUTH FERRATE AND SYSTEMS BASED ON
IT, AND SINGLE CRYSTAL LITHIUM NIOBATE)
KRAINIK NN + KHUCHUA NP + ZHDANOVA VV + MYLNIKOVA IE + PARFENOVA NN
P377-86, VOL 1 OF DVORAK V + FOUSKOVA A + GLOGAR P (EDS),
INT MEETING ON FERROELECTRICITY, PRAGUE, 1966, PROC.
INST PHYS CZECH ACAD SCI, PRAGUE, 1966.

    NONLINEAR EFFECTS IN FERROCERAMICS. (PZT-8, LITHIUM NIOBATE)
KRAUT EA + LIM TC
J ELECTROCHEM SOC 119: 98C (1972)

11-1750   PARAMETRIC LUMINESCENCE INTENSITY IN THE LITHIUM NIOBATE CRYSTAL.
KRINDACH DP + KRCL LM
OPT SPECTROSC 30: 73-5 (1971)

11-1751   PERTURBATION THEORY FOR ELECTROMAGNETIC COUPLING TO ELASTIC SURFACE WAVES ON
PIEZOELECTRIC SUBSTRATES (LITHIUM NIOBATE AND BISMUTH(12) GERMANIUM OXIDE(20)).
LAKIN KM
J APPL PHYS 42(3): 889-906 (1971)
    230

11-1752   OPTICAL PROBING OF ACOUSTIC SURFACE WAVE HARMONIC GENERATION (IN LITHIUM NIOBATE)
LEAN EGH + TSENG CC + POWELL CG
APPL PHYS LETT 16(1): 32-5 (1970)

11-1753   GENERATION AND PROPAGATION OF HYPERSONIC WAVES IN LITHIUM NIOBATE CRYSTALS. (80-950 DEG
K, 200-2000 MHZ)
LEMANOV VV + SMOLENSKII GA + SHERMAN AB + KLUEV VP
P303-5 OF INT MEETING ON FERROELECTRICITY, 2ND, 1969,, KYOTO, PROC,
(J PHYS SOC JAP, VOL 28 SUPPL). PHYS SOC JAP, 1970, 459P.

11. LITHIUM NIOBATE

11-1754   LIGHT SCATTERING BY ULTRASONIC WAVES IN LITHIUM NIOBATE CRYSTALS. (PHOTOELASTIC
           CONSTANT)
           LEMANOV VV + SHAKIN OV + SMOLENSKII GA
           SOV PHYS SOLID STATE 13(2): 426-8 (AUG 1971)
               245

11-1755   OPTIMUM CUT FOR A LITHIUM NIOBATE TRANSVERSE LIGHT MODULATOR.
           LIBRECHT FM + FRANCOIS GE
           IEEE J QUANTUM ELECTRON 7: 374-6 (1971)

11-1756   MICROWAVE PHONON ATTUNUATION IN X CUT LITHIUM NIOBATE.
           LIEKENS W + MICHIELS L + DE BOCK A
           PHYS LETT A 40: 309-10 (1972)

11-1757   EFFECT OF MASS LOADING ON THE PROPAGATION OF ACOUSTIC SURFACE WAVES ON LITHIUM NIOBATE
           AND QUARTZ.
           LIM TC + KRAUT EA
           IEEE TRANS SON ULTRASON 18: 67 (1970)

           COMPARISON OF FERROELECTRICITY IN ISOMORPHIC LITHIUM NICBATE AND LITHIUM TANTALATE
           USING AUTHOR'S STATISTICAL THEORY. (SPONTANEOUS POLARIZATION, BONDING)
           LINES ME
           PHYS REV B 2(3): 698-705 (1970)

11-1758   ACOUSTIC CONVOLUTION AND CORRELATION AND THE ASSOCIATED NONLINEARITY PARAMETERS IN
           LITHIUM NIOBATE.
           LUUKKALAA M + SURAKKA J
           J APPL PHYS 43: 2510-18 (1972)

11-1759   CRYSTAL FIELD IN CHROMIUM(3+) DOPED LITHIUM NIOBATE.
           MAJER M + SZYMCZAK H
           SOV PHYS SOLID STATE 13: 1027-8 (1971)

11-1760   THE MEASUREMENT OF HYPERSONIC ATTENUATION IN LITHIUM NIOBATE.
           MAJER J
           FYZ CAS 22(3): 137-42 (1972)
               SCI ABSTR A 75: 57050 (1972)
                   286

11-1761   FERMI RESONANCE OF POLARITON WITH BIPHONON IN AN LITHIUM NIOBATE CRYSTAL.
           MAVRIN BN + STERIN KHE
           JETP LETT 16: 187-8 (1972)

11-1762   TRANSVERSE POLARITONS IN A LITHIUM NIOBATE CRYSTAL. (RAMAN SPECTRUM, PERMITTIVITY)
           MAVRIN BN + ABRAMOVICH TE + STERIN KHE
           SOV PHYS SOLID STATE 14(6): 1562-3 (DEC 1972)
               279

11-1763   EPR OF GADOLINIUM(3+) IN LITHIUM NIOBATE.
           MCDONALD PF + TAM CP + MOK YW
           J CHEM PHYS 56: 1007-8 (1972)

11-1764   OPTIMUM SECOND HARMONIC GENERATION IN LITHIUM NIOBATE.
           MCGEOCH MW + SMITH RC
           IEEE J QUANTUM ELECTRON 6: 203-5 (1970)

11-1765   ELECTRON PARAMAGNETIC RESONANCE OF IRON(3+) DOPED LITHIUM NIOBATE.
           MEHRAN F + SCOTT BA
           SOLID STATE COMMUN 11: 15-19 (1972)

11-1766   EIGENFREQUENCIES AND EIGENVECTORS OF POLARITONS WITH APPLICATION TO LITHIUM NIOBATE.
           PART-2: POLARITONS OF SMALL WAVE VECTORS.
           MERTEN L + BORSTEL G
           Z NATURFORSCH A 27: 1073-81 (JUL 1972)

11-1767   BIREFRINGENCE INDUCED IN A LITHIUM NIOBATE CRYSTAL.
           MICHERON F + BISMUTH G
           OPT COMMUN 3: 390-4 (1971) (IN FRENCH)

11-1768   INVESTIGATION OF A SYSTEM FOR BEAM DEFLECTION BY MEANS OF LITHIUM NIOBATE CRYSTALS.
           MIKAELYAN AL + KOBLOVA MM + ZASOVIN EA
           SOV J QUANTUM ELECTRON 1: 87-90 (1971)

11. LITHIUM NIOBATE

11-1769   DEPENDENCE OF SECOND HARMONIC GENERATION COEFFICIENTS OF LITHIUM NIOBATE ON MELT
          COMPOSITION.
          MILLER RC + NORDLAND WA + BRIDENBAUGH PM
          J APPL PHYS 42(11): 4145-7 (OCT 1971)
              247

11-1770   TEMPERATURE DEPENDENCE OF THE OPTICAL PROPERTIES OF FERROELECTRIC LITHIUM NIOBATE AND
          TANTALATE.
          MILLER RC + SAVAGE A
          P405-12, VOL 1 OF DVORAK V + FOUSKOVA A + GLOGAR P (EDS),
          INT MEETING ON FERROELECTRICITY, PRAGUE, 1966, PROC.
          INST PHYS CZECH ACAD SCI, PRAGUE, 1966.
          APPL PHYS LETT 9: 169-71 (1966)

11-1771   SPONTANEOUS PARAMETRIC SCATTERING, IDLER ABSORPTION, AND DISPERSION IN THE NONLINEAR
          SUSCEPTIBILITY IN LITHIUM NIOBATE.
          MONTGOMERY GP + GIALLORENZI TG + HASS M
          IEEE J QUANTUM ELECTRON 8: 574 (1972)

11-1772   THEORETICAL STUDY OF THE GENERATION OF THE ACOUSTIC SECOND HARMONIC. (LITHIUM NIOBATE)
          MORLON B + COQUET E + DEVIN A
          COMPT REND B 270: 305-8 (1970) (IN FRENCH)

11-1773   EFFECT OF OPTICAL INHOMOGENEITIES ON PHASE MATCHING IN NONLINEAR CRYSTALS (LITHIUM
          NIOBATE).
          NASH FR + BOYD GD + SARGENT M + BRIDENBAUGH PM
          J APPL PHYS 41(6): 2564-76 (1970)
              197

11-1774   GROWTH AND PROPERTIES OF SINGLE DOMAIN CRYSTALS OF FERROELECTRIC LITHIUM NIOBATE.
          (CRACK-FREE, THROUGH ADDITION OF MAGNESIUM OXIDE TO MELT)
          NASSAU K
          P270-6, VOL 1 OF DVORAK V + FOUSKOVA A + GLOGAR P (EDS),
          INT MEETING ON FERROELECTRICITY, PRAGUE, 1966, PROC.
          INST PHYS CZECH ACAD SCI, PRAGUE, 1966.

11-1775   LITHIUM NIOBATE- A NEW TYPE OF FERROELECTRIC: GROWTH, STRUCTURE, AND PROPERTIES.
          (REVIEW OF WORK TO SEP 1966)
          NASSAU K
          P259-68 OF SYMP ON FERROELECTRICITY, 1966, GENERAL MOTORS
          RES LABS, WARREN, MICHIGAN, PROC, FERROELECTRICITY.
          WELLER EF(ED), ELSEVIER, 1967, 318P.

11-1776   STACKING FAULT MODEL FOR STOICHIOMETRY DEVIATIONS IN LITHIUM NIOBATE AND TANTALATE AND
          THE EFFECT ON THE CURIE TEMPERATURE.
          NASSAU K + LINES ME
          J APPL PHYS 41(2): 533-7 (1970)
              186

11-1777   STACKING FAULT MODEL FOR STOICHIOMETRY DEVIATIONS IN LITHIUM NIOBATE AND TANTALATE AND
          THE EFFECT ON THE CURIE TEMPERATURE.
          NASSAU K + LINES ME
          J APPL PHYS 41(2): 533-7 (1970)
              186

11-1778   POSSIBILITY OF SWITCHING AND STEERING SURFACE WAVES BY NONLINEAR MIXING IN ANISOTROPIC
          MEDIA. (LITHIUM NIOBATE, BISMUTH(12) GERMANIUM OXIDE(20))
          NEWHOUSE VL + CHEN CL + DAVIS KL
          J APPL PHYS 43: 2603-8 (1972)

11-1779   OBSERVATION OF THE SIMULTANEOUS GENERATION OF THE SECOND, THIRD, AND FOURTH HARMONICS
          OF 1.06 MICRON RADIATION IN BARIUM SODIUM NIOBATE, LITHIUM NIOBATE, AND LITHIUM IODATE
          CRYSTALS.
          NG WK + WOODBURY EJ
          APPL PHYS LETT (18(12): 550-1 (15 JUN 1971)
              240

11-1780   CHEMICAL COMPOSITION AND OPTICAL QUALITY OF FERROELECTRIC CRYSTALS, ESPECIALLY ON
          LITHIUM NIOBATE AND LITHIUM TANTALATE.
          NIIZEKI N
          J MATER SCI SOC JAP 9: 19-27 (1972) (IN JAPANESE)

11. LITHIUM NIOBATE

11-1781   GROWTH OF LITHIUM NIOBATE SINGLE DOMAIN CRYSTALS.
           NINOMIYA Y
           NHK TECH J 24: 215-33 (1972) (IN JAPANESE)

11-1782   LITHIUM NIOBATE LIGHT MODULATOR.
           NINOMIYA Y + MOTOKI T
           TRANS INST ELECTRON COMMUN ENG JAP C 55: 227-8 (1972) (IN JAPANESE)
           REV SCI INSTRUM 43: 519-24 (MAR 1972)

11-1783   LAPPING CHARACTERISTICS OF LITHIUM NIOBATE SINGLE CRYSTAL.
           NODA J + IDA I
           REV ELEC COMMUN LAB 20(1-2): 152-8 (1972)
                272

11-1784   LAPPING CHARACTERISTICS CF LITHIUM NIOBATE SINGLE CRYSTALS.
           NODA J + IDA I
           ELEC COMMUN LAB TECH J 19: 951-8 (1970) (IN JAPANESE)

11-1785   WAVELENGTH DEPENDENCE OF HOLOGRAPHIC PHENOMENA IN DOPED LITHIUM NIOBATE.
           NORMAN SL
           J OPT SOC AMER 62: 1396 (1972)

11-1786   BRILLOUIN SCATTERING IN LITHIUM NIOBATE.
           O'BRIEN RJ + ROSASCO GJ + WEBER A
           J OPT SOC AMER 60: 716 (1970)

11-1787   SPONTANEOUS RAMAN SCATTERING BY POLARITONS. PART-2: THE CASE OF CADMIUM SULFIDE AND
           LITHIUM NIOBATE CRYSTALS.
           OBUKHOVSKII VV + PONATH H + STRIZHEVSKII VL
           PHYS STATUS SOLIDI 41(2): 847-54 (1970)
                225

11-1788   LITHIUM NIOBATE SILICON SURFACE WAVE CONVOLUTER.
           OTTO OW + MOLL NJ
           ELECTRON LETT 8: 600-2 (1972)

11-1789   OPTICAL STUDIES OF REFRACTION AND REFLECTION OF ULTRASONIC SURFACE WAVES BY THIN GOLD
           FILM LAYERS ON LITHIUM NIOBATE SUBSTRATES.
           PEDINOFF ME + WALDNER M + JONES WR
           IEEE TRANS SON ULTRASON 18: 66 (1971)
           J APPL PHYS 42: 3025-34 (1971)

11-1790   REFRACTION AND REFLECTION OF ULTRASONIC SURFACE WAVES BY THIN GOLD FILM LAYERS ON
           LITHIUM NIOBATE SUBSTRATES.
           PEDINOFF ME + WALDNER M
           ELECTRON LETT 6: 533-4 (1970)

11-1791   CONTROL OF THE SUSCEPTIBILITY OF LITHIUM NIOBATE TO LASER INDUCED REFRACTIVE INDEX
           CHANGES.
           PETERSON GE + GLASS AM + NEGRAN TJ
           APPL PHYS LETT 19: 130-2 (1971)

11-1792   ELECTRON PARAMAGNETIC RESONANCE SPECTRUM IN (MANGANESE DOPED) LITHIUM NIOBATE AND
           (COPPER(2+) DOPED) ZINC BORACITE.
           PETROV MP + KIZHAEV SA + ANDREEVA GT + SMOLENSKII GA
           P128-30 OF INT MEETING ON FERROELECTRICITY, 2ND, 1969, KYOTO, PROC,
           (J PHYS SOC JAP, VOL 28 SUPPL). PHYS SOC JAP, 1970, 459P.

11-1793   LOW FREQUENCY (510 MHZ) PARAMAGNETIC RESONANCE INVESTIGATION OF LITHIUM NIOBATE.
           PETERSON GE + GLASS AM + CARNEVALE A + BRIDENBAUGH PM
           BULL AMER PHYS SOC 17: 48 (1972)

11-1794   NIOBIUM-93 NMR STUDY OF THE LITHIUM NIOBATE TANTALATE SOLID SOLUTION SYSTEM.
           PETERSON GE + CARRUTHERS JR + CARNEVALE A
           J CHEM PHYS 53(6): 2436-42 (1970)
                217

11-1795   NIOBIUM-93 NMR LINE WIDTHS IN NONSTOICHIOMETRIC LITHIUM NIOBATE.
           PETERSON GE + CARNEVALE A
           J CHEM PHYS 56: 4848-51 (1972)

11-1796 OPTICAL AND HOLOGRAPHIC STORAGE PROPERTIES OF TRANSITION METAL DOPED LITHIUM NIOBATE.
PHILLIPS W + AMODEI JJ + STAEBLER DL
RCA REV 33: 92-109 (1972)

11-1797 INVESTIGATION OF CERTAIN PARAMETERS OF LITHIUM NIOBATE SINGLE CRYSTALS USING SCATTERING
OF LASER RADIATION BY HYPERSCNIC VIBRATIONS.
PIROZHKOV VA + NAGIBAROV VB + SAMARTSEV RG + USMANOV RG + SHTYRKOV EI
OPT SPECTROSC 30: 178-9 (1971)
257

11-1798 INTERNAL FRICTION OF LITHIUM NIOBATE SINGLE CRYSTALS AT 1-40 HZ.
POSTNIKOV VS + KAVERIN LD + PAVLOV VS AND OTHERS
BULL ACAD SCI USSR, PHYS SER 35(9): 1741-3 (1971)

11-1799 DECORATION OF ION BOMBARDED LITHIUM NIOBATE SURFACES.
PRIMAK W
J ELECTROCHEM SOC 119: 228 (1972)

11-1800 EXPANSION, CRAZING AND EXFOLIATION OF LITHIUM NIOBATE ON ION BOMBARDMENT AND COMPARISON
RESULTS FOR SAPPHIRE.
PRIMAK W
J APPL PHYS 43: 4927-33 (1972)

11-1801 SECONDARY (ELECTRO) PHOTOELASTIC EFFECT IN LITHIUM NIOBATE (ON ELECTRON IRRADIATION).
PRIMAK W + ANDERSON TT
J APPL PHYS 41(11): 4745-6 (1970)
214

11-1802 NONLINEAR INTERACTION OF HYPERSONIC WAVES IN A LITHIUM NIOBATE CRYSTAL.
PUGOVKIN AV + SHANDAROV SM
SOV PHYS SOLID STATE 14: 201-2 (1972)

STIMULATED COMBINATION SCATTERING OF SOUND IN PIEZOELECTRICS. (THEORY, LITHIUM NIOBATE
EXAMPLE)
PUSHKINA NI + KHOKHLOV RV
SOV PHYS DOKL 17: 235-6 (1972)

PREPARATION AND PROPERTIES OF THIN FILM MATERIALS WITH PEROVSKITE TYPE STRUCTURE.
(MIXED OXIDES ABO_3) (BARIUM TITANATE AND LITHIUM NIOBATE FILMS ON PLATINUM OR
SILICON)
REIBER LM
J PHYS (PARIS) 33, SUPPL 4, COLLOQ 2: 265-7 (1972)

11-1803 ELECTRON SPIN RESONANCE STUDIES OF CRYSTAL FIELD PARAMETERS IN MANGANESE(2+) DOPED
LITHIUM NIOBATE.
REXFORD DG + KIM YM
J CHEM PHYS 57: 3094-8 (OCT 15, 1972)

11-1804 ELECTRON SPIN RESONANCE STUDIES OF CHROMIUM(3+) IN LITHIUM NIOBATE.
REXFORD OB + KIM YM + STORY HS
J CHEM PHYS 52(2): 860-3 (1970)

11-1805 FOREIDDEN HYPERFINE LINES IN ESR SPECTRA OF MANGANESE(2+) DOPED LITHIUM NIOBATE.
REXFORD DG + KIM YM
PHYS LETT A 35: 215-16 (1971)

11-1806 INTERACTION OF VOLUME ACOUSTIC MODES IN LITHIUM NIOBATE.
RISTIC VM
PHYS LETT A 35: 63-4 (1971)

11-1807 TEMPERATURE STUDY OF THE POLARITON ASSOCIATED WITH THE 248 CM(-1) SOFT MODE IN LITHIUM
NIOBATE.
ROKNI M + WALL LS + AMZALLAG E + CHANG TS
SOLID STATE COMMUN 10: 103-5 (1972)
260

11-1808 VIBRATIONAL SPECTRA OF LITHIUM NIOBATE, BARIUM SODIUM NIOBATE AND BARIUM SODIUM
TANTALATE (FORCE CONSTANTS, SPACE GROUP ASSIGNMENT).
ROSS SD
J PHYS C SOLID STATE PHYS 3: 1785-90 (1970)
213

11. LITHIUM NIOBATE

11-1809   DEFLECTED BEAM DIAMETER AND DEFLECTION CAPACITIES OF A DIGITAL LIGHT DEFLECTOR
          UTILIZING LITHIUM TANTALATE POLARIZATION SWITCHES.
          SAKAGUCHI M
          TRANS INST ELECTRON COMMUN ENG JAP B 54: 602-3 (1971) (IN JAPANESE)

11-1810   ZERO TEMPERATURE COEFFICIENT OF RESONANT FREQUENCY IN LITHIUM TANTALATE LENGTH EXPANDER
          BARS.
          SAWAMOTO K + NIIZEKI N
          PROC IEEE 58: 1289-90 (1970)

11-1811   NIOBIUM-93 NUCLEAR QUADRUPOLE RESONANCE INVESTIGATION OF (STOICHIOMETRIC) LITHIUM
          NIOBATE (TEMPERATURE DEPENDENCE OF COUPLING CONSTANT 21-515K).
          SCHEMPP E + PETERSON GE + CARRUTHERS JR
          J CHEM PHYS 53(1): 306-11 (1970)
              209

11-1812   TEMPERATURE DEPENDENCE OF RESONANT FREQUENCIES OF LITHIUM NIOBATE PLATE RESONATORS.
          (COMMENTS)
          SCHULZ BS + HOLLAND MG + FUKUMOTO A + WATANABE A
          PROC IEEE 58: 477 (1970)

11-1813   STOICHIOMETRY VARIATIONS IN LITHIUM NIOBATE AND LITHIUM TANTALATE BY RAMAN POWDER
          SPECTROSCOPY.
          SCOTT BA + BURNS G
          J AMER CERAM SOC 55: 225-30 (MAY 1972)

11-1814   SEPARATE EXCITATION OF LONGITUDINAL AND TRANSVERSE HYPERSONIC PULSES IN PIEZOELECTRIC
          LITHIUM NIOBATE AND BISMUTH BIGERMANATE.
          SEBER GA + BARANSKII KN
          BULL ACAD SCI USSR, PHYS SER 35(5): 864-6 (1971)

11-1815   INDEPENDENT EXCITATION OF LONGITUDINAL AND TRANSVERSE HYPERSONIC WAVE PULSES IN LITHIUM
          NIOBATE AND BISMUTH(12) GERMANIUM OXIDE(20) PIEZOELECTRIC CRYSTALS.
          SEVER GA + BARANSKII KN
          BULL ACAD SCI USSR, PHYS SER 35: 864-6 (1971)

11-1816   ABSORPTION OF ELASTIC WAVES IN REDUCED LITHIUM NIOBATE.
          SHERMAN AB + LEMANOV VV
          SOV PHYS SOLID STATE 13: 1413-15 (1971)

11-1817   MICROWAVE FREQUENCY ACOUSTIC SURFACE WAVE LOSS MECHANISMS ON LITHIUM NIOBATE
          (TEMPERATURE AND FREQUENCY DEPENDENCE, 0.5-5 GHZ).
          SLOBODNIK AJ + CARR PH + BUDREAU AJ
          J APPL PHYS 41(11): 4380-7 (1970)
              214

11-1818   NONLINEAR EFFECTS IN MICROWAVE ACOUSTIC LITHIUM NIOBATE SURFACE WAVE DELAY LINES.
          SLOBODNIK AJ
          J ACOUST SOC AMER 48: 203-10 (1970)
              206

11-1819   NEW HIGH FREQUENCY HIGH COUPLING LOW BEAM STEERING CUT FOR ACOUSTIC SURFACE WAVES ON
          LITHIUM NIOBATE.
          SLOBODNIK AJ + CONWAY ED
          ELECTRON LETT 6(6): 171-3 (MAR 19, 1970)
              214

11-1820   NEW HIGH COUPLING LOW DIFFRACTION CUT FOR ACOUSTIC SURFACE WAVES ON LITHIUM NIOBATE.
          SLOBODNIK AJ + SZABO TL
          ELECTRON LETT 7: 257-8 (1971)

11-1821   TEMPERATURE COEFFICIENTS OF ACOUSTIC SURFACE WAVE VELOCITY AND DELAY ON LITHIUM
          NIOBATE, LITHIUM TANTALATE, QUARTZ, AND TELLURIUM DIOXIDE.
          SLOBODNIK AJ
          AIR FORCE CAMBRIDGE RES LABS, MASS, 1972, 125P. TECH PAP 477.
              284

11-1822   GAMMA-RADIATION EFFECTS IN LITHIUM NIOBATE.
          SMITH RW
          PROC IEEE 59: 712-13 (1971)

11. LITHIUM NIOBATE

11-1823 TEMPERATURE DEPENDENCE OF THE ELASTIC, PIEZOELECTRIC, AND DIELECTRIC CONSTANTS OF LITHIUM TANTALATE AND LITHIUM NIOBATE.
SMITH RT + WELSH FS
J APPL PHYS 42: 2219-30 (MAY 1971)

11-1824 PIEZOELECTRIC SURFACE WAVES ON LITHIUM NIOBATE.
SPAIGHT RN + KOERBER GG
IEEE TRANS SON ULTRASON 18: 237-8 (1971)

11-1825 LOW POWER DENSITY GALLIUM ARSENIDE- LITHIUM NIOBATE SURFACE WAVE AMPLIFIER.
SPEARS DL + BURKE BE
IEEE TRANS SON ULTRASON 18: 54 (1971)

11-1826 COUPLED WAVE ANALYSIS OF HOLOGRAPHIC STORAGE IN LITHIUM NIOBATE.
STAEBLER DL + AMODEI JJ
J APPL PHYS 43: 1042-9 (MAR 1972)

11-1827 MULTIPLE STORAGE OF THICK PHASE HOLOGRAMS IN LITHIUM NIOBATE.
STAEBLER DL + AMODEI JJ + PHILLIPS W
IEEE J QUANTUM ELECTRON 8: 611 (1972)

11-1828 THERMALLY FIXED HOLOGRAMS IN LITHIUM NIOBATE.
STAEBLER DL + AMODEI JJ
FERROELECTRICS 3: 107-13 (1972);
IEEE TRANS SON ULTRASON 19: 107-13 (1972)
263

11-1829 TONE BURSTING OF QUADRUPOLAR NUCLEI. DETERMINATION OF THE SPIN LATTICE RELAXATION TIMES IN FERROELECTRIC LITHIUM TANTALATE AND LITHIUM NIOBATE.
STROUD JD + CREEL RB + SEGEL SL AND OTHERS
CAN J PHYS 50: 966-75 (1972)

11-1830 X-RAY TOPOGRAPHIC STUDIES ON LITHIUM NIOBATE AND LITHIUM TANTALATE SINGLE CRYSTALS.
SUGII K + MIYAZAWA S + NIIZEKI N
ACTA CRYSTALLOGR A 28: S169 (1972)

11-1831 CALCULATION OF THE (ABSORPTION) SPECTRA OF LITHIUM NIOBATE CRYSTALS CONTAINING CHROMIUM(3+) IONS (10 TO 30 THOUSAND/CM).
SVIRIDOV DT + SVIRIDOVA RK
SOV PHYS CRYSTALLOGR 15(4): 715-16 (1971)
221

11-1832 PIEZOELECTRIC LEAKY SURFACE WAVE IN LITHIUM NIOBATE (ELECTROMAGNETIC COUPLING).
TAKAYANAGI A + YAMANOUCHI K + SHIBAYAMA K
APPL PHYS LETT 17(5): 225-7 (1970)
201

11-1833 ELECTRON PARAMAGNETIC RESONANCE OF GADOLINIUM(3+) DOPED LITHIUM NIOBATE.
TAM CP + MCDONALD PF + MOK YW
BULL AMER PHYS SOC 15: 1306 (1970)

11-1834 EXPERIMENTAL STUDY OF LIGHT DIFFRACTION BY HIGH FREQUENCY ACOUSTIC WAVES. APPLICATION OF ACOUSTIC WAVES TO HARMONIC GENERATION LITHIUM NIOBATE.
THERY P + BRIDOUX E + DEROYON JP + MORIAMEZ M
COMPT REND B 270: 373-6 (1970) (IN FRENCH)

11-1835 GENERATION OF THE THIRD HARMONIC OF A LONGITUDINAL HIGH FREQUENCY (950 MHZ) ACOUSTIC WAVE ALONG THE (001) DIRECTION OF LITHIUM NIOBATE (LIQUID HELIUM TEMPERATURE).
THERY P + BRIDOUX E + DELANNOY M + MORIAMEZ M
COMPT REND B 270(1): 23-5 (1970) (IN FRENCH)
198

11-1836 INTERACTIONS BETWEEN HIGH FREQUENCY ACOUSTICAL WAVES IN LITHIUM NIOBATE.
THERY P + BRIDOUX E + MORIAMEZ M
J ACOUST SOC AMER 48: 772-3 (1970)

11-1837 (BACKWARD WAVE) ACOUSTIC PARAMETRIC OSCILLATIONS IN LITHIUM NIOBATE (60 DB GAIN).
THOMPSON RB + QUATE CF
APPL PHYS LETT 16(8): 295-8 (1970)
188

11. LITHIUM NIOBATE

11-1838 NONLINEAR INTERACTION OF MICROWAVE ELECTRIC FIELDS AND SOUND IN LITHIUM NIOBATE.
THOMPSON RB + QUATE CF
J APPL PHYS 42(3): 907-19 (1971)
230

11-1839 EPR STUDIES OF CRYSTAL FIELD PARAMETERS IN IRON(3+) DOPED LITHIUM NIOBATE.
TOWNER HH + KIM YM + STORY HS
J CHEM PHYS 56: 3676-9 (1972)

11-1840 MECHANISM AND KINETICS OF LITHIUM METANIOBATE FORMATION BY A SOLID PHASE REACTION.
TSIVILEV RP + FEDULOV SA + NEZAMAEVA NF
INORG MATER 6: 1358-9 (1970)

11-1841 DETERMINATION OF THE DISTRIBUTION COEFFICIENTS OF LITHIUM OXIDE INTO LITHIUM NIOBATE BY
PHASE MATCHING TEMPERATURES.
TSUYA H + FUJINO Y
JAP J APPL PHYS 10: 1509-12 (1971)

11-1842 DEPENDENCE OF LINEAR ELECTROOPTIC EFFECT AND DIELECTRIC CONSTANT ON MELT COMPOSITION
(STOICHIOMETRY) IN LITHIUM NIOBATE (VERY LITTLE EFFECT).
TURNER EH + NASH FR + BRIDENBAUGH PM
J APPL PHYS 41(13): 5278-81 (1970)
222

11-1843 FREQUENCY DEPENDENCES OF SOME PARAMETERS OF LITHIUM NIOBATE.
USMANOV RG + PIROZHKOV VA
UKR FIZ ZH 16: 1023-5 (JUN 1971) (IN RUSSIAN)

11-1844 ELECTROOPTIC EFFECT IN LITHIUM NIOBATE IN THE (SUB) MILLIMETER RANGE.
VINOGRADOV EA + PRISOVA NA + KOZLOV GV
SOV PHYS SOLID STATE 12(3): 605-7 (1970)
206

11-1845 DISPLAY OF TWINNING IN LITHIUM NIOBATE BY X-RAY DIFFRACTION TOPOGRAPHY.
WALLACE CA
J APPL CRYSTALLOGR 3: 546-7 (1970)

11-1846 FUNDAMENTAL AND HIGHER ORDER MODE SURFACE WAVE PROPAGATION ON LAYERED ANISOTROPIC
SUBSTRATES: GOLD ON LITHIUM NIOBATE.
WALDNER M + PEDINOFF ME + JONES WR + CAMPBELL JJ
J APPL PHYS 41: 2276-81 (1970)

11-1847 CONVOLUTION OF SURFACE WAVES IN A STRUCTURE OF SEMICONDUCTOR ON LITHIUM NIOBATE.
WANG WC
APPL PHYS LETT 20: 389-92 (1972)

11-1848 DIELECTRIC STUDIES OF THE CATION SUBSTITUTED MIXED CRYSTALS OF NIOBATES.
WANG FFY
STATE UNIV NEW YORK, STONY BROOK, COLLEGE OF ENG, 1969, 47P.
N70-12933
SCI TECH AEROSP REP 8(3): 410 (1970)
199

11-1849 DIFFRACTION SPREADING OF SURFACE WAVES ON LITHIUM NIOBATE.
WEGLEIN RD + PEDINOFF ME + WINSTON H
IEEE TRANS SON ULTRASON 18: 47 (1971)

11-1850 (FAR FIELD) DIFFRACTION SPREADING OF SURFACE (ACOUSTIC) WAVES ON (PIEZOELECTRIC)
LITHIUM NIOBATE.
WEGLEIN RD + PEDINOFF ME + WINSTON H
ELECTRON LETT 6(20): 654-5 (1970)
222

11-1851 MEDIUM FREQUENCY RESONATORS OF LITHIUM NIOBATE AND LITHIUM TANTALATE.
WELSH FS + HANNON JJ + LLOYD P + SMITH RT
IEEE TRANS SON ULTRASON 18: 55 (1971)

11-1852 SURFACE WAVE TEMPERATURE COEFFICIENTS ON LITHIUM TANTALATE.
WELSH FS
IEEE TRANS SON ULTRASON 18: 108-9 (1971)

11. LITHIUM NIOBATE

THEORY OF THE ELASTOOPTIC EFFECT IN NONMETALLIC CRYSTALS. (STRAIN DEPENDENT CURIE TEMPERATURE, POLARIZATION FLUCTUATIONS, LITHIUM NIOBATE, ALKALI HALIDES)
WEMPLE SH + DIDOMENICO M
PHYS REV B 1: 193-202 (1970)

11-1853 REFLECTION OF ELASTIC SURFACE WAVE FROM PERIODIC ARRAYS ON LITHIUM NIOBATE.
WILLIAMSON RC + SMITH HI
BULL AMER PHYS SOC 17: 777 (1972)

11-1854 OBSERVATION OF ORDINARY POLARITONS IN LITHIUM NIOBATE.
WINTER FX + CLAUS R
OPT COMMUN 6: 22-5 (1972)

11-1855 NONLINEAR OPTICAL SPECTROSCOPY OF THIRD ORDER SUSCEPTIBILITY IN LITHIUM NIOBATE.
WYNNE JJ
PHYS REV LETT 29(10): 650-3 (4 SEP 1972)
         278

11-1856 PROPAGATION AND AMPLIFICATION OF RAYLEIGH WAVES AND PIEZOELECTRIC LEAKY SURFACE WAVES IN LITHIUM NIOBATE.
YAMANOUCHI K + SHIBAYAMA K
J APPL PHYS 43(3): 856-62 (MAR 1972)
         260

11-1857 GENERATION OF FAR INFRARED RADIATION BY PICOSECOND LIGHT PULSES IN LITHIUM NIOBATE.
YANG KH + RICHARDS PL + SHEN YR
APPL PHYS LETT 19: 320-3 (1971)

11-1858 EFFECT TUNABLE OPTICAL EMISSION FROM LITHIUM NIOBATE WITHOUT A RESONATOR.
YARBOROUGH JM + SUSSMAN SS + PUTHOFF HE + PANTELL RH + JOHNSON BC
IEEE J QUANTUM ELECTRON 6: 6 (1970)

11-1859 SIMULTANEOUS OPTICAL PARAMETRIC OSCILLATION, SECOND HARMONIC GENERATION (IN LITHIUM NIOBATE).
YARBOROUGH JM + AMMANN EO
APPL PHYS LETT 18(4): 145-7 (1971)
         227

11-1860 GROWTH OF CRATERS ON ION ETCHED SURFACE OF LITHIUM NIOBATE AND ION ETCHING WITHOUT CRATERS.
YASUDA H + NAGAI K
JAP J APPL PHYS 11: 1713-6 (NOV 1972)

11-1861 NATURE OF INTERNAL ELECTRIC FIELD DURING OPTICAL DAMAGE PROCESS IN LITHIUM NIOBATE.
YASOJIMA Y + OHMORI Y + INUISHI Y
TECHNOL REP OSAKA UNIV 22: 575-83 (1972)

11-1862 THINNING OF LITHIUM NIOBATE FOR ACOUSTOOPTIC DEFLECTORS BY ION ETCHING.
YASUDA H + NAGAI K
JAP J APPL PHYS 11: 1216 (1972)
         278

11-1863 LONGITUDINAL ELECTROOPTICAL EFFECT IN OBLIQUE SECTIONS OF LITHIUM NIOBATE.
ZHASHKOV AA + PANKRATOV VM + MOIYA AM + SOMOV VG
SOV J QUANTUM ELECTRON 1(1): 111-12 (JUL-AUG 1971)
         289

11-1864 VISUALIZATION OF ACOUSTIC RADIATION IN LITHIUM NIOBATE.
ZULIANI M + RISTIC VM
PHYS LETT A 38: 87-8 (1972)

11-1865 AUTOFOCUSING OF LASER RADIATION IN ACTIVE MATERIALS AND NONLINEAR CRYSTALS. (KDP, ADP, RUBIDIUM DIHYDROGEN PHOSPHATE, LITHIUM NIOBATE, LITHIUM TANTALATE, QUARTZ)
ZVEREV GM + LEVCHUK EA + MALDUTIS EK + PASHKOV VA
SOV PHYS SOLID STATE 11(4): 865-6 (1969)
         176

11-1866 DAMAGE TO THE SURFACE OF LITHIUM NIOBATE BY LIGHT.
ZVEREV GM + LEVCHUK EA + PASHKOV VA
SOV PHYS JETP 35(1): 165-7 (1972)

12. LITHIUM TANTALATE

12-1867 NEW CRYSTAL FOR COMMUNICATIONS SYSTEMS ENTERS PRODUCTION STAGE. (LITHIUM TANTALATE)
BELL LAB REC 50: 296-7 (1972)

INHOMOGENEITY OF THE OPTICALLY INDUCED INDEX OF REFRACTION IN LITHIUM NIOBATE AND
LITHIUM TANTALATE.
ANGERT NB + PASHKOV VA + SOLOVEVA NM
SOV PHYS JETP 35(5): 867-9 (1972)

12-1868 TEMPERATURE DEPENDENCE OF AN X CUT LITHIUM TANTALATE CRYSTAL RESONATOR VIBRATING IN
THICKNESS SHEAR MODE OF MOTION.
ASHIDA T + SAWAMOTO K + NIIZEKI N
REV ELECTRON COMMUN LAB 18: 854-61 (1970)

12-1869 FERROELECTRIC DOMAIN REVERSAL IN LITHIUM METATANTALATE.
BALLMAN AA + BROWN H
FERROELECTRICS 4: 189-94 (1972)

12-1870 INFRARED (REFLECTIVITY) STUDY (20-10,000 PER CM, 300 DEG K) OF THE LATTICE VIBRATIONS
IN LITHIUM TANTALATE (DIELECTRIC CONSTANTS).
BARKER AS + BALLMAN AA + DITZENBERGER JA
PHYS REV B 2(10): 4233-9 (1970)
226

12-1871 LITHIUM TANTALATE SINGLE CRYSTAL STOICHIOMETRY (VERSUS MELT COMPOSITION, LATTICE AND
CURIE TEMPERATURE MEASUREMENTS).
BARNS RL + CARRUTHERS JR
J APPL CRYSTALLOGR 3(5): 395-9 (1970)
223

12-1872 FABRICATION OF A LITHIUM TANTALATE TEMPERATURE STABILIZED OPTICAL MODULATOR.
BIAZZO MR
APPL OPT 10: 1016-21 (1971)

12-1873 TEMPERATURE DEPENDENCE OF POLARITON DISPERSION IN LITHIUM TANTALATE. (OPTICAL
MODULATION VIA TEMPERATURE)
CHANG TS + JOHNSON BC + AMZALLAG E + PANTELL RH + ROKNI M + WALL LS
OPT COMMUN 4(1): 72-4 (SEP 1971)
247

12-1874 ELECTROOPTIC AND FERROELECTRIC PROPERTIES OF LITHIUM TANTALATE SINGLE CRYSTALS AS A
FUNCTION OF MELT COMPOSITION. (CURIE TEMPERATURE DEPENDENCE)
FUJINO Y + TSUYA H + SUGIBUCHI K
FERROELECTRICS 2: 113-17 (1971)
235

GROWTH AND PROPERTIES OF FERROELECTRIC POTASSIUM LITHIUM TANTALATE(X) NIOBATE(1-X).
FUKUDA T + HIRANO H + KOIDE S
J CRYST GROW 6: 293-6 (1970)

GROWTH AND PROPERTIES OF LITHIUM NIOBATE AND LITHIUM TANTALATE FILMS.
FUKUNISHI S + KAWANA A + UCHIDA N
ACTA CRYSTALLOGR A 28: S143 (1972)

12-1875 STUDY OF PIEZOELECTRIC OSCILLATIONS IN WIDEBAND PYROELECTRIC LITHIUM TANTALATE.
GLASS AM + ABRAMS RL
J APPL PHYS 41(11): 4455-9 (1970)
214

12. LITHIUM TANTALATE

LITHIUM TANTALATE AND LITHIUM NIOBATE PIEZOELECTRIC RESONATORS IN THE MEDIUM FREQUENCY RANGE WITH LOW RATIOS OF CAPACITANCE AND LOW TEMPERATURE COEFFICIENTS OF FREQUENCY (DEVICE DESIGN AND CHARACTERIZATION).
HANNON JJ + LLOYD P + SMITH RT
IEEE TRANS SON ULTRASON 17(4): 239-46 (1970)

12-1876 SINGLE CRYSTAL GROWTH AND PHYSICAL PROPERTIES OF LITHIUM TANTALATE. (DIELECTRIC CONSTANTS, PIEZOELECTRIC CONSTANTS, SPONTANEOUS POLARIZATION, THERMAL EXPANSION COEFFICIENTS, AND REFRACTIVE INDEX)
IWASAKI H + MIYAZAWA S + YAMADA T + UCHIDA N + NIIZEKI N
REV ELEC COMMUN LAB 20(1-2): 129-37 (1972)
   272

12-1877 BEHAVIOR OF LITHIUM TANTALATE SINGLE CRYSTAL NEAR ITS CURIE POINT. PART-1: PIEZOELECTRIC AND OPTICAL PROPERTIES.
IWASAKI H + YAMADA T + NIIZEKI N + TOYODA H
P306-8 OF INT MEETING ON FERROELECTRICITY, 2ND, 1969,, KYOTO, PROC,
(J PHYS SOC JAP, VOL 28 SUPPL). PHYS SOC JAP, 1970, 459P.

NONLINEAR OPTICAL POLARIZABILITY OF THE NIOBIUM- OXYGEN BOND. (LITHIUM NIOBATE, LITHIUM TANTALATE, POTASSIUM LITHIUM NIOBATE, BARIUM SODIUM NIOBATE, BARIUM STRONTIUM NIOBATE)
JEGGO CR + BOYD GD
J APPL PHYS 41: 2741-3 (1970)

SIMPLIFIED CHARACTERISTIC EQUATION FOR PLANE PIEZOELECTRIC VIBRATIONS OF LITHIUM NIOBATE AND LITHIUM TANTALATE CRYSTALS.
KALISKI S
BULL ACAD POLON SCI, SER SCI TECH 19: 127-32 (1971)

SOME PROPERTIES OF THE EQUATIONS OF PIEZOELECTRIC VIBRATION OF LITHIUM NIOBATE AND TANTALATE CRYSTALS.
KALISKI S
PROC VIBRATION PROBLEMS 12: 63-70 (1971)

SCATTERING OF LIGHT ON ACOUSTIC PHONONS IN LITHIUM NIOBATE AND LITHIUM TANTALATE.
KHASHKHOZHEV ZM + LEMANOV VV + PISAREV RV
BULL ACAD SCI USSR, PHYS SER 35: 911-15 (1971)

12-1878 SCATTERING OF LIGHT BY ACOUSTIC PHONONS AND POLARITONS IN LITHIUM TANTALATE.
KHASHKHOZHEV ZM + LEMANOV VV + PISAREV RV
SOV PHYS SOLID STATE 12(4): 941-5 (1970)
   212

COMPARISON OF FERROELECTRICITY IN ISOMORPHIC LITHIUM NIOBATE AND LITHIUM TANTALATE USING AUTHOR'S STATISTICAL THEORY. (SPONTANEOUS POLARIZATION, BONDING)
LINES ME
PHYS REV B 2(3): 698-705 (1970)

INTERCELL CORRECTIONS FOR IONIC MOTION IN DISPLACEMENT FERROELECTRICS (LITHIUM TANTALATE EXAMPLE).
LINES ME
PHYS REV B 2(3): 690-7 (1970)

12-1879 NATURE OF THE FERROELECTRIC- PARAELECTRIC PHASE TRANSITION IN LITHIUM TANTALATE.
LINES ME
SOLID STATE COMMUN 10: 793-6 (1972)
   267

12-1880 SIGNS OF THE ELECTROOPTIC COEFFICIENTS FOR LITHIUM TANTALATE.
LUTHER-DAVIES B + DAVIES PH + COUND VM + HULME KF
J PHYS C 3: L106-7 (1970)
   203

12-1881 OPTICAL BEAM STEERING USING A MULTICHANNEL LITHIUM TANTALATE CRYSTAL.
MEYER RA
APPL OPT 11: 613-16 (1972)

12. LITHIUM TANTALATE

TEMPERATURE DEPENDENCE OF THE OPTICAL PROPERTIES OF FERROELECTRIC LITHIUM NIOBATE AND TANTALATE.
MILLER RC + SAVAGE A
P405-12, VOL 1 OF DVORAK V + FOUSKOVA A + GLOGAR P (EDS),
INT MEETING ON FERROELECTRICITY, PRAGUE, 1966, PROC.
INST PHYS CZECH ACAD SCI, PRAGUE, 1966.
APPL PHYS LETT 9: 169-71 (1966)

12-1882   CONGRUENT MELTING COMPOSITION OF LITHIUM METATANTALATE.
MIYAZAWA S + IWASAKI H
J CRYST GROW 10: 276-8 (1971)

12-1883   STOICHIOMETRY AND OPTICAL QUALITY OF LITHIUM TANTALATE SINGLE CRYSTALS.
MIYAZAWA S + IWASAKI H
ELEC COMMUN LAB TECH J 21: 1739-51 (1972) (IN JAPANESE)
      284

CHEMICAL COMPOSITION AND OPTICAL QUALITY OF FERROELECTRIC CRYSTALS, ESPECIALLY ON LITHIUM NIOBATE AND LITHIUM TANTALATE.
NIIZEKI N
J MATER SCI SOC JAP 9: 19-27 (1972) (IN JAPANESE)

NIOBIUM-93 NMR STUDY OF THE LITHIUM NIOBATE TANTALATE SOLID SOLUTION SYSTEM.
PETERSON GE + CARRUTHERS JR + CARNEVALE A
J CHEM PHYS 53(6): 2436-42 (1970)

12-1884   PRIMARY AND SECONDARY PYROELECTRIC EFFECTS AND SPONTANEOUS POLARIZATION IN LITHIUM TANTALATE (THERMAL EXPANSION).
RABINOVICH AZ + ROITBERG MB
SOV PHYS SOLID STATE 12(9): 2178-9 (1971)
      227

12-1885   BEHAVIOR OF LITHIUM TANTALATE SINGLE CRYSTAL NEAR ITS CURIE POINT, PART-2, DIELECTRIC AND ULTRASONIC PROPERTIES.
SAWAMOTO K + ASHIDA T + OMACHI Y + UNO T
P309-11 OF INT MEETING ON FERROELECTRICITY, 2ND, 1969, KYOTO, PROC,
(J PHYS SOC JAP, VOL 28 SUPPL). PHYS SOC JAP, 1970, 459P.

12-1886   TEMPERATURE DEPENDENCE OF SURFACE ACOUSTIC WAVE VELOCITY IN LITHIUM TANTALATE.
SCHULZ MB + HOLLAND MG
IEEE TRANS SON ULTRASON 19: 381-4 (1972)

STOICHIOMETRY VARIATIONS IN LITHIUM NIOBATE AND LITHIUM TANTALATE BY RAMAN POWDER SPECTROSCOPY.
SCOTT BA + BURNS G
J AMER CERAM SOC 55: 225-30 (MAY 1972)

12-1887   CURIE TEMPERATURE OF FERROELECTRIC LITHIUM TANTALATE.
SHAPIRO ZI + FEDULOV SA + VENEVTSEV YUN
SOV PHYS SOLID STATE 6: 254-5 (1964)

TEMPERATURE COEFFICIENTS OF ACOUSTIC SURFACE WAVE VELOCITY AND DELAY ON LITHIUM NIOBATE, LITHIUM TANTALATE, QUARTZ, AND TELLURIUM DIOXIDE.
SLOBODNIK AJ
AIR FORCE CAMBRIDGE RES IABS, MASS, 1972, 125P. TECH PAP 477.

TEMPERATURE DEPENDENCE OF THE ELASTIC, PIEZOELECTRIC, AND DIELECTRIC CONSTANTS OF LITHIUM TANTALATE AND LITHIUM NIOBATE.
SMITH RT + WELSH FS
J APPL PHYS 42: 2219-30 (MAY 1971)

12-1888   PERFORMANCE OF AN 11 GHZ OPTICAL MODULATOR USING LITHIUM TANTALATE.
STANDLEY RD + MANDEVILLE GD
APPL OPT 10: 1022-3 (1971)

TONE BURSTING OF QUADRUPOLAR NUCLEI. DETERMINATION OF THE SPIN LATTICE RELAXATION TIMES IN FERROELECTRIC LITHIUM TANTALATE AND LITHIUM NIOBATE.
STROUD JD + CREEL RB + SEGEL SL AND OTHERS
CAN J PHYS 50: 966-75 (1972)

12. LITHIUM TANTALATE

X-RAY TOPOGRAPHIC STUDIES ON LITHIUM NIOBATE AND LITHIUM TANTALATE SINGLE CRYSTALS.
SUGII K + MIYAZAWA S + NIIZEKI N
ACTA CRYSTALLOGR A 28: S169 (1972)

12-1889 SELF INDUCED THERMAL EFFECTS ON LIGHT EXTINCTION OF LITHIUM TANTALATE.
TSUYA H + FUJINO Y + MATSUSHITA S
JAP J APPL PHYS 11: 1384-5 (1972)

12-1890 PIEZOELECTRIC PROPERTIES OF FERROELECTRIC POTASSIUM LITHIUM TANTALATE(0-0.5)
NIOBATE(1-0.5) SINGLE CRYSTALS.
UEMATSU Y + KOIDE S
JAP J APPL PHYS 9: 336-7 (1970)
190

12-1891 INTERACTION OF TWO HYPERSOUND WAVES IN LITHIUM TANTALATE.
VALITOVA NR + GONCHAROV KV
SOV PHYS SOLID STATE 12(11): 2499-502 (1971)
234

MEDIUM FREQUENCY RESONATORS OF LITHIUM NIOBATE AND LITHIUM TANTALATE.
WELSH FS + HANNON JJ + LLOYD P + SMITH RT
IEEE TRANS SON ULTRASON 18: 55 (1971)

12-1892 ANOMALOUS (LONGITUDINAL WAVE) ATTENUATION IN LITHIUM TANTALATE (EFFECT OF VACANCIES).
WORLEY JC + SMITH AB + KESTIGIAN M
J PHYS CHEM SOLIDS 31: 1857-61 (1970)
212

12-1893 PIEZOELECTRIC AND ELASTIC PROPERTIES OF LITHIUM TANTALATE CHARACTERISTICS.
YAMADA T + IWASAKI H + NIIZEKI N
JAP J APPL PHYS 8(9): 1127-32 (1969)
169

AUTOFOCUSING OF LASER RADIATION IN ACTIVE MATERIALS AND NONLINEAR CRYSTALS. (KDP, ADP,
RUBIDIUM DIHYDROGEN PHOSPHATE, LITHIUM NIOBATE, LITHIUM TANTALATE, QUARTZ)
ZVEREV GM + LEVCHUK EA + MALDUTIS EK + PASHKOV VA
SOV PHYS SOLID STATE 11(4): 865-6 (1969)

13. TUNGSTEN BRONZE TYPE
AND LAYER STRUCTURE OXIDES

13-1894   DILATOMETRIC STUDY OF THE ORTHORHOMBIC- TETRAGONAL PHASE TRANSITION IN BARIUM SODIUM
           NIOBATE. (THERMAL EXPANSION)
           ABELL JS + BARRACLOUGH KG + HARRIS IR + VERE AW + COCKAYNE B
           J MATER SCI 6: 1084-92 (1971)
                241

13-1895   FERROELECTRIC TUNGSTEN BRONZE TYPE CRYSTAL STRUCTURES. PART-3: POTASSIUM LITHIUM
           NIOBATE.
           ABRAHAMS SC + JAMIESON PB + BERNSTEIN JL
           J CHEM PHYS 54: 2355-64 (1971)

13-1896   FERROELECTRICS IN THE POTASSIUM OXIDE- STRONTIUM OXIDE- NIOBIUM PENTOXIDE.
           AINGER FW + BESWICK JA + PORTER SG + CLARKE R
           FERROELECTRICS 3: 321-5 (1972);
           IEEE TRANS SON ULTRASON 19: 321-5 (1972)
                263

13-1897   THE SEARCH FOR NEW FERROELECTRICS WITH THE TUNGSTEN BRONZE STRUCTURE.
           AINGER FW + BRICKLEY WP + SMITH GV
           P221-37 OF BRITISH CERAMIC SOC, ELECTRICAL MAGNETIC CERAMICS CONF,
           2ND, SEP 1969, WARWICK, ENGLAND, PROC. BRITISH CERAMIC SOC, 1970
                   SCI ABSTR A 74: 5664 (1971)
                284

           HOLOGRAPHIC PATTERN FIXING IN ELECTROOPTIC CRYSTALS. (LITHIUM NIOBATE, BARIUM SODIUM
           NIOBATE)
           AMODEI JJ + STAEBLER DL
           APPL PHYS LETT 18: 540-2 (1971)

13-1898   HOLOGRAPHIC STORAGE IN DOPED BARIUM SODIUM NIOBATE.
           AMODEI JJ + STAEBLER DL + STEPHENS AW
           APPL PHYS LETT 16(11): 507-9 (1 JUN 1971)
                240

13-1899   RAMAN SCATTERING BY BARIUM STRONTIUM NIOBATE.
           AMZALLAG E + CHANG TS + PANTELL RH
           J APPL PHYS 42(8): 3254-6 (JUL 1971)
                239

13-1900   ELECTROOPTIC EFFECTS OF HCT PRESSED CERAMICS IN THE SYSTEM LEAD (ZINCATE(0.333)
           NIOBATE(0.667)).
           ARIMA H + MING-CHONG L + NOMURA S
           JAP J APPL PHYS 11: 1225-6 (AUG 1972)

13-1901   GROWTH OF UNCRACKED BARIUM SODIUM NIOBATE CRYSTALS.
           BALLMAN AA + CARRUTHERS JR + O'BRYAN HM
           J CRYST GROW 6: 184-6 (1970)

13-1902   SOME EFFECTS OF MELT STOICHIOMETRY ON THE OPTICAL PROPERTIES BARIUM SODIUM NIOBATE.
           BALLMAN AA + KURTZ SK + BROWN H
           J CRYST GROW 10: 185-9 (1971)
                239

13-1903   SOME EFFECTS OF MELT STOICHIOMETRY ON THE OPTICAL PROPERTIES OF BARIUM SODIUM NIOBATE.
           BALLMAN AA + KURTZ SK + BROWN H
           J CRYST GROW 7: 1-5 (1971)

           SYNTHESIS OF OXIDE BRONZES. (REVIEW, 88 REFS)
           BANKS E + WOLD A
           P237-68 OF JOLLY WL, PREPARATIVE INORGANIC REACTIONS, VOL 4,
           INTERSCIENCE, NY, 1968.

13. TUNGSTEN BRONZE AND LAYER STRUCTURE OXIDES

13-1904　BARIUM SODIUM NIOBATE: CONSTITUTIONAL STUDIES AND (CZOCHRALSKI) CRYSTAL GROWTH.
BARRACLOUGH KG + HARRIS IR + COCKAYNE B + PLANT JG + VERE AW
J MATER SCI 5: 389-93 (1970)
　　197

13-1905　LOW FREQUENCY VIBRATIONAL MODES IN BARIUM SODIUM NIOBATE.
BOBB LC + DAHL J + LEFKOWITZ I + MULDAWER L
FERROELECTRICS 1: 247-51 (1970)

13-1906　RAMAN SPECTRA OF BARIUM SODIUM NIOBATE.
BOBB LC + LEFKOWITZ I + MULDAWER L
FERROELECTRICS 2: 217-23 (1971)
　　239

13-1907　EPITAXY OF PLATINUM ON BARIUM STRONTIUM NIOBATE.
BOHM J + STAHR S + WILKE KT
KRISTALL TECH 7: 179-82 (1972) (IN GERMAN)

13-1908　EFFECTS OF CHANGES IN MELT COMPOSITION ON CRYSTAL GROWTH OF BARIUM SODIUM NIOBATE.
BONNER WA + CARRUTHERS JR + O'BRYAN HM
MATER RES BULL 5: 243-52 (1970)
　　187

13-1909　GROWTH OF BARIUM SODIUM NIOBATE CRYSTALS BY THE KYROPOULOS TECHNIQUE.
BONNER WA
J ELECTROCHEM SOC 117: 195C (1970)

13-1910　TRANSIENT NMR STUDY OF PHASE TRANSITIONS IN METALLIC SODIUM TUNGSTEN BRONZES.
BONERA G + BORSA F + CRIPPA ML + RIGAMONTI A
PHYS REV B SOLID STATE 4(1): 52-9 (1 JUL 1971)
　　237

　　TEMPERATURE DEPENDENCE OF DIELECTRIC CONSTANTS OF CUBIC IONIC COMPOUNDS. (HALIDES AND
　　OXIDES, RELATION BETWEEN CURIE CONSTANT AND CURIE TEMPERATURE FOR FERROELECTRICS AND
　　ANTIFERROELECTRICS BARIUM TITANATE, COMPLEX LAYER OXIDES)
　　BOSMAN AJ + HAVINGA EE
　　PHYS REV 129: 1593-1600 (1963)

13-1911　CRYSTAL STRUCTURE AND PROPERTIES OF CALCIUM(2) NIOBIUM(2) OXIDE(7).
BRANDON JK + MEGAW HD
PHIL MAG 21(169): 189-94 (1970)
　　193

13-1912　ELECTRICAL PROPERTIES OF TERNARY BARIUM OXIDE- LANTHANUM OXIDE- TITANIUM DIOXIDE
COMPOUNDS.
BRADLEY FN + WANG T
BULL AMER CERAM SOC 50: 756 (1971)

13-1913　CZOCHRALSKI GROWTH OF BARIUM STRONTIUM NIOBATE CRYSTALS.
BRICE JC + HILL OF + WHIFFIN PAC + WILKINSON JA
J CRYST GROW 10: 133-8 (1971)
　　239

13-1914　MODIFICATION TO THE CZOCHRALSKI METHOD OF CRYSTAL PULLING. (CONTROL OF CRYSTAL DIAMETER
OF STRONTIUM BARIUM NIOBATE)
BRICE JC + HILL OF + WHIFFIN PAC
J CRYST GROW 6(3): 297-8 (1970)

13-1915　EFFECT OF OXYGEN STOICHIOMETRY AND HYDROXYL CONTENT ON THE PROPERTIES AND PROCESSING OF
SINGLE CRYSTAL BARIUM SODIUM NIOBATE.
BROWN RM + ARNOLD GL + KUNG R
J ELECTROCHEM SOC 117: 195C (1970)

13-1916　LOW FREQUENCY LINEAR ELECTROOPTIC EFFECT IN LEAD DITANATLATE.
BRUTON TM + WHITE EAD
J MATER SCI 7(11): 1233-4 (1972)
　　　SCI ABSTR A 76: 12115 (1973)
　　280

　　TUNGSTEN BRONZE TYPE FERROELECTRICS FOR OPTICAL USE (REVIEW, 22 REFS).
　　BURNS G
　　OYO BUTSURI 39(1): 3-10 (1970) (IN JAPANESE)

13. TUNGSTEN BRONZE AND LAYER STRUCTURE OXIDES

PROPERTIES OF TUNGSTEN BRONZE FERROELECTRICS. (REVIEW, 33 REFS)
BURNS G + GRESS EA + O'KANE DF + SCOTT BA + SMITH AW
P153-8 OF INT MEETING ON FERROELECTRICITY, 2ND, 1969, KYOTO, PROC,
(J PHYS SOC JAP, VOL 28 SUPPL). PHYS SOC JAP, 1970, 459P.

13-1917   BARIUM SODIUM NIOBATE FOR A TUNABLE STIMULATED RAMAN OSCILLATOR.
           BURNS G
           BULL AMER PHYS SOC 17: 245 (1972)

13-1918   POLARITON RESULTS IN FERROELECTRICS WITH THE TUNGSTEN BRONZE CRYSTAL STRUCTURE.
           BURNS G
           APPL PHYS LETT 20(6): 230-3 (15 MAR 1972)
               260

13-1919   PYROELECTRIC COEFFICIENT DIRECT MEASUREMENT TECHNIQUE AND APPLICATION TO A NANOSECOND
           RESPONSE TIME DETECTOR. (STRONTIUM BARIUM NIOBATE)
           BYER RL + ROUNDY CB
           FERROELECTRICS 3: 333-8 (1972);
           IEEE TRANS SON ULTRASON 19: 333-8 (1972)
               263

13-1920   PHASE EQUILIBRIA RELATIONS IN THE TERNARY SYSTEM BARIUM OXIDE- STRONTIUM OXIDE- NIOBIUM
           PENTOXIDE. (OXIDES, ESPECIALLY THE NIOBIUM-RICH REGION AND THE BINARY STRONTIUM OXIDE-
           NIOBIUM PENTOXIDE SYSTEM, DTA, X-RAY CURIE TEMPERATURE).
           CARRUTHERS JR + GRASSO M
           J ELECTROCHEM SOC 117: 1426-30 (1970)
               222

13-1921   EFFECT OF THE SUBSTITUTION OF TANTALUM FOR NIOBIUM ON THE CRYSTALLOGRAPHIC AND
           DIELECTRIC PROPERTIES OF STRONTIUM SODIUM NIOBATE AND BARIUM SODIUM NIOBATE.
           CHAMINADE JP + PERON A + RAVEZ J + HAGENMULLER P
           BULL SOC CHIM FR 1972(10): 3751-2 (IN FRENCH)
               290

13-1922   FERROELECTRICITY MEASUREMENTS ON STRONTIUM BARIUM NIOBATE.
           CHANG TS + AMZALLAG E
           FERROELECTRICS 3: 57-8 (1971)
               262

13-1923   FORMATION CONDITIONS AND PHASE TRANSITIONS IN CRYSTALS WITH POTASSIUM TUNGSTEN BRONZE
           STRUCTURE (POTASSIUM SODIUM LITHIUM NIOBATE, POTASSIUM STRONTIUM LITHIUM NIOBATE
           TITANATE, POTASSIUM LANTHANUM LITHIUM NIOBIUM TITANATE, BARIUM LITHIUM TITANIUM
           NIOBATE).
           CHAN VT + KRAINIK NN + MYLNIKOVA IE + ISMAILZADE IG + ISUPOV VA + AGAEV FA + CHMEL IS
           BULL ACAD SCI USSR, PHYS SER 35(9): 1661-4 (1971)

13-1924   USE OF PEROVSKITE PARAELECTRICS IN BEAM DEFLECTORS AND LIGHT MODULATORS. (POTASSIUM
           TANTALATE(0.65) NIOBATE(0.35))
           CHEN FS + GEUSIC JE + KURTZ SK + SKINNER JG + WEMPLE SH
           PROC IEEE 52: 1258-9 (1964)

13-1925   STUDIES ON THE DIELECTRIC AND FERROELECTRIC PROPERTIES OF THE BARIUM LEAD NIOBATE
           ZIRCONATE SYSTEM. (TETRAGONAL AND ORTHORHOMBIC PHASES, ABSTRACT ONLY)
           CHINCHOLKAR VS + MENEZES CA + BISWAS AB
           P324 OF INT MEETING ON FERROELECTRICITY, 2ND, 1969, KYOTO, PROC,
           (J PHYS SOC JAP, VOL 28 SUPPL). PHYS SOC JAP, 1970, 459P.

13-1926   TRAPPING SITES, OPTICAL DAMAGE, AND THERMALLY STIMULATED CURRENT IN FERROELECTRIC OXIDE
           CRYSTALS. (POTASSIUM LITHIUM NIOBATE, POTASSIUM STRONTIUM NIOBATE)
           CLARKE R + AINGER FW + BURFOOT JC
           J PHYS (PARIS) 33, SUPPL 4, COLLOQ 2: 143-5 (1972)
               294

13-1927   EFFECT OF IRON(4+) IN THE SYSTEM STRONTIUM FERRATE TITANATE.
           CLEVENGER TR
           J AMER CERAM SOC 46: 207-10 (1963)

13. TUNGSTEN BRONZE AND LAYER STRUCTURE OXIDES

13-1928   CRUCIBLE BASE COOLING: AN AID TO CZOCHRALSKI CRYSTAL GROWTH. (BARIUM SODIUM NIOBATE AND
          SODIUM LANTHANUM TANTALATE)
          COCKAYNE B + PLANT JG + VERE AW
          J CRYST GROW 15: 167-70 (1972)
               277

13-1929   ELECTROOPTIC POTASSIUM TANTALATE NIOBATE GRATINGS FOR LIGHT BEAM MODULATION AND
          DEFLECTION.
          COHEN MG + GORDON EI
          APPL PHYS LETT 5: 181-2 (1964)

13-1930   PHOTOCHROMIC EFFECT AND ELECTRON SPIN RESONANCE OF IRON AND MOLYBDENUM IN BARIUM SODIUM
          NIOBATE.
          CONSTANTINE JG + KLEIN PH + YANG GC
          BULL AMER PHYS SOC 15: 487 (1970)

13-1931   STRUCTURE OF THE BRONZE SODIUM(13) NIOBIUM(35) OXIDE(94) AND THE GEOMETRY OF
          FERROELECTRIC DOMAINS.
          CRAIG DC + STEPHENSON NC
          J SOLID STATE CHEM 3: 89-100 (1971)
               232

13-1932   FERROELECTRICITY IN BISMUTH OXIDE TYPE LAYER STRUCTURE COMPOUNDS.
          CROSS LE + POHANKA RC
          MATER RES BULL 6: 939-49 (OCT 1971)

13-1933   SOFT PHONON MODES IN THE MIXED CRYSTAL SYSTEMS POTASSIUM(X) SODIUM(1-X) TANTALATE AND
          POTASSIUM TANTALATE(X) NIOBATE(1-X).
          DAVIS TG
          P245-8 OF INT MEETING ON FERROELECTRICITY, 2ND, 1969, KYOTO, PROC,
          (J PHYS SOC JAP, VOL 28 SUPPL). PHYS SOC JAP, 1970, 459P.

13-1934   ELECTRONIC PROPERTIES (CONDUCTIVITY, MAGNETIC SUSCEPTIBILITY) OF HYDROGEN TUNGSTEN
          BRONZES, HYDROGEN(X) TUNGSTATE. (ANALOGOUS TO THE SODIUM COMPOUNDS)
          DICKENS PG + HURDITCH RJ
          P555-60 OF INST FOR ADVAN STUDY ON THE CHEM OF EXTENDED DEFECTS IN
          NONMETALLIC SOLIDS, 1969, EYRING L + O'KEEFE M(EDS), AMER ELSEVIER,
          1970, 669P.

13-1935   CRYSTAL STRUCTURE OF BISMUTH TITANIUM OXIDE.
          DORRIAN JF + NEWNHAM RE + SMITH DK + KAY MI
          FERROELECTRICS 3: 17-27 (1971)
               262

13-1936   PREPARATION AND STUDY OF BARIUM(X) STRONTIUM(1-X) NIOBATE SINGLE CRYSTALS (CZOCHRALSKI
          GROWTH, DIELECTRIC PROPERTIES).
          DUDNIK EF + GROMOV AK + KRAVCHENKO VB + KOPYLOV YUL + KUZNETSOV GF
          SOV PHYS CRYSTALLOGR 15(2): 330-32 (1970)
               208

13-1937   PREPARATION AND FABRICATION OF SODIUM POTASSIUM NIOBATE.
          EDWARDS DA
          ATOMIC WEAPONS RES ESTAB, ALDERMASTON, UK, AUG 1970, 10P.
          AWRE-0-68-70
               CHEM ABSTR 74: 6109S (1971)
               219

13-1938   TEMPERATURE DEPENDENCE OF THE DIELECTRIC PROPERTIES OF BARIUM STRONTIUM NIOBATE IN THE
          FREQUENCY RANGE OF 10 KILOHERTZ TO 10 MEGAHERTZ.
          ELSEN E + GRABMAIER J + GRAF P
          OPT LASER TECHNOL 3(4): 218-19 (NOV 1971)
               268

13-1939   VARIATION OF CRYSTAL GROWTH RATE WITH SUPERSATURATION IN FLUXED MELTS. (BARIUM
          STRONTIUM NIOBATE)
          ELWELL D + DAWSON RD
          J CRYST GROW 13-14: 555-9 (1972)
               266

13-1940 PHASE TRANSITION IN BARIUM BISMUTH(5+) BISMUTH(3+) OXIDE.
FESENKO EG + SHUVAEVA ET + GCLTSOV YUI
SOV PHYS CRYSTALLOGR 17(2): 362-3 (SEP-OCT 1972)
275

13-1941 COMPOSITION AND STRUCTURE OF SPUTTERED FILMS OF FERROELECTRIC NIOBATES.
FOSTER NF
J VACUUM SCI TECHNOL 8: 251-5 (1971)

13-1942 LONGITUDINAL ELECTROOPTIC EFFECT IN KTN.
FOX AJ + WHIPPS PW
ELECTRON LETT 7: 139-40 (1971)

13-1943 EFFECT OF ELECTRIC FIELD ON PHASE COMPOSITION OF (LEAD, BARIUM) NIOBATE SOLID
SOLUTIONS.
FREIMANIS VA + KRUCHAN YAYA
BULL ACAD SCI USSR, PHYS SER 35(9): 1795-7 (1971)

13-1944 ELECTROCHEMICAL DEPOSITICN AND DISSOLUTION OF TUNGSTEN OXIDE BRONZES. (SODIUM(X)
TUNGSTEN(1-X) Z(Y) OXIDE(3) WHERE Z IS NIOBIUM, TANTALUM, ZIRCONIUM, VANADIUM,
OR COBALT)
FREDLEIN RA + DAMJANOVIC A
J SOLID STATE CHEM 4: 94-102 (1972)

13-1945 PHASE COMPOSITION OF (LEAD, BARIUM) DINIOBATE. METHOD OF INVESTIGATION AND SOME
RESULTS.
FREIMANIS VA + KRUCHAN YAYA + GAVARS PE
P131-7 OF FAZOVYE PEREKHODY V SEGNETOELEKTRIKAKH (PHASE
TRANSFORMATIONS IN FERROELECTRICS), FRITSBERG VYA + ROLOV BN +
KRUCHAN YY(EDS), IZDATELSTVO ZINATNE RIGA, 1971, 206P. (IN RUSSIAN)

13-1946 ELASTIC AND PIEZOELECTRIC CONSTANTS OF STRONTIUM(4) POTASSIUM LITHIUM NIOBATE TYPE
FERROELECTRIC CRYSTAL.
FUKUMOTO A + WATANABE A
PROC IEEE 38(9): 1376 (1970)
218

13-1947 (X-RAY) STRUCTURAL (OPTICAL) AND DIELECTRIC STUDIES OF (SINGLE CRYSTALS IN THE TUNGSTEN
BRONZE TYPE) FERROELECTRIC (SYSTEM) POTASSIUM LITHIUM TANTALATE(X) NIOBATE(1-X).
FUKUDA T
JAP J APPL PHYS 9(6): 599-606 (1970)
196

13-1948 (TETRAGONAL) TUNGSTEN BRCNZE TYPE CRYSTALS GROWN IN MOLTEN SOLUTIONS OF THE
PSEUDOSYSTEM POTASSIUM LITHIUM BARIUM NIOBATE.
GIESS EA + SCOTT BA + BURNS G + SMITH AW + OLSON BL + O'KANE DF
MATER RES BULL 5: 109-16 (1970)
188

13-1949 (EXTENT OF) TUNGSTEN BRONZE (LIQUIDUS AND SOLIDUS) FIELD AND MELT GROWTH OF CRYSTALS IN
THE SODIUM OXIDE- BARIUM OXIDE NIOBIUM PENTOXIDE SYSTEM.
GIESS EA + SCOTT BA + OLSON BL + BURNS G + O'KANE DF
J AMER CERAM SOC 53: 14-17 (1970)
181

13-1950 PARAMETRIC FLUORESCENCE IN BARIUM SODIUM NIOBATE FOR NINE CW PUMP FREQUENCIES.
GLEASON TJ + KRUGER JS
BULL AMER PHYS SOC 15: 41 (1970)

NEUTRON SCATTERING BY A FERROELECTRIC SOLID SOLUTION. (THEORY, APPLIED TO BARIUM
STRONTIUM TITANATE AND POTASSIUM NIOBATE TANTALATE)
GLOGAR P
P92-9, VOL 1 OF DVORAK V + FOUSKOVA A + GLOGAR P (EDS),
INT MEETING ON FERROELECTRICITY, PRAGUE, 1966, PROC.
INST PHYS CZECH ACAD SCI, PRAGUE, 1966.

13-1951 ELECTROPHYSICAL PROPERTIES OF THE SOLID SOLUTION SYSTEM LEAD TITANATE- LEAD ZIRCONATE-
LEAD (ZINCATE(0.33) NIOBATE(0.67)).
GORELIK IF + ANANEVA AA + ZAIONTS LR + UGRYUMOVA MA
BULL ACAD SCI USSR, PHYS SER 35(9): 1798-800 (1971)

13. TUNGSTEN BRONZE AND LAYER STRUCTURE OXIDES

13-1952   NONLINEAR PROPERTIES AND PHASE TRANSITIONS OF STRONTIUM BISMUTH TITANATES.
           GUBKIN AN + KASHTANOVA AM + POTAPOV EV + SOLODUKHIN AV
           SOV PHYS SOLID STATE 4: 2411-16 (1963)

13-1953   MICROCHEMICAL ANALYSIS OF GADOLINIUM DOPED BARIUM SODIUM NIOBATES.
           GUPTA KP + WANG FFY
           BULL AMER CERAM SOC 49: 418 (1970)

13-1954   SOLIDIFICATION STUDY OF BARIUM STRONTIUM NIOBATE.
           GUPTA KP + WANG FFY
           BULL AMER CERAM SOC 51: 322 (1972)

13-1955   SCANNING ELECTRON MICROSCOPIC EXAMINATION OF SINTERED BARIUM SODIUM NIOBATES.
           GUPTA KP + SIEGEL RW + WANG FFY
           BULL AMER CERAM SOC 51: 361 (1972)

13-1956   SOLID SOLUTIONS IN ANTIFERROELECTRIC REGION OF THE SYSTEM LEAD HAFNATE TITANATE, LEAD
           STANNATE NIOBATE.
           HALL CA + DUNGAN RH + STARK AH
           J AMER CERAM SOC 47: 259-64 (1964)

13-1957   SOFT MODES AND STRUCTURE OF FERROELECTRIC TETRAGONAL POTASSIUM TANTALATE NIOBATE.
           HEWAT AW + ROUSE KD + ZACCAI G
           FERROELECTRICS 4(3): 153-7 (1972)

13-1958   NEW (UNIAXIAL) NONLINEAR OPTIC CRYSTAL BARIUM LITHIUM NIOBATE HAVING NO MICROTWINNING
           (AND RESISTANT TO LASER INDUCED REFRACTIVE INHOMOGENEITIES).
           HIRANO H + TAKEI H + KOIDE S
           JAP J APPL PHYS 9: 580 (1970)
                196

13-1959   CRYSTAL GROWTH OF LEAD (CADMATE(0.33) NIOBATE(0.67)) AND ITS DIELECTRIC PROPERTIES.
           (FERROELECTRIC- PARAELECTRIC PHASE TRANSITION)
           ICHINOSE N + TAKAHASHI T + YOKOMIZO Y
           J PHYS SOC JAP 31: 1848 (1971)
                268

13-1960   ELECTRICAL MAGNETIC AND X-RAY STUDIES OF PHASE TRANSITIONS IN LEAD COBALTATE(0.5)
           TUNGSTATE(0.5), TITANATE, ZIRCONATE SYSTEM.
           ICHINOSE N + KURIHARA K
           P321-3 OF INT MEETING ON FERROELECTRICITY, 2ND, 1969,, KYOTO, PROC,
           (J PHYS SOC JAP, VOL 28 SUPPL). PHYS SOC JAP, 1970, 459P.

13-1961   ANTIFERROELECTRIC PHASE IN THE SYSTEM LEAD TITANATE- LEAD ZIRCONATE- LANTHANUM FERRATE.
           IKEDA T + OKANO T
           JAP J APPL PHYS 3: 493-4 (1964)

13-1962   COMPLEX OXIDES WITH TUNGSTEN BRONZE TYPE STRUCTURES (POTASSIUM (BARIUM OR STRONTIUM)
           TITANATE NIOBATE, (BARIUM OR STRONTIUM) BISMUTH TITANATE NIOBATE, STRONTIUM TITANATE
           NIOBATE, POTASSIUM LANTHANUM TITANIUM NIOBATE, POTASSIUM STRONTIUM NIOBATE).
           IKEDA T + HARAGUCHI T
           JAP J APPL PHYS 9: 422-3 (1970)
                190

13-1963   STUDY OF SUBSOLIDUS EQUILERIA IN POTASSIUM OXIDE- LITHIUM OXIDE- NIOBIUM PENTOXIDE
           SYSTEM. (POTASSIUM LITHIUM NIOBATES)
           IKEDA T + KIYOHASHI K
           JAP J APPL PHYS 9: 1541-2 (1970)
                222

13-1964   SOME COMPOUNDS OF TUNGSTEN BRONZE TYPE A(6)B(10)O(30). (WHERE B IS (NIOBIUM, TITANIUM)
           OR (NIOBIUM, TUNGSTEN), COMPILATION, LATTICE CONSTANTS, CURIE TEMPERATURES)
           IKEDA T + HARAGUCHI T + ONODERA Y + SAITO T
           JAP J APPL PHYS 10(8): 987-94 (1971)
                267

13-1965   X-RAY STUDY OF HIGH TEMPERATURE PHASE TRANSITIONS OF SODIUM NIOBATE (TO 500C).
           ISHIDA K + HONJO G
           J PHYS SOC JAP 29: 249 (1970)
                202

13. TUNGSTEN BRONZE AND LAYER STRUCTURE OXIDES

13-1966   CRYSTAL CHEMISTRY OF PEROVSKITE LIKE LAYER TYPE FERROELECTRICS. (BISMUTH TUNGSTATE,
           BISMUTH NIOBIUM OXYFLUORIDE, BISMUTH TANTALUM OXYFLUORIDE)
           ISMAILZADE IH
           J PHYS (PARIS) 33, SUPPL 4, COLLOQ 2: 237-9 (1972)
               294

13-1967   ELASTIC AND PIEZOELECTRIC PROPERTIES OF CADMIUM PYRONIOBATE IN STRONG ELECTRIC FIELDS.
           ISUPOV VA + SKUBITSKII VN
           SOV PHYS SOLID STATE 5: 701-3 (1963)

13-1968   PHASE TRANSISITION IN THE ANTIFERROELECTRET LEAD COBALTATE(0.5) TUNGSTATE(0.5) AND LEAD
           COBALTATE(0.5) TUNGSTATE(0.5)- LEAD TITANATE SOLID SOLUTIONS.
           ISUPOV VA + BELOUS LP
           SOV PHYS CRYSTALLOGR 16: 129-32 (1971)

13-1969   DIELECTRIC PROPERTY OF BARIUM TITANATE NIOBATE SINGLE CRYSTAL (TETRAGONAL TUNGSTEN
           BRONZE).
           ITOH Y + MIYAZAWA S + YAMADA T + IWASAKI H
           JAP J APPL PHYS 9: 157-8 (1970)
               190

13-1970   TUNGSTEN BRONZE FIELD IN THE PSEUDOTERNARY SYSTEM SODIUM NIOBATE BARIUM TITANATE BARIUM
           DINIOBATE.
           ITOH Y + IWASAKI H
           MATER RES BULL 7: 663-72 (1972)
               286

13-1971   DIELECTRIC AND ELECTROOPTIC PROPERTIES OF BARIUM SODIUM YTTRIUM NIOBATE SINGLE CRYSTAL
           WITH TUNGSTEN BRONZE TYPE STRUCTURE.
           IWASAKI H + MIYAZAWA S
           JAP J APPL PHYS 10: 161-2 (1971)
               267

13-1972   SODIUM BARIUM RARE EARTH NIOBATES WITH THE TUNGSTEN BRONZE TYPE STRUCTURE.
           IWASAKI H
           MATER RES BULL 6: 251-60 (1971)

           NONLINEAR OPTICAL POLARIZABILITY OF THE NIOBIUM- OXYGEN BOND. (LITHIUM NIOBATE, LITHIUM
           TANTALATE, POTASSIUM LITHIUM NIOBATE, BARIUM SODIUM NIOBATE, BARIUM STRONTIUM NIOBATE)
           JEGGO CR + BOYD GD
           J APPL PHYS 41: 2741-3 (1970)

13-1973   OPTIMUM CRYSTAL ORIENTATION FOR A BARIUM SODIUM NIOBATE TRANSVERSE LIGHT MODULATOR.
           (HALF WAVE VOLTAGE, THEORY)
           JOHNSON DC
           COMMUN RES CENTER, OTTAWA, ONTARIO, 1972, 17P. CRC-1229, N73-13733
               286

13-1974   ANOMALOUS BEHAVIOR OF THE OPTICAL ASBSORPTION COEFFICIENT OF LEAD MAGNONIOBATE SINGLE
           CRYSTALS IN THE REGION OF A BROAD FERROELECTRIC TRANSITION.
           KAMZINA LS + KRAINIK NN + NESTEROVA NN
           SOV PHYS SOLID STATE 14(7): 1853-5 (JAN 1973)
               282

13-1975   NEODYMIUM: YTTRIUM ALUMINUM GARNET- BARIUM SODIUM NIOBATE 0.66 MILLIMICRON HARMONIC
           SOURCE.
           KARR MA
           J APPL PHYS 42: 4517-19 (1971)

13-1976   A NEW CLASS OF FERROELECTRIC CERAMICS OF THE TYPE AB(2) (XO(4)) (3).
           KEESTER KL + JACOBS JT
           BULL AMER CERAM SOC 50: 372 (1971)

13-1977   PHASE HOLOGRAMS IN A FERROELECTRIC PHOTOCONDUCTOR DEVICE. (BISMUTH(4) TITANIUM(3)
           OXIDE(12))
           KENEMAN SA + MILLER A + TAYLOR GW
           APPL OPT 9: 2279-82 (1970)

13-1978   BISMUTH(3) NIOBIUM(17) OXIDE(47): POTENTIALLY FERROELECTRIC CRYSTAL STRUCTURE OF THE
           TUNGSTEN BRONZE TYPE.
           KEVE ET + SKAPSKI AC
           J CHEM SOC LONDON PART A INORG PHYS THEOR NO 9: 1280-6 (1971)

13-1979   OPTICAL DAMAGE IN KTN.
          KING SR + HARTWICK TS + CHASE AB
          APPL PHYS LETT 21: 312-14 (1972)

13-1980   RELAXATION POLARIZATION OF A FERROELECTRIC LEAD MAGNESATE(0.33) NIOBATE(0.67) WITH A
          DIFFUSE PHASE TRANSITION.
          KIRILLOV VV + ISUPOV VA
          BULL ACAD SCI USSR, PHYS SER 35(12): 2360-4 (1971)

13-1981   TUNGSTEN BRONZE TYPE COMPLEX OXIDES.
          KOIDE S
          OYO BUTSURI 39: 997-1004 (1970) (IN JAPANESE)

13-1982   ELECTROOPTICAL EFFECT IN LEAD ZINCATE(0.33) NIOBATE(0.67). (DISPERSION OF QUADRATIC
          COEFFICIENTS)
          KRAUNIK NN + GOKHBERG LS + MYLNIKOVA IE
          SOV PHYS SOLID STATE 12(8): 1885-8 (1971)
              228

13-1983   PARAMETRIC FLUORESCENCE IN BARIUM SODIUM NIOBATE FOR 10 CW PUMP WAVELENGTHS (RANGE
          454.5-528.7 MICRONS).
          KRUGER JS + GLEASON TJ
          J APPL PHYS 41(9): 3903-04 (1970)
              207

13-1984   VISIBLE CW PARAMETRIC OSCILLATOR USING BARIUM SODIUM NIOBATE.
          LAURENCE C + TITTEL FK
          J APPL PHYS 42: 2137-8 (1971)

13-1985   NEW PHASES OF THE TETRAGONAL TUNGSTEN OXYGEN BRONZE TYPE: QUADRATIC LEAD OR BISMUTH
          TANTALATE NIOBATE, TANTALATE TUNGSTATE, OR TUNGSTATE NIOBATE.
          LE PARMENTIER L
          REV CHIM MIN 9(3): 519-38 (1972) (IN FRENCH)
              SCI ABSTR A 76: 7697 (1973)
              280

13-1986   FAR INFRARED DIELECTRIC DISPERSION IN BARIUM SODIUM NIOBATE.
          IURIO A
          J APPL PHYS 43: 3753-5 (1972)

13-1987   ABSORPTION EDGE SPLITTING IN POTASSIUM TANTALATE(X) NIOBATE(1-X).
          MANLIEF SK + FAN HY
          PHYS REV B SOLID STATE 6: 185-92 (1972)

13-1988   ABSORPTION EDGE SPLITTING IN POTASSIUM TANTALATE NIOBATE.
          MANLIEF SK + FAN HY
          BULL AMER PHYS SOC 16: 355 (1971)

13-1989   RAMAN SPECTRUM OF POTASSIUM TANTALUM(0.64) NIOBIUM(0.36) OXIDE.
          MANLIEF SK + FAN HY
          PHYS REV B 5(10): 4046-60 (15 MAY 1972)
              264

13-1990   GROWTH OF SINGLE CRYSTALS OF LEAD (ZINCATE NIOBATE) AND PROPERTIES OF THIS COMPOSITION.
          MATSUO Y
          J CERAM SOC JAP 78(899): 213-20 (1970) (IN JAPANESE)
                CHEM ABSTR 73: 70550K (1970)
                238

13-1991   GROWTH OF BARIUM LITHIUM NIOBATE SINGLE CRYSTALS BY THE CZOCHRALSKI METHOD.
          MATTHES H
          J CRYST GROW 15(2): 157-8 (1972)

13-1992   POLARITONS IN A BIAXIAL CRYSTAL OF BARIUM SODIUM NIOBATE.
          MAVRIN BN + STERIN KE
          SOV PHYS SOLID STATE 14(9): 2402-3 (MAR 1973)
              286

          FURTHER MEASUREMENTS OF ABSOLUTE SIGNS OF SECOND HARMONIC GENERATION COEFFICIENTS OF
          PIEZOELECTRIC CRYSTALS. (LEAD TITANATE, BARIUM STRONTIUM NIOBATE, LEAD NIOBIUM(4)
          OXIDE(11), SILVER GALLIUM DISULFIDE, SILICON CARBIDE)
          MILLER RC + NORDLAND WA
          PHYS REV B 5: 4931-4 (1972)

13. TUNGSTEN BRONZE AND LAYER STRUCTURE OXIDES

13-1993 SYNTHESES, X-RAY, AND DIELECTRIC STUDIES OF CERTAIN NEW LAYERED FERROELECTRICS.
(LANTHANUM BISMUTH TITANIUM IRON OXIDE, PRASEODYMIUM BISMUTH TITANIUM IRON OXIDE,
BISMUTH MOLYBDATE, BISMUTH FLUONIOBATE, BISMUTH FLUOTANTALATE)
MIRISHLI FA + ISMAILZADE IG
BULL ACAD SCI USSR, PHYS SER 35(9): 1668-71 (1971)
269

13-1994 PROPERTIES OF BISMUTH(12) TITANIUM OXIDE(20) AND THE SYSTEM BISMUTH OXIDE- TITANIUM
DIOXIDE.
MORRISON AD
FERROELECTRICS 2: 59-62 (JAN 1971)

13-1995 RESISTIVITY BEHAVIOR OF SEMICONDUCTING YTTRIUM BARIUM ZIRCONIUM TITANATE. (TEMPERATURE
DEPENDENCE)
MORATIS CJ + BRATTON RJ
JAP J APPL PHYS 10(4): 421-6 (1971)
267

13-1996 LINEAR THERMAL EXPANSIONS OF GADOLINIUM SUBSTITUTED BARIUM SODIUM NIOBATES.
MUKHERJEE JL + WANG FFY
BULL AMER CERAM SOC 49: 418 (1970)

13-1997 A NEW FERROELECTRIC SINGLE CRYSTAL STRONTIUM(2) NIOBIUM(2) OXIDE(7).
NANAMATSU S + KIMURA M + DOI K + TAKAHASHI M
NEC RES DEVELOP 27: 17-21 (1972)

13-1998 MEASUREMENTS OF SECOND HARMONIC GENERATION AND THE VARIATIONS IN THE FREE AND CLAMPED
VALUES OF THE DIELECTRIC CONSTANTS AND ELECTROOPTIC COEFFICIENTS IN BARIUM SODIUM
NIOBATE.
NASH FR + TURNER EH + BRIDENBAUGH PM + DZIEDZIC JM
J APPL PHYS 43(1): 1-9 (JAN 1972)
254

OBSERVATION OF THE SIMULTANEOUS GENERATION OF THE SECOND, THIRD, AND FOURTH HARMONICS
OF 1.06 MICRON RADIATION IN BARIUM SODIUM NIOBATE, LITHIUM NIOBATE, AND LITHIUM IODATE
CRYSTALS.
NG WK + WOODBURY EJ
APPL PHYS LETT (18(12): 550-1 (15 JUN 1971)

13-1999 X-RAY AND THERMAL EXPANSION STUDY OF A SODIUM(0.88) LITHIUM(0.12) NIOBATE CERAMIC.
NITTA T + MIYAZAWA T
J AMER CERAM SOC 54: 636-7 (1971)

13-2000 PTC EFFECT (POSITIVE TEMPERATURE COEFFICIENT OF RESISTIVITY) IN (DOPED AND UNDOPED
CERAMIC PEROVSKITE) LEAD FERRATE(0.5) NIOBATE(0.5).
NOMURA S + DOI K
JAP J APPL PHYS 9: 716 (1970)
196

13-2001 PREPARATION AND DIELECTRIC PROPERTIES OF POTASSIUM BISMUTH ZINC NIOBATE.
NOMURA S + KOJIMA F
J PHYS SOC JAP 31: 1286 (OCT 1971)

13-2002 SUPERLATTICE STRUCTURE OF STRONTIUM SODIUM LITHIUM NIOBATE (TUNGSTEN BRONZE) TYPE
FERROELECTRICS.
OHTA T + WATANABE A
JAP J APPL PHYS 9: 721-2 (1970)
196

13-2003 CRYSTAL GROWTH AND ELECTROOPTIC PROPERTIES OF TUNGSTEN BRONZE CRYSTALS FROM THE
POTASSIUM NIOBATE- LITHIUM NIOBATE- BARIUM DINIOBATE SYSTEM.
O'KANE DF + BURNS G + GIESS EA + SMITH AW
J ELECTROCHEM SOC 117: 195C (1970)

13-2004 EFFECTS OF GRAIN SIZE AND POROSITY ON THE ELECTRICAL AND OPTICAL PROPERTIES OF
TRANSPARENT FERROELECTRIC CERAMICS. ((LEAD, LANTHANUM) (ZIRCONATE, TANTALATE), (LEAD,
TUNGSTEN) (ZIRCONATE TANTALATE), LEAD (LANTHANATE, NIOBATE)- LEAD ZIRCONATE- LEAD
TITANATE)
OKAZAKI K + NAGATA K + SAEKI H
BULL AMER CERAM SOC 51: 355 (1972)

13. TUNGSTEN BRONZE AND LAYER STRUCTURE OXIDES

13-2005 FORMATION OF THIN LAYERS OF MIXED TITANATES BY SOLID SOLID REACTIONS.
OZDEMIR FS + SCHWARTZ RJ
BULL AMER CERAM SOC 51: 474-8 (1972)

13-2006 PROPERTIES OF FERROGLASS CERAMICS BASED ON LEAD BARIUM NIOBATE COMPOUNDS.
PAVLUSHKIN NM + ZHURAVLEV AK + EGOROVA LS + ZHABOTINSKII VA
INORG MATER 7: 735-8 (1971)
      256

13-2007 CONTINUOUS HOT PRESSING CF POTASSIUM SODIUM NIOBATE.
PERDUIJN DJ + VAREKAMP RFP + VERJANS HC
BER DEUT KERAM GESELL 47: 389 (1970) (IN GERMAN)
PROC BRIT CERAM SOC 18: 239-44 (1970)

13-2008 TEMPERATURE DEPENDENCE OF THE REFRACTIVE INDICES OF BARIUM SODIUM NIOBATE.
POMEROY WRM
OPTO-ELECTRON 3(3): 148-9 (AUG 1971)
      264

13-2009 FERROELECTRIC DIELECTRIC ANOMALY IN BISMUTH(4) TITANIUM(3) OXIDE(12) SINGLE CRYSTALS.
PULVARI CF
P347-60, VOL 1 OF DVORAK V + FOUSKOVA A + GLOGAR P (EDS),
INT MEETING ON FERROELECTRICITY, PRAGUE, 1966, PROC.
INST PHYS CZECH ACAD SCI, PRAGUE, 1966.

13-2010 CRYSTAL CHEMISTRY AND ELECTRICAL PROPERTIES OF TWO NEW PHASES OF QUADRATIC TUNGSTEN
BRONZES STRONTIUM POTASSIUM TANTALATE AND BARIUM POTASSIUM TANTALATE.
RAVEZ J + PERRON A + CHAMINADE JP
COMPT REND C 274: 1450-3 (1972) (IN FRENCH)

13-2011 FIRST FERROELECTRIC OXYFLUORIDE PHASES OF THE OXYGENATED QUADRATIC TUNGSTEN BRONZE
TYPE. (STRONTIUM POTASSIUM NIOBIUM OXYFLUORIDE)
RAVEZ J + TOURNEUR D + HAGENMULLER P
MATER RES BULL 7: 473-8 (1972 (IN FRENCH)
      265

13-2012 SODIUM ORDERING IN SODIUM TUNGSTEN BRONZES.
ROPER JG + KNOWLES HB
PHYS LETT A 38: 477-8 (1972)

VIBRATIONAL SPECTRA OF LITHIUM NIOBATE, BARIUM SODIUM NIOBATE AND BARIUM SODIUM
TANTALATE (FORCE CONSTANTS, SPACE GROUP ASSIGNMENT).
ROSS SD
J PHYS C SOLID STATE PHYS 3: 1785-90 (1970)

13-2013 DIELECTRIC AND CERAMIC PROPERTIES OF (BISMUTH(0.5) (SODIUM OR POTASSIUM) (0.5) (X)
LEAD(1-X) TITANATE SYSTEMS.
SAKATA K + MASUDA Y + OHARA G
SCI REP RES INST TOHOKU UNIV, SER B TECHNOL 15: 29-38 (1963)
      CHEM ABSTR 63: 5354 (1965)

13-2014 DIELECTRIC AND STRUCTURAL PROPERTIES OF FERROELECTRICS WITH DIFFUSE PHASE TRANSITIONS.
(MIXED LAYER OXIDES)
SALNIKOV VD + KUZMINOV YUS + VENEVTSEV YUN
INORG MATER 7: 1138-9 (JUL 1971)

RAMAN STUDY OF METALLIC TUNGSTEN BRONZES. (SODIUM AND RUBIDIUM TUNGSTATES)
SCOTT JF + LEHENY RF + SWEEDLER AR
PHYS REV B 2: 3883-7 (1970)

13-2015 CRYSTAL GROWTH AND NONLINEAR OPTICAL PROPERTIES OF POTASSIUM LITHIUM NIOBATES.
SCOTT BA + SMITH AW + BURNS G + EDMONDS HD + GIESS EA
J ELECTROCHEM SOC 117: 195C (1970)

13-2016 PHASE EQUILIBRIA IN THE POTASSIUM NIOBATE- STRONTIUM DINIOBATE AND POTASSIUM NIOBATE-
BARIUM DINIOBATE SYSTEMS.
SCOTT BA + GIESS EA + O'KANE DF + BURNS G
J AMER CERAM SOC 53: 106-9 (1970)
      183

13. TUNGSTEN BRONZE AND LAYER STRUCTURE OXIDES

13-2017   TUNGSTEN BRONZE FIELD IN THE (OXIDE) SYSTEM POTASSIA LITHIA NIOBIA (POTASSIUM LITHIUM
          NIOBATES, PHASE RELATIONSHIPS, SOLID SOLUTIONS).
          SCOTT BA + GIESS EA + OLSON BL + BURNS G + SMITH AW + O'KANE DF
          MATER RES BULL 5(1): 47-56 (1970)
              188

13-2018   PYROELECTRIC VOLTAGE RESPONSE TO SHORT INFRARED LASER PULSES IN TRIGLYCINE SULFATE AND
          BARIUM STRONTIUM NIOBATE.
          SHAULOV A + ROSENTHAL A + SIMHONY M
          J APPL PHYS 43: 4518-22 (NOV 1972)

13-2019   PYROELECTRIC VOLTAGE RESPONSE TO RECTANGULAR INFRARED SIGNALS IN TRIGLYCINE SULFATE AND
          STRONTIUM BARIUM NIOBATE.
          SHAULOV A + SIMHONY M
          J APPL PHYS 43: 1440-8 (1972)

13-2020   ANOMALIES OF ELASTICITY AND INTERNAL FRICTION IN THE REGION OF THE ANTIFERROELECTRIC
          CURIE POINT OF LEAD MAGNESATE(0.5) TUNGSTATE(0.5).
          SHUVALOV LA + MINAEVA KA
          SOV PHYS DOKL 7: 906-7 (1963)

13-2021   DEVELOPMENT OF SINGLE CRYSTAL BARIUM SODIUM NIOBATE AS A NONLINEAR OPTICAL MATERIAL.
          (GROWTH BY KYROPOULOS AND CZOCHRALSKI METHODS)
          SILVA WJ + NELSON WE + ROSENGREEN A
          CRYSTAL TECHNOLOGY INC, MOUNTAIN VIEW, CALIF, 1970, 83P. AD715987
              CHEM ABSTR 75: 41334D (1971) US GOVT RES DEVELOP REP 71(4): 180 (1971)
              220

13-2022   PYROELECTRIC VOLTAGE RESPONSE TO STEP SIGNALS OF INFRARED RADIATION IN TRIGLYCINE
          SULFATE AND STRONTIUM BARIUM NIOBATE.
          SIMHONY M + SHAULOV A
          J APPL PHYS 42(10): 3741-4 (SEP 1971)
              241

13-2023   NONLINEAR OPTICAL PROPERTIES OF FERROELECTRIC LEAD NIOBIUM(4) OXIDE(12).
          SINGH S + BONNER WA + POTOPOWICZ JR + VAN UITERT LG
          J ELECTROCHEM SOC 117: 195C (1970)

13-2024   OPTICAL AND FERROELECTRIC PROPERTIES OF BARIUM SODIUM NIOBATE. (OPTICAL ABSORPTION,
          REFRACTIVE INDEX, DIELECTRIC CONSTANT, NONLINEAR OPTICAL COEFFICIENTS, AND LINEAR
          ELECTROOPTIC COEFFICIENTS)
          SINGH S + DRAEGERT DA + GEUSIC JE
          PHYS REV B SOLID STATE 2(7): 2709-24 (1970)
              236

13-2025   ROLE OF HYDROGEN IN POLARIZATION REVERSAL OF FERROELECTRIC BARIUM SODIUM NIOBATE.
          SINGH S + LEVINSTEIN HJ + VAN UITERT LG
          APPL PHYS LETT 16(4): 176-8 (1970)
              181

13-2026   NONLINEAR OPTICAL PROPERTIES OF POTASSIUM LITHIUM NIOBATES (TRANSPARENT, TETRAGONAL
          TUNGSTEN BRONZE STRUCTURE, FERROELECTRIC).
          SMITH AW + BURNS G + SCOTT BA + EDMONDS HD
          J APPL PHYS 42(2): 684-6 (1971)

13-2027   NONLINEAR OPTICAL PROPERTIES OF POTASSIUM LITHIUM NIOBATE.
          SMITH AW + BURNS G + SCOTT BA + EDMONDS HD
          J OPT SOC AMER 60: 716 (1970)

13-2028   OPTICAL AND FERROELECTRIC PROPERTIES OF (THE TUNGSTEN BRONZE SYSTEM) POTASSIUM SODIUM
          BARIUM NIOBATE (SECOND HARMONIC GENERATION).
          SMITH AW + BURNS G + O'KANE DF
          J APPL PHYS 42(1): 250-5 (1971)
              224

13-2029   THERMOGRAPHIC STUDY OF THE REACTION OF BISMUTH(4) TITANIUM(3) OXIDE(12) WITH VANADIUM
          PENTOXIDE AND MOLYBDENUM AND TUNGSTEN TRIOXIDE.
          SMOLYANINOV NP + MOLOZOVA AP + BOCHKAREVA OB
          RUSS J INORG CHEM 15: 132-5 (1970)

13-2030   CHARACTERISTICS OF KTN (POTASSIUM TANTALATE NIOBATE) PYROELECTRIC DETECTORS.
          STAFSUDD OM + PINES MY
          J OPT SOC AMER 62: 1153-5 (OCT 1972)

13. TUNGSTEN BRONZE AND LAYER STRUCTURE OXIDES

13-2031 STRUCTURE OF TETRAGONAL TIN TUNGSTEN BRONZES.
         STEADMAN R
         MATER RES BULL 7: 1143-50 (1972)
            276

13-2032 CRYSTAL CHEMISTRY OF MIXED BISMUTH OXIDES WITH LAYER TYPE STRUCTURE.
         SUBBARAO EC
         J AMER CERAM SOC 45: 166-9 (1962)

13-2033 CRYSTAL GROWTH AND DIELECTRIC PROPERTIES OF BISMUTH NIOBATE.
         SUGAI T + WADA M
         JAP J APPL PHYS 11: 1863-4 (1972)

13-2034 SINGLE CRYSTAL GROWTH OF STRONTIUM POTASSIUM TANTALATE FROM MOLTEN SALTS.
         SUGAI T + WADA M
         JAP J APPL PHYS 10: 955-6 (1971)
            267

13-2035 MOSSBAUER STUDIES OF IRON-57 IN SOME PEROVSKITE LIKE LAYER TYPE FERROELECTRICS. (IN
         BISMUTH TITANATE FERRATES)
         SULTANOV GD + MIRISHLI FA + ISMAILZADE IH
         ACTA CRYSTALLOGR A 28: S241 (1972)

13-2036 ELECTROOPTIC LIGHT BEAM DEFLECTION WITH BARIUM STRONTIUM NIOBATE PRISM.
         TADA K + MURAI T + AOKI M + MUTO K + AWAZU K
         JAP J APPL PHYS 11: 1622-7 (1972)

13-2037 GROWTH OF FERROELECTRIC SINGLE CRYSTAL STRONTIUM(2) NIOBIUM(2) OXIDE(7) BY MEANS OF THE
         FLOATING ZONE TECHNIQUE.
         TAKAHASHI M + NANAMATSU S + KIMURA M
         J CRYST GROW 13-14: 681-5 (1972)
            266

13-2038 PREPARATION AND EPITAXY OF SPUTTERED FILMS OF FERROELECTRIC BISMUTH TITANIUM OXIDE.
         TAKEI NJ + FORMIGONI NP + FRANCOMBE MH
         J VACUUM SCI TECHNOL 7(3): 442-8 (1970)
            243

13-2039 ELECTROOPTIC G COEFFICIENTS (RELATING INDUCED BIREFRINGENCE AND POLARIZATION) OF LEAD
         CONTAINING OXYGEN OCTAHEDRA FERROELECTRICS: CERAMIC LEAD BARIUM ZIRCONATE TITANATE.
         THACHER PD
         J APPL PHYS 41(12): 4790-7 (1970)
            215

13-2040 SOFT PHONONS IN POTASSIUM TANTALATE NIOBATE.
         TODD LT + DAVID TG
         BULL AMER PHYS SOC 15: 810 (1970)

13-2041 PIEZOELECTRIC SURFACE WAVES IN CUBIC AND ORTHORHOMBIC CRYSTALS. (BISMUTH(12) GERMANIUM
         OXIDE(20) AND BARIUM SODIUM NIOBATE)
         TSENG CC
         APPL PHYS LETT 16: 253-5 (1970)

13-2042 MECHANISM AND KINETICS OF SOLID PHASE FORMATION OF LEAD METANIOBATE.
         TSIVILEV RP + FEDULOV SA + MARTYNOVA VF
         INORG MATER 7: 160-2 (JAN 1971)

13-2043 DEPENDENCE OF SECOND HARMONIC GENERATION ON CRYSTAL INHOMOGENEITY (THEORY AND TEST ON
         BARIUM SODIUM NIOBATE AND NEODYMIUM DOPED YAG LASER).
         TSUYA T + FUJINO Y + SUGIBUCHI K
         J APPL PHYS 41(6): 2557-63 (1970)
            197

13-2044 DIFFUSE TO SHARP FERROELECTRIC PHASE TRANSITION SYSTEMS LEAD (B(0.5) NIOBIUM(0.5))(1-X)
         B'(X) OXYGEN(3), WHERE B IS IRON OR SCANDIUM, AND B' IS ZIRCONIUM OR HAFNIUM.
         VALENTA MW + KUCHAR F + FRANKUS P
         J PHYS (PARIS) 33, SUPPL 4, COLLOQ 2: 247-8 (1972)
            294

13-2045   GROWTH OF BARIUM SODIUM NIOBATE SINGLE CRYSTALS FOR OPTICAL APPLICATIONS. (BY PULLING
          FROM THE MELT, STRUCTURE- PROPERTY CORRELATION WITH LITHIUM NIOBATE AND POTASSIUM
          LITHIUM NIOBATE)
          VAN UITERT LG + RUBIN JJ + BONNER WA
          IEEE J QUANTUM ELECTRON 4(10): 622-7 (1968)

          DIELECTRIC AND NONLINEAR MICROWAVE PROPERTIES OF POLYCRYSTALLINE SOLID SOLUTIONS BASED
          ON BARIUM TITANATE. (BARIUM STRONTIUM TITANATE, BARIUM TITANATE STANNATE, BARIUM
          TITANATE ZIRCONATE)
          VERBITSKAYA TN + IVANOV IV + MOROZOV NA
          SOV PHYS SOLID STATE 12(5): 1250-1 (1970)

13-2046   EFFECTS OF ROASTING CONDITIONS ON THE STRUCTURE AND PROPERTIES OF LEAD BARIUM NIOBATE
          PIEZOCERAMICS.
          VERIGINA ZS + DOLGAYA ZHA + RASTORGUEV LN
          INORG MATER 6: 806-9 (MAY 1970)

13-2047   PROGRESS TOWARDS THE GROWTH OF GOOD OPTICAL QUALITY BARIUM SODIUM NIOBATE (BY USING
          COMPOSITION WHERE LIQUIDUS AND SOLIDUS COINCIDE).
          VERE AW + PLANT JG + COCKAYNE B
          RRE NEWSLET AND RES REV NO 9, 1970, 3P.
              202

13-2048   CAUSES OF THERMAL EXPANSION OF GADOLINIUM SUBSTITUTED BARIUM SODIUM NIOBATES.
          WANG FFY + MUKHERJEE JL
          BULL AMER CERAM SOC 51: 329 (1972)

13-2049   EMPIRICAL FACTORS FOR CALCULATION OF THE FERROELECTRIC TRANSITION TEMPERATURES OF
          TUNGSTEN BRONZE TYPE NIOBATES (CORRELATION OF CURIE TEMPERATURE WITH CHEMICAL
          COMPOSITION).
          WANG FFY
          PHIL MAG 21(173): 903-6 (1970)
              199

13-2050   PHASE TRANSITIONS IN TANTALATES WITH THE STRUCTURE OF POTASSIUM TUNGSTEN BRONZE.
          (POTASSIUM BARIUM TANTALUM OXIDE, SODIUM BARIUM TANTALUM OXIDE, POTASSIUM STRONTIUM
          TANTALUM OXIDE, SODIUM STRONTIUM TANTALUM OXIDE, POTASSIUM CALCIUM TANTALUM OXIDE,
          SODIUM CALCIUM TANTALUM OXIDE, POTASSIUM LITHIUM TANTALUM OXIDE)
          WANG-TYAU C + KRAINIK NN + ISUPOV VA + ISMAILZADE IG + MYLNIKOVA IE + AGAEV FA
          + VOLKOVA LS
          SOV PHYS CRYSTALLOGR 17(1): 107-11 (JUL-AUG 1972)
              274

13-2051   ELECTROOPTICAL PROPERTIES OF STRONTIUM POTASSIUM LITHIUM NIOBATE. (TUNGSTEN BRONZE TYPE
          FERROELECTRICS).
          WATANABE A + SATO Y + YANO T + KITAHIRO I
          P93-5 OF INT MEETING ON FERROELECTRICITY, 2ND, 1969, KYOTO, PROC,
          (J PHYS SOC JAP, VOL 28 SUPPL). PHYS SOC JAP, 1970, 459P.

13-2052   CATASTROPHIC SURFACE DAMAGE PRODUCED IN BARIUM SODIUM NIOBATE CRYSTALS DURING
          INTRACAVITY FREQUENCY DOUBLING.
          WEBB R
          P98-103 OF DAMAGE IN LASER MATERIALS, 1971 PROC, GLASS AJ +
          GUENTHER AH(EDS), NBS SPEC PUB 356, 1971.

13-2053   IMPURITY EFFECTS IN SODIUM TUNGSTEN BRONZES: NIOBIUM AND TANTALUM DOPING.
          WELLER PF + TAYLOR BE
          J SOLID STATE CHEM 2: 9-15 (1970)

13-2054   KYROPOULOS SINGLE CRYSTAL GROWTH OF SODIUM TUNGSTEN BRONZES. (SODIUM TUNGSTATE)
          WELLER PF + GRANDITS DM
          J CRYST GROW 12: 63-5 (1972)

13-2055   GROWTH OF HIGH QUALITY CRYSTALS OF POTASSIUM TANTALATE NIOBATE.
          WHIPPS PW
          J CRYST GROW 12: 120-4 (1972)
              262

13-2056   STABILITY REGIONS FOR THE GROWTH OF BARIUM STRONTIUM NIOBATE CRYSTALS.
          WHIPPS PW
          J SOLID STATE CHEM 4: 281-5 (1972)
              278

13-2057 THERMAL CONDUCTIVITY OF CUBIC SODIUM TUNGSTEN BRONZES. (SODIUM TUNGSTATES)
         WHITEMAN AE + MARTIN JJ + SHANKS HR
         J PHYS CHEM SOLIDS 32: 2223-9 (1971)

13-2058 CRYSTAL STRUCTURE OF BISMUTH TITANIUM NIOBIUM OXIDE.
         WOLFE RW + NEWNHAM RE + SMITH DK
         FERROELECTRICS 3: 1-7 (1971)
              262

13-2059 NUCLEAR MAGNETIC RESONANCE STUDIES OF OXYGEN OCTAHEDRA FERROELECTRICS.
         WOLF FJ
         STATE UNIV NEW YORK ALBANY PHD THESIS, 1970
              DISS ABSTR INT B 32: 3622 (1971)
              203

13-2060 ELASTIC ANOMALY OF BARIUM SODIUM NIOBATE (ELASTIC, PIEZOELECTRIC, DIELECTRIC, AND
         OPTICAL CONSTANTS) ROOM TEMPERATURE TO THE CURIE POINT.
         YAMADA T + IWASAKI H + NIIZEKI N
         J APPL PHYS 41(10): 4141-47 (1970)
              209

13-2061 ELECTROMECHANICAL PROPERTIES CF OXYGEN OCTAHEDRAL FERROELECTRIC CRYSTALS.
         YAMADA T
         J APPL PHYS 43(2): 328-38 (1972)
              256

13-2062 ELECTROOPTIC PROPERTIES CF STRONTIUM LITHIUM NIOBATE TYPE FERROELECTRICS (SINGLE
         CRYSTAL).
         YANO T + OHTA T + WATANABE A
         JAP J APPL PHYS 9: 1008-9 (1970)
              207

13-2063 LONGITUDINAL ELECTROOPTIC EFFECT IN OBLIQUE CUT STRONTIUM BARIUM NIOBATE PLATES.
         YASAKI T + KANATANI K + SAKAMOTO S
         JAP J APPL PHYS 10(4): 522-3 (1971)
              254

13-2064 NEUTRON SCATTERING STUDY OF THE SOFT MODES IN CUBIC POTASSIUM TANTALATE NIOBATE.
         YELON WB + COCHRAN W + SHIRANE G AND OTHERS
         FERROELECTRICS 2(4): 261-9 (1971)

13-2065 DIELECTRIC AND OPTICAL PROPERTIES OF LEAD ZINCATE(0.33) NIOBATE(0.67) SINGLE CRYSTAL.
         YOKOMIZO Y + NOMURA S
         P150-2 OF INT MEETING ON FERROELECTRICITY, 2ND, 1969, KYOTO, PROC,
         (J PHYS SOC JAP, VOL 28 SUPPL). PHYS SOC JAP, 1970, 459P.

13-2066 FERROELECTRIC PROPERTIES OF LEAD ZINCATE(0.33) NIOBATE(0.67) (SINGLE CRYSTALS GROWN
         FROM LEAD OXIDE FLUX).
         YOKOMIZO Y + TAKAHASHI T + NOMURA S
         J PHYS SOC JAP 28(5): 1278-84 (1970)
              196

13-2067 CRYSTAL STRUCTURE AND PIEZOELECTRICITY OF THE SYSTEM LEAD (ZINCATE NIOBATE) - LEAD
         TITANATE.
         YONEZAWA M + DOI K + NANAMATSU S + TSUBOUCHI N + TAKAHASHI M + NOMURA S
         FUNTAI OYOBI FUNMATSU YAKIN (J JAP SOC POWDER MET) 16(6): 253-8
         (1969) (IN JAPANESE)
              CERAM ABSTR 1971: 156D
              236

13-2068 POTASSIUM TANTALATE NIOBATE, STRUCTURAL AND DYNAMICAL STUDIES.
         ZACCAI G + HEWAT AW + ROUSE KD
         J PHYS (PARIS) 33, SUPPL 4, COLLOQ 2: 133-4 (1972)
              294

13-2069 SUPPRESSION OF THE SPONTANEOUS MAGNETOELECTRIC MAGNETIZATION OF LEAD FERRATE(0.5)
         NIOBATE(0.5) AND LEAD MANGANATE(0.5) NIOBATE(0.5) BY A MAGNETIC FIELD.
         ZORIN RV + ALSHIN BI + ASTROV DN
         SOV PHYS SOLID STATE 13: 2862-3 (1972)

14. BORACITES

14-2070   MAGNETOELECTRIC PROPERTIES OF NICKEL IODINE BORACITE.
ALSHIN BI + ASTROV DN + GUFAN YUM
SOV PHYS SOLID STATE 12: 2143-6 (1971)

14-2071   DIELECTRIC PROPERTIES OF BORACITES AND EVIDENCE FOR FERROELECTRICITY.
ASCHER E + SCHMIDT H + TAR D
SOLID STATE COMMUN 2: 45-9 (1964)

14-2072   INTERACTIONS BETWEEN MAGNETIZATION AND POLARIZATION: PHENOMENOLOGICAL SYMMETRY
CONSIDERATIONS ON BORACITES.
ASCHER E
P7-14 OF INT MEETING ON FERROELECTRICITY, 2ND, 1969, KYOTO, PROC,
(J PHYS SOC JAP, VOL 28 SUPPL). PHYS SOC JAP, 1970, 459P.

14-2073   NEUTRON AND ELECTRON DIFFRACTION STUDY IN NICKEL IODINE BORACITE.
BECKER WJ + WILL G
Z KRISTALLOGR 13: 139-46 (1970) (IN GERMAN)

14-2074   OPTICAL ABSORPTION SPECTRA OF NICKEL(2+) IN NICKEL CHLORINE AND NICKEL IODINE BORACITES
(AND NICKEL DOPED CADMIUM CHLORINE AND CADMIUM IODINE BORACITES, 4.2-635 K)
DORMANN E
J PHYS CHEM SOLIDS 31: 199-214 (1970)
      187

14-2075   ATOMIC DISPLACEMENTS IN FERROELECTRIC TRIGONAL AND ORTHORHOMBIC BORACITE STRUCTURES.
(BORACITE, ERICAITE)
DOWTY E + CLARK JR
SOLID STATE COMMUN 10: 543-8 (1972)
      263

      THERMODYNAMIC THEORY OF THE CUBIC- ORTHORHOMBIC PHASE TRANSITION IN BORACITES.
DVORAK V
CZECH J PHYS B 21(12): 1250-61 (1971)

14-2076   BORACITES - AN EXAMPLE OF IMPROPER FERROELECTRICS.
DVORAK V
J PHYS (PARIS) 33, SUPPL 4, COLLOQ 2: 89-90 (1972)
      294

14-2077   SYMMETRY ASPECTS OF THE PHASE TRANSITIONS IN BORACITES.
DVORAK V
CZECH J PHYS B 21: 1141-52 (1971)
      257

14-2078   FERROELECTRIC TRANSITION IN COBALT- IODINE- BORACITE UNDER PRESSURE.
FOUSEK J + SMUTNY F + FRENZEL C + HEGENBARTH E
FERROELECTRICS 4: 23-8 (1972)
      267

      THEORY OF PHASE TRANSITIONS IN BORACITES.
GUFAN YUM + SAKHNENKO VP
SOV PHYS SOLID STATE 14(7): 1660-5 (JAN 1973)

14-2079   PREPARATION OF NICKEL BORACITE CRYSTALS BY HYDROTHERMAL SYNTHESIS.
JOUBERT JC + MULLER J + FOUASSIER C + LEVASSEUR A
KRISTALL TECH 6: 65-8 (1971)

14-2080 (OPTICAL) FREQUENCY DOUBLING IN (MAGNESIUM CHLORINE, ZINC CHLORINE, CADMIUM CHLORINE, CADMIUM IODINE, IRON CHLORINE IRON IODINE AND NICKEL CHLORINE) BORACITES (SECOND HARMONIC GENERATION INDEX MATCHING).
KELLER G + OPPELT A
Z ANGEW PHYS 29(3): 160-64 (1970) (IN GERMAN)
212

14-2081 LATTICE STRAINS OF BORACITE CRYSTALS.
KIKUCHI H + KOBAYASHI J
OYO BUTSURI 27: 597-601 (1972) (IN JAPANESE)

14-2082 ELECTROOPTICAL PROPERTIES OF FERROELECTRIC (IRON IODINE) BORACITE. (SPONTANEOUS LATTICE STRAINS)
KOBAYASHI J + MIZUTANI I + HARA H + YAMADA N + NAKADA O + KUMADA A + SCHMID H
P67-70 OF INT MEETING ON FERROELECTRICITY, 2ND, 1969,, KYOTO, PROC,
(J PHYS SOC JAP, VOL 28 SUPPL). PHYS SOC JAP, 1970, 459P.

14-2083 LATENT LATTICE STRAIN IN THE FERROELECTRIC STATE OF IRON- IODINE- BORACITE.
(TEMPERATURE DEPENDENCE OF PIEZOELECTRIC CONSTANT)
KOBAYASHI J + MIZUTANI I
PHYS STATUS SOLIDI A 2: K89-92 (1970)
268

14-2084 PHENOMENOLOGICAL THEORY OF DIELECTRIC AND MECHANICAL PROPERTIES OF IMPROPER FERROELECTRIC CRYSTALS. (AMMONIUM SULFATE, GADOLINIUM MOLYBDATE, IRON IODINE BORACITE)
KOBAYASHI J + ENOMOTO Y + SATO Y
PHYS STATUS SOLIDI B 50: 335-43 (1972)
262

14-2085 X-RAY STUDY ON THE LATTICE STRAINS OF (PARAELECTRIC AND) (IRON- IODINE- BORACITE).
KOBAYASHI J + MIZUTANI I + SCHMID H + SCHACHNER H
PHYS REV B 1(9): 3801-8 (1970)
213

14-2086 X-RAY STUDY ON PHASE TRANSITIONS OF FERROELECTRIC IRON IODINE BORACITE AT LOW TEMPERATURES. (MONOCLINIC AND RHOMBOHEDRAL PHASES)
KOBAYASHI J + SATO Y + SCHMID H
PHYS STATUS SOLIDI A 10: 259-70 (1972)
286

14-2087 RELATIONSHIP BETWEEN DIRECTIONS OF THE ELECTRIC AND MAGNETIC POLARIZATION IN BORACITE AND OTHER CRYSTALS.
KOVALEV OV
SOV PHYS SOLID STATE 14: 258-60 (1972)

14-2088 DOUBLETS CONCERNING THE STRUCTURE OF BORACITES.
KRIZ HM + BRAAY PJ
J PHYS CHEM SOLIDS 32: 302-4 (1971)

14-2089 NEW BORACITES OF FORMULA LITHIUM(4) BORON(7) OXYGEN(12) CHLORIDE OR BROMIDE (NOT EXAMINED FOR FERROELECTRICITY).
LEVASSEUR A + FOUASSIER C + HAGENMULLER P
MATER RES BULL 6: 15-22 (1971) (IN FRENCH)
223

14-2090 COEXISTENCE OF THE SPONTANEOUS ELECTRIC POLARIZATION AND MAGNETIZATION IN NICKEL IODIDE BORACITE (CURIE POINT 64 DEG K).
MIYASHITA T + MURAKAMI T
J PHYS SOC JAP 29: 1092 (1970)
217

14-2091 CRYSTAL GROWTH OF NICKEL IODINE BORACITE.
MIYASHITA T
ACTA CRYSTALLOGR A 28: S141 (1972)

14-2092 MODIFIED TECHNIQUE FOR THE GROWTH OF BORACITE CRYSTALS.
NASSAU K + SHIEVER JW
J CRYST GROW 16: 59-61 (1972)

14-2093   OPTICAL ANISOTROPY OF CUBIC NICKEL IODINE BORACITE DUE TO QUADRUPOLE TRANSITIONS.
          PASTRNAK J + CROSS LE
          PHYS STATUS SOLIDI B 44: 313-25 (1971)
               234

14-2094   OPTICAL ANISOTROPY OF THE CUBIC COBALT IODINE BORACITE.
          PASTRNAK J + CROSS LE
          PHYS STATUS SOLIDI B 43: K111-14 (1971)

          ELECTRON PARAMAGNETIC RESONANCE SPECTRUM IN (MANGANESE DOPED) LITHIUM NIOBATE AND
          (COPPER(2+) DOPED) ZINC BORACITE.
          PETROV MP + KIZHAEV SA + ANDREEVA GT + SMOLENSKII GA
          P128-30 OF INT MEETING ON FERROELECTRICITY, 2ND, 1969, KYOTO, PROC,
          (J PHYS SOC JAP, VOL 28 SUPPL). PHYS SOC JAP, 1970, 459P.

14-2095   CRYSTAL FIELD THEORY AND OPTICAL ABSORPTION (CALCULATIONS) OF FERROELECTRIC BORACITES.
          (IRON, NICKEL, OR COBALT, WITH CHLORIDE, BROMIDE, OR IODIDE)
          PISAREV RV + DRUZHININ VV + NESTEROVA NN + PROCHOROVA SD + ANDREEVA GT
          P71 OF INT MEETING ON FERROELECTRICITY, 2ND, 1969, KYOTO, PROC,
          (J PHYS SOC JAP, VOL 28 SUPPL). PHYS SOC JAP, 1970, 459P.

14-2096   CRYSTAL FIELD THEORY AND OPTICAL ABSORPTION OF COBALT AND NICKEL BORACITES.
          PISAREV RV + DRUZHININ VV + PROCHOROVA SD
          PHYS STATUS SOLIDI 35(1): 145-55 (1969)
               176

14-2097   OPTICAL ABSORPTION   (625-25 MICRONS, 77-650 DEG K) OF FERROELECTRIC IRON(2+) BORACITES.
          (CHLORIDE, BROMIDE AND IODIDE, AND RUBIDIUM IRON TRIFLUORIDE)
          PISAREV RV + DRUZHININ VV + NESTEROVA NN + PROCHOROVA SD + ANDREEVA GT
          PHYS STATUS SOLIDI 40: 503-12 (1970)
               216

14-2098   TRIGONAL BORACITES: A NEW TYPE OF FERROELECTRIC AND FERROMAGNETOELECTRIC MATERIAL THAT
          ALLOWS NO 180 DEGREE ELECTRIC POLARIZATION REVERSAL.
          SCHMID H
          P354-6 OF INT MEETING ON FERROELECTRICITY, 2ND, 1969, KYOTO, PROC,
          (J PHYS SOC JAP, VOL 28 SUPPL). PHYS SOC JAP, 1970, 459P.

          PHYS STATUS SOLIDI 37(1): 209-33 (1970)

14-2099   FERROELECTRIC TRANSITION IN COBALT IODINE BORACITE.
          SMUTNY F + FOUSEK J
          PHYS STATUS SOLIDI 40: K13-15 (1970)
               204

14-2100   PIEZOELECTRIC PROPERTIES OF COBALT IODINE BORACITE.
          SMUTNY F + ALBERS J + SMUTNY F + ALBERS J
          PHYS STATUS SOLIDI B, BASIC RES 49(2): K159-61 (1972)

14-2101   SPECIFIC HEAT ANOMALY OF FERROELECTRIC COBALT IODINE BORACITE.
          SMUTNY F
          PHYS STATUS SOLIDI A, APPL RES 9(2): K109-10 (1972)
               280

14-2102   FERROELECTRIC AND FERROELASTIC PROPERTIES OF MAGNESIUM CHLORINE BORACITE.
          TORRE LP + ABRAHAMS SC + BARNS RL
          FERROELECTRICS 4: 291-8 (1972)

14-2103   OBSERVATIONS OF FERROELECTRIC DOMAINS IN BORACITES.
          ZIMMERMANN A + BOLLMANN W + SCHMID H
          PHYS STATUS SOLIDI A 3: 707-20 (1970)

15. ANTIMONY SULFOIODIDE AND RELATED COMPOUNDS

15-2104   INFLUENCE OF NONEQUILIBRIUM CARRIERS ON THE DOMAIN STRUCTURE OF THE FERROELECTRIC
           SEMICONDUCTOR ANTIMONY SULFOIODIDE.
           AGARONOV BS + BEZDETNYI NM + ZEINALLY AKH + LEBEDEVA NN + SHEINKMAN MK
           SOV PHYS SOLID STATE 14(1): 205-6 (JUL 1972)
               270

15-2105   SHORT CIRCUIT PHOTOEMF IN FERROELECTRIC SEMICONDUCTORS. (ANTIMONY SULFOIODIDE CRYSTALS)
           AGARONOV BS + BEZDETNYI NM + ZEINALLY AKH + LEBEDEVA NN
           SOV PHYS CRYSTALLOGR 17(3): 591-2 (NOV-DEC 1972)
               280

15-2106   LONG WAVELENGTH OPTICAL PHONONS AND PHASE TRANSITIONS IN ANTIMONY SULFOIODIDE.
           (POLARIZED INFRARED AND RAMAN SPECTRA)
           AGRAWAL DK + PERRY CH
           P401-5 OF INT CONF ON LIGHT SCATTERING IN SOLIDS, 2ND PROC,
           BALKANSKI M(ED), PARIS, 1971, FLAMMARION, PARIS, 1971, 518P.

           PHYS REV B 4(6): 1893-902 (1971)
               246

15-2107   ELECTRICAL AND PHOTOELECTRIC PROPERTIES OF ANTIMONY SULFOIODIDE.
           ALEAKSEYEVA VG + LANDSBERG EG
           RADIO ENG ELECTRON PHYS 13(5): 770-4 (1968)
               141

15-2108   NUCLEAR GAMMA RESONANCE SPECTROSCOPIC STUDY OF SEMICONDUCTING COMPOUNDS OF THE TYPE
           A(2)(V)B(3)(VI)AND A(V)B(VI)C(VII). (ANTIMONY TRISULFIDE, TRISELENIDE, SULFOIODIDE,
           SELENOIODIDE, TELLUROIODIDE, TRITELLURIDE, TRIIODIDE, TRIBROMIDE, INDIUM ANTIMONIDE)
           ALEKSANDROV AYU + BALTRUNAS DI + BELYAEV LM + LYUBUTIN IS + LYAKHOVITSKAYA VA
           SOV PHYS CRYSTALLOGR 17(2): 281-3 (SEP-OCT 1972)
               275

15-2109   X-RAY CRYSTALLOGRAPHIC STUDIES OF ANTIMONY SULFOIODIDE.
           ARNDT R + NIGGLI A
           NATURWISSENSCHAFTEN 51(7): 158 (1964) (IN GERMAN)
               CHEM ABSTR 61: 1341E (1964)

15-2110   LATTICE MODES AND PHASE TRANSITION IN ANTIMONY SULFOIODIDE.
           BALKANSKI M + TENG MK + SHAPIRO SM + ZIOLKIEWICZ MK
           PHYS STATUS SOLIDI B 44: 355-68 (1971)

15-2111   RAMAN SCATTERING IN FERROELECTRIC SEMICONDUCTOR ANTIMONY SULFOIODIDE.
           BALKANSKI M + TENG MK + MASSOT M + SHAPIRO SM
           P392-5 OF INT CONF ON LIGHT SCATTERING IN SOLIDS, 2ND PROC,
           BALKANSKI M(ED), PARIS, 1971, FLAMMARION, PARIS, 1971, 518P.

15-2112   CRYSTAL STRUCTURE AND CONDUCTIVITY OF ANTIMONY SULFOIODIDE.
           BARBE M + BRULEBOIS D + DIMANE M + LAURENT M
           COMPT REND C 268: 2053-6 (1969) (IN FRENCH)
               CHEM ABSTR 71: 752040 (1969)
               165

           PHOTOFERROELECTRIC EFFECTS IN A(V)B(VI)C(VII) AND BARIUM TITANATE TYPE FERROELECTRICS.
           BELYAEV LM + FRIDKIN VM + GREKOV AA + KOSONOGOV NA + RODIN AI
           J PHYS (PARIS) 33, SUPPL 4, COLLOQ 2: 123-5 (1972)

15. ANTIMONY SULFOIODIDE AND RELATED COMPOUNDS

15-2113   AGING OF THE FERROELECTRIC SEMICONDUCTOR ANTIMONY SULFOIODIDE IN A STATIC ELECTRIC
          FIELD.
          BELYAEV AD + GROMASHEVSKII VL
          SOV PHYS SOLID STATE 12(11): 2693-4 (1971)
               234

15-2114   ABSORPTION OF ULTRASOUND IN A(V)B(VI)C(VII) FERROELECTRIC SEMICONDUCTORS AT
          TEMPERATURES NEAR A TYPE-1 PHASE TRANSITION. (ANTIMONY SULFOIODIDE, SELENOIODIDE,
          BISMUTH SULFOIODIDE, ANTIMONY SULFO SELENOIODIDE, CURIE TEMPERATURES)
          BELYAEV AD + GROMASHEVSKII VL + KRIVSHICH VV + OLIKH YAM + SLIVKA VYU + TURYANITSA ID
          BULL ACAD SCI USSR, PHYS SER 35(5): 915-8 (1971)
               248

15-2115   ABSORPTION OF ULTRASOUND IN AN ANTIMONY SULFOIODIDE FERROELECTRIC SEMICONDUCTOR NEAR
          ITS CURIE POINT.
          BELYAEV AD + GROMASHEVSKII VL + MISELYUK EG + NAKONECHNYI YUS + SLIVKA VYU
          + TURYANITSA ID
          SOV PHYS SOLID STATE 12: 1715-17 (1971)

15-2116   DETERMINATION OF THE LATTICE PARAMETERS AND CURIE POINTS OF SOLID SOLUTIONS OF
          A(V)B(VI)C(VII) COMPOUNDS. ANTIMONY(X) BISMUTH(1-X) SULFOIODIDE, ANTIMONY SULFO(X)
          SELENO(1-X) IODIDE, ANTIMONY(X) ARSENIC(1-X) SULFOIODIDE)
          BELYAEV AD + MISELYUK EG + SLIVKA VYU + TURYANITSA ID + CHEPUR DV
          UKR FIZ ZH 15(3): 499-502 (1970) (IN RUSSIAN)
               CHEM ABSTR 73: 49559K (1970)
               199

15-2117   DIELECTRIC DISPERSION IN ANTIMONY SULFOIODIDE CRYSTALS.
          BELYATSKAS RP + GRIGAS IP + ORLYUKAS AS + SHUGUROV VK
          LIET FIZ RINKINYS 11(6): 1029-38 (1971) (IN RUSSIAN)
               278

15-2118   EFFECT OF COMPOSITION ON GROWTH, DIELECTRIC, AND PHOTOELECTRIC PROPERTIES OF ANTIMONY
          SULFOIODIDE CRYSTALS.
          BELYAEV LM + LYAKHOVITSKAYA VA + SILVESTROVA IM
          INORG MATER 6(3): 377-80 (1970)
               191

15-2119   INFLUENCE OF ILLUMINATION ON THE FORM OF THE VOLTAGE AMPERE CHARACTERISTICS OF ANTIMONY
          SULFOIODIDE IN THE PARAELECTRIC REGION.
          BELYAEV AD + GROMASHEVSKII VL + TURYANITSA ID + CHEPUR DV + SHCHELKANOGOV VV
          IZV VUZ FIZ 1972(8): 144-5 (IN RUSSIAN)

15-2120   MOTION PICTURES OF THE PHASE TRANSITION IN ANTIMONY SULFOIODIDE CRYSTALS.
          BELYAEV LM + KLIYA MO + LYAKHOVITSKAYA VA
          SOV PHYS CRYSTALLOGR 16(2): 279-83 (SEP-OCT 1971)
               244

15-2121   PIEZOELECTRIC VIBRATIONS IN ANTIMONY SULFOIODIDE CRYSTALS NEAR THE CURIE POINT (EFFECTS
          OF TEMPERATURE, FIELD STRENGTH, SAMPLE ILLUMINATION).
          BELYAEV AD + GROMASHEVSKII VL + SLIVKA VYU + TURYANITSA ID
          UKR FIZ ZH 15(3): 517-9 (1970)
               223

15-2122   PHASE DIAGRAM AND ELECTRICAL CONDUCTIVITY OF BISMUTH SELENOIODIDE- ANTIMONY
          SELENOIODIDE SYSTEM ALLOYS.
          BELOTSKII DP + LAPSHIN VF
          INORG MATER 7(11): 1739-40 (1971)
               261

15-2123   POLARIZATION EFFECTS IN THE ANOMALOUS ABSORPTION OF ULTRASOUND IN ANTIMONY SULFOIODIDE
          CRYSTALS.
          BELYAEV AD + GROMASHEVSKII VL
          SOV PHYS SOLID STATE 12(10): 2264-6 (1971)
               232

15-2124   EFFECTS OF ILLUMINATION ON THE DOMAIN STRUCTURE OF ANTIMONY SULFOIODIDE. (ETCHING
          METHOD)
          PENDER VP + FRIDKIN VM
          SOV PHYS SOLID STATE 13(2): 501-2 (AUG 1971)
               240

15. ANTIMONY SULFOIODIDE AND RELATED COMPOUNDS

15-2125 ANISOTROPY OF THE REFLECTION SPECTRA OF ANTIMONY SULFOIODIDE CRYSTALS.
         BERCHA DM + SLIVKA VYU + SYRBU NN + TURYANITSA ID + CHEPUR DV
         SOV PHYS SOLID STATE 13(1): 217-18 (JUL 1971)
         IZV VUZ FIZ 1971(8): 13-19 (IN RUSSIAN)
         237

15-2126 CONDUCTIVITY, THERMOELECTRIC PROPERTIES, AND BAND STRUCTURE OF CRYSTALLINE BISMUTH
         SELENOIODIDE.
         BERCHA DM + ZAYACHKOVSKII MP
         SOV PHYS SOLID STATE 14(3): 766-7 (SEP 1972)
         276

15-2127 DOMAIN STRUCTURE AND LOCAL STATES IN ANTIMONY SULFOIODIDE TYPE SEMICONDUCTOR
         FERROELECTRICS.
         BERCHA DM + KIKINESHI AA + SEMAK DG + CHEPUR DV
         SOV PHYS SOLID STATE 14(5): 1359-60 (NOV 1972)
         278

15-2128 PIEZOELECTRIC RESISTANCE EFFECT IN BISMUTH SELENOIODIDE CRYSTALS.
         BERCHA DM + ZAYACHKOVSKII MP + SLIVKA VYU + LOVGA IV + TURYANITSA ID + CHEPUR DV
         POLUPROV ELEKTRON 1971: 53-8 (IN RUSSIAN)
             CHEM ABSTR 77: 26076U (1972)
             273

15-2129 CHARACTERISTICS OF THE PHOTOELECTRIC EMF OF ANTIMONY SULFOIODIDE.
         BEZDETNYI NM + GORBATOV GZ + ZEINALLY AKH + LEBEDEVA NN + SHEINKMAN MK
         SOV PHYS SOLID STATE 14(2): 477-8 (AUG 1972)
         274

15-2130 LIGHT INDUCED PYROCURRENTS IN FERROSEMICONDUCTOR ANTIMONY SULFOIODIDE.
         BEZDETNYI NM + ZEINALLY AKH + LEBEDEVA NN AND OTHERS
         UKR FIZ ZH 17: 1359-60 (AUG 1972) (IN RUSSIAN)

15-2131 PYROCURRENTS HIGHER THAN THE CURIE POINT IN ANTIMONY SULFOIODIDE.
         BEZDETNYI NM + ZEINALLY AKH + LEBEDEV NN AND OTHERS
         UKR FIZ ZH 16: 1377-8 (AUG 1971) (IN RUSSIAN)

15-2132 PHOTOEMF'S IN THE FERROELECTRIC SEMICONDUCTOR ANTIMONY SULFOIODIDE.
         BEZDETNYI NM + ZEINALLY AKH + LEBEDEVA NN + SHEINKMAN MK
         SOV PHYS SEMICOND 5(5): 904-5 (NOV 1971)
         251

15-2133 PHOTODOMAIN EFFECT IN NATURALLY POLARIZED ANTIMONY SULFOIODIDE CRYSTALS.
         BEZDETNYI NM + GORBATOV GZ + ZEINALLY AKH + LEBEDEVA NN + SHEINKMAN MK
         SOV PHYS SOLID STATE 14(3): 789-90 (SEP 1972)
         276

15-2134 PHOTOELECTRIC PROPERTIES (PHOTOCONDUCTIVITY) OF ANTIMONY SULFOIODIDE SINGLE CRYSTALS.
         BEZDETNYI NM + ZEINALLY AKH + LEBEDEVA NN + SHEINKMAN MK
         SOV PHYS SOLID STATE 12(8): 1990-1 (1971)
         227

15-2135 PHOTODOMAIN EFFECT IN NATURALLY POLARIZED ANTIMONY SULFOIODIDE CRYSTALS.
         BEZDETNYI NM + GORBATOV GZ + ZEINALLY AKH + LEBEDEVA NN + SHEINKMAN MK
         SOV PHYS SOLID STATE 14: 789-90 (1972)

15-2136 SOME FEATURES OF THE PHOTOEMF SPECTRUM OF ANTIMONY SULFOIODIDE.
         BEZDETNYI NM + GORBATOV GZ + ZEINALLY AKH + LEBEDEVA NN
         SOV PHYS SEMICOND 6: 1047-8 (1972)

15-2137 HIGH GAS PRESSURE CRYSTAL GROWTH OF ANTIMONY SULFOIODIDE, NICKEL CARBONATE AND BLACK
         PHOSPHORUS.
         BOKSHA SS
         J CRYST GROW 12: 113-19 (1972)

15-2138 PREPARATION OF ANTIMONY SULFOIODIDE TEXTURES FROM THE MELT UNDER PRESSURE AND
         INVESTIGATION OF THEIR ELECTRICAL PROPERTIES. (DIELECTRIC, ELASTIC, AND PIEZOELECTRIC
         CONSTANTS)
         BOKSHA SS + LJAKHOVICKJA VA + SILVESTROVA IM + TIKHOMIROVA NA
         INORG MATER 6(11): 1713-15 (1970)
             241

15. ANTIMONY SULFOIODIDE AND RELATED COMPOUNDS

15-2139   FREQUENCY DEPENDENCE OF ANOMALOUS ULTRASOUND DAMPING IN ANTIMONY SULFOIODIDE SINGLE
        CRYSTALS.
        BRAZDZIUNAS PP + KUNIGELIS VF + SAMULIONIS VI
        LIET FIZ RINKINYS 11(5): 831-5 (1971) (IN RUSSIAN)
            CHEM ABSTR 76: 158647V (1972)
            278

15-2140   DETERMINATION OF THE TEMPERATURE DEPENDENCE OF LATTICE CONSTANTS IN FERROELECTRIC
        ANTIMONY SULFOIODIDE.
        BRUHL HG + NEUMANN H + SCHMIDT W
        KRISTALL TECH 5: K15-17 (1970)

15-2141   PRODUCTION METHODS AND SOME OPTICAL PROPERTIES OF BISMUTH TELLUROIODIDE.
        CHEPUR DV + GORAK YAA + KOVACH DSH + TURYANITSA ID + BORETS AN + YATSKOVICH II
        INORG MATER 6(3): 336-7 (1970)
            191

15-2142   PRODUCTION AND SOME OPTICAL PROPERTIES OF ANTIMONY SULFOBROMIDE IN GLASSY AND
        CRYSTALLINE STATES.
        CHEPUR DV + TURYANITSA ID + GERZANICH EI + KOPERLES BM + SLIVKA VYU + PUGA PP
        IZV VUZ FIZ 1971(2): 114-16 (IN RUSSIAN)
        (TRANSLATION TO APPEAR IN SOV PHYS J)

15-2143   PHASE TRANSITION IN FERROELECTRIC ANTIMONY SULFOIODIDE CRYSTALS USING THE RAMAN
        SCATTERING SPECTRUM. (LATTICE VIBRATIONS, BONDING)
        CHISLER EV + SAVATINOVA II + FRIDKIN VM
        SOV PHYS SOLID STATE 12(10): 2327-31 (1971)
            232

15-2144   BOSON ECHOES. (ANTIMONY SULFOIODIDE)
        FRENOIS CH + JOFFRIN J + LEVELUT A + ZIOLKIEWICZ S
        SOLID STATE COMMUN 11: 327-31 (1972) (IN FRENCH)
            274

        EXTRINSIC PHOTOCONDUCTIVITY IN FERROELECTRICS DUE TO SURFACE LAYERS. (BARIUM TITANATE,
        ANTIMONY SULFOIODIDE)
        FRIDKIN VM + GREKOV AA + SAVICHENKO EA + VOLK TR
        J PHYS (PARIS) 33, SUPPL 4, COLLOQ 2: 127-9 (1972);
        PHYS STATUS SOLIDI A 8: K55-9 (1971)

15-2145   INFLUENCE OF NONEQUILIBRIUM CARRIERS ON THE PHASE TRANSITION IN FERROELECTRIC
        SEMICONDUCTOR (COMPOUNDS) $A(V)B(VI)C(VII)$.
        FRIDKIN VM + BELYAEV LM + GREKOV AA + RODIN AI
        P448-50 OF INT MEETING ON FERROELECTRICITY, 2ND, 1969, KYOTO, PROC,
        (J PHYS SOC JAP, VOL 28 SUPPL). PHYS SOC JAP, 1970, 459P.

15-2146   MECHANISM OF THE PHOTODOMAIN EFFECT IN ANTIMONY SULFOIODIDE.
        FRIDKIN VM
        FERROELECTRICS 2: 119-22 (1971)
            235

15-2147   TEMPERATURE DEPENDENCE OF THE CROSS SECTION OF CAPTURE CENTERS IN FERROELECTRIC
        SEMICONDUCTOR ANTIMONY SULFOIODIDE.
        FRIDKIN VM + GREKOV AA + RODIN AI
        PHYS LETT A 35: 59-60 (MAY 17 1971)

15-2148   FEATURES OF THE PHASE TRANSITION IN ANTIMONY SULFOIODIDE SINGLE CRYSTALS.
        GENE VV + GRZHEGORZHEVSKII OG
        IZV VUZ FIZ 1972(4): 155-6 (IN RUSSIAN)

15-2149   HOW THE DIELECTRIC PROPERTIES OF ANTIMONY SULFOIODIDE(X) BROMIDE(1-X) SOLID SOLUTIONS
        DEPEND ON TEMPERATURE AND HYDROSTATIC PRESSURE.
        GERZANICH EI
        INORG MATER 6: 1403-5 (SEP 1970)

15-2150   INDIRECT TRANSITIONS AND ABSORPTION IN THE MID-INFRARED IN ANTIMONY SULFOIODIDE
        CRYSTALS.
        GERZANICH EI + BORETS AN + KOVACH DSH
        OPT SPECTROSC 32: 618-21 (1972)
        IZV VUZ FIZ 1972(7): 85-9 (IN RUSSIAN)

15. ANTIMONY SULFOIODIDE AND RELATED COMPOUNDS

15-2151   ELECTROREFLECTION FROM SINGLE CRYSTALS OF FERROELECTRIC ANTIMONY SULFOIODIDE NEAR THE
PHASE TRANSITION TEMPERATURE.
GOLIK LL + ELINSON MI
SOV PHYS SOLID STATE 12: 2338-42 (1971)

15-2152   INSTABILITY OF FERROELECTRIC POLARIZATION IN ANTIMONY SULFOIODIDE SINGLE CRYSTALS.
GOLIK LL + ELINSON MI + CHENSKII EV
SOV PHYS SOLID STATE 13: 2978-81 (1972)

15-2153   CHARACTERISTICS OF THE PHOTOELECTRIC EMF OF ANTIMONY SULFOIODIDE.
GORBATOV GZ + ZEINALLY AKH + LEBEDEVA NN + SHEINKMAN MK
SOV PHYS SOLID STATE 14: 477-8 (1972)

15-2154   EFFECT OF THE ADSORPTION OF OXYGEN ON THE PROPERTIES OF ANTIMONY SULFOIODIDE SINGLE
CRYSTALS.
GORAK YAA + KOPINETS IF + MIKULANNINETS SV + TURYANITSA ID + CHEPUR DV
RUSS J PHYS CHEM 44: 1462-4 (1970)

15-2155   DILATOMETRIC INVESTIGATION OF ANTIMONY SULFOIODIDE SINGLE CRYSTALS.
GREKOV AA + ROGACH ED
IZV VUZ FIZ 1968(9): 138-9
     212

15-2156   DILATOMETRIC INVESTIGATION OF PHOTOFERROELECTRIC EFFECTS IN SINGLE CRYSTALS OF ANTIMONY
SULFOIODIDE.
GREKOV AA + ROGACH ED + SUKIYAZOV AG
SOV PHYS SOLID STATE 12(12): 2890-2 (JUN 1971)
     240

15-2157   FERROELECTRIC PHOTOELECTRETS BASED ON A(V)B(VI)C(VII) SOLID SOLUTIONS. (DIELECTRIC
CONSTANTS, PYROELECTRIC CURRENTS, ANTIMONY SULFOIODIDE(X) BROMIDE(1-X))
GREKOV AA + MALITSKAYA MA + FRIDKIN VM
SOV PHYS CRYSTALLOGR 17: 504-8 (NOV-DEC 1972)
     280

15-2158   INFLUENCE OF NONEQUILIBRIUM CARRIER SCREENING ON POLARIZATION PROCESSES IN
FERROELECTRIC ANTIMONY SULFOIODIDE.
GREKOV AA + LYAKHOVITSKAYA VA + RODIN AI + FRIDKIN VM
SOV PHYS SOLID STATE 10(7): 1762-3 (1969)
     140

15-2159   INDUCED IMPURITY PHOTOCONDUCTIVITY IN THE FERROELECTRIC SEMICONDUCTOR ANTIMONY
SULFOIODIDE.
GREKOV AA + RODIN AI + FRIDKIN VM
SOV PHYS SEMICOND 5(7): 1136-40 (JAN 1972)
     254

15-2160   METASTABLE OPTICAL CHARGE EXCHANGE AND "FROZEN" SHIFT OF THE CURIE TEMPERATURE OF A
FERROELECTRIC SEMICONDUCTOR. (ANTIMONY SULFOIODIDE BROMIDE)
GREKOV AA + RODIN AI + FRIDKIN VM
SOV PHYS SOLID STATE 12(12): 2966-7 (JUN 1971)
     240

15-2161   PHOTOFERROELECTRIC EFFECTS IN FERROELECTRIC SEMICONDUCTORS OF THE A(V)B(VI)C(VII) TYPE
WITH LOW TEMPERATURE PHASE CHANGES. (EFFECT OF PHOTOCONDUCTIVITY, ANTIMONY
SULFOIODIDE(X) BROMIDE(1-X))
GREKOV AA + MALITSKAYA MA + SPITSYNA VD + FRIDKIN VM
SOV PHYS CRYSTALLOGR 15(3): 423-30 (1970)
     217

15-2162   GROWING SINGLE CRYSTALS FROM THE VAPOR PHASE BY A DYNAMIC METHOD (ANTIMONY
SESQUISELENIDE OR SULFIDE, ANTIMONY SULFO OR SELENO IODIDE).
GRIGAS B + MIKALEVICIUS M
INORG MATER 6(1): 119-20 (1973)
     201

15. ANTIMONY SULFOIODIDE AND RELATED COMPOUNDS

15-2163   PYROELECTRIC EFFECT IN ANTIMONY SULFO AND SELENO IODIDE SINGLE CRYSTALS.
GRIGAS B
P178-80 OF NAUCH KONF MOLODYKH UCH LITOV SSR, RAB OBL FIZ,
MAT KIBERN, MATER (SCI CCNF CF YOUNG SCIENTISTS OF LITHUANIAN
SSR WORKING IN THE FIELDS OF PHYS, MATH AND CYBERNETICS,
MATERIALS). BRAZDZIUS P(ED), AKAD NAUK LITOV SSR, VILNIUS, 1967.
     CHEM ABSTR 69: 91082 (1968)
     142

15-2164   OPTICAL SECOND HARMONIC GENERATION IN ANTIMONY SULFOIODIDE.
HAFELE HG + WACHERNIG H + IRSLINGER C ET AL
PHYS STATUS SOLIDI 42(2): 31-41 (1970)

15-2165   ELECTROSTRICTION, PIEZOELECTRICITY AND ELASTICITY IN FERROELECTRIC ANTIMONY
SULFOIODIDE.
HAMANO K + SHINMI T
J PHYS SOC JAP 33(1): 118-24 (JUL 1972)
     278

15-2166   SWITCHING AND MEMORY EFFECT IN ANTIMONY SULFOIODIDE- TIN DIOXIDE HETEROSTRUCTURE THIN
FILM.
HAMAKAWA Y + YOSHIDA M
P165-78 OF INT CONF ON THE PHYSICS AND CHEMISTRY OF SEMICONDUCTOR
HETEROJUNCTION AND LAYER STRUCTURES, SZIEGTI G(ED), BUDAPEST,
1970, AKAD KIADO, BUDAPEST, 1971, 377P.

15-2167   SOFT PHONON MODE AND MODE COUPLING IN ANTIMONY SULFOIODIDE (BY RAMAN SCATTERING).
HARBEKE G + STEIGMEIER EF + WEHNER RK
SOLID STATE COMMUN 8: 1765-8 (1970)
     220

15-2168   CHALCOGENIDE BROMIDES OF ANTIMONY AND BISMUTH.
HORAK J + KOZAKOVA M + KLAZAR J
COLLECT CZECH CHEM COMMUN 37(7): 2309-16 (1972)
     288

15-2169   OPTICAL PROPERTIES OF THE SEMICONDUCTOR BISMUTH TELLUROIODIDE.
HORAK J
J PHYS (PARIS) 31: 121-3 (1970) (IN FRENCH)

15-2170   PHOTOELECTRIC PROPERTIES OF BISMUTH SULFOIODIDE CRYSTALS.
HORAK J + KOSEK F + FRUMAR M
COLLECT CZECH CHEM COMMUN 34(5): 1475-81 (1969) (IN GERMAN)
     CHEM ABSTR 70: 119338B (1969)
     286

15-2171   DIELECTRIC PROPERTIES CF ANTIMONY SULFOIODIDE AT MICROWAVE FREQUENCIES (3.3 AND 9.8
GHZ, 0-50 C).
HOSOYA M + NAKAMURA E
JAP J APPL PHYS 9(5): 552-6 (1970)
     195

15-2172   FERROELECTRIC BARKHAUSEN EFFECT IN ANTIMONY SULFOIODIDE.
IMAI K
MEM ISHIKAWA TECH COLL 3: 1-5 (1971)

15-2173   POTENTIAL DISTRIBUTION IN ANTIMONY SULFOIODIDE CRYSTAL (IN THE PARAELECTRIC PHASE,
UNDER DC FIELD ALONG C-AXIS).
IRIE K + OHI K
J PHYS SOC JAP 28: 1379 (1970)
     196

15-2174   PHOTOVOLTAIC EFFECT IN FERROELECTRIC AND PHOTOCONDUCTIVE ANTIMONY SULFOIODIDE.
IRIE K
J PHYS SOC JAP 30: 1506 (MAY 1971)

15-2175   RELATION BETWEEN THE ELECTRIC POLARIZATION AND THE ABSORPTION EDGE IN ANTIMONY
SULFOIODIDE.
ISHIKAWA K + TANAKA R + TOYODA K
PHYS LETT A 42(4): 289-90 (18 DEC 1972)
     286

15-2176   FERROELECTRIC TRANSITIONS WITH A GENUINE DIELECTRIC INSTABILITY (FERROELECTRIC
           TWINNING) TREATED AS CRITICAL POINTS IN CRYSTALS. (ANTIMONY SULFOIODIDE)
           JANOVEC V
           P178-80 OF INT MEETING ON FERROELECTRICITY, 2ND, 1969,, KYOTO, PROC,
           (J PHYS SOC JAP, VOL 28 SUPPL). PHYS SOC JAP, 1970, 459P.

15-2177   NOTES ON ABSORPTION BAND EDGE OF (PARAELECTRIC) ANTIMONY SULFOIODIDE.
           KAMIMURA H + SHAPIRO SM + BALKANSKI M
           PHYS LETT A 33(5): 277-8 (1970)
                223

15-2178   ELECTROOPTICAL AND ELECTROMECHANICAL EFFECT IN ANTIMONY SULFOIODIDE.
           KERN R
           J PHYS CHEM SOLIDS 23: 249-53 (1962)

15-2179   BAND STRUCTURE OF THE FERROELECTRIC SEMICONDUCTOR ANTIMONY SULFOIODIDE.
           KHASABOV AG + NIKIFOROV IYA
           BULL ACAD SCI USSR, PHYS SER 34(12): 2204-8 (1970)

15-2180   MOSSBAUER STUDIES OF ANTIMONY SULFOIODIDE TYPE CRYSTALS.
           KHIMICH TA + BELOV VF + ZHUKOV OK+ YURIN VA + KORABLIN LN + SHIPKO MN + LOBACHEV AN
           + POPOLITOV VI
           SOV PHYS SOLID STATE 13(5): 1265-6 (NOV 1971)
                252

15-2181   POLARIZATION NATURE IN SEMICONDUCTOR FERROELECTRICS OF THE ANTIMONY SULFOIODIDE TYPE.
           KIKINESHI AA + SEMAK DG
           UKR FIZ ZH 17(3): 498-500 (1972) (IN RUSSIAN)
                CHEM ABSTR 76: 132998K (1972)
                273

15-2182   SOME FEATURES OF TRAPPING LEVELS IN ANTIMONY SULFOIODIDE TYPE SEMICONDUCTING
           FERROELECTRICS.
           KIKINESHI AA + SEMAK DG
           SOV PHYS SEMICOND 6(3): 449-50 (SEP 1972)
                275

15-2183   DOMAIN STRUCTURE OF ANTIMONY SULFOIODIDE CRYSTALS (FROM SELECTIVE CRYSTALLIZATION AND
           ETCHING).
           KLIYA MO + LYAKHOVITSKAYA VA
           SOV PHYS CRYSTALLOGR 15(1): 59-62 (1970)
                206

15-2184   PHASE TRANSITION AND DOMAIN STRUCTURE IN ANTIMONY SULFOIODIDE CRYSTALS. (MICRO
           CINEMATOGRAPHY)
           KLIYA MO + LYAKHOVITSKAYA VA
           P217-19 OF INT MEETING ON FERROELECTRICITY, 2ND, 1969,, KYOTO, PROC,
           (J PHYS SOC JAP, VOL 28 SUPPL). PHYS SOC JAP, 1970, 459P.

15-2185   CRYSTAL FIELD GRADIENTS IN ANTIMONY SULFOIODIDE.
           KOIKOV SN + KRAINIK NN + MALININA VG AND OTHERS
           BULL ACAD SCI USSR, PHYS SER 35(9): 1628-32 (1971)

15-2186   EFFECT OF PHOTOELECTRET CHARGE ON VALUE OF FERROELECTRIC POLARIZATION IN ANTIMONY
           SULFOIODIDE.
           KOPINETS IF + RUBISH ID + SEMAK DG + CHEPUR DV
           UKR FIZ ZH 15: 505-7 (MAR 1970) (IN RUSSIAN)
                CHEM ABSTR 73: 71076X (1970)
                198

15-2187   EFFECT OF OXYGEN ADSORPTION ON THE FERROELECTRIC PHASE TRANSITION IN ANTIMONY
           SULFOIODIDE SINGLE CRYSTALS.
           KOPINETS IF + RUBISH ID + TURYANITSA ID
           UKR FIZ ZH 15(8): 1384-5 (1970) (IN RUSSIAN)
                CHEM ABSTR 74: 26019R (1971)
                225

15-2188   INJECTION CURRENTS AND ACCUMULATION OF POLARIZATION CHARGE IN ANTIMONY SULFOIODIDE.
           KOSMAN MS + SLEPTSOV AI
           SOV PHYS SOLID STATE 13(10): 2584-5 (APR 1972)
                263

15. ANTIMONY SULFOIODIDE AND RELATED COMPOUNDS

15-2189 NUCLEAR QUADRUPOLE AND ELECTROACOUSTIC ECHOES IN FERROELECTRIC ANTIMONY SULFOIODIDE.
KRAINIK NN + POPOV SN + MYLNIKOVA IE
J PHYS (PARIS) 33, SUPPL 4, COLLOQ 2: 179-82 (1972)
294

FERROELECTRIC SEMICONDUCTORS. (ANTIMONY SULFOIODIDE, REVIEW, 57 REFS)
KREHER K
WISS Z KARL MARX UNIV LEIPZIG, MATH NATURWISS REIH 20(2): 287-301
(1971)
CHEM ABSTR 75: 113587P (1971)

15-2190 RELAXATION PROCESSES OF ANTIMONY SULFOIODIDE SINGLE CRYSTALS.
KUDZIN AYU + SUKHINSKII AN + GENE VV
IZV VUZ FIZ 1972(1): 158-60 (IN RUSSIAN)

15-2191 TEMPERATURE DEPENDENCE OF THE SWITCHING PROCESS IN ANTIMONY SULFOIODIDE SINGLE
CRYSTALS.
KUDZIN AYU + SUKHINSKII AN + BUTSYCHENKO VA
SOV PHYS SOLID STATE 12(7): 1740-41 (1971)
221

15-2192 INTERACTION OF ULTRASONIC WAVES WITH MOVING DOMAIN WALLS IN FERROELECTRICS. (ANTIMONY
SULFOIODIDE)
KUNIGELIS VF
SOV PHYS SOLID STATE 13(8): 1986-8 (FEB 1972)
258

15-2193 TEMPERATURE DEPENDENCES OF THE HALL AND DRIFT MOBILITIES OF HOLES IN ANTIMONY
TRISULFIDE SINGLE CRYSTALS.
LIPSKIS KK + MIKALKYAVICHYUS MP + SAKALAS AP + YUSHKA GB
SOV PHYS SEMICOND 5(4): 608-11 (OCT 1971)
244

15-2194 HYDROTHERMAL METHOD FOR PREPARING A(V)B(VI)C(VII) COMPOUNDS. (USING ANTIMONY OR
BISMUTH, SULFUR, SELENIUM OR TELLURIUM, AND CHLORINE, BROMINE OR IODINE)
LITVIN BN + POPOLITOV VI
INORG MATER 6(3): 508-9 (1970)
205

15-2195 FERROELECTRIC TRANSITION OF ANTIMONY SULFOIODIDE.
MASSOT M
UNIV PARIS PHD THESIS, 49P, 1972. (IN FRENCH)
284

15-2196 ANOMALOUS PROPERTIES IN PHOTOFERROELECTRIC ANTIMONY SULFOIODIDE.
MIHAILOV MP
DOKL BOLG AKAD NAUK 21(5): 423-5 (1968)
PHYS ABSTR 71: 49373 (1968)
141

15-2197 EFFECT OF PHOTOCONDUCTION ON THE FERROELECTRIC PROPERTIES OF ANTIMONY SULFOIODIDE.
(PHASE BOUNDARY SHAPE STRIATIONS IN DC FIELD)
MORI T + TAMURA H + SAWAGUCHI E
P445-7 OF INT MEETING ON FERROELECTRICITY, 2ND, 1969, KYOTO, PROC,
(J PHYS SOC JAP, VOL 28 SUPPL). PHYS SOC JAP, 1970, 459P.

15-2198 ELECTROOPTIC EFFECTS OF THE ABSORPTION EDGE OF ANTIMONY SULFOIODIDE.
NAKAO K + BENNACEUR R + BALKANSKI M
PHYS LETT A 41: 219-20 (1972)

15-2199 PREPARATION AND PHYSICAL PROPERTIES OF NONSTOICHIOMETRIC ANTIMONY SULFOIODIDE SINGLE
CRYSTALS.
NAKONECHNYI YUS + FEDAK VV + RYZHAKOV AG + SEMAK DG + TURYANITSA ID + CHEPUR DV
INORG MATER 8: 1156-7 (1972)

15-2200 GROWTH OF LARGE ANTIMONY SULFOIODIDE CRYSTALS (TO 1 CM DIAMETER): CONTROL OF NEEDLE
MORPHOLOGY.
NASSAU K + SHIEVER JW + KOWALCHIK M
J CRYST GROW 7: 237-45 (1970)
211

15. ANTIMONY SULFOIODIDE AND RELATED COMPOUNDS

15-2201   CRYSTALLIZATION BEHAVIOR OF ANTIMONY SULFOIODIDE. (SINGLE CRYSTAL GROWTH, CHEMICAL
           TRANSPORT)
           NEELS H + SCHMITZ W + HOTTMANN H + ROSSNER R + TOPP W
           KRISTALL TECH 6(2): 225-43 (1971) (IN GERMAN)
                284

15-2202   RELAXATION PROCESSES IN ANTIMONY SULFOIODIDE SINGLE CRYSTALS EXCITED WITH MICROSECOND
           LIGHT PULSES.
           NESTERENKO PS + SAVCHENKO EA + TATARENKO LN
           SOV PHYS SOLID STATE 13(8): 2094-5 (FEB 1972)
                258

15-2203   FIELD EMISSION FROM ANTIMONY SULFOIODIDE.
           NEUMANN H
           ANN PHYS 23(1-2): 56-65 (1969) (IN GERMAN)
                CHEM ABSTR 71: 54687P (1969)
                    173

15-2204   ELECTRON STATE DENSITY AND OPTICAL PROPERTIES OF ANTIMONY SULFOIODIDE. (IN PARAELECTRIC
           AND FERROELECTRIC STATES)
           NIKIFOROV IYA + KHASABOV AG
           SOV PHYS SOLID STATE 13: 3030-2 (1972)

15-2205   NEW FERROELECTRIC A(V)B(VII)C(VII) COMPOUNDS OF ANTIMONY SULFOIODIDE TYPE. (PARTIAL OR
           TOTAL SUBSTITUTION OF ANTIMONY BY BISMUTH OR ARSENIC, SULFUR BY SELENIUM OR OXYGEN, AND
           IODINE BY BROMINE OR CHLORINE)
           NITSCHE R + ROETSCHI H + WOLD P
           APPL PHYS LETT 4(12): 210-11 (1964)

15-2206   REFRACTIVE INDICES OF ANTIMONY SULFOIODIDE. (DEPENDENCE ON TEMPERATURE, WAVELENGTH,
           APPLIED FIELD STRENGTH)
           OHI K
           P84-6 OF INT MEETING ON FERROELECTRICITY, 2ND, 1969, KYOTO, PROC,
           (J PHYS SOC JAP, VOL 28 SUPPL). PHYS SOC JAP, 1970, 459P.

           BULL SCI ENG RES LAB WASEDA UNIV NO 50: 81-91 (1970)

15-2207   (POLARIZED) RAMAN SPECTRUM OF (PARAELECTRIC AND) FERROELECTRIC ANTIMONY SULFOIODIDE
           (PHONON ASSIGNMENTS).
           PERRY CH + AGRAWAL DK
           SOLID STATE COMMUN 8: 225-30 (1970)
                186

15-2208   OPTICAL PHONONS AND PHASE TRANSITIONS IN SOME ORDER DISORDER AND DISPLACIVE
           FERROELECTRICS. (ANTIMONY SULFOIODIDE, KDP)
           PERRY CH + AGRAWAL DK
           P342-6 OF NUSIMOVICI M(ED), PHONONS, INT CONF, RENNES, 1971 PROC,
           FLAMMARION, PARIS, 1971, 502P.

15-2209   FAR INFRARED REFLECTIVITY OF ANTIMONY SULFOIODIDE.
           PETZELT J
           P287-92 OF EUROPEAN MEETING ON FERROELECTRICITY, 1969, SAARBRUCKEN,
           MUSER HE AND PETERSSON J(EDS), WISSENSCH VERLAGSGES, STUTTGART,
           1970, 396P.

15-2210   REFLECTION SPECTRA OF SOME FERROELECTRIC CRYSTALS OF GROUPS V, VI AND VII. (ANTIMONY
           AND BISMUTH SULFO HALIDES)
           PIKKA T
           TR TALLIN POLITEKH INST, SER A 251: 13-15 (1967) (IN RUSSIAN)
                CHEM ABSTR 69: 72341 (1968)
                    141

15-2211   ANOMALOUS ECHO IN FERROELECTRIC ANTIMONY SULFOIODIDE. (PIEZOELECTRIC DOMAIN
           OSCILLATIONS)
           POPOV SN + KRAINIK NN
           SOV PHYS SOLID STATE 12(10): 2440-4 (1971)
                232

15-2212   CRYSTALLIZATION OF BISMUTH SULFOIODIDE. (HYDROTHERMAL SYNTHESIS)
           POPOLITOV VI
           SOV PHYS CRYSTALLOGR 15(6): 1118-19 (MAY-JUN 1971)
                235

15. ANTIMONY SULFOIODIDE AND RELATED COMPOUNDS

15-2213   CRYSTALLIZATION OF SEMICONDUCTORS OF THE COMPOSITION A(V)B(VI)C(VII) UNDER HYDROTHERMAL
          CONDITIONS. (ANTIMONY AND BISMUTH SULFO, SELENO, AND TELLUROIODIDES AND BROMIDES)
          POPOLITOV VI + LOBACHEV AN
          INORG MATER 8: 1389-92 (1972)
              290

15-2214   HYDROTHERMAL CRYSTALLIZATION OF SEMICONDUCTING COMPOUNDS OF GROUP A(V)B(VI)C(VII).
          (HIGH GRADE SINGLE CRYSTALS, V IS ANTIMONY OR BISMUTH, VI IS SULFUR, SELENIUM OR
          TELLURIUM, VII IS IODINE, BROMINE OR CHLORINE)
          POPOLITOV VI + LITVIN BN + LOBACHEV AN
          PHYS STATUS SOLIDI A 3(1): K1-4 (1970)
              225

15-2215   KINETICS OF CRYSTALLIZATION OF ANTIMONY SULFOIODIDE (CRYSTAL GROWTH RATE, ANISOTROPY)
          POPOLITOV VI + LITVIN BN
          SOV PHYS CRYSTALLOGR 15(6): 1113-15 (MAY-JUN 1971)
              234

15-2216   NUCLEAR QUADRUPOLE RESONANCE IN FERROELECTRICS OF THE ANTIMONY SULFOIODIDE (AND
          BROMIDE) TYPE.
          POPOV SN + KRAINIK NN + MYLNIKOVA IE
          P120-2 OF INT MEETING ON FERROELECTRICITY, 2ND, 1969, KYOTO, PROC.
          (J PHYS SOC JAP, VOL 28 SUPPL). PHYS SOC JAP, 1970, 459P.

15-2217   ABSORPTION SPECTRA OF BISMUTH TELLUROBROMIDE AND BISMUTH TELLUROIODIDE CRYSTALS.
          PUGA GD + KOVACH DSH + TURYANITSA ID + BORETS AN + CHEPUR DV
          UKR FIZ ZH 16(2): 276-9 (1971) (IN RUSSIAN)
              243

15-2218   CRYSTAL GROWTH AND CHEMICAL SYNTHESIS UNDER HYDROTHERMAL CONDITIONS. (BISMUTH
          SULFOBROMIDE)
          RABENAU A + RAU H
          PHILLIPS TECH REV 30(4): 89-96 (1969)

15-2219   OPTICAL PROPERTIES OF ANTIMONY SELENOIODIDE IN THE INFRARED.
          RIEDE V
          PHYS STATUS SOLIDI A 2(3): K193-5 (1970)
              234

15-2220   EFFECT OF ILLUMINATION ON REPOLARIZATION PROCESS IN ANTIMONY SULFOIODIDE SINGLE
          CRYSTALS.
          RUDYAK VM + BOGOMOLOV AA + IVANOV VV
          BULL ACAD SCI USSR, PHYS SER 35(9): 1722-4 (1971)

15-2221   ANOMALOUS ABSORPTION OF SOUND NEAR THE FERROELECTRIC PHASE TRANSITION (ANTIMONY
          SULFOIODIDE).
          SAMULIONIS VI + KUNIGELIS VF
          JETP LETT 13: 207-8 (MAR 20, 1971)

15-2222   ANISOTROPY OF ANOMALOUS ABSORPTION OF ULTRASOUND IN SINGLE CRYSTALS OF ANTIMONY
          SULFOIODIDE.
          SAMULIONIS VI + KUNIGELIS VF
          SOV PHYS SOLID STATE 13(3): 611-14 (SEP 1971)
              244

15-2223   GENERATION OF THE SECOND ACOUSTIC HARMONIC IN ANTIMONY SULFOIODIDE SINGLE CRYSTALS.
          SAMULIONIS VI + KUNIGELIS VF + GIRSHOVICHUS MN
          SOV PHYS JETP 34(5): 1033-5 (MAY 1972)
                  PHYS ABSTR 75: 6666 (1972)
                  264

15-2224   GENERATION OF THE SECOND ACOUSTIC HARMONIC IN ANTIMONY SULFOIODIDE SINGLE CRYSTALS.
          SAMULIONIS VI + KUNIGELIS VF + GIRSHOVICHUS MN
          SOV PHYS JETP 34(5): 1033-5 (1972)

15-2225   (ELASTIC CONSTANTS FROM) BRILLOUIN SCATTERING STUDY OF ANTIMONY SULFOIODIDE USING A
          DOUBLE PASSED, STABILIZED SCANNING INTERFEROMETER.
          SANDERCOCK JR
          OPT COMMUN 2(2): 73-6 (1970)
              207

15-2226   PHOTOCONDUCTIVITY OF FERROELECTRIC PHOTOCONDUCTOR ANTIMONY SULFOIODIDE.
           SASAKI Y
           JAP J APPL PHYS 3: 558-9 (1964)

15-2227   PIEZOELECTRICITY AND ELECTROSTRICTION OF ANTIMONY SULFOIODIDE SINGLE CRYSTALS.
           SCHEIDING C + SCHMIDT G
           PHYS STATUS SOLIDI A, APPL RES 9(1): K77-80 (1972)

15-2228   INFLUENCE OF LIGHT ON SELF POLARIZATION OF SINGLE CRYSTALS OF ANTIMONY SULFOIODIDE.
           SHEINKMAN MK + ZEINALLY AKH + BEZDETNYI NM + LEBEDEVA NN + GORBATOV GZ
           UKR FIZ ZH 15: 1914-16 (1970) (IN RUSSIAN)

15-2229   LUMINESCENCE OF SINGLE CRYSTALS OF THE FERROELECTRIC SEMICONDUCTOR ANTIMONY
           SULFOIODIDE.
           SHEINKMAN MK + KROLEVETS NM + SAVCHENKO EA + TATARENKO LN
           SOV PHYS SOLID STATE 14(1): 253-4 (JUL 1972)
                270

15-2230   CRYSTAL GROWTH IN THE ANTIMONY SULFOIODIDE- ANTIMONY TRISULFIDE SYSTEM. (BRIDGMAN-
           STOCKBARGER METHOD)
           SPITSYNA VD + LYAKHOVITSKAYA VA + POPOVKIN BA
           SOV PHYS CRYSTALLOGR 16(1): 176-7 (JUL-AUG 1971)
                240

15-2231   CRITICAL PHENOMENA IN ANTIMONY SULFOIODIDE.
           STEIGMEIER EF + HARBEKE G
           J PHYS (PARIS) 33, SUPPL 4, COLLOQ 2: 55 (1972)
                294

15-2232   SOFT MODE COUPLING AND CRITICAL RAYLEIGH SCATTERING IN FERROELECTRIC ANTIMONY
           SULFOIODIDE.
           STEIGMEIER EF + HARBEKE G + WEHNER RK
           P396-418 OF INT CONF ON LIGHT SCATTERING IN SOLIDS, 2ND PROC,
           BALKANSKI M(ED), PARIS, 1971, FLAMMARION, PARIS, 1971, 518P.

15-2233   SOFT LATTICE VIBRATIONS AND VIBRATION COUPLING IN ANTIMONY SULFOIODIDE.
           STEIGMEIER EF + HARBEKE G + WEHNER RK
           HELV PHYS ACTA 43: 757 (1970) (IN GERMAN)

15-2234   FAR INFRARED REFLECTIVITY SPECTRA OF ANTIMONY SULFOIODIDE.
           SUGAWARA F + NAKAMURA T
           P221-2 OF INT MEETING ON FERROELECTRICITY, 2ND, 1969, KYOTO, PROC,
           (J PHYS SOC JAP, VOL 28 SUPPL). PHYS SOC JAP, 1970, 459P.

15-2235   FAR INFRARED REFLECTIVITY SPECTRA OF ANTIMONY SULFOIODIDE.
           SUGAWARA F + NAKAMURA T
           J PHYS CHEM SOLIDS 33: 1665-8 (1972)
                270

15-2236   PHOTODIELECTRIC EFFECT IN ANTIMONY SULFOIODIDE SINGLE CRYSTALS.
           SUGIURA A + IWASAKI H
           J RADIO RES LAB 18(95): 59-66 (JAN 1971)
                236

15-2237   CRITICAL POINT IN FERROELECTRIC ANTIMONY SULFOIODIDE.
           SYRKIN LN + POLANDOV IN + KACHALOV NP + GAMYNIN EV
           SOV PHYS SOLID STATE 14(2): 517-18 (AUG 1972)
                274

15-2238   HEAT CAPACTIY OF POLYCRYSTALLINE ANTIMONY SULFOIODIDE.
           TARASKIN SA + LYAKHOVITSKAYA VA + IVANOVSHITS AK
           SOV PHYS CRYSTALLOGR 17(3): 597-8 (NOV-DEC 1972)
                280

15-2239   STRAIN ALONG C-AXIS OF ANTIMONY SULFOIODIDE CAUSED BY ILLUMINATION IN DC ELECTRIC
           FIELD.
           TATSUZAKI I + ITOH K + UEDA S + SHINDO Y
           PHYS REV LETT 17(4): 198-200 (1966)

15. ANTIMONY SULFOIODIDE AND RELATED COMPOUNDS

15-2240 SOFT PHONON COUPLINGS IN THE PRESSURE INDUCED PHASE TRANSITION IN ANTIMONY SULFOIODIDE.
TENG MK + BALKANSKI M + MASSCT M
PHYS REV B SOLID STATE 5(3): 1031-4 (1 FEB 1972)
        258

15-2241 TRANSPORT PHENOMENA IN ANTIMONY SULFOIODIDE. (CONDUCTION MECHANISM FOR THERMOELECTRIC
CURRENTS)
TOYODA K + ISHIKAWA K
P451-3 OF INT MEETING ON FERROELECTRICITY, 2ND, 1969, KYOTO, PROC.
(J PHYS SOC JAP, VOL 28 SUPPL). PHYS SOC JAP, 1970, 459P.

15-2242 DIELECTRIC PROPERTIES OF ANTIMONY SULFOIODIDE- SULFOBROMIDE SOLID SOLUTIONS.
TURYANITSA ID + GROSHIK II + KOPERLES BM
UKR FIZ ZH 16: 505-6 (MAR 1971) (IN RUSSIAN)

15-2243 MODE FOR THE ANTIMONY SULFOICDIDE- TIN DIOXIDE HETEROSTRUCTURE SWITCH.
VAN DER ZIEL A
JAP J APPL PHYS 10: 1648-9 (NOV 1971)

15-2244 REFLECTIVITY ON ANTIMONY SULFOIODIDE.
VAZQUEZ F
BULL AMER PHYS SOC 17: 594 (1972)

15-2245 INFLUENCE OF ONE DIMENSIONAL MECHANICAL STRESS ON THE PHASE TRANSITION OF ANTIMONY
SULFOIODIDE SINGLE CRYSTALS.
VOLNYANSKII MD + KUDZIN AYU + SUKHINSKII AN
SOV PHYS CRYSTALLOGR 17(2): 364-6 (SEP-OCT 1972)
        275

15-2246 POLARIZATION REVERSAL OF ANTIMONY SULFOIODIDE SINGLE CRYSTALS IN THE REGION OF PHASE
COEXISTENCE.
VOLNYANSKII MD + KUDZIN AYU + SUKHINSKII AN
SOV PHYS SOLID STATE 13(11): 2905-7 (MAY 1972)
        265

15-2247 NONLINEAR EFFECTS AND THE LOW TEMPERATURE TRANSITION IN ANTIMONY SULFOIODIDE CRYSTALS.
ZAKS PL + PASYNKOV RE
BULL ACAD SCI USSR, PHYS SER 34(12): 2209-14 (1970)

15-2248 EFFECTS OF TEMPERATURE, ELECTRIC FIELD, AND ILLUMINATION ON THE ABSORPTION OF
ULTRASOUND IN ANTIMONY SULFOIODIDE NEAR THE PHASE TRANSITION TEMPERATURE.
ZAPOROZHETS OI + LYAKHOVITSKAYA VA + PEKAR SI + POLOTSKII IG + SILVESTROVA IM
SOV PHYS SOLID STATE 12(2): 523-4 (1970)
        206

15-2249 DIELECTRIC AND PIEZOELECTRIC PROPERTIES OF ANTIMONY SULFOIODIDE IN A CONSTANT ELECTRIC
FIELD.
ZAVYANLOVA AM + ZAKS PL + SYRKIN LN AND OTHERS
BULL ACAD SCI USSR, PHYS SER 35(9): 1718-21 (1971)

15-2250 INVESTIGATION OF THE PIEZORESISTANCE OF (POLYCRYSTALLINE) ANTIMONY SULFOIODIDE (DARK
CONDUCTIVITY NEAR TRANSITION POINT, 1-1000 ATM).
ZAVYANLOVA AM + ZAKS PL + SYRKIN LN
SOV PHYS SOLID STATE 12(5): 1252-3 (1970)

15-2251 ELECTRICAL AND PHOTOELECTRIC PROPERTIES OF THE CONTACTS BETWEEN ANTIMONY SULFOIODIDE
SINGLE CRYSTALS AND SOME METALS. (ANTIMONY, SILVER, ALUMINUM)
ZHDAN AG + ARTOBOLEVSKAYA ES
SOV PHYS SOLID STATE 13(4): 1040-1 (OCT 1971)
        252

15-2252 OPPOSING FERROELECTRIC DOMAINS IN SINGLE CRYSTALS OF ANTIMONY SULFOIODIDE.
ZHDAN AG + CHENSKII EV + ARTOBOLEVSKAYA ES AND OTHERS
JETP LETT 14(3): 105-7 (1971)

16. NITRATES AND NITRITES

16-2253   A RADICAL IN FERROELECTRIC DGN CRYSTAL.
ABE R + KAWAMURA A
REP TOYODA PHYS CHEM RES INST NO 25: 1-4 (1972) (IN JAPANESE)

16-2254   NUCLEAR QUADRUPOLE RELAXATION OF NITROGEN-14 IN SODIUM NITRITE.
ABE Y + OHNEDA Y + ABE S + KOJIMA S
J PHYS SOC JAP 33: 864 (1972)

16-2255   NUCLEAR MAGNETIC RESONANCE STUDIES OF THE PHASE TRANSITIONS IN FERROELECTRICS.
(POTASSIUM FERROCYANIDE TRIHYDRATE, SODIUM NITRITE)
ALEKSANDROV KS + HABUDA SP + LUNDIN AG + MIKHAJLOV GM
ACTA CRYSTALLOGR 16(SUPPL): A190-1 (1963) (ABSTRACT ONLY)

16-2256   ANGULAR DISPERSION OF LARGE WAVE VECTOR POLARITONS IN FERROELECTRIC SODIUM NITRITE.
ANDA E
SOLID STATE COMMUN 9: 1545-50 (1971)
     250

16-2257   RELATIONSHIP OF THE TEMPERATURE DEPENDENCE OF B1 PHONON AND DIELECTRIC RELAXATION IN
SODIUM NITRITE.
ANDRADE PR + PRASAD RAO AD + KATIYAR RS
BULL AMER PHYS SOC 17: 1182 (1972)

16-2258   ELECTRICAL RESISTIVITIES OF SODIUM NITRITE AND POTASSIUM NITRATE CRYSTALS.
ASAO Y + YOSHIDA I + ANDO R + SAWADA S
J PHYS SOC JAP 17: 442-6 (1962)

16-2259   MORPHOLOGY OF CRYSTAL GROWTH IN THE ALPHA TO BETA, BETA TO ALPHA, BETA TO GAMMA AND
GAMMA TO ALPHA TRANSITIONS IN POTASSIUM NITRATE.
ASADOV YUG + NASIROV VI
SOV PHYS CRYSTALLOGR 15(6): 1052-6 (MAY-JUN 1971)
     234

16-2260   MORPHOLOGY OF CRYSTAL GROWTH AT POLYMORPHIC TRANSFORMATIONS IN POTASSIUM NITRATE,
SILVER NITRATE, AND AMMONIUM NITRATE SINGLE CRYSTALS.
ASADOV YUG + NASIROV VI + JABRAILOVA GA
J CRYST GROW 15: 45-50 (1972)

16-2261   QUASI STATIC SWITCHING CURRENT IN SODIUM NITRITE.
ASAO Y + SAWADA S + HATTA I + SHIRAI Y
J PHYS SOC JAP 18: 1690-1 (1963)

16-2262   RAMAN SPECTRA AND MODE FREQUENCY SHIFTS OF FERROELECTRIC SODIUM NITRITE AT 77 AND 294
DEG K.
ASAWA CK + BARNOSKI MK
PHYS REV B 2(1): 205-13 (1970)
     209

16-2263   RHYTHMIC GROWTH OF NEW PHASE IN POTASSIUM NITRATE. (ALPHA TO BETA POLYMORPHIC
TRANSITION)
ASADOV YUG + NASIROV VI
SOV PHYS DOKL 15(4): 324-6 (1970)
     225

16-2264   NUCLEAR SPIN LATTICE RELAXATION OF SODIUM-23 IN THE PARAELECTRIC AND FERROELECTRIC
PHASE OF SODIUM NITRITE. (THERMODYNAMICS OF THE PHASE TRANSITION)
AVOGADRO A + CAVELIUS E + MULLER D + PETERSSON J
PHYS STATUS SOLIDI B 44: 639-46 (1971)
     235

16-2265   MELTING PROPERTIES OF POTASSIUM NITRATE UNDER HIGH PRESSURE.
          BABB SE + CHANEY PE + OWENS BB
          J CHEM PHYS 41: 2210-11 (1964)

16-2266   FERROELECTRIC PHASE TRANSITION IN SILVER SODIUM DINITRITE
          BELYAEV LM + VERKHOVSKAYA KA + VOLK TR + OGADZHANOVA VV + SOBOLEVA LV + TIKHOMIROVA NA
          + FRIDKIN VM
          BULL ACAD SCI USSR, PHYS SER 34(12): 2270-3 (1970)

16-2267   NMR STUDY OF THE FERROELECTRIC TRANSITIONS IN DIGLYCINE NITRATE AND TRIS-SARCOSINE
          CALCIUM CHLORIDE. (SHIFT OF CURIE TEMPERATURE)
          BLINC R + JAMSIK-VILFAN M + LAHAJNAR G + HAJDVKOVIC G
          J CHEM PHYS 52: 6407-11 (1970)
               238

16-2268   NUCLEAR QUADRUPOLE SPIN LATTICE RELAXATION AND CRITICAL DYNAMICS OF FERROELECTRIC
          CRYSTALS. (SODIUM NITRITE)
          BONERA G + BORSA F + RIGAMONTI A
          PHYS REV B 2(7): 2784-95 (1970)
               233

16-2269   PROGRESS TOWARD A FAST, NONVOLATILE, NONDESTRUCTIVE READOUT MEMORY ELEMENT UTILIZING
          POTASSIUM NITRATE. (FUSED THICK FILM OR VACUUM DEPOSITED THIN FILM)
          BORN RC + ROHRER GA + DHALL BS
          P149-54 OF ELECTRON COMPCN CONF, 20TH PROC, 1970,

16-2270   CRITICAL SCATTERING OF X-RAYS IN SODIUM NITRITE.
          CANUT ML + MENDIOLA J
          PHYS STATUS SOLIDI 5: 313-27 (1964)

16-2271   X-RAY ANALYSIS OF FERROELECTRIC DOMAINS IN THE PARAELECTRIC PHASE OF SODIUM NITRITE.
          CANUT ML + HOSEMANN R
          ACTA CRYSTALLOGR 17: 973-81 (1964)

16-2272   NONLINEAR OPTICAL PROPERTIES OF FERROELECTRIC SODIUM NITRITE.
          CHERN MJ + PHILLIPS RA
          BULL AMER PHYS SOC 16: 28 (1971)

16-2273   OPTICAL ACTIVITY IN NON-ENANTIOMORPHOUS BIAXIAL CRYSTALS: METHYL MESITYL OXIDE OXALATE
          AND SODIUM NITRITE.
          CHERN MJ + PHILLIPS RA
          J OPT SOC AMER 60(9): 1230-2, 1542 (1970)
               214

16-2274   TEMPERATURE DEPENDENCE OF NONLINEAR OPTICAL COEFFICIENTS IN FERROELECTRIC SODIUM
          NITRITE.
          CHERN MJ + PHILLIPS RA
          J APPL PHYS 43: 496-9 (1972)
               256

16-2275   INVESTIGATION OF FERROELECTRIC TRANSITION IN SODIUM NITRITE CRYSTAL BY OBSERVATION OF
          RAMAN SPECTRUM.
          CHISLER EV + SHUR MS
          BULL ACAD SCI USSR, PHYS SER 31(7): 1116-21 (1967)
               219

16-2276   PHASE CHANGE IN SODIUM NITRITE CRYSTALS AND COMBINATION SCATTER SPECTRUM IN THE REGION
          OF THE ANTISYMMETRICAL VIBRATION OF THE NITRITE ANION.
          CHISLER EV + GONCHARUK IN
          SOV PHYS SOLID STATE 13(5): 1157-9 (NOV 1971)
               251

16-2277   TWO-PHONON SCATTERING AND FERROELECTRIC TRANSITION IN A SODIUM NITRITE CRYSTAL.
          CHISLER EV + GONCHARUK IN
          SOV PHYS SOLID STATE 13(10): 2559-63 (APR 1972)
               263

16-2278   STUDIES OF PHASE TRANSFORMATIONS IN NITRATES AND NITRITES. PART-1: CHANGE IN
          ULTRAVIOLET ABSORPTION SPECTRA ON MELTING. PART-2: CHANGE IN ULTRAVIOLET ABSORPTION
          SPECTRA ACCOMPANYING THERMAL TRANSFORMATIONS IN THE CRYSTALS.
          CLEAVER B + RHODES E + UBBELOHDE AR
          PROC ROY SOC A 276: 437-52, 453-60 (1963)

16-2279   TRANSITION RATES OF POTASSIUM NITRATE HIGH PRESSURE POLYMORPHS.
           DAVIS BL + ADAMS LH
           J PHYS CHEM SOLIDS 24: 787-94 (1963)

16-2280   ATR INFRARED SPECTRA OF UNIAXIAL NITRATE CRYSTALS.
           DEVLIN JP + POLLARD G + FRECH R
           J CHEM PHYS 53: 4147-51 (1970)

16-2281   MULTIPHONON SELECTION RULES FOR INFRARED ABSORPTION AND RAMAN SCATTERING IN
           FERROELECTRIC AND PARAELECTRIC SODIUM NITRITE.
           DEVINE SD
           J PHYS C 4: 1036-48 (1971)

16-2282   SWITCHING PROPERTIES OF THICK FILM FERROELECTRIC POTASSIUM NITRATE.
           DHALL BS
           MICH TECHNOL UNIV MS THESIS, 1969.

16-2283   SPECIFIC HEAT, THERMAL DIFFUSIVITY AND THERMAL CONDUCTIVITY OF CALCIUM DOPED SODIUM
           NITRATE CRYSTALS NEAR THE TRANSITION POINT.
           DIKANT J
           CZECH J PHYS B 22(8): 697-703 (1972)

16-2284   CRYSTAL DYNAMICS OF SODIUM NITRITE. (NEUTRON SCATTERING)
           DOLLING G + SAKURAI J + COWLEY RA
           P258-60 OF INT MEETING ON FERROELECTRICITY, 2ND, 1969, KYOTO, PROC,
           (J PHYS SOC JAP, VOL 28 SUPPL). PHYS SOC JAP, 1970, 459P.

16-2285   SWITCHING BEHAVIOR IN FERROELECTRIC POTASSIUM NITRATE.
           DORK RA + SCHUBRING NW + NOLTA JP
           J APPL PHYS 35: 1984-5 (1964)

16-2286   ELECTRICAL PROPERTIES OF RUBIDIUM NITRATE AND CESIUM NITRATE.
           FERMOR JH + KJEKSHUS A
           ACTA CHEM SCAND 26: 2645-54 (1972)

16-2287   THE ORDER DISORDER TRANSITION OF SODIUM NITRITE. (TWO SUBLATTICE MODEL OF INTERACTING
           DIPOLES)
           FLUGGE S + MEYENN KV
           Z PHYS 253(5): 369-78 (1972)
                SCI ABSTR A 75: 61432 (1972)
                286

16-2288   ELECTRICAL BREAKDOWN IN FERROELECTRIC SODIUM NITRITE CRYSTALS.
           FOK J + HANSCOMB JR
           J APPL PHYS 43: 4824-5 (NOV 1972)

           PHENOMENOLOGICAL THEORY OF PHASE TRANSITIONS IN SODIUM NITRITE.
           GESI K
           J PHYS SOC JAP 20(10): 1764-72 (1965)

16-2289   DIELECTRIC STUDIES ON THE FERROELECTRIC PHASE TRANSITION IN SILVER SODIUM DINITRITE.
           GESI K
           J PHYS SOC JAP 28(2): 395-401 (1970)
                189

16-2290   DIELECTRIC RELAXATION IN FERROELECTRIC SILVER SODIUM DINITRITE.
           GESI K
           J PHYS SOC JAP 28: 1365 (1970)
                196

16-2291   DIELECTRIC RELAXATION MEASUREMENT ON FERROELECTRIC SILVER SODIUM DINITRITE BY STEP
           FUNCTION METHOD.
           GESI K
           JAP J APPL PHYS 11: 1745 (1972)

16-2292   DIELECTRIC RELAXATION IN FERROELECTRIC SILVER SODIUM DINITRITE.
           GESI K
           FERROELECTRICS 4: 245-52 (1972)

16-2293 FERROELECTRIC HYSTERESIS LOOP CF SILVER SODIUM DINITRITE BY QUASI STATIC ELECTRIC
         FIELD.
         GESI K
         J PHYS SOC JAP 32: 1679 (1972)
             278

16-2294 HYDROSTATIC PRESSURE EFFECT ON THE DIELECTRIC RELAXATION TIME IN FERROELECTRIC SILVER
         SODIUM DINITRITE.
         GESI K + OZAWA K
         J PHYS SOC JAP 33: 569 (1972)
         PHYS STATUS SOLIDI B 52(1): K45-8 (1972)
             278

16-2295 PYROELECTRIC STUDY ON THE SPONTANEOUS POLARIZATION IN SILVER SODIUM DINITRITE.
         GESI K
         J PHYS SOC JAP 33(1): 108-11 (JUL 1972)
             278

16-2296 SPECIFIC HEAT ANOMALY IN FERROELECTRIC SILVER SODIUM DINITRITE.
         GESI K
         J PHYS SOC JAP 28: 1377 (1970)
             196

16-2297 CRYSTAL STRUCTURE OF FERROELECTRIC GLYCINE SILVER NITRATE.
         GUHA S
         INDIAN J PHYS 46: 255-9 (1972)

16-2298 THERMAL EXPANSION OF SOME FERROELECTRICS. (POTASSIUM NITRATE, KDP, SODIUM NITRITE)
         GUTIERREZ M + CANUT ML + AMOROS JL
         ACTA CRYSTALLOGR 16(SUPPL): A166 (1963)

16-2299 ANALYSIS OF THE TEMPERATURE DEPENDENT PHONON STRUCTURE IN SODIUM NITRITE BY RAMAN
         SPECTROSCOPY.
         HARTWIG CM + WIENER E + PORTO SPS
         PHYS REV B 5(1): 79-91 (1 JAN 1972)
             255

16-2300 DIPOLE LATTICE RELAXATION TIME NEAR THE NEEL TEMPERATURE IN SODIUM NITRITE.
         HATTA I + IKUSHIMA A
         ACTA CRYSTALLOGR A 28: S176 (1972)

16-2301 FERROELECTRIC POLARIZATION SWITCHING IN SODIUM NITRITE.
         HATTA I + SAWADA S + ASAO Y + YANAGI T
         J PHYS SOC JAP 18: 1229-30 (1963)

16-2302 STATIC ELECTRIC SUSCEPTIBILITY AND DIELECTRIC RELAXATION TIME NEAR THE TRANSITION
         POINTS IN SODIUM NITRITE.
         HATTA I
         J PHYS SOC JAP 28(5): 1266-77 (1970)
             196

16-2303 SPECIFIC HEAT OF SODIUM NITRITE NEAR THE ANTIFERROELECTRIC TRANSITION POINT.
         HATTA I + IKUSHIMA A
         PHYS LETT A 37: 207-8 (1971)

16-2304 ULTRASONIC ATTENUATION NEAR THE TWO TRANSITION POINTS IN SODIUM NITRITE.
         HATTA I + ISHIGURO T + MIKOSHIBA N
         P211-13 OF INT MEETING ON FERROELECTRICITY, 2ND, 1969, KYOTO, PROC,
         (J PHYS SOC JAP, VOL 28 SUPPL). PHYS SOC JAP, 1970, 459P.

16-2305 BRILLOUIN SCATTERING IN SODIUM NITRITE AT ROOM TEMPERATURE.
         HAURET G + CHAPELLE JP + TAUREL L
         PHYS STATUS SOLIDI A 11: 255-61 (1972)

16-2306 MEASUREMENT OF THE ELASTIC CONSTANTS OF SODIUM NITRITE BY BRILLOUIN DIFFUSION AT
         ORDINARY TEMPERATURES.
         HAURET G + GHARBI A
         COMPT REND B 271: 1072-4 (1970) (IN FRENCH)

16-2307 SPLITTING OF THE SODIUM-23 NMR CENTRAL LINE IN FERROELECTRIC SILVER SODIUM DINITRITE.
         HIKITA T + KASAHARA M + TATSUZAKI I
         PHYS LETT A 37: 141-2 (1971)

16. NITRATES AND NITRITES

16-2308 FAR INFRARED SPECTRA AND FERROELECTRIC PHASE TRANSITION OF POTASSIUM NITRATE.
         HILL JC + MOHAN PV
         FERROELECTRICS 2: 201-7 (JUL 1971)

16-2309 ACCURATE FORCE CONSTANTS FROM ISOTOPIC SUBSTITUTION IN THE NITRITE RADICAL IN SODIUM
         NITRITE.
         HOLAH GD + HAPP H
         J PHYS C 3(8): 1807-14 (1970)
             218

16-2310 TWO PHONON ABSORPTION IN FERROELECTRIC SODIUM NITRITE.
         HOLAH GD
         J PHYS C 4: 2191-2201 (1971)

16-2311 TEMPERATURE DEPENDENCE OF THE FORCE CONSTANTS OF ION IN SODIUM NITRITE.
         HOLAH GD
         J PHYS C SOLID STATE PHYS 4: 2557-64 (1971)

16-2312 CRYSTAL STRUCTURE AND PHASE TRANSITION OF DIGLYCINE NITRATE.
         HOSHINO S + SATO S + TOYODA K
         JAP J APPL PHYS 2: 519-20 (1963)

16-2313 NUCLEAR QUADRUPOLE RESONANCE OF NITROGEN IN SODIUM NITRITE.
         IKEDA R + MIKAMI M + NAKAMURA D + KUBO M
         J MAGN RESON 1: 211-20 (1969)

16-2314 LONG RANGE ORDER IN SODIUM NITRITE BY OPTICAL SECOND HARMONIC GENERATION.
         INOUE K + ISHIDATE T
         REP FAC SCI, SHIZOPUOKA UNIV 7: 7-29 (1972)

16-2315 TEMPERATURE DEPENDENCE OF SECOND HARMONIC GENERATION IN SODIUM NITRITE.
         INOUE K
         JAP J APPL PHYS 9: 152 (1970)
             190

16-2316 RATE OF ROTATING CRYSTALS AND CHARACTERISTICS OF THEIR GROWTH. (AMMONIUM DIHYDROGEN
         PHOSPHATE, POTASSIUM FERROCYANIDE TRIHYDRATE, SODIUM NITRATE)
         INYUSHKIN GV + SHABALIN KN
         SOV PHYS CRYSTALLOGR 9: 242-4 (1964)

16-2317 LATTICE VIBRATION OF SODIUM NITRITE. (CALCULATIONS TO RELATE TO SPECIFIC
         ANTIFERROELECTRIC PHASE)
         ISHIBASHI Y + TAKAGI Y
         P261-3 OF INT MEETING ON FERROELECTRICITY, 2ND, 1969,, KYOTO, PROC,
         (J PHYS SOC JAP, VOL 28 SUPPL). PHYS SOC JAP, 1970, 459P.

16-2318 AN X-RAY STUDY OF THE PHASE TRANSITION IN SODIUM NITRITE.
         ISMAILZADE IG + ANNAGIEV MKH + ABDULLAEVA KHM
         SOV PHYS CRYSTALLOGR 6: 585-7 (1962)

16-2319 EFFECT OF ADDING TITANIUM(+) AND STRONTIUM(2+) IONS ON THE ELECTRICAL CONDUCTIVITY OF
         POTASSIUM NITRATE.
         KABANOV AA + LYKHIN VM + BOLDYREV VV + TRUBICYN AM
         IZV SIBIRSK OTD AKAD NAUK SSSR KKIM NAUK 1: 141-2 (1969) (IN RUSSIAN)
             CHEM ABSTR 71: 15590W
             189

16-2320 NITROGEN-14 NUCLEAR QUADRUPCLE RESONANCE IN FERROELECTRIC SODIUM NITRITE.
         KADABA PK + O'REILLY DE + BLINC R
         PHYS STATUS SOLIDI 42: 855-8 (1970)
             225

16-2321 ACOUSTIC NUCLEAR MAGNETIC RESONANCE IN FERROELECTRIC SODIUM NITRITE.
         KANASHIRO T + OHNO T + TAKI T + SATOH M
         BULL FAC ENG TAKUSHIMA UNIV 9: 15-22 (1972)

16-2322 X-RAY MEASUREMENTS ON THERMAL EXPANSION OF FERROELECTRIC POTASSIUM NITRATE.
         KANTOLA M + TARNA T
         ANN ACAD SCI FEN, SER A6, NO 335: 3-16 (1970)
             CHEM ABSTR 73: 59912J (1970)
             213

16-2323 PECULIARITIES IN THE VIBRATIONAL SPECTRA OF CESIUM AND RUBIDIUM NITRATE CRYSTALS IN ITS
LOW TEMPERATURE PHASES.
KARPOV SV + SHULTIN AA
PHYS STATUS SOLIDI 39: 33-38 (1970)
197

16-2324 NMR STUDY OF SODIUM-23 IN SILVER SODIUM DINITRITE.
KASAHARA M + HIKITA T + TATSUZAKI I
J PHYS SOC JAP 29: 240 (1970)
202

16-2325 DISORDERED STRUCTURE OF SODIUM NITRITE AT 185 DEG C.
KAY MI + FRAZER BC + UEDA R
ACTA CRYSTALLOGR 15: 506-8 (1962)

16-2326 NEUTRON DIFFRACTION STUDY OF SODIUM NITRITE.
KAY MI + GONZALO JA
P284-5 OF INT MEETING ON FERROELECTRICITY, 2ND, 1969,, KYOTO, PROC,
(J PHYS SOC JAP, VOL 28 SUPPL). PHYS SOC JAP, 1970, 459P.

16-2327 STRUCTURE OF SODIUM NITRITE AT 150, 185, AND 225 DEGREES C.
KAY MI
FERROELECTRICS 4: 235-43 (1972)

16-2328 X-RAY AND NEUTRON STUDY ON THE PHASE TRANSFORMATION OF SODIUM NITRITE.
KAY MI + FRAZER BC + UEDA R
J PHYS SOC JAP 17(SUPPL B-II): 389-91 (1962)
IN PART II P389-91 OF INT CONF ON MAGNETISM AND CRYSTALLOGRAPHY,
1961, KYOTO, PROC. PHYS SOC JAPAN, 1962

16-2329 METASTABLE POTASSIUM NITRATE(III) FROM SOLUTION. (GROWN IN WATER OR ALCOHOL)
KENNEDY SW
J CRYST GROW 16: 274-6 (1972)
283

16-2330 RUBIDIUM NITRATE NEAR THE MELTING POINT AND ITS POLYMORPHISM.
KENNEDY SW
PHYS STATUS SOLIDI A 2: 415-18 (1970)
234

16-2331 TEMPERATURE AND PRESSURE DEPENDENCE OF THE DIELECTRIC CONSTANT AND SPONTANEOUS
POLARIZATION OF FERROELECTRIC POTASSIUM NITRATE AND SODIUM NITRITE.
KEONG JT + EMRICK RM
J PHYS CHEM SOLIDS 32: 2593-603 (1971)

16-2332 INFLUENCE OF THE CONDITIONS OF CRYSTALLIZATION ON THE FORM OF CRYSTALS OF POTASSIUM
NITRATE. (CRYSTAL GROWTH)
KHAMSKIJ EV + DVEGUBSKIJ NS
J APPL CHEM USSR 44(3): 476-8 (1971)
254

16-2333 CRYSTAL STRUCTURE OF FERROELECTRIC GLYCINE SILVER NITRATE.
KRISHNAN RS + MOHANA RAO JK + VISWAMITRA MA
P298-300 OF INT MEETING ON FERROELECTRICITY, 2ND, 1969,, KYOTO, PROC,
(J PHYS SOC JAP, VOL 28 SUPPL). PHYS SOC JAP, 1970, 459P.

16-2334 PHONON DISPERSION IN FERROELECTRIC POTASSIUM NITRATE.
KRISHNAN RS + HARIDASAN TM
INDIAN J PURE APPL PHYS 10: 399-401 (1972)

16-2335 LUMINESCENCE OF POTASSIUM NITRATE IRRADIATED WITH GAMMA-RAYS.
KULYUPIN YUA + YATSENKO AF
SOV PHYS SOLID STATE 5: 2443-5 (1964)

16-2336 INFRARED ABSORPTION IN GAMMA-RAY IRRADIATED SINGLE CRYSTAL OF SODIUM NITRITE.
LEE MN + KWUN SI
J KOREAN PHYS SOC 3: 9-14 (1970)
230

16-2337   TEMPERATURE DEPENDENCE OF THE ELECTRIC FIELD GRADIENT IN FERROELECTRIC SODIUM NITRITE.
           LUNDIN AG + GABUDA SP
           SOV PHYS SOLID STATE 5: 1467-69 (1964)

16-2338   DIELECTRIC MEASUREMENTS OF AMMONIUM NITRATE.
           MAKOSZ JJ + GONSIOR A
           ACTA PHYS POLON A 39: 371-3 (MAR 1971)

16-2339   DIELECTRIC STUDIES OF PHASE 2 OF AMMONIUM NITRATE.
           MAKOSZ JJ
           ACTA PHYS POLON A 41: 63-75 (JUL 1972)

16-2340   DIELECTRIC AND ELECTRICAL CONDUCTIVITY STUDIES IN POTASSIUM NITRITE AND POTASSIUM
           NITRATE.
           MANSINGH A + SMITH AM
           J PHYS D APPL PHYS 4: 560-7 (APR 1971)

16-2341   DIELECTRIC DISPERSION IN THE PARAELECTRIC PHASE OF POTASSIUM NITRATE.
           MANSINGH A + SMITH AM
           J PHYS D 4: 1792-6 (1971)
                247

16-2342   POLARIZED INFRARED AND RAMAN SPECTRA OF SINGLE CRYSTAL CESIUM NITRITE.
           MELVEGER AJ + KHANNA RK + LIPPINCOTT ER
           J CHEM PHYS 52(5): 2747-51 (1970)
                198

16-2343   X-RAY CRITICAL SCATTERING OF SODIUM NITRITE AT 160 DEG C AND AT 215 DEG C.
           MENDIOLA J + CANUT ML + AMOROS JL
           ACTA CRYSTALLOGR 16(SUPPL): A190 (1963) (ABSTRACT)

16-2344   EFFECT OF HYDROSTATIC PRESSURE ON FERROELECTRICITY OF POTASSIUM NITRATE.
           MIDORIKAWA M + ISHIBASHI Y + TAKAGI Y
           J PHYS SOC JAP 30: 449-52 (FEB 1971)

16-2345   PHASE TRANSITIONS IN POTASSIUM NITRATE NITRITE MIXED CRYSTALS (GROWTH BY DOUBLE
           BRIDGMAN TECHNIQUE).
           MIDORIKAWA M + TAKAGI Y + ISHIBASHI Y
           J PHYS SOC JAP 28(4): 1001-5 (1970)
                189

16-2346   THERMAL EXPANSION IN FERROELECTRIC SILVER SODIUM DINITRITE.
           MIKI H + MAKITA Y + GESI K
           J PHYS SOC JAP 30: 1512 (MAY 1971)

16-2347   DIELECTRIC AND X-RAY STUDIES ON THE FERROELECTRIC PHASE TRANSITION OF GLYCINE SILVER
           NITRATE.
           MITANI S
           J PHYS SOC JAP 19: 481-6 (1964)

16-2348   CRYSTAL STRUCTURE OF GLYCINE SILVER NITRATE.
           MOHANA RAO JK + VISWAMITRA MA
           ACTA CRYSTALLOGR B 28: 1484-96 (1972)

16-2349   DOMAIN ETCH PATTERNS ON THE FACES PARALLEL TO THE FERROELECTRIC-AXIS OF SODIUM NITRATE.
           MORIMOTO S + HAMANO K
           JAP J APPL PHYS 9(3): 268-73 (1970)
                190

16-2350   INFLUENCE OF THE ELECTRICAL BOUNDARY CONDITIONS ON THE NMR RELAXATION RATE OF SODIUM-23
           IN SODIUM NITRITE.
           MULLER D + PETERSSON J
           J PHYS (PARIS) 33, SUPPL 4, COLLOQ 2: 193-4 (1972)
                294

16-2351   THERMAL EXPANSION IN SODIUM NITRITE CRYSTALS.
           MURUYAMA N + SAWADA S
           J PHYS SOC JAP 20(5): 811-16 (1965)

16-2352   INFRARED SPECTRA OF POTASSIUM NITRATE AND SODIUM NITRITE AT PHASE TRANSITIONS.
          MYASNIKOVA TP + EVSEEVA RYA
          IZV VUZ FIZ 1970(6): 103-5 (IN RUSSIAN)
          (TRANSLATION TO APPEAR IN SOV PHYS J)
               CA 73: 82225W (1970)
               231

16-2353   ANALYSIS OF LOW FREQUENCY ELECTROOPTIC RESPONSE IN SODIUM NITRITE.
          NAKAMURA T + JOHNSON AR
          P82-3 OF INT MEETING ON FERROELECTRICITY, 2ND, 1969, KYOTO, PROC,
          (J PHYS SOC JAP, VOL 28 SUPPL). PHYS SOC JAP, 1970, 459P.

16-2354   MEASUREMENT OF MICROWAVE DIELECTRIC CONSTANTS OF FERROELECTRICS. PART-2: DIELECTRIC
          CONSTANTS AND DIELECTRIC LOSSES OF SODIUM NITRITE AND GLYCINE SULFATE.
          NAKAMURA E
          J PHYS SOC JAP 17: 961-6 (1962)

16-2355   CONSTRAINED REFINEMENT TECHNIQUES APPLIED TO THE STRUCTURE OF AMMONIUM HYDROGEN SULFATE
          ABOVE THE FERROELECTRIC TRANSITION.
          NELMES RJ
          ACTA CRYSTALLOGR A 28: 445-54 (SEP 1972)

16-2356   DIELECTRIC BEHAVIOR OF (FERROELECTRIC) FILMS OF VACUUM DEPOSITED POTASSIUM NITRATE.
          NOLTA JP + SCHUBRING NW + DORK RA
          P237-53 OF VRATNY F(ED), THIN FILM DIELECTRICS, ELECTROCHEM SOC,
          NY, 1969, 669P.

16-2357   ELASTIC CONSTANTS AND ULTRASONIC ABSORPTION OF SODIUM NITRITE SINGLE CRYSTALS.
          OTA K + ISHIBASHI Y + TOKAGI Y
          J PHYS SOC JAP 29(6): 1545-51 (1970)
               222

16-2358   ELECTRON PARAMAGNETIC RESONANCE OF MANGANESE(2+) IN THE FERROELECTRIC PHASE OF SODIUM
          NITRITE.
          PANDEY SD + UPRETI GC
          FERROELECTRICS 2: 155-61 (1971)
               235

16-2359   FERROELECTRIC PHASE TRANSITION IN SODIUM NITRITE AS STUDIED BY EPR (PHASE TRANSITION AT
          163 DEG C).
          PANDEY SD + UPRETI GC
          PHYS STATUS SOLIDI A 1(2): K69-K72 (1970)
               223

16-2360   EFFECT OF HYDROSTATIC PRESSURE ON THE FERROELECTRIC PHASE TRANSITION OF AMMONIUM
          SULFATE.
          POLANDOV IN + LEVINA ME + MYLOV VP
          RUSS J PHYS CHEM 46: 281-2 (FEB 1972)

16-2361   PHASE DIAGRAMS OF SODIUM NITRITE AND POTASSIUM NITRITE TO 40 KBAR.
          RAPOPORT E
          J CHEM PHYS 45(8): 2721-8 (1966)

16-2362   QUADRUPOLE SPIN PHONON RELAXATION AND FERROELECTRIC TRANSITION. (IN SODIUM NITRITE)
          RIGAMONTI A
          PHYS REV LETT 19(8): 436-9 (1967)

16-2363   CALCULATION OF THE VARIATION OF THE RAMAN FREQUENCIES IN SODIUM NITRITE AT ORDINARY
          TEMPERATURES AS A FUNCTION OF CRYSTAL ORIENTATION
          SADOC A
          COMPT REND B 272: 1439-41 (1971) (IN FRENCH)

16-2364   CRYSTAL DYNAMICS AND THE FERROELECTRIC PHASE TRANSITION OF SODIUM NITRITE (STUDIED BY
          COHERENT NEUTRON INELASTIC SCATTERING).
          SAKURAI M + COWLEY RA + DOLLING G
          J PHYS SOC JAP 28(6): 1426-45 (1970)
               197

16. NITRATES AND NITRITES

16-2365   NEUTRON DIFFRACTION STUDIES OF FERROELECTRIC POLARIZATION REVERSALS IN SODIUM NITRITE.
           SHIBUYA I + IWATA Y + KOYANO N + FUKUI S + MITANI S + TOKUNAGA M
           P281-3 OF INT MEETING ON FERROELECTRICITY, 2ND, 1969, KYOTO, PROC,
           (J PHYS SOC JAP, VOL 28 SUPPL). PHYS SOC JAP, 1970, 459P.

16-2366   X-RAY STUDY ON THE DISORDERED STRUCTURE ABOVE THE FERROELECTRIC CURIE POINT IN
           POTASSIUM NITRATE.
           SHINNAKA Y
           J PHYS SOC JAP 17: 820-8 (1962)

16-2367   ELECTRICAL PROPERTIES OF RHOMBOHEDRAL POTASSIUM NITRATE.
           SIOUFFI JC + CERISIER P
           COMPT REND B 274: 754-7 (1972) (IN FRENCH)
                CHEM ABSTR 76: 145949 (1972)
                263

16-2368   TWO MODES OF ELECTRICAL CONDUCTIVITY IN POTASSIUM NITRATE.
           SIOUFFI JC
           COMPT REND B 275: 37-40 (1972) (IN FRENCH)

16-2369   PHASE TRANSFORMATIONS IN POTASSIUM NITRATE AND POTASSIUM NITRITE. (DOES NOT SEEM
           FERROELECTRIC- PARAELECTRIC, ABSTRACT)
           SMITH AM + MANSINGH A + MCLAY DB
           BULL AMER PHYS SOC 15(6): 774-5 (1970)
                191

16-2370   PROPERTIES OF SILVER SODIUM DINITRITE AND SILVER SODIUM NITRITE. (DIFFERENTIAL THERMAL
           ANALYSIS)
           SOBOLEVA LV + KOSYRBASOVA MG + OGADZHANOVA VV
           INORG MATER 8: 463-5 (1972)

16-2371   EVIDENCE OF ADDITIONAL DISORDER IN THE RHOMBOHEDRRAL FORM OF POTASSIUM NITRITE.
           STROMME KO
           ACTA CHEM SCAND 24: 1475-7 (1970)

16-2372   TOPOGRAPHIC STUDY ON FERROELECTRIC SODIUM NITRITE CRYSTALS. PART-1: STRUCTURE OF 180
           DEGREE DOMAIN WALLS.
           SUZUKI S + TAKAGI M
           J PHYS SOC JAP 30: 188-202 (JAN 1971)

16-2373   TEMPERATURE DEPENDENCE OF FAR INFRARED REFLECTIVITY SPECTRA OF SODIUM NITRITE CRYSTALS.
           (0.85 DEG K, 30-200 DEG C)
           SUZUKI K + SAWADA S + SUGAWARA F + NAKAMURA T
           P223-4 OF INT MEETING ON FERROELECTRICITY, 2ND, 1969, KYOTO, PROC,
           (J PHYS SOC JAP, VOL 28 SUPPL). PHYS SOC JAP, 1970, 459P.

16-2374   TOPOGRAPHIC STUDY OF FERROELECTRIC SODIUM NITRITE CRYSTALS. PART-2: MECHANISM OF
           POLARIZATION REVERSAL.
           SUZUKI S + TAKAGI M
           J PHYS SOC JAP 32(5): 1302-12 (1972)
                278

16-2375   APPLICATION OF THE X-RAY DIFFRACTION TOPOGRAPHIC METHOD TO FERROELECTRIC SODIUM NITRITE
           AND TGS CRYSTALS. (DOMAIN BOUNDARIES AND METASTABLE STRUCTURES)
           TAKAGI M + SUZUKI S + WATANABE H
           P369-72 OF INT MEETING ON FERROELECTRICITY, 2ND, 1969, KYOTO, PROC,
           (J PHYS SOC JAP, VOL 28 SUPPL). PHYS SOC JAP, 1970, 459P.

16-2376   DIELECTRIC ANOMALIES OF SODIUM NITRATE ABOVE THE FERROELECTRIC CURIE TEMPERATURE.
           TAKAGI Y + GESI K
           J PHYS SOC JAP 19: 142-3 (1964)

16-2377   LATTICE DYNAMICS MODEL OF POTASSIUM NITRATE IN ITS FERROELECTRIC PHASE.
           TENG MK
           PHYS STATUS SOLIDI 40: 639-46 (1970)
                215

16-2378   PRESSURE INDUCED FERROELECTRIC PHASE TRANSITION IN POTASSIUM NITRATE.
           TENG MK + BALKANSKI M + MOUREY JF
           SOLID STATE COMMUN 9: 465-9 (1971)

16-2379   STRUCTURE OF THE DISORDERED PHASE OF POTASSIUM NITRATE.
           TENG MK
           J PHYS (PARIS) 31: 771-7 (1970) (IN FRENCH)
               225

16-2380   X-RAY STUDY OF PHASE TRANSITION IN SODIUM NITRATE.
           TERAUCHI H + YAMADA Y
           J PHYS SOC JAP 33: 446-54 (AUG 1972)

16-2381   POLARIZATION REVERSAL IN A WIDE RANGE OF IMPRESSING RATE OF FIELD IN SODIUM NITRITE.
           TOKUGAWA Y
           J PHYS SOC JAP 33(2): 415-23 (AUG 1972)
               278

16-2382   STUDY OF THE CRITICAL SLOWING DOWN OF THE DIELECTRIC RELAXATION IN SILVER SODIUM
           DINITRITE BY THERMAL NOISE MEASUREMENTS.
           TOMINAGA Y + WADA S + IIDA S
           J PHYS SOC JAP 32: 1675 (1972)
               278

16-2383   ALLOWED AND FORBIDDEN HYPERFINE TRANSITIONS IN THE ELECTRON PARAMAGNETIC RESONANCE OF
           MANGANESE(2+) DOPED IN FERROELECTRIC SODIUM NITRITE.
           UPRETI GC
           PHYS STATUS SOLIDI B, BASIC RES 54(2): 387-92 (1972)

16-2384   OPTICAL SECOND HARMONIC GENERATION DURING FERROELECTRIC POLARIZATION REVERSAL. (SODIUM
           NITRITE)
           VOGT H + WEINMANN D
           PHYS STATUS SOLIDI A 14: 501-10 (1972)
               284

16-2385   SECOND HARMONIC GENERATION IN FERROELECTRIC SODIUM NITRITE.
           VOGT H + HAPP H
           P281-5 OF EUROPEAN MEETING ON FERROELECTRICITY, 1969, SAARBRUCKEN,
           MUSER HE AND PETERSSON J (EDS), WISSENSCH VERLAGSGES, STUTTGART,
           1970, 396P.
           PHYS STATUS SOLIDI A 1: 439-50 (1970)

16-2386   TEMPERATURE DEPENDENCE OF THE OPTICAL NONLINEAR SUSCEPTIBILITY OF SODIUM NITRITE.
           VOGT H + HAPP H
           PHYS STATUS SOLIDI B 44: 207-16 (1971)
               234

16-2387   TEMPERATURE DEPENDENCE OF ABSORPTION OF SODIUM NITRITE IN THE FAR INFRARED. (350-1100
           MICRONS)
           VOGT H + HAPP H
           PHYS STATUS SOLIDI 30(1): 67-72 (1968)

16-2388   ANHARMONICITY IN SODIUM NITRATE AND SODIUM NITRITE.
           WEGDAM GH + VAN DER ELSKEN J
           P469-72 OF NUSIMOVICI M(ED), PHONONS, INT CONF, RENNES, 1971 PROC,
           FLAMMARION, PARIS, 1971, 502P.

16-2389   DIPOLE CORRELATIONS IN MOLTEN ALKALI METAL NITRATES AND SODIUM NITRITE.
           WEGDAM GH + TEBEEK JB + VAN DER LINDEN H + VAN DER ELSKEN J
           J CHEM PHYS 55: 5207-14 (1971)

16-2390   GRUNEISEN PARAMETERS AROUND THE PHASE TRANSITIONS IN SODIUM NITRITE AND SODIUM NITRATE.
           WEGDAM GH + VAN DER ELSKEN J
           SOLID STATE COMMUN 9: 1867-9 (1971)
               255

16-2391   CHANGES IN ELECTRICAL CONDUCTIVITY ACCOMPANYING THE POLYMORPHIC TRANSITIONS IN
           POTASSIUM NITRATE AT 1 ATMOSPHERE.
           WEIDENTHALER P
           J PHYS CHEM SOLIDS 25: 1491-3 (1964)

16-2392   MODULATION OF THE MILLIMETER WAVE ABSORPTION IN SODIUM NITRITE BY POLARIZATION
           REVERSAL.
           WEINMANN D + VOGT H
           PHYS STATUS SOLIDI A, APPL RES 11(1): 75-80 (1972)

16-2393   TEMPERATURE DEPENDENCE OF THE SODIUM-23 NUCLEAR QUADRUPOLE COUPLING CONSTANT IN SODIUM
           NITRITE.
           WEISS A + BIEDENKAPP D
           Z NATURFORSCH A 17: 794-8 (1962) (IN GERMAN)

16-2394   THIN FILM NUCLEATION ON FERROELECTRIC SUBSTRATES. (TELLURIUM ON TRIGLYCINE SULFATE AND
           SODIUM NITRITE SURFACES) (DECORATION METHOD OF DOMAIN STUDY)
           WEIDMANN EJ + ANDERSON JC
           THIN SOLID FILMS 7: 27-39 (1971)
               232

16-2395   ELECTRIC FIELD GRADIENT AND LONG RANGE ORDER PARAMETER OF FERROELECTRIC SODIUM NITRITE.
           (AT THE SODIUM-23 NUCLEUS)
           YAGI T + TATSUZAKI I
           J PHYS SOC JAP 32(3): 750-6 (1972)
               278

16-2396   NUCLEAR MAGNETIC RESONANCE STUDY ON SODIUM-23 IN SODIUM NITRITE IN THE VICINITY OF THE
           PHASE TRANSITION TEMPERATURES.
           YAGI T + TATSUZAKI I + TODO I
           J PHYS SOC JAP 28(2): 321-6 (1970)
               189

16-2397   PHASE TRANSITION IN SODIUM NITRITE.
           YAMADA Y + SHIBUYA I + HOSHINO S
           J PHYS SOC JAP 18: 1594-1603 (1963)

16-2398   VACUUM ULTRAVIOLET ABSORPTION OF THE ALKALI NITRITES AND NITRATES.
           YAMASHITA H + KATO R
           J PHYS SOC JAP 26: 1561 (1969)
               157

16-2399   VACUUM ULTRAVIOLET ABSORPTION (5-23 EV AT ROOM AND LIQUID NITROGEN TEMPERATURES IN
           ALKALI NITRITES (SODIUM, POTASSIUM) AND ALKALI NITRATES. (LITHIUM, SODIUM, RUBIDIUM,
           CESIUM)
           YAMASHITA H + KATO R
           J PHYS SOC JAP 29(6): 1557-61 (1970)
               222

16-2400   X-RAY CRITICAL SCATTERING IN TRIGLYCINE SULFATE AND SODIUM NITRATE. (ORDER DISORDER
           TRANSITION, FERROELECTRIC AND NONFERROELECTRIC)
           YAMADA Y + FUJII Y + TERAUCHI H
           P274-7 OF INT MEETING ON FERROELECTRICITY, 2ND, 1969, KYOTO, PROC.
           (J PHYS SOC JAP, VOL 28 SUPPL). PHYS SOC JAP, 1970, 459P.

16-2401   FERROELECTRICITY IN AMMONIUM(X) POTASSIUM(1-X) NITRATE MIXED CRYSTAL.
           YANAGI T + SAWADA S
           J PHYS SOC JAP 18: 1228-9 (1963)

16-2402   SOME (DIELECTRIC MEASUREMENTS AND INFRARED ABSORPTION) STUDIES ON FERROELECTRICITY IN
           POTASSIUM NITRATE AND RELATED COMPOUNDS. (POTASSIUM(1-X) AMMONIUM(X) NITRATE)
           YANAGI T
           J PHYS SOC JAP 20: 1351-65 (1965)

16-2403   STUDY OF PHASE TRANSITION OF SODIUM NITRITE BY SECOND HARMONIC GENERATION. (TEMPERATURE
           DEPENDENCE)
           YANAGI T + IIO K + HANADATE H + SAWADA S
           P78-80 OF INT MEETING ON FERROELECTRICITY, 2ND, 1969, KYOTO, PROC.
           (J PHYS SOC JAP, VOL 28 SUPPL). PHYS SOC JAP, 1970, 459P.

16-2404   PARAMAGNETIC RESONANCE STUDY OF PRODUCTION OF NITROGEN DIOXIDE AND NITRITE ION IN
           IRRADIATED POTASSIUM NITRATE.
           ZELDES H + LIVINGSTON R
           J CHEM PHYS 37: 3017-19 (1962)

16-2405   PARAMAGNETIC SPECIES IN IRRADIATED NITRATE.
           ZELDES H
           P764-84 OF INT CONF ON PARAMAGNETIC RESONANCE, 1ST, 1962,
           HEBREW UNIV OF JERUSALEM, PROC. (PARAMAGNETIC RESONANCE), LOW W(ED),
           ACADEMIC PRESS, NY, 1963.

17. POTASSIUM DIHYDROGEN PHOSPHATE (KDP) AND RELATED PHOSPHATES

AUTOFOCUSING OF LASER RADIATION IN ACTIVE MATERIALS AND NONLINEAR CRYSTALS. (KDP, ADP, RUBIDIUM DIHYDROGEN PHOSPHATE, LITHIUM NIOBATE, LITHIUM TANTALATE, QUARTZ)
ZVEREV GM + LEVCHUK EA + MALDUTIS EK + PASHKOV VA
SOV PHYS SOLID STATE 11(4): 865-6 (1969)

17-2406   THE RADICAL IN GAMMA-IRRADIATED KDP.
ABE R + NAITO M
REP TOYODA PHYS CHEM RES INST 24: 1-4 (1971) (IN JAPANESE)

17-2407   DIFFRACTION OF POLARIZED LIGHT BY ULTRASONIC WAVES IN POTASSIUM DIHYDROGEN PHOSPHATE CRYSTALS.
ADRIANOVA II + VOLKONSKII VB + KOROLEV YUG
OPT SPECTROSC 30(6): 601-3 (JUN 1971)
264

17-2408   NMR PROTON SECOND MOMENTS IN POTASSIUM DIHYDROGEN PHOSPHATE TYPE CRYSTALS.
ADRIAENSSENS GJ + BJORKSTAM JL
J CHEM PHYS 55: 1137-9 (1971)

17-2409   TEMPERATURE DEPENDENT RAMAN SPECTRA OF POTASSIUM DIHYDROGEN PHOSPHATE, POTASSIUM DIDEUTERIUM PHOSPHATE, POTASSIUM DIHYDROGEN ARSENATE AND AMMONIUM DIHYDROGEN PHOSPHATE.
AGRAWAL DK + PERRY CH
P429-35 OF INT CONF ON LIGHT SCATTERING IN SOLIDS, 2ND PROC.
BALKANSKI M(ED), PARIS, 1971, FLAMMARION, PARIS, 1971, 518P.

17-2410   ELASTIC PROPERTIES OF AMMONIUM DIHYDROGEN PHOSPHATE AND THE LAVAL RAMAN ELASTICITY THEORY.
ALEKSANDROV KS + RYABINKIN LN
SOV PHYS DOKL 7: 99-101 (1962)

17-2411   GROWTH OF AMMONIUM DIHYDROGEN PHOSPHATE SINGLE CRYSTAL FROM SOLUTIONS.
ALEXANDRU HV + HANGEA N
AN UNIV BUCURESTI FIZ 18: 35-42 (1969)
CHEM ABSTR 74: 92031Q (1971)
238

17-2412   KINETICS OF GROWTH AND DISSOLUTION OF AMMONIUM DIHYDROGEN PHOSPHATE CRYSTALS IN SOLUTION.
ALEXANDRU HV
J CRYST GROW 10: 151-7 (1971)
239

17-2413   PRODUCTION OF LARGE SEED CRYSTALS OF KDP.
ALEKSANDROVA MV + MOKIEVSKII VA + SOBOLEEV CHS
INORG MATER 8: 341-2 (1972)

17-2414   EFFECTS OF DEUTERATION ON THE SPECIFIC HEAT OF ADP CRYSTALS.
AMIN M + STRUKOV BA
SOV PHYS SOLID STATE 12(7): 1616-18 (1971)
221

17-2415   HEAT CAPACITY OF RUBIDIUM DIHYDROGEN PHOSPHATE SINGLE CRYSTALS.
AMIN M + STRUKOV BA
SOV PHYS SOLID STATE 10(10): 2498-500 (1969)
141

FRACTURE OF NONLINEAR CRYSTALS (KDP AND LITHIUM NIOBATE) BY RADIATION FROM A RUBY LASER.
ANANIN OB + BYKOVSKII YUA + PETROVSKII AN + REZ IS
SOV PHYS TECH PHYS 17: 659-61 (1972)

17. POTASSIUM DIHYDROGEN PHOSPHATE (KDP) AND RELATED PHOSPHATES

17-2416   INCOHERENT NEUTRON SCATTERING FROM HYDROGEN BOND IN KDP AND ADP.
          ANTONINI M + SOSNOWSKA I + VALACCHINO M
          J PHYS (PARIS) 33, SUPPL 4, COLLOQ 2: 83-4 (1972) (IN FRENCH)
              294

17-2417   NEUTRON MEASUREMENT IN TWO SYSTEMS OF HYDROGEN BONDS. (KDP AND POTASSIUM BICARBONATE)
          ARSIC-ESKINJA M
          TECH HOCHSCHULE, AACHEN, PHD THESIS, 1972, 68P. (IN GERMAN)
              288

17-2418   NEUTRON MEASUREMENTS ON THE FERROELECTRIC PHASE TRANSFORMATION IN POTASSIUM DIHYDROGEN
          PHOSPHATE.
          ARSIC-ESKINJA M + GRIMM H + STILLER H
          P825-40 OF SYMP ON NEUTRON INELASTIC SCATTERING, 5TH PROC,
          GRENOBLE, 1972, IAEA, VIENNA, 1972.

17-2419   MACROSCOPIC STIMULATED POLARIZATION ECHO IN FERROELECTRICS. (POTASSIUM DIHYDROGEN
          PHOSPHATE)
          ASADULLIN YY + KOPVILLEM UKH + OSIPOV VN + SMOLYAKOV BP + SHARIPOV RZ
          SOV PHYS SOLID STATE 13(9): 2330-1 (MAR 1972)
              261

17-2420   GROWTH OF ELECTROOPTICAL CRYSTALS FROM AQUEOUS SOLUTIONS (APPARATUS, GROWTH OF ADP,
          KDP, HYDROIODIC ACID, AND AMMONIUM OXALATE MONOHYDRATE).
          ASCOLI A + KROSI M
          ENERG NUCL (ITALY) 17(9): 545-50 (1970)
              225

          X-RAY TOPOGRAPHICAL STUDY OF THE CONTRAST AT FERROELECTRIC DOMAIN BOUNDARIES (DOPED AND
          UNDOPED TGS, LITHIUM NIOBATE, KDP, ABSTRACT ONLY).
          AUTHIER A + PETROFF JF
          P373 OF INT MEETING ON FERROELECTRICITY, 2ND, 1969, KYOTO, PROC,
          (J PHYS SOC JAP, VOL 28 SUPPL). PHYS SOC JAP, 1970, 459P.

17-2421   LIGHT ABSORPTION IN POTASSIUM DIHYDROGEN PHOSPHATE SINGLE CRYSTALS.
          AVERBACH VS + BATYTEVA IA + BESPALOV VI
          IZV VUZ RADIOFIZ 13(2): 307-9 (1970) (IN RUSSIAN)
          (TRANSLATION TO APPEAR IN RADIOPHYS QUANTUM ELECTRON)
              198

17-2422   ULTRAVIOLET REFLECTION AND ABSORPTION OF POTASSIUM DIHYDROGEN PHOSPHATE AND AMMONIUM
          DIHYDROGEN PHOSPHATE CRYSTALS. (REFLECTANCE, REFRACTIVE INDICES)
          BALDINI G + COTTINI M + GRILLI E
          SOLID STATE COMMUN 11: 1257-60 (1972)
              281

17-2423   FAR INFRARED DIELECTRIC MEASUREMENTS ON POTASSIUM DIHYDROGEN PHOSPHATE, TRIGLYCINE
          SULFATE, AND RUTILE.
          BARKER AS + TINKHAM M
          J CHEM PHYS 38: 2257-64 (1963)

17-2424   INTERNAL STRESS AND THE FERROELECTRIC TRANSITION IN POTASSIUM DIHYDROGEN PHOSPHATE.
          BARTIS FJ
          PHYS STATUS SOLIDI B 45(2): K169-71 (1971)

          NEODYMIUM YAG LASER IRRADIATION INDUCED DAMAGE TO LITHIUM NIOBATE AND KDP.
          BASS M
          IEEE J QUANTUM ELECTRON 7: 350-9 (1971)

17-2425   RAMAN SPECTRA OF POLYMORPHIC MODIFICATIONS OF POTASSIUM DIHYDROGEN PHOSPHATE IN AN
          ELECTRIC FIELD.
          BATSANOV SS + ISUPOVA LA + SLYUDKIN OP
          J STRUCT CHEM 12: 305-6 (1971)

17-2426   CRYSTAL PERFECTION OF AMMONIUM DIHYDROGEN PHOSPHATE.
          BELT RF
          J APPL PHYS 35: 3063-4 (1964)

17-2427   EFFECT OF ANNEALING ON OPTICAL ANOMALIES OF POTASSIUM DIHYDROGEN PHOSPHATE CRYSTALS.
          BELYUSTIN AV + STEPANOVA NS + FRIDMAN SS
          INORG MATER 8: 1425-6 (1972)
              292

17. POTASSIUM DIHYDROGEN PHOSPHATE (KDP) AND RELATED PHOSPHATES

17-2428 TRAPPING OF PARAMETRICALLY AMPLIFIED LIGHT WAVES IN A KDP CRYSTAL.
BELYAEV YUN + FREIDMAN GI
JETP LETT 15: 165-8 (1972)

17-2429 ELECTRONIC STUDIES OF POTASSIUM DIHYDROGEN PHOSPHATE. (ELECTROCALORIC STUDIES,
SPONTANEOUS POLARIZATION)
BENEPE JW + REESE W
PHYS REV B SOLID STATE 3(9): 3032-9 (1 MAY 1971)
        235

17-2430 MEASUREMENTS OF THE SPONTANEOUS POLARIZATION IN POTASSIUM DIHYDROGEN PHOSPHATE.
BENEPE JW
NAVAL POSTGRAD SCHOOL, PHD THESIS, 1970, 91P. AD717598

17-2431 RAMAN SPECTRA OF ADP IN THE PARAELECTRIC AND ANTIFERROELECTRIC PHASES.
BENOIT JP
COMPT REND B 273: 483-5 (1971) (IN FRENCH)

17-2432 LOW TEMPERATURE INFRARED ABSORPTION SPECTRA OF AMMONIUM DIHYDROGEN ORTHOPHOSPHATE
(4000-10000 PER CM) AND HARMONIC VIBRATION FREQUENCIES OF THE HYDROGEN BONDS.
BERNARD MP
COMPT REND 256: 2812-5 (1963) (IN FRENCH)
        CHEM ABSTR 58: 13318G (1963)

17-2433 ORIGIN OF THE SPONTANEOUS POLARIZATION AND THE ISOTOPE EFFECTS IN FERROELECTRIC
POTASSIUM DIHYDROGEN PHOSPHATE. (INTERBAND INTERACTION AND DIPOLE INSTABILITY)
BERSUKER IB + VEKHTER BG + MUZALEVSKII AA
PHYS STATUS SOLIDI B 45: K25-7 (1971)
        235

17-2434 POLYTHERM OF SOLUBILITY OF THE AMMONIUM DIHYDROGEN PHOSPHATE- POTASSIUM DIHYDROGEN
PHOSPHATE- WATER SYSTEM.
BERGMAN AG + GLADKOVSKAYA AA + GALUSHKINA RA
RUSS J INORG CHEM 17: 1067-8 (1972)

17-2435 1.06 MICRON ABSORPTION COEFFICIENTS OF DEUTERATED KDP WITH 70-100% DEUTERATION.
BICHARD VM + DAVIES PH + HULME KF
ELECTRON LETT 8: 147-8 (1972)

RELATIVE MEASUREMENT OF THE OPTICAL NONLINEARITIES OF POTASSIUM DIHYDROGEN PHOSPHATE,
AMMONIUM DIHYDROGEN PHOSPHATE, LITHIUM NIOBATE, AND ALPHA- IODIC ACID.
BJORKHOLM JE
IEEE J QUANTUM ELECTRON 5(5): 260 (1969)

17-2436 ABSENCE OF A LINE WIDTH TRANSITION FOR THE PROTONS IN POTASSIUM DIHYDROGEN PHOSPHATE
TYPE CRYSTALS.
BJORKSTAM JL + JONES ED + SILSBEE HB + UEHLING EA
J CHEM PHYS 37: 469-70 (1962)

17-2437 EFFECT OF DEUTERON SUBSTITUTION ON DOMAIN DYNAMICS IN POTASSIUM DIHYDROGEN PHOSPHATE.
BJORKSTAM JL + OETTEL RE
P91-6, VOL 2 OF DVORAK V + FOUSKOVA A + GLOGAR P (EDS),
INT MEETING ON FERROELECTRICITY, PRAGUE, 1966, PROC.
INST PHYS CZECH ACAD SCI, PRAGUE, 1966.

PROTON LATTICE INTERACTIONS IN HYDROGEN BONDED FERROELECTRIC CRYSTALS. (THEORY, KDP
TYPE)
BLINC R + RIBARIC M
PHYS REV 130: 1816-21 (1963)

17-2438 DEUTERON MAGNETIC RESONANCE AND RELAXATION IN FERROELECTRIC POTASSIUM DIDEUTERIUM
PHOSPHATE, POTASSIUM DIDEUTERIUM ARSENATE, AND CESIUM DIDEUTERIUM ARSENATE.
BLINC R + STEPISNIK J + JAMSIK-VILFAN M + ZUMER S
J CHEM PHYS 54: 187-95 (1971)

17-2439 RUBIDIUM-87 AND ARSENIC-75 QUADRUPOLAR COUPLING IN FERROELECTRIC RUBIDIUM DIHYDROGEN
PHOSPHATE AND CESIUM DIHYDROGEN ARSENATE.
BLINC R + O'REILLY DE + PETERSON EM
PHYS REV B 1: 1953-7 (1970)

17-2440   SPIN LATTICE RELAXATION BY THE FERROELECTRIC MODE IN POTASSIUM DIHYDROGEN PHOSPHATE.
           BLINC R + ZUMER S
           PHYS REV LETT 21(14): 1004-6 (30 SEP 1968)
                143

17-2441   ULTRASLOW HYDROGEN MOTION IN POTASSIUM DIHYDROGEN PHOSPHATE AND SODIUM TRIHYDROGEN
           SELENATE TYPE CRYSTALS.
           BLINC R + PIRS J
           J CHEM PHYS 54: 1535-9 (1971)

17-2442   VANISHING OF FERROELECTRICITY IN POTASSIUM DIHYDROGEN PHOSPHATE AT HIGH PRESSURES.
           BLINC R + SVETINA S + ZEKS B
           SOLID STATE COMMUN 10: 387-9 (1972)
                262

17-2443   PHOTOELASTIC EFFECT IN 45 DEGREE X-CUTS OF POTASSIUM AND AMMONIUM DIHYDROGEN PHOSPHATE.
           BLOKH OG + LUSTIV-SHUMSKII LF
           SOV PHYS SOLID STATE 12(1): 256-57 (1970)
                202

17-2444   DIFFRACTION OF A CONTINUOUS LIGHT WAVE IN KDP.
           BOERSCH H + EICHLER H + FERY H
           PHYS LETT A 31: 468-9 (1970)

17-2445   THERMAL EXPANSION OF ADP AND DEUTERATED ADP IN THE REGION OF ANTIFERROELECTRIC PHASE
           TRANSITION.
           BOIKO AA + GOLOVNIN VA
           SOV PHYS CRYSTALLOGR 15(1): 153-55 (1970)
                206

17-2446   SOLUBILITY IN THE GROWTH OF ELECTROOPTIC CRYSTALS ADP AND KDP.
           BOLOGNESI GP + DROSI M
           ALTA FREQ 39(7): 641-3 (1970) (IN ITALIAN)
                219

17-2447   FERROELECTRIC DOMAINS OF POTASSIUM DIHYDROGEN PHOSPHATE. (93 REFS)
           BONARDEL J
           UNIV GRENOBLE PHD THESIS, 1971, 18P.
                254

17-2448   DISPLACEMENTS AND VELOCITIES OF THE DOMAIN WALLS IN KDP. EXISTENCE OF CRITICAL ELECTRIC
           FIELDS.
           BORNAREL J + LAJZEROWICZ J
           J PHYS (PARIS) 33, SUPPL 4, COLLOQ 2: 153-4 (1972)
                294

17-2449   ELECTRICAL AND OPTICAL INVESTIGATION OF FERROELECTRIC PROPERTIES OF POTASSIUM
           DIHYDROGEN PHOSPHATE.
           BORNAREL P + FOUSKOVA A + GUYON P + LAJZEROWICZ J
           P81-90, VOL 2 OF DVORAK V + FOUSKOVA A + GLOGAR P (EDS),
           INT MEETING ON FERROELECTRICITY, PRAGUE, 1966, PROC.
           INST PHYS CZECH ACAD SCI, PRAGUE, 1966.

17-2450   EXISTENCE OF DISLOCATIONS AT DOMAIN TIPS IN FERROELECTRIC CRYSTAL POTASSIUM DIHYDROGEN
           PHOSPHATE.
           BORNAREL J
           J APPL PHYS 43(3): 845-52 (MAR 1972)

17-2451   INTERDOMAIN AND DOMAIN EFFECT INTERACTIONS IN KDP (FERROELECTRIC DOMAINS).
           BORNAREL J + LAJZEROWICZ J
           FERROELECTRICS 4(3): 177-87 (1972)

17-2452   STRESS INDUCED BIREFRINGENCE IN AN ISOLATED AND A SHORTCIRCUITED POTASSIUM DIHYDROGEN
           PHOSPHATE CRYSTAL.
           BORNAREL J + FOUSEK J + GLOGAROVA M
           CZECH J PHYS B 22: 864-6 (1972)
                280

17-2453   VELOCITY OF PROPAGATION OF DOMAIN WALLS: INTERACTIONS BETWEEN DOMAINS. (IN KDP)
           BORNAREL J + LAJZEROWICZ J
           P360-1 OF INT MEETING ON FERROELECTRICITY, 2ND, 1969, KYOTO, PROC,
           (J PHYS SOC JAP, VOL 28 SUPPL). PHYS SOC JAP, 1970, 459P.

17. POTASSIUM DIHYDROGEN PHOSPHATE (KDP) AND RELATED PHOSPHATES

17-2454   MEASUREMENT OF THE ELASTIC CONSTANTS OF POTASSIUM DIHYDROGEN PHOSPHATE BY BRILLOUIN
DIFFUSION.
BOYER L + VACHER R
PHYS STATUS SOLIDI A 6: K105-8 (1971)
    244

17-2455   GROWTH OF POTASSIUM DIHYDROGEN PHOSPHATE SINGLE CRYSTALS IN GEL.
BREZINA B + HAVRANKOVA M
MATER RES BULL 6: 537-44 (1971)
    237

17-2456   INFLUENCE OF THE PHASE TRANSITION ON THE NORMAL VIBRATION MODES OF FERROELECTRIC
CRYSTALS AT LOW TEMPERATURES. (KDP TYPE CRYSTALS)
BREHAT F
UNIV NANCY PHD THESIS, 1972, 149P.
    299

17-2457   LOW FREQUENCY INFRARED ABSORPTION SPECTRA OF MONOCRYSTALS OF POTASSIUM DIHYDROGEN
PHOSPHATE AND RUBIDIUM DIHYDROGEN PHOSPHATE IN POLAR AND NONPOLAR PHASES.
BREHAT F + HADNI A + AUBRY A + ZANNE M
REV PHYS APPL, SUPPL J PHYS 7: 163-7 (SEP 1972) (IN FRENCH)
     SCI ABSTR A 76: 12136 (1973)
    286

17-2458   REGULAR BEHAVIOR OF SOLID SOLUTIONS IN POTASSIUM DIHYDROGEN(1-X) DIDEUTERIUM(X)
PHOSPHATE SINGLE CRYSTALS.
BREZINA B + FOUSKOVA A + SMUTNY F
PHYS STATUS SOLIDI A 11: K149-52 (1972)

17-2459   TRANSMISSION OF POTASSIUM DIHYDROGEN PHOSPHATE IN THE INFRARED REGION IN THE PARA- AND
FERROELECTRIC PHASES.
BREHAT F + HADNI A
COMPT REND B 271(23): 1137-40 (1970) (IN FRENCH)
    238

17-2460   CHARACTERISTICS OF AN ELECTROOPTICAL SHUTTER BASED ON Z CUT POTASSIUM DIDEUTERIUM
PHOSPHATE CRYSTALS.
BROVEEV SF + SAUKOV AI + UGODENKO AA
INSTRUM EXPER TECH 3: 829-30 (1972)

17-2461   LONG WAVELENGTH POLARIZATION FLUCTUATIONS IN ANTIFERROELECTRIC AMMONIUM DIHYDROGEN
PHOSPHATE.
BROBERG TW + SHE CY + WALL LS
PHYS REV B SOLID STATE 6(9): 3332-6 (1 NOV 1972)
    276

17-2462   LIGHT SCATTERING STUDY OF THE FERROELECTRIC TRANSITION IN KDP.
BRODY EM
BULL AMER PHYS SOC 15: 606 (1970)

17-2463   LIGHT SCATTERING FROM CRITICAL FLUCTUATIONS IN POTASSIUM DIDEUTERIUM PHOSPHATE.
BRODY EM
BULL AMER PHYS SOC 16: 62 (1971)

17-2464   LOW TEMPERATURE THERMAL EXPANSION MEASUREMENTS ON SINGLE CRYSTAL POTASSIUM DIHYDROGEN
PHOSPHATE.
BROWDER JS + BALLARD SS
J OPT SOC AMER 61: 683 (1971)

17-2465   MOSSBAUER EFFECT OBSERVATION OF A VERY LOW FREQUENCY MODE AT THE TRANSITION TEMPERATURE
IN KDP AND DEUTERATED KDP (DOPED WITH COBALT-57 AND IRON-57).
BRUNSTEIN M + GRINBERG J + PELAH I + WIENER E
SOLID STATE COMMUN 8: 1211-14 (1970)
    209

17-2466   DIELECTRIC PROPERTIES OF A (SMALL DOSE) GAMMA-IRRADIATED KDP CRYSTAL (TEMPERATURE
DEPENDENCE).
BURDANINA NA + KAMYSHEVA LN + ZHUKOV OK
SOV PHYS CRYSTALLOGR 15(4): 721-2 (1971)
    221

17. POTASSIUM DIHYDROGEN PHOSPHATE (KDP) AND RELATED PHOSPHATES

17-2467 DIELECTRIC PROPERTIES OF A GAMMA-IRRADIATED POTASSIUM DIHYDROGEN PHOSPHATE CRYSTAL.
BURDANINA NA + KAMYSHEVA LN + ZHUKOV OK
SOV PHYS CRYSTALLOGR 15: 721-2 (1971)

17-2468 PARAMETRIC LIGHT AMPLIFICATION AND OSCILLATION IN KDP WITH MODE LOCKED PUMP.
BURNEIKA K + IGNATAVICIUS M + KABELKA V + PISKARSKAS A + STABINIS A
IEEE J QUANTUM ELECTRON 8: 574 (1972)

17-2469 STRUCTURAL INHIBITION OF FERROELECTRIC SWITCHING IN TRIGLYCINE SULFATE. PART-2:
X-IRRADIATION TREATMENT.
BYE KL + KEVE ET
FERROELECTRICS 4: 87-95 (1972)

DEUTERON MAGNETIC RESONANCE STUDY OF SOME CRYSTALS CONTAINING O-D...O BOND. (INCLUDING
FULLY DEUTERATED AMMONIUM DIHYDROGEN PHOSPHATE)
CHIBA T
J CHEM PHYS 41: 1352-8 (1964)

17-2470 EFFECT OF TEMPERATURE AND PHASE TRANSITION ON THE HYDROXIDE VIBRATION SPECTRUM OF THE
FERROELECTRIC RUBIDIUM DIHYDROGEN PHOSPHATES.
CHISLER EV + DAVYDOV VYU + SAVATINOVA IT
SOV PHYS SOLID STATE 13(7): 1635-41 (JAN 1972)
255

17-2471 POLARIZATION OF THE RAMAN BANDS OF THE HYDROXYL VIBRATIONS AND STRUCTURE OF THE
HYDROGEN BONDS IN THE FERROELECTRIC CRYSTALS RUBIDIUM DIHYDROGEN PHOSPHATE AND RUBIDIUM
DIHYDROGEN ARSENATE.
CHISLER EV + SAVATINOVA II + DAVYDOV VYU
SOV PHYS SOLID STATE 13(6): 1339-41 (DEC 1971)
254

17-2472 VIBRATION SPECTRUM OF THE PHOSPHATE ANION IN A RUBIDIUM DIHYDROGEN PHOSPHATE CRYSTAL IN
PARAELECTRIC AND FERROELECTRIC PHASES.
CHISLER EV + SAVATINOVA II + DAVYDOV VYU
OPT SPECTROSC 32: 286-7 (1972)

17-2473 PHASE SYNCHRONIZATION IN NONLINEAR RUBIDIUM DIHYDROGEN PHOSPHATE.
CHMELA P
OPTIK 33: 312-14 (1971) (IN GERMAN)

17-2474 THERMAL VARIATION OF REFRACTIVE INDEX IN AMMONIUM DIHYDROGEN PHOSPHATE AND TRIGLYCINE
SULFATE CRYSTALS.
CHORVATOVA Z + VISNOVSKA M
ACTA FAC RERUM NATUR UNIV COMENIANAE, PHYS NO 12: 51-5 (1971)
(IN SLOVAK)
     CHEM ABSTR 76: 160255J (1972)
     273

17-2475 LASER INDUCED DAMAGE IN X DIHYDROGEN PHOSPHATE MATERIALS.
CHRISTMAS TM + LEY JM
ELECTRON LETT 7: 544-6 (1971)

17-2476 LASER INDUCED SELF FOCUSING IN AMMONIUM DIHYDROGEN PHOSPHATE AND POTASSIUM DIDEUTERIUM
PHOSPHATE.
CHRISTMAS TM
J PHYS D APPL PHYS 5: L13-16 (1972)

17-2477 PULSE MEASUREMENT OF R(63) IN KDP.
CHRISTMAS TM + WILDEY CG
ELECTRON LETT 6: 152-3 (1970)

17-2478 PHASE DIAGRAMS OF AMMONIUM DIHYDROGEN PHOSPHATE AND POTASSIUM DIHYDROGEN ARSENATE TO 52
KBAR AND 400 DEG C.
CLARK JB
HIGH TEMP HIGH PRESSURE 1: 553-9 (1969)
197

17-2479 SPECTRA OF INTERNAL VIBRATIONS OF THE PHOSPHATE ION IN THE PARAELECTRIC PHASE OF KDP
CRYSTALS (INFRARED ABSORPTION AND RAMAN SCATTERING).
COIGNAC JP
COMPT REND B 271(12): 583-4 (1970) (IN FRENCH)
227

17-2480   VIBRATION SPECTRA OF POTASSIUM DIHYDROGEN PHOSPHATE IN PARA- AND FERROELECTRIC PHASES.
           COIGNAC JP + POULET H
           J PHYS (PARIS) 32: 679-84 (1971) (IN FRENCH)
               256

17-2481   LASER LIGHT SCATTERING STUDIES OF PHASE TRANSITIONS. (TRIGLYCINE SULFATE, POTASSIUM
           DIHYDROGEN AND DIDEUTERIUM SULFATE)
           CUMMINS HZ
           JOHNS HOPKINS UNIV, 1972, 11P. AD744532

17-2482   VIBRATION SPECTRA OF HYDROGEN BONDS. (KDP)
           DE GENNES PG
           COMMENT SOLID STATE PHYS 1(3): 65-8 (1968)

17-2483   COLLECTIVE MOTIONS OF HYDROGEN BONDS. (CALCULATIONS FOR POTASSIUM DIHYDROGEN PHOSPHATE
           TYPE CRYSTALS)
           DE GENNES PG
           SOLID STATE COMMUN 1: 132-7 (1963)

17-2484   INDEX OF REFRACTION OF POTASSIUM DIHYDROGEN PHOSPHATE.
           DENNIS JH + KINGSTON RH
           APPL OPT 2: 1334-5 (1963)

17-2485   X-RAY DETERMINATION OF THE THERMAL EXPANSION OF AMMONIUM DIHYDROGEN PHOSPHATE.
           DESHPANDE VT + KHAN AA
           ACTA CRYSTALLOGR 16: 936-8 (1963)

17-2486   PARAMETRIC FLUORESCENCE IN AMMONIUM DIHYDROGEN PHOSPHATE AND POTASSIUM DIHYDROGEN
           PHOSPHATE EXCITED BY A 2573 ANGSTROM CW PUMP.
           DOWLEY MW
           OPTO-ELECTRON 1(4): 179-81 (1969)

17-2487   UNIT CELL VOLUME EFFECTS (RELATION WITH CURIE POINT) IN SOME ISOMORPHOUS FERROELECTRIC
           SYSTEMS. (DISCUSSED FOR POTASSIUM DIHYDROGEN PHOSPHATE SYSTEM)
           DUNNE TG + BURNS G
           J ELECTROCHEM SOC 109: 54-6 (1962)

           SOFT MODE BEHAVIOR IN IMPROPER FERROELECTRICS. (THEORY, KDP)
           DVORAK V + PETZELT J
           PHYS LETT A 35: 209-10 (1971)

           PHASE TRANSITION IN KDP AND THE ROLE OF THE DEPOLARIZING ENERGY IN IT. (THEORY)
           DVORAK V
           P252-4 OF INT MEETING ON FERROELECTRICITY, 2ND, 1969, KYOTO, PROC.
           (J PHYS SOC JAP, VOL 28 SUPPL). PHYS SOC JAP, 1970, 459P.

17-2488   PHASE TRANSITION IN KDP (MODIFICATION OF KOBAYASHI'S DYNAMICAL THEORY).
           DVORAK V
           CZECH J PHYS B 20(1): 1-8 (1970)
               198

17-2489   HIGH TEMPERATURE PHASE TRANSITION IN RUBIDIUM DIHYDROGEN PHOSPHATE BY SCATTERING OF
           COLD NEUTRONS.
           EFRON U + PELAH I + VULKAN U + ZAFRIR H
           J CHEM PHYS 55: 3599-601 (1971)
               255

17-2490   HYDROGEN DENSITY DISTRIBUTION IN PARAELECTRIC POTASSIUM DIHYDROGEN PHOSPHATE (NEUTRON
           SCATTERING).
           FELCHER G + PELAH I
           J CHEM PHYS 52(2): 905-10 (1970)
               225

17-2491   EFFECT OF DISLOCATIONS ON THE GROWTH OF POTASSIUM DIHYDROGEN PHOSPHATE CRYSTALS.
           FISHMAN YUM
           TER 8(12): 1962-3 (1972)

17-2492   X-RAY TOPOGRAPHIC STUDY OF THE DISLOCATIONS PRODUCED IN POTASSIUM DIHYDROGEN PHOSPHATE
           CRYSTALS BY GROWTH FROM SOLUTION.
           FISHMAN YUM
           SOV PHYS CRYSTALLOGR 17(3): 524-7 (NOV-DEC 1972)
               280

17. POTASSIUM DIHYDROGEN PHOSPHATE (KDP) AND RELATED PHOSPHATES

17-2493   OBSERVATION OF DOMAIN STRUCTURE IN POTASSIUM DIHYDROGEN PHOSPHATE CRYSTALS BY
POLARIZATION OPTICAL PROCEDURE.
FOMICHEV NN
BULL ACAD SCI USSR, PHYS SER 29(6): 965-6 (1965)

ORIENTATION OF COMPOSITION PLANES IN ANTIFERROELECTRICS. (APPLIED TO ADP, TUNGSTEN
TRIOXIDE, SODIUM NIOBATE)
FOUSEK J + JANOVEC V
P380-1 OF INT MEETING ON FERROELECTRICITY, 2ND, 1969, KYOTO, PROC,
(J PHYS SOC JAP, VOL 28 SUPPL). PHYS SOC JAP, 1970, 459P.

TRANSIENT ELECTROOPTIC EFFECTS AND Q SWITCHING PERFORMANCE IN LITHIUM NIOBATE AND
POTASSIUM DIDEUTERIUM PHOSPHATE POCKELS CELLS. (COMMENTS)
FOUNTAIN WD
APPL OPT 10: 972-3 (1971)

17-2494   OPTIMUM CUT IN X DIHYDROGEN PHOSPHATE CRYSTALS FOR TRANSVERSE LIGHT MODULATION
FRANCOIS GE + LIBRECHT FM + ENGELEN JJ
ELECTRON LETT 6: 778-9 (1970)

17-2495   INFLUENCE OF HYDROSTATIC PRESSURE ON THE PHASE TRANSITION TEMPERATURE OF FERROELECTRIC
CRYSTALS OF THE POTASSIUM DIHYDROGEN PHOSPHATE TYPE.
FRENZEL C + PIETRASS B + HEGENBARTH E
PHYS STATUS SOLIDI A 2: 273-9 (1970)
      234

17-2496   GROWTH LAYERS AND OTHER MACRODEFECTS IN POTASSIUM DIHYDROGEN PHOSPHATE CRYSTALS GROWN
FROM AN AQUEOUS SOLUTION.
FRIEDMANN SS + STEPANOVA NS + BELJUSTIN AW
KRISTALL TECH 6(1): 77-83 (1971) (IN GERMAN)
      SCI ABSTR A 74: 40428 (1971)
      278

17-2497   LIGHT SCATTERING STUDIES OF THE SOFT OPTIC AND ACOUSTIC MODES OF POTASSIUM DIHYDROGEN
PHOSPHATE AND POTASSIUM DIDEUTERIUM PHOSPHATE.
FRITZ IJ + REESE RL + BRODY EM + WILSON CM + CUMMINS HZ
P415-20 INT CONF ON LIGHT SCATTERING IN SOLIDS, 2ND PROC,
BALKANSKI M(ED), PARIS, 1971, FLAMMARION, PARIS, 1971, 518P.

17-2498   NOTE ON THE LATTICE DYNAMICAL ASPECT OF FERROELECTRIC MODES OF KDP. (CALCULATIONS)
FUJIWARA T
J PHYS SOC JAP 29(5): 1282-94 (1970)
      218

17-2499   ELECTROOPTIC CHARACTERISTICS OF DEUTERATED POTASSIUM DIHYDROGEN PHOSPHATE CRYSTAL AND
ITS APPLICATION FOR A HIGH SPEED SHUTTER.
FURUYA N + SHIMOMURA O
REP FAC ENG, YAMANASHI UNIV NO 21: 53-67 (1970) (IN JAPANESE)

17-2500   COMBINATION OF THE FREQUENCIES OF COHERENT AND INCOHERENT RADIATION IN KDP CRYSTAL.
GAINER AV + KRIVOSHCHEKOV GV + KRUGLOV SV + MARENNIKOV SI
J APPL SPECTROSC 13: 1245-6 (1970)

17-2501   TEMPERATURE DEPENDENCE OF THE DIELECTRIC CONSTANT OF POTASSIUM DIHYDROGEN PHOSPHATE AT
LAMBDA EQUAL 2.14 MM.
GAUSS KE + HAPP H
P63-7 OF EUROPEAN MEETING ON FERROELECTRICITY, 1969, SAARBRUCKEN,
MUSER HE AND PETERSSON J(EDS), WISSENSCH VERLAGSGES, STUTTGART,
1970, 396P.

ANALYTICAL PROPERTIES OF A CLASS OF PLANAR FERROELECTRIC MODELS. (THEORY, KDP TYPES)
GLASSER ML
P165-75 OF BATTELLE INST MATER SCI, COLLOQ, 5TH PROC, GENEVA AND
GSTAAD, SWITZERLAND, CRITICAL PHENOMENA IN ALLOYS, MAGNETS, AND
SUPERCONDUCTORS, MILLS RE + ASCHER E + JAFFEE RI(EDS),
MCGRAW-HILL, 1971, 661P.

17-2502 APPLICATIONS OF THE PIEZOELECTRIC NONLINEARITIES IN THE TEMPERATURE AUTOSTABILIZATION
REGIME OF A FERROELECTRIC CRYSTAL. (TGS TANDEL)
GLANC A
P334-6 OF INT MEETING ON FERROELECTRICITY, 2ND, 1969, KYOTO, PROC,
(J PHYS SOC JAP, VOL 28 SUPPL). PHYS SOC JAP, 1970, 459P.

OYO BUTSURI 39(5): 396-405 (1970) (IN JAPANESE)
CHEM ABSTR 73: 60278P

17-2503 DOUBLE DIELECTRIC HYSTERESIS LCOF OF POTASSIUM DIDEUTERIUM PHOSPHATE CRYSTALS.
GLADKII VV + SIDNENKO EV
SOV PHYS SOLID STATE 13(10): 2592-3 (APR 1972)
263

17-2504 POLARIZATION OF A RUBIDIUM DIHYDROGEN PHOSPHATE CRYSTAL IN THE PHASE TRANSITION REGION.
(COERCIVE FIELD)
GLADKII VV + SIDNENKO EV
SOV PHYS SOLID STATE 13(6): 1374 (DEC 1971)
253

17-2505 POLARIZATION (DOMAIN STRUCTURE) AND PHASE TRANSITION IN KDP SINGLE CRYSTAL. (ABSTRACT
ONLY)
GLADKII VV + ZHELUDEV IS + SIDNENKO EV
P206 OF INT MEETING ON FERROELECTRICITY, 2ND, 1969, KYOTO, PROC,
(J PHYS SOC JAP, VOL 28 SUPPL). PHYS SOC JAP, 1970, 459P.

17-2506 POLARIZATION OF A RUBIDIUM DIHYDROGEN PHOSPHATE CRYSTAL IN THE PHASE TRANSITION REGION.
GLADKII VV + SIDNENKO EV
SOV PHYS SOLID STATE 13: 1642-8 (1971)

17-2507 REAL-TIME HOLOGRAPHIC RECONSTRUCTION BY ELECTROOPTIC MODULATION (BY PASSING COHERENT
LIGHT THROUGH DEUTERATED KDP HAVING A CHARGE PATTERN).
GOETZ GG
APPL PHYS LETT 17(2): 63-6 (1970)
199

17-2508 NONLINEAR PROPERTIES OF A RUBIDIUM DIHYDROGEN PHOSPHATE CRYSTAL.
GOLOVEI MP + KALINKINA IN + KOSOUROV GI
OPT SPECTROSC 28: 535-6 (1970)

17-2509 (FINITE APERTURE LIGHT SWITCH USING) TRANSVERSE POCKELS EFFECT COMMUTATORS. (POTASSIUM
DIHYDROGEN PHOSPHATE)
GOUZERH J + HEPNER G
OPT COMMUN 1(9): 435-7 (APR 1970) (IN FRENCH)
SCI ABSTR A 73: 59371 (1970)
207

17-2510 EXPERIMENTAL PROOF FOR PROTON TUNNELING IN POTASSIUM DIHYDROGEN PHOSPHATE.
GRIMM H + STILLER H + PLESSER TH
P47-51 OF EUROPEAN MEETING ON FERROELECTRICITY, 1969, SAARBRUCKEN,
MUSER HE AND PETERSSON J (EDS), WISSENSCH VERLAGSGES, STUTTGART,
1970, 396P.

17-2511 NEUTRON MEASUREMENTS ON THE HYDROGEN BOND POTENTIAL IN PARAELECTRIC POTASSIUM
DIHYDROGEN PHOSPHATE.
GRIMM H + STILLER H + PLESSER TH
PHYS STATUS SOLIDI 42: 207-17 (1970)
226

17-2512 USE OF KDP TYPE CRYSTALS AS ELECTROOPTICAL REFLECTORS.
GRIB BN + KOROTKOV PA + TSYAHCHENKO YUP
UKR FIZ ZH 17: 546-50 (APR 1972) (IN RUSSIAN, ENGLISH ABSTRACT)

17-2513 HIGH TEMPERATURE PHASE TRANSITIONS AND METASTABILITY IN POTASSIUM DIHYDROGEN PHOSPHATE
TYPE CRYSTALS.
GRUNBERG J + LEVIN S + PELAH I + GERLICH D
PHYS STATUS SOLIDI B 49: 857-69 (1972)
262

THERMAL EXPANSION OF SOME FERROELECTRICS. (POTASSIUM NITRATE, KDP, SODIUM NITRITE)
GUTIERREZ M + CANUT ML + AMOROS JL
ACTA CRYSTALLOGR 16(SUPPL): A166 (1963)

17-2514 ELECTRICAL CONSTANTS OF TRIGLYCINE SULFATE, APPLICATIONS TO PYROELECTRICITY AND THE
DETECTION OF FAR INFRARED RADIATIONS.
HADNI A
RADIAT EFF 4: 195-206 (1970)

17-2515 RAMAN SPECTRA OF CRYSTALS OF THE POTASSIUM DIHYDROGEN PHOSPHATE TYPE IN THEIR
PARAELECTRIC AND LOW TEMPERATURE PHASES.
HAMMER H
P425-8 OF INT CONF ON LIGHT SCATTERING IN SOLIDS, 2ND PROC,
BALKANSKI M(ED), PARIS, 1971, FLAMMARION, PARIS, 1971, 518P.

17-2516 GROWING KINETIC OF THE AMMONIUM DIHYDROGEN PHOSPHATE SINGLE CRYSTAL (FROM SOLUTION) IN
STATIC REGIME AND AT NORMAL PH.
HANGEA N + ALEXANDRU HV
AN UNIV BUCURESTI FIZ 18: 27-34 (1969)
        CHEM ABSTR 74: 92273V (1971)
        238

17-2517 NEW METHOD OF MEASURING PYROELECTRIC COEFFICIENTS. (TGS EXAMPLE)
HARTLEY NP + SQUIRE PT + PUTLEY EH
J PHYS E 5 (8): 787-9 (1972)
        276

17-2518 POCKELS EFFECT IN AMMONIUM DIHYDROGEN PHOSPHATE ABOVE ROOM TEMPERATURE.
HARRIS LB + VELLA GJ
SOLID STATE COMMUN 10: 1229-31 (1972)
        270

17-2519 BRILLOUIN SCATTERING BY A SINGLE CRYSTAL OF RUBIDIUM DIHYDROGEN PHOSPHATE.
HAURET G
COMPT REND B 273(14): 627-30 (4 OCT 1971) (IN FRENCH)
        SCI ABSTR A 75: 7083 (1972)
        264

17-2520 BRILLOUIN DIFFUSION IN A SINGLE CRYSTAL OF RUBIDIUM DIHYDROGEN PHOSPHATE.
HAURET G + TAUREL L + CHAPELLE JP
COMPT REND B 273: 627-30 (1971) (IN FRENCH)
        256

17-2521 INTENSITY OF SOME BRILLOUIN AND RAYLEIGH LINES IN RUBIDIUM DIHYDROGEN PHOSPHATE.
HAURET G + TAUREL L + CHAPELLE JP
SOLID STATE COMMUN 10: 727-9 (1972)
        264

17-2522 STUDY OF THE BRILLOUIN EFFECT IN ADP (ALL ELASTIC CONSTANTS).
HAURET G + TAUREL L
J PHYS (PARIS) 31(7): 657-64 (1970) (IN FRENCH)
        223

17-2523 DISPLACEMENT OF THE CURIE POINTS IN KDP AND LITHIUM THALLIUM TARTRATE CRYSTALS BY
HYDROSTATIC PRESSURE.
HEGENBARTH E
SOV PHYS CRYSTALLOGR 15(2): 268-70 (1970)
        206

17-2524 STUDY OF THE LIMITATION OF THE ANGULAR FIELD BY THE COMMUTATOR OF POLARIZATION IN KDP,
IN A DIGITAL DEVIATOR.
HEPNER G + GOUZERH J + ROUSSEAU M
NOUV REPT OPT APPL 2: 3-13 (1971) (IN FRENCH)

17-2525 ELECTRON DELOCALIZATION WEIGHT OF THE HYDROGEN BOND IN POTASSIUM DIHYDROGEN PHOSPHATE.
HIDAKA T
J PHYS SOC JAP 33(3): 635-40 (SEP 1972)
        279

17-2526 HIGH FREQUENCY BEHAVIOR OF HYDROGEN BONDED FERROELECTRICS: TRIGLYCINE SULFATE AND
POTASSIUM DIHYDROGEN PHOSPHATE.
HILL RM + ICHIKI SK
PHYS REV 132: 1603-8 (1963)

17. POTASSIUM DIHYDROGEN PHOSPHATE (KDP) AND RELATED PHOSPHATES

17-2527   INFLUENCE OF TUNNELING ON DIELECTRIC BEHAVIOR OF POTASSIUM DIHYDROGEN PHOSPHATE.
HOLAKOVSKY J + BREZINA B + PACHEROVA O
PHYS STATUS SOLIDI B 53: K69-72 (1972)
276

17-2528   STUDY OF PHASE TRANSITION IN MIXED KDP- DEUTERATED KDP CRYSTALS BY MODIFIED MOLECULAR
FIELD APPROXIMATION.
HOLAKOVSKY J
CZECH J PHYS B 22(8): 651-65 (1972)
279

17-2529   LOSSLESS POTASSIUM DIDEUTERIUM PHOSPHATE POCKELS CELL FOR HIGH POWER Q SWITCHING.
HOOK WR + HILBERG RP
APPL OPT 10: 1179-80 (1971)

17-2530   NOVEL CUT OF POTASSIUM DIHYDROGEN PHOSPHATE TYPE CRYSTALS.
HOOKABE K + MATSUO Y
MEM INST SCI IND RES OSAKA UNIV 28: 23-7 (1971)
256

17-2531   NOVEL TYPE OF CUT FOR KDP CRYSTALS FOR LOW VOLTAGE LIGHT MODULATION.
HOOKABE K + MATSUO Y
ELECTRON LETT 6: 550-1 (1970)

17-2532   OPTIMUM CUT OF POTASSIUM DIHYDROGEN PHOSPHATE TYPE CRYSTALS FOR LIGHT MODULATION.
HOOKABE K + MATSUO Y
TRANS INST ELECTRON COMMUN ENG JAP B 54: 237-44, 371-2 (1971)
(IN JAPANESE)

QUANTUM FIELD THEORETICAL TREATMENT OF A DYNAMICAL MODEL OF THE FERROELECTRIC KDP.
PART-1. PART-2.
HOUSTON GD + BOLTON HC
J PHYS C SOLID STATE PHYS 4: 2097-108, 2894-902 (1971)

17-2533   ELECTRON SPIN RESONANCE OF IRRADIATED AND DEUTERATED POTASSIUM DIHYDROGEN PHOSPHATE.
HUGHES WE + MOULTON WG
J CHEM PHYS 39: 1359-60 (1963)

RATE OF ROTATING CRYSTALS AND CHARACTERISTICS OF THEIR GROWTH. (AMMONIUM DIHYDROGEN
PHOSPHATE, POTASSIUM FERROCYANIDE TRIHYDRATE, SODIUM NITRATE)
INYUSHKIN GV + SHABALIN KN
SOV PHYS CRYSTALLOGR 9: 242-4 (1964)

DYNAMIC MODEL OF THE SLATER KDP FERROELECTRIC.
IRVINE RD
AUSTRAL J PHYS 23: 833-45 (1970)

A THEORY OF THE PHASE TRANSITION IN AMMONIUM DIHYDROGEN PHOSPHATE.
ISHIBASHI Y + OHYA S + TAKAGI Y
J PHYS SOC JAP 33(6): 1545-50 (DEC 1972)

17-2534   DIELECTRIC DISPERSION OF TGS AND DEUTERATED KDP (THEORY, CALCULATIONS).
ISHIBASHI Y + SAWADA A + TAKAGI Y
J PHYS SOC JAP 28(6): 1488-94 (1970)
197

17-2535   STRUCTURE OF DEUTERATED KDP CRYSTAL: DEUTERIUM CONCENTRATION DETERMINATION.
ISHERWOOD BJ + JAMES JA
J PHYS (PARIS) 33, SUPPL 4, COLLOQ 2: 91-2 (1972)
294

17-2536   ELECTROCHEMISTRY ON THE POLYCRYSTAL OF POTASSIUM DIHYDROGEN PHOSPHATE.
ITOH K + ISHIHARA K + MATSUI N + YAMADA T
BULL NAGOYA INST TECHNOL 23: 113-17 (1971) (IN JAPANESE)

17-2537   DEPENDENCE OF THE ELASTIC COEFFICIENTS OF ADP ON THE ELECTRIC FIELD.
JANIK L + HRUSKA K
CZECH J PHYS B 20: 202-5 (1970)

17-2538   EFFICIENT SECOND HARMONIC GENERATION IN ADP WITH TWO NEW FLUORESCEIN DYE LASERS.
JENNINGS DA + VARGA AJ
J APPL PHYS 42: 5171-2 (1971)

17. POTASSIUM DIHYDROGEN PHOSPHATE (KDP) AND RELATED PHOSPHATES

17-2539 OPTICAL NONLINEAR SUSCEPTIBILITIES: ACCURATE RELATIVE VALUES FOR QUARTZ, AMMONIUM
DIHYDROGEN PHOSPHATE, AND POTASSIUM DIHYDROGEN PHOSPHATE.
JERPHAGNON J + KURTZ SK
PHYS REV B 1(4): 1739-44 (15 FEB 1970)
283

17-2540 DEPOLARIZATION OF LINEARLY POLARIZED LIGHT DUE TO HEATING OF PARALLEL PLANES OF
POTASSIUM DIHYDROGEN PHOSPHATE CRYSTALS.
JUNG B + RICHTER K
FEINGERATETECHNIK 21: 158-61 (1972) (IN GERMAN)

17-2541 OPTIMISATION OF ELECTROOPTIC CRYSTAL MODULATORS. (KDP)
KALYMNIOS D
ELECTRON LETT 6(25): 804-5 (1970)
231

17-2542 DIELECTRIC PROPERTIES OF GAMMA-IRRADIATED POTASSIUM DIHYDROGEN PHOSPHATE CRYSTALS.
KAMYSHEVA LN + BURDANINA NA + ZHUKOV OK + DARINSKII BM + SIZOVA LH
BULL ACAD SCI USSR, PHYS SER 34(12): 2327-31 (1970)

17-2543 NONLINEAR PROPERTIES OF POTASSIUM DIHYDROGEN PHOSPHATE. (DETERMINATION OF BETA IN THE P
EXPANSION OF FREE ENERGY F)
KAMYSHEVA LN + ZHUKOV OK
P200-3, VOL 2 OF DVORAK V + FOUSKOVA A + GLOGAR P (EDS),
INT MEETING ON FERROELECTRICITY, PRAGUE, 1966, PROC.
INST PHYS CZECH ACAD SCI, PRAGUE, 1966.

17-2544 TEMPERATURE DEPENDENCE OF INDEX MATCHING ANGLE OF THE SECOND HARMONIC GENERATION IN
POTASSIUM DIHYDROGEN PHOSPHATE CRYSTAL.
KAMIURA Y + KAWABE K
J PHYS SOC JAP 33(6): 1643-4 (DEC 1972)
282

17-2545 INVESTIGATION OF THE PHASE TRANSITION IN AMMONIUM DIHYDROGEN PHOSPHATE SINGLE CRYSTAL
BY NMR.
KATSURI SR + MORAN PR
BULL AMER PHYS SOC 16: 317 (1971)

17-2546 EPR OF CHROMIUM(3+) IONS IN ANTIFERROELECTRIC ADP CRYSTALS.
KAWANO T + NIIMORI K + HUKUDA K + FUJITA N
J PHYS SOC JAP 29(3): 633-42 (1970)
213

17-2547 ESR STUDY OF RADIATION DAMAGE CENTERS IN AMMONIUM DIHYDROGEN PHOSPHATE POTASSIUM
DIHYDROGEN ARSENATE MIXED CRYSTALS.
KAWANO T
J PHYS SOC JAP 33: 1492 (1972)
KOGOSHIMA UNIV, MATH PHYS CHEM 5: 39-48 (1972) (IN JAPANESE)
280

17-2548 EFFECT OF IONIC MOTION ON THE REFRACTIVE INDEX OF PARAELECTRIC KDP IN THE VICINITY OF
CURIE POINT.
KAWABE K + KAMIURA Y
TECHNOL REP OSAKA UNIV 22: 391-9 (1972)

17-2549 FAR INFRARED SPECTRA OF KDP AND ADP. (20-550 PER CM, 200 AND 300 DEG K)
KAWAMURA T + MITSUICHI A + YOSHINAGA H
P227-9 OF INT MEETING ON FERROELECTRICITY, 2ND, 1969,, KYOTO, PROC,
(J PHYS SOC JAP, VOL 28 SUPPL). PHYS SOC JAP, 1970, 459P.

17-2550 TEMPERATURE DEPENDENCE OF RAMAN SPECTRUM AND SECOND HARMONIC GENERATION IN KDP.
KAWABE K + SAKAMOTO A + KAMIURA Y + INUISHI Y
P225-6 OF INT MEETING ON FERROELECTRICITY, 2ND, 1969,, KYOTO, PROC,
(J PHYS SOC JAP, VOL 28 SUPPL). PHYS SOC JAP, 1970, 459P.

17-2551 ADP POLARIZER FOR LASERS.
KIEBURG H
FEINGERATETECHNIK 21: 151-5 (1972) (IN GERMAN)

17. POTASSIUM DIHYDROGEN PHOSPHATE (KDP) AND RELATED PHOSPHATES

17-2552 RAMAN SCATTERING FROM FERROELECTRIC CRYSTALS OF POTASSIUM DIHYDROGEN PHOSPHATE.
KIM JJ
UNIV CALIF PHD THESIS, 136P, 1970
DISS ABSTR INT B 32: 4820 (1972)
257

17-2553 MEASUREMENT OF THE REFRACTIVE INDEX OF ADP AND KDP CRYSTALS IN THE INFRARED RANGE BY PARAMETRIC SCATTERING OF LIGHT.
KLYSHKO DN + PENIN AN + POLKOVNIKOV BF
SOV J QUANTUM ELECTRON 1: 535-8 (1972)

17-2554 EFFECTS OF UNIAXIAL STRESS AND ELECTRIC FIELD ON PARAMAGNETIC SPECTRA OF CHROMIUM(3+) IN POTASSIUM DIHYDROGEN PHOSPHATE.
KOBAYASHI T + FURUKAWA K + HUKUDA K
J PHYS SOC JAP 32: 577 (FEB 1972)

17-2555 ORDER OF THE FERROELECTRIC PHASE TRANSFORMATION OF POTASSIUM DIHYDROGEN PHOSPHATE.
KOBAYASHI J + UESU Y + ENOMOTO Y
PHYS LETT A 34: 171-2 (1971)

17-2556 X-RAY STUDY ON THERMAL EXPANSION OF FERROELECTRIC POTASSIUM DIHYDROGEN PHOSPHATE. (NEAR CURIE TEMPERATURE)
KOBAYASHI J + UESU Y + MIZUTANI I + ENOMOTO Y
PHYS STATUS SOLIDI A 3: 63-9 (1970)
234

17-2557 X-RAY DILATOMETRIC STUDY OF THE FERROELECTRIC PHASE TRANSITION OF POTASSIUM DIHYDROGEN PHOSPHATE. PART-1: THE ORDER OF TRANSITION.
KOBAYASHI J + UESU Y + ENOMOTO Y
PHYS STATUS SOLIDI B 45: 293-304 (1971)

17-2558 ESR INVESTIGATIONS OF X-RAY IRRADIATED SINGLE CRYSTALS OF FERROELECTRIC POTASSIUM DIHYDROGEN PHOSPHATE.
KOCHERIL GP + LEE S + JACOBSMEYER VP
BULL AMER PHYS SOC 17: 309 (1972)

17-2559 CUTTING AND POLISHING IN UNIAXIAL CRYSTALS OF THE TYPE X DIHYDROGEN PHOSPHATE FOR LIGHT MODULATORS.
KOJIMA H + NOMURA S
PROC PHYS SOC JAP 25: 391-7 (1970) (IN JAPANESE)

17-2560 EFFECT OF REPLACEMENT OF POTASSIUM BY ALUMINUM ON THE FERROELECTRIC PROPERTIES OF POTASSIUM DIHYDROGEN PHOSPHATE.
KOLBENEVA GI + LEVINA ME
INORG MATER 7: 857-9 (JUN 1971)

17-2561 INVESTIGATION OF THE DOMAIN STRUCTURE OF A TRIGLYCINE SULFATE CRYSTAL DURING AGING.
KONSTANTINOVA VP + STANKOWSKA J
SOV PHYS CRYSTALLOGR 16: 123-8 (JUL-AUG 1971)

17-2562 CAN THE ISOTOPIC SHIFT IN KDP BE EXPLAINED BY TUNNELING ONLY.
KOPSKY V
PHYS STATUS SOLIDI A 5(1): K69-73 (1971)

17-2563 MOLECULAR FIELD STUDY OF THE ISOTOPIC EFFECT IN POTASSIUM DIHYDROGEN(X) DIDEUTERIUM(1-X) PHOSPHATE MIXED CRYSTALS.
KOPSKY V
CZECH J PHYS B 21(8): 896-916 (1971)

17-2564 POLARIZATION ECHO IN THE FERROELECTRIC SINGLE CRYSTAL POTASSIUM DIHYDROGEN PHOSPHATE.
KOPVILLEM UKH + SMOLYAKOV BP + SHARIPOV RZ
JETP LETT 13: 398-400 (1971)

17-2565 INVESTIGATION OF THE THERMAL EXPANSION OF POTASSIUM DIHYDROGEN PHOSPHATE CRYSTALS BY USING SEMICONDUCTOR FILMS.
KRAJEWSKI T
POZNAN TOW PRZYJ NAUK, PR KOM MAT-PRZYR 5(1): 73-81 (1969)
(IN POLISH)
CHEM ABSTR 72: 93912Y (1970)

17. POTASSIUM DIHYDROGEN PHOSPHATE (KDP) AND RELATED PHOSPHATES

17-2566 USE OF SPECIFIC HEAT (BY A DYNAMIC METHOD) OF POTASSIUM DIHYDROGEN PHOSPHATE CRYSTALS
FOR STUDY OF THE PYROELECTRIC PROPERTIES.
KRAJEWSKI T + DROZDOWSKI M
POZNAN TOW PRZYJ NAUK, PR KOM MAT-PRZYR 5(1): 83-7 (1969)
(IN POLISH)
      CHEM ABSTR 72: 115847K (1970)
      231

17-2567 GROWTH AND ETCHING OF POTASSIUM DIHYDROGEN PHOSPHATE CRYSTALS IN SOLUTIONS CONTAINING
ACETATE IONS.
KVAPIL J + KVAPIL J + SEDLAKOVA LN + MICHALEC R + CHALUPA B
J CRYST GROW 15: 126-8 (1972)
      272

17-2568 DYNAMICS OF PROTONS (AND DEUTERONS) IN FERROELECTRIC CRYSTALS WITH HYDROGEN BONDING.
(POTASSIUM DIHYDROGEN PHOSPHATE)
LAMOTTE B
CENTRE D'ETUDES NUCLEAIRES DE GRENOBLE, OCT 1970, 18P.
COMISSARIAT ENERGIE ATOMIQUE, FRANCE, CEA-BIB-153.
      234

FOURIER TRANSFORM FAR INFRARED REFLECTION SPECTRA OF PARAELECTRIC POTASSIUM DIHYDROGEN
PHOSPHATE AND POTASSIUM DIDEUTERIUM PHOSPHATE. (SOFT MADE THEORY)
LAVRENCIC B
FIZIKA 3(1): 67-9 (1971)
      SCI ABSTR A 74: 37536 (1971)

17-2569 LASER RAMAN STUDY OF QUASI SPIN WAVE HYDROGEN TUNNELING MODES IN KDP AND DEUTERATED
KDP.
LAVRENCIC B + LEVSTEK I + ZEKS B
CHEM PHYS LETT 5(7): 441-4 (1970)
      206

17-2570 PROTON LATTICE INTERACTIONS AND THE DYNAMICAL SUSCEPTIBILITY OF POTASSIUM DIHYDROGEN
PHOSPHATE TYPE FERROELECTRIC CRYSTALS.
LAVRENCIC B + LEVSTEK I + PIRC R + ZEKS B
P424 OF INT CONF ON LIGHT SCATTERING IN SOLIDS, 2ND PROC,
BALKANSKI M(ED), PARIS, 1971, FLAMMARION, PARIS, 1971, 518P.

17-2571 PHASE TRANSITION IN THALLIUM AND AMMONIUM THALLIUM (DIHYDROGEN) PHOSPHATES.
LE MONTAGNER S + LE DONCHE L
P244-9, VOL 1 OF DVORAK V + FOUSKOVA A + GLOGAR P (EDS),
INT MEETING ON FERROELECTRICITY, PRAGUE, 1966, PROC.
INST PHYS CZECH ACAD SCI, PRAGUE, 1966.

17-2572 DIELECTRIC PROPERTIES OF SINGLE CRYSTALS IN THE AMMONIUM DIHYDROGEN PHOSPHATE-
POTASSIUM DIHYDROGEN PHOSPHATE SYSTEM.
LEVINA ME + KOLBENEVA GI
VEST MOSK UNIV KHIM 13: 75-7 (1972) (IN RUSSIAN)

17-2573 ACOUSTIC PROPAGATION IN POTASSIUM DEUTERIUM PHOSPHATE AT THE FERROELECTRIC TRANSITION
TEMPERATURE.
LITOV E
UNIV WASHINGTON PHD THESIS, 1968, 145P.
      DISS ABSTR B 29: 3882 (APR 1969)
      185

17-2574 DIELECTRIC DISPERSION OF TGS.
LITHER G
J PHYS (PARIS) 33(2): 221-2 (1972)

17-2575 POLARIZATION RELAXATION AND SUSCEPTIBILITY IN THE FERROELECTRIC TRANSITION REGION OF
POTASSIUM DIDEUTERIUM PHOSPHATE (ANALYSIS AND THEORETICAL INTERPRETATION OF PREVIOUS
DATA).
LITOV E + UEHLING EA
PHYS REV B 1(9): 3713-24 (1970)
      221

17-2576 ULTRASONIC INVESTIGATION (15 MHZ SHEAR WAVES) OF THE FERROELECTRIC TRANSITION REGION IN
POTASSIUM DIHYDROGEN PHOSPHATE. (TEMPERATURE AND FIELD DEPENDENCE)
LITOV E + GARLAND CW
PHYS REV B 2(11): 4597-602 (1970)
228

17-2577 DTA STUDY OF THE FERROELECTRIC TRANSITION IN KDP TYPE CRYSTALS (CURIE TEMPERATURES OF
POTASSIUM, RUBIDIUM, CESIUM AND AMMONIUM DIHYDROGEN PHOSPHATES AND CESIUM AND AMMONIUM
DIHYDROGEN ARSENATES, STANDARD METHOD PROPOSED).
LOIACONO GM
MATER RES BULL 5: 775-82 (1970)
207

(EFFECT OF TEMPERATURE GRADIENT WITHIN THE SAMPLE ON THE) MICROWAVE DISPERSION OF
FERROELECTRICS NEAR THE TRANSITION POINT (THEORY, MEASUREMENTS ON TGS).
LUTHER G + MUESER HE
Z ANGEW PHYS 29(3): 237-40 (1970)

17-2578 X-RAY DIFFRACTION TOPOGRAPHY STUDY OF DEFECTS (DISLOCATIONS) IN KDP AND ADP SINGLE
CRYSTALS.
LUTSAU VG + FISHMAN YUM + RES IS
KRISTALL TECH 5(3): 445-58 (1970)
227

17-2579 EFFECTS OF DISPERSION AND FOCUSING ON THE PRODUCTION OF OPTICAL HARMONICS. (KDP)
MAKER PD + TERHUNE RW + NISENOFF M + SAVAGE CM
PHYS REV LETT 8: 21-2 (1962)

17-2580 RELAXATION OF FERROELECTRIC PHOSPHATES (AMMONIUM DIHYDROGEN PHOSPHATE AND POTASSIUM
DIHYDROGEN PHOSPHATE) IN THE EXTREME INFRARED.
MARTIN DH + STONE CD
PHYS LETT 5: 26-7 (1963)

17-2581 DIELECTRIC PROPERTIES OF DEUTERATED POTASSIUM DIHYDROGEN PHOSPHATE.
MAYER RJ + BJORKSTAM JL
J PHYS CHEM SOLIDS 23: 619-20 (1962)

QUANTITATIVE STUDIES OF OPTICAL HARMONIC GENERATION IN CADMIUM SULFIDE, BARIUM
TITANATE, AND POTASSIUM DIHYDROGEN PHOSPHATE TYPE CRYSTALS.
MILLER RC + KLEINMAN DA + SAVAGE A
PHYS REV LETT 11: 146-9 (1963)

17-2582 OPTICAL SECOND HARMONIC GENERATION IN PIEZOELECTRIC CRYSTALS. (KDP TYPE)
MILLER RC
APPL PHYS LETT 5: 17-19 (1964)

17-2583 CRYSTALLIZATION OF RUBIDIUM DIDEUTERIUM PHOSPHATE FROM SOLUTIONS IN HEAVY WATER.
MISHCHENKO AV + RASHKOVICH LN
SOV PHYS CRYSTALLOGR 16(5): 940-2 (MAR-APR 1972)
261

17-2584 EFFECT OF DEUTERATION ON THE ELASTIC PROPERTIES OF RUBIDIUM DIHYDROGEN PHOSPHATE (RDP).
MISHCHENKO AV + DOVCHENKO GV + RESHCHIKOVA IM
SOV PHYS CRYSTALLOGR 17(2): 360-1 (SEP-OCT 1972)
275

17-2585 ENTHALPIES OF SOLUTION OF RUBIDIUM DIHYDROGEN PHOSPHATE IN WATER AND RUBIDIUM
DIDEUTERIUM PHOSPHATE IN HEAVY WATER.
MONAENKOVA AS + KONKOVA TS + MISHCHENKO AV + VROBEV AF
J GEN CHEM 42: 2608-11 (1972)

17-2586 SYMMETRY OF ATOMIC VIBRATIONS IN POTASSIUM DIHYDROGEN PHOSPHATE.
MONTGOMERY H + PAUL GL
PROC ROY SOC EDINBURGH A 70: 107-24 (1972)

17-2587 FAR INFRARED ABSORPTION SPECTRA FOR AMMONIUM DIHYDROGEN PHOSPHATE.
MORLOT G + VILLERMAIN-LECOLLIER G + HADNI A
PHYS STATUS SOLIDI B, BASIC RES 49(1): K47-52 (1972) (IN FRENCH)

17-2588 PRESSURE EFFECTS ON THE LATTICE PARAMETERS AND STRUCTURE OF POTASSIUM DIHYDROGEN
PHOSPHATE TYPE CRYSTALS.
MOROSIN B + SAMARA GA
FERROELECTRICS 3: 49-56 (1971)
262

ONE DIMENSIONAL KDP MODEL IN STATISTICAL MECHANICS.
NAGLE JF
AMER J PHYS 36(12): 1114-7 (1968)

17-2589 OBSERVATION OF PHASE FRONT MOTION IN KDP CRYSTAL.
NAKAMURA T + IHARA H
P204-5 OF INT MEETING ON FERROELECTRICITY, 2ND, 1969, KYOTO, PROC,
(J PHYS SOC JAP, VOL 28 SUPPL). PHYS SOC JAP, 1970, 459P.

17-2590 STRUCTURAL STUDIES OF THE SYSTEM POTASSIUM DIHYDROGEN PHOSPHATE- POTASSIUM DIDEUTERIUM
PHOSPHATE.
NELMES RJ + EIRIKSSON VR
SOLID STATE COMMUN 11: 1261-4 (1972)
281

17-2591 THE CRYSTAL STRUCTURE OF MONOCLINIC POTASSIUM DIDEUTERIUM PHOSPHATE.
NELMES RJ
PHYS STATUS SOLIDI B 52: K89-93 (1972)
280

ANALYSIS OF PHONON TUNNELING MODES IN THE RAMAN SPECTRUM OF TETRAGONAL POTASSIUM
DIHYDROGEN PHOSPHATE. (THEORY)
NETTLETON RE
J PHYS SOC JAP 33(3): 641-6 (SEP 1972)

17-2592 INTERACTIONS BETWEEN PROTON TUNNELING AND OPTICAL PHONONS IN POTASSIUM DIHYDROGEN
PHOSPHATE.
NETTLETON RE
BULL AMER PHYS SOC 15: 336 (1970)

17-2593 INTERACTION BETWEEN PROTON TUNNELING AND OPTICAL PHONONS IN POTASSIUM DIHYDROGEN
PHOSPHATE.
NETTLETON RE
Z PHYS 248: 101-10 (1971)

17-2594 ABSENCE OF MOTIONAL NARROWING IN THE PROTON NMR IN POTASSIUM DIHYDROGEN PHOSPHATE AT
HIGH TEMPERATURES.
NICHOLSON JY + SOEST JF
BULL AMER PHYS SOC 15: 606 (1970)

17-2595 METHOD OF GROWTH IN SOLUTION OF SINGLE CRYSTALS BY THE GRADUAL ADDITION OF A REACTANT.
(POTASSIUM DIHYDROGEN PHOSPHATE)
NICOLAU IF + ITTU M + DABU R
J CRYST GROW 13-14: 462-6 (1972) (IN FRENCH)
266

17-2596 ELECTRON PARAMAGNETIC RESONANCE OF CHROMIUM(3+) IN ADP.
NIIMORI K + KAWANO T + HUKUDA K + FUJITA N
J PHYS SOC JAP 28: 801-2 (1970)
190

17-2597 ABSOLUTE SIGNS OF THE NONLINEAR OPTICAL COEFFICIENTS OF POTASSIUM DIHYDROGEN PHOSPHATE.
NORDLAND WA
FERROELECTRICS 2: 57-8 (JAN 1971)

17-2598 (PREPARATION AND STRUCTURE OF) RUBIDIUM AND CESIUM SEMIMETALLIC ORTHOPHOSPHATES.
NORBERT A + ANDRE D
COMPT REND C 270: 1718-20 (25 MAY 1970) (IN FRENCH)
CHEM ABSTR 73: 62087Z (1970)
218

17-2599 NUCLEAR DOUBLE RESONANCE STUDY OF POTASSIUM-39 IN FERROELECTRIC POTASSIUM DIHYDROGEN
PHOSPHATE.
NORDAL PE + HAHN EL
BULL AMER PHYS SOC 17: 129 (1972)

17. POTASSIUM DIHYDROGEN PHOSPHATE (KDP) AND RELATED PHOSPHATES

FREQUENCY SPECTRUM OF HYDROGEN BONDED FERROELECTRICS. (INFRARED ABSORPTION, NEUTRON
INELASTIC SCATTERING, POTASSIUM DIHYDROGEN PHOSPHATE, DEUTERATED POTASSIUM DIHYDROGEN
PHOSPHATE, THEORY)
NOVAKOVIC L
P127-39, VOL 1 OF DVORAK V + FOUSKOVA A + GLOGAR P (EDS),
INT MEETING ON FERROELECTRICITY, PRAGUE, 1966, PROC.
INST PHYS CZECH ACAD SCI, PRAGUE, 1966.

17-2600   DYNAMICAL BEHAVIOR OF FERROELECTRIC SPINS. (KDP TYPE CRYSTALS)
           NOVAKOVIC L
           J PHYS (PARIS) 33, SUPPL 4, COLLOQ 2: 73-4 (1972)
                294

17-2601   POSSIBLE THEORETICAL EXPLANATION OF SHIFT OF CURIE TEMPERATURE IN POTASSIUM DIDEUTERIUM
           PHOSPHATE AND DEUTERO POTASSIUM DIDEUTERIUM PHOSPHATE WITH HYDROSTATIC PRESSURE.
           NOVAKOVIC L
           J PHYS CHEM SOLIDS 29(6): 963-6 (1968)

17-2602   DIELECTRIC EVIDENCE OF A FIRST ORDER TRANSITION IN POTASSIUM DIHYDROGEN PHOSPHATE.
           OKADA K + SUGIE H
           PHYS LETT A 37: 337-8 (1971)

17-2603   EXPERIMENTS ON THE ORDER OF THE TRANSITION IN KDP.
           OKADA K
           KOTAI BUTSURI 7: 334-50 (1972) (IN JAPANESE)

17-2604   INTRACAVITY SECOND HARMONIC GENERATION WITH RUBY LASER BY KDP.
           OKADA K + TOMISHIMA K
           JAP J APPL PHYS 9: 153 (1970)

17-2605   LIGHT MODULATORS USING 45 DEGREE X CUT AND 45 DEGREE Y CUT ADP CRYSTALS.
           OKADA M + IEIRI S
           NHK TECH J 22: 490-500 (1970) (IN JAPANESE)

17-2606   EFFECT OF IMPURITIES ON THE STRUCTURE OF POTASSIUM DIHYDROGEN PHOSPHATE CRYSTALS.
           PAKHOMOV VI + SILNITSKAYA GB
           INORG MATER 7: 1060-3 (1971)

17-2607   STRUCTURE OF FERROELECTRIC CRYSTALS OF THE POTASSIUM DIHYDROGEN PHOSPHATE GROUP.
           PAKHOMOV VI + SILNITSKAYA GB
           BULL ACAD SCI USSR, PHYS SER 34(12): 2230-2 (1970)

17-2608   LOW FREQUENCY HYDROGEN VIBRATIONS IN POTASSIUM DIHYDROGEN PHOSPHATE.
           PALEVSKY H + OTNES K + WAKUTA Y
           P273-80, VOL 2 OF SYMP ON INELASTIC SCATTERING OF NEUTRONS IN
           SOLIDS AND LIQUIDS, CHALK RIVER, 1962, PROC. INT ATOMIC ENERGY AGENCY
           VIENNA, 1963.

17-2609   DETERMINATION OF THE COMPLEX PERMITTIVITY OF ADP AND KDP IN THE FAR INFRARED BY
           DISPERSIVE FOURIER TRANSFORM SPECTROSCOPY.
           PARKER TJ + BURFOOT JC + CHAMBERLAIN J
           P37-9 OF EUROPEAN MEETING ON FERROELECTRICITY SAARBRUCKEN MARCH 1969.
           MUESER HE + PETERSSON J (EDS), WISSENSCH VERLAGSGES STUTTGART, 1970.

           P230-2 OF INT MEETING ON FERROELECTRICITY, 2ND, 1969, KYOTO, PROC,
           (J PHYS SOC JAP, VOL 28 SUPPL). PHYS SOC JAP, 1970.

                205
                207

17-2610   FERROELECTRIC CRITICAL SCATTERING FROM DEUTERATED KDP. (NEUTRON INELASTIC SCATTERING)
           PAUL GL + COCHRAN W + BUYERS WJL + COWLEY RA
           P278-80 OF INT MEETING ON FERROELECTRICITY, 2ND, 1969, KYOTO, PROC,
           (J PHYS SOC JAP, VOL 28 SUPPL). PHYS SOC JAP, 1970, 459P.

17-2611   FERROELECTRIC TRANSITION IN POTASSIUM DIDEUTERIUM PHOSPHATE.
           PAUL GL + COCHRAN W + BUYERS WJL + COWLEY RA
           PHYS REV B 2: 4603-12 (1970)

17. POTASSIUM DIHYDROGEN PHOSPHATE (KDP) AND RELATED PHOSPHATES

MEASUREMENT OF THE RELATIVE NONLINEAR COEFFICIENTS OF POTASSIUM DIHYDROGEN PHOSPHATE, RUBIDIUM DIHYDROGEN PHOSPHATE, RUBIDIUM DIHYDROGEN ARSENATE, AND LITHIUM IODATE.
PEARSON JE + EVANS GA + YARIV A
OPT COMMUN 4(5): 366-7 (JAN 1972)

17-2612   DETECTION OF ATOM TUNNELING IN SOLIDS. (KDP)
PELAH I + IMRY Y
P223-6 OF MATER RES SYMP, 2ND, 1967, MOLECULAR DYNAMICS AND
STRUCTURE OF SOLIDS, RUSH JJ + CARTER RS(EDS), NBS SPEC PUB 301,
1969, 571P.
198

OPTICAL PHONONS AND PHASE TRANSITIONS IN SOME ORDER DISORDER AND DISPLACIVE
FERROELECTRICS. (ANTIMONY SULFOIODIDE, KDP)
PERRY CH + AGRAWAL DK
P342-6 OF NUSIMOVICI M(ED), PHONONS, INT CONF, RENNES, 1971 PROC,
FLAMMARION, PARIS, 1971, 502P.

17-2613   DIELECTRIC ANOMALIES IN CRYSTALS OF POTASSIUM DIHYDROGEN PHOSPHATE POTASSIUM
DIDEUTERIUM PHOSPHATE AND RUBIDIUM DIHYDROGEN PHOSPHATE AT HIGH TEMPERATURES.
PEREVERZEVA LP + POGOSSKAYA NZ + POPLAVKO YUM + PAKHOMOV VI + REZ IS + SILNITSKAYA GB
SOV PHYS SOLID STATE 13(11): 2690-2 (MAY 1972)
265

17-2614   EFFECT OF STRUCTURAL DEFECTS PRODUCED BY GAMMA AND REACTOR IRRADIATION ON PHASE
TRANSITIONS AND DIELECTRIC PROPERTIES OF KDP AND ADP CRYSTALS.
PESHIKOV EV + MUKHTAROVA II
BULL ACAD SCI USSR, PHYS SER 35(9): 1760-4 (1971)

17-2615   STRUCTURAL SENSITIVITY OF THE FERROELECTRIC PHASE TRANSITION AND DIELECTRIC PROPERTIES
OF POTASSIUM DIHYDROGEN PHOSPHATE CRYSTALS IRRADIATED WITH FAST NEUTRONS.
PESHIKOV EV
SOV PHYS CRYSTALLOGR 16(5): 820-3 (MAR-APR 1972)
261

17-2616   PARAMETRIC FREQUENCY CONVERSION OF COHERENT LIGHT BY THE ELECTROOPTIC EFFECT IN
POTASSIUM DIHYDROGEN PHOSPHATE.
PETERSON DG + YARIV A
APPL PHYS LETT 5: 184-6 (1964)

17-2617   ELECTRONIC EFFECT AND ELASTIC PROPERTIES OF RUBIDIUM DIHYDROGEN PHOSPHATE CRYSTALS.
PIERRE MC + DUFOUR JP + REMOISSENET M
SOLID STATE COMMUN 9: 1493-7 (SEP 1971)

IMPURITY MODES IN AN INDUCED MOMENT SYSTEM AND HYDROGEN BONDED FERROELECTRICS (KDP
TYPE, THEORY).
PINK DA
CAN J PHYS 49: 243-56 (1971)

17-2618   RESOLUTION OF A DISCREPANCY CONCERNING THE AVERAGE DIPOLE MOMENT IN KDP TYPE
FERROELECTRICS.
PINK DA
PHYS STATUS SOLIDI 37: K79-80 (1970)
185

17-2619   MEASUREMENT OF THE SPECIFIC HEAT OF KDP BY A DYNAMIC METHOD.
POMPE G + HEGENBARTH E
SOV PHYS SOLID STATE 12(2): 357-62 (1970)
206

17-2620   FERROELECTRIC TRANSITION IN RUBIDIUM DIHYDROGEN PHOSPHIDE USING RAMAN SCATTERING
SPECTRA.
POPOVA EA + SAVATINOVA IT + VELICHKO IA + POPEKO GS
SOV PHYS SOLID STATE 12(12): 2940-1 (JUN 1971)
240

17-2621   ISOTOPE EFFECT IN THE RAMAN SPECTRA OF A DEUTERATED KDP CRYSTAL.
POPOVA EA + SAVATINOVA IT + VELICHKO IA
SOV PHYS SOLID STATE 12(7): 1543-6 (1971)
221

17-2622   RAMAN SPECTRA OF THE FERROELECTRIC MODIFICATION OF A POTASSIUM DIHYDROGEN PHOSPHATE
           CRYSTAL.
           POPOVA EA
           SOV PHYS SOLID STATE 14(7): 1678-80 (JAN 1973)
               282

17-2623   SYMMETRY OF NORMAL OSCILLATIONS IN A FERROELECTRIC CRYSTAL OF KDP.
           POPOVA EA + STEKHANOV AI
           SOV PHYS SOLID STATE 12(1): 40-2 (1970)
               201

17-2624   STRUCTURE OF DIHYDROGEN PHOSPHATE ION IN THE FERROELECTRIC POTASSIUM SALT.
           POPOVA EA + STEKHANOV AI
           SOV PHYS SOLID STATE 10(12): 2948-9 (1969)
               146

17-2625   SPECTRUM OF RAMAN SCATTERED LASER LIGHT IN FERROELECTRICS OF THE KDP GROUP.
           POPOVA EA
           BULL ACAD SCI USSR, PHYS SER 35(9): 1648-51 (1971)

17-2626   TEMPERATURE DEPENDENCE OF RAMAN SCATTERING SPECTRUM OF FERROELECTRIC DEUTERATED KDP
           DURING THE PHASE TRANSITION.
           POPOVA EA + SAVATINOVA IT
           SOV PHYS SOLID STATE 12(9): 2075-7 (1971)
               227

17-2627   THERMOELASTIC PROPERTIES OF POTASSIUM DIHYDROGEN PHOSPHATE NEAR THE HIGH TEMPERATURE
           PHASE TRANSITION. (THERMAL EXPANSION)
           POPLAVKO YUM + REX IS + GORBOKON NV + DIMAROVA EN
           SOV PHYS CRYSTALLOGR 17(3): 595-6 (NOV-DEC 1972)
               280

17-2628   CONDITIONS FOR THE GROWTH OF POTASSIUM DIHYDROGEN PHOSPHATE WHISKERS.
           PORTNOV VN + SOZONTOVA GN + RYZHKOVA TM + BELYUSTIN AV
           INORG MATER 7: 297-8 (1971)

17-2629   PARAMAGNETIC RELAXATION AND DIVALENT COPPER IONS IN FERROELECTRIC KDP.
           PRATT TE + YANG GC + RUBIN B
           BULL AMER PHYS SOC 15: 336 (1970)

17-2630   PREPARATION (MORPHOLOGY) AND MECHANICAL PROPERTIES OF FILAMENTARY CRYSTALS OF ADP.
           PREDVODITELEV AA + PASTERNAK NA + ZAKHAROVA MV
           SOV PHYS CRYSTALLOGR 15(3): 465-7 (1970)
               217

17-2631   TRANSIENT PHENOMENA IN THE DISCHARGE OF CONDENSERS WITH POLAR DIELECTRICS. (POTASSIUM
           DIHYDROGEN PHOSPHATE AND POTASSIUM FERROCYANIDE)
           QUERROU M + LE MONTAGNER S
           COMPT REND 258(1): 106-8 (1964) (IN FRENCH)
               PHYS ABSTR 67: 22776 (1964)

17-2632   AN APPARATUS FOR POLISHING WATER SOLUBLE CRYSTALS (APPLICATION TO ADP AND KDP).
           RAMASWAMY V + DIVINO MD
           REV SCI INSTRUM 43: 1294-6 (SEP 1972)

17-2633   EPITAXIAL ELECTROOPTIC MIXED CRYSTAL AMMONIUM POTASSIUM DIHYDROGEN PHOSPHATE FILM
           WAVEGUIDE.
           RAMASWAMY V
           APPL PHYS LETT 21: 183-5 (1 SEP 1972)
               277

17-2634   PHASE TRANSFORMATIONS AND MELTING IN KDP TO 40 KBAR (BY DTA).
           RAPOPORT E
           J CHEM PHYS 53(1): 311-14 (1970)
               212

17-2635   PULSE RESPONSE OF ZERO DEGREE CUT ADP MODULATORS.
           RASHIDI K
           ELECTRON LETT 7: 114-15 (1971)

17-2636   SIMULTANEOUS OBSERVATION CF THE SOFT FERROELECTRIC AND ACOUSTIC MODES OF POTASSIUM
          DIDEUTERIUM PHOSPHATE.
          REESE RL + FRITZ IJ + CUMMINS HZ
          SOLID STATE COMMUN 9: 327-30 (1971)
              232

17-2637   CONTRIBUTION TO THE PREPARATION CN NON-WEDGE SHAPED RUBIDIUM DIHYDROGEN OR DIDEUTERIUM)
          PHOSPHATE CRYSTALS THEORY AND PRACTICAL CONSIDERATIONS).
          REMOISSENET M + DESVIGNES JM + MARECEK V
          KRISTALL TECH 5(4): 535-40 (1970)
              227

17-2638   FAR INFRARED SPECTRA OF RUBIDIUM DIHYDROGEN PHOSPHATE CRYSTALS.
          REMOISSENET M + GARD R
          J PHYS (PARIS) 33, SUPPL 4, COLLOQ 2: 106-7 (1972)
              294

          ELECTRICAL SUSCEPTIBILITY OF ADP (THEORY).
          RICHMOND P
          PHYS LETT A 32(1): 52-3 (1970)

17-2639   RAMAN SCATTERING FROM FERROELECTRIC MODES IN THE KDP ISOMORPHOUS PHOSPHATES AND
          ARSENATES. (KDP, AMMONIUM DIHYDROGEN PHOSPHATE)
          RYAN JF + KATIYAR RS + TAYLOR W
          J PHYS (PARIS) 33, SUPPL 4, COLLOQ 2: 49-51 (1972)
              294

17-2640   MICROWAVE MODULATOR OF RUBY LASER LIGHT USING POTASSIUM DIHYDROGEN PHOSPHATE CRYSTAL.
          SAITO S + KIMURA T
          JAP J APPL PHYS 2: 658-9 (1963)

          PRESSURE STUDIES OF FERROELECTRIC PROPERTIES. (PEROVSKITES, KDP TYPE)
          SAMARA GA + DRICKAMER HG
          COMMENT SOLID STATE PHYS 3: 1-7 (1970)

17-2641   PRESSURE DEPENDENCE OF THE FERROELECTRIC PROPERTIES OF POTASSIUM DIDEUTERIUM PHOSPHATE.
          SAMARA GA
          PHYS LETT A 25(9): 664-6 (1967)
              143

17-2642   VANISHING OF THE FERROELECTRIC AND ANTIFERROELECTRIC STATES IN POTASSIUM DIHYDROGEN
          PHOSPHATE TYPE CRYSTALS AT HIGH PRESSURE.
          SAMARA GA
          PHYS REV LETT 27: 103-6 (1971)

17-2643   MICROSCOPIC CBSERVATION CF ANTIFERROELECTRIC DOMAINS OF AMMONIUM DIHYDROGEN PHOSPHATE.
           (PHASE TRANSITION AT 148 DEG K)
          SARVADA A + MASE Y + TAKAGI Y + MIDORIKAWA M
          J PHYS SOC JAP 29(4): 969-72 (1970)
              217

17-2644   MOSSBAUER STUDY OF ANTIFERROELECTRIC TRANSITION IN IRON-57(3X) AMMONIUM DIHYDROGEN
          PHOSPHATE.
          SASTRY MD
          SOLID STATE COMMUN 11: 1671-4 (1972)
              283

17-2645   DEUTERON MOTIONS IN POTASSIUM DIDEUTERIUM PHOSPHATE (FROM POLARIZED REFLECTION SPECTRA,
          A TUNNELING MODEL).
          SATO Y
          PHYS LETT A 33(5): 156-7 (1970)
              222

17-2646   OPTICAL STUDY ON THE PHASE TRANSITIONS IN POTASSIUM DIHYDROGEN OR DEUTERIUM PHOSPHATE
           (INFRARED PHOSPHATE STRETCHING, WIDE TEMPERATURE RANGE).
          SATO Y
          J CHEM PHYS 53(3): 887-92 (1970)
              215

17-2647   RAMAN SPECTRUM OF POTASSIUM DIDEUTERIUM PHOSPHATE CRYSTAL IN PARA- AND FERROELECTRIC
          PHASE.
          SAVATINOVA IT + CHISLER EV + SUBEVA EV
          COMPT REND ACAD BULG SCI 25: 1475-8 (1972)

17. POTASSIUM DIHYDROGEN PHOSPHATE (KDP) AND RELATED PHOSPHATES

17-2648 STUDY OF THE (MECHANICAL) STRENGTH PROPERTIES OF POTASSIUM DIHYDROGEN PHOSPHATE
CRYSTALS.
SAVINKOV AI + BLISTANOV AA + GAIDUCHENYA VF + MARKOVSKII VYU + REZ IS + SHASKOLSKAYA MP
SOV PHYS CRYSTALLOGR 16(2): 380-1 (SEP-OCT 1971)
246

17-2649 TEMPERATURE DEPENDENCE OF THE MECHANICAL CHARACTERISTICS OF POTASSIUM DIHYDROGEN
PHOSPHATE SINGLE CRYSTALS.
SAVINKOV AI + SHASKOLSKAYA MP + MARKOVSKII VYU + REZ IS
BULL ACAD SCI USSR, PHYS SER 35(9): 1735-7 (1971)
269

17-2650 INFRARED VIBRATION FREQUENCIES OF THE HYDROGEN BOND IN KDP FERROELECTRIC
(CALCULATIONS).
SCHMIT J + LUCAS A
P461-70 OF HAIDEMENAKIS ED(ED), OPTICAL PROPERTIES OF SOLIDS,
GORDON AND BREACH, 1970, 520P. (CHANIA CONF, 5TH PROC,
CHANIA, CRETE, 1969)
224

17-2651 RANDOM MOTION OF DEUTERONS IN DEUTERATED POTASSIUM DIHYDROGEN PHOSPHATE.
SCHMIDT VH + UEHLING EA
PHYS REV 126: 447-57 (1962)

17-2652 THEORY OF DOMAIN WALL MOTION IN POTASSIUM DIHYDROGEN PHOSPHATE TYPE CRYSTALS.
SCHMIDT VH
P97-104, VOL 2 OF DVORAK V + FOUSKOVA A + GLOGAR P (EDS),
INT MEETING ON FERROELECTRICITY, PRAGUE, 1966, PROC.
INST PHYS CZECH ACAD SCI, PRAGUE, 1966.

17-2653 PROTON PHONON INTERACTIONS IN POTASSIUM DIHYDROGEN PHOSPHATE.
SCOTT JF
SOLID STATE COMMUN 10: 597-600 (1972)
265

17-2654 MEASUREMENT OF THE PRIMARY LINEAR ELECTROOPTIC EFFECT IN AMMONIUM DIHYDROGEN PHOSPHATE
CRYSTALS.
SHALDIN YUV
SOV PHYS CRYSTALLOGR 15(1): 146-7 (1970)
207

17-2655 EFFECT OF PROTON- PHONON COUPLING ON THE FERROELECTRIC MODE IN POTASSIUM DIHYDROGEN
PHOSPHATE. (COLLECTIVE PSEUDOSPIN MODEL OF KDP IS QUESTIONABLE)
SHE CY + BROBERG TW + WALL LS + EDWARDS DF
PHYS REV B SOLID STATE 6(5): 1847-50 (1 SEP 1972)
273

17-2656 LANDAU-KHALATNIKOV ATTENUATION IN POTASSIUM DIDEUTERIUM PHOSPHATE CRYSTALS.
SHERMAN AB + VAIDA D + VELICHKO IA + GUTNER OS + LEMANOV VV
SOV PHYS SOLID STATE 13: 3143-5 (1972)

17-2657 RAMAN SPECTRA OF TETRAGONAL POTASSIUM DIHYDROGEN PHOSPHATE.
SHE CY + BROBERG TW + EDWARDS DF
PHYS REV B 4: 1580-3 (1971)

17-2658 SOLUBILITY ISOBAR IN THE AMMONIUM DIHYDROGEN PHOSPHATE- POTASSIUM DIHYDROGEN PHOSPHATE-
WATER SYSTEM.
SHENKIN YAS + RUCHNOVA SA + RODIONOVA NA
INORG CHEM 17: 1769-70 (1972)

17-2659 ULTRASONIC STRESS WAVE INTERACTION WITH DOMAIN WALLS IN POTASSIUM DIDEUTERIUM
PHOSPHATE.
SHERMAN AB + VAJDA D
CZECH J PHYS B 22(9): 826-31 (1972)
288

17-2660 ANOMALOUS PROPAGATION OF LONGITUDINAL ULTRASONIC WAVES IN POTASSIUM DIDEUTERIUM
PHOSPHATE (TEMPERATURE DEPENDENCE OF THE YOUNGS MODULUS AND THE POLARIZATION RELAXATION
TIME).
SHIMSHONI M + HARNIK E
PHYS LETT A 32(5): 321-2 (1970)
222

17. POTASSIUM DIHYDROGEN PHOSPHATE (KDP) AND RELATED PHOSPHATES

17-2661 NEUTRON DIFFRACTION STUDIES OF FERROELECTRIC PHASE TRANSITIONS IN KDP AND POTASSIUM DIDEUTERIUM PHOSPHATE.
SHIBUYA I + IWATA Y + KOYANO N
ACTA CRYSTALLOGR A 28: S181 (1972)

17-2662 OSCILLATORY BEHAVIOR OF RAMAN SCATTERING INTENSITY WITH TEMPERATURE IN THE FERROELECTRIC PHASE POTASSIUM DIHYDROGEN PHOSPHATE. (DIFFRACTION BY DOMAIN WALLS)
SHIGENARI T + TAKAGI Y
SOLID STATE COMMUN 11: 481-4 (1972)
273

17-2663 RAMAN SPECTRUM OF FERROELECTRIC MODE IN A POTASSIUM DIHYDROGEN PHOSPHATE CRYSTAL.
SHIGENARI T + TAKAGI Y
J PHYS SOC JAP 31: 312 (1971)
256

17-2664 ULTRASONIC MEASUREMENT OF THE ELECTROCALORIC EFFECT IN KDP (NEAR THE CURIE TEMPERATURE).
SHIMSHONI M + HARNIK E
J PHYS CHEM SOLIDS 31: 1416-17 (1970)
197

17-2665 ANOMALIES IN THE INTERNAL STRESS FOR FERROELECTRIC PHOSPHATES NEAR THEIR CURIE POINTS. (POTASSIUM DIHYDROGEN PHOSPHATE, ALSO RUBIDIUM ANALOG AND POTASSIUM DIDEUTERIUM PHOSPHATE)
SHUVALOV LA + MNATSAKANYAN AV
BULL ACAD SCI USSR, PHYS SER 29(11): 1809-16 (1965)
CHEM ABSTR 64: 4393H (1966)

17-2666 PHASE DIAGRAM OF SODIUM TRIDEUTERIUM(X) TRIHYDROGEN(1-X) PHOSPHATE SYSTEM VERSUS CONCENTRATION, TEMPERATURE, FIELD, AND PRESSURE.
SHUVALOV LA + SHIROKOV AM + IVANOV NR + BARANOV AI + KIRIPICHNIKOVA LF + SCHAGINA NM
J PHYS (PARIS) 33, SUPPL 4, COLLOQ 2: 165-7 (1972)
294

DIELECTRIC RESONANCE AND RELAXATION IN ELECTRICALLY BIASED KDP AND DEUTERATED ISOMORPHS. (THEORY)
SILVERMAN BD
P255-7 OF INT MEETING ON FERROELECTRICITY, 2ND, 1969, KYOTO, PROC,
(J PHYS SOC JAP, VOL 28 SUPPL). PHYS SOC JAP, 1970, 459P.

17-2667 COUPLED ORDER DISORDER PHONON SYSTEM WITH DAMPING: KDP (FREQUENCY DEPENDENT SUSCEPTIBILITY).
SILVERMAN BD
PHYS REV LETT 25(2): 107-10 (1970)
206

17-2668 FERROELECTRIC (SOFT) MODE MOTION IN POTASSIUM DIDEUTERIUM PHOSPHATE . (BY NEUTRON TRIPLE AXIS SPECTROMETRY)
SKALYO J + FRAZER BC + SHIRANE G
PHYS REV B 1(1): 278-86 (1970)
195

17-2669 OPTICAL MIXING OF COHERENT AND INCOHERENT LIGHT. (KDP CRYSTAL FILTER)
SMITH AW + BRASLAU N
IBM J RES DEVELOP 6: 361-2 (1962)

DYNAMICAL THEORY OF AMMONIUM DIHYDROGEN PHOSPHATE TYPE ANTIFERROELECTRICS WITH HYDROGEN BONDS.
STASYUK IV + LEVITSKII RR
BULL ACAD SCI USSR, PHYS SER 35(9): 1616-19 (1971)

THEORY OF COHERENT NEUTRON SCATTERING BY HYDROGEN BONDED FERROELECTRICS AT LOW TEMPERATURES. PART-1: GENERAL EXPRESSION FOR INELASTIC COHERENT SCATTERING OF SLOW NEUTRONS AND EFFECTIVE THERMAL FACTORS. (KDP TYPE) PART-2: SCATTERING CHARACTERISTICS AND VARIOUS CONCEPTIONS OF THE TUNNELING QUASI SPIN MODEL.
STAMENKOVIC S
J LOW TEMP PHYS 9: 475-83, 485-94 (1972)

17. POTASSIUM DIHYDROGEN PHOSPHATE (KDP) AND RELATED PHOSPHATES

17-2670   COUPLED VIBRATIONS OF THE PROTON ION SYSTEM IN HYDROGEN BONDED FERROELECTRICS OF THE
           KDP TYPE (PHONON PROTON INTERACTION).
           STASYUK IV + LEVITSKII RR
           UKR FIZ ZH 15(3): 460-9 (1970)
                226

17-2671   INFLUENCE OF POLAR STATES OF PROTONS ON THE PHASE TRANSITION IN FERROELECTRICS WITH
           HYDROGEN BONDS. (KDP, DEUTERATED KDP)
           STASYUK IV + LVITSKII RR + LITVINOV VI
           UKR FIZ ZH 15: 470-6 (1970) (IN RUSSIAN)

17-2672   PROTON PHONON INTERACTION IN FERROELECTRICS WITH HYDROGEN BONDS. (KDP TYPE)
           STASYUK IV
           THEOR MAT FIZ 9: 431-9 (1971) (IN RUSSIAN)

17-2673   PROPERTIES OF THE KDP TYPE FERROELECTRICS WITH IMPURITIES.
           STAMENKOVIC S + ZEKOVIC S
           PHYS STATUS SOLIDI B 49: 277-85 (1972)

17-2674   ROLE OF PROTON PHONON INTERACTION IN THE PHASE TRANSITION OF FERROELECTRICS WITH
           HYDROGEN BONDS KDP TYPE.
           STASYUK IV + LEVITSKII RR
           PHYS STATUS SOLIDI 39(1): K35-K38 (1970)
                225

17-2675   TEMPERATURE INSTABILITY OF ELECTROOPTIC LASER MODULATORS WITH AMMONIUM DIHYDROGEN
           PHOSPHATE, POTASSIUM DIHYDROGEN PHOSPHATE AND PARTIALLY DEUTERATED POTASSIUM DIHYDROGEN
           PHOSPHATE CRYSTALS.
           STADNIK B
           ACTA TECH (PRAGUE) 15(1): 65-9 (1970)
                CHEM ABSTR 73: 9287P (1970)
                205

17-2676   CRITICAL BEHAVIOR OF KDP AND RUBIDIUM DIHYDROGEN PHOSPHATE CRYSTALS.
           STRUKOV BA + KORZHUEV MA + KOPTSIK VA
           BULL ACAD SCI USSR, PHYS SER 35(9): 1678-80 (1971)

17-2677   FIRST ORDER PHASE TRANSITION IN POTASSIUM DEUTERIUM PHOSPHATE CRYSTALS.
           STRUKOV BA + BADDUR A + VELICHKO IA
           SOV PHYS SOLID STATE 13(8): 2085-6 (FEB 1972)
                258

17-2678   SPONTANEOUS POLARIZATION OF A POTASSIUM DIHYDROGEN PHOSPHATE CRYSTAL NEAR THE CURIE
           POINT. (ELECTROCALORIC EFFECT)
           STRUKOV BA + KORZHUEV MA + BADDUR A + KOPTSIK VA
           SOV PHYS SOLID STATE 13(7): 1569-73 (JAN 1972)
                254

17-2679   SPECIFIC HEAT OF SOME KDP TYPE CRYSTALS. (POTASSIUM DIHYDROGEN PHOSPHATE, RUBIDIUM AND
           AMMONIUM ANALOGS SOME DEUTERATE)
           STRUKOV BA + SOLIMAN MA + KOPSIK VA
           P207-9 OF INT MEETING ON FERROELECTRICITY, 2ND, 1969, KYOTO, PROC.
           (J PHYS SOC JAP, VOL 28 SUPPL). PHYS SOC JAP, 1970, 459P.

17-2680   SOME THERMODYNAMICAL PROPERTIES OF POTASSIUM DIHYDROGEN PHOSPHATE- POTASSIUM
           DIHYDROGEN(1-X) DIDEUTERIUM(X) PHOSPHATE MIXED CRYSTALS.
           STRUKOV BA + BADDUR A + KOPTSIK VA
           J PHYS (PARIS) 33, SUPPL 4, COLLOQ 2: 155-7 (1972)
           SOV PHYS SOLID STATE 14: 885-9 (1972)
                294

17-2681   FAR INFRARED REFLECTIVITY SPECTRA OF KDP CRYSTAL (16-400 PER CM, 83-300K).
           SUGAWARA F + NAKAMURA T
           J PHYS SOC JAP 28(1): 158-60 (1970)
                190

17-2682   THERMAL HYSTERESIS OF THE FERROELECTRIC TRANSITION IN POTASSIUM DIHYDROGEN PHOSPHATE.
           SUGIE H + OKADA K + KANNO K
           J PHYS SOC JAP 33: 1727 (1972)
                282

17-2683 ESR STUDY OF ADP- KDA MIXED CRYSTALS.
SUZUKI I + ABE R
J PHYS SOC JAP 30: 1210 (APR 1971)

17-2684 ESR STUDY OF AMMONIUM IONS IN IRRADIATED AMMONIUM DIDEUTERIUM PHOSPHATE- 5 PERCENT
POTASSIUM DIDEUTERIUM ARSENATE.
SUZUKI I + ABE R
J PHYS SOC JAP 31: 951 (1971)

17-2685 ORDER DISORDER THEORY OF POTASSIUM DIHYDROGEN PHOSPHATE TYPE FERROELECTRICS.
SVETINA S + BLINC R
P119-26, VOL 1 OF DVORAK V + FOUSKOVA A + GLOGAR P (EDS),
INT MEETING ON FERROELECTRICITY, PRAGUE, 1966, PROC.
INST PHYS CZECH ACAD SCI, PRAGUE, 1966.

17-2686 CURIE-WEISS LAW FOR THE NONLINEAR SUSCEPTIBILITY OF A POTASSIUM DIHYDROGEN PHOSPHATE
CRYSTAL.
TALYANSKII VI + FILIMONOV AA + YASHCHIN EG
SOV PHYS SOLID STATE 12: 2224-5 (1971)

17-2687 PROTON MOTIONS IN HYDROGEN BONDED FERROELECTRIC AND ANTIFERROELECTRIC SOLIDS. PART-1:
ENERGY LOSS NEUTRON INCOHERENT SCATTERING SPECTRA OF POTASSIUM DIHYDROGEN PHOSPHATE AND
SILVER PERIODATE BELOW THE CRITICAL TEMPERATURE.
TEMME FP + WADDINGTON TC
J CHEM SOC FARADAY TRANS 68: 350-6 (1972)

17-2688 MEASUREMENT OF CHEMICAL SHIFT IN SOLID POTASSIUM DIHYDROGEN PHOSPHATE.
TERAO T + HASHI T
J MAGN RESON 7: 238-40 (1972)

17-2689 DIELECTRIC PERMITTIVITY OF FINELY DIVIDED POTASSIUM AND AMMONIUM DIHYDROGEN PHOSPHATE.
THOUY G + PALETTO J + CAPRON JP + OHANESSIAN H
COMPT REND B 273(21): 896-9 (22 NOV 1971) (IN FRENCH)
        CHEM ABSTR 76: 78022X (1972)
        264

17-2690 PULSE DISTORTIONS IN MISMATCHED SECOND HARMONIC GENERATION. (POTASSIUM DIHYDROGEN
PHOSPHATE)
THOMAS JM + TARAN JP
OPT COMMUN 4(5): 329-34 (JAN 1972)
        262

         MODELS OF THE PHASE TRANSITION IN POTASSIUM DIHYDROGEN PHOSPHATE TYPE FERROELECTRICS.
         (24 REFS)
TOKUNAGA M
FERROELECTRICS 1(4): 195-206 (OCT 1970)

17-2691 EFFECT OF GAMMA-IRRADIATION ON THE DIELECTRIC CONSTANTS OF KDP.
TOKUMARU Y + ABE R
JAP J APPL PHYS 9: 1548-9 (1970)
        222

17-2692 DYNAMICS OF POTASSIUM DIHYDROGEN PHOSPHATE TYPE FERROELECTRIC PHASE TRANSITIONS.
TSALLIS C
J PHYS (PARIS) 33: 1121-7 (NOV-DEC 1972)

17-2693 SECOND HARMONIC GENERATION OF LIGHT IN THE KDP CRYSTAL BY THE INDEX MATCHING METHOD.
TSUCHIYA A + SASAKI H + SUGAHARA M
REP FAC ENG YAMANASHI UNIV NO 23: 6-8 (1972) (IN JAPANESE)

17-2694 ELECTROOPTIC AMPLITUDE MODULATION OF LASER GENERATED SECOND HARMONICS IN POTASSIUM
DIHYDROGEN PHOSPHATE.
VAN DER ZIEL JP
APPL PHYS LETT 5: 27-9 (1964)

17-2695 TEMPERATURE DEPENDENCE OF OPTICAL (SECOND) HARMONIC GENERATION IN POTASSIUM DIHYDROGEN
PHOSPHATE (AND DEUTERATED POTASSIUM DIHYDROGEN PHOSPHATE) FERROELECTRICS.
VAN DER ZIEL JP + BLOEMBERGEN N
PHYS REV 135(6): 1662-9 (1964)

17. POTASSIUM DIHYDROGEN PHOSPHATE (KDP) AND RELATED PHOSPHATES

17-2696   RELATIONSHIP BETWEEN THE DIELECTRIC AND ELECTROOPTICAL PROPERTIES OF FERROELECTRIC
          CRYSTALS OF THE POTASSIUM DIHYDROGEN PHOSPHATE GROUP IN THE PARAELECTRIC PHASE.
          VASILEVSKAYA AS + SONIN AS
          SOV PHYS SOLID STATE 13(6): 1299-1304 (DEC 1971)
                254

          ENERGY PARAMETERS AND PROTON TUNNELING IN POTASSIUM DIHYDROGEN PHOSPHATE TYPE
          FERROELECTRICS. (THEORY)
          VILLAIN J + AUBRY S
          P41-5 OF EUROPEAN MEETING ON FERROELECTRICITY, 1969, SAARBRUCKEN,
          MUSER HE AND PETERSSON J(EDS), WISSENSCH VERLAGSGES, STUTTGART,
          1970, 396P.

17-2697   ELECTROOPTICAL AND PIEZOOPTICAL PROPERTIES OF POTASSIUM DIHYDROGEN PHOSPHATE AND
          AMMONIUM DIHYDROGEN PHOSPHATE CRYSTALS AT HIGH TEMPERATURES.
          VLOKH OG + LUSTIV-SHUMSKII LF + PYLYPYSHIN BP
          SOV PHYS CRYSTALLOGR 16(4): 717-18 (JAN-FEB 1972)
                254

17-2698   ELECTROOPTICAL AND OPTICAL PROPERTIES OF PARTIALLY DEUTERATED RUBIDIUM DIHYDROGEN
          PHOSPHATE CRYSTALS.
          VOLKOVA EN + BEREZHNOI BM + IZRAILENKO AN + MISHCHENKO AV + RASHKOVICH LN
          BULL ACAD SCI USSR, PHYS SER 35(9): 1690-3 (1971)

17-2699   LOW FREQUENCY HYDROGEN VIBRATIONS IN POTASSIUM DIHYDROGEN PHOSPHATE.
          WAKUTA Y
          J PHYS SOC JAP 18: 672-9 (1963)

17-2700   DETERMINATION OF THE EIGENVECTORS OF SOFT MODES IN POTASSIUM DIDEUTERIUM PHOSPHATE AND
          DEUTERATED AMMONIUM DIDEUTERIUM PHOSPHATE.
          WALLACE EA + COCHRAN W + STRINGFELLOW M
          J PHYS (PARIS) 33, SUPPL 4, COLLOQ 2: 59-61 (1972)
                294

17-2701   MODE INTERACTIONS IN POTASSIUM DIHYDROGEN PHOSPHATE.
          WALL LS + SHE CY
          OPT COMMUN 5(2): 123-5 (MAY 1972)
                265

17-2702   SIMULATION OF A DOUBLE 45 DEGREES, Z CUT POTASSIUM DIDEUTERIUM PHOSPHATE. ELECTROOPTIC
          Q SWITCH BY DESK-TOP COMPUTER.
          WARNER J
          OPT LASER TECHNOL 3: 215-17 (1971)

17-2703   (REVERSIBLE) OPTICAL DAMAGE IN KDP (WITH RELAXATION TIME OF SECONDS WHEN ILLUMINATED
          UNDER APPLIED ELECTRIC FIELD).
          WAX SI + CHODOROW M + PUTHOFF HE
          APPL PHYS LETT 16(4): 157-9 (1970)
                181

17-2704   GROWTH BEHAVIOR OF SMALL SEEDS OF AMMONIUM DIHYDROGEN PHOSPHATE IN A MODIFIED
          CIRCULATION APPARATUS.
          WEIS J
          KRISTALL TECH 6(1): 69-76 (1971) (IN GERMAN)
                CHEM ABSTR 74: 92029V (1971)
                278

17-2705   RAMAN SCATTERING BY THE TEMPERATURE DEPENDENT FERROELECTRIC MODE IN POTASSIUM
          DIDEUTERIUM PHOSPHATE (COMPARED WITH KDP AND WITH NEUTRON AND MICROWAVE EXPERIMENTS).
          WHITE KI + TAYLOR W + KATIYAR RS + KAY SM
          PHYS LETT A 33(3): 175-6 (1970)
                225

17-2706   ANTIFERROELECTRIC TRANSITIONS IN ADP AND AMMONIUM DIHYDROGEN ARSENATE STUDIED BY
          INFRARED ABSORPTION.
          WIENER E + LEVIN S + PELAH I
          J CHEM PHYS 52: 2891-2900 (1970)
                205

17. POTASSIUM DIHYDROGEN PHOSPHATE (KDP) AND RELATED PHOSPHATES

17-2707 CRYSTALLIZATION POTENTIALS IN AQUEOUS SOLUTIONS. (POTASSIUM BROMIDE, POTASSIUM CHLORIDE, SODIUM CHLORIDE, POTASSIUM SODIUM TARTRATE, POTASSIUM DIHYDROGEN PHOSPHATE)
WIEDERKEHR H
EIDGENOSSISCHE TECH HOCHSCHULE, ZURICH, 1970, 123P. N72-13712
(IN SWEDISH)
SCI TECH AEROSP REP 10(4): 528 (1972)
261

17-2708 PROTON DYNAMICS IN KDP TYPE FERROELECTRICS STUDIED BY INFRARED ABSORPTION.
WIENER E + LEVIN S + PELAH I
J CHEM PHYS 52(6): 2881-91 (1970)
200

17-2709 RAMAN SCATTERING IN POTASSIUM DIHYDROGEN PHOSPHATE.
WILSON CM + CUMMINS HZ
P420-3 OF INT CONF ON LIGHT SCATTERING IN SOLIDS, 2ND PROC,
BALKANSKI M(ED), PARIS, 1971, FLAMMARION, PARIS, 1971, 518P.

17-2710 MODIFIED POTASSIUM DIHYDROGEN PHOSPHATE MODEL IN A STAGGERED FIELD.
WU FY
PHYS REV B SOLID STATE 3: 3895-900 (1971)
242

17-2711 AN OBSERVATION OF THE CHANGE IN COOLED KDP TEMPERATURE DUE TO 337 MICRON INCIDENT RADIATION.
YAMANAKA M + KAWAMURA T + HINENO M
J PHYS D APPL PHYS 5: 1743-4 (SEP 1972)

17-2712 EFFICIENT HIGH GRAIN PARAMETRIC GENERATION IN ADP CONTINUOUSLY TUNABLE ACROSS THE VISIBLE SPECTRUM.
YARBOROUGH JM + MASSEY GA
APPL PHYS LETT 18: 438-40 (1971)

17-2713 FERROELECTRIC MODE AND NQR (NUCLEAR QUADRUPOLE RESONANCE) FREQUENCY. (POTASSIUM DIHYDROGEN ARSENATE)
YI PN
BULL AMER PHYS SOC 15: 56 (1970)

17-2714 RELATIVE PERMITTIVITY OF POTASSIUM DIHYDROGEN PHOSPHATE AND AMMONIUM DIHYDROGEN PHOSPHATE CRYSTALS AT MICROWAVE FREQUENCIES.
ZEHENTNER J
ELEKTROTECH CAS 21: 633-71 (1970) (IN CZECH)
256

17-2715 AMPLITUDE MODULATORS BASED ON THE MICHELSON INTERFEROMETER. (AMMONIUM DIHYDROGEN PHOSPHATE)
ZERNIKE F + WEBSTER JC
FERROELECTRICS 3: 163-7 (1972);
IEEE TRANS SON ULTRASON 19: 163-7 (1972)
263

17-2716 LIGHT AND CURRENT PULSES FROM X-RAYED POTASSIUM DIHYDROGEN PHOSPHATE CRYSTALS.
ZEREM JZ + HALPERIN A
J APPL PHYS 42: 5263-6 (DEC 1971)

17-2717 THERMAL SELF FOCUSING IN POTASSIUM DIHYDROGEN PHOSPHATE AND AMMONIUM DIHYDROGEN PHOSPHATE CRYSTALS OF LIGHT FROM A LASER OPERATING IN THE FREE GENERATION MODE.
ZVEREV GM + LEVCHUK EA + MALDUTIS EK
SOV PHYS JETP 31(5): 794-5 (NOV 1970)
208

18. POTASSIUM DIHYDROGEN ARSENATE (KDA) AND RELATED ARSENATES

18-2718   NMR AND DIELECTRIC RELAXATION STUDIES OF LONG RANGE CORRELATION IN FERROELECTRIC
          POTASSIUM DIHYDROGEN ARSENATE.
          ADRIAENSSENS GJ + BJORKSTAM JL + AIKINS J
          J MAGN RESON 7: 99-104 (1972)
               269

18-2719   NMR STUDIES IN SOME HYDROGEN BONDED FERROELECTRICS. (DEUTERATED ROCHELLE SALT,
          POTASSIUM OR RUBIDIUM OR AMMONIUM DIHYDROGEN ARSENATES, RUBIDIUM DIDEUTERIUM ARSENATE)
          BJORKSTAM JL
          P101-4 OF INT MEETING ON FERROELECTRICITY, 2ND, 1969, KYOTO, PROC,
          (J PHYS SOC JAP, VOL 28 SUPPL). PHYS SOC JAP, 1970, 459P.

          DEUTERON MAGNETIC RESONANCE AND RELAXATION IN FERROELECTRIC POTASSIUM DIDEUTERIUM
          PHOSPHATE, POTASSIUM DIDEUTERIUM ARSENATE, AND CESIUM DIDEUTERIUM ARSENATE.
          BLINC R + STEPISNIK J + JAMSIK-VILFAN M + ZUMER S
          J CHEM PHYS 54: 187-95 (1971)

18-2720   CESIUM-183 SPIN LATTICE RELAXATION AND RESONANCE IN FERROELECTRIC CESIUM DIDEUTERIUM
          ARSENATE AND CESIUM DIHYDROGEN ARSENATE.
          BLINC R + MALI M + SLAK J + STEPISNIK J + ZUMER S
          J CHEM PHYS 56: 3566-9 (1972)

18-2721   DEUTERON QUADRUPOLE COUPLING IN FERROELECTRIC CESIUM DIDEUTERIUM ARSENATE.
          BLINC R + MALI M + STEPISNIK J + JAMSIK-VILFAN M
          PHYS STATUS SOLIDI 41(2): K123-126 (1970)
               223

18-2722   STARK EFFECT IN THE NUCLEAR QUADRUPOLE RESONANCE SPECTRA OF THE FERROELECTRIC CRYSTAL
          CESIUM DIHYDROGEN ARSENATE.
          BOGUSLAVSKII AA + SEMIN GK
          BULL ACAD SCI USSR, PHYS SER 34(12): 2241-3 (1970)

          POLARIZATION OF THE RAMAN BANDS OF THE HYDROXYL VIBRATIONS AND STRUCTURE OF THE
          HYDROGEN BONDS IN THE FERROELECTRIC CRYSTALS RUBIDIUM DIHYDROGEN PHOSPHATE AND RUBIDIUM
          DIHYDROGEN ARSENATE.
          CHISLER EV + SAVATINOVA IT + DAVYDOV VYU
          SOV PHYS SOLID STATE 13(6): 1339-41 (DEC 1971)

          PHASE DIAGRAMS OF AMMONIUM DIHYDROGEN PHOSPHATE AND POTASSIUM DIHYDROGEN ARSENATE TO 52
          KBAR AND 400 DEG C.
          CLARK JB
          HIGH TEMP HIGH PRESSURE 1: 553-9 (1969)

18-2723   DIELECTRIC RESPONSE IN PIEZOELECTRIC CRYSTALS. (RAMAN SCATTERING IN POTASSIUM
          DIHYDROGEN ARSENATE AND CESIUM DIHYDROGEN ARSENATE)
          COWLEY RA + COOMBS GJ + KATIYAR RS + RYAN JF + SCOTT JF
          J PHYS C 4(10): L203-7 (1971)

18-2724   DETECTION BY ELECTRON PARAMAGNETIC RESONANCE AND ELECTRON NUCLEAR DOUBLE RESONANCE OF
          THE PROTON LATTICE COUPLED MODE IN POTASSIUM DIHYDROGEN ARSENATE, POTASSIUM DEUTERIUM
          ARSENATE, AND AMMONIUM DIHYDROGEN ARSENATE. (FERROELECTRICS AND ANTIFERROELECTRICS)
          DALAL NS + MCDOWELL CA
          PHYS REV B SOLID STATE 5(3): 1074-7 (1972)
               258

18-2725 ELECTRON PARAMAGNETIC RESONANCE STUDIES OF X-IRRADIATED POTASSIUM DIHYDROGEN ARSENATE, POTASSIUM DIDEUTERIUM ARSENATE, RUBIDIUM DIHYDROGEN ARSENATE, RUBIDIUM DIDEUTERIUM ARSENATE, CESIUM DIHYDROGEN ARSENATE, AMMONIUM DIHYDROGEN ARSENATE, AND DEUTERO AMMONIUM DIDEUTERIUM ARSENATE. (FERROELECTRICS AND ANTIFERROELECTRICS)
DALAL NS + DICKINSON JF + MCDOWELL CA
J CHEM PHYS 57: 4254-65 (NOV 15, 1972)

18-2726 ELECTRON PARAMAGNETIC RESONANCE AND ELECTRON NUCLEAR DOUBLE RESONANCE OBSERVATIONS OF DOMAINS IN THE IRRADIATED FERROELECTRICS POTASSIUM DIHYDROGEN ARSENATE AND POTASSIUM DIDEUTERIUM ARSENATE.
DALAL NS + MCDOWELL CA + SRINIVASAN R
PHYS REV LETT 25: 823-6 (1970)

18-2727 ELECTRON PARAMAGNETIC RESONANCE AND ELECTRON NUCLEAR DOUBLE RESONANCE OF THE ARSENATE CENTER IN X-IRRADIATED POTASSIUM DIHYDROGEN ARSENATE.
DALAL NS + MCDOWELL CA + SRINIVASAN R
MOL PHYS 24: 417-39 (1972)

18-2728 HYDROGEN BONDED FERROELECTRICS: EVIDENCE FOR SLATER CONFIGURATIONS FROM EPR STUDIES OF X-IRRADIATED POTASSIUM DIHYDROGEN ARSENATE AND AMMONIUM DIHYDROGEN ARSENATE.
DALAL NS + MCDOWELL CA + SRINIVASAN R
MOL PHYS 24(5): 1051-7 (NOV 1972)
    285

18-2729 OBSERVATION OF ELECTRON NUCLEAR TRIPLE RESONANCE FOR THE ARSENATE CENTER IN X-RAY IRRADIATED SINGLE CRYSTALS OF POTASSIUM DIHYDROGEN ARSENATE.
DALAL NS + MCDOWELL CA
CHEM PHYS LETT 6: 617-19 (1970)

18-2730 THERMODYNAMIC PROPERTIES OF POTASSIUM DIHYDROGEN ARSENATE AND POTASSIUM DIDEUTERIUM ARSENATE.
FAIRALL CW + REESE W
PHYS REV B SOLID STATE 6: 193-9 (1972)

18-2731 ESR OF FREE RADICALS IN FERROELECTRIC AND ANTIFERROELECTRIC ARSENATE SINGLE CRYSTALS.
GAILLARD J + CONSTANTINESCU O + LAMOTTE B
J CHEM PHYS 55: 5447-52 (1971)

18-2732 PURE QUADRUPOLE RESONANCE OF ARSENIC-75 IN AMMONIUM DIHYDROGEN ARSENATE.
GUPTA LC + VIJAYARAGHAVAN R
P237-9, VOL 3 OF NUCL PHYS SOLID STATE PHYS SYMP, 14TH PROC, 1969,
ROORKEE, INDIA, INDIAN DEPT ATOMIC ENERGY, BOMBAY, 1970.
35P OF CONF ON MAGNETIC RESONANCE AND RELATED PHENOMENA,
1970, BUCHAREST, ABSTRACTS. INST ATOMIC PHYS, BUCHAREST, 1971, 170P.
    CHEM ABSTR 75: 114569W (1971); SCI ABSTR A 75: 10589 (1972)
    206

18-2733 PROTON PHONON COUPLING IN CESIUM DIHYDROGEN ARSENATE AND POTASSIUM DIHYDROGEN ARSENATE.
KATIYAR RS + RYAN JF + SCOTT JF
P436-9 OF INT CONF ON LIGHT SCATTERING IN SOLIDS, 2ND PROC,
BALKANSKI M(ED), PARIS, 1971, FLAMMARION, PARIS, 1971, 518P.

PHYS REV B 4(8): 2635-8 (1971)
    246

18-2734 ESR IDENTIFICATION OF SLATER CONFIGURATIONS AND OF EXCHANGE OF THE ARSENATE RADICAL IN IRRADIATED FERROELECTRIC POTASSIUM DIHYDROGEN ARSENATE AND POTASSIUM DIDEUTERIUM ARSENATE AND ANTIFERROELECTRIC AMMONIUM DIHYDROGEN ARSENATE AND AMMONIUM DIDEUTERIUM ARSENATE.
LAMOTTE B + GAILLARD J + CONSTANTINESCU O
J CHEM PHYS 57: 3319-29 (OCT 15, 1972)

18-2735 NEW FERROELECTRIC: THE DOUBLE AMMONIUM THALLIUM MONOARSENATE OR ATLAS.
LE DONCHE L + LE MONTAGNER S
COMPT REND 256: 4406-8 (1963) (IN FRENCH)
    CHEM ABSTR 59: 2273H (1963)

THEORY OF HYDROGEN BONDED FERROELECTRICS. PART-1: (FORMALISM OF LOCAL MODE MODEL.)
PART-2: (DYNAMICAL PROPERTIES, TRANSVERSE ISING MODEL, SOFT MODE COUPLING TO TRANSVERSE OPTICAL PHONON). PART-3: (POTASSIUM DIHYDROGEN ARSENATE, CESIUM DIHYDROGEN ARSENATE)
MOORE MA + WILLIAMS HCWL
J PHYS C 5: 3168-84, 3185-221, 3222-44 (1972)

18. POTASSIUM DIHYDROGEN ARSENATE (KDA) AND RELATED ARSENATES

18-2736 ARSENIC-75 NMR RESONANCE IN AMMONIUM DIHYDROGEN ARSENATE AND DEUTERO AMMONIUM
DIDEUTERIUM ARSENATE.
NICHOLSON JY + SOEST JF
BULL AMER PHYS SOC 17: 246 (1972)

MEASUREMENT OF THE RELATIVE NONLINEAR COEFFICIENTS OF POTASSIUM DIHYDROGEN PHOSPHATE,
RUBIDIUM DIHYDROGEN PHOSPHATE, RUBIDIUM DIHYDROGEN ARSENATE, AND LITHIUM IODATE.
PEARSON JE + EVANS GA + YARIV A
OPT COMMUN 4(5): 366-7 (JAN 1972)

18-2737 PROTONIC CONDUCTION IN POTASSIUM DIHYDROGEN ARSENATE.
PERRINO CT + LAN B + ALSDORF R
INORG CHEM 11: 571-3 (MAR 1972)

18-2738 STRUCTURAL STUDIES OF HYDROGEN BONDED FERROELECTRICS USING POLARIZED IR RADIATION.
PART-2: INTERNAL FUNDAMENTAL VIBRATIONS OF POTASSIUM DIHYDROGEN ARSENATE IN THE
PARAELECTRIC PHASE.
RATAJCZAK H
J MOL STRUCT 11: 267-74 (1972)

SOFT MODE RAMAN SPECTROSCOPY: COUPLED MODES (FERROELECTRICS, CESIUM DIHYDROGEN
ARSENATE, STRONTIUM TITANATE)
SCOTT JF
P387-92 OF INT CONF ON LIGHT SCATTERING IN SOLIDS, 2ND, 1971.
BALKANSKI M(ED), FLAMMARION, PARIS, 1971.
    SCI ABSTR A 75: 15995 (1972)

18-2739 PHONON PHONON AND PROTON PHONON INTERACTIONS IN ALUMINUM PHOSPHATE AND CESIUM
DIHYDROGEN ARSENATE.
SCOTT JF
P478-82 OF NUSIMOVICI M(ED), PHONONS, INT CONF, RENNES, 1971 PROC,
FLAMMARION, PARIS, 1971, 502P.

18-2740 ARSENIC-75 NUCLEAR QUADRUPOLE RESONANCE SPECTRA OF PARTIALLY DEUTERATED FERROELECTRIC
CRYSTALS CESIUM DIHYDROGEN ARSENATE AND RUBIDIUM DIHYDROGEN ARSENATE.
SEMIN GK + GOLOVCHENKO LS + ZHUKOV AP
J STRUCT CHEM 13: 139-41 (1972)

ESR STUDY OF ADP- KDA MIXED CRYSTALS.
SUZUKI I + ABE R
J PHYS SOC JAP 30: 1210 (APR 1971)

ANTIFERROELECTRIC TRANSITIONS IN ADP AND AMMONIUM DIHYDROGEN ARSENATE STUDIED BY
INFRARED ABSORPTION.
WIENER E + LEVIN S + PELAH I
J CHEM PHYS 52: 2891-2900 (1970)

19. AMMONIUM SULFATE, AMMONIUM FLUOBERYLLATE, AND RELATED COMPOUNDS

MECHANISM OF THE FERROELECTRIC PHASE TRANSFORMATION OF DIAMMONIUM DICADMIUM TRISULFATE, A MODEL IN WHICH THE POLARIZATION VECTOR, P(S), ARISES PERPENDICULAR TO THE WAVE NORMAL OF THE SOFT WAVES. (THEORY)
AIZU K
J PHYS SOC JAP 32(4): 1033-6 (1972)

19-2741   CONJECTURED MECHANISM OF THE PHASE TRANSFORMATION OF A PECULIAR FERROELECTRIC-FERROELASTIC CRYSTAL: DIAMMONIUM DICADMIUM SULFATE.
AIZU K
J PHYS SOC JAP 32(1): 135-41 (1972)
          276

19-2742   INVESTIGATION OF FERROELECTRICITY IN SODIUM AMMONIUM SELENATE AND SODIUM AMMONIUM SULFATE.
ALEKSANDROV KS + ALEKSANDROVA IP + ZHEREBTSOVA LI AND OTHERS
FERROELECTRICS 2: 1-3 (JAN1971)

19-2743   INVESTIGATION OF FERROELECTRICS WITH GENERAL FORMULA METAL-1 METAL-2 SULFATE OR SELENATE OR FLUOBERYLLATE (METALS CHOSEN FROM SODIUM, POTASSIUM, RUBIDIUM, CESIUM, OR AMMONIUM, ABSTRACT ONLY).
ALEKSANDROV KS
P162 OF INT MEETING ON FERROELECTRICITY, 2ND, 1969, KYOTO, PROC,
(J PHYS SOC JAP, VOL 28 SUPPL). PHYS SOC JAP, 1970, 459P.

19-2744   NUCLEAR MAGNETIC RESONANCE OF DEUTERIUM IN THE PERDEUTERO AMMONIUM FLUOBERYLLATE CRYSTAL.
ALEKSANDROVA IP + SHCHERBAKOV VN
SOV PHYS CRYSTALLOGR 14(4): 608-10 (1970)

19-2745   FERROELECTRIC TRANSITION IN AMMONIUM SULFATE, THE DIELECTRIC, OPTICAL, AND ELECTROOPTICAL PROPERTIES IN THE NEIGHBORHOOD OF THE CURIE POINT (BIREFRINGENCE ANOMALIES).
ANISTRATOV AT + MARTYNOV VG
SOV PHYS CRYSTALLOGR 15(2): 256-60 (1970)
          207

19-2746   CRYSTAL STRUCTURE OF FERROELECTRIC RUBIDIUM HYDROGEN SULFATE. (ABSTRACT)
ASHMORE JP + PETCH HE
BULL AMER PHYS SOC 15(6): 775 (1970)
          191

19-2747   X-RAY TOPOGRAPHIC STUDY OF DISLOCATIONS AND FERROELECTRIC DOMAINS IN GLYCOL SULFATE.
AUTHIER A + PETROFF JF
COMPT REND 258(17): 4238-41 (1964) (IN FRENCH)
     CHEM ABSTR 61: 5026H (1964)

19-2748   ELECTRON SPIN RESONANCE OF GAMMA DEFECTS IN FERROELECTRIC AMMONIUM HYDROGEN SULFATE.
BARBUR I
PHYS STATUS SOLIDI B 45(2): K129-33 (1971)

19-2749   NITROGEN-14 QUADRUPOLE COUPLING IN PARAELECTRIC AMMONIUM SULFATE.
BLINC R + MALI M + OSREDKAR R + PRELESNIK A + SELIGER J + ZUPANCIC I
CHEM PHYS LETT 14(1): 49-51 (1972)
          269

19-2750   MICROWAVE DIELECTRIC MEASUREMENTS OF GAMMA-IRRADIATED FERROELECTRIC AMMONIUM SULFATE.
BODI A + BAICAN R + BARBUR I
ACTA PHYS POLON A 39: 39-44 (JAN 1971)

19. AMMONIUM SULFATE, AMMONIUM FLUOBERYLLATE AND RELATED COMPOUNDS

19-2751 RELAXATION PHENOMENA AND MOSSBAUER SPECTRA OF MAGNETIZED PARAMAGNETIC SUBSTANCES.
PART-1: IRON(X) ALUMINUM(1-X) AMMONIUM SULFATE DODECAHYDRATE.
BRUCKNER W + RITTER G + WEGENER H
Z PHYS 236: 52-69 (1970) (IN GERMAN)

19-2752 ROTATIONAL MOTIONS IN SOLIDS: THE FERROELECTRIC TRANSITION IN AMMONIUM SULFATE STUDIED
BY COLD NEUTRONS (SCATTERING).
DAHLBORG U + LARSSON KE + PIRKMAJER E
PHYSICA 49: 1-25 (1970)
220

19-2753 PREPARATION OF AMMONIUM SULFATE SINGLE CRYSTALS.
DESVIGNES JM + REMOISSENET M + MARECEK V
KRISTALL TECH 6: 203-11 (1971)

19-2754 CALCULATION OF CRYSTAL FIELD ENERGY LEVEL SPLITTINGS OF TITANIUM(3+) ION IN RUBIDIUM
ALUMINUM DISULFATE DODECAHYDRATE.
DIONNE GF
PHYS REV 137(3): A743-8 (1965)

19-2755 MOSSBAUER EFFECT STUDY OF THE FERROELECTRIC TRANSITIONS IN FERRIC AMMONIUM SULFATE
DODECAHYDRATE AND POTASSIUM FERROCYANIDE TRIHYDRATE.
GLEASON TJ
JOHNS HOPKINS UNIV PHD THESIS, 1968, 178P.
DISS ABSTR B 29: 4336
191

19-2756 PROPERTIES OF FERROELECTRIC CADMIUM AMMONIUM SULFATE.
GLOGAROVA M + FOUSEK J + BREZINA B
J PHYS (PARIS) 33, SUPPL 4, COLLOQ 2: 75-6 (1972)
294

19-2757 THE BEHAVIOR OF DIAMMONIUM DICADMIUM TRISULFATE UNDER PRESSURE.
GLOGAROVA M + FRENZEL C + HEGENBARTH E
PHYS STATUS SOLIDI B 53: 369-72 (1972)
276

19-2758 PREPARATION OF SINGLE CRYSTALS OF AMMONIUM SULFATE.
GODARD J + DRASKO G
KRISTALL TECH 7: K11-12 (1972)

19-2759 ELASTIC ANOMALY IN AMMONIUM SULFATE. (ELECTROSTRICTIVE CONSTANTS)
IKEDA T + FUJIBAYASHI K
J PHYS SOC JAP 33: 1487 (1972)
280

19-2760 EFFECT OF DEUTERATION IN AMMONIUM BISULFATE (CHANGE IN TRANSITION TEMPERATURES).
KASAHARA M + TATSUZAKI I
J PHYS SOC JAP 29: 1392 (1970)
218

19-2761 DEUTERON QUADRUPOLE COUPLING CONSTANT IN DEUTERATED AMMONIUM SULFATE.
KNISPEL RR + PETCH HE + PINTAR MM
J CHEM PHYS 56: 676 (1972)

PHENOMENOLOGICAL THEORY OF DIELECTRIC AND MECHANICAL PROPERTIES OF IMPROPER
FERROELECTRIC CRYSTALS. (AMMONIUM SULFATE, GADOLINIUM MOLYBDATE, IRON IODINE BORACITE)
KOBAYASHI J + ENOMOTO Y + SATO Y
PHYS STATUS SOLIDI B 50: 335-43 (1972)

19-2762 FERROELECTRIC PROPERTIES OF SOLID SOLUTIONS IN THE SYSTEM AMMONIUM SULFATE- AMMONIUM
FLUOBERYLLATE.
LEVINA ME + PARUZINA LYA
VEST MOSK UNIV KHIM 12(4): 454-6 (1971) (IN RUSSIAN)
CHEM ABSTR 75: 134531F (1971)
263

19-2763 FERROELECTRIC PROPERTIES OF SOLID SOLUTIONS IN THE AMMONIUM SULFATE- AMMONIUM
FLUOBERYLLATE SYSTEM.
LEVINA ME + KOLBENEVA GI
VEST MOSK UNIV KHIM 13(2): 211-4 (1972) (IN RUSSIAN)
CHEM ABSTR 77: 67541K (1972)
288

19. AMMONIUM SULFATE, AMMONIUM FLUOBERYLLATE AND RELATED COMPOUNDS

19-2764 FERROELECTRIC PROPERTIES OF SOLID SOLUTIONS IN THE AMMONIUM SULFATE- AMMONIUM
         FLUOBERYLLATE SYSTEM.
         LEVINA ME + KOLBENEVA GI
         VEST MOSK UNIV KHIM 12: 683-5 (1971) (IN RUSSIAN)

19-2765 (EXAMINATION OF) ANOMALIES OF DIELECTRIC PROPERTIES AT THE PHASE TRANSITIONS. (BASED ON
         LANDAU PHENOMENOLOGICAL THEORY, TESTED ON AMMONIUM FLUOBERYLLATE, ABSTRACT ONLY)
         LEVANYUK AP + SANNIKOV DG
         P188 OF INT MEETING ON FERROELECTRICITY, 2ND, 1969,, KYOTO, PROC,
         (J PHYS SOC JAP, VOL 28 SUPPL). PHYS SOC JAP, 1970, 459P.

19-2766 BRILLOUIN EFFECT AT ORDINARY TEMPERATURE IN AMMONIUM SULFATE.
         LUSPIN Y + HAURET G
         COMPT REND B 274: 995-7 (1972) (IN FRENCH)

19-2767 DIELECTRIC AND SOME OTHER STUDIES OF FERROELECTRIC PHASE TRANSITION IN SODIUM AMMONIUM
         SULFATE DIHYDRATE. (CRYSTALS)
         MAKITA Y
         P232-7, VOL 1 OF DVORAK V + FOUSKOVA A + GLOGAR P (EDS),
         INT MEETING ON FERROELECTRICITY, PRAGUE, 1966, PROC.
         INST PHYS CZECH ACAD SCI, PRAGUE, 1966.

19-2768 INFRARED SPECTRA OF RUBIDIUM HYDROGEN SULFATE AND RUBIDIUM DEUTERIUM SULFATE IN THE
         PARA- AND FERROELECTRIC PHASES.
         MIELKE Z + RATAJCZAK H
         BULL ACAD POLON SCI, SER SCI CHIM 20: 255-63 (1972)

19-2769 NUCLEAR SPIN LATTICE RELAXATION IN SOME FERROELECTRIC AMMONIUM SALTS. (AMMONIUM
         SULFATE, FLUOBERYLLATE, PERIODATE, BISULFATE)
         MILLER SR + BLINC R + BRENMAN M + WAUGH JS
         PHYS REV 126: 528-32 (1962)

19-2770 MOSSBAUER STUDY ON FERROELECTRIC PROPERTIES OF FERRIC AMMONIUM SULFATE DODECAHYDRATE
         SINGLE CRYSTALS.
         MONTANO PA + SHECHTER H + BIRAN A
         SOLID STATE COMMUN 9: 2029-32 (DEC 1, 1971)

19-2771 MOSSBAUER STUDY OF ELECTRONIC SPIN FLIP PROCESSES IN AMMONIUM IRON SULFATE
         DODECAHYDRATE.
         MORUP S + THRANE N
         PHYS REV B 4: 2087-91 (1971)

19-2772 CHANGES IN THE INFRARED SPECTRA OF AMMONIUM OR RUBIDIUM BISULFATE, AND AMMONIUM SULFATE
         AT FERROELECTRIC TRANSITIONS.
         MYASNIKOVA TP + YATSENKO AF
         SOV PHYS SOLID STATE 4: 475-8 (1962)

19-2773 AMMONIUM HYDROGEN SULFATE, ITS STRUCTURE AND THE FERROELECTRIC TRANSITION.
         NELMES RJ
         J PHYS (PARIS) 33, SUPPL 4, COLLOQ 2: 85-7 (1972)
              294

19-2774 STRUCTURE OF AMMONIUM SULFATE IN FERROELECTRIC PHASE AND TRANSITION.
         NELMES RJ
         FERROELECTRICS 4(3): 133-40 (1972)

19-2775 X-RAY DIFFRACTION DETERMINATION OF THE CRYSTAL STRUCTURE OF AMMONIUM HYDROGEN SULFATE
         ABOVE THE FERROELECTRIC TRANSITION.
         NELMES RJ
         ACTA CRYSTALLOGR B 27(2): 272-81 (1971)
              236

19-2776 STRUCTURE OF AMMONIUM FLUOBERYLLATE AT AMBIENT TEMPERATURE.
         NOZIK YUZ + TOKAR LF
         LATV PSR ZINDT AKAD VESTIS FIZ TEH ZINAT 1969(4): 75-7 (IN RUSSIAN)
              CHEM ABSTR 72: 48575
              198

19-2777 ON THE PHASE TRANSITION OF AMMONIUM HYDROGEN SULFATE AND THE STRUCTURE OF THE
         FERROELECTRIC PHASE.
         OGAWA K + YAMADA Y + TAGUCHI I + OSAKA K + WATANABE T + NITTA I
         SCI REP COLL GEN EDUC OSAKA UNIV 19(2): 1-5 (1970)

19. AMMONIUM SULFATE, AMMONIUM FLUOBERYLLATE AND RELATED COMPOUNDS

19-2778 DOMAIN STRUCTURE OF FERROELECTRIC SODIUM AMMONIUM SULFATE DIHYDRATE.
OSAKA T + MAKITA Y
J PHYS SOC JAP 28: 1378 (1970)

19-2779 THERMAL AND SOME OTHER STUDIES OF SODIUM AMMONIUM SULFATE DIHYDRATE, SODIUM TRIHYDROGEN
OR TRIDEUTERIUM DISELENITE. (SUMMARY ONLY)
OSAKA T + MIKI H + MAKITA Y
P202 OF INT MEETING ON FERROELECTRICITY, 2ND, 1969, KYOTO, PROC,
(J PHYS SOC JAP, VOL 28 SUPPL). PHYS SOC JAP, 1970, 459P.

19-2780 OPTICAL ABSORPTION, ELECTRIC CONDUCTIVITY AND PHOTOEMISSION OF GLYCOL SULFATE.
ROYAL G + MORLON B + GODEFROY G
COMPT REND B 275: 353-6 (4 SEP 1972) (IN FRENCH)
278

19-2781 ELECTRICAL CONDUCTIVITY OF AMMONIUM SULFATE SINGLE CRYSTALS.
SCHMIDT VH
J CHEM PHYS 38: 2783-4 (1963)

19-2782 LOW TEMPERATURE INFRARED STUDIES. PART-8: FERROELECTRIC AMMONIUM HYDROGEN SULFATE AND
AMMONIUM DEUTERIUM SULFATE.
SCHUTTE CJH
J MOL STRUCT 9(1-2): 77-90 (JUL 1971)
264

19-2783 LOW TEMPERATURE STUDIES. PART-4: PHASE TRANSITIONS OF AMMONIUM SULFATE: THE NATURE OF
HYDROGEN BONDING AND THE REORIENTATION OF THE AMMONIUM IONS.
SCHUTTE CJH + HEYNS AM
J CHEM PHYS 52: 864-71 (1970)

19-2784 (NMR STUDY OF) FERROELECTRIC TRANSITION IN (MANGANESE DOPED) AMMONIUM SULFATE.
SHRIVASTAVA KN
P105 OF INT MEETING ON FERROELECTRICITY, 2ND, 1969, KYOTO, PROC,
(J PHYS SOC JAP, VOL 28 SUPPL). PHYS SOC JAP, 1970, 459P.

19-2785 MANGANESE(2+) ABSORPTION NEAR THE CRITICAL REGION IN AMMONIUM SULFATE.
SHRIVASTAVA KN
PHYS STATUS SOLIDI A 1: K101-4 (1970)

19-2786 RAMAN SCATTERING IN A FERROELECTRIC AMMONIUM SULFATE CRYSTAL.
STEKHANOV AI + GABRICHIDZE ZA
SOV PHYS SOLID STATE 5: 2275-7 (1964)

19-2787 EXPERIMENTAL INVESTIGATION OF THE FERROELECTRIC PROPERTIES OF AMMONIUM BISULFATE IN THE
VICINITY OF THE HIGH TEMPERATURE PHASE TRANSITION.
STRUKOV BA + KOPTSIK VA + LIGASOVA VD
SOV PHYS SOLID STATE 4: 977-80 (1962)

19-2788 POLARIZATION REVERSAL CHARACTERISTICS OF AMMONIUM HYDROGEN SULFATE.
STRUKOV BA + MINAEVA KA + RODICHEVA EN
SOV PHYS SOLID STATE 6: 59-62 (1964)

19-2789 SPECIFIC HEAT OF AMMONIUM HYDROGEN SULFATE IN THE TEMPERATURE RANGE FROM -70 TO +14 DEG
C.
STRUKOV BA + DANILYCHEVA MN
SOV PHYS SOLID STATE 5: 1253-5 (1963)

19-2790 ESR OF GAMMA-IRRADIATED AMMONIUM SULFATE.
SUZUKI I + ABE R
J PHYS SOC JAP 30: 586 (FEB 1971)

19-2791 ELECTRON PARAMAGNETIC RESONANCE OF MANGANESE(2+) DOPED AMMONIUM CADMIUM SULFATE.
TATSUZAKI I
J PHYS SOC JAP 17: 582 (1962)

19-2792 TEMPERATURE DEPENDENCE OF THE ELECTRON PARAMAGNETIC RESONANCE OF AMMONIUM CADMIUM
SULFATE.
TATSUZAKI I
J PHYS SOC JAP 17: 1312-13 (1962)

19. AMMONIUM SULFATE, AMMONIUM FLUOBERYLLATE AND RELATED COMPOUNDS

19-2793 RAMAN AND INFRARED STUDIES OF THE FERROELECTRIC TRANSITION IN AMMONIUM SULFATE.
TORRIE BH + LIN CC + BINBREK OS + ANDERSON A
J PHYS CHEM SOLIDS 33: 697-709 (1972)
261

19-2794 DYNAMICS OF AMMONIUM IONS IN AMMONIUM SULFATE.
TREFLER M
CAN J PHYS 49(12): 1694-6 (1971)
263

19-2795 DEUTERON SPIN LATTICE RELAXATION IN FERROELECTRIC DEUTERATED AMMONIUM HYDROGEN SULFATE.
TRONTELJ Z + REBIC M
SOLID STATE COMMUN 11: 1337-9 (1972)
282

19-2796 EFFECT OF THE HYDROSTATIC PRESSURE ON THE TRANSITION TEMPERATURE IN AMMONIUM SULFATE.
TSUNEKAWA S + ISHIBASHI Y + TAKAGI Y
J PHYS SOC JAP 33: 8862 (1972)

19-2797 ELECTRON DIFFRACTION STUDY OF THE STRUCTURE OF AMMONIUM SULFATE.
UDALOVA VV + PINSKER ZG
SOV PHYS CRYSTALLOGR 8: 433-40 (1964)

19-2798 FERROELECTRIC TRANSITION OF AMMONIUM SULFATE.
UNRUH HG + RUDIGER U
J PHYS (PARIS) 33, SUPPL 4, COLLOQ 2: 77-8 (1972)
294

19-2799 (TEMPERATURE DEPENDENCE OF) THE SPONTANEOUS POLARIZATION OF AMMONIUM SULFATE.
UNRUH HG
SOLID STATE COMMUN 8: 1951-4 (1970)
220

19-2800 PROTON MAGNETIC RESONANCE STUDY OF THE SPIN SYMMETRY STATES OF AMMONIUM IONS IN SOLIDS.
(AMMONIUM SULFATE, SELENATE, VANADATE AND 15 OTHERS)
WATTON A + SHARP AR + PETCH HE + PINTAR MM
PHYS REV B 5: 4281-91 (1972)

20. ALUMS

20-2801 HEAT CAPACITY OF METHYL AMMONIUM CHROMIUM ALUM.
         BUNTING JG + STEEPLE H + ASHWORTH T
         PHYS LETT A 33: 37-8 (1970)

20-2802 NUCLEAR MAGNETIC RESONANCE (WIDE LINE SPECTRA AND SPIN LATTICE RELAXATION)
         INVESTIGATIONS OF THE PHASE TRANSITION IN ALUMS.
         GRANDE S
         P385-9, VOL 2 OF DVORAK V + FOUSKOVA A + GLOGAR P (EDS),
         INT MEETING ON FERROELECTRICITY, PRAGUE, 1966, PROC.
         INST PHYS CZECH ACAD SCI, PRAGUE, 1966.

20-2803 INFRARED AND RAMAN SPECTRA OF FERROELECTRIC ALUMS: METHYL AMMONIUM ALUMINUM (SULFATE OR
         SELENATE) DODECAHYDRATE. (AND THE DEUTERATED SULFATE)
         KRISHNAN RS + NARAYANAN PS + VENKATESH GM + SCHMIDT P
         P233-5 OF INT MEETING ON FERROELECTRICITY, 2ND, 1969,, KYOTO, PROC,
         (J PHYS SOC JAP, VOL 28 SUPPL). PHYS SOC JAP, 1970, 459P.

20-2804 DIELECTRIC RELAXATION IN FERROELECTRIC MASD, METHYL AMMONIUM ALUMINUM SULFATE
         DODECAHYDRATE. (COMPLEX DIELECTRIC CONSTANT)
         MAKITA Y + SUMITA M
         J PHYS SOC JAP 31(3): 792-6 (SEP 1971)
              252

20-2805 SPECIFIC HEAT AND SUSCEPTIBILITY IN CHROMIUM METHYL AMMONIUM ALUM ABOVE THE CRITICAL
         TEMPERATURE.
         MEIJER PH
         PHYS REV B 6: 214-22 (1972)

20-2806 POSSIBLE EXPLANATION OF THE UNUSUAL TEMPERATURE VARIATION OF THE MEAN MAGNETIC MOMENT
         OF CHROMIUM- POTASSIUM ALUM (UNSTABLE SOFT PHONON MODE, DISPLACIVE FERROELECTRICS).
         RAI R
         PHYSICA 57(1): 152-5 (1972)

20-2807 THERMAL CONDUCTIVITY OF METHYL AMMONIUM ALUMINUM ALUM.
         RECHOWICZ M + STEEPLE H
         CRYOGENICS 10: 331-2 (1970)

20-2808 SPECTROSCOPIC STUDIES OF FERROELECTRIC METHYL AMMONIUM ALUMS.
         VENKATESH GM + NARAYANAN PS
         J INDIAN INST SCI 52(4): 209-24 (1970)
              238

20-2809 INVESTIGATION OF PHASE TRANSITION IN FERROELECTRIC ALUMS (METHYL AMMONIUM ALUMINUM
         SULFATE, AMMONIUM ALUMINUM SULFATE AND A 36:64 RATIO MIXED CRYSTAL, PROTON MAGNETIC
         RESONANCE, DIELECTRIC CONSTANT).
         ZAITSEVA MP + ZHEREBTSOVA LI + VINOGRADOVA IS
         BULL ACAD SCI USSR, PHYS SER 29(6): 920-2 (1965)
              219

20-2810 PHYSICAL PROPERITES OF FERROELECTRIC ALUMS. (ESPECIALLY METHYL AMMONIUM ALUMINUM
         SULFATE)
         ZAITSEVA MP + ZHEREBTSOVA LI + VINOGRADOVA IS
         P341-6, VOL 1 OF DVORAK V + FOUSKOVA A + GLOGAR P (EDS),
         INT MEETING ON FERROELECTRICITY, PRAGUE, 1966, PROC.
         INST PHYS CZECH ACAD SCI, PRAGUE, 1966.

21. GUANIDINIUM ALUMINUM SULFATE HEXAHYDRATE (GASH) AND RELATED COMPOUNDS

21-2811   LOW TEMPERATURE INFRARED ABSORPTION SPECTRA (5200-4200 PER CM) OF GUANIDINIUM ALUMINUM
           SULFATE HEXAHYDRATE AND FREQUENCIES OF THE FUNDAMENTAL VIBRATIONS OF THE WATER
           MOLECULES.
           BERNARD MP
           COMPT REND 254: 450-2 (1962) (IN FRENCH)
               CHEM ABSTR 56: 15064A (1962)

21-2812   THERMOLUMINESCENCE STUDIES OF THE GAMMA-RAY IRRADIATED FERROELECTRICS ROCHELLE SALT AND
           GUANIDINE ALUMINUM SULFATE HEXAHYDRATE.
           GILLILAND JW + YOCKEY HP
           J PHYS CHEM SOLIDS 23: 367-74 (1962)

21-2813   DEFECTS AND THE FERROELECTRIC PROPERTIES OF GUANIDINIUM ALUMINUM SULFATE HEXAHYDRATE.
           (POLARIZATION VERSUS ELECTRIC FIELD, -80C TO ROOM TEMPERATURE)
           HILCZER B + PAWLACZYK C
           POZNAN TOW PRZYJ NAUK, PR KOM MAT-PRZYR 5(1): 55-64 (1969)
           (IN POLISH)
               CHEM ABSTR 72: 94419E (1970)
               231

21-2814   ZEEMAN EFFECT (STUDY) OF TWO FINE LINES IN (THE RED REGION OF) CHROMIUM GUANIDINIUM
           SULFATE HEXAHYDRATE.
           MARTIN-BRUNETIERE F + COUTURE L
           COMPT REND 256: 5327-30 (1963) (IN FRENCH)

21-2815   MAGNETIC SUSCEPTIBILITY OF GUANIDINIUM VANADIUM SULFATE HEXAHYDRATE AT LOW
           TEMPERATURES.
           MCELEARNEY JN + SCHWARTZ RW + MERCHANT S + CARLIN RL
           J CHEM PHYS 55: 466-7 (1971)

21-2816   OPTICAL AND INTERFEROMETRIC STUDIES ON FERROELECTRIC GUANIDINIUM ALUMINUM SULFATE
           HEXAHYDRATE CLEAVAGES AND THEIR ETCH PATTERNS.
           PATEL AR + DESAI CC
           J APPL CRYSTALLOGR 5: 286-95 (1972)
               274

21-2817   ELECTRON PARAMAGNETIC RESONANCE STUDY OF SOME METAL IONS IN GUANIDINIUM ALUMINUM
           SULFATE HEXAHYDRATE.
           SCHWARTZ RW + CARLIN RL
           J AMER CHEM SOC 92: 6763-71 (1970)

21-2818   FERROELECTRIC MOMENT IN GUANIDINIUM ALUMINUM SULFATE HEXAHYDRATE (AND MECHANISM OF
           REVERSAL).
           SCHEIN BJB + LINGAFELTER EC
           J CHEM PHYS 47(12): 5190-3 (1967)
               219

21-2819   REDETERMINATION OF STRUCTURE OF FERROELECTRIC CRYSTAL GUANIDINIUM ALUMINUM SULFATE
           HEXAHYDRATE AND ITS CHROMIUM ISOMORPH.
           SCHEIN BJB + LINGAFELTER EC + STEWART JM
           J CHEM PHYS 47(12): 5183-9 (1967)
               219

21-2820   EPR STUDY OF MAGNETIC COPPER(2+) COMPLEXES IN CRYSTALS OF GUANIDINIUM ALUMINUM SULFATE
           HEXAHYDRATE.
           SCZANIECKI B
           ACTA PHYS POLON A 38: 189-200 (1970)

21. GUANIDINIUM ALUMINUM SULFATE HEXAHYDRATE (GASH) AND RELATED COMPOUNDS

21-2821   ZEEMAN EFFECT AND DILUTION EFFECT OF ABSORPTION LINES IN GUANIDINIUM CHROMIUM SULFATE
          HEXAHYDRATE.
          TAMATANI M + BAN T + TSUJIKAWA I
          J PHYS SOC JAP 30: 481-90 (1971)

22. SELENITES AND SELENATES

22-2822   ANOMALOUS SPECIFIC HEAT OF SODIUM TRIHYDROGEN SELENITE ASSOCIATED COMBINATORIAL
          PROBLEM.
          ABRAHAM DB + LIEB EH
          J CHEM PHYS 54(4): 1446-50 (1971)
              234

22-2823   ANOMALOUS SPECIFIC HEAT OF SODIUM TRIHYDROGEN SELENITE.
          ABRAHAM DB + LIEB EH + OGUCHI T + YAMAMOTO T
          PROGR THEOR PHYS 44: 1114-16 (1970)

22-2824   DIELECTRIC AND THERMAL STUDY OF POTASSIUM SELENATE TRANSITIONS (FERROELECTRIC BELOW
          93K).
          AIKI K + HUKUDA K + KOGA H + KOBAYASHI T
          J PHYS SOC JAP 28(2): 389-94 (1970)
              189

22-2825   ESR STUDY OF GAMMA-IRRADIATED POTASSIUM SELENATE (STRUCTURAL PHASE CHANGES).
          AIKI K
          J PHYS SOC JAP 29(2): 379-88 (1970)
              208

          INVESTIGATION OF FERROELECTRICITY IN SODIUM AMMONIUM SELENATE AND SODIUM AMMONIUM
          SULFATE.
          ALEKSANDROV KS + ALEKSANDROVA IP + ZHEREBTSOVA LI AND OTHERS
          FERROELECTRICS 2: 1-3 (JAN1971)

22-2826   FERROELECTRICS WITH AMMONIUM GROUP ORIENTATION MOBILITY: METAL AMMONIUM (BX4) TYPE
          COMPOUNDS (PROTON MAGNETIC RESONANCE, 84 DEG K AND 120-300 DEG K, EXAMPLE SODIUM
          AMMONIUM SELENATE DIHYDRATE).
          ALEKSANDROVA IP + ALEKSANDROV KS + KRUPNAYA VP
          SOV PHYS SOLID STATE 12(4): 804-8 (1970)
              212

22-2827   MECHANISMS OF FERROELECTRIC PHASE TRANSITION IN CRYSTALS WITH REORIENTATING STRUCTURAL
          GROUPS. (SODIUM AMMONIUM SELENATE DIHYDRATE)
          ALEKSANDROVA IP + YUZVAK VI + SHABANOV VF
          J PHYS (PARIS) 33, SUPPL 4, COLLOQ 2: 63-5 (1972)
              294

22-2828   NUCLEAR MAGNETIC RESONANCE OF DEUTERIUM IN THE FERROELECTRIC SODIUM DEUTERO AMMONIUM
          SELENATE DIDEUTERATE.
          ALEKSANDROVA IP + YUZVAK VI + SHCHERBAKOV VN
          BULL ACAD SCI USSR, PHYS SER 35(9): 1644-7 (1971)

22-2829   LINEAR ELECTROOPTICAL EFFECT IN THE FERROELECTRIC SODIUM AMMONIUM SELENATE DIHYDRATE.
          ANISTRATOV AT + MELNIKOVA SV
          SOV PHYS CRYSTALLOGR 17(1): 119-21 (JUL-AUG 1972)
              274

22-2830   DOMAIN STRUCTURE OF THE BETA PHASE OF SODIUM TRIHYDROGEN SELENATE IN ELECTRIC FIELDS.
          ASKOCHENSKII AA + KIRIKOV VA + SHUVALOV LA
          BULL ACAD SCI USSR, PHYS SER 35(9): 1705-8 (1971)

          ULTRASLOW HYDROGEN MOTION IN POTASSIUM DIHYDROGEN PHOSPHATE AND SODIUM TRIHYDROGEN
          SELENATE TYPE CRYSTALS.
          BLINC R + PIRS J
          J CHEM PHYS 54: 1535-9 (1971)

22-2831   DOUBLE HYSTERESIS LOOPS AT THE UPPER CURIE POINT OF SODIUM TRIHYDROGEN SELENITE.
          BLINC R + LEVSTIK A
          PHYS STATUS SOLIDI A 2: K131-3 (1970)

22-2832   NUCLEAR MAGNETIC RESONANCE STUDY OF THE FERROELECTRIC TRANSITIONS IN SODIUM TRIHYDROGEN
           AND TRIDEUTERIUM SELENATE.
           BLINC R + STEPISNIK J + ZUPANCIC I
           PHYS REV 176: 732-9 (1968)
               146

22-2833   ESR STUDY OF GAMMA-IRRADIATED CESIUM TRIHYDROGEN SELENITE AT ROOM TEMPERATURE.
           CESANI FA + FARACH HA + POOLE CP + FENRICK HW
           BULL AMER PHYS SOC 15: 1307 (1970)

22-2834   IMPROPER FERROELECTRIC PHASE TRANSITION IN RUBIDIUM TRIHYDROGEN SELENITE.
           DVORAK V
           PHYS STATUS SOLIDI B 51: K129-32 (1972)
               270

22-2835   RESIDUAL ENTROPY OF SODIUM TRIHYDROGEN SELENATE.
           FUCHIKAMI N + OGUCHI T
           PROGR THEOR PHYS 44(6): 1500-8 (1970)
               253

22-2836   INFRARED SPECTRA OF ISOTOPIC NONISOMORPHISM OF SODIUM TRIHYDROGEN SELENITE (SHS) OR
           SODIUM TRIDEUTERIUM SELENITE (SDS).
           GALANOV EK + IVANOV NR + SHUVALOV LA
           BULL ACAD SCI USSR, PHYS SER 34(12): 2233-41 (1970)

22-2837   REVERSAL MECHANISM IN FERROELECTRICS: LITHIUM OR SODIUM TRIHYDROGEN DISELENITE, NUCLEAR
           MAGNETIC RESONANCE INVESTIGATION.
           GAVRILOVA-PODOLSKAYA GV
           P390-9, VOL 2 OF DVORAK V + FOUSKOVA A + GLOGAR P (EDS),
           INT MEETING ON FERROELECTRICITY, PRAGUE, 1966, PROC.
           INST PHYS CZECH ACAD SCI, PRAGUE, 1966.

22-2838   CRYSTAL STRUCTURES OF POTASSIUM TRIHYDROGEN SELENITE AND SODIUM TRIHYDROGEN SELENITE.
           GORBATYI LV + PONOMAREV VI + KHEIKER DM
           SOV PHYS CRYSTALLOGR 16(5): 781-5 (MAR-APR 1972)
               261

22-2839   BEHAVIOR OF THE FUNDAMENTAL ABSORPTION EDGE IN TRIHYDROGEN SELENITE CRYSTALS AT THEIR
           PHASE TRANSITIONS.
           IVANOV NR + VERKHOVSKAYA KA
           BULL ACAD SCI USSR, PHYS SER 34(12): 2281-4 (1970)

22-2840   PHASE TRANSFORMATION AND THERMOOPTICAL AND ELECTROOPTICAL EFFECT IN RUBIDIUM
           TRIHYDROGEN OR DEUTERIUM DISELENITE CRYSTALS.
           IVANOV NR + TUKHTASUNOV IT + SHUVALOV LA
           SOV PHYS CRYSTALLOGR 15(4): 647-50 (1971)
               221

22-2841   THERMOOPTIC AND ELECTROOPTIC PROPERTIES OF TRIHYDROGEN SELENITES.
           IVANOV NR + SHUVALOV LA
           P72-4 OF INT MEETING ON FERROELECTRICITY, 2ND, 1969,, KYOTO, PROC,
           (J PHYS SOC JAP, VOL 28 SUPPL). PHYS SOC JAP, 1970, 459P.

22-2842   ELECTRON PARAMAGNETIC RESONANCE IN GAMMA-IRRADIATED SINGLE CRYSTAL OF FERROELECTRIC
           LITHIUM TRIHYDROGEN DISELENITE.
           IWASAKI H
           J RADIO RES LAB 11(55): 153-6 (1964)

22-2843   PARAMAGNETIC SPECIES IN GAMMA-IRRADIATED SINGLE CRYSTAL OF FERROELECTRIC LITHIUM
           TRIHYDROGEN DISELENITE.
           IWASAKI H
           J RADIO RES LAB 12(60): 127-39 (1965)

22-2844   COERCIVE FIELD IN LITHIUM TRIHYDROGEN DISELENITE. (EXPERIMENTAL)
           JIMENEZ B
           P251-4 OF EUROPEAN MEETING ON FERROELECTRICITY, 1969, SAARBRUCKEN,
           MUSER HE AND PETERSSON J (EDS), WISSENSCH VERLAGSGES, STUTTGART,
           1970, 396P.
           ANN FIS 66(5-6): 171-7 (1970) (IN SPANISH)
               CHEM ABSTR 73: 103270Z
               216

22. SELENITES AND SELENATES

22-2845   CRYSTAL STRUCTURE OF POTASSIUM SELENATE.
          KALMAN A + STEPHENS JS + CRUICKSHANK DWJ
          ACTA CRYSTALLOGR B 26: 1451-4 (1970)

22-2846   NEUTRON DIFFRACTION STUDY ON PARAELECTRIC SODIUM TRIHYDROGEN DISELENITE.
          KAPLAN SF + KAY MI + MOROSIN B
          FERROELECTRICS 1: 31-6 (1970)
               191

22-2847   NUCLEAR SPIN LATTICE RELAXATION TIME MEASUREMENTS IN SODIUM DEUTERIUM SELENATE.
          KINSPEL RR
          MONTANA STATE UNIV PHD THESIS, 148P, 1970
               DISS ABSTR INT B 31: 1486-7 (1970)
               261

22-2848   (FERROELECTRIC AND OPTICAL PROPERTIES OF) CRYSTALS OF THE ALKALINE TRIHYDROGEN
          DISELENITE FAMILY. (ESPECIALLY THE SODIUM MEMBER, AND EFFECTS OF DEUTERATION)
          KIRPICHNIKOVA LF + SHIROKCV AM + SCHAGINA NM
          P75-7 OF INT MEETING ON FERROELECTRICITY, 2ND, 1969, KYOTO, PROC,
          (J PHYS SOC JAP, VOL 28 SUPPL). PHYS SOC JAP, 1970, 459P.

22-2849   NMR STUDY OF DEUTERIUM DIFFUSION AND DEUTERIUM AND SODIUM-23 RELAXATION IN SODIUM
          TRIDEUTERIUM SELENATE.
          KNISPEL RR + PARKER RS + SCHMIDT VH
          BULL AMER PHYS SOC 15: 104 (1970)

22-2850   THE ELECTRON PARAMAGNETIC RESONANCE OF POTASSIUM SELENATE.
          KOBAYASHI T + YAKABE M + HUKUDA K
          J PHYS SOC JAP 32: 578 (1972)

22-2851   RELAXATION BETWEEN THE STRUCTURE OF FERROELECTRIC SODIUM TRIHYDROGEN DISELENITE AND ITS
          THERMAL EXPANSION.
          KRISHNAN RS + NARAYANAN PS + DEVANARAYANAN S
          P163-5 OF INT MEETING ON FERROELECTRICITY, 2ND, 1969,, KYOTO, PROC,
          (J PHYS SOC JAP, VOL 28 SUPPL). PHYS SOC JAP, 1970, 459P.

22-2852   PROTON SPIN LATTICE RELAXATION IN FERROELECTRIC SODIUM TRIHYDROGEN SELENATE.
          KURODA N + TABATA Y + KAUAMORI A
          J PHYS SOC JAP 31: 609 (1971)
               254

22-2853   PROTON SPIN LATTICE RELAXATION IN IRRADIATED SODIUM TRIHYDROGEN SELENATE.
          LAVRENCIC B
          PHYS STATUS SOLIDI A, APPL RES 5, NO 2: K133-6 (1971)

22-2854   THE HYDROGEN BOND SYSTEM IN POTASSIUM TRIHYDROGEN SELENITE, AND IN POTASSIUM
          TRIDEUTERIUM SELENITE, AS DETERMINED BY NEUTRON DIFFRACTION.
          LEHMANN MS + LARSEN FK
          ACTA CHEM SCAND 25: 3859-71 (1971)

22-2855   DIELECTRIC PROPERTIES OF HYDROGEN BONDED FERROELECTRICS. (SODIUM TRIHYDROGEN SELENATE
          AND SODIUM TRIDEUTERIUM SELENATE)
          LEVSTIK A
          INST JOZEF STEFAN, 1970, 35P. P-259
               CHEM ABSTR 74: 117242Z (1971)
               252

22-2856   ANOMALOUS SPECIFIC HEAT AND CONFIGURATIONAL ENTROPY CHANGE IN SODIUM TRIHYDROGEN
          DISELENITE.
          MAKITA Y + MIKI H
          J PHYS SOC JAP 28(5): 1221-7 (1970)
               196

22-2857   THERMAL EXPANSION OF HYDROGEN BONDED FERROELECTRICS. LITHIUM TRIHYDROGEN SELENITE.
          MENDIOLA J + BRAVO C + ALEMANY C
          ELECTRON FIS APLIC 15: 81-3 (1972) (IN SPANISH)

22. SELENITES AND SELENATES

22-2858   THERMAL EXPANSION IN SODIUM TRIHYDROGEN(1-X) TRIDEUTERIUM(X) DISELENITE AND ISOTOPE
          EFFECT.
          MIKI H + MAKITA Y
          J PHYS SOC JAP 29(1): 143-9 (1970)
              202

22-2859   REFINEMENT OF THE CRYSTAL STRUCTURE OF FERROELECTRIC ACID LITHIUM SELENITE, POSITION OF
          THE LITHIUM ION.
          MOHANA RAO JK + VISWAMITRA MA
          ACTA CRYSTALLOGR B 27(9): 1765-75 (1971)
              255

22-2860   REFINEMENT OF THE CRYSTAL STRUCTURE OF FERROELECTRIC LITHIUM HYDROGEN SELENITE:
          POSITION OF THE LITHIUM ION.
          MOHANA RAO JK + VISWAMITRA MA
          ACTA CRYSTALLOGR B 27: 1765-75 (1971)

22-2861   ROOM TEMPERATURE CRYSTAL STRUCTURE OF THE FERROELECTRIC SODIUM TRIDEUTERO SELENITE.
          MOHANA RAO JK
          SOV PHYS CRYSTALLOGR 17(3): 432-8 (1972)
              280

22-2862   MODELS FOR THE ORDER DISORDER TRANSITION IN SODIUM TRIHYDROGEN SELENATE.
          NAGLE JF + ALLEN GR
          J CHEM PHYS 55: 2708 (1971)
              257

22-2863   PREPARATION AND ANALYSIS OF SODIUM AMMONIUM SELENATE DIHYDRATE CRYSTALS.
          ORLYANSKII YUN + ROSTUNTSEVA AI + LEIBOVICH TA + BEZNOSIKOVA NV
          INORG MATER 8: 510-11 (1972)

          THERMAL AND SOME OTHER STUDIES OF SODIUM AMMONIUM SULFATE DIHYDRATE, SODIUM TRIHYDROGEN
          OR TRIDEUTERIUM DISELENITE. (SUMMARY ONLY)
          OSAKA T + MIKI H + MAKITA Y
          P202 OF INT MEETING ON FERROELECTRICITY, 2ND, 1969, KYOTO, PROC,
          (J PHYS SOC JAP, VOL 28 SUPPL). PHYS SOC JAP, 1970, 459P.

22-2864   RAMAN SPECTRUM OF SODIUM TRIHYDROGEN DISELENITE (PARAELECTRIC PHASE, COMPARISON WITH
          FERROELECTRIC).
          PEERCY PS
          OPT COMMUN 2(6): 270-2 (1970)
              222

22-2865   RADIATION EFFECTS IN SODIUM TRIHYDROGEN SELENATE.
          PESHIKOV EV
          BULL ACAD SCI USSR, PHYS SER 35(9): 1757-9 (1971)

22-2866   A NEUTRON DIFFRACTION STUDY OF POTASSIUM TRIHYDROGEN SELENITE.
          PRELESNIK B + HERAK R + MANOJLOVIC-MUR LJ + MUIR KW
          ACTA CRYSTALLOGR B 28: 2104 (1972)

22-2867   HEAT OF FORMATION OF LITHIUM TRIHYDROGEN DISELENITE.
          ROSHCHINA ZV + SELIVANOVA NM
          IZV VUZ KHIM KHIM TEKHNOL 1970(1): 3-5 (IN RUSSIAN)
              205

22-2868   THERMAL STABILITY OF LITHIUM HYDROGEN SELENITES
          ROSHCHINA ZV + SELIVANOVA NM
          TR MOSK KHIM TEKHNOL INST 1969(62): 37-9 (IN RUSSIAN)
              216

22-2869   NEUTRON SCATTERING ANALYSIS OF CESIUM TRIHYDROGEN SELENITE AT ROOM TEMPERATURE.
          SATO S
          J PHYS SOC JAP 32: 1670 (1972)
              278

22-2870   DIELECTRIC PROPERTIES AND PRESSURE EFFECTS ON SELENITE (ION). RADIATION DAMAGE ELECTRON
          SPIN RESONANCE SPECTRA IN FERROELECTRIC SODIUM TRIHYDROGEN DISELENITE.
          SCHARA M + LEVSTIK A + CEVC P
          PHYS STATUS SOLIDI A 1(2): 323-5 (1970)
              223

22-2871   ESR STUDY OF THE FERROELECTRIC PHASE TRANSITIONS IN SODIUM TRIHYDROGEN DISELENITE.
SCHARA M + CEVC P + MARUSIC M
P309-11 OF EUROPEAN MEETING ON FERROELECTRICITY, 1969, SAARBRUCKEN,
MUSER HE AND PETERSSON J (EDS), WISSENSCH VERLAGSGES, STUTTGART,
1970, 396P.

22-2872   THE STRUCTURE AND PHASE TRANSITION IN POTASSIUM SELENATE.
SHIMAOKA K + TSUDA N + YOSHIMURA Y
ACTA CRYSTALLOGR A 28: S187 (1972)

22-2873   ISOTOPIC NONISOMORPHISM OF RUBIDIUM TRIHYDROGEN(1-X) TRIDEUTERIUM(X) DISELENATE
CRYSTALS.
SHUVALOV LA + IVANOV NR + KIRPICHNIKOVA LF + GORDEYEVA NV
PHYS LETT A 33(8): 490-1 (1970)
234

22-2874   THE PECULIAR DOMAIN STRUCTURE AND ITS UNUSUAL BEHAVIOR IN THE BETA PHASE OF THE
FERROELECTRIC SODIUM TRIHYDROGEN SELENATE.
SHUVALOV LA + ASKOCHENSKII AA + KIRIKOV VA
J PHYS (PARIS) 33, SUPPL 4, COLLOQ 2: 163-4 (1972)
294

22-2875   UNUSUAL EFFECT OF AN ELECTRIC FIELD ON THE PERMITTIVITY OF A NEW DELTA PHASE OF SODIUM
TRIDEUTERIUM(X) TRIHYDROGEN(1-X) SELENITE CRYSTALS. (PRESSURE EFFECTS)
SHUVALOV LA + SHIROKOV AM + IVANOV NR + BARANOV AI
SOV PHYS CRYSTALLOGR 15(5): 879-84 (1971)
232

22-2876   PROTON MAGNETIC RESONANCE STUDY OF SODIUM TRIHYDROGEN SELENATE AND POTASSIUM
TRIHYDROGEN SELENATE.
SILVIDI AA + WORKMAN DT
J CHEM PHYS 55: 4672-3 (1971)

22-2877   FAR INFRARED REFLECTIVITY SPECTRA OF SODIUM TRIHYDROGEN DISELENITE CRYSTAL.
SUGAWARA F + NAKAMURA T
J PHYS SOC JAP 29(1): 162-3 (1970)
202

22-2878   HYDROGEN BOND STUDIES. PART-54: A NEUTRON DIFFRACTION STUDY OF THE FERROELECTRIC
LITHIUM TRIHYDROGEN SELENITE. (CRYSTAL STRUCTURE)
TELLGREN R + LIMINGA R
J SOLID STATE CHEM 4: 255-61 (1972)
263

22-2879   HYDROGEN BOND STUDIES. PART-58: CRYSTAL STRUCTURE OF AMMONIUM TRIHYDROGEN SELENITE.
TELLGREN R + DABIR A + LIMINGA R
CHEM SCR 2(5): 215-19 (1972)
297

22-2880   ANOMALOUS THERMAL EXPANSION OF POTASSIUM TRIHYDROGEN SELENATE CRYSTALS.
TIVARI HW + NARAYANAN PS
SOV PHYS CRYSTALLOGR 16(5): 824-6 (MAR-APR 1972)
261

22-2881   CRYSTAL STRUCTURE OF RUBIDIUM TRIHYDROGEN SELENITE.
TOVBIS AB + DAVYDOVA TS + SIMONOV VI
SOV PHYS CRYSTALLOGR 17(1): 81-4 (JUL-AUG 1972)
274

22-2882   EFFECT OF HIGH PRESSURE ON PHASE TRANSITIONS OF ALKALI TRIHYDROGEN SELENITE CRYSTALS.
WHIROKOV AM + BARANOV AI + SHUVALOV LA
BULL ACAD SCI USSR, PHYS SER 35(9): 1727-31 (1971)

22-2883   NUCLEAR MAGNETIC RESONANCE OF SODIUM-23 IN THE SODIUM AMMONIUM SELENATE DIHYDRATE
CRYSTAL.
YUZVAK VI + ALEKSANDROVA IP + SHCHERBAKOV VN
SOV PHYS SOLID STATE 13(7): 1620-3 (JAN 1972)
255

22-2884   ELECTROMECHANICAL PROPERTIES OF THE FERROELECTRIC SODIUM AMMONIUM SELENATE DIHYDRATE.
ZAITSEVA MP + ZHEREBTSOVA LI + SHABANOVA LA
BULL ACAD SCI USSR, PHYS SER 35(9): 1715-18 (1971)

23. COLEMANITE

23-2885   PROTON SPIN LATTICE RELAXATION IN FERROELECTRIC COLEMANITE.
BLINC R + BRENMAN M + MILLER SR + WAUGH JS
J PHYS CHEM SOLIDS 23: 156-7 (1962)

23-2886   BIAS FIELDS IN FERROELECTRIC COLEMANITE.
DAVISSON JW + MOLNAR B
MATER RES BULL 6: 951-8 (OCT 1971)

23-2887   PROTON DYNAMICS IN FERROELECTRIC COLEMANITE.
WATTON A + PETCH HE + PINTAR MM
CAN J PHYS 48: 1081-5 (1970)

23-2888   SWITCHING PROPERTIES OF BIASED FERROELECTRIC COLEMANITE.
WIEDER HH + CLAWSON AR
SOLID STATE ELECTRON 6: 255-60 (1963)

24. POTASSIUM FERROCYANIDE

NUCLEAR MAGNETIC RESONANCE STUDIES OF THE PHASE TRANSITIONS IN FERROELECTRICS. (POTASSIUM FERROCYANIDE TRIHYDRATE, SODIUM NITRITE)
ALEKSANDROV KS + HABUDA SP + LUNDIN AG + MIKHAJLOV GM
ACTA CRYSTALLOGR 16(SUPPL): A190-1 (1963) (ABSTRACT ONLY)

24-2889 STUDY OF HYDROGEN-1 IN PARAELECTRIC AND FERROELECTRIC POTASSIUM FERROCYANIDE TRIHYDRATE BY PULSED NMR.
AVOGADRO A + CAVELIUS E + MULLER D + PETERSSON J
PHYS STATUS SOLIDI B 48: 247-53 (1971)
259

24-2890 MOSSBAUER RESONANCE OF SINGLE CRYSTAL POTASSIUM FERROCYANIDE TRIHYDRATE.
CARTHEY L + GROSKREUTZ HE
BULL AMER PHYS SOC 15: 206 (1970)

24-2891 MOSSBAUER EFFECT STUDY OF FERROELECTRIC POTASSIUM FERROCYANIDE TRIHYDRATE (NEAR THE TRANSITION TEMPERATURE, NO ANOMALY BUT SOFT PHONON MODE STILL POSSIBLE).
CLAUSER MJ
PHYS REV B 1(1): 357-9 (1970)
195

INTERNAL FIELD IN GENERAL DIPOLE LATTICES. (THEORY APPLIED TO POTASSIUM FERROCYANIDE TRIHYDRATE)
DE WETLE FW + SCHACHER GE
PHYS REV 137(1): 78-91 (1965)

MOSSBAUER EFFECT STUDY OF THE FERROELECTRIC TRANSITIONS IN FERRIC AMMONIUM SULFATE DODECAHYDRATE AND POTASSIUM FERROCYANIDE TRIHYDRATE.
GLEASON TJ
JOHNS HOPKINS UNIV PHD THESIS, 1968, 178P.
DISS ABSTR B 29: 4336

24-2892 MOSSBAUER EFFECT STUDY OF THE FERROELECTRIC TRANSITIONS IN FERRIC AMMONIUM SULFATE DODECAHYDRATE AND POTASSIUM FERROCYANIDE TRIHYDRATE.
GLEASON TG + WALKER JC
PHYS REV 188(2): 893-8 (1969)
CHEM ABSTR 72: 83968G (1970)

24-2893 ORDERING OF DIPOLE MOMENTS OF WATER MOLECULES AND FERROELECTRIC PHASE TRANSITION IN POTASSIUM FERROCYANIDE AND ISOMORPHOUS CRYSTALS. (NEW MODEL BASED ON NMR OF UNTWINNED SAMPLES)
HABUDA SP + ZEER EP + LUNDIN AG
P203 OF INT MEETING ON FERROELECTRICITY, 2ND, 1969, KYOTO, PROC, (J PHYS SOC JAP, VOL 28 SUPPL). PHYS SOC JAP, 1970, 459P.

FERROELECTRICS 1: 71-4 (1970)

RATE OF ROTATING CRYSTALS AND CHARACTERISTICS OF THEIR GROWTH. (AMMONIUM DIHYDROGEN PHOSPHATE, POTASSIUM FERROCYANIDE TRIHYDRATE, SODIUM NITRATE)
INYUSHKIN GV + SHABALIN KN
SOV PHYS CRYSTALLOGR 9: 242-4 (1964)

24-2894 NUCLEAR MAGNETIC RESONANCE AND X-RAY STUDIES OF POTASSIUM FERROCYANIDE TRIHYDRATE CRYSTALS.
KIRIYAMA R + KIRIYAMA H + WADA T + NIIZEKI N + HIRABAYASHI H
J PHYS SOC JAP 19: 540-9 (1964)

24-2895 DOMAIN STRUCTURE IN THE FERROELECTRIC POTASSIUM FERROCYANIDE.
KRASNIKOVA AYA + TOSHEV SD + KOPTSIK VA
BULL ACAD SCI USSR, PHYS SER 35(9): 1709-12 (1971)

24. POTASSIUM FERROCYANIDE

24-2896 FERROELECTRIC TRANSITION IN A POTASSIUM FERROCYANIDE TRIHYDRATE CRYSTAL UNDER HIGH
         HYDROSTATIC PRESSURE (TO 5600 KG/SQ CM, TEMPERATURE DEPENDENCE OF DIELECTRIC CONSTANT).
         KRASNIKOVA AYA + POLANDOV IN
         SOV PHYS SOLID STATE 11(7): 1421-4 (1970)
             181

24-2897 FERROELECTRIC BEHAVIOR OF POTASSIUM FERROCYANIDE TRIHYDRATE CRYSTALS AT HIGH PRESSURE.
         (TO 5100 KG/SQ CM)
         KRASNIKOVA AYA + POLANDOV IN
         P361-9, VOL 1 OF DVORAK V + FOUSKOVA A + GLOGAR P (EDS),
         INT MEETING ON FERROELECTRICITY, PRAGUE, 1966, PROC.
         INST PHYS CZECH ACAD SCI, PRAGUE, 1966.

24-2898 POLYTYPE AND TWINNING OF THE FERROELECTRIC CRYSTALS POTASSIUM FERROCYANIDE TRIHYDRATE
         BY THE NMR METHOD.
         KRASNIKOVA AYA + ZEER EP + KOPTSIK VA
         SOV PHYS CRYSTALLOGR 17(2): 287-91 (SEP-OCT 1972)
             275

24-2899 ANOMALOUS SPIN LATTICE RELAXATION BY QUASI SPIN WAVES IN POTASSIUM FERROCYANIDE
         TRIHYDRATE.
         KUBAREV YUG + MOSKVICH YUN + ZEER EP
         PHYS STATUS SOLIDI B 53: 41-5 (1972)
             276

24-2900 RADIO SPECTROSCOPIC INVESTIGATION OF POTASSIUM FERROCYANIDE TRIHYDRATE AND ISOMORPHOUS
         COMPOUNDS.
         LUNDIN AG + ZEER EP
         P400-9, VOL 2 OF DVORAK V + FOUSKOVA·A + GLOGAR P (EDS),
         INT MEETING ON FERROELECTRICITY, PRAGUE, 1966, PROC.
         INST PHYS CZECH ACAD SCI, PRAGUE, 1966.

24-2901 MOSSBAUER STUDY OF THE FERROELECTRIC PHASE TRANSITION IN POTASSIUM FERROCYANIDE
         TRIHYDRATE.
         MONTANO PA + SHECHTER H + SHIMONY U
         PHYS REV B 3(3): 858-62 (1971)
             234

24-2902 ANOMALOUS SPECIFIC HEAT OF FERROELECTRIC THIOUREA AND POTASSIUM FERROCYANIDE
         TRIHYDRATE.
         NAKAGAWA T + SAWADA S + KAWAKUBO T + NOMURA S
         J PHYS SOC JAP 18: 1227 (1963)

         TRANSIENT PHENOMENA IN THE DISCHARGE OF CONDENSERS WITH POLAR DIELECTRICS. (POTASSIUM
         DIHYDROGEN PHOSPHATE AND POTASSIUM FERROCYANIDE)
         QUERROU M + LE MONTAGNER S
         COMPT REND 258(1): 106-8 (1964) (IN FRENCH)
             PHYS ABSTR 67: 22776 (1964)

24-2903 NEUTRON DIFFRACTION STUDY OF FERROELECTRIC POTASSIUM FERROCYANIDE TRIDEUTERO HYDRATE
         ABOVE THE CURIE TEMPERATURE (LATTICE CONSTANTS, SPACE GROUP).
         TAYLOR JC + MUELLER MH + HITTERMAN RL
         ACTA CRYSTALLOGR A 26(5): 559-67 (1970)
             222

24-2904 APPLICATION OF THE DEW METHOD FOR REVEALING THE DOMAIN STRUCTURE IN POTASSIUM
         FERROCYANIDE TRIHYDRATE SINGLE CRYSTALS, P-N JUNCTIONS AND P- AND N- RANGES IN
         SEMICONDUCTORS. (SILICON INTEGRATED CIRCUITS)
         TOSHEV SD + AMOV IG
         P241-4 OF EUROPEAN MEETING ON FERROELECTRICITY, 1969, SAARBRUCKEN,
         MUSER HE AND PETERSSON J (EDS), WISSENSCH VERLAGSGES, STUTTGART,
         1970, 396P.

25. THIOUREA

25-2905　LATTICE VIBRATIONAL SPECTRA OF FIVE CRYSTAL MODIFICATIONS OF THIOUREA.
BANDY A + CESSAC GL + LIPPINCOTT ER
SPECTROCHIM ACTA A 28: 1807-12 (1972)

25-2906　LOW FREQUENCY RAMAN SPECTRUM OF A THIOUREA CRYSTAL AS A FUNCTION OF TEMPERATURE.
BENOIT JP + DENIAU M + CHAPELLE JP
COMPT REND B 275: 665-8 (1972) (IN FRENCH)

25-2907　MOLECULAR AND LATTICE VIBRATIONS OF ORTHORHOMBIC AND FERROELECTRIC THIOUREA. PART-2:
CALCULATION OF THE INTERMOLECULAR AND INTRAMOLECULAR FORCE CONSTANTS.
BLECKMANN P + SCHRADER B + MEIER W + TAKAHASHI H
BER BUNSEN GESELL PHYSIK CHEM 75(12): 1279-87 (1971) (IN GERMAN)
　　　CHEM ABSTR 76: 52067R (1972)
　　　278

25-2908　INFRARED ABSORPTION SPECTRUM OF THIOUREA IN THE PARA- AND FERROELECTRIC PHASES.
BREHAT F + HADNI A
J PHYS (PARIS) 32: 759-62 (1971) (IN FRENCH)

25-2909　STARK EFFECT ON THE NUCLEAR QUADRUPOLE RESONANCE OF NITROGEN-14 AND THE
FERROELECTRICITY OF THIOUREA.
COLOT JL
SOLID STATE COMMUN 10: 207-9 (1972)
　　　260

25-2910　INFRARED ABSORPTION SPECTRUM OF A SINGLE CRYSTAL OF THIOUREA IN THE PARA- AND
FERROELECTRIC PHASES.
DELAHAIGUE A + KHELIFA B + JOUVE P
J PHYS (PARIS) 33: 507-12 (1972) (IN FRENCH)

25-2911　LONG PERIOD STRUCTURE OF THIOUREA CRYSTAL. (CALCULATIONS FROM SATELLITE SCATTERING
DATA)
FUTAMA H
P295-7 OF INT MEETING ON FERROELECTRIC, 2ND, 1969, KYOTO, PROC,
(J PHYS SOC JAP, VOL 28 SUPPL). PHYS SOC JAP, 1970, 459P.

25-2912　DIELECTRIC DISPERSION OF UREA AND THIOUREA.
GRANT EH + KEEFE S + SHACK R
ADVAN MOL RELAXATION PROCESSES 4: 217-28 (1972)

25-2913　OPTICAL PROPERTIES OF THIOUREA.
JEFFERY JW
J APPL CRYSTALLOGR 4: 334 (1971)

25-2914　X-RAY TOPOGRAPHY OF THE DEFECT STRUCTURE OF THIOUREA.
KLAPPER H
J CRYST GROW 15: 281-7 (1972) (IN GERMAN)

25-2915　VIBRATIONAL LINE WIDTHS OF THIOUREA.
LAULICHT I + PELLACH E + BRITH M
J CHEM PHYS 57: 2857-61 (1972)

25-2916　INFRARED AND RAMAN STUDIES IN THE FERROELECTRIC TRANSITION OF THIOUREA.
LEWIS JE + SIAPKAS D + WILKINSON GR
P347 OF NUSIMOVICI M(ED), PHONONS, INT CONF, RENNES, 1971 PROC,
FLAMMARION, PARIS, 1971, 502P.

25-2917　FAR INFRARED TRANSMISSION SPECTRUM OF THIOUREA.
MCKENZIE DR + HAM NS + WHITFIELD HJ
SOLID STATE COMMUN 8: 2059-61 (1970)

25. THIOUREA

ANOMALOUS SPECIFIC HEAT OF FERROELECTRIC THIOUREA AND POTASSIUM FERROCYANIDE TRIHYDRATE.
NAKAGAWA T + SAWADA S + KAWAKUBO T + NOMURA S
J PHYS SOC JAP 18: 1227 (1963)

25-2918   DEUTERON MAGNETIC RESONANCE OF THE HIGH TEMPERATURE PHASE OF FERROELECTRIC THIOUREA.
O'REILLY DE + PETERSON EM + ELSAFFAR ZM
BULL AMER PHYS SOC 15: 276 (1970)

25-2919   STUDIES OF FERROELECTRIC SOLIDS BY MAGNETIC RESONANCE PART-18: PROTON AND DEUTERON RESONANCE OF THIOUREA.
O'REILLY DE + PETERSON EM + ELSAFFAR ZM
J CHEM PHYS 54(3): 1304-12 (1 FEB 1971)

25-2920   MOLECULAR AND LATTICE VIBRATIONS OF ORTHORHOMBIC AND FERROELECTRIC THIOUREA. PART-1: TEMPERATURE DEPENDENCE OF THE INFRARED AND RAMAN SPECTRA AND THE CALCULATED ENTROPY.
SCHRADER B + MEIER W + GOTTLIEB K + AGATHA H + BARENTZEN H + BLECKMANN P
BER BUNSEN GESELL PHYSIK CHEM 75: 1263-78 (1971) (IN GERMAN)

25-2921   SATELLITE X-RAY SCATTERING AND STRUCTURAL MODULATION OF THIOUREA.
SHIOZAKI Y
FERROELECTRICS 2(4): 245-60 (1971)

25-2922   STUDIES ON THE MECHANISM OF THE PHASE TRANSITIONS IN THIOUREA. (SATTELITE SCATTERING)
SHIOZAKI Y + HOSOYA M
P290-2 OF INT MEETING ON FERROELECTRICITY, 2ND, 1969, KYOTO, PROC,
(J PHYS SOC JAP, VOL 28 SUPPL). PHYS SOC JAP, 1970, 459P.

25-2923   CRYSTAL STRUCTURE OF THE IV PHASE OF THIOUREA. (SATELLITE SCATTERING)
TANISAKI S + NAKAMURA N
P293-4 OF INT MEETING ON FERROELECTRICITY, 2ND, 1969, KYOTO, PROC,
(J PHYS SOC JAP, VOL 28 SUPPL). PHYS SOC JAP, 1970, 459P.

25-2924   SEMICONDUCTIVITY OF THIOUREA.
YOGANARASIMHAN SR + SOOD RK
PHIL MAG 22: 1075-80 (1970)

26. FORMATES AND PROPIONATES

26-2925 ANTIPOLARIZATION IN COPPER FORMATE CRYSTALS BY THE ELECTROOPTICAL METHOD.
(ANTIFERROELECTRIC)
APKARYANTS PA + IZRAILENKO AN + SONIN AS
SOV PHYS SOLID STATE 12(11): 2580-3 (1971)
234

26-2926 VIBRATION SPECTRUM OF THE FORMATE ION IN COPPER FORMATE TETRAHYDRATE AND COPPER DEUTERO
FORMATE TETRADEUTERATE.
BERGER J
COMPT REND B 273: 927-9 (1971) (IN FRENCH)

26-2927 UNIT CELL AND SPACE GROUP OF THE ANTIFERROELECTRIC PHASE OF COPPER FORMATE
TETRAHYDRATE.
BIRD MJ + LOMER TR
ACTA CRYSTALLOGR B 27: 859-60 (1971)

26-2928 SPIN LATTICE RELAXATION BY QUASI SPIN WAVES IN ORDER DISORDER TYPE FERROELECTRICS:
POLARIZATION FLUCTUATIONS IN DICALCIUM STRONTIUM PROPIONATE.
BLINC R + ZUMER S + LAHAJNAR G
PHYS REV B 1: 4456-63 (1970)

26-2929 ZERO POINT SPIN DEVIATION AND SPONTANEOUS SUBLATTICE MAGNETIZATION IN THE TWO
DIMENSIONAL ANTIFERROMAGNET COPPER FORMATE TETRADEUTERATE.
DUPAS A + RENARD JP
PHYS LETT A 33: 470-1 (1970)

26-2930 MAGNETISM AND THE PHASE TRANSITION OF COPPER FORMATE TETRAHYDRATE.
FURUKAWA M + HIRAKAWA K
TECHNOL REP KYUSHU UNIV 44: 71-4 (1971) (IN JAPANESE)

26-2931 NMR INVESTIGATIONS OF FERROELECTRIC DICALCIUM STRONTIUM PROPIONATE.
GRANDE S + LIPPOLD B
P137-8 OF INT MEETING ON FERROELECTRICITY, 2ND, 1969, KYOTO, PROC,
(J PHYS SOC JAP, VOL 28 SUPPL). PHYS SOC JAP, 1970, 459P.

26-2932 THERMAL CONDUCTION IN A TWO DIMENSIONAL ANTIFERROMAGNETIC COPPER FORMATE TETRAHYDRATE.
HIRAKAWA K + HAYASHI H + MIIKE H
J PHYS SOC JAP 32: 1667 (1972)

26-2933 DIELECTRIC RESPONSE FUNCTION OF COPPER FORMATE TETRAHYDRATES.
ISHIBASHI Y + TAKAGI Y
MATER RES BULL 6: 999-1004 (1971)
247

26-2934 TEMPERATURE DEPENDENCE OF SOUND VELOCITY IN DICALCIUM STRONTIUM PROPIONATE.
KAMEYAMA H + ISHIBASHI Y + TAKAGI Y
J PHYS SOC JAP 33: 861 (1972)
278

26-2935 APPROXIMATE STRUCTURE FOR ANTIFERROELECTRIC PHASE OF COPPER FORMATE TETRAHYDRATE BY
NEUTRON DIFFRACTION.
KAY MI + KLEINBERG R
FERROELECTRICS 4(3): 147-52 (1972)

26-2936 RAMAN AND INFRARED SPECTRA OF COPPER FORMATE TETRAHYDRATE.
KRISHNAN RS + RAMANUJAM PS
SPECTROCHIM ACTA A 28: 2227-31 (1972)

26. FORMATES, PROPIONATES

26-2937 RAMAN AND INFRARED SPECTRA OF INORGANIC FORMATES.
         KRISHNAN RS + RAMANUJAM PS
         P277-80 OF INT CONF ON LIGHT SCATTERING IN SOLIDS, 2ND PROC,
         BALKANSKI M(ED), PARIS, 1971, FLAMMARION, PARIS, 1971, 518P.

26-2938 X-RAY DIFFRACTION STUDY OF PHASE TRANSITION IN COPPER FORMATE.
         MAKITA Y + SUZUKI S
         ACTA CRYSTALLOGR A 28: S186 (1972)

26-2939 MEASUREMENTS OF THE THERMAL CONDUCTIONS IN THE LOW DIMENSIONAL ANTIFERROMAGNETS
         POTASSIUM COPPER TRIFLUORIDE AND COPPER FORMATE TETRAHYDRATE.
         MIIKE H + HAYASHI H + HIRAKAWA K
         TECHNOL REP KYUSHU UNIV 45: 130-5 (1972) (IN JAPANESE)

26-2940 CRYSTAL STRUCTURE OF PIEZOELECTRIC LITHIUM FORMATE MONOHYDRATE.
         MOHANA RAO JK + VISWAMITRA MA
         FERROELECTRICS 2: 209-16 (1971)

26-2941 ANOMALOUS SHIFT OF ESR LINES IN THE TWO DIMENSIONAL ANTIFERROMAGNET COPPER FORMATE
         TETRAHYDRATE.
         MORIMOTO Y + DATE M
         J PHYS SOC JAP 29: 1093 (1970)

26-2942 THE ANOMALOUS BEHAVIOR OF PHONON NEAR THE 60 DEGREES C PHASE TRANSITION IN LEAD
         DICALCIUM PROPIONATE.
         QUILICHINI M + POULET H
         SOLID STATE COMMUN 10: 239-42 (1972)
              260

26-2943 STUDY OF THE ORDERED MAGNETIC STATE OF COPPER FORMATE TETRAHYDRATE BY ANTIFERROMAGNETIC
         RESONANCE.
         SEEHRA MS + CASTNER TG
         PHYS REV B 1: 2289-303 (1970)

26-2944 PHASE TRANSITION IN DIVALENT METAL DICALCIUM PROPIONATES SUBSTITUTED PARTIALLY BY
         ACETATE IONS.
         SHIRAKI H + TATSUZAKI I + YAGI T
         PHYS STATUS SOLIDI A, APPL RES 7(1): 227-32 (1971)

26-2945 NONLINEAR OPTICAL SUSCEPTIBILITY OF LITHIUM FORMATE MONOHYDRATE.
         SINGH S + BONNER WA + POTOPOWICZ JR + VAN UITERT LG
         APPL PHYS LETT 17: 292-4 (1970)

         ELECTRON MIRROR MICROSCOPIC ASPECTS OF FERROELECTRIC DOMAINS OF BARIUM TITANATE AND
         DICALCIUM STRONTIUM PROPIONATE.
         SOMEYA T + KOBAYASHI J
         P478-9 OF ANNU MEET ELECTRON MICROSC SOC AMER, 28TH PROC, 1970,
         HOUSTON, CONDENSED PAP, CLAITORS PUB DIV, BATON ROUGE, 1970.
              SCI ABSTR A 75: 10120 (1972)

26-2946 ELECTRON MIRROR MACROSCOPIC OBSERVATION OF FERROELECTRIC DOMAINS OF DICALCIUM STRONTIUM
         PROPIONATE.
         SOMEYA T + KOBAYASHI J
         PHYS STATUS SOLIDI A 4: K161-5 (1971)

26-2947 ANOMALY OF THE PROTON SPIN LATTICE RELAXATION TIME NEAR THE CRITICAL TEMPERATURE OF
         DICALCIUM LEAD PROPIONATE.
         TATSUZAKI I + SAKATA K + TODO I + TOKUNAGA M
         J PHYS SOC JAP 33(2): 438-43 (AUG 1972)
              278

26-2948 ULTRASONIC ATTENUATION NEAR THE CURIE TEMPERATURE IN DICALCIUM STRONTIUM PROPIONATE.
         TODO I + TATSUZAKI I
         J PHYS SOC JAP 31: 1479-82 (NOV 1971)

26-2949 ULTRASONIC ATTENUATION NEAR THE PHASE TRANSITION IN DICALCIUM LEAD PROPIONATE
         TODO I + TATSUZAKI I
         PHYS LETT A 38: 41-2 (1972)

26. FORMATES, PROPIONATES

26-2950   PROTON RESONANCE STUDY OF SUBLATTICE ROTATIONS AND SPIN DEVIATION IN COPPER FORMATE
         TETRADEUTERATE.
         YAMAGATA K + HAYAMA M + ODAKA T
         J PHYS SOC JAP 31: 1279 (1971)

27. TRIGLYCINE SULFATE (TGS) AND RELATED COMPOUNDS

27-2951 A NEW SECOND HARMONIC TYPE FERROELECTRIC MODULATOR FOR ELECTROMETER. (WITH A TGS SINGLE
CRYSTAL)
ABE Z + KATO Y + FURUHATA Y
REV SCI INSTRUM 42: 805-9 (1971)

27-2952 (POLARIZATION CHANGE) AFTER EFFECTS IN TRIGLYCINE SULFATE. (DIELECTRIC DISPLACEMENT AND
COERCIVE FIELD)
ALBERS J
P207-17 OF EUROPEAN MEETING ON FERROELECTRICITY, 1969, SAARBRUCKEN,
MUSER HE AND PETERSSON J (EDS), WISSENSCH VERLAGSGES, STUTTGART,
1970, 396P.

27-2953 TIME DEPENDENCE OF MATERIAL CONSTANTS OF TGS AFTER POLARIZATION REVERSAL.
ALBERS J
J PHYS (PARIS) 33, SUPPL 4, COLLOQ 2: 199-200 (1972)
294

THERMODYNAMIC THEORY OF IRRADIATED TRIGLYCINE SULFATE.
ALEMANY C + MENDIOLA J + KIMENEZ B + MAURER E
ACTA CRYSTALLOGR A 28: S233 (1972)

27-2954 MEASUREMENT OF THE DIELECTRIC CONSTANT OF TRIGLYCINE SULFATE AT MICROWAVE FREQUENCIES.
ANDERSON RL
NAVAL POSTGRAD SCHOOL, JUN 1970, 44P. AD709930
SCI ABSTR A 74: 28574 (1971)
238

X-RAY TOPOGRAPHICAL STUDY OF THE CONTRAST AT FERROELECTRIC DOMAIN BOUNDARIES (DOPED AND
UNDOPED TGS, LITHIUM NIOBATE, KDP, ABSTRACT ONLY).
AUTHIER A + PETROFF JF
P373 OF INT MEETING ON FERROELECTRICITY, 2ND, 1969, KYOTO, PROC,
(J PHYS SOC JAP, VOL 28 SUPPL). PHYS SOC JAP, 1970, 459P.

27-2955 INFLUENCE OF THE ELECTRIC FIELD AND THE DOMAIN STRUCTURE ON THE ULTRASONIC ABSORPTION
IN TRIGLYCINE SULFATE.
BAJAK IL + FOUSEK J + KEJST J
FYZ CAS 21: 99-108 (1971)

27-2956 THE INFLUENCE OF THE ELECTRIC FIELD ON THE ULTRASONIC ATTENUATION IN FERROELECTRIC
TRIGLYCINE SULFATE.
BAJAK IL
FYZ CAS 21(2-3): 90-8 (1971)
CHEM ASTR 75: 123815A (1971)
253

27-2957 RAMAN SPECTRUM OF TRIGLYCINE SELENATE.
BALASUBRAMANIAN K + KRISHNAN RS
PROC INDIAN ACAD SCI A 58: 209-15 (1963)

FAR INFRARED DIELECTRIC MEASUREMENTS ON POTASSIUM DIHYDROGEN PHOSPHATE, TRIGLYCINE
SULFATE, AND RUTILE.
BARKER AS + TINKHAM M
J CHEM PHYS 38: 2257-64 (1963)

27-2958 OPTICAL MIXING. (OF PULSED RUBY LASER EMISSIONS BY TRIGLYCINE SULFATE CRYSTAL)
BASS M + FRANKEN PA + HILL AE + PETERS CW + WEINREICH G
PHYS REV LETT 8: 18 (1962)

27-2959 THERMODYNAMIC STABILITY CF THIN FERROELECTRIC FILMS. (TRIGLYCINE SULFATE)
BATRA IP + SILVERMAN BD
SOLID STATE COMMUN 11: 291-4 (1972)
274

27-2960 SPECIFIC HEAT UNDER APPLIED ELECTRIC FIELD AND ELECTROCALORIC EFFECT IN TRIGLYCINE
SULFATE.
BAUMBERGER C + DURPAIRE JP + GODEFROY L
P199-204, VOL 1 OF DVORAK V + FOUSKOVA A + GLOGAR P (EDS),
INT MEETING ON FERROELECTRICITY, PRAGUE, 1966, PROC.
INST PHYS CZECH ACAD SCI, PRAGUE, 1966.

27-2961 INFLUENCE OF THE SPONTANEOUS POLARIZATION ON THE PHOTOSTIMULATED EXOELECTRON EMISSION
YIELD OF TRIGLYCINE SULFATE CRYSTALS.
BELYAEV LM + BENDRIKOVA GG
SOV PHYS SOLID STATE 6: 506-7 (1964)

27-2962 SPECIFIC HEAT ANOMALY IN THE NEIGHBORHOOD OF A FERRO- PARAELECTRIC TRANSITION. (TGS)
BERNARD M + PERRIGOT J + RICHARD M + EYRAUD L
J PHYS (PARIS) 33, SUPPL 4, COLLOQ 2: 94-6 (1972)
294

27-2963 SUPERSONIC DOMAIN WALL MOTION IN TRIGLYCINE SULFATE.
BINGGELI B + FATUZZO E
J APPL PHYS 36(4): 1431-5 (1965)

27-2964 THERMAL ANALYSIS OF PYROELECTRIC DETECTORS. (TRIGLYCINE SULFATE EXAMPLE)
BLACKBURN H + WRIGHT HC
INFRARED PHYS 10: 191-7 (1970)
232

27-2965 BERYLLIUM-9 QUADRUPOLE PERTURBED NMR STUDY OF THE FERROELECTRIC TRANSITION IN
DEUTERATED TRIGLYCINE FLUOBERYLLATE.
BLINC R + SLAK J + STEPISNIK J
J CHEM PHYS 55: 4848-50 (1971)

27-2966 CRITICAL BEHAVIOR OF FERROELECTRIC TGS AND DEUTERATED TGS (HYSTERESIS LOOPS JUST BELOW
CURIE POINT CORRECTED FOR ELECTROCALORIC EFFECT).
BLINC R + BURGAR M + LEVSTIK A
SOLID STATE COMMUN 8: 317-21 (1970)
190

27-2967 NUCLEAR SPIN LATTICE RELAXATION IN FERROELECTRIC TRIGLYCINE SULFATE.
BLINC R + LAHAJNAR G + PINTAR MM + ZUPANCIC I
J CHEM PHYS 44(5): 1784-7 (1966)

27-2968 PULSED NITROGEN PROTON DOUBLE RESONANCE STUDY OF THE FERROELECTRIC TRANSITION IN
TRIGLYCINE SULFATE.
BLINC R + MALI M + EHRENBERG L
J CHEM PHYS 55: 4843-8 (1971)

27-2969 PULSED DOUBLE RESONANCE STUDY OF THE NUCLEAR QUADRUPOLE INTERACTIONS OF NITROGEN-14 IN
PARAELECTRIC TRIGLYCINE SULFATE.
BLINC R + MALI M + OSREDKAR R + PRELESNIK A + ZUPANCIC I + EHRENBERG L
ACTA CHEM SCAND 25: 2403-8 (1971)

27-2970 ELECTROGYRATION AND FERROELECTRIC PHASE TRANSITIONS (TGS).
BLOKH OG + KUTNYI IV + LAZKO LA AND OTHERS
BULL ACAD SCI USSR, PHYS SER 35(9): 1683-7 (1971)

27-2971 CONCENTRATION DEPENDENCE OF SOME FERROELECTRIC PROPERTIES OF SOLID SOLUTIONS OF
TRIGLYCINE SULFATE WITH ISOMORPHOUS SUBSTANCES.
BREZINA B
P176-81, VOL 1 OF DVORAK V + FOUSKOVA A + GLOGAR P (EDS),
INT MEETING ON FERROELECTRICITY, PRAGUE, 1966, PROC.
INST PHYS CZECH ACAD SCI, PRAGUE, 1966.

27-2972 GROWTH AND CHARACTERIZATION OF SOLID SOLUTIONS OF FERROELECTRIC TRIGLYCINE SULFATE
SINGLE CRYSTALS WITH ISOMORPHOUS COMPOUNDS. (TRIGLYCINE SELENATE, FLUOBERYLLATE, AND
THE DEUTERATED SOLID SOLUTIONS)
BREZINA B
MATER RES BULL 6: 401-12 (JUN 1971)
234

27. TRIGLYCINE SULFATE (TGS) AND RELATED COMPOUNDS

27-2973 PROPERTIES OF DEUTERATED TRIGLYCINE FLUOBERYLLATE SINGLE CRYSTALS.
BREZINA B + SUMUTNY F
P182-4, VOL 1 OF DVORAK V + FOUSKOVA A + GLOGAR P (EDS),
INT MEETING ON FERROELECTRICITY, PRAGUE, 1966, PROC.
INST PHYS CZECH ACAD SCI, PRAGUE, 1966.

27-2974 CRITICAL FLUCTUATIONS IN TRIGLYCINE SULFATE.
BROPHY JJ + WEBB SL
PHYS REV 128: 584-8 (1962)

27-2975 TEMPERATURE DEPENDENCE OF THE DIELECTRIC CONSTANTS OF TRIGLYCINE SULFATE ORTHOGONAL TO
THE FERROELECTRIC AXIS.
BROSOWSKI G + LUTHER G + MUSER HE
PHYS STATUS SOLIDI A 14: K15-7 (1972)
      283

      ACOUSTIC EMISSION FROM FERROELECTRIC CRYSTALS. (BARIUM TITANATE, TGS)
      BUCHMAN P
      SOLID STATE ELECTRON 15: 142-4 (JAN 1972)

27-2976 DIELECTRIC LOSSES OF TRIGLYCINE SULFATE CRYSTALS EXPOSED TO SMALL DOSES OF X- AND
GAMMA-RAYS.
BURDANINA NA + ZOLOTOTRUBOV YUS + KAMYSHEVA LN + ZHUKOV OV + KOVALENKO AN
BULL ACAD SCI USSR, PHYS SER 35(9): 1764-7 (1971)

27-2977 HIGH INTERNAL BIAS FIELDS IN L-ALANINE SUBSTITUTED TGS.
BYE KL + WHIPPS PW + KEVE ET
FERROELECTRICS 4: 253-6 (1972)

27-2978 PERMITTIVITY OF SINGLE CRYSTALS OF TRIGLYCINE SULFATE IN A STRONG ELECTRIC FIELD.
TSEDRIK MS + MARGOLIN LN
DOKL AKAD NAUK BELORUS SSR 14(9): 802-5 (1970) (IN RUSSIAN)
      238

      PHASE CHANGES IN TRIGLYCINE SULFATE AND BARIUM TITANATE FERROELECTRICS. (26 REFS)
      CHANUSSOT G
      P125-41 OF SUCHET JP(ED), INFLUENCE DES CHANGEMENTS DE PHASE SUR LES
      PROPERIETES PHYSIQUES DES CORPS SOLIDS. MASSON, PARIS, 1970.
      (IN FRENCH)
            CHEM ABSTR 72: 137545N (1970)

      EXPERIMENTAL AND THEORETICAL STUDY OF DEFECTS IN TRIGLYCINE SULFATE AND BARIUM
      TITANATE.
      CHANUSSOT G
      PHD THESIS, DIJON, 1970, 197P. ARCH ORIG CENTRE DOCUMENT CNRS, 4706,
      23 JUN 1970, (IN FRENCH)

27-2979 PYROELECTRIC RESPONSE OF TGS IN THE FERROELECTRIC REGION NEAR THE CURIE POINT (IMPROVED
METHOD OF MEASUREMENT).
CHANUSSOT G + MALEK Z
COMPT REND B 270: 844-7 (1970) (IN FRENCH)
      195

27-2980 POLARIZATION REVERSAL IN TGS SINGLE CRYSTALS.
CHABIN M + GILLETTA F
J PHYS (PARIS) 33, SUPPL 4, COLLOQ 2: 211-13 (1972)
COMPT REND B 272: 243-6 (1971) (IN FRENCH)
      294

27-2981 EFFECT OF SEVERAL PARAMETERS ON THE FERROELECTRIC- PARAELECTRIC TRANSITION IN
TRIGLYCINE SULFATE.
CORROCHANO F + JIMENEZ B + MAURER E
ELECTRON FIS APLIC 13: 83-7 (1970) (IN SPANISH)

27-2982 ORDER DISORDER TRANSITION IN FERROELECTRICS. (TRIGLYCINE SULFATE, HIGHLY ACCURATE
TEMPERATURE CONTROL)
CORROCHANO F + JIMENEZ B + MAURER E
ELECTRON FIS APPLIC 13(1): 15-20 (1970) (IN SPANISH)
      CHEM ABSTR 73: 60277N (1970)
            213

27. TRIGLYCINE SULFATE (TGS) AND RELATED COMPOUNDS

27-2983   THERMAL CONDUCTIVITY OF TRIGLYCINE SULFATE NEAR THE CURIE POINT.
CORONEL G + GONZALO JA
FERROELECTRICS 4: 19-22 (1972)
    267

LASER LIGHT SCATTERING STUDIES OF PHASE TRANSITIONS. (TRIGLYCINE SULFATE, POTASSIUM
DIHYDROGEN AND DIDEUTERIUM SULFATE)
CUMMINS HZ
JOHNS HOPKINS UNIV, 1972, 11P. AD744532

27-2984   DYNAMIC DETERMINATION OF (ELASTIC AND PIEZOELECTRIC) MATERIAL CONSTANTS OF TRIGLYCINE
SULFATE.
DADOUREK K + HAJICEK P + TICHY J + ZELENKA J
P442-7, VOL 1 OF DVORAK V + FOUSKOVA A + GLOGAR P (EDS),
INT MEETING ON FERROELECTRICITY, PRAGUE, 1966, PROC.
INST PHYS CZECH ACAD SCI, PRAGUE, 1966.

27-2985   CRITICAL REGION IN FERROELECTRIC TRIGLYCINE SULFATE.
DEGUCHI K + NAKAMURA E
PHYS REV B SOLID STATE 5(3): 1072-3 (1 FEB 1972)
    258

27-2986   DEUTERATION EFFECT ON THERMAL AND ELASTIC PROPERTIES OF FERROELECTRIC CRYSTALS THE
TRIGLYCINE SULFATE GROUP.
DIMAROVA EN + GORBOKON NV + VARIKASH VM
UKR FIZ ZH 17(7): 1189-92 (1972) (IN RUSSIAN)

27-2987   ELECTRICAL PROPERTIES OF THE SURFACES OF IONIC CRYSTALS (TGS AND OTHERS, BY SELECTIVE
DEPOSITION OF CHARGED POLYSTYRENE LATEX PARTICLES).
DISTLER GI + TOKMAKOVA EI
SOV PHYS CRYSTALLOGR 14(6): 913-15 (1970)
    194

27-2988   SPECIAL ELECTRICAL PROPERTIES OF REGIONS NEAR DOMAIN WALLS IN TRIGLYCINE SULFATE
CRYSTALS.
DISTLER GI + KOBZAREVA SA
SOV PHYS SOLID STATE 13(9): 2366-7 (MAR 1972)
    261

27-2989   EFFECTS OF DOMAIN SHAPES ON SECOND HARMONIC SCATTERING IN TRIGLYCINE SULFATE.
DOLINO G
PHYS REV B SOLID STATE 6(10): 4025-35 (15 NOV 1972)
    277

27-2990   GENERATION OF THE SECOND HARMONIC BY TRIGLYCINE SULFATE. (84 REFS)
DOLINO G
UNIV GRENOBLE PHD THESIS, 1971, 108P.
    254

27-2991   SECOND HARMONIC LIGHT SCATTERING BY DOMAINS IN FERROELECTRIC TRIGLYCINE SULFATE.
DOLINO G + LAJZEROWICZ J + VALLADE M
PHYS REV B 2(6): 2194-2200 (1970)
    222

27-2992   STUDY OF THE PHASE TRANSITION OF TGS BY SECOND HARMONIC SCATTERING.
DOLINO G + LAJZEROWICZ J + VALLADE M
P439-42 OF INT CONF ON LIGHT SCATTERING IN SOLIDS, 2ND PROC,
BALKANSKI M(ED), PARIS, 1971, FLAMMARION, PARIS, 1971, 518P.

      SCI ABSTR A 75: 16294
    264

27-2993   TIGLYCINE SULFATE AN INTERESTING NEW DIELECTRIC CRYSTAL SPECIES.
DOMINQUEZ E + JIMENEZ B + MENDIOLA J + VIVAS E
J MATER SCI 7: 363-4 (1972)
    261

27-2994   METHODS OF DETERMINING SPONTANEOUS POLARIZATION NEAR THE PHASE TRANSITION POINT OF
TRIGLYCINE SELENATE CRYSTALS.
DUDEK J
ACTA PHYS POLON A 39(6): 675-86 (1971)
    248

27. TRIGLYCINE SULFATE (TGS) AND RELATED COMPOUNDS

27-2995 NEW MEASUREMENTS OF DIVERSE THERMODYNAMICAL COEFFICIENTS OF ROCHELLE SALT AND TRIGLYCINE SULFATE.
EHSES KH + MUSER HE + FORSCH K + SCHMITT H + TOPF KJ
P195-206 OF EUROPEAN MEETING ON FERROELECTRICITY, 1969, SAARBRUCKEN,
MUSER HE AND PETERSSON J (EDS), WISSENSCH VERLAGSGES, STUTTGART,
1970, 396P.

27-2996 CRYSTALLOGRAPHIC STUDIES OF IRRADIATION FIELD TREATED TRIGLYCINE SULFATE: NEW STRUCTURE FORM.
FLETCHER SR + SKAPSKI AC + KEVE ET
J PHYS C SOLID STATE PHYS 4: L255-8 (1971)

27-2997 FERROELECTRIC DOMAIN WALL IN TRIGLYCINE SULFATE. (THEORY AND CALCULATIONS OF WALL THICKNESS AND ENERGY DENSITY)
FOUSEK J
JAP J APPL PHYS 6(8): 950-3 (1967)
     219

27-2998 X-RAY CRITICAL SCATTERING IN FERROELECTRIC TRIGLYCINE SULFATE.
FUJII Y + YAMADA Y
J PHYS SOC JAP 30: 1676-85 (JUN 1971)

27-2999 GROWTH REGIONS IN FERROELECTRIC TRIGLYCINE SULFATE CRYSTALS. (TOPOGRAPHY OF IMPERFECTIONS, DOMAIN STRUCTURE)
FURUHATA Y
P425-7 OF INT MEETING ON FERROELECTRICITY, 2ND, 1969, KYOTO, PROC,
(J PHYS SOC JAP, VOL 28 SUPPL). PHYS SOC JAP, 1970, 459P.

27-3000 THERMAL EXPANSION OF TRIGLYCINE SULFATE.
GANESAN S
ACTA CRYSTALLOGR 15: 81-7 (1962)

STUDY OF THE PYROELECTRIC EFFECT IN THIN FILMS OF TRIGLYCINE SULFATE AND BARIUM TITANATE.
GAVRILOVA ND + ZVIRGZD YUA + NOVIK VK + POSHIN VG
SOV PHYS SOLID STATE 13(6): 1506-7 (DEC 1971)

27-3001 DIELECTRIC RELAXATION IN MULTIDOMAIN TGS SINGLE CRYSTALS.
GILLETTA F
PHYS STATUS SOLIDI A, APPL RES 12(1): 143-51 (1972)

27-3002 DIELECTRIC RELAXATION IN MULTIDOMAIN CRYSTALS OF TRIGLYCINE SULFATE.
GILLETTA F + LAUGINIE P + TAUREL L
COMPT REND B 270: 94-6 (1970) (IN FRENCH)

27-3003 EVOLUTION OF FERROELECTRIC DOMAINS IN TGS SINGLE CRYSTALS.
GILLETTA F
PHYS STATUS SOLIDI A 11: 721-7 (1972)

27-3004 INFLUENCE OF DISLOCATIONS, SURFACE LAYER AND X-RAY IRRADIATION ON DOMAIN STRUCTURE IN TRIGLYCINE SULFATE.
GILLETTA F + TAUREL L + LAUGINIE P
P225-30 OF EUROPEAN MEETING ON FERROELECTRICITY, 1969, SAARBRUCKEN,
MUSER HE AND PETERSSON J (EDS), WISSENSCH VERLAGSGES, STUTTGART,
1970, 396P.

27-3005 TEMPERATURE AUTOSTABILIZATION EFFECT OF TRIGLYCINE SULFATE SINGLE CRYSTALS IN AN AC ELECTRIC FIELD.
GLANC A + DVORAK V + JANOVEC V + RECHZIEGEL E + JANOUSEK V
PHYS LETT 7: 106-7 (1963)

NOISE MEASUREMENTS IN FERROELECTRICS. (TGS, ROCHELLE SALT, BARIUM TITANATE)
GODEFROY L
J PHYS (PARIS) 33, SUPPL 4, COLLOQ 2: 44-8 (1972)

STATISTICAL THEORY FOR FERROELECTRICITY IN TRIGLYCINE SULFATE.
GONZALO JA + LOPEZ-ALONSO JR
J PHYS CHEM SOLIDS 25(3): 303-10 (1964)

27-3006   EQUATION OF STATE FOR THE COOPERATIVE TRANSITION OF TRIGLYCINE SULFATE NEAR CURIE
          TEMPERATURE.
          GONZALO JA
          PHYS REV B 1(7): 3125-32 (1970)
               234

27-3007   OPTICAL CONSTANTS OF TRIGLYCINE SELENATE IN THE NEAR INFRARED TO FAR INFRARED AND IN
          THE RADIO FREQUENCY REGION.
          GRANDJEAN D + CLAUDEL J + BREHAT F + HADNI A + STRIMER P + THOMAS R
          J PHYS (PARIS) 31(5-6): 471-6 (1970) (IN FRENCH)
               223

27-3008   BLOCKING OF SPONTANEOUS POLARIZATION IN TRIGLYCINE SULFATE.
          HADNI A + PERRIN J + THOMAS R + SCHOUMACHER P
          COMPT REND B 273: 537-40 (1971) (IN FRENCH)

27-3009   DISPERSION OF THE DIELECTRIC CONSTANT OF TGS IN THE FAR INFRARED (TEMPERATURE AND
          FREQUENCY DEPENDENCE).
          HADNI A + GRANDJEAN D + CLAUDEL J + GERBAUX X
          J PHYS (PARIS) 31(10): 899-902 (1970) (IN FRENCH)
               226

27-3010   IRREVERSIBLE AND SPONTANEOUS POLARIZATION REVERSAL IN TRIGLYCINE SULFATE BY MEANS OF A
          LASER BEAM. APPLICATION TO FERROELECTRIC MEMORIES.
          HADNI A + THOMAS R
          OPT COMMUN 6: 314-16 (1972) (IN FRENCH)

27-3011   LASER STUDY OF REVERSIBLE NUCLEATION SITES IN TRIGLYCINE SULFATE AND APPLICATIONS TO
          PYROELECTRIC DETECTORS.
          HADNI A + THOMAS R
          FERROELECTRICS 4: 39-49 (MAY 1972)

27-3012   REVERSIBLE DOMAIN REVERSAL IN TRIGLYCINE SULFATE CAUSED BY A LASER BEAM.
          HADNI A + THOMAS R
          J PHYS (PARIS) 33, SUPPL 4, COLLOQ 2: 202 (1972)
               294

27-3013   CHANGES OF THE DOMAIN STRUCTURE IN VANADIUM DOPED TRIGLYCINE SULFATE OBSERVED BY EPR.
          HARTMANN E + WINDSCH W
          PHYS STATUS SOLIDI A 13: 119-25 (1972)
               280

27-3014   EFFECT OF PRESSURE ON THE SWITCHING RATE OF TRIGLYCINE SULFATE.
          HAYASHI M
          J PHYS SOC JAP 31: 1450-4 (NOV 1971)

27-3015   TEMPERATURE DEPENDENCE OF THE SWITCHING RATE OF TRIGLYCINE SULFATE.
          HAYASHI M
          J PHYS SOC JAP 33(3): 739-42 (SEP 1972)
               278

27-3016   THERMAL CONDUCTIVITY OF TRIGLYCINE SULFATE NEAR THE CURIE POINT.
          HELWIG J + ALBERS J
          PHYS STATUS SOLIDI A 7: 151-4 (1971)
               252

27-3017   OPTICAL ACTIVITY OF SOME FERROELECTRIC CRYSTALS (PYROELECTRICITY, TGS, ROCHELLE SALT).
          HERMELBRACHT K + UNRUH HG
          Z ANGEW PHYS 28(5): 285-8 (1970)
               195

27-3018   TEMPERATURE DEPENDENCE OF THE FERROELECTRIC FIELD EFFECT IN TELLURIUM FILMS ON
          TRIGLYCINE SUBSTRATES.
          HETZLER U + WURFEL P + RUPPEL W
          PHYS STATUS SOLIDI B 50: K85-7 (1972)
               280

          HIGH FREQUENCY BEHAVIOR OF HYDROGEN BONDED FERROELECTRICS: TRIGLYCINE SULFATE AND
          POTASSIUM DIHYDROGEN PHOSPHATE.
          HILL RM + ICHIKI SK
          PHYS REV 132: 1603-8 (1963)

27. TRIGLYCINE SULFATE (TGS) AND RELATED COMPOUNDS

27-3019   EFFECT OF X- AND GAMMA-RADIATION ON THE SWITCHING PROCESS IN TRIGLYCINE SULFATE.
HILCZER B
P155-8, VOL 2 OF DVORAK V + FOUSKOVA A + GLOGAR P (EDS),
INT MEETING ON FERROELECTRICITY, PRAGUE, 1966, PROC.
INST PHYS CZECH ACAD SCI, PRAGUE, 1966.

27-3020   SWITCHING IN FERROELECTRIC TRIGLYCINE SULFATE, PURE AND DOPED WITH PARAMAGNETIC IONS.
HILCZER B
FIZ DIELEK RADIOSPEKTROSK 2: 261-72 (1972) (IN POLISH)

27-3021   (OPTICAL) ENERGY GAP (AND DIELECTRIC CONSTANT) TEMPERATURE CHARACTERISTICS OF
FERROELECTRIC TRIGLYCINE SELENATE.
HONEYMAN WN + LAND DN
J PHYS D 3(8): 129-30 (1970)
213

27-3022   PROPERTIES OF MIXED CRYSTALS OF TRIGLYCINE SULFATE AND SELENATE.
HONEYMAN WN + LEE MK
J PHYS D APPL PHYS 5: 188-92 (JAN 1972)

27-3023   DIELECTRIC DISPERSION OF TRIGLYCINE SULFATE AT LOW FREQUENCIES.
IDA M + KAWADA S
SCI REP KANAZAWA UNIV 8(1): 39-44 (1962)
CHEM ABSTR 59: 4622B (1963)

DIELECTRIC DISPERSION OF TGS AND DEUTERATED KDP (THEORY, CALCULATIONS).
ISHIBASHI Y + SAWADA A + TAKAGI Y
J PHYS SOC JAP 28(6): 1488-94 (1970)

27-3024   REFINEMENT OF CRYSTAL STRUCTURE OF TRIGLYCINE SULFATE.
ITOH K + MITSUI T
FERROELECTRICS 2: 225-6 (1971)
239

27-3025   X-RAY TOPOGRAPHIC STUDY OF GROWTH DEFECTS IN TRIGLYCINE SULFATE CRYSTALS IN RELATION TO
THEIR GROWTH CONDITIONS.
IZRAEL A + PETROFF JF + AUTHIER A
J CRYST GROW 16: 131-41 (1972)
281

27-3026   X-RAY TOPOGRAPHIC STUDY OF DOMAIN WALL MOVEMENT IN TGS.
IZRAEL A + PETROFF JF + AUTHIER A
J PHYS (PARIS) 33, SUPPL 4, COLLOQ 2: 206-8 (1972)
294

27-3027   IMPEDANCE OF FERROELECTRIC TRIGLYCINE FLUOBERYLLATE CRYSTALS DURING SWITCHING IN PULSE
ELECTRIC FIELD.
JANOUSEK V + FOUSKOVA A
CZECH J PHYS B 13: 549-50 (1963)

27-3028   IMPEDANCE MEASUREMENT OF A DIELECTRIC FERROELECTRIC CONDENSER UNDER CONTINUOUS HEATING.
(TGS SINGLE CRYSTAL)
JANNIN M
COMPT REND B 270: 411-14 (1970) (IN FRENCH)

27-3029   LARGE SIGNAL PERMITTIVITY OF TRIGLYCINE SULFATE AT 100 MHZ. (NEAR THE CURIE
TEMPERATURE)
JANTA J + VELVARSKY J
P214-19, VOL 2 OF DVORAK V + FOUSKOVA A + GLOGAR P (EDS),
INT MEETING ON FERROELECTRICITY, PRAGUE, 1966, PROC.
INST PHYS CZECH ACAD SCI, PRAGUE, 1966.

27-3030   EXPERIMENTAL DETAILED STUDY OF THE PHASE TRANSITION IN TRIGLYCINE SULFATE. (CLOSE
THERMAL CONTROL)
JIMENEZ B + MAURER E + CORROCHANO F
P375-9 OF EUROPEAN MEETING ON FERROELECTRICITY, 1969, SAARBRUCKEN,
MUSER HE AND PETERSSON J (EDS), WISSENSCH VERLAGSGES, STUTTGART,
1970, 396P.

27-3031   A NEW FERROELECTRIC TRANSDUCER FOR USE IN HEAT TRANSFER AND FLOW STUDIES. (TRIGLYCINE
SULFATE)
JOLLS KR + SFORZA PM
BROOKLYN POLYTECHNIC INST, 1970, 9P. AD707738

27-3032   DIELECTRIC DISPERSION IN FERROELECTRIC TRIGLYCINE SULFATE (SEVERAL KHZ TO 23 KMHZ, DOMAIN WALL RESONANCE).
KACZMAREK F
ACTA PHYS POLON A 38(3): 393-403 (1970)
223

27-3033   DISPLACIVE MOVEMENT OF THE RADICAL IN TRIGLYCINE SULFATE NEAR THE CURIE POINT. (ESR OF GAMMA-IRRADIATED SPECIMENS)
KATO T + ABE R
J PHYS SOC JAP 32(3): 717-22 (1972)
278

27-3034   ESR OF GAMMA-IRRADIATED TGS. PART-2: INTERNAL MOTION OF HYDROGEN AND THE ACTIVATION ENERGY IN GLYCINE.
KATO T + ABE R
J PHYS SOC JAP 29(2): 389-93 (1970)
208

27-3035   ESR IN GAMMA-IRRADIATED TGS.
KATO T + ABE R + SUZUKI I
P123-4 OF INT MEETING ON FERROELECTRICITY, 2ND, 1969,, KYOTO, PROC,
(J PHYS SOC JAP, VOL 28 SUPPL). PHYS SOC JAP, 1970, 459P.

27-3036   THICKNESS DEPENDENCE OF THE NUCLEATION FIELD OF TRIGLYCINE SULFATE.
KAY HF + DUNN JW
PHIL MAG 7(8): 2027-34 (1962)

27-3037   EFFECTS OF ADDITIVES ON THE PYROELECTRIC PROPERTIES OF TGS.
KEVE ET + BYE KL + WHIPPS PW + ANNIS AD
J PHYS (PARIS) 33, SUPPL 4, COLLOQ 2: 229-31 (1972)
294

27-3038   STRUCTURAL INHIBITION OF FERROELECTRIC SWITCHING IN TRIGLYCINE SULFATE. PART-1: ADDITIVES.
KEVE ET + BYE KL + WHIPPS PW + ANNIS AD
FERROELECTRICS 3: 39-48 (1971)
262

27-3039   ELECTRICAL STRUCTURE OF THE SURFACE OF REAL CRYSTAL SUBSTRATES AS THE DETERMINING FACTOR OF THE GROWTH STAGE IN EPITAXY. (TRIGLYCINE SULFATE)
KOBZAREVA SA + DISTLER GI
J CRYST GROW 10: 269-75 (1971)
241

27-3040   OPTICAL ACTIVITY OF FERROELECTRIC DICALCIUM STRONTIUM PROPIONATE. (TRANSITION AT 209 DEGREES K)
KOBAYASHI J + BOUILLOT J + KINOSHITA K
PHYS STATUS SOLIDI B 47: 619-28 (1971)
250

27-3041   REVEALING THE DOMAIN STRUCTURE OF TGS CRYSTALS BY SELECTIVE CRYSTALLIZATION OF ANTHRAQUINONE (SUBLIMED ONTO TGS).
KOBZAREVA SA + DISTLER GI + KONSTANTINOVA VP
SOV PHYS CRYSTALLOGR 15(3): 431-4 (1970)
217

27-3042   APPLICATION OF SELECTIVE ETCHING TO THE STUDY OF TWIN AND DISLOCATION STRUCTURES IN TRIGLYCINE SULFATE.
KONSTANTINOVA VP
SOV PHYS CRYSTALLOGR 7: 605-10 (1963)

27-3043   DOMAIN STRUCTURE OF TRIGLYCINE SELENATE CRYSTAL (DURING COOLING THROUGH CURIE POINT).
KONSTANTINOVA VP + STANKOVSKAYA Y
SOV PHYS CRYSTALLOGR 15(2): 325-6 (1970)
207

27-3044   DEFECT STRUCTURE OF TRIGLYCINE SULFATE. (CORRELATION WITH DOMAIN STRUCTURES)
KONSTANTINOVA VP + DISTLER GI
P428-9 OF INT MEETING ON FERROELECTRICITY, 2ND, 1969,, KYOTO, PROC,
(J PHYS SOC JAP, VOL 28 SUPPL). PHYS SOC JAP, 1970, 459P.

27-3045   EFFECT OF A POLARIZING FIELD AND OF THE ADDITION OF A MIXTURE OF COPPER AND CHROMIUM
          IONS ON THE DIELECTRIC PROPERTIES OF TRIGLYCINE SULFATE.
          KRAJEWSKI T + NAWROCIK W + SZCZEPANIAK B
          POZNAN TOW PRZYJ NAUK, PR KOM MAT-PRZYR 5(1): 35-45 (1969)
          (IN POLISH)
               CHEM ABSTR 72: 94360D (1970)
               231

27-3046   FIELD EFFECT IN TELLURIUM THIN FILMS ON A SUBSTRATE OF FERROELECTRIC TRIGLYCINE SULFATE
          AND SELENATE CRYSTALS. (POLARIZATION CHARGE)
          KRAJEWSKI T + KILARSKA J
          POZNAN TOW PRZYJ NAUK, PR KOM MAT-PRZYR 5(2): 281-91 (1972)
          (IN POLISH)
               CHEM ABSTR 77: 10989H (1972)
               273

27-3047   THERMAL EXPANSION OF TRIGLYCINE SULFATE CRYSTALS STUDIED WITH TELLURIUM THIN FILMS.
          KRAJEWSKI T + PLOKARZ H
          POZNAN TOW PRZYJ NAUK, PR KOM MAT-PRZYR 5(2): 273-9 (1972)
          (IN POLISH)
               CHEM ABSTR 76: 145984C (1972)
               273

27-3048   ANOMALIES OF THERMAL PROPERTIES OF TRIGLYCINE SULFIDE.
          KUBICAR L + DIKANT J + AMBROVIC P
          FYZ CAS 20(3): 185-7 (1970)
               CHEM ABSTR 74: 57956Y (1971)
               225

27-3049   DIELECTRIC CONSTANT OF TRIGLYCINE SULFATE.
          LAUGINIE P
          P76-80, VOL 2 OF DVORAK V + FOUSKOVA A + GLOGAR P (EDS),
          INT MEETING ON FERROELECTRICITY, PRAGUE, 1966, PROC.
          INST PHYS CZECH ACAD SCI, PRAGUE, 1966.

27-3050   DIELECTRIC RELAXATION OF MULTIDOMAIN TRIGLYCINE SULFATE.
          LAUGINIE P + GILLETTA F
          P219-24 OF EUROPEAN MEETING ON FERROELECTRICITY, 1969, SAARBRUCKEN,
          MUSER HE AND PETERSSON J (EDS), WISSENSCH VERLAGSGES, STUTTGART,
          1970, 396P.

27-3051   DIRECT OBSERVATION OF FERROELECTRIC DOMAINS IN TRIGLYCINE SULFATE USING THE SCANNING
          ELECTRON MICROSCOPE.
          LE BIHAN R + MAUSSION M
          J PHYS (PARIS) 33, SUPPL 4, COLLOQ 2: 217-19 (1972)
               294

27-3052   SCANNING ELECTRON MICROSCOPE STUDY OF FERROELECTRIC DOMAINS IN TGS.
          LE BIHAN R + MAUSSION M
          COMPT REND B 272: 1010-3 (1971) (IN FRENCH)

27-3053   EFFECT OF HYDROSTATIC PRESSURE ON THE UNIPOLAR PROPERTIES OF TRIGLYCINE SULFATE.
          LEONIDOVA GG + NETESOVA NP + MELESHINA VA + GULISH OK
          VEST MOSK UNIV KHIM 12: 436-40 (1971) (IN RUSSIAN)

27-3054   INVESTIGATION OF THE PHASE TRANSITION IN TRIGLYCINE SULFATE.
          LEONIDOVA GG + BUZIN VN + ALIKHANOV RA
          SOV PHYS DOKL 16: 9-11 (1971)
               256

27-3055   ANISOTROPY OF ULTRASONIC ATTENUATION IN UNIAXIAL FERROELECTRICS. (TGS, ABSTRACT ONLY)
          LEVANYUK AP + STRUKOV BA + MINAEVA KA
          P210 OF INT MEETING ON FERROELECTRICITY, 2ND, 1969,, KYOTO, PROC,
          (J PHYS SOC JAP, VOL 28 SUPPL). PHYS SOC JAP, 1970, 459P.

27-3056   CRITICAL PROPERTIES OF TGS AND ROCHELLE SALT AS DETERMINED BY AN ELECTRIC FIELD METHOD.
          LEVSTIK A + BURGAR M + BLINC R
          J PHYS (PARIS) 33, SUPPL 4, COLLOQ 2: 235-6 (1972)
               294

27-3057 DOPED TRIGLYCINE SULFATE FOR PYROELECTRIC APPLICATIONS.
LOCK PJ
APPL PHYS LETT 19: 390-1 (1971)

27-3058 POLARIZED TRIGLYCINE SULFATES.
LOCK PJ + KEVE ET
NV PHILIPS GLOELAMPENFABRIEKEN
GER PAT 2,118,823, BRIT APPL 24 APR 1970, PUBL 18 NOV 1971, 13P.
    CHEM ABSTR 76: 39031E (1972)
      261

27-3059 NUCLEAR MAGNETIC RESONANCE INVESTIGATION OF TRIGLYCINE SULFATE.
LOESCHE A
P378-84, VOL 2 OF DVORAK V + FOUSKOVA A + GLOGAR P (EDS),
INT MEETING ON FERROELECTRICITY, PRAGUE, 1966, PROC.
INST PHYS CZECH ACAD SCI, PRAGUE, 1966.

27-3060 DIELECTRIC DISPERSION OF TGS.
LUTHER G
J PHYS (PARIS) 33, SUPPL 4, COLLOQ 2: 221-2 (1972)
      294

27-3061 TEMPERATURE DEPENDENCE OF DIELECTRIC PROPERTIES OF FERROELECTRIC TRIGLYCINE SULFATE IN
THE FAR INFRARED.
LUTHER G
PHYS STATUS SOLIDI B 52: K41-4 (1972)
      275

INFLUENCE OF THE QUALITY CF TRIGLYCINE SULFATE SINGLE CRYSTALS ON THEIR DIELECTRIC
PROPERTIES. (REVIEW)
MALEK Z + MORAVEC F + STRAJBLOVA J + NOVOTNY J + JANTA J
MASTNER J
P181-9 OF EUROPEAN MEETING ON FERROELECTRICITY, 1969, SAARBRUCKEN,
MUSER HE AND PETERSSON J (EDS), WISSENSCH VERLAGSGES, STUTTGART,
1970, 396P.

27-3062 FREQUENCY DEPENDENCE OF THE COERCIVE FIELD OF TRIGLYCINE SULFATE CRYSTALS.
MALEK Z + FOUSEK J + AL ALI NS + SALIM AJ
SOV PHYS SOLID STATE 5: 705-7 (1963)

27-3063 INFLUENCE OF DEFECTS ON SOME DIELECTRIC PROPERTIES OF TGS SINGLE CRYSTALS.
MALEK Z + MORAVEC F + STRAJBLOVA J + NOVOTNY J
P430-3 OF INT MEETING ON FERROELECTRICITY, 2ND, 1969, KYOTO, PROC,
(J PHYS SOC JAP, VOL 28 SUPPL). PHYS SOC JAP, 1970, 459P.

27-3064 INFLUENCE OF SURFACE EFFECTS ON THE PYROELECTRIC BEHAVIOR OF TGS CLOSE TO THE PHASE
TRANSITION.
MALEK Z + JANTA J + CHANUSSOT G
J PHYS (PARIS) 33, SUPPL 4, COLLOQ 2: 233-4 (1972)
      294

27-3065 MATERIAL RESEARCH OF FERROELECTRIC TRIGLYCINE SULFATE FROM THE VIEWPOINT OF PRACTICAL
APPLICATIONS.
MALEK Z
SLABOPROUDY OBZOR 31: 281-5 (1970) (IN CZECH)

27-3066 PYROELECTRIC BEHAVIOR OF TRIGLYCINE SULFATE IN THE PARAELECTRIC REGION.
MALEK Z + CHANUSSOT G
COMPT REND B 270: 1297-300 (1970) (IN FRENCH)

27-3067 THE EFFECT OF GROWTH RATE ON THE DEFECT STRUCTURE AND DIELECTRIC PROPERTIES OF TGS
SINGLE CRYSTALS.
MALEK Z + POLCAROVA M + STRAJBLOVA J AND OTHERS
PHYS STATUS SOLIDI A, APPL RES 11(1): 195-206 (1972)

27-3068 TEMPERATURE AND FIELD STRENGTH DEPENDENCE OF LARGE SIGNAL PERMITTIVITY OF TGS. (10
KHZ-1 MHZ)
MALEK Z + MASTNER J + HRDLICKA J + STRAJBLOVA J
P204-13, VOL 2 OF DVORAK V + FOUSKOVA A + GLOGAR P (EDS),
INT MEETING ON FERROELECTRICITY, PRAGUE, 1966, PROC.
INST PHYS CZECH ACAD SCI, PRAGUE, 1966.

27-3069   INFLUENCE OF GROWTH CONDITIONS OF TRIGLYCINE SULFATE AND TRIGLYCINE SELENATE CRYSTALS
          ON CHANGES IN SPECIFIC RESISTANCE.
          MARGOLIN LN + TSEDRIK MS
          IZV VUZ FIZ 1972(12): 143-6 (IN RUSSIAN)

27-3070   NONLINEARITY OF THE DIELECTRIC CONSTANT OF FERROELECTRIC TRIGLYCINE SULFATE.
          MATSUDA T + ABE R + SAWADA A
          J PHYS SOC JAP 32(4): 999-1002 (1972)
                278

27-3071   MOVEMENT OF DOMAIN WALLS AND THE NUCLEATION OF DOMAINS IN CRYSTALS OF TRIGLYCINE
          SULFATE.
          MELESHINA VA
          SOV PHYS CRYSTALLOGR 16: 471-5 (NOV-DEC 1971)

27-3072   WALL MOTION AND NUCLEATION OF DOMAINS IN TGS CRYSTALS. (SUMMARY, WITH PHOTOS)
          MELESHINA VA
          P357 OF INT MEETING ON FERROELECTRICITY, 2ND, 1969, KYOTO, PROC,
          (J PHYS SOC JAP, VOL 28 SUPPL). PHYS SOC JAP, 1970, 459P.

27-3073   DAMAGE PRODUCED BY X-RAYS ON TRIGLYCINE SULFATE.
          MENDIOLA J + ALEMANY C
          NAT LENDING LIBRARY SCI TECHNOL, BOSTON SPA, ENGLAND, 1971, 11P.
          (TRANSLATED FROM ELECTRON FIS APPL 13: 237-42 (1970))
                SCI TECH AEROSP REP 10(7): 945 ([972)
                268

27-3074   X-RAY DAMAGE IN TRIGLYCINE SULFATE.
          MENDIOLA J + ALEMANY C
          ELECTRON FIS APLIC 13: 237-42 (1970) (IN SPANISH)

27-3075   CRITICAL BEHAVIOR OF TRIGLYCINE FLUOBERYLLATE.
          MERCADO A + GONZALO JA
          BULL AMER PHYS SOC 17: 497 (1972)

27-3076   INFLUENCE OF AN ELECTRIC FIELD ON THE NOISE AND THE PYROELECTRIC EFFECT IN TGS.
          MICHERON F + GODEFROY L
          COMPT REND B 273: 143-6 (1971) (IN FRENCH)

27-3077   POLARIZATION FLUCTUATION OF TRIGLYCINE SULFATE NEAR THE CURIE POINT.
          MICHERON F + BAUMBERGER C + GODEFROY L
          P185-90, VOL 1 OF DVORAK V + FOUSKOVA A + GLOGAR P (EDS),
          INT MEETING ON FERROELECTRICITY, PRAGUE, 1966, PROC.
          INST PHYS CZECH ACAD SCI, PRAGUE, 1966.

27-3078   ANISOTROPY OF SOUND ABSORPTION IN TRIGLYCINE SULFATE SINGLE CRYSTALS.   (THEORY OF
          ANISOTROPIC RELAXATION ATTENUATION OF QUASI LONGITUDINAL PHONONS)
          MINAEVA KA + STRUKOV BA + VARNSTORFF K
          SOV PHYS SOLID STATE 10(7): 1665-7 (1969)
                139

27-3079   FREQUENCY DEPENDENCE OF THE FLUCTUATION ABSORPTION OF ULTRASOUND IN TRIGLYCINE SULFATE
          SINGLE CRYSTALS (AT 10, 30, 50 MHZ).
          MINAEVA KA + STRUKOV BA + THU HC
          SOV PHYS SOLID STATE 12(5): 1256-7 (1970)
                215

27-3080   MEASUREMENT OF INTERNAL FRICTION IN SINGLE CRYSTALS OF FERROELECTRIC SUBSTANCES BY THE
          COMPOSITE RESONATOR METHOD. (TGS)
          MINAEVA KA
          SOV PHYS CRYSTALLOGR 7: 335-7 (1962)

          TRANSFORMATION OF ELECTROMAGNETIC RADIATION INTO MOTION. (PYROELECTRIC EFFECT IN TGS,
          ETHYL DIAMINE TARTRATE, BARIUM TITANATE, SALICYLIDENE- 4-BROMO ANILINE.
          MOCKEL P
          NATURWISSENSCHAFTEN 57(5): 240 (1970) (IN GERMAN)

27-3081   AGING PROCESSES IN TRIGLYCINE SULFATE. (TANDEL OR CAPACITANCE PROPERTIES DOMAIN
          STRUCTURE)
          MORAVEC F + NOVOTNY J
          P294-302, VOL 2 OF DVORAK V + FOUSKOVA A + GLOGAR P (EDS),
          INT MEETING ON FERROELECTRICITY, PRAGUE, 1966, PROC.
          INST PHYS CZECH ACAD SCI, PRAGUE, 1966.

27-3082   EFFECT OF A TWO DIMENSIONAL PRESSURE ON THE CURIE POINTS OF TRIGLYCINE SULFATE AND
          ROCHELLE SALT.
          MORI K + HAYASHI M
          J PHYS SOC JAP 33(5): 1396-400 (NOV 1972)
               280

27-3083   GROWTH OF TRIGLYCINE SULFATE SINGLE CRYSTALS.
          MORAVEC F + NOVOTNY J
          KRISTALL TECH 7: 891-902 (1972)

27-3084   INFLUENCE OF IMPURITIES ON THE GROWTH AND SOME PHYSICAL PROPERTIES OF TGS SINGLE
          CRYSTALS.
          MORAVEC F + NOVOTNY J
          KRISTALL TECH 6: 335-42 (1971)

27-3085   PREPARATION OF PURE GLYCINE USED FOR GROWING OF TRIGLYCINE SULFATE SINGLE CRYSTALS.
          MORAVEC F + SULCEK Z
          COLLECT CZECH CHEM COMMUN 36: 3374-7 (SEP 1971)

27-3086   PREPARATION OF PURE TGS SINGLE CRYSTALS AND SOME INVESTIGATIONS OF THEIR QUALITY.
          (VOLUME DISTRIBUTION OF DEFECTS)
          MORAVEC F + MALEK Z + SULCEK Z + HRDLICKA J
          P434-6 OF INT MEETING ON FERROELECTRICITY, 2ND, 1969, KYOTO, PROC,
          (J PHYS SOC JAP, VOL 28 SUPPL). PHYS SOC JAP, 1970, 459P.

27-3087   PREPARATION OF PURE TRIGLYCINE SULFATE. (USING ION EXCHANGE)
          MORAVEC F + SULCEK Z
          P191-4 OF EUROPEAN MEETING ON FERROELECTRICITY, 1969, SAARBRUCKEN,
          MUSER HE AND PETERSSON J (EDS), WISSENSCH VERLAGSGES, STUTTGART,
          1970, 396P.

27-3088   MICROWAVE BEHAVIOR OF ROCHELLE SALT AND TGS (18 REFS).
          MUESER HE + GREGORIUS P + LUTHER G + POTTHARST J
          P91-101 OF EUROPEAN MEETING CN FERROELECTRICITY SAARBRUCKEN MARCH
          1969, MUESER HE + PETERSSON J (EDS), WISSENSCH VERLAGSGES, STUTTGART,
          1970.
               204

27-3089   INVESTIGATIONS OF SURFACES OF FERROELECTRICS (TGS) WITH SEMICONDUCTING ELEMENTS.
          MUSER HE
          J PHYS (PARIS) 33, SUPPL 4, COLLOQ 2: 17-19 (1972)
               294

27-3090   NEW HIGH PRESSURE PHASE TRANSITION IN TRIGLYCINE SELENATE.
          MYLOV VP + CHURAGULOV BR + LEONIDOVA GG
          SOV PHYS SOLID STATE 12(4): 1012-13 (1970)
               212

          MEASUREMENT OF MICROWAVE DIELECTRIC CONSTANTS OF FERROELECTRICS. PART-2: DIELECTRIC
          CONSTANTS AND DIELECTRIC LOSSES OF SODIUM NITRITE AND GLYCINE SULFATE.
          NAKAMURA E
          J PHYS SOC JAP 17: 961-6 (1962)

27-3091   DOMAIN WALL CAUGHT IN DISLOCATIONS IN FERROELECTRIC TRIGLYCINE SULFATE CRYSTALS.
          NAKAMURA T + NAKAMURA H
          JAP J APPL PHYS 1: 253-9 (1962)

27-3092   DEPENDENCE OF THE COERCIVE FIELD OF TRIGLYCINE SULFATE ON FREQUENCY, AMPLITUDE AND
          TEMPERATURE.
          NAKATANI N
          J PHYS SOC JAP 32(6): 1556-9 (1972)
               278

27-3093 STUDIES ON THE MECHANISM OF THE PHASE TRANSITION IN TGS. (STATIC DIELECTRIC CONSTANT,
         X-RAY SCATTERING)
         NAKAMURA E + NAGAI T + ISHIDA K + ITOH K + MITSUI T
         P271-3 OF INT MEETING ON FERROELECTRICITY, 2ND, 1969, KYOTO, PROC,
         (J PHYS SOC JAP, VOL 28 SUPPL). PHYS SOC JAP, 1970, 459P.

         NONLINEAR PROPERTIES OF FERROELECTRICS AT MICROWAVE FREQUENCIES. (TRIGLYCINE SULFATE,
         POLYCRYSTALLINE BARIUM TITANATE)
         NEKRASOV MM
         P193-9, VOL 2 OF DVORAK V + FOUSKOVA A + GLOGAR P (EDS),
         INT MEETING ON FERROELECTRICITY, PRAGUE, 1966, PROC.
         INST PHYS CZECH ACAD SCI, PRAGUE, 1966.

27-3094 DIELECTRIC DISPERSION IN TGS CRYSTALS. (PERMITTIVITY AND LOSS FROM 10 GHZ TO 100 GHZ)
         NEKRASOV MM
         P145-6 OF INT MEETING ON FERROELECTRICITY, 2ND, 1969, KYOTO, PROC,
         (J PHYS SOC JAP, VOL 28 SUPPL). PHYS SOC JAP, 1970, 459P.

27-3095 STUDY OF SWITCHING PARAMETERS IN TRIGLYCINE SULFATE.
         NEVOT L
         P112-16, VOL 2 OF DVORAK V + FOUSKOVA A + GLOGAR P (EDS),
         INT MEETING ON FERROELECTRICITY, PRAGUE, 1966, PROC.
         INST PHYS CZECH ACAD SCI, PRAGUE, 1966.

27-3096 GROWTH OF TRIGLYCINE SULFATE FROM SLIGHTLY SUPERSATURATED SOLUTIONS.
         NOVOTNY J + MORAVEC F
         J CRYST GROW 11: 329-35 (1971)
             262

27-3097 ULTRASONIC RELAXATION NEAR THE CURIE TEMPERATURE OF FERROELECTRIC TRIGLYCINE SULFATE.
         O'BRIEN EJ + LITOVITZ TA
         J APPL PHYS 35: 180-6 (1964)

27-3098 FERROELECTRIC BEHAVIOR OF RADIATION DAMAGED TRIGLYCINE SULFATE AND ROCHELLE SALT UP TO
         HIGH DOSAGE (DESTRUCTION OR MASKING OF FERROELECTRICITY).
         OKADA K + GONZALO JA + RIVERA JM
         J PHYS CHEM SOLIDS 28(4): 689-95 (1967)
             219

27-3099 NEUTRON DIFFRACTION STUDY OF TRIGLYCINE SULFATE.
         PADMANABHAN VM + YADAV VS
         CURR SCI 40: 60-1 (1971)

27-3100 EFFECT OF EVAPORATED METALLIC ELECTRODES ON THE TRANSITION TEMPERATURE OF TRIGLYCINE
         SULFATE.
         PELL RF
         NAVAL POSTGRAD SCHOOL, 1971, 46P. AD728572

27-3101 RESULTS OF EXPERIMENTAL STUDY OF GAMMA-RADIATION EFFECTS IN ROCHELLE SALT AND
         TRIGLYCINE SULFATE. (ON ELECTRONIC PROPERTIES.)
         PESHIKOV EV + STARODUBTSEV SV
         P267-76, VOL 2 OF DVORAK V + FOUSKOVA A + GLOGAR P (EDS),
         INT MEETING ON FERROELECTRICITY, PRAGUE, 1966, PROC.
         INST PHYS CZECH ACAD SCI, PRAGUE, 1966.

27-3102 UNUSUAL EFFECT OF ISOTOPIC AND ISOMORPHOUS SUBSTITUTION IN FERROELECTRIC CRYSTALS OF
         THE TRIGLYCINE SULFATE GROUP. (NORMAL AND DEUTERATED)
         PESHIKOV EV
         SOV PHYS SOLID STATE 14(6): 1377-80 (DEC 1972)
             279

27-3103 RELAXATION OF DOMAIN WALLS IN TRIGLYCINE SULFATE. (TEMPERATURE DEPENDENCES OF THE
         COMPLEX DIELECTRIC CONSTANTS)
         PETROV VM + KOGAN OI
         SOV PHYS CRYSTALLOGR 15(5): 885-7 (1971)
             232

27-3104 INVESTIGATIONS OF NUCLEAR RELAXATION OF TRIGLYCINE FLUOBERYLLATE.
         PISLEWSKI N + GROSESCU R
         BULL ACAD POLON SCI, SER SCI MATH PHYS ASTRON 20: 1027-32 (1972)

27-3105   SECOND HARMONIC GENERATION IN THE TRIGLYCINE CRYSTAL IN THE PARAELECTRIC PHASE.
           PLESHAKOV IA + SUVOROV VS + FILIMONOV AA
           BULL ACAD SCI USSR, PHYS SER 35(9): 1687-9 (1971)

27-3106   X-RAY TOPOGRAPHIC OBSERVATIONS OF RADIATION DAMAGE IN TRIGLYCINE SULFATE SINGLE
           CRYSTALS.
           POLCAROVA M + BRADLER J + JANTA J
           PHYS STATUS SOLIDI A 2: K137-9 (1970)

27-3107   PYROELECTRIC THERMAL IMAGING DEVICES. (TGS)
           PUTLEY EH + WATTON R + LUDLOW JH
           IEEE TRANS SON ULTRASON 19: 263-8 (1972)

           REMARKS ABOUT SWITCHING IN BARIUM TITANATE AND TRIGLYCINE SULFATE SINGLE CRYSTALS.
           REIBER M + GODEFROY L
           P117-21, VOL 2 OF DVORAK V + FOUSKOVA A + GLOGAR P (EDS),
           INT MEETING ON FERROELECTRICITY, PRAGUE, 1966, PROC.
           INST PHYS CZECH ACAD SCI, PRAGUE, 1966.

27-3108   IONIZATION OF THE (X-RADIATION AND COPPER AND IRON IMPURITY) ABSORPTION BANDS OF DEFECT
           ROCHELLE SALT AND TRIGLYCINE SULFATE CRYSTALS.
           ROMANYUK NA + VIBLYI IF
           SOV PHYS CRYSTALLOGR 15(4): 642-6 (1971)
                221

           EFFECT OF ULTRASOUND ON THE POLARIZATION AND REPOLARIZATION PROCESSES IN TRIGLYCINE
           SULFATE CRYSTALS AND BARIUM TITANATE CERAMIC.
           RUDYAK VM + BARANOV AI
           BULL ACAD SCI USSR, PHYS SER 29(6): 955-9 (1965)
                CHEM ABSTR 63: 10801H (1965)

27-3109   BARKHAUSEN JUMPS AND SWITCHING CURRENT IN TGS CRYSTALS.
           RUDYAK VM + GORNOSTAEV VF
           IZV VUZ FIZ 1970(5): 84-8 (IN RUSSIAN)
           (TRANSLATION TO APPEAR IN SOV PHYS J)
                CHEM ABSTR 73: 49692Y (1970)
                216

27-3110   DISTINCTIVE FEATURES OF BARKHAUSEN EFFECT IN ROCHELLE SALT AND TRIGLYCINE SULFATE
           CRYSTALS.
           RUDYAK VM + SHUVALOV LA + KAMAEV VE
           BULL ACAD SCI USSR, PHYS SER 29(6): 947-51 (1965)
                219

27-3111   INVESTIGATION OF BARKHAUSEN EFFECT IN TRIGLYCINE SULFATE CRYSTALS.
           RUDYAK VM + KAMAEV VE
           BULL ACAD SCI USSR, PHYS SER 29(6): 942-6 (1965)
                219

27-3112   EQUILIBRIUM DOMAIN STRUCTURE OF TRIGLYCINE SULFATE IN THE VICINITY OF THE CURIE POINT.
           SAFRANKOVA M
           P231-4 OF EUROPEAN MEETING ON FERROELECTRICITY, 1969, SAARBRUCKEN,
           MUSER HE AND PETERSSON J (EDS), WISSENSCH VERLAGSGES, STUTTGART,
           1970, 396P.
           CZECH J PHYS B 20: 797-802 (1970)
                216

27-3113   LOW FREQUENCY RAMAN SCATTERING BY FERROELECTRIC TRIGLYCINE FLUOBERYLLATE CRYSTALS.
           SAVATINOVA IT + SIMOVA PD + MARKOV M
           J APPL SPECTROSC 10: 339-41 (MAR 1969)

27-3114   LOW FREQUENCY RAMAN SPECTRA OF FERROELECTRIC CRYSTALS OF THE TRIGLYCINE SULFATE TYPE.
           SAVATINOVA IT + SIMOVA P + MARKOV M
           IZV FIZ INST ANEB 21: 251-6 (1971) (IN BULGARIAN)

27-3115   VARIATION OF THE UNIPOLARITY OF TRIGLYCINE SULFATE CRYSTALS WITH TIME.
           SAVVINOV AM + GAVRILOVA ND + NOVIK VK
           BULL ACAD SCI USSR, PHYS SER 34(12): 2317-19 (1970)

27. TRIGLYCINE SULFATE (TGS) AND RELATED COMPOUNDS

27-3116   DIELECTRIC STUDY OF CRITICAL BEHAVIOR OF FERROELECTRIC TRIGLYCINE SULFATE BY A DIGITAL
          TECHNIQUE. (MINUTE CHANGES IN DIELECTRIC CONSTANT)
          SAWADA A + ISHIBASHI Y + TAKAGI Y
          J PHYS SOC JAP 31(3): 823-7 (SEP 1971)
               252

27-3117   LIGHT EMISSION FROM TGS. (CRYSTALS DURING HEATING)
          SCHMIDT G + PETERSSON J + MUESER HE
          P147-8 OF INT MEETING ON FERROELECTRICITY, 2ND, 1969, KYOTO, PROC,
          (J PHYS SOC JAP, VOL 28 SUPPL). PHYS SOC JAP, 1970, 459P.

27-3118   PIEZOELECTRICITY AND ELECTROSTRICTION OF TRIGLYCINE SULFATE.
          SCHMIDT G + PFANNSCHMIDT F
          PHYS STATUS SOLIDI 3: 2215-20 (1963) (IN GERMAN)
               CHEM ABSTR 60: 13966H (1964)

27-3119   PERFORMANCE CHARACTERISTICS OF A SMALL TRIGLYCINE SULFATE DETECTOR OPERATED IN THE
          PYROELECTRIC MODE.
          SCHWARZ F + POOLE RR
          APPL OPT 9: 1940-1 (AUG 1970)
               216

27-3120   SHAPE DEPENDENCE OF PROPERTIES (CURIE TEMPERATURE, CURIE CONSTANT) OF FERROELECTRIC
          CRYSTALS. (5 COMPOUNDS MEASURED, TGS RESULTS GIVEN)
          SCHACHER GE
          J APPL PHYS 37(7): 2736-7 (1966)

          PYROELECTRIC VOLTAGE RESPONSE TO SHORT INFRARED LASER PULSES IN TRIGLYCINE SULFATE AND
          BARIUM STRONTIUM NIOBATE.
          SHAULOV A + ROSENTHAL A + SIMHONY M
          J APPL PHYS 43: 4518-22 (NOV 1972)

          PYROELECTRIC VOLTAGE RESPONSE TO RECTANGULAR INFRARED SIGNALS IN TRIGLYCINE SULFATE AND
          STRONTIUM BARIUM NIOBATE.
          SHAULOV A + SIMHONY M
          J APPL PHYS 43: 1440-8 (1972)

27-3121   REEXAMINATION OF THE THERMAL EXPANSION OF THE FERROELECTRIC TRIGLYCINE SULFATE.
          SHIBUYA I + HOSHINO S
          JAP J APPL PHYS 1: 249-52 (1962)

27-3122   AMPLITUDE DEPENDENCE OF INTERNAL FRICTION IN SINGLE CRYSTALS OF FERROELECTRICS.
          (ROCHELLE SALT AND TRIGLYCINE SULFATE)
          SHUVALOV LA + SHIROKOV AM
          SOV PHYS CRYSTALLOGR 9(6): 746-9 (1965)

27-3123   INFLUENCE OF GAMMA-IRRADIATION ON BARKHAUSEN EFFECT IN FERROELECTRIC MATERIALS (TGS AND
          ROCHELLE SALT).
          SHUVALOV LA + RUDYAK VM + KOMLYAKOVA NS + KAMAEV VE
          BULL ACAD SCI USSR, PHYS SER 29(11): 1844-8 (1965)
               219

          PYROELECTRIC VOLTAGE RESPONSE TO STEP SIGNALS OF INFRARED RADIATION IN TRIGLYCINE
          SULFATE AND STRONTIUM BARIUM NIOBATE.
          SIMHONY M + SHAULOV A
          J APPL PHYS 42(10): 3741-4 (SEP 1971)

27-3124   BAND GAP OF TRIGLYCINE SELENATE, TRIGLYCINE FLUOBERYLLATE, TRIGLYCINE SULFATE AND ITS
          VARIATION UNDER X-IRRADIATION.
          SIROTA NN + KORINA RV
          KRISTALL TECH 6: 387-93 (1971)

27-3125   DIELECTRIC PERMEABILITY AND DIELECTRIC LOSS ANGLE TANGENT OF TRIGLYCINE SELENATE
          DEPENDING ON THE CONDITIONS OF GROWING.
          SIROTA NN + TSEDRIK MS + MARGOLIN LN
          KRISTALL TECH 6: 245-54 (1971)

27-3126   THERMAL HYSTERESIS OF THE DIELECTRIC CONSTANT OF TRIGLYCINE SULFATE AS A FUNCTION OF
          THE CONDITIONS OF GROWTH.
          SIROTA NN + TSEDRIK MS + MARGOLIN LN
          DOKL AKAD NAUK BELORUS SSR 14(8): 693-6 (1970) (IN RUSSIAN)
               238

27-3127   LIP ETCH PATTERNS ON THE FACES OF TRIGLYCINE SULFATE CRYSTALS.
          SOEYA T + YAHATA Y + NOZAKI K
          JAP J APPL PHYS 10(2): 279-80 (FEB 1971)
              238

          DETECTION OF ELECTROMAGNETIC RADIATION USING PYROELECTRIC EFFECT. (BARIUM TITANATE,
          TRIGLYCINE SULFATE, AT 5.6 AND 10.5 GHZ)
          STANFORD AL
          SOLID STATE ELECTRON 8(9): 747-55 (1965)

          SUSCEPTIBILITY INCREASE CURING POLARIZATION REVERSAL. (IN BARIUM TITANATE AND
          TRIGLYCINE SULFATE)
          STADLER HL & NAKAMURA T & ISHIBASHI Y & KUROKI H
          P138-43, VOL 2 OF DVORAK V + FOUSKOVA A + GLOGAR P (EDS),
          INT MEETING ON FERROELECTRICITY, PRAGUE, 1966, PROC.
          INST PHYS CZECH ACAD SCI, PRAGUE, 1966.

27-3128   AGING PROCESS IN TRIGLYCINE SULFATE. (HYSTERESIS LAPS, DOMAIN STRUCTURE)
          STANKOWSKA J
          P288-93, VOL 2 OF DVORAK V + FOUSKOVA A + GLOGAR P (EDS),
          INT MEETING ON FERROELECTRICITY, PRAGUE, 1966, PROC.
          INST PHYS CZECH ACAD SCI, PRAGUE, 1966.

27-3129   DEPENDENCE OF THE DOMAIN STRUCTURE OF TRIGLYCINE SULFATE CRYSTALS ON TEMPERATURE AND
          APPLIED ELECTRIC FIELD STRENGTH.
          STANKOWSKA J + KRYSINKA M
          ACTA PHYS POLON A 40(2): 239-49 (1971)
              263

27-3130   DOMAIN STRUCTURE OF PURE AND DOPED TRIGLYCINE SULFATE CRYSTALS GROWN BELOW AND ABOVE
          THE CURIE POINT.
          STANKOWSKI J + KONSTANTINOVA B
          POZNAN TOW PRZYJ NAUK PR KOM MAT-PRZYR, FIZ DIELEK RADIOSPECTROSK
          2: 251-9 (1972) (IN POLISH)
              CHEM ABSTR 77: 10988G

27-3131   EFFECT OF A DC FIELD ON THE DIELECTRIC PROPERTIES OF TRIGLYCINE SELENATE. (AGING AND
          PHASE CHANGES NEAR THE CURIE POINT)
          STANKOWSKA J + JACKOWIAK I
          POZNAN TOW PRZYJ NAUK, PR KOM MAT-PRZYR 5(1): 47-53 (1969)
          (IN POLISH)
              CHEM ABSTR 72: 94368N (1970)
              231

27-3132   ELECTRON PARAMAGNETIC RESONANCE INVESTIGATIONS OF (COPPER) DOPED FERROELECTRIC
          CRYSTALS. (TRIGLYCINE SULFATE AND ROCHELLE SALT)
          STANKOWSKA J
          P364-8, VOL 2 OF DVORAK V + FOUSKOVA A + GLOGAR P (EDS),
          INT MEETING ON FERROELECTRICITY, PRAGUE, 1966, PROC.
          INST PHYS CZECH ACAD SCI, PRAGUE, 1966.

27-3133   EPR OF THE CHROMIUM COMPLEX IN TRIGLYCINE FLUOBERYLLATE.
          STANKOWSKI J + WAPLAK S
          BULL ACAD POLON SCI, SER SCI CHIM 19: 243-7 (1971)

27-3134   EPR OF THE FOUR COPPER ION COMPLEX WITH SPIN S=2 IN TRIGLYCINE FLUOBERYLLATE
          MONOCRYSTAL.
          STANKOWSKI J + MACKOWIAK M
          PHYS STATUS SOLIDI B 51: 449-56 (1972)

27-3135   EPR STUDIES OF CHROMIUM(3+), COPPER(2+), AND MANGANESE(2+) PARAMAGNETIC ION COMPLEXES
          IN GLYCINE CRYSTALS.
          STANKOWSKI J + WAPLAK S
          FIZ DIELEK RADIOSPEKTROSK 2: 317-21 (1972) (IN POLISH)

27-3136   INFLUENCE OF SPONTANEOUS POLARIZATION ON THE EPR SPECTRA OF CHROMIUM(3+) IONS IN
          TRIGLYCINE FLUOBERYLLATE.
          STANKOWSKI J + WAPLAK S
          J PHYS (PARIS) 33, SUPPL 4, COLLOQ 2: 177-8 (1972)
              294

27. TRIGLYCINE SULFATE (TGS) AND RELATED COMPOUNDS

27-3137   ANOMALY OF THERMAL PROPERTIES OF FERROELECTRICS (SINGLE CRYSTAL TRIGLYCINE SULFATE AND
           TRIGLYCINE FLUOBERYLLATE) AT PHASE TRANSITIONS OF THE SECOND ORDER.
           STRUKOV PA
           P191-8, VOL 1 OF DVORAK V + FOUSKOVA A + GLOGAR P (EDS),
           INT MEETING ON FERROELECTRICITY, PRAGUE, 1966, PROC.
           INST PHYS CZECH ACAD SCI, PRAGUE, 1966.

27-3138   SECOND HARMONIC GENERATION IN TRIGLYCINE SULFATE CRYSTALS.
           SUVOROV VS + SONIN AS
           SOV PHYS JETP 27(4): 557-60 (1968)
                143

27-3139   ELECTRON SPIN RESONANCE OF GAMMA-IRRADIATED TRIGLYCINE SELENATE.
           SUZUKI I + ABE R
           J PHYS SOC JAP 31: 179-83 (JUL 1971)

           APPLICATION OF THE X-RAY DIFFRACTION TOPOGRAPHIC METHOD
           TO FERROELECTRIC SODIUM NITRITE AND TGS CRYSTALS. (DOMAIN
           BOUNDARIES AND METASTABLE STRUCTURES)
           TAKAGI M + SUZUKI S + WATANABE H
           P369-72 OF INT MEETING ON FERROELECTRICITY, 2ND, 1969, KYOTO, PROC,
           (J PHYS SOC JAP, VOL 28 SUPPL). PHYS SOC JAP, 1970, 459P.

27-3140   EFFECTS ON ELECTRICAL BOUNDARY CONDITIONS ON THE THERMAL CAPACITY OF SINGLE CRYSTALS OF
           TRIGLYCINE SULFATE (NEAR THE FERROELECTRIC TRANSITION).
           TARASKIN SA
           SOV PHYS SOLID STATE 12(6): 1443-4 (1970)
                221

27-3141   THERMAL AND ELECTRICAL PROPERTIES OF TRIGLYCINE SULFATE SINGLE CRYSTALS (HEAT CAPACITY
           AND ELECTRIC PERMITTIVITY).
           TARASKIN SA + STRUKOV BA + MELESHINA VA
           SOV PHYS SOLID STATE 12(5): 1089-94 (1970)
                215

27-3142   CORROSION FIGURES DUE TO THE FERROELECTRIC DOMAIN BOUNDARIES OF TRIGLYCINE SULFATE.
           TAUREL L + EIMER M
           COMPT REND 256: 642-5 (1963) (IN FRENCH)

27-3143   ETCHING AND ULTRAMICROSCOPIC STUDIES OF FERROELECTRIC DOMAINS IN TRIGLYCINE SULFATE.
           TAUREL L + GILLETTA F
           P43-50, VOL 2 OF DVORAK V + FOUSKOVA A + GLOGAR P (EDS),
           INT MEETING ON FERROELECTRICITY, PRAGUE, 1966, PROC.
           INST PHYS CZECH ACAD SCI, PRAGUE, 1966.

27-3144   UTILIZATION OF THE (CRITICAL PULSE WIDTH) PARTIAL SWITCHING PROPERTIES OF
           FERROELECTRICS IN MEMORY DEVICES. (EXAMPLE TRIGLYCINE SULFATE)
           TAYLOR GW
           IEEE TRANS ELECTRON COMPUT 14: 881-6 (1965)

27-3145   ELECTROMECHANICAL PROPERTIES OF TRIGLYCINE SULFATE. (MEASUREMENTS VERIFY THEORY
           APPROXIMATELY)
           TELLE F + CHAPELLE JP
           P448-56, VOL 1 OF DVORAK V + FOUSKOVA A + GLOGAR P (EDS),
           INT MEETING ON FERROELECTRICITY, PRAGUE, 1966, PROC.
           INST PHYS CZECH ACAD SCI, PRAGUE, 1966.

27-3146   FERROELECTRIC SPECIFIC HEAT OF TRIGLYCINE SULFATE.
           TELLO MJ + GONZALO JA
           P199-201 OF INT MEETING ON FERROELECTRICITY, 2ND, 1969, KYOTO, PROC,
           (J PHYS SOC JAP, VOL 28 SUPPL). PHYS SOC JAP, 1970, 459P.

27-3147   STUDY OF THE CRITICAL BEHAVIOR OF TRIGLYCINE SULFATE BY THERMAL NOISE MEASUREMENTS.
           TOMINAGA Y + IIDA S
           J PHYS SOC JAP 32: 1437 (1972)
                278

27-3148   EFFECT OF SAMPLE TREATMENT AND ELECTRODE NATURE ON SOME TRIGLYCINE SULFATE PARAMETERS.
           TOSHEV SD + AMOV IG + KIROV KI
           COMPT REND ACAD BULG SCI 25: 1479-81 (1972)

27. TRIGLYCINE SULFATE (TGS) AND RELATED COMPOUNDS

27-3149   DIELECTRIC BEHAVIOR OF GAMMA-RAY IRRADIATED TRIGLYCINE SULFATE.
TOYODA K + KAWABATA A + TANAKA T
JAP J APPL PHYS 2: 311 (1963)

27-3150   DIELECTRIC CONSTANT OF TRIGLYCINE SULFATE SINGLE CRYSTALS IN A STRONG ELECTRIC FIELD.
TSEDRIK MS + MARGOLIN LN
DOKL AKAD NAUK BELORUS SSR 14: 802-5 (1970) (IN RUSSIAN)

27-3151   EFFECT OF TRIGLYCINE FLUOBERYLLATE SINGLE CRYSTAL GROWTH CONDITIONS ON A CHANGE IN THE
DIELECTRIC CONSTANT AND TANGENT OF THE ANGLE OF DIELECTRIC LOSS IN A STRONG ELECTRIC
FIELD.
TSEDRIK MS + MARGOLIN LN
DOKL AKAD NAUK BELORUS SSR 14: 681-4 (1971) (IN RUSSIAN)

27-3152   EFFECT OF GROWTH CONDITIONS ON DIELECTRIC LOSSES IN TRIGLYCINE SULFATE SINGLE CRYSTALS.
TSEDRIK MS + MARGOLIN LN
VEST AKAD NAVUK BELARUS SSR, SER FIZ MAT NAVUK NO 3: 120-2 (1970)
(IN RUSSIAN)

27-3153   EFFECT OF GROWTH CONDITIONS OF TRIGLYCINE SULFATE SINGLE CRYSTALS ON A CHANGE IN THE
TANGENT OF THE ANGLE OF DIELECTRIC LOSSES IN A STRONG ELECTRIC FIELD.
TSEDRIK MS + MARGOLIN LN
VEST AKAD NAVUK BELARUS SSR, SER FIZ MAT NAVUK NO 5: 130-2 (1970)
(IN RUSSIAN)

27-3154   MAXIMUM REPOLARIZATION CURRENT FOR TRIGLYCINE SULFATE CRYSTALS DEPENDENT ON GROWTH
CONDITIONS.
TSEDRIK MS + DEMIDOVICH NP
VEST AKAD NAVUK BELARUS SSR, SER FIZ MAT NAVUK 2: 127-30 (1972)
(IN RUSSIAN)

27-3155   CRITICAL SLOWING DOWN AT FERROELECTRIC TRANSITIONS (TRIGLYCINE SULFATE).
UNRUH HG + WAHL HJ
PHYS STATUS SOLIDI A, APPL RES 9(1): 119-24 (1972)

27-3156   SHORT LIVED FERROELECTRIC "AFTER EFFECT" PHENOMENA. (HYSTERESIS LOOP CONSTRICTIONS,
ROCHELLE SALT, TGS)
UNRUH HG + MUESER HE
Z ANGEW PHYS 14: 121-5 (1962) (IN GERMAN)

27-3157   APPLICABILITY OF THE PIPPARD RELATIONS TO THE PHASE TRANSITION IN TRIGLYCINE SULFATE.
VARIKASH VM + ZAREMBOVSKAYA TA + PUPKEVICH PA
VEST AKAD NAUK BELORUS SSR, SER FIZ MAT NAVUK 1972(6): 120-2
(IN RUSSIAN)
           297

27-3158   NONLINEAR PROPERTIES OF TRIGLYCINE SELENATE. (DIELECTRIC CONSTANT VERSUS ALTERNATING
FIELD VOLTAGE, TEMPERATURE AND POLARIZATION)
VARIKASH VM + PUPKEVICH PA
IZV VUZ FIZ 1970(2): 151-2 (IN RUSSIAN)
(TRANSLATION TO APPEAR IN SOV PHYS J)
       CHEM ABSTR 73: 8339V (1970)
           205

27-3159   NATURE OF RADIATION INDUCED DEFECTS IN CRYSTALS OF ROCHELLE SALT AND TRIGLYCINE
SULFATE.
VIBLYI IF + ROMANYUK NA
SOV PHYS CRYSTALLOGR 15: 274-7 (1970)

27-3160   POSITION OF THE FUNDAMENTAL ABSORPTION EDGE AS A FUNCTION OF TEMPERATURE FOR ROCHELLE
SALT AND TRIGLYCINE SULFATE CRYSTALS.
VIBLYI IF + ROMANYUK NA
OPT SPECTROSC 28(2): 165-7 (1970)
           216

27-3161   EPR STUDIES OF CHROMIUM- AND VANADYL- DOPED TRIGLYCINE SULFATE SINGLE CRYSTALS.
WARTEWIG S + WINDSCH W
ANN PHYS 24(5): 243-55 (1970) (IN GERMAN)
       CHEM ABSTR 73: 61108V (1970)
           216

27. TRIGLYCINE SULFATE (TGS) AND RELATED COMPOUNDS

THIN FILM NUCLEATION ON FERROELECTRIC SUBSTRATES. (TELLURIUM ON TRIGLYCINE SULFATE AND SODIUM NITRITE SURFACES) (DECORATION METHOD OF DOMAIN STUDY)
WEIDMANN EJ + ANDERSON JC
THIN SOLID FILMS 7: 27-39 (1971)

27-3162   INDUCED GROWTH ANISOTROPY IN TGS CRYSTALS.
WEIDMANN EJ + WHITE EAD + WOOD VM
J MATER SCI 7: 719-20 (1972)

27-3163   THERMODYNAMIC STABILITY OF POLARIZATION IN THIN FERROELECTRIC FILMS ON SEMICONDUCTING SUBSTRATES. (TGS ON SILICON)
WURFEL P + BATRA IP
BULL AMER PHYS SOC 17: 1182 (1972)

X-RAY CRITICAL SCATTERING IN TRIGLYCINE SULFATE AND SODIUM NITRATE. (ORDER DISORDER TRANSITION, FERROELECTRIC AND NONFERROELECTRIC)
YAMADA Y + FUJII Y + TERAUCHI H
P274-7 OF INT MEETING ON FERROELECTRICITY, 2ND, 1969, KYOTO, PROC,
(J PHYS SOC JAP, VOL 28 SUPPL). PHYS SOC JAP, 1970, 459P.

27-3164   DOMAIN STRUCTURE OF GAMMA-IRRADIATED TGS CRYSTALS.
YURIN VA + BELUGINA NV + MELESHINA VA + ANKUDINOV MA + ZHELUDEV IS
BULL ACAD SCI USSR, PHYS SER 35(9): 1749-52 (1971)

27-3165   PHASE STABILIZATION (HINDERED TRANSITION) IN GAMMA-IRRADIATED FERROELECTRIC MATERIALS.
(ROCHELLE SALT, TRIGLYCINE SULFATE)
YURIN VA
BULL ACAD SCI USSR, PHYS SER 29(11): 1834-9 (1965)
219

27-3166   DIELECTRIC PERMEABILITY OF TRIGLYCINE SELENATE NEAR THE PHASE TRANSITION.
ZAREMBOVSKAYA TA + VARIKASH VM
SOV PHYS SOLID STATE 13: 2989-92 (1972)

27-3167   EFFECT OF CONSTANT ELECTRICAL FIELD ON PHASE TRANSITIONS IN TRIGLYCINE FLUOBERYLLATE.
ZAREMBOVSKAYA TA + VARIKASH VM + LAGUTINA ZP + PUPKEVICH PA
VEST AKAD NAVUK BELARUS SSR SER FIZ MAT NAVUK 3: 97-104 (1971)
(IN RUSSIAN)

27-3168   RELATION BETWEEN ELECTRICAL AND THERMAL PROPERTIES OF TRIGLYCINE SELENATE NEAR THE CURIE POINT.
ZAREMBOVSKAYA TA + VARIKASH VM
FIZ MAT NAVUK 6: 122-5 (1971) (IN RUSSIAN)

27-3169   THERMAL EXPANSION OF TRIGLYCINE FLUOBERYLLATE CRYSTALS IN THE FERROELECTRIC TRANSITION RANGE.
ZAREMBOVSKAYA TA + VARIKASH VM + PUPKEVICH PA
IZV VUZ FIZ 1972(6): 153-5 (IN RUSSIAN)

27-3170   AFTER EFFECTS IN SEMICONDUCTOR ELECTRODES EVAPORATED ON INSULATORS ESPECIALLY ON TRIGLYCINE SULFATE.
ZIEBERT V
FERROELECTRICS 4: 77-82 (1972)

27-3171   AFTER EFFECTS IN THE FIELD EFFECT AT VERY LARGE GATE CHARGES APPLIED TO TELLURIUM FILMS ON UHV CLEAVED TGS.
ZIEBERT V
FERROELECTRICS 4: 83-6 (1972)

27-3172   FIELD EFFECT AND HALL MOBILITY OF TELLURIUM AND LEAD TELLURIDE ON UHV CLEAVED TRIGLYCINE SULFATE.
ZIEBERT V
FERROELECTRICS 4: 71-6 (1972)

27-3173   THIN FILMS ON TGS CLEAVED IN ULTRAHIGH VACUUM.
ZIEBERT V
J PHYS (PARIS) 33, SUPPL 4, COLLOQ 2: 223-5 (1972)
294

27. TRIGLYCINE SULFATE (TGS) AND RELATED COMPOUNDS

27-3174 ELECTROMECHANICAL COUPLING NEAR THE FERROELECTRIC PHASE TRANSITION OF TRIGLYCINE
         SULFATE.
         ZUCKER J
         J APPL PHYS 43(9): 3656-62 (1972)
             274

27-3175 CRYSTAL STRUCTURE OF TRIGLYCINE FLUOBERYLLATE IN THE FERRO- AND PARAELECTRIC PHASE.
         ZUKASZEWICZ K + WARKUSZ F
         ACTA CRYSTALLOGR A 28: S186 (1972)

28. ROCHELLE SALT AND RELATED COMPOUNDS

28-3176   ELECTRON PARAMAGNETIC RESCNANCE CF COPPER IONS IN ROCHELLE SALT.
ABDULSABIROV RYU + GREZNEV YUS + ZAITOV MM + STEPANOV VG
SOV PHYS SOLID STATE 12: 2204-5 (1971)

28-3177   ANOMALOUS DISPERSION OF CCERSIVE FIELD AT VERY LOW FREQUENCIES AND JERKY WALL MOTION IN
ROCHELLE SALT.
ABE R
JAP J APPL PHYS 3(5): 243-9 (1964)

28-3178   COERCIVE FIELDS OF ROCHELLE SALT HEAVILY IRRADIATED WITH GAMMA-RAYS.
ABE R + MIZUNO O
JAP J APPL PHYS 10: 1122 (AUG 1971)

28-3179   DIELECTRIC CONSTANT AND CIELECTRIC RELAXATION TIME OF ROCHELLE SALT.
ABE R + TOKUMARU Y
J PHYS SOC JAP 31(6): 1748-53 (DEC 1971)
      268

28-3180   ELECTRON PARAMAGNETIC RESONANCE OF GAMMA-IRRADIATED ROCHELLE SALT.
ABE R + SUZUKI I
P358-63, VOL 2 OF DVORAK V + FOUSKOVA A + GLOGAR P (EDS),
INT MEETING ON FERROELECTRICITY, PRAGUE, 1966, PROC.
INST PHYS CZECH ACAD SCI, PRAGUE, 1966.

28-3181   REARRANGEMENT OF THE DOMAIN STRUCTURE IN ROCHELLE SALT NEAR THE PHASE TRANSITION.
ABOLINSH Y + BIELIS IYA
P149-53 OF FAZOVYE PEREKHODY V SEGNETOELEKTRIKAKH (PHASE
TRANSFORMATICNS IN FERROELECTRICS), FRITSBERG VYA + ROLOV BN +
KRUCHAN YY (EDS), IZDATELSTVO ZINATNE RIGA, 1971, 206P. (IN RUSSIAN)

28-3182   CIELECTRIC PHENOMENA IN ROCHELLE SALT.
BAE P + GUILLIEN R
P272-80 OF COLLOQUE AMPERE, 11TH, 1962, EINDHOVEN, MAGNETIC AND
ELECTRIC RESCNANCE AND RELAXATION. SMIDT J (ED), NORTH HOLLAND, 1963,
789P.   (IN FRENCH)
     CHEM ABSTR 59: 13427 (1963)

28-3183   FREQUENCY DEPENDENCE OF THE ABSORPTION OF SOUND IN ROCHELLE SALT NEAR ITS UPPER CURIE
POINT.
BARANSKII KN + SHUSTIN OA + VELICHKINA TS + YAKOVLEV IA
SOV PHYS JETP 16(2): 518-19 (1963)

28-3184   MEASUREMENT OF THE COMPLEX CIELECTRIC CONSTANT OF ROCHELLE SALT AT 10 GHZ AS A FUNCTION
OF TEMPERATURE.
BAUMBER P + BLUM W + DEYDA H
Z PHYS 180(1): 96-104 (1964) (IN GERMAN)
     CHEM ABSTR 61: 10136G (1964)

     NMR STUDIES IN SOME HYDROGEN BONDED FERROELECTRICS. (DEUTERATED ROCHELLE SALT,
POTASSIUM OR RUBIDIUM OR AMMONIUM DIHYDROGEN ARSENATES, RUBIDIUM DIDEUTERIUM ARSENATE)
BJORKSTAM JL
P101-4 OF INT MEETING ON FERROELECTRICITY, 2ND, 1969, KYOTO, PROC,
(J PHYS SOC JAP, VOL 28 SUPPL). PHYS SOC JAP, 1970, 459P.

28-3185   EFFECT OF APPLIED ELECTRIC FIELD ON EPR SPECTRA OF (COPPER(2+)) DOPED FERROELECTRIC
ROCHELLE SALT.
BLINC R + SENTJURC M
PHYS REV LETT 19(21): 1231-3 (1967)

28-3186 SECOND ORDER QUADRUPOLE SHIFTS OF THE SODIUM-23 MAGNETIC RESONANCE IN ROCHELLE SALT.
BLINC R + MALI M + ZPANCIC I
PHYS LETT 5: 309-10 (1963)

28-3187 SPECIFIC HEAT ANOMALY IN FERROELECTRIC ROCHELLE SALT (CALCULATION OF TEMPERATURE
DEPENDENCE FOR FERROELECTRIC CONTRIBUTION).
BLINC R + ZEKS B
PHYS LETT A 39: 167-8 (1972)

28-3188 STRUCTURAL EFFECTS OF (DAMAGE FROM) IONIZING RADIATION IN FERROELECTRIC ROCHELLE SALT.
(NEUTRON DIFFRACTION MEASUREMENTS)
BOUTIN H + FRAZER BC + JONA F
J PHYS CHEM SOLIDS 24: 1341-7 (1963)

28-3189 TEMPERATURE DEPENDENCE OF THE INTENSITY OF HYDROGEN BOND BANDS IN RAMAN SCATTERING
SPECTRA OF GYPSUM AND ROCHELLE SALT CRYSTALS.
CHISLER EV
SOV PHYS SOLID STATE 5: 1789-92 (1964)

28-3190 EFFECT OF ELECTRIC FIELD ON DOMAIN STRUCTURE FORMATION IN SEIGNETTES SALT (ROCHELLE
SALT). (TEMPERATURE ANHYSTERESIS).
CHYLA A + POCHABA AM
ACTA PHYS POLON A 40: 109-10 (JUL 1971)

28-3191 INFLUENCE OF THE DOMAIN STRUCTURE ON POSITRON (TWO-PHONON) ANNIHILATION IN ROCHELLE
SALT.
DWORAKOWSKI J + KOLODZIEJ HB
ACTA PHYS POLON A 38(5): 809-11 (1970)
    231

NEW MEASUREMENTS OF DIVERSE THERMODYNAMICAL COEFFICIENTS OF ROCHELLE SALT AND
TRIGLYCINE SULFATE.
EHSES KH + MUSER HE + FORSCH K + SCHMITT H + TOPF KJ
P195-206 OF EUROPEAN MEETING ON FERROELECTRICITY, 1969, SAARBRUCKEN,
MUSER HE AND PETERSSON J (EDS), WISSENSCH VERLAGSGES, STUTTGART,
1970, 396P.

28-3192 ELECTRONIC EFFECTS (OF X-RAY AND GAMMA-RAY IRRADIATION) IN ROCHELLE SALT CRYSTALS.
EISNER IYA
BULL ACAD SCI USSR, PHYS SER 29(11): 1848-50 (1965)
    219

28-3193 INFLUENCE OF THE CRYSTALLIZATION TEMPERATURE OF ROCHELLE SALT ON ITS FERROELECTRIC
PROPERTIES.
EISNER J
FERROELECTRICS 4: 213-19 (1972)

28-3194 NMR STUDIES OF SPONTANEOUS POLARIZATION IN ROCHELLE SALT.
FITZGERALD ME
RENSSELAER POLYTECH INST PHD THESIS, 1969, 135P.
    DISS ABSTR INT B 30: 5651 (1970)
    229

28-3195 NMR STUDY OF PHASE TRANSITIONS IN ROCHELLE SALT.
FITZGERALD ME + CASABELLA PA
PHYS REV B 2: 1350-4 (1970)

28-3196 PERMITTIVITY OF ROCHELLE SALT DURING SWITCHING.
FOUSKOVA A + JANOUSEK V
CZECH J PHYS B 12: 413-15 (1962)

NEUTRON DIFFRACTION STUDIES OF FERROELECTRICS. (BARIUM TITANATE AND ROCHELLE SALT AS
EXAMPLES)
FRAZER BC
J PHYS SOC JAP 17(SUPPL B-II): 376-82 (1962)
IN PART II P376-82 OF INT CONF ON MAGNETISM AND CRYSTALLOGR.
1961, KYOTO, PROC. PHYS SOC JAPAN, 1962

THERMOLUMINESCENCE STUDIES OF THE GAMMA-RAY IRRADIATED FERROELECTRICS ROCHELLE SALT AND
GUANIDINE ALUMINUM SULFATE HEXAHYDRATE.
GILLILAND JW + YOCKEY HP
J PHYS CHEM SOLIDS 23: 367-74 (1962)

28-3197   ULTRASONIC ANOMALIES IN ROCHELLE SALT.
          GOBRAN NK + YOUSSEF H + MUHMOUD SA
          ACUSTICA 27: 1-7 (1972)

          NOISE MEASUREMENTS IN FERROELECTRICS. (TGS, ROCHELLE SALT, BARIUM TITANATE)
          GODEFROY L
          J PHYS (PARIS) 33, SUPPL 4, COLLOQ 2: 44-8 (1972)

28-3198   CONTRIBUTION OF DOMAIN WALL MOTION TO THE PERMITTIVITY OF ROCHELLE SALT.
          GURK P
          PHYS STATUS SOLIDI A, APPL RES 10(2): 407-14 (1972)

28-3199   LOW FREQUENCY DISPERSIONS OF THE PERMITTIVITY OF ROCHELLE SALT CAUSED BY DOMAIN WALLS.
          (ABSTRACT ONLY)
          GURK P
          P239 OF EUROPEAN MEETING ON FERROELECTRICITY, 1969, SAARBRUCKEN,
          MUSER HE AND PETERSSON J (EDS), WISSENSCH VERLAGSGES, STUTTGART,
          1970, 396P.

28-3200   CRYSTALLIZATION OF ROCHELLE SALT IN ELECTRIC FIELDS (KINETICS OF CRYSTAL GROWTH)
          HANANI M + SCHIEBER M
          J APPL PHYS 42(1): 206-11 (1971)
                224

          OPTICAL ACTIVITY OF SOME FERROELECTRIC CRYSTALS (PYROELECTRICITY, TGS, ROCHELLE SALT).
          HERMELBRACHT K + UNRUH HG
          Z ANGEW PHYS 28(5): 285-8 (1970)

28-3201   HIGHLY SENSITIVE, SIMPLE CALORIMETER FOR PHASE CHANGE DETECTION (INCLUDING
          FERROELECTRIC AND MAGNETIC TRANSITIONS, EXAMPLE ROCHELLE SALT)
          HIRAKAWA K + FURUKAWA M
          JAP J APPL PHYS 9(8): 971-5 (1970)
                207

28-3202   MEASUREMENT OF THERMAL EXPANSION OF ROCHELLE SALT BY USING THE SHIFT OF RESONANCE
          FREQUENCY IN CAVITY.
          HORIOKA M + ABE R
          MEM KANAZAWA INST TECHNOL 1: 55-62 (1971)

          THE ROLE OF LORENTZ FACTORS IN ROCHELLE SALT. (THEORY)
          ISHIBASHI Y + TAKAGI Y
          J PHYS SOC JAP 32(3): 723-8 (1972)

28-3203   NUCLEAR MAGNETIC RESONANCE IN ROCHELLE SALT, DETERMINATION OF THE ORIENTATIONS OF
          PROTON PAIRS (IN THE WATER MOLECULES).
          KATO T + MIZUNO O + ABE R
          J PHYS SOC JAP 29(2): 393-7 (1970)
                208

28-3204   ATTENUATION OF SOUND WAVES IN DEUTERATED ROCHELLE SALT AT THE PHASE TRANSITION POINT.
          KESSENIKH GG + SHIROKOV AM + SHUVALOV LA + SHCHAGINA NM
          SOV PHYS CRYSTALLOGR 15(6): 1097-8 (MAY-JUN 1971)
                234

28-3205   ATTENUATION OF SOUND WAVES IN DEUTERATED ROCHELLE SALT AT THE PHASE TRANSITION POINT.
          KESSENIKH GG + SHIROKOV AM + SHUVALOV LA + SHCHAGINA NM
          SOV PHYS CRYSTALLOGR 15(6): 1097-8 (MAY-JUN 1971)
                234

28-3206   FAST AFTEREFFECT OF SEIGNETTE SALT.
          KLEIN G + LUTHER G
          Z NATURFORSCH A 25(7): 1159-60 (1970) (IN GERMAN)
                234

28-3207   RADIATION DAMAGE AND THE FERROELECTRIC EFFECT IN ROCHELLE SALT.
          KRUEGER HHA + COOK WR + SARTAIN CC + YOCKEY HP
          J APPL PHYS 34: 218-24 (1963)

28-3208   CORRELATION BETWEEN BARKHAUSEN NOISE AND SURFACE MICRODAMAGES ON SINGLE CRYSTALS OF
          SEIGNETTE SALT.
          KUSZ J + SUJAK B
          ACTA PHYS POLON A 37: 221-5 (1970)

28. ROCHELLE SALT AND RELATED COMPOUNDS

PHENOMENOLOGICAL THEORY OF DIELECTRIC ANOMALIES IN FERROELECTRIC MATERIAL WITH SEVERAL
PHASE TRANSITIONS AT TEMPERATURES CLOSE TOGETHER. (ROCHELLE SALT)
LEVANYUK AP + SANNIKOV DG
SOV PHYS JETP 33: 600-4 (1971)

CRITICAL PROPERTIES OF TGS AND ROCHELLE SALT AS DETERMINED BY AN ELECTRIC FIELD METHOD.
LEVSTIK A + BURGAR M + BLINC R
J PHYS (PARIS) 33, SUPPL 4, COLLOQ 2: 235-6 (1972)

28-3209   SPONTANEOUS ELECTROOPTICAL EFFECT IN ROCHELLE SALT CRYSTALS.
          LOMOVA LG + SONIN AS
          SOV PHYS SOLID STATE 12(11): 2699-700 (1971)
               234

28-3210   FAR INFRARED ABSORPTION SPECTRA OF ROCHELLE SALT.
          MALINEAU M + STRIMER P + HADNI A
          J CHIM PHYS 69: 343-5 (FEB 1972) (IN FRENCH)

28-3211   ABSORPTION OF LONGITUDINAL ULTRASONIC WAVES IN ROCHELLE SALT CRYSTALS NEAR THE CURIE
          POINTS.
          MINAEVA KA + CHUNG TK + STRUKOV BA + KOPTSIK VA
          BULL ACAD SCI USSR, PHYS SER 35(9): 1738-40 (1971)
               269

          EFFECT OF A TWO DIMENSIONAL PRESSURE ON THE CURIE POINTS OF TRIGLYCINE SULFATE AND
          ROCHELLE SALT.
          MORI K + HAYASHI M
          J PHYS SOC JAP 33(5): 1396-400 (NOV 1972)

          MICROWAVE BEHAVIOR OF ROCHELLE SALT AND TGS (18 REFS).
          MUESER HE + GREGORIUS P + LUTHER G + POTTHARST J
          P91-101 OF EUROPEAN MEETING ON FERROELECTRICITY SAARBRUCKEN MARCH
          1969, MUESER HE + PETERSSON J (EDS), WISSENSCH VERLAGSGES, STUTTGART,
          1970.

28-3212   ANALYSIS OF DOUBLE HYSTERESIS LOOPS. (ROCHELLE SALT)
          MUSER HE + BERNDES G
          P149-54, VOL 2 OF DVORAK V + FOUSKOVA A + GLOGAR P (EDS),
          INT MEETING ON FERROELECTRICITY, PRAGUE, 1966, PROC.
          INST PHYS CZECH ACAD SCI, PRAGUE, 1966.

28-3213   MEASUREMENTS OF COMPLEX PIEZOELECTRIC CONSTANTS IN FERROELECTRIC CRYSTALS. (ROCHELLE
          SALT)
          MUSER HE + SCHMITT H
          J PHYS (PARIS) 33, SUPPL 4, COLLOQ 2: 103-4 (1972)
               294

28-3214   HIGH PRESSURE PHASE TRANSITIONS OF THE FIRST KIND IN ROCHELLE SALT.
          MYLOV VP + LEONIDOVA GG + CHURAGULOV BR
          SOV PHYS SOLID STATE 14(7): 1893-4 (JAN 1973)
               282

28-3215   DOMAIN WALLS CAUGHT IN (FINE LINE) "SUDARES" IN ROCHELLE SALT CRYSTAL.
          OHI K + NAKAMURA T
          J PHYS SOC JAP 17: 1195 (1962)

          FERROELECTRIC BEHAVIOR OF RADIATION DAMAGED TRIGLYCINE SULFATE AND ROCHELLE SALT UP TO
          HIGH DOSAGE (DESTRUCTION OR MASKING OF FERROELECTRICITY).
          OKADA K + GONZALO JA + RIVERA JM
          J PHYS CHEM SOLIDS 28(4): 689-95 (1967)

          RESULTS OF EXPERIMENTAL STUDY OF GAMMA-RADIATION EFFECTS IN ROCHELLE SALT AND
          TRIGLYCINE SULFATE. (ON ELECTRONIC PROPERTIES.)
          PESHIKOV EV + STARODUBTSEV SV
          P267-76, VOL 2 OF DVORAK V + FOUSKOVA A + GLOGAR P (EDS),
          INT MEETING ON FERROELECTRICITY, PRAGUE, 1966, PROC.
          INST PHYS CZECH ACAD SCI, PRAGUE, 1966.

28-3216   ZERO SHIFT IN PIEZOELECTRIC ACCELEROMETERS. (POLARIZATION SWITCHING IN POLYCRYSTALLINE
          FERROELECTRICS AT VERY LOW FIELDS AND STRESSES, ROCHELLE SALT)
          PLUMLEE RH
          SANDIA LABS, 1971, 61P. SC-RR-70-755
               NUCL SCI ABSTR 25: 34477 (1971)
               243

          UHF (WAVEGUIDE RESONATOR) METHOD OF STUDYING TEMPERATURE CHARACTERISTICS OF ROCHELLE
          TYPE CERAMICS. (BARIUM CALCIUM TITANATE CERAMIC)
          POPLAVKO YUM + YAZYTSKII BYA
          INSTRUM EXPER TECH 1967(4): 841-3

28-3217   GROWTH RATE OF THE (010) PLANE OF ROCHELLE SALT IN THE PRESENCE OF PURE DIRECT SKY BLUE
          DYE.
          PORTNOV VN + RYKHLYUK AV + BELYUSTIN AV
          SOV PHYS CRYSTALLOGR 14(4): 643-4 (1970)

28-3218   FERROELECTRICITY IN ROCHELLE SALT.
          POTTHARST J
          UNIV SAARLAND PHD THESIS, INSTITUTE FOR EXPERIMENTAL PHYSICS, 1969,
          72P. (IN GERMAN) N70-27303
               SCI TECH AEROSP REP 8(13): 2466 (1970)

28-3219   INFLUENCE OF CRYSTAL DEFECTS ON THE DIELECTRIC BEHAVIOR OF ROCHELLE SALT IN MICROWAVE
          RANGE.
          POTTHARST J
          P439-40 OF INT MEETING ON FERROELECTRICITY, 2ND, 1969,
          KYOTO, PROC (J PHYS SOC JAP, VOL 28 SUPPL). PHYS SOC JAP,
          1970.
               205

28-3220   TEMPERATURE AND FREQUENCY DEPENDENCE OF THE LOSS FACTOR IN FERROELECTRIC MATERIALS
          (ROCHELLE SALT NEAR THE UPPER CURIE POINT AS EXAMPLE).
          POTTHARST J
          CZECH J PHYS B 20: 752-4 (1970)
               204

28-3221   EFFECT OF THE CUPRIC ION ON DELAYED PHENOMENA IN ROCHELLE SALT.
          REWAJ T
          ACTA PHYS POLON 24: 45-50 (1963)

28-3222   ON THE EFFECTS OF X-RAY IRRADIATION ON TRANSIENT PHENOMENA IN ROCHELLE SALT.
          REWAJ T
          PHYS STATUS SOLIDI 2: 1151-7 (1962)

          IONIZATION OF THE (X-RADIATION AND COPPER AND IRON IMPURITY) ABSORPTION BANDS OF DEFECT
          ROCHELLE SALT AND TRIGLYCINE SULFATE CRYSTALS.
          ROMANYUK NA + VIBLYI IF
          SOV PHYS CRYSTALLOGR 15(4): 642-6 (1971)

          DISTINCTIVE FEATURES OF BARKHAUSEN EFFECT IN ROCHELLE SALT AND TRIGLYCINE SULFATE
          CRYSTALS.
          RUDYAK VM + SHUVALOV LA + KAMAEV VE
          BULL ACAD SCI USSR, PHYS SER 29(6): 947-51 (1965)

          ELECTROMAGNETIC AND ACOUSTIC WAVES IN FERROELECTRICS. (THEORY, BARIUM TITANATE,
          ROCHELLE SALT)
          SANNIKOV DG
          SOV PHYS SOLID STATE 4: 1187-92 (1962)

28-3223   MECHANISM OF FERROELECTRIC PHASE TRANSITION IN AMMONIUM ROCHELLE SALT. (THEORY)
          SAWADA A + TAKAGI Y
          J PHYS SOC JAP 33(4): 1071-5 (OCT 1972)
               278

28-3224   SUPERSTRUCTURE IN THE FERROELECTRIC PHASE OF AMMONIUM ROCHELLE SALT.
          SAWASA A + TAKAGI Y
          J PHYS SOC JAP 31: 952 (1971)
               251

28-3225   EFFECT OF MECHANICAL STRESS ON DIELECTRIC PROPERTIES OF PARAELECTRIC ROCHELLE SALT.
          SCHMIDT G + NEUMANN KH
          Z PHYS 166: 207-15 (1962) (IN GERMAN)

28-3226   THERMAL CONDUCTIVITY OF ROCHELLE SALT.
           SCHAFER H + ALBERS J + HEIWIG J
           PHYS LETT A 39: 159-60 (1972)

28-3227   INTERNAL FRICTION AS A FUNCTION OF TEMPERATURE FOR ROCHELLE SALT.
           SHIROKOV AM + SHUVALOV LA
           SOV PHYS CRYSTALLOGR 8: 586-8 (1964)

28-3228   LOW FREQUENCY DIELECTRIC DISPERSION IN ROCHELLE SALT CRYSTALS.
           SHILNIKOV AV + POPOV ES + RAPOPORT SL AND OTHERS
           SOV PHYS CRYSTALLOGR 15: 1027-31 (MAY-JUN 1971)

           INFLUENCE OF GAMMA-IRRADIATION ON BARKHAUSEN EFFECT IN FERROELECTRIC MATERIALS (TGS AND
           ROCHELLE SALT).
           SHUVALOV LA + RUDYAK VM + KOMLYAKOVA NS + KAMAEV VE
           BULL ACAD SCI USSR, PHYS SER 29(11): 1844-8 (1965)

           AMPLITUDE DEPENDENCE OF INTERNAL FRICTION IN SINGLE CRYSTALS OF FERROELECTRICS.
           (ROCHELLE SALT AND TRIGLYCINE SULFATE)
           SHUVALOV LA + SHIROKOV AM
           SOV PHYS CRYSTALLOGR 9(6): 746-9 (1965)

28-3229   PIEZOELECTRIC EFFECT AND INTERNAL FRICTION IN GAMMA-IRRADIATED ROCHELLE SALT CRYSTALS.
           SILVESTROVA IM + YURIN VA + SHUVALOV LA + PODLESSKAYA AF
           BULL ACAD SCI USSR, PHYS SER 29(11): 1839-43 (1965)
                219

           ELECTRON PARAMAGNETIC RESONANCE INVESTIGATIONS OF (COPPER) DOPED FERROELECTRIC
           CRYSTALS. (TRIGLYCINE SULFATE AND ROCHELLE SALT)
           STANKOWSKA J
           P364-8, VOL 2 OF DVORAK V + FOUSKOVA A + GLOGAR P (EDS),
           INT MEETING ON FERROELECTRICITY, PRAGUE, 1966, PROC.
           INST PHYS CZECH ACAD SCI, PRAGUE, 1966.

28-3230   EFFECT OF A DC FIELD ON DOMAIN STRUCTURE IN SEIGNETTE SALT. (EFFECT OF AGING AND
           COPPER(2+) DOPING)
           STANKOWSKA J + KASPRZAK J
           POZNAN TOW PRZYJ NAUK, PR KOM MAT-PRZYR 5(1): 27-34 (1969)
           (IN POLISH)
                CHEM ABSTR 72: 84047T (1969)
                231

28-3231   RAMAN LIGHT SCATTERING IN ROCHELLE SALT.
           STEKHANOV AI + GABRICHIDZE ZA
           SOV PHYS SOLID STATE 5: 972-4 (1963)

           MICROSCOPIC THEORY OF THE PHASE TRANSITIONS IN ROCHELLE SALT.
           TAKAGI Y + ISHIBASHI Y
           J PHYS SOC JAP 33(5): 1381-9 (NOV 1972)

28-3232   STUDY OF SLOW MOTION OF WATER MOLECULES IN ROCHELLE SALT AND AMMONIUM ROCHELLE SALT BY
           SPIN LATTICE RELAXATION OF PROTONS IN ROTATING FRAME.
           TRONTELJ Z
           J PHYS (PARIS) 33, SUPPL 4, COLLOQ 2: 189-91 (1972)
                294

           SHORT LIVED FERROELECTRIC "AFTER EFFECT" PHENOMENA. (HYSTERESIS LOOP CONSTRICTIONS,
           ROCHELLE SALT, TGS)
           UNRUH HG + MUESER HE
           Z ANGEW PHYS 14: 121-5 (1962) (IN GERMAN)

           POSITION OF THE FUNDAMENTAL ABSORPTION EDGE AS A FUNCTION OF TEMPERATURE FOR ROCHELLE
           SALT AND TRIGLYCINE SULFATE CRYSTALS.
           VIBLYI IF + ROMANYUK NA
           OPT SPECTROSC 28(2): 165-7 (1970)

           NATURE OF RADIATION INDUCED DEFECTS IN CRYSTALS OF ROCHELLE SALT AND TRIGLYCINE
           SULFATE.
           VIBLYI IF + ROMANYUK NA
           SOV PHYS CRYSTALLOGR 15: 274-7 (1970)

28-3233   EPR INVESTIGATION OF THE INTERNAL ELECTRIC FIELD IN COPPER DOPED ROCHELLE SALT.
          VOLKEL G + WINDSCH W
          PHYS STATUS SOLIDI 48: 263-77 (1971)
               226

28-3234   SURFACE STRUCTURES AND CORRESPONDING LIGHT FIGURES REVEALED BY WATER ETCHED ROCHELLE
          SALT CRYSTALS.
          WATANABE J + SATO T + YAMAMOTO M
          SCI REP RES INST, TOHOKU UNIV, SER-A 22(4): 165-71 (FEB 1971)
               245

28-3235   GRAVIMETRIC MEASUREMENTS CF THE CHANGES OF WATER OF CRYSTALLIZATION OF ROCHELLE SALT.
          WEISSBACH G
          PHYS STATUS SOLIDI A 3: K113-15 (1970)
               234

          PHASE STABILIZATION (HINDERED TRANSITION) IN GAMMA-IRRADIATED FERROELECTRIC MATERIALS.
          (ROCHELLE SALT, TRIGLYCINE SULFATE)
          YURIN VA
          BULL ACAD SCI USSR, PHYS SER 29(11): 1834-9 (1965)

28-3236   DYNAMICS OF FERROELECTRIC ROCHELLE SALT.
          ZEKS B + SHUKLA GC + BLINC R
          PHYS REV B SOLID STATE 3: 2306-9 (1971)

28-3237   DYNAMICS OF FERROELECTRIC ROCHELLE SALT.
          ZEKS B + SHULKA GC + BLINC R
          J PHYS (PARIS) 33, SUPPL 4, COLLOQ 2: 67-8 (1972)
               294

29. TARTRATES

29-3238 INFERRED MECHANISM OF THE PHASE TRANSFORMATION OF PECULIAR FERROELECTRIC- FERROELASTIC CRYSTAL: SODIUM AMMONIUM TARTRATE.
AIZU K
J PHYS SOC JAP 31: 1521-6 (NOV 1971)

29-3239 SOME PROPERTIES OF NONDEUTERATED AND DEUTERATED LITHIUM THALLIUM TARTRATE MONOHYDRATE. (EFFECT ON LATTICE AND TRANSITION)
BREZINA B + JANOUSEK V + MARECEK V + SMUTNY F
P369-73 OF EUROPEAN MEETING ON FERROELECTRICITY, 1969, SAARBRUCKEN, MUSER HE AND PETERSSON J(EDS), WISSENSCH VERLAGSGES, STUTTGART, 1970, 396P.

29-3240 NUCLEAR MAGNETIC RESONANCE OF FERROELECTRIC LITHIUM AMMONIUM TARTRATE.
ELSAFFAR ZM + O'REILLY DE + PETERSON EM + FLICK C
J CHEM PHYS 57(4): 2372-9 (15 SEP 1972)
285

29-3241 NUCLEAR MAGNETIC RESONANCE OF FERROELECTRIC LITHIUM AMMONIUM TARTRATE.
ELSAFFAR ZM + O'REILLY DE + PETERSON EM
BULL AMER PHYS SOC 16: 94 (1970)

29-3242 SOME PROPERTIES OF THE FERROELECTRIC LITHIUM THALLIUM TARTRATE.
FOUSEK J + CROSS LE + SEELY K
FERROELECTRICS 1: 63-70 (1970)

29-3243 EFFECT OF TEMPERATURE AND PROPAGATION DIRECTION UPON OPTICAL PHONONS IN CUBIC SODIUM CHLORIDE AND EPITAXIAL FERROELECTRIC SODIUM TARTRATE.
HARTING CM
UNIV SOUTHERN CALIF PHD THESIS, 1971, 154P.
DISS ABSTR INT B 32: 503-4 (1971)
238

DISPLACEMENT OF THE CURIE POINTS IN KDP AND LITHIUM THALLIUM TARTRATE CRYSTALS BY HYDROSTATIC PRESSURE.
HEGENBARTH E
SOV PHYS CRYSTALLOGR 15(2): 268-70 (1970)

29-3244 X-RAY REFINEMENT OF THE CRYSTAL STRUCTURE OF LITHIUM AMMONIUM TARTRATE MONOHYDRATE.
HINAZUMI H + MITSUI T
ACTA CRYSTALLOGR B 28: 3299-3305 (1972)

29-3245 IS SODIUM RUBIDIUM TARTRATE TETRAHYDRATE A FERROELECTRIC. (ANSWER-NO)
KRISHNAN RS + TIWARY HV + NARAYANAN PS
P167-8 OF INT MEETING ON FERROELECTRICITY, 2ND, 1969,, KYOTO, PROC, (J PHYS SOC JAP, VOL 28 SUPPL). PHYS SOC JAP, 1970, 459P.

29-3246 SOME CHEMICAL REACTIONS IN SILICA GELS. PART-3: FORMATION OF POTASSIUM ACID TARTRATE CRYSTALS.
KURZ PF
OHIO J SCI 69(5): 296-304 (1969)
CHEM ABSTR 71: 117174S (1969)
198

29-3247 CRYSTAL DATA FOR LITHIUM THALLIUM TARTRATE MONOHYDRATE.
MCCARTHY GJ + SCHLEGEL LH + SAWAGUCHI E
J APPL CRYSTALLOGR 4: 180-1 (1971)

29-3248 PROTON SPIN LATTICE RELAXATION IN FERROELECTRIC SODIUM AMMONIUM TARTRATE.
MORIMOTO K
PHYS LETT A 35: 472-3 (1971)

29-3249   CONTROL OF THE ELASTIC COMPLIANCE OF FERROELECTRIC LITHIUM THALLIUM TARTRATE HYDRATE
          (LTT) BY AN ELECTRIC FIELD.
          SAWAGUCHI E + CROSS LE
          FERROELECTRICS 3: 327-32 (1972);
          IEEE TRANS SON ULTRACON 19: 327-32 (1972)
                263

29-3250   ELECTRIC FIELD CONTROL OF THE ELASTIC COMPLIANCE IN LITHIUM THALLIUM TARTRATE (TUNING
          AND DELAY DEVICES).
          SAWAGUCHI E + CROSS LE
          APPL PHYS LETT 18(1): 1-2 (1971)
                224

29-3251   ELECTROMECHANICAL COUPLING EFFECTS ON THE DIELECTRIC PROPERTIES AND FERROELECTRIC PHASE
          TRANSITION ON LITHIUM THALLATE TARTRATE.
          SAWAGUCHI E + CROSS LE
          FERROELECTRICS 2: 37-46 (JAN 1971)

29-3252   ESR OF COPPER(2+) DOPED SODIUM AMMONIUM TARTRATE.
          SUZUKI I + MAEDA M + ABE R
          J PHYS SOC JAP 33: 860 (1972)

          CRYSTALLIZATION POTENTIALS IN AQUEOUS SOLUTIONS. (POTASSIUM BROMIDE, POTASSIUM
          CHLORIDE, SODIUM CHLORIDE, POTASSIUM SODIUM TARTRATE, POTASSIUM DIHYDROGEN PHOSPHATE)
          WIEDERKEHR H
          EIDGENOSSISCHE TECH HOCHSCHULE, ZURICH, 1970, 123P. N72-13712
          (IN SWEDISH)
                SCI TECH AEROSP REP 10(4): 528 (1972)

30. MOLYBDATES

30-3253   INFERRED TEMPERATURE DEPENDENCES OF ELECTRICAL, MECHANICAL AND OPTICAL PROPERTIES OF
           FERROELECTRIC FERROELASTIC GADOLINIUM MOLYBDATE.
           AIZU K
           J PHYS SOC JAP 31(3): 802-11 (SEP 1971)
                253

30-3254   MECHANISM OF THE FERROELECTRIC PHASE TRANSFORMATION IN RARE EARTH MOLYBDATES. (TERBIUM
           MOLYBDATE)
           AXE JD + DORNER B + SHIRANE G
           PHYS REV LETT 26(9): 519-23 (1 MAR 1971)
                233

30-3255   CERTAIN GROWTH, LASING AND SPECTROSCOPIC PROPERTIES OF NEODYMIUM(3+) DOPED GADOLINIUM
           MOLYBDATE CRYSTALS.
           EAGDASAROV KHS + BOGOMOLOVA GA + KAMINSKII AA + MELESHINA VA + PROKHORTSEVA TM
           + SHUVALOV LA
           BULL ACAD SCI USSR, PHYS SER 35(9): 1681-3 (1971)

30-3256   CONTROL AND APPLICATION OF DOMAIN WALL MOTION IN GADOLINIUM MOLYBDATE.
           BARKLEY JR + BRIXNER LH + HOGAN EM + WARING RK
           FERROELECTRICS 3: 191-7 (1972);
           IEEE TRANS SON ULTRASON 19: 191-7 (1972)
                263

30-3257   CONTROL AND APPLICATION OF DOMAIN WALL MOTION IN GADOLINIUM MOLYBDATE.
           BARKLEY JR + BRIXNER LH + HOGAN EM AND OTHERS
           IEEE TRANS SON ULTRASON 19: 115-23 (1972)

30-3258   DISLOCATION ETCHANTS FOR BETA GADOLINIUM MOLYBDATE.
           BHALLA AS
           J ELECTROCHEM SOC 119: 1602-3 (NOV 1972)

30-3259   GROWTH AND MORPHOLOGY OF SINGLE CRYSTALS OF GADOLINIUM MOLYBDATE, A FERROELECTRIC AND
           FERROELASTIC SUBSTANCE.
           BOHM J + KURSTEN HD
           KRISTALL TECH 6(2): 213-17 (1971) (IN GERMAN)
                284

30-3260   FERROELECTRIC DOMAINS AND THE CONDENSED SOFT MODE IN GADOLINIUM MOLYBDATE.
           BOYER L + HARDY JR
           SOLID STATE COMMUN 11: 555-8 (1972)
                276

30-3261   PHASE TRANSITIONS IN COMPLEX PEROVSKITES OF THE TYPE BARIUM LANTHANON MOLYBDATE.
           BRANDLE CD + STEINFINK H
           P235-45, VOL 1 OF CONF ON RARE EARTH RESEARCH, 8TH PROC, 1970, RENO,
           HENRIE TA + LINDSTROM RE(EDS), US BUR MINES, 1971.
                255

30-3262   PRECISION PARAMETERS OF SOME LANTHANON MOLYBDATE TYPE RARE EARTH MOLYBDATES. (CRYSTAL
           STRUCTURE)
           BRIXNER LH + BIERSTEDT PE + SLEIGHT AW + LICIS MS
           MATER RES BULL 6: 545-54 (1971)
                237

30-3263   PI- GMO: ANOTHER MODIFICATION OF GADOLINIUM MOLYBDATE.
           BRIXNER LH
           MATER RES BULL 7: 879-82 (1972)
                275

30-3264 PRECISION PARAMETERS OF THE FERROELECTRIC RARE EARTH MOLYBDATES: LANTHANUM MOLYBDATE.
BRIXNER LH + BIERSTEDT PE + SLEIGHT AW + LICIS MS
P437-43 OF MATER RES SYMP, 5TH PROC, 1971, SOLID STATE CHEMISTRY,
ROTH RS + SCHNEIDER SJ (EDS), NBS SPEC PUB 364, 1972, 783P.

271

30-3265 ABNORMAL ABSORPTION OF ELASTIC WAVES NEAR THE PHASE TRANSITION OF GADOLINIUM MOLYBDATE.
CHIZHIKOV SI + SOROKIN NG + OSTROVSKII BI + MELESHINA VA
JETP LETT 14(9): 336-8 (5 NOV 1971)
273

30-3266 FERROELECTRICITY IN GADOLINIUM MOLYBDATE.
CROSS LE
ACTA CRYSTALLOGR A 25: S233 (1969)

30-3267 ELECTRICAL, OPTICAL, AND MECHANICAL BEHAVIOR OF FERROELECTRIC GADOLINIUM MOLYBDATE.
CUMMINS SE
FERROELECTRICS 1: 11-17 (1970)
191

30-3268 NEUTRON SCATTERING STUDY OF THE FERROELECTRIC PHASE TRANSFORMATION IN TERBIUM
MOLYBDATE.
DORNER B + AXE JD + SHIRANE G
PHYS REV B SOLID STATE 6(5): 1950-63 (1 SEP 1972)
273

30-3269 ZONE BOUNDARY SOFT MODE IN TERBIUM MOLYBDATE.
DORNER B + AXE JD + SHIRANE G
BULL AMER PHYS SOC 16: 393 (1971)

FERROELECTRIC RARE EARTH MOLYBDATES. (REVIEW, FORMATION, CRYSTAL GROWTH, PHASE
TRANSITIONS, STRUCTURE, 49 REFS)
DROBYSHEV LA + RABINOVICH AZ + VENEVTSEV YUN
BULL ACAD SCI USSR, PHYS SER 34(12): 2250-61 (1970)

30-3270 FORMATION AND PHASE TRANSFORMATIONS OF THE M MODIFICATION OF SAMARIUM MOLYBDATE (SOLID
PHASE SYNTHESIS, NEW MODIFICATION).
DROBYSHEV LA + FROLKINA IT
SOV PHYS CRYSTALLOGR 15(4): 686-9 (1971)
221

30-3271 PHASE TRANSITIONS OF GADOLINIUM MOLYBDATE AND ISOSTRUCTURAL COMPOUNDS (EUROPIUM
TERBIUM, DYSPROSIUM AND HOLMIUM MOLYBDATES).
DROBYSHEV LA + FROLKINA IT + PONOMAREV VI + TOMASHPOLSKII YUYA + VENEVTSEV YUN
+ ZHDANOV GS
SOV PHYS CRYSTALLOGR 15(1): 53-8 (1970)
206

30-3272 PHASE TRANSITIONS OF ANHYDROUS MODIFICATIONS OF RARE EARTH MOLYBDATES.
DROBYSHEV LA
SOV PHYS CRYSTALLOGR 15(5): 835-7 (1971)
232

PREPARATION OF SODIUM AND BARIUM NIOBATES AND LEAD MOLYBDATE.
DUMAS J + VILLELA G + BONIORT JY
DGRST, 1971, 13P. FINAL REPT CONTRACT 7002170.

THERMODYNAMIC THEORY OF GADOLINIUM MOLYBDATE.
DVORAK V
PHYS STATUS SOLIDI B 46: 763-72 (1971)

30-3273 ORIGIN OF THE STRUCTURAL PHASE TRANSITION IN GADOLINIUM MOLYBDATE. (SYMMETRY CHANGES)
DVORAK V
PHYS STATUS SOLIDI B 45: 147-52 (1971)
235

GROWTH AND EVALUATION OF SINGLE CRYSTAL BISMUTH TITANATE AND RELATED COMPOUNDS.
(GADOLINIUM MOLYBDATE, GROWTH BY RF CZOCHRALSKI AND FROM TOP SEEDED SOLUTION)
EPSTEIN DJ
MIT, AUG 1970, 96P. AD715312
CHEM ABSTR 75: 54390U (1971) US GOVT RES DEVELOP REP 71(2): 169 (1971)

30. MOLYBDATES

30-3274 ELASTIC CONSTANTS OF GADOLINIUM MOLYBDATE (NEAR THE CURIE POINT).
EPSTEIN DJ + HERRICK V + TUREK RF
SOLID STATE COMMUN 8: 1491-3 (1970)
214

30-3275 MAGNETOTHERMODYNAMICS OF ANTIFERROMAGNETIC, FERROELECTRIC BETA GADOLINIUM MOLYBDATE.
PARTS 1-3: HEAT CAPACITY, ENTROPY, MAGNETIC MOMENT OF THE ELECTRICALLY POLARIZED FORM
FROM 0.4 TO 4.2 DEGREES K WITH FIELDS TO 90 KILOGAUSS ALONG THE A, B, AND C CRYSTAL
AXES.
FISHER RA + HORNUNG EW + BRODALE GE + GIAUQUE WF
J CHEM PHYS 56: 193-212, 5007-18, 6118-25 (1972)
257

30-3276 ANOMALOUS PHONON BEHAVIOR NEAR THE PHASE TRANSITION IN FERROELASTIC FERROELECTRIC
GADOLINIUM MOLYBDATE.
FLEURY PA
SOLID STATE COMMUN 8: 601-5 (1970)
192

30-3277 ELECTROOPTICAL PROPERTIES OF FERROELECTRIC GADOLINIUM MOLYBDATE.
FOUSEK J + KONAK C
CZECH J PHYS B 22: 995-1006 (1972)
281

30-3278 INDUCED AND SPONTANEOUS BIREFRINGENCE IN IMPROPER FERROELECTRIC GADOLINIUM MOLYBDATE.
FOUSEK J + KONAK C
PHYS STATUS SOLIDI B 52: K13-16 (1972)
275

30-3279 PYROELECTRIC BEHAVIOR OF FERROELECTRIC GADOLINIUM MOLYBDATE.
GANGULY BN + ULLMAN FG
BULL AMER PHYS SOC 16: 415 (1971)

30-3280 FERROELECTRIC DOMAIN SHIFTING DEVICES. (GADOLINIUM MOLYBDATE)
GEUSIC JE + NELSON TJ + SCHINKE DP
BELL TELEPHONE LABORATORIES
US PAT 3,701,122, 24 OCT 1972.
285

30-3281 LATTICE DYNAMICS OF A RIGID ION MODEL OF GADOLINIUM MOLYBDATE.
HARDY JR + BOYER L
BULL AMER PHYS SOC 17: 246 (1972)

30-3282 PRODUCTION OF SPONTANEOUS POLARIZATION BY ELASTIC INSTABILITIES IN PIEZOELECTRIC
MATERIALS. (GADOLINIUM MOLYBDATE)
HARDY JR + ULLMAN FG + BOYER L
UNIV NEBRASKA, 1971, 160P. AD729924

30-3283 ELASTIC PROPERTIES OF GADOLINIUM MOLYBDATE AT ITS FERROELECTRIC PHASE TRANSITION.
HOCHLI UT + MULLER F
P61 OF EUROPEAN CONF ON CONDENSED MATTER, 1ST, 14-17 SEP 1971,
EUROPEAN PHYS SOC, 1971
SCI PHYS A 75: 16009 (1972)
264

30-3284 ELASTIC CONSTANTS AND SOFT OPTICAL MODES IN GADOLINIUM MOLYBDATE.
HOCHLI UT
PHYS REV B SOLID STATE 6(5): 1814-23 (1 SEP 1972)
273

30-3285 TEMPERATURE DEPENDENCE OF THE RAMAN SPECTRUM OF GADOLINIUM MOLYBDATE.
HOLDEN BJ + ULLMAN FG + HARDY JR
BULL AMER PHYS SOC 17: 246 (1972)

30-3286 SECOND HARMONIC GENERATION IN MOLYBDATES (BETA GADOLINIUM MOLYBDATE AND TERBIUM
MOLYBDATE.
JEGGO CR
J PHYS C SOLID STATE PHYS 5(11): L133-5 (1972)

30-3287   COMPREHENSIVE X-RAY STUDY OF THE FERROELECTRIC- FERROELASTIC AND PARAELECTRIC-
           PARAELASTIC PHASES OF GADOLINIUM MOLYBDATE.
           JEITSCHKO W
           ACTA CRYSTALLOGR B 28: 60-76 (1972)
               272

30-3288   THE CRYSTAL STRUCTURE OF FERROELECTRIC GADOLINIUM MOLYBDATE.
           JEITSCHKO W
           NATURWISSENSCHAFTEN 57(11): 544 (NOV 1970)
               238

           LASER AND SPECTROSCOPIC PROPERTIES OF ACTIVATED FERROELECTRICS. (LITHIUM NIOBATE AND
           GADOLINIUM MOLYBDATE)
           KAMINSKII AA
           SOV PHYS CRYSTALLOGR 17(1): 194-207 (JUL-AUG 1972)

           THALLIUM PEROVSKITES. THALLIUM A(1-X) TUNGSTATE OR MOLYBDATE. (WHERE A IS ZIRCONIUM,
           HAFNIUM, CERIUM, THORIUM, TITANIUM, SODIUM, MAGNESIUM, CADMIUM, YTTRIUM, GADOLINIUM,
           DYSPROSIUM, IRON, SELENIUM, NIOBIUM, COBALT, SCANDIUM, OR COPPER)
           KAPYSHEV AG + VENEVTSEV YUN
           BULL ACAD SCI USSR, PHYS SER 35(9): 1675-7 (1971)

30-3289   FERROELECTRIC FERROELASTIC PARAMAGNETIC BETA GADOLINIUM MOLYBDATE. CRYSTAL STRUCTURE OF
           THE TRANSITION METAL MOLYBDATES AND TUNGSTATES. PART-6.
           KEVE ET + ABRAHAMS SC + BERNSTEIN JL
           J CHEM PHYS 54: 3185-94 (1971)

30-3290   (ROOM TEMPERATURE) FERROELECTRIC FERROELASTIC PARAMAGNETIC (BETA) TERBIUM MOLYBDATE
           (CZOCHRALSKI SINGLE CRYSTALS).
           KEVE ET + ABRAHAMS SC + NASSAU K + GLASS AM
           SOLID STATE COMMUN 8: 1517-20 (1970)
               215

           THICKNESS OF DOMAIN WALLS IN FERROELECTRIC AND FERROELASTIC CRYSTALS. (BARIUM TITANATE,
           GADOLINIUM MOLYBDATE)
           KITTEL C
           SOLID STATE COMMUN 10: 119-21 (1972)

30-3291   ORIENTATION AND THICKNESS DEPENDENCE OF CONTRAST AND BRIGHTNESS IN GADOLINIUM MOLYBDATE
           LIGHT VALVES.
           KMETZ AR
           IEEE TRANS ELECTRON DEV 18: 756-61 (1971)

           PHENOMENOLOGICAL THEORY OF DIELECTRIC AND MECHANICAL PROPERTIES OF IMPROPER
           FERROELECTRIC CRYSTALS. (AMMONIUM SULFATE, GADOLINIUM MOLYBDATE, IRON IODINE BORACITE)
           KOBAYASHI J + ENOMOTO Y + SATO Y
           PHYS STATUS SOLIDI B 50: 335-43 (1972)

30-3292   AN EXPLANATION OF ANOMALOUS OPTICAL BEHAVIOR OF THE IMPROPER FERROELECTRIC GADOLINIUM
           MOLYBDATE.
           KOBAYASHI J + ASCHER E
           PHYS LETT A 38(1): 47-8 (3 JAN 1972)
               260

30-3293   APPLICATION OF ELECTRON MIRROR MICROSCOPY TO DIRECT OBSERVATION OF MOVING FERROELECTRIC
           DOMAINS OF GADOLINIUM MOLYBDATE.
           KOBAYASHI J + SOMEYA T + FURUHATA Y
           PHYS LETT A 38(5): 309-10 (28 FEB 1972)
               261

30-3294   X-RAY STUDY OF THERMAL EXPANSION OF FERROELECTRIC GADOLINIUM MOLYBDATE.
           KOBAYASHI J + SATO Y + NAKAMURA T
           PHYS STATUS SOLIDI A 14: 259-64 (1972)
               283

30-3295   OPTICAL DEVICE ELEMENT OF FERROELECTRIC GADOLINIUM MOLYBDATE.
           KOGA M
           J INST ELECTRON COMMUN ENG JAP 53: 33-41 (1970) (IN JAPANESE)

30-3296   RAMAN SPECTRUM AND STRUCTURE OF TERBIUM MOLYBDATE.
           KONINGSTEIN JA + PREUDHOMME JM
           J CHEM PHYS 55: 461-3 (1971)

30-3297 FERROELECTRIC FERROELASTIC CRYSTAL GADOLINIUM MOLYBDATE.
         KUMADA A + YUMOTO H + ASHIDA S
         P351-3 OF INT MEETING ON FERROELECTRICITY, 2ND, 1969,, KYOTO, PROC,
         (J PHYS SOC JAP, VOL 28 SUPPL). PHYS SOC JAP, 1970, 459P.

30-3298 (PROPERTIES,) FUNCTION AND APPLICATIONS OF GADOLINIUM MOLYBDATE (OPTICAL SHUTTER, COLOR
         MODULATOR).
         KUMADA A
         P258-63 OF CONF ON SOLID STATE DEVICES, FIRST, TOKYO 1969. JAPAN SOC
         APPL PHYS, TOKYO, 1970. OYO BUTSURI SUPPLEMENT 1970.
             196

30-3299 OPTICAL PROPERTIES OF GADOLINIUM MOLYBDATE AND THEIR DEVICE APPLICATIONS.
         KUMADA A
         FERROELECTRICS 3: 115-23 (1972);
         IEEE TRANS SON ULTRASON 19: 115-23 (1972)
             263

30-3300 OPTICAL PROPERTIES OF GADOLINIUM MOLYBDATE AND THEIR DEVICE APPLICATIONS.
         KUMADA A
         IEEE TRANS SON ULTRASON 19: 191-7 (APR 1972)

30-3301 DOMAIN STRUCTURE OF GADOLINIUM MOLYBDATE.
         KURSTEN HD + BOHM J
         KRISTALL TECH 7: 957-63 (1972) (IN GERMAN)

30-3302 STRUCTURAL STUDY OF GADOLINIUM MOLYBDATE CRYSTALS.
         KVAPIL J + JOHN V
         PHYS STATUS SOLIDI 39: K15-K17 (1970)
             202

30-3303 PHENOMENOLOGICAL THEORY OF THE FERROELECTRIC PHASE TRANSITION IN GADOLINIUM MOLYBDATE.
         (DIELECTRIC AND ELASTIC PROPERTIES)
         LEVANYUK AP + SANNIKOV DG
         SOV PHYS SOLID STATE 12(10): 2418-21 (1971)
             232

30-3304 STUDY OF FERROELECTRIC DOMAINS IN GADOLINIUM MOLYBDATE.
         MALGRANGE C + GLOGAROVA M
         J PHYS (PARIS) 33, SUPPL 4, COLLOQ 2: 159-61 (1972)
             294

30-3305 NONLINEAR OPTICAL PROPERTIES OF GADOLINIUM MOLYBDATE AND TERBIUM MOLYBDATE.
         MILLER RC + NORDLAND WA + NASSAU K
         FERROELECTRICS 2: 97-9 (1971)

         SYNTHESES, X-RAY, AND DIELECTRIC STUDIES OF CERTAIN NEW LAYERED FERROELECTRICS.
         (LANTHANUM BISMUTH TITANIUM IRON OXIDE, PRASEODYMIUM BISMUTH TITANIUM IRON OXIDE,
         BISMUTH MOLYBDATE, BISMUTH FLUONIOBATE, BISMUTH FLUOTANTALATE)
         MIRISHLI FA + ISMAILZADE IG
         BULL ACAD SCI USSR, PHYS SER 35(9): 1668-71 (1971)

30-3306 OBSERVATION OF PHASE BOUNDARIES BETWEEN FERRO- AND PARAELECTRIC PHASES IN GADOLINIUM
         MOLYBDATE CRYSTALS.
         NAKAMURA T + KONDO T + KUMADA A
         SOLID STATE COMMUN 9: 2265-8 (1971)
             257

30-3307 SPONTANEOUS BIREFRINGENCE AND ELECTROOPTIC RESPONSE IN GADOLINIUM MOLYBDATE.
         NAKAMURA T + KONDO T + KUMADA A
         PHYS LETT A 36: 141-2 (1971)

         STRUCTURAL AND PHASE RELATIONSHIPS AMONG TRIVALENT TUNGSTATES AND MOLYBDATES.
         NASSAU K + SHIEVER JW
         P445-56 OF MATER RES SYMP, 5TH PROC, 1971, SOLID STATE CHEMISTRY,
         ROTH RS + SCHNEIDER SJ (EDS), NBS SPEC PUB 364, 1972, 783P.

         STRUCTURAL AND PHASE RELATIONSHIPS AMONG TRIVALENT TUNGSTATES AND MOLYBDATES.
         NASSAU K + SHIEVER JW + KEVE ET
         J SOLID STATE CHEM 3: 411-19 (1971)

30-3308   FERROELECTRIC AND FERROELASTIC CRYSTAL, GADOLINIUM MOLYBDATE.
          OHSUMI K + ASHIDA S
          J CRYSTALLOGR SOC JAP 14: 132-40 (1972) (IN JAPANESE)

30-3309   TEMPERATURE DEPENDENCE OF DIELECTRIC CONSTANT IN THE SYSTEM POTASSIUM YTTRIUM
          MOLYBDATE- POTASSIUM DYSPROSIUM MOLYBDATE.
          PELIKH LN + KOBETS MI + ZVYAGIN AI
          UKR FIZ ZH 16: 333-5 (FEB 1971) (IN RUSSIAN)

30-3310   NEW TYPE OF FAR INFRARED SOFT MODE IN FERROELECTRIC GADOLINIUM MOLYBDATE.
          PETZELT J
          SOLID STATE COMMUN 9: 1485-8 (1971)
               246

30-3311   NEW TYPE OF FERROELECTRIC SOFT MODE IN GADOLINIUM MOLYBDATE.
          PETZELT J + DVORAK V
          PHYS STATUS SOLIDI B 46: 413-23 (1971)

30-3312   X-RAY DIFFRACTION STUDY OF GADOLINIUM MOLYBDATE (CRYSTALLOGRAPHIC PHASE CHANGE AT CURIE
          TEMPERATURE).
          PREWITT CT
          SOLID STATE COMMUN 8: 2037-40 (1970)
               220

30-3313   MODEL FOR FERROELECTRIC GADOLINIUM MOLYBDATE.
          PYTTE E
          SOLID STATE COMMUN 8: 2101-4 (1970)

30-3314   PYROELECTRIC EFFECT AND SPONTANEOUS POLARIZATION OF GADOLINIUM MOLYBDATE (TEMPERATURE
          DEPENDENCE 150-165 C).
          RABINOVICH AZ + SAFONOV AI
          SOV PHYS CRYSTALLOGR 15(1): 148-9 (1970)
               207

30-3315   PYROELECTRIC EFFECT AND DOMAIN STRUCTURE OF GADOLINIUM MOLYBDATE.
          RABINOVICH AZ + ROITBERG MB
          SOV PHYS CRYSTALLOGR 15: 1023-6 (MAY-JUN 1971)

30-3316   PHASE TRANSITION IN GADOLINIUM MOLYBDATE.
          SAKUDO T
          KOTAI BUTSURI (SOLID STATE PHYS) 6(11): 679-84 (1971) (IN JAPANESE)
               SCI ABSTR A 75: 19763 (1972)
               256

30-3317   PIEZOELECTRIC COEFFICIENT OF GADOLINIUM MOLYBDATE.
          SCHEIDING C + SCHMIDT G
          PHYS STATUS SOLIDI B 53: K95-8 (1972)
               280

30-3318   DOMAIN WALL STRUCTURE IN GADOLINIUM MOLYBDATE BY RAMAN SCATTERING.
          SHEPHERD IW + BARKLEY JR
          SOLID STATE COMMUN 10: 123-6 (1972)
               260

30-3319   RAMAN SCATTERING TECHNIQUES APPLIED TO PROBLEMS IN SOLID STATE PHYSICS (DOMAIN WALL
          THICKNESS IN GADOLINIUM MOLYBDATE).
          SHEPHERD IW
          APPL OPT 11: 1924-7 (SEP 1972)

30-3320   WAVEVECTOR DEPENDENT RELAXATION OF AN OPTICAL PHONON IN GADOLINIUM MOLYBDATE.
          SHEPHERD IW
          SOLID STATE COMMUN 9: 1857-60 (1971)
               255

30-3321   EFFECT OF HYDROSTATIC PRESSURE ON PHASE TRANSITION IN GADOLINIUM MOLYBDATE.
          SHIROKOV AM + MYLOV VP + BARANOV AI + PROKHORTSEVA TM
          SOV PHYS SOLID STATE 13(10): 2610-11 (APR 1972)
               263

30-3322   OBSERVATION OF DIFFRACTED LIGHT FROM DOMAIN WALLS IN GADOLINIUM MOLYBDATE.
          SUZUKI K
          SOLID STATE COMMUN 11: 937-9 (1972)
               279

30. MOLYBDATES

30-3323 PYROELECTRIC EFFECT IN GADOLINIUM MOLYBDATE.
ULLMAN FG + GANGULY BN + HARDY JR
FERROELECTRICS 2(4): 303-6 (1971)

30-3324 PYROELECTRIC DETECTION PROPERTIES OF GADOLINIUM MOLYBDATE.
ULLMAN FG + GANGULY BN + ZEIDLER JR
J ELECTRON MATER 1(3): 425-34 (1972)
274

31. HYDROGEN HALIDES

31-3325   PARAELECTRIC RESONANCE OF LITHIUM ION IN POTASSIUM CHLORIDE.
BLUMENSTOCK D + OSSWALD R + WOLF HC
Z PHYS 231: 333-46 (1970)

31-3326   LONGITUDINAL MODES IN HYDROGEN AND DEUTERIUM HALIDE CRYSTALS. (RAMAN SPECTRA, DIPOLAR
COUPLING MODEL COMPUTATICN)
FRIEDRICH HB + CARLSON RE
J CHEM PHYS 53(12): 4441-3 (1970)
          231

31-3327   PHASE TRANSITIONS OF SCLID HYDROGEN HALIDES.
FUJII Y + HOSHINO S
J CRYSTALLOGR SOC JAP 14: 326-34 (1972) (IN JAPANESE)

31-3328   EFFECT OF CHARGE TRANSFER ON RAMAN SCATTERING OF HYDROGEN HALIDES.
HANAMURA E
J CHEM PHYS 52: 797-802 (1970)

31-3329   PHASE TRANSITIONS OF HYDROGEN HALIDE CRYSTALS (IMPORTANCE OF HYDROGEN BONDING IN THE
FERROELECTRIC MECHANISM).
HANAMURA E
P192-4 OF INT MEETING ON FERROELECTRICITY, 2ND, 1969,
KYOTO, PROC (J PHYS SOC JAP, VOL 28 SUPPL). PHYS SOC JAP,
1970.
          210

31-3330   RAMAN SPECTRA OF ORTHORHCMBIC AND CUBIC HYDROGEN AND DEUTERIUM CHLORIDES AT
TEMPERATURES ABOVE 80 DEGREES K.
HEASTIE R
CHEM PHYS LETT 15: 613-16 (1972)

31-3331   FERROELECTRICITY AND PHASE TRANSITIONS IN SOLID HYDROGEN HALIDES (SINGLE CRYSTAL X-RAY
AND NEUTRON DIFFRACTION).
HOSHINO S + SHIMAVKA K + NIMURA N + MOTEGI H + MARUYAMA N
P189-91 OF INT MEETING ON FERROELECTRICITY, 2ND, 1969,
KYOTO, PROC (J PHYS SOC JAP, VCL 28 SUPPL). PHYS SOC JAP,
1970.
          210

31-3332   BROMINE NUCLEAR QUADRUPOLE RESONANCE IN FERROELECTRIC HYDROGEN BROMIDE.
KADABA PK + O'REILLY DE
J CHEM PHYS 52(5): 2403-6 (1970)
          232

31-3333   STRESS FREE SINGLE CRYSTAL GROWTH OF FERROELECTRIC HYDROGEN HALIDES.
MARUYAMA N + HOSHINO S
ACTA CRYSTALLOGR A 28: S140 (1972)

31-3334   CRYSTAL STRUCTURE AND PHASE TRANSITION OF HYDROGEN CHLORIDE.
NIIMURA N + SHIMAOKA K + MOTEGI H + HOAHINO S
J PHYS SOC JAP 32(4): 1019-26 (1972)
          278

31-3335   NUCLEAR QUADRUPOLE RESONANCE OF IODINE-127 IN SOLID HYDROGEN IODIDE.
O'REILLY DE + PETERSON EM + KADABA PK
J CHEM PHYS 52(12): 6444-5 (1970)
          232

31-3336   NUCLEAR QUADRUPOLE FREQUENCIES AND RELAXATION TIMES IN FERROELECTRIC HYDROGEN CHLORIDE.
O'REILLY DE
J CHEM PHYS 52: 2396-402 (1970)

31-3337   STRUCTURE AND PHASE TRANSITION IN HYDROGEN CHLORIDE.
           SHIMAOKA K + NIIMURA N + HOSHINO S
           ACTA CRYSTALLOGR A 25: S50 (1969)

31-3338   CRYSTAL SIZE EFFECTS IN THE RAMAN SPECTRA OF ORTHORHOMBIC HYDROGEN CHLORIDE CRYSTALS.
           SUN TS + ANDERSON A
           SPECTROSC LETT 4: 377-83 (1971)

31-3339   INTERMOLECULAR POTENTIAL AND FERROELECTRIC TRANSITIONS IN HYDROGEN HALIDES.
           TSANG T + SHAW EL
           J CHEM PHYS 55: 2337-42 (1971)

           CRYSTALLIZATION POTENTIALS IN AQUEOUS SOLUTIONS. (POTASSIUM BROMIDE, POTASSIUM
           CHLORIDE, SODIUM CHLORIDE, POTASSIUM SODIUM TARTRATE, POTASSIUM DIHYDROGEN PHOSPHATE)
           WIEDERKEHR H
           EIDGENOSSISCHE TECH HOCHSCHULE, ZURICH, 1970, 123P. N72-13712
           (IN SWEDISH)
                SCI TECH AEROSP REP 10(4): 528 (1972)

31-3340   MOLECULAR MOTION IN THE FERROELECTRIC PHASE OF HYDROGEN CHLORIDE. (CALCULATED IN THE
           HARMONIC APPROXIMATION)
           YI PN + GAVRIELIDES AT
           J CHEM PHYS 54: 3777-84 (1971)
                235

32. LITHIUM HYDRAZINIUM SULFATE AND FLUOBERYLLATE

32-3341   ANOMALOUS NEUTRON SCATTERING AND THE QUESTION OF FERROELECTRICITY IN LITHIUM
          HYDRAZINIUM SULFATE.
          ANDERSON MR + BROWN ID
          ACTA CRYSTALLOGR A 28: 663-5 (1972)

32-3342   CRYSTAL STRUCTURE OF LITHIUM HYDRAZINIUM SULFATE.
          BROWN ID
          ACTA CRYSTALLOGR 17: 654-60 (1964)

32-3343   NUCLEAR MAGNETIC RESONANCE STUDY OF FERROELECTRIC LITHIUM HYDRAZINIUM SULFATE.
          CUTHBERT JD + PETCH HE
          CAN J PHYS 41: 1629-50 (1963)

32-3344   DIELECTRIC PROPERTIES OF LITHIUM HYDRAZINIUM SULFATE.
          FUKUTOMI K + SHIRAISHI T
          MATH NAT SCI 19: 37-47 (1972)

32-3345   NITROGEN-14 NUCLEAR QUADRUPOLE RESONANCE STUDY OF LITHIUM HYDRAZINIUM SULFATE.
          HASTINGS RN + OJA T
          J CHEM PHYS 57: 2139-46 (1972)

32-3346   HYDROGEN BONDING IN LITHIUM HYDRAZINIUM SULFATE.
          HOWELL FL
          MONTANA STATE UNIV PHD THESIS, 1970, 122P. N71-33864
              DISS ABSTR INT B 31(3): 1485 (1970) SCI TECH AEROSP REP 9(20): 3335 (1971)
              250

32-3347   PROTON ROTATING FRAME RELAXATION IN LITHIUM HYDRAZINIUM SULFATE.
          KNISPEL RR + PETCH HE
          CAN J PHYS 49: 870-5 (1971)

32-3348   FERROELECTRIC-LIKE BEHAVIOR OF LITHIUM HYDRAZINIUM SULFATE.
          NIIZEKI N + KOIZUMI H
          J PHYS SOC JAP 19: 132-3 (1964)

32-3349   DETERMINATION OF FERROELECTRIC PROPERTIES OF LITHIUM HYDRAZINIUM FLUOBERYLLATE.
          PALAU JM + LASSABATERE L
          COMPT REND B 273(16): 714-7 (1971) (IN FRENCH)
              CHEM ABSTR 76: 51527D (1972)
              263

32-3350   DIPOLAR RELAXATION OF LITHIUM-7 BY HINDERED ROTATORS IN LITHIUM HYDRAZINIUM SULFATE.
          PARKER RD + SCHMIDT VH
          J MAGN RESON 6: 507-15 (1972)

32-3351   ANALYSIS OF HYSTERESIS LOOPS IN LITHIUM HYDRAZINIUM SULFATE.
          SCHMIDT VH + PARKER RS
          J PHYS (PARIS) 33, SUPPL 4, COLLOQ 2: 109-11 (1972)
              294

32-3352   DIELECTRIC PROPERTIES OF LITHIUM HYDRAZINIUM SULFATE.
          SCHMIDT VH + DRUMHELLER JE
          PHYS REV B SOLID STATE 4(12): 4592-7 (15 DEC 1971)
              253

32-3353   DIELECTRIC PROPERTIES OF LITHIUM HYDRAZINIUM SULFATE.
          SCHMIDT VH + DRUMHELLER JE + HOWELL FL
          PHYS REV B SOLID STATE 4: 4582-97 (1971)

32-3354 DEUTERON NMR STUDIES OF LITHIUM HYDRAZINIUM SULFATE.
         SCHMIDT VH + HOWELL FL
         P106-8 OF INT MEETING ON FERROELECTRICITY, 2ND, 1969, KYOTO, PROC,
         (J PHYS SOC JAP, VOL 28 SUPPL). PHYS SOC JAP, 1970, 459P.

32-3355 CRYSTALLOGRAPHIC STUDY OF SEVERAL HYDRAZINIUM ORTHO FLUOBERYLLATES.
         TEDENAC JC + VILMINOT S + COT L + NORBERT A + MAURIN M
         MATER RES BULL 6: 183-8 (1971) (IN FRENCH)

32-3356 CRYSTAL STRUCTURE OF FERROELECTRIC LITHIUM HYDRAZINIUM SULFATE.
         VAN DEN HENDE JH + BOUTIN H
         ACTA CRYSTALLOGR 17: 660-3 (1964)

32-3357 MICROWAVE DIELECTRIC STUDY OF LITHIUM HYDRAZINIUM SULFATE.
         WAN P + HOWELL FL + SCHMIDT VH + DRUMHELLER JE
         BULL AMER PHYS SOC 15: 606 (1970)

33. CESIUM LEAD TRICHLORIDE, POTASSIUM MANGANESE TRIFLUORIDE, AND RELATED HALIDES

33-3358 FERROELASTIC EFFECT IN RUBIDIUM IRON TETRAFLUORIDE AND CESIUM IRON TETRAFLUORIDE.
ABRAHAMS SC + BERNSTEIN JL
MATER RES BULL 7: 715-20 (1972)

33-3359 PREPARATION AND PROPERTIES OF CESIUM LEAD HALIDES. (FERROELECTRICITY INDETERMINITE)
AINGER FW + CLARK CC + MARSH A + WATERWORTH P
P295-9, VOL 1 OF DVORAK V + FOUSKOVA A + GLOGAR P (EDS),
INT MEETING ON FERROELECTRICITY, PRAGUE, 1966, PROC.
INST PHYS CZECH ACAD SCI, PRAGUE, 1966.

33-3360 ELASTIC PROPERTIES OF CESIUM LEAD TRICHLORIDE.
ALEKSANDROV KS + KRUPNYI AI + ZINENKO VI + BEZNOSIKOV BV
SOV PHYS CRYSTALLOGR 17: 515-17 (1972)

33-3361 LOW FREQUENCY DIELECTRIC CONSTANTS OF THE ALKALINE EARTH FLUORIDES BY THE METHOD OF
SUBSTITUTION.
ANDEEN C + FONTANELLA J + SCHUELE D
J APPL PHYS 42: 2216-9 (MAY 1971)

33-3362 INFRARED ACTIVE MODE BELOW THE STRUCTURAL TRANSITION IN POTASSIUM MANGANESE
TRIFLUORIDE.
BALTES HP + KNEUBUHL FK
SOLID STATE COMMUN 8: 1029-30 (1970)

33-3363 TWO PHONON PROCESSES AND PHASE TRANSITIONS IN POTASSIUM MANGANESE TRIFLUORIDE:
TEMPERATURE DISCONTINUITIES IN THE INFRARED SPECTRUM AND GROUP THEORETICAL ANALYSIS.
BALTES HP + TOSI M + KNEUBUHL FK
J PHYS CHEM SOLIDS 31: 321-9 (1970)

X-RAY DETERMINATION OF POLARITY SENSE BY ANOMALOUS SCATTERING AT AN ABSORPTION EDGE.
(LITHIUM NIOBATE, BARIUM MANGANESE TETRAFLUORIDE)
BARNS RL + KEVE ET + ABRAHAMS SC
J APPL CRYSTALLOGR 3: 27-32 (1970)

33-3364 FINE STRUCTURE OF EXCITON- MAGNON ABSORPTION OF LIGHT IN POTASSIUM MANGANESE
TRIFLUORIDE.
BELYAEVA AI + EREMENKO VV + BEZNOSIKOV BV
SOV PHYS JETP 31: 429-33 (1970)

33-3365 ELECTRICAL CONDUCTION AND (NONFERROELECTRIC) PHASE TRANSITIONS IN CESIUM LEAD
TRICHLORIDE.
BUSMUNDRUD O + FEDER J
SOLID STATE COMMUN 9: 1575-7 (1971)
249

33-3366 ANTIFERROELECTRICITY IN (PURE AND) COBALT DOPED POTASSIUM MANGANESE TRIFLUORIDE. (20
PERCENT MANGANESE REPLACED BY COBALT, COHERENT NEUTRON SCATTERING)
BUYERS WJL + COWLEY RA + PAUL GL
P242-4 OF INT MEETING ON FERROELECTRICITY, 2ND, 1969, KYOTO, PROC,
(J PHYS SOC JAP, VOL 28 SUPPL). PHYS SOC JAP, 1970, 459P.

33-3367 ELECTRIC FIELD GRADIENTS IN IONIC CHLORIDES: CESIUM LEAD TRICHLORIDE.
CARLSON EH
J CHEM PHYS 55: 4662-3 (1971)

33-3368 PHASE TRANSITIONS IN CESIUM LEAD TRICHLORIDE. (LOSS OF A CENTER OF SYMMETRY AT 194
DEGREES K)
COHEN MI + YOUNG KF + CHANG TT + BROWER WS
J APPL PHYS 42: 5267-72 (1971)
250

33. CESIUM LEAD TRICHLCRIDE, POTASSIUM MANGANESE TRIFLUORIDE AND RELATED HALIDES

33-3369 CRITICAL ANISOTROPIC FLUCTUATIONS AT THE 184 DEG K TRANSITION OF POTASSIUM MANGANESE
TRIFLUORIDE SINGLE CRYSTALS, (X-RAY DIFFUSE SCATTERING).
COMES R + DENOYER F + DESCHAMPS I + LAMBERT M
PHYS LETT 34A: 65-6 (1971)
220

33-3370 ATTENUATION AND DISPERSICN OF ULTRASONIC WAVES IN POTASSIUM MANGANESE TRIFLUORIDE IN
THE NEIGHBORHOOD OF THE STRUCTURAL TRANSITION.
COURDILLE JM + DUMAS J
SOLID STATE COMMUN 9: 609-12 (1971) (IN FRENCH)

ANISOTROPIC CRITICAL X-RAY DIFFUSE SCATTERING FROM POTASSIUM MANGANESE TRIFLUORIDE AND
SODIUM NIOBATE SINGLE CRYSTALS.
DENOYER F + LAMBERT M
J PHYS (PARIS) 33, SUPPL 4, COLLOQ 2: 132 (1972)

33-3371 PARAMAGNETIC SCATTERING OF NEUTRONS FROM POTASSIUM MANGANESE TRIFLUORIDE IN THE SHORT
RANGE ORDERED REGION.
DENIZ KU + GOYAL PS
J PHYS (PARIS) 32, COLLOQ C1: 619-21 (1971)

33-3372 RAMAN SCATTERING AND PHASE TRANSITIONS IN POTASSIUM MANGANESE TRIFLUORIDE CRYSTALS.
EREMENKO VV + POPKOV YUA + FOMIN VI
P372-6 OF INT CONF ON LIGHT SCATTERING IN SOLIDS, 2ND PROC,
BALKANSKI M(ED), PARIS, 1971, FLAMMARION, PARIS, 1971, 518P.

THEORY OF SUPEREXCHANGE REACTION. PART-1: POTASSIUM MANGANESE TRIFLUORIDE.
FUCHIKAMI N
J PHYS SOC JAP 28: 871-90 (1970)

33-3373 ULTRASONIC ATTENUATION NEAR THE SOFT MODE TRANSITION POINT IN POTASSIUM MANGANESE
TRIFLUORIDE (ANOMALOUS HEAT CAPACITY, CRITICAL TEMPERATURE ABOUT 186-65 DEG K).
FURUKAWA M + FUJIMORI Y + HIRAKAWA K
J PHYS SOC JAP 29(6): 1528-32 (1970)
222

33-3374 DISPERSION AND DAMPING OF SOFT ZONE BONDARY PHONONS IN POTASSIUM MANGANESE TRIFLUORIDE.
GESI K + AXE JD + SHIRANE G + LINZ A
PHYS REV B SOLID STATE 5(5): 1933-41 (1 MAR 1972)
260

33-3375 THERMAL EXPANSIVITY OF RUBIDIUM MANGANESE FLUORIDE NEAR THE NEEL TEMPERATURE.
GOLDING B
J APPL PHYS 42: 1381-2 (1971)

33-3376 EXCHANGE NARROWING: MAGNETIC RESONANCE LINE SHAPES AND SPIN CORRELATION IN PARAMAGNETIC
POTASSIUM MANGANESE TRIFLUORIDE, RUBIDIUM MAGANESE TRIFLUORIDE, AND MANGANESE
DIFLUORIDE.
GULLEY JE + HONE D + SCALAPINO DJ + SILBERNAGEL BG
PHYS REV B 1: 1020-30 (1970)

33-3377 SHIFT OF NEEL TEMPERATURE AND EPR LINE WIDTH OF POTASSIUM MANGANESE TRIFLUORIDE WITH
MAGNESIUM DOFING.
GUPTA KP + SEEHRA MS + VEHSE WE
PHYS REV B 5: 92-5 (1972)

DETERMINATION OF THE NORMAL VIBRATIONAL DISPLACEMENTS OF SEVERAL PEROVSKITES (POTASSIUM
TANTALATE, STRONTIUM TITANATE, RUBIDIUM MANGANESE TRIFLUORIDE) BY INELASTIC NEUTRON
SCATTERING.
HARADA J + AXE JD + SHIRANE G
BULL AMER PHYS SOC 15: 102 (1970)

33-3378 EXPERIMENTAL STUDIES OF STRUCTURAL PHASE TRANSITIONS IN CESIUM LEAD TRICHLORIDE.
HIROTSU S
J PHYS SOC JAP 31: 552-60 (1971)

33-3379 FAR INFRARED REFLECTIVITY SPECTRA OF CESIUM LEAD TRICHLORIDE.
HIROTSU S
PHYS LETT A 41: 55-6 (1972)

33. CESIUM LEAD TRICHLORIDE, POTASSIUM MANGANESE TRIFLUORIDE AND RELATED HALIDES

33-3380   (MECHANISM OF) STRUCTURAL PHASE TRANSITIONS IN CESIUM LEAD TRICHLORIDE.
HIROTSU S
P185-7 OF INT MEETING ON FERROELECTRICITY, 2ND, 1969,
KYOTO, PROC (J PHYS SOC JAP, VOL 28 SUPPL). PHYS SOC JAP,
1970
197

33-3381   LUMINESCENCE OF EUROPIUM(3+) ION IN ANTIFERROMAGNETIC POTASSIUM MANGANESE TRIFLUORIDE.
HIRANO M + SHIONOYA S
J PHYS SOC JAP 28: 926-34 (1970)

33-3382   OBSERVATION OF THE ANOMALY OF THE THERMAL CONDUCTIVITY OF POTASSIUM MANGANESE
TRIFLUORIDE.
HIRAKAWA K + HAMAZAKI K + MLIKE H + HAYASHI H
J PHYS SOC JAP 33: 268 (1972)

33-3383   MICROSCOPIC OBSERVATIONS OF PHASE TRANSITIONS IN CESIUM LEAD TRICHLORIDE.
IMAOKA K + MIDORIKAWA M + ISHIBASHI Y + TAKAGI Y
JAP J APPL PHYS 11: 120-1 (1972)

33-3384   INFRARED ABSORPTION SPECTRA OF POTASSIUM MANGANESE TRIFLUORIDE, POTASSIUM COBALT
TRIFLUORIDE, AND RUBIDIUM COBALT TRIFLUORIDE.
KARAMYAN AA
OPT SPECTROSC 30: 314-15 (1971)

33-3385   VIBRATIONAL SPECTRUM AND DISPERSION OF THE OPTICAL CONSTANTS OF ANTIFERROMAGNETIC
POTASSIUM MANGANESE TRIFLUORIDE AND POTASSIUM COBALT TRIFLUORIDE CRYSTALS.
KARAMYAN AA
OPT SPECTROSC 33: 97-8 (1972)

33-3386   FERROELECTRIC PARAELASTIC PARAMAGNETIC BARIUM COBALT TETRAFLUORIDE CRYSTAL STRUCTURE
(SPACE GROUP AND LATTICE CONSTANTS).
KEVE ET + ABRAHAMS SC + BERNSTEIN JL
J CHEM PHYS 53(8): 3279-87 (1970)
225

33-3387   SPECIFIC HEAT OF POTASSIUM MANGANESE TRIFLUORIDE.
KHLYUSTOV VG + FLEROV IN + SILIN AT + SALNIKOV AN
SOV PHYS SOLID STATE 14: 139-41 (1972)

33-3388   A PHASE TRANSITION IN A COMPOUND WITH HELICAL ELECTRIC DIPOLE STRUCTURE: CESIUM COPPER
TRICHLORIDE.
KROESE CJ + TINDERMANS-VAN EYNDHOVEN JCM + MAASKANT WJA
SOLID STATE COMMUN 9: 1707-9 (1971)

33-3389   MAGNETOELASTIC EFFECTS IN POTASSIUM MANGANESE TRIFLUORIDE.
MAARTENSE I + SEARLE CW
PHYS REV B 6: 894-901 (1972)

33-3390   ACOUSTIC AND MAGNETIC EFFECTS INVOLVING THE FLUORINE-19 NUCLEI IN ANTIFERROMAGNETIC
POTASSIUM MANGANESE TRIFLUORIDE.
MAHLER RJ + JAMES LW
J APPL PHYS 41: 1633-6 (1970)

33-3391   COVALENCY IN POTASSIUM MANGANESE TRIFLUORIDE.
MATSUOKA O + KUNII TL
J PHYS SOC JAP 28: 1296-302 (1970); 30: 1771 (1971)

33-3392   ANOMALOUS ELASTIC BEHAVIOR OF POTASSIUM MANGANESE TRIFLUORIDE NEAR A STRUCTURAL PHASE
TRANSITION.
MELCHER RL + PLOVNICK RH
P348-52 OF NUSIMOVICI M(ED), PHONONS, INT CONF, RENNES, 1971 PROC,
FLAMMARION, PARIS, 1971, 502P.

33-3393   PRESSURE DEPENDENCE OF CUBIC TETRAGONAL TRANSITION TEMPERATURE OF CESIUM LEAD
TRICHLORIDE.
MIDORIKAWA M + ISHIBASHI Y + TAKAGI Y
J PHYS SOC JAP 32: 1672 (1972)

MEASUREMENTS OF THE THERMAL CONDUCTIONS IN THE LOW DIMENSIONAL ANTIFERROMAGNETS
POTASSIUM COPPER TRIFLUORIDE AND COPPER FORMATE TETRAHYDRATE.
MIIKE H + HAYASHI H + HIRAKAWA K
TECHNOL REP KYUSHU UNIV 45: 130-5 (1972) (IN JAPANESE)

33. CESIUM LEAD TRICHLORIDE, POTASSIUM MANGANESE TRIFLUORIDE AND RELATED HALIDES

33-3394   X-RAY SCATTERING AND THE PHASE TRANSITION OF POTASSIUM MANGANESE TRIFLUORIDE AT 184
           DEGREES K.
           MINKIEWICZ VJ + FUJII Y + YAMADA Y
           J PHYS SOC JAP 28: 443-50 (1970)

33-3395   IMPURITY CONCENTRATION DEPENDENCE OF RAMAN SCATTERING BY MAGNONS IN NICKEL DOPED
           RUBIDIUM MANGANATE AND POTASSIUM MANGANESE TRIFLUORIDE.
           PARISOT G + DIETZ RE + GUGGENHEIM HJ + MOCH P + DUGAUTIER C
           J PHYS (PARIS) 32, COLLOQ C1: 803-5 (1971)

33-3396   ANTIFERROMAGNETIC RESONANCE SPECTRUM IN BARIUM MANGANESE TRIFLUORIDE.
           PETROV SV + POPOV MA + PROZOROVA LA
           SOV PHYS JETP 35: 981-3 (1972)

33-3397   PREPARATION AND DISLOCATION DENSITY DETERMINATION OF LARGE, PURE RUBIDIUM MANGANESE
           TRIFLUORIDE AND POTASSIUM MANGANESE TRIFLUORIDE SINGLE CRYSTALS.
           PLOVNICK RH + CAMOBRECO SJ
           MATER RES BULL 7: 573-82 (1972)

33-3398   MANDELSTAM BRILLOUIN SCATTERING OF LIGHT IN MANGANESE FLUORIDE CRYSTALS (POTASSIUM
           TRIFLUORIDE, RUBIDIUM MANGANESE TRIFLUORIDE, MANGANESE DIFLUORIDE).
           POPKOV YUA + FOMIN VI + KHARCHENKO LT
           SOV PHYS SOLID STATE 13: 1360-4 (1971)

33-3399   MANDELSTAM BRILLOUIN SCATTERING OF LIGHT IN MANGANESE FLUORIDE, COBALT FLUORIDE,
           POTASSIUM MANGANESE TRIFLUORIDE AND RUBIDIUM MANGANESE TRIFLUORIDE CRYSTALS.
           POPKOV YUA + FOMIN VI
           P502-7 OF INT CONF ON LIGHT SCATTERING IN SOLIDS, 2ND PROC,
           BALKANSKI M(ED), PARIS, 1971, FLAMMARION, PARIS, 1971, 518P.

33-3400   RAMAN SCATTERING AND PHASE TRANSITIONS IN A POTASSIUM MANGANESE TRIFLUORIDE CRYSTAL.
           POPKOV YUA + EREMENKO VV + FOMIN VI
           SOV PHYS SOLID STATE 13: 1701-8 (1972)

33-3401   TWO MAGNON SCATTERING OF LIGHT IN ANTIFERROMAGNETIC POTASSIUM MANGANESE TRIFLUORIDE.
           POPKOV YUA + FOMIN VI + BEZNOSIKOV BV
           JETP LETT 11: 264-6 (1970)

33-3402   SUPERSTRUCTURE ARISING DURING THE PHASE TRANSITION IN POTASSIUM MANGANESE TRIFLUORIDE.
           POZDNYAKOVA LA + KRUGLIK AI + ALEKSANDROV KS
           SOV PHYS CRYSTALLOGR 17(3): 284-6 (SEP-OCT 1972)
                275

33-3403   RESONANCE BEHAVIOR IN CANTED ANTIFERROMAGNETIC POTASSIUM MANGANESE TRIFLUORIDE.
           SASAKI K
           J PHYS SOC JAP 33: 1284-91 (1972)

33-3404   ELECTRONIC HALL MOBILITY IN THE ALKALINE EARTH FLUORIDES.
           SEAGER CH
           PHYS REV B SOLID STATE 3: 3479-84 (1971)

           CRITICAL NEUTRON SCATTERING IN STRONTIUM TITANATE AND POTASSIUM MANGANESE TRIFLUORIDE.
           SHAPIRO SM + AXE JD + SHIRANE G
           PHYS REV B 6(12): 4332-41 (1 DEC 1972)

33-3405   NEUTRON SCATTERING STUDY OF THE LATTICE DYNAMICAL PHASE TRANSITIONS IN POTASSIUM
           MANGANESE TRIFLUORIDE.
           SHIRANE G + MINKIEWICZ VJ + LINZ A
           SOLID STATE COMMUN 8: 1941-4 (1970)
                220

33-3406   BARIUM MANGANESE FLUORIDE, A NEW CRYSTAL FOR MICROWAVE ULTRASONICS.
           SPENCER EG + GUGGENHEIM HJ + KOMINIAK GJ
           APPL PHYS LETT 17: 300-1 (1970)

33-3407   MAGNETIC EXCITATIONS IN NICKEL DOPED POTASSIUM MANGANESE TRIFLUORIDE.
           SVENSSON EC + HOLDEN TM + COWLEY RA + BUYERS WJL + STEVENSON RWH
           BULL AMER PHYS SOC 15: 823 (1970)

33-3408   DETERMINATION OF TRANSITION TEMPERATURES IN CESIUM LEAD TRICHLORIDE USING EPR.
           YOUNG KF + COHEN MI + CHANG TT + BROWER WS
           BULL AMER PHYS SOC 16: 394 (1971)

33-3409   LOW FREQUENCY MAGNETOELECTRIC RESONANCES IN BARIUM COBALT TETRAFLUORIDE.
          ZORIN RV + ALSHIN BI + ASTROV DN
          SOV PHYS JETP 35: 634-6 (1972)

34-3410   LOW ENTROPY FORM OF ICE I(H) OBTAINED FROM A LINEAR STEP GROWTH MODEL.
AUVERT G + BULLEMER B + KAHANE A
SOLID STATE COMMUN 11: 1031-4 (1972)

34-3411   EQUILIBRIUM STRUCTURE OF POLARIZED ICE.
BABCOCK RV + LONGINI RL
J CHEM PHYS 56(1): 344-53 (1972)
      256

34-3412   ELECTRIC POLARIZATION EFFECTS IN PURE AND DOPED ICE AT LOW TEMPERATURES.
BISHOP PB + GLEN JW
P492-501 OF INT SYMP ON PHYSICS OF ICE, 3RD PROC, 1968,
RIEHL N + BULLEMER B + ENGELHARDT H(EDS), PLENUM, NY,
1969, 642P.

34-3413   ENDOR (ELECTRON NUCLEAR DOUBLE RESONANCE) STUDY OF X-IRRADIATED SINGLE CRYSTALS OF ICE.
BOX HC + BUDZINSKI EE + LILGA KT + FREUND HG
J CHEM PHYS 53: 1059-65 (1970)

34-3414   PROTONIC CONDUCTION OF ICE. PART-1: HIGH TEMPERATURE REGION. PART-2: LOW TEMPERATURE
REGION.
BULLEMER B + ENGELHARDT H + RIEHL N
P416-29, 430-42 OF INT SYMP ON PHYSICS OF ICE, 3RD PROC, 1968,
RIEHL N + BULLEMER B + ENGELHARDT H(EDS), PLENUM, NY,
1969, 642P.

34-3415   ELECTRICAL CONDUCTION IN ICE.
CAMP PR + KISZENICK W + ARNOLD D
P450-70 OF INT SYMP ON PHYSICS OF ICE, 3RD PROC, 1968,
RIEHL N + BULLEMER B + ENGELHARDT H(EDS), PLENUM, NY,
1969, 642P.

34-3416   MOSSBAUER EFFECT OF FERROUS IONS IN CUBIC ICE.
CAMERON JA + KESZTHELYI L + NAGY G + KACSON L
CHEM PHYS LETT 8: 628-30 (1971)

34-3417   THERMODIELECTRIC EFFECT AND FREEZING POTENTIAL IN GROWING ICE.
CASSETTARI M + SALVETTI G
NUOVO CIMENTO B 12: 95-100 (NOV 1972)

34-3418   LOW TEMPERATURE POLARIZATION EFFECTS IN ICE. (CURRENT REVERSAL DUE TO DIPOLE
RELAXATION)
CHAMBERLAIN J + FLETCHER NH
PHYS KONDENS MATER, PHYS MATER CONDENSEE 12(3): 193-209 (1971)
      234

34-3419   DIELECTRIC PROPERTIES CF ICE I.
COLE RH + WORZ O
P546-54 OF INT SYMP ON PHYSICS OF ICE, 3RD PROC, 1968,
RIEHL N + BULLEMER B + ENGELHARDT H(EDS), PLENUM, NY,
1969, 642P.

34-3420   STUDY OF THE SURFACE OF ICE WITH A SCANNING ELECTRON MICROSCOPE.
CROSS JD
P81-94 OF INT SYMP ON PHYSICS OF ICE, 3RD PROC, 1968,
RIEHL N + BULLEMER B + ENGELHARDT H(EDS), PLENUM, NY,
1969, 642P.

34-3421   ELASTIC MODULI OF ICE.
          DANTL G
          P223-30 OF INT SYMP ON PHYSICS OF ICE, 3RD PROC, 1968,
          RIEHL N + BULLEMER B + ENGELHARDT H(EDS), PLENUM, NY,
          1969, 642P.

34-3422   TENSILE AND FLEXURE PROPERTIES OF SALINE ICE.
          DYKINS JE
          P251-70 OF INT SYMP ON PHYSICS OF ICE, 3RD PROC, 1968,
          RIEHL N + BULLEMER B + ENGELHARDT H(EDS), PLENUM, NY,
          1969, 642P.

34-3423   IRRADIATION PRODUCED SOLVATED ELECTRONS IN ICE.
          EIBEN K
          P184-94 OF INT SYMP ON PHYSICS OF ICE, 3RD PROC, 1968,
          RIEHL N + BULLEMER B + ENGELHARDT H(EDS), PLENUM, NY,
          1969, 642P.

34-3424   DYNAMIC POLAR MODEL OF MONOCRYSTALLINE ICE I(H).
          FAURE P + KAHANE A
          P243-7 OF NUSIMOVICI M(ED), PHONONS, INT CONF, RENNES, 1971 PROC,
          FLAMMARION, PARIS, 1971, 502P.

34-3425   LOW FREQUENCY RAMAN SPECTRUM OF ICE I(H). TENTATIVE INTERPRETATION WITH A MIXED COULOMB
          VALENCE DYNAMICAL MODEL.
          FAURE P + CHOSSON A
          P272-7 OF INT CONF ON LIGHT SCATTERING IN SOLIDS, 2ND PROC,
          BALKANSKI M(ED), PARIS, 1971, FLAMMARION, PARIS, 1971, 518P.

          THEORY OF THE MOBILITY OF STRUCTURAL DEFECTS IN ICE.
          FISCHER SF + HOFACKER GL + RATNER MA
          P385-400 OF INT SYMP ON PHYSICS OF ICE, 3RD PROC, 1968,
          RIEHL N + BULLEMER B + ENGELHARDT H(EDS), PLENUM, NY,
          1969, 642P.

34-3426   X-RAY DIFFRACTION TOPOGRAPHIC STUDIES OF THE DEFORMATION BEHAVIOR OF ICE SINGLE
          CRYSTALS.
          FUKUDA A + HIGASHI A
          P219-50 OF INT SYMP ON PHYSICS OF ICE, 3RD PROC, 1968,
          RIEHL N + BULLEMER B + ENGELHARDT H(EDS), PLENUM, NY,
          1969, 642P.

34-3427   ANALYTIC PROPERTIES OF THE FREE ENERGY FOR THE "ICE" MODELS (FERROELECTRICS).
          GLASSER ML + ABRAHAM DB + LIEB EH
          J MATH PHYS 13: 887-900 (JUN 1972)

          STRUCTURE AND POINT DEFECTS OF ICE: THEIR EFFECT ON THE ELECTRICAL AND MECHANICAL
          PROPERTIES. (REVIEW, 56 REFS)
          GLEN JW
          SCI PROGR 57: 1-21 (1969)

34-3428   CONDUCTION ANOMALIES AND POLARIZATION IN ICE AT LOW TEMPERATURES.
          GLOCKMANN HP
          P502-13 OF INT SYMP ON PHYSICS OF ICE, 3RD PROC, 1968,
          RIEHL N + BULLEMER B + ENGELHARDT H(EDS), PLENUM, NY,
          1969, 642P.

34-3429   CALCULATION OF THE DIELECTRIC CORRELATION FACTOR OF CUBIC ICE.
          GOBUSH W + HOEVE CAJ
          J CHEM PHYS 57: 3416-21 (1972)

34-3430   PROTON- PROTON AND PROTON- LATTICE INTERACTIONS IN ICE.
          GOSAR P
          P401-15 OF INT SYMP ON PHYSICS OF ICE, 3RD PROC, 1968,
          RIEHL N + BULLEMER B + ENGELHARDT H(EDS), PLENUM, NY,
          1969, 642P.

34-3431   DIELECTRIC BEHAVIOR OF CUBIC AND HEXAGONAL ICES AT LOW TEMPERATURES.
          GOUGH SR + DAVIDSON DW
          J CHEM PHYS 52(10): 5442-9 (1970)
               205

34-3432 LOW TEMPERATURE DIELECTRIC CELL AND PERMITTIVITY OF HEXAGONAL ICE TO 2 K.
         GOUGH SR
         CAN J CHEM 50: 3046-51 (1972)

         REVIEW ON THE PROBLEMS OF THE PHYSICS OF ICE.
         GRANICHER H
         P1-18 OF INT SYMP ON PHYSICS OF ICE, 3RD PROC, 1968,
         RIEHL N + BULLEMER B + ENGELHARDT H(EDS), PLENUM, NY,
         1969, 642P.

34-3433 EVALUATION OF DIELECTRIC DISPERSION DATA. (ICE)
         GRANICHER H
         P527-33 OF INT SYMP ON PHYSICS OF ICE, 3RD PROC, 1968,
         RIEHL N + BULLEMER B + ENGELHARDT H(EDS), PLENUM, NY,
         1969, 642P.

34-3434 INTERPRETATION OF THE PRESSURE DEPENDENCE OF PROPERTIES CAUSED BY LATTICE DEFECTS.
         (ICE)
         GRANICHER H
         P534-40 OF INT SYMP ON PHYSICS OF ICE, 3RD PROC, 1968,
         RIEHL N + BULLEMER B + ENGELHARDT H(EDS), PLENUM, NY,
         1969, 642P.

34-3435 SOME EXPERIMENTS ON THE REGELATION OF ICE.
         HAHNE E + GRUGULL U
         P320-8 OF INT SYMP ON PHYSICS OF ICE, 3RD PROC, 1968,
         RIEHL N + BULLEMER B + ENGELHARDT H(EDS), PLENUM, NY,
         1969, 642P.

34-3436 RELAXATIONAL PROTON ORDERING AND GLASSY CRYSTALLINE STATE IN HEXAGONAL ICE.
         HAIDA O + MATSUO T + SUGA H + SEKI S
         PROC JAP ACAD 48: 489-94 (1972)

34-3437 DIFFUSION OF HYDROGEN FLUORIDE IN ICE.
         HALTENORTH H + KLINGER J
         P579-84 OF INT SYMP ON PHYSICS OF ICE, 3RD PROC, 1968,
         RIEHL N + BULLEMER B + ENGELHARDT H(EDS), PLENUM, NY,
         1969, 642P.

34-3438 DEUTERON ARRANGEMENTS IN THE HIGH PRESSURE FORMS OF ICE.
         HAMILTON WC + KAMB B + LAPLACA SJ + PRAKASH A
         P19-43 OF INT SYMP ON PHYSICS OF ICE, 3RD PROC, 1968,
         RIEHL N + BULLEMER B + ENGELHARDT H(EDS), PLENUM, NY,
         1969, 642P.

34-3439 ELASTIC ANOMALIES OF ICE AT LOW TEMPERATURES.
         HELMREICH D
         P231-8 OF INT SYMP ON PHYSICS OF ICE, 3RD PROC, 1968,
         RIEHL N + BULLEMER B + ENGELHARDT H(EDS), PLENUM, NY,
         1969, 642P.

34-3440 MOLECULAR FORCES OF HEAVY AND LIGHT ICE.
         HELMREICH D
         P279-83 OF NUSIMOVICI M(ED), PHONONS, INT CONF, RENNES, 1971 PROC,
         FLAMMARION, PARIS, 1971, 502P.

34-3441 MECHANICAL PROPERTIES OF ICE SINGLE CRYSTALS.
         HIGASHI A
         P197-212 OF INT SYMP ON PHYSICS OF ICE, 3RD PROC, 1968,
         RIEHL N + BULLEMER B + ENGELHARDT H(EDS), PLENUM, NY,
         1969, 642P.

34-3442 PHASE TRANSITION IN ICE.
         HIGASHI A
         J CRYSTALLOGR SOC JAP 14: 274-85 (1972) (IN JAPANESE)

34-3443 A STUDY OF BROUT'S MODEL FOR FERROELECTRICS. INVESTIGATIONS ON HEXAGONAL ICE.
         HIPOLOTO O + LOBO R
         FERROELECTRICS 1: 169-75 (1970)

34-3444   PLANAR GROWTH OF ICE FROM THE PURE MELT.
          HOBBS PV
          P95-112 OF INT SYMP ON PHYSICS OF ICE, 3RD PROC, 1968,
          RIEHL N + BULLEMER B + ENGELHARDT H(EDS), PLENUM, NY,
          1969, 642P.

34-3445   ON THE SYMMETRY OF THE HYDROGEN BONDS IN ICE-VII.
          HOLZAPFEL WB
          J CHEM PHYS 56(2): 712-15
              258

34-3446   THERMOELECTRIC EFFECT IN ICE.
          JACCARD C
          P348-62 OF INT SYMP ON PHYSICS OF ICE, 3RD PROC, 1968,
          RIEHL N + BULLEMER B + ENGELHARDT H(EDS), PLENUM, NY,
          1969, 642P.

34-3447   DIPOLAR RELAXATION AT LOW TEMPERATURE OF ICE SINGLE CRYSTAL.
          JENEVEAU A + SIXOU P
          SOLID STATE COMMUN 10: 191-4 (1972)

34-3448   ELECTRICAL BEHAVIOR OF ICE AT LOW TEMPERATURES, FERROELECTRICITY AND SPACE CHARGE.
          JENEVEAU A + SIXOU P + DANSAS P
          PHYS KONDENS MATER, PHYS MATER CONDENSEE 14: 252-64 (1972)
          (IN FRENCH)
              273

34-3449   IMPURITY EFFECTS ON THE PLASTICITY OF ICE AND THEIR EXPLANATION IN TERMS OF HYDROGEN
          REORIENTATION.
          JONES SJ + GLEN JW
          P217-22 OF INT SYMP ON PHYSICS OF ICE, 3RD PROC, 1968,
          RIEHL N + BULLEMER B + ENGELHARDT H(EDS), PLENUM, NY,
          1969, 642P.

34-3450   EXPERIMENTAL AND THEORETICAL STUDIES ON THE DC CONDUCTIVITY OF ICE.
          KAHANE A
          P443-9 OF INT SYMP ON PHYSICS OF ICE, 3RD PROC, 1968,
          RIEHL N + BULLEMER B + ENGELHARDT H(EDS), PLENUM, NY,
          1969, 642P.

34-3451   STRUCTURES OF THE FORMS OF ICES. (ICES II, VIII AND IX ARE FULLY ORDERED
          (ANTIFERROELECTRIC STRUCTURES))
          KAMB B
          INT CONGR CRYSTALLOGR, 8TH, 1969, NY, ABSTR. AMER INST PHYS, 1969.
              SCI ABSTR B 73: 65244 (1970)
              219

34-3452   DIELECTRIC DISPERSION AND PHASE TRANSITION OF POTASSIUM HYDROXIDE DOPED ICE.
          KAWADA S
          J PHYS SOC JAP 32: 1442 (1972)
              278

34-3453   ICE, FERRO- AND ANTIFERROELECTRICS.
          LIEB EH
          P21-8 OF BOWCOCK JE(ED), METHODS AND PROBLEMS THEOR PHYS,
          NORTH HOLLAND, 1970, 440P.

34-3454   DIELECTRIC PROPERTIES OF POTASSIUM CHLORIDE ICE.
          MAENO N
          J APPL PHYS 43(2): 312-16 (FEB 1972)
              257

34-3455   TRANSFER OF PROTONS THROUGH "PURE" ICE I(H), SINGLE CRYSTALS. PART-2: MOLECULAR MODELS
          FOR POLARIZATION AND CONDUCTION. PART-3: EXTRINSIC VERSUS INTRINSIC POLARIZATION,
          SURFACE VERSUS VOLUME CONDUCTION (INCLUDES ELECTRODE EFFECTS).
          MAIDIQUE MA + VON HIPPEL AR + WESTPHAL WB
          J CHEM PHYS 54(1): 145-9, 150-60 (1971)
              230

34-3456 CHARGE AND POLARIZATION STORAGE IN ICE CRYSTALS.
         MASCARENHAS S
         P483-91 OF INT SYMP ON PHYSICS OF ICE, 3RD PROC, 1968,
         RIEHL N + BULLEMER B + ENGELHARDT H(EDS), PLENUM, NY,
         1969, 642P.

34-3457 ANGULAR CORRELATION OF ANNIHILATION PHOTONS IN ICE SINGLE CRYSTALS.
         MOGENSEN O + KVAJIC G + ELDRUP M + MILOSEVIC-KVAJIC M
         PHYS REV B 4: 71-3 (1971)

34-3458 STUDY OF CONDUCTIVITY AND DIPOLAR RELAXATION IN DOPED ICE SINGLE CRYSTALS.
         MOUNIER S + SIXOU P
         P562-70 OF INT SYMP ON PHYSICS OF ICE, 3RD PROC, 1968,
         RIEHL N + BULLEMER B + ENGELHARDT H(EDS), PLENUM, NY,
         1969, 642P.

34-3459 INFLUENCE OF THE SURFACE LAYER ON THE PLASTIC DEFORMATION OF ICE SINGLE CRYSTALS.
         MUGURUMA J
         P213-16 OF INT SYMP ON PHYSICS OF ICE, 3RD PROC, 1968,
         RIEHL N + BULLEMER B + ENGELHARDT H(EDS), PLENUM, NY,
         1969, 642P.

34-3460 HARDNESS ANISOTROPY OF SINGLE CRYSTALS OF ICE I(H).
         OFFENBACHER EL + ROSELMAN I
         BULL AMER PHYS SOC 16: 1424 (1971)

34-3461 PROTONIC SEMICONDUCTORS. (ICE)
         ONSAGER L
         P363-8 OF INT SYMP ON PHYSICS OF ICE, 3RD PROC, 1968,
         RIEHL N + BULLEMER B + ENGELHARDT H(EDS), PLENUM, NY,
         1969, 642P.

34-3462 SYMMETRY ANALYSIS AND ELECTRONIC STATES IN CUBIC ICE.
         PASTORI-PARRAVICINI G + RESCA L
         J PHYS C 4: L314-17 (1971)

34-3463 SPECIFIC HEAT OF PURE AND DOPED ICE NEAR 120 DEGREES K.
         PICK MA + WENZL H + ENGELHARDT H
         Z NATURFORSCH A 26: 810-14 (1971)

34-3464 SPECIFIC HEAT OF ICE I(H).
         PICK MA
         P344-7 OF INT SYMP ON PHYSICS OF ICE, 3RD PROC, 1968,
         RIEHL N + BULLEMER B + ENGELHARDT H(EDS), PLENUM, NY,
         1969, 642P.

34-3465 ICE I: LATTICE DYNAMICS AND INCOHERENT NEUTRON SCATTERING.
         PRASK HJ + TREVINO SF + GAULT JD + LOGAN KW
         J CHEM PHYS 56: 3217-25 (1972)

34-3466 STRUCTURAL STUDIES OF ICE POLYMORPHS BY NEUTRON DIFFRACTION, PROTON AND DEUTERON
         NUCLEAR MAGNETIC RESONANCE.
         RABIDEAU SW + FINCH ED
         P59-80 OF INT SYMP ON PHYSICS OF ICE, 3RD PROC, 1968,
         RIEHL N + BULLEMER B + ENGELHARDT H(EDS), PLENUM, NY,
         1969, 642P.

34-3467 LATTICE DYNAMICS OF ICE.
         RENKER KB + BLANCKENHAGEN PV
         P287-304 OF INT SYMP ON PHYSICS OF ICE, 3RD PROC, 1968,
         RIEHL N + BULLEMER B + ENGELHARDT H(EDS), PLENUM, NY,
         1969, 642P.

34-3468 LATTICE DYNAMICS OF ICE I(H).
         RENKER KB
         P167-70 OF NUSIMOVICI M(ED), PHONONS, INT CONF, RENNES, 1971 PROC,
         FLAMMARION, PARIS, 1971, 502P.

34-3469 DIELECTRIC RELAXATION, BULK AND SURFACE CONDUCTIVITY OF ICE SINGLE CRYSTALS.
         RUEPP R + KASS M
         P555-61 OF INT SYMP ON PHYSICS OF ICE, 3RD PROC, 1968,
         RIEHL N + BULLEMER B + ENGELHARDT H(EDS), PLENUM, NY,
         1969, 642P.

34-3470   DIFFUSION AND RELAXATION PHENOMENA IN ICE.
           RUNNELS LK
           P514-26 OF INT SYMP ON PHYSICS OF ICE, 3RD PROC, 1968,
           RIEHL N + BULLEMER B + ENGELHARDT H(EDS), PLENUM, NY,
           1969, 642P.

34-3471   NMR INVESTIGATIONS OF ICE CRYSTALS. PART-1: PROTON MAGNETIC RESONANCE IN WATER ICE.
           PART-2: DEUTERON SPIN LATTICE RELAXATION IN DEUTERATED ICE.
           SCHMIDT VH + GRANICHER H
           P223-31, 232-7 OF AMPERE INT SUMMER SCHOOL, 2ND PROC, BLINC R(ED),
           1971.

34-3472   IMPURITY STATISTICS IN ICE.
           SEIDENSTICKER RG + LONGINI RL
           P471-82 OF INT SYMP ON PHYSICS OF ICE, 3RD PROC, 1968,
           RIEHL N + BULLEMER B + ENGELHARDT H(EDS), PLENUM, NY,
           1969, 642P.

34-3473   INFLUENCE OF IMPURITIES ON THE ORDER DISORDER REACTION IN HEXAGONAL ICE I(H).
           SESSELMANN I + HELMREICH D
           Z NATURFORSCH A 26: 803-9 (1971) (IN GERMAN)

34-3474   ATOMIC VIBRATIONS IN ORIENTATIONALLY DISORDERED SYSTEMS. PART-1: A TWO DIMENSIONAL
           MODEL. (ICE I)
           SHAWYER RE + DEAN P
           J PHYS C SOLID STATE PHYS 5: 1017-27 (1972)

34-3475   INTERPRETATION OF THE PROTON SPIN LATTICE RELAXATION IN HEXAGONAL ICE.
           SIEGLE G + WEITHASE M
           P5771-8 OF INT SYMP ON PHYSICS OF ICE, 3RD PROC, 1968,
           RIEHL N + BULLEMER B + ENGELHARDT H(EDS), PLENUM, NY,
           1969, 642P.

34-3476   CALORIMETRIC STUDY OF GLASS TRANSITION OF THE AMORPHOUS ICE AND OF THE PHASE
           TRANSFORMATION BETWEEN THE CUBIC AND HEXAGONAL ICES.
           SUGISAKI M + SUGA H + SEKI S
           P329-43 OF INT SYMP ON PHYSICS OF ICE, 3RD PROC, 1968,
           RIEHL N + BULLEMER B + ENGELHARDT H(EDS), PLENUM, NY,
           1969, 642P.

34-3477   ELECTRIC RESONANCE: APPLICATION TO THE HYDROGEN BOND. (ICE)
           SUSSMANN JA
           P541-5 OF INT SYMP ON PHYSICS OF ICE, 3RD PROC, 1968,
           RIEHL N + BULLEMER B + ENGELHARDT H(EDS), PLENUM, NY,
           1969, 642P.

34-3478   DIELECTRIC CONSTANT OF PURE ICE I(H) SINGLE CRYSTALS.
           TAUBENBERGER R
           HELV PHYS ACTA 45: 881 (1972) (IN GERMAN)

34-3479   PRESSURE DEPENDENCE OF THE COMPLEX DISPERSION COEFFICIENT OF ICE I(H) SINGLE CRYSTALS.
           TAUBENBERGER R + HUBMANN M + GRANICHER H
           HELV PHYS ACTA 44(5): 567 (1971) (IN GERMAN)
               SCI ABSTR A 74: 80251 (1971)
               256

34-3480   STRONG COLLISION LIMIT OF SPIN LATTICE RELAXATION IN HEXAGONAL ICE.
           VALIC MI + GORNOSTANSKY S + PINTAR MM
           CHEM PHYS LETT 9: 362-4 (1971)

34-3481   THE THREE DIMENSIONAL EIGHT VERTEX MODEL AND THE PROTON PROTON CORRELATION FUNCTIONS IN
           ICE.
           VILLAIN J
           SOLID STATE COMMUN 10: 967-70 (1972)

34-3482   DIELECTRIC RELAXATION SPECTRA OF WATER, ICE, AND AQUEOUS SOLUTIONS AND THEIR
           INTERPRETATION. PART-8: TRANSFER OF PROTONS THROUGH "PURE" ICE I(H) SINGLE CRYSTALS.
           SECT-1: POLARIZATION SPECTRA OF ICE I(H). SECT-2: MOLECULAR MODELS FOR POLARIZATION AND
           CONDUCTION. SECT-3: EXTRINSIC VERSUS INTRINSIC POLARIZATION, SURFACE VERSUS VOLUME
           CONDUCTION.
           VON HIPPEL AR + KNOLL DB + MAIDIQUE MA + WESTPHAL WB
           J CHEM PHYS 54(1): 150-60 (1971)
               186

34. ICE

34-3483 DIELECTRIC AND MECHANICAL RESPONSE OF ICE I(H) SINGLE CRYSTALS AND ITS INTERPRETATION.
VON HIPPEL AR + MYKOLAJEWYCZ R + RUNCK AH + WESTPHAL WB
J CHEM PHYS 57: 2560-71 (1972)

34-3484 MOLECULAR UNDERSTANDING OF ELECTROCHEMICAL PROCESSES BY ICE RESEARCH.
VON HIPPEL AR
J ELECTROCHEM SOC 119(2): C45-54 (1972)
261

34-3485 MOLECULAR PHENOMENA IN WATER SYSTEMS. PART-1: MOLECULAR INTERPRETATION OF THE PHASE
DIAGRAM OF ICE. PART-2: DIELECTRIC AND MECHANICAL RESPONSE OF ICE I(H) SINGLE CRYSTALS
AND ITS INTERPRETATION.
VON HIPPEL AR + FARRELL EF
MIT LAB INSULATION RES, 1971, 27P. AD724731. TECH REPT 10, 1971, 33P.
GOVT REP ANNOUNCE 71(15): 235 (1971)
232
253

34-3486 TRANSFER OF PROTONS THROUGH PURE ICE I(H) SINGLE CRYSTALS. PART-1: POLARIZATION SPECTRA
OF ICE I(H). PART-2: MOLECULAR MODELS FOR POLARIZATION AND CONDUCTION.
VON HIPPEL AR + KNOLL DB + WESTPHAL WB
J CHEM PHYS 54: 134-44, 145-9 (1971)

34-3487 FORMATION OF COLOR CENTERS IN IRRADIATED ICE.
WEISS JJ
P195-6 OF INT SYMP ON PHYSICS OF ICE, 3RD PROC, 1968,
RIEHL N + BULLEMER B + ENGELHARDT H(EDS), PLENUM, NY,
1969, 642P.

34-3488 PROTEIN SPIN RELAXATION IN HEXAGONAL ICE. PART-2: THE T(1 RHO) MINIMUM.
WEITHASE M + NOACK F + SHUTZ J
Z PHYS 246: 91-6 (1971)
256

34-3489 DIPOLE MOMENT DERIVATIVE OF THE HYDROGEN BOND IN ICE.
WHALLEY E
CAN J CHEM 50: 310-14 (1972)

34-3490 INFRARED SPECTRUM OF ICE I(H) IN THE RANGE 4000 TO 15 CM(-1).
WHALLEY E
P271-86 OF INT SYMP ON PHYSICS OF ICE, 3RD PROC, 1968,
RIEHL N + BULLEMER B + ENGELHARDT H(EDS), PLENUM, NY,
1969, 642P.

34-3491 VAPOR PRESSURE ISOTOPE EFFECT OF ICE AND ITS ISOMERS.
WOLFF H
P305:19 OF INT SYMP ON PHYSICS OF ICE, 3RD PROC, 1968,
RIEHL N + BULLEMER B + ENGELHARDT H(EDS), PLENUM, NY,
1969, 642P.

34-3492 OPTICAL CONSTANTS OF ICE I OVER THE ENTIRE INFRARED REGION.
ZOLOTARYOV VM
OPT SPECTROSC 29: 599-601 (1970)

35. LEAD GERMANATE AND BISMUTH GERMANATE

35-3493   CRYSTAL STRUCTURE OF PIEZOELECTRIC BISMUTH(12) GERMANIUM OXIDE(20).
ABRAHAMS SC + JAMIESON PB + BERNSTEIN JL
J CHEM PHYS 47(10): 4034-41 (15 NOV 1967)

TOP SEEDED SOLUTION GROWTH OF OXIDE CRYSTALS FROM NONSTOICHIOMETRIC MELTS. (BARIUM
TITANATE, STRONTIUM TITANATE, GERMANATES)
BELRUSS V + KALNAJS J + LINZ A + FOLWEILER RC
MATER RES BULL 6: 899-906 (1971)

35-3494   ELECTROOPTIC MEASUREMENTS ON LITHIUM GERMANATE AND LEAD GERMANATE. (LITHIUM GERMANATE
NOT FERROELECTRIC)
BICHARD VM + DAVIES PH + HULME KF + JONES GR + ROBERTSON DS
J PHYS D 5: 2124-8 (1972)
     279

35-3495   SOFT OPTIC PHONON MODE IN FERROELECTRIC LEAD(5) GERMANIUM(3) OXIDE(11).
BURNS G + SCOTT BA
PHYS LETT A 39: 177-8 (1972)

ACOUSTIC SURFACE WAVE CONVOLUTION ON CRYSTALS OF CADMIUM SULFIDE, LITHIUM NIOBATE AND
BISMUTH(12) GERMANIUM OXIDE(20).
CHAMBERS J + MASON IM + TURNER CW
ELECTRON LETT 8: 314-16 (1972)

35-3496   CZOCHRALSKI SYNTHESIS AND PROPERTIES OF RARE EARTH DOPED BISMUTH GERMANATE.
DICKINSON SK + HILTON RM + LIPSON HG
MATER RES BULL 7: 181-91 (MAR 1972)

BISMUTH(12) GERMANIUM OXIDE(20) VERSUS LITHIUM NIOBATE IN THE SURFACE WAVE AMPLIFIER. A
THEORETICAL AND EXPERIMENTAL COMPARISON.
HAGON PJ + SALLEE CF + LAKIN KM
IEEE TRANS SON ULTRASON 18: 54 (1971)

35-3497   QUASI ELASTIC SCATTERING IN LEAD GERMANATE.
HISANO K + RYAN JF
SOLID STATE COMMUN 11: 1745-9 (1972)
     283

35-3498   RAMAN SCATTERING FROM THE FERROELECTRIC MODE IN LEAD(5) GERMANIUM(3) OXIDE(11).
HISANO K + RYAN JF
SOLID STATE COMMUN 11: 119-21 (1972)
     274

35-3499   FERROELECTRIC PROPERTY AND ENANTIOMORPHISM OF LEAD(5) GERMANIUM(3) OXIDE(11) AND THE
RELATED MATERIALS.
IWASAKI H + FUSHIMI S
ACTA CRYSTALLOGR A 28: S182 (1972)

35-3500   FERROELECTRIC LEAD(5) GERMANIUM(3) OXIDE(11) CRYSTAL.
IWASAKI H
OYO BUTSURI 41: 415-20 (1972) (IN JAPANESE)

35-3501   FERROELECTRIC AND OPTICAL PROPERTIES OF LEAD(5) GERMANIUM(3) OXIDE(11) AND ITS
ISOMORPHOUS COMPOUND LEAD GERMANIUM SILICON OXIDE. (DIELECTRIC CONSTANTS, SPONTANEOUS
POLARIZATION, THERMAL EXPANSION, INDICES OF REFRACTION, OPTICAL ROTATORY POWER)
IWASAKI H + MIYAZAWA S + KOIZUMI H + SUGII K + NIIZEKI N
J APPL PHYS 43(12): 4907-15 (1972)
     282

35-3502 LEAD(5) GERMANIUM(3) OXIDE(11) CRYSTAL, A NEW FERROELECTRIC. (DIELECTRIC CONSTANTS AND REFRACTIVE INDICES)
IWASAKI H + SUGII K + YAMADA T + NIIZEKI N
APPL PHYS LETT 18(10): 444-5 (15 MAY 1971)
234

35-3503 OPTICAL ACTIVITY OF FERROELECTRIC LEAD(5) GERMANIUM(3) OXIDE(11) SINGLE CRYSTALS.
IWASAKI H + SUGII K
APPL PHYS LETT 19: 92-3 (1971)

35-3504 CRYSTAL STRUCTURE OF FERROELECTRIC LEAD(5) GERMANIUM(3) OXIDE(11).
KOIZUMI H + NIIZEKI N
ACTA CRYSTALLOGR A 28: S60 (1972)

PERTURBATION THEORY FOR ELECTROMAGNETIC COUPLING TO ELASTIC SURFACE WAVES ON PIEZOELECTRIC SUBSTRATES (LITHIUM NIOBATE AND BISMUTH(12) GERMANIUM OXIDE(20)).
LAKIN KM
J APPL PHYS 42(3): 889-906 (1971)

35-3505 LIGHT AND ELECTRIC FIELD DEPENDENT OSCILLATION OF SPACE CHARGE LIMITED CURRENT IN BISMUTH(12) GERMANIUM OXIDE(20).
LENZO PV
J APPL PHYS 43: 1107-12 (MAR 1972)

35-3506 NONLINEAR OPTICAL PROPERTIES OF FERROELECTRIC LEAD(5) GERMANIUM(3) OXIDE(11).
MILLER RC + NORDLAND WA + BALLMAN AA
OPT COMMUN 6(2): 210-12 (OCT 1972)
281

35-3507 GROWTH OF LARGE SINGLE CRYSTAL BISMUTH(4) TITANIUM(3) OXIDE(12).
MORRISON AD + LEWIS FA + MILLER A
FERROELECTRICS 1: 75-8 (1970)

35-3508 FERROELECTRICITY IN LEAD(5) GERMANIUM(3) OXIDE(11).
NANAMATSU S + SUGIYAMA H + DOI K AND OTHERS
J PHYS SOC JAP 31: 616 (AUG 1971)

POSSIBILITY OF SWITCHING AND STEERING SURFACE WAVES BY NONLINEAR MIXING IN ANISOTROPIC MEDIA. (LITHIUM NIOBATE, BISMUTH(12) GERMANIUM OXIDE(20))
NEWHOUSE VL + CHEN CL + DAVIS KL
J APPL PHYS 43: 2603-8 (1972)

35-3509 ACOUSTOOPTIC PROPERTY OF SINGLE CRYSTAL LEAD(5) GERMANIUM(3) OXIDE(11).
OHMACHI Y + UCHIDA N
J APPL PHYS 43(8): 3583-4 (1972)
270

35-3510 ULTRASONIC ATTENUATION IN SINGLE CRYSTALS OF BISMUTH(12) GERMANIUM OXIDE(20) AT LOW TEMPERATURES.
OMOTOSO EA
TUFTS UNIV PHD THESIS, 1972, 144P. N-7325798

SEPARATE EXCITATION OF LONGITUDINAL AND TRANSVERSE HYPERSONIC PULSES IN PIEZOELECTRIC LITHIUM NIOBATE AND BISMUTH BIGERMANATE.
SEBER GA + BARANSKII KN
BULL ACAD SCI USSR, PHYS SER 35(5): 864-6 (1971)

INDEPENDENT EXCITATION OF LONGITUDINAL AND TRANSVERSE HYPERSONIC WAVE PULSES IN LITHIUM NIOBATE AND BISMUTH(12) GERMANIUM OXIDE(20) PIEZOELECTRIC CRYSTALS.
SEVER GA + BARANSKII KN
BULL ACAD SCI USSR, PHYS SER 35: 864-6 (1971)

35-3511 ACOUSTIC SURFACE WAVE LOSS MECHANISMS ON BISMUTH(12) GERMANIUM OXIDE(20) AT MICROWAVE FREQUENCIES.
SLOBODNIK AJ + BUDREAU AJ
J APPL PHYS 43(8): 3278-83 (1972)

35-3512 ELASTIC, PIEZOELECTRIC, AND DIELECTRIC CONSTANTS OF BISMUTH(12) GERMANIUM OXIDE(20).
SLOBODNIK AJ + SETHARES JC
J APPL PHYS 43: 247-8 (JAN 1972)

35-3513   CRYSTAL GROWTH AND PROPERTIES OF LEAD(5) GERMANIUM(3) OXIDE(11) SINGLE CRYSTALS.
           SUGII K + IWASAKI H + MIYAZAWA S
           MATER RES BULL 6: 503-12 (1971)
                234

35-3514   GROWTH OF SINGLE CRYSTALS IN LEAD MONOXIDE- GERMANIUM DIOXIDE BINARY SYSTEM.
           SUGII K + MIYAZAWA S + IWASAKI H
           REV ELECTRON COMMUN LAB 20: 886-96 (1972)
           ELEC COMMUN LAB TECH J 21: 359-66 (1972) (IN JAPANESE)

35-3515   X-RAY TOPOGRAPHIC OBSERVATION OF 180 DEG DOMAINS IN FERROELECTRIC LEAD GERMANATE SINGLE
           CRYSTALS.
           SUGH K + IWASAKI H + ITOH Y + NIIZEKI N
           J CRYST GROW 16: 291-3 (1972)
                283

35-3516   X-RAY TCPOGRAPHIC OBSERVATION OF 180 DEGREE DOMAINS IN FERROELECTRIC LEAD(5)
           GERMANIUM(3) OXIDE(11).
           SUGII K + IWASAKI H + ITOH Y + NIIZEKI N
           J CRYST GROW 16: 291-3 (1972)

35-3517   ELECTRICAL AND OPTICAL SWITCHING PROPERTIES OF SINGLE CRYSTAL BISMUTH(4) TITANIUM(3)
           OXIDE(12).
           TAYLOR GW
           FERROELECTRICS 1: 79-86 (1970)

           PIEZOELECTRIC SURFACE WAVES IN CUBIC AND ORTHORHOMBIC CRYSTALS. (BISMUTH(12) GERMANIUM
           OXIDE(20) AND BARIUM SODIUM NIOBATE)
           TSENG CC
           APPL PHYS LETT 16: 253-5 (1970)

35-3518   ELECTROOPTIC PROPERTIES OF FERROELECTRIC LEAD GERMANATE SINGLE CRYSTAL.
           UCHIDA N + SAKU T + IWASAKI H + ONUKI K
           J APPL PHYS 43(12): 4933-6 (1972)
                282

35-3519   RAMAN SPECTRUM OF BISMUTH(12) GERMANIUM OXIDE(20).
           VENUGOPALAN S + RAMDAS AK
           PHYS LETT A 34(1): 9-10 (25 JAN 1971)
                238

35-3520   ELASTIC AND PIEZOELECTRIC PROPERTIES OF FERROELECTRIC LEAD(5) GERMANIUM(3) OXIDE(11)
           CRYSTALS.
           YAMADA T + IWASAKI H + NIZEKI N
           J APPL PHYS 43(3): 771-5 (MAR 1972)
                260

36. OTHER MATERIALS

36-3521   ANISOTROPY OF THE ELECTROOPTICAL EFFECT IN CRYSTALS OF LEAD MAGNONIOBATE.
ADRIANOVA II + BEREZHNOI AA + KAMZINA LS + KRAINIK NN
SOV PHYS SOLID STATE 13(11): 2813-15 (MAY 1972)
     265

36-3522   USE OF ELECTRON MICROSCOPY IN THE STUDY OF EXTENDED DEFECTS RELATED TO NONSTOICHIOMETRY
(AND ANTIFERROELECTRIC DOMAINS IN TRANSITION METAL OXIDES).
AMELINCKX S + VAN LANDUYT J
P295-322 OF INST FOR ADVAN STUDY ON THE CHEMISTRY OF EXTENDED
DEFECTS IN NONMETALLIC SOLIDS, EYRING L + O'KEEFE M(EDS), 1969,
AMER ELSEVIER, 1960, 669P.
     188

36-3523   GEL GROWTH OF SILVER PERIODATE CRYSTALS.
AREND H + PERISON J
MATER RES BULL 6: 1205-10 (1971)
     250

36-3524   THE CRYSTAL STRUCTURE OF TRIS-SARCOSINE CALCIUM CHLORIDE.
ASHIDA T + BANDO S + KAKUDO M
ACTA CRYSTALLOGR B 28: 1560-5 (1972)

36-3525   EPR INVESTIGATIONS OF MANGANESE(2+) COMPLEXES IN FERROELECTRIC TRIS-SARCOSINE CALCIUM
CHLORIDE (TSCC) SINGLE CRYSTALS.
BARTUCH H + WINDSCH W
PHYS STATUS SOLIDI A 14: K51-3 (1972)
     283

36-3526   INFRARED OPTICAL PROPERTIES OF VANADIUM DIOXIDE ABOVE AND BELOW TRANSITION TEMPERATURE.
(NO COCHRAN TYPE FERROELECTRIC MODE)
BARKER AS + VERLEUR HW + GUGGENHEIM HJ
PHYS REV LETT 17(26): 1286-9 (1966)

36-3527   ELECTRICAL PROPERTIES AND STRUCTURES OF METANIOBATES AND METATANTALATES OF TRANSITION
METALS OF THE 3D SERIES.
BAZUEV GV + KRYLOV EI
INORG MATER 7: 1072-4 (JUL 1971)

36-3528   X-RAY PHASE STUDY OF THE CADMIUM BORATE- ABO(3) SYSTEMS. (WHERE A IS STRONTIUM OR
BARIUM, AND B IS HAFNIUM, TITANIUM, OR TIN)
BELYAEV IN + AVERYANOVA LN + SOLOVEV AA
RUSS J INORG CHEM 15: 1475-8 (1970)

36-3529   NEW FERROELECTRIC (PYROCHLORE TYPE) COMPOUND (AMMONO THIO CADMIUM NIOBATE, WITH 4
ELECTRIC PHASES, 2 FERROELECTRIC).
BERNARD D + LE MONTAGNER S + PANNETIER J + LUCAS J
MATER RES BULL 6: 75-80 (1971) (IN FRENCH)
     223

36-3530   PYROELECTRICITY AND OPTICAL SECOND HARMONIC GENERATION IN POLYVINYLIDENE FLUORIDE FILMS
(AFTER COOLING STRETCHED FILM IN ELECTRIC FIELD, THERMOELECTRETS).
BERGMAN JG + MCFEE JH + CRANE GR
APPL PHYS LETT 18(5): 203-5 (1971)
     220

36-3531   RADIATIONLESS TRANSITIONS IN THE EUROPIUM(3+) CENTER IN LANTHANUM ALUMINATE.
BLASSE G + BRIL A + DE POORTER JA
J CHEM PHYS 53: 4450-3 (1970)

36. OTHER MATERIALS

NMR STUDY OF THE FERROELECTRIC TRANSITIONS IN DIGLYCINE NITRATE AND TRIS-SARCOSINE CALCIUM CHLORIDE. (SHIFT OF CURIE TEMPERATURE)
BLINC R + JAMSIK-VILFAN M + LAHAJNAR G + HAJDVKOVIC G
J CHEM PHYS 52: 6407-11 (1970)

36-3532 STARK EFFECT AND FERROELECTRICITY IN THE ORGANIC SEMICONDUCTOR COPPER PHTHALOCYANINE.
BLINOV LM + KIRICHENKO NA
SOV PHYS SOLID STATE 12: 1246-7 (1970)

36-3533 NMR STUDY OF THE STRUCTURAL PHASE TRANSITION IN LANTHANUM ALUMINATE.
BORSA F + CRIPPA ML + DERIGHETTI B
PHYS LETT A 34: 5-6 (1971)

36-3534 NEW FERROELECTRIC LANGBEINITE THALLIUM CADMIUM SULFATE.
BREZINA B + GLOGAROVA M
PHYS STATUS SOLIDI A, APPL RES 11(1): K39-42 (1972)

36-3535 POLYMORPHISM OF STRONTIUM METANIOBATE (TWO REVERSIBLE PHASES).
BRUSSET H
MATER RES BULL 6: 5-14 (1971) (IN FRENCH)
        223

36-3536 FABRICATION (BY FLUX GROWTH) OF HEXAGONAL SINGLE CRYSTALS OF RARE EARTH MANGANATES
(PHASE DIAGRAMS, CONVECTION EFFECTS).
BUISSON G
P93-106 OF SUCHET JP(ED), CROISSANCE DE COMPOSES MINERAUX
CRYSTALLINS: GROWTH OF SINGLE CRYSTAL MINERAL COMPOUNDS, MASSON,
1969, 169P. (IN FRENCH)
        CHEM ABSTR 71: 129530T
        178

36-3537 PREPARATION AND CRYSTALLOGRAPHIC PROPERTIES OF A(2+)B(3+)(2)O(4) TYPE CALCIUM AND
STRONTIUM SCANDATES.
CARTER JR + FEIGELSON RS
J AMER CERAM SOC 47: 141-4 (1964)

36-3538 FERROELECTRICITY AND PHASE TRANSITION IN AMMONIUM DICHLORO ACETATE AND DEUTERATED
AMMONIUM DICHLORO ACETATE.
CHIHARA H + INABA A + NAKAMURA N + OKUMA H + SODA G + YAMAOTO T
ACTA CRYSTALLOGR A 28: S182 (1972)

36-3539 FIRST ORDER NEGATIVE PHOTODIELECTRIC EFFECT IN PHTHALOCYANINE IN THE FERROELECTRIC
STATE.
CHISTYAKOV EA + VIDADI YUA + ROZENSHTEIN LD
SOV PHYS SOLID STATE 11: 2751-2 (1970)

36-3540 PYROELECTRIC EFFECT IN METAL-FREE PHTHALOCYANINE. (IN THE POLAR, OR FERROELECTRIC
STATE)
CHISTYAKOV EA + VIDADI YUA + ROZENSHTEIN LD
SOV PHYS SOLID STATE 12(9): 2241-2 (1971)
        225

36-3541 LATTICE DYNAMICS OF DIATOMIC CRYSTALS OF HIGH DIELECTRIC CONSTANT. (POTASSIUM BROMIDE,
LEAD TELLURIDE, TIN TELLURIDE, LEAD SULFIDE, GERMANIUM TELLURIDE)
COCHRAN W
P62-71 OF SYMP ON FERROELECTRICITY, 1966, GENERAL MOTORS
RES LABS, WARREN, MICHIGAN, PROC, FERROELECTRICITY.
WELLER EF(ED), ELSEVIER, 1967, 318P.

36-3542 FERROELECTRIC PROPERTIES OF HEXAGONAL ORTHOMANGANITES OF YTTRIUM AND RARE EARTHS.
(YTTERBIUM, HOLMIUM, ERBIUM)
COEURE PH + GUINET P + PEUZIN JC + BUISSON G + BERTAUT EF
P332-40, VOL 1 OF DVORAK V + FOUSKOVA A + GLOGAR P (EDS),
INT MEETING ON FERROELECTRICITY, PRAGUE, 1966, PROC.
INST PHYS CZECH ACAD SCI, PRAGUE, 1966.

36-3543 DIRECT PIEZOELECTRIC EFFECT IN POLYVINYL CHLORIDE FILMS.
COHEN J + EDELMAN S
J APPL PHYS 42: 893-4 (1971)
        226

36-3544   PYROELECTRICITY AND PIEZOELECTRICITY IN ORIENTED FILMS OF POLYVINYL FLUORIDE AND
           POLYVINYLIDENE FLUORIDE.
           COHEN J + EDELMAN S + VEZZETTI CF
           J ELECTROCHEM SOC 119: 227C (1972)

36-3545   MAGNETIC PROPERTIES OF HEAVY RARE EARTH ORTHOMANGANATES.
           COLLINGS EW + AUSTIN AE + BROG KC + WOOD VE
           BULL AMER PHYS SOC 16: 325 (1971)

36-3546   NEUTRON SCATTERING ANALYSIS OF THE LINEAR DISPLACEMENT CORRELATIONS IN POTASSIUM
           TANTALATE.
           COMES R + SHIRANE G
           PHYS REV B SOLID STATE 5(5): 1886-92 (1 MAR 1972)
                260

36-3547   DIELECTRIC ANOMALIES IN SOLID METHANE AT LOW TEMPERATURES AND HIGH PRESSURES. (POSSIBLY
           FERROELECTRIC)
           COSTANTINO MS + DANIELS WB + CRAWFORD RK
           PHYS REV LETT 29: 1098-100 (1972)
                292

36-3548   VIBRATIONAL SPECTRA OF IODIC ACID SINGLE CRYSTALS.
           COUTURE L + KRAUZMAN M + MATHIEU JP
           COMPT REND B 269: 1278-80 (1969) (IN FRENCH)
                195

36-3549   THE VIBRATIONAL SPECTRUM AND DIELECTRIC BEHAVIOR OF SODIUM CHLORATE. (NOT
           FERROELECTRIC)
           DAWSON P
           PHYS STATUS SOLIDI B 50: 571-6 (1972)
                280

           ELECTRICAL PROPERTIES OF THE SURFACES OF IONIC CRYSTALS (TGS AND OTHERS, BY SELECTIVE
           DEPOSITION OF CHARGED POLYSTYRENE LATEX PARTICLES).
           DISTLER GI + TOKMAKOVA EI
           SOV PHYS CRYSTALLOGR 14(6): 913-15 (1970)

36-3550   FERROELECTRIC OPTICAL ROTATION DOMAINS IN SINGLE CRYSTAL LEAD(5) GERMANIUM(3)
           OXIDE(11).
           DOUGHERTY JP + SAWAGUCHI E + CROSS LE
           APPL PHYS LETT 20(9): 364-5 (1 MAY 1972)
                263

36-3551   NMR INVESTIGATIONS OF PHTHALOCYANINE.
           DUDREVA B + GRANDE S
           J PHYS (PARIS) 33, SUPPL 4, COLLOQ 2: 183-5 (1972)
                294

36-3552   STRUCTURAL PHASE TRANSITIONS IN LANGBEINITES.
           DVORAK V
           PHYS STATUS SOLIDI B 52: 93-8 (1972)

           THE SYSTEM LEAD TITANATE- STRONTIUM COPPER NIOBATE. (CERAMIC SYNTHESIS)
           DZHMUKHADZE DF + VENEVTSEV YUN + ZHDANOV GS
           SOV PHYS CRYSTALLOGR 16(1): 136-9 (JUL-AUG 1971)

36-3553   LATTICE DYNAMICS OF IONIC CRYSTALS WITH HIGH DIELECTRIC CONSTANTS.   (LEAD AND TIN
           TELLURIDES)
           ELCOMBE MM + PAWLEY GS
           P100-3, VOL 1 OF DVORAK V + FOUSKOVA A + GLOGAR P (EDS),
           INT MEETING ON FERROELECTRICITY, PRAGUE, 1966, PROC.
           INST PHYS CZECH ACAD SCI, PRAGUE, 1966.

36-3554   MICROSTRUCTURE OF TSTS-19 TYPE FERROELECTRIC CERAMICS.
           FEDCHENKO ED + VERIGINA ZS + FOMICHEV VV
           INORG MATER 8: 785-7 (1972)

36-3555   (WEAK FIELD) DIELECTRIC (PERMITTIVITY) PROPERTIES OF BISMUTH TITANATE (EFFECT OF
           TEMPERATURE).
           FOUSKOVA A + CROSS LE
           J APPL PHYS 41(7): 2834-8 (1970)
                202

36-3556   FERROELECTRIC FILMS AND THEIR DEVICE APPLICATIONS. (POLYVINYLIDENE FLUORIDE)
           FRANCOMBE MH
           THIN SOLID FILMS 13: 413-33 (1972)
               281

36-3557   PIEZOELECTRICITY IN POLARIZED POLYVINYLIDENE FLUORIDE FILMS.
           FUKADA E + SAKURAI T
           POLYM J (JAPAN) 2(5): 656-62 (1971)

36-3558   FLUX GROWTH OF MAGNESIUM OXIDE AND LANTHANUM ALUMINATE CRYSTALS DOPED WITH ISOTOPE 170.
           GARTON G + HANN BF + WANKLYN BM + SMITH SH
           J CRYST GROW 12: 66-8 (1972)

36-3559   FERROELECTRIC BEHAVIOR OF THE MINERAL STIBOTANTALITE.
           GAVRILOVA ND + KARYAKINA NF + KOPTSIK VA + NOVIK VK
           SOV PHYS DOKL 15: 1075-7 (1971)

36-3560   PYROELECTRIC PROPERTIES OF POLYVINYLIDENE FLUORIDE AND ITS USE FOR INFRARED DETECTION.
           GLASS AM + MCFEE JH + BERGMAN JG
           J APPL PHYS 42: 5219-22 (DEC 1971)

36-3561   PECULIARITIES OF THE DIELECTRIC POLARIZATION OF CADMIUM PYRONIOBATE.
           GOLOVSHCHIKOVA GI + ISUPOV VA + MYLNIKOVA IE
           SOV PHYS SOLID STATE 13: 1967-9 (1972)
               258

36-3562   TWO COMPONENTS OF THE CRYSTALLOGRAPHIC TRANSITION IN VANADIUM DIOXIDE.
           (ANTIFERROELECTRIC TEMPERATURE 340 DEG K)
           GOODENOUGH JB
           J SOLID STATE CHEM 3(4): 490-500 (NOV 1971)
               257

36-3563   MAGNETOELECTRIC EFFECT IN THE ANTIFERROMAGNET IRON ANTIMONY(2) OXIDE(4) (MAY BE
           FERROELECTRIC DUE TO ANTIFERROMAGNETIC ORDERING).
           GORODETSKY G + SAYAR M + SHTRIKMAN S
           MATER RES BULL 5: 253-6 (1970)
               187

36-3564   EXPERIMENTAL RELATIONSHIP BETWEEN THE SUPERCONDUCTING AND THE FERROELECTRIC PHASE
           TRANSITIONS (TIN TELLURIDE PURE AND ALLOYED WITH SMALL AMOUNTS OF GERMANIUM TELLURIDE
           OR LEAD TELLURIDE).
           GRASSIE ADC + BENYON A
           PHYS LETT A 39: 199-201 (1972)

36-3565   SPECTRUM AND ENERGY LEVELS OF CHROMIUM(3+) IONS AND EXCHANGE COUPLED CHROMIUM(3+) PAIRS
           IN LANTHANUM ALUMINATE.
           HEBER J + HELLWEGE KH + LEUTLOFF S + PLATZ W
           Z PHYS 246: 261-80 (1971)

36-3566   RESONANCE ABSORPTION OF GAMMA QUANTA IN BARIUM, STRONTIUM, AND CALCIUM STANNATE.
           HIEN PZ + SHPINEL VS + VISKOV AS + VENEVTSEV YUN
           SOV PHYS JETP 17: 1271-5 (1963)

36-3567   THE CRYSTAL STRUCTURE AND PHASE TRANSITION OF AMMONIUM HYDROGEN DICHLORO ACETATE.
           PART-1: THE CRYSTAL STRUCTURE OF THE PARAELECTRIC PHASE.
           ICHIKAWA M
           ACTA CRYSTALLOGR B 28: 755-60 (1972)

36-3568   MAGNETOELECTRIC EFFECT IN FERROELECTRIC- ANTIFERROMAGNETIC LITHIUM IRON(0.5)
           TANTALUM(0.5) OXY FLUORIDE.
           ISMAILZADE IH + YAKUPOV RG + MELIK-SHANAZAROVA TA
           PHYS STATUS SOLIDI A 8: K63-4 (1971)
               259

36-3569   X-RAY AND ELECTRIC INVESTIGATIONS OF THE SYSTEMS YTTRIUM MANGANESE(1-X) (B(X))
           OXYGEN(3) (B=TRIVALENT IRON, CHROMIUM OR ALUMINUM).
           ISMAILZADE IH + SMOLENSKII GA + NESTERENKO VI + AGAEV FA
           PHYS STATUS SOLIDI A 5: 83-9 (1971)
               237

36-3570   SWITCHING OF OPTICAL ROTATORY POWER IN FERROELECTRIC LEAD(5) GERMANIUM(3) OXIDE(11)
           SINGLE CRYSTAL.
           IWASAKI H + SUGII K + NIIZEKI N + TOYODA H
           FERROELECTRICS 3: 157-61 (1972);
           IEEE TRANS SON ULTRASON 19: 157-61 (1972)
                263

36-3571   YTTRIUM MANGANATE SWITCHING BEHAVIOR.
           JIMENEZ B
           P127-30, VOL 2 OF DVORAK V + FOUSKOVA A + GLOGAR P (EDS),
           INT MEETING ON FERROELECTRICITY, PRAGUE, 1966, PROC.
           INST PHYS CZECH ACAD SCI, PRAGUE, 1966.

36-3572   DIELECTRIC ABSORPTION IN ORIENTED POLYVINYLIDENE FLUORIDE.
           KAKUTANI H
           J POLYM SCI A-2 8(7): 1177-86 (1970)
                223

36-3573   PIEZOELECTRICITY OF POLYVINYLIDENE FLUORIDE.
           KAWAI H
           JAP J AFPL PHYS 8: 975-6 (1969)

36-3574   PYROELECTRIC AMMONIUM IODATE, A POTENTIAL FERROELASTIC: CRYSTAL STRUCTURE.
           KEVE ET + ABRAHAMS SC + BERNSTEIN JL
           J CHEM PHYS 54: 2556-63 (1971)

36-3575   BAND STRUCTURE OF THE FERROELECTRIC SEMICONDUCTOR ANTIMONY TRISULFIDE.
           KHASABOV AG + NIKIFOROV IYA
           SOV PHYS CRYSTALLOGR 16(1): 28-31 (JUL-AUG 1971)
                240

36-3576   ELECTROOPTIC AND PIEZOELECTRIC PROPERTIES OF LANTHANUM TITANATE SINGLE CRYSTAL. (FOUND
           NOT FERROELECTRIC)
           KIMURA M + NANAMATSU S + DOI K + MATSUSHITA S + TAKAHASHI M
           JAP J APPL PHYS 11: 904 (1972)
                278

36-3577   LEVEL CROSSING (CROSS RELAXATION) EXPERIMENTS IN AMMONIUM TRIHYDROGEN PERIODATE
           CRYSTALS, PULSED MAGNETIC AND OPTICAL RESONANCE.
           KIND R + GRANICHER H
           P239-40 OF AMPERE INT SUMMER SCHOOL, 2ND PROC, BLINC R(ED), 1971.

36-3578   NUCLEAR QUADRUPOLE INTERACTION IN ANTIFERROELECTRIC DIAMMONIUM TRIHYDROGEN PERIODATE
           PROTON CROSSOVER RELAXATION.
           KIND R + GRANICHER H
           P109-11 OF INT MEETING ON FERROELECTRICITY, 2ND, 1969,, KYOTO, PROC,
           (J PHYS SOC JAP, VOL 28 SUPPL). PHYS SOC JAP, 1970, 459P.

36-3579   NUCLEAR QUADRUPOLE INTERACTION IN ANTIFERROELECTRIC DIAMMONIUM TRIHYDROGEN PERIODATE BY
           IODINE PROTON CROSSOVER RELAXATION.
           KIND R + GRANICHER H
           P305-8 OF EUROPEAN MEETING ON FERROELECTRICITY, 1969, SAARBRUCKEN,
           MUSER HE AND PETERSSON J(EDS), WISSENSCH VERLAGSGES, STUTTGART,
           1970, 396P.

36-3580   STUDY OF THE STRUCTURE OF ANTIFERROELECTRIC AMMONIUM PERIODATE BY NUCLEAR RESONANCE
           METHODS.
           KIND R
           PHYS KONDENS MATER, PHYS MATER CONDENSEE 13: 217-45 (1971)
           (IN GERMAN)
                257

36-3581   STATISTICAL CONSIDERATIONS OF THE ORDER DISORDER REACTION IN AMMONIUM PERIODATE.
           KIND R + GRANICHER H
           HELV PHYS ACTA 43: 483 (1970) (IN GERMAN)

36-3582   DIELECTRIC AND NMR STUDIES OF THE PHASE TRANSITION IN STANNOUS CHLORIDE DIHYDRATE.
           KIRIYAMA H + KIRIYAMA R
           P114-16 OF INT MEETING ON FERROELECTRICITY, 2ND, 1969,, KYOTO, PROC,
           (J PHYS SOC JAP, VOL 28 SUPPL). PHYS SOC JAP, 1970, 459P.

36. OTHER MATERIALS

36-3583 ELECTRON SPIN RESONANCE SPECTRA OF CHROMIUM(3+) AND GADOLINIUM(3+) IN LANTHANUM
ALUMINATE.
KIRO D + LOW W + ZUSMAN A
P44-50 OF INT CONF ON PARAMAGNETIC RESONANCE, 2ND PROC,
LOW W(ED), HEBREW UNIV JERUSALEM, 1963.

36-3584 X-RAY AND OPTICAL STUDIES ON THE PHASE TRANSITION OF FERROELECTRIC DICALCIUM STRONTIUM
HEXAACETATE.
KOBAYASHI J
P225-31, VOL 1 OF DVORAK V + FOUSKOVA A + GLOGAR P (EDS),
INT MEETING ON FERROELECTRICITY, PRAGUE, 1966, PROC.
INST PHYS CZECH ACAD SCI, PRAGUE, 1966.

36-3585 THERMALLY STIMULATED CURRENTS AND LUMINESCENCE IN BISMUTH(12) SILICON OXIDE(20) AND
BISMUTH(12) GERMANIUM OXIDE(20).
LAUER RB
J APPL PHYS 42: 2147-9 (APR 1971)

36-3586 TRANSITION IN TIN TELLURIDE- GERMANIUM TELLURIDE ALLOYS. (NEUTRON SCATTERING)
LEFKOWITZ I + SHIELDS M + DOLLING G + BUYERS WJL + COWLEY RA
P249-51 OF INT MEETING ON FERROELECTRICITY, 2ND, 1969,, KYOTO, PROC,
(J PHYS SOC JAP, VOL 28 SUPPL). PHYS SOC JAP, 1970, 459P.

TWO PHASE FERROELECTRIC SYSTEMS. PART-2: FERROELECTRICS- SILVER (BARIUM TITANATE AND
BARIUM TITANATE- MAGNESIUM STANNATE).
LEILBLER K + BRANSKI W
ACTA PHYS POLON 23: 279-85 (1963)

36-3587 PREPARATION OF THE TERNARY TITANATE BARIUM(0.8) LEAD(0.12) CALCIUM(0.08) TITANATE.
LIMAR TF + NAKHODNOVA AF + KAGAN YUA + SAVENKOVA GE + STOLSHTEIN DI
INORG MATER 6: 1145-8 (JUL 1970)

36-3588 SURFACE EFFECT IN FERROELECTRIC YTTRIUM MANGANATE.
LISSALDE FC + PEUZIN JC
FERROELECTRICS 4(3): 159-68 (1972)

36-3589 NEW FERROELECTRIC: ANTIMONY ORTHONIOBATE. (CURIE TEMPERATURE)
LOBACHEV AN + PESKIN VF + POPOLITOV VI + SYRKIN LN + FEOKTISTOVA NN
SOV PHYS SOLID STATE 14(2): 509-10 (AUG 1972)
274

36-3590 FAR INFRARED SPECTRA OF PIEZOELECTRIC POLYVINYLIDENE FLUORIDE.
LUONGO JP
J POLYM SCI A-2 10: 1119-23 (1972)

36-3591 MONOCLINIC STRUCTURE OF SYNTHETIC CALCIUM CHLORAPATITE. (DEVELOPMENT OF FERROELECTRIC
CHARACTER IN AN APPLIED ELECTRIC FIELD)
MACKIE PE + ELLIOTT JC + YOUNG RA
ACTA CRYSTALLOGR B 28: 1840-8 (1972)
270

36-3592 SYSTEM LANTHANA TITANIA PHASE EQUILIBRIA AND ELECTRICAL PROPERTIES (LANTHANUM
TITANATE).
MACCHESNEY JB + SAUER HA
J AMER CERAM SOC 45: 416-22 (1962)

36-3593 OPTICAL PROPERTIES OF POTASSIUM TANTALATE NIOBATE.
MANLIEF SK
PURDUE UNIV PHD THESIS, 1971, 155P.
DISS ABSTR INT B 32: 4812 (1972)
287

36-3594 GROWTH OF SINGLE CRYSTALS OF LEAD ZINCATE NIOBATE AND ITS PROPERTIES.
MATSUO Y
YOGYO KYOKAI SHI 78(899): 213-20 (1970) (IN JAPANESE)
CHEM ABSTR 73: 70550K (1970)
261

ELECTRET EFFECT IN A MIXTURE OF POLY METHACRYLATE AND BARIUM TITANATE
MAZUR K + HANDEREK J + PIECH T
ACTA PHYS POLON A 37: 31-40 (1970)

36-3595   PYROELECTRIC AND NONLINEAR OPTICAL PROPERTIES OF POLED POLYVINYLIDENE FLUORIDE FILMS.
MCFEE JH + BERGMAN JG + CRANE GR
FERROELECTRICS 3: 305-13 (1972);
IEEE TRANS SON ULTRASON 19: 305-13 (1972)
        263

NUCLEAR SPIN LATTICE RELAXATION IN SOME FERROELECTRIC AMMONIUM SALTS. (AMMONIUM
SULFATE, FLUOBERYLLATE, PERIODATE, BISULFATE)
MILLER SR + BLINC R + BRENMAN M + WAUGH JS
PHYS REV 126: 528-32 (1962)

PARAMAGNETIC RESONANCE AND OPTICAL ABSORPTION OF TRANSITION ELEMENT IONS IN STRONTIUM
TITANATE AND LANTHANUM ALUMINATE.
MUELLER KA
P17-43 OF INT CONF ON PARAMAGNETIC RESONANCE, 1ST PROC, HEBREW
UNIV JERUSALEM, 1962, 128P.

36-3596   MOSSBAUER EFFECT AND FERROELECTRIC PROPERTIES OF IONIC CRYSTALS.
MUZIKAR C + JANOVEC V + DVORAK V
PHYS STATUS SOLIDI 3(1): K9-12 (1963)

36-3597   PHASE RELATIONS AND SUPERSTRUCTURES OF PYRROHOTITE, IRON(1-X) SULFIDE.
NAKAZAWA H + MORIMOTO N
MATER RES BULL 6: 345-57 (1971)

36-3598   PIEZOELECTRICITY, PYROELECTRICITY, AND THE ELECTROSTRICTION CONSTANT OF POLYVINYLIDENE
FLUORIDE.
NAKAMURA K + WADA Y
J POLYM SCI A-2 9: 161-73 (1971)

36-3599   POSSIBLE METHOD FOR ESTIMATING THE PARAMETER X IN COMPOUNDS OF THE PYROCHLORE TYPE WITH
GENERAL FORMULA A(2)B(2)O(7).
NIKIFOROV LG
SOV PHYS CRYSTALLOGR 17(2): 347-9 (SEP-OCT 1972)
        275

36-3600   ANOMALIES IN THE (1 KHZ) DIELECTRIC CONSTANT (5-300K) OF THE PRASEODYMIUM(1-X)
NEODYMIUM(X) (ORTHO) ALUMINATE SYSTEM (NOT FERROELECTRIC BY HYSTERESIS LOOP TESTS).
NORDLAND WA + VAN UITERT LG
J PHYS CHEM SOLIDS 31: 1257-62 (1970)
        196

36-3601   DOMAIN STRUCTURE AND TEMPERATURE DEPENDENCE OF THE PIEZOELECTRIC CONSTANTS OF
STIBOTANTALITE. (ANTIMONY (TANTALUM, NIOBIUM) OXIDE(4))
NOVIK VK + KARYAKINA NF + BOCHKOV BG + KOPTSIK VA + GAVRILOVA ND
BULL ACAD SCI USSR, PHYS SER 35(9): 1703-4 (1971)

36-3602   NEW DESCRIPTION OF THE PYROCHLORE STRUCTURE (CADMIUM(2) NIOBIUM(2) OXIDE(6) SULFIDE).
PANNETIER J + LUCAS J
MATER RES BULL 5: 797-806 (1970) (IN FRENCH)
        207

36-3603   NEW FERROELECTRIC: SEMICARBAZIDE HYDROCHLORIDE.
ROCARIES JC + BOLDRINI P
APPL PHYS LETT 20: 49-51 (1972)

36-3604   STRUCTURE AND MAGNETIC PROPERTIES OF FERROELECTRIC FERROMAGNETIC SOLID SOLUTIONS IN THE
LEAD COBALT TUNGSTATE- CADMIUM MANGANATE SYSTEM (BOTH CURIE POINTS STUDIED, VARIOUS
COMPOSITIONS)
ROGINSKAYA YUE + VENEVTSEV YUN + ZHDANOV GS
BULL ACAD SCI USSR, PHYS SER 29(6): 1021-4 (1965)

INERTIAL ELECTRICAL POLARIZATION OF CRYSTALS (ALKALI HALIDES, STRONTIUM TITANATE).
ROST A + SCHMIDT G
ANN PHYS 28(2): 147-53 (1972) (IN GERMAN)

36-3605   DIELECTRIC PROPERTIES OF LITHIUM IODATE. (FOUND NOT FERROELECTRIC)
SAILER E
PHYS STATUS SOLIDI A 4: K173-5 (1971)
        234

36-3606   A NOTE ON THE APPARENT CCNTACT FERROELECTRIC EFFECT OF MANGANESE DIOXIDE.
           SATYANARAYANA BS
           J SCI IND RES (INDIA) B 21: 139-40 (1962)
           CHEM ABSTR 57: 1733

36-3607   SYNTHESIS AND X-RAY INVESTIGATION OF THE PEROVSKITES BARIUM OR STRONTIUM METAPLUMBATES
           (NOT FERROELECTRICS).
           SHUVAEVA ET + FESENKO EG
           SOV PHYS CRYSTALLOGR 15(2): 321-22 (1970)
                207

36-3608   ELECTRICAL TRANSPORT IN (SEMICONDUCTING AND MAGNETIC) RARE EARTH ORTHOCHROMITES,
           MANGANITES AND FERRITES
           SUBBARAO GV + WANKLYN BM + RAO CNR
           J PHYS CHEM SOLIDS 32: 345-58 (1971)
                230

36-3609   STRUCTURAL AND DIELECTRIC PRCPERTIES OF BARIUM ZINCATE(0.33) NIOBATE(0.67) (NOT
           FERROELECTRIC -160 TO +250C).
           TENNERY VJ + HANG KW
           J AMER CERAM SOC 53(2): 118 (1970)
                183

36-3610   SILVER CITHORIUM PHOSPHATE- A NEW CASE OF THE NON-HYDROGEN BONDED PHOSPHATE
           FERROELECTRIC (CRYSTAL GROWTH AND STRUCTURE HYSTERESIS LOOP RESULTS).
           TOPIC M + KOJIE-PRODIC B + POPOVIC S
           CZECH J PHYS B 20: 1003-6 (1970)
                221

36-3611   TEMPERATURE DEPENDENCE OF SOME FROPERTIES OF SODIUM DITHORIUM PHOSPHATE FERROELECTRIC
           CRYSTALS.
           TOPIC M + NAPIJALO M + POPVUC S + ZELJIC Z
           PHYS STATUS SOLIDI A 11: 787-90 (1972)

36-3612   DIELECTRIC PROPERTIES OF DIAMMONIUM HEXABROMO STANNATE (DIELECTRIC TRANSITIONS NOT
           FERROELECTRIC).
           TSUNEKAWA S + ISHIBASHI Y + TAKAGI Y
           JAP J APPL PHYS 9: 1530 (1970)
                222

36-3613   SYNTHESIS AND X-RAY DIFFRACTION STUDY OF SINGLE CRYSTALS OF A NEW PYROCHLORE CONTAINING
           LITHIUM, STRONTIUM LANTHANUM TANTALUM OXYFLUORIDE.
           TUTOV AG + BADER VI + MYLNIKOVA IE
           SOV PHYS CRYSTALLOGR 17(3): 345-6 (SEP-OCT 1972)
                275

36-3614   FERROELECTRICITY IN THE IRON DEFICIENT FERROUS SULFIDE SYSTEM. (EXISTS TO DEFICIENCIES
           OF 0.04)
           VAN DEN BERG CB
           PHYS STATUS SOLIDI 40: K65-8 (1970)
                211

36-3615   MODEL FOR THE SEMICONDUCTING UNIAXIAL FERROELECTRIC IRON(1-X) SULFIDE. PART-1. PART-2.
           VAN DEN BERG CB
           FERROELECTRICS 4: 103-16, 195-212 (1972)

36-3616   OPTICAL SPECTRA OF CHROMIUM(3+) PAIRS IN LANTHANUM ALUMINATE.
           VAN DER ZIEL JP
           PHYS REV B 4: 2888-905 (1971)

36-3617   PHASE CIAGRAM OF THE SEMICONDUCTING UNIAXIAL FERROELECTRIC IRON(1-X) SULFIDE.
           VAN DEN BERG CB
           FERROELECTRICS 4: 117-20 (1971)

36-3618   A NEW MAGNETOFERROELECTRIC: CADMIUM IRON NIOBATE.
           VENEVTSEV YUN + IVANOVA VV + GOLCVNIN VA + KASHLINSKII AI + KOZLOVA LA + KUZMIN RN
           BULL ACAD SCI USSR, PHYS SER 35(9): 1664-7 (1971)
                269

36-3619 NEW FERROELECTRICS. (BARIUM(2) BISMUTH (NIOBIUM OR TANTALUM OR VANADIUM) OXIDE(6), AND
         BARIUM(3) BISMUTH(2) (TUNGSTEN OR MOLYBDENUM) OXIDE(6), (BARIUM OR STRONTIUM)(2) COPPER
         TUNGSTEN OXIDE(6), AND (BARIUM OR STRONTIUM)(3) COPPER (NIOBIUM OR TANTALUM)(2)
         OXIDE(9), AND VARIOUS TUNGSTEN SUBSTITUTED LEAD TUNGSTATES)
         VENEVTSEV YUN + KAPYSHEV AG + VISKOV AS + LEBEDEV VM + ZHDANOV GS
         P261-9, VOL 1 OF DVORAK V + FOUSKOVA A + GLOGAR P (EDS),
         INT MEETING ON FERROELECTRICITY, PRAGUE, 1966, PROC.
         INST PHYS CZECH ACAD SCI, PRAGUE, 1966.

36-3620 CRYSTALLOGRAPHIC, OPTICAL AND MAGNETIC PROPERTIES OF (MONOCLINIC) EUROPIUM SILICATE.
         (UNSTABLE FERROELECTRIC PHASE)
         VERREAULT R
         PHYS KONDENS MATER, PHYS MATER CONDENSEE 14(1): 37-54 (1971)
             CHEM ABSTR 76: 51085H (1972)
             272

36-3621 DEBYE TYPE ABSORPTION IN POLAR PHTHALOCYANINE.
         VIDADI YUA + ROZENSHTEIN LD + CHISTYAKOV EA
         SOV PHYS SOLID STATE 11: 2183-4 (1970)

36-3622 ELECTRICAL PROPERTIES OF PHTHALOCYANINE AS A SEMICONDUCTOR IN THE FERROELECTRIC STATE.
         VIDADI YUA + ROZENSHTEIN LD + CHISTYAKOV EA
         SOV PHYS SOLID STATE 12: 486-7 (1970)

36-3623 LOW TEMPERATURE PHASE TRANSITION IN THE POLAR STATE OF THE ORGANIC SEMICONDUCTOR
         PHTHALOCYANINE.
         VIDADI YUA + CHISTYAKOV EA + ROZENSHTEIN LD
         SOV PHYS SOLID STATE 11: 1945-6 (1970)

36-3624 ORTHORHOMBIC DIGLYCINE SULFATE. (NOT PIEZOELECTRIC, NOT FERROELECTRIC)
         WHIPPS PW + COSIER RS + BYE KL
         J MATER SCI 7: 1476-7 (1972)
             282

36-3625 NUCLEAR QUADRUPOLE RESONANCE STUDY OF PHASE TRANSITION IN AMMONIUM MONOCHLORO ACETATE,
         MONOCHLORO ACETIC ACID MIXED CRYSTAL.
         YAMAMOTO T + NAKAMURA N + CHIHARA H
         P112-13 OF INT MEETING ON FERROELECTRICITY, 2ND, 1969, KYOTO, PROC,
         (J PHYS SOC JAP, VOL 28 SUPPL). PHYS SOC JAP, 1970, 459P.

36-3626 NEW FERROELECTRIC COMPOUND STRONTIUM TELLURATE.
         YAMADA T + IWASAKI H
         APPL PHYS LETT 21(3): 89-90 (1972)
             273

36-3627 PHASE TRANSITION OF A NEW FERROELECTRIC OXIDE, STRONTIUM OXIDE- TELLURIUM DIOXIDE.
         YAMADA T + IWASAKI H
         ACTA CRYSTALLOGR A 28: S181 (1972)

37. DEVICE APPLICATIONS OF FERROELECTRICS

NEW CRYSTAL FOR COMMUNICATIONS SYSTEMS ENTERS PRODUCTION STAGE. (LITHIUM TANTALATE)

BELL LAB REC 50: 296-7 (1972)

A NEW SECOND HARMONIC TYPE FERROELECTRIC MODULATOR FOR ELECTROMETER. (WITH A TGS SINGLE CRYSTAL)
ABE Z + KATO Y + FURUHATA Y
REV SCI INSTRUM 42: 805-9 (1971)

ELECTRON THERMAL DIFFUSION EFFECTS (INTERNAL GENERATION OF STRONG FIELDS) DURING HOLOGRAM RECORDING IN CRYSTALS (IRON DOPED STRONTIUM TITANATE).
AMODEI JJ
APPL PHYS LETT 18(1): 22-4 (1971)

HOLOGRAPHIC RECORDING IN LITHIUM NIOBATE.
AMODEI JJ + STAEBLER DL
RCA REV 33: 71-93 (1972)

FERROELECTRICS IN OPTICAL MEMORIES AND DISPLAYS: A CRITICAL APPRAISAL.
ANDERSON LK
IEEE TRANS SON ULTRASON 19: 69-79 (1972)

37-3628 FERROELECTRICS IN OPTICAL MEMORIES AND DISPLAYS: A CRITICAL APPRAISAL.
ANDERSON LK
FERROELECTRICS 3: 69-79 (1972);
IEEE TRANS SON ULTRASON 19: 69-79 (1972)
263

LOW TEMPERATURE COEFFICIENT OF LITHIUM NIOBATE RESONATORS BY USE OF TEMPERATURE SENSITIVE CAPACITANCES.
ASHIDA T
PROC IEEE 58: 146-7 (1970)

37-3629 PERFORMANCE OF SPUTTERED LEAD(0.92) BISMUTH(0.07) LANTHANUM(0.01) (IRON(0.405) NIOBIUM(0.325) ZIRCONIUM(0.27)) OXYGEN(3) FERROELECTRIC MEMORY FILMS.
ATKIN RB
FERROELECTRICS 3: 213-15 (1972);
IEEE TRANS SON ULTRASON 19: 213-15 (1972)
263

37-3630 EXPERIMENTAL METHODS FOR INVESTIGATING STRAIN WAVE PROPAGATION AND ASSOCIATED CHARGE RELEASE TO FERROELECTRIC MATERIALS.
BEADLE CW + DALLY JW
EXP MECH 4(3): 70-6 (1964)

IMPROVEMENT IN THE PYROELECTRIC INFRARED RADIATION DETECTOR. (TRIGLYCINE SULFATE, TRIGLYCINE FLUOROBERYLLATE, LITHIUM SULFATE, BARIUM STRONTIUM NIOBATE)
BEERMAN HP
FERROELECTRICS 2: 123-8 (1971)

THERMAL ANALYSIS OF PYROELECTRIC DETECTORS. (TRIGLYCINE SULFATE EXAMPLE)
BLACKBURN H + WRIGHT HC
INFRARED PHYS 10: 191-7 (1970)

PROGRESS TOWARD A FAST, NONVOLATILE, NONDESTRUCTIVE READOUT MEMORY ELEMENT UTILIZING POTASSIUM NITRATE. (FUSED THICK FILM OR VACUUM DEPOSITED THIN FILM)
BORN RC + ROHRER GA + DHALL BS
P149-54 OF ELECTRON COMPON CONF, 20TH PROC, 1970,

PARAMETRIC LIGHT AMPLIFICATION AND OSCILLATION IN KDP WITH MODE LOCKED PUMP.
BURNEIKA K + IGNATAVICIUS M + KABELKA V + PISKARSKAS A
STABINIS A
IEEE J QUANTUM ELECTRON 8: 574 (1972)

BARIUM SODIUM NIOBATE FOR A TUNABLE STIMULATED RAMAN OSCILLATOR.
BURNS G
BULL AMER PHYS SOC 17: 245 (1972)

PYROELECTRIC COEFFICIENT DIRECT MEASUREMENT TECHNIQUE AND APPLICATION TO A NANOSECOND
RESPONSE TIME DETECTOR. (STRONTIUM BARIUM NIOBATE)
BYER RL + ROUNDY CB
FERROELECTRICS 3: 333-8 (1972);
IEEE TRANS SON ULTRASON 19: 333-8 (1972)

37-3631   MESOMORPHIC MATERIALS FOR ELECTROOPTICAL APPLICATION.
CASTELLANO JA
FERROELECTRICS 3: 29-38 (1971)

THIN FILM FERROELECTRIC PHOTOCONDUCTOR MEMORY DEVICE. (PLZT)
CHAPMAN DW
J VACUUM SCI TECHNOL 9: 425-31 (1972)

37-3632   DESIGN AND PERFORMANCE OF A THIN FILM FERROELECTRIC PHOTOCONDUCTOR STORAGE DEVICE.
CHAPMAN DW + MEHTA RR
FERROELECTRICS 3: 101-6 (1972);
IEEE TRANS SON ULTRASON 19: 101-6 (1972)
263

(HIGH SPEED, SMALL APERTURE) MODULATORS FOR OPTICAL COMMUNICATIONS (REVIEW EMPHASIZING
ELECTROOPTIC MODULATION, LITHIUM TANTALATE, LITHIUM NIOBATE, KDP, 113 REFS).
CHEN FS
PROC IEEE 58(10): 1440-57 (1970)

USE OF PEROVSKITE PARAELECTRICS IN BEAM DEFLECTORS AND LIGHT MODULATORS. (POTASSIUM
TANTALATE(0.65) NIOBATE(0.35))
CHEN FS + GEUSIC JE + KURTZ SK + SKINNER JG + WEMPLE SH
PROC IEEE 52: 1258-9 (1964)

EVALUATION OF LEAD LANTHANATE ZIRCONATE CERAMICS FOR APPLICATIONS IN OPTICAL
COMMUNICATIONS.
CHEN FS
OPT COMMUN 6(3): 297-300 (NOV 1972)

ELECTROOPTIC POTASSIUM TANTALATE NIOBATE GRATINGS FOR LIGHT BEAM MODULATION AND
DEFLECTION.
COHEN MG + GORDON EI
APPL PHYS LETT 5: 181-2 (1964)

37-3633   (COMPACT, SURFACE WAVE) FERROELECTRIC PHASE SHIFTERS FOR VHF AND UHF. (BASED ON
LEAD(0.35) STRONTIUM(0.65) TITANATE)
COHN M + EIKENBERG AF
IRE TRANS MICROW THEORY TECH 10: 536-48 (1962)

MINIMUM DETECTABLE POWER OF A PYROELECTRIC THERMAL RECEIVER. (USING BARIUM TITANATE
CRYSTAL)
COOPER J
REV SCI INSTRUM 33: 92-5 (1962)

37-3634   AN ANALYSIS OF MULTIPLE LIGHT SCATTERING BY RANDOM OPTICAL INHOMOGENEITIES WITH
APPLICATIONS TO FERROELECTRIC CERAMIC DISPLAY DEVICES.
COQUIN GA
J OPT SOC AMER 62: 1363 (1972)

A FERROELECTRIC PIEZOELECTRIC RANDOM ACCESS MEMORY. (LEAD ZIRCONATE- LEAD TITANATE
CERAMIC MEMORY PLATE)
CRAWFORD JC
IEEE TRANS ELECTRON DEV 18: 951-8 (1971)

37. DEVICE APPLICATIONS OF FERROELECTRICS

37-3635   PIEZOELECTRIC RESPONSE OF A FERROELECTRIC MEMORY ARRAY.
CRAWFORD JC
FERROELECTRICS 3: 139-46 (1972);
IEEE TRANS SON ULTRASON 19: 139-46 (1972)
263

ELECTRON BEAM WRITING OF FERROELECTRIC DOMAINS IN BISMUTH TITANIUM OXIDE SINGLE
CRYSTALS.
CUMMINS SE + HILL BH
PROC IEEE 58: 938-9 (JUN 1970)

A NEW METHOD OF OPTICALLY READING DOMAINS IN BISMUTH TITANATE AND MEMORY APPLICATIONS.
CUMMINS SE + LUKE TE
IEEE TRANS ELECTRON DEV 18: 761-8 (1971)

MEASURED ELECTRICAL CHARACTERISTICS OF INTERDIGITAL SURFACE WAVE TRANSDUCERS ON LITHIUM
NIOBATE.
DE KLERK J + DANIEL MR
APPL PHYS LETT 16: 219-21 (1970)

37-3636   APPLICATION OF FERROELECTRIC MATERIALS IN ELECTRICAL FILTERS.
DE JONG M
J PHYS (PARIS) 33, SUPPL 4, COLLOQ 2: 33-7 (1972)
294

FERROELECTRIC FILMS AND THEIR DEVICE APPLICATIONS. (POLYVINYLIDENE FLUORIDE)
FRANCOMBE MH
THIN SOLID FILMS 13: 413-33 (1972)

FERROELECTRIC DISPLAYS. (BISMUTH TITANATE, LEAD ZIRCONATE TITANATE)
FRANCOMBE MH + WU SY + TAKEI WJ
WESTINGHOUSE RES LAB, 1972, 89P. AD755746

OPTIMUM CUT IN X DIHYDROGEN PHOSPHATE CRYSTALS FOR TRANSVERSE LIGHT MODULATION
FRANCOIS GE + LIBRECHT FM + ENGELEN JJ
ELECTRON LETT 6: 778-9 (1970)

37-3637   RESEARCH STATUS AND DEVICE POTENTIAL OF FERROELECTRIC THIN FILMS.
FRANCOMBE MH
FERROELECTRICS 3: 199-211 (1972);
IEEE TRANS SON ULTRASON 19: 199-211 (1972)
263

37-3638   OPTICALLY ERASABLE AND REWRITABLE SOLID STATE HOLOGRAMS.
GAYLORD TK + RABSON TA + TITTEL FK
APPL PHYS LETT 20(1): 47-9 (1 JAN 1972)
257

FERROELECTRIC DOMAIN SHIFTING DEVICES. (GADOLINIUM MOLYBDATE)
GEUSIC JE + NELSON TJ + SCHINKE DP
BELL TELEPHONE LABORATORIES
US PAT 3,701,122, 24 OCT 1972.

CUTS OF LITHIUM NIOBATE CRYSTALS SUITABLE FOR LIGHT MODULATORS WITH A LOW HALF WAVE
VOLTAGE.
GISIN BV
SOV J QUANTUM ELECTRON 1: 686-7 (1972)

REAL-TIME HOLOGRAPHIC RECONSTRUCTION BY ELECTROOPTIC MODULATION (BY PASSING COHERENT
LIGHT THROUGH DEUTERATED KDP HAVING A CHARGE PATTERN).
GOETZ GG
APPL PHYS LETT 17(2): 63-6 (1970)

(FINITE APERTURE LIGHT SWITCH USING) TRANSVERSE POCKELS EFFECT COMMUTATORS. (POTASSIUM
DIHYDROGEN PHOSPHATE)
GOUZERH J + HEPNER G
OPT COMMUN 1(9): 435-7 (APR 1970) (IN FRENCH)
    SCI ABSTR A 73: 59371 (1970)

USE OF KDP TYPE CRYSTALS AS ELECTROOPTICAL REFLECTORS.
GRIB BN + KOROTKOV PA + TSYAHCHENKO YUP
UKR FIZ ZH 17: 546-50 (APR 1972) (IN RUSSIAN, ENGLISH ABSTRACT)

37. DEVICE APPLICATIONS OF FERROELECTRICS

37-3639   FERROELECTRIC MICROTHERMOSTATS.
GUSKOV VP + ERMACHENKOV NS + IVANOV IV
INSTRUM EXPER TECH 15(2): 579-81 (1972)

LASER STUDY OF REVERSIBLE NUCLEATION SITES IN TRIGLYCINE SULFATE AND APPLICATIONS TO
PYROELECTRIC DETECTORS.
HADNI A + THOMAS R
FERROELECTRICS 4: 39-49 (MAY 1972)

HOT PRESSED (LEAD, LANTHANUM) (ZIRCONATE, TITANATE) FERROELECTRIC CERAMICS FOR
ELECTROOPTIC APPLICATIONS.
HAERTLING GH + LAND CE
J AMER CERAM SOC 54(1): 1-11 (1971)

A NEW LONGITUDINAL DISPLAY MODE FOR CERAMIC ELECTROOPTIC DEVICES. (PLZT)
HAERTLING GH + MCCAMPBELL CB
PROC IEEE 60: 450-1 (1972)

BISMUTH (12) GERMANIUM OXIDE (20) VERSUS LITHIUM NIOBATE IN THE SURFACE WAVE AMPLIFIER. A
THEORETICAL AND EXPERIMENTAL COMPARISON.
HAGON PJ + SALLEE CF + LAKIN KM
IEEE TRANS SON ULTRASON 18: 54 (1971)

LITHIUM TANTALATE AND LITHIUM NIOBATE PIEZOELECTRIC RESONATORS IN THE MEDIUM FREQUENCY
RANGE WITH LOW RATIOS OF CAPACITANCE AND LOW TEMPERATURE COEFFICIENTS OF FREQUENCY
(DEVICE DESIGN AND CHARACTERIZATION).
HANNON JJ + LLOYD P + SMITH RT
IEEE TRANS SON ULTRASON 17(4): 239-46 (1970)

NEW METHOD OF MEASURING PYROELECTRIC COEFFICIENTS. (TGS EXAMPLE)
HARTLEY NP + SQUIRE PT + PUTLEY EH
J PHYS E 5(8): 787-9 (1972)

USE OF FERROELECTRIC CERAMICS IN COMPUTERS AND OTHER DEVICES. (LEAD ZIRCONATE- LEAD
TITANATE AND PLZT)
HARRISON WB
BULL AMER CERAM SOC 50: 805 (1971)

POLARIZATION REFLECTION TYPE LIGHT MODULATOR USING FERROELECTRICS. (BARIUM TITANATE)
HASEGAWA T + SATO H
FERROELECTRICS 3: 183-90 (1972;
IEEE TRANS SON ULTRASON 19: 183-90 (1972)

DOMAIN WALL MOTION IN FERROELECTRIC SWITCHING.
HAYASHI M
MEM FAC ENG, NAGOYA UNIV 24: 216-69 (1972)

HIGHLY SENSITIVE, SIMPLE CALORIMETER FOR PHASE CHANGE DETECTION (INCLUDING
FERROELECTRIC AND MAGNETIC TRANSITIONS, EXAMPLE ROCHELLE SALT)
HIRAKAWA K + FURUKAWA M
JAP J APPL PHYS 9(8): 971-5 (1970)

37-3640   VARIOUS HYSTERESIS LOOP TRACERS FOR A WIDE FREQUENCY RANGE. (CIRCUITRY DESCRIBED)
HRDLICKA J + MASTNER J + VELVARSKY J
P255-8 OF EUROPEAN MEETING ON FERROELECTRICITY, 1969, SAARBRUCKEN,
MUSER HE AND PETERSSON J (EDS), WISSENSCH VERLAGSGES, STUTTGART,
1970, 396P.

37-3641   ON THE SWITCHING PROPERTIES OF THE FERROELECTRIC SURFACES. (NEW MODEL REPRODUCES
EXPERIMENTAL PULSES)
IBEAS JG + LOPEZ V + PEINADO F
P245-9 OF EUROPEAN MEETING ON FERROELECTRICITY, 1969, SAARBRUCKEN,
MUSER HE AND PETERSSON J (EDS), WISSENSCH VERLAGSGES, STUTTGART,
1970, 396P.

ACOUSTOOPTICAL MEASUREMENTS OF ACOUSTIC WAVE GENERATION IN PIEZOELECTRICS COUPLED TO A
GUNN OSCILLATOR. (LITHIUM NIOBATE)
ISHIDA A + SUMI M
APPL PHYS LETT 18: 252-3 (1971)

37. DEVICE APPLICATIONS OF FERROELECTRICS

RH DOPED LITHIUM NIOBATE AS AN IMPROVED NEW MATERIAL FOR REVERSIBLE HOLOGRAPHIC
STORAGE.
ISHIDA A + MIKAMI O + MIYAZAWA S + SUMI M
APPL PHYS LETT 21: 192-3 (1972)

37-3642   FERROELECTRIC DOMAIN SWITCHING.
          ISHIBASHI Y + TAKAGI Y
          J PHYS SOC JAP 31: 506-10 (AUG 1971)

37-3643   THIN FILM MICROWAVE FERROELECTRIC CAPACITORS AND NONLINEAR MICROWAVE PROPERTIES OF
          FERROELECTRIC CERAMICS.
          IVANOV IV + MOROZOV NA
          P180-92, VOL 2 OF DVORAK V + FOUSKOVA A + GLOGAR P (EDS),
          INT MEETING ON FERROELECTRICITY, PRAGUE, 1966, PROC.
          INST PHYS CZECH ACAD SCI, PRAGUE, 1966.

          SWITCHING OF OPTICAL RCTATORY POWER IN FERROELECTRIC LEAD(5) GERMANIUM(3) OXIDE(11)
          SINGLE CRYSTAL.
          IWASAKI H + SUGII K + NIIZEKI N + TOYODA H
          FERROELECTRICS 3: 157-61 (1972);
          IEEE TRANS SON ULTRASON 19: 157-61 (1972)

37-3644   EQUIVALENT CIRCUIT ANALYSIS OF A FERROELECTRIC PHOTOCONDUCTOR MEMORY DEVICE.
          JACOBS JT + SILVERMAN BD + BATRA IP
          FERROELECTRICS 3: 177-82 (1972);
          IEEE TRANS SON ULTRASON 19: 177-82 (1972)
                263

37-3645   NOISE AND IMPEDANCE MEASUREMENTS OF A FERROELECTRIC CONDENSER UNDER A DC FIELD.
          JANNIN M
          J PHYS (PARIS) 32: 981-91 (NOV-DEC 1971) (IN FRENCH)

          OPTIMUM CRYSTAL ORIENTATION FOR A BARIUM SODIUM NIOBATE TRANSVERSE LIGHT MODULATOR.
          (HALF WAVE VOLTAGE, THEORY)
          JOHNSON DC
          COMMUN RES CENTER, OTTAWA, ONTARIO, 1972, 17P. CRC-1229, N73-13733

          A NEW FERROELECTRIC TRANSDUCER FOR USE IN HEAT TRANSFER AND FLOW STUDIES. (TRIGLYCINE
          SULFATE)
          JOLLS KR + SFORZA PM
          BROOKLYN POLYTECHNIC INST, 1970, 9P. AD707738

          ELECTRON BEAM SENSING OF SURFACE ELASTIC WAVES FOR SIGNAL PROCESSING PURPOSES.
          (TECHNIQUE APPLIED TO LITHIUM NIOBATE AND QUARTZ)
          JOSHI SG + EPSTEIN M + SERAFIN RJ + VAN DEN HEUVEL AP
          P72-5 OF EUROPEAN MICROWAVE CONF, 1969 PROC. IEE, LONDON, 1970, 570P.

          OPTIMISATION OF ELECTROOPTIC CRYSTAL MODULATORS. (KDP)
          KALYMNIOS D
          ELECTRON LETT 6(25): 804-5 (1970)

          BISMUTH TITANATE FERROELECTRIC- PHOTOCONDUCTOR OPTICAL STORAGE MEDIUM FURTHER
          DEVELOPMENTS.
          KENEMAN SA + MILLER A + TAYLOR GW
          FERROELECTRICS 3: 131-7 (1972);
          IEEE TRANS SON ULTRASON 19: 131-7 (1972)

          FERROELECTRIC PHOTOCONDUCTOR OPTICAL STORAGE MEDIUM UTILIZING BISMUTH TITANATE.
          KENEMAN SA + TAYLOR GW + MILLER A
          FERROELECTRICS 1: 227-41 (1970)

37-3646   STORAGE CF HOLOGRAMS IN A FERROELECTRIC PHOTOCONDUCTOR DEVICE (ERASABLE, NONDESTRUCTIVE
          READOUT, BASED ON BISMUTH TITANATE AND ZINC SELENIDE).
          KENEMAN SA + TAYLOR GW + MILLER A + FONGER WH
          APPL PHYS LETT 17(4): 173-5 (1970)
                217

          ADP POLARIZER FOR LASERS.
          KIEBURG H
          FEINGERATETECHNIK 21: 151-5 (1972) (IN GERMAN)

37. DEVICE APPLICATIONS OF FERROELECTRICS

STUDIES OF FLAME SPRAYED BARIUM TITANATE CAPACITORS. PART-10: STRUCTURE OF
SEMICONDUCTIVE TITANIUM DIOXIDE AND TITANIUM DIOXIDE- BARIUM TITANATE COATINGS PRODUCED
BY FLAME SPRAYING. PART-11: ELECTRICAL CHARACTERISTICS OF SEMICONDUCTIVE BARIUM
TITANATE COATINGS PRODUCED BY FLAME SPRAYING. PART-12: STRUCTURE OF SEMICONDUCTIVE
BARIUM TITANATE COATINGS PRODUCED BY FLAME SPRAYING.
KIMURA S
KINZOKU HYOMEN GIJUTSU 21: 14-19, 327-34, 363-7 (1970) (IN JAPANESE)

FLAME SPRAYED BARIUM TITANATE AS A CAPACITOR DIELECTRIC.
KIMURA S
IEEE TRANS PARTS MATER PACKAG 6: 3-11 (1970)

ORIENTATION AND THICKNESS DEPENDENCE OF CONTRAST AND BRIGHTNESS IN GADOLINIUM MOLYBDATE
LIGHT VALVES.
KMETZ AR
IEEE TRANS ELECTRON DEV 18: 756-61 (1971)

CUTTING AND POLISHING IN UNIAXIAL CRYSTALS OF THE TYPE X DIHYDROGEN PHOSPHATE FOR LIGHT
MODULATORS.
KOJIMA H + NOMURA S
PROC PHYS SOC JAP 25: 391-7 (1970) (IN JAPANESE)

37-3647 APPLICATION OF NONLINEAR INTERACTIONS IN FERROELECTRIC CERAMICS TO MICROWAVE SIGNAL
         PROCESSING.
         KRAUT EA + LIM TC + TITTMANN BR
         FERROELECTRICS 3: 247-55 (1972);
         TRANS SON ULTRASON 19: 247-55 (1972)
              263

DISPERSION STUDY OF DOMAIN WALL MOBILITY AND MECHANICAL DAMPING IN LEAD TITANATE
ZIRCONATE TRANSDUCER CERAMICS.
KRUEGER HHA
CLEVITE CORP, CLEVELAND, 1970, 23P, AD706793; 1971, 59P, AD717964

OPTICAL PROPERTIES OF GADOLINIUM MOLYBDATE AND THEIR DEVICE APPLICATIONS.
KUMADA A
FERROELECTRICS 3: 115-23 (1972);
IEEE TRANS SON ULTRASON 19: 115-23 (1972)

37-3648 ELECTROOPTIC EFFECT IN FERROELECTRICS.
         KUMADA A + NOMURA S
         CERAM JAP 6: 590-9 (1971) (IN JAPANESE)

ELECTROOPTIC CERAMICS- NEW MATERIALS FOR INFORMATION STORAGE AND DISPLAY (HOT PRESSED
LEAD ZIRCONATE TITANATE, TUTORIAL, 17 REFS).
LAND CE + THACHER PD
W ELEC ENG 14(1): 13-24 (JAN, 1970)

FERROELECTRIC CERAMICS FOR INFORMATION STORAGE AND DISPLAY. (PZT)
LAND CE
MET TRANS 2: 781-8 (1971)

37-3649 DEPENDENCE OF THE SMALL SIGNAL PARAMETERS OF FERROELECTRIC CERAMIC RESONATORS UPON
         STATE OF POLARIZATION.
         LAND CE + SMITH GW + WESTGATE CR
         IEEE TRANS SON ULTRASON 11: 8-19 (1964)

37-3650 ELECTROOPTIC CERAMIC (INFORMATION) STORAGE (PROCESSING) AND DISPLAY DEVICES. (BASED ON
         ELECTRICALLY CONTROLLED LIGHT SCATTERING AND BIREFRINGENCE, REVIEW)
         LAND CE
         OYO BUTSURI 39(1): 18-29 (1970) (IN JAPANESE)
              CHEM ABSTR 72: 83966E (1970)

VISIBLE CW PARAMETRIC OSCILLATOR USING BARIUM SODIUM NIOBATE.
LAURENCE C + TITTEL FK
J APPL PHYS 42: 2137-8 (1971)

37-3651 OPEN CIRCUIT SENSITIVITY OF RADIALLY POLARIZED FERROELECTRIC CERAMIC CYLINDERS.
         (HYDROPHONE ELEMENTS)
         LE BLANC CL
         NAVAL UNDERWATER SYST CENTER, NEW LONDON, CONN, 1970, 28P. AD712766

37-3652 OPEN CIRCUIT SENSITIVITY OF AXIALLY POLARIZED FERROELECTRIC CERAMIC CYLINDERS.
(HYDROPHONE ELEMENTS)
LE BLANC CL
NAVAL UNDERWATER SYST CENTER, NEW LONDON, CONN, 1970, 40P. AD715786

37-3653 OPEN CIRCUIT SENSITIVITY OF RADIALLY POLARIZED FERROELECTRIC CERAMIC HOLLOW SPHERES.
(HYDROPHONE ELEMENTS)
LE BLANC CL
NAVAL UNDERWATER SYST CENTER, NEW LONDON, CONN, 1970, 34P. AD722401

INPUT ADMITTANCE AND FIELDS RADIATED BY AN OPEN WAVEGUIDE FILLED WITH HIGH PERMITTIVITY
DIELECTRIC. (THEORY)
LEFEUVRE S + JONGEJANS A
J PHYS (PARIS) 33, SUPPL 4, COLLOQ 2: 99-101 (1972)

NEW COMPOSITE CERAMIC PIEZOELECTRIC TRANSDUCER MATERIAL. (ALUMINA WHISKERS IN LEAD
ZIRCONATE TITANATE)
LESTER WW
OFF NAV RES 1970, 11P. AD714494
    GOV REP ANNOUNCE 71(1): 97 (1971)

STABILIZED BARIUM TITANATE CERAMICS FOR CAPACITOR DIELECTRICS.
MACCHESNEY JB + GALLAGHER PK + DI MARCELLO FV
J AMER CERAM SOC 46: 197-202 (1963)

HIGH D*, FAST, LEAD ZIRCONATE TITANATE PYROELECTRIC DETECTORS.
MAHLER RJ + PHELAN RJ + COOK AR
INFRARED PHYS 12: 57-9 (1972)

STRAIN BIASED FERROELECTRIC- PHOTOCONDUCTOR IMAGE STORAGE AND DISPLAY DEVICES OPERATED
IN A REFLECTION MODE. (PLZT)
MALDONADO JR + ANDERSON LK
IEEE TRANS ELECTRON DEV 18: 774-7 (1971)

ELECTROOPTIC DEVICES USING STRAIN BIASED PLZT FERROELECTRIC CERAMICS.
MALDONADO JR
WESCON TECH PAP 15 (SESSION 31, PAPER 3): 1-8 (1971)

DISPLAY APPLICATIONS OF PLZT FERROELECTRIC CERAMICS.
MALDONADO JR
IEEE INT CONV DIGEST 1972: 190-1

37-3654 FERROELECTRIC CERAMIC LIGHT GATES OPERATED IN A VOLTAGE CONTROLLED MODE.
MALDONADO JR + MEITZLER AH
IEEE TRANS ELECTRON DEV 17: 148-57 (1970)

37-3655 TEMPERATURE AUTOSTABILIZATION EFFECT IN FERROELECTRICS. (CAPACITOR PROPERTIES, TANDEL)
MALEK Z + JANTA J
P220-8, VOL 2 OF DVORAK V + FOUSKOVA A + GLOGAR P (EDS),
INT MEETING ON FERROELECTRICITY, PRAGUE, 1966, PROC.
INST PHYS CZECH ACAD SCI, PRAGUE, 1966.

37-3656 FREQUENCY SPECTRUM OF FERROCERAMIC TRANSDUCERS.
MAZAK E
FYZ CAS 21: 136-50 (1971)

37-3657 RADIALLY VIBRATING FERROCERAMIC TRANSDUCER AS A SOURCE OF ULTRASONIC VIBRATIONS
RADIATING IN THE DIRECTION OF THE TRANSDUCER AXIS.
MAZAK E
FYZ CAS 21: 151-3 (1971)

AN ADAPTIVE FERROELECTRIC TRANSFORMER - A SOLID STATE ANALOG MEMORY DEVICE. (USING, FOR
EXAMPLE, PZT)
MCCUSKER JH + PERLMAN SS
IEEE TRANS ELECTRON DEV 17: 534-40 (1970)

FRACTURE TOUGHNESS OF COMMERCIAL BARIUM TITANATE TRANSDUCER MATERIAL.
MCKINNEY KR + SMITH HL
BULL AMER CERAM SOC 51: 650 (1972)

37-3658 MEASUREMENT OF POISSON'S RATIO IN POLED FERROELECTRIC CERAMIC DISKS.
MCMAHON GW
IEEE TRANS ULTRASON ENG 10: 102-3 (1963)

37. DEVICE APPLICATIONS OF FERROELECTRICS

SOME THEORETICAL CONSIDERATICNS RELATING TO SWITCHING AND REMANENCE IN FERROELECTRIC
PHOTOCONDUCTOR MEMORY DEVICES.
MEHTA RR
FERROELECTRICS 4: 5-18 (1972)

FERROELECTRIC- PHOTOCONDUCTOR MEMORY DEVICE. (PLZT)
MEHTA RR
J APPL PHYS 42: 1842-5 (1971)

37-3659 DEPOLARIZATION FIELD IN THIN FERROELECTRIC FILMS.
MEHTA RR + SILVERMAN BD + JACOBS JT
BULL AMER PHYS SOC 17: 103 (1972)

DEVICE APPLICATIONS OF PLZT CERAMICS.
MEITZLER AH
IEEE TRANS SON ULTRASON 19: 339 (1972)

ELECTRICAL CONTROL OF FIXATICN AND ERASURE OF HOLOGRAPHIC PATTERNS IN FERROELECTRIC
MATERIALS. (BARIUM TITANATE, POTASSIUM NIOBATE)
MICHERON F + BISMUTH G
APPL PHYS LETT 20: 79-81 (1972)

HOLOGRAPHIC STORAGE, ELECTRICAL FIXING AND ERASING IN DOPED BARIUM TITANATE CRYSTALS.
MICHERON F + BISMUTH G
J PHYS (PARIS) 33, SUPPL 4, COLLOQ 2: 149-50 (1972)

POTENTIAL BARRIERS ON SEMICONDUCTING BARIUM TITANATE. (PTCR DEVICES)
MILLER CA
J PHYS D APPL PHYS 4: 690-6 (MAY 1971)

NONLINEAR OPTICAL DEVICES. (BASED IN SECOND HARMONIC GENERATION AND OPTICAL PARAMETRIC
OSCILLATICN IN FERROELECTRIC CRYSTALS, REVIEW, 41 REFS)
MILLER RC
OYO BUTSURI 39(1): 11-17 (1970) (IN JAPANESE)
    CHEM ABSTR 72: 84356T (1970)

37-3660 NONLINEAR DIELECTRICS FOR HIGH POWER ELECTRONIC TUNING. (WITH ELECTRODES REDUCING
POLARIZATION SWITCHING TIME, LEAD(0.35) STRONTIUM(0.65) TITANATE, BARIUM(0.65 OR 0.75)
STRONTIUM(0.35 OR 0.25) TITANATES)
MURDOCH FJ
P140-8 OF ELECTRONIC COMPONENTS CONF, 20TH PROC, 1970, IEEE, 1970.

        242

LEAD TITANATE CERAMIC RESONATORS OPERATING IN VHF BAND.
NAGATA T + NAKAJIMA Y + SASAKI R
TRANS INST ELECTRON COMMUN ENG JAP C 55: 345-50 (1972) (IN JAPANESE)

LIGHT MODULATORS USING 45 DEGREE X CUT AND 45 DEGREE Y CUT ADP CRYSTALS.
OKADA M + IEIRI S
NHK TECH J 22: 490-500 (1970) (IN JAPANESE)

37-3661 USE OF CUBIC PHOTOSENSITIVE CRYSTALS AS POCKELS EFFECT LIGHT VALVES IN DISPLAY AND
MODULATOR APPLICATIONS.
OLIVER DS
FERROELECTRICS 3: 339 (1972)
IEEE TRANS SON ULTRASON 19: 339 (1972)

DYNAMIC SUSCEPTIBILITY OF CLASSICAL ANHARMONIC OSCILLATOR. A UNIFIED OSCILLATOR MODEL
FOR ORDER DISORDER AND DISPLACIVE FERROELECTRICS.
ONODERA Y
PROGR THEOR PHYS 44(6): 1477-99 (1970)

37-3662 COLOR IMAGING WITH FERROELECTRIC CERAMICS.
PARISI AJ
PROD ENG 41: 16-18 (22 JUN 1970)
        204

37-3663   ANALOG MEMORY DEVICES EMPLOYING PIEZOELECTRIC- FERROELECTRIC INTERACTIONS FOR ADAPTIVE
           CONTROL VOLTAGE MODULES.
           PERLMAN SS + MCCUSKER JH + BOARDMAN SM
           FERROELECTRICS 3: 239-45 (1972);
           IEEE TRANS SON ULTRASON 19: 239-45 (1972)
               263

37-3664   POSSIBLE REALIZATION OF ULTRA LOW NOISE LEVEL PARAELECTRIC MICROWAVE AMPLIFIER (BASED
           ON FERROELECTRICS IN THE PARAELECTRIC STATE).
           PETROV VM
           BULL ACAD SCI USSR, PHYS SER 29(11): 1954-8 (1965)
               219

           ZERO SHIFT IN PIEZOELECTRIC ACCELEROMETERS. (POLARIZATION SWITCHING IN POLYCRYSTALLINE
           FERROELECTRICS AT VERY LOW FIELDS AND STRESSES, ROCHELLE SALT)
           PLUMLEE RH
           SANDIA LABS, 1971, 61P. SC-RR-70-755
               NUCL SCI ABSTR 25: 34477 (1971)

           MICROWAVE DISPERSION MECHANISM IN BARIUM TITANATE TYPE FERROELECTRICS.
           POPLAVKO YUM
           BULL ACAD SCI USSR, PHYS SER 29(11): 1856-61 (1965)

37-3665   FERROELECTRICS AND THEIR APPLICATION IN SOLID STATE DEVICES AS ADAPTIVE CONTROL.
           PULVARI CF
           IEEE TRANS MIL ELECTRON 7(2-3): 254-60 (1963)

           PYROELECTRIC THERMAL IMAGING DEVICES. (TGS)
           PUTLEY EH + WATTON R + LUDLOW JH
           IEEE TRANS SON ULTRASON 19: 263-8 (1972)

           EPITAXIAL ELECTROOPTIC MIXED CRYSTAL AMMONIUM POTASSIUM DIHYDROGEN PHOSPHATE FILM
           WAVEGUIDE.
           RAMASWAMY V
           APPL PHYS LETT 21: 183-5 (1 SEP 1972)

           ULTRASONIC INTERFEROMETER WITH LEAD ZIRCONATE TITANATE AS TRANSDUCER.
           RAMACHANDRA RS
           INDIAN J PURE APPL PHYS 8: 848-9 (1970)

           PULSE RESPONSE OF 0 DEGREE CUT ADP MODULATORS.
           RASHIDI K
           ELECTRON LETT 7: 114-15 (1971)

37-3666   NEW TYPE OF LOOP TRACER FOR FERROELECTRICS.
           ROETSCHI H
           J SCI INSTRUM 39: 152-3 (1962)

           A NEW PTC THERMISTOR MATERIAL. (BARIUM TITANATE)
           SASAKI H + MATSUO Y + HAYAKAWA S
           BULL AMER CERAM SOC 47(4): 385 (APR 1968)

           ZERO TEMPERATURE COEFFICIENT OF RESONANT FREQUENCY IN LITHIUM TANTALATE LENGTH EXPANDER
           BARS.
           SAWAMOTO K + NIIZEKI N
           PROC IEEE 58: 1289-90 (1970)

           A SEMICONDUCTOR- FERROELECTRIC MEMORY DEVICE. (PLZT)
           SAWYER DE + SANDSTROM DB
           PROC IEEE 59: 87-8 (1971)

           INCORPORATION OF ANTIMONY INTO THE BARIUM TITANATE LATTICE. (STRUCTURE, AND PCT
           RESISTORS)
           SCHMELZ H
           P351-6 OF EUROPEAN MEETING ON FERROELECTRICITY, 1969, SAARBRUCKEN,
           MUSER HE AND PETERSSON J (EDS), WISSENSCH VERLAGSGES, STUTTGART,
           1970, 396P.

           TEMPERATURE DEPENDENCE OF RESONANT FREQUENCIES OF LITHIUM NIOBATE PLATE RESONATORS.
           (COMMENTS)
           SCHULZ BS + HOLLAND MG + FUKUMOTO A + WATANABE A
           PROC IEEE 58: 477 (1970)

37-3667 A SELF SCANNED FERROELECTRIC IMAGE SENSOR.
         SCHLOSSER PA + GLOWER DD
         IEEE TRANS SON ULTRASON 19: 257-62 (1972)

37-3668 FERROELECTRIC HYSTERESIS TRACER FEATURING COMPENSATION AND SAMPLE GROUNDING.
         SCHUBRING NW + NOLTA JP + DORK RA
         REV SCI INSTRUM 35: 1517-21 (1964)

37-3669 THEORY AND APPLICATION OF FERROELECTRIC RADIATION DETECTORS. (PYROELECTRICS)
         SCHLOSSER PA + MILLER DW + GLOWER DD
         INT J NONDESTRUCT TEST 2(1): 19-29 (1970)
              SCI ABSTR A 74(878): 5709 (1971)
              234

37-3670 EFFECT OF GAUSSIAN IRRADIANCE PROFILE ON THE SWITCHING CHARACTERISTICS AND THE BIT SIZE
         OF THE FE-PC MEMORY DEVICE.
         SHARMA BS
         FERROELECTRICS 4: 283-9 (1972)

37-3671 PHOTOCONDUCTOR AND ELECTRODE REQUIREMENTS FOR THIN FILM FERROELECTRIC PHOTOCONDUCTOR
         MEMORY DEVICE.
         SHARMA BS + MEHTA RR
         FERROELECTRICS 3: 225-9 (1972);
         IEEE TRANS SON ULTRASON 19: 225-9 (1972)
              263

37-3672 ACOUSTIC TRANSIENTS GENERATED BY ANTIFERROELECTRIC- FERROELECTRIC TRANSDUCERS.
         SINGAL SP + CAROME EF
         J ACOUST SOC AMER 44: 1211-15 (1968)
              143

         OPTICAL MIXING OF COHERENT AND INCOHERENT LIGHT. (KDP CRYSTAL FILTER)
         SMITH AW + BRASLAU N
         IBM J RES DEVELOP 6: 361-2 (1962)

         SCATTERING MODE FERROELECTRIC PHOTOCONDUCTOR IMAGE STORAGE AND DISPLAY DEVICES. (PLZT
         CERAMICS)
         SMITH WD + LAND CE
         APPL PHYS LETT 20: 169-71 (1972)

         LOW POWER DENSITY GALLIUM ARSENIDE- LITHIUM NIOBATE SURFACE WAVE AMPLIFIER.
         SPEARS DL + BURKE BE
         IEEE TRANS SON ULTRASON 18: 54 (1971)

         TEMPERATURE INSTABILITY OF ELECTROOPTIC LASER MODULATORS WITH AMMONIUM DIHYDROGEN
         PHOSPHATE, POTASSIUM DIHYDROGEN PHOSPHATE AND PARTIALLY DEUTERATED POTASSIUM DIHYDROGEN
         PHOSPHATE CRYSTALS.
         STADNIK B
         ACTA TECH (PRAGUE) 15(1): 65-9 (1970)
              CHEM ABSTR 73: 9287P (1970)

         MULTIPLE STORAGE OF THICK PHASE HOLOGRAMS IN LITHIUM NIOBATE.
         STAEBLER DL + AMODEI JJ + PHILLIPS W
         IEEE J QUANTUM ELECTRON 8: 611 (1972)

37-3673 GENERATOR OF BIPOLAR PULSES FOR INVESTIGATION OF FERROELECTRIC CRYSTALS.
         TAMBOVTSEV DA + NOVOSELOV AS
         INSTRUM EXPER TECH 1963(5): 906-8

37-3674 AN APPROXIMATIVE ANALYSIS OF A FERROELECTRIC RECORDING SYSTEM.
         TANAKA H + SATO R
         TRANS INST ELECTRON COMMUN ENG JAP A 53: 343-9 (1970) (IN JAPANESE)

         UTILIZATION OF THE (CRITICAL PULSE WIDTH) PARTIAL SWITCHING PROPERTIES OF
         FERROELECTRICS IN MEMORY DEVICES. (EXAMPLE TRIGLYCINE SULFATE)
         TAYLOR GW
         IEEE TRANS ELECTRON COMPUT 14: 881-6 (1965)

         BISMUTH TITANATE PHOTOCONDUCTOR OPTICAL STORAGE MEDIUM. (HIGH RESOLUTION READ WRITE)
         TAYLOR GW + MILLER A + KENEMAN SA
         RCA CORP, 1970, 180P. AFAL-TR-70-61

37. DEVICE APPLICATIONS OF FERROELECTRICS

FEASIBILITY OF ELECTROOPTIC DEVICES UTILIZING FERROELECTRIC BISMUTH TITANATE.
TAYLOR GW + MILLER A
PROC IEEE 58: 1220-9 (1970)

37-3675 ACTIVE COMPENSATORS FOR FERROELECTRIC OPTICAL CIRCUITS.
TAYLOR GW + KENEMAN SA + STEWART WC
FERROELECTRICS 2: 101-12 (1971)
         235

37-3676 FERROELECTRIC LIGHT VALVE ARRAYS FOR OPTICAL MEMORIES.
TAYLOR GW + KOSONOCKY WF
FERROELECTRICS 3: 81-99 (1972);
IEEE TRANS SON ULTRASON 19: 81-99 (1972)
         263

SURFACE WAVE DELAY LINES WITH INTERDIGITAL TRANSDUCERS ON UNPOLARIZED PZT CERAMIC
PLATES.
TODA K + KAWABATA A + TANAKA T
JAP J APPL PHYS 10: 671-7 (1971)

APPLICATION OF THE DEW METHOD FOR REVEALING THE DOMAIN STRUCTURE IN POTASSIUM
FERROCYANIDE TRIHYDRATE SINGLE CRYSTALS, P-N JUNCTIONS AND P- AND N- RANGES IN
SEMICONDUCTORS. (SILICON INTEGRATED CIRCUITS)
TOSHEV SD + AMOV IG
P241-4 OF EUROPEAN MEETING ON FERROELECTRICITY, 1969, SAARBRUCKEN,
MUSER HE AND PETERSSON J (EDS), WISSENSCH VERLAGSGES, STUTTGART,
1970, 396P.

ELECTROOPTIC CERAMICS AS WAVELENGTH SELECTION DEVICES IN DYE LASERS (PLZT).
VARNADO SG + SMITH WD
IEEE J QUANTUM ELECTRON 8: 88-9 (1972)

37-3677 SOME PROPERTIES AND APPLICATION OF FERROELECTRICS AT MICROWAVES.
VENDIK OG + MIRONENKO IG + TERMARTIROSYAN LT
J PHYS (PARIS) 33, SUPPL 4, COLLOQ 2: 277-80 (1972)
         294

37-3678 HIGH DIELECTRIC CONSTANT MATERIALS AND FERROELECTRICITY AS CAPACITOR DIELECTRICS, A
STUDY IN DIELECTRIC SPECTROSCOPY.
VON HIPPEL AR + WESTPHAL WB
MIT LAB INSULATION RES, DEC 1959, 76P. TECH REP 145,
AECU-4458
         CHEM ABSTR 57: 2962 (1962) NUCL SCI ABSTR 14(9): 8731 (1960)

37-3679 OBSERVATION OF THE SURFACES OF FERROELECTRIC CRYSTALS IN A PHOTOEMISSION ELECTRON
MICROSCOPE.
WEGMANN L
HELV PHYS ACTA 44: 581 (1971) (IN GERMAN)

LEAD TITANATE PYROELECTRIC INFRARED DETECTOR.
YAMAKA E + HAYASHI T + MATSUMOTO M
INFRARED PHYS 11: 247-8 (1971)

37-3680 ACOUSTIC SURFACE WAVE GUIDE, AVAILING FERROELECTRIC POLARIZATIONS.
YAMANISHI M + YOSHIDA K
OYO BUTSURI 40: 630-8 (1971) (IN JAPANESE)

37-3681 MANUFACTURING METHOD AND PROPERTIES OF FERROELECTRIC THIN FILM.
YAMANAKA S + INOKUMA T + HOSOKAI M
ELECTROTECH J JAP 7: 30-4 (1962)

EFFECT TUNABLE OPTICAL EMISSION FROM LITHIUM NIOBATE WITHOUT A RESONATOR.
YARBOROUGH JM + SUSSMAN SS + PUTHOFF HE + PANTELL RH + JOHNSON BC
IEEE J QUANTUM ELECTRON 6: 6 (1970)

EFFICIENT HIGH GRAIN PARAMETRIC GENERATION IN ADP CONTINUOUSLY TUNABLE ACROSS THE
VISIBLE SPECTRUM.
YARBOROUGH JM + MASSEY GA
APPL PHYS LETT 18: 438-40 (1971)

37. DEVICE APPLICATIONS OF FERROELECTRICS

AMPLITUDE MODULATORS BASED ON THE MICHELSON INTERFEROMETER. (AMMONIUM DIHYDROGEN PHOSPHATE)
ZERNIKE F + WEBSTER JC
FERROELECTRICS 3: 163-7 (1972);
IEEE TRANS SON ULTRASON 19: 163-7 (1972)

PERMUTED TITLE INDEX

PERMUTED TITLE INDEX

A

PHASE TRANSITION IN FERROELECTRIC SEMICONDUCTOR (COMPOUNDS) A(V)B(VI)C(VII). /NEQUILIBRIUM CARRIERS ON THE 15-2145
ELECTRICS.   PHOTOFERROELECTRIC EFFECTS IN A(V)B(VI)C(VII) AND BARIUM TITANATE TYPE FERRO 04-0552
/ OF SEMICONDUCTING COMPOUNDS OF THE TYPE A(2)(V)B(3)(VI)AND A(V)B(VI)C(VII). (ANTIMONY TRISULFIDE, TRISEL/ 15-2108
/E LATTICE PARAMETERS AND CURIE POINTS OF SOLID SOLUTIONS OF A(V)B(VI)C(VII) COMPOUNDS. ANTIMONY(X) BISMUT/ 15-2116
BISMUTH, SULFUR, SELENIUM/ HYDROTHERMAL METHOD FOR PREPARING A(V)B(VI)C(VII) COMPOUNDS. (USING ANTIMONY OR 15-2194
T TEMPERATURES NEAR A TYPE-1 PH/ ABSORPTION OF ULTRASOUND IN A(V)B(VI)C(VII) FERROELECTRIC SEMICONDUCTORS A 15-2114
/HERMAL CRYSTALLIZATION OF SEMICONDUCTING COMPOUNDS OF GROUP A(V)B(VI)C(VII). (HIGH GRADE SINGLE CRYSTALS,/ 15-2214
ONSTANTS, PYROELECTRI/ FERROELECTRIC PHOTOELECTRETS BASED ON A(V)B(VI)C(VII) SOLID SOLUTIONS. (DIELECTRIC C 15-2157
/ERROELECTRIC EFFECTS IN FERROELECTRIC SEMICONDUCTORS OF THE A(V)B(VI)C(VII) TYPE WITH LOW TEMPERATURE PHA/ 15-2161
(ANTI/ CRYSTALLIZATION OF SEMICONDUCTORS OF THE COMPOSITION A(V)B(VI)C(VII) UNDER HYDROTHERMAL CONDITIONS. 15-2213
DIDE TYPE. (PARTIAL OR TOTAL SUBSTITUTION/ NEW FERROELECTRIC A(V)B(VII)C(VII) COMPOUNDS OF ANTIMONY SULFOIO 15-2205
/MPT TO CORRELATE THE FERROELECTRIC PROPERTIES OF ABO(3) AND A(0.5)BO(3) TYPE PHASES AND THE IONIC MASS OF/ 03-0203
CANDATES.   PREPARATION AND CRYSTALLOGRAPHIC PROPERTIES OF A(2+)B(3+)(2)O(4) TYPE CALCIUM AND STRONTIUM S 36-3537
/SPECTROSCOPIC STUDY OF SEMICONDUCTING COMPOUNDS OF THE TYPE A(2)(V)B(3)(VI)AND A(V)B(VI)C(VII). (ANTIMONY/ 15-2108
R X IN COMPOUNDS OF THE PYROCHLORE TYPE WITH GENERAL FORMULA A(2)B(2)O(7). /HOD FOR ESTIMATING THE PARAMETE 36-3599
) OR (NIOBIUM, TUNGS/ SOME COMPOUNDS OF TUNGSTEN BRONZE TYPE A(6)B(10)O(30). (WHERE B IS (NIOBIUM, TITANIUM 13-1964
CITY ORIGINS)   FERROELECTRICALLY ACTIVE A-IONS IN AEO(3) STRUCTURES. (ANTIFERROELECTRI 08-1364
CTURE INVESTIGATION OF LANTHANUM MODIFIED LEAD TITANATE WITH A-SITE AND B-SITE VACANCIES.   X-RAY STRU 05-1059
M) (TITANATE, ZIRCONATE) CERAMICS.   DISTRIBUTION OF A-SITE AND B-SITE VACANCIES IN (LEAD, LANTHANU 09-1530
     A NEW CLASS OF FERROELECTRIC CERAMICS OF THE TYPE AB(2)(XO(4))(3).   13-1976
PHASE TRANSITION OF GADOLINIUM MOLYBDATE.   ABNORMAL ABSORPTION OF ELASTIC WAVES NEAR THE 30-3265
     GROUP ANALYSIS OF LATTICE VIBRATIONS OF CUBIC PEROVSKITE ABO(3).   08-1321
IC / AN ATTEMPT TO CORRELATE THE FERROELECTRIC PROPERTIES OF ABO(3) AND A(0.5)BO(3) TYPE PHASES AND THE ION 03-0203
/ILM MATERIALS WITH PEROVSKITE TYPE STRUCTURE. (MIXED OXIDES ABO(3)) (BARIUM TITANATE AND LITHIUM NIOBATE / 04-0914
E TYPE STRUCTURE.   CHARACTER OF THE CHEMICAL BONDS IN ABO(3) FERROELECTRIC CRYSTALS WITH A PEROVSKIT 08-1341
/PROPERTIES (REFLECTANCE SPECTRA AND BAND STRUCTURE) OF SOME ABO(3) FERROELECTRICS. (PHENOMENOLOGICAL MODE/ 08-1383
/AND INVESTIGATION OF FERROELECTRICS OF ILMENITE STRUCTURE. (ABO(3) NIOBATES, TANTALATES, AND MIXTURES WIT/ 08-1438
/CULATION OF INTERNAL ELECTRIC FIELDS AND THEIR GRADIENTS IN ABO(3) PEROVSKITE COMPOUNDS WITH DISTINCTIVE / 08-1455
(POTASSIUM TANTALATE AND TA/ ELECTROOPTIC PROPERTIES OF SOME ABO(3) PEROVSKITES IN THE PARAELECTRIC PHASE. 04-0690
ELECTRIC CRITICAL TEMPERATURE (ROTATIONS OF BO6 OCTAHEDRA IN ABO(3) PEROVSKITES, STRONTIUM TITANATE). /ERRO 04-1205
NS)   FERROELECTRICALLY ACTIVE A-IONS IN ABO(3) STRUCTURES. (ANTIFERROELECTRICITY ORIGI 08-1364
TRIC/ DIPOLE PATTERNS IN ORTHORHOMBIC AND TRIGONAL PHASES OF ABO(3) SUBSTANCE. (WHEN THE B ION IS FERROELEC 08-1363
M, AND B IS HAFNIU/ X-RAY PHASE STUDY OF THE CADMIUM BORATE- ABO(3) SYSTEMS. (WHERE A IS STRONTIUM OR BARIU 36-3528
CAL PHONONS IN PARA- AND FERROELECTRIC PEROVSKITES. (THEORY, ABO(3) TYPE)   DIELECTRIC PROPERTIES AND OPTI 03-0413
UM TITANATE.   SURFACE DOMAIN REORIENTATION PRODUCED BY ABRASION AND ANNEALING OF POLYCRYSTALLINE BARI 04-0639
    MOSSBAUER EFFECT AT THE PHASE TRANSITION IN FERROELECTRICS. ABRUPT (STEPWISE) CHANGE IN PROBABILITY OF THE 03-0276
TONS IN POTASSIUM DIHYDROGEN PHOSPHATE TYPE CRYSTALS.   ABSENCE OF A LINE WIDTH TRANSITION FOR THE PRO 17-2436
R IN POTASSIUM DIHYDROGEN PHOSPHATE AT HIGH TEMPERATURES.   ABSENCE OF MOTIONAL NARROWING IN THE PROTON NM 17-2594
BILITIES. (BARIUM TITANATE, LITHIUM TANTAL/ DETERMINATION OF ABSOLUTE SIGNS OF MICROWAVE NONLINEAR SUSCEPTI 02-0107
RMONIC GENERATION) COEFFICIENTS OF POLAR (PYROELECTRIC) CRY/ ABSOLUTE SIGNS OF NONLINEAR OPTICAL (SECOND HA 02-0095
OEFFICIENTS OF PIEZOELECTRIC CRYSTALS.   ABSOLUTE SIGNS OF SECOND HARMONIC GENERATION C 02-0096
OEFFICIENTS OF PIEZOELECTRIC CRYSTA/ FURTHER MEASUREMENTS OF ABSOLUTE SIGNS OF SECOND HARMONIC GENERATION C 05-1066
CIENTS OF POTASSIUM DIHYDROGEN PHOSPHATE.   ABSOLUTE SIGNS OF THE NONLINEAR OPTICAL COEFFI 17-2597
     ABSOLUTE STRUCTURE OF LITHIUM IODATE CRYSTALS. 08-1290
IN ADP AND AMMONIUM DIHYDROGEN ARSENATE STUDIED BY INFRARED ABSORPTION.   ANTIFERROELECTRIC TRANSITIONS 17-2706
OTON DYNAMICS IN KDP TYPE FERROELECTRICS STUDIED BY INFRARED ABSORPTION.   PR 17-2708
BARIUM TITANATE. (RESISTIVITY, SEEBECK COEFFICIENT, OPTICAL ABSORPTION) SOME SEMICONDUCTING PROPERTIES OF 04-0919
A TWO-LAYER SYSTEM CONSISTING OF A PIEZOELECTRIC CERAMIC (B/ ABSORPTION AND AMPLIFICATION OF ULTRASOUND IN 04-1037
SCEPTIBILITY IN LI/ SPONTANEOUS PARAMETRIC SCATTERING, IDLER ABSORPTION, AND DISPERSION IN THE NONLINEAR SU 11-1771
LEVELS OF NEODYMIUM(3+) AND ERBIUM(3+) IONS IN LITHIUM NIO/ ABSORPTION AND LUMINESCENCE SPECTRA AND ENERGY 11-1773
/INTERACTION IN FERROELECTRIC CRYSTALS (CYCLOTRON RESONANCE, ABSORPTION AND LUMINESCENCE SPECTRA, INFRARED/ 03-0427
HATE ION IN THE PARAELECTRIC PHASE OF KDP CRYSTALS (INFRARED ABSORPTION AND RAMAN SCATTERING). /F THE PHOSP 17-2479
IC AND PARAELECTRI/ MULTIPHONON SELECTION RULES FOR INFRARED ABSORPTION AND RAMAN SCATTERING IN FERROELECTR 16-2281
TANATE IN THE LONG WAVELENGTH INFRARED REGION.   ABSORPTION AND REFLECTION SPECTRA OF BARIUM TI 04-0644
Y SULFOIODIDE.   NOTES ON ABSORPTION BAND EDGE OF (PARAELECTRIC) ANTIMON 15-2177
/ONIZATION OF THE (X-RADIATION AND COPPER AND IRON IMPURITY) ABSORPTION BANDS OF DEFECT ROCHELLE SALT AND / 27-3108
ARED. (APPLIED TO BARIUM TITANATE)   FIELD DEPENDENCE OF ABSORPTION BANDS OF FERROELECTRICS IN FAR INFR 04-0665
UENCY RANGE OF DOUBLE ELECTRONIC EXC/ ON THE THEORY OF LIGHT ABSORPTION BY ANTIFERRODIELECTRICS IN THE FREQ 03-0419
ACITES. (IRON, NICKEL, OR / CRYSTAL FIELD THEORY AND OPTICAL ABSORPTION (CALCULATIONS) OF FERROELECTRIC BOR 14-2095
ONUNIFORM THIN SAMPLES (STRONTIUM TI/ DETERMINATION OF X-RAY ABSORPTION COEFFICIENTS AND THE THICKNESS OF N 06-1179
70-100% DEUTERATION.   1.06 MICRON ABSORPTION COEFFICIENTS OF DEUTERATED KDP WITH 17-2435
N IN STRONTIUM TITANATE.   ABSORPTION EDGE AND STRUCTURAL PHASE TRANSITIO 06-1107
OR ROCHELLE SALT AND TRIGLYCINE/ POSITION OF THE FUNDAMENTAL ABSORPTION EDGE AS A FUNCTION OF TEMPERATURE F 27-3160
     RELATION BETWEEN THE ELECTRIC POLARIZATION AND THE ABSORPTION EDGE IN ANTIMONY SULFOIODIDE. 15-2175
     POLARIZATION FLUCTUATIONS AND THE OPTICAL ABSORPTION EDGE IN BARIUM TITANATE. 04-1028
LS AT THEIR PHASE TRANSITIONS.   BEHAVIOR OF THE FUNDAMENTAL ABSORPTION EDGE IN TRIHYDROGEN SELENITE CRYSTA 22-2839
/TERMINATION OF POLARITY SENSE BY ANOMALOUS SCATTERING AT AN ABSORPTION EDGE. (LITHIUM NIOBATE, BARIUM MAN/ 11-1673
     ELECTROOPTIC EFFECTS OF THE ABSORPTION EDGE OF ANTIMONY SULFOIODIDE. 15-2198
     FUNDAMENTAL ABSORPTION EDGE OF STRONTIUM TITANATE. 06-1186
     OPTICAL ABSORPTION EDGE OF STRONTIUM TITANATE. 06-1187
HE 105 DEGREES K PHASE TRANSITION.   OPTICAL ABSORPTION EDGE OF STRONTIUM TITANATE AROUND T 06-1099
TE(X) NIOBATE(1-X).   ABSORPTION EDGE SPLITTING IN POTASSIUM TANTALA 13-1987
TE NIOBATE.   ABSORPTION EDGE SPLITTING IN POTASSIUM TANTALA 13-1988
SSION OF GLYCOL SULFATE.   OPTICAL ABSORPTION, ELECTRIC CONDUCTIVITY AND PHOTOEMI 19-2780
     MICROWAVE ABSORPTION IN CUBIC STRONTIUM TITANATE. 06-1212
     TWO PHONON ABSORPTION IN FERROELECTRIC SODIUM NITRITE. 16-2310
TAL OF SODIUM NITRITE.   INFRARED ABSORPTION IN GAMMA-RAY IRRADIATED SINGLE CRYS 16-2336
.   FREE CARRIER OPTICAL ABSORPTION IN NIOBIUM DOPED STRONTIUM TITANATE 06-1214
.   DIELECTRIC ABSORPTION IN ORIENTED POLYVINYLIDENE FLUORIDE 36-3572
     DEBYE TYPE ABSORPTION IN POLAR PHTHALOCYANINE. 36-3621
INGLE CRYSTALS.   LIGHT ABSORPTION IN POTASSIUM DIHYDROGEN PHOSPHATE S 17-2421

EVERSAL.   MODULATION OF THE MILLIMETER WAVE ABSORPTION IN SODIUM NITRITE BY POLARIZATION R 16-2392
TO TETRAGONAL TRANSITION.   DISTORTION ENHANCED OPTICAL ABSORPTION IN STRONTIUM TITANATE AT THE CUBIC   06-1109
STING PHASES.   REINTERPRETATION OF WAVELENGTH MODULATED ABSORPTION IN STRONTIUM TITANATE WITHOUT COEXI 06-1188
FOIODIDE CRYSTALS.   INDIRECT TRANSITIONS AND ABSORPTION IN THE MID-INFRARED IN ANTIMONY SUL 15-2150
HASES OF BARIUM TITANATE.   DIELECTRIC ABSORPTION IN THE ORTHORHOMBIC AND QUADRATIC P   04-0900
HASES OF BARIUM TITANATE.   DIELECTRIC ABSORPTION IN THE ORTHORHOMBIC AND QUADRATIC P   04-0899
TITANATE.   DIELECTRIC ABSORPTION IN THE RHOMOHEDRAL PHASE OF BAPIUM   04-0809
HE ELECTRIC FIELD AND THE DOMAIN STRUCTURE ON THE ULTRASONIC ABSORPTION IN TRIGLYCINE SULFATE. /LUENCE OF T 27-2955
LS. (THEORY OF ANISOTROPIC RELAXATION / ANISOTROPY OF SOUND ABSORPTION IN TRIGLYCINE SULFATE SINGLE CRYSTA 27-3078
TE HEXAHYDRATE.   ZEEMAN EFFECT AND DILUTION EFFECT OF ABSORPTION LINES IN GUANIDINIUM CHROMIUM SULFA   21-2821
M SULFATE.   MANGANESE(2+) ABSORPTION NEAR THE CRITICAL REGION IN AMMONIU   19-2785
/UENCY SPECTRUM OF HYDROGEN BONDED FERROELECTRICS. (INFRARED ABSORPTION, NEUTRON INELASTIC SCATTERING, POT/ 03-0401
CRYSTAL FIELD THEORY AND OPTICAL ABSORPTION OF COBALT AND NICKEL BORACITES.   14-2096
NIOBATE.   ABSORPTION OF ELASTIC WAVES IN REDUCED LITHIUM   11-1816
NSITION OF GADOLINIUM MOLYBDATE.   ABNORMAL ABSORPTION OF ELASTIC WAVES NEAR THE PHASE TRA   30-3265
M, AND CALCIUM STANNATE.   RESONANCE ABSORPTION OF GAMMA QUANTA IN BARIUM, STRONTIU 36-3566
FLUORIDE.   FINE STRUCTURE OF EXCITON- MAGNON ABSORPTION OF LIGHT IN POTASSIUM MANGANESE TRI   33-3364
ROCHELLE SALT CRYSTALS NEAR THE CURIE POINTS.   ABSORPTION OF LONGITUDINAL ULTRASONIC WAVES IN   28-3211
ND AMMONIUM DIHYDROGEN PHOSPHATE/ ULTRAVIOLET REFLECTION AND ABSORPTION OF POTASSIUM DIHYDROGEN PHOSPHATE A 17-2422
ALS AND (POTASSIUM SODIUM/ DIELECTRIC PROPERTIES AND OPTICAL ABSORPTION OF POTASSIUM TANTALATE SINGLE CRYST 08-1272
ED. (350-1100 MICRONS)   TEMPERATURE DEPENDENCE OF ABSORPTION OF SODIUM NITRITE IN THE FAR INFRAR   06-1208
ELASTIC CONSTANTS AND ULTRASONIC ABSORPTION OF SODIUM NITRITE SINGLE CRYSTALS.   16-2357
UPPER CURIE POINT.   FREQUENCY DEPENDENCE OF THE ABSORPTION OF SOUND IN ROCHELLE SALT NEAR ITS   28-3183
RROELECTRIC SEMICONDUCTORS.   RELAXATION ABSORPTION OF SOUND NEAR THE CURIE POINT OF FE 03-0492
SE TRANSITION (ANTIMONY SULFOIODIDE).   ANOMALOUS ABSORPTION OF SOUND NEAR THE FERROELECTRIC PHA 15-2221
.   VACUUM ULTRAVIOLET ABSORPTION OF THE ALKALI NITRITES AND NITRATES   16-2398
TIUM TITANATE AND LANTHA/ PARAMAGNETIC RESONANCE AND OPTICAL ABSORPTION OF TRANSITION ELEMENT IONS IN STRON 06-1165
RROELECTRIC SEMICONDUCTORS AT TEMPERATURES NEAR A TYPE-1 PH/ ABSORPTION OF ULTRASOUND IN A(V)B(VI)C(VII) FE 15-2114
ODIDE FERROELECTRIC SEMICONDUCTOR NEAR ITS CURIE POINT.   ABSORPTION OF ULTRASOUND IN AN ANTIMONY SULFOI 15-2115
DE CRYSTALS.   POLARIZATION EFFECTS IN THE ANOMALOUS ABSORPTION OF ULTRASOUND IN ANTIMONY SULFOIODI 15-2123
/CTS OF TEMPERATURE, ELECTRIC FIELD, AND ILLUMINATION ON THE ABSORPTION OF ULTRASOUND IN ANTIMONY SULFOIOD/ 15-2248
BARIUM TITANATE IN THE PRESENCE OF A STEADY ELE/ RELAXATION ABSORPTION OF ULTRASOUND IN SINGLE CRYSTALS OF 04-0850
ANTIMONY SULFOIODIDE.   ANISOTROPY OF ANOMALOUS ABSORPTION OF ULTRASOUND IN SINGLE CRYSTALS OF   15-2222
SINGLE CRYSTALS (A/ FREQUENCY DEPENDENCE OF THE FLUCTUATION ABSORPTION OF ULTRASOUND IN TRIGLYCINE SULFATE 27-3079
XIDE AND LITHIUM NIOBATE SIN/ TEMPERATURE DEPENDENCES OF THE ABSORPTION OF 9.4 GHZ HYPERSOUND IN ALUMINUM O 11-1719
/FERROELECTRIC PROPERTIES OF BARIUM SODIUM NIOBATE. (OPTICAL ABSORPTION, REFRACTIVE INDEX, DIELECTRIC CONS/ 13-2024
SPHATE.   FAR INFRARED ABSORPTION SPECTRA FOR AMMONIUM DIHYDROGEN PHO   17-2587
OPHOSPHATE (4000-10000 PER CM) AND/ LOW TEMPERATURE INFRARED ABSORPTION SPECTRA OF AMMONIUM DIHYDROGEN ORTH 17-2432
NTIUM TIT/ EFFECT OF PARTICLE SIZE AND SHAPE ON THE INFRARED ABSORPTION SPECTRA OF BARIUM TITANATE AND STRO 04-0823
ND BISMUTH TELLUROIODIDE CRYSTALS.   ABSORPTION SPECTRA OF BISMUTH TELLUROBROMIDE A 15-2217
ONTIUM OR CALCIUM TITANATE/ EXTENDED FINE STRUCTURE IN X-RAY ABSORPTION SPECTRA OF CERTAIN PEROVSKITES (STR 03-0414
S CONTAINING CHROMIUM(3+) IONS (10 TO 3/ CALCULATION OF THE (ABSORPTION) SPECTRA OF LITHIUM NIOBATE CRYSTAL 11-1831
M DIHYDROGEN PHOSPHATE AND RUBIDIUM / LOW FREQUENCY INFRARED ABSORPTION SPECTRA OF MONOCRYSTALS OF POTASSIU 17-2457
ORINE AND NICKEL IODINE BORACITES (AND NICKEL DOPED/ OPTICAL ABSORPTION SPECTRA OF NICKEL(2+) IN NICKEL CHL 14-2074
LUORIDE, POTASSIUM COBALT TRIFLUORIDE, AND RUBIDIU/ INFRARED ABSORPTION SPECTRA OF POTASSIUM MANGANESE TRIF 33-3384
TE.   OPTICAL ABSORPTION SPECTRA OF REDUCED STRONTIUM TITANA   06-1245
FAR INFRARED ABSORPTION SPECTRA OF ROCHELLE SALT.   28-3210
/L PROPERTIES (CONDUCTIVITY AND MOBILITY VERSUS TEMPERATURE, ABSORPTION SPECTRA) OF SEMICONDUCTING STRONTI/ 06-1198
IN-119 DURING PHASE TRANSISTIONS OF THE/ CHANGE OF RESONANCE ABSORPTION SPECTRA OF 23.8 KEV GAMMA-RAYS OF T 08-1400
/ONS IN NITRATES AND NITRITES. PART-1: CHANGE IN ULTRAVIOLET ABSORPTION SPECTRA ON MELTING. PART-2: CHANGE/ 16-2278
DINIUM ALUMINUM SULFATE HEXAHYDRAT/ LOW TEMPERATURE INFRARED ABSORPTION SPECTRA (5200-4200 PER CM) OF GUANI 21-2811
ENT METAL NIOBATES.   INFRARED ABSORPTION SPECTROPHOTOMETRIC STUDY OF MONOVAL   02-0113
OUREA IN THE PARA- AND FERROELECTRIC PHASES.   INFRARED ABSORPTION SPECTRUM OF A SINGLE CRYSTAL OF THE   25-2910
ND FERROELECTRIC PHASES.   INFRARED ABSORPTION SPECTRUM OF THIOUREA IN THE PARA- A 25-2908
ASSIUM NITRATE A/ SOME (DIELECTRIC MEASUREMENTS AND INFRARED ABSORPTION) STUDIES ON FERROELECTRICITY IN POT 16-2402
N TEMPERATURES IN ALKALI NITRITES (SODIU/ VACUUM ULTRAVIOLET ABSORPTION (5-23 EV AT ROOM AND LIQUID NITROGE 16-2393
FERROELECTRIC IRON(2+) BORACITES. (CHLORIDE, BROMID/ OPTICAL ABSORPTION   (625-25 MICRONS, 77-650 DEG K) OF 14-2097
VSKITES.   CRYSTAL CHEMISTRY OF THE ABX(3) FAMILY WITH PARTICULAR EMPHASIS ON PERO   08-1403
AC SEE ALTERNATING CURRENT
YCRYSTALLINE FERROELECTRICS AT / ZERO SHIFT IN PIEZOELECTRIC ACCELEROMETERS. (POLARIZATION SWITCHING IN POL 28-3216
CERAMIC MEMORY PLATE) A FERROELECTRIC PIEZOELECTRIC RANDOM ACCESS MEMORY. (LEAD ZIRCONATE- LEAD TITANATE FE 09-1504
RRITES AND FERROELECTRI/ NATURE OF DISSIPATIVE EFFECTS WHICH ACCOMPANY DISACCOMODATION TYPE PHENOMENA IN FE 03-0447
NS MODEL FOR DIFFUSE FERROELECTRIC PHASE TRANSITIONS.   ACCOUNTING FOR INTERACTION IN THE KANZIG REGIO 03-0438
Y SULFOIODIDE.   INJECTION CURRENTS AND ACCUMULATION OF POLARIZATION CHARGE IN ANTIMON   15-2188
UTION IN THE NITRITE RADICAL IN SODIUM NITRITE.   ACCURATE FORCE CONSTANTS FROM ISOTOPIC SUBSTIT   16-2303
IES OF FERROELECTRICS AT MICROWAVE FREQUENCI/ PROBLEM OF THE ACCURATE MEASUREMENT OF THE DIELECTRIC PROPERT 03-0429
DIHYDROGEN PHOSPHATE, A/ OPTICAL NONLINEAR SUSCEPTIBILITIES: ACCURATE RELATIVE VALUES FOR QUARTZ, AMMONIUM 17-2539
IN CONNECTION WITH THEIR PHASE TRANSITIONS. PART-1: TGS. PA/ ACCURATE STRUCTURE ANALYSIS OF FERROELECTRICS 02-0077
ER TRANSITION IN FERROELECTRICS. (TRIGLYCINE SULFATE, HIGHLY ACCURATE TEMPERATURE CONTROL)   ORDER DISORD   27-2982
YDROGEN AMMONIUM MONOCHLORO ACETATE, DICALCIUM STRONTIUM HEXAACETATE. /IDE, THIOUREA, TRIGLYCINE SULFATE, H 02-0099
HE PHASE TRANSITION OF FERROELECTRIC DICALCIUM STRONTIUM HEXAACETATE.   X-RAY AND OPTICAL STUDIES ON T 36-3544
/ FERROELECTRICITY AND PHASE TRANSITION IN AMMONIUM DICHLORO ACETATE AND DEUTERATED AMMONIUM DICHLORO ACETA 36-3538
, THIOUREA, TRIGLYCINE SULFATE, HYDROGEN AMMONIUM MONOCHLORO ACETATE, DICALCIUM STRONTIUM HEXAACETATE. /IDE 02-0099
ASSIUM DIHYDROGEN PHOSPHATE CRYSTALS IN SOLUTIONS CONTAINING ACETATE IONS.   GROWTH AND ETCHING OF POT 17-2567
IVALENT METAL DICALCIUM PROPIONATES SUBSTITUTED PARTIALLY BY ACETATE IONS.   PHASE TRANSITION IN D   26-2944
/ RESONANCE STUDY OF PHASE TRANSITION IN AMMONIUM MONOCHLORO ACETATE, MONOCHLORO ACETIC ACID MIXED CRYSTAL. 36-3625
/TRUCTURE AND PHASE TRANSITION OF AMMONIUM HYDROGEN DICHLORO ACETATE. PART-1: THE CRYSTAL STRUCTURE OF THE/ 36-3567
PHASE TRANSITION IN AMMONIUM MONOCHLORO ACETATE, MONOCHLORO ACETIC ACID MIXED CRYSTAL. /RESONANCE STUDY OF 36-3625
STABLE FERROELECTRICS. (BARIUM TITANATE, STRONT/ METHOD FOR ACHIEVING HIGH DIELECTRIC CONSTANT TEMPERATURE 04-0551
NIUM DIHYDROGEN PHOSPHATE, LITHIUM NIOBATE, AND ALPHA- IODIC ACID. /OF POTASSIUM DIHYDROGEN PHOSPHATE, AMMO 11-1677
AQUEOUS SOLUTIONS (APPARATUS, GROWTH OF ADP, KDP, HYDROIODIC ACID, AND AMMONIUM OXALATE MONOHYDRATE). /ROM 17-2420
METHOD.   GROWTH OF IODIC ACID AND IODATE CRYSTALS BY A SOLUTION GROWTH   08-1392
ION.   REFINEMENT OF THE CRYSTAL STRUCTURE OF FERROELECTRIC ACID LITHIUM SELENITE, POSITION OF THE LITHIUM 22-2859
TRANSITION IN AMMONIUM MONOCHLORO ACETATE, MONOCLORO ACETIC ACID MIXED CRYSTAL. /RESONANCE STUDY OF PHASE   36-3625
VIBRATIONAL SPECTRA OF IODIC ACID SINGLE CRYSTALS.   36-3548
CAL REACTIONS IN SILICA GELS. PART-3: FORMATION OF POTASSIUM ACID TARTRATE CRYSTALS.   SOME CHEMI   29-3246
UORINE-19 NUCLEI IN ANTIFERROMAGNETIC POTASSIUM MANGANESE T/ ACOUSTIC AND MAGNETIC EFFECTS INVOLVING THE FL 33-3390
OELECTRIC AND FERROELECTRIC SEMICONDUCTORS.   ACOUSTIC ATTENUATION AND AMPLIFICATION IN PIEZ   03-0275
SSOCIATED NONLINEARITY PARAMETERS IN LITHIUM NIOBATE.   ACOUSTIC CONVOLUTION AND CORRELATION AND THE A   11-1758
SITIONS IN PLZT CERAMICS.   ACOUSTIC DETECTION OF FERROELECTRIC PHASE TRAN   09-1545

/ CERAMICS WITH HIGH DIELECTRIC CONSTANT. PART-1: EFFECTS OF ADDITION OF GLASSES CONTAINING BARIUM OXIDE A/ 04-0829
RGE EFFECT IN LEAD ZIRCONATE TITANATE CERAMICS CAUSED BY THE ADDITION OF IMPURITIES.   SPACE CHA 09-1609
STALS OF FERROELECTRIC LITHIUM NIOBATE. (CRACK-FREE, THROUGH ADDITION OF MAGNESIUM OXIDE TO MELT) /MAIN CRY 11-1774
/ LEAD ZIRCONATE(0.52) TITANATE(0.48) CERAMICS CAUSED BY THE ADDITION OF NIOBIUM PENTOXIDE (BELOW 0.2 PERC/ 09-1606
APPLICATION TO QUAR/ WIDE RANGE TEMPERATURE COMPENSATION BY ADDITION OF TWO CRYSTAL RESONATOR FREQUENCIES: 11-1686
/ETWEEN LEAD MONOXIDE AND PZT. PART-4: EFFECT OF BORON OXIDE ADDITION ON THE PRODUCT OF PZT. PART-5: EFFEC/ 09-1624
IC PROPERTIES OF SINGLE CRYSTAL BARIUM TITANATE WITH NIOBIUM ADDITIONS.   OPTICAL AND DIELECTR 04-0875
STRUCTURE AND SINTERABILITY.   EFFECT OF CLAY AND DOLOMITE ADDITIONS ON BARIUM TITANATE CERAMICS. CRYSTAL 04-0739
DOPED BARIUM TITANATE.   EFFECT OF SILICA ADDITIONS ON THE SEMICONDUCTING PROPERTIES OF 04-0660
COMPACT   EFFECT OF OXIDE ADDITIONS ON THE SINTERING OF BARIUM TITANATE 04-0996
NG LEAD ZIRCONATE TITANATES CONTAINING BISMUTH AND STRONTIUM ADDITIVES. CHROMATOGRAPHIC METHOD FOR ANALYZI 09-1625
TIES OF BARIUM TITANATE CONTAINING NIOBIUM ANC THE EFFECT OF ADDITIVES.   DIELECTRIC PROPER 04-0575
BARIUM TITANATE WITH NIOBIUM PENTOXIDE AND MANGANESE DIOXIDE ADDITIVES. /ICIENT OF RESISTANCE) BEHAVIOR OF 04-0837
ON OF FERROELECTRIC SWITCHING IN TRIGLYCINE SULFATE. PART-1: ADDITIVES.   STRUCTURAL INHIBITI 27-3038
PIEZOCERAMICS BASED ON LEAD ZIRCONATE TITANATE WITH COMPLEX ADDITIVES CONTAINING GERMANIUM.   09-1510
INTERED AN PRESSURE SINTERED BARIUM TITANATE.   INFLUENCE OF ADDITIVES ON DENSIFICATION AND PROPERTIES OF S 04-1026
IES OF LEAD TITANATE CERAMICS. (DIELECTRIC CONST/ EFFECTS OF ADDITIVES ON PIEZOELECTRIC AND RELATED PROPERT 05-1084
OF LEAD ZIRCONA/ EFFECT OF BISMUTH AND MANGANESE CONTAINING ADDITIVES ON THE PROPERTIES OF SOLID SOLUTIONS 09-1588
S.   EFFECTS OF ADDITIVES ON THE PYROELECTRIC PROPERTIES OF TG 27-3037
GUIDE FILLED WITH HIGH PERMITTIVITY DIELECTRIC. (THEO/ INPUT ADMITTANCE AND FIELDS RADIATED BY AN OPEN WAVE 03-0359
BARIUM TITANATE SINGLE CRYSTALS. (LATTICE PA/ EFFECT OF SOME ADMIXED NICKEL ON SOME PHYSICAL PROPERTIES OF 04-0909
ATE OF THE MANGANESE(2+) ION IN A CUBIC ANTIFERRODIELECTRIC/ ADMIXTURE OF THE 4T1G STATE TO THE 6A1G(6S) ST 03-0317
ELECTRON PARAMAGNETIC RESONANCE OF CHROMIUM(3+) IN ADP.   17-2596
INCOHERENT NEUTRON SCATTERING FROM HYDROGEN BOND IN KDP AND ADP.   17-2416
HOLOGY) AND MECHANICAL PROPERTIES OF FILAMENTARY CRYSTALS OF ADP.   PREPARATION (MORP 17-2630
STUDY CF THE BRILLOUIN EFFECT IN ADP (ALL ELASTIC CONSTANTS).   17-2522
Y INFRARED ABSORPTION.   ANTIFERROELECTRIC TRANSITIONS IN ADP AND AMMONIUM DIHYDROGEN ARSENATE STUDIED B 17-2706
RROELECTRIC PHASE TRANSITION.   THERMAL EXPANSION OF ADP AND DEUTERATED ADP IN THE REGION OF ANTIFE 17-2445
PARATUS FOR POLISHING WATER SOLUBLE CRYSTALS (APPLICATION TO ADP AND KDP).   AN AP 17-2632
SOLUBILITY IN THE GRCWTH OF ELECTROOPTIC CRYSTALS ADP AND KDP.   17-2446
PARAMETRIC SCATTERIN/ MEASUREMENT OF THE REFRACTIVE INDEX OF ADP AND KDP CRYSTALS IN THE INFRARED RANGE BY 17-2553
FOURIER TRANSF/ DETERMINATION OF THE COMPLEX PERMITTIVITY OF ADP AND KDP IN THE FAR INFRARED BY DISPERSIVE 17-2609
TE, LITHIUM TANTALATE, LITHIUM NICBATE, LITHIUM IODATE, KDP, ADP, AND OTHER NONFERROELECTRIC MATERIALS) /NA 02-0107
ECTRUM.   EFFICIENT HIGH GRAIN PARAMETRIC GENERATION IN ADP CONTINUOUSLY TUNABLE ACROSS THE VISIBLE SP 17-2712
ON ON PHASE TRANSITIONS AND DIELECTRIC PROPERTIES OF KDP AND ADP CRYSTALS. / BY GAMMA AND REACTOR IRRADIATI 17-2614
EFFECTS OF DEUTERATION ON THE SPECIFIC HEAT OF ADP CRYSTALS.   17-2414
EPR OF CHROMIUM(3+) IONS IN ANTIFERROELECTRIC ADP CRYSTALS.   17-2546
LIGHT MODULATORS USING 45 DEGREE X CUT AND 45 DEGREE Y CUT ADP CRYSTALS.   17-2605
PHASES.   RAMAN SPECTRA OF ADP IN THE PARAELECTRIC AND ANTIFERROELECTRIC 17-2431
/TICAL CRYSTALS FROM AQUEOUS SOLUTIONS (APPARATUS, GROWTH OF ADP, KDP, HYDROIODIC ,ACID, AND AMMONIUM OXALA/ 17-2420
PULSE RESPONSE OF ZERO DEGREE CUT ADP MODULATORS.   17-2635
DEPENDENCE OF THE ELASTIC COEFFICIENTS OF ADP ON THE ELECTRIC FIELD.   17-2537
ADP POLARIZER FOR LASERS.   17-2551
/RADIATION IN ACTIVE MATERIALS AND NONLINEAR CRYSTALS. (KDP, ADP, RUBIDIUM DIHYDROGEN PHOSPHATE, LITHIUM N/ 11-1865
ACTION TOPOGRAPHY STUDY OF DEFECTS (DISLOCATIONS) IN KDP AND ADP SINGLE CRYSTALS.   X-RAY DIFFR 17-2578
ELECTRICAL SUSCEPTIBILITY OF ADP (THEORY).   03-0436
ION OF COMPOSITION PLANES IN ANTIFERROELECTRICS. (APPLIED TO ADP, TUNGSTEN TRIOXIDE, SODIUM NIOBATE) /ENTAT 08-1327
EFFICIENT SECOND HARMONIC GENERATION IN ADP WITH TWO NEW FLUORESCEIN DYE LASERS.   17-2538
FAR INFRARED SPECTRA OF KDP AND ADP. (20-550 PER CM, 200 AND 300 DEG K)   17-2549
ESR STUDY OF ADP- KDA MIXED CRYSTALS.   17-2683
MONY SULFOIODIDE SINGLE CRYSTALS.   EFFECT OF THE ADSORPTION OF OXYGEN ON THE PROPERTIES OF ANTI 15-2154
ON IN ANTIMONY SULFOIODIDE SINGLE CRYSTALS. EFFECT OF OXYGEN ADSORPTION ON THE FERROELECTRIC PHASE TRANSITI 15-2187
ERAMICS.   THE ROLE OF IRON OXIDE ADULTERANT IN A LEAD ZIRCONATE LEAD TITANATE C 09-1592
FAST AFTEREFFECT OF SEIGNETTE SALT.   28-3206
THE DOMAIN STRUCTURE OF A TRIGLYCINE SULFATE CRYSTAL DURING AGING.   INVESTIGATION OF 28-2561
A DC FIELD ON DOMAIN STRUCTURE IN SEIGNETTE SALT. (EFFECT OF AGING AND COPPER(2+) DOPING)   EFFECT OF 28-3230
/FIELD ON THE DIELECTRIC PROPERTIES OF TRIGLYCINE SELENATE. (AGING AND PHASE CHANGES NEAR THE CURIE POINT) 27-3131
TANATE CERAMICS WITH INDIUM ELECTRODES.   IMPROVED AGING AND SWITCHING OF LEAD ZIRCONATE- LEAD TI 09-1516
TANATE CERAMICS WITH INDIUM ELECTRCDES.   IMPROVED AGING AND SWITCHING OF LEAD ZIRCONATE- LEAD TI 09-1517
NATE.   DEMONSTRATION OF AGING EFFECTS IN BARIUM TITANATE- BARIUM ZIRCO 09-1543
NATE. (CAPACITANCE, PERMITTIVITY)   AGING IN COMPLEX CERAMICS BASED ON BARIUM TITA 04-0724
ATE, LEAC ZIRCONATE TITANATE)   NATURE OF AGING IN FEFROELECTRIC CERAMICS. (BARIUM TITAN 04-0757
ATE.   AGING IN TETRAGONAL FERROELECTRIC BARIUM TITAN 04-0579
TITANATE IN BARIUM TITANATE.   DIELECTRIC AGING IN TETRAGONAL SOLID SOLUTIONS OF CALCIUM 04-1295
TANATE FERROELECTRIC CERAMICS. (REVIEW, 31 REFERENCES)   AGING OF BARIUM TITANATE AND LEAD ZIRCONATE TI 02-0062
SPACE CHARGE POLARIZATION AND AGING OF BARIUM TITANATE CERAMICS.   04-0880
ONY SULFOIODIDE IN A STATIC ELECTRIC FIELD.   AGING OF THE FERROELECTRIC SEMICONDUCTOR ANTIM 15-2113
STALS.   STUDY OF THE AGING PHENOMENON IN BARIUM TITANATE SINGLE CRY 04-0545
IS LAPS, DOMAIN STRUCTURE)   AGING PROCESS IN TRIGLYCINE SULFATE. (HYSTERES 27-3128
N NUMBER OF 180 DEG DOMAINS).   AGING PROCESSES IN BARIUM TITANATE (DECREASE I 04-0911
OR CAPACITANCE PROPERTIES DOMAIN STRUCTURE)   AGING PROCESSES IN TRIGLYCINE SULFATE. (TANDEL 27-3081
COND HARMONIC GENERATED LIGHT (FROM LITHIUM NIOBATE CRYSTAL, AGREEMENT WITH THEORY). /ON CORRELATIONS OF SE 11-1701
REPARATION OF TUNGSTIC OXIDE SINGLE CRYSTALS (SUBLIMATION IN AIR OF TUNGSTEN TRIOXIDE).   P 10-1644
HIGH INTERNAL BIAS FIELDS IN L-ALANINE SUBSTITUTED TGS.   27-2977
BLE POTASSIUM NITRATE(III) FROM SOLUTION. (GROWN IN WATER OR ALCOHOL)   METASTA 16-2329
D OTHER CERAMIC POWDERS BY COPRECIPITATION OF CITRATES IN AN ALCOHOL.   PREPARATION OF BARIUM TITANATE AN 04-0861
NTHANUM OXIDE (P/ TEMPERATURE DEPENDENT FERROELECTRIC DOMAIN ALIGNMENT IN LEAD ZIRCONATE- LEAD TITANATE- LA 09-1586
RIE TEMPERATURE, POLARIZATION FLUCTUATIONS, LITHIUM NIOBATE, ALKALI HALIDES) /RYSTALS. (STRAIN DEPENDENT CU 03-0513
INERTIAL ELECTRICAL POLARIZATION OF CRYSTALS (ALKALI HALIDES, STRONTIUM TITANATE).   06-1197
LE RESONANCE, PHASE TRANSFORMATIONS, AND FERROELECTRICITY OF ALKALI IODATES.   NUCLEAR QUADRUPO 08-1350
DIPOLE CORRELATIONS IN MOLTEN ALKALI METAL NITRATES AND SODIUM NITRITE.   16-2389
)   CHEMISTRY AND PHYSICS OF (ALKALI) METAL TANTALATES. (PREPARATION, REVIEW 02-0048
VACUUM ULTRAVIOLET ABSORPTION OF THE ALKALI NITRITES AND NITRATES.   16-2398
/RPTION (5-23 EV AT ROOM AND LIQUID NITROGEN TEMPERATURES IN ALKALI NITRITES (SODIUM, POTASSIUM) AND ALKAL/ 16-2399
EFFECT OF HIGH PRESSURE ON PHASE TRANSITIONS OF ALKALI TRIHYDROGEN SELENITE CRYSTALS. 22-2882
ELECTRONIC HALL MOBILITY IN THE ALKALINE EARTH FLUORIDES.   33-3404
TITUTION.   LOW FREQUENCY DIELECTRIC CONSTANTS OF THE ALKALINE EARTH FLUORIDES BY THE METHOD OF SUBS 33-3361
I/ (FERROELECTRIC AND OPTICAL PROPERTIES OF) CRYSTALS OF THE ALKALINE TRIHYDROGEN DISELENITE FAMILY. (ESPEC 22-2848
COLD PRESSING AND LOW TEMPERATURE SINTERING OF ALKOXY-DERIVED PLZT AND PLZT.   09-1493
UM TITANATE (OF HIGH PU/ PREPARATION AND CHARACTERIZATION OF ALKOXY-DERIVED STRONTIUM ZIRCONATE AND STRONTI 06-1217

1970, VORONEZH, USSR, PROC, PARTS 1-3.   ALL-UNION CONFERENCE ON FERROELECTRICITY, 7TH, 01-0001
TIVITY OF BISMUTH SELENOIODIDE- ANTIMONY SELENOIODIDE SYSTEM ALLOYS.   PHASE DIAGRAM AND ELECTRICAL CONDUC 15-2122
TRANSITION IN TIN TELLURIDE- GERMANIUM TELLURIDE ALLOYS. (NEUTRON SCATTERING)   36-3586
THERMODYNAMICS, THEORY, MAGNETISM, FERROELECTRICITY, ORDERED ALLOYS, SUPERFLUIDITY, SUPERCONDUCTIVITY) /. (03-0183
/THE FERROELECTRIC PHASE TRANSITIONS (TIN TELLURIDE PURE AND ALLOYED WITH SMALL AMOUNTS OF GERMANIUM TELLU/ 36-3564
ITHIUM NIOBATE (CRYSTAL GROWTH BY ADD/ SPATIALLY UNIFORM AND ALTERABLE SHG PHASE MATCHING TEMPERATURES IN L 11-1684
ILIZATION EFFECT OF TRIGLYCINE SULFATE SINGLE CRYSTALS IN AN ALTERNATING CURRENT ELECTRIC FIELD. / AUTOSTAB 27-3005
C HYSTERESIS NEAR THE TRANSITION POINT OF DIFFUSE FERROELEC/ ALTERNATING CURRENT POLARIZATION AND DIELECTRI 03-0437
/PERTIES OF TRIGLYCINE SELENATE. (DIELECTRIC CONSTANT VERSUS ALTERNATING FIELD VOLTAGE, TEMPERATURE AND PO/ 27-3158
HEAT CAPACITY OF METHYL AMMONIUM CHROMIUM ALUM.   20-2801
THERMAL CONDUCTIVITY OF METHYL AMMONIUM ALUMINUM ALUM.   20-2807
SPECIFIC HEAT AND SUSCEPTIBILITY IN CHROMIUM METHYL AMMONIUM ALUM ABOVE THE CRITICAL TEMPERATURE.   20-2805
/ARIATION OF THE MEAN MAGNETIC MOMENT OF CHROMIUM- POTASSIUM ALUM (UNSTABLE SOFT PHONON MODE, DISPLACIVE F/ 20-2806
ODIUM TARTRATE, TGS, AMMONIUM FLUOBERYLLATE, SODIUM NITRITE, ALUMS) /CS. (BARIUM TITANATE, KDP, POTASSIUM S 02-0134
ATTICE RELAXATION) INVESTIGATIONS OF THE PHASE TRANSITION IN ALUMS. /ESONANCE (WIDE LINE SPECTRA AND SPIN L 20-2802
SPECTROSCOPIC STUDIES OF FERROELECTRIC METHYL AMMONIUM ALUMS.   20-2808
LFATE)   PHYSICAL PROPERITES OF FERROELECTRIC ALUMS. (ESPECIALLY METHYL AMMONIUM ALUMINUM SU 20-2810
IUM ALUM/ INVESTIGATION OF PHASE TRANSITION IN FERROELECTRIC ALUMS (METHYL AMMONIUM ALUMINUM SULFATE, AMMON 20-2809
/E. (COPPER FORMATE TETRAHYDRATE, AMMONIUM IRON AND CHROMIUM ALUMS, METHYL AMMONIUM ALUMINUM SULFATE DODEC/ 02-0087
LENATE) DODECAH/ INFRARED AND RAMAN SPECTRA OF FERROELECTRIC ALUMS: METHYL AMMONIUM ALUMINUM (SULFATE OR SE 20-2803
DIELECTRIC PROPERTIES OF TITANATE- ALUMINA CERAMICS.   08-1414
/NTAINING BARIUM OXIDE AND TITANIUM DIOXIDE (WITH SILICA AND ALUMINA) ON PROPERTIES OF BARIUM TITANATE CER/ 04-0829
IUM OR LEAD TITANATE, FROM SYSTEMS BARIA OR LITHIA, TITANIA, ALUMINA, SILICA). /FERROELECTRIC CRYSTALS (BAR 04-0783
NEW COMPOSITE CERAMIC PIEZOELECTRIC TRANSDUCER MATERIAL. (ALUMINA WHISKERS IN LEAD ZIRCONATE TITANATE) 09-1556
ANCE SPECTRA OF CHROMIUM (3+) AND GADOLINIUM (3+) IN LANTHANUM ALUMINATE.   ELECTRON SPIN RESON 36-3583
NMR STUDY OF THE STRUCTURAL PHASE TRANSITION IN LANTHANUM ALUMINATE.   36-3533
OPTICAL SPECTRA OF CHROMIUM (3+) PAIRS IN LANTHANUM ALUMINATE.   36-3616
TRANSITION ELEMENT IONS IN STRONTIUM TITANATE AND LANTHANUM ALUMINATE. /ESONANCE AND OPTICAL ABSORPTION OF 06-1165
TIONLESS TRANSITIONS IN THE EUROPIUM (3+) CENTER IN LANTHANUM ALUMINATE.   RADIA 36-3531
IN PEROVSKITE TYPE CRYSTALS. (STRONTIUM TITANATE, LANTHANUM ALUMINATE) /RY OF STRUCTURAL PHASE TRANSITIONS 03-0424
+) IONS AND EXCHANGE COUPLED CHROMIUM (3+) PAIRS IN LANTHANUM ALUMINATE. /UM AND ENERGY LEVELS OF CHROMIUM (3 36-3565
ACTION WITH ELASTIC STRAIN (STRONTIUM TITANATE AND LANTHANUM ALUMINATE). /ITE- TYPE CRYSTALS, PART-2, INTER 03-0241
FLUX GROWTH OF MAGNESIUM OXIDE AND LANTHANUM ALUMINATE CRYSTALS DOPED WITH ISOTOPE 170.   36-3558
NIOBATE, OR BISMUTHATE, OR BY BISMUTH OR LANTHANUM FERRATE, ALUMINATE OR CHROMATE. / POTASSIUM ANTIMONATE, 09-1539
/ANT (5-300K) OF THE PRASEODYMIUM (1-X) NEODYMIUM (X) (ORTHO) ALUMINATE SYSTEM (NOT FERROELECTRIC BY HYSTER/ 36-3600
TROLYTE FUEL CELL. ((CALCIUM, STRONTIUM) LANTHANUM TITANATE- ALUMINATES) /ITS APPLICATION TO THE SOLID ELEC 08-1444
NATURE OF THE HIGH PERMITTIVITY OF CERIUM ALUMINATES WITH PEROVSKITE TYPE STRUCTURE.   08-1355
FOIODIDE SINGLE CRYSTALS AND SOME METALS. (ANTIMONY, SILVER, ALUMINUM) /F THE CONTACTS BETWEEN ANTIMONY SUL 15-2251
NGANESE (1-X) (B(X)) OXYGEN (3) (B=TRIVALENT IRON, CHROMIUM OR ALUMINUM). /IGATIONS OF THE SYSTEMS YTTRIUM MA 36-3569
THERMAL CONDUCTIVITY OF METHYL AMMONIUM ALUMINUM ALUM.   20-2807
IELD ENERGY LEVEL SPLITTINGS OF TITANIUM (3+) ION IN RUBIDIUM ALUMINUM DISULFATE DODECAHYDRATE. /F CRYSTAL F 19-2754
PIEZOELECTRIC EFFECT IN ALUMINUM DOPED BARIUM TITANATE CERAMIC.   04-0536
LLIMICRON HARMONIC SOURCE.   NEODYMIUM: YTTRIUM ALUMINUM GARNET- BARIUM SODIUM NIOBATE 0.66 MI 13-1975
SOLUBILITY OF CHROMIUM (3+) IN LEAD ZIRCONATE TITANATE.   09-1479
TASSIUM DIHYDROGEN PH/ EFFECT OF REPLACEMENT OF POTASSIUM BY ALUMINUM ON THE FERROELECTRIC PROPERTIES OF PO 17-2560
/TURE DEPENDENCES OF THE ABSORPTION OF 9.4 GHZ HYPERSOUND IN ALUMINUM OXIDE AND LITHIUM NIOBATE SINGLE CRY/ 11-1719
ATE.   PHONON PHONON AND PROTON PHONON INTERACTIONS IN ALUMINUM PHOSPHATE AND CESIUM DIHYDROGEN ARSEN 18-2739
OPERITES OF FERROELECTRIC ALUMS. (ESPECIALLY METHYL AMMONIUM ALUMINUM SULFATE)   PHYSICAL PR 20-2810
/OF PHASE TRANSITION IN FERROELECTRIC ALUMS (METHYL AMMONIUM ALUMINUM SULFATE, AMMONIUM ALUMINUM SULFATE A/ 20-2809
/AHYDRATE, AMMONIUM IRON AND CHROMIUM ALUMS, METHYL AMMONIUM ALUMINUM SULFATE DODECAHYDRATE, COLEMANITE, S/ 02-0087
/IELECTRIC RELAXATION IN FERROELECTRIC MASD, METHYL AMMONIUM ALUMINUM SULFATE DODECAHYDRATE. (COMPLEX DIEL/ 20-2804
RAMAGNETIC RESONANCE STUDY OF SOME METAL IONS IN GUANIDINIUM ALUMINUM SULFATE HEXAHYDRATE.   ELECTRON PA 21-2817
OF MAGNETIC COPPER (2+) COMPLEXES IN CRYSTALS OF GUANIDINIUM ALUMINUM SULFATE HEXAHYDRATE.   EPR STUDY 21-2820
EORY. (AMMONIUM SULFATE, AMMONIUM FLUOBERYLLATE, GUANIDINIUM ALUMINUM SULFATE HEXAHYDRATE) /OMENOLOGICAL TH 02-0040
SEE BOTH GASH AND GUANIDINIUM ALUMINUM SULFATE HEXAHYDRATE.
MA-RAY IRRADIATED FERROELECTRICS POCHELLE SALT AND GUANIDINE ALUMINUM SULFATE HEXAHYDRATE. /DIES OF THE GAM 21-2812
/FRARED ABSORPTION SPECTRA (5200-4200 PER CM) OF GUANIDINIUM ALUMINUM SULFATE HEXAHYDRATE AND FREQUENCIES / 21-2811
/RMINATION OF STRUCTURE OF FERROELECTRIC CRYSTAL GUANIDINIUM ALUMINUM SULFATE HEXAHYDRATE AND ITS CHROMIUM/ 21-2819
REVERSAL).   FERROELECTRIC MOMENT IN GUANIDINIUM ALUMINUM SULFATE HEXAHYDRATE (AND MECHANISM OF 21-2818
/AL AND INTERFEROMETRIC STUDIES ON FERROELECTRIC GUANIDINIUM ALUMINUM SULFATE HEXAHYDRATE CLEAVAGES AND TH/ 21-2816
RSU/ DEFECTS AND THE FERROELECTRIC PROPERTIES OF GUANIDINIUM ALUMINUM SULFATE HEXAHYDRATE. (POLARIZATION VE 21-2813
/D AND RAMAN SPECTRA OF FERROELECTRIC ALUMS: METHYL AMMONIUM ALUMINUM (SULFATE OR SELENATE) DODECAHYDRATE./ 20-2803
/CTRA OF MAGNETIZED PARAMAGNETIC SUBSTANCES. PART-1: IRON (X) ALUMINUM (1-X) AMMONIUM SULFATE DODECAHYDRATE. 19-2751
STRUCTURE OF AMMONIUM FLUOBERYLLATE AT AMBIENT TEMPERATURE.   19-2776
TERNATIONAL MEETING ON FERROELECTRICITY. A EUROPEAN VIEW.   AN AMERICAN VIEW.   REPORT ON THE SECOND IN 01-0008
/RADIATION INTO MOTION. (PYROELECTRIC EFFECT IN TGS, ETHYL DIAMINE TARTRATE, BARIUM TITANATE, SALICYLIDENE/ 04-0854
(METALS CHOSEN FROM SODIUM, POTASSIUM, RUBIDIUM, CESIUM, OR AMMONIUM, ABSTRACT ONLY). /TE OR FLUOBERYLLATE 19-2743
SPECIFIC HEAT AND SUSCEPTIBILITY IN CHROMIUM METHYL AMMONIUM ALUM ABOVE THE CRITICAL TEMPERATURE. 20-2805
SPECTROSCOPIC STUDIES OF FERROELECTRIC METHYL AMMONIUM ALUMS.   20-2808
THERMAL CONDUCTIVITY OF METHYL AMMONIUM ALUMINUM ALUM.   20-2807
YSICAL PROPERITES OF FERROELECTRIC ALUMS. (ESPECIALLY METHYL AMMONIUM ALUMINUM SULFATE)   PH 20-2810
/TIGATION OF PHASE TRANSITION IN FERROELECTRIC ALUMS (METHYL AMMONIUM ALUMINUM SULFATE, AMMONIUM ALUMINUM / 20-2809
LEX DIE/ DIELECTRIC RELAXATION IN FERROELECTRIC MASD, METHYL AMMONIUM ALUMINUM SULFATE DODECAHYDRATE. (COMP 20-2804
H/ INFRARED AND RAMAN SPECTRA OF FERROELECTRIC ALUMS: METHYL AMMONIUM ALUMINUM (SULFATE OR SELENATE) DODECA 20-2803
TYPE CRYSTALS. (POTASSIUM DIHYDROGEN PHOSPHATE, RUBIDIUM AND AMMONIUM ANALOGS SOME DEUTERATE) /OF SOME KDP 17-2679
RATURES).   EFFECT OF DEUTERATION IN AMMONIUM BISULFATE (CHANGE IN TRANSITION TEMPE 19-2760
/PERIMENTAL INVESTIGATION OF THE FERROELECTRIC PROPERTIES OF AMMONIUM BISULFATE IN THE VICINITY OF THE HIG/ 19-2787
ELECTRON PARAMAGNETIC RESONANCE OF MANGANESE (2+) DOPED AMMONIUM CADMIUM SULFATE.   19-2791
ERATURE DEPENDENCE OF THE ELECTRON PARAMAGNETIC RESONANCE OF AMMONIUM CADMIUM SULFATE.   TEMP 19-2792
HEAT CAPACITY OF METHYL AMMONIUM CHROMIUM ALUM.   20-2801
RMATION OF A PECULIAR FERROELECTRIC- FERROELASTIC CRYSTAL: DIAMMONIUM DICADMIUM SULFATE. /THE PHASE TRANSFO 19-2741
H / MECHANISM OF THE FERROELECTRIC PHASE TRANSFORMATION OF DIAMMONIUM DICADMIUM TRISULFATE, A MODEL IN WHIC 03-0141
THE BEHAVIOR OF DIAMMONIUM DICADMIUM TRISULFATE UNDER PRESSURE.   19-2757
IUM DICHLORO ACETA/ FERROELECTRICITY AND PHASE TRANSITION IN AMMONIUM DICHLORO ACETATE AND DEUTERATED AMMON 36-3538
SOFT MODES IN POTASSIUM DIDEUTERIUM PHOSPHATE AND DEUTERATED AMMONIUM DIDEUTERIUM PHOSPHATE. /ENVECTORS OF 17-2700
PR STUDIES OF X-IRRADIATED POTASSIUM DIHYDROGEN ARSENATE AND AMMONIUM DIHYDROGEN ARSENATE. /URATIONS FROM E 18-2728
PURE QUADRUPOLE RESONANCE OF ARSENIC-75 IN AMMONIUM DIHYDROGEN ARSENATE.   18-2732
SEE BOTH ADA AND AMMONIUM DIHYDROGEN ARSENATE.
/TE AND POTASSIUM DIDEUTERIUM ARSENATE AND ANTIFERROELECTRIC AMMONIUM DIHYDROGEN ARSENATE AND AMMONIUM DID/ 18-2734
/ RUBIDIUM DIDEUTERIUM ARSENATE, CESIUM DIHYDROGEN ARSENATE, AMMONIUM DIHYDROGEN ARSENATE, AND DEUTERO AMM/ 18-2725

IUM DIDEUTERIUM ARSENATE.   ARSENIC-75 NMR RESONANCE IN AMMONIUM DIHYDROGEN ARSENATE AND DEUTERO AMMON 18-2736
/SIUM DIHYDROGEN ARSENATE, POTASSIUM DEUTERIUM ARSENATE, AND AMMONIUM DIHYDROGEN ARSENATE. (FERROELECTRICS/ 18-2724
ED ABSORPTION.   ANTIFERROELECTRIC TRANSITIONS IN ADP AND AMMONIUM DIHYDROGEN ARSENATE STUDIED BY INFRAR 17-2706
/CTRICS. (DEUTERATED ROCHELLE SALT, POTASSIUM OR RUBIDIUM OR AMMONIUM DIHYDROGEN ARSENATES, RUBIDIUM DIDEU/ 18-2719
PER CM) AND/ LOW TEMPERATURE INFRARED ABSORPTION SPECTRA OF AMMONIUM DIHYDROGEN ORTHOPHOSPHATE (4000-10000 17-2432
              A THEORY OF THE PHASE TRANSITION IN AMMONIUM DIHYDROGEN PHOSPHATE.   03-0286
AMPLITUDE MODULATORS BASED ON THE MICHELSON INTERFEROMETER. (AMMONIUM DIHYDROGEN PHOSPHATE)   17-2715
              CRYSTAL PERFECTION OF AMMONIUM DIHYDROGEN PHOSPHATE.   17-2426
RYSTALS CONTAINING O-D...O BOND. (INCLUDING FULLY DEUTERATED AMMONIUM DIHYDROGEN PHOSPHATE) /TUDY OF SOME C 03-0204
   DIELECTRIC PERMITTIVITY OF FINELY DIVIDED POTASSIUM AND AMMONIUM DIHYDROGEN PHOSPHATE.   17-2689
              FAR INFRARED ABSORPTION SPECTRA FOR AMMONIUM DIHYDROGEN PHOSPHATE.   17-2587
NG WAVELENGTH POLARIZATION FLUCTUATIONS IN ANTIFERROELECTRIC AMMONIUM DIHYDROGEN PHOSPHATE.   LO 17-2461
SSBAUER STUDY OF ANTIFERROELECTRIC TRANSITION IN IRON-57(3X) AMMONIUM DIHYDROGEN PHOSPHATE.   MO 17-2644
   PHOTOELASTIC EFFECT IN 45 DEGREE X-CUTS OF POTASSIUM AND AMMONIUM DIHYDROGEN PHOSPHATE.   17-2443
MODES IN THE KDP ISOMORPHOUS PHOSPHATES AND ARSENATES. (KDP, AMMONIUM DIHYDROGEN PHOSPHATE) /FERROELECTRIC 17-2639
IUM DIDEUTERIUM PHOSPHATE, POTASSIUM DIHYDROGEN ARSENATE AND AMMONIUM DIHYDROGEN PHOSPHATE. /SPHATE, POTASS 17-2409
   X-RAY DETERMINATION OF THE THERMAL EXPANSION OF AMMONIUM DIHYDROGEN PHOSPHATE.   17-2485
RATURE.   POCKELS EFFECT IN AMMONIUM DIHYDROGEN PHOSPHATE ABOVE ROOM TEMPE 17-2518
DEUTERIUM PHOSPHATE.   LASER INDUCED SELF FOCUSING IN AMMONIUM DIHYDROGEN PHOSPHATE AND POTASSIUM DI 17-2476
HYDROGEN PHOSPHATE)/ RELAXATION OF FERROELECTRIC PHOSPHATES (AMMONIUM DIHYDROGEN PHOSPHATE AND POTASSIUM DI 17-2580
HYDROGEN ARSENATE TO 52 KBAR AND 400 DEG / PHASE DIAGRAMS OF AMMONIUM DIHYDROGEN PHOSPHATE AND POTASSIUM DI 17-2478
/NEAR SUSCEPTIBILITIES: ACCURATE RELATIVE VALUES FOR QUARTZ, AMMONIUM DIHYDROGEN PHOSPHATE, AND POTASSIUM / 17-2539
HYDROGEN PHOSPHATE EXCITED BY A / PARAMETRIC FLUORESCENCE IN AMMONIUM DIHYDROGEN PHOSPHATE AND POTASSIUM DI 17-2486
MAN ELASTICITY THEORY.   ELASTIC PROPERTIES OF AMMONIUM DIHYDROGEN PHOSPHATE AND THE LAVAL RA 17-2410
ULFATE CRYSTALS.   THERMAL VARIATION OF REFRACTIVE INDEX IN AMMONIUM DIHYDROGEN PHOSPHATE AND TRIGLYCINE S 17-2474
   MEASUREMENT OF THE PRIMARY LINEAR ELECTROOPTIC EFFECT IN AMMONIUM DIHYDROGEN PHOSPHATE CRYSTALS.   17-2654
/EZOOPTICAL PROPERTIES OF POTASSIUM DIHYDROGEN PHOSPHATE AND AMMONIUM DIHYDROGEN PHOSPHATE CRYSTALS AT HIG/ 17-2697
   RELATIVE PERMITTIVITY OF POTASSIUM DIHYDROGEN PHOSPHATE AND AMMONIUM DIHYDROGEN PHOSPHATE CRYSTALS AT MIC/ 17-2714
TION.   KINETICS OF GROWTH AND DISSOLUTION OF AMMONIUM DIHYDROGEN PHOSPHATE CRYSTALS IN SOLU 17-2412
   THERMAL SELF FOCUSING IN POTASSIUM DIHYDROGEN PHOSPHATE AND AMMONIUM DIHYDROGEN PHOSPHATE CRYSTALS OF LIG/ 17-2422
/ECTION AND ABSORPTION OF POTASSIUM DIHYDROGEN PHOSPHATE AND AMMONIUM DIHYDROGEN PHOSPHATE CRYSTALS. (REFL/ 17-2422
RCULATION APPARATUS.   GROWTH BEHAVIOR OF SMALL SEEDS OF AMMONIUM DIHYDROGEN PHOSPHATE IN A MODIFIED CI 17-2704
/E OPTICAL NONLINEARITIES OF POTASSIUM DIHYDROGEN PHOSPHATE, AMMONIUM DIHYDROGEN PHOSPHATE, LITHIUM NIOBAT/ 01-0021
ASSI/ LINEAR ELECTROOPTIC MODULATOR MATERIALS. (POTASSIUM OR AMMONIUM DIHYDROGEN PHOSPHATE OR ARSENATE, POT 01-0021
ON / MICROSCOPIC OBSERVATION OF ANTIFERROELECTRIC DOMAINS OF AMMONIUM DIHYDROGEN PHOSPHATE. (PHASE TRANSITI 17-2643
OGEN ARSENATE MIXE/ ESR STUDY OF RADIATION DAMAGE CENTERS IN AMMONIUM DIHYDROGEN PHOSPHATE POTASSIUM DIHYDR 17-2547
/MPERATURE INSTABILITY OF ELECTROOPTIC LASER MODULATORS WITH AMMONIUM DIHYDROGEN PHOSPHATE, POTASSIUM DIHY/ 17-2675
/ OF ROTATING CRYSTALS AND CHARACTERISTICS OF THEIR GROWTH. (AMMONIUM DIHYDROGEN PHOSPHATE, POTASSIUM FERR/ 16-2316
Y NMR.   INVESTIGATION OF THE PHASE TRANSITION IN AMMONIUM DIHYDROGEN PHOSPHATE SINGLE CRYSTAL B 17-2545
FROM SOLUTION) IN STATIC REGIME AND / GROWING KINETIC OF THE AMMONIUM DIHYDROGEN PHOSPHATE SINGLE CRYSTAL (17-2516
ROM SOLUTIONS.   GROWTH OF AMMONIUM DIHYDROGEN PHOSPHATE SINGLE CRYSTAL F 17-2411
ECTRICS WITH HYDROGEN BONDS.   DYNAMICAL THEORY OF AMMONIUM DIHYDROGEN PHOSPHATE TYPE ANTIFERROEL 03-0476
/TALS (CURIE TEMPERATURES OF POTASSIUM, RUBIDIUM, CESIUM AND AMMONIUM DIHYDROGEN PHOSPHATES AND CESIUM AND/ 17-2577
ROGEN PHOSPHATE- WATER SYSTEM.   SOLUBILITY ISOBAR IN THE AMMONIUM DIHYDROGEN PHOSPHATE- POTASSIUM DIHYD 17-2658
ROGEN PHOSPHATE- WATER SYSTE/ POLYTHERM OF SOLUBILITY OF THE AMMONIUM DIHYDROGEN PHOSPHATE- POTASSIUM DIHYD 17-2434
ROGEN PHOSP/ DIELECTRIC PROPERTIES OF SINGLE CRYSTALS IN THE AMMONIUM DIHYDROGEN PHOSPHATE- POTASSIUM DIHYD 17-2572
SITIONS. (BASED ON LANDAU PHENOMENOLOGICAL THEORY, TESTED ON AMMONIUM FLUOBERYLLATE, ABSTRACT ONLY) /E TRAN 19-2765
              STRUCTURE OF AMMONIUM FLUOBERYLLATE AT AMBIENT TEMPERATURE. 19-2776
   NUCLEAR MAGNETIC RESONANCE OF DEUTERIUM IN THE PERDEUTERO AMMONIUM FLUOBERYLLATE CRYSTAL.   19-2744
/ICS. (BARIUM TITANATE, KDP, POTASSIUM SODIUM TARTRATE, TGS, AMMONIUM FLUOBERYLLATE, SODIUM NITRITE, ALUMS) 02-0134
ONIUM (BX4) TYPE COMPOUNDS (PROTON MAGN/ FERROELECTRICS WITH AMMONIUM GROUP ORIENTATION MOBILITY: METAL AMM 22-2826
TIONS NOT FERROELECTRIC).   DIELECTRIC PROPERTIES OF DIAMMONIUM HEXABROMO STANNATE (DIELECTRIC TRANSI 36-3612
E CRYSTAL STR/ THE CRYSTAL STRUCTURE AND PHASE TRANSITION OF AMMONIUM HYDROGEN DICHLORO ACETATE. PART-1: TH 36-3567
DEUTERON SPIN LATTICE RELAXATION IN FERROELECTRIC DEUTERATED AMMONIUM HYDROGEN SULFATE.   19-2795
   ELECTRON SPIN RESONANCE OF GAMMA DEFECTS IN FERROELECTRIC AMMONIUM HYDROGEN SULFATE.   19-2748
              POLARIZATION REVERSAL CHARACTERISTICS OF AMMONIUM HYDROGEN SULFATE.   19-2788
/NSTRAINED REFINEMENT TECHNIQUES APPLIED TO THE STRUCTURE OF AMMONIUM HYDROGEN SULFATE ABOVE THE FERROELEC/ 16-2355
   X-RAY DIFFRACTION DETERMINATION OF THE CRYSTAL STRUCTURE OF AMMONIUM HYDROGEN SULFATE ABOVE THE FERROELEC/ 19-2775
UM / LOW TEMPERATURE INFRARED STUDIES. PART-8: FERROELECTRIC AMMONIUM HYDROGEN SULFATE AND AMMONIUM DEUTERI 19-2782
THE FERROELECTRIC PHASE.   ON THE PHASE TRANSITION OF AMMONIUM HYDROGEN SULFATE AND THE STRUCTURE OF 19-2777
ANGE FROM -70 TO +14 DEG C.   SPECIFIC HEAT OF AMMONIUM HYDROGEN SULFATE IN THE TEMPERATURE R 19-2789
HE FERROELECTRIC TRANSITION.   AMMONIUM HYDROGEN SULFATE, ITS STRUCTURE AND T 19-2773
STAL STRUCTURE.   PYROELECTRIC AMMONIUM IODATE, A POTENTIAL FERROELASTIC: CRY 36-3574
              DYNAMICS OF AMMONIUM IONS IN AMMONIUM SULFATE.   19-2794
UM PHOSPHATE- 5 PERCENT POTASSIUM DIDEUTERIUM / ESR STUDY OF AMMONIUM IONS IN IRRADIATED AMMONIUM DIDEUTERI 17-2684
/TON MAGNETIC RESONANCE STUDY OF THE SPIN SYMMETRY STATES OF AMMONIUM IONS IN SOLIDS. (AMMONIUM SULFATE, S/ 19-2800
/STALS OF ORDER DISORDER TYPE. (COPPER FORMATE TETRAHYDRATE, AMMONIUM IRON AND CHROMIUM ALUMS, METHYL AMMO/ 02-0087
   MOSSBAUER STUDY OF ELECTRONIC SPIN FLIP PROCESSES IN AMMONIUM IRON SULFATE DODECAHYDRATE.   19-2771
/NTIMONY SULFOIODIDE, THIOUREA, TRIGLYCINE SULFATE, HYDROGEN AMMONIUM MONOCHLORO ACETATE, DICALCIUM STRONT/ 02-0099
/ NUCLEAR QUADRUPOLE RESONANCE STUDY OF PHASE TRANSITION IN AMMONIUM MONOCHLORO ACETATE, MONOCHLORO ACETIC 36-3625
              DIELECTRIC MEASUREMENTS OF AMMONIUM NITRATE.   16-2338
              DIELECTRIC STUDIES OF PHASE 2 OF AMMONIUM NITRATE.   16-2339
IC TRANSFORMATIONS IN POTASSIUM NITRATE, SILVER NITRATE, AND AMMONIUM NITRATE SINGLE CRYSTALS. /T POLYMORPH 16-2260
ULFATE AT FERROELECTRIC / CHANGES IN THE INFRARED SPECTRA OF AMMONIUM OR RUBIDIUM BISULFATE, AND AMMONIUM S 19-2772
LUTIONS (APPARATUS, GROWTH OF ADP, KDP, HYDROIODIC ACID, AND AMMONIUM OXALATE MONOHYDRATE). /ROM AQUEOUS SO 17-2420
STATISTICAL CONSIDERATIONS OF THE ORDER DISORDER REACTION IN AMMONIUM PERIODATE.   36-3581
S.   STUDY OF THE STRUCTURE OF ANTIFERROELECTRIC AMMONIUM PERIODATE BY NUCLEAR RESONANCE METHOD 36-3580
AVEGUIDE.   EPITAXIAL ELECTROOPTIC MIXED CRYSTAL AMMONIUM POTASSIUM DIHYDROGEN PHOSPHATE FILM W 17-2633
   SUPERSTRUCTURE IN THE FERROELECTRIC PHASE OF AMMONIUM ROCHELLE SALT.   28-3224
/TUDY OF SLOW MOTION OF WATER MOLECULES IN ROCHELLE SALT AND AMMONIUM ROCHELLE SALT BY SPIN LATTICE RELAXA/ 28-3232
   MECHANISM OF FERROELECTRIC PHASE TRANSITION IN AMMONIUM ROCHELLE SALT. (THEORY)   28-3223
TE, P/ NUCLEAR SPIN LATTICE RELAXATION IN SOME FERROELECTRIC AMMONIUM SALTS. (AMMONIUM SULFATE, FLUOBERYLLA 19-2769
   INVESTIGATION OF FERROELECTRICITY IN SODIUM AMMONIUM SELENATE AND SODIUM AMMONIUM SULFATE.   19-2742
C RESONANCE OF DEUTERIUM IN THE FERROELECTRIC SODIUM DEUTERO AMMONIUM SELENATE DIDEUTERATE. NUCLEAR MAGNETI 22-2828
   ELECTROMECHANICAL PROPERTIES OF THE FERROELECTRIC SODIUM AMMONIUM SELENATE DIHYDRATE.   22-2884
   LINEAR ELECTROOPTICAL EFFECT IN THE FERROELECTRIC SODIUM AMMONIUM SELENATE DIHYDRATE.   22-2829
ON IN CRYSTALS WITH REORIENTATING STRUCTURAL GROUPS. (SODIUM AMMONIUM SELENATE DIHYDRATE) /C PHASE TRANSITI 22-2827
   NUCLEAR MAGNETIC RESONANCE OF SODIUM-23 IN THE SODIUM AMMONIUM SELENATE DIHYDRATE CRYSTAL.   22-2883
   PREPARATION AND ANALYSIS OF SODIUM AMMONIUM SELENATE DIHYDRATE CRYSTALS.   22-2863
   BRILLOUIN EFFECT AT ORDINARY TEMPERATURE IN AMMONIUM SULFATE.   19-2766

DEUTERON QUADRUPOLE COUPLING CONSTANT IN DEUTERATED AMMONIUM SULFATE. 19-2761
YDROSTATIC PRESSURE ON THE FERROELECTRIC PHASE TRANSITION OF AMMONIUM SULFATE. EFFECT OF H 16-2360
OF THE HYDROSTATIC PRESSURE ON THE TRANSITION TEMPERATURE IN AMMONIUM SULFATE. EFFECT 19-2796
ELECTRON DIFFRACTION STUDY OF THE STRUCTURE OF AMMONIUM SULFATE. 19-2797
ESR OF GAMMA-IRRADIATED AMMONIUM SULFATE. 19-2790
FERROELECTRIC TRANSITION OF AMMONIUM SULFATE. 19-2798
MANGANESE(2+) ABSORPTION NEAR THE CRITICAL REGION IN AMMONIUM SULFATE. 19-2785
VE DIELECTRIC MEASUREMENTS OF GAMMA-IRRADIATED FERROELECTRIC AMMONIUM SULFATE. MICROWA 19-2750
NITROGEN-14 QUADRUPOLE COUPLING IN PARAELECTRIC AMMONIUM SULFATE. 19-2749
(NMR STUDY OF) FERROELECTRIC TRANSITION IN (MANGANESE DOPED) AMMONIUM SULFATE. 19-2784
PREPARATION OF SINGLE CRYSTALS OF AMMONIUM SULFATE. 19-2758
PROPERTIES OF FERROELECTRIC CADMIUM AMMONIUM SULFATE. 19-2756
AMAN AND INFRARED STUDIES OF THE FERROELECTRIC TRANSITION IN AMMONIUM SULFATE. R 19-2793
(TEMPERATURE DEPENDENCE OF) THE SPONTANEOUS POLARIZATION IN AMMONIUM SULFATE. 19-2799
/ FERROELECTRICS. DIGRESSIONS FROM PHENOMENOLOGICAL THEORY. (AMMONIUM SULFATE, AMMONIUM FLUOBERYLLATE, GUA/ 02-0040
RAMAN SCATTERING IN A FERROELECTRIC AMMONIUM SULFATE CRYSTAL. 19-2786
DOMAIN STRUCTURE OF FERROELECTRIC SODIUM AMMONIUM SULFATE DIHYDRATE. 19-2778
ME OTHER STUDIES OF FERROELECTRIC PHASE TRANSITION IN SODIUM AMMONIUM SULFATE DIHYDRATE. (CRYSTALS) /AND SO 19-2767
OR TRIDEUTERIUM D/ THERMAL AND SOME OTHER STUDIES OF SODIUM AMMONIUM SULFATE DIHYDRATE, SODIUM TRIHYDROGEN 19-2779
TIZED PARAMAGNETIC SUBSTANCES. PART-1: IRON(X) ALUMINUM(1-X) AMMONIUM SULFATE DODECAHYDRATE. /CTRA OF MAGNE 19-2751
/UER EFFECT STUDY OF THE FERROELECTRIC TRANSITIONS IN FERRIC AMMONIUM SULFATE DODECAHYDRATE AND POTASSIUM / 19-2755
/UER EFFECT STUDY OF THE FERROELECTRIC TRANSITIONS IN FERRIC AMMONIUM SULFATE DODECAHYDRATE AND POTASSIUM / 24-2892
.   MOSSBAUER STUDY ON FERROELECTRIC PROPERTIES OF FERRIC AMMONIUM SULFATE DODECAHYDRATE SINGLE CRYSTALS 19-2770
ELASTIC ANOMALY IN AMMONIUM SULFATE. (ELECTROSTRICTIVE CONSTANTS) 19-2759
/ MECHANICAL PROPERTIES OF IMPROPER FERROELECTRIC CRYSTALS. (AMMONIUM SULFATE, GADOLINIUM MOLYBDATE, IRON / 14-2084
ANSITION.   STRUCTURE OF AMMONIUM SULFATE IN FERROELECTRIC PHASE AND TR 19-2774
/YSTALS (TRIGLYCINE SULFATE, POTASSIUM DIHYDROGEN PHOSPHATE, AMMONIUM SULFATE, ROCHELLE SALT) AT THE TRANS/ 02-0125
ELECTRICAL CONDUCTIVITY OF AMMONIUM SULFATE SINGLE CRYSTALS. 19-2781
PREPARATION OF AMMONIUM SULFATE SINGLE CRYSTALS. 19-2753
/TATIONAL MOTIONS IN SOLIDS: THE FERROELECTRIC TRANSITION IN AMMONIUM SULFATE STUDIED BY COLD NEUTRONS (SC/ 19-2752
ELECTROOPTICAL PROPERTIES IN T/ FERROELECTRIC TRANSITION IN AMMONIUM SULFATE, THE DIELECTRIC, OPTICAL, AND 19-2745
NG AN/ LOW TEMPERATURE STUDIES. PART-4: PHASE TRANSITIONS OF AMMONIUM SULFATE: THE NATURE OF HYDROGEN BONDI 19-2783
FERROELECTRIC PROPERTIES OF SOLID SOLUTIONS IN THE SYSTEM AMMONIUM SULFATE- AMMONIUM FLUOBERYLLATE. 19-2762
M.   FERROELECTRIC PROPERTIES OF SOLID SOLUTIONS IN THE AMMONIUM SULFATE- AMMONIUM FLUOBERYLLATE SYSTE 19-2764
M.   FERROELECTRIC PROPERTIES OF SOLID SOLUTIONS IN THE AMMONIUM SULFATE- AMMONIUM FLUOBERYLLATE SYSTE 19-2763
ESR OF COPPER(2+) DOPED SODIUM AMMONIUM TARTRATE. 29-3252
TION OF PECULIAR FERROELECTRIC- FERROELASTIC CRYSTAL: SODIUM AMMONIUM TARTRATE. /SM OF THE PHASE TRANSFORMA 29-3238
NUCLEAR MAGNETIC RESONANCE OF FERROELECTRIC LITHIUM AMMONIUM TARTRATE. 29-3240
NUCLEAR MAGNETIC RESONANCE OF FERROELECTRIC LITHIUM AMMONIUM TARTRATE. 29-3241
PROTON SPIN LATTICE RELAXATION IN FERROELECTRIC SODIUM AMMONIUM TARTRATE. 29-3248
/SODIUM, POTASSIUM(2), OR LITHIUM THALLIUM TARTRATE, LITHIUM AMMONIUM TARTRATE, AND MIXED CRYSTALS, BARIUM/ 02-0089
VER SODIUM DINITRITE. PART-3: ROCHELLE SALT. PART-4: LITHIUM AMMONIUM TARTRATE HYDRATE. /: TGS. PART-2: SIL 02-0077
/E RESONANCE OF LITHIUM IN FERROELECTRIC COMPOUNDS. (LITHIUM AMMONIUM TARTRATE, LITHIUM THALLIUM TARTRATE,/ 02-0052
X-RAY REFINEMENT OF THE CRYSTAL STRUCTURE OF LITHIUM AMMONIUM TARTRATE MONOHYDRATE. 29-3244
PHASE TRANSITION IN THALLIUM AND AMMONIUM THALLIUM (DIHYDROGEN) PHOSPHATES. 17-2571
NEW FERROELECTRIC: THE DOUBLE AMMONIUM THALLIUM MONOARSENATE OR ATLAS. 18-2735
N CRO/ NUCLEAR QUADRUPOLE INTERACTION IN ANTIFERROELECTRIC DIAMMONIUM TRIHYDROGEN PERIODATE BY IODINE PROTO 36-3579
D MAGNETIC/ LEVEL CROSSING (CROSS RELAXATION) EXPERIMENTS IN AMMONIUM TRIHYDROGEN PERIODATE CRYSTALS, PULSE 36-3577
R REL/ NUCLEAR QUADRUPOLE INTERACTION IN ANTIFERROELECTRIC DIAMMONIUM TRIHYDROGEN PERIODATE PROTON CROSSOVE 36-3578
HYDROGEN BOND STUDIES. PART-58: CRYSTAL STRUCTURE OF AMMONIUM TRIHYDROGEN SELENITE. 22-2879
IN POTASSIUM NITRATE AND RELATED COMPOUNDS. (POTASSIUM(1-X) AMMONIUM(X) NITRATE) /DIES ON FERROELECTRICITY 16-2402
AL.   FERROELECTRICITY IN AMMONIUM(1-X) POTASSIUM(1-X) NITRATE MIXED CRYST 16-2401
HASES, 2 FERR/ NEW FERROELECTRIC (PYROCHLORE TYPE) COMPOUND (AMMONO THIO CADMIUM NIOBATE, WITH 4 ELECTRIC P 36-3529
IN DEDUCING MEAN DAMAGE RANGES. (STUDI/ BOMBARDMENT INDUCED AMORPHIZATION IN TUNGSTEN TRIOXIDE AND ITS USE 10-1642
BETWEEN THE C/ CALORIMETRIC STUDY OF GLASS TRANSITION OF THE AMORPHOUS ICE AND OF THE PHASE TRANSFORMATION 34-3476
IN TH/ INFLUENCE OF ILLUMINATION ON THE FORM OF THE VOLTAGE AMPERE CHARACTERISTICS OF ANTIMONY SULFOIODIDE 15-2119
ERS ON BARIUM TITANATE- STRONTIUM TITAN/ SATURATION OF VOLT- AMPERE CHARACTERISTICS OF CADMIUM SELENIDE LAY 04-0592
LOCKED PUMP.   PARAMETRIC LIGHT AMPLIFICATION AND OSCILLATION IN KDP WITH MODE 17-2468
IUM NIOBATE STRUCTURE.   LAYER WAVE AMPLIFICATION IN A CADMIUM SULFIDE FLUID- LITH 11-1667
IC SEMICONDUCTORS.   ACOUSTIC ATTENUATION AND AMPLIFICATION IN PIEZOELECTRIC AND FERROELECTR 03-0275
A LAYERED STRUCTURE CONSISTING OF LITHIUM NIOBATE AND P-TYP/ AMPLIFICATION OF RAYLEIGH ULTRASONIC WAVES IN 11-1747
RIC LEAKY SURFACE WAVES IN LITHIUM NIOBATE. PROPAGATION AND AMPLIFICATION OF RAYLEIGH WAVES AND PIEZOELECT 11-1856
ADMIUM SELENIDE FILM ON LITHIUM NIOBATE.   ACOUSTOELECTRIC AMPLIFICATION OF SURFACE WAVE STRUCTURE OF A C 11-1723
TEM CONSISTING OF A PIEZOELECTRIC CERAMIC (B/ ABSORPTION AND AMPLIFICATION OF ULTRASOUND IN A TWO-LAYER SYS 04-1037
TRAPPING OF PARAMETRICALLY AMPLIFIED LIGHT WAVES IN A KDP CRYSTAL. 17-2428
POWER DENSITY GALLIUM ARSENIDE- LITHIUM NIOBATE SURFACE WAVE AMPLIFIER.   LOW 11-1825
/MANIUM OXIDE(20) VERSUS LITHIUM NIOBATE IN THE SURFACE WAVE AMPLIFIER. A THEORETICAL AND EXPERIMENTAL COM/ 11-1720
/REALIZATION OF ULTRA LOW NOISE LEVEL PARAELECTRIC MICROWAVE AMPLIFIER (BASED ON FERROELECTRICS IN THE PAR/ 37-3664
CE OF THE COERCIVE FIELD OF TRIGLYCINE SULFATE ON FREQUENCY, AMPLITUDE AND TEMPERATURE.   DEPENDEN 27-3092
INGLE CRYSTALS OF FERROELECTRICS. (ROCHELLE SALT AND TRIGLY/ AMPLITUDE DEPENDENCE OF INTERNAL FRICTION IN S 27-3122
IN BARIUM TITANATE SINGLE CRYSTALS. (PIEZOELECTRIC MODULUS) AMPLITUDE DEPENDENCE OF THE INTERNAL FRICTION 04-0791
YSTALS.   NONLINEAR EFFECTS IN THE PROPAGATION OF FINITE AMPLITUDE HYPERSONIC WAVES IN FERROELECTRIC CR 03-0498
HARMONICS IN POTASSIUM DIHYDROGEN PHOSPHATE.   ELECTROOPTIC AMPLITUDE MODULATION OF LASER GENERATED SECOND 17-2694
TERFEROMETER. (AMMONIUM DIHYDROGEN PHOSPHATE)   AMPLITUDE MODULATORS BASED ON THE MICHELSON IN 17-2715
TY OF BARIUM TITANATE SINGLE CRYSTALS.   INFLUENCE OF THE AMPLITUDE OF FREE VIBRATIONS ON THE PERMITTIVI 04-0768
TITANATE.   VIBRATIONAL AMPLITUDES IN FERROELECTRIC TETRAGONAL BARIUM 04-0726
CURIE POINTS. (POTASSIUM DIHYDROGEN PHOSPHATE, ALSO RUBIDIUM ANALOG AND POTASSIUM DIDEUTERIUM PHOSPHATE) / 17-2665
)   AN ADAPTIVE FERROELECTRIC TRANSFORMER - A SOLID STATE ANALOG MEMORY DEVICE. (USING, FOR EXAMPLE, PZT 09-1566
FERROELECTRIC INTERACTIONS FOR ADAPTIVE CONTROL VOLTAGE MO/ ANALOG MEMORY DEVICES EMPLOYING PIEZOELECTRIC- 37-3663
LINE FERROELECTRICS. (THEORY)   MACROSCOPIC ANALOG OF THE SPIN- ECHO EFFECT IN POLYCRYSTAL 03-0310
TALS. (POTASSIUM DIHYDROGEN PHOSPHATE, RUBIDIUM AND AMMONIUM ANALOGS SOME DEUTERATE) /OF SOME KDP TYPE CRYS 17-2679
PEROVSKITE) STRONTIUM IRON TUNGSTATE. (MAGNETIC G STRUCTURE ANALOGOUS TO ANTIFERROELECTRIC STRUCTURE) /IC 10-1655
ILITY) OF HYDROGEN TUNGSTEN BRONZES, HYDROGEN(X) TUNGSTATE. (ANALOGOUS TO THE SODIUM COMPOUNDS) / SUSCEPTIB 13-1934
ROELECTRIC MODELS. (THEORY, KDP TYPES)   ANALYTICAL PROPERTIES OF A CLASS OF PLANAR FER 03-0258
/S ON A CHANGE IN THE DIELECTRIC CONSTANT AND TANGENT OF THE ANGLE OF DIELECTRIC LOSS IN A STRONG ELECTRIC/ 27-3151
/E SULFATE SINGLE CRYSTALS ON A CHANGE IN THE TANGENT OF THE ANGLE OF DIELECTRIC LOSSES IN A STRONG ELECTR/ 27-3153
ASSIUM DIHYDROGEN / TEMPERATURE DEPENDENCE OF INDEX MATCHING ANGLE OF THE SECOND HARMONIC GENERATION IN POT 17-2544
ON THE CONDITI/ DIELECTRIC PERMEABILITY AND DIELECTRIC LOSS ANGLE TANGENT OF TRIGLYCINE SELENATE DEPENDING 27-3125
CRYSTALS OF DIFFERENT STOICHIOMETRIES.   PHASE MATCHING ANGLES AND TEMPERATURE OF LITHIUM METANIOBATE 11-1665
LIED TO LEAD(0.99) NIOBI/ MEASUREMENT OF PIEZOELECTRIC PHASE ANGLES IN A FERROELECTRIC CERAMIC (THEORY, APP 09-1537

ICE SINGLE CRYSTALS.	ANGULAR CORRELATION OF ANNIHILATION PHOTONS IN 34-3457
/KITE TYPE COMPOUNDS OF HAFNIUM-181 STUDIED BY THE PERTURBED	ANGULAR CORRELATION TECHNIQUE. (LEAD HAFNATE,/ 08-1276
ENT IN ANTIFERROELECTRIC LEAD ZIRCONATE STUDIED BY PERTURBED	ANGULAR CORRELATIONS. /AL ELECTRIC FIELD GRADI 07-1255
TONS IN FERROELECTRIC SODIUM NITRITE.	ANGULAR DISPERSION OF LARGE WAVE VECTOR POLARI 16-2256
LL IN BARIUM TITANATE. (THERMODYNAMIC CALCULATIONS)	ANGULAR ENERGY DEPENDENCE OF 180 DEG DOMAIN WA 04-0658
N IN KDP, IN A DIGITAL DEVIA/ STUDY OF THE LIMITATION CF THE	ANGULAR FIELD BY THE COMMUTATOR OF POLARIZATIO 17-2524
THEORY) DIELECTRIC PROPERTIES OF AN	ANHARMONIC CRYSTAL. (EXTENSION OF "SOFT MODE" 03-0212
/CROSCOPIC AND THERMODYNAMIC THEORY CF FERROELECTRICITY AND	ANHARMONIC EFFECTS IN CRYSTALS. (AND CALCULAT/ 03-0213
ON TO THE THEORY OF FERROELECTR/ VIBRATIONAL APPROACH TO THE	ANHARMONIC LATTICE PROBLEM. (WITH AN APPLICATI 03-0221
. (LATTICE DYNAMICS)	ANHARMONIC MODE MIXING IN IONIC FERROELECTRICS 03-0234
DEL FCR ORDER DISORDER / DYNAMIC SUSCEPTIBILITY OF CLASSICAL	ANHARMONIC OSCILLATOR. A UNIFIED OSCILLATOR MO 03-0405
IN FERROELECTRIC PHASE TRANSITION IN TERMS OF A SINGLE MODE	ANHARMONIC OSCILLATOR MODEL. /ND SPECIFIC HEAT 03-0406
PHASE TRANSITION IN PEROVSKITE LIKE FERROELECTRICS, USING AN	ANHARMONIC OSCILLATOR MODEL. /EARITY NEAR THE 03-0249
IN FERROELECTRICS: A SELF CONSISTENT POLARIZATION SCHEME FOR	ANHARMONIC POTENTIALS. (THEORY) / TRANSITIONS 03-0355
NONSTATIONARY POLARIZATION IN THE	ANHARMONIC VIBRATOR MODEL. (FERROELECTRICS) 03-0345
RITE.	ANHARMONICITY IN SODIUM NITRATE AND SODIUM NIT 16-2388
F STRONTIUM TITANATE. EFFECT OF THE QUARTIC	ANHARMONICITY ON THE SOFT PHONON INSTABILITY O 06-1211
ES. PHASE TRANSITIONS OF	ANHYDROUS MODIFICATIONS OF RARE EARTH MOLYBDAT 30-3272
HYL DIAMINE TARTRATE, BARIUM TITANATE, SALICYLIDENE- 4-BROMO	ANILINE. /ION. (PYROELECTRIC EFFECT IN TGS, ET 04-0854
N THE REGION OF THE ANTISYMMETRICAL VIBRATION OF THE NITRITE	ANION. /ALS AND COMBINATION SCATTER SPECTRUM I 16-2276
AL IN PARAELECTRIC AND / VIBRATION SPECTRUM OF THE PHOSPHATE	ANION IN A RUBIDIUM DIHYDROGEN PHOSPHATE CRYST 17-2472
FROM SODIUM NIOBATE CRYSTALS (NEAR THE 640 DEG C PHASE TRAN/	ANISOTROPIC CRITICAL X-RAY DIFFUSE SCATTERING 08-1311
FROM POTASSIUM MANGANESE TRIFLUCRIDE AND SODIUM NIOBATE SIN/	ANISOTROPIC CRITICAL X-RAY DIFFUSE SCATTERING 08-1312
/NE BOUNDARY MODES IN PEROVSKITE FERROELECTRICS INDICATED BY	ANISOTROPIC DEBYE- WALLER FACTORS. (POTASSIUM/ 08-1351
TRANSITIONS IN SODIUM NIOBATE.	ANISOTROPIC DIFFUSE X-RAY SCATTERING AND PHASE 08-1357
FOR STRONTIUM TITANATE.	ANISOTROPIC ENERGY BANDS (THEORY, CALCULATIONS 03-0305
/R AND POLAR- NEUTRAL CRYSTAL STRUCTURES. (CLASSIFICATION OF	ANISOTROPIC FERROELECTRIC AND ANTIFERROELECTR/ 03-0336
SITION OF POTASSIUM MANGANESE TRIFLUORIDE SINGLE C/ CRITICAL	ANISOTROPIC FLUCTUATIONS AT THE 184 DEG K TRAN 33-3369
/SWITCHING AND STEERING SURFACE WAVES BY NONLINEAR MIXING IN	ANISOTROPIC MEDIA. (LITHIUM NIOBATE, BISMUTH(/ 11-1778
/SORPTION IN TRIGLYCINE SULFATE SINGLE CRYSTALS. (THEORY OF	ANISOTROPIC RELAXATION ATTENUATION OF QUASI 27-3078
ICITY. (106 REFS)	ANISOTROPIC, SOLID DIELECTRICS AND FERROELEC 02-0065
/L AND HIGHER ORDER MODE SURFACE WAVE PROPAGATION ON LAYERED	ANISOTROPIC SUBSTRATES: GOLD ON LITHIUM NIOBA 11-1846
NTIUM TITANATE AT LOW TEMPERATURES. (TEMPERATURE DEPENDENCE,	ANISOTROPY) DIELECTRIC PROPERTIES OF STRO 06-1202
RYSTALLIZATION OF ANTIMONY SULFOIODIDE (CRYSTAL GROWTH RATE,	ANISOTROPY) KINETICS OF C 15-2215
/NTROL OF ELASTIC COMPLIANCE, NICKEL IODINE BORACITE OPTICAL	ANISOTROPY, BISMUTH TITANATE, POSSIBLE FERROE/ 02-0060
INDUCED GROWTH	ANISOTROPY IN TGS CRYSTALS. 27-3162
ROELECTRICS, SUBJECTED BEFOREHAND TO AN EL/ INVESTIGATION OF	ANISOTROPY IN THE DIELECTRIC PROPERTIES OF FER 04-0661
C. (BARIUM TITANATE, BARIUM / ULTRAHIGH FREQUENCY DIELECTRIC	ANISOTROPY OF A POLARIZED FERROELECTRIC CERAMI 04-0767
ND IN SINGLE CRYSTALS OF ANTIMONY SULFOIODIDE.	ANISOTROPY OF ANOMALOUS ABSORPTION OF ULTRASOU 15-2222
TO QUADRUPOLE TRANSITIONS.	ANISOTROPY OF CUBIC NICKEL IODINE BORACITE DUE 14-2093
S IN FERROELECTRIC BARIUM TITANATE. EFFECTS OF	ANISOTROPY OF ELASTIC CONSTANTS ON DISLOCATION 04-0674
HARDNESS	ANISOTROPY OF SINGLE CRYSTALS OF ICE I(H). 34-3460
ULFATE SINGLE CRYSTALS. (THEORY OF ANISOTROPIC RELAXATION /	ANISOTROPY OF SOUND ABSORPTION IN TRIGLYCINE S 27-3078
OPTICAL	ANISOTROPY OF THE CUBIC COBALT IODINE BORACITE 14-2094
ARIUM TITANATE SINGLE CRYSTALS GROWN FRCM SOLUTION IN A POT/	ANISOTROPY OF THE ELECTRICAL CONDUCTIVITY OF B 04-1035
STALS OF LEAD MAGNONIOBATE.	ANISOTROPY OF THE ELECTROOPTICAL EFFECT IN CRY 36-3521
NY SULFOIODIDE CRYSTALS.	ANISOTROPY OF THE REFLECTION SPECTRA OF ANTIMO 15-2125
AL FERROELECTRICS. (TGS, ABSTRACT ONLY)	ANISOTROPY OF ULTRASONIC ATTENUATION IN UNIAXI 27-3055
5-300 DEG K). DIELECTRIC PROPERTIES OF STRONTIUM TITANATE	ANNEALED SINGLE CRYSTALS) AT LOW TEMPERATURE (06-1201
D, TWIN-FREE CRYSTALS OF FERROELECTRIC BISMUTH TITANATE, (BY	ANNEALING). PREPARATION OF POLE 08-1352
SURFACE DOMAIN REORIENTATICN PRODUCED BY ABRASION AND	ANNEALING OF POLYCRYSTALLINE BARIUM TITANATE. 04-0639
HYDROGEN PHOSPHATE CRYSTALS.	ANNEALING ON OPTICAL ANOMALIES OF POTASSIUM DI 17-2427
INFLUENCE OF THE DOMAIN STRUCTURE ON POSITRON (TWO-PHONON)	ANNIHILATION IN ROCHELLE SALT. 28-3191
ANGULAR CORRELATION OF	ANNIHILATION PHOTONS IN ICE SINGLE CRYSTALS. 34-3457
TANATE SINGLE CRYSTALS HEATED IN OXYGEN (EFFECT TIED TO GOLD	ANODES). /ELECTRICAL CONDUCTIVITY OF BARIUM TI 04-0997
ERTIES IN THE NEIGHBORHOOD OF THE CURIE PCINT (BIREFRINGENCE	ANOMALIES). / OPTICAL, AND ELECTROOPTICAL PROP 19-2765
RATURES. CONDUCTION	ANOMALIES AND POLARIZATION IN ICE AT LOW TEMPE 34-3428
ARIZATION IN FERROELECTRICS. (ABSTRACT ON/ THERMAL EXPANSION	ANOMALIES AND THE DIRECTION OF SPONTANEOUS POL 03-0350
F STRONTIUM TITANATE NEAR 105 DEG K. ANALYSIS OF THE ELASTIC	ANOMALIES AT THE STRUCTURAL PHASE TRANSITION O 06-1183
PHOSPHATE POTASSIUM DIDEUTERIUM FHOSPHATE AND RU/ DIELECTRIC	ANOMALIES IN CRYSTALS OF POTASSIUM DIHYDROGEN 17-2613
OPTICAL	ANOMALIES IN FERROELECTRIC CRYSTALS. 02-0082
AL PHASE TRANSITIONS / PHENOMENOLOGICAL THEORY OF DIELECTRIC	ANOMALIES IN FERROELECTRIC MATERIAL WITH SEVER 03-0363
ULTRASONIC	ANOMALIES IN ROCHELLE SALT. 28-3197
AND HIGH PRESSURES. (POSSIBLY FERROELECTRIC) DIELECTRIC	ANOMALIES IN SOLID METHANE AT LOW TEMPERATURES 36-3547
ELASTIC CONSTANT	ANOMALIES IN STRONTIUM TITANATE. 06-1177
FERROELECTRIC SEMICONDUCTOR NEAR THE CUBIC POINT.	ANOMALIES IN THE ELECTRICAL CONDUCTIVITY OF A 03-0268
TRIC PHOSPHATES NEAR THEIR CURIE POINTS. (POTASSIUM DIHYDRO/	ANOMALIES IN THE INTERNAL STRESS FOR FERROELEC 17-2665
(5-300K) OF THE PRASEODYMIUM(1-X) NEODYMIUM(X) (ORTHO) ALUM/	ANOMALIES IN THE (1 KHZ) DIELECTRIC CONSTANT 36-3600
THE POINT OF VIEW OF THE THEORY OF CRITICAL POINTS.	ANOMALIES NEAR FERROELECTRIC TRANSITIONS FROM 03-0298
E TRANSITIONS. (BASED ON LANDAU PHENOMENOL/ EXAMINATION OF)	ANOMALIES OF DIELECTRIC PROPERTIES AT THE PHAS 19-2765
IN THE REGION OF THE ANTIFERROELECTRIC CURIE POINT OF LEAD /	ANOMALIES OF ELASTICITY AND INTERNAL FRICTION 13-2020
ELASTIC	ANOMALIES OF ICE AT LOW TEMPERATURES. 34-3439
YSTALS. EFFECT OF ANNEALING ON OPTICAL	ANOMALIES OF POTASSIUM DIHYDROGEN PHOSPHATE CR 17-2427
CTRIC CURIE TEMPERATURE. DIELECTRIC	ANOMALIES OF SODIUM NITRATE ABOVE THE FERROELE 16-2376
SULFIDE.	ANOMALIES OF THERMAL PROPERTIES OF TRIGLYCINE 27-3048
OND ORDER TRANSITION POINTS IN F/ PHENOMENOLOGICAL THEORY OF	ANOMALIES OF THERMODYNAMIC PARAMETERS NEAR SEC 03-0362
ECTRIC PHASE TRANSITICN (ANTIMONY SULFOIODIDE).	ANOMALOUS ABSORPTION OF SOUND NEAR THE FERROEL 15-2221
SULFOIODIDE CRYSTALS. POLARIZATION EFFECTS IN THE	ANOMALOUS ABSORPTION OF ULTRASOUND IN ANTIMONY 15-2123
RYSTALS CF ANTIMONY SULFOIODIDE. ANISOTROPY OF	ANOMALOUS ABSORPTION OF ULTRASOUND IN SINGLE C 15-2222
ES C PHASE TRANSITION IN LEAD DICALCIUM PROPIONATE. THE	ANOMALOUS BEHAVIOR OF PHONON NEAR THE 60 DEGRE 26-2942
ITY COEFFICIENT IN STRONTIUM TITANATE NEAR THE PHASE TRANSI/	ANOMALOUS BEHAVIOR OF THE DIELECTRIC NONLINEAR 06-1105
COEFFICIENT OF LEAD MAGNONIOBATE SINGLE CRYSTALS IN THE REG/	ANOMALOUS BEHAVIOR OF THE OPTICAL ASBSORPTION 13-1974
BARIUM TITANATE PARTICLES. ON THE	ANOMALOUS CRYSTALLOGRAPHIC PROPERTIES OF SMALL 04-0935
UM TITANATE.	ANOMALOUS DECAY EFFECT IN POLYCRYSTALLINE BARI 04-0752
YSTALS OF ORDER DISORDER / CRITICAL SLOWING DOWN PROCESS OF	(ANOMALOUS) DIELECTRIC RELAXATION IN SEVERAL CR 02-0087
LOW FREQUENCIES AND JERKY WALL MOTION IN ROCHELLE SALT.	ANOMALOUS DISPERSION OF COERSIVE FIELD AT VERY 28-3177
IODIDE. (PIEZOELECTRIC DOMAIN OSCILLATIONS)	ANOMALOUS ECHO IN FERROELECTRIC ANTIMONY SULFO 15-2211
ESE TRIFLUORIDE NEAR A STRUCTURAL PHASE TRANSITION.	ANOMALOUS ELASTIC BEHAVIOR OF POTASSIUM MANGAN 33-3392
ICS IN THE PHASE TRANSITION REGION. NATURE OF THE	ANOMALOUS ELECTRIC CONDUCTIVITY OF FERROELECTR 03-0508
D STRONTIUM TITANATE)	ANOMALOUS FERROELECTRIC HYSTERESIS LOOPS. (LEA 05-1067

/T MODE TRANSITION POINT IN POTASSIUM MANGANESE TRIFLUORIDE (ANOMALOUS HEAT CAPACITY, CRITICAL TEMPERATURE/ 33-3373
0.002) BARIUM(0.998) TITANATE- BISMUTH OXIDE COMPO/ TWO STEP ANOMALOUS INCREASE OF RESISTANCE ON LANTHANUM(04-0801
DOPED BARIUM TITANATE- BISMUTH TRIOXIDE COMPOSITE/ TWO-STEP ANOMALOUS INCREASE OF RESISTIVITY ON LANTHANUM 04-0800
RADIATION AND ACOUSTIC EXCITATION ON ANOMALOUS LAYERS OF PEROVSKITE FERROELECTRICS. 08-1388
ITHIUM TANTALATE (EFFECT OF VACANCIES). ANOMALOUS (LONGITUDINAL WAVE) ATTENUATION IN L 12-1892
OF FERROELECTRICITY IN LITHIUM HYDRAZINIUM SULFATE. ANOMALOUS NEUTRON SCATTERING AND THE QUESTION 32-3341
ROELECTRIC GADOLINIUM MOLYBDATE. AN EXPLANATION OF ANOMALOUS OPTICAL BEHAVIOR OF THE IMPROPER FER 30-3292
ITION IN FERROELASTIC FERROELECTRIC GADOLINIUM MOLYBDATE. ANOMALOUS PHONON BEHAVIOR NEAR THE PHASE TRANS 30-3276
IC WAVES IN POTASSIUM DIDEUTERIUM PHOSPHATE (TEMPERATURE DE/ ANOMALOUS PROPAGATION OF LONGITUDINAL ULTRASON 17-2660
IMONY SULFOIODIDE. ANOMALOUS PROPERTIES IN PHOTOFERROELECTRIC ANT 15-2196
ANOMALOUS RESONANCE OF STRONTIUM TITANATE. 06-1173
ITHIUM NIOBATE, BA/ X-RAY DETERMINATION OF POLARITY SENSE BY ANOMALOUS SCATTERING AT AN ABSORPTION EDGE. (L 11-1673
IONAL ANTIFERROMAGNET COPPER FORMATE TETRAHYDRATE. ANOMALOUS SHIFT OF ESR LINES IN THE TWO DIMENS 26-2941
TROPY CHANGE IN SODIUM TRIHYDROGEN DISELENITE. ANOMALOUS SPECIFIC HEAT AND CONFIGURATIONAL EN 22-2856
REA AND POTASSIUM FERROCYANIDE TRIHYDRATE. ANOMALOUS SPECIFIC HEAT OF FERROELECTRIC THIOU 24-2902
SELENITE. ANOMALOUS SPECIFIC HEAT OF SODIUM TRIHYDROGEN 22-2823
SELENITE ASSOCIATED COMBINATORIAL PROBLEM. ANOMALOUS SPECIFIC HEAT OF SODIUM TRIHYDROGEN 22-2822
ANOMALOUS SPECIFIC HEAT OF STRONTIUM TITANATE. 06-1159
N WAVES IN POTASSIUM FERROCYANIDE TRIHYDRATE. ANOMALOUS SPIN LATTICE RELAXATION BY QUASI SPI 24-2899
DROGEN SELENATE CRYSTALS. ANOMALOUS THERMAL EXPANSION OF POTASSIUM TRIHY 22-2880
RANSITION IN STRONTIUM TITANATE. ANOMALOUS ULTRASONIC ATTENUATION AT THE 105K T 06-1189
IODIDE SINGLE CRYSTALS. FREQUENCY DEPENDENCE OF ANOMALOUS ULTRASOUND DAMPING IN ANTIMONY SULFO 15-2139
RIC PHASE TRANSITION AROUND 47 DEG K IN S/ PHOTOCONDUCTIVITY ANOMALY AS AN EVIDENCE OF THE LOCAL FERROELECT 06-1249
ANSITION IN STRONTIUM TITANATE. (ABSTRACT) SPECIFIC HEAT ANOMALY ASSOCIATED WITH THE 110 DEG K PHASE TR 06-1226
ELECTRICAL PROPERTIES OF LITHIUM NIOBATE. (ANOMALY AT 263 DEG K IS STRUCTURAL TRANSITION) 11-1709
/ERROCYANIDE TRIHYDRATE (NEAR THE TRANSITION TEMPERATURE, NO ANOMALY BUT SOFT PHONON MODE STILL POSSIBLE). 24-2891
CONSTANTS) ELASTIC ANOMALY IN AMMONIUM SULFATE. (ELECTROSTRICTIVE 19-2759
NGLE CRYSTALS. FERROELECTRIC DIELECTRIC ANOMALY IN BISMUTH(4) TITANIUM(3) OXIDE(12) SI 13-2009
ULTRASONIC ATTENUATION ANOMALY IN DISPLACIVE TYPE FERROELECTRICS. 03-0393
QUASI PHENOMENOLOGICAL THEORY OF ELASTIC ANOMALY IN FERROELECTRIC CRYSTALS. 03-0374
ATION OF TEMPERATURE DEPENDENCE FOR FERROELEC/ SPECIFIC HEAT ANOMALY IN FERROELECTRIC ROCHELLE SALT (CALCUL 28-3187
TE. SPECIFIC HEAT ANOMALY IN FERROELECTRIC SILVER SODIUM DINITRI 16-2296
ANATE. HIGH L/F NOISE ANOMALY IN SEMICONDUCTING BARIUM STRONTIUM TIT 04-0866
SITIVE TEMPERATURE COEFFICIENT OF RESISTIVITY) RESISTIVITY ANOMALY IN SEMICONDUCTING BARIUM TITANATE. (PO 04-0584
MICS. EFFECT OF GRAIN SIZE ON THE RESISTIVITY ANOMALY IN SEMICONDUCTIVE BARIUM TITANATE CERA 04-0735
NEW ESR EVIDENCE FOR THE 65 DEGREES K ANOMALY IN STRONTIUM TITANATE. 06-1230
ECTRIC TRANSITION. (TGS) SPECIFIC HEAT ANOMALY IN THE NEIGHBORHOOD OF A FERRO- PARAEL 27-2962
ZOELECTRIC, DIELECTRIC, AND OPTICAL CONSTANTS) ROOM/ ELASTIC ANOMALY OF BARIUM SODIUM NIOBATE (ELASTIC, PIE 13-2060
E. SPECIFIC HEAT ANOMALY OF FERROELECTRIC COBALT IODINE BORACIT 14-2101
TE CRYSTAL. ANOMALY OF RESISTANCE IN REDUCED BARIUM TITANA 04-0988
TIME NEAR THE CRITICAL TEMPERATURE OF DICALCIUM LEAD PROPIO/ ANOMALY OF THE PROTON LATTICE RELAXATION 26-2947
AR THE TRANSITION TEMPERATURE IN THE ORDER DISORDER TYPE FE/ ANOMALY OF THE SPIN LATTICE RELAXATION TIME NE 03-0488
UM MANGANESE TRIFLUORIDE. OBSERVATION OF THE ANOMALY OF THE THERMAL CONDUCTIVITY OF POTASSI 33-3382
S (SINGLE CRYSTAL TRIGLYCINE SULFATE AND TRIGLYCINE FLUOBER/ ANOMALY OF THERMAL PROPERTIES OF FERROELECTRIC 27-3137
IN STRUCTURE OF TGS CRYSTALS BY SELECTIVE CRYSTALLIZATION OF ANTHRAQUINONE (SUBLIMED ONTO TGS). /G THE DOMA 27-3041
ECT OF IRRADIATION OF DIELECTRIC PROPERTIES OF SOME FERRO-, ANTIFERRO- AND PARAELECTRIC PEROVSKITES. EFF 08-1474
EXCITON MODEL OF SPIN AND OPTICAL EXCITATIONS OF AN ANTIFERRODIELECTRIC. 03-0416
/E TO THE 6A1G(6S) STATE OF THE MANGANESE(2+) ION IN A CUBIC ANTIFERRODIELECTRIC AN EXTERNAL MAGNETIC FIEL/ 03-0317
NTS OF OPTICAL TRANSITIO/ ELECTRONIC AND SPIN EXCITATIONS IN ANTIFERRODIELECTRIC ELECTRIC AND MAGNETIC MOME 03-0418
DOUBLE ELECTRONIC EXC/ ON THE THEORY OF LIGHT ABSORPTION BY ANTIFERRODIELECTRICS IN THE FREQUENCY RANGE OF 03-0419
LECT/ CONSIDERATIONS OF PARTIALLY FERROELASTIC AND PARTIALLY ANTIFERROELASTIC CRYSTALS AND PARTIALLY FERROE 03-0136
E(0.5) AND LEAD COBALTATE(0.5) / PHASE TRANSITION IN THE ANTIFERROELECTRET LEAD COBALTATE(0.5) TUNGSTAT 13-1968
ON IN COPPER FORMATE CRYSTALS BY THE ELECTROOPTICAL METHOD. (ANTIFERROELECTRIC) ANTIPOLARIZATI 26-2925
EPR OF CHROMIUM(3+) IONS IN ANTIFERROELECTRIC ADP CRYSTALS. 17-2546
/ DIHYDROGEN ARSENATE AND POTASSIUM DIDEUTERIUM ARSENATE AND ANTIFERROELECTRIC AMMONIUM DIHYDROGEN ARSENAT/ 18-2734
E. LONG WAVELENGTH POLARIZATION FLUCTUATIONS IN ANTIFERROELECTRIC AMMONIUM DIHYDROGEN PHOSPHAT 17-2461
R RESONANCE METHODS. STUDY OF THE STRUCTURE OF ANTIFERROELECTRIC AMMONIUM PERIODATE BY NUCLEA 36-3580
ESR OF FREE RADICALS IN FERROELECTRIC AND ANTIFERROELECTRIC ARSENATE SINGLE CRYSTALS. 18-2731
N AND PHASE T/ EFFECT OF ELECTRIC FIELD ON FERROELECTRIC AND ANTIFERROELECTRIC CERAMICS. PART-1: DEFORMATIO 03-0169
XATION. FERRO- AND ANTIFERROELECTRIC COUPLING AND DIELECTRIC RELA 03-0354
PHENOMENOLOGICAL THEORY OF FERROELECTRIC AND ANTIFERROELECTRIC CRYSTALS. 03-0270
ROELASTIC CRYSTALS AND PARTIALLY FERROELECTRIC AND PARTIALLY ANTIFERROELECTRIC CRYSTALS (THEORY). / ANTIFER 03-0136
/ES OF ELASTICITY AND INTERNAL FRICTION IN THE REGION OF THE ANTIFERROELECTRIC CURIE POINT OF LEAD MAGNESA/ 13-2020
DATE PROTON CROSSOVER REL/ NUCLEAR QUADRUPOLE INTERACTION IN ANTIFERROELECTRIC DIAMMONIUM TRIHYDROGEN PERIO 36-3578
DATE BY IODINE PROTON CRO/ NUCLEAR QUADRUPOLE INTERACTION IN ANTIFERROELECTRIC DIAMMONIUM TRIHYDROGEN PERIO 36-3579
D, AND ACTIVATION ENERGY. APPLICATION TO LEAD/ RELAXATION OF ANTIFERROELECTRIC DOMAIN WALLS, EFFECTIVE FIEL 07-1257
/ STUDY OF EXTENDED DEFECTS RELATED TO NONSTOICHIOMETRY (AND ANTIFERROELECTRIC DOMAINS IN TRANSITION METAL/ 36-3522
EN PHOSPHATE. (PHASE TRANSITION / MICROSCOPIC OBSERVATION OF ANTIFERROELECTRIC DOMAINS OF AMMONIUM DIHYDROG 17-2643
IN PLZT CERAMICS. TEMPERATURE AND FIELD DEPENDENCE OF THE ANTIFERROELECTRIC FERROELECTRIC PHASE BOUNDARY 09-1491
TRIC L/ MOSSBAUER EFFECT STUDIES OF THE PHASE TRANSITIONS IN ANTIFERROELECTRIC LEAD ZIRCONATE AND FERROELEC 07-1253
UER EFFECT. TEMPERATURE DEPENDENT OPTICAL MODE IN ANTIFERROELECTRIC LEAD ZIRCONATE BY THE MOSSBA 07-1260
RTURBED ANGULAR COR/ THE INTERNAL ELECTRIC FIELD GRADIENT IN ANTIFERROELECTRIC LEAD ZIRCONATE STUDIED BY PE 07-1255
ITES) TEMPERATURE DEPENDENT SPECTRUM OF AN ANTIFERROELECTRIC LINEAR CHAIN MODEL. (PEROVSK 03-0468
NATURE OF PHASE TRANSITIONS IN FERROELECTRIC AND ANTIFERROELECTRIC MATERIALS. 03-0311
MPERATURES) ANTIFERROELECTRICITY AND ANTIFERROELECTRIC MATERIALS. (TABLE OF NEEL TE 01-0020
TICE VIBRATIONS) ANTIFERROELECTRIC MODE IN LEAD ZIRCONATE. (LAT 17-1262
AL ENTROPY OF ICE, PART-2, HYDROGEN BONDED FERROELECTRIC AND ANTIFERROELECTRIC MODELS. / PART-1, THE RESIDU 03-0388
/E OF THE DIELECTRIC PROPERTIES AND PHASE TRANSITIONS OF THE ANTIFERROELECTRIC PEROVSKITES LEAD ZIRCONATE / 08-1433
ATION OF SODIUM NITRITE. (CALCULATIONS TO RELATE TO SPECIFIC ANTIFERROELECTRIC PHASE) LATTICE VIBR 16-2317
ANATE- LEAD ZIRCONATE- LANTHANUM FERRATE. ANTIFERROELECTRIC PHASE IN THE SYSTEM LEAD TIT 13-1961
AHYDRATE. UNIT CELL AND SPACE GROUP OF THE ANTIFERROELECTRIC PHASE OF COPPER FORMATE TETR 26-2927
AHYDRATE BY NEUTRON DIFFRACTION. APPROXIMATE STRUCTURE FOR ANTIFERROELECTRIC PHASE OF COPPER FORMATE TETR 26-2935
THERMAL EXPANSION OF ADP AND DEUTERATED ADP IN THE REGION OF ANTIFERROELECTRIC PHASE TRANSITION. 17-2445
D SCATTERING EXPERIMENTS. (REVIEW) FERRO- AND ANTIFERROELECTRIC PHASE TRANSITIONS AND RELATE 02-0132
ECTRIC OR FERROELECTRIC, THEORY). SOME REMARKS ABOUT ANTIFERROELECTRIC PHASE TRANSITIONS (TO PARAEL 03-0171
RAMAN SPECTRA OF ADP IN THE PARAELECTRIC AND ANTIFERROELECTRIC PHASES. 17-2431
M DOPED LEA/ ELECTROSTRICTIVE CONSTANTS IN FERROELECTRIC AND ANTIFERROELECTRIC PHASES. (LANTHANUM OR NIOBIU 09-1487
FNATE TITANATE, LEAD STANNATE NIOBATE. SOLID SOLUTIONS IN ANTIFERROELECTRIC REGION OF THE SYSTEM LEAD HA 13-1956
NEUTRON/ PROTON MOTIONS IN HYDROGEN BONDED FERROELECTRIC AND ANTIFERROELECTRIC SOLIDS. PART-1: ENERGY LOSS 17-2687

FERROELECTRICS--PERMUTED TITLE INDEX

EN PHOSPHATE TYPE CRYSTA/ VANISHING OF THE FERROELECTRIC AND ANTIFERROELECTRIC STATES IN POTASSIUM DIHYDROG 17-2642
TRONTIUM IRON TUNGSTATE . (MAGNETIC G STRUCTURE ANALOGOUS TO ANTIFERROELECTRIC STRUCTURE) /IC PEROVSKITE) S 10-1655
STRUCTURES. (CLASSIFICATION OF ANISOTROPIC FERROELECTRIC AND ANTIFERROELECTRIC STRUCTURES) /EUTRAL CRYSTAL 03-0336
THE FORMS OF ICES. (ICES II, VIII AND IX ARE FULLY ORDERED (ANTIFERROELECTRIC STRUCTURES)) STRUCTURES OF 34-3451
LANDOLT-BORNSTEIN TABELLEN, NEUE SERIE. VOL 3: FERRO- AND ANTIFERROELECTRIC SUBSTANCES. 01-0002
COMPARISON, APPLICATION TO BARIUM TITANAT/ FERROELECTRIC AND ANTIFERROELECTRIC SYMMETRY GROUPS. (AND THEIR 04-0933
NTS OF THE CRYSTALLOGRAPHIC TRANSITION IN VANADIUM DIOXIDE. (ANTIFERROELECTRIC TEMPERATURE 340 DEG K) /PONE 36-3562
MONIUM DIHYDROGEN PHOSPHATE. MOSSBAUER STUDY OF ANTIFERROELECTRIC TRANSITION IN IRON-57(3X) AM 17-2644
DER TRANSITION. .PHENOMENOLOGICAL THEORY OF ANTIFERROELECTRIC TRANSITION. PART-2: FIRST OR 03-0404
SPECIFIC HEAT OF SODIUM NITRITE NEAR THE ANTIFERROELECTRIC TRANSITION POINT. 16-2303
IUM DIHYDROGEN ARSENATE STUDIED BY INFRARED ABSORPTION. ANTIFERROELECTRIC TRANSITIONS IN ADP AND AMMON 17-2706
RANSITIONS AND CRITICAL POINTS. SECTION-5: FERROELECTRIC AND ANTIFERROELECTRIC TRANSITIONS. (REVIEW) /ASE T 02-0070
THE STRUCTURAL CHANGES INVOLVED IN PHASE TRANSFORMATIONS IN ANTIFERROELECTRICS. CHARACTERISTICS OF 03-0526
ENATE, AND AMMONIUM DIHYDROGEN ARSENATE. (FERROELECTRICS AND ANTIFERROELECTRICS) /, POTASSIUM DEUTERIUM ARS 18-2724
DYNAMICS OF ORDER DISORDER TYPE FERROELECTRICS AND ANTIFERROELECTRICS 03-0179
DYNAMICAL THEORY OF RIGID LATTICES OF ANTIFERROELECTRICS. 03-0344
RTIES. (SUMMARY WITH NEW RESULTS ON LEAD ZIRCONATE AND OTHER ANTIFERROELECTRICS) /RE ON FERROELECTRIC PROPE 07-1263
D DEUTERO AMMONIUM DIDEUTERIUM ARSENATE. (FERROELECTRICS AND ANTIFERROELECTRICS) /M DIHYDROGEN ARSENATE, AN 18-2725
FERROELECTRICS AND ANTIFERROELECTRICS. 01-0029
FERROELECTRICS AND ANTIFERROELECTRICS. 01-0030
ICE, FERRO- AND ANTIFERROELECTRICS. 34-3453
LATTICE DYNAMICS OF PHASE TRANSITIONS IN PEROVSKITE TYPE ANTIFERROELECTRICS. 08-1475
TRIOXIDE, SODIUM NIOBA/ ORIENTATION OF COMPOSITION PLANES IN ANTIFERROELECTRICS. (APPLIED TO ADP, TUNGSTEN 08-1327
/CURIE CONSTANT AND CURIE TEMPERATURE FOR FERROELECTRICS AND ANTIFERROELECTRICS BARIUM TITANATE, COMPLEX L/ 03-0188
/IPOLE PATTERNS IN THE STRUCTURES OF SOME FERROELECTRICS AND ANTIFERROELECTRICS. (BARIUM TITANATE, KDP, PO/ 02-0134
CRITICAL PHENOMENA IN ANTIFERROELECTRICS. (EQUATION OF STATE) 03-0343
/OF THE PEROVSKITE MODIFICATION OF CADMIUM TITANATE (AND THE ANTIFERROELECTRIC- FERROELECTRIC PHASE CHANGE/ 08-1387
ACOUSTIC TRANSIENTS GENERATED BY ANTIFERROELECTRIC- FERROELECTRIC TRANSDUCERS. 37-3672
THE LEAD (ZIRCONATE, STANNATE, TITANATE) SOLID SOLUTIONS. / ANTIFERROELECTRIC- FERROELECTRIC TRANSITION IN 09-1486
YSTALLIN/ MICROWAVE INVESTIGATION OF SOME FERROELECTRICS AND ANTIFERROELECTRICS OF PEROVSKITE TYPE. (POLYCR 08-1375
LIED, 18 REFS) COLLINEAR AND NONCOLLINEAR FERRO- AND ANTIFERROELECTRICS. (REPRESENTATION THEORY APP 03-0173
/PROPERTIES AND THERMAL EXPANSION OF SOME FERROELECTRICS AND ANTIFERROELECTRICS WITH HIGH CURIE TEMPERATUR/ 08-1378
DYNAMICAL THEORY OF AMMONIUM DIHYDROGEN PHOSPHATE TYPE ANTIFERROELECTRICS WITH HYDROGEN BONDS. 03-0476
/ONSTANT ABOVE THE CURIE POINT IN VARIOUS FERROELECTRICS AND ANTIFERROELECTRICS. (ZIRCONATES, NIOBATES, TU/ 02-0124
MAGNETIC RESONANCE AND THE PROBLEM OF THE DEFINITION OF ANTIFERROELECTRICITY. 03-0266
ERIALS. (TABLE OF NEEL TEMPERATURES) ANTIFERROELECTRICITY AND ANTIFERROELECTRIC MAT 01-0020
IED TO SODIUM NIOBATE AND THE SY/ PHENOMENOLOGICAL THEORY OF ANTIFERROELECTRICITY AND FERROELECTRICITY APPL 03-0218
/NCTION FOR THE SLATER MODEL OF FERROELECTRICITY (KDP MODEL, ANTIFERROELECTRICITY F MODEL, WU KDP MODEL, C/ 03-0308
ITE STRUCTURE. (EXAMINATION OF PUBLISHED DATA, CLASSIFICATI/ ANTIFERROELECTRICITY IN COMPOUNDS WITH PEROVSK 01-0018
SOME ASPECTS OF THE PROBLEM OF FERROELECTRICITY AND ANTIFERROELECTRICITY IN PEROVSKITES. 03-0504
Y, LEAD ZIRCONATE) FERROELECTRICITY AND ANTIFERROELECTRICITY IN POLAR CRYSTALS. (THEOR 03-0372
D POTASSIUM MANGANESE TRIFLUORIDE. (20 PERCENT MANGANESE RE/ ANTIFERROELECTRICITY IN (PURE AND) COBALT DOPE 33-3366
LEAD ZIRCONATE. FERROELECTRICITY AND ANTIFERROELECTRICITY IN PURE AND NIOBIUM DOPED 07-1251
CERTAIN ASPECTS OF THE PROBLEM OF FERROELECTRICITY AND ANTIFERROELECTRICITY IN THE PEROVSKITES. 08-1456
(TUNGSTEN TRIOXIDE) ON FERROELECTRICITY AND ANTIFERROELECTRICITY OF THE AO3 TYPE CRYSTAL. 10-1641
FERROELECTRICALLY ACTIVE A-IONS IN ABO(3) STRUCTURES. (ANTIFERROELECTRICITY ORIGINS) 08-1364
/SPONTANEOUS SUBLATTICE MAGNETIZATION IN THE TWO DIMENSIONAL ANTIFERROMAGNET COPPER FORMATE TETRADEUTERATE. 26-2929
ANOMALOUS SHIFT OF ESR LINES IN THE TWO DIMENSIONAL ANTIFERROMAGNET COPPER FORMATE TETRAHYDRATE. 26-2941
BE FERROELECTRIC DUE TO ANTI/ MAGNETOELECTRIC EFFECT IN THE ANTIFERROMAGNET IRON ANTIMONY(2) OXIDE(4) (MAY 36-3563
INSTABILITY OF SPIN WAVES IN FERROELECTRIC ANTIFERROMAGNETS IN EXTERNAL FIELDS. (THEORY) 03-0149
/SUREMENTS OF THE THERMAL CONDUCTIONS IN THE LOW DIMENSIONAL ANTIFERROMAGNETS POTASSIUM COPPER TRIFLUORIDE/ 26-2939
ECTRIC BISMUTH FERRATE X-RAY AND NEUTRON DIFFRACTION STUDY. (ANTIFERROMAGNETIC) FERROEL 08-1402
RTIES IN THE BISMUTH- LANTHANUM FERRATE SYST/ COEXISTENCE OF ANTIFERROMAGNETIC AND SPECIAL DIELECTRIC PROPE 08-1425
MAGNETOELECTRIC EFFECT IN FERROELECTRIC- ANTIFERROMAGNETIC BISMUTH TITANIUM FERRATE. 08-1360
THERMAL CONDUCTION IN A TWO DIMENSIONAL ANTIFERROMAGNETIC COPPER FORMATE TETRAHYDRATE. 26-2932
UM MOLYBDATE. PARTS 1-3: HEAT CAPA/ MAGNETOTHERMODYNAMICS OF ANTIFERROMAGNETIC, FERROELECTRIC BETA GADOLINI 30-3275
.5) OXY FLUORIDE. MAGNETOELECTRIC EFFECT IN FERROELECTRIC- ANTIFERROMAGNETIC LITHIUM IRON(0.5) TANTALUM(0 36-3568
WITH AN ACOUSTIC OR ELECTROMAGNETIC WAVE IN A FERROELECTRIC ANTIFERROMAGNETIC MATERIAL. /EQUENCY SPIN WAVE 03-0284
AGNET IRON ANTIMONY(2) OXIDE(4) (MAY BE FERROELECTRIC DUE TO ANTIFERROMAGNETIC ORDERING). /N THE ANTIFERROM 36-3563
IDE. LUMINESCENCE OF EUROPIUM(3+) ION IN ANTIFERROMAGNETIC POTASSIUM MANGANESE TRIFLUOR 33-3381
IDE. TWO MAGNON SCATTERING OF LIGHT IN ANTIFERROMAGNETIC POTASSIUM MANGANESE TRIFLUOR 33-3401
/IC AND MAGNETIC EFFECTS INVOLVING THE FLUORINE-19 NUCLEI IN ANTIFERROMAGNETIC POTASSIUM MANGANESE TRIFLUO/ 33-3390
/ATIONAL SPECTRUM AND DISPERSION OF THE OPTICAL CONSTANTS OF ANTIFERROMAGNETIC POTASSIUM MANGANESE TRIFLUO/ 33-3385
IDE. RESONANCE BEHAVIOR IN CANTED ANTIFERROMAGNETIC POTASSIUM MANGANESE TRIFLUOR 33-3403
ES. (BISMUTH FERRATE) ANTIFERROMAGNETIC PROPERTIES OF SOME PEROVSKIT 08-1441
THE ORDERED MAGNETIC STATE OF COPPER FORMATE TETRAHYDRATE BY ANTIFERROMAGNETIC RESONANCE. STUDY OF 26-2943
MANGANESE TRIFLUORIDE. ANTIFERROMAGNETIC RESONANCE SPECTRUM IN BARIUM 33-3396
/F LEAD ZIRCONATE TITANATE MODIFIED WITH SODIUM OR POTASSIUM ANTIMONATE, NIOBATE, OR BISMUTHATE, OR BY BIS/ 09-1539
, TELLURCIODIDE, TRITELLURIDE, TRIIODIDE, TRIBROMIDE, INDIUM ANTIMONIDE) /LENIDE, SULFOIODIDE, SELENOIODIDE 15-2108
BARIUM TITANATE FILMS PREPARED BY RF SPUTTERING ONTO INDIUM ANTIMONIDE OR GALLIUM ARSENIDE. 04-0741
/TALS. (GADOLINIUM MOLYBDATE, KDP, ORTHOFERRITES, DYSPROSIUM ANTIMONIDE, RUBIDIUM IRON TRIFLUORIDE, BORACI/ 04-0036
CHALCOGENIDE BROMIDES OF ANTIMONY AND BISMUTH. 15-2168
TRA OF SOME FERROELECTRIC CRYSTALS OF GROUPS V, VI AND VII. (ANTIMONY AND BISMUTH SULFO HALIDES) /TION SPEC 15-2210
/COMPOSITION A(V)B(VI)C(VII) UNDER HYDROTHERMAL CONDITIONS. (ANTIMONY AND BISMUTH SULFO, SELENO, TELLU/ 15-2213
TORS. (ORIGIN AND EFFECTS OF GRAIN BOUNDARY BLOCKING LAYERS, ANTIMONY DOPANT, REVIEW) /OELECTRIC SEMICONDUC 02-0074
HIGH EFFECTIVE D/ GRAIN BOUNDARY BARRIER LAYERS IN (N TYPE, ANTIMONY DOPED) BARIUM TITANATE CERAMIC WITH A 04-0581
RUCTURE, AND PCT RESISTORS) INCORPORATION OF ANTIMONY INTO THE BARIUM TITANATE LATTICE. (ST 04-0934
/RMAL METHOD FOR PREPARING A(V)B(VI)C(VII) COMPOUNDS. (USING ANTIMONY OR BISMUTH, SULFUR, SELENIUM OR TELL/ 15-2194
/OF GROUP A(V)B(VI)C(VII). (HIGH GRADE SINGLE CRYSTALS, V IS ANTIMONY OR BISMUTH, VI IS SULFUR, SELENIUM O/ 15-2214
NEW FERROELECTRIC: ANTIMONY ORTHONIOBATE. (CURIE TEMPERATURE) 36-3589
OPTICAL PROPERTIES OF ANTIMONY SELENOIODIDE IN THE INFRARED. 15-2219
DIAGRAM AND ELECTRICAL CONDUCTIVITY OF BISMUTH SELENOIODIDE- ANTIMONY SELENOIODIDE SYSTEM ALLOYS. PHASE 15-2122
/G SINGLE CRYSTALS FROM THE VAPOR PHASE BY A DYNAMIC METHOD (ANTIMONY SESQUISELENIDE OR SULFIDE, ANTIMONY / 15-2162
LS. PYROELECTRIC EFFECT IN ANTIMONY SULFO AND SELENO IODIDE SINGLE CRYSTA 15-2163
/(VI)C(VII) COMPOUNDS. ANTIMONY(X) BISMUTH(1-X) SULFOIODIDE, ANTIMONY SULFO(X) SELENO(1-X) IODIDE, ANTIMON/ 15-2116
E STATES. PRODUCTION AND SOME OPTICAL PROPERTIES OF ANTIMONY SULFOBROMIDE IN GLASSY AND CRYSTALLIN 15-2142
OF ANOMALOUS ABSORPTION OF ULTRASOUND IN SINGLE CRYSTALS OF ANTIMONY SULFOIODIDE. ANISOTROPY 15-2222
ABSORPTION OF SOUND NEAR THE FERROELECTRIC PHASE TRANSITION (ANTIMONY SULFOIODIDE). ANOMALOUS 15-2221
ANOMALOUS PROPERTIES IN PHOTOFERROELECTRIC ANTIMONY SULFOIODIDE. 15-2196
BAND STRUCTURE OF THE FERROELECTRIC SEMICONDUCTOR ANTIMONY SULFOIODIDE. 15-2179

PAGE 338

BOSON ECHOES. (ANTIMONY SULFOIODIDE)   15-2144
CHARACTERISTICS OF THE PHOTOELECTRIC EMF OF ANTIMONY SULFOIODIDE.   15-2129
CHARACTERISTICS OF THE PHOTOELECTRIC EMF OF ANTIMONY SULFOIODIDE.   15-2153
CRITICAL PHENOMENA IN ANTIMONY SULFOIODIDE.   15-2231
CRITICAL POINT IN FERROELECTRIC ANTIMONY SULFOIODIDE.   15-2237
CRYSTAL FIELD GRADIENTS IN ANTIMONY SULFOIODIDE.   15-2185
CRYSTAL STRUCTURE AND CONDUCTIVITY OF ANTIMONY SULFOIODIDE.   15-2112
TEMPERATURE DEPENDENCE OF LATTICE CONSTANTS IN FERROELECTRIC ANTIMONY SULFOIODIDE.   DETERMINATION OF THE 15-2140
TIGATION OF PHOTOFERROELECTRIC EFFECTS IN SINGLE CRYSTALS OF ANTIMONY SULFOIODIDE.   DILATOMETRIC INVES 15-2156
OTOELECTRET CHARGE ON VALUE OF FERROELECTRIC POLARIZATION IN ANTIMONY SULFOIODIDE.   EFFECT OF PH 15-2186
ELECTRICAL AND PHOTOELECTRIC PROPERTIES OF ANTIMONY SULFOIODIDE.   15-2107
ELECTROOPTIC EFFECTS OF THE ABSORPTION EDGE OF ANTIMONY SULFOIODIDE.   15-2198
ELECTROOPTICAL AND ELECTROMECHANICAL EFFECT IN ANTIMONY SULFOIODIDE.   15-2178
OSTRICTION, PIEZOELECTRICITY AND ELASTICITY IN FERROELECTRIC ANTIMONY SULFOIODIDE.   ELECTR 15-2165
Y IN FERROELECTRICS DUE TO SURFACE LAYERS. (BARIUM TITANATE, ANTIMONY SULFOIODIDE) /RINSIC PHOTOCONDUCTIVIT 04-0681
FAR INFRARED REFLECTIVITY OF ANTIMONY SULFOIODIDE.   15-2209
FAR INFRARED REFLECTIVITY SPECTRA OF ANTIMONY SULFOIODIDE.   15-2234
FAR INFRARED REFLECTIVITY SPECTRA OF ANTIMONY SULFOIODIDE.   15-2235
FERROELECTRIC BARKHAUSEN EFFECT IN ANTIMONY SULFOIODIDE.   15-2172
OELECTRIC TWINNING) TREATED AS CRITICAL POINTS IN CRYSTALS. (ANTIMONY SULFOIODIDE) /CTRIC INSTABILITY (FERR 15-2176
FERROELECTRIC TRANSITION OF ANTIMONY SULFOIODIDE.   15-2195
FIELD EMISSION FROM ANTIMONY SULFOIODIDE.   15-2203
HEAT CAPACTIV OF POLYCRYSTALLINE ANTIMONY SULFOIODIDE.   15-2238
MPURITY PHOTOCONDUCTIVITY IN THE FERROELECTRIC SEMICONDUCTOR ANTIMONY SULFOIODIDE.   INDUCED I 15-2159
NFLUENCE OF LIGHT ON SELF POLARIZATION OF SINGLE CRYSTALS OF ANTIMONY SULFOIODIDE.   I 15-2228
S ON THE DOMAIN STRUCTURE OF THE FERROELECTRIC SEMICONDUCTOR ANTIMONY SULFOIODIDE. / NONEQUILIBRIUM CARRIER 15-2104
CARRIER SCREENING ON POLARIZATION PROCESSES IN FERROELECTRIC ANTIMONY SULFOIODIDE. /ENCE OF NONEQUILIBRIUM 15-2158
NJECTION CURRENTS AND ACCUMULATION OF POLARIZATION CHARGE IN ANTIMONY SULFOIODIDE.   I 15-2188
LTRASONIC WAVES WITH MOVING DOMAIN WALLS IN FERROELECTRICS. (ANTIMONY SULFOIODIDE)   INTERACTION OF U 15-2192
LATTICE MODES AND PHASE TRANSITION IN ANTIMONY SULFOIODIDE.   15-2110
LIGHT INDUCED PYROCURRENTS IN FERROSEMICONDUCTOR ANTIMONY SULFOIODIDE.   15-2130
SCENCE OF SINGLE CRYSTALS OF THE FERROELECTRIC SEMICONDUCTOR ANTIMONY SULFOIODIDE.   LUMINE 15-2229
MECHANISM OF THE PHOTODOMAIN EFFECT IN ANTIMONY SULFOIODIDE.   15-2146
NOTES ON ABSORPTION BAND EDGE OF (PARAELECTRIC) ANTIMONY SULFOIODIDE.   15-2177
CLEAR QUADRUPOLE AND ELECTROACOUSTIC ECHOES IN FERROELECTRIC ANTIMONY SULFOIODIDE.   NU 15-2189
OPPOSING FERROELECTRIC DOMAINS IN SINGLE CRYSTALS OF ANTIMONY SULFOIODIDE.   15-2252
OPTICAL SECOND HARMONIC GENERATION IN ANTIMONY SULFOIODIDE.   15-2164
PHOTOCONDUCTIVITY OF FERROELECTRIC PHOTOCONDUCTOR ANTIMONY SULFOIODIDE.   15-2226
PHOTOEMF'S IN THE FERROELECTRIC SEMICONDUCTOR ANTIMONY SULFOIODIDE.   15-2132
PHOTOVOLTAIC EFFECT IN FERROELECTRIC AND PHOTOCONDUCTIVE ANTIMONY SULFOIODIDE.   15-2174
PYROCURRENTS HIGHER THAN THE CURIE POINT IN ANTIMONY SULFOIODIDE.   15-2131
RAMAN SCATTERING IN FERROELECTRIC SEMICONDUCTOR ANTIMONY SULFOIODIDE.   15-2111
REFLECTIVITY ON ANTIMONY SULFOIODIDE.   15-2244
BETWEEN THE ELECTRIC POLARIZATION AND THE ABSORPTION EDGE IN ANTIMONY SULFOIODIDE.   RELATION 15-2175
SOFT LATTICE VIBRATIONS AND VIBRATION COUPLING IN ANTIMONY SULFOIODIDE.   15-2233
E COUPLING AND CRITICAL RAYLEIGH SCATTERING IN FERROELECTRIC ANTIMONY SULFOIODIDE.   SOFT MOD 15-2232
PHONON COUPLINGS IN THE PRESSURE INDUCED PHASE TRANSITION IN ANTIMONY SULFOIODIDE.   SOFT 15-2240
SOME FEATURES OF THE PHOTOEMF SPECTRUM OF ANTIMONY SULFOIODIDE.   15-2136
SS SECTION OF CAPTURE CENTERS IN FERROELECTRIC SEMICONDUCTOR ANTIMONY SULFOIODIDE. /E DEPENDENCE OF THE CRO 15-2147
X-RAY CRYSTALLOGRAPHIC STUDIES OF ANTIMONY SULFOIODIDE.   15-2109
NUCLEAR QUADRUPOLE RESONANCE IN FERROELECTRICS OF THE ANTIMONY SULFOIODIDE (AND BROMIDE) TYPE.   15-2216
(3.3 AND 9.8 GHZ, 0-50 C).   DIELECTRIC PROPERTIES OF ANTIMONY SULFOIODIDE AT MICROWAVE FREQUENCIES 15-2171
OF THE CURIE TEMPERATURE OF A FERROELECTRIC SEMICONDUCTOR. (ANTIMONY SULFOIODIDE BROMIDE) / "FROZEN" SHIFT 15-2160
SOFT PHONON MODE AND MODE COUPLING IN ANTIMONY SULFOIODIDE (BY RAMAN SCATTERING).   15-2167
DC ELECTRIC FIELD.   STRAIN ALONG C-AXIS OF ANTIMONY SULFOIODIDE CAUSED BY ILLUMINATION IN 15-2239
R THERMOELECTRIC CURRENTS)   TRANSPORT PHENOMENA IN ANTIMONY SULFOIODIDE. (CONDUCTION MECHANISM FO 15-2241
SOTROPY)   KINETICS OF CRYSTALLIZATION OF ANTIMONY SULFOIODIDE (CRYSTAL GROWTH RATE, ANI 15-2215
RIC PHASE, UNDER DC FIELD ALONG C/ POTENTIAL DISTRIBUTION IN ANTIMONY SULFOIODIDE CRYSTAL (IN THE PARAELECT 15-2173
ANISOTROPY OF THE REFLECTION SPECTRA OF ANTIMONY SULFOIODIDE CRYSTALS.   15-2125
DIELECTRIC DISPERSION IN ANTIMONY SULFOIODIDE CRYSTALS.   15-2117
ITION ON GROWTH, DIELECTRIC, AND PHOTOELECTRIC PROPERTIES OF ANTIMONY SULFOIODIDE CRYSTALS. /FECT OF COMPOS 15-2118
INDIRECT TRANSITIONS AND ABSORPTION IN THE MID-INFRARED IN ANTIMONY SULFOIODIDE CRYSTALS.   15-2150
MOTION PICTURES OF THE PHASE TRANSITION IN ANTIMONY SULFOIODIDE CRYSTALS.   15-2120
NONLINEAR EFFECTS AND THE LOW TEMPERATURE TRANSITION IN ANTIMONY SULFOIODIDE CRYSTALS.   15-2247
PHOTODOMAIN EFFECT IN NATURALLY POLARIZED ANTIMONY SULFOIODIDE CRYSTALS.   15-2133
PHOTODOMAIN EFFECT IN NATURALLY POLARIZED ANTIMONY SULFOIODIDE CRYSTALS.   15-2135
IZATION EFFECTS IN THE ANOMALOUS ABSORPTION OF ULTRASOUND IN ANTIMONY SULFOIODIDE CRYSTALS.   POLAR 15-2123
SHORT CIRCUIT PHOTOEMF IN FERROELECTRIC SEMICONDUCTORS. (ANTIMONY SULFOIODIDE CRYSTALS)   15-2105
CRYSTALLIZATION AND ETCHING).   DOMAIN STRUCTURE OF ANTIMONY SULFOIODIDE CRYSTALS (FROM SELECTIVE 15-2183
GRAPHY)   PHASE TRANSITION AND DOMAIN STRUCTURE IN ANTIMONY SULFOIODIDE CRYSTALS. (MICRO CINEMATO 15-2184
OINT (EFFECTS OF TEMPERATURE, F/ FIEZOELECTRIC VIBRATIONS IN ANTIMONY SULFOIODIDE CRYSTALS NEAR THE CURIE P 15-2121
R): CONTROL OF NEEDLE MORPHOLOGY.   GROWTH OF LARGE ANTIMONY SULFOIODIDE CRYSTALS (TO 1 CM DIAMETE 15-2200
SCATTERING SPECTRUM. (LAT/ PHASE TRANSITION IN ANTIMONY SULFOIODIDE CRYSTALS USING THE RAMAN 15-2143
R/ INVESTIGATION OF THE PIEZORESISTANCE OF (POLYCRYSTALLINE) ANTIMONY SULFOIODIDE (DARK CONDUCTIVITY NEAR T 15-2250
RE, WAVELENGTH, APPLIED FIELD STRENGT/ REFRACTIVE INDICES OF ANTIMONY SULFOIODIDE. (DEPENDENCE ON TEMPERATU 15-2206
EFFECTS OF ILLUMINATION CN THE DOMAIN STRUCTURE OF ANTIMONY SULFOIODIDE. (ETCHING METHOD) 15-2124
OR NEAR ITS CURIE POINT.   ABSORPTION OF ULTRASOUND IN AN ANTIMONY SULFOIODIDE FERROELECTRIC SEMICONDUCT 15-2115
ELD.   DIELECTRIC AND PIEZOELECTRIC PROPERTIES OF ANTIMONY SULFOIODIDE IN A CONSTANT ELECTRIC FI 15-2249
D.   AGING OF THE FERROELECTRIC SEMICONDUCTOR ANTIMONY SULFOIODIDE IN A STATIC ELECTRIC FIEL 15-2113
ROELECTRIC/ ELECTRON STATE DENSITY AND OPTICAL PROPERTIES OF ANTIMONY SULFOIODIDE. (IN PARAELECTRIC AND FER 15-2204
/NATION ON THE FORM OF THE VOLTAGE AMPERE CHARACTERISTICS OF ANTIMONY SULFOIODIDE IN THE PARAELECTRIC REGI/ 15-2119
TIONS IN SOME ORDER DISORDER AND DISPLACIVE FERROELECTRICS. (ANTIMONY SULFOIODIDE, KDP) /S AND PHASE TRANSI 15-2208
TE/ ELECTROREFLECTION FROM SINGLE CRYSTALS OF FERROELECTRIC ANTIMONY SULFOIODIDE NEAR THE PHASE TRANSITION 15-2151
/ FIELD, AND ILLUMINATION ON THE ABSORPTION OF ULTRASOUND IN ANTIMONY SULFOIODIDE NEAR THE PHASE TRANSITIO/ 15-2248
CK PHOSPHORUS.   HIGH GAS PRESSURE CRYSTAL GROWTH OF ANTIMONY SULFOIODIDE, NICKEL CARBONATE AND BLA 15-2137
/FFECT OF PHOTOCONDUCTION ON THE FERROELECTRIC PROPERTIES OF ANTIMONY SULFOIODIDE. (PHASE BOUNDARY SHAPE S/ 15-2197
OLARIZED) RAMAN SPECTRUM OF (PARAELECTRIC AND) FERROELECTRIC ANTIMONY SULFOIODIDE (PHONON ASSIGNMENTS). (P 15-2207
CILIATIONS)   ANOMALOUS ECHO IN FERROELECTRIC ANTIMONY SULFOIODIDE. (PIEZOELECTRIC DOMAIN OS 15-2211
RA/ LONG WAVELENGTH OPTICAL PHONONS AND PHASE TRANSITIONS IN ANTIMONY SULFOIODIDE. (POLARIZED INFRARED AND 15-2106
FERROELECTRIC SEMICONDUCTORS. (ANTIMONY SULFOIODIDE, REVIEW, 57 REFS)   02-0084

/CONDUCTORS AT TEMPERATURES NEAR A TYPE-1 PHASE TRANSITION. (ANTIMONY SULFOIODIDE, SELENOIODIDE, BISMUTH S/ 15-2114
CHEMICAL TRANSPORT)   CRYSTALLIZATION BEHAVIOR OF ANTIMONY SULFOIODIDE. (SINGLE CRYSTAL GROWTH, 15-2201
DILATOMETRIC INVESTIGATION OF ANTIMONY SULFOIODIDE SINGLE CRYSTALS. 15-2155
F OXYGEN ADSORPTION ON THE FERROELECTRIC PHASE TRANSITION IN ANTIMONY SULFOIODIDE SINGLE CRYSTALS. EFFECT O 15-2187
EFFECT OF ILLUMINATION ON REPOLARIZATION PROCESS IN ANTIMONY SULFOIODIDE SINGLE CRYSTALS. 15-2220
EFFECT OF THE ADSORPTION OF OXYGEN ON THE PROPERTIES OF ANTIMONY SULFOIODIDE SINGLE CRYSTALS. 15-2154
FEATURES OF THE PHASE TRANSITION IN ANTIMONY SULFOIODIDE SINGLE CRYSTALS. 15-2148
FREQUENCY DEPENDENCE OF ANOMALOUS ULTRASOUND DAMPING IN ANTIMONY SULFOIODIDE SINGLE CRYSTALS. 15-2139
GENERATION OF THE SECOND ACOUSTIC HARMONIC IN ANTIMONY SULFOIODIDE SINGLE CRYSTALS. 15-2224
GENERATION OF THE SECOND ACOUSTIC HARMONIC IN ANTIMONY SULFOIODIDE SINGLE CRYSTALS. 15-2223
ONE DIMENSIONAL MECHANICAL STRESS ON THE PHASE TRANSITION IN ANTIMONY SULFOIODIDE SINGLE CRYSTALS. /NCE OF 15-2245
INSTABILITY OF FERROELECTRIC POLARIZATION IN ANTIMONY SULFOIODIDE SINGLE CRYSTALS. 15-2152
PHOTODIELECTRIC EFFECT IN ANTIMONY SULFOIODIDE SINGLE CRYSTALS. 15-2236
PHOTOELECTRIC PROPERTIES (PHOTOCONDUCTIVITY) OF ANTIMONY SULFOIODIDE SINGLE CRYSTALS. 15-2134
PIEZOELECTRICITY AND ELECTROSTRICTION OF ANTIMONY SULFOIODIDE SINGLE CRYSTALS. 15-2227
PREPARATION AND PHYSICAL PROPERTIES OF NONSTOICHIOMETRIC ANTIMONY SULFOIODIDE SINGLE CRYSTALS. 15-2199
RELAXATION PROCESSES OF ANTIMONY SULFOIODIDE SINGLE CRYSTALS. 15-2190
TEMPERATURE DEPENDENCE OF THE SWITCHING PROCESS IN ANTIMONY SULFOIODIDE SINGLE CRYSTALS. 15-2191
/TRICAL AND PHOTOELECTRIC PROPERTIES OF THE CONTACTS BETWEEN ANTIMONY SULFOIODIDE SINGLE CRYSTALS AND SOME/ 15-2251
ITH MICROSECOND LIGHT PULSES.   RELAXATION PROCESSES IN ANTIMONY SULFOIODIDE SINGLE CRYSTALS EXCITED W 15-2202
GION OF PHASE COEXISTENCE.   POLARIZATION REVERSAL OF ANTIMONY SULFOIODIDE SINGLE CRYSTALS IN THE RE 15-2246
DER PRESSURE AND INVESTIGATION OF THEIR ELEC/ PREPARATION OF ANTIMONY SULFOIODIDE TEXTURES FROM THE MELT UN 15-2138
/OF THE PHASE TRANSITIONS IN ROCHELLE SALT, BARIUM TITANATE, ANTIMONY SULFOIODIDE, THIOUREA, TRIGLYCINE SU/ 02-0099
POLARIZATION NATURE IN SEMICONDUCTOR FERROELECTRICS OF THE ANTIMONY SULFOIODIDE TYPE. 15-2181
MOSSBAUER STUDIES OF ANTIMONY SULFOIODIDE TYPE CRYSTALS. 15-2180
UBSTITUTION/ NEW FERROELECTRIC A(V)B(VII)C(VII) COMPOUNDS OF ANTIMONY SULFOIODIDE TYPE. (PARTIAL OR TOTAL S 15-2205
ELECTRICS.   SOME FEATURES OF TRAPPING LEVELS IN ANTIMONY SULFOIODIDE TYPE SEMICONDUCTING FERRO 15-2182
LECTRICS.   DOMAIN STRUCTURE AND LOCAL STATES IN ANTIMONY SULFOIODIDE TYPE SEMICONDUCTOR FERROE 15-2127
ABIL/ (ELASTIC CONSTANTS FROM) BRILLOUIN SCATTERING STUDY OF ANTIMONY SULFOIODIDE USING A DOUBLE PASSED, ST 15-2225
LID SOLUTIONS. (DIELECTRIC CONSTANTS, PYROELECTRIC CURRENTS, ANTIMONY SULFOIODIDE(X) BROMIDE(1-X)) /VII) SO 15-2157
LOW TEMPERATURE PHASE CHANGES. (EFFECT OF PHOTOCONDUCTIVITY, ANTIMONY SULFOIODIDE(X) BROMIDE(1-X)) /E WITH 15-2161
UTIONS DEPEND ON TEMPERATU/ HOW THE DIELECTRIC PROPERTIES OF ANTIMONY SULFOIODIDE(X) BROMIDE(1-X) SOLID SOL 15-2149
EM. (BRIDGMAN- STOCKBARGER METHOD)   CRYSTAL GROWTH IN THE ANTIMONY SULFOIODIDE- ANTIMONY TRISULFIDE SYST 15-2230
IONS.   DIELECTRIC PROPERTIES OF ANTIMONY SULFOIODIDE- SULFOBROMIDE SOLID SOLUT 15-2242
URE THIN FILM.   SWITCHING AND MEMORY EFFECT IN ANTIMONY SULFOIODIDE- TIN DIOXIDE HETEROSTRUCT 15-2166
URE SWITCH.   MODE FOR THE ANTIMONY SULFOIODIDE- TIN DIOXIDE HETEROSTRUCT 15-2243
EPENDENCE OF THE PIEZOELECTRIC CONSTANTS OF STIBOTANTALITE. (ANTIMONY (TANTALUM, NIOBIUM) OXIDE(4)) /TURE D 36-3601
BAND STRUCTURE OF THE FERROELECTRIC SEMICONDUCTOR ANTIMONY TRISULFIDE. 36-3575
URE DEPENDENCES OF THE HALL AND DRIFT MOBILITIES OF HOLES IN ANTIMONY TRISULFIDE SINGLE CRYSTALS. TEMPERAT 15-2193
/ COMPOUNDS OF THE TYPE A(2)(V)B(3)(VI) AND A(V)B(VI)C(VII). (ANTIMONY TRISULFIDE, TRISELENIDE, SULFOIODIDE/ 15-2108
/RIE POINTS OF SOLID SOLUTIONS OF A(V)B(VI)C(VII) COMPOUNDS. ANTIMONY(X) BISMUTH(1-X) SULFOIODIDE, ANTIMON/ 15-2116
TO ANTI/ MAGNETOELECTRIC EFFECT IN THE ANTIFERROMAGNET IRON ANTIMONY(2) OXIDE(4) (MAY BE FERROELECTRIC DUE 36-3563
ON FERROELECTRICITY AND ANTIFERROELECTRICITY OF THE AO3 TYPE CRYSTAL. (TUNGSTEN TRIOXIDE) 10-1641
S EFFECT COMMUTATORS. (POTASSIUM DIHYDROGEN PHOSPHA/ (FINITE APERTURE LIGHT SWITCH USING) TRANSVERSE POCKEL 02-0056
S (REVIEW EMPHASIZING ELECTROOPTIC MODUL/ (HIGH SPEED, SMALL APERTURE) MODULATORS FOR OPTICAL COMMUNICATION 02-0056
PHASE TRANSITION IN TRIGLYCINE SULFATE.   APPLICABILITY OF THE PIPPARD RELATIONS TO THE 27-3157
CTS IN POTASSIUM DIHYDROGEN PHOSPHATE CRYSTALS GROWN FROM AN AQUEOUS SOLUTION. / LAYERS AND OTHER MACRODEFE 17-2496
RT-8: TRAN/ DIELECTRIC RELAXATION SPECTRA OF WATER, ICE, AND AQUEOUS SOLUTIONS AND THEIR INTERPRETATION. PA 34-3482
DP, HYDROIODIC ACID,/ GROWTH OF ELECTROOPTICAL CRYSTALS FROM AQUEOUS SOLUTIONS (APPARATUS, GROWTH OF ADP, K 17-2420
UM CHLORIDE, SODIUM CHLORIDE,/ CRYSTALLIZATION POTENTIALS IN AQUEOUS SOLUTIONS. (POTASSIUM BROMIDE, POTASSI 17-2707
ATE CRYSTALS.   ANTIPARALLEL SWITCHING IN ALPHA DOMAIN BARIUM TITAN 04-0655
THREE APPLICATION AREAS FOR STRONTIUM TITANATE GLASS CERAMICS. 06-1150
DEUTERON ARRANGEMENTS IN THE HIGH PRESSURE FORMS OF ICE 34-3438
PIEZOELECTRIC RESPONSE OF A FERROELECTRIC MEMORY ARRAY. 37-3635
FERROELECTRIC LIGHT VALVE ARRAYS FOR OPTICAL MEMORIES. 37-3676
REFLECTION OF ELASTIC SURFACE WAVE FROM PERIODIC ARRAYS ON LITHIUM NIOBATE. 11-1853
ERON QUADRUPOLE COUPLING IN FERROELECTRIC CESIUM DIDEUTERIUM ARSENATE.   DEUT 18-2721
ONIUM DIDEUTERIUM PHOSPHATE- 5 PERCENT POTASSIUM DIDEUTERIUM ARSENATE. / OF AMMONIUM IONS IN IRRADIATED AMM 17-2684
CLEAR QUADRUPOLE RESONANCE) FREQUENCY. (POTASSIUM DIHYDROGEN ARSENATE)   FERROELECTRIC MODE AND NQR (NU 17-2713
LONG RANGE CORRELATION IN FERROELECTRIC POTASSIUM DIHYDROGEN ARSENATE. /D DIELECTRIC RELAXATION STUDIES OF 18-2718
IDIUM OR AMMONIUM ARSENATES, RUBIDIUM DIDEUTERIUM ARSENATE) /TED ROCHELLE SALT, POTASSIUM OR RUB 18-2719
NON INTERACTIONS IN ALUMINUM PHOSPHATE AND CESIUM DIHYDROGEN ARSENATE.   PHONON PHONON AND PROTON PHO 18-2739
YSTALS RUBIDIUM DIHYDROGEN PHOSPHATE AND RUBIDIUM DIHYDROGEN ARSENATE. /ROGEN BONDS IN THE FERROELECTRIC CR 17-2471
PROTONIC CONDUCTION IN POTASSIUM DIHYDROGEN ARSENATE. 18-2732
RE QUADRUPOLE RESONANCE OF ARSENIC-75 IN AMMONIUM DIHYDROGEN ARSENATE.   PU 18-2732
ELECTRIC RUBIDIUM DIHYDROGEN PHOSPHATE AND CESIUM DIHYDROGEN ARSENATE. /IC-75 QUADRUPOLAR COUPLING IN FERRO 17-2439
.   SEE BOTH ADA AND AMMONIUM DIHYDROGEN ARSENATE.
SEE BOTH KDA AND POTASSIUM DIHYDROGEN ARSENATE.
NANCE SPECTRA OF THE FERROELECTRIC CRYSTAL CESIUM DIHYDROGEN ARSENATE. /FECT IN THE NUCLEAR QUADRUPOLE RESO 18-2722
ATIONS FROM EPR STUDIES OF X-IRRADIATED POTASSIUM DIHYDROGEN ARSENATE AND AMMONIUM DIHYDROGEN ARSENATE. /UR 18-2728
PHATE, POTASSIUM DIDEUTERIUM PHOSPHATE, POTASSIUM DIHYDROGEN ARSENATE AND AMMONIUM DIHYDROGEN PHOSPHATE. /S 17-2409
CTRIC POTASSIUM DIDEUTERIUM PHOSPHATE, POTASSIUM DIDEUTERIUM ARSENATE, AND CESIUM DIDEUTERIUM ARSENATE. /LE 17-2438
RELAXATION AND RESONANCE IN FERROELECTRIC CESIUM DIDEUTERIUM ARSENATE AND CESIUM DIHYDROGEN ARSENATE. /ICE 18-2720
ELECTRIC CRYSTALS. (RAMAN SCATTERING IN POTASSIUM DIHYDROGEN ARSENATE AND CESIUM DIHYDROGEN ARSENATE) /IEZO 18-2723
NATE.   ARSENIC-75 NMR RESONANCE IN AMMONIUM DIHYDROGEN ARSENATE AND DEUTERO AMMONIUM DIDEUTERIUM ARSE 18-2736
HOSPHATE, RUBIDIUM DIHYDROGEN PHOSPHATE, RUBIDIUM DIHYDROGEN ARSENATE, AND LITHIUM IODATE. /UM DIHYDROGEN P 08-1413
OMAINS IN THE IRRADIATED FERROELECTRICS POTASSIUM DIHYDROGEN ARSENATE AND POTASSIUM DIDEUTERIUM ARSENATE. /D 18-2726
THERMODYNAMIC PROPERTIES OF POTASSIUM DIHYDROGEN ARSENATE AND POTASSIUM DIDEUTERIUM ARSENATE. 18-2730
PROTON PHONON COUPLING IN CESIUM DIHYDROGEN ARSENATE AND CESIUM DIHYDROGEN ARSENATE. 18-2733
ARTIALLY DEUTERATED FERROELECTRIC CRYSTALS CESIUM DIHYDROGEN ARSENATE AND RUBIDIUM DIHYDROGEN ARSENATE. / P 18-2740
/ETIC RESONANCE AND ELECTRON NUCLEAR DOUBLE RESONANCE OF THE ARSENATE CENTER IN X-IRRADIATED POTASSIUM DIH/ 18-2727
ST/ OBSERVATION OF ELECTRON NUCLEAR TRIPLE RESONANCE FOR THE ARSENATE CENTER IN X-RAY IRRADIATED SINGLE CRY 18-2729
TO TRANSVERSE OPTICAL PHONON). PART-3: (POTASSIUM DIHYDROGEN ARSENATE, CESIUM DIHYDROGEN ARSENATE) /UPLING 03-0384
T-2: INTERNAL FUNDAMENTAL VIBRATIONS OF POTASSIUM DIHYDROGEN ARSENATE IN THE PARAELECTRIC PHASE. /TION. PAR 18-2738
ENTERS IN AMMONIUM DIHYDROGEN PHOSPHATE POTASSIUM DIHYDROGEN ARSENATE MIXED CRYSTALS. /F RADIATION DAMAGE C 17-2547
NEW FERROELECTRIC: THE DOUBLE AMMONIUM THALLIUM MONOARSENATE OR ATLAS. 18-2735
/ OF THE PROTON LATTICE COUPLED MODE IN POTASSIUM DIHYDROGEN ARSENATE, POTASSIUM DEUTERIUM ARSENATE, AND A/ 18-2724
/ETIC RESONANCE STUDIES OF X-IRRADIATED POTASSIUM DIHYDROGEN ARSENATE, POTASSIUM DIDEUTERIUM ARSENATE, RUB/ 18-2725
/R MATERIALS. (POTASSIUM OR AMMONIUM DIHYDROGEN PHOSPHATE OR ARSENATE, POTASSIUM DIDEUTERIUM PHOSPHATE, PO/ 01-0021
/NTIFICATION OF SLATER CONFIGURATIONS AND OF EXCHANGE OF THE ARSENATE RADICAL IN IRRADIATED FERROELECTRIC / 18-2734

ESR OF FREE RADICALS IN FERROELECTRIC AND ANTIFERROELECTRIC ARSENATE SINGLE CRYSTALS. 18-2731
ECTROSCOPY: COUPLED MODES (FERROELECTRICS, CESIUM DIHYDROGEN ARSENATE, STRONTIUM TITANATE) /T MODE RAMAN SP 06-1207
ANTIFERROELECTRIC TRANSITIONS IN ADP AND AMMONIUM DIHYDROGEN ARSENATE STUDIED BY INFRARED ABSORPTION. 17-2706
MS OF AMMONIUM DIHYDROGEN PHOSPHATE AND POTASSIUM DIHYDROGEN ARSENATE TO 52 KBAR AND 400 DEG C. /ASE DIAGRA 17-2478
/M FERROELECTRIC MODES IN THE KDP ISOMORPHOUS PHOSPHATES AND ARSENATES. (KDP, AMMONIUM DIHYDROGEN PHOSPHAT/ 17-2639
ROCHELLE SALT, POTASSIUM OR RUBIDIUM OR AMMONIUM DIHYDROGEN ARSENATES, RUBIDIUM DIDEUTERIUM ARSENATE) /TED 18-2719
IUM DIHYDROGEN PHOSPHATES AND CESIUM AND AMMONIUM ARSENATES, STANDARD METHOD PROPOSED). /D AMMON 17-2577
/E. (PARTIAL OR TOTAL SUBSTITUTION OF ANTIMONY BY BISMUTH OR ARSENIC, SULFUR BY SELENIUM OR OXYGEN, AND IO/ 15-2205
LFOIODIDE, ANTIMONY SULFO(X) SELENC(1-X) IODIDE, ANTIMONY(X) ARSENIC(1-X) SULFOIODIDE) /(X) BISMUTH(1-X) SU 15-2116
PURE QUADRUPOLE RESONANCE OF ARSENIC-75 IN AMMONIUM DIHYDROGEN ARSENATE. 18-2732
N ARSENATE AND DEUTERO AMMONIUM DIDEUTERIUM ARSENATE. ARSENIC-75 NMR RESONANCE IN AMMONIUM DIHYDROGE 18-2736
A OF PARTIALLY DEUTERATED FERROELECTRIC CRYSTALS CESIUM DIH/ ARSENIC-75 NUCLEAR QUADRUPOLE RESONANCE SPECTR 18-2740
IC RUBIDIUM DIHYDROGEN PHOSPHATE AND CESIUM/ RUBIDIUM-87 AND ARSENIC-75 QUADRUPOLAR COUPLING IN FERROELECTR 17-2439
PREPARED BY RF SPUTTERING ONTO INDIUM ANTIMONIDE OR GALLIUM ARSENIDE. BARIUM TITANATE FILMS 04-0741
ER. LOW POWER DENSITY GALLIUM ARSENIDE- LITHIUM NIOBATE SURFACE WAVE AMPLIFI 11-1825
INGLE CRYSTALS IN THE REG/ ANOMALOUS BEHAVIOR OF THE OPTICAL ASBSORPTION COEFFICIENT OF LEAD MAGNONIOBATE S 13-1974
TE AND BARIUM SODIUM TANTALATE (FORCE CONSTANTS, SPACE GROUP ASSIGNMENT). /IUM NIOBATE, BARIUM SODIUM NIOBA 11-1808
TE. DIRECTIONAL DISPERSION AND ASSIGNMENT OF OPTICAL PHONONS IN LITHIUM NIOBA 11-1695
PARAELECTRIC AND) FERROELECTRIC ANTIMONY SULFOIODIDE (PHONON ASSIGNMENTS). (POLARIZED) RAMAN SPECTRUM OF (15-2207
TEMPERATURE DEPENDENT POLARIZABILITIES AND COCHRAN MODE ASSIGNMENTS IN BARIUM AND STRONTIUM TITANATES. 04-0805
NIOBATE. ASSIGNMENTS OF OPTICAL PHONON MODES IN LITHIUM 11-1694
COMPUTATIONS FOR FERROELECTRICS. ASYMMETRIC MINOR LOOP AND TRANSIENT OPEN LOOP 03-0296
ITANATE ABOVE THE CRITICAL TEMPERATURE. CRITICAL ASYMMETRY IN LOCAL FLUCTUATIONS IN STRONTIUM T 06-1166
ATHERMAL PHASE TRANSITION IN LEAD ZIRCONATE. 07-1258
FERROELECTRIC: THE DOUBLE AMMONIUM THALLIUM MONOARSENATE OR ATLAS. NEW 18-2735
YSTAL BARIUM TITANATE FILMS GROWN FROM THE MELT IN AN OXYGEN ATMOSPHERE. SINGLE CR 04-0612
/TERING. PART-6: FORMATION OF PZT AND SINTERING UNDER OXYGEN ATMOSPHERE. PART-7: EFFECT OF CALCINATION ON / 09-1624
FABRICATION OF ELECTROOPTIC PLZT CERAMICS BY ATMOSPHERE SINTERING. 09-1600
DETECTION OF ATOM TUNNELING IN SOLIDS. (KDP) 17-2612
UM TITANATE) COMPUTER EXPERIMENT ON LINEAR CHAIN OF ATOMS LYING IN DOUBLE MINIMUM POTENTIAL. (BARI 03-0287
ON IN THE FERROELECTRIC PHASES OF / RELATIONSHIP BETWEEN THE ATOMIC DISPLACEMENTS AND SPONTANEOUS DEFORMATI 01-1377
AND ORTHORHOMBIC BORACITE STRUCTURES. (BORACITE, ERICAITE) ATOMIC DISPLACEMENTS IN FERROELECTRIC TRIGONAL 14-2075
ITANATE (AT OPTICAL FREQUENCIES, CALCULATIONS). ATOMIC DISPLACEMENTS IN PEROVSKITE STRONTIUM T 06-1142
TEMPERATURE DEPENDENCE OF THE LITHIUM NMR SPECTRUM AND ATOMIC MOTION IN LITHIUM NIOBATE. 11-1722
ATOMIC STRUCTURE IN THE PEROVSKITIC FERRATE. 08-1398
IEW, 244 REFS) NEUTRON DIFFRACTION INVESTIGATIONS OF THE ATOMIC STRUCTURES OF INORGANIC MATERIALS. (REV 02-0133
DS LEAD ZIRCONATE(X) TITANATE(1-X) FOR TWO RHOMBOHEDRAL PHA/ ATOMIC STRUCTURES OF THE FERROELECTRIC COMPOUN 09-1574
D SYSTEMS. PART-1: A TWO DIMENSIONAL MODEL. (ICE I) ATOMIC VIBRATIONS IN ORIENTATIONALLY DISORDERE 34-3474
PHATE. SYMMETRY OF ATOMIC VIBRATIONS IN POTASSIUM DIHYDROGEN PHOS 17-2586
ALS. ATR INFRARED SPECTRA OF UNIAXIAL NITRATE CRYST 16-2280
AND FERROELECTRIC SEMICONDUCTORS. ACOUSTIC ATTENUATION AND AMPLIFICATION IN PIEZOELECTRIC 03-0275
IN POTASSIUM MANGANESE TRIFLUORIDE IN THE NEIGHBORHOOD OF / ATTENUATION AND DISPERSION OF ULTRASONIC WAVES 33-3370
CIVE TYPE FERROELECTRICS. (THEORY, BAR/ HIGH FREQUENCY SOUND ATTENUATION AND THERMAL CONDUCTIVITY IN DISPLA 03-0160
OELECTRICS / INFLUENCE OF LONG RANGE DIPOLE FORCES ON ATTENUATION AND VELOCITY OF ULTRASOUND IN FERR 03-0253
ECTRICS. ULTRASONIC ATTENUATION ANOMALY IN DISPLACIVE TYPE FERROEL 03-0393
M TITANATE. ANOMALOUS ULTRASONIC ATTENUATION AT THE 105K TRANSITION IN STRONTIU 06-1189
E. THE INFLUENCE OF THE ELECTRIC FIELD ON THE ULTRASONIC ATTENUATION IN FERROELECTRIC TRIGLYCINE SULFAT 27-2956
PHENOMENOLOGICAL THEORY OF ULTRASONIC ATTENUATION IN FERROELECTRICS. 03-0228
THE MEASUREMENT OF HYPERSONIC ATTENUATION IN LITHIUM NIOBATE. 11-1760
CANCIES). ANOMALOUS (LONGITUDINAL WAVE) ATTENUATION IN LITHIUM TANTALATE (EFFECT OF VA 12-1892
CRYSTALS. LANDAU-KHALATNIKOV ATTENUATION IN POTASSIUM DIDEUTERIUM PHOSPHATE 17-2656
GERMANIUM OXIDE(20) AT LOW TEMPERATURES. ULTRASONIC ATTENUATION IN SINGLE CRYSTALS OF BISMUTH(12) 35-3510
BSTRACT ONLY) ANISOTROPY OF ULTRASONIC ATTENUATION IN UNIAXIAL FERROELECTRICS.(TGS, A 27-3055
ICIUM STRONTIUM PROPIONATE. ULTRASONIC ATTENUATION NEAR THE CURIE TEMPERATURE IN DICA 26-2948
CIUM LEAD PROPIONATE · ULTRASONIC ATTENUATION NEAR THE PHASE TRANSITION IN DICAL 26-2949
T IN POTASSIUM MANGANESE TRIFLUORIDE (ANOMALOUS / ULTRASONIC ATTENUATION NEAR THE SOFT MODE TRANSITION POIN 33-3373
SODIUM NITRITE. ULTRASONIC ATTENUATION NEAR THE TWO TRANSITION POINTS IN 16-2304
RIC LOSSES IN DISPLACIVE TYPE FERROELECTRICS (THEORY). ATTENUATION OF CRITICAL VIBRATIONS AND DIELECT 03-0159
SULFATE SINGLE CRYSTALS. (THEORY OF ANISOTROPIC RELAXATION ATTENUATION OF QUASI LONGITUDINAL PHONONS) /F 27-3078
LLE SALT AT THE PHASE TRANSITION POINT. ATTENUATION OF SOUND WAVES IN DEUTERATED ROCHE 28-3204
LLE SALT AT THE PHASE TRANSITION POINT. ATTENUATION OF SOUND WAVES IN DEUTERATED ROCHE 28-3205
MICROWAVE PHONON ATTUNUATION IN X CUT LITHIUM NIOBATE. 11-1756
ALS (TRIGLYCINE SULFA/ BEHAVIOR OF THE COMPLEX PERMITTIVITY (AUDIO FREQUENCIES) OF SOME FERROELECTRIC CRYST 02-0125
TERING FROM ONE AND TWO PHONON STAGES OF BARIUM TITANATE. AUGER-LIKE RESONANT INTERFERENCE IN RAMAN SCAT 04-0923
RIALS AND NONLINEAR CRYSTALS. (KDF, ADP, RUBIDIUM DIHYDROGE/ AUTOFOCUSING OF LASER RADIATION IN ACTIVE MATE 11-1865
NOISE ANALYSIS (APPLICATION TO BARIUM TITANATE). AUTOMATIC IMPEDANCE MEASUREMENTS USING THERMAL 04-0844
ECTRIC OSCILLATORY SYSTEMS. EFFECT OF AUTOMATIC TEMPERATURE STABILIZATION IN FERROEL 03-0294
ON THE FREQUENCY DEPENDENCE OF THE TEMPERATURE AUTOSTABILIZATION EFFECT IN FERROELECTRICS. 03-0479
APACITOR PROPERTIES, TANDEL) TEMPERATURE AUTOSTABILIZATION EFFECT IN FERROELECTRICS. (C 37-3655
SINGLE CRYSTALS IN AN AC ELECTRIC FIELD. TEMPERATURE AUTOSTABILIZATION EFFECT OF TRIGLYCINE SULFATE 27-3005
IELECTRIC HEATING AND POLARIZ/ INTERPRETATION OF TEMPERATURE AUTOSTABILIZATION OF FERROELECTRIC CRYSTAL. (D 03-0245
/IONS OF THE PIEZOELECTRIC NONLINEARITIES IN THE TEMPERATURE AUTOSTABILIZATION REGIME OF A FERROELECTRIC C/ 17-2502
ACOUSTIC SURFACE WAVE GUIDE, AVAILING FERROELECTRIC POLARIZATIONS. 37-3680
CS. RESOLUTION OF A DISCREPANCY CONCERNING THE AVERAGE DIPOLE MOMENT IN KDP TYPE FERROELECTRI 17-2618
/G BETWEEN LEAD MONOXIDE AND PZT. PART-1. PART-2: EFFECTS OF AVERAGE PARTICLE SIZE OF RAW POWDER ON SINTER/ 09-1624
K WITH FIELDS TO 90 KILOGAUSS ALONG THE A, B, AND C CRYSTAL AXES. / POLARIZED FORM FROM 0.4 TO 4.2 DEGREES 30-3275
ONIC VIBRATIONS RADIATING IN THE DIRECTION OF THE TRANSDUCER AXIS. /ERAMIC TRANSDUCER AS A SOURCE OF ULTRAS 37-3657
STANTS OF TRIGLYCINE SULFATE ORTHOGONAL TO THE FERROELECTRIC AXIS. /RATURE DEPENDENCE OF THE DIELECTRIC CON 27-2975
TION IN POTASSIUM DIDEUTERIUM PHOSPHATE . (BY NEUTRON TRIPLE AXIS SPECTROMETRY) /RROELECTRIC (SOFT) MODE MO 17-2668
OF IRON(3+) IN STRONTIUM TITANATE DUE TO NEAREST NE/ STRONG AXIAL ELECTRON PARAMAGNETIC RESONANCE SPECTRUM 06-1144
ERS. (HYDROPHONE ELEMENTS) OPEN CIRCUIT SENSITIVITY OF AXIALLY POLARIZED FERROELECTRIC CERAMIC CYLIND 37-3652

B

.4 TO 4.2 DEGREES K WITH FIELDS TC 90 KILOGAUSS ALONG THE A,	B, AND C CRYSTAL AXES. / POLARIZED FORM FROM O	30-3275
HORHOMBIC AND TRIGONAL PHASES OF ABO(3) SUBSTANCE. (WHEN THE	B ION IS FERROELECTRICALLY ACTIVE) /RNS IN ORT	08-1363
ABO(3) AND A(0.5)BO(3) TYPE PHASES ANC THE IONIC MASS OF THE	B IONS. /LATE THE FERROELECTRIC PROPERTIES OF	03-0203
TIGATION OF LANTHANUM MODIFIED LEAD TITANATE WITH A-SITE AND	B-SITE VACANCIES. X-RAY STRUCTURE INVES	05-1059
E, ZIRCONATE) CERAMICS. DISTRIBUTION OF A-SITE AND	B-SITE VACANCIES IN (LEAD, LANTHANUM) (TITANAT	09-1530
S IN LITHIUM NIOBATE (60 DB GAIN).	(BACKWARD WAVE) ACOUSTIC PARAMETRIC OSCILLATION	11-1837
SPERSION IN BARIUM TITANATE TYPE FERROELECTRICS IN THE METER	BAND. DIELECTRIC DI	04-1009
LEAD TITANATE CERAMIC RESONATORS OPERATING IN VHF	BAND.	05-1069
TE. THREE LAYER BROAD	BAND ANTIREFLECTION COATINGS FOR LITHIUM NIOBA	11-1676
TUNGSTATE, AND POTASSIUM TANTALATE. TIGHT BONDING	BAND CALCULATIONS ON RHENIUM TRIOXIDE, SODIUM	08-1334
ARIZATION. (LCAO ANALYSIS OF BAND SCHEME, STRONTIUM / ENERGY	BAND CHANGES IN PEROVSKITES DUE TO LATTICE POL	06-1101
DE. NOTES ON ABSORPTION	BAND EDGE OF (PARAELECTRIC) ANTIMONY SULFOIODI	15-2177
UOBERYLIATE, TRIGLYCINE SULFATE AND ITS VARIATION UNDER X-I/	BAND GAP OF TRIGLYCINE SELENATE, TRIGLYCINE FL	27-3124
/EMPERATURE ELECTRICAL CONDUCTIVITY MEASUREMENTS. (INTRINSIC	BAND GAP, PTCR (POSITIVE TEMPERATURE COEFFICI/	04-0822
DIDE. CONDUCTIVITY, THERMOELECTRIC PROPERTIES, AND	BAND STRUCTURE OF CRYSTALLINE BISMUTH SELENOIO	15-2126
TITANATE.	BAND STRUCTURE OF CUBIC AND TETRAGONAL BARIUM	04-0846
TANATE AND (CALCIUM DOPED) POTASSIUM/ ELECTRON TUNNELING AND	BAND STRUCTURE OF (NIOBIUM DOPED) STRONTIUM TI	06-1222
/AND ULTRAVIOLET OPTICAL PROPERTIES (REFLECTANCE SPECTRA AND	BAND STRUCTURE) OF SOME ABO(3) FERROELECTRICS/	08-1383
ED BY X-RAY PHOTOELECTRON SPECTROSCOPY (ESCA).	BAND STRUCTURE OF STRONTIUM TITANATE AS MEASUR	06-1100
LF CONSISTENT FIELD TIGHT BINDING CALCULATION. ENERGY	BAND STRUCTURE OF STRONTIUM TITANATE FROM A SE	06-1221
TANATE. (IMAGINARY CIELECTRIC CONSTANT)	BAND STRUCTURE OF THE CUBIC PHASE OF BARIUM TI	04-0845
TOR ANTIMONY SULFOIODICE.	BAND STRUCTURE OF THE FERROELECTRIC SEMICONDUC	15-2179
TOR ANTIMONY TRISULFIDE.	BAND STRUCTURE OF THE FERROELECTRIC SEMICONDUC	36-3575
GREE K PHASE TRANSITION CN THE STRONTIUM TITANATE CONDUCTION	BANDS. EFFECT OF THE 110 DE	06-1158
IUM TITANATE, POTASSIUM MOLYBDATE, AND POTASSIUM TAN/ ENERGY	BANDS FOR POTASSIUM NICKEL TRIFLUORIDE, STRONT	22-0091
D / TEMPERATURE DEPENDENCE OF THE INTENSITY OF HYDROGEN BOND	BANDS IN RAMAN SCATTERING SPECTRA OF GYPSUM AN	28-3189
/F THE (X-RADIATION AND COPPER AND IRON IMPURITY) ABSORPTION	BANDS OF DEFECT ROCHELLE SALT AND TRIGLYCINE /	27-3108
IED TO BARIUM TITANATE. FIELD DEPENDENCE CF ABSORPTION	BANDS OF FERROELECTRICS IN FAR INFRARED. (APPL	04-0665
ENERGY	BANDS OF POTASSIUM NIOBATE.	08-1372
OF THE HYDROGEN PONDS IN THE FER/ POLARIZATION OF THE RAMAN	BANDS OF THE HYDROXYL VIBRATIONS AND STRUCTURE	17-2471
NATE). POLARCNS IN ANISOTROPIC ENERGY	BANDS (THEOPY, CALCULATIONS FOR STRONTIUM TITA	03-0305
T OF RESONANT FREQUENCY IN LITHIUM TANTALATE LENGTH EXPANDER	BARS. ZERO TEMPERATURE COEFFICIEN	11-1810
ERROELECTRIC CRYSTALS (BARIUM OR LEAD TITANATE, FROM SYSTEMS	BARIA OR LITHIA, TITANIA, ALUMINA, SILICA). /F	04-0783
THE CADMIUM BORATE- ABO(3) SYSTEMS. (WHERE A IS STRONTIUM OR	BARIUM, AND B IS HAFNIUM, TITANIUM, OR TIN) /	36-3528
/ FERROELECTRICS WITH PEROVSKITE AND PYROCHLORE STRUCTURES.	(BARIUM AND BISMUTH(0.5)) (NIOBIUM, TANTALUM, /	08-1462
/ THE ABC NIOBATES (WHERE A AND B ARE CALCIUM, STRONTIUM, OR	BARIUM, AND C IS SODIUM OR POTASSIUM) PHASES /	04-0912
VIBRATIONAL SPECTRA OF STRONTIUM,	BARIUM AND CALCIUM TITANATES.	04-0864
VIBRATIONAL SPECTRA OF THE TITANATES OF STRONTIUM,	BARIUM, AND CALCIUM. (75 REFS)	04-0865
CTRIC QUADRUPOLE INTERACTION OF TANTALUM AND HAFNIUM IONS IN	BARIUM AND LEAD TITANATE. STATIC ELE	04-0937
PPLICATICN TO THE CALCULATION OF TETRAGONAL BIREFRINGENCE IN	BARIUM AND LEAD TITANATES. /C IONIC CRYSTAL. A	04-0578
STATIC SHELL MODEL FOR	BARIUM AND LEAD TITANATES.	04-0826
CULATING THE WALL ENERGY AND T/ THEORETICAL STATIC MODEL FOR	BARIUM AND LEAD TITANATES - APPLICATION TO CAL	04-0189
/AND PROPERTIES OF HIGH FERMITTIVITY THIN FILM DIELECTRICS.	(BARIUM AND LEAD TITANATES, EFFECT OF FERROELE/	04-0670
EUTRON) RADIATION INDUCEL PHASE TRANSFORMATIONS IN (CERAMIC)	BARIUM AND LEAD TITANATES. (SUMMARY ONLY) /T N	04-0971
AND PSEUDOBINARY PHASES BETWEEN NIOBIUM FENTOXIDE AND LEAD,	BARIUM, AND STRONTIUM OXIDES. BINARY	04-0590
SSIUM LIHYDROGEN PHOSPHATE PRESENTED, TRIGLYCINE SULFATE AND	BARIUM AND STRONTIUM TITANATE DISCUSSED) /POTA	03-0180
AELECTRIC PHASE. (POTASSIUM TANTALATE AND TANTALATE NIOBATE,	BARIUM AND STRONTIUM TITANATES) /ES IN THE PAR	04-0690
VELOCITY OF ULTRASOUND IN POLYCRYSTALLINE SOLID SOLUTIONS OF	BARIUM AND STRONTIUM TITANATES. /DENCE OF THE	04-0894
E DEPENDENT FOLARIZABILITIES AND COCHRAN MODE ASSIGNMENTS IN	BARIUM AND STRONTIUM TITANATES. TEMPERATUR	04-0805
/OUPLED CPTICAL PHONON MODE THEORY OF INFRARED DISPERSION IN	BARIUM AND STRONTIUM TITANATES AND POTASSIUM /	04-0547
OXIDE. FAR INFRARED DIELECTRIC DISPERSION IN	BARIUM AND STRONTIUM TITANATES AND TITANIUM DI	04-0975
ICRON/ TEMPERATURE STUDIES OF INFRAREE REFLECTION SPECTRA OF	BARIUM AND STRONTIUM TITANATES IN THE 2-1000 M	04-0863
/T MODES. (REVIEW OF BROOKHAVEN WCRK CN POTASSIUM TANTALATE,	BARIUM AND STRONTIUM TITANATES, KDP, 22 REFS)	02-0115
EMPERATURE COEFFICIENTS OF RESISTIVITY. SEMICONDUCTING	BARIUM AND STRONTIUM TITANATES WITH POSITIVE T	04-0995
PHASE TRANSITION IN	BARIUM BISMUTH(5+) BISMUTH(3+) OXIDE.	13-1940
YING TEMPERATURE CHARACTERISTICS OF ROCHELLE TYPE CERAMICS.	(BARIUM CALCIUM TITANATE CERAMIC) /THOD OF STUD	04-0905
LCW FREQUENCY MAGNETOELECTRIC RESONANCES IN	BARIUM COBALT TETRAFLUORIDE.	33-3409
(SPACE GROUP AND LAT/ FERROELECTRIC PARAELASTIC PARAMAGNETIC	BARIUM COBALT TETRAFLUORIDE CRYSTAL STRUCTURE	33-3386
SOME RESULTS. PHASE COMPOSITION OF (LEAD,	BARIUM) DINIOBATE. METHOD OF INVESTIGATION AND	13-1945
BRONZE CRYSTALS FROM THE POTASSIUM NIOBATE- LITHIUM NICBATE-	BARIUM DINIOBATE SYSTEM. /PERTIES OF TUNGSTEN	13-2003
OELECTRIC AND PYROELECTRIC PROPERTIES OF THE LEAD DINIOBATE-	BARIUM DINIOBATE SYSTEM. DIELECTRIC, PIEZ	08-1384
OTASSIUM NIOBATE- STRONTIUM DINIOBATE AND POTASSIUM NIOBATE-	BARIUM DINIOBATE SYSTEMS. /EQUILIBRIA IN THE P	13-2016
PHASE TRANSITIONS IN COMPLEX PEROVSKITES OF THE TYPE	BARIUM LANTHANON MOLYBDATE.	30-3261
PERATURES) FERROELECTRIC PROPERTIES OF TETRAGONAL	(BARIUM, LEAD) (NIOBATE, ZIRCONATE). (CURIE TEM	08-1293
/UDIES ON THE DIELECTRIC AND FERROELECTRIC PROPERTIES OF THE	BARIUM LEAD NIOBATE ZIRCONATE SYSTEM. (TETRAG/	13-1925
URE. EMISSION OF LIGHT BY SURFACE MICROPLASMA OF CERAMIC	(BARIUM, LEAD) TITANATE NEAR THE CURIE TEMPERAT	04-0982
CIELECTRIC AND ELASTIC PROPERTIES OF SINGLE CRYSTALS OF	BARIUM LEAD TITANATE SOLID SOLUTIONS.	04-0597
(AND RESISTANT TO L/ NEW (UNIAXIAL) NONLINEAR OPTIC CRYSTAL	BARIUM LITHIUM NIOBATE HAVING NO MICROTWINNING	13-1958
CZOCHRALSKI METHOD. GROWTH OF	BARIUM LITHIUM NIOBATE SINGLE CRYSTALS BY THE	13-1991
BATE TITANATE, POTASSIUM LANTHANUM LITHIUM NIOBIUM TITANATE,	BARIUM LITHIUM TITANIUM NIOBATE). /LITHIUM NIO	13-1923
/CHING PROPERTIES IN FERROELECTRICS OF THE FAMILY BISMUTH(4)	BARIUM(M-2) TITANIUM(M+1) OXIDE(3) (M=2,/	02-0066
ICROWAVE ULTRASONICS.	BARIUM MANGANESE FLUORIDE, A NEW CRYSTAL FOR M	33-3406
NOMALOUS SCATTERING AT AN ABSORPTION EDGE. (LITHIUM NIOBATE,	BARIUM MANGANESE TETRAFLUORIDE) /TY SENSE BY A	11-1673
ANTIFERROMAGNETIC RESONANCE SPECTRUM IN	BARIUM MANGANESE TRIFLUORIDE.	33-3396
SPECTRA AND STATIC MAGNETIC SUSCEPTIBILITY OF CHROMIUM DOPED	BARIUM METATITANATE. /NCY FROM STUDIES OF EPR	04-0542
STATE OF CHROMIUM ADDED TO	BARIUM METATITANATE.	04-0543
RNACE. REACTION SEQUENCE IN FORMING	BARIUM METATITANATE IN AN INDUSTRIAL ROTARY FU	04-1003
FERROELECTRICITY MEASUREMENTS ON STRONTIUM	BARIUM NIOBATE.	13-1922
F CRYSTAL PULLING. (CONTROL OF CRYSTAL DIAMETER OF STRONTIUM	BARIUM NIOBATE) /N TO THE CZOCHRALSKI METHOD O	13-1914
PLICATION TO A NANOSECOND RESPONSE TIME DETECTOR. (STRONTIUM	BARIUM NIOBATE) / MEASUREMENT TECHNIQUE AND AP	13-1919
ANGULAR INFRARED SIGNALS IN TRIGLYCINE SULFATE AND STRONTIUM	BARIUM NIOBATE. /TRIC VOLTAGE RESPONSE TO RECT	13-2019
LS OF INFRARED RADIATION IN TRIGLYCINE SULFATE AND STRONTIUM	BARIUM NIOBATE. /OLTAGE RESPONSE TO STEP SIGNA	13-2022
WN IN MOLTEN SOLUTIONS OF THE PSEUDOSYSTEM POTASSIUM LITHIUM	BARIUM NIOBATE. /STEN BRONZE TYPE CRYSTALS GRO	13-1948
PROPERTIES OF FERROGLASS CERAMICS BASED ON LEAD	BARIUM NIOBATE COMPOUNDS.	13-2006
ROASTING CONDITIONS ON THE STRUCTURE AND PROPERTIES OF LEAD	BARIUM NIOBATE PIEZOCERAMICS. EFFECTS OF	13-2046
LONGITUDINAL ELECTROOPTIC EFFECT IN OBLIQUE CUT STRONTIUM	BARIUM NIOBATE PLATES.	08-1472
LONGITUDINAL ELECTROOPTIC EFFECT IN OBLIQUE CUT STRONTIUM	BARIUM NIOBATE PLATES.	13-2063

PHOTOMIXING AT 10.6 MICRONS WITH STRONTIUM BARIUM NIOBATE PYROELECTRIC DETECTORS. 09-1477
PROPERTIES OF (THE TUNGSTEN BRONZE SYSTEM) POTASSIUM SODIUM BARIUM NIOBATE (SECOND HARMONIC GENERATION). / 13-2028
EFFECT OF ELECTRIC FIELD ON PHASE COMPOSITION OF (LEAD, BARIUM) NIOBATE SOLID SOLUTIONS. 13-1943
LECTRIC AND PYROELECTRIC PROPERTIES OF THE LEAD METANIOBATE- BARIUM NIOBATE SYSTEM. DIELECTRIC, PIEZOE 08-1385
PREPARATION OF SODIUM AND BARIUM NIOBATES AND LEAD MOLYBDATE. 04-0657
/VIOR OF NONLINEAR DIELECTRICS (EXPERIMENT AND DISCUSSION OF BARIUM (OR LEAD OR CALCIUM) STRONTIUM TITANAT/ 08-1281
/RTIES OF GLASS- CERAMICS CONTAINING FERROELECTRIC CRYSTALS (BARIUM OR LEAD TITANATE, FROM SYSTEMS BARIA O/ 04-0783
M TITANATE SYSTEMS./ DIELECTRIC NONLINEARITY AND LOSS IN THE BARIUM (OR LEAD, 4 COMPOSITIONS EACH) STRONTIU 09-1522
/ND BARIUM(3) BISMUTH(2) (TUNGSTEN OR MOLYBDENUM) OXIDE(6), (BARIUM OR STRONTIUM) (2) COPPER TUNGSTEN OXIDE/ 36-3619
ECTRIC/ SYNTHESIS AND X-RAY INVESTIGATION OF THE PEROVSKITES BARIUM OR STRONTIUM METAPLUMBATES (NOT FERROEL 36-3607
/LEX OXIDES WITH TUNGSTEN BRONZE TYPE STRUCTURES (POTASSIUM BARIUM OR STRONTIUM) TITANATE NIOBATE, (BARIU 13-1962
IDUS) FIELD AND MELT GROWTH OF CRYSTALS IN THE SODIUM OXIDE- BARIUM OXIDE NIOBIUM PENTOXIDE SYSTEM. /ND SOL 13-1949
E COMPOUNDS. ELECTRICAL PROPERTIES OF TERNARY BARIUM OXIDE- LANTHANUM OXIDE- TITANIUM DIOXID 13-1912
DE. (OXIDE/ PHASE EQUILIBRIA RELATIONS IN THE TERNARY SYSTEM BARIUM OXIDE- STRONTIUM OXIDE- NIOBIUM PENTOXI 13-1920
QUADRATIC TUNGSTEN BRONZES STRONTIUM POTASSIUM TANTALATE AND BARIUM POTASSIUM TANTALATE. /WO NEW PHASES OF 13-2010
RONZE TYPE STRUCTURE. SODIUM BARIUM RARE EARTH NIOBATES WITH THE TUNGSTEN B 13-1972
L CONTENT ON THE PROPERTIES AND PROCESSING OF SINGLE CRYSTAL BARIUM SODIUM NIOBATE. /ICHIOMETRY AND HYDROXY 13-1915
IC AND DIELECTRIC PROPERTIES OF STRONTIUM SODIUM NIOBATE AND BARIUM SODIUM NIOBATE. / ON THE CRYSTALLOGRAPH 13-1921
EFFECTS OF CHANGES IN MELT COMPOSITION ON CRYSTAL GROWTH OF BARIUM SODIUM NIOBATE. 13-1908
FAR INFRARED DIELECTRIC DISPERSION IN BARIUM SODIUM NIOBATE. 13-1986
C PATTERN FIXING IN ELECTROOPTIC CRYSTALS. (LITHIUM NIOBATE, BARIUM SODIUM NIOBATE) HOLOGRAPHI 11-1661
HOLOGRAPHIC STORAGE IN DOPED BARIUM SODIUM NIOBATE. 13-1898
LOW FREQUENCY VIBRATIONAL MODES IN BARIUM SODIUM NIOBATE. 13-1905
OF THE DIELECTRIC CONSTANTS AND ELECTROOPTIC COEFFICIENTS IN BARIUM SODIUM NIOBATE. /EE AND CLAMPED VALUES 13-1998
EFFECT AND ELECTRON SPIN RESONANCE OF IRON AND MOLYBDENUM IN BARIUM SODIUM NIOBATE. PHOTOCHROMIC 13-1930
ORTHORHOMBIC CRYSTALS. (BISMUTH(12) GERMANIUM OXIDE(20) AND BARIUM SODIUM NIOBATE) /ACE WAVES IN CUBIC AND 13-2041
POLARITONS IN A BIAXIAL CRYSTAL OF BARIUM SODIUM NIOBATE. 13-1992
RAMAN SPECTRA OF BARIUM SODIUM NIOBATE. 13-1906
ROLE OF HYDROGEN IN POLARIZATION REVERSAL OF FERROELECTRIC BARIUM SODIUM NIOBATE. 13-2025
SOME EFFECTS OF MELT STOICHIOMETRY ON THE OPTICAL PROPERTIES BARIUM SODIUM NIOBATE. 13-1902
E EFFECTS OF MELT STOICHIOMETRY ON THE OPTICAL PROPERTIES OF BARIUM SODIUM NIOBATE. SOM 13-1903
TEMPERATURE DEPENDENCE OF THE REFRACTIVE INDICES OF BARIUM SODIUM NIOBATE. 13-2008
VISIBLE CW PARAMETRIC OSCILLATOR USING BARIUM SODIUM NIOBATE. 13-1984
ATE (FORCE CONSTANT/ VIBRATIONAL SPECTRA OF LITHIUM NIOBATE, BARIUM SODIUM NIOBATE AND BARIUM SODIUM TANTAL 11-1808
/NIC GENERATION ON CRYSTAL INHOMOGENEITY (THEORY AND TEST ON BARIUM SODIUM NIOBATE AND NEODYMIUM DOPED YAG/ 13-2043
/UCIBLE EASE COOLING: AN AID TO CZOCHRALSKI CRYSTAL GROWTH. (BARIUM SODIUM NIOBATE AND SODIUM LANTHANUM TA/ 13-1928
ATERIAL. (GROWTH BY KYROPOULO/ DEVELOPMENT OF SINGLE CRYSTAL BARIUM SODIUM NIOBATE AS A NONLINEAR OPTICAL M 13-2021
/HIUM NIOBATE, LITHIUM TANTALATE, POTASSIUM LITHIUM NIOBATE, BARIUM SODIUM NIOBATE, BARIUM STRONTIUM NIOBAT 11-1734
/ROWAVE ACOUSTICS HANDBOOK. VOL 1: SURFACE WAVE VELOCITIES. (BARIUM SODIUM NIOBATE, BISMUTH(12) GERMANIUM / 01-0028
ERE LIQ/ PROGRESS TOWARDS THE GROWTH OF GOOD OPTICAL QUALITY BARIUM SODIUM NIOBATE (BY USING COMPOSITION WH 13-2047
AND (CZOCHRALSKI) CRYSTAL GROWTH. BARIUM SODIUM NIOBATE: CONSTITUTIONAL STUDIES 13-1904
GROWTH OF UNCRACKED BARIUM SODIUM NIOBATE CRYSTALS. 13-1901
OS TECHNIQUE. GROWTH OF BARIUM SODIUM NIOBATE CRYSTALS BY THE KYROPOUL 13-1909
ITY FREQUENCY DOUBL/ CATASTROPHIC SURFACE DAMAGE PRODUCED IN BARIUM SODIUM NIOBATE CRYSTALS DURING INTRACAV 13-2052
/TASSIUM TANTALATE OR NIOBATE, LITHIUM NIOBATE OR TANTALATE, BARIUM SODIUM NIOBATE, CUPROUS CHLORIDE, PROU/ 01-0021
DIELECTRIC, AND OPTICAL CONSTANTS) ROOM/ ELASTIC ANOMALY OF BARIUM SODIUM NIOBATE (ELASTIC, PIEZOELECTRIC, 13-2060
RAMAN OSCILLATOR. BARIUM SODIUM NIOBATE FOR A TUNABLE STIMULATED 13-1917
CIES. PARAMETRIC FLUORESCENCE IN BARIUM SODIUM NIOBATE FOR NINE CW PUMP FREQUEN 13-1950
HS (RANGE 454.5-528.7 MICRONS). PARAMETRIC FLUORESCENCE IN BARIUM SODIUM NIOBATE FOR 10 CW PUMP WAVELENGT 13-1983
/ND, THIRD, AND FOURTH HARMONICS OF 1.06 MICRON RADIATION IN BARIUM SODIUM NIOBATE, LITHIUM NIOBATE, AND L/ 11-1779
FRACTIVE INDEX, DIE/ OPTICAL AND FERROELECTRIC PROPERTIES OF BARIUM SODIUM NIOBATE. (OPTICAL ABSORPTION, RE 13-2024
CAL APPLICATIONS. (BY PULLING FROM THE MELT, STRU/ GROWTH OF BARIUM SODIUM NIOBATE SINGLE CRYSTALS FOR OPTI 13-2045
IC STUDY OF THE ORTHORHOMBIC- TETRAGONAL PHASE TRANSITION IN BARIUM SODIUM NIOBATE. (THERMAL EXPANSION) /TR 13-1894
OR. (HALF WAVE VOLTAGE, T/ OPTIMUM CRYSTAL ORIENTATION FOR A BARIUM SODIUM NIOBATE TRANSVERSE LIGHT MODULAT 13-1973
C SOURCE. NEODYMIUM: YTTRIUM ALUMINUM GARNET- BARIUM SODIUM NIOBATE 0.66 MILLIMICRON HARMONI 13-1975
CAUSES OF THERMAL EXPANSION OF GADOLINIUM SUBSTITUTED BARIUM SODIUM NIOBATES. 13-2048
LINEAR THERMAL EXPANSIONS OF GADOLINIUM SUBSTITUTED BARIUM SODIUM NIOBATES. 13-1996
MICROCHEMICAL ANALYSIS OF GADOLINIUM DOPED BARIUM SODIUM NIOBATES. 13-1953
SCANNING ELECTRON MICROSCOPIC EXAMINATION OF SINTERED BARIUM SODIUM NIOBATES. 13-1955
ITH TUNGSTEN BRON/ DIELECTRIC AND ELECTROOPTIC PROPERTIES OF BARIUM SODIUM YTTRIUM NIOBATE SINGLE CRYSTAL W 13-1971
RESONANCE ABSORPTION OF GAMMA QUANTA IN BARIUM, STRONTIUM, AND CALCIUM STANNATE. 36-3566
S. ELECTRICAL AND OPTICAL PROPERTIES OF (BARIUM, STRONTIUM) MIXED TITANATE THIN SECTION 04-0603
EPITAXY OF PLATINUM ON BARIUM STRONTIUM NIOBATE. 13-1907
LYCINE SULFATE, TRIGLYCINE FLUOROBERYLLATE, LITHIUM SULFATE, BARIUM STRONTIUM NIOBATE) /ION DETECTOR. (TRIG 02-0046
NSE TO SHORT INFRARED LASER PULSES IN TRIGLYCINE SULFATE AND BARIUM STRONTIUM NIOBATE. /CTRIC VOLTAGE RESPO 13-2018
RAMAN SCATTERING BY BARIUM STRONTIUM NIOBATE. 13-1899
SOLIDIFICATION STUDY OF BARIUM STRONTIUM NIOBATE. 13-1954
F CRYSTAL GROWTH RATE WITH SUPERSATURATION IN FLUXED MELTS. (BARIUM STRONTIUM NIOBATE) VARIATION O 13-1939
CZOCHRALSKI GROWTH OF BARIUM STRONTIUM NIOBATE CRYSTALS. 13-1913
STABILITY REGIONS FOR THE GROWTH OF BARIUM STRONTIUM NIOBATE CRYSTALS. 13-2056
E OF/ TEMPERATURE DEPENDENCE OF THE DIELECTRIC PROPERTIES OF BARIUM STRONTIUM NIOBATE IN THE FREQUENCY RANG 13-1938
/ICN COEFFICIENTS OF PIEZOELECTRIC CRYSTALS. (LEAD TITANATE, BARIUM STRONTIUM NIOBATE, LEAD NIOBIUM(4) OXI/ 05-1066
ELECTROOPTIC LIGHT BEAM DEFLECTION WITH BARIUM STRONTIUM NIOBATE PRISM. 13-2036
TH BOUNDARY LAYER STRUCTURE FOR HIGH FREQUENCY APPLICATION. (BARIUM, STRONTIUM TITANATE) /CTRIC CERAMICS WI 04-0833
HIGH L/F NOISE ANOMALY IN SEMICONDUCTING BARIUM STRONTIUM TITANATE. 04-0866
TE, POLYCRYSTALLINE, STRONTIUM TITANATE, AND SOLID SOLUTIONS BARIUM STRONTIUM TITANATE) /L STRONTIUM TITANA 04-0721
TERMINATIONS FROM WEIGHT MEASUREMENTS IN SOLID SOLUTIONS OF (BARIUM, STRONTIUM) TITANATE. MOBILITY DE 04-0689
PREPARATION AND STUDY OF THIN FILMS OF (BARIUM, STRONTIUM) TITANATE. 04-1000
/RING BY A FERROELECTRIC SOLID SOLUTION. (THEORY, APPLIED TO BARIUM STRONTIUM TITANATE AND POTASSIUM NIOBA/ 03-0260
IC DIELECTRICS. STUDIES ON THE BARIUM STRONTIUM TITANATE BOUNDARY LAYER CERAM 04-1025
ZT. ATTEMPT TO PREPARE HIGH DOPED BARIUM STRONTIUM TITANATE SEMICONDUCTOR WITH P 04-0717
PREPARATION AND PROPERTIES OF (BARIUM, STRONTIUM) TITANATE SINGLE CRYSTALS. 04-0560
CRYSTAL CHEMICAL PROBLEMS OF (BARIUM, STRONTIUM) TITANATE SOLID SOLUTIONS. 04-0633
AR THE PHASE TRANSI/ DIELECTRIC PROPERTIES OF FERROELECTRIC (BARIUM, STRONTIUM) TITANATE SOLID SOLUTIONS NE 04-0680
W FREQUENCY DIELECTRIC PROPE/ SINGLE CRYSTALS OF THE SYSTEM (BARIUM, STRONTIUM) TITANATE. (SYNTHESIS AND LO 04-0561
IE-WEISS CONSTANT IN POLYCRYSTALLINE SOLID SOLUTIONS IN THE (BARIUM, STRONTIUM) TITANATE SYSTEM. /F THE CUR 04-0586
/WITH THE STRUCTURE OF POTASSIUM TUNGSTEN BRONZE. (POTASSIUM BARIUM TANTALUM OXIDE, SODIUM BARIUM TANTALUM/ 13-2050
A FLOATING ZONE SINGLE CRYSTAL GROWING APPARATUS. (BARIUM TITANATE) 04-0998
A NEW PTC THERMISTOR MATERIAL. (BARIUM TITANATE) 04-0931
ACOUSTIC PHONON SOFT OPTIC PHONON INTERACTION IN BARIUM TITANATE. 04-0806

ACOUSTIC SOFT OPTIC MODE INTERACTIONS IN FERROELECTRIC BARIUM TITANATE. 04-0671
AGING IN TETRAGONAL FERROELECTRIC BARIUM TITANATE. 04-0579
ANOMALOUS DECAY EFFECT IN POLYCRYSTALLINE BARIUM TITANATE. 04-0752
ERENCE IN RAMAN SCATTERING FROM ONE AND TWO PHONON STATES OF BARIUM TITANATE. AUGER-LIKE RESONANT INTERF 04-0923
CE MEASUREMENTS USING THERMAL NOISE ANALYSIS (APPLICATION TO BARIUM TITANATE). AUTOMATIC IMPEDAN 04-0844
BAND STRUCTURE OF CUBIC AND TETRAGONAL BARIUM TITANATE. 04-0846
BINARY LIGHT BEAM DEFLECTION IN SINGLE CRYSTAL BARIUM TITANATE. 04-0653
BINARY LIGHT BEAM DETECTION IN SINGLE CRYSTAL BARIUM TITANATE. 04-0654
CALCULATION OF (DOMAIN) WALL ENERGY IN BARIUM TITANATE. 04-0673
RACTIONS IN FINE GRAINED PEROVSKITE FERROELECTRIC CERAMICS. (BARIUM TITANATE) /ON OF THE ELECTROSTATIC INTE 04-0642
CALCULATIONS OF ELECTRONIC CONDUCTIVITY IN NONSTOICHIOMETRIC BARIUM TITANATE. 04-0821
CATALYTIC OXIDATION OF CARBON MONOXIDE ON SEMICONDUCTIVE BARIUM TITANATE. 04-0945
CATION NONSTOICHIOMETRY IN TERNARY OXIDES. (BARIUM TITANATE) 04-0969
CERAMIC ACOUSTIC DETECTORS. (BARIUM TITANATE) 01-0003
E CUBIC AND TETRAGONAL (FERROELECTRICALLY POLARIZED PHASE OF BARIUM TITANATE. /ANNELING OF LIGHT IONS IN TH 04-0688
ICS OBTAINED BY PROJECTION WITH A PLASMA TORCH OR FRITTING. (BARIUM TITANATE) /ILAYERED FERROELECTRIC CERAM 04-0778
CHARACTERIZATION OF SURFACE AND GRAIN BOUNDARY LAYER OF BARIUM TITANATE. 04-0701
TS ON THE NEUTRON SPECTRUM OF FERROELECTRIC AND PARAELECTRIC BARIUM TITANATE. COMMEN 04-0813
ON LINEAR CHAIN OF ATOMS LYING IN DOUBLE MINIMUM POTENTIAL. (BARIUM TITANATE) COMPUTER EXPERIMENT 03-0287
CONDUCTIVITY INJECTION AND EXTRACTION IN POLYCRYSTALLINE BARIUM TITANATE. 04-0814
TION BETWEEN RHEOLOGICAL AND ELECTRICAL BEHAVIOR IN SINTERED BARIUM TITANATE. CORRELA 04-1027
UCTURE AND ELECTRIC PROPERTIES OF FERROELECTRIC SUBSTANCES. (BARIUM TITANATE) CORRELATION BETWEEN STR 04-0728
TYPE PHASE TRANSITIONS IN FERROELECTRICS. (THEORY APPLIED TO BARIUM TITANATE) /ION EFFECTS IN DISPLACEMENT 03-0497
COUPLED POLARITONS OF A1 SYMMETRY IN BARIUM TITANATE. 04-0619
CRITICAL FIELD IN CERAMIC BARIUM TITANATE. 04-0951
CURRENT EXCITATION IN GAMMA-IRRADIATED BARIUM TITANATE. 04-0884
DIELECTRIC ABSORPTION IN THE RHOMOHEDRAL PHASE OF BARIUM TITANATE. 04-0809
CTRIC ABSORPTION IN THE ORTHORHOMBIC AND QUADRATIC PHASES OF BARIUM TITANATE. DIELE 04-0900
CTRIC ABSORPTION IN THE ORTHORHOMBIC AND QUADRATIC PHASES OF BARIUM TITANATE. DIELE 04-0899
C AGING IN TETRAGONAL SOLID SOLUTIONS OF CALCIUM TITANATE IN BARIUM TITANATE. DIELECTRI 08-1295
DIELECTRIC BREAKDOWN IN BARIUM TITANATE. 04-0580
DIELECTRIC BREAKDOWN OF POLYCRYSTALLINE BARIUM TITANATE. 04-1015
DIELECTRIC DISPERSION IN BARIUM TITANATE. 04-0646
DIELECTRIC PROPERTIES OF SINGLE DOMAIN MELT GROWN BARIUM TITANATE. 04-0616
ITIONS AND THE STABLE DOMAIN STRUCTURE OF (HIGH SENSITIVITY) BARIUM TITANATE. DIFFUSIONLESS PHASE TRANS 04-0888
DISPERSION IN FERROELECTRICS. (THEORY, BARIUM TITANATE) 03-0454
DISPERSION OF THE ELECTROOPTIC EFFECT IN BARIUM TITANATE. 04-0756
DOMAIN INTERACTIONS DURING PHASE TRANSITIONS IN BARIUM TITANATE. 04-0651
DYNAMIC FATIGUE OF SINGLE CRYSTALS OF BARIUM TITANATE. 04-0792
EFFECT OF "A" DOMAINS UPON "C" DOMAIN SWITCHING IN BARIUM TITANATE. 04-0615
CRYSTALLINE GRAIN SIZE ON THE PHYSICAL PROPERTIES OF CERAMIC BARIUM TITANATE. EFFECT OF 04-0785
F GRAIN SIZE ON THE FERROELECTRIC PARAELECTRIC TRANSITION OF BARIUM TITANATE. EFFECT O 04-0917
OF GRAIN SIZE ON THE PIEZOELECTRIC AND ELASTIC PROPERTIES OF BARIUM TITANATE. EFFECT 04-1011
FECT OF HEAT TREATMENT ON THE PTCR ANOMALY IN SEMICONDUCTING BARIUM TITANATE. EF 04-0762
IMPURITIES ON THE ELECTRICAL RESISTIVITY OF LANTHANUM DOPED BARIUM TITANATE. EFFECT OF OXIDE 04-0587
ECT OF PRESSURE (TO ABOUT 2KBAR) ON POLARIZATION REVERSAL IN BARIUM TITANATE. EFF 04-0718
F SILICA ADDITIONS ON THE SEMICONDUCTING PROPERTIES OF DOPED BARIUM TITANATE. EFFECT O 04-0660
OTROPY OF ELASTIC CONSTANTS ON DISLOCATIONS IN FERROELECTRIC BARIUM TITANATE. EFFECTS OF ANIS 04-0674
ELECTRET EFFECT IN A MIXTURE OF POLY METHACRYLATE AND BARIUM TITANATE 04-0838
OF CERTAIN TWO PHASE FERROELECTRIC FERROMAGNETIC MIXTURES. (BARIUM TITANATE) /IC AND MAGNETIC MEASUREMENTS 04-0565
ELECTRICAL CONDUCTIVITY OF SINGLE CRYSTALLINE BARIUM TITANATE. 04-0663
SS. (60-90 VOL PERCENT, FERROELECTRIC, COMPARED WITH CERAMIC BARIUM TITANATE) /ERIALS CRYSTALLIZED FROM GLA 04-0725
ELECTROCALORIC EFFECT IN POLYCRYSTALLINE BARIUM TITANATE. 04-0766
SOLUTIONS OF ISO- AND HETEROVALENT CATIONIC SUBSTITUTIONS IN BARIUM TITANATE. /ECT IN SEMICONDUCTING SOLID 04-0943
ICAL THEORY. PART-2: OBSERVATIONS. (FERROELECTRIC DOMAINS IN BARIUM TITANATE) /IN BOUNDARIES. PART-1: DYNAM 03-0255
F PEROVSKITE TYPE OXIDES. (LEAD TITANATE, POTASSIUM NIOBATE, BARIUM TITANATE) /S ON FERROELECTRIC DOMAINS O 04-0993
CROSCOPIC OBSERVATION OF FERROELECTRIC DOMAINS. (GOLD COATED BARIUM TITANATE) ELECTRON MIRROR MI 04-0974
ELECTRON PARAMAGNETIC RESONANCE IN NONSTOICHIOMETRIC BARIUM TITANATE. 04-0640
ELECTRON SPIN RESONANCE OF MANGANESE(2+) IN BARIUM TITANATE. 04-0747
ENERGY STORAGE IN CERAMIC DIELECTRICS. (STRONTIUM TITANATE, BARIUM TITANATE) 04-0602
GE IN FERROELECTRIC CERAMICS: SURVEY OF MATERIALS BASED UPON BARIUM TITANATE. ENERGY STORA 02-0075
IN HIGH PERMITTIVITY CERAMICS: SURVEY OF MATERIALS BASED ON BARIUM TITANATE. ENERGY STORAGE 04-0733
EXAGGERATED GRAIN GROWTH IN LIQUID PHASE SINTERING OF BARIUM TITANATE. 04-0835
L AND THEORETICAL STUDY OF DEFECTS IN TRIGLYCINE SULFATE AND BARIUM TITANATE. EXPERIMENTA 04-0620
MATION IN THE THEORY OF OPTICAL LATTICE VIBRATIONS. (QUARTZ, BARIUM TITANATE) /F THE POULET- LOUDON APPROXI 03-0380
FAR INFRARED REFLECTIVITY OF BARIUM TITANATE. 04-0744
CATTERING) SHADOW EFFECT IN (DISPLACIVE) PHASE TRANSITIONS. (BARIUM TITANATE) (FAST ION S 04-0533
CTRIC SYMMETRY GROUPS. (AND THEIR COMPARISON, APPLICATION TO BARIUM TITANATE) /RROELECTRIC AND ANTIFERROELE 04-0933
FERROELECTRIC FAMILY OF BARIUM TITANATE. 04-1018
FERROELECTRIC PHASE TRANSITION OF FIRST ORDER IN BARIUM TITANATE. 04-1030
ORPTION BANDS OF FERROELECTRICS IN FAR INFRARED. (APPLIED TO BARIUM TITANATE) FIELD DEPENDENCE OF ABS 04-0665
CED NUCLEATION AT THE (CUBIC TETRAGONAL) PHASE TRANSITION IN BARIUM TITANATE. FIELD INDU 04-0753
H EVAPORATION OF FERROELECTRIC THIN FILMS OF POLYCRYSTALLINE BARIUM TITANATE. FLAS 04-0968
M OF THE ELECTRON PARAMAGNETIC RESONANCE OF MANGANESE(+2) IN BARIUM TITANATE. /NES IN THE HYPERFINE SPECTRU 04-0877
FORMATION OF FLUORINE CONTAINING SOLID SOLUTIONS BASED ON BARIUM TITANATE. 04-0932
GENERATION OF SECOND OPTICAL HARMONIC IN THE CUBIC PHASE OF BARIUM TITANATE. 04-0538
E DEPENDENCE OF DIELECTRIC AND STRUCTURAL CHARACTERISTICS OF BARIUM TITANATE. GRAIN SIZ 04-0582
MAL CONDUCTIVITY IN DISPLACIVE TYPE FERROELECTRICS. (THEORY, BARIUM TITANATE) /Y SOUND ATTENUATION AND THER 03-0160
ERMODYNAMIC AND X-RAY DIFFRACTION STUDY OF NONSTOICHIOMETRIC BARIUM TITANATE. HIGH TEMPERATURE TH 04-0885
HYDROXYL AS DEFECT IN BARIUM TITANATE. 04-0636
ECTRIC FIELD ON THE DIELECTRIC PROPERTIES OF POLYCRYSTALLINE BARIUM TITANATE. INFLUENCE OF A CONSTANT EL 04-1012
ENSIFICATION AND PROPERTIES OF SINTERED AN PRESSURE SINTERED BARIUM TITANATE. INFLUENCE OF ADDITIVES ON D 04-1026
AL POLARIZATION FLUCTUATIONS ON THE PHOTOELASTIC BEHAVIOR OF BARIUM TITANATE. INFLUENCE OF CRITIC 04-0628
CE OF GRAIN GROWTH ON DIELECTRIC PROPERTIES OF NIOBIUM DOPED BARIUM TITANATE. INFLUEN 04-0763
UCTURES (PREPARATION CONDITIONS) ON DIELECTRIC PROPERTIES OF BARIUM TITANATE. INFLUENCE OF MICROSTR 04-0798
E OF OXYGEN PARTIAL PRESSURE ON PROPERTIES OF SEMICONDUCTING BARIUM TITANATE. INFLUENC 04-0999
INFLUENCE OF PHASE TRANSITIONS ON ELECTRICAL RESISTANCE OF BARIUM TITANATE. 04-0599
INFRARED SPECTRA OF FLUORINE CONTAINING BARIUM TITANATE. 04-0983
INTERACTION BETWEEN THE 90 DEG AND 180 DEG DOMAINS IN BARIUM TITANATE. 04-0585
INVERSION OF ELECTRICAL PROPERTIES OF SEMICONDUCTING BARIUM TITANATE. 04-0572

UCTURE AND THE PYROELECTRIC PROPERTIES OF SINGLE CRYSTALS OF BARIUM TITANATE. / THICKNESS ON THE DOMAIN STR 04-1039
MOSSBAUER ISOMER SHIFTS. (DEGREE CF COVALENCY IN IRON DOPED BARIUM TITANATE) IONIC CHARACTERS FROM 04-0563
KINETICS OF THE PHASE TRANSITIONS IN BARIUM TITANATE. 04-0650
LEED STUDY OF SURFACE STRUCTURES ON THE (001) FACE OF BARIUM TITANATE. 04-0529
R DISORDER AND TEMPERATURE DEPENDENCE OF RAMAN SCATTERING IN BARIUM TITANATE. LINEA 04-0675
LOWERING THE CURIE TEMPERATURE IN REDUCED BARIUM TITANATE. 04-0711
NTANEOUS POLARIZATION IN THIN CRYSTALS AND FINE PARTICLES OF BARIUM TITANATE. MAGNITUDE OF SPO 04-0607
N DIFFUSION CONSTANT IN SOME TITANATES. (STRONTIUM TITANATE, BARIUM TITANATE) MEASUREMENT OF OXYGE 04-1033
MECHANISM OF ELECTROCONDUCTIVITY IN BARIUM TITANATE. 04-0908
HANISM OF MICROWAVE DIELECTRIC DISPERSION IN POLYCRYSTALLINE BARIUM TITANATE. MEC 04-0745
SCANDIUM TRIOXIDE DOPED, RARE EARTH OXIDE DOPED, AND UNDOPED BARIUM TITANATE. /ND ELECTRICAL PROPERTIES OF 04-0839
CTURE OF DOMAINS AND DOMAIN WALLS IN SINGLE CRYSTAL FILMS OF BARIUM TITANATE. MICROSTRU 04-1031
MOSSBAUER EFFECT OF COBALT-57 IN BARIUM TITANATE. 04-0723
MOVING STRIPE PATTERN IN BARIUM TITANATE. 04-0643
NEUTRON SCATTERING STUDY OF SOFT MODES IN CUBIC BARIUM TITANATE. 04-0712
NOISE MEASUREMENTS IN FERROELECTRICS. (TGS, ROCHELLE SALT, BARIUM TITANATE) 04-0698
MICROWAVE FREQUENCIES. (TRIGLYCINE SULFATE, POLYCRYSTALLINE BARIUM TITANATE) /PERTIES OF FERROELECTRICS AT 04-0870
NONSTOICHIOMETRY IN BARIUM TITANATE. 04-0631
DELS FOR A FIRST ORDER TRANSITION FRONT, AND OBSERVATIONS IN BARIUM TITANATE. NUCLEATION MO 04-0608
OHMIC CONTACTS TO SEMICONDUCTING BARIUM TITANATE. 04-0764
ON THE INFLUENCE OF DEFECTS ON THE TEMPERATURE HYSTERESIS OF BARIUM TITANATE. 04-0987
TATION OF INFRARED SPECTRA AND THE LATTICE DYNAMICS OF CUBIC BARIUM TITANATE. ON THE INTERPRE 04-0659
ON THE LATTICE VIBRATION OF CUBIC BARIUM TITANATE. 04-0775
ON THE PHASE COMPOSITION OF REDUCED AND REOXIDIZED BARIUM TITANATE. 04-0541
TICAL EVIDENCE FOR POLARIZATION FLUCTUATIONS IN PARAELECTRIC BARIUM TITANATE. OP 04-0737
OPTICALLY INDUCED REFRACTIVE INDEX CHANGES IN BARIUM TITANATE. 04-1005
OXIDATION OF CARBON MONOXIDE ON LANTHANUM OXIDE DOPED BARIUM TITANATE. 04-0947
PARAMAGNETIC RESONANCE OF AN F CENTER IN NONSTOICHIOMETRIC BARIUM TITANATE. 04-0976
RRELATION OF SCANDIUM-44 WITHIN THE LATTICE OF FERROELECTRIC BARIUM TITANATE. PERTURBED DIRECTIONAL CO 04-0691
AL CORRELATION OF SCANDIUM-44 IN THE FERROELECTRIC PHASES OF BARIUM TITANATE. PERTURBED DIRECTION 04-0558
ENA IN CRYSTALS OF PEROVSKITE STRUCTURE (WITH APPLICATION TO BARIUM TITANATE). /X AND ELECTROOPTICAL PHENOM 03-0518
PHONON TRANSPORT IN BARIUM TITANATE. 04-0830
PHOTOCONDUCTIVITY AND PHOTOFERROELECTRIC PHENOMENA IN BARIUM TITANATE. 04-0574
PHOTODOMAIN EFFECT IN BARIUM TITANATE. 04-0682
PHOTON PHONON INTERACTION IN THE NEAR INFRARED. (BARIUM TITANATE) 04-0925
POLARITON DISPERSION RELATION IN CUBIC BARIUM TITANATE. 04-0878
POLARIZATION FLUCTUATIONS AND THE OPTICAL ABSORPTION EDGE IN BARIUM TITANATE. 04-1028
ATION REFLECTION TYPE LIGHT MODULATOR USING FERROELECTRICS. (BARIUM TITANATE) POLARIZ 04-0714
POLARIZED STATES OF FERROELECTRIC CERAMICS. (BARIUM TITANATE) 04-0666
POLARON MECHANISM OF CONDUCTIVITY IN BARIUM TITANATE. 04-0609
POWDER RAMAN STUDY FOR BARIUM TITANATE. 04-0610
PREPARATION AND PROPERTIES OF THIN FILMS OF BARIUM TITANATE. 04-0843
PREPARATION OF FLUORINE CONTAINING BARIUM TITANATE. 04-0984
SMALL GRAINED AND LARGE GRAINED CERAMICS FROM NIOBIUM DOPED BARIUM TITANATE. PREPARATION OF 04-0765
ACOUSTIC MODE- SOFT OPTIC MODE INTERACTIONS IN FERROELECTRIC BARIUM TITANATE. PRESSURE DEPENDENCE OF 04-0893
TRANSITION TEMPERATURES AND ELECTROSTRICTIONS IN PEROVSKITE BARIUM TITANATE. PRESSURE DEPENDENCE OF 04-0853
PROPERTIES OF BLACK SEMICONDUCTING BARIUM TITANATE. 04-0702
RAMAN SCATTERING IN THE FERROELECTRIC SYSTEM LEAD BARIUM TITANATE. 04-0611
RAMAN SCATTERING LINE SHAPE OF THE SOFT E POLARITON MODE IN BARIUM TITANATE. 04-0803
RAMAN SPECTRUM OF BARIUM TITANATE. 04-0746
RELATION BETWEEN ELECTRICAL PROPERTIES AND MICROSTRUCTURE OF BARIUM TITANATE. 04-0980
RESISTIVITY AND SINTERING OF BARIUM TITANATE. 04-0819
URRENCE OF ORDERED PHASE IN PEROVSKITE TYPE FERROELECTRICS. (BARIUM TITANATE) /GIN OF FERROELECTRICITY, OCC 04-0964
SHELL MODEL OF INTERIONIC INTERACTIONS FOR BARIUM TITANATE. 04-0716
SHOCK INDUCED TRANSITION IN BARIUM TITANATE. 04-0589
SHOCK WAVE STUDIES OF SURFACE LAYERS IN BARIUM TITANATE. 04-0588
SIZE EFFECTS IN FERROELECTRICS. (BARIUM TITANATE) 04-0656
SOLID SOLUTIONS BASED ON BARIUM TITANATE. 04-0786
SOLUBILITY LIMIT OF LANTHANUM(0.5) LITHIUM(0.5) TITANATE IN BARIUM TITANATE. 04-0558
SOME ASPECTS OF SEMICONDUCTING BARIUM TITANATE. 04-0758
SPECIFIC HEAT OF BARIUM TITANATE. 04-0715
SPONTANEOUS DESTRUCTION OF FERROCERAMICS. (BARIUM TITANATE) 04-0787
STATIC QUADRUPLE INTERACTION OF SCANDIUM-44 IN BARIUM TITANATE. 04-0692
FERROELECTRIC PROPERTIES OF X POTASSIUM DINIOBATE AND (1-X) BARIUM TITANATE. STRUCTURAL DI- AND 04-0623
TURE OF A MOVING DOMAIN WALL. (MECHANISM OF A WALL MOTION IN BARIUM TITANATE) STRUC 04-0678
STRUCTURE OF 90 DEGREE DOMAIN WALLS IN BARIUM TITANATE. 04-0641
CRYSTALS WITH A PHOTOEMISSION MICROSCOPE (TUNGSTEN TRIOXIDE, BARIUM TITANATE). /F DOMAINS IN FERROELECTRIC 04-0811
PYROELECTRIC EFFECT IN THIN FILMS OF TRIGLYCINE SULFATE AND BARIUM TITANATE. STUDY OF THE 04-0686
AL LATTICE BASED ON THE SHADOW EFFECT. (NEUTRON DIFFRACTION, BARIUM TITANATE) /MALL DEFORMATIONS OF A CRYST 04-0534
SURFACE BREAKDOWN IN VACUUM ON BARIUM TITANATE. 04-0596
TATION PRODUCED BY ABRASION AND ANNEALING OF POLYCRYSTALLINE BARIUM TITANATE. SURFACE DOMAIN REORIEN 04-0639
TEMPERATURE DEPENDENCE OF CARRIER MOBILITY IN BARIUM TITANATE. 04-0614
URE DEPENDENCE OF ELECTRONIC POLARIZABILITY OF STRONTIUM AND BARIUM TITANATE. TEMPERAT 04-0738
TEMPERATURE DEPENDENCE OF THE BRILLOUIN- RAMAN SPECTRUM OF BARIUM TITANATE. 04-0807
TEMPERATURE FACTORS OF TETRAGONAL AND CUBIC BARIUM TITANATE. 04-0532
THEORETICAL DYNAMIC STUDY OF CUBIC PHASE BARIUM TITANATE. 03-0261
THEORY OF BARIUM TITANATE. 03-0483
RY OF ELECTROSTRICTION IN CUBIC CRYSTALS WITH APPLICATION TO BARIUM TITANATE. THEO 03-0267
THEORY OF THE DIELECTRIC CONSTANTS OF BARIUM TITANATE. 04-0776
THERMAL HYSTERESIS IN THE TETRAGONAL CUBIC TRANSITION OF BARIUM TITANATE. 04-0921
THERMOSTABILIZING PROPERTIES OF SEMICONDUCTING BARIUM TITANATE. 04-0570
DEPENDENCE OF THE DIELECTRIC PERMITTIVITY OF HYDROXYL DOPED BARIUM TITANATE. TIME 04-0550
ULTRASONIC STUDIES OF FIELD INDUCED PHASE TRANSITION IN BARIUM TITANATE. 04-0549
VACUUM EVAPORATION OF BARIUM TITANATE. 04-1001
VARIATION OF THE DECAY CONSTANT OF STRONTIUM-85 IN BARIUM TITANATE. 04-0876
VIBRATIONAL AMPLITUDES IN FERROELECTRIC TETRAGONAL BARIUM TITANATE. 04-0726
FFECT IN THE TEMPERATURE DEPENDENCE OF THE SOFT FREQUENCY OF BARIUM TITANATE. VIBRONIC E 04-0662
WET SYNTHETIC METHOD OF MAKING PURE, EXACTLY STOICHIOMETRIC BARIUM TITANATE. 04-0790
X-RAY AND NEUTRON DIFFRACTION STUDY OF TETRAGONAL BARIUM TITANATE. 04-0713
X-RAY AND NEUTRON DIFFRACTION STUDY OF TETRAGONAL BARIUM TITANATE. 04-0740

```
                      X-RAY AND THERMOLUMINESCENCE OF BARIUM TITANATE.                                           04-0750
ON OF THE EFFECT OF HYDROSTATIC PRESSURE ON THE STRUCTURE OF BARIUM TITANATE.              X-RAY INVESTIGATI 04-0760
                      SEMICONDUCTING AND NONSTOICHIOMETRIC BARIUM TITANATE. A REFERENCE GUIDE.                   01-0022
              DOUBLE SPACE CHARGE INJECTION IN SOLIDS. (USING BARIUM TITANATE ABOVE CURIE POINT)                 04-0638
TEMPERATURE, AS SH/ CRITICAL (POLARIZATION) FLUCTUATIONS IN BARIUM TITANATE (ABOVE THE FERROELECTRIC CURIE 04-1032
          DISPERSION AT MILLIMETER WAVELENGTHS IN BARIUM TITANATE ABOVE THE TRANSITION POINT.                04-0601
AND DECIMETER WAVELENGTHS EMPLOYING A SYMMETRIC STRIPLINE. (BARIUM TITANATE AND BARIUM CALCIUM TITANATE) / 04-0855
/UM TARTRATE, LITHIUM AMMONIUM TARTRATE, AND MIXED CRYSTALS, BARIUM TITANATE, AND BARIUM STRONTIUM TITANAT/ 02-0089
/HASE FERROELECTRIC SYSTEMS. PART-2: FERROELECTRICS- SILVER (BARIUM TITANATE AND BARIUM TITANATE- MAGNESIU/ 04-0816
/IC FIELD, IN STRONG ELECTRIC FIELDS. (ANISOTROPY DEVELOPED, BARIUM TITANATE AND BARIUM(0.95) CALCIUM(0.05/ 04-0661
R CENTERS)    RELATIONSHIP BETWEEN THE BREAKDOWN MECHANISM IN BARIUM TITANATE AND COLORATION. (INJECTED COLO 04-0944
/TRON MIRROR MICROSCOPIC ASPECTS OF FERROELECTRIC DOMAINS IN BARIUM TITANATE AND DICALCIUM STRONTIUM PROPI/ 04-0973
    HETEROVALENT SUBSTITUTION AND OXYGEN NONSTOICHIOMETRY OF BARIUM TITANATE AND HEMATITE SINGLE CRYSTALS.    04-0874
) NIOBATE(0.5), BISMUTH FERRATE AND ITS SOLID SOLUTIONS WITH BARIUM TITANATE, AND LANTHANUM FERRATE) /E(0.5 08-1368
    TWO WAVE SHOCK STRUCTURES IN THE FERROELECTRIC CERAMICS BARIUM TITANATE AND LEAD ZIRCONATE TITANATE.      04-0916
RROELECTRIC CERAMICS. (REVIEW, 31 REFERENCES)     AGING OF BARIUM TITANATE AND LEAD ZIRCONATE TITANATE FE 02-0062
/IALS WITH PEROVSKITE TYPE STRUCTURE. (MIXED OXIDES ABO(3)) (BARIUM TITANATE AND LITHIUM NIOBATE FILMS ON / 04-0914
ALLY ORDERED "NUCLEI", PHASE STRUCTU/ DIELECTRIC SPECTRUM OF BARIUM TITANATE. (AND MICROSCOPIC MODEL OF LOC 04-0647
OPRECIPITATION OF CITRATES IN AN ALCOHOL.     PREPARATION OF BARIUM TITANATE AND OTHER CERAMIC POWDERS BY C 04-0861
/ STUDIES OF OPTICAL HARMONIC GENERATION IN CADMIUM SULFIDE, BARIUM TITANATE, AND POTASSIUM DIHYDROGEN PHO/ 04-0852
LS (CASE OF SILICON, QUARTZ AND FERROELECTRIC PEROVSKITES). (BARIUM TITANATE AND POTASSIUM NIOBATE) /CRYSTA 04-0629
              STRUCTURE DISORDER AND RAMAN SCATTERING IN BARIUM TITANATE AND POTASSIUM NIOBATE.              04-0676
    NEUTRON DIFFRACTION STUDIES OF FERROELECTRICS. (BARIUM TITANATE AND ROCHELLE SALT AS EXAMPLES) 04-0679
/ELECTRIC PROPERTIES OF SOME POLYCRYSTALLINE FERROELECTRICS (BARIUM TITANATE AND SAME WITH 1 PERCENT SILIC/ 04-0818
                          FORCE CONSTANTS IN BARIUM TITANATE AND STRONTIUM TITANATE.              04-0748
    TEMPERATURE DEPENDENCE OF THE RAMAN CROSS SECTIONS IN BARIUM TITANATE AND STRONTIUM TITANATE.            04-0548
/RTICLE SIZE AND SHAPE ON THE INFRARED ABSORPTION SPECTRA OF BARIUM TITANATE AND STRONTIUM TITANATE POWDER/ 04-0823
S.                         DIFFUSION POTENTIALS IN BARIUM TITANATE AND THE THEORY OF PTC MATERIAL 03-0352
/ DIELECTRIC DISPERSION IN FERROELECTRICS OF VARIOUS TYPES. (BARIUM TITANATE AND TITANATE ZIRCONATE, TRIGL/ 02-0108
YSTALS.          SOME PROPERTIES OF THE THERMOCURRENTS IN BARIUM TITANATE AND TITANIUM DIOXIDE SINGLE CR 04-0622
OMETRIC COMPENSATION)    OXYGEN STOICHIOMETRY OF DONOR DOPED BARIUM TITANATE AND TITANIUM DIOXIDE. (STOICHI 04-0664
    SUSCEPTIBILITY INCREASE DURING POLARIZATION REVERSAL. (IN BARIUM TITANATE AND TRIGLYCINE SULFATE) 04-0981
CRYSTALS.                REMARKS ABOUT SWITCHING IN BARIUM TITANATE AND TRIGLYCINE SULFATE SINGLE 04-0915
OOPTICAL PROPERTIES OF SINGLE CRYSTALS OF SOLID SOLUTIONS OF BARIUM TITANATE AND ZINC OXIDE.        ELECTR 04-0953
PHOTOCONDUCTIVITY IN FERROELECTRICS DUE TO SURFACE LAYERS. (BARIUM TITANATE, ANTIMONY SULFOIODIDE) /RINSIC 04-0681
/ND X-RAY STUDIES OF THE PHASE TRANSITIONS IN ROCHELLE SALT, BARIUM TITANATE, ANTIMONY SULFOIODIDE, THIOUR/ 02-0099
                        FLAME SPRAYED BARIUM TITANATE AS A CAPACITOR DIELECTRIC.              04-0773
VARIATION IN CERTAIN ELECTRICAL PROPERTIES OF SINTERED BARIUM TITANATE AS A FUNCTION OF GRAIN SIZE.      04-0627
NSITIONS.                      NUCLEATION PROCESS IN BARIUM TITANATE AT CUBIC- TETRAGONAL PHASE TRA 04-0754
        NUCLEATION AND GROWTH OF FERROELECTRIC DOMAINS IN BARIUM TITANATE AT FIELDS FROM 2 TO 450 KV/CM. 04-0979
      ELECTRON PARAMAGNETIC RESONANCE OF IRON(3+) IN BARIUM TITANATE AT LOW TEMPERATURES.            04-0927
N PRESENCE / CHANGE IN THE DECAY CONSTANT OF ZIRCONIUM-89 IN BARIUM TITANATE (BARIUM CHLORIDE FLUX GROWTH I 04-0685
STEN BRONZE FIELD IN THE PSEUDOTERNARY SYSTEM SODIUM NIOBATE BARIUM TITANATE BARIUM DINIOBATE.       TUNG 13-1970
/WAVE PROPERTIES OF POLYCRYSTALLINE SOLID SOLUTIONS BASED ON BARIUM TITANATE. (BARIUM STRONTIUM TITANATE, / 04-1019
/IELECTRIC ANISOTROPY OF A POLARIZED FERROELECTRIC CERAMIC. (BARIUM TITANATE, BARIUM TIN TITANATE, LEAD BA/ 04-0767
              SINGLE FIRE PROCESS FOR BARIUM TITANATE BARRIER LAYER DIELECTRICS.              04-1029
ACTIONS ON THE DOMAIN STRUCTURE OF SINGLE CRYSTALS FORMED BY BARIUM TITANATE BASE SOLID SOLUTION. /XTERNAL 04-0954
/RE DEPENDENCE OF LEED INTENSITIES FROM THE (001) SURFACE OF BARIUM TITANATE BETWEEN 20 DEGREES C AND 650 / 04-0531
ICES)          DISPERSION OF ELECTROOPTIC EFFECT IN BARIUM TITANATE. (BIREFRINGENCE, PRINCIPAL IND 04-0755
180 / TEMPERATURE DEPENDENCE OF THE DEBYE- WALLER FACTORS OF BARIUM TITANATE (BRAGG REFLECTIONS C-AXIS, 20- 04-0892
CHNIQUE.              DIFFUSION COEFFICIENT OF NIOBIUM IN BARIUM TITANATE BY A DIELECTRIC MEASUREMENT TE 04-0926
FFRACTION.                        BARIUM TITANATE BY CONVERGENT BEAM ELECTRON DI 04-0991
              CURIE SHIFT IN BARIUM TITANATE BY HETEROSUBSTITUTION.              04-0727
SCOPE.        DIRECT OBSERVATION OF 180 DEG DOMAINS IN BARIUM TITANATE BY MEANS OF A POLARIZING MICRO 04-0952
/F THE CHANGE IN THE POLARIZATION OF SINGLE CRYSTAL FILMS OF BARIUM TITANATE BY MEANS OF A STROBOSCOPIC TR/ 04-0942
N.          CHANGE IN THE DECAY CONSTANT OF TECHNETIUM-99 IN BARIUM TITANATE BY THE FERROELECTRIC TRANSITIO 04-0871
              HYDROTHERMAL PREPARATION OF BARIUM TITANATE BY TRANSPORT REACTIONS.              04-0626
ING TITANIUM ESTERS TO YIELD SM/ HYDROTHERMAL PREPARATION OF BARIUM TITANATE BY TRANSPORT REACTIONS (INVOLV 04-0625
THIN FILMS: PREPARATION OF FERROELECTRIC THIN FILMS OF BARIUM TITANATE BY VACUUM EVAPORATION.          04-0834
    POLARITON DISPERSION OF TETRAGONAL FERROELECTRIC BARIUM TITANATE. (CALCULATIONS)              04-0842
/ DEG DOMAIN WALL CONTRIBUTION TO THE DIELECTRIC CONSTANT OF BARIUM TITANATE. (CALCULATIONS, SUMMARY ONLY,/ 04-0804
          AGING IN COMPLEX CERAMICS BASED ON BARIUM TITANATE. (CAPACITANCE, PERMITTIVITY) 04-0722
OF SEMICONDUCTIVE TITANIUM DIOXID/ STUDIES OF FLAME SPRAYED BARIUM TITANATE CAPACITORS. PART-10: STRUCTURE 04-0774
              DOUBLE INJECTION IN BARIUM TITANATE CERAMIC.              04-0554
REPOLARIZATION PROCESSES IN TRIGLYCINE SULFATE CRYSTALS AND BARIUM TITANATE CERAMIC. /THE POLARIZATION AND 04-0924
              FERROELECTRIC LOSSES IN (PURE) BARIUM TITANATE CERAMIC.              04-0555
          PIEZOELECTRIC EFFECT IN ALUMINUM DOPED BARIUM TITANATE CERAMIC.              04-0536
R.                        BARIUM TITANATE CERAMIC: A NEW ELECTRON EMITTE 04-0802
MANGANESE IONS AND ITS APPLICATIONS.     A SEMICONDUCTING BARIUM TITANATE CERAMIC DOPED WITH NIOBIUM AND 04-0734
IVEN AN IMPACT.        PIEZOELECTRIC CHARACTERISTIC OF BARIUM TITANATE CERAMIC EXHIBITED WHEN IT IS G 04-0881
Y DIFFUSION.              FORMATION OF BARIUM TITANATE CERAMIC SEMICONDUCTING LAYER B 04-0684
    TWO SETS OF ELECTRON SPIN RESONANCE OF GADOLINIUM(3+) IN BARIUM TITANATE CERAMIC SEMICONDUCTOR.          04-0990
L/ GRAIN BOUNDARY BARRIER LAYERS IN (N TYPE, ANTIMONY DOPED) BARIUM TITANATE CERAMIC WITH A HIGH EFFECTIVE 04-0581
                DC BIAS EFFECTS IN BARIUM TITANATE CERAMICS.              04-0566
EXCESS TITANIA ON THE ELECTRIC RESISTANCE OF SEMICONDUCTIVE BARIUM TITANATE CERAMICS.        EFFECT OF 04-0966
T OF GRAIN SIZE ON THE RESISTIVITY ANOMALY IN SEMICONDUCTIVE BARIUM TITANATE CERAMICS.        EFFEC 04-0735
FFECT OF INTERNAL BOUNDARIES ON THE DIELECTRIC PROPERTIES OF BARIUM TITANATE CERAMICS.        E 04-0891
CT OF NEUTRON IRRADIATION ON THE FERROELECTRIC PROPERTIES OF BARIUM TITANATE CERAMICS.        EFFE 04-0732
    EFFECTS OF GRAIN GROWTH ON THE DISTRIBUTION OF NIOBIUM IN BARIUM TITANATE CERAMICS.              04-0595
      ELECTRICAL PROPERTIES OF RUBIDIUM AND CESIUM DOPED BARIUM TITANATE CERAMICS.              04-0860
      ELECTROSTRICTIVELY GENERATED AND MODULATED FRACTURE IN BARIUM TITANATE CERAMICS.              04-0700
IN THE EMISSION ELECTRON MICROSCOPE (BARRIER LAYERS IN DOPED BARIUM TITANATE CERAMICS). /F SOLID SURFACES) 04-0913
    INFLUENCE OF VIBRATIONS ON THE DIELECTRIC PROPERTIES OF BARIUM TITANATE CERAMICS.              04-0882
              INFLUENCE OF WATER VAPOR ON BARIUM TITANATE CERAMICS.              04-0880
CTROMECHANICAL AND MECHANICAL PROPERTIES OF FERROELECTRICS. (BARIUM TITANATE CERAMICS)    NONLINEARITY OF ELE 04-0986
PRODUCTION AND STUDY OF THE INNER BARRIERS IN SEMICONDUCTING BARIUM TITANATE CERAMICS.              04-0583
          SPACE CHARGE POLARIZATION AND AGING OF BARIUM TITANATE CERAMICS.              04-0880
NCE OF THE DIELECTRIC CONSTANT OF FERROELECTRIC SUBSTANCES. (BARIUM TITANATE CERAMICS)    TEMPERATURE DEPENDE 04-0591
INFLUENCE OF HUMIDITY ON THE POTENTIAL DISTRIBUTION IN BARIUM TITANATE CERAMICS AND CRYSTALS.         04-0883
EMPERATURES FOR MAXIMUM DIELECTRIC CON/ INTERNAL STRESSES IN BARIUM TITANATE CERAMICS. (AND DIFFERENCE IN T 04-0849
```

FERROELECTRIC PROPERTIES OF (INSULATING, BOUNDARY LAYER) BARIUM TITANATE CERAMICS AT GRAIN BOUNDARY.   04-1024
WITH CADMIUM TR/ CHANGES IN THE FERROELECTRIC PROPERTIES OF BARIUM TITANATE CERAMICS CAUSED BY IRRADIATION   04-0731
D SINTERABILITY.   EFFECT OF CLAY AND DOLOMITE ADDITIONS ON BARIUM TITANATE CERAMICS. CRYSTAL STRUCTURE AN   04-0739
RICS.   STABILIZED BARIUM TITANATE CERAMICS FOR CAPACITOR DIELECT   04-0825
CAL NATURE OF THE SO/ DEPENDENCE CF DIELECTRIC PROPERTIES OF BARIUM TITANATE CERAMICS ON THE PHYSICO- CHEMI   04-0789
TRANSVERSE COMPRESSI/ PIEZOELECTRIC PROPERTIES OF POLARIZED BARIUM TITANATE CERAMICS SUBJECTED TO UNIAXIAL   04-0962
ECTRON MICROSCOPE.   SURFACE STUDIES OF BARIUM TITANATE CERAMICS USING THE SCANNING EL   04-0567
A GRAIN BOUNDARY ELEMENT.   LANTHANUM DOPED BARIUM TITANATE CERAMICS WITH BISMUTH OXIDE AS   04-0799
CONSTANT. PART-1: EFFECTS OF ADDITION OF GLASSES CONTAINING/ BARIUM TITANATE CERAMICS WITH HIGH DIELECTRIC   04-0829
TES AND INTERCRYSTALLITE LAYERS IN THE SEMICONDUCTOR CERAMIC BARIUM TITANATE. (CERIUM DOPED) / OF CRYSTALLI   04-0568
                EFFECT OF OXIDE ADDITIONS ON THE SINTERING OF BARIUM TITANATE COMPACT   04-0996
/ONGENERGETIC) ELECTRON AND (COBALT-60) GAMMA-IRRADIATION ON BARIUM TITANATE, (COMPARED TO STUDY DAMAGE ME/   04-0970
CURIE TEMPERATURE FOR FERROELECTRICS AND ANTIFERROELECTRICS BARIUM TITANATE, COMPLEX LAYER OXIDES /NT AND   03-0188
ETHOD.   IONIC CONTRIBUTION TO INJECTION AND CONDUCTION IN BARIUM TITANATE (CONDENSER CHARGE- DISCHARGE M   04-0571
ECT OF ADDITIVES.   DIELECTRIC PROPERTIES OF BARIUM TITANATE CONTAINING NIOBIUM AND THE EFF   04-0575
            ANOMALY OF RESISTANCE IN REDUCED BARIUM TITANATE CRYSTAL.   04-0988
DETECTABLE POWER OF A PYROELECTRIC THERMAL RECEIVER. (USING BARIUM TITANATE CRYSTAL)   MINIMUM 04-0632
            THEORY OF POLARIZATION REVERSAL OF BARIUM TITANATE CRYSTAL.   03-0326
)   90 DEGREE TYPE DOMAIN WALL OF BARIUM TITANATE CRYSTAL. (THEORY, CALCULATIONS   03-0324
            ANTIPARALLEL SWITCHING IN ALPHA DOMAIN BARIUM TITANATE CRYSTALS.   04-0655
CHEMICAL PROCESSES DURING HYDROGEN REDUCTION OF BARIUM TITANATE CRYSTALS.   04-0540
DIELECTRIC PROPERTIES AND DOMAIN TEXTURE OF BARIUM TITANATE CRYSTALS.   04-0537
HOLOGRAPHIC STORAGE, ELECTRICAL FIXING AND ERASING IN DOPED BARIUM TITANATE CRYSTALS.   04-0848
            LEED STUDIES ON THE (001) FACE OF BARIUM TITANATE CRYSTALS.   04-0530
PHOTOEMISSION MICROSCOPIC OBSERVATION OF THE SURFACE OF BARIUM TITANATE CRYSTALS.   04-0810
SEM OBSERVATION OF THE FERROELECTRIC SURFACE OF BARIUM TITANATE CRYSTALS.   04-0812
            SMALL POLARONS IN CONDUCTING BARIUM TITANATE CRYSTALS.   04-0613
            SURFACE CONDUCTIVITY IN POLARIZED BARIUM TITANATE CRYSTALS.   04-0569
AND BELOW THE FERROELECTRIC / CHANNELING OF PROTONS IN THIN BARIUM TITANATE CRYSTALS AT TEMPERATURES ABOVE   04-0687
MICROSCOPE.   DIRECT OBSERVATION OF BARIUM TITANATE CRYSTALS BY SCANNING ELECTRON   04-0808
LD (1.5-6 KV/CM) POLARIZATION REVERSALS IN LIQUID ELECTRODED BARIUM TITANATE CRYSTALS (CALCULATIONS). / FIE   04-0994
SIUM FLUORIDE).   INVESTIGATION OF THE FINE STRUCTURE OF BARIUM TITANATE CRYSTALS FLUX GROWN FROM POTAS   04-0708
/ECT OF IMPURITIES AND GROWTH CONDITIONS ON THE STRUCTURE OF BARIUM TITANATE CRYSTALS GROWN FROM SOLUTION /   04-0598
/ CONDUCTIVITY (SEEBECK COEFFICIENT, CURIE POINT) OF REDUCED BARIUM TITANATE CRYSTALS (POLARON HOPPING INT/   04-0918
STRESS PULSES.   LIGHT EMISSION FROM BARIUM TITANATE CRYSTALS SUBJECTED TO UNIAXIAL   04-0672
ICS, 150-37/ SPACE CHARGE LIMITED CURRENT IN POLYCRYSTALLINE BARIUM TITANATE. (CURRENT VOLTAGE CHARACTERIST   04-0557
DOMAINS).   AGING PROCESSES IN BARIUM TITANATE (DECREASE IN NUMBER OF 180 DEG   04-0911
IVITY) ELECTRICAL PROPERTIES OF HIGH PURITY POLYCRYSTALLINE BARIUM TITANATE. (DIELECTRIC CONSTANT, CONDUCT   04-0704
RESISTANCE)   SEMICONDUCTING BARIUM TITANATE. (DIELECTRIC CONSTANT, CONTACT   04-0730
SPRAYING (CAPACITORS). ELECTRICAL PROPERTIES OF THICK FILM BARIUM TITANATE DIELECTRICS PRODUCED BY FLAME   04-0772
            DIELECTRIC PROPERTIES OF BARIUM TITANATE DOPED WITH TUNGSTEN OXIDE.   04-0696
TANT AND LATTICE DYNAMICS OF A THIN FILM OF A FERROELECTRIC (BARIUM TITANATE DOWN TO 3 MICRONS). /TRIC CONS   04-0600
COMPARED WITH DA/ (STEP OR SMOOTH) DOMAIN WALL STRUCTURE IN BARIUM TITANATE DURING SWITCHING (CALCULATIONS   04-0677
THE ORTHORHOMBIC PHASE IN MODIFIED BARIUM TITANATE. (EFFECT OF SUBSTITUTIONS)   04-0617
            FLUORINE CONTAINING BARIUM TITANATE. (ELECTRICAL PROPERTIES)   04-0869
OMAIN BOUNDARIES DURING SWITCHING IN SINGLE CRYSTAL FILMS OF BARIUM TITANATE (ELECTRON MICROSCOPY). /N OF D   04-0940
            GROWTH AND STRUCTURE OF BARIUM TITANATE EVAPORATED FILMS.   04-0604
HE OPTICAL DA/ OPTICALLY INDUCED REFRACTIVE INDEX CHANGES IN BARIUM TITANATE (EXPERIMENTS TO CHARACTERIZE T   04-1004
FHASE CHANGES IN TRIGLYCINE SULFATE AND BARIUM TITANATE FERROELECTRICS. (26 REFS)   04-0621
            PREPARATION OF BARIUM TITANATE FILM BY EVAPORATION.   04-0743
ECTRON PROBE X-RAY MICROANALYSIS ON IMPURITIES IN EVAPORATED BARIUM TITANATE FILMS.   EL 04-0742
            EXPERIMENTAL STUDY OF EVAPORATED BARIUM TITANATE FILMS.   04-0939
            PREPARATION AND PROPERTIES OF THIN BARIUM TITANATE FILMS.   04-0669
ONTO PLATINUM OR FUSED QUARTZ).   PREPARATION OF THIN BARIUM TITANATE FILMS BY DC DIODE SPUTTERING (04-0948
N OXYGEN ATMOSPHERE.   SINGLE CRYSTAL BARIUM TITANATE FILMS GROWN FROM THE MELT IN A   04-0612
            CHARACTERISTICS OF RF SPUTTERED BARIUM TITANATE FILMS ON SILICON.   04-0930
G ONTO INDIUM ANTIMONIDE OR GALLIUM ARSENIDE.   BARIUM TITANATE FILMS PREPARED BY RF SPUTTERIN   04-0741
TERING.   STRUCTURE AND PROPERTIES OF BARIUM TITANATE FILMS PRODUCED BY CATHODE SPUT   04-0832
ESENCE OF CARBON.   CHEMICAL CHANGES IN BARIUM TITANATE FIRED UNDER NITROGEN IN THE PR   04-0872
UM DIOXIDE IN THE PRESENCE OF SODIUM CARB/ THE PRODUCTION OF BARIUM TITANATE FROM BARIUM SULFATE AND TITANI   04-0624
OF DOMAIN WALLS IN FERROELECTRIC AND FERROELASTIC CRYSTALS. (BARIUM TITANATE, GADOLINIUM MOLYBDATE) /KNESS   04-0777
LUO/ FERROELECTRIC SINGLE CRYSTALS WITH LARGE NONLINEARITY. (BARIUM TITANATE HAFNATE GROWN FROM POTASSIUM F   08-1373
RINGES.   STUDY OF CRYSTALS OF BARIUM TITANATE HAVING PARALLEL INTERFERENCE F   04-0897
T)   BAND STRUCTURE OF THE CUBIC PHASE OF BARIUM TITANATE. (IMAGINARY DIELECTRIC CONSTAN   04-0845
ELECTRON MICROSCOPIC STUDY CF FERROELECTRIC DOMAINS OF BARIUM TITANATE IN LOW TEMPERATURE PHASES.   04-1034
ERHIGH FREQUENCIES (MICROWAVES). FERROELECTRIC PROPERTIES OF BARIUM TITANATE IN THE CENTIMETER RANGE OF SUP   04-0901
C REGION.   ABSORPTION AND REFLECTION SPECTRA OF BARIUM TITANATE IN THE LONG WAVELENGTH INFRARE   04-0644
            DIELECTRIC DISPERSION OF IRRADIATED BARIUM TITANATE IN THE PHASE TRANSITION REGION   04-0652
E/ RELAXATION ABSORPTION OF ULTRASOUND IN SINGLE CRYSTALS OF BARIUM TITANATE IN THE PRESENCE OF A STEADY EL   04-0850
EIC PHASES. ELECTRON PARAMAGNETIC RESONANCE OF IRON(3+) IN BARIUM TITANATE IN THE RHOMBOHEDRAL AND THE CU   04-0929
/ STRUCTURES OF SOME FERROELECTRICS AND ANTIFERROELECTRICS. (BARIUM TITANATE, KDP, POTASSIUM SODIUM TARTRA/   02-0134
INELASTIC NEUTRON SCATTERING FROM SINGLE DOMAIN BARIUM TITANATE. (LATTICE DYNAMICS)   04-0946
MOSSBAUER STUDIES OF LOWER TRANSITIONS IN BARIUM TITANATE. (LATTICE INSTABILITIES)   04-0564
ESISTORS)   INCORPORATION OF ANTIMONY INTO THE BARIUM TITANATE LATTICE. (STRUCTURE, AND PCT R   04-0934
/ ON THE PROPERTIES OF A SERIES OF FERROELECTRIC MATERIALS. (BARIUM TITANATE, LEAD FERRATE(0.5) NIOBATE(0./   04-0780
/UBIC PEROVSKITES. (STRONTIUM TITANATE, POTASSIUM TANTALATE, BARIUM TITANATE, LEAD TITANATE, LEAD ZIRCONAT/   02-0114
NATURE OF AGING IN FERROELECTRIC CERAMICS. (BARIUM TITANATE, LEAD ZIRCONATE TITANATE)   04-0757
/ OF CERAMIC FERROELECTRICS ON THEIR MECHANICAL PROPERTIES. (BARIUM TITANATE, LEAD ZIRCONATE TITANATE, LEA/   04-0985
/OF ABSOLUTE SIGNS OF MICROWAVE NONLINEAR SUSCEPTIBILITIES. (BARIUM TITANATE, LITHIUM TANTALATE, LITHIUM N/   02-0107
ON BEHAVIOR.   ORIGIN OF LATTICE DEFECTS IN (BARIUM) TITANATE MATERIALS AND THEIR INFLUENCE   04-0784
DESCRIPTION)   FERROELECTRIC TRANSFORMATION IN BARIUM TITANATE. (MODEL GIVES NEARLY COMPLETE   04-0836
CONDUCTIVITY OF STRONTIUM TITANATE, POTASSIUM TANTALATE AND BARIUM TITANATE NEAR THE CURIE POINTS. THERMAL   04-0989
CRITICAL X-RAY INCOHERENT SCATTERING BY BARIUM TITANATE NEAR THE CURIE TEMPERATURE.   04-0535
ELECTRON PARAMAGNETIC RESONANCE IN BARIUM TITANATE NEAR THE CURIE TEMPERATURE.   04-0928
DIELECTRIC DISPERSION OF IRRADIATED BARIUM TITANATE NEAR THE PHASE TRASNITION.   04-0648
ONAL TUNGSTEN BRONZE).   DIELECTRIC PROPERTY OF BARIUM TITANATE NIOBATE SINGLE CRYSTAL (TETRAG   13-1969
SPONTANEOUS INDUCTION.   DEPENDENCE OF THE PIEZOEFFECT IN BARIUM TITANATE ON THE MEANS OF SCREENING THE   04-1021
, DOPANT NOT NAMED)   PIEZORESISTIVITY OF DOPED BARIUM TITANATE. (OR BARIUM STRONTIUM TITANATE)   04-0729
TRICAL CONDUCTIVITY MEASURE/ TITANIUM RICH NONSTOICHIOMETRIC BARIUM TITANATE. PART-1: HIGH TEMPERATURE ELEC   04-0822
MODYNAMIC AND X-RAY DIFFRAC/ TITANIUM RICH NONSTOICHIOMETRIC BARIUM TITANATE. PART-3: HIGH TEMPERATURE THER   04-0886

ON THE ANOMALOUS CRYSTALLOGRAPHIC PROPERTIES OF SMALL BARIUM TITANATE PARTICLES. 04-0935
OF SURFACE MOISTURE ON THE DIELECTRIC BEHAVIOR OF ULTRAFINE BARIUM TITANATE PARTICULATES. EFFECT 04-0859
TUDYING ULTRASONIC FIELDS. BARIUM TITANATE PIEZOELECTRIC MICROPROBE FOR S 04-0938
NIUM OR TELLURIUM) SEMICONDUCTOR F/ EFFECT OF FERROELECTRIC (BARIUM TITANATE) POLARIZATION FIELDS ON (GERMA 04-0594
IENT OF RESISTIVITY) RESISTIVITY ANOMALY IN SEMICONDUCTING BARIUM TITANATE. (POSITIVE TEMPERATURE COEFFIC 04-0584
ERASURE OF HOLOGRAPHIC PATTERNS IN FERROELECTRIC MATERIALS. (BARIUM TITANATE, POTASSIUM NIOBATE) /TION AND 04-0847
/HASE TRANSITIONS. (STRONTIUM TITANATE, POTASSIUM TANTALATE, BARIUM TITANATE, POTASSIUM NIOBATE, SODIUM NI/ 02-0059
FAR INFRARED DIELECTRIC DISPERSION OF PEROVSKITES. (BARIUM TITANATE, POTASSIUM TANTALATE) 04-0546
ILICON SUPS/ PHYSICAL AND ELECTRICAL PROPERTIES OF THIN FILM BARIUM TITANATE PREPARED BY RF SPUTTERING ON S 04-0828
FRINGES. CRYSTAL STRUCTURE OF BARIUM TITANATE PRODUCING SQUARE INTERFERENCE 04-0706
POTENTIAL BARRIERS ON SEMICONDUCTING BARIUM TITANATE. (PTCR DEVICES) 04-0851
D (IRON, HYDROXYL) SINGLE CRYSTALS. CONDUCTIVITY OF BARIUM TITANATE, PURE SINGLE CRYSTALS AND DOPE 04-0695
CAL AND INFRARED RADIATION (0.5 TO 2 MIC/ CHARACTERISTICS OF BARIUM TITANATE PYROELECTRIC DETECTOR FOR OPTI 04-0751
IENT, OPTICAL ABSORPTION) SOME SEMICONDUCTING PROPERTIES OF BARIUM TITANATE. (RESISTIVITY, SEEBECK COEFFIC 04-0919
ECTROMAGNETIC AND ACOUSTIC WAVES IN FERROELECTRICS. (THEORY, BARIUM TITANATE, ROCHELLE SALT) EL 03-0455
GY AND COVALENCY ON THE STRUCTURE OF AB(5) OXIDE COMPOUNDS. (BARIUM TITANATE, RUBIDIUM NIOBATE) /ELUNG ENER 04-0761
/OTION. (PYROELECTRIC EFFECT IN TGS, ETHYL DIAMINE TARTRATE, BARIUM TITANATE, SALICYLIDENE- 4-BROMO ANILIN/ 04-0854
ISM FOR FORMING INSULATION LAYERS ON THE GRAIN BOUNDARIES OF BARIUM TITANATE SEMICONDUCTIVE CERAMICS. /CHAN 04-0867
THERMAL NEUTRON IRRADIATION ON THE RELATIVE PERMITTIVITY OF BARIUM TITANATE SINGLE CRYSTAL. EFFECT OF 04-0781
PARAMAGNETIC RESONANCE OF PLATINUM IONS IN NIOBIUM DOPED BARIUM TITANATE SINGLE CRYSTAL. 04-0977
PRODUCTION AND STUDY OF PURE AND DOPED BARIUM TITANATE SINGLE CRYSTAL 04-0667
X-RAY STUDY OF THE SURFACE LAYER ON BARIUM TITANATE SINGLE CRYSTAL. 04-0858
COMPENSATION OF HETEROVALENT FERRIC IONS IN THE LATTICE OF BARIUM TITANATE SINGLE CRYSTALS. 04-0539
DETERMINATION OF THE DEVIATION FROM OXYGEN STOICHIOMETRY OF BARIUM TITANATE SINGLE CRYSTALS. 04-0736
DIELECTRIC MEASUREMENTS ON PILE-IRRADIATED BARIUM TITANATE SINGLE CRYSTALS. 04-0910
DISLOCATIONS IN BARIUM TITANATE SINGLE CRYSTALS. 04-1007
EFFECT OF ADDING NICKEL ON THE PROPERTIES OF BARIUM TITANATE SINGLE CRYSTALS. 04-0793
EFFECT OF EXTERNAL ACTIONS ON THE REPOLARIZATION OF BARIUM TITANATE SINGLE CRYSTALS. 04-0957
EFFECT OF FERRIC OXIDE ON THE 120 DEGREE PHASE TRANSITION IN BARIUM TITANATE SINGLE CRYSTALS. 04-0544
CT OF LINEAR COMPRESSION ON THE UHF DIELECTRIC PROPERTIES OF BARIUM TITANATE SINGLE CRYSTALS. EFFE 04-0896
ION OF TITANIUM(4+) IONS ON THE ELECTROOPTICAL PROPERTIES OF BARIUM TITANATE SINGLE CRYSTALS. /NT SUBSTITUT 04-0959
EFFECT OF TRANSVERSE COMPRESSION OF DIELECTRIC PROPERTIES OF BARIUM TITANATE SINGLE CRYSTALS. 04-0956
ETCH PITS CORRESPONDING TO DISLOCATIONS IN BARIUM TITANATE SINGLE CRYSTALS. 04-1023
ETCH PITS IN BARIUM TITANATE SINGLE CRYSTALS. 04-0922
E OF THE AMPLITUDE OF FREE VIBRATIONS ON THE PERMITTIVITY OF BARIUM TITANATE SINGLE CRYSTALS. INFLUENC 04-0768
TRESSES ON THE ELECTRICAL PROPERTIES AND DOMAIN STRUCTURE OF BARIUM TITANATE SINGLE CRYSTALS. /AL TENSILE S 04-0961
INTERNAL FRICTION IN BARIUM TITANATE SINGLE CRYSTALS. 04-0794
KERR EFFECT STUDIES OF ELECTRIC CONDUCTION IN IRON DOPED BARIUM TITANATE SINGLE CRYSTALS. 04-0703
LATTICE DYNAMICAL THEORY OF SWITCHING IN BARIUM TITANATE SINGLE CRYSTALS. 03-0398
MICROWAVE DYNAMIC NONLINEARITY OF BARIUM TITANATE SINGLE CRYSTALS. 04-0749
HASE TRANSITIONS AND) THERMOCURRENTS IN TITANIUM DIOXIDE AND BARIUM TITANATE SINGLE CRYSTALS. (P 04-0697
PREPARATION AND PROPERTIES OF PURE AND DOPED BARIUM TITANATE SINGLE CRYSTALS. 04-0668
PROBLEM OF THE SURFACE LAYER IN BARIUM TITANATE SINGLE CRYSTALS. 04-1013
STABILIZATION OF SPONTANEOUS POLARIZATION OF BARIUM TITANATE SINGLE CRYSTALS. 04-0795
STUDY OF THE AGING PHENOMENON IN BARIUM TITANATE SINGLE CRYSTALS. 04-0545
TEMPERATURE DEPENDENCE OF INTERNAL FRICTION OF BARIUM TITANATE SINGLE CRYSTALS. 04-0796
THICKNESS DEPENDENCE OF THE PERMITTIVITY OF BARIUM TITANATE SINGLE CRYSTALS. 04-0637
THICKNESS DEPENDENCE OF POLARIZATION REVERSAL IN BARIUM TITANATE SINGLE CRYSTALS. 04-0949
/AL (MARTENSITIC TYPE) PHASE TRANSITIONS IN HIGH RESISTIVITY BARIUM TITANATE SINGLE CRYSTALS. (AND TWINNIN/ 04-0887
REQUENCIES. MEASUREMENT OF THE PERMITTIVITY OF BARIUM TITANATE SINGLE CRYSTALS AT MICROWAVE F 04-0759
XIDE GROUPS. DIELECTRIC PROPERTIES OF BARIUM TITANATE SINGLE CRYSTALS DOPED BY HYDRO 04-0634
UTION IN A POT/ ANISOTROPY OF THE ELECTRICAL CONDUCTIVITY OF BARIUM TITANATE SINGLE CRYSTALS GROWN FROM SOL 04-1035
EN (EFFEC/ TIME DEPENDENCE OF THE ELECTRICAL CONDUCTIVITY OF BARIUM TITANATE SINGLE CRYSTALS HEATED IN OXYG 04-0997
C AND RHOMBOHEDRAL PHASES. REPOLARIZATION PROCESSES OF BARIUM TITANATE SINGLE CRYSTALS IN ORTHORHOMBI 04-0963
LUORIDE FLUX. DISCUSSION ON THE CRYSTAL GROWTH OF BARIUM TITANATE SINGLE CRYSTALS IN POTASSIUM F 04-1022
CTRIC PHASE. PIEZOELECTRIC ACTIVITY OF BARIUM TITANATE SINGLE CRYSTALS IN THE PARAELE 04-0771
RANSITION REGION. DIELECTRIC PERMEABILITY OF BARIUM TITANATE SINGLE CRYSTALS IN THE PHASE T 04-0645
OF THE CU/ POSITIVE TEMPERATURE COEFFICIENT OF RESISTANCE OF BARIUM TITANATE SINGLE CRYSTALS IN THE REGION 04-1036
/FFECT OF SOME ADMIXED NICKEL ON SOME PHYSICAL PROPERTIES OF BARIUM TITANATE SINGLE CRYSTALS. (LATTICE PAR/ 04-0909
NT. PHASE BOUNDARIES IN BARIUM TITANATE SINGLE CRYSTALS NEAR CURIE POI 04-0577
N STRUCTURE. DEPENDENCE OF DIELECTRIC DISPERSION IN BARIUM TITANATE SINGLE CRYSTALS ON THEIR DOMAI 04-0895
NESS. DEPENDENCE OF COERCIVE FIELD OF BARIUM TITANATE SINGLE CRYSTALS ON THEIR THICK 04-0693
C MODULUS) AMPLITUDE DEPENDENCE OF THE INTERNAL FRICTION IN BARIUM TITANATE SINGLE CRYSTALS. (PIEZOELECTRI 04-0791
ECHANICAL IMPULSES. BARIUM TITANATE SINGLE CRYSTALS SUBJECTED TO M 04-1006
STERESIS LOOP. PECULIARITIES OF POLARIZATION OF BARIUM TITANATE SINGLE CRYSTALS WITH DOUBLE HY 04-0709
OMAIN S/ DIELECTRIC, ELASTIC AND PIEZOELECTRIC PROPERTIES OF BARIUM TITANATE SINGLE CRYSTALS WITH LAMINAR D 04-1010
NOLOGICAL PROPERTIES OF FERROELECTRI/ INFLUENCE OF FIRING ON BARIUM TITANATE SINTERS ON ELECTRICAL AND TECH 04-1002
PED BY TUNGSTEN TRIOXIDE. SEMICONDUCTOR PROPERTIES OF BARIUM (TITANATE, STANNATE) SOLID SOLUTIONS DO 04-0965
E / CHANGE OF THE MOSSBAUER EFFECT PROBABILITY ON TIN-119 IN BARIUM TITANATE STANNATE SOLID SOLUTIONS IN TH 04-0573
HIGH DIELECTRIC CONSTANT TEMPERATURE STABLE FERROELECTRICS. (BARIUM TITANATE, STRONTIUM STANNATE) /HIEVING 04-0551
CERAMIC DIELECTRICS. (BARIUM TITANATE, STRONTIUM TITANATE) 04-0683
F DISPLACIVE TYPE FERROELECTRICS NEAR THE TRANSITION POINT. (BARIUM TITANATE, STRONTIUM TITANATE) /HAVIOR O 04-0992
NCE SPECTRA IN A FERROELECTRIC CERAMIC CONTAINING TITANIUM. (BARIUM TITANATE, STRONTIUM TITANATE) /C RESONA 04-1020
ENCE OF THE DIELECTRIC CONSTANTS OF PARAELECTRIC MATERIALS. (BARIUM TITANATE, STRONTIUM TITANATE) /E DEPEND 04-0856
/ PERMITTIVITY OF SOME TITANATES BETWEEN 20 AND 1000 DEG C. (BARIUM TITANATE, STRONTIUM TITANATE AND OTHER/ 04-0820
/ION GROWTH OF OXIDE CRYSTALS FROM NONSTOICHIOMETRIC MELTS. (BARIUM TITANATE, STRONTIUM TITANATE, GERMANAT/ 04-0553
C FIELD IN A PSEUDOCUBIC IONIC STRUCTURE. APPLICATION TO ALL BARIUM TITANATE STRUCTURE TYPES. ELECTROSTATI 04-1008
CHELLE SALT, TRIGLYCINE SULFATE, SELENATE, FLUOBERYLLATE AND BARIUM TITANATE, SUMMARY ONLY) /ELECTRICS. (RO 02-0083
ACOUSTIC EMISSION FROM FERROELECTRIC CRYSTALS. (BARIUM TITANATE, TGS) 04-0593
PERMITTIVITY OF THE STATIONARY (100)-180 DEG DOMAIN WALL IN BARIUM TITANATE (THEORY, CALCULATIONS). /IGNAL 03-0358
E. POLARITONS IN TETRAGONAL BARIUM TITANATE: THEORY OF THE RAMAN LINE SHAP 04-0556
ANGULAR ENERGY DEPENDENCE OF 180 DEG DOMAIN WALL IN BARIUM TITANATE. (THERMODYNAMIC CALCULATIONS). 04-0658
FORMATION PROCESS OF BARIUM TITANATE THIN FILM EVAPORATED IN VACUO. 04-0862
CHARACTERISTICS OF RF SPUTTERED BARIUM TITANATE THIN FILMS. 04-0906
FABRICATION OF RF SPUTTERED BARIUM TITANATE THIN FILMS. 04-0907
/CROSTRUCTURE) AND ELECTRICAL PROPERTIES OF FLASH EVAPORATED BARIUM TITANATE THIN FILMS (SEPARATED FROM SU/ 04-0605
NCE) ELECTRICAL PROPERTIES OF FLASH EVAPORATED FERROELECTRIC BARIUM TITANATE THIN FILMS. (THICKNESS DEPENDE 04-0967
LOGICAL MECHANISMS, MEMORY, ORGANICS) FROM BARIUM TITANATE TO DNA. (FERROELECTRICITY, BIO 02-0092
FRACTURE TOUGHNESS OF COMMERCIAL BARIUM TITANATE TRANSDUCER MATERIAL. 04-0841
ION. (CONCEPTS) INTERPRETATION OF THE DYNAMICS OF THE BARIUM TITANATE TRANSITION IN TERMS OF NUCLEAT 04-0606

/ON OF ELECTROMAGNETIC RADIATION USING PYROELECTRIC EFFECT. (BARIUM TITANATE, TRIGLYCINE SULFATE, AT 5.6 A/ 04-0978
CALCULATION OF ELECTRIC FIELDS IN IONIC CRYSTALS. (BARIUM TITANATE TYPE) 04-0824
CE IMPERFECTIONS ON THE DIELECTRIC PROPERTIES OF MATERIALS. (BARIUM TITANATE TYPE) EFFECT OF CHARGED LATTI 04-1017
RADIATION PHYSICS OF FERROELECTRICS OF BARIUM TITANATE TYPE. 04-0972
MICROWAVE PERMI/ THERMAL CHARACTERISTICS OF A FERROELECTRIC (BARIUM TITANATE TYPE) CERAMIC WITH CONTROLLED 04-0904
O DENSI/ ELECTRICAL PROPERTIES OF HOT PRESSED FERROELECTRIC (BARIUM TITANATE TYPE) CERAMICS. (IN RELATION T 04-0879
UHF REVERSAL PROPERTIES OF BARIUM TITANATE TYPE FERROELECTRIC CERAMICS. 04-1014
THE FORM OF THICK PLATES. MACROSCOPIC THEORY OF BARIUM TITANATE TYPE FERROELECTRIC CRYSTALS IN 04-0770
OF THE THEORY OF MICROWAVE DISPERSION OF THE PERMITTIVITY OF BARIUM TITANATE TYPE FERROELECTRICS. /L CHECK 04-0649
MICROWAVE DISPERSION MECHANISM IN BARIUM TITANATE TYPE FERROELECTRICS. 04-0902
PHOTOFERROELECTRIC EFFECTS IN A (V) B (VI) C (VII) AND BARIUM TITANATE TYPE FERROELECTRICS. 04-0552
THERMODYNAMICAL THEORY OF BARIUM TITANATE TYPE FERROELECTRICS. 03-0322
NEW TREATMENT AND INTERPRETATION OF X/ STRUCTURE DISORDER OF BARIUM TITANATE TYPE FERROELECTRICS. BASED ON 04-0630
ER BAND. DIELECTRIC DISPERSION IN BARIUM TITANATE TYPE FERROELECTRICS IN THE MET 04-1009
RYSTALS AND CERAMICS) PERMITTIVITY DISPERSION OF BARIUM TITANATE TYPE FERROELECTRICS. (SINGLE C 04-0903
FERROELECTRIC CERAMICS) WITH GREAT DIELECTRIC PERMITTIVITY. (BARIUM TITANATE TYPE MATERIALS) LOW LOSS (04-0779
IATION EFFECT ON DIELECTRIC PROPERTIES OF SOLID SOLUTIONS OF BARIUM TITANATE TYPE MATERIALS. X-RAY IRRAD 04-0782
YNAMICS. FERROELECTRICITY (IN BARIUM TITANATE TYPE STRUCTURES) AND LATTICE D 04-0950
LD. PERMITTIVITY CHANGE OF BARIUM TITANATE UNDER VERY SLOWLY CHANGING FIE 04-0868
AND THE / INVESTIGATION OF THE FERROELECTRIC PHASE CHANGE IN BARIUM TITANATE USING ELECTRON SPIN RESONANCE 04-0827
(4+) ION/ GROWING SINGLE CRYSTALS OF SOLID SOLUTION BASED ON BARIUM TITANATE, WHERE BARIUM (2+) AND TITANIUM 04-0960
YSTAL CHEMICAL AND PHYSICAL PROPERTIES OF SINGLE CRYSTALS OF BARIUM TITANATE WITH COBALT ADDED. SOME CR 04-0618
PERTIES. DEVELOPMENT OF BARIUM TITANATE WITH CONSISTENT DIELECTRIC PRO 04-0797
RTIES. (DIELECTRIC CON/ SYNTHESIS OF NIOBIUM PENTOXIDE DOPED BARIUM TITANATE WITH IMPROVED ELECTRICAL PROPE 04-0840
/N A TWO-LAYER SYSTEM CONSISTING OF A PIEZOELECTRIC CERAMIC (BARIUM TITANATE) WITH LARGE DIELECTRIC CONSTA/ 04-1037
EFFECT OF DOPING BARIUM TITANATE WITH MANGANESE DIOXIDE. 04-0562
OPTICAL AND DIELECTRIC PROPERTIES OF SINGLE CRYSTAL BARIUM TITANATE WITH NIOBIUM ADDITIONS. 04-0875
/POSITIVE TEMPERATURE COEFFICIENT OF RESISTANCE) BEHAVIOR OF BARIUM TITANATE WITH NIOBIUM PENTOXIDE AND MA/ 04-0837
NGES. CRYSTAL STUDY OF BARIUM TITANATE WITH PARALLEL INTERFERENCE FRI 04-0705
FFECT, PARAELECTRIC AND FERROELECTRIC PHAS/ SURFACE LAYER IN BARIUM TITANATE (X-RAY DIFFRACTION, DC FIELD E 04-0857
/EFFICIENTS OF POLAR CRYSTALS (LITHIUM NIOBATE OR TANTALATE, BARIUM TITANATE, ZINC OXIDE, AND LITHIUM GALL/ 02-0098
/LECTRIC) CRYSTALS (LITHIUM NIOBATE OR TANTALATE OR GALLATE, BARIUM TITANATE, ZINC OXIDE, CADMIUM SULFIDE). 02-0095
FI/ ELECTROMECHANICAL PROPERTIES OF CERAMIC FERROELECTRICS (BARIUM TITANATE ZIRCONATES) IN STRONG ELECTRIC 09-1605
. SPACE CHARGE LIMITED CURRENTS IN HEXAGONAL BARIUM TITANATE (0.75) PLATINATE (0.25) CRYSTALS 04-0936
PLA/ PROPAGATION CHARACTERISTICS OF SURFACE ELASTIC WAVES ON BARIUM TITANATE (0.95) ZIRCONATE (0.05) CERAMIC 09-1613
TZ, LISSAJOUS FI/ ENERGY LOSS NONPOLAR STATE POLYCRYSTALLINE BARIUM TITANATE. (150-500 DEG C, 20-50 KILOHER 04-0831
ETIC RESONANCE OF TRIVALENT GADOLINIUM IONS IN STRONTIUM AND BARIUM TITANATES. ELECTRON PARAMAGN 04-0920
PHASE EQUILIBRIA IN THE SYSTEM BARIUM TITANATE- BARIUM GERMANATE. 04-0710
DEMONSTRATION OF AGING EFFECTS IN BARIUM TITANATE, BARIUM ZIRCONATE. 09-1543
/DIES OF FERROELECTRICS AND FERROELECTRIC MAGNETS. (LEAD AND BARIUM TITANATES, BISMUTH FERRATE, POTASSIUM / 04-1038
/O-STEP ANOMALOUS INCREASE OF RESISTIVITY ON LANTHANUM DOPED BARIUM TITANATE- BISMUTH TRIOXIDE COMPOSITE C/ 04-0800
TWO-PHASE FERROELECTRIC SYSTEM. PART-1: BARIUM TITANATE- METAL. 04-0815
/D FERROELECTRIC PROPERTIES OF SOLID SOLUTIONS IN THE SYSTEM BARIUM TITANATE- STRONTIUM NIOBIUM (0.8) OXIDE/ 04-0707
/ VOLT- AMPERE CHARACTERISTICS OF CADMIUM SELENIDE LAYERS ON BARIUM TITANATE- STRONTIUM TITANATE SUBSTRATE/ 04-0592
TION SINGLE CRYSTALS. ELECTROOPTICAL PROPERTIES OF BARIUM TITANATE- TANTALUM PENTOXIDE SOLID SOLU 04-0955
ELECTROOPTICAL EFFECT IN SINGLE CRYSTALS OF BARIUM TITANATE- ZINC TITANATE SOLID SOLUTIONS 04-0958
/IDE (3) (M+2) (M=2,3,4). (DATA FOR THE TITANIUM (3) OXIDE (12), BARIUM TITANIUM (4) OXIDE (15), AND BARIUM (2) T/ 02-0066
LS (CZOCHRALSKI GROWTH, DIELECTRIC/ PREPARATION AND STUDY OF BARIUM (X) STRONTIUM (1-X) NIOBATE SINGLE CRYSTA 13-1936
BRATING WIRE" STUDY OF POTENTIAL DISTRIBUTION (STABILITY) IN BARIUM (X) STRONTIUM (1-X) TITANATE CERAMICS. /I 04-0699
TWEEN 4 AND 100 DEGREES K. POLARIZATION OF SOLID BARIUM (X) STRONTIUM (1-X) TITANATE COMPOUNDS BE 04-0719
/E TRANSITIONS AT LOW TEMPERATURES: STRONTIUM TITANATE AND ((BARIUM (X) STRONTIUM (1-X) TITANATE (RECIPROCAL/ 04-0898
/T LOW TEMPERATURES. (POLYCRYSTALLINE STRONTIUM TITANATE AND BARIUM (X) STRONTIUM (1-X) TITANATE SOLID SOLUT/ 04-0720
PHASE TRANSITIONS IN THE MIXED CRYSTAL SYSTEM STRONTIUM (1-X) BARIUM (X) TITANATE. 04-1016
/IC PROPERTIES IN THE TERNARY SYSTEM LEAD (ZINCATE NIOBATE)- BARIUM (ZINCATE NIOBATE)- LEAD TITANATE. (PHA/ 05-1001
LECTRIC -160 TO +25/ STRUCTURAL AND DIELECTRIC PROPERTIES OF BARIUM ZINCATE (0.33) NIOBATE (0.67) (NOT FERROE 36-3609
ERAMICS. BARIUM ZIRCONATE LEAD TITANATE FERROELECTRIC C 09-1492
EAD CONTAINING OXYGEN OCTAHEDRA FERROELECTRICS: CERAMIC LEAD BARIUM ZIRCONATE TITANATE. /POLARIZATION) OF L 13-2039
ENCE) RESISTIVITY BEHAVIOR OF SEMICONDUCTING YTTRIUM BARIUM ZIRCONIUM TITANATE. (TEMPERATURE DEPEND 13-1995
/ZATION SWITCHING TIME, LEAD (0.35) STRONTIUM (0.65) TITANATE, BARIUM (0.65 OR 0.75) STRONTIUM (0.35 OR 0.25) / 37-3660
REJUVENATION OF A BARIUM (0.72) STRONTIUM (0.28) TITANATE CERAMIC. 04-0817
PREPARATION OF THE TERNARY TITANATE BARIUM (0.8) LEAD (0.12) CALCIUM (0.08) TITANATE. 36-3587
ELECTRIC FIELDS. (ANISOTROPY DEVELOPED, BARIUM TITANATE AND BARIUM (0.95) CALCIUM (0.05) TITANATE USED) /ONG 04-0661
/O STEP ANOMALOUS INCREASE OF RESISTANCE ON LANTHANUM (0.002) BARIUM (0.998) TITANATE- BISMUTH OXIDE COMPOSI/ 04-0801
SOLUTIONS BY HYDR/ SHIFTING OF THE TRANSITION TEMPERATURE OF BARIUM (0-0.1) STRONTIUM (0.9-1) TITANATE SOLID 04-0722
YGEN OCTAHEDRON FERROELECTRICS ABOVE THE CURIE TEMPERATURE. (BARIUM (0-1.0) STRONTIUM (1.0-0) TITANATE) /N OX 04-0950
ALS GROWN FROM A POTASSIUM FLUORID/ DIELECTRIC PROPERTIES OF BARIUM (1-X) STRONTIUM (X) TITANATE SINGLE CRYST 04-0635
TALS FROM A POTASSIUM FLUORIDE MELT. DEFECT STRUCTURE OF (BARIUM (1-X) STRONTIUM (X)) TITANATE- MIXED CRYS 04-0873
THERMAL STUDIES OF CALCIUM (X) BARIUM (1-X) TITANATE. 04-0694
/5) TITANATES AND PHASE DIAGRAMS FOR BISMUTH (X) POTASSIUM (X) BARIUM (1-2X) TITANATE, BISMUTH (0.5) POTASSIUM/ 08-1289
/ CRYSTALS OF SOLID SOLUTION BASED ON BARIUM TITANATE, WHERE BARIUM (2+) AND TITANIUM (4+) IONS ARE REPLACED/ 04-0960
DIUM) OXIDE (6), AND BARIUM (3) BISMUTH (/ NEW FERROELECTRICS. (BARIUM (2) BISMUTH (NIOBIUM OR TANTALUM OR VANA 36-3619
THE TITANIUM (3) OXIDE (12), BARIUM TITANIUM (4) OXIDE (15), AND BARIUM (2) TITANIUM (5) OXYGEN (18) COMPOUNDS) / 02-0066
/ (2) BISMUTH (NIOBIUM OR TANTALUM OR VANADIUM) OXIDE (6), AND BARIUM (3) BISMUTH (2) (TUNGSTEN OR MOLYBDENUM)/ 36-3619
FERROELECTRIC BARKHAUSEN EFFECT IN ANTIMONY SULFOIODIDE. 15-2172
TGS AND ROCHELLE SALT). INFLUENCE OF GAMMA-IRRADIATION ON BARKHAUSEN EFFECT IN FERROELECTRIC MATERIALS (27-3123
INUOUS X-RAY DIFFRACTION. BARKHAUSEN EFFECT IN FERROELECTRICS UNDER CONT 03-0337
INE SULFATE CRYSTALS. DISTINCTIVE FEATURES OF BARKHAUSEN EFFECT IN ROCHELLE SALT AND TRIGLYC 27-3110
LS. INVESTIGATION OF BARKHAUSEN EFFECT IN TRIGLYCINE SULFATE CRYSTA 27-3111
N EFFECT IN FERROELECTRIC CRYSTALS. MECHANISMS FOR BARKHAUSEN JUMPS AND BEHAVIOR OF THE BARKHAUSE 03-0448
CRYSTALS. BARKHAUSEN JUMPS AND SWITCHING CURRENT IN TGS 27-3109
INGLE CRYSTALS OF SEIGNETTE SALT. CORRELATION BETWEEN BARKHAUSEN NOISE AND SURFACE MICRODAMAGES ON S 28-3208
SURFACE BARRIER ENERGIES ON STRONTIUM TITANATE. 06-1175
ELECTRON TUNNELING INTO POTASSIUM TANTALATE SCHOTTKY BARRIER JUNCTIONS. 08-1365
M TITANATE- BISMUTH TRIOXIDE COMPOSITE CERAMICS WITH SURFACE BARRIER LAYER. /IVITY ON LANTHANUM DOPED BARIU 04-0800
SINGLE FIRE PROCESS FOR BARIUM TITANATE BARRIER LAYER DIELECTRICS. 04-1029
/DS (OF SOLID SURFACES) IN THE EMISSION ELECTRON MICROSCOPE (BARRIER LAYERS IN DOPED BARIUM TITANATE CERAM/ 04-0913
IUM TITANATE CERAMIC WITH A HIGH EFFECTIVE D/ GRAIN BOUNDARY BARRIER LAYERS IN (N TYPE, ANTIMONY DOPED) BAR 04-0581
IC PROPERTIES OF ZINC OXIDE AND STRONTIUM TITANATE. (SURFACE BARRIER SYSTEMS, METAL CONTACTS) SOME ELECTRON 06-1176
WEIGHT OF SHOTTKY BARRIERS IN FERROELECTRICS. 03-0459
AMICS. PRODUCTION AND STUDY OF THE INNER BARRIERS IN SEMICONDUCTING BARIUM TITANATE CER 04-0583

EORY OF TRAN/ TUNNELING RESULTS ON INDIUM STRONTIUM TITANATE BARRIERS INTERPRETED AS CONFIRMATION OF THE TH 03-0233
TCR DEVICES) POTENTIAL BARRIERS ON SEMICONDUCTING BARIUM TITANATE. (P 04-0851
M TITANATE AND (CALCIUM DOPED) POTASSIUM TANTALATE (SCHOTTKY BARRIERS WITH INDIUM). /IOBIUM DOPED) STRONTIU 06-1222
OWTH. (BARIUM SODIUM NIOBATE AND SODIUM LANTHANUM / CRUCIBLE BASE COOLING: AN AID TO CZOCHRALSKI CRYSTAL GR 13-1928
OMAIN STRUCTURE OF SINGLE CRYSTALS FORMED BY BARIUM TITANATE BASE SOLID SOLUTION. /XTERNAL ACTIONS ON THE D 04-0954
IBLE DOMAIN REVERSAL IN TRIGLYCINE SULFATE CAUSED BY A LASER BEAM. REVERS 27-3012
ARIZATION REVERSAL IN TRIGLYCINE SULFATE BY MEANS OF A LASER BEAM. APPLICATION TO FERROELECTRIC MEMORIES. / 27-3010
YSTALS. INVESTIGATION OF A SYSTEM FOR BEAM DEFLECTION BY MEANS OF LITHIUM NIOBATE CR 11-1768
D MAGNESIC NIOBATE CRYSTALS. LIGHT BEAM DEFLECTION CONTROLLED WITH A PRISM OF LEA 08-1280
ATE. BINARY LIGHT BEAM DEFLECTION IN SINGLE CRYSTAL BARIUM TITAN 04-0653
PRISM. ELECTROOPTIC LIGHT BEAM DEFLECTION WITH BARIUM STRONTIUM NIOBATE 13-2036
UM TANTALATE(0.65) NIOBA/ USE OF PEROVSKITE PARAELECTRICS IN BEAM DEFLECTORS AND LIGHT MODULATORS. (POTASSI 13-1924
TE. BINARY LIGHT BEAM DETECTION IN SINGLE CRYSTAL BARIUM TITANA 04-0654
IGITAL LIGHT DEFLECTOR UTILIZING LITHIUM TANTALAT/ DEFLECTED BEAM DIAMETER AND DEFLECTION CAPACITIES OF A D 11-1809
BARIUM TITANATE BY CONVERGENT BEAM ELECTRON DIFFRACTION. 04-0991
ELECTROOPTIC POTASSIUM TANTALATE NIOBATE GRATINGS FOR LIGHT BEAM MODULATION AND DEFLECTION. 13-1929
AL PROCESSING PURPOSES. (TECHNIQUE APPLIED TO LITH/ ELECTRON BEAM SENSING OF SURFACE ELASTIC WAVES FOR SIGN 11-1738
N LITHIUM NIOBATE. NEW HIGH FREQUENCY HIGH COUPLING LOW BEAM STEERING CUT FOR ACOUSTIC SURFACE WAVES O 11-1819
TALATE CRYSTAL. OPTICAL BEAM STEERING USING A MULTICHANNEL LITHIUM TAN 12-1881
TH TITANIUM OXIDE SINGLE CRYSTALS. ELECTRON BEAM WRITING OF FERROELECTRIC DOMAINS IN BISMU 08-1300
HASE TRANSITION FERROELECTRICS. THE BEHAVIORS OF THE MICRODOMAINS ON THE DIFFUSE P 03-0283
C FIELD. (OBSERVATION/ POLARIZATION OF FERROELECTRIC FILM BY BENDING AND BENDING DEFORMATION IN THE ELECTRI 03-0195
THE FERROELECTRIC TRANSITION IN DEUTERATED TRIGLYCINE FLUOB/ BERYLLIUM-9 QUADRUPOLE PERTURBED NMR STUDY OF 27-2965
BI... SEE THE PREFIXED WORD.
DC BIAS EFFECTS IN BARIUM TITANATE CERAMICS. 04-0566
CS. DC BIAS EFFECTS IN LEAD ZIRCONATE TITANATE CERAMI 09-1490
BIAS FIELDS IN FERROELECTRIC COLEMANITE. 23-2886
HIGH INTERNAL BIAS FIELDS IN L-ALANINE SUBSTITUTED TGS. 27-2977
SWITCHING PROPERTIES OF BIASED FERROELECTRIC COLEMANITE. 23-2888
AGE AND DISPLAY DEVICES. (PLZT) STRAIN BIASED FERROELECTRIC PHOTOCONDUCTOR IMAGE STOR 09-1565
RAGE AND DISPLAY DEVICES OPERATED IN A REFLECTION MO/ STRAIN BIASED FERROELECTRIC- PHOTOCONDUCTOR IMAGE STO 09-1564
DIELECTRIC RESONANCE AND RELAXATION IN ELECTRICALLY BIASED KDP AND DEUTERATED ISOMORPHS. (THEORY) 03-0467
ELECTROOPTIC DEVICES USING STRAIN BIASED PLZT FERROELECTRIC CERAMICS. 09-1562
ELECTRIC CONSTANT OF POTASSIUM NIOBATE SINGLE CRYSTALS UNDER BIASING CONDITIONS. DI 08-1381
POLARITONS IN A BIAXIAL CRYSTAL OF BARIUM SODIUM NIOBATE. 13-1992
AND SODIUM NITRITE. OPTICAL ACTIVITY IN NON-ENANTIOMORPHOUS BIAXIAL CRYSTALS: METHYL MESITYL OXIDE OXALATE 16-2273
ERROELECTRICITY AND FERROELECTRIC MATERIALS: A COMPREHENSIVE BIBLIOGRAPHY. F 01-0004
BIBLIOGRAPHY OF FERROELECTRICS. 01-0032
48 REFERENCES, SUBJECT AND AUTHOR INDEXES) BIBLIOGRAPHY ON FERROELECTRICS. (1949-1964, 30 01-0033
PENTOXIDE AND LEAD, BARIUM, AND STRONTIUM OXIDES. BINARY AND PSEUDOBINARY PHASES BETWEEN NIOBIUM 04-0590
. STUDY OF DIVALENT BINARY AND TERNARY NIOBATES IN THE SOLID STATE 08-1286
BARIUM TITANATE. BINARY LIGHT BEAM DEFLECTION IN SINGLE CRYSTAL 04-0653
BARIUM TITANATE. BINARY LIGHT BEAM DETECTION IN SINGLE CRYSTAL 04-0654
ETIC, ELECTRONIC AND OPTICAL PROPERTIES OF EACH OF THE II-VI BINARY SEMICONDUCTING COMPOUNDS) /THERMAL MAGN 01-0025
/TOXIDE. (OXIDES, ESPECIALLY THE NIOBIUM-RICH REGION AND THE BINARY STRONTIUM OXIDE- NIOBIUM PENTOXIDE SYS/ 13-1920
ROWTH OF SINGLE CRYSTALS IN LEAD MONOXIDE- GERMANIUM DIOXIDE BINARY SYSTEM. G 35-3514
TUNGSTATE(0.5). ELECTROMECHANICAL PROPERTIES IN THE BINARY SYSTEM LEAD TITANATE- LEAD CADMATE(0.5) 10-1638
ERMINATION OF THE GROWTH RATE OF FERROELECTRIC CRYSTALS FROM BINARY SYSTEMS. (THEORY) DET 03-0178
URE OF STRONTIUM TITANATE FROM A SELF CONSISTENT FIELD TIGHT BINDING CALCULATION. ENERGY BAND STRUCT 06-1221
FROM BARIUM TITANATE TO DNA. (FERROELECTRICITY, BIOLOGICAL MECHANISMS, MEMORY, ORGANICS) 02-0092
RIC CRYSTALS. GENERATOR OF BIPOLAR PULSES FOR INVESTIGATION OF FERROELECT 37-3673
OLINIUM MOLYBDATE. SPONTANEOUS BIREFRINGENCE AND ELECTROOPTIC RESPONSE IN GAD 30-3307
NING OXYGEN O/ ELECTROOPTIC G COEFFICIENTS (RELATING INDUCED BIREFRINGENCE AND POLARIZATION) OF LEAD CONTAI 13-2039
SIUM NIOBATE (LINEAR ELECTROOPTICS). OPTICAL PROPERTIES (BIREFRINGENCE AND REFRACTIVE INDICES) OF POTAS 08-1468
ROOPTICAL PROPERTIES IN THE NEIGHBORHOOD OF THE CURIE POINT (BIREFRINGENCE ANOMALIES). / OPTICAL, AND ELECT 19-2745
TED POTASSIUM DIHYDROGEN PHOSPHATE CRYSTAL. STRESS INDUCED BIREFRINGENCE IN AN ISOLATED AND A SHORTCIRCUI 17-2452
IONIC CRYSTAL. APPLICATION TO THE CALCULATION OF TETRAGONAL BIREFRINGENCE IN BARIUM AND LEAD TITANATES. /C 04-0578
NIUM MOLYBDATE. INDUCED AND SPONTANEOUS BIREFRINGENCE IN IMPROPER FERROELECTRIC GADOLI 30-3278
STAL. BIREFRINGENCE INDUCED IN A LITHIUM NIOBATE CRY 11-1767
METER WAVELENGTH RANGE. (LITHIUM NIOBATE) BIREFRINGENCE OF CERTAIN CRYSTALS IN THE MILLI 11-1730
CERAMICS. (PZT) OPTICAL PROPERTIES (ELECTRICALLY VARIABLE BIREFRINGENCE) OF (HOT PRESSED) FERROELECTRIC 09-1554
Y THE 105 DEGREE K STRUCTURAL PHASE TRANSITION. BIREFRINGENCE OF STRONTIUM TITANATE PRODUCED B 11-1116
DISPERSION OF ELECTROOPTIC EFFECT IN BARIUM TITANATE. (BIREFRINGENCE, PRINCIPAL INDICES) 04-0755
ICES. (BASED ON ELECTRICALLY CONTROLLED LIGHT SCATTERING AND BIREFRINGENCE, REVIEW) /SSING) AND DISPLAY DEV 37-3650
/ITY LITHIUM NIOBATE CRYSTALS (FOR NONLINEAR OPTICS WITH LOW BIREFRINGENCE VARIATIONS) FROM THE CONGRUENT / 11-1688
ARIZATION PROCESS IN STRONTIUM TITANATE AT LOW TEMPERATURE. (BIREFRINGENCE, 20-275 DEG K) /UDIES OF THE POL 06-1117
CHALCOGENIDE BROMIDES OF ANTIMONY AND BISMUTH. 15-2168
HOT PRESSED LEAD ZIRCONATE TITANATE CERAMICS CONTAINING BISMUTH. 09-1525
THE PROPERTIES OF SOLID SOLUTIONS OF LEAD ZIRCONA/ EFFECT OF BISMUTH AND MANGANESE CONTAINING ADDITIVES ON 09-1588
3) TITANATE(0.47) (EXPERIMENT AND STRUCTURA/ SUBSTITUTION OF BISMUTH AND NIOBIUM IONS IN LEAD ZIRCONATE(0.5 09-1480
HIC METHOD FOR ANALYZING LEAD ZIRCONATE TITANATES CONTAINING BISMUTH AND STRONTIUM ADDITIVES. CHROMATOGRAP 09-1625
VERSE HYPERSONIC PULSES IN PIEZOELECTRIC LITHIUM NIOBATE AND BISMUTH BIGERMANATE. /F LONGITUDINAL AND TRANS 11-1814
PEROVSKITE STRUCTURE (PEROVSKITE COMPOUNDS CONTAINING EITHER BISMUTH, COPPER, OR LEAD). /IC MATERIALS WITH 08-1460
ANTIFERROMAGNETIC PROPERTIES OF SOME PEROVSKITES. (BISMUTH FERRATE) 08-1441
TRIC PHASE TRANSITIONS IN THE LEAD ZIRCONATE- LEAD TITANATE- BISMUTH FERRATE. / EFFECT STUDIES OF FERROELEC 09-1498
TICE PARAMETERS AND THE COEFFICIENTS OF THERMAL EXPANSION OF BISMUTH FERRATE. /ION DETERMINATION OF THE LAT 08-1287
/RRATE(0.67) TUNGSTATE(0.33) LEAD FERRATE(0.5) NIOBATE(0.5), BISMUTH FERRATE AND ITS SOLID SOLUTIONS WITH / 08-1368
/TRICS AND ANTIFERROELECTRICS WITH HIGH CURIE TEMPERATURES. (BISMUTH FERRATE AND SYSTEMS BASED ON IT, AND / 08-1378
). DIELECTRIC HYSTERESIS IN (FLUX GROWN) SINGLE CRYSTAL BISMUTH FERRATE (CONFIRMS FERROELECTRIC NATURE 08-1449
SOLID SOLUTIONS (ROOM TEMPERATURE TO 6/ MOSSBAUER STUDIES OF BISMUTH FERRATE LEAD TITANATE PEROVSKITE TYPE 08-1471
/PHASES. (X=0.9 AND 0.58) (CORRELATION OF CATION SHIFTS WITH BISMUTH FERRATE, LITHIUM NIOBATE AND LITHIUM / 09-1574
TRICS AND FERROELECTRIC MAGNETS. (LEAD AND BARIUM TITANATES, BISMUTH FERRATE, POTASSIUM NIOBATE) /FERROELEC 04-1038
D RELATED MATERIALS (SYSTEM WIT/ FERROELECTRIC PROPERTIES OF BISMUTH FERRATE (PREPARATION AND STRUCTURE) AN 08-1333
/ KEV GAMMA-RAYS OF TIN-119 DURING PHASE TRANSITIONS OF THE BISMUTH FERRATE STRONTIUM STANNATE(0.33) MANG/ 08-1400
STUDY. (ANTIFERROMAGNETIC) FERROELECTRIC BISMUTH FERRATE X-RAY AND NEUTRON DIFFRACTION 08-1402
TE SYSTEMS. MOSSBAUER MEASUREMENTS OF BISMUTH FERRATE- BISMUTH FERRATE- LEAD ZIRCONA 08-1282
CZOCHRALSKI SYNTHESIS AND PROPERTIES OF RARE EARTH DOPED BISMUTH GERMANATE. 35-3496
XIDE(20). BISMUTH GERMANATE, SEE BISMUTH(12) GERMANIUM O
TAL IS SCANDIUM, CHROMIUM, MANGANESE, IRO/ SUPERSTRUCTURE OF BISMUTH METAL OXYGEN(3) PEROVSKITES. (WHERE ME 08-1451
CRYSTAL GROWTH AND DIELECTRIC PROPERTIES OF BISMUTH NIOBATE. 13-2033

E(6), AND BARIUM(3) BISMUTH(/ NEW FERROELECTRICS. (BARIUM(2) BISMUTH (NIOBIUM OR TANTALUM OR VANADIUM) OXID 36-3619
/OIODIDE TYPE. (PARTIAL OR TOTAL SUBSTITUTION OF ANTIMONY BY BISMUTH OR ARSENIC, SULFUR BY SELENIUM OR OXY/ 15-2205
/DIUM OR POTASSIUM ANTIMONATE, NIOBATE, OR BISMUTHATE, OR BY BISMUTH OR LANTHANUM FERRATE, ALUMINATE OR CH/ 09-1539
LANTHANUM DOPED BARIUM TITANATE CERAMICS WITH BISMUTH OXIDE AS A GRAIN BOUNDARY ELEMENT. 04-0799
SE OF RESISTANCE ON LANTHANUM(0.002) TITANATE- BISMUTH OXIDE COMPOSITE CERAMICS. /LOUS INCREA 04-0801
X-RAY STUDY OF THE SYSTEM BISMUTH OXIDE FERRIC OXIDE. 08-1376
FERROELECTRICITY IN BISMUTH OXIDE TYPE LAYER STRUCTURE COMPOUNDS. 13-1932
PROPERTIES OF BISMUTH(12) TITANIUM OXIDE(20) AND THE SYSTEM BISMUTH OXIDE- TITANIUM DIOXIDE. 13-1994
THE BISMUTH OXIDE- TUNGSTEN TRIOXIDE SYSTEM. 10-1656
CRYSTAL CHEMISTRY OF MIXED BISMUTH OXIDES WITH LAYER TYPE STRUCTURE. 13-2032
ATE, BISMUT/ (CRYSTALLOGRAPHIC AND DIELECTRIC) PROPERTIES OF BISMUTH PEROVSKITES. (AND THE FERROELECTRIC ST 08-1289
THERMOELECTRIC PROPERTIES, AND BAND STRUCTURE OF CRYSTALLINE BISMUTH SELENOIODIDE. CONDUCTIVITY, 15-2126
PIEZOELECTRIC RESISTANCE EFFECT IN BISMUTH SELENOIODIDE CRYSTALS. 15-2128
STEM ALLOYS. PHASE DIAGRAM AND ELECTRICAL CONDUCTIVITY OF BISMUTH SELENOIODIDE- ANTIMONY SELENOIODIDE SY 15-2122
ERROELECTRIC CRYSTALS OF GROUPS V, VI AND VII. (ANTIMONY AND BISMUTH SULFO HALIDES) /TION SPECTRA OF SOME F 15-2210
/ (V) B (VI) C (VII) UNDER HYDROTHERMAL CONDITIONS. (ANTIMONY AND BISMUTH SULFO, SELENO, AND TELLUROIODIDES AND/ 15-2213
ROWTH AND CHEMICAL SYNTHESIS UNDER HYDROTHERMAL CONDITIONS. (BISMUTH SULFOBROMIDE) CRYSTAL G 15-2205
/PE-1 PHASE TRANSITION. (ANTIMONY SULFOIODIDE, SELENOIODIDE, BISMUTH SULFOIODIDE, ANTIMONY SULFO SELENOIOD/ 15-2114
PHOTOELECTRIC PROPERTIES OF BISMUTH SULFOIODIDE CRYSTALS. 15-2170
CRYSTALLIZATION OF BISMUTH SULFOIODIDE. (HYDROTHERMAL SYNTHESIS) 15-2212
/FOR PREPARING A (V) B (VI) C (VII) COMPOUNDS. (USING ANTIMONY OR BISMUTH, SULFUR, SELENIUM OR TELLURIUM, AND C/ 15-2194
/E TETRAGONAL TUNGSTEN OXYGEN BRONZE TYPE: QUADRATIC LEAD OR BISMUTH TANTALATE NIOBATE, TANTALATE TUNGSTAT/ 13-1985
DE CRYSTALS. ABSORPTION SPECTRA OF BISMUTH TELLUROBROMIDE AND BISMUTH TELLUROIODI 15-2217
OPTICAL PROPERTIES OF THE SEMICONDUCTOR BISMUTH TELLUROIODIDE. 15-2169
PRODUCTION METHODS AND SOME OPTICAL PROPERTIES OF BISMUTH TELLUROIODIDE. 15-2141
COEXISTENCE OF THE ANTIPOLARIZED AND POLARIZED STATE IN BISMUTH TITANATE. 08-1422
DEPOLING OF SINGLE DOMAIN BISMUTH TITANATE. 08-1446
TRUCTURE AND POLARIZATION REVERSAL IN FILMS OF FERROELECTRIC BISMUTH TITANATE. DOMAIN S 08-1470
EFFICIENT WHITE LIGHT READING OF DOMAIN PATTERNS IN BISMUTH TITANATE. 08-1299
FEASIBILITY OF ELECTROOPTIC DEVICES UTILIZING FERROELECTRIC BISMUTH TITANATE. 08-1447
ERROELECTRIC PHOTOCONDUCTOR OPTICAL STORAGE MEDIUM UTILIZING BISMUTH TITANATE. F 04-0769
OPTICAL STUDY OF DOMAIN WALLS IN BISMUTH TITANATE. 08-1301
SITIONS OF THE FERROELECTRIC PEROVSKITES: LEAD TITANATE AND BISMUTH TITANATE. /C PROPERTIES AND PHASE TRAN 08-1434
SELF REVERSAL AND WAITING TIME EFFECTS IN SINGLE CRYSTAL BISMUTH TITANATE. 08-1448
RYSTALS OF POTASSIUM TANTALATE NIOBATE, LITHIUM NIOBATE, AND BISMUTH TITANATE) /ROOPTIC MATERIALS, SINGLE C 02-0073
THERMODYNAMICS OF FERROELECTRICITY IN BISMUTH TITANATE. 08-1297
THERMODYNAMICS OF FERROELECTRICITY IN BISMUTH TITANATE. 08-1419
A NEW METHOD OF OPTICALLY READING DOMAINS IN BISMUTH TITANATE AND MEMORY APPLICATIONS. 08-1298
INIUM MOLYBDATE, GR/ GROWTH AND EVALUATION OF SINGLE CRYSTAL BISMUTH TITANATE AND RELATED COMPOUNDS. (GADOL 08-1323
CONDUCTOR DEVICE (ERASABLE, NONDESTRUCTIVE READOUT, BASED ON BISMUTH TITANATE AND ZINC SELENIDE). /IC PHOTO 37-3646
/ (PERMITTIVITY, SPONTANEOUS POLARIZATION, COERCIVE FIELD) OF BISMUTH TITANATE AT LOW TEMPERATURE (5-350K). 08-1436
PREPARATION OF POLED, TWIN-FREE CRYSTALS OF FERROELECTRIC BISMUTH TITANATE, (BY ANNEALING). 08-1352
NONLINEAR OPTICAL PROPERTIES OF LEAD TITANATE AND BISMUTH TITANATE CRYSTALS. 04-0941
(WEAK FIELD) DIELECTRIC (PERMITTIVITY) PROPERTIES OF BISMUTH TITANATE (EFFECT OF TEMPERATURE). 36-3555
GROWTH AND PROPERTIES OF EPITAXIAL FILMS OF FERROELECTRIC BISMUTH TITANATE. (ELECTROOPTICS) 08-1329
STRUCTURAL BASIS OF FERROELECTRICITY IN THE BISMUTH TITANATE FAMILY. 08-1406
ON-57 IN SOME PEROVSKITE LIKE LAYER TYPE FERROELECTRICS. (IN BISMUTH TITANATE FERRATES) /AUER STUDIES OF IR 13-2035
OPTICAL STORAGE MEDIUM FURTHER DEVELOPMENTS. BISMUTH TITANATE FERROELECTRIC- PHOTOCONDUCTOR 08-1371
FERROELECTRIC DISPLAYS. (BISMUTH TITANATE, LEAD ZIRCONATE TITANATE) 09-1515
FERROELECTRICITY AND CONDUCTION IN FERROELECTRIC CRYSTALS. (BISMUTH TITANATE, LITHIUM THALLIUM TARTRATE M/ 02-0061
/ARIUM OR STRONTIUM) TITANATE NIOBATE, (BARIUM OR STRONTIUM) BISMUTH TITANATE NIOBATE, STRONTIUM TITANATE / 13-1962
E MEDIUM. (HIGH RESOLUTION READ WRITE) BISMUTH TITANATE PHOTOCONDUCTOR OPTICAL STORAG 08-1445
/STIC COMPLIANCE, NICKEL IODINE BORACITE OPTICAL ANISOTROPY, BISMUTH TITANATE, POSSIBLE FERROELECTRIC SILI/ 02-0060
BISMUTH TITANATE SOLID SOLUTIONS. 08-1277
HYDROTHERMAL CRYSTALLIZATION AND SOME PROPERTIES OF BISMUTH TITANATES 08-1279
NONLINEAR PROPERTIES AND PHASE TRANSITIONS OF STRONTIUM BISMUTH TITANATES. 13-1952
TRUCTURE OF POLYCRYSTALLINE SOLID SOLUTIONS OF STRONTIUM AND BISMUTH TITANATES. S 06-1145
MAGNETOELECTRIC EFFECT IN FERROELECTRIC- ANTIFERROMAGNETIC BISMUTH TITANIUM FERRATE. 08-1360
/C STUDIES OF CERTAIN NEW LAYERED FERROELECTRICS. (LANTHANUM BISMUTH TITANIUM IRON OXIDE, PRASEODYMIUM BIS/ 13-1993
CRYSTAL STRUCTURE OF BISMUTH TITANIUM NIOBIUM OXIDE. 13-2058
CRYSTAL STRUCTURE OF BISMUTH TITANIUM OXIDE. 13-1935
PREPARATION AND EPITAXY OF SPUTTERED FILMS OF FERROELECTRIC BISMUTH TITANIUM OXIDE. 13-2038
ELECTRON BEAM WRITING OF FERROELECTRIC DOMAINS IN BISMUTH TITANIUM OXIDE SINGLE CRYSTALS. 08-1300
/INCREASE OF RESISTIVITY ON LANTHANUM DOPED BARIUM TITANATE- BISMUTH TRIOXIDE COMPOSITE CERAMICS WITH SURF/ 04-0800
GROWTH, CRYSTALLOGRAPHY AND DIELECTRIC PROPERTIES OF BISMUTH TUNGSTATE. 10-1650
/AL CHEMISTRY OF PEROVSKITE LIKE LAYER TYPE FERROELECTRICS. (BISMUTH TUNGSTATE, BISMUTH NIOBIUM OXYFLUORID/ 13-1966
/) B (VI) C (VII). (HIGH GRADE SINGLE CRYSTALS, V IS ANTIMONY OR BISMUTH, VI IS SULFUR, SELENIUM OR TELLURIUM,/ 15-2214
/ (POTASSIUM OR SODIUM(0.5) TITANATES AND PHASE DIAGRAMS FOR BISMUTH(X) POTASSIUM(X) BARIUM(1-2X) TITANATE/ 08-1289
PREPARATION AND DIELECTRIC PROPERTIES OF POTASSIUM BISMUTH ZINC NIOBATE. 13-2001
/RIC FIELD EFFECT STUDIES. (ON TIN OXIDE FILMS ON LEAD(0.97) BISMUTH(0.02) TITANATE (0.35) ZIRCONATE(0.65)/ 09-1506
BIUM(0.325) ZIRCONIUM(0/ PERFORMANCE OF SPUTTERED LEAD(0.92) BISMUTH(0.07) LANTHANUM(0.01) (IRON(0.405) NIO 37-3629
/RICS WITH PEROVSKITE AND PYROCHLORE STRUCTURES. (BARIUM AND BISMUTH(0.5)) (NIOBIUM, TANTALUM, OR VANADIUM/ 08-1462
/RTIES OF BISMUTH PEROVSKITES. (AND THE FERROELECTRIC STATE, BISMUTH(0.5) (POTASSIUM OR SODIUM(0.5) TITANATE/ 08-1289
D (1-X) TITANATE SYSTE/ DIELECTRIC AND CERAMIC PROPERTIES OF (BISMUTH(0.5) (SODIUM OR POTASSIUM) (0.5) (X) LEA 13-2013
/Y OF THE FERROELECTRIC COMPOUNDS (POTASSIUM OR SODIUM) (0.5) BISMUTH(0.5) TITANATE AND THE HIGH TEMPERATUR/ 08-1362
/IOBIUM, TANTALUM, OR VANADIUM) (0.5) OXIDE, (BARIUM OR LEAD) BISMUTH(0.67) (TUNGSTEN OR MOLYBDENUM) (0.33) / 08-1462
/F SOLID SOLUTIONS OF A (V) B (VI) C (VII) COMPOUNDS. ANTIMONY(X) BISMUTH(1-X) SULFOIODIDE, ANTIMONY SULFO(X) S/ 15-2116
VOLUTION ON CRYSTALS OF CADMIUM SULFIDE, LITHIUM NIOBATE AND BISMUTH(12) GERMANIUM OXIDE(20). /ACE WAVE CON 11-1692
CRYSTAL STRUCTURE OF PIEZOELECTRIC BISMUTH(12) GERMANIUM OXIDE(20). 35-3493
ELASTIC, PIEZOELECTRIC, AND DIELECTRIC CONSTANTS OF BISMUTH(12) GERMANIUM OXIDE(20). 35-3512
ELD DEPENDENT OSCILLATION OF SPACE CHARGE LIMITED CURRENT IN BISMUTH(12) GERMANIUM OXIDE(20). / ELECTRIC FI 35-3505
RFACE WAVES ON PIEZOELECTRIC SUBSTRATES (LITHIUM NIOBATE AND BISMUTH(12) GERMANIUM OXIDE(20)). / ELASTIC SU 11-1751
BY NONLINEAR MIXING IN ANISOTROPIC MEDIA. (LITHIUM NIOBATE, BISMUTH(12) GERMANIUM OXIDE(20)) /URFACE WAVES 11-1778
RAMAN SPECTRUM OF BISMUTH(12) GERMANIUM OXIDE(20). 35-3519
/ELECTRIC SURFACE WAVES IN CUBIC AND ORTHORHOMBIC CRYSTALS. (BISMUTH(12) GERMANIUM OXIDE(20) AND BARIUM SO/ 13-2041
TURES. ULTRASONIC ATTENUATION IN SINGLE CRYSTALS OF BISMUTH(12) GERMANIUM OXIDE(20) AT LOW TEMPERA 35-3510
REQUENCIES. ACOUSTIC SURFACE WAVE LOSS MECHANISMS ON BISMUTH(12) GERMANIUM OXIDE(20) AT MICROWAVE F 35-3511
/OK. VOL 1: SURFACE WAVE VELOCITIES. (BARIUM SODIUM NIOBATE, BISMUTH(12) GERMANIUM OXIDE(20), LITHIUM NIOB/ 01-0028
/ND TRANSVERSE HYPERSONIC WAVE PULSES IN LITHIUM NIOBATE AND BISMUTH(12) GERMANIUM OXIDE(20) PIEZOELECTRIC/ 11-1815
OBATE, CUPROUS CHLORIDE, PROUSTITE, CALCIUM PYRONIOBATE, AND BISMUTH(12) GERMANIUM OXIDE(20). (REVIEW) / NI 01-0021

NIOBATE IN THE SURFACE WAVE AMPLIFIER. A THEORETICAL AND E/ BISMUTH(12) GERMANIUM OXIDE(20) VERSUS LITHIUM 11-1720
GERMANIUM/ THERMALLY STIMULATED CURRENTS AND LUMINESCENCE IN BISMUTH(12) SILICON OXIDE(20) AND BISMUTH(12) 36-3585
BISMUTH OXIDE- TITANIUM DIOXIDE. PROPERTIES OF BISMUTH(12) TITANIUM OXIDE(20) AND THE SYSTEM 13-1994
/H (NIOBIUM OR TANTALUM OR VANADIUM) OXIDE(6), AND BARIUM(3) BISMUTH(2) (TUNGSTEN OR MOLYBDENUM) OXIDE(6),/ 36-3619
PHASE TRANSITION IN BARIUM BISMUTH(5+) BISMUTH(3+) OXIDE. 13-1940
FERROELECTRIC CRYSTAL STRUCTURE OF THE TUNGSTEN BRONZE TYP/ BISMUTH(3) NIOBIUM(17) OXIDE(47): POTENTIALLY 13-1978
M+2) (/ SWITCHING PROPERTIES IN FERROELECTRICS OF THE FAMILY BISMUTH(4) BARIUM(M-2) TITANIUM(M+1) OXIDE(3) (02-0066
LECTRICAL AND OPTICAL SWITCHING PROPERTIES OF SINGLE CRYSTAL BISMUTH(4) TITANIUM(3) OXIDE(12). E 35-3517
GROWTH OF LARGE SINGLE CRYSTAL BISMUTH(4) TITANIUM(3) OXIDE(12). 35-3507
PHASE HOLOGRAMS IN A FERROELECTRIC PHOTOCONDUCTOR DEVICE. (BISMUTH(4) TITANIUM(3) OXIDE(12)) 13-1977
/YSPROSIUM ANTIMONIDE, RUBIDIUM IRON TRIFLUORIDE, BORACITES, BISMUTH(4) TITANIUM(3) OXIDE(12), HAFNIUM DIV/ 02-0036
LS. FERROELECTRIC DIELECTRIC ANOMALY IN BISMUTH(4) TITANIUM(3) OXIDE(12) SINGLE CRYSTA 13-2009
PENTOXIDE AND MOLYB/ THERMOGRAPHIC STUDY OF THE REACTION OF BISMUTH(4) TITANIUM(3) OXIDE(12) WITH VANADIUM 13-2029
PHASE TRANSITION IN BARIUM BISMUTH(5+) BISMUTH(3+) OXIDE. 13-1940
F ANTIFERROMAGNETIC AND SPECIAL DIELECTRIC PROPERTIES IN THE BISMUTH- LANTHANUM FERRATE SYSTEM. /XISTENCE O 08-1425
/E MODIFIED WITH SODIUM OR POTASSIUM ANTIMONATE, NIOBATE, OR BISMUTHATE, OR BY BISMUTH OR LANTHANUM FERRAT/ 09-1539
IRRADIANCE PROFILE ON THE SWITCHING CHARACTERISTICS AND THE BIT SIZE OF THE FE-PC MEMORY DEVICE. /GAUSSIAN 37-3670
CRYSTAL GROWTH OF ANTIMONY SULFOIODIDE, NICKEL CARBONATE AND BLACK PHOSPHORUS. HIGH GAS PRESSURE 15-2137
PROPERTIES OF BLACK SEMICONDUCTING BARIUM TITANATE. 04-0702
NSIENTS) IN PHOTOCHROMIC STRONTIUM TITANATE (ON COLORING AND BLEACHING). PHOTOCONDUCTIVITY (TRA 06-1240
OELECTRICS. RANDOM PHASE APPROXIMATION FOR THE BLINC-DE GENNES MODEL FOR HYDROGEN BONDED FERR 03-0338
ECTRIC SEMICONDUCTORS. (ORIGIN AND EFFECTS OF GRAIN BOUNDARY BLOCKING LAYERS, ANTIMONY DOPANT, REVIEW) /OEL 02-0074
CINE SULFATE. BLOCKING OF SPONTANEOUS POLARIZATION IN TRIGLY 27-3008
0) PLANE OF ROCHELLE SALT IN THE PRESENCE OF PURE DIRECT SKY BLUE DYE. GROWTH RATE OF THE (01 28-3217
D BY SHEARING DUE TO POLARIZA/ OSCILLATIONS OF FERROELECTRIC BODIES. (WITH FREQUENCY OF APPLIED FIELD CAUSE 03-0197
/ION ON SINTERING. PART-8: MECHANOCHEMISTRY OF THE SINTERING BODY. PART-9: EFFECT OF MOTHER SALT OF LEAD M/ 09-1624
DECORATION OF ION BOMBARDED LITHIUM NIOBATE SURFACES. 11-1799
RATION INDUCED IN PEROVSKITE TYPE SINGLE CRYSTALS BY NEUTRON BOMBARDMENT. SPLIT INTERSTITIAL CONFIGU 08-1467
/XPANSION, CRAZING AND EXFOLIATION OF LITHIUM NIOBATE ON ION BOMBARDMENT AND COMPARISON RESULTS FOR SAPPHI/ 11-1800
TRIOXIDE AND ITS USE IN DEDUCING MEAN DAMAGE RANGES. (STUDI/ BOMBARDMENT INDUCED AMORPHIZATION IN TUNGSTEN 10-1642
UM AND / TEMPERATURE DEPENDENCE OF THE INTENSITY OF HYDROGEN BOND BANDS IN RAMAN SCATTERING SPECTRA OF GYPS 28-3189
LECTRIC ORDERING AND STABILI/ PEROVSKITE FAMILY AND CHEMICAL BOND. (COVALENCY, DIRECTIVITY CONDITIONS FOR E 03-0243
ELECTRIC RESONANCE: APPLICATION TO THE HYDROGEN BOND. (ICE) 34-3477
DIPOLE MOMENT DERIVATIVE OF THE HYDROGEN BOND IN ICE. 34-3489
INCOHERENT NEUTRON SCATTERING FROM HYDROGEN BOND IN KDP AND ADP. 17-2416
INFRARED VIBRATION FREQUENCIES OF THE HYDROGEN BOND IN KDP FERROELECTRIC (CALCULATIONS). 17-2650
FERROELECTRICITY AND THE CHEMICAL BOND IN PEROVSKITE TYPE OXIDES. (43 REFS) 02-0102
ELECTRON DELOCALIZATION WEIGHT OF THE HYDROGEN BOND IN POTASSIUM DIHYDROGEN PHOSPHATE. 17-2525
/AGNETIC RESONANCE STUDY OF SOME CRYSTALS CONTAINING O-D...O BOND. (INCLUDING FULLY DEUTERATED AMMONIUM DI/ 03-0204
ASS/ NONLINEAR OPTICAL POLARIZABILITY OF THE NIOBIUM- OXYGEN BOND. (LITHIUM NIOBATE, LITHIUM TANTALATE, POT 11-1734
ROGEN PHOSPHATE. NEUTRON MEASUREMENTS ON THE HYDROGEN BOND POTENTIAL IN PARAELECTRIC POTASSIUM DIHYD 17-2511
TUDY OF THE FERROELECTRIC LITHIUM TRIHYDROGEN SELE/ HYDROGEN BOND STUDIES. PART-54: A NEUTRON DIFFRACTION S 22-2878
MONIUM TRIHYDROGEN SELENITE. HYDROGEN BOND STUDIES. PART-58: CRYSTAL STRUCTURE OF AM 22-2879
AND IN POTASSIUM TRIDEUTERIUM SELENITE, AS DE/ THE HYDROGEN BOND SYSTEM IN POTASSIUM TRIHYDROGEN SELENITE, 22-2854
M DIHYDROGEN PHOSPHATE TYPE ANTIFERROELECTRICS WITH HYDROGEN BONDS. DYNAMICAL THEORY OF AMMONIU 03-0476
0 PER CM) AND HARMONIC VIBRATION FREQUENCIES OF THE HYDROGEN BONDS. /M DIHYDROGEN ORTHOPHOSPHATE (4000-1000 17-2432
QUANTUM EFFECTS IN FERROELECTRIC MATERIALS WITH HYDROGEN BONDS. 03-0347
PHOSPHATE TYPE CRYSTALS) COLLECTIVE MOTIONS OF HYDROGEN BONDS. (CALCULATIONS FOR POTASSIUM DIHYDROGEN 17-2483
PEROVSKITE TYPE STRUCTURE. CHARACTER OF THE CHEMICAL BONDS IN ABO(3) FERROELECTRIC CRYSTALS WITH A 08-1341
ON THE SYMMETRY OF THE HYDROGEN BONDS IN ICE-VII. 34-3445
/DS OF THE HYDROXYL VIBRATIONS AND STRUCTURE OF THE HYDROGEN BONDS IN THE FERROELECTRIC CRYSTALS RUBIDIUM / 17-2471
VIBRATION SPECTRA OF HYDROGEN BONDS. (KDP) 17-2482
NEUTRON MEASUREMENT IN TWO SYSTEMS OF HYDROGEN BONDS. (KDP AND POTASSIUM BICARBONATE) 17-2417
TONS ON THE PHASE TRANSITION IN FERROELECTRICS WITH HYDROGEN BONDS. (KDP, DEUTERATED KDP) /AR STATES OF PRO 17-2671
PROTON PHONON INTERACTION IN FERROELECTRICS WITH HYDROGEN BONDS. (KDP TYPE) 17-2672
TION IN THE PHASE TRANSITION OF FERROELECTRICS WITH HYDROGEN BONDS KDP TYPE. ROLE OF PROTON PHONON INTERAC 17-2674
ITY ON THE PHASE TRANSITION IN A FERROELECTRIC WITH HYDROGEN BONDS (THEORY, WU MODEL). /OF ELECTRIC NEUTRAL 03-0449
RIDE. DISPERSION AND DAMPING OF SOFT ZONE BONDARY PHONONS IN POTASSIUM MANGANESE TRIFLUO 33-3374
OF ICE, PART-2, HYDROGEN BON/ LATTICE STATISTICS OF HYDROGEN BONDED CRYSTALS, PART-1, THE RESIDUAL ENTROPY 03-0388
IDS. PART-1: ENERGY LOSS NEUTRON/ PROTON MOTIONS IN HYDROGEN BONDED FERROELECTRIC AND ANTIFERROELECTRIC SOL 17-2687
PE) PROTON LATTICE INTERACTIONS IN HYDROGEN BONDED FERROELECTRIC CRYSTALS. (THEORY, KDP TY 03-0181
CRITICAL BEHAVIOR OF HYDROGEN BONDED FERROELECTRICS. 03-0516
EFFECT OF POINT DEFECTS ON THE ENERGY SPECTRUM OF HYDROGEN BONDED FERROELECTRICS. 03-0486
CITY, PART-1: SPECTRAL ANALYSIS (PROTON DYNAMICS IN HYDROGEN BONDED FERROELECTRICS). /HEORY OF FERROELECTRI 03-0402
ASE APPROXIMATION FOR THE BLINC-DE GENNES MODEL FOR HYDROGEN BONDED FERROELECTRICS. RANDOM PH 03-0338
URE. PROTON DYNAMICS IN HYDROGEN BONDED FERROELECTRICS ABOVE THE CURIE TEMPERAT 03-0206
T-1: GENE/ THEORY OF COHERENT NEUTRON SCATTERING BY HYDROGEN BONDED FERROELECTRICS AT LOW TEMPERATURES. PAR 03-0478
LT, POTASSIUM OR RUBIDIUM OR A/ NMR STUDIES IN SOME HYDROGEN BONDED FERROELECTRICS: (DEUTERATED ROCHELLE SA 18-2719
FIGURATIONS FROM EPR STUDIES OF X-IRRADIATED POTAS/ HYDROGEN BONDED FERROELECTRICS: EVIDENCE FOR SLATER CON 18-2728
EUTRON INELASTIC SCATTERING./ FREQUENCY SPECTRUM OF HYDROGEN BONDED FERROELECTRICS. (INFRARED ABSORPTION, N 03-0401
IMPURITY MODES IN AN INDUCED MOMENT SYSTEM AND HYDROGEN BONDED FERROELECTRICS (KDP TYPE, THEORY). 03-0426
ENITE. THERMAL EXPANSION OF HYDROGEN BONDED FERROELECTRICS. LITHIUM TRIHYDROGEN SEL 22-2857
PRO/ COUPLED VIBRATIONS OF THE PROTON ION SYSTEM IN HYDROGEN BONDED FERROELECTRICS OF THE KDP TYPE (PHONON 17-2670
REQUENCY) SHIFTS (OF PROTONIC / DYNAMICAL THEORY OF HYDROGEN BONDED FERROELECTRICS. PART-1: DAMPINGS AND (F 03-0475
OCAL MODE MODEL.) PART-2: (DYNAMICAL PRO/ THEORY OF HYDROGEN BONDED FERROELECTRICS. PART-1: (FORMALISM OF L 03-0384
LUTIONS TO THE KINETIC EQUATION/ PROTON DYNAMICS IN HYDROGEN BONDED FERROELECTRICS. PART-2: APPROXIMATED SO 03-0205
ENATE AND SODIUM TRIDEUTE/ DIELECTRIC PROPERTIES OF HYDROGEN BONDED FERROELECTRICS. (SODIUM TRIHYDROGEN SEL 22-2855
POTASSIUM DIHYDROGEN PH/ HIGH FREQUENCY BEHAVIOR OF HYDROGEN BONDED FERROELECTRICS: TRIGLYCINE SULFATE AND 17-2526
TION. PART-2: INTERNAL FUNDA/ STRUCTURAL STUDIES OF HYDROGEN BONDED FERROELECTRICS USING POLARIZED IR RADIA 18-2738
EG C BELOW THE CUR/ FIELD ENFORCED FERROELECTRICITY IN GLASS BONDED LEAD ZIRCONATE. (PHASE TRANSITION 200 D 07-1252
/ SILVER DITHORIUM PHOSPHATE- A NEW CASE OF THE NON-HYDROGEN BONDED PHOSPHATE FERROELECTRIC (CRYSTAL GROWTH 36-3610
SING AUTHOR'S STATISTICAL THEORY. (SPONTANEOUS POLARIZATION, BONDING) /HIUM NIOBATE AND LITHIUM TANTALATE U 03-0366
LS USING THE RAMAN SCATTERING SPECTRUM. (LATTICE VIBRATIONS, BONDING) /ELECTRIC ANTIMONY SULFOIODIDE CRYSTA 15-2143
/ASE TRANSITIONS OF AMMONIUM SULFATE: THE NATURE OF HYDROGEN BONDING AND THE REORIENTATION OF THE AMMONIUM/ 19-2783
SODIUM TUNGSTATE, AND POTASSIUM TANTALATE. TIGHT BONDING BAND CALCULATIONS ON RHENIUM TRIOXIDE, 08-1334
HYDROGEN BONDING IN LITHIUM HYDRAZINIUM SULFATE. 32-3346
NSITIONS OF HYDROGEN HALIDE CRYSTALS (IMPORTANCE OF HYDROGEN BONDING IN THE FERROELECTRIC MECHANISM). / TRA 31-3329
TONS (AND DEUTERONS) IN FERROELECTRIC CRYSTALS WITH HYDROGEN BONDING. (POTASSIUM DIHYDROGEN PHOSPHATE) /PRO 17-2568
CRYSTAL GROWTH OF NICKEL IODINE BORACITE. 14-2091
MANGANESE DOPED) LITHIUM NIOBATE AND (COPPER(2+) DOPED) ZINC BORACITE. /ARAMAGNETIC RESONANCE SPECTRUM IN (11-1792

ROELECTRIC AND FERROELASTIC PROPERTIES OF MAGNESIUM CHLORINE BORACITE.   FER 14-2102
                       FERROELECTRIC TRANSITION IN COBALT IODINE BORACITE.   14-2099
                  MAGNETOELECTRIC PROPERTIES OF NICKEL IODINE BORACITE.   14-2070
      NEUTRON AND ELECTRON DIFFRACTION STUDY IN NICKEL IODINE BORACITE.   14-2073
                  OPTICAL ANISOTROPY OF THE CUBIC COBALT IODINE BORACITE.   14-2094
YSTALS. (AMMONIUM SULFATE, GADOLINIUM MOLYBDATE, IRON IODINE BORACITE) /ERTIES OF IMPROPER FERROELECTRIC CR   14-2084
                    PIEZOELECTRIC PROPERTIES OF COBALT IODINE BORACITE.   14-2100
         SPECIFIC HEAT ANOMALY OF FERROELECTRIC COBALT IODINE BORACITE.   14-2101
ON THE LATTICE STRAINS OF (PARAELECTRIC AND) (IRON- IODINE- BORACITE).   X-RAY STUDY 14-2085
WEEN DIRECTIONS OF THE ELECTRIC AND MAGNETIC POLARIZATION IN BORACITE AND OTHER CRYSTALS.   RELATIONSHIP BET   14-2087
/RAY STUDY ON PHASE TRANSITIONS OF FERROELECTRIC IRON IODINE BORACITE AT LOW TEMPERATURES. (MONOCLINIC AND/   14-2086
                                           LATTICE STRAINS OF BORACITE CRYSTALS.   14-2081
             MODIFIED TECHNIQUE FOR THE GROWTH OF BORACITE CRYSTALS.   14-2092
          PREPARATION OF NICKEL BORACITE CRYSTALS BY HYDROTHERMAL SYNTHESIS.   14-2079
OUS ELECTRIC POLARIZATION AND MAGNETIZATION IN NICKEL IODIDE BORACITE (CURIE POINT 64 DEG K). /THE SPONTANE   14-2090
         OPTICAL ANISOTROPY OF CUBIC NICKEL IODINE BORACITE DUE TO QUADRUPOLE TRANSITIONS.   14-2093
/RATE, ELECTRIC CONTROL OF ELASTIC COMPLIANCE, NICKEL IODINE BORACITE OPTICAL ANISOTROPY, BISMUTH TITANATE/   02-0060
     ELECTROOPTICAL PROPERTIES OF FERROELECTRIC (IRON IODINE- BORACITE. (SPONTANEOUS LATTICE STRAINS)   14-2082
MIC DISPLACEMENTS IN FERROELECTRIC TRIGONAL AND ORTHORHOMBIC BORACITE STRUCTURES. (BORACITE, ERICAITE) . ATO   14-2075
/ LATTICE STRAIN IN THE FERROELECTRIC STATE OF IRON- IODINE- BORACITE. (TEMPERATURE DEPENDENCE OF PIEZOELE/   14-2083
           FERROELECTRIC TRANSITION IN COBALT- IODINE- BORACITE UNDER PRESSURE.   14-2078
TAL FIELD THEORY AND OPTICAL ABSORPTION OF COBALT AND NICKEL BORACITES.   CRYS 14-2096
                DOUBLETS CONCERNING THE STRUCTURE OF BORACITES.   14-2088
ND POLARIZATION: PHENOMENOLOGICAL SYMMETRY CONSIDERATIONS ON BORACITES. /TERACTIONS BETWEEN MAGNETIZATION A   14-2072
              OBSERVATIONS OF FERROELECTRIC DOMAINS IN BORACITES.   14-2103
           SYMMETRY ASPECTS OF THE PHASE TRANSITIONS IN BORACITES.   14-2077
                     THEORY OF PHASE TRANSITIONS IN BORACITES.   03-0269
YNAMIC THEORY OF THE CUBIC- ORTHORHOMBIC PHASE TRANSITION IN BORACITES.   THERMOD 03-0231
ICS.   BORACITES - AN EXAMPLE OF IMPROPER FERROELECTR   14-2076
ROMAGNETOELECTRIC MATERIAL THAT ALLOWS NO 180 DEGR/ TRIGONAL BORACITES: A NEW TYPE OF FERROELECTRIC AND FER   14-2098
                    DIELECTRIC PROPERTIES OF BORACITES AND EVIDENCE FOR FERROELECTRICITY.   14-2071
/ SPECTRA OF NICKEL(2+) IN NICKEL CHLORINE AND NICKEL IODINE BORACITES (AND NICKEL DOPED CADMIUM CHLORINE /   14-2074
/FERRITES, DYSPROSIUM ANTIMONIDE, RUBIDIUM IRON TRIFLUORIDE, BORACITES, BISMUTH(4) TITANIUM(3) OXIDE(12), /   02-0036
/N (625-25 MICRONS, 77-650 DEG K) OF FERROELECTRIC IRON(2+) BORACITES. (CHLORIDE, BROMIDE AND IODIDE, AND/   14-2097
/EORY AND OPTICAL ABSORPTION (CALCULATIONS) OF FERROELECTRIC BORACITES. (IRON, NICKEL, OR COBALT, WITH CHL/   14-2095
TES)   NEW CLASSES OF FERROELECTRICS OF DISPLACEMENT TYPE. (BORACITES, MOLYBDATES, TUNGSTEN BRONZES, NIOBA   02-0118
N(12) CHLORIDE OR BROMIDE (NOT EXAMINED FOR FERROELECTR/ NEW BORACITES OF FORMULA LITHIUM(4) BORON(7) OXYGE   14-2089
/MIUM IODIDE, IRON CHLORINE IRON IODINE AND NICKEL CHLORINE) BORACITES (SECOND HARMONIC GENERATION INDEX M/   14-2080
OR BARIUM, AND B IS HAFNIU/ X-RAY PHASE STUDY OF THE CADMIUM BORATE- ABO(3) SYSTEMS. (WHERE A IS STRONTIUM   36-3528
/E REACTION BETWEEN LEAD MONOXIDE AND PZT. PART-4: EFFECT OF BORON OXIDE ADDITION ON THE PRODUCT OF PZT. P/   09-1624
XAMINED FOR FERROELECTR/ NEW BORACITES OF FORMULA LITHIUM(4) BORON(7) OXYGEN(12) CHLORIDE OR BROMIDE (NOT E   14-2089
                                                             BOSON ECHOES. (ANTIMONY SULFOIODIDE)   15-2144
OD TO FERROELECTRIC SODIUM NITRITE AND TGS CRYSTALS. (DOMAIN BOUNDARIES AND METASTABLE STRUCTURES) /IC METH   16-2375
/H AND WITHOUT NIOBIA ADDITION) DUE TO INTERACTION OF DOMAIN BOUNDARIES AND POINT DEFECTS (TORSION PENDULU/   09-1584
SES IN GADOLINIUM MOLYBDATE CRYSTALS.   OBSERVATION OF PHASE BOUNDARIES BETWEEN FERRO- AND PARAELECTRIC PHA   30-3306
/TOPOGRAPHICAL STUDY OF THE CONTRAST AT FERROELECTRIC DOMAIN BOUNDARIES (DOPED AND UNDOPED TGS, LITHIUM NI/   11-1671
FILMS OF BARIUM TITANATE (ELECTRON MICROSC/ MOTION OF DOMAIN BOUNDARIES DURING SWITCHING IN SINGLE CRYSTAL   04-0940
            SURFACE ELASTOPOLARIZATION WAVES AT DOMAIN BOUNDARIES IN A FERROELECTRIC.   03-0315
NEAR CURIE POINT.   PHASE BOUNDARIES IN BARIUM TITANATE SINGLE CRYSTALS   04-0577
ON THE REFLECTION OF A TRANSVERSE ACOUSTIC WAVE FROM DOMAIN BOUNDARIES IN FERROELECTRICS. /ELECTRIC EFFECT   03-0312
LONGITUDINAL AND QUASI TRANSVERSE SOUND WAVES BY 180 DOMAIN BOUNDARIES IN FERROELECTRICS. (ACTION OF QUASI   03-0314
EFLECTION AND REFRACTION OF TRANSVERSE SOUND WAVES AT DOMAIN BOUNDARIES IN FERROELECTRICS. (THEORY)   R   03-0313
                  MOVEMENT OF 180 DEGREE DOMAIN BOUNDARIES IN PEROVSKITE FERROELECTRICS.   08-1305
ERAMIC/ MECHANISM FOR FORMING INSULATION LAYERS ON THE GRAIN BOUNDARIES OF BARIUM TITANATE SEMICONDUCTIVE C   04-0867
      CORROSION FIGURES DUE TO THE FERROELECTRIC DOMAIN BOUNDARIES OF TRIGLYCINE SULFATE.   27-3142
IUM TITANATE CERAMICS.   EFFECT OF INTERNAL BOUNDARIES ON THE DIELECTRIC PROPERTIES OF BAR   04-0891
/ ELECTRON MICROSCOPE TRANSMISSION IMAGES OF COHERENT DOMAIN BOUNDARIES. PART-1: DYNAMICAL THEORY. PART-2:   03-0255
CONATE(X) TITANATE(1-X) CERAMICS NEAR THE MORPHOTROPIC PHASE BOUNDARY. /CTROMECHANICAL ACTIVITY IN LEAD ZIR   09-1499
D ZIRCONATE(X) TITANATE(1-X) CERAMICS NEAR THE MORPHIC PHASE BOUNDARY. /E ELECTROMECHANICAL ACTIVITY IN LEA   09-1500
PHOTOEFFECTS AT THE PHOTOCONDUCTOR- FERROELECTRIC BOUNDARY.   03-0519
CERAMICS.   PHASE BOUNDARY AND DEFECT STRUCTURE STUDIES IN PLZT   09-1590
OPED) BARIUM TITANATE CERAMIC WITH A HIGH EFFECTIVE D/ GRAIN BOUNDARY BARRIER LAYERS IN (N TYPE, ANTIMONY D   04-0581
/ FERROELECTRIC SEMICONDUCTORS. (ORIGIN AND EFFECTS OF GRAIN BOUNDARY BLOCKING LAYERS, ANTIMONY DOPANT, RE/   02-0074
OF SODIUM-23 IN SODIUM NITRITE. INFLUENCE OF THE ELECTRICAL BOUNDARY CONDITIONS ON THE NMR RELAXATION RATE   16-2350
SINGLE CRYSTALS OF TRIGLYCINE SULFAT/ EFFECTS ON BOUNDARY CONDITIONS ON THE THERMAL CAPACITY OF   27-3140
DOPED BARIUM TITANATE CERAMICS WITH BISMUTH OXIDE AS A GRAIN BOUNDARY ELEMENT.   LANTHANUM 04-0799
IELD DEPENDENCE OF THE ANTIFERROELECTRIC FERROELECTRIC PHASE BOUNDARY IN PLZT CERAMICS.   TEMPERATURE AND F   09-1491
AIN BOUNDARY.   FERROELECTRIC PROPERTIES OF (INSULATING, BOUNDARY LAYER) BARIUM TITANATE CERAMICS AT GR   04-1024
           STUDIES ON THE BARIUM STRONTIUM TITANATE BOUNDARY LAYER CERAMIC DIELECTRICS.   04-1025
        STUDIES ON THE (STRONTIUM, CALCIUM) TITANATE BOUNDARY LAYER CERAMIC DIELECTRICS.   06-1247
              STUDIES ON STRONTIUM TITANATE BOUNDARY LAYER DIELECTRICS.   06-1248
           CHARACTERIZATION OF SURFACE AND GRAIN BOUNDARY LAYER OF BARIUM TITANATE.   04-0701
HE (STRONTIUM, CALCIUM) TITANATE SOLID SOLUTION CERAMIC WITH BOUNDARY LAYER STRUCTURE.   STUDIES ON T 06-1246
PLICATION. (BARIUM, STRONTIUM TITA/ DIELECTRIC CERAMICS WITH BOUNDARY LAYER STRUCTURE FOR HIGH FREQUENCY AP   04-0833
LICATED BY ANISOTROPIC DEBYE- WALLER FAC/ LOW FREQUENCY ZONE BOUNDARY MODES IN PEROVSKITE FERROELECTRICS IN   08-1351
SCOPIC CALCULATION OF THE RESPONSE FUNCTION OF THE SOFT ZONE BOUNDARY PHONON IN STRONTIUM TITANATE.   MICRO   06-1213
THE FERROELECTRIC PROPERTIES OF ANTIMONY SULFOIODIDE. (PHASE BOUNDARY SHAPE STRIATIONS IN DC FIELD) /ON ON   15-2197
                     ZONE BOUNDARY SOFT MODE IN TERBIUM MOLYBDATE.   30-3269
S (THERMODYNAMIC EXISTENCE CONDITIONS FOR MORPHOTROPIC PHASE BOUNDARY, STOICHIOMETRY DEPENDENCE). /SOLUTION   09-1541
/IONS ABOVE FERROELECTRIC CRITICAL TEMPERATURE (ROTATIONS OF BO6 OCTAHEDRA IN ABO(3) PEROVSKITES, STRONTIU/   06-1205
/DEPENDENCE OF THE DEBYE- WALLER FACTORS OF BARIUM TITANATE (BRAGG REFLECTIONS C-AXIS, 20-180 DEG C, CALCU/   04-0892
                                         DIELECTRIC BREAKDOWN IN BARIUM TITANATE.   04-0580
TALS.   ELECTRICAL BREAKDOWN IN FERROELECTRIC SODIUM NITRITE CRYS   16-2288
                               SURFACE BREAKDOWN IN VACUUM ON BARIUM TITANATE.   04-0596
ORATION. (INJECTED COLOR CENTERS)   RELATIONSHIP BETWEEN THE BREAKDOWN MECHANISM IN BARIUM TITANATE AND COL   04-0944
                   DIELECTRIC BREAKDOWN OF POLYCRYSTALLINE BARIUM TITANATE.   04-1015
N POTASSIUM NITRATE NITRITE MIXED CRYSTALS (GROWTH BY DOUBLE BRIDGMAN TECHNIQUE).   PHASE TRANSITIONS I 16-2345
TH IN THE ANTIMONY SULFOIODIDE- ANTIMONY TRISULFIDE SYSTEM. (BRIDGMAN- STOCKBARGER METHOD)   CRYSTAL GROW   15-2230
S.   ORIENTATION AND THICKNESS DEPENDENCE OF CONTRAST AND BRIGHTNESS IN GADOLINIUM MOLYBDATE LIGHT VALVE   30-3291
ROGEN PHOSPHATE.   INTENSITY OF SOME BRILLOUIN AND RAYLEIGH LINES IN RUBIDIUM DIHYD   17-2521

F THE ELASTIC CONSTANTS OF POTASSIUM DIHYDROGEN PHOSPHATE BY BRILLOUIN DIFFUSION.   MEASUREMENT O 17-2454
   MEASUREMENT OF THE ELASTIC CONSTANTS OF SODIUM NITRITE BY BRILLOUIN DIFFUSION AT ORDINARY TEMPERATURES.   16-2306
IDIUM DIHYDROGEN PHOSPHATE.   BRILLOUIN DIFFUSION IN A SINGLE CRYSTAL OF RUB 17-2520
MONIUM SULFATE.   BRILLOUIN EFFECT AT ORDINARY TEMPERATURE IN AM 19-2766
).   STUDY OF THE BRILLOUIN EFFECT IN ADP (ALL ELASTIC CONSTANTS 17-2522
BIDIUM DIHYDROGEN PHOSPHATE.   BRILLOUIN SCATTERING BY A SINGLE CRYSTAL OF RU 17-2519
                                                           BRILLOUIN SCATTERING IN LITHIUM NIOBATE.   11-1786
   TEMPERATURE.   BRILLOUIN SCATTERING IN SODIUM NITRITE AT ROOM 16-2305
GLE CRYSTALS IN THE TEMPERATURE RANGE 5 DEG K-300 DEG K.   BRILLOUIN SCATTERING IN STRONTIUM TITANATE SIN 06-1149
.   THEORY OF RAYLEIGH AND BRILLOUIN SCATTERING NEAR THE PHASE TRANSITION 03-0334
TE (FROM TRANSVERSE (MORE INTENSE) AND LONGITUDI/ MANDELSTAM BRILLOUIN SCATTERING OF LIGHT IN LITHIUM NIOBA 11-1743
ORIDE, COBALT FLUORIDE, POTASSIUM MANGANESE TRIF/ MANDELSTAM BRILLOUIN SCATTERING OF LIGHT IN MANGANESE FLU 33-3399
ORIDE CRYSTALS (POTASSIUM TRIFLUORIDE, RUBIDIUM / MANDELSTAM BRILLOUIN SCATTERING OF LIGHT IN MANGANESE FLU 33-3398
DIDE USING A DOUBLE PASSED, STABIL/ (ELASTIC CONSTANTS FROM) BRILLOUIN SCATTERING STUDY OF ANTIMONY SULFOIO 15-2225
                      TEMPERATURE DEPENDENCE OF THE BRILLOUIN- RAMAN SPECTRUM OF BARIUM TITANATE.   04-0807
NIOBATE.   THREE LAYER BROAD BAND ANTIREFLECTION COATINGS FOR LITHIUM 11-1676
IENT OF LEAD MAGNONIOBATE SINGLE CRYSTALS IN THE REGION OF A BROAD FERROELECTRIC TRANSITION. /PTION COEFFIC 13-1974
ECTRIC POLARIZATION RELAXATION I/ CAUSES OF PHASE TRANSITION BROADENING (DIFFUSNESS) AND THE NATURE OF DIEL 03-0291
OMINE NUCLEAR QUADRUPOLE RESONANCE IN FERROELECTRIC HYDROGEN BROMIDE.   BR 31-3332
TURE OF A FERROELECTRIC SEMICONDUCTOR. (ANTIMONY SULFOIODIDE BROMIDE) / "FROZEN" SHIFT OF THE CURIE TEMPERA 15-2160
/-650 DEG K) OF FERROELECTRIC IRON(2+) BORACITES. (CHLORIDE, BROMIDE AND IODIDE, AND RUBIDIUM IRON TRIFLUO/ 14-2097
DE, SELENOIODIDE, TELLUROIODIDE, TRITELLURIDE, TRIIODIDE, TRIBROMIDE, INDIUM ANTIMONIDE) /LENIDE, SULFOIODI 15-2108
/F DIATOMIC CRYSTALS OF HIGH DIELECTRIC CONSTANT. (POTASSIUM BROMIDE, LEAD TELLURIDE, TIN TELLURIDE, LEAD / 36-3541
ACITES OF FORMULA LITHIUM(4) BORON(7) OXYGEN(12) CHLORIDE OR BROMIDE (NOT EXAMINED FOR FERROELECTRICITY). / 14-2089
ELECTRIC BORACITES. (IRON, NICKEL, OR COBALT, WITH CHLORIDE, BROMIDE, OR IODIDE /N (CALCULATIONS) OF FERRO 14-2095
CRYSTALLIZATION POTENTIALS IN AQUEOUS SOLUTIONS. (POTASSIUM BROMIDE, POTASSIUM CHLORIDE, SODIUM CHLORIDE,/ 17-2707
RESONANCE IN FERROELECTRICS OF THE ANTIMONY SULFOIODIDE (AND BROMIDE) TYPE.   NUCLEAR QUADRUPOLE 15-2216
IC CONSTANTS, PYROELECTRIC CURRENTS, ANTIMONY SULFOIODIDE(X) BROMIDE(1-X) /VII) SOLID SOLUTIONS. (DIELECTR 15-2157
ANGES. (EFFECT OF PHOTOCONDUCTIVITY, ANTIMONY SULFOIODIDE(X) BROMIDE(1-X) /E WITH LOW TEMPERATURE PHASE CH 15-2161
TU/ HOW THE DIELECTRIC PROPERTIES OF ANTIMONY SULFOIODIDE(X) BROMIDE(1-X) SOLID SOLUTIONS DEPEND ON TEMPERA 15-2149
(ANTIMONY AND BISMUTH SULFO, SELENO, AND TELLUROIODIDES AND BROMIDES) /VII) UNDER HYDROTHERMAL CONDITIONS. 15-2213
                                 CHALCOGENIDE BROMIDES OF ANTIMONY AND BISMUTH.   15-2168
LECTRIC HYDROGEN BROMIDE.   BROMINE NUCLEAR QUADRUPOLE RESONANCE IN FERROE 31-3332
BISMUTH, VI IS SULFUR, SELENIUM OR TELLURIUM, VII IS IODINE, BROMINE OR CHLORINE) /STALS, V IS ANTIMONY OR 15-2214
MUTH OR ARSENIC, SULFUR BY SELENIUM OR OXYGEN, AND IODINE BY BROMINE OR CHLORINE) /UTION OF ANTIMONY BY BIS 15-2205
ONY OR BISMUTH, SULFUR, SELENIUM OR TELLURIUM, AND CHLORINE, BROMINE OR IODINE) /I) COMPOUNDS. (USING ANTIM 15-2194
GS, ETHYL DIAMINE TARTRATE, BARIUM TITANATE, SALICYLIDENE- 4-BROMO ANILINE. /ION. (PYROELECTRIC EFFECT IN T 04-0854
ROELECTRIC).   DIELECTRIC PROPERTIES OF DIAMMONIUM HEXABROMO STANNATE (DIELECTRIC TRANSITIONS NOT FER 36-3612
BARIUM TITANATE NIOBATE SINGLE CRYSTAL (TETRAGONAL TUNGSTEN BRONZE).   DIELECTRIC PROPERTY OF 13-1969
   POLARITON RESULTS IN FERROELECTRICS WITH THE TUNGSTEN BRONZE CRYSTAL STRUCTURE.   13-1918
THIU/ CRYSTAL GROWTH AND ELECTROOPTIC PROPERTIES OF TUNGSTEN BRONZE CRYSTALS FROM THE POTASSIUM NIOBATE- LI 13-2003
                      PROPERTIES OF TUNGSTEN BRONZE FERROELECTRICS. (REVIEW, 33 REFS)   02-0053
THIA NIOBIA (POTASSIUM LITHIUM NIOBATES, PHASE REL/ TUNGSTEN BRONZE FIELD IN THE (OXIDE) SYSTEM POTASSIA LI 13-2017
M NIOBATE BARIUM DINIOBATE.   TUNGSTEN BRONZE FIELD IN THE PSEUDOTERNARY SYSTEM SODIU 13-1970
ROWTH OF CRYSTALS IN THE SODIUM OXIDE-/ (EXTENT OF) TUNGSTEN BRONZE (LIQUIDUS AND SOLIDUS) FIELD AND MELT G 13-1949
C IS SODIUM OR POTASSIUM) PHASES OF THE TETRAGONAL TUNGSTEN BRONZE OXIDE TYPE. / STRONTIUM, OR BARIUM, AND 04-0912
/IONS IN TANTALATES WITH THE STRUCTURE OF POTASSIUM TUNGSTEN BRONZE. (POTASSIUM BARIUM TANTALUM OXIDE, SOD/ 13-2050
E GEOMETRY OF FERROELECTRIC DOMAINS.   STRUCTURE OF THE BRONZE SODIUM(13) NIOBIUM(35) OXIDE(94) AND TH 13-1931
      THE SEARCH FOR NEW FERROELECTRICS WITH THE TUNGSTEN BRONZE STRUCTURE.   13-1897
POTASSIUM LITHIUM NIOBATES (TRANSPARENT, TETRAGONAL TUNGSTEN BRONZE STRUCTURE, FERROELECTRIC). /PERTIES OF 13-2026
/S AND PHASE TRANSITIONS IN CRYSTALS WITH POTASSIUM TUNGSTEN BRONZE STRUCTURE (POTASSIUM SODIUM LITHIUM NI/ 13-1923
(SEC/ OPTICAL AND FERROELECTRIC PROPERTIES OF (THE TUNGSTEN BRONZE SYSTEM) POTASSIUM SODIUM BARIUM NIOBATE 13-2028
POTENTIALLY FERROELECTRIC CRYSTAL STRUCTURE OF THE TUNGSTEN BRONZE TYPE. /SMUTH(3) NIOBIUM(17) OXIDE(47): 13-1978
UM, TITANIUM) OR (NIOBIUM, TUNGS/ SOME COMPOUNDS OF TUNGSTEN BRONZE TYPE A(6)B(10)O(30). (WHERE B IS (NIOBI 13-1964
                                         TUNGSTEN BRONZE TYPE COMPLEX OXIDES.   13-1981
IUM LITHIUM NIOBATE.   FERROELECTRIC TUNGSTEN BRONZE TYPE CRYSTAL STRUCTURES. PART-3: POTASS 13-1895
OF THE PSEUDOSYSTEM POTASSIUM LITHIU/ (TETRAGONAL) TUNGSTEN BRONZE TYPE CRYSTALS GROWN IN MOLTEN SOLUTIONS 13-1948
/ AND DIELECTRIC STUDIES OF (SINGLE CRYSTALS IN THE TUNGSTEN BRONZE TYPE) FERROELECTRIC (SYSTEM) POTASSIUM/ 13-1947
PROPERTIES OF STRONTIUM POTASSIUM LITHIUM NIOBATE (TUNGSTEN BRONZE TYPE FERROELECTRICS).   ELECTROOPTICAL 13-2051
TICE STRUCTURE OF STRONTIUM SODIUM LITHIUM NIOBATE (TUNGSTEN BRONZE) TYPE FERROELECTRICS.   SUPERLAT 13-2002
VIEW, 22 REFS).   TUNGSTEN BRONZE TYPE FERROELECTRICS FOR OPTICAL USE (RE 02-0054
/ON OF THE FERROELECTRIC TRANSITION TEMPERATURES OF TUNGSTEN BRONZE TYPE NIOBATES (CORRELATION OF CURIE TE/ 13-2049
TE NIOBATE, TA/ NEW PHASES OF THE TETRAGONAL TUNGSTEN OXYGEN BRONZE TYPE: QUADRATIC LEAD OR BISMUTH TANTALA 13-1985
/RIC OXYFLUORIDE PHASES OF THE OXYGENATED QUADRATIC TUNGSTEN BRONZE TYPE. (STRONTIUM POTASSIUM NIOBIUM OXY/ 13-2011
F BARIUM SODIUM YTTRIUM NIOBATE SINGLE CRYSTAL WITH TUNGSTEN BRONZE TYPE STRUCTURE. /CTROOPTIC PROPERTIES O 13-1971
      SODIUM BARIUM RARE EARTH NIOBATES WITH THE TUNGSTEN BRONZE TYPE STRUCTURE.   13-1972
TRONTIUM) TITANATE NIOBATE, (B/ COMPLEX OXIDES WITH TUNGSTEN BRONZE TYPE STRUCTURES (POTASSIUM (BARIUM OR S 13-1962
            SODIUM ORDERING IN SODIUM TUNGSTEN BRONZES.   13-2012
            STRUCTURE OF TETRAGONAL TIN TUNGSTEN BRONZES.   13-2031
T NMR STUDY OF PHASE TRANSITIONS IN METALLIC SODIUM TUNGSTEN BRONZES.   TRANSIEN 13-1910
/CONDUCTIVITY, MAGNETIC SUSCEPTIBILITY) OF HYDROGEN TUNGSTEN BRONZES, HYDROGEN(X) TUNGSTATE. (ANALOGOUS TO/ 13-1934
TRICS OF DISPLACEMENT TYPE. (BORACITES, MOLYBDATES, TUNGSTEN BRONZES, NIOBATES)   NEW CLASSES OF FERROELEC 02-0118
            IMPURITY EFFECTS IN SODIUM TUNGSTEN BRONZES: NIOBIUM AND TANTALUM DOPING.   13-2053
                      SYNTHESIS OF OXIDE BRONZES. (REVIEW, 88 REFS)   02-0042
      RAMAN STUDY OF METALLIC TUNGSTEN BRONZES. (SODIUM AND RUBIDIUM TUNGSTATES)   10-1654
KYROPOULOS SINGLE CRYSTAL GROWTH OF SODIUM TUNGSTEN BRONZES. (SODIUM TUNGSTATE)   13-2054
      THERMAL CONDUCTIVITY OF CUBIC SODIUM TUNGSTEN BRONZES. (SODIUM TUNGSTATES)   13-2057
/LECTROCHEMICAL DEPOSITION AND DISSOLUTION OF TUNGSTEN OXIDE BRONZES. (SODIUM(X) TUNGSTEN(1-X) Z(Y) OXIDE(/ 13-1944
/ECTRICAL PROPERTIES OF TWO NEW PHASES OF QUADRATIC TUNGSTEN BRONZES STRONTIUM POTASSIUM TANTALATE AND BAR/ 13-2010
NS ON HEXAGONAL ICE.   A STUDY OF BROUT'S MODEL FOR FERROELECTRICS. INVESTIGATIO 34-3443
YSTALS.   DIELECTRIC RELAXATION, BULK AND SURFACE CONDUCTIVITY OF ICE SINGLE CR 34-3469
MEASUREMENT OF ELECTRON ENERGY LOSS FUNCTION IN BULK STRONTIUM TITANATE.   06-1138
OF THE SPIN LATTICE RELAXATION TIMES IN FERROELECTRIC/ TONE BURSTING OF QUADRUPOLAR NUCLEI.   DETERMINATION 11-1829
/S WITH AMMONIUM GROUP ORIENTATION MOBILITY: METAL AMMONIUM (BX4) TYPE COMPOUNDS (PROTON MAGNETIC RESONANC/ 22-2826

C

.2 DEGREES K WITH FIELDS TO 90 KILOGAUSS ALONG THE A, B, AND C CRYSTAL AXES. / POLARIZED FORM FROM 0.4 TO 4 30-3275
IDE CRYSTAL (IN THE PARAELECTRIC PHASE, UNDER DC FIELD ALONG C-AXIS). /AL DISTRIBUTION IN ANTIMONY SULFOIOD 15-2173
ETHOD).   GROWTH TWINNING (UNDER DIRECT CURRENT) IN C-AXIS LITHIUM NIOBATE CRYSTALS (CZOCHRALSKI M 11-1672
INATION IN DC ELECTRIC FIELD.   STRAIN ALONG C-AXIS OF ANTIMONY SULFOIODIDE CAUSED BY ILLUM 15-2239
DEBYE- WALLER FACTORS OF BARIUM TITANATE (BRAGG REFLECTIONS C-AXIS, 20-180 DEG C, CALCULATIONS). /E OF THE 04-0892
U/ FIELD EFFECT AT THE CONTACT BETWEEN A SEMICONDUCTOR AND A C-DOMAIN FERROELECTRIC (THEORY, TITANATES DISC 03-0511
INFLUENCE OF COMPENSATING CHARGES ON THE C-DOMAIN STRUCTURE OF FERROELECTRICS.   03-0458
PIEZOELECTRIC SURFACE WAVES ON A C-DOMAIN WALL OF A FERROELECTRIC (CLASS 6MM).   03-0421
C PROPERTIES. (FERROELECTRIC- PARAE/ CRYSTAL GROWTH OF LEAD (CADMATE(0.33) NIOBATE(0.67)) AND ITS DIELECTRI 13-1959
CHANICAL PROPERTIES IN THE BINARY SYSTEM LEAD TITANATE- LEAD CADMATE(0.5) TUNGSTATE(0.5).   ELECTROME 10-1638
PROPERTIES OF FERROELECTRIC CADMIUM AMMONIUM SULFATE.   19-2756
TITANATES, HAFNATES/ CRYSTAL STRUCTURE AND PROPERTIES OF NEW CADMIUM AND THALLIUM CONTAINING PEROVSKITES. (08-1458
RONTIUM OR BARIUM, AND B IS HAFNIU/ X-RAY PHASE STUDY OF THE CADMIUM BORATE- ABO(3) SYSTEMS. (WHERE A IS ST 36-3528
/CKEL CHLORINE AND NICKEL IODINE BORACITES (AND NICKEL DOPED CADMIUM CHLORINE AND CADMIUM IODINE BORACITES/ 14-2074
/) FREQUENCY DOUBLING IN (MAGNESIUM CHLORINE, ZINC CHLORINE, CADMIUM CHLORINE, CADMIUM IODINE, IRON CHLORI/ 14-2080
GROWTH AND STUDY OF NEW SINGLE CRYSTAL CADMIUM COMPOUNDS WITH PEROVSKITE STRUCTURE.   08-1429
A NEW MAGNETOFERROELECTRIC: CADMIUM IRON NIOBATE.   36-3618
/FERROMAGNETIC SOLID SOLUTIONS IN THE LEAD COBALT TUNGSTATE- CADMIUM MANGANATE SYSTEM (BOTH CURIE POINTS S/ 36-3604
R/ NEW FERROELECTRIC (PYROCHLORE TYPE) COMPOUND (AMMONO THIO CADMIUM NIOPATE, WITH 4 ELECTRIC PHASES, 2 FER 36-3529
PECULIARITIES OF THE DIELECTRIC POLARIZATION OF CADMIUM PYRONIOBATE.   36-3561
ELASTIC AND PIEZOELECTRIC PROPERTIES OF CADMIUM PYRONIOBATE IN STRONG ELECTRIC FIELDS. 13-1967
ACOUSTOELECTRIC AMPLIFICATION OF SURFACE WAVE STRUCTURE OF A CADMIUM SELENIDE FILM ON LITHIUM NIOBATE.   11-1723
RONTIUM TITAN/ SATURATION OF VOLT- AMPERE CHARACTERISTICS OF CADMIUM SELENIDE LAYERS ON BARIUM TITANATE- ST 04-0592
A PECULIAR FERROELECTRIC- FERROELASTIC CRYSTAL: DIAMMONIUM DICADMIUM SULFATE. /THE PHASE TRANSFORMATION OF 19-2741
CTRON PARAMAGNETIC RESONANCE OF MANGANESE(2+) DOPED AMMONIUM CADMIUM SULFATE.   ELE 19-2791
NEW FERROELECTRIC LANGBEINITE THALLIUM CADMIUM SULFATE.   36-3534
EPENDENCE OF THE ELECTRON PARAMAGNETIC RESONANCE OF AMMONIUM CADMIUM SULFATE.   TEMPERATURE D 19-2792
IOBATE OR TANTALATE OR GALLATE, BARIUM TITANATE, ZINC OXIDE, CADMIUM SULFIDE). /ECTRIC) CRYSTALS (LITHIUM N 02-0095
REPORT IS CONCERNED WITH ION IMPLANTATION OF SINGLE CRYSTAL CADMIUM SULFIDE) /. (THE MAJOR PORTION OF THIS 02-0057
/TANEOUS RAMAN SCATTERING BY POLARITONS. PART-2: THE CASE OF CADMIUM SULFIDE AND LITHIUM NIOBATE CRYSTALS. 11-1787
NONCOLLINEAR PARAMETRIC 4 PHOTON RECIPROCAL EFFECT IN CADMIUM SULFIDE AND STRONTIUM TITANATE.   06-1122
M DI/ QUANTITATIVE STUDIES OF OPTICAL HARMONIC GENERATION IN CADMIUM SULFIDE, BARIUM TITANATE, AND POTASSIU 04-0852
E.   SURFACE WAVE CONVOLUTION USING A CADMIUM SULFIDE FLUID LITHIUM NIOBATE STRUCTUR 11-1700
RE.   LAYER WAVE AMPLIFICATION IN A CADMIUM SULFIDE FLUID- LITHIUM NIOBATE STRUCTU 11-1667
2) GERMANI/ ACOUSTIC SURFACE WAVE CONVOLUTION ON CRYSTALS OF CADMIUM SULFIDE, LITHIUM NIOBATE AND BISMUTH(1 11-1692
INTERNAL ELECTRIC FIELDS IN CRYSTALS OF SODIUM TANTALATE AND CADMIUM TITANATE.   08-1389
/ATURE X-RAY INVESTIGATION OF THE PEROVSKITE MODIFICATION OF CADMIUM TITANATE (AND THE ANTIFERROELECTRIC- / 08-1387
/TURE OF DIELECTRIC PROPERTIES IN PEROVSKITE MODIFICATION OF CADMIUM TITANATE. (SOLID SOLUTIONS WITH STRON/ 08-1424
RTIES OF BARIUM TITANATE CERAMICS CAUSED BY IRRADIATION WITH CADMIUM TRANSMITTED NEUTRONS. /OELECTRIC PROPE 04-0731
/M OF THE FERROELECTRIC PHASE TRANSFORMATION OF DIAMMONIUM DICADMIUM TRISULFATE, A MODEL IN WHICH THE POLA/ 03-0141
THE BEHAVIOR OF DIAMMONIUM DICADMIUM TRISULFATE UNDER PRESSURE.   19-2757
INVESTIGATION OF PHASE TRANSITION IN LEAD CADMIUM TUNGSTATE.   10-1633
/IUM, HAFNIUM, CERIUM, THORIUM, TITANIUM, SODIUM, MAGNESIUM, CADMIUM, YTTRIUM, GADOLINIUM, DYSPROSIUM, IRO/ 10-1642
NEW DESCRIPTION OF THE PYROCHLORE STRUCTURE (CADMIUM(2) NIOBIUM(2) OXIDE(6) SULFIDE).   36-3602
/RAY ANALYSIS, AND STUDY OF THE DIELECTRIC PROPERTIES OF NEW CADMIUM-CONTAINING PEROVSKITES AND PYROCHLORE/ 08-1309
/ZT AND SINTERING UNDER OXYGEN ATMOSPHERE. PART-7: EFFECT OF CALCINATION ON SINTERING. PART-8: MECHANOCHEM/ 09-1624
ATE CERAMICS.   EFFECTS OF CALCINING ON SINTERING OF LEAD ZIRCONATE TITAN 09-1494
NATE CERAMIC   EFFECTS OF CALCINING ON THE FIRING OF LEAD ZIRCONATE TITA 09-1622
ON AND CRYSTALLOGRAPHIC PROPERTIES OF A(2+)B(3+)2O(4) TYPE CALCIUM AND STRONTIUM SCANDATES.   PREPARATI 36-3574
CTRIC CHARACTER IN AN APF/ MONOCLINIC STRUCTURE OF SYNTHETIC CALCIUM CHLORAPATITE. (DEVELOPMENT OF FERROELE 36-3591
THE CRYSTAL STRUCTURE OF TRIS-SARCOSINE CALCIUM CHLORIDE.   36-3524
/LECTRIC TRANSITIONS IN DIGLYCINE NITRATE AND TRIS-SARCOSINE CALCIUM CHLORIDE. (SHIFT OF CURIE TEMPERATURE) 16-2267
S OF MANGANESE(2+) COMPLEXES IN FERROELECTRIC TRIS-SARCOSINE CALCIUM CHLORIDE (TSCC) SINGLE CRYSTALS. /TION 36-3525
CENTRATION IN SINGLE CRYSTAL POTASSIUM TANTALATE. (CONDUCTI/ CALCIUM CONCENTRATION VS NET IONIZED DONOR CON 08-1437
/D BAND STRUCTURE OF (NIOBIUM DOPED) STRONTIUM TITANATE AND (CALCIUM DOPED) POTASSIUM TANTALATE (SCHOTTKY / 06-1122
/CIFIC HEAT, THERMAL DIFFUSIVITY AND THERMAL CONDUCTIVITY OF CALCIUM DOPED SODIUM NITRATE CRYSTALS NEAR TH/ 16-2283
/ MATERIALS RESEARCH. (STRONTIUM TITANATE PHOTOCONDUCTIVITY, CALCIUM FLUORIDE PHOTOCONDUCTIVITY AND OPTICA/ 06-1140
IUM TI/ ERASE MODE RECORDING CHARACTERISTICS OF PHOTOCHROMIC CALCIUM FLUORIDE, STRONTIUM TITANATE, AND CALC 06-1119
/S UNDER HIGH ELECTRIC FIELDS. (SUBSTITUTION OF STRONTIUM OR CALCIUM FOR LEAD, IRON FOR ZIRCONIUM AND TITA/ 09-1484
D ANGULAR CORRELATION TECHNIQUE. (LEAD HAFNATE, TIN HAFNATE, CALCIUM HAFNATE, STRONTIUM HAFNATE) / PERTURBE 08-1276
N LATTICE RELAXATION TIME NEAR THE CRITICAL TEMPERATURE OF LEAD PROPIONATE. /LY OF THE PROTON SPI 26-2947
ULTRASONIC ATTENUATION NEAR THE PHASE TRANSITION IN DICALCIUM LEAD PROPIONATE   26-2949
R OF PHONON NEAR THE 60 DEGREES C PHASE TRANSITION IN LEAD DICALCIUM PROPIONATE.   THE ANOMALOUS BEHAVIO 26-2942
CETATE IONS.   PHASE TRANSITION IN DIVALENT METAL DICALCIUM PROPIONATES SUBSTITUTED PARTIALLY BY A 26-2944
/TALATE, BARIUM SODIUM NIOBATE, CUPROUS CHLORIDE, PROUSTITE, CALCIUM PYRONIOBATE, AND BISMUTH(12) GERMANIU/ 01-0021
SONANCE ABSORPTION OF GAMMA QUANTA IN BARIUM, STRONTIUM, AND CALCIUM STANNATE.   RE 36-3566
TRIGLYCINE SULFATE, HYDROGEN AMMONIUM MONOCHLORO ACETATE, DICALCIUM STRONTIUM HEXAACETATE. /IDE, THIOUREA, 02-0009
D OPTICAL STUDIES ON THE PHASE TRANSITION OF FERROELECTRIC DICALCIUM STRONTIUM HEXAACETATE.   X-RAY AN 36-3584
/N AND ITS APPLICATION TO THE SOLID ELECTROLYTE FUEL CELL. ((CALCIUM, STRONTIUM) LANTHANUM TITANATE- ALUMI/ 08-1444
/R OPTICAL PROPERTIES OF THE ABC NIOBATES (WHERE A AND B ARE CALCIUM, STRONTIUM, OR BARIUM, AND C IS SODIU/ 04-0912
IC ASPECTS OF FERROELECTRIC DOMAINS OF BARIUM TITANATE AND DICALCIUM STRONTIUM PROPIONATE. /IRROR MICROSCOP 04-0973
MIRROR MACROSCOPIC OBSERVATION OF FERROELECTRIC DOMAINS OF DICALCIUM STRONTIUM PROPIONATE.   ELECTRON 26-2946
NMR INVESTIGATIONS OF FERROELECTRIC DICALCIUM STRONTIUM PROPIONATE.   26-2931
DISORDER TYPE FERROELECTRICS: POLARIZATION FLUCTUATIONS IN DICALCIUM STRONTIUM PROPIONATE. /WAVES IN ORDER 26-2928
TEMPERATURE DEPENDENCE OF SOUND VELOCITY IN DICALCIUM STRONTIUM PROPIONATE.   26-2934
ULTRASONIC ATTENUATION NEAR THE CURIE TEMPERATURE IN DICALCIUM STRONTIUM PROPIONATE.   26-2948
09 DEGREES K)   OPTICAL ACTIVITY OF FERROELECTRIC DICALCIUM STRONTIUM PROPIONATE. (TRANSITION AT 2 27-3040
/IELECTRICS (EXPERIMENT AND DISCUSSION OF BARIUM (OR LEAD OR CALCIUM) STRONTIUM TITANATE, BARIUM TITANATE / 08-1281
/ TANTALUM OXIDE, SODIUM STRONTIUM TANTALUM OXIDE, POTASSIUM CALCIUM TANTALUM OXIDE, SODIUM CALCIUM TANTAL/ 13-2050
EMPLOYING A SYMMETRIC STRIPLINE. (BARIUM TITANATE AND BARIUM CALCIUM TITANATE) / AND DECIMETER WAVELENGTHS 04-0855
CTRICS.   STUDIES ON THE (STRONTIUM, CALCIUM) TITANATE BOUNDARY LAYER CERAMIC DIELE 06-1247
MPERATURE CHARACTERISTICS OF ROCHELLE TYPE CERAMICS. (BARIUM CALCIUM TITANATE CERAMIC) /THOD OF STUDYING TE 04-0905
DIELECTRIC AGING IN TETRAGONAL SOLID SOLUTIONS OF CALCIUM TITANATE IN BARIUM TITANATE.   08-1295
IONIC CONDUCTION IN SINTERED OXIDES BASED ON CALCIUM TITANATE OR STRONTIUM TITANATE.   06-1141
/RAY ABSORPTION SPECTRA OF CERTAIN PEROVSKITES (STRONTIUM OR CALCIUM TITANATE OR ZIRCONATE, TEST OF SHORT / 03-0414
BOUNDARY LAYER STRUCTURE.   STUDIES ON THE (STRONTIUM, CALCIUM) TITANATE SOLID SOLUTION CERAMIC WITH 06-1246
VIBRATIONAL SPECTRA OF STRONTIUM, BARIUM AND CALCIUM TITANATES.   04-0864

THERMAL STUDIES OF CALCIUM(X) BARIUM(1-X) TITANATE. 04-0694
/CAL PROPERTIES OF PEROVSKITE TYPE COMPOUNDS, LANTHANUM(1-X) CALCIUM(X) MANGANATE (PHASE TRANSFORMATIONS C/ 08-1356
SOFT MODE SUPERCONDUCTIVITY IN STRONTIUM TITANATE(3-X) AND CALCIUM(Y) STRONTIUM(1-Y) TITANATE(3-X). 06-1093
LDS. (ANISOTROPY DEVELOPED, BARIUM TITANATE AND BARIUM(0.95) CALCIUM(0.05) TITANATE USED) /ONG ELECTRIC FIE 04-0661
PREPARATION OF THE TERNARY TITANATE BARIUM(0.8) LEAD(0.12) CALCIUM(0.08) TITANATE. 36-3587
CRYSTAL STRUCTURE AND PROPERTIES OF CALCIUM(2) NIOBIUM(2) OXIDE(7). 13-1911
BRATIONAL SPECTRA OF THE TITANATES OF STRONTIUM, BARIUM, AND CALCIUM. (75 REFS) VI 04-0865
ERATURE DEPENDENCE OF THE INFRARED AND RAMAN SPECTRA AND THE CALCULATED ENTROPY. /IC THIOUREA. PART-1: TEMP 25-2920
LAR MOTION IN THE FERROELECTRIC PHASE OF HYDROGEN CHLORIDE. (CALCULATED IN THE HARMONIC APPROXIMATION) /ECU 31-3340
/STATIC MODEL FOR BARIUM AND LEAD TITANATES - APPLICATION TO CALCULATING THE WALL ENERGY AND THE DIELECTRI/ 03-0189
N FERROELECTRICS POSSESSING FIRST ORDER TRANSITION. (THEORY, CALCULATION) /POINT SHIFT WITH APPLIED FIELD I 03-0327
TRONTIUM TITANATE FROM A SELF CONSISTENT FIELD TIGHT BINDING CALCULATION. ENERGY BAND STRUCTURE OF S 06-1221
TRIC HEATING AND POLARIZATION REVERSAL, COMPLEX PERMITTIVITY CALCULATION) /F FERROELECTRIC CRYSTAL. (DIELEC 03-0245
TTINGS OF TITANIUM(3+) ION IN RUBIDIUM ALUMINUM DISULFATE D/ CALCULATION OF CRYSTAL FIELD ENERGY LEVEL SPLI 19-2754
TITANATE. CALCULATION OF (DOMAIN) WALL ENERGY IN BARIUM 04-0673
LS. (BARIUM TITANATE TYPE) CALCULATION OF ELECTRIC FIELDS IN IONIC CRYSTA 04-0824
OPIC THEORY OF SURFACE PHENOMENA IN FERROELECTRIC CRYSTALS. (CALCULATION OF FIELD NEAR SURFACE) MICROSC 03-0501
EIR GRADIENTS IN ABO(3) PEROVSKITE COMPOUNDS WITH DISTINCTI/ CALCULATION OF INTERNAL ELECTRIC FIELDS AND TH 08-1455
OELEC/ SPECIFIC HEAT ANOMALY IN FERROELECTRIC ROCHELLE SALT (CALCULATION OF TEMPERATURE DEPENDENCE FOR FERR 28-3187
/N A SLIGHTLY ORTHORHOMBIC IONIC CRYSTAL. APPLICATION TO THE CALCULATION OF TETRAGONAL BIREFRINGENCE IN BA/ 04-0578
HIUM NIOBATE CRYSTALS CONTAINING CHROMIUM(3+) IONS (10 TO 3/ CALCULATION OF THE (ABSORPTION) SPECTRA OF LIT 11-1831
OR OF CUBIC ICE. CALCULATION OF THE DIELECTRIC CORRELATION FACT 34-3429
IN FINE GRAINED PEROVSKITE FERROELECTRIC CERAMICS. (BARIUM / CALCULATION OF THE ELECTROSTATIC INTERACTIONS 04-0642
MPERATURES OF TUNGSTEN BRONZE TYPE NI/ EMPIRICAL FACTORS FOR CALCULATION OF THE FERROELECTRIC TRANSITION TE 12-2049
/RATIONS OF ORTHORHOMBIC AND FERROELECTRIC THIOUREA. PART-2: CALCULATION OF THE INTERMOLECULAR AND INTRAMO/ 25-2907
FT ZONE BOUNDARY PHONON IN STRONTIUM TITANATE. MICROSCOPIC CALCULATION OF THE RESPONSE FUNCTION OF THE SO 06-1213
UENCIES IN SODIUM NITRITE AT ORDINARY TEMPERATURES AS A FUN/ CALCULATION OF THE VARIATION OF THE RAMAN FREQ 16-2363
CE OF 180 DEG DOMAIN WALL IN BARIUM TITANATE. (THERMODYNAMIC CALCULATIONS) ANGULAR ENERGY DEPENDEN 04-0658
TS IN PEROVSKITE STRONTIUM TITANATE (AT OPTICAL FREQUENCIES, CALCULATIONS). ATOMIC DISPLACEMEN 06-1142
DIELECTRIC DISPERSION OF TGS AND DEUTERATED KDP (THEORY, CALCULATIONS). 17-2534
CITY (KDP MODEL, ANTIFERROELECTRICITY F MODEL, WU KDP MODEL, CALCULATIONS). /E SLATER MODEL OF FERROELECTRI 03-0308
ION REVERSALS IN LIQUID ELECTRODED BARIUM TITANATE CRYSTALS (CALCULATIONS). / FIELD (1.5-6 KV/CM) POLARIZAT 04-0994
ATION FREQUENCIES OF THE HYDROGEN BOND IN KDP FERROELECTRIC (CALCULATIONS). INFRARED VIBR 17-2650
THE LATTICE DYNAMICAL ASPECT OF FERROELECTRIC MODES OF KDP. (CALCULATIONS) NOTE ON 17-2498
TON DISPERSION OF TETRAGONAL FERROELECTRIC BARIUM TITANATE. (CALCULATIONS) POLARI 04-0842
IONARY (100)-180 DEG DOMAIN WALL IN BARIUM TITANATE (THEORY, CALCULATIONS). /IGNAL PERMITTIVITY OF THE STAT 03-0358
OF BARIUM TITANATE (BRAGG REFLECTIONS C-AXIS, 20-180 DEG C, CALCULATIONS). /E OF THE DEBYE- WALLER FACTORS 04-0892
DEGREE TYPE DOMAIN WALL OF BARIUM TITANATE CRYSTAL. (THEORY, CALCULATIONS) 90 03-0324
/F FERROELECTRICITY AND ANHARMONIC EFFECTS IN CRYSTALS. (AND CALCULATIONS BASED ON STRONTIUM TITANATE MODE/ 03-0213
) DOMAIN WALL STRUCTURE IN BARIUM TITANATE DURING SWITCHING (CALCULATIONS COMPARED WITH DATA). /P OR SMOOTH 04-0677
EFFECT (THEORY, PROPAGATION OF RADIATION AND FIRST HARMONIC, CALCULATIONS FOR LITHIUM NIOBATE. /SADOVSKII 03-0186
E TYPE CRYSTALS) COLLECTIVE MOTIONS OF HYDROGEN BONDS. (CALCULATIONS FOR POTASSIUM DIHYDROGEN PHOSPHAT 17-2483
POLARONS IN ANISOTROPIC ENERGY BANDS (THEORY, CALCULATIONS FOR STRONTIUM TITANATE). 03-0305
LONG PERIOD STRUCTURE OF THIOUREA CRYSTAL. (CALCULATIONS FROM SATELLITE SCATTERING DATA) 25-2911
FREQUENCY) SHIFTS (OF PROTONIC COLLECTIVE MOTION, TUNNELING, CALCULATIONS, INTERPRETATION). /DAMPINGS AND (03-0475
STOICHIOMETRIC BARIUM TITANATE. CALCULATIONS OF ELECTRONIC CONDUCTIVITY IN NON 04-0821
N, NICKEL, OR / CRYSTAL FIELD THEORY AND OPTICAL ABSORPTION (CALCULATIONS) OF FERROELECTRIC BORACITES. (IRO 14-2095
THEORY OF POLARIZATION REVERSAL IN FERROELECTRICS. (CALCULATIONS OF SWITCHING CURRENT) 03-0422
ND ELECTRIC FIELD GRADIENTS IN THE PEROVSKITE TYPE COMPOUND/ CALCULATION OF THE INTERNAL ELECTRIC FIELDS A 08-1442
/ERROELECTRIC DOMAIN WALL IN TRIGLYCINE SULFATE. (THEORY AND CALCULATIONS OF WALL THICKNESS AND ENERGY DEN/ 27-2997
TATE, AND POTASSIUM TANTALATE. TIGHT BONDING BAND CALCULATIONS ON RHENIUM TRIOXIDE, SODIUM TUNGS 08-1334
CONTRIBUTION TO THE DIELECTRIC CONSTANT OF BARIUM TITANATE. (CALCULATIONS, SUMMARY ONLY, DISCUSSION) /WALL 04-0804
ECTRIC PHASE) LATTICE VIBRATION OF SODIUM NITRITE. (CALCULATIONS TO RELATE TO SPECIFIC ANTIFERROEL 16-2317
ING FERROELECTRIC AND MAGNETIC TRA/ HIGHLY SENSITIVE, SIMPLE CALORIMETER FOR PHASE CHANGE DETECTION (INCLUD 28-3201
RROELECTRIC TRANSITIONS. CALORIMETRIC INVESTIGATIONS OF SECOND ORDER FE 04-0435
AMORPHOUS ICE AND OF THE PHASE TRANSFORMATION BETWEEN THE C/ CALORIMETRIC STUDY OF GLASS TRANSITION OF THE 34-3476
RIFLUORIDE. RESONANCE BEHAVIOR IN CANTED ANTIFERROMAGNETIC POTASSIUM MANGANESE T 33-3403
/RESONATORS IN THE MEDIUM FREQUENCY RANGE WITH LOW RATIOS OF CAPACITANCE AND LOW TEMPERATURE COEFFICIENTS / 11-1724
TERS) RADIATION DAMAGE IN LITHIUM NIOBATE. (CAPACITANCE, OPTICAL SPECTRA, ESR, CELL PARAME 11-1705
AGING IN COMPLEX CERAMICS BASED ON BARIUM TITANATE. (CAPACITANCE, PERMITTIVITY) 04-0724
AGING PROCESSES IN TRIGLYCINE SULFATE. (TANDEL OR CAPACITANCE PROPERTIES DOMAIN STRUCTURE) 27-2860
F LITHIUM NIOBATE RESONATORS BY USE OF TEMPERATURE SENSITIVE CAPACITANCES. LOW TEMPERATURE COEFFICIENT O 11-1670
ING LITHIUM TANTALAT/ DEFLECTED BEAM DIAMETER AND DEFLECTION CAPACITIES OF A DIGITAL LIGHT DEFLECTOR UTILIZ 11-1809
FLAME SPRAYED BARIUM TITANATE AS A CAPACITOR DIELECTRIC. 04-0773
STABILIZED BARIUM TITANATE CERAMICS FOR CAPACITOR DIELECTRICS. 04-0825
/ HIGH DIELECTRIC CONSTANT MATERIALS AND FERROELECTRICITY AS CAPACITOR DIELECTRICS, A STUDY IN DIELECTRIC S 37-3678
TEMPERATURE AUTOSTABILIZATION EFFECT IN FERROELECTRICS. (CAPACITOR PROPERTIES, TANDEL) 37-3655
ONS EACH) STRONTIUM TITANATE SYSTEMS. (FOR VOLTAGE DEPENDENT CAPACITORS) / THE BARIUM (OR LEAD, 4 COMPOSITI 09-1522
FILM BARIUM TITANATE DIELECTRICS PRODUCED BY FLAME SPRAYING (CAPACITORS). ELECTRICAL PROPERTIES OF THICK 04-0772
OF FERROELECTRIC CERAMICS. THIN FILM MICROWAVE FERROELECTRIC CAPACITORS AND NONLINEAR MICROWAVE PROPERTIES 37-3643
VE TITANIUM DIOXID/ STUDIES OF FLAME SPRAYED BARIUM TITANATE CAPACITORS. PART-10: STRUCTURE OF SEMICONDUCTI 04-0774
RICAL PROPERTIES OF TRIGLYCINE SULFATE SINGLE CRYSTALS (HEAT CAPACITY AND ELECTRIC PERMITTIVITY). /ND ELECT 27-3141
/ON POINT IN POTASSIUM MANGANESE TRIFLUORIDE (ANOMALOUS HEAT CAPACITY, CRITICAL TEMPERATURE ABOUT 186-65 D/ 33-3373
/C, FERROELECTRIC BETA GADOLINIUM MOLYBDATE. PARTS 1-3: HEAT CAPACITY, ENTROPY, MAGNETIC MOMENT OF THE ELE/ 30-3275
N (THEORY, RELATION TO KAENZIG DOMAIN MODEL). HEAT CAPACITY IN THE FERROELECTRIC TRANSITION REGIO 04-0443
HEAT CAPACITY OF METHYL AMMONIUM CHROMIUM ALUM. 20-2801
LE CRYSTALS. HEAT CAPACITY OF RUBIDIUM DIHYDROGEN PHOSPHATE SING 17-2415
AT/ EFFECTS ON ELECTRICAL BOUNDARY CONDITIONS ON THE THERMAL CAPACITY OF SINGLE CRYSTALS OF TRIGLYCINE SULF 27-3140
DE. HEAT CAPACTIY OF POLYCRYSTALLINE ANTIMONY SULFOIODI 15-2238
ANTIMONY SU/ TEMPERATURE DEPENDENCE OF THE CROSS SECTION OF CAPTURE CENTERS IN FERROELECTRIC SEMICONDUCTOR 15-2147
LEAD NIOBIUM(4) OXIDE(11), SILVER GALLIUM DISULFIDE, SILICON CARBIDE) /TITANATE, BARIUM STRONTIUM NIOBATE, 05-1066
S IN BARIUM TITANATE FIRED UNDER NITROGEN IN THE PRESENCE OF CARBON. CHEMICAL CHANGE 04-0872
M TITANATE. OXIDATION OF CARBON MONOXIDE ON LANTHANUM OXIDE DOPED BARIU 04-0947
ATE. CATALYTIC OXIDATION OF CARBON MONOXIDE ON SEMICONDUCTIVE BARIUM TITAN 04-0945
EMENT IN TWO SYSTEMS OF HYDROGEN BONDS. (KDP AND POTASSIUM BICARBONATE) NEUTRON MEASUR 17-2417
GAS PRESSURE CRYSTAL GROWTH OF ANTIMONY SULFOIODIDE, NICKEL CARBONATE AND BLACK PHOSPHORUS. HIGH 15-2137
ARIUM SULFATE AND TITANIUM DIOXIDE IN THE PRESENCE OF SODIUM CARBONATE AT HIGH TEMPERATURE. /ITANATE FROM B 04-0624
TEMPERATURE DEPENDENCE OF CARRIER MOBILITY IN BARIUM TITANATE. 04-0743
RONTIUM TITANATE. FREE CARRIER OPTICAL ABSORPTION IN NIOBIUM DOPED ST 06-1214
FERROELECTRIC ANTIMONY SULFOIO/ INFLUENCE OF NONEQUILIBRIUM CARRIER SCREENING ON POLARIZATION PROCESSES IN 15-2158

TITANATE SUBSTRATES AT SUPERSONIC DRIFT VELOCITY OF CURRENT CARRIERS. /AYERS ON BARIUM TITANATE- STRONTIUM 04-0592
LECTRIC SEMICONDUCTOR ANTIMONY / INFLUENCE OF NONEQUILIBRIUM CARRIERS ON THE DOMAIN STRUCTURE OF THE FERROE 15-2104
ECTRIC. (THEORY, ABSTRACT ONLY) EFFECT OF FREE CARRIERS ON THE INFRARED SPECTRUM OF A FERROEL 03-0408
RIC SEMICONDUCTOR (COMPOUNDS) A/ INFLUENCE OF NONEQUILIBRIUM CARRIERS ON THE PHASE TRANSITION IN FERROELECT 15-2145
MAGNETIC PROPERTIES OF FERRODIELECTRIC CATALYSTS. 03-0235
EN TRIOXIDE. EPR STUDY OF THE CATALYTIC ACTIVITY OF NONSTOICHIOMETRIC TUNGST 10-1628
CONDUCTIVE BARIUM TITANATE. CATALYTIC OXIDATION OF CARBON MONOXIDE ON SEMI 04-0945
SODIUM NIOBATE CRYSTALS DURING INTRACAVITY FREQUENCY DOUBL/ CATASTROPHIC SURFACE DAMAGE PRODUCED IN BARIUM 13-2052
TRUCTURE AND PROPERTIES OF BARIUM TITANATE FILMS PRODUCED BY CATHODE SPUTTERING. S 04-0832
/ELAXATION IN SOME FERROELECTRICS (SOLID SOLUTIONS AND MIXED CATION COMPOUNDS, KANZIG REGIONS, PEROVSKITES) 03-0291
RIUM TITANATE) CATION NONSTOICHIOMETRY IN TERNARY OXIDES. (BA 04-0969
/R TWC RHOMBOHEDRAL PHASES. (X=0.9 AND 0.58) (CORRELATION OF CATION SHIFTS WITH BISMUTH FERRATE, LITHIUM N/ 09-1574
DIELECTRIC STUDIES OF THE CATION SUBSTITUTED MIXED CRYSTALS OF NIOBATES. 11-1848
TITANIUM(4+) IONS ARE REPLACED BY ISOVALENT AND NONISOVALENT CATIONS. /RIUM TITANATE, WHERE BARIUM(2+) AND 04-0960
T IN SEMICONDUCTING SOLID SOLUTIONS OF ISO- AND HETEROVALENT CATIONIC SUBSTITUTIONS IN BARIUM TITANATE. /EC 04-0943
F ROCHELLE SALT BY USING THE SHIFT OF RESONANCE FREQUENCY IN CAVITY. MEASUREMENT OF THERMAL EXPANSION O 28-3202
ON PZT FERROELECTRICS. OXYGEN CONCENTRATION CELL AND ELECTRICAL CONDUCTIVITY MEASUREMENTS 09-1495
LOW TEMPERATURE DIELECTRIC CELL AND PERMITTIVITY OF HEXAGONAL ICE TO 2 K. 34-3432
PHASE CF COPPER FORMATE TETRAHYDRATE. UNIT CELL AND SPACE GROUP OF THE ANTIFERROELECTRIC 26-2927
/ SOLUTION AND ITS APPLICATION TO THE SOLID ELECTROLYTE FUEL CELL. ((CALCIUM, STRONTIUM) LANTHANUM TITANAT/ 08-1444
LOSSLESS POTASSIUM DIDEUTERIUM PHOSPHATE POCKELS CELL FOR HIGH POWER Q SWITCHING. 17-2529
N PZT (LEAD ZIRCONATE TITANATE) FERROE/ OXYGEN CONCENTRATION CELL MEASUREMENTS OF IONIC TRANSPORT NUMBERS I 09-1512
D FERROELASTICS HAVING THE INDEX OF FAINTNESS UNEQUAL TO THE CELL MULTIPLICITY. /MPLES OF FERROELECTRICS AN 03-0145
MAGE IN LITHIUM NIOBATE. (CAPACITANCE, OPTICAL SPECTRA, ESR, CELL PARAMETERS) RADIATION DA 11-1705
) IN SOME ISOMORPHOUS FERROELECTRIC SYSTEMS. (DISCUSSE/ UNIT CELL VOLUME EFFECTS (RELATION WITH CURIE POINT 17-2487
D Q SWITCHING PERFORMANCE IN LITHIUM NIOBATE AND KDP POCKELS CELLS. TRANSIENT ELASTOOPTIC EFFECTS AN 11-1727
LITHIUM NIOBATE AND POTASSIUM DIDEUTERIUM PHOSPHATE POCKELS CELLS. (COMMENTS) / Q SWITCHING PERFORMANCE IN 11-1710
IBRATIONS IN FERROELECTRIC FERROMAGNETICS WITHOUT A SYMMETRY CENTER. QUANTUM THEORY OF V 03-0166
RADIATIONLESS TRANSITIONS IN THE EUROPIUM(3+) CENTER IN LANTHANUM ALUMINATE. 36-3531
PARAMAGNETIC RESONANCE OF AN F CENTER IN NONSTOICHIOMETRIC BARIUM TITANATE. 04-0976
TANATE. ANALYSIS OF THE IRON OXYGEN VACANCY CENTER IN THE TETRAGONAL PHASE OF STRONTIUM TI 06-1231
TANATE. THE IRON(3+)- VACANCY CENTER IN THE TETRAGONAL PHASE OF STRONTIUM TI 06-1234
/NANCE AND ELECTRON NUCLEAR DOUBLE RESONANCE OF THE ARSENATE CENTER IN X-IRRADIATED POTASSIUM DIHYDROGEN A/ 18-2727
/ATION OF ELECTRON NUCLEAR TRIPLE RESONANCE FOR THE ARSENATE CENTER IN X-RAY IRRADIATED SINGLE CRYSTALS OF/ 18-2729
PHASE TRANSITIONS IN CESIUM LEAD TRICHLORIDE. (LOSS OF A CENTER OF SYMMETRY AT 194 DEGREES K) 33-3368
MECHANISM IN BARIUM TITANATE AND COLORATION. (INJECTED COLOR CENTERS) RELATIONSHIP BETWEEN THE BREAKDOWN 04-0944
SIUM DIHYDROGEN ARSENATE MIXE/ ESR STUDY OF RADIATION DAMAGE CENTERS IN AMMONIUM DIHYDROGEN PHOSPHATE POTAS 17-2547
Y SU/ TEMPERATURE DEPENDENCE OF THE CROSS SECTION OF CAPTURE CENTERS IN FERROELECTRIC SEMICONDUCTOR ANTIMON 15-2147
FORMATION OF COLOR CENTERS IN IRRADIATED ICE. 34-3487
EFFECTS OF RADIATION INDUCED DAMAGE CENTERS IN LEAD ZIRCONATE TITANATE CERAMICS. 09-1520
MOSSBAUER STUDY OF COLOR CENTERS IN STRONTIUM TITANATE. 06-1153
/MENT OF DIELECTRIC PROPERTIES OF FERROELECTRIC MATERIALS AT CENTIMETER AND DECIMETER WAVELENGTHS EMPLOYIN/ 04-0855
ROWAVES). FERROELECTRIC PROPERTIES OF BARIUM TITANATE IN THE CENTIMETER RANGE OF SUPERHIGH FREQUENCIES (MIC 04-0901
NITRITE. SPLITTING OF THE SODIUM-23 NMR CENTRAL LINE IN FERROELECTRIC SILVER SODIUM DI 16-2307
DOUBLE INJECTION IN BARIUM TITANATE CERAMIC. 04-0554
PROCESSES IN TRIGLYCINE SULFATE CRYSTALS AND BARIUM TITANATE CERAMIC. /THE POLARIZATION AND REPOLARIZATION 04-0924
FFECTS OF CALCINING ON THE FIRING OF LEAD ZIRCONATE TITANATE CERAMIC E 09-1622
(WITH TIN OXIDE FILMS ON LEAD ZIRCONATE(0.65) TITANATE(0.35) CERAMIC). /EFFECT STUDIES AT LOW TEMPERATURES 09-1505
FERROELECTRIC LOSSES IN (PURE) BARIUM TITANATE CERAMIC. 04-0555
PIEZOELECTRIC EFFECT IN ALUMINUM DOPED BARIUM TITANATE CERAMIC. 04-0536
REJUVENATION OF A BARIUM(0.72) STRONTIUM(0.28) TITANATE CERAMIC. 04-0817
FATIGUING EFFECTS IN 8/65/35 PLZT FINE GRAINED FERROELECTRIC CERAMIC. SOME 08-1432
G CHARACTERISTICS OF A LEAD ZIRCONATE TITANATE FERROELECTRIC CERAMIC. SOME PHYSICAL AND ELECTRICAL SWITCHIN 09-1601
ENT IN LEAD ZIRCONATE- LEAD TITANATE- LANTHANUM OXIDE (PLZT) CERAMIC. /EPENDENT FERROELECTRIC DOMAIN ALIGNM 09-1586
TEMPERATURE DEPENDENCES OF THE PARAMETERS OF A FERROELECTRIC CERAMIC. TIME AND 09-1514
ERISTICS OF ROCHELLE TYPE CERAMICS. (BARIUM CALCIUM TITANATE CERAMIC) /THOD OF STUDYING TEMPERATURE CHARACT 04-0905
RMAL EXPANSION STUDY OF A SODIUM(0.88) LITHIUM(0.12) NIOBATE CERAMIC. X-RAY AND THE 13-1999
BARIUM TITANATE CERAMIC: A NEW ELECTRON EMITTER. 04-0802
CERAMIC ACOUSTIC DETECTORS. (BARIUM TITANATE) 01-0003
/, FAST NEUTRON) RADIATION INDUCED PHASE TRANSFORMATIONS IN (CERAMIC) BARIUM AND LEAD TITANATES. (SUMMARY / 04-0971
TEMPERATURE. EMISSION OF LIGHT BY SURFACE MICROPLASMA OF CERAMIC (BARIUM, LEAD) TITANATE NEAR THE CURIE 04-0982
CRITICAL FIELD IN CERAMIC BARIUM TITANATE. 04-0951
FECT OF CRYSTALLINE GRAIN SIZE ON THE PHYSICAL PROPERTIES OF CERAMIC BARIUM TITANATE. EF 04-0785
FROM GLASS. (60-90 VOL PERCENT, FERROELECTRIC, COMPARED WITH CERAMIC BARIUM TITANATE) /ERIALS CRYSTALLIZED 04-0725
/REQUENCY DIELECTRIC ANISOTROPY OF A POLARIZED FERROELECTRIC CERAMIC. (BARIUM TITANATE, BARIUM TIN TITANAT/ 04-0767
RYSTALLITES AND INTERCRYSTALLITE LAYERS IN THE SEMICONDUCTOR CERAMIC BARIUM TITANATE. (CERIUM DOPED) / OF C 04-0568
/RASOUND IN A TWO-LAYER SYSTEM CONSISTING OF A PIEZOELECTRIC CERAMIC (BARIUM TITANATE) WITH LARGE DIELECTR/ 04-1037
EQUATIONS OF STATE: ISOTROPIC AND ORTHOTROPIC (FERROELECTRIC CERAMIC) CASES. /R ELASTIC AND ELECTROELASTIC 03-0277
/ ELECTRON PARAMAGNETIC RESONANCE SPECTRA IN A FERROELECTRIC CERAMIC CONTAINING TITANIUM. (BARIUM TITANATE/ 04-1020
OPEN CIRCUIT SENSITIVITY OF RADIALLY POLARIZED FERROELECTRIC CERAMIC CYLINDERS. (HYDROPHONE ELEMENTS) 37-3651
OPEN CIRCUIT SENSITIVITY OF AXIALLY POLARIZED FERROELECTRIC CERAMIC CYLINDERS. (HYDROPHONE ELEMENTS) 37-3652
D LANTHANUM ZIRCONATE TITANATE) TRANSPARENT CERAMIC DEVELOPED BY SANDIA LABORATORIES. (LEA 09-1476
STUDIES ON THE BARIUM STRONTIUM TITANATE BOUNDARY LAYER CERAMIC DIELECTRICS. 04-1025
STUDIES ON THE (STRONTIUM, CALCIUM) TITANATE BOUNDARY LAYER CERAMIC DIELECTRICS. 06-1247
UM TITANATE) CERAMIC DIELECTRICS. (BARIUM TITANATE, STRONTI 04-0683
UM TITANATE) ENERGY STORAGE IN CERAMIC DIELECTRICS. (STRONTIUM TITANATE, BARI 04-0602
MEASUREMENT OF POISSON'S RATIO IN POLED FERROELECTRIC CERAMIC DISKS. 37-3658
M OPTICAL INHOMOGENEITIES WITH APPLICATIONS TO FERROELECTRIC CERAMIC DISPLAY DEVICES. / SCATTERING BY RANDO 37-3634
AND ITS APPLICATIONS. A SEMICONDUCTING BARIUM TITANATE CERAMIC DOPED WITH NIOBIUM AND MANGANESE IONS 04-0734
MOSSBAUER STUDIES OF LEAD ZIRCONATE TITANATE STANNATE. (CERAMIC DOPED WITH NIOBIUM AND TITANIUM) 09-1604
A NEW LONGITUDINAL DISPLAY MODE FOR CERAMIC ELECTROOPTIC DEVICES. (PLZT) 09-1523
PIEZOELECTRIC CHARACTERISTIC OF BARIUM TITANATE CERAMIC EXHIBITED WHEN IT IS GIVEN AN IMPACT. 04-0881
ETHOD OF MEASURING POLARIZATION VIBRATIONS OF A PREPOLARIZED CERAMIC FERROELECTRIC. (PZT) M 09-1615
CERAMIC FERROELECTRICS. 02-0130
ATES) IN STRONG ELECTRIC FI/ ELECTROMECHANICAL PROPERTIES OF CERAMIC FERROELECTRICS (BARIUM TITANATE ZIRCON 09-1605
UM TITANA/ INSCRIPTION OF HOLOGRAPHIC RESULTS IN TRANSPARENT CERAMIC FERROELECTRICS. (LEAD LANTHANUM ZIRCON 09-1575
PERTIES. (BARIUM TITANATE,/ INFLUENCE OF DOMAIN STRUCTURE OF CERAMIC FERROELECTRICS ON THEIR MECHANICAL PRO 04-0985
OPEN CIRCUIT SENSITIVITY OF RADIALLY POLARIZED FERROELECTRIC CERAMIC HOLLOW SPHERES. (HYDROPHONE ELEMENTS) 37-3653
DISPLAY DEVICES. (BASED ON ELECTRICALLY CONTR/ ELECTROOPTIC CERAMIC (INFORMATION) STORAGE (PROCESSING) AND 37-3650
IZATION) OF LEAD CONTAINING OXYGEN OCTAHEDRA FERROELECTRICS: CERAMIC LEAD BARIUM ZIRCONATE TITANATE. /POLAR 13-2039

/ CONDITIONS ON DIELECTRIC AND PIEZOELECTRIC PROPERTIES OF A CERAMIC (LEAD STRONTIUM TITANATE ZIRCONATE DO/ 09-1591
ISOTHERMAL PHASE TRANSITIONS IN CERAMIC LEAD ZIRCONATE. 08-1466
T NIOBIUM PENTOXIDE) SURFACE / PROPERTIES OF OXYGEN DEPLETED CERAMIC LEAD ZIRCONATE TITANATE (PLUS 1 PERCEN 09-1576
ROLLED MODE. FERROELECTRIC CERAMIC LIGHT GATES OPERATED IN A VOLTAGE CONT 37-3654
ITANATE ZIRCO/ PREPARATION AND ELECTRICAL PROPERTIES OF SOME CERAMIC MATERIALS AT MICROWAVE FREQUENCIES. (T 09-1602
LECTRIC RANDOM ACCESS MEMORY. (LEAD ZIRCONATE- LEAD TITANATE CERAMIC MEMORY PLATE) A FERROELECTRIC PIEZOE 09-1504
FERROELECTRIC CERAMIC OXIDES. 02-0038
/MPERATURE COEFFICIENT OF RESISTIVITY) IN (DOPED AND UNDOPED CERAMIC PEROVSKITE) LEAD FERRATE(0.5) NIOBATE/ 13-2000
UMINA WHISKERS IN LEAD ZIRCONATE TITANATE) NEW COMPOSITE CERAMIC PIEZOELECTRIC TRANSDUCER MATERIAL. (AL 09-1556
ELASTIC SURFACE WAVE ON THE PZT CERAMIC PLATE. 09-1612
RFACE ELASTIC WAVES ON BARIUM TITANATE(0.95) ZIRCONATE(0.05) CERAMIC PLATES. /AGATION CHARACTERISTICS OF SU 09-1613
DELAY LINES WITH INTERDIGITAL TRANSDUCERS ON UNPOLARIZED PZT CERAMIC PLATES. SURFACE WAVE 09-1614
IN AN ALCOHOL. PREPARATION OF BARIUM TITANATE AND OTHER CERAMIC POWDERS BY COPRECIPITATION OF CITRATES 04-0861
POTASSIUM) (0.5) (X) LEAD(1-X) TITANATE SYSTE/ DIELECTRIC AND CERAMIC PROPERTIES OF (BISMUTH(0.5) (SODIUM OR 13-2013
LEAD TITANATE CERAMIC RESONATORS OPERATING IN VHF BAND. 05-1069
DEPENDENCE OF THE SMALL SIGNAL PARAMETERS OF FERROELECTRIC CERAMIC RESONATORS UPON STATE OF POLARIZATION. 37-3649
FORMATION OF BARIUM TITANATE CERAMIC SEMICONDUCTING LAYER BY DIFFUSION. 04-0684
ELECTRON SPIN RESONANCE OF GADOLINIUM(3+) IN BARIUM TITANATE CERAMIC SEMICONDUCTOR. TWO SETS OF 04-0990
DYNAMIC PIEZOELECTRIC THEORY FOR FERROELECTRIC CERAMIC SHELLS. 03-0225
LENT) RARE EARTH IONS. DIELECTRIC RELAXATION IN (CERAMIC) STRONTIUM TITANATES CONTAINING (TRIVA 06-1143
PIEZOELECTRIC SUBSTANCES. (REVIEW, 23 REFS) 02-0105
THE SYSTEM LEAD TITANATE- STRONTIUM COPPER NIOBATE. (CERAMIC SYNTHESIS) 05-1044
ION OF DIELECTRIC MIXTURE FORMULAS TO (LEAD TITANATE) GLASS- CERAMIC SYSTEMS. APPLICAT 05-1054
IC AND OPTICAL PROPERTIES OF TRANSPARENT FERROELECTRIC GLASS CERAMIC SYSTEMS. DIELECTR 04-0576
AMIC PHENOMENOLOGY OF FERROELECTRICITY IN SINGLE CRYSTAL AND CERAMIC SYSTEMS. THERMODYN 03-0214
/EASUREMENT OF PIEZOELECTRIC PHASE ANGLES IN A FERROELECTRIC CERAMIC (THEORY, APPLIED TO LEAD(0.99) NIOBIU/ 09-1537
/ BARRIER LAYERS IN (N TYPE, ANTIMONY DOPED) BARIUM TITANATE CERAMIC WITH A HIGH EFFECTIVE DIELECTRIC CONS/ 04-0581
STUDIES ON THE (STRONTIUM, CALCIUM) TITANATE SOLID SOLUTION CERAMIC WITH BOUNDARY LAYER STRUCTURE. 06-1246
/L CHARACTERISTICS OF A FERROELECTRIC (BARIUM TITANATE TYPE) CERAMIC WITH CONTROLLED MICROWAVE PERMITTIVIT/ 04-0904
COUSTIC DETECTION OF FERROELECTRIC PHASE TRANSITIONS IN PLZT CERAMICS. A 09-1545
BARIUM ZIRCONATE LEAD TITANATE FERROELECTRIC CERAMICS. 09-1492
COLOR IMAGING WITH FERROELECTRIC CERAMICS. 37-3662
DC BIAS EFFECTS IN BARIUM TITANATE CERAMICS. 04-0566
DC BIAS EFFECTS IN LEAD ZIRCONATE TITANATE CERAMICS. 09-1490
DEVICE APPLICATIONS OF PLZT CERAMICS. 09-1569
TATE(0.33) NIOBATE(0.67)- TITANATE- ZIRCONATE SOLID SOLUTION CERAMICS. /ZOELECTRIC PROPERTIES OF LEAD COBAL 09-1549
DIELECTRIC PROPERTIES OF TITANATE- ALUMINA CERAMICS. DI 08-1414
SPERSION IN STRONTIUM TITANATE- LANTHANUM FERRATE DIELECTRIC CERAMICS. 06-1120
AND MECHANICAL DAMPING IN LEAD TITANATE ZIRCONATE TRANSDUCER CERAMICS. /SION STUDY OF DOMAIN WALL MOBILITY 09-1547
DISPLAY APPLICATIONS OF PLZT FERROELECTRIC CERAMICS. 09-1561
B-SITE VACANCIES IN (LEAD, LANTHANUM) (TITANATE, ZIRCONATE) CERAMICS. DISTRIBUTION OF A-SITE AND 09-1530
DYNAMIC SHELL THEORY FOR FERROELECTRIC CERAMICS. 03-0224
ON THE ELECTRIC RESISTANCE OF SEMICONDUCTIVE BARIUM TITANATE CERAMICS. EFFECT OF EXCESS TITANIA 04-0966
ON THE RESISTIVITY ANOMALY IN SEMICONDUCTIVE BARIUM TITANATE CERAMICS. EFFECT OF GRAIN SIZE 04-0735
I BOUNDARIES ON THE DIELECTRIC PROPERTIES OF BARIUM TITANATE CERAMICS. EFFECT OF INTERNA 04-0891
RADIATION ON THE FERROELECTRIC PROPERTIES OF BARIUM TITANATE CERAMICS. EFFECT OF NEUTRON IR 04-0732
EFFECTS OF CALCINING ON SINTERING OF LEAD ZIRCONATE TITANATE CERAMICS. 09-1494
AIN GROWTH ON THE DISTRIBUTION OF NIOBIUM IN BARIUM TITANATE CERAMICS. EFFECTS OF GR 04-0595
ON THE MECHANICAL QUALITY FACTOR OF LEAD ZIRCONATE TITANATE CERAMICS. EFFECTS OF IMPURITIES 09-1608
RADIATION INDUCED DAMAGE CENTERS IN LEAD ZIRCONATE TITANATE CERAMICS. EFFECTS OF 09-1520
EFFECTS OF SILICA ON MANGANESE DOPED STRONTIUM TITANATE CERAMICS. 06-1172
ICAL PROPERTIES OF RUBIDIUM AND CESIUM DOPED BARIUM TITANATE CERAMICS. ELECTR 04-0860
ELECTROELASTIC CONSTANT MEASUREMENTS FOR PLZT CERAMICS. 09-1598
LECTROMECHANICAL PROPERTIES AND APPLICATION OF LEAD TITANATE CERAMICS. E 05-1085
ELECTROOPTIC DEVICES USING STRAIN BIASED PLZT FERROELECTRIC CERAMICS. 09-1562
ICTIVELY GENERATED AND MODULATED FRACTURE IN BARIUM TITANATE CERAMICS. ELECTROSTR 04-0700
ENHANCED ORDERING IN PLZT CERAMICS. 09-1570
ENHANCED ORDERING OF FERROELECTRIC DOMAINS IN PLZT CERAMICS. 09-1579
ENHANCED ORDERING OF FERROELECTRIC DOMAINS IN PLZT CERAMICS. 09-1580
FERROELECTRIC CERAMICS. 01-0006
FERROELECTRIC DOMAIN SWITCHING IN RHOMBOHEDRAL PHASE PLZT CERAMICS. 09-1563
ELECTRON MICROSCOPE (BARRIER LAYERS IN DOPED BARIUM TITANATE CERAMICS). /F SOLID SURFACES) IN THE EMISSION 04-0913
F VIBRATIONS ON THE DIELECTRIC PROPERTIES OF BARIUM TITANATE CERAMICS. INFLUENCE O 04-0890
INFLUENCE OF WATER VAPOR ON BARIUM TITANATE CERAMICS. 04-0882
MAL GRAIN GROWTH AND ELECTROMECHANICAL BEHAVIOR OF FULLY DENSE PLZT CERAMICS. ISOTHER 09-1553
OCIATED WITH THE ELECTROMECHANICAL COUPLING IN PIEZOELECTRIC CERAMICS. MEASUREMENT OF THE DISSIPATION ASS 03-0514
RS ON THE GRAIN BOUNDARIES OF BARIUM TITANATE SEMICONDUCTIVE CERAMICS. /CHANISM FOR FORMING INSULATION LAYE 04-0867
MICROSTRUCTURE OF TSTS-19 TYPE FERROELECTRIC CERAMICS. 36-3554
ND MECHANICAL PROPERTIES OF FERROELECTRICS. (BARIUM TITANATE CERAMICS) NONLINEARITY OF ELECTROMECHANICAL A 04-0986
OF BARIUM TITANATE TYPE FERROELECTRICS. (SINGLE CRYSTALS AND CERAMICS) PERMITTIVITY DISPERSION 04-0903
PHASE BOUNDARY AND DEFECT STRUCTURE STUDIES IN PLZT CERAMICS. 09-1590
PIEZOELECTRIC PROPERTIES OF LEAD TITANATE CERAMICS. 05-1062
ND PIEZOELECTRIC ACTIVITY OF LEAD ZIRCONATE(X) TITANATE(1-X) CERAMICS. POLARIZATION, ELECTROSTRICTION, A 09-1501
TUDY OF THE INNER BARRIERS IN SEMICONDUCTING BARIUM TITANATE CERAMICS. PRODUCTION AND S 04-0583
LECTRIC PROPERTIES OF THE LANTHANUM DOPED FERROELECTRIC PLZT CERAMICS. PYROE 09-1558
ROVEMENTS IN THE OPTICAL AND ELECTROOPTIC PROPERTIES OF PLZT CERAMICS. RECENT IMP 09-1529
TRIC PHOTOCONDUCTOR IMAGE STORAGE AND DISPLAY DEVICES. (PLZT CERAMICS) SCATTERING MODE FERROELEC 09-1599
RING AND FERROELECTRIC PROPERTIES OF LEAD ZIRCONATE TITANATE CERAMICS. SINTE 09-1481
SPACE CHARGE POLARIZATION AND AGING OF BARIUM TITANATE CERAMICS. 04-0880
ANSITION TETRAGONAL- RHOMBOHEDRAL IN LEAD ZIRCONATE TITANATE CERAMICS. /ABILIZATION EFFECTS AT THE PHASE TR 09-1611
IGATING (DYNAMICS OF) POLARIZATION REVERSAL IN FERROELECTRIC CERAMICS. STATISTICAL METHOD FOR INVEST 03-0493
ANICAL PROPERTIES OF LANTHANUM DOPED LEAD TITANATE ZIRCONATE CERAMICS. STRUCTURE AND ELECTROMECH 09-1502
ANICAL PROPERTIES OF LANTHANUM DOPED LEAD TITANATE ZIRCONATE CERAMICS. STRUCTURE AND ELECTROMECH 09-1531
F THE ANTIFERROELECTRIC FERROELECTRIC PHASE BOUNDARY IN PLZT CERAMICS. TEMPERATURE AND FIELD DEPENDENCE O 09-1491
CTRIC CONSTANT OF FERROELECTRIC SUBSTANCES. (BARIUM TITANATE CERAMICS) TEMPERATURE DEPENDENCE OF THE DIELE 04-0591
E OF IRON OXIDE ADULTERANT IN A LEAD ZIRCONATE TITANATE CERAMICS. THE ROL 09-1592
PACITORS AND NONLINEAR MICROWAVE PROPERTIES OF FERROELECTRIC CERAMICS. THIN FILM MICROWAVE FERROELECTRIC CA 37-3643
THREE APPLICATION AREAS FOR STRONTIUM TITANATE GLASS CERAMICS. 06-1150
HANUM(0.002) BARIUM(0.998) TITANATE- BISMUTH OXIDE COMPOSITE CERAMICS. /LOUS INCREASE OF RESISTANCE ON LANT 04-0801
HF REVERSAL PROPERTIES OF BARIUM TITANATE TYPE FERROELECTRIC CERAMICS. U 04-1014

ISTRIBUTION (STABILITY) IN BARIUM(X) STRONTIUM(1-X) TITANATE CERAMICS. /IBRATING WIRE" STUDY OF POTENTIAL D 04-0699
X-RAY ANALYSIS OF MODIFIED LEAD ZIRCONATE- LEAD TITANATE CERAMICS. 09-1619
RGY. (LEAD ZIRCONATE TITANATE) FERROELECTRIC CERAMICS AND CONVERSION OF MECHANOELECTRIC ENE 09-1483
OF HUMIDITY ON THE POTENTIAL DISTRIBUTION IN BARIUM TITANATE CERAMICS AND CRYSTALS. INFLUENCE 04-0883
MAXIMUM DIELECTRIC CON/ INTERNAL STRESSES IN BARIUM TITANATE CERAMICS. (AND DIFFERENCE IN TEMPERATURES FOR 04-0849
E TITANATE) FERROELECTRIC CERAMICS AND ENERGY CONVERSION. (LEAD ZIRCONAT 09-1508
/AL AND STRUCTURAL PARAMETERS OF FERROELECTRIC PIEZOELECTRIC CERAMICS AND METHODS OF THEIR CHANGING. (SPON/ 03-0217
ADVANCES IN PRODUCTION CF (COMMERCIAL) PIEZOELECTRIC CERAMICS AND (TABLES OF) THEIR PROPERTIES. 02-0110
GY CONVERSION). PIEZOELECTRIC EFFECTS IN FERROELECTRIC CERAMICS (AND THE THEORY OF PIEZOELECTRIC ENER 03-0278
E LASERS (PLZT). ELECTROOPTIC CERAMICS AS WAVELENGTH SELECTION DEVICES IN DY 09-1618
C PROPERTIES OF (INSULATING, BOUNDARY LAYER) BARIUM TITANATE CERAMICS AT GRAIN BOUNDARY. FERROELECTRI 04-1024
LECTRIC AND DIELECTRIC PROPERTIES OF LEAD TITANATE ZIRCONATE CERAMICS AT LOW TEMPERATURES. PIEZOE 09-1519
/CIFIC HEAT, ELECTROCALORIC EFFECT) STUDIES OF FERROELECTRIC CERAMICS AT LOW TEMPERATURES. (POLYCRYSTALLIN/ 04-0720
HOD OF STUDYING TEMPERATURE CHARACTERISTICS OF ROCHELLE TYPE CERAMICS. (BARIUM CALCIUM TITANATE CERAMIC) /T 04-0905
STATIC INTERACTIONS IN FINE GRAINED PEROVSKITE FERROELECTRIC CERAMICS. (BARIUM TITANATE) /ON OF THE ELECTRO 04-0642
POLARIZED STATES OF FERROELECTRIC CERAMICS. (BARIUM TITANATE) 04-0666
TANATE. TWO WAVE SHOCK STRUCTURES IN THE FERROELECTRIC CERAMICS BARIUM TITANATE AND LEAD ZIRCONATE TI 04-0916
ANATE) NATURE OF AGING IN FERROELECTRIC CERAMICS. (BARIUM TITANATE, LEAD ZIRCONATE TIT 04-0757
E, PERMITTIVITY) AGING IN COMPLEX CERAMICS BASED ON BARIUM TITANATE. (CAPACITANC 04-0724
S. PROPERTIES OF FERROGLASS CERAMICS BASED ON LEAD BARIUM NIOBATE COMPOUND 13-2006
FABRICATION OF ELECTROOPTIC PLZT CERAMICS BY ATMOSPHERE SINTERING. 09-1600
/ CHANGES IN THE FERROELECTRIC PROPERTIES OF BARIUM TITANATE CERAMICS CAUSED BY IRRADIATION WITH CADMIUM TR 04-0731
SPACE CHARGE EFFECT IN LEAD ZIRCONATE TITANATE CERAMICS CAUSED BY THE ADDITION OF IMPURITIES. 09-1609
/ECHANICAL PROPERTIES OF LEAD ZIRCONATE(0.52) TITANATE(0.48) CERAMICS CAUSED BY THE ADDITION OF NIOBIUM PE/ 09-1606
HOT PRESSED LEAD ZIRCONATE TITANATE CERAMICS CONTAINING BISMUTH. 09-1525
RIUM OR LEAD TITANATE,/ PREPARATION AND PROPERTIES OF GLASS- CERAMICS CONTAINING FERROELECTRIC CRYSTALS (BA 04-0783
ELECTRICAL RESISTIVITY OF LEAD ZIRCONATE TITANATE CERAMICS CONTAINING IMPURITIES. 09-1607
ELECTROMECHANICAL PROPERTIES OF LEAD TITANATE CERAMICS CONTAINING LANTANUM AND MANGANESE. 05-1060
/ POLARIZATION AND PIEZOELECTRIC PROPERTIES OF LEAD TITANATE CERAMICS. (CONTAINING 1 MOLE PERCENT MANGANESE 05-1061
EFFECT OF CLAY AND DOLOMITE ADDITIONS ON BARIUM TITANATE CERAMICS. CRYSTAL STRUCTURE AND SINTERABILITY, 04-0739
VES ON PIEZOELECTRIC AND RELATED PROPERTIES OF LEAD TITANATE CERAMICS. (DIELECTRIC CONSTANT, RESISTIVITY) / 05-1084
TIONS. EVALUATION OF LEAD LANTHANATE ZIRCONATE CERAMICS FOR APPLICATIONS IN OPTICAL COMMUNICA 07-1254
STABILIZED BARIUM TITANATE CERAMICS FOR CAPACITOR DIELECTRICS. 04-0825
RESSED (LEAD, LANTHANUM) (ZIRCONATE, TITANATE) FERROELECTRIC CERAMICS FOR ELECTROOPTIC APPLICATIONS. HOT P 09-1526
HOT PRESSED FERROELECTRIC LEAD ZIRCONATE TITANATE CERAMICS FOR ELECTROOPTICAL APPLICATIONS. 09-1524
D ZIRCONATE TITANATE AND SODIUM F/ HOT PRESSED FERROELECTRIC CERAMICS FOR ELECTROOPTICAL APPLICATIONS. (LEA 09-1527
(PZT) FERROELECTRIC CERAMICS FOR INFORMATION STORAGE AND DISPLAY. 09-1552
GH ELECTRIC FIELDS ON MODIFIED LEAD ZIRCONATE- LEAD TITANATE CERAMICS FOR PIEZOELECTRIC APPLICATIONS. /F HI 09-1485
PREPARATION OF SMALL GRAINED AND LARGE GRAINED CERAMICS FROM NIOBIUM DOPED BARIUM TITANATE. 04-0765
EFFECT OF GRAIN SIZE, / RECENT DEVELOPMENTS IN ELECTROOPTIC CERAMICS (HOT PRESSED LEAD ZIRCONATE TITANATE, 02-0086
PREPARATION OF TRANSPARENT FERROELECTRIC CERAMICS. (HOT PRESSING) 02-0103
IES ON PROPAGATION OF ELASTIC SURFACE WAVES ON FERROELECTRIC CERAMICS. (IMPROVEMENT) /ALLY DIFFUSED IMPURIT 03-0487
ZIRCONATE- LEAD TITANATE AND PLZT) USE OF FERROELECTRIC CERAMICS IN COMPUTERS AND OTHER DEVICES. (LEAD 09-1540
/PERTIES OF HOT PRESSED FERROELECTRIC (BARIUM TITANATE TYPE) CERAMICS. (IN RELATION TO DENSIFICATION AND G/ 04-0879
ATE SYSTEM. IMPROVED HOT PRESSED ELECTROOPTIC CERAMICS IN THE LEAD LANTHANUM ZIRCONATE TITAN 09-1528
/TRIC SYSTEM POLARIZATION AND THERMAL EXPANSION OF POLARIZED CERAMICS IN THE LEAD TITANATE ZIRCONATE MAGNE/ 09-1603
OBATE(0.667)). ELECTROOPTIC EFFECTS OF HOT PRESSED CERAMICS IN THE SYSTEM LEAD (ZINCATE(0.333) NI 13-1900
/CTRICAL AND OPTICAL PROPERTIES OF TRANSPARENT FERROELECTRIC CERAMICS. ((LEAD, LANTHANUM) (ZIRCONATE, TANT/ 13-2004
OBATE(0.5), LEAD TITANATE- LEAD ZIRCONATE) NEW PIEZOELECTRIC CERAMICS. (LEAD TITANATE- LEAD FERRATE(0.5) NI 09-1544
ELECTROOPTIC EFFECTS IN FERROELECTRIC CERAMICS. (LEAD ZIRCONATE TITANATE) 09-1551
E STORAGE AND DISPLAY DEVICES USING FINE GRAIN FERROELECTRIC CERAMICS. (LEAD ZIRCONATE- LEAD TITANATE) IMAG 09-1572
PROPERTIES OF NEW TRANSPARENT FERROELECTRIC CERAMICS. (LEAD ZIRCONATE- LEAD TITANATE TYPE) 09-1518
LEAD TITANATE PIEZOELECTRIC CERAMICS. (MATSUSHITA LEAD TITANATE) 09-1616
LECTROMECHANICAL ACTIVITY IN LEAD ZIRCONATE(X) TITANATE(1-X) CERAMICS NEAR THE MORPHIC PHASE BOUNDARY. /E E 09-1500
/ECTROMECHANICAL ACTIVITY IN LEAD ZIRCONATE(X) TITANATE(1-X) CERAMICS NEAR THE MORPHOTROPIC PHASE BOUNDARY. 09-1499
PIEZOELECTRIC CERAMICS. (NONMETALLIC SOLIDS, VOL-3, REVIEW) 01-0015
TORCH O/ CHARACTERISTICS OF THIN, MULTILAYERED FERROELECTRIC CERAMICS OBTAINED BY PROJECTION WITH A PLASMA 04-0778
PRECIPITATION OF ZIRCONIA PHASE IN NIOBIUM MODIFIED CERAMICS OF LEAD ZIRCONATE TITANATE. 09-1539
ITH SODIUM OR POTASSIUM ANTIMONATE, NIOBATE, / PIEZOELECTRIC CERAMICS OF LEAD ZIRCONATE TITANATE MODIFIED W 09-1539
A NEW CLASS OF FERROELECTRIC CERAMICS OF THE TYPE AB(2) (XO(4)) (3). 13-1976
E SO/ DEPENDENCE OF DIELECTRIC PROPERTIES OF BARIUM TITANATE CERAMICS ON THE PHYSICO- CHEMICAL NATURE OF TH 04-0789
/CT OF ELECTRIC FIELD ON FERROELECTRIC AND ANTIFERROELECTRIC CERAMICS. PART-1: DEFORMATION AND PHASE TRANS/ 03-0169
VOLTAGES) LINEAR ELECTROOPTIC EFFECT IN FERROELECTRIC CERAMICS: PLZT 12-40-60. (USED FOR SENSING LOW 09-1610
DISTRIBUTION OF POTENTIAL ALONG TITANATE CERAMICS PREVIOUSLY POLARIZED. 04-0788
INTERNAL FRICTION OF MODIFIED LEAD ZIRCONATE- LEAD TITANATE CERAMICS. (PROPRIETARY COMPOSITIONS) 09-1623
CALLY VARIABLE BIREFRINGENCE) OF (HOT PRESSED) FERROELECTRIC CERAMICS. (PZT) OPTICAL PROPERTIES (ELECTRI 09-1554
OF BARIUM TITANATE AND LEAD ZIRCONATE TITANATE FERROELECTRIC CERAMICS. (REVIEW, 31 REFERENCES) AGING 02-0062
DEFECTS IN OXIDE CERAMICS. (STRONTIUM TITANATE) 06-1163
RESSI/ PIEZOELECTRIC PROPERTIES OF POLARIZED BARIUM TITANATE CERAMICS SUBJECTED TO UNIAXIAL TRANSVERSE COMP 04-0962
TITANATE. ENERGY STORAGE IN HIGH PERMITTIVITY CERAMICS: SURVEY OF MATERIALS BASED ON BARIUM 04-0733
M TITANATE. ENERGY STORAGE IN HIGH PERMITTIVITY CERAMICS: SURVEY OF MATERIALS BASED UPON BARIU 02-0075
APPLICATION OF NONLINEAR INTERACTIONS IN FERROELECTRIC CERAMICS TO MICROWAVE SIGNAL PROCESSING. 37-3647
/VIOR OF MODIFIED LEAD ZIRCONATE LEAD TITANATE PIEZOELECTRIC CERAMICS UNDER HIGH ELECTRIC FIELDS. (SUBSTIT/ 09-1484
E. SURFACE STUDIES OF BARIUM TITANATE CERAMICS USING THE SCANNING ELECTRON MICROSCOP 04-0567
/ FERROELASTIC BEHAVIOR OF LEAD LANTHANUM ZIRCONATE TITANATE CERAMICS WHEN SUBJECTED TO LARGE TENSILE STRAI 09-1571
/ DEPENDENCE OF ELASTIC AND PIEZOELECTRIC PROPERTIES OF PLZT CERAMICS WITH A ZIRCONIUM TO TITANIUM RATIO O/ 09-1573
Y ELEMENT. LANTHANUM DOPED BARIUM TITANATE CERAMICS WITH BISMUTH OXIDE AS A GRAIN BOUNDAR 04-0799
H FREQUENCY APPLICATION. (BARIUM, STRONTIUM TITA/ DIELECTRIC CERAMICS WITH BOUNDARY LAYER STRUCTURE FOR HIG 04-0833
(BARIUM TITANATE TYPE MATERIALS) LOW LOSS (FERROELECTRIC CERAMICS) WITH GREAT DIELECTRIC PERMITTIVITY. 04-0779
: EFFECTS OF ADDITION OF GLASSES CONTAINING/ BARIUM TITANATE CERAMICS WITH HIGH DIELECTRIC CONSTANT. PART-1 04-0829
MPROVED AGING AND SWITCHING OF LEAD ZIRCONATE- LEAD TITANATE CERAMICS WITH INDIUM ELECTRODES. I 09-1516
MPROVED AGING AND SWITCHING OF LEAD ZIRCONATE- LEAD TITANATE CERAMICS WITH INDIUM ELECTRODES. I 09-1517
LANTHANUM DOPED BARIUM TITANATE- BISMUTH TRIOXIDE COMPOSITE CERAMICS WITH SURFACE BARRIER LAYER. /IVITY ON 04-0800
ELECTRICAL PROPERTIES OF NONLINEAR FERROELECTRIC CERAMICS WITH THE PEROVSKITE STRUCTURE. 08-1461
E AND DISPLAY (HOT PRESSED LEAD ZIRCONATE TITANATE) ELECTROOPTIC CERAMICS- NEW MATERIALS FOR INFORMATION STORAG 09-1550
RE. NATURE OF THE HIGH PERMITTIVITY OF CERIUM ALUMINATES WITH PEROVSKITE TYPE STRUCTU 08-1355
ALLITE LAYERS IN THE SEMICONDUCTOR CERAMIC BARIUM TITANATE. (CERIUM DOPED) / OF CRYSTALLITES AND INTERCRYST 04-0568
UCTING STRONTIUM TITANATE SINGLE CRYSTALS (HYDROGEN REDUCED, CERIUM OR NIOBIUM DOPED). /PECTRA) OF SEMICOND 06-1198
/-X) TUNGSTATE OR MOLYBDATE. (WHERE A IS ZIRCONIUM, HAFNIUM, CERIUM, THORIUM, TITANIUM, SODIUM, MAGNESIUM,/ 10-1640
POTASSIUM) AND ALKALI NITRATES. (LITHIUM, SODIUM, RUBIDIUM, CESIUM) /PERATURES IN ALKALI NITRITES (SODIUM, 16-2399

/P TYPE CRYSTALS (CURIE TEMPERATURES OF POTASSIUM, RUBIDIUM, CESIUM AND AMMONIUM DIHYDROGEN PHOSPHATES AND/ 17-2577
W TEMPERATURE P/ PECULIARITIES IN THE VIBRATIONAL SPECTRA OF CESIUM AND RUBIDIUM NITRATE CRYSTALS IN ITS LO 16-2323
SITION IN A COMPOUND WITH HELICAL ELECTRIC DIPOLE STRUCTURE: CESIUM COPPER TRICHLORIDE.   A PHASE TRAN 33-3388
M DIDEUTERIUM PHOSPHATE, POTASSIUM DIDEUTERIUM ARSENATE, AND CESIUM DIDEUTERIUM ARSENATE. /LECTRIC POTASSIU 17-2438
            DEUTERON QUADRUPOLE COUPLING IN FERROELECTRIC CESIUM DIDEUTERIUM ARSENATE.   18-2721
/-183 SPIN LATTICE RELAXATION AND RESONANCE IN FERROELECTRIC CESIUM DIDEUTERIUM ARSENATE AND CESIUM DIHYDR/ 18-2720
TALS. (RAMAN SCATTERING IN POTASSIUM DIHYDROGEN ARSENATE AND CESIUM DIHYDROGEN ARSENATE) /IEZOELECTRIC CRYS 18-2723
NON AND PROTON PHONON INTERACTIONS IN ALUMINUM PHOSPHATE AND CESIUM DIHYDROGEN ARSENATE.   PHONON PHO 18-2739
  COUPLING IN FERROELECTRIC RUBIDIUM DIHYDROGEN PHOSPHATE AND CESIUM DIHYDROGEN ARSENATE. /IC-75 QUADRUPOLAR 17-2439
AR QUADRUPOLE RESONANCE SPECTRA OF THE FERROELECTRIC CRYSTAL CESIUM DIHYDROGEN ARSENATE. /FECT IN THE NUCLE 18-2722
RSE OPTICAL PHONON). PART-3: (POTASSIUM DIHYDROGEN ARSENATE, CESIUM DIHYDROGEN ARSENATE) /UPLING TO TRANSVE 03-0384
/UBIDIUM DIHYDROGEN ARSENATE, RUBIDIUM DIDEUTERIUM ARSENATE, CESIUM DIHYDROGEN ARSENATE, AMMONIUM DIHYDROG/ 18-2725
ROGEN ARSENATE.   PROTON PHONON COUPLING IN CESIUM DIHYDROGEN ARSENATE AND POTASSIUM DIHYD 18-2733
/ANCE SPECTRA OF PARTIALLY DEUTERATED FERROELECTRIC CRYSTALS CESIUM DIHYDROGEN ARSENATE AND RUBIDIUM DIHYD/ 18-2740
/OFT MODE RAMAN SPECTROSCOPY: COUPLED MODES (FERROELECTRICS, CESIUM DIHYDROGEN ARSENATE, STRONTIUM TITANAT/ 06-1207
            ELECTRICAL PROPERTIES OF RUBIDIUM AND CESIUM DOPED BARIUM TITANATE CERAMICS.   04-0860
FERROELASTIC EFFECT IN RUBIDIUM IRON TETRAFLUORIDE AND CESIUM IRON TETRAFLUORIDE.   33-3358
MINITE)   PREPARATION AND PROPERTIES OF CESIUM LEAD HALIDES. (FERROELECTRICITY INDETER 33-3359
            ELASTIC PROPERTIES OF CESIUM LEAD TRICHLORIDE.   33-3360
            ELECTRIC FIELD GRADIENTS IN IONIC CHLORIDES: CESIUM LEAD TRICHLORIDE.   33-3367
RICAL CONDUCTION AND (NONFERROELECTRIC) PHASE TRANSITIONS IN CESIUM LEAD TRICHLORIDE.   ELECT 33-3365
    EXPERIMENTAL STUDIES OF STRUCTURAL PHASE TRANSITIONS IN CESIUM LEAD TRICHLORIDE.   33-3378
            FAR INFRARED REFLECTIVITY SPECTRA OF CESIUM LEAD TRICHLORIDE.   33-3379
            (MECHANISM OF) STRUCTURAL PHASE TRANSITIONS IN CESIUM LEAD TRICHLORIDE.   33-3380
            MICROSCOPIC OBSERVATIONS OF PHASE TRANSITIONS IN CESIUM LEAD TRICHLORIDE.   33-3383
URE DEPENDENCE OF CUBIC TETRAGONAL TRANSITION TEMPERATURE OF CESIUM LEAD TRICHLORIDE.   PRESS 33-3393
SYMMETRY AT 194 DEGREES K)   PHASE TRANSITIONS IN CESIUM LEAD TRICHLORIDE. (LOSS OF A CENTER OF 33-3368
    DETERMINATION OF TRANSITION TEMPERATURES IN CESIUM LEAD TRICHLORIDE USING EPR.   33-3408
            ELECTRICAL PROPERTIES OF RUBIDIUM NITRATE AND CESIUM NITRATE.   16-2286
    POLARIZED INFRARED AND RAMAN SPECTRA OF SINGLE CRYSTAL CESIUM NITRITE.   16-2342
UOBERYLLATE (METALS CHOSEN FROM SODIUM, POTASSIUM, RUBIDIUM, CESIUM, OR AMMONIUM, ABSTRACT ONLY). /TE OR FL 19-2743
    (PREPARATION AND STRUCTURE OF) RUBIDIUM AND CESIUM SEMIMETALLIC ORTHOPHOSPHATES.   17-2598
E.   ESR STUDY OF GAMMA-IRRADIATED CESIUM TRIHYDROGEN SELENITE AT ROOM TEMPERATUR 22-2833
E.   NEUTRON SCATTERING ANALYSIS OF CESIUM TRIHYDROGEN SELENITE AT ROOM TEMPERATUR 22-2869
CE IN FERROELECTRIC CESIUM DIDEUTERIUM ARSENATE AND CESIUM / CESIUM-183 SPIN LATTICE RELAXATION AND RESONAN 18-2720
EMPERATURE DEPENDENT SPECTRUM OF AN ANTIFERROELECTRIC LINEAR CHAIN MODEL. (PEROVSKITES)   T 03-0468
AL. (BARIUM TITANATE)   COMPUTER EXPERIMENT ON LINEAR CHAIN OF ATOMS LYING IN DOUBLE MINIMUM POTENTI 03-0287
            CHALCOGENIDE BROMIDES OF ANTIMONY AND BISMUTH. 15-2168
AGONAL (FERROELECTRICALLY POLARIZED PHASE OF BARIUM TITANAT/ CHANNELING OF LIGHT IONS IN THE CUBIC AND TETR 04-0688
CRYSTALS AT TEMPERATURES ABOVE AND BELOW THE FERROELECTRIC / CHANNELING OF PROTONS IN THIN BARIUM TITANATE 04-0687
YNTHETIC CALCIUM CHLORAPATITE. (DEVELOPMENT OF FERROELECTRIC CHARACTER IN AN APPLIED ELECTRIC FIELD) / OF S 36-3591
POTASSIUM TANTALATE. (SECOND ORDER)   CHARACTER OF PHASE TRANSITION OF FERROELECTRIC 08-1307
OELECTRIC CRYSTALS WITH A PEROVSKITE TYPE STRUCTURE.   CHARACTER OF THE CHEMICAL BONDS IN ABO(3) FERR 08-1341
EE OF COVALENCY IN IRON DOPED BARIUM TITANATE)   IONIC CHARACTERS FROM MOSSBAUER ISOMER SHIFTS. (DEGR 04-0563
/ASS) FOR ACTUAL FERROELECTRICS TOGETHER WITH EXAMINATION OF CHARACTERS OF THEIR PARAELECTRIC PHASE TRANSF/ 03-0138
LOW TEMPERATURE.COEFFICIENTS OF FREQUENCY (DEVICE DESIGN AND CHARACTERIZATION). /RATIOS OF CAPACITANCE AND 11-1724
IRCONATE AND STRONTIUM TITANATE (OF HIGH PU/ PREPARATION AND CHARACTERIZATION OF ALKOXY-DERIVED STRONTIUM Z 06-1217
    (TOP SEEDED FLUX) CRYSTAL GROWTH AND NEUTRON CHARACTERIZATION OF POTASSIUM NIOBATE.   08-1353
ECTRIC TRIGLYCINE SULFATE SINGLE CRYSTALS WITH I/ GROWTH AND CHARACTERIZATION OF SOLID SOLUTIONS OF FERROEL 27-2972
LAYER OF BARIUM TITANATE.   CHARACTERIZATION OF SURFACE AND GRAIN BOUNDARY 04-0701
ICONDUCTING FILMS. (THE MAJOR PORTION OF THIS/ SYNTHESIS AND CHARACTERIZATION OF THIN FERROELECTRIC AND SEM 02-0057
REFRACTIVE INDEX CHANGES IN BARIUM TITANATE (EXPERIMENTS TO CHARACTERIZE THE OPTICAL DAMAGE). /LLY INDUCED 04-1004
AVIOR OF ICE AT LOW TEMPERATURES, FERROELECTRICITY AND SPACE CHARGE.   ELECTRICAL BEH 34-3448
S.   CHARGE) /HIN FILMS ON A SUBSTRATE OF FERROELEC 27-3046
TRIC TRIGLYCINE SULFATE AND SELENATE CRYSTALS. (POLARIZATION CHARGE AND POLARIZATION STORAGE IN ICE CRYSTAL 34-3456
UM OF IRON(3+) IN STRONTIUM TITANATE DUE TO NEAREST NEIGHBOR CHARGE COMPENSATION. /AGNETIC RESONANCE SPECTR 06-1144
CHROMIUM(3+) DOPED STRONTIUM TITANATE (ESR MEASUREMENTS, M/ CHARGE COMPENSATION BY OXYGEN(2-) VACANCIES IN 06-1161
            SHARED HOLES TRAPPED BY CHARGE DEFECTS IN STRONTIUM TITANATE.   06-1123
ICS CAUSED BY THE ADDITION OF IMPURITIES.   SPACE CHARGE EFFECT IN LEAD ZIRCONATE TITANATE CERAM 09-1609
E TEMPERATURE OF A FERROELECTRIC SEMICON/ METASTABLE OPTICAL CHARGE EXCHANGE AND "FROZEN" SHIFT OF THE CURI 15-2160
INJECTION CURRENTS AND ACCUMULATION OF POLARIZATION CHARGE IN ANTIMONY SULFOIODIDE.   15-2188
NATE ABOVE CURIE POINT)   DOUBLE SPACE CHARGE INJECTION IN SOLIDS. (USING BARIUM TITA 04-0638
TE CRYSTALS.   SURFACE CHARGE LAYERS ON FERROELECTRIC POTASSIUM NIOBA 08-1369
M O/ LIGHT AND ELECTRIC FIELD DEPENDENT OSCILLATION OF SPACE CHARGE LIMITED CURRENT IN BISMUTH(12) GERMANIU 35-3505
UM TITANATE. (CURRENT VOLTAGE CHARACTERISTICS, 150-37/ SPACE CHARGE LIMITED CURRENT IN POLYCRYSTALLINE BARI 04-0557
IC METAL SYSTEM (THEORY).   SPACE CHARGE LIMITED CURRENTS IN A METAL FERROELECTR 03-0348
            THEORY OF SPACE CHARGE LIMITED CURRENTS IN FERROELECTRICS.   03-0251
TANATE(0.75) PLATINATE(0.25) CRYSTALS.   SPACE CHARGE LIMITED CURRENTS IN HEXAGONAL BARIUM TI 04-0936
SSIUM NIOBATE CRYSTALS.   SPACE CHARGE LIMITED CURRENTS IN SEMICONDUCTING POTA 08-1412
ONATE TITANATE, STOICHIOMETRY EFFECTS)   CHARGE MECHANISM IN FERROELECTRICS. (LEAD ZIRC 03-0259
ESONANCE OF MANGANESE(4+) IN STRONTIUM TITANATE.   EFFECTIVE CHARGE OF TITANIUM(4+) FROM THE PARAMAGNETIC R 06-1167
IN ANTIMONY SULFOIODIDE.   EFFECT OF PHOTOELECTRET CHARGE ON VALUE OF FERROELECTRIC POLARIZATION 15-2186
N (BY PASSING COHERENT LIGHT THROUGH DEUTERATED KDP HAVING A CHARGE PATTERN). /ON BY ELECTROOPTIC MODULATIO 17-2507
TE CERAMICS.   SPACE CHARGE POLARIZATION AND AGING OF BARIUM TITANA 04-0880
ODS FOR INVESTIGATING STRAIN WAVE PROPAGATION AND ASSOCIATED CHARGE RELEASE TO FERROELECTRIC MATERIALS. /TH 37-3630
N HALIDES.   EFFECT OF CHARGE TRANSFER ON RAMAN SCATTERING OF HYDROGE 31-3328
VED TG/ AFTER EFFECTS IN THE FIELD EFFECT AT VERY LARGE GATE CHARGES APPLIED TO TELLURIUM FILMS ON UHV CLEA 27-3171
ON TO INJECTION AND CONDUCTION IN BARIUM TITANATE (CONDENSER CHARGE- DISCHARGE METHOD).   IONIC CONTRIBUTI 04-0571
TRICS.   INFLUENCE OF COMPENSATING CHARGES ON THE C-DOMAIN STRUCTURE OF FERROELEC 03-0458
C PROPERTIES OF MATERIALS. (BARIUM TITANATE TYPE)   EFFECT OF CHARGED LATTICE IMPERFECTIONS ON THE DIELECTRI 04-1017
F IONIC CRYSTALS (TGS AND OTHERS, BY SELECTIVE DEPOSITION OF CHARGED POLYSTYRENE LATEX PARTICLES. /FACES O 27-2987
THE PERMITTIVITY OF BARIUM TITANATE TYPE FERR/ EXPERIMENTAL CHECK OF THE THEORY OF MICROWAVE DISPERSION OF 04-0649
UM IN LEAD TITANATE AND SIMILAR COMPOUNDS. STUDIES O/ DIRECT CHELATOMETRIC DETERMINATION OF LEAD AND TITANI 05-1068
STALS OF BARIUM TITANATE WITH COBALT ADDED.   SOME CRYSTAL CHEMICAL AND PHYSICAL PROPERTIES OF SINGLE CRY 04-0618
ONS FOR ELECTRIC ORDERING AND STABILI/ PEROVSKITE FAMILY AND CHEMICAL BOND. (COVALENCY, DIRECTIVITY CONDITI 03-0243
EFS)   FERROELECTRICITY AND THE CHEMICAL BOND IN PEROVSKITE TYPE OXIDES. (43 R 02-0102
S WITH A PEROVSKITE TYPE STRUCTURE.   CHARACTER OF THE CHEMICAL BONDS IN ABO(3) FERROELECTRIC CRYSTAL 08-1341
R NITROGEN IN THE PRESENCE OF CARBON.   CHEMICAL CHANGES IN BARIUM TITANATE FIRED UNDE 04-0872
BRONZE TYPE NIOBATES (CORRELATION OF CURIE TEMPERATURE WITH CHEMICAL COMPOSITION). /MPERATURES OF TUNGSTEN 13-2049
RROELECTRIC CRYSTALS, ESPECIALLY ON LITHIUM NIOBATE AND LIT/ CHEMICAL COMPOSITION AND OPTICAL QUALITY OF FE 11-1780

/TRIC PROPERTIES OF BARIUM TITANATE CERAMICS ON THE PHYSICO- CHEMICAL NATURE OF THE SOLID SOLUTION FORMED / 04-0789
CHEMICAL PHYSICS OF ICE. 01-0009
ATE SOLID SOLUTIONS. CRYSTAL CHEMICAL PROBLEMS OF (BARIUM, STRONTIUM) TITAN 04-0633
F BARIUM TITANATE CRYSTALS. CHEMICAL PROCESSES DURING HYDROGEN REDUCTION O 04-0540
MATION OF POTASSIUM ACID TARTRATE CRYSTALS. SOME CHEMICAL REACTIONS IN SILICA GELS. PART-3: FOR 29-3246
HOSPHATE. MEASUREMENT OF CHEMICAL SHIFT IN SOLID POTASSIUM DIHYDROGEN P 17-2688
NS. (BISMUTH SULFOBROMIDE) CRYSTAL GROWTH AND CHEMICAL SYNTHESIS UNDER HYDROTHERMAL CONDITIO 15-2218
ON BEHAVIOR OF ANTIMONY SULFOICDIIE. (SINGLE CRYSTAL GROWTH, CHEMICAL TRANSPORT) CRYSTALLIZATI 15-2201
PHASES OF QUADRATIC TUNGSTEN BRONZES STRONTIUM POT/ CRYSTAL CHEMISTRY AND ELECTRICAL PROPERTIES OF TWO NEW 13-2010
ATES. (PREPARATION, REVIEW) CHEMISTRY AND PHYSICS OF (ALKALI) METAL TANTAL 02-0048
YPE STRUCTURE. CRYSTAL CHEMISTRY OF MIXED BISMUTH OXIDES WITH LAYER T 13-2032
LECTRICS. (BISMUTH TUNGSTATE, BISMUTH NIOBIUM OXYFL/ CRYSTAL CHEMISTRY OF PEROVSKITE LIKE LAYER TYPE FERROE 13-1966
EMPHASIS ON PEROVSKITES. CRYSTAL CHEMISTRY OF THE ABX(3) FAMILY WITH PARTICULAR 08-1403
ARACTER IN AN APP/ MONOCLINIC STRUCTURE OF SYNTHETIC CALCIUM CHLORAPATITE. (DEVELOPMENT OF FERROELECTRIC CH 36-3591
THE VIBRATIONAL SPECTRUM AND DIELECTRIC BEHAVIOR OF SODIUM CHLORATE. (NOT FERROELECTRIC) 36-3549
UND WITH HELICAL ELECTRIC DIPOLE STRUCTURE: CESIUM COPPER TRICHLORIDE. A PHASE TRANSITION IN A COMPO 31-3388
CRYSTAL STRUCTURE AND PHASE TRANSITION OF HYDROGEN CHLORIDE. 31-3334
ELASTIC PROPERTIES OF CESIUM LEAD CHLORIDE. 33-3360
ELECTRIC FIELD GRADIENTS IN IONIC CHLORIDES: CESIUM LEAD TRICHLORIDE. 33-3367
N AND (NONFERROELECTRIC) PHASE TRANSITIONS IN CESIUM LEAD TRICHLORIDE. ELECTRICAL CONDUCTIO 33-3365
AL STUDIES OF STRUCTURAL PHASE TRANSITIONS IN CESIUM LEAD TRICHLORIDE. EXPERIMENT 33-3378
FAR INFRARED REFLECTIVITY SPECTRA OF CESIUM LEAD TRICHLORIDE. 33-3379
MECHANISM OF) STRUCTURAL PHASE TRANSITIONS IN CESIUM LEAD TRICHLORIDE. (33-3380
ROSCOPIC OBSERVATIONS OF PHASE TRANSITIONS IN CESIUM LEAD TRICHLORIDE. MIC 33-3383
E FREQUENCIES AND RELAXATION TIMES IN FERROELECTRIC HYDROGEN CHLORIDE. NUCLEAR QUADRUPOL 31-3336
PARAELECTRIC RESONANCE OF LITHIUM ION IN POTASSIUM CHLORIDE. 31-3325
OF CUBIC TETRAGONAL TRANSITION TEMPERATURE OF CESIUM LEAD TRICHLORIDE. PRESSURE DEPENDENCE 33-3393
STRUCTURE AND PHASE TRANSITION IN HYDROGEN CHLORIDE. 31-3337
THE CRYSTAL STRUCTURE OF TRIS-SARCOSINE CALCIUM CHLORIDE. 36-3524
/ PROPAGATION DIRECTION UPON OPTICAL PHONONS IN CUBIC SODIUM CHLORIDE AND EPITAXIAL FERROELECTRIC SODIUM T/ 29-3243
/ICRONS, 77-650 DEG K) OF FERROELECTRIC IRON(2+) BORACITES. (CHLORIDE, BROMIDE AND IODIDE, AND RUBIDIUM IR/ 14-2097
) OF FERROELECTRIC BORACITES. (IRON, NICKEL, OR COBALT, WITH CHLORIDE, BROMIDE, OR IODIDE) /N (CALCULATIONS 14-2095
ATI/ MOLECULAR MOTION IN THE FERROELECTRIC PHASE OF HYDROGEN CHLORIDE. (CALCULATED IN THE HARMONIC APPROXIM 31-3340
L SIZE EFFECTS IN THE RAMAN SPECTRA OF ORTHORHOMBIC HYDROGEN CHLORIDE CRYSTALS. CRYSTA 31-3338
ELECTRIC AND NMR STUDIES OF THE PHASE TRANSITION IN STANNOUS CHLORIDE DIHYDRATE. DI 36-3582
DIRECT PIEZOELECTRIC EFFECT IN POLYVINYL CHLORIDE FILMS. 36-3543
/E DECAY CONSTANT OF ZIRCONIUM-89 IN BARIUM TITANATE (BARIUM CHLORIDE FLUX GROWTH IN PRESENCE OF ZIRCONIUM/ 04-0685
DIELECTRIC PROPERTIES OF POTASSIUM CHLORIDE ICE. 34-3454
/ SYSTEMS (SPIN WAVES IN HYDROXYL DIPOLE SYSTEM IN POTASSIUM CHLORIDE LATTICE, FERROELECTRIC STATE AT LOW / 03-0452
DEGREES K) PHASE TRANSITIONS IN CESIUM LEAD TRICHLORIDE. (LOSS OF A CENTER OF SYMMETRY AT 194 33-3368
CTR/ NEW BORACITES OF FORMULA LITHIUM(4) BORON(7) OXYGEN(12) CHLORIDE OR BROMIDE (NOT EXAMINED FOR FERROELE 14-2089
/ITHIUM NIOBATE OR TANTALATE, BARIUM SODIUM NIOBATE, CUPROUS CHLORIDE, PROUSTITE, CALCIUM PYRONIOBATE, AND/ 15-0021
TRANSITIONS IN DIGLYCINE NITRATE AND TRIS-SARCOSINE CALCIUM CHLORIDE. (SHIFT OF CURIE TEMPERATURE) /ECTRIC 16-2267
/ENTIALS IN AQUEOUS SOLUTIONS. (POTASSIUM BROMIDE, POTASSIUM CHLORIDE, SODIUM CHLORIDE, POTASSIUM SODIUM T/ 17-2707
GANESE(2+) COMPLEXES IN FERROELECTRIC TRIS-SARCOSINE CALCIUM CHLORIDE (TSCC) SINGLE CRYSTALS. /TIONS OF MAN 36-3525
DETERMINATION OF TRANSITION TEMPERATURES IN CESIUM LEAD TRICHLORIDE USING EPR. 33-3408
/AN SPECTRA OF ORTHORHOMBIC AND CUBIC HYDROGEN AND DEUTERIUM CHLORIDES AT TEMPERATURES ABOVE 80 DEGREES K. 31-3330
ELECTRIC FIELD GRADIENTS IN IONIC CHLORIDES: CESIUM LEAD TRICHLORIDE. 33-3367
IS SULFUR, SELENIUM OR TELLURIUM, VII IS IODINE, BROMINE OR CHLORINE) /STALS, V IS ANTIMONY OR BISMUTH, VI 15-2214
ENIC, SULFUR BY SELENIUM OR OXYGEN, AND IODINE BY BROMINE OR CHLORINE) /UTION OF ANTIMONY BY BISMUTH OR ARS 15-2205
EL DOPED/ OPTICAL ABSORPTION SPECTRA OF NICKEL(2+) IN NICKEL CHLORINE AND NICKEL IODINE BORACITES (AND NICK 14-2074
FERROELECTRIC AND FERROELASTIC PROPERTIES OF MAGNESIUM CHLORINE BORACITE. 14-2102
SING ANTIMONY OR BISMUTH, SULFUR, SELENIUM OR TELLURIUM, AND CHLORINE, BROMINE OR IODINE) /I) COMPOUNDS. (U 15-2194
MIUM IODINE, IRO/ (OPTICAL) FREQUENCY DOUBLING IN (MAGNESIUM CHLORINE, ZINC CHLORINE, CADMIUM CHLORINE, CAD 14-2080
O ACETA/ FERROELECTRICITY AND PHASE TRANSITION IN AMMONIUM DICHLORO ACETATE AND DEUTERATED AMMONIUM DICHLOR 36-3538
/IODIDE, THIOUREA, TRIGLYCINE SULFATE, HYDROGEN AMMONIUM MONOCHLORO ACETATE, DICALCIUM STRONTIUM HEXAACETA/ 02-0099
/DRUPOLE RESONANCE STUDY OF PHASE TRANSITION IN AMMONIUM MONOCHLORO ACETATE, MONOCHLORO ACETIC ACID MIXED / 36-3625
/YSTAL STRUCTURE AND PHASE TRANSITION OF AMMONIUM HYDROGEN DICHLORO ACETATE. PART-1: THE CRYSTAL STRUCTURE/ 36-3567
BISMUTHATE, OR BY BISMUTH OR LANTHANUM FERRATE, ALUMINATE OR CHROMATE. / POTASSIUM ANTIMONATE, NIOBATE, OR 01-1539
XIDE MATERIALS. (NIOBATES, TANTALATES, STANNATES, TITANATES, CHROMATES, COBALTATES) /OWTH OF SOME COMPLEX O 02-0129
ONATE TITANATES CONTAINING BISMUTH AND STRONTIUM ADDITIVES. CHROMATOGRAPHIC METHOD FOR ANALYZING LEAD ZIRC 09-1625
L TRANSPORT IN (SEMICONDUCTING AND MAGNETIC) RARE EARTH ORTHOCHROMITES, MANGANITES AND FERRITES ELECTRICA 36-3608
STATE OF CHROMIUM ADDED TO BARIUM METATITANATE. 04-0543
HEAT CAPACITY OF METHYL AMMONIUM CHROMIUM ALUM. 20-2801
/ORDER TYPE. (COPPER FORMATE TETRAHYDRATE, AMMONIUM IRON AND CHROMIUM ALUMS, METHYL AMMONIUM ALUMINUM SULF/ 02-0087
EPR OF THE CHROMIUM COMPLEX IN TRIGLYCINE FLUOBERYLLATE. 27-3133
ALS. ESR OF CHROMIUM(3+) IN CHROMIUM DOPED STRONTIUM TITANATE SINGLE CRYST 06-1147
EMAN EFFECT (STUDY) OF TWO FINE LINES IN (THE RED REGION OF) CHROMIUM GUANIDINIUM SULFATE HEXAHYDRATE. ZE 21-2814
/RIZING FIELD AND OF THE ADDITION OF A MIXTURE OF COPPER AND CHROMIUM IONS ON THE DIELECTRIC PROPERTIES OF/ 27-3045
RIC CRYSTAL GUANIDINIUM ALUMINUM SULFATE HEXAHYDRATE AND ITS CHROMIUM ISOMORPH. /OF STRUCTURE OF FERROELECT 21-2819
/MUTH METAL OXYGEN(3) PEROVSKITES. (WHERE METAL IS SCANDIUM, CHROMIUM, MANGANESE, IRON, COBALT, NICKEL, YT/ 08-1451
AL TEMPERATURE. SPECIFIC HEAT AND SUSCEPTIBILITY IN CHROMIUM METHYL AMMONIUM ALUM ABOVE THE CRITIC 20-2805
S YTTRIUM MANGANESE(1-X) (B(X)) OXYGEN(3) (B=TRIVALENT IRON, CHROMIUM OR ALUMINUM). /IGATIONS OF THE SYSTEM 36-3569
S OF A CERAMIC (LEAD STRONTIUM TITANATE ZIRCONATE DOPED WITH CHROMIUM OXIDE). / AND PIEZOELECTRIC PROPERTIE 09-1591
FFECT AND DILUTION EFFECT OF ABSORPTION LINES IN GUANIDINIUM CHROMIUM SULFATE HEXAHYDRATE. ZEEMAN E 21-2821
ND STATIC MAGNETIC SUSCEPTIBILITY OF CHROM/ DETERMINATION OF CHROMIUM VALENCY FROM STUDIES OF EPR SPECTRA A 04-0542
LUMINATE. ELECTRON SPIN RESONANCE SPECTRA OF CHROMIUM(3+) AND GADOLINIUM(3+) IN LANTHANUM A 36-3583
RAMAGNETIC ION COMPLEXES IN GLYCINE CRYSTALS. EPR STUDIES OF CHROMIUM(3+), COPPER(2+), AND MANGANESE(2+) PA 27-3135
CRYSTAL FIELD IN CHROMIUM(3+) DOPED LITHIUM NIOBATE. 11-1759
YSIS OF STRESS AND TEMPERATURE DEPENDENCE OF FLUORESCENCE IN CHROMIUM(3+) DOPED STRONTIUM TITANATE. ANAL 06-1215
EFFECT OF DOMAIN STRUCTURE ON THE FLUORESCENCE OF CHROMIUM(3+) DOPED STRONTIUM TITANATE. 06-1114
SUREMENTS, M/ CHARGE COMPENSATION BY OXYGEN(2-) VACANCIES IN CHROMIUM(3+) DOPED STRONTIUM TITANATE (ESR MEA 06-1161
ELECTRON PARAMAGNETIC RESONANCE OF CHROMIUM(3+) IN ADP. 17-2596
ATE SINGLE CRYSTALS. ESR OF CHROMIUM(3+) IN CHROMIUM DOPED STRONTIUM TITAN 06-1147
ELECTRON SPIN RESONANCE STUDIES OF CHROMIUM(3+) IN LITHIUM NIOBATE. 11-1804
/IAXIAL STRESS AND ELECTRIC FIELD ON PARAMAGNETIC SPECTRA OF CHROMIUM(3+) IN POTASSIUM DIHYDROGEN PHOSPHAT/ 17-2554
M(3+) PAIRS IN LANTHANUM ALUM/ SPECTRUM AND ENERGY LEVELS OF CHROMIUM(3+) IONS AND EXCHANGE COUPLED CHROMIU 36-3565
STALS. EPR OF CHROMIUM(3+) IONS IN ANTIFERROELECTRIC ADP CRY 17-2546
INFLUENCE OF SPONTANEOUS POLARIZATION ON THE EPR SPECTRA OF CHROMIUM(3+) IONS IN TRIGLYCINE FLUOBERYLLATE. 27-3136
(ABSORPTION) SPECTRA OF LITHIUM NIOBATE CRYSTALS CONTAINING CHROMIUM(3+) IONS (10 TO 30 THOUSAND/CM). /THE 11-1831

OPTICAL SPECTRA OF CHROMIUM(3+) PAIRS IN LANTHANUM ALUMINATE. 36-3616
ANSITION IN STRONTIUM TITAN/ POLARIZED FLUORESCENCE STUDY OF CHROMIUM(3+) THROUGH A STRESS INDUCED PHASE TR 06-1115
E SINGLE CRYSTALS. EPR STUDIES OF CHROMIUM- AND VANADYL- DOPED TRIGLYCINE SULFAT 27-3161
/NUSUAL TEMPERATURE VARIATION OF THE MEAN MAGNETIC MOMENT OF CHROMIUM- POTASSIUM ALUM (UNSTABLE SOFT PHONO/ 20-2806
ND DOMAIN STRUCTURE IN ANTIMONY SULFOIODIDE CRYSTALS. (MICRO CINEMATOGRAPHY) PHASE TRANSITION A 15-2184
LS.
CTOR MEMORY DEVICE. CIRCLE THEOREM FOR ICE RULE FERROELECTRIC MODE 03-0200
RS. (ANTIMONY SULFOIODIDE CRYSTALS) EQUIVALENT CIRCUIT ANALYSIS OF A FERROELECTRIC PHOTOCONDU 37-3644
ELECTRIC CERAMIC CYLINDERS. (HYDROPHONE ELEMENTS) SHORT CIRCUIT PHOTOEMF IN FERROELECTRIC SEMICONDUCTO 15-2105
OELECTRIC CERAMIC CYLINDERS. (HYDROPHONE ELEMENTS) OPEN CIRCUIT SENSITIVITY OF AXIALLY POLARIZED FERRO- 37-3652
OELECTRIC CERAMIC HOLLOW SPHERES. (HYDROPHONE ELEMENTS) OPEN CIRCUIT SENSITIVITY OF RADIALLY POLARIZED FERR 37-3651
OPEN CIRCUIT SENSITIVITY OF RADIALLY POLARIZED FERR 37-3653
ACTIVE COMPENSATORS FOR FERROELECTRIC OPTICAL CIRCUITS. 37-3675
AND P- AND N- RANGES IN SEMICONDUCTORS. (SILICON INTEGRATED CIRCUITS) /RATE SINGLE CRYSTALS, P-N JUNCTIONS 24-2904
VARIOUS HYSTERESIS LOOP TRACERS FOR A WIDE FREQUENCY RANGE. (CIRCUITRY DESCRIBED) 37-3640
F SMALL SEEDS OF AMMONIUM DIHYDROGEN PHOSPHATE IN A MODIFIED CIRCULATION APPARATUS. GROWTH BEHAVIOR O 17-2704
IUM TITANATE AND OTHER CERAMIC POWDERS BY COPRECIPITATION OF CITRATES IN AN ALCOHOL. PREPARATION OF BAR 04-0861
/COND HARMONIC GENERATION AND THE VARIATIONS IN THE FREE AND CLAMPED VALUES OF THE DIELECTRIC CONSTANTS AN/ 13-1998
EXAMINATION OF CHARACTERS OF / DETERMINATION OF KIND (TYPE, CLASS) FOR ACTUAL FERROELECTRICS TOGETHER WITH 03-0138
(2)(XO(4))(3). A NEW CLASS OF FERROELECTRIC CERAMICS OF THE TYPE AB 13-1976
KDP TYPES) ANALYTICAL PROPERTIES OF A CLASS OF PLANAR FERROELECTRIC MODELS. (THEORY, 03-0258
LECTRIC SURFACE WAVES ON A C-DOMAIN WALL OF A FERROELECTRIC (CLASS 6MM). PIEZOE 03-0421
CAL PROPERTIES OF FERROELECTRICS OF DIFFERENT CLASSIFICATION CLASSES. /S OF DOMAIN STRUCTURE AND SOME PHYSI 03-0480
RMATIONS. (AND SUGGESTIONS FOR DISCOVERING MEMBERS OF VACANT CLASSES) / OF THEIR PARAELECTRIC PHASE TRANSFO 03-0138
. (BORACITES, MOLYBDATES, TUNGSTEN BRONZES, NIOBATES) NEW CLASSES OF FERROELECTRICS OF DISPLACEMENT TYPE 02-0118
ILLATOR MODEL FOR ORDER DISORDER / DYNAMIC SUSCEPTIBILITY OF CLASSICAL ANHARMONIC OSCILLATOR. A UNIFIED OSC 03-0405
CRYSTALLOCHEMISTRY OF PEROVSKITES. (EXISTENCE CONDITIONS AND CLASSIFICATION) 08-1324
/ WITH PEROVSKITE STRUCTURE. (EXAMINATION OF PUBLISHED DATA, CLASSIFICATION, LISTING AND DISCUSSION OF STR/ 01-0018
D ANTI- POLAR- POLAR AND POLAR- NEUTRAL CRYSTAL STRUCTURES. (CLASSIFICATION OF ANISOTROPIC FERROELECTRIC AN 03-0336
MAIN STRUCTURE AND ITS APPLICATIONS) CRYSTALLOPHYSICAL CLASSIFICATION OF FERROELECTRICS. (BASED ON DO 03-0463
ISS CONSTANT) A NOTE ON THE CLASSIFICATION OF FERROELECTRICS. (BY CURIE-WE 01-0024
C PHASE TRANSITIONS AND SPECIAL FEATURES/ CRYSTALLOGRAPHICAL CLASSIFICATION OF FERROELECTRICS, FERROELECTRI 03-0480
HASE TRANSITIONS. THERMODYNAMIC CLASSIFICATION OF SECOND ORDER FERROELECTRIC P 03-0303
CERAMICS. CRYSTAL STRUCTURE AND SINTERABILITY. EFFECT OF CLAY AND DOLOMITE ADDITIONS ON BARIUM TITANATE 04-0739
IN POTASSIUM NIOBATE SINGLE CRYSTALS FROM PSEUDOCUBIC (001) CLEAVAGE PLANES. DOMAINS 08-1317
ES ON FERROELECTRIC GUANIDINIUM ALUMINUM SULFATE HEXAHYDRATE CLEAVAGES AND THEIR ETCH PATTERNS. /TRIC STUDI 21-2816
THIN FILMS ON TGS CLEAVED IN ULTRAHIGH VACUUM. 27-3173
AT VERY LARGE GATE CHARGES APPLIED TO TELLURIUM FILMS ON UHV CLEAVED TGS. /TER EFFECTS IN THE FIELD EFFECT 27-3171
ECT AND HALL MOBILITY OF TELLURIUM AND LEAD TELLURIDE ON UHV CLEAVED TRIGLYCINE SULFATE. FIELD EFF 27-3172
TAILED STUDY OF THE PHASE TRANSITION IN TRIGLYCINE SULFATE. (CLOSE THERMAL CONTROL) EXPERIMENTAL DE 27-3030
UENCE OF SURFACE EFFECTS ON THE PYROELECTRIC BEHAVIOR OF TGS CLOSE TO THE PHASE TRANSITION. INFL 27-3064
TRIC MATERIAL WITH SEVERAL PHASE TRANSITIONS AT TEMPERATURES CLOSE TOGETHER. (ROCHELLE SALT) / IN FERROELEC 03-0363
RROR MICROSCOPIC OBSERVATION OF FERROELECTRIC DOMAINS. (GOLD COATED BARIUM TITANATE) ELECTRON MI 04-0974
THREE LAYER BROAD BAND ANTIREFLECTION COATINGS FOR LITHIUM NIOBATE. 11-1676
/TIVE TITANIUM DIOXIDE AND TITANIUM DIOXIDE- BARIUM TITANATE COATINGS PRODUCED BY FLAME SPRAYING. PART-11:/ 04-0774
FERROELECTRIC TUBES. VIBRATIONS OF COAXIALLY SEGMENTED, LONGITUDINALLY POLARIZED 03-0376
IDE(3) WHERE Z IS NIOBIUM, TANTALUM, ZIRCONIUM, VANADIUM, OR COBALT) /ZES. (SODIUM(X) TUNGSTEN(1-X) Z(Y) OX 13-1944
YSICAL PROPERTIES OF SINGLE CRYSTALS OF BARIUM TITANATE WITH COBALT ADDED. SOME CRYSTAL CHEMICAL AND PH 04-0618
CRYSTAL FIELD THEORY AND OPTICAL ABSORPTION OF COBALT AND NICKEL BORACITES. 14-2096
(20 PERCENT MANGANESE RE/ ANTIFERROELECTRICITY IN (PURE AND) COBALT DOPED POTASSIUM MANGANESE TRIFLUORIDE. 33-3366
/ELSTAM BRILLOUIN SCATTERING OF LIGHT IN MANGANESE FLUORIDE, COBALT FLUORIDE, POTASSIUM MANGANESE TRIFLUOR/ 33-3399
FERROELECTRIC TRANSITION IN COBALT IODINE BORACITE. 14-2099
OPTICAL ANISOTROPY OF THE CUBIC COBALT IODINE BORACITE. 14-2094
PIEZOELECTRIC PROPERTIES OF COBALT IODINE BORACITE. 14-2100
SPECIFIC HEAT ANOMALY OF FERROELECTRIC COBALT IODINE BORACITE. 14-2101
SKITES. (WHERE METAL IS SCANDIUM, CHROMIUM, MANGANESE, IRON, COBALT, NICKEL, YTTRIUM) /ETAL OXYGEN(3) PEROV 08-1451
M, YTTRIUM, GADOLINIUM, DYSPROSIUM, IRON, SELENIUM, NIOBIUM, COBALT, SCANDIUM, OR COPPER) /AGNESIUM, CADMIU 08-1451
LOW FREQUENCY MAGNETOELECTRIC RESONANCES IN BARIUM COBALT TETRAFLUORIDE. 33-3409
GROUP AND LAT/ FERROELECTRIC PARAELASTIC PARAMAGNETIC BARIUM COBALT TETRAFLUORIDE CRYSTAL STRUCTURE (SPACE 33-3386
/PTICN SPECTRA OF POTASSIUM MANGANESE TRIFLUORIDE, POTASSIUM COBALT TRIFLUORIDE, AND RUBIDIUM COBALT TRIFL/ 33-3384
IFERROMAGNETIC POTASSIUM MANGANESE TRIFLUORIDE AND POTASSIUM COBALT TRIFLUORIDE CRYSTALS. /CONSTANTS OF ANT 33-3385
/ OF FERROELECTRIC FERROMAGNETIC SOLID SOLUTIONS IN THE LEAD COBALT TUNGSTATE- CADMIUM MANGANATE SYSTEM (B/ 36-3604
(CALCULATIONS) OF FERROELECTRIC BORACITES. (IRON, NICKEL, OR COBALT, WITH CHLORIDE, BROMIDE, OR IODIDE) /N 14-2095
EPR OF COBALT(2+) IN POTASSIUM TANTALATE. 08-1342
FERROELECTRIC TRANSITION IN COBALT- IODINE- BORACITE UNDER PRESSURE. 14-2078
TRANSITION TEMPERATURE IN KDP AND DEUTERATED KDP (DOPED WITH COBALT-57 AND IRON-57). /REQUENCY MODE AT THE 14-2465
MOSSBAUER EFFECT OF COBALT-57 IN BARIUM TITANATE. 04-0723
E, (COMPARED TO ST/ EFFECTS OF (MONOENERGETIC) ELECTRON AND (COBALT-60) GAMMA-IRRADIATION ON BARIUM TITANAT 04-0970
NATE SOLID / DIELECTRIC AND PIEZOELECTRIC PROPERTIES OF LEAD COBALTATE(0.33) NIOBATE(0.67)- TITANATE- ZIRCO 09-1549
TE(0.5) TU/ PHASE TRANSITION IN THE ANTIFERROELECTRET LEAD COBALTATE(0.5) TUNGSTATE(0.5) AND LEAD COBALTA 13-1968
/CAL MAGNETIC AND X-RAY STUDIES OF PHASE TRANSITIONS IN LEAD COBALTATE(0.5) TUNGSTATE(0.5), TITANATE, ZIRC/ 13-1960
ALS. (NIOBATES, TANTALATES, STANNATES, TITANATES, CHROMATES, COBALTATES) /OWTH OF SOME COMPLEX OXIDE MATERI 02-0129
UM TITANATES. TEMPERATURE DEPENDENT POLARIZABILITIES AND COCHRAN MODE ASSIGNMENTS IN BARIUM AND STRONTI 04-0805
VANADIUM DIOXIDE ABOVE AND BELOW TRANSITION TEMPERATURE. (NO COCHRAN TYPE FERROELECTRIC MODE) /OPERTIES OF 36-3526
ECTRICS. SIMULTANEOUS MEASUREMENT OF PYROELECTRIC COEFFICIENT AND DIELECTRIC CONSTANT OF FERROEL 03-0333
TRIC RESPONSE TO STEP RADIA/ MEASUREMENT OF THE PYROELECTRIC COEFFICIENT AND PERMITTIVITY FROM THE PYROELEC 03-0469
GENERATION IN LITHIUM IODATE. LARGE NONLINEAR OPTICAL COEFFICIENT AND PHASE MATCHED SECOND HARMONIC 08-1405
SEE BOTH POSITIVE TEMPERATURE COEFFICIENT AND PTC.
TANATE CRYSTALS (POLARON H/ ELECTRICAL CONDUCTIVITY (SEEBECK COEFFICIENT, CURIE POINT) OF REDUCED BARIUM TI 04-0918
PPLICATION TO A NANOSECOND RESPONSE TIME DETEC/ PYROELECTRIC COEFFICIENT DIRECT MEASUREMENT TECHNIQUE AND A 13-1919
SE TRANSI/ ANOMALOUS BEHAVIOR OF THE DIELECTRIC NONLINEARITY COEFFICIENT IN STRONTIUM TITANATE NEAR THE PHA 06-1105
PIEZOELECTRIC COEFFICIENT OF GADOLINIUM MOLYBDATE. 30-3317
PRESSURE DEPENDENCE OF THE COMPLEX DISPERSION COEFFICIENT OF ICE I(H) SINGLE CRYSTALS. 34-3479
LS IN THE REG/ ANOMALOUS BEHAVIOR OF THE OPTICAL ABSORPTION COEFFICIENT OF LEAD MAGNONIOBATE SINGLE CRYSTA 13-1974
SE OF TEMPERATURE SENSITIVE CAPACITANCES. LOW TEMPERATURE COEFFICIENT OF LITHIUM NIOBATE RESONATORS BY U 11-1670
DIELECTRIC MEASUREMENT TECHNIQUE. DIFFUSION COEFFICIENT OF NIOBIUM IN BARIUM TITANATE BY A 04-0926
SEE BOTH POSITIVE TEMPERATURE COEFFICIENT OF RESISTANCE AND PTCR.
TITANATE WITH NIOBIUM PENTOXIDE / PTCR (POSITIVE TEMPERATURE COEFFICIENT OF RESISTANCE) BEHAVIOR OF BARIUM 04-0837
TE. POSITIVE TEMPERATURE COEFFICIENT OF RESISTANCE IN DOPED LEAD TITANA 05-1051
INGLE CRYSTALS IN THE REGION OF THE CU/ POSITIVE TEMPERATURE COEFFICIENT OF RESISTANCE OF BARIUM TITANATE S 04-1036
ALY IN SEMICONDUCTING BARIUM TITANATE. (POSITIVE TEMPERATURE COEFFICIENT OF RESISTIVITY) RESISTIVITY ANOM 04-0584

/ASUREMENTS. (INTRINSIC BAND GAP, PTCR (POSITIVE TEMPERATURE COEFFICIENT OF RESISTIVITY) DEVICES) PART-2: / 04-0822
PED CERAMIC PEROVSKITE) LE/ PTC EFFECT (POSITIVE TEMPERATURE COEFFICIENT OF RESISTIVITY) IN (DOPED AND UNDO 13-2000
ANTALATE LENGTH EXPANDER BARS.   ZERO TEMPERATURE COEFFICIENT OF RESONANT FREQUENCY IN LITHIUM T 11-1810
NATE)   MEASUREMENT OF THE PYROELECTRIC COEFFICIENT OF 65-35 PZT. (LEAD ZIRCONATE TITA 09-1567
DUCTING PROPERTIES OF BARIUM TITANATE. (RESISTIVITY, SEEBECK COEFFICIENT, OPTICAL ABSORPTION) SOME SEMICON 04-0919
N LEAD ZINCATE(0.33) NIOBATE(0.67). (DISPERSION OF QUADRATIC COEFFICIENTS)   ELECTROOPTICAL EFFECT I 13-1982
OOPTIC CRYSTALS. (NONLINEAR OPTICAL PROPERTIES, ELECTROOPTIC COEFFICIENTS) / LUMINOUS WAVES IN CUBIC ELECTR 02-0044
/N, REFRACTIVE INDEX, DIELECTRIC CONSTANT, NONLINEAR OPTICAL COEFFICIENTS, AND LINEAR ELECTROOPTIC COEFFIC/ 13-2024
CTRIC CONSTANTS, SPONTANEOUS POLARIZATION, THERMAL EXPANSION COEFFICIENTS, AND REFRACTIVE INDEX) / PIEZOELE 12-1876
HIN SAMPLES (STRONTIUM TI/ DETERMINATION OF X-RAY ABSORPTION COEFFICIENTS AND THE THICKNESS OF NONUNIFORM T 06-1179
                                            SIGNS OF THE ELECTROOPTIC COEFFICIENTS FOR LITHIUM TANTALATE.   12-1880
             FERROELECTRIC FREE ENERGY EXPANSION COEFFICIENTS FROM DOUBLE HYSTERESIS LOOPS.   03-0262
CLAMPED VALUES OF THE DIELECTRIC CONSTANTS AND ELECTROOPTIC COEFFICIENTS IN BARIUM SODIUM NIOBATE. /EE AND 13-1998
TEMPERATURE DEPENDENCE OF NONLINEAR OPTICAL COEFFICIENTS IN FERROELECTRIC SODIUM NITRITE.   16-2274
INTERFEROMETRIC DETERMINATION OF THE QUADRATIC ELECTROOPTIC COEFFICIENTS IN STRONTIUM TITANATE (SINGLE) C/ 06-1130
EG / ELECTRONIC TRANSPORT (CONDUCTIVITY AND HALL AND SEEBECK COEFFICIENTS) IN STRONTIUM TITANATE (4.2-300 D 06-1129
AND DELAY ON LITHIUM NIOBATE, LITHIUM TANTALAT/ TEMPERATURE COEFFICIENTS OF ACOUSTIC SURFACE WAVE VELOCITY 11-1821
                 DEPENDENCE OF THE ELASTIC COEFFICIENTS OF ADP ON THE ELECTRIC FIELD.   17-2537
UTERATION.   1.06 MICRON ABSORPTION COEFFICIENTS OF DEUTERATED KDP WITH 70-100% DE 17-2435
/CY RANGE WITH LOW RATIOS OF CAPACITANCE AND LOW TEMPERATURE COEFFICIENTS OF FREQUENCY (DEVICE DESIGN AND / 11-1724
ITION.   DEPENDENCE OF SECOND HARMONIC GENERATION COEFFICIENTS OF LITHIUM NIOBATE ON MELT COMPOS 11-1769
BATE BY PHASE MATCHING TE/ DETERMINATION OF THE DISTRIBUTION COEFFICIENTS OF LITHIUM OXIDE INTO LITHIUM NIO 11-1841
             ABSOLUTE SIGNS OF SECOND HARMONIC GENERATION COEFFICIENTS OF PIEZOELECTRIC CRYSTALS.   02-0096
/EASUREMENTS OF ABSOLUTE SIGNS OF SECOND HARMONIC GENERATION COEFFICIENTS OF PIEZOELECTRIC CRYSTALS. (LEAD/ 05-1066
E OR TANTALATE, BARIUM / RELATIVE SIGNS OF NONLINEAR OPTICAL COEFFICIENTS OF POLAR CRYSTALS (LITHIUM NIOBAT 02-0098
/UTE SIGNS OF NONLINEAR OPTICAL (SECOND HARMONIC GENERATION) COEFFICIENTS OF POLAR (PYROELECTRIC) CRYSTALS/ 02-0095
, RUBIDIUM DIHYDROGEN/ MEASUREMENT OF THE RELATIVE NONLINEAR COEFFICIENTS OF POTASSIUM DIHYDROGEN PHOSPHATE 08-1413
.   ABSOLUTE SIGNS OF THE NONLINEAR OPTICAL COEFFICIENTS OF POTASSIUM DIHYDROGEN PHOSPHATE 17-2597
ING BARIUM AND STRONTIUM TITANATES WITH POSITIVE TEMPERATURE COEFFICIENTS OF RESISTIVITY.   SEMICONDUCT 04-0995
ULFATE.   NEW MEASUREMENTS OF DIVERSE THERMODYNAMICAL COEFFICIENTS OF ROCHELLE SALT AND TRIGLYCINE S 27-2995
E/ PRECISION DETERMINATION OF THE LATTICE PARAMETERS AND THE COEFFICIENTS OF THERMAL EXPANSION OF BISMUTH F 08-1287
                       SURFACE WAVE COEFFICIENTS ON LITHIUM TANTALATE.   11-1852
ND POLARIZATION) OF LEAD CONTAINING OXYGEN O/ ELECTROOPTIC G COEFFICIENTS (RELATING INDUCED BIREFRINGENCE A 13-2039
         NEW METHOD OF MEASURING PYROELECTRIC COEFFICIENTS. (TGS EXAMPLE)   17-2517
EFFECTS IN TRIGLYCINE SULFATE. (DIELECTRIC DISPLACEMENT AND COERCIVE FIELD)   (POLARIZATION CHANGE) AFTER 27-2952
IHYDROGEN PHOSPHATE CRYSTAL IN THE PHASE TRANSITION REGION. (COERCIVE FIELD)   POLARIZATION OF A RUBIDIUM D 17-2504
TE. (EXPERIMENTAL)   COERCIVE FIELD IN LITHIUM TRIHYDROGEN DISELENI 22-2844
ALS ON THEIR THICKNESS.   DEPENDENCE OF COERCIVE FIELD OF BARIUM TITANATE SINGLE CRYST 04-0693
/IELECTRIC BEHAVIOR (PERMITTIVITY, SPONTANEOUS POLARIZATION, COERCIVE FIELD) OF BISMUTH TITANATE AT LOW TE/ 08-1436
NGLE CRYSTALS).   SPONTANEOUS POLARIZATION AND COERCIVE FIELD OF LEAD TITANATE (FLUX GROWN SI 05-1050
          FREQUENCY DEPENDENCE OF THE COERCIVE FIELD OF TRIGLYCINE SULFATE CRYSTALS. 27-3062
NCY, AMPLITUDE AND TEMPERATURE.   DEPENDENCE OF THE COERCIVE FIELD OF TRIGLYCINE SULFATE ON FREQUE 27-3092
IATED WITH GAMMA-RAYS.   COERCIVE FIELDS OF ROCHELLE SALT HEAVILY IRRAD 28-3178
M LEAD ZI/ THERMAL VARIATION OF SPONTANEOUS POLARIZATION AND COERCIVE FORCE OF THE PEROVSKITES IN THE SYSTE 09-1597
KY WALL MOTION IN ROCHELLE SALT.   ANOMALOUS DISPERSION OF COERSIVE FIELD AT VERY LOW FREQUENCIES AND JER 28-3177
ANTIMONY SULFOIODIDE SINGLE CRYSTALS IN THE REGION OF PHASE COEXISTENCE.   POLARIZATION REVERSAL OF 15-2246
IELECTRIC PROPERTIES IN THE BISMUTH- LANTHANUM FERRATE SYST/ COEXISTENCE OF ANTIFERROMAGNETIC AND SPECIAL D 08-1425
STATE IN BISMUTH TITANATE.   COEXISTENCE OF THE ANTIPOLARIZED AND POLARIZED 08-1422
ZATION AND MAGNETIZATION IN NICKEL IODIDE BORACITE (CURIE P/ COEXISTENCE OF THE SPONTANEOUS ELECTRIC POLARI 14-2090
AVELENGTH MODULATED ABSORPTION IN STRONTIUM TITANATE WITHOUT COEXISTING PHASES.   REINTERPRETATION OF W 06-1188
LTER)   OPTICAL MIXING OF COHERENT AND INCOHERENT LIGHT. (KDP CRYSTAL FI 17-2669
AL.   COMBINATION OF THE FREQUENCIES OF COHERENT AND INCOHERENT RADIATION IN KDP CRYST 17-2500
THEORY. PART-2: / ELECTRON MICROSCOPE TRANSMISSION IMAGES OF COHERENT DOMAIN BOUNDARIES. PART-1: DYNAMICAL 03-0255
OTASSIUM DIHYDROGEN PHOS/ PARAMETRIC FREQUENCY CONVERSION OF COHERENT LIGHT BY THE ELECTROOPTIC EFFECT IN P 17-2616
/APHIC RECONSTRUCTION BY ELECTROOPTIC MODULATION (BY PASSING COHERENT LIGHT THROUGH DEUTERATED KDP HAVING / 17-2507
FERROELECTRIC PHASE TRANSITION OF SODIUM NITRITE (STUDIED BY COHERENT NEUTRON INELASTIC SCATTERING). / THE 16-2364
ANESE TRIFLUORIDE. (20 PERCENT MANGANESE REPLACED BY COBALT, COHERENT NEUTRON SCATTERING) /D POTASSIUM MANG 33-3366
FERROELECTRICS AT LOW TEMPERATURES. PART-1: GENE/ THEORY OF COHERENT NEUTRON SCATTERING BY HYDROGEN BONDED 03-0478
TRANSITION IN RUBIDIUM DIHYDROGEN PHOSPHATE BY SCATTERING OF COLD NEUTRONS.   HIGH TEMPERATURE PHASE 17-2489
THE FERROELECTRIC TRANSITION IN AMMONIUM SULFATE STUDIED BY COLD NEUTRONS (SCATTERING). /OTIONS IN SOLIDS: 19-2752
ALKOXY-DERIVED PLZT AND PLZT.   COLD PRESSING AND LOW TEMPERATURE SINTERING OF 09-1493
                      BIAS FIELDS IN FERROELECTRIC COLEMANITE.   23-2886
                      PROTON DYNAMICS IN FERROELECTRIC COLEMANITE.   23-2887
        PROTON SPIN LATTICE RELAXATION IN FERROELECTRIC COLEMANITE.   23-2885
          SWITCHING PROPERTIES OF BIASED FERROELECTRIC COLEMANITE.   23-2888
OMIUM ALUMS, METHYL AMMONIUM ALUMINUM SULFATE DODECAHYDRATE, COLEMANITE, SODIUM TRIDEUTERIUM DISELENITE) /R 02-0087
NONLINEAR OPTICAL MATERIALS (REVIEW, DATA COLLECTION, MATERIALS NEEDS, 118 REFS)   02-0049
E OF DIELECTRIC DISPERSION IN THE FERROELECTRIC P/ EFFECT OF COLLECTIVE INTERACTION OF DIPOLES ON THE NATUR 03-0417
/TRICS. PART-1: DAMPINGS AND (FREQUENCY) SHIFTS (OF PROTONIC COLLECTIVE MOTION, TUNNELING, CALCULATIONS, I/ 03-0475
TIONS FOR POTASSIUM DIHYDROGEN PHOSPHATE TYPE CRYSTALS) COLLECTIVE MOTIONS OF HYDROGEN BONDS. (CALCULA 17-2483
ERIALS (INCLUDING FERROELEC/ STUDY OF OPTICAL PROPERTIES AND COLLECTIVE OSCILLATIONS IN NEW SOLID STATE MAT 02-0104
/ THE FERROELECTRIC MODE IN POTASSIUM DIHYDROGEN PHOSPHATE. (COLLECTIVE PSEUDOSPIN MODEL OF KDP IS QUESTIO/ 17-2655
OELECTRICS. (REPRESENTATION THEORY APPLIED, 18 REFS) COLLINEAR AND NONCOLLINEAR FERRO- AND ANTIFERRO 03-0173
COUSTIC WAVE AND TWO LUMINOUS WAVES ALONG THE (100) DIRECTI/ COLLINEAR INTERACTION BETWEEN A LONGITUDINAL A 11-1680
WAVES IN LITHIUM NIOBATE CRYSTALS.   COLLINEAR INTERACTION OF LONGITUDINAL ELASTIC 11-1678
HEXAGONAL ICE.   STRONG COLLISION LIMIT OF SPIN LATTICE RELAXATION IN 34-3480
KDOWN MECHANISM IN BARIUM TITANATE AND COLORATION. (INJECTED COLOR CENTERS)   RELATIONSHIP BETWEEN THE BREA 04-0944
                   FORMATION OF COLOR CENTERS IN IRRADIATED ICE.   34-3487
                MOSSBAUER STUDY OF COLOR CENTERS IN STRONTIUM TITANATE.   06-1153
                       COLOR IMAGING WITH FERROELECTRIC CERAMICS.   37-3662
N AND APPLICATIONS OF GADOLINIUM MOLYBDATE (OPTICAL SHUTTER, COLOR MODULATOR).   (PROPERTIES,) FUNCTIO 30-3298
CONDUCTIVITY.   THE COLOR PROBLEM OF TUNGSTEN TRIOXIDE ELECTRICAL 10-1636
IES.   THE COLOR PROBLEM OF TUNGSTEN TRIOXIDE. X-RAY STUD 10-1632
CE WAVE TRANSDUCERS ON LEAD ZIRCONA/ REALIZATION OF A 40 MHZ COLOR TELEVISION INFRARED RESPONSE USING SURFA 09-1509
NSHIP BETWEEN THE BREAKDOWN MECHANISM IN BARIUM TITANATE AND COLORATION. (INJECTED COLOR CENTERS)   RELATIO 04-0944
UCTIVITY (TRANSIENTS) IN PHOTOCHROMIC STRONTIUM TITANATE (ON COLORING AND BLEACHING).   PHOTOCOND 06-1240
LOUS SPECIFIC HEAT OF SODIUM TRIHYDROGEN SELENITE ASSOCIATED COMBINATORIAL PROBLEM.   ANOMA 22-2822
.   FRACTURE TOUGHNESS OF COMMERCIAL BARIUM TITANATE TRANSDUCER MATERIAL 04-0841
OF) THEIR PROPERTIES.   ADVANCES IN PRODUCTION OF (COMMERCIAL) PIEZOELECTRIC CERAMICS AND (TABLES 02-0110
AD LANTHANATE ZIRCONATE CERAMICS FOR APPLICATIONS IN OPTICAL COMMUNICATIONS.   EVALUATION OF LE 07-1254

C MODUL/ (HIGH SPEED, SMALL APERTURE) MODULATORS FOR OPTICAL COMMUNICATIONS (REVIEW EMPHASIZING ELECTROOPTI 02-0056
. (LITHIUM TANTALATE)   NEW CRYSTAL FOR COMMUNICATIONS SYSTEMS ENTERS PRODUCTION STAGE 12-1867
L DEVIA/ STUDY OF THE LIMITATION OF THE ANGULAR FIELD BY THE COMMUTATOR OF POLARIZATION IN KDP, IN A DIGITA 17-2524
/NITE APERTURE LIGHT SWITCH USING) TRANSVERSE POCKELS EFFECT COMMUTATORS. (POTASSIUM DIHYDROGEN PHOSPHATE)   17-2509
FFECT OF OXIDE ADDITIONS ON THE SINTERING OF BARIUM TITANATE COMPACT   E 04-0996
FTERS FOR VHF AND UHF. (BASED ON LEAD(0.35) STRONTIUM(0.65/ (COMPACT, SURFACE WAVE) FERROELECTRIC PHASE SHI 37-3633
OF FERROELECTRICS.   INFLUENCE OF COMPENSATING CHARGES ON THE C-DOMAIN STRUCTURE 03-0458
DOPED BARIUM TITANATE AND TITANIUM DIOXIDE. (STOICHIOMETRIC COMPENSATION)   OXYGEN STOICHIOMETRY OF DONOR 04-0664
RON(3+) IN STRONTIUM TITANATE DUE TO NEAREST NEIGHBOR CHARGE COMPENSATION. /AGNETIC RESONANCE SPECTRUM OF I 06-1144
FERROELECTRIC HYSTERESIS TRACER FEATURING COMPENSATION AND SAMPLE GROUNDING.   37-3668
TOR FREQUENCIES: APPLICATION TO QUAR/ WIDE RANGE TEMPERATURE COMPENSATION BY ADDITION OF TWO CRYSTAL RESONA 11-1686
UM(3+) DOPED STRONTIUM TITANATE (ESR MEASUREMENTS, M/ CHARGE COMPENSATION BY OXYGEN(2-) VACANCIES IN CHROMI 06-1161
E LATTICE OF BARIUM TITANATE SINGLE CRYSTALS.   COMPENSATION OF HETEROVALENT FERRIC IONS IN TH 04-0539
S.   ACTIVE COMPONENTS FOR FERROELECTRIC OPTICAL CIRCUIT   37-3675
/0). (WHERE B IS (NIOBIUM, TITANIUM) OR (NIOBIUM, TUNGSTEN), COMPILATION, LATTICE CONSTANTS, CURIE TEMPERA/ 13-1964
NG THE LINEAR ELASTIC AND ELECTROELASTIC EQUATIONS OF STATE/ COMPILATION OF CONVERSION FORMULAS INTERRELATI 03-0277
PIEZOCERAMICS BASED ON LEAD ZIRCONATE TITANATE WITH COMPLEX ADDITIVES CONTAINING GERMANIUM.   09-1510
PACITANCE, PERMITTIVITY)   AGING IN COMPLEX CERAMICS BASED ON BARIUM TITANATE. (CA 04-0724
TRUCT/ GROWTH OF SINGLE CRYSTALS OF VARIOUS PEROVSKITES WITH COMPLEX COMPOSITIONS, AND THE STUDY OF THEIR S 08-1428
ATE(0.5) AND LEAD MANGANATE(0.33) N/ SYNTHESIS OF PEROVSKITE COMPLEX COMPOSITIONS: LEAD MANGANATE(0.5) NIOB 08-1453
CTRIC MASD, METHYL AMMONIUM ALUMINUM SULFATE DODECAHYDRATE. (COMPLEX DIELECTRIC CONSTANT) /TION IN FERROELE 20-2804
NEAR THE CURIE/ FREQUENCY AND TEMPERATURE DEPENDENCES OF THE COMPLEX DIELECTRIC CONSTANT OF FERROELECTRICS 03-0361
20-520 DEG K.   COMPLEX DIELECTRIC CONSTANT OF LEAD HAFNATE, 3 08-1339
UNDER POLARIZING FIELD, 370-520 DEG K.   COMPLEX DIELECTRIC CONSTANT OF LEAD ZIRCONATE   07-1256
T 10 GHZ AS A FUNCTION OF TEMPERATURE.   MEASUREMENT OF THE COMPLEX DIELECTRIC CONSTANT OF ROCHELLE SALT A 28-3184
WALLS IN TRIGLYCINE SULFATE. (TEMPERATURE DEPENDENCES OF THE COMPLEX DIELECTRIC CONSTANTS) /TION OF DOMAIN 27-3103
GLE CRYSTALS.   PRESSURE DEPENDENCE OF THE COMPLEX DISPERSION COEFFICIENT OF ICE I(H) SIN 34-3479
EPR OF THE CHROMIUM COMPLEX IN TRIGLYCINE FLUOBERYLLATE.   27-3133
E FOR FERROELECTRICS AND ANTIFERROELECTRICS BARIUM TITANATE, COMPLEX LAYER OXIDES) /NT AND CURIE TEMPERATUR 03-0188
, STANNATES, TITANATES, CHROMATES, COBA/ FLUX GROWTH OF SOME COMPLEX OXIDE MATERIALS. (NIOBATES, TANTALATES 02-0129
TUNGSTEN BRONZE TYPE COMPLEX OXIDES.   13-1981
TURES (POTASSIUM (BARIUM OR STRONTIUM) TITANATE NIOBATE, (B/ COMPLEX OXIDES WITH TUNGSTEN BRONZE TYPE STRUC 13-1962
ME FERROELECTRIC CRYSTALS (TRIGLYCINE SULFA/ BEHAVIOR OF THE COMPLEX PERMITTIVITY (AUDIO FREQUENCIES) OF SO 02-0125
TRIC CRYSTAL. (DIELECTRIC HEATING AND POLARIZATION REVERSAL, COMPLEX PERMITTIVITY CALCULATION) /F FERROELEC 03-0245
INFRARED BY DISPERSIVE FOURIER TRANSF/ DETERMINATION OF THE COMPLEX PERMITTIVITY OF ADP AND KDP IN THE FAR 17-2609
20 AND 1000 DEG C. (BARIUM T/ TEMPERATURE DEPENDENCE OF THE COMPLEX PERMITTIVITY OF SOME TITANATES BETWEEN 04-0820
TITANATE ZIRCONATES, 2 SUBSTITUTED LEAD ZIRCONATE STANNATES, COMPLEX PEROVSKITES) /CS. (8 SUBSTITUTED LEAD 08-1418
ON MOLYBDATE.   PHASE TRANSITIONS IN COMPLEX PEROVSKITES OF THE TYPE BARIUM LANTHAN 30-3261
TION TO SINGLE DOMAIN FERROELECTRICS BELOW CURIE POINT).   COMPLEX PHASE TRANSITIONS IN CRYSTALS (APPLICA 03-0319
IC CRYSTALS. (ROCHELLE SALT)   MEASUREMENTS OF COMPLEX PIEZOELECTRIC CONSTANTS IN FERROELECTR 28-3213
ATE MONOCRYSTAL.   EPR OF THE FOUR COPPER ION COMPLEX WITH SPIN S=2 IN TRIGLYCINE FLUOBERYLL 27-3134
SULFATE HEXAHYDRATE.   EPR STUDY OF MAGNETIC COPPER(2+) COMPLEXES IN CRYSTALS OF GUANIDINIUM ALUMINUM 21-2820
EPR STUDY OF PARAMAGNETIC ION COMPLEXES IN FERROELECTRIC CRYSTALS.   03-0477
IUM CHLORIDE (TSCC) SIN/ EPR INVESTIGATIONS OF MANGANESE(2+) COMPLEXES IN FERROELECTRIC TRIS-SARCOSINE CALC 36-3525
CHROMIUM(3+), COPPER(2+), AND MANGANESE(2+) PARAMAGNETIC ION COMPLEXES IN GLYCINE CRYSTALS. EPR STUDIES OF 27-3135
G AND DELAY DEVICES).   ELECTRIC FIELD CONTROL OF THE ELASTIC COMPLIANCE IN LITHIUM THALLIUM TARTRATE (TUNIN 29-3250
/ROELASTICS SUCH THAT THE ELECTRIC SUSCEPTIBILITY OR ELASTIC COMPLIANCE IS TEMPERATURE INDEPENDENT IN THE / 03-0140
/ THALLIUM TARTRATE MONOHYDRATE, ELECTRIC CONTROL OF ELASTIC COMPLIANCE, NICKEL IODINE BORACITE OPTICAL AN/ 02-0060
ARTRATE HYDRATE (LTT) BY AN ELECTRIC/ CONTROL OF THE ELASTIC COMPLIANCE OF FERROELECTRIC LITHIUM THALLIUM T 29-3249
SITION TEMPERATURE.   PRESSURE DEPENDENCE OF THE ELASTIC COMPLIANCE OF STRONTIUM TITANATE NEAR THE TRAN 06-1219
LS IN THE TEMPERATURE/ TEMPERATURE DEPENDENCE OF THE ELASTIC COMPLIANCE OF STRONTIUM TITANATE SINGLE CRYSTA 06-1220
IN VANADIUM DIOXIDE. (ANTIFERROELECTRIC TEMPERATURE 340/ TWO COMPONENTS OF THE CRYSTALLOGRAPHIC TRANSITION 36-3562
ERIAL. (ALUMINA WHISKERS IN LEAD ZIRCONATE TITANATE)   NEW COMPOSITE CERAMIC PIEZOELECTRIC TRANSDUCER MAT 09-1556
CE ON LANTHANUM(0.002) BARIUM(0.998) TITANATE- BISMUTH OXIDE COMPOSITE CERAMICS. /LOUS INCREASE OF RESISTAN 04-0801
/TIVITY ON LANTHANUM DOPED BARIUM- BISMUTH TRIOXIDE COMPOSITE CERAMICS WITH SURFACE BARRIER LAYER.   04-0800
ICTION IN SINGLE CRYSTALS OF FERROELECTRIC SUBSTANCES BY THE COMPOSITE RESONATOR METHOD. (TGS) /INTERNAL FR 27-3076
HARMONIC GENERATION COEFFICIENTS OF LITHIUM NIOBATE ON MELT COMPOSITION.   DEPENDENCE OF SECOND 11-1769
YPE NIOBATES (CORRELATION OF CURIE TEMPERATURE WITH CHEMICAL COMPOSITION). /MPERATURES OF TUNGSTEN BRONZE T 13-2049
LE CRYSTALS OF LEAD (ZINCATE NIOBATE) AND PROPERTIES OF THIS COMPOSITION.   GROWTH OF SING 13-1990
IC CRYSTALS, ESPECIALLY ON LITHIUM NIOBATE AND LIT/ CHEMICAL COMPOSITION AND OPTICAL QUALITY OF FERROELECTR 11-1780
LID SOLUTIONS OF THE PE/ INVESTIGATING THE RELATIONS BETWEEN COMPOSITION AND PROPERTIES IN FERROELECTRIC SO 08-1330
F FERROELECTRIC NIOBATES.   COMPOSITION AND STRUCTURE OF SPUTTERED FILMS O 13-1941
S OF LITHIUM TANTALATE SINGLE CRYSTALS AS A FUNCTION OF MELT COMPOSITION. (CURIE TEMPERATURE DEPENDENCE) /E 12-1874
TRANSFORMATION IN FERROELECTRIC SOLID SOLUTIONS.   COMPOSITION DEPENDENCE OF PHASE TRANSITION) /E 03-0250
LITHIUM TANTALATE SINGLE CRYSTAL STOICHIOMETRY (VERSUS MELT COMPOSITION, LATTICE AND CURIE TEMPERATURE ME/ 12-1871
N GRO/ EFFECT OF CRYSTALLIZATION CONDITIONS ON THE SHAPE AND COMPOSITION OF FERROELECTRIC CRYSTALS (SOLUTIO 02-0089
D OF INVESTIGATION AND SOME RESULTS.   PHASE COMPOSITION OF (LEAD, BARIUM) DINIOBATE. METHO 13-1945
LUTIONS.   EFFECT OF ELECTRIC FIELD ON PHASE COMPOSITION OF (LEAD, BARIUM) NIOBATE SOLID SO 13-1943
CONGRUENT MELTING COMPOSITION OF LITHIUM METATANTALATE.   12-1882
ITANATE.   ON THE PHASE COMPOSITION OF REDUCED AND REOXIDIZED BARIUM T 04-0541
NIOBATE.   EFFECTS OF CHANGES IN MELT COMPOSITION ON CRYSTAL GROWTH OF BARIUM SODIUM 13-1908
ECTRIC PROPERTIES OF ANTIMONY SULFOIODIDE CRYSTAL/ EFFECT OF COMPOSITION ON GROWTH, DIELECTRIC, AND PHOTOEL 15-2118
LIED TO ADP, TUNGSTEN TRIOXIDE, SODIUM NIOBA/ ORIENTATION OF COMPOSITION PLANES IN ANTIFERROELECTRICS. (APP 08-1327
ELS EFFECT IN LITHIUM NIOBATE CRYSTALS.   MELT COMPOSITION (STOICHIOMETRY) DEPENDENCE OF POCK 11-1728
/ LINEAR ELECTROOPTIC EFFECT AND DIELECTRIC CONSTANT ON MELT COMPOSITION (STOICHIOMETRY) IN LITHIUM NIOBAT/ 11-1842
/WTH OF GOOD OPTICAL QUALITY BARIUM SODIUM NIOBATE (BY USING COMPOSITION WHERE LIQUIDUS AND SOLIDUS COINCI/ 13-2047
ODIFIED LEAD ZIRCONATE- LEAD TITANATE CERAMICS. (PROPRIETARY COMPOSITIONS)   INTERNAL FRICTION OF M 09-1623
CADMIUM MANGANATE SYSTEM (BOTH CURIE POINTS STUDIED, VARIOUS COMPOSITIONS) / IN THE LEAD COBALT TUNGSTATE- 36-3604
/OWTH OF SINGLE CRYSTALS OF VARIOUS PEROVSKITES WITH COMPLEX COMPOSITIONS, AND THE STUDY OF THEIR STRUCTUR/ 08-1428
LEAD ZIRCONATE- LEAD TITANATE MATERIALS. PART-2: HOT PRESSED COMPOSITIONS AT NEAR THEORETICAL DENSITY. /OF 09-1578
AND LEAD MANGANATE(0.33) N/ SYNTHESIS OF PEROVSKITE COMPLEX COMPOSITIONS: LEAD MANGANATE(0.5) NIOBATE(0.5) 08-1453
/N THE PROPERTIES OF POLYCRYSTALLINE LEAD TITANATE ZIRCONATE COMPOSITIONS NEAR THE MORPHOTROPIC PHASE TRAN/ 09-1557
- FERROELASTIC AND PARAELECTRIC- PARAELASTIC PHASES OF GADO/ COMPREHENSIVE X-RAY STUDY OF THE FERROELECTRIC 30-3287
ONTIUM TITANATE.   HIGH PRESSURE COMPRESSIBILITY AND GRUNEISEN PARAMETER OF STR 06-1121
ED BARIUM TITANATE CERAMICS SUBJECTED TO UNIAXIAL TRANSVERSE COMPRESSION. /ZOELECTRIC PROPERTIES OF POLARIZ 04-0962
TITANATE SINGLE CRYSTALS.   EFFECT OF TRANSVERSE COMPRESSION OF DIELECTRIC PROPERTIES OF BARIUM 04-0956
F BARIUM TITANATE SINGLE CRYSTALS.   EFFECT OF LINEAR COMPRESSION ON THE UHF DIELECTRIC PROPERTIES O 04-0896
RIUM HALIDE CRYSTALS. (RAMAN SPECTRA, DIPOLAR COUPLING MODEL COMPUTATION) /INAL MODES IN HYDROGEN AND DEUTE 31-3326
ASYMMETRIC MINOR LOOP AND TRANSIENT OPEN LOOP COMPUTATIONS FOR FERROELECTRICS.   03-0296
IUM DIDEUTERIUM PHOSPHATE. ELECTROOPTIC Q SWITCH BY DESK-TOP COMPUTER. /F A DOUBLE 45 DEGREES, Z CUT POTASS 17-2702

YING IN DOUBLE MINIMUM POTENTIAL. (BARIUM TITANATE) COMPUTER EXPERIMENT ON LINEAR CHAIN OF ATOMS L 03-0287
LEAD TITANATE AND PLZT) USE OF FERROELECTRIC CERAMICS IN COMPUTERS AND OTHER DEVICES. (LEAD ZIRCONATE- 09-1532
MEASUREMENTS ON PZT FERROELECTRICS. OXYGEN CONCENTRATION CELL AND ELECTRICAL CONDUCTIVITY 09-1495
PORT NUMBERS IN PZT (LEAD ZIRCONATE TITANATE) FERROE/ OXYGEN CONCENTRATION CELL MEASUREMENTS OF IONIC TRANS 09-1512
Y MAGNONS IN NICKEL DOPED RUBIDIUM MANGANATE AND P/ IMPURITY CONCENTRATION DEPENDENCE OF RAMAN SCATTERING B 33-3395
PROPERTIES OF SOLID SOLUTIONS OF TRIGLYCINE SULFATE WITH I/ CONCENTRATION DEPENDENCE OF SOME FERROELECTRIC 27-2971
NSTANT IN POLYCRYSTALLINE SOLID SOLUTIONS IN THE (BARIUM, S/ CONCENTRATION DEPENDENCE OF THE CURIE-WEISS CO 04-0586
STRUCTURE OF DEUTERATED KDP CRYSTAL: DEUTERIUM CONCENTRATION DETERMINATION. 17-2535
CONSTANT OF FERROELECTRIC SOLID SOLUTIONS WITH ALLOWANCE FOR CONCENTRATION FLUCTUATIONS. DIELECTRIC 03-0440
FRICTION IN POLYCRYSTALLINE LEAD ZIRCONAT/ INFLUENCE OF THE CONCENTRATION OF POINT DEFECTS ON THE INTERNAL 09-1582
/UM TRIDEUTERIUM(X) TRIHYDROGEN(1-X) PHOSPHATE SYSTEM VERSUS CONCENTRATION, TEMPERATURE, FIELD, AND PRESSU/ 17-2666
ON IN SINGLE CRYSTAL POTASSIUM TANTALATE. (CONDUCTI/ CALCIUM CONCENTRATION VS NET IONIZED DONOR CONCENTRATI 08-1437
/. (KDP TYPE) PART-2: SCATTERING CHARACTERISTICS AND VARIOUS CONCEPTIONS OF THE TUNNELING QUASI SPIN MODEL. 03-0478
FERROELECTRIC DOMAINS AND THE CONDENSED SOFT MODE IN GADOLINIUM MOLYBDATE. 30-3260
CONTRIBUTION TO INJECTION AND CONDUCTION IN BARIUM TITANATE (CONDENSER CHARGE- DISCHARGE METHOD). IONIC 04-0571
NOISE AND IMPEDANCE MEASUREMENTS OF A FERROELECTRIC CONDENSER UNDER A DC FIELD. 37-3645
E CRYST/ IMPEDANCE MEASUREMENT OF A DIELECTRIC FERROELECTRIC CONDENSER UNDER CONTINUOUS HEATING. (TGS SINGL 27-3028
DIHYDROGEN PHOSPHAT/ TRANSIENT PHENOMENA IN THE DISCHARGE OF CONDENSERS WITH POLAR DIELECTRICS. (POTASSIUM 17-2631
SMALL POLARONS IN CONDUCTING BARIUM TITANATE CRYSTALS. 04-0613
A OF ICE I(H). PART-2: MOLECULAR MODELS FOR POLARIZATION AND CONDUCTION. /TALS. PART-1: POLARIZATION SPECTR 34-3486
TIONS IN CESIUM LEAD TRICHLORIDE. ELECTRICAL CONDUCTION AND (NONFERROELECTRIC) PHASE TRANSI 33-3365
T LOW TEMPERATURES. CONDUCTION ANOMALIES AND POLARIZATION IN ICE A 34-3428
THE 110 DEGREE K PHASE TRANSITION ON THE STRONTIUM TITANATE CONDUCTION BANDS. EFFECT OF 06-1158
DINAL OPTICAL PHONONS IN IONIC FERROELECTRIC / SCATTERING OF CONDUCTION ELECTRONS BY TRANSVERSE AND LONGITU 03-0292
TIC COPPER FORMATE TETRAHYDRATE. THERMAL CONDUCTION IN A TWO DIMENSIONAL ANTIFERROMAGNE 26-2932
E- DISCHARGE METHOD). IONIC CONTRIBUTION TO INJECTION AND CONDUCTION IN BARIUM TITANATE (CONDENSER CHARG 04-0571
TITANATE, LITHIUM THALLIUM TARTRATE M/ FERROELECTRICITY AND CONDUCTION IN FERROELECTRIC CRYSTALS. (BISMUTH 02-0061
THALLIUM TARTRATE MONOHYDRATE, ELECTR/ FERROELECTRICITY AND CONDUCTION IN FERROELECTRIC CRYSTALS. (LITHIUM 02-0060
ELECTRICAL CONDUCTION IN ICE. 34-3415
E CRYSTALS. KERR EFFECT STUDIES OF ELECTRIC CONDUCTION IN IRON DOPED BARIUM TITANATE SINGL 04-0703
ELECTRICAL CONDUCTION IN METAL OXIDES. (REVIEW, 319 REFS) 02-0109
TION AND ITS APPLICATION TO THE SOLID ELECTROLYTE FUE/ IONIC CONDUCTION IN PEROVSKITE TYPE OXIDE SOLID SOLU 08-1444
PROTONIC CONDUCTION IN POTASSIUM DIHYDROGEN ARSENATE. 18-2737
TITANATE OR STRONTIUM TITANATE. IONIC CONDUCTION IN SINTERED OXIDES BASED ON CALCIUM 06-1141
ATE AT LOW TEMPERATURES. ELECTRONIC CONDUCTION IN SLIGHTLY REDUCED STRONTIUM TITAN 06-1151
RT-1: DEFORMATION AND PHASE TRANSFORMATION. PART-2: ELECTRIC CONDUCTION IN THE INDUCTION REGION. /AMICS. PA 03-0169
TS) TRANSPORT PHENOMENA IN ANTIMONY SULFOIODIDE. (CONDUCTION MECHANISM FOR THERMOELECTRIC CURREN 15-2241
GION. PART-2: LOW TEMPERATURE REGION. PROTONIC CONDUCTION OF ICE. PART-1: HIGH TEMPERATURE RE 34-3414
/GLE CRYSTALS. PART-2: MOLECULAR MODELS FOR POLARIZATION AND CONDUCTION. PART-3: EXTRINSIC VERSUS INTRINSI/ 34-3455
/ OF ICE I(H). SECT-2: MOLECULAR MODELS FOR POLARIZATION AND CONDUCTION. SECT-3: EXTRINSIC VERSUS INTRINSI/ 34-3482
GNETS POTASSIUM COPPER TRIFLUOR/ MEASUREMENTS OF THE THERMAL CONDUCTIONS IN THE LOW DIMENSIONAL ANTIFERROMA 26-2939
DONOR CONCENTRATION IN SINGLE CRYSTAL POTASSIUM TANTALATE. (CONDUCTIVE, KYROPOULOS GROWN) / VS NET IONIZED 08-1437
/AN SCATTERING. PART-5: MOSSBAUER EFFECT. PART-6: ELECTRICAL CONDUCTIVE, SURFACE, AND ELECTROOPTIC PROPERT/ 03-0397
URITY POLYCRYSTALLINE BARIUM TITANATE. (DIELECTRIC CONSTANT, CONDUCTIVITY) ELECTRICAL PROPERTIES OF HIGH P 04-0704
THE COLOR PROBLEM OF TUNGSTEN TRIOXIDE ELECTRICAL CONDUCTIVITY. 10-1636
SITIONS IN POTASSIUM NITRATE AT 1 ATM/ CHANGES IN ELECTRICAL CONDUCTIVITY ACCOMPANYING THE POLYMORPHIC TRAN 16-2391
CE SINGLE CRYSTALS. STUDY OF CONDUCTIVITY AND DIPOLAR RELAXATION IN DOPED I 34-3458
) IN STRONTIUM TITANATE (4.2-300 DEG / ELECTRONIC TRANSPORT (CONDUCTIVITY AND HALL AND SEEBECK COEFFICIENTS 06-1129
ABSORPTION SPECTRA) OF S/ ELECTRICAL AND OPTICAL PROPERTIES (CONDUCTIVITY AND MOBILITY VERSUS TEMPERATURE, 06-1198
TE. OPTICAL ABSORPTION, ELECTRIC CONDUCTIVITY AND PHOTOEMISSION OF GLYCOL SULFA 19-2780
POLARON MECHANISM OF CONDUCTIVITY IN BARIUM TITANATE. 04-0609
PROPERTIES OF SOME FERROELECT/ INVESTIGATION OF THE THERMAL CONDUCTIVITY IN CONNECTION WITH OTHER PHYSICAL 04-0721
. (THEORY, BAR/ HIGH FREQUENCY SOUND ATTENUATION AND THERMAL CONDUCTIVITY IN DISPLACIVE TYPE FERROELECTRICS 03-0160
ATE. CALCULATIONS OF ELECTRONIC CONDUCTIVITY IN NONSTOICHIOMETRIC BARIUM TITAN 04-0821
ION PROPOSED FOR STRONTIUM TITANATE AND POTASSIUM T/ THERMAL CONDUCTIVITY IN PEROVSKITES (3 PHONON INTERACT 06-1095
TALS. SURFACE CONDUCTIVITY IN POLARIZED BARIUM TITANATE CRYS 04-0569
TWO MODES OF ELECTRICAL CONDUCTIVITY IN POTASSIUM NITRATE. 16-2368
TALS. THERMALLY STIMULATED CONDUCTIVITY IN STRONTIUM TITANATE SINGLE CRYS 06-1227
RYSTALLINE BARIUM TITANATE. CONDUCTIVITY INJECTION AND EXTRACTION IN POLY 04-0814
OGEN TUNGSTEN BRONZES, HYDROGEN(X) T/ ELECTRONIC PROPERTIES (CONDUCTIVITY, MAGNETIC SUSCEPTIBILITY) OF HYDR 13-1934
/METRIC BARIUM TITANATE. PART-1: HIGH TEMPERATURE ELECTRICAL CONDUCTIVITY MEASUREMENTS. (INTRINSIC BAND GA/ 04-0822
S. OXYGEN CONCENTRATION CELL AND ELECTRICAL CONDUCTIVITY MEASUREMENTS ON PZT FERROELECTRIC 09-1495
/ORESISTANCE OF (POLYCRYSTALLINE) ANTIMONY SULFOIODIDE (DARK CONDUCTIVITY NEAR TRANSITION POINT, 1-1000 AT/ 15-2250
NEAR THE CUBIC POINT. ANOMALIES IN THE ELECTRICAL CONDUCTIVITY OF A FERROELECTRIC SEMICONDUCTOR 03-0268
LS. ELECTRICAL CONDUCTIVITY OF AMMONIUM SULFATE SINGLE CRYSTA 19-2781
CRYSTAL STRUCTURE AND CONDUCTIVITY OF ANTIMONY SULFOIODIDE. 15-2112
RYSTALS AND DOPED (IRON, HYDROXYL) SINGLE CRYSTALS. CONDUCTIVITY OF BARIUM TITANATE, PURE SINGLE C 04-0695
S HEATED IN OXYGEN (EFFEC/ TIME DEPENDENCE OF THE ELECTRICAL CONDUCTIVITY OF BARIUM TITANATE SINGLE CRYSTAL 04-0997
S GROWN FROM SOLUTION IN A POT/ ANISOTROPY OF THE ELECTRICAL CONDUCTIVITY OF BARIUM TITANATE SINGLE CRYSTAL 04-1035
SELENOIODIDE SYSTEM ALLOYS. PHASE DIAGRAM AND ELECTRICAL CONDUCTIVITY OF BISMUTH SELENOIODIDE- ANTIMONY 15-2122
RYSTALS NEAR/ SPECIFIC HEAT, THERMAL DIFFUSIVITY AND THERMAL CONDUCTIVITY OF CALCIUM DOPED SODIUM NITRATE C 16-2283
(SODIUM TUNGSTATES) THERMAL CONDUCTIVITY OF CUBIC SODIUM TUNGSTEN BRONZES. 13-2057
MEASUREMENT OF THE TEMPERATURE DEPENDENCE OF THE ELECTRICAL CONDUCTIVITY OF FERROELECTRICS. 03-0525
NDUCTIVITY AND RELATED PROPERTIES OF 60 COMPOUNDS,/ CONDUCTIVITY OF FERROELECTRICS. (ELECTRICAL CO 01-0014
ANSITION REGION. NATURE OF THE ANOMALOUS ELECTRIC CONDUCTIVITY OF FERROELECTRICS IN THE PHASE TR 03-0508
EXPERIMENTAL AND THEORETICAL STUDIES ON THE DC CONDUCTIVITY OF ICE. 34-3450
DIELECTRIC RELAXATION, BULK AND SURFACE CONDUCTIVITY OF ICE SINGLE CRYSTALS. 34-3469
S. ELECTRICAL CONDUCTIVITY OF LITHIUM NIOBATE SINGLE CRYSTAL 11-1733
THERMAL CONDUCTIVITY OF METHYL AMMONIUM ALUMINUM ALUM. 20-2807
E. OBSERVATION OF THE ANOMALY OF THE THERMAL CONDUCTIVITY OF POTASSIUM MANGANESE TRIFLUORID 33-3382
ADDING TITANIUM(+) AND STRONTIUM(2+) IONS ON THE ELECTRICAL CONDUCTIVITY OF POTASSIUM NITRATE. EFFECT OF 16-2319
THERMAL CONDUCTIVITY OF ROCHELLE SALT. 28-3226
NATE. ELECTRICAL CONDUCTIVITY OF SINGLE CRYSTALLINE BARIUM TITA 04-0663
TANTALATE AND BARIUM TITANATE NEAR THE CURIE POINTS. THERMAL CONDUCTIVITY OF STRONTIUM TITANATE, POTASSIUM 04-0989
E AT HIGH TEMPERATURES. ELECTRICAL CONDUCTIVITY OF STRONTIUM ZIRCONATE AND HAFNAT 08-1417
RIE POINT. THERMAL CONDUCTIVITY OF TRIGLYCINE SULFATE NEAR THE CU 27-2983
RIE POINT. THERMAL CONDUCTIVITY OF TRIGLYCINE SULFATE NEAR THE CU 27-3016
ELECTRICAL CONDUCTIVITY OF TUNGSTEN TRIOXIDE. 10-1637
/ART-1: INTRODUCTION AND MAGNETIC RESONANCE. PART-2: THERMAL CONDUCTIVITY. PART-3: NEUTRON SCATTERING. PAR/ 03-0397
METAL- LEAD TITANATE- METAL. (METHOD OF STUDYING THE TYPE OF CONDUCTIVITY, PLATINUM ELECTRODES) /HE SYSTEM 05-1057

) OF REDUCED BARIUM TITANATE CRYSTALS (POLARON H/ ELECTRICAL CONDUCTIVITY (SEEBECK COEFFICIENT, CURIE POINT 04-0918
POTASSIUM NITRATE.   DIELECTRIC AND ELECTRICAL CONDUCTIVITY STUDIES IN POTASSIUM NITRITE AND 16-2340
AND STRUCTURE OF CRYSTALLINE BISMUTH SELENOIODIDE.   CONDUCTIVITY, THERMOELECTRIC PROPERTIES, AND B 15-2126
ONEZH, USSR, PROC, PARTS 1-3.   ALL-UNION CONFERENCE ON FERROELECTRICITY, 7TH, 1970, VOR 01-0001
E CRYSTALS BY NEUTRON BOMBARDMENT.   SPLIT INTERSTITIAL CONFIGURATION INDUCED IN PEROVSKITE TYPE SINGL 08-1467
RADICAL IN IRRADIATED FERROEL/ ESR IDENTIFICATION OF SLATER CONFIGURATIONS AND OF EXCHANGE OF THE ARSENATE 18-2734
D POTAS/ HYDROGEN BONDED FERROELECTRICS: EVIDENCE FOR SLATER CONFIGURATIONS FROM EPR STUDIES OF X-IRRADIATE 18-2728
WITH SPECIAL DIELECTRIC PROPERTIES.   ORIGIN OF DIPOLE CONFIGURATIONS IN SOME (PEROVSKITE) STRUCTURES 08-1390
, LANTHANUM(1-X) CALCIUM(X) MANGANATE (PHASE TRANSFORMATIONS CONFIGURATIONAL ENERGY). /SKITE TYPE COMPOUNDS 01-1356
DROGEN DISELENITE.   ANOMALOUS SPECIFIC HEAT AND CONFIGURATIONAL ENTROPY CHANGE IN SODIUM TRIHY 22-2856
/ESULTS ON INDIUM STRONTIUM TITANATE BARRIERS INTERPRETED AS CONFIRMATION OF THE THEORY OF TRANSITIONS FRO/ 03-0233
NONLINEAR OPTICS WITH LOW BIREFRINGENCE VARIATIONS) FROM THE CONGRUENT MELT. /ITHIUM NIOBATE CRYSTALS (FOR 11-1688
THIUM NIOBATE (CRYSTAL GROWTH BY ADDING MAGNESIUM OXIDE TO A CONGRUENT MELT). / MATCHING TEMPERATURES IN LI 11-1684
ANTALATE.   CONGRUENT MELTING COMPOSITION OF LITHIUM METAT 12-1882
ION OF A PECULIAR FERROELECTRIC- FERROELASTIC CRYSTAL: DIAM/ CONJECTURED MECHANISM OF THE PHASE TRANSFORMAT 19-2741
HAUSEN STEPS AND THE CRITICAL POLARIZATION AND REPOLARIZATI/ CONNECTION BETWEEN THE STARTING FIELD FOR PARK 03-0318
OME FERROELECT/ INVESTIGATION OF THE THERMAL CONDUCTIVITY IN CONNECTION WITH OTHER PHYSICAL PROPERTIES OF S 04-0721
1: TGS. FA/ ACCURATE STRUCTURE ANALYSIS OF FERROELECTRICS IN CONNECTION WITH THEIR PHASE TRANSITIONS. PART- 02-0077
OGEN OR DIDEUTERIUM) PHOSPHATE CRYSTALS THEORY AND PRACTICAL CONSIDERATIONS). /WEDGE SHAPED RUBIDIUM DIHYDR 17-2637
ARTIALLY ANTIFERROELASTIC CRYSTALS AND PARTIALLY FERROELECT/ CONSIDERATIONS OF PARTIALLY FERROELASTIC AND P 03-0136
IN AMMONIUM PERIODATE.   STATISTICAL CONSIDERATIONS OF THE ORDER DISORDER REACTION 36-3581
EN MAGNETIZATION AND POLARIZATION: PHENOMENOLOGICAL SYMMETRY CONSIDERATIONS ON BORACITES. /TERACTIONS BETWE 14-2072
ENCE IN FERROELECTRIC PHOTOCONDUCTOR MEMOR/ SOME THEORETICAL CONSIDERATIONS RELATING TO SWITCHING AND REMAN 03-0378
N FROM PIEZO TO FERROELECTRICITY (THEORY).   SELF CONSISTENCY OF LANDAU'S MODEL IN THE TRANSITIO 03-0507
DEVELOPMENT OF BARIUM TITANATE WITH CONSISTENT DIELECTRIC PROPERTIES.   04-0797
ENERGY BAND STRUCTURE OF STRONTIUM TITANATE FROM A SELF CONSISTENT FIELD TIGHT BINDING CALCULATION. 06-1221
RAL PHASE TRANSITIONS IN PEROVSKITE TYPE CRYSTALS. (ST/ SELF CONSISTENT LATTICE DYNAMICAL THEORY OF STRUCTU 03-0424
/ND SECOND ORDER PHASE TRANSITIONS IN FERROELECTRICS: A SELF CONSISTENT POLARIZATION SCHEME FOR ANHARMONIC/ 03-0355
OTE ON THE CLASSIFICATION OF FERROELECTRICS. (BY CURIE-WEISS CONSTANT)   A N 01-0024
OF THE CUBIC PHASE OF BARIUM TITANATE. (IMAGINARY DIELECTRIC CONSTANT)   BAND STRUCTURE 04-0845
AMMONIUM ALUMINUM SULFATE DODECAHYDRATE. (COMPLEX DIELECTRIC CONSTANT) /TION IN FERROELECTRIC MASD, METHYL 20-2804
ULFATE BY A DIGITAL TECHNIQUE. (MINUTE CHANGES IN DIELECTRIC CONSTANT) /AVIOR OF FERROELECTRIC TRIGLYCINE S 27-3116
SURE (TO 5600 KG/SQ CM, TEMPERATURE DEPENDENCE OF DIELECTRIC CONSTANT). /RYSTAL UNDER HIGH HYDROSTATIC PRES 24-2896
4 RATIO MIXED CRYSTAL, PROTON MAGNETIC RESONANCE, DIELECTRIC CONSTANT). /MONIUM ALUMINUM SULFATE AND A 36:6 20-2809
- IODINE- BORACITE. (TEMPERATURE DEPENDENCE OF PIEZOELECTRIC CONSTANT) / IN THE FERROELECTRIC STATE OF IRON 14-2083
ULTRASONIC WAVES IN LITHIUM NIOBATE CRYSTALS. (PHOTOELASTIC CONSTANT)   LIGHT SCATTERING BY 11-1754
OELECTRI/ MILLIMETER WAVELENGTH DISPERSION IN THE DIELECTRIC CONSTANT ABOVE THE CURIE POINT IN VARIOUS FERR 02-0124
IEZOELECTRIC CERAMIC (BARIUM TITANATE) WITH LARGE DIELECTRIC CONSTANT AND A SEMICONDUCTOR. /NSISTING OF A P 04-1037
/ONIC COMPOUNDS. (HALIDES AND OXIDES, RELATION BETWEEN CURIE CONSTANT AND CURIE TEMPERATURE FOR FERROELECT/ 03-0188
HELLE SALT.   DIELECTRIC CONSTANT AND DIELECTRIC RELAXATION TIME OF ROC 28-3179
F A FERROELECTRIC (BARIUM TITANATE DOWN TO 3 MIC/ DIELECTRIC CONSTANT AND LATTICE DYNAMICS OF A THIN FILM O 04-0600
MICS. (AND DIFFERENCE IN TEMPERATURES FOR MAXIMUM DIELECTRIC CONSTANT AND LOSS) /ES IN BARIUM TITANATE CERA 04-0849
UM TITANATE SINGLE CRYSTALS (40 HZ-10 MHZ, EFFE/ DIELECTRIC (CONSTANT AND LOSS) BEHAVIOR OF REDUCED STRONTI 06-1148
ELECT/ TEMPERATURE AND PRESSURE DEPENDENCE OF THE DIELECTRIC CONSTANT AND SPONTANEOUS POLARIZATION OF FERRO 16-2331
/GLE CRYSTAL GROWTH CONDITIONS ON A CHANGE IN THE DIELECTRIC CONSTANT AND TANGENT OF THE ANGLE OF DIELECTR/ 27-3151
ELASTIC CONSTANT ANOMALIES IN STRONTIUM TITANATE.   06-1177
HASE TRANSITION IN CHARACTERISTICALLY FERROELECT/ DIELECTRIC CONSTANT BEHAVIOR IN THE NEIGHBORHOOD OF THE P 03-0522
OF HIGH PURITY POLYCRYSTALLINE BARIUM TITANATE. (DIELECTRIC CONSTANT, CONDUCTIVITY)   ELECTRICAL PROPERTIES 04-0704
SEMICONDUCTING BARIUM TITANATE. (DIELECTRIC CONSTANT, CONTACT RESISTANCE)   04-0730
UM TITANATE WITH IMPROVED ELECTRICAL PROPERTIES. (DIELECTRIC CONSTANT, DISSIPATION FACTOR) /XIDE DOPED BARI 04-0840
/D) BARIUM TITANATE CERAMIC WITH A HIGH EFFECTIVE DIELECTRIC CONSTANT (DUE TO ADDING CUPROUS OR FERRIC OXI/ 04-0581
IC AND PIEZOELECTRIC PROPERTIES OF ANTIMONY SULFOIODIDE IN A CONSTANT ELECTRIC FIELD.   DIELECTR 15-2249
ERTIES OF POLYCRYSTALLINE BARIUM TITANATE.   INFLUENCE OF A CONSTANT ELECTRIC FIELD ON THE DIELECTRIC PROP 04-1012
IN TRIGLYCINE FLUOBERYLLATE.   EFFECT OF CONSTANT ELECTRICAL FIELD ON PHASE TRANSITIONS 27-3167
D ((BARIUM(X) STRONTIUM(1-X) TITANATE (RECIPROCAL DIELECTRIC CONSTANT FOLLOWS QUADRATIC TEMPERATURE LAW). / 04-0898
DEUTERON QUADRUPOLE COUPLING CONSTANT IN DEUTERATED AMMONIUM SULFATE.   19-2761
THE (BARIUM, S/ CONCENTRATION DEPENDENCE OF THE CURIE-WEISS CONSTANT IN POLYCRYSTALLINE SOLID SOLUTIONS IN 04-0586
TURE DEPENDENCE OF THE SODIUM-23 NUCLEAR QUADRUPOLE COUPLING CONSTANT IN SODIUM NITRITE.   TEMPERA 16-2393
E, BARIUM TITANATE)   MEASUREMENT OF OXYGEN DIFFUSION CONSTANT IN SOME TITANATES. (STRONTIUM TITANAT 04-1033
DATE- POTASSIUM DYSPRO/ TEMPERATURE DEPENDENCE OF DIELECTRIC CONSTANT IN THE SYSTEM POTASSIUM YTTRIUM MOLYB 30-3309
ACITOR DIELECTRICS, A STUDY IN DIELECTRIC S/ HIGH DIELECTRIC CONSTANT MATERIALS AND FERROELECTRICITY AS CAP 37-3678
ETIC RESONANCE. (ABSTRACT ON/ INFORMATION ON HIGH DIELECTRIC CONSTANT MATERIALS GAINED BY ELECTRON PARAMAGN 02-0101
ELECTROELASTIC CONSTANT MEASUREMENTS FOR PLZT CERAMICS.   09-1598
/ NIOBATE. (OPTICAL ABSORPTION, REFRACTIVE INDEX, DIELECTRIC CONSTANT, NONLINEAR OPTICAL COEFFICIENTS, AND/ 13-2024
MMARY ON/ 180 DEG DOMAIN WALL CONTRIBUTION TO THE DIELECTRIC CONSTANT OF BARIUM TITANATE. (CALCULATIONS, SU 04-0804
ND/ SHAPE DEPENDENCE OF PROPERTIES (CURIE TEMPERATURE, CURIE CONSTANT) OF FERROELECTRIC CRYSTALS. (5 COMPOU 27-3120
ALLOWANCE FOR CONCENTRATION FLUCTUATIONS.   DIELECTRIC CONSTANT OF FERROELECTRIC SOLID SOLUTIONS WITH 03-0440
TITANATE CERAMICS) TEMPERATURE DEPENDENCE OF THE DIELECTRIC CONSTANT OF FERROELECTRIC SUBSTANCES. (BARIUM 04-0591
NONLINEARITY OF THE DIELECTRIC CONSTANT OF FERROELECTRIC TRIGLYCINE SULFATE. 27-3070
NEOUS MEASUREMENT OF PYROELECTRIC COEFFICIENT AND DIELECTRIC CONSTANT OF FERROELECTRICS.   SIMULTA 03-0333
/UENCY AND TEMPERATURE DEPENDENCES OF THE COMPLEX DIELECTRIC CONSTANT OF FERROELECTRICS NEAR THE CURIE POI/ 03-0361
ELD, 370-520 DEG K.   COMPLEX DIELECTRIC CONSTANT OF LEAD HAFNATE, 320-520 DEG K.   08-1339
X-RAY DIFFUSE SCATTERING.   COMPLEX DIELECTRIC CONSTANT OF LEAD ZIRCONATE UNDER POLARIZING FI 07-1256
PIEZOELECTRICITY, PYROELECTRICITY, AND THE ELECTROSTRICTION ELASTIC CONSTANT OF ORTHORHOMBIC POTASSIUM NIOBATE BY 08-1416
LAMBDA EQUAL 2.14 / TEMPERATURE DEPENDENCE OF THE DIELECTRIC CONSTANT OF POLYVINYLIDENE FLUORIDE.   36-3598
UNDER BIASING CONDITIONS.   DIELECTRIC CONSTANT OF POTASSIUM DIHYDROGEN PHOSPHATE AT 17-2501
EFFECT OF PRESSURE ON THE STATIC DIELECTRIC CONSTANT OF POTASSIUM NIOBATE SINGLE CRYSTALS 08-1381
DIELECTRIC CONSTANT OF POTASSIUM TANTALATE.   08-1269
DIELECTRIC CONSTANT OF PURE ICE I(H) SINGLE CRYSTALS.   34-3478
ION OF TEMPERATURE.   MEASUREMENT OF THE COMPLEX DIELECTRIC CONSTANT OF ROCHELLE SALT AT 10 GHZ AS A FUNCT 28-3184
PRESSURE DEPENDENCE OF THE ELASTIC CONSTANT OF STRONTIUM TITANATE.   06-1097
TEMPERATURE DEPENDENCE OF THE DIELECTRIC CONSTANT OF STRONTIUM TITANATE.   06-1134
VARIATION OF THE DECAY CONSTANT OF STRONTIUM-85 IN BARIUM TITANATE.   04-0876
Y THE FERROELECTRIC TRANSITION.   CHANGE IN THE DECAY CONSTANT OF TECHNETIUM-99 IN BARIUM TITANATE B 04-0871
RE AND FREQUENCY DEPENDENCE).   DISPERSION OF THE DIELECTRIC CONSTANT OF TGS IN THE FAR INFRARED (TEMPERATU 27-3009
DIELECTRIC CONSTANT OF TRIGLYCINE SULFATE.   27-3049
F THE CONDITIONS OF GR/ THERMAL HYSTERESIS OF THE DIELECTRIC CONSTANT OF TRIGLYCINE SULFATE AS A FUNCTION O 27-3126
EQUENCIES.   MEASUREMENT OF THE DIELECTRIC CONSTANT OF TRIGLYCINE SULFATE AT MICROWAVE FR 27-2954
IN A STRONG ELECTRIC FIELD.   DIELECTRIC CONSTANT OF TRIGLYCINE SULFATE SINGLE CRYSTALS 27-3150
ARIUM CHLORIDE FLUX GROWTH IN PRESENCE / CHANGE IN THE DECAY CONSTANT OF ZIRCONIUM-89 IN BARIUM TITANATE (B 04-0685

N L/ DEPENDENCE OF LINEAR ELECTROOPTIC EFFECT AND DIELECTRIC CONSTANT ON MELT COMPOSITION (STOICHIOMETRY) I 11-1842
DATE - A DISORDERED FERROELECTRIC. (DEPENDENCE OF DIELECTRIC CONSTANT ON THERMAL AND FIELD HISTORY) /IUM IO 08-1348
ES CONTAINING/ BARIUM TITANATE CERAMICS WITH HIGH DIELECTRIC CONSTANT. PART-1: EFFECTS OF ADDITION OF GLASS 04-0829
TI/ LATTICE DYNAMICS OF DIATOMIC CRYSTALS OF HIGH DIELECTRIC CONSTANT. (POTASSIUM BROMIDE, LEAD TELLURIDE, 36-3541
ND RELATED PROPERTIES OF LEAD TITANATE CERAMICS. (DIELECTRIC CONSTANT, RESISTIVITY) /VES ON PIEZOELECTRIC A 05-1084
ELECTRIC TRIGLYCINE SE/ (OPTICAL) ENERGY GAP (AND DIELECTRIC CONSTANT) TEMPERATURE CHARACTERISTICS OF FERRO 27-3021
ARIUM TITANATE, STRONT/ METHOD FOR ACHIEVING HIGH DIELECTRIC CONSTANT TEMPERATURE STABLE FERROELECTRICS. (B 04-0551
PE/ NONLINEAR PROPERTIES OF TRIGLYCINE SELENATE. (DIELECTRIC CONSTANT VERSUS ALTERNATING FIELD VOLTAGE, TEM 27-3158
MECHANISM OF THE PHASE TRANSITION IN TGS. (STATIC DIELECTRIC CONSTANT, X-RAY SCATTERING)   STUDIES ON THE 27-3093
OMETRIC) LITHIUM NIOBATE (TEMPERATURE DEPENDENCE OF COUPLING CONSTANT 21-515K). / INVESTIGATION OF (STOICHI 11-1811
ODYMIUM(X) (ORTHO) ALUM/ ANOMALIES IN THE (1 KHZ) DIELECTRIC CONSTANT (5-300K) OF THE PRASEODYMIUM(1-X) NE 36-3600
NATE, STANNATE, TITANATE) SOLID SOLUTIONS. (ELECTROSTRICTION CONSTANTS) /TRIC TRANSITION IN THE LEAD (ZIRCO 09-1486
UM (ZINCATE NIOBATE)- LEAD TITANATE. (PHASE DIAGRAM, LATTICE CONSTANTS) /YSTEM LEAD (ZINCATE NIOBATE)- BARI 05-1070
      ELASTIC ANOMALY IN AMMONIUM SULFATE. (ELECTROSTRICTIVE CONSTANTS) 19-2759
ALT TETRAFLUORIDE CRYSTAL STRUCTURE (SPACE GROUP AND LATTICE CONSTANTS). /RAELASTIC PARAMAGNETIC BARIUM COB 33-3386
) OF THE LATTICE VIBRATIONS IN LITHIUM TANTALATE (DIELECTRIC CONSTANTS). /TUDY (20-10,000 PER CM, 300 DEG K 12-1870
: CALCULATION OF THE INTERMOLECULAR AND INTRAMOLECULAR FORCE CONSTANTS. /AND FERROELECTRIC THIOUREA. PART-2 25-2907
ECTRICAL PROPERTIES. (DIELECTRIC, ELASTIC, AND PIEZOELECTRIC CONSTANTS) /SURE AND INVESTIGATION OF THEIR EL 15-2138
SULFATE. (TEMPERATURE DEPENDENCES OF THE COMPLEX DIELECTRIC CONSTANTS) /TION OF DOMAIN WALLS IN TRIGLYCINE 27-3103
      STUDY OF THE BRILLOUIN EFFECT IN ADP (ALL ELASTIC CONSTANTS). 17-2522
PPLICATION TO CALCULATING THE WALL ENERGY AND THE DIELECTRIC CONSTANTS. / FOR BARIUM AND LEAD TITANATES - A 03-0189
/VARIATIONS IN THE FREE AND CLAMPED VALUES OF THE DIELECTRIC CONSTANTS AND ELECTROOPTIC COEFFICIENTS IN BA/ 13-1998
ANIUM(3) OXIDE(11) CRYSTAL, A NEW FERROELECTRIC. (DIELECTRIC CONSTANTS AND REFRACTIVE INDICES) LEAD(5) GERM 35-3502
MOLYBDATE.   ELASTIC CONSTANTS AND SOFT OPTICAL MODES IN GADOLINIUM 30-3284
NITRITE SINGLE CRYSTALS.   ELASTIC CONSTANTS AND ULTRASONIC ABSORPTION OF SODIUM 16-2357
BIUM, TITANIUM) OR (NIOBIUM, TUNGSTEN), COMPILATION, LATTICE CONSTANTS, CURIE TEMPERATURES) /HERE B IS (NIO 13-1964
LOW TEMPERATURES.   MEASUREMENT OF DIELECTRIC CONSTANTS FOR A FEW FERROELECTRIC MATERIALS AT 02-0035
ANTIMONY SULFOIODIDE USING A DOUBLE PASSED, STABIL/ (ELASTIC CONSTANTS FROM) BRILLOUIN SCATTERING STUDY OF 15-2225
TRITE RADICAL IN SODIUM NITRITE.   ACCURATE FORCE CONSTANTS FROM ISOTOPIC SUBSTITUTION IN THE NI 16-2309
ANATE.   FORCE CONSTANTS IN BARIUM TITANATE AND STRONTIUM TIT 04-0748
IC PHASES. (LANTHANUM OR NIOBIUM DOPED LEA/ ELECTROSTRICTIVE CONSTANTS IN FERROELECTRIC AND ANTIFERROELECTR 09-1487
E.   DETERMINATION OF THE TEMPERATURE DEPENDENCE OF LATTICE CONSTANTS IN FERROELECTRIC ANTIMONY SULFOIODID 15-2140
SALT)   MEASUREMENTS OF COMPLEX PIEZOELECTRIC CONSTANTS IN FERROELECTRIC CRYSTALS. (ROCHELLE 28-3213
LATTICE DYNAMICS OF IONIC CRYSTALS WITH HIGH DIELECTRIC CONSTANTS. (LEAD AND TIN TELLURIDES) 36-3553
NESE TRI/ VIBRATIONAL SPECTRUM AND DISPERSION OF THE OPTICAL CONSTANTS OF ANTIFERROMAGNETIC POTASSIUM MANGA 33-3385
THEORY OF THE DIELECTRIC CONSTANTS OF BARIUM TITANATE. 04-0776
ELASTIC, PIEZOELECTRIC, AND DIELECTRIC CONSTANTS OF BISMUTH(12) GERMANIUM OXIDE(20). 35-3512
ND OXIDES, RELATION BE/ TEMPERATURE DEPENDENCE OF DIELECTRIC CONSTANTS OF CUBIC IONIC COMPOUNDS. (HALIDES A 03-0188
ELASTIC CONSTANTS OF FERROCERAMICS. 09-1513
ELECTRICAL AND OPTICAL CONSTANTS OF FERROELECTRIC POTASSIUM IODATE. 08-1427
C CONSTANTS AND DIELECT/ MEASUREMENT OF MICROWAVE DIELECTRIC CONSTANTS OF FERROELECTRICS. PART-2: DIELECTRI 16-2354
RIE POINT).   ELASTIC CONSTANTS OF GADOLINIUM MOLYBDATE (NEAR THE CU 30-3274
GION.   OPTICAL CONSTANTS OF ICE I OVER THE ENTIRE INFRARED RE 34-3492
TEMPERATURE DEPENDENCE OF THE FORCE CONSTANTS OF ION IN SODIUM NITRITE. 16-2311
EFFECT OF GAMMA-IRRADIATION ON THE DIELECTRIC CONSTANTS OF KDP. 17-2691
PHOTOELASTIC CONSTANTS OF LITHIUM NIOBATE CRYSTALS. 11-1745
DETERMINATION OF ELASTIC AND PIEZOELECTRIC CONSTANTS OF LITHIUM NIOBATE SINGLE CRYSTALS. 11-1749
/RE DEPENDENCE OF THE ELASTIC, PIEZOELECTRIC, AND DIELECTRIC CONSTANTS OF LITHIUM TANTALATE AND LITHIUM NI/ 11-1823
ITANATE, STRONTIUM TI/ PRESSURE DEPENDENCE OF THE DIELECTRIC CONSTANTS OF PARAELECTRIC MATERIALS. (BARIUM T 04-0856
CONATE WITH THOROUGHNESS OF POL/ VIBRATION OF ELECTROELASTIC CONSTANTS OF POLYCRYSTALLINE LEAD TITANATE ZIR 09-1489
BRILLOUIN DIFFUSION.   MEASUREMENT OF THE ELASTIC CONSTANTS OF POTASSIUM DIHYDROGEN PHOSPHATE BY 17-2454
DETERMINATION OF THE ELASTIC CONSTANTS OF POTASSIUM NITRATE. 08-1399
SION AT ORDINARY TEMPERATURES.   MEASUREMENT OF THE ELASTIC CONSTANTS OF SODIUM NITRITE BY BRILLOUIN DIFFU 16-2306
/N STRUCTURE AND TEMPERATURE DEPENDENCE OF THE PIEZOELECTRIC CONSTANTS OF STIBOTANTALITE. (ANTIMONY (TANTA/ 36-3601
PRESSURE DEPENDENCE OF THE ELASTIC CONSTANTS OF STRONTIUM TITANATE. 06-1096
TEMPERATURE DEPENDENCE OF THE FIRST ORDER ELASTIC CONSTANTS OF STRONTIUM TITANATE. 06-1160
OBATE TYPE FERROELECTRIC CRYSTAL. ELASTIC AND PIEZOELECTRIC CONSTANTS OF STRONTIUM(4) POTASSIUM LITHIUM NI 13-1946
TIME DEPENDENCE OF MATERIAL CONSTANTS OF TGS AFTER POLARIZATION REVERSAL. 27-2953
HE METHOD OF SUBSTITUTION.   LOW FREQUENCY DIELECTRIC CONSTANTS OF THE ALKALINE EARTH FLUORIDES BY T 33-3361
HANIUM NIOBATE.   DIELECTRIC CONSTANTS OF THE SYSTEM SAMARIUM NIOBATE- LANT 08-1268
NFRARED TO FAR INFRARED AND IN THE RADIO FREQUENCY / OPTICAL CONSTANTS OF TRIGLYCINE SELENATE IN THE NEAR I 27-3007
YNAMIC DETERMINATION OF (ELASTIC AND PIEZOELECTRIC) MATERIAL CONSTANTS OF TRIGLYCINE SULFATE.   D 27-2984
TO PYROELECTRICITY AND THE DETECTION OF FAR INFR/ ELECTRICAL CONSTANTS OF TRIGLYCINE SULFATE, APPLICATIONS 17-2514
THE FERROELECTRIC / TEMPERATURE DEPENDENCE OF THE DIELECTRIC CONSTANTS OF TRIGLYCINE SULFATE ORTHOGONAL TO 27-2975
IUM TITANATE.   EFFECTS OF ANISOTROPY OF ELASTIC CONSTANTS ON DISLOCATIONS IN FERROELECTRIC BAR 04-0674
/H AND PHYSICAL PROPERTIES OF LITHIUM TANTALATE. (DIELECTRIC CONSTANTS, PIEZOELECTRIC CONSTANTS, SPONTANEO/ 12-1876
/TRETS BASED ON A(V)B(VI)C(VII) SOLID SOLUTIONS. (DIELECTRIC CONSTANTS, PYROELECTRIC CURRENTS, ANTIMONY SU/ 15-2157
/UM NIOBATE (ELASTIC, PIEZOELECTRIC, DIELECTRIC, AND OPTICAL CONSTANTS) ROOM TEMPERATURE TO THE CURIE POIN/ 13-2060
NIDE TRIDEUTERO HYDRATE ABOVE THE CURIE TEMPERATURE (LATTICE CONSTANTS, SPACE GROUP). /C POTASSIUM FERROCYA 24-2903
TE, BARIUM SODIUM NIOBATE AND BARIUM SODIUM TANTALATE (FORCE CONSTANTS, SPACE GROUP ASSIGNMENT). /IUM NIOBA 11-1808
/MORPHOUS COMPOUND LEAD GERMANIUM SILICON OXIDE. (DIELECTRIC CONSTANTS, SPONTANEOUS POLARIZATION, THERMAL / 35-3501
AL GROWTH.   BARIUM SODIUM NIOBATE: CONSTITUTIONAL STUDIES AND (CZOCHRALSKI) CRYST 13-1904
HE STRUCTURE OF AMMONIUM HYDROGEN SULFATE ABOVE THE FERROEL/ CONSTRAINED REFINEMENT TECHNIQUES APPLIED TO T 16-2355
HENOMENOLOGICAL MODEL/ EFFECT OF THE CHANGE OF ELECTROSTATIC CONSTRAINTS ON THE PHASE TRANSITIONS OF SOME P 03-0280
VED FERROELECTRIC "AFTER EFFECT" PHENOMENA. (HYSTERESIS LOOP CONSTRICTIONS, ROCHELLE SALT, TGS)   SHORT LI 27-3156
FERROELECTRIC (THEORY, TITANATES DISCU/ FIELD EFFECT AT THE CONTACT BETWEEN A SEMICONDUCTOR AND A C-DOMAIN 03-0511
IDE.   A NOTE ON THE APPARENT CONTACT FERROELECTRIC EFFECT OF MANGANESE DIOX 36-3606
SEMICONDUCTING BARIUM TITANATE. (DIELECTRIC CONSTANT, CONTACT RESISTANCE) 04-0730
SEEBECK EFFECT IN METAL LEAD TITANATE METAL CONTACT SYSTEMS. 05-1056
ELECTRICAL CONTACT TO N- AND P-TYPE FERROELECTRIC OXIDES. 03-0512
XIDE AND STRONTIUM TITANATE. (SURFACE BARRIER SYSTEMS, METAL CONTACTS) SOME ELECTRONIC PROPERTIES OF ZINC O 06-1176
RYSTALS AND / ELECTRICAL AND PHOTOELECTRIC PROPERTIES OF THE CONTACTS BETWEEN ANTIMONY SULFOIODIDE SINGLE C 15-2251
OHMIC CONTACTS TO SEMICONDUCTING BARIUM TITANATE. 04-0764
CE MEASUREMENT OF A DIELECTRIC FERROELECTRIC CONDENSER UNDER CONTINUOUS HEATING. (TGS SINGLE CRYSTAL) /EDAN 27-3028
OBATE.   CONTINUOUS HOT PRESSING OF POTASSIUM SODIUM NI 13-2007
DIFFRACTION OF A CONTINUOUS LIGHT WAVE IN KDP. 17-2444
RING OF LIGHT IN LITHIUM IODATE (INDEX OF REFRACTION 1.2-7/ (CONTINUOUS WAVE) SPONTANEOUS PARAMETRIC SCATTE 08-1291
BARKHAUSEN EFFECT IN FERROELECTRICS UNDER CONTINUOUS X-RAY DIFFRACTION. 03-0337
UM.   EFFICIENT HIGH GRAIN PARAMETRIC GENERATION IN ADP CONTINUOUSLY TUNABLE ACROSS THE VISIBLE SPECTR 17-2712
E LIGHT VALVES.   ORIENTATION AND THICKNESS DEPENDENCE OF CONTRAST AND BRIGHTNESS IN GADOLINIUM MOLYBDAT 30-3291

OPED AND UNDOPED TGS, LITH/ X-RAY TOPOGRAPHICAL STUDY OF THE CONTRAST AT FERROELECTRIC DOMAIN BOUNDARIES (D 11-1671
F THE PHASE TRANSITION IN TRIGLYCINE SULFATE. (CLOSE THERMAL CONTROL).   EXPERIMENTAL DETAILED STUDY O 27-3030
ICS AND THEIR APPLICATION IN SOLID STATE DEVICES AS ADAPTIVE CONTROL.   FERROELECTR 37-3665
OELECTRICS. (TRIGLYCINE SULFATE, HIGHLY ACCURATE TEMPERATURE CONTROL)   ORDER DISORDER TRANSITION IN FERR 27-2982
IN GADOLINIUM MOLYBDATE.   CONTROL AND APPLICATION OF DOMAIN WALL MOTION 30-3257
IN GADOLINIUM MOLYBDATE.   CONTROL AND APPLICATION OF DOMAIN WALL MOTION 30-3256
 MODIFICATION TO THE CZOCHRALSKI METHOD OF CRYSTAL PULLING. (CONTROL OF CRYSTAL DIAMETER OF STRONTIUM BARI/ 13-1914
/ CRYSTALS. (LITHIUM THALLIUM TARTRATE MONOHYDRATE, ELECTRIC CONTROL OF ELASTIC COMPLIANCE, NICKEL IODINE / 02-0060
PATTERNS IN FERROELECTRIC MATERIALS. (BARIUM TI/ ELECTRICAL CONTROL OF FIXATION AND ERASURE OF HOLOGRAPHIC 04-0847
                                                             CONTROL OF LASER DAMAGE IN LITHIUM NIOBATE. 11-1716
H OF LARGE ANTIMONY SULFOIODIDE CRYSTALS (TO 1 CM DIAMETER): CONTROL OF NEEDLE MORPHOLOGY.   GROWT 15-2200
HALLIUM TARTRATE (TUNING AND DELAY DEVICES).   ELECTRIC FIELD CONTROL OF THE ELASTIC COMPLIANCE IN LITHIUM T 29-3250
TRIC LITHIUM THALLIUM TARTRATE HYDRATE (LTT) BY AN ELECTRIC/ CONTROL OF THE ELASTIC COMPLIANCE OF FERROELEC 29-3249
TE TO LASER INDUCED REFRACTIVE INDEX CHANGES.   CONTROL OF THE SUSCEPTIBILITY OF LITHIUM NIOBA 11-1791
OYING PIEZOELECTRIC- FERROELECTRIC INTERACTIONS FOR ADAPTIVE CONTROL VOLTAGE MODULES. / MEMORY DEVICES EMPL 37-3663
/GE (PROCESSING) AND DISPLAY DEVICES. (BASED ON ELECTRICALLY CONTROLLED LIGHT SCATTERING AND BIREFRINGENCE/ 37-3650
STICS OF A FERROELECTRIC (BARIUM TITANATE TYPE) CERAMIC WITH CONTROLLED MICROWAVE PERMITTIVITY. /CHARACTERI 04-0904
      FERROELECTRIC CERAMIC LIGHT GATES OPERATED IN A VOLTAGE CONTROLLED MODE.   37-3654
TE CRYSTALS.   LIGHT BEAM DEFLECTION CONTROLLED WITH A PRISM OF LEAD MAGNESIO NIOBA 08-1280
AL SINGLE CRYSTALS OF RARE EARTH MANGANATES (PHASE DIAGRAMS, CONVECTION EFFECTS). / FLUX GROWTH) OF HEXAGON 36-3536
                  BARIUM TITANATE BY CONVERGENT BEAM ELECTRON DIFFRACTION.   04-0991
RROELECTRIC CERAMICS (AND THE THEORY OF PIEZOELECTRIC ENERGY CONVERSION).   PIEZOELECTRIC EFFECTS IN FE 03-0278
LASTIC AND ELECTROELASTIC EQUATIONS OF STATE/ COMPILATION OF CONVERSION FORMULAS INTERRELATING THE LINEAR E 03-0277
                 SHOCK WAVE INDUCED ELECTROMECHANICAL CONVERSION IN FERROELECTRIC MATERIALS.   03-0491
                   FERROELECTRIC CERAMICS AND ENERGY CONVERSION. (LEAD ZIRCONATE TITANATE)   09-1508
IC EFFECT IN POTASSIUM DIHYDROGEN PHOS/ PARAMETRIC FREQUENCY CONVERSION OF COHERENT LIGHT BY THE ELECTROOPT 17-2616
RCONATE TITANATE)   FERROELECTRIC CERAMICS AND CONVERSION OF MECHANOELECTRIC ENERGY. (LEAD ZI 09-1483
                    LITHIUM NIOBATE SILICON SURFACE WAVE CONVOLUTER.   11-1788
NONLINEARITY PARAMETERS IN LITHIUM NIOBATE.   ACOUSTIC CONVOLUTION AND CORRELATION AND THE ASSOCIATED 11-1758
RMINATION OF A NONLINEAR PARAMETER FOR ACOUSTIC SURFACE WAVE CONVOLUTION IN LITHIUM NIOBATE.   DETE 11-1681
SEMICONDUCTOR ON LITHIUM NIOBATE.   CONVOLUTION OF SURFACE WAVES IN A STRUCTURE OF 11-1847
THIUM NIOBATE AND BISMUTH(12) GERMANI/ ACOUSTIC SURFACE WAVE CONVOLUTION ON CRYSTALS OF CADMIUM SULFIDE, LI 11-1692
IUM NIOBATE STRUCTURE.   SURFACE WAVE CONVOLUTION USING A CADMIUM SULFIDE FLUID LITH 11-1700
ENT RADIATION.   AN OBSERVATION OF THE CHANGE IN COOLED KDP TEMPERATURE DUE TO 337 MICRON INCID 17-2711
(BARIUM SODIUM NIOBATE AND SODIUM LANTHANUM / CRUCIBLE BASE COOLING: AN AID TO CZOCHRALSKI CRYSTAL GROWTH. 13-1928
/HARMONIC GENERATION IN POLYVINYLIDENE FLUORIDE FILMS (AFTER COOLING STRETCHED FILM IN ELECTRIC FIELD, THE/ 36-3530
       DOMAIN STRUCTURE OF TRIGLYCINE SELENATE CRYSTAL (DURING COOLING THROUGH CURIE POINT).   27-3043
EAR CURIE TEMPERATURE.   EQUATION OF STATE FOR THE COOPERATIVE TRANSITION OF TRIGLYCINE SULFATE N 27-3006
/TONIAN IN TERMS OF LOCALIZED D-STRAIN AND SOFT- NORMAL MODE COORDINATES AND TEMPERATURE INDEPENDENT PARAM/ 03-0432
M, DYSPROSIUM, IRON, SELENIUM, NIOBIUM, COBALT, SCANDIUM, CR COPPER) /AGNESIUM, CADMIUM, YTTRIUM, GADOLINIU 10-1640
/T OF A POLARIZING FIELD AND OF THE ADDITION OF A MIXTURE OF COPPER AND CHROMIUM IONS ON THE DIELECTRIC PR/ 27-3045
DEFECT ROCHELLE SALT AND/ IONIZATION OF THE (X-RADIATION AND COPPER AND IRON IMPURITY) ABSORPTION BANDS OF 27-3108
INE SULF/ ELECTRON PARAMAGNETIC RESONANCE INVESTIGATIONS OF (COPPER) DOPED FERROELECTRIC CRYSTALS. (TRIGLYC 27-3132
       EPR INVESTIGATION OF THE INTERNAL ELECTRIC FIELD IN COPPER DOPED ROCHELLE SALT.   28-3233
            X-RAY DIFFRACTION STUDY OF PHASE TRANSITION IN COPPER FORMATE.   26-2938
METHOD. (ANTIFERROELECTRIC)   ANTIPOLARIZATION IN COPPER FORMATE CRYSTALS BY THE ELECTROOPTICAL 26-2925
ESONANCE STUDY OF SUBLATTICE ROTATIONS AND SPIN DEVIATION IN COPPER FORMATE TETRADEUTERATE.   PROTON R 26-2950
LATTICE MAGNETIZATION IN THE TWO DIMENSIONAL ANTIFERROMAGNET COPPER FORMATE TETRADEUTERATE. /PONTANEOUS SUB 26-2929
US SHIFT OF ESR LINES IN THE TWO DIMENSIONAL ANTIFERROMAGNET COPPER FORMATE TETRAHYDRATE.   ANOMALO 26-2941
               MAGNETISM AND THE PHASE TRANSITION OF COPPER FORMATE TETRAHYDRATE.   26-2930
                RAMAN AND INFRARED SPECTRA OF COPPER FORMATE TETRAHYDRATE.   26-2936
THERMAL CONDUCTION IN A TWO DIMENSIONAL ANTIFERROMAGNETIC COPPER FORMATE TETRAHYDRATE.   26-2932
UNIT CELL AND SPACE GROUP OF THE ANTIFERROELECTRIC PHASE OF COPPER FORMATE TETRAHYDRATE.   26-2927
/RIC RELAXATION IN SEVERAL CRYSTALS OF ORDER DISORDER TYPE. (COPPER FORMATE TETRAHYDRATE, AMMONIUM IRON AN/ 02-0087
FORMATE TETRADEUT/ VIBRATION SPECTRUM OF THE FORMATE ION IN COPPER FORMATE TETRAHYDRATE AND COPPER DEUTERO 26-2926
IC RESONANCE.   STUDY OF THE ORDERED MAGNETIC STATE OF COPPER FORMATE TETRAHYDRATE BY ANTIFERROMAGNET 26-2943
TION.   APPROXIMATE STRUCTURE FOR ANTIFERROELECTRIC PHASE OF COPPER FORMATE TETRAHYDRATE BY NEUTRON DIFFRAC 26-2935
          DIELECTRIC RESPONSE FUNCTION OF COPPER FORMATE TETRAHYDRATES.   26-2933
FLUOBERYLLATE MONOCRYSTAL.   EPR OF THE FOUR COPPER ION COMPLEX WITH SPIN S=2 IN TRIGLYCINE 27-3134
        PARAMAGNETIC RELAXATION AND DIVALENT COPPER IONS IN FERROELECTRIC KDP.   17-2629
     ELECTRON PARAMAGNETIC RESONANCE OF COPPER IONS IN ROCHELLE SALT.   28-3176
            THE SYSTEM LEAD TITANATE- STRONTIUM COPPER NIOBATE. (CERAMIC SYNTHESIS)   05-1044
E STRUCTURE (PEROVSKITE COMPOUNDS CONTAINING EITHER BISMUTH, COPPER, OR LEAD). /IC MATERIALS WITH PEROVSKIT 08-1460
ARK EFFECT AND FERROELECTRICITY IN THE ORGANIC SEMICONDUCTOR COPPER PHTHALOCYANINE.   ST 36-3532
IN A COMPOUND WITH HELICAL ELECTRIC DIPOLE STRUCTURE: CESIUM COPPER TRICHLORIDE.   A PHASE TRANSITION 33-3388
/NDUCTIONS IN THE LOW DIMENSIONAL ANTIFERROMAGNETS POTASSIUM COPPER TRIFLUORIDE AND COPPER FORMATE TETRAHY/ 26-2939
/(TUNGSTEN OR MOLYBDENUM) OXIDE(6), (BARIUM OR STRONTIUM) (2) COPPER TUNGSTEN OXIDE(6), AND (BARIUM OR STRO/ 36-3619
COMPLEXES IN GLYCINE CRYSTALS. EPR STUDIES OF CHROMIUM(3+), COPPER(2+), AND MANGANESE(2+) PARAMAGNETIC ION 27-3135
M ALUMINUM SULFATE HEXAHYDRATE.   EPR STUDY OF MAGNETIC COPPER(2+) COMPLEXES IN CRYSTALS OF GUANIDINIU 21-2820
     EFFECT OF APPLIED ELECTRIC FIELD ON EPR SPECTRA OF (COPPER(2+) DOPED FERROELECTRIC ROCHELLE SALT. 28-3185
                        ESR OF COPPER(2+) DOPED SODIUM AMMONIUM TARTRATE. 29-3252
RESONANCE SPECTRUM IN (MANGANESE DOPED) LITHIUM NIOBATE AND (COPPER(2+) DOPED) ZINC BORACITE. /ARAMAGNETIC 11-1792
ON DOMAIN STRUCTURE IN SEIGNETTE SALT. (EFFECT OF AGING AND COPPER(2+) DOPING)   EFFECT OF A DC FIELD 28-3230
PREPARATION OF BARIUM TITANATE AND OTHER CERAMIC POWDERS BY COPRECIPITATION OF CITRATES IN AN ALCOHOL. 04-0861
AND DEUTERATED TGS (HYSTERESIS LOOPS JUST BELOW CURIE POINT CORRECTED FOR ELECTROCALORIC EFFECT). /RIC TGS 27-2966
THEORY OF ORDER DISORDER FERROELECTRICS: FIRST PERTURBATIVE CORRECTION.   STATISTICAL PERTURBATION 03-0399
ERROELECTRICS (LITHIUM TANTALATE EXAMPLE).   INTERCELL CORRECTIONS FOR IONIC MOTION IN DISPLACEMENT F 03-0367
3) AND A(0.5)BO(3) TYPE PHASES AND THE IONIC / AN ATTEMPT TO CORRELATE THE FERROELECTRIC PROPERTIES OF ABO(03-0203
RAMETERS IN LITHIUM NIOBATE.   ACOUSTIC CONVOLUTION AND CORRELATION AND THE ASSOCIATED NONLINEARITY PA 11-1758
CE MICRODAMAGES ON SINGLE CRYSTALS OF SEIGNETTE SALT.   CORRELATION BETWEEN BARKHAUSEN NOISE AND SURFA 28-3208
BEHAVIOR IN SINTERED BARIUM TITANATE.   CORRELATION BETWEEN RHEOLOGICAL AND ELECTRICAL 04-1027
PERTIES OF FERROELECTRIC SUBSTANCES. (BARIUM TITANATE) CORRELATION BETWEEN STRUCTURE AND ELECTRIC PRO 04-0728
TRANSITIONS IN FERROELECTRICS. (THEORY APPLIED TO BARIUM T/ CORRELATION EFFECTS IN DISPLACEMENT TYPE PHASE 03-0497
              CALCULATION OF THE DIELECTRIC CORRELATION FACTOR OF CUBIC ICE.   34-3429
E THREE DIMENSIONAL EIGHT VERTEX MODEL AND THE PROTON PROTON CORRELATION FUNCTIONS IN ICE.   TH 34-3481
GEN ARS/ NMR AND DIELECTRIC RELAXATION STUDIES OF LONG RANGE CORRELATION IN FERROELECTRIC POTASSIUM DIHYDRO 18-2718
EXCHANGE NARROWING: MAGNETIC RESONANCE LINE SHAPES AND SPIN CORRELATION IN PARAMAGNETIC POTASSIUM MANGANE/ 33-3376
GLE CRYSTALS.   ANGULAR CORRELATION OF ANNIHILATION PHOTONS IN ICE SIN 34-3457
/ITANATE(1-X) FOR TWO RHOMBOHEDRAL PHASES. (X=0.9 AND 0.58) (CORRELATION OF CATION SHIFTS WITH BISMUTH FER/ 09-1574
/C TRANSITION TEMPERATURES OF TUNGSTEN BRONZE TYPE NIOBATES (CORRELATION OF CURIE TEMPERATURE WITH CHEMICA/ 13-2049

PERTURBED DIRECTIONAL CORRELATION OF SCANDIUM-44 IN LEAD TITANATE. 05-1055
C PHASES OF BARIUM TITANATE. PERTURBED DIRECTIONAL CORRELATION OF SCANDIUM-44 IN THE FERROELECTRI 04-0558
OF FERROELECTRIC BARIUM TITANATE. PERTURBED DIRECTIONAL CORRELATION OF SCANDIUM-44 WITHIN THE LATTICE 04-0691
/E COMPOUNDS OF HAFNIUM-181 STUDIED BY THE PERTURBED ANGULAR CORRELATION TECHNIQUE. (LEAD HAFNATE, TIN HAF/ 08-1276
DEFECT STRUCTURE OF TRIGLYCINE SULFATE. (CORRELATION WITH DOMAIN STRUCTURES) 27-3044
/PPLICATIONS. (BY PULLING FROM THE MELT, STRUCTURE- PROPERTY CORRELATION WITH LITHIUM NIOBATE AND POTASSIU/ 13-2045
NTIFERROELECTRIC LEAD ZIRCONATE STUDIED BY PERTURBED ANGULAR CORRELATIONS. /AL ELECTRIC FIELD GRADIENT IN A 07-1255
ND SODIUM NITRITE. DIPOLE CORRELATIONS IN MOLTEN ALKALI METAL NITRATES A 16-2389
NEUTRON SCATTERING ANALYSIS OF THE LINEAR DISPLACEMENT CORRELATIONS IN POTASSIUM TANTALATE. 36-3546
RITICAL TEMPERATURE. FLUCTUATIONS AND CORRELATIONS IN STRONTIUM TITANATE ABOVE THE C 06-1233
T (FROM LITHIUM NIOBATE CRYSTAL, AGR/ MEASUREMENTS OF PHOTON CORRELATIONS OF SECOND HARMONIC GENERATED LIGH 11-1701
ETCHED ROCHELLE SALT CRYSTALS. SURFACE STRUCTURES AND CORRESPONDING LIGHT FIGURES REVEALED BY WATER 28-3234
TE SINGLE CRYSTALS. ETCH PITS CORRESPONDING TO DISLOCATIONS IN BARIUM TITANA 04-1023
AIN BOUNDARIES OF TRIGLYCINE SULFATE. CORROSION FIGURES DUE TO THE FERROELECTRIC DOM 27-3142
SPECTRUM OF ICE I(H). TENTATIVE INTERPRETATION WITH A MIXED COULOMB VALENCE DYNAMICAL MODEL. /QUENCY RAMAN 34-3425
/PECTRUM AND ENERGY LEVELS OF CHROMIUM(3+) IONS AND EXCHANGE COUPLED CHROMIUM(3+) PAIRS IN LANTHANUM ALUMI/ 36-3565
ALLY ORDERED FERROELECTRICS. COUPLED ELECTROMAGNETIC SPIN WAVES IN MAGNETIC 03-0148
YEDATE, KDP, ORTHOFERRITES, DYSPROSIUM ANTIMONIDE, RUBIDIUM/ COUPLED FERROELASTIC CRYSTALS. (GADOLINIUM MOL 02-0036
/AND ELECTRON NUCLEAR DOUBLE RESONANCE OF THE PROTON LATTICE COUPLED MODE IN POTASSIUM DIHYDROGEN ARSENATE/ 18-2724
EN ARSENATE, STRONTIUM TITANA/ SOFT MODE RAMAN SPECTROSCOPY: COUPLED MODES (FERROELECTRICS, CESIUM DIHYDROG 06-1207
C) IN THE PHONON SPECTRA OF STRONTIUM TITANATE AND POTASSIU/ COUPLED MODES (OPTICAL, FERROELECTRIC- ACOUSTI 06-1157
DISPERSION IN BARIUM AND STRONTIUM TITANATES AND POTASSIUM/ COUPLED OPTICAL PHONON MODE THEORY OF INFRARED 04-0547
ING: KDP (FREQUENCY DEPENDENT SUSCEPTIBILITY). COUPLED ORDER DISORDER PHONON SYSTEM WITH DAMP 17-2667
ITANATE. COUPLED POLARIZATIONS OF A1 SYMMETRY IN BARIUM T 04-0619
/ MEASUREMENTS OF ACOUSTIC WAVE GENERATION IN PIEZOELECTRICS COUPLED TO A GUNN OSCILLATOR. (LITHIUM NIOBAT/ 11-1731
HYDROGEN BONDED FERROELECTRICS OF THE KDP TYPE (PHONON PRO/ COUPLED VIBRATIONS OF THE PROTON ION SYSTEM IN 17-2670
N LITHIUM NIOBATE. COUPLED WAVE ANALYSIS OF HOLOGRAPHIC STORAGE I 11-1826
CTRIC LEAKY SURFACE WAVE IN LITHIUM NIOBATE (ELECTROMAGNETIC COUPLING). PIEZOELE 11-1832
ERROELECTRIC ANTIMONY SULFOIODIDE. SOFT MODE COUPLING AND CRITICAL RAYLEIGH SCATTERING IN F 15-2232
FERRO- AND ANTIFERROELECTRIC COUPLING AND DIELECTRIC RELAXATION. 03-0354
TE. DEUTERON QUADRUPOLE COUPLING CONSTANT IN DEUTERATED AMMONIUM SULFA 19-2761
TEMPERATURE DEPENDENCE OF THE SODIUM-23 NUCLEAR QUADRUPOLE COUPLING CONSTANT IN SODIUM NITRITE. 16-2393
(STOICHIOMETRIC) LITHIUM NIOBATE (TEMPERATURE DEPENDENCE OF COUPLING CONSTANT 21-515K). / INVESTIGATION OF 11-1811
AND FERROELECTRIC PHASE TRANSITION ON LIT/ ELECTROMECHANICAL COUPLING EFFECTS ON THE DIELECTRIC PROPERTIES 29-3251
SOFT LATTICE VIBRATIONS AND VIBRATION COUPLING IN ANTIMONY SULFOIODIDE. 15-2233
TTERING). SOFT PHONON MODE AND MODE COUPLING IN ANTIMONY SULFOIODIDE (BY RAMAN SCA 15-2167
ASSIUM DIHYDROGEN ARSENATE. PROTON PHONON COUPLING IN CESIUM DIHYDROGEN ARSENATE AND POT 18-2733
RSENATE. DEUTERON QUADRUPOLE COUPLING IN FERROELECTRIC CESIUM DIDEUTERIUM A 18-2721
PHOSPHATE AND CESIUM/ RUBIDIUM-87 AND ARSENIC-75 QUADRUPOLAR COUPLING IN FERROELECTRIC RUBIDIUM DIHYDROGEN 17-2439
NITROGEN-14 QUADRUPOLE COUPLING IN PARAELECTRIC AMMONIUM SULFATE. 19-2749
ENT OF THE DISSIPATION ASSOCIATED WITH THE ELECTROMECHANICAL COUPLING IN PIEZOELECTRIC CERAMICS. MEASUREM 03-0514
RFACE WAVES ON LITHIUM NIOBATE. NEW HIGH FREQUENCY HIGH COUPLING LOW BEAM STEERING CUT FOR ACOUSTIC SU 11-1819
ACE WAVES ON LITHIUM NIOBATE. NEW HIGH COUPLING LOW DIFFRACTION CUT FOR ACOUSTIC SURF 11-1820
ROGEN AND DEUTERIUM HALIDE CRYSTALS. (RAMAN SPECTRA, DIPOLAR COUPLING MODEL COMPUTATION) /INAL MODES IN HYD 31-3326
ON OF TRIGLYCINE SULFATE. ELECTROMECHANICAL COUPLING NEAR THE FERROELECTRIC PHASE TRANSITI 27-3114
M DIHYDROGEN PHOSPHATE. (COLLECTIV/ EFFECT OF PROTON- PHONON COUPLING ON THE FERROELECTRIC MODE IN POTASSIU 17-2655
TRIC SUBSTRATES (LI/ PERTURBATION THEORY FOR ELECTROMAGNETIC COUPLING TO ELASTIC SURFACE WAVES ON PIEZOELEC 11-1751
/2: (DYNAMICAL PROPERTIES, TRANSVERSE ISING MODEL, SOFT MODE COUPLING TO TRANSVERSE OPTICAL PHONON). PART-/ 03-0384
TION IN ANTIMONY SULFOIODIDE. SOFT PHONON COUPLINGS IN THE PRESSURE INDUCED PHASE TRANSI 15-2240
ORDERING AND STABILI/ PEROVSKITE FAMILY AND CHEMICAL BOND. (COVALENCY, DIRECTIVITY CONDITIONS FOR ELECTRIC 03-0243
IONIC CHARACTERS FROM MOSSBAUER ISOMER SHIFTS. (DEGREE OF COVALENCY IN IRON DOPED BARIUM TITANATE) 04-0563
COVALENCY IN POTASSIUM MANGANESE TRIFLUORIDE. 33-3391
OUNDS. (BARIUM TITANATE, R/ INFLUENCE OF MADELUNG ENERGY AND COVALENCY ON THE STRUCTURE OF AB(5) OXIDE COMP 04-0761
/F SINGLE DOMAIN CRYSTALS OF FERROELECTRIC LITHIUM NIOBATE. (CRACK-FREE, THROUGH ADDITION OF MAGNESIUM OXI/ 11-1774
TE AND ION ETCHING WITHOUT CRATERS. GROWTH OF CRATERS ON ION ETCHED SURFACE OF LITHIUM NIOBA 11-1860
ION BOMBARDMENT AND COMPARISON RESULTS FOR SAPPH/ EXPANSION, CRAZING AND EXFOLIATION OF LITHIUM NIOBATE ON 11-1800
IC THEORY. CONCERNING THE CRITERION OF FERROELECTRICITY IN THE MICROSCOP 03-0293
SE TRANSITIONS) EXPERIMENTAL CRITERION OF FERROELECTRICITY. (SHARP OR DIFFU 03-0184
EG K TRANSITION OF POTASSIUM MANGANESE TRIFLUORIDE SINGLE C/ CRITICAL ANISOTROPIC FLUCTUATIONS AT THE 184 D 33-3369
FERROELECTRICS IN OPTICAL MEMORIES AND DISPLAYS: A CRITICAL APPRAISAL. 37-3628
RONTIUM TITANATE ABOVE THE CRITICAL TEMPERATURE. CRITICAL ASYMMETRY IN LOCAL FLUCTUATIONS IN ST 06-1166
TERATED TGS (HYSTERESIS LOOPS JUST BELOW CURIE POINT CORREC/ CRITICAL BEHAVIOR OF FERROELECTRIC TGS AND DEU 27-2966
SULFATE BY A DIGITAL TECHNIQUE. (MINUTE/ DIELECTRIC STUDY OF CRITICAL BEHAVIOR OF FERROELECTRIC TRIGLYCINE 27-3116
TRICS. CRITICAL BEHAVIOR OF HYDROGEN BONDED FERROELEC 03-0516
EN PHOSPHATE CRYSTALS. CRITICAL BEHAVIOR OF KDP AND RUBIDIUM DIHYDROG 17-2676
HE 105 K PHASE TRANSITION. CRITICAL BEHAVIOR OF STRONTIUM TITANATE NEAR T 06-1194
ORDER DISORDER TYPE FERROELECTRICS WITH A SHORT RANGE INTER/ CRITICAL BEHAVIOR OF THE PSEUDOSPIN MODEL FOR 03-0423
CRITICAL BEHAVIOR OF TRIGLYCINE FLUOBERYLLATE. 27-3115
RMAL NOISE MEASUREMENTS. STUDY OF THE CRITICAL BEHAVIOR OF TRIGLYCINE SULFATE BY THE 27-3147
SODIUM NITRI/ NUCLEAR QUADRUPOLE SPIN LATTICE RELAXATION AND CRITICAL DYNAMICS OF FERROELECTRIC CRYSTALS. (16-2268
THE CRITICAL TEMPERATURE. CRITICAL DYNAMICS OF STRONTIUM TITANATE ABOVE 06-1204
ORMATION IN FERROELECTRICS. (THEORY) CRITICAL DYNAMICS OF THE ORDER DISORDER TRANSF 03-0281
ENTS AND VELOCITIES OF THE DOMAIN WALLS IN KDP. EXISTENCE OF CRITICAL ELECTRIC FIELDS. DISPLACEM 17-2448
CRITICAL FIELD IN CERAMIC BARIUM TITANATE. 04-0951
PHOSPHATE. LIGHT SCATTERING FROM CRITICAL FLUCTUATIONS IN POTASSIUM DIDEUTERIUM 17-2463
CRITICAL FLUCTUATIONS IN TRIGLYCINE SULFATE. 27-2974
ANATE. CRITICAL NEUTRON SCATTERING FROM STRONTIUM TIT 06-1195
ATE AND POTASSIUM MANGANESE TRIFLUORIDE. CRITICAL NEUTRON SCATTERING IN STRONTIUM TITAN 06-1208
ATION OF STATE) CRITICAL PHENOMENA IN ANTIFERROELECTRICS. (EQU 03-0343
CRITICAL PHENOMENA IN ANTIMONY SULFOIODIDE. 15-2231
CTING STRONTIUM TITANATE. (ABSTRACT) ULTRASONIC STUDIES OF CRITICAL PHENOMENA IN INSULATING AND SEMICONDU 06-1118
IODIDE. CRITICAL POINT IN FERROELECTRIC ANTIMONY SULFO 15-2237
DIELECTRIC RELAXATION PROCESSES NEAR THE CRITICAL POINT IN FERROELECTRICS. 03-0272
RROMAGNETIC PHASE TRANSITIONS. CRITICAL POINT INDICES OF FERROELECTRIC AND FE 03-0264
ELECTRIC TRANSITIONS FROM THE POINT OF VIEW OF THE THEORY OF CRITICAL POINTS. ANOMALIES NEAR FERRO 03-0298
/ DIELECTRIC INSTABILITY (FERROELECTRIC TWINNING) TREATED AS CRITICAL POINTS IN CRYSTALS. (ANTIMONY SULFOI/ 15-2176
ANTIFERRO/ ULTRASONIC INVESTIGATION OF PHASE TRANSITIONS AND CRITICAL POINTS. SECTION-5: FERROELECTRIC AND 02-0070
/ION BETWEEN THE STARTING FIELD FOR BARKHAUSEN STEPS AND THE CRITICAL POLARIZATION AND REPOLARIZATION FIEL/ 03-0318
TITANATE (ABOVE THE FERROELECTRIC CURIE TEMPERATURE, AS SH/ CRITICAL (POLARIZATION) FLUCTUATIONS IN BARIUM 04-1032
OELASTIC BEHAVIOR OF BARIUM TITANATE. INFLUENCE OF CRITICAL POLARIZATION FLUCTUATIONS ON THE PHOT 04-0628

S DETERMINED BY AN ELECTRIC FIELD METHOD.   CRITICAL PROPERTIES OF TGS AND ROCHELLE SALT A 27-3056
TIES OF FERROELECTRICS IN MEMORY DEVICE/ UTILIZATION OF THE (CRITICAL PULSE WIDTH) PARTIAL SWITCHING PROPER 27-3144
ANTIMONY SULFOIODIDE.   SOFT MODE COUPLING AND CRITICAL RAYLEIGH SCATTERING IN FERROELECTRIC 15-2232
     MANGANESE(2+) ABSORPTION NEAR THE CRITICAL REGION IN AMMONIUM SULFATE. 19-2785
LFATE.   CRITICAL REGION IN FERROELECTRIC TRIGLYCINE SU 27-2985
RON INELASTIC SCATTERING)   FERROELECTRIC CRITICAL SCATTERING FROM DEUTERATED KDP. (NEUT 17-2610
E SULFATE.   X-RAY CRITICAL SCATTERING IN FERROELECTRIC TRIGLYCIN 27-2998
SODIUM NITRATE. (ORDER DISORDER TRANSITION, FERROELEC/ X-RAY CRITICAL SCATTERING IN TRIGLYCINE SULFATE AND 16-2400
EG C AND AT 215 DEG C.   X-RAY CRITICAL SCATTERING OF SODIUM NITRITE AT 160 D 16-2343
E.   CRITICAL SCATTERING OF X-RAYS IN SODIUM NITRIT 16-2270
IONS (TRIGLYCINE SULFATE).   CRITICAL SLOWING DOWN AT FERROELECTRIC TRANSIT 27-3155
TION IN SILVER SODIUM DINITRITE BY THERMAL NOI/ STUDY OF THE CRITICAL SLOWING DOWN OF THE DIELECTRIC RELAXA 16-2382
IELECTRIC RELAXATION IN SEVERAL CRYSTALS OF ORDER DISORDER / CRITICAL SLOWING DOWN PROCESS OF (ANOMALOUS) D 02-0087
     CRITICAL STATES OF FERROELECTRICS. (THEORY) 03-0238
LUCTUATIONS AND CORRELATIONS IN STRONTIUM TITANATE ABOVE THE CRITICAL TEMPERATURE.   F 06-1233
OTASSIUM DIHYDROGEN PHOSPHATE AND SILVER PERIODATE BELOW THE CRITICAL TEMPERATURE. /SCATTERING SPECTRA OF P 17-2687
ND SUSCEPTIBILITY IN CHROMIUM METHYL AMMONIUM ALUM ABOVE THE CRITICAL TEMPERATURE.   SPECIFIC HEAT A 20-2805
IN POTASSIUM MANGANESE TRIFLUORIDE (ANOMALOUS HEAT CAPACITY, CRITICAL TEMPERATURE ABOUT 186-65 DEG K). /NT 33-3373
ANOMALY OF THE PROTON SPIN LATTICE RELAXATION TIME NEAR THE CRITICAL TEMPERATURE OF DICALCIUM LEAD PROPIO/ 26-2947
RA I/ DYNAMICS AT STRUCTURAL TRANSITIONS ABOVE FERROELECTRIC CRITICAL TEMPERATURE (ROTATIONS OF BO6 OCTAHED 06-1205
ISPLACIVE TYPE FERROELECTRICS (THEORY).   ATTENUATION OF CRITICAL VIBRATIONS AND DIELECTRIC LOSSES IN D 03-0159
UM MANGANESE TRIFLUORIDE AND SODIUM NIOBATE SIN/ ANISOTROPIC CRITICAL X-RAY DIFFUSE SCATTERING FROM POTASSI 08-1312
NIOBATE CRYSTALS (NEAR THE 640 DEG C PHASE TRAN/ ANISOTROPIC CRITICAL X-RAY DIFFUSE SCATTERING FROM SODIUM 08-1311
TITANATE NEAR THE CURIE TEMPERATURE.   CRITICAL X-RAY INCOHERENT SCATTERING BY BARIUM 04-0535
YDROGEN PERIODATE CRYSTALS, PULSED MAGNETIC/ LEVEL CROSSING (CROSS RELAXATION) EXPERIMENTS IN AMMONIUM TRIH 36-3577
RIC SEMICONDUCTOR ANTIMONY SU/ TEMPERATURE DEPENDENCE OF THE CROSS SECTION OF CAPTURE CENTERS IN FERROELECT 15-2147
     INELASTIC SCATTERING CROSS SECTION OF NEUTRONS IN FERROELECTRICS. 03-0297
M TITANATE.   TEMPERATURE DEPENDENCE OF THE RAMAN CROSS SECTIONS IN BARIUM TITANATE AND STRONTIU 04-0548
ONIUM TRIHYDROGEN PERIODATE CRYSTALS, PULSED MAGNETIC/ LEVEL CROSSING (CROSS RELAXATION) EXPERIMENTS IN AMM 36-3577
IN ANTIFERROELECTRIC DIAMMONIUM TRIHYDROGEN PERIODATE PROTON CROSSOVER RELAXATION. /QUADRUPOLE INTERACTION 36-3578
ROELECTRIC DIAMMONIUM TRIHYDROGEN PERIODATE BY IODINE PROTON CROSSOVER RELAXATION. / INTERACTION IN ANTIFER 36-3579
RYSTAL GROWTH. (BARIUM SODIUM NIOBATE AND SODIUM LANTHANUM / CRUCIBLE BASE COOLING: AN AID TO CZOCHRALSKI C 13-1928
     ELECTRICAL CONDUCTIVITY OF SINGLE CRYSTALLINE BARIUM TITANATE. 04-0663
NDUCTIVITY, THERMOELECTRIC PROPERTIES, AND BAND STRUCTURE OF CRYSTALLINE BISMUTH SELENOIODIDE.   CO 15-2126
IES OF CERAMIC BARIUM TITANATE.   EFFECT OF CRYSTALLINE GRAIN SIZE ON THE PHYSICAL PROPERT 04-0785
ERTIES OF THE IRON TRANSITION GROUP IN THE LATTICE OF SINGLE CRYSTALLINE LITHIUM NIOBATE.   PROP 11-1669
EFFECT OF INTENSE LIGHT ON THE REFRACTIVE INDEX OF A CRYSTALLINE POTASSIUM NIOBATE. 08-1410
RELAXATIONAL PROTON ORDERING AND GLASSY CRYSTALLINE STATE IN HEXAGONAL ICE. 34-3436
ME OPTICAL PROPERTIES OF ANTIMONY SULFOBROMIDE IN GLASSY AND CRYSTALLINE STATES.   PRODUCTION AND SO 15-2142
E SEMICONDUCTOR CERAMIC BARIUM TIT/ ELECTRICAL PROPERTIES OF CRYSTALLITES AND INTERCRYSTALLITE LAYERS IN TH 04-0568
N STRUCTURE OF ANTIMONY SULFOIODIDE CRYSTALS (FROM SELECTIVE CRYSTALLIZATION AND ETCHING).   DOMAI 15-2183
TITANATES   HYDROTHERMAL CRYSTALLIZATION AND SOME PROPERTIES OF BISMUTH 08-1279
DE. (SINGLE CRYSTAL GROWTH, CHEMICAL TRANSPORT)   CRYSTALLIZATION BEHAVIOR OF ANTIMONY SULFOIODI 15-2201
MPOSITION OF FERROELECTRIC CRYSTALS (SOLUTION GRO/ EFFECT OF CRYSTALLIZATION CONDITIONS ON THE SHAPE AND CO 02-0089
REVEALING THE DOMAIN STRUCTURE OF TGS CRYSTALS BY SELECTIVE CRYSTALLIZATION OF ANTHRAQUINONE (SUBLIMED ON/ 27-3041
AL GROWTH RATE, ANISOTROPY)   KINETICS OF CRYSTALLIZATION OF ANTIMONY SULFOIODIDE (CRYST 15-2215
THERMAL SYNTHESIS)   CRYSTALLIZATION OF BISMUTH SULFOIODIDE. (HYDRO 15-2212
.   CRYSTALLIZATION OF LEAD METANIOBATE FROM GLASS 08-1275
ID SOLUTIONS.   HYDROTHERMAL CRYSTALLIZATION OF LEAD TITANATE ZIRCONATE SOL 09-1482
     CRYSTALLIZATION OF POTASSIUM IODATE. 08-1367
GRAVIMETRIC MEASUREMENTS OF THE CHANGES OF WATER OF CRYSTALLIZATION OF ROCHELLE SALT. 28-3235
IELDS (KINETICS OF CRYSTAL GROWTH)   CRYSTALLIZATION OF ROCHELLE SALT IN ELECTRIC F 28-3200
ATE FROM SOLUTIONS IN HEAVY WATER.   CRYSTALLIZATION OF RUBIDIUM DIDEUTERIUM PHOSPH 17-2583
GROUP A(V)B(VI)C(VII). (HIGH GRADE SINGLE CRY/ HYDROTHERMAL CRYSTALLIZATION OF SEMICONDUCTING COMPOUNDS OF 15-2214
SITION A(V)B(VI)C(VII) UNDER HYDROTHERMAL CONDITIONS. (ANTI/ CRYSTALLIZATION OF SEMICONDUCTORS OF THE COMPO 15-2213
IOBATE FROM GLASS.   CRYSTALLIZATION OF SOLID SOLUTIONS OF SODIUM N 08-1386
ASSIUM NITRATE. (CRYSTAL GRO/ INFLUENCE OF THE CONDITIONS OF CRYSTALLIZATION ON THE FORM OF CRYSTALS OF POT 16-2332
S. (POTASSIUM BROMIDE, POTASSIUM CHLORIDE, SODIUM CHLORIDE,/ CRYSTALLIZATION POTENTIALS IN AQUEOUS SOLUTION 17-2707
N ITS FERROELECTRIC PROPERTIES.   INFLUENCE OF THE CRYSTALLIZATION TEMPERATURE OF ROCHELLE SALT O 28-3193
/ICAL PROPERTIES OF MICROCRYSTALLINE FERROELECTRIC MATERIALS CRYSTALLIZED FROM GLASS. (60-90 VOL PERCENT, / 04-0725
CONDITIONS AND CLASSIFICATION)   CRYSTALLOCHEMISTRY OF PEROVSKITES. (EXISTENCE 08-1324
S/ EFFECT OF THE SUBSTITUTION OF TANTALUM FOR NIOBIUM ON THE CRYSTALLOGRAPHIC AND DIELECTRIC PROPERTIES OF 13-1921
BISMUTH PEROVSKITES. (AND THE FERROELECTRIC STATE, BISMUT/ (CRYSTALLOGRAPHIC AND DIELECTRIC) PROPERTIES OF 08-1289
TICAL PROPERTIES OF THE ABO NIOBAT/ COMPARATIVE STUDY OF THE CRYSTALLOGRAPHIC, DIELECTRIC, AND NONLINEAR OP 04-0912
IES OF (MONOCLINIC) EUROPIUM SILICATE. (UNSTABLE FERROELEC/ CRYSTALLOGRAPHIC, OPTICAL AND MAGNETIC PROPERT 36-3620
TURE).   X-RAY DIFFRACTION STUDY OF GADOLINIUM MOLYBDATE (CRYSTALLOGRAPHIC PHASE CHANGE AT CURIE TEMPERA 30-3312
/ II-VI SEMICONDUCTING COMPOUNDS - DATA TABLES. (MECHANICAL, CRYSTALLOGRAPHIC, PHYSICAL, THERMAL MAGNETIC, 01-0025
4) TYPE CALCIUM AND STRONTIUM SCANDATES.   PREPARATION AND CRYSTALLOGRAPHIC PROPERTIES OF A(2+)B(3+)(2)O(36-3537
TANATE PARTICLES.   ON THE ANOMALOUS CRYSTALLOGRAPHIC PROPERTIES OF SMALL BARIUM TI 04-0935
THE RHENIUM TRIOXIDE STRUCTURE TYPE: AN ELECTRON MICRO/ CS (CRYSTALLOGRAPHIC SHEAR) FAMILIES DERIVED FROM 10-1630
DE.   X-RAY CRYSTALLOGRAPHIC STUDIES OF ANTIMONY SULFOIODI 15-2109
TREATED TRIGLYCINE SULFATE: NEW STRUCTURE FORM.   CRYSTALLOGRAPHIC STUDIES OF IRRADIATION FIELD 27-2996
ORTHO FLUOBERYLLATES.   CRYSTALLOGRAPHIC STUDY OF SEVERAL HYDRAZINIUM 32-3355
E. (ANTIFERROELECTRIC TEMPERATURE 340/ TWO COMPONENTS OF THE CRYSTALLOGRAPHIC TRANSITION IN VANADIUM DIOXID 36-3562
EORY OF STATIC DOMAIN STRUCTURES OF FERROELECTRIC CRYSTALS. (CRYSTALLOGRAPHIC VIEWPOINT)   MICROSCOPIC TH 03-0340
TRICS, FERROELECTRIC PHASE TRANSITIONS AND SPECIAL FEATURES/ CRYSTALLOGRAPHICAL CLASSIFICATION OF FERROELEC 03-0480
ISMUTH TUNGSTATE.   GROWTH, CRYSTALLOGRAPHY AND DIELECTRIC PROPERTIES OF B 10-1650
RICS. (BASED ON DOMAIN STRUCTURE AND ITS APPLICATIONS)   CRYSTALLOPHYSICAL CLASSIFICATION OF FERROELECT 03-0463
FROM THE RHENIUM TRIOXIDE STRUCTURE TYPE: AN ELECTRON MICRO/ CS (CRYSTALLOGRAPHIC SHEAR) FAMILIES DERIVED 10-1630
HE AMORPHOUS ICE AND OF THE PHASE TRANSFORMATION BETWEEN THE CUBIC AND HEXAGONAL ICES. /ASS TRANSITION OF T 34-3476
     DIELECTRIC BEHAVIOR OF CUBIC AND HEXAGONAL ICES AT LOW TEMPERATURES. 34-3431
GERMANIUM OXIDE(20) AND BARI/ PIEZOELECTRIC SURFACE WAVES IN CUBIC AND ORTHORHOMBIC CRYSTALS. (BISMUTH(12) 13-2041
     BAND STRUCTURE OF CUBIC AND TETRAGONAL BARIUM TITANATE. 04-0846
ZED PHASE OF BARIUM TITANAT/ CHANNELING OF LIGHT IONS IN THE CUBIC AND TETRAGONAL (FERROELECTRICALLY POLARI 04-0688
/G STATE TO THE 6A1G(6S) STATE OF THE MANGANESE(2+) ION IN A CUBIC ANTIFERRODIELECTRIC AN EXTERNAL MAGNETI/ 03-0317
     NEUTRON SCATTERING STUDY OF SOFT MODES IN CUBIC BARIUM TITANATE. 04-0712
TERPRETATION OF INFRARED SPECTRA AND THE LATTICE DYNAMICS OF CUBIC BARIUM TITANATE.   ON THE IN 04-0659
     ON THE LATTICE VIBRATION OF CUBIC BARIUM TITANATE. 04-0775
     POLARITON DISPERSION RELATION IN CUBIC BARIUM TITANATE. 04-0878
     TEMPERATURE FACTORS OF TETRAGONAL AND CUBIC BARIUM TITANATE. 04-0532

```
                       OPTICAL ANISOTROPY OF THE CUBIC COBALT IODINE BORACITE.                                14-2094
NATE.                            THEORY OF ELECTROSTRICTION IN CUBIC CRYSTALS WITH APPLICATION TO BARIUM TITA   03-0267
L PROPERTIES, / PHASES AND POLARIZATION OF LUMINOUS WAVES IN CUBIC ELECTROOPTIC CRYSTALS. (NONLINEAR OPTICA   02-0044
ERATURES ABOVE 80 DEGREES/ RAMAN SPECTRA OF ORTHORHOMBIC AND CUBIC HYDROGEN AND DEUTERIUM CHLORIDES AT TEMP   31-3330
        CALCULATION OF THE DIELECTRIC CORRELATION FACTOR OF CUBIC ICE.                                        34-3429
                             MOSSBAUER EFFECT OF FERROUS IONS IN CUBIC ICE.                                        34-3416
                       SYMMETRY ANALYSIS AND ELECTRONIC STATES IN CUBIC ICE.                                        34-3462
IATION BE/ TEMPERATURE DEPENDENCE OF DIELECTRIC CONSTANTS OF CUBIC IONIC COMPOUNDS. (HALIDES AND OXIDES, RE   03-0188
TRANSITIONS.                        OPTICAL ANISOTROPY OF CUBIC NICKEL IODINE BORACITE DUE TO QUADRUPOLE     14-2093
            GROUP ANALYSIS OF LATTICE VIBRATIONS OF CUBIC PEROVSKITE ABO(3).                               08-1321
MAGNETIC G STRUCT/ NEUTRON DIFFRACTION STUDY OF (TETRAGONAL- CUBIC PEROVSKITE) STRONTIUM IRON TUNGSTATE . (   10-1655
/PENDENCE OF OPTICAL PHONON FREQUENCIES (RAMAN SCATTERING IN CUBIC PEROVSKITE STRONTIUM TITANATE, 8-250K,     06-1241
/F MAGNESIUM AND NIOBIUM IN THE OCTAHEDRAL POSITIONS OF THE "CUBIC" PEROVSKITE STRUCTURE OF LEAD MAGNESIUM/   08-1379
ONTIUM TITANATE.                      LATTICE DYNAMICS OF CUBIC PEROVSKITE STRUCTURES, IN PARTICULAR STR    06-1185
/E GRUNEISEN PARAMETER OF THE SOFT FERROELECTRIC MODE IN THE CUBIC PEROVSKITES. (STRONTIUM TITANATE, POTAS/   02-0114
               THEORETICAL DYNAMIC STUDY OF CUBIC PHASE BARIUM TITANATE.                               03-0261
        GENERATION OF SECOND OPTICAL HARMONIC IN THE CUBIC PHASE OF BARIUM TITANATE.                        04-0538
LECTRIC CONSTANT)                    BAND STRUCTURE OF THE CUBIC PHASE OF BARIUM TITANATE. (IMAGINARY DIE     04-0845
E OF IRON(3+) IN BARIUM TITANATE IN THE RHOMBOHEDRAL AND THE CUBIC PHASES.     ELECTRON PARAMAGNETIC RESONANC   04-0929
T LIGHT VALVES IN DISPLAY AND MODULATOR APPLICATIONS. USE OF CUBIC PHOTOSENSITIVE CRYSTALS AS POCKELS EFFEC   37-3661
RICAL CONDUCTIVITY OF A FERROELECTRIC SEMICONDUCTOR NEAR THE CUBIC POINT.          ANOMALIES IN THE ELECT     03-0268
             NEUTRON STUDY OF DIFFUSE SCATTERING IN CUBIC POTASSIUM NIOBATE.                            08-1407
      NEUTRON SCATTERING STUDY OF THE SOFT MODES IN CUBIC POTASSIUM TANTALATE NIOBATE.                 13-2064
/MPERATURE AND PROPAGATION DIRECTION UPON OPTICAL PHONONS IN CUBIC SODIUM CHLORIDE AND EPITAXIAL FERROELEC/   29-3243
TES)                    THERMAL CONDUCTIVITY OF CUBIC SODIUM TUNGSTEN BRONZES. (SODIUM TUNGSTA   13-2057
RAMAGNETIC RESONANCE SPECTRA OF SUBSTITUTIONAL IMPURITIES IN CUBIC STRONTIUM TITANATE. / ON THE ELECTRON PA   06-1193
                 MICROWAVE ABSORPTION IN CUBIC STRONTIUM TITANATE.                             06-1212
SITIONS IN HIGH RESISTIVITY BARIUM TITANATE SINGLE CRYSTALS/ CUBIC TETRAGONAL (MARTENSITIC TYPE) PHASE TRAN   04-0887
ITANATE.             FIELD INDUCED NUCLEATION AT THE (CUBIC TETRAGONAL) PHASE TRANSITION IN BARIUM T   04-0753
IUM LEAD TRICHLORIDE.            PRESSURE DEPENDENCE OF CUBIC TETRAGONAL TRANSITION TEMPERATURE OF CES   33-3393
IUM TIT/ LIGHT SCATTERING STUDIES OF THE SOFT MODES NEAR THE CUBIC TO TETRAGONAL PHASE TRANSITION IN STRONT   06-1242
ION ENHANCED OPTICAL ABSORPTION IN STRONTIUM TITANATE AT THE CUBIC TO TETRAGONAL PHASE TRANSITION.   DISTORT   06-1109
ION IN STRONTIUM TITANATE.                CUBIC TO TRIGONAL STRESS INDUCED PHASE TRANSIT    06-1235
    THERMAL HYSTERESIS IN THE TETRAGONAL CUBIC TRANSITION OF BARIUM TITANATE.                 04-0921
TES.               THERMODYNAMIC THEORY OF THE CUBIC- ORTHORHOMBIC PHASE TRANSITION IN BORACI   03-0231
    NUCLEATION PROCESS IN BARIUM TITANATE AT CUBIC- TETRAGONAL PHASE TRANSITIONS.           04-0754
/IN STRONTIUM TITANATE (OCCURANCE IN THIN (110) PLATES BELOW CUBIC- TETRAGONAL TRANSITION, AND PRODUCTION /   06-1164
/ THE STRUCTURE OF SODIUM NIOBATE I(2) AT 600 DEG C, AND THE CUBIC- TETRAGONAL TRANSITION IN RELATION TO SO   08-1337
        THE SYSTEM LEAD TITANATE- STRONTIUM CUPRATE(0.33) NIOBATE(0.67).                       05-1045
LT.                            EFFECT OF THE CUPRIC ION ON DELAYED PHENOMENA IN ROCHELLE SA   28-3221
/OBATE, LITHIUM NIOBATE OR TANTALATE, BARIUM SODIUM NIOBATE, CUPROUS CHLORIDE, PROUSTITE, CALCIUM PYRONIOB/   01-0021
MIC WITH A HIGH EFFECTIVE DIELECTRIC CONSTANT (DUE TO ADDING CUPROUS OR FERRIC OXIDE). /ARIUM TITANATE CERA   04-0581
/UBIC IONIC COMPOUNDS. (HALIDES AND OXIDES, RELATION BETWEEN CURIE CONSTANT AND CURIE TEMPERATURE FOR FERR/   03-0188
AN ANTIMONY SULFOIODIDE FERROELECTRIC SEMICONDUCTOR NEAR ITS CURIE POINT.       ABSORPTION OF ULTRASOUND IN   15-2115
E CRYSTALS AT TEMPERATURES ABOVE AND BELOW THE FERROELECTRIC CURIE POINT. /F PROTONS IN THIN BARIUM TITANAT   04-0687
 CRYSTALS (APPLICATION TO SINGLE DOMAIN FERROELECTRICS BELOW CURIE POINT).       COMPLEX PHASE TRANSITIONS IN   03-0319
DOPED TRIGLYCINE SULFATE CRYSTALS GROWN BELOW AND ABOVE THE CURIE POINT.        DOMAIN STRUCTURE OF PURE AND   27-3130
CTURE OF TRIGLYCINE SELENATE CRYSTAL (DURING COOLING THROUGH CURIE POINT).         DOMAIN STRU   27-3043
ACE CHARGE INJECTION IN SOLIDS. (USING BARIUM TITANATE ABOVE CURIE POINT)          DOUBLE SP   04-0638
ES OF TRIGLYCINE SELENATE. (AGING AND PHASE CHANGES NEAR THE CURIE POINT) /FIELD ON THE DIELECTRIC PROPERTI   27-3131
THE REFRACTIVE INDEX OF PARAELECTRIC KDP IN THE VICINITY OF CURIE POINT.       EFFECT OF IONIC MOTION ON    17-2548
, DIELECTRIC, AND OPTICAL CONSTANTS) ROOM TEMPERATURE TO THE CURIE POINT. / NIOBATE (ELASTIC, PIEZOELECTRIC   13-2060
       ELASTIC CONSTANTS OF GADOLINIUM MOLYBDATE (NEAR THE CURIE POINT).                       30-3274
OMAIN STRUCTURE OF TRIGLYCINE SULFATE IN THE VICINITY OF THE CURIE POINT.           EQUILIBRIUM D   27-3112
BONDED LEAD ZIRCONATE. (PHASE TRANSITION 200 DEG C BELOW THE CURIE POINT) /ORCED FERROELECTRICITY IN GLASS   07-1252
E OF THE ABSORPTION OF SOUND IN ROCHELLE SALT NEAR ITS UPPER CURIE POINT.         FREQUENCY DEPENDENC   28-3183
    PHASE BOUNDARIES IN BARIUM TITANATE SINGLE CRYSTALS NEAR CURIE POINT.                       04-0577
CROWAVE LOSSES IN FERROELECTRICS. (AT TEMPERATURES ABOVE THE CURIE POINT)       PHENOMENOLOGICAL THEORY OF MI   03-0503
       POLARIZATION FLUCTUATION OF TRIGLYCINE SULFATE NEAR THE CURIE POINT.                   27-3077
RROELECTRIC CRYSTAL WITH A SECOND ORDER TRANSITION ABOVE THE CURIE POINT. / GENERATION OF HARMONICS IN A FE   03-0331
RICAL AND THERMAL PROPERTIES OF TRIGLYCINE SELENATE NEAR THE CURIE POINT.       RELATION BETWEEN ELECT   27-3168
S, SPONTANEOUS POLARIZATION OF FERROELECTRIC PLATES NEAR THE CURIE POINT. /ORY OF DOMAINLESS, NONHOMOGENEOU   03-0321
        THERMAL CONDUCTIVITY OF TRIGLYCINE SULFATE NEAR THE CURIE POINT.                       27-2983
        THERMAL CONDUCTIVITY OF TRIGLYCINE SULFATE NEAR THE CURIE POINT.                       27-3016
S IN PEROVSKITE MODIFICATION OF CADMIUM TITANATE./ ABOUT THE CURIE POINT AND NATURE OF DIELECTRIC PROPERTIE   08-1424
TOR IN FERROELECTRIC MATERIALS (ROCHELLE SALT NEAR THE UPPER CURIE POINT AS EXAMPLE). /ENCE OF THE LOSS FAC   28-3220
AL, AND ELECTROOPTICAL PROPERTIES IN THE NEIGHBORHOOD OF THE CURIE POINT (BIREFRINGENCE ANOMALIES). / OPTIC   19-2745
/LECTRIC TGS AND DEUTERATED TGS (HYSTERESIS LOOPS JUST BELOW CURIE POINT CORRECTED FOR ELECTROCALORIC EFFE/   27-2966
/ECTRIC VIBRATIONS IN ANTIMONY SULFOIODIDE CRYSTALS NEAR THE CURIE POINT (EFFECTS OF TEMPERATURE, FIELD ST/   15-2121
IZATION OF A POTASSIUM DIHYDROGEN PHOSPHATE CRYSTAL NEAR THE CURIE POINT. (ELECTROCALORIC EFFECT) /US POLAR   17-2678
/CIVE MOVEMENT OF THE RADICAL IN TRIGLYCINE SULFATE NEAR THE CURIE POINT. (ESR OF GAMMA-IRRADIATED SPECIME/   27-3033
/ECTRIC RESPONSE OF TGS IN THE FERROELECTRIC REGION NEAR THE CURIE POINT (IMPROVED METHOD OF MEASUREMENT).   27-2979
                  PYROCURRENTS HIGHER THAN THE CURIE POINT IN ANTIMONY SULFOIODIDE.           15-2131
AY STUDY ON THE DISORDERED STRUCTURE ABOVE THE FERROELECTRIC CURIE POINT IN POTASSIUM NITRATE.      X-R   16-2366
SYSTEMS. (DISCUSSE/ UNIT CELL VOLUME EFFECTS (RELATION WITH CURIE POINT) IN SOME ISOMORPHOUS FERROELECTRIC   17-2487
/ WAVELENGTH DISPERSION IN THE DIELECTRIC CONSTANT ABOVE THE CURIE POINT IN VARIOUS FERROELECTRICS AND ANT/   02-0124
            RELAXATION ABSORPTION OF SOUND NEAR THE CURIE POINT OF FERROELECTRIC SEMICONDUCTORS.   03-0492
/ND INTERNAL FRICTION IN THE REGION OF THE ANTIFERROELECTRIC CURIE POINT OF LEAD MAGNESATE(0.5) TUNGSTATE(/   13-2020
IS (POLARON H/ ELECTRICAL CONDUCTIVITY (SEEBECK COEFFICIENT, CURIE POINT) OF REDUCED BARIUM TITANATE CRYSTA   04-0918
            DOUBLE HYSTERESIS LOOPS AT THE UPPER CURIE POINT OF SODIUM TRIHYDROGEN SELENITE.   22-2831
PROP/ BEHAVIOR OF LITHIUM TANTALATE SINGLE CRYSTAL NEAR ITS CURIE POINT. PART-1: PIEZOELECTRIC AND OPTICAL   12-1877
PROP/ BEHAVIOR OF LITHIUM TANTALATE SINGLE CRYSTAL NEAR ITS CURIE POINT. PART-2, DIELECTRIC AND ULTRASONIC   12-1885
/IC MODEL OF LOCALLY ORDERED "NUCLEI", PHASE STRUCTURE ABOVE CURIE POINT, RAMAN SPECTRUM OF STRONTIUM TITA/   04-0647
LECTRICS POSSESSING FIRST ORDER TRANSITION. (THEORY, CALCUL/ CURIE POINT SHIFT WITH APPLIED FIELD IN FERROE   03-0327
F THE COMPLEX DIELECTRIC CONSTANT OF FERROELECTRICS NEAR THE CURIE POINT. (THEORY) /MPERATURE DEPENDENCES O   03-0361
CTOR).          INFLUENCE OF SCREENING ON THE FERROELECTRIC CURIE POINT (UNIAXIAL FERROELECTRIC PHOTOCONDU   03-0392
IC POLARIZATION AND MAGNETIZATION IN NICKEL IODIDE BORACITE (CURIE POINT 64 DEG K). /THE SPONTANEOUS ELECTR   14-2090
ITUDINAL ULTRASONIC WAVES IN ROCHELLE SALT CRYSTALS NEAR THE CURIE POINTS.          ABSORPTION OF LONG   28-3211
E TEMPERATURE RANGE 4.2-300 DEG K (LATTICE PARAMETERS VERSUS CURIE POINTS). /ATE LEAD TITANATE SYSTEM IN TH   08-1308
M TITANATE, POTASSIUM TANTALATE AND BARIUM TITANATE NEAR THE CURIE POINTS. THERMAL CONDUCTIVITY OF STRONTIU   04-0989
```

ATE CRYSTALS BY HYDROSTATIC PRESSURE.   DISPLACEMENT OF THE CURIE POINTS IN KDP AND LITHIUM THALLIUM TARTR 17-2523
VII) COMPOUNDS./ DETERMINATION OF THE LATTICE PARAMETERS AND CURIE POINTS OF SOLID SOLUTIONS OF A(V)B(VI)C(15-2116
E SALT.   EFFECT OF A TWO DIMENSIONAL PRESSURE ON THE CURIE POINTS OF TRIGLYCINE SULFATE AND ROCHELL 27-3082
/THE INTERNAL STRESS FOR FERROELECTRIC PHOSPHATES NEAR THEIR CURIE POINTS. (POTASSIUM DIHYDROGEN PHOSPHATE/ 17-2665
IN THE LEAD COBALT TUNGSTATE- CADMIUM MANGANATE SYSTEM (BOTH CURIE POINTS STUDIED, VARIOUS COMPOSITIONS/ 36-3604
TUTION.   CURIE SHIFT IN BARIUM TITANATE BY HETEROSUBSTI 04-0727
ICAL X-RAY INCOHERENT SCATTERING BY BARIUM TITANATE NEAR THE CURIE TEMPERATURE.   CRIT 04-0535
ELECTRIC ANOMALIES OF SODIUM NITRATE ABOVE THE FERROELECTRIC CURIE TEMPERATURE.   DI 16-2376
 ELECTRON PARAMAGNETIC RESONANCE IN BARIUM TITANATE NEAR THE CURIE TEMPERATURE.   04-0928
FACE MICROPLASMA OF CERAMIC (BARIUM, LEAD) TITANATE NEAR THE CURIE TEMPERATURE.   EMISSION OF LIGHT BY SUR 04-0982
TE FOR THE COOPERATIVE TRANSITION OF TRIGLYCINE SULFATE NEAR CURIE TEMPERATURE.   EQUATION OF STA 27-3006
NAL PERMITTIVITY OF TRIGLYCINE SULFATE AT 100 MHZ. (NEAR THE CURIE TEMPERATURE)   LARGE SIG 27-3029
              NEW FERROELECTRIC: ANTIMONY ORTHONIOBATE. (CURIE TEMPERATURE)   36-3589
YCINE NITRATE AND TRIS-SARCOSINE CALCIUM CHLORIDE. (SHIFT OF CURIE TEMPERATURE) /ECTRIC TRANSITIONS IN DIGL 16-2267
BINARY STRONTIUM OXIDE- NIOBIUM PENTOXIDE SYSTEM, DTA, X-RAY CURIE TEMPERATURE). /BIUM-RICH REGION AND THE 13-1920
ANCE OF BARIUM TITANATE SINGLE CRYSTALS IN THE REGION OF THE CURIE TEMPERATURE. /TURE COEFFICIENT OF RESIST 04-1036
PROTON DYNAMICS IN HYDROGEN BONDED FERROELECTRICS ABOVE THE CURIE TEMPERATURE.   03-0206
TIONS IN LITHIUM NIOBATE AND TANTALATE AND THE EFFECT ON THE CURIE TEMPERATURE. /EL FOR STOICHIOMETRY DEVIA 11-1776
TIONS IN LITHIUM NIOBATE AND TANTALATE AND THE EFFECT ON THE CURIE TEMPERATURE. /EL FOR STOICHIOMETRY DEVIA 11-1777
IC MEASUREMENT OF THE ELECTROCALORIC EFFECT IN KDP (NEAR THE CURIE TEMPERATURE).   ULTRASON 17-2664
DY OF GADOLINIUM MOLYBDATE (CRYSTALLOGRAPHIC PHASE CHANGE AT CURIE TEMPERATURE).   X-RAY DIFFRACTION STU 30-3312
NSION OF FERROELECTRIC POTASSIUM DIHYDROGEN PHOSPHATE. (NEAR CURIE TEMPERATURE) X-RAY STUDY ON THERMAL EXPA 17-2556
IN DISPLAC/ RELATIONSHIP BETWEEN IONIC DISPLACEMENT AND THE CURIE TEMPERATURE AND SPONTANEOUS POLARIZATION 03-0453
F SPONTANEOUS PO/ X-RAY DIFFRACTOMETRIC DETERMINATION OF THE CURIE TEMPERATURE AND TEMPERATURE DEPENDENCE O 03-0290
/N) FLUCTUATIONS IN BARIUM TITANATE (ABOVE THE FERROELECTRIC CURIE TEMPERATURE, AS SHOWN) BY NEUTRON SCATT/ 04-1032
/ECTRIC LOSSES IN OXYGEN OCTAHEDRON FERROELECTRICS ABOVE THE CURIE TEMPERATURE. (BARIUM(0-1.0) STRONTIUM(1/ 04-0559
CTRIC CRYSTALS. (5 COMPOUND/ SHAPE DEPENDENCE OF PROPERTIES (CURIE TEMPERATURE, CURIE CONSTANT) OF FERROELE 27-3120
ANTALATE SINGLE CRYSTALS AS A FUNCTION OF MELT COMPOSITION. (CURIE TEMPERATURE DEPENDENCE) /ES OF LITHIUM T 12-1874
TANTALATE NIOBATE. (POTASSIUM TANTALATE(0.65) NIOBATE(0.35), CURIE TEMPERATURE EQUALS 10 DEG C, ABSTRACT) 08-1391
ONATE.   ULTRASONIC ATTENUATION NEAR THE CURIE TEMPERATURE IN DICALCIUM STRONTIUM PROPI 26-2948
SPHATE AND DEU/ POSSIBLE THEORETICAL EXPLANATION OF SHIFT OF CURIE TEMPERATURE IN POTASSIUM DIDEUTERIUM PHO 17-2601
                         LOWERING THE CURIE TEMPERATURE IN REDUCED BARIUM TITANATE. 04-0711
RACT)   RAMAN STUDIES OF STRESS INDUCED PHASE TRANSITION NEAR CURIE TEMPERATURE IN STRONTIUM TITANATE. (ABST 06-1236
/LECTRIC POTASSIUM FERROCYANIDE TRIDEUTERO HYDRATE ABOVE THE CURIE TEMPERATURE (LATTICE CONSTANTS, SPACE G/ 24-2903
CRYSTAL STOICHIOMETRY (VERSUS MELT COMPOSITION, LATTICE AND CURIE TEMPERATURE MEASUREMENTS. /ALATE SINGLE 12-1871
/ETASTABLE OPTICAL CHARGE EXCHANGE AND "FROZEN" SHIFT OF THE CURIE TEMPERATURE OF A FERROELECTRIC SEMICOND/ 15-2160
TALATE.   CURIE TEMPERATURE OF FERROELECTRIC LITHIUM TAN 12-1887
SULFATE.   ULTRASONIC RELAXATION NEAR THE CURIE TEMPERATURE OF FERROELECTRIC TRIGLYCINE 27-3097
/ASTOOPTIC EFFECT IN NONMETALLIC CRYSTALS. (STRAIN DEPENDENT CURIE TEMPERATURE, POLARIZATION FLUCTUATIONS,/ 03-0513
/ DISTRIBUTION OF RELAXATION TIME IN FERROELECTRICS NEAR THE CURIE TEMPERATURE (POLY VERSUS MONODISPERSIVE 03-0521
/MPERATURES OF TUNGSTEN BRONZE TYPE NIOBATES (CORRELATION OF CURIE TEMPERATURE WITH CHEMICAL COMPOSITION). 13-2049
ENOIODIDE, BISMUTH SULFOIODIDE, ANTIMONY SULFO SELENOIODIDE, CURIE TEMPERATURES) /ANTIMONY SULFOIODIDE, SEL 15-2114
            DEFECT LEAD TITANATES WITH DIVERSE CURIE TEMPERATURES.   05-1076
OPERTIES OF TETRAGONAL (BARIUM, LEAD) (NIOBATE, ZIRCONATE). (CURIE TEMPERATURES)   FERROELECTRIC PR 08-1293
IUM) OR (NIOBIUM, TUNGSTEN), COMPILATION, LATTICE CONSTANTS, CURIE TEMPERATURES) /HERE B IS (NIOBIUM, TITAN 13-1964
/ION OF SOME FERROELECTRICS AND ANTIFERROELECTRICS WITH HIGH CURIE TEMPERATURES. (BISMUTH FERRATE AND SYST/ 08-1378
E) FERROELECTRICS (WITH AND WITHOUT / INTERNAL FRICTION (AND CURIE TEMPERATURES) IN (LEAD ZIRCONATE TITANAT 09-1584
/STUDY OF THE FERROELECTRIC TRANSITION IN KDP TYPE CRYSTALS (CURIE TEMPERATURES OF POTASSIUM, RUBIDIUM, CE/ 17-2577
D ZIRCONATE- LEAD TITANATE- LEAD METANIOBATE TERNARY SYSTEM. CURIE TEMPERATURES OF SOLID SOLUTIONS IN A LEA 09-1595
/POUNDS, 58 SOLID SOLUTIONS, AND 10 TECHNICAL FERROCERAMICS. CURIE TEMPERATURES OF 290 COMPOUNDS TABULATED) 01-0014
         A NOTE ON THE CLASSIFICATION OF FERROELECTRICS. (BY CURIE-WEISS CONSTANT)   01-0024
SOLUTIONS IN THE (BARIUM, S/ CONCENTRATION DEPENDENCE OF THE CURIE-WEISS CONSTANT IN POLYCRYSTALLINE SOLID 04-0586
TY OF A POTASSIUM DIHYDROGEN PHOSPHATE CRYSTAL.   CURIE-WEISS LAW FOR THE NONLINEAR SUSCEPTIBILI 17-2686
ATION REVERSAL IN FERROELECTRICS. (CALCULATIONS OF SWITCHING CURRENT)   THEORY OF POLARIZ 03-0422
CS.   DIRECT CURRENT BIAS EFFECTS IN BARIUM TITANATE CERAMI 04-0566
E CERAMICS.   DIRECT CURRENT BIAS EFFECTS IN LEAD ZIRCONATE TITANAT 09-1490
TRONTIUM TITANATE SUBSTRATES AT SUPERSONIC DRIFT VELOCITY OF CURRENT CARRIERS. /AYERS ON BARIUM TITANATE- S 04-0592
     EXPERIMENTAL AND THEORETICAL STUDIES ON THE DIRECT CURRENT CONDUCTIVITY OF ICE.   34-3450
ED QUAR/ PREPARATION OF THIN BARIUM TITANATE FILMS BY DIRECT CURRENT DIODE SPUTTERING (ONTO PLATINUM OR FUS 04-0948
XIS OF ANTIMONY SULFOIODIDE CAUSED BY ILLUMINATION IN DIRECT CURRENT ELECTRIC FIELD.   STRAIN ALONG C-A 15-2239
FECT OF TRIGLYCINE SULFATE SINGLE CRYSTALS IN AN ALTERNATING CURRENT ELECTRIC FIELD. / AUTOSTABILIZATION EF 27-3005
TITANATE.   CURRENT EXCITATION IN GAMMA-IRRADIATED BARIUM 04-0884
MONY SULFOIODIDE. (PHASE BOUNDARY SHAPE STRIATIONS IN DIRECT CURRENT FIELD) /RROELECTRIC PROPERTIES OF ANTI 15-2197
NCE MEASUREMENTS IN A FERROELECTRIC CONDENSER UNDER A DIRECT CURRENT FIELD.   NOISE AND IMPEDA 37-3645
SULFOIODIDE CRYSTAL (IN THE PARAELECTRIC PHASE, UNDER DIRECT CURRENT FIELD ALONG C-AXIS. /ION IN ANTIMONY 15-2173
 SURFACE LAYER IN BARIUM TITANATE (X-RAY DIFFRACTION, DIRECT CURRENT FIELD EFFECT, PARAELECTRIC AND FERROE/ 04-0857
SALT. (EFFECT OF AGING AND COPPER(2+) D/ EFFECT OF A DIRECT CURRENT FIELD ON DOMAIN STRUCTURE IN SEIGNETTE 28-3230
TRIGLYCINE SELENATE. (AGING AND PHASE CH/ EFFECT OF A DIRECT CURRENT FIELD ON THE DIELECTRIC PROPERTIES OF 27-3131
TUNGSTATE(0.5)), EFFECTS OF POSITIVE OR NEGATIVE HIGH DIRECT CURRENT FIELDS SUMMARY ONLY) / MAGNESATE(0.5) 05-1063
ENT ON GROWTH CONDITIONS.   MAXIMUM REPOLARIZATION CURRENT FOR TRIGLYCINE SULFATE CRYSTALS DEPEND 27-3154
ELECTRIC FIELD DEPENDENT OSCILLATION OF SPACE CHARGE LIMITED CURRENT IN BISMUTH(12) GERMANIUM OXIDE(20). / 35-3505
ZOCHRALSKI METHOD).   GROWTH TWINNING (UNDER DIRECT CURRENT) IN C-AXIS LITHIUM NIOBATE CRYSTALS (C 11-1672
SS/ TRAPPING SITES, OPTICAL DAMAGE, AND THERMALLY STIMULATED CURRENT IN FERROELECTRIC OXIDE CRYSTALS. (POTA 13-1926
URRENT VOLTAGE CHARACTERISTICS, 150-37/ SPACE CHARGE LIMITED CURRENT IN POLYCRYSTALLINE BARIUM TITANATE. (C 04-0557
         QUASI STATIC SWITCHING CURRENT IN SODIUM NITRITE.   16-2261
     BARKHAUSEN JUMPS AND SWITCHING CURRENT IN TGS CRYSTALS.   27-3109
NEAR THE TRANSITION POINT OF DIFFUSE FERROELEC/ ALTERNATING CURRENT POLARIZATION AND DIELECTRIC HYSTERESIS 03-0437
EN PHOSPHATE CRYSTALS.   LIGHT AND CURRENT PULSES FROM X-RAYED POTASSIUM DIHYDROG 17-2716
     LOW TEMPERATURE POLARIZATION EFFECTS IN ICE.   CURRENT REVERSAL DUE TO DIPOLE RELAXATION) 34-3418
MPORTANT PROPERTIES OF FERROELECTRICS.   SOME ASPECTS OF THE CURRENT STUDY AND USE OF SEVERAL TECHNICALLY I 02-0111
TIMONY SULFOIODIDE. (CONDUCTION MECHANISM FOR THERMOELECTRIC CURRENTS)   TRANSPORT PHENOMENA IN AN 15-2241
GE IN ANTIMONY SULFOIODIDE.   INJECTION CURRENTS AND ACCUMULATION OF POLARIZATION CHAR 15-2188
ON OXIDE(20) AND BISMUTH(12) GERMANIUM/ THERMALLY STIMULATED CURRENTS AND LUMINESCENCE IN BISMUTH(12) SILIC 36-3585
/C(VII) SOLID SOLUTIONS. (DIELECTRIC CONSTANTS, PYROELECTRIC CURRENTS, ANTIMONY SULFOIODIDE(X) BROMIDE(1-X/ 15-2157
RIC MATERIAL. (THEORY)   VOLTAGES AND CURRENTS DURING DEPOLARIZATION OF A FERROELECT 03-0177
(THEORY).   SPACE CHARGE LIMITED CURRENTS IN A METAL FERROELECTRIC METAL SYSTEM 03-0348
     THEORY OF SPACE CHARGE LIMITED CURRENTS IN FERROELECTRICS.   03-0251
ATINATE(0.25) CRYSTALS.   SPACE CHARGE LIMITED CURRENTS IN HEXAGONAL BARIUM TITANATE(0.75) PL 04-0936
RYSTALS.   SPACE CHARGE LIMITED CURRENTS IN SEMICONDUCTING POTASSIUM NIOBATE C 08-1412
     PULSE RESPONSE OF ZERO DEGREE CUT ADP MODULATORS.   17-2635

```
                LIGHT MODULATORS USING 45 DEGREE X CUT AND 45 DEGREE Y CUT ADP CRYSTALS.             17-2605
ULATOR.                                  OPTIMUM CUT FOR A LITHIUM NIOBATE TRANSVERSE LIGHT MOD       11-1755
ATE.                NEW HIGH COUPLING LOW DIFFRACTION CUT FOR ACOUSTIC SURFACE WAVES ON LITHIUM NIOB  11-1820
ATE.      NEW HIGH FREQUENCY HIGH COUPLING LOW BEAM STEERING CUT FOR ACOUSTIC SURFACE WAVES ON LITHIUM NIOB  11-1819
ULATION.                                   NOVEL TYPE OF CUT FOR KDP CRYSTALS FOR LOW VOLTAGE LIGHT MOD  17-2531
NSVERSE LIGHT MODULATION                         OPTIMUM CUT IN X DIHYDROGEN PHOSPHATE CRYSTALS FOR TRA  17-2494
               MICROWAVE PHONON ATTUNUATION IN X CUT LITHIUM NIOBATE.                                 11-1756
ING IN THICKNESS SHEAR MODE / TEMPERATURE DEPENDENCE OF AN X CUT LITHIUM TANTALATE CRYSTAL RESONATOR VIBRAT  12-1868
STALS.                                           NOVEL CUT OF POTASSIUM DIHYDROGEN PHOSPHATE TYPE CRY  17-2530
STALS FOR LIGHT MODULATION.                    OPTIMUM CUT OF POTASSIUM DIHYDROGEN PHOSPHATE TYPE CRY  17-2532
    CHARACTERISTICS OF AN ELECTROOPTICAL SHUTTER BASED ON Z CUT POTASSIUM DIDEUTERIUM PHOSPHATE CRYSTALS.    17-2460
TIC Q SWITCH BY DESK-T/ SIMULATION OF A DOUBLE 45 DEGREES, Z CUT POTASSIUM DIDEUTERIUM PHOSPHATE. ELECTROOP  17-2702
       LONGITUDINAL ELECTROOPTIC EFFECT IN OBLIQUE CUT STRONTIUM BARIUM NIOBATE PLATES.               08-1472
       LONGITUDINAL ELECTROOPTIC EFFECT IN OBLIQUE CUT STRONTIUM BARIUM NIOBATE PLATES.               13-2063
LIGHT MODULATORS WITH A LOW HALF WAVE VOLTAGE.      CUTS OF LITHIUM NIOBATE CRYSTALS SUITABLE FOR     11-1715
THE TYPE X DIHYDROGEN PHOSPHATE FOR LIGHT MODULATORS.  CUTTING AND POLISHING IN UNIAXIAL CRYSTALS OF  17-2559
IOBATE.                                VISIBLE CW PARAMETRIC OSCILLATOR USING BARIUM SODIUM N         13-1984
ND POTASSIUM DIHYDROGEN PHOSPHATE EXCITED BY A 2573 ANGSTROM CW PUMP. /E IN AMMONIUM DIHYDROGEN PHOSPHATE A  17-2486
PARAMETRIC FLUORESCENCE IN BARIUM SODIUM NIOBATE FOR NINE CW PUMP FREQUENCIES.                        13-1950
/S OF ELECTRON PHONON INTERACTION IN FERROELECTRIC CRYSTALS (CYCLOTRON RESONANCE, ABSORPTION AND LUMINESCE/ 03-0427
CUIT SENSITIVITY OF RADIALLY POLARIZED FERROELECTRIC CERAMIC CYLINDERS. (HYDROPHONE ELEMENTS)    OPEN CIR   37-3651
RCUIT SENSITIVITY OF AXIALLY POLARIZED FERROELECTRIC CERAMIC CYLINDERS. (HYDROPHONE ELEMENTS)   OPEN CI    37-3652
    VIBRATIONS OF LONGITUDINALLY POLARIZED FERROELECTRIC CYLINDRICAL TUBES.                           03-0375
E AND RELATED COMPOUNDS. (GADOLINIUM MOLYBDATE, GROWTH BY RF CZOCHRALSKI AND FROM TOP SEEDED SOLUTION) /NAT 08-1323
     BARIUM SODIUM NIOBATE: CONSTITUTIONAL STUDIES AND CZOCHRALSKI CRYSTAL GROWTH.                    13-1904
BATE AND SODIUM LANTHANUM / CRUCIBLE BASE COOLING: AN AID TO CZOCHRALSKI CRYSTAL GROWTH. (BARIUM SODIUM NIO  13-1928
D STUDY OF BARIUM(X) STRONTIUM(1-X) NIOBATE SINGLE CRYSTALS (CZOCHRALSKI GROWTH, DIELECTRIC PROPERTIES). /N 13-1936
CRYSTALS.                                               CZOCHRALSKI GROWTH OF BARIUM STRONTIUM NIOBATE 13-1913
     GROWTH OF BARIUM LITHIUM NIOBATE SINGLE CRYSTALS BY THE CZOCHRALSKI METHOD.                      13-1991
G (UNDER DIRECT CURRENT) IN C-AXIS LITHIUM NIOBATE CRYSTALS (CZOCHRALSKI METHOD).            GROWTH TWINNIN 11-1672
L OF CRYSTAL DIAMETER OF STRONTIUM BARI/ MODIFICATION TO THE CZOCHRALSKI METHOD OF CRYSTAL PULLING. (CONTRO 13-1914
E AS A NONLINEAR OPTICAL MATERIAL. (GROWTH BY KYROPOULOS AND CZOCHRALSKI METHODS) /TAL BARIUM SODIUM NIOBAT 13-2021
ELECTRIC FERROELASTIC PARAMAGNETIC (BETA) TERBIUM MOLYBDATE (CZOCHRALSKI SINGLE CRYSTALS). /PERATURE) FERRO 30-3290
ARTH DOPED BISMUTH GERMANATE.                           CZOCHRALSKI SYNTHESIS AND PROPERTIES OF RARE E 35-3496

                                                   D

DETECTORS.                                        HIGH D*, FAST, LEAD ZIRCONATE TITANATE PYROELECTRIC 09-1560
/EROVSKITE FERROELECTRICS (HAMILTONIAN IN TERMS OF LOCALIZED D-STRAIN AND SOFT- NORMAL MODE COORDINATES AN/ 03-0432
IN BARIUM TITANATE (EXPERIMENTS TO CHARACTERIZE THE OPTICAL DAMAGE). /LLY INDUCED REFRACTIVE INDEX CHANGES 04-1004
E SALT.                                       RADIATION DAMAGE AND THE FERROELECTRIC EFFECT IN ROCHELL 28-3207
RROELECTRIC OXIDE CRYSTALS. (POTASS/ TRAPPING SITES, OPTICAL DAMAGE, AND THERMALLY STIMULATED CURRENT IN FE 13-1926
E POTASSIUM DIHYDROGEN ARSENATE MIXE/ ESR STUDY OF RADIATION DAMAGE CENTERS IN AMMONIUM DIHYDROGEN PHOSPHAT 17-2547
MICS.              EFFECTS OF RADIATION INDUCED DAMAGE CENTERS IN LEAD ZIRCONATE TITANATE CERA       09-1520
/ROPERTIES AND PRESSURE EFFECTS ON SELENITE (ION). RADIATION DAMAGE ELECTRON SPIN RESONANCE SPECTRA IN FER/ 22-2870
IC ROCHELLE SALT. (NEUTRON DIFFRACTI/ STRUCTURAL EFFECTS OF (DAMAGE FROM) IONIZING RADIATION IN FERROELECTR 28-3188
WHEN ILLUMINATED UNDER APPLIED ELECTR/ (REVERSIBLE) OPTICAL DAMAGE IN KDP (WITH RELAXATION TIME OF SECONDS  17-2703
                                            OPTICAL DAMAGE IN KTN.                                    13-1979
              CONTROL OF LASER DAMAGE IN LITHIUM NIOBATE.                                             11-1716
IC INSULATORS.          (MECHANISM OF) OPTICAL INDEX DAMAGE IN LITHIUM NIOBATE AND OTHER PYROELECTR     11-1735
AL SPECTRA, ESR, CELL PARAMETERS)        RADIATION DAMAGE IN LITHIUM NIOBATE. (CAPACITANCE, OPTIC        11-1705
   X-RAY TOPOGRAPHIC OBSERVATIONS OF RADIATION DAMAGE IN TRIGLYCINE SULFATE.                          27-3074
              X-RAY TOPOGRAPHIC OBSERVATIONS OF RADIATION DAMAGE IN TRIGLYCINE SULFATE SINGLE CRYSTALS.     27-3106
                   LASER INDUCED DAMAGE IN X DIHYDROGEN PHOSPHATE MATERIALS.                          17-2475
60) GAMMA-IRRADIATION ON BARIUM TITANATE, (COMPARED TO STUDY DAMAGE MECHANISM) /ETIC) ELECTRON AND (COBALT- 04-0970
     NATURE OF INTERNAL ELECTRIC FIELD DURING OPTICAL DAMAGE PROCESS IN LITHIUM NIOBATE.              11-1861
E.                                               DAMAGE PRODUCED BY X-RAYS ON TRIGLYCINE SULFAT        27-3073
ALS DURING INTRACAVITY FREQUENCY DOUBL/ CATASTROPHIC SURFACE DAMAGE PRODUCED IN BARIUM SODIUM NIOBATE CRYST 13-2052
/PHIZATION IN TUNGSTEN TRIOXIDE AND ITS USE IN DEDUCING MEAN DAMAGE RANGES. (STUDIES ON BOMBARDMENT INDUCE/ 10-1642
          NEODYMIUM YAG LASER IRRADIATION INDUCED DAMAGE TO LITHIUM NIOBATE AND KDP.                  11-1674
IUM IODATE.        OPTICALLY INDUCED PHYSICAL DAMAGE TO LITHIUM NIOBATE, PROUSTITE, AND LITH           08-1328
GHT.                                             DAMAGE TO THE SURFACE OF LITHIUM NIOBATE BY LI        11-1866
               EPR STUDY OF X-RAY DAMAGED FERROELECTRICS.                                             02-0079
P TO HIGH DOSAGE (DESTR/ FERROELECTRIC BEHAVIOR OF RADIATION DAMAGED TRIGLYCINE SULFATE AND ROCHELLE SALT U 27-3098
        RAMAN STUDIES (TEMPERATURE DEPENDENCE) OF DAMPED SOFT MODES IN LEAD TITANATE.                 05-1041
S.        FREQUENCY DEPENDENCE OF ANOMALOUS ULTRASOUND DAMPING IN ANTIMONY SULFOIODIDE SINGLE CRYSTAL  15-2139
CER/ DISPERSION STUDY OF DOMAIN WALL MOBILITY AND MECHANICAL DAMPING IN LEAD TITANATE ZIRCONATE TRANSDUCER  09-1547
EE CENTIMETER RAD/ TEMPERATURE DEPENDENCE OF HYPERSONIC WAVE DAMPING IN RUBY AND LITHIUM NIOBATE IN THE THR 11-1718
TY).            COUPLED ORDER DISORDER PHONON SYSTEM WITH DAMPING: KDP (FREQUENCY DEPENDENT SUSCEPTIBILI   17-2667
IUM MANGANESE TRIFLUORIDE.          DISPERSION AND DAMPING OF SOFT ZONE BOUNDARY PHONONS IN POTASS     33-3374
DYNAMICAL THEORY OF HYDROGEN BONDED FERROELECTRICS. PART-1: DAMPINGS AND (FREQUENCY) SHIFTS (OF PROTONIC /  03-0475
/ PIEZORESISTANCE OF (POLYCRYSTALLINE) ANTIMONY SULFOIODIDE (DARK CONDUCTIVITY NEAR TRANSITION POINT, 1-10/ 15-2250
OF THIOUREA CRYSTAL. (CALCULATIONS FROM SATELLITE SCATTERING DATA)            LONG PERIOD STRUCTURE      25-2911
DES OF THE PEROVSKITE STRUCTURE. (THEORY, STRONTIUM TITANATE DATA)                  OPTICAL MO         03-0304
OSPHATE (ANALYSIS AND THEORETICAL INTERPRETATION OF PREVIOUS DATA). /ION REGION OF POTASSIUM DIDEUTERIUM PH 17-2575
BARIUM TITANATE DURING SWITCHING (CALCULATIONS COMPARED WITH DATA). /P OR SMOOTH) DOMAIN WALL STRUCTURE IN  04-0677
LECTRICS. BASED ON NEW TREATMENT AND INTERPRETATION OF X-RAY DATA) /DISORDER OF BARIUM TITANATE TYPE FERROE 04-0630
/POUNDS WITH PEROVSKITE STRUCTURE. (EXAMINATION OF PUBLISHED DATA, CLASSIFICATION, LISTING AND DISCUSSION / 01-0018
         NONLINEAR OPTICAL MATERIALS (REVIEW, DATA COLLECTION, MATERIALS NEEDS, 118 REFS)             02-0049
.                                          CRYSTAL DATA FOR LITHIUM THALLIUM TARTRATE MONOHYDRATE      29-3247
/MUTH(4) BARIUM(M-2) TITANIUM(M+1) OXIDE(3) (M+2) (M=2,3,4). (DATA FOR THE TITANIUM(3) OXIDE(12), BARIUM TI/ 02-0066
ONS IN SODIUM NIOBATE.              NEW DATA FROM AN X-RAY STUDY OF PHASE TRANSFORMATI                 08-1361
YSICAL, THERMAL MAGNETIC, / II-VI SEMICONDUCTING COMPOUNDS - DATA TABLES. (MECHANICAL, CRYSTALLOGRAPHIC, PH 01-0025
                                                        DC SEE DIRECT CURRENT
IN A FERROELECTRIC.                              DEBYE SCREENING OF A NONUNIFORM ELECTRIC FIELD        03-0499
                                                 DEBYE TYPE ABSORPTION IN POLAR PHTHALOCYANINE.       36-3621
G REFLECTIONS C-AXIS, 20-180 / TEMPERATURE DEPENDENCE OF THE DEBYE- WALLER FACTORS OF BARIUM TITANATE (BRAG 04-0892
/MODES IN PEROVSKITE FERROELECTRICS INDICATED BY ANISOTROPIC DEBYE- WALLER FACTORS. (POTASSIUM TANTALATE N/ 08-1351
ATE.                         VARIATION OF THE DECAY CONSTANT OF STRONTIUM-85 IN BARIUM TITAN          04-0876
NATE BY THE FERROELECTRIC TRANSITION.   CHANGE IN THE DECAY CONSTANT OF TECHNETIUM-99 IN BARIUM TITA  04-0871
```

FERROELECTRICS--PERMUTED TITLE INDEX

ATE (BARIUM CHLORIDE FLUX GROWTH IN PRESENCE / CHANGE IN THE DECAY CONSTANT OF ZIRCONIUM-89 IN BARIUM TITAN 04-0685
E. ANOMALOUS DECAY EFFECT IN POLYCRYSTALLINE BARIUM TITANAT 04-0752
/RIC PROPERTIES OF FERROELECTRIC MATERIALS AT CENTIMETER AND DECIMETER WAVELENGTHS EMPLOYING A SYMMETRIC S/ 04-0855
TITANATE (OF HIGH PURITY AND SURFACE ACTIVITY VIA HYDROLYTIC DECOMPOSITION). /TIUM ZIRCONATE AND STRONTIUM 06-1217
ERTIES OF FERROELECTRIC MATERIALS. (FINE POWDERS PREPARED BY DECOMPOSITION OF ORGANIC COMPOUNDS) /ICAL PROP 02-0093
ELLURIUM ON TRIGLYCINE SULFATE AND SODIUM NITRITE SURFACES) (DECORATION METHOD OF DOMAIN STUDY) /TRATES. (T 16-2394
RFACES. DECORATION OF ION BOMBARDED LITHIUM NIOBATE SU 11-1799
AGING PROCESSES IN BARIUM TITANATE (DECREASE IN NUMBER OF 180 DEG DOMAINS). 04-0911
/T INDUCED AMORPHIZATION IN TUNGSTEN TRIOXIDE AND ITS USE IN DEDUCING MEAN DAMAGE RANGES. (STUDIES ON BOMB/ 10-1642
HYDROXYL AS DEFECT IN BARIUM TITANATE. 04-0636
CHARACTERISTICS OF DEFECT LEAD TITANATE. 05-1075
KATURES. DEFECT LEAD TITANATES WITH DIVERSE CURIE TEMPE 05-1076
/RADIATION AND COPPER AND IRON IMPURITY) ABSORPTION BANDS OF DEFECT ROCHELLE SALT AND TRIGLYCINE SULFATE C/ 27-3108
SIC NONSTOICHIOMETRY IN THE LEAD ZIRCONATE TITANATE SYSTEM. (DEFECT STRUCTURE) INTRINSIC AND EXTRIN 09-1536
URE COEFFICIENT OF RESISTIVITY) DEVICES) PART-2: ANALYSIS OF DEFECT STRUCTURE. /AP, PTCR (POSITIVE TEMPERAT 04-0822
TGS SINGLE CRYSTALS. THE EFFECT OF GROWTH RATE ON THE DEFECT STRUCTURE AND DIELECTRIC PROPERTIES OF 27-3067
TITANATE- MIXED CRYSTALS FROM A POTASSIUM FLUORIDE MELT. DEFECT STRUCTURE OF (BARIUM(1-X) STRONTIUM(X)) 04-0873
ELECTRICAL PROPERTIES AND DEFECT STRUCTURE OF POTASSIUM TANTALATE. 08-1314
X-RAY TOPOGRAPHY OF THE DEFECT STRUCTURE OF THIOUREA. 25-2914
LATION WITH DOMAIN STRUCTURES) DEFECT STRUCTURE OF TRIGLYCINE SULFATE. (CORRE 27-3044
PHASE BOUNDARY AND DEFECT STRUCTURE STUDIES IN PLZT CERAMICS. 09-1590
DEFECT STUDIES IN STRONTIUM TITANATE. 06-1244
OME INVESTIGATIONS OF THEIR QUALITY. (VOLUME DISTRIBUTION OF DEFECTS) /ON OF PURE TGS SINGLE CRYSTALS AND S 27-3086
TE. POINT DEFECTS AND SINTERING OF LEAD ZIRCONATE TITANA 09-1478
ANIDINIUM ALUMINUM SULFATE HEXAHYDRATE. (POLARIZATION VERSU/ DEFECTS AND THE FERROELECTRIC PROPERTIES OF GU 21-2813
RYSTALS. X-RAY DIFFRACTION TOPOGRAPHY STUDY OF DEFECTS (DISLOCATIONS) IN KDP AND ADP SINGLE C 17-2578
N OF THE PRESSURE DEPENDENCE OF PROPERTIES CAUSED BY LATTICE DEFECTS. (ICE) INTERPRETATIO 34-3434
IR INFLUENCE ON BEHAVIOR. ORIGIN OF LATTICE DEFECTS IN (BARIUM) TITANATE MATERIALS AND THE 04-0784
YCINE SULFATE. NATURE OF RADIATION INDUCED DEFECTS IN CRYSTALS OF ROCHELLE SALT AND TRIGL 27-3159
FATE. ELECTRON SPIN RESONANCE OF GAMMA DEFECTS IN FERROELECTRIC AMMONIUM HYDROGEN SUL 19-2748
ELECTRON SPIN RESONANCE STUDIES AT DEFECTS IN FLUORITES. (LEAD TITANATE) 05-1086
THEORY OF THE MOBILITY OF STRUCTURAL DEFECTS IN ICE. 03-0244
HEAT TREATMENT INDUCED POINT DEFECTS IN LITHIUM METANIOBATE SINGLE CRYSTALS 11-1668
D IN LASER MODULATION. POINT DEFECTS IN LITHIUM NIOBATE SINGLE CRYSTALS USE 11-1737
) DEFECTS IN OXIDE CERAMICS. (STRONTIUM TITANATE 06-1163
VIEW, 77 REFS) POINT DEFECTS IN PEROVSKITE TYPE FERROELECTRICS. (RE 02-0041
SHARED HOLES TRAPPED BY CHARGE DEFECTS IN STRONTIUM TITANATE. 06-1123
ATE. EXPERIMENTAL AND THEORETICAL STUDY OF DEFECTS IN TRIGLYCINE SULFATE AND BARIUM TITAN 04-0620
TION TO THEIR GROWTH COND/ X-RAY TOPOGRAPHIC STUDY OF GROWTH DEFECTS IN TRIGLYCINE SULFATE CRYSTALS IN RELA 27-3025
AND MECHANICAL PROPERTIES. (REVIEW, 56/ STRUCTURE AND POINT DEFECTS OF ICE: THEIR EFFECT ON THE ELECTRICAL 02-0071
ICS, AND PHASE TRANSITIONS OF FERROELECT/ INFLUENCE OF POINT DEFECTS ON DIELECTRIC PROPERTIES, DOMAIN DYNAM 03-0495
INGLE CRYSTALS. INFLUENCE OF DEFECTS ON SOME DIELECTRIC PROPERTIES OF TGS S 27-3063
SALT IN MICROWAVE RANGE. INFLUENCE OF CRYSTAL DEFECTS ON THE DIELECTRIC BEHAVIOR OF ROCHELLE 28-3219
DED FERROELECTRICS. EFFECT OF DEFECTS ON THE ENERGY SPECTRUM OF HYDROGEN BON 03-0486
LLINE LEAD ZIRCONAT/ INFLUENCE OF THE CONCENTRATION OF POINT DEFECTS ON THE INTERNAL FRICTION IN POLYCRYSTA 09-1582
M TITANATE. ON THE INFLUENCE OF DEFECTS ON THE TEMPERATURE HYSTERESIS OF BARIU 04-0987
ION ON PHASE TRANSITIONS AND DIELECTRI/ EFFECT OF STRUCTURAL DEFECTS PRODUCED BY GAMMA AND REACTOR IRRADIAT 17-2614
ERROELE/ USE OF ELECTRON MICROSCOPY IN THE STUDY OF EXTENDED DEFECTS RELATED TO NONSTOICHIOMETRY (AND ANTIF 36-3522
/ADDITION) DUE TO INTERACTION OF DOMAIN BOUNDARIES AND POINT DEFECTS (TORSION PENDULUM AND RESONANCE TECHN/ 09-1584
SION, RELAXATION TIME AND STREN/ ON THE INFLUENCE OF CRYSTAL DEFECTS UPON FERROELECTRIC PROPERTIES. (DISPER 03-0496
ITY IN THE IRON DEFICIENT FERROUS SULFIDE SYSTEM. (EXISTS TO DEFICIENCIES OF 0.04) FERROELECTRIC 36-3614
OF TUNGSTEN TRIOXIDE CRYSTALS. EFFECT OF OXYGEN DEFICIENCY ON ELECTRICAL TRANSPORT PROPERTIES 10-1629
EFICIENCIES OF 0.04) FERROELECTRICITY IN THE IRON DEFICIENT FERROUS SULFIDE SYSTEM. (EXISTS TO D 36-3614
MAGNETIC RESONANCE AND THE PROBLEM OF THE DEFINITION OF ANTIFERROELECTRICITY. 03-0266
IES OF A DIGITAL LIGHT DEFLECTOR UTILIZING LITHIUM TANTALAT/ DEFLECTED BEAM DIAMETER AND DEFLECTION CAPACIT 11-1809
IUM TANTALATE NIOBATE GRATINGS FOR LIGHT BEAM MODULATION AND DEFLECTION. ELECTROOPTIC POTASS 13-1929
S. INVESTIGATION OF A SYSTEM FOR BEAM DEFLECTION BY MEANS OF LITHIUM NIOBATE CRYSTAL 11-1768
CTOR UTILIZING LITHIUM TANTALAT/ DEFLECTED BEAM DIAMETER AND DEFLECTION CAPACITIES OF A DIGITAL LIGHT DEFLE 11-1809
NESIO NIOBATE CRYSTALS. LIGHT BEAM DEFLECTION CONTROLLED WITH A PRISM OF LEAD MAG 08-1267
BINARY LIGHT BEAM DEFLECTION IN SINGLE CRYSTAL BARIUM TITANATE. 04-0653
ELECTROOPTIC LIGHT BEAM DEFLECTION WITH BARIUM STRONTIUM NIOBATE PRISM 13-2036
/ BEAM DIAMETER AND DEFLECTION CAPACITIES OF A DIGITAL LIGHT DEFLECTOR UTILIZING LITHIUM TANTALATE POLARIZ/ 11-1809
NTALATE(0.65) NIOBA/ USE OF PEROVSKITE PARAELECTRICS IN BEAM DEFLECTORS AND LIGHT MODULATORS. (POTASSIUM TA 13-1924
THINNING OF LITHIUM NIOBATE FOR ACOUSTOOPTIC DEFLECTORS BY ION ETCHING. 11-1862
/LD ON FERROELECTRIC AND ANTIFERROELECTRIC CERAMICS. PART-1: DEFORMATION AND PHASE TRANSFORMATION. PART-2:/ 03-0169
ECTRIC POTASSIUM NIOBATE SINGLE CRYSTALS. DEFORMATION AND RELAXATION BEHAVIOR IN FERROEL 08-1410
X-RAY DIFFRACTION TOPOGRAPHIC STUDIES OF THE DEFORMATION BEHAVIOR OF ICE SINGLE CRYSTALS. 34-3426
N/ POLARIZATION OF FERROELECTRIC FILM BY BENDING AND BENDING DEFORMATION IN THE ELECTRIC FIELD. (OBSERVATIO 03-0195
/LATIONSHIP BETWEEN THE ATOMIC DISPLACEMENTS AND SPONTANEOUS DEFORMATION IN THE FERROELECTRIC PHASES OF OX/ 08-1377
INFLUENCE OF THE SURFACE LAYER ON THE PLASTIC DEFORMATION OF ICE SINGLE CRYSTALS. 34-3459
/ECTRIC CERAMICS AND METHODS OF THEIR CHANGING. (SPONTANEOUS DEFORMATION RELATED TO THEIR CHARACTERISTICS) 03-0217
SHADOW EFFECT. (NEUTRON DIFFRACTION, BARIUM/ STUDYING SMALL DEFORMATIONS OF A CRYSTAL LATTICE BASED ON THE 04-0534
ELASTIC COMPLIANCE IN LITHIUM THALLIUM TARTRATE (TUNING AND DELAY DEVICES). ELECTRIC FIELD CONTROL OF THE 29-3250
R EFFECTS IN MICROWAVE ACOUSTIC LITHIUM NIOBATE SURFACE WAVE DELAY LINES. NONLINEA 11-1818
NPOLARIZED PZT CERAMIC PLATES. SURFACE WAVE DELAY LINES WITH INTERDIGITAL TRANSDUCERS ON U 09-1614
/PERATURE COEFFICIENTS OF ACOUSTIC SURFACE WAVE VELOCITY AND DELAY ON LITHIUM NIOBATE, LITHIUM TANTALATE, / 11-1821
EFFECT OF THE CUPRIC ION ON DELAYED PHENOMENA IN ROCHELLE SALT. 28-3221
POTASSIUM DIHYDROGEN PHOSPHATE. ELECTRON DELOCALIZATION WEIGHT OF THE HYDROGEN BOND IN 17-2525
ISOTHERMAL GRAIN GROWTH AND ELECTRICAL BEHAVIOR OF FULLY DENSE PLZT CERAMICS. 09-1553
RROELECTRIC (BARIUM TITANATE TYPE) CERAMICS. (IN RELATION TO DENSIFICATION AND GRAIN GROWTH) /OT PRESSED FE 04-0879
ESSURE SINTERED BARIUM TITANATE. INFLUENCE OF ADDITIVES ON DENSIFICATION AND PROPERTIES OF SINTERED AN PR 04-1026
LFATE. (THEORY AND CALCULATIONS OF WALL THICKNESS AND ENERGY DENSITY) /LECTRIC DOMAIN WALL IN TRIGLYCINE SU 27-2997
ERIALS. PART-2: HOT PRESSED COMPOSITIONS AT NEAR THEORETICAL DENSITY. /OF LEAD ZIRCONATE- LEAD TITANATE MAT 09-1578
FOIODIDE. (IN PARAELECTRIC AND FERROELECTRIC/ ELECTRON STATE DENSITY AND OPTICAL PROPERTIES OF ANTIMONY SUL 15-2204
MANGANESE TRIFLUORIDE AND POTAS/ PREPARATION AND DISLOCATION DENSITY DETERMINATION OF LARGE, PURE RUBIDIUM 33-3397
DIHYDROGEN PHOSPHATE (NEUTRON SCATTERING). HYDROGEN DENSITY DISTRIBUTION IN PARAELECTRIC POTASSIUM 17-2490
ACE WAVE AMPLIFIER. LOW POWER DENSITY GALLIUM ARSENIDE- LITHIUM NIOBATE SURF 11-1825
/IES OF ANTIMONY SULFOIODIDE(X) BROMIDE(1-X) SOLID SOLUTIONS DEPEND ON TEMPERATURE AND HYDROSTATIC PRESSUR/ 15-2149
CAL PROPERTIES OF FERROELECTRIC FERRO/ INFERRED TEMPERATURE DEPENDENCES OF ELECTRICAL, MECHANICAL AND OPTI 30-3253
ATE. FREQUENCY DEPENDENCES OF SOME PARAMETERS OF LITHIUM NIOB 11-1843

PAGE 374

SOUND IN ALUMINUM OXIDE AND LITHIUM NIOBATE SIN/ TEMPERATURE DEPENDENCES OF THE ABSORPTION OF 9.4 GHZ HYPER 11-1719
/AXATION OF DOMAIN WALLS IN TRIGLYCINE SULFATE. (TEMPERATURE DEPENDENCES OF THE COMPLEX DIELECTRIC CONSTAN/ 27-3103
OF FERROELECTRICS NEAR THE CURIE/ FREQUENCY AND TEMPERATURE DEPENDENCES OF THE COMPLEX DIELECTRIC CONSTANT 03-0361
THIUM METANIOBATE. TEMPERATURE AND FREQUENCY DEPENDENCES OF THE ELECTRICAL PROPERTIES OF LI 11-1717
F HOLES IN ANTIMONY TRISULFIDE SINGLE CRYSTALS. TEMPERATURE DEPENDENCES OF THE HALL AND DRIFT MOBILITIES O 15-2193
IC CERAMIC. TIME AND TEMPERATURE DEPENDENCES OF THE PARAMETERS OF A FERROELECTR 09-1514
COMPOSITIONS EACH) STRONTIUM TITANATE SYSTEMS. (FOR VOLTAGE DEPENDENT CAPACITORS) / THE BARIUM (OR LEAD, 4 09-1522
/ OF THE ELASTOOPTIC EFFECT IN NONMETALLIC CRYSTALS. (STRAIN DEPENDENT CURIE TEMPERATURE, POLARIZATION FLU/ 03-0513
AD ZIRCONATE- LEAD TITANATE- LANTHANUM OXIDE (P/ TEMPERATURE DEPENDENT FERROELECTRIC DOMAIN ALIGNMENT IN LE 09-1586
UTERIUM PHOSPHATE (COMP/ RAMAN SCATTERING BY THE TEMPERATURE DEPENDENT FERROELECTRIC MODE IN POTASSIUM DIDE 17-2705
XIMUM REPOLARIZATION CURRENT FOR TRIGLYCINE SULFATE CRYSTALS DEPENDENT ON GROWTH CONDITIONS. MA 27-3154
AD ZIRCONATE BY THE MOSSBAUER EFFECT. TEMPERATURE DEPENDENT OPTICAL MODE IN ANTIFERROELECTRIC LE 07-1260
TEMPERATURE DEPENDENT OPTICAL PHONONS IN LEAD TITANATE. 05-1083
CURRENT IN BISMUTH(12) GERMANIUM O/ LIGHT AND ELECTRIC FIELD DEPENDENT OSCILLATION OF SPACE CHARGE LIMITED 35-3505
Y RAMAN SPECTROSCOPY. ANALYSIS OF THE TEMPERATURE DEPENDENT PHONON STRUCTURE IN SODIUM NITRITE B 16-2299
SIGNMENTS IN BARIUM AND STRONTIUM TITANATES. TEMPERATURE DEPENDENT POLARIZABILITIES AND COCHRAN MODE AS 04-0805
CRYSTAL GROWTH AND TEMPERATURE DEPENDENT RAMAN SPECTRA OF LEAD TITANATE. 05-1074
N PHOSPHATE, POTASSIUM DIDEUTERIUM PHOSPHATE, P/ TEMPERATURE DEPENDENT RAMAN SPECTRA OF POTASSIUM DIHYDROGE 17-2409
ADOLINIUM MOLYEDATE. WAVEVECTOR DEPENDENT RELAXATION OF AN OPTICAL PHONON IN G 30-3320
EAR CHAIN MODEL. (PEROVSKITES) TEMPERATURE DEPENDENT SPECTRUM OF AN ANTIFERROELECTRIC LIN 03-0468
ED ORDER DISORDER PHONON SYSTEM WITH DAMPING: KDP (FREQUENCY DEPENDENT SUSCEPTIBILITY). COUPL 17-2667
1 PERCENT NIOBIUM PENTOXIDE) SURFACE / PROPERTIES OF OXYGEN DEPLETED CERAMIC LEAD ZIRCONATE TITANATE (PLUS 09-1576
MS. DEPOLARIZATION FIELD IN THIN FERROELECTRIC FIL 37-3659
HEORY) VOLTAGES AND CURRENTS DURING DEPOLARIZATION OF A FERROELECTRIC MATERIAL. (T 03-0177
TO HEATING OF PARALLEL PLANES OF POTASSIUM DIHYDROGEN PHOS/ DEPOLARIZATION OF LINEARLY POLARIZED LIGHT DUE 17-2540
PHASE TRANSITION IN KDP AND THE ROLE OF THE DEPOLARIZING ENERGY IN IT. (THEORY) 03-0229
DEPOLING OF SINGLE DOMAIN BISMUTH TITANATE. 08-1446
DIELECTRIC BEHAVIOR OF (FERROELECTRIC) FILMS OF VACUUM DEPOSITED POTASSIUM NITRATE. 16-2356
ENT UTILIZING POTASSIUM NITRATE. (FUSED THICK FILM OR VACUUM DEPOSITED THIN FILM) /TIVE READOUT MEMORY ELEM 16-2269
RONZES. (SODIUM(X) TUNGSTEN(1-X) Z(Y) OXIDE/ ELECTROCHEMICAL DEPOSITION AND DISSOLUTION OF TUNGSTEN OXIDE B 13-1944
/HE SURFACES OF IONIC CRYSTALS (TGS AND OTHERS, BY SELECTIVE DEPOSITION OF CHARGED POLYSTYRENE LATEX PARTI/ 27-2987
DIPOLE MOMENT DERIVATIVE OF THE HYDROGEN BOND IN ICE. 34-3489
(MACROSCOPIC) SYMMETRY ASPECTS OF FERROELECTRICITY. (DESCRIPTIVE AND PREDICTIVE POSSIBILITIES) 03-0464
UT POTASSIUM DIDEUTERIUM PHOSPHATE. ELECTROOPTIC Q SWITCH BY DESK-TOP COMPUTER. /F A DOUBLE 45 DEGREES, Z C 17-2702
) SPONTANEOUS DESTRUCTION OF FERROCERAMICS. (BARIUM TITANATE 04-0782
AGED TRIGLYCINE SULFATE AND ROCHELLE SALT UP TO HIGH DOSAGE (DESTRUCTION OR MASKING OF FERROELECTRICITY). / 27-3098
EIVER. (USING BARIUM TITANATE CRYSTAL) MINIMUM DETECTABLE POWER OF A PYROELECTRIC THERMAL REC 04-0632
OPERTIES OF POLYVINYLIDENE FLUORIDE AND ITS USE FOR INFRARED DETECTION. PYROELECTRIC PR 36-3560
ND ELECTRON NUCLEAR DOUBLE RESONANCE OF THE PROTON LATTICE / DETECTION BY ELECTRON PARAMAGNETIC RESONANCE A 18-2724
BINARY LIGHT BEAM DETECTION IN SINGLE CRYSTAL BARIUM TITANATE. 04-0654
C TRA/ HIGHLY SENSITIVE, SIMPLE CALORIMETER FOR PHASE CHANGE DETECTION (INCLUDING FERROELECTRIC AND MAGNETI 28-3201
DETECTION OF ATOM TUNNELING IN SOLIDS. (KDP) 17-2612
YROELECTRIC EFFECT. (BARIUM TITANATE, TRIGLYCINE SULFATE, A/ DETECTION OF ELECTROMAGNETIC RADIATION USING P 04-0978
TRIGLYCINE SULFATE, APPLICATIONS TO PYROELECTRICITY AND THE DETECTION OF FAR INFRARED RADIATIONS. /ANTS OF 17-2514
N PLZT CERAMICS. ACOUSTIC DETECTION OF FERROELECTRIC PHASE TRANSITIONS I 09-1545
PYROELECTRIC DETECTION PROPERTIES OF GADOLINIUM MOLYBDATE. 30-3324
LEAD TITANATE PYROELECTRIC INFRARED DETECTOR. 05-1087
.5 TO 2 MIC/ CHARACTERISTICS OF BARIUM TITANATE PYROELECTRIC DETECTOR FOR OPTICAL AND INFRARED RADIATION (0 04-0751
PERFORMANCE CHARACTERISTICS OF A SMALL TRIGLYCINE SULFATE DETECTOR OPERATED IN THE PYROELECTRIC MODE. 27-3119
MENT TECHNIQUE AND APPLICATION TO A NANOSECOND RESPONSE TIME DETECTOR. (STRONTIUM BARIUM NIOBATE) / MEASURE 13-1919
ROBERYLL/ IMPROVEMENT IN THE PYROELECTRIC INFRARED RADIATION DETECTOR. (TRIGLYCINE SULFATE, TRIGLYCINE FLUO 02-0046
CTERISTICS OF KTN (POTASSIUM TANTALATE NIOBATE) PYROELECTRIC DETECTORS. CHARA 13-2030
HIGH D*, FAST, LEAD ZIRCONATE TITANATE PYROELECTRIC DETECTORS. 09-1560
SITES IN TRIGLYCINE SULFATE AND APPLICATIONS TO PYROELECTRIC DETECTORS. /ER STUDY OF REVERSIBLE NUCLEATION 27-3011
G AT 10.6 MICRONS WITH STRONTIUM BARIUM NIOBATE PYROELECTRIC DETECTORS. PHOTOMIXIN 09-1477
CERAMIC ACOUSTIC DETECTORS. (BARIUM TITANATE) 01-0003
THEORY AND APPLICATION OF FERROELECTRIC RADIATION DETECTORS. (PYROELECTRICS) 37-3669
THERMAL ANALYSIS OF PYROELECTRIC DETECTORS. (TRIGLYCINE SULFATE EXAMPLE) 27-2964
IES OF LEAD ZIRCONATE(0.52) TITANATE(0.48) CERAMICS CAUSED / DETERIORATION OF THE ELECTROMECHANICAL PROPERT 09-1606
ID SOLUTIONS OF (BARIUM, STRONTIUM) TITANATE. MOBILITY DETERMINATIONS FROM WEIGHT MEASUREMENTS IN SOL 04-0689
RIUM IN THE FERROELECTRIC SODIUM DEUTERO AMMONIUM SELENATE DIDEUTERATE. NUCLEAR MAGNETIC RESONANCE OF DEUTE 22-2828
BLATTICE ROTATIONS AND SPIN DEVIATION IN COPPER FORMATE TETRADEUTERATE. PROTON RESONANCE STUDY OF SU 26-2950
IUM DIHYDROGEN PHOSPHATE, RUBIDIUM AND AMMONIUM ANALOGS SOME DEUTERATE) /OF SOME KDP TYPE CRYSTALS. (POTASS 17-2679
COPPER FORMATE TETRAHYDRATE AND COPPER DEUTERO FORMATE TETRADEUTERATE. /ION SPECTRUM OF THE FORMATE ION IN 26-2926
N IN THE TWO DIMENSIONAL ANTIFERROMAGNET COPPER FORMATE TETRADEUTERATE. /PONTANEOUS SUBLATTICE MAGNETIZATIO 26-2929
ECTRIC CRYSTALS OF THE TRIGLYCINE SULFATE GROUP. (NORMAL AND DEUTERATED) /OMORPHOUS SUBSTITUTION IN FERROEL 27-3102
RIC PHASE TRANSITION. THERMAL EXPANSION OF ADP AND DEUTERATED ADP IN THE REGION OF ANTIFERROELECT 17-2445
RICITY AND PHASE TRANSITION IN AMMONIUM DICHLORO ACETATE AND DEUTERATED AMMONIUM DICHLORO ACETATE. /ROELECT 36-3538
VECTORS OF SOFT MODES IN POTASSIUM DIDEUTERIUM PHOSPHATE AND DEUTERATED AMMONIUM DIDEUTERIUM PHOSPHATE. /EN 17-2700
Y OF SOME CRYSTALS CONTAINING O-D...O BOND. (INCLUDING FULLY DEUTERATED AMMONIUM DIHYDROGEN PHOSPHATE) /TUD 03-0204
DEUTERON SPIN LATTICE RELAXATION IN FERROELECTRIC DEUTERATED AMMONIUM HYDROGEN SULFATE. 19-2795
DEUTERON QUADRUPOLE COUPLING CONSTANT IN DEUTERATED AMMONIUM SULFATE. 19-2761
/RSENIC-75 NUCLEAR QUADRUPOLE RESONANCE SPECTRA OF PARTIALLY DEUTERATED FERROELECTRIC CRYSTALS CESIUM DIHY/ 18-2740
CE IN WATER ICE. PART-2: DEUTERON SPIN LATTICE RELAXATION IN DEUTERATED ICE. /RT-1: PROTON MAGNETIC RESONAN 34-3471
TRIC RESONANCE AND RELAXATION IN ELECTRICALLY BIASED KDP AND DEUTERATED ISOMORPHS. (THEORY) DIELEC 03-0467
HASE TRANSITION IN FERROELECTRICS WITH HYDROGEN BONDS. (KDP, DEUTERATED KDP) /AR STATES OF PROTONS ON THE P 17-2671
STUDY OF QUASI SPIN WAVE HYDROGEN TUNNELING MODES IN KDP AND DEUTERATED KDP. LASER RAMAN 17-2569
ISOTOPE EFFECT IN THE RAMAN SPECTRA OF A DEUTERATED KDP CRYSTAL. 17-2621
N DETERMINATION. STRUCTURE OF DEUTERATED KDP CRYSTAL: DEUTERIUM CONCENTRATIO 17-2535
FIELD APPROXIMATION. STUDY OF PHASE TRANSITION IN MIXED KDP- DEUTERATED KDP CRYSTALS BY MODIFIED MOLECULAR 17-2528
/LOW FREQUENCY MODE AT THE TRANSITION TEMPERATURE IN KDP AND DEUTERATED KDP (DOPED WITH COBALT-57 AND IRON/ 17-2465
URE DEPENDENCE OF RAMAN SCATTERING SPECTRUM OF FERROELECTRIC DEUTERATED KDP DURING THE PHASE TRANSITION. /T 17-2626
Y ELECTROOPTIC MODULATION (BY PASSING COHERENT LIGHT THROUGH DEUTERATED KDP HAVING A CHARGE PATTERN). /ON B 17-2507
FERROELECTRIC CRITICAL SCATTERING FROM DEUTERATED KDP. (NEUTRON INELASTIC SCATTERING) 17-2610
DIELECTRIC DISPERSION OF TGS AND DEUTERATED KDP (THEORY, CALCULATIONS). 17-2534
1.06 MICRON ABSORPTION COEFFICIENTS OF DEUTERATED KDP WITH 70-100% DEUTERATION. 17-2435
TE. (EFFECT ON LATTICE/ SOME PROPERTIES OF NONDEUTERATED AND DEUTERATED LITHIUM THALLIUM TARTRATE MONOHYDRA 29-3239
DIELECTRIC PROPERTIES OF DEUTERATED POTASSIUM DIHYDROGEN PHOSPHATE. 17-2581
ELECTRON SPIN RESONANCE OF IRRADIATED AND DEUTERATED POTASSIUM DIHYDROGEN PHOSPHATE. 17-2533
RANDOM MOTION OF DEUTERONS IN DEUTERATED POTASSIUM DIHYDROGEN PHOSPHATE. 17-2651

TAL AND ITS APPLICATION FOR/ ELECTROOPTIC CHARACTERISTICS OF DEUTERATED POTASSIUM DIHYDROGEN PHOSPHATE CRYS 17-2499
/GEN PHOSPHATE, POTASSIUM DIHYDROGEN PHOSPHATE AND PARTIALLY DEUTERATED POTASSIUM DIHYDROGEN PHOSPHATE CRY/ 17-2675
/ HARMONIC GENERATION IN POTASSIUM DIHYDROGEN PHOSPHATE (AND DEUTERATED POTASSIUM DIHYDROGEN PHOSPHATE) FE/ 17-2695
/UTRON INELASTIC SCATTERING, POTASSIUM DIHYDROGEN PHOSPHATE, DEUTERATED POTASSIUM DIHYDROGEN PHOSPHATE, TH/ 03-0401
ON POINT.   ATTENUATION OF SOUND WAVES IN DEUTERATED ROCHELLE SALT AT THE PHASE TRANSITI 28-3204
ON POINT.   ATTENUATION OF SOUND WAVES IN DEUTERATED ROCHELLE SALT AT THE PHASE TRANSITI 28-3205
M OR A/ NMR STUDIES IN SOME HYDROGEN BONDED FERROELECTRICS. (DEUTERATED ROCHELLE SALT, POTASSIUM OR RUBIDIU 18-2719
ALS.   ELECTROOPTICAL AND OPTICAL PROPERTIES OF PARTIALLY DEUTERATED RUBIDIUM DIHYDROGEN PHOSPHATE CRYST 17-2698
HOUS COMPOUNDS. (TRIGLYCINE SELENATE, FLUOBERYLLATE, AND THE DEUTERATED SOLID SOLUTIONS) /TALS WITH ISOMORP 27-2972
ONIUM ALUMINUM (SULFATE OR SELENATE) DODECAHYDRATE. (AND THE DEUTERATED SULFATE) /LECTRIC ALUMS: METHYL AMM 20-2803
RIE POINT CORREC/ CRITICAL BEHAVIOR OF FERROELECTRIC TGS AND DEUTERATED TGS (HYSTERESIS LOOPS JUST BELOW CU 27-2966
UPLE PERTURBED NMR STUDY OF THE FERROELECTRIC TRANSITION IN DEUTERATED TRIGLYCINE FLUOBERYLLATE. /-9 QUADR 27-2965
STALS.   PROPERTIES OF DEUTERATED TRIGLYCINE FLUOBERYLLATE SINGLE CRY 27-2973
LENITE FAMILY. (ESPECIALLY THE SODIUM MEMBER, AND EFFECTS OF DEUTERATION) /OF THE ALKALINE TRIHYDROGEN DISE 22-2848
ICRON ABSORPTION COEFFICIENTS OF DEUTERATED KDP WITH 70-100% DEUTERATION.   1.06 M 17-2435
ERTIES OF FERROELECTRIC CRYSTALS THE TRIGLYCINE SULFATE GRO/ DEUTERATION EFFECT ON THERMAL AND ELASTIC PROP 27-2986
RANSITION TEMPERATURES).   EFFECT OF DEUTERATION IN AMMONIUM BISULFATE (CHANGE IN T 19-2760
IUM DIHYDROGEN PHOSPHATE (RDP).   EFFECT OF DEUTERATION ON THE ELASTIC PROPERTIES OF RUBID 17-2584
LS.   EFFECTS OF DEUTERATION ON THE SPECIFIC HEAT OF ADP CRYSTA 17-2414
NANCE IN AMMONIUM DIHYDROGEN ARSENATE AND DEUTERO AMMONIUM DIDEUTERIUM ARSENATE.   ARSENIC-75 NMR RESO 18-2736
LEUTERON QUADRUPOLE COUPLING IN FERROELECTRIC CESIUM DIDEUTERIUM ARSENATE.   18-2721
FERROELECTRICS POTASSIUM DIHYDROGEN ARSENATE AND POTASSIUM DIDEUTERIUM ARSENATE. /OMAINS IN THE IRRADIATED 18-2726
IUM OR RUBIDIUM OR AMMONIUM DIHYDROGEN ARSENATES, RUBIDIUM DIDEUTERIUM ARSENATE) /TED ROCHELLE SALT, POTASS 18-2719
PROPERTIES OF POTASSIUM DIHYDROGEN ARSENATE AND POTASSIUM DIDEUTERIUM ARSENATE.   THERMODYNAMIC 18-2730
/CE COUPLED MODE IN POTASSIUM DIHYDROGEN ARSENATE, POTASSIUM DEUTERIUM ARSENATE, AND AMMONIUM DIHYDROGEN A/ 18-2724
/FERROELECTRIC POTASSIUM DIHYDROGEN ARSENATE AND POTASSIUM DIDEUTERIUM ARSENATE AND ANTIFERROELECTRIC AMMO/ 18-2734
/ LATTICE RELAXATION AND RESONANCE IN FERROELECTRIC CESIUM DIDEUTERIUM ARSENATE AND CESIUM DIHYDROGEN ARSE/ 18-2720
/ OF X-IRRADIATED POTASSIUM DIHYDROGEN ARSENATE, POTASSIUM DIDEUTERIUM ARSENATE, RUBIDIUM DIHYDROGEN ARSEN/ 18-2725
EGREES/ RAMAN SPECTRA OF ORTHORHOMBIC AND CUBIC HYDROGEN AND DEUTERIUM CHLORIDES AT TEMPERATURES ABOVE 80 D 31-3330
STRUCTURE OF DEUTERATED KDP CRYSTAL: DEUTERIUM CONCENTRATION DETERMINATION.   17-2535
3 RELAXATION IN SODIUM TRIDEUTERIUM SELENATE.   NMR STUDY OF DEUTERIUM DIFFUSION AND DEUTERIUM AND SODIUM-2 22-2849
MONIUM ALUMINUM SULFATE DODECAHYDRATE, COLEMANITE, SODIUM TRIDEUTERIUM DISELENITE /ROMIUM ALUMS, METHYL AM 02-0087
OPTICAL AND ELECTROOPTICAL EFFECT IN RUBIDIUM TRIHYDROGEN OR DEUTERIUM DISELENITE CRYSTALS. /ION AND THERMO 22-2840
SODIUM AMMONIUM SULFATE DIHYDRATE, SODIUM TRIHYDROGEN OR TRIDEUTERIUM DISELENITE. (SUMMARY ONLY) /UDIES OF 19-2779
OLAR COUPLING MODEL COMP/ LONGITUDINAL MODES IN HYDROGEN AND DEUTERIUM HALIDE CRYSTALS. (RAMAN SPECTRA, DIP 31-3326
AMMONIUM SELENATE DIDEUTERATE. NUCLEAR MAGNETIC RESONANCE OF DEUTERIUM IN THE FERROELECTRIC SODIUM DEUTERO 22-2828
LATE CRYSTAL.   NUCLEAR MAGNETIC RESONANCE OF DEUTERIUM IN THE PERDEUTERO AMMONIUM FLUOBERYL 19-2744
M DIHYDROGEN PHOSPHATE, ALSO RUBIDIUM ANALOG AND POTASSIUM DIDEUTERIUM PHOSPHATE) / CURIE POINTS. (POTASSIU 17-2665
FERROELECTRIC TRANSITION IN POTASSIUM DIDEUTERIUM PHOSPHATE.   17-2611
LF FOCUSING IN AMMONIUM DIHYDROGEN PHOSPHATE AND POTASSIUM DIDEUTERIUM PHOSPHATE.   LASER INDUCED SE 17-2659
LIGHT SCATTERING FROM CRITICAL FLUCTUATIONS IN POTASSIUM DIDEUTERIUM PHOSPHATE.   17-2463
STIC MODES OF POTASSIUM DIHYDROGEN PHOSPHATE AND POTASSIUM DIDEUTERIUM PHOSPHATE. / THE SOFT OPTIC AND ACOU 17-2497
ES OF FERROELECTRIC PHASE TRANSITIONS IN KDP AND POTASSIUM DIDEUTERIUM PHOSPHATE. NEUTRON DIFFRACTION STUDI 17-2661
IUM TITANATE, POTASSIUM NIOBATE, SODIUM NITRITE, POTASSIUM DIDEUTERIUM PHOSPHATE) /POTASSIUM TANTALATE, BAR 02-0059
RE DEPENDENCE OF THE FERROELECTRIC PROPERTIES OF POTASSIUM DIDEUTERIUM PHOSPHATE.   PRESSU 17-2641
OF THE SOFT FERROELECTRIC AND ACOUSTIC MODES OF POTASSIUM DIDEUTERIUM PHOSPHATE. SIMULTANEOUS OBSERVATION 17-2636
ES OF THE SYSTEM POTASSIUM DIHYDROGEN PHOSPHATE- POTASSIUM DIDEUTERIUM PHOSPHATE. STRUCTURAL STUDI 17-2590
THE CRYSTAL STRUCTURE OF MONOCLINIC POTASSIUM DIDEUTERIUM PHOSPHATE.   17-2591
NIC STRESS WAVE INTERACTION WITH DOMAIN WALLS IN POTASSIUM DIDEUTERIUM PHOSPHATE.   ULTRASO 17-2659
/ILITY IN THE FERROELECTRIC TRANSITION REGION OF POTASSIUM DIDEUTERIUM PHOSPHATE (ANALYSIS AND THEORETICAL/ 17-2575
/ERMINATION OF THE EIGENVECTORS OF SOFT MODES IN POTASSIUM DIDEUTERIUM PHOSPHATE AND DEUTERATED AMMONIUM D/ 17-2700
/AL EXPLANATION OF SHIFT OF CURIE TEMPERATURE IN POTASSIUM DIDEUTERIUM PHOSPHATE AND DEUTERO POTASSIUM DID/ 17-2601
/S IN CRYSTALS OF POTASSIUM DIHYDROGEN PHOSPHATE POTASSIUM DIDEUTERIUM PHOSPHATE AND RUBIDIUM DIHYDROGEN P/ 17-2613
ITION TEMPERATURE.   ACOUSTIC PROPAGATION IN POTASSIUM DEUTERIUM PHOSPHATE AT THE FERROELECTRIC TRANS 17-2573
SPECTROMETR/ FERROELECTRIC (SOFT) MODE MOTION IN POTASSIUM DIDEUTERIUM PHOSPHATE . (BY NEUTRON TRIPLE AXIS 17-2668
/THE TEMPERATURE DEPENDENT FERROELECTRIC MODE IN POTASSIUM DIDEUTERIUM PHOSPHATE (COMPARED WITH KDP AND WI/ 17-2705
ELECTRIC PHASE.   RAMAN SPECTRUM OF POTASSIUM DIDEUTERIUM PHOSPHATE CRYSTAL IN PARA- AND FERRO 17-2647
TICS OF AN ELECTROOPTICAL SHUTTER BASED ON Z CUT POTASSIUM DIDEUTERIUM PHOSPHATE CRYSTALS.   CHARACTERIS 17-2460
DOUBLE DIELECTRIC HYSTERESIS LOOP OF POTASSIUM DEUTERIUM PHOSPHATE CRYSTALS.   17-2503
FIRST ORDER PHASE TRANSITION IN POTASSIUM DEUTERIUM PHOSPHATE CRYSTALS.   17-2677
LANDAU-KHALATNIKOV ATTENUATION IN POTASSIUM DIDEUTERIUM PHOSPHATE CRYSTALS.   17-2656
/HE PREPARATION ON NON-WEDGE SHAPED RUBIDIUM DIHYDROGEN OR DIDEUTERIUM) PHOSPHATE CRYSTALS THEORY AND PRAC/ 17-2637
DESK-T/ SIMULATION OF A DOUBLE 45 DEGREES, Z CUT POTASSIUM DIDEUTERIUM PHOSPHATE. ELECTROOPTIC Q SWITCH BY 17-2702
SPECTRA, A TUNNELING MODEL/ DEUTERON MOTIONS IN POTASSIUM DIDEUTERIUM PHOSPHATE (FROM POLARIZED REFLECTION 17-2645
TER.   CRYSTALLIZATION OF RUBIDIUM DIDEUTERIUM PHOSPHATE FROM SOLUTIONS IN HEAVY WA 17-2583
ION OF RUBIDIUM DIHYDROGEN PHOSPHATE IN WATER AND RUBIDIUM DIDEUTERIUM PHOSPHATE IN HEAVY WATER. / OF SOLUT 17-2585
/L STUDY ON THE PHASE TRANSITIONS IN POTASSIUM DIHYDROGEN OR DEUTERIUM PHOSPHATE (INFRARED PHOSPHATE STRET/ 17-2646
R Q SWITCHING.   LOSSLESS POTASSIUM DIDEUTERIUM PHOSPHATE POCKELS CELL FOR HIGH POWE 17-2529
/ Q SWITCHING PERFORMANCE IN LITHIUM NIOBATE AND POTASSIUM DIDEUTERIUM PHOSPHATE POCKELS CELLS. (COMMENTS) 11-1710
/NETIC RESONANCE AND RELAXATION IN FERROELECTRIC POTASSIUM DIDEUTERIUM PHOSPHATE, POTASSIUM DIDEUTERIUM AR/ 17-2438
/AMAN SPECTRA OF POTASSIUM DIHYDROGEN PHOSPHATE, POTASSIUM DIDEUTERIUM PHOSPHATE, POTASSIUM DIHYDROGEN ARS/ 17-2409
/M OR AMMONIUM DIHYDROGEN PHOSPHATE OR ARSENATE, POTASSIUM DIDEUTERIUM PHOSPHATE, POTASSIUM TANTALATE OR N/ 01-0021
PARAELECTRIC POTASSIUM DIHYDROGEN PHOSPHATE AND POTASSIUM DIDEUTERIUM PHOSPHATE. (SOFT MADE THEORY) /RA OF 03-0356
/PROPAGATION OF LONGITUDINAL ULTRASONIC WAVES IN POTASSIUM DIDEUTERIUM PHOSPHATE (TEMPERATURE DEPENDENCE O/ 17-2660
TERIUM / ESR STUDY OF AMMONIUM IONS IN IRRADIATED AMMONIUM DIDEUTERIUM PHOSPHATE- 5 PERCENT POTASSIUM DIDEU 17-2684
D FERROELECTRICS. (SODIUM TRIHYDROGEN SELENATE AND SODIUM TRIDEUTERIUM SELENATE) /PERTIES OF HYDROGEN BONDE 22-2855
IFFUSION AND DEUTERIUM AND SODIUM-23 RELAXATION IN SODIUM TRIDEUTERIUM SELENATE.   NMR STUDY OF DEUTERIUM D 22-2849
F THE FERROELECTRIC TRANSITIONS IN SODIUM TRIHYDROGEN AND TRIDEUTERIUM SELENATE. /AGNETIC RESONANCE STUDY O 22-2832
NUCLEAR SPIN LATTICE RELAXATION TIME MEASUREMENTS IN SODIUM DEUTERIUM SELENATE.   22-2847
/STEM IN POTASSIUM TRIHYDROGEN SELENITE, AND IN POTASSIUM TRIDEUTERIUM SELENITE, AS DETERMINED BY NEUTRON / 22-2854
SOMORPHISM OF SODIUM TRIHYDROGEN SELENITE (SHS) OR SODIUM TRIDEUTERIUM SELENITE (SDS). /RA OF ISOTOPIC NONI 22-2836
TRANSITIONS. (TRIGLYCINE SULFATE, POTASSIUM DIHYDROGEN AND DIDEUTERIUM SULFATE) /ATTERING STUDIES OF PHASE 17-2481
PART-8: FERROELECTRIC AMMONIUM HYDROGEN SULFATE AND AMMONIUM DEUTERIUM SULFATE. /ERATURE INFRARED STUDIES. 19-2782
/ INFRARED SPECTRA OF RUBIDIUM HYDROGEN SULFATE AND RUBIDIUM DEUTERIUM SULFATE IN THE PARA- AND FERROELECTR 19-2768
ISOTOPIC NONISOMORPHISM OF RUBIDIUM TRIHYDROGEN(1-X) TRIDEUTERIUM(X) DISELENATE CRYSTALS.   22-2873
THERMAL EXPANSION IN SODIUM TRIHYDROGEN(1-X) TRIDEUTERIUM(X) DISELENITE AND ISOTOPE EFFECT.   22-2858
POTASSIUM DIHYDROGEN PHOSPHATE- POTASSIUM DIHYDROGEN(1-X) DIDEUTERIUM(X) PHOSPHATE MIXED CRYSTALS. /IES OF 17-2680
R BEHAVIOR OF SOLID SOLUTIONS IN POTASSIUM DIHYDROGEN(1-X) DIDEUTERIUM(X) PHOSPHATE SINGLE CRYSTALS. REGULA 17-2458
VERSUS CONCENTRATION, TEMPERATU/ PHASE DIAGRAM OF SODIUM TRIDEUTERIUM(X) TRIHYDROGEN(1-X) PHOSPHATE SYSTEM 17-2666
/FIELD ON THE PERMITTIVITY OF A NEW DELTA PHASE OF SODIUM TRIDEUTERIUM(X) TRIHYDROGEN(1-X) SELENITE CRYSTA/ 22-2875

LD STUDY OF THE ISOTOPIC EFFECT IN POTASSIUM DIHYDROGEN(X) DIDEUTERIUM(1-X) PHOSPHATE MIXED CRYSTALS. / FIE 17-2563
ARSENIC-75 NMR RESONANCE IN AMMONIUM DIHYDROGEN ARSENATE AND DEUTERO AMMONIUM DIDEUTERIUM ARSENATE. 18-2736
/SIUM DIHYDROGEN ARSENATE, AMMONIUM DIHYDROGEN ARSENATE, AND DEUTERO AMMONIUM DIDEUTERIUM ARSENATE. (FERRO/ 18-2725
NUCLEAR MAGNETIC RESONANCE OF DEUTERIUM IN THE PERDEUTERO AMMONIUM FLUOBERYLLATE CRYSTAL. 19-2744
MAGNETIC RESONANCE OF DEUTERIUM IN THE FERROELECTRIC SODIUM DEUTERO AMMONIUM SELENATE DIDEUTERATE. NUCLEAR 22-2828
OF THE FORMATE ION IN COPPER FORMATE TETRAHYDRATE AND COPPER DEUTERO FORMATE TETRADEUTERATE. /ION SPECTRUM 26-2926
/IFFRACTION STUDY OF FERROELECTRIC POTASSIUM FERROCYANIDE TRIDEUTERO HYDRATE ABOVE THE CURIE TEMPERATURE (/ 24-2903
/OF CURIE TEMPERATURE IN POTASSIUM DIDEUTERIUM PHOSPHATE AND DEUTERO POTASSIUM DIDEUTERIUM PHOSPHATE WITH / 17-2601
TEMPERATURE CRYSTAL STRUCTURE OF THE FERROELECTRIC SODIUM TRIDEUTERO SELENITE. ROOM 22-2861
MS OF ICE. DEUTERON ARRANGEMENTS IN THE HIGH PRESSURE FOR 34-3438
FERROELECTRIC POTASSIUM DIDEUTERIUM PHOSPHATE, POTASSIUM DI/ DEUTERON MAGNETIC RESONANCE AND RELAXATION IN 17-2438
ATURE PHASE OF FERROELECTRIC THIOUREA. DEUTERON MAGNETIC RESONANCE OF THE HIGH TEMPER 25-2918
TALS CONTAINING O-D...O BOND. (INCLUDING FULLY DEUTERATED A/ DEUTERON MAGNETIC RESONANCE STUDY OF SOME CRYS 03-0204
PHATE (FROM POLARIZED REFLECTION SPECTRA, A TUNNELING MODEL/ DEUTERON MOTIONS IN POTASSIUM DIDEUTERIUM PHOS 17-2645
LFATE. DEUTERON NMR STUDIES OF LITHIUM HYDRAZINIUM SU 32-3354
STUDIES OF ICE POLYMORPHS BY NEUTRON DIFFRACTION, PROTON AND DEUTERON NUCLEAR MAGNETIC RESONANCE. /UCTURAL 34-3466
RATED AMMONIUM SULFATE. DEUTERON QUADRUPOLE COUPLING CONSTANT IN DEUTE 19-2761
CESIUM DIDEUTERIUM ARSENATE. DEUTERON QUADRUPOLE COUPLING IN FERROELECTRIC 18-2721
RROELECTRIC SOLIDS BY MAGNETIC RESONANCE PART-18: PROTON AND DEUTERON RESONANCE OF THIOUREA. STUDIES OF FE 25-2919
/LS. PART-1: PROTON MAGNETIC RESONANCE IN WATER ICE. PART-2: DEUTERON SPIN LATTICE RELAXATION IN DEUTERATE/ 34-3471
RIC DEUTERATED AMMONIUM HYDROGEN SULFATE. DEUTERON SPIN LATTICE RELAXATION IN FERROELECT 19-2795
TASSIUM DIHYDROGEN PHOSPHATE. EFFECT OF DEUTERON SUBSTITUTION ON DOMAIN DYNAMICS IN PO 17-2437
HOSPHATE. RANDOM MOTION OF DEUTERONS IN DEUTERATED POTASSIUM DIHYDROGEN P 17-2651
OGEN BONDING. (POTASSIUM DIHYDROGE/ DYNAMICS OF PROTONS (AND DEUTERONS) IN FERROELECTRIC CRYSTALS WITH HYDR 17-2568
TION IN THE TWO DIMENSIONAL ANTIFERROMAGNET/ ZERO POINT SPIN DEVIATION AND SPONTANEOUS SUBLATTICE MAGNETIZA 26-2929
TITANATE SINGLE CRYSTALS. DETERMINATION OF THE DEVIATION FROM OXYGEN STOICHIOMETRY OF BARIUM 04-0736
OPTICAL GAP OF STRONTIUM TITANATE. (DEVIATION FROM URBACH TAIL BEHAVIOR) 06-1111
PROTON RESONANCE STUDY OF SUBLATTICE ROTATIONS AND SPIN DEVIATION IN COPPER FORMATE TETRADEUTERATE. 26-2950
D THE EFFECT ON THE / STACKING FAULT MODEL FOR STOICHIOMETRY DEVIATIONS IN LITHIUM NIOBATE AND TANTALATE AN 11-1776
D THE EFFECT ON THE / STACKING FAULT MODEL FOR STOICHIOMETRY DEVIATIONS IN LITHIUM NIOBATE AND TANTALATE AN 11-1777
FIELD BY THE COMMUTATOR OF POLARIZATION IN KDP, IN A DIGITAL DEVIATOR. /Y OF THE LIMITATION OF THE ANGULAR 17-2524
A NEW TYPE OF SECOND ORDER PHASE TRANSITION DERIVED FROM DEVONSHIRE'S THEORY OF FERROELECTRICS. 03-0226
IN POTASSIUM FERROCYANIDE TRIHYDRATE SIN/ APPLICATION OF THE DEW METHOD FOR REVEALING THE DOMAIN STRUCTURE 24-2904
DOMAIN STRUCTURE IN LOW TEMPERATURE FERROELECTRICS BY SOLID DEW TECHNIQUE. OBSERVATION OF 03-0341
A RADICAL IN FERROELECTRIC DGN CRYSTAL. 16-2253
DI... SEE THE PREFIXED WORD.
DIELECTRIC STUDY OF THE LEAD ZIRCONATE LEAD HAFNATE (PHASE) DIAGRAM. 08-1340
SELENOIODIDE- ANTIMONY SELENOIODIDE SYSTEM ALLOYS. PHASE DIAGRAM AND ELECTRICAL CONDUCTIVITY OF BISMUTH 15-2122
TE NIOBATE)- BARIUM (ZINCATE NIOBATE)- LEAD TITANATE. (PHASE DIAGRAM, LATTICE CONSTANTS) /YSTEM LEAD (ZINCA 05-1070
/ATER SYSTEMS. PART-1: MOLECULAR INTERPRETATION OF THE PHASE DIAGRAM OF ICE. PART-2: DIELECTRIC AND MECHAN/ 34-3485
S 0.8 PERCENT TUNGSTEN TRIOXIDE. PHASE DIAGRAM OF LEAD ZIRCONATE(1-X) TITANATE(X) PLU 09-1581
1-X) PHOSPHATE SYSTEM VERSUS CONCENTRATION, TEMPERATU/ PHASE DIAGRAM OF SODIUM TRIDEUTERIUM(X) TRIHYDROGEN(17-2666
ECTRIC IRON(1-X) SULFIDE. PHASE DIAGRAM OF THE SEMICONDUCTING UNIAXIAL FERROEL 36-3617
OF HEXAGONAL SINGLE CRYSTALS OF RARE EARTH MANGANATES (PHASE DIAGRAMS, CONVECTION EFFECTS). / FLUX GROWTH) 36-3536
/ BISMUTH(0.5) (POTASSIUM OR SODIUM(0.5) TITANATES AND PHASE DIAGRAMS FOR BISMUTH(X) POTASSIUM(X) BARIUM(1/ 08-1289
POTASSIUM DIHYDROGEN ARSENATE TC 52 KBAR AND 400 DEG / PHASE DIAGRAMS OF AMMONIUM DIHYDROGEN PHOSPHATE AND 17-2478
TE TO 40 KBAR. PHASE DIAGRAMS OF SODIUM NITRITE AND POTASSIUM NITRI 16-2361
(POTASSIUM BROMIDE, LEAD TELLURIDE, TI/ LATTICE DYNAMICS OF DIATOMIC CRYSTALS OF HIGH DIELECTRIC CONSTANT. 36-3541
FLAME SPRAYED BARIUM TITANATE AS A CAPACITOR DIELECTRIC. 04-0773
NE FLUORIDE. DIELECTRIC ABSORPTION IN ORIENTED POLYVINYLIDE 36-3572
QUADRATIC PHASES OF BARIUM TITANATE. DIELECTRIC ABSORPTION IN THE ORTHORHOMBIC AND 04-0900
QUADRATIC PHASES OF BARIUM TITANATE. DIELECTRIC ABSORPTION IN THE ORTHORHOMBIC AND 04-0899
OF BARIUM TITANATE. DIELECTRIC ABSORPTION IN THE RHOMOHEDRAL PHASE 04-0809
OF CALCIUM TITANATE IN BARIUM TITANATE. DIELECTRIC AGING IN TETRAGONAL SOLID SOLUTIONS 08-1295
0.5) (SODIUM OR POTASSIUM)(0.5)(X) LEAD(1-X) TITANATE SYSTE/ DIELECTRIC AND CERAMIC PROPERTIES OF (BISMUTH(13-2013
THE FERROELECTRIC PHASE TRANSITION IN GADOLINIUM MOLYBDATE. (DIELECTRIC AND ELASTIC PROPERTIES) /THEORY OF 30-3303
YSTALS OF BARIUM LEAD TITANATE SOLID SOLUTIONS. DIELECTRIC AND ELASTIC PROPERTIES OF SINGLE CR 04-0597
IN POTASSIUM NITRITE AND POTASSIUM NITRATE. DIELECTRIC AND ELECTRICAL CONDUCTIVITY STUDIES 16-2340
UM SODIUM YTTRIUM NIOBATE SINGLE CRYSTAL WITH TUNGSTEN BRON/ DIELECTRIC AND ELECTROOPTIC PROPERTIES OF BARI 13-1971
ARIZATION PROCESS IN STRONTIUM TITANATE AT LOW TEMPERATURE./ DIELECTRIC AND ELECTROOPTIC STUDIES OF THE POL 06-1117
RROELECTRIC CRYSTALS OF THE POTASS/ RELATIONSHIP BETWEEN THE DIELECTRIC AND ELECTROOPTICAL PROPERTIES OF FE 17-2696
BARIUM LEAD NIOBATE ZIRCONATE SYSTEM. (TETR/ STUDIES ON THE DIELECTRIC AND FERROELECTRIC PROPERTIES OF THE 13-1925
ER FERROELECTRIC CRYSTALS. (AMMO/ PHENOMENOLOGICAL THEORY OF DIELECTRIC AND MECHANICAL PROPERTIES OF IMPROP 14-2084
/LECULAR INTERPRETATION OF THE PHASE DIAGRAM OF ICE. PART-2: DIELECTRIC AND MECHANICAL RESPONSE OF ICE I(H/ 34-3485
SINGLE CRYSTALS AND ITS INTERPRETATION. DIELECTRIC AND MECHANICAL RESPONSE OF ICE I(H) 34-3483
TION IN STANNOUS CHLORIDE DIHYDRATE. DIELECTRIC AND NMR STUDIES OF THE PHASE TRANSI 36-3582
OF POLYCRYSTALLINE SOLID SOLUTIONS BASED ON BARIUM TITANATE/ DIELECTRIC AND NONLINEAR MICROWAVE PROPERTIES 04-1019
F THE ABC NIOBAT/ COMPARATIVE STUDY OF THE CRYSTALLOGRAPHIC, DIELECTRIC, AND NONLINEAR OPTICAL PROPERTIES O 04-0912
/C ANOMALY OF BARIUM SODIUM NIOBATE (ELASTIC, PIEZOELECTRIC, DIELECTRIC, AND OPTICAL CONSTANTS) ROOM TEMPE/ 13-2060
ATE(0.33) NIOBATE(0.67) SINGLE CRYSTAL. DIELECTRIC AND OPTICAL PROPERTIES OF LEAD ZINC 13-2065
NT FERROELECTRIC GLASS CERAMIC SYSTEMS. DIELECTRIC AND OPTICAL PROPERTIES OF TRANSPARE 04-0576
TIMONY SULFOIODIDE CRYSTAL/ EFFECT OF COMPOSITION ON GROWTH, DIELECTRIC, AND PHOTOELECTRIC PROPERTIES OF AN 15-2118
PZT-5 UNDER PRESSURE, DIELECTRIC AND PIEZOELECTRIC PROPERTIES. 09-1496
TERNARY SYSTEM LEAD (ZINCATE NIOBATE)- BARIUM (ZINCATE NIO/ DIELECTRIC AND PIEZOELECTRIC PROPERTIES IN THE 05-1070
ERAM/ EFFECT OF FERROELECTRIC PHASE TRANSITION CONDITIONS ON DIELECTRIC AND PIEZOELECTRIC PROPERTIES OF A C 09-1591
IMONY SULFOIODIDE IN A CONSTANT ELECTRIC FIELD. DIELECTRIC AND PIEZOELECTRIC PROPERTIES OF ANT 15-2249
C COBALTATE(0.33) NIOBATE(0.67)- TITANATE- ZIRCONATE SOLID / DIELECTRIC AND PIEZOELECTRIC PROPERTIES OF LEA 09-1549
ELECTRICS. (THEORY) DYNAMICAL (DIELECTRIC) AND SCATTERING PROPERTIES OF FERRO 03-0208
TRIC PHASE TRANSITION IN SODIUM AMMONIUM SULFATE DIHYDRATE./ DIELECTRIC AND SOME OTHER STUDIES OF FERROELEC 19-2767
ARIUM TITANATE. GRAIN SIZE DEPENDENCE OF DIELECTRIC AND STRUCTURAL CHARACTERISTICS OF B 04-0582
LECTRICS WITH DIFFUSE PHASE TRANSITIONS. (MIXED LAYER OXIDE/ DIELECTRIC AND STRUCTURAL PROPERTIES OF FERROE 13-2014
/L CONDUCTIVE, SURFACE, AND ELECTROOPTIC PROPERTIES. PART-7: DIELECTRIC AND SWITCHING MEASUREMENTS. PART-8/ 03-0397
F SOLID SOLUTIONS OF TUNGSTEN OXIDE. DIELECTRIC (AND TEMPERATURE) CHARACTERISTICS O 10-1657
CALORIC EFFECT) STUDIES OF FERROELECTRIC CERAMICS AT LOW TE/ DIELECTRIC AND THERMAL (SPECIFIC HEAT, ELECTRO 04-0720
NATE TRANSITIONS (FERROELECTRIC BELOW 93K). DIELECTRIC AND THERMAL STUDY OF POTASSIUM SELE 22-2824
THIUM TANTALATE SINGLE CRYSTAL NEAR ITS CURIE POINT, PART-2, DIELECTRIC AND ULTRASONIC PROPERTIES. /R OF LI 12-1885
SITIONS IN ROCHELLE SALT, BARIUM TITANATE, ANTIMONY SULFOIO/ DIELECTRIC AND X-RAY STUDIES OF THE PHASE TRAN 02-0099
RIC PHASE TRANSITION OF GLYCINE SILVER NITRATE. DIELECTRIC AND X-RAY STUDIES ON THE FERROELECT 16-2347
TRIC CERAMIC. (BARIUM TITANATE, BARIUM / ULTRAHIGH FREQUENCY DIELECTRIC ANISOTROPY OF A POLARIZED FERROELEC 04-0767
DIHYDROGEN PHOSPHATE POTASSIUM DIDEUTERIUM PHOSPHATE AND RU/ DIELECTRIC ANOMALIES IN CRYSTALS OF POTASSIUM 17-2613

WITH SEVERAL PHASE TRANSITIONS / PHENOMENOLOGICAL THEORY OF DIELECTRIC ANOMALIES IN FERROELECTRIC MATERIAL 03-0363
EMPERATURES AND HIGH PRESSURES. (POSSIBLY FERROELECTRIC) DIELECTRIC ANOMALIES IN SOLID METHANE AT LOW T 36-3547
HE FERROELECTRIC CURIE TEMPERATURE. DIELECTRIC ANOMALIES OF SODIUM NITRATE ABOVE T 16-2376
XIDE(12) SINGLE CRYSTALS. FERROELECTRIC DIELECTRIC ANOMALY IN BISMUTH(4) TITANIUM(3) O 13-2009
THEORETICAL AND PRACTICAL STUDIES OF DIELECTRIC APPLICATIONS OF FERROELECTRICS. 03-0462
S AT LOW TEMPERATURES. DIELECTRIC BEHAVIOR OF CUBIC AND HEXAGONAL ICE 34-3431
F VACUUM DEPOSITED POTASSIUM NITRATE. DIELECTRIC BEHAVIOR OF (FERROELECTRIC) FILMS O 16-2356
IGLYCINE SULFATE. DIELECTRIC BEHAVIOR OF GAMMA-RAY IRRADIATED TR 27-3149
OSPHATE. INFLUENCE OF TUNNELING ON DIELECTRIC BEHAVIOR OF POTASSIUM DIHYDROGEN PH 17-2527
AVE RANGE. INFLUENCE OF CRYSTAL DEFECTS ON THE DIELECTRIC BEHAVIOR OF ROCHELLE SALT IN MICROW 28-3219
ERROELECTRIC) THE VIBRATIONAL SPECTRUM AND DIELECTRIC BEHAVIOR OF SODIUM CHLORATE. (NOT F 36-3549
TE PARTICULATES. EFFECT OF SURFACE MOISTURE ON THE DIELECTRIC BEHAVIOR OF ULTRAFINE BARIUM TITANA 04-0859
POLARIZATION, COERCIVE FIELD) OF BISMUTH TITANATE AT LOW T/ DIELECTRIC BEHAVIOR (PERMITTIVITY, SPONTANEOUS 08-1436
DIELECTRIC BREAKDOWN IN BARIUM TITANATE. 04-0580
TITANATE. DIELECTRIC BREAKDOWN OF POLYCRYSTALLINE BARIUM 04-1015
ICE TO 2 K. LOW TEMPERATURE DIELECTRIC CELL AND PERMITTIVITY OF HEXAGONAL 34-3432
DISPERSION IN STRONTIUM TITANATE- LANTHANUM FERRATE DIELECTRIC CERAMICS. 06-1120
URE FOR HIGH FREQUENCY APPLICATION. (BARIUM, STRONTIUM TITA/ DIELECTRIC CERAMICS WITH BOUNDARY LAYER STRUCT 04-0833
STRUCTURE OF THE CUBIC PHASE OF BARIUM TITANATE. (IMAGINARY DIELECTRIC CONSTANT) BAND 04-0845
STATIC PRESSURE (TO 5600 KG/SQ CM, TEMPERATURE DEPENDENCE OF DIELECTRIC CONSTANT). /RYSTAL UNDER HIGH HYDRO 24-2896
AND A 36:64 RATIO MIXED CRYSTAL, PROTON MAGNETIC RESONANCE, DIELECTRIC CONSTANT). /MONIUM ALUMINUM SULFATE 20-2809
ARIOUS FERROELECTRI/ MILLIMETER WAVELENGTH DISPERSION IN THE DIELECTRIC CONSTANT ABOVE THE CURIE POINT IN V 02-0124
TING OF A PIEZOELECTRIC CERAMIC (BARIUM TITANATE) WITH LARGE DIELECTRIC CONSTANT AND A SEMICONDUCTOR. /NSIS 04-1037
TIME OF ROCHELLE SALT. DIELECTRIC CONSTANT AND DIELECTRIC RELAXATION 28-3179
THIN FILM OF A FERROELECTRIC (BARIUM TITANATE DOWN TO 3 MIC/ DIELECTRIC CONSTANT AND LATTICE DYNAMICS OF A 04-0600
TANATE CERAMICS. (AND DIFFERENCE IN TEMPERATURES FOR MAXIMUM DIELECTRIC CONSTANT AND LOSS) /ES IN BARIUM TI 04-0849
UCED STRONTIUM TITANATE SINGLE CRYSTALS (40 HZ-10 MHZ, EFFE/ DIELECTRIC (CONSTANT AND LOSS) BEHAVIOR OF RED 06-1148
ON OF FERROELECT/ TEMPERATURE AND PRESSURE DEPENDENCE OF THE DIELECTRIC CONSTANT AND SPONTANEOUS POLARIZATI 16-2331
/RYLLATE SINGLE CRYSTAL GROWTH CONDITIONS ON A CHANGE IN THE DIELECTRIC CONSTANT AND TANGENT OF THE ANGLE / 27-3151
OD OF THE PHASE TRANSITION IN CHARACTERISTICALLY FERROELECT/ DIELECTRIC CONSTANT BEHAVIOR IN THE NEIGHBORHO 03-0522
PROPERTIES OF HIGH PURITY POLYCRYSTALLINE BARIUM TITANATE. (DIELECTRIC CONSTANT, CONDUCTIVITY) ELECTRICAL 04-0704
SEMICONDUCTING BARIUM TITANATE. (DIELECTRIC CONSTANT, CONTACT RESISTANCE) 04-0730
DOPED BARIUM TITANATE WITH IMPROVED ELECTRICAL PROPERTIES. (DIELECTRIC CONSTANT, DISSIPATION FACTOR) /XIDE 04-0840
/TIMONY DOPED) BARIUM TITANATE CERAMIC WITH A HIGH EFFECTIVE DIELECTRIC CONSTANT (DUE TO ADDING CUPROUS OR/ 04-0581
/ITANATE AND ((BARIUM(X) STRONTIUM(1-X) TITANATE (RECIPROCAL DIELECTRIC CONSTANT FOLLOWS QUADRATIC TEMPERA/ 04-0898
TRIUM MOLYBDATE- POTASSIUM DYSPRO/ TEMPERATURE DEPENDENCE OF DIELECTRIC CONSTANT IN THE SYSTEM POTASSIUM YT 30-3309
CITY AS CAPACITOR DIELECTRICS, A STUDY IN DIELECTRIC S/ HIGH DIELECTRIC CONSTANT MATERIALS AND FERROELECTRI 37-3678
ON PARAMAGNETIC RESONANCE. (ABSTRACT ON/ INFORMATION ON HIGH DIELECTRIC CONSTANT MATERIALS GAINED BY ELECTR 02-0101
/RIUM SODIUM NIOBATE. (OPTICAL ABSORPTION, REFRACTIVE INDEX, DIELECTRIC CONSTANT, NONLINEAR OPTICAL COEFFI/ 13-2024
LATIONS, SUMMARY ON/ 180 DEG DOMAIN WALL CONTRIBUTION TO THE DIELECTRIC CONSTANT OF BARIUM TITANATE. (CALCU 04-0804
UTIONS WITH ALLOWANCE FOR CONCENTRATION FLUCTUATIONS. DIELECTRIC CONSTANT OF FERROELECTRIC SOLID SOL 03-0440
S. (BARIUM TITANATE CERAMICS) TEMPERATURE DEPENDENCE OF THE DIELECTRIC CONSTANT OF FERROELECTRIC SUBSTANCE 04-0591
E SULFATE. NONLINEARITY OF THE DIELECTRIC CONSTANT OF FERROELECTRIC TRIGLYCIN 27-3070
SIMULTANEOUS MEASUREMENT OF PYROELECTRIC COEFFICIENT AND DIELECTRIC CONSTANT OF FERROELECTRICS. 03-0333
CURIE/ FREQUENCY AND TEMPERATURE DEPENDENCES OF THE COMPLEX DIELECTRIC CONSTANT OF FERROELECTRICS NEAR THE 03-0361
EG K. COMPLEX DIELECTRIC CONSTANT OF LEAD HAFNATE, 320-520 D 08-1339
LARIZING FIELD, 370-520 DEG K. COMPLEX DIELECTRIC CONSTANT OF LEAD ZIRCONATE UNDER PO 07-1256
OSPHATE AT LAMBDA EQUAL, 2.14 / TEMPERATURE DEPENDENCE OF THE DIELECTRIC CONSTANT OF POTASSIUM DIHYDROGEN PH 17-2501
E CRYSTALS UNDER BIASING CONDITIONS. DIELECTRIC CONSTANT OF POTASSIUM NIOBATE SINGL 08-1381
EFFECT OF PRESSURE ON THE STATIC DIELECTRIC CONSTANT OF POTASSIUM TANTALATE. 08-1269
YSTALS. DIELECTRIC CONSTANT OF PURE ICE I(H) SINGLE CR 34-3478
AS A FUNCTION OF TEMPERATURE. MEASUREMENT OF THE COMPLEX DIELECTRIC CONSTANT OF ROCHELLE SALT AT 10 GHZ 28-3184
TEMPERATURE DEPENDENCE OF THE DIELECTRIC CONSTANT OF STRONTIUM TITANATE. 06-1134
(TEMPERATURE AND FREQUENCY DEPENDENCE). DISPERSION OF THE DIELECTRIC CONSTANT OF TGS IN THE FAR INFRARED 27-3009
DIELECTRIC CONSTANT OF TRIGLYCINE SULFATE. 27-3049
FUNCTION OF THE CONDITIONS OF GR/ THERMAL HYSTERESIS OF THE DIELECTRIC CONSTANT OF TRIGLYCINE SULFATE AS A 27-3126
ICROWAVE FREQUENCIES. MEASUREMENT OF THE DIELECTRIC CONSTANT OF TRIGLYCINE SULFATE AT M 27-2954
LE CRYSTALS IN A STRONG ELECTRIC FIELD. DIELECTRIC CONSTANT OF TRIGLYCINE SULFATE SING 27-3150
HIOMETRY) IN L/ DEPENDENCE OF LINEAR ELECTROOPTIC EFFECT AND DIELECTRIC CONSTANT ON MELT COMPOSITION (STOIC 11-1842
/TASSIUM IODATE - A DISORDERED FERROELECTRIC. (DEPENDENCE OF DIELECTRIC CONSTANT ON THERMAL AND FIELD HIST/ 01-1348
ON OF GLASSES CONTAINING/ BARIUM TITANATE CERAMICS WITH HIGH DIELECTRIC CONSTANT. PART-1: EFFECTS OF ADDITI 04-0829
TELLURIDE, TI/ LATTICE DYNAMICS OF DIATOMIC CRYSTALS OF HIGH DIELECTRIC CONSTANT. (POTASSIUM BROMIDE, LEAD 36-3541
OELECTRIC AND RELATED PROPERTIES OF LEAD TITANATE CERAMICS. (DIELECTRIC CONSTANT, RESISTIVITY) /VES ON PIEZ 05-1084
CS OF FERROELECTRIC TRIGLYCINE SE/ (OPTICAL) ENERGY GAP (AND DIELECTRIC CONSTANT) TEMPERATURE CHARACTERISTI 27-3021
ECTRICS. (BARIUM TITANATE, STRONT/ METHOD FOR ACHIEVING HIGH DIELECTRIC CONSTANT TEMPERATURE STABLE FERROEL 04-0551
OLTAGE, TEMPE/ NONLINEAR PROPERTIES OF TRIGLYCINE SELENATE. (DIELECTRIC CONSTANT VERSUS ALTERNATING FIELD V 37-3158
IES ON THE MECHANISM OF THE PHASE TRANSITION IN TGS. (STATIC DIELECTRIC CONSTANT, X-RAY SCATTERING) STUD 27-3093
IUM(1-X) NEODYMIUM(X) (ORTHO) ALUM/ ANOMALIES IN THE (1 KHZ) DIELECTRIC CONSTANT (5-300K) OF THE PRASEODYM 36-3600
, 300 DEG K) OF THE LATTICE VIBRATIONS IN LITHIUM TANTALATE (DIELECTRIC CONSTANTS) /TUDY (20-10,000 PER CM 12-1870
TRIGLYCINE SULFATE. (TEMPERATURE DEPENDENCES OF THE COMPLEX DIELECTRIC CONSTANTS) /TION OF DOMAIN WALLS IN 27-3103
TANATES - APPLICATION TO CALCULATING THE WALL ENERGY AND THE DIELECTRIC CONSTANTS. / FOR BARIUM AND LEAD TI 03-0189
/ON AND THE VARIATIONS IN THE FREE AND CLAMPED VALUES OF THE DIELECTRIC CONSTANTS AND ELECTROOPTIC COEFFIC/ 13-1998
EAD(5) GERMANIUM(3) OXIDE(11) CRYSTAL, A NEW FERROELECTRIC. (DIELECTRIC CONSTANTS AND REFRACTIVE INDICES) L 35-3502
ATERIALS AT LOW TEMPERATURES. MEASUREMENT OF DIELECTRIC CONSTANTS FOR A FEW FERROELECTRIC M 02-0035
S) LATTICE DYNAMICS OF IONIC CRYSTALS WITH HIGH DIELECTRIC CONSTANTS. (LEAD AND TIN TELLURIDE 36-3553
THEORY OF THE DIELECTRIC CONSTANTS OF BARIUM TITANATE. 04-0776
OXIDE(20). ELASTIC, PIEZOELECTRIC, AND DIELECTRIC CONSTANTS OF BISMUTH(12) GERMANIUM 35-3512
(HALIDES AND OXIDES, RELATION BE/ TEMPERATURE DEPENDENCE OF DIELECTRIC CONSTANTS OF CUBIC IONIC COMPOUNDS. 03-0188
: DIELECTRIC CONSTANTS AND DIELECT/ MEASUREMENT OF MICROWAVE DIELECTRIC CONSTANTS OF FERROELECTRICS. PART-2 16-2354
EFFECT OF GAMMA-IRRADIATION ON THE DIELECTRIC CONSTANTS OF KDP. 17-2691
L/ TEMPERATURE DEPENDENCE OF THE ELASTIC, PIEZOELECTRIC, AND DIELECTRIC CONSTANTS OF LITHIUM TANTALATE AND 11-1823
. (BARIUM TITANATE, STRONTIUM TI/ PRESSURE DEPENDENCE OF THE DIELECTRIC CONSTANTS OF PARAELECTRIC MATERIALS 04-0856
ORIDES BY THE METHOD OF SUBSTITUTION. LOW FREQUENCY DIELECTRIC CONSTANTS OF THE ALKALINE EARTH FLU 33-3361
OBATE- LANTHANUM NIOBATE. DIELECTRIC CONSTANTS OF THE SYSTEM SAMARIUM NI 08-1268
HOGONAL TO THE FERROELECTRIC / TEMPERATURE DEPENDENCE OF THE DIELECTRIC CONSTANTS OF TRIGLYCINE SULFATE ORT 27-2975
/YSTAL GROWTH AND PHYSICAL PROPERTIES OF LITHIUM TANTALATE. DIELECTRIC CONSTANTS, PIEZOELECTRIC CONSTANTS/ 12-1876
/C PHOTOELECTRETS BASED ON A(V)B(VI)C(VII) SOLID SOLUTIONS. (DIELECTRIC CONSTANTS, PYROELECTRIC CURRENTS, / 15-2157
/AND ITS ISOMORPHOUS COMPOUND LEAD GERMANIUM SILICON OXIDE. (DIELECTRIC CONSTANTS, SPONTANEOUS POLARIZATIO/ 35-3501
CALCULATION OF THE DIELECTRIC CORRELATION FACTOR OF CUBIC ICE. 34-3429
DIGLYCINE SULFATE AN INTERESTING NEW DIELECTRIC CRYSTAL SPECIES. 27-2993

POTASSIUM HYDROXIDE DOPEL ICE.   DIELECTRIC DISPERSION AND PHASE TRANSITION OF   34-3452

REVERSIBLE PROCESSES. (APPLIED TO FERROELECTRIC TRANSITIONS/ DIELECTRIC DISPERSION AND THERMODYNAMICS OF IR   03-0415

EVALUATION OF DIELECTRIC DISPERSION DATA. (ICE)   34-3433

CRYSTALS.   DIELECTRIC DISPERSION IN ANTIMONY SULFOIODIDE   15-2117

TITANATES AND TITANIUM DIOXIDE.   FAR INFRARED DIELECTRIC DISPERSION IN BARIUM AND STRONTIUM   04-0975

.   FAR INFRARED DIELECTRIC DISPERSION IN BARIUM SODIUM NIOBATE   13-1986

DIELECTRIC DISPERSION IN BARIUM TITANATE.   04-0646

E CRYSTALS ON THEIR DOMAIN STRUCTURE.   DEPENDENCE OF DIELECTRIC DISPERSION IN BARIUM TITANATE SINGL   04-0895

FERROELECTRICS IN THE METER BAND.   DIELECTRIC DISPERSION IN BARIUM TITANATE TYPE   04-1009

INE SULFATE (SEVERAL KHZ TO 23 KMHZ, DOMAIN WALL RESONANCE). DIELECTRIC DISPERSION IN FERROELECTRIC TRIGLYC   27-3032

DIELECTRIC DISPERSION IN FERROELECTRICS.   03-0428

IOUS TYPES. (BARIUM TITANATE AND TITANAT/ STUDY OF MICROWAVE DIELECTRIC DISPERSION IN FERROELECTRICS OF VAR   02-0108

LECTRICS. (THEORY) DIELECTRIC DISPERSION IN ORDER DISORDER FERROE   03-0373

M TITANATE.   MECHANISM OF MICROWAVE DIELECTRIC DISPERSION IN POLYCRYSTALLINE BARIU   04-0745

S.   LCW FREQUENCY DIELECTRIC DISPERSION IN ROCHELLE SALT CRYSTAL   28-3228

TIVITY AND LOSS FROM 10 GHZ TO 100 GHZ) DIELECTRIC DISPERSION IN TGS CRYSTALS. (PERMIT   27-3094

/FFECT OF COLLECTIVE INTERACTION CF DIPOLES CN THE NATURE OF DIELECTRIC DISPERSION IN THE FERROELECTRIC PH/   04-0417

E OF POTASSIUM NITRATE.   DIELECTRIC DISPERSION IN THE PARAELECTRIC PHAS   16-2341

ANATE NEAR THE PHASE TRASNITION.   DIELECTRIC DISPERSION OF IRRADIATED BARIUM TIT   04-0648

ANATE IN THE PHASE TRANSITION REGICN.   DIELECTRIC DISPERSION OF IRRADIATED BARIUM TIT   04-0652

TITANATE, POTASSIUM TANTALATE)   FAR INFRARED DIELECTRIC DISPERSION OF PEROVSKITES. (BARIUM   04-0546

DIELECTRIC DISPERSION OF TGS.   17-2574

DIELECTRIC DISPERSION OF TGS.   27-3060

P (THEORY, CALCULATIONS).   DIELECTRIC DISPERSION OF TGS AND DEUTERATED KD   17-2534

LOW FREQUENCIES.   DIELECTRIC DISPERSION OF TRIGLYCINE SULFATE AT   27-3023

DIELECTRIC DISPERSION OF UREA AND THIOUREA.   25-2912

(POLARIZATION CHANGE) AFTER EFFECTS IN TRIGLYCINE SULFATE. (DIELECTRIC DISPLACEMENT AND COERCIVE FIELD)   27-2952

METRY, AND PHASE TRANSITION REPRESENTATICN. (18 REFS) DIELECTRIC, ELASTIC AND MAGNETIC PROPERTY, SYM   03-0316

/PRESSURE AND INVESTIGATICN OF THEIR ELECTRICAL PROPERTIES. (DIELECTRIC, ELASTIC, AND PIEZOELECTRIC CONSTA/   15-2138

ES OF BARIUM TITANATE SINGLE CRYSTALS WITH LAMINAR DOMAIN S/ DIELECTRIC, ELASTIC AND PIEZOELECTRIC PROPERTI   04-1010

N IN POTASSIUM DIHYDROGEN PHOSPHATE.   DIELECTRIC EVIDENCE OF A FIRST ORDER TRANSITIO   17-2602

NUOUS HEATING. (TGS SINGLE CRYSI/ IMPEDANCE MEASUREMENT OF A DIELECTRIC FERROELECTRIC CONDENSER UNDER CONTI   27-3028

/OF TEMPERATURE AUTOSTABILIZATION OF FERROELECTRIC CRYSTAL. (DIELECTRIC HEATING AND POLARIZATION REVERSAL,/   03-0245

RYSTAL BISMUTH FERRATE (CONFIRMS FERROELECTRIC NATURE).   DIELECTRIC HYSTERESIS IN (FLUX GROWN) SINGLE C   08-1449

ERIUM FHCSPHATE CRYSTALS.   DOUBLE DIELECTRIC HYSTERESIS IN POTASSIUM DIDEUT   17-2503

T OF DIFFUSE FERROELECTRIC PHASE. (KAEN/ AC POLARIZATION AND DIELECTRIC HYSTERESIS NEAR THE TRANSITION POIN   03-0437

NSHIP TO FERROELECTRICITY) DIELECTRIC INSTABILITIES. (THEORETICAL RELATIO   02-0068

) TREATED AS CRITI/ FERROELECTRIC TRANSITIONS WITH A GENUINE DIELECTRIC INSTABILITY. (FERROELECTRIC TWINNING   15-2176

/ SINGLE CRYSTALS ON A CHANGE IN THE TANGENT OF THE ANGLE OF DIELECTRIC LOSSES IN A STRONG ELECTRIC FIELD.   27-3153

TRICS (THEORY).   ATTENUATION OF CRITICAL VIERATIONS AND DIELECTRIC LOSSES IN DISPLACIVE TYPE FERROELEC   03-0159

ECTRICS ABOVE THE CURIE TEMPERATURE. (BARIUM(0-1./ MICROWAVE DIELECTRIC LOSSES IN OXYGEN OCTAHEDRON FERROEL   04-0559

CRYSTALS.   EFFECT OF GROWTH CONDITIONS ON DIELECTRIC LOSSES IN TRIGLYCINE SULFATE SINGLE   27-3152

SYNTHESIZED AT HIGH PRESSURES.   STRUCTURE AND DIELECTRIC LOSSES OF FERROELECTRIC PEROVSKITES   08-1452

LS EXPOSED TO SMALL DOSES OF X- AND GAMMA-RAYS.   DIELECTRIC LOSSES OF TRIGLYCINE SULFATE CRYSTA   27-2976

DIFFUSICN COEFFICIENT OF NIOBIUM IN BARIUM TITANATE BY A DIELECTRIC MEASUREMENT TECHNIQUE.   04-0926

N) STUDIES ON FERROELECTRICITY IN POTASSIUM NITRATE A/ SOME (DIELECTRIC MEASUREMENTS AND INFRARED ABSORPTIO   16-2402

DIELECTRIC MEASUREMENTS OF AMMONIUM NITRATE.   16-2338

RROELECTRIC AMMONIUM SULFATE.   MICROWAVE DIELECTRIC MEASUREMENTS OF GAMMA-IRRADIATED FE   19-2750

IUM TITANATE SINGLE CRYSTALS.   DIELECTRIC MEASUREMENTS ON PILE-IRRADIATED BAR   04-0910

N PHOSPHATE, TRIGLYCINE SULFATE, AND RUTILE.   FAR INFRARED DIELECTRIC MEASUREMENTS ON POTASSIUM DIHYDROGE   17-2423

GLASS- CERAMIC SYSTEMS.   APPLICATION OF DIELECTRIC MIXTURE FORMULAS TO (LEAD TITANATE)   05-1054

DUE TC NEUTRON IRRADIATION.   STRUCTURAL AND DIELECTRIC MODIFICATIONS OF TUNGSTEN TRIOXIDE   10-1651

(OR LEAD, 4 COMPOSITIONS EACH) STRONTIUM TITANATE SYSTEMS./ DIELECTRIC NONLINEARITY AND LOSS IN THE BARIUM   09-1522

UM TITANATE NEAR THE PHASE TRANSI/ ANCMALOUS BEHAVIOR OF THE DIELECTRIC NONLINEARITY COEFFICIENT IN STRONTI   06-1105

SINGLE CRYSTALS NEAR PHASE TRANSITION AT 110 DEGREES K. DIELECTRIC NONLINEARITY OF STRONTIUM TITANATE   06-1106

TIES IN T/ FERROELECTRIC TRANSITION IN AMMONIUM SULFATE, THE DIELECTRIC, OPTICAL, AND ELECTROOPTICAL PROPER   19-2745

GLE TANGENT OF TRIGLYCINE SELENATE DEPENDING ON THE CONDITI/ DIELECTRIC PERMEABILITY AND DIELECTRIC LOSS AN   27-3125

GLE CRYSTALS IN THE PHASE TRANSITION REGION.   DIELECTRIC PERMEABILITY OF BARIUM TITANATE SIN   04-0565

NEAR THE PHASE TRANSITION.   DIELECTRIC PERMEABILITY OF TRIGLYCINE SELENATE   27-3166

MATERIALS)   IOW LOSS (FERROELECTRIC CERAMICS) WITH GREAT DIELECTRIC PERMITTIVITY. (BARIUM TITANATE TYPE   04-0779

ES IN FERROELECTRICS WITH ORDER DISORDER TYPE PHASE TRANSIT/ DIELECTRIC PERMITTIVITY DISPERSION PECULIARITI   03-0412

SSIUM AND AMMONIUM DIHYDROGEN PHOSPHATE.   DIELECTRIC PERMITTIVITY OF FINELY DIVIDED POTA   17-2689

UM TITANATE.   TIME DEPENDENCE OF THE DIELECTRIC PERMITTIVITY OF HYDROXYL DOPED BARI   04-0550

H TITANATE (EFFECT OF TEMPERATURE).   (WEAK FIELD) DIELECTRIC (PERMITTIVITY) PROPERTIES OF BISMUT   36-3555

DIELECTRIC PHENOMENA IN ROCHELLE SALT.   28-3182

IES OF LITHIUM NIOBATE SINGLE CRYSTAL/ THERMAL DEPENDENCE OF DIELECTRIC, PIEZOELECTRIC, AND ELASTIC PROPERT   11-1693

PERTIES OF THE LEAD DINIOBATE- BARIUM DINIOBATE SYSTEM. DIELECTRIC, PIEZOELECTRIC AND PYROELECTRIC PRO   08-1384

PERTIES OF THE LEAD METANIOBATE- EARIUM NIOBATE SYSTEM. DIELECTRIC, PIEZOELECTRIC AND PYROELECTRIC PRO   08-1385

ERTIES OF MATERIALS.   NONLINEARITY OF DIELECTRIC POLARIZATION AND FERROELECTRIC PROP   03-0185

.   PECULIARITIES OF THE DIELECTRIC POLARIZATION OF CADMIUM PYRONIOBATE   36-3561

NATE SOLID SOLUTIONS (THERMODYNAMIC EXISTENCE CONDITIONS FO/ DIELECTRIC POLARIZATION OF LEAD TITANATE ZIRCO   09-1541

/ PHASE TRANSITION BROADENING (DIFFUSNESS) AND THE NATURE OF DIELECTRIC POLARIZATION RELAXATION IN SOME FE/   03-0291

IR GRADIENTS IN ABO(3) PEROVSKITE COMPOUNDS WITH DISTINCTIVE DIELECTRIC PROPERTIES. /LECTRIC FIELDS AND THE   08-1455

IELD GRADIENTS IN THE PEROVSKITE TYPE COMPCUNDS WITH SPECIAL DIELECTRIC PROPERTIES. / FIELDS AND ELECTRIC F   08-1442

DEVELOPMENT OF BARIUM TITANATE WITH CONSISTENT DIELECTRIC PROPERTIES.   04-0797

COMPLEX COMPOSITIONS, AND THE STUDY OF THEIR STRUCTURES AND DIELECTRIC PROPERTIES. /RIOUS PEROVSKITES WITH   08-1428

CONFIGURATIONS IN SOME (PEROVSKITE) STRUCTURES WITH SPECIAL DIELECTRIC PROPERTIES.   ORIGIN OF DIPOLE   08-1390

STRONTIUM(1-X) NIOBATE SINGLE CRYSTALS (CZOCHRALSKI GROWTH, DIELECTRIC PROPERTIES). /ND STUDY OF BARIUM(X)   13-1936

M (BARIUM, STRONTIUM) TITANATE. (SYNTHESIS AND LOW FREQUENCY DIELECTRIC PROPERTIES) / CRYSTALS OF THE SYSTE   04-0561

RIUM TITANATE CRYSTALS.   DIELECTRIC PROPERTIES AND DOMAIN TEXTURE OF BA   04-0537

TRANSITION ON LIT/ ELECTROMECHANICAL COUPLING EFFECTS ON THE DIELECTRIC PROPERTIES AND FERROELECTRIC PHASE   29-3251

F POTASSIUM TANTALATE SINGLE CRYSTALS AND (POTASSIUM SODIUM/ DIELECTRIC PROPERTIES AND OPTICAL ABSORPTION O   08-1272

ARA- AND FERROELECTRIC PEROVSKITES. (THEORY, ABO(3) TYPE) DIELECTRIC PROPERTIES AND OPTICAL PHONONS IN P   03-0413

THE FERROELECTR/ PRESSURE AND TEMPERATURE DEPENDENCE OF THE DIELECTRIC PROPERTIES AND PHASE TRANSITIONS OF   08-1434

THE ANTIFERROEL/ PRESSURE AND TEMPERATURE DEPENDENCE OF THE DIELECTRIC PROPERTIES AND PHASE TRANSITIONS OF   08-1433

SELENITE (ION). RADIATION DAMAGE ELECTRON SPIN RESONANCE SP/ DIELECTRIC PROPERTIES AND PRESSURE EFFECTS ON   22-2870

RROELECTRIC MIXED CRYSTAIS POTASSIUM(1-X) SODIUM(X) TANTALA/ DIELECTRIC PROPERTIES AND SOFT MODES IN THE FE   08-1306

SOME FERROELECTRICS AND ANTIFERRCELECTRICS WITH HARD CURIE/ DIELECTRIC PROPERTIES AND THERMAL EXPANSION OF   08-1378

. (BASED ON LANDAU PHENOMENOL/ (EXAMINATION OF) ANOMALIES OF DIELECTRIC PROPERTIES AT THE PHASE TRANSITIONS   19-2765

ASE TRANSITIONS OF FERROELECT/ INFLUENCE OF POINT DEFECTS ON DIELECTRIC PROPERTIES, DOMAIN DYNAMICS, AND PH   03-0495

/RYSTAL GROWTH OF LEAD (CADMATE(0.33) NIOBATE(0.67)) AND ITS DIELECTRIC PROPERTIES. (FERROELECTRIC- PARAEL/ 13-1959
ON OF CADMIUM TITANATE./ ABOUT THE CURIE POINT AND NATURE OF DIELECTRIC PROPERTIES IN PEROVSKITE MODIFICATI 08-1424
M FERRATE SYST/ COEXISTENCE OF ANTIFERROMAGNETIC AND SPECIAL DIELECTRIC PROPERTIES IN THE BISMUTH- LANTHANU 08-1425
TASSIUM DIHYDROGEN PHOSPHATE CRYSTAL. DIELECTRIC PROPERTIES OF A GAMMA-IRRADIATED PO 17-2467
AL. (POTASSIUM TANTALATE)   EFFECT OF PRESSURE ON THE DIELECTRIC PROPERTIES OF A PARAELECTRIC MATERI 08-1335
IRRADIATED KDP CRYSTAL (TEMPERATURE DEPENDENCE). DIELECTRIC PROPERTIES OF A (SMALL DOSE) GAMMA- 17-2466
. (EXTENSION OF "SOFT MODE" THEORY) DIELECTRIC PROPERTIES OF AN ANHARMONIC CRYSTAL 03-0212
AT MICROWAVE FREQUENCIES (3.3 AND 9.8 GHZ, 0-50 C). DIELECTRIC PROPERTIES OF ANTIMONY SULFOIODIDE 15-2171
X) BROMIDE(1-X) SOLID SOLUTIONS DEPEND ON TEMPERATU/ HOW THE DIELECTRIC PROPERTIES OF ANTIMONY SULFOIODIDE(15-2149
SULFOBROMIDE SOLID SOLUTIONS. DIELECTRIC PROPERTIES OF ANTIMONY SULFOIODIDE- 15-2242
ATE IN THE FREQUENCY RANGE OF/ TEMPERATURE DEPENDENCE OF THE DIELECTRIC PROPERTIES OF BARIUM STRONTIUM NIOB 13-1938
    INFLUENCE OF MICROSTRUCTURES (PREPARATION CONDITIONS) ON DIELECTRIC PROPERTIES OF BARIUM TITANATE. 04-0798
ICS ON THE PHYSICO- CHEMICAL NATURE OF THE SO/ DEPENDENCE OF DIELECTRIC PROPERTIES OF BARIUM TITANATE CERAM 04-0789
ICS.   EFFECT OF INTERNAL BOUNDARIES ON THE DIELECTRIC PROPERTIES OF BARIUM TITANATE CERAM 04-0891
ICS.   INFLUENCE OF VIBRATIONS ON THE DIELECTRIC PROPERTIES OF BARIUM TITANATE CERAM 04-0890
INING NIOBIUM AND THE EFFECT OF ADDITIVES. DIELECTRIC PROPERTIES OF BARIUM TITANATE CONTA 04-0575
WITH TUNGSTEN OXIDE. DIELECTRIC PROPERTIES OF BARIUM TITANATE DOPED 04-0696
E CRYSTALS DOPED BY HYDROXIDE GROUPS. DIELECTRIC PROPERTIES OF BARIUM TITANATE SINGL 04-0634
E CRYSTALS.   EFFECT OF LINEAR COMPRESSION ON THE UHF DIELECTRIC PROPERTIES OF BARIUM TITANATE SINGL 04-0896
E CRYSTALS.   EFFECT OF TRANSVERSE COMPRESSION OF DIELECTRIC PROPERTIES OF BARIUM TITANATE SINGL 04-0956
NIOBATE(0.67) (NOT FERROELECTRIC -160 TO +25/ STRUCTURAL AND DIELECTRIC PROPERTIES OF BARIUM ZINCATE(0.33) 36-3609
(X) TITANATE SINGLE CRYSTALS GROWN FROM A POTASSIUM FLUORID/ DIELECTRIC PROPERTIES OF BARIUM(1-X) STRONTIUM 04-0635
                  CRYSTAL GROWTH AND DIELECTRIC PROPERTIES OF BISMUTH NIOBATE. 13-2033
(AND THE FERROELECTRIC STATE, BISMUT/ (CRYSTALLOGRAPHIC AND DIELECTRIC) PROPERTIES OF BISMUTH PEROVSKITES. 08-1289
          GROWTH, CRYSTALLOGRAPHY AND DIELECTRIC PROPERTIES OF BISMUTH TUNGSTATE. 10-1650
E FOR FERROELECTRICITY. DIELECTRIC PROPERTIES OF BORACITES AND EVIDENC 14-2071
DIHYDROGEN PHOSPHATE. DIELECTRIC PROPERTIES OF DEUTERATED POTASSIUM 17-2581
STANNATE (DIELECTRIC TRANSITIONS NOT FERROELECTRIC). DIELECTRIC PROPERTIES OF DIAMMONIUM HEXABROMO 36-3612
, STRONTIUM TITANATE SOLID SOLUTIONS NEAR THE PHASE TRANSI/ DIELECTRIC PROPERTIES OF FERROELECTRIC (BARIUM 04-0680
LS AT CENTIMETER AND DECIMETER WAVELENGTHS E/ MEASUREMENT OF DIELECTRIC PROPERTIES OF FERROELECTRIC MATERIA 04-0855
INE SULFATE IN THE FAR INFRARED.   TEMPERATURE DEPENDENCE OF DIELECTRIC PROPERTIES OF FERROELECTRIC TRIGLYC 27-3061
ROWAVE FREQUENCI/ PROBLEM OF THE ACCURATE MEASUREMENT OF THE DIELECTRIC PROPERTIES OF FERROELECTRICS AT MIC 03-0429
MILLIMETER WAVELENGTHS. (STRONTIUM TITANATE) MEASUREMENT OF DIELECTRIC PROPERTIES OF FERROELECTRICS AT SUB 06-1089
CTED BEFOREHAND TO AN EL/ INVESTIGATION OF ANISOTROPY IN THE DIELECTRIC PROPERTIES OF FERROELECTRICS, SUBJE 04-0661
SSIUM DIHYDROGEN PHOSPHATE CRYSTALS. DIELECTRIC PROPERTIES OF GAMMA-IRRADIATED POTA 17-2542
ELECTRICS. (SODIUM TRIHYDROGEN SELENATE AND SODIUM TRIDEUTE/ DIELECTRIC PROPERTIES OF HYDROGEN BONDED FERRO 22-2855
DIELECTRIC PROPERTIES OF ICE I. 34-3419
/D BY GAMMA AND REACTOR IRRADIATION ON PHASE TRANSITIONS AND DIELECTRIC PROPERTIES OF KDP AND ADP CRYSTALS. 17-2614
E.   A MICROSCOPIC THEORY FOR THE DIELECTRIC PROPERTIES OF LEAD MAGNESIUM NIOBAT 08-1440
TE CERAMICS AT LOW TEMPERATURES.   PIEZOELECTRIC AND DIELECTRIC PROPERTIES OF LEAD TITANATE ZIRCONA 09-1519
FERRATE NIOBATE) SOLID SOLUTIONS AT HIGH PRESSURES. DIELECTRIC PROPERTIES OF LEAD TITANATE- LEAD (15-1071
          EFFECT OF PILE NEUTRON IRRADIATION ON THE DIELECTRIC PROPERTIES OF LEAD ZIRCONATE. 07-1259
ROVSKITES. (STRONTIUM LITHIUM NIOBATE, STRONT/ STRUCTURE AND DIELECTRIC PROPERTIES OF LITHIUM CONTAINING PE 08-1292
ULFATE. DIELECTRIC PROPERTIES OF LITHIUM HYDRAZINIUM S 32-3344
ULFATE. DIELECTRIC PROPERTIES OF LITHIUM HYDRAZINIUM S 32-3352
D NOT FERROELECTRIC) DIELECTRIC PROPERTIES OF LITHIUM HYDRAZINIUM S 32-3353
TANATE TYPE) EFFECT OF CHARGED LATTICE IMPERFECTIONS ON THE DIELECTRIC PROPERTIES OF LITHIUM IODATE. (FOUN 36-3605
G PEROVSKITES A/ SYNTHESIS, X-RAY ANALYSIS, AND STUDY OF THE DIELECTRIC PROPERTIES OF MATERIALS. (BARIUM TI 04-1017
ES. DIELECTRIC PROPERTIES OF NEW CADMIUM-CONTAININ 08-1309
TITANATE.   INFLUENCE OF GRAIN GROWTH ON DIELECTRIC PROPERTIES OF NIOBATES AND TUNGSTAT 08-1322
SALT.   EFFECT OF MECHANICAL STRESS ON DIELECTRIC PROPERTIES OF NIOBIUM DOPED BARIUM 04-0763
M TITANATE.   INFLUENCE OF A CONSTANT ELECTRIC FIELD ON THE DIELECTRIC PROPERTIES OF PARAELECTRIC ROCHELLE 28-3225
C NIOBATE.   PREPARATION AND DIELECTRIC PROPERTIES OF POLYCRYSTALLINE BARIU 04-1012
E.   DIELECTRIC PROPERTIES OF POTASSIUM BISMUTH ZIN 13-2001
/TURAL SENSITIVITY OF THE FERROELECTRIC PHASE TRANSITION AND DIELECTRIC PROPERTIES OF POTASSIUM CHLORIDE IC 34-3454
TITANATE WITH NIOBIUM ADDITIONS.   OPTICAL AND DIELECTRIC PROPERTIES OF POTASSIUM DIHYDROGEN/ 17-2615
IRCONATE   GROWTH AND DIELECTRIC PROPERTIES OF SINGLE CRYSTAL BARIUM 04-0875
E AMMONIUM DIHYDROGEN PHOSPHATE- POTASSIUM DIHYDROGEN PHOSP/ DIELECTRIC PROPERTIES OF SINGLE CRYSTAL LEAD Z 07-1265
OWN BARIUM TITANATE. DIELECTRIC PROPERTIES OF SINGLE CRYSTALS IN TH 17-2572
ON THE FERROELECTRIC POTASSIUM TANTALAT/ X-RAY ANALYSIS AND DIELECTRIC PROPERTIES OF SINGLE DOMAIN MELT GR 04-0616
RIUM TITANATE TYPE MATERIALS.   X-RAY IRRADIATION EFFECT ON DIELECTRIC PROPERTIES OF SOLID SOLUTIONS BASED 04-1310
RO- AND PARAELECTRIC PEROVSKITES.   EFFECT OF IRRADIATION OF DIELECTRIC PROPERTIES OF SOLID SOLUTIONS OF BA 04-0782
FERROELECTRICS (BARIUM TITANATE AND SAME WITH 1 PERCENT/ UHF DIELECTRIC PROPERTIES OF SOME FERRO-, ANTIFER 08-1474
/ITUTION OF TANTALUM FOR NIOBIUM ON THE CRYSTALLOGRAPHIC AND DIELECTRIC PROPERTIES OF SOME POLYCRYSTALLINE 04-0818
NNEALED SINGLE CRYSTALS) AT LOW TEMPERATURE (5-300 DEG K). DIELECTRIC PROPERTIES OF STRONTIUM SODIUM NIO/ 13-1921
LOW TEMPERATURES. DIELECTRIC PROPERTIES OF STRONTIUM TITANATE (A 06-1201
LOW TEMPERATURES. (TEMPERATURE DEPENDENCE, ANISOTROPY) DIELECTRIC PROPERTIES OF STRONTIUM TITANATE AT 06-1200
                INFLUENCE OF DEFECTS ON SOME DIELECTRIC PROPERTIES OF STRONTIUM TITANATE AT 06-1202
THE EFFECT OF GROWTH RATE ON THE DEFECT STRUCTURE AND DIELECTRIC PROPERTIES OF TGS SINGLE CRYSTALS. 27-3063
AMICS. DIELECTRIC PROPERTIES OF TGS SINGLE CRYSTALS. 27-3067
(AGING AND PHASE CHANGES NEAR T/ EFFECT OF A DC FIELD ON THE DIELECTRIC PROPERTIES OF TITANATE- ALUMINA CER 08-1414
THE ADDITION OF A MIXTURE OF COPPER AND CHROMIUM IONS ON THE DIELECTRIC PROPERTIES OF TRIGLYCINE SELENATE. 27-3131
F THE QUALITY OF TRIGLYCINE SULFATE SINGLE CRYSTALS ON THEIR DIELECTRIC PROPERTIES OF TRIGLYCINE SULFATE. 27-3045
SINGLE CRYSTAL (TETRAGONAL TUNGSTEN BRONZE). DIELECTRIC PROPERTIES. (REVIEW)   INFLUENCE O 02-0088
            FERRO- AND ANTIFERROELECTRIC COUPLING AND DIELECTRIC PROPERTY OF BARIUM TITANATE NIOBATE 13-1969
TIVITY OF ICE SINGLE CRYSTALS. DIELECTRIC RELAXATION. 03-0354
ITANATES CONTAINING (TRIVALENT) RARE EARTH IONS. DIELECTRIC RELAXATION, BULK AND SURFACE CONDUC 34-3469
ETHYL AMMONIUM ALUMINUM SULFATE DODECAHYDRATE. (COMPLEX DIE/ DIELECTRIC RELAXATION IN (CERAMIC) STRONTIUM T 06-1143
SODIUM DINITRITE. DIELECTRIC RELAXATION IN FERROELECTRIC MASD, M 20-2804
SODIUM DINITRITE. DIELECTRIC RELAXATION IN FERROELECTRIC SILVER 16-2290
ETICAL DISCREPANCY).   NOTE ON THE DIELECTRIC RELAXATION IN FERROELECTRIC SILVER 16-2292
OF TRIGLYCINE SULFATE. DIELECTRIC RELAXATION IN FERROELECTRICS (THEOR 03-0391
E CRYSTALS. DIELECTRIC RELAXATION IN MULTIDOMAIN CRYSTALS 27-3002
RDER DISORDER / CRITICAL SLOWING DOWN PROCESS OF (ANOMALOUS) DIELECTRIC RELAXATION IN MULTIDOMAIN TGS SINGL 27-3001
TE BY THERMAL NOI/ STUDY OF THE CRITICAL SLOWING DOWN OF THE DIELECTRIC RELAXATION IN SEVERAL CRYSTALS OF O 02-0087
RELATIONSHIP OF THE TEMPERATURE DEPENDENCE OF B1 PHONON AND DIELECTRIC RELAXATION IN SILVER SODIUM DINITRI 16-2382
TRIC SILVER SODIUM DINITRITE BY STEP FUNCTION METHOD. DIELECTRIC RELAXATION IN SODIUM NITRITE. 16-2257
E SULFATE. DIELECTRIC RELAXATION MEASUREMENT ON FERROELEC 16-2291
CAL POINT IN FERROELECTRICS. DIELECTRIC RELAXATION OF MULTIDOMAIN TRIGLYCIN 27-3050
DIELECTRIC RELAXATION PROCESSES NEAR THE CRITI 03-0272

ND AQUEOUS SOLUTIONS AND THEIR INTERPRETATION. PART-8: TRAN/ DIELECTRIC RELAXATION SPECTRA OF WATER, ICE, A 34-3482
RRELATION IN FERROELECTRIC POTASSIUM DIHYDROGEN ARS/ NMR AND DIELECTRIC RELAXATION STUDIES OF LONG RANGE CO 18-2718
IVER SODIUM DINITRITE.   HYDROSTATIC PRESSURE EFFECT ON THE DIELECTRIC RELAXATION TIME IN FERROELECTRIC SI 16-2294
  POINTS IN SODIUM NITRIT/ STATIC ELECTRIC SUSCEPTIBILITY AND DIELECTRIC RELAXATION TIME NEAR THE TRANSITION 16-2302
CALLY BIASED KDP AND DEUTERATED ISOMORPHS. (THEORY)   DIELECTRIC RESONANCE AND RELAXATION IN ELECTRI 03-0467
  TETRAHYDRATES.   DIELECTRIC RESPONSE FUNCTION OF COPPER FORMATE 26-2933
  (RAMAN SCATTERING IN POTASSIUM DIHYDROGEN ARSENATE AND CES/ DIELECTRIC RESPONSE IN PIEZOELECTRIC CRYSTALS. 18-2723
C FLUCTUATIONS)   DIELECTRIC RESPONSE OF CRYSTALS. (FERROELECTRI 03-0210
ICROSCCPIC MODEL OF LOCALLY ORDERED "NUCLEI", PHASE STRUCTU/ DIELECTRIC SPECTRUM OF BARIUM TITANATE. (AND M 04-0647
OELECTRICS. (LANTHANUM BISMUTH TITANI/ SYNTHESES, X-RAY, AND DIELECTRIC STUDIES OF CERTAIN NEW LAYERED FERR 13-1993
PE COMPOUNDS.   X-RAY AND DIELECTRIC STUDIES OF IRRADIATED PEROVSKITE TY 08-1443
ATE.   DIELECTRIC STUDIES OF PHASE 2 OF AMMONIUM NITR 16-2339
TUNGSTEN BRONZE TYPE) FERR/ (X-RAY) STRUCTURAL (OPTICAL) AND DIELECTRIC STUDIES OF (SINGLE CRYSTALS IN THE 13-1947
IXED CRYSTALS OF NIOBATES.   DIELECTRIC STUDIES OF THE CATION SUBSTITUTED M 11-1848
TRANSITION IN SILVER SODIUM DINITRITE.   DIELECTRIC STUDIES ON THE FERROELECTRIC PHASE 16-2289
ELECTRIC TRIGLYCINE SULFATE BY A DIGITAL TECHNIQUE. (MINUTE/ DIELECTRIC STUDY OF CRITICAL BEHAVIOR OF FERRO 27-3116
E.   MICROWAVE DIELECTRIC STUDY OF LITHIUM HYDRAZINIUM SULFAT 32-3357
FNATE (PHASE) DIAGRAM.   DIELECTRIC STUDY OF THE LEAD ZIRCONATE LEAD HA 08-1340
ROELECTRIC PHASE TRANSITIONS.   DIELECTRIC SUSCEPTIBILITY AND THE ORDER OF FER 03-0223
IC STRONTIUM TITANATE.   DIELECTRIC SUSCEPTIBILITY OF QUASI FERROELECTR 06-1133
ANSION OF POLARIZED CERAMICS IN THE LEAD TITANATE ZIRCONATE/ DIELECTRIC SYSTEM POLARIZATION AND THERMAL EXP 09-1603
             EFFECT OF THE SURFACE ON POLARIZATION OF A DIELECTRIC (THEORY).   03-0271
RADIATED BY AN OPEN WAVEGUIDE FILLED WITH HIGH PERMITTIVITY DIELECTRIC. (THEORY) /UT ADMITTANCE AND FIELDS 03-0359
SINGLE FIRE PROCESS FOR BARIUM TITANATE BARRIER LAYER DIELECTRICS.   04-1029
STABILIZED BARIUM TITANATE CERAMICS FOR CAPACITOR DIELECTRICS.   04-0825
STUDIES ON STRONTIUM TITANATE BOUNDARY LAYER DIELECTRICS.   06-1248
DIES ON THE BARIUM STRONTIUM TITANATE BOUNDARY LAYER CERAMIC DIELECTRICS.   STU 04-1025
ON THE (STRONTIUM, CALCIUM) TITANATE BOUNDARY LAYER CERAMIC DIELECTRICS.   STUDIES 06-1247
/ECTRIC CONSTANT MATERIALS AND FERROELECTRICITY AS CAPACITOR DIELECTRICS, A STUDY IN DIELECTRIC SPECTROSCO/ 37-3678
EFERENCE/ NONLINEAR OPTICAL AND ELECTROOPTICAL PROPERTIES OF DIELECTRICS AND FERROELECTRICS. (REVIEW, 195 R 02-0081
                 ANISOTROPIC, SOLID DIELECTRICS AND FERROELECTRICITY. (106 REFS) 02-0065
T/ PREPARATION AND PROPERTIES OF HIGH PERMITTIVITY THIN FILM DIELECTRICS. (BARIUM AND LEAD TITANATES, EFFEC 04-0670
ATE)   CERAMIC DIELECTRICS. (BARIUM TITANATE, STRONTIUM 04-0683
UM (OR LEAD OR CALCIUM) STR/ MICROWAVE BEHAVIOR OF NONLINEAR DIELECTRICS (EXPERIMENT AND DISCUSSION OF BARI 08-1281
(WITH ELECTRODES REDUCING POLARIZATION SWITCHING / NONLINEAR DIELECTRICS FOR HIGH POWER ELECTRONIC TUNING. 37-3660
/ANSIENT PHENOMENA IN THE DISCHARGE OF CONDENSERS WITH POLAR DIELECTRICS. (POTASSIUM DIHYDROGEN PHOSPHATE / 17-2631
TORS).   ELECTRICAL PROPERTIES OF THICK FILM BARIUM TITANATE DIELECTRICS PRODUCED BY FLAME SPRAYING (CAPACI 04-0772
ATE)   ENERGY STORAGE IN CERAMIC DIELECTRICS. (STRONTIUM TITANATE, BARIUM TITAN 04-0602
ETHOD)   SPONTANEOUS POLARIZATION NOISE IN POLAR DIELECTRICS. (THEORY, EINSTEIN THERMODYNAMIC M 03-0219
ES AND ZIRCONATES). FORBIDDEN ZONE WIDTHS IN POLYCRYSTALLINE DIELECTRICS WITH PEROVSKITE STRUCTURE (TITANAT 08-1288
FACE REGIONS IN FERROELECTR/ ELECTRIC FIELD DISTRIBUTIONS IN DIELECTRICS, WITH SPECIAL EMPHASIS ON NEAR SUR 03-0182
RIC CON/ INTERNAL STRESSES IN BARIUM TITANATE CERAMICS. (AND DIFFERENCE IN TEMPERATURES FOR MAXIMUM DIELECT 04-0849
RTIES OF SILVER SODIUM DINITRITE AND SILVER SODIUM NITRITE. (DIFFERENTIAL THERMAL ANALYSIS)   PROPE 16-2370
UM MOLYBDATE.   OBSERVATION OF DIFFRACTED LIGHT FROM DOMAIN WALLS IN GADOLINI 30-3322
ERROELECTRIC PHASE OF COPPER FORMATE TETRAHYDRATE BY NEUTRON DIFFRACTION.   APPROXIMATE STRUCTURE FOR ANTIF 26-2935
BARIUM TITANATE BY CONVERGENT BEAM ELECTRON DIFFRACTION.   04-0991
  BARKHAUSEN EFFECT IN FERROELECTRICS UNDER CONTINUOUS X-RAY DIFFRACTION.   03-0337
IN SOLID HYDROGEN HALIDES (SINGLE CRYSTAL X-RAY AND NEUTRON DIFFRACTION). /ECTRICITY AND PHASE TRANSITIONS 31-3331
575 DEG C, AND THEIR RELEVANCE TO SOFT PHONON MODES. (X-RAY DIFFRACTION) /F SODIUM NIOBATE BETWEEN 480 AND 08-1273
IN POTASSIUM TRIDEUTERIUM SELENITE, AS DETERMINED BY NEUTRON DIFFRACTION. /SSIUM TRIHYDROGEN SELENITE, A 22-2854
NS OF A CRYSTAL LATTICE BASED ON THE SHADOW EFFECT. (NEUTRON DIFFRACTION, BARIUM TITANATE) /MALL DEFORMATIO 04-0534
IN THE FERROELECTRIC PHASE POTASSIUM DIHYDROGEN PHOSPHATE. (DIFFRACTION BY DOMAIN WALLS) /WITH TEMPERATURE 17-2662
APPLICATION OF ACOUSTIC WAVES T/ EXPERIMENTAL STUDY OF LIGHT DIFFRACTION BY HIGH FREQUENCY ACOUSTIC WAVES. 11-1834
LITHIUM NIOBATE.   NEW HIGH COUPLING LOW DIFFRACTION CUT FOR ACOUSTIC SURFACE WAVES ON 11-1820
FERROELECTRIC PHAS/ SURFACE LAYER IN BARIUM TITANATE (X-RAY DIFFRACTION, DC FIELD EFFECT, PARAELECTRIC AND 04-0857
TURE OF AMMONIUM HYDROGEN SULFATE ABOVE THE FERROELEC/ X-RAY DIFFRACTION DETERMINATION OF THE CRYSTAL STRUC 19-2775
SURFACE WAVE DIFFRACTION IN LITHIUM NIOBATE.   11-1698
TURES OF INORGANIC MATERIALS. (REVIEW, 244 REFS)   NEUTRON DIFFRACTION INVESTIGATIONS OF THE ATOMIC STRUC 02-0133
ORMATION IN FERROELECTRICS.   THEORY OF X-RAY DIFFRACTION LINE SHAPE EFFECTS DUE TO DOMAIN F 03-0236
IONIZING RADIATION IN FERROELECTRIC ROCHELLE SALT. (NEUTRON DIFFRACTION MEASUREMENTS) /TS OF (DAMAGE FROM) 28-3188
M TITANATE. PART-3: HIGH TEMPERATURE THERMODYNAMIC AND X-RAY DIFFRACTION MEASUREMENTS. /TOICHIOMETRIC BARIU 04-0886
DIFFRACTION OF A CONTINUOUS LIGHT WAVE IN KDP. 17-2444
AVES IN POTASSIUM DIHYDROGEN PHOSPHATE CRYSTALS.   DIFFRACTION OF POLARIZED LIGHT BY ULTRASONIC W 17-2407
TIC RESONAN/ STRUCTURAL STUDIES OF ICE POLYMORPHS BY NEUTRON DIFFRACTION, PROTON AND DEUTERON NUCLEAR MAGNE 34-3466
VES ON (PIEZOELECTRIC) LITHIUM NIOBATE.   (FAR FIELD) DIFFRACTION SPREADING OF SURFACE (ACOUSTIC) WA 11-1850
IUM NIOBATE.   DIFFRACTION SPREADING OF SURFACE WAVES ON LITH 11-1849
N FERROELECTRICS.   X-RAY DIFFRACTION STUDIES OF ELECTRON POLARIZATION I 03-0199
NSITIONS IN KDP AND POTASSIUM DIDEUTERIUM PHOSPHATE. NEUTRON DIFFRACTION STUDIES OF FERROELECTRIC PHASE TRA 17-2661
ION REVERSALS IN SODIUM NITRITE.   NEUTRON DIFFRACTION STUDIES OF FERROELECTRIC POLARIZAT 16-2365
TITANATE AND ROCHELLE SALT AS EXAMPLES)   NEUTRON DIFFRACTION STUDIES OF FERROELECTRICS. (BARIUM 04-0679
FERROELECTRIC BISMUTH FERRATE X-RAY AND NEUTRON DIFFRACTION STUDY. (ANTIFERROMAGNETIC)   08-1402
NEUTRON AND ELECTRON DIFFRACTION STUDY IN NICKEL IODINE BORACITE. 14-2073
GNETIC SUBSTANCES. (LEAD TITANATE) STRUCTURAL MICROELECTRON DIFFRACTION STUDY OF FERROELECTRIC AND FERROMA 05-1082
ERROCYANIDE TRIDEUTERO HYDRATE ABOVE THE CURIE TEMP/ NEUTRON DIFFRACTION STUDY OF FERROELECTRIC POTASSIUM F 24-2903
STALLOGRAPHIC PHASE CHANGE AT CURIE TEMPERATURE.   X-RAY DIFFRACTION STUDY OF GADOLINIUM MOLYBDATE (CRY 30-3312
TITANATE.   HIGH TEMPERATURE THERMODYNAMIC AND X-RAY DIFFRACTION STUDY OF NONSTOICHIOMETRIC BARIUM 04-0885
R FORMATE.   X-RAY DIFFRACTION STUDY OF PHASE TRANSITION IN COPPE 26-2938
ENITE.   A NEUTRON DIFFRACTION STUDY OF POTASSIUM TRIHYDROGEN SEL 22-2866
PYROCHLORE CONTAINING LITHIUM, STRONTIU/ SYNTHESIS AND X-RAY DIFFRACTION STUDY OF SINGLE CRYSTALS OF A NEW 36-3613
NEUTRON DIFFRACTION STUDY OF SODIUM NITRITE.   16-2326
E.   X-RAY AND NEUTRON DIFFRACTION STUDY OF TETRAGONAL BARIUM TITANAT 04-0713
E.   X-RAY AND NEUTRON DIFFRACTION STUDY OF TETRAGONAL BARIUM TITANAT 04-0740
KITE) STRONTIUM IRON TUNGSTATE . (MAGNETIC G STRUCT/ NEUTRON DIFFRACTION STUDY OF (TETRAGONAL- CUBIC PEROVS 10-1655
TRIHYDROGEN SELE/ HYDROGEN BOND STUDIES. PART-54: A NEUTRON DIFFRACTION STUDY OF THE FERROELECTRIC LITHIUM 22-2878
SULFATE.   ELECTRON DIFFRACTION STUDY OF THE STRUCTURE OF AMMONIUM 19-2797
NEUTRON DIFFRACTION STUDY OF TRIGLYCINE SULFATE. 27-3099
DROGEN DISELENITE.   NEUTRON DIFFRACTION STUDY ON PARAELECTRIC SODIUM TRIHY 22-2846
C SODIUM NITRITE AND TGS CRYSTALS./ APPLICATION OF THE X-RAY DIFFRACTION TOPOGRAPHIC METHOD TO FERROELECTRI 16-2375
TION BEHAVIOR OF ICE SINGLE CRYSTALS.   X-RAY DIFFRACTION TOPOGRAPHIC STUDIES OF THE DEFORMA 34-3426
DISPLAY OF TWINNING IN LITHIUM NIOBATE BY X-RAY DIFFRACTION TOPOGRAPHY. 11-1845

LS.   X-RAY DIFFRACTION TOPOGRAPHY OF FERROELECTRIC CRYSTA 02-0121
CATIONS) IN KDP AND ADP SINGLE CRYSTALS.   X-RAY DIFFRACTION TOPOGRAPHY STUDY OF DEFECTS (DISLO 17-2578
PERATURE AND TEMPERATURE DEPENDENCE OF SPONTANEOUS PO/ X-RAY DIFFRACTOMETRIC DETERMINATION OF THE CURIE TEM 03-0290
ORMATIONS IN STRONTIUM TITANATE.   X-RAY DIFFRACTOMETRY OF LOW TEMPERATURE PHASE TRANSF 06-1156
ATION AND DIELECTRIC HYSTERESIS NEAR THE TRANSITION POINT OF DIFFUSE FERROELECTRIC PHASE. (KAENZIG MODEL) / 03-0437
G INTO ACCOUNT TH/ EFFECT OF AN EXTERNAL ELECTRIC FIELD ON A DIFFUSE FERROELECTRIC PHASE TRANSITION, TALKIN 03-0441
ACCOUNTING FOR INTERACTION IN THE KANZIG REGIONS MODEL FOR DIFFUSE FERROELECTRIC PHASE TRANSITIONS. 03-0438
EFFECT OF HYDROSTATIC PRESSURE ON THE NATURE OF DIFFUSE FERROELECTRIC PHASE TRANSITIONS. 03-0442
THERMODYNAMICS AND STATISTICAL MECHANICS OF DIFFUSE FERROELECTRIC PHASE TRANSITIONS. 03-0445
ELASTOOPTICAL EFFECT IN FERROELECTRIC WITH DIFFUSE PHASE TRANSITION. 03-0307
OF A FERROELECTRIC LEAD MAGNESATE(0.33) NIOBATE(0.67) WITH A DIFFUSE PHASE TRANSITION. /ATION POLARIZATION 13-1980
THE BEHAVIORS OF THE MICRODOMAINS ON THE DIFFUSE PHASE TRANSITION FERROELECTRICS. 03-0283
DOMAIN STRUCTURE OF POLYCRYSTALLINE FERROELECTRICS WITH DIFFUSE PHASE TRANSITIONS. 03-0505
DIELECTRIC AND STRUCTURAL PROPERTIES OF FERROELECTRICS WITH DIFFUSE PHASE TRANSITIONS. (MIXED LAYER OXIDE/ 13-2014
N OF POTASSIUM MANGANESE TRIFLUORIDE SINGLE CRYSTALS, (X-RAY DIFFUSE SCATTERING). / THE 184 DEG K TRANSITIO 33-3369
ELASTIC CONSTANT OF ORTHORHOMBIC POTASSIUM NIOBATE BY X-RAY DIFFUSE SCATTERING. 08-1416
IFLUORIDE AND SODIUM NIOBATE SIN/ ANISOTROPIC CRITICAL X-RAY DIFFUSE SCATTERING FROM POTASSIUM MANGANESE TR 08-1312
NCTION OF TEMPERATURE.   X-RAY DIFFUSE SCATTERING FROM SODIUM NIOBATE AS A FU 08-1313
S (NEAR THE 640 DEG C PHASE TRAN/ ANISOTROPIC CRITICAL X-RAY DIFFUSE SCATTERING FROM SODIUM NIOBATE CRYSTAL 08-1311
NEUTRON STUDY OF DIFFUSE SCATTERING IN CUBIC POTASSIUM NIOBATE. 08-1407
EMPERATURE.   DIFFUSE SCATTERING IN SODIUM NIOBATE AT ROOM T 13-1304
N SYSTEMS LEAD (B(0.5) NIOBIUM(0.5))(1-X) B'(X) OXYGEN(3), / DIFFUSE TO SHARP FERROELECTRIC PHASE TRANSITIO 13-2044
EXPERIMENTAL CRITERION OF FERROELECTRICITY. (SHARP OR DIFFUSE TRANSITIONS) 03-0184
IN SODIUM NIOBATE.   ANISOTROPIC DIFFUSE X-RAY SCATTERING AND PHASE TRANSITIONS 08-1357
NSITION IN SODIUM NIOBATE.   DIFFUSE X-RAY STUDY OF 640 DEGREES C PHASE TRA 08-1358
SURFACE WAVES ON FERROELECTRIC CERAMIC/ EFFECTS OF THERMALLY DIFFUSED IMPURITIES ON PROPAGATION OF ELASTIC 03-0487
NISMS, ELECTROOPT/ PHYSICAL PHENOMENA IN FERROELECTRICS WITH DIFFUSED PHASE TRANSITION. (POLARIZATION MECHA 02-0119
FORMATION OF BARIUM TITANATE CERAMIC SEMICONDUCTING LAYER BY DIFFUSION. 04-0684
TIC CONSTANTS OF POTASSIUM DIHYDROGEN PHOSPHATE BY BRILLOUIN DIFFUSION.   MEASUREMENT OF THE ELAS 17-2454
ON IN SODIUM TRIDEUTERIUM SELENATE.   NMR STUDY OF DEUTERIUM DIFFUSION AND DEUTERIUM AND SODIUM-23 RELAXATI 22-2849
DIFFUSION AND RELAXATION PHENOMENA IN ICE. 34-3470
MENT OF THE ELASTIC CONSTANTS OF SODIUM NITRITE BY BRILLOUIN DIFFUSION AT ORDINARY TEMPERATURES.   MEASURE 16-2306
ANATE BY A DIELECTRIC MEASUREMENT TECHNIQUE.   DIFFUSION COEFFICIENT OF NIOBIUM IN BARIUM TIT 04-0926
UM TITANATE, BARIUM TITANATE)   MEASUREMENT OF OXYGEN DIFFUSION CONSTANT IN SOME TITANATES. (STRONTI 04-1033
NG FIELDS) DURING HOLOGRAM RECORDING IN CR/ ELECTRON THERMAL DIFFUSION EFFECTS (INTERNAL GENERATION OF STRO 06-1090
DROGEN PHOSPHATE.   BRILLOUIN DIFFUSION IN A SINGLE CRYSTAL OF RUBIDIUM DIHY 17-2520
S OF DISPLACIVE STRUCTURAL PHASE TRANSFORMATIONS. (INELASTIC DIFFUSION IN FERROELECTRICS)   NEUTRON STUDIE 03-0155
DIFFUSION OF HYDROGEN FLUORIDE IN ICE. 34-3437
E THEORY OF PTC MATERIALS.   DIFFUSION POTENTIALS IN BARIUM TITANATE AND TH 03-0352
DOMAIN STRUCTURE OF (HIGH SENSITIVITY) BARIUM TITANATE. DIFFUSIONLESS PHASE TRANSITIONS AND THE STABLE 04-0888
M DOPED SODIUM NITRATE CRYSTALS NEAR/ SPECIFIC HEAT, THERMAL DIFFUSIVITY AND THERMAL CONDUCTIVITY OF CALCIU 16-2283
IZATION RELAXATION I/ CAUSES OF PHASE TRANSITION BROADENING (DIFFUSNESS) AND THE NATURE OF DIELECTRIC POLAR 03-0291
ANGULAR FIELD BY THE COMMUTATOR OF POLARIZATION IN KDP, IN A DIGITAL DEVIATOR. /Y OF THE LIMITATION OF THE 17-2524
ALAT/ DEFLECTED BEAM DIAMETER AND DEFLECTION CAPACITIES OF A DIGITAL LIGHT DEFLECTOR UTILIZING LITHIUM TANT 11-1809
/ CRITICAL BEHAVIOR OF FERROELECTRIC TRIGLYCINE SULFATE BY A DIGITAL TECHNIQUE. (MINUTE CHANGES IN DIELECT 27-3116
ONIUM SULFATE, AMMONI/ OPTICAL PROPERTIES OF FERROELECTRICS. DIGRESSIONS FROM PHENOMENOLOGICAL THEORY. (AMM 02-0040
SEE BOTH ADA AND AMMONIUM DIHYDROGEN ARSENATE.
SEE BOTH KDA AND POTASSIUM DIHYDROGEN ARSENATE.
SEE BOTH KDP AND POTASSIUM DIHYDROGEN PHOSPHATE.
DIDE SINGLE CRYSTALS.   DILATOMETRIC INVESTIGATION OF ANTIMONY SULFOIO 15-2155
IC EFFECTS IN SINGLE CRYSTALS OF ANTIMONY SULFOIODIDE. DILATOMETRIC INVESTIGATION OF PHOTOFERROELECTR 15-2156
TRANSITION OF POTASSIUM DIHYDROGEN PHOSPHATE. PART-1:/ X-RAY DILATOMETRIC STUDY OF THE FERROELECTRIC PHASE 17-2557
ONAL PHASE TRANSITION IN BARIUM SODIUM NIOBATE. (THERMAL EX/ DILATOMETRIC STUDY OF THE ORTHORHOMBIC- TETRAG 13-1894
NIUM CHROMIUM SULFATE HEXAHYDRATE.   ZEEMAN EFFECT AND DILUTION EFFECT OF ABSORPTION LINES IN GUANIDI 21-2821
/VIATION AND SPONTANEOUS SUBLATTICE MAGNETIZATION IN THE TWO DIMENSIONAL ANTIFERROMAGNET COPPER FORMATE TE/ 26-2929
RAHYDRATE.   ANOMALOUS SHIFT OF ESR LINES IN THE TWO DIMENSIONAL ANTIFERROMAGNET COPPER FORMATE TET 26-2941
TRIFLUOR/ MEASUREMENTS OF THE THERMAL CONDUCTIONS IN THE LOW DIMENSIONAL ANTIFERROMAGNETS POTASSIUM COPPER 26-2939
ETRAHYDRATE.   THERMAL CONDUCTION IN A TWO DIMENSIONAL ANTIFERROMAGNETIC COPPER FORMATE T 26-2932
PROTON CORRELATION FUNCTIONS IN ICE.   THE THREE DIMENSIONAL EIGHT VERTEX MODEL AND THE PROTON 34-3481
TYPE-II PHASE CHANGE IN THREE DIMENSIONAL FERROELECTRIC MODEL. (THEORY) 03-0451
TWO DIMENSIONAL FERROELECTRIC MODELS. 03-0365
AL ELECTRIC FIELD.   EXACT SOLUTION OF MODEL OF TWO DIMENSIONAL FERROELECTRICS IN ARBITRARY EXTERN 03-0520
.   ONE DIMENSIONAL KDP MODEL IN STATISTICAL MECHANICS 03-0389
NSITION OF ANTIMONY SULFOIODIDE SINGLE CRY/ INFLUENCE OF ONE DIMENSIONAL MECHANICAL STRESS ON THE PHASE TRA 15-2245
RATIONS IN ORIENTATIONALLY DISORDERED SYSTEMS. PART-1: A TWO DIMENSIONAL MODEL. (ICE I)   ATOMIC VIB 34-3474
SECOND ORDER PHASE TRANSITION IN THE THREE DIMENSIONAL MODEL OF A FERROELECTRIC. 03-0450
IGLYCINE SULFATE AND ROCHELLE SALT.   EFFECT OF A TWO DIMENSIONAL PRESSURE ON THE CURIE POINTS OF TR 27-3082
Z).   PREPARATION OF THIN BARIUM TITANATE FILMS BY DC DIODE SPUTTERING (ONTO PLATINUM OR FUSED QUART 04-0940
ULFATE CRYSTALS.   DIP ETCH PATTERNS ON THE FACES OF TRIGLYCINE S 27-3127
S IN HYDROGEN AND DEUTERIUM HALIDE CRYSTALS. (RAMAN SPECTRA, DIPOLAR COUPLING MODEL COMPUTATION) /INAL MODE 31-3326
INGLE CRYSTAL.   DIPOLAR RELAXATION AT LOW TEMPERATURE OF ICE S 34-3447
S.   STUDY OF CONDUCTIVITY AND DIPOLAR RELAXATION IN DOPED ICE SINGLE CRYSTAL 34-3458
TATORS IN LITHIUM HYDRAZINIUM SULFATE.   DIPOLAR RELAXATION OF LITHIUM-7 BY HINDERED RO 32-3350
M-7 NMR SIGNALS IN LITHIUM NIOBATE.   ACOUSTIC DIPOLE AND QUADRUPOLE SATURATION OF THE LITHIU 11-1666
UCTURES WITH SPECIAL DIELECTRIC PROPERTIES.   ORIGIN OF DIPOLE CONFIGURATIONS IN SOME (PEROVSKITE) STR 08-1390
RATES AND SODIUM NITRITE.   DIPOLE CORRELATIONS IN MOLTEN ALKALI METAL NIT 16-2389
TY OF ULTRASOUND IN FERROELECTRICS / INFLUENCE OF LONG RANGE DIPOLE DIPOLE FORCES ON ATTENUATION AND VELOCI 03-0253
LEAD TITANATE.   TESTING A POINT DIPOLE ELECTROOPTIC MODEL BY AN APPLICATION TO 05-1073
C POTASSIUM DIHYDROGEN PHOSPHATE. (INTERBAND INTERACTION AND DIPOLE INSTABILITY) /E EFFECTS IN FERROELECTRI 17-2433
EMPERATURE IN SODIUM NITRITE.   DIPOLE LATTICE RELAXATION TIME NEAR THE NEEL T 16-2300
FERROCYANIDE TRIHYDRATE.   INTERNAL FIELD IN GENERAL DIPOLE LATTICES. (THEORY APPLIED TO POTASSIUM 03-0220
IN ICE.   DIPOLE MOMENT DERIVATIVE OF THE HYDROGEN BOND 34-3489
RESOLUTION OF A DISCREPANCY CONCERNING THE AVERAGE DIPOLE MOMENT IN KDP TYPE FERROELECTRICS. 17-2618
CTRIC PHASE TRANSITION IN POTASSIUM FERROCYANID/ ORDERING OF DIPOLE MOMENTS OF WATER MOLECULES AND FERROELE 24-2893
HASES OF ABO(3) SUBSTANCE. (WHEN THE B ION IS FERROELECTRIC/ DIPOLE PATTERNS IN ORTHORHOMBIC AND TRIGONAL P 08-1363
OELECTRICS AND ANTIFERROELECTRICS. (BARIUM TITANATE, KDP, P/ DIPOLE PATTERNS IN THE STRUCTURES OF SOME FERR 02-0134
RATURE POLARIZATION EFFECTS IN ICE. (CURRENT REVERSAL DUE TO DIPOLE RELAXATION)   LOW TEMPE 34-3418
IN LEAD ZIRCONATE.   DIPOLE STRUCTURE AND INTERNAL ELECTRIC FIELDS 07-1261
A PHASE TRANSITION IN A COMPOUND WITH HELICAL ELECTRIC DIPOLE STRUCTURE: CESIUM COPPER TRICHLORIDE. 33-3388
/N THE QUANTUM THEORY OF THE FERROELECTRIC STATE IN ELECTRIC DIPOLE SYSTEMS (SPIN WAVES IN HYDROXYL DIPOLE/ 03-0452

TION OF SODIUM NITRITE. (TWO SUBLATTICE MODEL OF INTERACTING DIPOLES) THE ORDER DISORDER TRANSI 16-2287
IN THE FERROELECTRIC P/ EFFECT OF COLLECTIVE INTERACTION OF DIPOLES ON THE NATURE OF DIELECTRIC DISPERSION 03-0417
TITANIUM IN LEAD TITANATE AND SIMILAR COMPOUNDS. STUDIES O/ DIRECT CHELATOMETRIC DETERMINATION OF LEAD AND 05-1068
CERAMICS. DIRECT CURRENT BIAS EFFECTS IN BARIUM TITANATE 04-0566
TITANATE CERAMICS. DIRECT CURRENT BIAS EFFECTS IN LEAD ZIRCONATE 09-1490
EXPERIMENTAL AND THEORETICAL STUDIES ON THE DIRECT CURRENT CONDUCTIVITY OF ICE. 34-3450
OR FUSED QUAR/ PREPARATION OF THIN BARIUM TITANATE FILMS BY DIRECT CURRENT DIODE SPUTTERING (ONTO PLATINUM 04-0948
ONG C-AXIS OF ANTIMONY SULFOIODIDE CAUSED BY ILLUMINATION IN DIRECT CURRENT ELECTRIC FIELD. STRAIN AL 15-2239
OF ANTIMONY SULFOIODIDE. (PHASE BOUNDARY SHAPE STRIATIONS IN DIRECT CURRENT FIELD) /RROELECTRIC PROPERTIES 15-2197
IMPEDANCE MEASUREMENTS OF A FERROELECTRIC CONDENSER UNDER A DIRECT CURRENT FIELD. NOISE AND 37-3645
TIMONY SULFOIODIDE CRYSTAL (IN THE PARAELECTRIC PHASE, UNDER DIRECT CURRENT FIELD ALONG C-AXIS). /ION IN AN 15-2173
FERROE/ SURFACE LAYER IN BARIUM TITANATE (X-RAY DIFFRACTION, DIRECT CURRENT FIELD EFFECT, PARAELECTRIC AND 04-0857
IGNETTE SALT. (EFFECT OF AGING AND COPPER(2+) D/ EFFECT OF A DIRECT CURRENT FIELD ON DOMAIN STRUCTURE IN SE 28-3230
IES OF TRIGLYCINE SELENATE. (AGING AND PHASE CH/ EFFECT OF A DIRECT CURRENT FIELD ON THE DIELECTRIC PROPERT 27-3131
E(0.5) TUNGSTATE(0.5)), EFFECTS OF POSITIVE OR NEGATIVE HIGH DIRECT CURRENT FIELDS SUMMARY ONLY) / MAGNESAT 05-1063
TALS (CZOCHRALSKI METHOD). GROWTH TWINNING (UNDER DIRECT CURRENT) IN C-AXIS LITHIUM NIOBATE CRYS 11-1672
O A NANOSECOND RESPONSE TIME DETEC/ PYROELECTRIC COEFFICIENT DIRECT MEASUREMENT TECHNIQUE AND APPLICATION T 13-1919
BY SCANNING ELECTRON MICROSCOPE. DIRECT OBSERVATION OF BARIUM TITANATE CRYSTALS 04-0808
TRIGLYCINE SULFATE USING THE SCANNING ELECTRON MICROSCOPE. DIRECT OBSERVATION OF FERROELECTRIC DOMAINS IN 27-3051
AINS OF GADOLI/ APPLICATION OF ELECTRON MIRROR MICROSCOPY TO DIRECT OBSERVATION OF MOVING FERROELECTRIC DOM 30-3293
TITANATE (OPTICAL STUDY UNDER UNIAXIAL STRESS). DIRECT OBSERVATION OF SINGLE DOMAIN STRONTIUM 06-1112
M TITANATE BY MEANS OF A POLARIZING MICROSCOPE. DIRECT OBSERVATION OF 180 DEG DOMAINS IN BARIU 04-0952
IDE FILMS. DIRECT PIEZOELECTRIC EFFECT IN POLYVINYL CHLOR 36-3543
OF THE (010) PLANE OF ROCHELLE SALT IN THE PRESENCE OF PURE DIRECT SKY BLUE DYE. GROWTH RATE 28-3217
TUDINAL ACOUSTIC WAVE AND TWO LUMINOUS WAVES ALONG THE (100) DIRECTION OF LITHIUM NIOBATE. /BETWEEN A LONGI 11-1680
Y A TRANSVERSE ACOUSTIC WAVE PROPAGATING ITSELF IN THE (001) DIRECTION OF LITHIUM NIOBATE. /IRD HARMONICS B 11-1682
/INAL HIGH FREQUENCY (950 MHZ) ACOUSTIC WAVE ALONG THE (001) DIRECTION OF LITHIUM NIOBATE (LIQUID HELIUM T/ 11-1835
ELECTRICS. (ABSTRACT ON/ THERMAL EXPANSION ANOMALIES AND THE DIRECTION OF SPONTANEOUS POLARIZATION IN FERRO 03-0350
SDUCER AS A SOURCE OF ULTRASONIC VIBRATIONS RADIATING IN THE DIRECTION OF THE TRANSDUCER AXIS. /ERAMIC TRAN 37-3657
CHLORIDE AND EPITAXI/ EFFECT OF TEMPERATURE AND PROPAGATION DIRECTION UPON OPTICAL PHONONS IN CUBIC SODIUM 29-3243
ZATION IN BORACITE AND OTHER CRYSTALS. RELATIONSHIP BETWEEN DIRECTIONS OF THE ELECTRIC AND MAGNETIC POLARI 14-2087
TITANATE. PERTURBED DIRECTIONAL CORRELATION OF SCANDIUM-44 IN LEAD 05-1055
FERROELECTRIC PHASES OF BARIUM TITANATE. PERTURBED DIRECTIONAL CORRELATION OF SCANDIUM-44 IN THE 04-0558
THE LATTICE OF FERROELECTRIC BARIUM TITANATE. PERTURBED DIRECTIONAL CORRELATION OF SCANDIUM-44 WITHIN 04-0691
AL PHONONS IN LITHIUM NIOBATE. DIRECTIONAL DISPERSION AND ASSIGNMENT OF OPTIC 11-1695
ND POLARITON SCATTERING IN LITHIUM NIOBATE. DIRECTIONAL DISPERSION OF PHONON FREQUENCIES A 11-1696
ND STAEILI/ PEROVSKITE FAMILY AND CHEMICAL BOND. (COVALENCY, DIRECTIVITY CONDITIONS FOR ELECTRIC ORDERING A 03-0243
FERROELECTRI/ NATURE OF DISSIPATIVE EFFECTS WHICH ACCOMPANY DISACCOMODATION TYPE PHENOMENA IN FERRITES AND 03-0447
JECTION AND CONDUCTION IN BARIUM TITANATE (CONDENSER CHARGE- DISCHARGE METHOD). IONIC CONTRIBUTION TO IN 04-0571
. (POTASSIUM DIHYDROGEN PHOSPHAT/ TRANSIENT PHENOMENA IN THE DISCHARGE OF CONDENSERS WITH POLAR DIELECTRICS 17-2631
/TRANSITIONS IN POTASSIUM MANGANESE TRIFLUORIDE: TEMPERATURE DISCONTINUITIES IN THE INFRARED SPECTRUM AND / 33-3363
EIR PARAELECTRIC PHASE TRANSFORMATIONS. (AND SUGGESTIONS FOR DISCOVERING MEMBERS OF VACANT CLASSES) / OF TH 03-0138
ON THE DIELECTRIC RELAXATION IN FERROELECTRICS (THEORETICAL DISCREPANCY). NOTE 03-0391
NT IN KDP TYPE FERROELECTRICS. RESOLUTION OF A DISCREPANCY CONCERNING THE AVERAGE DIPOLE MOME 17-2618
EASUREMENT OF POISSON'S RATIO IN POLED FERROELECTRIC CERAMIC DISKS. M 37-3658
RE RUBIDIUM MANGANESE TRIFLUORIDE AND POTAS/ PREPARATION AND DISLOCATION DENSITY DETERMINATION OF LARGE, PU 33-3397
DATE. DISLOCATION ETCHANTS FOR BETA GADOLINIUM MOLYB 30-3258
APPLICATION OF SELECTIVE ETCHING TO THE STUDY OF TWIN AND DISLOCATION STRUCTURES IN TRIGLYCINE SULFATE. 27-3042
NERGY SPECTRUM OF AN ORDER DISORDER FERROELECTRIC CONTAINING DISLOCATIONS. E 03-0485
OL SULFATE. X-RAY TOPOGRAPHIC STUDY OF DISLOCATIONS AND FERROELECTRIC DOMAINS IN GLYC 19-2747
RYSTAL POTASSIUM DIHYDROGEN PHOSPHATE. EXISTENCE OF DISLOCATIONS AT DOMAIN TIPS IN FERROELECTRIC C 17-2450
/OELECTRIC DOMAIN STRUCTURE AND ITS INTERACTION WITH CRYSTAL DISLOCATIONS. (DOMAIN ORIGIN, SHAPES, MOVEMEN/ 03-0390
S. DISLOCATIONS IN BARIUM TITANATE SINGLE CRYSTAL 04-1007
S. ETCH PITS CORRESPONDING TO DISLOCATIONS IN BARIUM TITANATE SINGLE CRYSTAL 04-1023
EFFECTS OF ANISOTROPY OF ELASTIC CONSTANTS ON DISLOCATIONS IN FERROELECTRIC BARIUM TITANATE. 04-0674
TE CRYSTALS. DOMAIN WALL CAUGHT IN DISLOCATIONS IN FERROELECTRIC TRIGLYCINE SULFA 27-3091
X-RAY DIFFRACTION TOPOGRAPHY STUDY OF DEFECTS (DISLOCATIONS) IN KDP AND ADP SINGLE CRYSTALS. 17-2578
OGEN PHOSPHATE CRYSTALS. EFFECT OF DISLOCATIONS ON THE GROWTH OF POTASSIUM DIHYDR 17-2491
PHOSPHATE CRYSTALS BY GROWTH/ X-RAY TOPOGRAPHIC STUDY OF THE DISLOCATIONS PRODUCED IN POTASSIUM DIHYDROGEN 17-2492
ION ON DOMAIN STRUCTURE IN TRIGLYCINE SULFATE. INFLUENCE OF DISLOCATIONS, SURFACE LAYER AND X-RAY IRRADIAT 27-3004
ANHARMONIC OSCILLATOR. A UNIFIED OSCILLATOR MODEL FOR ORDER DISORDER AND DISPLACIVE FERROELECTRICS. /SICAL 03-0405
ONY SUL/ OPTICAL PHONONS AND PHASE TRANSITIONS IN SOME ORDER DISORDER AND DISPLACIVE FERROELECTRICS. (ANTIM 15-2208
TE AND POTASSIUM NIOBATE. STRUCTURE DISORDER AND RAMAN SCATTERING IN BARIUM TITANA 04-0676
CATTERING IN BARIUM TITANATE. LINEAR DISORDER AND TEMPERATURE DEPENDENCE OF RAMAN S 04-0675
. ENERGY SPECTRUM OF AN ORDER DISORDER FERROELECTRIC CONTAINING DISLOCATIONS 03-0485
RRECTION. STATISTICAL PERTURBATION THEORY OF ORDER DISORDER FERROELECTRICS: FIRST PERTURBATIVE CO 03-0399
DIELECTRIC DISPERSION IN ORDER DISORDER FERROELECTRICS. (THEORY) 03-0373
AND FERROELECTRIC PEROVSKITES). (BARIUM TITANATE AND/ LINEAR DISORDER IN CRYSTALS (CASE OF SILICON, QUARTZ 04-0629
M NITRITE. EVIDENCE OF ADDITIONAL DISORDER IN THE RHOMBOHEDRAL FORM OF POTASSIU 16-2371
S. BASED ON NEW TREATMENT AND INTERPRETATION OF X/ STRUCTURE DISORDER OF BARIUM TITANATE TYPE FERROELECTRIC 04-0630
DISORDER OF LITHIUM NIOBATE CRYSTALS. 11-1679
DEDUCING MEAN DAMAGE RANGES. (STUDIES ON BOMBARDMENT INDUCED DISORDER. PART-6.) /N TRIOXIDE AND ITS USE IN 10-1642
UENCY DEPENDENT SUSCEPTIBILITY). COUPLED DISORDER PHONON SYSTEM WITH DAMPING: KDP (FREQ 17-2667
SILVER PERIODATE, / FERROELECTRICS. (THEORY, DYNAMICAL ORDER DISORDER PROPERTIES, KDP, TGS, ROCHELLE SALT, 02-0050
STATISTICAL CONSIDERATIONS OF THE ORDER DISORDER REACTION IN AMMONIUM PERIODATE. 36-3581
INFLUENCE OF IMPURITIES ON THE ORDER DISORDER REACTION IN HEXAGONAL ICE I(H). 34-3473
ATE TYPE FERROELECTRICS. ORDER DISORDER THEORY OF POTASSIUM DIHYDROGEN PHOSPH 17-2685
EORY) CRITICAL DYNAMICS OF THE ORDER DISORDER TRANSFORMATION IN FERROELECTRICS. (TH 03-0281
/SCATTERING IN TRIGLYCINE SULFATE AND SODIUM NITRATE. (ORDER DISORDER TRANSITION, FERROELECTRIC AND NONFER/ 16-2400
NISM) ENTROPY OF AN ORDER DISORDER TRANSITION. (IN FERROELECTRICS, MECHA 03-0192
CINE SULFATE, HIGHLY ACCURATE TEMPERATURE CONTROL) ORDER DISORDER TRANSITION IN FERROELECTRICS. (TRIGLY 27-2982
NATE. MODELS FOR THE ORDER DISORDER TRANSITION IN SODIUM TRIHYDROGEN SELE 22-2862
BLATTICE MODEL OF INTERACTING DIPOLES) THE ORDER DISORDER TRANSITION OF SODIUM NITRITE. (TWO SU 16-2287
/OMALOUS) DIELECTRIC RELAXATION IN SEVERAL CRYSTALS OF ORDER DISORDER TYPE. (COPPER FORMATE TETRAHYDRATE, / 02-0087
TEMPERATURE DEPENDENCE OF NMR SPECTRA SECOND MOMENT IN ORDER DISORDER TYPE FERROELECTRICS. NATURE OF THE 03-0460
TRICS DYNAMICS OF ORDER DISORDER TYPE FERROELECTRICS AND ANTIFERROELEC 03-0179
CTUATI/ SPIN LATTICE RELAXATION BY QUASI SPIN WAVES IN ORDER DISORDER TYPE FERROELECTRICS: POLARIZATION FLU 26-2928
RELAXATION TIME NEAR THE TRANSITION TEMPERATURE IN THE ORDER DISORDER TYPE FERROELECTRICS. (THEORY) /TTICE 03-0488
E INTER/ CRITICAL BEHAVIOR OF THE PSEUDOSPIN MODEL FOR ORDER DISORDER TYPE FERROELECTRICS WITH A SHORT RANG 03-0423
TIVITY DISPERSION PECULIARITIES IN FERROELECTRICS WITH ORDER DISORDER TYPE PHASE TRANSITION. /ECTRIC PERMIT 03-0412

CTRIC CONSTANT ON THERMAL AND FIELD HI/ POTASSIUM IODATE - A DISORDERED FERROELECTRIC. (DEPENDENCE OF DIELE 08-1348
STRUCTURE OF THE DISORDERED FHASE OF POTASSIUM NITRATE. 16-2379
URIE POINT IN POTASSIUM NITRATE. X-RAY STUDY ON THE DISORDERED STRUCTURE ABOVE THE FERROELECTRIC C 16-2366
DEG C. DISORDERED STRUCTURE OF SODIUM NITRITE AT 185 16-2325
MODEL. (ICE I) ATOMIC VIERATIONS IN ORIENTATIONALLY DISORDERED SYSTEMS. PART-1: A TWO DIMENSIONAL 34-3474
N LITHIUM NIOBATE. DIRECTIONAL DISPERSION AND ASSIGNMENT OF OPTICAL PHONONS I 11-1695
ONONS IN POTASSIUM MANGANESE TRIFLUORIDE. DISPERSION AND DAMPING OF SOFT ZONE BONDARY PH 33-3374
PTICAL HARMONICS. (KDP) EFFECTS OF DISPERSION AND FOCUSING ON THE PRODUCTION O 17-2579
YDROXIDE DOPED ICE. DIELECTRIC DISPERSION AND PHASE TRANSITION OF POTASSIUM H 34-3452
CRYSTALS. (THEORY) FAR INFRARED DISPERSION AND RAMAN SPECTRA OF FERROELECTRIC 03-0165
PROCESSES. (APPLIED TC FERROELECTFIC TRANSITIONS/ DIELECTRIC DISPERSION AND THERMODYNAMICS OF IRREVERSIBLE 03-0415
TITANATE ABOVE THE TRANSITION POINT. DISPERSION AT MILLIMETER WAVELENGTHS IN BARIUM 04-0601
TALS. PRESSURE DEPENDENCE OF THE COMPLEX DISPERSION COEFFICIENT OF ICE I(H) SINGLE CRYS 34-3479
EVALUATION OF DIELECTRIC DISPERSION DATA. (ICE) 34-3433
DIELECTRIC DISPERSION IN ANTIMONY SULFOIODIDE CRYSTALS. 15-2117
ND POTASSIUM/ COUPLED OPTICAL PHONCN MODE THEORY OF INFRARED DISPERSION IN BARIUM AND STRONTIUM TITANATES A 04-0547
NC TITANIUM CIOXIDE. FAR INFRARED DIELECTRIC DISPERSION IN BARIUM AND STRONTIUM TITANATES A 04-0975
FAR INFRARED DIELECTRIC DISPERSION IN BARIUM SODIUM NIOBATE. 13-1986
DIELECTRIC DISPERSION IN BARIUM TITANATE. 04-0646
ON THEIR DOMAIN STRUCTURE. DEPENDENCE OF DIELECTRIC DISPERSION IN BARIUM TITANATE SINGLE CRYSTALS 04-0895
ICS IN THE METER BAND. DIELECTRIC DISPERSION IN BARIUM TITANATE TYPE FERROELECTR 04-1009
PHONON DISPERSION IN FERROELECTRIC POTASSIUM NITRATE. 16-2334
(SEVERAL KHZ TO 23 KMHZ, DOMAIN WALL RESONANCE). CIELECTRIC DISPERSION IN FERROELECTRIC TRIGLYCINE SULFATE 27-3032
DIELECTRIC DISPERSION IN FERROELECTRICS. 03-0428
(BARIUM TITANATE AND TITANAT/ STUDY OF MICROWAVE DIELECTRIC DISPERSION IN FERROELECTRICS OF VARIOUS TYPES. 02-0108
TITANATE) DISPERSION IN FERROELECTRICS. (THEORY, BARIUM 03-0454
LATION VIA TEMPERATURE) TEMPERATURE DEPENDENCE OF POLARITON DISPERSION IN LITHIUM TANTALATE. (OPTICAL MODU 12-1873
THEORY) DIELECTRIC DISPERSION IN ORDER DISORDER FERROELECTRICS. (03-0373
MECHANISM OF MICROWAVE DIELECTRIC DISPERSION IN POLYCRYSTALLINE BARIUM TITANATE. 04-0745
LOW FREQUENCY DIELECTRIC DISPERSION IN ROCHELLE SALT CRYSTALS. 28-3228
ULTRASONIC DISPERSION IN STRONTIUM TITANATE. 06-1239
RRATE DIELECTRIC CERAMICS. DISPERSION IN STRONTIUM TITANATE- LANTHANUM FE 06-1120
LOSS FROM 10 GHZ TO 100 GHZ) DIELECTRIC DISPERSION IN TGS CRYSTALS. (PERMITTIVITY AND 27-3094
E CURIE POINT IN VARIOUS FERROELECTRI/ MILLIMETER WAVELENGTH DISPERSION IN THE DIELECTRIC CONSTANT ABOVE TH 02-0124
/LLECTIVE INTERACTION OF DIPOLES CN THE NATURE OF DIELECTRIC DISPERSION IN THE FERROELECTRIC PHASE TRANSIT/ 03-0417
LI/ SPONTANEOUS PARAMETRIC SCATTERING, IDLER ABSORPTION, AND DISPERSION IN THE NONLINEAR SUSCEPTIBILITY IN 11-1771
IUM NITRATE. CIELECTRIC DISPERSION IN THE PARAELECTRIC PHASE OF POTASS 16-2341
ERROELECTRICS. MICROWAVE DISPERSION MECHANISM IN BARIUM TITANATE TYPE F 04-0902
ICS. (SINGLE CRYSTALS ANC CERAMICS) PERMITTIVITY DISPERSION OF BARIUM TITANATE TYPE FERROELECTR 04-0903
ENCIES AND JERKY WAIL MOTION IN ROCHELLE SALT. ANOMALOUS DISPERSION OF COERSIVE FIELD AT VERY LOW FREQU 03-3177
(LITHIUM NIOBATE) DISPERSION OF ELASTIC WAVES IN FERROELECTRICS. 11-1706
TANATE. (BIREFRINGENCE, PRINCIPAL INDICES) DISPERSION OF ELECTROOPTIC EFFECT IN BARIUM TI 04-0755
/OF TEMPERATURE GRADIENT WITHIN THE SAMPLE ON THE) MICROWAVE DISPERSION OF FERROELECTRICS NEAR THE TRANSIT/ 03-0370
E PHASE TRANSITION REGION. DIELECTRIC DISPERSION OF IRRADIATED BARIUM TITANATE IN TH 04-0652
THE PHASE TRASNITION. DIELECTRIC DISPERSION OF IRRADIATED BARIUM TITANATE NEAR 04-0648
FERROELECTRIC SODIUM NITRITE. ANGULAR DISPERSION OF LARGE WAVE VECTOR POLARITONS IN 11-2256
OTASSIUM TANTALATE) FAR INFRARED CIELECTRIC DISPERSION OF PEROVSKITES. (BARIUM TITANATE, P 04-0546
SCATTERING IN LITHIUM NIOBATE. DIRECTIONAL DISPERSION OF PHONON FREQUENCIES AND POLARITON 11-1696
ELECTROOPTICAL EFFECT IN LEAD ZINCATE(0.33) NIOBATE(0.67). (DISPERSION OF QUADRATIC COEFFICIENTS) 13-1982
TITANATE. (CALCULATIONS) POLARITON DISPERSION OF TETRAGONAL FERROELECTRIC BARIUM 04-0842
DIELECTRIC DISPERSION OF TGS. 17-2574
DIELECTRIC DISPERSION OF TGS. 27-3060
CALCULATIONS). DIELECTRIC DISPERSION OF TGS AND DEUTERATED KDP (THEORY, 17-2534
N THE FAR INFRARED (TEMPERATURE AND FREQUENCY DEPENDENCE). DISPERSION OF THE DIELECTRIC CONSTANT OF TGS I 27-3009
M TITANATE. DISPERSION OF THE ELECTROOPTIC EFFECT IN 04-0756
ROMAGNETIC POTASSIUM MANGANESE TRI/ VIBRATIONAL SPECTRUM AND DISPERSION OF THE OPTICAL CONSTANTS OF ANTIFER 33-3385
ATE TYPE FERR/ EXPERIMENTAL CHECK OF THE THEORY OF MICROWAVE DISPERSION OF THE PERMITTIVITY OF BARIUM TITAN 04-0649
NCIES. DIELECTRIC DISPERSION OF TRIGLYCINE SULFATE AT LOW FREQUE 27-3023
NGANESE TRIFLUORIDE IN THE NEIGHBORHOOD OF / ATTENUATION AND DISPERSION OF ULTRASONIC WAVES IN POTASSIUM MA 33-3370
DIELECTRIC DISPERSION OF UREA AND THIOUREA. 25-2912
H ORDER DISORDER TYPE PHASE TRANSIT/ DIELECTRIC PERMITTIVITY DISPERSION PECULIARITIES IN FERROELECTRICS WIT 03-0412
THERMODYNAMIC THEORY OF DISPERSION PHENOMENA. (FERROELECTRICITY) 03-0420
POLARITON DISPERSION RELATION IN CUBIC BARIUM TITANATE. 04-0878
/NFLUENCE OF CRYSTAL DEFECTS UPON FERROELECTRIC PROPERTIES. (DISPERSION, RELAXATION TIME AND STRENGTH, HYS/ 03-0496
ECHANICAI DAMPING IN LEAD TITANATE ZIRCONATE TRANSDUCER CER/ DISPERSION STUDY OF DOMAIN WALL MOBILITY AND M 09-1547
LT CAUSED BY DOMAIN WALLS. (ABSTRACT ONLY) LCW FREQUENCY DISPERSIONS OF THE PERMITTIVITY OF ROCHELLE SA 28-3199
E COMPLEX PERMITTIVITY OF ACP ANE KDP IN THE FAR INFRARED BY DISPERSIVE FOURIER TRANSFORM SPECTROSCOPY. /TH 17-2609
ION CHANGE) AFTER EFFECTS IN TRIGLYCINE SULFATE. (DIELECTRIC DISPLACEMENT AND COERCIVE FIELD) (POLARIZAT 27-2952
NTANEOUS POLARIZATION IN DISPLAC/ FELATICNSHIP BETWEEN IONIC DISPLACEMENT AND THE CURIE TEMPERATURE AND SPO 03-0453
TE. NEUTRON SCATTERING ANALYSIS OF THE LINEAR DISPLACEMENT CORRELATIONS IN POTASSIUM TANTALA 36-3546
STATISTICAL THEORY FOR DISPLACEMENT FERROELECTRICS. 03-0369
EXAMPLE). INTERCELL CCRRECTIONS FOR IONIC MOTION IN DISPLACEMENT FERROELECTRICS (LITHIUM TANTALATE 03-0367
THIUM THALLIUM TARTRATE CRYSTALS BY HYDROSTATIC PRESSURE. DISPLACEMENT OF THE CURIE POINTS IN KDP AND LI 17-2523
DISPLACEMENT PHASE TRANSITIONS IN PEROVSKITES. 08-1274
GSTEN BRONZES, NIOBATES) NEW CLASSES OF FERROELECTRICS OF DISPLACEMENT TYPE. (BORACITES, MOLYBDATES, TUN 02-0118
. (THEORY, FERROELECTRICS, PIEZCELECTRICS) LOW TEMPERATURE DISPLACEMENT TYPE PHASE TRANSITION IN CRYSTALS 03-0320
ECTRICS. (THEORY APPLIED TO BARIUM T/ CORRELATION EFFECTS IN DISPLACEMENT TYPE PHASE TRANSITIONS IN FERROEL 03-0497
HE FERROELECTRIC PHASES OF / RELATIONSHIP BETWEEN THE ATOMIC DISPLACEMENTS AND SPONTANEOUS DEFORMATION IN T 08-1377
LS IN KDP. EXISTENCE OF CRITICAL ELECTRIC FIELDS. DISPLACEMENTS AND VELOCITIES OF THE DOMAIN WAL 17-2448
THORHCMBIC BORACITE STRUCTURES. (BORACITE, ERICAITE) ATOMIC DISPLACEMENTS IN FERROELECTRIC TRIGONAL AND OR 14-2075
(AT OPTICAL FREQUENCIES, CALCULATIONS). ATOMIC DISPLACEMENTS IN PEROVSKITE STRONTIUM TITANATE 06-1142
M TANTALATE, STRONT/ DETERMINATION OF THE NORMAL VIBRATIONAL DISPLACEMENTS OF SEVERAL PEROVSKITES (POTASSIU 06-1139
SCILLATOR. A UNIFIED OSCILLATOR MODEL FOR ORDER DISORDER AND DISPLACIVE FERROELECTRICS. /SICAL ANHARMONIC O 03-0405
MENT OF CHROMIUM- POTASSIUM ALUM (UNSTABLE SOFT PHONCN MODE, DISPLACIVE FERROELECTRICS). / MEAN MAGNETIC MO 20-2806
NT AND THE CURIE TEMPERATURE AND SPONTANEOUS POLARIZATION IN DISPLACIVE FERROELECTRICS. /N IONIC DISPLACEME 03-0453
/AL PHONONS AND PHASE TRANSITIONS IN SOME ORDER DISORDER AND DISPLACIVE FERROELECTRICS. (ANTIMONY SULFOIOD/ 15-2208
) POWDER RAMAN SPECTRA: APPLICATION TO DISPLACIVE FERROELECTRICS. (LATTICE VIBRATIONS 03-0194
NE SULFATE NEAR THE CURIE POINT. (ESR OF GAMMA-IRRADIATED S/ DISPLACIVE MOVEMENT OF THE RADICAL IN TRIGLYCI 27-3033
E) (FAST ION SCATTERING) SHADOW EFFECT IN (DISPLACIVE) PHASE TRANSITIONS. (BARIUM TITANAT 04-0533
INELASTIC DIFFUSION IN FERROELECTRICS) NEUTRON STUDIES OF DISPLACIVE STRUCTURAL PHASE TRANSFORMATIONS. (03-0155

IN FERROELECTRICS)   NEUTRON STUDIES OF DISPLACIVE STRUCTURAL PHASE TRANSFORMATIONS. (03-0156
OPAGATION, STRESS EFFECTS, AND INTERACTION PARAMETERS AT THE DISPLACIVE TRANSITION IN STRONTIUM TITANATE. / 06-1127
CE VIBRATIONS (REDEVELOPED FROM FIRST PRINCIPLES, APPLIED TO DISPLACIVE TYPE FERROELECTRICS). /G FROM LATTI 03-0484
                        ULTRASONIC ATTENUATION ANOMALY IN DISPLACIVE TYPE FERROELECTRICS.   03-0393
TION POINT. (BARIUM TITANATE, STRONTI/ DYNAMICAL BEHAVIOR OF DISPLACIVE TYPE FERROELECTRICS NEAR THE TRANSI 04-0992
ATTENUATION OF CRITICAL VIBRATIONS AND DIELECTRIC LOSSES IN DISPLACIVE TYPE FERROELECTRICS (THEORY).   03-0159
/IGH FREQUENCY SOUND ATTENUATION AND THERMAL CONDUCTIVITY IN DISPLACIVE TYPE FERROELECTRICS. (THEORY, BARI 03-0160
IC PHOTOSENSITIVE CRYSTALS AS POCKELS EFFECT LIGHT VALVES IN DISPLAY AND MODULATOR APPLICATIONS. USE OF CUB 37-3661
AMICS.   DISPLAY APPLICATIONS OF PLZT FERROELECTRIC CER 09-1561
L INHOMOGENEITIES WITH APPLICATIONS TO FERROELECTRIC CERAMIC DISPLAY DEVICES. / SCATTERING BY RANDOM OPTICA 37-3634
ELECTROOPTIC CERAMIC (INFORMATION) STORAGE (PROCESSING) AND DISPLAY DEVICES. (BASED ON ELECTRICALLY CONTR/ 37-3650
/RAIN EIASED FERROELECTRIC- PHOTOCONDUCTOR IMAGE STORAGE AND DISPLAY DEVICES OPERATED IN A REFLECTION MODE/ 09-1564
STRAIN EIASED FERROELECTRIC PHOTOCCCNDUCTOR IMAGE STORAGE AND DISPLAY DEVICES. (PLZT)   09-1565
ATTERING MODE FERROELECTRIC PHOTOCONDUCTOR IMAGE STORAGE AND DISPLAY DEVICES. (PLZT CERAMICS)   SC 09-1599
CERAMICS. (LEAD ZIRCCNATE- LEAD TITANATE) IMAGE STORAGE AND DISPLAY DEVICES USING FINE GRAIN FERROELECTRIC 09-1572
/ROOPTIC CERAMICS- NEW MATERIALS FCR INFORMATION STORAGE AND DISPLAY (HOT PRESSED LEAD ZIRCONATE TITANATE,/ 09-1550
(PLZT)   A NEW LONGITUDINAL DISPLAY MODE FOR CERAMIC ELECTROOPTIC DEVICES. 09-1523
Y DIFFRACTION TOPOGRAPHY.   DISPLAY OF TWINNING IN LITHIUM NIOBATE BY X-RA 11-1845
       FERROELECTRIC CERAMICS FOR INFORMATION STORAGE AND   DISPLAY. (PZT)   09-1552
              FERROELECTRICS IN OPTICAL MEMORIES AND   DISPLAYS: A CRITICAL APPRAISAL.   02-0039
              FERROELECTRICS IN OPTICAL MEMORIES AND   DISPLAYS: A CRITICAL APPRAISAL.   37-3628
TANATE)   FERROELECTRIC DISPLAYS. (BISMUTH TITANATE, LEAD ZIRCONATE TI 09-1515
CAL COUPLING IN PIEZOELECTRIC CERAMICS.   MEASUREMENT OF THE DISSIPATION ASSOCIATED WITH THE ELECTROMECHANI 03-0514
E WITH IMPROVED ELECTRICAL PROPERTIES. (DIELECTRIC CONSTANT, DISSIPATION FACTOR) /XIDE DOPED BARIUM TITANAT 04-0840
ATION TYPE PHENOMENA IN FERRITES AND FERROELECTRI/ NATURE OF DISSIPATIVE EFFECTS WHICH ACCOMPANY DISACCOMOD 03-0447
RYSTALS IN SOLUTION.   KINETICS OF GROWTH AND   DISSOLUTION OF AMMONIUM DIHYDROGEN PHOSPHATE C 17-2412
(X) TUNGSTEN(1-X) Z(Y) OXIDE/ ELECTROCHEMICAL DEPOSITION AND DISSOLUTION OF TUNGSTEN OXIDE BRONZES. (SODIUM 13-1944
URE.   INTERATOMIC DISTANCES IN OXIDES WITH THE PEROVSKITE STRUCT 08-1439
ELDS AND THEIR GRADIENTS IN ABO(3) PEROVSKITE COMPOUNDS WITH DISTINCTIVE DIELECTRIC PROPERTIES. /LECTRIC FI 08-1455
OCHELLE SALT AND TRIGLYCINE SULFATE CRYSTALS.   DISTINCTIVE FEATURES OF BARKHAUSEN EFFECT IN R 27-3110
NTIUM TITANATE AT THE CUBIC TO TETRAGONAL TRANSITION.   DISTORTION ENHANCED OPTICAL ABSORPTION IN STRO 06-1109
STAL ELECTRET.   LATTICE DISTORTIONS IN A STRONTIUM TITANATE SINGLE CRY 06-1238
RATION. (POTASSIUM DIHYDROGEN PHOSPHATE)   PULSE   DISTORTIONS IN MISMATCHED SECOND HARMONIC GENE 17-2690
O LITHIUM NIOBATE BY PHASE MATCHING TE/ DETERMINATION OF THE DISTRIBUTION COEFFICIENTS OF LITHIUM OXIDE INT 11-1841
IN THE PARAELECTRIC PHASE, UNDER DC FIELD ALONG C/ POTENTIAL DISTRIBUTION IN ANTIMONY SULFOIODIDE CRYSTAL (15-2173
RYSTALS.   INFLUENCE OF HUMIDITY ON THE POTENTIAL DISTRIBUTION IN BARIUM TITANATE CERAMICS AND C 04-0883
GEN PHOSPHATE (NEUTRON SCATTERING).   HYDROGEN DENSITY   DISTRIBUTION IN PARAELECTRIC POTASSIUM DIHYDRO 17-2490
C SEMICONDUCTOR INTERFACE.   POLARIZATION   DISTRIBUTION IN THE VICINITY OF A FERROELECTRI 03-0167
(LEAD, LANTHANUM) (TITANATE, ZIRCONATE) CERAMICS.   DISTRIBUTION OF A-SITE AND B-SITE VACANCIES IN 09-1530
E CRYSTALS AND SOME INVESTIGATIONS OF THEIR QUALITY. (VOLUME DISTRIBUTION OF DEFECTS) /ON OF PURE TGS SINGL 27-3086
AMICS.   EFFECTS OF GRAIN GROWTH ON THE DISTRIBUTION OF NIOBIUM IN BARIUM TITANATE CER 04-0595
ICS PREVIOUSLY POLARIZED.   DISTRIBUTION OF POTENTIAL ALONG TITANATE CERAM 04-0788
ICS NEAR THE CURIE TEMPERATURE (POLY VERSUS MONODISPERSIVE / DISTRIBUTION OF RELAXATION TIME IN FERROELECTR 03-0521
ELECTRICS.   DISTRIBUTION OF RELAXATION TIMES IN SOME FERRO 03-0288
AD TITANATE.   DISTRIBUTION OF VACANCIES IN LANTHANA DOPED LE 05-1058
ON FOR THE SLATER MODEL OF FERROELECTRICITY (KDP MODEL, ANT/ DISTRIBUTION OF ZEROES OF THE PARTITION FUNCTI 03-0308
M(1-X) TITANATE CERAMIC/ "VIBRATING WIRE" STUDY OF POTENTIAL DISTRIBUTION (STABILITY) IN BARIUM(X) STRONTIU 04-0699
HASIS ON NEAR SURFACE REGIONS IN FERROELECTR/ ELECTRIC FIELD DISTRIBUTIONS IN DIELECTRICS, WITH SPECIAL EMP 03-0182
        LOW FREQUENCY LINEAR ELECTROOPTIC EFFECT IN LEAD   DITANATLATE.   13-1916
LID STATE.   STUDY OF   DIVALENT BINARY AND TERNARY NIOBATES IN THE SO 08-1286
                          PARAMAGNETIC RELAXATION AND   DIVALENT COPPER IONS IN FERROELECTRIC KDP.   17-2629
ED PARTIALLY BY ACETATE IONS.   PHASE TRANSITION IN   DIVALENT METAL DICALCIUM PROPIONATES SUBSTITUT 26-2944
                      APPARENT SOFT MODE LINE WIDTH   DIVERGENCE IN LEAD TITANATE.   05-1079
                      DEFECT LEAD TITANATES WITH   DIVERSE CURIE TEMPERATURES.   05-1076
LE SALT AND TRIGLYCINE SULFATE.   NEW MEASUREMENTS OF   DIVERSE THERMODYNAMICAL COEFFICIENTS OF ROCHEL 27-2995
PHATE.   DIELECTRIC PERMITTIVITY OF FINELY   DIVIDED POTASSIUM AND AMMONIUM DIHYDROGEN PHOS 17-2689
MEMORY, ORGANICS)   FROM BARIUM TITANATE TO   DNA. (FERROELECTRICITY, BIOLOGICAL MECHANISMS, 02-0092
                                                             DODECA... SEE THE PREFIXED WORD
. CRYSTAL STRUCTURE AND SINTERABILITY.   EFFECT OF CLAY AND DOLOMITE ADDITIONS ON BARIUM TITANATE CERAMICS 04-0739
ATE- LANTHANUM OXIDE (P/ TEMPERATURE DEPENDENT FERROELECTRIC DOMAIN ALIGNMENT IN LEAD ZIRCONATE- LEAD TITAN 09-1586
                  ANTIPARALLEL SWITCHING IN ALPHA   DOMAIN BARIUM TITANATE CRYSTALS.   04-0655
           INELASTIC NEUTRON SCATTERING FROM SINGLE   DOMAIN BARIUM TITANATE. (LATTICE DYNAMICS) 04-0946
                         DEPOLING OF SINGLE   DOMAIN BISMUTH TITANATE.   08-1446
IC METHOD TO FERROELECTRIC SODIUM NITRITE AND TGS CRYSTALS. (DOMAIN BOUNDARIES AND METASTABLE STRUCTURES) / 16-2375
/CS (WITH AND WITHOUT NIOBIA ADDITION) DUE TO INTERACTION OF DOMAIN BOUNDARIES AND POINT DEFECTS (TORSION / 09-1564
/ X-RAY TOPOGRAPHICAL STUDY OF THE CONTRAST AT FERROELECTRIC DOMAIN BOUNDARIES (DOPED AND UNDOPED TGS, LITH 11-1671
RYSTAL FILMS OF BARIUM TITANATE (ELECTRON MICROSC/ MOTION OF DOMAIN BOUNDARIES DURING SWITCHING IN SINGLE C 04-0940
             SURFACE ELASTCPOLARIZATION WAVES AT   DOMAIN BOUNDARIES IN A FERROELECTRIC. 03-0315
EFFECT ON THE REFLECTION OF A TRANSVERSE ACOUSTIC WAVE FROM DOMAIN BOUNDARIES IN FERROELECTRICS. /ELECTRIC 03-0312
F QUASI LONGITUDINAL AND QUASI TRANSVERSE SOUND WAVES BY 180 DOMAIN BOUNDARIES IN FERROELECTRICS. /ACTION O 03-0314
REFLECTION AND REFRACTION CF TRANSVERSE SOUND WAVES AT   DOMAIN BOUNDARIES IN FERROELECTRICS. (THEORY) 03-0313
.   MOVEMENT OF 180 DEGREE   DOMAIN BOUNDARIES IN PEROVSKITE FERROELECTRICS 08-1305
           CORROSION FIGURES DUE TO THE FERROELECTRIC   DOMAIN BOUNDARIES OF TRIGLYCINE SULFATE. 27-3142
ART-2: / ELECTRON MICROSCOPE TRANSMISSION IMAGES OF COHERENT DOMAIN BOUNDARIES. PART-1: DYNAMICAL THEORY. P 03-0255
           GROWTH OF LITHIUM NIOBATE SINGLE   DOMAIN CRYSTALS.   11-1781
TE. (CRACK-FREE, THROUGH AD/ GROWTH AND PROPERTIES OF SINGLE DOMAIN CRYSTALS OF FERROELECTRIC LITHIUM NIOBA 11-1774
OELECT/ INFLUENCE CF POINT DEFECTS ON DIELECTRIC PROPERTIES, DOMAIN DYNAMICS, AND PHASE TRANSITIONS OF FERR 03-0495
ATE.   EFFECT OF DEUTERON SUBSTITUTION ON   DOMAIN DYNAMICS IN POTASSIUM DIHYDROGEN PHOSPH 17-2437
IC DOMAINS).   INTERDOMAIN AND   DOMAIN EFFECT INTERACTIONS IN KDP (FERROELECTR 17-2451
THE FERROELECTRIC-AXIS OF SODIUM NITRATE.   DOMAIN ETCH PATTERNS ON THE FACES PARALLEL TO 16-2349
MECHANICAL MODEL OF LONG WAVELENGTH VIBRATIONS IN A (SINGLE DOMAIN) FERROELECTRIC TYPE CRYSTAL.   03-0342
COMPLEX PHASE TRANSITIONS IN CRYSTALS (APPLICATION TO SINGLE DOMAIN FERROELECTRICS BELOW CURIE POINT). 03-0319
    THEORY OF X-RAY DIFFRACTION LINE SHAPE EFFECTS DUE TO   DOMAIN FORMATION IN FERROELECTRICS.   03-0236
SOME RESULTS ABOUT PHOTOEMISSION ON FERROELECTRIC CRYSTALS. (DOMAIN IMAGING, ABSTRACT ONLY)   03-0386
N BARIUM TITANATE.   DOMAIN INTERACTIONS DURING PHASE TRANSITIONS I 04-0651
         APPARATUS FOR GROWTH OF SINGLE CRYSTAL, SINGLE   DOMAIN LITHIUM NIOBATE.   11-1726
/TION CF SUPERHIGH FREQUENCY ACOUSTIC HARMONICS IN A (SINGLE DOMAIN) LITHIUM NIOBATE CRYSTAL (NONLINEAR EL/ 11-1707
       DIELECTRIC PROPERTIES OF SINGLE   DOMAIN MELT GROWN BARIUM TITANATE.   04-0616
FERROELECTRIC TRANSITION REGION (THEORY, RELATION TO KAENZIG DOMAIN MODEL).   HEAT CAPACITY IN THE 03-0443
S ECHO IN FERROELECTRIC ANTIMONY SULFOIODIDE. (PIEZOELECTRIC DOMAIN OSCILLATIONS)   ANOMALOU 15-2211

EFFICIENT WHITE LIGHT READING OF DOMAIN PATTERNS IN BISMUTH TITANATE. 08-1299
A SECCND ORDER PHASE TRANSITION (THEORY). SINGLE DOMAIN POLARIZATION OF A FERROELECTRIC HAVING 03-0202
ANNEALING CF POLYCRYSTALLINE BARIUM TITANATE. SURFACE DOMAIN REORIENTATION PRODUCED BY ABRASION AND 04-0639
FERROELECTRIC DOMAIN REVERSAL IN LITHIUM METATANTALATE. 12-1869
Y A LASER BEAM. REVERSIBLE DOMAIN REVERSAL IN TRIGLYCINE SULFATE CAUSED B 27-3012
TRIGLYCINE SULFATE. EFFECTS OF DOMAIN SHAPES ON SECOND HARMONIC SCATTERING IN 27-2989
) FERROELECTRIC DOMAIN SHIFTING DEVICES. (GADOLINIUM MOLYBDATE 30-3280
RECENT PROBLEMS IN DOMAIN STATICS AND DYNAMICS. (56 REFS) 03-0248
DOMAIN STATISTICS AND FERROELECTRIC TRANSIENTS 03-0216
UNIAXIAL STRESS). DIRECT CBSERVATION OF SINGLE DOMAIN STRONTIUM TITANATE (OPTICAL STUDY UNDER 06-1112
AGING PROCESS IN TRIGLYCINE SULFATE. (HYSTERESIS LAPS, DOMAIN STRUCTURE) 27-3128
SES IN TRIGLYCINE SULFATE. (TANDEL OR CAPACITANCE PROPERTIES DOMAIN STRUCTURE) AGING PROCES 27-3081
CTRIC CISPERSION IN BARIUM TITANATE SINGLE CRYSTALS ON THEIR DOMAIN STRUCTURE. DEPENDENCE OF DIELE 04-0895
C PROPERTIES OF BARIUM TITANATE SINGLE CRYSTALS WITH LAMINAR DOMAIN STRUCTURE. /C, ELASTIC AND PIEZOELECTRI 04-1010
GROWTH OF LEAC TITANATE CRYSTALS AND EXAMINATION OF THEIR DOMAIN STRUCTURE. 05-1047
C TRIGLYCINE SULFATE CRYSTALS. (TCPOGRAPHY OF IMPERFECTIONS, DOMAIN STRUCTURE) /WTH REGIONS IN FERROELECTRI 27-2999
AGNETICALLY EQUIAXIAL FERRODIELECTRIC SINGLE CRYSTALS HAVING DOMAIN STRUCTURE. RESONANCE EFFECTS IN M 03-0509
RYSTALLOPHYSICAL CLASSIFICATION OF FERROELECTRICS. (BASED ON DOMAIN STRUCTURE AND ITS APPLICATIONS) C 03-0463
TAL DISLOCATIONS. (DOMAIN ORIGIN, SHAPES, MOV/ FERROELECTRIC DOMAIN STRUCTURE AND ITS INTERACTION WITH CRYS 03-0390
HE BETA FHASE OF THE FERROELECTRIC SODIUM TRIB/ THE PECULIAR DOMAIN STRUCTURE AND ITS UNUSUAL BEHAVIOR IN T 22-2874
SULFOIOCIDE TYPE SEMICONDUCTOR FERROELECTRICS. DOMAIN STRUCTURE AND LOCAL STATES IN ANTIMONY 15-2127
SINGLE CRYSTAL. (AESTRACT ONLY) POLARIZATION (DOMAIN STRUCTURE) AND PHASE TRANSITION IN KDP 17-2505
FILMS OF FERROELECTRIC BISMUTH TITANATE. DOMAIN STRUCTURE AND POLARIZATION REVERSAL IN 08-1470
/CS, FERROELECTRIC PHASE TRANSITIONS AND SPECIAL FEATURES OF DOMAIN STRUCTURE AND SOME PHYSICAL PROPERTIES/ 04-0480
THE PIEZOELECTRIC CONSTANTS OF STIBICTANTALITE. (ANTIMONY (/ DOMAIN STRUCTURE AND TEMPERATURE DEPENDENCE OF 36-3601
ES O/ INVESTIGATION OF THE EFFECT OF SAMPLE THICKNESS ON THE DOMAIN STRUCTURE AND THE PYROELECTRIC PROPERTI 04-1039
(ROCHELLE SALT). (TEMPERATURE A/ EFFECT OF ELECTRIC FIELD ON DOMAIN STRUCTURE FORMATION IN SEIGNETTES SALT 28-3190
ALS. (MICRO CINEMATOGRAPHY) PHASE TRANSITION AND DOMAIN STRUCTURE IN ANTIMONY SULFOIODIDE CRYST 15-2184
STALS. DOMAIN STRUCTURE IN LITHIUM NIOBATE SINGLE CRY 11-1703
RICS BY SOLID DEW TECHNIQUE. OBSERVATION OF DOMAIN STRUCTURE IN LOW TEMPERATURE FERROELECT 03-0341
HATE CRYSTALS BY POLARIZATION OPTICAL PROCED/ OBSERVATION OF DOMAIN STRUCTURE IN POTASSIUM DIHYDROGEN PHOSP 17-2493
HYDRATE SIN/ APPLICATION OF THE DEW METHCD FOR REVEALING THE DOMAIN STRUCTURE IN POTASSIUM FERROCYANIDE TRI 24-2904
RYSTALS. INTERFEROMETRIC STUDIES OF DOMAIN STRUCTURE IN POTASSIUM NIOBATE SINGLE C 08-1319
SE TRANSITION. REARRANGEMENT OF THE DOMAIN STRUCTURE IN ROCHELLE SALT NEAR THE PHA 28-3181
AGING AND COPPER(2+) DOPING) EFFECT OF A DC FIELD ON DOMAIN STRUCTURE IN SEIGNETTE SALT. (EFFECT OF 28-3230
M FERROCYANIDE. DOMAIN STRUCTURE IN THE FERROELECTRIC POTASSIU 24-2895
ENCE OF DISLOCATIONS, SURFACE LAYER AND X-RAY IRRADIATION ON DOMAIN STRUCTURE IN TRIGLYCINE SULFATE. INFLU 27-3004
SULFATE OBSERVED BY EPR. CHANGES OF THE DOMAIN STRUCTURE IN VANADIUM DOPED TRIGLYCINE 27-3013
AL DURING AGING. INVESTIGATION OF THE DOMAIN STRUCTURE OF A TRIGLYCINE SULFATE CRYST 17-2561
ALS (FROM SELECTIVE CRYSTALLIZATION AND ETCHING). DOMAIN STRUCTURE OF ANTIMONY SULFOIODIDE CRYST 15-2183
HING METHOD) EFFECTS OF ILLUMINATION ON THE DOMAIN STRUCTURE OF ANTIMONY SULFOIODIDE. (ETC 15-2124
/ UNIAXIAL TENSILE STRESSES ON THE ELECTRICAL PROPERTIES AND DOMAIN STRUCTURE OF BARIUM TITANATE SINGLE CR/ 04-0961
THEIR MECHANICAL PRCPERTIES. (BARIUM TITANATE,/ INFLUENCE OF DOMAIN STRUCTURE OF CERAMIC FERROELECTRICS ON 04-0985
IUM SULFATE DIHYCRATE. DOMAIN STRUCTURE OF FERROELECTRIC SODIUM AMMON 19-2778
DOMAIN STRUCTURE OF GADOLINIUM MOLYBDATE. 30-3301
PYROELECTRIC EFFECT AND DOMAIN STRUCTURE OF GADOLINIUM MOLYBDATE. 30-3315
ALS. DOMAIN STRUCTURE OF GAMMA-IRRADIATED TGS CRYST 27-3164
TITANATE. DIFFUSICNLESS PHASE TRANSITIONS AND THE STABLE DOMAIN STRUCTURE OF (HIGH SENSITIVITY) BARIUM 04-0888
LS. DOMAIN STRUCTURE OF LITHIUM METANIOBATE CRYSTA 11-1708
RICS WITH DIFFUSE PHASE TRANSITIONS. DOMAIN STRUCTURE OF POLYCRYSTALLINE FERROELECT 03-0505
RYSTALS. OPTICAL STUDIES OF DOMAIN STRUCTURE OF POTASSIUM NIOBATE SINGLE C 08-1320
SULFATE CRYSTALS GROWN BELOW AND ABOVE THE CURIE POINT. DOMAIN STRUCTURE OF PURE AND DOPED TRIGLYCINE 27-3130
BARIUM TITANATE BASE SOLI/ EFFECT OF EXTERNAL ACTIONS ON THE DOMAIN STRUCTURE OF SINGLE CRYSTALS FORMED BY 04-0954
NIAXIAL STRESSES. DOMAIN STRUCTURE OF STRONTIUM TITANATE UNDER U 06-1113
CRYSTALLIZATION OF ANTHRAQUINONE (SUBLIMED CN/ REVEALING THE DOMAIN STRUCTURE OF TGS CRYSTALS BY SELECTIVE 27-3041
RIHYDROGEN SELENATE IN ELECTRIC FIELDS. DOMAIN STRUCTURE OF THE BETA PHASE OF SODIUM T 22-2830
UCTOR ANTIMONY / INFLUENCE OF NONEQUIIIBRIUM CARRIERS ON THE DOMAIN STRUCTURE OF THE FERROELECTRIC SEMICOND 15-2104
L (DURING COOLING THROUGH CURIE PCINT). DOMAIN STRUCTURE OF TRIGLYCINE SELENATE CRYSTA 27-3043
S ON TEMPERATURE AND APPLIED ELECTRIC FIE/ DEPENDENCE OF THE DOMAIN STRUCTURE OF TRIGLYCINE SULFATE CRYSTAL 27-3129
VICINITY OF THE CURIE POINT. EQUILIBRIUM DOMAIN STRUCTURE OF TRIGLYCINE SULFATE IN THE 27-3112
HILATION IN ROCHELLE SALT. INFLUENCE OF THE DOMAIN STRUCTURE ON POSITRON (TWO-PHONON) ANNI 28-3191
UM(3+) DOPED STRONTIUM TITANATE. EFFECT OF DOMAIN STRUCTURE ON THE FLUORESCENCE OF CHROMI 06-1114
TALLINE LEAD TITANATE ZIRCONATE COMPCSITIONS N/ INFLUENCE OF DOMAIN STRUCTURE ON THE PROPERTIES OF POLYCRYS 09-1557
IN TRIGLYCINE SULFA/ INFLUENCE OF THE ELECTRIC FIELD AND THE DOMAIN STRUCTURE ON THE ULTRASONIC ABSORPTION 27-2955
(ROCHELLE SALT, TRIGLYCI/ PYROELECTRIC INVESTIGATIONS OF THE DOMAIN STRUCTURE STABILITY IN FERROELECTRICS. 02-0083
DEFECT STRUCTURE OF TRIGLYCINE SULFATE. (CORRELATION WITH DOMAIN STRUCTURES) 27-3044
FACTORS INFLUENCING) TYPE AND ORIENTATION OF THE EQUILIBRIUM DOMAIN STRUCTURES IN FERROELECTRICS. (03-0246
S. CONTRIBUTION TO THE THEORY OF DOMAIN STRUCTURES IN MAGNETS AND FERROELECTRIC 03-0430
CRYSTALS ASSOCIATED WITH STEP LADDERS ON PSEUDOCUBIC (001) / DOMAIN STRUCTURES IN POTASSIUM NIOBATE SINGLE 08-1316
CRYSTALS. INTERFEROMETRIC STUDIES OF DOMAIN STRUCTURES IN POTASSIUM NIOBATE SINGLE 08-1318
TION, IN A SCANNING ELECTRCN M/ OBSERVATION OF FERROELECTRIC DOMAIN STRUCTURES, MODULATED BY A PHASE TRANSI 03-0340
CRYSTALLCGRAPHIC VIEWPOINT) MICROSCOPIC THEORY OF STATIC DOMAIN STRUCTURES OF FERROELECTRIC CRYSTALS. (03-0340
E SULFATE AND SODIUM NITRITE SURFACES) (DECORATION METHOD OF DOMAIN STUDY) /TRATES. (TELLURIUM ON TRIGLYCIN 16-2394
FERROELECTRIC DOMAIN SWITCHING. 37-3642
EFFECT OF "A" DOMAINS UPON "C" DOMAIN SWITCHING IN BARIUM TITANATE. 04-0615
RAMICS. FERROELECTRIC DOMAIN SWITCHING IN RHOMBOHEDRAL PHASE PLZT CE 09-1563
DIELECTRIC PROPERTIES AND DOMAIN TEXTURE OF BARIUM TITANATE CRYSTALS. 04-0537
DIHYDROGEN PHOSPHATE. EXISTENCE OF DISLOCATIONS AT DOMAIN TIPS IN FERROELECTRIC CRYSTAL POTASSIUM 17-2450
CTRIC TRIGLYCINE SULFATE CRYSTALS. DOMAIN WALL CAUGHT IN DISLOCATIONS IN FERROELE 27-3091
STANT OF BARIUM TITANATE. (CALCULATIONS, SUMMARY ON/ 180 DEG DOMAIN WALL CONTRIBUTION TO THE DIELECTRIC CON 04-0804
CALCULATION OF (DOMAIN) WALL ENERGY IN BARIUM TITANATE. 04-0673
A/ SMALL SIGNAL PERMITTIVITY OF THE STATIONARY (100)-180 DEG DOMAIN WALL IN BARIUM TITANATE (THEORY, CALCUL 03-0358
CALCULATIONS) ANGULAR ENERGY DEPENDENCE OF 180 DEG DOMAIN WALL IN BARIUM TITANATE. (THERMODYNAMIC 04-0658
CS. TEMPERATURE DEPENDENCE OF THE SHAPE OF THE DOMAIN WALL IN FERROMAGNETICS AND FERROELECTRI 03-0191
CALCULATIONS OF WALL THICKNESS AND ENERGY DE/ FERROELECTRIC DOMAIN WALL IN TRIGLYCINE SULFATE. (THEORY AND 27-2997
RIUM TITANATE). STRUCTURE OF A MOVING DOMAIN WALL. (MECHANISM OF A WALL MOTION IN BA 04-0672
LEAD TITANATE ZIRCONATE TRANSDUCER CER/ DISPERSION STUDY OF DOMAIN WALL MOBILITY AND MECHANICAL DAMPING IN 09-1547
ERESIS LOOP OF FERROELECTRICS. SIDEWAYS DOMAIN WALL MOTION AND (MECHANISM OF THE) HYST 03-0301
DOMAIN WALL MOTION IN FERROELECTRIC SWITCHING. 03-0273
PART-1: GENERAL FORMULATION. KINETICS OF DOMAIN WALL MOTION IN FERROELECTRIC SWITCHING. 03-0274

CONTROL AND APPLICATION OF DOMAIN WALL MOTION IN GADOLINIUM MOLYBDATE. 30-3257
CONTROL AND APPLICATION OF DOMAIN WALL MOTION IN GADOLINIUM MOLYBDATE. 30-3256
SPHATE TYPE CRYSTALS.   THEORY OF DOMAIN WALL MOTION IN POTASSIUM DIHYDROGEN PHO 17-2652
SUPERSONIC DOMAIN WALL MOTION IN TRIGLYCINE SULFATE. 27-2963
ELLE SALT.   CONTRIBUTION OF DOMAIN WALL MOTION TO THE PERMITTIVITY OF ROCH 28-3198
X-RAY TOPOGRAPHIC STUDY OF DOMAIN WALL MOVEMENT IN TGS. 27-3026
Y, CALCULATIONS)   90 DEGREE TYPE DOMAIN WALL OF BARIUM TITANATE CRYSTAL. (THEOR 03-0324
IN FERROELECTRIC TRIGLYCINE SULFATE (SEVERAL KHZ TO 23 KMHZ, DOMAIN WALL RESONANCE). DIELECTRIC DISPERSION 27-3032
G SWITCHING (CALCULATIONS COMPARED WITH DA/ (STEP OR SMOOTH) DOMAIN WALL STRUCTURE IN BARIUM TITANATE DURIN 04-0677
BY RAMAN SCATTERING.   DOMAIN WALL STRUCTURE IN GADOLINIUM MOLYBDATE 30-3318
/RING TECHNIQUES APPLIED TO PROBLEMS IN SOLID STATE PHYSICS (DOMAIN WALL THICKNESS IN GADOLINIUM MOLYBDATE/ 30-3319
ECTRIC PHASE POTASSIUM DIHYDROGEN PHOSPHATE. (DIFFRACTION BY DOMAIN WALLS) /WITH TEMPERATURE IN THE FERROEL 17-2662
IC SODIUM NITRITE CRYSTALS.   PART-1: STRUCTURE OF 180 DEGREE DOMAIN WALLS. TOPOGRAPHIC STUDY ON FERROELECTR 16-2372
Y DISPERSIONS OF THE PERMITTIVITY OF ROCHELLE SALT CAUSED BY DOMAIN WALLS. (ABSTRACT ONLY)   LOW FREQUENC 28-3199
CRYSTALS OF TRIGLYCINE SULFATE.   MOVEMENT OF DOMAIN WALLS AND THE NUCLEATION OF DOMAINS IN 27-3071
N ROCHELLE SALT CRYSTAL.   DOMAIN WALLS CAUGHT IN (FINE LINE) "SUDARES" I 28-3215
ENERGY. APPLICATION TO LEAD/ RELAXATION OF ANTIFERROELECTRIC DOMAIN WALLS, EFFECTIVE FIELD, AND ACTIVATION 07-1257
STRUCTURE OF 90 DEGREE DOMAIN WALLS IN BARIUM TITANATE. 04-0641
OPTICAL STUDY OF DOMAIN WALLS IN BISMUTH TITANATE. 08-1301
CRYSTALS. (BARIUM TITANATE, GADOLINIUM MOLYBD/ THICKNESS OF DOMAIN WALLS IN FERROELECTRIC AND FERROELASTIC 04-0777
HE PEROVSKITE STRUCTURE.   THERMODYNAMIC THEORY OF DOMAIN WALLS IN FERROELECTRIC MATERIALS WITH T 03-0190
E SINGLE CRYSTALS.   MECHANICAL SWITCHING OF 60 DEGREE DOMAIN WALLS IN FERROELECTRIC POTASSIUM NIOBAT 08-1421
AND TABULATION)   PERMISSIBLE DOMAIN WALLS IN FERROELECTRIC SPECIES (THEORY 03-0247
OIODIDE)   INTERACTION OF ULTRASONIC WAVES WITH MOVING DOMAIN WALLS IN FERROELECTRICS. (ANTIMONY SULF 15-2192
OBSERVATION OF DIFFRACTED LIGHT FROM DOMAIN WALLS IN GADOLINIUM MOLYBDATE. 30-3322
CTRIC FIELDS.   DISPLACEMENTS AND VELOCITIES OF THE DOMAIN WALLS IN KDP. EXISTENCE OF CRITICAL ELE 17-2448
F.   ULTRASONIC STRESS WAVE INTERACTION WITH DOMAIN WALLS IN POTASSIUM DIDEUTERIUM PHOSPHAT 17-2659
TITANATE.   MICROSTRUCTURE OF DOMAINS AND DOMAIN WALLS IN SINGLE CRYSTAL FILMS OF BARIUM 04-1031
SPECIAL ELECTRICAL PROPERTIES OF REGIONS NEAR DOMAIN WALLS IN TRIGLYCINE SULFATE CRYSTALS. 27-2988
RE DEPENDENCES OF THE COMPLEX DIELECTRIC CONS/ RELAXATION OF DOMAIN WALLS IN TRIGLYCINE SULFATE. (TEMPERATU 27-3103
N KDP)   VELOCITY OF PROPAGATION OF DOMAIN WALLS: INTERACTIONS BETWEEN DOMAINS. (I 17-2453
LOCAL ELECTRON STATES IN DOMAIN WALLS OF FERROELECTRICS. 03-0161
HASE TRANSFORMATIONS IN FERROELECTRIC CRYSTAL IN PRESENCE OF DOMAIN WALLS. (THEORY)   P 03-0346
PROCESSES IN BARIUM TITANATE (DECREASE IN NUMBER OF 180 DEG DOMAINS).   AGING 04-0911
ELECTRON MIRROR MICROSCOPIC OBSERVATION OF FERROELECTRIC DOMAINS.   INTE 03-0473
RDOMAIN AND DOMAIN EFFECT INTERACTIONS IN KDP (FERROELECTRIC DOMAINS).   INTE 17-2451
(13) NIOBIUM(35) OXIDE(94) AND THE GEOMETRY OF FERROELECTRIC DOMAINS.   STRUCTURE OF THE BRONZE SODIUM 13-1931
MS OF BARIUM TITANATE.   MICROSTRUCTURE OF DOMAINS AND DOMAIN WALLS IN SINGLE CRYSTAL FIL 04-1031
IUM MOLYBDATE.   FERROELECTRIC DOMAINS AND THE CONDENSED SOFT MODE IN GADOLIN 30-3260
ELECTRON MIRROR MICROSCOPIC OBSERVATION OF FERROELECTRIC DOMAINS. (GOLD COATED BARIUM TITANATE) 04-0974
RT-1: DYNAMICAL THEORY. PART-2: OBSERVATIONS. (FERROELECTRIC DOMAINS IN BARIUM TITANATE) /IN BOUNDARIES. PA 03-0255
INTERACTION BETWEEN THE 90 DEG AND 180 DEG DOMAINS IN BARIUM TITANATE. 04-0585
450 KV/CM.   NUCLEATION AND GROWTH OF FERROELECTRIC DOMAINS IN BARIUM TITANATE AT FIELDS FROM 2 TO 04-0979
IZING MICROSCOPE.   DIRECT OBSERVATION OF 180 DEG DOMAINS IN BARIUM TITANATE BY MEANS OF A POLAR 04-0952
TIONS.   A NEW METHOD OF OPTICALLY READING DOMAINS IN BISMUTH TITANATE AND MEMORY APPLICA 08-1298
ALS.   ELECTRON BEAM WRITING OF DOMAINS IN BISMUTH TITANIUM OXIDE SINGLE CRYST 08-1300
OBSERVATIONS OF FERROELECTRIC DOMAINS IN BORACITES. 14-2103
MOVEMENT OF DOMAIN WALLS AND THE NUCLEATION OF DOMAINS IN CRYSTALS OF TRIGLYCINE SULFATE. 27-3071
EMISSION MICROSCOPE (TUNGSTEN TRIOXIDE, BARIUM TIT/ STUDY OF DOMAINS IN FERROELECTRIC CRYSTALS WITH A PHOTO 04-0811
CRYSTALS.   X-RAY TOPOGRAPHIC OBSERVATION OF 180 DEG DOMAINS IN FERROELECTRIC LEAD GERMANATE SINGLE 35-3515
OXIDE(11).   X-RAY TOPOGRAPHIC OBSERVATION OF 180 DEGREE DOMAINS IN FERROELECTRIC LEAD(5) GERMANIUM(3) 35-3516
GLE CRYSTALS.   DOMAINS IN FERROELECTRIC POTASSIUM NIOBATE SIN 08-1315
SECOND HARMONIC LIGHT SCATTERING BY DOMAINS IN FERROELECTRIC TRIGLYCINE SULFATE. 27-2991
STUDY OF FERROELECTRIC DOMAINS IN GADOLINIUM MOLYBDATE. 30-3304
X-RAY TOPOGRAPHIC STUDY OF DISLOCATIONS AND FERROELECTRIC DOMAINS IN GLYCOL SULFATE. 19-2747
ELOCITY OF PROPAGATION OF DOMAIN WALLS: INTERACTIONS BETWEEN DOMAINS. (IN KDP)   V 17-2453
RECIPROCAL DOMAINS IN LEAD TITANATE CRYSTALS. 05-1048
ENHANCED ORDERING OF FERROELECTRIC DOMAINS IN PLZT CERAMICS. 09-1579
ENHANCED ORDERING OF FERROELECTRIC DOMAINS IN PLZT CERAMICS. 09-1580
ROM PSEUDOCUBIC (001) CLEAVAGE PLANES.   DOMAINS IN POTASSIUM NIOBATE SINGLE CRYSTALS F 08-1317
OXIDE(11).   FERROELECTRIC OPTICAL ROTATION DOMAINS IN SINGLE CRYSTAL LEAD(5) GERMANIUM(3) 36-3550
DIDE.   OPPOSING FERROELECTRIC DOMAINS IN SINGLE CRYSTALS OF ANTIMONY SULFOIO 15-2252
RES.   FIELD INDUCED SWITCHING OF TETRAGONAL DOMAINS IN STRONTIUM TITANATE AT LOW TEMPERATU 06-1184
SCANNING ELECTRON MICROSCOPE STUDY OF FERROELECTRIC DOMAINS IN TGS. 27-3053
)   WALL MOTION AND NUCLEATION OF DOMAINS IN TGS CRYSTALS. (SUMMARY, WITH PHOTOS 27-3072
EVOLUTION OF FERROELECTRIC DOMAINS IN TGS SINGLE CRYSTALS. 27-3003
/NANCE AND ELECTRON NUCLEAR DOUBLE RESONANCE OBSERVATIONS OF DOMAINS IN THE IRRADIATED FERROELECTRICS POTA/ 18-2726
TRITE.   X-RAY ANALYSIS OF FERROELECTRIC DOMAINS IN THE PARAELECTRIC PHASE OF SODIUM NI 16-2271
D DEFECTS RELATED TO NONSTOICHIOMETRY (AND ANTIFERROELECTRIC DOMAINS IN TRANSITION METAL OXIDES). / EXTENDE 36-3522
ETCHING AND ULTRAMICROSCOPIC STUDIES OF FERROELECTRIC DOMAINS IN TRIGLYCINE SULFATE. 27-3143
NG ELECTRON MICROSCOPE.   DIRECT OBSERVATION OF FERROELECTRIC DOMAINS IN TRIGLYCINE SULFATE USING THE SCANNI 27-3051
SE TRANSITION / MICROSCOPIC OBSERVATION OF ANTIFERROELECTRIC DOMAINS OF AMMONIUM DIHYDROGEN PHOSPHATE. (PHA 17-2643
TIUM P/ ELECTRON MIRROR MICROSCOPIC ASPECTS OF FERROELECTRIC DOMAINS OF BARIUM TITANATE AND DICALCIUM STRON 04-0973
PHASES.   ELECTRON MICROSCOPIC STUDY OF FERROELECTRIC DOMAINS OF BARIUM TITANATE IN LOW TEMPERATURE 04-1034
ELECTRON MIRROR MACROSCOPIC OBSERVATION OF FERROELECTRIC DOMAINS OF DICALCIUM STRONTIUM PROPIONATE. 26-2946
G ELECTRON MICROSCOPE. (TUNGSTEN TRIOXIDE SINGLE C/ STUDY OF DOMAINS OF FERROELECTRIC CRYSTALS WITH SCANNIN 10-1646
ROR MICROSCOPY TO DIRECT OBSERVATION OF MOVING FERROELECTRIC DOMAINS OF GADOLINIUM MOLYBDATE. /ELECTRON MIR 30-3293
ATE, POTASSIU/ ELECTRON MICROSCOPIC STUDIES ON FERROELECTRIC DOMAINS OF PEROVSKITE TYPE OXIDES. (LEAD TITAN 04-0993
REFS)   FERROELECTRIC DOMAINS OF POTASSIUM DIHYDROGEN PHOSPHATE. (93 17-2447
INFLUENCE OF THE SHAPE OF DOMAINS ON THE FERROELECTRIC HYSTERESIS LOOP. 03-0300
TANATE.   EFFECT OF "A" DOMAINS UPON "C" DOMAIN SWITCHING IN BARIUM TI 04-0615
ZATION OF FERROELECTRIC PLATES NEAR THE CURIE POI/ THEORY OF DOMAINLESS, NONHOMOGENEOUS, SPONTANEOUS POLARI 03-0321
M TANTALATE. (CONDUCTI/ CALCIUM CONCENTRATION VS NET IONIZED DONOR CONCENTRATION IN SINGLE CRYSTAL POTASSIU 08-1437
DE. (STOICHIOMETRIC COMPENSATION)   OXYGEN STOICHIOMETRY OF DONOR DOPED BARIUM TITANATE AND TITANIUM DIOXI 04-0664
ITY OF DOPED BARIUM TITANATE. (OR BARIUM STRONTIUM TITANATE, DOPANT NOT NAMED)   PIEZORESISTIV 04-0729
IGIN AND EFFECTS OF GRAIN BOUNDARY BLOCKING LAYERS, ANTIMONY DOPANT, REVIEW) /OELECTRIC SEMICONDUCTORS. (OR 02-0074
(HOT PRESSED LEAD ZIRCONATE TITANATE, EFFECT OF GRAIN SIZE, DOPANTS, REVIEW). /TS IN ELECTROOPTIC CERAMICS 02-0086
ITANATE SINGLE CRYSTALS (HYDROGEN REDUCED, CERIUM OR NIOBIUM DOPED). /PECTRA) OF SEMICONDUCTING STRONTIUM T 06-1198
LAYERS IN THE SEMICONDUCTOR CERAMIC BARIUM TITANATE. (CERIUM DOPED) / OF CRYSTALLITES AND INTERCRYSTALLITE 04-0568
ELECTRON PARAMAGNETIC RESONANCE OF MANGANESE(2+) DOPED AMMONIUM CADMIUM SULFATE. 19-2791

(NMR STUDY OF) FERROELECTRIC TRANSITION IN (MANGANESE DOPED) AMMONIUM SULFATE.   19-2784
/FFECT (POSITIVE TEMPERATURE COEFFICIENT OF RESISTIVITY) IN (DOPED AND UNDOPED CERAMIC PEROVSKITE) LEAD FE/   13-2000
/L STUDY OF THE CONTRAST AT FERROELECTRIC DOMAIN BOUNDARIES (DOPED AND UNDOPED TGS, LITHIUM NIOBATE, KDP, /   11-1671
F EPR SPECTRA AND STATIC MAGNETIC SUSCEPTIBILITY OF CHROMIUM DOPED BARIUM METATITANATE. /NCY FROM STUDIES O   04-0542
HOLOGRAPHIC STORAGE IN DOPED BARIUM SODIUM NIOBATE.   13-1898
MICROCHEMICAL ANALYSIS OF GADOLINIUM DOPED BARIUM SODIUM NIOBATES.   13-1953
WITH PZT.   ATTEMPT TO PREPARE HIGH DOPED BARIUM STRONTIUM TITANATE SEMICONDUCTOR   04-0717
OXIDE IMPURITIES ON THE ELECTRICAL RESISTIVITY OF LANTHANUM DOPED BARIUM TITANATE.   EFFECT OF 04-0587
FECT OF SILICA ADDITIONS ON THE SEMICONDUCTING PROPERTIES OF DOPED BARIUM TITANATE.   EF 04-0660
NFLUENCE OF GRAIN GROWTH ON DIELECTRIC PROPERTIES OF NIOBIUM DOPED BARIUM TITANATE.   I 04-0763
S FROM MOSSBAUER ISOMER SHIFTS. (DEGREE CF COVALENCY IN IRON DOPED BARIUM TITANATE)   IONIC CHARACTER 04-0563
OXIDATION OF CARBON MONOXIDE ON LANTHANUM OXIDE DOPED BARIUM TITANATE.   04-0947
ION OF SMALL GRAINED AND LARGE GRAINED CERAMICS FROM NIOBIUM DOPED BARIUM TITANATE.   PREPARAT 04-0765
TIME DEPENDENCE OF THE DIELECTRIC PERMITTIVITY OF HYDROXYL DOPED BARIUM TITANATE.   04-0550
TOICHIOMETRIC COMPENSATION)   OXYGEN STOICHIOMETRY OF DONOR DOPED BARIUM TITANATE AND TITANIUM DIOXIDE. (S 04-0664
PIEZOELECTRIC EFFECT IN ALUMINUM DOPED BARIUM TITANATE CERAMIC.   04-0536
ECTIVE D/ GRAIN BOUNDARY BARRIER LAYERS IN (N TYPE, ANTIMONY DOPED) BARIUM TITANATE CERAMIC WITH A HIGH EFF 04-0581
ELECTRICAL PROPERTIES OF RUBIDIUM AND CESIUM DOPED BARIUM TITANATE CERAMICS.   04-0860
ACES) IN THE EMISSION ELECTRON MICROSCOPE (BARRIER LAYERS IN DOPED BARIUM TITANATE CERAMICS). /F SOLID SURF 04-0913
IDE AS A GRAIN BOUNDARY ELEMENT.   LANTHANUM DOPED BARIUM TITANATE CERAMICS WITH BISMUTH OX 04-0799
HOLOGRAPHIC STORAGE, ELECTRICAL FIXING AND ERASING IN DOPED BARIUM TITANATE CRYSTALS.   04-0848
TANATE, DOPANT NOT NAMED)   PIEZORESISTIVITY OF DOPED BARIUM TITANATE. (OR BARIUM STRONTIUM TI 04-0729
PARAMAGNETIC RESONANCE OF PLATINUM IONS IN DOPED BARIUM TITANATE SINGLE CRYSTAL.   04-0977
PRODUCTION AND STUDY OF PURE AND DOPED BARIUM TITANATE SINGLE CRYSTAL   04-0667
KERR EFFECT STUDIES OF ELECTRIC CONDUCTION IN IRON DOPED BARIUM TITANATE SINGLE CRYSTALS.   04-0703
PREPARATION AND PROPERTIES OF PURE AND DOPED BARIUM TITANATE SINGLE CRYSTALS.   04-0668
PROPERTIES. (DIELECTRIC CON/ SYNTHESIS OF NIOBIUM PENTOXIDE DOPED BARIUM TITANATE WITH IMPROVED ELECTRICAL 04-0840
ITE/ TWO-STEP ANOMALOUS INCREASE OF RESISTIVITY ON LANTHANUM DOPED BARIUM TITANATE- BISMUTH TRIOXIDE COMPOS 04-0800
CZOCHRALSKI SYNTHESIS AND PROPERTIES OF RARE EARTH DOPED BISMUTH GERMANATE.   35-3496
DIELECTRIC PROPERTIES OF BARIUM TITANATE SINGLE CRYSTALS DOPED BY HYDROXIDE GROUPS.   04-0634
OR PROPERTIES OF BARIUM (TITANATE, STANNATE) SOLID SOLUTIONS DOPED BY TUNGSTEN TRIOXIDE.   SEMICONDUCT 04-0965
/ IN NICKEL CHLORINE AND NICKEL IODINE BORACITES (AND NICKEL DOPED CADMIUM CHLORINE AND CADMIUM IODINE BOR/ 14-2074
/ ELECTRON PARAMAGNETIC RESONANCE INVESTIGATIONS OF (COPPER) DOPED FERROELECTRIC CRYSTALS. (TRIGLYCINE SULF 27-3132
PYROELECTRIC PROPERTIES OF THE LANTHANUM DOPED FERROELECTRIC PLZT CERAMICS.   09-1558
ECT OF APPLIED ELECTRIC FIELD ON EPR SPECTRA OF (COPPER(2+)) DOPED FERROELECTRIC ROCHELLE SALT.   EFF 28-3185
GROWTH, LASING AND SPECTROSCOPIC PROPERTIES OF NEODYMIUM(3+) DOPED GADOLINIUM MOLYBDATE CRYSTALS.   CERTAIN 30-3255
CTRIC DISPERSION AND PHASE TRANSITION OF POTASSIUM HYDROXIDE DOPED ICE.   DIELE 34-3452
ELECTRIC POLARIZATION EFFECTS IN PURE AND DOPED ICE AT LOW TEMPERATURES.   34-3412
SPECIFIC HEAT OF PURE AND DOPED ICE NEAR 120 DEGREES K.   34-3463
STUDY OF CONDUCTIVITY AND DIPOLAR RELAXATION IN DOPED ICE SINGLE CRYSTALS.   34-3458
IONS IN THE ELECTRON PARAMAGNETIC RESONANCE OF MANGANESE(2+) DOPED IN FERROELECTRIC SODIUM NITRITE. /RANSIT 16-2383
CONDUCTIVITY OF BARIUM TITANATE, PURE SINGLE CRYSTALS AND DOPED (IRON, HYDROXYL) SINGLE CRYSTALS.   04-0695
DISTRIBUTION OF VACANCIES IN LANTHANA DOPED LEAD TITANATE.   05-1058
POSITIVE TEMPERATURE COEFFICIENT OF RESISTANCE IN DOPED LEAD TITANATE.   05-1051
STRUCTURE AND ELECTROMECHANICAL PROPERTIES OF LANTHANUM DOPED LEAD TITANATE ZIRCONATE CERAMICS.   09-1502
STRUCTURE AND ELECTROMECHANICAL PROPERTIES OF LANTHANUM DOPED LEAD TITANATE ZIRCONATE CERAMICS.   09-1531
ERROELECTRICITY AND ANTIFERROELECTRICITY IN PURE AND NIOBIUM DOPED LEAD ZIRCONATE.   F 07-1251
ELECTRIC AND ANTIFERROELECTRIC PHASES. (LANTHANUM OR NIOBIUM DOPED LEAD ZIRCONATE STANNATE TITANATE) /FERRO 09-1487
CRYSTAL FIELD IN CHROMIUM(3+) DOPED LITHIUM NIOBATE.   11-1759
ELECTRON PARAMAGNETIC RESONANCE OF IRON(3+) DOPED LITHIUM NIOBATE.   11-1765
ELECTRON PARAMAGNETIC RESONANCE OF GADOLINIUM(3+) DOPED LITHIUM NIOBATE.   11-1833
SONANCE STUDIES OF CRYSTAL FIELD PARAMETERS IN MANGANESE(2+) DOPED LITHIUM NIOBATE.   ELECTRON SPIN RE 11-1803
EPR STUDIES OF CRYSTAL FIELD PARAMETERS IN IRON(3+) DOPED LITHIUM NIOBATE.   11-1839
FORBIDDEN HYPERFINE LINES IN ESR SPECTRA OF MANGANESE(2+) DOPED LITHIUM NIOBATE.   11-1805
TICAL AND HOLOGRAPHIC STORAGE PROPERTIES OF TRANSITION METAL DOPED LITHIUM NIOBATE.   OP 11-1796
WAVELENGTH DEPENDENCE OF HOLOGRAPHIC PHENOMENA IN DOPED LITHIUM NIOBATE.   11-1785
ZINC/ ELECTRON PARAMAGNETIC RESONANCE SPECTRUM IN (MANGANESE DOPED) LITHIUM NIOBATE AND (COPPER(2+) DOPED) 11-1792
IAL FOR REVERSIBLE HOLOGRAPHIC STORAGE.   RH DOPED LITHIUM NIOBATE AS AN IMPROVED NEW MATER 08-1359
MAGNETIC EXCITATIONS IN NICKEL DOPED POTASSIUM MANGANESE TRIFLUORIDE.   33-3407
CENT MANGANESE RE/ ANTIFERROELECTRICITY IN (PURE AND) COBALT DOPED POTASSIUM MANGANESE TRIFLUORIDE. (20 PER 33-3366
/CROSTRUCTURE AND ELECTRICAL PROPERTIES OF SCANDIUM TRIOXIDE DOPED, RARE EARTH OXIDE DOPED, AND UNDOPED BA/ 04-0839
EPR INVESTIGATION OF THE INTERNAL ELECTRIC FIELD IN COPPER DOPED ROCHELLE SALT.   28-3233
/TRATION DEPENDENCE OF RAMAN SCATTERING BY MAGNONS IN NICKEL DOPED RUBIDIUM MANGANATE AND POTASSIUM MANGAN/ 33-3395
ESR OF COPPER(2+) DOPED SODIUM AMMONIUM TARTRATE.   29-3252
/AT, THERMAL DIFFUSIVITY AND THERMAL CONDUCTIVITY OF CALCIUM DOPED SODIUM NITRATE CRYSTALS NEAR THE TRANSI/ 16-2283
S AND TEMPERATURE DEPENDENCE OF FLUORESCENCE IN CHROMIUM(3+) DOPED STRONTIUM TITANATE.   ANALYSIS OF STRES 06-1215
FECT OF DOMAIN STRUCTURE ON THE FLUORESCENCE OF CHROMIUM(3+) DOPED STRONTIUM TITANATE.   EF 06-1114
ELECTRIC PROPERTIES OF HYDROGEN DOPED STRONTIUM TITANATE.   06-1088
F STRONG FIELDS) DURING HOLOGRAM RECORDING IN CRYSTALS (IRON DOPED STRONTIUM TITANATE). /ERNAL GENERATION O 06-1090
FREE CARRIER OPTICAL ABSORPTION IN NIOBIUM DOPED STRONTIUM TITANATE.   06-1214
PRETATION OF ELECTRIC FIELD EFFECTS IN EPR OF GADOLINIUM(3+) DOPED STRONTIUM TITANATE.   INTER 06-1168
PHOTOCHROISM IN TRANSITION METAL DOPED STRONTIUM TITANATE.   06-1126
POTASSIUM/ ELECTRON TUNNELING AND BAND STRUCTURE OF (NIOBIUM DOPED) STRONTIUM TITANATE AND (CALCIUM DOPED) 06-1222
EFFECTS OF SILICA ON MANGANESE DOPED STRONTIUM TITANATE CERAMICS.   06-1172
CHARGE COMPENSATION BY OXYGEN(2-) VACANCIES IN CHROMIUM(3+) DOPED STRONTIUM TITANATE (ESR MEASUREMENTS, M/ 06-1161
ELASTIC PROPERTIES OF YTTRIUM DOPED STRONTIUM TITANATE SINGLE CRYSTALS.   06-1180
ESR OF CHROMIUM(3+) IN CHROMIUM DOPED STRONTIUM TITANATE SINGLE CRYSTALS.   06-1147
AND ABOVE THE CURIE POINT.   DOMAIN STRUCTURE OF PURE AND DOPED TRIGLYCINE SULFATE CRYSTALS GROWN BELOW 27-3130
ICATIONS.   DOPED TRIGLYCINE SULFATE FOR PYROELECTRIC APPL 27-3057
CHANGES OF THE DOMAIN STRUCTURE IN VANADIUM DOPED TRIGLYCINE SULFATE OBSERVED BY EPR.   27-3013
EPR STUDIES OF CHROMIUM- AND VANADYL- DOPED TRIGLYCINE SULFATE SINGLE CRYSTALS.   27-3161
C PROPERTIES OF A CERAMIC (LEAD STRONTIUM TITANATE ZIRCONATE DOPED WITH CHROMIUM OXIDE). / AND PIEZOELECTRI 09-1591
ODE AT THE TRANSITION TEMPERATURE IN KDF AND DEUTERATED KDP (DOPED WITH COBALT-57 AND IRON-57). /REQUENCY M 17-2465
LD EFFECT OF ELECTRON SPIN RESONANCE IN STRONTIUM TITANATE. (DOPED WITH GADOLINIUM(3+) OR IRON(3+)) /IC FIE 06-1229
X GROWTH OF MAGNESIUM OXIDE AND LANTHANUM ALUMINATE CRYSTALS DOPED WITH ISOTOPE 170.   FLU 36-3558
APPLICATIONS.   A SEMICONDUCTING BARIUM TITANATE CERAMIC DOPED WITH NIOBIUM AND MANGANESE IONS AND ITS 04-0734
SBAUER STUDIES OF LEAD ZIRCONATE TITANATE STANNATE. (CERAMIC DOPED WITH NIOBIUM AND TITANIUM)   MOS 09-1604
SWITCHING IN FERROELECTRIC TRIGLYCINE SULFATE, PURE AND DOPED WITH PARAMAGNETIC IONS.   27-3020
DIELECTRIC PROPERTIES OF BARIUM TITANATE DOPED WITH TUNGSTEN OXIDE.   04-0696
EITY (THEORY AND TEST ON BARIUM SODIUM NIOBATE AND NEODYMIUM DOPED YAG LASER). /RATION ON CRYSTAL INHOMOGEN 13-2043

Y AT FERROELECTRIC-1, FERROELECTRIC-2 TRANSITION IN (NIOBIUM DOPED) ZIRCONATE(0.95) TITANATE(0.05). /ELOCIT 09-1511
STRUCTURE IN SEIGNETTE SALT. (EFFECT OF AGING AND COPPER(2+) DOPING)   EFFECT OF A DC FIELD ON DOMAIN 28-3230
ITY EFFECTS IN SODIUM TUNGSTEN BRONZES: NIOBIUM AND TANTALUM DOPING.   IMPUR 13-2053
LINE WIDTH OF POTASSIUM MANGANESE TRIFLUORIDE WITH MAGNESIUM DOPING.   SHIFT OF NEEL TEMPERATURE AND EPR 33-3377
                                            EFFECT OF DOPING BARIUM TITANATE WITH MANGANESE DIOXIDE.   04-0562
MICONDUCTORS. (ORIGIN AND EFFECTS OF GRAIN BOUNDARY BLOCKIN/ DOPING PROBLEMS IN (TITANATE) FERROELECTRIC SE 02-0074
/ION DAMAGED TRIGLYCINE SULFATE AND ROCHELLE SALT UP TO HIGH DOSAGE (DESTRUCTION OR MASKING OF FERROELECTR/ 27-3098
E DEPENDENCE).   DIELECTRIC PROPERTIES OF A (SMALL DOSE) GAMMA-IRRADIATED KDP CRYSTAL (TEMPERATUR 17-2466
CTRIC LOSSES OF TRIGLYCINE SULFATE CRYSTALS EXPOSED TO SMALL DOSES OF X- AND GAMMA-RAYS.   DIELE 27-2976
.   NEW FERROELECTRIC: THE DOUBLE AMMONIUM THALLIUM MONOARSENATE OR ATLAS 18-2735
TIONS IN POTASSIUM NITRATE NITRITE MIXED CRYSTALS (GROWTH BY DOUBLE BRIDGMAN TECHNIQUE).   PHASE TRANSI 16-2345
DIDEUTERIUM PHOSPHATE CRYSTALS.   DOUBLE DIELECTRIC HYSTERESIS LOOP OF POTASSIUM 17-2503
/BSORPTION BY ANTIFERRODIELECTRICS IN THE FREQUENCY RANGE OF DOUBLE ELECTRONIC EXCITATIONS OF MOLECULES (I/ 03-0419
TIES CF POLARIZATION OF BARIUM TITANATE SINGLE CRYSTALS WITH DOUBLE HYSTERESIS LOOP.   PECULIARI 04-0709
         FERROELECTRIC FREE ENERGY EXPANSION COEFFICIENTS FROM DOUBLE HYSTERESIS LOOPS.   03-0262
TRANSITIONS IN FERROELECTRIC CRYSTALS.   DOUBLE HYSTERESIS LOOPS AND FIRST ORDER PHASE 03-0456
NT OF SODIUM TRIHYDROGEN SELENITE.   DOUBLE HYSTERESIS LOOPS AT THE UPPER CURIE POI 22-2831
                                                ANALYSIS OF DOUBLE HYSTERESIS LOOPS. (ROCHELLE SALT)   28-3212
                                                             DOUBLE INJECTION IN BARIUM TITANATE CERAMIC. 04-0554
        COMPUTER EXPERIMENT ON LINEAR CHAIN OF ATOMS LYING IN DOUBLE MINIMUM POTENTIAL. (BARIUM TITANATE) 03-0287
/ BRILLOUIN SCATTERING STUDY OF ANTIMONY SULFOIODIDE USING A DOUBLE PASSED, STABILIZED SCANNING INTERFEROM/ 15-2225
E IRRA/ ELECTRON PARAMAGNETIC RESONANCE AND ELECTRON NUCLEAR DOUBLE RESONANCE OBSERVATIONS OF DOMAINS IN TH 18-2726
RRADIA/ ELECTRON PARAMAGNETIC RESONANCE AND ELECTRON NUCLEAR DOUBLE RESONANCE OF THE ARSENATE CENTER IN X-I 18-2727
/ION BY ELECTRON PARAMAGNETIC RESONANCE AND ELECTRON NUCLEAR DOUBLE RESONANCE OF THE PROTON LATTICE COUPLE/ 18-2724
OELECTRIC POTASSIUM DIHYDROGEN PHOSPHATE.   NUCLEAR DOUBLE RESONANCE STUDY OF POTASSIUM-39 IN FERR 17-2599
ANSITION IN TRIGLYCINE SULFATE.   PULSED NITROGEN PROTON DOUBLE RESONANCE STUDY OF THE FERROELECTRIC TR 27-2968
LE INTERACTIONS OF NITROGEN-14 IN PARAELECTRIC TRIGL/ PULSED DOUBLE RESONANCE STUDY OF THE NUCLEAR QUADRUPO 27-2969
CRYSTALS OF ICE.   ENDOR (ELECTRON NUCLEAR DOUBLE RESONANCE) STUDY OF X-IRRADIATED SINGLE 34-3413
G BARIUM TITANATE ABOVE CURIE POINT)   DOUBLE SPACE CHARGE INJECTION IN SOLIDS. (USIN 04-0638
PHOSPHATE. ELECTROOPTIC Q SWITCH BY DESK-T/ SIMULATION OF A DOUBLE 45 DEGREES, Z CUT POTASSIUM DIDEUTERIU 17-2702
                                                             DOUBLETS CONCERNING THE STRUCTURE OF BORACITES 14-2088
BARIUM SODIUM NIOBATE CRYSTALS DURING INTRACAVITY FREQUENCY DOUBLING. /STROPHIC SURFACE DAMAGE PRODUCED IN 13-2052
, CADMIUM CHLORINE, CADMIUM IODINE, IRO/ (OPTICAL) FREQUENCY DOUBLING IN (MAGNESIUM CHLORINE, ZINC CHLORINE 14-2080
TALS.   ELECTROCOLORATION IN STRONTIUM TITANATE: VACANCY DRIFT AND OXIDATION REDUCTION OF TRANSITION ME 06-1098
DE SINGLE CRYSTALS.   TEMPERATURE DEPENDENCES OF THE HALL AND DRIFT MOBILITIES OF HOLES IN ANTIMONY TRISULFI 15-2193
BARIUM TITANATE- STRONTIUM TITANATE SUBSTRATES AT SUPERSONIC DRIFT VELOCITY OF CURRENT CARRIERS. /AYERS ON 04-0592
PHASE TRANSFORMATIONS AND MELTING IN KDP TO 40 KBAR (BY DTA).   17-2634
DP TYPE CRYSTALS (CURIE TEMPERATURES OF POTASSIUM, RUBIDIUM/ DTA STUDY OF THE FERROELECTRIC TRANSITION IN K 27-2577
ON AND THE BINARY STRONTIUM OXIDE- NIOBIUM PENTOXIDE SYSTEM, DTA, X-RAY CURIE TEMPERATURE). /BIUM-RICH REGI 13-1920
ANE OF ROCHELLE SALT IN THE PRESENCE OF PURE DIRECT SKY BLUE DYE.   GROWTH RATE OF THE (010) PL 28-3217
T SECOND HARMONIC GENERATION IN ADP WITH TWO NEW FLUORESCEIN DYE LASERS.   EFFICIEN 17-2538
ELECTROOPTIC CERAMICS AS WAVELENGTH SELECTION DEVICES IN DYE LASERS (PLZT).   09-1618
SIUM IODATE CRYSTALS.   POLAR AND LATTICE DYNAMIC CHARACTERISTICS OF FERROELECTRIC POTAS 08-1430
CTRIC) MATERIAL CONSTANTS OF TRIGLYCINE SULFATE.   DYNAMIC DETERMINATION OF (ELASTIC AND PIEZOELE 27-2984
ITANATE.   DYNAMIC FATIGUE OF SINGLE CRYSTALS OF BARIUM T 04-0792
M OF MOLYBDENUM(5+) IN STRONTIUM TITANATE: AN EXAMPLE OF THE DYNAMIC JAHN TELLER EFFECT. /RESONANCE SPECTRU 06-1125
           MEASUREMENT OF THE SPECIFIC HEAT OF KDP BY A DYNAMIC METHOD.   17-2619
FIDE, ANT/ GROWING SINGLE CRYSTALS FROM THE VAPOR PHASE BY A DYNAMIC METHOD (ANTIMONY SESQUISELENIDE OR SUL 15-2162
FERROELECTRICS.   A MORE EXACT DYNAMIC METHOD OF DETERMINING POLARIZATION IN 03-0381
ATE CRYSTALS FOR STUDY OF THE PY/ USE OF SPECIFIC HEAT (BY A DYNAMIC METHOD) OF POTASSIUM DIHYDROGEN PHOSPH 17-2566
                                                             DYNAMIC MODEL OF THE SLATER KDP FERROELECTRIC. 03-0285
CRYSTALS.   MICROWAVE DYNAMIC NONLINEARITY OF BARIUM TITANATE SINGLE 04-0749
CERAMIC SHELLS.   DYNAMIC PIEZOELECTRIC THEORY FOR FERROELECTRIC 03-0225
).   DYNAMIC POLAR MODEL OF MONOCRYSTALLINE ICE I(H 34-3424
S.   DYNAMIC SHELL THEORY FOR FERROELECTRIC CERAMIC 03-0224
                                            THEORETICAL DYNAMIC STUDY OF CUBIC PHASE BARIUM TITANATE.   03-0261
OSCILLATOR. A UNIFIED OSCILLATOR MODEL FOR ORDER DISORDER / DYNAMIC SUSCEPTIBILITY OF CLASSICAL ANHARMONIC 03-0405
TION OF THE DIMENSIONS OF KANZIG REGIONS ON THE BASIS OF THE DYNAMIC THEORY OF CRYSTAL LATTICES.   ESTIMA 03-0446
ANHARMONIC MODE MIXING IN IONIC FERROELECTRICS. (LATTICE DYNAMICS)   03-0234
ECTRA OF STRONTIUM TITANATE AND POTASSIUM TANTALATE (LATTICE DYNAMICS). /ECTRIC- ACOUSTIC) IN THE PHONON SP 06-1157
ELECTRICITY (IN BARIUM TITANATE TYPE STRUCTURES) AND LATTICE DYNAMICS.   FERRO 04-0950
TRON SCATTERING FROM SINGLE DOMAIN BARIUM TITANATE. (LATTICE DYNAMICS)   INELASTIC NEU 04-0946
TRONTIUM TITANATE.   LATTICE DYNAMICS ABOVE STRUCTURAL PHASE TRANSITIONS: S 06-1124
                         ICE I: LATTICE DYNAMICS AND INCOHERENT NEUTRON SCATTERING.   34-3465
INFLUENCE OF POINT DEFECTS ON DIELECTRIC PROPERTIES, DOMAIN DYNAMICS, AND PHASE TRANSITIONS OF FERROELECT/ 03-0495
N OF SODIUM NITRITE (STUDIED BY COHERENT NEUTRON IN/ CRYSTAL DYNAMICS AND THE FERROELECTRIC PHASE TRANSITIO 16-2364
ELECTRIC CRITICAL TEMPERATURE (ROTATIONS OF BO6 OCTAHEDRA I/ DYNAMICS AT STRUCTURAL TRANSITIONS ABOVE FERRO 06-1205
(REVIEW)   LATTICE DYNAMICS BY NEUTRON SCATTERING. PARTS 1 AND 2. 02-0051
                               PROTON DYNAMICS IN FERROELECTRIC COLEMANITE.   23-2887
HEORY OF FERROELECTRICITY, PART-1: SPECTRAL ANALYSIS (PROTON DYNAMICS IN HYDROGEN BONDED FERROELECTRICS). / 03-0402
VE THE CURIE TEMPERATURE.   PROTON DYNAMICS IN HYDROGEN BONDED FERROELECTRICS ABO 03-0206
RT-2: APPROXIMATED SOLUTIONS TO THE KINETIC EQUATION/ PROTON DYNAMICS IN HYDROGEN BONDED FERROELECTRICS. PA 03-0205
INFRARED ABSORPTION.   PROTON DYNAMICS IN KDP TYPE FERROELECTRICS STUDIED BY 17-2708
          EFFECT OF DEUTERON SUBSTITUTION ON DOMAIN DYNAMICS IN POTASSIUM DIHYDROGEN PHOSPHATE. 17-2437
HOOD OF THE PHASE TRANSFORMATION.   EPR STUDIES AND DYNAMICS IN STRONTIUM TITANATE IN THE NEIGHBOR 06-1232
ROELECTRIC PHASE.   LATTICE DYNAMICS MODEL OF POTASSIUM NITRATE IN ITS FER 16-2377
LYEDATE.   LATTICE DYNAMICS OF A RIGID ION MODEL OF GADOLINIUM MO 30-3281
RIUM TITANATE DOWN TO 3 MIC/ DIELECTRIC CONSTANT AND LATTICE DYNAMICS OF A THIN FILM OF A FERROELECTRIC (BA 04-0600
                                                             DYNAMICS OF AMMONIUM IONS IN AMMONIUM SULFATE. 19-2794
ON THE INTERPRETATION OF INFRARED SPECTRA AND THE LATTICE DYNAMICS OF CUBIC BARIUM TITANATE.   04-0659
RTICULAR STRONTIUM TITANATE.   LATTICE DYNAMICS OF CUBIC PEROVSKITE STRUCTURES, IN PA 06-1185
IC CONSTANT. (POTASSIUM BROMIDE, LEAD TELLURIDE, TI/ LATTICE DYNAMICS OF DIATOMIC CRYSTALS OF HIGH DIELECTR 36-3541
TRI/ NUCLEAR QUADRUPOLE SPIN LATTICE RELAXATION AND CRITICAL DYNAMICS OF FERROELECTRIC CRYSTALS. (SODIUM NI 16-2268
                                                             DYNAMICS OF FERROELECTRIC ROCHELLE SALT.   28-3237
                                                             DYNAMICS OF FERROELECTRIC ROCHELLE SALT.   28-3236
                                              LATTICE DYNAMICS OF ICE.   34-3467
                                              LATTICE DYNAMICS OF ICE I(H).   34-3468
C CONSTANTS. (LEAD AND TIN TELLURIDES)   LATTICE DYNAMICS OF IONIC CRYSTALS WITH HIGH DIELECTRI 36-3553
AND ANTIFERROELECTRICS   DYNAMICS OF ORDER DISORDER TYPE FERROELECTRICS 03-0179
PE ANTIFERROELECTRICS.   LATTICE DYNAMICS OF PHASE TRANSITIONS IN PEROVSKITE TY 08-1475

M TITANATE MATERIALS.   STUDY OF THE DYNAMICS OF PHOTOCHROMIC SWITCHING IN STRONTIU 06-1092
D SOLUTIONS OF THE PEROVSKITE TYPE WIT/ INVESTIGATION OF THE DYNAMICS OF POLARIZATION OF FERROELECTRIC SOLI 08-1380
CTRIC CERAMICS.   STATISTICAL METHOD FOR INVESTIGATING (DYNAMICS OF) POLARIZATION REVERSAL IN FERROELE 03-0493
E FERROELECTRIC PHASE TRANSITIONS.   DYNAMICS OF POTASSIUM DIHYDROGEN PHOSPHATE TYP 17-2692
ECTRIC CRYSTALS WITH HYDROGEN BONDING. (POTASSIUM DIHYDROGE/ DYNAMICS OF PROTONS (AND DEUTERONS) IN FERROEL 17-2568
G)   CRYSTAL DYNAMICS OF SODIUM NITRITE. (NEUTRON SCATTERIN 16-2284
                                                 LATTICE DYNAMICS OF STRONTIUM TITANATE.   06-1224
CAL TEMPERATURE.   CRITICAL DYNAMICS OF STRONTIUM TITANATE ABOVE THE CRITI 06-1204
S.   NEUTRON INELASTIC SCATTERING STUDY OF THE LATTICE DYNAMICS OF STRONTIUM TITANATE: HARMONIC MODEL 06-1225
TERMS OF NUCLEATION. (CONCEPTS)   INTERPRETATION OF THE DYNAMICS OF THE BARIUM TITANATE TRANSITION IN 04-0606
IN FERROELECTRICS. (THEORY)   CRITICAL DYNAMICS OF THE ORDER DISORDER TRANSFORMATION 03-0281
ROVSKITE STRUCTURE. (RELATION OF FERROELECTRICITY TO LATTICE DYNAMICS, REVIEW, 68 REFS) /OF MATERIALS OF PE 02-0043
            RECENT PROBLEMS IN DOMAIN STATICS AND DYNAMICS. (56 REFS)   03-0248
. (CALCULATIONS)   NOTE ON THE LATTICE DYNAMICAL ASPECT OF FERROELECTRIC MODES OF KDP 17-2498
CTRICS NEAR THE TRANSITION POINT. (BARIUM TITANATE, STRONTI/ DYNAMICAL BEHAVIOR OF DISPLACIVE TYPE FERROELE 04-0992
P TYPE CRYSTALS.   DYNAMICAL BEHAVIOR OF FERROELECTRIC SPINS. (KD 17-2600
ES OF FERROELECTRICS. (THEORY)   DYNAMICAL (DIELECTRIC) AND SCATTERING PROPERTI 03-0208
RMATIONS. (FERROELECTRICS)   DYNAMICAL EFFECTS IN SOLID STATE PHASE TRANSFO 03-0154
I(H). TENTATIVE INTERPRETATION WITH A MIXED COULOMB VALENCE DYNAMICAL MODEL. /QUENCY RAMAN SPECTRUM OF ICE 34-3425
-1. PART-2.   QUANTUM FIELD THEORETICAL TREATMENT OF A DYNAMICAL MODEL OF THE FERROELECTRIC KDP. PART 03-0279
ROCHELLE SALT, SILVER PERIODATE, / FERROELECTRICS. (THEORY, DYNAMICAL ORDER DISORDER PROPERTIES, KDP, TGS, 02-0050
NESE TRIFLUORIDE.   NEUTRON SCATTERING STUDY OF THE LATTICE DYNAMICAL PHASE TRANSITIONS IN POTASSIUM MANGA 33-3405
. (THEORY)   NUCLEAR QUADRUPOLE SPIN LATTICE RELAXATION AND DYNAMICAL PROPERTIES OF FERROELECTRIC CRYSTALS 03-0187
/LECTRICS. PART-1: (FORMALISM OF LOCAL MODE MODEL.) PART-2: (DYNAMICAL PROPERTIES, TRANSVERSE ISING MODEL,/ 03-0384
            POTASSIUM TANTALATE NIOBATE, STRUCTURAL AND DYNAMICAL STUDIES.   13-2068
EN PHOSPHATE TYPE FERRO/ PROTON LATTICE INTERACTIONS AND THE DYNAMICAL SUSCEPTIBILITY OF POTASSIUM DIHYDROG 17-2570
    PHASE TRANSITION IN KDP (MODIFICATION OF KOBAYASHI'S DYNAMICAL THEORY).   17-2488
ATE TYPE ANTIFERROELECTRICS WITH HYDROGEN BONDS.   DYNAMICAL THEORY OF AMMONIUM DIHYDROGEN PHOSPH 03-0476
            PHENOMENOLOGICAL LATTICE DYNAMICAL THEORY OF FERROELASTICITY.   03-0144
RICS. PART-1: DAMPINGS AND (FREQUENCY) SHIFTS (OF PROTONIC / DYNAMICAL THEORY OF HYDROGEN BONDED FERROELECT 03-0475
OELECTRICS.   DYNAMICAL THEORY OF RIGID LATTICES OF ANTIFERR 03-0344
NS IN PEROVSKITE TYPE CRYSTALS. (ST/ SELF CONSISTENT LATTICE DYNAMICAL THEORY OF STRUCTURAL PHASE TRANSITIO 03-0424
TE SINGLE CRYSTALS.   LATTICE DYNAMICAL THEORY OF SWITCHING IN BARIUM TITANA 03-0398
/ TRANSMISSION IMAGES OF COHERENT DOMAIN BOUNDARIES. PART-1: DYNAMICAL THEORY. PART-2: OBSERVATIONS. (FERR/ 03-0255
IUM MOLYBDATE AND ISOSTRUCTURAL COMPOUNDS (EUROPIUM TERBIUM, DYSPROSIUM AND HOLMIUM MOLYBDATES). /F GADOLIN 30-3271
/LASTIC CRYSTALS. (GADOLINIUM MOLYBDATE, KDP, ORTHOFERRITES, DYSPROSIUM ANTIMONIDE, RUBIDIUM IRON TRIFLUOR/ 03-0036
/ TITANIUM, SODIUM, MAGNESIUM, CADMIUM, YTTRIUM, GADOLINIUM, DYSPROSIUM, IRON, SELENIUM, NIOBIUM, COBALT, / 10-1640
ONSTANT IN THE SYSTEM POTASSIUM YTTRIUM MOLYBDATE- POTASSIUM DYSPROSIUM MOLYBDATE. /ENDENCE OF DIELECTRIC C 30-3309

E

       RAMAN SCATTERING LINE SHAPE OF THE SOFT E POLARITON MODE IN BARIUM TITANATE.   04-0803
                                          EARLY HISTORY OF FERROELECTRICITY (1920-1943). 02-0126
RIC ORDERING)   PREPARATION AND PROPERTIES OF RARE EARTH COMPOUNDS. (WITH MAGNETIC AND FERROELECT 02-0131
        CZOCHRALSKI SYNTHESIS AND PROPERTIES OF RARE EARTH DOPED BISMUTH GERMANATE.   35-3496
            ELECTRONIC HALL MOBILITY IN THE ALKALINE EARTH FLUORIDES.   33-3404
    LOW FREQUENCY DIELECTRIC CONSTANTS OF THE ALKALINE EARTH FLUORIDES BY THE METHOD OF SUBSTITUTION. 33-3361
        ELECTRON PARAMAGNETIC RESONANCE OF SOME RARE EARTH IMPURITIES IN STRONTIUM TITANATE.   06-1192
IN (CERAMIC) STRONTIUM TITANATES CONTAINING (TRIVALENT) RARE EARTH IONS.   DIELECTRIC RELAXATION 06-1143
/ATION (BY FLUX GROWTH) OF HEXAGONAL SINGLE CRYSTALS OF RARE EARTH MANGANATES (PHASE DIAGRAMS, CONVECTION / 36-3536
    PHASE TRANSITIONS OF ANHYDROUS MODIFICATIONS OF RARE EARTH MOLYBDATES.   30-3272
    PRECISION PARAMETERS OF SOME LANTHANON MOLYBDATE TYPE RARE EARTH MOLYBDATES. (CRYSTAL STRUCTURE) 30-3262
        PRECISION PARAMETERS OF THE FERROELECTRIC RARE EARTH MOLYBDATES: LANTHANUM MOLYBDATE. 30-3264
GROWTH, PHASE TRANSITIONS, STRUCTURE, 49/ FERROELECTRIC RARE EARTH MOLYBDATES. (REVIEW, FORMATION, CRYSTAL 02-0064
    MECHANISM OF THE FERROELECTRIC PHASE TRANSFORMATION IN RARE EARTH MOLYBDATES. (TERBIUM MOLYBDATE) 30-3254
TRUCTURE.   SODIUM BARIUM RARE EARTH NIOBATES WITH THE TUNGSTEN BRONZE TYPE S 13-1972
    ELECTRICAL TRANSPORT IN (SEMICONDUCTING AND MAGNETIC) RARE EARTH ORTHOCHROMITES, MANGANITES AND FERRITES 36-3608
            MAGNETIC PROPERTIES OF HEAVY RARE EARTH ORTHOMANGANATES.   36-3545
/ AND ELECTRICAL PROPERTIES OF SCANDIUM TRIOXIDE DOPED, RARE EARTH OXIDE DOPED, AND UNDOPED BARIUM TITANAT/ 04-0839
(THEORY)   MACROSCOPIC ANALOG OF THE SPIN- ECHO EFFECT IN POLYCRYSTALLINE FERROELECTRICS. 03-0310
IEZOELECTRIC DOMAIN OSCILLATIONS)   ANOMALOUS ECHO IN FERROELECTRIC ANTIMONY SULFOIODIDE. (P 15-2211
PHOSPHATE)   MACROSCOPIC STIMULATED POLARIZATION ECHO IN FERROELECTRICS. (POTASSIUM DIHYDROGEN 17-2419
SIUM DIHYDROGEN PHOSPHATE.   POLARIZATION ECHO IN THE FERROELECTRIC SINGLE CRYSTAL POTAS 17-2564
                            BOSON ECHOES. (ANTIMONY SULFOIODIDE)   15-2144
            NUCLEAR QUADRUPOLE AND ELECTROACOUSTIC ECHOES IN FERROELECTRIC ANTIMONY SULFOIODIDE. 15-2169
IUM TITANATE.   ABSORPTION EDGE AND STRUCTURAL PHASE TRANSITION IN STRONT 06-1107
SALT AND TRIGLYCINE/ POSITION OF THE FUNDAMENTAL ABSORPTION EDGE AS A FUNCTION OF TEMPERATURE FOR ROCHELLE 27-3160
ELATION BETWEEN THE ELECTRIC POLARIZATION AND THE ABSORPTION EDGE IN ANTIMONY SULFOIODIDE.   R 15-2175
        POLARIZATION FLUCTUATIONS AND THE OPTICAL ABSORPTION EDGE IN BARIUM TITANATE.   04-1028
PHASE TRANSITIONS.   BEHAVIOR OF THE FUNDAMENTAL ABSORPTION EDGE IN TRIHYDROGEN SELENITE CRYSTALS AT THEIR 22-2839
/ OF POLARITY SENSE BY ANOMALOUS SCATTERING AT AN ABSORPTION EDGE. (LITHIUM NIOBATE, BARIUM MANGANESE TETR/ 11-1673
        ELECTROOPTIC EFFECTS OF THE ABSORPTION EDGE OF ANTIMONY SULFOIODIDE.   15-2198
        NOTES ON ABSORPTION BAND EDGE OF (PARAELECTRIC) ANTIMONY SULFOIODIDE. 15-2177
            FUNDAMENTAL ABSORPTION EDGE OF STRONTIUM TITANATE.   06-1186
            OPTICAL ABSORPTION EDGE OF STRONTIUM TITANATE.   06-1187
EES K PHASE TRANSITION.   OPTICAL ABSORPTION EDGE OF STRONTIUM TITANATE AROUND THE 105 DEGR 06-1099
            ABSORPTION EDGE SPLITTING IN POTASSIUM TANTALATE NIOBATE. 13-1987
TE(1-X).   ABSORPTION EDGE SPLITTING IN POTASSIUM TANTALATE(X) NIOBA 13-1987
MAGNETIC RESONANCE OF MANGANESE(4+) IN STRONTIUM TITANATE. EFFECTIVE CHARGE OF TITANIUM(4+) FROM THE PARA 06-1167
/N TYPE, ANTIMONY DOPED) BARIUM TITANATE CERAMIC WITH A HIGH EFFECTIVE DIELECTRIC CONSTANT (DUE TO ADDING / 04-0581
ATION TO LEAD/ RELAXATION OF ANTIFERROELECTRIC DOMAIN WALLS, EFFECTIVE FIELD, AND ACTIVATION ENERGY. APPLIC 07-1257
E INTENSITY OF THE ELECTRIC FIELD.   DEPENDENCE OF EFFECTIVE POLARIZATION IN FERROELECTRICS ON TH 03-0382
/SION FOR INELASTIC COHERENT SCATTERING OF SLOW NEUTRONS AND EFFECTIVE THERMAL FACTORS. (KDP TYPE) PART-2:/ 03-0478
            THREE LAYER BROAD BAND ANTIREFLECTION COATINGS FOR LITHIUM NIOBATE.   11-1676
S WITH APPLICATION TO LITHIUM NIOBATE. PART-2: POLARITONS O/ EIGENFREQUENCIES AND EIGENVECTORS OF POLARITON 11-1766
S WITH APPLICATION TO LEAD NIOBATE. PART-1: LONG OPTICAL PH/ EIGENFREQUENCIES AND EIGENVECTORS OF POLARITON 08-1284
LITHIUM NIOBATE. PART-2: POLARITONS C/ EIGENFREQUENCIES AND EIGENVECTORS OF POLARITONS WITH APPLICATION TO 11-1766
LEAD NIOBATE. PART-1: LONG OPTICAL PH/ EIGENFREQUENCIES AND EIGENVECTORS OF POLARITONS WITH APPLICATION TO 08-1284
ERIUM PHOSPHATE AND DEUTERATED AMMONIU/ DETERMINATION OF THE EIGENVECTORS OF SOFT MODES IN POTASSIUM DIDEUT 17-2700
PONTANEOUS POLARIZATION NOISE IN POLAR DIELECTRICS. (THEORY, EINSTEIN THERMODYNAMIC METHOD)   S 03-0219

COMPILATION OF CONVERSION FORMULAS INTERRELATING THE LINEAR ELASTIC AND ELECTROELASTIC EQUATIONS OF STATE/ 03-0277
TRIC CRYSTALS.   ELASTIC AND ELECTROMAGNETIC WAVES IN PIEZOELEC 03-0371
HASE TRANSITION REPRESENTATION. (18 REFS)   DIELECTRIC, ELASTIC AND MAGNETIC PROPERTY, SYMMETRY, AND P 03-0316
D INVESTIGATION OF THEIR ELECTRICAL PROPERTIES. (DIELECTRIC, ELASTIC, AND PIEZOELECTRIC CONSTANTS) /SURE AN 15-2138
NIOBATE SINGLE CRYSTALS.   DETERMINATION OF ELASTIC AND PIEZOELECTRIC CONSTANTS OF LITHIUM 11-1749
UM(4) POTASSIUM LITHIUM NIOBATE TYPE FERROELECTRIC CRYSTAL. ELASTIC AND PIEZOELECTRIC CONSTANTS OF STRONTI 13-1946
OF TRIGLYCINE SULFATE.   DYNAMIC DETERMINATION OF (ELASTIC AND PIEZOELECTRIC) MATERIAL CONSTANTS 27-2984
TITANATE SINGLE CRYSTALS WITH LAMINAR DOMAIN S/ DIELECTRIC, ELASTIC AND PIEZOELECTRIC PROPERTIES OF BARIUM 04-1010
M PYRONIOBATE IN STRONG ELECTRIC FIELDS.   ELASTIC AND PIEZOELECTRIC PROPERTIES OF CADMIU 13-1967
LECTRIC LEAD(5) GERMANIUM(3) OXIDE(11) CRYSTALS.   ELASTIC AND PIEZOELECTRIC PROPERTIES OF FERROE 35-3520
ERAMICS WITH A ZIRCONIUM TO TITANIU/ LANTHANUM DEPENDENCE OF ELASTIC AND PIEZOELECTRIC PROPERTIES OF PLZT C 09-1573
SITION OF STRONTIUM TITANATE NEAR 105 DEG K. ANALYSIS OF THE ELASTIC ANOMALIES AT THE STRUCTURAL PHASE TRAN 06-1183
                                                       ELASTIC ANOMALIES OF ICE AT LOW TEMPERATURES. 34-3439
TRICTIVE CONSTANTS)   ELASTIC ANOMALY IN AMMONIUM SULFATE. (ELECTROS 19-2759
                            QUASI PHENOMENOLOGICAL THEORY OF ELASTIC ANOMALY IN FERROELECTRIC CRYSTALS. 03-0374
TIC, PIEZOELECTRIC, DIELECTRIC, AND OPTICAL CONSTANTS) ROOM/ ELASTIC ANOMALY OF BARIUM SODIUM NIOBATE (ELAS 13-2060
ORIDE NEAR A STRUCTURAL PHASE TRANSITION.   ANOMALOUS ELASTIC BEHAVIOR OF POTASSIUM MANGANESE TRIFLU 33-3392
ELD.   DEPENDENCE OF THE ELASTIC COEFFICIENTS OF ADP ON THE ELECTRIC FI 17-2537
E (TUNING AND DELAY DEVICES). ELECTRIC FIELD CONTROL OF THE ELASTIC COMPLIANCE IN LITHIUM THALLIUM TARTRAT 29-3250
/ AND FERROELASTICS SUCH THAT THE ELECTRIC SUSCEPTIBILITY OR ELASTIC COMPLIANCE IS TEMPERATURE INDEPENDENT/ 03-0140
/(LITHIUM THALLIUM TARTRATE MONOHYDRATE, ELECTRIC CONTROL OF ELASTIC COMPLIANCE, NICKEL IODINE BORACITE OP/ 02-0060
ALLIUM TARTRATE HYDRATE (LTT) BY AN ELECTRIC/ CONTROL OF THE ELASTIC COMPLIANCE OF FERROELECTRIC LITHIUM TH 29-3249
THE TRANSITION TEMPERATURE.   PRESSURE DEPENDENCE OF THE ELASTIC COMPLIANCE OF STRONTIUM TITANATE NEAR 06-1219
E CRYSTALS IN THE TEMPERATURE/ TEMPERATURE DEPENDENCE OF THE ELASTIC COMPLIANCE OF STRONTIUM TITANATE SINGL 06-1220
TE.   ELASTIC CONSTANT ANOMALIES IN STRONTIUM TITANA 06-1177
BATE BY X-RAY DIFFUSE SCATTERING.   ELASTIC CONSTANT OF ORTHORHOMBIC POTASSIUM NIO 08-1416
                           PRESSURE DEPENDENCE OF THE ELASTIC CONSTANT OF STRONTIUM TITANATE. 06-1097
                           STUDY OF THE BRILLOUIN EFFECT IN ADP (ALL ELASTIC CONSTANTS). 17-2522
DOLINIUM MOLYBDATE.   ELASTIC CONSTANTS AND SOFT OPTICAL MODES IN GA 30-3284
SODIUM NITRITE SINGLE CRYSTALS.   ELASTIC CONSTANTS AND ULTRASONIC ABSORPTION OF 16-2357
TUDY OF ANTIMONY SULFOIODIDE USING A DOUBLE PASSED, STABIL/ (ELASTIC CONSTANTS FROM) BRILLOUIN SCATTERING S 15-2225
                                                       ELASTIC CONSTANTS OF FERROCERAMICS. 09-1513
R THE CURIE POINT).   ELASTIC CONSTANTS OF GADOLINIUM MOLYBDATE (NEA 30-3274
PHATE BY BRILLOUIN DIFFUSION.   MEASUREMENT OF THE ELASTIC CONSTANTS OF POTASSIUM DIHYDROGEN PHOS 17-2454
                           DETERMINATION OF THE ELASTIC CONSTANTS OF POTASSIUM NITRATE. 08-1399
IN DIFFUSION AT ORDINARY TEMPERATURES.   MEASUREMENT OF THE ELASTIC CONSTANTS OF SODIUM NITRITE BY BRILLOU 16-2306
                    PRESSURE DEPENDENCE OF THE ELASTIC CONSTANTS OF STRONTIUM TITANATE. 06-1096
        TEMPERATURE DEPENDENCE OF THE FIRST ORDER ELASTIC CONSTANTS OF STRONTIUM TITANATE. 06-1160
TRIC BARIUM TITANATE.   EFFECTS OF ANISOTROPY OF ELASTIC CONSTANTS ON DISLOCATIONS IN FERROELEC 04-0674
RIC TRANSITION.   ELASTIC ENERGY AND THE FERROELECTRIC PARAELECT 03-0237
                            SOLUTION OF AN ELASTIC FERROELECTRIC MODEL. 03-0211
                    THERMODYNAMICS OF AN ELASTIC FERROELECTRIC MODEL. 03-0240
STAL.   ELASTIC GIBBS FUNCTION FOR A FERROELECTRIC CRY 03-0243
LS. (GADOLINIUM M/ PRODUCTION OF SPONTANEOUS POLARIZATION BY ELASTIC INSTABILITIES IN PIEZOELECTRIC MATERIA 30-3282
                                                       ELASTIC MODULI OF ICE. 34-3421
0 DEG K).   LOW TEMPERATURE ELASTIC MODULI OF STRONTIUM TITANATE (10 TO 13 06-1190
TS OF LITHIUM TANTALATE AND L/ TEMPERATURE DEPENDENCE OF THE ELASTIC, PIEZOELECTRIC, AND DIELECTRIC CONSTAN 11-1823
TS OF BISMUTH(12) GERMANIUM OXIDE(20).   ELASTIC, PIEZOELECTRIC, AND DIELECTRIC CONSTAN 35-3512
NICS IN A (SINGLE DOMAIN) LITHIUM NIOBATE CRYSTAL (NONLINEAR ELASTIC PROPERTIES). /FREQUENCY ACOUSTIC HARMO 11-1707
IC PHASE TRANSITION IN GADOLINIUM MOLYBDATE. (DIELECTRIC AND ELASTIC PROPERTIES) /THEORY OF THE FERROELECTR 30-3303
GENERATION IN STRONTIUM TITANATE SINGLE CRYSTALS (NONLINEAR ELASTIC PROPERTIES). ULTRASONIC THIRD HARMONIC 06-1181
PHATE AND THE LAVAL RAMAN ELASTICITY THEORY.   ELASTIC PROPERTIES OF AMMONIUM DIHYDROGEN PHOS 17-2410
                       EFFECT OF GRAIN SIZE ON THE PIEZOELECTRIC AND ELASTIC PROPERTIES OF BARIUM TITANATE. 04-1011
                                                       ELASTIC PROPERTIES OF CESIUM LEAD TRICHLORIDE. 33-3360
HE TRIGLYCINE SULFATE GRO/ DEUTERATION EFFECT ON THERMAL AND ELASTIC PROPERTIES OF FERROELECTRIC CRYSTALS T 27-2986
ITS FERROELECTRIC PHASE TRANSITION.   ELASTIC PROPERTIES OF GADOLINIUM MOLYBDATE AT 30-3283
RYSTAL/ THERMAL DEPENDENCE OF DIELECTRIC, PIEZOELECTRIC, AND ELASTIC PROPERTIES OF LITHIUM NIOBATE SINGLE C 11-1693
TERISTICS.   PIEZOELECTRIC AND ELASTIC PROPERTIES OF LITHIUM TANTALATE CHARAC 12-1893
PHATE CRYSTALS.   ELECTRONIC EFFECT AND ELASTIC PROPERTIES OF RUBIDIUM DIHYDROGEN PHOS 17-2617
PHATE (RDP).   EFFECT OF DEUTERATION ON THE ELASTIC PROPERTIES OF RUBIDIUM DIHYDROGEN PHOS 17-2584
M LEAD TITANATE SOLID SOLUTIONS.   DIELECTRIC AND ELASTIC PROPERTIES OF SINGLE CRYSTALS OF BARIU 04-0574
TITANATE SINGLE CRYSTALS.   ELASTIC PROPERTIES OF YTTRIUM DOPED STRONTIUM 06-1180
                                      QUASI ELASTIC SCATTERING IN LEAD GERMANATE. 35-3497
/TION IN PEROVSKITE- TYPE CRYSTALS, PART-2, INTERACTION WITH ELASTIC STRAIN (STRONTIUM TITANATE AND LANTHA/ 03-0241
OF STRONTIUM TITANATE.   INTERACTION OF ELASTIC STRAIN WITH THE STRUCTURAL TRANSITION 06-1216
ONS WITH FERROELECTRIC PROPERTIES.   EFFECT OF ELASTIC STRESSES ON PARAMETERS OF SOLID SOLUTI 03-0265
ITHIUM NIOBATE.   REFLECTION OF ELASTIC SURFACE WAVE FROM PERIODIC ARRAYS ON L 11-1853
                                                       ELASTIC SURFACE WAVE ON THE PZT CERAMIC PLATE. 09-1612
/ EFFECTS OF THERMALLY DIFFUSED IMPURITIES ON PROPAGATION OF ELASTIC SURFACE WAVES ON FERROELECTRIC CERAMIC 03-0487
MIXING OF NONCOLLINEAR ELASTIC SURFACE WAVES ON LITHIUM NIOBATE. 11-1710
TES (LI/ PERTURBATION THEORY FOR ELECTROMAGNETIC COUPLING TO ELASTIC SURFACE WAVES ON PIEZOELECTRIC SUBSTRA 11-1751
(TECHNIQUE APPLIED TO LITH/ ELECTRON BEAM SENSING OF SURFACE ELASTIC WAVES FOR SIGNAL PROCESSING PURPOSES. 11-1738
ATE)   DISPERSION OF ELASTIC WAVES IN FERROELECTRICS. (LITHIUM NIOB 11-1706
         COLLINEAR INTERACTION OF LONGITUDINAL ELASTIC WAVES IN LITHIUM NIOBATE CRYSTALS. 11-1678
                            ABSORPTION OF ELASTIC WAVES IN REDUCED LITHIUM NIOBATE. 11-1816
OLINIUM MOLYBDATE.   ABNORMAL ABSORPTION OF ELASTIC WAVES NEAR THE PHASE TRANSITION OF GAD 30-3265
TE(0.05) CERAMIC FLA/ PROPAGATION CHARACTERISTICS OF SURFACE ELASTIC WAVES ON BARIUM TITANATE(0.95) ZIRCONA 09-1613
OF THE ANTIFERROELECTRIC CURIE POINT OF LEAD / ANOMALIES OF ELASTICITY AND INTERNAL FRICTION IN THE REGION 13-2020
DE.   ELECTROSTRICTION, PIEZOELECTRICITY AND ELASTICITY IN FERROELECTRIC ANTIMONY SULFOIODI 15-2165
PERTIES OF AMMONIUM DIHYDROGEN PHOSPHATE AND THE LAVAL RAMAN ELASTICITY THEORY.   ELASTIC PRO 17-2410
TRAIN DEPENDENT CURIE TEMPERATURE, POLARIZATI/ THEORY OF THE ELASTOOPTIC EFFECT IN NONMETALLIC CRYSTALS. (S 03-0513
E IN LITHIUM NIOBATE AND KDP POCKELS CELLS.   TRANSIENT ELASTOOPTIC EFFECTS AND Q SWITCHING PERFORMANC 11-1727
                                                       ELASTOOPTICAL EFFECT IN FERROELECTRIC WITH DIF 03-0307
IN A FERROELECTRIC.   SURFACE ELASTOPOLARIZATION WAVES AT DOMAIN BOUNDARIES 03-0315
LATTICE DISTORTIONS IN A STRONTIUM TITANATE SINGLE CRYSTAL ELECTRET. 06-1238
ATE AND BARIUM TITANATE   ELECTRET EFFECT IN A MIXTURE OF POLY METHACRYL 04-0838
TWO PHASE FERROELECTRIC FERROMAGNETIC MIXTURES. (BARIUM TIT/ ELECTRIC AND MAGNETIC MEASUREMENTS OF CERTAIN 04-0565
ITIO/ ELECTRONIC AND SPIN EXCITATIONS IN ANTIFERRODIELECTRIC ELECTRIC AND MAGNETIC MOMENTS OF OPTICAL TRANS 04-0418
AND OTHER CRYSTALS.   RELATIONSHIP BETWEEN DIRECTIONS OF THE ELECTRIC AND MAGNETIC POLARIZATION IN BORACITE 14-2087
ATE SINGLE CRYSTALS.   KERR EFFECT STUDIES OF ELECTRIC CONDUCTION IN IRON DOPED BARIUM TITAN 04-0703

COL SULFATE.   OPTICAL ABSORPTION, ELECTRIC CONDUCTIVITY AND PHOTOEMISSION OF GLY 19-2780
TRICAL CONDUCTIVITY AND RELATED PROPERTIES OF 60 COMPOUNDS,/ ELECTRIC CONDUCTIVITY OF FERROELECTRICS. (ELEC 01-0014
 PHASE TRANSITION REGION.   NATURE OF THE ANOMALOUS ELECTRIC CONDUCTIVITY OF FERROELECTRICS IN THE 03-0508
/OELECTRIC CRYSTALS. (LITHIUM THALLIUM TARTRATE MONOHYDRATE, ELECTRIC CONTROL OF ELASTIC COMPLIANCE, NICKE/ 02-0060
LORIDE.   A PHASE TRANSITION IN A COMPOUND WITH HELICAL ELECTRIC DIPOLE STRUCTURE: CESIUM COPPER TRICH 33-3388
L DIPOL/ ON THE QUANTUM THEORY OF THE FERROELECTRIC STATE IN ELECTRIC DIPOLE SYSTEMS (SPIN WAVES IN HYDROXY 03-0452
NATE STANNATE BY TEMPERATURE AND PRESSURE ENFORC/ RELEASE OF ELECTRIC ENERGY IN LEAD NIOBIUM ZIRCONATE TITA 09-1488
FERROELECTRIC SEMICONDUCTOR ANTIMONY SULFOIODIDE IN A STATIC ELECTRIC FIELD.   AGING OF THE 15-2113
MAGNETICALLY ORDERED FERROELECTRICS IN A RAPIDLY OSCILLATING ELECTRIC FIELD.   BEHAVIOR OF 03-0147
 FERROELECTRIC LITHIUM THALLIUM TARTRATE HYDRATE (LTT) BY AN ELECTRIC FIELD. / OF THE ELASTIC COMPLIANCE OF 29-3249
CTIVE POLARIZATION IN FERROELECTRICS ON THE INTENSITY OF THE ELECTRIC FIELD.   DEPENDENCE OF EFFE 03-0382
          DEPENDENCE OF THE ELASTIC COEFFICIENTS OF ADP ON THE ELECTRIC FIELD.   17-2537
EZOELECTRIC PROPERTIES OF ANTIMONY SULFOIODIDE IN A CONSTANT ELECTRIC FIELD.   DIELECTRIC AND PI 15-2249
C CONSTANT OF TRIGLYCINE SULFATE SINGLE CRYSTALS IN A STRONG ELECTRIC FIELD.   DIELECTRI 27-3150
IN THE TANGENT OF THE ANGLE OF DIELECTRIC LOSSES IN A STRONG ELECTRIC FIELD. / SINGLE CRYSTALS ON A CHANGE 27-3153
TANT AND TANGENT OF THE ANGLE OF DIELECTRIC LOSS IN A STRONG ELECTRIC FIELD. /CHANGE IN THE DIELECTRIC CONS 27-3151
ONLINEARITY OF A FERRCELECTRIC SEMICONDUCTOR IN A NONUNIFORM ELECTRIC FIELD.   ELECTRICAL N 03-0500
ODEL OF TWO DIMENSIONAL FERROELECTRICS IN ARBITRARY EXTERNAL ELECTRIC FIELD.   EXACT SOLUTION OF M 03-0520
C HYSTERESIS LOOP OF SILVER SODIUM DINITRITE BY QUASI STATIC ELECTRIC FIELD.   FERROELECTRI 16-2293
TRIGLYCINE FLUOBERYLLATE CRYSTALS DURING SWITCHING IN PULSE ELECTRIC FIELD)   IMPEDANCE OF FERROELECTRIC 27-3027
ATITE. (DEVELOPMENT OF FERROELECTRIC CHARACTER IN AN APPLIED ELECTRIC FIELD) / OF SYNTHETIC CALCIUM CHLORAP 36-3591
          PARAELECTRIC SEMICONDUCTOR IN A NONUNIFORM ELECTRIC FIELD.   03-0502
TTIVITY OF SINGLE CRYSTALS OF TRIGLYCINE SULFATE IN A STRONG ELECTRIC FIELD.   PERMI 27-2978
 STABLE POSITIONS, SPONTANEOUS) POLARIZATION (ON APPLYING AN ELECTRIC FIELD). /VERSIBLE, BUT TILTABLE (TO 3 08-1346
ORPHIC MODIFICATIONS OF POTASSIUM DIHYDROGEN PHOSPHATE IN AN ELECTRIC FIELD.   RAMAN SPECTRA OF POLYM 17-2425
NGLE CRYSTALS OF BARIUM TITANATE IN THE PRESENCE OF A STEADY ELECTRIC FIELD. /BSORPTION OF ULTRASOUND IN SI 04-0850
TH RELAXATION TIME OF SECONDS WHEN ILLUMINATED UNDER APPLIED ELECTRIC FIELD). /E) OPTICAL DAMAGE IN KDP (WI 17-2703
C-AXIS OF ANTIMONY SULFOIODIDE CAUSED BY ILLUMINATION IN DC ELECTRIC FIELD.   STRAIN ALONG 15-2239
ZATION EFFECT OF TRIGLYCINE SULFATE SINGLE CRYSTALS IN AN AC ELECTRIC FIELD.   TEMPERATURE AUTOSTABILI 27-3005
ANIDINIUM ALUMINUM SULFATE HEXAHYDRATE. (POLARIZATION VERSUS ELECTRIC FIELD, -80C TO ROOM TEMPERATURE) / GU 21-2813
IGLYCINE SULFATE.   SPECIFIC HEAT UNDER APPLIED ELECTRIC FIELD AND ELECTROCALORIC EFFECT IN TR 27-2960
TION OF ULTRASOUND IN ANTIMONY SULF/ EFFECTS OF TEMPERATURE, ELECTRIC FIELD, AND ILLUMINATION ON THE ABSORP 15-2248
 ULTRASONIC ABSORPTION IN TRIGLYCINE SULFA/ INFLUENCE OF THE ELECTRIC FIELD AND THE DOMAIN STRUCTURE ON THE 27-2955
CE IN LITHIUM THALLIUM TARTRATE (TUNING AND DELAY DEVICES). ELECTRIC FIELD CONTROL OF THE ELASTIC COMPLIAN 29-3250
TY MODE IN STRONTIUM TITANATE.   (TEMPERATURE AND) ELECTRIC FIELD DEPENDENCE OF A HYDROGEN IMPURI 06-1146
EQUENCIES (RAMAN SCATTERING IN CUBIC PEROVSKITE STRONTIUM T/ ELECTRIC FIELD DEPENDENCE OF OPTICAL PHONON FR 06-1241
CHARGE LIMITED CURRENT IN BISMUTH(12) GERMANIUM O/ LIGHT AND ELECTRIC FIELD DEPENDENT OSCILLATION OF SPACE 35-3505
ITH SPECIAL EMPHASIS ON NEAR SURFACE REGIONS IN FERROELECTR/ ELECTRIC FIELD DISTRIBUTIONS IN DIELECTRICS, W 03-0182
N LITHIUM NIOBATE.   NATURE OF INTERNAL ELECTRIC FIELD DURING OPTICAL DAMAGE PROCESS I 11-1861
CE IN STRONTIUM TITANATE. (DOPED WITH GADOLINIUM(3+) OR IRO/ ELECTRIC FIELD EFFECT OF ELECTRON SPIN RESONAN 06-1229
) DOPED STRONTIUM TITANATE.   INTERPRETATION OF ELECTRIC FIELD EFFECTS IN EPR OF GADOLINIUM(3+ 06-1168
ARAMETER OF FERROELECTRIC SODIUM NITRITE. (AT THE SODIUM-23/ ELECTRIC FIELD GRADIENT AND LONG RANGE ORDER P 16-2395
EAD ZIRCONATE STUDIED BY PERTURBED ANGULAR COR/ THE INTERNAL ELECTRIC FIELD GRADIENT IN ANTIFERROELECTRIC L 07-1255
M NITRITE.   TEMPERATURE DEPENDENCE OF THE ELECTRIC FIELD GRADIENT IN FERROELECTRIC SODIU 16-2337
OMAG/ MOSSBAUER STUDY AND THEORETICAL ESTIMATION OF INTERNAL ELECTRIC FIELD GRADIENTS IN FERROELECTRIC FERR 08-1401
ESIUM LEAD TRICHLORIDE.   ELECTRIC FIELD GRADIENTS IN IONIC CHLORIDES: C 33-3367
          DEBYE SCREENING OF A NONUNIFORM ELECTRIC FIELD IN A FERROELECTRIC. 03-0499
          EPR INVESTIGATION OF THE INTERNAL ELECTRIC FIELD IN COPPER DOPED ROCHELLE SALT. 28-3233
/IC PROPERTIES OF FERROELECTRICS, SUBJECTED BEFOREHAND TO AN ELECTRIC FIELD, IN STRONG ELECTRIC FIELDS. (A/ 04-0661
TION IN STRONTIUM TITANATE AND POTASSIUM TANTALATE. ELECTRIC FIELD INDUCED OPTICAL HARMONIC GENERA 06-1203
CTRIC CRYSTALS. (POTASSIUM TANTALATE) ELECTRIC FIELD INDUCED RAMAN EFFECT IN PARAELE 08-1326
/THE PEROVSKITE TYPE WITHIN A WIDE RANGE OF TEMPERATURES AND ELECTRIC FIELD INTENSITIES AND TIMES OF APPLI/ 08-1380
ICAL PROPERTIES OF TGS AND ROCHELLE SALT AS DETERMINED BY AN ELECTRIC FIELD METHOD.   CRIT 27-3056
FERROELECTRIC FILM BY BENDING AND BENDING DEFORMATION IN THE ELECTRIC FIELD. (OBSERVATION REPORTED) /ON OF 03-0195
E TRANSITION, TALKING INTO ACCOUNT TH/ EFFECT OF AN EXTERNAL ELECTRIC FIELD ON A DIFFUSE FERROELECTRIC PHAS 03-0441
N SEIGNETTES SALT (ROCHELLE SALT). (TEMPERATURE A/ EFFECT OF ELECTRIC FIELD ON DOMAIN STRUCTURE FORMATION I 28-3190
DOPED FERROELECTRIC ROCHELLE SALT.   EFFECT OF APPLIED ELECTRIC FIELD ON EPR SPECTRA OF (COPPER(2+)) 28-3185
LECTRIC CERAMICS. PART-1: DEFORMATION AND PHASE T/ EFFECT OF ELECTRIC FIELD ON FERROELECTRIC AND ANTIFERROE 03-0169
MIUM(3+) IN POTASSIUM DIHYDR/ EFFECTS OF UNIAXIAL STRESS AND ELECTRIC FIELD ON PARAMAGNETIC SPECTRA OF CHRO 17-2554
BARIUM) NIOBATE SOLID SOLUTIONS.   EFFECT OF ELECTRIC FIELD ON PHASE COMPOSITION OF (LEAD, 13-1943
 POLYCRYSTALLINE BARIUM TITANATE.   INFLUENCE OF A CONSTANT ELECTRIC FIELD ON THE DIELECTRIC PROPERTIES OF 04-1012
IC EFFECT IN TGS.   INFLUENCE OF AN ELECTRIC FIELD ON THE NOISE AND THE PYROELECTR 27-3076
LTA PHASE OF SODIUM TRIDEUTERIUM(X) TR/ UNUSUAL EFFECT OF AN ELECTRIC FIELD ON THE PERMITTIVITY OF A NEW DE 22-2875
N FERROELECTRIC TRIGLYCINE SULFATE.   THE INFLUENCE OF THE ELECTRIC FIELD ON THE ULTRASONIC ATTENUATION I 27-2956
RE OF TRIGLYCINE SULFATE CRYSTALS ON TEMPERATURE AND APPLIED ELECTRIC FIELD STRENGTH. /F THE DOMAIN STRUCTU 27-3129
LYVINYLIDENE FLUORIDE FILMS (AFTER COOLING STRETCHED FILM IN ELECTRIC FIELD, THERMOELECTRETS). /ATION IN PO 36-3530
VELOCITIES OF THE DOMAIN WALLS IN KDP. EXISTENCE OF CRITICAL ELECTRIC FIELDS.   DISPLACEMENTS AN 17-2448
TRUCTURE OF THE BETA PHASE OF SODIUM TRIHYDROGEN SELENATE IN ELECTRIC FIELDS.   DOMAIN S 22-2830
ND PIEZOELECTRIC PROPERTIES OF CADMIUM PYRONIOBATE IN STRONG ELECTRIC FIELDS.   ELASTIC A 13-1967
ERAMIC FERROELECTRICS (BARIUM TITANATE ZIRCONATES) IN STRONG ELECTRIC FIELDS AND AT HIGH PRESSURE. /ES OF C 09-1605
N THE PEROVSKITE TYPE COMPOUND/ CALCULATIONS OF THE INTERNAL ELECTRIC FIELDS AND ELECTRIC FIELD GRADIENTS I 08-1442
          NONLINEAR INTERACTION OF MICROWAVE ELECTRIC FIELDS AND SOUND IN LITHIUM NIOBATE. 11-1838
PEROVSKITE COMPOUNDS WITH DISTINCTI/ CALCULATION OF INTERNAL ELECTRIC FIELDS AND THEIR GRADIENTS IN ABO(3) 08-1455
E AND CADMIUM TITANATE.   INTERNAL ELECTRIC FIELDS IN CRYSTALS OF SODIUM TANTALATE 08-1389
ANATE TYPE)   CALCULATION OF ELECTRIC FIELDS IN IONIC CRYSTALS. (BARIUM TIT 04-0824
          DIPOLE STRUCTURE AND INTERNAL ELECTRIC FIELDS IN LEAD ZIRCONATE. 07-1261
          CRYSTALLIZATION OF ROCHELLE SALT IN ELECTRIC FIELDS (KINETICS OF CRYSTAL GROWTH) 28-3200
AD TITANATE CERAMICS FOR PIEZOELECTRIC APPL/ EFFECTS OF HIGH ELECTRIC FIELDS ON MODIFIED LEAD ZIRCONATE- LE 09-1485
/D ZIRCONATE LEAD TITANATE PIEZOELECTRIC CERAMICS UNDER HIGH ELECTRIC FIELDS. (SUBSTITUTION OF STRONTIUM O/ 09-1484
CONDUCTOR CRYSTALS.   THEORY OF ELECTRIC INDUCTION WAVES IN FERROELECTRIC SEMI 03-0466
MANGANESE(1-X) (B(X)) OXYGEN(3) (B=TRIVALENT IRO/ X-RAY AND ELECTRIC INVESTIGATIONS OF THE SYSTEMS YTTRIUM 36-3569
HE EMISSION ELECTRON MICROSCOPE (BARRIER LAYERS IN / IMAGING ELECTRIC MICRO FIELDS (OF SOLID SURFACES) IN T 04-0913
 A FERROELECTRIC WITH HYDROGEN BONDS/ EFFECT OF VIOLATION OF ELECTRIC NEUTRALITY ON THE PHASE TRANSITION IN 03-0449
LY AND CHEMICAL BOND. (COVALENCY, DIRECTIVITY CONDITIONS FOR ELECTRIC ORDERING AND STABILITY) /OVSKITE FAMI 03-0243
IES OF TRIGLYCINE SULFATE SINGLE CRYSTALS (HEAT CAPACITY AND ELECTRIC PERMITTIVITY). /ND ELECTRICAL PROPERT 27-3141
ROCHLORE TYPE) COMPOUND (AMMONO THIO CADMIUM NIOBATE, WITH 4 ELECTRIC PHASES, 2 FERROELECTRIC). /ECTRIC (PY 36-3529
KEL IODIDE BORACITE (CURIE P/ COEXISTENCE OF THE SPONTANEOUS ELECTRIC POLARIZATION AND MAGNETIZATION IN NIC 14-2090
IN ANTIMONY SULFOIODIDE.   RELATION BETWEEN THE ELECTRIC POLARIZATION AND THE ABSORPTION EDGE 15-2175
D ICE AT LOW TEMPERATURES.   ELECTRIC POLARIZATION EFFECTS IN PURE AND DOPE 34-3412
AND FERROMAGNETOELECTRIC MATERIAL THAT ALLOWS NO 180 DEGREE ELECTRIC POLARIZATION REVERSAL. /FERROELECTRIC 14-2098

D LEAD TITANATES, EFFECT OF FERROELECTRIC CHARACTERISTICS ON ELECTRIC PROPERTIES) / DIELECTRICS. (BARIUM AN 04-0670
S. (BARIUM TITANATE)   CORRELATION BETWEEN STRUCTURE AND ELECTRIC PROPERTIES OF FERROELECTRIC SUBSTANCE 04-0728
M TITANATE.   ELECTRIC PROPERTIES OF HYDROGEN DOPED STRONTIU 06-1088
D HAFNIUM IONS IN BARIUM AND LEAD TITANATE.   STATIC ELECTRIC QUADRUPOLE INTERACTION OF TANTALUM AN 04-0937
ITANATE CERAMICS.   EFFECT OF EXCESS TITANIA ON THE ELECTRIC RESISTANCE OF SEMICONDUCTIVE BARIUM T 04-0966
N BOND. (ICE)   ELECTRIC RESONANCE: APPLICATION TO THE HYDROGE 34-3477
ION TIME NEAR THE TRANSITION POINTS IN SODIUM NITRIT/ STATIC ELECTRIC SUSCEPTIBILITY AND DIELECTRIC RELAXAT 16-2302
/IDERATION OF FERROELECTRICS AND FERROELASTICS SUCH THAT THE ELECTRIC SUSCEPTIBILITY OR ELASTIC COMPLIANCE/ 03-0140
56/ STRUCTURE AND POINT DEFECTS OF ICE: THEIR EFFECT ON THE ELECTRICAL AND MECHANICAL PROPERTIES. (REVIEW, 02-0071
RIC POTASSIUM IODATE.   ELECTRICAL AND OPTICAL CONSTANTS OF FERROELECT 08-1427
LECTRIC PROPERTIES OF POTASSIUM DIHYDROGEN PHOSPHATE.   ELECTRICAL AND OPTICAL INVESTIGATION OF FERROE 17-2449
Y AND MOBILITY VERSUS TEMPERATURE, ABSORPTION SPECTRA) OF S/ ELECTRICAL AND OPTICAL PROPERTIES (CONDUCTIVIT 06-1198
STRONTIUM) MIXED TITANATE THIN SECTIONS.   ELECTRICAL AND OPTICAL PROPERTIES OF (BARIUM, 04-0603
NT FERROELECTRIC / EFFECTS OF GRAIN SIZE AND POROSITY ON THE ELECTRICAL AND OPTICAL PROPERTIES OF TRANSPARE 13-2004
XIDE AT LOW TEMPERATURE.   (PREPARATION AND ELECTRICAL AND OPTICAL) STUDY OF TUNGSTEN TRIO 10-1643
SINGLE CRYSTAL BISMUTH(4) TITANIUM(3) OXIDE(12).   ELECTRICAL AND OPTICAL SWITCHING PROPERTIES OF 35-3517
IMONY SULFOIODIDE.   ELECTRICAL AND PHOTOELECTRIC PROPERTIES OF ANT 15-2107
CONTACTS BETWEEN ANTIMONY SULFOIODIDE SINGLE CRYSTALS AND / ELECTRICAL AND PHOTOELECTRIC PROPERTIES OF THE 15-2251
ROELECTRI/ INFLUENCE OF FIRING OF BARIUM TITANATE SINTERS ON ELECTRICAL AND TECHNOLOGICAL PROPERTIES OF FER 04-1002
E SELENATE NEAR THE CURIE POINT.   RELATION BETWEEN ELECTRICAL AND THERMAL PROPERTIES OF TRIGLYCIN 27-3168
E.   CORRELATION BETWEEN RHEOLOGICAL AND ELECTRICAL BEHAVIOR IN SINTERED BARIUM TITANAT 04-1027
CS.   ISOTHERMAL GRAIN GROWTH AND ELECTRICAL BEHAVIOR OF FULLY DENSE PLZT CERAMI 09-1553
, FERROELECTRICITY AND SPACE CHARGE.   ELECTRICAL BEHAVIOR OF ICE AT LOW TEMPERATURES 34-3448
XATION RATE OF SODIUM-23 IN SODIUM NITRITE. INFLUENCE OF THE ELECTRICAL BOUNDARY CONDITIONS ON THE NMR RELA 16-2350
CAPACITY OF SINGLE CRYSTALS OF TRIGLYCINE SULFAT/ EFFECTS ON ELECTRICAL BOUNDARY CONDITIONS ON THE THERMAL 27-3140
ITRITE CRYSTALS.   ELECTRICAL BREAKDOWN IN FERROELECTRIC SODIUM N 16-2288
FACE WAVE TRANSDUCERS ON LITHIUM NIOBATE.   MEASURED ELECTRICAL CHARACTERISTICS OF INTERDIGITAL SUR 11-1702
/RIUM TITANATE COATINGS PRODUCED BY FLAME SPRAYING. PART-11: ELECTRICAL CHARACTERISTICS OF SEMICONDUCTIVE / 04-0774
HASE TRANSITIONS IN CESIUM LEAD TRICHLORIDE.   ELECTRICAL CONDUCTION AND (NONFERROELECTRIC) P 33-3365
                                                             ELECTRICAL CONDUCTION IN ICE. 34-3415
, 319 REFS)   ELECTRICAL CONDUCTION IN METAL OXIDES. (REVIEW 02-0109
/RED AND RAMAN SCATTERING. PART-5: MOSSBAUER EFFECT. PART-6: ELECTRICAL CONDUCTIVE, SURFACE, AND ELECTROOP/ 03-0397
THE COLOR PROBLEM OF TUNGSTEN TRIOXIDE ELECTRICAL CONDUCTIVITY. 10-1636
ORPHIC TRANSITIONS IN POTASSIUM NITRATE AT 1 ATM/ CHANGES IN ELECTRICAL CONDUCTIVITY ACCOMPANYING THE POLYM 16-2391
OF 60 COMPOUNDS,/ ELECTRIC CONDUCTIVITY OF FERROELECTRICS. (ELECTRICAL CONDUCTIVITY AND RELATED PROPERTIES 01-0014
TWO MODES OF ELECTRICAL CONDUCTIVITY IN POTASSIUM NITRATE. 16-2368
/NONSTOICHIOMETRIC BARIUM TITANATE. PART-1: HIGH TEMPERATURE ELECTRICAL CONDUCTIVITY MEASUREMENTS. (INTRIN/ 04-0822
RROELECTRICS.   OXYGEN CONCENTRATION CELL AND ELECTRICAL CONDUCTIVITY MEASUREMENTS ON PZT FE 09-1495
ICONDUCTOR NEAR THE CUBIC POINT.   ANOMALIES IN THE ELECTRICAL CONDUCTIVITY OF A FERROELECTRIC SEM 03-0268
NGLE CRYSTALS.   ELECTRICAL CONDUCTIVITY OF AMMONIUM SULFATE SI 19-2781
GLE CRYSTALS HEATED IN OXYGEN (EFFEC/ TIME DEPENDENCE OF THE ELECTRICAL CONDUCTIVITY OF BARIUM TITANATE SIN 04-0997
GLE CRYSTALS GROWN FROM SOLUTION IN A POT/ ANISOTROPY OF THE ELECTRICAL CONDUCTIVITY OF BARIUM TITANATE SIN 04-1035
E- ANTIMONY SELENOIODIDE SYSTEM ALLOYS.   PHASE DIAGRAM AND ELECTRICAL CONDUCTIVITY OF BISMUTH SELENOIODID 15-2122
MEASUREMENT OF THE TEMPERATURE DEPENDENCE OF THE ELECTRICAL CONDUCTIVITY OF FERROELECTRICS. 03-0525
GLE CRYSTALS.   ELECTRICAL CONDUCTIVITY OF LITHIUM NIOBATE SIN 11-1733
EFFECT OF ADDING TITANIUM(+) AND STRONTIUM(2+) IONS ON THE ELECTRICAL CONDUCTIVITY OF POTASSIUM NITRATE. 16-2319
BARIUM TITANATE.   ELECTRICAL CONDUCTIVITY OF SINGLE CRYSTALLINE 04-0663
AND HAFNATE AT HIGH TEMPERATURES.   ELECTRICAL CONDUCTIVITY OF STRONTIUM ZIRCONATE 08-1417
                                                             ELECTRICAL CONDUCTIVITY OF TUNGSTEN TRIOXIDE. 10-1637
CURIE POINT) OF REDUCED BARIUM TITANATE CRYSTALS (POLARON H/ ELECTRICAL CONDUCTIVITY (SEEBECK COEFFICIENT, 04-0918
ITRITE AND POTASSIUM NITRATE.   DIELECTRIC AND ELECTRICAL CONDUCTIVITY STUDIES IN POTASSIUM N 16-2340
PLICATIONS TO PYROELECTRICITY AND THE DETECTION OF FAR INFR/ ELECTRICAL CONSTANTS OF TRIGLYCINE SULFATE, AP 17-2514
RIC OXIDES.   ELECTRICAL CONTACT TO N- AND P-TYPE FERROELECT 03-0512
HOLOGRAPHIC PATTERNS IN FERROELECTRIC MATERIALS. (BARIUM TI/ ELECTRICAL CONTROL OF FIXATION AND ERASURE OF 04-0847
YCINE FLUOBERYLLATE.   EFFECT OF CONSTANT ELECTRICAL FIELD ON PHASE TRANSITIONS IN TRIGL 27-3167
APPLICATION OF FERROELECTRIC MATERIALS IN ELECTRICAL FILTERS. 37-3636
TITANATE CRYSTALS.   HOLOGRAPHIC STORAGE, ELECTRICAL FIXING AND ERASING IN DOPED BARIUM 04-0848
TRANSITIONS IN LEAD COBALTATE(0.5) TUNGSTATE(0.5), TITANAT/ ELECTRICAL MAGNETIC AND X-RAY STUDIES OF PHASE 13-1960
OF FERROELECTRIC FERROE/ INFERRED TEMPERATURE DEPENDENCES OF ELECTRICAL, MECHANICAL AND OPTICAL PROPERTIES 33-3253
ICONDUCTOR IN A NONUNIFORM ELECTRIC FIELD.   ELECTRICAL NONLINEARITY OF A FERROELECTRIC SEM 03-0500
F FERROELECTRIC GADOLINIUM MOLYBDATE.   ELECTRICAL, OPTICAL, AND MECHANICAL BEHAVIOR O 30-3267
LIDES, STRONTIUM TITANATE).   INERTIAL ELECTRICAL POLARIZATION OF CRYSTALS (ALKALI HA 06-1197
FLUORINE CONTAINING BARIUM TITANATE. (ELECTRICAL PROPERTIES) 04-0869
FLUX GROWN KTN SINGLE CRYSTALS AND THEIR ELECTRICAL PROPERTIES. 09-1533
POTASSIUM TANTALATE.   ELECTRICAL PROPERTIES AND DEFECT STRUCTURE OF 08-1314
BARIUM TITANA/ INFLUENCE OF UNIAXIAL TENSILE STRESSES ON THE ELECTRICAL PROPERTIES AND DOMAIN STRUCTURE OF 04-0961
RIUM TITANATE.   RELATION BETWEEN ELECTRICAL PROPERTIES AND MICROSTRUCTURE OF BA 04-0980
OBATES AND METATANTALATES OF TRANSITION METALS OF THE 3D SE/ ELECTRICAL PROPERTIES AND STRUCTURES OF METANI 04-0840
/IS OF NIOBIUM PENTOXIDE DOPED BARIUM TITANATE WITH IMPROVED ELECTRICAL PROPERTIES. (DIELECTRIC CONSTANT, / 04-0840
/RES FROM THE MELT UNDER PRESSURE AND INVESTIGATION OF THEIR ELECTRICAL PROPERTIES. (DIELECTRIC, ELASTIC, / 15-2138
SYSTEM LANTHANA TITANIA PHASE EQUILIBRIA AND ELECTRICAL PROPERTIES (LANTHANUM TITANATE). 36-3592
RCRYSTALLITE LAYERS IN THE SEMICONDUCTOR CERAMIC BARIUM TIT/ ELECTRICAL PROPERTIES OF CRYSTALLITES AND INTE 04-0568
IS. (FINE POWDE/ QUANTITATIVE ANALYSIS OF MICROSTRUCTURE AND ELECTRICAL PROPERTIES OF FERROELECTRIC MATERIA 02-0093
AN APPARATUS FOR INVESTIGATING THE ELECTRICAL PROPERTIES OF FERROELECTRICS. 03-0153
UM TITANATE THIN FILM/ GROWTH STRUCTURE (MICROSTRUCTURE) AND ELECTRICAL PROPERTIES OF FLASH EVAPORATED BARI 04-0605
OELECTRIC BARIUM TITANATE THIN FILMS. (THICKNESS DEPENDENCE) ELECTRICAL PROPERTIES OF FLASH EVAPORATED FERR 04-0967
ALLINE BARIUM TITANATE. (DIELECTRIC CONSTANT, CONDUCTIVITY) ELECTRICAL PROPERTIES OF HIGH PURITY POLYCRYST 04-0704
TRIC (BARIUM TITANATE TYPE) CERAMICS. (IN RELATION TO DENSI/ ELECTRICAL PROPERTIES OF HOT PRESSED FERROELEC 04-0879
TEMPERATURE AND FREQUENCY DEPENDENCES OF THE ELECTRICAL PROPERTIES OF LITHIUM METANIOBATE. 11-1717
MALY AT 263 DEG K IS STRUCTURAL TRANSITION)   ELECTRICAL PROPERTIES OF LITHIUM NIOBATE. (ANO 11-1709
OELECTRIC MATERIALS CRYSTALLIZED FROM GLASS. (60-90 VOL PER/ ELECTRICAL PROPERTIES OF MICROCRYSTALLINE FERR 04-0725
IC CERAMICS WITH THE PEROVSKITE STRUCTURE.   ELECTRICAL PROPERTIES OF NONLINEAR FERROELECTR 08-1461
EMICONDUCTOR IN. THE FERROELECTRIC STATE.   ELECTRICAL PROPERTIES OF PHTHALOCYANINE AS A S 36-3622
ALLS IN TRIGLYCINE SULFATE CRYSTALS.   SPECIAL ELECTRICAL PROPERTIES OF REGIONS NEAR DOMAIN W 27-2988
M NITRATE.   ELECTRICAL PROPERTIES OF RHOMBOHEDRAL POTASSIU 16-2367
OPED BARIUM TITANATE CERAMICS.   ELECTRICAL PROPERTIES OF RUBIDIUM AND CESIUM D 04-0860
CESIUM NITRATE.   ELECTRICAL PROPERTIES OF RUBIDIUM NITRATE AND 16-2286
ED, RARE EARTH OXIDE DOPED, AND UNDOPED / MICROSTRUCTURE AND ELECTRICAL PROPERTIES OF SCANDIUM TRIOXIDE DOP 04-0839
TITANATE.   INVERSION OF ELECTRICAL PROPERTIES OF SEMICONDUCTING BARIUM 04-0572
ATE AS A FUNCTION OF GRAIN SIZE.   VARIATION IN CERTAIN ELECTRICAL PROPERTIES OF SINTERED BARIUM TITAN 04-0627

S AT MICROWAVE FREQUENCIES. (TITANATE ZIRCC/ PREPARATION AND ELECTRICAL PROPERTIES OF SOME CERAMIC MATERIAL 09-1602

LANTHANUM OXIDE- TITANIUM DIOXIDE COMPOUNDS. ELECTRICAL PROPERTIES OF TERNARY BARIUM OXIDE- 13-1912

CRYSTALS (TGS AND OTHERS, BY SELECTIVE DEPOSITION OF CHARG/ ELECTRICAL PROPERTIES OF THE SURFACES OF IONIC 27-2987

ANATE DIELECTRICS PRODUCED BY FLAME SPRAYING (CAPACITORS). ELECTRICAL PROPERTIES OF THICK FILM BARIUM TIT 04-0772

NATE PREPARED BY RF SPUTTERING CN SILICON SUBS/ PHYSICAL AND ELECTRICAL PROPERTIES OF THIN FILM BARIUM TITA 04-0828

NGLE CRYSTALS (HEAT CAPACITY AND ELECTRIC PERMI/ THERMAL AND ELECTRICAL PROPERTIES OF TRIGLYCINE SULFATE SI 27-3141

DRATIC TUNGSTEN BRONZES STRONTIUM POT/ CRYSTAL CHEMISTRY AND ELECTRICAL PROPERTIES OF TWO NEW PHASES OF QUA 13-2010

INFLUENCE CF PHASE TRANSITICNS ON ELECTRICAL RESISTANCE OF BARIUM TITANATE. 04-0599

POTASSIUM NITRATE CRYSTALS. ELECTRICAL RESISTIVITIES OF SODIUM NITRITE AND 16-2258

UM TITANATE. EFFECT OF OXIDE IMPURITIES ON THE ELECTRICAL RESISTIVITY OF LANTHANUM DOPED BARI 04-0587

ATE CERAMICS CONTAINING IMPURITIES. ELECTRICAL RESISTIVITY OF LEAD ZIRCONATE TITAN 09-1607

YSTAL SUBSTRATES AS THE DETERMINING FACTOR OF THE GROWTH ST/ ELECTRICAL STRUCTURE OF THE SURFACE OF REAL CR 27-3039

ELECTRICAL SUSCEPTIBILITY OF ADP (THEORY). 03-0436

ZIRCONATE TITANATE FERROELECTRIC CERAMIC. SOME PHYSICAL AND ELECTRICAL SWITCHING CHARACTERISTICS OF A LEAD 09-1601

GNETIC) RARE EARTH ORTHOCHROMITES, MANGANITES AND FERRITES ELECTRICAL TRANSPORT IN (SEMICONDUCTING AND MA 36-3608

IOXIDE CRYSTALS. EFFECT CF OXYGEN DEFICIENCY ON ELECTRICAL TRANSPORT PROPERTIES OF TUNGSTEN TR 10-1629

HS. (THECRY) DIELECTRIC RESONANCE AND RELAXATION IN ELECTRICALLY BIASED KDP AND DEUTERATED ISOMORP 03-0467

/MATION) STORAGE (PROCESSING) AND DISPLAY DEVICES. (BASED ON ELECTRICALLY CONTROLLED LIGHT SCATTERING AND / 37-3650

/. PARTS 1-3: HEAT CAPACITY, ENTROPY, MAGNETIC MCMENT OF THE ELECTRICALLY POLARIZED FORM FROM 0.4 TO 4.2 D/ 30-3275

RESSEL) FERROELECTRIC CERAMICS. (PZT) OPTICAL PROPERTIES (ELECTRICALLY VARIABLE BIREFRINGENCE) OF (HOT P 09-1554

E (ON ELECTRON IRRADIATION). SECONDARY (ELECTRO) PHOTOELASTIC EFFECT IN LITHIUM NIOBAT 11-1801

NY SULFOIODIDE. NUCLEAR QUADRUPOLE AND ELECTROACOUSTIC ECHOES IN FERROELECTRIC ANTIMO 15-2189

D TGS (HYSTERESIS LOOPS JUST BELOW CURIE POINT CCRRECTED FOR ELECTROCALORIC EFFECT). /RIC TGS AND DEUTERATE 27-2966

OTASSIUM DIHYDROGEN PHOSPHATE CRYSTAL NEAR THE CURIE POINT. (ELECTROCALORIC EFFECT) /US POLARIZATION OF A P 17-2678

EMPERATURE). ULTRASONIC MEASUREMENT OF THE ELECTROCALORIC EFFECT IN KDP (NEAR THE CURIE T 17-2664

M TITANATE. ELECTROCALORIC EFFECT IN POLYCRYSTALLINE BARIU 04-0766

SPECIFIC HEAT UNDER APPLIED ELECTRIC FIELD AND ELECTROCALORIC EFFECT IN TRIGLYCINE SULFATE. 27-2960

C CERAMICS AT LOW TE/ DIELECTRIC AND THERMAL (SPECIFIC HEAT, ELECTROCALORIC EFFECT) STUDIES OF FERROELECTRI 04-0720

ON) ELECTRONIC STUDIES OF POTASSIUM DIHYDROGEN PHOSPHATE. (ELECTROCALORIC STUDIES, SPONTANEOUS POLARIZATI 17-2429

/LEAD TITANATE AND SIMILAR COMPOUNDS. STUDIES ON ANALYSIS OF ELECTROCERAMICS AND ITS RAW MATERIALS. PART-1/ 05-1068

TUNGSTEN OXIDE BRONZES. (SODIUM(X) TUNGSTEN(1-X) Z(Y) OXIDE/ ELECTROCHEMICAL DEPOSITION AND DISSOLUTION OF 13-1944

MOLECULAR UNDERSTANDING OF ELECTROCHEMICAL PROCESSES BY ICE RESEARCH. 34-3484

UM DIHYDROGEN PHOSPHATE. ELECTROCHEMISTRY ON THE POLYCRYSTAL OF POTASSI 17-2536

CY DRIFT AND OXIDATION REDUCTION CF TRANSITION METALS. ELECTROCOLORATION IN STRONTIUM TITANATE: VACAN 06-1098

DUCTING SOLID SOLUTIONS OF ISO- AND HETEROVALENT CATIONIC S/ ELECTROCONDUCTIVITY AND HALL EFFECT IN SEMICON 04-0943

MECHANISM OF ELECTROCONDUCTIVITY IN BARIUM TITANATE. 04-0908

SIC POLARIZATION, SURFACE VERSUS VOLUME CONDUCTION (INCLUDES ELECTRODE EFFECTS). /: EXTRINSIC VERSUS INTRIN 34-3455

/L- FERROELECTRIC- METAL JUNCTION SYSTEMS. (INFLUENCE OF THE ELECTRODE MATERIAL ON THE PYROELECTRIC EFFECT) 02-0127

RAMETERS. EFFECT OF SAMPLE TREATMENT AND ELECTRODE NATURE ON SOME TRIGLYCINE SULFATE PA 27-3148

TRIC PHOTOCONDUCTOR MEMORY DEVICE. PHOTOCONDUCTOR AND ELECTRODE REQUIREMENTS FOR THIN FILM FERROELEC 37-3671

TCHING OF LEAD ZIRCONATE- LEAD TITANATE CERAMICS WITH INDIUM ELECTRODES. IMPROVED AGING AND SWI 09-1516

TCHING OF LEAD ZIRCONATE- LEAD TITANATE CERAMICS WITH INDIUM ELECTRODES. IMPROVED AGING AND SWI 09-1531

ETAL. (METHOD OF STUDYING THE TYPE OF CONDUCTIVITY, PLATINUM ELECTRODES) /HE SYSTEM METAL- LEAD TITANATE- M 05-1057

ON TRIGLYCINE SULFATE. AFTER EFFECTS IN SEMICONDUCTOR ELECTRODES EVAPORATED ON INSULATORS ESPECIALLY 27-3170

IGLYCINE SULFATE. EFFECT OF EVAPORATED METALLIC ELECTRODES ON THE TRANSITION TEMPERATURE OF TR 27-3100

/NLINEAR DIELECTRICS FOR HIGH POWER ELECTRONIC TUNING. (WITH ELECTRODES REDUCING POLARIZATION SWITCHING TI/ 37-3660

O/ HIGH FIELD (1.5-6 KV/CM) POLARIZATION REVERSALS IN LIQUID ELECTRODED BARIUM TITANATE CRYSTALS (CALCULATI 04-0994

CERAMICS. ELECTROELASTIC CONSTANT MEASUREMENTS FOR PLZT 09-1598

AD TITANATE ZIRCONATE WITH THOROUGHNESS OF POL/ VIBRATION OF ELECTROELASTIC CONSTANTS OF POLYCRYSTALLINE LE 09-1489

/OF CONVERSION FORMULAS INTERRELATING THE LINEAR ELASTIC AND ELECTROELASTIC EQUATIONS OF STATE: ISOTROPIC / 03-0277

TIONS (TGS). ELECTROGYRATION AND FERROELECTRIC PHASE TRANSI 27-2970

/ TYPE OXIDE SOLID SOLUTICN AND ITS APPLICATION TO THE SOLID ELECTROLYTE FUEL CELL. ((CALCIUM, STRONTIUM) / 08-1444

CTRICS. (THEORY, BARIUM TITANATE, ROCHELLE SALT) ELECTROMAGNETIC AND ACOUSTIC WAVES IN FERROELE 03-0455

PIEZOELECTRIC LEAKY SURFACE WAVE IN LITHIUM NIOBATE (ELECTROMAGNETIC COUPLING). 11-1832

VES ON PIEZOELECTRIC SUBSTRATES (LI/ PERTURBATION THEORY FOR ELECTROMAGNETIC COUPLING TO ELASTIC SURFACE WA 11-1751

ECTRIC EFFECT IN TGS, ETHYL DIAMINE TARTR/ TRANSFORMATION OF ELECTROMAGNETIC RADIATION INTO MOTION. (PYROEL 04-0854

FFECT. (BARIUM TITANATE, TRIGLYCINE SULFATE, A/ DETECTION OF ELECTROMAGNETIC RADIATION USING PYROELECTRIC E 04-0978

ERED FERROELECTRICS. COUPLED ELECTROMAGNETIC SPIN WAVES IN MAGNETICALLY ORD 03-0148

/ RESONANCE OF A LOW FREQUENCY SPIN WAVE WITH AN ACOUSTIC OR ELECTROMAGNETIC WAVE IN A FERROELECTRIC DOMAIN 03-0284

S. ELASTIC AND ELECTROMAGNETIC WAVES IN PIEZOELECTRIC CRYSTAL 03-0371

) TITANATE(1-X) CERAMICS NEAR / ORIGIN OF THE MAXIMUM IN THE ELECTROMECHANICAL ACTIVITY IN LEAD ZIRCONATE(X 09-1499

) TITANATE(1-X) CERAMICS NEAR / ORIGIN OF THE MAXIMUM IN THE ELECTROMECHANICAL ACTIVITY IN LEAD ZIRCONATE(X 09-1500

FERROELECTRICS. (BARIUM TITANATE CERAMICS) NONLINEARITY OF ELECTROMECHANICAL AND MECHANICAL PROPERTIES OF 04-0986

MATERIALS. SHOCK WAVE INDUCED ELECTROMECHANICAL CONVERSION IN FERROELECTRIC 03-0491

ECTRIC PROPERTIES AND FERROELECTRIC PHASE TRANSITION ON LIT/ ELECTROMECHANICAL COUPLING EFFECTS ON THE DIEL 29-3251

RAMICS. MEASUREMENT OF THE DISSIPATION ASSOCIATED WITH THE ELECTROMECHANICAL COUPLING IN PIEZOELECTRIC CE 03-0514

RIC PHASE TRANSITION CF TRIGLYCINE SULFATE. ELECTROMECHANICAL COUPLING NEAR THE FERROELECT 27-3174

DE. ELECTROOPTICAL AND ELECTROMECHANICAL EFFECT IN ANTIMONY SULFOIODI 15-2178

F LEAD TITANATE CERAMICS. ELECTROMECHANICAL PROPERTIES AND APPLICATION O 05-1085

TEM LEAD TITANATE- LEAD CADMATE(0.5) TUNGSTATE(0.5). ELECTROMECHANICAL PROPERTIES IN THE BINARY SYS 10-1638

LECTRICS (BARIUM TITANATE ZIRCONATES) IN STRONG ELECTRIC FI/ ELECTROMECHANICAL PROPERTIES OF CERAMIC FERROE 09-1605

D LEAD TITANATE ZIRCONATE CERAMICS. STRUCTURE AND ELECTROMECHANICAL PROPERTIES OF LANTHANUM DOPE 09-1502

D LEAD TITANATE ZIRCONATE CERAMICS. STRUCTURE AND ELECTROMECHANICAL PROPERTIES OF LANTHANUM DOPE 09-1531

CERAMICS CONTAINING LANTANUM AND MANGANESE. ELECTROMECHANICAL PROPERTIES OF LEAD TITANATE 05-1060

(0.52) TITANATE(0.48) CERAMICS CAUSED / DETERIORATION OF THE ELECTROMECHANICAL PROPERTIES OF LEAD ZIRCONATE 09-1606

RAL FERROELECTRIC CRYSTALS. ELECTROMECHANICAL PROPERTIES OF OXYGEN OCTAHED 13-2061

NATE SINGLE CRYSTALS. ELECTROMECHANICAL PROPERTIES OF STRONTIUM TITA 06-1204

RIC SODIUM AMMONIUM SELENATE DIHYDRATE. ELECTROMECHANICAL PROPERTIES OF THE FERROELECT 22-2884

FATE. (MEASUREMENTS VERIFY THEORY APPROXIMATELY) ELECTROMECHANICAL PROPERTIES OF TRIGLYCINE SUL 27-3145

A NEW SECOND HARMONIC TYPE FERROELECTRIC MODULATOR FOR ELECTROMETER. (WITH A TGS SINGLE CRYSTAL) 27-2951

BARIUM TITANATE, (COMPARED TO ST/ EFFECTS OF (MONOENERGETIC) ELECTRON AND (COBALT-60) GAMMA-IRRADIATION ON 04-0970

FOR SIGNAL PROCESSING PURPOSES. (TECHNIQUE APPLIED TO LITH/ ELECTRON BEAM SENSING OF SURFACE ELASTIC WAVES 11-1738

IN BISMUTH TITANIUM OXIDE SINGLE CRYSTALS. ELECTRON BEAM WRITING OF FERROELECTRIC DOMAINS 08-1300

BOND IN POTASSIUM DIHYDROGEN PHOSPHATE. ELECTRON DELOCALIZATION WEIGHT OF THE HYDROGEN 17-2525

BARIUM TITANATE BY CONVERGENT BEAM ELECTRON DIFFRACTION. 04-0991

RACITE. NEUTRON AND ELECTRON DIFFRACTION STUDY IN NICKEL IODINE BO 14-2073

AMMONIUM SULFATE. ELECTRON DIFFRACTION STUDY OF THE STRUCTURE OF 19-2797

BARIUM TITANATE CERAMIC: A NEW ELECTRON EMITTER. 04-0802

M TITANATE. MEASUREMENT OF ELECTRON ENERGY LOSS FUNCTION IN BULK STRONTIU 06-1138

CONDARY (ELECTRO) PHOTOELASTIC EFFECT IN LITHIUM NIOBATE (ON ELECTRON IRRADIATION). SE 11-1801

TRICS AND FERROELECTRIC MAGNETS. (LEAD AND BARIU/ STRUCTURAL ELECTRON MICRODIFFRACTION STUDIES OF FERROELEC 04-1038
CTURAL STUDIES OF FERROELECTRICS AND FERROELECTRIC MAGNETS. (ELECTRON MICRODIFFRACTION, SUMMARY ONLY) /STRU 03-0490
  DIRECT OBSERVATION OF BARIUM TITANATE CRYSTALS BY SCANNING ELECTRON MICROSCOPE.   04-0808
RROELECTRIC DOMAINS IN TRIGLYCINE SULFATE USING THE SCANNING ELECTRON MICROSCOPE.   DIRECT OBSERVATION OF FE 27-3051
N STRUCTURES, MODULATED BY A PHASE TRANSITION, IN A SCANNING ELECTRON MICROSCOPE. /N OF FERROELECTRIC DOMAI 03-0474
S OF BARIUM TITANATE BY MEANS OF A STROBOSCOPIC TRANSMISSION ELECTRON MICROSCOPE. /N OF SINGLE CRYSTAL FILM 04-0942
OF THE SURFACES OF FERROELECTRIC CRYSTALS IN A PHOTOEMISSION ELECTRON MICROSCOPE.   OBSERVATION 37-3679
         STUDY OF THE SURFACE OF ICE WITH A SCANNING ELECTRON MICROSCOPE.   34-3420
RFACE STUDIES OF BARIUM TITANATE CERAMICS USING THE SCANNING ELECTRON MICROSCOPE.   SU 04-0567
/G ELECTRIC MICRO FIELDS (OF SOLID SURFACES) IN THE EMISSION ELECTRON MICROSCOPE (BARRIER LAYERS IN DOPED / 04-0913
AINS IN TGS.   SCANNING ELECTRON MICROSCOPE STUDY OF FERROELECTRIC DOM 27-3052
/MILIES DERIVED FROM THE RHENIUM TRIOXIDE STRUCTURE TYPE: AN ELECTRON MICROSCOPE STUDY OF REDUCED TUNGSTEN- 10-1630
ERENT DOMAIN BOUNDARIES. PART-1: DYNAMICAL THEORY. PART-2: / ELECTRON MICROSCOPE TRANSMISSION IMAGES OF COH 03-0255
C/ STUDY OF DOMAINS OF FERROELECTRIC CRYSTALS WITH SCANNING ELECTRON MICROSCOPE. (TUNGSTEN TRIOXIDE SINGLE 10-1646
ARIUM SODIUM NIOBATES.   SCANNING ELECTRON MICROSCOPIC EXAMINATION OF SINTERED B 13-1955
DOMAINS OF PEROVSKITE TYPE OXIDES. (LEAD TITANATE, POTASSIU/ ELECTRON MICROSCOPIC STUDIES ON FERROELECTRIC 04-0993
MAINS OF BARIUM TITANATE IN LOW TEMPERATURE PHASES.   ELECTRON MICROSCOPIC STUDY OF FERROELECTRIC DO 04-1034
INATION OF SUBSTOICHIOMETRIC TUNGSTEN OXIDE(3-X) CRYSTALS BY ELECTRON MICROSCOPY.   EXAM 10-1627
DURING SWITCHING IN SINGLE CRYSTAL FILMS OF BARIUM TITANATE (ELECTRON MICROSCOPY). /N OF DOMAIN BOUNDARIES 04-0940
EFECTS RELATED TO NONSTOICHIOMETRY (AND ANTIFERROELE/ USE OF ELECTRON MICROSCOPY IN THE STUDY OF EXTENDED D 36-3522
ROELECTRIC DOMAINS OF DICALCIUM STRONTIUM PROFIONATE.   ELECTRON MIRROR MACROSCOPIC OBSERVATION OF FER 26-2946
ECTRIC DOMAINS OF BARIUM TITANATE AND DICALCIUM STRONTIUM P/ ELECTRON MIRROR MICROSCOPIC ASPECTS OF FERROEL 04-0973
ROELECTRIC DOMAINS.   ELECTRON MIRROR MICROSCOPIC OBSERVATION OF FER 03-0473
ROELECTRIC DOMAINS. (GOLD COATED BARIUM TITANATE)   ELECTRON MIRROR MICROSCOPIC OBSERVATION OF FER 04-0974
ON OF MOVING FERROELECTRIC DOMAINS OF GADOLI/ APPLICATION OF ELECTRON MIRROR MICROSCOPY TO DIRECT OBSERVATI 30-3293
IRRADIATED SINGLE CRYSTALS OF ICE.   ENDOR (ELECTRON NUCLEAR DOUBLE RESONANCE) STUDY OF X- 34-3413
NATE CENTER IN X-RAY IRRADIATED SINGLE CRYST/ OBSERVATION OF ELECTRON NUCLEAR TRIPLE RESONANCE FOR THE ARSE 18-2729
N STRUCTURE IN VANADIUM DOPED TRIGLYCINE SULFATE OBSERVED BY ELECTRON PARAMAGNETIC RESONANCE /OF THE DOMAI 27-3013
OF TRANSITION TEMPERATURES IN CESIUM LEAD TRICHLORIDE USING ELECTRON PARAMAGNETIC RESONANCE DETERMINATION 33-3408
INFORMATION ON HIGH DIELECTRIC CONSTANT MATERIALS GAINED BY ELECTRON PARAMAGNETIC RESONANCE. (ABSTRACT ON/ 02-0101
UCLEAR DOUBLE RESONANCE OF THE PROTON LATTICE / DETECTION BY ELECTRON PARAMAGNETIC RESONANCE AND ELECTRON N 18-2724
UCLEAR DOUBLE RESONANCE OF THE ARSENATE CENTER IN X-IRRADIA/ ELECTRON PARAMAGNETIC RESONANCE AND ELECTRON N 18-2727
UCLEAR DOUBLE RESONANCE OBSERVATIONS OF DOMAINS IN THE IRRA/ ELECTRON PARAMAGNETIC RESONANCE AND ELECTRON N 18-2726
NATE NEAR THE CURIE TEMPERATURE.   ELECTRON PARAMAGNETIC RESONANCE IN BARIUM TITA 04-0928
IATED SINGLE CRYSTAL OF FERROELECTRIC LITHIUM TRIHYDROGEN D/ ELECTRON PARAMAGNETIC RESONANCE IN GAMMA-IRRAD 22-2842
METRIC BARIUM TITANATE.   ELECTRON PARAMAGNETIC RESONANCE IN NONSTOICHIO 04-0640
OF IRON(3+) AND MANGANESE(2+) IN LITHIUM NIOBATE.   AN ELECTRON PARAMAGNETIC RESONANCE INVESTIGATION 11-1725
OF THE INTERNAL ELECTRIC FIELD IN COPPER DOPED ROCHELLE SALT ELECTRON PARAMAGNETIC RESONANCE INVESTIGATION 28-3233
OF POLYCRYSTALLINE STRONTIUM TITANATE CONTAINING MANGANESE / ELECTRON PARAMAGNETIC RESONANCE INVESTIGATION 06-1162
OF MANGANESE(2+) COMPLEXES IN FERROELECTRIC TRIS-SARCOSINE/ ELECTRON PARAMAGNETIC RESONANCE INVESTIGATIONS 36-3525
OF (COPPER) DOPED FERROELECTRIC CRYSTALS. (TRIGLYCINE SULF/ ELECTRON PARAMAGNETIC RESONANCE INVESTIGATIONS 27-3132
POTASSIUM MANGANESE TRIFLUORI/ SHIFT OF NEEL TEMPERATURE AND ELECTRON PARAMAGNETIC RESONANCE LINE WIDTH OF 33-3377
DMIUM SULFATE.   TEMPERATURE DEPENDENCE OF THE ELECTRON PARAMAGNETIC RESONANCE OF AMMONIUM CA 19-2792
) IONS IN ANTIFERROELECTRIC ADP CRYSTALS.   ELECTRON PARAMAGNETIC RESONANCE OF CHROMIUM(3+ 17-2546
) IN ADP.   ELECTRON PARAMAGNETIC RESONANCE OF CHROMIUM(3+ 17-2596
IN POTASSIUM TANTALATE.   ELECTRON PARAMAGNETIC RESONANCE OF COBALT(2+) 08-1342
IN ROCHELLE SALT.   ELECTRON PARAMAGNETIC RESONANCE OF COPPER IONS 28-3176
3+) DOPED STRON/ INTERPRETATION OF ELECTRIC FIELD EFFECTS IN ELECTRON PARAMAGNETIC RESONANCE OF GADOLINIUM(06-1168
3+) IN LITHIUM NIOBATE.   ELECTRON PARAMAGNETIC RESONANCE OF GADOLINIUM(11-1763
3+) DOPED LITHIUM NIOBATE.   ELECTRON PARAMAGNETIC RESONANCE OF GADOLINIUM(11-1833
IATED ROCHELLE SALT.   ELECTRON PARAMAGNETIC RESONANCE OF GAMMA-IRRAD 28-3180
PED LITHIUM NIOBATE.   ELECTRON PARAMAGNETIC RESONANCE OF IRON(3+) DO 11-1161
BARIUM TITANATE AT LOW TEMPERATURES.   ELECTRON PARAMAGNETIC RESONANCE OF IRON(3+) IN 04-0927
BARIUM TITANATE IN THE RHOMBOHEDRAL AND THE CUBIC PHASES.   ELECTRON PARAMAGNETIC RESONANCE OF IRON(3+) IN 04-0929
/ETER AND PHASE TRANSITIONS OF STRESSED STRONTIUM TITANATE. (ELECTRON PARAMAGNETIC RESONANCE OF IRON- VANA/ 06-1170
2) IN BARI/ FORBIDDEN LINES IN THE HYPERFINE SPECTRUM OF THE ELECTRON PARAMAGNETIC RESONANCE OF MANGANESE(+ 04-0877
+) IN POTASSIUM TANTALATE.   ELECTRON PARAMAGNETIC RESONANCE OF MANGANESE(2 08-1343
+) IN THE FERROELECTRIC PHASE OF SODIUM NITRITE.   ELECTRON PARAMAGNETIC RESONANCE OF MANGANESE(2 16-2358
+) DOPED/ ALLOWED AND FORBIDDEN HYPERFINE TRANSITIONS IN THE ELECTRON PARAMAGNETIC RESONANCE OF MANGANESE(2 16-2383
+) DOPED AMMONIUM CADMIUM SULFATE.   ELECTRON PARAMAGNETIC RESONANCE OF MANGANESE(2 19-2791
ELENATE.   THE ELECTRON PARAMAGNETIC RESONANCE OF POTASSIUM S 22-2850
ARTH IMPURITIES IN STRONTIUM TITANATE.   ELECTRON PARAMAGNETIC RESONANCE OF SOME RARE E 06-1192
M COMPLEX IN TRIGLYCINE FLUOBERYLLATE.   ELECTRON PARAMAGNETIC RESONANCE OF THE CHROMIU 27-3133
PPER ION COMPLEX WITH SPIN S=2 IN TRIGLYCINE FLUOBERYLLATE / ELECTRON PARAMAGNETIC RESONANCE OF THE FOUR CO 27-3134
ADOLINIUM IONS IN STRONTIUM AND BARIUM TITANATES.   ELECTRON PARAMAGNETIC RESONANCE OF TRIVALENT G 04-0920
/ROELECTRIC PHASE TRANSITION IN SODIUM NITRITE AS STUDIED BY ELECTRON PARAMAGNETIC RESONANCE (PHASE TRANSI/ 16-2359
ATIC MAGN/ DETERMINATION OF CHROMIUM VALENCY FROM STUDIES OF ELECTRON PARAMAGNETIC RESONANCE SPECTRA AND ST 04-0542
ERROELECTRIC CERAMIC CONTAINING TITANI/ PECULIARITIES OF THE ELECTRON PARAMAGNETIC RESONANCE SPECTRA IN A F 04-1020
OMIUM(3+) IONS/ INFLUENCE OF SPONTANEOUS POLARIZATION ON THE ELECTRON PARAMAGNETIC RESONANCE SPECTRA OF CHR 27-3136
PPER(2+)) DOPED FERROEL/ EFFECT OF APPLIED ELECTRIC FIELD ON ELECTRON PARAMAGNETIC RESONANCE SPECTRA OF (CO 28-3185
STITUTIONAL IMPUR/ EFFECT OF TEMPERATURE AND PRESSURE ON THE ELECTRON PARAMAGNETIC RESONANCE SPECTRA OF SUB 06-1193
ANGANESE DOPED) LITHIUM NIOBATE AND (COPPER(2+) DOPED) ZINC/ ELECTRON PARAMAGNETIC RESONANCE SPECTRUM IN (M 11-1792
ON(3+) IN STRONTIUM TITANATE DUE TO NEAREST NE/ STRONG AXIAL ELECTRON PARAMAGNETIC RESONANCE SPECTRUM OF IR 06-1144
LYBDENUM(5+) IN STRONTIUM TITANATE: AN EXAMPLE OF THE DYNAM/ ELECTRON PARAMAGNETIC RESONANCE SPECTRUM OF MO 06-1125
NAMICS IN STRONTIUM TITANATE IN THE NEIGHBORHOOD OF THE PHA/ ELECTRON PARAMAGNETIC RESONANCE STUDIES AND DY 06-1232
OMIUM- AND VANADYL- DOPED TRIGLYCINE SULFATE SINGLE CRYSTALS ELECTRON PARAMAGNETIC RESONANCE STUDIES OF CHR 27-3161
OMIUM(3+), COPPER(2+), AND MANGANESE(2+) PARAMAGNETIC ION C/ ELECTRON PARAMAGNETIC RESONANCE STUDIES OF CHR 27-3135
STAL FIELD PARAMETERS IN IRON(3+) DOPED LITHIUM NIOBATE.   ELECTRON PARAMAGNETIC RESONANCE STUDIES OF CRY 11-1839
/DED FERROELECTRICS: EVIDENCE FOR SLATER CONFIGURATIONS FROM ELECTRON PARAMAGNETIC RESONANCE STUDIES OF X-/ 18-2728
RRADIATED POTASSIUM DIHYDROGEN ARSENATE, POTASSIUM DIDEUTER/ ELECTRON PARAMAGNETIC RESONANCE STUDIES OF X-I 18-2725
TIC COPPER(2+) COMPLEXES IN CRYSTALS OF GUANIDINIUM ALUMINU/ ELECTRON PARAMAGNETIC RESONANCE STUDY OF MAGNE 21-2820
AGNETIC ION COMPLEXES IN FERROELECTRIC CRYSTALS.   ELECTRON PARAMAGNETIC RESONANCE STUDY OF PARAM 03-0477
METAL IONS IN GUANIDINIUM ALUMINUM SULFATE HEXAHYDRATE.   ELECTRON PARAMAGNETIC RESONANCE STUDY OF SOME 21-2817
ATALYTIC ACTIVITY OF NONSTOICHIOMETRIC TUNGSTEN TRIOXIDE.   ELECTRON PARAMAGNETIC RESONANCE STUDY OF THE C 10-1628
DAMAGED FERROELECTRICS.   ELECTRON PARAMAGNETIC RESONANCE STUDY OF X-RAY 02-0079
RYSTALS (CYCLOTRON RESONANCE, ABSORPTION AND LUM/ EFFECTS OF ELECTRON PHONON INTERACTION IN FERROELECTRIC C 03-0427
                    X-RAY DIFFRACTION STUDIES OF ELECTRON POLARIZATION IN FERROELECTRICS.   03-0199
ES IN EVAPORATED BARIUM TITANATE FILMS.   ELECTRON PROBE X-RAY MICROANALYSIS ON IMPURITI 04-0742
/ OF THE FERROELECTRIC PHASE CHANGE IN BARIUM TITANATE USING ELECTRON SPIN RESONANCE AND THE MOSSBAUER EFF/ 04-0827
ON DAMAGE IN LITHIUM NIOBATE. (CAPACITANCE, OPTICAL SPECTRA, ELECTRON SPIN RESONANCE CELL PARAMETERS) /ATI 11-1705

GREES K ANOMALY IN STRONTIUM TITANATE.   NEW ELECTRON SPIN RESONANCE EVIDENCE FOR THE 65 DE 06-1230
ER CONFIGURATIONS AND OF EXCHANGE OF THE ARSENATE RADICAL I/   ELECTRON SPIN RESONANCE IDENTIFICATION OF SLAT 18-2734
S.   ELECTRON SPIN RESONANCE IN GAMMA-IRRADIATED TG 27-3035
  (DOPED WITH GADOLINIUM(3+) CR IRC/ ELECTRIC FIELD EFFECT OF   ELECTRON SPIN RESONANCE IN STRONTIUM TITANATE. 06-1229
Y IRRADIATED SINGLE CRYSTALS OF FERROELECTRIC POTASSIUM DIH/   ELECTRON SPIN RESONANCE INVESTIGATIONS OF X-RA 17-2558
SIONAL ANTIFERROMAGNET CCPPER FORMATE TE/ ANOMALOUS SHIFT OF   ELECTRON SPIN RESONANCE LINES IN THE TWO DIMEN 26-2941
/GEN (2-) VACANCIES IN CHRCMIUM(3+) DOPED STRONTIUM TITANATE (ELECTRON SPIN RESONANCE MEASUREMENTS, MODEL P/ 06-1161
OMIUM DOPED STRONTIUM TITANATE SINGLE CRYSTALS.   ELECTRON SPIN RESONANCE OF CHROMIUM(3+) IN CHR 06-1147
DIUM AMMONIUM TARTRATE.   ELECTRON SPIN RESONANCE OF COPPER(2+) DOPED SO 29-3252
RROELECTRIC AND ANTIFERROELECTRIC ARSENATE SINGLE CRYSTALS.   ELECTRON SPIN RESONANCE OF FREE RADICALS IN FE 18-2731
ARIUM TITANATE CERAMIC SEMICONDUCTOR.   TWO SETS OF ELECTRON SPIN RESONANCE OF GADOLINIUM(3+) IN B 04-0990
RROELECTRIC AMMONIUM HYDROGEN SULFATE.   ELECTRON SPIN RESONANCE OF GAMMA DEFECTS IN FE 29-2748
MONIUM SULFATE.   ELECTRON SPIN RESONANCE OF GAMMA-IRRADIATED AM 19-2790
/OF THE RADICAL IN TRIGLYCINE SULFATE NEAR THE CURIE POINT. (ELECTRON SPIN RESONANCE OF GAMMA-IRRADIATED S/ 27-3033
S. PART-2: INTERNAL MCTICN OF HYDROGEN AND THE ACTIVATION E/   ELECTRON SPIN RESONANCE OF GAMMA-IRRADIATED TG 27-3034
IGLYCINE SELENATE.   ELECTRON SPIN RESONANCE OF GAMMA-IRRADIATED TR 27-3139
IN BARIUM SODIUM NIOBATE.   PHOTOCHROMIC EFFECT AND   ELECTRON SPIN RESONANCE OF IRON AND MOLYBDENUM 13-1930
ERATED POTASSIUM DIHYDROGEN PHOSPHATE.   ELECTRON SPIN RESONANCE OF IRRADIATED AND DEUT 17-2533
RIUM TITANATE.   ELECTRON SPIN RESONANCE OF MANGANESE(2+) IN BA 04-0747
/ES AND PRESSURE EFFECTS ON SELENITE (ION). RADIATION DAMAGE   ELECTRON SPIN RESONANCE SPECTRA IN FERROELECT/ 22-2870
) AND GADOLINIUM(3+) IN LANTHANUM ALUMINATE.   ELECTRON SPIN RESONANCE SPECTRA OF CHROMIUM(3+ 36-3583
+) DOPED LITHIUM NIOBATE.   FCRBIDDEN HYPERFINE LINES IN   ELECTRON SPIN RESONANCE SPECTRA OF MANGANESE(2 11-1805
DEPENDENCE) IN SOME FERROELECTRICS AND FERROELECTRIC FERROM/   ELECTRON SPIN RESONANCE (SPECTRA, TEMPERATURE 08-1368
) IN LITHIUM NIOBATE.   ELECTRON SPIN RESONANCE STUDIES OF CHROMIUM(3+ 11-1804
LD PARAMETERS IN MANGANESE(2+) DOPED LITHIUM NIOBATE.   ELECTRON SPIN RESONANCE STUDIES OF CRYSTAL FIE 11-1803
ANATE.   ELECTRON SPIN RESONANCE STUDY IN STRONTIUM TIT 06-1228
D CRYSTALS.   ELECTRON SPIN RESONANCE STUDY OF ADP- KDA MIXE 17-2683
IN IRRADIATED AMMONIUM DIDEUTERIUM PHOSPHATE- 5 PERCENT PO/   ELECTRON SPIN RESONANCE STUDY OF AMMONIUM IONS 17-2684
TED CESIUM TRIHYDROGEN SELENITE AT ROCM TEMPERATURE.   ELECTRON SPIN RESONANCE STUDY OF GAMMA-IRRADIA 22-2833
TED POTASSIUM SELENATE (STRUCTURAL PHASE CHANGES).   ELECTRON SPIN RESONANCE STUDY OF GAMMA-IRRADIA 22-2825
AGE CENTERS IN AMMONIUM DIHYDROGEN PHOSPHATE POTASSIUM DIHY/   ELECTRON SPIN RESONANCE STUDY OF RADIATION DAM 17-2547
TRIC PHASE TRANSITIONS IN SODIUM TRIHYDROGEN DISELENITE.   ELECTRON SPIN RESONANCE STUDY OF THE FERROELEC 22-2771
LUORITES. (LEAD TITANATE)   ELECTRON SPIN RESOURCE STUDIES AT DEFECTS IN F 05-1086
OF ANTIMONY SULFOIODIDE. (IN PARAELECTRIC AND FERROELECTRIC/   ELECTRON STATE DENSITY AND OPTICAL PROPERTIES 15-2204
ICS.   LOCAL ELECTRON STATES IN DOMAIN WALLS OF FERROELECTR 03-0161
ENERATION OF STRONG FIELDS) DURING HOLOGRAM RECORDING IN CR/   ELECTRON THERMAL DIFFUSION EFFECTS (INTERNAL G 06-1090
IUM DOPED) STRONTIUM TITANATE AND (CALCIUM DOPED) POTASSIUM/   ELECTRON TUNNELING AND BAND STRUCTURE OF (NIOB 06-1222
HOTTKY BARRIER JUNCTIONS.   ELECTRON TUNNELING INTO POTASSIUM TANTALATE SC 08-1365
ALS.   PHONONS, ELECTRONS, AND PROTONS IN FERROELECTRIC MATERI 02-0116
AL PHONONS IN IONIC FERROELECTRIC / SCATTERING OF CONDUCTION ELECTRONS BY TRANSVERSE AND LONGITUDINAL OPTIC 03-0292
                               IRRADIATION PRODUCED SOLVATED ELECTRONS IN ICE.   34-3423
/ (MECHANICAL, CRYSTALLOGRAPHIC, PHYSICAL, THERMAL MAGNETIC,   ELECTRONIC AND OPTICAL PROPERTIES OF EACH OF / 01-0025
ELECTRIC ELECTRIC AND MAGNETIC MOMENTS OF OPTICAL TRANSITIO/   ELECTRONIC AND SPIN EXCITATIONS IN ANTIFERRODI 03-0414
NTIUM TITANATE AT LOW TEMPERATURES.   ELECTRONIC CONDUCTION IN SLIGHTLY REDUCED STRO 06-1151
ARIUM TITANATE.   CALCULATIONS OF   ELECTRONIC CONDUCTIVITY IN NONSTOICHIOMETRIC B 04-0821
BIDIUM DIHYDROGEN PHOSPHATE CRYSTALS.   ELECTRONIC EFFECT AND ELASTIC PROPERTIES OF RU 17-2617
ADIATION) IN ROCHELLE SALT CRYSTALS.   ELECTRONIC EFFECTS (OF X-RAY AND GAMMA-RAY IRR 28-3192
ION BY ANTIFERRODIELECTRICS IN THE FREQUENCY RANGE OF DOUBLE   ELECTRONIC EXCITATIONS OF MOLECULES (IONS). /T 03-0419
FLUORIDES.   ELECTRONIC HALL MOBILITY IN THE ALKALINE EARTH 33-3404
IUM TITANATE.   TEMPERATURE DEPENDENCE OF   ELECTRONIC POLARIZABILITY OF STRONTIUM AND BAR 04-0738
RY OF ENERGY STATES INVOLVED)   STATISTICAL AND   ELECTRONIC PROBLEMS IN FERROELECTRICITY. (THEO 03-0325
DIATION EFFECTS IN ROCHELLE SALT AND TRIGLYCINE SULFATE. (ON   ELECTRONIC PROPERTIES.) /TAL STUDY OF GAMMA-RA 27-3101
SUSCEPTIBILITY) OF HYDROGEN TUNGSTEN BRONZES, HYDROGEN(X) T/   ELECTRONIC PROPERTIES (CONDUCTIVITY, MAGNETIC 13-1934
IUM TITANATE. (SURFACE BARRIER SYSTEMS, METAL CONTACTS) SOME   ELECTRONIC PROPERTIES OF ZINC OXIDE AND STRONT 06-1176
N SULFATE DODECAHYDRATE.   MOSSBAUER STUDY OF   ELECTRONIC SPIN FLIP PROCESSES IN AMMONIUM IRO 29-2771
                             SYMMETRY ANALYSIS AND   ELECTRONIC STATES IN CUBIC ICE.   34-3462
SPHATE. (ELECTROCALORIC STUDIES, SPONTANEOUS POLARIZATION)   ELECTRONIC STUDIES OF POTASSIUM DIHYDROGEN PHO 17-2429
D SEEBECK COEFFICIENTS) IN STRONTIUM TITANATE (4.2-300 DEG /   ELECTRONIC TRANSPORT (CONDUCTIVITY AND HALL AN 06-1129
OLARIZATION SWITCHING / NONLINEAR DIELECTRICS FOR HIGH POWER   ELECTRONIC TUNING. (WITH ELECTRODES REDUCING P 37-3660
ERATED SECOND HARMONICS IN POTASSIUM DIHYDROGEN PHOSPHATE.   ELECTROOPTIC AMPLITUDE MODULATION OF LASER GEN 17-2694
ITHIUM TANTALATE SINGLE CRYSTALS AS A FUNCTION OF MELT COMP/   ELECTROOPTIC AND FERROELECTRIC PROPERTIES OF L 12-1874
ANTHANUM TITANATE SINGLE CRYSTAL. (FOUND NOT FERROELECTRIC)   ELECTROOPTIC AND PIEZOELECTRIC PROPERTIES OF L 36-3576
LANTHANUM) (ZIRCONATE, TITANATE) FERROELECTRIC CERAMICS FOR   ELECTROOPTIC APPLICATIONS. HOT PRESSED (LEAD, 09-1526
OCESSING) AND DISPLAY DEVICES. (BASED ON ELECTRICALLY CONTR/   ELECTROOPTIC CERAMIC (INFORMATION) STORAGE (PR 37-3650
DEVICES IN DYE LASERS (PLZT).   ELECTROOPTIC CERAMICS AS WAVELENGTH SELECTION 09-1618
ATE TITANATE, EFFECT OF GRAIN SIZE, / RECENT DEVELOPMENTS IN   ELECTROOPTIC CERAMICS (HOT PRESSED LEAD ZIRCON 02-0086
RCONATE TITANATE SYSTEM.   IMPROVED HOT PRESSED   ELECTROOPTIC CERAMICS IN THE LEAD LANTHANUM ZI 09-1528
MATION STORAGE AND DISPLAY (HOT PRESSED LEAD ZIRCONATE TITA/   ELECTROOPTIC CERAMICS- NEW MATERIALS FOR INFOR 09-1550
ASSIUM DIHYDROGEN PHOSPHATE CRYSTAL AND ITS APPLICATION FOR/   ELECTROOPTIC CHARACTERISTICS OF DEUTERATED POT 17-2499
LECTRIC CONSTANT, NONLINEAR OPTICAL COEFFICIENTS, AND LINEAR   ELECTROOPTIC COEFFICIENTS) /RACTIVE INDEX, DIE 13-2024
E.   SIGNS OF THE   ELECTROOPTIC COEFFICIENTS FOR LITHIUM TANTALAT 12-1880
/THE FREE AND CLAMPED VALUES OF THE DIELECTRIC CONSTANTS AND   ELECTROOPTIC COEFFICIENTS IN BARIUM SODIUM NI/ 13-1998
E (SINGLE) C/ INTERFEROMETRIC DETERMINATION OF THE QUADRATIC   ELECTROOPTIC COEFFICIENTS IN STRONTIUM TITANAT 06-1130
                                        OPTIMISATION OF   ELECTROOPTIC CRYSTAL MODULATORS. (KDP)   17-2541
                            SOLUBILITY IN THE GROWTH OF   ELECTROOPTIC CRYSTALS ADP AND KDP.   17-2446
M SODIUM NIOBATE.   HOLOGRAPHIC PATTERN FIXING IN   ELECTROOPTIC CRYSTALS. (LITHIUM NIOBATE, BARIU 11-1651
ERTIES, / PHASES AND POLARIZATION OF LUMINOUS WAVES IN CUBIC   ELECTROOPTIC CRYSTALS. (NONLINEAR OPTICAL PROP 02-0044
                        A NEW LONGITUDINAL DISPLAY MODE FOR CERAMIC ELECTROOPTIC DEVICES. (PLZT)   09-1523
FERROELECTRIC CERAMICS.   ELECTROOPTIC DEVICES USING STRAIN BIASED PLZT 09-1562
ISMUTH TITANATE.   FEASIBILITY OF   ELECTROOPTIC DEVICES UTILIZING FERROELECTRIC B 08-1447
MELT COMPOSITION (STOICHIOMETRY) IN L/ DEPENDENCE OF LINEAR   ELECTROOPTIC EFFECT AND DIELECTRIC CONSTANT ON 11-1842
SPHATE CRYSTALS.   MEASUREMENT OF THE PRIMARY LINEAR   ELECTROOPTIC EFFECT IN AMMONIUM DIHYDROGEN PHO 17-2654
                                  DISPERSION OF THE   ELECTROOPTIC EFFECT IN BARIUM TITANATE.   04-0756
RINGENCE, PRINCIPAL INDICES)   DISPERSION OF   ELECTROOPTIC EFFECT IN BARIUM TITANATE. (BIREF 04-0755
PLZT 12-40-60. (USED FOR SENSING ICW VOLTAGES)   LINEAR ELECTROOPTIC EFFECT IN FERROELECTRIC CERAMICS: 09-1610
                                                                ELECTROOPTIC EFFECT IN FERROELECTRICS.   37-3648
                                           LONGITUDINAL   ELECTROOPTIC EFFECT IN KTN.   13-1942
                            LOW FREQUENCY LINEAR   ELECTROOPTIC EFFECT IN LEAD DITANATLATE.   13-1916
(SUB) MILLIMETER RANGE.   ELECTROOPTIC EFFECT IN LITHIUM NIOBATE IN THE 11-1844
ARIUM NIOBATE PLATES.   LONGITUDINAL   ELECTROOPTIC EFFECT IN OBLIQUE CUT STRONTIUM B 08-1472

ARIUM NIOBATE PLATES.   LONGITUDINAL ELECTROOPTIC EFFECT IN OBLIQUE CUT STRONTIUM B 13-2063
   CRYSTALS (THEORY AND METHOD OF DETERMINING HALF WAVE VOLTA/ ELECTROOPTIC EFFECT IN OPTICALLY HETEROGENEOUS 03-0461
OS/ PARAMETRIC FREQUENCY CONVERSION OF COHERENT LIGHT BY THE ELECTROOPTIC EFFECT IN POTASSIUM DIHYDROGEN PH 17-2616
CE IN LITHIUM NIOBATE AND POTASSIUM DIDEUTERIUM P/ TRANSIENT ELECTROOPTIC EFFECTS AND Q SWITCHING PERFORMAN 11-1710
. (LEAD ZIRCONATE TITANATE)   ELECTROOPTIC EFFECTS IN FERROELECTRIC CERAMICS 09-1551
N THE SYSTEM LEAD (ZINCATE(0.333) NIOBATE(0.667)).   ELECTROOPTIC EFFECTS OF HOT PRESSED CERAMICS I 13-1900
. ANTIMONY SULFOIODIDE.   ELECTROOPTIC EFFECTS OF THE ABSORPTION EDGE OF 15-2198
              PRECISION MACHINING OF LITHIUM NIOBATE CRYSTALS FOR ELECTROOPTIC ELEMENTS.   11-1742
BIREFRINGENCE AND POLARIZATION) OF LEAD CONTAINING OXYGEN O/ ELECTROOPTIC G COEFFICIENTS (RELATING INDUCED 17-2039
HYDROGEN PHOSPHATE, POTASSIUM DI/ TEMPERATURE INSTABILITY OF ELECTROOPTIC LASER MODULATORS WITH AMMONIUM DI 17-2675
STRONTIUM NIOBATE PRISM.   ELECTROOPTIC LIGHT BEAM DEFLECTION WITH BARIUM 13-2036
                  FERROELECTRIC, PIEZOELECTRIC, AND ELECTROOPTIC MATERIALS. (REVIEW, 991 REFS)   02-0058
ASSIUM TANTALATE N/ SOME RECENT ADVANCES IN FERROELECTRICS. (ELECTROOPTIC MATERIALS, SINGLE CRYSTALS OF POT 02-0073
AND LEAD GERMANATE. (LITHIUM GERMANATE NOT FERROELECTRIC)   ELECTROOPTIC MEASUREMENTS ON LITHIUM GERMANATE 35-3494
DIHYDROGEN PHOSPHATE FILM WAVEGUIDE.   EPITAXIAL ELECTROOPTIC MIXED CRYSTAL AMMONIUM POTASSIUM 17-2633
ITANATE.   TESTING A POINT DIPOLE ELECTROOPTIC MODEL BY AN APPLICATION TO LEAD T 05-1073
/) MODULATORS FOR OPTICAL COMMUNICATIONS (REVIEW EMPHASIZING ELECTROOPTIC MODULATION, LITHIUM TANTALATE, L/ 02-0056
R AMMONIUM DIHYDROGEN PHOSPHATE OF ARSENATE, POTASSI/ LINEAR ELECTROOPTIC MODULATOR MATERIALS. (POTASSIUM O 01-0021
IODATE.   PIEZOELECTRIC AND ELECTROOPTIC MODULI OF FERROELECTRIC POTASSIUM 08-1431
RING.   FABRICATION OF ELECTROOPTIC PLZT CERAMICS BY ATMOSPHERE SINTE 09-1600
NGS FOR LIGHT BEAM MODULATION AND DEFLECTION.   ELECTROOPTIC POTASSIUM TANTALATE NIOBATE GRATI 13-1929
UM NIOBATE SINGLE CRYSTAL WITH TUNGSTEN BRON/ DIELECTRIC AND ELECTROOPTIC PROPERTIES OF BARIUM SODIUM YTTRI 13-1971
GERMANATE SINGLE CRYSTAL.   ELECTROOPTIC PROPERTIES OF FERROELECTRIC LEAD 35-3518
              RECENT IMPROVEMENTS IN THE OPTICAL AND ELECTROOPTIC PROPERTIES OF PLZT CERAMICS.   09-1529
ITES IN THE PARAELECTRIC PHASE. (POTASSIUM TANTALATE AND TA/ ELECTROOPTIC PROPERTIES OF SOME ABO(3) PEROVSK 04-0690
IOBATE TYPE FERROELECTRICS (SINGLE CRYSTAL).   ELECTROOPTIC PROPERTIES OF STRONTIUM LITHIUM N 13-2062
M) (HAFNATE, TITANATE) SYSTEM.   MATERIAL AND ELECTROOPTIC PROPERTIES OF THE (LEAD, LANTHANU 08-1302
ES.   THERMOOPTIC AND ELECTROOPTIC PROPERTIES OF TRIHYDROGEN SELENIT 22-2841
STALS FROM THE POTASSIUM NIOBATE- LITHIU/ CRYSTAL GROWTH AND ELECTROOPTIC PROPERTIES OF TUNGSTEN BRONZE CRY 13-2003
/SSBAUER EFFECT. PART-6: ELECTRICAL CONDUCTIVE, SURFACE, AND ELECTROOPTIC PROPERTIES. PART-7: DIELECTRIC A/ 03-0397
A DOUBLE 45 DEGREES, Z CUT POTASSIUM DIDEUTERIUM PHOSPHATE. ELECTROOPTIC Q SWITCH BY DESK-TOP COMPUTER. /F 17-2702
         SPONTANEOUS BIREFRINGENCE AND ELECTROOPTIC RESPONSE IN GADOLINIUM MOLYBDATE. 30-3307
               ANALYSIS OF LOW FREQUENCY ELECTROOPTIC RESPONSE IN SODIUM NITRITE.   16-2353
SS IN STRONTIUM TITANATE AT LOW TEMPERATURE./ DIELECTRIC AND ELECTROOPTIC STUDIES OF THE POLARIZATION PROCE 06-1117
ITHIUM NIOBATE.   AN ELECTROOPTIC SWITCH USING ROTATED Y PLATE OF L 11-1732
RTIES OF EPITAXIAL FILMS OF FERROELECTRIC BISMUTH TITANATE. (ELECTROOPTICS).   GROWTH AND PROPE 08-1329
INGENCE AND REFRACTIVE INDICES) OF POTASSIUM NIOBATE (LINEAR ELECTROOPTICS).   OPTICAL PROPERTIES (BIREFR 08-1468
ANTIMONY SULFOIODIDE.   ELECTROOPTICAL AND ELECTROMECHANICAL EFFECT IN 15-2178
ALLY DEUTERATED RUBIDIUM DIHYDROGEN PHOSPHATE CRYSTALS.   ELECTROOPTICAL AND OPTICAL PROPERTIES OF PARTI 17-2698
POTASSIUM DIHYDROGEN PHOSPHATE AND AMMONIUM DIHYDROGEN PHOS/ ELECTROOPTICAL AND PIEZOOPTICAL PROPERTIES OF 17-2697
           MESOMORPHIC MATERIALS FOR ELECTROOPTICAL APPLICATION.   37-3631
T PRESSED FERROELECTRIC LEAD ZIRCONATE TITANATE CERAMICS FOR ELECTROOPTICAL APPLICATIONS.   HO 09-1524
ITANATE AND SODIUM P/ HOT PRESSED FERROELECTRIC CERAMICS FOR ELECTROOPTICAL APPLICATIONS. (LEAD ZIRCONATE T 09-1527
(APPARATUS, GROWTH OF ADP, KDP, HYDROIODIC ACID,/ GROWTH OF ELECTROOPTICAL CRYSTALS FROM AQUEOUS SOLUTIONS 17-2509
E) OF SOME ABO(3) FERROELECTRICS. (PHENOMENOLOGICAL MODEL OF ELECTROOPTICAL EFFECT) /CTRA AND BAND STRUCTUR 08-1383
ONIOBATE.   ANISOTROPY OF THE ELECTROOPTICAL EFFECT IN CRYSTALS OF LEAD MAGN 36-3521
S HAVING A SMEARED PHASE TRANSITION.   ELECTROOPTICAL EFFECT IN FERROELECTRIC CRYSTAL 03-0174
OBATE(0.67). (DISPERSION OF QUADRATIC COEFFICIENTS)   ELECTROOPTICAL EFFECT IN LEAD ZINCATE(0.33) NI 13-1982
ITHIUM NIOBATE.   LONGITUDINAL ELECTROOPTICAL EFFECT IN OBLIQUE SECTIONS OF L 11-1863
S.   SPONTANEOUS ELECTROOPTICAL EFFECT IN ROCHELLE SALT CRYSTAL 28-3209
OR DEUTERIUM DIS/ PHASE TRANSFORMATION AND THERMOOPTICAL AND ELECTROOPTICAL EFFECT IN RUBIDIUM TRIHYDROGEN 22-2840
RIUM TITANATE- ZINC TITANATE SOLID SOLUTIONS.   ELECTROOPTICAL EFFECT IN SINGLE CRYSTALS OF BA 04-0958
IUM AMMONIUM SELENATE DIHYDRATE.   LINEAR ELECTROOPTICAL EFFECT IN THE FERROELECTRIC SOD 22-2829
                                     ELECTROOPTICAL MATERIALS.   02-0106
                             LINEAR ELECTROOPTICAL MATERIALS. (REVIEW, 138 REFS)   01-0016
         ANTIPOLARIZATION IN COPPER FORMATE CRYSTALS BY THE ELECTROOPTICAL METHOD. (ANTIFERROELECTRIC)   26-2925
KITE STRUCT/ PHENOMENOLOGICAL THEORY OF REFRACTIVE INDEX AND ELECTROOPTICAL PHENOMENA IN CRYSTALS OF PEROVS 03-0518
/RANSITION IN AMMONIUM SULFATE, THE DIELECTRIC, OPTICAL, AND ELECTROOPTICAL PROPERTIES IN THE NEIGHBORHOOD/ 19-2745
/CT OF NONISOVALENT SUBSTITUTION OF TITANIUM(4+) IONS ON THE ELECTROOPTICAL PROPERTIES OF BARIUM TITANATE / 04-0959
TANTALUM PENTOXIDE SOLID SOLUTION SINGLE CRYSTALS.   ELECTROOPTICAL PROPERTIES OF BARIUM TITANATE- 04-0955
ERROELECTRICS. (REVIEW, 195 REFERENCE/ NONLINEAR OPTICAL AND ELECTROOPTICAL PROPERTIES OF DIELECTRICS AND F 02-0081
STALS OF THE POTASS/ RELATIONSHIP BETWEEN THE DIELECTRIC AND ELECTROOPTICAL PROPERTIES OF FERROELECTRIC CRY 17-2696
OLINIUM MOLYBDATE.   ELECTROOPTICAL PROPERTIES OF FERROELECTRIC GAD 30-3277
ON IODINE) BORACITE. (SPONTANEOUS LATTICE STRAINS)   ELECTROOPTICAL PROPERTIES OF FERROELECTRIC (IR 14-2082
NGLE CRYSTALS.   ELECTROOPTICAL PROPERTIES OF LITHIUM IODATE SI 08-1423
F SOLID SOLUTIONS OF BARIUM TITANATE AND ZINC OXIDE.   ELECTROOPTICAL PROPERTIES OF SINGLE CRYSTALS O 04-0953
UM LITHIUM NIOBATE. (TUNGSTEN BRONZE TYPE FERROELECTRICS). ELECTROOPTICAL PROPERTIES OF STRONTIUM POTASSI 13-2051
/S WITH DIFFUSED PHASE TRANSITION. (POLARIZATION MECHANISMS, ELECTROOPTICAL PROPERTIES, REVIEW OF RUSSIAN / 02-0119
              USE OF KDP TYPE CRYSTALS AS ELECTROOPTICAL REFLECTORS.   17-2512
M DIDEUTERIUM PHOSPHATE CRYSTALS.   CHARACTERISTICS OF AN ELECTROOPTICAL SHUTTER BASED ON Z CUT POTASSIU 17-2460
ERROELECTRIC PIEZOELECTRIC CERAMI/ RELATION BETWEEN THE MAIN ELECTROPHYSICAL AND STRUCTURAL PARAMETERS OF F 03-0217
ON SYSTEM LEAD TITANATE- LEAD ZIRCONATE- LEAD (ZINCATE(0.33/ ELECTROPHYSICAL PROPERTIES OF THE SOLID SOLUTI 13-1951
OELECTRIC ANTIMONY SULFOIODIDE NEAR THE PHASE TRANSITION TE/ ELECTROREFLECTION FROM SINGLE CRYSTALS OF FERR 15-2151
IONS OF SOME PHENOMENOLOGICAL MODEL/ EFFECT OF THE CHANGE OF ELECTROSTATIC CONSTRAINTS ON THE PHASE TRANSIT 03-0280
UCTURE. APPLICATION TO ALL BARIUM TITANATE STRUCTURE TYPES. ELECTROSTATIC FIELD IN A PSEUDOCUBIC IONIC STR 04-1008
IONIC CRYSTAL. APPLICATION TO THE CALCULATION OF TETRAGONA/ ELECTROSTATIC FIELD IN A SLIGHTLY ORTHORHOMBIC 04-0578
OVSKITE FERROELECTRIC CERAMICS. (BARIUM / CALCULATION OF THE ELECTROSTATIC INTERACTIONS IN FINE GRAINED PER 04-0642
F LEAD ZIRCONATE(X) TITANATE(1-X) CERAMICS.   POLARIZATION, ELECTROSTRICTION, AND PIEZOELECTRIC ACTIVITY O 09-1501
UORIDE.   PIEZOELECTRICITY, PYROELECTRICITY, AND THE ELECTROSTRICTION CONSTANT OF POLYVINYLIDENE FL 36-3598
N THE LEAD (ZIRCONATE, STANNATE, TITANATE) SOLID SOLUTIONS. (ELECTROSTRICTION CONSTANTS) /TRIC TRANSITION I 09-1486
ATION TO BARIUM TITANATE.   THEORY OF ELECTROSTRICTION IN CUBIC CRYSTALS WITH APPLIC 03-0267
E CRYSTALS.   PIEZOELECTRICITY AND ELECTROSTRICTION OF ANTIMONY SULFOIODIDE SINGL 15-2227
               PIEZOELECTRICITY AND ELECTROSTRICTION OF TRIGLYCINE SULFATE.   27-3118
ITY IN FERROELECTRIC ANTIMONY SULFOIODIDE.   ELECTROSTRICTION, PIEZOELECTRICITY AND ELASTIC 15-2165
E.   PRESSURE DEPENDENCE OF TRANSITION TEMPERATURES AND ELECTROSTRICTIONS IN PEROVSKITE BARIUM TITANAT 04-0853
            ELASTIC ANOMALY IN AMMONIUM SULFATE. (ELECTROSTRICTIVE CONSTANTS)   19-2759
D ANTIFERROELECTRIC PHASES. (LANTHANUM OR NIOBIUM DOPED LEA/ ELECTROSTRICTIVE CONSTANTS IN FERROELECTRIC AN 09-1487
CTURE IN BARIUM TITANATE CERAMICS.   ELECTROSTRICTIVELY GENERATED AND MODULATED FRA 04-0700
IUM TITANATE CERAMICS WITH BISMUTH OXIDE AS A GRAIN BOUNDARY ELEMENT.   LANTHANUM DOPED BAR 04-0799

PARAMAGNETIC RESONANCE AND OPTICAL ABSORPTION OF TRANSITION ELEMENT IONS IN STRONTIUM TITANATE AND LANTHA/ 06-1165
OPTICAL DEVICE ELEMENT OF FERROELECTRIC GADOLINIUM MOLYBDATE. 30-3295
/S TOWARD A FAST, NONVOLATILE, NONDESTRUCTIVE READOUT MEMORY ELEMENT UTILIZING POTASSIUM NITRATE. (FUSED T/ 16-2269
IONS OF SURFACES OF FERROELECTRICS (TGS) WITH SEMICONDUCTING ELEMENTS. INVESTIGAT 27-3089
ISION MACHINING OF LITHIUM NIOBATE CRYSTALS FOR ELECTROOPTIC ELEMENTS. PREC 11-1742
CHARACTERISTICS OF THE PHOTOELECTRIC EMF OF ANTIMONY SULFOIODIDE. 15-2129
CHARACTERISTICS OF THE PHOTOELECTRIC EMF OF ANTIMONY SULFOIODIDE. 15-2153
N / IMAGING ELECTRIC MICRO FIELDS (OF SOLID SURFACES) IN THE EMISSION ELECTRON MICROSCOPE (BARRIER LAYERS I 04-0913
FIELD EMISSION FROM ANTIMONY SULFOIODIDE. 15-2203
ED TO UNIAXIAL STRESS PULSES. LIGHT EMISSION FROM BARIUM TITANATE CRYSTALS SUBJECT 04-0672
TITANATE, TGS) ACOUSTIC EMISSION FROM FERROELECTRIC CRYSTALS. (BARIUM 04-0593
TOR. EFFECT TUNABLE OPTICAL EMISSION FROM LITHIUM NIOBATE WITHOUT A RESONA 11-1858
LIGHT EMISSION FROM TGS. (CRYSTALS DURING HEATING) 27-3117
POWER AND LINE WIDTH OF TUNABLE STIMULATED FAR INFRARED EMISSION IN LITHIUM NIOBATE. 11-1736
RAMIC (BARIUM, LEAD) TITANATE NEAR THE CURIE TEMPERATURE. EMISSION OF LIGHT BY SURFACE MICROPLASMA OF CE 04-0982
/SPONTANEOUS POLARIZATION ON THE PHOTOSTIMULATED EXOELECTRON EMISSION YIELD OF TRIGLYCINE SULFATE CRYSTALS. 27-2961
OPTICAL MIXING. (OF PULSED RUBY LASER EMISSIONS BY TRIGLYCINE SULFATE CRYSTAL) 27-2958
BARIUM TITANATE CERAMIC: A NEW ELECTRON EMITTER. 04-0802
ELECTRIC TRANSITION TEMPERATURES OF TUNGSTEN BRONZE TYPE NI/ EMPIRICAL FACTORS FOR CALCULATION OF THE FERRO 13-2049
11) AND THE RELATED MATERIALS. FERROELECTRIC PROPERTY AND ENANTIOMORPHISM OF LEAD (5) GERMANIUM (3) OXIDE (35-3499
Y OF X-IRRADIATED SINGLE CRYSTALS OF ICE. ENDOR (ELECTRON NUCLEAR DOUBLE RESONANCE) STUD 34-3413
SURFACE BARRIER ENERGIES ON STRONTIUM TITANATE. 06-1175
CALCIUM(X) MANGANATE (PHASE TRANSFORMATIONS CONFIGURATIONAL ENERGY). /SKITE TYPE COMPOUNDS, LANTHANUM(1-X) 08-1356
OXIDE COMPOUNDS. (BARIUM TITANATE, R/ INFLUENCE OF MADELUNG ENERGY AND COVALENCY ON THE STRUCTURE OF AB(5) 04-0761
E TRANSITION IN TERMS OF A SINGLE MODE ANHARMONIC OSCI/ FREE ENERGY AND SPECIFIC HEAT IN FERROELECTRIC PHAS 03-0406
IUM AND LEAD TITANATES - APPLICATION TO CALCULATING THE WALL ENERGY AND THE DIELECTRIC CONSTANTS. / FOR BAR 03-0189
SITION. ELASTIC ENERGY AND THE FERROELECTRIC PARAELECTRIC TRAN 03-0237
IFERROELECTRIC DOMAIN WALLS, EFFECTIVE FIELD, AND ACTIVATION ENERGY. APPLICATION TO LEAD ZIRCONATE. /OF ANT 07-1257
ICE POLARIZATION. (LCAO ANALYSIS OF BAND SCHEME, STRONTIUM / ENERGY BAND CHANGES IN PEROVSKITES DUE TO LATT 06-1101
OM A SELF CONSISTENT FIELD TIGHT BINDING CALCULATION. ENERGY BAND STRUCTURE OF STRONTIUM TITANATE FR 06-1221
STRONTIUM TITANATE, POTASSIUM MOLYBDATE, AND POTASSIUM TAN/ ENERGY BANDS FOR POTASSIUM NICKEL TRIFLUORIDE. 02-0091
ENERGY BANDS OF POTASSIUM NIOBATE. 08-1372
UM TITANATE). POLARONS IN ANISOTROPIC ENERGY BANDS (THEORY, CALCULATIONS FOR STRONTI 03-0305
S IN FERROELECTRIC CERAMICS (AND THE THEORY OF PIEZOELECTRIC ENERGY CONVERSION). PIEZOELECTRIC EFFECT 03-0278
FERROELECTRIC CERAMICS AND ENERGY CONVERSION. (LEAD ZIRCONATE TITANATE) 09-1508
CINE SULFATE. (THEORY AND CALCULATIONS OF WALL THICKNESS AND ENERGY DENSITY) /LECTRIC DOMAIN WALL IN TRIGLY 27-2997
RIUM TITANATE. (THERMODYNAMIC CALCULATIONS) ANGULAR ENERGY DEPENDENCE OF 180 DEG DOMAIN WALL IN BA 04-0658
ERESIS LOOPS. FERROELECTRIC FREE ENERGY EXPANSION COEFFICIENTS FROM DOUBLE HYST 03-0262
PHOSPHATE. (DETERMINATION OF BETA IN THE P EXPANSION OF FREE ENERGY F) /PROPERTIES OF POTASSIUM DIHYDROGEN 27-2543
ANALYTIC PROPERTIES OF THE FREE ENERGY FOR THE "ICE" MODELS (FERROELECTRICS). 34-3427
(MACROSCOPIC AND MICROSCOPIC) SYMMETRY OF FERROELECTRIC ENERGY FUNCTIONS. (SOME COMMENTS) 03-0193
RE CHARACTERISTICS OF FERROELECTRIC TRIGLYCINE SE/ (OPTICAL) ENERGY GAP (AND DIELECTRIC CONSTANT) TEMPERATU 27-3021
(OR FERROELECTRIC) MATERIALS. ENERGY GENERATION FROM IMPACT ON PIEZOELECTRIC 03-0222
CALCULATION OF (DOMAIN) WALL ENERGY IN BARIUM TITANATE. 04-0673
TGS. PART-2: INTERNAL MOTION OF HYDROGEN AND THE ACTIVATION ENERGY IN GLYCINE. ESR OF GAMMA-IRRADIATED 27-3034
PHASE TRANSITION IN KDP AND THE ROLE OF THE DEPOLARIZING ENERGY IN IT. (THEORY) 03-0229
NATE BY TEMPERATURE AND PRESSURE ENFORC/ RELEASE OF ELECTRIC ENERGY IN LEAD NIOBIUM ZIRCONATE TITANATE STAN 09-1488
FERROELECTRIC CERAMICS AND CONVERSION OF MECHANOELECTRIC ENERGY. (LEAD ZIRCONATE TITANATE) 09-1483
RUBIDIUM ALUMINUM DISULFATE D/ CALCULATION OF CRYSTAL FIELD ENERGY LEVEL SPLITTINGS OF TITANIUM(3+) ION IN 19-2754
E COUPLED CHROMIUM(3+) PAIRS IN LANTHANUM ALUM/ SPECTRUM AND ENERGY LEVELS OF CHROMIUM(3+) IONS AND EXCHANG 36-3565
IONS IN LITHIUM NIO/ ABSORPTION AND LUMINESCENCE SPECTRA AND ENERGY LEVELS OF NEODYMIUM(3+) AND ERBIUM(3+) 11-1713
E. MEASUREMENT OF ELECTRON ENERGY LOSS FUNCTION IN BULK STRONTIUM TITANAT 06-1138
/ BONDED FERROELECTRIC AND ANTIFERROELECTRIC SOLIDS. PART-1: ENERGY LOSS NEUTRON INCOHERENT SCATTERING SPE/ 17-2687
IUM TITANATE. (150-500 DEG C, 20-50 KILOHERTZ, LISSAJOUS FI/ ENERGY LOSS NONPOLAR STATE POLYCRYSTALLINE BAR 04-0831
SSIUM DIHYDROGEN PHOSPHATE TYPE FERROELECTRICS. (THEORY) ENERGY PARAMETERS AND PROTON TUNNELING IN POTA 03-0506
TRIC CONTAINING DISLOCATIONS. ENERGY SPECTRUM OF AN ORDER DISORDER FERROELEC 03-0485
ICS. EFFECT OF POINT DEFECTS ON THE ENERGY SPECTRUM OF HYDROGEN BONDED FERROELECTR 03-0486
ICAL AND ELECTRONIC PROBLEMS IN FERROELECTRICITY. (THEORY OF ENERGY STATES INVOLVED) STATIST 03-0325
IUM TITANATE, BARIUM TITANATE) ENERGY STORAGE IN CERAMIC DIELECTRICS. (STRONT 04-0602
EY OF MATERIALS BASED UPON BARIUM TITANATE. ENERGY STORAGE IN FERROELECTRIC CERAMICS: SURV 04-0075
SURVEY OF MATERIALS BASED ON BARIUM TITANATE. ENERGY STORAGE IN HIGH PERMITTIVITY CERAMICS: 04-0733
ZIRCONATE. (PHASE TRANSITION 200 DEG C BELOW THE CUR/ FIELD ENFORCED FERROELECTRICITY IN GLASS BONDED LEAD 07-1252
BIUM ZIRCONATE TITANATE STANNATE BY TEMPERATURE AND PRESSURE ENFORCED PHASE TRANSITIONS. /NERGY IN LEAD NIO 09-1488
ATE AT THE CUBIC TO TETRAGONAL TRANSITION. DISTORTION ENHANCED OPTICAL ABSORPTION IN STRONTIUM TITAN 06-1109
ENHANCED ORDERING IN PLZT CERAMICS. 09-1570
PLZT CERAMICS. ENHANCED ORDERING OF FERROELECTRIC DOMAINS IN 09-1579
PLZT CERAMICS. ENHANCED ORDERING OF FERROELECTRIC DOMAINS IN 09-1580
PHOSPHATE IN WATER AND RUBIDIUM DIDEUTERIUM PHOSPHATE IN HE/ ENTHALPIES OF SOLUTION OF RUBIDIUM DIHYDROGEN 17-2585
ENDENCE OF THE INFRARED AND RAMAN SPECTRA AND THE CALCULATED ENTROPY. /IC THIOUREA. PART-1: TEMPERATURE DEP 25-2920
E. ANOMALOUS SPECIFIC HEAT AND CONFIGURATIONAL ENTROPY CHANGE IN SODIUM TRIHYDROGEN DISELENIT 22-2856
R STEP GROWTH MODEL. LOW ENTROPY FORM OF ICE I(H) OBTAINED FROM A LINEA 34-3410
/ECTRIC BETA GADOLINIUM MOLYBDATE. PARTS 1-3: HEAT CAPACITY, ENTROPY, MAGNETIC MOMENT OF THE ELECTRICALLY / 30-3275
ERROELECTRICS, MECHANISM) ENTROPY OF AN ORDER DISORDER TRANSITION. (IN F 03-0192
/TATISTICS OF HYDROGEN BONDED CRYSTALS, PART-1, THE RESIDUAL ENTROPY OF ICE, PART-2, HYDROGEN BONDED FERRO/ 03-0388
RESIDUAL ENTROPY OF SODIUM TRIHYDROGEN SELENATE. 22-2835
POTASSIUM DIHYDROGEN PHOSPHATE FILM WAVEGUIDE. EPITAXIAL ELECTROOPTIC MIXED CRYSTAL AMMONIUM 17-2633
DIRECTION UPON OPTICAL PHONONS IN CUBIC SODIUM CHLORIDE AND EPITAXIAL FERROELECTRIC SODIUM TARTRATE. /TION 29-3243
ATE. (ELECTROOPTICS) GROWTH AND PROPERTIES OF EPITAXIAL FILMS OF FERROELECTRIC BISMUTH TITAN 08-1329
E. EPITAXY OF PLATINUM ON BARIUM STRONTIUM NIOBAT 13-1907
SMUTH TITANIUM OXIDE. PREPARATION AND EPITAXY OF SPUTTERED FILMS OF FERROELECTRIC BI 13-2038
SUBSTRATES AS THE DETERMINING FACTOR OF THE GROWTH STAGE IN EPITAXY. (TRIGLYCINE SULFATE) /OF REAL CRYSTAL 27-3039
EPR SEE ELECTRON PARAMAGNETIC RESONANCE
LITHIUM NIOBATE AND LITHIUM TANT/ SIMPLIFIED CHARACTERISTIC EQUATION FOR PLANE PIEZOELECTRIC VIBRATIONS OF 11-1739
TRICS NEAR THE PHASE TRANSITION POINT. KINETIC EQUATION FOR QUADRUPOLE RESONANCE IN FERROELEC 03-0309
CRITICAL PHENOMENA IN ANTIFERROELECTRICS. (EQUATION OF STATE) 03-0343
ON OF TRIGLYCINE SULFATE NEAR CURIE TEMPERATURE. EQUATION OF STATE FOR THE COOPERATIVE TRANSITI 27-3006
ERROELECTRICS. PART-2: APPROXIMATED SOLUTIONS TO THE KINETIC EQUATIONS. /OTON DYNAMICS IN HYDROGEN BONDED F 03-0205
M NIOBATE AND TANTALATE CRYSTALS. SOME PROPERTIES OF THE EQUATIONS OF PIEZOELECTRIC VIBRATION OF LITHIU 11-1740
/ORMULAS INTERRELATING THE LINEAR ELASTIC AND ELECTROELASTIC EQUATIONS OF STATE: ISOTROPIC AND ORTHOTROPIC/ 03-0277
NG DOMAIN STRUCTURE. RESONANCE EFFECTS IN MAGNETICALLY EQUIAXIAL FERRODIELECTRIC SINGLE CRYSTALS HAVI 03-0509

IOBIUM PENTOXIDE SYSTEM. (POTASSIUM LIT/ STUDY OF SUBSOLIDUS EQUILIBRIA IN POTASSIUM OXIDE- LITHIUM OXIDE- N 13-1963
XIDE- NIOBIUM PENTOXIDE SYSTEM.   PHASE EQUILIBRIA AND CRYSTAL GROWTH IN THE LITHIUM O 11-1683
M TITANATE).   SYSTEM LANTHANA TITANIA PHASE EQUILIBRIA AND ELECTRICAL PROPERTIES (LANTHANU 36-3592
EM.   HIGH TEMPERATURE PHASE EQUILIBRIA IN THE LEAD ZIRCONATE TITANATE SYST 09-1577
DINIOBATE AND POTASSIUM NIOBATE- BARIUM DINIOBATE SY/ PHASE EQUILIBRIA IN THE POTASSIUM NIOBATE- STRONTIUM 13-2016
UM GERMANATE.   PHASE EQUILIBRIA IN THE SYSTEM BARIUM TITANATE- BARI 04-0710
IUM OXIDE- STRONTIUM OXIDE- NIOBIUM PENTOXIDE. (OXIDE/ PHASE EQUILIBRIA RELATIONS IN THE TERNARY SYSTEM BAR 13-1920
FATE IN THE VICINITY OF THE CURIE POINT.   EQUILIBRIUM DOMAIN STRUCTURE OF TRIGLYCINE SUL 27-3112
S.   (FACTORS INFLUENCING) TYPE AND ORIENTATION OF THE EQUILIBRIUM DOMAIN STRUCTURES IN FERROELECTRIC 03-0246
                                                            EQUILIBRIUM STRUCTURE OF POLARIZED ICE.   34-3411
PHOTOCONDUCTOR MEMORY DEVICE.   EQUIVALENT CIRCUIT ANALYSIS OF A FERROELECTRIC 37-3644
                                                  OPTICALLY ERASABLE AND REWRITABLE SOLID STATE HOLOGRAMS. 37-3638
/RAGE OF HOLOGRAMS IN A FERROELECTRIC PHOTOCONDUCTOR DEVICE (ERASABLE, NONDESTRUCTIVE READOUT, BASED ON BI/ 37-3646
HROMIC CALCIUM FLUORIDE, STRONTIUM TITANATE, AND CALCIUM TI/ ERASE MODE RECORDING CHARACTERISTICS OF PHOTOC 06-1119
             HOLOGRAPHIC STORAGE, ELECTRICAL FIXING AND ERASING IN DOPED BARIUM TITANATE CRYSTALS.   04-0848
IC MATERIALS. (BARIUM TI/ ELECTRICAL CONTROL OF FIXATION AND ERASURE OF HOLOGRAPHIC PATTERNS IN FERROELECTR 04-0847
CMANGANITES OF YTTRIUM AND RARE EARTHS. (YTTERBIUM, HOLMIUM, ERBIUM) /ELECTRIC PROPERTIES OF HEXAGONAL ORTH 36-3542
LUMINESCENCE SPECTRA AND ENERGY LEVELS OF NEODYMIUM(3+) AND ERBIUM(3+) IONS IN LITHIUM NIOBATE CRYSTALS. / 11-1713
IC TRIGONAL AND ORTHORHOMBIC BORACITE STRUCTURES. (BORACITE, ERICAITE) ATOMIC DISPLACEMENTS IN FERROELECTR 14-2075
UM TITANATE AS MEASURED BY X-RAY PHOTOELECTRON SPECTROSCOPY (ESCA).   BAND STRUCTURE OF STRONTI 06-1100
                                                            ESR SEE ELECTRON SPIN RESONANCE
F BARIUM TITANATE BY TRANSPORT REACTIONS (INVOLVING TITANIUM ESTERS TO YIELD SMALL CRYSTAL PRISMS). /TION O 04-0625
GUANIDINIUM ALUMINUM SULFATE HEXAHYDRATE CLEAVAGES AND THEIR ETCH PATTERNS. /TRIC STUDIES ON FERROELECTRIC 21-2816
TE CRYSTALS.   DIP ETCH PATTERNS ON THE FACES OF TRIGLYCINE SULFA 27-3127
ROELECTRIC-AXIS OF SODIUM NITRATE.   DOMAIN ETCH PATTERNS ON THE FACES PARALLEL TO THE FER 16-2349
IUM TITANATE SINGLE CRYSTALS.   ETCH PITS CORRESPONDING TO DISLOCATIONS IN BAR 04-1023
                                                            ETCH PITS IN BARIUM TITANATE SINGLE CRYSTALS. 04-0922
                                               DISLOCATION ETCHANTS FOR BETA GADOLINIUM MOLYBDATE.   30-3258
STRUCTURES AND CORRESPONDING LIGHT FIGURES REVEALED BY WATER ETCHED ROCHELLE SALT CRYSTALS.   SURFACE 28-3234
ING WITHOUT CRATERS.   GROWTH OF CRATERS ON ION ETCHED SURFACE OF LITHIUM NIOBATE AND ION ETCH 11-1860
ONY SULFOIODIDE CRYSTALS (FROM SELECTIVE CRYSTALLIZATION AND ETCHING).   DOMAIN STRUCTURE OF ANTIM 15-2183
INNING OF LITHIUM NIOBATE FOR ACOUSTOOPTIC DEFLECTORS BY ION ETCHING.   TH 11-1862
LECTRIC DOMAINS IN TRIGLYCINE SULFATE.   ETCHING AND ULTRAMICROSCOPIC STUDIES OF FERROE 27-3143
LUMINATION ON THE DOMAIN STRUCTURE OF ANTIMONY SULFOIODIDE. (ETCHING METHOD)   EFFECTS OF IL 15-2124
TALS IN SOLUTIONS CONTAINING ACETATE IONS.   GROWTH AND ETCHING OF POTASSIUM DIHYDROGEN PHOSPHATE CRYS 17-2567
ATE, SILICON DIOXIDE, AND SILICON.   MEASUREMENT OF SPUTTER ETCHING RATES FOR MONOCRYSTALLINE LITHIUM NIOB 11-1714
TRUCTURES IN TRIGLYCINE SULFATE.   APPLICATION OF SELECTIVE ETCHING TO THE STUDY OF TWIN AND DISLOCATION S 27-3042
OF CRATERS ON ION ETCHED SURFACE OF LITHIUM NIOBATE AND ION ETCHING WITHOUT CRATERS.   GROWTH 11-1860
/AGNETIC RADIATION INTO MOTION. (PYROELECTRIC EFFECT IN TGS, ETHYL DIAMINE TARTRATE, BARIUM TITANATE, SALI/ 04-0854
ROELECTRICS AND RELATED MATERIALS: REVIEW OF DEVELOPMENTS IN EUROPE.   APPLICATIONS OF FER 02-0122
                                                            EUROPEAN MEETING ON FERROELECTRICITY, PROC.   01-0023
, DIJON, 1971.   EUROPEAN MEETING ON FERROELECTRICITY, 2ND PROC 01-0012
T ON THE SECOND INTERNATIONAL MEETING ON FERROELECTRICITY. A EUROPEAN VIEW. AN AMERICAN VIEW.   REPOR 01-0008
/LLOGRAPHIC, OPTICAL AND MAGNETIC PROPERTIES OF (MONOCLINIC) EUROPIUM SILICATE. (UNSTABLE FERROELECTRIC PH/ 36-3620
/ITIONS OF GADOLINIUM MOLYBDATE AND ISOSTRUCTURAL COMPOUNDS (EUROPIUM TERBIUM, DYSPROSIUM AND HOLMIUM MOLY/ 30-3271
                      RADIATIONLESS TRANSITIONS IN THE EUROPIUM(3+) CENTER IN LANTHANUM ALUMINATE. 36-3531
M MANGANESE TRIFLUORIDE.   LUMINESCENCE OF EUROPIUM(3+) ION IN ANTIFERROMAGNETIC POTASSIU 33-3381
     ELECTRON PROBE X-RAY MICROANALYSIS ON IMPURITIES IN EVAPORATED BARIUM TITANATE FILMS.   04-0742
             EXPERIMENTAL STUDY OF EVAPORATED BARIUM TITANATE FILMS.   04-0939
/RUCTURE (MICROSTRUCTURE) AND ELECTRICAL PROPERTIES OF FLASH EVAPORATED BARIUM TITANATE THIN FILMS (SEPARA/ 04-0605
FILMS. (THICKNESS DEPENDENCE) ELECTRICAL PROPERTIES OF FLASH EVAPORATED FERROELECTRIC BARIUM TITANATE THIN 04-0967
           GROWTH AND STRUCTURE OF BARIUM TITANATE EVAPORATED FILMS.   04-0604
                                                            EVAPORATED FILMS OF TUNGSTEN OXIDE.   10-1647
       FORMATION PROCESS OF BARIUM TITANATE THIN FILM EVAPORATED IN VACUO.   04-0862
ON TEMPERATURE OF TRIGLYCINE SULFATE.   EFFECT OF EVAPORATED METALLIC ELECTRODES ON THE TRANSITI 27-3100
INE SULFATE.   AFTER EFFECTS IN SEMICONDUCTOR ELECTRODES EVAPORATED ON INSULATORS ESPECIALLY ON TRIGLYC 27-3170
          PREPARATION OF BARIUM TITANATE FILM BY EVAPORATION.   04-0743
ION OF FERROELECTRIC THIN FILMS OF BARIUM TITANATE BY VACUUM EVAPORATION.   THIN FILMS: PREPARAT 04-0834
                             VACUUM EVAPORATION OF BARIUM TITANATE.   04-1001
YCRYSTALLINE BARIUM TITANATE.   FLASH EVAPORATION OF FERROELECTRIC THIN FILMS OF POL 04-0968
ION TO THIN FILMS OF PEROVSKITES AND III-V COMPOUNDS,/ FLASH EVAPORATION. (REVIEW OF TECHNIQUE AND APPLICAT 02-0112
LE CRYSTALS.   EVOLUTION OF FERROELECTRIC DOMAINS IN TGS SING 27-3003
ON IN FERROELECTRICS.   A MORE EXACT DYNAMIC METHOD OF DETERMINING POLARIZATI 03-0381
ROELECTRICS IN ARBITRARY EXTERNAL ELECTRIC FIELD.   EXACT SOLUTION OF MODEL OF TWO DIMENSIONAL FER 03-0520
                   WET SYNTHETIC METHOD OF MAKING PURE, EXACTLY STOICHIOMETRIC BARIUM TITANATE.   04-0790
RING OF BARIUM TITANATE.   EXAGGERATED GRAIN GROWTH IN LIQUID PHASE SINTE 04-0835
TIES AT THE PHASE TRANSITIONS. (BASED ON LANDAU PHENOMENOL/ (EXAMINATION OF) ANOMALIES OF DIELECTRIC PROPER 19-2765
/ KIND (TYPE, CLASS) FOR ACTUAL FERROELECTRICS TOGETHER WITH EXAMINATION OF CHARACTERS OF THEIR PARAELECTR/ 03-0138
/TIFERROELECTRICITY IN COMPOUNDS WITH PEROVSKITE STRUCTURE. (EXAMINATION OF PUBLISHED DATA, CLASSIFICATION/ 01-0018
.   SCANNING ELECTRON MICROSCOPIC EXAMINATION OF SINTERED BARIUM SODIUM NIOBATES 13-1955
E(3-X) CRYSTALS BY ELECTRON MICROSCOPY.   EXAMINATION OF SUBSTOICHIOMETRIC TUNGSTEN OXID 10-1627
          GROWTH OF LEAD TITANATE CRYSTALS AND EXAMINATION OF THEIR DOMAIN STRUCTURE.   05-1047
EMICONDUCTIVE BARIUM TITANATE CERAMICS.   EFFECT OF EXCESS TITANIA ON THE ELECTRIC RESISTANCE OF S 04-0966
       PREPARATION OF PURE TRIGLYCINE SULFATE. (USING ION EXCHANGE)   27-3087
RATURE OF A FERROELECTRIC SEMICON/ METASTABLE OPTICAL CHARGE EXCHANGE AND "FROZEN" SHIFT OF THE CURIE TEMPE 15-2160
UM ALUM/ SPECTRUM AND ENERGY LEVELS OF CHROMIUM(3+) IONS AND EXCHANGE COUPLED CHROMIUM(3+) PAIRS IN LANTHAN 36-3565
APES AND SPIN CORRELATION IN PARAMAGNETIC POTASSIUM MANGANE/ EXCHANGE NARROWING: MAGNETIC RESONANCE LINE SH 33-3376
FERROEL/ ESR IDENTIFICATION OF SLATER CONFIGURATIONS AND OF EXCHANGE OF THE ARSENATE RADICAL IN IRRADIATED 18-2734
                                                    CURRENT EXCITATION IN GAMMA-IRRADIATED BARIUM TITANATE 04-0884
RSONIC PULSES IN PIEZOELECTRIC LITHIUM NIOBATE AND/ SEPARATE EXCITATION OF LONGITUDINAL AND TRANSVERSE HYPE 11-1814
RSONIC WAVE PULSES IN LITHIUM NIOBATE AND BISMU/ INDEPENDENT EXCITATION OF LONGITUDINAL AND TRANSVERSE HYPE 11-1815
ED FERROELECTRICS. (THEORY)   POSSIBILITY OF EXCITATION OF SPIN WAVES IN MAGNETICALLY ORDER 03-0150
ERROELECTRICS.   RADIATION AND ACOUSTIC EXCITATION ON ANOMALOUS LAYERS OF PEROVSKITE F 08-1388
D MAGNETIC MOMENTS OF OPTICAL TRANSITIO/ ELECTRONIC AND SPIN EXCITATIONS IN ANTIFERRODIELECTRIC ELECTRIC AN 03-0418
E TRIFLUORIDE.   MAGNETIC EXCITATIONS IN NICKEL DOPED POTASSIUM MANGANES 33-3407
       EXCITON MODEL OF SPIN AND OPTICAL EXCITATIONS OF AN ANTIFERRODIELECTRIC.   03-0416
FERRODIELECTRICS IN THE FREQUENCY RANGE OF DOUBLE ELECTRONIC EXCITATIONS OF MOLECULES (IONS). /TION BY ANTI 03-0419
NIUM DIHYDROGEN PHOSPHATE AND POTASSIUM DIHYDROGEN PHOSPHATE EXCITED BY A 2573 ANGSTROM CW PUMP. /E IN AMMO 17-2486
RELAXATION PROCESSES IN ANTIMONY SULFOIODIDE SINGLE CRYSTALS EXCITED WITH MICROSECOND LIGHT PULSES.   15-2202
OF AN ANTIFERRODIELECTRIC.   EXCITON MODEL OF SPIN AND OPTICAL EXCITATIONS 03-0416

UM MANGANESE TRIFLUORIDE.   FINE STRUCTURE OF EXCITON- MAGNON ABSORPTION OF LIGHT IN POTASSI 33-3364
ENT AND COMPARISON RESULTS FOR SAPPH/ EXPANSION, CRAZING AND EXFOLIATION OF LITHIUM NIOBATE ON ION BOMBARDM 11-1800
                    CRYSTALLOCHEMISTRY OF PEROVSKITES. (EXISTENCE CONDITIONS AND CLASSIFICATION)   08-1324
/N OF LEAD TITANATE ZIRCONATE SOLID SOLUTICNS (THERMODYNAMIC EXISTENCE CONDITIONS FOR MORPHOTROPIC PHASE B/ 09-1541
DISPLACEMENTS AND VELOCITIES OF THE DOMAIN WALLS IN KDP. EXISTENCE OF CRITICAL ELECTRIC FIELDS.   17-2448
RROELECTRIC CRYSTAL POTASSIUM DIHYDROGEN PHOSPHATE.   EXISTENCE OF DISLOCATIONS AT DOMAIN TIPS IN FE   17-2450
EAD OXIDE- TITANIUM DIOXIDE- LANTHANUM TRIOXIDE.   RANGE OF EXISTENCE OF PEROVSKITE PHASES IN THE SYSTEM L 08-1349
/ENCE OF THE SPONTANEOUS POLARIZATION ON THE PHOTOSTIMULATED EXOELECTRON EMISSION YIELD OF TRIGLYCINE SULF/ 27-2961
OEFFICIENT OF RESONANT FREQUENCY IN LITHIUM TANTALATE LENGTH EXPANDER BARS.   ZERO TEMPERATURE C   11-1810
TRAGONAL PHASE TRANSITION IN BARIUM SODIUM NIOBATE. (THERMAL EXPANSION) /TRIC STUDY OF THE ORTHORHOMBIC- TE 13-1894
S AND SPONTANEOUS POLARIZATION IN LITHIUM TANTALATE (THERMAL EXPANSION). /AND SECONDARY PYROELECTRIC EFFECT 12-1884
 FERROELECTRIC SODIUM TRIHYDROGEN DISELENITE AND ITS THERMAL EXPANSION. RELAXATION BETWEEN THE STRUCTURE OF 22-2851
OSPHATE NEAR THE HIGH TEMPERATURE PHASE TRANSITION. (THERMAL EXPANSION) /PERTIES OF POTASSIUM DIHYDROGEN PH 17-2627
NSITION CF THE TUNGSTEN TRIOXIDE PHASE.   THERMAL EXPANSION AND THE ORTHORHOMBIC- TETRAGONAL TRA 10-1626
ANEOUS POLARIZATION IN FERROELECTRICS. (ABSTRACT ON/ THERMAL EXPANSION ANOMALIES AND THE DIRECTION OF SPONT 03-0350
/ PIEZOELECTRIC CONSTANTS, SPONTANEOUS POLARIZATION, THERMAL EXPANSION COEFFICIENTS, AND REFRACTIVE INDEX) 12-1876
LOOPS.   FERROELECTRIC FREE ENERGY EXPANSION COEFFICIENTS FROM DOUBLE HYSTERESIS   13-0262
NIOBATE ON ION BOMBARDMENT AND COMPARISON RESULTS FOR SAPPH/ EXPANSION, CRAZING AND EXFOLIATION OF LITHIUM 11-1800
RITE.   THERMAL EXPANSION IN FERROELECTRIC SILVER SODIUM DINIT 16-2346
                                                   THERMAL EXPANSION IN SODIUM NITRITE CRYSTALS.   16-2351
IUM(X) DISELENITE AND ISOTOPE EFFECT.   THERMAL EXPANSION IN SODIUM TRIHYDROGEN(1-X) TRIDEUTER 22-2858
/E. (DIELECTRIC CONSTANTS, SPONTANEOUS POLARIZATION, THERMAL EXPANSION, INDICES OF REFRACTION, OPTICAL ROT/ 35-3501
SIUM DIHYDROGEN PHOSPHATE.   LOW TEMPERATURE THERMAL EXPANSION MEASUREMENTS ON SINGLE CRYSTAL POTAS 17-2464
ION OF ANTIFERROELECTRIC PHASE TRANSITION.   THERMAL EXPANSION OF ADP AND DEUTERATED ADP IN THE REG 17-2445
          X-RAY DETERMINATION OF THE THERMAL EXPANSION OF AMMONIUM DIHYDROGEN PHOSPHATE.   17-2485
ON OF THE LATTICE PARAMETERS AND THE COEFFICIENTS OF THERMAL EXPANSION OF BISMUTH FERRATE. /ION DETERMINATI 08-1287
E.   X-RAY STUDY OF THERMAL EXPANSION OF FERROELECTRIC GADOLINIUM MOLYBDAT 30-3294
N PHOSPHATE. (NEAR CURIE TEMPERATURE) X-RAY STUDY ON THERMAL EXPANSION OF FERROELECTRIC POTASSIUM DIHYDROGE 17-2556
          X-RAY MEASUREMENTS ON THERMAL EXPANSION OF FERROELECTRIC POTASSIUM NITRATE.   16-2322
ASSIUM DIHYDROGEN PHOSPHATE. (DETERMINATION OF BETA IN THE P EXPANSION OF FREE ENERGY F) /PROPERTIES OF POT 17-2543
IUM NIOBATES.   CAUSES OF THERMAL EXPANSION OF GADOLINIUM SUBSTITUTED BARIUM SOD 13-2048
ITHIUM TRIHYDROGEN SELENITE.   THERMAL EXPANSION OF HYDROGEN BONDED FERROELECTRICS. L 22-2857
E RANGE OF 70-300 K.   THERMAL EXPANSION OF LITHIUM NIOBATE IN THE TEMPERATUR 11-1685
TANATE ZIRCONATE/ DIELECTRIC SYSTEM POLARIZATION AND THERMAL EXPANSION OF POLARIZED CERAMICS IN THE LEAD TI 09-1603
YSTALS BY USING SEMICONDUCTOR / INVESTIGATION OF THE THERMAL EXPANSION OF POTASSIUM DIHYDROGEN PHOSPHATE CR 17-2565
YSTALS.   ANOMALOUS THERMAL EXPANSION OF POTASSIUM TRIHYDROGEN SELENATE CR 22-2880
OF RESONANCE FREQUENCY IN CAVITY.   MEASUREMENT OF THERMAL EXPANSION OF ROCHELLE SALT BY USING THE SHIFT 28-3202
ELECTRICS WITH HIGH CURIE/ DIELECTRIC PROPERTIES AND THERMAL EXPANSION OF SOME FERROELECTRICS AND ANTIFERRO 08-1378
ITRATE, KDP, SODIUM NITRITE)   THERMAL EXPANSION OF SOME FERROELECTRICS. (POTASSIUM N 16-2298
ATE.   REEXAMINATION OF THE THERMAL EXPANSION OF THE FERROELECTRIC TRIGLYCINE SULF 27-3121
 IN THE FERROELECTRIC TRANSITION RANGE.   THERMAL EXPANSION OF TRIGLYCINE FLUOBERYLLATE CRYSTALS 27-3169
                                                   THERMAL EXPANSION OF TRIGLYCINE SULFATE.   27-3000
ED WITH TELLURIUM THIN FILMS.   THERMAL EXPANSION OF TRIGLYCINE SULFATE CRYSTALS STUDI 27-3047
) NIOBATE CERAMIC.   X-RAY AND THERMAL EXPANSION STUDY OF A SODIUM(0.88) LITHIUM(0.12 13-1999
DIUM NIOBATES.   LINEAR THERMAL EXPANSIONS OF GADOLINIUM SUBSTITUTED BARIUM SO 13-1996
E STRUCTURAL TRANSITICN OF STRONTIUM TITANATE.   THERMAL EXPANSIVITY AND ULTRASONIC PROPAGATION NEAR TH 06-1153
R THE NEEL TEMPERATURE.   THERMAL EXPANSIVITY OF RUBIDIUM MANGANESE FLUORIDE NEA 33-3375
UCTURAL TRANSITION.   THERMAL EXPANSIVITY OF STRONTIUM TITANATE NEAR THE STR 06-1136
          DIELECTRIC LOSSES CF TRIGLYCINE SULFATE CRYSTALS EXPOSED TO SMALL DOSES OF X- AND GAMMA-RAYS.   03-2976
TEMPERATURES.   RADIATION EXPOSURE OF A LITHIUM NIOBATE CRYSTAL AT HIGH   11-1721
/ BONDED FERROELECTRICS AT LOW TEMPERATURES. PART-1: GENERAL EXPRESSION FOR INELASTIC COHERENT SCATTERING / 03-0478
AND ANTIFERROELE/ USE OF ELECTRON MICROSCCPY IN THE STUDY OF EXTENDED DEFECTS RELATED TO NONSTOICHIOMETRY / 36-3522
ECTRA OF CERTAIN PEROVSKITES (STRONTIUM OR CALCIUM TITANATE/ EXTENDED FINE STRUCTURE IN X-RAY ABSORPTION SP 03-0414
/F FERROMAGNETIC, AND FERROELASTIC CRYSTALS (PREVIOUS THEORY EXTENDED TO MAGNETIC CRYSTALS AND TO INCLUDE / 03-0142
          DIELECTRIC PROPERTIES OF AN ANHARMONIC CRYSTAL. (EXTENSION OF "SOFT MODE" THEORY)   03-0212
IN THE THEORY OF OPTICAL LATTICE VIBRATIONS. (QUARTZ, BARIU/ EXTENSION OF THE POULET- LOUDON APPROXIMATION 03-0380
US) FIELD AND MELT GROWTH OF CRYSTALS IN THE SODIUM OXIDE-/ (EXTENT OF) TUNGSTEN BRONZE (LIQUIDUS AND SOLID 13-1949
NGLE CRYSTALS FORMED BY BARIUM TITANATE BASE SOLI/ EFFECT OF EXTERNAL ACTIONS ON THE DOMAIN STRUCTURE OF SI 04-0594
UM TITANATE SINGLE CRYSTALS.   EFFECT OF EXTERNAL ACTIONS ON THE REPOLARIZATION OF BARI 04-0957
TION OF MODEL OF TWO DIMENSIONAL FERROELECTRICS IN ARBITRARY EXTERNAL ELECTRIC FIELD.   EXACT SOLU 03-0520
TRIC PHASE TRANSITION, TALKING INTO ACCOUNT TH/ EFFECT OF AN EXTERNAL ELECTRIC FIELD ON A DIFFUSE FERROELEC 03-0441
STABILITY OF SPIN WAVES IN FERROELECTRIC ANTIFERROMAGNETS IN EXTERNAL FIELDS. (THEORY)   IN 03-0149
E OF THE MANGANESE(2+) ION IN A CUBIC ANTIFERRODIELECTRIC AN EXTERNAL MAGNETIC FIELD. /TO THE 6A1G(6S) STAT 03-0317
          SELF INDUCED THERMAL EFFECTS ON LIGHT EXTINCTION OF LITHIUM TANTALATE.   12-1889
          CONDUCTIVITY INJECTION AND EXTRACTION IN POLYCRYSTALLINE BARIUM TITANATE.   04-0814
YDROGEN PHOSPHATE AND POTASSIUM DIHYDROGEN PHOSPHATE) IN THE EXTREME INFRARED. /IC PHOSPHATES (AMMONIUM DIH 17-2580
TE TITANATE SYSTEM. (DEFECT STRUCTURE)   INTRINSIC EXTRINSIC NONSTOICHIOMETRY IN THE LEAD ZIRCONA 09-1536
DUE TO SURFACE LAYERS. (BARIUM TITANATE, ANTIMONY SULFOIODI/ EXTRINSIC PHOTOCONDUCTIVITY IN FERROELECTRIC 04-0681
/: MOLECULAR MODELS FOR POLARIZATION AND CONDUCTION. PART-3: EXTRINSIC VERSUS INTRINSIC POLARIZATION, SURF/ 34-3455
/: MOLECULAR MODELS FOR POLARIZATION AND CONDUCTION. SECT-3: EXTRINSIC VERSUS INTRINSIC POLARIZATION, SURF/ 34-3482

F

          PARAMAGNETIC RESONANCE OF AN F CENTER IN NONSTOICHIOMETRIC BARIUM TITANATE. 04-0976
R MODEL OF FERROELECTRICITY (KDP MODEL, ANTIFERROELECTRICITY F MODEL, WU KDP MODEL, CALCULATIONS). /E SLATE 03-0308
LE CRYSTALS OF RARE EARTH MANGANATES (PHASE DIAGRAMS, CONVE/ FABRICATION (BY FLUX GROWTH) OF HEXAGONAL SING 36-3536
STABILIZED OPTICAL MODULATOR.   FABRICATION OF A LITHIUM TANTALATE TEMPERATURE 12-1872
TMOSPHERE SINTERING.   FABRICATION OF ELECTROOPTIC PLZT CERAMICS BY A 09-1600
IN FILMS.   FABRICATION OF RF SPUTTERED BARIUM TITANATE TH 04-0907
          PREPARATION AND FABRICATION OF SODIUM POTASSIUM NIOBATE.   13-1937
LEED STUDY OF SURFACE STRUCTURES ON THE (001) FACE OF BARIUM TITANATE.   04-0529
LEED STUDIES ON THE (001) FACE OF BARIUM TITANATE CRYSTALS.   04-0530
DIP ETCH PATTERNS ON THE FACES OF TRIGLYCINE SULFATE CRYSTALS.   27-3127
DIUM NITRATE.   DOMAIN ETCH PATTERNS ON THE FACES PARALLEL TO THE FERROELECTRIC-AXIS OF SO 16-2349
VEL ELECTRICAL PROPERTIES. (DIELECTRIC CCNSTANT, DISSIPATION FACTOR) /XIDE DOPED BARIUM TITANATE WITH IMPRO 04-0840
LT NEAR TH/ TEMPERATURE AND FREQUENCY DEPENDENCE OF THE LOSS FACTOR IN FERROELECTRIC MATERIALS (ROCHELLE SA 28-3220
          CALCULATION OF THE DIELECTRIC CORRELATION FACTOR OF CUBIC ICE.   34-3429
EFFECTS OF IMPURITIES ON THE MECHANICAL QUALITY FACTOR OF LEAD ZIRCONATE TITANATE CERAMICS.   09-1608
/F THE SURFACE OF REAL CRYSTAL SUBSTRATES AS THE DETERMINING FACTOR OF THE GROWTH STAGE IN EPITAXY. (TRIGL/ 27-3039
PLES CF FERROELECTRICS AND FERROELASTICS HAVING THE INDEX OF FAINTNESS UNEQUAL TO THE CELL MULTIPLICITY. /M 03-0145

UCTURE TYPE: AN ELECTRON MICRO/ CS (CRYSTALLOGRAPHIC SHEAR) FAMILIES DERIVED FROM THE RHENIUM TRIOXIDE STR 10-1630
STRUCTURAL BASIS OF FERROELECTRICITY IN THE BISMUTH TITANATE FAMILY. 08-1406
ITY CONDITIONS FOR ELECTRIC ORDERING AND STABILI/ PEROVSK FAMILY AND CHEMICAL BOND. (COVALENCY, DIRECTIV 03-0243
IDE(3)(M+2) (/ SWITCHING PROPERTIES IN FERROELECTRICS OF THE FAMILY BISMUTH(4) BARIUM (M-2) TITANIUM(M+1) OX 02-0066
/PERTIES OF) CRYSTALS OF THE ALKALINE TRIHYDROGEN DISELENITE FAMILY. (ESPECIALLY THE SODIUM MEMBER, AND EF/ 22-2848
FERROELECTRIC FAMILY OF BARIUM TITANATE. 04-1018
CRYSTAL CHEMISTRY OF THE ABX(3) FAMILY WITH PARTICULAR EMPHASIS ON PEROVSKITES 08-1403
DYNAMIC FATIGUE OF SINGLE CRYSTALS OF BARIUM TITANATE. 04-0792
FERROELECTRIC CERAMIC. SOME FATIGUING EFFECTS IN 8/65/35 PLZT FINE GRAINED 08-1432
THIUM NIOBATE AND TANTALATE AND THE EFFECT ON THE / STACKING FAULT MODEL FOR STOICHIOMETRY DEVIATIONS IN LI 11-1776
THIUM NIOBATE AND TANTALATE AND THE EFFECT ON THE / STACKING FAULT MODEL FOR STOICHIOMETRY DEVIATIONS IN LI 11-1777
ZATION IN PEROVSKITES. (ABSTRACT ONLY) CONDITIONS FAVORING THE ORIGINATION OF SPONTANEOUS POLARI 08-1457
ILE ON THE SWITCHING CHARACTERISTICS AND THE BIT SIZE OF THE FE-PC MEMORY DEVICE. /GAUSSIAN IRRADIANCE PROF 37-3670
FERROELECTRIC BISMUTH TITANATE. FEASIBILITY OF ELECTROOPTIC DEVICES UTILIZING 08-1447
OPTICAL TRANSITION TO THE FERMI LEVEL IN SODIUM TUNGSTATE. 10-1634
AN LITHIUM NIOBATE CRYSTAL. FERMI RESONANCE OF POLARITON WITH BIPHONON IN 11-1761
PHASE IN THE SYSTEM LEAD TITANATE- LEAD ZIRCONATE- LANTHANUM FERRATE. ANTIFERROELECTRIC 13-1961
ANTIFERROMAGNETIC PROPERTIES OF SOME PEROVSKITES. (BISMUTH FERRATE) 08-1441
ATOMIC STRUCTURE IN THE PEROVSKITIC FERRATE. 08-1398
EFFECT IN FERROELECTRIC- ANTIFERROMAGNETIC BISMUTH TITANIUM FERRATE. MAGNETOELECTRIC 08-1360
SE TRANSITIONS IN THE LEAD ZIRCONATE- LEAD TITANATE- BISMUTH FERRATE. / EFFECT STUDIES OF FERROELECTRIC PHA 09-1498
AMETERS AND THE COEFFICIENTS OF THERMAL EXPANSION OF BISMUTH FERRATE. /ION DETERMINATION OF THE LATTICE PAR 08-1287
TIMONATE, NIOBATE, OR BISMUTHATE, OR BY BISMUTH OR LANTHANUM FERRATE, ALUMINATE OR CHROMATE. / POTASSIUM AN 09-1539
/67) TUNGSTATE(0.33) LEAD FERRATE(0.5) NIOBATE(0.5), FERRATE AND ITS SOLID SOLUTIONS WITH BARIUM T/ 08-1457
/D ANTIFERROELECTRICS WITH HIGH CURIE TEMPERATURES. (BISMUTH FERRATE AND SYSTEMS BASED ON IT, AND SINGLE C/ 08-1378
DIELECTRIC HYSTERESIS IN (FLUX GROWN) SINGLE CRYSTAL BISMUTH FERRATE (CONFIRMS FERROELECTRIC NATURE). 08-1449
DISPERSION IN STRONTIUM TITANATE- LANTHANUM FERRATE DIELECTRIC CERAMICS. 06-1120
LUTIONS (ROOM TEMPERATURE TO 6/ MOSSBAUER STUDIES OF BISMUTH FERRATE LEAD TITANATE PEROVSKITE TYPE SOLID SO 08-1471
/(X=0.9 AND 0.58) (CORRELATION OF CATION SHIFTS WITH BISMUTH FERRATE, LITHIUM NIOBATE AND LITHIUM TANTALAT/ 09-1574
URES. DIELECTRIC PROPERTIES OF LEAD TITANATE- LEAD (FERRATE NIOBATE) SOLID SOLUTIONS AT HIGH PRESS 05-1071
D FERROELECTRIC MAGNETS. (LEAD AND BARIUM TITANATES, BISMUTH FERRATE, POTASSIUM NIOBATE) /FERROELECTRICS AN 04-1038
D MATERIALS (SYSTEM WIT/ FERROELECTRIC PROPERTIES OF BISMUTH FERRATE (PREPARATION AND STRUCTURE) AND RELATE 08-1333
/MA-RAYS OF TIN-119 DURING PHASE TRANSISTIONS OF THE BISMUTH FERRATE STRONTIUM STANNATE(0.33) MANGANATE(0./ 08-1400
AND SPECIAL DIELECTRIC PROPERTIES IN THE BISMUTH- LANTHANUM FERRATE SYSTEM. /XISTENCE OF ANTIFERROMAGNETIC 08-1425
AN X-RAY STUDY OF THE PHASE TRANSITION IN STRONTIUM FERRATE TANTALATE 08-1382
EFFECT OF IRON(4+) IN THE SYSTEM STRONTIUM FERRATE TITANATE. 13-1927
ANTIFERROMAGNETIC) FERROELECTRIC BISMUTH FERRATE X-RAY AND NEUTRON DIFFRACTION STUDY. (08-1402
A SERIES OF FERROELECTRIC MATERIALS. (BARIUM TITANATE, LEAD FERRATE(0.5) NIOBATE(0.5)) / THE PROPERTIES OF 04-0780
SINGLE CRYSTAL LEAD MAGNESATE(0.33) NIOBATE(0.67), AND LEAD FERRATE(0.5) NIOBATE(0.5)) /OLYCRYSTALLINE AND 08-1375
RESISTIVITY) IN (DOPED AND UNDOPED CERAMIC PEROVSKITE) LEAD FERRATE(0.5) NIOBATE(0.5). /URE COEFFICIENT OF 13-2000
/ON OF THE SPONTANEOUS MAGNETOELECTRIC MAGNETIZATION OF LEAD FERRATE(0.5) NIOBATE(0.5) AND LEAD MANGANATE(/ 13-2069
/TRIC FERROMAGNETS. (LEAD FERRATE(0.67) TUNGSTATE(0.33) LEAD FERRATE(0.5) NIOBATE(0.5), BISMUTH FERRATE AN/ 08-1368
ZIRCONATE) NEW PIEZOELECTRIC CERAMICS. (LEAD TITANATE- LEAD FERRATE(0.5) NIOBATE(0.5), LEAD TITANATE- LEAD 09-1544
YSTEM. SEIGNETTO MAGNETIC SOLID SOLUTIONS OF THE LEAD FERRATE(0.67) TUNGSTATE(0.33)- LEAD TITANATE S 05-1065
/N SOME FERROELECTRICS AND FERROELECTRIC FERROMAGNETS. (LEAD FERRATE(0.67) TUNGSTATE(0.33). LEAD FERRATE(0./ 08-1368
OVSKITE LIKE LAYER TYPE FERROELECTRICS. (IN BISMUTH TITANATE FERRATES) /AUER STUDIES OF IRON-57 IN SOME PER 13-2035
MS. MOSSBAUER MEASUREMENTS OF BISMUTH FERRATE- BISMUTH FERRATE- LEAD ZIRCONATE SYSTE 08-1282
/ MOSSBAUER EFFECT STUDY OF THE FERROELECTRIC TRANSITIONS IN FERRIC AMMONIUM SULFATE DODECAHYDRATE AND POTA 19-2755
/ MOSSBAUER EFFECT STUDY OF THE FERROELECTRIC TRANSITIONS IN FERRIC AMMONIUM SULFATE DODECAHYDRATE AND POTA 24-2892
RYSTALS. MOSSBAUER STUDY ON FERROELECTRIC PROPERTIES OF FERRIC AMMONIUM SULFATE DODECAHYDRATE SINGLE C 19-2770
SINGLE CRYSTALS. COMPENSATION OF HETEROVALENT FERRIC IONS IN THE LATTICE OF BARIUM TITANATE 04-0539
HIGH EFFECTIVE DIELECTRIC CONSTANT (DUE TO ADDING CUPROUS OR FERRIC OXIDE). /ARIUM TITANATE CERAMIC WITH A 04-0581
X-RAY STUDY OF THE SYSTEM BISMUTH OXIDE FERRIC OXIDE. 08-1376
N IN BARIUM TITANATE SINGLE CRYSTALS. EFFECT OF FERRIC OXIDE ON THE 120 DEGREE PHASE TRANSITIO 04-0544
FERROELASTIC EFFECT IN LANTHANUM ORTHOFERRITE. 08-1270
TING AND MAGNETIC) RARE EARTH ORTHOCHROMITES, MANGANITES AND FERRITES ELECTRICAL TRANSPORT IN (SEMICONDUC 36-3608
VE EFFECTS WHICH ACCOMPANY DISACCOMODATION TYPE PHENOMENA IN FERRITES AND FERROELECTRICS. /URE OF DISSIPATI 03-0447
/LED FERROELASTIC CRYSTALS. (GADOLINIUM MOLYBDATE, KDP, ORTHOFERRITES, DYSPROSIUM ANTIMONIDE, RUBIDIUM IRO/ 02-0036
STRUCTURE AND TRANSITIONS IN PEROVSKITES. (ORTHOFERRITES, SODIUM NIOBATE) 08-1396
NIC VIBRATIONS RADIATING IN THE DIRECTIO/ RADIALLY VIBRATING FERROCERAMIC TRANSDUCER AS A SOURCE OF ULTRASO 37-3657
FREQUENCY SPECTRUM OF FERROCERAMIC TRANSDUCERS. 37-3656
ELASTIC CONSTANTS OF FERROCERAMICS. 09-1513
SPONTANEOUS DESTRUCTION OF FERROCERAMICS. (BARIUM TITANATE) 04-0787
/RTIES OF 60 COMPOUNDS, 58 SOLID SOLUTIONS, AND 10 TECHNICAL FERROCERAMICS, CURIE TEMPERATURES OF 290 COMP/ 01-0014
NONLINEAR EFFECTS IN FERROCERAMICS. (PZT-8, LITHIUM NIOBATE) 09-1546
DOMAIN STRUCTURE IN THE FERROELECTRIC POTASSIUM FERROCYANIDE. 24-2895
R DIELECTRICS. (POTASSIUM DIHYDROGEN PHOSPHATE AND POTASSIUM FERROCYANIDE) /SCHARGE OF CONDENSERS WITH POLA 17-2631
/R MOLECULES AND FERROELECTRIC PHASE TRANSITION IN POTASSIUM FERROCYANIDE AND ISOMORPHOUS CRYSTALS. (NEW M/ 24-2893
E TEMP/ NEUTRON DIFFRACTION STUDY OF FERROELECTRIC POTASSIUM FERROCYANIDE TRIDEUTERO HYDRATE ABOVE THE CURI 24-2903
MALOUS SPECIFIC HEAT OF FERROELECTRIC THIOUREA AND POTASSIUM FERROCYANIDE TRIHYDRATE. ANO 24-2902
OUS SPIN LATTICE RELAXATION BY QUASI SPIN WAVES IN POTASSIUM FERROCYANIDE TRIHYDRATE. ANOMAL 24-2891
LD IN GENERAL DIPOLE LATTICES. (THEORY APPLIED TO POTASSIUM FERROCYANIDE TRIHYDRATE) INTERNAL FI 03-0220
TIONS IN FERRIC AMMONIUM SULFATE DODECAHYDRATE AND POTASSIUM FERROCYANIDE TRIHYDRATE. /FERROELECTRIC TRANSI 19-2755
TIONS IN FERRIC AMMONIUM SULFATE DODECAHYDRATE AND POTASSIUM FERROCYANIDE TRIHYDRATE. /FERROELECTRIC TRANSI 24-2892
MOSSBAUER RESONANCE OF SINGLE CRYSTAL POTASSIUM FERROCYANIDE TRIHYDRATE. 24-2890
UER STUDY OF THE FERROELECTRIC PHASE TRANSITION IN POTASSIUM FERROCYANIDE TRIHYDRATE. MOSSBA 24-2901
NDS. RADIO SPECTROSCOPIC INVESTIGATION OF POTASSIUM FERROCYANIDE TRIHYDRATE AND ISOMORPHOUS COMPOU 24-2900
DY OF HYDROGEN-1 IN PARAELECTRIC AND FERROELECTRIC POTASSIUM FERROCYANIDE TRIHYDRATE BY PULSED NMR. STU 24-2889
OLYTYPE AND TWINNING OF THE FERROELECTRIC CRYSTALS POTASSIUM FERROCYANIDE TRIHYDRATE BY THE NMR METHOD. P 24-2898
ROSTATIC PRESSURE (/ FERROELECTRIC TRANSITION IN A POTASSIUM FERROCYANIDE TRIHYDRATE CRYSTAL UNDER HIGH HYD 24-2897
NUCLEAR MAGNETIC RESONANCE AND X-RAY STUDIES OF POTASSIUM FERROCYANIDE TRIHYDRATE CRYSTALS. 24-2894
URE. (TO 5100 KG/SQ CM) FERROELECTRIC BEHAVIOR OF POTASSIUM FERROCYANIDE TRIHYDRATE CRYSTALS AT HIGH PRESS 24-2897
EMPERATUR/ MOSSBAUER EFFECT STUDY OF FERROELECTRIC POTASSIUM FERROCYANIDE TRIHYDRATE (NEAR THE TRANSITION T 24-2891
/ DEW METHOD FOR REVEALING THE DOMAIN STRUCTURE IN POTASSIUM FERROCYANIDE TRIHYDRATE SINGLE CRYSTALS, P-N / 24-2904
S OF THEIR GROWTH. (AMMONIUM DIHYDROGEN PHOSPHATE, POTASSIUM FERROCYANIDE TRIHYDRATE, SODIUM NITRATE) /STIC 16-2316
UCIES OF THE PHASE TRANSITIONS IN FERROELECTRICS. (POTASSIUM FERROCYANIDE TRIHYDRATE, SODIUM NITRITE) /E ST 16-2255
MAGNETIC PROPERTIES OF FERRODIELECTRIC CATALYSTS. 03-0235
STRUCTURE. RESONANCE EFFECTS IN MAGNETICALLY EQUIAXIAL FERRODIELECTRIC SINGLE CRYSTALS HAVING DOMAIN 03-0509
INTERACTION OF LIGHT WITH MAGNETOACOUSTIC WAVES IN A FERRODIELECTRIC (THEORY). 03-0157
MMET/ SPIN WAVE PHENOMENOLOGICAL THEORY IN FINITE SAMPLES OF FERRODIELECTRICS WITH AN ARBITRARY MAGNETIC SY 03-0164

SES OF GADO/ COMPREHENSIVE X-RAY STUDY OF THE FERROELECTRIC- FERROELASTIC AND PARAELECTRIC- PARAELASTIC PHA 30-3287
YSTALS AND PARTIALLY FERROELECT/ CONSIDERATIONS OF PARTIALLY FERROELASTIC AND PARTIALLY ANTIFERROELASTIC CR 03-0136
ATE TITANATE CERAMICS WHEN SUBJECTED TO LARGE TENSILE STRAI/ FERROELASTIC BEHAVIOR OF LEAD LANTHANUM ZIRCON 09-1571
/SM OF THE PHASE TRANSFORMATION OF A PECULIAR FERROELECTRIC- FERROELASTIC CRYSTAL: DIAMMONIUM DICADMIUM SU/ 19-2741
FERROELECTRIC AND FERROELASTIC CRYSTAL, GADOLINIUM MOLYBDATE. 30-3308
FERROELECTRIC FERROELASTIC CRYSTAL GADOLINIUM MOLYBDATE. 30-3297
/NISM OF THE PHASE TRANSFORMATION OF PECULIAR FERROELECTRIC- FERROELASTIC CRYSTAL: SODIUM AMMONIUM TARTRAT/ 29-3238
PYROELECTRIC AMMONIUM IODATE, A POTENTIAL FERROELASTIC: CRYSTAL STRUCTURE. 36-3574
INIUM MOLYBD/ THICKNESS OF DOMAIN WALLS IN FERROELECTRIC AND FERROELASTIC CRYSTALS. (BARIUM TITANATE, GADOL 04-0777
KDP, ORTHOFERRITES, DYSPROSIUM ANTIMONIDE, RUBIDIUM/ COUPLED FERROELASTIC CRYSTALS. (GADOLINIUM MOLYBDATE, 02-0036
D TO MAGNETIC CRYSTA/ POSSIBLE SPECIES OF FERROMAGNETIC, AND FERROELASTIC CRYSTALS (PREVIOUS THEORY EXTENDE 03-0142
FERROELASTIC EFFECT IN LANTHANUM ORTHOFERRITE. 08-1270
RIDE AND CESIUM IRON TETRAFLUORIDE. FERROELASTIC EFFECT IN RUBIDIUM IRON TETRAFLUO 33-3358
E.   ANOMALOUS PHONON BEHAVIOR NEAR THE PHASE TRANSITION IN FERROELASTIC FERROELECTRIC GADOLINIUM MOLYBDAT 30-3276
ECTRICAL, MECHANICAL AND OPTICAL PROPERTIES OF FERROELECTRIC FERROELASTIC GADOLINIUM MOLYBDATE. /NCES OF EL 30-3253
BDATE. CRYSTAL STRUCTURE OF THE TRANSITION ME/ FERROELECTRIC FERROELASTIC PARAMAGNETIC BETA GADOLINIUM MOLY 30-3289
DATE (CZOCHRALSKI SINGLE C/ (ROOM TEMPERATURE) FERROELECTRIC FERROELASTIC PARAMAGNETIC (BETA) TERBIUM MOLYB 30-3290
BORACITE.   FERROELECTRIC AND FERROELASTIC PROPERTIES OF MAGNESIUM CHLORINE 14-2102
SINGLE CRYSTALS OF GADOLINIUM MOLYBDATE, A FERROELECTRIC AND FERROELASTIC SUBSTANCE. /TH AND MORPHOLOGY OF 30-3259
SUPPLEMENT TO A PREVIOUS THEORY OF WEAK FERROELASTICS AND WEAK FERROELECTRICS. 03-0146
/ESENTATION AND DISCUSSION OF EXAMPLES OF FERROELASTICS AND FERROELASTICS HAVING THE INDEX OF FAINTNESS U/ 03-0145
ILITY OR ELASTI/ GENERAL CONSIDERATION OF FERROELECTRICS AND FERROELASTICS SUCH THAT THE ELECTRIC SUSCEPTIB 03-0140
E STATE PARAMETERS AND FORMULATION OF SPONTANEOUS STRAIN FOR FERROELASTICS (THEORY).   DETERMINATION OF TH 03-0137
FERROELASTICITY. 02-0037
PHENOMENOLOGICAL LATTICE DYNAMICAL THEORY OF FERROELASTICITY. 03-0144
STRUCTURAL BASIS OF FERROELECTRICITY AND FERROELASTICITY. (SYMMETRY RELATIONS) 03-0135
FERROELECT.... IS NOT INDEXED.
ANISOTROPY OF ULTRASONIC ATTENUATION IN UNIAXIAL FERROELECTRICS. (TGS, ABSTRACT ONLY) 27-3055
TE COMPOUNDS.   PROPERTIES OF FERROGLASS CERAMICS BASED ON LEAD BARIUM NIOBA 13-2006
EXTENDED TO MAGNETIC CRYSTALS AND TO INCLUDE PARTIAL CASES, "FERROIC CRYSTALS"). /RYSTALS (PREVIOUS THEORY 03-0142
FERROMAGNETIC RESONANCE LINE WIDTH IN FERROELECTRIC FERROMAGNETS. 03-0207
/RATURE DEPENDENCE) IN SOME FERROELECTRICS AND FERROELECTRIC FERROMAGNETS. (LEAD FERRATE(0.67) TUNGSTATE(0/ 08-1368
MATION OF INTERNAL ELECTRIC FIELD GRADIENTS IN FERROELECTRIC FERROMAGNETS. (PEROVSKITES) / THEORETICAL ESTI 08-1401
IOUS THEORY EXTENDED TO MAGNETIC CRYSTA/ POSSIBLE SPECIES OF FERROMAGNETIC, AND FERROELASTIC CRYSTALS (PREV 03-0142
EAR SIMULTANEOUSLY.   SYMMETRY OF CRYSTALS IN WHICH FERROMAGNETIC AND FERROELECTRIC PROPERTIES APP 03-0465
ULTS.   NEW TWO- AND THREE PHASE FERROELECTRIC FERROMAGNETIC MATERIALS. SOME PROBLEMS AND RES 03-0360
AND MAGNETIC MEASUREMENTS OF CERTAIN TWO PHASE FERROELECTRIC FERROMAGNETIC MIXTURES. (BARIUM TITANATE) /IC 04-0565
CRITICAL POINT INDICES OF FERROELECTRIC AND FERROMAGNETIC PHASE TRANSITIONS. 03-0264
HERMODYNAMIC THEORY OF CRYSTALS POSSESSING FERROELECTRIC AND FERROMAGNETIC PROPERTIES.   T 03-0472
CTRIC FERROMAGNETS. FERROMAGNETIC RESONANCE LINE WIDTH IN FERROELE 03-0207
LT TUNGS/ STRUCTURE AND MAGNETIC PROPERTIES OF FERROELECTRIC FERROMAGNETIC SOLID SOLUTIONS IN THE LEAD COBA 36-3604
UCTURAL MICROELECTRON DIFFRACTION STUDY OF FERROELECTRIC AND FERROMAGNETIC SUBSTANCES. (LEAD TITANATE)   STR 05-1082
ECTRIC STA/ THEORY OF SECOND ORDER PHASE TRANSITION FROM THE FERROMAGNETIC TO THE FERROMAGNETIC AND FERROEL 03-0395
CRYSTALS DURING THE PHASE TRANSITIONS IN FERROELECTRICS AND FERROMAGNETICS. /E OF COMPLETE SYMMETRY OF THE 03-0527
TEMPERATURE DEPENDENCE OF THE SHAPE OF THE DOMAIN WALL IN FERROMAGNETICS AND FERROELECTRICS. 03-0191
QUANTUM THEORY OF VIBRATIONS IN FERROELECTRIC FERROMAGNETICS WITHOUT A SYMMETRY CENTER. 03-0166
80 DEGR/ TRIGONAL BORACITES: A NEW TYPE OF FERROELECTRIC AND FERROMAGNETOELECTRIC MATERIAL THAT ALLOWS NO 1 14-2098
LIGHT INDUCED PYROCURRENTS IN FERROSEMICONDUCTOR ANTIMONY SULFOIODIDE. 15-2130
MOSSBAUER EFFECT OF FERROUS IONS IN CUBIC ICE. 34-3416
S OF 0.04)   FERROELECTRICITY IN THE IRON DEFICIENT FERROUS SULFIDE SYSTEM. (EXISTS TO DEFICIENCIE 36-3614
(2+) ION IN A CUBIC ANTIFERRODIELECTRIC AN EXTERNAL MAGNETIC FIELD. /TO THE 6A1G(6S) STATE OF THE MANGANESE 03-0317
TRIC SEMICONDUCTOR ANTIMONY SULFOIODICE IN A STATIC ELECTRIC FIELD.   AGING OF THE FERROELEC 15-2113
ILY ORDERED FERROELECTRICS IN A RAPIDLY OSCILLATING ELECTRIC FIELD.   BEHAVIOR OF MAGNETICA 03-0147
CTRIC LITHIUM THALLIUM TARTRATE HYDRATE (LTT) BY AN ELECTRIC FIELD. / OF THE ELASTIC COMPLIANCE OF FERROELE 29-3249
ARIZATION IN FERROELECTRICS ON THE INTENSITY OF THE ELECTRIC FIELD.   DEPENDENCE OF EFFECTIVE POL 03-0382
EPENDENCE OF THE ELASTIC COEFFICIENTS OF ADP ON THE ELECTRIC FIELD.   D 17-2537
IC PROPERTIES OF ANTIMONY SULFOIODIDE IN A CONSTANT ELECTRIC FIELD.   DIELECTRIC AND PIEZOELECTR 15-2249
T OF TRIGLYCINE SULFATE SINGLE CRYSTALS IN A STRONG ELECTRIC FIELD.   DIELECTRIC CONSTAN 27-3150
NGENT OF THE ANGLE OF DIELECTRIC LOSSES IN A STRONG ELECTRIC FIELD. / SINGLE CRYSTALS ON A CHANGE IN THE TA 27-3153
ANTIMONY SULFOIODIDE. (PHASE BOUNDARY SHAPE STRIATIONS IN DC FIELD) /ON ON THE FERROELECTRIC PROPERTIES OF 15-2197
TANGENT OF THE ANGLE OF DIELECTRIC LOSS IN A STRONG ELECTRIC FIELD. /CHANGE IN THE DIELECTRIC CONSTANT AND 27-3151
TY OF A FERROELECTRIC SEMICONDUCTOR IN A NONUNIFORM ELECTRIC FIELD.   ELECTRICAL NONLINEARI 03-0500
WO DIMENSIONAL FERROELECTRICS IN ARBITRARY EXTERNAL ELECTRIC FIELD.   EXACT SOLUTION OF MODEL OF T 03-0520
SIS LOOP OF SILVER SODIUM DINITRITE BY QUASI STATIC ELECTRIC FIELD.   FERROELECTRIC HYSTERE 16-2293
NE FLUOBERYLLATE CRYSTALS DURING SWITCHING IN PULSE ELECTRIC FIELD.   IMPEDANCE OF FERROELECTRIC TRIGLYCI 27-3027
MODIFIED POTASSIUM DIHYDROGEN PHOSPHATE MODEL IN A STAGGERED FIELD.   17-2710
EVELOPMENT OF FERROELECTRIC CHARACTER IN AN APPLIED ELECTRIC FIELD) / OF SYNTHETIC CALCIUM CHLORAPATITE. (D 36-3591
PEDANCE MEASUREMENTS OF A FERROELECTRIC CONDENSER UNDER A DC FIELD.   NOISE AND IM 37-3645
PARAELECTRIC SEMICONDUCTOR IN A NONUNIFORM ELECTRIC FIELD.   03-0502
TTIVITY CHANGE OF BARIUM TITANATE UNDER VERY SLOWLY CHANGING FIELD.   PERMI 04-0868
F SINGLE CRYSTALS OF TRIGLYCINE SULFATE IN A STRONG ELECTRIC FIELD.   PERMITTIVITY O 27-2978
IN TRIGLYCINE SULFATE. (DIELECTRIC DISPLACEMENT AND COERCIVE FIELD)   (POLARIZATION CHANGE) AFTER EFFECTS 27-2952
PHOSPHATE CRYSTAL IN THE PHASE TRANSITION REGION. (COERCIVE FIELD)   POLARIZATION OF A RUBIDIUM DIHYDROGEN 17-2504
OSITIONS, SPONTANEOUS) POLARIZATION (ON APPLYING AN ELECTRIC FIELD). /VERSIBLE, BUT TILTABLE (TO 3 STABLE P 08-1346
DIFICATIONS OF POTASSIUM DIHYDROGEN PHOSPHATE IN AN ELECTRIC FIELD.   RAMAN SPECTRA OF POLYMORPHIC MO 17-2425
TALS OF BARIUM TITANATE IN THE PRESENCE OF A STEADY ELECTRIC FIELD. /BSORPTION OF ULTRASOUND IN SINGLE CRYS 04-0850
TION TIME OF SECONDS WHEN ILLUMINATED UNDER APPLIED ELECTRIC FIELD). /E) OPTICAL DAMAGE IN KDP (WITH RELAXA 17-2703
F ANTIMONY SULFOIODIDE CAUSED BY ILLUMINATION IN DC ELECTRIC FIELD.   STRAIN ALONG C-AXIS O 15-2239
BATE(0.5) AND LEAD MANGANATE(0.5) NIOBATE(0.5) BY A MAGNETIC FIELD. /MAGNETIZATION OF LEAD FERRATE(0.5) NIO 13-2069
FECT OF TRIGLYCINE SULFATE SINGLE CRYSTALS IN AN AC ELECTRIC FIELD.   TEMPERATURE AUTOSTABILIZATION EF 27-3005
ALUMINUM SULFATE HEXAHYDRATE. (POLARIZATION VERSUS ELECTRIC FIELD, -80C TO ROOM TEMPERATURE) / GUANIDINIUM 21-2813
ONY SULFOIODIDE CRYSTAL (IN THE PARAELECTRIC PHASE, UNDER DC FIELD ALONG C-AXIS). /AL DISTRIBUTION IN ANTIM 15-2173
EAD/ RELAXATION OF ANTIFERROELECTRIC DOMAIN WALLS, EFFECTIVE FIELD, AND ACTIVATION ENERGY. APPLICATION TO L 07-1257
SULFATE.   SPECIFIC HEAT UNDER APPLIED ELECTRIC FIELD AND ELECTROCALORIC EFFECT IN TRIGLYCINE 27-2960
LTRASOUND IN ANTIMONY SULF/ EFFECTS OF TEMPERATURE, ELECTRIC FIELD, AND ILLUMINATION ON THE ABSORPTION OF U 15-2248
M OXIDE-/ (EXTENT OF) TUNGSTEN BRONZE (LIQUIDUS AND SOLIDUS) FIELD AND MELT GROWTH OF CRYSTALS IN THE SODIU 13-1949
ER AND CHROMIUM IONS ON THE DIELECTR/ EFFECT OF A POLARIZING FIELD AND THE ADDITION OF A MIXTURE OF COPP 27-3045
GEN(1-X) PHOSPHATE SYSTEM VERSUS CONCENTRATION, TEMPERATURE, FIELD, AND PRESSURE. /TRIDEUTERIUM(X) TRIHYDRO 17-2666
IC ABSORPTION IN TRIGLYCINE SULFA/ INFLUENCE OF THE ELECTRIC FIELD AND THE DOMAIN STRUCTURE ON THE ULTRASON 27-2955
IN MIXED KDP- DEUTERATED KDP CRYSTALS BY MODIFIED MOLECULAR FIELD APPROXIMATION. STUDY OF PHASE TRANSITION 17-2528

OTION IN ROCHELLE SALT.   ANOMALOUS DISPERSION OF COERSIVE FIELD AT VERY LOW FREQUENCIES AND JERKY WALL M 28-3177
, IN A DIGITAL DEVIA/ STUDY OF THE LIMITATION OF THE ANGULAR FIELD BY THE COMMUTATOR OF POLARIZATION IN KDP 17-2524
/LATIONS OF FERROELECTRIC BODIES. (WITH FREQUENCY OF APPLIED FIELD CAUSED BY SHEARING DUE TO POLARIZATION / 03-0197
HIUM THALLIUM TARTRATE (TUNING AND DELAY DEVICES).   ELECTRIC FIELD CONTROL OF THE ELASTIC COMPLIANCE IN LIT 29-3250
N REGICN IN POTASSIUM DIHYDROGEN PHOSPHATE. (TEMPERATURE AND FIELD DEPENDENCE) /THE FERROELECTRIC TRANSITIO 17-2576
N STRONTIUM TITANATE.   (TEMPERATURE AND) ELECTRIC FIELD DEPENDENCE OF A HYDROGEN IMPURITY MODE I 06-1146
LECTRICS IN FAR INFRARED. (APPLIED TO BARIUM TITANATE)   FIELD DEPENDENCE OF ABSORPTION BANDS OF FERROE 04-0665
(RAMAN SCATTERING IN CUBIC PEROVSKITE STRONTIUM T/ ELECTRIC FIELD DEPENDENCE OF OPTICAL PHONON FREQUENCIES 06-1241
OELECTRIC PHASE BOUNDARY IN PLZT CERAMICS.   TEMPERATURE AND FIELD DEPENDENCE OF THE ANTIFERROELECTRIC FERR 09-1491
MITEC CURRENT IN BISMUTH(12) GERMANIUM O/ LIGHT AND ELECTRIC FIELD DEPENDENT OSCILLATION OF SPACE CHARGE LI 35-3505
BISMUTH TITANATE (EFFECT OF TEMPERATURE).   (WEAK FIELD) DIELECTRIC (PERMITTIVITY) PROPERTIES OF 36-3555
TIC) WAVES ON (PIEZOELECTRIC) LITHIUM NIOBATE.   (FAR FIELD) DIFFRACTION SPREADING OF SURFACE (ACOUS 11-1850
AL EMPHASIS ON NEAR SURFACE REGIONS IN FERROELECTR/ ELECTRIC FIELD DISTRIBUTIONS IN DIELECTRICS, WITH SPECI 03-0182
NIOBATE.   NATURE OF INTERNAL ELECTRIC FIELD DURING OPTICAL DAMAGE PROCESS IN LITHIUM 11-1861
D LEAD TELLURIDE ON UHV CLEAVED TRIGLYCINE SULFATE.   FIELD EFFECT AND HALL MOBILITY OF TELLURIUM AN 27-3172
UCTOR AND A C-DOMAIN FERROELECTRIC (THEORY, TITANATES DISCU/ FIELD EFFECT AT THE CONTACT BETWEEN A SEMICOND 03-0511
D TO TELLURIUM FILMS CN UHV CLEAVED TG/ AFTER EFFECTS IN THE FIELD EFFECT AT VERY LARGE GATE CHARGES APPLIE 27-3171
SUBSTRATES.   TEMPERATURE DEPENDENCE OF THE FERROELECTRIC FIELD EFFECT IN TELLURIUM FILMS ON TRIGLYCINE 27-3018
TRATE OF FERROELECTRIC TRIGLYCINE SULFATE AND SELENATE CRYS/ FIELD EFFECT IN TELLURIUM THIN FILMS ON A SUBS 27-3046
ONTIUM TITANATE. (DOPED WITH GADCLINIUM(3+) CR IRC/ ELECTRIC FIELD EFFECT OF ELECTRON SPIN RESONANCE IN STR 06-1229
HAS/ SURFACE LAYER IN BARIUM TITANATE (X-RAY DIFFRACTION, DC FIELD EFFECT, PARAELECTRIC AND FERROELECTRIC P 04-0857
TIN OXIDE FILMS ON LEAD ZIRCONATE(0.65) TITA/ FERROELECTRIC FIELD EFFECT STUDIES AT LOW TEMPERATURES (WITH 09-1505
EAD(0.97) BISMUTH(0.02) TITANATE (0.35) ZIRCO/ FERROELECTRIC FIELD EFFECT STUDIES. (ON TIN OXIDE FILMS ON L 09-1506
TRONTIUM TITANATE.   INTERPRETATION OF ELECTRIC FIELD EFFECTS IN EPR OF GADOLINIUM(3+) DOPED S 06-1168
                                                             FIELD EMISSION FROM ANTIMONY SULFOIODIDE.   15-2203
ION IN RUBIDIUM ALUMINUM DISULFATE D/ CALCULATION OF CRYSTAL FIELD ENERGY LEVEL SPLITTINGS OF TITANIUM(3+) 19-2754
D LEAD ZIRCONATE. (PHASE TRANSITION 200 DEG C BELOW THE CUR/ FIELD ENFORCED FERROELECTRICITY IN GLASS BONDE 07-1252
LARIZATION AND REPOLARIZATI/ CONNECTION BETWEEN THE STARTING FIELD FOR PARKHAUSEN STEPS AND THE CRITICAL PO 03-0318
OF FERROELECTRIC SODIUM NITRITE. (AT THE SODIUM-23/ ELECTRIC FIELD GRADIENT AND LONG RANGE ORDER PARAMETER 16-2395
NATE STUDIED BY PERTURBED ANGULAR COR/ THE INTERNAL ELECTRIC FIELD GRADIENT IN ANTIFERROELECTRIC LEAD ZIRCO 07-1255
.   TEMPERATURE DEPENDENCE OF THE ELECTRIC FIELD GRADIENT IN FERROELECTRIC SCDIUM NITRITE 16-2337
                                                    CRYSTAL FIELD GRADIENTS IN ANTIMONY SULFOIODIDE.   15-2185
/BAUER STUDY AND THEORETICAL ESTIMATION OF INTERNAL ELECTRIC FIELD GRADIENTS IN FERROELECTRIC FERROMAGNETS/ 08-1401
D TRICHLORIDE.   ELECTRIC FIELD GRADIENTS IN IONIC CHLORIDES: CESIUM LEA 33-3367
D/ CALCULATIONS OF THE INTERNAL ELECTRIC FIELDS AND ELECTRIC FIELD GRADIENTS IN THE PEROVSKITE TYPE COMPOUN 08-1442
OELECTRIC. (DEPENDENCE OF DIELECTRIC CONSTANT ON THERMAL AND FIELD HISTORY) /IUM IODATE - A DISORDERED FERR 08-1348
DEBYE SCREENING OF A NCNUNIFORM ELECTRIC FIELD IN A FERROELECTRIC.   03-0499
ATION TC ALL BARIUM TITANATE STRUCTURE TYPES.   ELECTROSTATIC FIELD IN A PSEUDOCUBIC IONIC STRUCTURE. APPLIC 04-1008
. APPLICATION TO THE CALCULATION OF TETRAGONA/ ELECTROSTATIC FIELD IN A SLIGHTLY ORTHORHOMBIC IONIC CRYSTAL 04-0578
                                          CRITICAL FIELD IN CERAMIC BARIUM TITANATE.   04-0951
                                           CRYSTAL FIELD IN CHROMIUM(3+) DOPED LITHIUM NIOBATE. 11-1759
EPR INVESTIGATION OF THE INTERNAL ELECTRIC FIELD IN COPPER DOPED ROCHELLE SALT.   28-3233
TRANSITION. (THEORY, CALCUL/ CURIE PCINT SHIFT WITH APPLIED FIELD IN FERROELECTRICS POSSESSING FIRST ORDER 03-0327
IED TO POTASSIUM FERROCYANIDE TRIHYDRATE)   INTERNAL FIELD IN GENERAL DIPOLE LATTICES. (THEORY APPL 03-0220
RIMENTAL)   COERCIVE FIELD IN LITHIUM TRIHYDROGEN DISELENITE. (EXPE 22-2844
POLARIZATION REVERSAL IN A WIDE RANGE OF IMPRESSING RATE OF FIELD IN SODIUM NITRITE.   16-2381
/TIES CF FERROELECTRICS, SUBJECTED BEFOREHAND TO AN ELECTRIC FIELD, IN STRONG ELECTRIC FIELDS. (ANISOTROPY/ 04-0661
OBIA (PCTASSIUM LITHIUM NICBATES, PHASE REL/ TUNGSTEN BRONZE FIELD IN THE (OXIDE) SYSTEM POTASSIA LITHIA NI 13-2017
TE BARIUM TITANATE BARIUM DINICBATE.   TUNGSTEN BRONZE FIELD IN THE PSEUDOTERNARY SYSTEM SODIUM NIOBA 13-1970
                                       DEPOLARIZATION FIELD IN THIN FERROELECTRIC FILMS.   37-3659
NAL) PHASE TRANSITION IN BARIUM TITANATE.   FIELD INDUCED NUCLEATION AT THE (CUBIC TETRAGO 04-0753
TRONTIUM TITANATE AND POTASSIUM TANTALATE.   ELECTRIC FIELD INDUCED OPTICAL HARMONIC GENERATION IN S 06-1203
ATE.   ULTRASONIC STUDIES OF FIELD INDUCED PHASE TRANSITION IN BARIUM TITAN 04-0509
STALS. (POTASSIUM TANTALATE)   ELECTRIC FIELD INDUCED RAMAN EFFECT IN PARAELECTRIC CRY 08-1326
IN STRONTIUM TITANATE AT LOW TEMPERATURES.   FIELD INDUCED SWITCHING OF TETRAGONAL DOMAINS 06-1184
VSKITE TYPE WITHIN A WIDE RANGE OF TEMPERATURES AND ELECTRIC FIELD INTENSITIES AND TIMES OF APPLICATION. /O 08-1380
ERTIES OF TGS AND ROCHELLE SALT AS DETERMINED BY AN ELECTRIC FIELD METHOD.   CRITICAL PROP 27-3056
SURFACE PHENOMENA IN FERROELECTRIC CRYSTALS. (CALCULATION OF FIELD NEAR SURFACE)   MICROSCOPIC THEORY OF 03-0501
TRIC FILM BY BENDING AND BENDING DEFORMATION IN THE ELECTRIC FIELD. (OBSERVATION REPORTED) /ON OF FERROELEC 03-0195
EIR THICKNESS.   DEPENDENCE OF COERCIVE FIELD OF BARIUM TITANATE SINGLE CRYSTALS ON TH 04-0693
/ BEHAVICR (PERMITTIVITY, SPONTANEOUS POLARIZATION, COERCIVE FIELD) OF BISMUTH TITANATE AT LOW TEMPERATURE/ 08-1436
TALS).   SPONTANEOUS POLARIZATION AND COERCIVE FIELD OF LEAD TITANATE (FLUX GROWN SINGLE CRYS 05-1050
THICKNESS DEPENDENCE OF THE NUCLEATION FIELD OF TRIGLYCINE SULFATE.   27-3036
FREQUENCY DEPENDENCE OF THE COERCIVE FIELD OF TRIGLYCINE SULFATE CRYSTALS.   27-3062
ITUDE AND TEMPERATURE.   DEPENDENCE OF THE COERCIVE FIELD OF TRIGLYCINE SULFATE ON FREQUENCY, AMPL 27-3092
ION, TALKING INTO ACCOUNT TH/ EFFECT OF AN EXTERNAL ELECTRIC FIELD ON A DIFFUSE FERROELECTRIC PHASE TRANSIT 03-0441
TES SALT (ROCHELLE SALT). (TEMPERATURE A/ EFFECT OF ELECTRIC FIELD ON DOMAIN STRUCTURE FORMATION IN SEIGNET 03-3190
EFFECT OF AGING AND CCPPER(2+) DOPING)   EFFECT OF A DC FIELD ON DOMAIN STRUCTURE IN SEIGNETTE SALT. (28-3230
ROELECTRIC ROCHELLE SALT.   EFFECT OF APPLIED ELECTRIC FIELD ON EPR SPECTRA OF (COPPER(2+)) DOPED FER 28-3185
ERAMICS. PART-1: DEFORMATION AND PHASE T/ EFFECT OF ELECTRIC FIELD ON FERROELECTRIC AND ANTIFERROELECTRIC C 03-0169
IN POTASSIUM DIHYDR/ EFFECTS OF UNIAXIAL STRESS AND ELECTRIC FIELD ON PARAMAGNETIC SPECTRA OF CHROMIUM(3+) 17-2554
IOBATE SOLID SOLUTIONS.   EFFECT OF ELECTRIC FIELD ON PHASE COMPOSITION OF (LEAD, BARIUM) N 13-1943
ERYLLATE.   EFFECT OF CONSTANT ELECTRICAL FIELD ON PHASE TRANSITIONS IN TRIGLYCINE FLUOB 27-3167
TALLINE BARIUM TITANATE.   INFLUENCE CF A CONSTANT ELECTRIC FIELD ON THE DIELECTRIC PROPERTIES OF POLYCRYS 04-1012
NE SELENATE. (AGING AND PHASE CHANGES NEAR T/ EFFECT OF A DC FIELD ON THE DIELECTRIC PROPERTIES OF TRIGLYCI 27-3131
IN TGS.   INFLUENCE OF AN ELECTRIC FIELD ON THE NOISE AND THE PYROELECTRIC EFFECT 27-3076
OF SODIUM TRIDEUTERIUM(X) TR/ UNUSUAL EFFECT OF AN ELECTRIC FIELD ON THE PERMITTIVITY OF A NEW DELTA PHASE 22-2875
ECTRIC TRIGLYCINE SULFATE.   THE INFLUENCE OF THE ELECTRIC FIELD ON THE ULTRASONIC ATTENUATION IN FERROEL 27-2956
EFFECTS OF LONGITUDINAL MAGNETIC FIELD ON ULTRASOUND IN STRONTIUM TITANATE.   06-1243
BATE.   EPR STUDIES OF CRYSTAL FIELD PARAMETERS IN IRON(3+) DOPED LITHIUM NIO 11-1839
M NIOBATE.   ELECTRON SPIN RESONANCE STUDIES OF CRYSTAL FIELD PARAMETERS IN MANGANESE(2+) DOPED LITHIU 11-1803
FERROELECTRICS. (8 SUBSTITUTED LEAD TITANATE ZIRCONAT/ HIGH FIELD POLARIZATION REVERSAL IN POLYCRYSTALLINE 08-1418
GLYCINE SULFATE CRYSTALS ON TEMPERATURE AND APPLIED ELECTRIC FIELD STRENGTH. /F THE DOMAIN STRUCTURE OF TRI 27-3129
SULFOIODIDE. (DEPENDENCE ON TEMPERATURE, WAVELENGTH, APPLIED FIELD STRENGTH) /FRACTIVE INDICES OF ANTIMONY 15-2206
ITTIVITY OF TGS. (10 KHZ-1 MHZ)   TEMPERATURE AND FIELD STRENGTH DEPENDENCE OF LARGE SIGNAL PERM 27-3068
ODIDE CRYSTALS NEAR THE CURIE PCINT (EFFECTS OF TEMPERATURE, FIELD STRENGTH, SAMPLE ILLUMINATION). / SULFOI 15-2121
M DIHYDROGEN(X) DIDEUTERIUM(1-X) FHOSPHATE MIXED / MOLECULAR FIELD STUDY OF THE ISOTOPIC EFFECT IN POTASSIU 17-2563
EL OF THE FERROELECTRIC KDP. PART-1. PART-2.   QUANTUM FIELD THEORETICAL TREATMENT OF A DYNAMICAL MOD 03-0279
ONS) OF FERROELECTRIC BORACITES. (IRON, NICKEL, OR / CRYSTAL FIELD THEORY AND OPTICAL ABSORPTION (CALCULATI 14-2095
ANC NICKEL BORACITES.   CRYSTAL FIELD THEORY AND OPTICAL ABSORPTION OF COBALT 14-2096

ENE FLUORIDE FILMS (AFTER COOLING STRETCHED FILM IN ELECTRIC FIELD, THERMOELECTRETS). /ATION IN POLYVINYLID 36-3530
BAND STRUCTURE OF STRONTIUM TITANATE FROM A SELF CONSISTENT FIELD TIGHT BINDING CALCULATION.   ENERGY 06-1221
RE FORM.   CRYSTALLOGRAPHIC STUDIES OF IRRADIATION FIELD TREATED TRIGLYCINE SULFATE: NEW STRUCTU 27-2996
TRIGLYCINE SELENATE. (DIELECTRIC CONSTANT VERSUS ALTERNATING FIELD VOLTAGE, TEMPERATURE AND POLARIZATION) / 27-3158
LIQUID ELECTRODED BARIUM TITANATE CRYSTALS (CALCULATIO/ HIGH FIELD (1.5-6 KV/CM) POLARIZATION REVERSALS IN 04-0994
MPLEX DIELECTRIC CONSTANT OF LEAD ZIRCONATE UNDER POLARIZING FIELD, 370-520 DEG K.   CO 07-1256
UM TITANATE PIEZOELECTRIC MICROPROBE FOR STUDYING ULTRASONIC FIELDS.   BARI 04-0938
S OF THE DOMAIN WALLS IN KDP. EXISTENCE OF CRITICAL ELECTRIC FIELDS.   DISPLACEMENTS AND VELOCITIE 17-2448
OF THE BETA PHASE OF SODIUM TRIHYDROGEN SELENATE IN ELECTRIC FIELDS.   DOMAIN STRUCTURE 22-2830
LECTRIC PROPERTIES OF CADMIUM PYRONIOBATE IN STRONG ELECTRIC FIELDS.   ELASTIC AND PIEZOE 13-1967
RROELECTRICS (BARIUM TITANATE ZIRCONATES) IN STRONG ELECTRIC FIELDS AND AT HIGH PRESSURE. /ES OF CERAMIC FE 09-1605
OVSKITE TYPE COMPOUND/ CALCULATIONS OF THE INTERNAL ELECTRIC FIELDS AND ELECTRIC FIELD GRADIENTS IN THE PER 08-1442
NONLINEAR INTERACTION OF MICROWAVE ELECTRIC FIELDS AND SOUND IN LITHIUM NIOBATE.   11-1838
TION SWITCHING IN POLYCRYSTALLINE FERROELECTRICS AT VERY LOW FIELDS AND STRESSES, ROCHELLE SALT) /(POLARIZA 28-3216
E COMPOUNDS WITH DISTINCTI/ CALCULATION OF INTERNAL ELECTRIC FIELDS AND THEIR GRADIENTS IN ABO(3) PEROVSKIT 08-1455
/BJECTED BEFOREHAND TO AN ELECTRIC FIELD, IN STRONG ELECTRIC FIELDS. (ANISOTROPY DEVELOPED, BARIUM TITANAT/ 04-0661
/ON AND DIFFUSION EFFECTS (INTERNAL GENERATION OF STRONG FIELDS) DURING HOLOGRAM RECORDING IN CRYSTALS/ 06-1090
ON AND GROWTH OF FERROELECTRIC DOMAINS IN BARIUM TITANATE AT FIELDS FROM 2 TO 450 KV/CM.   NUCLEATI 04-0979
MIUM TITANATE.   INTERNAL ELECTRIC FIELDS IN CRYSTALS OF SODIUM TANTALATE AND CAD 08-1389
BIAS FIELDS IN FERROELECTRIC COLEMANITE.   23-2886
AUSEN STEPS AND THE CRITICAL POLARIZATION AND REPOLARIZATION FIELDS IN FERROELECTRICS /TING FIELD FOR PARKH 03-0318
E)   CALCULATION OF ELECTRIC FIELDS IN IONIC CRYSTALS. (BARIUM TITANATE TYP 04-0824
HIGH INTERNAL BIAS FIELDS IN L-ALANINE SUBSTITUTED TGS.   27-2977
DIPOLE STRUCTURE AND INTERNAL ELECTRIC FIELDS IN LEAD ZIRCONATE.   07-1261
CRYSTALLIZATION OF ROCHELLE SALT IN ELECTRIC FIELDS (KINETICS OF CRYSTAL GROWTH)   28-3200
H GAMMA-RAYS.   COERCIVE FIELDS OF ROCHELLE SALT HEAVILY IRRADIATED WIT 28-3178
CTRON MICROSCOPE (BARRIER LAYERS IN / IMAGING ELECTRIC MICRO FIELDS (OF SOLID SURFACES) IN THE EMISSION ELE 04-0913
OR F/ EFFECT OF FERROELECTRIC (BARIUM TITANATE) POLARIZATION FIELDS ON (GERMANIUM OR TELLURIUM) SEMICONDUCT 04-0594
TE CERAMICS FOR PIEZOELECTRIC APPL/ EFFECTS OF HIGH ELECTRIC FIELDS ON MODIFIED LEAD ZIRCONATE- LEAD TITANA 09-1485
TH HIGH PERMITTIVITY DIELECTRIC. (THEO/ INPUT ADMITTANCE AND FIELDS RADIATED BY AN OPEN WAVEGUIDE FILLED WI 03-0359
/TE LEAD TITANATE PIEZOELECTRIC CERAMICS UNDER HIGH ELECTRIC FIELDS. (SUBSTITUTION OF STRONTIUM OR CALCIUM/ 09-1484
.5) TUNGSTATE(0.5)), EFFECTS OF POSITIVE OR NEGATIVE HIGH DC FIELDS SUMMARY ONLY) /LUTIONS LEAD MAGNESATE(0 05-1063
OF SPIN WAVES IN FERROELECTRIC ANTIFERROMAGNETS IN EXTERNAL FIELDS. (THEORY)   INSTABILITY 03-0149
/ ELECTRICALLY POLARIZED FORM FROM 0.4 TO 4.2 DEGREES K WITH FIELDS TO 90 KILOGAUSS ALONG THE A, B, AND C / 30-3275
HE PROCEEDINGS OF THE SECOND INTERNATIONAL MEETING ON FERRO/ FIFTY YEARS OF FERROELECTRICITY: A REVIEW OF T 01-0010
FIFTY YEARS OF FERROELECTRICITY (REVIEW).   02-0090
BARIUM TITANATE. (150-500 DEG C, 20-50 KILOHERTZ, LISSAJOUS FIGURE) /Y LOSS NONPOLAR STATE POLYCRYSTALLINE 04-0831
RIES OF TRIGLYCINE SULFATE.   CORROSION FIGURES DUE TO THE FERROELECTRIC DOMAIN BOUNDA 27-3142
CRYSTALS.   SURFACE STRUCTURES AND CORRESPONDING LIGHT FIGURES REVEALED BY WATER ETCHED ROCHELLE SALT 28-3234
PREPARATION (MORPHOLOGY) AND MECHANICAL PROPERTIES OF FILAMENTARY CRYSTALS OF ADP.   17-2630
O/ INPUT ADMITTANCE AND FIELDS RADIATED BY AN OPEN WAVEGUIDE FILLED WITH HIGH PERMITTIVITY DIELECTRIC. (THE 03-0359
MANUFACTURING METHOD AND PROPERTIES OF FERROELECTRIC THIN FILM.   37-3681
CT IN ANTIMONY SULFOIODIDE- TIN DIOXIDE HETEROSTRUCTURE THIN FILM.   SWITCHING AND MEMORY EFFE 15-2166
LAME SPRAYING (CAPACITORS).   ELECTRICAL PROPERTIES OF THICK FILM BARIUM TITANATE DIELECTRICS PRODUCED BY F 04-0772
ON SILICON SUBS/ PHYSICAL AND ELECTRICAL PROPERTIES OF THIN FILM BARIUM TITANATE PREPARED BY RF SPUTTERING 04-0828
ELECTRIC FIELD. (OBSERVATION/ POLARIZATION OF FERROELECTRIC FILM BY BENDING AND BENDING DEFORMATION IN THE 03-0195
PREPARATION OF BARIUM TITANATE FILM BY EVAPORATION.   04-0743
EFFECT/ PREPARATION AND PROPERTIES OF HIGH PERMITTIVITY THIN FILM DIELECTRICS. (BARIUM AND LEAD TITANATES, 04-0670
FORMATION PROCESS OF BARIUM TITANATE THIN FILM EVAPORATED IN VACUO.   04-0862
E. (PLZT)   THIN FILM FERROELECTRIC PHOTOCONDUCTOR MEMORY DEVIC 09-1503
E.   PHOTOCONDUCTOR AND ELECTRODE REQUIREMENTS FOR THIN FILM FERROELECTRIC PHOTOCONDUCTOR MEMORY DEVIC 37-3671
CE.   DESIGN AND PERFORMANCE OF A THIN FILM FERROELECTRIC PHOTOCONDUCTOR STORAGE DEVI 37-3632
SWITCHING PROPERTIES OF THICK FILM FERROELECTRIC POTASSIUM NITRATE.   16-2282
ON IN POLYVINYLIDENE FLUORIDE FILMS (AFTER COOLING STRETCHED FILM IN ELECTRIC FIELD, THERMOELECTRETS). /ATI 36-3530
TION AND REFLECTION OF ULTRASONIC SURFACE WAVES BY THIN GOLD FILM LAYERS ON LITHIUM NIOBATE SUBSTRATES. /AC 11-1789
TION AND REFLECTION OF ULTRASONIC SURFACE WAVES BY THIN GOLD FILM LAYERS ON LITHIUM NIOBATE SUBSTRATES. /AC 11-1790
(MIXED OXIDES ABO(3)) (/ PREPARATION AND PROPERTIES OF THIN FILM MATERIALS WITH PEROVSKITE TYPE STRUCTURE. 04-0914
NLINEAR MICROWAVE PROPERTIES OF FERROELECTRIC CERAMICS. THIN FILM MICROWAVE FERROELECTRIC CAPACITORS AND NO 37-3643
TELLURIUM ON TRIGLYCINE SULFATE AND SODIUM NITRITE SUR/ THIN FILM NUCLEATION ON FERROELECTRIC SUBSTRATES. (16-2394
TO 3 MIC/ DIELECTRIC CONSTANT AND LATTICE DYNAMICS OF A THIN FILM OF A FERROELECTRIC (BARIUM TITANATE DOWN 04-0600
MPLIFICATION OF SURFACE WAVE STRUCTURE OF A CADMIUM SELENIDE FILM ON LITHIUM NIOBATE.   ACOUSTOELECTRIC A 11-1723
OUT MEMORY ELEMENT UTILIZING POTASSIUM NITRATE. (FUSED THICK FILM OR VACUUM DEPOSITED THIN FILM) /TIVE READ 16-2269
OOPTIC MIXED CRYSTAL AMMONIUM POTASSIUM DIHYDROGEN PHOSPHATE FILM WAVEGUIDE.   EPITAXIAL ELECTR 17-2633
CHARACTERISTICS OF RF SPUTTERED BARIUM TITANATE THIN FILMS.   04-0906
DEPOLARIZATION FIELD IN THIN FERROELECTRIC FILMS.   37-3659
DIRECT PIEZOELECTRIC EFFECT IN POLYVINYL CHLORIDE FILMS.   36-3543
AY MICROANALYSIS ON IMPURITIES IN EVAPORATED BARIUM TITANATE FILMS.   ELECTRON PROBE X-R 04-0742
EXPERIMENTAL STUDY OF EVAPORATED BARIUM TITANATE FILMS.   04-0939
FABRICATION OF RF SPUTTERED BARIUM TITANATE THIN FILMS.   04-0907
ECTIVITY AND TRANSMISSION MEASUREMENTS ON STRONTIUM TITANATE FILMS.   FAR INFRARED REFL 06-1154
FERROELECTRICITY IN SPUTTERED LEAD TITANATE ZIRCONATE FILMS.   09-1559
OWTH AND PROPERTIES OF LITHIUM NIOBATE AND LITHIUM TANTALATE FILMS.   GR 11-1712
GROWTH AND STRUCTURE OF BARIUM TITANATE EVAPORATED FILMS.   04-0604
TASSIUM DIHYDROGEN PHOSPHATE CRYSTALS BY USING SEMICONDUCTOR FILMS. /IGATION OF THE THERMAL EXPANSION OF PO 17-2565
OBIUM(0.325) ZIRCONIUM(0.27)) OXYGEN(3) FERROELECTRIC MEMORY FILMS. /(0.07) LANTHANUM(0.01) (IRON(0.405) NI 37-3629
PIEZOELECTRICITY IN POLARIZED POLYVINYLIDENE FLUORIDE FILMS.   36-3557
PREPARATION AND PROPERTIES OF THIN BARIUM TITANATE FILMS.   04-0669
ONLINEAR OPTICAL PROPERTIES OF POLED POLYVINYLIDENE FLUORIDE FILMS.   PYROELECTRIC AND N 36-3595
RESEARCH STATUS AND DEVICE POTENTIAL OF FERROELECTRIC THIN FILMS.   37-3637
N OF TRIGLYCINE SULFATE CRYSTALS STUDIED WITH TELLURIUM THIN FILMS.   THERMAL EXPANSIO 27-3047
/TICAL SECOND HARMONIC GENERATION IN POLYVINYLIDENE FLUORIDE FILMS (AFTER COOLING STRETCHED FILM IN ELECTR/ 36-3530
LIDENE FLUORIDE)   FERROELECTRIC FILMS AND THEIR DEVICE APPLICATIONS. (POLYVINY 36-3556
/LARIZATION FIELDS ON (GERMANIUM OR TELLURIUM) SEMICONDUCTOR FILMS. (AND USE OF THESE FILMS TO STUDY SURFA/ 04-0594
FUSED QUARTZ).   PREPARATION OF THIN BARIUM TITANATE FILMS BY DC DIODE SPUTTERING (ONTO PLATINUM OR 04-0948
ERE.   SINGLE CRYSTAL BARIUM TITANATE FILMS GROWN FROM THE MELT IN AN OXYGEN ATMOSPH 04-0612
PREPARATION AND STUDY OF THIN FILMS OF (BARIUM, STRONTIUM) TITANATE.   04-1000
MICROSTRUCTURE OF DOMAINS AND DOMAIN WALLS IN SINGLE CRYSTAL FILMS OF BARIUM TITANATE.   04-1031
PREPARATION AND PROPERTIES OF THIN FILMS OF BARIUM TITANATE.   04-0843
/RVATION OF THE CHANGE IN THE POLARIZATION OF SINGLE CRYSTAL FILMS OF BARIUM TITANATE BY MEANS OF A STROBO- 04-0942
/ION OF DOMAIN BOUNDARIES DURING SWITCHING IN SINGLE CRYSTAL FILMS OF BARIUM TITANATE (ELECTRON MICROSCOPY/ 04-0940

DOMAIN STRUCTURE AND POLARIZATION REVERSAL IN FILMS OF FERROELECTRIC BISMUTH TITANATE. 08-1470
TROOPTICS) GROWTH AND PROPERTIES OF EPITAXIAL FILMS OF FERROELECTRIC BISMUTH TITANATE. (ELEC 08-1329
PREPARATION AND EPITAXY OF SPUTTERED FILMS OF FERROELECTRIC BISMUTH TITANIUM OXIDE. 13-2038
COMPOSITION AND STRUCTURE OF SPUTTERED FILMS OF FERROELECTRIC NIOBATES. 13-1941
/H EVAPORATION. (REVIEW OF TECHNIQUE AND APPLICATION TO THIN FILMS OF PEROVSKITES AND III-V COMPOUNDS, 8 R/ 02-0112
FLASH EVAPORATION OF FERROELECTRIC THIN FILMS OF POLYCRYSTALLINE BARIUM TITANATE. 04-0968
FLUORIDE. PYROELECTRICITY AND PIEZOELECTRICITY IN ORIENTED FILMS OF POLYVINYL FLUORIDE AND POLYVINYLIDENE 36-3544
E. STUDY OF THE PYROELECTRIC EFFECT IN THIN FILMS OF TRIGLYCINE SULFATE AND BARIUM TITANAT 04-0686
EVAPORATED FILMS OF TUNGSTEN OXIDE. 10-1647
DIELECTRIC BEHAVIOR OF (FERROELECTRIC) FILMS OF VACUUM DEPOSITED POTASSIUM NITRATE. 16-2356
NE SULFATE AND SELENATE CRYS/ FIELD EFFECT IN TELLURIUM THIN FILMS ON A SUBSTRATE OF FERROELECTRIC TRIGLYCI 27-3046
/IC FIELD EFFECT STUDIES AT LOW TEMPERATURES (WITH TIN OXIDE FILMS ON LEAD ZIRCONATE(0.65) TITANATE(0.35) / 09-1505
35) ZIRCO/ FERROELECTRIC FIELD EFFECT STUDIES. (ON TIN OXIDE FILMS ON LEAD(0.97) BISMUTH(0.02) TITANATE (0. 09-1506
. (MIXED OXIDES ABO(3)) (BARIUM TITANATE AND LITHIUM NIOBATE FILMS ON PLATINUM OR SILICON) / TYPE STRUCTURE 04-0914
/ERMODYNAMIC STABILITY OF POLARIZATION IN THIN FERROELECTRIC FILMS ON SEMICONDUCTING SUBSTRATES. (TGS ON S/ 27-3163
CHARACTERISTICS OF RF SPUTTERED BARIUM TITANATE FILMS ON SILICON. 04-0930
THIN FILMS ON TGS CLEAVED IN ULTRAHIGH VACUUM. 27-3173
RE DEPENDENCE OF THE FERROELECTRIC FIELD EFFECT IN TELLURIUM FILMS ON TRIGLYCINE SUBSTRATES. TEMPERATU 27-3018
FIELD EFFECT AT VERY LARGE GATE CHARGES APPLIED TO TELLURIUM FILMS ON UHV CLEAVED TGS. /TER EFFECTS IN THE 27-3171
RETENTION IN THE THIN FERROELECTRIC FILMS. (PLZT) 09-1593
OF BARIUM TITANATE BY VACUUM EVAPORATION. THIN FILMS: PREPARATION OF FERROELECTRIC THIN FILMS 09-1593
TIMONIDE OR GALLIUM ARSENIDE. BARIUM TITANATE FILMS PREPARED BY RF SPUTTERING ONTO INDIUM AN 04-0741
STRUCTURE AND PROPERTIES OF BARIUM TITANATE FILMS PRODUCED BY CATHODE SPUTTERING. 04-0832
ECTRICAL PROPERTIES OF FLASH EVAPORATED BARIUM TITANATE THIN FILMS (SEPARATED FROM SUBSTRATE). /URE) AND EL 04-0605
/D CHARACTERIZATION OF THIN FERROELECTRIC AND SEMICONDUCTING FILMS. (THE MAJOR PORTION OF THIS REPORT IS C/ 02-0057
RTIES OF FLASH EVAPORATED FERROELECTRIC BARIUM TITANATE THIN FILMS. (THICKNESS DEPENDENCE) ELECTRICAL PROPE 04-0967
THERMODYNAMIC STABILITY OF THIN FERROELECTRIC FILMS. (TRIGLYCINE SULFATE) 27-2959
PTICAL MIXING OF COHERENT AND INCOHERENT LIGHT. (KDP CRYSTAL FILTER) O 17-2669
APPLICATION OF FERROELECTRIC MATERIALS IN ELECTRICAL FILTERS. 37-3636
NATE- LEAD TITANATE) IMAGE STORAGE AND DISPLAY DEVICES USING FINE GRAIN FERROELECTRIC CERAMICS. (LEAD ZIRCO 09-1572
SOME FATIGUING EFFECTS IN 8/65/35 PLZT FINE GRAINED FERROELECTRIC CERAMIC. 08-1432
. (BARIUM / CALCULATION OF THE ELECTROSTATIC INTERACTIONS IN FINE GRAINED PEROVSKITE FERROELECTRIC CERAMICS 04-0642
DOMAIN WALLS CAUGHT IN (FINE LINE) "SUDARES" IN ROCHELLE SALT CRYSTAL. 28-3215
NIDINIUM SULFATE HEXAHYDRATE. ZEEMAN EFFECT (STUDY) OF TWO FINE LINES IN (THE RED REGION OF) CHROMIUM GUA 21-2814
MAGNITUDE OF SPONTANEOUS POLARIZATION IN THIN CRYSTALS AND FINE PARTICLES OF BARIUM TITANATE. 04-0607
/TURE AND ELECTRICAL PROPERTIES OF FERROELECTRIC MATERIALS. (FINE POWDERS PREPARED BY DECOMPOSITION OF ORG/ 02-0093
CERTAIN PEROVSKITES (STRONTIUM OR CALCIUM TITANATE/ EXTENDED FINE STRUCTURE IN X-RAY ABSORPTION SPECTRA OF 03-0414
X GROWN FROM POTASSIUM FLUORIDE. INVESTIGATION OF THE FINE STRUCTURE OF BARIUM TITANATE CRYSTALS FLU 04-0708
F LIGHT IN POTASSIUM MANGANESE TRIFLUORIDE. FINE STRUCTURE OF EXCITON- MAGNON ABSORPTION O 33-3364
EN PHOSPHATE. DIELECTRIC PERMITTIVITY OF FINELY DIVIDED POTASSIUM AND AMMONIUM DIHYDROG 17-2689
TRIC CRYSTALS. NONLINEAR EFFECTS IN THE PROPAGATION OF FINITE AMPLITUDE HYPERSONIC WAVES IN FERROELEC 03-0498
POCKELS EFFECT COMMUTATORS. (POTASSIUM DIHYDROGEN PHOSPHA/ (FINITE APERTURE LIGHT SWITCH USING) TRANSVERSE 17-2509
ITRARY MAGNETIC SYMMET/ SPIN WAVE PHENOMENOLOGICAL THEORY IN FINITE SAMPLES OF FERRODIELECTRICS WITH AN ARB 03-0164
DIELECTRICS. SINGLE FIRE PROCESS FOR BARIUM TITANATE BARRIER LAYER 04-1029
. CHEMICAL CHANGES IN BARIUM TITANATE FIRED UNDER NITROGEN IN THE PRESENCE OF CARBON 04-0872
L AND TECHNOLOGICAL PROPERTIES OF FERROELECTRI/ INFLUENCE OF FIRING OF BARIUM TITANATE SINTERS ON ELECTRICA 04-1002
EFFECTS OF CALCINING ON THE FIRING OF LEAD ZIRCONATE TITANATE CERAMIC 09-1622
RROELECTRICS: A SELF CONSISTENT POLARIZATION SCHEME FOR ANH/ FIRST AND SECOND ORDER PHASE TRANSITIONS IN FE 03-0355
OXYGENATED QUADRATIC TUNGSTEN BRONZE TYPE. (STRONTIUM POTAS/ FIRST FERROELECTRIC OXYFLUORIDE PHASES OF THE 13-2011
/NEAR SADOVSKII EFFECT (THEORY, PROPAGATION OF RADIATION AND FIRST HARMONIC, CALCULATIONS FOR LITHIUM NIOB/ 03-0186
HIGH PRESSURE PHASE TRANSITIONS OF THE FIRST KIND IN ROCHELLE SALT. 28-3214
ANATE. TEMPERATURE DEPENDENCE OF THE FIRST ORDER ELASTIC CONSTANTS OF STRONTIUM TIT 06-1160
RAMAN SPECTROSCOPY OF PHONON INSTABILITY IN FIRST ORDER FERROELECTRIC PHASE TRANSITIONS. 03-0163
FERROELECTRIC PHASE TRANSITION OF FIRST ORDER IN BARIUM TITANATE. 04-1030
PHTHALOCYANINE IN THE FERROELECTRIC STATE. FIRST ORDER NEGATIVE PHOTODIELECTRIC EFFECT IN 36-3539
ERIUM PHOSPHATE CRYSTALS. FIRST ORDER PHASE TRANSITION IN POTASSIUM DEUT 17-2677
CRYSTALS. DOUBLE HYSTERESIS LOOPS AND FIRST ORDER PHASE TRANSITIONS IN FERROELECTRIC 03-0456
OMENOLOGICAL THEORY OF ANTIFERROELECTRIC TRANSITION. PART-2: FIRST ORDER TRANSITION. PHEN 03-0404
IN BARIUM TITANATE. NUCLEATION MODELS FOR A FIRST ORDER TRANSITION FRONT, AND OBSERVATIONS 04-0608
PHOSPHATE. DIELECTRIC EVIDENCE OF A FIRST ORDER TRANSITION IN POTASSIUM DIHYDROGEN 17-2602
/POINT SHIFT WITH APPLIED FIELD IN FERROELECTRICS POSSESSING FIRST ORDER TRANSITION. (THEORY, CALCULATION) 03-0327
STICAL PERTURBATION THEORY OF ORDER DISORDER FERROELECTRICS: FIRST PERTURBATIVE CORRECTION. STATI 03-0399
/EUTRON SCATTERING FROM LATTICE VIBRATIONS (REDEVELOPED FROM FIRST PRINCIPLES, APPLIED TO DISPLACIVE TYPE / 03-0484
N FERROELECTRIC MATERIALS. (BARIUM TI/ ELECTRICAL CONTROL OF FIXATION AND ERASURE OF HOLOGRAPHIC PATTERNS I 04-0847
THERMALLY FIXED HOLOGRAMS IN LITHIUM NIOBATE. 11-1828
YSTALS. HOLOGRAPHIC STORAGE, ELECTRICAL FIXING AND ERASING IN DOPED BARIUM TITANATE CR 04-0848
ATE, BARIUM SODIUM NIOBATE) HOLOGRAPHIC PATTERN FIXING IN ELECTROOPTIC CRYSTALS. (LITHIUM NIOB 11-1661
IELECTRIC. FLAME SPRAYED BARIUM TITANATE AS A CAPACITOR D 04-0773
-10: STRUCTURE OF SEMICONDUCTIVE TITANIUM DIOXID/ STUDIES OF FLAME SPRAYED BARIUM TITANATE CAPACITORS. PART 04-0774
ERTIES OF THICK FILM BARIUM TITANATE DIELECTRICS PRODUCED BY FLAME SPRAYING (CAPACITORS). ELECTRICAL PROP 04-0772
/WTH STRUCTURE (MICROSTRUCTURE) AND ELECTRICAL PROPERTIES OF FLASH EVAPORATED BARIUM TITANATE THIN FILMS (/ 04-0605
THIN FILMS. (THICKNESS DEPENDENCE) ELECTRICAL PROPERTIES OF FLASH EVAPORATED FERROELECTRIC BARIUM TITANATE 04-0967
OF POLYCRYSTALLINE BARIUM TITANATE. FLASH EVAPORATION OF FERROELECTRIC THIN FILMS 04-0968
PLICATION TO THIN FILMS OF PEROVSKITES AND III-V COMPOUNDS,/ FLASH EVAPORATION. (REVIEW OF TECHNIQUE AND AP 02-0112
TENSILE AND FLEXURE PROPERTIES OF SALINE ICE. 34-3422
HYDRATE. MOSSBAUER STUDY OF ELECTRONIC SPIN FLIP PROCESSES IN AMMONIUM IRON SULFATE DODECA 19-2771
. (BARIUM TITANATE) A FLOATING ZONE SINGLE CRYSTAL GROWING APPARATUS 04-0998
GLE CRYSTAL STRONTIUM(2) NIOBIUM(2) OXIDE(7) BY MEANS OF THE FLOATING ZONE TECHNIQUE. /OF FERROELECTRIC SIN 13-2037
A NEW FERROELECTRIC TRANSDUCER FOR USE IN HEAT TRANSFER AND FLOW STUDIES. (TRIGLYCINE SULFATE) 27-3031
CINE SULFATE SINGLE CRYSTALS (A/ FREQUENCY DEPENDENCE OF THE FLUCTUATION ABSORPTION OF ULTRASOUND IN TRIGLY 27-3079
IE POINT. POLARIZATION FLUCTUATION OF TRIGLYCINE SULFATE NEAR THE CUR 27-3077
RROELECTRIC SOLID SOLUTIONS WITH ALLOWANCE FOR CONCENTRATION FLUCTUATIONS. DIELECTRIC CONSTANT OF FE 03-0440
DIELECTRIC RESPONSE OF CRYSTALS. (FERROELECTRIC FLUCTUATIONS) 03-0210
ANATE ABOVE THE CRITICAL TEMPERATURE. FLUCTUATIONS AND CORRELATIONS IN STRONTIUM TIT 06-1233
N BARIUM TITANATE. POLARIZATION FLUCTUATIONS AND THE OPTICAL ABSORPTION EDGE I 04-1028
TASSIUM MANGANESE TRIFLUORIDE SINGLE C/ CRITICAL ANISOTROPIC FLUCTUATIONS AT THE 184 DEG K TRANSITION OF PO 33-3369
YDROGEN PHOSPHATE. LONG WAVELENGTH POLARIZATION FLUCTUATIONS IN ANTIFERROELECTRIC AMMONIUM DIH 17-2461
ROELECTRIC CURIE TEMPERATURE, AS SH/ CRITICAL (POLARIZATION) FLUCTUATIONS IN BARIUM TITANATE (ABOVE THE FER 04-1032
/N WAVES IN ORDER DISORDER TYPE FERROELECTRICS: POLARIZATION FLUCTUATIONS IN DICALCIUM STRONTIUM PROPIONAT/ 26-2928
OPTICAL EVIDENCE FOR POLARIZATION FLUCTUATIONS IN PARAELECTRIC BARIUM TITANATE. 04-0737

E.   LIGHT SCATTERING FROM CRITICAL FLUCTUATIONS IN POTASSIUM DIDEUTERIUM PHOSPHAT 17-2463
RITICAL TEMPERATURE.   CRITICAL ASYMMETRY IN LOCAL FLUCTUATIONS IN STRONTIUM TITANATE ABOVE THE C 06-1166
                                             CRITICAL FLUCTUATIONS IN TRIGLYCINE SULFATE. 27-2974
/CRYSTALS. (STRAIN DEPENDENT CURIE TEMPERATURE, POLARIZATION FLUCTUATIONS, LITHIUM NIOBATE, ALKALI HALIDES) 03-0513
TION.   POLARIZATION FLUCTUATIONS NEAR A FERROELECTRIC PHASE TRANSI 03-0368
ARIUM TITANATE.   INFLUENCE OF CRITICAL POLARIZATION FLUCTUATIONS ON THE PHOTOELASTIC BEHAVIOR OF B 04-0628
        SURFACE WAVE CONVOLUTION USING A CADMIUM SULFIDE FLUID LITHIUM NIOBATE STRUCTURE. 11-1700
          LAYER WAVE AMPLIFICATION IN A CADMIUM SULFIDE FLUID- LITHIUM NIOBATE STRUCTURE. 11-1667
UDY OF THE FERROELECTRIC TRANSITION IN DEUTERATED TRIGLYCINE FLUOBERYLLATE. /-9 QUADRUPOLE PERTURBED NMR ST 27-2965
                                    CRITICAL BEHAVIOR OF TRIGLYCINE FLUOBERYLLATE. 27-3075
RMINATION OF FERROELECTRIC PROPERTIES OF LITHIUM HYDRAZINIUM FLUOBERYLLATE.   DETE 32-3349
CONSTANT ELECTRICAL FIELD ON PHASE TRANSITIONS IN TRIGLYCINE FLUOBERYLLATE.   EFFECT OF 27-3167
                    EPR OF THE CHROMIUM COMPLEX IN TRIGLYCINE FLUOBERYLLATE. 27-3133
OF SOLID SOLUTIONS IN THE SYSTEM AMMONIUM SULFATE- AMMONIUM FLUOBERYLLATE.   FERROELECTRIC PROPERTIES 19-2762
ZATION ON THE EPR SPECTRA OF CHROMIUM(3+) IONS IN TRIGLYCINE FLUOBERYLLATE. INFLUENCE OF SPONTANEOUS POLARI 27-3136
         INVESTIGATIONS OF NUCLEAR RELAXATION OF TRIGLYCINE FLUOBERYLLATE. 27-3104
(BASED ON LANDAU PHENOMENOLOGICAL THEORY, TESTED ON AMMONIUM FLUOBERYLLATE, ABSTRACT ONLY) /E TRANSITIONS. 19-2765
/RROELECTRICS. (ROCHELLE SALT, TRIGLYCINE SULFATE, SELENATE, FLUOBERYLLATE AND BARIUM TITANATE, SUMMARY ON/ 02-0083
/ CRYSTALS WITH ISOMORPHOUS COMPOUNDS. (TRIGLYCINE SELENATE, FLUOBERYLLATE, AND THE DEUTERATED SOLID SOLUT/ 27-2972
                         STRUCTURE OF AMMONIUM FLUOBERYLLATE AT AMBIENT TEMPERATURE. 19-2776
/ELECTRICS (SINGLE CRYSTAL TRIGLYCINE SELENATE AND TRIGLYCINE FLUOBERYLLATE) AT PHASE TRANSITIONS OF THE SE/ 27-3137
R MAGNETIC RESONANCE OF DEUTERIUM IN THE PERDEUTERO AMMONIUM FLUOBERYLLATE CRYSTAL.   NUCLEA 19-2744
   LOW FREQUENCY RAMAN SCATTERING BY FERROELECTRIC TRIGLYCINE FLUOBERYLLATE CRYSTALS. 27-3113
SE ELECTRIC FIELD.   IMPEDANCE OF FERROELECTRIC TRIGLYCINE FLUOBERYLLATE CRYSTALS DURING SWITCHING IN PUL 27-3027
ANSITION RANGE.   THERMAL EXPANSION OF TRIGLYCINE FLUOBERYLLATE CRYSTALS IN THE FERROELECTRIC TR 27-3169
/QUASI SYMMETRICAL HARD IONS (PHOSPHATE, TITANATE, SULFATE, FLUOBERYLLATE, ETC) OF SOME MONOCLINIC CRYSTA/ 03-0330
/S FROM PHENOMENOLOGICAL THEORY. (AMMONIUM SULFATE, AMMONIUM FLUOBERYLLATE, GUANIDINIUM ALUMINUM SULFATE H/ 02-0040
HASE.   CRYSTAL STRUCTURE OF TRIGLYCINE FLUOBERYLLATE IN THE FERRO- AND PARAELECTRIC P 27-3175
/WITH GENERAL FORMULA METAL-1 METAL-2 SULFATE OR SELENATE OR FLUOBERYLLATE (METALS CHOSEN FROM SODIUM, POT/ 19-2743
R OF THE FOUR COPPER ION COMPLEX WITH SPIN S=2 IN TRIGLYCINE FLUOBERYLLATE MONOCRYSTAL.   EP 27-3134
ION IN SOME FERROELECTRIC AMMONIUM SALTS. (AMMONIUM SULFATE, FLUOBERYLLATE, PERIODATE, BISULFATE) / RELAXAT 19-2769
/TRIC CRYSTALS (SOLUTION GROWTH, TGS, TRIGLYCINE SELENATE OR FLUOBERYLLATE, POTASSIUM OR RUBIDIUM DIHYDROG/ 02-0089
ON A CHANGE IN THE DIELECTRIC CONSTAN/ EFFECT OF TRIGLYCINE FLUOBERYLLATE SINGLE CRYSTAL GROWTH CONDITIONS 27-3151
           PROPERTIES OF DEUTERATED TRIGLYCINE FLUOBERYLLATE SINGLE CRYSTALS. 27-2973
RIUM TITANATE, KDP, POTASSIUM SODIUM TARTRATE, TGS, AMMONIUM FLUOBERYLLATE, SODIUM NITRITE, ALUMS) /CS. (BA 02-0134
PERTIES OF SOLID SOLUTIONS IN THE AMMONIUM SULFATE- AMMONIUM FLUOBERYLLATE SYSTEM.   FERROELECTRIC PRO 19-2764
PERTIES OF SOLID SOLUTIONS IN THE AMMONIUM SULFATE- AMMONIUM FLUOBERYLLATE SYSTEM.   FERROELECTRIC PRO 19-2763
ATION UNDER X-I/ BAND GAP OF TRIGLYCINE SELENATE, TRIGLYCINE FLUOBERYLLATE, TRIGLYCINE SULFATE AND ITS VARI 27-3124
         CRYSTALLOGRAPHIC STUDY OF SEVERAL HYDRAZINIUM ORTHO FLUOBERYLLATES. 32-3355
MIUM BISMUTH TITANIUM IRON OXIDE, BISMUTH MOLYBDATE, BISMUTH FLUONIOBATE, BISMUTH FLUOTANTALATE) / PRASEODY 13-1993
EFFICIENT SECOND HARMONIC GENERATION IN ADP WITH TWO NEW FLUORESCEIN DYE LASERS. 17-2538
AND POTASSIUM DIHYDROGEN PHOSPHATE EXCITED BY A / PARAMETRIC FLUORESCENCE IN AMMONIUM DIHYDROGEN PHOSPHATE 17-2486
CW PUMP FREQUENCIES.   PARAMETRIC FLUORESCENCE IN BARIUM SODIUM NIOBATE FOR NINE 13-1950
W PUMP WAVELENGTHS (RANGE 454.5-528.7 MICRONS).   PARAMETRIC FLUORESCENCE IN BARIUM SODIUM NIOBATE FOR 10 C 13-1983
ITANATE.   ANALYSIS OF STRESS AND TEMPERATURE DEPENDENCE OF FLUORESCENCE IN CHROMIUM(3+) DOPED STRONTIUM T 06-1215
ITANATE.   EFFECT OF DOMAIN STRUCTURE ON THE FLUORESCENCE OF CHROMIUM(3+) DOPED STRONTIUM T 06-1114
TRESS INDUCED PHASE TRANSITION IN STRONTIUM TITAN/ POLARIZED FLUORESCENCE STUDY OF CHROMIUM(3+) THROUGH A S 06-1115
UOPINE-19 NUCLEI IN ANTIFERROMAGNETIC POTASSIUM MANGANESE TRIFLUORIDE. /D MAGNETIC EFFECTS INVOLVING THE FL 33-3390
ANTIFERROMAGNETIC RESONANCE SPECTRUM IN BARIUM MANGANESE TRIFLUORIDE. 33-3396
                COVALENCY IN POTASSIUM MANGANESE TRIFLUORIDE.   33-3391
SCATTERING IN STRONTIUM TITANATE AND POTASSIUM MANGANESE TRIFLUORIDE.   CRITICAL NEUTRON 06-1208
        DIELECTRIC ABSORPTION IN ORIENTED POLYVINYLIDENE FLUORIDE. 36-3572
MPING OF SOFT ZONE BONDARY PHONONS IN POTASSIUM MANGANESE TRIFLUORIDE.   DISPERSION AND DA 33-3374
TRIFLUORIDE, RUBIDIUM MAGANESE TRIFLUORIDE, AND MANGANESE DIFLUORIDE. /IN PARAMAGNETIC POTASSIUM MANGANESE 33-3376
      FAR INFRARED SPECTRA OF PIEZOELECTRIC POLYVINYLIDENE FLUORIDE. 36-3590
LECTRIC FILMS AND THEIR DEVICE APPLICATIONS. (POLYVINYLIDENE FLUORIDE)   FERROE 36-3556
XCITON- MAGNON ABSORPTION OF LIGHT IN POTASSIUM MANGANESE TRIFLUORIDE.   FINE STRUCTURE OF E 33-3364
N NICKEL DOPED RUBIDIUM MANGANATE AND POTASSIUM MANGANESE TRIFLUORIDE. /CE OF RAMAN SCATTERING BY MAGNONS I 33-3395
CE BELOW THE STRUCTURAL TRANSITION IN POTASSIUM MANGANESE TRIFLUORIDE.   INFRARED ACTIVE MO 33-3382
UCTURE OF BARIUM TITANATE CRYSTALS FLUX GROWN FROM POTASSIUM FLUORIDE).   INVESTIGATION OF THE FINE STR 04-0708
W FREQUENCY MAGNETOELECTRIC RESONANCES IN BARIUM COBALT TETRAFLUORIDE.   LO 33-3409
EUROPIUM(3+) ION IN ANTIFERROMAGNETIC POTASSIUM MANGANESE TRIFLUORIDE.   LUMINESCENCE OF 33-3381
MAGNETIC EXCITATIONS IN NICKEL DOPED POTASSIUM MANGANESE TRIFLUORIDE.   33-3407
         MAGNETOELASTIC EFFECTS IN POTASSIUM MANGANESE TRIFLUORIDE.   33-3389
CTRIC- ANTIFERROMAGNETIC LITHIUM IRON(0.5) TANTALUM(0.5) OXY FLUORIDE.   MAGNETOELECTRIC EFFECT IN FERROELE 36-3568
IUM TRIFLUORIDE, RUBIDIUM MANGANESE TRIFLUORIDE, MANGANESE DIFLUORIDE). /ANGANESE FLUORIDE CRYSTALS (POTASS 33-3398
ATTICE DYNAMICAL PHASE TRANSITIONS IN POTASSIUM MANGANESE TRIFLUORIDE.   NEUTRON SCATTERING STUDY OF THE L 33-3405
NOMALY OF THE THERMAL CONDUCTIVITY OF POTASSIUM MANGANESE TRIFLUORIDE.   OBSERVATION OF THE 33-3382
RACTICES. (CHLORIDE, BROMIDE AND IODIDE, AND RUBIDIUM IRON TRIFLUORIDE) /DEG K) OF FERROELECTRIC IRON(2+) BO 14-2097
              PIEZOELECTRICITY OF POLYVINYLIDENE FLUORIDE. 36-3573
TRICITY, AND THE ELECTROSTRICTION CONSTANT OF POLYVINYLIDENE FLUORIDE.   PIEZOELECTRICITY, PYROELEC 36-3598
BEHAVIOR IN CANTED ANTIFERROMAGNETIC POTASSIUM MANGANESE TRIFLUORIDE.   RESONANCE 33-3403
          SPECIFIC HEAT OF POTASSIUM MANGANESE TRIFLUORIDE.   33-3387
RISING DURING THE PHASE TRANSITION IN POTASSIUM MANGANESE TRIFLUORIDE.   SUPERSTRUCTURE A 33-3402
ROCHLORE CONTAINING LITHIUM, STRONTIUM LANTHANUM TANTALUM OXYFLUORIDE. /TUDY OF SINGLE CRYSTALS OF A NEW PY 36-3613
RY OF SUPEREXCHANGE REACTION. PART-1: POTASSIUM MANGANESE TRIFLUORIDE.   THEO 03-0252
TTERING OF LIGHT IN ANTIFERROMAGNETIC POTASSIUM MANGANESE TRIFLUORIDE.   TWO MAGNON SCA 33-3401
AN ABSORPTION EDGE. (LITHIUM NIOBATE, BARIUM MANGANESE TETRAFLUORIDE) /TY SENSE BY ANOMALOUS SCATTERING AT 11-1673
ICS.   BARIUM MANGANESE FLUORIDE, A NEW CRYSTAL FOR MICROWAVE ULTRASON 33-3406
FERROELASTIC EFFECT IN RUBIDIUM IRON TETRAFLUORIDE AND CESIUM IRON TETRAFLUORIDE. 33-3358
IN THE LOW DIMENSIONAL ANTIFERROMAGNETS POTASSIUM COPPER TRIFLUORIDE AND COPPER FORMATE TETRAHYDRATE. /ONS 26-2939
PYROELECTRIC PROPERTIES OF POLYVINYLIDENE FLUORIDE AND ITS USE FOR INFRARED DETECTION. 36-3560
CTRICITY AND PIEZOELECTRICITY IN ORIENTED FILMS OF POLYVINYL FLUORIDE AND POLYVINYLIDENE FLUORIDE. PYROELE 36-3544
/TICAL CONSTANTS OF ANTIFERROMAGNETIC POTASSIUM MANGANESE TRIFLUORIDE AND POTASSIUM COBALT TRIFLUORIDE CRY/ 33-3385
/ DENSITY DETERMINATION OF LARGE, PURE RUBIDIUM MANGANESE TRIFLUORIDE AND POTASSIUM MANGANESE TRIFLUORIDE / 33-3397
/MANGANESE FLUORIDE, COBALT FLUORIDE, POTASSIUM MANGANESE TRIFLUORIDE AND RUBIDIUM MANGANESE TRIFLUORIDE C/ 33-3399
RITICAL X-RAY DIFFUSE SCATTERING FROM POTASSIUM MANGANESE TRIFLUORIDE AND SODIUM NIOBATE SINGLE CRYSTALS. / 08-1312
/AR THE SOFT MODE TRANSITION POINT IN POTASSIUM MANGANESE TRIFLUORIDE (ANOMALOUS HEAT CAPACITY, CRITICAL T/ 33-3373
CATTERING AND THE PHASE TRANSITION OF POTASSIUM MANGANESE TRIFLUORIDE AT 184 DEGREES K.   X-RAY S 33-3394
TYPE FERROELECTRICS. (BISMUTH TUNGSTATE, BISMUTH NIOBIUM OXYFLUORIDE, BISMUTH TANTALUM OXYFLUORIDE) /LAYER 13-1966

/KDP, ORTHOFERRITES, DYSPROSIUM ANTIMONIDE, RUBIDIUM IRON TRIFLUORIDE, BORACITES, BISMUTH(4) TITANIUM(3) O/ 02-0036
TASSIUM TANTALATE, STRONTIUM TITANATE, RUBIDIUM MANGANESE TRIFLUORIDE) BY INELASTIC NEUTRON SCATTERING. /PO 06-1139
TRIF/ MANDELSTAM BRILLOUIN SCATTERING OF LIGHT IN MANGANESE FLUORIDE, COBALT FLUORIDE, POTASSIUM MANGANESE 33-3399
SCATTERING AND PHASE TRANSITIONS IN A POTASSIUM MANGANESE TRIFLUORIDE CRYSTAL.   RAMAN 33-3400
T/ FERROELECTRIC PARAELASTIC PARAMAGNETIC BARIUM COBALT TETRAFLUORIDE CRYSTAL STRUCTURE (SPACE GROUP AND LA 33-3386
N SCATTERING AND PHASE TRANSITIONS IN POTASSIUM MANGANESE TRIFLUORIDE CRYSTALS.   RAMA 33-3372
DIUM / MANDELSTAM BRILLOUIN SCATTERING OF LIGHT IN MANGANESE FLUORIDE CRYSTALS (POTASSIUM TRIFLUORIDE, RUBI 33-3398
                    PIEZOELECTRICITY IN POLARIZED POLYVINYLIDENE FLUORIDE FILMS.   36-3557
RIC AND NONLINEAR OPTICAL PROPERTIES OF POLED POLYVINYLIDENE FLUORIDE FILMS.   PYROELECT 36-3595
/TY AND OPTICAL SECOND HARMONIC GENERATION IN POLYVINYLIDENE FLUORIDE FILMS (AFTER COOLING STRETCHED FILM / 36-3530
STRONTIUM(X) TITANATE SINGLE CRYSTALS GROWN FROM A POTASSIUM FLUORIDE FLUX. /RIC PROPERTIES OF BARIUM(1-X) 04-0635
YSTAL GROWTH OF BARIUM TITANATE SINGLE CRYSTALS IN POTASSIUM FLUORIDE FLUX.   DISCUSSION ON THE CR 04-1022
NONLINEARITY. (BARIUM TITANATE HAFNATE GROWN FROM POTASSIUM FLUORIDE FLUX) /RIC SINGLE CRYSTALS WITH LARGE 08-1373
                    DIFFUSION OF HYDROGEN FLUORIDE IN ICE.   34-3437
/ND DISPERSION OF ULTRASONIC WAVES IN POTASSIUM MANGANESE TRIFLUORIDE IN THE NEIGHBORHOOD OF THE STRUCTURA/ 33-3370
AMAGNETIC SCATTERING OF NEUTRONS FROM POTASSIUM MANGANESE TRIFLUORIDE IN THE SHORT RANGE ORDERED REGION. /R 33-3371
TITANATE SINGLE CRYSTALS GROWN FROM SOLUTION IN A POTASSIUM FLUORIDE MELT. /CTRICAL CONDUCTIVITY OF BARIUM 04-1035
1-X) STRONTIUM(X) TITANATE- MIXED CRYSTALS FROM A POTASSIUM FLUORIDE MELT.   DEFECT STRUCTURE OF (BARIUM(04-0873
OF BARIUM TITANATE CRYSTALS GROWN FROM SOLUTION IN POTASSIUM FLUORIDE MELT. /H CONDITIONS ON THE STRUCTURE 04-0598
        ANOMALOUS ELASTIC BEHAVIOR OF POTASSIUM MANGANESE TRIFLUORIDE NEAR A STRUCTURAL PHASE TRANSITION. 33-3392
        THERMAL EXPANSIVITY OF RUBIDIUM MANGANESE FLUORIDE NEAR THE NEEL TEMPERATURE.   33-3375
NGSTEN BRONZE TYPE. (STRONTIUM POTAS/ FIRST FERROELECTRIC OXYFLUORIDE PHASES OF THE OXYGENATED QUADRATIC TU 13-2011
/LS RESEARCH. (STRONTIUM TITANATE PHOTOCONDUCTIVITY, CALCIUM FLUORIDE PHOTOCONDUCTIVITY AND OPTICAL PROPER/ 06-1140
BIDIU/ INFRARED ABSORPTION SPECTRA OF POTASSIUM MANGANESE TRIFLUORIDE, POTASSIUM COBALT TRIFLUORIDE, AND RU 33-3384
/AND SPIN CORRELATION IN PARAMAGNETIC POTASSIUM MANGANESE TRIFLUORIDE, RUBIDIUM MAGANESE TRIFLUORIDE, AND / 33-3376
/ERING OF LIGHT IN MANGANESE FLUORIDE CRYSTALS (POTASSIUM TRIFLUORIDE, RUBIDIUM MANGANESE TRIFLUORIDE, MAN/ 33-3398
/TIONS AT THE 184 DEG K TRANSITION OF POTASSIUM MANGANESE TRIFLUORIDE SINGLE CRYSTALS, (X-RAY DIFFUSE SCAT/ 33-3369
/RASE MODE RECORDING CHARACTERISTICS OF PHOTOCHROMIC CALCIUM FLUORIDE, STRONTIUM TITANATE, AND CALCIUM TIT/ 06-1119
ATE, AND POTASSIUM TAN/ ENERGY BANDS FOR POTASSIUM NICKEL TRIFLUORIDE, STRONTIUM TITANATE, POTASSIUM MOLYBD 02-0091
/N PROCESSES AND PHASE TRANSITIONS IN POTASSIUM MANGANESE TRIFLUORIDE: TEMPERATURE DISCONTINUITIES IN THE / 33-3363
EEL TEMPERATURE AND EPR LINE WIDTH OF POTASSIUM MANGANESE TRIFLUORIDE WITH MAGNESIUM DOPING.   SHIFT OF N 33-3377
/ECTRICITY IN (PURE AND) COBALT DOPED POTASSIUM MANGANESE TRIFLUORIDE. (20 PERCENT MANGANESE REPLACED BY C/ 33-3366
                    ELECTRONIC HALL MOBILITY IN THE ALKALINE EARTH FLUORIDES.   33-3404
LOW FREQUENCY DIELECTRIC CONSTANTS OF THE ALKALINE EARTH FLUORIDES BY THE METHOD OF SUBSTITUTION.   33-3361
                    INFRARED SPECTRA OF FLUORINE CONTAINING BARIUM TITANATE.   04-0983
                    PREPARATION OF FLUORINE CONTAINING BARIUM TITANATE.   04-0984
AL PROPERTIES)   FLUORINE CONTAINING BARIUM TITANATE. (ELECTRIC 04-0869
ARIUM TITANATE.   FORMATION OF FLUORINE CONTAINING SOLID SOLUTIONS BASED ON B 04-0932
IUM MANGANESE T/ ACOUSTIC AND MAGNETIC EFFECTS INVOLVING THE FLUORINE-19 NUCLEI IN ANTIFERROMAGNETIC POTASS 33-3390
        ELECTRON SPIN RESOURCE STUDIES AT DEFECTS IN FLUORITES. (LEAD TITANATE)   05-1086
/NFRARED RADIATION DETECTOR. (TRIGLYCINE SULFATE, TRIGLYCINE FLUOROBERYLLATE, LITHIUM SULFATE, BARIUM STRO/ 02-0046
IRON OXIDE, BISMUTH MOLYBDATE, BISMUTH FLUONIOBATE, BISMUTH FLUOTANTALATE) / PRASEODYMIUM BISMUTH TITANIUM 13-1993
(X) TITANATE SINGLE CRYSTALS GROWN FROM A POTASSIUM FLUORIDE FLUX. /RIC PROPERTIES OF BARIUM(1-X) STRONTIUM 04-0635
WTH OF BARIUM TITANATE SINGLE CRYSTALS IN POTASSIUM FLUORIDE FLUX.   DISCUSSION ON THE CRYSTAL GRO 04-1022
E(0.33) NIOBATE(0.67) (SINGLE CRYSTALS GROWN FROM LEAD OXIDE FLUX). FERROELECTRIC PROPERTIES OF LEAD ZINCAT 13-2066
RITY. (BARIUM TITANATE HAFNATE GROWN FROM POTASSIUM FLUORIDE FLUX) /RIC SINGLE CRYSTALS WITH LARGE NONLINEA 08-1373
ION OF POTASSIUM NIOBATE.   (TOP SEEDED FLUX) CRYSTAL GROWTH AND NEUTRON CHARACTERIZAT 08-1353
ESTIGATION OF THE FINE STRUCTURE OF BARIUM TITANATE CRYSTALS FLUX GROWN FROM POTASSIUM FLUORIDE).   INV 04-0708
RICAL PROPERTIES.   FLUX GROWN KTN SINGLE CRYSTALS AND THEIR ELECT 09-1533
NFIRMS FERROELECTRIC NATURE).   DIELECTRIC HYSTERESIS IN (FLUX GROWN) SINGLE CRYSTAL BISMUTH FERRATE (CO 08-1449
PONTANEOUS POLARIZATION AND COERCIVE FIELD OF LEAD TITANATE (FLUX GROWN SINGLE CRYSTALS).   S 05-1050
IGH RESISTIVITY SINGLE CRYSTALS OF LEAD TITANATE.   (FLUX) GROWTH AND FERROELECTRIC PROPERTIES OF H 05-1072
CONSTANT OF ZIRCONIUM-89 IN BARIUM TITANATE (BARIUM CHLORIDE FLUX GROWTH IN PRESENCE OF ZIRCONIUM-89). /AY 04-0685
ARE EARTH MANGANATES (PHASE DIAGRAMS, CONVE/ FABRICATION (BY FLUX GROWTH) OF HEXAGONAL SINGLE CRYSTALS OF R 36-3536
LUMINATE CRYSTALS DOPED WITH ISOTOPE 170.   FLUX GROWTH OF MAGNESIUM OXIDE AND LANTHANUM A 36-3558
NIOBATES, TANTALATES, STANNATES, TITANATES, CHROMATES, COBA/ FLUX GROWTH OF SOME COMPLEX OXIDE MATERIALS. (02-0129
VARIATION OF CRYSTAL GROWTH RATE WITH SUPERSATURATION IN FLUXED MELTS. (BARIUM STRONTIUM NIOBATE) 13-1939
POTASSIUM DIDEUTERIUM PHOSPHATE.   LASER INDUCED SELF FOCUSING IN AMMONIUM DIHYDROGEN PHOSPHATE AND 17-2476
AMMONIUM DIHYDROGEN PHOSPHATE CRYSTALS OF LIG/ THERMAL SELF FOCUSING IN POTASSIUM DIHYDROGEN PHOSPHATE AND 17-2717
S. (KDP)   EFFECTS OF DISPERSION AND FOCUSING ON THE PRODUCTION OF OPTICAL HARMONIC 17-2579
NGANESE(2+) DOPED LITHIUM NIOBATE.   FORBIDDEN HYPERFINE LINES IN ESR SPECTRA OF MA 11-1805
N PARAMAGNETIC RESONANCE OF MANGANESE(2+) DOPED/ ALLOWED AND FORBIDDEN HYPERFINE TRANSITIONS IN THE ELECTRO 16-2383
HE ELECTRON PARAMAGNETIC RESONANCE OF MANGANESE(+2) IN BARI/ FORBIDDEN LINES IN THE HYPERFINE SPECTRUM OF T 04-0877
CTRICS WITH PEROVSKITE STRUCTURE (TITANATES AND ZIRCONATES)/ FORBIDDEN ZONE WIDTHS IN POLYCRYSTALLINE DIELE 08-1288
PART-2: CALCULATION OF THE INTERMOLECULAR AND INTRAMOLECULAR FORCE CONSTANTS. /AND FERROELECTRIC THIOUREA. 25-2907
THE NITRITE RADICAL IN SODIUM NITRITE.   ACCURATE FORCE CONSTANTS FROM ISOTOPIC SUBSTITUTION IN 16-2309
UM TITANATE.   FORCE CONSTANTS IN BARIUM TITANATE AND STRONTI 04-0748
TEMPERATURE DEPENDENCE OF THE FORCE CONSTANTS OF ION IN SODIUM NITRITE.   16-2311
NIOBATE, BARIUM SODIUM NIOBATE AND BARIUM SODIUM TANTALATE (FORCE CONSTANTS, SPACE GROUP ASSIGNMENT). /IUM 11-1808
/ THERMAL VARIATION OF SPONTANEOUS POLARIZATION AND COERCIVE FORCE OF THE PEROVSKITES IN THE SYSTEM LEAD ZI 09-1597
                    MOLECULAR FORCES OF HEAVY AND LIGHT ICE.   34-3440
ND IN FERROELECTRICS / INFLUENCE OF LONG RANGE DIPOLE DIPOLE FORCES ON ATTENUATION AND VELOCITY OF ULTRASOU 03-0253
IRRADIATION FIELD TREATED TRIGLYCINE SULFATE: NEW STRUCTURAL FORM.   CRYSTALLOGRAPHIC STUDIES OF 27-2996
RIC (LEAD DINIOBATE INTO THE NONFERROELECTRIC (RHOMBOHEDRIC) FORM. /SITION OF THE (ORTHORHOMBIC) FERROELECT 08-1285
/ITY, ENTROPY, MAGNETIC MOMENT OF THE ELECTRICALLY POLARIZED FORM FROM 0.4 TO 4.2 DEGREES K WITH FIELDS TO/ 30-3275
L GRO/ INFLUENCE OF THE CONDITIONS OF CRYSTALLIZATION ON THE FORM OF CRYSTALS OF POTASSIUM NITRATE. (CRYSTA 16-2332
ROWTH MODEL.   LOW ENTROPY FORM OF ICE I(H) OBTAINED FROM A LINEAR STEP G 34-3410
EVIDENCE OF ADDITIONAL DISORDER IN THE RHOMBOHEDRRAL FORM OF POTASSIUM NITRITE.   16-2371
ANTIMONY SULFOIODIDE IN TH/ INFLUENCE OF ILLUMINATION ON THE FORM OF THE VOLTAGE AMPERE CHARACTERISTICS OF 15-2119
THEORY OF BARIUM TITANATE TYPE FERROELECTRIC CRYSTALS IN THE FORM OF THICK PLATES.   MACROSCOPIC 04-0770
                    A NEW FORM OF TUNGSTEN TRIOXIDE.   10-1652
                    DEUTERON ARRANGEMENTS IN THE HIGH PRESSURE FORMS OF ICE.   34-3438
ORDERED (ANTIFERROELECTRIC STRUCTURES))   STRUCTURES OF THE FORMS OF ICES. (ICES II, VIII AND IX ARE FULLY 34-3451
ICAL PRO/ THEORY OF HYDROGEN BONDED FERROELECTRICS. PART-1: (FORMALISM OF LOCAL MODE MODEL.) PART-2: (DYNAM 03-0384
X-RAY DIFFRACTION STUDY OF PHASE TRANSITION IN COPPER FORMATE.   26-2938
(ANTIFERROELECTRIC)   ANTIPOLARIZATION IN COPPER FORMATE CRYSTALS BY THE ELECTROOPTICAL METHOD. 26-2925
COPPER DEUTERO FORMATE TETRADEUT/ VIBRATION SPECTRUM OF THE FORMATE ION IN COPPER FORMATE TETRAHYDRATE AND 26-2926
CRYSTAL STRUCTURE OF PIEZOELECTRIC LITHIUM FORMATE MONOHYDRATE.   26-2940
NONLINEAR OPTICAL SUSCEPTIBILITY OF LITHIUM FORMATE MONOHYDRATE.   26-2945
E STUDY OF SUBLATTICE ROTATIONS AND SPIN DEVIATION IN COPPER FORMATE TETRADEUTERATE.   PROTON RESONANC 26-2950

MAGNETIZATION IN THE TWO DIMENSIONAL ANTIFERROMAGNET COPPER FORMATE TETRADEUTERATE. /PONTANEOUS SUBLATTICE 26-2929
T OF ESR LINES IN THE TWO DIMENSIONAL ANTIFERROMAGNET COPPER FORMATE TETRAHYDRATE.   ANOMALOUS SHIF 26-2941
           MAGNETISM AND THE PHASE TRANSITION OF COPPER FORMATE TETRAHYDRATE.   26-2930
NAL ANTIFERROMAGNETS POTASSIUM COPPER TRIFLUORIDE AND COPPER FORMATE TETRAHYDRATE. /ONS IN THE LOW DIMENSIO 26-2939
           RAMAN AND INFRARED SPECTRA OF COPPER FORMATE TETRAHYDRATE.   26-2936
MAL CONDUCTION IN A TWO DIMENSIONAL ANTIFERROMAGNETIC COPPER FORMATE TETRAHYDRATE.   THER 26-2932
ELL AND SPACE GROUP OF THE ANTIFERROELECTRIC PHASE OF COPPER FORMATE TETRAHYDRATE.   UNIT C 26-2927
/AXATION IN SEVERAL CRYSTALS OF ORDER DISORDER TYPE. (COPPER FORMATE TETRAHYDRATE, AMMONIUM IRON AND CHROM/ 02-0087
NANCE.   STUDY OF THE ORDERED MAGNETIC STATE OF COPPER FORMATE TETRAHYDRATE BY ANTIFERROMAGNETIC RESO 26-2943
APPROXIMATE STRUCTURE FOR ANTIFERROELECTRIC PHASE OF COPPER FORMATE TETRAHYDRATE BY NEUTRON DIFFRACTION.   26-2935
        DIELECTRIC RESPONSE FUNCTION OF COPPER FORMATE TETRAHYDRATES.   26-2933
           RAMAN AND INFRARED SPECTRA OF INORGANIC FORMATES.   26-2937
ODIFICATION OF SAMARIUM MOLYBDATE (SOLID PHASE SYNTHESIS, N/ FORMATION AND PHASE TRANSFORMATIONS OF THE M M 30-3270
          MECHANISM AND KINETICS OF LITHIUM METANIOBATE FORMATION BY A SOLID PHASE REACTION.   11-1840
CRYSTALS WITH POTASSIUM TUNGSTEN BRONZE STRUCTURE (POTASSIU/ FORMATION CONDITIONS AND PHASE TRANSITIONS IN 13-1923
STRUCTURE, 49/ FERROELECTRIC RARE EARTH MOLYBDATES. (REVIEW, FORMATION, CRYSTAL GROWTH, PHASE TRANSITIONS, 02-0064
THEORY OF X-RAY DIFFRACTION LINE SHAPE EFFECTS DUE TO DOMAIN FORMATION IN FERROELECTRICS.   03-0236
(TEMPERATURE A/ EFFECT OF ELECTRIC FIELD ON DOMAIN STRUCTURE FORMATION IN SEIGNETTES SALT (ROCHELLE SALT). 28-3190
CTING LAYER BY DIFFUSION.   FORMATION OF BARIUM TITANATE CERAMIC SEMICONDU 04-0684
                                                              FORMATION OF COLOR CENTERS IN IRRADIATED ICE. 34-3487
NS BASED ON BARIUM TITANATE.   FORMATION OF FLUORINE CONTAINING SOLID SOLUTIO 04-0932
                MECHANISM AND KINETICS OF SOLID PHASE FORMATION OF LEAD METANIOBATE.   13-2042
                                                 HEAT OF FORMATION OF LITHIUM TRIHYDROGEN DISELENITE.   22-2867
        SOME CHEMICAL REACTIONS IN SILICA GELS. PART-3: FORMATION OF POTASSIUM ACID TARTRATE CRYSTALS.   29-3246
/ART-5: EFFECT OF THE FORMING PRESSURE ON SINTERING. PART-6: FORMATION OF PZT AND SINTERING UNDER OXYGEN A/ 09-1624
TRIC TUNGSTEN TRIOXIDE.   FORMATION OF SHEAR STRUCTURES IN SUBSTOICHIOME 10-1659
SOLID SOLID REACTIONS.   FORMATION OF THIN LAYERS OF MIXED TITANATES BY 13-2005
EVAPORATED IN VACUO.   FORMATION PROCESS OF BARIUM TITANATE THIN FILM 04-0862
N THE SOLID STATE REACTION.   FORMATION PROCESSES OF LEAD MAGNESIO NIOBATE I 08-1354
/EXTERNAL ACTIONS ON THE DOMAIN STRUCTURE OF SINGLE CRYSTALS FORMED BY BARIUM TITANATE BASE SOLID SOLUTION. 04-0954
RAMICS ON THE PHYSICO- CHEMICAL NATURE OF THE SOLID SOLUTION FORMED FROM TWO TITANATES. /BARIUM TITANATE CE 04-0789
OTARY FURNACE.   REACTION SEQUENCE IN FORMING BARIUM METATITANATE IN AN INDUSTRIAL R 04-1003
IES OF BARIUM TITANATE SEMICONDUCTIVE CERAMIC/ MECHANISM FOR FORMING INSULATION LAYERS ON THE GRAIN BOUNDAR 04-0867
/OXIDE ADDITION ON THE PRODUCT OF PZT. PART-5: EFFECT OF THE FORMING PRESSURE ON SINTERING. PART-6: FORMAT/ 09-1624
PARAMETER X IN COMPOUNDS OF THE PYROCHLORE TYPE WITH GENERAL FORMULA A(2)B(2)O(7). /HOD FOR ESTIMATING THE 36-3599
E OR BROMIDE (NOT EXAMINED FOR FERROELECTR/ NEW BORACITES OF FORMULA LITHIUM(4) BORON(7) OXYGEN(12) CHLORID 14-2089
FLUOBERYLLATE/ INVESTIGATION OF FERROELECTRICS WITH GENERAL FORMULA METAL-1 METAL-2 SULFATE OR SELENATE OR 19-2743
ELECTROELASTIC EQUATIONS OF STATE/ COMPILATION OF CONVERSION FORMULAS INTERRELATING THE LINEAR ELASTIC AND 03-0277
TEMS.   APPLICATION OF DIELECTRIC MIXTURE FORMULAS TO (LEAD TITANATE) GLASS- CERAMIC SYS 05-1054
MAIN WALL MOTION IN FERROELECTRIC SWITCHING. PART-1: GENERAL FORMULATION.   KINETICS OF DO 03-0274
STICS (THEORY).   DETERMINATION OF THE STATE PARAMETERS AND FORMULATION OF SPONTANEOUS STRAIN FOR FERROELA 03-0137
TRA OF PARAELECTRIC POTASSIUM DIHYDROGEN PHOSPHATE AND POTA/ FOURIER TRANSFORM FAR INFRARED REFLECTION SPEC 03-0356
ERMITTIVITY OF ADP AND KDP IN THE FAR INFRARED BY DISPERSIVE FOURIER TRANSFORM SPECTROSCOPY. /THE COMPLEX P 27-2609
/ON OF THE SIMULTANEOUS GENERATION OF THE SECOND, THIRD, AND FOURTH HARMONICS OF 1.06 MICRON RADIATION IN / 11-1779
          ELECTROSTRICTIVELY GENERATED AND MODULATED FRACTURE IN BARIUM TITANATE CERAMICS.   · 04-0700
M NIOBATE) BY RADIATION FROM A RUBY LASER.   FRACTURE OF NONLINEAR CRYSTALS (KDP AND LITHIU 11-1663
TE TRANSDUCER MATERIAL.   FRACTURE TOUGHNESS OF COMMERCIAL BARIUM TITANA 04-0841
HELLE SALT BY SPIN LATTICE RELAXATION OF PROTONS IN ROTATING FRAME. /ULES IN ROCHELLE SALT AND AMMONIUM ROC 28-3232
E.   PROTON ROTATING FRAME RELAXATION IN LITHIUM HYDRAZINIUM SULFAT 32-3347
/NTS OF SECOND HARMONIC GENERATION AND THE VARIATIONS IN THE FREE AND CLAMPED VALUES OF THE DIELECTRIC CON/ 13-1998
ED STRONTIUM TITANATE.   FREE CARRIER OPTICAL ABSORPTION IN NIOBIUM DOP 06-1214
RROELECTRIC. (THEORY, ABSTRACT ONLY)   EFFECT OF FREE CARRIERS ON THE INFRARED SPECTRUM OF A FE 03-0408
PHASE TRANSITION IN TERMS OF A SINGLE MODE ANHARMONIC OSCI/ FREE ENERGY AND SPECIFIC HEAT IN FERROELECTRIC 03-0406
HYSTERESIS LOOPS.   FERROELECTRIC FREE ENERGY EXPANSION COEFFICIENTS FROM DOUBLE 03-0262
OGEN PHOSPHATE. (DETERMINATION OF EETA IN THE P EXPANSION OF FREE ENERGY F) /PROPERTIES OF POTASSIUM DIHYDR 17-2543
CS).   ANALYTIC PROPERTIES OF THE FREE ENERGY FOR THE "ICE" MODELS (FERROELECTRI 34-3427
EN PHOSPHATE CRYSTALS OF LIGHT FROM A LASER OPERATING IN THE FREE GENERATION MODE. /E AND AMMONIUM DIHYDROG 17-2717
ECTRIC ARSENATE SINGLE CRYSTALS.   ESR OF FREE RADICALS IN FERROELECTRIC AND ANTIFERROEL 18-2731
DROGEN HALIDES.   STRESS FREE SINGLE CRYSTAL GROWTH OF FERROELECTRIC HY 31-3333
TITANATE SINGLE CRYSTALS.   INFLUENCE OF THE AMPLITUDE OF FREE VIBRATIONS ON THE PERMITTIVITY OF BARIUM 04-0768
                 THERMODIELECTRIC EFFECT AND FREEZING POTENTIAL IN GROWING ICE.   34-3417
S MECHANISMS ON BISMUTH(12) GERMANIUM OXIDE(20) AT MICROWAVE FREQUENCIES.   ACOUSTIC SURFACE WAVE LOS 35-3511
          DIELECTRIC DISPERSION OF TRIGLYCINE SULFATE AT LOW FREQUENCIES.   27-3023
F THE DIELECTRIC CONSTANT OF TRIGLYCINE SULFATE AT MICROWAVE FREQUENCIES.   MEASUREMENT O 27-2954
PERMITTIVITY OF BARIUM TITANATE SINGLE CRYSTALS AT MICROWAVE FREQUENCIES.   MEASUREMENT OF THE 04-0759
ETRIC FLUORESCENCE IN BARIUM SODIUM NIOBATE FOR NINE CW PUMP FREQUENCIES.   PARAM 13-1950
OF THE DIELECTRIC PROPERTIES OF FERROELECTRICS AT MICROWAVE FREQUENCIES. /BLEM OF THE ACCURATE MEASUREMENT 03-0429
HATE AND AMMONIUM DIHYDROGEN PHOSPHATE CRYSTALS AT MICROWAVE FREQUENCIES. /TY OF POTASSIUM DIHYDROGEN PHOSP 17-2714
SALT.   ANOMALOUS DISPERSION OF COERSIVE FIELD AT VERY LOW FREQUENCIES AND JERKY WALL MOTION IN ROCHELLE 28-3177
M NIOBATE.   DIRECTIONAL DISPERSION OF PHONON FREQUENCIES AND POLARITON SCATTERING IN LITHIU 11-1696
RIC HYDROGEN CHLORIDE.   NUCLEAR QUADRUPOLE FREQUENCIES AND RELAXATION TIMES IN FERROELECT 31-3336
/MPERATURE COMPENSATION BY ADDITION OF TWO CRYSTAL RESONATOR FREQUENCIES: APPLICATION TO QUARTZ AND LITHIU/ 11-1686
C DISPLACEMENTS IN PEROVSKITE STRONTIUM TITANATE (AT OPTICAL FREQUENCIES, CALCULATIONS).   ATOMI 06-1142
ERATURES AS A FUN/ CALCULATION OF THE VARIATION OF THE RAMAN FREQUENCIES IN SODIUM NITRITE AT ORDINARY TEMP 16-2363
TIES OF BARIUM TITANATE IN THE CENTIMETER RANGE OF SUPERHIGH FREQUENCIES (MICROWAVES). FERROELECTRIC PROPER 04-0901
ON IN KDP CRYSTAL.   COMBINATION OF THE FREQUENCIES OF COHERENT AND INCOHERENT RADIATI 17-2500
S. (COMMENTS)   TEMPERATURE DEPENDENCE OF RESONANT FREQUENCIES OF LITHIUM NIOBATE PLATE RESONATOR 11-1812
RIGLYCINE SULFA/ BEHAVIOR OF THE COMPLEX PERMITTIVITY (AUDIO FREQUENCIES) OF SOME FERROELECTRIC CRYSTALS (T 02-0125
/200 PER CM) OF GUANIDINIUM ALUMINUM SULFATE HEXAHYDRATE AND FREQUENCIES OF THE FUNDAMENTAL VIBRATIONS OF / 21-2811
LECTRIC (CALCULATIONS).   INFRARED VIBRATION FREQUENCIES OF THE HYDROGEN BOND IN KDP FERROE 17-2650
EN ORTHOPHOSPHATE (4000-10000 PER CM) AND HARMONIC VIBRATION FREQUENCIES OF THE HYDROGEN BONDS. /M DIHYDROG 17-2432
                                              VIBRATION FREQUENCIES OF TITANATES.   08-1394
ITE STRONTIUM T/ ELECTRIC FIELD DEPENDENCE OF OPTICAL PHONON FREQUENCIES (RAMAN SCATTERING IN CUBIC PEROVSK 06-1241
/LECTRICAL PROPERTIES OF SOME CERAMIC MATERIALS AT MICROWAVE FREQUENCIES. (TITANATE ZIRCONATE GARNET MIXTU/ 09-1602
INE BAR/ NONLINEAR PROPERTIES OF FERROELECTRICS AT MICROWAVE FREQUENCIES. (TRIGLYCINE SULFATE, POLYCRYSTALL 04-0870
DIELECTRIC PROPERTIES OF ANTIMONY SULFOIODIDE AT MICROWAVE FREQUENCIES (3.3 AND 9.8 GHZ, 0-50 C).   15-2171
IN) LITHIUM NIOBATE CRYSTAL (NCNLIN/ GENERATION OF SUPERHIGH FREQUENCY ACOUSTIC HARMONICS IN A (SINGLE DOMA 11-1707
S ON LITHIUM NIOBATE (TEMPERATURE AND FREQUENCY D/ MICROWAVE FREQUENCY ACOUSTIC SURFACE WAVE LOSS MECHANISM 11-1817
TIC WAVES T/ EXPERIMENTAL STUDY OF LIGHT DIFFRACTION BY HIGH FREQUENCY ACOUSTIC WAVES. APPLICATION OF ACOUS 11-1834
          INTERACTIONS BETWEEN HIGH FREQUENCY ACOUSTICAL WAVES IN LITHIUM NIOBATE. 11-1836

DEPENDENCE OF THE COERCIVE FIELD OF TRIGLYCINE SULFATE ON FREQUENCY, AMPLITUDE AND TEMPERATURE.   27-3092
OMPLEX DIELECTRIC CONSTANT OF FERROELECTRICS NEAR THE CURIE/ FREQUENCY AND TEMPERATURE DEPENDENCES OF THE C   03-0361
/ DIELECTRIC CERAMICS WITH BOUNDARY LAYER STRUCTURE FOR HIGH FREQUENCY APPLICATION. (BARIUM, STRONTIUM TITA   04-0833
/CE WAVE) FERROELECTRIC PHASE SHIFTERS FOR VHF AND ULTRAHIGH FREQUENCY   (BASED ON LEAD(0.35) STRONTIUM(0.6/   37-3633
CTRICS: TRIGLYCINE SULFATE AND POTASSIUM DIHYDROGEN PH/ HIGH FREQUENCY BEHAVIOR OF HYDROGEN BONDED FERROELE   17-2526
ELECTROOPTIC EFFECT IN POTASSIUM DIHYDROGEN PHOS/ PARAMETRIC FREQUENCY CONVERSION OF COHERENT LIGHT BY THE   17-2616
LECTRIC CONSTANT OF TGS IN THE FAR INFRARED (TEMPERATURE AND FREQUENCY DEPENDENCE).   DISPERSION OF THE DIE   27-3009
AMPING IN ANTIMONY SULFOIODIDE SINGLE CRYSTALS.   FREQUENCY DEPENDENCE OF ANOMALOUS ULTRASOUND D   15-2139
D IN ROCHELLE SALT NEAR ITS UPPER CURIE POINT.   FREQUENCY DEPENDENCE OF THE ABSORPTION OF SOUN   28-3183
TRIGLYCINE SULFATE CRYSTALS.   FREQUENCY DEPENDENCE OF THE COERCIVE FIELD OF   27-3062
TION OF ULTRASOUND IN TRIGLYCINE SULFATE SINGLE CRYSTALS (A/ FREQUENCY DEPENDENCE OF THE FLUCTUATION ABSORP   27-3079
ROELECTRIC MATERIALS (ROCHELLE SALT NEAR TH/ TEMPERATURE AND FREQUENCY DEPENDENCE OF THE LOSS FACTOR IN FER   28-3220
ABILIZATION EFFECT IN FERROELECTRICS.   ON THE FREQUENCY DEPENDENCE OF THE TEMPERATURE AUTOST   03-0479
THIUM NIOBATE.   FREQUENCY DEPENDENCES OF SOME PARAMETERS OF LI   11-1843
TIES OF LITHIUM METANIOBATE.   TEMPERATURE AND FREQUENCY DEPENDENCES OF THE ELECTRICAL PROPER   11-1717
     COUPLED ORDER DISORDER PHONON SYSTEM WITH DAMPING: KDP (FREQUENCY DEPENDENT SUSCEPTIBILITY).   17-2667
FERROELECTRIC CERAMIC. (BARIUM TITANATE, BARIUM / ULTRAHIGH FREQUENCY DIELECTRIC ANISOTROPY OF A POLARIZED   04-0767
EARTH FLUORIDES BY THE METHOD OF SUBSTITUTION.   LOW FREQUENCY DIELECTRIC CONSTANTS OF THE ALKALINE   33-3361
LT CRYSTALS.   LOW FREQUENCY DIELECTRIC DISPERSION IN ROCHELLE SA   28-3228
THE SYSTEM (BARIUM, STRONTIUM) TITANATE. (SYNTHESIS AND LOW FREQUENCY DIELECTRIC PROPERTIES) / CRYSTALS OF   04-0561
NATE SINGLE C/ EFFECT OF LINEAR COMPRESSION ON THE ULTRAHIGH FREQUENCY DIELECTRIC PROPERTIES OF BARIUM TITA   04-0896
YSTALLINE FERROELECTRICS (BARIUM TITANATE AND SAM/ ULTRAHIGH FREQUENCY DIELECTRIC PROPERTIES OF SOME POLYCR   04-0818
OCHELLE SALT CAUSED BY DOMAIN WALLS. (ABSTRACT ONLY)   LOW FREQUENCY DISPERSIONS OF THE PERMITTIVITY OF R   28-3199
RODUCED IN BARIUM SODIUM NIOBATE CRYSTALS DURING INTRACAVITY FREQUENCY DOUBLING. /STROPHIC SURFACE DAMAGE P   13-2052
C CHLORINE, CADMIUM CHLORINE, CADMIUM IODINE, IRO/ (OPTICAL) FREQUENCY DOUBLING IN (MAGNESIUM CHLORINE, ZIN   14-2080
ITE.   ANALYSIS OF LOW FREQUENCY ELECTROOPTIC RESPONSE IN SODIUM NITR   16-2353
FOR ACOUSTIC SURFACE WAVES ON LITHIUM NIOBATE.   NEW HIGH FREQUENCY HIGH COUPLING LOW BEAM STEERING CUT   11-1819
YDROGEN PHOSPHATE.   LOW FREQUENCY HYDROGEN VIBRATIONS IN POTASSIUM DIH   17-2699
YDROGEN PHOSPHATE.   LOW FREQUENCY HYDROGEN VIBRATIONS IN POTASSIUM DIH   17-2608
L EXPANSION OF ROCHELLE SALT BY USING THE SHIFT OF RESONANCE FREQUENCY IN CAVITY.   MEASUREMENT OF THERMA   28-3202
BARS.   ZERO TEMPERATURE COEFFICIENT OF RESONANT FREQUENCY IN LITHIUM TANTALATE LENGTH EXPANDER   11-1810
RYSTALS OF POTASSIUM DIHYDROGEN PHOSPHATE AND RUBIDIUM / LOW FREQUENCY INFRARED ABSORPTION SPECTRA OF MONOC   17-2457
ITANATLATE.   LOW FREQUENCY LINEAR ELECTROOPTIC EFFECT IN LEAD D   13-1916
COBALT TETRAFLUORIDE.   LOW FREQUENCY MAGNETOELECTRIC RESONANCES IN BARIUM   33-3409
N KDP AND DEUTER/ MOSSBAUER EFFECT OBSERVATION OF A VERY LOW FREQUENCY MODE AT THE TRANSITION TEMPERATURE I   17-2465
DUE TO POLARIZA/ OSCILIATIONS OF FERROELECTRIC BODIES. (WITH FREQUENCY OF APPLIED FIELD CAUSED BY SHEARING   03-0197
VIBRONIC EFFECT IN THE TEMPERATURE DEPENDENCE OF THE SOFT FREQUENCY OF BARIUM TITANATE.   04-0662
FERROELECTRIC MODE AND NQR (NUCLEAR QUADRUPOLE RESONANCE) FREQUENCY. (POTASSIUM DIHYDROGEN ARSENATE)   17-2713
IGLYCINE FLUOBERYLLATE CRYSTALS.   LOW FREQUENCY RAMAN SCATTERING BY FERROELECTRIC TR   27-3113
ALS OF THE TRIGLYCINE SULFATE TYPE.   LOW FREQUENCY RAMAN SPECTRA OF FERROELECTRIC CRYST   27-3114
AS A FUNCTION OF TEMPERATURE.   LOW FREQUENCY RAMAN SPECTRUM OF A THIOUREA CRYSTAL   25-2906
E INTERPRETATION WITH A MIXED COULOMB VALENCE DYNAMICAL/ LOW FREQUENCY RAMAN SPECTRUM OF ICE I(H). TENTATIV   34-3425
NG IN RUBY AND LITHIUM NIOBATE IN THE THREE CENTIMETER RADIO FREQUENCY RANGE. /NCE OF HYPERSONIC WAVE DAMPI   11-1718
     VARIOUS HYSTERESIS LOOP TRACERS FOR A WIDE FREQUENCY RANGE. (CIRCUITRY DESCRIBED)   37-3640
/E THEORY OF LIGHT ABSORPTION BY ANTIFERRODIELECTRICS IN THE FREQUENCY RANGE OF DOUBLE ELECTRONIC EXCITATI/   03-0419
/HE DIELECTRIC PROPERTIES OF BARIUM STRONTIUM NIOBATE IN THE FREQUENCY RANGE OF 10 KILOHERTZ TO 10 MEGAHER/   13-1938
/ AND LITHIUM NIOBATE PIEZOELECTRIC RESONATORS IN THE MEDIUM FREQUENCY RANGE WITH LOW RATIOS OF CAPACITANC/   11-1724
LENATE IN THE NEAR INFRARED TO FAR INFRARED AND IN THE RADIO FREQUENCY REGION. / CONSTANTS OF TRIGLYCINE SE   27-3007
THIUM TANTALATE.   MEDIUM FREQUENCY RESONATORS OF LITHIUM NIOBATE AND LI   11-1851
TE TYPE FERROELECTRIC CERAMICS.   ULTRAHIGH FREQUENCY REVERSAL PROPERTIES OF BARIUM TITANA   04-1014
TE AT 77 AND 294 DEG K.   RAMAN SPECTRA AND MODE FREQUENCY SHIFTS OF FERROELECTRIC SODIUM NITRI   16-2262
/RY OF HYDROGEN BONDED FERROELECTRICS. PART-1: DAMPINGS AND (FREQUENCY) SHIFTS (OF PROTONIC COLLECTIVE MOT/   03-0475
     ROTATING WAVEPLATE OPTICAL FREQUENCY SHIFTING IN LITHIUM NIOBATE.   11-1689
TIVITY IN DISPLACIVE TYPE FERROELECTRICS. (THEORY, BAR/ HIGH FREQUENCY SOUND ATTENUATION AND THERMAL CONDUC   03-0160
.   FREQUENCY SPECTRUM OF FERROCERAMIC TRANSDUCERS   37-3656
CTRICS. (INFRARED ABSORPTION, NEUTRON INELASTIC SCATTERING,/ FREQUENCY SPECTRUM OF HYDROGEN BONDED FERROELE   03-0401
OMAGNETIC WAVE IN A FERROELECTRIC ANTIFE/ RESONANCE OF A LOW FREQUENCY SPIN WAVE WITH AN ACOUSTIC OR ELECTR   03-0284
IOBATE.   LOW FREQUENCY VIBRATIONAL MODES IN BARIUM SODIUM N   13-1905
YING TEMPERATURE CHARACTERISTICS OF ROCHELLE TYPE/ ULTRAHIGH FREQUENCY (WAVEGUIDE RESONATOR) METHOD OF STUD   04-0905
RROELECTRICS INDICATED BY ANISOTROPIC DEBYE- WALLER FAC/ LOW FREQUENCY ZONE BOUNDARY MODES IN PEROVSKITE FE   08-1351
ESTIGATION OF LITHIUM NIOBATE.   LOW FREQUENCY (510 MHZ) PARAMAGNETIC RESONANCE INV   11-1793
01)/ GENERATION OF THE THIRD HARMONIC OF A LONGITUDINAL HIGH FREQUENCY (950 MHZ) ACOUSTIC WAVE ALONG THE (0   11-1835
CONATE TITANATE) FERROELECTRICS (WITH AND WITHOUT / INTERNAL FRICTION (AND CURIE TEMPERATURES) IN (LEAD ZIR   09-1584
ELLE SALT.   INTERNAL FRICTION AS A FUNCTION OF TEMPERATURE FOR ROCH   28-3227
     INTERNAL FRICTION IN BARIUM TITANATE SINGLE CRYSTALS.   04-0794
PIEZOELECTRIC MODULUS)   AMPLITUDE DEPENDENCE OF THE INTERNAL FRICTION IN BARIUM TITANATE SINGLE CRYSTALS. (04-0791
STALS.   PIEZOELECTRIC EFFECT AND INTERNAL FRICTION IN GAMMA-IRRADIATED ROCHELLE SALT CRY   28-3229
/UENCE OF THE CONCENTRATION OF POINT DEFECTS ON THE INTERNAL FRICTION IN POLYCRYSTALLINE LEAD ZIRCONATE TI/   9-1553
UBSTANCES BY THE COMPOSITE RESONATC/ MEASUREMENT OF INTERNAL FRICTION IN SINGLE CRYSTALS OF FERROELECTRIC S   27-3080
(ROCHELLE SALT AND TRIGLY/ AMPLITUDE DEPENDENCE OF INTERNAL FRICTION IN SINGLE CRYSTALS OF FERROELECTRICS.   27-3122
C CURIE POINT OF LEAD / ANOMALIES OF ELASTICITY AND INTERNAL FRICTION IN THE REGION OF THE ANTIFERROELECTRI   13-2020
     TEMPERATURE DEPENDENCE OF INTERNAL FRICTION OF BARIUM TITANATE SINGLE CRYSTALS.   04-0796
1-40 HZ.   INTERNAL FRICTION OF LITHIUM NIOBATE SINGLE CRYSTALS AT   11-1798
NATE CERAMICS. (PROPRIETARY COMPOSITIONS)   INTERNAL FRICTION OF MODIFIED LEAD ZIRCONATE- LEAD TITA   09-1623
L STRUCTURE OF BARIUM TITANATE PRODUCING SQUARE INTERFERENCE FRINGES.   CRYSTA   04-0706
CRYSTAL STUDY OF BARIUM TITANATE WITH PARALLEL INTERFERENCE FRINGES.   04-0705
OF CRYSTALS OF BARIUM TITANATE HAVING PARALLEL INTERFERENCE FRINGES.   STUDY   04-0897
CTRIC CERAMICS OBTAINED BY PROJECTION WITH A PLASMA TORCH OR FRITTING. (BARIUM TITANATE) /ILAYERED FERROELE   04-0778
     NUCLEATION MODELS FOR A FIRST ORDER TRANSITION FRONT, AND OBSERVATIONS IN BARIUM TITANATE.   04-0608
     OBSERVATION OF PHASE FRONT MOTION IN KDP CRYSTAL.   17-2589
RROELECTRIC SEMICON/ METASTABLE OPTICAL CHARGE EXCHANGE AND "FROZEN" SHIFT OF THE CURIE TEMPERATURE OF A FE   15-2160
/SOLID SOLUTION AND ITS APPLICATION TO THE SOLID ELECTROLYTE FUEL CELL. ((CALCIUM, STRONTIUM) LANTHANUM TI/   08-1444
     ISOTHERMAL GRAIN GROWTH AND ELECTRICAL BEHAVIOR OF FULLY DENSE PLZT CERAMICS.   09-1553
/ STUDY OF SOME CRYSTALS CONTAINING O-D...O BOND. (INCLUDING FULLY DEUTERATED AMMONIUM DIHYDROGEN PHOSPHAT/   03-0204
STRUCTURES OF THE FORMS OF ICES. (ICES II, VIII AND IX ARE FULLY ORDERED (ANTIFERROELECTRIC STRUCTURES))   34-3451
ATE (OPTICAL SHUTTER, COLOR MODULATOR).   (PROPERTIES), FUNCTION AND APPLICATIONS OF GADOLINIUM MOLYBD   30-3298
     ELASTIC GIBBS FUNCTION FOR A FERROELECTRIC CRYSTAL.   03-0328
D GENERAL ISING MODELS.   ZEROES OF THE PARTITION FUNCTION FOR THE HEISENBERG, FERROELECTRIC, AN   03-0481
ITY (KDP MODEL, ANT/ DISTRIBUTION OF ZEROES OF THE PARTITION FUNCTION FOR THE SLATER MODEL OF FERROELECTRIC   03-0308
     MEASUREMENT OF ELECTRON ENERGY LOSS FUNCTION IN BULK STRONTIUM TITANATE.   06-1138

MEASUREMENT ON FERROELECTRIC SILVER SODIUM DINITRITE BY STEP FUNCTION METHOD.   DIELECTRIC RELAXATION 16-2291
                              DIELECTRIC RESPONSE FUNCTION OF COPPER FORMATE TETRAHYDRATES.   26-2933
  FREQUENCIES IN SODIUM NITRITE AT ORDINARY TEMPERATURES AS A FUNCTION OF CRYSTAL ORIENTATION / OF THE RAMAN 16-2363
RTAIN ELECTRICAL PROPERTIES CF SINTERED BARIUM TITANATE AS A FUNCTION OF GRAIN SIZE.   VARIATION IN CE 04-0627
/ECTRIC PROPERTIES OF LITHIUM TANTALATE SINGLE CRYSTALS AS A FUNCTION OF MELT COMPOSITION. (CURIE TEMPERAT/ 12-1874
       LCW FREQUENCY RAMAN SPECTRUM OF A THIOUREA CRYSTAL AS A FUNCTION OF TEMPERATURE.   25-2906
  COMPLEX DIELECTRIC CONSTANT OF ROCHELLE SALT AT 10 GHZ AS A FUNCTION OF TEMPERATURE.   MEASUREMENT OF THE 28-3184
       X-RAY DIFFUSE SCATTERING FROM SODIUM NIOBATE AS A FUNCTION OF TEMPERATURE.   08-1313
                            INTERNAL FRICTION AS A FUNCTION OF TEMPERATURE FOR ROCHELLE SALT.   28-3227
TRIGLYCINE/ POSITION OF THE FUNDAMENTAL ABSORPTION EDGE AS A FUNCTION OF TEMPERATURE FOR ROCHELLE SALT AND 27-3160
/N NEW SOLID STATE MATERIALS (INCLUDING FERROELECTRICS) AS A FUNCTION OF TEMPERATURE USING INFRARED AND RA/ 02-0104
ERESIS OF THE DIELECTRIC CONSTANT CF TRIGLYCINE SULFATE AS A FUNCTION OF THE CONDITIONS OF GROWTH. /AL HYST 27-3126
TRONTIUM TITANATE.   MICROSCOPIC CALCULATION OF THE RESPONSE FUNCTION OF THE SOFT ZONE BOUNDARY PHONON IN S 06-1213
/RAY PHOTOELECTRON SPECTRUM OF TUNGSTEN- TUNGSTEN OXIDE AS A FUNCTION OF THICKNESS OF THE SURFACE OXIDE LA/ 10-1631
NSIONAL EIGHT VERTEX MODEL AND THE PROTON PROTON CORRELATION FUNCTIONS IN ICE.   THE THREE DIME 34-3481
ACROSCOPIC AND MICROSCOPIC) SYMMETRY OF FERROELECTRIC ENERGY FUNCTIONS. (SOME COMMENTS)   (M 03-0193
EMPERATURE FOR ROCHELLE SALT AND TRIGLYCINE/ POSITION OF THE FUNDAMENTAL ABSORPTION EDGE AS A FUNCTION OF T 27-3160
ENITE CRYSTALS AT THEIR PHASE TRANSITIONS.   BEHAVIOR OF THE FUNDAMENTAL ABSORPTION EDGE IN TRIHYDROGEN SEL 22-2839
ATE.   FUNDAMENTAL ABSORPTION EDGE OF STRONTIUM TITAN 06-1186
   PROPAGATION ON LAYERED ANISOTROPIC SUBSTRATES: GOLD ON LIT/ FUNDAMENTAL AND HIGHER ORDER MODE SURFACE WAVE 11-1846
- METAL JUNCTION SYSTEMS. (INFLUENCE OF THE ELECTRODE MATER/ FUNDAMENTAL PROPERTIES OF METAL- FERROELECTRIC 02-0127
/RROELECTRICS USING POLARIZED IR RADIATION. PART-2: INTERNAL FUNDAMENTAL VIBRATIONS OF POTASSIUM DIHYDROGE/ 18-2738
/IDINIUM ALUMINUM SULFATE HEXAHYDRATE AND FREQUENCIES OF THE FUNDAMENTAL VIBRATIONS OF THE WATER MOLECULES. 21-2811
UENCE IN FORMING BARIUM METATITANATE IN AN INDUSTRIAL ROTARY FURNACE.   REACTION SEQ 41-1003
RIUM TITANATE FILMS BY DC DIODE SPUTTERING (ONTO PLATINUM OR FUSED QUARTZ).   PREPARATION OF THIN BA 04-0948
/UCTIVE READOUT MEMORY ELEMENT UTILIZING POTASSIUM NITRATE. (FUSED THICK FILM OR VACUUM DEPOSITED THIN FIL/ 16-2269
                                                            FUTURE PROBLEMS ON FERROELECTRICS.   02-0100

                                                  G

  AND POLARIZATION) OF LEAD CONTAINING OXYGEN O/ ELECTROOPTIC G COEFFICIENTS (RELATING INDUCED BIREFRINGENCE 13-2039
/NAL- CUBIC PEROVSKITE) STRONTIUM IRON TUNGSTATE . (MAGNETIC G STRUCTURE ANALOGOUS TO ANTIFERROELECTRIC ST/ 10-1655
                MICROCHEMICAL ANALYSIS OF GADOLINIUM DOPED BARIUM SODIUM NIOBATES.   13-1953
/UM, THORIUM, TITANIUM, SODIUM, MAGNESIUM, CADMIUM, YTTRIUM, GADOLINIUM, DYSPROSIUM, IRON, SELENIUM, NIOBI/ 10-1640
TES.   ELECTRON PARAMAGNETIC RESONANCE OF TRIVALENT GADOLINIUM IONS IN STRONTIUM AND BARIUM TITANA 04-0920
MAL ABSORPTION OF ELASTIC WAVES NEAR THE PHASE TRANSITION OF GADOLINIUM MOLYBDATE.   ABNOR 30-3265
OF ANOMALOUS OPTICAL BEHAVIOR OF THE IMPROPER FERROELECTRIC GADOLINIUM MOLYBDATE.   AN EXPLANATION 30-3292
VIOR NEAR THE PHASE TRANSITION IN FERROELASTIC FERROELECTRIC GADOLINIUM MOLYBDATE.   ANOMALOUS PHONON BEHA 30-3276
OPY TO DIRECT OBSERVATION CF MOVING FERROELECTRIC DOMAINS OF GADOLINIUM MOLYBDATE. /ELECTRON MIRROR MICROSC 30-3293
ECTRIC- FERROELASTIC AND PARAELECTRIC- PARAELASTIC PHASES OF GADOLINIUM MOLYBDATE. /AY STUDY OF THE FERROEL 30-3287
        CONTROL AND APPLICATION OF DOMAIN WALL MOTION IN GADOLINIUM MOLYBDATE.   30-3257
        CONTROL AND APPLICATION CF DOMAIN WALL MOTION IN GADOLINIUM MOLYBDATE.   30-3256
              DISLOCATION ETCHANTS FOR BETA GADOLINIUM MOLYBDATE.   30-3258
                    DOMAIN STRUCTURE OF GADOLINIUM MOLYBDATE.   30-3301
   EFFECT OF HYDROSTATIC PRESSURE ON PHASE TRANSITION IN GADOLINIUM MOLYBDATE.   30-3321
        ELASTIC CONSTANTS AND SOFT OPTICAL MODES IN GADOLINIUM MOLYBDATE.   30-3284
LECTRICAL, OPTICAL, AND MECHANICAL BEHAVIOR OF FERROELECTRIC GADOLINIUM MOLYBDATE.   E 30-3267
        ELECTROOPTICAL PROPERTIES OF FERROELECTRIC GADOLINIUM MOLYBDATE.   30-3277
         FERROELECTRIC AND FERROELASTIC CRYSTAL, GADOLINIUM MOLYBDATE.   30-3308
            FERROELECTRIC DOMAIN SHIFTING DEVICES. (GADOLINIUM MOLYBDATE)   30-3280
  FERROELECTRIC DOMAINS AND THE CONDENSED SOFT MODE IN GADOLINIUM MOLYBDATE.   30-3260
            FERROELECTRIC FERROELASTIC CRYSTAL GADOLINIUM MOLYBDATE.   30-3297
                        FERROELECTRICITY IN GADOLINIUM MOLYBDATE.   30-3266
UCED AND SPONTANEOUS BIREFRINGENCE IN IMPROPER FERROELECTRIC GADOLINIUM MOLYBDATE.   IND 30-3278
HANICAL AND OPTICAL PROPERTIES OF FERROELECTRIC FERROELASTIC GADOLINIUM MOLYBDATE. /NCES OF ELECTRICAL, MEC 30-3253
PROPERTIES OF ACTIVATED FERROELECTRICS. (LITHIUM NIOBATE AND GADOLINIUM MOLYBDATE) LASER AND SPECTROSCOPIC 11-1741
        LATTICE DYNAMICS OF A RIGID ION MODEL OF GADOLINIUM MOLYBDATE.   30-3281
                   MODEL FOR FERROELECTRIC GADOLINIUM MOLYBDATE.   30-3313
  NEW TYPE OF FAR INFRARED SOFT MODE IN FERROELECTRIC GADOLINIUM MOLYBDATE.   30-3310
             NEW TYPE OF FERROELECTRIC SOFT MODE IN GADOLINIUM MOLYBDATE.   30-3311
  OBSERVATION OF DIFFRACTED LIGHT FROM DOMAIN WALLS IN GADOLINIUM MOLYBDATE.   30-3322
        OPTICAL DEVICE ELEMENT OF FERROELECTRIC GADOLINIUM MOLYBDATE.   30-3295
                       PHASE TRANSITION IN GADOLINIUM MOLYBDATE.   30-3316
        PI- GMO: ANOTHER MODIFICATION OF GADOLINIUM MOLYBDATE.   30-3263
               PIEZOELECTRIC COEFFICIENT OF GADOLINIUM MOLYBDATE.   30-3317
ZATION BY ELASTIC INSTABILITIES IN PIEZOELECTRIC MATERIALS. (GADOLINIUM MOLYBDATE) /N OF SPONTANEOUS POLARI 30-3282
      PYROELECTRIC BEHAVIOR OF FERROELECTRIC GADOLINIUM MOLYBDATE.   30-3279
          PYROELECTRIC DETECTION PROPERTIES OF GADOLINIUM MOLYBDATE.   30-3324
  PYROELECTRIC EFFECT AND DOMAIN STRUCTURE OF GADOLINIUM MOLYBDATE.   30-3315
                 PYROELECTRIC EFFECT IN GADOLINIUM MOLYBDATE.   30-3323
TO PROBLEMS IN SOLID STATE PHYSICS (DOMAIN WALL THICKNESS IN GADOLINIUM MOLYBDATE). /NG TECHNIQUES APPLIED 30-3319
 SPONTANEOUS BIREFRINGENCE AND ELECTROOPTIC RESPONSE IN GADOLINIUM MOLYBDATE.   30-3307
              STUDY CF FERROELECTRIC DOMAINS IN GADOLINIUM MOLYBDATE.   30-3304
  TEMPERATURE DEPENDENCE OF THE RAMAN SPECTRUM OF GADOLINIUM MOLYBDATE.   30-3285
           THE CRYSTAL STRUCTURE OF FERROELECTRIC GADOLINIUM MOLYBDATE.   30-3288
                 THERMODYNAMIC THEORY OF GADOLINIUM MOLYBDATE.   03-0232
N FERROELECTRIC AND FERROELASTIC CRYSTALS. (BARIUM TITANATE, GADOLINIUM MOLYBDATE) /KNESS OF DOMAIN WALLS I 04-0777
  WAVEVECTOR DEPENDENT RELAXATION CF AN OPTICAL PHONON IN GADOLINIUM MOLYBDATE.   30-3320
    X-RAY STUDY OF THERMAL EXPANSION OF FERROELECTRIC GADOLINIUM MOLYBDATE.   30-3294
OELASTIC SUBSTA/ GROWTH AND MORPHOLOGY OF SINGLE CRYSTALS OF GADOLINIUM MOLYBDATE, A FERROELECTRIC AND FERR 30-3259
LS (EUROPIUM TERBIUM, DYSPROSIUM AND B/ PHASE TRANSITIONS OF GADOLINIUM MOLYBDATE AND ISOSTRUCTURAL COMPOUN 30-3271
    SECOND HARMONIC GENERATION IN MOLYBDATES (BETA GADOLINIUM MOLYBDATE AND TERBIUM MOLYBDATE.   30-3286
           NONLINEAR OPTICAL PROPERTIES OF GADOLINIUM MOLYBDATE AND TERBIUM MOLYBDATE.   30-3305
IONS.   OPTICAL PROPERTIES OF GADOLINIUM MOLYBDATE AND THEIR DEVICE APPLICAT 30-3299
IONS.   OPTICAL PROPERTIES OF GADOLINIUM MOLYBDATE AND THEIR DEVICE APPLICAT 30-3300
E TRANSITION.   ELASTIC PROPERTIES OF GADOLINIUM MOLYBDATE AT ITS FERROELECTRIC PHAS 30-3283
        DOMAIN WALL STRUCTURE IN GADOLINIUM MOLYBDATE BY RAMAN SCATTERING.   30-3318
TRANSITION ME/ FERROELECTRIC FERROELASTIC PARAMAGNETIC BETA GADOLINIUM MOLYBDATE. CRYSTAL STRUCTURE OF THE 30-3289
, LASING AND SPECTROSCOPIC PROPERTIES OF NEODYMIUM(3+) DOPED GADOLINIUM MOLYBDATE CRYSTALS.   CERTAIN GROWTH 30-3255
F PHASE BOUNDARIES BETWEEN FERRO- AND PARAELECTRIC PHASES IN GADOLINIUM MOLYBDATE CRYSTALS.   OBSERVATION O 30-3306

STRUCTURAL STUDY OF GADOLINIUM MOLYBDATE CRYSTALS.		30-3302
HANGE AT CURIE TEMPERATURE). X-RAY DIFFRACTION STUDY OF GADOLINIUM MOLYBDATE (CRYSTALLOGRAPHIC PHASE C		30-3312
/MENOLOGICAL THEORY OF THE FERROELECTRIC PHASE TRANSITION IN GADOLINIUM MOLYBDATE. (DIELECTRIC AND ELASTIC/		30-3303
/ OF SINGLE CRYSTAL BISMUTH TITANATE AND RELATED COMPOUNDS. (GADOLINIUM MOLYBDATE, GROWTH BY RF CZOCHRALSK/		08-1323
RTIES OF IMPROPER FERROELECTRIC CRYSTALS. (AMMONIUM SULFATE, GADOLINIUM MOLYBDATE, IRON IODINE BORACITE) /E		14-2084
ROSIUM ANTIMONIDE, RUBIDIUM/ COUPLED FERROELASTIC CRYSTALS. (GADOLINIUM MOLYBDATE, KDP, ORTHOFERRITES, DYSP		02-0036
ATION AND THICKNESS DEPENDENCE OF CONTRAST AND BRIGHTNESS IN GADOLINIUM MOLYBDATE LIGHT VALVES. ORIENT		30-3291
ELASTIC CONSTANTS OF GADOLINIUM MOLYBDATE (NEAR THE CURIE POINT).		30-3274
ODULATOR). (PROPERTIES,) FUNCTION AND APPLICATIONS OF GADOLINIUM MOLYBDATE (OPTICAL SHUTTER, COLOR M		30-3298
/NETOTHERMODYNAMICS OF ANTIFERROMAGNETIC, FERROELECTRIC BETA GADOLINIUM MOLYBDATE. PARTS 1-3: HEAT CAPACIT/		30-3275
ORIGIN OF THE STRUCTURAL PHASE TRANSITION IN GADOLINIUM MOLYBDATE. (SYMMETRY CHANGES)		30-3273
50-165 / PYROELECTRIC EFFECT AND SPONTANEOUS POLARIZATION IN GADOLINIUM MOLYBDATE (TEMPERATURE DEPENDENCE 1		30-3314
LINEAR THERMAL EXPANSIONS OF GADOLINIUM SUBSTITUTED BARIUM SODIUM NIOBATES.		13-1996
CAUSES OF THERMAL EXPANSION OF GADOLINIUM SUBSTITUTED BARIUM SODIUM NIOBATES.		13-2048
ELECTRON PARAMAGNETIC RESONANCE OF GADOLINIUM(3+) DOPED LITHIUM NIOBATE.		11-1833
INTERPRETATION OF ELECTRIC FIELD EFFECTS IN EPR OF GADOLINIUM(3+) DOPED STRONTIUM TITANATE.		06-1168
CONDUCTOR. TWO SETS OF ELECTRON SPIN RESONANCE OF GADOLINIUM(3+) IN BARIUM TITANATE CERAMIC SEMI		04-0990
ELECTRON SPIN RESONANCE SPECTRA OF CHROMIUM(3+) AND GADOLINIUM(3+) IN LANTHANUM ALUMINATE.		36-3583
EPR OF GADOLINIUM(3+) IN LITHIUM NIOBATE.		11-1763
F ELECTRON SPIN RESONANCE IN STRONTIUM TITANATE. (DOPED WITH GADOLINIUM(3+) OR IRON(3+)) /IC FIELD EFFECT O		06-1229
) ACOUSTIC PARAMETRIC OSCILLATIONS IN LITHIUM NIOBATE (60 DB GAIN). (BACKWARD WAVE		11-1837
STRACT ON/ INFORMATION ON HIGH DIELECTRIC CONSTANT MATERIALS GAINED BY ELECTRON PARAMAGNETIC RESONANCE. (AB		02-0101
/AR (PYROELECTRIC) CRYSTALS (LITHIUM NIOBATE OR TANTALATE GALLATE, BARIUM TITANATE, ZINC OXIDE, CADMIUM/		02-0095
TE FILMS PREPARED BY RF SPUTTERING ONTO INDIUM ANTIMONIDE OR GALLIUM ARSENIDE. BARIUM TITANA		04-0741
AMPLIFIER. LOW POWER DENSITY GALLIUM ARSENIDE- LITHIUM NIOBATE SURFACE WAVE		11-1825
BARIUM STRONTIUM NIOBATE, LEAD NIOBIUM(4) OXIDE(11), SILVER GALLIUM DISULFIDE, SILICON CARBIDE) /TITANATE,		05-1066
OBATE OR TANTALATE, BARIUM TITANATE, ZINC OXIDE, AND LITHIUM GALLIUM OXIDE). /OF POLAR CRYSTALS (LITHIUM NI		02-0098
ESR OF GAMMA-IRRADIATED AMMONIUM SULFATE.		19-2790
CURRENT EXCITATION IN GAMMA-IRRADIATED BARIUM TITANATE.		04-0884
T ROOM TEMPERATURE. ESR STUDY OF GAMMA-IRRADIATED CESIUM TRIHYDROGEN SELENITE A		22-2833
E. MICROWAVE DIELECTRIC MEASUREMENTS OF GAMMA-IRRADIATED FERROELECTRIC AMMONIUM SULFAT		19-2750
HELLE SALT, TR/ PHASE STABILIZATION (HINDERED TRANSITION) IN GAMMA-IRRADIATED FERROELECTRIC MATERIALS. (ROC		27-3165
THE RADICAL IN GAMMA-IRRADIATED KDP.		17-2406
NDENCE). DIELECTRIC PROPERTIES OF A (SMALL DOSE) GAMMA-IRRADIATED KDP CRYSTAL (TEMPERATURE DEPE		17-2466
E CRYSTAL. DIELECTRIC PROPERTIES OF A GAMMA-IRRADIATED POTASSIUM DIHYDROGEN PHOSPHAT		17-2467
E CRYSTALS. DIELECTRIC PROPERTIES OF GAMMA-IRRADIATED POTASSIUM DIHYDROGEN PHOSPHAT		17-2542
L PHASE CHANGES). ESR STUDY OF GAMMA-IRRADIATED POTASSIUM SELENATE (STRUCTURA		22-2825
ELECTRON PARAMAGNETIC RESONANCE OF GAMMA-IRRADIATED ROCHELLE SALT.		28-3180
PIEZOELECTRIC EFFECT AND INTERNAL FRICTION IN GAMMA-IRRADIATED ROCHELLE SALT CRYSTALS.		28-3229
IC LITHIUM TRIHYDROGEN DISELENITE. PARAMAGNETIC SPECIES IN GAMMA-IRRADIATED SINGLE CRYSTAL OF FERROELECTR		22-2843
IC LITHIUM TRIHYDROGEN D/ ELECTRON PARAMAGNETIC RESONANCE IN GAMMA-IRRADIATED SINGLE CRYSTAL OF FERROELECTR		22-2842
RADICAL IN TRIGLYCINE SULFATE NEAR THE CURIE POINT. (ESR OF GAMMA-IRRADIATED SPECIMENS) /E MOVEMENT OF THE		27-3033
ESR IN GAMMA-IRRADIATED TGS.		27-3035
DOMAIN STRUCTURE OF GAMMA-IRRADIATED TGS CRYSTALS.		27-3164
OF HYDROGEN AND THE ACTIVATION ENERGY IN GLYCINE. ESR OF GAMMA-IRRADIATED TGS. PART-2: INTERNAL MOTION		27-3034
ELECTRON SPIN RESONANCE OF GAMMA-IRRADIATED TRIGLYCINE SELENATE.		27-3139
D TO ST/ EFFECTS OF (MONOENERGETIC) ELECTRON AND (COBALT-60) GAMMA-IRRADIATION ON BARIUM TITANATE. (COMPARE		04-0970
OELECTRIC MATERIALS (TGS AND ROCHELLE SALT). INFLUENCE OF GAMMA-IRRADIATION ON BARKHAUSEN EFFECT IN FERR		27-3123
OF KDP. EFFECT OF GAMMA-IRRADIATION ON THE DIELECTRIC CONSTANTS		17-2691
GAMMA-RADIATION EFFECTS IN LITHIUM NIOBATE.		11-1822
RIGLYCINE SULFATE. (ON ELE/ RESULTS OF EXPERIMENTAL STUDY OF GAMMA-RADIATION EFFECTS IN ROCHELLE SALT AND T		27-3101
IGLYCINE SULFATE. EFFECT OF X- AND GAMMA-RADIATION ON THE SWITCHING PROCESS IN TR		27-3019
ALT AND GUANIDINE ALUMINU/ THERMOLUMINESCENCE STUDIES OF THE GAMMA-RAY IRRADIATED FERROELECTRICS ROCHELLE S		21-2812
NITRITE. INFRARED ABSORPTION IN GAMMA-RAY IRRADIATED SINGLE CRYSTAL OF SODIUM		16-2336
DIELECTRIC BEHAVIOR OF GAMMA-RAY IRRADIATED TRIGLYCINE SULFATE.		27-3149
LS. ELECTRONIC EFFECTS (OF X-RAY AND GAMMA-RAY IRRADIATION) IN ROCHELLE SALT CRYSTA		28-3192
COERCIVE FIELDS OF ROCHELLE SALT HEAVILY IRRADIATED WITH GAMMA-RAYS.		28-3178
TRIGLYCINE SULFATE CRYSTALS EXPOSED TO SMALL DOSES OF X- AND GAMMA-RAYS. DIELECTRIC LOSSES OF		27-2976
LUMINESCENCE OF POTASSIUM NITRATE IRRADIATED WITH GAMMA-RAYS.		16-2335
S OF THE/ CHANGE OF RESONANCE ABSORPTION SPECTRA OF 23.8 KEV GAMMA-RAYS OF TIN-119 DURING PHASE TRANSISTION		08-1400
ACTERISTICS OF FERROELECTRIC TRIGLYCINE SE/ (OPTICAL) ENERGY GAP (AND DIELECTRIC CONSTANT) TEMPERATURE CHAR		27-3021
THEORY OF VIBRONIC PHASE TRANSFORMATIONS IN WIDE GAP FERROELECTRICS.		03-0349
DETERMINATION OF THE NATURE OF THE OPTICAL GAP OF STRONTIUM TITANATE.		06-1108
ACH TAIL BEHAVIOR) OPTICAL GAP OF STRONTIUM TITANATE. (DEVIATION FROM URB		06-1111
YLLATE, TRIGLYCINE SULFATE AND ITS VARIATION UNDER X-I/ BAND GAP OF TRIGLYCINE SELENATE, TRIGLYCINE FLUOBER		27-3124
/ATURE ELECTRICAL CONDUCTIVITY MEASUREMENTS. (INTRINSIC BAND GAP, PTCR (POSITIVE TEMPERATURE COEFFICIENT O/		04-0822
AMIC MATERIALS AT MICROWAVE FREQUENCIES. (TITANATE ZIRCONATE GARNET MIXTURES) /RICAL PROPERTIES OF SOME CER		09-1602
HARMONIC SOURCE. NEODYMIUM: YTTRIUM ALUMINUM GARNET- BARIUM SODIUM NIOBATE 0.66 MILLIMICRON		13-1975
ODIDE, NICKEL CARBONATE AND BLACK PHOSPHORUS. HIGH GAS PRESSURE CRYSTAL GROWTH OF ANTIMONY SULFOI		15-2137
ATE. SEE BOTH GASH AND GUANIDINIUM ALUMINUM SULFATE HEXAHYDR		
CLEAVED TG/ AFTER EFFECTS IN THE FIELD EFFECT AT VERY LARGE GATE CHARGES APPLIED TO TELLURIUM FILMS ON UHV		27-3171
FERROELECTRIC CERAMIC LIGHT GATES OPERATED IN A VOLTAGE CONTROLLED MODE.		37-3654
HARACTERISTICS AND THE BIT SIZE OF THE FE-PC MEMO/ EFFECT OF GAUSSIAN IRRADIANCE PROFILE ON THE SWITCHING C		37-3670
GROWTH OF POTASSIUM DIHYDROGEN PHOSPHATE SINGLE CRYSTALS IN GEL.		17-2455
GEL GROWTH OF SILVER PERIODATE CRYSTALS.		36-3523
RATE CRYSTALS. SOME CHEMICAL REACTIONS IN SILICA GELS. PART-3: FORMATION OF POTASSIUM ACID TART		29-3243
S AND THEIR APPLICATION TO FERROELECTRIC (LAMBDA) PHASE TRA/ GENERALIZATION OF THE PIPPARD-GARLAND RELATION		03-0299
TTICE. GENERALIZED FERROELECTRIC MODEL ON A SQUARE LA		03-0168
PROPERTIES OF FERROELECTRICS. (ABSTRACT ONLY, DISCUSSION) GENERALIZED SUSCEPTIBILITY AND THE SCATTERING		03-0209
ANATE CERAMICS. ELECTROSTRICTIVELY GENERATED AND MODULATED FRACTURE IN BARIUM TIT		04-0700
TRANSDUCERS. ACOUSTIC TRANSIENTS GENERATED BY ANTIFERROELECTRIC- FERROELECTRIC		37-3672
AGR/ MEASUREMENTS OF PHOTON CORRELATIONS OF SECOND HARMONIC GENERATED LIGHT (FROM LITHIUM NIOBATE CRYSTAL,		11-1701
OGEN PHOSPHATE. ELECTROOPTIC AMPLITUDE MODULATION OF LASER GENERATED SECOND HARMONICS IN POTASSIUM DIHYDR		17-2694
REVERSAL OF A FERROELECTRIC. POSSIBILITY OF GENERATING INFRARED RADIATION BY POLARIZATION L		03-0162
CNG RANGE ORDER IN SODIUM NITRITE BY OPTICAL SECOND HARMONIC GENERATION. L		16-2314
NZE SYSTEM) POTASSIUM SODIUM BARIUM NIOBATE (SECOND HARMONIC GENERATION). / PROPERTIES OF (THE TUNGSTEN BRO		13-2028
IN FER/ NONLINEAR OPTICAL DEVICES. (BASED IN SECOND HARMONIC GENERATION AND OPTICAL PARAMETRIC OSCILLATION		02-0097
IN LITHIUM NIOBATE CRYSTALS. (80-950 DEG K, 200-2000 MHZ) GENERATION AND PROPAGATION OF HYPERSONIC WAVES		11-1753
CLAMPED VALUES OF THE DIELE/ MEASUREMENTS OF SECOND HARMONIC GENERATION AND THE VARIATIONS IN THE FREE AND		13-1998
MELT COMPOSITION. DEPENDENCE OF SECOND HARMONIC GENERATION COEFFICIENTS OF LITHIUM NIOBATE ON		11-1769

ALS.   ABSOLUTE SIGNS OF SECOND HARMONIC GENERATION COEFFICIENTS OF PIEZOELECTRIC CRYST 02-0096
A/ FURTHER MEASUREMENTS OF ABSOLUTE SIGNS OF SECOND HARMONIC GENERATION COEFFICIENTS OF PIEZOELECTRIC CRYST 05-1066
C) CRY/ ABSOLUTE SIGNS OF NONLINEAR OPTICAL (SECOND HARMONIC GENERATION) COEFFICIENTS OF POLAR (PYROELECTRI 02-0095
EVERSAL. (SODIUM NITRITE)   OPTICAL SECOND HARMONIC GENERATION DURING FERROELECTRIC POLARIZATION R 16-2384
RROELECTRIC) MATERIALS.   ENERGY GENERATION FROM IMPACT ON PIEZOELECTRIC (OR FE 03-0222
THE VISIBLE SPECTRUM.   EFFICIENT HIGH GRAIN PARAMETRIC GENERATION IN ADP CONTINUOUSLY TUNABLE ACROSS 17-2712
LASERS.   EFFICIENT SECOND HARMONIC GENERATION IN ADP WITH TWO NEW FLUORESCEIN DYE 17-2538
                            OPTICAL SECOND HARMONIC GENERATION IN ANTIMONY SULFOIODIDE. 15-2164
, AND POTASSIUM DI/ QUANTITATIVE STUDIES OF OPTICAL HARMONIC GENERATION IN CADMIUM SULFIDE, BARIUM TITANATE 04-0852
                                           HARMONICS GENERATION IN FERROELECTRIC MEDIA. 03-0215
                            SECOND HARMONIC GENERATION IN FERROELECTRIC SODIUM NITRITE. 16-2385
TEMPERATURE DEPENDENCE OF RAMAN SPECTRUM AND SECOND HARMONIC GENERATION IN KDP. 17-2550
LINEAR OPTICAL COEFFICIENT AND PHASE MATCHED SECOND HARMONIC GENERATION IN LITHIUM IODATE.   LARGE NON 08-1405
OPTICAL PROBING OF ACOUSTIC SURFACE WAVE HARMONIC GENERATION (IN LITHIUM NIOBATE) 11-1752
                           OPTIMUM SECOND HARMONIC GENERATION IN LITHIUM NIOBATE. 11-1764
SIMULTANEOUS OPTICAL PARAMETRIC OSCILLATION, SECOND HARMONIC GENERATION (IN LITHIUM NIOBATE). 11-1859
BDATE AND TERBIUM MOLYBDATE.   SECOND HARMONIC GENERATION IN MOLYBDATES (BETA GADOLINIUM MOLY 30-3286
E)   OPTICAL SECOND HARMONIC GENERATION IN PIEZOELECTRIC CRYSTALS. (KDP TYP 17-2582
OSCILLATOR. (/ ACOUSTOOPTICAL MEASUREMENTS OF ACOUSTIC WAVE GENERATION IN PIEZOELECTRICS COUPLED TO A GUNN 11-1731
FTER COOLING ST/ PYROELECTRICITY AND OPTICAL SECOND HARMONIC GENERATION IN POLYVINYLIDENE FLUORIDE FILMS (A 36-3530
AND DEU/ TEMPERATURE DEPENDENCE OF OPTICAL (SECOND) HARMONIC GENERATION IN POTASSIUM DIHYDROGEN PHOSPHATE (17-2695
/E DEPENDENCE OF INDEX MATCHING ANGLE OF THE SECOND HARMONIC GENERATION IN POTASSIUM DIHYDROGEN PHOSPHATE / 17-2544
       TEMPERATURE DEPENDENCE OF SECOND HARMONIC GENERATION IN SODIUM NITRITE. 16-2315
ONTIUM TITANATE.   SECOND HARMONIC GENERATION IN STRESS INDUCED FERROELECTRIC STR 06-1131
TANTALATE.   ELECTRIC FIELD INDUCED OPTICAL HARMONIC GENERATION IN STRONTIUM TITANATE AND POTASSIUM 06-1203
LS (NONLINEAR ELASTIC PROPERTIES). ULTRASONIC THIRD HARMONIC GENERATION IN STRONTIUM TITANATE SINGLE CRYSTA 06-1181
RAELECTRIC PHASE.   SECOND HARMONIC GENERATION IN THE TRIGLYCINE CRYSTAL IN THE PA 27-3105
                           SECOND HARMONIC GENERATION IN TRIGLYCINE SULFATE CRYSTALS. 27-3138
IRON IODINE AND NICKEL CHLORINE) BORACITES (SECOND HARMONIC GENERATION INDEX MATCHING). /NE, IRON CHLORINE 14-2080
CY ACOUSTIC WAVES. APPLICATION OF ACOUSTIC WAVES TO HARMONIC GENERATION LITHIUM NIOBATE. /N BY HIGH FREQUEN 11-1834
E REFRACTIVE INDEX OF A NONLINEAR CRYSTAL ON SECOND HARMONIC GENERATION. (LITHIUM NIOBATE) /MOGENEITY OF TH 11-1687
OSPHATE CRYSTALS OF LIGHT FROM A LASER OPERATING IN THE FREE GENERATION MODE. /E AND AMMONIUM DIHYDROGEN PH 17-2717
COND LIGHT PULSES IN LITHIUM NIOBATE.   GENERATION OF FAR INFRARED RADIATION BY PICOSE 11-1857
STAL WITH A SECOND ORDER TRANSITION ABOVE THE C/ QUASISTATIC GENERATION OF HARMONICS IN A FERROELECTRIC CRY 03-0331
INDEX MATCHING METHOD.   SECOND HARMONIC GENERATION OF LIGHT IN THE KDP CRYSTAL BY THE 17-2693
TRANSVERSE ACOUSTIC WAVE PROPAGATING ITSELF IN THE (001) DI/ GENERATION OF SECOND AND THIRD HARMONICS BY A 11-1682
UBIC PHASE OF BARIUM TITANATE.   GENERATION OF SECOND OPTICAL HARMONIC IN THE C 04-0538
ECORDING IN CR/ ELECTRON THERMAL DIFFUSION EFFECTS (INTERNAL GENERATION OF STRONG FIELDS) DURING HOLOGRAM R 06-1090
MONICS IN A (SINGLE DOMAIN) LITHIUM NIOBATE CRYSTAL (NONLIN/ GENERATION OF SUPERHIGH FREQUENCY ACOUSTIC HAR 11-1707
ITHIUM NIOBATE)   THEORETICAL STUDY OF THE GENERATION OF THE ACOUSTIC SECOND HARMONIC. (L 11-1772
ANTIMONY SULFOIODIDE SINGLE CRYSTALS.   GENERATION OF THE SECOND ACOUSTIC HARMONIC IN 15-2224
ANTIMONY SULFOIODIDE SINGLE CRYSTALS.   GENERATION OF THE SECOND ACOUSTIC HARMONIC IN 15-2223
CTRICITY. (THEORY)   GENERATION OF THE SECOND HARMONIC AND FERROELE 02-0063
E SULFATE. (84 REFS)   GENERATION OF THE SECOND HARMONIC BY TRIGLYCIN 27-2990
RMONICS OF 1.06 MICRON RADI/ OBSERVATION OF THE SIMULTANEOUS GENERATION OF THE SECOND, THIRD, AND FOURTH HA 11-1779
INAL HIGH FREQUENCY (950 MHZ) ACOUSTIC WAVE ALONG THE (001)/ GENERATION OF THE THIRD HARMONIC OF A LONGITUD 11-1835
D TEST ON BARIUM SODIUM NIOBA/ DEPENDENCE OF SECOND HARMONIC GENERATION ON CRYSTAL INHOMOGENEITY (THEORY AN 13-2043
TICAL PROPERTIES OF FERROELECTRIC CRYSTALS. (SECOND HARMONIC GENERATION, OPTICAL PARAMETRIC OSCILLATION) /P 03-0383
      PULSE DISTORTIONS IN MISMATCHED SECOND HARMONIC GENERATION. (POTASSIUM DIHYDROGEN PHOSPHATE) 17-2690
UDY OF PHASE TRANSITION OF SODIUM NITRITE BY SECOND HARMONIC GENERATION. (TEMPERATURE DEPENDENCE)   ST 16-2403
      INTRACAVITY SECOND HARMONIC GENERATION WITH RUBY LASER BY KDP. 17-2604
OF FERROELECTRIC CRYSTALS.   GENERATOR OF BIPOLAR PULSES FOR INVESTIGATION 37-3673
CTURE OF THE BRONZE SODIUM(13) NIOBIUM(35) OXIDE(94) AND THE GEOMETRY OF FERROELECTRIC DOMAINS.   STRU 13-1931
HRALSKI SYNTHESIS AND PROPERTIES OF RARE EARTH DOPED BISMUTH GERMANATE.   CZOC 35-3496
FHASE EQUILIBRIA IN THE SYSTEM BARIUM TITANATE- BARIUM GERMANATE. 04-0710
QUASI ELASTIC SCATTERING IN LEAD GERMANATE. 35-3497
RSONIC PULSES IN PIEZOELECTRIC LITHIUM NIOBATE AND BISMUTH BIGERMANATE. /F LONGITUDINAL AND TRANSVERSE HYPE 11-1814
TE NOT FERROELECTRIC) ELECTROOPTIC MEASUREMENTS ON LITHIUM GERMANATE AND LEAD GERMANATE. (LITHIUM GERMANA 35-3494
.   BISMUTH GERMANATE, SEE BISMUTH(12) GERMANIUM OXIDE(20)
                   LEAD GERMANATE, SEE LEAD(5) GERMANIUM(3) OXIDE(11).
ELECTROOPTIC PROPERTIES OF FERROELECTRIC LEAD GERMANATE SINGLE CRYSTAL. 35-3518
GRAPHIC OBSERVATION OF 180 DEG DOMAINS IN FERROELECTRIC LEAD GERMANATE SINGLE CRYSTALS.   X-RAY TOPO 35-3515
THE STRUCTURE AND PROPERTIES OF FERROELECTRIC LEAD TITANATE (GERMANATE SOLID SOLUTIONS TO 7 OR 8 MO1). /ON 05-1043
NSTOICHIOMETRIC MELTS. (BARIUM TITANATE, STRONTIUM TITANATE, GERMANATES) / GROWTH OF OXIDE CRYSTALS FROM NO 04-0553
ON LEAD ZIRCONATE TITANATE WITH COMPLEX ADDITIVES CONTAINING GERMANIUM.   PIEZOCERAMICS BASED 09-1510
GROWTH OF SINGLE CRYSTALS IN LEAD MONOXIDE- GERMANIUM DIOXIDE BINARY SYSTEM. 35-3514
ERROELECTRIC LEAD TITANATE (GERMANATE SOLID SOLUT/ EFFECT OF GERMANIUM ON THE STRUCTURE AND PROPERTIES OF F 05-1043
/ OF FERROELECTRIC (BARIUM TITANATE) POLARIZATION FIELDS ON (GERMANIUM OR TELLURIUM) SEMICONDUCTOR FILMS. / 04-0594
CRYSTALS OF CADMIUM SULFIDE, LITHIUM NIOBATE AND BISMUTH(12) GERMANIUM OXIDE(20). /ACE WAVE CONVOLUTION ON 11-1692
CRYSTAL STRUCTURE OF PIEZOELECTRIC BISMUTH(12) GERMANIUM OXIDE(20). 35-3493
STIC, PIEZOELECTRIC, AND DIELECTRIC CONSTANTS OF BISMUTH(12) GERMANIUM OXIDE(20).   ELA 35-3512
T OSCILLATION OF SPACE CHARGE LIMITED CURRENT IN BISMUTH(12) GERMANIUM OXIDE(20). / ELECTRIC FIELD DEPENDEN 35-3505
ON PIEZOELECTRIC SUBSTRATES (LITHIUM NIOBATE AND BISMUTH(12) GERMANIUM OXIDE(20)). / ELASTIC SURFACE WAVES 11-1751
R MIXING IN ANISOTROPIC MEDIA. (LITHIUM NIOBATE, BISMUTH(12) GERMANIUM OXIDE(20)) /URFACE WAVES BY NONLINEA 11-1778
RAMAN SPECTRUM OF BISMUTH(12) GERMANIUM OXIDE(20). 35-3519
UMINESCENCE IN BISMUTH(12) SILICON OXIDE(20) AND BISMUTH(12) GERMANIUM OXIDE(20). /TIMULATED CURRENTS AND L 36-3585
/FACE WAVES IN CUBIC AND ORTHORHOMBIC CRYSTALS. (BISMUTH(12) GERMANIUM OXIDE(20) AND BARIUM SODIUM NIOBATE) 13-2041
ULTRASONIC ATTENUATION IN SINGLE CRYSTALS OF BISMUTH(12) GERMANIUM OXIDE(20) AT LOW TEMPERATURES. 35-3510
ACOUSTIC SURFACE WAVE LOSS MECHANISMS ON BISMUTH(12) GERMANIUM OXIDE(20) AT MICROWAVE FREQUENCIES. 35-3511
/URFACE WAVE VELOCITIES. (BARIUM SODIUM NIOBATE, BISMUTH(12) GERMANIUM OXIDE(20), LITHIUM NIOBATE, LITHIUM/ 01-0028
SE HYPERSONIC WAVE PULSES IN LITHIUM NIOBATE AND BISMUTH(12) GERMANIUM OXIDE(20) PIEZOELECTRIC CRYSTALS. /R 11-1815
US CHLORIDE, PROUSTITE, CALCIUM PYRONIOBATE, AND BISMUTH(12) GERMANIUM OXIDE(20). (REVIEW) / NIOBATE, CUPRO 01-0021
THE SURFACE WAVE AMPLIFIER. A THEORETICAL AND E/ BISMUTH(12) GERMANIUM OXIDE(20) VERSUS LITHIUM NIOBATE IN 11-1720
/5) GERMANIUM(3) OXIDE(11) AND ITS ISOMORPHOUS COMPOUND LEAD GERMANIUM SILICON OXIDE. (DIELECTRIC CONSTANT/ 35-3501
ASSIUM BROMIDE, LEAD TELLURIDE, TIN TELLURIDE, LEAD SULFIDE, GERMANIUM TELLURIDE. /IELECTRIC CONSTANT. (POT 36-3541
G)   TRANSITION IN TIN TELLURIDE- GERMANIUM TELLURIDE ALLOYS. (NEUTRON SCATTERIN 36-3586
ITIONS (TIN TELLURIDE PURE AND ALLOYED WITH SMALL AMOUNTS OF GERMANIUM TELLURIDE OR LEAD TELLURIDE). /TRANS 36-3564
ACOUSTOOPTIC PROPERTY OF SINGLE CRYSTAL LEAD(5) GERMANIUM(3) OXIDE(11). 35-3509
CRYSTAL STRUCTURE OF FERROELECTRIC LEAD(5) GERMANIUM(3) OXIDE(11). 35-3504
OELECTRIC OPTICAL ROTATION DOMAINS IN SINGLE CRYSTAL LEAD(5) GERMANIUM(3) OXIDE(11).   FERR 36-3550

UTH TITANATE, LITHIUM THALLIUM TARTRATE MONOHYDRATE, LEAD(5) GERMANIUM(3) OXIDE(11)) /CTRIC CRYSTALS. (BISM 02-0061
            FERROELECTRICITY IN LEAD(5) GERMANIUM(3) OXIDE(11).   35-3508
    NONLINEAR OPTICAL PROPERTIES OF FERROELECTRIC LEAD(5) GERMANIUM(3) OXIDE(11).   35-3506
  RAMAN SCATTERING FROM THE FERROELECTRIC MODE IN LEAD(5) GERMANIUM(3) OXIDE(11).   35-3498
      SOFT OPTIC PHONON MODE IN FERROELECTRIC LEAD(5) GERMANIUM(3) OXIDE(11).   35-3495
C OBSERVATION OF 180 DEGREE DOMAINS IN FERROELECTRIC LEAD(5) GERMANIUM(3) OXIDE(11).   X-RAY TOPOGRAPHI 35-3516
POUND LEAD / FERROELECTRIC AND OPTICAL PROPERTIES OF LEAD(5) GERMANIUM(3) OXIDE(11) AND ITS ISOMORPHOUS COM 35-3501
LS. FERROELECTRIC PROPERTY AND ENANTIOMORPHISM OF LEAD(5) GERMANIUM(3) OXIDE(11) AND THE RELATED MATERIA 35-3499
            FERROELECTRIC LEAD(5) GERMANIUM(3) OXIDE(11) CRYSTAL.   35-3500
CTRIC. (DIELECTRIC CONSTANTS AND REFRACTIVE INDICES) LEAD(5) GERMANIUM(3) OXIDE(11) CRYSTAL, A NEW FERROELE 35-3502
LASTIC AND PIEZOELECTRIC PROPERTIES OF FERROELECTRIC LEAD(5) GERMANIUM(3) OXIDE(11) CRYSTALS. E 35-3520
SWITCHING OF OPTICAL ROTATORY POWER IN FERROELECTRIC LEAD(5) GERMANIUM(3) OXIDE(11) SINGLE CRYSTAL. 36-3570
     CRYSTAL GROWTH AND PROPERTIES OF LEAD(5) GERMANIUM(3) OXIDE(11) SINGLE CRYSTALS. 35-3513
     OPTICAL ACTIVITY OF FERROELECTRIC LEAD(5) GERMANIUM(3) OXIDE(11) SINGLE CRYSTALS. 35-3503
            ELASTIC GIBBS FUNCTION FOR A FERROELECTRIC CRYSTAL.   03-0328
       CRYSTALLIZATION OF LEAD METANIOBATE FROM GLASS.   08-1275
   CRYSTALLIZATION OF SOLID SOLUTIONS OF SODIUM NIOBATE FROM GLASS.   08-1386
200 DEG C BELOW THE CUR/ FIELD ENFORCED FERROELECTRICITY IN GLASS BONDED LEAD ZIRCONATE. (PHASE TRANSITION 07-1252
ELECTRIC AND OPTICAL PROPERTIES OF TRANSPARENT FERROELECTRIC GLASS CERAMIC SYSTEMS.   DI 04-0576
   THREE APPLICATION AREAS FOR STRONTIUM TITANATE GLASS CERAMICS.   06-1150
/ARIUM TITANATE CERAMICS. PART-2: EFFECTS OF THE ADDITION OF GLASS POWDERS CONTAINING BARIUM OXIDE AND TIT/ 04-0829
HE PHASE TRANSFORMATION BETWEEN THE C/ CALORIMETRIC STUDY OF GLASS TRANSITION OF THE AMORPHOUS ICE AND OF T 34-3476
/ MICROCRYSTALLINE FERROELECTRIC MATERIALS CRYSTALLIZED FROM GLASS. (60-90 VOL PERCENT, FERROELECTRIC, COM/ 04-0725
PPLICATION OF DIELECTRIC MIXTURE FORMULAS TO (LEAD TITANATE) GLASS- CERAMIC SYSTEMS.   A 05-1054
ALS (BARIUM OR LEAD TITANATE,/ PREPARATION AND PROPERTIES OF GLASS- CERAMICS CONTAINING FERROELECTRIC CRYST 04-0783
/TH HIGH DIELECTRIC CONSTANT. PART-1: EFFECTS OF ADDITION OF GLASSES CONTAINING BARIUM OXIDE AND TITANIUM / 04-0829
TION AND SOME OPTICAL PROPERTIES OF ANTIMONY SULFOBROMIDE IN GLASSY AND CRYSTALLINE STATES. PRODUC 15-2142
     RELAXATIONAL PROTON ORDERING AND GLASSY CRYSTALLINE STATE IN HEXAGONAL ICE. 34-3436
-2: INTERNAL MOTION OF HYDROGEN AND THE ACTIVATION ENERGY IN GLYCINE. ESR OF GAMMA-IRRADIATED TGS. PART 27-3034
COPPER(2+), AND MANGANESE(2+) PARAMAGNETIC ION COMPLEXES IN GLYCINE CRYSTALS. EPR STUDIES OF CHROMIUM(3+), 27-3135
   CRYSTAL STRUCTURE AND PHASE TRANSITION OF DIGLYCINE NITRATE.   16-2312
ORIDE. (SHI/ NMR STUDY OF THE FERROELECTRIC TRANSITIONS IN DIGLYCINE NITRATE AND TRIS-SARCOSINE CALCIUM CHL 16-2267
    CRYSTAL STRUCTURE OF FERROELECTRIC GLYCINE SILVER NITRATE.   16-2297
      CRYSTAL STRUCTURE OF FERROELECTRIC GLYCINE SILVER NITRATE.   16-2348
    CRYSTAL STRUCTURE OF FERROELECTRIC GLYCINE SILVER NITRATE.   16-2333
C AND X-RAY STUDIES ON THE FERROELECTRIC PHASE TRANSITION OF GLYCINE SILVER NITRATE.   DIELECTRI 16-2347
ECTRIC CONSTANTS AND DIELECTRIC LOSSES OF SODIUM NITRITE AND GLYCINE SULFATE. /FERROELECTRICS. PART-2: DIEL 16-2354
CRYSTAL SPECIES.   DIGLYCINE SULFATE AN INTERESTING NEW DIELECTRIC 27-2993
ELECTRIC)   ORTHORHOMBIC DIGLYCINE SULFATE. (NOT PIEZOELECTRIC, NOT FERRO 36-3624
SINGLE CRYSTALS.   PREPARATION OF PURE GLYCINE USED FOR GROWING OF TRIGLYCINE SULFATE 27-3085
TICAL ABSORPTION, ELECTRIC CONDUCTIVITY AND PHOTOEMISSION OF GLYCOL SULFATE.   OP 19-2780
POGRAPHIC STUDY OF DISLOCATIONS AND FERROELECTRIC DOMAINS IN GLYCOL SULFATE.   X-RAY TO 19-2747
ATE.   PI- GMO: ANOTHER MODIFICATION OF GADOLINIUM MOLYBD 30-3263
UM TITANATE SINGLE CRYSTALS HEATED IN OXYGEN (EFFECT TIED TO GOLD ANODES). /ELECTRICAL CONDUCTIVITY OF BARI 04-0907
ON MIRROR MICROSCOPIC OBSERVATION OF FERROELECTRIC DOMAINS. (GOLD COATED BARIUM TITANATE) ELECTR 04-0974
/FRACTION AND REFLECTION OF ULTRASONIC SURFACE WAVES BY THIN GOLD FILM LAYERS ON LITHIUM NIOBATE SUBSTRATE/ 11-1789
/FRACTION AND REFLECTION OF ULTRASONIC SURFACE WAVES BY THIN GOLD FILM LAYERS ON LITHIUM NIOBATE SUBSTRATE/ 11-1790
SURFACE WAVE PROPAGATION ON LAYERED ANISOTROPIC SUBSTRATES: GOLD ON LITHIUM NIOBATE. /ND HIGHER ORDER MODE 11-1846
/OF SEMICONDUCTING COMPOUNDS OF GROUP A(V)B(VI)C(VII). (HIGH GRADE SINGLE CRYSTALS, V IS ANTIMONY OR BISMU/ 15-2214
ROELECTRIC SODIUM NITRITE. (AT THE SODIUM-23/ ELECTRIC FIELD GRADIENT AND LONG RANGE ORDER PARAMETER OF FER 16-2395
TUDIED BY PERTURBED ANGULAR COR/ THE INTERNAL ELECTRIC FIELD GRADIENT IN ANTIFERROELECTRIC LEAD ZIRCONATE S 07-1255
  TEMPERATURE DEPENDENCE OF THE ELECTRIC FIELD GRADIENT IN FERROELECTRIC SODIUM NITRITE. 16-2337
ISPERSION OF FERROELECTRICS NEAR THE/ (EFFECT OF TEMPERATURE GRADIENT WITHIN THE SAMPLE ON THE) MICROWAVE D 03-0370
DISTINCTI/ CALCULATION OF INTERNAL ELECTRIC FIELDS AND THEIR GRADIENTS IN ABO(3) PEROVSKITE COMPOUNDS WITH 08-1455
       CRYSTAL FIELD GRADIENTS IN ANTIMONY SULFOIODIDE.   15-2185
/STUDY AND THEORETICAL ESTIMATION OF INTERNAL ELECTRIC FIELD GRADIENTS IN FERROELECTRIC FERROMAGNETS. (PER/ 08-1401
HLORIDE.   ELECTRIC FIELD GRADIENTS IN IONIC CHLORIDES: CESIUM LEAD TRIC 33-3367
/ULATIONS OF THE INTERNAL ELECTRIC FIELDS AND ELECTRIC FIELD GRADIENTS IN THE PEROVSKITE TYPE COMPOUNDS WI/ 08-1442
YDRO/ METHOD OF GROWTH IN SOLUTION OF SINGLE CRYSTALS BY THE GRADUAL ADDITION OF A REACTANT. (POTASSIUM DIH 17-2595
TIVE CERAMIC/ MECHANISM FOR FORMING INSULATION LAYERS ON THE GRAIN BOUNDARIES OF BARIUM TITANATE SEMICONDUC 04-0867
OF (INSULATING, BOUNDARY LAYER) BARIUM TITANATE CERAMICS AT GRAIN BOUNDARY. FERROELECTRIC PROPERTIES 04-1024
MONY DOPED) BARIUM TITANATE CERAMIC WITH A HIGH EFFECTIVE D/ GRAIN BOUNDARY BARRIER LAYERS IN (N TYPE, ANTI 04-0581
/ANATE) FERROELECTRIC SEMICONDUCTORS. (ORIGIN AND EFFECTS OF GRAIN BOUNDARY BLOCKING LAYERS, ANTIMONY DOPA/ 02-0074
HANUM DOPED BARIUM TITANATE CERAMICS WITH BISMUTH OXIDE AS A GRAIN BOUNDARY ELEMENT.   LANT 04-0799
    CHARACTERIZATION OF SURFACE AND GRAIN BOUNDARY LAYER OF BARIUM TITANATE. 04-0701
LEAD TITANATE) IMAGE STORAGE AND DISPLAY DEVICES USING FINE GRAIN FERROELECTRIC CERAMICS. (LEAD ZIRCONATE- 09-1572
M TITANATE TYPE) CERAMICS. (IN RELATION TO DENSIFICATION AND GRAIN GROWTH) /OT PRESSED FERROELECTRIC (BARIU 04-0879
DENSE PLZT CERAMICS.   ISOTHERMAL GRAIN GROWTH AND ELECTRICAL BEHAVIOR OF FULLY 09-1553
UM TITANATE.   EXAGGERATED GRAIN GROWTH IN LIQUID PHASE SINTERING OF BARI 04-0835
UM DOPED BARIUM TITANATE.   INFLUENCE OF GRAIN GROWTH ON DIELECTRIC PROPERTIES OF NIOBI 04-0763
BARIUM TITANATE CERAMICS.   EFFECTS OF GRAIN GROWTH ON THE DISTRIBUTION OF NIOBIUM IN 04-0595
Y TUNABLE ACROSS THE VISIBLE SPECTRUM. EFFICIENT HIGH GRAIN PARAMETRIC GENERATION IN ADP CONTINUOUSL 17-2712
ICAL PROPERTIES OF SINTERED BARIUM TITANATE AS A FUNCTION OF GRAIN SIZE. VARIATION IN CERTAIN ELECTR 04-0627
OPTICAL PROPERTIES OF TRANSPARENT FERROELECTRIC / EFFECTS OF GRAIN SIZE AND POROSITY ON THE ELECTRICAL AND 13-2004
URAL CHARACTERISTICS OF BARIUM TITANATE. GRAIN SIZE DEPENDENCE OF DIELECTRIC AND STRUCT 04-0582
TIC CERAMICS (HOT PRESSED LEAD ZIRCONATE TITANATE, EFFECT OF GRAIN SIZE, DOPANTS, REVIEW). /TS IN ELECTROOP 02-0086
RANSITION OF BARIUM TITANATE.   EFFECT OF GRAIN SIZE ON THE FERROELECTRIC PARAELECTRIC T 04-0997
IC BARIUM TITANATE. EFFECT OF CRYSTALLINE GRAIN SIZE ON THE PHYSICAL PROPERTIES OF CERAM 04-0785
OPERTIES OF BARIUM TITANATE.   EFFECT OF GRAIN SIZE ON THE PIEZOELECTRIC AND ELASTIC PR 04-1011
ONDUCTIVE BARIUM TITANATE CERAMICS. EFFECT OF GRAIN SIZE ON THE RESISTIVITY ANOMALY IN SEMIC 04-0735
M DOPED BARIUM TITANATE. PREPARATION OF SMALL GRAINED AND LARGE GRAINED CERAMICS FROM NIOBIU 04-0765
  SOME FATIGUING EFFECTS IN 8/65/35 PLZT FINE GRAINED FERROELECTRIC CERAMIC.   08-1432
RIUM / CALCULATION OF THE ELECTROSTATIC INTERACTIONS IN FINE GRAINED PEROVSKITE FERROELECTRIC CERAMICS. (BA 04-0642
ION. ELECTROOPTIC POTASSIUM TANTALATE NIOBATE GRATINGS FOR LIGHT BEAM MODULATION AND DEFLECT 13-1929
ER OF CRYSTALLIZATION OF ROCHELLE SALT. GRAVIMETRIC MEASUREMENTS OF THE CHANGES OF WAT 28-3235
E TYPE MATERIALS) LOW LOSS (FERROELECTRIC CERAMICS) WITH GREAT DIELECTRIC PERMITTIVITY. (BARIUM TITANAT 04-0779
FECT OF HEAT TREATMENT ON TUNGSTEN TRIOXIDE CRYSTAL. PART-2: GREEN STRIPES.   AN EF 10-1639
  ACOUSTIC RADIATION BY INTERDIGITAL GRIDS ON LITHIUM NIOBATE.   11-1699
ELECTRIC HYSTERESIS TRACER FEATURING COMPENSATION AND SAMPLE GROUNDING.   FERRO 37-3668
PROPERTIES OF FERROELECTRIC CRYSTALS THE TRIGLYCINE SULFATE GROUP. /TERATION EFFECT ON THERMAL AND ELASTIC 27-2986

YDRATE ABOVE THE CURIE TEMPERATURE (LATTICE CONSTANTS, SPACE GROUP). /C POTASSIUM FERROCYANIDE TRIDEUTERO H 24-2903
OF RAMAN SCATTERED LASER LIGHT IN FERROELECTRICS OF THE KDP GROUP. SPECTRUM 17-2625
FERROELECTRIC CRYSTALS OF THE POTASSIUM DIHYDROGEN PHOSPHATE GROUP. STRUCTURE OF 17-2607
HYDROTHERMAL CRYSTALLIZATION OF SEMICONDUCTING COMPOUNDS OF GROUP A(V)B(VI)C(VII). (HIGH GRADE SINGLE CRY/ 15-2214
PEROVSKITE ABO(3). GROUP ANALYSIS OF LATTICE VIBRATIONS OF CUBIC 08-1321
AGNETIC BARIUM COBALT TETRAFLUORIDE CRYSTAL STRUCTURE (SPACE GROUP AND LATTICE CONSTANTS). /RAELASTIC PARAM 33-3386
NIOBATE AND BARIUM SODIUM TANTALATE (FORCE CONSTANTS, SPACE GROUP ASSIGNMENT). /IUM NIOBATE, BARIUM SODIUM 11-1808
HIUM NIOBATE. PROPERTIES OF THE IRON TRANSITION GROUP IN THE LATTICE OF SINGLE CRYSTALLINE LIT 11-1669
FERROELECTRIC CRYSTALS OF THE POTASSIUM DIHYDROGEN PHOSPHATE GROUP IN THE PARAELECTRIC PHASE. /OPERTIES OF 27-2696
TITUTION IN FERROELECTRIC CRYSTALS OF THE TRIGLYCINE SULFATE GROUP. (NORMAL AND DEUTERATED) /OMORPHOUS SUBS 27-3102
NOTE ON THE SPACE GROUP OF POTASSIUM HYDROGEN IODATE. 08-1370
FORMATE TETRAHYDRATE. UNIT CELL AND SPACE GROUP OF THE ANTIFERROELECTRIC PHASE OF COPPER 26-2927
4) TYPE COMPOUNDS (FROTON MAGN/ FERROELECTRICS WITH AMMONIUM GROUP ORIENTATION MOBILITY: METAL AMMONIUM (BX 22-2826
DE: TEMPERATURE DISCONTINUITIES IN THE INFRARED SPECTRUM AND GROUP THEORETICAL ANALYSIS. /NGANESE TRIFLUORI 33-3363
C NEUTRON SCATTERING WITH APPLICATIONS TO LITHIUM NIOBATE. GROUP THEORETICAL SELECTION RULES FOR INELASTI 11-1704
ERTIES OF BARIUM TITANATE SINGLE CRYSTALS DOPED BY HYDROXIDE GROUPS. DIELECTRIC PROP 04-0634
BARIUM TITANAT/ FERROELECTRIC AND ANTIFERROELECTRIC SYMMETRY GROUPS. (AND THEIR COMPARISON, APPLICATION TO 04-0933
C PHASE TRANSITION IN CRYSTALS WITH REORIENTATING STRUCTURAL GROUPS. (SODIUM AMMONIUM SELENATE DIHYDRATE) / 22-2827
LFO HA/ REFLECTION SPECTRA OF SOME FERROELECTRIC CRYSTALS OF GROUPS V, VI AND VII. (ANTIMONY AND BISMUTH SU 15-2210
ANGENT OF TRIGLYCINE SELENATE DEPENDING ON THE CONDITIONS OF GROWING. /EABILITY AND DIELECTRIC LOSS ANGLE T 27-3125
A FLOATING ZONE SINGLE CRYSTAL GROWING APPARATUS. (BARIUM TITANATE) 04-0998
THERMODIELECTRIC EFFECT AND FREEZING POTENTIAL IN GROWING ICE. 34-3417
SPHATE SINGLE CRYSTAL (FROM SOLUTION) IN STATIC REGIME AND / GROWING KINETIC OF THE AMMONIUM DIHYDROGEN PHO 17-2516
PREPARATION OF PURE GLYCINE USED FOR GROWING OF TRIGLYCINE SULFATE SINGLE CRYSTALS. 27-3085
Y A DYNAMIC METHOD (ANTIMONY SESQUISELENIDE OR SULFIDE, ANT/ GROWING SINGLE CRYSTALS FROM THE VAPOR PHASE B 15-2162
D ON BARIUM TITANATE, WHERE BARIUM(2+) AND TITANIUM(4+) ION/ GROWING SINGLE CRYSTALS OF SOLID SOLUTION BASE 04-0960
SINGLE CRYSTAL POTASSIUM TANTALATE. (CONDUCTIVE, KYROPOULOS GROWN) / VS NET IONIZED DONOR CONCENTRATION IN 04-1437
DIELECTRIC PROPERTIES OF SINGLE DOMAIN MELT GROWN BARIUM TITANATE. 04-0616
MAIN STRUCTURE OF PURE AND DOPED TRIGLYCINE SULFATE CRYSTALS GROWN BELOW AND ABOVE THE CURIE POINT. DO 27-3130
PERTIES OF BARIUM(1-X) STRONTIUM(X) TITANATE SINGLE CRYSTALS GROWN FROM A POTASSIUM FLUORIDE FLUX. /RIC PRO 04-0635
THER MACRODEFECTS IN POTASSIUM DIHYDROGEN PHOSPHATE CRYSTALS GROWN FROM AN AQUEOUS SOLUTION. / LAYERS AND O 17-2496
PERTIES OF LEAD ZINCATE(0.33) NIOBATE(0.67) (SINGLE CRYSTALS GROWN FROM LEAD OXIDE FLUX). FERROELECTRIC PRO 13-2066
ATION OF THE FINE STRUCTURE OF BARIUM TITANATE CRYSTALS FLUX GROWN FROM POTASSIUM FLUORIDE. INVESTIG 04-0708
E CRYSTALS WITH LARGE NONLINEARITY. (BARIUM TITANATE HAFNATE GROWN FROM POTASSIUM FLUORIDE FLUX) /RIC SINGL 08-1373
/ ELECTRICAL CONDUCTIVITY OF BARIUM TITANATE SINGLE CRYSTALS GROWN FROM SOLUTION IN A POTASSIUM FLUORIDE M/ 04-1035
/WTH CONDITIONS ON THE STRUCTURE OF BARIUM TITANATE CRYSTALS GROWN FROM SOLUTION IN POTASSIUM FLUORIDE MEL/ 04-0598
SINGLE CRYSTAL BARIUM TITANATE FILMS GROWN FROM THE MELT IN AN OXYGEN ATMOSPHERE. 04-0612
POTASSIUM LITHIU/ (TETRAGONAL) TUNGSTEN BRONZE TYPE CRYSTALS GROWN IN MOLTEN SOLUTIONS OF THE PSEUDOSYSTEM 13-1948
METASTABLE POTASSIUM NITRATE(III) FROM SOLUTION. (GROWN IN WATER OR ALCOHOL) 16-2329
PROPERTIES. FLUX GROWN KTN SINGLE CRYSTALS AND THEIR ELECTRICAL 09-1533
S FERROELECTRIC NATURE). DIELECTRIC HYSTERESIS IN (FLUX GROWN) SINGLE CRYSTAL BISMUTH FERRATE (CONFIRM 08-1449
NEOUS POLARIZATION AND COERCIVE FIELD OF LEAD TITANATE (FLUX GROWN SINGLE CRYSTALS). SPONTA 05-1050
UM NIOBATE: CONSTITUTIONAL STUDIES AND (CZOCHRALSKI) CRYSTAL GROWTH. BARIUM SODI 13-1904
ION OF ROCHELLE SALT IN ELECTRIC FIELDS (KINETICS OF CRYSTAL GROWTH) CRYSTALLIZAT 28-3200
NATE TYPE) CERAMICS. (IN RELATION TO DENSIFICATION AND GRAIN GROWTH) /OT PRESSED FERROELECTRIC (BARIUM TITA 04-0879
ATION ON THE FORM OF CRYSTALS OF POTASSIUM NITRATE. (CRYSTAL GROWTH) /UENCE OF THE CONDITIONS OF CRYSTALLIZ 16-2332
SINGLE CRYSTAL TITANATES AND ZIRCONATES. (METHODS OF GROWTH) 02-0045
ANT OF TRIGLYCINE SULFATE AS A FUNCTION OF THE CONDITIONS OF GROWTH. /AL HYSTERESIS OF THE DIELECTRIC CONST 27-3126
IUM / RATE OF ROTATING CRYSTALS AND CHARACTERISTICS OF THEIR GROWTH. (AMMONIUM DIHYDROGEN PHOSPHATE, POTASS 16-2316
OF FERROELECTRIC TRIGLYCINE SULFATE SINGLE CRYSTALS WITH I/ GROWTH AND CHARACTERIZATION OF SOLID SOLUTIONS 27-2972
AL CONDITIONS. (BISMUTH SULFOBROMIDE) CRYSTAL GROWTH AND CHEMICAL SYNTHESIS UNDER HYDROTHERM 15-2218
OBATE. CRYSTAL GROWTH AND DIELECTRIC PROPERTIES OF BISMUTH NI 13-2033
STAL LEAD ZIRCONATE GROWTH AND DIELECTRIC PROPERTIES OF SINGLE CRY 07-1265
PHOSPHATE CRYSTALS IN SOLUTION. KINETICS OF GROWTH AND DISSOLUTION OF AMMONIUM DIHYDROGEN 17-2412
PLZT CERAMICS. ISOTHERMAL GRAIN GROWTH AND ELECTRICAL BEHAVIOR OF FULLY DENSE 09-1553
BRONZE CRYSTALS FROM THE POTASSIUM NIOBATE- LITHIU/ CRYSTAL GROWTH AND ELECTROOPTIC PROPERTIES OF TUNGSTEN 13-2003
SPHATE CRYSTALS IN SOLUTIONS CONTAINING ACETATE IONS. GROWTH AND ETCHING OF POTASSIUM DIHYDROGEN PHO 17-2567
H TITANATE AND RELATED COMPOUNDS. (GADOLINIUM MOLYBDATE, GR/ GROWTH AND EVALUATION OF SINGLE CRYSTAL BISMUT 08-1323
SISTIVITY SINGLE CRYSTALS OF LEAD TITANATE. (FLUX) GROWTH AND FERROELECTRIC PROPERTIES OF HIGH RE 05-1072
DOLINIUM MOLYBDATE, A FERROELECTRIC AND FERROELASTIC SUBSTA/ GROWTH AND MORPHOLOGY OF SINGLE CRYSTALS OF GA 30-3259
UM NIOBATE. (TOP SEEDED FLUX) CRYSTAL GROWTH AND NEUTRON CHARACTERIZATION OF POTASSI 08-1353
ASSIUM LITHIUM NIOBATES. CRYSTAL GROWTH AND NONLINEAR OPTICAL PROPERTIES OF POT 13-2015
ASE OF LEAD ZIRCONATE. CRYSTAL GROWTH AND OBSERVATION OF THE FERROELECTRIC PH 07-1264
ALATE. (DIELECTRIC CONSTANTS, PIEZOELECTRIC / SINGLE CRYSTAL GROWTH AND PHYSICAL PROPERTIES OF LITHIUM TANT 12-1876
RROELECTRIC BISMUTH TITANATE. (ELECTROOPTICS) GROWTH AND PROPERTIES OF EPITAXIAL FILMS OF FE 08-1329
UM LITHIUM TANTALATE(X) NIOBATE(1-X). GROWTH AND PROPERTIES OF FERROELECTRIC POTASSI 11-1711
OXIDE(11) SINGLE CRYSTALS. CRYSTAL GROWTH AND PROPERTIES OF LEAD(5) GERMANIUM(3) 35-3513
ITHIUM TANTALATE FILMS. GROWTH AND PROPERTIES OF LITHIUM NIOBATE AND L 11-1712
S OF FERROELECTRIC LITHIUM NIOBATE. (CRACK-FREE, THROUGH AD/ GROWTH AND PROPERTIES OF SINGLE DOMAIN CRYSTAL 11-1774
GLE CRYSTALS. INFLUENCE OF IMPURITIES ON THE GROWTH AND SOME PHYSICAL PROPERTIES OF TGS SIN 27-3084
/OF THE NON-HYDROGEN BONDED PHOSPHATE FERROELECTRIC (CRYSTAL GROWTH AND STRUCTURE HYSTERESIS LOOP RESULTS). 36-3610
ATED FILMS. GROWTH AND STRUCTURE OF BARIUM TITANATE EVAPOR 04-0604
COMPOUNDS WITH PEROVSKITE STRUCTURE. GROWTH AND STUDY OF NEW SINGLE CRYSTAL CADMIUM 08-1429
OF LEAD TITANATE. CRYSTAL GROWTH AND TEMPERATURE DEPENDENT RAMAN SPECTRA 05-1074
INDUCED GROWTH ANISOTROPY IN TGS CRYSTALS. 27-3162
SIUM NITRATE, SILVER NITRATE, AND AMM/ MORPHOLOGY OF CRYSTAL GROWTH AT POLYMORPHIC TRANSFORMATIONS IN POTAS 16-2260
HANUM / CRUCIBLE BASE COOLING: AN AID TO CZOCHRALSKI CRYSTAL GROWTH. (BARIUM SODIUM NIOBATE AND SODIUM LANT 13-1928
YDROGEN PHOSPHATE IN A MODIFIED CIRCULATION APPARATUS. GROWTH BEHAVIOR OF SMALL SEEDS OF AMMONIUM DIH 17-2704
/SHG PHASE MATCHING TEMPERATURES IN LITHIUM NIOBATE (CRYSTAL GROWTH BY ADDING MAGNESIUM OXIDE TO A CONGRUE/ 11-1684
ASE TRANSITIONS IN POTASSIUM NITRATE NITRITE MIXED CRYSTALS. (GROWTH BY DOUBLE BRIDGMAN TECHNIQUE). PH 16-2345
/TAL BARIUM SODIUM NIOBATE AS A NONLINEAR OPTICAL MATERIAL. (GROWTH BY KYROPOULOS AND CZOCHRALSKI METHODS) 13-2021
ALLIZATION BEHAVIOR OF ANTIMONY SULFOIODIDE. (SINGLE CRYSTAL GROWTH, CHEMICAL TRANSPORT) CRYST 15-2201
IZATION CURRENT FOR TRIGLYCINE SULFATE CRYSTALS DEPENDENT ON GROWTH CONDITIONS. MAXIMUM REPOLAR 27-3154
IGLYCINE SELENATE CRYSTALS ON CHANGES IN SPECI/ INFLUENCE OF GROWTH CONDITIONS OF TRIGLYCINE SULFATE AND TR 27-3069
CRYSTALS ON A CHANGE IN THE TANGENT OF THE ANGLE/ EFFECT OF GROWTH CONDITIONS OF TRIGLYCINE SULFATE SINGLE 27-3153
C CONSTAN/ EFFECT OF TRIGLYCINE FLUOBERYLLATE SINGLE CRYSTAL GROWTH CONDITIONS ON A CHANGE IN THE DIELECTRI 27-3151
LYCINE SULFATE SINGLE CRYSTALS. EFFECT OF GROWTH CONDITIONS ON DIELECTRIC LOSSES IN TRIG 27-3152
ITANATE CRYSTALS GROWN FROM SOLUTI/ EFFECT OF IMPURITIES AND GROWTH CONDITIONS ON THE STRUCTURE OF BARIUM T 04-0598
GROWTH, CRYSTALLOGRAPHY AND DIELECTRIC PROPERT 10-1650
IN RELATION TO THEIR GROWTH COND/ X-RAY TOPOGRAPHIC STUDY OF GROWTH DEFECTS IN TRIGLYCINE SULFATE CRYSTALS 27-3025

ES OF ANTIMONY SULFOIODIDE CRYSTAL/ EFFECT OF COMPOSITION ON GROWTH, DIELECTRIC, AND PHOTOELECTRIC PROPERTI 15-2118
ARIUM(X) STRONTIUM(1-X) NIOBATE SINGLE CRYSTALS (CZOCHRALSKI GROWTH, DIELECTRIC PROPERTIES). /ND STUDY OF B 13-1936
TIONS PRODUCED IN POTASSIUM DIHYDROGEN PHOSPHATE CRYSTALS BY GROWTH FROM SOLUTION. /IC STUDY OF THE DISLOCA 17-2492
ANATE.   EXAGGERATED GRAIN GROWTH IN LIQUID PHASE SINTERING OF BARIUM TIT 04-0835
ANT OF ZIRCONIUM-89 IN BARIUM TITANATE (BARIUM CHLORIDE FLUX GROWTH IN PRESENCE OF ZIRCONIUM-89). /AY CONST 04-0685
RADUAL ADDITION OF A REACTANT. (POTASSIUM DIHYDRO/ METHOD OF GROWTH IN SOLUTION OF SINGLE CRYSTALS BY THE G 17-2595
TA TO GAMMA AND GAMMA TO ALPHA TRANSI/ MORPHOLOGY OF CRYSTAL GROWTH IN THE ALPHA TO BETA, BETA TO ALPHA, BE 16-2259
RISULFIDE SYSTEM. (BRIDGMAN- STOCKBARGER METHOD)   CRYSTAL GROWTH IN THE ANTIMONY SULFOIODIDE- ANTIMONY T 15-2230
SYSTEM.   PHASE EQUILIBRIA AND CRYSTAL GROWTH IN THE LITHIUM OXIDE- NIOBIUM PENTOXIDE 11-1683
NEODYMIUM(3+) DOPED GADOLINIUM MOLYBDATE CRYSTALS.   CERTAIN GROWTH, LASING AND SPECTROSCOPIC PROPERTIES OF 30-3255
IUM DIHYDROGEN PHOSPHATE CRYSTALS GROWN FROM AN AQUEOUS SOL/ GROWTH LAYERS AND OTHER MACRODEFECTS IN POTASS 17-2496
LOW ENTROPY FORM OF ICE I(B) OBTAINED FROM A LINEAR STEP GROWTH MODEL.   34-3410
CRYSTAL FROM SOLUTIONS.   GROWTH OF AMMONIUM DIHYDROGEN PHOSPHATE SINGLE 17-2411
TE AND BLACK PHOSPHORUS.   HIGH GAS PRESSURE CRYSTAL GROWTH OF ANTIMONY SULFOIODIDE, NICKEL CARBONA 15-2137
LS BY THE CZOCHRALSKI METHOD.   GROWTH OF BARIUM LITHIUM NIOBATE SINGLE CRYSTA 13-1991
EFFECTS OF CHANGES IN MELT COMPOSITION ON CRYSTAL GROWTH OF BARIUM SODIUM NIOBATE.   13-1908
E KYROPOULOS TECHNIQUE.   GROWTH OF BARIUM SODIUM NIOBATE CRYSTALS BY TH 13-1909
S FOR OPTICAL APPLICATIONS. (BY PULLING FROM THE MELT, STRU/ GROWTH OF BARIUM SODIUM NIOBATE SINGLE CRYSTAL 13-2045
CZOCHRALSKI GROWTH OF BARIUM STRONTIUM NIOBATE CRYSTALS.   13-1913
STABILITY REGIONS FOR THE GROWTH OF BARIUM STRONTIUM NIOBATE CRYSTALS.   13-2056
OTASSIUM FLUORIDE FLUX.   DISCUSSION ON THE CRYSTAL GROWTH OF BARIUM TITANATE SINGLE CRYSTALS IN P 04-1022
MODIFIED TECHNIQUE FOR THE GROWTH OF BORACITE CRYSTALS.   14-2092
HIUM NIOBATE AND ION ETCHING WITHOUT CRATERS.   GROWTH OF CRATERS ON ION ETCHED SURFACE OF LIT 11-1860
/T OF) TUNGSTEN BRONZE (LIQUIDUS AND SOLIDUS) FIELD AND MELT GROWTH OF CRYSTALS IN THE SODIUM OXIDE- BARIU 13-1949
SOLUBILITY IN THE GROWTH OF ELECTROOPTIC CRYSTALS ADP AND KDP.   17-2446
SOLUTIONS (APPARATUS, GROWTH OF ADP, KDP, HYDROIODIC ACID,/ GROWTH OF ELECTROOPTICAL CRYSTALS FROM AQUEOUS 17-2420
ITANATE, LEAD(X) STRONTIUM(1-X) TITANATE, LEAD/ SOLVENT ZONE GROWTH OF FERROELECTRIC CRYSTALS. (STRONTIUM T 05-1042
NATE AT FIELDS FROM 2 TO 450 KV/CM.   NUCLEATION AND GROWTH OF FERROELECTRIC DOMAINS IN BARIUM TITA 04-0979
STRESS FREE SINGLE CRYSTAL GROWTH OF FERROELECTRIC HYDROGEN HALIDES.   31-3333
UM(2) NIOBIUM(2) OXIDE(7) BY MEANS OF THE FLOATING ZONE TEC/ GROWTH OF FERROELECTRIC SINGLE CRYSTAL STRONTI 13-2037
IOBATE (BY USING COMPOSITION WHERE LIQ/ PROGRESS TOWARDS THE GROWTH OF GOOD OPTICAL QUALITY BARIUM SODIUM N 13-2047
ARTH MANGANATES (PHASE DIAGRAMS, CONVE/ FABRICATION (BY FLUX GROWTH) OF HEXAGONAL SINGLE CRYSTALS OF RARE E 36-3536
ANTALATE NIOBATE.   GROWTH OF HIGH QUALITY CRYSTALS OF POTASSIUM T 13-2055
S (FOR NONLINEAR OPTICS WITH LOW BIREFRINGENCE VARIATIONS) / GROWTH OF HIGH QUALITY LITHIUM NIOBATE CRYSTAL 11-1688
PLANAR GROWTH OF ICE FROM THE PURE MELT.   34-3444
SOLUTION GROWTH METHOD.   GROWTH OF IODIC ACID AND IODATE CRYSTALS BY A 08-1392
(TO 1 CM DIAMETER): CONTROL OF NEEDLE MORPHOLOGY.   GROWTH OF LARGE ANTIMONY SULFOIODIDE CRYSTALS 15-2200
NIUM(3) OXIDE(12).   GROWTH OF LARGE SINGLE CRYSTAL BISMUTH(4) TITA 35-3507
ND ITS DIELECTRIC PROPERTIES. (FERROELECTRIC- PARAE/ CRYSTAL GROWTH OF LEAD (CADMATE(0.33) NIOBATE(0.67)) A 13-1959
ON OF THEIR DOMAIN STRUCTURE.   GROWTH OF LEAD TITANATE CRYSTALS AND EXAMINATI 05-1047
NONSTOICHIOMETRY AND CRYSTAL GROWTH OF LITHIUM NIOBATE.   11-1691
LS.   GROWTH OF LITHIUM NIOBATE SINGLE DOMAIN CRYSTA 11-1781
ATE CRYSTALS DOPED WITH ISOTOPE 170.   FLUX GROWTH OF MAGNESIUM OXIDE AND LANTHANUM ALUMIN 36-3558
HA TO BETA POLYMORPHIC TRANSITION)   RHYTHMIC GROWTH OF NEW PHASE IN POTASSIUM NITRATE. (ALP 16-2263
CRYSTAL GROWTH OF NICKEL IODINE BORACITE.   14-2091
C MELTS. (BARIUM TITANATE, STRONTIUM TI/ TOP SEEDED SOLUTION GROWTH OF OXIDE CRYSTALS FROM NONSTOICHIOMETRI 04-0553
ALS.   EFFECT OF DISLOCATIONS ON THE GROWTH OF POTASSIUM DIHYDROGEN PHOSPHATE CRYST 17-2491
E CRYSTALS IN GEL.   GROWTH OF POTASSIUM DIHYDROGEN PHOSPHATE SINGL 17-2455
ERS.   CONDITIONS FOR THE GROWTH OF POTASSIUM DIHYDROGEN PHOSPHATE WHISK 17-2628
GEL GROWTH OF SILVER PERIODATE CRYSTALS.   36-3523
M NIOBATE.   APPARATUS FOR GROWTH OF SINGLE CRYSTAL, SINGLE DOMAIN LITHIU 11-1726
RMANIUM DIOXIDE BINARY SYSTEM.   GROWTH OF SINGLE CRYSTALS IN LEAD MONOXIDE- GE 35-3514
BATE) AND PROPERTIES OF THIS COMPOSITION.   GROWTH OF SINGLE CRYSTALS OF LEAD (ZINCATE NIO 13-1990
ATE AND ITS PROPERTIES.   GROWTH OF SINGLE CRYSTALS OF LEAD ZINCATE NIOB 36-3594
ES WITH COMPLEX COMPOSITIONS, AND THE STUDY OF THEIR STRUCT/ GROWTH OF SINGLE CRYSTALS OF VARIOUS PEROVSKIT 08-1428
CRYSTAL GROWTH OF SODIUM NIOBATE.   08-1409
GSTATE)   KYROPOULOS SINGLE CRYSTAL GROWTH OF SODIUM TUNGSTEN BRONZES. (SODIUM TUN 13-2054
TES, TANTALATES, STANNATES, TITANATES, CHROMATES, COBA/ FLUX GROWTH OF SOME COMPLEX OXIDE MATERIALS. (NIOBA 02-0129
OLTEN SALTS.   SINGLE CRYSTAL GROWTH OF STRONTIUM POTASSIUM TANTALATE FROM M 13-2034
ERSATURATED SOLUTIONS.   GROWTH OF TRIGLYCINE SULFATE FROM SLIGHTLY SUP 27-3096
GROWTH OF TRIGLYCINE SULFATE SINGLE CRYSTALS. 27-3083
TALS.   GROWTH OF UNCRACKED BARIUM SODIUM NIOBATE CRYS 13-1901
ED BARIUM TITANATE.   INFLUENCE OF GRAIN GROWTH ON DIELECTRIC PROPERTIES OF NIOBIUM DOP 04-0763
M TITANATE CERAMICS.   EFFECTS OF GRAIN GROWTH ON THE DISTRIBUTION OF NIOBIUM IN BARIU 04-0595
/ELECTRIC RARE EARTH MOLYBDATES. (REVIEW, FORMATION, CRYSTAL GROWTH, PHASE TRANSITIONS, STRUCTURE, 49 REFS) 02-0064
KINETICS OF CRYSTALLIZATION OF ANTIMONY SULFOIODIDE (CRYSTAL GROWTH RATE, ANISOTROPY)   15-2215
ARY SYSTEMS. (THEORY)   DETERMINATION OF THE GROWTH RATE OF FERROELECTRIC CRYSTALS FROM BIN 03-0178
T IN THE PRESENCE OF PURE DIRECT SKY BLUE DYE.   GROWTH RATE OF THE (010) PLANE OF ROCHELLE SAL 28-3217
TRIC PROPERTIES OF TGS SINGLE CRYSTALS.   THE EFFECT OF GROWTH RATE ON THE DEFECT STRUCTURE AND DIELEC 27-3067
TS. (BARIUM STRONTIUM NIOBATE)   VARIATION OF CRYSTAL GROWTH RATE WITH SUPERSATURATION IN FLUXED MEL 13-1939
FATE CRYSTALS. (TOPOGRAPHY OF IMPERFECTIONS, DOMAIN STRUCTU/ GROWTH REGIONS IN FERROELECTRIC TRIGLYCINE SUL 27-2999
/OF REAL CRYSTAL SUBSTRATES AS THE DETERMINING FACTOR OF THE GROWTH STAGE IN EPITAXY. (TRIGLYCINE SULFATE) 27-3039
WORK TO SEP 1/ LITHIUM NIOBATE- A NEW TYPE OF FERROELECTRIC/ GROWTH, STRUCTURE, AND PROPERTIES. (REVIEW OF 11-1775
AL PROPERTIES OF FLASH EVAPORATED BARIUM TITANATE THIN FILM/ GROWTH STRUCTURE (MICROSTRUCTURE) AND ELECTRIC 04-0605
/E SHAPE AND COMPOSITION OF FERROELECTRIC CRYSTALS (SOLUTION GROWTH, TGS, TRIGLYCINE SELENATE OR FLUOBERYL/ 02-0089
IS LITHIUM NIOBATE CRYSTALS (CZOCHRALSKI METHOD).   GROWTH TWINNING (UNDER DIRECT CURRENT) IN C-AX 11-1672
HIGH PRESSURE COMPRESSIBILITY AND GRUNEISEN PARAMETER OF STRONTIUM TITANATE.   06-1121
MODE IN THE CUBIC PEROVSKITES. (STRONTIUM TITANATE, POT/ THE GRUNEISEN PARAMETER OF THE SOFT FERROELECTRIC 02-0114
ONS IN SODIUM NITRITE AND SODIUM NITRATE.   GRUNEISEN PARAMETERS AROUND THE PHASE TRANSITI 16-2390
OF THE GAMMA-RAY IRRADIATED FERROELECTRICS ROCHELLE SALT AND GUANIDINE ALUMINUM SULFATE HEXAHYDRATE. /DIES 21-2812
ELECTRON PARAMAGNETIC RESONANCE STUDY OF SOME METAL IONS IN GUANIDINIUM ALUMINUM SULFATE HEXAHYDRATE.   21-2817
EPR STUDY OF MAGNETIC COPPER(2+) COMPLEXES IN CRYSTALS OF GUANIDINIUM ALUMINUM SULFATE HEXAHYDRATE.   21-2820
NOLOGICAL THEORY. (AMMONIUM SULFATE, AMMONIUM FLUOBERYLLATE, GUANIDINIUM ALUMINUM SULFATE HEXAHYDRATE) /OME 02-0040
SEE BOTH GASH AND GUANIDINIUM ALUMINUM SULFATE HEXAHYDRATE.
/MPERATURE INFRARED ABSORPTION SPECTRA (5200-4200 PER CM) OF GUANIDINIUM ALUMINUM SULFATE HEXAHYDRATE AND / 21-2811
TS CH/ REDETERMINATION OF STRUCTURE OF FERROELECTRIC CRYSTAL GUANIDINIUM ALUMINUM SULFATE HEXAHYDRATE AND I 21-2819
MECHANISM OF REVERSAL).   FERROELECTRIC MOMENT IN GUANIDINIUM ALUMINUM SULFATE HEXAHYDRATE (AND 21-2818
AGES A/ OPTICAL AND INTERFEROMETRIC STUDIES ON FERROELECTRIC GUANIDINIUM ALUMINUM SULFATE HEXAHYDRATE CLEAV 21-2816
ARIZATION VERSU/ DEFECTS AND THE FERROELECTRIC PROPERTIES OF GUANIDINIUM ALUMINUM SULFATE HEXAHYDRATE. (POL 21-2813
ZEEMAN EFFECT AND DILUTION EFFECT OF ABSORPTION LINES IN GUANIDINIUM CHROMIUM SULFATE HEXAHYDRATE.   21-2821

CT (STUDY) OF TWO FINE LINES IN (THE RED REGION OF) CHROMIUM GUANIDINIUM SULFATE HEXAHYDRATE.   ZEEMAN EFFE 21-2814
W TEMPERATURES.   MAGNETIC SUSCEPTIBILITY OF GUANIDINIUM VANADIUM SULFATE HEXAHYDRATE AT LO 21-2815
                          ACOUSTIC SURFACE WAVE GUIDE, AVAILING FERROELECTRIC POLARIZATIONS.   37-3680
MATERIALS.   GUIDE LINES FOR THE SELECTION OF ACOUSTOOPTIC 03-0425
S OF ACOUSTIC WAVE GENERATION IN PIEZOELECTRICS COUPLED TO A GUNN OSCILLATOR. (LITHIUM NIOBATE) /EASUREMENT 11-1731
ENSITY OF HYDROGEN BOND BANDS IN RAMAN SCATTERING SPECTRA OF GYPSUM AND ROCHELLE SALT CRYSTALS. /OF THE INT 28-3189

H

OF THE ANTIFERROELECTRIC PEROVSKITES LEAD ZIRCONATE AND LEAD HAFNATE. /IC PROPERTIES AND PHASE TRANSITIONS 08-1433
TALATE, BARIUM TITANATE, LEAD TITANATE, LEAD ZIRCONATE, LEAD HAFNATE) /. (STRONTIUM TITANATE, POTASSIUM TAN 02-0114
        ELECTRICAL CONDUCTIVITY OF STRONTIUM ZIRCONATE AND HAFNATE AT HIGH TEMPERATURES.   08-1417
IC SINGLE CRYSTALS WITH LARGE NONLINEARITY. (BARIUM TITANATE HAFNATE GROWN FROM POTASSIUM FLUORIDE FLUX) /R 08-1373
        DIELECTRIC STUDY OF THE LEAD ZIRCONATE LEAD HAFNATE (PHASE) DIAGRAM.   08-1340
/UDIED BY THE PERTURBED ANGULAR CORRELATION TECHNIQUE. (LEAD HAFNATE, TIN HAFNATE, CALCIUM HAFNATE, STRONT/ 08-1276
LID SOLUTIONS IN ANTIFERROELECTRIC REGION OF THE SYSTEM LEAD HAFNATE TITANATE, LEAD STANNATE NIOBATE.   SO 13-1956
TERIAL AND ELECTROOPTIC PROPERTIES OF THE (LEAD, LANTHANUM) (HAFNATE, TITANATE) SYSTEM.   MA 08-1302
                COMPLEX DIELECTRIC CONSTANT OF LEAD HAFNATE, 320-520 DEG K.   08-1339
NEW CADMIUM AND THALLIUM CONTAINING PEROVSKITES. (TITANATES, HAFNATES, MIXED NIOBATES AND TANTALATES) / OF 08-1458
YGEN(3), WHERE B IS IRON OR SCANDIUM, AND B' IS ZIRCONIUM OR HAFNIUM. / (B(0.5) NIOBIUM(0.5)) (1-X) B'(X) OX 13-2044
/ILLIUM A(1-X) TUNGSTATE OR MOLYBDATE. (WHERE A IS ZIRCONIUM, HAFNIUM, CERIUM, THORIUM, TITANIUM, SODIUM, M/ 10-1640
ON TRIFLUORIDE, BORACITES, BISMUTH(4) TITANIUM(3) OXIDE(12), HAFNIUM DIVANADIDE) /M ANTIMONIDE, RUBIDIUM IR 02-0036
        STATIC ELECTRIC QUADRUPOLE INTERACTION OF TANTALUM AND HAFNIUM IONS IN BARIUM AND LEAD TITANATE.   04-0937
E- ABO(3) SYSTEMS. (WHERE A IS STRONTIUM OR BARIUM, AND B IS HAFNIUM, TITANIUM, OR TIN) / THE CADMIUM BORAT 36-3528
/EAR QUADRUPOLE INTERACTIONS IN PEROVSKITE TYPE COMPOUNDS OF HAFNIUM-181 STUDIED BY THE PERTURBED ANGULAR / 08-1276
UM NIOBATE CRYSTALS SUITABLE FOR LIGHT MODULATORS WITH A LOW HALF WAVE VOLTAGE.   CUTS OF LITHI 11-1715
LLY HETEROGENEOUS CRYSTALS (THEORY AND METHOD OF DETERMINING HALF WAVE VOLTAGE). /TROOPTIC EFFECT IN OPTICA 03-0461
ION FOR A BARIUM SODIUM NIOBATE TRANSVERSE LIGHT MODULATOR. (HALF WAVE VOLTAGE, THEORY) /M CRYSTAL ORIENTAT 13-1973
G IN THE FERROELECTRIC MECHAN/ PHASE TRANSITIONS OF HYDROGEN HALIDE CRYSTALS (IMPORTANCE OF HYDROGEN BONDIN 31-3329
ING MODEL COMP/ LONGITUDINAL MODES IN HYDROGEN AND DEUTERIUM HALIDE CRYSTALS. (RAMAN SPECTRA, DIPOLAR COUPL 31-3326
        EFFECT OF CHARGE TRANSFER ON RAMAN SCATTERING OF HYDROGEN HALIDES.   31-3328
OLECULAR POTENTIAL AND FERROELECTRIC TRANSITIONS IN HYDROGEN HALIDES.   INTERM 31-3339
                PHASE TRANSITIONS OF SOLID HYDROGEN HALIDES.   31-3327
RYSTALS OF GROUPS V, VI AND VII. (ANTIMONY AND BISMUTH SULFO HALIDES) /TION SPECTRA OF SOME FERROELECTRIC C 15-2210
        STRESS FREE SINGLE CRYSTAL GROWTH OF FERROELECTRIC HYDROGEN HALIDES.   31-3333
PERATURE, POLARIZATION FLUCTUATIONS, LITHIUM NIOBATE, ALKALI HALIDES) /RYSTALS. (STRAIN DEPENDENT CURIE TEM 03-0513
/PENDENCE OF DIELECTRIC CONSTANTS OF CUBIC IONIC COMPOUNDS. (HALIDES AND OXIDES, RELATION BETWEEN CURIE CO/ 03-0188
        PREPARATION AND PROPERTIES OF CESIUM LEAD HALIDES. (FERROELECTRICITY INDETERMINITE)   33-3359
RA/ FERROELECTRICITY AND PHASE TRANSITIONS IN SOLID HYDROGEN HALIDES (SINGLE CRYSTAL X-RAY AND NEUTRON DIFF 31-3331
        INERTIAL ELECTRICAL POLARIZATION OF CRYSTALS (ALKALI HALIDES, STRONTIUM TITANATE).   06-1101
TRISULFIDE SINGLE CRYSTALS. TEMPERATURE DEPENDENCES OF THE HALL AND DRIFT MOBILITIES OF HOLES IN ANTIMONY 15-2193
TANATE (4.2-300 DEG / ELECTRONIC TRANSPORT (CONDUCTIVITY AND HALL AND SEEBECK COEFFICIENTS) IN STRONTIUM TI 06-1129
OF ISO- AND HETEROVALENT CATIONIC S/ ELECTROCONDUCTIVITY AND HALL EFFECT IN SEMICONDUCTING SOLID SOLUTIONS 04-0943
                ELECTRONIC HALL MOBILITY IN THE ALKALINE EARTH FLUORIDES.   33-3404
ON UHV CLEAVED TRIGLYCINE SULFATE.   FIELD EFFECT AND HALL MOBILITY OF TELLURIUM AND LEAD TELLURIDE 27-3172
SOFT- NORMAL MODE COU/ THEORY OF PEROVSKITE FERROELECTRICS (HAMILTONIAN IN TERMS OF LOCALIZED D-STRAIN AND 03-0432
IUM SODIUM NIOBATE, BISMUTH(12) GERMANI/ MICROWAVE ACOUSTICS HANDBOOK. VOL 1: SURFACE WAVE VELOCITIES. (BAR 01-0028
ERYLLATE, ETC/ (EFFECT OF TEMPERATURE ON) QUASI SYMMETRICAL HARD IONS (PHOSPHATE, TITANATE, SULFATE, FLUOB 03-0330
I(H).   HARDNESS ANISOTROPY OF SINGLE CRYSTALS OF ICE 34-3460
                GENERATION OF THE SECOND HARMONIC AND FERROELECTRICITY. (THEORY)   02-0063
FERROELECTRIC PHASE OF HYDROGEN CHLORIDE. (CALCULATED IN THE HARMONIC APPROXIMATION) /ECULAR MOTION IN THE 31-3340
                GENERATION OF THE SECOND HARMONIC BY TRIGLYCINE SULFATE. (84 REFS)   27-2990
SADOVSKII EFFECT (THEORY, PROPAGATION OF RADIATION AND FIRST HARMONIC CALCULATIONS FOR LITHIUM NIOBATE). / 03-0186
CRYSTAL, AGR/ MEASUREMENTS OF PHOTON CORRELATIONS OF SECOND HARMONIC GENERATED LIGHT (FROM LITHIUM NIOBATE 11-1701
        LONG RANGE ORDER IN SODIUM NITRITE BY OPTICAL SECOND HARMONIC GENERATION.   16-2314
GSTEN BRONZE SYSTEM) POTASSIUM SODIUM BARIUM NIOBATE (SECOND HARMONIC GENERATION). / PROPERTIES OF (THE TUN 13-2028
ILLATION IN FER/ NONLINEAR OPTICAL DEVICES. (BASED IN SECOND HARMONIC GENERATION AND OPTICAL PARAMETRIC OSC 02-0097
FREE AND CLAMPED VALUES OF THE DIELE/ MEASUREMENTS OF SECOND HARMONIC GENERATION AND THE VARIATIONS IN THE 11-1998
OBATE ON MELT COMPOSITION.   DEPENDENCE OF SECOND HARMONIC GENERATION COEFFICIENTS OF LITHIUM NI 11-1769
RIC CRYSTALS.   ABSOLUTE SIGNS OF SECOND HARMONIC GENERATION COEFFICIENTS OF PIEZOELECT 02-0096
RIC CRYSTA/ FURTHER MEASUREMENTS OF ABSOLUTE SIGNS OF SECOND HARMONIC GENERATION COEFFICIENTS OF PIEZOELECT 05-1066
ROELECTRIC) CRY/ ABSOLUTE SIGNS OF NONLINEAR OPTICAL (SECOND HARMONIC GENERATION) COEFFICIENTS OF POLAR (PY 02-0095
IZATION REVERSAL. (SODIUM NITRITE)   OPTICAL SECOND HARMONIC GENERATION DURING FERROELECTRIC POLAR 16-2384
SCEIN DYE LASERS.   EFFICIENT SECOND HARMONIC GENERATION IN ADP WITH TWO NEW FLUORE 15-2164
        OPTICAL SECOND HARMONIC GENERATION IN ANTIMONY SULFOIODIDE.   15-2164
TITANATE, AND POTASSIUM DI/ QUANTITATIVE STUDIES OF OPTICAL HARMONIC GENERATION IN CADMIUM SULFIDE, BARIUM 04-0852
TRITE.   SECOND HARMONIC GENERATION IN FERROELECTRIC SODIUM NI 16-2385
        TEMPERATURE DEPENDENCE OF RAMAN SPECTRUM AND SECOND HARMONIC GENERATION IN KDP.   17-2550
LARGE NONLINEAR OPTICAL COEFFICIENT AND PHASE MATCHED SECOND HARMONIC GENERATION IN LITHIUM IODATE.   08-1405
        OPTICAL PROBING OF ACOUSTIC SURFACE WAVE HARMONIC GENERATION (IN LITHIUM NIOBATE).   11-1752
                OPTIMUM SECOND HARMONIC GENERATION (IN LITHIUM NIOBATE).   11-1764
        SIMULTANEOUS OPTICAL PARAMETRIC OSCILLATION, SECOND HARMONIC GENERATION (IN LITHIUM NIOBATE).   11-1859
NIUM MOLYBDATE AND TERBIUM MOLYBDATE.   SECOND HARMONIC GENERATION IN MOLYBDATES (BETA GADOLI 30-3208
(KDP TYPE)   OPTICAL SECOND HARMONIC GENERATION IN PIEZOELECTRIC CRYSTALS. 17-2582
FILMS (AFTER COOLING ST/ PYROELECTRICITY AND OPTICAL SECOND HARMONIC GENERATION IN POLYVINYLIDENE FLUORIDE 36-3530
/EMPERATURE DEPENDENCE OF INDEX MATCHING ANGLE OF THE SECOND HARMONIC GENERATION IN POTASSIUM DIHYDROGEN P/ 17-2544
OSPHATE (AND DEU/ TEMPERATURE DEPENDENCE OF OPTICAL (SECOND) HARMONIC GENERATION IN POTASSIUM DIHYDROGEN PH 17-2695
        TEMPERATURE DEPENDENCE OF SECOND HARMONIC GENERATION IN SODIUM NITRITE.   16-2315
CTRIC STRONTIUM TITANATE.   SECOND HARMONIC GENERATION IN STRESS INDUCED FERROELE 06-1131
POTASSIUM TANTALATE.   ELECTRIC FIELD INDUCED OPTICAL HARMONIC GENERATION IN STRONTIUM TITANATE AND 06-1203
LE CRYSTALS (NONLINEAR ELASTIC PROPERTIES). ULTRASONIC THIRD HARMONIC GENERATION IN STRONTIUM TITANATE SING 06-1181
IN THE PARAELECTRIC PHASE.   SECOND HARMONIC GENERATION IN THE TRIGLYCINE CRYSTAL 27-3105
TALS.   SECOND HARMONIC GENERATION IN TRIGLYCINE SULFATE CRYS 27-3138
CHLORINE IRON IODINE AND NICKEL CHLORINE) BORACITES (SECOND HARMONIC GENERATION INDEX MATCHING). /NE, IRON 14-2080
H FREQUENCY ACOUSTIC WAVES. APPLICATION OF ACOUSTIC WAVES TO HARMONIC GENERATION LITHIUM NIOBATE. /N BY HIG 11-1834
ITY OF THE REFRACTIVE INDEX OF A NONLINEAR CRYSTAL ON SECOND HARMONIC GENERATION. (LITHIUM NIOBATE) /MOGENE 11-1687
L BY THE INDEX MATCHING METHOD.   SECOND HARMONIC GENERATION OF LIGHT IN THE KDP CRYSTA 17-2693
THEORY AND TEST ON BARIUM SODIUM NIOBA/ DEPENDENCE OF SECOND HARMONIC GENERATION ON CRYSTAL INHOMOGENEITY / 13-2043
/INEAR OPTICAL PROPERTIES OF FERROELECTRIC CRYSTALS. (SECOND HARMONIC GENERATION, OPTICAL PARAMETRIC OSCIL/ 03-0383
SPHATE)   PULSE DISTORTIONS IN MISMATCHED SECOND HARMONIC GENERATION. (POTASSIUM DIHYDROGEN PHO 17-2690

STUDY OF PHASE TRANSITION CF SODIUM NITRITE BY SECOND HARMONIC GENERATION. (TEMPERATURE DEPENDENCE) 16-2403
INTRACAVITY SECOND HARMONIC GENERATION WITH RUBY LASER BY KDP. 17-2604
LS. GENERATION OF THE SECOND ACOUSTIC HARMONIC IN ANTIMONY SULFOIODIDE SINGLE CRYSTA 15-2224
LS. GENERATION OF THE SECOND ACOUSTIC HARMONIC IN ANTIMONY SULFOIODIDE SINGLE CRYSTA 15-2223
. GENERATION OF SECOND OPTICAL HARMONIC IN THE CUBIC PHASE OF BARIUM TITANATE 04-0538
LECTRIC TRIGLYCINE SULFATE. SECOND HARMONIC LIGHT SCATTERING BY DOMAINS IN FERROE 27-2991
THEORETICAL STUDY OF THE GENERATION CF THE ACOUSTIC SECOND HARMONIC. (LITHIUM NIOBATE) 11-1772
TTERING STUDY OF THE LATTICE DYNAMICS OF STRONTIUM TITANATE: HARMONIC MODELS. NEUTRON INELASTIC SCA 06-1225
MHZ) ACOUSTIC WAVE ALONG THE (001) GENERATION OF THE THIRD HARMONIC OF A LONGITUDINAL HIGH FREQUENCY (950 11-1835
STUDY OF THE PHASE TRANSITION OF TGS BY SECOND HARMONIC SCATTERING. 27-2992
EFFECTS OF DCMAIN SHAPES ON SECOND HARMONIC SCATTERING IN TRIGLYCINE SULFATE. 27-2989
RIUM ALUMINUM GARNET- BARIUM SODIUM NIOBATE 0.66 MILLIMICRON HARMONIC SOURCE. NEODYMIUM: YTT 13-1975
TROMETER. (WITH A TGS SINGLE CRYSTAL) A NEW SECOND HARMONIC TYPE FERROELECTRIC MODULATOR FOR ELEC 27-2951
/ AMMCNIUM DIHYDROGEN ORTHOPHOSPHATE (4000-10000 PER CM) AND HARMONIC VIBRATION FREQUENCIES OF THE HYDROGE/ 17-2432
VIEW) OPTICAL HARMONICS AND NONLINEAR PHENOMENA. (THEORY, RE 02-0067
ATING ITSELF IN THE (001) DI/ GENERATION CF SECOND AND THIRD HARMONICS BY A TRANSVERSE ACOUSTIC WAVE PROPAG 11-1682
HARMONICS GENERATION IN FERROELECTRIC MEDIA. 03-0215
COND ORDER TRANSITION ABOVE THE C/ QUASISTATIC GENERATION OF HARMONICS IN A FERROELECTRIC CRYSTAL WITH A SE 03-0331
CRYSTAL (NONLIN/ GENERATION OF SUPERHIGH FREQUENCY ACCUSTIC HARMONICS IN A (SINGLE DOMAIN) LITHIUM NIOBATE 11-1707
ELECTROOPTIC AMPLITUDE MODULATION OF LASER GENERATED SECOND HARMONICS IN POTASSIUM DIHYDROGEN PHOSPHATE. 17-2694
ECTS OF DISPERSICN AND FOCUSING CN THE PRODUCTION OF OPTICAL HARMONICS. (KDP) EFF 17-2579
/HE SIMULTANEOUS GENERATICN OF THE SECOND, THIRD, AND FOURTH HARMONICS OF 1.06 MICRON RADIATION IN BARIUM / 11-1779
IUM TRIHYDROGEN DISELENITE. ANOMALOUS SPECIFIC HEAT AND CONFIGURATIONAL ENTROPY CHANGE IN SOD 22-2856
ONIUM ALUM ABOVE THE CRITICAL TEMPERATURE. SPECIFIC HEAT AND SUSCEPTIBILITY IN CHROMIUM METHYL AMM 20-2805
SE TRANSITION IN STRONTIUM TITANATE. (ABSTRACT) SPECIFIC HEAT ANOMALY ASSOCIATED WITH THE 110 DEG K PHA 06-1226
ALCULATION OF TEMPERATURE DEPENDENCE FOR FERRCELEC/ SPECIFIC HEAT ANOMALY IN FERROELECTRIC ROCHELLE SALT (C 28-3187
NITRITE. SPECIFIC HEAT ANOMALY IN FERROELECTRIC SILVER SODIUM DI 16-2296
ARAELECTRIC TRANSITION. (TGS) SPECIFIC HEAT ANOMALY IN THE NEIGHBORHOOD OF A FERRO- P 27-2962
RACITE. SPECIFIC HEAT ANOMALY OF FERROELECTRIC COBALT IODINE BO 14-2101
OGEN PHOSPHATE CRYSTALS FCR STUDY CF THE FY/ USE OF SPECIFIC HEAT (BY A DYNAMIC METHOD) OF POTASSIUM DIHYDR 17-2566
ELECTRICAL PROPERTIES OF TRIGLYCINE SULFATE SINGLE CRYSTALS (HEAT CAPACITY AND ELECTRIC PERMITTIVITY). /ND 27-3141
/NSITION POINT IN POTASSIUM MANGANESE TRIFLUORIDE (ANOMALOUS HEAT CAPACITY, CRITICAL TEMPERATURE ABOUT 186/ 33-3373
/GNETIC, FERROELECTRIC BETA GADOLINIUM MOLYBDATE. PARTS 1-3: HEAT CAPACITY, ENTROPY, MAGNETIC MOMENT OF TH/ 30-3275
REGION (THEORY, RELATION TO KAENZIG DOMAIN MODEL). HEAT CAPACITY IN THE FERROELECTRIC TRANSITION 03-0443
. HEAT CAPACITY OF METHYL AMMONIUM CHROMIUM ALUM 20-2801
SINGLE CRYSTALS. HEAT CAPACITY OF RUBIDIUM DIHYDROGEN PHOSPHATE 17-2415
OIODIDE. HEAT CAPACTIY OF POLYCRYSTALLINE ANTIMONY SULF 15-2238
LECTRIC CERAMICS AT LOW TE/ DIELECTRIC AND THERMAL (SPECIFIC HEAT, ELECTROCALORIC EFFECT) STUDIES OF FERROE 04-0720
S OF A SINGLE MODE ANHARMONIC OSCI/ FREE ENERGY AND SPECIFIC HEAT IN FERROELECTRIC PHASE TRANSITION IN TERM 03-0406
EFFECTS OF DEUTERATION ON THE SPECIFIC HEAT OF ADP CRYSTALS. 17-2414
RATURE RANGE FROM -70 TO +14 DEG C. SPECIFIC HEAT OF AMMONIUM HYDROGEN SULFATE IN THE TEMPE 19-2789
SPECIFIC HEAT OF BARIUM TITANATE. 04-0715
ERROCYANIDE TRIHYDRATE. ANOMALOUS SPECIFIC HEAT OF FERROELECTRIC THIOUREA AND POTASSIUM F 24-2902
ENITE. HEAT OF FORMATION OF LITHIUM TRIHYDROGEN DISEL 22-2867
SPECIFIC HEAT OF ICE I(H). 34-3464
MEASUREMENT OF THE SPECIFIC HEAT OF KDP BY A DYNAMIC METHOD. 17-2619
SPECIFIC HEAT OF POTASSIUM MANGANESE TRIFLUORIDE. 33-3387
SPECIFIC HEAT OF PURE AND DOPED ICE NEAR 120 DEGREES K. 34-3463
RIC TRANSITION POINT. SPECIFIC HEAT OF SODIUM NITRITE NEAR THE ANTIFERROELECT 16-2303
ANOMALOUS SPECIFIC HEAT OF SODIUM TRIHYDROGEN SELENITE. 22-2823
COMBINATORIAL PROBLEM. ANOMALOUS SPECIFIC HEAT OF SODIUM TRIHYDROGEN SELENITE ASSOCIATED 22-2822
YDROGEN PHOSPHATE, RUBIDIUM AND AMMONIUM ANALOGS S/ SPECIFIC HEAT OF SOME KDP TYPE CRYSTALS. (POTASSIUM DIH 17-2679
ANCMALOUS SPECIFIC HEAT OF STRONTIUM TITANATE. 06-1159
TRANSITION. SPECIFIC HEAT OF STRONTIUM TITANATE NEAR THE STRUCTURAL 06-1132
FERROELECTRIC SPECIFIC HEAT OF TRIGLYCINE SULFATE. 27-3146
/SUREMENTS. PART-8: ULTRASOUND PROPAGATION. PART-9: SPECIFIC HEAT. PART-10: PHOTON RENORMALIZATION AND THE/ 03-0337
VITY OF CALCIUM DOPED SODIUM NITRATE CRYSTALS NEAR/ SPECIFIC HEAT, THERMAL DIFFUSIVITY AND THERMAL CONDUCTI 16-2283
LFATE) A NEW FERROELECTRIC TRANSDUCER FOR USE IN HEAT TRANSFER AND FLOW STUDIES. (TRIGLYCINE SU 27-3031
M METANIOBATE SINGLE CRYSTALS. HEAT TREATMENT INDUCED POINT DEFECTS IN LITHIU 11-1668
UCTING BARIUM TITANATE. EFFECT OF HEAT TREATMENT ON THE PTCR ANOMALY IN SEMICOND 04-0762
ART-2: GREEN STRIPES. AN EFFECT OF HEAT TREATMENT ON TUNGSTEN TRIOXIDE CRYSTAL. P 10-1639
ALORIC EFFECT IN TRIGLYCINE SULFATE. SPECIFIC HEAT UNDER APPLIED ELECTRIC FIELD AND ELECTROC 27-2960
/ ELECTRICAL CONDUCTIVITY OF BARIUM TITANATE SINGLE CRYSTALS HEATED IN OXYGEN (EFFECT TIED TO GOLD ANODES). 04-0997
LIGHT EMISSICN FRCM TGS. (CRYSTALS DURING HEATING) 27-3117
/URE AUTOSTABILIZATION OF FERROELECTRIC CRYSTAL. (DIELECTRIC HEATING AND POLARIZATION REVERSAL, COMPLEX PE/ 03-0245
OGEN PHOS/ DEPOLARIZATION OF LINEARLY POLARIZED LIGHT DUE TO HEATING OF PARALLEL PLANES OF POTASSIUM DIHYDR 17-2540
ENT OF A DIELECTRIC FERROELECTRIC CONDENSER UNDER CONTINUOUS HEATING. (TGS SINGLE CRYSTAL) /EDANCE MEASUREM 27-3028
MOLECULAR FORCES OF HEAVY AND LIGHT ICE. 34-3440
MAGNETIC PROPERTIES OF HEAVY RARE EARTH ORTHOMANGANATES. 36-3545
LIZATION OF RUBIDIUM DIDEUTERIUM PHOSPHATE FROM SOLUTIONS IN HEAVY WATER. CRYSTAL 17-2583
GEN PHOSPHATE IN WATER AND RUBIDIUM DIDEUTERIUM PHOSPHATE IN HEAVY WATER. / OF SOLUTION OF RUBIDIUM DIHYDRO 17-2585
ODELS. ZEROES OF THE PARTITICN FUNCTION FOR THE HEISENBERG, FERROELECTRIC, AND GENERAL ISING M 03-0481
ER TRICHLORIDE. A PHASE TRANSITION IN A COMPOUND WITH HELICAL ELECTRIC DIPOLE STRUCTURE: CESIUM COPP 33-3388
IC WAVE ALONG THE (001) DIRECTION OF LITHIUM NIOBATE (LIQUID HELIUM TEMPERATURE). /EQUENCY (950 MHZ) ACOUST 11-1835
STITUTION AND OXYGEN NONSTOICHICMETRY OF BARIUM TITANATE AND HEMATITE SINGLE CRYSTALS. HETEROVALENT SUB 04-0874
ETERMINING HALF WAVE VOLTA/ ELECTRCCPTIC EFFECT IN OPTICALLY HETEROGENEOUS CRYSTALS (THEORY AND METHOD OF D 03-0461
MODE FOR THE ANTIMONY SULFOIODIDE- TIN DIOXIDE HETEROSTRUCTURE SWITCH. 15-2243
CHING AND MEMORY EFFECT IN ANTIMONY SULFOIODIDE- TIN DIOXIDE HETEROSTRUCTURE THIN FILM. SWIT 15-2166
CURIE SHIFT IN BARIUM TITANATE BY HETEROSUBSTITUTION. 04-0727
STRUCTURE OF SODIUM NIOBATE HETEROTYPES. 08-1395
OCTAHEDRON AND LARGER POLYHEDRA IN FERROELECTRIC MATERIALS. (HETEROTYPES, REVIEW) ROLE OF THE REGULAR 02-0094
/D HALL EFFECT IN SEMICONLUCTING SCLID SCLUTIONS OF ISO- AND HETEROVALENT CATIONIC SUBSTITUTIONS IN BARIUM/ 04-0943
IUM TITANATE SINGLE CRYSTALS. COMPENSATION OF HETEROVALENT FERRIC IONS IN THE LATTICE OF BAR 04-0539
IOMETRY OF BARIUM TITANATE AND HEMATITE SINGLE CRYSTALS. HETEROVALENT SUBSTITUTION AND OXYGEN NONSTOICH 04-0874
HEXA... SEE THE PREFIXED WORD.
) CRYSTALS. SPACE CHARGE LIMITED CURRENTS IN HEXAGONAL BARIUM TITANATE(0.75) PLATINATE(0.25 04-0936
STUDY OF BROUT'S MODEL FCR FERROELECTRICS. INVESTIGATIONS ON HEXAGONAL ICE. A 34-3443
INTERPRETATION OF THE PROTCN SPIN LATTICE RELAXATION IN HEXAGONAL ICE. 34-3475
RELAXATICNAL PROTON ORDERING AND GLASSY CRYSTALLINE STATE IN HEXAGONAL ICE. 34-3436
STRONG COLLISION LIMIT OF SPIN LATTICE RELAXATION IN HEXAGONAL ICE. 34-3480

INFLUENCE OF IMPURITIES ON THE ORDER DISORDER REACTION IN HEXAGONAL ICE I(H).   34-3473
                    PROTEIN SPIN RELAXATION IN HEXAGONAL ICE. PART-2: THE T(1 RHO) MINIMUM.   34-3488
        LOW TEMPERATURE DIELECTRIC CELL AND PERMITTIVITY OF HEXAGONAL ICE TO 2 K.   34-3432
US ICE AND OF THE PHASE TRANSFORMATION BETWEEN THE CUBIC AND HEXAGONAL ICES. /ASS TRANSITION OF THE AMORPHO 34-3476
                        DIELECTRIC BEHAVIOR OF CUBIC AND HEXAGONAL ICES AT LOW TEMPERATURES.   34-3431
EARTHS. (YTTERBIUM, HOLMIUM, ER/ FERROELECTRIC PROPERTIES OF HEXAGONAL ORTHOMANGANITES OF YTTRIUM AND RARE   36-3542
RE AND TEMPERATURE DEPENDENCE OF SPONTANEOUS POLARIZATION OF HEXAGONAL (RHOMBOHEDRAL) FERROELECTRICS. /RATU 03-0290
ATES (PHASE DIAGRAMS, CONVE/ FABRICATION (BY FLUX GROWTH) OF HEXAGONAL SINGLE CRYSTALS OF RARE EARTH MANGAN 36-3536
              SEE BOTH GASH AND GUANIDINIUM ALUMINUM SULFATE HEXAHYDRATE.
ROPERTIES OF BARIUM TITANATE IN THE CENTIMETER RANGE OF SUPERHIGH FREQUENCIES (MICROWAVES). FERROELECTRIC P 04-0901
  DOMAIN) LITHIUM NIOBATE CRYSTAL (NONLIN/ GENERATION OF SUPERHIGH FREQUENCY ACOUSTIC HARMONICS IN A (SINGLE 11-1707
TE.   DIPOLAR RELAXATION OF LITHIUM-7 BY HINDERED ROTATORS IN LITHIUM HYDRAZINIUM SULFA   32-3350
ELECTRIC MATERIALS. (ROCHELLE SALT, TR/ PHASE STABILIZATION (HINDERED TRANSITION) IN GAMMA-IRRADIATED FERRO 27-3165
RIC. (DEPENDENCE OF DIELECTRIC CONSTANT ON THERMAL AND FIELD HISTORY) /IUM IODATE - A DISORDERED FERROELECT 08-1348
                              EARLY HISTORY OF FERROELECTRICITY (1920-1943).   02-0126
     TEMPERATURE DEPENDENCES OF THE HALL AND DRIFT MOBILITIES OF HOLES IN ANTIMONY TRISULFIDE SINGLE CRYSTALS. 15-2193
ITANATE.   SHARED HOLES TRAPPED BY CHARGE DEFECTS IN STRONTIUM T   06-1123
CUIT SENSITIVITY OF RADIALLY POLARIZED FERROELECTRIC CERAMIC HOLLOW SPHERES. (HYDROPHONE ELEMENTS) OPEN CIR 37-3653
ONAL ORTHOMANGANITES OF YTTRIUM AND RARE EARTHS. (YTTERBIUM, HOLMIUM, ERBIUM) /ELECTRIC PROPERTIES OF HEXAG 36-3542
ND ISOSTRUCTURAL COMPOUNDS (EUROPIUM TERBIUM, DYSPROSIUM AND HOLMIUM MOLYBDATES). /F GADOLINIUM MOLYBDATE A 30-3271
/USION EFFECTS (INTERNAL GENERATION OF STRONG FIELDS) DURING HOLOGRAM RECORDING IN CRYSTALS (IRON DOPED ST/ 06-1090
         OPTICALLY ERASABLE AND REWRITABLE SOLID STATE HOLOGRAMS.   37-3638
VICE. (BISMUTH(4) TITANIUM(3) OXIDE(12))   PHASE HOLOGRAMS IN A FERROELECTRIC PHOTOCONDUCTOR DE   13-1977
VICE (ERASABLE, NONDESTRUCTIVE READOUT, BASED ON/ STORAGE OF HOLOGRAMS IN A FERROELECTRIC PHOTOCONDUCTOR DE 37-3646
          MULTIPLE STORAGE OF THICK PHASE HOLOGRAMS IN LITHIUM NIOBATE.   11-1827
                  THERMALLY FIXED HOLOGRAMS IN LITHIUM NIOBATE.   11-1828
STALS. (LITHIUM NIOBATE, BARIUM SODIUM NIOBATE)   HOLOGRAPHIC PATTERN FIXING IN ELECTROOPTIC CRY   11-1661
S. (BARIUM TI/ ELECTRICAL CONTROL OF FIXATION AND ERASURE OF HOLOGRAPHIC PATTERNS IN FERROELECTRIC MATERIAL 04-0847
.   WAVELENGTH DEPENDENCE OF HOLOGRAPHIC PHENOMENA IN DOPED LITHIUM NIOBATE   11-1785
ULATION (BY PASSING COHERENT LIGHT THROUGH DEUTER/ REAL-TIME HOLOGRAPHIC RECONSTRUCTION BY ELECTROOPTIC MOD 17-2507
                                  HOLOGRAPHIC RECORDING IN LITHIUM NIOBATE.   11-1662
ROELECTRICS. (LEAD LANTHANUM ZIRCONUM TITANA/ INSCRIPTION OF HOLOGRAPHIC RESULTS IN TRANSPARENT CERAMIC FER 09-1575
D LITHIUM NIOBATE AS AN IMPROVED NEW MATERIAL FOR REVERSIBLE HOLOGRAPHIC STORAGE.   RH DOPE   08-1359
SING IN DOPED BARIUM TITANATE CRYSTALS.   HOLOGRAPHIC STORAGE, ELECTRICAL FIXING AND ERA   04-0848
BATE.   HOLOGRAPHIC STORAGE IN DOPED BARIUM SODIUM NIO   13-1898
                  COUPLED WAVE ANALYSIS OF HOLOGRAPHIC STORAGE IN LITHIUM NIOBATE.   11-1826
ETAL DOPED LITHIUM NIOBATE.   OPTICAL AND HOLOGRAPHIC STORAGE PROPERTIES OF TRANSITION M   11-1796
T, CURIE POINT) OF REDUCED BARIUM TITANATE CRYSTALS (POLARON HOPPING INTERPRETATION). / (SEEBECK COEFFICIEN 04-0918
TE(0.333) NIOBATE(0.667)).   ELECTROOPTIC EFFECTS OF HOT PRESSED CERAMICS IN THE SYSTEM LEAD (ZINCA 13-1900
/IZATION OF LEAD ZIRCONATE- LEAD TITANATE MATERIALS. PART-2: HOT PRESSED COMPOSITIONS AT NEAR THEORETICAL / 09-1578
LANTHANUM ZIRCONATE TITANATE SYSTEM.   IMPROVED HOT PRESSED ELECTROOPTIC CERAMICS IN THE LEAD   09-1528
E) CERAMICS. (IN RELATION TO DENSI/ ELECTRICAL PROPERTIES OF HOT PRESSED FERROELECTRIC (BARIUM TITANATE TYP 04-0879
OPTICAL APPLICATIONS. (LEAD ZIRCONATE TITANATE AND SODIUM P/ HOT PRESSED FERROELECTRIC CERAMICS FOR ELECTRO 09-1527
OPTICAL PROPERTIES (ELECTRICALLY VARIABLE BIREFRINGENCE) OF (HOT PRESSED) FERROELECTRIC CERAMICS. (PZT)   09-1554
ATE CERAMICS FOR ELECTROOPTICAL APPLICATIONS.   HOT PRESSED FERROELECTRIC LEAD ZIRCONATE TITAN   09-1524
NATE) FERROELECTRIC CERAMICS FOR ELECTROOPTIC APPLICATIONS. HOT PRESSED (LEAD, LANTHANUM) (ZIRCONATE, TITA 09-1526
ONTAINING BISMUTH.   HOT PRESSED LEAD ZIRCONATE TITANATE CERAMICS C   09-1525
  GRAIN SIZE, / RECENT DEVELOPMENTS IN ELECTROOPTIC CERAMICS (HOT PRESSED LEAD ZIRCONATE TITANATE, EFFECT OF 02-0086
/ERAMICS- NEW MATERIALS FOR INFORMATION STORAGE AND DISPLAY (HOT PRESSED LEAD ZIRCONATE TITANATE, TUTORIAL/ 09-1550
       PREPARATION OF TRANSPARENT FERROELECTRIC CERAMICS. (HOT PRESSING)   02-0103
                      CONTINUOUS HOT PRESSING OF POTASSIUM SODIUM NIOBATE.   13-2007
                                  HOT PRESSING OF POTASSIUM- SODIUM NIOBATE.   09-1542
UM TITANATE CERAMICS AND CRYSTALS.   INFLUENCE OF HUMIDITY ON THE POTENTIAL DISTRIBUTION IN BARI   04-0883
TE. PART-3: ROCHELLE SALT. PART-4: LITHIUM AMMONIUM TARTRATE HYDRATE. /: TGS. PART-2: SILVER SODIUM DINITRI 02-0077
S IN THE TWO DIMENSIONAL ANTIFERROMAGNET COPPER FORMATE TETRAHYDRATE.   ANOMALOUS SHIFT OF ESR LINE   26-2941
HEAT OF FERROELECTRIC THIOUREA AND POTASSIUM FERROCYANIDE TRIHYDRATE.   ANOMALOUS SPECIFIC   24-2902
RELAXATION BY QUASI SPIN WAVES IN POTASSIUM FERROCYANIDE TRIHYDRATE.   ANOMALOUS SPIN LATTICE   24-2899
NGS OF TITANIUM(3+) ION IN RUBIDIUM ALUMINUM DISULFATE DODECAHYDRATE. /F CRYSTAL FIELD ENERGY LEVEL SPLITTI 19-2754
           CRYSTAL DATA FOR LITHIUM THALLIUM TARTRATE MONOHYDRATE.   29-3247
       CRYSTAL STRUCTURE OF PIEZOELECTRIC LITHIUM FORMATE MONOHYDRATE.   26-2940
D NMR STUDIES OF THE PHASE TRANSITION IN STANNOUS CHLORIDE DIHYDRATE.   DIELECTRIC AN   36-3582
   DOMAIN STRUCTURE OF FERROELECTRIC SODIUM AMMONIUM SULFATE DIHYDRATE.   19-2778
L PROPERTIES OF THE FERROELECTRIC SODIUM AMMONIUM SELENATE DIHYDRATE.   ELECTROMECHANICA   22-2884
STUDY OF SOME METAL IONS IN GUANIDINIUM ALUMINUM SULFATE HEXAHYDRATE.   ELECTRON PARAMAGNETIC RESONANCE 21-2817
+) COMPLEXES IN CRYSTALS OF GUANIDINIUM ALUMINUM SULFATE HEXAHYDRATE.   EPR STUDY OF MAGNETIC COPPER (2 21-2812
DEG K AND 120-300 DEG K, EXAMPLE SODIUM AMMONIUM SELENATE DIHYDRATE). /UNDS (PROTON MAGNETIC RESONANCE, 84 22-2826
ROWTH OF ADP, KDP, HYDROIODIC ACID, AND AMMONIUM OXALATE MONOHYDRATE). /ROM AQUEOUS SOLUTIONS (APPARATUS, G 17-2420
IPOLE LATTICES. (THEORY APPLIED TO POTASSIUM FERROCYANIDE TRIHYDRATE)   INTERNAL FIELD IN GENERAL D   03-0220
TICAL EFFECT IN THE FERROELECTRIC SODIUM AMMONIUM SELENATE DIHYDRATE.   LINEAR ELECTROOP   22-2829
     MAGNETISM AND THE PHASE TRANSITION OF COPPER FORMATE TETRAHYDRATE.   26-2930
MAGNETS POTASSIUM COPPER TRIFLUORIDE AND COPPER FORMATE TETRAHYDRATE. /ONS IN THE LOW DIMENSIONAL ANTIFERRO 26-2939
REORIENTATING STRUCTURAL GROUPS. (SODIUM AMMONIUM SELENATE DIHYDRATE) /C PHASE TRANSITION IN CRYSTALS WITH 22-2827
AMMONIUM SULFATE DODECAHYDRATE AND POTASSIUM FERROCYANIDE TRIHYDRATE. /FERROELECTRIC TRANSITIONS IN FERRIC 19-2755
AMMONIUM SULFATE DODECAHYDRATE AND POTASSIUM FERROCYANIDE TRIHYDRATE. /FERROELECTRIC TRANSITIONS IN FERRIC 24-2892
SBAUER RESONANCE OF SINGLE CRYSTAL POTASSIUM FERROCYANIDE TRIHYDRATE.   MOS   24-2890
LECTRONIC SPIN FLIP PROCESSES IN AMMONIUM IRON SULFATE DODECAHYDRATE.   MOSSBAUER STUDY OF E   19-2771
FERROELECTRIC PHASE TRANSITION IN POTASSIUM FERROCYANIDE TRIHYDRATE.   MOSSBAUER STUDY OF THE   24-2901
NONLINEAR OPTICAL SUSCEPTIBILITY OF LITHIUM FORMATE MONOHYDRATE.   26-2945
TE, AMMONIUM FLUOBERYLLATE, GUANIDINIUM ALUMINUM SULFATE HEXAHYDRATE) /OMENOLOGICAL THEORY. (AMMONIUM SULFA 02-0040
       RAMAN AND INFRARED SPECTRA OF COPPER FORMATE TETRAHYDRATE.   26-2936
TANCES. PART-1: IRON(X) ALUMINUM(1-X) AMMONIUM SULFATE DODECAHYDRATE. /CTRA OF MAGNETIZED PARAMAGNETIC SUBS 19-2751
N IN A TWO DIMENSIONAL ANTIFERROMAGNETIC COPPER FORMATE TETRAHYDRATE.   THERMAL CONDUCTIO   26-2932
ROELECTRICS ROCHELLE SALT AND GUANIDINE ALUMINUM SULFATE HEXAHYDRATE. /DIES OF THE GAMMA-RAY IRRADIATED FER 21-2812
GROUP OF THE ANTIFERROELECTRIC PHASE OF COPPER FORMATE TETRAHYDRATE.   UNIT CELL AND SPACE   26-2927
NT OF THE CRYSTAL STRUCTURE OF LITHIUM AMMONIUM TARTRATE MONOHYDRATE.   X-RAY REFINEME   29-3244
FECT OF ABSORPTION LINES IN GUANIDINIUM CHROMIUM SULFATE HEXAHYDRATE.   ZEEMAN EFFECT AND DILUTION EF 21-2821
INES IN (THE RED REGION OF) CHROMIUM GUANIDINIUM SULFATE HEXAHYDRATE. ZEEMAN EFFECT (STUDY) OF TWO FINE L 21-2814
        IS SODIUM RUBIDIUM TARTRATE TETRAHYDRATE A FERROELECTRIC. (ANSWER-NO)   29-3245
/ON STUDY OF FERROELECTRIC POTASSIUM FERROCYANIDE TRIDEUTERO HYDRATE ABOVE THE CURIE TEMPERATURE (LATTICE / 24-2903
/VERAL CRYSTALS OF ORDER DISORDER TYPE. (COPPER FORMATE TETRAHYDRATE, AMMONIUM IRON AND CHROMIUM ALUMS, ME/ 02-0087

/IBRATION SPECTRUM OF THE FORMATE ION IN COPPER FORMATE TETRAHYDRATE AND COPPER DEUTERO FORMATE TETRADEUTE/ 26-2926
/CTRA (5200-4200 PER CM) OF GUANIDINIUM ALUMINUM SULFATE HEXAHYDRATE AND FREQUENCIES OF THE FUNDAMENTAL VI/ 21-2811
DIO SPECTROSCOPIC INVESTIGATION OF POTASSIUM FERROCYANIDE TRIHYDRATE AND ISOMORPHOUS COMPOUNDS.   RA 24-2900
RE OF FERROELECTRIC CRYSTAL GUANIDINIUM ALUMINUM SULFATE HEXAHYDRATE AND ITS CHROMIUM ISOMORPH. /OF STRUCTU 21-2819
    FERROELECTRIC MOMENT IN GUANIDINIUM ALUMINUM SULFATE HEXAHYDRATE (AND MECHANISM OF REVERSAL).   21-2818
/ FERROELECTRIC TRANSITIONS IN FERRIC AMMONIUM SULFATE DODECAHYDRATE AND POTASSIUM FERROCYANIDE TRIHYDRATE. 19-2755
/ FERROELECTRIC TRANSITIONS IN FERRIC AMMONIUM SULFATE DODECAHYDRATE AND POTASSIUM FERROCYANIDE TRIHYDRATE. 24-2892
ALUMS: METHYL AMMONIUM ALUMINUM (SULFATE OR SELENATE) DODECAHYDRATE. (AND THE DEUTERATED SULFATE) /LECTRIC 20-2803
MAGNETIC SUSCEPTIBILITY OF GUANIDINIUM VANADIUM SULFATE HEXAHYDRATE AT LOW TEMPERATURES.   21-2815
    STUDY OF THE ORDERED MAGNETIC STATE OF COPPER FORMATE TETRAHYDRATE BY ANTIFERROMAGNETIC RESONANCE. 26-2943
STRUCTURE FOR ANTIFERROELECTRIC PHASE OF COPPER FORMATE TETRAHYDRATE BY NEUTRON DIFFRACTION. APPROXIMATE 26-2935
    IN PARAELECTRIC AND FERROELECTRIC POTASSIUM FERROCYANIDE TRIHYDRATE BY PULSED NMR.   STUDY OF HYDROGEN-1 24-2889
NING OF THE FERROELECTRIC CRYSTALS POTASSIUM FERROCYANIDE TRIHYDRATE BY THE NMR METHOD.   POLYTYPE AND TWIN 24-2898
IC STUDIES ON FERROELECTRIC GUANIDINIUM ALUMINUM SULFATE HEXAHYDRATE CLEAVAGES AND THEIR ETCH PATTERNS. /TR 21-2816
/ AND CHROMIUM ALUMS, METHYL AMMONIUM ALUMINUM SULFATE DODECAHYDRATE, COLEMANITE, SODIUM TRIDEUTERIUM DISE/ 02-0087
N FERROELECTRIC MASD, METHYL AMMONIUM ALUMINUM SULFATE DODECAHYDRATE. (COMPLEX DIELECTRIC CONSTANT) /TION I 20-2804
TIC RESONANCE OF SODIUM-23 IN THE SODIUM AMMONIUM SELENATE DIHYDRATE CRYSTAL.   NUCLEAR MAGNE 22-2883
E (/ FERROELECTRIC TRANSITION IN A POTASSIUM FERROCYANIDE TRIHYDRATE CRYSTAL UNDER HIGH HYDROSTATIC PRESSUR 24-2896
    FERROELECTRIC PHASE TRANSITION IN SODIUM AMMONIUM SULFATE DIHYDRATE. (CRYSTALS) /AND SOME OTHER STUDIES OF 19-2767
TIC RESONANCE AND X-RAY STUDIES OF POTASSIUM FERROCYANIDE TRIHYDRATE CRYSTALS.   NUCLEAR MAGNE 24-2894
    PREPARATION AND ANALYSIS OF SODIUM AMMONIUM SELENATE DIHYDRATE CRYSTALS.   22-2863
/SQ CM)   FERROELECTRIC BEHAVIOR OF POTASSIUM FERROCYANIDE TRIHYDRATE CRYSTALS AT HIGH PRESSURE. (TO 5100 KG 24-2897
F NONDEUTERATED AND DEUTERATED LITHIUM THALLIUM TARTRATE MONOHYDRATE. (EFFECT ON LATTICE AND TRANSITION) /O 29-3239
/N IN FERROELECTRIC CRYSTALS. (LITHIUM THALLIUM TARTRATE MONOHYDRATE, ELECTRIC CONTROL OF ELASTIC COMPLIAN/ 02-0060
C CRYSTALS. (BISMUTH TITANATE, LITHIUM THALLIUM TARTRATE MONOHYDRATE, LEAD(5) GERMANIUM(3) OXIDE(11)) /CTRI 02-0061
LASTIC COMPLIANCE OF FERROELECTRIC LITHIUM THALLIUM TARTRATE HYDRATE (LTT) BY AN ELECTRIC FIELD. / OF THE E 29-3249
/UER EFFECT STUDY OF FERROELECTRIC POTASSIUM FERROCYANIDE TRIHYDRATE (NEAR THE TRANSITION TEMPERATURE, NO / 24-2891
/ERROELECTRIC PROPERTIES OF GUANIDINIUM ALUMINUM SULFATE HEXAHYDRATE. (POLARIZATION VERSUS ELECTRIC FIELD,/ 21-2813
ON FERROELECTRIC PROPERTIES OF FERRIC AMMONIUM SULFATE DODECAHYDRATE SINGLE CRYSTALS.   MOSSBAUER STUDY 19-2770
/REVEALING THE DOMAIN STRUCTURE IN POTASSIUM FERROCYANIDE TRIHYDRATE SINGLE CRYSTALS, P-N JUNCTIONS AND P-/ 24-2904
H. (AMMONIUM DIHYDROGEN PHOSPHATE, POTASSIUM FERROCYANIDE TRIHYDRATE, SODIUM NITRATE) /STICS OF THEIR GROWT 16-2316
SE TRANSITIONS IN FERROELECTRICS. (POTASSIUM FERROCYANIDE TRIHYDRATE, SODIUM NITRITE) /E STUDIES OF THE PHA 16-2255
THERMAL AND SOME OTHER STUDIES OF SODIUM AMMONIUM SULFATE DIHYDRATE, SODIUM TRIHYDROGEN OR TRIDEUTERIUM D/ 19-2779
    DIELECTRIC RESPONSE FUNCTION OF COPPER FORMATE TETRAHYDRATES.   26-2933
    DETERMINATION OF FERROELECTRIC PROPERTIES OF LITHIUM HYDRAZINIUM FLUOBERYLLATE.   32-3349
    CRYSTALLOGRAPHIC STUDY OF SEVERAL HYDRAZINIUM ORTHO FLUOBERYLLATES.   32-3355
    ANALYSIS OF HYSTERESIS LOOPS IN LITHIUM HYDRAZINIUM SULFATE.   32-3351
N SCATTERING AND THE QUESTION OF FERROELECTRICITY IN LITHIUM HYDRAZINIUM SULFATE.   ANOMALOUS NEUTRO 32-3341
    CRYSTAL STRUCTURE OF FERROELECTRIC LITHIUM HYDRAZINIUM SULFATE.   32-3356
    CRYSTAL STRUCTURE OF LITHIUM HYDRAZINIUM SULFATE.   32-3342
    DEUTERON NMR STUDIES OF LITHIUM HYDRAZINIUM SULFATE.   32-3354
    DIELECTRIC PROPERTIES OF LITHIUM HYDRAZINIUM SULFATE.   32-3344
    DIELECTRIC PROPERTIES OF LITHIUM HYDRAZINIUM SULFATE.   32-3352
    DIELECTRIC PROPERTIES OF LITHIUM HYDRAZINIUM SULFATE.   32-3353
OLAR RELAXATION OF LITHIUM-7 BY HINDERED ROTATORS IN LITHIUM HYDRAZINIUM SULFATE.   DIP 32-3350
    FERROELECTRIC-LIKE BEHAVIOR OF LITHIUM HYDRAZINIUM SULFATE.   32-3348
    HYDROGEN BONDING IN LITHIUM HYDRAZINIUM SULFATE.   32-3346
    MICROWAVE DIELECTRIC STUDY OF LITHIUM HYDRAZINIUM SULFATE.   32-3357
NITROGEN-14 NUCLEAR QUADRUPOLE RESONANCE STUDY OF LITHIUM HYDRAZINIUM SULFATE.   32-3345
NUCLEAR MAGNETIC RESONANCE STUDY OF FERROELECTRIC LITHIUM HYDRAZINIUM SULFATE.   32-3343
ALLIUM TARTRATE, LITHIUM TANTALATE, LITHIUM NIOBATE, LITHIUM HYDRAZINIUM SULFATE) /IUM TARTRATE, LITHIUM TH 02-0052
    FROTON ROTATING FRAME RELAXATION IN LITHIUM HYDRAZINIUM SULFATE.   32-3347
    NEW FERROELECTRIC: SEMICARBAZIDE HYDROCHLORIDE.   36-3603
/TANATE, ANTIMONY SULFOIODIDE, THIOUREA, TRIGLYCINE SULFATE, HYDROGEN AMMONIUM MONOCHLORO ACETATE, DICALCI/ 02-0099
ES ABOVE 80 DEGREES/ RAMAN SPECTRA OF ORTHORHOMBIC AND CUBIC HYDROGEN AND DEUTERIUM CHLORIDES AT TEMPERATUR 31-3330
SPECTRA, DIPOLAR COUPLING MODEL COMP/ LONGITUDINAL MODES IN HYDROGEN AND DEUTERIUM HALIDE CRYSTALS. (RAMAN 31-3326
UDIES OF PHASE TRANSITIONS. (TRIGLYCINE SULFATE, POTASSIUM DIHYDROGEN AND DIDEUTERIUM SULFATE) /ATTERING ST 17-2481
    ESR OF GAMMA-IRRADIATED TGS. PART-2: INTERNAL MOTION OF HYDROGEN AND THE ACTIVATION ENERGY IN GLYCINE. 27-3034
ESONANCE STUDY OF THE FERROELECTRIC TRANSITIONS IN SODIUM TRIHYDROGEN AND TRIDEUTERIUM SELENATE. /AGNETIC R 22-2832
CE IN FERROELECTRIC CESIUM DIDEUTERIUM ARSENATE AND CESIUM DIHYDROGEN ARSENATE. /ICE RELAXATION AND RESONAN 18-2720
RESONANCE OF THE ARSENATE CENTER IN X-IRRADIATED POTASSIUM DIHYDROGEN ARSENATE. /D ELECTRON NUCLEAR DOUBLE 18-2727
D NQR (NUCLEAR QUADRUPOLE RESONANCE) FREQUENCY. (POTASSIUM DIHYDROGEN ARSENATE)   FERROELECTRIC MODE AN 17-2713
UDIES OF LONG RANGE CORRELATION IN FERROELECTRIC POTASSIUM DIHYDROGEN ARSENATE. /D DIELECTRIC RELAXATION ST 18-2718
TE CENTER IN X-RAY IRRADIATED SINGLE CRYSTALS OF POTASSIUM DIHYDROGEN ARSENATE. /E RESONANCE FOR THE ARSENA 18-2729
ROTON PHONON INTERACTIONS IN ALUMINUM PHOSPHATE AND CESIUM DIHYDROGEN ARSENATE.   PHONON PHONON AND P 18-2739
    PROTONIC CONDUCTION IN POTASSIUM DIHYDROGEN ARSENATE.   18-2737
    PURE QUADRUPOLE RESONANCE OF ARSENIC-75 IN AMMONIUM DIHYDROGEN ARSENATE.   18-2732
POLE RESONANCE SPECTRA OF THE FERROELECTRIC CRYSTAL CESIUM DIHYDROGEN ARSENATE. /FECT IN THE NUCLEAR QUADRU 18-2722
/CONFIGURATIONS FROM EPR STUDIES OF X-IRRADIATED POTASSIUM DIHYDROGEN ARSENATE AND AMMONIUM DIHYDROGEN ARS/ 18-2728
/IN PIEZOELECTRIC CRYSTALS. (RAMAN SCATTERING IN POTASSIUM DIHYDROGEN ARSENATE AND CESIUM DIHYDROGEN ARSEN/ 18-2723
RIUM ARSENATE.   ARSENIC-75 NMR RESONANCE IN AMMONIUM DIHYDROGEN ARSENATE AND DEUTERO AMMONIUM DIDEUTE 18-2736
/HE ARSENATE RADICAL IN IRRADIATED FERROELECTRIC POTASSIUM DIHYDROGEN ARSENATE AND POTASSIUM DIDEUTERIUM A/ 18-2734
/ONS OF DOMAINS IN THE IRRADIATED FERROELECTRICS POTASSIUM DIHYDROGEN ARSENATE AND POTASSIUM DIDEUTERIUM A/ 18-2726
SENATE.   THERMODYNAMIC PROPERTIES OF POTASSIUM DIHYDROGEN ARSENATE AND POTASSIUM DIDEUTERIUM AR 18-2730
ENATE.   PROTON PHONON COUPLING IN CESIUM DIHYDROGEN ARSENATE AND POTASSIUM DIHYDROGEN ARS 18-2733
/TRA OF PARTIALLY DEUTERATED FERROELECTRIC CRYSTALS CESIUM DIHYDROGEN ARSENATE AND RUBIDIUM DIHYDROGEN ARS/ 18-2740
/OUPLING TO TRANSVERSE OPTICAL PHONON). PART-3: (POTASSIUM DIHYDROGEN ARSENATE, CESIUM DIHYDROGEN ARSENATE) 03-0384
TION. PART-2: INTERNAL FUNDAMENTAL VIBRATIONS OF POTASSIUM DIHYDROGEN ARSENATE IN THE PARAELECTRIC PHASE. / 18-2738
/RESONANCE OF THE PROTON LATTICE COUPLED MODE IN POTASSIUM DIHYDROGEN ARSENATE, POTASSIUM DEUTERIUM ARSENA/ 18-2724
/ PARAMAGNETIC RESONANCE STUDIES OF X-IRRADIATED POTASSIUM DIHYDROGEN ARSENATE, POTASSIUM DIDEUTERIUM ARSE/ 18-2725
RAMAN SPECTROSCOPY: COUPLED MODES (FERROELECTRICS, CESIUM DIHYDROGEN ARSENATE, STRONTIUM TITANATE) /T MODE 06-1207
ON.   ANTIFERROELECTRIC TRANSITIONS IN ADP AND AMMONIUM DIHYDROGEN ARSENATE STUDIED BY INFRARED ABSORPTI 17-2706
/UTERATED ROCHELLE SALT, POTASSIUM OR RUBIDIUM OR AMMONIUM DIHYDROGEN ARSENATES, RUBIDIUM DIDEUTERIUM ARSE/ 18-2719
A OF GYPSUM AND / TEMPERATURE DEPENDENCE OF THE INTENSITY OF HYDROGEN BOND BANDS IN RAMAN SCATTERING SPECTR 28-3189
    ELECTRIC RESONANCE: APPLICATION TO THE HYDROGEN BOND. (ICE)   34-3477
    DIPOLE MOMENT DERIVATIVE OF THE HYDROGEN BOND IN ICE.   34-3489
    INCOHERENT NEUTRON SCATTERING FROM HYDROGEN BOND IN KDP AND ADP.   17-2416
NS).   INFRARED VIBRATION FREQUENCIES OF THE HYDROGEN BOND IN KDP FERROELECTRIC (CALCULATIO 17-2650
E.   ELECTRON DELOCALIZATION WEIGHT OF THE HYDROGEN BOND IN POTASSIUM DIHYDROGEN PHOSPHAT 17-2525
IUM DIHYDROGEN PHOSPHATE.   NEUTRON MEASUREMENTS ON THE HYDROGEN BOND POTENTIAL IN PARAELECTRIC POTASS 17-2511

RACTION STUDY OF THE FERROELECTRIC LITHIUM TRIHYDROGEN SELE/ HYDROGEN BOND STUDIES. PART-54: A NEUTRON DIFF 22-2878
URE OF AMMONIUM TRIHYDROGEN SELENITE. HYDROGEN BOND STUDIES. PART-58: CRYSTAL STRUCT 22-2879
SELENITE, AND IN POTASSIUM TRIDEUTERIUM SELENITE, AS DE/ THE HYDROGEN BOND SYSTEM IN POTASSIUM TRIHYDROGEN 22-2854
F AMMONIUM DIHYDROGEN PHOSPHATE TYPE ANTIFERROELECTRICS WITH HYDROGEN BONDS. DYNAMICAL THEORY O 03-0476
4000-10000 PER CM) AND HARMONIC VIBRATION FREQUENCIES OF THE HYDROGEN BONDS. /M DIHYDROGEN ORTHOPHOSPHATE (17-2432
QUANTUM EFFECTS IN FERROELECTRIC MATERIALS WITH HYDROGEN BONDS. 03-0347
HYDROGEN PHOSPHATE TYPE CRYSTALS) COLLECTIVE MOTIONS OF HYDROGEN BONDS. (CALCULATIONS FOR POTASSIUM DI 17-2483
ON THE SYMMETRY OF THE HYDROGEN BONDS IN ICE-VII. 34-3445
/RAMAN BANDS OF THE HYDROXYL VIBRATIONS AND STRUCTURE OF THE HYDROGEN BONDS IN THE FERROELECTRIC CRYSTALS / 17-2471
VIBRATION SPECTRA OF HYDROGEN BONDS. (KDP) 17-2482
) NEUTRON MEASUREMENT IN TWO SYSTEMS OF HYDROGEN BONDS. (KDP AND POTASSIUM BICARBONATE 17-2417
ES OF PROTONS ON THE PHASE TRANSITION IN FERROELECTRICS WITH HYDROGEN BONDS. (KDP, DEUTERATED KDP) /AR STAT 17-2671
PROTON PHONON INTERACTION IN FERROELECTRICS WITH HYDROGEN BONDS. (KDP TYPE) 17-2672
N INTERACTION IN THE PHASE TRANSITION OF FERROELECTRICS WITH HYDROGEN BONDS KDP TYPE. ROLE OF PROTON PHONO 17-2674
C NEUTRALITY ON THE PHASE TRANSITION IN A FERROELECTRIC WITH HYDROGEN BONDS (THEORY, WU MODEL). /OF ELECTRI 03-0449
ENTROPY OF ICE, PART-2, HYDROGEN BON/ LATTICE STATISTICS OF HYDROGEN BONDED CRYSTALS, PART-1, THE RESIDUAL 03-0388
CTRIC SOLIDS. PART-1: ENERGY LOSS NEUTRON/ PROTON MOTIONS IN HYDROGEN BONDED FERROELECTRIC AND ANTIFERROELE 17-2687
Y, KDP TYPE) PROTON LATTICE INTERACTIONS IN HYDROGEN BONDED FERROELECTRIC CRYSTALS. (THEOR 03-0181
CRITICAL BEHAVIOR OF HYDROGEN BONDED FERROELECTRICS. 03-0516
EFFECT OF POINT DEFECTS ON THE ENERGY SPECTRUM IN HYDROGEN BONDED FERROELECTRICS. 03-0486
ROELECTRICITY, PART-1: SPECTRAL ANALYSIS (PROTON DYNAMICS IN HYDROGEN BONDED FERROELECTRICS). /HEORY OF FER 03-0402
RANDOM PHASE APPROXIMATION FOR THE BLINC-DE GENNES MODEL FOR HYDROGEN BONDED FERROELECTRICS. 03-0338
TEMPERATURE. PROTON DYNAMICS IN HYDROGEN BONDED FERROELECTRICS ABOVE THE CURIE 03-0206
URES. PART-1: GENE/ THEORY OF COHERENT NEUTRON SCATTERING BY HYDROGEN BONDED FERROELECTRICS AT LOW TEMPERAT 03-0478
CHELLE SALT, POTASSIUM OR RUBIDIUM OR A/ NMR STUDIES IN SOME HYDROGEN BONDED FERROELECTRICS. (DEUTERATED RO 18-2719
LATER CONFIGURATIONS FROM EPR STUDIES OF X-IRRADIATED POTAS/ HYDROGEN BONDED FERROELECTRICS: EVIDENCE FOR S 18-2728
RPTION, NEUTRON INELASTIC SCATTERING,/ FREQUENCY SPECTRUM OF HYDROGEN BONDED FERROELECTRICS. (INFRARED ABSO 03-0401
RY). IMPURITY MODES IN AN INDUCED MOMENT SYSTEM AND HYDROGEN BONDED FERROELECTRICS (KDP TYPE, THEO 03-0426
ROGEN SELENITE. THERMAL EXPANSION OF HYDROGEN BONDED FERROELECTRICS. LITHIUM TRIHYD 22-2857
(PHONON FRO/ COUPLED VIBRATIONS OF THE PROTON ION SYSTEM IN HYDROGEN BONDED FERROELECTRICS OF THE KDP TYPE 17-2670
GS AND (FREQUENCY) SHIFTS (OF PROTONIC / DYNAMICAL THEORY OF HYDROGEN BONDED FERROELECTRICS. PART-1: DAMPIN 03-0475
LISM OF LOCAL MODE MODEL.) PART-2: (DYNAMICAL PRO/ THEORY OF HYDROGEN BONDED FERROELECTRICS. PART-1: (FORMA 03-0384
IMATED SOLUTIONS TO THE KINETIC EQUATION/ PROTON DYNAMICS IN HYDROGEN BONDED FERROELECTRICS. PART-2: APPROX 03-0205
ROGEN SELENATE AND SODIUM TRIDEUTE/ DIELECTRIC PROPERTIES OF HYDROGEN BONDED FERROELECTRICS. (SODIUM TRIHYD 22-2855
FATE AND POTASSIUM DIHYDROGEN PH/ HIGH FREQUENCY BEHAVIOR OF HYDROGEN BONDED FERROELECTRICS: TRIGLYCINE SUL 17-2526
IR RADIATION. PART-2: INTERNAL FUNDA/ STRUCTURAL STUDIES OF HYDROGEN BONDED FERROELECTRICS USING POLARIZED 18-2738
/ART-4: PHASE TRANSITIONS OF AMMONIUM SULFATE: THE NATURE OF HYDROGEN BONDING AND THE REORIENTATION OF THE/ 19-2783
E. HYDROGEN BONDING IN LITHIUM HYDRAZINIUM SULFAT 32-3346
/S OF PROTONS (AND DEUTERONS) IN FERROELECTRIC CRYSTALS WITH HYDROGEN BONDING. (POTASSIUM DIHYDROGEN PHOSP/ 17-2568
BROMINE NUCLEAR QUADRUPOLE RESONANCE IN FERROELECTRIC HYDROGEN BROMIDE. 31-3332
CRYSTAL STRUCTURE AND PHASE TRANSITION OF HYDROGEN CHLORIDE. 31-3334
QUADRUPOLE FREQUENCIES AND RELAXATION TIMES IN FERROELECTRIC HYDROGEN CHLORIDE. NUCLEAR 31-3336
STRUCTURE AND PHASE TRANSITION IN HYDROGEN CHLORIDE. 31-3337
APPROXIMATI/ MOLECULAR MOTION IN THE FERROELECTRIC PHASE OF HYDROGEN CHLORIDE. (CALCULATED IN THE HARMONIC 31-3340
CRYSTAL SIZE EFFECTS IN THE RAMAN SPECTRA OF ORTHORHOMBIC HYDROGEN CHLORIDE CRYSTALS. 31-3338
POTASSIUM DIHYDROGEN PHOSPHATE (NEUTRON SCATTERING). HYDROGEN DENSITY DISTRIBUTION IN PARAELECTRIC 17-2490
STR/ THE CRYSTAL STRUCTURE AND PHASE TRANSITION OF AMMONIUM HYDROGEN DICHLORO ACETATE. PART-1: THE CRYSTAL 36-3567
PECIFIC HEAT AND CONFIGURATIONAL ENTROPY CHANGE IN SODIUM TRIHYDROGEN DISELENITE. ANOMALOUS S 22-2856
E ELECTRON SPIN RESONANCE SPECTRA IN FERROELECTRIC SODIUM TRIHYDROGEN DISELENITE. /E (ION). RADIATION DAMAG 22-2870
GAMMA-IRRADIATED SINGLE CRYSTAL OF FERROELECTRIC LITHIUM TRIHYDROGEN DISELENITE. /ARAMAGNETIC RESONANCE IN 22-2842
SR STUDY OF THE FERROELECTRIC PHASE TRANSITIONS IN SODIUM TRIHYDROGEN DISELENITE. E 22-2871
HEAT OF FORMATION OF LITHIUM TRIHYDROGEN DISELENITE. 22-2867
NEUTRON DIFFRACTION STUDY ON PARAELECTRIC SODIUM TRIHYDROGEN DISELENITE. 22-2846
GAMMA-IRRADIATED SINGLE CRYSTAL OF FERROELECTRIC LITHIUM TRIHYDROGEN DISELENITE. PARAMAGNETIC SPECIES IN 22-2843
RELAXATION BETWEEN THE STRUCTURE OF FERROELECTRIC SODIUM TRIHYDROGEN DISELENITE AND ITS THERMAL EXPANSION. 22-2851
FAR INFRARED REFLECTIVITY SPECTRA OF SODIUM TRIHYDROGEN DISELENITE CRYSTAL. 22-2877
COERCIVE FIELD IN LITHIUM TRIHYDROGEN DISELENITE. (EXPERIMENTAL) 22-2844
/TRIC AND OPTICAL PROPERTIES OF) CRYSTALS OF THE ALKALINE TRIHYDROGEN DISELENITE FAMILY. (ESPECIALLY THE S/ 22-2848
/ REVERSAL MECHANISM IN FERROELECTRICS: LITHIUM OR SODIUM TRIHYDROGEN DISELENITE, NUCLEAR MAGNETIC RESONANC 22-2837
RISON WITH FERROELECTRIC). RAMAN SPECTRUM OF SODIUM TRIHYDROGEN DISELENITE (PARAELECTRIC PHASE, COMPA 22-2864
ELECTRIC PROPERTIES OF HYDROGEN DOPED STRONTIUM TITANATE. 06-1088
DIFFUSION OF HYDROGEN FLUORIDE IN ICE. 34-3437
EN BONDING IN THE FERROELECTRIC MECHAN/ PHASE TRANSITIONS OF HYDROGEN HALIDE CRYSTALS (IMPORTANCE OF HYDROG 31-3329
EFFECT OF CHARGE TRANSFER ON RAMAN SCATTERING OF HYDROGEN HALIDES. 31-3328
INTERMOLECULAR POTENTIAL AND FERROELECTRIC TRANSITIONS IN HYDROGEN HALIDES. 31-3339
PHASE TRANSITIONS OF SOLID HYDROGEN HALIDES. 31-3327
STRESS FREE SINGLE CRYSTAL GROWTH OF FERROELECTRIC HYDROGEN HALIDES. 31-3333
TRON DIFFRA/ FERROELECTRICITY AND PHASE TRANSITIONS IN SOLID HYDROGEN HALIDES (SINGLE CRYSTAL X-RAY AND NEU 31-3331
(TEMPERATURE AND) ELECTRIC FIELD DEPENDENCE OF A HYDROGEN IMPURITY MODE IN STRONTIUM TITANATE. 06-1146
TRIC BARIUM SODIUM NIOBATE. ROLE OF HYDROGEN IN POLARIZATION REVERSAL OF FERROELEC 13-2025
NOTE ON THE SPACE GROUP OF POTASSIUM HYDROGEN IODATE. 08-1370
NUCLEAR QUADRUPOLE RESONANCE OF IODINE-127 IN SOLID HYDROGEN IODIDE. 31-3335
ATE AND SODIUM TRIHYDROGEN SELENATE TYPE CRYSTALS. ULTRASLOW HYDROGEN MOTION IN POTASSIUM DIHYDROGEN PHOSPH 17-2441
N AND THERMOOPTICAL AND ELECTROOPTICAL EFFECT IN RUBIDIUM TRIHYDROGEN OR DEUTERIUM DISELENITE CRYSTALS. /IO 22-2840
PHATE/ OPTICAL STUDY ON THE PHASE TRANSITIONS IN POTASSIUM DIHYDROGEN OR DEUTERIUM PHOSPHATE (INFRARED PHOS 17-2646
/TRIBUTION TO THE PREPARATION ON NON-WEDGE SHAPED RUBIDIUM DIHYDROGEN OR DIDEUTERIUM) PHOSPHATE CRYSTALS T/ 17-2637
/HER STUDIES OF SODIUM AMMONIUM SULFATE DIHYDRATE, SODIUM TRIHYDROGEN OR TRIDEUTERIUM DISELENITE. (SUMMARY/ 19-2779
D/ LOW TEMPERATURE INFRARED ABSORPTION SPECTRA OF AMMONIUM DIHYDROGEN ORTHOPHOSPHATE (4000-10000 PER CM) AN 17-2432
/R QUADRUPOLE INTERACTION IN ANTIFERROELECTRIC DIAMMONIUM TRIHYDROGEN PERIODATE BY IODINE PROTON CROSSOVER/ 36-3579
/EVEL CROSSING (CROSS RELAXATION) EXPERIMENTS IN AMMONIUM TRIHYDROGEN PERIODATE CRYSTALS, PULSED MAGNETIC / 36-3577
/R QUADRUPOLE INTERACTION IN ANTIFERROELECTRIC DIAMMONIUM TRIHYDROGEN PERIODATE PROTON CROSSOVER RELAXATIO/ 36-3578
A THEORY OF THE PHASE TRANSITION IN AMMONIUM DIHYDROGEN PHOSPHATE. 03-0286
E SIGNS OF THE NONLINEAR OPTICAL COEFFICIENTS OF POTASSIUM DIHYDROGEN PHOSPHATE. ABSOLUT 17-2597
ODULATORS BASED ON THE MICHELSON INTERFEROMETER. (AMMONIUM DIHYDROGEN PHOSPHATE) AMPLITUDE M 17-2715
BRILLOUIN DIFFUSION IN A SINGLE CRYSTAL OF RUBIDIUM DIHYDROGEN PHOSPHATE. 17-2520
BRILLOUIN SCATTERING BY A SINGLE CRYSTAL OF RUBIDIUM DIHYDROGEN PHOSPHATE. 17-2519
CRYSTAL PERFECTION OF AMMONIUM DIHYDROGEN PHOSPHATE. 17-2426
IDE, SODIUM CHLORIDE, POTASSIUM SODIUM TARTRATE, POTASSIUM DIHYDROGEN PHOSPHATE) / BROMIDE, POTASSIUM CHLOR 17-2707
TAINING O-D...O BOND. (INCLUDING FULLY DEUTERATED AMMONIUM DIHYDROGEN PHOSPHATE) /TUDY OF SOME CRYSTALS CON 03-0204
ELECTRIC EVIDENCE OF A FIRST ORDER TRANSITION IN POTASSIUM DIHYDROGEN PHOSPHATE. DI 17-2602

TRIC PERMITTIVITY OF FINELY DIVIDED POTASSIUM AND AMMONIUM DIHYDROGEN PHOSPHATE.   DIELEC 17-2689
                 DIELECTRIC PROPERTIES OF DEUTERATED POTASSIUM DIHYDROGEN PHOSPHATE.   17-2581
N FERROELECTRIC CRYSTALS WITH HYDROGEN BONDING. (POTASSIUM DIHYDROGEN PHOSPHATE) /PROTONS (AND DEUTERONS) I 17-2568
T OF DEUTERON SUBSTITUTION ON DOMAIN DYNAMICS IN POTASSIUM DIHYDROGEN PHOSPHATE.   EFFEC 17-2437
M BY ALUMINUM ON THE FERROELECTRIC PROPERTIES OF POTASSIUM DIHYDROGEN PHOSPHATE. /F REPLACEMENT OF POTASSIU 17-2560
FIELD ON PARAMAGNETIC SPECTRA OF CHROMIUM(3+) IN POTASSIUM DIHYDROGEN PHOSPHATE. /XIAL STRESS AND ELECTRIC 17-2554
CAL INVESTIGATION OF FERROELECTRIC PROPERTIES OF POTASSIUM DIHYDROGEN PHOSPHATE.   ELECTRICAL AND OPTI 17-2449
                 ELECTROCHEMISTRY ON THE POLYCRYSTAL OF POTASSIUM DIHYDROGEN PHOSPHATE.   17-2536
ON DELOCALIZATION WEIGHT OF THE HYDROGEN BOND IN POTASSIUM DIHYDROGEN PHOSPHATE.   ELECTR 17-2525
TRON SPIN RESONANCE OF IRRADIATED AND DEUTERATED POTASSIUM DIHYDROGEN PHOSPHATE.   ELEC 17-2533
ODULATION OF LASER GENERATED SECOND HARMONICS IN POTASSIUM DIHYDROGEN PHOSPHATE.   ELECTROOPTIC AMPLITUDE M 17-2694
-RAY IRRADIATED SINGLE CRYSTALS OF FERROELECTRIC POTASSIUM DIHYDROGEN PHOSPHATE.   ESR INVESTIGATIONS OF X 17-2558
OCATIONS AT DOMAIN TIPS IN FERROELECTRIC CRYSTAL POTASSIUM DIHYDROGEN PHOSPHATE.   EXISTENCE OF DISL 17-2450
        EXPERIMENTAL PROOF FOR PROTON TUNNELING IN POTASSIUM DIHYDROGEN PHOSPHATE.   17-2510
             FAR INFRARED ABSORPTION SPECTRA FOR AMMONIUM DIHYDROGEN PHOSPHATE.   17-2587
H USING) TRANSVERSE POCKELS EFFECT COMMUTATORS. (POTASSIUM DIHYDROGEN PHOSPHATE) /NITE APERTURE LIGHT SWITC 17-2509
EN BONDED FERROELECTRICS: TRIGLYCINE SULFATE AND POTASSIUM DIHYDROGEN PHOSPHATE. /QUENCY BEHAVIOR OF HYDROG 17-2526
                                 INDEX OF REFRACTION OF POTASSIUM DIHYDROGEN PHOSPHATE.   17-2484
INFLUENCE OF TUNNELING ON DIELECTRIC BEHAVIOR OF POTASSIUM DIHYDROGEN PHOSPHATE.   17-2527
INTENSITY OF SOME BRILLOUIN AND RAYLEIGH LINES IN RUBIDIUM DIHYDROGEN PHOSPHATE.   17-2521
    BETWEEN PROTON TUNNELING AND OPTICAL PHONONS IN POTASSIUM DIHYDROGEN PHOSPHATE.   INTERACTION 17-2593
    BETWEEN PROTON TUNNELING AND OPTICAL PHONONS IN POTASSIUM DIHYDROGEN PHOSPHATE.   INTERACTIONS 17-2592
ERNAL STRESS AND THE FERROELECTRIC TRANSITION IN POTASSIUM DIHYDROGEN PHOSPHATE.   INT 17-2424
TH POLARIZATION FLUCTUATIONS IN ANTIFERROELECTRIC AMMONIUM DIHYDROGEN PHOSPHATE.   LONG WAVELENG 17-2461
                LOW FREQUENCY HYDROGEN VIBRATIONS IN AMMONIUM DIHYDROGEN PHOSPHATE.   17-2699
                LOW FREQUENCY HYDROGEN VIBRATIONS IN POTASSIUM DIHYDROGEN PHOSPHATE.   17-2608
THERMAL EXPANSION MEASUREMENTS ON SINGLE CRYSTAL POTASSIUM DIHYDROGEN PHOSPHATE.   LOW TEMPERATURE 17-2464
STIMULATED POLARIZATION ECHO IN FERROELECTRICS. (POTASSIUM DIHYDROGEN PHOSPHATE)   MACROSCOPIC 17-2419
        MEASUREMENT OF CHEMICAL SHIFT IN SOLID POTASSIUM DIHYDROGEN PHOSPHATE.   17-2688
        MEASUREMENTS OF THE SPONTANEOUS POLARIZATION IN POTASSIUM DIHYDROGEN PHOSPHATE.   17-2430
CRYSTALS BY THE GRADUAL ADDITION OF A REACTANT. (POTASSIUM DIHYDROGEN PHOSPHATE) /TH IN SOLUTION OF SINGLE 17-2595
                       MODE INTERACTIONS IN POTASSIUM DIHYDROGEN PHOSPHATE.   17-2701
DY OF ANTIFERROELECTRIC TRANSITION IN IRON-57(3X) AMMONIUM DIHYDROGEN PHOSPHATE.   MOSSBAUER STU 17-2644
S ON THE HYDROGEN BOND POTENTIAL IN PARAELECTRIC POTASSIUM DIHYDROGEN PHOSPHATE.   NEUTRON MEASUREMENT 17-2511
NTS ON THE FERROELECTRIC PHASE TRANSFORMATION IN POTASSIUM DIHYDROGEN PHOSPHATE.   NEUTRON MEASUREME 17-2418
RESONANCE STUDY OF POTASSIUM-39 IN FERROELECTRIC POTASSIUM DIHYDROGEN PHOSPHATE.   NUCLEAR DOUBLE 17-2599
DER OF THE FERROELECTRIC PHASE TRANSFORMATION OF POTASSIUM DIHYDROGEN PHOSPHATE.   OR 17-2555
OF COHERENT LIGHT BY THE ELECTROOPTIC EFFECT IN POTASSIUM DIHYDROGEN PHOSPHATE. /TRIC FREQUENCY CONVERSION 17-2616
        PHASE SYNCHRONIZATION IN NONLINEAR RUBIDIUM DIHYDROGEN PHOSPHATE.   17-2473
ASTIC EFFECT IN 45 DEGREE X-CUTS OF POTASSIUM AND AMMONIUM DIHYDROGEN PHOSPHATE.   PHOTOEL 17-2443
IZATION ECHO IN THE FERROELECTRIC SINGLE CRYSTAL POTASSIUM DIHYDROGEN PHOSPHATE.   POLAR 17-2564
                 PROTON PHONON INTERACTIONS IN POTASSIUM DIHYDROGEN PHOSPHATE.   17-2653
TIONS IN MISMATCHED SECOND HARMONIC GENERATION. (POTASSIUM DIHYDROGEN PHOSPHATE)   PULSE DISTOR 17-2690
RAMAN SCATTERING FROM FERROELECTRIC CRYSTALS OF POTASSIUM DIHYDROGEN PHOSPHATE.   17-2552
E KDP ISOMORPHOUS PHOSPHATES AND ARSENATES. (KDP, AMMONIUM DIHYDROGEN PHOSPHATE) /FERROELECTRIC MODES IN TH 17-2639
                               RAMAN SCATTERING IN POTASSIUM DIHYDROGEN PHOSPHATE.   17-2709
                    RAMAN SPECTRA OF TETRAGONAL POTASSIUM DIHYDROGEN PHOSPHATE.   17-2657
        RANDOM MOTION OF DEUTERONS IN DEUTERATED POTASSIUM DIHYDROGEN PHOSPHATE.   17-2651
LATTICE RELAXATION BY THE FERROELECTRIC MODE IN POTASSIUM DIHYDROGEN PHOSPHATE.   SPIN 17-2440
        SYMMETRY OF ATOMIC VIBRATIONS IN POTASSIUM DIHYDROGEN PHOSPHATE.   17-2586
AL HYSTERESIS OF THE FERROELECTRIC TRANSITION IN POTASSIUM DIHYDROGEN PHOSPHATE.   THERM 17-2682
X-RAY DETERMINATION OF THE THERMAL EXPANSION OF AMMONIUM DIHYDROGEN PHOSPHATE.   17-2485
                     POCKELS EFFECT IN AMMONIUM DIHYDROGEN PHOSPHATE ABOVE ROOM TEMPERATURE.   17-2518
/ROELECTRIC PHOSPHATES NEAR THEIR CURIE POINTS. (POTASSIUM DIHYDROGEN PHOSPHATE, ALSO RUBIDIUM ANALOG AND / 17-2665
/VE MEASUREMENT OF THE OPTICAL NONLINEARITIES OF POTASSIUM DIHYDROGEN PHOSPHATE, AMMONIUM DIHYDROGEN PHOSP/ 11-1677
/OME FERROELECTRIC CRYSTALS (TRIGLYCINE SULFATE, POTASSIUM DIHYDROGEN PHOSPHATE, AMMONIUM SULFATE, ROCHELL/ 02-0125
S/ ELECTROOPTICAL AND PIEZOOPTICAL PROPERTIES OF POTASSIUM DIHYDROGEN PHOSPHATE AND AMMONIUM DIHYDROGEN PHO 17-2697
SPHATE CRYSTALS OF LIG/ THERMAL SELF FOCUSING IN POTASSIUM DIHYDROGEN PHOSPHATE AND AMMONIUM DIHYDROGEN PHO 17-2717
SPHATE CRYSTALS AT MIC/ RELATIVE PERMITTIVITY OF POTASSIUM DIHYDROGEN PHOSPHATE AND AMMONIUM DIHYDROGEN PHO 17-2714
SPHATE/ ULTRAVIOLET REFLECTION AND ABSORPTION OF POTASSIUM DIHYDROGEN PHOSPHATE AND AMMONIUM DIHYDROGEN PHO 17-2422
/ARSENIC-75 QUADRUPOLAR COUPLING IN FERROELECTRIC RUBIDIUM DIHYDROGEN PHOSPHATE AND CESIUM DIHYDROGEN ARSE/ 17-2439
/ENCE OF OPTICAL (SECOND) HARMONIC GENERATION IN POTASSIUM DIHYDROGEN PHOSPHATE (AND DEUTERATED POTASSIUM / 17-2695
/FAR INFRARED REFLECTION SPECTRA OF PARAELECTRIC POTASSIUM DIHYDROGEN PHOSPHATE AND POTASSIUM DIDEUTERIUM / 03-0356
HOSPHATE.   LASER INDUCED SELF FOCUSING IN AMMONIUM DIHYDROGEN PHOSPHATE AND POTASSIUM DIDEUTERIUM P 17-2476
/STUDIES OF THE SOFT OPTIC AND ACOUSTIC MODES OF POTASSIUM DIHYDROGEN PHOSPHATE AND POTASSIUM DIDEUTERIUM / 17-2497
SENATE TC 52 KBAR AND 400 DEG / PHASE DIAGRAMS OF AMMONIUM DIHYDROGEN PHOSPHATE AND POTASSIUM DIHYDROGEN AR 17-2478
OSPHATE EXCITED BY A / PARAMETRIC FLUORESCENCE IN AMMONIUM DIHYDROGEN PHOSPHATE AND POTASSIUM DIHYDROGEN PH 17-2486
/TIBILITIES: ACCURATE RELATIVE VALUES FOR QUARTZ, AMMONIUM DIHYDROGEN PHOSPHATE, AND POTASSIUM DIHYDROGEN / 17-2539
OSPHATE)/ RELAXATION OF FERROELECTRIC PHOSPHATES (AMMONIUM DIHYDROGEN PHOSPHATE AND POTASSIUM DIHYDROGEN PH 17-2580
/ISCHARGE OF CONDENSERS WITH POLAR DIELECTRICS. (POTASSIUM DIHYDROGEN PHOSPHATE AND POTASSIUM FERROCYANIDE) 17-2631
/THE HYDROGEN BONDS IN THE FERROELECTRIC CRYSTALS RUBIDIUM DIHYDROGEN PHOSPHATE AND RUBIDIUM DIHYDROGEN AR/ 17-2471
/ INFRARED ABSORPTION SPECTRA OF MONOCRYSTALS OF POTASSIUM DIHYDROGEN PHOSPHATE AND RUBIDIUM DIHYDROGEN PH/ 17-2457
/Y LOSS NEUTRON INCOHERENT SCATTERING SPECTRA OF POTASSIUM DIHYDROGEN PHOSPHATE AND SILVER PERIODATE BELOW/ 17-2687
NATE TYPE CRYSTALS. ULTRASLOW HYDROGEN MOTION IN POTASSIUM DIHYDROGEN PHOSPHATE AND SODIUM TRIHYDROGEN SELE 17-2441
ITY THEORY.   ELASTIC PROPERTIES OF AMMONIUM DIHYDROGEN PHOSPHATE AND THE LAVAL RAMAN ELASTIC 17-2410
TALS.   THERMAL VARIATION OF REFRACTIVE INDEX IN AMMONIUM DIHYDROGEN PHOSPHATE AND TRIGLYCINE SULFATE CRYS 17-2474
             VANISHING OF FERROELECTRICITY IN POTASSIUM DIHYDROGEN PHOSPHATE AT HIGH PRESSURES.   17-2442
SENCE OF MOTIONAL NARROWING IN THE PROTON NMR IN POTASSIUM DIHYDROGEN PHOSPHATE AT HIGH TEMPERATURES.   AB 17-2594
ERATURE DEPENDENCE OF THE DIELECTRIC CONSTANT OF POTASSIUM DIHYDROGEN PHOSPHATE AT LAMBDA EQUAL 2.14 MM. /P 17-2501
        MEASUREMENT OF THE ELASTIC CONSTANTS OF POTASSIUM DIHYDROGEN PHOSPHATE BY BRILLOUIN DIFFUSION.   17-2454
ONS.   HIGH TEMPERATURE PHASE TRANSITION IN RUBIDIUM DIHYDROGEN PHOSPHATE BY SCATTERING OF COLD NEUTR 17-2489
/N- PHONON COUPLING ON THE FERROELECTRIC MODE IN POTASSIUM DIHYDROGEN PHOSPHATE. (COLLECTIVE PSEUDOSPIN MO/ 17-2655
-WEISS LAW FOR THE NONLINEAR SUSCEPTIBILITY OF A POTASSIUM DIHYDROGEN PHOSPHATE CRYSTAL.   CURIE 17-2686
        DIELECTRIC PROPERTIES OF A GAMMA-IRRADIATED POTASSIUM DIHYDROGEN PHOSPHATE CRYSTAL.   17-2467
MICROWAVE MODULATOR OF RUBY LASER LIGHT USING POTASSIUM DIHYDROGEN PHOSPHATE CRYSTAL.   17-2640
                 NONLINEAR PROPERTIES OF A RUBIDIUM DIHYDROGEN PHOSPHATE CRYSTAL.   17-2508
N SPECTRA OF THE FERROELECTRIC MODIFICATION OF A POTASSIUM DIHYDROGEN PHOSPHATE CRYSTAL.   RAMA 17-2622
        RAMAN SPECTRUM OF FERROELECTRIC MODE IN A POTASSIUM DIHYDROGEN PHOSPHATE CRYSTAL.   17-2663
IREFRINGENCE IN AN ISOLATED AND A SHORTCIRCUITED POTASSIUM DIHYDROGEN PHOSPHATE CRYSTAL.   STRESS INDUCED B 17-2452
CHING ANGLE OF THE SECOND HARMONIC GENERATION IN POTASSIUM DIHYDROGEN PHOSPHATE CRYSTAL. /ENCE OF INDEX MAT 17-2544

FOR/ ELECTROOPTIC CHARACTERISTICS OF DEUTERATED POTASSIUM DIHYDROGEN PHOSPHATE CRYSTAL AND ITS APPLICATION 17-2499
/ VIBRATION SPECTRUM OF THE PHOSPHATE ANION IN A RUBIDIUM DIHYDROGEN PHOSPHATE CRYSTAL IN PARAELECTRIC AND 17-2472
TION REGICN. (COERCIVE FIELD)   POLARIZATION OF A RUBIDIUM DIHYDROGEN PHOSPHATE CRYSTAL IN THE PHASE TRANSI 17-2504
TION REGION.   POLARIZATION OF A RUBIDIUM DIHYDROGEN PHOSPHATE CRYSTAL IN THE PHASE TRANSI 17-2506
T. (ELECTROCALORI/ SPONTANEOUS POLARIZATION OF A POTASSIUM DIHYDROGEN PHOSPHATE CRYSTAL NEAR THE CURIE POIN 17-2678
CRITICAL BEHAVIOR OF KDP AND RUBIDIUM DIHYDROGEN PHOSPHATE CRYSTALS. 17-2676
RIZED LIGHT DUE TO HEATING OF PARALLEL PLANES OF POTASSIUM DIHYDROGEN PHOSPHATE CRYSTALS. /OF LINEARLY POLA 17-2540
DIELECTRIC PROPERTIES OF GAMMA-IRRADIATED POTASSIUM DIHYDROGEN PHOSPHATE CRYSTALS. 17-2542
ACTION OF POLARIZED LIGHT BY ULTRASONIC WAVES IN POTASSIUM DIHYDROGEN PHOSPHATE CRYSTALS.   DIFFR 17-2407
EFFECT OF ANNEALING ON OPTICAL ANOMALIES OF POTASSIUM DIHYDROGEN PHOSPHATE CRYSTALS. 17-2427
EFFECT OF DISLOCATIONS ON THE GROWTH OF POTASSIUM DIHYDROGEN PHOSPHATE CRYSTALS. 17-2491
EFFECT OF IMPURITIES ON THE STRUCTURE OF POTASSIUM DIHYDROGEN PHOSPHATE CRYSTALS. 17-2606
ELECTRONIC EFFECT AND ELASTIC PROPERTIES OF RUBIDIUM DIHYDROGEN PHOSPHATE CRYSTALS. 17-2617
AL AND OPTICAL PROPERTIES OF PARTIALLY DEUTERATED RUBIDIUM DIHYDROGEN PHOSPHATE CRYSTALS.   ELECTROOPTIC 17-2698
FAR INFRARED SPECTRA OF RUBIDIUM DIHYDROGEN PHOSPHATE CRYSTALS. 17-2638
LIGHT AND CURRENT PULSES FROM X-RAYED POTASSIUM DIHYDROGEN PHOSPHATE CRYSTALS. 17-2716
MENT OF THE PRIMARY LINEAR ELECTROOPTIC EFFECT IN AMMONIUM DIHYDROGEN PHOSPHATE CRYSTALS.   MEASURE 17-2654
STUDY OF THE (MECHANICAL) STRENGTH PROPERTIES OF POTASSIUM DIHYDROGEN PHOSPHATE CRYSTALS. 17-2648
/POGRAPHIC STUDY OF THE DISLOCATIONS PRODUCED IN POTASSIUM DIHYDROGEN PHOSPHATE CRYSTALS BY GROWTH FROM SO/ 17-2492
TICAL PROCEED/ OBSERVATION OF DOMAIN STRUCTURE IN POTASSIUM DIHYDROGEN PHOSPHATE CRYSTALS BY POLARIZATION OP 17-2493
CTOR / INVESTIGATION OF THE THERMAL EXPANSION OF POTASSIUM DIHYDROGEN PHOSPHATE CRYSTALS BY USING SEMICONDU 17-2565
Y/ USE OF SPECIFIC HEAT (BY A DYNAMIC METHOD) OF POTASSIUM DIHYDROGEN PHOSPHATE CRYSTALS FOR STUDY OF THE P 17-2566
HT MODULATION   OPTIMUM CUT IN X DIHYDROGEN PHOSPHATE CRYSTALS FOR TRANSVERSE LIG 17-2494
OUS SOL/ GROWTH LAYERS AND OTHER MACRODEFECTS IN POTASSIUM DIHYDROGEN PHOSPHATE CRYSTALS GROWN FROM AN AQUE 17-2496
KINETICS OF GROWTH AND DISSOLUTION OF AMMONIUM DIHYDROGEN PHOSPHATE CRYSTALS IN SOLUTION. 17-2412
INING ACETATE IONS.   GROWTH AND ETCHING OF POTASSIUM DIHYDROGEN PHOSPHATE CRYSTALS IN SOLUTIONS CONTA 17-2567
/C PHASE TRANSITION AND DIELECTRIC PROPERTIES OF POTASSIUM DIHYDROGEN PHOSPHATE CRYSTALS IRRADIATED WITH F/ 17-2615
THE P EXPANSION OF FREE/ NONLINEAR PROPERTIES OF POTASSIUM DIHYDROGEN PHOSPHATE. (DETERMINATION OF BETA IN 17-2543
/RARED ABSORPTION, NEUTRON INELASTIC SCATTERING, POTASSIUM DIHYDROGEN PHOSPHATE, DEUTERATED POTASSIUM DIHY/ 03-0401
/ITY WITH TEMPERATURE IN THE FERROELECTRIC PHASE POTASSIUM DIHYDROGEN PHOSPHATE. (DIFFRACTION BY DOMAIN WA/ 17-2662
PONTANEOUS POLARIZATION)   ELECTRONIC STUDIES OF POTASSIUM DIHYDROGEN PHOSPHATE. (ELECTROCALORIC STUDIES, S 17-2429
EPITAXIAL ELECTROOPTIC MIXED CRYSTAL AMMONIUM POTASSIUM DIHYDROGEN PHOSPHATE FILM WAVEGUIDE. 17-2633
CUTTING AND POLISHING IN UNIAXIAL CRYSTALS OF THE TYPE X DIHYDROGEN PHOSPHATE FOR LIGHT MODULATORS. 17-2559
STRUCTURE OF FERROELECTRIC CRYSTALS OF THE POTASSIUM DIHYDROGEN PHOSPHATE GROUP. 17-2607
/CAL PROPERTIES OF FERROELECTRIC CRYSTALS OF THE POTASSIUM DIHYDROGEN PHOSPHATE GROUP IN THE PARAELECTRIC / 17-2696
PPARATUS.   GROWTH BEHAVIOR OF SMALL SEEDS OF AMMONIUM DIHYDROGEN PHOSPHATE IN A MODIFIED CIRCULATION A 17-2704
RAMAN SPECTRA OF POLYMORPHIC MODIFICATIONS OF POTASSIUM DIHYDROGEN PHOSPHATE IN AN ELECTRIC FIELD. 17-2425
PHASES.   VIBRATION SPECTRA OF POTASSIUM DIHYDROGEN PHOSPHATE IN PARA- AND FERROELECTRIC 17-2480
HE PARA- AND FERROELECTRIC PHAS/ TRANSMISSION OF POTASSIUM DIHYDROGEN PHOSPHATE IN THE INFRARED REGION IN T 17-2459
TERIUM PHOSPHATE IN HE/ ENTHALPIES OF SOLUTION OF RUBIDIUM DIHYDROGEN PHOSPHATE IN WATER AND RUBIDIUM DIDEU 17-2585
/ZATION AND THE ISOTOPE EFFECTS IN FERROELECTRIC POTASSIUM DIHYDROGEN PHOSPHATE. (INTERBAND INTERACTION AN/ 17-2433
TASSIUM SALT.   STRUCTURE OF DIHYDROGEN PHOSPHATE ION IN THE FERROELECTRIC PO 17-2624
/IVE MATERIALS AND NONLINEAR CRYSTALS. (KDP, ADP, RUBIDIUM DIHYDROGEN PHOSPHATE, LITHIUM NIOBATE, LITHIUM / 11-1865
LASER INDUCED DAMAGE IN X DIHYDROGEN PHOSPHATE MATERIALS. 17-2475
MODIFIED POTASSIUM DIHYDROGEN PHOSPHATE MODEL IN A STAGGERED FIELD. 17-2710
-RAY STUDY ON THERMAL EXPANSION OF FERROELECTRIC POTASSIUM DIHYDROGEN PHOSPHATE. (NEAR CURIE TEMPERATURE) X 17-2556
HASE TRANSITION. (T/ THERMOELASTIC PROPERTIES OF POTASSIUM DIHYDROGEN PHOSPHATE NEAR THE HIGH TEMPERATURE P 17-2627
HYDROGEN DENSITY DISTRIBUTION IN PARAELECTRIC POTASSIUM DIHYDROGEN PHOSPHATE (NEUTRON SCATTERING). 17-2648
/ ELECTROOPTIC MODULATOR MATERIALS. (POTASSIUM OR AMMONIUM DIHYDROGEN PHOSPHATE OR ARSENATE, POTASSIUM DID/ 01-0021
/ STUDY OF THE FERROELECTRIC PHASE TRANSITION OF POTASSIUM DIHYDROGEN PHOSPHATE. PART-1: THE ORDER OF TRAN/ 17-2557
/OPIC OBSERVATION OF ANTIFERROELECTRIC DOMAINS OF AMMONIUM DIHYDROGEN PHOSPHATE. (PHASE TRANSITION AT 148 / 17-2643
HATE AND RU/ DIELECTRIC ANOMALIES IN CRYSTALS OF POTASSIUM DIHYDROGEN PHOSPHATE POTASSIUM DIDEUTERIUM PHOSP 17-2613
PHATE, P/ TEMPERATURE DEPENDENT RAMAN SPECTRA OF POTASSIUM DIHYDROGEN PHOSPHATE, POTASSIUM DIDEUTERIUM PHOS 17-2409
TE MIXE/ ESR STUDY OF RADIATION DAMAGE CENTERS IN AMMONIUM DIHYDROGEN PHOSPHATE POTASSIUM DIHYDROGEN ARSENA 17-2547
/NSTABILITY OF ELECTROOPTIC LASER MODULATORS WITH AMMONIUM DIHYDROGEN PHOSPHATE, POTASSIUM DIHYDROGEN PHOS/ 17-2675
/G CRYSTALS AND CHARACTERISTICS OF THEIR GROWTH. (AMMONIUM DIHYDROGEN PHOSPHATE, POTASSIUM FERROCYANIDE TR/ 16-2316
/AGNETIC RESONANCE IN FERROELECTRICS. (MODEL FOR POTASSIUM DIHYDROGEN PHOSPHATE PRESENTED, TRIGLYCINE SULF/ 03-0180
FFECT OF DEUTERATION ON THE ELASTIC PROPERTIES OF RUBIDIUM DIHYDROGEN PHOSPHATE (RDP).   E 17-2584
OGS S/ SPECIFIC HEAT OF SOME KDP TYPE CRYSTALS. (POTASSIUM DIHYDROGEN PHOSPHATE, RUBIDIUM AND AMMONIUM ANAL 17-2679
/EMENT OF THE RELATIVE NONLINEAR COEFFICIENTS OF POTASSIUM DIHYDROGEN PHOSPHATE, RUBIDIUM DIHYDROGEN PHOSP/ 08-1413
INVESTIGATION OF THE PHASE TRANSITION IN AMMONIUM DIHYDROGEN PHOSPHATE SINGLE CRYSTAL BY NMR. 17-2545
ON) IN STATIC REGIME AND / GROWING KINETIC OF THE AMMONIUM DIHYDROGEN PHOSPHATE SINGLE CRYSTAL (FROM SOLUTI 17-2516
NS.   GROWTH OF AMMONIUM DIHYDROGEN PHOSPHATE SINGLE CRYSTAL FROM SOLUTIO 17-2411
HEAT CAPACITY OF RUBIDIUM DIHYDROGEN PHOSPHATE SINGLE CRYSTALS. 17-2415
LIGHT ABSORPTION IN POTASSIUM DIHYDROGEN PHOSPHATE SINGLE CRYSTALS. 17-2421
DEPENDENCE OF THE MECHANICAL CHARACTERISTICS OF POTASSIUM DIHYDROGEN PHOSPHATE SINGLE CRYSTALS. /MPERATURE 17-2649
GROWTH OF POTASSIUM DIHYDROGEN PHOSPHATE SINGLE CRYSTALS IN GEL. 17-2455
/IGLYCINE SELENATE OR FLUOBERYLLATE, POTASSIUM OR RUBIDIUM DIHYDROGEN PHOSPHATE, SODIUM, POTASSIUM(2), OR / 02-0089
SOMORPHOUS FERROELECTRIC SYSTEMS. (DISCUSSED FOR POTASSIUM DIHYDROGEN PHOSPHATE SYSTEM) /E POINT) IN SOME I 17-2487
/AVES) OF THE FERROELECTRIC TRANSITION REGION IN POTASSIUM DIHYDROGEN PHOSPHATE. (TEMPERATURE AND FIELD DE/ 17-2576
NELING MODES IN THE RAMAN SPECTRUM OF TETRAGONAL POTASSIUM DIHYDROGEN PHOSPHATE. (THEORY) /IS OF PHONON TUN 03-0396
TILE.   FAR INFRARED DIELECTRIC MEASUREMENTS ON POTASSIUM DIHYDROGEN PHOSPHATE, TRIGLYCINE SULFATE, AND RU 17-2423
ION TEMPERATURE OF FERROELECTRIC CRYSTALS OF THE POTASSIUM DIHYDROGEN PHOSPHATE TYPE. /ON THE PHASE TRANSIT 17-2495
H HYDROGEN BONDS.   DYNAMICAL THEORY OF AMMONIUM DIHYDROGEN PHOSPHATE TYPE ANTIFERROELECTRICS WIT 03-0476
CE OF A LINE WIDTH TRANSITION FOR THE PROTONS IN POTASSIUM DIHYDROGEN PHOSPHATE TYPE CRYSTALS.   ABSEN 17-2436
IVE MOTIONS OF HYDROGEN BONDS. (CALCULATIONS FOR POTASSIUM DIHYDROGEN PHOSPHATE TYPE CRYSTALS)   COLLECT 17-2483
MPERATURE PHASE TRANSITIONS AND METASTABILITY IN POTASSIUM DIHYDROGEN PHOSPHATE TYPE CRYSTALS.   HIGH TE 17-2513
NMR PROTON SECOND MOMENTS IN POTASSIUM DIHYDROGEN PHOSPHATE TYPE CRYSTALS. 17-2408
NOVEL CUT OF POTASSIUM DIHYDROGEN PHOSPHATE TYPE CRYSTALS. 17-2530
FECTS ON THE LATTICE PARAMETERS AND STRUCTURE OF POTASSIUM DIHYDROGEN PHOSPHATE TYPE CRYSTALS.   PRESSURE EF 17-2588
ERATION IN CADMIUM SULFIDE, BARIUM TITANATE, AND POTASSIUM DIHYDROGEN PHOSPHATE TYPE CRYSTALS. /ARMONIC GEN 04-0852
THEORY OF DOMAIN WALL MOTION IN POTASSIUM DIHYDROGEN PHOSPHATE TYPE CRYSTALS. 17-2652
/E FERROELECTRIC AND ANTIFERROELECTRIC STATES IN POTASSIUM DIHYDROGEN PHOSPHATE TYPE CRYSTALS AT HIGH PRES/ 17-2642
ULATION.   OPTIMUM CUT OF POTASSIUM DIHYDROGEN PHOSPHATE TYPE CRYSTALS FOR LIGHT MOD 17-2532
/NTERACTIONS AND THE DYNAMICAL SUSCEPTIBILITY OF POTASSIUM DIHYDROGEN PHOSPHATE TYPE FERROELECTRIC CRYSTAL/ 17-2570
ANSITIONS.   DYNAMICS OF POTASSIUM DIHYDROGEN PHOSPHATE TYPE FERROELECTRIC PHASE TR 17-2692
ORDER DISORDER THEORY OF POTASSIUM DIHYDROGEN PHOSPHATE TYPE FERROELECTRICS. 17-2685
Y)   ENERGY PARAMETERS AND PROTON TUNNELING IN POTASSIUM DIHYDROGEN PHOSPHATE TYPE FERROELECTRICS. (THEOR 03-0506
FS)   MODELS OF THE PHASE TRANSITION IN POTASSIUM DIHYDROGEN PHOSPHATE TYPE FERROELECTRICS. (24 RE 03-0489
AND LOW TEMPER/ RAMAN SPECTRA OF CRYSTALS OF THE POTASSIUM DIHYDROGEN PHOSPHATE TYPE IN THEIR PARAELECTRIC 17-2515

CONDITIONS FOR THE GROWTH OF POTASSIUM DIHYDROGEN PHOSPHATE WHISKERS.   17-2628
FERROELECTRIC DOMAINS OF POTASSIUM DIHYDROGEN PHOSPHATE. (93 REFS)   17-2447
HYDROXIDE VIBRATION SPECTRUM OF THE FERROELECTRIC RUBIDIUM DIHYDROGEN PHOSPHATES. /PHASE TRANSITION ON THE   17-2470
PHASE TRANSITION IN THALLIUM AND AMMONIUM THALLIUM (DIHYDROGEN) PHOSPHATES.   17-2571
/ TEMPERATURES OF POTASSIUM, RUBIDIUM, CESIUM AND AMMONIUM DIHYDROGEN PHOSPHATES AND CESIUM AND AMMONIUM D/   17-2577
PHATE.   STRUCTURAL STUDIES OF THE SYSTEM POTASSIUM DIHYDROGEN PHOSPHATE- POTASSIUM DIDEUTERIUM PHOS   17-2590
HATE- WATER SYSTEM.   SOLUBILITY ISOBAR IN THE AMMONIUM DIHYDROGEN PHOSPHATE- POTASSIUM DIHYDROGEN PHOSP   17-2658
HATE- WATER SYSTE/ POLYTHERM OF SOLUBILITY OF THE AMMONIUM DIHYDROGEN PHOSPHATE- POTASSIUM DIHYDROGEN PHOSP   17-2434
/ DIELECTRIC PROPERTIES OF SINGLE CRYSTALS IN THE AMMONIUM DIHYDROGEN PHOSPHATE- POTASSIUM DIHYDROGEN PHOSP   17-2572
DIDEUTERIUM(/ SOME THERMODYNAMICAL PROPERTIES OF POTASSIUM DIHYDROGEN PHOSPHATE- POTASSIUM DIHYDROGEN(1-X)   17-2680
TRA.   FERROELECTRIC TRANSITION IN RUBIDIUM DIHYDROGEN PHOSPHIDE USING RAMAN SCATTERING SPEC   17-2620
ECTRA) OF SEMICONDUCTING STRONTIUM TITANATE SINGLE CRYSTALS (HYDROGEN REDUCED, CERIUM OR NIOBIUM DOPED). /P   06-1198
.   CHEMICAL PROCESSES DURING HYDROGEN REDUCTION OF BARIUM TITANATE CRYSTALS   04-0540
S ON THE PLASTICITY OF ICE AND THEIR EXPLANATION IN TERMS OF HYDROGEN REORIENTATION.   IMPURITY EFFECT   34-3449
MODELS FOR THE ORDER DISORDER TRANSITION IN SODIUM TRIHYDROGEN SELENATE.   22-2862
PROTON SPIN LATTICE RELAXATION IN IRRADIATED SODIUM TRIHYDROGEN SELENATE.   22-2853
PROTON SPIN LATTICE RELAXATION IN FERROELECTRIC SODIUM TRIHYDROGEN SELENATE.   22-2852
RADIATION EFFECTS IN SODIUM TRIHYDROGEN SELENATE.   22-2865
RESIDUAL ENTROPY OF SODIUM TRIHYDROGEN SELENATE.   22-2835
AL BEHAVIOR IN THE BETA PHASE OF THE FERROELECTRIC SODIUM TRIHYDROGEN SELENATE. /IN STRUCTURE AND ITS UNUSU   22-2874
LENATE.   PROTON MAGNETIC RESONANCE STUDY OF SODIUM TRIHYDROGEN SELENATE AND POTASSIUM TRIHYDROGEN SE   22-2876
/IC PROPERTIES OF HYDROGEN BONDED FERROELECTRICS. (SODIUM TRIHYDROGEN SELENATE AND SODIUM TRIDEUTERIUM SEL/   22-2855
ANOMALOUS THERMAL EXPANSION OF POTASSIUM TRIHYDROGEN SELENATE CRYSTALS.   22-2880
DOMAIN STRUCTURE OF THE BETA PHASE OF SODIUM TRIHYDROGEN SELENATE IN ELECTRIC FIELDS.   22-2830
ROGEN MOTION IN POTASSIUM DIHYDROGEN PHOSPHATE AND SODIUM TRIHYDROGEN SELENATE TYPE CRYSTALS. ULTRASLOW HYD   17-2441
A NEUTRON DIFFRACTION STUDY OF POTASSIUM TRIHYDROGEN SELENITE.   22-2866
ANOMALOUS SPECIFIC HEAT OF SODIUM TRIHYDROGEN SELENITE.   22-2823
CRYSTAL STRUCTURE OF RUBIDIUM TRIHYDROGEN SELENITE.   22-2881
OUBLE HYSTERESIS LOOPS AT THE UPPER CURIE POINT OF SODIUM TRIHYDROGEN SELENITE.   D 22-2831
ERTIES, KDP, TGS, ROCHELLE SALT, SILVER PERIODATE, SODIUM TRIHYDROGEN SELENITE) /AMICAL ORDER DISORDER PROP   02-0050
OGEN BOND STUDIES. PART-58: CRYSTAL STRUCTURE OF AMMONIUM TRIHYDROGEN SELENITE.   HYDR 22-2879
IMPROPER FERROELECTRIC PHASE TRANSITION IN RUBIDIUM TRIHYDROGEN SELENITE.   22-2834
RMAL EXPANSION OF HYDROGEN BONDED FERROELECTRICS. LITHIUM TRIHYDROGEN SELENITE.   THE 22-2857
UM SELENITE, AS DE/ THE HYDROGEN BOND SYSTEM IN POTASSIUM TRIHYDROGEN SELENITE, AND IN POTASSIUM TRIDEUTERI   22-2854
ITE.   CRYSTAL STRUCTURES OF POTASSIUM TRIHYDROGEN SELENITE AND SODIUM TRIHYDROGEN SELEN   22-2838
BLEM.   ANOMALOUS SPECIFIC HEAT OF SODIUM TRIHYDROGEN SELENITE ASSOCIATED COMBINATORIAL PRO   22-2822
ESR STUDY OF GAMMA-IRRADIATED CESIUM TRIHYDROGEN SELENITE AT ROOM TEMPERATURE.   22-2833
NEUTRON SCATTERING ANALYSIS OF CESIUM TRIHYDROGEN SELENITE AT ROOM TEMPERATURE.   22-2869
A NEUTRON DIFFRACTION STUDY OF THE FERROELECTRIC LITHIUM TRIHYDROGEN SELENITE. (CRYSTAL STRUCTURE) /RT-54:   22-2878
EFFECT OF HIGH PRESSURE ON PHASE TRANSITIONS OF ALKALI TRIHYDROGEN SELENITE CRYSTALS.   22-2882
SITIONS.   BEHAVIOR OF THE FUNDAMENTAL ABSORPTION EDGE IN TRIHYDROGEN SELENITE CRYSTALS AT THEIR PHASE TRAN   22-2839
/EFINEMENT OF THE CRYSTAL STRUCTURE OF FERROELECTRIC LITHIUM HYDROGEN SELENITE: POSITION OF THE LITHIUM IO/   22-2860
S/ INFRARED SPECTRA OF ISOTOPIC NONISOMORPHISM OF SODIUM TRIHYDROGEN SELENITE (SHS) OR SODIUM TRIDEUTERIUM   22-2836
THERMAL STABILITY OF LITHIUM HYDROGEN SELENITES   22-2868
THERMOOPTIC AND ELECTROOPTIC PROPERTIES OF TRIHYDROGEN SELENITES.   22-2841
SPIN LATTICE RELAXATION IN FERROELECTRIC DEUTERATED AMMONIUM HYDROGEN SULFATE.   DEUTERON 19-2795
ON SPIN RESONANCE OF GAMMA DEFECTS IN FERROELECTRIC AMMONIUM HYDROGEN SULFATE.   ELECTR 19-2748
POLARIZATION REVERSAL CHARACTERISTICS OF AMMONIUM HYDROGEN SULFATE.   19-2788
/ REFINEMENT TECHNIQUES APPLIED TO THE CRYSTAL STRUCTURE OF AMMONIUM HYDROGEN SULFATE ABOVE THE FERROELECTRIC TRAN/   16-2355
/FRACTION DETERMINATION OF THE CRYSTAL STRUCTURE OF AMMONIUM HYDROGEN SULFATE ABOVE THE FERROELECTRIC TRAN/   19-2775
CRYSTAL STRUCTURE OF FERROELECTRIC RUBIDIUM HYDROGEN SULFATE. (ABSTRACT)   19-2746
/EMPERATURE INFRARED STUDIES. PART-8: FERROELECTRIC AMMONIUM HYDROGEN SULFATE AND AMMONIUM DEUTERIUM SULFA/   19-2782
E IN THE PARA- AND FERROELECTR/ INFRARED SPECTRA OF RUBIDIUM HYDROGEN SULFATE AND RUBIDIUM DEUTERIUM SULFAT   19-2768
OELECTRIC PHASE.   ON THE PHASE TRANSITION OF AMMONIUM HYDROGEN SULFATE AND THE STRUCTURE OF THE FERR   19-2777
-70 TO +14 DEG C.   SPECIFIC HEAT OF AMMONIUM HYDROGEN SULFATE IN THE TEMPERATURE RANGE FROM   19-2789
LECTRIC TRANSITION.   AMMONIUM HYDROGEN SULFATE, ITS STRUCTURE AND THE FERROE   19-2773
/RONIC PROPERTIES (CONDUCTIVITY, MAGNETIC SUSCEPTIBILITY) OF HYDROGEN TUNGSTEN BRONZES, HYDROGEN(X) TUNGST/   13-1934
KDP.   LASER RAMAN STUDY OF QUASI SPIN WAVE HYDROGEN TUNNELING MODES IN KDP AND DEUTERATED   17-2569
OSPHATE.   LOW FREQUENCY HYDROGEN VIBRATIONS IN POTASSIUM DIHYDROGEN PH   17-2699
OSPHATE.   LOW FREQUENCY HYDROGEN VIBRATIONS IN POTASSIUM DIHYDROGEN PH   17-2608
MOLECULAR FIELD STUDY OF THE ISOTOPIC EFFECT IN POTASSIUM DIHYDROGEN(X) DIDEUTERIUM(1-X) PHOSPHATE MIXED /   17-2563
/ITY, MAGNETIC SUSCEPTIBILITY) OF HYDROGEN TUNGSTEN BRONZES, HYDROGEN(X) TUNGSTATE. (ANALOGOUS TO THE SODI/   13-1934
/L PROPERTIES OF POTASSIUM DIHYDROGEN PHOSPHATE- POTASSIUM DIHYDROGEN(1-X) DIDEUTERIUM(X) PHOSPHATE MIXED /   17-2680
CRYSTALS. REGULAR BEHAVIOR OF SOLID SOLUTIONS IN POTASSIUM DIHYDROGEN(1-X) DIDEUTERIUM(X) PHOSPHATE SINGLE   17-2458
ATION, TEMPERATU/ PHASE DIAGRAM OF SODIUM TRIDEUTERIUM(X) TRIHYDROGEN(1-X) PHOSPHATE SYSTEM VERSUS CONCENTR   17-2666
/MITTIVITY OF A NEW DELTA PHASE OF SODIUM TRIDEUTERIUM(X) TRIHYDROGEN(1-X) SELENITE CRYSTALS. (PRESSURE EF/   22-2875
ALS.   ISOTOPIC NONISOMORPHISM OF RUBIDIUM TRIHYDROGEN(1-X) TRIDEUTERIUM(X) DISELENATE CRYST   22-2873
SOTOPE EFFECT.   THERMAL EXPANSION IN SODIUM TRIHYDROGEN(1-X) TRIDEUTERIUM(X) DISELENITE AND I   22-2858
OTASSIUM FERROCYANIDE TRIHYDRATE BY PULSED NMR.   STUDY OF HYDROGEN-1 IN PARAELECTRIC AND FERROELECTRIC P   24-2889
/TALS FROM AQUEOUS SOLUTIONS (APPARATUS, GROWTH OF ADP, KDP, HYDROIODIC ACID, AND AMMONIUM OXALATE MONOHYD/   17-2420
STRONTIUM TITANATE (OF HIGH PURITY AND SURFACE ACTIVITY VIA HYDROLYTIC DECOMPOSITION). /TIUM ZIRCONATE AND   06-1217
VITY OF RADIALLY POLARIZED FERROELECTRIC CERAMIC CYLINDERS. (HYDROPHONE ELEMENTS)   OPEN CIRCUIT SENSITI 37-3651
IVITY OF AXIALLY POLARIZED FERROELECTRIC CERAMIC CYLINDERS. (HYDROPHONE ELEMENTS)   OPEN CIRCUIT SENSIT 37-3652
OF RADIALLY POLARIZED FERROELECTRIC CERAMIC HOLLOW SPHERES. (HYDROPHONE ELEMENTS) OPEN CIRCUIT SENSITIVITY 37-3653
URIE POINTS IN KDP AND LITHIUM THALLIUM TARTRATE CRYSTALS BY HYDROSTATIC PRESSURE.   DISPLACEMENT OF THE C 17-2523
DE(X) BROMIDE(1-X) SOLID SOLUTIONS DEPEND ON TEMPERATURE AND HYDROSTATIC PRESSURE. /S OF ANTIMONY SULFOIODI 15-2149
M PHOSPHATE AND DEUTERO POTASSIUM DIDEUTERIUM PHOSPHATE WITH HYDROSTATIC PRESSURE. /IN POTASSIUM DIDEUTERIU 17-2601
RELAXATION TIME IN FERROELECTRIC SILVER SODIUM DINITRITE.   HYDROSTATIC PRESSURE EFFECT ON THE DIELECTRIC 16-2294
AL. (PZT)   PHASE CHANGES INDUCED BY HYDROSTATIC PRESSURE IN A FERROELECTRIC MATERI 09-1521
ES. (SUMMARY WITH NEW RESULTS ON LEAD ZIRCONATE / EFFECTS OF HYDROSTATIC PRESSURE ON FERROELECTRIC PROPERTI 07-1263
TASSIUM NITRATE.   EFFECT OF HYDROSTATIC PRESSURE ON FERROELECTRICITY OF PO 16-2344
DOLINIUM MOLYBDATE.   EFFECT OF HYDROSTATIC PRESSURE ON PHASE TRANSITION IN GA 30-3321
E TRANSITION OF AMMONIUM SULFATE.   EFFECT OF HYDROSTATIC PRESSURE ON THE FERROELECTRIC PHAS 16-2360
FERROELECTRIC PHASE TRANSITIONS.   EFFECT OF HYDROSTATIC PRESSURE ON THE NATURE OF DIFFUSE 03-0442
EMPERATURE OF FERROELECTRIC CRYSTALS OF THE PO/ INFLUENCE OF HYDROSTATIC PRESSURE ON THE PHASE TRANSITION T 17-2495
M TITANATE.   X-RAY INVESTIGATION OF THE EFFECT OF HYDROSTATIC PRESSURE ON THE STRUCTURE OF BARIU 04-0760
TITANATE.   AN X-RAY STUDY OF THE EFFECT OF HYDROSTATIC PRESSURE ON THE STRUCTURE OF LEAD 05-1064
TURE IN AMMONIUM SULFATE.   EFFECT OF THE HYDROSTATIC PRESSURE ON THE TRANSITION TEMPERA 19-2796
S OF TRIGLYCINE SULFATE.   EFFECT OF HYDROSTATIC PRESSURE ON THE UNIPOLAR PROPERTIE 27-3053
F BARIUM(0-0.1) STRONTIUM(0.9-1) TITANATE SOLID SOLUTIONS BY HYDROSTATIC PRESSURE. (TO 1500 ATM) /ERATURE O 04-0722

/N IN A POTASSIUM FERROCYANIDE TRIHYDRATE CRYSTAL UNDER HIGH HYDROSTATIC PRESSURE (TO 5600 KG/SQ CM, TEMPE/ 24-2896
/ OF SEMICONDUCTORS OF THE COMPOSITION A(V)B(VI)C(VII) UNDER HYDROTHERMAL CONDITIONS. (ANTIMONY AND BISMUT/ 15-2213
)   CRYSTAL GROWTH AND CHEMICAL SYNTHESIS UNDER HYDROTHERMAL CONDITIONS. (BISMUTH SULFOBROMIDE 15-2218
ES OF BISMUTH TITANATES   HYDROTHERMAL CRYSTALLIZATION AND SOME PROPERTI 08-1279
ZIRCONATE SOLID SOLUTIONS.   HYDROTHERMAL CRYSTALLIZATION OF LEAD TITANATE 09-1482
COMPOUNDS OF GROUP A(V)B(VI)C(VII). (HIGH GRADE SINGLE CRY/ HYDROTHERMAL CRYSTALLIZATION OF SEMICONDUCTING 15-2214
II) COMPOUNDS. (USING ANTIMONY OR BISMUTH, SULFUR, SELENIUM/ HYDROTHERMAL METHOD FOR PREPARING A(V)B(VI)C(V 15-2194
TRANSPORT REACTIONS (INVOLVING TITANIUM ESTERS TO YIELD SM/ HYDROTHERMAL PREPARATION OF BARIUM TITANATE BY 04-0625
TRANSPORT REACTIONS.   HYDROTHERMAL PREPARATION OF BARIUM TITANATE BY 04-0626
CRYSTALLIZATION OF BISMUTH SULFOIODIDE. (HYDROTHERMAL SYNTHESIS) 15-2212
PREPARATION OF NICKEL BORACITE CRYSTALS BY HYDROTHERMAL SYNTHESIS. 14-2079
DIELECTRIC DISPERSION AND PHASE TRANSITION OF POTASSIUM HYDROXIDE DOPED ICE. 34-3452
CTRIC PROPERTIES OF BARIUM TITANATE SINGLE CRYSTALS DOPED BY HYDROXIDE GROUPS.   DIELE 04-0634
D MONOXIDE ON THE SINTERING. PART-10: PROPERTIES OF LEAD OXY HYDROXIDE POWDER. /FFECT OF MOTHER SALT OF LEA 09-1624
RIC RUBID/ EFFECT OF TEMPERATURE AND PHASE TRANSITION ON THE HYDROXIDE VIBRATION SPECTRUM OF THE FERROELECT 17-2470
                                         HYDROXYL AS DEFECT IN BARIUM TITANATE. 04-0636
ING OF SINGLE CRYSTAL BA/ EFFECT OF OXYGEN STOICHIOMETRY AND HYDROXYL CONTENT ON THE PROPERTIES AND PROCESS 13-1915
/RROELECTRIC STATE IN ELECTRIC DIPOLE SYSTEMS (SPIN WAVES IN HYDROXYL DIPOLE SYSTEM IN POTASSIUM CHLORIDE / 03-0452
TIME DEPENDENCE OF THE DIELECTRIC PERMITTIVITY OF HYDROXYL DOPED BARIUM TITANATE. 04-0550
TY OF BARIUM TITANATE, PURE SINGLE CRYSTALS AND DOPED (IRON, HYDROXYL) SINGLE CRYSTALS.   CONDUCTIVI 04-0695
ANATE A/ EFFECTS OF TEMPERATURE AND PHASE TRANSITIONS ON THE HYDROXYL STRETCHING VIBRATION IN STRONTIUM TIT 06-1209
GEN BONDS IN THE FER/ POLARIZATION OF THE RAMAN BANDS OF THE HYDROXYL VIBRATIONS AND STRUCTURE OF THE HYDRO 17-2471
) DOPED LITHIUM NIOBATE.   FORBIDDEN LINES IN THE HYPERFINE LINES IN ESR SPECTRA OF MANGANESE (2+ 11-1805
C RESONANCE OF MANGANESE(+2) IN BARI/ FORBIDDEN LINES IN THE HYPERFINE SPECTRUM OF THE ELECTRON PARAMAGNETI 04-0877
ETIC RESONANCE OF MANGANESE(2+) DOPED/ ALLOWED AND FORBIDDEN HYPERFINE TRANSITIONS IN THE ELECTRON PARAMAGN 16-2383
THE MEASUREMENT OF HYPERSONIC ATTENUATION IN LITHIUM NIOBATE. 11-1760
BATE AND/ SEPARATE EXCITATION OF LONGITUDINAL AND TRANSVERSE HYPERSONIC PULSES IN PIEZOELECTRIC LITHIUM NIO 11-1814
OBATE SINGLE CRYSTALS USING SCATTERING OF LASER RADIATION BY HYPERSONIC VIBRATIONS. /RAMETERS OF LITHIUM NI 11-1797
OBATE IN THE THREE CENTIMETER RAD/ TEMPERATURE DEPENDENCE OF HYPERSONIC WAVE DAMPING IN RUBY AND LITHIUM NI 11-1718
BISMU/ INDEPENDENT EXCITATION OF LONGITUDINAL AND TRANSVERSE HYPERSONIC WAVE PULSES IN LITHIUM NIOBATE AND 11-1815
NONLINEAR INTERACTION OF HYPERSONIC WAVES IN A LITHIUM NIOBATE CRYSTAL. 11-1802
NONLINEAR EFFECTS IN THE PROPAGATION OF FINITE AMPLITUDE HYPERSONIC WAVES IN FERROELECTRIC CRYSTALS. 03-0498
(80-950 DEG K, 200-2000 MHZ)   GENERATION AND PROPAGATION OF HYPERSONIC WAVES IN LITHIUM NIOBATE CRYSTALS. 11-1753
TE SIN/ TEMPERATURE DEPENDENCES OF THE ABSORPTION OF 9.4 GHZ HYPERSOUND IN ALUMINUM OXIDE AND LITHIUM NIOBA 11-1719
INTERACTION OF TWO HYPERSOUND WAVES IN LITHIUM TANTALATE. 12-1891
SPIN MODEL OF STRONTIUM TITANATE.   STATIC SCALING HYPOTHESIS AND ITS IMPLICATIONS FOR THE PLANAR 06-1223
ITION.   APPARENT VIOLATION OF THE UNIVERSALITY HYPOTHESIS IN A LATTICE MODEL OF A PHASE TRANS 03-0515
FORMATION IN SEIGNETTES SALT (ROCHELLE SALT). (TEMPERATURE ANHYSTERESIS). /CTRIC FIELD ON DOMAIN STRUCTURE 28-3190
UTH FERRATE (CONFIRMS FERROELECTRIC NATURE).   DIELECTRIC HYSTERESIS IN (FLUX GROWN) SINGLE CRYSTAL BISM 08-1449
OF BARIUM TITANATE.   THERMAL HYSTERESIS IN THE TETRAGONAL CUBIC TRANSITION 04-0921
AGING PROCESS IN TRIGLYCINE SULFATE. (HYSTERESIS LAPS, DOMAIN STRUCTURE) 27-3128
INFLUENCE OF THE SHAPE OF DOMAINS ON THE FERROELECTRIC HYSTERESIS LOOP. 03-0300
POLARIZATION OF BARIUM TITANATE SINGLE CRYSTALS WITH DOUBLE HYSTERESIS LOOP.   PECULIARITIES OF 04-0709
TGS)   SHORT LIVED FERROELECTRIC "AFTER EFFECT" PHENOMENA. (HYSTERESIS LOOP CONSTRICTIONS, ROCHELLE SALT, 27-3156
SIDEWAYS DOMAIN WALL MOTION AND (MECHANISM OF THE) HYSTERESIS LOOP OF FERROELECTRICS. 03-0301
TIME EFFECTS IN THE HYSTERESIS LOOP OF (LEAD, STRONTIUM) TITANATE. 05-1052
TIME EFFECTS IN THE HYSTERESIS LOOP OF LEAD STRONTIUM TITANATE. 05-1053
HATE CRYSTALS.   DOUBLE DIELECTRIC HYSTERESIS LOOP OF POTASSIUM DIDEUTERIUM PHOSP 17-2503
QUASI STATIC ELECTRIC FIELD.   FERROELECTRIC HYSTERESIS LOOP OF SILVER SODIUM DINITRITE BY 16-2293
BONDED PHOSPHATE FERROELECTRIC (CRYSTAL GROWTH AND STRUCTURE HYSTERESIS LOOP RESULTS). /F THE NON-HYDROGEN 36-3601
NEODYMIUM(X) (ORTHO) ALUMINATE SYSTEM (NOT FERROELECTRIC BY HYSTERESIS LOOP TESTS). /THE PRASEODYMIUM(1-X) 36-3600
ANGE. (CIRCUITRY DESCRIBED)   VARIOUS HYSTERESIS LOOP TRACERS FOR A WIDE FREQUENCY R 37-3640
FERROELECTRIC FREE ENERGY EXPANSION COEFFICIENTS FROM DOUBLE HYSTERESIS LOOPS. 03-0262
IONS IN FERROELECTRIC CRYSTALS.   DOUBLE HYSTERESIS LOOPS AND FIRST ORDER PHASE TRANSIT 03-0456
ODIUM TRIHYDROGEN SELENITE.   DOUBLE HYSTERESIS LOOPS AT THE UPPER CURIE POINT OF S 22-2831
E.   ANALYSIS OF HYSTERESIS LOOPS IN LITHIUM HYDRAZINIUM SULFAT 32-3351
/ CRITICAL BEHAVIOR OF FERROELECTRIC TGS AND DEUTERATED TGS (HYSTERESIS LOOPS JUST BELOW CURIE POINT CORREC 27-2966
ANOMALOUS FERROELECTRIC HYSTERESIS LOOPS. (LEAD STRONTIUM TITANATE) 05-1067
ANALYSIS OF DOUBLE HYSTERESIS LOOPS. (ROCHELLE SALT) 28-3212
E FERROELECTRIC PHASE. (KAEN/ AC POLARIZATION AND DIELECTRIC HYSTERESIS NEAR THE TRANSITION POINT OF DIFFUS 03-0437
ON THE INFLUENCE OF DEFECTS ON THE TEMPERATURE HYSTERESIS OF BARIUM TITANATE. 04-0987
THEORY OF POLARIZATION AND HYSTERESIS OF FERROELECTRICS. 03-0494
YCINE SULFATE AS A FUNCTION OF THE CONDITIONS OF GR/ THERMAL HYSTERESIS OF THE DIELECTRIC CONSTANT OF TRIGL 27-3126
POTASSIUM DIHYDROGEN PHOSPHATE.   THERMAL HYSTERESIS OF THE FERROELECTRIC TRANSITION IN 17-2682
AMPLE GROUNDING.   FERROELECTRIC HYSTERESIS TRACER FEATURING COMPENSATION AND S 37-3668
CTRIC PROPERTIES. (DISPERSION, RELAXATION TIME AND STRENGTH, HYSTERESIS, 26 REFS) /AL DEFECTS UPON FERROELE 03-0496

I

ROUT'S MODEL FOR FERROELECTRICS. INVESTIGATIONS ON HEXAGONAL ICE.   A STUDY OF B 34-3443
CALCULATION OF THE DIELECTRIC CORRELATION FACTOR OF CUBIC ICE. 34-3429
CHEMICAL PHYSICS OF ICE. 01-0009
DEUTERON ARRANGEMENTS IN THE HIGH PRESSURE FORMS OF ICE. 34-3438
DISPERSION AND PHASE TRANSITION OF POTASSIUM HYDROXIDE DOPED ICE.   DIELECTRIC 34-3452
DIELECTRIC PROPERTIES OF POTASSIUM CHLORIDE ICE. 34-3454
DIFFUSION AND RELAXATION PHENOMENA IN ICE. 34-3470
DIFFUSION OF HYDROGEN FLUORIDE IN ICE. 34-3437
DIPOLE MOMENT DERIVATIVE OF THE HYDROGEN BOND IN ICE. 34-3489
ELASTIC MODULI OF ICE. 34-3421
ELECTRIC RESONANCE: APPLICATION TO THE HYDROGEN BOND. (ICE) 34-3477
ELECTRICAL CONDUCTION IN ICE. 34-3415
R DOUBLE RESONANCE) STUDY OF X-IRRADIATED SINGLE CRYSTALS OF ICE.   ENDOR (ELECTRON NUCLEA 34-3413
EQUILIBRIUM STRUCTURE OF POLARIZED ICE. 34-3411
EVALUATION OF DIELECTRIC DISPERSION DATA. (ICE) 34-3433
PERIMENTAL AND THEORETICAL STUDIES ON THE DC CONDUCTIVITY OF ICE.   EX 34-3450
FORMATION OF COLOR CENTERS IN IRRADIATED ICE. 34-3487
IMPURITY STATISTICS IN ICE. 34-3472
RESSURE DEPENDENCE OF PROPERTIES CAUSED BY LATTICE DEFECTS. (ICE)   INTERPRETATION OF THE P 34-3434
PRETATION OF THE PROTON SPIN LATTICE RELAXATION IN HEXAGONAL ICE.   INTER 34-3475
IRRADIATION PRODUCED SOLVATED ELECTRONS IN ICE. 34-3423

LATTICE DYNAMICS OF ICE. 34-3467
MOLECULAR FORCES OF HEAVY AND LIGHT ICE. 34-3440
MOSSBAUER EFFECT OF FERROUS IONS IN CUBIC ICE. 34-3416
PHASE TRANSITION IN ICE. 34-3442
PROTON- PROTON AND PROTON- LATTICE INTERACTIONS IN ICE. 34-3430
PROTONIC SEMICONDUCTORS. (ICE) 34-3461
AL PROTON ORDERING AND GLASSY CRYSTALLINE STATE IN HEXAGONAL ICE. RELAXATION 34-3436
REVIEW ON THE PROBLEMS OF THE PHYSICS OF ICE. 02-0072
SOME EXPERIMENTS ON THE REGELATION OF ICE. 34-3435
RONG COLLISION LIMIT OF SPIN LATTICE RELAXATION IN HEXAGONAL ICE. ST 34-3480
SYMMETRY ANALYSIS AND ELECTRONIC STATES IN CUBIC ICE. 34-3462
TENSILE AND FLEXURE PROPERTIES OF SALINE ICE. 34-3422
VERTEX MODEL AND THE PROTON PROTON CORRELATION FUNCTIONS IN ICE. THE THREE DIMENSIONAL EIGHT 34-3481
THEORY OF THE MOBILITY OF STRUCTURAL DEFECTS IN ICE. 03-0244
THERMODIELECTRIC EFFECT AND FREEZING POTENTIAL IN GROWING ICE. 34-3417
THERMOELECTRIC EFFECT IN ICE. 34-3446
ATION. PART-8: TRAN/ DIELECTRIC RELAXATION SPECTRA OF WATER, ICE, AND AQUEOUS SOLUTIONS AND THEIR INTERPRET 34-3482
VAPOR PRESSURE ISOTOPE EFFECT OF ICE AND ITS ISOMERS. 34-3491
E C/ CALORIMETRIC STUDY OF GLASS TRANSITION OF THE AMORPHOUS ICE AND OF THE PHASE TRANSFORMATION BETWEEN TH 34-3476
REORIENTATION. IMPURITY EFFECTS ON THE PLASTICITY OF ICE AND THEIR EXPLANATION IN TERMS OF HYDROGEN 34-3449
CONDUCTION ANOMALIES AND POLARIZATION IN ICE AT LOW TEMPERATURES. 34-3428
ELASTIC ANOMALIES OF ICE AT LOW TEMPERATURES. 34-3439
ELECTRIC POLARIZATION EFFECTS IN PURE AND DOPED ICE AT LOW TEMPERATURES. 34-3412
SPACE CHARGE. ELECTRICAL BEHAVIOR OF ICE AT LOW TEMPERATURES, FERROELECTRICITY AND 34-3448
CHARGE AND POLARIZATION STORAGE IN ICE CRYSTALS. 34-3456
E IN WATER ICE. PART-2: DEUTERON SPIN/ NMR INVESTIGATIONS OF ICE CRYSTALS. PART-1: PROTON MAGNETIC RESONANC 34-3471
N) LOW TEMPERATURE POLARIZATION EFFECTS IN ICE. (CURRENT REVERSAL DUE TO DIPOLE RELAXATIO 34-3418
ICE, FERRO- AND ANTIFERROELECTRICS. 34-3453
PLANAR GROWTH OF ICE FROM THE PURE MELT. 34-3444
ONALLY DISORDERED SYSTEMS. PART-1: A TWO DIMENSIONAL MODEL. (ICE I) ATOMIC VIBRATIONS IN ORIENTATI 34-3474
DIELECTRIC PROPERTIES OF ICE I. 34-3419
DYNAMIC POLAR MODEL OF MONOCRYSTALLINE ICE I(H). 34-3424
HARDNESS ANISOTROPY OF SINGLE CRYSTALS OF ICE I(H). 34-3460
CE OF IMPURITIES ON THE ORDER DISORDER REACTION IN HEXAGONAL ICE I(H). INFLUEN 34-3473
LATTICE DYNAMICS OF ICE I(H). 34-3468
SPECIFIC HEAT OF ICE I(H). 34-3464
INFRARED SPECTRUM OF ICE I(H) IN THE RANGE 4000 TO 15 CM(-1). 34-3490
DEL. LOW ENTROPY FORM OF ICE I(H) OBTAINED FROM A LINEAR STEP GROWTH MO 34-3410
DIELECTRIC CONSTANT OF PURE ICE I(H) SINGLE CRYSTALS. 34-3478
PRESSURE DEPENDENCE OF THE COMPLEX DISPERSION COEFFICIENT OF ICE I(H) SINGLE CRYSTALS. 34-3479
N. DIELECTRIC AND MECHANICAL RESPONSE OF ICE I(H) SINGLE CRYSTALS AND ITS INTERPRETATIO 34-3483
SPECTRA OF ICE I(H). PART/ TRANSFER OF PROTONS THROUGH PURE ICE I(H) SINGLE CRYSTALS. PART-1: POLARIZATION 34-3455
ODELS FOR POLARIZATION A/ TRANSFER OF PROTONS THROUGH "PURE" ICE I(H), SINGLE CRYSTALS. PART-2: MOLECULAR M 34-3425
D COULOMB VALENCE DYNAMICAL/ LOW FREQUENCY RAMAN SPECTRUM OF ICE I(H). TENTATIVE INTERPRETATION WITH A MIXE 34-3425
SCATTERING. ICE I: LATTICE DYNAMICS AND INCOHERENT NEUTRON 34-3465
OPTICAL CONSTANTS OF ICE I OVER THE ENTIRE INFRARED REGION. 34-3492
ANALYTIC PROPERTIES OF THE FREE ENERGY FOR THE "ICE" MODELS (FERROELECTRICS). 34-3427
SPECIFIC HEAT OF PURE AND DOPED ICE NEAR 120 DEGREES K. 34-3463
LOW TEMPERATURE REGION. PROTONIC CONDUCTION OF ICE. PART-1: HIGH TEMPERATURE REGION. PART-2: 34-3414
/S. PART-1: MOLECULAR INTERPRETATION OF THE PHASE DIAGRAM OF ICE. PART-2: DIELECTRIC AND MECHANICAL RESPON/ 34-3485
/F HYDROGEN BONDED CRYSTALS, PART-1, THE RESIDUAL ENTROPY OF ICE, PART-2, HYDROGEN BONDED FERROELECTRIC AN/ 03-0388
PROTEIN SPIN RELAXATION IN HEXAGONAL ICE. PART-2: THE T(1 RHO) MINIMUM. 34-3488
AND DEUTERON NUCLEAR MAGNETIC RESONAN/ STRUCTURAL STUDIES OF ICE POLYMORPHS BY NEUTRON DIFFRACTION, PROTON 34-3466
PHYSICS OF ICE. (PROC OF INT SYMP, 9-14 SEP 1968 MUNICH) 01-0026
MOLECULAR UNDERSTANDING OF ELECTROCHEMICAL PROCESSES BY ICE RESEARCH. 34-3484
CIRCLE THEOREM FOR ICE RULE FERROELECTRIC MODELS. 03-0200
DIPOLAR RELAXATION AT LOW TEMPERATURE IN ICE SINGLE CRYSTAL. 34-3447
ANGULAR CORRELATION OF ANNIHILATION PHOTONS IN ICE SINGLE CRYSTALS. 34-3457
DIELECTRIC RELAXATION, BULK AND SURFACE CONDUCTIVITY OF ICE SINGLE CRYSTALS. 34-3469
INFLUENCE OF THE SURFACE LAYER ON THE PLASTIC DEFORMATION OF ICE SINGLE CRYSTALS. 34-3459
MECHANICAL PROPERTIES OF ICE SINGLE CRYSTALS. 34-3441
STUDY OF CONDUCTIVITY AND DIPOLAR RELAXATION IN DOPED ICE SINGLE CRYSTALS. 34-3458
FFRACTION TOPOGRAPHIC STUDIES OF THE DEFORMATION BEHAVIOR OF ICE SINGLE CRYSTALS. X-RAY DI 34-3426
ICAL PROPERTIES. (REVIEW, 56/ STRUCTURE AND POINT DEFECTS OF ICE: THEIR EFFECT ON THE ELECTRICAL AND MECHAN 02-0071
OW TEMPERATURE DIELECTRIC CELL AND PERMITTIVITY OF HEXAGONAL ICE TO 2 K. L 34-3432
STUDY OF THE SURFACE OF ICE WITH A SCANNING ELECTRON MICROSCOPE. 34-3420
OF THE PHASE TRANSFORMATION BETWEEN THE CUBIC AND HEXAGONAL ICES. /ASS TRANSITION OF THE AMORPHOUS ICE AND 34-3476
DIELECTRIC BEHAVIOR OF CUBIC AND HEXAGONAL ICES AT LOW TEMPERATURES. 34-3431
(ANTIFERROELECTRIC STRUCTURES)) STRUCTURES OF THE FORMS OF ICES. (ICES II, VIII AND IX ARE FULLY ORDERED 34-3451
ON THE SYMMETRY OF THE HYDROGEN BONDS IN ICE-VII. 34-3445
EXCHANGE OF THE ARSENATE RADICAL IN IRRADIATED FERROEL/ ESR IDENTIFICATION OF SLATER CONFIGURATIONS AND OF 18-2734
EAR SUSCEPTIBILITY IN LI/ SPONTANEOUS PARAMETRIC SCATTERING, IDLER ABSORPTION, AND DISPERSION IN THE NONLIN 11-1771
(MECHANICAL, CRYSTALLOGRAPHIC, PHYSICAL, THERMAL MAGNETIC, / II-VI SEMICONDUCTING COMPOUNDS - DATA TABLES. 01-0025
F TECHNIQUE AND APPLICATION TO THIN FILMS OF PEROVSKITES AND III-V COMPOUNDS, 8 REFS) /PORATION. (REVIEW O 02-0112
OPTICAL DAMAGE IN KDP (WITH RELAXATION TIME OF SECONDS WHEN ILLUMINATED UNDER APPLIED ELECTRIC FIELD). /E) 17-2703
CURIE POINT (EFFECTS OF TEMPERATURE, FIELD STRENGTH, SAMPLE ILLUMINATION). / SULFOIODIDE CRYSTALS NEAR THE 15-2121
STRAIN ALONG C-AXIS OF ANTIMONY SULFOIODIDE CAUSED BY ILLUMINATION IN DC ELECTRIC FIELD. 15-2239
MONY SULFOIODIDE SINGLE CRYSTALS. EFFECT OF ILLUMINATION ON REPOLARIZATION PROCESS IN ANTI 15-2220
N ANTIMONY SULF/ EFFECTS OF TEMPERATURE, ELECTRIC FIELD, AND ILLUMINATION ON THE ABSORPTION OF ULTRASOUND I 15-2248
NY SULFOIODIDE. (ETCHING METHOD) EFFECTS OF ILLUMINATION ON THE DOMAIN STRUCTURE OF ANTIMO 15-2124
CHARACTERISTICS OF ANTIMONY SULFOIODIDE IN TH/ INFLUENCE OF ILLUMINATION ON THE FORM OF THE VOLTAGE AMPERE 15-2119
ES, AND / PREPARATION AND INVESTIGATION OF FERROELECTRICS OF ILMENITE STRUCTURE. (ABO(3) NIOBATES, TANTALAT 08-1438
A SELF SCANNED FERROELECTRIC IMAGE SENSOR. 37-3667
A REFLECTION MO/ STRAIN BIASED FERROELECTRIC- PHOTOCONDUCTOR IMAGE STORAGE AND DISPLAY DEVICES OPERATED IN 09-1564
STRAIN BIASED FERROELECTRIC PHOTOCONDUCTOR IMAGE STORAGE AND DISPLAY DEVICES. (PLZT) 09-1565
ICS) SCATTERING MODE FERROELECTRIC PHOTOCONDUCTOR IMAGE STORAGE AND DISPLAY DEVICES. (PLZT CERAM 09-1599
RAIN FERROELECTRIC CERAMICS. (LEAD ZIRCONATE- LEAD TITANATE) IMAGE STORAGE AND DISPLAY DEVICES USING FINE G 09-1572
DYNAMICAL THEORY. PART-2: / ELECTRON MICROSCOPE TRANSMISSION IMAGES OF COHERENT DOMAIN BOUNDARIES. PART-1: 03-0255
BAND STRUCTURE OF THE CUBIC PHASE OF BARIUM TITANATE. (IMAGINARY DIELECTRIC CONSTANT) 04-0845
SULTS ABOUT PHOTOEMISSION ON FERROELECTRIC CRYSTALS. (DOMAIN IMAGING, ABSTRACT ONLY) SOME RE 03-0386

PYROELECTRIC THERMAL IMAGING DEVICES. (TGS) 27-3107
ES) IN THE EMISSION ELECTRON MICRCSCCPE (BARRIER LAYERS IN / IMAGING ELECTRIC MICRO FIELDS (OF SOLID SURFAC 04-0913
COLOR IMAGING WITH FERROELECTRIC CERAMICS. 37-3662
TIC OF BARIUM TITANATE CERAMIC EXHIBITED WHEN IT IS GIVEN AN IMPACT. PIEZOELECTRIC CHARACTERIS 04-0881
ERIALS. ENERGY GENERATION FROM IMPACT ON PIEZOELECTRIC (OR FERROELECTRIC) MAT 03-0222
CTRIC CONDENSER UNDER CONTINUOUS HEATING. (TGS SINGLE CRYST/ IMPEDANCE MEASUREMENT OF A DIELECTRIC FERROELE 27-3028
ENSER UNDER A DC FIELD. NOISE AND IMPEDANCE MEASUREMENTS OF A FERROELECTRIC COND 37-3645
LYSIS (APPLICATICN TO BARIUM TITANATE). AUTOMATIC IMPEDANCE MEASUREMENTS USING THERMAL NOISE ANA 04-0844
LLATE CRYSTALS DURING SWITCHING IN PULSE ELECTRIC FIELD. IMPEDANCE OF FERROELECTRIC TRIGLYCINE FLUOBERY 27-3027
IN FERROELECTRIC TRIGLYCINE SULFATE CRYSTALS. (TOPOGRAPHY OF IMPERFECTIONS, DOMAIN STRUCTURE) /WTH REGIONS 27-2999
MATERIALS. (BARIUM TITANATE TYPE) EFFECT OF CHARGED LATTICE IMPERFECTIONS ON THE DIELECTRIC PROPERTIES OF 04-1017
/MS. (THE MAJOR PORTICN OF THIS REPORT IS CONCERNED WITH ION IMPLANTATION OF SINGLE CRYSTAL CADMIUM SULFID/ 02-0057
POLARIZATION REVERSAL IN A WIDE RANGE OF IMPRESSING RATE OF FIELD IN SODIUM NITRITE. 16-2381
/NOLOGICAL THEORY OF DIELECTRIC AND MECHANICAL PROPERTIES OF IMPROPER FERROELECTRIC CRYSTALS. (AMMONIUM SU/ 14-2084
INDUCED AND SPCNTANECUS BIREFRINGENCE IN IMPROPER FERROELECTRIC GADOLINIUM MOLYBDATE. 30-3278
AN EXPLANATION OF ANOMALCUS OPTICAL BEHAVIOR OF THE IMPROFER FERROELECTRIC GADOLINIUM MOLYBDATE. 30-3292
IDIUM TRIHYDROGEN SELENITE. IMPROPER FERROELECTRIC PHASE TRANSITION IN RUB 22-2834
BORACITES - AN EXAMPLE OF IMPROPER FERROELECTRICS. 14-2076
PHASE TRANSFORMATIONS OF IMPROPER FERROELECTRICS. 03-0335
SOFT MODE BEHAVIOR IN IMPROPER FERROELECTRICS. (THEORY, KDP) 03-0230
BARIUM TITANATE SINGLE CRYSTALS SUBJECTED TO MECHANICAL IMPULSES. 04-1006
L RESISTIVITY OF LEAD ZIRCONATE TITANATE CERAMICS CONTAINING IMPURITIES. ELECTRICA 09-1607
PROPERTIES OF THE KDP TYPE FERROELECTRICS WITH IMPURITIES. 17-2673
N LEAD ZIRCONATE TITANATE CERAMICS CAUSED BY THE ADDITION OF IMPURITIES. SPACE CHARGE EFFECT I 09-1609
URE OF BARIUM TITANATE CRYSTALS GROWN FRCM SOLUTI/ EFFECT OF IMPURITIES AND GROWTH CONDITIONS ON THE STRUCT 04-0598
HE ELECTRON PARAMAGNETIC RESONANCE SPECTRA OF SUBSTITUTIONAL IMPURITIES IN CUBIC STRONTIUM TITANATE. / ON T 06-1193
. ELECTRON PROBE X-RAY MICROANALYSIS ON IMPURITIES IN EVAPORATED BARIUM TITANATE FILMS 04-0742
ELECTRON PARAMAGNETIC RESONANCE CF SOME RARE EARTH IMPURITIES IN STRONTIUM TITANATE. 06-1192
AVES ON FERROELECTRIC CERAMIC/ EFFECTS OF THERMALLY DIFFUSED IMPURITIES ON PROPAGATION OF ELASTIC SURFACE W 03-0487
NTHANUM DOFED BARIUM TITANATE. EFFECT OF OXIDE IMPURITIES ON THE ELECTRICAL RESISTIVITY OF LA 04-0587
PERTIES OF TGS SINGLE CRYSTALS. INFLUENCE OF IMPURITIES ON THE GROWTH AND SOME PHYSICAL PRO 27-3084
LEAD ZIRCONATE TITANATE CERAMICS. EFFECTS OF IMPURITIES ON THE MECHANICAL QUALITY FACTOR OF 09-1608
EXAGONAL ICE I (H). INFLUENCE OF IMPURITIES ON THE ORDER DISORDER REACTION IN H 34-3473
ROGEN PHCSPHATE CRYSTALS. EFFECT OF IMPURITIES ON THE STRUCTURE OF POTASSIUM DIHYD 17-2606
SALT AND/ IONIZATION OF THE (X-RADIATICN AND COPPER AND IRON IMPURITY) ABSORPTION BANDS OF DEFECT ROCHELLE 27-3108
TTERING BY MAGNONS IN NICKEL DOPED RUBIDIUM MANGANATE AND P/ IMPURITY CONCENTRATION DEPENDENCE OF RAMAN SCA 33-3395
IOBIUM AND TANTALUM DOPING. IMPURITY EFFECTS IN SODIUM TUNGSTEN BRONZES: N 13-2053
THEIR EXPLANATION IN TERMS OF HYDRCGEN REORIENTATION. IMPURITY EFFECTS ON THE PLASTICITY OF ICE AND 34-3449
(TEMPERATURE AND) ELECTRIC FIELD DEPENDENCE OF A HYDROGEN IMPURITY MODE IN STRONTIUM TITANATE. 06-1146
HYDROGEN BONDED FERROELECTRICS (KDP TYPE, THEORY). IMPURITY MODES IN AN INDUCED MOMENT SYSTEM AND 03-0426
C SEMICONDUCTOR ANTIMONY SULFOIODIDE. INDUCED IMPURITY PHOTOCONDUCTIVITY IN THE FERROELECTRI 15-2159
IMPURITY STATISTICS IN ICE. 34-3472
ON OF THE CHANGE IN COOLED KDP TEMPERATURE DUE TO 337 MICRON INCIDENT RADIATION. AN OBSERVATI 17-2711
OPTICAL MIXING OF COHERENT AND INCOHERENT LIGHT. (KDP CRYSTAL FILTER) 17-2669
ICE I: LATTICE DYNAMICS AND INCOHERENT NEUTRON SCATTERING. 34-3465
ND IN KDP AND ADP. INCOHERENT NEUTRON SCATTERING FROM HYDROGEN BO 17-2416
COMBINATION OF THE FREQUENCIES OF COHERENT AND INCOHERENT RADIATION IN KDP CRYSTAL. 17-2500
THE CURIE TEMPERATURE. CRITICAL X-RAY INCOHERENT SCATTERING BY BARIUM TITANATE NEAR 04-0535
/C AND ANTIFERROELECTRIC SOLIDS. PART-1: ENERGY LOSS NEUTRON INCOHERENT SCATTERING SPECTRA OF POTASSIUM DI/ 17-2687
NATE LATTICE. (STRUCTURE, AND PCT RESISTORS) INCORPORATION OF ANTIMONY INTO THE BARIUM TITA 04-0934
NSVERSE HYPERSONIC WAVE PULSES IN LITHIUM NIOBATE AND BISMU/ INDEPENDENT EXCITATION OF LONGITUDINAL AND TRA 11-1815
ELECTRIC SUSCEPTIBILITY OR ELASTIC COMPLIANCE IS TEMPERATURE INDEPENDENT IN THE PROTOTYPIC PHASE. /HAT THE 03-0140
D D-STRAIN AND SOFT- NORMAL MODE COORDINATES AND TEMPERATURE INDEPENDENT PARAMETERS). /IN TERMS OF LOCALIZE 03-0432
ION AND PROPERTIES OF CESIUM LEAD HALIDES. (FERROELECTRICITY INDETERMINITE) PREPARAT 33-3359
POLARIZATION, THERMAL EXPANSION COEFFICIENTS, AND REFRACTIVE INDEX) / PIEZOELECTRIC CONSTANTS, SPONTANEOUS 12-1876
OF PEROVSKITE STRUCT/ PHENOMENOLOGICAL THEORY OF REFRACTIVE INDEX AND ELECTROOPTICAL PHENOMENA IN CRYSTALS 03-0518
USCEPTIBILITY OF LITHIUM NIOBATE TO LASER INDUCED REFRACTIVE INDEX CHANGES. CONTROL OF THE S 11-1791
OPTICALLY INDUCED REFRACTIVE INDEX CHANGES IN BARIUM TITANATE. 04-1005
TO CHARACTERIZE THE OPTICAL DA/ OPTICALLY INDUCED REFRACTIVE INDEX CHANGES IN BARIUM TITANATE (EXPERIMENTS 04-1004
ELECTRIC INSULATORS. (MECHANISM OF) OPTICAL INDEX DAMAGE IN LITHIUM NIOBATE AND OTHER PYRO 11-1735
/S OF BARIUM SODIUM NIOBATE. (OPTICAL ABSORPTION, REFRACTIVE INDEX, DIELECTRIC CONSTANT, NONLINEAR OPTICAL/ 13-2024
GLYCINE SULFATE CRYSTALS. THERMAL VARIATION OF REFRACTIVE INDEX IN AMMONIUM DIHYDROGEN PHOSPHATE AND TRI 17-2474
E AND NICKEL CHLORINE) BORACITES (SECOND HARMONIC GENERATION INDEX MATCHING). /NE, IRON CHLORINE IRON IODIN 14-2080
NERATION IN POTASSIUM DIHYDROGEN / TEMPERATURE DEPENDENCE OF INDEX MATCHING ANGLE OF THE SECOND HARMONIC GE 17-2544
ECOND HARMONIC GENERATION OF LIGHT IN THE KDP CRYSTAL BY THE INDEX MATCHING METHOD. S 17-2693
EFFECT OF INTENSE LIGHT ON THE REFRACTIVE INDEX OF A CRYSTALLINE POTASSIUM NIOBATE. 08-1410
C/ INFLUENCE OF A TRANSVERSE INHOMOGENEITY OF THE REFRACTIVE INDEX OF A NONLINEAR CRYSTAL ON SECOND HARMONI 11-1687
RANGE BY PARAMETRIC SCATTERIN/ MEASUREMENT OF THE REFRACTIVE INDEX OF ADP AND KDP CRYSTALS IN THE INFRARED 17-2553
/ OF EXAMPLES OF FERROELECTRICS AND FERROELASTICS HAVING THE INDEX OF FAINTNESS UNEQUAL TO THE CELL MULTIP/ 03-0145
URIE POINT. EFFECT OF IONIC MOTION ON THE REFRACTIVE INDEX OF PARAELECTRIC KDP IN THE VICINITY OF C 17-2548
HIUM TANTALATE. INHOMOGENEITY OF THE OPTICALLY INDUCED INDEX OF REFRACTION IN LITHIUM NIOBATE AND LIT 11-1664
OSPHATE. INDEX OF REFRACTION OF POTASSIUM DIHYDROGEN PH 17-2484
PONTANEOUS PARAMETRIC SCATTERING OF LIGHT IN LITHIUM IODATE (INDEX OF REFRACTION 1.2-7 MICRONS). /S WAVE) S 08-1291
/ FREQUENCY ZONE BOUNDARY MODES IN PERCVSKITE FERROELECTRICS INDICATED BY ANISOTROPIC DEBYE- WALLER FACTOR/ 08-1351
ROOPTIC EFFECT IN BARIUM TITANATE. (BIREFRINGENCE, PRINCIPAL INDICES) DISPERSION OF ELECT 04-0755
L, A NEW FERROELECTRIC. (DIELECTRIC CONSTANTS AND REFRACTIVE INDICES) LEAD(5) GERMANIUM(3) OXIDE(11) CRYSTA 35-3502
PROPERTIES OF POTASSIUM NIOBATE SINGLE CRYSTALS. (REFRACTIVE INDICES) NONLINEAR OPTICAL 08-1454
E SYMMETRY OF POTASSIUM IODATE. (NONLINEAR OPTIC, REFRACTIVE INDICES) RELATION OF PHYSICAL PROPERTIES TO TH 08-1296
NIUM DIHYDROGEN PHOSPHATE CRYSTALS. (REFLECTANCE, REFRACTIVE INDICES) /ASSIUM DIHYDROGEN PHOSPHATE AND AMMO 17-2422
N TEMPERATURE, WAVELENGTH, APPLIED FIELD STRENGT/ REFRACTIVE INDICES OF ANTIMONY SULFOIODIDE. (DEPENDENCE O 15-2206
TEMPERATURE DEPENDENCE OF THE REFRACTIVE INDICES OF BARIUM SODIUM NIOBATE. 13-2008
SE TRANSITIONS. CRITICAL POINT INDICES OF FERROELECTRIC AND FERROMAGNETIC PHA 03-0264
PTICS). OPTICAL PROPERTIES (BIREFRINGENCE AND REFRACTIVE INDICES) OF POTASSIUM NIOBATE (LINEAR ELECTRO 08-1468
/RIC CONSTANTS, SPONTANEOUS POLARIZATION, THERMAL EXPANSION, INDICES OF REFRACTION, OPTICAL ROTATORY POWER) 35-3501
-INFRARED IN ANTIMONY SULFOIODIDE CRYSTALS. INDIRECT TRANSITIONS AND ABSORPTION IN THE MID 15-2150
(CALCIUM DOFED) POTASSIUM TANTALATE (SCHOTTKY BARRIERS WITH INDIUM). /IOBIUM DOPED) STRONTIUM TITANATE AND 06-1222
OIODIDE, TELLUROIODIDE, TRITELLURIDE, TRIIODIDE, TRIBRCMIDE, INDIUM ANTIMONIDE /LENIDE, SULFOIODIDE, SELEN 15-2108
BARIUM TITANATE FILMS PREPARED BY RF SPUTTERING ONTO INDIUM ANTIMONIDE OR GALLIUM ARSENIDE. 04-0741
AND SWITCHING OF LEAD ZIRCONATE- LEAD TITANATE CERAMICS WITH INDIUM ELECTRODES. IMPROVED AGING 09-1516
AND SWITCHING OF LEAD ZIRCONATE- LEAD TITANATE CERAMICS WITH INDIUM ELECTRODES. IMPROVED AGING 09-1517

AS CONFIRMATION OF THE THEORY OF TRAN/ TUNNELING RESULTS ON INDIUM STRONTIUM TITANATE BARRIERS INTERPRETED 03-0233
IN BARIUM TITANATE ON THE MEANS OF SCREENING THE SPONTANEOUS INDUCTION.   DEPENDENCE OF THE PIEZOEFFECT 04-1021
AND PHASE TRANSFORMATION. PART-2: ELECTRIC CONDUCTION IN THE INDUCTION REGION. /AMICS. PART-1: DEFORMATION 03-0169
CRYSTALS.   THEORY OF ELECTRIC INDUCTION WAVES IN FERROELECTRIC SEMICONDUCTOR 03-0466
REACTION SEQUENCE IN FORMING BARIUM METATITANATE IN AN INDUSTRIAL ROTARY FURNACE. 04-1003
/ECTRICS AT LOW TEMPERATURES. PART-1: GENERAL EXPRESSION FOR INELASTIC COHERENT SCATTERING OF SLOW NEUTRON/ 03-0478
RON STUDIES OF DISPLACIVE STRUCTURAL PHASE TRANSFORMATIONS. (INELASTIC DIFFUSION IN FERROELECTRICS)   NEUT 03-0155
LATE, STRONTIUM TITANATE, RUBIDIUM MANGANESE TRIFLUORIDE) BY INELASTIC NEUTRON SCATTERING. /POTASSIUM TANTA 06-1139
N BARIUM TITANATE. (LATTICE DYNAMICS) INELASTIC NEUTRON SCATTERING FROM SINGLE DOMAI 04-0946
TO LITHIUM NIOBATE.   GROUP THEORETICAL SELECTION RULES FOR INELASTIC NEUTRON SCATTERING WITH APPLICATIONS 11-1704
SE TRANSITION OF SODIUM NITRITE (STUDIED BY COHERENT NEUTRON INELASTIC SCATTERING). / THE FERROELECTRIC PHA 16-2364
ROELECTRIC CRITICAL SCATTERING FROM DEUTERATED KDP. (NEUTRON INELASTIC SCATTERING)   FER 17-2610
SOFT FERROELECTRIC MODES IN LEAD TITANATE (NEUTRON INELASTIC SCATTERING). 05-1077
ITIONS. (STRONTIUM TITANATE, POTASSIUM TANTALATE, B/ NEUTRON INELASTIC SCATTERING AT STRUCTURAL PHASE TRANS 02-0059
IN FERROELECTRICS. INELASTIC SCATTERING CROSS SECTION OF NEUTRONS 03-0297
/DROGEN BONDED FERROELECTRICS. (INFRARED ABSORPTION, NEUTRON INELASTIC SCATTERING, POTASSIUM DIHYDROGEN PH/ 03-0401
MICS OF STRONTIUM TITANATE: HARMONIC MODELS.   NEUTRON INELASTIC SCATTERING STUDY OF THE LATTICE DYNA 06-1225
IEW OF BROOKHAVEN WORK ON POTASSIUM TANTALATE, BARI/ NEUTRON INELASTIC SCATTERING STUDY ON SOFT MODES. (REV 02-0115
ALKALI HALIDES, STRONTIUM TITANATE). INERTIAL ELECTRICAL POLARIZATION OF CRYSTALS (06-1197
OF PECULIAR FERROELECTRIC- FERROELASTIC CRYSTAL: SODIUM AM/ INFERRED MECHANISM OF THE PHASE TRANSFORMATION 29-3238
, MECHANICAL AND OPTICAL PROPERTIES OF FERROELECTRIC FERROE/ INFERRED TEMPERATURE DEPENDENCES OF ELECTRICAL 30-3253
ALS GAINED BY ELECTRON PARAMAGNETIC RESONANCE. (ABSTRACT ON/ INFORMATION ON HIGH DIELECTRIC CONSTANT MATERI 02-0101
EAD ZIRCONATE TITA/ ELECTROOPTIC CERAMICS- NEW MATERIALS FOR INFORMATION STORAGE AND DISPLAY (HOT PRESSED L 09-1550
FERROELECTRIC CERAMICS FOR INFORMATION STORAGE AND DISPLAY. (PZT) 09-1552
DEVICES. (BASED ON ELECTRICALLY CONTR/ ELECTROOPTIC CERAMIC (INFORMATION) STORAGE (PROCESSING) AND DISPLAY 37-3650
OPTICAL PROPERTIES OF ANTIMONY SELENOIODIDE IN THE INFRARED. 15-2219
PHOSPHATE AND POTASSIUM DIHYDROGEN PHOSPHATE) IN THE EXTREME INFRARED. /IC PHOSPHATES (AMMONIUM DIHYDROGEN 17-2580
IC PROPERTIES OF FERROELECTRIC TRIGLYCINE SULFATE IN THE FAR INFRARED.   TEMPERATURE DEPENDENCE OF DIELECTR 27-3061
ANSITIONS IN ADP AND AMMONIUM DIHYDROGEN ARSENATE STUDIED BY INFRARED ABSORPTION.   ANTIFERROELECTRIC TR 17-2706
FROTON DYNAMICS IN KDP TYPE FERROELECTRICS STUDIED BY INFRARED ABSORPTION. 17-2708
THE PHOSPHATE ION IN THE PARAELECTRIC PHASE OF KDP CRYSTALS (INFRARED ABSORPTION AND RAMAN SCATTERING). /F 17-2479
RROELECTRIC AND PARAELECTRI/ MULTIPHONON SELECTION RULES FOR INFRARED ABSORPTION AND RAMAN SCATTERING IN FE 16-2281
NGLE CRYSTAL OF SODIUM NITRITE.   INFRARED ABSORPTION IN GAMMA-RAY IRRADIATED SI 16-2336
ING,/ FREQUENCY SPECTRUM OF HYDROGEN BONDED FERROELECTRICS. (INFRARED ABSORPTION, NEUTRON INELASTIC SCATTER 03-0401
ROGEN PHOSPHATE.   FAR INFRARED ABSORPTION SPECTRA FOR AMMONIUM DIHYD 17-2587
OGEN ORTHOPHOSPHATE (4000-10000 PER CM) AND/ LOW TEMPERATURE INFRARED ABSORPTION SPECTRA OF AMMONIUM DIHYDR 17-2432
AND STRONTIUM TIT/ EFFECT OF PARTICLE SIZE AND SHAPE ON THE INFRARED ABSORPTION SPECTRA OF BARIUM TITANATE 04-0823
POTASSIUM DIHYDROGEN PHOSPHATE AND RUBIDIUM / LOW FREQUENCY INFRARED ABSORPTION SPECTRA OF MONOCRYSTALS OF 17-2457
NESE TRIFLUORIDE, POTASSIUM COBALT TRIFLUORIDE, AND RUBIDIU/ INFRARED ABSORPTION SPECTRA OF POTASSIUM MANGA 33-3384
FAR INFRARED ABSORPTION SPECTRA OF ROCHELLE SALT. 28-3210
OF GUANIDINIUM ALUMINUM SULFATE HEXAHYDRAT/ LOW TEMPERATURE INFRARED ABSORPTION SPECTRA (5200-4200 PER CM) 21-2841
F MONOVALENT METAL NIOBATES.   INFRARED ABSORPTION SPECTROPHOTOMETRIC STUDY O 02-0113
AL OF THIOUREA IN THE PARA- AND FERROELECTRIC PHASES.   INFRARED ABSORPTION SPECTRUM OF A SINGLE CRYST 25-2910
E PARA- AND FERROELECTRIC PHASES.   INFRARED ABSORPTION SPECTRUM OF THIOUREA IN TH 25-2908
TY IN POTASSIUM NITRATE A/ SOME (DIELECTRIC MEASUREMENTS AND INFRARED ABSORPTION) STUDIES ON FERROELECTRICI 16-2402
SITION IN POTASSIUM MANGANESE TRIFLUORIDE.   INFRARED ACTIVE MODE BELOW THE STRUCTURAL TRAN 33-3362
INFRARED ACTIVE MODES IN POTASSIUM TANTALATE. 08-1366
/: THERMAL CONDUCTIVITY. PART-3: NEUTRON SCATTERING. PART-4: INFRARED AND RAMAN SCATTERING. PART-5: MOSSBA/ 03-0397
NS AND PHASE TRANSITIONS IN ANTIMONY SULFOIODIDE. (POLARIZED INFRARED AND RAMAN SPECTRA) /GTH OPTICAL PHONO 15-2106
/RROELECTRIC THIOUREA. PART-1: TEMPERATURE DEPENDENCE OF THE INFRARED AND RAMAN SPECTRA AND THE CALCULATED/ 25-2920
UMS: METHYL AMMONIUM ALUMINUM (SULFATE OR SELENATE) DODECAH/ INFRARED AND RAMAN SPECTRA OF FERROELECTRIC AL 20-2803
TE ZIRCONATES.   INFRARED AND RAMAN SPECTRA OF SEVERAL PEROVSKI 14-1411
ESIUM NITRITE.   POLARIZED INFRARED AND RAMAN SPECTRA OF SINGLE CRYSTAL C 16-2342
C TRANSITION OF THIOUREA.   INFRARED AND RAMAN SPECTROSCOPY OF POTASSIUM M 08-1303
INCLUDING FERROELECTRICS) AS A FUNCTION OF TEMPERATURE USING INFRARED AND RAMAN STUDIES IN THE FERROELECTRI 25-2916
IELD DEPENDENCE OF ABSORPTION BANDS OF FERROELECTRICS IN FAR INFRARED AND RAMAN TECHNIQUES. /TE MATERIALS (02-0104
PHOTON PHONON INTERACTION IN THE NEAR INFRARED. (APPLIED TO BARIUM TITANATE)   F 04-0665
/ATION OF THE COMPLEX PERMITTIVITY OF ADP AND KDP IN THE FAR INFRARED. (BARIUM TITANATE) 04-0909
ECTRIC PROPERTIES OF POLYVINYLIDENE FLUORIDE AND ITS USE FOR INFRARED BY DISPERSIVE FOURIER TRANSFORM SPEC/ 17-2609
LEAD TITANATE PYROELECTRIC INFRARED DETECTION.   PYROEL 36-3560
TRONTIUM TITANATES AND TITANIUM DIOXIDE.   FAR INFRARED DETECTOR. 05-1087
M NIOBATE.   FAR INFRARED DIELECTRIC DISPERSION IN BARIUM AND S 04-0975
(BARIUM TITANATE, POTASSIUM TANTALATE)   FAR INFRARED DIELECTRIC DISPERSION IN BARIUM SODIU 13-1986
DIHYDROGEN PHOSPHATE, TRIGLYCINE SULFATE, AND RUTILE.   FAR INFRARED DIELECTRIC DISPERSION OF PEROVSKITES. 04-0546
ELECTRIC CRYSTALS. (THEORY)   FAR INFRARED DIELECTRIC MEASUREMENTS ON POTASSIUM 17-2423
TANATES AND POTASSIUM/ COUPLED OPTICAL PHONON MODE THEORY OF INFRARED DISPERSION AND RAMAN SPECTRA OF FERRO 03-0165
POWER AND LINE WIDTH OF TUNABLE STIMULATED FAR INFRARED DISPERSION IN BARIUM AND STRONTIUM TI 04-0547
TIUM TITANATE.   INFRARED EMISSION IN LITHIUM NIOBATE. 11-1736
INDIRECT TRANSITIONS AND ABSORPTION IN THE MID-INFRARED FERROELECTRIC VIBRATION MODE IN STRON 06-1094
D BARIUM STRONTIUM N/ PYROELECTRIC VOLTAGE RESPONSE TO SHORT INFRARED IN ANTIMONY SULFOIODIDE CRYSTALS. 15-2150
S (CYCLOTRON RESONANCE, ABSORPTION AND LUMINESCENCE SPECTRA, INFRARED LASER PULSES IN TRIGLYCINE SULFATE AN 13-2018
E ABOVE AND BELOW TRANSITION TEMPERATURE. (NO COCHRAN TYPE / INFRARED, OPTICAL, AND ACOUSTIC SCATTERING). / 03-0427
/TRANSITIONS IN POTASSIUM DIHYDROGEN OR DEUTERIUM PHOSPHATE (INFRARED OPTICAL PROPERTIES OF VANADIUM DIOXID 36-3526
IN LITHIUM NIOBATE.   GENERATION OF FAR INFRARED PHOSPHATE STRETCHING, WIDE TEMPERATU/ 17-2646
A FERROELECTRIC.   POSSIBILITY OF GENERATING INFRARED RADIATION BY PICOSECOND LIGHT PULSES 11-1857
TE, TRIGLYCINE FLUOROBERYLL/ IMPROVEMENT IN THE PYROELECTRIC INFRARED RADIATION BY POLARIZATION REVERSAL OF 03-0162
TRONTIUM B/ PYROELECTRIC VOLTAGE RESPONSE TO STEP SIGNALS OF INFRARED RADIATION DETECTOR. (TRIGLYCINE SULFA 02-0046
ICS OF BARIUM TITANATE PYROELECTRIC DETECTOR FOR OPTICAL AND INFRARED RADIATION IN TRIGLYCINE SULFATE AND S 13-2022
TE, APPLICATIONS TO PYROELECTRICITY AND THE DETECTION OF FAR INFRARED RADIATION (0.5 TO 2 MICRONS). /TERIST 04-0751
/MENT OF THE REFRACTIVE INDEX OF ADP AND KDP CRYSTALS IN THE INFRARED RADIATIONS. /ANTS OF TRIGLYCINE SULFA 17-2514
NTIUM TITANATES IN THE 2-1000 MICRON/ TEMPERATURE STUDIES OF INFRARED RANGE BY PARAMETRIC SCATTERING OF LI/ 17-2553
TASSIUM DIHYDROGEN PHOSPHATE AND POTA/ FOURIER TRANSFORM FAR INFRARED REFLECTION SPECTRA OF BARIUM AND STRO 04-0863
MENTS ON STRONTIUM TITANATE FILMS.   INFRARED REFLECTION SPECTRA OF PARAELECTRIC PO 03-0356
FAR INFRARED REFLECTIVITY AND TRANSMISSION MEASURE 06-1154
FAR INFRARED REFLECTIVITY OF ANTIMONY SULFOIODIDE. 15-2209
FAR INFRARED REFLECTIVITY OF BARIUM TITANATE. 04-0744
OIODIDE.   FAR INFRARED REFLECTIVITY SPECTRA OF ANTIMONY SULF 15-2234
OIODIDE.   FAR INFRARED REFLECTIVITY SPECTRA OF ANTIMONY SULF 15-2235
RICHLORIDE.   FAR INFRARED REFLECTIVITY SPECTRA OF CESIUM LEAD T 33-3379
16-400 PER CM, 83-300K).   FAR INFRARED REFLECTIVITY SPECTRA OF KDP CRYSTAL (17-2681

E CRYSTALS. (0.85 DEG K, 30-2/ TEMPERATURE DEPENDENCE OF FAR INFRARED REFLECTIVITY SPECTRA OF SODIUM NITRIT 16-2373
ROGEN DISELENITE CRYSTAL.   FAR INFRARED REFLECTIVITY SPECTRA OF SODIUM TRIHYD 22-2877
M, 300 DEG K) OF THE LATTICE VIBRATIONS IN LITHIUM TANTALAT/ INFRARED (REFLECTIVITY) STUDY (20-10,000 PER C 12-1870
REFLECTICN SPECTRA CF BARIUM TITANATE IN THE LONG WAVELENGTH INFRARED REGION.   ABSORPTION AND 04-0644
          OPTICAL CONSTANTS CF ICE I OVER THE ENTIRE INFRARED REGION.   34-3492
PHAS/ TRANSMISSION OF POTASSIUM DIHYDROGEN PHOSPHATE IN THE INFRARED REGION IN THE PARA- AND FERROELECTRIC 17-2459
RS ON LEAD ZIRCONA/ REALIZATION OF A 40 MHZ COLOR TELEVISION INFRARED RESPONSE USING SURFACE WAVE TRANSDUCE 09-1509
ONTIUM BARIUM / PYROELECTRIC VOLTAGE RESPCNSE TO RECTANGULAR INFRARED SIGNALS IN TRIGLYCINE SULFATE AND STR 13-2019
MOLYBDATE.   NEW TYPE OF FAR INFRARED SOFT MODE IN FERROELECTRIC GADOLINIUM 30-3310
ITION OF POTASSIUM NITRATE.   FAR INFRARED SPECTRA AND FERROELECTRIC PHASE TRANS 16-2308
UBIC BARIUM TITANATE.   ON THE INTERPRETATION OF INFRARED SPECTRA AND THE LATTICE DYNAMICS OF C 04-0659
FATE, AND AMMONIUM SULFATE AT FERROELECTRIC / CHANGES IN THE INFRARED SPECTRA OF AMMONIUM OR RUBIDIUM BISUL 19-2772
E.   RAMAN AND INFRARED SPECTRA OF COPPER FORMATE TETRAHYDRAT 26-2936
TITANATE.   INFRARED SPECTRA OF FLUORINE CONTAINING BARIUM 04-0983
                                          RAMAN AND INFRARED SPECTRA OF INORGANIC FORMATES.   26-2937
SODIUM TRIHYDROGEN SELENITE (SHS) OR SODIUM TRIDEUTERIUM S/ INFRARED SPECTRA OF ISOTOPIC NONISOMORPHISM OF 22-2836
M, 200 AND 300 DEG K)   FAR INFRARED SPECTRA OF KDP AND ADP. (20-550 PER C 17-2549
NE FLUORIDE.   FAR INFRARED SPECTRA OF PIEZOELECTRIC POLYVINYLIDE 36-3590
UM NITRITE AT PHASE TRANSITIONS.   INFRARED SPECTRA OF POTASSIUM NITRATE AND SODI 16-2352
ATE CRYSTALS.   FAR INFRARED SPECTRA OF RUBIDIUM DIHYDROGEN PHOSPH 17-2638
AND RUBIDIUM DEUTERIUM SULFATE IN THE PARA- AND FERROELECTR/ INFRARED SPECTRA OF RUBIDIUM HYDROGEN SULFATE 19-2768
STRONTIUM TITANATE.   INFRARED SPECTRA OF SINGLE AND POLYCRYSTALLINE 06-1250
                                                   ATR INFRARED SPECTRA OF UNIAXIAL NITRATE CRYSTALS. 16-2280
ESONANCE.   POTASSIUM IODATE CRYSTAL STRUCTURE STUDIED BY INFRARED SPECTROSCOPY AND NUCLEAR QUADRUPOLE R 08-1278
/M MANGANESE TRIFLUCRIDE: TEMPERATURE DISCONTINUITIES IN THE INFRARED SPECTRUM AND GROUP THEORETICAL ANALY/ 33-3363
          ABSTRACT ONLY)   EFFECT CF FREE CARRIERS ON THE INFRARED SPECTRUM OF A FERROELECTRIC. (THEORY, 03-0408
0 TO 15 CM(-1).   INFRARED SPECTRUM OF ICE I(H) IN THE RANGE 400 34-3490
ON IN AMMONIUM SULFATE.   RAMAN AND INFRARED STUDIES OF THE FERROELECTRIC TRANSITI 19-2793
UM HYDROGEN SULFATE AND AMMCNIUM DEUTERIUM / LOW TEMPERATURE INFRARED STUDIES. PART-8: FERROELECTRIC AMMONI 17-2782
).   DISPERSION OF THE DIELECTRIC CONSTANT OF TGS IN THE FAR INFRARED (TEMPERATURE AND FREQUENCY DEPENDENCE 27-3009
UENCY / OPTICAL CONSTANTS OF TRIGLYCINE SELENATE IN THE NEAR INFRARED TO FAR INFRARED AND IN THE RADIO FREQ 27-3007
                                                   FAR INFRARED TRANSMISSION SPECTRUM OF THIOUREA.   25-2917
BOND IN KDP FERROELECTRIC (CALCULATIONS).   INFRARED VIBRATION FREQUENCIES OF THE HYDROGEN 17-2650
RATURE DEPENDENCE OF ABSORPTION OF SODIUM NITRITE IN THE FAR INFRARED. (350-1100 MICRONS)   TEMPE 16-2387
YCINE SULFATE. PART-2: X-IRRADIATION TREATMENT. STRUCTURAL INHIBITION OF FERROELECTRIC SWITCHING IN TRIGL 17-2469
YCINE SULFATE. PART-1: ADDITIVES.   STRUCTURAL INHIBITION OF FERROELECTRIC SWITCHING IN TRIGL 27-3038
NO MICROTWINNING (AND RESISTANT TO LASER INDUCED REFRACTIVE INHOMOGENEITIES). /RIUM LITHIUM NIOBATE HAVING 13-1958
CRYSTALS (LITHIUM NICBATE).   EFFECT OF OPTICAL INHOMOGENEITIES ON PHASE MATCHING IN NONLINEAR 11-1773
/ AN ANALYSIS OF MULTIPLE LIGHT SCATTERING BY RANDOM OPTICAL INHOMOGENEITIES WITH APPLICATIONS TO FERROELEC 37-3634
F REFRACTION IN LITHIUM NIOBATE AND LITHIUM TANTALATE. INHOMOGENEITY OF THE OPTICALLY INDUCED INDEX O 11-1664
LINEAR CRYSTAL ON SECOND HARMONIC/ INFLUENCE OF A TRANSVERSE INHOMOGENEITY OF THE REFRACTIVE INDEX OF A NON 11-1687
M NIOBA/ DEPENDENCE OF SECOND HARMONIC GENERATION ON CRYSTAL INHOMOGENEITY (THEORY AND TEST ON BARIUM SODIU 13-2043
THE BREAKDOWN MECHANISM IN BARIUM TITANATE AND COLORATION. (INJECTED COLOR CENTERS) RELATIONSHIP BETWEEN 04-0944
ONDENSER CHARGE- DISCHARGE METHOD).   IONIC CONTRIBUTION TO INJECTION AND CONDUCTION IN BARIUM TITANATE (C 04-0571
RIUM TITANATE.   CONDUCTIVITY INJECTION AND EXTRACTION IN POLYCRYSTALLINE BA 04-0814
ATION CHARGE IN ANTIMONY SULFOIODIDE.   INJECTION CURRENTS AND ACCUMULATION OF POLARIZ 15-2188
                                          DOUBLE INJECTION IN BARIUM TITANATE CERAMIC.   04-0554
OVE CURIE POINT)   DOUBLE SPACE CHARGE INJECTION IN SOLIDS. (USING BARIUM TITANATE AB 04-0638
TE CERAMICS.   PRODUCTION AND STUDY OF THE INNER BARRIERS IN SEMICONDUCTING BARIUM TITANA 04-0583
                        RAMAN AND INFRARED SPECTRA OF INORGANIC FORMATES.   26-2937
UTRON DIFFRACTION INVESTIGATIONS CF THE ATOMIC STRUCTURES OF INORGANIC MATERIALS. (REVIEW, 244 REFS)   NE 02-0133
N WAVEGUIDE FILLED WITH HIGH PERMITTIVITY DIELECTRIC. (THEO/ INPUT ADMITTANCE AND FIELDS RADIATED BY AN OPE 03-0359
ENT CERAMIC FERROELECTRICS. (LEAD LANTHANUM ZIRCONUM TITANA/ INSCRIPTION OF HOLOGRAPHIC RESULTS IN TRANSPAR 09-1575
ER STUDIES OF LOWER TRANSITIONS IN BARIUM TITANATE. (LATTICE INSTABILITIES)   MOSSBAU 04-0564
OLINIUM M/ PRODUCTION OF SPONTANEOUS POLARIZATION BY ELASTIC INSTABILITIES IN PIEZOELECTRIC MATERIALS. (GAD 30-3282
RROELECTRICITY)   DIELECTRIC INSTABILITIES. (THEORETICAL RELATIONSHIP TO FE 02-0068
SIUM DIHYDROGEN PHOSPHATE. (INTERBAND INTERACTION AND DIPOLE INSTABILITY) /E EFFECTS IN FERROELECTRIC POTAS 17-2433
S CRITI/ FERROELECTRIC TRANSITIONS WITH A GENUINE DIELECTRIC INSTABILITY (FERROELECTRIC TWINNING) TREATED A 15-2176
TRANSITIONS.   RAMAN SPECTROSCOPY OF PHONON INSTABILITY IN FIRST ORDER FERROELECTRIC PHASE 03-0163
ITH AMMONIUM DIHYDROGEN PHOSPHATE, POTASSIUM DI/ TEMPERATURE INSTABILITY OF ELECTROOPTIC LASER MODULATORS W 17-2675
NTIMONY SULFOIODIDE SINGLE CRYSTALS.   INSTABILITY OF FERROELECTRIC POLARIZATION IN A 15-2152
IFERROMAGNETS IN EXTERNAL FIELDS. (THEORY)   INSTABILITY OF SPIN WAVES IN FERROELECTRIC ANT 03-0149
          EFFECT OF THE QUARTIC ANHARMONICITY ON THE SOFT PHONON INSTABILITY OF STRONTIUM TITANATE.   06-1211
DUCTORS. (THEORY)   INSTABILITY PHENOMENA IN FERROELECTRIC SEMICON 03-0201
TE. (ABSTRACT)   ULTRASONIC STUDIES OF CRITICAL PHENOMENA IN INSULATING AND SEMICONDUCTING STRONTIUM TITANA 06-1118
RAMICS AT GRAIN BOUNDARY.   FERROELECTRIC PROPERTIES OF (INSULATING, BOUNDARY LAYER) BARIUM TITANATE CE 04-1024
ARIUM TITANATE SEMICONDUCTIVE CERAMIC/ MECHANISM FOR FORMING INSULATION LAYERS ON THE GRAIN BOUNDARIES OF B 04-0867
TICAL INDEX DAMAGE IN LITHIUM NIOBATE AND OTHER PYROELECTRIC INSULATORS.   (MECHANISM OF) OP 11-1735
          AFTER EFFECTS IN SEMICONDUCTOR ELECTRODES EVAPORATED ON INSULATORS ESPECIALLY ON TRIGLYCINE SULFATE. 27-3170
FERROELECTRIC SWITCHING AND THE SIEVERT INTEGRAL.   03-0239
N JUNCTIONS AND P- AND N- RANGES IN SEMICONDUCTORS. (SILICON INTEGRATED CIRCUITS) /RATE SINGLE CRYSTALS, P- 24-2904
CATTERING OF LIGHT IN LITHIUM NIOBATE (FROM TRANSVERSE (MORE INTENSE) AND LONGITUDINAL PHONONS). /ILLOUIN S 11-1743
STALLINE POTASSIUM NIOBATE.   EFFECT OF INTENSE LIGHT ON THE REFRACTIVE INDEX OF A CRY 08-1410
TYPE WITHIN A WIDE RANGE OF TEMPERATURES AND ELECTRIC FIELD INTENSITIES AND TIMES OF APPLICATION. /OVSKITE 08-1380
ITANATE BETWEEN/ STUDY OF THE TEMPERATURE DEPENDENCE OF LEED INTENSITIES FROM THE (001) SURFACE OF BARIUM T 04-0531
          PARAMETRIC LUMINESCENCE INTENSITY IN THE LITHIUM NIOBATE CRYSTAL.   11-1750
TERING SPECTRA OF GYPSUM AND / TEMPERATURE DEPENDENCE OF THE INTENSITY OF HYDROGEN BOND BANDS IN RAMAN SCAT 28-3189
IN RUBIDIUM DIHYDROGEN PHOSPHATE.   INTENSITY OF SOME BRILLOUIN AND RAYLEIGH LINES 17-2521
EPENDENCE OF EFFECTIVE POLARIZATION IN FERROELECTRICS ON THE INTENSITY OF THE ELECTRIC FIELD.   D 03-0382
C PHASE POTASSIUM / OSCILLATORY BEHAVIOR OF RAMAN SCATTERING INTENSITY WITH TEMPERATURE IN THE FERROELECTRI 17-2662
ORDER TRANSITION OF SODIUM NITRITE. (TWO SUBLATTICE MODEL OF INTERACTING DIPOLES)   THE ORDER DIS 16-2287
USE FERROELECTRIC PHASE TRANSITION, TALKING INTO ACCOUNT THE INTERACTING KANZIG REGIONS. /C FIELD ON A DIFF 03-0441
YDROGEN BONDED FERROELECTRICS OF THE KDP TYPE (PHONON PROTON INTERACTION). /S OF THE PROTON ION SYSTEM IN H 17-2670
IN FERROELECTRIC POTASSIUM DIHYDROGEN PHOSPHATE. (INTERBAND INTERACTION AND DIPOLE INSTABILITY) /E EFFECTS 17-2433
VE AND TWO LUMINOUS WAVES ALONG THE (100) DIRECTI/ COLLINEAR INTERACTION BETWEEN A LONGITUDINAL ACOUSTIC WA 11-1680
AL PHONONS IN POTASSIUM DIHYDROGEN PHOSPHATE.   INTERACTION BETWEEN PROTON TUNNELING AND OPTIC 17-2593
AINS IN BARIUM TITANATE.   INTERACTION BETWEEN THE 90 DEG AND 180 DEG DOM 04-0585
IHYDROGEN PERIODATE PROTON CROSSOVER REL/ NUCLEAR QUADRUPOLE INTERACTION IN ANTIFERROELECTRIC DIAMMONIUM TR 36-3578
IHYDROGEN PERIODATE BY IODINE PROTON CRO/ NUCLEAR QUADRUPOLE INTERACTION IN ANTIFERROELECTRIC DIAMMONIUM TR 36-3579
          ACOUSTIC PHONON SOFT OPTIC PHONON INTERACTION IN BARIUM TITANATE.   04-0806

ON RESONANCE, ABSORPTION AND LUM/ EFFECTS OF ELECTRON PHONON INTERACTION IN FERROELECTRIC CRYSTALS (CYCLOTR 03-0427
        ACOUSTOELECTRIC INTERACTION IN FERROELECTRIC SUBSTANCES. 03-0353
NDS. (KDP TYPE)   PROTON PHONON INTERACTION IN FERROELECTRICS WITH HYDROGEN BO 17-2672
FFUSE FERROELECTRIC PHASE TRANSITIONS. ACCOUNTING FOR INTERACTION IN THE KANZIG REGIONS MODEL FOR DI 03-0438
NATE)   PHOTON PHONON INTERACTION IN THE NEAR INFRARED. (BARIUM TITA 04-0925
ECTRICS WITH HYDROGEN BONDS KDP TYPE. ROLE OF PROTON INTERACTION IN THE PHASE TRANSITION OF FERROEL 17-2674
TRIC DISPERSION IN THE FERROELECTRIC P/ EFFECT OF COLLECTIVE INTERACTION OF DIPOLES ON THE NATURE OF DIELEC 03-0417
/E) FERROELECTRICS (WITH AND WITHOUT NIOBIA ADDITION) DUE TO INTERACTION OF DOMAIN BOUNDARIES AND POINT DE/ 09-1584
RAL TRANSITION OF STRONTIUM TITANATE.   INTERACTION OF ELASTIC STRAIN WITH THE STRUCTU 06-1216
IOBATE CRYSTAL.   NONLINEAR INTERACTION OF HYPERSONIC WAVES IN A LITHIUM N 11-1802
S IN A FERROELECTRIC (THEORY).   INTERACTION OF LIGHT WITH MAGNETOACOUSTIC WAVE 03-0157
LECTRICS. (THEORY).   INTERACTION OF LIGHT WITH SPIN WAVES IN FERROE 03-0158
OUND IN LITHIUM NIOBATE.   COLLINEAR INTERACTION OF LONGITUDINAL ELASTIC WAVES IN L 11-1678
       NONLINEAR INTERACTION OF MICROWAVE ELECTRIC FIELDS AND S 11-1838
    STATIC QUADRUPOLE INTERACTION OF SCANDIUM-44 IN BARIUM TITANATE. 04-0692
RIUM AND LEAD TITANATE. STATIC ELECTRIC QUADRUPOLE INTERACTION OF TANTALUM AND HAFNIUM IONS IN BA 04-0937
TANTALATE.   INTERACTION OF TWO HYPERSOUND WAVES IN LITHIUM 12-1891
MAIN WALLS IN FERROELECTRICS. (ANTIMONY SULFOIODIDE) INTERACTION OF ULTRASONIC WAVES WITH MOVING DO 15-2192
M NIOBATE.   INTERACTION OF VOLUME ACOUSTIC MODES IN LITHIU 11-1806
ITION IN STRONT/ ULTRASONIC PROPAGATION, STRESS EFFECTS, AND INTERACTION PARAMETERS AT THE DISPLACIVE TRANS 06-1127
D POTASSIUM T/ THERMAL CONDUCTIVITY IN PEROVSKITES (3 PHONON INTERACTION PROPOSED FOR STRONTIUM TITANATE AN 06-1095
EL FOR ORDER DISORDER TYPE FERROELECTRICS WITH A SHORT RANGE INTERACTION. (THEORY) /R OF THE PSEUDOSPIN MOD 03-0423
ORIGIN, SHAPES, MOV/ FERROELECTRIC DOMAIN STRUCTURE AND ITS INTERACTION WITH CRYSTAL DISLOCATIONS. (DOMAIN 03-0390
EUTERIUM PHOSPHATE.   ULTRASONIC STRESS WAVE INTERACTION WITH DOMAIN WALLS IN POTASSIUM DID 17-2659
/URAL PHASE TRANSITION IN PEROVSKITE- TYPE CRYSTALS, PART-2, INTERACTION WITH ELASTIC STRAIN (STRONTIUM TI/ 03-0241
OF POTASSIUM DIHYDROGEN PHOSPHATE TYPE FERRO/ PROTON LATTICE INTERACTIONS AND THE DYNAMICAL SUSCEPTIBILITY 17-2570
  VELOCITY OF PROPAGATION OF DOMAIN WALLS: INTERACTIONS BETWEEN DOMAINS. (IN KDP) 17-2453
WAVES IN LITHIUM NIOBATE.   INTERACTIONS BETWEEN HIGH FREQUENCY ACOUSTICAL 11-1836
ATION: PHENOMENOLOGICAL SYMMETRY CONSIDERATIONS ON BORACITE/ INTERACTIONS BETWEEN MAGNETIZATION AND POLARIZ 14-2072
CAL PHONONS IN POTASSIUM DIHYDROGEN PHOSPHATE. INTERACTIONS BETWEEN PROTON TUNNELING AND OPTI 17-2592
M TITANATE.   DOMAIN INTERACTIONS DURING PHASE TRANSITIONS IN BARIU 04-0651
/NALOG MEMORY DEVICES EMPLOYING PIEZOELECTRIC- FERROELECTRIC INTERACTIONS FOR ADAPTIVE CONTROL VOLTAGE MOD/ 37-3663
  SHELL MODEL OF INTERIONIC INTERACTIONS FOR BARIUM TITANATE. 04-0716
·DIHYDROGEN ARSENATE. PHONON PHONON AND PROTON PHONON INTERACTIONS IN ALUMINUM PHOSPHATE AND CESIUM 18-2739
PRESSURE DEPENDENCE OF ACOUSTIC MODE- SOFT OPTIC MODE INTERACTIONS IN FERROELECTRIC BARIUM TITANATE. 04-0893
    ACOUSTIC SOFT OPTIC MODE INTERACTIONS IN FERROELECTRIC BARIUM TITANATE. 04-0671
OWAVE SIGNAL PROCESSING. APPLICATION OF NONLINEAR INTERACTIONS IN FERROELECTRIC CERAMICS TO MICR 37-3647
LECTRIC CERAMICS. (BARIUM / CALCULATION OF THE ELECTROSTATIC INTERACTIONS IN FINE GRAINED PEROVSKITE FERROE 04-0642
CRYSTALS. (THEORY, KDP TYPE)   PROTON LATTICE INTERACTIONS IN HYDROGEN BONDED FERROELECTRIC 04-0181
  PROTON- PROTON AND PROTON- LATTICE INTERACTIONS IN ICE. 34-3430
   INTERDOMAIN AND DOMAIN EFFECT INTERACTIONS IN KDP (FERROELECTRIC DOMAINS). 17-2451
AFNIUM-181 STUDIED BY THE PERTURBED ANGU/ NUCLEAR QUADRUPOLE INTERACTIONS IN PEROVSKITE TYPE COMPOUNDS OF H 08-1276
·   PROTON PHONON INTERACTIONS IN POTASSIUM DIHYDROGEN PHOSPHATE 17-2653
        MODE INTERACTIONS IN POTASSIUM DIHYDROGEN PHOSPHATE 17-2701
IGL/ PULSED DOUBLE RESONANCE STUDY OF THE NUCLEAR QUADRUPOLE INTERACTIONS OF NITROGEN-14 IN PARAELECTRIC TR 27-2969
SKITE STRUCTURE.   INTERATOMIC DISTANCES IN OXIDES WITH THE PEROV 08-1439
/E EFFECTS IN FERROELECTRIC POTASSIUM DIHYDROGEN PHOSPHATE. (INTERBAND INTERACTION AND DIPOLE INSTABILITY) 17-2433
RANSITIONS.   INTERBAND MECHANISMS FOR FERROELECTRIC PHASE T 02-0085
LACEMENT FERROELECTRICS (LITHIUM TANTALATE EXAMPLE). INTERCELL CORRECTIONS FOR IONIC MOTION IN DISP 03-0367
ERAMIC BARIUM TIT/ ELECTRICAL PROPERTIES OF CRYSTALLITES AND INTERCRYSTALLITE LAYERS IN THE SEMICONDUCTOR C 04-0568
ACOUSTIC RADIATION BY INTERDIGITAL GRIDS ON LITHIUM NIOBATE. 11-1699
UM NIOBATE. MEASURED ELECTRICAL CHARACTERISTICS OF INTERDIGITAL SURFACE WAVE TRANSDUCERS ON LITHI 11-1702
RAMIC PLATES. SURFACE WAVE DELAY LINES WITH INTERDIGITAL TRANSDUCERS ON UNPOLARIZED PZT CE 09-1614
KDP (FERROELECTRIC DOMAINS).   INTERDOMAIN AND DOMAIN EFFECT INTERACTIONS IN 17-2451
ISTRIBUTION IN THE VICINITY OF A FERROELECTRIC SEMICONDUCTOR INTERFACE. POLARIZATION D 03-0167
CRYSTAL STRUCTURE OF BARIUM TITANATE PRODUCING SQUARE INTERFERENCE FRINGES. 04-0706
CRYSTAL STUDY OF BARIUM TITANATE WITH PARALLEL INTERFERENCE FRINGES. 04-0705
STUDY OF CRYSTALS OF BARIUM TITANATE HAVING PARALLEL INTERFERENCE FRINGES. 04-0897
TWO PHONON STATES OF BARIUM TITANATE. AUGER-LIKE RESONANT INTERFERENCE IN RAMAN SCATTERING FROM ONE AND 04-0923
IMONY SULFOIODIDE USING A DOUBLE PASSED, STABILIZED SCANNING INTERFEROMETER. /LOUIN SCATTERING STUDY OF ANT 15-2225
) AMPLITUDE MODULATORS BASED ON THE MICHELSON INTERFEROMETER. (AMMONIUM DIHYDROGEN PHOSPHATE 17-2715
TRANSDUCER.   ULTRASONIC INTERFEROMETER WITH LEAD ZIRCONATE TITANATE AS 09-1585
ELECTROOPTIC COEFFICIENTS IN STRONTIUM TITANATE (SINGLE) C/ INTERFEROMETRIC DETERMINATION OF THE QUADRATIC 06-1130
POTASSIUM NIOBATE SINGLE CRYSTALS.   INTERFEROMETRIC STUDIES OF DOMAIN STRUCTURE IN 08-1319
N POTASSIUM NIOBATE SINGLE CRYSTALS.   INTERFEROMETRIC STUDIES OF DOMAIN STRUCTURES I 08-1318
DINIUM ALUMINUM SULFATE HEXAHYDRATE CLEAVAGES A/ OPTICAL AND INTERFEROMETRIC STUDIES ON FERROELECTRIC GUANI 21-2816
  SHELL MODEL OF INTERIONIC INTERACTIONS FOR BARIUM TITANATE. 04-0716
/MBIC AND FERROELECTRIC THIOUREA. PART-2: CALCULATION OF THE INTERMOLECULAR AND INTRAMOLECULAR FORCE CONST/ 25-2907
NSITIONS IN HYDROGEN HALIDES.   INTERMOLECULAR POTENTIAL AND FERROELECTRIC TRA 31-3339
TGS.   HIGH INTERNAL BIAS FIELDS IN L-ALANINE SUBSTITUTED 27-2977
ES OF BARIUM TITANATE CERAMICS. EFFECT OF INTERNAL BOUNDARIES ON THE DIELECTRIC PROPERTI 04-0891
PROCESS IN LITHIUM NIOBATE.   NATURE OF INTERNAL ELECTRIC FIELD DURING OPTICAL DAMAGE 11-1861
LECTRIC LEAD ZIRCONATE STUDIED BY PERTURBED ANGULAR COR/ THE INTERNAL ELECTRIC FIELD GRADIENT IN ANTIFERROE 07-1255
TRIC FERROMAG/ MOSSBAUER STUDY AND THEORETICAL ESTIMATION OF INTERNAL ELECTRIC FIELD GRADIENTS IN FERROELEC 08-1401
LE SALT.   EPR INVESTIGATION OF THE INTERNAL ELECTRIC FIELD IN COPPER DOPED ROCHEL 28-3233
ADIENTS IN THE PEROVSKITE TYPE COMPOUND/ CALCULATIONS OF THE INTERNAL ELECTRIC FIELDS AND ELECTRIC FIELD GR 08-1442
N ABO(3) PEROVSKITE COMPOUNDS WITH DISTINCTI/ CALCULATION OF INTERNAL ELECTRIC FIELDS AND THEIR GRADIENTS I 08-1455
TANTALATE AND CADMIUM TITANATE.   INTERNAL ELECTRIC FIELDS IN CRYSTALS OF SODIUM 08-1389
  DIPOLE STRUCTURE AND INTERNAL ELECTRIC FIELDS IN LEAD ZIRCONATE. 07-1261
EORY APPLIED TO POTASSIUM FERROCYANIDE TRIHYDRATE) INTERNAL FIELD IN GENERAL DIPOLE LATTICES. (TH 03-0220
(LEAD ZIRCONATE TITANATE) FERROELECTRICS (WITH AND WITHOUT / INTERNAL FRICTION (AND CURIE TEMPERATURES) IN 09-1584
FOR ROCHELLE SALT.   INTERNAL FRICTION AS A FUNCTION OF TEMPERATURE 28-3227
YSTALS. (PIEZOELECTRIC MODULUS) AMPLITUDE DEPENDENCE OF THE INTERNAL FRICTION IN BARIUM TITANATE SINGLE CR 04-0791
YSTALS.   INTERNAL FRICTION IN BARIUM TITANATE SINGLE CR 04-0794
SALT CRYSTALS. PIEZOELECTRIC EFFECT AND INTERNAL FRICTION IN GAMMA-IRRADIATED ROCHELLE 28-3229
ONAT/ INFLUENCE OF THE CONCENTRATION OF POINT DEFECTS ON THE INTERNAL FRICTION IN POLYCRYSTALLINE LEAD ZIRC 09-1582
LECTRIC SUBSTANCES BY THE COMPOSITE RESONATO/ MEASUREMENT OF INTERNAL FRICTION IN SINGLE CRYSTALS OF FERROE 27-3080
LECTRICS. (ROCHELLE SALT AND TRIGLY/ AMPLITUDE DEPENDENCE OF INTERNAL FRICTION IN SINGLE CRYSTALS OF FERROE 27-3122
ROELECTRIC CURIE POINT OF LEAD / ANOMALIES OF ELASTICITY AND INTERNAL FRICTION IN THE REGION OF THE ANTIFER 13-2020
YSTALS.   TEMPERATURE DEPENDENCE OF INTERNAL FRICTION OF BARIUM TITANATE SINGLE CR 04-0796

YSTALS AT 1-40 HZ.
LEAD TITANATE CERAMICS. (PROPRIETARY COMPOSITIONS)
/BONDED FERROELECTRICS USING POLARIZED IR RADIATION. PART-2:
OLOGRAM RECORDING IN CR/ ELECTRON THERMAL DIFFUSION EFFECTS
  ENERGY IN GLYCINE.   ESR OF GAMMA-IRRADIATED TGS. PART-2:
ON IN POTASSIUM DIHYDROGEN PHOSPHATE.
EAR THEIR CURIE POINTS. (POTASSIUM DIHYDRO/ ANOMALIES IN THE
  (AND DIFFERENCE IN TEMPERATURES FOR MAXIMUM DIELECTRIC CON/
E PARAELECTRIC PHASE OF KDP CRYSTALS (INFRARED A/ SPECTRA OF
FERROELECTRICITY: A REVIEW OF THE PROCEEDINGS OF THE SECOND
UROPEAN VIEW. AN AMERICAN VIEW.   REPORT ON THE SECOND
GUE, PROC.
OND PROC.
  AND MECHANICAL RESPONSE OF ICE I(H) SINGLE CRYSTALS AND ITS
FTS (OF PROTONIC COLLECTIVE MOTION, TUNNELING, CALCULATIONS,
  POINT/ OF REDUCED BARIUM TITANATE CRYSTALS (POLARON HOPPING
R OF GADOLINIUM(3+) DOPED STRONTIUM TITANATE.
TICE DYNAMICS OF CUBIC BARIUM TITANATE.   ON THE
OF POTASSIUM DIDEUTERIUM PHOSPHATE (ANALYSIS AND THEORETICAL
N OF FERROELECTRIC CRYSTAL. (DIELECTRIC HEATING AND POLARIZ/
ITANATE TRANSITION IN TERMS OF NUCLEATION. (CONCEPTS)
RT-/ MOLECULAR PHENOMENA IN WATER SYSTEMS. PART-1: MOLECULAR
ROPERTIES CAUSED BY LATTICE DEFECTS. (ICE)
XATION IN HEXAGONAL ICE.
IUM TITANATE TYPE FERROELECTRICS. BASED ON NEW TREATMENT AND
/TION SPECTRA OF WATER, ICE, AND AQUEOUS SOLUTIONS AND THEIR
NAMICAL/ LOW FREQUENCY RAMAN SPECTRUM OF ICE I(H). TENTATIVE
RAN/ TUNNELING RESULTS ON INDIUM STRONTIUM TITANATE BARRIERS
ASTIC EQUATIONS OF STATE/ COMPILATION OF CONVERSION FORMULAS
TE TYPE SINGLE CRYSTALS BY NEUTRON BOMBARDMENT.   SPLIT
ACE DAMAGE PRODUCED IN BARIUM SODIUM NIOBATE CRYSTALS DURING
BY LASER BY KDP.
TRIC THIOUREA. PART-2: CALCULATION OF THE INTERMOLECULAR AND
E LEAD ZIRCONATE TITANATE SYSTEM. (DEFECT STRUCTURE)
/-1: HIGH TEMPERATURE ELECTRICAL CONDUCTIVITY MEASUREMENTS. (
D ZIRCONATE TITANATE.
TE TITANATE SYSTEM.
/S FOR POLARIZATION AND CONDUCTION. PART-3: EXTRINSIC VERSUS
/S FOR POLARIZATION AND CONDUCTION. SECT-3: EXTRINSIC VERSUS
TION PROCESS OF FERROELECTRICS WITH THE PEROVSKITE STRU/ THE
UCTING BARIUM TITANATE.
SAL IN FERROELECTRIC CERAMICS.   STATISTICAL METHOD FOR
CIATED CHARGE RELEASE TO FERROELEC/ EXPERIMENTAL METHODS FOR
ROELECTRICS.   AN APPARATUS FOR
N AND PROPERTIES IN FERROELECTRIC SOLID SOLUTIONS OF THE PE/
     CRYSTALLIZATION OF POTASSIUM
  ELECTRICAL AND OPTICAL CONSTANTS OF FERROELECTRIC POTASSIUM
IENT AND PHASE MATCHED SECOND HARMONIC GENERATION IN LITHIUM
YDROGEN PHOSPHATE, RUBIDIUM DIHYDROGEN ARSENATE, AND LITHIUM
     NOTE ON THE SPACE GROUP OF POTASSIUM HYDROGEN
     OPTICAL PARAMETRIC OSCILLATION IN LITHIUM
D PHYSICAL DAMAGE TO LITHIUM NIOBATE, PROUSTITE, AND LITHIUM
     PECULIAR FERROELECTRIC BEHAVIOR OF POTASSIUM
OELECTRIC AND ELECTROOPTIC MODULI OF FERROELECTRIC POTASSIUM
     PIEZOELECTRIC AND PHOTOELASTIC PROPERTIES OF LITHIUM
     PIEZOELECTRIC AND PHOTOELASTIC PROPERTIES OF LITHIUM
     POLARITONS IN LITHIUM
CE OF DIELECTRIC CONSTANT ON THERMAL AND FIELD HI/ POTASSIUM
T TILTABLE (TO 3 STABLE POSITIONS, SPONTANEOUS) P/ POTASSIUM
CTURE.   PYROELECTRIC AMMONIUM
PECTROSCOPY AND NUCLEAR QUADRUPOLE RESONANCE.   POTASSIUM
     ABSOLUTE STRUCTURE OF POTASSIUM
ATION IN BARIUM SODIUM NIOBATE, LITHIUM NIOBATE, AND LITHIUM
D LATTICE DYNAMIC CHARACTERISTICS OF FERROELECTRIC POTASSIUM
     GROWTH OF IODIC ACID AND
     DIELECTRIC PROPERTIES OF LITHIUM
WAVE) SPONTANEOUS PARAMETRIC SCATTERING OF LIGHT IN LITHIUM
/ARIUM TITANATE, LITHIUM TANTALATE, LITHIUM NIOBATE, LITHIUM
RELATION OF PHYSICAL PROPERTIES TO THE SYMMETRY OF POTASSIUM
     ELECTROOPTICAL PROPERTIES OF LITHIUM
     FERROELECTRICITY OF POTASSIUM
NANCE, PHASE TRANSFORMATIONS, AND FERROELECTRICITY OF ALKALI
, AMMONIUM DIHYDROGEN PHOSPHATE, LITHIUM NIOBATE, AND ALPHA-
ROWTH METHOD.   GROWTH
     VIBRATIONAL SPECTRA OF
ACITES. (IRON, NICKEL, OR COBALT, WITH CHLORIDE, BROMIDE, OR
ANTIMONY SESQUISELENIDE OR SULFIDE, ANTIMONY SULFO OR SELENO
NUCLEAR QUADRUPOLE RESONANCE OF IODINE-127 IN SOLID HYDROGEN
OF FERROELECTRIC IRON(2+) BORACITES. (CHLORIDE, BROMIDE AND
/(X) BISMUTH(1-X) SULFOIODIDE, ANTIMONY SULFO(X) SELENO(1-X)
PONTANEOUS ELECTRIC POLARIZATION AND MAGNETIZATION IN NICKEL
     PYROELECTRIC EFFECT IN ANTIMONY SULFO AND SELENO
, SULFOIODIDE, SELENOIODIDE, TELLUROIODIDE, TRITELLURIDE, TRIIODIDE, TRIBROMIDE, INDIUM ANTIMONIDE) /LENIDE
UTH, SULFUR, SELENIUM OR TELLURIUM, AND CHLORINE, BROMINE OR
     CRYSTAL GROWTH OF NICKEL
     FERROELECTRIC TRANSITION IN COBALT
     MAGNETOELECTRIC PROPERTIES OF NICKEL
    NEUTRON AND ELECTRON DIFFRACTION STUDY IN NICKEL
     OPTICAL ANISOTROPY OF THE CUBIC COBALT
TRIC CRYSTALS. (AMMONIUM SULFATE, GADOLINIUM MOLYBDATE, IRON

INTERNAL FRICTION OF LITHIUM NIOBATE SINGLE CR 11-1798
INTERNAL FRICTION OF MODIFIED LEAD ZIRCONATE- 09-1623
INTERNAL FUNDAMENTAL VIBRATIONS OF POTASSIUM / 18-2738
(INTERNAL GENERATION OF STRONG FIELDS) DURING H 06-1090
INTERNAL MOTION OF HYDROGEN AND THE ACTIVATION 27-3034
INTERNAL STRESS AND THE FERROELECTRIC TRANSITI 17-2424
INTERNAL STRESS FOR FERROELECTRIC PHOSPHATES N 17-2665
INTERNAL STRESSES IN BARIUM TITANATE CERAMICS. 04-0849
INTERNAL VIBRATIONS OF THE PHOSPHATE ION IN TH 17-2479
INTERNATIONAL MEETING ON FERROELECTRICITY. /OF 01-0010
INTERNATIONAL MEETING ON FERROELECTRICITY. A E 01-0008
INTERNATIONAL MEETING ON FERROELECTRICITY, PRA 01-0007
INTERNATIONAL MEETING ON FERROELECTRICITY, SEC 01-0031
INTERPRETATION.   DIELECTRIC 34-3483
INTERPRETATION). /DAMPINGS AND (FREQUENCY) SHI 03-0475
INTERPRETATION). / (SEEBECK COEFFICIENT, CURIE 04-0918
INTERPRETATION OF ELECTRIC FIELD EFFECTS IN EP 06-1168
INTERPRETATION OF INFRARED SPECTRA AND THE LAT 04-0659
INTERPRETATION OF PREVIOUS DATA). /ION REGION 17-2575
INTERPRETATION OF TEMPERATURE AUTOSTABILIZATIO 03-0245
INTERPRETATION OF THE DYNAMICS OF THE BARIUM T 04-0606
INTERPRETATION OF THE PHASE DIAGRAM OF ICE. PA 34-3485
INTERPRETATION OF THE PRESSURE DEPENDENCE OF P 34-3434
INTERPRETATION OF THE PROTON SPIN LATTICE RELA 34-3475
INTERPRETATION OF X-RAY DATA) /DISORDER OF BAR 04-0630
INTERPRETATION. PART-8: TRANSFER OF PROTONS T/ 34-3482
INTERPRETATION WITH A MIXED COULOMB VALENCE DY 34-3425
INTERPRETED AS CONFIRMATION OF THE THEORY OF T 03-0233
INTERRELATING THE LINEAR ELASTIC AND ELECTROEL 03-0277
INTERSTITIAL CONFIGURATION INDUCED IN PEROVSKI 08-1467
INTRACAVITY FREQUENCY DOUBLING. /STROPHIC SURF 13-2052
INTRACAVITY SECOND HARMONIC GENERATION WITH RU 17-2604
INTRAMOLECULAR FORCE CONSTANTS. /AND FERROELEC 25-2907
INTRINSIC AND EXTRINSIC NONSTOICHIOMETRY IN TH 09-1536
(INTRINSIC BAND GAP, PTCR (POSITIVE TEMPERATUR/ 04-0822
INTRINSIC NONSTOICHIOMETRY IN SINGLE PHASE LEA 09-1534
INTRINSIC NONSTOICHIOMETRY IN THE LEAD ZIRCONA 09-1535
INTRINSIC POLARIZATION, SURFACE VERSUS VOLUME/ 34-3455
INTRINSIC POLARIZATION, SURFACE VERSUS VOLUME/ 34-3482
INVERSE PIEZOELECTRIC EFFECT IN THE REPOLARIZA 08-1374
INVERSION OF ELECTRICAL PROPERTIES OF SEMICOND 04-0572
INVESTIGATING (DYNAMICS OF) POLARIZATION REVER 03-0493
INVESTIGATING STRAIN WAVE PROPAGATION AND ASSO 37-3630
INVESTIGATING THE ELECTRICAL PROPERTIES OF FER 03-0153
INVESTIGATING THE RELATIONS BETWEEN COMPOSITIO 08-1330
IODATE.   08-1367
IODATE.   08-1427
IODATE.   LARGE NONLINEAR OPTICAL COEFFIC 08-1405
IODATE. /UM DIHYDROGEN PHOSPHATE, RUBIDIUM DIH 08-1413
IODATE.   08-1370
IODATE.   08-1338
IODATE.   OPTICALLY INDUCE 08-1328
IODATE.   08-1347
IODATE.   PIEZ 08-1431
IODATE.   08-1465
IODATE.   08-1464
IODATE.   08-1294
IODATE - A DISORDERED FERROELECTRIC. (DEPENDEN 08-1348
IODATE, A FERROELECTRIC WITH NONREVERSIBLE, BU 08-1346
IODATE, A POTENTIAL FERROELASTIC: CRYSTAL STRU 36-3574
IODATE CRYSTAL STRUCTURE STUDIED BY INFRARED S 08-1278
IODATE CRYSTALS.   08-1290
IODATE CRYSTALS. /ARMONICS OF 1.06 MICRON RADI 11-1779
IODATE CRYSTALS.   POLAR AN 08-1430
IODATE CRYSTALS BY A SOLUTION GROWTH METHOD. 08-1392
IODATE. (FOUND NOT FERROELECTRIC) 36-3605
IODATE (INDEX OF REFRACTION 1.2-7 MICRONS). /S 08-1291
IODATE, KDP, ADP, AND OTHER NONFERROELECTRIC / 02-0107
IODATE. (NONLINEAR OPTIC, REFRACTIVE INDICES) 08-1296
IODATE SINGLE CRYSTALS. 08-1423
IODATE. (UNUSUAL FERROELECTRIC BEHAVIOR) 08-1345
IODATES.   NUCLEAR QUADRUPOLE RESO 08-1350
IODIC ACID. /OF POTASSIUM DIHYDROGEN PHOSPHATE 11-1677
IODIC ACID AND IODATE CRYSTALS BY A SOLUTION G 08-1392
IODIC ACID AND IODATE CRYSTALS BY A SOLUTION G 36-3548
IODIDE) /N (CALCULATIONS) OF FERROELECTRIC BOR 14-2095
IODIDE). /HE VAPOR PHASE BY A DYNAMIC METHOD (31-3335
IODIDE.   14-2097
IODIDE, AND RUBIDIUM IRON TRIFLUORIDE) /DEG K) 15-2116
IODIDE, ANTIMONY(X) ARSENIC(1-X) SULFOIODIDE) 15-2090
IODIDE BORACITE (CURIE POINT 64 DEG K). /THE S 15-2163
IODIDE SINGLE CRYSTALS. 15-2108
IODIDE, TRIBROMIDE, INDIUM ANTIMONIDE) /LENIDE 15-2194
IODINE) /I) COMPOUNDS. (USING ANTIMONY OR BISM 14-2091
IODINE BORACITE. 14-2099
IODINE BORACITE. 14-2070
IODINE BORACITE. 14-2073
IODINE BORACITE. 14-2094
IODINE BORACITE) /ERTIES OF IMPROPER FERROELEC 14-2084

PIEZOELECTRIC PROPERTIES OF COBALT IODINE BORACITE. 14-2100
SPECIFIC HEAT ANOMALY OF FERROELECTRIC COBALT IODINE BORACITE. 14-2101
IC A/ X-RAY STUDY ON PHASE TRANSITIONS OF FERROELECTRIC IRON IODINE BORACITE AT LOW TEMPERATURES. (MONOCLIN 14-2086
OPTICAL ANISOTROPY OF CUBIC NICKEL IODINE BORACITE DUE TO QUADRUPOLE TRANSITIONS. 14-2093
/MONOHYDRATE, ELECTRIC CONTROL OF ELASTIC COMPLIANCE, NICKEL IODINE BORACITE OPTICAL ANISOTROPY, BISMUTH T/ 02-0060
) ELECTROOPTICAL PROPERTIES OF FERROELECTRIC (IRON IODINE) BORACITE. (SPONTANEOUS LATTICE STRAINS 14-2082
/ORPTION SPECTRA OF NICKEL(2+) IN NICKEL CHLORINE AND NICKEL IODINE BORACITES (AND NICKEL DOPED CADMIUM CH/ 14-2074
MONY OR BISMUTH, VI IS SULFUR, SELENIUM OR TELLURIUM, VII IS IODINE, BROMINE OR CHLORINE) /STALS, V IS ANTI 15-2214
ONY BY BISMUTH OR ARSENIC, SULFUR BY SELENIUM OR OXYGEN, AND IODINE BY BROMINE OR CHLORINE) /UTION OF ANTIM 15-2205
/AGNESIUM CHLORINE, ZINC CHLORINE, CADMIUM CHLORINE, CADMIUM IODINE, IRON CHLORINE IRON IODINE AND NICKEL / 14-2080
ION IN ANTIFERROELECTRIC DIAMMONIUM TRIHYDROGEN PERIODATE BY IODINE PROTON CROSSOVER RELAXATION. / INTERACT 36-3579
AY STUDY ON THE LATTICE STRAINS OF (PARAELECTRIC AND) (IRON- IODINE- BORACITE. X-R 14-2085
I/ LATENT LATTICE STRAIN IN THE FERROELECTRIC STATE OF IRON- IODINE- BORACITE. (TEMPERATURE DEPENDENCE OF P 14-2083
FERROELECTRIC TRANSITION IN COBALT- IODINE- BORACITE UNDER PRESSURE. 14-2078
NUCLEAR QUADRUPOLE RESONANCE OF IODINE-127 IN SILVER PERIODATE. 08-1426
NUCLEAR QUADRUPOLE RESONANCE OF IODINE-127 IN SOLID HYDROGEN IODIDE. 31-3335
FERROELECTRIC ACID LITHIUM SELENITE, POSITION OF THE LITHIUM ION. REFINEMENT OF THE CRYSTAL STRUCTURE OF 22-2859
OELECTRIC LITHIUM HYDROGEN SELENITE: POSITION OF THE LITHIUM ION. /INEMENT OF THE CRYSTAL STRUCTURE OF FERR 22-2860
DECORATION OF ION BOMBARDED LITHIUM NIOBATE SURFACES. 11-1799
PH/ EXPANSION, CRAZING AND EXFOLIATION OF LITHIUM NIOBATE ON ION BOMBARDMENT AND COMPARISON RESULTS FOR SAP 11-1800
RYLLATE MONOCRYSTAL. EPR OF THE FOUR COPPER ION COMPLEX WITH SPIN S=2 IN TRIGLYCINE FLUOBE 27-3134
EPR STUDY OF PARAMAGNETIC ION COMPLEXES IN FERROELECTRIC CRYSTALS. 03-0477
OF CHROMIUM(3+), COPPER(2+), AND MANGANESE(2+) PARAMAGNETIC ION COMPLEXES IN GLYCINE CRYSTALS. EPR STUDIES 27-3135
ETCHING WITHOUT CRATERS. GROWTH OF CRATERS ON ION ETCHED SURFACE OF LITHIUM NIOBATE AND ION 11-1860
THINNING OF LITHIUM NIOBATE FOR ACOUSTOOPTIC DEFLECTORS BY ION ETCHING. 11-1862
PREPARATION OF PURE TRIGLYCINE SULFATE. (USING ION EXCHANGE) 27-3087
/ FILMS. (THE MAJOR PORTION OF THIS REPORT IS CONCERNED WITH ION IMPLANTATION OF SINGLE CRYSTAL CADMIUM SU/ 02-0057
/F THE 4T1G STATE TO THE 6A1G(6S) STATE OF THE MANGANESE(2+) ION IN A CUBIC ANTIFERRODIELECTRIC AN EXTERNA/ 03-0317
RIFLUORIDE. LUMINESCENCE OF EUROPIUM(3+) ION IN ANTIFERROMAGNETIC POTASSIUM MANGANESE T 33-3381
DEUTERO FORMATE TETRADEUT/ VIBRATION SPECTRUM OF THE FORMATE ION IN COPPER FORMATE TETRAHYDRATE AND COPPER 26-2926
ESONANCE STUDY OF PRODUCTION OF NITROGEN DIOXIDE AND NITRITE ION IN IRRADIATED POTASSIUM NITRATE. /GNETIC R 16-2404
PARAELECTRIC RESONANCE OF LITHIUM ION IN POTASSIUM CHLORIDE. 31-3325
/ON OF CRYSTAL FIELD ENERGY LEVEL SPLITTINGS OF TITANIUM(3+) ION IN RUBIDIUM ALUMINUM DISULFATE DODECAHYDR/ 19-2754
TEMPERATURE DEPENDENCE OF THE FORCE CONSTANTS OF ION IN SODIUM NITRITE. 16-2311
STRUCTURE OF DIHYDROGEN PHOSPHATE ION IN THE FERROELECTRIC POTASSIUM SALT. 17-2624
(INFRARED A/ SPECTRA OF INTERNAL VIBRATIONS OF THE PHOSPHATE ION IN THE PARAELECTRIC PHASE OF KDP CRYSTALS 17-2479
RHOMBIC AND TRIGONAL PHASES OF ABO(3) SUBSTANCE. (WHEN THE B ION IS FERROELECTRICALLY ACTIVE) /RNS IN ORTHO 08-1363
LATTICE DYNAMICS OF A RIGID ION MODEL OF GADOLINIUM MOLYBDATE. 30-3281
EFFECT OF THE CUPRIC ION ON DELAYED PHENOMENA IN ROCHELLE SALT. 28-3221
SP/ DIELECTRIC PROPERTIES AND PRESSURE EFFECTS ON SELENITE (ION). RADIATION DAMAGE ELECTRON SPIN RESONANCE 22-2870
PHASE TRANSITIONS. (BARIUM TITANATE) (FAST ION SCATTERING) SHADOW EFFECT IN (DISPLACIVE) 04-0533
F THE KDP TYPE (PHONON PRO/ COUPLED VIBRATIONS OF THE PROTON ION SYSTEM IN HYDROGEN BONDED FERROELECTRICS O 17-2670
O(3) AND A(0.5)BO(3) TYPE PHASES AND THE IONIC MASS OF THE B IONS. /LATE THE FERROELECTRIC PROPERTIES OF AB 03-0203
RAMIC) STRONTIUM TITANATES CONTAINING (TRIVALENT) RARE EARTH IONS. DIELECTRIC RELAXATION IN (CE 06-1143
IHYDROGEN PHOSPHATE CRYSTALS IN SOLUTIONS CONTAINING ACETATE IONS. GROWTH AND ETCHING OF POTASSIUM D 17-2567
RE OF HYDROGEN BONDING AND THE REORIENTATION OF THE AMMONIUM IONS. /ANSITIONS OF AMMONIUM SULFATE: THE NATU 19-2783
EQUENCY RANGE OF DOUBLE ELECTRONIC EXCITATIONS OF MOLECULES (IONS). /TION BY ANTIFERRODIELECTRICS IN THE PR 03-0419
METAL DICALCIUM PROPIONATES SUBSTITUTED PARTIALLY BY ACETATE IONS. PHASE TRANSITION IN DIVALENT 26-2944
LECTRIC TRIGLYCINE SULFATE, PURE AND DOPED WITH PARAMAGNETIC IONS. SWITCHING IN FERROE 27-3020
N LANTHANUM ALUM/ SPECTRUM AND ENERGY LEVELS OF CHROMIUM(3+) IONS AND EXCHANGE COUPLED CHROMIUM(3+) PAIRS I 36-3565
ING BARIUM TITANATE CERAMIC DOPED WITH NIOBIUM AND MANGANESE IONS AND ITS APPLICATIONS. A SEMICONDUCT 04-0734
/BASED ON BARIUM TITANATE, WHERE BARIUM(2+) AND TITANIUM(4+) IONS ARE REPLACED BY ISOVALENT AND NONISOVALE/ 04-0960
DYNAMICS OF AMMONIUM IONS IN AMMONIUM SULFATE. 19-2794
EPR OF CHROMIUM(3+) IONS IN ANTIFERROELECTRIC ADP CRYSTALS. 17-2546
ATIC ELECTRIC QUADRUPOLE INTERACTION OF TANTALUM AND HAFNIUM IONS IN BARIUM AND LEAD TITANATE. ST 04-0937
MOSSBAUER EFFECT OF FERROUS IONS IN CUBIC ICE. 34-3416
PARAMAGNETIC RELAXATION AND DIVALENT COPPER IONS IN FERROELECTRIC KDP. 17-2629
TE. ELECTRON PARAMAGNETIC RESONANCE STUDY OF SOME METAL IONS IN GUANIDINIUM ALUMINUM SULFATE HEXAHYDRA 21-2817
ATE- 5 PERCENT POTASSIUM DIDEUTERIUM / ESR STUDY OF AMMONIUM IONS IN IRRADIATED AMMONIUM DIDEUTERIUM PHOSPH 17-2684
SOLUTION KINETICS OF LUTETIUM AND SCANDIUM IONS IN LEAD TITANATE(0.5) ZIRCONATE(0.5). 09-1538
XPERIMENT AND STRUCTURA/ SUBSTITUTION OF BISMUTH AND NIOBIUM IONS IN LEAD ZIRCONATE(0.53) TITANATE(0.47) (E 09-1480
CE SPECTRA AND ENERGY LEVELS OF NEODYMIUM(3+) AND ERBIUM(3+) IONS IN LITHIUM NIOBATE CRYSTALS. / LUMINESCEN 11-1713
RYSTAL. PARAMAGNETIC RESONANCE OF PLATINUM IONS IN NIOBIUM DOPED BARIUM TITANATE SINGLE C 04-0977
ELECTRON PARAMAGNETIC RESONANCE OF COPPER IONS IN ROCHELLE SALT. 28-3176
/TIC RESONANCE STUDY OF THE SPIN SYMMETRY STATES OF AMMONIUM IONS IN SOLIDS. (AMMONIUM SULFATE, SELENATE, / 19-2800
ELECTRON PARAMAGNETIC RESONANCE OF TRIVALENT GADOLINIUM IONS IN STRONTIUM AND BARIUM TITANATES. 04-0920
/ETIC RESONANCE AND OPTICAL ABSORPTION OF TRANSITION ELEMENT IONS IN STRONTIUM TITANATE AND LANTHANUM ALUM/ 06-1145
CALLY POLARIZED PHASE OF BARIUM TITANAT/ CHANNELING OF LIGHT IONS IN THE CUBIC AND TETRAGONAL (FERROELECTRI 04-0688
CRYSTALS. COMPENSATION OF HETEROVALENT FERRIC IONS IN THE LATTICE OF BARIUM TITANATE SINGLE 04-0539
SPONTANEOUS POLARIZATION ON THE EPR SPECTRA OF CHROMIUM(3+) IONS IN TRIGLYCINE FLUOBERYLLATE. INFLUENCE OF 27-3136
/ELD AND OF THE ADDITION OF A MIXTURE OF COPPER AND CHROMIUM IONS ON THE DIELECTRIC PROPERTIES OF TRIGLYCI/ 27-3045
UM NITRATE. EFFECT OF ADDING TITANIUM(+) AND STRONTIUM(2+) IONS ON THE ELECTRICAL CONDUCTIVITY OF POTASSI 16-2319
M TITAN/ EFFECT OF NONISOVALENT SUBSTITUTION OF TITANIUM(4+) IONS ON THE ELECTROOPTICAL PROPERTIES OF BARIU 04-0959
ATE, ETC/ (EFFECT OF TEMPERATURE ON) QUASI SYMMETRICAL HARD IONS (PHOSPHATE, TITANATE, SULFATE, FLUOBERYLL 03-0330
SPECTRA OF LITHIUM NIOBATE CRYSTALS CONTAINING CHROMIUM(3+) IONS (10 TO 30 THOUSAND/CM). /THE (ABSORPTION) 11-1831
(DEGREE OF COVALENCY IN IRON DOPED BARIUM TITANATE) IONIC CHARACTERS FROM MOSSBAUER ISOMER SHIFTS. 04-0563
ELECTRIC FIELD GRADIENTS IN IONIC CHLORIDES: CESIUM LEAD TRICHLORIDE. 33-3367
BE/ TEMPERATURE DEPENDENCE OF DIELECTRIC CONSTANTS OF CUBIC IONIC COMPOUNDS. (HALIDES AND OXIDES, RELATION 03-0188
D SOLUTION AND ITS APPLICATION TO THE SOLID ELECTROLYTE FUE/ IONIC CONDUCTION IN PEROVSKITE TYPE OXIDE SOLI 08-1444
ALCIUM TITANATE OR STRONTIUM TITANATE. IONIC CONDUCTION IN SINTERED OXIDES BASED ON C 06-1141
IN BARIUM TITANATE (CONDENSER CHARGE- DISCHARGE METHOD). IONIC CONTRIBUTION TO INJECTION AND CONDUCTION 04-0571
OF TETRAGONA/ ELECTROSTATIC FIELD IN A SLIGHTLY ORTHORHOMBIC IONIC CRYSTAL. APPLICATION TO THE CALCULATION 04-0578
MOSSBAUER EFFECT AND FERROELECTRIC PROPERTIES OF IONIC CRYSTALS. 36-3596
CALCULATION OF ELECTRIC FIELDS IN IONIC CRYSTALS. (BARIUM TITANATE TYPE) 04-0824
EPOSITION OF CHARG/ ELECTRICAL PROPERTIES OF THE SURFACES OF IONIC CRYSTALS (TGS AND OTHERS, BY SELECTIVE D 27-2987
POLARIZATION, PYROELECTRICTY, AND FERROELECTRICITY OF IONIC CRYSTALS. (THEORY, NO REFS) 03-0143
(LEAD AND TIN TELLURIDES) LATTICE DYNAMICS OF IONIC CRYSTALS WITH HIGH DIELECTRIC CONSTANTS. 36-3553
ND SPONTANEOUS POLARIZATION IN DISPLAC/ RELATIONSHIP BETWEEN IONIC DISPLACEMENT AND THE CURIE TEMPERATURE A 03-0453
ELECTRONS BY TRANSVERSE AND LONGITUDINAL OPTICAL PHONONS IN IONIC FERROELECTRIC SEMICONDUCTORS. /ONDUCTION 03-0292
ANHARMONIC MODE MIXING IN IONIC FERROELECTRICS. (LATTICE DYNAMICS) 03-0234

RIC PROPERTIES OF AEO(3) AND A(0.5)BO(3) TYPE PHASES AND THE IONIC MASS OF THE B IONS. /LATE THE FERROELECT 03-0203
ITHIUM TANTALATE EXAMPLE).   INTERCELL CORRECTIONS FOR IONIC MOTION IN DISPLACEMENT FERROELECTRICS (L 03-0367
ECTRIC KDP IN THE VICINITY OF CURIE POINT.   EFFECT OF IONIC MOTION ON THE REFRACTIVE INDEX OF PARAEL 17-2548
ANATE STRUCTURE TYPES. ELECTROSTATIC FIELD IN A PSEUDOCUBIC IONIC STRUCTURE. APPLICATION TO ALL BARIUM TIT 04-1008
TITANATE) FERROE/ OXYGEN CONCENTRATION CELL MEASUREMENTS OF IONIC TRANSPORT NUMBERS IN PZT (LEAD ZIRCONATE 09-1512
IRON IMPURITY) ABSORPTION BANDS OF DEFECT ROCHELLE SALT AND/ IONIZATION OF THE (X-RADIATION AND COPPER AND 27-3108
POTASSIUM TANTALATE. (CONDUCTI/ CALCIUM CONCENTRATION VS NET IONIZED DONOR CONCENTRATION IN SINGLE CRYSTAL 08-1437
ALT. (NEUTRON DIFFRACTI/ STRUCTURAL EFFECTS OF (DAMAGE FROM) IONIZING RADIATION IN FERROELECTRIC ROCHELLE S 28-3188
/ORDER DISORDER TYPE. (COPPER FORMATE TETRAHYDRATE, AMMONIUM IRON AND CHROMIUM ALUMS, METHYL AMMONIUM ALUM/ 02-0087
        PHOTOCHROMIC EFFECT AND ELECTRON SPIN RESONANCE OF IRON AND MOLYBDENUM IN BARIUM SODIUM NIOBATE. 13-1930
C DUE TO ANTI/ MAGNETOELECTRIC EFFECT IN THE ANTIFERROMAGNET IRON ANTIMONY(2) OXIDE(4) (MAY BE FERROELECTRI 36-3563
/ CHLORINE, ZINC CHLORINE, CADMIUM CHLORINE, CADMIUM IODINE IRON CHLORINE IRON IODINE AND NICKEL CHLORINE/ 14-2080
SYSTEMS YTTRIUM MANGANESE(1-X) (B(X)) OXYGEN(3) (B=TRIVALENT IRON, CHROMIUM OR ALUMINUM). /IGATIONS OF THE 36-3569
PEROVSKITES. (WHERE METAL IS SCANDIUM, CHROMIUM, MANGANESE, IRON, COBALT, NICKEL, YTTRIUM) /ETAL OXYGEN(3) 08-1451
TO DEFICIENCIES OF 0.04)   FERROELECTRICITY IN THE IRON DEFICIENT FERROUS SULFIDE SYSTEM. (EXISTS 36-3614
ACTERS FROM MOSSBAUER ISOMER SHIFTS. (DEGREE OF COVALENCY IN IRON DOPED BARIUM TITANATE)   IONIC CHAR 04-0563
            KERR EFFECT STUDIES OF ELECTRIC CONDUCTION IN IRON DOPED BARIUM TITANATE SINGLE CRYSTALS. 04-0703
ION OF STRONG FIELDS DURING HOLOGRAM RECORDING IN CRYSTALS (IRON DOPED STRONTIUM TITANATE). /ERNAL GENERAT 06-1090
/RIC FIELDS. (SUBSTITUTION OF STRONTIUM OR CALCIUM FOR LEAD, IRON FOR ZIRCONIUM AND TITANIUM, EFFECT ON AC/ 09-1484
UCTIVITY OF BARIUM TITANATE, PURE SINGLE CRYSTALS AND DOPED (IRON, HYDROXYL) SINGLE CRYSTALS.   COND 04-0695
ELLE SALT AND/ IONIZATION OF THE (X-RADIATION AND COPPER AND IRON IMPURITY) ABSORPTION BANDS OF DEFECT ROCH 27-3108
AMAGNETIC RESONANCE.   PHOTOCHROMIC IRON IN STRONTIUM TITANATE. EVIDENCE FROM PAR 06-1171
OELECTRIC CRYSTALS. (AMMONIUM SULFATE, GADOLINIUM MOLYBDATE, IRON IODINE BORACITE) /ERTIES OF IMPROPER FERR 14-2084
OCLINIC A/ X-RAY STUDY ON PHASE TRANSITIONS OF FERROELECTRIC IRON IODINE BORACITE AT LOW TEMPERATURES. (MON 14-2086
RAINS)   ELECTROOPTICAL PROPERTIES OF FERROELECTRIC (IRON IODINE) BORACITE. (SPONTANEOUS LATTICE ST 14-2082
/ICAL ABSORPTION (CALCULATIONS) OF FERROELECTRIC BORACITES. (IRON, NICKEL, OR COBALT, WITH CHLORIDE, BROMI/ 14-2095
          A NEW MAGNETOFERROELECTRIC: CADMIUM IRON NIOBATE. 36-3618
/LEAD (B(0.5) NIOBIUM(0.5)) (1-X) B'(X) OXYGEN(3), WHERE B IS IRON OR SCANDIUM, AND B' IS ZIRCONIUM OR HAFN/ 13-2044
TITANATE CERAMICS.   THE ROLE OF IRON OXIDE ADULTERANT IN A LEAD ZIRCONATE LEAD 09-1592
/AIN NEW LAYERED FERROELECTRICS. (LANTHANUM BISMUTH TITANIUM IRON OXIDE, PRASEODYMIUM BISMUTH TITANIUM IRO/ 13-1993
HASE OF STRONTIUM TITANATE.   ANALYSIS OF THE IRON OXYGEN VACANCY CENTER IN THE TETRAGONAL P 06-1231
/ODIUM, MAGNESIUM, CADMIUM, YTTRIUM, GADOLINIUM, DYSPROSIUM, IRON, SELENIUM, NIOBIUM, COBALT, SCANDIUM, OR/ 10-1640
OSSBAUER STUDY OF ELECTRONIC SPIN FLIP PROCESSES IN AMMONIUM IRON SULFATE DODECAHYDRATE.   M 19-2771
DE.   FERROELASTIC EFFECT IN RUBIDIUM IRON TETRAFLUORIDE AND CESIUM IRON TETRAFLUORI 33-3358
CRYSTALLINE LITHIUM NIOBATE.   PROPERTIES OF THE IRON TRANSITION GROUP IN THE LATTICE OF SINGLE 11-1669
N(2+) BORACITES. (CHLORIDE, BROMIDE AND IODIDE, AND RUBIDIUM IRON TRIFLUORIDE) /DEG K) OF FERROELECTRIC IRO 14-2097
/YBDATE, KDP, ORTHOFERRITES, DYSPROSIUM ANTIMONIDE, RUBIDIUM IRON TRIFLUORIDE, BORACITES, BISMUTH(4) TITAN/ 02-0036
/FFRACTION STUDY OF (TETRAGONAL- CUBIC PEROVSKITE) STRONTIUM IRON TUNGSTATE . (MAGNETIC G STRUCTURE ANALOG/ 10-1655
/AUER SPECTRA OF MAGNETIZED PARAMAGNETIC SUBSTANCES. PART-1: IRON(X) ALUMINUM(1-X) AMMONIUM SULFATE DODECA/ 19-2751
/ANCE OF SPUTTERED LEAD(0.92) BISMUTH(0.07) LANTHANUM(0.01) (IRON(0.405) NIOBIUM(0.325) ZIRCONIUM(0.27)) O/ 37-3629
OELECTRIC EFFECT IN FERROELECTRIC- ANTIFERROMAGNETIC LITHIUM IRON(0.5) TANTALUM(0.5) OXY FLUORIDE. MAGNET 36-3568
  PHASE DIAGRAM OF THE SEMICONDUCTING UNIAXIAL FERROELECTRIC IRON(1-X) SULFIDE. 36-3617
     PHASE RELATIONS AND SUPERSTRUCTURES OF PYRRHOTITE, IRON(1-X) SULFIDE. 36-3597
     MODEL FOR THE SEMICONDUCTING UNIAXIAL FERROELECTRIC IRON(1-X) SULFIDE. PART-1. PART-2. 36-3615
/ABSORPTION (625-25 MICRONS, 77-650 DEG K) OF FERROELECTRIC IRON(2+) BORACITES. (CHLORIDE, BROMIDE AND IO/ 14-2097
SONANCE IN STRONTIUM TITANATE. (DOPED WITH GADOLINIUM(3+) OR IRON(3+)) /IC FIELD EFFECT OF ELECTRON SPIN RE 06-1120
SE OF STRONTIUM TITANATE.   THE IRON(3+)- VACANCY CENTER IN THE TETRAGONAL PHA 06-1234
        AN EPR INVESTIGATION OF IRON(3+) AND MANGANESE(2+) IN LITHIUM NIOBATE. 11-1725
        ELECTRON PARAMAGNETIC RESONANCE OF IRON(3+) DOPED LITHIUM NIOBATE. 11-1765
      EPR STUDIES OF CRYSTAL FIELD PARAMETERS IN IRON(3+) DOPED LITHIUM NIOBATE. 11-1839
S.   ELECTRON PARAMAGNETIC RESONANCE OF IRON(3+) IN BARIUM TITANATE AT LOW TEMPERATURE 04-0927
L AND THE CUBIC PHASES.   ELECTRON PARAMAGNETIC RESONANCE OF IRON(3+) IN BARIUM TITANATE IN THE RHOMBOHEDRA 04-0929
NE/ STRONG AXIAL ELECTRON PARAMAGNETIC RESONANCE SPECTRUM OF IRON(3+) IN STRONTIUM TITANATE DUE TO NEAREST 06-1144
ATE.   EFFECT OF IRON(4+) IN THE SYSTEM STRONTIUM FERRATE TITAN 13-1927
X-RAY STUDY ON THE LATTICE STRAINS OF (PARAELECTRIC AND) (IRON- IODINE- BORACITE). 14-2085
E OF PI/ LATENT LATTICE STRAIN IN THE FERROELECTRIC STATE OF (IRON- IODINE- BORACITE. (TEMPERATURE DEPENDENC 14-2083
ND PHASE TRANSITIONS OF STRESSED STRONTIUM TITANATE. (EPR OF IRON- VANADIUM PAIRS)   ORDER PARAMETER A 06-1170
PERATURE IN KDP AND DEUTERATED KDP (DOPED WITH COBALT-57 AND IRON-57). /REQUENCY MODE AT THE TRANSITION TEM 17-2465
            MOSSBAUER EFFECT FOR IRON-57 IN FERROELECTRIC LEAD TITANATE. 05-1040
ROELECTRICS. (IN BISMUTH TITANATE FERR/ MOSSBAUER STUDIES OF IRON-57 IN SOME PEROVSKITE LIKE LAYER TYPE FER 13-2035
    MOSSBAUER STUDY OF ANTIFERROELECTRIC TRANSITION IN IRON-57(3X) AMMONIUM DIHYDROGEN PHOSPHATE. 17-2644
STICS AND THE BIT SIZE OF THE FE-FC MEMO/ EFFECT OF GAUSSIAN IRRADIANCE PROFILE ON THE SWITCHING CHARACTERI 37-3670
ERCENT POTASSIUM DIDEUTERIUM / ESR STUDY OF AMMONIUM IONS IN IRRADIATED AMMONIUM DIDEUTERIUM PHOSPHATE- 5 P 17-2684
PHOSPHATE.   ELECTRON SPIN RESONANCE OF IRRADIATED AND DEUTERATED POTASSIUM DIHYDROGEN 17-2533
TION REGION.   DIELECTRIC DISPERSION OF IRRADIATED BARIUM TITANATE IN THE PHASE TRANSI 04-0652
NITION.   DIELECTRIC DISPERSION OF IRRADIATED BARIUM TITANATE NEAR THE PHASE TRAS 04-0648
/R CONFIGURATIONS AND OF EXCHANGE OF THE ARSENATE RADICAL IN IRRADIATED FERROELECTRIC POTASSIUM DIHYDROGEN/ 18-2734
/RON NUCLEAR DOUBLE RESONANCE OBSERVATIONS OF DOMAINS IN THE IRRADIATED FERROELECTRICS POTASSIUM DIHYDROGE/ 18-2726
ANIDINE ALUMINU/ THERMOLUMINESCENCE STUDIES OF THE GAMMA-RAY IRRADIATED FERROELECTRICS ROCHELLE SALT AND GU 21-2812
          FORMATION OF COLOR CENTERS IN IRRADIATED ICE. 34-3407
          X-RAY INVESTIGATION OF IRRADIATED LEAD TITANATE. 05-1081
          PARAMAGNETIC SPECIES IN IRRADIATED NITRATE. 16-2405
      X-RAY AND DIELECTRIC STUDIES OF IRRADIATED PEROVSKITE TYPE COMPOUNDS. 08-1443
E STUDY OF PRODUCTION OF NITROGEN DIOXIDE AND NITRITE ION IN IRRADIATED POTASSIUM NITRATE. /GNETIC RESONANC 16-2404
      INFRARED ABSORPTION IN GAMMA-RAY IRRADIATED SINGLE CRYSTAL OF SODIUM NITRITE. 16-2336
TASSIUM DIHYDROGEN PHOSPHATE. ESR INVESTIGATIONS OF X-RAY IRRADIATED SINGLE CRYSTALS OF FERROELECTRIC PO 17-2558
/N NUCLEAR TRIPLE RESONANCE FOR THE ARSENATE CENTER IN X-RAY IRRADIATED SINGLE CRYSTALS OF POTASSIUM DIHYD/ 18-2729
    PROTON SPIN LATTICE RELAXATION IN IRRADIATED SODIUM TRIHYDROGEN SELENATE. 22-2853
      DIELECTRIC BEHAVIOR OF GAMMA-RAY IRRADIATED TRIGLYCINE SULFATE. 27-3149
          THERMODYNAMIC THEORY OF IRRADIATED TRIGLYCINE SULFATE. 03-0152
ECTRIC PROPERTIES OF POTASSIUM DIHYDROGEN PHOSPHATE CRYSTALS IRRADIATED WITH FAST NEUTRONS. /ITION* AND DIEL 17-2615
      COERCIVE FIELDS OF ROCHELLE SALT HEAVILY IRRADIATED WITH GAMMA-RAYS. 28-3178
      LUMINESCENCE OF POTASSIUM NITRATE IRRADIATED WITH GAMMA-RAYS. 16-2335
ELECTRO) PHOTOELASTIC EFFECT IN LITHIUM NIOBATE (ON ELECTRON IRRADIATION).   SECONDARY (11-1801
DIELECTRIC MODIFICATIONS OF TUNGSTEN TRIOXIDE DUE TO NEUTRON IRRADIATION.   STRUCTURAL AND 10-1651
SOLID SOLUTIONS OF BARIUM TITANATE TYPE MATERIALS.   X-RAY IRRADIATION EFFECT ON DIELECTRIC PROPERTIES OF 04-0782
NEW STRUCTURE FORM.   CRYSTALLOGRAPHIC STUDIES OF IRRADIATION FIELD TREATED TRIGLYCINE SULFATE: 27-2996
ELECTRONIC EFFECTS (OF X-RAY AND GAMMA-RAY IRRADIATION) IN ROCHELLE SALT CRYSTALS. 28-3192
AND KDP.   NEODYMIUM YAG LASER IRRADIATION INDUCED DAMAGE TO LITHIUM NIOBATE 11-1674
ERRO-, ANTIFERRO- AND PARAELECTRIC PEROVSKITES. EFFECT OF IRRADIATION OF DIELECTRIC PROPERTIES OF SOME F 08-1474

SULFATE. INFLUENCE OF DISLOCATIONS, SURFACE LAYER AND X-RAY IRRADIATION ON DOMAIN STRUCTURE IN TRIGLYCINE 27-3004
/ EFFECT OF STRUCTURAL DEFECTS PRODUCED BY GAMMA AND REACTOR IRRADIATION ON PHASE TRANSITIONS AND DIELECTRI 17-2614
AD ZIRCONATE. EFFECT OF PILE NEUTRON IRRADIATION ON THE DIELECTRIC PROPERTIES OF LE 07-1259
  BARIUM TITANATE CERAMICS. EFFECT OF NEUTRON IRRADIATION ON THE FERROELECTRIC PROPERTIES OF 04-0732
RIUM TITANATE SINGLE CRYSTAL. EFFECT OF THERMAL NEUTRON IRRADIATION ON THE RELATIVE PERMITTIVITY OF BA 04-0781
  SALT. ON THE EFFECTS OF X-RAY IRRADIATION ON TRANSIENT PHENOMENA IN ROCHELLE 28-3222
                                              IRRADIATION PRODUCED SOLVATED ELECTRONS IN ICE 34-3423
/ROELECTRIC PROPERTIES OF BARIUM TITANATE CERAMICS CAUSED BY IRRADIATION WITH CADMIUM TRANSMITTED NEUTRONS. 04-0731
RSAL IN TRIGLYCINE SULFATE BY MEANS OF A LASER BEAM. APPLIC/ IRREVERSIBLE AND SPONTANEOUS POLARIZATION REVE 27-3010
RIC TRANSITIONS/ DIELECTRIC DISPERSION AND THERMODYNAMICS OF IRREVERSIBLE PROCESSES. (APPLIED TO FERROELECT 03-0415
/OCAL MODE MODEL.) PART-2: (DYNAMICAL PROPERTIES, TRANSVERSE ISING MODEL, SOFT MODE COUPLING TO TRANSVERSE/ 03-0384
TION FUNCTION FOR THE HEISENBERG, FERROELECTRIC, AND GENERAL ISING MODELS. ZEROES OF THE PARTI 03-0481
/TIVITY AND HALL EFFECT IN SEMICONDUCTING SOLID SOLUTIONS OF ISO- AND HETEROVALENT CATIONIC SUBSTITUTIONS / 04-0943
OTASSIUM DIHYDROGEN PHOSPHATE- WATER SYSTEM. SOLUBILITY ISOBAR IN THE AMMONIUM DIHYDROGEN PHOSPHATE- P 17-2658
OGEN PHOSPHATE CRYSTAL. STRESS INDUCED BIREFRINGENCE IN AN ISOLATED AND A SHORTCIRCUITED POTASSIUM DIHYDR 17-2452
PED BARIUM TITANATE) IONIC CHARACTERS FROM MOSSBAUER ISOMER SHIFTS. (DEGREE OF COVALENCY IN IRON DO 04-0563
                VAPOR PRESSURE ISOTOPE EFFECT OF ICE AND ITS ISOMERS. 34-3491
AL GUANIDINIUM ALUMINUM SULFATE HEXAHYDRATE AND ITS CHROMIUM ISOMORPH. /OF STRUCTURE OF FERROELECTRIC CRYST 21-2819
NCE AND RELAXATION IN ELECTRICALLY BIASED KDP AND DEUTERATED ISOMORPHS. (THEORY) DIELECTRIC RESONA 03-0467
TE USING AUTHOR'S STATIST/ COMPARISON OF FERROELECTRICITY IN ISOMORPHIC LITHIUM NIOBATE AND LITHIUM TANTALA 03-0366
/PTICAL PROPERTIES OF LEAD(5) GERMANIUM(3) OXIDE(11) AND ITS ISOMORPHOUS COMPOUND LEAD GERMANIUM SILICON O/ 35-3501
COPIC INVESTIGATION OF POTASSIUM FERROCYANIDE TRIHYDRATE AND ISOMORPHOUS COMPOUNDS. RADIO SPECTROS 24-2900
/NS OF FERROELECTRIC TRIGLYCINE SULFATE SINGLE CRYSTALS WITH ISOMORPHOUS COMPOUNDS. (TRIGLYCINE SELENATE, / 27-2972
/ERROELECTRIC PHASE TRANSITION IN POTASSIUM FERROCYANIDE AND ISOMORPHOUS CRYSTALS. (NEW MODEL BASED ON NMR/ 24-2893
/NIT CELL VOLUME EFFECTS (RELATION WITH CURIE POINT) IN SOME ISOMORPHOUS FERROELECTRIC SYSTEMS. (DISCUSSED/ 17-2487
             THERMODYNAMIC THEORY OF SOLID SOLUTIONS OF ISOMORPHOUS FERROELECTRICS. 03-0302
MONIUM/ RAMAN SCATTERING FROM FERROELECTRIC MODES IN THE KDP ISOMORPHOUS PHOSPHATES AND ARSENATES. (KDP, AM 17-2639
RIC PROPERTIES OF SOLID SOLUTIONS OF TRIGLYCINE SULFATE WITH ISOMORPHOUS SUBSTANCES. /CE OF SOME FERROELECT 27-2971
TALS OF THE TRIGLYCINE SULFA/ UNUSUAL EFFECT OF ISOTOPIC AND ISOMORPHOUS SUBSTITUTION IN FERROELECTRIC CRYS 27-3102
PROSIUM AND H/ PHASE TRANSITIONS OF GADOLINIUM MOLYBDATE AND ISOSTRUCTURAL COMPOUNDS (EUROPIUM TERBIUM, DYS 30-3271
R OF FULLY DENSE PLZT CERAMICS. ISOTHERMAL GRAIN GROWTH AND ELECTRICAL BEHAVIO 09-1553
IRCONATE. ISOTHERMAL PHASE TRANSITIONS IN CERAMIC LEAD Z 08-1466
ON IN SODIUM TRIHYDROGEN(1-X) TRIDEUTERIUM(X) DISELENITE AND ISOTOPE EFFECT. THERMAL EXPANSI 22-2858
RATED KDP CRYSTAL. ISOTOPE EFFECT IN THE RAMAN SPECTRA OF A DEUTE 17-2621
                VAPOR PRESSURE ISOTOPE EFFECT OF ICE AND ITS ISOMERS. 34-3491
YDROGEN PHOS/ ORIGIN OF THE SPONTANEOUS POLARIZATION AND THE ISOTOPE EFFECTS IN FERROELECTRIC POTASSIUM DIH 17-2433
  MAGNESIUM OXIDE AND LANTHANUM ALUMINATE CRYSTALS DOPED WITH ISOTOPE 170. FLUX GROWTH OF 36-3558
ELECTRIC CRYSTALS OF THE TRIGLYCINE SULFA/ UNUSUAL EFFECT OF ISOTOPIC AND ISOMORPHOUS SUBSTITUTION IN FERRO 27-3102
EUTERIUM(1-X) PHOSPHATE MIXED / MOLECULAR FIELD STUDY OF THE ISOTOPIC EFFECT IN POTASSIUM DIHYDROGEN(X) DID 17-2563
N(1-X) TRIDEUTERIUM(X) DISELENATE CRYSTALS. ISOTOPIC NONISOMORPHISM OF RUBIDIUM TRIHYDROGE 22-2873
SELENITE (SHS) OR SODIUM TRIDEUTERIUM S/ INFRARED SPECTRA OF ISOTOPIC NONISOMORPHISM OF SODIUM TRIHYDROGEN 22-2836
G ONLY. CAN THE ISOTOPIC SHIFT IN KDP BE EXPLAINED BY TUNNELIN 17-2562
N SODIUM NITRITE. ACCURATE FORCE CONSTANTS FROM ISOTOPIC SUBSTITUTION IN THE NITRITE RADICAL I 16-2309
/G THE LINEAR ELASTIC AND ELECTROELASTIC EQUATIONS OF STATE: ISOTROPIC AND ORTHOTROPIC (FERROELECTRIC CERA/ 03-0277
NATE, WHERE BARIUM(2+) AND TITANIUM(4+) IONS ARE REPLACED BY ISOVALENT AND NONISOVALENT CATIONS. /RIUM TITA 04-0960

J

YBDENUM(5+) IN STRONTIUM TITANATE: AN EXAMPLE OF THE DYNAMIC JAHN TELLER EFFECT. /RESONANCE SPECTRUM OF MOL 06-1125
ITY, OCCURRENCE OF ORDERED PHASE IN PEROVSKITE TYPE/ ROLE OF JAHN TELLER EFFECTS IN ORIGIN OF FERROELECTRIC 04-0964
OUS DISPERSION OF COERSIVE FIELD AT VERY LOW FREQUENCIES AND JERKY WALL MOTION IN ROCHELLE SALT. ANOMAL 28-3177
  FERROELECTRIC CRYSTALS. MECHANISMS FOR BARKHAUSEN JUMPS AND BEHAVIOR OF THE BARKHAUSEN EFFECT IN 03-0448
               BARKHAUSEN JUMPS AND SWITCHING CURRENT IN TGS CRYSTALS. 27-3109
MATER/ FUNDAMENTAL PROPERTIES OF METAL- FERROELECTRIC- METAL JUNCTION SYSTEMS. (INFLUENCE OF THE ELECTRODE 02-0127
ELECTRON TUNNELING INTO POTASSIUM TANTALATE SCHOTTKY BARRIER JUNCTIONS. 08-1365
/E IN POTASSIUM FERROCYANIDE TRIHYDRATE SINGLE CRYSTALS, P-N JUNCTIONS AND P- AND N- RANGES IN SEMICONDUCT/ 24-2904

K

IN THE FERROELECTRIC TRANSITION REGION (THEORY, RELATION TO KAENZIG DOMAIN MODEL). HEAT CAPACITY 03-0443
S NEAR THE TRANSITION POINT OF DIFFUSE FERROELECTRIC PHASE. (KAENZIG MODEL) /ATION AND DIELECTRIC HYSTERESI 03-0437
CTRIC PHASE TRANSITION, TALKING INTO ACCOUNT THE INTERACTING KANZIG REGIONS. /C FIELD ON A DIFFUSE FERROELE 03-0441
       SOME POSSIBILITIES OF DETERMINING THE VOLUME OF KANZIG REGIONS IN FERROELECTRICS. 03-0444
  PHASE TRANSITIONS. ACCOUNTING FOR INTERACTION IN THE KANZIG REGIONS MODEL FOR DIFFUSE FERROELECTRIC 03-0438
ORY OF CRYSTAL LATTICES. ESTIMATION OF THE DIMENSIONS OF KANZIG REGIONS ON THE BASIS OF THE DYNAMIC TH 03-0440
  FERROELECTRICS (SOLID SOLUTIONS AND MIXED CATION COMPOUNDS, KANZIG REGIONS, PEROVSKITES) /LAXATION IN SOME 03-0291
RIENTATIONS OF THE SPONTANEOUS POLARIZATION. A MODEL OF KANZIG REGIONS TAKING INTO ACCOUNT DIFFERENT O 03-0439
                    SEE BOTH KDA AND POTASSIUM DIHYDROGEN ARSENATE.
         ESR STUDY OF ADP- KDA MIXED CRYSTALS. 17-2683
FOR POLISHING WATER SOLUBLE CRYSTALS (APPLICATION TO ADP AND KDP). AN APPARATUS 17-2632
    DETECTION OF ATOM TUNNELING IN SOLIDS. (KDP) 17-2612
    DIFFRACTION OF A CONTINUOUS LIGHT WAVE IN KDP. 17-2444
  EFFECT OF GAMMA-IRRADIATION ON THE DIELECTRIC CONSTANTS OF KDP. 17-2691
ERSION AND FOCUSING ON THE PRODUCTION OF OPTICAL HARMONICS. (KDP) EFFECTS OF DISP 17-2579
    EXPERIMENTS ON THE ORDER OF THE TRANSITION IN KDP. 17-2603
   INTRACAVITY SECOND HARMONIC GENERATION WITH RUBY LASER BY KDP. 17-2604
    LIGHT SCATTERING STUDY OF THE FERROELECTRIC TRANSITION IN KDP. 17-2462
YAG LASER IRRADIATION INDUCED DAMAGE TO LITHIUM NIOBATE AND KDP. NEODYMIUM 11-1674
SORDER AND DISPLACIVE FERROELECTRICS. (ANTIMONY SULFOIODIDE, KDP) /S AND PHASE TRANSITIONS IN SOME ORDER DI 15-2208
    OPTIMISATION OF ELECTROOPTIC CRYSTAL MODULATORS. (KDP) 17-2541
AGNETIC RELAXATION AND DIVALENT COPPER IONS IN FERROELECTRIC KDP. PARAM 17-2629
    PRODUCTION OF LARGE SEED CRYSTALS OF KDP. 17-2413
    PULSE MEASUREMENT OF R(63) IN KDP. 17-2477
    SOFT MODE BEHAVIOR IN IMPROPER FERROELECTRICS. (THEORY, KDP) 03-0230
    SOLUBILITY IN THE GROWTH OF ELECTROOPTIC CRYSTALS ADP AND KDP. 17-2446
YMMETRY OF NORMAL OSCILLATIONS IN A FERROELECTRIC CRYSTAL OF KDP. S 17-2623
PENDENCE OF RAMAN SPECTRUM AND SECOND HARMONIC GENERATION IN KDP. TEMPERATURE DE 17-2550
    THE RADICAL IN GAMMA-IRRADIATED KDP. 17-2406
OPAGATION OF DOMAIN WALLS: INTERACTIONS BETWEEN DOMAINS. (IN KDP) VELOCITY OF PR 17-2453
    VIBRATION SPECTRA OF HYDROGEN BONDS. (KDP) 17-2482

C DOMAIN BOUNDARIES (DOPED AND UNDOPED TGS, LITHIUM NIOBATE, KDP, ABSTRACT ONLY). /CONTRAST AT FERROELECTRI 11-1671
/TANATE, LITHIUM TANTALATE, LITHIUM NIOBATE, LITHIUM IODATE, KDP, ADP, AND OTHER NONFERROELECTRIC MATERIAL/ 02-0107
/ASER RADIATION IN ACTIVE MATERIALS AND NONLINEAR CRYSTALS. (KDP, ADP, RUBIDIUM DIHYDROGEN PHOSPHATE, LITH/ 11-1865
            INCOHERENT NEUTRON SCATTERING FROM HYDROGEN BOND IN KDP AND ADP.   17-2416
RRADIATION ON PHASE TRANSITIONS AND DIELECTRIC PROPERTIES OF KDP AND ADP CRYSTALS. / BY GAMMA AND REACTOR I 17-2614
AY DIFFRACTION TOPOGRAPHY STUDY OF DEFECTS (DISLOCATIONS) IN KDP AND ADP SINGLE CRYSTALS.   X-R 17-2578
)   FAR INFRARED SPECTRA OF KDP AND ADP. (20-550 PER CM, 200 AND 300 DEG K 17-2549
   DIELECTRIC RESONANCE AND RELAXATION IN ELECTRICALLY BIASED KDP AND DEUTERATED ISOMORPHS. (THEORY)   03-0467
R RAMAN STUDY OF QUASI SPIN WAVE HYDROGEN TUNNELING MODES IN KDP AND DEUTERATED KDP.   LASE 17-2569
/ A VERY LOW FREQUENCY MODE AT THE TRANSITION TEMPERATURE IN KDP AND DEUTERATED KDP (DOPED WITH COBALT-57 / 17-2465
UBY LASER.   FRACTURE OF NONLINEAR CRYSTALS (KDP AND LITHIUM NIOBATE) BY RADIATION FROM A R 11-1663
HYDROSTATIC PRESSURE.   DISPLACEMENT OF THE CURIE POINTS IN KDP AND LITHIUM THALLIUM TARTRATE CRYSTALS BY 17-2523
         NEUTRON MEASUREMENT IN TWO SYSTEMS OF HYDROGEN BONDS. (KDP AND POTASSIUM BICARBONATE)   17-2417
ON DIFFRACTION STUDIES OF FERROELECTRIC PHASE TRANSITIONS IN KDP AND POTASSIUM DIDEUTERIUM PHOSPHATE. NEUTR 17-2661
         SEE BOTH KDP AND POTASSIUM DIHYDROGEN PHOSPHATE.
                                       CRITICAL BEHAVIOR OF KDP AND RUBIDIUM DIHYDROGEN PHOSPHATE CRYSTALS 17-2676
IT. (THEORY)   PHASE TRANSITION IN KDP AND THE ROLE OF THE DEPOLARIZING ENERGY IN 03-0229
/TRIC MODE IN POTASSIUM DIDEUTERIUM PHOSPHATE (COMPARED WITH KDP AND WITH NEUTRON AND MICROWAVE EXPERIMENT/ 17-2705
                     CAN THE ISOTOPIC SHIFT IN KDP BE EXPLAINED BY TUNNELING ONLY.   17-2562
                MEASUREMENT OF THE SPECIFIC HEAT OF KDP BY A DYNAMIC METHOD.   17-2619
TE ON THE LATTICE DYNAMICAL ASPECT OF FERROELECTRIC MODES OF KDP. (CALCULATIONS)   NO 17-2498
N OF THE FREQUENCIES OF COHERENT AND INCOHERENT RADIATION IN KDP CRYSTAL.   COMBINATIO 17-2500
          ISOTOPE EFFECT IN THE RAMAN SPECTRA OF A DEUTERATED KDP CRYSTAL.   17-2621
     TRAPPING OF PARAMETRICALLY AMPLIFIED LIGHT WAVES IN A KDP CRYSTAL.   17-2589
             SECOND HARMONIC GENERATION OF LIGHT IN THE KDP CRYSTAL BY THE INDEX MATCHING METHOD.   17-2428
TION.   STRUCTURE OF DEUTERATED KDP CRYSTAL: DEUTERIUM CONCENTRATION DETERMINA 17-2693
        OPTICAL MIXING OF COHERENT AND INCOHERENT LIGHT. (KDP CRYSTAL FILTER)   17-2535
  DIELECTRIC PROPERTIES OF A (SMALL DOSE) GAMMA-IRRADIATED KDP CRYSTAL (TEMPERATURE DEPENDENCE).   17-2669
            FAR INFRARED REFLECTIVITY SPECTRA OF KDP CRYSTAL (16-400 PER CM, 83-300K).   17-2466
XIMATION. STUDY OF PHASE TRANSITION IN MIXED KDP- DEUTERATED KDP CRYSTALS BY MODIFIED MOLECULAR FIELD APPRO 17-2681
                NOVEL TYPE OF CUT FOR KDP CRYSTALS FOR LOW VOLTAGE LIGHT MODULATION.   17-2528
IC SCATTERIN/ MEASUREMENT OF THE REFRACTIVE INDEX OF ADP AND KDP CRYSTALS IN THE INFRARED RANGE BY PARAMETR 17-2531
/IBRATIONS OF THE PHOSPHATE ION IN THE PARAELECTRIC PHASE OF KDP CRYSTALS (INFRARED ABSORPTION AND RAMAN S/ 17-2553
THE PHASE TRANSITION IN FERROELECTRICS WITH HYDROGEN BONDS. (KDP, DEUTERATED KDP) /AR STATES OF PROTONS ON 17-2479
NCE OF RAMAN SCATTERING SPECTRUM OF FERROELECTRIC DEUTERATED KDP DURING THE PHASE TRANSITION. /TURE DEPENDE 17-2671
     DISPLACEMENTS AND VELOCITIES OF THE DOMAIN WALLS IN KDP. EXISTENCE OF CRITICAL ELECTRIC FIELDS. 17-2626
               DYNAMIC MODEL OF THE SLATER KDP FERROELECTRIC.   17-2448
INFRARED VIBRATION FREQUENCIES OF THE HYDROGEN BOND IN KDP FERROELECTRIC (CALCULATIONS).   03-0285
     INTERDOMAIN AND DOMAIN EFFECT INTERACTIONS IN KDP (FERROELECTRIC DOMAINS).   17-2650
COUPLED ORDER DISORDER PHONON SYSTEM WITH DAMPING IN KDP (FREQUENCY DEPENDENT SUSCEPTIBILITY).   17-2451
TRUM OF RAMAN SCATTERED LASER LIGHT IN FERROELECTRICS OF THE KDP GROUP.   SPEC 17-2667
TIC MODULATION (BY PASSING COHERENT LIGHT THROUGH DEUTERATED KDP HAVING A CHARGE PATTERN). /ON BY ELECTROOP 17-2625
/ CRYSTALS FROM AQUEOUS SOLUTIONS (APPARATUS, GROWTH OF ADP, KDP, HYDROIODIC ACID, AND AMMONIUM OXALATE MO/ 17-2507
ON OF THE ANGULAR FIELD BY THE COMMUTATOR OF POLARIZATION IN KDP, IN A DIGITAL DEVIATOR. /Y OF THE LIMITATI 17-2420
TRANSF/ DETERMINATION OF THE COMPLEX PERMITTIVITY OF ADP AND KDP IN THE FAR INFRARED BY DISPERSIVE FOURIER 17-2524
FECT OF IONIC MOTION ON THE REFRACTIVE INDEX OF PARAELECTRIC KDP IN THE VICINITY OF CURIE POINT.   EF 17-2609
ASSIUM DIHYDROGEN PHOSPHATE. (COLLECTIVE PSEUDOSPIN MODEL OF KDP IS QUESTIONABLE) /ERROELECTRIC MODE IN POT 17-2548
, AMMONIUM/ RAMAN SCATTERING FROM FERROELECTRIC MODES IN THE KDP ISOMORPHOUS PHOSPHATES AND ARSENATES. (KDP 17-2655
/ARTITION FUNCTION FOR THE SLATER MODEL OF FERROELECTRICITY (KDP MODEL, ANTIFERROELECTRICITY F MODEL, WU K/ 17-2639
                ONE DIMENSIONAL KDP MODEL IN STATISTICAL MECHANICS.   03-0308
ORY).   PHASE TRANSITION IN KDP (MODIFICATION OF KOBAYASHI'S DYNAMICAL THE 03-0389
   ULTRASONIC MEASUREMENT OF THE ELECTROCALORIC EFFECT IN KDP (NEAR THE CURIE TEMPERATURE).   17-2488
      FERROELECTRIC CRITICAL SCATTERING FROM DEUTERATED KDP. (NEUTRON INELASTIC SCATTERING)   17-2664
IDIUM/ COUPLED FERROELASTIC CRYSTALS. (GADOLINIUM MOLYBDATE, KDP, ORTHOFERRITES, DYSPROSIUM ANTIMONIDE, RUB 17-2610
ORETICAL TREATMENT OF A DYNAMICAL MODEL OF THE FERROELECTRIC KDP. PART-1. PART-2.   QUANTUM FIELD THE 02-0036
C EFFECTS AND Q SWITCHING PERFORMANCE IN LITHIUM NIOBATE AND KDP POCKELS CELLS.   TRANSIENT ELASTOOPTI 11-1727
/ME FERROELECTRICS AND ANTIFERROELECTRICS. (BARIUM TITANATE, KDP, POTASSIUM SODIUM TARTRATE, TGS, AMMONIUM/ 02-0134
        POLARIZATION (DOMAIN STRUCTURE) AND PHASE TRANSITION IN KDP SINGLE CRYSTAL. (ABSTRACT ONLY)   02-0505
HERMAL EXPANSION OF SOME FERROELECTRICS. (POTASSIUM NITRATE, KDP, SODIUM NITRITE)   T 16-2298
IATION.   AN OBSERVATION OF THE CHANGE IN COOLED KDP TEMPERATURE DUE TO 337 MICRON INCIDENT RAD 17-2711
/RROELECTRICS. (THEORY, DYNAMICAL ORDER DISORDER PROPERTIES, KDP, TGS, ROCHELLE SALT, SILVER PERIODATE, SO/ 02-0050
       DIELECTRIC DISPERSION OF TGS AND DEUTERATED KDP (THEORY, CALCULATIONS).   17-2534
           PHASE TRANSFORMATIONS AND MELTING IN KDP TO 40 KBAR (BY DTA).   17-2634
TICAL SECOND HARMONIC GENERATION IN PIEZOELECTRIC CRYSTALS. (KDP TYPE)   OP 17-2582
  PRESSURE STUDIES OF FERROELECTRIC PROPERTIES. (PEROVSKITES, KDP TYPE)   08-1435
RACTIONS IN HYDROGEN BONDED FERROELECTRIC CRYSTALS. (THEORY, KDP TYPE)   PROTON LATTICE INTE 03-0181
N PHONON INTERACTION IN FERROELECTRICS WITH HYDROGEN BONDS. (KDP TYPE)   PROTO 17-2672
N THE PHASE TRANSITION OF FERROELECTRICS WITH HYDROGEN BONDS KDP TYPE.   ROLE OF PROTON PHONON INTERACTION I 17-2674
            DYNAMICAL BEHAVIOR OF FERROELECTRIC SPINS. (KDP TYPE CRYSTALS)   17-2600
RATION MODES OF FERROELECTRIC CRYSTALS AT LOW TEMPERATURES. (KDP TYPE CRYSTALS) /ANSITION ON THE NORMAL VIB 17-2456
.   USE OF KDP TYPE CRYSTALS AS ELECTROOPTICAL REFLECTORS 17-2512
SIUM, RUBIDIUM/ DTA STUDY OF THE FERROELECTRIC TRANSITION IN KDP TYPE CRYSTALS (CURIE TEMPERATURES OF POTAS 17-2577
HATE, RUBIDIUM AND AMMONIUM ANALOGS S/ SPECIFIC HEAT OF SOME KDP TYPE CRYSTALS. (POTASSIUM DIHYDROGEN PHOSP 17-2629
ION OF A DISCREPANCY CONCERNING THE AVERAGE DIPOLE MOMENT IN KDP TYPE FERROELECTRICS.   RESOLUT 17-2618
SORPTION.   PROTON DYNAMICS IN KDP TYPE FERROELECTRICS STUDIED BY INFRARED AB 17-2708
         PROPERTIES OF THE KDP TYPE FERROELECTRICS WITH IMPURITIES.   17-2670
/SCATTERING OF SLOW NEUTRONS AND EFFECTIVE THERMAL FACTORS. (KDP TYPE) PART-2: SCATTERING CHARACTERISTICS / 03-0478
E PROTON ION SYSTEM IN HYDROGEN BONDED FERROELECTRICS OF THE KDP TYPE (PHONON PROTON INTERACTION). /S OF TH 17-2670
AN INDUCED MOMENT SYSTEM AND HYDROGEN BONDED FERROELECTRICS (KDP TYPE, THEORY).   IMPURITY MODES IN 03-0426
OPER.IES OF A CLASS OF PLANAR FERROELECTRIC MODELS. (THEORY, KDP TYPES)   ANALYTICAL PR 03-0258
       PARAMETRIC LIGHT AMPLIFICATION AND OSCILLATION IN KDP WITH MODE LOCKED PUMP.   17-2468
MINATED UNDER APPLIED ELECTR/ (REVERSIBLE) OPTICAL DAMAGE IN KDP (WITH RELAXATION TIME OF SECONDS WHEN ILLU 17-2703
   1.06 MICRON ABSORPTION COEFFICIENTS OF DEUTERATED KDP WITH 70-100% DEUTERATION.   17-2435
ELECTROOPTIC MODULATION, LITHIUM TANTALATE, LITHIUM NIOBATE, KDP, 113 REFS). /ICATIONS (REVIEW EMPHASIZING 02-0056
WORK ON POTASSIUM TANTALATE, BARIUM AND STRONTIUM TITANATES, KDP, 22 REFS) /T MODES. (REVIEW OF BROOKHAVEN 02-0115
ULAR FIELD APPROXIMATION. STUDY OF PHASE TRANSITION IN MIXED KDP- DEUTERATED KDP CRYSTALS BY MODIFIED MOLEC 17-2528
IRON DOPED BARIUM TITANATE SINGLE CRYSTALS.   KERR EFFECT STUDIES OF ELECTRIC CONDUCTION IN 04-0703
ERROELECTRICS NEAR THE PHASE TRANSITION POINT.   KINETIC EQUATION FOR QUADRUPOLE RESONANCE IN F 03-0309
BONDED FERROELECTRICS. PART-2: APPROXIMATED SOLUTIONS TO THE KINETIC EQUATIONS. /OTON DYNAMICS IN HYDROGEN 03-0205

INGLE CRYSTAL (FROM SOLUTION) IN STATIC REGIME AND / GROWING KINETIC OF THE AMMONIUM DIHYDROGEN PHOSPHATE S 17-2516
CRYSTALLIZATION OF ROCHELLE SALT IN ELECTRIC FIELDS (KINETICS OF CRYSTAL GROWTH) 28-3200
ODIDE (CRYSTAL GROWTH RATE, ANISOTROPY) KINETICS OF CRYSTALLIZATION OF ANTIMONY SULFOI 15-2215
C SWITCHING. PART-1: GENERAL FORMULATION. KINETICS OF DOMAIN WALL MOTION IN FERROELECTRI 03-0274
DIHYDROGEN PHOSPHATE CRYSTALS IN SOLUTION. KINETICS OF GROWTH AND DISSOLUTION OF AMMONIUM 17-2412
SOLID PHASE REACTION. MECHANISM AND KINETICS OF LITHIUM METANIOBATE FORMATION BY A 11-1840
TITANATE (0.5) ZIRCONATE (0.5). SOLUTION KINETICS OF LUTETIUM AND SCANDIUM IONS IN LEAD 09-1538
NIOBATE. MECHANISM AND KINETICS OF SOLID PHASE FORMATION OF LEAD META 13-2042
TANATE. KINETICS OF THE PHASE TRANSITIONS OF BARIUM TI 04-0650
PHASE TRANSITION IN KDP (MODIFICATION OF KOBAYASHI'S DYNAMICAL THEORY). 17-2488
KTN SEE POTASSIUM TANTALATE NIOBATE
M SODIUM NIOBATE AS A NONLINEAR OPTICAL MATERIAL. (GROWTH BY KYROPOULOS AND CZOCHRALSKI METHODS) /TAL BARIU 13-2021
NTRATION IN SINGLE CRYSTAL POTASSIUM TANTALATE. (CONDUCTIVE, KYROPOULOS GROWN) / VS NET IONIZED DONOR CONCE 08-1437
GSTEN BRONZES. (SODIUM TUNGSTATE) KYROPOULOS SINGLE CRYSTAL GROWTH OF SODIUM TUN 13-2054
GROWTH OF BARIUM SODIUM NIOBATE CRYSTALS BY THE KYROPOULOS TECHNIQUE. 13-1909

L

ES IN POTASSIUM NIOBATE SINGLE CRYSTALS ASSOCIATED WITH STEP LADDERS ON PSEUDOCUBIC (001) PLANES. /STRUCTUR 08-1316
THE DIELECTRIC CONSTANT OF POTASSIUM DIHYDROGEN PHOSPHATE AT LAMBDA EQUAL 2.14 MM. /PERATURE DEPENDENCE OF 17-2501
RD-GARLAND RELATIONS AND THEIR APPLICATION TO FERROELECTRIC (LAMBDA) PHASE TRANSITIONS. /ATION OF THE PIPPA 03-0299
OELECTRIC PROPERTIES OF BARIUM TITANATE SINGLE CRYSTALS WITH LAMINAR DOMAIN STRUCTURE. /C, ELASTIC AND PIEZ 04-1010
FERROELECTRICITY (THEORY). SELF CONSISTENCY OF LANDAU'S MODEL IN THE TRANSITION FROM PIEZO T 03-0507
/F DIELECTRIC PROPERTIES AT THE PHASE TRANSITIONS. (BASED ON LANDAU PHENOMENOLOGICAL THEORY, TESTED ON AMM/ 19-2765
DEUTERIUM PHOSPHATE CRYSTALS. LANDAU-KHALATNIKOV ATTENUATION IN POTASSIUM DI 17-2656
FERRO- AND ANTIFERROELECTRIC SUBSTANCES. LANDOLT-BORNSTEIN TABELLEN, NEUE SERIE. VOL 3: 01-0002
NEW FERROELECTRIC LANGBEINITE THALLIUM CADMIUM SULFATE. 36-3534
STRUCTURAL PHASE TRANSITIONS IN LANGBEINITES. 36-3552
ROMECHANICAL PROPERTIES OF LEAD TITANATE CERAMICS CONTAINING LANTANUM AND MANGANESE. ELECT 05-1060
DISTRIBUTION OF VACANCIES IN LANTHANA DOPED LEAD TITANATE. 05-1058
AL PROPERTIES (LANTHANUM TITANATE). SYSTEM LANTHANA TITANIA PHASE EQUILIBRIA AND ELECTRIC 36-3592
/, TANTALATE), (LEAD, TUNGSTEN) (ZIRCONATE TANTALATE), LEAD (LANTHANATE, NIOBATE)- LEAD ZIRCONATE- LEAD TI 13-2004
IN OPTICAL COMMUNICATIONS. EVALUATION OF LEAD LANTHANATE ZIRCONATE CERAMICS FOR APPLICATIONS 07-1254
PHASE TRANSITIONS IN COMPLEX PEROVSKITES OF THE TYPE BARIUM LANTHANON MOLYBDATE. 30-3261
. (CRYSTAL STRUCTURE) PRECISION PARAMETERS OF SOME LANTHANON MOLYBDATE TYPE RARE EARTH MOLYBDATES 30-3262
SPIN RESONANCE SPECTRA OF CHROMIUM(3+) AND GADOLINIUM(3+) IN LANTHANUM ALUMINATE. ELECTRON 36-3583
NMR STUDY OF THE STRUCTURAL PHASE TRANSITION IN LANTHANUM ALUMINATE. 36-3533
OPTICAL SPECTRA OF CHROMIUM(3+) PAIRS IN LANTHANUM ALUMINATE. 36-3616
ORPTION OF TRANSITION ELEMENT IONS IN STRONTIUM TITANATE AND LANTHANUM ALUMINATE. /ESONANCE AND OPTICAL ABS 06-1165
RADIATIONLESS TRANSITIONS IN THE EUROPIUM(3+) CENTER IN LANTHANUM ALUMINATE. 36-3531
RANSITIONS IN PEROVSKITE TYPE CRYSTALS. (STRONTIUM TITANATE, LANTHANUM ALUMINATE) /RY OF STRUCTURAL PHASE T 03-0424
CHROMIUM(3+) IONS AND EXCHANGE COUPLED CHROMIUM(3+) PAIRS IN LANTHANUM ALUMINATE. /UM AND ENERGY LEVELS OF 36-3565
T-2, INTERACTION WITH ELASTIC STRAIN (STRONTIUM TITANATE AND LANTHANUM ALUMINATE). /ITE- TYPE CRYSTALS, PAR 03-0241
E 170. FLUX GROWTH OF MAGNESIUM OXIDE AND LANTHANUM ALUMINATE CRYSTALS DOPED WITH ISOTOP 36-3558
/ DIELECTRIC STUDIES OF CERTAIN NEW LAYERED FERROELECTRICS. (LANTHANUM BISMUTH TITANIUM IRON OXIDE, PRASEO/ 13-1993
RIC PROPERTIES OF PLZT CERAMICS WITH A ZIRCONIUM TO TITANIU/ LANTHANUM DEPENDENCE OF ELASTIC AND PIEZOELECT 09-1573
EFFECT OF OXIDE IMPURITIES ON THE ELECTRICAL RESISTIVITY OF LANTHANUM DOPED BARIUM TITANATE. 04-0587
BISMUTH OXIDE AS A GRAIN BOUNDARY ELEMENT. LANTHANUM DOPED BARIUM TITANATE CERAMICS WITH 04-0799
IDE COMPOSITE/ TWO-STEP ANOMALOUS INCREASE OF RESISTIVITY IN LANTHANUM DOPED BARIUM TITANATE- BISMUTH TRIOX 04-0800
PYROELECTRIC PROPERTIES OF THE LANTHANUM DOPED FERROELECTRIC PLZT CERAMICS. 09-1558
CS. STRUCTURE AND ELECTROMECHANICAL PROPERTIES OF LANTHANUM DOPED LEAD TITANATE ZIRCONATE CERAMI 09-1502
CS. STRUCTURE AND ELECTROMECHANICAL PROPERTIES OF LANTHANUM DOPED LEAD TITANATE ZIRCONATE CERAMI 09-1531
OELECTRIC PHASE IN THE SYSTEM LEAD TITANATE- LEAD ZIRCONATE- LANTHANUM FERRATE. ANTIFERR 13-1961
TH FERRATE AND ITS SOLID SOLUTIONS WITH BARIUM TITANATE, AND LANTHANUM FERRATE) /E (0.5) NIOBATE (0.5), BISMU 08-1368
TASSIUM ANTIMONATE, NIOBATE, OR BISMUTHATE, OR BY BISMUTH OR LANTHANUM FERRATE, ALUMINATE OR CHROMATE. / PO 08-1539
DISPERSION IN STRONTIUM TITANATE- LANTHANUM FERRATE DIELECTRIC CERAMICS. 06-1120
ROMAGNETIC AND SPECIAL DIELECTRIC PROPERTIES IN THE BISMUTH- LANTHANUM FERRATE SYSTEM. /XISTENCE OF ANTIFER 08-1425
MATERIAL AND ELECTROOPTIC PROPERTIES OF THE (LEAD, LANTHANUM) (HAFNATE, TITANATE) SYSTEM. 08-1302
/TE, POTASSIUM STRONTIUM LITHIUM NIOBATE TITANATE, POTASSIUM LANTHANUM LITHIUM NIOBIUM TITANATE, BARIUM LI/ 13-1923
ND B-SITE VACANCIES. X-RAY STRUCTURE INVESTIGATION OF LANTHANUM MODIFIED LEAD TITANATE WITH A-SITE A 05-1059
ISION PARAMETERS OF THE FERROELECTRIC RARE EARTH MOLYBDATES: LANTHANUM MOLYBDATE. PREC 30-3264
DIELECTRIC CONSTANTS OF THE SYSTEM SAMARIUM NIOBATE- LANTHANUM NIOBATE. 08-1268
/E CONSTANTS IN FERROELECTRIC AND ANTIFERROELECTRIC PHASES. (LANTHANUM OR NIOBIUM DOPED LEAD ZIRCONATE STA/ 09-1487
FERROELASTIC EFFECT IN LANTHANUM ORTHOFERRITE. 08-1270
OXIDATION OF CARBON MONOXIDE ON LANTHANUM OXIDE DOPED BARIUM TITANATE. 04-0947
OELECTRIC DOMAIN ALIGNMENT IN LEAD ZIRCONATE- LEAD TITANATE- LANTHANUM OXIDE (PLZT) CERAMIC. /EPENDENT FERR 09-1586
ELECTRICAL PROPERTIES OF TERNARY BARIUM LANTHANUM OXIDE- TITANIUM DIOXIDE COMPOUNDS. 13-1912
ZOCHRALSKI CRYSTAL GROWTH. (BARIUM SODIUM NIOBATE AND SODIUM LANTHANUM TANTALATE. /ASE COOLING: AN AID TO C 13-1928
E CRYSTALS OF A NEW PYROCHLORE CONTAINING LITHIUM, STRONTIUM LANTHANUM TANTALUM OXYFLUORIDE. /TUDY OF SINGL 36-3613
LANTHANA TITANIA PHASE EQUILIBRIA AND ELECTRICAL PROPERTIES (LANTHANUM TITANATE). SYSTEM 36-3592
FERROELECTRIC) ELECTROOPTIC AND PIEZOELECTRIC PROPERTIES OF LANTHANUM TITANATE SINGLE CRYSTAL. (FOUND NOT 36-3576
OXIDATION REDUCTION PHENOMENA IN THE LEAD TITANATE- LANTHANUM TITANATE SYSTEM. 09-1587
DISTRIBUTION OF A-SITE AND B-SITE VACANCIES IN (LEAD, LANTHANUM) (TITANATE, ZIRCONATE) CERAMICS. 09-1530
ON TO THE SOLID ELECTROLYTE FUEL CELL. ((CALCIUM, STRONTIUM) LANTHANUM TITANATE- ALUMINATES) /ITS APPLICATI 08-1444
/UTH TITANATE NIOBATE, STRONTIUM TITANATE NIOBATE, POTASSIUM LANTHANUM TITANIUM NIOBATE, POTASSIUM STRONTI/ 13-1962
EROVSKITE PHASES IN THE SYSTEM LEAD OXIDE- TITANIUM DIOXIDE- LANTHANUM TRIOXIDE. RANGE OF EXISTENCE OF P 08-1349
/L PROPERTIES OF TRANSPARENT FERROELECTRIC CERAMICS. ((LEAD, LANTHANUM) (ZIRCONATE, TANTALATE), (LEAD, TUN/ 13-2004
TRANSPARENT CERAMIC DEVELOPED BY SANDIA LABORATORIES. (LEAD LANTHANUM ZIRCONATE TITANATE) 09-1476
SEE BOTH LEAD LANTHANUM ZIRCONATE TITANATE AND PLZT.
JECTED TO LARGE TENSILE STRAI/ FERROELASTIC BEHAVIOR OF LEAD LANTHANUM ZIRCONATE TITANATE CERAMICS WHEN SUB 09-1571
CERAMICS FOR ELECTROOPTIC APPLICATIONS. HOT PRESSED (LEAD, LANTHANUM) (ZIRCONATE, TITANATE) FERROELECTRIC 09-1526
IMPROVED HOT PRESSED ELECTROOPTIC CERAMICS IN THE LEAD LANTHANUM ZIRCONATE TITANATE SYSTEM. 09-1529
GRAPHIC RESULTS IN TRANSPARENT CERAMIC FERROELECTRICS. (LEAD LANTHANUM ZIRCONUM TITANATE) /CRIPTION OF HOLO 09-1575
TH OXIDE COMPO/ TWO STEP ANOMALOUS INCREASE OF RESISTANCE ON LANTHANUM(0.002) BARIUM(0.998) TITANATE- BISMU 04-0801
RCONIUM(0/ PERFORMANCE OF SPUTTERED LEAD(0.92) BISMUTH(0.07) LANTHANUM(0.01) (IRON(0.405) NIOBIUM(0.325) ZI 37-3609
TITANATE. SOLUBILITY LIMIT OF LANTHANUM(0.5) LITHIUM(0.5) TITANATE IN BARIUM 04-0889
/ATION AND PHYSICAL PROPERTIES OF PEROVSKITE TYPE COMPOUNDS, LANTHANUM(1-X) CALCIUM(X) MANGANATE (PHASE TR/ 08-1356
AGING PROCESS IN TRIGLYCINE SULFATE. (HYSTERESIS LAPS, DOMAIN STRUCTURE) 27-3128
GLE CRYSTAL. LAPPING CHARACTERISTICS OF LITHIUM NIOBATE SIN 11-1783
GLE CRYSTALS. LAPPING CHARACTERISTICS OF LITHIUM NIOBATE SIN 11-1784

RY AND TEST ON BARIUM SODIUM NIOBATE AND NEODYMIUM DOPED YAG LASER). /RATION ON CRYSTAL INHOMOGENEITY (THEO 13-2043
CRYSTALS (KDP AND LITHIUM NIOBATE) BY RADIATION FROM A RUBY LASER.   FRACTURE OF NONLINEAR 11-1663
D FERROELECTRICS. (LITHIUM NIOBATE AND GADOLINIUM MOLYBDATE) LASER AND SPECTROSCOPIC PROPERTIES OF ACTIVATE 11-1741
REVERSIBLE DOMAIN REVERSAL IN TRIGLYCINE SULFATE CAUSED BY A LASER BEAM.   27-3012
/S POLARIZATION REVERSAL IN TRIGLYCINE SULFATE BY MEANS OF A LASER BEAM. APPLICATION TO FERROELECTRIC MEMO/ 27-3010
        INTRACAVITY SECOND HARMONIC GENERATION WITH RUBY LASER BY KDP.   •   17-2604
                            CONTROL OF LASER DAMAGE IN LITHIUM NIOBATE.   11-1716
            OPTICAL MIXING. (OF PULSED RUBY LASER EMISSIONS BY TRIGLYCINE SULFATE CRYSTAL) 27-2958
DIHYDROGEN PHOSPHATE.   ELECTROOPTIC AMPLITUDE MODULATION OF LASER GENERATED SECOND HARMONICS IN POTASSIUM 17-2694
MATERIALS.   LASER INDUCED DAMAGE IN X DIHYDROGEN PHOSPHATE 17-2475
        CONTROL OF THE SUSCEPTIBILITY OF LITHIUM NIOBATE TO LASER INDUCED REFRACTIVE INDEX CHANGES.   11-1791
UM LITHIUM NIOBATE HAVING NO MICROTWINNING (AND RESISTANT TO LASER INDUCED REFRACTIVE INHOMOGENEITIES). /RI 13-1958
OGEN PHOSPHATE AND POTASSIUM DIDEUTERIUM PHOSPHATE.   LASER INDUCED SELF FOCUSING IN AMMONIUM DIHYDR 17-2476
OBATE AND KDP.   LASER IRRADIATION INDUCED DAMAGE TO LITHIUM NI 11-1674
                        SPECTRUM OF RAMAN SCATTERED NEODYMIUM YAG LASER LIGHT IN FERROELECTRICS OF THE KDP GROUP 17-2625
TIONS. (TRIGLYCINE SULFATE, POTASSIUM DIHYDROGEN AND DIDEUT/ LASER LIGHT SCATTERING STUDIES OF PHASE TRANSI 17-2481
TE CRYSTAL.   MICROWAVE MODULATOR OF RUBY LASER LIGHT USING POTASSIUM DIHYDROGEN PHOSPHA 17-2640
        POINT DEFECTS IN LITHIUM NIOBATE SINGLE CRYSTALS USED IN LASER MODULATION.   11-1737
PHATE, POTASSIUM DI/ TEMPERATURE INSTABILITY OF ELECTROOPTIC LASER MODULATORS WITH AMMONIUM DIHYDROGEN PHOS 17-2675
E AND AMMONIUM DIHYDROGEN PHOSPHATE CRYSTALS OF LIGHT FROM A LASER OPERATING IN THE FREE GENERATION MODE. / 17-2717
STRONTIUM N/ PYROELECTRIC VOLTAGE RESPONSE TO SHORT INFRARED LASER PULSES IN TRIGLYCINE SULFATE AND BARIUM 13-2018
ETERS OF LITHIUM NIOBATE SINGLE CRYSTALS USING SCATTERING OF LASER RADIATION BY HYPERSONIC VIBRATIONS. /RAM 11-1797
EAR CRYSTALS. (KDP, ADP, RUBIDIUM DIHYDROG/ AUTOFOCUSING OF LASER RADIATION IN ACTIVE MATERIALS AND NONLIN 11-1865
TUNNELING MODES IN KDP AND DEUTERATED KDP.   LASER RAMAN STUDY OF QUASI SPIN WAVE HYDROGEN 17-2569
TRIGLYCINE SULFATE AND APPLICATIONS TO PYROELECTRIC DETECTO/ LASER STUDY OF REVERSIBLE NUCLEATION SITES IN 27-3011
                              ADP POLARIZER FOR LASERS.   17-2551
COND HARMONIC GENERATION IN ADP WITH TWO NEW FLUORESCEIN DYE LASERS.   EFFICIENT SE 17-2538
ELECTROOPTIC CERAMICS AS WAVELENGTH SELECTION DEVICES IN DYE LASERS (PLZT).   09-1618
UM(3+) DOPED GADOLINIUM MOLYBDATE CRYSTALS.   CERTAIN GROWTH, LASING AND SPECTROSCOPIC PROPERTIES OF NEODYMI 30-3255
TE OF IRON- IODINE- BORACITE. (TEMPERATURE DEPENDENCE OF PI/ LATENT LATTICE STRAIN IN THE FERROELECTRIC STA 14-2083
S AND OTHERS, BY SELECTIVE DEPOSITION OF CHARGED POLYSTYRENE LATEX PARTICLES). /FACES OF IONIC CRYSTALS (TG 27-2987
            GENERALIZED FERROELECTRIC MODEL ON A SQUARE LATTICE.   03-0168
            NORMAL VIBRATIONS OF STRONTIUM TITANATE LATTICE.   06-1128
ALATE SINGLE CRYSTAL STOICHIOMETRY (VERSUS MELT COMPOSITION, LATTICE AND CURIE TEMPERATURE MEASUREMENTS). / 12-1871
DEUTERATED LITHIUM THALLIUM TARTRATE MONOHYDRATE. (EFFECT ON LATTICE AND TRANSITION) /OF NONDEUTERATED AND 29-3239
IFFRACTION, BARIUM/ STUDYING SMALL DEFORMATIONS OF A CRYSTAL LATTICE BASED ON THE SHADOW EFFECT. (NEUTRON D 04-0534
E)- BARIUM (ZINCATE NIOBATE)- LEAD TITANATE. (PHASE DIAGRAM, LATTICE CONSTANTS) /YSTEM LEAD (ZINCATE NIOBAT 05-1070
RIUM COBALT TETRAFLUORIDE CRYSTAL STRUCTURE (SPACE GROUP AND LATTICE CONSTANTS). /RAELASTIC PARAMAGNETIC BA 33-3386
IS (NIOBIUM, TITANIUM) OR (NIOBIUM, TUNGSTEN), COMPILATION, LATTICE CONSTANTS, CURIE TEMPERATURES) /HERE B 13-1964
LFOIODIDE.   DETERMINATION OF THE TEMPERATURE DEPENDENCE OF LATTICE CONSTANTS IN FERROELECTRIC ANTIMONY SU 15-2140
FERROCYANIDE TRIDEUTERO HYDRATE ABOVE THE CURIE TEMPERATURE (LATTICE CONSTANTS, SPACE GROUP). /C POTASSIUM 24-2903
/SONANCE AND ELECTRON NUCLEAR DOUBLE RESONANCE OF THE PROTON LATTICE COUPLED MODE IN POTASSIUM DIHYDROGEN / 18-2724
PRETATION OF THE PRESSURE DEPENDENCE OF PROPERTIES CAUSED BY LATTICE DEFECTS. (ICE)   INTER 34-3434
AND THEIR INFLUENCE ON BEHAVIOR.   ORIGIN OF LATTICE DEFECTS IN (BARIUM) TITANATE MATERIALS 04-0784
NGLE CRYSTAL ELECTRET.   LATTICE DISTORTIONS IN A STRONTIUM TITANATE SI 06-1238
IC POTASSIUM IODATE CRYSTALS.   POLAR AND LATTICE DYNAMIC CHARACTERISTICS OF FERROELECTR 08-1430
        ANHARMONIC MODE MIXING IN IONIC FERROELECTRICS. (LATTICE DYNAMICS)   03-0234
HONON SPECTRA OF STRONTIUM TITANATE AND POTASSIUM TANTALATE (LATTICE DYNAMICS). /ECTRIC- ACOUSTIC) IN THE P 06-1157
    FERROELECTRICITY (IN BARIUM TITANATE TYPE STRUCTURES) AND LATTICE DYNAMICS.   04-0950
STIC NEUTRON SCATTERING FROM SINGLE DOMAIN BARIUM TITANATE. (LATTICE DYNAMICS)   INELA 04-0946
TIONS: STRONTIUM TITANATE.   LATTICE DYNAMICS ABOVE STRUCTURAL PHASE TRANSI 06-1124
RING.   ICE I: LATTICE DYNAMICS AND INCOHERENT NEUTRON SCATTE 34-3465
1 AND 2. (REVIEW)   LATTICE DYNAMICS BY NEUTRON SCATTERING. PARTS 02-0051
ITS FERROELECTRIC PHASE.   LATTICE DYNAMICS MODEL OF POTASSIUM NITRATE IN 16-2377
INIUM MOLYBDATE.   LATTICE DYNAMICS OF A RIGID ION MODEL OF GADOL 30-3281
TRIC (BARIUM TITANATE DOWN TO 3 MIC/ DIELECTRIC CONSTANT AND LATTICE DYNAMICS OF A THIN FILM OF A FERROELEC 04-0600
    ON THE INTERPRETATION OF INFRARED SPECTRA AND THE LATTICE DYNAMICS OF CUBIC BARIUM TITANATE.   04-0659
S, IN PARTICULAR STRONTIUM TITANATE.   LATTICE DYNAMICS OF CUBIC PEROVSKITE STRUCTURE 06-1185
DIELECTRIC CONSTANT. (POTASSIUM BROMIDE, LEAD TELLURIDE, TI/ LATTICE DYNAMICS OF DIATOMIC CRYSTALS OF HIGH 36-3541
                                                             LATTICE DYNAMICS OF ICE.   34-3467
                                                             LATTICE DYNAMICS OF ICE I(H).   34-3468
IELECTRIC CONSTANTS.   (LEAD AND TIN TELLURIDES)   LATTICE DYNAMICS OF IONIC CRYSTALS WITH HIGH D 36-3553
SKITE TYPE ANTIFERROELECTRICS.   LATTICE DYNAMICS OF PHASE TRANSITIONS IN PEROV 08-1475
                                                             LATTICE DYNAMICS OF STRONTIUM TITANATE.   06-1224
IC MODELS.   NEUTRON INELASTIC SCATTERING STUDY OF THE LATTICE DYNAMICS OF STRONTIUM TITANATE: HARMON 06-1225
LS OF PEROVSKITE STRUCTURE. (RELATION OF FERROELECTRICITY TO LATTICE DYNAMICS, REVIEW, 68 REFS) /OF MATERIA 02-0043
S OF KDP. (CALCULATIONS)   NOTE ON THE LATTICE DYNAMICAL ASPECT OF FERROELECTRIC MODE 17-2498
UM MANGANESE TRIFLUORIDE.   NEUTRON SCATTERING STUDY OF THE LATTICE DYNAMICAL PHASE TRANSITIONS IN POTASSI 33-3405
            PHENOMENOLOGICAL LATTICE DYNAMICAL THEORY OF FERROELASTICITY.   03-0144
RANSITIONS IN PEROVSKITE TYPE CRYSTALS. (ST/ SELF CONSISTENT LATTICE DYNAMICAL THEORY OF STRUCTURAL PHASE T 03-0424
M TITANATE SINGLE CRYSTALS.   LATTICE DYNAMICAL THEORY OF SWITCHING IN BARIU 03-0398
/(SPIN WAVES IN HYDROXYL DIPOLE SYSTEM IN POTASSIUM CHLORIDE LATTICE, FERROELECTRIC STATE AT LOW TEMPERATU/ 03-0452
TIES OF MATERIALS. (BARIUM TITANATE TYPE) EFFECT OF CHARGED LATTICE IMPERFECTIONS ON THE DIELECTRIC PROPER 04-1017
MOSSBAUER STUDIES OF LOWER TRANSITIONS IN BARIUM TITANATE. (LATTICE INSTABILITIES)   04-0564
IBILITY OF POTASSIUM DIHYDROGEN PHOSPHATE TYPE FERRO/ PROTON LATTICE INTERACTIONS AND THE DYNAMICAL SUSCEP 17-2570
LECTRIC CRYSTALS. (THEORY, KDP TYPE)   PROTON LATTICE INTERACTIONS IN HYDROGEN BONDED FERROE 03-0181
            PROTON- PROTON AND PROTON- LATTICE INTERACTIONS IN ICE.   34-3430
SULFOIODIDE.   LATTICE MODES AND PHASE TRANSITION IN ANTIMONY 15-2110
        PHASE TRANSITION IN A SIXTEEN VERTEX LATTICE MODEL (FERROELECTRIC SQUARE LATTICE). 03-0517
APPARENT VIOLATION OF THE UNIVERSALITY HYPOTHESIS IN A LATTICE MODEL OF A PHASE TRANSITION.   03-0515
    COMPENSATION OF HETEROVALENT FERRIC IONS IN THE LATTICE OF BARIUM TITANATE SINGLE CRYSTALS. 04-0539
PERTURBED DIRECTIONAL CORRELATION OF SCANDIUM-44 WITHIN THE LATTICE OF FERROELECTRIC BARIUM TITANATE.   04-0691
PROPERTIES OF THE IRON TRANSITION GROUP IN THE LATTICE OF SINGLE CRYSTALLINE LITHIUM NIOBATE. 11-1669
OME PHYSICAL PROPERTIES OF BARIUM TITANATE SINGLE CRYSTALS. (LATTICE PARAMETERS) / SOME ADMIXED NICKEL ON S 04-0909
OLUTIONS OF A(V)B(VI)C(VII) COMPOUNDS./ DETERMINATION OF THE LATTICE PARAMETERS AND CURIE POINTS OF SOLID S 15-2116
DIHYDROGEN PHOSPHATE TYPE CRYSTALS. PRESSURE EFFECTS ON THE LATTICE PARAMETERS AND STRUCTURE OF POTASSIUM 17-2588
RMAL EXPANSION OF BISMUTH FE/ PRECISION DETERMINATION OF THE LATTICE PARAMETERS AND THE COEFFICIENTS OF THE 08-1287
LEAD TITANATE SYSTEM IN THE TEMPERATURE RANGE 4.2-300 DEG K (LATTICE PARAMETERS VERSUS CURIE POINTS). /ATE 08-1308
CHEME, STRONTIUM / ENERGY BAND CHANGES IN PEROVSKITES DUE TO LATTICE POLARIZATION. (LCAO ANALYSIS OF BAND S 06-1101
HEORY OF FERROELECTR/ VIBRATIONAL APPROACH TO THE ANHARMONIC LATTICE PROBLEM. (WITH AN APPLICATION TO THE T 03-0221

RROELECTRIC CRYSTALS. (SODIUM NITRI/ NUCLEAR QUADRUPOLE SPIN LATTICE RELAXATION AND CRITICAL DYNAMICS OF FE 16-2268
FERROELECTRIC CRYSTALS. (THEORY)   NUCLEAR QUADRUPOLE SPIN LATTICE RELAXATION AND DYNAMICAL PROPERTIES OF 03-0187
RIC CESIUM DIDEUTERIUM ARSENATE AND CESIUM / CESIUM-183 SPIN LATTICE RELAXATION AND RESONANCE IN FERROELECT 18-2720
R DISORDER TYPE FERROELECTRICS: POLARIZATION FLUCTUATI/ SPIN LATTICE RELAXATION BY QUASI SPIN WAVES IN ORDE 26-2928
SSIUM FERROCYANIDE TRIHYDRATE.   ANOMALOUS SPIN LATTICE RELAXATION BY QUASI SPIN WAVES IN POTA 24-2899
N POTASSIUM DIHYDROGEN PHOSPHATE.   SPIN LATTICE RELAXATION BY THE FERROELECTRIC MODE I 17-2440
ROTON MAGNETIC RESONANCE IN WATER ICE. PART-2: DEUTERON SPIN LATTICE RELAXATION IN DEUTERATED ICE. /RT-1: P 34-3471
                               PROTON SPIN LATTICE RELAXATION IN FERROELECTRIC COLEMANITE 23-2885
AMMONIUM HYDROGEN SULFATE.   DEUTERON SPIN LATTICE RELAXATION IN FERROELECTRIC DEUTERATED 19-2795
ONIUM TARTRATE.   PROTON SPIN LATTICE RELAXATION IN FERROELECTRIC SODIUM AMM 29-3248
HYDROGEN SELENATE.   PROTON SPIN LATTICE RELAXATION IN FERROELECTRIC SODIUM TRI 22-2852
SULFATE.   NUCLEAR SPIN LATTICE RELAXATION IN FERROELECTRIC TRIGLYCINE 27-2967
          INTERPRETATION OF THE PROTON SPIN LATTICE RELAXATION IN HEXAGONAL ICE.   34-3475
          STRONG COLLISION LIMIT OF SPIN LATTICE RELAXATION IN HEXAGONAL ICE.   34-3480
ROGEN SELENATE.   PROTON SPIN LATTICE RELAXATION IN IRRADIATED SODIUM TRIHYD 22-2853
IUM SALTS. (AMMONIUM SULFATE, FLUOBERYLLATE, P/ NUCLEAR SPIN LATTICE RELAXATION IN SOME FERROELECTRIC AMMON 19-2769
E TR/ NUCLEAR MAGNETIC RESONANCE (WIDE LINE SPECTRA AND SPIN LATTICE RELAXATION) INVESTIGATIONS OF THE PHAS 20-2802
/LECULES IN ROCHELLE SALT AND AMMONIUM ROCHELLE SALT BY SPIN LATTICE RELAXATION OF PROTONS IN ROTATING FRA/ 28-3232
CTRIC AND FERROELECTRIC PHASE OF SODIUM NITRIT/ NUCLEAR SPIN LATTICE RELAXATION OF SODIUM-23 IN THE PARAELE 16-2264
DEUTERIUM SELENATE.   NUCLEAR SPIN LATTICE RELAXATION TIME MEASUREMENTS IN SODIUM 22-2847
ERATURE OF DICALCIUM LEAD PROPIO/ ANOMALY OF THE PROTON SPIN LATTICE RELAXATION TIME NEAR THE CRITICAL TEMP 26-2947
URE IN SODIUM NITRITE.   DIPOLE LATTICE RELAXATION TIME NEAR THE NEEL TEMPERAT 16-2300
MPERATURE IN THE ORDER DISORDER TYPE FE/ ANOMALY OF THE SPIN LATTICE RELAXATION TIME NEAR THE TRANSITION TE 03-0488
/ BURSTING OF QUADRUPOLAR NUCLEI. DETERMINATION OF THE SPIN LATTICE RELAXATION TIMES IN FERROELECTRIC LIT/ 11-1829
, PART-1, THE RESIDUAL ENTROPY OF ICE, PART-2, HYDROGEN BON/ LATTICE STATISTICS OF HYDROGEN BONDED CRYSTALS 03-0388
RON- IODINE- BORACITE. (TEMPERATURE DEPENDENCE OF PI/ LATENT LATTICE STRAIN IN THE FERROELECTRIC STATE OF I 14-2083
ERTIES OF FERROELECTRIC (IRON IODINE) BORACITE. (SPONTANEOUS LATTICE STRAINS)   ELECTROOPTICAL PROP 14-2082
                                        LATTICE STRAINS OF BORACITE CRYSTALS.   14-2081
ODINE- BORACITE).   X-RAY STUDY ON THE LATTICE STRAINS OF (PARAELECTRIC AND) (IRON- I 14-2085
           INCORPORATION OF ANTIMONY INTO THE BARIUM TITANATE LATTICE. (STRUCTURE, AND PCT RESISTORS) 04-0934
              ON THE LATTICE VIBRATION OF CUBIC BARIUM TITANATE. 04-0775
IONS TO RELATE TO SPECIFIC ANTIFERROELECTRIC PHASE)   LATTICE VIBRATION OF SODIUM NITRITE. (CALCULAT 16-2317
         ANTIFERROELECTRIC MODE IN LEAD ZIRCONATE. (LATTICE VIBRATIONS) 07-1262
ER RAMAN SPECTRA: APPLICATION TO DISPLACIVE FERROELECTRICS. (LATTICE VIBRATIONS)   POWD 03-0194
NTIMONY SULFOIODIDE.   SOFT LATTICE VIBRATIONS AND VIBRATION COUPLING IN A 15-2233
Y SULFOIODIDE CRYSTALS USING THE RAMAN SCATTERING SPECTRUM. (LATTICE VIBRATIONS, BONDING) /ELECTRIC ANTIMON 15-2143
/D (REFLECTIVITY) STUDY (20-10,000 PER CM, 300 DEG K) OF THE LATTICE VIBRATIONS IN LITHIUM TANTALATE (DIEL/ 12-1870
              GROUP ANALYSIS OF LATTICE VIBRATIONS OF CUBIC PEROVSKITE ABO(3). 08-1321
ECTRIC THIOUREA. PART-1: TEMPERATURE DEPENDEN/ MOLECULAR AND LATTICE VIBRATIONS OF ORTHORHOMBIC AND FERROEL 25-2920
ECTRIC THIOUREA. PART-2: CALCULATION OF THE I/ MOLECULAR AND LATTICE VIBRATIONS OF ORTHORHOMBIC AND FERROEL 25-2907
/F THE POULET- LOUDON APPROXIMATION IN THE THEORY OF OPTICAL LATTICE VIBRATIONS. (QUARTZ, BARIUM TITANATE) 03-0380
NCIPLES, APPLIED TO DISPL/ THEORY OF NEUTRON SCATTERING FROM LATTICE VIBRATIONS (REDEVELOPED FROM FIRST PRI 03-0484
DIFICATIONS OF THIOUREA.   LATTICE VIBRATIONAL SPECTRA OF FIVE CRYSTAL MO 25-2905
KANZIG REGIONS ON THE BASIS OF THE DYNAMIC THEORY OF CRYSTAL LATTICES.   ESTIMATION OF THE DIMENSIONS OF 03-0446
.   DYNAMICAL THEORY OF RIGID LATTICES OF ANTIFERROELECTRICS. 03-0344
ANIDE TRIHYDRATE)   INTERNAL FIELD IN GENERAL DIPOLE LATTICES. (THEORY APPLIED TO POTASSIUM FERROCY 03-0220
ELASTIC PROPERTIES OF AMMONIUM DIHYDROGEN PHOSPHATE AND THE LAVAL RAMAN ELASTICITY THEORY. 17-2410
RECIPROCAL DIELECTRIC CONSTANT FOLLOWS QUADRATIC TEMPERATURE LAW). /D ((BARIUM(X) STRONTIUM(1-X) TITANATE (04-0898
SSIUM DIHYDROGEN PHOSPHATE CRYSTAL.   CURIE-WEISS LAW FOR THE NONLINEAR SUSCEPTIBILITY OF A POTA 17-2686
NGSTEN OXIDE AS A FUNCTION OF THICKNESS OF THE SURFACE OXIDE LAYER. /PHOTOELECTRON SPECTRUM OF TUNGSTEN- TU 10-1631
TE- BISMUTH TRIOXIDE COMPOSITE CERAMICS WITH SURFACE BARRIER LAYER. /IVITY ON LANTHANUM DOPED BARIUM TITANA 04-0800
E IN TRIGLYCINE SULFATE. INFLUENCE OF DISLOCATIONS, SURFACE LAYER AND X-RAY IRRADIATION ON DOMAIN STRUCTUR 27-3004
ARY.   FERROELECTRIC PROPERTIES OF (INSULATING, BOUNDARY LAYER) BARIUM TITANATE CERAMICS AT GRAIN BOUND 04-1024
ITHIUM NIOBATE.   THREE LAYER BROAD BAND ANTIREFLECTION COATINGS FOR L 11-1676
FORMATION OF BARIUM TITANATE CERAMIC SEMICONDUCTING LAYER BY DIFFUSION.   04-0684
STUDIES ON THE BARIUM STRONTIUM TITANATE BOUNDARY LAYER CERAMIC DIELECTRICS.   04-1025
STUDIES ON THE (STRONTIUM, CALCIUM) TITANATE BOUNDARY LAYER CERAMIC DIELECTRICS.   06-1247
SINGLE FIRE PROCESS FOR BARIUM TITANATE BARRIER LAYER DIELECTRICS.   04-1029
STUDIES ON STRONTIUM TITANATE BOUNDARY LAYER DIELECTRICS.   06-1248
PROBLEM OF THE SURFACE LAYER IN BARIUM TITANATE SINGLE CRYSTALS.   04-1013
C FIELD EFFECT, PARAELECTRIC AND FERROELECTRIC PHAS/ SURFACE LAYER IN BARIUM TITANATE (X-RAY DIFFRACTION, D 04-0857
CHARACTERIZATION OF SURFACE AND GRAIN BOUNDARY LAYER OF BARIUM TITANATE.   04-0701
X-RAY STUDY OF THE SURFACE LAYER ON BARIUM TITANATE SINGLE CRYSTAL.   04-0858
CRYSTALS.   INFLUENCE OF THE SURFACE LAYER ON THE PLASTIC DEFORMATION OF ICE SINGLE 34-3459
IES OF FERROELECTRICS WITH DIFFUSE PHASE TRANSITIONS. (MIXED LAYER OXIDES) /ELECTRIC AND STRUCTURAL PROPERT 13-2014
RROELECTRICS AND ANTIFERROELECTRICS BARIUM TITANATE, COMPLEX LAYER OXIDES) /NT AND CURIE TEMPERATURE FOR FE 03-0188
TIUM, CALCIUM) TITANATE SOLID SOLUTION CERAMIC WITH BOUNDARY LAYER STRUCTURE.   STUDIES ON THE (STRON 06-1246
FERROELECTRICITY IN BISMUTH OXIDE TYPE LAYER STRUCTURE COMPOUNDS. 13-1932
. (BARIUM, STRONTIUM TITA/ DIELECTRIC CERAMICS WITH BOUNDARY LAYER STRUCTURE FOR HIGH FREQUENCY APPLICATION 04-0833
BISMUTH NIOBIUM OXYFL/ CRYSTAL CHEMISTRY OF PEROVSKITE LIKE LAYER TYPE FERROELECTRICS. (BISMUTH TUNGSTATE, 13-1966
E FERR/ MOSSBAUER STUDIES OF IRON-57 IN SOME PEROVSKITE LIKE LAYER TYPE FERROELECTRICS. (IN BISMUTH TITANAT 13-2035
CRYSTAL CHEMISTRY OF MIXED BISMUTH OXIDES WITH LAYER TYPE STRUCTURE. 13-2032
FLUID- LITHIUM NIOBATE STRUCTURE.   LAYER WAVE AMPLIFICATION IN A CADMIUM SULFIDE 11-1667
IRCONATE TITANATE (PLUS 1 PERCENT NIOBIUM PENTOXIDE) SURFACE LAYERS. /IES OF OXYGEN DEPLETED CERAMIC LEAD Z 09-1576
YDROGEN PHOSPHATE CRYSTALS GROWN FROM AN AQUEOUS SOL/ GROWTH LAYERS AND OTHER MACRODEFECTS IN POTASSIUM DIH 17-2496
MICONDUCTORS. (ORIGIN AND EFFECTS OF GRAIN BOUNDARY BLOCKING LAYERS, ANTIMONY DOPANT, REVIEW) /OELECTRIC SE 02-0074
/XTRINSIC PHOTOCONDUCTIVITY IN FERROELECTRICS DUE TO SURFACE LAYERS. (BARIUM TITANATE, ANTIMONY SULFOIODID/ 04-0681
SHOCK WAVE STUDIES OF SURFACE LAYERS IN BARIUM TITANATE. 04-0588
SOLID SURFACES) IN THE EMISSION ELECTRON MICROSCOPE (BARRIER LAYERS IN DOPED BARIUM TITANATE CERAMICS). /F 04-0913
NATE CERAMIC WITH A HIGH EFFECTIVE D/ GRAIN BOUNDARY BARRIER LAYERS IN (N TYPE, ANTIMONY DOPED) BARIUM TITA 04-0581
/ ELECTRICAL PROPERTIES OF CRYSTALLITES AND INTERCRYSTALLITE LAYERS IN THE SEMICONDUCTOR CERAMIC BARIUM TIT 04-0568
IONS.   FORMATION OF THIN LAYERS OF MIXED TITANATES BY SOLID SOLID REACT 13-2005
RADIATION AND ACOUSTIC EXCITATION ON ANOMALOUS LAYERS OF PEROVSKITE FERROELECTRICS. 08-1388
/URATION OF VOLT- AMPERE CHARACTERISTICS OF CADMIUM SELENIDE LAYERS ON BARIUM TITANATE- STRONTIUM TITANATE/ 04-0592
TALS.   SURFACE CHARGE LAYERS ON FERROELECTRIC POTASSIUM NIOBATE CRYS 08-1369
AND REFLECTION OF ULTRASONIC SURFACE WAVES BY THIN GOLD FILM LAYERS ON LITHIUM NIOBATE SUBSTRATES. /ACTION 11-1789
AND REFLECTION OF ULTRASONIC SURFACE WAVES BY THIN GOLD FILM LAYERS ON LITHIUM NIOBATE SUBSTRATES. /ACTION 11-1790
ATE SEMICONDUCTIVE CERAMIC/ MECHANISM FOR FORMING INSULATION LAYERS ON THE GRAIN BOUNDARIES OF BARIUM TITAN 04-0867
/NDAMENTAL AND HIGHER ORDER MODE SURFACE WAVE PROPAGATION ON LAYERED ANISOTROPIC SUBSTRATES: GOLD ON LITHI/ 11-1846
ANI/ SYNTHESES, X-RAY, AND DIELECTRIC STUDIES OF CERTAIN NEW LAYERED FERROELECTRICS. (LANTHANUM BISMUTH TIT 13-1993

TRANSVERSE ACOUSTOELECTRIC EFFECT IN A LAYERED LITHIUM NIOBATE- SILICON STRUCTURE.   11-1748
E AND P-TYP/ AMPLIFICATION OF RAYLEIGH ULTRASONIC WAVES IN A LAYERED STRUCTURE CONSISTING OF LITHIUM NIOBAT 11-1747
/Y BAND CHANGES IN PEROVSKITES DUE TO LATTICE POLARIZATION. (LCAO ANALYSIS OF BAND SCHEME, STRONTIUM TITAN/ 06-1101
(PEROVSKITE COMPOUNDS CONTAINING EITHER BISMUTH, COPPER, OR LEAD). /IC MATERIALS WITH PEROVSKITE STRUCTURE 08-1460
/CTION STUDIES OF FERROELECTRICS AND FERROELECTRIC MAGNETS. (LEAD AND BARIUM TITANATES, BISMUTH FERRATE, P/ 04-1038
DYNAMICS OF IONIC CRYSTALS WITH HIGH DIELECTRIC CONSTANTS. (LEAD AND TIN TELLURIDES)   LATTICE 36-3553
COMPCUNDS. STUDIES O/ DIRECT CHELATOMETRIC DETERMINATION OF LEAD AND TITANIUM IN LEAD TITANATE AND SIMILAR 05-1068
), / DIFFUSE TO SHARP FERROELECTRIC PHASE TRANSITION SYSTEMS LEAD (B(0.5) NIOBIUM(0.5)) (1-X) B'(X) OXYGEN(3 13-2044
BINARY AND PSEUDOBINARY PHASES BETWEEN NIOBIUM PENTOXIDE AND LEAD, BARIUM, AND STRONTIUM OXIDES. 04-0590
ERROELECTRIC CERAMIC. (BARIUM TITANATE, BARIUM TIN TITANATE, LEAD BARIUM DINIOBATE) /TROPY OF A POLARIZED F 04-0767
ON AND SOME RESULTS.   PHASE COMPOSITION OF (LEAD, BARIUM) DINIOBATE. METHOD OF INVESTIGATI 13-1945
PROPERTIES OF FERROGLASS CERAMICS BASED ON LEAD BARIUM NIOBATE COMPOUNDS. 13-2006
TS OF ROASTING CONDITIONS ON THE STRUCTURE AND PROPERTIES OF LEAD BARIUM NIOBATE PIEZOCERAMICS.   EFFEC 13-2046
EFFECT OF ELECTRIC FIELD ON PHASE COMPOSITION OF (LEAD, BARIUM) NIOBATE SOLID SOLUTIONS. 13-1943
RAMAN SCATTERING IN THE FERROELECTRIC SYSTEM BARIUM TITANATE. 04-0611
/5)) (NIOBIUM, TANTALUM, OR VANADIUM)(0.5) OXIDE, (BARIUM OR LEAD) BISMUTH(0.67) (TUNGSTEN OR MOLYBDENUM)(/ 08-1462
LECTRIC FROPERTIES. (FERROELECTRIC- PARAE/ CRYSTAL GROWTH OF LEAD (CADMATE(0.33) NIOBATE(0.67)) AND ITS DIE 13-1959
INVESTIGATICN CF FHASE TRANSITICN IN LEAD CADMIUM TUNGSTATE. 10-1633
/RTIES OF FERROELECTRIC FERROMAGNETIC SOLID SOLUTIONS IN THE LEAD COBALT TUNGSTATE- CADMIUM MANGANATE SYST/ 36-3604
ZIRCONATE SOLID / DIELECTRIC AND PIEZOELECTRIC PROPERTIES OF LEAD COBALTATE(0.33) NIOBATE(0.67)- TITANATE- 09-1549
BALTATE(0.5) TU/ PHASE TRANSISITICN IN THE ANTIFERROELECTRET LEAD COBALTATE(0.5) TUNGSTATE(0.5) AND LEAD CO 13-1968
/ECTRICAL MAGNETIC AND X-RAY STUDIES OF PHASE TRANSITIONS IN LEAD COBALTATE(0.5) TUNGSTATE(0.5), TITANATE/ 13-1960
/CIENTS (RELATING INDUCED BIREFRINGENCE AND POLARIZATION) OF LEAD CONTAINING OXYGEN OCTAHEDRA FERROELECTRI/ 13-2039
BEHAVIOR OF PHONON NEAR THE 60 DEGREES C PHASE TRANSITION IN LEAD DICALCIUM PROPIONATE.   THE ANOMALOUS 26-2942
/TUDY OF THE TRANSITICN OF THE (ORTHORHOMBIC) FERROELECTRIC (LEAD DINIOBATE INTO THE NONFERROELECTRIC (RHO/ 08-1285
DIELECTRIC, PIEZOELECTRIC AND PYROELECTRIC PROPERTIES OF THE LEAD DINIOBATE- BARIUM DINIOBATE SYSTEM. 08-1384
LOW FREQUENCY LINEAR ELECTROOPTIC EFFECT IN LEAD DITANATLATE. 13-1916
ES OF A SERIES OF FERROELECTRIC MATERIALS. (BARIUM TITANATE, LEAD FERRATE(0.5) NIOBATE(0.5)) / THE PROPERTI 04-0780
NT OF RESISTIVITY) IN (DOPED AND UNDOPED CERAMIC PEROVSKITE) LEAD FERRATE(0.5) NIOBATE(0.5). /URE COEFFICIE 13-2000
/RESSION OF THE SPCNTANEOUS MAGNETOELECTRIC MAGNETIZATION OF THE LEAD FERRATE(0.5) NIOBATE(0.5) AND LEAD MANGA/ 13-2069
ATE SYSTEM.   SEIGNETTO MAGNETIC SOLID SOLUTIONS OF THE LEAD FERRATE(0.67) TUNGSTATE(0.33)- LEAD TITAN 05-1065
/CE) IN SOME FERROELECTRICS AND FERROELECTRIC FERROMAGNETS. (LEAD FERRATE(0.67) TUNGSTATE(0.33) LEAD FERRA/ 08-1368
QUASI ELASTIC SCATTERING IN LEAD GERMANATE. 35-3497
ECTRIC) ELECTROOPTIC MEASUREMENTS ON LITHIUM GERMANATE AND LEAD GERMANATE. (LITHIUM GERMANATE NOT FERROEL 35-3494
(11).   LEAD GERMANATE, SEE LEAD(5) GERMANIUM(3) OXIDE
ELECTROOPTIC FROPERTIES OF FERROELECTRIC LEAD GERMANATE SINGLE CRYSTAL. 35-3518
TOPOGRAFHIC OBSERVATION OF 180 DEG DOMAINS IN FERROELECTRIC LEAD GERMANATE SINGLE CRYSTALS.   X-RAY 35-3515
/LEAD(5) GERMANIUM(3) OXIDE(11) AND ITS ISOMORPHOUS COMPOUND LEAD GERMANIUM SILICON OXIDE. (DIELECTRIC CON/ 35-3501
/81 STUDIED BY THE PERTURBED ANGULAR CORRELATION TECHNIQUE. (LEAD HAFNATE, TIN HAFNATE, CALCIUM HAFNATE, S/ 08-1276
SOLID SOLUTIONS IN ANTIFERROELECTRIC REGION OF THE SYSTEM LEAD HAFNATE TITANATE, LEAD STANNATE NIOBATE. 13-1956
COMPLEX DIELECTRIC CONSTANT OF LEAD HAFNATE, 320-520 DEG K. 08-1339
PREPARATION AND PROPERTIES OF CESIUM LEAD HALIDES. (FERROELECTRICITY INDETERMINITE) 33-3359
TIONS IN OPTICAL COMMUNICATIONS.   EVALUATION OF LEAD LANTHANATE ZIRCONATE CERAMICS FOR APPLICA 07-1254
MATERIAL AND ELECTROOPTIC PROPERTIES OF THE (LEAD, LANTHANATE) (HAFNATE, TITANATE) SYSTEM. 08-1302
S.   DISTRIBUTION OF A-SITE AND B-SITE VACANCIES IN (LEAD, LANTHANUM) (TITANATE, ZIRCONATE) CERAMIC 09-1530
/OPTICAL PROPERTIES OF TRANSPARENT FERROELECTRIC CERAMICS. ((LEAD, LANTHANUM) (ZIRCONATE, TANTALATE), (LEA/ 13-2004
TRANSPARENT CERAMIC DEVELOPED BY SANDIA LABORATORIES. (LEAD LANTHANUM ZIRCONATE TITANATE) 09-1476
SEE BOTH LEAD LANTHANUM ZIRCONATE TITANATE AND PLZT.
N SUBJECTED TO LARGE TENSILE STRAI/ FERROELASTIC BEHAVIOR OF LEAD LANTHANUM ZIRCONATE TITANATE CERAMICS WHE 09-1571
ECTRIC CERAMICS FOR ELECTROOPTIC APPLICATIONS. HOT PRESSED (LEAD, LANTHANUM) (ZIRCONATE, TITANATE) FERROEL 09-1526
IMPROVED HOT PRESSED ELECTROOPTIC CERAMICS IN THE LEAD LANTHANUM ZIRCONATE TITANATE SYSTEM. 09-1528
HOLOGRAFHIC RESULTS IN TRANSPARENT CERAMIC FERROELECTRICS. (LEAD LANTHANUM ZIRCONUM TITANATE) /CRIPTION OF 09-1575
/ICS OF PEROVSKITE TYPE. (POLYCRYSTALLINE AND SINGLE CRYSTAL LEAD MAGNESATE(0.33) NIOBATE(0.67), AND LEAD / 08-1375
USE PHASE TRANSI/ RELAXATION POLARIZATION OF A FERROELECTRIC LEAD MAGNESATE(0.33) NIOBATE(0.67) WITH A DIFF 13-1980
ICTION IN THE REGION CF THE ANTIFERROELECTRIC CURIE POINT OF LEAD MAGNESATE(0.5) TUNGSTATE(0.5). /TERNAL FR 13-2020
LIGHT BEAM LEFLECTICN CCNTROLLED WITH A PRISM OF LEAD MAGNESIO NIOBATE CRYSTALS. 08-1280
ION.   FORMATION PROCESSES OF LEAD MAGNESIO NIOBATE IN THE SOLID STATE REACT 08-1354
A MICROSCOPIC THEORY FOR THE DIELECTRIC PROPERTIES OF LEAD MAGNESIUM NIOBATE. 08-1440
OCTAHEDRAL POSITIONS OF THE "CUBIC" FEROVSKITE STRUCTURE OF LEAD MAGNESIUM NIOBATE. /UM AND NIOBIUM IN THE 08-1379
ANISOTROPY OF THE ELECTROOPTICAL EFFECT IN CRYSTALS OF LEAD MAGNONIOBATE. 36-3521
/NOMALOUS BEHAVIOR OF THE OPTICAL ASBSORPTION COEFFICIENT OF LEAD MAGNONIOBATE SINGLE CRYSTALS IN THE REGI/ 13-1974
ANATE(0.33) N/ SYNTHESIS OF PEROVSKITE CCMFLEX COMPOSITIONS: LEAD MANGANATE(0.5) NIOBATE(0.5) AND LEAD MANG 13-1453
HIGH PRESSURE SYNTHESIS OF LEAD METAL- OXYGEN(3) TYPE PEROVSKITE. 08-1331
MECHANISM AND KINETICS OF SOLID PHASE FORMATION OF LEAD METANIOBATE. 13-2042
CRYSTALLIZATION OF LEAD METANIOBATE FROM GLASS. 08-1275
DIELECTRIC, PIEZOELECTRIC AND PYROELECTRIC PROPERTIES OF THE LEAD METANIOBATE- BARIUM NIOBATE SYSTEM. 08-1385
PREPARATION OF SODIUM AND BARIUM NIOBATES AND LEAD MOLYBDATE. 04-0657
OF AVERAGE PARTICLE SIZE / STUDIES ON THE SINTERING BETWEEN LEAD MONOXIDE AND PZT. PART-1. PART-2: EFFECTS 09-1624
.   GROWTH OF SINGLE CRYSTALS IN LEAD MONOXIDE- GERMANIUM DIOXIDE BINARY SYSTEM 35-3514
EQUENCIES ANC EIGENVECTORS OF POLARITONS WITH APPLICATION TO LEAD NIOBATE. PART-1: LONG OPTICAL PHONONS. /R 08-1284
S)   FERROELECTRIC PROPERTIES CF TETRAGONAL (BARIUM, LEAD) (NIOBATE, ZIRCONATE). (CURIE TEMPERATURE 08-1293
/N THE DIELECTRIC AND FERROELECTRIC PROPERTIES OF THE BARIUM LEAD NIOBATE ZIRCONATE SYSTEM. (TETRAGONAL AN/ 13-1925
MPERATURE AND PRESSURE ENFORC/ RELEASE OF ELECTRIC ENERGY IN LEAD NIOBIUM ZIRCONATE TITANATE STANNATE BY TE 09-1488
NONLINEAR CPTICAL PROPERTIES OF FERROELECTRIC LEAD NIOBIUM(4) OXIDE(12). 13-2023
/ES OF THE TETRAGONAL TUNGSTEN OXYGEN BRONZE TYPE: QUADRATIC LEAD OR BISMUTH TANTALATE NIOBATE, TANTALATE / 13-1985
/LINEAR CIELECTRICS (EXPERIMENT AND DISCUSSION OF BARIUM (OR LEAD OR CALCIUM) STRONTIUM TITANATE, BARIUM T/ 08-1281
DE.   RANGE OF EXISTENCE OF PEROVSKITE PHASES IN THE SYSTEM LEAD OXIDE- TITANIUM DIOXIDE- LANTHANUM TRIOXI 13-1349
E RELAXATION TIME NEAR THE CRITICAL TEMPERATURE OF DICALCIUM LEAD PROPIONATE. /LY OF THE PROTON SPIN LATTIC 26-2947
LTRASONIC ATTENUATION NEAR THE PHASE TRANSITION IN DICALCIUM LEAD PROPIONATE   U 26-2949
SYNTHESIS AND STUDY OF LEAD SCANDIUM NIOBATE SINGLE CRYSTALS. 08-1325
TE(X) PEROVSKITE TYPE SOLID SOL/ FERROELECTRIC PROPERTIES OF LEAD (STANNATE(0.5) NIOBATE(0.5)) (1-X) ZIRCONA 09-1548
ANCMALOUS FERROELECTRIC HYSTERESIS LOOPS. (LEAD STRONTIUM TITANATE) 05-1067
TIME EFFECTS IN THE HYSTERESIS LOOP OF (LEAD, STRONTIUM) TITANATE. 05-1052
TIME EFFECTS IN THE HYSTERESIS LOOP OF (LEAD, STRONTIUM) TITANATE. 05-1053
/NS ON LIELECTRIC AND PIEZOELECTRIC PROPERTIES OF A CERAMIC (LEAD STRONTIUM TITANATE ZIRCONATE DOPED WITH / 09-1591
URE AND ALLOYED WITH SMALL AMCUNTS OF GERMANIUM TELLURIDE ON LEAD TELLURIDE). /TRANSITIONS (TIN TELLURIDE P 36-3564
TE.   FIELD EFFECT AND HALL MOBILITY OF TELLURIUM AND LEAD TELLURIDE ON UHV CLEAVED TRIGLYCINE SULFA 27-3172
/C CRYSTALS OF HIGH DIELECTRIC CONSTANT. (POTASSIUM BROMIDE, LEAD TELLURIDE, TIN TELLURIDE, LEAD SULFIDE, / 36-3541
DY OF THE EFFECT OF HYDROSTATIC PRESSURE CN THE STRUCTURE OF LEAD TITANATE.   AN X-RAY STU 05-1064
APPARENT SOFT MODE LINE WIDTH DIVERGENCE IN LEAD TITANATE. 05-1078

CHARACTERISTICS OF DEFECT LEAD TITANATE. 05-1075
CRYSTAL GROWTH AND TEMPERATURE DEPENDENT RAMAN SPECTRA OF LEAD TITANATE. 05-1074
DISTRIBUTION OF VACANCIES IN LANTHANA DOPED LEAD TITANATE. 05-1058
ELECTRON SPIN RESOURCE STUDIES AT DEFECTS IN FLUORITES. (LEAD TITANATE) 05-1086
ROELECTRIC PROPERTIES OF HIGH RESISTIVITY SINGLE CRYSTALS OF LEAD TITANATE. (FLUX) GROWTH AND FER 05-1072
MOSSBAUER EFFECT FOR IRON-57 IN FERROELECTRIC LEAD TITANATE. 05-1040
NONLINEAR OPTICAL PROPERTIES OF FERROELECTRIC LEAD TITANATE. 05-1080
NONLINEAR OPTICAL PROPERTIES OF FERROELECTRIC LEAD TITANATE. 05-1079
PERTURBED DIRECTIONAL CORRELATION OF SCANDIUM-44 IN LEAD TITANATE. 05-1055
POSITIVE TEMPERATURE COEFFICIENT OF RESISTANCE IN DOPED LEAD TITANATE. 05-1051
UDIES (TEMPERATURE DEPENDENCE) OF UNDER DAMPED SOFT MODES IN LEAD TITANATE. RAMAN ST 05-1041
UPOLE INTERACTION OF TANTALUM AND HAFNIUM IONS IN BARIUM AND LEAD TITANATE. STATIC ELECTRIC QUADR 04-0937
ACTION STUDY OF FERROELECTRIC AND FERROMAGNETIC SUBSTANCES. (LEAD TITANATE) STRUCTURAL MICROELECTRON DIFFR 05-1082
TEMPERATURE DEPENDENT OPTICAL PHONONS IN LEAD TITANATE. 05-1083
STING A POINT DIPOLE ELECTROOPTIC MODEL BY AN APPLICATION TO LEAD TITANATE. TE 05-1073
X-RAY INVESTIGATION OF IRRADIATED LEAD TITANATE. 05-1081
IES AND PHASE TRANSITIONS OF THE FERROELECTRIC PEROVSKITES: LEAD TITANATE AND BISMUTH TITANATE. /C PROPERT 08-1434
NONLINEAR OPTICAL PROPERTIES OF LEAD TITANATE AND BISMUTH TITANATE CRYSTALS. 04-0941
LEAD MAGNESATE/ PHYSICAL PHENOMENA IN PIEZOCERAMICS BASED ON LEAD TITANATE AND ZIRCONATE. (SOLID SOLUTIONS 05-1063
/ARMONIC GENERATION COEFFICIENTS OF PIEZOELECTRIC CRYSTALS. (LEAD TITANATE, BARIUM STRONTIUM NIOBATE, LEAD/ 05-1066
VHF BAND. LEAD TITANATE CERAMIC RESONATORS OPERATING IN 05-1069
ELECTROMECHANICAL PROPERTIES AND APPLICATION OF LEAD TITANATE CERAMICS. 05-1085
PIEZOELECTRIC PROPERTIES OF LEAD TITANATE CERAMICS. 05-1062
MANGANESE. ELECTROMECHANICAL PROPERTIES OF LEAD TITANATE CERAMICS CONTAINING LANTANUM AND 05-1060
CENT MANGANESE/ POLARIZATION AND PIEZOELECTRIC PROPERTIES OF LEAD TITANATE CERAMICS. (CONTAINING 1 MOLE PER 05-1061
/CTS OF ADDITIVES ON PIEZOELECTRIC AND RELATED PROPERTIES OF LEAD TITANATE CERAMICS. (DIELECTRIC CONSTANT,/ 05-1084
RECIPROCAL DOMAINS IN LEAD TITANATE CRYSTALS. 05-1048
R DOMAIN STRUCTURE. GROWTH OF LEAD TITANATE CRYSTALS AND EXAMINATION OF THEI 05-1047
BARIUM ZIRCONATE LEAD TITANATE FERROELECTRIC CERAMICS. 09-1492
SPONTANEOUS POLARIZATION AND COERCIVE FIELD OF LEAD TITANATE (FLUX GROWN SINGLE CRYSTALS). 05-1050
/LASS- CERAMICS CONTAINING FERROELECTRIC CRYSTALS (BARIUM OR LEAD TITANATE, FROM SYSTEMS BARIA OR LITHIA, / 04-0783
/ GERMANIUM ON THE STRUCTURE AND PROPERTIES OF FERROELECTRIC LEAD TITANATE (GERMANATE SOLID SOLUTIONS TO 7/ 05-1043
APPLICATION OF DIELECTRIC MIXTURE FORMULAS TO (LEAD TITANATE) GLASS- CERAMIC SYSTEMS. 05-1054
. (STRONTIUM TITANATE, POTASSIUM TANTALATE, BARIUM TITANATE, LEAD TITANATE, LEAD ZIRCONATE, LEAD HAFNATE) / 02-0114
SEEBECK EFFECT IN METAL LEAD TITANATE METAL CONTACT SYSTEMS. 05-1056
EMISSION OF LIGHT BY SURFACE MICROPLASMA OF CERAMIC (BARIUM, LEAD) TITANATE NEAR THE CURIE TEMPERATURE. 04-0982
SOFT FERROELECTRIC MODES IN LEAD TITANATE (NEUTRON INELASTIC SCATTERING). 05-1077
(ROOM TEMPERATURE TO 6/ MOSSBAUER STUDIES OF BISMUTH FERRATE LEAD TITANATE PEROVSKITE TYPE SOLID SOLUTIONS 18-1471
ITA LEAD TITANATE) LEAD TITANATE PIEZOELECTRIC CERAMICS. (MATSUSH 09-1616
/TUDIES ON FERROELECTRIC DOMAINS OF PEROVSKITE TYPE OXIDES. (LEAD TITANATE, POTASSIUM NIOBATE, BARIUM TITA/ 04-0993
LEAD TITANATE PYROELECTRIC INFRARED DETECTOR. 05-1087
FERROELECTRIC PROPERTIES OF LEAD TITANATE SINGLE CRYSTALS. 05-1046
PIEZOELECTRIC EFFECT IN LEAD TITANATE SINGLE CRYSTALS. 05-1049
ELECTRIC AND ELASTIC PROPERTIES OF SINGLE CRYSTALS OF BARIUM LEAD TITANATE SOLID SOLUTIONS. DI 04-0597
4.2-300 DEG K (LATTICE PARAMETERS VERSU/ POTASSIUM TANTALATE LEAD TITANATE SYSTEM IN THE TEMPERATURE RANGE 08-1308
. X-RAY STRUCTURE INVESTIGATION OF LANTHANUM MODIFIED LEAD TITANATE WITH A-SITE AND B-SITE VACANCIES 05-1059
TRUCTURE AND ELECTROMECHANICAL PROPERTIES OF LANTHANUM DOPED LEAD TITANATE ZIRCONATE CERAMICS. S 09-1502
TRUCTURE AND ELECTROMECHANICAL PROPERTIES OF LANTHANUM DOPED LEAD TITANATE ZIRCONATE CERAMICS. S 09-1531
ATURES. PIEZOELECTRIC AND DIELECTRIC PROPERTIES OF LEAD TITANATE ZIRCONATE CERAMICS AT LOW TEMPER 09-1519
/CE OF DOMAIN STRUCTURE ON THE PROPERTIES OF POLYCRYSTALLINE LEAD TITANATE ZIRCONATE COMPOSITIONS NEAR THE/ 09-1557
FERROELECTRICITY IN SPUTTERED LEAD TITANATE ZIRCONATE FILMS. 09-1559
/RIZATION AND THERMAL EXPANSION OF POLARIZED CERAMICS IN THE LEAD TITANATE ZIRCONATE MAGNESATE(0.5) TUNGST/ 09-1603
HYDROTHERMAL CRYSTALLIZATION OF LEAD TITANATE ZIRCONATE SOLID SOLUTIONS. 09-1482
ODYNAMIC EXISTENCE CONDITIONS FO/ DIELECTRIC POLARIZATION OF LEAD TITANATE ZIRCONATE SOLID SOLUTIONS (THERM 09-1541
RAMAN SPECTRA OF POLYCRYSTALLINE SOLIDS: APPLICATION TO THE LEAD TITANATE ZIRCONATE SYSTEM (MEASUREMENT O/ 09-1497
SION STUDY OF DOMAIN WALL MOBILITY AND MECHANICAL DAMPING IN LEAD TITANATE ZIRCONATE TRANSDUCER CERAMICS. / 09-1547
CL/ VIBRATION OF ELECTROELASTIC CONSTANTS OF POLYCRYSTALLINE LEAD TITANATE ZIRCONATE WITH THOROUGHNESS OF P 09-1489
/ REVERSAL IN POLYCRYSTALLINE FERROELECTRICS. (8 SUBSTITUTED LEAD TITANATE ZIRCONATES, 2 SUBSTITUTED LEAD / 18-1418
SOLUTION KINETICS OF LUTETIUM AND SCANDIUM IONS IN LEAD TITANATE(0.5) ZIRCONATE(0.5). 09-1538
TO THE CALCULATION OF TETRAGONAL BIREFRINGENCE IN BARIUM AND LEAD TITANATES. /C IONIC CRYSTAL. APPLICATION 04-0578
STATIC SHELL MODEL FOR BARIUM AND LEAD TITANATES. 04-0826
E WALL ENERGY AND T/ THEORETICAL STATIC MODEL FOR BARIUM AND LEAD TITANATES - APPLICATION TO CALCULATING TH 03-0189
/IES OF HIGH PERMITTIVITY THIN FILM DIELECTRICS. (BARIUM AND LEAD TITANATES, EFFECT OF FERROELECTRIC CHARA/ 04-0670
OXIDATION REDUCTION PHENOMENA IN THE LEAD TITANATE- LANTHANUM TITANATE SYSTEM. 09-1587
). ELECTROMECHANICAL PROPERTIES IN THE BINARY SYSTEM LEAD TITANATE- LEAD CADMATE(0.5) TUNGSTATE(0.5 10-1638
LUTIONS AT HIGH PRESSURES. DIELECTRIC PROPERTIES OF LEAD TITANATE- LEAD (FERRATE NIOBATE) SOLID SO 05-1071
LEAD TITANATE- LEAD ZIRCONATE) NEW PIEZOELECTRIC CERAMICS. (LEAD TITANATE- LEAD FERRATE(0.5) NIOBATE(0.5), 09-1544
TE. ANTIFERROELECTRIC PHASE IN THE SYSTEM LEAD TITANATE- LEAD ZIRCONATE- LANTHANUM FERRA 13-1961
67) ZINCATE(0.33). LEAD TITANATE- LEAD ZIRCONATE- LEAD NIOBATE(0. 09-1507
.33) NIOBAT/ FERROELECTRIC AND PIEZOELECTRIC PROPERTIES IN A LEAD TITANATE- LEAD ZIRCONATE- LEAD (ZINCATE(0 09-1617
.33/ ELECTROPHYSICAL PROPERTIES OF THE SOLID SOLUTION SYSTEM LEAD TITANATE- LEAD ZIRCONATE- LEAD (ZINCATE(0 13-1951
TYPE OF CONDUCT/ THERMOELECTRIC EFFECTS IN THE SYSTEM METAL- LEAD TITANATE- METAL. (METHOD OF STUDYING THE 05-1057
MIC SYNTHESIS) THE SYSTEM LEAD TITANATE- STRONTIUM COPPER NIOBATE. (CERA 05-1044
(0.67). THE SYSTEM LEAD TITANATE- STRONTIUM CUPRATE(0.33) NIOBATE 05-1045
IATION INDUCED PHASE TRANSFORMATIONS IN (CERAMIC) BARIUM AND LEAD TITANATES. (SUMMARY ONLY) /T NEUTRON) RAD 04-0971
. DEFECT LEAD TITANATES WITH DIVERSE CURIE TEMPERATURES 05-1076
ELASTIC PROPERTIES OF CESIUM LEAD TRICHLORIDE. 33-3360
ELECTRIC FIELD GRADIENTS IN IONIC CHLORIDES: CESIUM LEAD TRICHLORIDE. 33-3367
ONDUCTION AND (NONFERROELECTRIC) PHASE TRANSITIONS IN CESIUM LEAD TRICHLORIDE. ELECTRICAL C 33-3365
PERIMENTAL STUDIES OF STRUCTURAL PHASE TRANSITIONS IN CESIUM LEAD TRICHLORIDE. EX 33-3378
FAR INFRARED REFLECTIVITY SPECTRA OF CESIUM LEAD TRICHLORIDE. 33-3379
(MECHANISM OF) STRUCTURAL PHASE TRANSITIONS IN CESIUM LEAD TRICHLORIDE. 33-3380
MICROSCOPIC OBSERVATIONS OF PHASE TRANSITIONS IN CESIUM LEAD TRICHLORIDE. 33-3383
ENDENCE OF CUBIC TETRAGONAL TRANSITION TEMPERATURE OF CESIUM LEAD TRICHLORIDE. PRESSURE DEP 33-3393
Y AT 194 DEGREES K) PHASE TRANSITIONS IN CESIUM LEAD TRICHLORIDE. (LOSS OF A CENTER OF SYMMETR 33-3368
DETERMINATION OF TRANSITION TEMPERATURES IN CESIUM LEAD TRICHLORIDE USING EPR. 33-3408
M OR TANTALUM) (2) OXIDE(9), AND VARIOUS TUNGSTEN SUBSTITUTED LEAD TUNGSTATES) /STRONTIUM)(3) COPPER (NIOBIU 36-3619
LITHIUM NIOBATE, STRONTIUM LITHIUM TANTALATE, STRONTIUM(1-X) LEAD(X) LITHIUM NIOBATE) /VSKITES. (STRONTIUM 08-1292
/ZONE GROWTH OF FERROELECTRIC CRYSTALS. (STRONTIUM TITANATE, LEAD(X) STRONTIUM(1-X) TITANATE, LEAD(X) TITA/ 05-1042
/ELECTRIC AND PIEZOELECTRIC PROPERTIES IN THE TERNARY SYSTEM LEAD (ZINCATE NIOBATE)- BARIUM (ZINCATE NIOBA/ 05-1070

CRYSTAL STRUCTURE AND PIEZOELECTRICITY OF THE SYSTEM LEAD (ZINCATE NIOBATE)- LEAD TITANATE. 13-2067
GROWTH OF SINGLE CRYSTALS OF LEAD ZINCATE NIOBATE AND ITS PROPERTIES. 36-3594
COMPOSITION. GROWTH OF SINGLE CRYSTALS OF LEAD (ZINCATE NIOBATE) AND PROPERTIES OF THIS 13-1990
STATE REACTION, SINTERING) AND PROPERTIES OF PEROVSKITE TYPE LEAD ZINCATE(0.33) NIOBATE(0.67). /SIS (SOLID 08-1393
THE EFFECT OF PRESSURE ON THE TRANSITION OF LEAD ZINCATE(0.33) NIOBATE(0.67). 08-1473
OF QUADRATIC COEFFICIENTS) ELECTROOPTICAL EFFECT IN LEAD ZINCATE(0.33) NIOBATE(0.67). (DISPERSION 13-1982
L. DIELECTRIC AND OPTICAL PROPERTIES OF LEAD ZINCATE(0.33) NIOBATE(0.67) SINGLE CRYSTA 13-2065
ALS GROWN FROM LEAD OXIDE FLUX). FERROELECTRIC PROPERTIES OF LEAD ZINCATE(0.33) NIOBATE(0.67) (SINGLE CRYST 13-2066
ELECTROOPTIC EFFECTS OF HOT PRESSED CERAMICS IN THE SYSTEM LEAD (ZINCATE(0.333) NIOBATE(0.667)). 13-1900
ATHERMAL PHASE TRANSITION IN LEAD ZIRCONATE. 07-1258
CRYSTAL GROWTH AND OBSERVATION OF THE FERROELECTRIC PHASE OF LEAD ZIRCONATE. 07-1264
DIPOLE STRUCTURE AND INTERNAL ELECTRIC FIELDS IN LEAD ZIRCONATE. 07-1261
OF PILE NEUTRON IRRADIATION ON THE DIELECTRIC PROPERTIES OF LEAD ZIRCONATE. EFFECT 07-1259
TRICITY AND ANTIFERROELECTRICITY IN POLAR CRYSTALS. (THEORY, LEAD ZIRCONATE) FERROELEC 03-0372
ECTRICITY AND ANTIFERROELECTRICITY IN PURE AND NIOBIUM DOPED LEAD ZIRCONATE. FERROEL 07-1251
GROWTH AND DIELECTRIC PROPERTIES OF SINGLE CRYSTAL LEAD ZIRCONATE 07-1265
EMPERATURE (TO 236 DEG C) PHASE TRANSITIONS IN (HIGH PURITY) LEAD ZIRCONATE. HIGH T 08-1466
ISOTHERMAL PHASE TRANSITIONS IN CERAMIC LEAD ZIRCONATE. 08-1466
ALLS, EFFECTIVE FIELD, AND ACTIVATION ENERGY. APPLICATION TO LEAD ZIRCONATE. /OF ANTIFERROELECTRIC DOMAIN W 07-1257
THERMOELECTRIC EFFECT IN SEMICONDUCTIVE LEAD ZIRCONATE. 07-1267
/FFECT STUDIES OF THE PHASE TRANSITIONS IN ANTIFERROELECTRIC LEAD ZIRCONATE AND FERROELECTRIC LEAD TITANAT/ 07-1253
S AND PHASE TRANSITIONS OF THE ANTIFERROELECTRIC PEROVSKITES LEAD ZIRCONATE AND LEAD HAFNATE. /IC PROPERTIE 08-1433
RE ON FERROELECTRIC PROPERTIES. (SUMMARY WITH NEW RESULTS ON LEAD ZIRCONATE AND OTHER ANTIFERROELECTRICS) / 07-1263
/EPARATION AND STRUCTURE) AND RELATED MATERIALS (SYSTEM WITH LEAD ZIRCONATE AND WITH LEAD ZIRCONATE TITANA/ 08-1333
TEMPERATURE DEPENDENT OPTICAL MODE IN ANTIFERROELECTRIC LEAD ZIRCONATE BY THE MOSSBAUER EFFECT. 07-1260
ANTIFERROELECTRIC MODE IN LEAD ZIRCONATE. (LATTICE VIBRATIONS) 07-1262
DIELECTRIC STUDY OF THE LEAD ZIRCONATE LEAD HAFNATE (PHASE) DIAGRAM. 08-1340
THE ROLE OF IRON OXIDE ADULTERANT IN A LEAD ZIRCONATE LEAD TITANATE CERAMICS. 09-1592
AMICS UNDER HIGH ELECTRIC FIELDS. (SUB/ BEHAVIOR OF MODIFIED LEAD ZIRCONATE LEAD TITANATE PIEZOELECTRIC CER 09-1484
TERNARY SYSTE/ CRYSTAL STRUCTURE AND PHASE RELATIONS IN THE LEAD ZIRCONATE LEAD TITANATE- LEAD METANIOBATE 09-1596
LOW THE CUR/ FIELD ENFORCED FERROELECTRICITY IN GLASS BONDED LEAD ZIRCONATE. (PHASE TRANSITION 200 DEG C BE 07-1252
IC AND ANTIFERROELECTRIC PHASES. (LANTHANUM OR NIOBIUM DOPED LEAD ZIRCONATE STANNATE TITANATE) /FERROELECTR 09-1487
UTIONS. / ANTIFERROELECTRIC- FERROELECTRIC TRANSITION IN THE LEAD (ZIRCONATE, STANNATE, TITANATE) SOLID SOL 09-1486
R/ THE INTERNAL ELECTRIC FIELD GRADIENT IN ANTIFERROELECTRIC LEAD ZIRCONATE STUDIED BY PERTURBED ANGULAR CO 07-1255
MOSSBAUER MEASUREMENTS OF BISMUTH FERRATE- BISMUTH FERRATE- LEAD ZIRCONATE SYSTEMS. 08-1282
CONTAINING ADDITIVES ON THE PROPERTIES OF SOLID SOLUTIONS OF LEAD ZIRCONATE TITANATE. /SMUTH AND MANGANESE 09-1588
ELECTROOPTIC EFFECTS IN FERROELECTRIC CERAMICS. (LEAD ZIRCONATE TITANATE) 09-1551
ELECTRIC CERAMICS AND CONVERSION OF MECHANOELECTRIC ENERGY. (LEAD ZIRCONATE TITANATE) FERRO 09-1483
FERROELECTRIC CERAMICS AND ENERGY CONVERSION. (LEAD ZIRCONATE TITANATE) 09-1508
FERROELECTRIC DISPLAYS. (BISMUTH TITANATE, LEAD ZIRCONATE TITANATE) 09-1515
FERROELECTRIC MEMORY DEVICE PARAMETER STUDY. (LEAD ZIRCONATE TITANATE) 09-1594
OF POINT DEFECTS ON THE INTERNAL FRICTION IN POLYCRYSTALLINE LEAD ZIRCONATE TITANATE. /F THE CONCENTRATION 09-1582
INTRINSIC NONSTOICHIOMETRY IN SINGLE PHASE LEAD ZIRCONATE TITANATE. 09-1534
MEASUREMENT OF THE PYROELECTRIC COEFFICIENT OF 65-35 PZT. (LEAD ZIRCONATE TITANATE) 09-1567
NATURE OF AGING IN FERROELECTRIC CERAMICS. (BARIUM TITANATE, LEAD ZIRCONATE TITANATE) 04-0757
AMIC PIEZOELECTRIC TRANSDUCER MATERIAL. (ALUMINA WHISKERS IN LEAD ZIRCONATE TITANATE) NEW COMPOSITE CER 09-1556
POINT DEFECTS AND SINTERING OF LEAD ZIRCONATE TITANATE. 09-1478
CIPITATION OF ZIRCONIA PHASE IN NIOBIUM MODIFIED CERAMICS OF LEAD ZIRCONATE TITANATE. PRE 09-1540
LEVISION INFRARED RESPONSE USING SURFACE WAVE TRANSDUCERS ON LEAD ZIRCONATE TITANATE. /OF A 40 MHZ COLOR TE 09-1509
SINTERING SCANDIUM AND NIOBIUM MODIFIED LEAD ZIRCONATE TITANATE. 09-1555
SOLUBILITY OF ALUMINUM IN LEAD ZIRCONATE TITANATE. 09-1479
STRUCTURES IN THE FERROELECTRIC CERAMICS BARIUM TITANATE AND LEAD ZIRCONATE TITANATE. TWO WAVE SHOCK 04-0916
SEE BOTH LEAD ZIRCONATE TITANATE AND PZT.
/ED FERROELECTRIC CERAMICS FOR ELECTROOPTICAL APPLICATIONS. (LEAD ZIRCONATE TITANATE AND SODIUM POTASSIUM / 09-1527
ULTRASONIC INTERFEROMETER WITH LEAD ZIRCONATE TITANATE AS TRANSDUCER. 09-1585
EFFECTS OF CALCINING ON THE FIRING OF LEAD ZIRCONATE TITANATE CERAMIC 09-1622
DC BIAS EFFECTS IN LEAD ZIRCONATE TITANATE CERAMICS. 09-1490
EFFECTS OF RADIATION INDUCED DAMAGE CENTERS IN LEAD ZIRCONATE TITANATE CERAMICS. 09-1520
EFFECTS OF CALCINING ON SINTERING OF LEAD ZIRCONATE TITANATE CERAMICS. 09-1494
EFFECTS OF IMPURITIES ON THE MECHANICAL QUALITY FACTOR OF LEAD ZIRCONATE TITANATE CERAMICS. 09-1608
SINTERING AND FERROELECTRIC PROPERTIES OF LEAD ZIRCONATE TITANATE CERAMICS. 09-1481
EFFECTS AT THE PHASE TRANSITION TETRAGONAL- RHOMBOHEDRAL IN LEAD ZIRCONATE TITANATE CERAMICS. /ABILIZATION 09-1611
ADDITION OF IMPURITIES. SPACE CHARGE EFFECT IN LEAD ZIRCONATE TITANATE CERAMICS CAUSED BY THE 09-1609
SMUTH. HOT PRESSED LEAD ZIRCONATE TITANATE CERAMICS CONTAINING BI 09-1525
PURITIES. ELECTRICAL RESISTIVITY OF LEAD ZIRCONATE TITANATE CERAMICS CONTAINING IM 09-1607
TICAL APPLICATIONS. HOT PRESSED FERROELECTRIC LEAD ZIRCONATE TITANATE CERAMICS FOR ELECTROOP 09-1524
/ RECENT DEVELOPMENTS IN ELECTROOPTIC CERAMICS (HOT PRESSED LEAD ZIRCONATE TITANATE, EFFECT OF GRAIN SIZE, 02-0086
SOME PHYSICAL AND ELECTRICAL SWITCHING CHARACTERISTICS OF A LEAD ZIRCONATE TITANATE FERROELECTRIC CERAMIC. 09-1601
. (REVIEW, 31 REFERENCES) AGING OF BARIUM TITANATE AND LEAD ZIRCONATE TITANATE FERROELECTRIC CERAMICS 02-0062
TRATION CELL MEASUREMENTS OF IONIC TRANSPORT NUMBERS IN PZT (LEAD ZIRCONATE TITANATE) FERROELECTRICS. /NCEN 09-1512
AND WITHOUT / INTERNAL FRICTION (AND CURIE TEMPERATURES) IN (LEAD ZIRCONATE TITANATE) FERROELECTRICS (WITH 09-1584
/ELECTRICS ON THEIR MECHANICAL PROPERTIES. (BARIUM TITANATE, LEAD ZIRCONATE TITANATE, LEAD METANIOBATE, 20/ 04-0985
RIZATION AND COERCIVE FORCE OF THE PEROVSKITES IN THE SYSTEM LEAD ZIRCONATE TITANATE METANIOBATE. /OUS POLA 09-1597
R POTASSIUM ANTIMONATE, NIOBATE, / PIEZOELECTRIC CERAMICS OF LEAD ZIRCONATE TITANATE MODIFIED WITH SODIUM O 09-1539
M PENTOXIDE) SURFACE / PROPERTIES OF OXYGEN DEPLETED CERAMIC LEAD ZIRCONATE TITANATE (PLUS 1 PERCENT NIOBIU 09-1576
. HIGH D*, FAST, LEAD ZIRCONATE TITANATE PYROELECTRIC DETECTORS 09-1560
ED WITH NIOBIUM AND TITANIUM) MOSSBAUER STUDIES OF LEAD ZIRCONATE TITANATE STANNATE. (CERAMIC DOP 09-1604
) CHARGE MECHANISM IN FERROELECTRICS. (LEAD ZIRCONATE TITANATE, STOICHIOMETRY EFFECTS 03-0259
HIGH TEMPERATURE PHASE EQUILIBRIA IN THE LEAD ZIRCONATE TITANATE SYSTEM. 09-1577
INTRINSIC NONSTOICHIOMETRY IN THE LEAD ZIRCONATE TITANATE SYSTEM. 09-1535
STUDY OF PHASE TRANSITIONS IN THE LEAD ZIRCONATE TITANATE SYSTEM. 09-1620
URE) INTRINSIC AND EXTRINSIC NONSTOICHIOMETRY IN THE LEAD ZIRCONATE TITANATE SYSTEM. (DEFECT STRUCT 09-1536
W MATERIALS FOR INFORMATION STORAGE AND DISPLAY (HOT PRESSED LEAD ZIRCONATE TITANATE, TUTORIAL, 17 REFS). / 09-1550
CONTAINING GERMANIUM. PIEZOCERAMICS BASED ON LEAD ZIRCONATE TITANATE WITH COMPLEX ADDITIVES 09-1510
D STRONTIUM ADDITIVES. CHROMATOGRAPHIC METHOD FOR ANALYZING LEAD ZIRCONATE TITANATES CONTAINING BISMUTH AN 09-1625
DEG K. COMPLEX DIELECTRIC CONSTANT OF LEAD ZIRCONATE UNDER POLARIZING FIELD, 370-520 07-1256
OLARIZATION, ELECTROSTRICTION, AND PIEZOELECTRIC ACTIVITY OF LEAD ZIRCONATE(X) TITANATE(1-X) CERAMICS. P 09-1501
/ ORIGIN OF THE MAXIMUM IN THE ELECTROMECHANICAL ACTIVITY IN LEAD ZIRCONATE(X) TITANATE(1-X) CERAMICS NEAR 09-1499
/ ORIGIN OF THE MAXIMUM IN THE ELECTROMECHANICAL ACTIVITY IN LEAD ZIRCONATE(X) TITANATE(1-X) CERAMICS NEAR 09-1500
HEDRAL PHA/ ATOMIC STRUCTURES OF THE FERROELECTRIC COMPOUNDS LEAD ZIRCONATE(X) TITANATE(1-X) FOR TWO RHOMBO 09-1574

AUSED / DETERIORATION OF THE ELECTROMECHANICAL PROPERTIES OF LEAD ZIRCONATE(0.52) TITANATE(0.48) CERAMICS C 09-1606
T AND STRUCTURA/ SUBSTITUTION OF BISMUTH AND NIOBIUM IONS IN LEAD ZIRCONATE(0.53) TITANATE(0.47) (EXPERIMEN 09-1480
/EFFECT STUDIES AT LOW TEMPERATURES (WITH TIN OXIDE FILMS ON LEAD ZIRCONATE(0.65) TITANATE(0.35) CERAMIC). 09-1505
NT TUNGSTEN TRIOXIDE.   PHASE DIAGRAM OF LEAD ZIRCONATE(1-X) TITANATE(X) PLUS 0.8 PERCE 09-1581
ND DISPLAY DEVICES USING FINE GRAIN FERROELECTRIC CERAMICS. (LEAD ZIRCONATE- LEAD TITANATE) IMAGE STORAGE A 09-1572
E OF FERROELECTRIC CERAMICS IN COMPUTERS AND OTHER DEVICES. (LEAD ZIRCONATE- LEAD TITANATE AND PLZT)   US 09-1532
LATE)   A FERROELECTRIC PIEZOELECTRIC RANDOM ACCESS MEMORY. (LEAD ZIRCONATE- LEAD TITANATE CERAMIC MEMORY P 09-1504
                     X-RAY ANALYSIS OF MODIFIED LEAD ZIRCONATE- LEAD TITANATE CERAMICS.   09-1619
ZOELECTRIC APPL/ EFFECTS OF HIGH ELECTRIC FIELDS ON MODIFIED LEAD ZIRCONATE- LEAD TITANATE CERAMICS FOR PIE 09-1485
IETARY COMPOSITIONS)   INTERNAL FRICTION OF MODIFIED LEAD ZIRCONATE- LEAD TITANATE CERAMICS. (PROPR 09-1623
DIUM ELECTRODES.   IMPROVED AGING AND SWITCHING OF LEAD ZIRCONATE- LEAD TITANATE CERAMICS WITH IN 09-1516
DIUM ELECTRODES.   IMPROVED AGING AND SWITCHING OF LEAD ZIRCONATE- LEAD TITANATE CERAMICS WITH IN 09-1517
2: HOT PRESSED COMPOSITIONS AT NEAR THEORET/ VAPORIZATION OF LEAD ZIRCONATE- LEAD TITANATE MATERIALS. PART- 09-1578
            STUDY OF PHASE TRANSITIONS IN THE LEAD ZIRCONATE- LEAD TITANATE SYSTEM.   09-1621
PROPERTIES OF NEW TRANSPARENT FERROELECTRIC CERAMICS. (LEAD ZIRCONATE- LEAD TITANATE TYPE)   09-1518
/ER EFFECT STUDIES OF FERROELECTRIC PHASE TRANSITIONS IN THE LEAD ZIRCONATE- LEAD TITANATE- BISMUTH FERRAT/ 09-1498
(P/ TEMPERATURE DEPENDENT FERROELECTRIC DOMAIN ALIGNMENT IN LEAD ZIRCONATE- LEAD TITANATE- LANTHANUM OXIDE 09-1586
E TERNARY SYSTEM. CURIE TEMPERATURES OF SOLID SOLUTIONS IN A LEAD ZIRCONATE- LEAD TITANATE- LEAD METANIOBAT 09-1595
PREPARATION OF THE TERNARY TITANATE BARIUM(0.8) LEAD(0.12) CALCIUM(0.08) TITANATE.   36-3587
AVE) FERROELECTRIC PHASE SHIFTERS FOR VHF AND UHF. (BASED ON LEAD(0.35) STRONTIUM(0.65) TITANATE) /URFACE W 37-3633
/ING. (WITH ELECTRODES REDUCING POLARIZATION SWITCHING TIME, LEAD(0.35) STRONTIUM(0.65) TITANATE, BARIUM(0/ 37-3660
(0.405) NIOBIUM(0.325) ZIRCONIUM(0/ PERFORMANCE OF SPUTTERED LEAD(0.92) BISMUTH(0.07) LANTHANUM(0.01) (IRON 37-3629
/ FERROELECTRIC FIELD EFFECT STUDIES. (ON TIN OXIDE FILMS ON LEAD(0.97) BISMUTH(0.02) TITANATE (0.35) ZIRCO 09-1506
/PHASE ANGLES IN A FERROELECTRIC CERAMIC (THEORY, APPLIED TO LEAD(0.99) NIOBIUM(0.02) ZIRCONATE(0.65) TITA/ 09-1537
IC PROPERTIES OF (BISMUTH(0.5) (SODIUM OR POTASSIUM) (0.5) (X) LEAD(1-X) TITANATE SYSTEMS. /LECTRIC AND CERAM 13-2013
SODIUM(0.5-X) TITANATE, BISMUTH(X) (POTASSIUM OR SODIUM) (X) LEAD (1-2X) ZIRCONATES). /TH(0.5) POTASSIUM(X) 08-1289
SYSTEMS./ DIELECTRIC NONLINEARITY AND LOSS IN THE BARIUM (OR LEAD, 4 COMPOSITIONS EACH) STRONTIUM TITANATE 09-1522
ACOUSTOOPTIC PROPERTY OF SINGLE CRYSTAL LEAD(5) GERMANIUM(3) OXIDE(11).   35-3509
CRYSTAL STRUCTURE OF FERROELECTRIC LEAD(5) GERMANIUM(3) OXIDE(11).   35-3504
FERROELECTRIC OPTICAL ROTATION DOMAINS IN SINGLE CRYSTAL LEAD(5) GERMANIUM(3) OXIDE(11).   36-3550
S. (BISMUTH TITANATE, LITHIUM THALLIUM TARTRATE MONOHYDRATE, LEAD(5) GERMANIUM(3) OXIDE(11)) /CTRIC CRYSTAL 02-0061
FERROELECTRICITY IN LEAD(5) GERMANIUM(3) OXIDE(11).   35-3508
NONLINEAR OPTICAL PROPERTIES OF FERROELECTRIC LEAD(5) GERMANIUM(3) OXIDE(11).   35-3506
RAMAN SCATTERING FROM THE FERROELECTRIC MODE IN LEAD(5) GERMANIUM(3) OXIDE(11).   35-3498
SOFT OPTIC PHONON MODE IN FERROELECTRIC LEAD(5) GERMANIUM(3) OXIDE(11).   35-3495
POGRAPHIC OBSERVATION OF 180 DEGREE DOMAINS IN FERROELECTRIC LEAD(5) GERMANIUM(3) OXIDE(11).   X-RAY TO 35-3516
HOUS COMPOUND LEAD / FERROELECTRIC AND OPTICAL PROPERTIES OF LEAD(5) GERMANIUM(3) OXIDE(11) AND ITS ISOMORP 35-3501
MATERIALS.   FERROELECTRIC PROPERTY AND ENANTIOMORPHISM OF LEAD(5) GERMANIUM(3) OXIDE(11) AND THE RELATED 35-3499
FERROELECTRIC LEAD(5) GERMANIUM(3) OXIDE(11) CRYSTAL.   35-3500
FERROELECTRIC. (DIELECTRIC CONSTANTS AND REFRACTIVE INDICES) LEAD(5) GERMANIUM(3) OXIDE(11) CRYSTAL, A NEW 35-3502
ELASTIC AND PIEZOELECTRIC PROPERTIES OF FERROELECTRIC LEAD(5) GERMANIUM(3) OXIDE(11) CRYSTALS.   35-3520
SWITCHING OF OPTICAL ROTATORY POWER IN FERROELECTRIC LEAD(5) GERMANIUM(3) OXIDE(11) SINGLE CRYSTAL. 36-3570
.   OPTICAL ACTIVITY OF FERROELECTRIC LEAD(5) GERMANIUM(3) OXIDE(11) SINGLE CRYSTALS 35-3503
.   CRYSTAL GROWTH AND PROPERTIES OF LEAD(5) GERMANIUM(3) OXIDE(11) SINGLE CRYSTALS 35-3513
MAGNETIC COUPLING).   PIEZOELECTRIC LEAKY SURFACE WAVE IN LITHIUM NIOBATE (ELECTRO 11-1832
GATION AND AMPLIFICATION OF RAYLEIGH WAVES AND PIEZOELECTRIC LEAKY SURFACE WAVES IN LITHIUM NIOBATE. PROPA 11-1856
IUM TITANATE BETWEEN/ STUDY OF THE TEMPERATURE DEPENDENCE OF LEED INTENSITIES FROM THE (001) SURFACE OF BAR 04-0531
ATE CRYSTALS.   LEED STUDIES ON THE (001) FACE OF BARIUM TITAN 04-0530
FACE OF BARIUM TITANATE.   LEED STUDY OF SURFACE STRUCTURES ON THE (001) 04-0529
ATURE COEFFICIENT OF RESONANT FREQUENCY IN LITHIUM TANTALATE LENGTH EXPANDER BARS.   ZERO TEMPER 11-1810
IN AMMONIUM TRIHYDROGEN PERIODATE CRYSTALS, PULSED MAGNETIC/ LEVEL CROSSING (CROSS RELAXATION) EXPERIMENTS 36-3577
OPTICAL TRANSITION TO THE FERMI LEVEL IN SODIUM TUNGSTATE.   10-1634
ON FERROELECTRICS I/ POSSIBLE REALIZATION OF ULTRA LOW NOISE LEVEL PARAELECTRIC MICROWAVE AMPLIFIER (BASED 37-3664
UM ALUMINUM DISULFATE D/ CALCULATION OF CRYSTAL FIELD ENERGY LEVEL SPLITTINGS OF TITANIUM(3+) ION IN RUBIDI 19-2754
TING FERROELECTRICS.   SOME FEATURES OF TRAPPING LEVELS IN ANTIMONY SULFOIODIDE TYPE SEMICONDUC 15-2182
ED CHROMIUM(3+) PAIRS IN LANTHANUM ALUM/ SPECTRUM AND ENERGY LEVELS OF CHROMIUM(3+) IONS AND EXCHANGE COUPL 36-3565
LITHIUM NIO/ ABSORPTION AND LUMINESCENCE SPECTRA AND ENERGY LEVELS OF NEODYMIUM(3+) AND ERBIUM(3+) IONS IN 11-1713
PEROVSKITE CRYSTALS (STRONTIUM TITANATE, POTASSIUM TANTALA/ LIFETIME OF THE "FERROELECTRIC" PHONON IN SOME 03-0227
DAMAGE TO THE SURFACE OF LITHIUM NIOBATE BY LIGHT.   11-1866
P CRYSTALS IN THE INFRARED RANGE BY PARAMETRIC SCATTERING OF LIGHT. / OF THE REFRACTIVE INDEX OF ADP AND KD 17-2553
E FREQUENCY RANGE OF DOUBLE ELECTRONIC EXC/ ON THE THEORY OF LIGHT ABSORPTION BY ANTIFERRODIELECTRICS IN TH 03-0419
HATE SINGLE CRYSTALS.   LIGHT ABSORPTION IN POTASSIUM DIHYDROGEN PHOSP 17-2421
H MODE LOCKED PUMP.   PARAMETRIC LIGHT AMPLIFICATION AND OSCILLATION IN KDP WIT 17-2468
M DIHYDROGEN PHOSPHATE CRYSTALS.   LIGHT AND CURRENT PULSES FROM X-RAYED POTASSIU 17-2716
OF SPACE CHARGE LIMITED CURRENT IN BISMUTH(12) GERMANIUM O/ LIGHT AND ELECTRIC FIELD DEPENDENT OSCILLATION 35-3505
OF LEAD MAGNESIO NIOBATE CRYSTALS.   LIGHT BEAM DEFLECTION CONTROLLED WITH A PRISM 08-1280
TITANATE.   BINARY LIGHT BEAM DEFLECTION IN SINGLE CRYSTAL BARIUM 04-0653
OBATE PRISM.   ELECTROOPTIC LIGHT BEAM DEFLECTION WITH BARIUM STRONTIUM NI 13-2036
TITANATE.   BINARY LIGHT BEAM DETECTION IN SINGLE CRYSTAL BARIUM 04-0654
ELECTROOPTIC POTASSIUM TANTALATE NIOBATE GRATINGS FOR LIGHT BEAM MODULATION AND DEFLECTION.   13-1929
THIUM TANTALATE.   SCATTERING OF LIGHT BY ACOUSTIC PHONONS AND POLARITONS IN LI 12-1876
M, LEAD) TITANATE NEAR THE CURIE TEMPERATURE. EMISSION OF LIGHT BY SURFACE MICROPLASMA OF CERAMIC (BARIU 04-0982
DIHYDROGEN PHOS/ PARAMETRIC FREQUENCY CONVERSION OF COHERENT LIGHT BY THE ELECTROOPTIC EFFECT IN POTASSIUM 17-2616
GEN PHOSPHATE CRYSTALS.   DIFFRACTION OF POLARIZED LIGHT BY ULTRASONIC WAVES IN POTASSIUM DIHYDRO 17-2407
/LECTED BEAM DIAMETER AND DEFLECTION CAPACITIES OF A DIGITAL LIGHT DEFLECTOR UTILIZING LITHIUM TANTALATE P/ 11-1809
AVES. APPLICATION OF ACOUSTIC WAVES T/ EXPERIMENTAL STUDY OF LIGHT DIFFRACTION BY HIGH FREQUENCY ACOUSTIC W 11-1834
ASSIUM DIHYDROGEN PHOS/ DEPOLARIZATION OF LINEARLY POLARIZED LIGHT DUE TO HEATING OF PARALLEL PLANES OF POT 17-2540
UBJECTED TO UNIAXIAL STRESS PULSES.   LIGHT EMISSION FROM BARIUM TITANATE CRYSTALS S 04-0672
ING)   LIGHT EMISSION FROM TGS. (CRYSTALS DURING HEAT 27-3117
SELF INDUCED THERMAL EFFECTS ON LIGHT EXTINCTION OF LITHIUM TANTALATE.   12-1889
E SALT CRYSTALS.   SURFACE STRUCTURES AND CORRESPONDING LIGHT FIGURES REVEALED BY WATER ETCHED ROCHELL 28-3234
/GEN PHOSPHATE AND AMMONIUM DIHYDROGEN PHOSPHATE CRYSTALS OF LIGHT FROM A LASER OPERATING IN THE FREE GENE/ 17-2717
E.   OBSERVATION OF DIFFRACTED LIGHT FROM DOMAIN WALLS IN GADOLINIUM MOLYBDAT 30-3322
/REMENTS OF PHOTON CORRELATIONS OF SECOND HARMONIC GENERATED LIGHT (FROM LITHIUM NIOBATE CRYSTAL, AGREEMEN/ 11-1701
ODE.   FERROELECTRIC CERAMIC LIGHT GATES OPERATED IN A VOLTAGE CONTROLLED M 37-3654
                                  34-3440
TRIFLUORIDE.   MOLECULAR FORCES OF HEAVY AND LIGHT ICE.
TWO MAGNON SCATTERING OF LIGHT IN ANTIFERROMAGNETIC POTASSIUM MANGANESE 33-3401
SPECTRUM OF RAMAN SCATTERED LASER LIGHT IN FERROELECTRICS OF THE KDP GROUP.   17-2625
.2-7/ (CONTINUOUS WAVE) SPONTANEOUS PARAMETRIC SCATTERING OF LIGHT IN LITHIUM IODATE (INDEX OF REFRACTION 1 08-1291
E INTENSE) AND LONGITUDI/ MANDELSTAM BRILLOUIN SCATTERING OF LIGHT IN LITHIUM NIOBATE (FROM TRANSVERSE (MOR 11-1743

POTASSIUM MANGANESE TRIF/ MANDELSTAM BRILLCUIN SCATTERING OF LIGHT IN MANGANESE FLUORIDE, COBALT FLUORIDE, 33-3399
M TRIFLUORIDE, RUBIDIUM / MANDELSTAM BRILLOUIN SCATTERING OF LIGHT IN MANGANESE FLUORIDE CRYSTALS (POTASSIU 33-3398
FINE STRUCTURE OF EXCITON- MAGNON ABSORPTION OF LIGHT IN POTASSIUM MANGANESE TRIFLUORIDE. 33-3364
METHOD.   SECOND HARMONIC GENERATION OF LIGHT IN THE KDP CRYSTAL BY THE INDEX MATCHING 17-2693
OR ANTIMONY SULFOIODIDE.   LIGHT INDUCED PYROCURRENTS IN FERROSEMICONDUCT 15-2130
LECTRICALLY POLARIZED PHASE OF BARIUM TITANAT/ CHANNELING OF LIGHT IONS IN THE CUBIC AND TETRAGONAL (FERROE 04-0688
OPTICAL MIXING CF COHERENT AND INCOHERENT LIGHT. (KDP CRYSTAL FILTER) 17-2669
NOVEL TYPE OF CUT FOR KLP CRYSTALS FOR LOW VOLTAGE LIGHT MODULATION. 17-2531
PTIMUM CUT IN X DIHYDROGEN FHOSPHATE CRYSTALS FOR TRANSVERSE LIGHT MODULATION   O 17-2494
IMUM CUT OF POTASSIUM DIHYDROGEN FHOSPHATE TYPE CRYSTALS FOR LIGHT MODULATION.   OPT 17-2532
LITHIUM NIOBATE LIGHT MODULATOR. 11-1782
OPTIMUM CUT FOR A LITHIUM NIOBATE TRANSVERSE LIGHT MODULATOR. 11-1755
M CRYSTAL ORIENTATION FOR A BARIUM SODIUM NICBATE TRANSVERSE LIGHT MODULATOR. (HALF WAVE VOLTAGE, THEORY) / 13-1973
TITANATE)   POLARIZATION REFLECTION TYPE LIGHT MODULATOR USING FERROELECTRICS. (BARIUM 04-0714
IN UNIAXIAL CRYSTALS OF THE TYPE X DIHYDROGEN FHOSPHATE FOR LIGHT MODULATORS.   CUTTING AND POLISHING 17-2559
IOBA/ USE OF PEROVSKITE PARAELECTRICS IN BEAM DEFLECTORS AND LIGHT MODULATORS. (POTASSIUM TANTALATE(0.65) N 13-1924
LEGREE Y CUT ADP CRYSTALS.   LIGHT MODULATORS USING 45 DEGREE X CUT AND 45 17-2605
CUTS OF LITHIUM NICBATE CRYSTALS SUITABLE FOR LIGHT MODULATORS WITH A LOW HALF WAVE VOLTAGE. 11-1715
ND LITHIUM TANTALATE.   SCATTERING OF LIGHT ON ACOUSTIC PHONONS IN LITHIUM NIOBATE A 11-1744
OF ANTIMONY SULFOIODIDE.   INFLUENCE OF LIGHT ON SELF POLARIZATION OF SINGLE CRYSTALS 15-2228
POTASSIUM NIOBATE.   EFFECT OF INTENSE LIGHT ON THE REFRACTIVE INDEX OF A CRYSTALLINE 08-1410
NTIMONY SULFOIODIDE SINGLE CRYSTALS EXCITED WITH MICROSECOND LIGHT PULSES.   RELAXATION PROCESSES IN A 15-2202
GENERATION OF FAR INFRARED RADIATION BY PICOSECOND LIGHT PULSES IN LITHIUM NIOBATE. 11-1857
TANATE.   EFFICIENT WHITE LIGHT READING OF DOMAIN PATTERNS IN BISMUTH TI 08-1299
SING) ANC DISPLAY DEVICES. (BASED CN ELECTRICALLY CONTROLLED LIGHT SCATTERING AND BIREFRINGENCE, REVIEW) /S 37-3650
LECTRIC AND PIEZOELECTRIC CRYSTALS.   LIGHT SCATTERING BY ACOUSTIC PHONONS IN FERROE 03-0471
RIGLYCINE SULFATE.   SECOND HARMONIC LIGHT SCATTERING BY DOMAINS IN FERROELECTRIC T 27-2991
ORDINARY PHOTONS IN LITHIUM NICBATE.   LIGHT SCATTERING BY POLARITONS ASSOCIATED WITH 11-1697
TE.   PARAMETRIC LUMINESCENCE AND LIGHT SCATTERING BY POLARITONS. (LITHIUM NIOBA 11-1746
TIES WITH APPLICATIONS TC FERROELEC/ AN ANALYSIS OF MULTIPLE LIGHT SCATTERING BY RANDOM OPTICAL INHOMOGENEI 37-3634
M NIOBATE CRYSTALS. (PHOTCELASTIC CONSTANT)   LIGHT SCATTERING BY ULTRASONIC WAVES IN LITHIU 11-1754
POTASSIUM DIDEUTERIUM PHOSPHATE.   LIGHT SCATTERING FROM CRITICAL FLUCTUATIONS IN 17-2463
RAMAN LIGHT SCATTERING IN ROCHELLE SALT. 28-3231
(TRIGLYCINE SULFATE, POTASSIUM DIHYDROGEN AND DIDEUT/ LASER LIGHT SCATTERING STUDIES OF PHASE TRANSITIONS. 17-2481
R THE CUBIC TO TETRAGONAL PHASE TRANSITION IN STRONTIUM TIT/ LIGHT SCATTERING STUDIES OF THE SOFT MODES NEA 06-1242
ACOUSTIC MODES CF POTASSIUM DIHYCROGEN FHOSPHATE AND POTAS/ LIGHT SCATTERING STUDIES OF THE SOFT OPTIC AND 17-2497
ANSITION IN KDP.   LIGHT SCATTERING STUDY OF THE FERROELECTRIC TR 17-2462
COMMUTATCRS. (POTASSIUM DIHYDROGEN PHOSPHA/ (FINITE APERTURE LIGHT SWITCH USING) TRANSVERSE POCKELS EFFECT 17-2509
/ONSTRUCTION BY ELECTROOPTIC MODULATION (BY PASSING COHERENT LIGHT THROUGH DEUTERATED KDP HAVING A CHARGE / 17-2507
STAL.   MICROWAVE MODULATOR OF RUBY LASER LIGHT USING POTASSIUM DIHYDROGEN PHOSPHATE CRY 17-2640
FERROELECTRIC LIGHT VALVE ARRAYS FOR OPTICAL MEMORIES. 37-3676
EPENDENCE OF CONTRAST AND BRIGHTNESS IN GADOLINIUM MOLYBDATE LIGHT VALVES.   ORIENTATION AND THICKNESS D 30-3291
IONS. USE OF CUBIC PHCTOSENSITIVE CRYSTALS AS POCKELS EFFECT LIGHT VALVES IN DISPLAY AND MODULATOR APPLICAT 37-3661
DIFFRACTION OF A CONTINUOUS LIGHT WAVE IN KDP. 17-2444
TRAPPING OF PARAMETRICALLY AMPLIFIED LIGHT WAVES IN A KDP CRYSTAL. 17-2428
LECTRIC (THEORY).   INTERACTION OF LIGHT WITH MAGNETOACOUSTIC WAVES IN A FERRODIE 03-0157
RY)   INTERACTION OF LIGHT WITH SPIN WAVES IN FERROELECTRICS. (THEO 03-0158
TOR OF POLARIZATION IN KDP, IN A DIGITAL DEVIA/ STUDY OF THE LIMITATION OF THE ANGULAR FIELD BY THE COMMUTA 17-2524
PHONCN SPECTRUM OF FERROELECTRICS STUDIED BY LIMITATION OF THE CRYSTAL SIZE. 03-0196
FERROELECTRICS.   LIMITATIONS TO POLARIZATION OF POLYCRYSTALLINE 03-0434
/HT AND ELECTRIC FIELD DEPENDENT OSCILLATION OF SPACE CHARGE LIMITED CURRENT IN BISMUTH(12) GERMANIUM OXID/ 35-3505
NATE. (CURRENT VOLTAGE CHARACTERISTICS, 150-37/ SPACE CHARGE LIMITED CURRENT IN POLYCRYSTALLINE BARIUM TITA 04-0557
L SYSTEM (THEORY).   SPACE CHARGE LIMITED CURRENTS IN A METAL FERROELECTRIC META 03-0348
THEORY OF SPACE CHARGE LIMITED CURRENTS IN FERROELECTRICS. 03-0251
0.75) PLATINATE(0.25) CRYSTALS.   SPACE CHARGE LIMITED CURRENTS IN HEXAGONAL BARIUM TITANATE(04-0936
IOBATE CRYSTALS.   SPACE CHARGE LIMITED CURRENTS IN SEMICONDUCTING POTASSIUM N 08-1412
SPLITTING OF THE SODIUM-23 NMR CENTRAL LINE IN FERROELECTRIC SILVER SODIUM DINITRITE. 16-2307
OLARITONS IN TETRAGONAL BARIUM TITANATE: THEORY OF THE RAMAN LINE SHAPE.   P 04-0556
FERROELECTRICS.   THEORY OF X-RAY DIFFRACTION LINE SHAPE EFFECTS DUE TO DOMAIN FORMATION IN 03-0236
IUM TITANATE.   RAMAN SCATTERING LINE SHAPE OF THE SOFT E POLARITON MODE IN BAR 04-0803
IC POTASSIUM MANGANE/ EXCHANGE NARROWING: MAGNETIC RESCNANCE LINE SHAPES AND SPIN CORRELATION IN PARAMAGNET 33-3376
STIGATIONS OF THE PHASE TR/ NUCLEAR MAGNETIC RESONANCE (WIDE LINE SPECTRA AND SPIN LATTICE RELAXATION) INVE 20-2802
DCMAIN WALLS CAUGHT IN (FINE LINE) "SUDARES" IN ROCHELLE SALT CRYSTAL. 28-3215
APPARENT SOFT MODE LINE WIDTH DIVERGENCE IN LEAD TITANATE. 05-1078
FERROMAGNETIC RESONANCE LINE WIDTH IN FERROELECTRIC FERROMAGNETS. 03-0207
WITH MAGNESIUM DOPING.   SHIFT OF NEEL TEMPERATURE AND EPR LINE WIDTH OF POTASSIUM MANGANESE TRIFLUORIDE 33-3377
EMISSION IN LITHIUM NIOBATE.   POWER AND LINE WIDTH OF TUNABLE STIMULATED FAR INFRARED 11-1736
SIUM CIHYDROGEN PHOSPHATE TYPE CRYSTALS.   ABSENCE OF A LINE WIDTH TRANSITION FOR THE PROTONS IN POTAS 17-2436
TE.   NIOBIUM-93 NMR LINE WIDTHS IN NONSTOICHIOMETRIC LITHIUM NIOBA 11-1795
VIBRATIONAL LINE WIDTHS OF THIOUREA. 25-2915
CTS IN MICROWAVE ACOUSTIC LITHIUM NICBATE SURFACE WAVE DELAY LINES.   NONLINEAR EFFE 11-1818
ALS.   GUIDE LINES FOR THE SELECTION OF ACOUSTOOPTIC MATERI 03-0425
THIUM NIOBATE.   FORBIDDEN HYPERFINE LINES IN ESR SPECTRA OF MANGANESE(2+) DOPED LI 11-1805
ATE.   ZEEMAN EFFECT AND DILUTION EFFECT OF ABSORPTION LINES IN GUANIDINIUM CHROMIUM SULFATE HEXAHYDR 21-2821
INTENSITY OF SCME BRILLOUIN AND RAYLEIGH LINES IN RUBIDIUM DIHYDROGEN PHOSPHATE. 17-2521
N PARAMAGNETIC RESONANCE OF MANGANESE(+2) IN BARI/ FORBIDDEN LINES IN THE HYPERFINE SPECTRUM OF THE ELECTRO 04-0877
IUM SULFATE HEXAHYDRATE.   ZEEMAN EFFECT (STUDY) OF TWO FINE LINES IN (THE RED REGION OF) CHROMIUM GUANIDIN 21-2814
OPPER FORMATE TETRAHYDRATE.   ANOMALOUS SHIFT OF ESR LINES IN THE TWO DIMENSIONAL ANTIFERROMAGNET C 26-2941
IZED PZT CERAMIC PLATES.   SURFACE WAVE DELAY LINES WITH INTERDIGITAL TRANSDUCERS ON UNPOLAR 09-1614
TEMPERATURE DEPENDENT SPECTRUM OF AN ANTIFERROELECTRIC LINEAR CHAIN MODEL. (PEROVSKITES) 03-0468
POTENTIAL. (BARIUM TITANATE)   COMPUTER EXPERIMENT ON LINEAR CHAIN OF ATOMS LYING IN DOUBLE MINIMUM 03-0287
RTIES OF BARIUM TITANATE SINGLE CRYSTALS.   EFFECT OF LINEAR COMPRESSION ON THE UHF DIELECTRIC PROPE 04-0896
RAMAN SCATTERING IN BARIUM TITANATE.   LINEAR DISORDER AND TEMPERATURE DEPENDENCE OF 04-0675
QUARTZ AND FERROELECTRIC PERCVSKITES). (BARIUM TITANATE AND/ LINEAR DISORDER IN CRYSTALS (CASE OF SILICON, 04-0629
TANTALATE.   NEUTRON SCATTERING ANALYSIS OF THE LINEAR DISPLACEMENT CORRELATIONS IN POTASSIUM 36-3546
STATE/ COMPILATION OF CONVERSION FORMULAS INTERRELATING THE LINEAR ELASTIC AND ELECTROELASTIC EQUATIONS OF 03-0277
EX, DIELECTRIC CONSTANT, NONLINEAR OPTICAL COEFFICIENTS, AND LINEAR ELECTROOPTIC COEFFICIENTS) /RACTIVE IND 13-2024
TANT ON MELT COMPCSITION (STOICHIOMETRY) IN L/ DEPENDENCE OF LINEAR ELECTROOPTIC EFFECT AND DIELECTRIC CONS 11-1842
GEN PHOSPHATE CRYSTALS.   MEASUREMENT OF THE PRIMARY LINEAR ELECTROOPTIC EFFECT IN AMMONIUM DIHYDRO 17-2654
RAMICS: PLZT 12-40-60. (USED FOR SENSING ICW VOLTAGES) LINEAR ELECTROOPTIC EFFECT IN FERROELECTRIC CE 09-1610

LOW FREQUENCY LINEAR ELECTROOPTIC EFFECT IN LEAD DITANATLATE 13-1916
SSIUM OR AMMONIUM DIHYDROGEN PHOSPHATE OR ARSENATE, POTASSI/ LINEAR ELECTROOPTIC MODULATOR MATERIALS. (POTA 01-0021
(BIREFRINGENCE AND REFRACTIVE INDICES) OF POTASSIUM NIOBATE (LINEAR ELECTROOPTICS). OPTICAL PROPERTIES 08-1468
RIC SODIUM AMMONIUM SELENATE DIHYDRATE. LINEAR ELECTROOPTICAL EFFECT IN THE FERROELECT 22-2829
REFS) LINEAR ELECTROOPTICAL MATERIALS. (REVIEW, 138 01-0016
LOW ENTROPY FORM OF ICE I(H) OBTAINED FROM A LINEAR STEP GROWTH MODEL. 34-3410
TUTED BARIUM SODIUM NIOBATES. LINEAR THERMAL EXPANSIONS OF GADOLINIUM SUBSTI 13-1996
ALLEL PLANES OF POTASSIUM DIHYDROGEN PHOS/ DEPOLARIZATION OF LINEARLY POLARIZED LIGHT DUE TO HEATING OF PAR 17-2540
LCULATIO/ HIGH FIELD (1.5-6 KV/CM) POLARIZATION REVERSALS IN LIQUID ELECTRODED BARIUM TITANATE CRYSTALS (CA 04-0994
ACOUSTIC WAVE ALONG THE (001) DIRECTION OF LITHIUM NIOBATE (LIQUID HELIUM TEMPERATURE). /EQUENCY (950 MHZ) 11-1835
S (SODIU/ VACUUM ULTRAVIOLET ABSORPTION (5-23 EV AT ROOM AND LIQUID NITROGEN TEMPERATURES IN ALKALI NITRITE 16-2399
EXAGGERATED GRAIN GROWTH IN LIQUID PHASE SINTERING OF BARIUM TITANATE. 04-0835
AL QUALITY BARIUM SODIUM NIOBATE (BY USING COMPOSITION WHERE LIQUIDUS AND SOLIDUS COINCIDE). /OF GOOD OPTIC 13-2047
CRYSTALS IN THE SODIUM OXIDE-/ (EXTENT OF) TUNGSTEN BRONZE (LIQUIDUS AND SOLIDUS) FIELD AND MELT GROWTH OF 13-1949
RYSTALLINE BARIUM TITANATE. (150-500 DEG C, 20-50 KILOHERTZ, LISSAJOUS FIGURE) /Y LOSS NONPOLAR STATE POLYC 04-0831
SE REL/ TUNGSTEN BRONZE FIELD IN THE (OXIDE) SYSTEM POTASSIA LITHIA NIOBIA (POTASSIUM LITHIUM NIOBATES, PHA 13-2017
RIC CRYSTALS (BARIUM OR LEAD TITANATE, FROM SYSTEMS BARIA OR LITHIA, TITANIA, ALUMINA, SILICA). /FERROELECT 04-0783
NUCLEAR MAGNETIC RESONANCE OF FERROELECTRIC LITHIUM AMMONIUM TARTRATE. 29-3240
NUCLEAR MAGNETIC RESONANCE OF FERROELECTRIC LITHIUM AMMONIUM TARTRATE. 29-3241
T-2: SILVER SODIUM DINITRITE. PART-3: ROCHELLE SALT. PART-4: LITHIUM AMMONIUM TARTRATE HYDRATE. /: TGS. PAR 02-0077
X-RAY REFINEMENT OF THE CRYSTAL STRUCTURE OF LITHIUM AMMONIUM TARTRATE MONOHYDRATE. 29-3244
TALS GROWN IN MOLTEN SOLUTIONS OF THE PSEUDOSYSTEM POTASSIUM LITHIUM BARIUM NIOBATE. /STEN BRONZE TYPE CRYS 13-1948
HIUM NIOBATE, STRON/ STRUCTURE AND DIELECTRIC PROPERTIES OF LITHIUM CONTAINING PEROVSKITES. (STRONTIUM LIT 08-1292
CRYSTAL STRUCTURE OF PIEZOELECTRIC LITHIUM FORMATE MONOHYDRATE. 26-2940
NONLINEAR OPTICAL SUSCEPTIBILITY OF LITHIUM FORMATE MONOHYDRATE. 26-2945
GERMANATE NOT FERROELECTRIC) ELECTROOPTIC MEASUREMENTS ON LITHIUM GERMANATE AND LEAD GERMANATE. (LITHIUM 35-3494
DETERMINATION OF FERROELECTRIC PROPERTIES OF LITHIUM HYDRAZINIUM FLUOBERYLLATE. 32-3349
ANALYSIS OF HYSTERESIS LOOPS IN LITHIUM HYDRAZINIUM SULFATE. 32-3351
S NEUTRON SCATTERING AND THE QUESTION OF FERROELECTRICITY IN LITHIUM HYDRAZINIUM SULFATE. ANOMALOU 32-3341
CRYSTAL STRUCTURE OF LITHIUM HYDRAZINIUM SULFATE. 32-3342
CRYSTAL STRUCTURE OF FERROELECTRIC LITHIUM HYDRAZINIUM SULFATE. 32-3356
DEUTERON NMR STUDIES OF LITHIUM HYDRAZINIUM SULFATE. 32-3354
DIELECTRIC PROPERTIES OF LITHIUM HYDRAZINIUM SULFATE. 32-3344
DIELECTRIC PROPERTIES OF LITHIUM HYDRAZINIUM SULFATE. 32-3352
DIELECTRIC PROPERTIES OF LITHIUM HYDRAZINIUM SULFATE. 32-3353
DIPOLAR RELAXATION OF LITHIUM-7 BY HINDERED ROTATORS IN LITHIUM HYDRAZINIUM SULFATE. 32-3350
FERROELECTRIC-LIKE BEHAVIOR OF LITHIUM HYDRAZINIUM SULFATE. 32-3348
HYDROGEN BONDING IN LITHIUM HYDRAZINIUM SULFATE. 32-3346
MICROWAVE DIELECTRIC STUDY OF LITHIUM HYDRAZINIUM SULFATE. 32-3357
NITROGEN-14 NUCLEAR QUADRUPOLE RESONANCE STUDY OF LITHIUM HYDRAZINIUM SULFATE. 32-3345
NUCLEAR MAGNETIC RESONANCE STUDY OF FERROELECTRIC LITHIUM HYDRAZINIUM SULFATE. 32-3343
PROTON ROTATING FRAME RELAXATION IN LITHIUM HYDRAZINIUM SULFATE. 32-3347
HIUM I/ REFINEMENT OF THE CRYSTAL STRUCTURE OF FERROELECTRIC LITHIUM HYDROGEN SELENITE: POSITION OF THE LIT 22-2860
THERMAL STABILITY OF LITHIUM HYDROGEN SELENITES 22-2868
MMONIUM TARTRATE, LITHIUM T/ NUCLEAR QUADRUPOLE RESONANCE OF LITHIUM IN FERROELECTRIC COMPOUNDS. (LITHIUM A 02-0052
COEFFICIENT AND PHASE MATCHED SECOND HARMONIC GENERATION IN LITHIUM IODATE. LARGE NONLINEAR OPTICAL 08-1405
DIUM DIHYDROGEN PHOSPHATE, RUBIDIUM DIHYDROGEN ARSENATE, AND LITHIUM IODATE. /UM DIHYDROGEN PHOSPHATE, RUBI 08-1413
OPTICAL PARAMETRIC OSCILLATION IN LITHIUM IODATE. 08-1338
PIEZOELECTRIC AND PHOTOELASTIC PROPERTIES OF LITHIUM IODATE. 08-1465
PIEZOELECTRIC AND PHOTOELASTIC PROPERTIES OF LITHIUM IODATE. 08-1464
POLARITONS IN LITHIUM IODATE. 08-1294
ABSOLUTE STRUCTURE OF LITHIUM IODPTE CRYSTALS. 08-1290
DIELECTRIC PROPERTIES OF LITHIUM IODATE. (FOUND NOT FERROELECTRIC) 36-3605
/TINUOUS WAVE) SPONTANEOUS PARAMETRIC SCATTERING OF LIGHT IN LITHIUM IODATE (INDEX OF REFRACTION 1.2-7 MIC/ 08-1291
ELECTROOPTICAL PROPERTIES OF LITHIUM IODATE SINGLE CRYSTALS. 08-1423
PARAELECTRIC RESONANCE OF LITHIUM ION IN POTASSIUM CHLORIDE. 31-3325
MAGNETOELECTRIC EFFECT IN FERROELECTRIC- ANTIFERROMAGNETIC LITHIUM IRON(0.5) TANTALUM(0.5) OXY FLUORIDE. 36-3568
RE AND FREQUENCY DEPENDENCES OF THE ELECTRICAL PROPERTIES OF LITHIUM METANIOBATE. TEMPERATU 11-1717
DOMAIN STRUCTURE OF LITHIUM METANIOBATE CRYSTALS. 11-1708
CHIOMETRIES. PHASE MATCHING ANGLES AND TEMPERATURE OF LITHIUM METANIOBATE CRYSTALS OF DIFFERENT STOI 11-1665
REACTION. MECHANISM AND KINETICS OF LITHIUM METANIOBATE FORMATION BY A SOLID PHASE 11-1840
HEAT TREATMENT INDUCED POINT DEFECTS IN LITHIUM METANIOBATE SINGLE CRYSTALS. 11-1668
CONGRUENT MELTING COMPOSITION OF LITHIUM METATANTALATE. 12-1882
FERROELECTRIC DOMAIN REVERSAL IN LITHIUM METATANTALATE. 12-1869
ABSORPTION OF ELASTIC WAVES IN REDUCED LITHIUM NIOBATE. 11-1816
ND CORRELATION AND THE ASSOCIATED NONLINEARITY PARAMETERS IN LITHIUM NIOBATE. ACOUSTIC CONVOLUTION A 11-1758
LE AND QUADRUPOLE SATURATION OF THE LITHIUM-7 NMR SIGNALS IN LITHIUM NIOBATE. ACOUSTIC DIPO 11-1666
ACOUSTIC RADIATION BY INTERDIGITAL GRIDS ON LITHIUM NIOBATE. 11-1699
TION OF SURFACE WAVE STRUCTURE OF A CADMIUM SELENIDE FILM ON LITHIUM NIOBATE. ACOUSTOELECTRIC AMPLIFICA 11-1723
GENERATION IN PIEZOELECTRICS COUPLED TO A GUNN OSCILLATOR. (LITHIUM NIOBATE) /EASUREMENTS OF ACOUSTIC WAVE 11-1731
AN ELECTROPTIC SWITCH USING ROTATED Y PLATE OF LITHIUM NIOBATE. 11-1732
AN EPR INVESTIGATION OF IRON(3+) AND MANGANESE(2+) IN LITHIUM NIOBATE. 11-1725
APPARATUS FOR GROWTH OF SINGLE CRYSTAL, SINGLE DOMAIN LITHIUM NIOBATE. 11-1726
ASSIGNMENTS OF OPTICAL PHONON MODES IN LITHIUM NIOBATE. 11-1694
NCE OF CERTAIN CRYSTALS IN THE MILLIMETER WAVELENGTH RANGE. (LITHIUM NIOBATE) BIREFRINGE 11-1730
BRILLOUIN SCATTERING IN LITHIUM NIOBATE. 11-1786
TIC WAVE AND TWO LUMINOUS WAVES ALONG THE (100) DIRECTION OF LITHIUM NIOBATE. /BETWEEN A LONGITUDINAL ACOUS 11-1680
CONTROL OF LASER DAMAGE IN LITHIUM NIOBATE. 11-1716
VOLUTION OF SURFACE WAVES IN A STRUCTURE OF SEMICONDUCTOR ON LITHIUM NIOBATE. CON 11-1847
COUPLED WAVE ANALYSIS OF HOLOGRAPHIC STORAGE IN LITHIUM NIOBATE. 11-1826
CRYSTAL FIELD IN CHROMIUM(3+) DOPED LITHIUM NIOBATE. 11-1759
NONLINEAR PARAMETER FOR ACOUSTIC SURFACE WAVE CONVOLUTION IN LITHIUM NIOBATE. DETERMINATION OF A 11-1681
(BISMUTH FERRATE AND SYSTEMS BASED ON IT, AND SINGLE CRYSTAL LITHIUM NIOBATE) /TH HIGH CURIE TEMPERATURES. 08-1378
DIFFRACTION SPREADING OF SURFACE WAVES ON LITHIUM NIOBATE. 11-1849
DIRECTIONAL DISPERSION AND ASSIGNMENT OF OPTICAL PHONONS IN LITHIUM NIOBATE. 11-1695
DISPERSION OF PHONON FREQUENCIES AND POLARITON SCATTERING IN LITHIUM NIOBATE. DIRECTIONAL 11-1696
DISPERSION OF ELASTIC WAVES IN FERROELECTRICS. (LITHIUM NIOBATE) 11-1706
CAL INHOMOGENEITIES ON PHASE MATCHING IN NONLINEAR CRYSTALS (LITHIUM NIOBATE). EFFECT OF OPTI 11-1773
ELECTRON PARAMAGNETIC RESONANCE OF IRON(3+) DOPED LITHIUM NIOBATE. 11-1765
ELECTRON PARAMAGNETIC RESONANCE OF GADOLINIUM(3+) DOPED LITHIUM NIOBATE. 11-1833

```
                    ELECTRON SPIN RESONANCE STUDIES OF CHROMIUM(3+) IN LITHIUM NIOBATE.                                          11-1804
E STUDIES OF CRYSTAL FIELD PARAMETERS IN MANGANESE(2+) DOPED LITHIUM NIOBATE.       ELECTRON SPIN RESONANC 11-1803
                                  EPR OF GADOLINIUM(3+) IN LITHIUM NIOBATE.                                11-1763
        EPR STUDIES OF CRYSTAL FIELD PARAMETERS IN IRON(3+) DOPED LITHIUM NIOBATE.                                 11-1839
WAVES. APPLICATION OF ACOUSTIC WAVES TO HARMONIC GENERATION IN LITHIUM NIOBATE. /N BY HIGH FREQUENCY ACOUSTIC 11-1834
ION SPREADING OF SURFACE (ACOUSTIC) WAVES ON (PIEZOELECTRIC) LITHIUM NIOBATE.       (FAR FIELD) DIFFRACT 11-1850
C TUNGSTEN BRONZE TYPE CRYSTAL STRUCTURES. PART-3: POTASSIUM LITHIUM NIOBATE.       FERROELECTRI 13-1895
EIDEN HYPERFINE LINES IN ESR SPECTRA OF MANGANESE(2+) DOPED LITHIUM NIOBATE.       FOR 11-1805
               FREQUENCY DEPENDENCES OF SOME PARAMETERS OF LITHIUM NIOBATE.                                11-1843
WAVE PROPAGATION ON LAYERED ANISOTROPIC SUBSTRATES: GOLD ON LITHIUM NIOBATE. /ND HIGHER ORDER MODE SURFACE 11-1846
                                  GAMMA-RADIATION EFFECTS IN LITHIUM NIOBATE.                                11-1822
TION OF FAR INFRARED RADIATION BY PICOSECOND LIGHT PULSES IN LITHIUM NIOBATE.       GENERA 11-1857
E ACOUSTIC WAVE PROPAGATING ITSELF IN THE (001) DIRECTION OF LITHIUM NIOBATE. /IRD HARMONICS BY A TRANSVERS 11-1682
RULES FOR INELASTIC NEUTRON SCATTERING WITH APPLICATIONS TO LITHIUM NIOBATE. GROUP THEORETICAL SELECTION 11-1704
                                  HOLOGRAPHIC RECORDING IN LITHIUM NIOBATE.                                11-1662
INDEX OF A NONLINEAR CRYSTAL ON SECOND HARMONIC GENERATION. (LITHIUM NIOBATE) /MOGENEITY OF THE REFRACTIVE 11-1687
                      INTERACTION OF VOLUME ACOUSTIC MODES IN LITHIUM NIOBATE.                                11-1806
               INTERACTIONS BETWEEN HIGH FREQUENCY ACOUSTICAL WAVES IN LITHIUM NIOBATE.                                11-1836
SCATTERING BY POLARITONS ASSOCIATED WITH ORDINARY PHOTONS IN LITHIUM NIOBATE.       LIGHT 11-1697
        LONGITUDINAL ELECTROOPTICAL EFFECT IN OBLIQUE SECTIONS OF LITHIUM NIOBATE.                                11-1863
FREQUENCY (510 MHZ) PARAMAGNETIC RESONANCE INVESTIGATION OF LITHIUM NIOBATE.       LOW 11-1793
                                  LUMINESCENCE FROM LITHIUM NIOBATE.                                11-1729
CHARACTERISTICS OF INTERDIGITAL SURFACE WAVE TRANSDUCERS ON LITHIUM NIOBATE.       MEASURED ELECTRICAL 11-1702
                        MICROWAVE PHONON ATTUNUATION IN X CUT LITHIUM NIOBATE.                                11-1756
               MIXING OF NONCOLLINEAR ELASTIC SURFACE WAVES ON LITHIUM NIOBATE.                                11-1690
                   MULTIPLE STORAGE OF THICK PHASE HOLOGRAMS IN LITHIUM NIOBATE.                                11-1827
OF INTERNAL ELECTRIC FIELD DURING OPTICAL DAMAGE PROCESS IN LITHIUM NIOBATE.       NATURE 11-1861
H COUPLING LOW DIFFRACTION CUT FOR ACOUSTIC SURFACE WAVES ON LITHIUM NIOBATE.       NEW HIG 11-1820
COUPLING LOW BEAM STEERING CUT FOR ACOUSTIC SURFACE WAVES ON LITHIUM NIOBATE. NEW HIGH FREQUENCY HIGH 11-1819
               NIOBIUM-93 NMR LINE WIDTHS IN NONSTOICHIOMETRIC LITHIUM NIOBATE.                                11-1795
                      NONCOLLINEAR PHASE MATCHING EFFECTS IN LITHIUM NIOBATE.                                11-1675
                        NONLINEAR EFFECTS IN FERROCERAMICS. (PZT-8, LITHIUM NIOBATE)                             09-1546
LINEAR INTERACTION OF MICROWAVE ELECTRIC FIELDS AND SOUND IN LITHIUM NIOBATE.       NON 11-1838
               NONLINEAR OPTICAL PROPERTIES OF POTASSIUM LITHIUM NIOBATE.                             13-2027
LINEAR OPTICAL SPECTROSCOPY OF THIRD ORDER SUSCEPTIBILITY IN LITHIUM NIOBATE.       NON 11-1855
ROPAGATION OF RADIATION AND FIRST HARMONIC, CALCULATIONS FOR LITHIUM NIOBATE). /SADOVSKII EFFECT (THEORY, P 03-0186
               NONSTOICHIOMETRY AND CRYSTAL GROWTH OF LITHIUM NIOBATE.                                11-1691
               OBSERVATION OF ORDINARY POLARITONS IN LITHIUM NIOBATE.                                11-1854
AND HOLOGRAPHIC STORAGE PROPERTIES OF TRANSITION METAL DOPED LITHIUM NIOBATE.       OPTICAL 11-1796
CAL PROBING OF ACOUSTIC SURFACE WAVE HARMONIC GENERATION (IN LITHIUM NIOBATE)       OPTI 11-1752
               OPTIMUM SECOND HARMONIC GENERATION IN LITHIUM NIOBATE.                                11-1764
PARAMETRIC LUMINESCENCE AND LIGHT SCATTERING BY POLARITONS. (LITHIUM NIOBATE)                                11-1746
                      PIEZOELECTRIC SURFACE WAVES IN LITHIUM NIOBATE.                                11-1824
ND LINE WIDTH OF TUNABLE STIMULATED FAR INFRARED EMISSION IN LITHIUM NIOBATE.       POWER A 11-1736
N OF RAYLEIGH WAVES AND PIEZOELECTRIC LEAKY SURFACE WAVES IN LITHIUM NIOBATE. PROPAGATION AND AMPLIFICATIO 11-1856
E IRON TRANSITION GROUP IN THE LATTICE OF SINGLE CRYSTALLINE LITHIUM NIOBATE. PROPERTIES OF TH 11-1669
REFLECTION OF ELASTIC SURFACE WAVE FROM PERIODIC ARRAYS ON LITHIUM NIOBATE.                                11-1853
               ROTATING WAVEPLATE OPTICAL FREQUENCY SHIFTING IN LITHIUM NIOBATE.                                11-1689
TICAL PARAMETRIC OSCILLATION, SECOND HARMONIC GENERATION (IN LITHIUM NIOBATE). SIMULTANEOUS OP 11-1859
ESORPTION, AND DISPERSION IN THE NONLINEAR SUSCEPTIBILITY IN LITHIUM NIOBATE. /RAMETRIC SCATTERING, IDLER A 11-1771
                                  SURFACE WAVE DIFFRACTION IN LITHIUM NIOBATE.                                11-1698
OF THE POLARITON ASSOCIATED WITH THE 248 CM(-1) SOFT MODE IN LITHIUM NIOBATE. TEMPERATURE STUDY 11-1807
               THE MEASUREMENT OF HYPERSONIC ATTENUATION IN LITHIUM NIOBATE.                                11-1760
AL STUDY OF THE GENERATION OF THE ACOUSTIC SECOND HARMONIC. (LITHIUM NIOBATE)       THEORETIC 11-1772
                                  THERMALLY FIXED HOLOGRAMS IN LITHIUM NIOBATE.                                11-1828
THREE LAYER BROAD BAND ANTIREFLECTION COATINGS FOR LITHIUM NIOBATE.                                11-1676
                   VISUALIZATION OF ACOUSTIC RADIATION IN LITHIUM NIOBATE.                                11-1864
WAVELENGTH DEPENDENCE OF HOLOGRAPHIC PHENOMENA IN DOPED LITHIUM NIOBATE.                                11-1785
RAIN DEPENDENT CURIE TEMPERATURE, POLARIZATION FLUCTUATIONS, LITHIUM NIOBATE, ALKALI HALIDES) /RYSTALS. (ST 03-0513
TASSIUM DIHYDROGEN PHOSPHATE, AMMONIUM DIHYDROGEN PHOSPHATE, LITHIUM NIOBATE, AND ALPHA- IODIC ACID. /OF PO 11-1677
NGITUDINAL AND TRANSVERSE HYPERSONIC PULSES IN PIEZOELECTRIC LITHIUM NIOBATE AND BISMUTH BIGERMANATE. /F LO 11-1814
C MATERIALS, SINGLE CRYSTALS OF POTASSIUM TANTALATE NIOBATE, LITHIUM NIOBATE, AND BISMUTH TITANATE) /ROOPTI 02-0073
/PLING TO ELASTIC SURFACE WAVES ON PIEZOELECTRIC SUBSTRATES (LITHIUM NIOBATE AND BISMUTH(12) GERMANIUM OXI/ 11-1751
/IC SURFACE WAVE CONVOLUTION ON CRYSTALS OF CADMIUM SULFIDE, LITHIUM NIOBATE AND BISMUTH(12) GERMANIUM OXI/ 11-1692
/ON OF LONGITUDINAL AND TRANSVERSE HYPERSONIC WAVE PULSES IN LITHIUM NIOBATE AND BISMUTH(12) GERMANIUM OXI/ 11-1815
/ECTRON PARAMAGNETIC RESONANCE SPECTRUM IN (MANGANESE DOPED) LITHIUM NIOBATE AND (COPPER(2+) DOPED) ZINC B/ 11-1792
R AND SPECTROSCOPIC PROPERTIES OF ACTIVATED FERROELECTRICS. (LITHIUM NIOBATE AND GADOLINIUM MOLYBDATE) LASE 11-1741
S.              GROWTH OF CRATERS ON ION ETCHED SURFACE OF LITHIUM NIOBATE AND ION ETCHING WITHOUT CRATER 11-1860
NEODYMIUM YAG LASER IRRADIATION INDUCED DAMAGE TO LITHIUM NIOBATE AND KDP.                                11-1674
TRANSIENT ELASTOOPTIC EFFECTS AND Q SWITCHING PERFORMANCE IN LITHIUM NIOBATE AND KDP POCKELS CELLS.         11-1727
/ARMONICS OF 1.06 MICRON RADIATION IN BARIUM SODIUM NIOBATE, LITHIUM NIOBATE, AND LITHIUM IODATE CRYSTALS. 11-1779
ND 0.58) (CORRELATION OF CATION SHIFTS WITH BISMUTH FERRATE, LITHIUM NIOBATE, AND LITHIUM TANTALATE) /=0.9 A 09-1574
AND OPTICAL QUALITY OF FERROELECTRIC CRYSTALS, ESPECIALLY IN LITHIUM NIOBATE AND LITHIUM TANTALATE. /ITION 11-1780
NHOMOGENEITY OF THE OPTICALLY INDUCED INDEX OF REFRACTION IN LITHIUM NIOBATE AND LITHIUM TANTALATE.       I 11-1664
               MEDIUM FREQUENCY RESONATORS OF LITHIUM NIOBATE AND LITHIUM TANTALATE.                                11-1851
        SCATTERING OF LIGHT ON ACOUSTIC PHONONS IN LITHIUM NIOBATE AND LITHIUM TANTALATE.                                11-1744
POWDER SPECTROSCOPY.        STOICHIOMETRY VARIATIONS IN LITHIUM NIOBATE AND LITHIUM TANTALATE BY RAMAN 11-1813
/ARACTERISTIC EQUATION FOR PLANE PIEZOELECTRIC VIBRATIONS OF LITHIUM NIOBATE AND LITHIUM TANTALATE CRYSTAL/ 11-1739
               GROWTH AND PROPERTIES OF LITHIUM NIOBATE AND LITHIUM TANTALATE FILMS.                                11-1712
RYSTALS.        X-RAY TOPOGRAPHIC STUDIES ON LITHIUM NIOBATE AND LITHIUM TANTALATE SINGLE C 11-1830
THOR'S STATIST/ COMPARISON OF FERROELECTRICITY IN ISOMORPHIC LITHIUM NIOBATE AND LITHIUM TANTALATE USING AU 03-0366
ORS.              (MECHANISM OF) OPTICAL INDEX DAMAGE IN LITHIUM NIOBATE AND OTHER PYROELECTRIC INSULAT 11-1735
YLEIGH ULTRASONIC WAVES IN A LAYERED STRUCTURE CONSISTING OF LITHIUM NIOBATE AND P-TYPE SILICON. /ION OF RA 11-1747
/ANSIENT ELECTROOPTIC EFFECTS AND Q SWITCHING PERFORMANCE IN LITHIUM NIOBATE AND POTASSIUM DIDEUTERIUM PHO/ 11-1710
/PULLING FROM THE MELT, STRUCTURE- PROPERTY CORRELATION WITH LITHIUM NIOBATE AND POTASSIUM LITHIUM NIOBATE) 13-2045
MASS LOADING ON THE PROPAGATION OF ACOUSTIC SURFACE WAVES ON LITHIUM NIOBATE AND QUARTZ.       EFFECT OF 11-1757
WAVES FOR SIGNAL PROCESSING PURPOSES. (TECHNIQUE APPLIED TO LITHIUM NIOBATE AND QUARTZ) /F SURFACE ELASTIC 11-1738
RATURE DEPENDENCE OF THE OPTICAL PROPERTIES OF FERROELECTRIC LITHIUM NIOBATE AND TANTALATE.       TEMPE 11-1770
N THE / STACKING FAULT MODEL FOR STOICHIOMETRY DEVIATIONS IN LITHIUM NIOBATE AND TANTALATE AND THE EFFECT O 11-1776
N THE / STACKING FAULT MODEL FOR STOICHIOMETRY DEVIATIONS IN LITHIUM NIOBATE AND TANTALATE AND THE EFFECT O 11-1777
```

ME PROPERTIES OF THE EQUATIONS OF PIEZOELECTRIC VIBRATION OF LITHIUM NIOBATE AND TANTALATE CRYSTALS.   SO 11-1740
CTURAL TRANSITION)   ELECTRICAL PROPERTIES OF LITHIUM NIOBATE. (ANOMALY AT 263 DEG K IS STRU 11-1709
R REVERSIBLE HOLOGRAPHIC STORAGE.   RH DOPED LITHIUM NIOBATE AS AN IMPROVED NEW MATERIAL FO 08-1359
/ARITY SENSE BY ANOMALOUS SCATTERING AT AN ABSORPTION EDGE. (LITHIUM NIOBATE, BARIUM MANGANESE TETRAFLUORI/ 11-1673
         HOLOGRAPHIC PATTERN FIXING IN ELECTROOPTIC CRYSTALS. (LITHIUM NIOBATE, BARIUM SODIUM NIOBATE) 11-1661
IUM SODIUM TANTALATE (FORCE CONSTANT/ VIBRATIONAL SPECTRA OF LITHIUM NIOBATE, BARIUM SODIUM NIOBATE AND BAR 11-1808
/NG SURFACE WAVES BY NONLINEAR MIXING IN ANISOTROPIC MEDIA. (LITHIUM NIOBATE, BISMUTH(12) GERMANIUM OXIDE(/ 11-1778
                   DAMAGE TO THE SURFACE OF LITHIUM NIOBATE BY LIGHT.   11-1866
R.   FRACTURE OF NONLINEAR CRYSTALS (KDP AND LITHIUM NIOBATE) BY RADIATION FROM A RUBY LASE 11-1663
Y.   DISPLAY OF TWINNING IN LITHIUM NIOBATE BY X-RAY DIFFRACTION TOPOGRAPH 11-1845
, ESR, CELL PARAMETERS)   RADIATION DAMAGE IN LITHIUM NIOBATE. (CAPACITANCE, OPTICAL SPECTRA 11-1705
/H AND PROPERTIES OF SINGLE DOMAIN CRYSTALS OF FERROELECTRIC LITHIUM NIOBATE. (CRACK-FREE, THROUGH ADDITIO/ 11-1774
                BIREFRINGENCE INDUCED IN A LITHIUM NIOBATE CRYSTAL.   11-1767
FERMI RESONANCE OF POLARITON WITH BIPHONON IN AN LITHIUM NIOBATE CRYSTAL.   11-1761
NONLINEAR INTERACTION OF HYPERSONIC WAVES IN A LITHIUM NIOBATE CRYSTAL.   11-1802
PARAMETRIC LUMINESCENCE INTENSITY IN THE LITHIUM NIOBATE CRYSTAL.   11-1750
/HOTON CORRELATIONS OF SECOND HARMONIC GENERATED LIGHT (FROM LITHIUM NIOBATE CRYSTAL, AGREEMENT WITH THEOR/ 11-1701
        RADIATION EXPOSURE OF A LITHIUM NIOBATE CRYSTAL AT HIGH TEMPERATURES.   11-1721
/LY UNIFORM AND ALTERABLE SHG PHASE MATCHING TEMPERATURES IN LITHIUM NIOBATE (CRYSTAL GROWTH BY ADDING MAG/ 11-1684
/SUPERHIGH FREQUENCY ACOUSTIC HARMONICS IN A (SINGLE DOMAIN) LITHIUM NIOBATE CRYSTAL (NONLINEAR ELASTIC PR/ 11-1707
ITTIVITY)   TRANSVERSE POLARITONS IN A LITHIUM NIOBATE CRYSTAL. (RAMAN SPECTRUM, PERM 11-1762
RA AND ENERGY LEVELS OF NEODYMIUM(3+) AND ERBIUM(3+) IONS IN LITHIUM NIOBATE CRYSTALS. / LUMINESCENCE SPECT 11-1713
COLLINEAR INTERACTION OF LONGITUDINAL ELASTIC WAVES IN LITHIUM NIOBATE CRYSTALS.   11-1678
                           DISORDER OF LITHIUM NIOBATE CRYSTALS.   11-1679
INVESTIGATION OF A SYSTEM FOR BEAM DEFLECTION BY MEANS OF LITHIUM NIOBATE CRYSTALS.   11-1768
COMPOSITION (STOICHIOMETRY) DEPENDENCE OF POCKELS EFFECT IN LITHIUM NIOBATE CRYSTALS.   MELT 11-1728
                  PHOTOELASTIC CONSTANTS OF LITHIUM NIOBATE CRYSTALS.   11-1745
ERING BY POLARITONS. PART-2: THE CASE OF CADMIUM SULFIDE AND LITHIUM NIOBATE CRYSTALS. /TANEOUS RAMAN SCATT 11-1787
+) IONS (10 TO 3/ CALCULATION OF THE (ABSORPTION) SPECTRA OF LITHIUM NIOBATE CRYSTALS CONTAINING CHROMIUM(3 11-1831
       GROWTH TWINNING (UNDER DIRECT CURRENT) IN C-AXIS LITHIUM NIOBATE CRYSTALS (CZOCHRALSKI METHOD). 11-1672
ENTS.   PRECISION MACHINING OF LITHIUM NIOBATE CRYSTALS FOR ELECTROOPTIC ELEM 11-1742
WITH LOW BIREFRINGENCE VARIATIONS / GROWTH OF HIGH QUALITY LITHIUM NIOBATE CRYSTALS (FOR NONLINEAR OPTICS 11-1688
NT)   LIGHT SCATTERING BY ULTRASONIC WAVES IN LITHIUM NIOBATE CRYSTALS. (PHOTOELASTIC CONSTA 11-1754
DULATORS WITH A LOW HALF WAVE VOLTAGE.   CUTS OF LITHIUM NIOBATE CRYSTALS SUITABLE FOR LIGHT MO 11-1715
000 MHZ)   GENERATION AND PROPAGATION OF HYPERSONIC WAVES IN LITHIUM NIOBATE CRYSTALS. (80-950 DEG K, 200-2 11-1753
            PIEZOELECTRIC LEAKY SURFACE WAVE IN LITHIUM NIOBATE (ELECTROMAGNETIC COUPLING).   11-1832
COMBINATION SCATTERING OF SOUND IN PIEZOELECTRICS. (THEORY, LITHIUM NIOBATE EXAMPLE)   STIMULATED 03-0431
/ TYPE STRUCTURE. (MIXED OXIDES ABC(3)) (BARIUM TITANATE AND LITHIUM NIOBATE FILMS ON PLATINUM OR SILICON) 04-0914
ION ETCHING.   THINNING OF LITHIUM NIOBATE FOR ACOUSTOOPTIC DEFLECTORS BY 11-1862
) AND LONGITUDI/ MANDELSTAM BRILLOUIN SCATTERING OF LIGHT IN LITHIUM NIOBATE (FROM TRANSVERSE (MORE INTENSE 11-1743
ESISTANT TO L/ NEW (UNIAXIAL) NONLINEAR OPTIC CRYSTAL BARIUM LITHIUM NIOBATE HAVING NO MICROTWINNING (AND R 13-1958
       ELECTROOPTIC EFFECT IN LITHIUM NIOBATE IN THE (SUB) MILLIMETER RANGE.   11-1844
A THEORETICAL AND E/ BISMUTH(12) GERMANIUM OXIDE(20) VERSUS LITHIUM NIOBATE IN THE SURFACE WAVE AMPLIFIER. 11-1720
-300 K.   THERMAL EXPANSION OF LITHIUM NIOBATE IN THE TEMPERATURE RANGE OF 70 11-1685
/MPERATURE DEPENDENCE OF HYPERSONIC WAVE DAMPING IN RUBY AND LITHIUM NIOBATE IN THE THREE CENTIMETER RADIO/ 11-1718
T AT FERROELECTRIC DOMAIN BOUNDARIES (DOPED AND UNDOPED TGS, LITHIUM NIOBATE, KDP, ABSTRACT ONLY). /CONTRAS 11-1671
                   LITHIUM NIOBATE LIGHT MODULATOR.   11-1782
EQUENCY (950 MHZ) ACOUSTIC WAVE ALONG THE (001) DIRECTION OF LITHIUM NIOBATE (LIQUID HELIUM TEMPERATURE). / 11-1835
/S. (BARIUM SODIUM NIOBATE, BISMUTH(12) GERMANIUM OXIDE(20), LITHIUM NIOBATE, LITHIUM TANTALATE, AND OTHER/ 01-0028
/LINEAR OPTICAL POLARIZABILITY OF THE NIOBIUM- OXYGEN BOND. (LITHIUM NIOBATE, LITHIUM TANTALATE, POTASSIU/ 11-1734
ONLINEAR CRYSTALS. (KDP, ADP, RUBIDIUM DIHYDROGEN PHOSPHATE, LITHIUM NIOBATE, LITHIUM TANTALATE, QUARTZ/ /N 11-1865
/COEFFICIENTS OF ACOUSTIC SURFACE WAVE VELOCITY AND DELAY ON LITHIUM NIOBATE, LITHIUM TANTALATE, QUARTZ, A/ 11-1821
      SECONDARY (ELECTRO) PHOTOELASTIC EFFECT IN LITHIUM NIOBATE (ON ELECTRON IRRADIATION).   11-1801
SON RESULTS FOR SAPPH/ EXPANSION, CRAZING AND EXFOLIATION OF LITHIUM NIOBATE ON ION BOMBARDMENT AND COMPARI 11-1800
DEPENDENCE OF SECOND HARMONIC GENERATION COEFFICIENTS OF LITHIUM NIOBATE ON MELT COMPOSITION.   11-1769
/SIUM DIDEUTERIUM PHOSPHATE, POTASSIUM TANTALATE OR NIOBATE, LITHIUM NIOBATE OR TANTALATE, BARIUM SODIUM N/ 01-0021
/ SIGNS OF NONLINEAR OPTICAL COEFFICIENTS OF POLAR CRYSTALS (LITHIUM NIOBATE OR TANTALATE, BARIUM TITANATE/ 02-0098
/ GENERATION) COEFFICIENTS OF POLAR (PYROELECTRIC) CRYSTALS (LITHIUM NIOBATE OR TANTALATE OR GALLATE, BARI/ 02-0095
/QUENCIES AND EIGENVECTORS OF POLARITONS WITH APPLICATION TO LITHIUM NIOBATE. PART-2: POLARITONS OF SMALL / 11-1766
       TEMPERATURE DEPENDENCE OF RESONANT FREQUENCIES OF LITHIUM NIOBATE PLATE RESONATORS. (COMMENTS) 11-1812
/ WITH POTASSIUM TUNGSTEN BRONZE STRUCTURE (POTASSIUM SODIUM LITHIUM NIOBATE, POTASSIUM STRONTIUM LITHIUM / 13-1923
/MULATED CURRENT IN FERROELECTRIC OXIDE CRYSTALS. (POTASSIUM LITHIUM NIOBATE, POTASSIUM STRONTIUM NIOBATE) 13-1926
.   OPTICALLY INDUCED PHYSICAL DAMAGE TO LITHIUM NIOBATE, PROUSTITE, AND LITHIUM IODATE 08-1328
RE SENSITIVE CAPACITANCES.   LOW TEMPERATURE COEFFICIENT OF LITHIUM NIOBATE RESONATORS BY USE OF TEMPERATU 11-1670
MEASUREMENT OF SPUTTER ETCHING RATES FOR MONOCRYSTALLINE LITHIUM NIOBATE, SILICON DIOXIDE, AND SILICON. 11-1714
R.   LITHIUM NIOBATE SILICON SURFACE WAVE CONVOLUTE 11-1788
               LAPPING CHARACTERISTICS OF LITHIUM NIOBATE SINGLE CRYSTAL.   11-1783
DETERMINATION OF ELASTIC AND PIEZOELECTRIC CONSTANTS OF LITHIUM NIOBATE SINGLE CRYSTALS.   11-1749
             DOMAIN STRUCTURE IN LITHIUM NIOBATE SINGLE CRYSTALS.   11-1703
          ELECTRICAL CONDUCTIVITY OF LITHIUM NIOBATE SINGLE CRYSTALS.   11-1733
             LAPPING CHARACTERISTICS OF LITHIUM NIOBATE SINGLE CRYSTALS.   11-1784
F THE ABSORPTION OF 9.4 GHZ HYPERSOUND IN ALUMINUM OXIDE AND LITHIUM NIOBATE SINGLE CRYSTALS. /EPENDENCES O 11-1719
ENCE OF DIELECTRIC, PIEZOELECTRIC, AND ELASTIC PROPERTIES OF LITHIUM NIOBATE SINGLE CRYSTALS. /ERMAL DEPEND 11-1693
               INTERNAL FRICTION OF LITHIUM NIOBATE SINGLE CRYSTALS AT 1-40 HZ.   11-1798
LSKI METHOD.   GROWTH OF BARIUM LITHIUM NIOBATE SINGLE CRYSTALS BY THE CZOCHRA 13-1991
MODULATION.   POINT DEFECTS IN LITHIUM NIOBATE SINGLE CRYSTALS USED IN LASER 11-1737
NG OF LASER RADIATIO/ INVESTIGATION OF CERTAIN PARAMETERS OF LITHIUM NIOBATE SINGLE CRYSTALS USING SCATTERI 11-1797
          GROWTH OF LITHIUM NIOBATE SINGLE DOMAIN CRYSTALS.   11-1781
LAYER WAVE AMPLIFICATION IN A CADMIUM SULFIDE FLUID- LITHIUM NIOBATE STRUCTURE.   11-1667
SURFACE WAVE CONVOLUTION USING A CADMIUM SULFIDE FLUID LITHIUM NIOBATE STRUCTURE.   11-1700
TION OF ULTRASONIC SURFACE WAVES BY THIN GOLD FILM LAYERS ON LITHIUM NIOBATE SUBSTRATES. /ACTION AND REFLEC 11-1789
TION OF ULTRASONIC SURFACE WAVES BY THIN GOLD FILM LAYERS ON LITHIUM NIOBATE SUBSTRATES. /ACTION AND REFLEC 11-1790
LOW POWER DENSITY GALLIUM ARSENIDE- LITHIUM NIOBATE SURFACE WAVE AMPLIFIER.   11-1825
NONLINEAR EFFECTS IN MICROWAVE ACOUSTIC LITHIUM NIOBATE SURFACE WAVE DELAY LINES.   11-1818
DECORATION OF ION BOMBARDED LITHIUM NIOBATE SURFACES.   11-1799
M.   NIOBIUM-93 NMR STUDY OF THE LITHIUM NIOBATE TANTALATE SOLID SOLUTION SYSTE 11-1794
/ICROWAVE FREQUENCY ACOUSTIC SURFACE WAVE LOSS MECHANISMS ON LITHIUM NIOBATE (TEMPERATURE AND FREQUENCY DE/ 11-1817
/LEAR QUADRUPOLE RESONANCE INVESTIGATION OF (STOICHIOMETRIC) LITHIUM NIOBATE (TEMPERATURE DEPENDENCE OF CO/ 11-1811
DEX CHANGES.   CONTROL OF THE SUSCEPTIBILITY OF LITHIUM NIOBATE TO LASER INDUCED REFRACTIVE IN 11-1791
        OPTIMUM CUT FOR A LITHIUM NIOBATE TRANSVERSE LIGHT MODULATOR.   11-1755

ECTRICS.   SUPERLATTICE STRUCTURE OF STRONTIUM SODIUM LITHIUM NIOBATE (TUNGSTEN BRONZE) TYPE FERROEL 13-2002
ECTRICS).   ELECTROOPTICAL PROPERTIES OF STRONTIUM POTASSIUM LITHIUM NIOBATE. (TUNGSTEN BRONZE TYPE FERROEL 13-2051
LASTIC AND PIEZOELECTRIC CONSTANTS OF STRONTIUM(4) POTASSIUM LITHIUM NIOBATE TYPE FERROELECTRIC CRYSTAL. E 13-1946
YSTAL).   ELECTROOPTIC PROPERTIES OF STRONTIUM LITHIUM NIOBATE TYPE FERROELECTRICS (SINGLE CR 13-2062
D DIELECTRIC CONSTANT ON MELT COMPOSITION (STOICHIOMETRY) IN LITHIUM NIOBATE (VERY LITTLE EFFECT). /FECT AN 11-1842
                   EFFECT TUNABLE OPTICAL EMISSION FROM LITHIUM NIOBATE WITHOUT A RESONATOR.   11-1858
           (BACKWARD WAVE) ACOUSTIC PARAMETRIC OSCILLATIONS IN LITHIUM NIOBATE (60 DB GAIN).   11-1837
CRYSTAL GROWTH AND NONLINEAR OPTICAL PROPERTIES OF POTASSIUM LITHIUM NIOBATES.   13-2015
GROWTH, STRUCTURE, AND PROPERTIES. (REVIEW OF WORK TO SEP 1/ LITHIUM NIOBATE- A NEW TYPE OF FERROELECTRIC: 11-1775
TIES OF TUNGSTEN BRONZE CRYSTALS FROM THE POTASSIUM NIOBATE- LITHIUM NIOBATE- BARIUM DINIOBATE SYSTEM. /PER 13-2003
/ELD IN THE (OXIDE) SYSTEM POTASSIA LITHIA NIOBIA (POTASSIUM LITHIUM NIOBATES, PHASE RELATIONSHIPS, SOLID / 13-2017
          TRANSVERSE ACOUSTOELECTRIC EFFECT IN A LAYERED LITHIUM NIOBATE- SILICON STRUCTURE.   11-1748
STEN BRONZE STRUC/ NONLINEAR OPTICAL PROPERTIES OF POTASSIUM LITHIUM NIOBATES (TRANSPARENT, TETRAGONAL TUNG 13-2026
IUM NIOBATE.   TEMPERATURE DEPENDENCE OF THE LITHIUM NMR SPECTRUM AND ATOMIC MOTION IN LITH 11-1722
EAR MAGNETIC RESONANC/ REVERSAL MECHANISM IN FERROELECTRICS: LITHIUM OR SODIUM TRIHYDROGEN DISELENITE, NUCL 22-2837
TCHING TE/ DETERMINATION OF THE DISTRIBUTION COEFFICIENTS OF LITHIUM OXIDE INTO LITHIUM NIOBATE BY PHASE MA 11-1841
FERROELECTRIC MATERIALS. (BARIUM TITANATE, LEAD / EFFECT OF LITHIUM OXIDE ON THE PROPERTIES OF A SERIES OF 04-0780
          PHASE EQUILIBRIA AND CRYSTAL GROWTH IN LITHIUM OXIDE- NIOBIUM PENTOXIDE SYSTEM.   11-1683
SSIUM LIT/ STUDY OF SUBSOLIDUS EQUILIBRIA IN POTASSIUM OXIDE- LITHIUM OXIDE- NIOBIUM PENTOXIDE SYSTEM. (POTA 13-1963
          FERROELECTRICS IN THE LITHIUM POTASSIUM NIOBATE SYSTEM.   11-1660
REFINEMENT OF THE CRYSTAL STRUCTURE OF FERROELECTRIC ACID LITHIUM SELENITE, POSITION OF THE LITHIUM ION. 22-2859
IN ALKALI NITRITES (SODIUM, POTASSIUM) AND ALKALI NITRATES. (LITHIUM, SODIUM, RUBIDIUM, CESIUM) /PERATURES 16-2399
/ION STUDY OF SINGLE CRYSTALS OF A NEW PYROCHLORE CONTAINING LITHIUM, STRONTIUM LANTHANUM TANTALUM OXYFLUO/ 36-3613
N DETECTOR. (TRIGLYCINE SULFATE, TRIGLYCINE FLUOROBERYLLATE, LITHIUM SULFATE, BARIUM STRONTIUM NIOBATE) /IO 02-0046
          CURIE TEMPERATURE OF FERROELECTRIC LITHIUM TANTALATE.   12-1887
          INTERACTION OF TWO HYPERSOUND WAVES IN LITHIUM TANTALATE.   12-1891
ATURE OF THE FERROELECTRIC- PARAELECTRIC PHASE TRANSITION IN LITHIUM TANTALATE.   N 12-1879
CRYSTAL FOR COMMUNICATIONS SYSTEMS ENTERS PRODUCTION STAGE. (LITHIUM TANTALATE)   NEW 12-1867
PERFORMANCE OF AN 11 GHZ OPTICAL MODULATOR USING LITHIUM TANTALATE.   12-1888
SCATTERING OF LIGHT BY ACOUSTIC PHONONS AND POLARITONS IN LITHIUM TANTALATE.   12-1878
SELF INDUCED THERMAL EFFECTS ON LIGHT EXTINCTION OF LITHIUM TANTALATE.   12-1889
          SIGNS OF THE ELECTROOPTIC COEFFICIENTS FOR LITHIUM TANTALATE.   12-1880
STUDY OF PIEZOELECTRIC OSCILLATIONS IN WIDEBAND PYROELECTRIC LITHIUM TANTALATE.   12-1875
          SURFACE WAVE TEMPERATURE COEFFICIENTS ON LITHIUM TANTALATE.   11-1852
TEMPERATURE DEPENDENCE OF SURFACE ACOUSTIC WAVE VELOCITY IN LITHIUM TANTALATE.   12-1886
TWO CRYSTAL RESONATOR FREQUENCIES: APPLICATION TO QUARTZ AND LITHIUM TANTALATE. /MPENSATION BY ADDITION OF 11-1686
E OF THE ELASTIC, PIEZOELECTRIC, AND DIELECTRIC CONSTANTS OF LITHIUM TANTALATE AND LITHIUM NIOBATE. /ENDENC 11-1823
NATION OF THE SPIN LATTICE RELAXATION TIMES IN FERROELECTRIC LITHIUM TANTALATE AND LITHIUM NIOBATE. /ETERMI 11-1829
CTRIC RESONATORS IN THE MEDIUM FREQUENCY RANGE WITH LOW RAT/ LITHIUM TANTALATE AND LITHIUM NIOBATE PIEZOELE 11-1724
          PIEZOELECTRIC AND ELASTIC PROPERTIES OF LITHIUM TANTALATE CHARACTERISTICS.   12-1893
          OPTICAL BEAM STEERING USING A MULTICHANNEL LITHIUM TANTALATE CRYSTAL.   12-1881
IN THICKNESS SHEAR MODE / TEMPERATURE DEPENDENCE OF AN X CUT LITHIUM TANTALATE CRYSTAL RESONATOR VIBRATING 12-1868
Y (20-10,000 PER CM, 300 DEG K) OF THE LATTICE VIBRATIONS IN LITHIUM TANTALATE (DIELECTRIC CONSTANTS). /TUD 12-1870
OELECTRIC / SINGLE CRYSTAL GROWTH AND PHYSICAL PROPERTIES OF LITHIUM TANTALATE. (DIELECTRIC CONSTANTS, PIEZ 12-1876
          ANOMALOUS (LONGITUDINAL WAVE) ATTENUATION IN LITHIUM TANTALATE (EFFECT OF VACANCIES).   12-1892
CORRECTIONS FOR IONIC MOTION IN DISPLACEMENT FERROELECTRICS (LITHIUM TANTALATE EXAMPLE).   INTERCELL 03-0367
ZERO TEMPERATURE COEFFICIENT OF RESONANT FREQUENCY IN LITHIUM TANTALATE LENGTH EXPANDER BARS.   11-1810
F CADMIUM TITANATE. (SOLID SOLUTIONS WITH STRONTIUM TITANATE LITHIUM TANTALATE, LITHIUM NIOBATE) /ICATION O 08-1424
/COMMUNICATIONS (REVIEW EMPHASIZING ELECTROOPTIC MODULATION, LITHIUM TANTALATE, LITHIUM NIOBATE, KDP, 113 / 02-0056
/ OF MICROWAVE NONLINEAR SUSCEPTIBILITIES. (BARIUM TITANATE, LITHIUM TANTALATE, LITHIUM NIOBATE, LITHIUM I/ 02-0107
PERATURE) TEMPERATURE DEPENDENCE OF POLARITON DISPERSION IN LITHIUM TANTALATE. (OPTICAL MODULATION VIA TEM 12-1873
DEFLECTION CAPACITIES OF A DIGITAL LIGHT DEFLECTOR UTILIZING LITHIUM TANTALATE POLARIZATION SWITCHES. /AND 11-1809
E POINT. PART-1: PIEZOELECTRIC AND OPTICAL PROP/ BEHAVIOR OF LITHIUM TANTALATE SINGLE CRYSTAL NEAR ITS CURI 12-1877
E POINT, PART-2, DIELECTRIC AND ULTRASONIC PROP/ BEHAVIOR OF LITHIUM TANTALATE SINGLE CRYSTAL NEAR ITS CURI 12-1885
(VERSUS MELT COMPOSITION, LATTICE AND CURIE TEMPERATURE ME/ LITHIUM TANTALATE SINGLE CRYSTAL STOICHIOMETRY 12-1871
          STOICHIOMETRY AND OPTICAL QUALITY OF LITHIUM TANTALATE SINGLE CRYSTALS.   12-1883
N OF MELT COMP/ ELECTROOPTIC AND FERROELECTRIC PROPERTIES OF LITHIUM TANTALATE SINGLE CRYSTALS AS A FUNCTIO 12-1874
AL MODULATOR.   FABRICATION OF A LITHIUM TANTALATE TEMPERATURE STABILIZED OPTIC 12-1872
CONDARY PYROELECTRIC EFFECTS AND SPONTANEOUS POLARIZATION IN LITHIUM TANTALATE (THERMAL EXPANSION). /AND SE 12-1884
          GROWTH AND PROPERTIES OF FERROELECTRIC POTASSIUM LITHIUM TANTALATE(X) NIOBATE(1-X).   11-1711
N THE TUNGSTEN BRONZE TYPE) FERROELECTRIC (SYSTEM) POTASSIUM LITHIUM TANTALATE(X) NIOBATE(1-X). /CRYSTALS I 13-1947
CRYSTA/ PIEZOELECTRIC PROPERTIES OF FERROELECTRIC POTASSIUM LITHIUM TANTALATE(0-0.5) NIOBATE(1-0.5) SINGLE 12-1890
IUM TANTALUM OXIDE, SODIUM CALCIUM TANTALUM OXIDE, POTASSIUM LITHIUM TANTALUM OXIDE. /OXIDE, POTASSIUM CALC 13-2050
DIELECTRIC PROPERTIES AND FERROELECTRIC PHASE TRANSITION ON LITHIUM THALLATE TARTRATE. /ING EFFECTS ON THE 29-3251
          SOME PROPERTIES OF THE FERROELECTRIC LITHIUM THALLIUM TARTRATE.   29-3242
TIC PRESSURE.   DISPLACEMENT OF THE CURIE POINTS IN KDP AND LITHIUM THALLIUM TARTRATE CRYSTALS BY HYDROSTA 17-2523
ELECTRIC/ CONTROL OF THE ELASTIC COMPLIANCE OF FERROELECTRIC LITHIUM THALLIUM TARTRATE HYDRATE (LTT) BY AN 29-3249
/ OR RUBIDIUM DIHYDROGEN PHOSPHATE, SODIUM, POTASSIUM(2), OR LITHIUM THALLIUM TARTRATE, LITHIUM AMMONIUM T/ 02-0089
          CRYSTAL DATA FOR LITHIUM THALLIUM TARTRATE MONOHYDRATE.   29-3247
ON LATTICE/ SOME PROPERTIES OF NONDEUTERATED AND DEUTERATED LITHIUM THALLIUM TARTRATE MONOHYDRATE. (EFFECT 29-3239
FERROELECTRICITY AND CONDUCTION IN FERROELECTRIC CRYSTALS. (LITHIUM THALLIUM TARTRATE MONOHYDRATE, ELECTR/ 02-0060
/ND CONDUCTION IN FERROELECTRIC CRYSTALS. (BISMUTH TITANATE, LITHIUM THALLIUM TARTRATE MONOHYDRATE, LEAD(5/ 02-0061
VICES). ELECTRIC FIELD CONTROL OF THE ELASTIC COMPLIANCE IN LITHIUM THALLIUM TARTRATE (TUNING AND DELAY DE 29-3250
ESONANCE IN GAMMA-IRRADIATED SINGLE CRYSTAL OF FERROELECTRIC LITHIUM TRIHYDROGEN DISELENITE. /ARAMAGNETIC R 22-2842
          HEAT OF FORMATION OF LITHIUM TRIHYDROGEN DISELENITE.   22-2867
SPECIES IN GAMMA-IRRADIATED SINGLE CRYSTAL OF FERROELECTRIC LITHIUM TRIHYDROGEN DISELENITE. PARAMAGNETIC 22-2843
          COERCIVE FIELD IN LITHIUM TRIHYDROGEN DISELENITE. (EXPERIMENTAL) 22-2844
THERMAL EXPANSION OF HYDROGEN BONDED FERROELECTRICS. LITHIUM TRIHYDROGEN SELENITE.   22-2857
/. PART-5A: A NEUTRON DIFFRACTION STUDY OF THE FERROELECTRIC LITHIUM TRIHYDROGEN SELENITE. (CRYSTAL STRUCT/ 22-2878
X-RAY AND THERMAL EXPANSION STUDY OF A SODIUM(0.88) LITHIUM(0.12) NIOBATE CERAMIC.   13-1999
          SOLUBILITY LIMIT OF LANTHANUM(0.5) LITHIUM(0.5) TITANATE IN BARIUM TITANATE.   04-0889
MIDE (NOT EXAMINED FOR FERROELECTR/ NEW BORACITES OF FORMULA LITHIUM(4) BORON(7) OXYGEN(12) CHLORIDE OR BRO 14-2089
AZINIUM SULFATE.   DIPOLAR RELAXATION OF LITHIUM-7 BY HINDERED ROTATORS IN LITHIUM HYDR 32-3350
ACOUSTIC DIPOLE AND QUADRUPOLE SATURATION OF THE LITHIUM-7 NMR SIGNALS IN LITHIUM NIOBATE. 11-1666
(HYSTERESIS LOOP CONSTRICTIONS, ROCHELLE SALT, TGS)   SHORT LIVED FERROELECTRIC "AFTER EFFECT" PHENOMENA. 27-3156
WAVES ON LITHIUM NIOBATE AND QUARTZ.   EFFECT OF MASS LOADING ON THE PROPAGATION OF ACOUSTIC SURFACE 11-1757
ELECTRICS.   LOCAL ELECTRON STATES IN DOMAIN WALLS OF FERRO 03-0161
DEG K IN S/ PHOTOCONDUCTIVITY ANOMALY AS AN EVIDENCE OF THE LOCAL FERROELECTRIC PHASE TRANSITION AROUND 47 06-1249
THE CRITICAL TEMPERATURE.   CRITICAL ASYMMETRY IN LOCAL FLUCTUATIONS IN STRONTIUM TITANATE ABOVE 06-1166
/RY OF HYDROGEN BONDED FERROELECTRICS. PART-1: (FORMALISM OF LOCAL MODE MODEL.) PART-2: (DYNAMICAL PROPERT/ 03-0384

CONDUCTOR FERROELECTRICS.   DOMAIN STRUCTURE AND LOCAL STATES IN ANTIMONY SULFOIODIDE TYPE SEMI 15-2127
/HEORY OF PEROVSKITE FERROELECTRICS (HAMILTONIAN IN TERMS OF LOCALIZED D-STRAIN AND SOFT- NORMAL MODE COOR/ 03-0432
/TRIC SPECTRUM OF BARIUM TITANATE. (AND MICROSCOPIC MODEL OF LOCALLY ORDERED "NUCLEI", PHASE STRUCTURE ABO/ 04-0647
AMETRIC LIGHT AMPLIFICATION AND OSCILLATION IN KDP WITH MODE LOCKED PUMP.   PAR 17-2468
TORS OF POLARITONS WITH APPLICATION TO LEAD NIOBATE. PART-1: LONG OPTICAL PHONONS. /REQUENCIES AND EIGENVEC 08-1284
LCULATIONS FROM SATELLITE SCATTERING DATA)   LONG PERIOD STRUCTURE OF THIOUREA CRYSTAL. (CA 25-2911
IUM DIHYDROGEN ARS/ NMR AND DIELECTRIC RELAXATION STUDIES OF LONG RANGE CORRELATION IN FERROELECTRIC POTASS 18-2718
AND VELOCITY OF ULTRASOUND IN FERROELECTRICS / INFLUENCE OF LONG RANGE DIPOLE DIPOLE FORCES ON ATTENUATION 03-0253
SECOND HARMONIC GENERATION.   LONG RANGE ORDER IN SODIUM NITRITE BY OPTICAL 16-2314
DIUM NITRITE. (AT THE SODIUM-23/ ELECTRIC FIELD GRADIENT AND LONG RANGE ORDER PARAMETER OF FERROELECTRIC SO 16-2395
STRONTIUM OR CALCIUM TITANATE OF ZIRCONATE, TEST OF SHORT OR LONG RANGE ORDER THEORIES). /AIN PEROVSKITES (03-0414
ABSORPTION AND REFLECTION SPECTRA OF BARIUM TITANATE IN THE LONG WAVELENGTH INFRARED REGION.   04-0644
SITIONS IN ANTIMONY SULFOIODIDE. (POLARIZED INFRARED AND RA/ LONG WAVELENGTH OPTICAL PHONONS AND PHASE TRAN 15-2106
NTIFERROELECTRIC AMMONIUM DIHYDROGEN PHOSPHATE.   LONG WAVELENGTH POLARIZATION FLUCTUATIONS IN A 17-2461
) FERROELECTRIC TYPE CRYSTAL.   MECHANICAL MODEL OF LONG WAVELENGTH VIBRATIONS IN A (SINGLE DOMAIN 03-0342
VES ALONG THE (100) DIRECTI/ COLLINEAR INTERACTION BETWEEN A LONGITUDINAL ACOUSTIC WAVE AND TWO LUMINOUS WA 11-1680
BY 180 DOMAIN BOUNDARIES/ REFLECTION AND REFRACTION OF QUASI LONGITUDINAL AND QUASI TRANSVERSE SOUND WAVES 03-0314
IN PIEZOELECTRIC LITHIUM NIOBATE AND/ SEPARATE EXCITATION OF LONGITUDINAL AND TRANSVERSE HYPERSONIC PULSES 11-1814
LSES IN LITHIUM NIOBATE AND BISMU/ INDEPENDENT EXCITATION OF LONGITUDINAL AND TRANSVERSE HYPERSONIC WAVE PU 11-1815
PTIC DEVICES. (PLZT)   A NEW LONGITUDINAL DISPLAY MODE FOR CERAMIC ELECTROO 09-1523
CRYSTALS.   COLLINEAR INTERACTION OF LONGITUDINAL ELASTIC WAVES IN LITHIUM NIOBATE 11-1678
                                                             LONGITUDINAL ELECTROOPTIC EFFECT IN KTN.   13-1942
T STRONTIUM BARIUM NIOBATE PLATES.   LONGITUDINAL ELECTROOPTIC EFFECT IN OBLIQUE CU 08-1472
T STRONTIUM BARIUM NIOBATE PLATES.   LONGITUDINAL ELECTROOPTIC EFFECT IN OBLIQUE CU 13-2063
SECTIONS OF LITHIUM NIOBATE.   LONGITUDINAL ELECTROOPTICAL EFFECT IN OBLIQUE 11-1863
WAVE ALONG THE (001)/ GENERATION OF THE THIRD HARMONIC OF A LONGITUDINAL HIGH FREQUENCY (950 MHZ) ACOUSTIC 11-1835
TRONTIUM TITANATE.   EFFECTS OF LONGITUDINAL MAGNETIC FIELD ON ULTRASOUND IN S 06-1243
ALIDE CRYSTALS. (RAMAN SPECTRA, DIPOLAR COUPLING MODEL COMP/ LONGITUDINAL MODES IN HYDROGEN AND DEUTERIUM H 31-3326
CTRIC / SCATTERING OF CONDUCTION ELECTRONS BY TRANSVERSE AND LONGITUDINAL OPTICAL PHONONS IN IONIC FERROELE 03-0292
LS. (THEORY OF ANISOTROPIC RELAXATION ATTENUATION OF QUASI LONGITUDINAL PHONONS /E SULFATE SINGLE CRYSTA 27-3078
LIGHT IN LITHIUM NIOBATE (FROM TRANSVERSE (MORE INTENSE) AND LONGITUDINAL PHONONS). /ILLOUIN SCATTERING OF 11-1743
EUTERIUM PHOSPHATE (TEMPERATURE DE/ ANOMALOUS PROPAGATION OF LONGITUDINAL ULTRASONIC WAVES IN POTASSIUM DID 17-2660
CRYSTALS NEAR THE CURIE POINTS.   ABSORPTION OF LONGITUDINAL ULTRASONIC WAVES IN ROCHELLE SALT 28-3211
ALATE (EFFECT OF VACANCIES).   ANOMALOUS (LONGITUDINAL WAVE) ATTENUATION IN LITHIUM TANT 12-1892
ICAL TUBES.   VIBRATIONS OF LONGITUDINALLY POLARIZED FERROELECTRIC CYLINDR 03-0375
VIBRATIONS OF COAXIALLY SEGMENTED, LONGITUDINALLY POLARIZED FERROELECTRIC TUBES. 03-0376
ENCE OF THE SHAPE OF DOMAINS ON THE FERROELECTRIC HYSTERESIS LOOP.   INFLU 03-0300
ON OF BARIUM TITANATE SINGLE CRYSTALS WITH DOUBLE HYSTERESIS LOOP.   PECULIARITIES OF POLARIZATI 04-0709
FERROELECTRICS.   ASYMMETRIC MINOR LOOP AND TRANSIENT OPEN LOOP COMPUTATIONS FOR 03-0296
RT LIVED FERROELECTRIC "AFTER EFFECT" PHENOMENA. (HYSTERESIS LOOP CONSTRICTIONS, ROCHELLE SALT, TGS) SHO 27-3156
IDEWAYS DOMAIN WALL MOTION AND (MECHANISM OF THE) HYSTERESIS LOOP OF FERROELECTRICS.   S 03-0301
TIME EFFECTS IN THE HYSTERESIS LOOP OF (LEAD, STRONTIUM) TITANATE.   05-1052
TIME EFFECTS IN THE HYSTERESIS LOOP OF LEAD STRONTIUM TITANATE.   05-1053
LS.   DOUBLE DIELECTRIC HYSTERESIS LOOP OF POTASSIUM DIDEUTERIUM PHOSPHATE CRYSTA 17-2503
C ELECTRIC FIELD.   FERROELECTRIC HYSTERESIS LOOP OF SILVER SODIUM DINITRITE BY QUASI STATI 16-2293
PHATE FERROELECTRIC (CRYSTAL GROWTH AND STRUCTURE HYSTERESIS LOOP RESULTS). /F THE NON-HYDROGEN BONDED PHOS 36-3610
X) (ORTHO) ALUMINATE SYSTEM (NOT FERROELECTRIC BY HYSTERESIS LOOP TESTS). /THE PRASEODYMIUM(1-X) NEODYMIUM(36-3600
NEW TYPE OF LOOP TRACER FOR FERROELECTRICS.   37-3666
UITRY DESCRIBED)   VARIOUS HYSTERESIS LOOP TRACERS FOR A WIDE FREQUENCY RANGE. (CIRC 37-3640
IC FREE ENERGY EXPANSION COEFFICIENTS FROM DOUBLE HYSTERESIS LOOPS.   FERROELECTR 03-0262
ROELECTRIC CRYSTALS.   DOUBLE HYSTERESIS LOOPS AND FIRST ORDER PHASE TRANSITIONS IN FER 03-0456
DROGEN SELENITE.   DOUBLE HYSTERESIS LOOPS AT THE UPPER CURIE POINT OF SODIUM TRIHY 22-2831
ANALYSIS OF HYSTERESIS LOOPS IN LITHIUM HYDRAZINIUM SULFATE.   32-3351
/EHAVIOR OF FERROELECTRIC TGS AND DEUTERATED TGS (HYSTERESIS LOOPS JUST BELOW CURIE POINT CORRECTED FOR EL/ 27-2966
ANOMALOUS FERROELECTRIC HYSTERESIS LOOPS. (LEAD STRONTIUM TITANATE)   05-1067
ANALYSIS OF DOUBLE HYSTERESIS LOOPS. (ROCHELLE SALT)   28-3212
THE RCLE OF LORENTZ FACTORS IN ROCHELLE SALT. (THEORY) 03-0289
FFERENCE IN TEMPERATURES FOR MAXIMUM DIELECTRIC CONSTANT AND LOSS) /ES IN BARIUM TITANATE CERAMICS. (AND DI 04-0849
NDING ON THE CONDITI/ DIELECTRIC PERMEABILITY AND DIELECTRIC LOSS ANGLE TANGENT OF TRIGLYCINE SELENATE DEPE 27-3125
INGLE CRYSTALS (40 HZ-10 MHZ, EFFE/ DIELECTRIC (CONSTANT AND LOSS) BEHAVIOR OF REDUCED STRONTIUM TITANATE S 06-1148
LE SALT NEAR TH/ TEMPERATURE AND FREQUENCY DEPENDENCE OF THE LOSS FACTOR IN FERROELECTRIC MATERIALS (ROCHEL 28-3220
CTRIC PERMITTIVITY. (BARIUM TITANATE TYPE MATERIALS)   LOW LOSS (FERROELECTRIC CERAMICS) WITH GREAT DIELE 04-0779
DIELECTRIC DISPERSION IN TGS CRYSTALS. (PERMITTIVITY AND LOSS FROM 10 GHZ TO 100 GHZ)   27-3094
MEASUREMENT OF ELECTRON ENERGY LOSS FUNCTION IN BULK STRONTIUM TITANATE.   06-1138
E DIELECTRIC CONSTANT AND TANGENT OF THE ANGLE OF DIELECTRIC LOSS IN A STRONG ELECTRIC FIELD. /CHANGE IN TH 27-3151
CH) STRONTIUM TITANATE SYSTEMS./ DIELECTRIC NONLINEARITY AND LOSS IN THE BARIUM (OR LEAD, 4 COMPOSITIONS EA 09-1522
(20) AT MICROWAVE FREQUENCIES.   ACOUSTIC SURFACE WAVE LOSS MECHANISMS ON BISMUTH(12) GERMANIUM OXIDE 35-3511
E AND FREQUENCY D/ MICROWAVE FREQUENCY ACOUSTIC SURFACE WAVE LOSS MECHANISMS ON LITHIUM NIOBATE (TEMPERATUR 11-1817
/ FERROELECTRIC AND ANTIFERROELECTRIC SOLIDS. PART-1: ENERGY LOSS NEUTRON INCOHERENT SCATTERING SPECTRA OF/ 17-2687
ANATE. (150-500 DEG C, 20-50 KILCHERTZ, LISSAJOUS FI/ ENERGY LOSS NONPOLAR STATE POLYCRYSTALLINE BARIUM DID 04-0831
PHASE TRANSITIONS IN CESIUM LEAD TRICHLORIDE. (LOSS OF A CENTER OF SYMMETRY AT 194 DEGREES K) 33-3368
YSTALS ON A CHANGE IN THE TANGENT OF THE ANGLE OF DIELECTRIC LOSSES IN A STRONG ELECTRIC FIELD. / SINGLE CR 27-3153
RY)   ATTENUATION OF CRITICAL VIBRATIONS AND DIELECTRIC LOSSES IN DISPLACIVE TYPE FERROELECTRICS (THEO 03-0159
VE THE CURIE POINT)   PHENOMENOLOGICAL THEORY OF MICROWAVE LOSSES IN FERROELECTRICS. (AT TEMPERATURES ABO 03-0503
VE THE CURIE TEMPERATURE. (BARIUM(0-1./ MICROWAVE DIELECTRIC LOSSES IN OXYGEN OCTAHEDRON FERROELECTRICS ABO 04-0559
FERROELECTRIC LOSSES IN (PURE) BARIUM TITANATE CERAMIC.   04-0555
RANSITION.   MICROWAVE LOSSES IN STRONTIUM TITANATE ABOVE THE PHASE T 06-1199
EFFECT OF GROWTH CONDITIONS ON DIELECTRIC LOSSES IN TRIGLYCINE SULFATE SINGLE CRYSTALS. 27-3152
D AT HIGH PRESSURES.   STRUCTURE AND DIELECTRIC LOSSES OF FERROELECTRIC PEROVSKITES SYNTHESIZE 08-1452
/FERROELECTRICS. PART-2: DIELECTRIC CONSTANTS AND DIELECTRIC LOSSES OF SODIUM NITRITE AND GLYCINE SULFATE. 16-2354
TO SMALL DOSES OF X- AND GAMMA-RAYS.   DIELECTRIC LOSSES OF TRIGLYCINE SULFATE CRYSTALS EXPOSED 27-2976
LS CELL FOR HIGH POWER Q SWITCHING.   LOSSLESS POTASSIUM DIDEUTERIUM PHOSPHATE POCKE 17-2529
LATTICE VIBRATIONS. (QUARTZ, BARIU/ EXTENSION OF THE POULET- LOUDON APPROXIMATION IN THE THEORY OF OPTICAL 03-0380
UM TITANATE.   LOWERING THE CURIE TEMPERATURE IN REDUCED BARI 04-0711
MPLIANCE OF FERROELECTRIC LITHIUM THALLIUM TARTRATE HYDRATE (LTT) BY AN ELECTRIC FIELD. / OF THE ELASTIC CO 29-3249
S. (LITHIUM NIOBATE)   PARAMETRIC LUMINESCENCE AND LIGHT SCATTERING BY POLARITON 11-1746
LUMINESCENCE FROM LITHIUM NIOBATE.   11-1729
AND BISMUTH(12) GERMANIUM/ THERMALLY STIMULATED CURRENTS AND LUMINESCENCE IN BISMUTH(12) SILICON OXIDE(20) 36-3585
CRYSTAL.   PARAMETRIC LUMINESCENCE INTENSITY IN THE LITHIUM NIOBATE 11-1750
AGNETIC POTASSIUM MANGANESE TRIFLUORIDE.   LUMINESCENCE OF EUROPIUM(3+) ION IN ANTIFERROM 33-3381

ITH GAMMA-RAYS.
ECTRIC SEMICONDUCTOR ANTIMONY SULFOIODIDE.
YMIUM(3+) AND ERBIUM(3+) IONS IN LITHIUM NIO/ ABSORPTION AND
/FERROELECTRIC CRYSTALS (CYCLOTRON RESONANCE, ABSORPTION AND
/AR INTERACTION BETWEEN A LONGITUDINAL ACOUSTIC WAVE AND TWO
 (NONLINEAR OPTICAL PROPERTIES, / PHASES AND POLARIZATION OF
5) ZIRCONATE(0.5).   SOLUTION KINETICS OF
ANATE)   COMPUTER EXPERIMENT ON LINEAR CHAIN OF ATOMS

LUMINESCENCE OF POTASSIUM NITRATE IRRADIATED W 16-2335
LUMINESCENCE OF SINGLE CRYSTALS OF THE FERROEL 15-2229
LUMINESCENCE SPECTRA AND ENERGY LEVELS OF NEOD 11-1713
LUMINESCENCE SPECTRA, INFRARED, OPTICAL, AND / 03-0427
LUMINOUS WAVES ALONG THE (100) DIRECTION OF L/ 11-1680
LUMINOUS WAVES IN CUBIC ELECTROOPTIC CRYSTALS. 02-0044
LUTETIUM AND SCANDIUM IONS IN LEAD TITANATE(0. 09-1538
LYING IN DOUBLE MINIMUM POTENTIAL. (BARIUM TIT 03-0287

M

ASE SYNTHESIS, N/ FORMATION AND PHASE TRANSFORMATIONS OF THE
TROOPTIC ELEMENTS.
   CRYSTALS GROWN FROM AN AQUEOUS SOL/ GROWTH LAYERS AND OTHER
   POLYCRYSTALLINE FERROELECTRICS. (THEORY)
ELECTRIC ENERGY FUNCTIONS. (SOME COMMENTS)
NS OF DICALCIUM STRONTIUM PROPIONATE.   ELECTRON MIRROR
RROELECTRICS. (POTASSIUM DIHYDROGEN PHOSPHATE)
ITY. (DESCRIPTIVE AND PREDICTIVE POSSIBILITIES)
ROELECTRIC CRYSTALS IN THE FORM OF THICK PLATES.
                                               MACROSO

M MODIFICATION OF SAMARIUM MOLYBDATE (SOLID PH 30-3270
PRECISION MACHINING OF LITHIUM NIOBATE CRYSTALS FOR ELEC 11-1742
MACRODEFECTS IN POTASSIUM DIHYDROGEN PHOSPHATE 17-2496
MACROSCOPIC ANALOG OF THE SPIN- ECHO EFFECT IN 03-0310
(MACROSCOPIC AND MICROSCOPIC) SYMMETRY OF FERRO 03-0193
MACROSCOPIC OBSERVATION OF FERROELECTRIC DOMAI 26-2946
MACROSCOPIC STIMULATED POLARIZATION ECHO IN FE 17-2419
(MACROSCOPIC) SYMMETRY ASPECTS OF FERROELECTRIC 03-0464
MACROSCOPIC THEORY OF BARIUM TITANATE TYPE FER 04-0770
MACROSCOPIC THEORY OF FERROELECTRICS. 03-0295
MACROSO MACROSCOPIC THEORY OF FERROELECTRICS. 03-0295

OF AB(5) OXIDE COMPOUNDS. (BARIUM TITANATE, R/ INFLUENCE OF
/N IN PARAMAGNETIC POTASSIUM MANGANESE TRIFLUORIDE, RUBIDIUM
/F PEROVSKITE TYPE. (POLYCRYSTALLINE AND SINGLE CRYSTAL LEAD
HASE TRANSI/ RELAXATION POLARIZATION OF A FERROELECTRIC LEAD
N IN THE REGION OF THE ANTIFERROELECTRIC CURIE POINT OF LEAD
PANSION OF POLARIZED CERAMICS IN THE LEAD TITANATE ZIRCONATE
/BASED ON LEAD TITANATE AND ZIRCONATE. (SOLID SOLUTIONS LEAD
   LIGHT BEAM DEFLECTION CONTROLLED WITH A PRISM OF LEAD
           FORMATION PROCESSES OF LEAD
ONS OF THE "CUBIC" PEROVSKITE STRUCTURE OF LEAD/ ORDERING OF
/A IS ZIRCONIUM, HAFNIUM, CERIUM, THORIUM, TITANIUM, SODIUM,
           FERROELECTRIC AND FERROELASTIC PROPERTIES OF
ORINE, CADMIUM IODINE, IRO/ (OPTICAL) FREQUENCY DOUBLING IN
E AND EPR LINE WIDTH OF POTASSIUM MANGANESE TRIFLUORIDE WITH
   A MICROSCOPIC THEORY FOR THE DIELECTRIC PROPERTIES OF LEAD
LS DOPED WITH ISOTOPE 170.   FLUX GROWTH OF
NG TEMPERATURES IN LITHIUM NIOBATE (CRYSTAL GROWTH BY ADDING
ROELECTRIC LITHIUM NIOBATE. (CRACK-FREE, THROUGH ADDITION OF
FERROELECTRICS- SILVER (BARIUM TITANATE AND BARIUM TITANATE-
           PEROVSKITE TYPE SEIGNETTO
   CONTRIBUTION TO THE THEORY OF DOMAIN STRUCTURES IN
/ OF) STRUCTURAL STUDIES OF FERROELECTRICS AND FERROELECTRIC
/ICRODIFFRACTION STUDIES OF FERROELECTRICS AND FERROELECTRIC
SYNTHESIS AND STUDY OF TH/ NEW FERROELECTRICS AND SEIGNETTO
   PREPARATION AND PROPERTIES OF RARE EARTH COMPOUNDS. (WITH
PERIMENTS IN AMMONIUM TRIHYDROGEN PERIODATE CRYSTALS, PULSED
S IN LEAD COBALTATE(0.5) TUNGSTATE(0.5), TITANAT/ ELECTRICAL
UANIDINIUM ALUMINUM SULFATE HEXAHYDRATE.   EPR STUDY OF
/TIC, AND FERROELASTIC CRYSTALS (PREVIOUS THEORY EXTENDED TO
LEI IN ANTIFERROMAGNETIC POTASSIUM MANGANESE T/ ACOUSTIC AND
/TA TABLES. (MECHANICAL, CRYSTALLOGRAPHIC, PHYSICAL, THERMAL
   MANGANESE TRIFLUORIDE.
MANGANESE(2+) ION IN A CUBIC ANTIFERRODIELECTRIC AN EXTERNAL
(0.5) NIOBATE(0.5) AND LEAD MANGANATE(0.5) NIOBATE(0.5) BY A
NATE.   EFFECTS OF LONGITUDINAL
/ (TETRAGONAL- CUBIC PEROVSKITE) STRONTIUM IRON TUNGSTATE .
ROELECTRIC FERROMAGNETIC MIXTURES. (BARIUM TIT/ ELECTRIC AND
/XPLANATION OF THE UNUSUAL TEMPERATURE VARIATION OF THE MEAN
/TA GADOLINIUM MOLYBDATE. PARTS 1-3: HEAT CAPACITY, ENTROPY,
NIC AND SPIN EXCITATIONS IN ANTIFERRODIELECTRIC ELECTRIC AND
YSTALS. RELATIONSHIP BETWEEN DIRECTIONS OF THE ELECTRIC AND
TS.
ETIC SOLID SOLUTIONS IN THE LEAD COBALT TUNGS/ STRUCTURE AND
ANGANATES.
ILICATE. (UNSTABLE FERROELECT/ CRYSTALLOGRAPHIC, OPTICAL AND
ION REPRESENTATION. (18 REFS)   DIELECTRIC, ELASTIC AND
S AND FERRITES   ELECTRICAL TRANSPORT IN (SEMICONDUCTING AND
N IN AMMONIUM DIHYDROGEN PHOSPHATE SINGLE CRYSTAL BY NUCLEAR
LYMORPHS BY NEUTRON DIFFRACTION, PROTON AND DEUTERON NUCLEAR
ELECTRIC POTASSIUM FERROCYANIDE TRIHYDRATE BY PULSED NUCLEAR
TUDIES OF LONG RANGE CORRELATION IN FERROELECTRIC P/ NUCLEAR
TRIC POTASSIUM DIDEUTERIUM PHOSPHATE, POTASSIUM DI/ DEUTERON
NITION OF ANTIFERROELECTRICITY.
IUM FERROCYANIDE TRIHYDRATE CRYSTALS.   NUCLEAR
IC SILVER SODIUM DINITRI/ SPLITTING OF THE SODIUM-23 NUCLEAR
IUM ALUMINUM SULFATE AND A 36:64 RATIO MIXED CRYSTAL, PROTON
RITE.   ACOUSTIC NUCLEAR
OR POTASSIUM DIHYDROGEN PHOSPHATE PRESENTED, TRIGLYCINE SUL/
SPHATE / ABSENCE OF MOTIONAL NARROWING IN THE PROTON NUCLEAR
   SECOND ORDER QUADRUPOLE SHIFTS OF THE SODIUM-23
TION OF THE ORIENTATIONS OF PROTON PAIRS (IN THE WA/ NUCLEAR
RON SPIN/ NMR INVESTIGATIONS OF ICE CRYSTALS. PART-1: PROTON
ELECTRICS: LITHIUM OR SODIUM TRIHYDROGEN DISELENITE, NUCLEAR
   SULFATE.
TRIC DICALCIUM STRONTIUM PROPIONATE.   NUCLEAR
ALS. PART-1: PROTON MAGNETIC RESONANCE IN WATER ICE/ NUCLEAR
ANINE.   NUCLEAR
ATION IN PARAMAGNETIC POTASSIUM MANGANE/ EXCHANGE NARROWING:

MADELUNG ENERGY AND COVALENCY ON THE STRUCTURE 04-0761
MAGANESE TRIFLUORIDE, AND MANGANESE DIFLUORID/ 33-3376
MAGNESATE(0.33) NIOBATE(0.67), AND LEAD FERRA/ 08-1375
MAGNESATE(0.33) NIOBATE(0.67) WITH A DIFFUSE P 13-1980
MAGNESATE(0.5) TUNGSTATE(0.5) ./TERNAL FRICTIO 13-2020
MAGNESATE(0.5) TUNGSTATE(0.5). /AND THERMAL EX 09-1603
MAGNESATE(0.5) TUNGSTATE(0.5)), EFFECTS OF PO/ 05-1063
MAGNESIO NIOBATE CRYSTALS. 08-1280
MAGNESIO NIOBATE IN THE SOLID STATE REACTION. 08-1354
MAGNESIUM AND NIOBIUM IN THE OCTAHEDRAL POSITI 08-1379
MAGNESIUM, CADMIUM, YTTRIUM, GADOLINIUM, DYSP/ 10-1640
MAGNESIUM CHLORINE BORACITE. 14-2102
(MAGNESIUM CHLORINE, ZINC CHLORINE, CADMIUM CHL 14-2080
MAGNESIUM DOPING.   SHIFT OF NEEL TEMPERATUR 33-3377
MAGNESIUM NIOBATE. 08-1440
MAGNESIUM OXIDE AND LANTHANUM ALUMINATE CRYSTA 36-3558
MAGNESIUM OXIDE TO A CONGRUENT MELT). / MATCHI 11-1684
MAGNESIUM OXIDE TO MELT) /MAIN CRYSTALS OF FER 11-1774
MAGNESIUM STANNATE). /ECTRIC SYSTEMS. PART-2: 04-0816
MAGNETS. 08-1459
MAGNETS AND FERROELECTRICS. 03-0430
MAGNETS. (ELECTRON MICRODIFFRACTION, SUMMARY / 04-0490
MAGNETS. (LEAD AND BARIUM TITANATES, BISMUTH / 04-1038
MAGNETS OF THE PEROVSKITE AND PYROCHLORE TYPE. 02-0128
MAGNETIC AND FERROELECTRIC ORDERING) 02-0131
MAGNETIC AND OPTICAL RESONANCE. /ELAXATION) EX 36-3577
MAGNETIC AND X-RAY STUDIES OF PHASE TRANSITION 13-1960
MAGNETIC COPPER(2+) COMPLEXES IN CRYSTALS OF G 21-2820
MAGNETIC CRYSTALS AND TO INCLUDE PARTIAL CASE/ 03-0142
MAGNETIC EFFECTS INVOLVING THE FLUORINE-19 NUC 33-3390
MAGNETIC, ELECTRONIC AND OPTICAL PROPERTIES O/ 01-0025
MAGNETIC EXCITATIONS IN NICKEL DOPED POTASSIUM 33-3407
MAGNETIC FIELD. /TO THE 6A1G(6S) STATE OF THE 03-0317
MAGNETIC FIELD. /MAGNETIZATION OF LEAD FERRATE 13-2069
MAGNETIC FIELD ON ULTRASOUND IN STRONTIUM TITA 06-1243
(MAGNETIC G STRUCTURE ANALOGOUS TO ANTIFERROEL/ 10-1655
MAGNETIC MEASUREMENTS OF CERTAIN TWO PHASE FER 04-0565
MAGNETIC MOMENT OF CHROMIUM- POTASSIUM ALUM (/ 20-2806
MAGNETIC MOMENT OF THE ELECTRICALLY POLARIZED/ 30-3275
MAGNETIC MOMENTS OF OPTICAL TRANSITIONS. /CTRO 03-0418
MAGNETIC POLARIZATION IN BORACITE AND OTHER CR 14-2087
MAGNETIC PROPERTIES OF FERRODIELECTRIC CATALYS 03-0235
MAGNETIC PROPERTIES OF FERROELECTRIC FERROMAGN 36-3604
MAGNETIC PROPERTIES OF HEAVY RARE EARTH ORTHOM 36-3545
MAGNETIC PROPERTIES OF (MONOCLINIC) EUROPIUM S 36-3620
MAGNETIC PROPERTY, SYMMETRY, AND PHASE TRANSIT 03-0316
MAGNETIC) RARE EARTH ORTHOCHROMITES, MANGANITE 36-3608
MAGNETIC RESONANCE /ON OF THE PHASE TRANSITIO 17-2545
MAGNETIC RESONANCE. /UCTURAL STUDIES OF ICE PO 34-3466
MAGNETIC RESONANCE /IN PARAELECTRIC AND FERRO 24-2889
MAGNETIC RESONANCE AND DIELECTRIC RELAXATION S 18-2718
MAGNETIC RESONANCE AND RELAXATION IN FERROELEC 17-2438
MAGNETIC RESONANCE AND THE PROBLEM OF THE DEFI 03-0266
MAGNETIC RESONANCE AND X-RAY STUDIES OF POTASS 24-2894
MAGNETIC RESONANCE CENTRAL LINE IN FERROELECTR 16-2307
MAGNETIC RESONANCE, DIELECTRIC CONSTANT). /MON 20-2809
MAGNETIC RESONANCE IN FERROELECTRIC SODIUM NIT 16-2321
MAGNETIC RESONANCE IN FERROELECTRICS. (MODEL F 03-0180
MAGNETIC RESONANCE IN POTASSIUM DIHYDROGEN PHO 17-2594
MAGNETIC RESONANCE IN ROCHELLE SALT. 28-3186
MAGNETIC RESONANCE IN ROCHELLE SALT, DETERMINA 28-3203
MAGNETIC RESONANCE IN WATER ICE. PART-2: DEUTE 34-3471
MAGNETIC RESONANCE INVESTIGATION. /SM IN FERRO 22-2837
MAGNETIC RESONANCE INVESTIGATION OF TRIGLYCINE 27-3059
MAGNETIC RESONANCE INVESTIGATIONS OF FERROELEC 26-2931
MAGNETIC RESONANCE INVESTIGATIONS OF ICE CRYST 34-3471
MAGNETIC RESONANCE INVESTIGATIONS OF PHTHALOCY 36-3551
MAGNETIC RESONANCE LINE SHAPES AND SPIN CORREL 33-3376

ETRIC LITHIUM NIOBATE.   NIOBIUM-93 NUCLEAR MAGNETIC RESONANCE LINE WIDTHS IN NONSTOICHIOM 11-1795
IC CRYSTALS POTASSIUM FERROCYANIDE TRIHYDRATE BY THE NUCLEAR MAGNETIC RESONANCE METHOD. /OF THE FERROELECTR 24-2898
ECTRIC SODIUM DEUTERO AMMONIUM SELENATE DIDEUTERATE. NUCLEAR MAGNETIC RESONANCE OF DEUTERIUM IN THE FERROEL 22-2828
ERO AMMONIUM FLUOBERYLLATE CRYSTAL.   NUCLEAR MAGNETIC RESONANCE OF DEUTERIUM IN THE PERDEUT 19-2744
MONIUM TARTRATE.   NUCLEAR MAGNETIC RESONANCE OF FERROELECTRIC LITHIUM AM 29-3240
MONIUM TARTRATE.   NUCLEAR MAGNETIC RESONANCE OF FERROELECTRIC LITHIUM AM 29-3241
AMMONIUM SELENATE DIHYDRATE CRYSTAL.   NUCLEAR MAGNETIC RESONANCE OF SODIUM-23 IN THE SODIUM 22-2883
SE OF FERROELECTRIC THIOUREA.   DEUTERON MAGNETIC RESONANCE OF THE HIGH TEMPERATURE PHA 25-2918
YANIDE AND ISOMORPHOUS CRYSTALS. (NEW MODEL BASED ON NUCLEAR MAGNETIC RESONANCE OF UNTWINNED SAMPLES) /RROC 24-2893
N RESONANCE OF THIOUREA. STUDIES OF FERROELECTRIC SOLIDS BY MAGNETIC RESONANCE PART-18: PROTON AND DEUTERO 25-2919
/A REVIEW OF THEORY AND EXPERIMENT. PART-1: INTRODUCTION AND MAGNETIC RESONANCE. PART-2: THERMAL CONDUCTIV/ 03-0397
TASSIUM DIHYDROGEN PHOSPHATE TYPE CRYSTALS.   NUCLEAR MAGNETIC RESONANCE PROTON SECOND MOMENTS IN PO 17-2408
/LUENCE OF THE ELECTRICAL BOUNDARY CONDITIONS ON THE NUCLEAR MAGNETIC RESONANCE RELAXATION RATE OF SODIUM-/ 16-2350
ROGEN ARSENATE AND DEUTERO AMMONIUM DIDE/ ARSENIC-75 NUCLEAR MAGNETIC RESONANCE RESONANCE IN AMMONIUM DIHYD 18-2736
/C DIPOLE AND QUADRUPOLE SATURATION OF THE LITHIUM-7 NUCLEAR MAGNETIC RESONANCE SIGNALS IN LITHIUM NIOBATE. 11-1666
DER DISORDE/ NATURE OF THE TEMPERATURE DEPENDENCE OF NUCLEAR MAGNETIC RESONANCE SPECTRA SECOND MOMENT IN OR 03-0460
IN LITHIUM NI/ TEMPERATURE DEPENDENCE OF THE LITHIUM NUCLEAR MAGNETIC RESONANCE SPECTRUM AND ATOMIC MOTION 11-1722
NDED FERROELECTRICS. (DEUTERATED ROCHELLE SALT, POT/ NUCLEAR MAGNETIC RESONANCE STUDIES IN SOME HYDROGEN BO 18-2719
IUM SULFATE.   DEUTERON NUCLEAR MAGNETIC RESONANCE STUDIES OF LITHIUM HYDRAZIN 32-3354
 FERROELECTRICS.   NUCLEAR MAGNETIC RESONANCE STUDIES OF OXYGEN OCTAHEDRA 13-2059
RIZATION IN ROCHELLE SALT.   NUCLEAR MAGNETIC RESONANCE STUDIES OF SPONTANEOUS POLA 28-3194
TIONS IN FERROELECTRICS. (POTASSIUM FERROCYANIDE TR/ NUCLEAR MAGNETIC RESONANCE STUDIES OF THE PHASE TRANSI 16-2255
TION IN STANNOUS CHLORIDE DIHYDRATE. DIELECTRIC AND NUCLEAR MAGNETIC RESONANCE STUDIES OF THE PHASE TRANSI 36-3582
N AND DEUTERIUM AND SODIUM-23 RELAXATION IN SODIUM / NUCLEAR MAGNETIC RESONANCE STUDY OF DEUTERIUM DIFFUSIO 22-2849
IUM HYDRAZINIUM SULFATE.   NUCLEAR MAGNETIC RESONANCE STUDY OF FERROELECTRIC LITH 32-3343
NSITION IN (MANGANESE DOPED) AMMONIUM SULFATE.   (NUCLEAR MAGNETIC RESONANCE STUDY OF) FERROELECTRIC TRA 19-2784
IN PEROVSKITES CRYSTALS. (NIOBIUM-93 AND SODIUM-23 / NUCLEAR MAGNETIC RESONANCE STUDY OF PHASE TRANSITIONS 08-1283
IN METALLIC SODIUM TUNGSTEN BRONZES.   TRANSIENT NUCLEAR MAGNETIC RESONANCE STUDY OF PHASE TRANSITIONS 13-1910
IN ROCHELLE SALT.   NUCLEAR MAGNETIC RESONANCE STUDY OF PHASE TRANSITIONS 28-3195
SELENATE AND POTASSIUM TRIHYDROGEN SELENATE.   PROTON MAGNETIC RESONANCE STUDY OF SODIUM TRIHYDROGEN 22-2876
R SODIUM DINITRITE.   NUCLEAR MAGNETIC RESONANCE STUDY OF SODIUM-23 IN SILVE 16-2324
AINING O-D...O BOND. (INCLUDING FULLY DEUTERATED A/ DEUTERON MAGNETIC RESONANCE STUDY OF SOME CRYSTALS CONT 03-0204
TRANSITIONS IN DIGLYCINE NITRATE AND TRIS-SARCOSINE/ NUCLEAR MAGNETIC RESONANCE STUDY OF THE FERROELECTRIC 16-2267
TRANSITIONS IN SODIUM TRIHYDROGEN AND TRIDEUTERIUM / NUCLEAR MAGNETIC RESONANCE STUDY OF THE FERROELECTRIC 22-2832
TRANSITION IN DEUT/ BERYLLIUM-9 QUADRUPOLE PERTURBED NUCLEAR MAGNETIC RESONANCE STUDY OF THE FERROELECTRIC 27-2965
E TANTALATE SOLID SOLUTION SYSTEM.   NIOBIUM-93 NUCLEAR MAGNETIC RESONANCE STUDY OF THE LITHIUM NIOBAT 11-1794
ON IN STRONTIUM TITANATE.   NUCLEAR MAGNETIC RESONANCE STUDY OF THE PHASE TRANSITI 06-1237
STATES OF AMMONIUM IONS IN SOLIDS. (AMMONIUM SULFATE/ PROTON MAGNETIC RESONANCE STUDY OF THE SPIN SYMMETRY 19-2800
SE TRANSITION IN LANTHANUM ALUMINATE.   NUCLEAR MAGNETIC RESONANCE STUDY OF THE STRUCTURAL PHA 36-3533
M NITRITE IN THE VICINITY OF THE PHASE TRANSITION T/ NUCLEAR MAGNETIC RESONANCE STUDY ON SODIUM-23 IN SODIU 16-2396
 LATTICE RELAXATION) INVESTIGATIONS OF THE PHASE TR/ NUCLEAR MAGNETIC RESONANCE, (WIDE LINE SPECTRA AND SPIN 20-2802
/ATION MOBILITY: METAL AMMONIUM (BX4) TYPE COMPOUNDS PROTON MAGNETIC RESONANCE, 84 DEG K AND 120-300 DEG / 22-2826
.67) TUNGSTATE(0.33)- LEAD TITANATE SYSTEM.   SEIGNETTO MAGNETIC SOLID SOLUTIONS OF THE LEAD FERRATE(0 05-1065
BY ANTIFERROMAGNETIC RESONANCE.   STUDY OF THE ORDERED MAGNETIC STATE OF COPPER FORMATE TETRAHYDRATE 26-2943
/ OF CHROMIUM VALENCY FROM STUDIES OF EPR SPECTRA AND STATIC MAGNETIC SUSCEPTIBILITY OF CHROMIUM DOPED BAR/ 04-0542
M SULFATE HEXAHYDRATE AT LOW TEMPERATURES.   MAGNETIC SUSCEPTIBILITY OF GUANIDINIUM VANADIU 21-2815
BRONZES, HYDROGEN(X) T/ ELECTRONIC PROPERTIES (CONDUCTIVITY, MAGNETIC SUSCEPTIBILITY) OF HYDROGEN TUNGSTEN 13-1934
EORY IN FINITE SAMPLES OF FERRODIELECTRICS WITH AN ARBITRARY MAGNETIC SYMMETRY. /N WAVE PHENOMENOLOGICAL TH 03-0164
ETER FOR PHASE CHANGE DETECTION (INCLUDING FERROELECTRIC AND MAGNETIC TRANSITIONS, EXAMPLE ROCHELLE SALT) / 28-3201
CRYSTALS HAVING DOMAIN STRUCTURE.   RESONANCE EFFECTS IN MAGNETICALLY EQUIAXIAL FERRODIELECTRIC SINGLE 03-0509
REVIEW, 251 REFS).   MAGNETICALLY ORDERED FERROELECTRIC MATERIALS (02-0117
EORY)   PHASE TRANSITION FROM NORMAL FERROELECTRIC TO MAGNETICALLY ORDERED FERROELECTRIC STATE. (TH 03-0394
COUPLED ELECTROMAGNETIC SPIN WAVES IN MAGNETICALLY ORDERED FERROELECTRICS. 03-0148
LY OSCILLATING ELECTRIC FIELD.   BEHAVIOR OF MAGNETICALLY ORDERED FERROELECTRICS IN A RAPID 03-0147
POSSIBILITY OF EXCITATION OF SPIN WAVES IN MAGNETICALLY ORDERED FERROELECTRICS. (THEORY) 03-0150
ORMATE TETRAHYDRATE.   MAGNETISM AND THE PHASE TRANSITION OF COPPER F 26-2930
UPERFLUIDI/ PHYSICS OF TRANSITIONS. (THERMODYNAMICS, THEORY, MAGNETISM, FERROELECTRICITY, ORDERED ALLOYS, S 03-0183
AL SYMMETRY CONSIDERATIONS ON BORACITE/ INTERACTIONS BETWEEN MAGNETIZATION AND POLARIZATION: PHENOMENOLOGIC 14-2072
P/ COEXISTENCE OF THE SPONTANEOUS ELECTRIC POLARIZATION AND MAGNETIZATION IN NICKEL IODIDE BORACITE (CURIE 14-2090
MAGNET/ ZERO POINT SPIN DEVIATION AND SPONTANEOUS SUBLATTICE MAGNETIZATION IN THE TWO DIMENSIONAL ANTIFERRO 26-2929
) AND LEAD M/ SUPPRESSION OF THE SPONTANEOUS MAGNETOELECTRIC MAGNETIZATION OF LEAD FERRATE(0.5) NIOBATE(0.5 13-2069
ON(X) ALUMINU/ RELAXATION PHENOMENA AND MOSSBAUER SPECTRA OF MAGNETIZED PARAMAGNETIC SUBSTANCES. PART-1: IR 19-2751
EORY).   INTERACTION OF LIGHT WITH MAGNETOACOUSTIC WAVES IN A FERRODIELECTRIC (TH 03-0157
TRIFLUORIDE.   MAGNETOELASTIC EFFECTS IN POTASSIUM MANGANESE 33-3389
ERROMAGNETIC BISMUTH TITANIUM FERRATE.   MAGNETOELECTRIC EFFECT IN FERROELECTRIC- ANTIF 08-1360
ERROMAGNETIC LITHIUM IRON(0.5) TANTALUM(0.5) OXY FLUORIDE. MAGNETOELECTRIC EFFECT IN FERROELECTRIC- ANTIF 36-3568
IRON ANTIMONY(2) OXIDE(4) (MAY BE FERROELECTRIC DUE TO ANTI/ MAGNETOELECTRIC EFFECT IN THE ANTIFERROMAGNET 36-3563
0.5) NIOBATE(0.5) AND LEAD M/ SUPPRESSION OF THE SPONTANEOUS MAGNETOELECTRIC MAGNETIZATION OF LEAD FERRATE(13-2069
RACITE.   MAGNETOELECTRIC PROPERTIES OF NICKEL IODINE BO 14-2070
TRAFLUORIDE.   LOW FREQUENCY MAGNETOELECTRIC RESONANCES IN BARIUM COBALT TE 33-3409
A NEW MAGNETOFERROELECTRIC: CADMIUM IRON NIOBATE. 36-3618
RROELECTRIC BETA GADOLINIUM MOLYBDATE. PARTS 1-3: HEAT CAPA/ MAGNETOTHERMODYNAMICS OF ANTIFERROMAGNETIC, FE 30-3275
CRYSTALS AND FINE PARTICLES OF BARIUM TITANATE.   MAGNITUDE OF SPONTANEOUS POLARIZATION IN THIN 04-0607
ESE TRIFLUORIDE.   FINE STRUCTURE OF EXCITON- MAGNON ABSORPTION OF LIGHT IN POTASSIUM MANGAN 33-3404
C POTASSIUM MANGANESE TRIFLUORIDE.   TWO MAGNON SCATTERING OF LIGHT IN ANTIFERROMAGNETI 33-3401
P/ IMPURITY CONCENTRATION DEPENDENCE OF RAMAN SCATTERING BY MAGNONS IN NICKEL DOPED RUBIDIUM MANGANATE AND 33-3395
ANISOTROPY OF THE ELECTROOPTICAL EFFECT IN CRYSTALS OF LEAD MAGNONIOBATE. 36-3521
/OUS BEHAVIOR OF THE OPTICAL ABSORPTION COEFFICIENT OF LEAD MAGNONIOBATE SINGLE CRYSTALS IN THE REGION OF/ 13-1974
OF FERROELECTRIC PIEZOELECTRIC CERAMI/ RELATION BETWEEN THE MAIN ELECTROPHYSICAL AND STRUCTURAL PARAMETERS 03-0217
THIUM NIOBATE (FROM TRANSVERSE (MORE INTENSE) AND LONGITUDI/ MANDELSTAM BRILLOUIN SCATTERING OF LIGHT IN LI 11-1743
NGANESE FLUORIDE, COBALT FLUORIDE, POTASSIUM MANGANESE TRIF/ MANDELSTAM BRILLOUIN SCATTERING OF LIGHT IN MA 33-3399
NGANESE FLUORIDE CRYSTALS (POTASSIUM TRIFLUORIDE, RUBIDIUM / MANDELSTAM BRILLOUIN SCATTERING OF LIGHT IN MA 33-3398
SURFACE EFFECT IN FERROELECTRIC YTTRIUM MANGANATE. 36-3588
/NCE OF RAMAN SCATTERING BY MAGNONS IN NICKEL DOPED RUBIDIUM MANGANATE AND POTASSIUM MANGANESE TRIFLUORIDE. 33-3395
/IES OF PEROVSKITE TYPE COMPOUNDS, LANTHANUM(1-X) CALCIUM(X) MANGANATE (PHASE TRANSFORMATIONS CONFIGURATIO/ 08-1356
YTTRIUM MANGANATE SWITCHING BEHAVIOR. 36-3571
/NETIC SOLID SOLUTIONS IN THE LEAD COBALT TUNGSTATE- CADMIUM MANGANATE SYSTEM (BOTH CURIE POINTS STUDIED, / 36-3604
PLEX COMPOSITIONS: LEAD MANGANATE(0.5) NIOBATE(0.5) AND LEAD MANGANATE(0.33) NIOBATE(0.67). /PEROVSKITE COM 08-1453
(0.33) N/ SYNTHESIS OF PEROVSKITE COMPLEX COMPOSITIONS: LEAD MANGANATE(0.5) NIOBATE(0.5) AND LEAD MANGANATE 08-1453
/IC MAGNETIZATION OF LEAD FERRATE(0.5) NIOBATE(0.5) AND LEAD MANGANATE(0.5) NIOBATE(0.5) BY A MAGNETIC FIE/ 13-2069

TRANSISTICNS OF THE BISMUTH FERRATE STRONTIUM STANNATE(0.33) MANGANATE(0.67) SYSTEM. /TIN-119 DURING PHASE 08-1400
MAGNETIC PROPERTIES OF HEAVY RARE EARTH ORTHOMANGANATES. 36-3545
/(BY FLUX GROWTH) OF HEXAGONAL SINGLE CRYSTALS OF RARE EARTH MANGANATES (PHASE DIAGRAMS, CONVECTION EFFECT/ 36-3536
PROPERTIES OF LEAD TITANATE CERAMICS CONTAINING LANTANUM AND MANGANESE. ELECTROMECHANICAL 05-1060
HYPERFINE SPECTRUM OF THE ELECTRON PARAMAGNETIC RESONANCE OF MANGANESE(+2) IN BARIUM TITANATE. /NES IN THE 04-0877
ES OF SCLID SOLUTIONS OF LEAD ZIRCONA/ EFFECT OF BISMUTH AND MANGANESE CONTAINING ADDITIVES ON THE PROPERTI 09-1588
A NOTE ON THE APPARENT CONTACT FERROELECTRIC EFFECT OF MANGANESE DIOXIDE. 36-3606
EFFECT CF DCPING BARIUM TITANATE WITH MANGANESE DIOXIDE. 04-0562
VESTIGATION OF PCLYCRYSTALLINE STRONTIUM TITANATE CONTAINING MANGANESE DIOXIDE. EPR IN 06-1162
ERTIES OF LEAD TITANATE CERAMICS. (CONTAINING 1 MOLE PERCENT MANGANESE DIOXIDE) /ION AND PIEZOELECTRIC PROP 05-1061
ANCE) BEHAVIOR OF BARIUM TITANATE WITH NIOBIUM PENTOXIDE AND MANGANESE DIOXIDE ADDITIVES. /ICIENT OF RESIST 04-0837
(NMR STUDY OF) FERROELECTRIC TRANSITION IN (MANGANESE DOPED) AMMONIUM SULFATE. 19-2784
+) DOPED) ZINC/ ELECTRON PARAMAGNETIC RESCNANCE SPECTRUM IN (MANGANESE DOPED) LITHIUM NIOBATE AND (COPPER(2 11-1792
EFFECTS OF SILICA ON MANGANESE DOPED STRONTIUM TITANATE CERAMICS. 06-1172
E ULTRASCNICS. BARIUM MANGANESE FLUORIDE, A NEW CRYSTAL FOR MICROWAV 33-3406
MANGANESE TRIF/ MANDELSTAM BRILLCUIN SCATTERING OF LIGHT IN MANGANESE FLUORIDE, COBALT FLUORIDE, POTASSIUM 33-3399
RIDE, RUBIDIUM / MANDELSTAM BRILLCUIN SCATTERING OF LIGHT IN MANGANESE FLUORIDE CRYSTALS (POTASSIUM TRIFLUO 33-3398
THERMAL EXPANSIVITY OF RUBIDIUM MANGANESE FLUORIDE NEAR THE NEEL TEMPERATURE. 33-3375
EMICONDUCTING BARIUM TITANATE CERAMIC DOPED WITH NIOBIUM AND MANGANESE IONS AND ITS APPLICATIONS. A S 04-0734
L OXYGEN(3) PEROVSKITES. (WHERE METAL IS SCANDIUM, CHROMIUM, MANGANESE, IRON, COBALT, NICKEL, YTTRIUM) /ETA 08-1451
S SCATTERING AT AN ABSORPTION EDGE. (LITHIUM NIOBATE, BARIUM MANGANESE TETRAFLUORIDE) /TY SENSE BY ANOMALOU 11-1673
OLVING THE FLUORINE-19 NUCLEI IN ANTIFERRCMAGNETIC POTASSIUM MANGANESE TRIFLUORIDE. /D MAGNETIC EFFECTS INV 33-3390
ANTIFERROMAGNETIC RESONANCE SPECTRUM IN BARIUM MANGANESE TRIFLUORIDE. 33-3396
COVALENCY IN POTASSIUM MANGANESE TRIFLUORIDE. 33-3391
TICAL NEUTRON SCATTERING IN STRONTIUM TITANATE AND POTASSIUM MANGANESE TRIFLUORIDE. CRI 06-1208
ERSION AND DAMPING OF SOFT ZCNE BCNDARY PHONONS IN POTASSIUM MANGANESE TRIFLUORIDE. DISP 33-3374
TRUCTURE OF EXCITON- MANGNON ABSORPTION OF LIGHT IN POTASSIUM MANGANESE TRIFLUORIDE. FINE S 33-3364
BY MAGNONS IN NICKEL DOPED RUBIDIUM MANGANATE AND POTASSIUM MANGANESE TRIFLUORIDE. /CE OF RAMAN SCATTERING 33-3395
RED ACTIVE MODE BELOW THE STRUCTURAL TRANSITION IN POTASSIUM MANGANESE TRIFLUORIDE. INFRA 33-3362
INESCENCE OF EUROPIUM(3+) ION IN ANTIFERROMAGNETIC PCTASSIUM MANGANESE TRIFLUORIDE. LUM 33-3381
MAGNETIC EXCITATIONS IN NICKEL DOPED POTASSIUM MANGANESE TRIFLUORIDE. 33-3407
MAGNETOELASTIC EFFECTS IN PCTASSIUM MANGANESE TRIFLUORIDE. 33-3389
TUDY OF THE LATTICE DYNAMICAL PHASE TRANSITIONS IN POTASSIUM MANGANESE TRIFLUORIDE. NEUTRON SCATTERING S 33-3405
TION OF THE ANOMALY OF THE THERMAL CONDUCTIVITY OF POTASSIUM MANGANESE TRIFLUORIDE. OBSERVA 33-3382
RESONANCE BEHAVIOR IN CANTED ANTIFERRCMAGNETIC POTASSIUM MANGANESE TRIFLUORIDE. 33-3403
SPECIFIC HEAT OF POTASSIUM MANGANESE TRIFLUORIDE. 33-3387
ERSTRUCTURE ARISING DURING THE PHASE TRANSITION IN POTASSIUM MANGANESE TRIFLUORIDE. SUP 33-3402
THEORY OF SUPEREXCHANGE REACTION. PART-1: POTASSIUM MANGANESE TRIFLUORIDE. 03-0252
WO MAGNCN SCATTERING OF LIGHT IN ANTIFERRCMAGNETIC POTASSIUM MANGANESE TRIFLUORIDE. T 33-3401
/ION OF THE OPTICAL CCNSTANTS OF ANTIFERRCMAGNETIC POTASSIUM MANGANESE TRIFLUORIDE AND POTASSIUM COBALT TR/ 33-3385
/D DISLOCATION DENSITY DETERMINATICN CF LARGE, PURE RUBIDIUM MANGANESE TRIFLUORIDE AND POTASSIUM MANGANESE/ 33-3397
/NISCTROPIC CRITICAL X-RAY DIFFUSE SCATTERING FROM POTASSIUM MANGANESE TRIFLUORIDE AND SODIUM NIOBATE SING/ 08-1312
/TTENUATION NEAR THE SOFT MODE TRANSITICN PCINT IN POTASSIUM MANGANESE TRIFLUORIDE (ANOMALOUS HEAT CAPACIT/ 33-3373
X-RAY SCATTERING AND THE PHASE TRANSITION OF PCTASSIUM MANGANESE TRIFLUORIDE AT 184 DEGREES K. 33-3394
/OVSKITES (POTASSIUM TANTALATE, STRONTIUM TITANATE, RUBIDIUM MANGANESE TRIFLUORIDE) BY INELASTIC NEUTRON S/ 06-1139
RAMAN SCATTERING AND PHASE TRANSITICNS IN A POTASSIUM MANGANESE TRIFLUORIDE CRYSTAL. 33-3400
RAMAN SCATTERING AND PHASE TRANSITIONS IN POTASSIUM MANGANESE TRIFLUORIDE CRYSTALS. 33-3372
ATTENUATION AND DISPERSION OF ULTRASONIC WAVES IN POTASSIUM MANGANESE TRIFLUORIDE IN THE NEIGHBORHOOD OF / 33-3370
ED REGIO/ PARAMAGNETIC SCATTERING OF NEUTRONS FRCM POTASSIUM MANGANESE TRIFLUORIDE IN THE SHORT RANGE ORDER 33-3371
TRANSITION. ANCMALOUS ELASTIC BEHAVIOR OF POTASSIUM MANGANESE TRIFLUORIDE NEAR A STRUCTURAL PHASE 33-3392
ORIDE, AND RUBIDIU/ INFRARED ABSORPTION SPECTRA CF POTASSIUM MANGANESE TRIFLUORIDE, POTASSIUM COBALT TRIFLU 33-3384
/ LINE SHAPES AND SPIN CCRRELATION IN PARAMAGNETIC POTASSIUM MANGANESE TRIFLUORIDE, RUBIDIUM MAGANESE TRIF/ 33-3376
/ROPIC FLUCTUATIONS AT THE 184 DEG K TRANSITION OF POTASSIUM MANGANESE TRIFLUORIDE SINGLE CRYSTALS, (X-RAY/ 33-3369
TIE/ TWO PHONON PROCESSES AND PHASE TRANSITIONS IN POTASSIUM MANGANESE TRIFLUORIDE: TEMPERATURE DISCONTINUI 33-3363
SHIFT OF NEEL TEMPERATURE AND EPR LINE WIDTH OF POTASSIUM MANGANESE TRIFLUORIDE WITH MAGNESIUM DOPING. 33-3377
E/ ANTIFERROELECTRICITY IN (PURE AND) COBALT DOPED POTASSIUM MANGANESE TRIFLUORIDE. (20 PERCENT MANGANESE R 33-3366
RO/ X-RAY AND ELECTRIC INVESTIGATIONS OF THE SYSTEMS YTTRIUM MANGANESE(1-X) (B(X)) OXYGEN(3) (B=TRIVALENT / 36-3569
ION IN AMMONIUM SULFATE. MANGANESE(2+) ABSORPTION NEAR THE CRITICAL REG 19-2785
SARCOSINE CALCIUM CHLORIDE (TSCC) SIN/ EPR INVESTIGATIONS OF MANGANESE(2+) COMPLEXES IN FERROELECTRIC TRIS- 36-3525
ELECTRON PARAMAGNETIC RESONANCE OF MANGANESE(2+) DOPED AMMONIUM CADMIUM SULFATE. 19-2791
/RFINE TRANSITIONS IN THE ELECTRON PARAMAGNETIC RESONANCE OF MANGANESE(2+) DOPED IN FERROELECTRIC SODIUM N/ 16-2383
ECTRON SPIN RESONANCE STUDIES OF CRYSTAL FIELD PARAMETERS IN MANGANESE(2+) DOPED LITHIUM NIOBATE. EL 11-1803
FORBIDDEN HYPERFINE LINES IN ESR SPECTRA OF MANGANESE(2+) DOPED LITHIUM NIOBATE. 11-1805
ELECTRON SPIN RESONANCE OF MANGANESE(2+) IN BARIUM TITANATE. 04-0747
AN EPR INVESTIGATION CF IRON(3+) AND MANGANESE(2+) IN LITHIUM NIOBATE. 11-1725
ELECTRON PARAMAGNETIC RESONANCE OF MANGANESE(2+) IN POTASSIUM TANTALATE. 08-1343
DIUM NITRITE. ELECTRON PARAMAGNETIC RESONANCE OF MANGANESE(2+) IN THE FERROELECTRIC PHASE OF SO 16-2358
IC/ ADMIXTURE OF THE 4T1G STATE TC THE 6A1G(6S) STATE OF THE MANGANESE(2+) ION IN A CUBIC ANTIFERRODIELECTR 03-0317
YCINE CRYSTALS. EPR STUDIES OF CHROMIUM(3+), COPPER(2+), AND MANGANESE(2+) PARAMAGNETIC ION COMPLEXES IN GL 27-3135
VE CHARGE OF TITANIUM(4+) FROM THE PARAMAGNETIC RESONANCE OF MANGANESE(4+) IN STRONTIUM TITANATE. EFFECTI 06-1167
IN (SEMICONDUCTING AND MAGNETIC) RARE EARTH ORTHOCHROMITES, MANGANITES AND FERRITES ELECTRICAL TRANSPORT 36-3608
IUM, HOLMIUM, ER/ FERROELECTRIC PROPERTIES OF HEXAGONAL ORTHOMANGANITES OF YTTRIUM AND RARE EARTHS. (YTTERB 36-3542
ECTRIC THIN FILM. MANUFACTURING METHOD AND PROPERTIES OF FERROEL 37-3681
SISTIVITY BARIUM TITANATE SINGLE CRYSTALS/ CUBIC TETRAGONAL (MARTENSITIC TYPE) PHASE TRANSITIONS IN HIGH RE 04-0887
RFACE WAVES ON LITHIUM NICBATE AND QUARTZ. EFFECT OF MASS LOADING ON THE PROPAGATION OF ACOUSTIC SU 11-1757
OPERTIES OF ABO(3) AND A(0.5)BO(3) TYPE PHASES AND THE IONIC MASS OF THE B IONS. /LATE THE FERROELECTRIC PR 03-0203
SULFATE AND ROCHELLE SALT UP TO HIGH DOSAGE (DESTRUCTION OR MASKING OF FERROELECTRICITY). /AGED TRIGLYCINE 27-3098
IODATE. LARGE NONLINEAR CPTICAL COEFFICIENT AND PHASE MATCHED SECOND HARMONIC GENERATION IN LITHIUM 08-1405
NICKEL CHLORINE) BORACITES (SECOND HARMONIC GENERATION INDEX MATCHING). /NE, IRON CHLORINE IRON IODINE AND 14-2080
ON IN POTASSIUM DIHYDROGEN / TEMPERATURE DEPENDENCE OF INDEX MATCHING ANGLE OF THE SECOND HARMONIC GENERATI 17-2544
ANIOBATE CRYSTALS CF DIFFERENT STCICHICMETRIES. PHASE MATCHING ANGLES AND TEMPERATURE OF LITHIUM MET 11-1665
NONCOLLINEAR PHASE MATCHING EFFECTS IN LITHIUM NIOBATE. 11-1675
E). EFFECT OF OPTICAL INHOMOGENEITIES ON PHASE MATCHING IN NONLINEAR CRYSTALS (LITHIUM NIOBAT 11-1773
HARMONIC GENERATION OF LIGHT IN THE KDP CRYSTAL BY THE INDEX MATCHING METHOD. SECOND 17-2693
COEFFICIENTS OF LITHIUM OXIDE INTO LITHIUM NIOBATE BY PHASE MATCHING TEMPERATURES. /ON OF THE DISTRIBUTION 11-1841
TAL GROWTH BY ADD/ SPATIALLY UNIFORM AND ALTERABLE SHG PHASE MATCHING TEMPERATURES IN LITHIUM NIOBATE (CRYS 11-1684
/ AMORPHIZATION IN TUNGSTEN TRIOXIDE AND ITS USE IN DEDUCING MEAN DAMAGE RANGES. (STUDIES ON BOMBARDMENT I/ 10-1642
/BLE EXPLANATION OF THE UNUSUAL TEMPERATURE VARIATION OF THE MEAN MAGNETIC MOMENT OF CHROMIUM- POTASSIUM A/ 20-2806
ESCA). BAND STRUCTURE OF STRONTIUM TITANATE AS MEASURED BY X-RAY PHOTOELECTRON SPECTROSCOPY (06-1100
GITAL SURFACE WAVE TRANSDUCERS ON LITHIUM NIOBATE. MEASURED ELECTRICAL CHARACTERISTICS OF INTERDI 11-1702

URE, CURIE CONSTANT) OF FERROELECTRIC CRYSTALS. (5 COMPOUNDS MEASURED, TGS RESULTS GIVEN) / (CURIE TEMPERAT 27-3120
ERROELECTRIC REGICN NEAR ThE CURIE POINT (IMPROVED METHOD OF MEASUREMENT). /ECTRIC RESPONSE OF TGS IN THE F 27-2979
(KDP AND POTASSIUM BICARBONATE)   NEUTRON MEASUREMENT IN TWO SYSTEMS OF HYDROGEN BONDS.   17-2417
ENSER UNDER CONTINUOUS HEATING. (TGS SINGLE CRYST/ IMPEDANCE MEASUREMENT OF A DIELECTRIC FERROELECTRIC COND 27-3028
UM DIHYDROGEN PHOSPHATE.   MEASUREMENT OF CHEMICAL SHIFT IN SOLID POTASSI 17-2688
FERROELECTRIC MATERIALS AT LOW TEMPERATURES.   MEASUREMENT OF DIELECTRIC CONSTANTS FOR A FEW   02-0035
LECTRIC MATERIALS AT CENTIMETER AND DECIMETER WAVELENGTHS E/ MEASUREMENT OF DIELECTRIC PROPERTIES OF FERROE 04-0855
LECTRICS AT SUBMILLIMETER WAVELENGTHS. (STRONTIUM TITANATE)   MEASUREMENT OF DIELECTRIC PROPERTIES OF FERROE 06-1089
N BULK STRONTIUM TITANATE.   MEASUREMENT OF ELECTRON ENERGY LOSS FUNCTION I 06-1138
UM NIOBATE.   THE MEASUREMENT OF HYPERSONIC ATTENUATION IN LITHI 11-1760
STALS OF FERROELECTRIC SUBSTANCES BY THE COMPOSITE RESONATO/ MEASUREMENT OF INTERNAL FRICTION IN SINGLE CRY 27-3080
OF FERROELECTRICS. PART-2: DIELECTRIC CONSTANTS AND DIELECT/ MEASUREMENT OF MICROWAVE DIELECTRIC CONSTANTS   16-2354
ME TITANATES. (STRONTIUM TITANATE, BARIUM TITANATE)   MEASUREMENT OF OXYGEN DIFFUSION CONSTANT IN SO 04-1033
FERROELECTRIC CERAMIC (THEORY, APPLIED TO LEAD(0.99) NIOBI/ MEASUREMENT OF PIEZOELECTRIC PHASE ANGLES IN A 09-1537
LECTRIC CERAMIC DISKS.   MEASUREMENT OF POISSON'S RATIO IN POLED FERROE 37-3658
ELECTRIC CONSTANT OF FERROELECTRICS.   SIMULTANEOUS MEASUREMENT OF PYROELECTRIC COEFFICIENT AND DI 03-0333
                                                        PULSE MEASUREMENT OF R(63) IN KDP.   17-2477
/ SOLIDS: APPLICATION TO THE LEAD TITANATE ZIRCONATE SYSTEM (MEASUREMENT OF SOFT OPTICAL PHONON MODES IN T/ 09-1497
RYSTALLINE LITHIUM NIOBATE, SILICON DIOXIDE, AND SILICON.   MEASUREMENT OF SPUTTER ETCHING RATES FOR MONOC 11-1714
OF ROCHELLE SALT AT 10 GHZ AS A FUNCTION OF TEMPERATURE.   MEASUREMENT OF THE COMPLEX DIELECTRIC CONSTANT 28-3184
LYCINE SULFATE AT MICROWAVE FREQUENCIES.   MEASUREMENT OF THE DIELECTRIC CONSTANT OF TRIG 27-2954
RROELECTRICS AT MICROWAVE FREQUENCI/ PROBLEM OF THE ACCURATE MEASUREMENT OF THE DIELECTRIC PROPERTIES OF FE 03-0429
THE ELECTROMECHANICAL COUPLING IN PIEZOELECTRIC CERAMICS.   MEASUREMENT OF THE DISSIPATION ASSOCIATED WITH 03-0514
IUM DIHYDROGEN PHOSPHATE BY BRILLOUIN DIFFUSION.   MEASUREMENT OF THE ELASTIC CONSTANTS OF POTASS 17-2454
NITRITE BY BRILLOUIN DIFFUSION AT ORDINARY TEMPERATURES.   MEASUREMENT OF THE ELASTIC CONSTANTS OF SODIUM 16-2306
P (NEAR ThE CURIE TEMPERATURE).   ULTRASONIC MEASUREMENT OF THE ELECTROCALORIC EFFECT IN KD 17-2664
OTASSIUM DIHYDROGEN PHOSPHATE, AMMONIUM DIHYDROGEN/ RELATIVE MEASUREMENT OF THE OPTICAL NONLINEARITIES OF P 11-1677
NATE SINGLE CRYSTALS AT MICROWAVE FREQUENCIES.   MEASUREMENT OF THE PERMITTIVITY OF BARIUM TITA 04-0759
EFFECT IN AMMONIUM DIHYDROGEN PHOSPHATE CRYSTALS.   MEASUREMENT OF THE PRIMARY LINEAR ELECTROOPTIC 17-2654
D PERMITTIVITY FROM THE PYROELECTRIC RESPONSE TO STEP RADIA/ MEASUREMENT OF THE PYROELECTRIC COEFFICIENT AN 03-0469
65-35 PZT. (LEAD ZIRCONATE TITANATE)   MEASUREMENT OF THE PYROELECTRIC COEFFICIENT OF 09-1567
KDP CRYSTALS IN THE INFRARED RANGE BY PARAMETRIC SCATTERIN/ MEASUREMENT OF THE REFRACTIVE INDEX OF ADP AND 17-2553
ENTS OF POTASSIUM DIHYDROGEN PHOSPHATE, RUBIDIUM DIHYDROGEN/ MEASUREMENT OF THE RELATIVE NONLINEAR COEFFICI 08-1413
YNAMIC METHOD.   MEASUREMENT OF THE SPECIFIC HEAT OF KDP BY A D 17-2619
HE ELECTRICAL CONDUCTIVITY OF FERROELECTRICS.   MEASUREMENT OF THE TEMPERATURE DEPENDENCE OF T 03-0525
ALT BY USING THE SHIFT OF RESONANCE FREQUENCY IN CAVITY.   MEASUREMENT OF THERMAL EXPANSION OF ROCHELLE S 28-3202
ITRITE BY STEP FUNCTION METHOD.   DIELECTRIC RELAXATION MEASUREMENT ON FERROELECTRIC SILVER SODIUM DIN 16-2291
CN COEFFICIENT OF NIOBIUM IN BARIUM TITANATE BY A DIELECTRIC MEASUREMENT TECHNIQUE.   DIFFUSI 04-0926
OSECOND RESPONSE TIME DETEC/ PYROELECTRIC COEFFICIENT DIRECT MEASUREMENT TECHNIQUE AND APPLICATION TO A NAN 13-1919
ETRY (VERSUS MELT COMPOSITION, LATTICE AND CURIE TEMPERATURE MEASUREMENTS). /ALATE SINGLE CRYSTAL STOICHIOM 12-1871
DIATION IN FERROELECTRIC ROCHELLE SALT. (NEUTRON DIFFRACTION MEASUREMENTS) /TS OF (DAMAGE FROM) IONIZING RA 28-3188
THE CRITICAL BEHAVIOR OF TRIGLYCINE SULFATE BY THERMAL NOISE MEASUREMENTS.   STUDY OF 27-3147
CTRIC RELAXATION IN SILVER SODIUM DINITRITE BY THERMAL NOISE MEASUREMENTS. /TICAL SLOWING DOWN OF THE DIELE 16-2382
PART-3: HIGH TEMPERATURE THERMODYNAMIC AND X-RAY DIFFRACTION MEASUREMENTS. /TOICHIOMETRIC BARIUM TITANATE. 04-0886
ON FERROELECTRICITY IN POTASSIUM NITRATE A/ SOME (DIELECTRIC MEASUREMENTS AND INFRARED ABSORPTION) STUDIES 16-2402
                    ELECTROELASTIC CONSTANT MEASUREMENTS FOR PLZT CERAMICS.   09-1598
SALT, BARIUM TITANATE)   NOISE MEASUREMENTS IN FERROELECTRICS. (TGS, ROCHELLE 04-0698
                NUCLEAR SPIN LATTICE RELAXATION TIME MEASUREMENTS IN SODIUM DEUTERIUM SELENATE.   22-2847
RONTIUM) TITANATE.   MOBILITY DETERMINATIONS FROM WEIGHT MEASUREMENTS IN SOLID SOLUTIONS OF (BARIUM, ST 06-1161
/ TITANATE. PART-1: HIGH TEMPERATURE ELECTRICAL CONDUCTIVITY MEASUREMENTS. (INTRINSIC BAND GAP, PTCR (POSI/ 04-0822
(2-) VACANCIES IN CHROMIUM(3+) DOPED STRONTIUM TITANATE (ESR MEASUREMENTS, MODEL PROPOSED). /TION BY OXYGEN 06-1161
R A DC FIELD.   NOISE AND IMPEDANCE MEASUREMENTS OF A FERROELECTRIC CONDENSER UNDE 37-3645
NIC GENERATION COEFFICIENTS OF PIEZOELECTRIC CRYSTA/ FURTHER MEASUREMENTS OF ABSOLUTE SIGNS OF SECOND HARMO 05-1066
EZOELECTRICS COUPLED TO A GUNN OSCILLATOR. (/ ACOUSTOOPTICAL MEASUREMENTS OF ACOUSTIC WAVE GENERATION IN PI 11-1731
                                           DIELECTRIC MEASUREMENTS OF AMMONIUM SULFATE.   16-2338
TE- LEAD ZIRCONATE SYSTEMS.   MOSSBAUER MEASUREMENTS OF BISMUTH FERRATE- BISMUTH FERRA 08-1282
C FERROMAGNETIC MIXTURES. (BARIUM TIT/ ELECTRIC AND MAGNETIC MEASUREMENTS OF CERTAIN TWO PHASE FERROELECTRI 04-0565
S IN FERROELECTRIC CRYSTALS. (ROCHELLE SALT)   MEASUREMENTS OF COMPLEX PIEZOELECTRIC CONSTANT 28-3213
CIENTS CF ROCHELLE SALT AND TRIGLYCINE SULFATE.   NEW MEASUREMENTS OF DIVERSE THERMODYNAMICAL COEFFI 27-2995
AMMONIUM SULFATE.   MICROWAVE DIELECTRIC MEASUREMENTS OF GAMMA-IRRADIATED FERROELECTRIC 19-2750
(LEAD ZIRCONATE TITANATE) FERRCE/ OXYGEN CONCENTRATION CELL MEASUREMENTS OF IONIC TRANSPORT NUMBERS IN PZT 09-1512
HARMONIC GENERATED LIGHT (FROM LITHIUM NIOBATE CRYSTAL, AGR/ MEASUREMENTS OF PHOTON CORRELATIONS OF SECOND 11-1701
THE VARIATIONS IN ThE FREE AND CLAMPED VALUES OF THE DIELE/ MEASUREMENTS OF SECOND HARMONIC GENERATION AND 13-1998
LLIZATION OF ROCHELLE SALT.   GRAVIMETRIC MEASUREMENTS OF THE CHANGES OF WATER OF CRYSTA 28-3235
N POTASSIUM DIHYDROGEN PHOSPHATE.   MEASUREMENTS OF THE SPONTANEOUS POLARIZATION I 17-2430
LOW DIMENSIONAL ANTIFERROMAGNETS POTASSIUM COPPER TRIFLUOR/ MEASUREMENTS OF THE THERMAL CONDUCTIONS IN THE 26-2939
MANATE. (LITHIUM GERMANATE NOT FERROELECTRIC)   ELECTROOPTIC MEASUREMENTS ON LITHIUM GERMANATE AND LEAD GER 35-3494
E SINGLE CRYSTALS.   DIELECTRIC MEASUREMENTS ON PILE-IRRADIATED BARIUM TITANAT 04-0910
, TRIGLYCINE SULFATE, AND RUTILE.   FAR INFRARED DIELECTRIC MEASUREMENTS ON POTASSIUM DIHYDROGEN PHOSPHATE 17-2423
OXYGEN CONCENTRATION CELL AND ELECTRICAL CONDUCTIVITY MEASUREMENTS ON PZT FERROELECTRICS.   09-1495
ROGEN PHOSPHATE.   LOW TEMPERATURE THERMAL EXPANSION MEASUREMENTS ON SINGLE CRYSTAL POTASSIUM DIHYD 17-2464
            FERROELECTRICITY AND TRANSMISSION MEASUREMENTS ON STRONTIUM BARIUM NIOBATE.   13-1922
FAR INFRARED REFLECTIVITY AND TRANSMISSION MEASUREMENTS ON STRONTIUM TITANATE FILMS.   06-1038
PERSION OF FERROELECTRICS NEAR THE TRANSITION POINT (THEORY, MEASUREMENTS ON TGS). /E ON THE) MICROWAVE DIS 03-0370
ORMATION IN POTASSIUM DIHYDROGEN PHOSPHATE.   NEUTRON MEASUREMENTS ON THE FERROELECTRIC PHASE TRANSF 17-2418
PARAELECTRIC POTASSIUM DIHYDROGEN PHOSPHATE.   NEUTRON MEASUREMENTS ON THE HYDROGEN BOND POTENTIAL IN 17-2511
TRIC POTASSIUM NITRATE.   X-RAY MEASUREMENTS ON THERMAL EXPANSION OF FERROELEC 16-2322
/D ELECTROOPTIC PROPERTIES. PART-7: DIELECTRIC AND SWITCHING MEASUREMENTS. PART-8: ULTRASOUND PROPAGATION./ 03-0397
LICATION TO BARIUM TITANATE).   AUTOMATIC IMPEDANCE MEASUREMENTS USING THERMAL NOISE ANALYSIS (APP 03-0389
ELECTROMECHANICAL PROPERTIES OF TRIGLYCINE SULFATE. (MEASUREMENTS VERIFY THEORY APPROXIMATELY) 27-3145
RIZED CERAMIC FERROELECTRIC. (PZT)   METHOD OF MEASURING POLARIZATION VIBRATIONS OF A PREPOLA 09-1615
PLE)   NEW METHOD OF MEASURING PYROELECTRIC COEFFICIENTS. (TGS EXAM 17-2517
        ONE DIMENSIONAL KDP MODEL IN STATISTICAL MECHANICS.   03-0389
ITIONS.   THERMODYNAMICS AND STATISTICAL MECHANICS OF DIFFUSE FERROELECTRIC PHASE TRANS 03-0445
          STATISTICAL MECHANICS OF FERROELECTRIC PHASE TRANSITIONS.   03-0198
TRIC FERROE/ INFERRED TEMPERATURE DEPENDENCES OF ELECTRICAL, MECHANICAL AND OPTICAL PROPERTIES OF FERROELEC 30-3253
M MOLYBDATE.   ELECTRICAL, OPTICAL, AND MECHANICAL BEHAVIOR OF FERROELECTRIC GADOLINIU 30-3267
OGEN PHOSPHATE SINGLE CRYSTAL/ TEMPERATURE DEPENDENCE OF THE MECHANICAL CHARACTERISTICS OF POTASSIUM DIHYDR 17-2649
L MAGNETIC, / II-VI SEMICONDUCTING COMPOUNDS - DATA TABLES. (MECHANICAL, CRYSTALLOGRAPHIC, PHYSICAL, THERMA 01-0025
TRANSDUCER CER/ DISPERSION STUDY OF DOMAIN WALL MOBILITY AND MECHANICAL DAMPING IN LEAD TITANATE ZIRCONATE 09-1547

BARIUM TITANATE SINGLE CRYSTALS SUBJECTED TO MECHANICAL IMPULSES. 04-1006
IN A (SINGLE DOMAIN) FERROELECTRIC TYPE CRYSTAL. MECHANICAL MODEL OF LONG WAVELENGTH VIBRATIONS 03-0342
A SIMPLE MECHANICAL MODEL WITH FERROELECTRIC BEHAVIOR. 03-0151
/ENCE OF DOMAIN STRUCTURE OF CERAMIC FERROELECTRICS ON THEIR MECHANICAL PROPERTIES. (BARIUM TITANATE, LEAD/ 04-0985
UM TITANATE CERAMICS) NONLINEARITY OF ELECTROMECHANICAL AND MECHANICAL PROPERTIES OF FERROELECTRICS. (BARI 04-0986
OF ADP. PREPARATION (MORPHOLOGY) AND MECHANICAL PROPERTIES OF FILAMENTARY CRYSTALS 17-2630
MECHANICAL PROPERTIES OF ICE SINGLE CRYSTALS. 34-3441
C CRYSTALS. (AMMO/ PHENOMENOLOGICAL THEORY OF DIELECTRIC AND MECHANICAL PROPERTIES OF IMPROPER FERROELECTRI 14-2084
AND POINT DEFECTS OF ICE: THEIR EFFECT ON THE ELECTRICAL AND MECHANICAL PROPERTIES. (REVIEW, 56 REFS) /URE 02-0071
TANATE CERAMICS. EFFECTS OF IMPURITIES ON THE MECHANICAL QUALITY FACTOR OF LEAD ZIRCONATE TI 09-1608
NTIUM TITANATE SINGLE CRYSTALS. MECHANICAL RELAXATION AND NONLINEARITY IN STRO 06-1218
/ETATION OF THE PHASE DIAGRAM OF ICE. PART-2: DIELECTRIC AND MECHANICAL RESPONSE OF ICE I(H) SINGLE CRYSTA/ 34-3485
S AND ITS INTERPRETATION. DIELECTRIC AND MECHANICAL RESPONSE OF ICE I(H) SINGLE CRYSTAL 34-3483
IHYDROGEN PHOSPHATE CRYSTALS. STUDY OF THE (MECHANICAL) STRENGTH PROPERTIES OF POTASSIUM D 17-2648
PARAELECTRIC ROCHELLE SALT. EFFECT OF MECHANICAL STRESS ON DIELECTRIC PROPERTIES OF 28-3225
NTIMONY SULFOIODIDE SINGLE CRY/ INFLUENCE OF ONE DIMENSIONAL MECHANICAL STRESS ON THE PHASE TRANSITION OF A 15-2245
IN FERROELECTRIC POTASSIUM NIOBATE SINGLE CRYSTALS. MECHANICAL SWITCHING OF 60 DEGREE DOMAIN WALLS 08-1421
OF SOME MONOCLINIC CRYSTALS IN THE FERROELECTRIC TRANSITION MECHANISM. /NATE, SULFATE, FLUOBERYLLATE, ETC) 03-0330
MA-IRRADIATION ON BARIUM TITANATE, (COMPARED TO STUDY DAMAGE MECHANISM) /ETIC) ELECTRON AND (COBALT-60) GAM 04-0970
ENTROPY OF AN ORDER DISORDER TRANSITION. (IN FERROELECTRICS, MECHANISM) 03-0192
RYSTALS (IMPORTANCE OF HYDROGEN BONDING IN THE FERROELECTRIC MECHANISM). / TRANSITIONS OF HYDROGEN HALIDE C 31-3329
THEORY OF FERROELECTRIC PHASE TRANSITIONS WITH VIBRONIC MECHANISM. 03-0351
FORMATION BY A SOLID PHASE REACTION. MECHANISM AND KINETICS OF LITHIUM METANIOBATE 11-1840
N OF LEAD METANIOBATE. MECHANISM AND KINETICS OF SOLID PHASE FORMATIO 13-2042
GRAIN BOUNDARIES OF BARIUM TITANATE SEMICONDUCTIVE CERAMIC/ MECHANISM FOR FORMING INSULATION LAYERS ON THE 04-0867
TRANSPORT PHENOMENA IN ANTIMONY SULFOIODIDE. (CONDUCTION MECHANISM FOR THERMOELECTRIC CURRENTS) 15-2241
ROVSKITE STRUCTURED TITANATES. PROPOSED MECHANISM FOR THERMOPHOTOTROPIC BEHAVIOR IN PE 03-0400
INJECTED COLOR CENTERS) RELATIONSHIP BETWEEN THE BREAKDOWN MECHANISM IN BARIUM TITANATE AND COLORATION. (04-0944
CS. MICROWAVE DISPERSION MECHANISM IN BARIUM TITANATE TYPE FERROELECTRI 04-0902
ITANATE, STOICHIOMETRY EFFECTS) CHARGE MECHANISM IN FERROELECTRICS. (LEAD ZIRCONATE T 03-0259
TRIHYDROGEN DISELENITE, NUCLEAR MAGNETIC RESONANC/ REVERSAL MECHANISM IN FERROELECTRICS: LITHIUM OR SODIUM 22-2837
STRUCTURE OF A MOVING DOMAIN WALL. (MECHANISM OF A WALL MOTION IN BARIUM TITANATE) 04-0678
PCLARON MECHANISM OF CONDUCTIVITY IN BARIUM TITANATE. 04-0609
ANATE. MECHANISM OF ELECTROCONDUCTIVITY IN BARIUM TIT 04-0908
AMMONIUM ROCHELLE SALT. (THEORY) MECHANISM OF FERROELECTRIC PHASE TRANSITION IN 28-3223
N POLYCRYSTALLINE BARIUM TITANATE. MECHANISM OF MICROWAVE DIELECTRIC DISPERSION I 04-0745
NIOBATE AND OTHER PYROELECTRIC INSULATORS. (MECHANISM OF) OPTICAL INDEX DAMAGE IN LITHIUM 11-1735
PHIC STUDY OF FERROELECTRIC SODIUM NITRITE CRYSTALS. PART-2: MECHANISM OF POLARIZATION REVERSAL. TOPOGRA 16-2374
TRIC MOMENT IN GUANIDINIUM ALUMINUM SULFATE HEXAHYDRATE (AND MECHANISM OF REVERSAL). FERROELEC 21-2818
CESIUM LEAD TRICHLORIDE. (MECHANISM OF) STRUCTURAL PHASE TRANSITIONS IN 33-3380
ATION OF DIAMMONIUM DICADMIUM TRISULFATE, A MODEL IN WHICH / MECHANISM OF THE FERROELECTRIC PHASE TRANSFORM 03-0141
ATION IN RARE EARTH MOLYBDATES. (TERBIUM MOLYBDATE) MECHANISM OF THE FERROELECTRIC PHASE TRANSFORM 30-3254
TRICS. SIDEWAYS DOMAIN WALL MOTION AND (MECHANISM OF THE) HYSTERESIS LOOP OF FERROELEC 03-0301
ULIAR FERROELECTRIC- FERROELASTIC CRYSTAL: DIAM/ CONJECTURED MECHANISM OF THE PHASE TRANSFORMATION OF A PEC 19-2741
IAR FERROELECTRIC- FERROELASTIC CRYSTAL: SODIUM AM/ INFERRED MECHANISM OF THE PHASE TRANSFORMATION OF PECUL 29-3238
TIC DIELECTRIC CONSTANT, X-RAY SCATTERING) STUDIES ON THE MECHANISM OF THE PHASE TRANSITION IN TGS. (STA 27-3093
. (SATTELITE SCATTERING) STUDIES ON THE MECHANISM OF THE PHASE TRANSITIONS IN THIOUREA 25-2922
Y SULFOIODIDE. MECHANISM OF THE PHOTODOMAIN EFFECT IN ANTIMON 15-2146
/E OF RAW POWDER ON SINTERING. PART-3: SINTERING PROCESS AND MECHANISM OF THE REACTION BETWEEN LEAD MONOXI/ 09-1624
TITANATE, POTASSIUM TITANATE) IS THERE A UNIFIED MECHANISM UNDERLYING URBACH'S RULE. (STRONTIUM 06-1110
/ERROELECTRICS WITH DIFFUSED PHASE TRANSITION. (POLARIZATION MECHANISMS, ELECTROOPTICAL PROPERTIES, REVIEW/ 02-0119
F THE BARKHAUSEN EFFECT IN FERROELECTRIC CRYSTALS. MECHANISMS FOR BARKHAUSEN JUMPS AND BEHAVIOR O 03-0448
. INTERBAND MECHANISMS FOR FERROELECTRIC PHASE TRANSITIONS 02-0085
SOME STRUCTURAL MECHANISMS IN FERROELECTRICITY. 03-0411
FROM BARIUM TITANATE TO DNA. (FERROELECTRICITY, BIOLOGICAL MECHANISMS, MEMORY, ORGANICS) 03-0092
N CRYSTALS WITH REORIENTATING STRUCTURAL GROUPS. (SODIUM AM/ MECHANISMS OF FERROELECTRIC PHASE TRANSITION I 22-2827
AT MICROWAVE FREQUENCIES. ACOUSTIC SURFACE WAVE LOSS MECHANISMS ON BISMUTH(12) GERMANIUM OXIDE(20) 35-3511
FREQUENCY D/ MICROWAVE FREQUENCY ACOUSTIC SURFACE WAVE LOSS MECHANISMS ON LITHIUM NIOBATE (TEMPERATURE AND 11-1617
/SPHERE. PART-7: EFFECT OF CALCINATION ON SINTERING. PART-8: MECHANOCHEMISTRY OF THE SINTERING BODY. PART-/ 09-1624
TE) FERROELECTRIC CERAMICS AND CONVERSION OF MECHANOELECTRIC ENERGY. (LEAD ZIRCONATE TITANA 09-1483
HARMONICS GENERATION IN FERROELECTRIC MEDIA. 03-0215
/D STEERING SURFACE WAVES BY NONLINEAR MIXING IN ANISOTROPIC MEDIA. (LITHIUM NIOBATE, BISMUTH(12) GERMANIU/ 11-1778
/NTALATE AND LITHIUM NIOBATE PIEZOELECTRIC RESONATORS IN THE MEDIUM FREQUENCY RANGE WITH LOW RATIOS OF CAP/ 11-1724
AND LITHIUM TANTALATE. MEDIUM FREQUENCY RESONATORS OF LITHIUM NIOBATE 11-1851
SMUTH TITANATE FERROELECTRIC- PHOTOCONDUCTOR OPTICAL STORAGE MEDIUM FURTHER DEVELOPMENTS. BI 08-1371
BISMUTH TITANATE PHOTOCONDUCTOR OPTICAL STORAGE MEDIUM. (HIGH RESOLUTION READ WRITE) 08-1445
FERROELECTRIC PHOTOCONDUCTOR OPTICAL STORAGE MEDIUM UTILIZING BISMUTH TITANATE. 04-0769
SINGLE CRYSTALS GROWN FROM SOLUTION IN A POTASSIUM FLUORIDE MELT. /CTRICAL CONDUCTIVITY OF BARIUM TITANATE 04-1035
NTIUM(X)) TITANATE- MIXED CRYSTALS FROM A POTASSIUM FLUORIDE MELT. DEFECT STRUCTURE OF (BARIUM(1-X) STRO 04-0873
TITANATE CRYSTALS GROWN FROM SOLUTION IN POTASSIUM FLUORIDE MELT. /H CONDITIONS ON THE STRUCTURE OF BARIUM 04-0598
NIOBATE. (CRACK-FREE, THROUGH ADDITION OF MAGNESIUM OXIDE TO MELT) /MAIN CRYSTALS OF FERROELECTRIC LITHIUM 11-1774
OPTICS WITH LOW BIREFRINGENCE VARIATIONS) FROM THE CONGRUENT MELT. /ITHIUM NIOBATE CRYSTALS (FOR NONLINEAR 11-1688
PLANAR GROWTH OF ICE FROM THE PURE MELT. 34-3444
ATE (CRYSTAL GROWTH BY ADDING MAGNESIUM OXIDE TO A CONGRUENT MELT). / MATCHING TEMPERATURES IN LITHIUM NIOB 11-1684
ECOND HARMONIC GENERATION COEFFICIENTS OF LITHIUM NIOBATE ON MELT COMPOSITION. DEPENDENCE OF S 11-1769
/RTIES OF LITHIUM TANTALATE SINGLE CRYSTALS AS A FUNCTION OF MELT COMPOSITION. (CURIE TEMPERATURE DEPENDEN/ 12-1874
E ME/ LITHIUM TANTALATE SINGLE CRYSTAL STOICHIOMETRY (VERSUS MELT COMPOSITION, LATTICE AND CURIE TEMPERATUR 12-1871
ODIUM NIOBATE. EFFECTS OF CHANGES IN MELT COMPOSITION ON CRYSTAL GROWTH OF BARIUM S 13-1908
POCKELS EFFECT IN LITHIUM NIOBATE CRYSTALS. MELT COMPOSITION (STOICHIOMETRY) DEPENDENCE OF 11-1728
/CE OF LINEAR ELECTROOPTIC EFFECT AND DIELECTRIC CONSTANT ON MELT COMPOSITION (STOICHIOMETRY) IN LITHIUM N/ 11-1842
DIELECTRIC PROPERTIES OF SINGLE DOMAIN MELT GROWN BARIUM TITANATE. 04-0616
/EXTENT OF) TUNGSTEN BRONZE (LIQUIDUS AND SOLIDUS) FIELD AND MELT GROWTH OF CRYSTALS IN THE SODIUM OXIDE- / 13-1949
SINGLE CRYSTAL BARIUM TITANATE FILMS GROWN FROM THE MELT IN AN OXYGEN ATMOSPHERE. 04-0612
ARIUM SODIUM NIOBATE. SOME EFFECTS OF MELT STOICHIOMETRY ON THE OPTICAL PROPERTIES B 13-1902
F BARIUM SODIUM NIOBATE. SOME EFFECTS OF MELT STOICHIOMETRY ON THE OPTICAL PROPERTIES O 13-1903
/GLE CRYSTALS FOR OPTICAL APPLICATIONS. (BY PULLING FROM THE MELT, STRUCTURE- PROPERTY CORRELATION WITH LI/ 13-2045
ELEC/ PREPARATION OF ANTIMONY SULFOIODIDE TEXTURES FROM THE MELT UNDER PRESSURE AND INVESTIGATION OF THEIR 15-2138
IATION OF CRYSTAL GROWTH RATE WITH SUPERSATURATION IN FLUXED MELTS. (BARIUM STRONTIUM NIOBATE) VAR 13-1939
/ED SOLUTION GROWTH OF OXIDE CRYSTALS FROM NONSTOICHIOMETRIC MELTS. (BARIUM TITANATE, STRONTIUM TITANATE, / 04-0553
CONGRUENT MELTING COMPOSITION OF LITHIUM METATANTALATE. 12-1882

PHASE TRANSFORMATIONS AND MELTING IN KDP TO 40 KBAR (BY DTA). 17-2634
/TRITES. PART-1: CHANGE IN ULTRAVIOLET ABSORPTION SPECTRA ON MELTING. PART-2: CHANGE IN ULTRAVIOLET ABSORP/ 16-2278
RUBIDIUM NITRATE NEAR THE MELTING POINT AND ITS POLYMORPHISM. 16-2330
HIGH PRESSURE. MELTING PROPERTIES OF POTASSIUM NITRATE UNDER 16-2265
KALINE TRIHYDROGEN DISELENITE FAMILY. (ESPECIALLY THE SODIUM MEMBER, AND EFFECTS OF DEUTERATION) /OF THE AL 22-2848
TRIC PHASE TRANSFORMATIONS. (AND SUGGESTIONS FOR DISCOVERING MEMBERS OF VACANT CLASSES) / OF THEIR PARAELEC 03-0138
FERROELECTRIC LIGHT VALVE ARRAYS FOR OPTICAL MEMORIES. 37-3676
LFATE BY MEANS OF A LASER BEAM. APPLICATION TO FERROELECTRIC MEMORIES. /ARIZATION REVERSAL IN TRIGLYCINE SU 27-3010
FERROELECTRICS IN OPTICAL MEMORIES AND DISPLAYS: A CRITICAL APPRAISAL. 02-0039
FERROELECTRICS IN OPTICAL MEMORIES AND DISPLAYS: A CRITICAL APPRAISAL. 37-3628
METHOD OF OPTICALLY READING DOMAINS IN BISMUTH TITANATE AND MEMORY APPLICATIONS. A NEW 08-1298
PIEZOELECTRIC RESPONSE OF A FERROELECTRIC MEMORY ARRAY. 37-3635
THE SWITCHING CHARACTERISTICS AND THE BIT SIZE OF THE FE-PC MEMORY DEVICE. /GAUSSIAN IRRADIANCE PROFILE ON 37-3670
QUIVALENT CIRCUIT ANALYSIS OF A FERROELECTRIC PHOTOCONDUCTOR MEMORY DEVICE. E 37-3644
RODE REQUIREMENTS FOR THIN FILM FERROELECTRIC PHOTOCONDUCTOR MEMORY DEVICE. PHOTOCONDUCTOR AND ELECT 37-3671
TITANATE) FERROELECTRIC MEMORY DEVICE PARAMETER STUDY. (LEAD ZIRCONATE 09-1594
A SEMICONDUCTOR- FERROELECTRIC MEMORY DEVICE. (PLZT) 09-1589
FERROELECTRIC- PHOTOCONDUCTOR MEMORY DEVICE. (PLZT) 09-1568
THIN FILM FERROELECTRIC PHOTOCONDUCTOR MEMORY DEVICE. (PLZT) 09-1503
AN ADAPTIVE FERROELECTRIC TRANSFORMER - A SOLID STATE ANALOG MEMORY DEVICE. (USING, FOR EXAMPLE, PZT) 09-1566
G TO SWITCHING AND REMANENCE IN FERROELECTRIC PHOTOCONDUCTOR MEMORY DEVICES. /ETICAL CONSIDERATIONS RELATIN 03-0378
LECTRIC INTERACTIONS FOR ADAPTIVE CONTROL VOLTAGE MO/ ANALOG MEMORY DEVICES EMPLOYING PIEZOELECTRIC- FERROE 37-3663
LSE WIDTH) PARTIAL SWITCHING PROPERTIES OF FERROELECTRICS IN MEMORY DEVICES. (EXAMPLE TRIGLYCINE SULFATE) / 27-3144
XIDE HETEROSTRUCTURE THIN FILM. SWITCHING AND MEMORY EFFECT IN ANTIMONY SULFOIODIDE- TIN DIO 15-2166
PROGRESS TOWARD A FAST, NONVOLATILE, NONDESTRUCTIVE READOUT MEMORY ELEMENT UTILIZING POTASSIUM NITRATE. (/ 16-2269
405) NIOBIUM(0.325) ZIRCONIUM(0.27)) OXYGEN(3) FERROELECTRIC MEMORY FILMS. /(0.07) LANTHANUM(0.01) (IRON(0. 37-3629
MEMORY PLATE) A FERROELECTRIC PIEZOELECTRIC RANDOM ACCESS MEMORY. (LEAD ZIRCONATE- LEAD TITANATE CERAMIC 09-1504
M TITANATE TO DNA. (FERROELECTRICITY, BIOLOGICAL MECHANISMS, MEMORY, ORGANICS) FROM BARIU 02-0092
CAL ACTIVITY IN NON-ENANTIOMORPHOUS BIAXIAL CRYSTALS: METHYL MESITYL OXIDE OXALATE AND SODIUM NITRITE. OPTI 16-2273
CATION. MESOMORPHIC MATERIALS FOR ELECTROOPTICAL APPLI 37-3631
META... SEE THE PREFIXED WORD.
TWO-PHASE FERROELECTRIC SYSTEM. PART-1: BARIUM TITANATE- METAL. 04-0815
GN/ FERROELECTRICS WITH AMMONIUM GROUP ORIENTATION MOBILITY: METAL AMMONIUM (BX4) TYPE COMPOUNDS (PROTON MA 22-2826
ZINC OXIDE AND STRONTIUM TITANATE. (SURFACE BARRIER SYSTEMS, METAL CONTACTS) SOME ELECTRONIC PROPERTIES OF 06-1176
LLY BY ACETATE IONS. PHASE TRANSITION IN DIVALENT METAL DICALCIUM PROPIONATES SUBSTITUTED PARTIA 26-2944
OPTICAL AND HOLOGRAPHIC STORAGE PROPERTIES OF TRANSITION METAL DOPED LITHIUM NIOBATE. 11-1796
PHOTOCHROMISM IN TRANSITION METAL DOPED STRONTIUM TITANATE. 06-1126
SPACE CHARGE LIMITED CURRENTS IN A METAL FERROELECTRIC METAL SYSTEM (THEORY). 03-0348
AHYDRATE. ELECTRON PARAMAGNETIC RESONANCE STUDY OF SOME METAL IONS IN GUANIDINIUM ALUMINUM SULFATE HEX 21-2817
TRODE MATER/ FUNDAMENTAL PROPERTIES OF METAL- FERROELECTRIC- METAL JUNCTION SYSTEMS. (INFLUENCE OF THE ELEC 02-0127
SEEBECK EFFECT IN METAL LEAD TITANATE METAL CONTACT SYSTEMS. 05-1056
/ THERMOELECTRIC EFFECTS IN THE SYSTEM METAL- LEAD TITANATE- METAL. (METHOD OF STUDYING THE TYPE OF CONDUCT 05-1057
TA GADOLINIUM MOLYBDATE. CRYSTAL STRUCTURE OF THE TRANSITION METAL MOLYBDATES AND TUNGSTATES. PART-6. /C BE 30-3289
INFRARED ABSORPTION SPECTROPHOTOMETRIC STUDY OF MONOVALENT METAL NIOBATES. 02-0113
DIPOLE CORRELATIONS IN MOLTEN ALKALI METAL NITRATES AND SODIUM NITRITE. 16-2389
ONSTOICHIOMETRY (AND ANTIFERROELECTRIC DOMAINS IN TRANSITION METAL OXIDES). / EXTENDED DEFECTS RELATED TO N 36-3522
ELECTRICAL CONDUCTION IN METAL OXIDES. (REVIEW, 319 REFS) 02-0109
CHEMISTRY AND PHYSICS OF (ALKALI) METAL TANTALATES. (PREPARATION, REVIEW) 02-0048
ITANATE: VACANCY DRIFT AND OXIDATION REDUCTION OF TRANSITION METALS. ELECTROCOLORATION IN STRONTIUM T 06-1098
NTACTS BETWEEN ANTIMONY SULFOIODIDE SINGLE CRYSTALS AND SOME METALS. (ANTIMONY, SILVER, ALUMINUM) /F THE CO 15-2251
/RMULA METAL-1 METAL-2 SULFATE OR SELENATE OR FLUOBERYLLATE (METALS CHOSEN FROM SODIUM, POTASSIUM, RUBIDIU 19-2743
(INFLUENCE OF THE ELECTRODE MATER/ FUNDAMENTAL PROPERTIES OF METAL- FERROELECTRIC- METAL JUNCTION SYSTEMS. 02-0127
NG THE TYPE OF CONDUCT/ THERMOELECTRIC EFFECTS IN THE SYSTEM METAL- LEAD TITANATE- METAL. (METHOD OF STUDYI 05-1057
STRUCTURES OF METANIOBATES AND METATANTALATES OF TRANSITION METALS OF THE 3D SERIES. /RICAL PROPERTIES AND 36-3527
HIGH PRESSURE SYNTHESIS OF LEAD METAL- OXYGEN(3) TYPE PEROVSKITE. 08-1331
ERROELECTRIC STATE) PYROELECTRIC EFFECT IN METAL-FREE PHTHALOCYANINE. (IN THE POLAR, OR F 36-3540
YLLATE/ INVESTIGATION OF FERROELECTRICS WITH GENERAL FORMULA METAL-1 METAL-2 SULFATE OR SELENATE OR FLUOBER 19-2743
/NVESTIGATION OF FERROELECTRICS WITH GENERAL FORMULA METAL-1 METAL-2 SULFATE OR SELENATE OR FLUOBERYLLATE / 19-2743
URE OF TRIGLYCINE SULFATE. EFFECT OF EVAPORATED METALLIC ELECTRODES ON THE TRANSITION TEMPERAT 27-3100
TRANSIENT NMR STUDY OF PHASE TRANSITIONS IN METALLIC SODIUM TUNGSTEN BRONZES. 13-1910
M TUNGSTATES) RAMAN STUDY OF METALLIC TUNGSTEN BRONZES. (SODIUM AND RUBIDIU 10-1654
E TYPE CRYSTALS. HIGH TEMPERATURE PHASE TRANSITIONS AND METASTABILITY IN POTASSIUM DIHYDROGEN PHOSPHAT 17-2513
" SHIFT OF THE CURIE TEMPERATURE OF A FERROELECTRIC SEMICON/ METASTABLE OPTICAL CHARGE EXCHANGE AND "FROZEN 15-2160
N. (GROWN IN WATER OR ALCOHOL) METASTABLE POTASSIUM NITRATE(III) FROM SOLUTIO 16-2329
TRIC SODIUM NITRITE AND TGS CRYSTALS. (DOMAIN BOUNDARIES AND METASTABLE STRUCTURES) /IC METHOD TO FERROELEC 16-2315
RIC DISPERSION IN BARIUM TITANATE TYPE FERROELECTRICS IN THE METER BAND. DIELECT 04-1009
ELECTRET EFFECT IN A MIXTURE OF POLY METHACRYLATE AND BARIUM TITANATE 04-0838
. (POSSIBLY FERROELECTRIC) DIELECTRIC ANOMALIES IN SOLID METHANE AT LOW TEMPERATURES AND HIGH PRESSURES 36-3547
ATURE. SPECIFIC HEAT AND SUSCEPTIBILITY IN CHROMIUM METHYL AMMONIUM ALUM ABOVE THE CRITICAL TEMPER 20-2805
SPECTROSCOPIC STUDIES OF FERROELECTRIC METHYL AMMONIUM ALUMS. 20-2808
THERMAL CONDUCTIVITY OF METHYL AMMONIUM ALUMINUM ALUM. 20-2807
PHYSICAL PROPERTIES OF FERROELECTRIC ALUMS. (ESPECIALLY METHYL AMMONIUM ALUMINUM SULFATE) 20-2810
M/ INVESTIGATION OF PHASE TRANSITION IN FERROELECTRIC ALUMS (METHYL AMMONIUM ALUMINUM SULFATE, AMMONIUM ALU 20-2809
/PER FORMATE TETRAHYDRATE, AMMONIUM IRON AND CHROMIUM ALUMS, METHYL AMMONIUM ALUMINUM SULFATE DODECAHYDRAT/ 02-0087
. (COMPLEX DIE/ DIELECTRIC RELAXATION IN FERROELECTRIC MASD, METHYL AMMONIUM ALUMINUM SULFATE DODECAHYDRATE 20-2804
DODECAH/ INFRARED AND RAMAN SPECTRA OF FERROELECTRIC ALUMS: METHYL AMMONIUM ALUMINUM (SULFATE OR SELENATE) 20-2803
HEAT CAPACITY OF METHYL AMMONIUM CHROMIUM ALUM. 20-2801
E. OPTICAL ACTIVITY IN NON-ENANTIOMORPHOUS BIAXIAL CRYSTALS: METHYL MESITYL OXIDE OXALATE AND SODIUM NITRIT 16-2273
PHOSPHATE) AMPLITUDE MODULATORS BASED ON THE MICHELSON INTERFEROMETER. (AMMONIUM DIHYDROGEN 17-2715
TION AND DOMAIN STRUCTURE IN ANTIMONY SULFOIODIDE CRYSTALS. (MICRO CINEMATOGRAPHY) PHASE TRANSI 15-2184
ON ELECTRON MICROSCOPE (BARRIER LAYERS IN / IMAGING ELECTRIC MICRO FIELDS (OF SOLID SURFACES) IN THE EMISSI 04-0913
UM TITANATE FILMS. ELECTRON PROBE X-RAY MICROANALYSIS ON IMPURITIES IN EVAPORATED BARI 04-0742
IUM SODIUM NIOBATES. MICROCHEMICAL ANALYSIS OF GADOLINIUM DOPED BAR 13-1953
ALLIZED FROM GLASS. (60-90 VOL PER/ ELECTRICAL PROPERTIES OF MICROCRYSTALLINE FERROELECTRIC MATERIALS CRYST 04-0725
ALT. CORRELATION BETWEEN BARKHAUSEN NOISE AND SURFACE MICRODAMAGES ON SINGLE CRYSTALS OF SEIGNETTE S 28-3208
FERROELECTRIC MAGNETS. (LEAD AND BARIU/ STRUCTURAL MICRODIFFRACTION STUDIES OF FERROELECTRICS AND 04-1038
UDIES OF FERROELECTRICS AND FERROELECTRIC MAGNETS. (ELECTRON MICRODIFFRACTION, SUMMARY ONLY) /STRUCTURAL ST 03-0490
ERROELECTRICS. THE BEHAVIORS OF THE MICRODOMAINS ON THE DIFFUSE PHASE TRANSITION F 03-0283
IC AND FERROMAGNETIC SUBSTANCES. (LEAD TITANATE) STRUCTURAL MICROELECTRON DIFFRACTION STUDY OF FERROELECTR 05-1082
SERVATION OF THE CHANGE IN COOLED KDP TEMPERATURE DUE TO 337 MICRON INCIDENT RADIATION. AN OB 17-2711

NEAR THE CURIE TEMPERATURE.   EMISSION OF LIGHT BY SURFACE MICROPLASMA OF CERAMIC (BARIUM, LEAD) TITANATE 04-0982
                            BARIUM TITANATE PIEZOELECTRIC MICROPROBE FOR STUDYING ULTRASONIC FIELDS.   04-0938
OBSERVATION OF BARIUM TITANATE CRYSTALS BY SCANNING ELECTRON MICROSCOPE.   DIRECT 04-0808
IC DOMAINS IN TRIGLYCINE SULFATE USING THE SCANNING ELECTRON MICROSCOPE.   DIRECT OBSERVATION OF FERROELECTR 27-3051
180 DEG DOMAINS IN BARIUM TITANATE BY MEANS OF A POLARIZING MICROSCOPE.   DIRECT OBSERVATION OF 04-0952
RES, MODULATED BY A PHASE TRANSITION, IN A SCANNING ELECTRON MICROSCOPE.   /N OF FERROELECTRIC DOMAIN STRUCTU 03-0474
UM TITANATE BY MEANS OF A STROBOSCOPIC TRANSMISSION ELECTRON MICROSCOPE.   /N OF SINGLE CRYSTAL FILMS OF BARI 04-0942
RFACES OF FERROELECTRIC CRYSTALS IN A PHOTOEMISSION ELECTRON MICROSCOPE.   OBSERVATION OF THE SU 37-3679
              STUDY OF THE SURFACE OF ICE WITH A SCANNING ELECTRON MICROSCOPE.   34-3420
DIES OF BARIUM TITANATE CERAMICS USING THE SCANNING ELECTRON MICROSCOPE.   SURFACE STU 04-0567
/C MICRO FIELDS (OF SOLID SURFACES) IN THE EMISSION ELECTRON MICROSCOPE (BARRIER LAYERS IN DOPED BARIUM TI/ 04-0913
GS.   SCANNING ELECTRON MICROSCOPE STUDY OF FERROELECTRIC DOMAINS IN T 27-3052
/RIVED FROM THE RHENIUM TRIOXIDE STRUCTURE TYPE: AN ELECTRON MICROSCOPE STUDY OF REDUCED TUNGSTEN TRIOXIDE/ 10-1630
AIN BOUNDARIES. PART-1: DYNAMICAL THEORY. PART-2: / ELECTRON MICROSCOPE TRANSMISSION IMAGES OF COHERENT DOM 03-0255
/Y OF DOMAINS IN FERROELECTRIC CRYSTALS WITH A PHOTOEMISSION MICROSCOPE (TUNGSTEN TRIOXIDE, BARIUM TITANAT/ 04-0811
/OF DOMAINS OF FERROELECTRIC CRYSTALS WITH SCANNING ELECTRON MICROSCOPE. (TUNGSTEN TRIOXIDE SINGLE CRYSTAL/ 10-1646
ELECTRICITY AND ANHARMONIC EFFECTS IN CRYSTALS. (AND CALCU/ (MICROSCOPIC AND THERMODYNAMIC) THEORY OF FERRO 03-0213
F BARIUM TITANATE AND DICALCIUM STRONTIUM P/ ELECTRON MIRROR MICROSCOPIC ASPECTS OF FERROELECTRIC DOMAINS O 04-0973
ON OF THE SOFT ZONE BOUNDARY PHONON IN STRONTIUM TITANATE.   MICROSCOPIC CALCULATION OF THE RESPONSE FUNCTI 06-1213
IUM NIOBATES.   SCANNING ELECTRON MICROSCOPIC EXAMINATION OF SINTERED BARIUM SOD 13-1955
PHASE STRUCTU/ DIELECTRIC SPECTRUM OF BARIUM TITANATE. (AND MICROSCOPIC MODEL OF LOCALLY ORDERED "NUCLEI", 04-0647
OMAINS OF AMMONIUM DIHYDROGEN PHOSPHATE. (PHASE TRANSITION / MICROSCOPIC OBSERVATION OF ANTIFERROELECTRIC D 17-2643
NS.   ELECTRON MIRROR MICROSCOPIC OBSERVATION OF FERROELECTRIC DOMAI 03-0473
NS. (GOLD COATED BARIUM TITANATE)   ELECTRON MIRROR MICROSCOPIC OBSERVATION OF FERROELECTRIC DOMAI 04-0974
UM TITANATE CRYSTALS.   PHOTOEMISSION MICROSCOPIC OBSERVATION OF THE SURFACE OF BARI 04-0810
STEN TRIOXIDE CRYSTALS.   PHOTOEMISSION MICROSCOPIC OBSERVATION OF THE SURFACE OF TUNG 10-1645
IN CESIUM LEAD TRICHLORIDE.   MICROSCOPIC OBSERVATIONS OF PHASE TRANSITIONS 33-3383
F PEROVSKITE TYPE OXIDES. (LEAD TITANATE, POTASSIU/ ELECTRON MICROSCOPIC STUDIES ON FERROELECTRIC DOMAINS O 04-0993
BARIUM TITANATE IN LOW TEMPERATURE PHASES.   ELECTRON MICROSCOPIC STUDY OF FERROELECTRIC DOMAINS OF 04-1034
FUNCTIONS. (SOME COMMENTS)   (MACROSCOPIC AND MICROSCOPIC) SYMMETRY OF FERROELECTRIC ENERGY 03-0193
          CONCERNING THE CRITERION OF FERROELECTRICITY IN THE MICROSCOPIC THEORY.   03-0293
ES OF LEAD MAGNESIUM NIOBATE.   A MICROSCOPIC THEORY FOR THE DIELECTRIC PROPERTI 08-1440
OF FERROELECTRIC CRYSTALS. (CRYSTALLOGRAPHIC VIEWPOINT)   MICROSCOPIC THEORY OF STATIC DOMAIN STRUCTURES 03-0340
ROELECTRIC CRYSTALS. (CALCULATION OF FIELD NEAR SURFACE)   MICROSCOPIC THEORY OF SURFACE PHENOMENA IN FER 04-0501
ROCHELLE SALT.   MICROSCOPIC THEORY OF THE PHASE TRANSITIONS IN 03-0482
F SUBSTOICHIOMETRIC TUNGSTEN OXIDE(3-X) CRYSTALS BY ELECTRON MICROSCOPY.   EXAMINATION O 10-1627
ITCHING IN SINGLE CRYSTAL FILMS OF BARIUM TITANATE (ELECTRON MICROSCOPY). /N OF DOMAIN BOUNDARIES DURING SW 04-0940
LATED TO NONSTOICHIOMETRY (AND ANTIFERROELE/ USE OF ELECTRON MICROSCOPY IN THE STUDY OF EXTENDED DEFECTS RE 36-3522
ROELECTRIC DOMAINS OF GADOLI/ APPLICATION OF ELECTRON MIRROR MICROSCOPY TO DIRECT OBSERVATION OF MOVING FER 30-3293
OCESSES IN ANTIMONY SULFOIODIDE SINGLE CRYSTALS EXCITED WITH MICROSECOND LIGHT PULSES.   RELAXATION PR 15-2202
RROELECTRIC MATERIALS. (FINE POWDE/ QUANTITATIVE ANALYSIS OF MICROSTRUCTURE AND ELECTRICAL PROPERTIES OF FE 02-0093
LASH EVAPORATED BARIUM TITANATE THIN FILM/ GROWTH STRUCTURE (MICROSTRUCTURE) AND ELECTRICAL PROPERTIES OF F 04-0605
ANDIUM TRIOXIDE DOPED, RARE EARTH OXIDE DOPED, AND UNDOPED / MICROSTRUCTURE AND ELECTRICAL PROPERTIES OF SC 04-0839
            RELATION BETWEEN ELECTRICAL PROPERTIES AND MICROSTRUCTURE OF BARIUM TITANATE.   04-0980
SINGLE CRYSTAL FILMS OF BARIUM TITANATE.   MICROSTRUCTURE OF DOMAINS AND DOMAIN WALLS IN 04-1031
ERAMICS.   MICROSTRUCTURE OF TSTS-19 TYPE FERROELECTRIC C 36-3554
ELECTRIC PROPERTIES OF BARIUM TITANATE.   INFLUENCE OF MICROSTRUCTURES (PREPARATION CONDITIONS) ON DI 04-0798
ASE TRANSITIONS IN CRYSTALS OF PEROVSKITE TYPE AND OF ONES / MICROTHEORY OF SPONTANEOUS POLARIZATION AND PH 03-0175
                    FERROELECTRIC MICROTHERMOSTATS.   37-3639
/L) NONLINEAR OPTIC CRYSTAL BARIUM LITHIUM NIOBATE HAVING NO MICROTWINNING (AND RESISTANT TO LASER INDUCED/ 13-1958
TE.   MICROWAVE ABSORPTION IN CUBIC STRONTIUM TITANA 06-1212
E DELAY LINES.   NONLINEAR EFFECTS IN MICROWAVE ACOUSTIC LITHIUM NIOBATE SURFACE WAV 11-1818
AVE VELOCITIES. (BARIUM SODIUM NIOBATE, BISMUTH(12) GERMANI/ MICROWAVE ACOUSTICS HANDBOOK. VOL 1: SURFACE W 01-0028
/ POSSIBLE REALIZATION OF ULTRA LOW NOISE LEVEL PARAELECTRIC MICROWAVE AMPLIFIER (BASED ON FERROELECTRICS I 37-3664
XPERIMENT AND DISCUSSION OF BARIUM (OR LEAD OR CALCIUM) STR/ MICROWAVE BEHAVIOR OF NONLINEAR DIELECTRICS (E 08-1281
8 REFS)   MICROWAVE BEHAVIOR OF ROCHELLE SALT AND TGS (1 27-3088
CS. PART-2: DIELECTRIC CONSTANTS AND DIELECT/ MEASUREMENT OF MICROWAVE DIELECTRIC CONSTANTS OF FERROELECTRI 16-2354
ICS OF VARIOUS TYPES. (BARIUM TITANATE AND TITANAT/ STUDY OF MICROWAVE DIELECTRIC DISPERSION IN FERROELECTR 02-0108
LINE BARIUM TITANATE.   MECHANISM OF MICROWAVE DIELECTRIC DISPERSION IN POLYCRYSTAL 04-0745
ON FERROELECTRICS ABOVE THE CURIE TEMPERATURE. (BARIUM(0-1./ MICROWAVE DIELECTRIC LOSSES IN OXYGEN OCTAHEDR 04-0559
ADIATED FERROELECTRIC AMMONIUM SULFATE.   MICROWAVE DIELECTRIC MEASUREMENTS OF GAMMA-IRR 19-2750
IUM SULFATE.   MICROWAVE DIELECTRIC STUDY OF LITHIUM HYDRAZIN 32-3357
ATE TYPE FERROELECTRICS.   MICROWAVE DISPERSION MECHANISM IN BARIUM TITAN 04-0902
E/ (EFFECT OF TEMPERATURE GRADIENT WITHIN THE SAMPLE ON THE) MICROWAVE DISPERSION OF FERROELECTRICS NEAR TH 03-0370
RIUM TITANATE TYPE FERR/ EXPERIMENTAL CHECK OF THE THEORY OF MICROWAVE DISPERSION OF THE PERMITTIVITY OF BA 04-0649
ATE SINGLE CRYSTALS.   MICROWAVE DYNAMIC NONLINEARITY OF BARIUM TITAN 04-0749
NIOBATE.   NONLINEAR INTERACTION OF MICROWAVE ELECTRIC FIELDS AND SOUND IN LITHIUM 11-1838
ICEUTERIUM PHOSPHATE (COMPARED WITH KDP AND WITH NEUTRON AND MICROWAVE EXPERIMENTS). /C MODE IN POTASSIUM D 17-2705
AR MICROWAVE PROPERTIES OF FERROELECTRIC CERAMICS. THIN FILM MICROWAVE FERROELECTRIC CAPACITORS AND NONLINE 37-3643
E WAVE LOSS MECHANISMS ON BISMUTH(12) GERMANIUM OXIDE(20) AT MICROWAVE FREQUENCIES.   ACOUSTIC SURFAC 35-3511
SUREMENT OF THE DIELECTRIC CONSTANT OF TRIGLYCINE SULFATE AT MICROWAVE FREQUENCIES.   MEA 27-2954
NT OF THE PERMITTIVITY OF BARIUM TITANATE SINGLE CRYSTALS AT MICROWAVE FREQUENCIES.   MEASUREME 04-0759
EASUREMENT OF THE DIELECTRIC PROPERTIES OF FERROELECTRICS AT MICROWAVE FREQUENCIES. /BLEM OF THE ACCURATE M 03-0429
OGEN PHOSPHATE AND AMMONIUM DIHYDROGEN PHOSPHATE CRYSTALS AT MICROWAVE FREQUENCIES. /TY OF POTASSIUM DIHYDR 17-2714
/TION AND ELECTRICAL PROPERTIES OF SOME CERAMIC MATERIALS AT MICROWAVE FREQUENCIES. (TITANATE ZIRCONATE GA/ 09-1602
LYCRYSTALLINE BAR/ NONLINEAR PROPERTIES OF FERROELECTRICS AT MICROWAVE FREQUENCIES. (TRIGLYCINE SULFATE, PO 04-0870
).   DIELECTRIC PROPERTIES OF ANTIMONY SULFOIODIDE AT MICROWAVE FREQUENCIES (3.3 AND 9.8 GHZ, 0-50 C 15-2171
MECHANISMS ON LITHIUM NIOBATE (TEMPERATURE AND FREQUENCY D/ MICROWAVE FREQUENCY ACOUSTIC SURFACE WAVE LOSS 11-1817
AND ANTIFERROELECTRICS OF PEROVSKITE TYPE. (POLYCRYSTALLIN/ MICROWAVE INVESTIGATION OF SOME FERROELECTRICS 08-1375
ATURES ABOVE THE CURIE POINT)   PHENOMENOLOGICAL THEORY OF MICROWAVE LOSSES IN FERROELECTRICS. (AT TEMPER 03-0503
HE PHASE TRANSITION.   MICROWAVE LOSSES IN STRONTIUM TITANATE ABOVE T 06-1199
POTASSIUM DIHYDROGEN PHOSPHATE CRYSTAL.   MICROWAVE MODULATOR OF RUBY LASER LIGHT USING 17-2640
TITANATE, LITHIUM TANTAL/ DETERMINATION OF ABSOLUTE SIGNS OF MICROWAVE NONLINEAR SUSCEPTIBILITIES. (BARIUM 02-0107
FERROELECTRIC (BARIUM TITANATE TYPE) CERAMIC WITH CONTROLLED MICROWAVE PERMITTIVITY. /CHARACTERISTICS OF A 04-0904
NIOBATE.   MICROWAVE PHONON ATTENUATION IN X CUT LITHIUM 11-1756
SOLUTIONS BASED ON BARIUM TITANATE/ DIELECTRIC AND NONLINEAR MICROWAVE PROPERTIES OF POLYCRYSTALLINE SOLID 04-1019
YSTAL DEFECTS ON THE DIELECTRIC BEHAVIOR OF ROCHELLE SALT IN MICROWAVE RANGE.   INFLUENCE OF CR 28-3219
ATION OF NONLINEAR INTERACTIONS IN FERROELECTRIC CERAMICS TO MICROWAVE SIGNAL PROCESSING.   APPLIC 37-3647
            BARIUM MANGANESE FLUORIDE, A NEW CRYSTAL FOR MICROWAVE ULTRASONICS.   33-3406
M TITANATE IN THE CENTIMETER RANGE OF SUPERHIGH FREQUENCIES (MICROWAVES). FERROELECTRIC PROPERTIES OF BARIU 04-0901

SOME PROPERTIES AND APPLICATION OF FERROELECTRICS AT MICROWAVES. 37-3677
ELECTROOPTIC EFFECT IN LITHIUM NIOBATE IN THE (SUB) MILLIMETER RANGE. 11-1844
Y POLARIZATION REVERSAL. MODULATION OF THE MILLIMETER WAVE ABSORPTION IN SODIUM NITRITE B 16-2392
TRIC CONSTANT ABOVE THE CURIE POINT IN VARIOUS FERROELECTRI/ MILLIMETER WAVELENGTH DISPERSION IN THE DIELEC 02-0124
BIREFRINGENCE OF CERTAIN CRYSTALS IN THE MILLIMETER WAVELENGTH RANGE. (LITHIUM NIOBATE) 11-1730
E THE TRANSITION POINT. DISPERSION AT MILLIMETER WAVELENGTHS IN BARIUM TITANATE ABOV 04-0601
MEASUREMENT OF DIELECTRIC PROPERTIES OF FERROELECTRICS AT SUBMILLIMETER WAVELENGTHS. (STRONTIUM TITANATE) 06-1089
FERROELECTRIC BEHAVIOR OF THE MINERAL STIBOTANTALITE. 36-3559
C DOMAINS OF DICALCIUM STRONTIUM PROFIONATE. ELECTRON MIRROR MACROSCOPIC OBSERVATION OF FERROELECTRI 26-2946
MAINS OF BARIUM TITANATE AND DICALCIUM STRONTIUM P/ ELECTRON MIRROR MICROSCOPIC ASPECTS OF FERROELECTRIC DO 04-0973
C DOMAINS. ELECTRON MIRROR MICROSCOPIC OBSERVATION OF FERROELECTRI 03-0473
C DOMAINS. (GOLD COATED BARIUM TITANATE) ELECTRON MIRROR MICROSCOPIC OBSERVATION OF FERROELECTRI 04-0974
ING FERROELECTRIC DOMAINS OF GADOLI/ APPLICATION OF ELECTRON MIRROR MICROSCOPY TO DIRECT OBSERVATION OF MOV 30-3293
IUM DIHYDROGEN PHOSPHATE) PULSE DISTORTIONS IN MISMATCHED SECOND HARMONIC GENERATION. (POTASS 17-2690
CRYSTAL CHEMISTRY OF MIXED BISMUTH OXIDES WITH LAYER TYPE STRUCTURE 13-2032
/TION RELAXATION IN SOME FERROELECTRICS (SOLID SOLUTIONS AND MIXED CATION COMPOUNDS, KANZIG REGIONS, PEROV/ 03-0291
RAMAN SPECTRUM OF ICE I (H). TENTATIVE INTERPRETATION WITH A MIXED COULOMB VALENCE DYNAMICAL MODEL. /QUENCY 34-3425
FERROELECTRICITY IN AMMONIUM(X) POTASSIUM(1-X) NITRATE MIXED CRYSTAL. 16-2401
ITION IN AMMONIUM MONOCHLORO ACETATE, MONOCHLORO ACETIC ACID MIXED CRYSTAL. /RESONANCE STUDY OF PHASE TRANS 36-3625
OSPHATE FILM WAVEGUIDE. EPITAXIAL ELECTROOPTIC MIXED CRYSTAL AMMONIUM POTASSIUM DIHYDROGEN PH 17-2633
/UMINUM SULFATE, AMMONIUM ALUMINUM SULFATE AND A 36:64 RATIO MIXED CRYSTAL, PROTON MAGNETIC RESONANCE, DIE/ 20-2809
TITANATE. PHASE TRANSITIONS IN THE MIXED CRYSTAL SYSTEM STRONTIUM(1-X) BARIUM(X) 04-1016
TANTALATE AND POTASSIUM TANTALATE/ SOFT PHONON MODES IN THE MIXED CRYSTAL SYSTEMS POTASSIUM(X) SODIUM(1-X) 13-1933
ESR STUDY OF ADP- KDA MIXED CRYSTALS. 17-2683
AMMONIUM DIHYDROGEN PHOSPHATE POTASSIUM DIHYDROGEN ARSENATE MIXED CRYSTALS. /F RADIATION DAMAGE CENTERS IN 17-2547
EFFECT IN POTASSIUM DIHYDROGEN(X) DIDEUTERIUM(1-X) PHOSPHATE MIXED CRYSTALS. / FIELD STUDY OF THE ISOTOPIC 17-2563
HOSPHATE- POTASSIUM DIHYDROGEN(1-X) DIDEUTERIUM(X) PHOSPHATE MIXED CRYSTALS. /IES OF POTASSIUM DIHYDROGEN P 17-2680
/R LITHIUM THALLIUM TARTRATE, LITHIUM AMMONIUM TARTRATE, AND MIXED CRYSTALS, BARIUM TITANATE, AND BARIUM S/ 02-0089
DEFECT STRUCTURE OF (BARIUM(1-X) STRONTIUM(X)) TITANATE- MIXED CRYSTALS FROM A POTASSIUM FLUORIDE MELT. 04-0873
NIQUE). PHASE TRANSITIONS IN POTASSIUM NITRATE NITRITE MIXED CRYSTALS (GROWTH BY DOUBLE BRIDGMAN TECH 16-2345
DIELECTRIC STUDIES OF THE CATION SUBSTITUTED MIXED CRYSTALS OF NIOBATES. 11-1848
ATE. PROPERTIES OF MIXED CRYSTALS OF TRIGLYCINE SULFATE AND SELEN 27-3022
A/ DIELECTRIC PROPERTIES AND SOFT MODES IN THE FERROELECTRIC MIXED CRYSTALS POTASSIUM(1-X) SODIUM(X) TANTAL 08-1306
UM TANTALATE SINGLE CRYSTALS AND (POTASSIUM SODIUM TANTALATE MIXED CRYSTALS (100 HZ-10 MHZ, 30-350 C). /SSI 08-1272
MOLECULAR FIELD APPROXIMATION. STUDY OF PHASE TRANSITION IN MIXED KDP- DEUTERATED KDP CRYSTALS BY MODIFIED 17-2528
ROPERTIES OF FERROELECTRICS WITH DIFFUSE PHASE TRANSITIONS. (MIXED LAYER OXIDES) /ELECTRIC AND STRUCTURAL P 13-2014
M AND THALLIUM CONTAINING PEROVSKITES. (TITANATES, HAFNATES, MIXED NIOBATES AND TANTALATES) / OF NEW CADMIU 08-1458
/IES OF THIN FILM MATERIALS WITH PEROVSKITE TYPE STRUCTURE. (MIXED OXIDES ABO(3)) (BARIUM TITANATE AND LIT/ 04-0914
ELECTRICAL AND OPTICAL PROPERTIES OF (BARIUM, STRONTIUM) MIXED TITANATE THIN SECTIONS. 04-0603
FORMATION OF THIN LAYERS OF MIXED TITANATES BY SOLID SOLID REACTIONS. 13-2005
/BILITY OF SWITCHING AND STEERING SURFACE WAVES BY NONLINEAR MIXING IN ANISOTROPIC MEDIA. (LITHIUM NIOBATE/ 11-1778
MICS) ANHARMONIC MODE MIXING IN IONIC FERROELECTRICS. (LATTICE DYNA 03-0234
CRYSTAL FILTER) OPTICAL MIXING OF COHERENT AND INCOHERENT LIGHT. (KDP 17-2669
N LITHIUM NIOBATE. MIXING OF NONCOLLINEAR ELASTIC SURFACE WAVES O 11-1690
GLYCINE SULFATE CRYSTAL) OPTICAL MIXING. (OF PULSED RUBY LASER EMISSIONS BY TRI 27-2958
AMIC SYSTEMS. APPLICATION OF DIELECTRIC MIXTURE FORMULAS TO (LEAD TITANATE) GLASS- CER 05-1054
LECTR/ EFFECT OF A POLARIZING FIELD AND OF THE ADDITION OF A MIXTURE OF COPPER AND CHROMIUM IONS ON THE DIE 27-3045
TE ELECTRET EFFECT IN A MIXTURE OF POLY METHACRYLATE AND BARIUM TITANA 04-0838
TERIALS AT MICROWAVE FREQUENCIES. (TITANATE ZIRCONATE GARNET MIXTURES) /RICAL PROPERTIES OF SOME CERAMIC MA 09-1490
EASUREMENTS OF CERTAIN TWO PHASE FERROELECTRIC FERROMAGNETIC MIXTURES. (BARIUM TITANATE) /IC AND MAGNETIC M 04-0565
ICS OF ILMENITE STRUCTURE. (ABO(3) NIOBATES, TANTALATES, AND MIXTURES WITH THEM) /ESTIGATION OF FERROELECTR 08-1438
GLE CRYSTALS. TEMPERATURE DEPENDENCES OF THE HALL AND DRIFT MOBILITIES OF HOLES IN ANTIMONY TRISULFIDE SIN 15-2193
TE ZIRCONATE TRANSDUCER CER/ DISPERSION STUDY OF DOMAIN WALL MOBILITY AND MECHANICAL DAMPING IN LEAD TITANA 09-1547
TS IN SOLID SOLUTIONS OF (BARIUM, STRONTIUM) TITANATE. MOBILITY DETERMINATIONS FROM WEIGHT MEASUREMEN 04-0689
TEMPERATURE DEPENDENCE OF CARRIER MOBILITY IN BARIUM TITANATE. 04-0614
ELECTRONIC HALL MOBILITY IN THE ALKALINE EARTH FLUORIDES. 33-3404
(PROTON MAGN/ FERROELECTRICS WITH AMMONIUM GROUP ORIENTATION MOBILITY: METAL AMMONIUM (BX4) TYPE COMPOUNDS 22-2826
THEORY OF THE MOBILITY OF STRUCTURAL DEFECTS IN ICE. 03-0244
V CLEAVED TRIGLYCINE SULFATE. FIELD EFFECT AND HALL MOBILITY OF TELLURIUM AND LEAD TELLURIDE ON UH 27-3172
A) OF S/ ELECTRICAL AND OPTICAL PROPERTIES (CONDUCTIVITY AND MOBILITY VERSUS TEMPERATURE, ABSORPTION SPECTR 06-1198
LECTRIC CERAMIC LIGHT GATES OPERATED IN A VOLTAGE CONTROLLED MODE. FERROE 37-3654
BELOW TRANSITION TEMPERATURE. (NO COCHRAN TYPE FERROELECTRIC MODE) /OPERTIES OF VANADIUM DIOXIDE ABOVE AND 36-3526
MODEL OF THE FERROELECTRIC MODE. 03-0170
ALL TRIGLYCINE SULFATE DETECTOR OPERATED IN THE PYROELECTRIC MODE. PERFORMANCE CHARACTERISTICS OF A SM 27-3119
STALS OF LIGHT FROM A LASER OPERATING IN THE FREE GENERATION MODE. /E AND AMMONIUM DIHYDROGEN PHOSPHATE CRY 17-2717
(BY RAMAN SCATTERING). SOFT PHONON MODE AND MODE COUPLING IN ANTIMONY SULFOIODIDE 15-2167
EQUENCY. (POTASSIUM DIHYDROGEN ARSENATE) FERROELECTRIC MODE AND NQR (NUCLEAR QUADRUPOLE RESONANCE) FR 17-2713
HEAT IN FERROELECTRIC PHASE TRANSITION IN TERMS OF A SINGLE MODE ANHARMONIC OSCILLATOR MODEL. /ND SPECIFIC 03-0406
ATES. TEMPERATURE DEPENDENT POLARIZABILITIES AND COCHRAN MODE ASSIGNMENTS IN BARIUM AND STRONTIUM TITAN 04-0805
DEUTER/ MOSSBAUER EFFECT OBSERVATION OF A VERY LOW FREQUENCY MODE AT THE TRANSITION TEMPERATURE IN KDP AND 17-2465
ORY, KDP) SOFT MODE BEHAVIOR IN IMPROPER FERROELECTRICS. (THE 03-0230
IUM MANGANESE TRIFLUORIDE. INFRARED ACTIVE MODE BELOW THE STRUCTURAL TRANSITION IN POTASS 33-3362
/HAMILTONIAN IN TERMS OF LOCALIZED D-STRAIN AND SOFT- NORMAL MODE COORDINATES AND TEMPERATURE INDEPENDENT / 03-0402
IN FERROELECTRIC ANTIMONY SULFOIODIDE. SOFT MODE COUPLING AND CRITICAL RAYLEIGH SCATTERING 15-2232
TIC MOMENT OF CHROMIUM- POTASSIUM ALUM (UNSTABLE SOFT PHONON MODE, DISPLACIVE FERROELECTRICS). / MEAN MAGNE 20-2806
E AND DISPLAY DEVICES. (PLZT CERAMICS) SCATTERING MODE FERROELECTRIC PHOTOCONDUCTOR IMAGE STORAG 09-1599
A NEW LONGITUDINAL DISPLAY MODE FOR CERAMIC ELECTROOPTIC DEVICES. (PLZT) 09-1523
HETEROSTRUCTURE SWITCH. MODE FOR THE ANTIMONY SULFOIODIDE- TIN DIOXIDE 15-2243
NITRITE AT 77 AND 294 DEG K. RAMAN SPECTRA AND MODE FREQUENCY SHIFTS OF FERROELECTRIC SODIUM 16-2262
AL. RAMAN SPECTRUM OF FERROELECTRIC MODE IN A POTASSIUM DIHYDROGEN PHOSPHATE CRYST 17-2663
E MOSSBAUER EFFECT. TEMPERATURE DEPENDENT OPTICAL MODE IN ANTIFERROELECTRIC LEAD ZIRCONATE BY TH 07-1260
RAMAN SCATTERING LINE SHAPE OF THE SOFT E POLARITON MODE IN BARIUM TITANATE. 04-0803
NEW TYPE OF FAR INFRARED SOFT MODE IN FERROELECTRIC GADOLINIUM MOLYBDATE. 30-3310
DE(11). SOFT OPTIC PHONON MODE IN FERROELECTRIC LEAD(5) GERMANIUM(3) OXI 35-3495
FERROELECTRIC DOMAINS AND THE CONDENSED SOFT MODE IN GADOLINIUM MOLYBDATE. 30-3260
NEW TYPE OF FERROELECTRIC SOFT MODE IN GADOLINIUM MOLYBDATE. 30-3311
ANTIFERROELECTRIC MODE IN LEAD ZIRCONATE. (LATTICE VIBRATIONS) 07-1262
RAMAN SCATTERING FROM THE FERROELECTRIC MODE IN LEAD(5) GERMANIUM(3) OXIDE(11). 35-3498
E STUDY OF THE POLARITON ASSOCIATED WITH THE 248 CM(-1) SOFT MODE IN LITHIUM NIOBATE. TEMPERATUR 11-1807
RAMAN SCATTERING BY THE TEMPERATURE DEPENDENT FERROELECTRIC MODE IN POTASSIUM DIDEUTERIUM PHOSPHATE (COMP/ 17-2705

/TRON NUCLEAR DOUBLE RESONANCE OF THE PROTON LATTICE COUPLED MODE IN POTASSIUM DIHYDROGEN ARSENATE, POTASS/ 18-2724
SPIN LATTICE RELAXATION BY THE FERROELECTRIC MODE IN POTASSIUM DIHYDROGEN PHOSPHATE. 17-2440
CTIV/ EFFECT OF PROTON- PHONON COUPLING ON THE FERROELECTRIC MODE IN POTASSIUM DIHYDROGEN PHOSPHATE. (COLLE 17-2655
FAR INFRARED FERROELECTRIC VIBRATION MODE IN STRONTIUM TITANATE. 06-1094
RATURE AND) ELECTRIC FIELD DEPENDENCE OF A HYDROGEN IMPURITY MODE IN STRONTIUM TITANATE. (TEMPE 06-1146
TRANSITION. TEMPERATURE DEPENDENCE OF THE SOFT MODE IN STRONTIUM TITANATE ABOVE THE 105 DEG K 06-1178
ZONE BOUNDARY SOFT MODE IN TERBIUM MOLYBDATE. 30-3269
NATE, POT/ THE GRUNEISEN PARAMETER OF THE SOFT FERROELECTRIC MODE IN THE CUBIC PEROVSKITES. (STRONTIUM TITA 02-0114
NATE. PRESSURE DEPENDENCE OF ACOUSTIC MODE- SOFT OPTIC MODE INTERACTIONS IN FERROELECTRIC BARIUM TITA 04-0893
NATE. ACOUSTIC SOFT OPTIC MODE INTERACTIONS IN FERROELECTRIC BARIUM TITA 04-0671
PHATE. MODE INTERACTIONS IN POTASSIUM DIHYDROGEN PHOS 17-2701
APPARENT SOFT MODE LINE WIDTH DIVERGENCE IN LEAD TITANATE. 05-1078
PARAMETRIC LIGHT AMPLIFICATION AND OSCILLATION IN KDP WITH MODE LOCKED PUMP. 17-2468
DYNAMICS) ANHARMONIC MODE MIXING IN IONIC FERROELECTRICS. (LATTICE 03-0234
/HYDROGEN BONDED FERROELECTRICS. PART-1: (FORMALISM OF LOCAL MODE MODEL.) PART-2: (DYNAMICAL PROPERTIES, T/ 03-0384
. (BY NEUTRON TRIPLE AXIS SPECTROMETR/ FERROELECTRIC (SOFT) MODE MOTION IN POTASSIUM DIDEUTERIUM PHOSPHATE 17-2668
IUM TANTALATE CRYSTAL RESONATOR VIBRATING IN THICKNESS SHEAR MODE OF MOTION. /E DEPENDENCE OF AN X CUT LITH 12-1868
R IMAGE STORAGE AND DISPLAY DEVICES OPERATED IN A REFLECTION MODE. (PLZT) /SED FERROELECTRIC- PHOTOCONDUCTO 09-1564
LECTRICS, CESIUM DIHYDROGEN ARSENATE, STRONTIUM TITANA/ SOFT MODE RAMAN SPECTROSCOPY: COUPLED MODES (FERROE 06-1207
CALCIUM FLUORIDE, STRONTIUM TITANATE, AND CALCIUM TI/ ERASE MODE RECORDING CHARACTERISTICS OF PHOTOCHROMIC 06-1119
(NEAR THE TRANSITION TEMPERATURE, NO ANOMALY BUT SOFT PHONON MODE STILL POSSIBLE). /ERROCYANIDE TRIHYDRATE 24-2891
IONS NEAR THE TRANSITION TEMPERATURE IN STRONTIUM TITA/ SOFT MODE STUDY OF THE STRESS INDUCED PHASE TRANSIT 06-1196
-X) AND CALCIUM(Y) STRONTIUM(1-Y) TITANATE(3-X). SOFT MODE SUPERCONDUCTIVITY IN STRONTIUM TITANATE(3 06-1093
TROPIC SUBSTRATES: GOLD ON LIT/ FUNDAMENTAL AND HIGHER ORDER MODE SURFACE WAVE PROPAGATION ON LAYERED ANISO 11-1846
RIC PROPERTIES OF AN ANHARMONIC CRYSTAL. (EXTENSION OF "SOFT MODE" THEORY) DIELECT 03-0212
ND STRONTIUM TITANATES AND POTASSIUM/ COUPLED OPTICAL PHONON MODE THEORY OF INFRARED DISPERSION IN BARIUM A 04-0547
RIFLUORIDE (ANOMALOUS / ULTRASONIC ATTENUATION NEAR THE SOFT MODE TRANSITION POINT IN POTASSIUM MANGANESE T 33-3373
STRUCTURAL PHASE TRANSITIONS AND SOFT MODES. 03-0176
THE CUBIC- TETRAGONAL TRANSITION IN RELATION TO SOFT PHONON MODES. / SODIUM NIOBATE T(2) AT 600 DEG C, AND 08-1337
DIDE. LATTICE MODES AND PHASE TRANSITION IN ANTIMONY SULFOIO 15-2110
L POTASSIUM TANTALATE NIOBATE. SOFT MODES AND STRUCTURE OF FERROELECTRIC TETRAGONA 13-1957
ATE, STRONTIUM TITANA/ SOFT MODE RAMAN SPECTROSCOPY: COUPLED MODES (FERROELECTRICS, CESIUM DIHYDROGEN ARSEN 06-1207
BONDED FERROELECTRICS (KDP TYPE, THEORY). IMPURITY MODES IN AN INDUCED MOMENT SYSTEM AND HYDROGEN 03-0426
LOW FREQUENCY VIBRATIONAL MODES IN BARIUM SODIUM NIOBATE. 13-1905
NEUTRON SCATTERING STUDY OF SOFT MODES IN CUBIC BARIUM TITANATE. 04-0712
NEUTRON SCATTERING STUDY OF THE SOFT MODES IN CUBIC POTASSIUM TANTALATE NIOBATE. 13-2064
ELASTIC CONSTANTS AND SOFT OPTICAL MODES IN GADOLINIUM MOLYBDATE. 30-3284
S. (RAMAN SPECTRA, DIPOLAR COUPLING MODEL COMP/ LONGITUDINAL MODES IN HYDROGEN AND DEUTERIUM HALIDE CRYSTAL 31-3326
LASER RAMAN STUDY OF QUASI SPIN WAVE HYDROGEN TUNNELING MODES IN KDP AND DEUTERATED KDP. 17-2569
RAMAN STUDIES (TEMPERATURE DEPENDENCE) OF UNDER DAMPED SOFT MODES IN LEAD TITANATE. 05-1041
TERING). SOFT FERROELECTRIC MODES IN LEAD TITANATE (NEUTRON INELASTIC SCAT 05-1077
ASSIGNMENTS OF OPTICAL PHONON MODES IN LITHIUM NIOBATE. 11-1694
INTERACTION OF VOLUME ACOUSTIC MODES IN LITHIUM NIOBATE. 11-1806
Y ANISOTROPIC DEBYE- WALLER FAC/ LOW FREQUENCY ZONE BOUNDARY MODES IN PEROVSKITE FERROELECTRICS INDICATED B 08-1351
EUTERATED AMMONIU/ DETERMINATION OF THE EIGENVECTORS OF SOFT MODES IN POTASSIUM DIDEUTERIUM PHOSPHATE AND D 17-2700
INFRARED ACTIVE MODES IN POTASSIUM TANTALATE. 08-1366
ITANATE ZIRCONATE SYSTEM (MEASUREMENT OF SOFT OPTICAL PHONON MODES IN TETRAGONAL FERROELECTRIC REGION). / T 09-1497
SSIUM(1-X) SODIUM(X) TANTALA/ DIELECTRIC PROPERTIES AND SOFT MODES IN THE FERROELECTRIC MIXED CRYSTALS POTA 08-1306
ON THE SYMMETRY OF THE SOFT MODES IN THE FERROELECTRIC PHASE TRANSITIONS. 03-0357
SENATES. (KDP, AMMONIUM/ RAMAN SCATTERING FROM FERROELECTRIC MODES IN THE KDP ISOMORPHOUS PHOSPHATES AND AR 17-2639
) SODIUM(1-X) TANTALATE AND POTASSIUM TANTALATE/ SOFT PHONON MODES IN THE MIXED CRYSTAL SYSTEMS POTASSIUM(X 13-1933
SSIUM DIHYDROGEN PHOSPHATE. (T/ ANALYSIS OF PHONON TUNNELING MODES IN THE RAMAN SPECTRUM OF TETRAGONAL POTA 03-0396
ITION IN STRONTIUM TIT/ LIGHT SCATTERING STUDIES OF THE SOFT MODES NEAR THE CUBIC TO TETRAGONAL PHASE TRANS 06-1242
NITRATE. TWO MODES OF ELECTRICAL CONDUCTIVITY IN POTASSIUM 16-2368
T/ INFLUENCE OF THE PHASE TRANSITION ON THE NORMAL VIBRATION MODES OF FERROELECTRIC CRYSTALS AT LOW TEMPERA 17-2456
NOTE ON THE LATTICE DYNAMICAL ASPECT OF FERROELECTRIC MODES OF KDP. (CALCULATIONS) 17-2498
ULTANEOUS OBSERVATION OF THE SOFT FERROELECTRIC AND ACOUSTIC MODES OF POTASSIUM DIDEUTERIUM PHOSPHATE. SIM 17-2636
TAS/ LIGHT SCATTERING STUDIES OF THE SOFT OPTIC AND ACOUSTIC MODES OF POTASSIUM DIHYDROGEN PHOSPHATE AND PO 17-2497
RONTIUM TITANATE DATA) OPTICAL MODES OF THE PEROVSKITE STRUCTURE. (THEORY, ST 03-0304
E PHONON SPECTRA OF STRONTIUM TITANATE AND POTASSIU/ COUPLED MODES (OPTICAL, FERROELECTRIC- ACOUSTIC) IN TH 06-1157
TANTALATE, BARI/ NEUTRON INELASTIC SCATTERING STUDY ON SOFT MODE. (REVIEW OF BROOKHAVEN WORK ON POTASSIUM 02-0115
CTRIC BARIUM TITANATE. PRESSURE DEPENDENCE OF ACOUSTIC MODE- SOFT OPTIC MODE INTERACTIONS IN FERROELE 04-0893
ETWEEN 480 AND 575 DEG C, AND THEIR RELEVANCE TO SOFT PHONON MODES. (X-RAY DIFFRACTION) /F SODIUM NIOBATE B 08-1273
HE TRANSITION POINT OF DIFFUSE FERROELECTRIC PHASE. (KAENZIG MODEL) /ATION AND DIELECTRIC HYSTERESIS NEAR T 03-0437
UM PHOSPHATE (FROM POLARIZED REFLECTION SPECTRA, A TUNNELING MODEL). /UTERON MOTIONS IN POTASSIUM DIDEUTERI 17-2645
RANSITION IN A FERROELECTRIC WITH HYDROGEN BONDS (THEORY, WU MODEL). /OF ELECTRIC NEUTRALITY ON THE PHASE T 03-0449
E TRANSITION IN TERMS OF A SINGLE MODE ANHARMONIC OSCILLATOR MODEL. /ND SPECIFIC HEAT IN FERROELECTRIC PHAS 03-0406
ECTRIC TRANSITION REGION (THEORY, RELATION TO KAENZIG DOMAIN MODEL). HEAT CAPACITY IN THE FERROEL 03-0443
ROVSKITE LIKE FERROELECTRICS, USING AN ANHARMONIC OSCILLATOR MODEL. /EARITY NEAR THE PHASE TRANSITION IN PE 03-0249
ENTROPY FORM OF ICE I(H) OBTAINED FROM A LINEAR STEP GROWTH MODEL. LOW 34-3410
TATIVE INTERPRETATION WITH A MIXED COULOMB VALENCE DYNAMICAL MODEL. /QUENCY RAMAN SPECTRUM OF ICE I(H). TEN 34-3425
S IN CRYSTALS. (AND CALCULATIONS BASED ON STRONTIUM TITANATE MODEL) /FERROELECTRICITY AND ANHARMONIC EFFECT 03-0213
SOLUTION OF AN ELASTIC FERROELECTRIC MODEL. 03-0211
ONAL FEATURES OF FERROELECTRIC PHASE TRANSITIONS IN VIBRONIC MODEL. SOME ADDITI 03-0339
ERISTICS AND VARIOUS CONCEPTIONS OF THE TUNNELING QUASI SPIN MODEL. / (KDP TYPE) PART-2: SCATTERING CHARACT 03-0478
THERMODYNAMICS OF AN ELASTIC FERROELECTRIC MODEL. 03-0240
ONS IN ICE. THE THREE DIMENSIONAL EIGHT VERTEX MODEL AND THE PROTON PROTON CORRELATION FUNCTI 34-3481
ION IN POTASSIUM FERROCYANIDE AND ISOMORPHOUS CRYSTALS. (NEW MODEL BASED ON NMR OF UNTWINNED SAMPLES) /NSIT 24-2893
TESTING A POINT DIPOLE ELECTROOPTIC MODEL BY AN APPLICATION TO LEAD TITANATE. 05-1073
DEUTERIUM HALIDE CRYSTALS. (RAMAN SPECTRA, DIPOLAR COUPLING MODEL COMPUTATION) /INAL MODES IN HYDROGEN AND 31-3326
PHASE TRANSITIONS IN A MODEL FERROELECTRIC. 03-0257
PHASE TRANSITIONS IN A MODEL FERROELECTRIC. 03-0256
PHASE TRANSITION IN A SIXTEEN VERTEX LATTICE MODEL (FERROELECTRIC SQUARE LATTICE). 03-0517
NONSTATIONARY POLARIZATION IN THE ANHARMONIC VIBRATOR MODEL. (FERROELECTRICS) 03-0345
STATIC SHELL MODEL FOR BARIUM AND LEAD TITANATES. 04-0826
ION TO CALCULATING THE WALL ENERGY AND T/ THEORETICAL STATIC MODEL FOR BARIUM AND LEAD TITANATES - APPLICAT 03-0189
ONS. ACCOUNTING FOR INTERACTION IN THE KANZIG REGIONS MODEL FOR DIFFUSE FERROELECTRIC PHASE TRANSITI 03-0438
MODEL FOR FERROELECTRIC GADOLINIUM MOLYBDATE. 30-3313
XAGONAL ICE. A STUDY OF BROUT'S MODEL FOR FERROELECTRICS. INVESTIGATIONS ON HE 34-3443
RANDOM PHASE APPROXIMATION FOR THE BLINC-DE GENNES MODEL FOR HYDROGEN BONDED FERROELECTRICS. 03-0338

/TY OF CLASSICAL ANHARMONIC OSCILLATOR. A UNIFIED OSCILLATOR MODEL FOR ORDER DISORDER AND DISPLACIVE FERRO/ 03-0405
ITH A SHORT RANGE INTER/ CRITICAL BEHAVIOR OF THE PSEUDOSPIN MODEL FOR ORDER DISORDER TYPE FERROELECTRICS W 03-0423
NTED, TRIGLYCINE SUL/ MAGNETIC RESONANCE IN FERROELECTRICS. (MODEL FOR POTASSIUM DIHYDROGEN PHOSPHATE PRESE 03-0180
NIOBATE AND TANTALATE AND THE EFFECT ON THE / STACKING FAULT MODEL FOR STOICHIOMETRY DEVIATIONS IN LITHIUM 11-1776
NIOBATE AND TANTALATE AND THE EFFECT ON THE / STACKING FAULT MODEL FOR STOICHIOMETRY DEVIATIONS IN LITHIUM 11-1777
CTRIC IRON (1-X) SULFIDE. PART-1. PART-2.   MODEL FOR THE SEMICONDUCTING UNIAXIAL FERROELE 36-3615
        FERROELECTRIC TRANSFORMATION IN BARIUM TITANATE. (MODEL GIVES NEARLY COMPLETE DESCRIPTION)   04-0836
RIENTATIONALLY DISORDERED SYSTEMS. PART-1: A TWO DIMENSIONAL MODEL. (ICE I)   ATOMIC VIBRATIONS IN O 34-3474
            MODIFIED POTASSIUM DIHYDROGEN PHOSPHATE MODEL IN A STAGGERED FIELD.   17-2710
                ONE DIMENSIONAL KDP MODEL IN STATISTICAL MECHANICS.   03-0389
CTRICITY (THEORY).   SELF CONSISTENCY OF LANDAU'S MODEL IN THE TRANSITION FROM PIEZO TO FERROELE 03-0507
/ PHASE TRANSFORMATION OF DIAMMONIUM DICADMIUM TRISULFATE, A MODEL IN WHICH THE POLARIZATION VECTOR, P(S),/ 03-0141
        SECOND ORDER PHASE TRANSITION IN THE THREE DIMENSIONAL MODEL OF A FERROELECTRIC.   03-0450
PARENT VIOLATION OF THE UNIVERSALITY HYPOTHESIS IN A LATTICE MODEL OF A PHASE TRANSITION.   AP 03-0515
STRUCTURE) OF SOME ABO(3) FERROELECTRICS. (PHENOMENOLOGICAL MODEL OF ELECTROOPTICAL EFFECT) /CTRA AND BAND 08-1383
            INDUCED POLARIZATION AND A ONE POLARIZATION MODEL OF FERROELECTRICITY.   03-0329
/RIBUTION OF ZEROES OF THE PARTITION FUNCTION FOR THE SLATER MODEL OF FERROELECTRICITY (KDP MODEL, ANTIFER/ 03-0308
            LATTICE DYNAMICS OF A RIGID ION MODEL OF GADOLINIUM MOLYBDATE.   30-3281
ORDER DISORDER TRANSITION OF SODIUM NITRITE. (TWO SUBLATTICE MODEL OF INTERACTING DIPOLES)   THE 16-2287
TANATE.   SHELL MODEL OF INTERIONIC INTERACTIONS FOR BARIUM TI 04-0716
FFERENT ORIENTATIONS OF THE SPONTANEOUS POLARIZATION.   A MODEL OF KANZIG REGIONS TAKING INTO ACCOUNT DI 03-0439
DE IN POTASSIUM DIHYDROGEN PHOSPHATE. (COLLECTIVE PSEUDOSPIN MODEL OF KDP IS QUESTIONABLE) /ERROELECTRIC MO 17-2655
TU/ DIELECTRIC SPECTRUM OF BARIUM TITANATE. (AND MICROSCOPIC MODEL OF LOCALLY ORDERED "NUCLEI", PHASE STRUC 04-0647
LE DOMAIN) FERROELECTRIC TYPE CRYSTAL.   MECHANICAL MODEL OF LONG WAVELENGTH VIBRATIONS IN A (SING 03-0342
                          DYNAMIC POLAR MODEL OF MONOCRYSTALLINE ICE I (H).   34-3424
C PHASE.   LATTICE DYNAMICS MODEL OF POTASSIUM NITRATE IN ITS FERROELECTRI 16-2377
TIFERRODIELECTRIC.   EXCITON MODEL OF SPIN AND OPTICAL EXCITATIONS OF AN AN 03-0416
SCALING HYPOTHESIS AND ITS IMPLICATIONS FOR THE PLANAR SPIN MODEL OF STRONTIUM TITANATE.   STATIC 06-1223
        QUANTUM FIELD THEORETICAL TREATMENT OF A DYNAMICAL MODEL OF THE FERROELECTRIC KDP. PART-1. PART-2 03-0279
                          MODEL OF THE FERROELECTRIC MODE.   03-0170
                    DYNAMIC MODEL OF THE SLATER KDP FERROELECTRIC.   03-0285
ITRARY EXTERNAL ELECTRIC FIELD.   EXACT SOLUTION OF MODEL OF TWO DIMENSIONAL FERROELECTRICS IN ARB 03-0520
                    GENERALIZED FERROELECTRIC MODEL ON A SQUARE LATTICE.   03-0168
/GEN BONDED FERROELECTRICS. PART-1: (FORMALISM OF LOCAL MODE MODEL.) PART-2: (DYNAMICAL PROPERTIES, TRANSV/ 03-0384
TURE DEPENDENT SPECTRUM OF AN ANTIFERROELECTRIC LINEAR CHAIN MODEL. (PEROVSKITES)   TEMPERA 03-0468
IN CHROMIUM (3+) DOPED STRONTIUM TITANATE (ESR MEASUREMENTS, MODEL PROPOSED). /TION BY OXYGEN (2-) VACANCIES 06-1161
THE SWITCHING PROPERTIES OF THE FERROELECTRIC SURFACES. (NEW MODEL REPRODUCES EXPERIMENTAL PULSES)   ON 37-3641
TYPE-II PHASE CHANGE IN THREE DIMENSIONAL FERROELECTRIC MODEL. (THEORY)   03-0451
                TWO POLARIZATION MODEL THEORY OF FERROELECTRICITY.   03-0332
            A SIMPLE MECHANICAL MODEL WITH FERROELECTRIC BEHAVIOR.   03-0151
        CIRCLE THEOREM FOR ICE RULE FERROELECTRIC MODELS.   03-0200
PART-2, HYDROGEN BONDED FERROELECTRIC AND ANTIFERROELECTRIC MODELS. / PART-1, THE RESIDUAL ENTROPY OF ICE, 03-0388
TUDY OF THE LATTICE DYNAMICS OF STRONTIUM TITANATE: HARMONIC MODELS.   NEUTRON INELASTIC SCATTERING S 06-1225
            TWO DIMENSIONAL FERROELECTRIC MODELS.   03-0365
UNCTION FOR THE HEISENBERG, FERROELECTRIC, AND GENERAL ISING MODELS.   ZEROES OF THE PARTITION F 03-0481
ANALYTIC PROPERTIES OF THE FREE ENERGY FOR THE "ICE" MODELS (FERROELECTRICS).   34-3427
OBSERVATIONS IN BARIUM TITANATE.   NUCLEATION MODELS FOR A FIRST ORDER TRANSITION FRONT, AND 04-0608
PART-1: POLARIZATION SPECTRA OF ICE I (H). PART-2: MOLECULAR MODELS FOR POLARIZATION AND CONDUCTION. /TALS. 34-3486
/THROUGH "PURE" ICE I (H), SINGLE CRYSTALS. PART-2: MOLECULAR MODELS FOR POLARIZATION AND CONDUCTION. PART-/ 34-3455
/SECT-1: POLARIZATION SPECTRA OF ICE I (H). SECT-2: MOLECULAR MODELS FOR POLARIZATION AND CONDUCTION. SECT-/ 34-3482
DIUM TRIHYDROGEN SELENATE.   MODELS FOR THE ORDER DISORDER TRANSITION IN SO 22-2862
ONSTRAINTS ON THE PHASE TRANSITIONS OF SOME PHENOMENOLOGICAL MODELS OF FERROELECTRICS. / OF ELECTROSTATIC C 03-0280
HYDROGEN PHOSPHATE TYPE FERROELECTRICS. (24 REFS)   MODELS OF THE PHASE TRANSITION IN POTASSIUM DI 03-0489
ANALYTICAL PROPERTIES OF A CLASS OF PLANAR FERROELECTRIC MODELS. (THEORY, KDP TYPES)   03-0258
                          MODERN APPLICATIONS OF FERROELECTRICITY.   02-0080
CTRICITY)   MODERN ASPECTS OF PHYSICS OF SOLIDS. (FERROELE 01-0019
HOUT COEXISTING PHASES.   REINTERPRETATION OF WAVELENGTH MODULATED ABSORPTION IN STRONTIUM TITANATE WIT 06-1188
ELECTRON M/ OBSERVATION OF FERROELECTRIC DOMAIN STRUCTURES, MODULATED BY A PHASE TRANSITION, IN A SCANNING 03-0474
.   ELECTROSTRICTIVELY GENERATED AND MODULATED FRACTURE IN BARIUM TITANATE CERAMICS 04-0700
NOVEL TYPE OF CUT FOR KDP CRYSTALS FOR LOW VOLTAGE LIGHT MODULATION.   17-2531
CUT IN X DIHYDROGEN PHOSPHATE CRYSTALS FOR TRANSVERSE LIGHT MODULATION.   OPTIMUM 17-2494
UT OF POTASSIUM DIHYDROGEN PHOSPHATE TYPE CRYSTALS FOR LIGHT MODULATION.   OPTIMUM C 17-2532
INT DEFECTS IN LITHIUM NIOBATE SINGLE CRYSTALS USED IN LASER MODULATION.   PO 11-1737
TROOPTIC POTASSIUM TANTALATE NIOBATE GRATINGS FOR LIGHT BEAM MODULATION AND DEFLECTION.   ELEC 13-1929
DEUTER/ REAL-TIME HOLOGRAPHIC RECONSTRUCTION BY ELECTROOPTIC MODULATION (BY PASSING COHERENT LIGHT THROUGH 17-2507
/FOR OPTICAL COMMUNICATIONS (REVIEW EMPHASIZING ELECTROOPTIC MODULATION, LITHIUM TANTALATE, LITHIUM NIOBAT/ 02-0056
IN POTASSIUM DIHYDROGEN PHOSPHATE.   ELECTROOPTIC AMPLITUDE MODULATION OF LASER GENERATED SECOND HARMONICS 17-2694
N SODIUM NITRITE BY POLARIZATION REVERSAL.   MODULATION OF THE MILLIMETER WAVE ABSORPTION I 16-2392
SATELLITE X-RAY SCATTERING AND STRUCTURAL MODULATION OF THIOUREA.   25-2921
DENCE OF POLARITON DISPERSION IN LITHIUM TANTALATE. (OPTICAL MODULATION VIA TEMPERATURE)   TEMPERATURE DEPEN 12-1873
CATION OF A LITHIUM TANTALATE TEMPERATURE STABILIZED OPTICAL MODULATOR.   FABRI 12-1872
            LITHIUM NIOBATE LIGHT MODULATOR.   11-1782
OPTIMUM CUT FOR A LITHIUM NIOBATE TRANSVERSE LIGHT MODULATOR.   11-1755
APPLICATIONS OF GADOLINIUM MOLYBDATE (OPTICAL SHUTTER, COLOR MODULATOR).   (PROPERTIES,) FUNCTION AND 30-3298
ITIVE CRYSTALS AS POCKELS EFFECT LIGHT VALVES IN DISPLAY AND MODULATOR APPLICATIONS. USE OF CUBIC PHOTOSENS 37-3661
CRYSTAL)   A NEW SECOND HARMONIC TYPE FERROELECTRIC MODULATOR FOR ELECTROMETER. (WITH A TGS SINGLE 27-2951
TAL ORIENTATION FOR A BARIUM SODIUM NIOBATE TRANSVERSE LIGHT MODULATOR. (HALF WAVE VOLTAGE, THEORY) /M CRYS 13-1973
HYDROGEN PHOSPHATE OR ARSENATE, POTASSI/ LINEAR ELECTROOPTIC MODULATOR MATERIALS. (POTASSIUM OR AMMONIUM DI 01-0021
DIHYDROGEN PHOSPHATE CRYSTAL.   MICROWAVE MODULATOR OF RUBY LASER LIGHT USING POTASSIUM 17-2640
TE)   POLARIZATION REFLECTION TYPE LIGHT MODULATOR USING FERROELECTRICS. (BARIUM TITANA 04-0714
PERFORMANCE OF AN 11 GHZ OPTICAL MODULATOR USING LITHIUM TANTALATE.   12-1888
IAXIAL CRYSTALS OF THE TYPE X DIHYDROGEN PHOSPHATE FOR LIGHT MODULATORS.   CUTTING AND POLISHING IN UN 17-2559
            PULSE RESPONSE OF ZERO DEGREE CUT ADP MODULATORS.   17-2635
ER. (AMMONIUM DIHYDROGEN PHOSPHATE)   AMPLITUDE MODULATORS BASED ON THE MICHELSON INTERFEROMET 17-2715
EMPHASIZING ELECTROOPTIC MODUL/ (HIGH SPEED, SMALL APERTURE) MODULATORS FOR OPTICAL COMMUNICATIONS (REVIEW 02-0056
            OPTIMISATION OF ELECTROOPTIC CRYSTAL MODULATORS. (KDP)   17-2541
/SE OF PEROVSKITE PARAELECTRICS IN BEAM DEFLECTORS AND LIGHT MODULATORS. (POTASSIUM TANTALATE (0.65) NIOBAT/ 13-1924
Y CUT ADP CRYSTALS.   LIGHT MODULATORS USING 45 DEGREE X CUT AND 45 DEGREE 17-2605
CUTS OF LITHIUM NIOBATE CRYSTALS SUITABLE FOR LIGHT MODULATORS WITH A LOW HALF WAVE VOLTAGE.   11-1715
POTASSIUM DI/ TEMPERATURE INSTABILITY OF ELECTROOPTIC LASER MODULATORS WITH AMMONIUM DIHYDROGEN PHOSPHATE, 17-2675

RIC- FERROELECTRIC INTERACTIONS FOR ADAPTIVE CONTROL VOLTAGE MODULES. / MEMORY DEVICES EMPLOYING PIEZOELECT 37-3663
PIEZOELECTRIC AND ELECTROOPTIC MODULI OF FERROELECTRIC POTASSIUM IODATE. 08-1431
ELASTIC MODULI OF ICE. 34-3421
LOW TEMPERATURE ELASTIC MODULI OF STRONTIUM TITANATE (10 TO 130 DEG K) 06-1190
FRICTION IN BARIUM TITANATE SINGLE CRYSTALS. (PIEZOELECTRIC MODULUS) AMPLITUDE DEPENDENCE OF THE INTERNAL 04-0791
/DIDEUTERIUM PHOSPHATE (TEMPERATURE DEPENDENCE OF THE YOUNGS MODULUS AND THE POLARIZATION RELAXATION TIME). 17-2660
NE BARIUM TITANATE PARTICULATES. EFFECT OF SURFACE MOISTURE ON THE DIELECTRIC BEHAVIOR OF ULTRAFI 04-0859
IC AND FERROELECTRIC THIOUREA. PART-1: TEMPERATURE DEPENDEN/ MOLECULAR AND LATTICE VIBRATIONS OF ORTHORHOMB 25-2920
IC AND FERROELECTRIC THIOUREA. PART-2: CALCULATION OF THE I/ MOLECULAR AND LATTICE VIBRATIONS OF ORTHORHOMB 25-2907
NONLINEAR OPTICAL PROPERTIES OF MOLECULAR CRYSTALS. 02-0078
TRANSITION IN MIXED KDP- DEUTERATED KDP CRYSTALS BY MODIFIED MOLECULAR FIELD APPROXIMATION. STUDY OF PHASE 17-2528
N POTASSIUM DIHYDROGEN(X) DIDEUTERIUM(1-X) PHOSPHATE MIXED / MOLECULAR FIELD STUDY OF THE ISOTOPIC EFFECT I 17-2563
MOLECULAR FORCES OF HEAVY AND LIGHT ICE. 34-3440
/F PROTONS THROUGH "PURE" ICE I(H), SINGLE CRYSTALS. PART-2: MOLECULAR MODELS FOR POLARIZATION AND CONDUCT/ 34-3455
/CRYSTALS. SECT-1: POLARIZATION SPECTRA OF ICE I(H). SECT-2: MOLECULAR MODELS FOR POLARIZATION AND CONDUCT/ 34-3482
/CRYSTALS. PART-1: POLARIZATION SPECTRA OF ICE I(H). PART-2: MOLECULAR MODELS FOR POLARIZATION AND CONDUCT/ 34-3486
HYDROGEN CHLORIDE. (CALCULATED IN THE HARMONIC APPROXIMATI/ MOLECULAR MOTION IN THE FERROELECTRIC PHASE OF 31-3340
MOLECULAR INTERPRETATION OF THE PHASE DIAGRAM OF ICE. PART-/ MOLECULAR PHENOMENA IN WATER SYSTEMS. PART-1: 34-3485
CESSES BY ICE RESEARCH. MOLECULAR UNDERSTANDING OF ELECTROCHEMICAL PRO 34-3484
E AND FREQUENCIES OF THE FUNDAMENTAL VIBRATIONS OF THE WATER MOLECULES. /DINIUM ALUMINUM SULFATE HEXAHYDRAT 21-2811
ERMINATION OF THE ORIENTATIONS OF PROTON PAIRS (IN THE WATER MOLECULES). /C RESONANCE IN ROCHELLE SALT, DET 28-3203
N POTASSIUM FERROCYANID/ ORDERING OF DIPOLE MOMENTS OF WATER MOLECULES AND FERROELECTRIC PHASE TRANSITION I 24-2893
LE SALT BY SPIN LATTICE RELAX/ STUDY OF SLOW MOTION OF WATER MOLECULES IN ROCHELLE SALT AND AMMONIUM ROCHEL 28-3232
S IN THE FREQUENCY RANGE OF DOUBLE ELECTRONIC EXCITATIONS OF MOLECULES (IONS). /TION BY ANTIFERRODIELECTRIC 03-0419
E. DIPOLE CORRELATIONS IN MOLTEN ALKALI METAL NITRATES AND SODIUM NITRIT 16-2389
SINGLE CRYSTAL GROWTH OF STRONTIUM POTASSIUM TANTALATE FROM MOLTEN SALTS. 13-2034
LITHIU/ (TETRAGONAL) TUNGSTEN BRONZE TYPE CRYSTALS GROWN IN MOLTEN SOLUTIONS OF THE PSEUDOSYSTEM POTASSIUM 13-1948
ION OF ELASTIC WAVES NEAR THE PHASE TRANSITION OF GADOLINIUM MOLYBDATE. ABNORMAL ABSORPT 30-3265
US OPTICAL BEHAVIOR OF THE IMPROPER FERROELECTRIC GADOLINIUM MOLYBDATE. AN EXPLANATION OF ANOMALO 30-3292
HE PHASE TRANSITION IN FERROELASTIC FERROELECTRIC GADOLINIUM MOLYBDATE. ANOMALOUS PHONON BEHAVIOR NEAR T 30-3276
CT OBSERVATION OF MOVING FERROELECTRIC DOMAINS OF GADOLINIUM MOLYBDATE. /ELECTRON MIRROR MICROSCOPY TO DIRE 30-3293
ROELASTIC AND PARAELECTRIC- PARAELASTIC PHASES OF GADOLINIUM MOLYBDATE. /AY STUDY OF THE FERROELECTRIC- FER 30-3287
CONTROL AND APPLICATION OF DOMAIN WALL MOTION IN GADOLINIUM MOLYBDATE. 30-3257
CONTROL AND APPLICATION OF DOMAIN WALL MOTION IN GADOLINIUM MOLYBDATE. 30-3256
DISLOCATION ETCHANTS FOR BETA GADOLINIUM MOLYBDATE. 30-3258
DOMAIN STRUCTURE OF GADOLINIUM MOLYBDATE. 30-3301
CT OF HYDROSTATIC PRESSURE ON PHASE TRANSITION IN GADOLINIUM MOLYBDATE. EFFE 30-3321
ELASTIC CONSTANTS AND SOFT OPTICAL MODES IN GADOLINIUM MOLYBDATE. 30-3284
OPTICAL, AND MECHANICAL BEHAVIOR OF FERROELECTRIC GADOLINIUM MOLYBDATE. ELECTRICAL, 30-3267
ELECTROOPTICAL PROPERTIES OF FERROELECTRIC GADOLINIUM MOLYBDATE. 30-3277
FERROELECTRIC AND FERROELASTIC CRYSTAL, GADOLINIUM MOLYBDATE. 30-3308
FERROELECTRIC DOMAIN SHIFTING DEVICES. (GADOLINIUM MOLYBDATE) 30-3280
ROELECTRIC DOMAINS AND THE CONDENSED SOFT MODE IN GADOLINIUM MOLYBDATE. FER 30-3260
FERROELECTRIC FERROELASTIC CRYSTAL GADOLINIUM MOLYBDATE. 30-3297
FERROELECTRICITY IN GADOLINIUM MOLYBDATE. 30-3266
ONTANEOUS BIREFRINGENCE IN IMPROPER FERROELECTRIC GADOLINIUM MOLYBDATE. INDUCED AND SP 30-3278
OPTICAL PROPERTIES OF FERROELECTRIC FERROELASTIC GADOLINIUM MOLYBDATE. /NCES OF ELECTRICAL, MECHANICAL AND 30-3253
OF ACTIVATED FERROELECTRICS. (LITHIUM NIOBATE AND GADOLINIUM MOLYBDATE) LASER AND SPECTROSCOPIC PROPERTIES 11-1741
LATTICE DYNAMICS OF A RIGID ION MODEL OF GADOLINIUM MOLYBDATE. 30-3281
TRIC PHASE TRANSFORMATION IN RARE EARTH MOLYBDATES. (TERBIUM MOLYBDATE) MECHANISM OF THE FERROELEC 30-3254
MODEL FOR FERROELECTRIC GADOLINIUM MOLYBDATE. 30-3313
G STUDY OF THE FERROELECTRIC PHASE TRANSFORMATION IN TERBIUM MOLYBDATE. NEUTRON SCATTERIN 30-3268
W TYPE OF FAR INFRARED SOFT MODE IN FERROELECTRIC GADOLINIUM MOLYBDATE. NE 30-3310
NEW TYPE OF FERROELECTRIC SOFT MODE IN GADOLINIUM MOLYBDATE. 30-3311
ERVATION OF DIFFRACTED LIGHT FROM DOMAIN WALLS IN GADOLINIUM MOLYBDATE. OBS 30-3322
OPTICAL DEVICE ELEMENT OF FERROELECTRIC GADOLINIUM MOLYBDATE. 30-3295
PHASE TRANSITION IN GADOLINIUM MOLYBDATE. 30-3316
NSITIONS IN COMPLEX PEROVSKITES OF THE TYPE BARIUM LANTHANON MOLYBDATE. PHASE TRA 30-3261
PI- GMO: ANOTHER MODIFICATION OF GADOLINIUM MOLYBDATE. 30-3263
PIEZOELECTRIC COEFFICIENT OF GADOLINIUM MOLYBDATE. 30-3317
METERS OF THE FERROELECTRIC RARE EARTH MOLYBDATES: LANTHANUM MOLYBDATE. PRECISION PARA 30-3264
PREPARATION OF SODIUM AND BARIUM NIOBATES AND LEAD MOLYBDATE. 04-0657
LASTIC INSTABILITIES IN PIEZOELECTRIC MATERIALS. (GADOLINIUM MOLYBDATE) /N OF SPONTANEOUS POLARIZATION BY E 30-3282
PYROELECTRIC BEHAVIOR OF FERROELECTRIC GADOLINIUM MOLYBDATE. 30-3279
PYROELECTRIC DETECTION PROPERTIES OF GADOLINIUM MOLYBDATE. 30-3324
PYROELECTRIC EFFECT AND DOMAIN STRUCTURE OF GADOLINIUM MOLYBDATE. 30-3315
PYROELECTRIC EFFECT IN GADOLINIUM MOLYBDATE. 30-3323
IN SOLID STATE PHYSICS (DOMAIN WALL THICKNESS IN GADOLINIUM MOLYBDATE). /NG TECHNIQUES APPLIED TO PROBLEMS 30-3319
RAMAN SPECTRUM AND STRUCTURE OF TERBIUM MOLYBDATE. 30-3296
ANEOUS BIREFRINGENCE AND ELECTROOPTIC RESPONSE IN GADOLINIUM MOLYBDATE. SPONT 30-3307
STUDY OF FERROELECTRIC DOMAINS IN GADOLINIUM MOLYBDATE. 30-3304
THE SYSTEM POTASSIUM YTTRIUM MOLYBDATE- POTASSIUM DYSPROSIUM MOLYBDATE. /ENDENCE OF DIELECTRIC CONSTANT IN 30-3309
TEMPERATURE DEPENDENCE OF THE RAMAN SPECTRUM OF GADOLINIUM MOLYBDATE. 30-3285
THE CRYSTAL STRUCTURE OF FERROELECTRIC GADOLINIUM MOLYBDATE. 30-3288
THERMODYNAMIC THEORY OF GADOLINIUM MOLYBDATE. 03-0232
TRIC AND FERROELASTIC CRYSTALS. (BARIUM TITANATE, GADOLINIUM MOLYBDATE) /KNESS OF DOMAIN WALLS IN FERROELEC 04-0777
CTOR DEPENDENT RELAXATION OF AN OPTICAL PHONON IN GADOLINIUM MOLYBDATE. WAVEVE 30-3320
X-RAY STUDY OF THERMAL EXPANSION OF FERROELECTRIC GADOLINIUM MOLYBDATE. 30-3294
ZONE BOUNDARY SOFT MODE IN TERBIUM MOLYBDATE. 30-3269
BSTA/ GROWTH AND MORPHOLOGY OF SINGLE CRYSTALS OF GADOLINIUM MOLYBDATE, A FERROELECTRIC AND FERROELASTIC SU 30-3259
M TERBIUM, DYSPROSIUM AND H/ PHASE TRANSITIONS OF GADOLINIUM MOLYBDATE AND ISOSTRUCTURAL COMPOUNDS (EUROPIU 30-3271
POTASSIUM NICKEL TRIFLUORIDE, STRONTIUM TITANATE, POTASSIUM MOLYBDATE, AND POTASSIUM TANTALATE. /BANDS FOR 02-0091
NONLINEAR OPTICAL PROPERTIES OF GADOLINIUM MOLYBDATE AND TERBIUM MOLYBDATE. 30-3305
SECOND HARMONIC GENERATION IN MOLYBDATES (BETA GADOLINIUM MOLYBDATE AND TERBIUM MOLYBDATE. 30-3286
OPTICAL PROPERTIES OF GADOLINIUM MOLYBDATE AND THEIR DEVICE APPLICATIONS. 30-3299
OPTICAL PROPERTIES OF GADOLINIUM MOLYBDATE AND THEIR DEVICE APPLICATIONS. 30-3300
N. ELASTIC PROPERTIES OF GADOLINIUM MOLYBDATE AT ITS FERROELECTRIC PHASE TRANSITIO 30-3283
/ON OXIDE, PRASEODYMIUM BISMUTH TITANIUM IRON OXIDE, BISMUTH MOLYBDATE, BISMUTH FLUONIOBATE, BISMUTH FLUOT/ 13-1993
DOMAIN WALL STRUCTURE IN GADOLINIUM MOLYBDATE BY RAMAN SCATTERING. 30-3318
ME/ FERROELECTRIC FERROELASTIC PARAMAGNETIC BETA GADOLINIUM MOLYBDATE. CRYSTAL STRUCTURE OF THE TRANSITION 30-3289

D SPECTROSCOPIC PROPERTIES OF NEODYMIUM(3+) DOPED GADOLINIUM MOLYBDATE CRYSTALS.   CERTAIN GROWTH, LASING AN 30-3255
NDARIES BETWEEN FERRO- AND PARAELECTRIC PHASES IN GADOLINIUM MOLYBDATE CRYSTALS.   OBSERVATION OF PHASE BOU 30-3306
                                 STRUCTURAL STUDY OF GADOLINIUM MOLYBDATE CRYSTALS.   30-3302
RIE TEMPERATURE).   X-RAY DIFFRACTION STUDY OF GADOLINIUM MOLYBDATE (CRYSTALLOGRAPHIC PHASE CHANGE AT CU 30-3312
TURE) FERROELECTRIC FERROELASTIC PARAMAGNETIC (BETA) TERBIUM MOLYBDATE (CZOCHRALSKI SINGLE CRYSTALS). /PERA 30-3290
/ THEORY OF THE FERROELECTRIC PHASE TRANSITION IN GADOLINIUM MOLYBDATE. (DIELECTRIC AND ELASTIC PROPERTIES) 30-3303
/CRYSTAL BISMUTH TITANATE AND RELATED COMPOUNDS. (GADOLINIUM MOLYBDATE, GROWTH BY RF CZOCHRALSKI AND FROM / 08-1323
PROPER FERROELECTRIC CRYSTALS. (AMMONIUM SULFATE, GADOLINIUM MOLYBDATE, IRON IODINE BORACITE) /ERTIES OF IM 14-2084
MONIDE, RUBIDIUM/ COUPLED FERROELASTIC CRYSTALS. (GADOLINIUM MOLYBDATE, KDP, ORTHOFERRITES, DYSPROSIUM ANTI 02-0036
HICKNESS DEPENDENCE OF CONTRAST AND BRIGHTNESS IN GADOLINIUM MOLYBDATE LIGHT VALVES.   ORIENTATION AND T 30-3291
                     ELASTIC CONSTANTS OF GADOLINIUM MOLYBDATE (NEAR THE CURIE POINT).   30-3274
      (PROPERTIES,) FUNCTION AND APPLICATIONS OF GADOLINIUM MOLYBDATE (OPTICAL SHUTTER, COLOR MODULATOR).   30-3298
/YNAMICS OF ANTIFERROMAGNETIC, FERROELECTRIC BETA GADOLINIUM MOLYBDATE. PARTS 1-3: HEAT CAPACITY, ENTROPY,/ 30-3275
/AND PHASE TRANSFORMATIONS OF THE M MODIFICATION OF SAMARIUM MOLYBDATE (SOLID PHASE SYNTHESIS, NEW MODIFIC/ 30-3270
       ORIGIN OF THE STRUCTURAL PHASE TRANSITION IN GADOLINIUM MOLYBDATE. (SYMMETRY CHANGES)   30-3273
/OELECTRIC EFFECT AND SPONTANEOUS POLARIZATION OF GADOLINIUM MOLYBDATE (TEMPERATURE DEPENDENCE 150-165 C).   30-3314
STRUCTURE)   PRECISION PARAMETERS OF SOME LANTHANON MOLYBDATE TYPE RARE EARTH MOLYBDATES. (CRYSTAL 30-3262
IUM, THO/ THALLIUM PEROVSKITES. THALLIUM A(1-X) TUNGSTATE OR MOLYBDATE. (WHERE A IS ZIRCONIUM, HAFNIUM, CER 10-1640
   PHASE TRANSITIONS OF ANHYDROUS MODIFICATIONS OF RARE EARTH MOLYBDATES.   30-3272
RUCTURAL COMPOUNDS (EUROPIUM TERBIUM, DYSPROSIUM AND HOLMIUM MOLYBDATES). /F GADOLINIUM MOLYBDATE AND ISOST 30-3271
TURAL AND PHASE RELATIONSHIPS AMONG TRIVALENT TUNGSTATES AND MOLYBDATES.   STRUC 10-1648
TURAL AND PHASE RELATIONSHIPS AMONG TRIVALENT TUNGSTATES AND MOLYBDATES.   STRUC 10-1649
OLINIUM MOLYBDATE. CRYSTAL STRUCTURE OF THE TRANSITION METAL MOLYBDATES AND TUNGSTATES. PART-6. /C BETA GAD 30-3289
IUM MOLYBDATE.   SECOND HARMONIC GENERATION IN MOLYBDATES (BETA GADOLINIUM MOLYBDATE AND TERB 30-3286
ISION PARAMETERS OF SOME LANTHANON MOLYBDATE TYPE RARE EARTH MOLYBDATES. (CRYSTAL STRUCTURE)   PREC 30-3262
       PRECISION PARAMETERS OF THE FERROELECTRIC RARE EARTH MOLYBDATES: LANTHANUM MOLYBDATE.   30-3254
DENCE OF DIELECTRIC CONSTANT IN THE SYSTEM POTASSIUM YTTRIUM MOLYBDATE- POTASSIUM DYSPROSIUM MOLYBDATE. /EN 30-3309
, PHASE TRANSITIONS, STRUCTURE, 49/ FERROELECTRIC RARE EARTH MOLYBDATES. (REVIEW, FORMATION, CRYSTAL GROWTH 02-0064
NISM OF THE FERROELECTRIC PHASE TRANSITION IN RARE EARTH MOLYBDATES. (TERBIUM MOLYBDATE)   MECHA 30-3250
CLASSES OF FERROELECTRICS OF DISPLACEMENT TYPE. (BORACITES, MOLYBDATES, TUNGSTEN BRONZES, NIOBATES)   NEW 02-0118
IUM (0.5) OXIDE, (BARIUM OR LEAD) BISMUTH(0.67) (TUNGSTEN OR MOLYBDENUM) (0.33) OXIDE) /, TANTALUM, OR VANAD 08-1462
BISMUTH(4) TITANIUM(3) OXIDE(12) WITH VANADIUM PENTOXIDE AND MOLYBDENUM AND TUNGSTEN TRIOXIDE. /EACTION OF 13-2029
PHOTOCHROMIC EFFECT AND ELECTRON SPIN RESONANCE OF IRON AND MOLYBDENUM IN BARIUM SODIUM NIOBATE.   13-1930
/R VANADIUM) OXIDE(6), AND BARIUM(3) BISMUTH(2) (TUNGSTEN OR MOLYBDENUM) OXIDE(6), (BARIUM OR STRONTIUM)(2/ 36-3619
LE OF THE DYNAM/ ELECTRON PARAMAGNETIC RESONANCE SPECTRUM OF MOLYBDENUM(5+) IN STRONTIUM TITANATE: AN EXAMP 06-1125
                         DIPOLE MOMENT DERIVATIVE OF THE HYDROGEN BOND IN ICE. 34-3489
RATE (AND MECHANISM OF REVERSAL).   FERROELECTRIC MOMENT IN GUANIDINIUM ALUMINUM SULFATE HEXAHYD 21-2818
   RESOLUTION OF A DISCREPANCY CONCERNING THE AVERAGE DIPOLE MOMENT IN KDP TYPE FERROELECTRICS.   17-2618
   NATURE OF THE TEMPERATURE DEPENDENCE OF NMR SPECTRA SECOND MOMENT IN ORDER DISORDER TYPE FERROELECTRICS. 03-0460
/N OF THE UNUSUAL TEMPERATURE VARIATION OF THE MEAN MAGNETIC MOMENT OF CHROMIUM- POTASSIUM ALUM (UNSTABLE / 20-2806
/NIUM MOLYBDATE. PARTS 1-3: HEAT CAPACITY, ENTROPY, MAGNETIC MOMENT OF THE ELECTRICALLY POLARIZED FORM FRO/ 30-3275
CS (KDP TYPE, THEORY).   IMPURITY MODES IN AN INDUCED MOMENT SYSTEM AND HYDROGEN BONDED FERROELECTRI 03-0426
CRYSTALS.   NMR PROTON SECOND MOMENTS IN POTASSIUM DIHYDROGEN PHOSPHATE TYPE 17-2408
PIN EXCITATIONS IN ANTIFERRODIELECTRIC ELECTRIC AND MAGNETIC MOMENTS OF OPTICAL TRANSITIONS. /CTRONIC AND S 03-0418
HASE TRANSITION IN POTASSIUM FERROCYANID/ ORDERING OF DIPOLE MOMENTS OF WATER MOLECULES AND FERROELECTRIC P 24-2893
                         MONO... SEE THE PREFIXED WORD.
OF FERROELECTRIC IRON IODINE BORACITE AT LOW TEMPERATURES. (MONOCLINIC AND RHOMBOHEDRAL PHASES) /ANSITIONS 14-2086
/ (PHOSPHATE, TITANATE, SULFATE, FLUOBERYLLATE, ETC) OF SOME MONOCLINIC CRYSTALS IN THE FERROELECTRIC TRAN/ 03-0330
ELECT/ CRYSTALLOGRAPHIC, OPTICAL AND MAGNETIC PROPERTIES OF (MONOCLINIC) EUROPIUM SILICATE. (UNSTABLE FERRO 36-3620
            THE CRYSTAL STRUCTURE OF MONOCLINIC POTASSIUM DIDEUTERIUM PHOSPHATE. 17-2591
RAPATITE. (DEVELOPMENT OF FERROELECTRIC CHARACTER IN AN APP/ MONOCLINIC STRUCTURE OF SYNTHETIC CALCIUM CHLO 36-3591
COPPER ION COMPLEX WITH SPIN S=2 IN TRIGLYCINE FLUOBERYLLATE MONOCRYSTAL.   EPR OF THE FOUR 27-3174
AND RUBIDIUM / LOW FREQUENCY INFRARED ABSORPTION SPECTRA OF MONOCRYSTALS OF POTASSIUM DIHYDROGEN PHOSPHATE 17-2457
            DYNAMIC POLAR MODEL OF MONOCRYSTALLINE ICE I(H).   34-3424
DE, AND SILICON.   MEASUREMENT OF SPUTTER ETCHING RATES FOR MONOCRYSTALLINE LITHIUM NIOBATE, SILICON DIOXI 11-1714
ME IN FERROELECTRICS NEAR THE CURIE TEMPERATURE (POLY VERSUS MONODISPERSIVE BEHAVIOR). /ON OF RELAXATION TI 03-0521
                         MONODOMAIN STRONTIUM TITANATE. 06-1169
IN (110) PLATES BELOW CUBIC- TETRAGONAL TRANSITION, AND PRO/ MONODOMAIN STRONTIUM TITANATE (OCCURANCE IN TH 06-1164
IRRADIATION ON BARIUM TITANATE, (COMPARED TO ST/ EFFECTS OF (MONOENERGETIC) ELECTRON AND (COBALT-60) GAMMA- 04-0970
      INFRARED ABSORPTION SPECTROPHOTOMETRIC STUDY OF MONOVALENT METAL NIOBATES.   02-0113
CTIVITY IN LEAD ZIRCONATE(X) TITANATE(1-X) CERAMICS NEAR THE MORPHIC PHASE BOUNDARY. /E ELECTROMECHANICAL A 09-1500
Y SULFOIODIDE CRYSTALS (TO 1 CM DIAMETER): CONTROL OF NEEDLE MORPHOLOGY.   GROWTH OF LARGE ANTIMON 15-2200
ENTARY CRYSTALS OF ADP.   PREPARATION (MORPHOLOGY) AND MECHANICAL PROPERTIES OF FILAM 17-2630
ANSFORMATIONS IN POTASSIUM NITRATE, SILVER NITRATE, AND AMM/ MORPHOLOGY OF CRYSTAL GROWTH AT POLYMORPHIC TR 16-2260
ETA, BETA TO ALPHA, BETA TO GAMMA AND GAMMA TO ALPHA TRANSI/ MORPHOLOGY OF CRYSTAL GROWTH IN THE ALPHA TO B 16-2259
LYBDATE, A FERROELECTRIC AND FERROELASTIC SUBSTA/ GROWTH AND MORPHOLOGY OF SINGLE CRYSTALS OF GADOLINIUM MO 30-3259
CTIVITY IN LEAD ZIRCONATE(X) TITANATE(1-X) CERAMICS NEAR THE MORPHOTROPIC PHASE BOUNDARY. /CTROMECHANICAL A 09-1499
/ATE SOLID SOLUTIONS (THERMODYNAMIC EXISTENCE CONDITIONS FOR MORPHOTROPIC PHASE BOUNDARY, STOICHIOMETRY DE/ 09-1541
OLYCRYSTALLINE LEAD TITANATE ZIRCONATE COMPOSITIONS NEAR THE MORPHOTROPIC PHASE TRANSITION. /ROPERTIES OF P 09-1557
BOHEDRAL IN THE PEROVSKITE FERR/ THERMODYNAMIC THEORY OF THE MORPHOTROPIC PHASE TRANSITION TETRAGONAL- RHOM 03-0172
NGE IN BARIUM TITANATE USING ELECTRON SPIN RESONANCE AND THE MOSSBAUER EFFECT. /THE FERROELECTRIC PHASE CHA 04-0827
DENT OPTICAL MODE IN ANTIFERROELECTRIC LEAD ZIRCONATE BY THE MOSSBAUER EFFECT.   TEMPERATURE DEPEN 07-1260
OF IONIC CRYSTALS.   MOSSBAUER EFFECT AND FERROELECTRIC PROPERTIES 36-3596
RROELECTRICS. ABRUPT (STEPWISE) CHANGE IN PROBABILITY OF THE MOSSBAUER EFFECT AT THE PHASE TRANSITION IN FE 03-0276
LEAD TITANATE.   MOSSBAUER EFFECT FOR IRON-57 IN FERROELECTRIC 05-1040
QUENCY MODE AT THE TRANSITION TEMPERATURE IN KDP AND DEUTER/ MOSSBAUER EFFECT OBSERVATION OF A VERY LOW FRE 17-2465
TE.   MOSSBAUER EFFECT OF COBALT-57 IN BARIUM TITANA 04-0723
            MOSSBAUER EFFECT OF FERROUS IONS IN CUBIC ICE. 34-3416
/ SCATTERING. PART-4: INFRARED AND RAMAN SCATTERING. PART-5: MOSSBAUER EFFECT. PART-6: ELECTRICAL CONDUCTI/ 03-0397
IUM TITANATE STANNATE SOLID SOLUTIONS IN THE / CHANGE OF THE MOSSBAUER EFFECT PROBABILITY ON TIN-119 IN BAR 04-0573
E TRANSITIONS IN THE LEAD ZIRCONATE- LEAD TITANATE- BISMUTH/ MOSSBAUER EFFECT STUDIES OF FERROELECTRIC PHAS 09-1498
ONS IN ANTIFERROELECTRIC LEAD ZIRCONATE AND FERROELECTRIC L/ MOSSBAUER EFFECT STUDIES OF THE PHASE TRANSITI 07-1253
IUM FERROCYANIDE TRIHYDRATE (NEAR THE TRANSITION TEMPERATUR/ MOSSBAUER EFFECT STUDY OF FERROELECTRIC POTASS 24-2891
ANSITIONS IN FERRIC AMMONIUM SULFATE DODECAHYDRATE AND POTA/ MOSSBAUER EFFECT STUDY OF THE FERROELECTRIC TR 19-2755
ANSITIONS IN FERRIC AMMONIUM SULFATE DODECAHYDRATE AND POTA/ MOSSBAUER EFFECT STUDY OF THE FERROELECTRIC TR 24-2892
IN IRON DOPED BARIUM TITANATE)   IONIC CHARACTERS FROM MOSSBAUER ISOMER SHIFTS. (DEGREE OF COVALENCY 04-0563
MUTH FERRATE- LEAD ZIRCONATE SYSTEMS.   MOSSBAUER MEASUREMENTS OF BISMUTH FERRATE- BIS 08-1282
M FERROCYANIDE TRIHYDRATE.   MOSSBAUER RESONANCE OF SINGLE CRYSTAL POTASSIU 24-2890
UBSTANCES. PART-1: IRON(X) ALUMINU/ RELAXATION PHENOMENA AND MOSSBAUER SPECTRA OF MAGNETIZED PARAMAGNETIC S 19-2751
(SYSTEM WITH LEAD ZIRCONATE AND WITH LEAD ZIRCONATE TITANATE MOSSBAUER SPECTROSCOPY. /ND RELATED MATERIALS 08-1333

CRYSTALS.   MOSSBAUER STUDIES OF ANTIMONY SULFOIODIDE TYPE 15-2180
NATE PEROVSKITE TYPE SOLID SOLUTIONS (ROOM TEMPERATURE TO 6/ MOSSBAUER STUDIES OF BISMUTH FERRATE LEAD TITA 08-1471
E LIKE LAYER TYPE FERROELECTRICS. (IN BISMUTH TITANATE FERR/ MOSSBAUER STUDIES OF IRON-57 IN SOME PEROVSKIT 13-2035
TANNATE. (CERAMIC DOPED WITH NIOBIUM AND TITANIUM)   MOSSBAUER STUDIES OF LEAD ZIRCONATE TITANATE S 09-1604
UM TITANATE. (LATTICE INSTABILITIES)   MOSSBAUER STUDIES OF LOWER TRANSITIONS IN BARI 04-0564
INTERNAL ELECTRIC FIELD GRADIENTS IN FERROELECTRIC FERROMAG/ MOSSBAUER STUDY AND THEORETICAL ESTIMATION OF 08-1401
N IN IRON-57(3X) AMMONIUM DIHYDROGEN PHOSPHATE.   MOSSBAUER STUDY OF ANTIFERROELECTRIC TRANSITIO 17-2644
TITANATE.   MOSSBAUER STUDY OF COLOR CENTERS IN STRONTIUM 06-1153
SES IN AMMONIUM IRON SULFATE DODECAHYDRATE.   MOSSBAUER STUDY OF ELECTRONIC SPIN FLIP PROCES 19-2771
NSITION IN POTASSIUM FERROCYANIDE TRIHYDRATE.   MOSSBAUER STUDY OF THE FERROELECTRIC PHASE TRA 24-2901
 FERRIC AMMONIUM SULFATE DODECAHYDRATE SINGLE CRYSTALS.   MOSSBAUER STUDY ON FERROELECTRIC PROPERTIES OF 19-2770
/: MECHANOCHEMISTRY OF THE SINTERING BODY. PART-9: EFFECT OF MOTHER SALT OF LEAD MONOXIDE ON THE SINTERING/ 09-1624
ALATE CRYSTAL RESONATOR VIBRATING IN THICKNESS SHEAR MODE OF MOTION. /E DEPENDENCE OF AN X CUT LITHIUM TANT 12-1868
OF FERROELECTRICS.   SIDEWAYS DOMAIN WALL MOTION AND (MECHANISM OF THE) HYSTERESIS LOOP 03-0301
LS. (SUMMARY, WITH PHOTOS)   WALL MOTION AND NUCLEATION OF DOMAINS IN TGS CRYSTA 27-3072
   STRUCTURE OF A MOVING DOMAIN WALL. (MECHANISM OF A WALL MOTION IN BARIUM TITANATE) 04-0678
TANTALATE EXAMPLE).   INTERCELL CORRECTIONS FOR IONIC MOTION IN DISPLACEMENT FERROELECTRICS (LITHIUM 03-0367
                              DOMAIN WALL MOTION IN FERROELECTRIC SWITCHING. 03-0273
ERAL FORMULATION.   KINETICS OF DOMAIN WALL MOTION IN FERROELECTRIC SWITCHING. PART-1: GEN 03-0274
       CONTROL AND APPLICATION OF DOMAIN WALL MOTION IN GADOLINIUM MOLYBDATE. 30-3257
       CONTROL AND APPLICATION OF DOMAIN WALL MOTION IN GADOLINIUM MOLYBDATE. 30-3256
           OBSERVATION OF PHASE FRONT MOTION IN KDP CRYSTAL. 17-2589
EMPERATURE DEPENDENCE OF THE LITHIUM NMR SPECTRUM AND ATOMIC MOTION IN LITHIUM NIOBATE.   T 11-1722
Y NEUTRON TRIPLE AXIS SPECTROMETR/ FERROELECTRIC (SOFT) MODE MOTION IN POTASSIUM DIDEUTERIUM PHOSPHATE . (B 17-2668
ODIUM TRIHYDROGEN SELENATE TYPE CRYSTALS. ULTRASLOW HYDROGEN MOTION IN POTASSIUM DIHYDROGEN PHOSPHATE AND S 17-2441
CRYSTALS.   THEORY OF DOMAIN WALL MOTION IN POTASSIUM DIHYDROGEN PHOSPHATE TYPE 17-2652
ION OF COERSIVE FIELD AT VERY LOW FREQUENCIES AND JERKY WALL MOTION IN ROCHELLE SALT.   ANOMALOUS DISPERS 28-3177
CHLORIDE. (CALCULATED IN THE HARMONIC APPROXIMATI/ MOLECULAR MOTION IN THE FERROELECTRIC PHASE OF HYDROGEN 31-3340
             SUPERSONIC DOMAIN WALL MOTION IN TRIGLYCINE SULFATE. 27-2963
HYDROGEN PHOSPHATE.   RANDOM MOTION OF DEUTERONS IN DEUTERATED POTASSIUM DI 17-2651
N SINGLE CRYSTAL FILMS OF BARIUM TITANATE (ELECTRON MICROSC/ MOTION OF DOMAIN BOUNDARIES DURING SWITCHING I 04-0940
N GLYCINE.   ESR OF GAMMA-IRRADIATED TGS. PART-2: INTERNAL MOTION OF HYDROGEN AND THE ACTIVATION ENERGY I 27-3034
 AMMONIUM ROCHELLE SALT BY SPIN LATTICE RELAX/ STUDY OF SLOW MOTION OF WATER MOLECULES IN ROCHELLE SALT AND 28-3232
KDP IN THE VICINITY OF CURIE POINT.   EFFECT OF IONIC MOTION ON THE REFRACTIVE INDEX OF PARAELECTRI 17-2548
IMONY SULFOIODIDE CRYSTALS.   MOTION PICTURES OF THE PHASE TRANSITION IN ANT 15-2120
MINE TARTR/ TRANSFORMATION OF ELECTROMAGNETIC RADIATION INTO MOTION. (PYROELECTRIC EFFECT IN TGS, ETHYL DIA 04-0854
              CONTRIBUTION OF DOMAIN WALL MOTION TO THE PERMITTIVITY OF ROCHELLE SALT. 28-3194
/-1: DAMPINGS AND (FREQUENCY) SHIFTS (OF PROTONIC COLLECTIVE MOTION, TUNNELING, CALCULATIONS, INTERPRETATI/ 03-0475
NTIFERROELECTRIC SOLIDS. PART-1: ENERGY LOSS NEUTRON/ PROTON MOTIONS IN HYDROGEN BONDED FERROELECTRIC AND A 17-2687
OM POLARIZED REFLECTION SPECTRA, A TUNNELING MODEL/ DEUTERON MOTIONS IN POTASSIUM DIDEUTERIUM PHOSPHATE (FR 17-2645
N IN AMMONIUM SULFATE STUDIED BY COLD NEUTRONS (/ ROTATIONAL MOTIONS IN SOLIDS: THE FERROELECTRIC TRANSITIO 19-2752
OTASSIUM DIHYDROGEN PHOSPHATE TYPE CRYSTALS)   COLLECTIVE MOTIONS OF HYDROGEN BONDS. (CALCULATIONS FOR P 17-2483
IUM DIHYDROGEN PHOSPHATE AT HIGH TEMPERATURES.   ABSENCE OF MOTIONAL NARROWING IN THE PROTON NMR IN POTASS 17-2594
      X-RAY TOPOGRAPHIC STUDY OF DOMAIN WALL MOVEMENT IN TGS. 27-3026
DOMAINS IN CRYSTALS OF TRIGLYCINE SULFATE.   MOVEMENT OF DOMAIN WALLS AND THE NUCLEATION OF 27-3071
NEAR THE CURIE POINT. (ESR OF GAMMA-IRRADIATED S/ DISPLACIVE MOVEMENT OF THE RADICAL IN TRIGLYCINE SULFATE 27-3033
ROVSKITE FERROELECTRICS.   MOVEMENT OF 180 DEGREE DOMAIN BOUNDARIES IN PE 08-1305
TERACTION WITH CRYSTAL DISLOCATIONS. (DOMAIN ORIGIN, SHAPES, MOVEMENT, 34 REFS) /OMAIN STRUCTURE AND ITS IN 03-0390
N IN BARIUM TITANATE)   STRUCTURE OF A MOVING DOMAIN WALL. (MECHANISM OF A WALL MOTIO 04-0678
NY SULFOIODIDE)   INTERACTION OF ULTRASONIC WAVES WITH MOVING DOMAIN WALLS IN FERROELECTRICS. (ANTIMO 15-2192
/TION OF ELECTRON MIRROR MICROSCOPY TO DIRECT OBSERVATION OF MOVING FERROELECTRIC DOMAINS OF GADOLINIUM MO/ 30-3293
                                   MOVING STRIPE PATTERN IN BARIUM TITANATE. 04-0643
OPTICAL BEAM STEERING USING A MULTICHANNEL LITHIUM TANTALATE CRYSTAL. 12-1881
DIELECTRIC RELAXATION IN MULTIDOMAIN CRYSTALS OF TRIGLYCINE SULFATE. 27-3002
DIELECTRIC RELAXATION IN MULTIDOMAIN TGS SINGLE CRYSTALS. 27-3001
DIELECTRIC RELAXATION OF MULTIDOMAIN TRIGLYCINE SULFATE. 27-3050
Y PROJECTION WITH A PLASMA TORCH O/ CHARACTERISTICS OF THIN, MULTILAYERED FERROELECTRIC CERAMICS OBTAINED B 04-0778
PTION AND RAMAN SCATTERING IN FERROELECTRIC AND PARAELECTRI/ MULTIPHONON SELECTION RULES FOR INFRARED ABSOR 16-2281
HOMOGENEITIES WITH APPLICATIONS TO FERROELEC/ AN ANALYSIS OF MULTIPLE LIGHT SCATTERING BY RANDOM OPTICAL IN 37-3634
ITHIUM NIOBATE.   MULTIPLE STORAGE OF THICK PHASE HOLOGRAMS IN L 11-1827
ROELASTICS HAVING THE INDEX OF FAINTNESS UNEQUAL TO THE CELL MULTIPLICITY. /MPLES OF FERROELECTRICS AND FER 03-0145

N

C WITH A HIGH EFFECTIVE D/ GRAIN BOUNDARY BARRIER LAYERS IN (N TYPE, ANTIMONY DOPED) BARIUM TITANATE CERAMI 04-0581
/YANIDE TRIHYDRATE SINGLE CRYSTALS, P-N JUNCTIONS AND P- AND N- RANGES IN SEMICONDUCTORS. (SILICON INTEGRA/ 24-2904
/EFFICIENT DIRECT MEASUREMENT TECHNIQUE AND APPLICATION TO A NANOSECOND RESPONSE TIME DETECTOR. (STRONTIUM 13-1919
ROGEN PHOSPHATE AT HIGH TEMPERATURES.   ABSENCE OF MOTIONAL NARROWING IN THE PROTON NMR IN POTASSIUM DIHYD 17-2594
SPIN CORRELATION IN PARAMAGNETIC POTASSIUM MANGANE/ EXCHANGE NARROWING: MAGNETIC RESONANCE LINE SHAPES AND 33-3376
ROWN) SINGLE CRYSTAL BISMUTH FERRATE (CONFIRMS FERROELECTRIC NATURE).   DIELECTRIC HYSTERESIS IN (FLUX G 08-1449
O WE REALLY UNDERSTAND (THE CAUSE OF) FERROELECTRICITY. (ITS NATURE AND THE PREDICTION OF ITS OCCURANCE)   D 03-0510
ANTIMONY SULFOIODIDE TYPE.   POLARIZATION NATURE IN SEMICONDUCTOR FERROELECTRICS OF THE 15-2181
RIUM TITANATE, LEAD ZIRCONATE TITANATE)   NATURE OF AGING IN FERROELECTRIC CERAMICS. (BA 04-0757
ECTRIC P/ EFFECT OF COLLECTIVE INTERACTION OF DIPOLES ON THE NATURE OF DIELECTRIC DISPERSION IN THE FERROEL 03-0417
/ CAUSES OF PHASE TRANSITION BROADENING (DIFFUSNESS) AND THE NATURE OF DIELECTRIC POLARIZATION RELAXATION I 03-0291
MODIFICATION OF CADMIUM TITANATE./ ABOUT THE CURIE POINT AND NATURE OF DIELECTRIC PROPERTIES IN PEROVSKITE 08-1424
ONS.   EFFECT OF HYDROSTATIC PRESSURE ON THE NATURE OF DIFFUSE FERROELECTRIC PHASE TRANSITI 03-0442
DISACCOMODATION TYPE PHENOMENA IN FERRITES AND FERROELECTRI/ NATURE OF DISSIPATIVE EFFECTS WHICH ACCOMPANY 03-0447
/STUDIES. PART-4: PHASE TRANSITIONS OF AMMONIUM SULFATE: THE NATURE OF HYDROGEN BONDING AND THE REORIENTAT/ 19-2783
AL DAMAGE PROCESS IN LITHIUM NIOBATE.   NATURE OF INTERNAL ELECTRIC FIELD DURING OPTIC 11-1861
ND ANTIFERROELECTRIC MATERIALS.   NATURE OF PHASE TRANSITIONS IN FERROELECTRIC A 03-0311
S OF ROCHELLE SALT AND TRIGLYCINE SULFATE.   NATURE OF RADIATION INDUCED DEFECTS IN CRYSTAL 27-3159
OF FERROELECTRICS IN THE PHASE TRANSITION REGION.   NATURE OF THE ANOMALOUS ELECTRIC CONDUCTIVITY 03-0508
E TRANSITION IN LITHIUM TANTALATE.   NATURE OF THE FERROELECTRIC- PARAELECTRIC PHAS 12-1879
INATES WITH PEROVSKITE TYPE STRUCTURE.   NATURE OF THE HIGH PERMITTIVITY OF CERIUM ALUM 08-1355
E.   DETERMINATION OF THE NATURE OF THE OPTICAL GAP OF STRONTIUM TITANAT 06-1108
/ERTIES OF BARIUM TITANATE CERAMICS ON THE PHYSICO- CHEMICAL NATURE OF THE SOLID SOLUTION FORMED FROM TWO / 04-0789
ECTRA SECOND MOMENT IN ORDER DISORDER TYPE FERROELECTRICS. NATURE OF THE TEMPERATURE DEPENDENCE OF NMR SP 03-0460
        EFFECT OF SAMPLE TREATMENT AND ELECTRODE NATURE ON SOME TRIGLYCINE SULFATE PARAMETERS. 27-3148
RESONANCE SPECTRUM OF IRON(3+) IN STRONTIUM TITANATE DUE TO NEAREST NEIGHBOR CHARGE COMPENSATION. /AGNETIC 06-1144

ANTIMONY SULFOIODIDE CRYSTALS (TO 1 CM DIAMETER): CONTROL OF NEEDLE MORPHOLOGY.   GROWTH OF LARGE 15-2200
THERMAL EXPANSIVITY OF RUBIDIUM MANGANESE FLUORIDE NEAR THE NEEL TEMPERATURE.   33-3375
UM MANGANESE TRIFLUCRIDE WITH MAGNESIUM DOPING.   SHIFT OF NEEL TEMPERATURE AND EPR LINE WIDTH OF POTASSI 33-3377
                    DIPOLE LATTICE RELAXATION TIME NEAR THE NEEL TEMPERATURE IN SODIUM NITRITE.   16-2300
IFERROELECTRICITY AND ANTIFERROELECTRIC MATERIALS. (TABLE OF NEEL TEMPERATURES)   ANT 01-0020
LEAD MAGNESATE(0.5) TUNGSTATE(0.5)), EFFECTS OF POSITIVE OR NEGATIVE HIGH DC FIELDS SUMMARY ONLY) /LUTIONS 05-1063
INE IN THE FERROELECTRIC STATE.   FIRST ORDER NEGATIVE PHOTODIELECTRIC EFFECT IN PHTHALOCYAN 36-3539
CE SPECTRUM OF IRON(3+) IN STRONTIUM TITANATE DUE TO NEAREST NEIGHBOR CHARGE COMPENSATION. /AGNETIC RESONAN 06-1144
INHOMOGENEITY (THEORY AND TEST ON BARIUM SODIUM NIOBATE AND NEODYMIUM DOPED YAG LASER). /RATION ON CRYSTAL 13-2043
/KHZ) CIELECTRIC CONSTANT   (5-300K) CF THE PRASEODYMIUM(1-X) NEODYMIUM(X) (ORTHO) ALUMINATE SYSTEM (NOT FE/ 36-3600
TO LITHIUM NIOBATE AND KDP.   NEODYMIUM YAG LASER IRRADIATION INDUCED DAMAGE 11-1674
IUM NIOBATE 0.66 MILLIMICRON HARMONIC SOURCE.   NEODYMIUM: YTTRIUM ALUMINUM GARNET- BARIUM SOD 13-1975
IO/ ABSORPTION AND LUMINESCENCE SPECTRA AND ENERGY LEVELS OF NEODYMIUM(3+) AND ERBIUM(3+) IONS IN LITHIUM N 11-1713
ALS.   CERTAIN GROWTH, LASING AND SPECTROSCOPIC PROPERTIES OF NEODYMIUM(3+) DOPED GADOLINIUM MOLYBDATE CRYST 30-3255
TAL POTASSIUM TANTALATE. (CONDUCTI/ CALCIUM CONCENTRATION VS NET IONIZED DONOR CONCENTRATION IN SINGLE CRYS 08-1437
ANISOTROPIC FERROELECTRIC AND ANTI/ PCLAR- POLAR AND POLAR- NEUTRAL CRYSTAL STRUCTURES. (CLASSIFICATION OF 03-0336
LECTRIC WITH HYDROGEN BONDS/ EFFECT OF VIOLATION OF ELECTRIC NEUTRALITY ON THE PHASE TRANSITION IN A FERROE 03-0449
EL IODINE EORACITE.   NEUTRON AND ELECTRON DIFFRACTION STUDY IN NICK 13-2073
POTASSIUM DIDEUTERIUM PHOSPHATE (CCMPARED WITH KDP AND WITH NEUTRON AND MICROWAVE EXPERIMENTS). /C MODE IN 17-2705
CONFIGURATICN INDUCED IN PEROVSKITE TYPE SINGLE CRYSTALS BY NEUTRON BOMBARDMENT.   SPLIT INTERSTITIAL 08-1467
          (TOP SEEDED FLUX) CRYSTAL GROWTH AND NEUTRON CHARACTERIZATION OF POTASSIUM NIOBATE. 13-1353
OR ANTIFERROELECTRIC PHASE OF COPPER FORMATE TETRAHYDRATE BY NEUTRON DIFFRACTION.   APPROXIMATE STRUCTURE F 26-2935
NSITIONS IN SOLID HYDROGEN HALIDES (SINGLE CRYSTAL X-RAY AND NEUTRON DIFFRACTION). /ECTRICITY AND PHASE TRA 31-3331
TE, AND IN POTASSIUM TRIDEUTERIUM SELENITE, AS DETERMINED BY NEUTRON DIFFRACTION. /SSIUM TRIHYDROGEN SELENI 22-2854
FORMATIONS OF A CRYSTAL LATTICE BASED ON THE SHADOW EFFECT. (NEUTRON DIFFRACTION, BARIUM TITANATE) /MALL DE 04-0534
IC STRUCTURES OF INORGANIC MATERIALS. (REVIEW, 244 REFS) NEUTRON DIFFRACTION INVESTIGATIONS OF THE ATOM 02-0133
GE FROM) IONIZING RADIATION IN FERROELECTRIC ROCHELLE SALT. (NEUTRON DIFFRACTION MEASUREMENTS) /TS OF (DAMA 28-3188
AR MAGNETIC RESONAN/ STRUCTURAL STUDIES OF ICE POLYMORPHS BY NEUTRON DIFFRACTION, PROTON AND DEUTERON NUCLE 34-3466
OLARIZATION REVERSALS IN SODIUM NITRITE.   NEUTRON DIFFRACTION STUDIES OF FERROELECTRIC P 16-2365
HASE TRANSITIONS IN KDP AND POTASSIUM DIDEUTERIUM PHOSPHATE. NEUTRON DIFFRACTION STUDIES OF FERROELECTRIC P 17-2661
(BARIUM TITANATE AND ROCHELLE SALT AS EXAMPLES)   NEUTRON DIFFRACTION STUDIES OF FERROELECTRICS. 04-0679
           FERROELECTRIC BISMUTH FERRATE X-RAY AND NEUTRON DIFFRACTION STUDY. (ANTIFERROMAGNETIC) 08-1402
ASSIUM FERROCYANIDE TRIDEUTERO HYDRATE ABOVE THE CURIE TEMP/ NEUTRON DIFFRACTION STUDY OF FERROELECTRIC POT 24-2903
OGEN SELENITE.   A NEUTRON DIFFRACTION STUDY OF POTASSIUM TRIHYDR 22-2866
                                                NEUTRON DIFFRACTION STUDY OF SODIUM NITRITE. 16-2326
TITANATE.   X-RAY AND NEUTRON DIFFRACTION STUDY OF TETRAGONAL BARIUM 04-0713
TITANATE.   X-RAY AND NEUTRON DIFFRACTION STUDY OF TETRAGONAL BARIUM 04-0740
C PEROVSKITE) STRCNTIUM IRON TUNGSTATE . (MAGNETIC G STRUCT/ NEUTRON DIFFRACTION STUDY OF (TETRAGONAL- CUBI 10-1655
LITHIUM TRIHYDROGEN SELE/ HYDROGEN BOND STUDIES. PART-54: A NEUTRON DIFFRACTION STUDY OF THE FERROELECTRIC 22-2878
E.   NEUTRON DIFFRACTION STUDY OF TRIGLYCINE SULFAT 27-3099
UM TRIHYDROGEN DISELENITE.   NEUTRON DIFFRACTION STUDY ON PARAELECTRIC SODI 22-2846
/OELECTRIC AND ANTIFERROELECTRIC SOLIDS. PART-1: ENERGY LOSS NEUTRON INCOHERENT SCATTERING SPECTRA OF POTA/ 17-2687
TRIC PHASE TRANSITION CF SODIUM NITRITE (STUDIED BY COHERENT NEUTRON INELASTIC SCATTERING). / THE FERROELEC 16-2364
   FERROELECTRIC CRITICAL SCATTERING FRCM DEUTERATED KDP. (NEUTRON INELASTIC SCATTERING)   17-2610
       SOFT FERROELECTRIC MODES IN LEAD TITANATE (NEUTRON INELASTIC SCATTERING) .   05-1077
SE TRANSITIONS. (STRONTIUM TITANATE, POTASSIUM TANTALATE, B/ NEUTRON INELASTIC SCATTERING AT STRUCTURAL PHA 02-0059
/UM OF HYDROGEN ECNDED FERROELECTRICS. (INFRARED ABSORPTION, NEUTRON INELASTIC SCATTERING, POTASSIUM DIHYD/ 03-0401
ICE DYNAMICS OF STRONTIUM TITANATE: HARMONIC MODELS.   NEUTRON INELASTIC SCATTERING STUDY OF THE LATT 06-1225
RAL ANC CIELECTRIC MODIFICATIONS OF TUNGSTEN TRIOXIDE DUE TO NEUTRON INELASTIC SCATTERING STUDY ON SOFT MOD 02-0115
ES. (REVIEW OF BROOKHAVEN WORK ON POTASSIUM TANTALATE, BARI/ NEUTRON IRRADIATION.   STRUCTU 10-1615
ES OF LEAD ZIRCONATE.   EFFECT OF PILE NEUTRON IRRADIATION ON THE DIELECTRIC PROPERTI 07-1259
RTIES OF BARIUM TITANATE CERAMICS.   EFFECT OF NEUTRON IRRADIATION ON THE FERROELECTRIC PROPE 04-0732
TY OF BARIUM TITANATE SINGLE CRYSTAL.   EFFECT OF THERMAL NEUTRON IRRADIATION ON THE RELATIVE PERMITTIVI 04-0781
BONDS. (KDP AND POTASSIUM BICARBCNATE)   NEUTRON MEASUREMENT IN TWO SYSTEMS OF HYDROGEN 17-2417
E TRANSFORMATION IN POTASSIUM DIHYDROGEN FHOSPHATE. . NEUTRON MEASUREMENTS ON THE FERROELECTRIC PHAS 17-2418
NTIAL IN PARAELECTRIC POTASSIUM DIHYDROGEN PHOSPHATE. NEUTRON MEASUREMENTS ON THE HYDROGEN BOND POTE 17-2511
NS IN (CERAMIC) BARIUM AND LEAD TITANATES. (/ (IN PILE, FAST NEUTRON) RADIATION INDUCED PHASE TRANSFORMATIO 04-0971
FLUORIDE. (20 PERCENT MANGANESE REPLACED BY COBALT, COHERENT NEUTRON SCATTERING) /D POTASSIUM MANGANESE TRI 33-3366
ATE (ABOVE THE FERROELECTRIC CURIE TEMPERATURE, AS SHOWN) BY NEUTRON SCATTERING. /CTUATIONS IN BARIUM TITAN 04-1032
          CRYSTAL DYNAMICS CF SODIUM NITRITE. (NEUTRON SCATTERING)   16-2284
NTIUM IITANATE, RUBICIUM MANGANESE TRIFLUCRIDE) BY INELASTIC NEUTRON SCATTERING. /POTASSIUM TANTALATE, STRO 06-1139
DISTRIBUTION IN PARAELECTRIC POTASSIUM DIHYDROGEN PHOSPHATE (NEUTRON SCATTERING).   HYDROGEN DENSITY 17-2490
           ICE I: LATTICE DYNAMICS AND INCOHERENT NEUTRON SCATTERING.   17-2490
     TRANSITION IN TIN TELLURIDE- GERMANIUM TELLURIDE ALLOYS. (NEUTRON SCATTERING)   36-3586
GEN SELENITE AT ROOM TEMPERATURE.   NEUTRON SCATTERING ANALYSIS OF CESIUM TRIHYDRO 22-2869
LACEMENT CORRELATIONS IN POTASSIUM TANTALATE.   NEUTRON SCATTERING ANALYSIS OF THE LINEAR DISP 36-3546
ECTRICITY IN LITHIUM HYDRAZINIUM SULFATE.   ANOMALOUS NEUTRON SCATTERING AND THE QUESTION OF FERROEL 32-3341
LUTION. (THEORY, APPLIED TO BARIUM STRONTIUM TITANATE AND P/ NEUTRON SCATTERING BY A FERROELECTRIC SOLID SO 03-0260
CTRICS AT LOW TEMPERATURES. PART-1: GENE/ THEORY OF COHERENT NEUTRON SCATTERING BY HYDROGEN BONDED FERROELE 03-0478
NC ADP.   INCOHERENT NEUTRON SCATTERING FROM HYDROGEN BOND IN KDP A 17-2416
DEVELOPED FROM FIRST PRINCIPLES, AFPLIED TO DISPL/ THEORY OF NEUTRON SCATTERING FROM LATTICE VIBRATIONS (RE 03-0484
ITANATE. (LATTICE DYNAMICS)   INELASTIC NEUTRON SCATTERING FROM SINGLE DOMAIN BARIUM T 04-0946
                               CRITICAL NEUTRON SCATTERING FROM STRONTIUM TITANATE. 06-1195
                               CRITICAL NEUTRON SCATTERING IN STRONTIUM TITANATE AND P 06-1208
                        LATTICE DYNAMICS BY NEUTRON SCATTERING. PARTS 1 AND 2. (REVIEW) 02-0051
/D MAGNETIC RESONANCE. PART-2: THERMAL CONDUCTIVITY. PART-3: NEUTRON SCATTERING. PART-4: INFRARED AND RAMA/ 03-0397
HASE TRANSITION IN STRONTIUM TITANATE. NEW OBSERVATIONS FROM NEUTRON SCATTERING STUDIES OF THE STRUCTURAL P 06-1210
C BARIUM TITANATE.   NEUTRON SCATTERING STUDY OF SOFT MODES IN CUBI 04-0712
PHASE TRANSFORMATION IN TERBIUM MCLYEDATE.   NEUTRON SCATTERING STUDY OF THE FERROELECTRIC 30-3268
CAL PHASE TRANSITIONS IN POTASSIUM MANGANESE TRIFLUORIDE. NEUTRON SCATTERING STUDY OF THE LATTICE DYNAMI 33-3405
CUBIC POTASSIUM TANTALATE NIOBATE.   NEUTRON SCATTERING STUDY OF THE SOFT MODES IN 13-2064
M NIOBATE.   GROUP THEORETICAL SELECTION RULES FOR INELASTIC NEUTRON SCATTERING WITH APPLICATIONS TO LITHIU 11-1704
TRIC BARIUM TITANATE.   COMMENTS ON THE NEUTRON SPECTRUM OF FERROELECTRIC AND PARAELEC 04-0813
TRANSFORMATIONS. (INELASTIC DIFFUSION IN FERROELECTRICS) NEUTRON STUDIES OF DISPLACIVE STRUCTURAL PHASE 03-0155
TRANSFORMATIONS. (IN FERROELECTRICS)   NEUTRON STUDIES OF DISPLACIVE STRUCTURAL PHASE 03-0156
OTASSIUM NIOBATE.   NEUTRON STUDY OF DIFFUSE SCATTERING IN CUBIC P 08-1407
ODIUM NITRITE.   X-RAY AND NEUTRON STUDY ON THE PHASE TRANSFORMATION OF S 16-2328
(SOFT) MODE MOTION IN POTASSIUM CIEEUTERIUM FHOSPHATE . (BY NEUTRON TRIPLE AXIS SPECTROMETRY) /RROELECTRIC 17-2668
NATE CERAMICS CAUSED BY IRRADIATION WITH CADMIUM TRANSMITTED NEUTRONS. /OELECTRIC PROPERTIES OF BARIUM TITA 04-0731
ITION IN RUBIDIUM DIHYDRCGEN PHCSFHATE BY SCATTERING OF COLD NEUTRONS.   HIGH TEMPERATURE PHASE TRANS 17-2489

POTASSIUM DIHYDROGEN PHOSPHATE CRYSTALS IRRADIATED WITH FAST NEUTRONS. /ITION AND DIELECTRIC PROPERTIES OF 17-2615
/ENERAL EXPRESSION FOR INELASTIC COHERENT SCATTERING OF SLOW NEUTRONS AND EFFECTIVE THERMAL FACTORS. (KDP / 03-0478
IN THE SHORT RANGE ORDERED REGIC/ PARAMAGNETIC SCATTERING OF NEUTRONS FROM POTASSIUM MANGANESE TRIFLUORIDE 33-3371
INELASTIC SCATTERING CROSS SECTION OF NEUTRONS IN FERROELECTRICS. 03-0297
FERROELECTRIC TRANSITION IN AMMONIUM SULFATE STUDIED BY COLD NEUTRONS (SCATTERING). /OTIONS IN SOLIDS: THE 19-2752
ESIS. PREPARATION OF NICKEL BORACITE CRYSTALS BY HYDROTHERMAL SYNTH 14-2079
CRYSTAL FIELD THEORY AND OPTICAL ABSORPTION CF COBALT AND NICKEL BORACITES. 14-2096
HIGH GAS PRESSURE CRYSTAL GROWTH OF ANTIMONY SULFOIODIDE, NICKEL CARBONATE AND BLACK PHOSPHORUS. 15-2137
ND NICKEL DOPED/ OPTICAL ABSORPTION SPECTRA OF NICKEL(2+) IN NICKEL CHLORINE AND NICKEL IODINE BORACITES (A 14-2074
/IUM CHLORINE, CADMIUM IODINE, IRON CHLORINE IRON IODINE AND NICKEL CHLORINE) BORACITES (SECOND HARMONIC G/ 14-2080
MAGNETIC EXCITATIONS IN NICKEL DOPED POTASSIUM MANGANESE TRIFLUORIDE. 33-3407
/ CONCENTRATION DEPENDENCE OF RAMAN SCATTERING BY MAGNONS IN NICKEL DOPED RUBIDIUM MANGANATE AND POTASSIUM/ 33-3395
/ THE SPONTANEOUS ELECTRIC POLARIZATION AND MAGNETIZATION IN NICKEL IODIDE BORACITE (CURIE POINT 64 DEG K). 14-2090
CRYSTAL GROWTH OF NICKEL IODINE BORACITE. 14-2091
MAGNETOELECTRIC PROPERTIES OF NICKEL IODINE BORACITE. 14-2070
NEUTRON AND ELECTRON DIFFRACTION STUDY IN NICKEL IODINE BORACITE. 14-2073
ITIONS. OPTICAL ANISOTROPY OF CUBIC NICKEL IODINE BORACITE DUE TO QUADRUPOLE TRANS 14-2093
/RTRATE MONOHYDRATE, ELECTRIC CONTROL OF ELASTIC COMPLIANCE, NICKEL IODINE BORACITE OPTICAL ANISOTROPY, BI/ 02-0060
ITANATE SINGLE CRYSTALS. (LATTICE PA/ EFFECT OF SOME ADMIXED NICKEL ON SOME PHYSICAL PROPERTIES OF BARIUM T 04-0909
NGLE CRYSTALS. EFFECT OF ADDING NICKEL ON THE PROPERTIES OF BARIUM TITANATE SI 04-0793
/BSORPTION (CALCULATIONS) OF FERROELECTRIC BORACITES. (IRON, NICKEL, OR COBALT, WITH CHLORIDE, BROMIDE, OR/ 14-2095
IUM MOLYBDATE, AND POTASSIUM TAN/ ENERGY BANDS FOR POTASSIUM NICKEL TRIFLUORIDE, STRONTIUM TITANATE, POTASS 02-0091
(WHERE METAL IS SCANDIUM, CHROMIUM, MANGANESE, IRON, COBALT, NICKEL, YTTRIUM) /ETAL OXYGEN(3) PEROVSKITES. 08-1451
E BORACITES (AND NICKEL DOPED/ OPTICAL ABSORPTION SPECTRA OF NICKEL(2+) IN NICKEL CHLORINE AND NICKEL IODIN 14-2074
COPIC THEORY FOR THE DIELECTRIC PROPERTIES OF LEAD MAGNESIUM NIOBATE. A MICROS 08-1440
A NEW MAGNETOFERROELECTRIC: CADMIUM IRON NIOBATE. 36-3618
SOLUTIONS WITH STRONTIUM TITANATE LITHIUM TANTALATE, LITHIUM NIOBATE) /ICATION OF CADMIUM TITANATE. (SOLID 13-1424
ABSORPTION EDGE SPLITTING IN POTASSIUM TANTALATE NIOBATE. 13-1988
ABSORPTION OF ELASTIC WAVES IN REDUCED LITHIUM NIOBATE. 11-1816
LATION AND THE ASSOCIATED NONLINEARITY PARAMETERS IN LITHIUM NIOBATE. ACOUSTIC CONVOLUTION AND CORRE 11-1758
UADRUPOLE SATURATION OF THE LITHIUM-7 NMR SIGNALS IN LITHIUM NIOBATE. ACOUSTIC DIPOLE AND Q 11-1666
ACOUSTIC RADIATION BY INTERDIGITAL GRIDS ON LITHIUM NIOBATE. 11-1699
SURFACE WAVE STRUCTURE OF A CADMIUM SELENIDE FILM ON LITHIUM NIOBATE. ACOUSTOELECTRIC AMPLIFICATION OF 11-1723
ION IN PIEZOELECTRICS COUPLED TO A GUNN OSCILLATOR. (LITHIUM NIOBATE) /EASUREMENTS OF ACOUSTIC WAVE GENERAT 11-1731
AN ELECTROOPTIC SWITCH USING ROTATED Y PLATE OF LITHIUM NIOBATE. 11-1732
N EPR INVESTIGATION OF IRON(3+) AND MANGANESE(2+) IN LITHIUM NIOBATE. A 11-1725
PIC DIFFUSE X-RAY SCATTERING AND PHASE TRANSITIONS IN SODIUM NIOBATE. ANISOTRO 08-1357
PPARATUS FOR GROWTH OF SINGLE CRYSTAL, SINGLE DOMAIN LITHIUM NIOBATE. A 11-1726
ASSIGNMENTS OF OPTICAL PHONON MODES IN LITHIUM NIOBATE. 11-1694
ERTAIN CRYSTALS IN THE MILLIMETER WAVELENGTH RANGE. (LITHIUM NIOBATE) BIREFRINGENCE OF C 11-1730
BRILLOUIN SCATTERING IN LITHIUM NIOBATE. 11-1786
AND TWO LUMINOUS WAVES ALONG THE (100) DIRECTION OF LITHIUM NIOBATE. /BETWEEN A LONGITUDINAL ACOUSTIC WAVE 11-1680
CONTINUOUS HOT PRESSING OF POTASSIUM SODIUM NIOBATE. 13-2007
CONTROL OF LASER DAMAGE IN LITHIUM NIOBATE. 11-1716
OF SURFACE WAVES IN A STRUCTURE OF SEMICONDUCTOR ON LITHIUM NIOBATE. CONVOLUTION 11-1847
COUPLED WAVE ANALYSIS OF HOLOGRAPHIC STORAGE IN LITHIUM NIOBATE. 11-1826
CRYSTAL FIELD IN CHROMIUM(3+) DOPED LITHIUM NIOBATE. 11-1759
CRYSTAL GROWTH AND DIELECTRIC PROPERTIES OF BISMUTH NIOBATE. 13-2033
CRYSTAL GROWTH OF SODIUM NIOBATE. 08-1409
R PARAMETER FOR ACOUSTIC SURFACE WAVE CONVOLUTION IN LITHIUM NIOBATE. DETERMINATION OF A NONLINEA 11-1681
ELECTRIC CONSTANTS OF THE SYSTEM SAMARIUM NIOBATE- LANTHANUM NIOBATE. DI 08-1268
FERRATE AND SYSTEMS BASED ON IT, AND SINGLE CRYSTAL LITHIUM NIOBATE) /TH HIGH CURIE TEMPERATURES. (BISMUTH 08-1378
DIFFRACTION SPREADING OF SURFACE WAVES ON LITHIUM NIOBATE. 11-1849
FUSE X-RAY STUDY OF 640 DEGREES C PHASE TRANSITION IN SODIUM NIOBATE. DIF 08-1358
ONAL DISPERSION AND ASSIGNMENT OF OPTICAL PHONONS IN LITHIUM NIOBATE. DIRECTI 11-1695
ON OF PHONON FREQUENCIES AND POLARITON SCATTERING IN LITHIUM NIOBATE. DIRECTIONAL DISPERSI 11-1696
DISPERSION OF ELASTIC WAVES IN FERROELECTRICS. (LITHIUM NIOBATE) 11-1706
NSE LIGHT ON THE REFRACTIVE INDEX OF A CRYSTALLINE POTASSIUM NIOBATE. EFFECT OF INTE 08-1410
MOGENEITIES ON PHASE MATCHING IN NONLINEAR CRYSTALS (LITHIUM NIOBATE). EFFECT OF OPTICAL INHO 11-1773
HE PROPERTIES AND PROCESSING OF SINGLE CRYSTAL BARIUM SODIUM NIOBATE. /ICHIOMETRY AND HYDROXYL CONTENT ON T 13-1915
ANGES IN MELT COMPOSITION ON CRYSTAL GROWTH OF BARIUM SODIUM NIOBATE. EFFECTS OF CH 13-1908
ERNS IN FERROELECTRIC MATERIALS. (BARIUM TITANATE, POTASSIUM NIOBATE) /TION AND ERASURE OF HOLOGRAPHIC PATT 04-0761...

NIOBATE. ELECTRON PARAMAGNETIC RESONANCE OF IRON(3+) DOPED LITHIUM NIOBATE. 11-1765
CTRON PARAMAGNETIC RESONANCE OF GADOLINIUM(3+) DOPED LITHIUM NIOBATE. ELE 11-1833
ELECTRON SPIN RESONANCE STUDIES OF CHROMIUM(3+) IN LITHIUM NIOBATE. 11-1804
S OF CRYSTAL FIELD PARAMETERS IN MANGANESE(2+) DOPED LITHIUM NIOBATE. ELECTRON SPIN RESONANCE STUDIE 11-1803
ENERGY BANDS OF POTASSIUM NIOBATE. 08-1372
EPITAXY OF PLATINUM ON BARIUM STRONTIUM NIOBATE. 13-1907
EPR OF GADOLINIUM(3+) IN LITHIUM NIOBATE. 11-1763
TUDIES OF CRYSTAL FIELD PARAMETERS IN IRON(3+) DOPED LITHIUM NIOBATE. EPR S 11-1839
APPLICATION OF ACOUSTIC WAVES TO HARMONIC GENERATION LITHIUM NIOBATE. /N BY HIGH FREQUENCY ACOUSTIC WAVES. 11-1834
ADING OF SURFACE (ACOUSTIC) WAVES ON (PIEZOELECTRIC) LITHIUM NIOBATE. (FAR FIELD) DIFFRACTION SPRE 11-1850
FAR INFRARED DIELECTRIC DISPERSION IN BARIUM SODIUM NIOBATE. 13-1986
EN BRONZE TYPE CRYSTAL STRUCTURES. PART-3: POTASSIUM LITHIUM NIOBATE. FERROELECTRIC TUNGST 13-1895
FERROELECTRICITY MEASUREMENTS ON STRONTIUM BARIUM NIOBATE. 13-1922
YPERFINE LINES IN ESR SPECTRA OF MANGANESE(2+) DOPED LITHIUM NIOBATE. FORBIDDEN H 11-1805
FREQUENCY DEPENDENCES OF SOME PARAMETERS OF LITHIUM NIOBATE. 11-1843
OPAGATION ON LAYERED ANISOTROPIC SUBSTRATES: GOLD ON LITHIUM NIOBATE. /ND HIGHER ORDER MODE SURFACE WAVE PR 11-1846
GAMMA-RADIATION EFFECTS IN LITHIUM NIOBATE. 11-1822
FAR INFRARED RADIATION BY PICOSECOND LIGHT PULSES IN LITHIUM NIOBATE. GENERATION OF 11-1857
IC WAVE PROPAGATING ITSELF IN THE (001) DIRECTION OF LITHIUM NIOBATE. /IED HARMONICS BY A TRANSVERSE ACOUST 11-1682
OR INELASTIC NEUTRON SCATTERING WITH APPLICATIONS TO LITHIUM NIOBATE. GROUP THEORETICAL SELECTION RULES F 11-1704
GROWTH OF HIGH QUALITY CRYSTALS OF POTASSIUM TANTALATE NIOBATE. 13-2055
HOLOGRAPHIC RECORDING IN LITHIUM NIOBATE. 11-1662
HOLOGRAPHIC STORAGE IN DOPED BARIUM SODIUM NIOBATE. 13-1898
APPLICATIONS. (LEAD ZIRCONATE TITANATE AND SODIUM POTASSIUM NIOBATE) /ELECTRIC CERAMICS FOR ELECTROOPTICAL 09-1527
HOT PRESSING OF POTASSIUM- SODIUM NIOBATE. 09-1542
RIGLYCINE FLUOROBERYLLATE, LITHIUM SULFATE, BARIUM STRONTIUM NIOBATE) /ION DETECTOR. (TRIGLYCINE SULFATE, T 02-0046
A NONLINEAR CRYSTAL ON SECOND HARMONIC GENERATION. (LITHIUM NIOBATE) /MOGENEITY OF THE REFRACTIVE INDEX OF 11-1687
RUCTURE OF AB(5) OXIDE COMPOUNDS. (BARIUM TITANATE, RUBIDIUM NIOBATE) /ELUNG ENERGY AND COVALENCY ON THE ST 04-0761

INFRARED AND RAMAN SPECTROSCOPY OF POTASSIUM METANIOBATE. 08-1303
INTERACTION OF VOLUME ACOUSTIC MODES IN LITHIUM NIOBATE. 11-1806
ERACTIONS BETWEEN HIGH FREQUENCY ACOUSTICAL WAVES IN LITHIUM NIOBATE. INT 11-1836
NG BY POLARITONS ASSOCIATED WITH ORDINARY PHOTONS IN LITHIUM NIOBATE. LIGHT SCATTERI 11-1697
D FERROELECTRIC PEROVSKITES). (BARIUM TITANATE AND POTASSIUM NIOBATE) /CRYSTALS (CASE OF SILICON, QUARTZ AN 04-0629
LONGITUDINAL ELECTROOPTIC EFFECT IN POTASSIUM TANTALATE NIOBATE 13-1942
TUDINAL ELECTROOPTICAL EFFECT IN OBLIQUE SECTIONS OF LITHIUM NIOBATE. LONGI 11-1863
LOW FREQUENCY VIBRATIONAL MODES IN BARIUM SODIUM NIOBATE. 13-1905
D BY ANISOTROPIC DEBYE- WALLER FACTORS. (POTASSIUM TANTALATE NIOBATE) /N PEROVSKITE FERROELECTRICS INDICATE 08-1351
CY (510 MHZ) PARAMAGNETIC RESONANCE INVESTIGATION OF LITHIUM NIOBATE. LOW FREQUEN 11-1793
LUMINESCENCE FROM LITHIUM NIOBATE. 11-1729
ERISTICS OF INTERDIGITAL SURFACE WAVE TRANSDUCERS ON LITHIUM NIOBATE. MEASURED ELECTRICAL CHARACT 11-1702
RIC CONSTANTS AND ELECTROOPTIC COEFFICIENTS IN BARIUM SODIUM NIOBATE. /EE AND CLAMPED VALUES OF THE DIELECT 13-1998
MECHANISM AND KINETICS OF SOLID PHASE FORMATION OF LEAD METANIOBATE. 13-2042
MICROWAVE PHONON ATTENUATION IN X CUT LITHIUM NIOBATE. 11-1756
MIXING OF NONCOLLINEAR ELASTIC SURFACE WAVES ON LITHIUM NIOBATE. 11-1690
AL PULLING. (CONTROL OF CRYSTAL DIAMETER OF STRONTIUM BARIUM NIOBATE) /N TO THE CZOCHRALSKI METHOD OF CRYST 13-1914
MULTIPLE STORAGE OF THICK PHASE HOLOGRAMS IN LITHIUM NIOBATE. 11-1827
RNAL ELECTRIC FIELD DURING OPTICAL DAMAGE PROCESS IN LITHIUM NIOBATE. NATURE OF INTE 11-1861
TTERING STUDY OF THE SOFT MODES IN CUBIC POTASSIUM TANTALATE NIOBATE. NEUTRON SCA 13-2064
NEUTRON STUDY OF DIFFUSE SCATTERING IN CUBIC POTASSIUM NIOBATE. 08-1407
DATA FROM AN X-RAY STUDY OF PHASE TRANSFORMATIONS IN SODIUM NIOBATE. NEW 08-1361
NG LOW DIFFRACTION CUT FOR ACOUSTIC SURFACE WAVES ON LITHIUM NIOBATE. NEW HIGH COUPLI 11-1820
LOW BEAM STEERING CUT FOR ACOUSTIC SURFACE WAVES ON LITHIUM NIOBATE. NEW HIGH FREQUENCY HIGH COUPLING 11-1819
NIOBIUM-93 AND SODIUM-23 IN POLYCRYSTALLINE SODIUM NIOBATE. 08-1469
NIOBIUM-93 NMR LINE WIDTHS IN NONSTOICHIOMETRIC LITHIUM NIOBATE. 11-1795
NONCOLLINEAR PHASE MATCHING EFFECTS IN LITHIUM NIOBATE. 11-1675
NONLINEAR EFFECTS IN FERROCERAMICS. (PZT-8, LITHIUM NIOBATE) 09-1546
NTERACTION OF MICROWAVE ELECTRIC FIELDS AND SOUND IN LITHIUM NIOBATE. NONLINEAR I 11-1838
NONLINEAR OPTICAL PROPERTIES OF POTASSIUM LITHIUM NIOBATE. 13-2027
PTICAL SPECTROSCOPY OF THIRD ORDER SUSCEPTIBILITY IN LITHIUM NIOBATE. NONLINEAR O 11-1855
ON OF RADIATION AND FIRST HARMONIC, CALCULATIONS FOR LITHIUM NIOBATE). /SADOVSKII EFFECT (THEORY, PROPAGATI 03-0186
NONSTOICHIOMETRY AND CRYSTAL GROWTH OF LITHIUM NIOBATE. 11-1691
OBSERVATION OF ORDINARY POLARITONS IN LITHIUM NIOBATE. 11-1854
GRAPHIC STORAGE PROPERTIES OF TRANSITION METAL DOPED LITHIUM NIOBATE. OPTICAL AND HOLO 11-1796
OPTICAL DAMAGE IN POTASSIUM TANTALATE NIOBATE 13-1979
ING OF ACOUSTIC SURFACE WAVE HARMONIC GENERATION (IN LITHIUM NIOBATE) OPTICAL PROB 11-1752
OPTICAL PROPERTIES OF POTASSIUM TANTALATE NIOBATE. 36-3593
OPTIMUM SECOND HARMONIC GENERATION IN LITHIUM NIOBATE. 11-1764
ITIONS OF THE "CUBIC" PEROVSKITE STRUCTURE OF LEAD MAGNESIUM NIOBATE. /UM AND NIOBIUM IN THE OCTAHEDRAL POS 08-1379
TIFERROELECTRICS. (APPLIED TO ADP, TUNGSTEN TRIOXIDE, SODIUM NIOBATE) /ENTATION OF COMPOSITION PLANES IN AN 08-1327
IC LUMINESCENCE AND LIGHT SCATTERING BY POLARITONS. (LITHIUM NIOBATE) PARAMETR 11-1746
CTRON SPIN RESONANCE OF IRON AND MOLYBDENUM IN BARIUM SODIUM NIOBATE. PHOTOCHROMIC EFFECT AND ELE 13-1930
CRYSTALS. (BISMUTH(12) GERMANIUM OXIDE(20) AND BARIUM SODIUM NIOBATE) /ACE WAVES IN CUBIC AND ORTHORHOMBIC 13-2041
PIEZOELECTRIC SURFACE WAVES ON LITHIUM NIOBATE. 11-1824
POLARITONS IN A BIAXIAL CRYSTAL OF BARIUM SODIUM NIOBATE. 13-1992
WIDTH OF TUNABLE STIMULATED FAR INFRARED EMISSION IN LITHIUM NIOBATE. POWER AND LINE 11-1736
PARATION AND DIELECTRIC PROPERTIES OF POTASSIUM BISMUTH ZINC NIOBATE. PRE 13-2001
PREPARATION AND FABRICATION OF SODIUM POTASSIUM NIOBATE. 13-1937
LEIGH WAVES AND PIEZOELECTRIC LEAKY SURFACE WAVES IN LITHIUM NIOBATE. PROPAGATION AND AMPLIFICATION OF RAY 11-1856
RANSITION GROUP IN THE LATTICE OF SINGLE CRYSTALLINE LITHIUM NIOBATE. PROPERTIES OF THE IRON T 11-1669
ON TO A NANOSECOND RESPONSE TIME DETECTOR. (STRONTIUM BARIUM NIOBATE) / MEASUREMENT TECHNIQUE AND APPLICATI 13-1919
INFRARED SIGNALS IN TRIGLYCINE SULFATE AND STRONTIUM BARIUM NIOBATE. /TRIC VOLTAGE RESPONSE TO RECTANGULAR 13-2019
ARED LASER PULSES IN TRIGLYCINE SULFATE AND BARIUM STRONTIUM NIOBATE. /CTRIC VOLTAGE RESPONSE TO SHORT INFR 13-2018
NFRARED RADIATION IN TRIGLYCINE SULFATE AND STRONTIUM BARIUM NIOBATE. /OLTAGE RESPONSE TO STEP SIGNALS OF I 13-2022
RAMAN SCATTERING BY BARIUM STRONTIUM NIOBATE. 13-1899
RAMAN SPECTRA OF BARIUM SODIUM NIOBATE. 13-1906
TION OF ELASTIC SURFACE WAVE FROM PERIODIC ARRAYS ON LITHIUM NIOBATE. REFLEC 11-1853
OGEN IN POLARIZATION REVERSAL OF FERROELECTRIC BARIUM SODIUM NIOBATE. ROLE OF HYDR 13-2025
ROTATING WAVEPLATE OPTICAL FREQUENCY SHIFTING IN LITHIUM NIOBATE. 11-1689
RAMETRIC OSCILLATION, SECOND HARMONIC GENERATION IN LITHIUM NIOBATE). SIMULTANEOUS OPTICAL PA 11-1859
ND STRUCTURE OF FERROELECTRIC TETRAGONAL POTASSIUM TANTALATE NIOBATE. SOFT MODES A 13-1957
SOFT PHONONS IN POTASSIUM TANTALATE NIOBATE. 13-2040
IC REGION OF THE SYSTEM LEAD HAFNATE TITANATE, LEAD STANNATE NIOBATE. SOLID SOLUTIONS IN ANTIFERROELECTR 13-1956
SOLIDIFICATION STUDY OF BARIUM STRONTIUM NIOBATE. 13-1954
F MELT STOICHIOMETRY ON THE OPTICAL PROPERTIES BARIUM SODIUM NIOBATE. SOME EFFECTS O 13-1902
ELT STOICHIOMETRY ON THE OPTICAL PROPERTIES OF BARIUM SODIUM NIOBATE. SOME EFFECTS OF M 13-1903
N, AND DISPERSION IN THE NONLINEAR SUSCEPTIBILITY IN LITHIUM NIOBATE. /RAMETRIC SCATTERING, IDLER ABSORPTIO 11-1771
NETS. (LEAD AND BARIUM TITANATES, BISMUTH FERRATE, POTASSIUM NIOBATE) /FERROELECTRICS AND FERROELECTRIC MAG 04-1038
CTURE AND TRANSITIONS IN PEROVSKITES. (ORTHOFERRITES, SODIUM NIOBATE) STRU 08-1396
SORDER AND RAMAN SCATTERING IN BARIUM TITANATE AND POTASSIUM NIOBATE. STRUCTURE DI 04-0676
SURFACE WAVE DIFFRACTION IN LITHIUM NIOBATE. 11-1698
ENCY DEPENDENCES OF THE ELECTRICAL PROPERTIES OF LITHIUM METANIOBATE. TEMPERATURE AND FREQU 11-1717
C, AND DIELECTRIC CONSTANTS OF LITHIUM TANTALATE AND LITHIUM NIOBATE. /ENDENCE OF THE ELASTIC, PIEZOELECTRI 11-1823
NCE OF THE LITHIUM NMR SPECTRUM AND ATOMIC MOTION IN LITHIUM NIOBATE. TEMPERATURE DEPENDE 11-1722
RATURE DEPENDENCE OF THE REFRACTIVE INDICES OF BARIUM SODIUM NIOBATE. TEMPE 13-2008
OLARITON ASSOCIATED WITH THE 248 CM(-1) SOFT MODE IN LITHIUM NIOBATE. TEMPERATURE STUDY OF THE P 11-1807
OLTEN SOLUTIONS OF THE PSEUDOSYSTEM POTASSIUM LITHIUM BARIUM NIOBATE. /STEN BRONZE TYPE CRYSTALS GROWN IN M 13-1948
THE MEASUREMENT OF HYPERSONIC ATTENUATION IN LITHIUM NIOBATE. 11-1760
OF THE GENERATION OF THE ACOUSTIC SECOND HARMONIC. (LITHIUM NIOBATE) THEORETICAL STUDY 11-1772
OF THE PEROVSKITES IN THE SYSTEM LEAD ZIRCONATE TITANATE METANIOBATE. /OUS POLARIZATION AND COERCIVE FORCE 09-1597
THERMALLY FIXED HOLOGRAMS IN LITHIUM NIOBATE. 11-1828
THREE LAYER BROAD BAND ANTIREFLECTION COATINGS FOR LITHIUM NIOBATE. 11-1676
AXATION TIMES IN FERROELECTRIC LITHIUM TANTALATE AND LITHIUM NIOBATE. /ETERMINATION OF THE SPIN LATTICE REL 11-1829
UX) CRYSTAL GROWTH AND NEUTRON CHARACTERIZATION OF POTASSIUM NIOBATE. (TOP SEEDED FL 08-1353
TRANSFORMATIONS IN SODIUM NIOBATE. 08-1397
PSEUDOTERNARY SYSTEM SODIUM NIOBATE BARIUM TITANATE BARIUM DINIOBATE. TUNGSTEN BRONZE FIELD IN THE 13-1970
ERAMIC. (BARIUM TITANATE, BARIUM TIN TITANATE, LEAD BARIUM DINIOBATE) /TROPY OF A POLARIZED FERROELECTRIC C 04-0767
RATE WITH SUPERSATURATION IN FLUXED MELTS. (BARIUM STRONTIUM NIOBATE) VARIATION OF CRYSTAL GROWTH 13-1939
VISIBLE CW PARAMETRIC OSCILLATOR USING BARIUM SODIUM NIOBATE. 13-1984

VISUALIZATION OF ACOUSTIC RADIATION IN LITHIUM NIOBATE. 11-1864
ELENGTH DEPENDENCE OF HOLOGRAPHIC PHENOMENA IN DOPED LITHIUM NIOBATE. WAV 11-1785
/IEZOELECTRIC PROPERTIES IN THE TERNARY SYSTEM LEAD (ZINCATE NIOBATE)- BARIUM (ZINCATE NIOBATE)- LEAD TITA/ 05-1070
L STRUCTURE AND PIEZOELECTRICITY OF THE SYSTEM LEAD (ZINCATE NIOBATE)- LEAD TITANATE. CRYSTA 13-2067
), (LEAD, TUNGSTEN) (ZIRCONATE TANTALATE), LEAD (LANTHANATE, NIOBATE)- LEAD ZIRCONATE- LEAD TITANATE) /LATE 13-2004
ENDENT CURIE TEMPERATURE, POLARIZATION FLUCTUATIONS, LITHIUM NIOBATE, ALKALI HALIDES) /RYSTALS. (STRAIN DEP 03-0513
DIHYDROGEN PHOSPHATE, AMMONIUM DIHYDROGEN PHOSPHATE, LITHIUM NIOBATE, AND ALPHA- IODIC ACID. /OF POTASSIUM 11-1677
YSTALLOGRAPHIC AND DIELECTRIC PROPERTIES OF STRONTIUM NIOBATE AND BARIUM SODIUM NIOBATE. / ON THE CR 13-1921
AL AND TRANSVERSE HYPERSONIC PULSES IN PIEZOELECTRIC LITHIUM NIOBATE AND BISMUTH BIGERMANATE. /F LONGITUDIN 11-1814
ACE WAVE CONVOLUTION ON CRYSTALS OF CADMIUM SULFIDE, LITHIUM NIOBATE AND BISMUTH(12) GERMANIUM OXIDE(20). / 11-1692
/ ELASTIC SURFACE WAVES ON PIEZOELECTRIC SUBSTRATES (LITHIUM NIOBATE AND BISMUTH(12) GERMANIUM OXIDE(20)). 11-1751
/NGITUDINAL AND TRANSVERSE HYPERSONIC WAVE PULSES IN LITHIUM NIOBATE AND BISMUTH(12) GERMANIUM OXIDE(20) P/ 11-1815
/ARAMAGNETIC RESONANCE SPECTRUM IN (MANGANESE DOPED) LITHIUM NIOBATE AND (COPPER(2+) DOPED) ZINC BORACITE. 11-1792
ECTROSCOPIC PROPERTIES OF ACTIVATED FERROELECTRICS. (LITHIUM NIOBATE AND GADOLINIUM MOLYBDATE) LASER AND SP 11-1741
GROWTH OF CRATERS ON ION ETCHED SURFACE OF LITHIUM NIOBATE AND ION ETCHING WITHOUT CRATERS. 11-1860
GROWTH OF SINGLE CRYSTALS OF LEAD ZINCATE NIOBATE AND ITS PROPERTIES. 36-3594
NEODYMIUM YAG LASER IRRADIATION INDUCED DAMAGE TO LITHIUM NIOBATE AND KDP. 11-1674
T ELASTOOPTIC EFFECTS AND Q SWITCHING PERFORMANCE IN LITHIUM NIOBATE AND KDP POCKELS CELLS. TRANSIEN 11-1727
(CORRELATION OF CATION SHIFTS WITH BISMUTH FERRATE, LITHIUM NIOBATE AND LITHIUM TANTALATE) /=0.9 AND 0.58) 09-1574
CAL QUALITY OF FERROELECTRIC CRYSTALS, ESPECIALLY IN LITHIUM NIOBATE AND LITHIUM TANTALATE. /ITION AND OPTI 11-1780
EITY OF THE OPTICALLY INDUCED INDEX OF REFRACTION IN LITHIUM NIOBATE AND LITHIUM TANTALATE. INHOMOGEN 11-1664
MEDIUM FREQUENCY RESONATORS OF LITHIUM NIOBATE AND LITHIUM TANTALATE. 11-1851
SCATTERING OF LIGHT ON ACOUSTIC PHONONS IN LITHIUM NIOBATE AND LITHIUM TANTALATE. 11-1744
SPECTROSCOPY. STOICHIOMETRY VARIATIONS IN LITHIUM NIOBATE AND LITHIUM TANTALATE BY RAMAN POWDER 11-1813
ISTIC EQUATION FOR PLANE PIEZOELECTRIC VIBRATIONS OF LITHIUM NIOBATE AND LITHIUM TANTALATE CRYSTALS. /ACTER 11-1739
GROWTH AND PROPERTIES OF LITHIUM NIOBATE AND LITHIUM TANTALATE FILMS. 11-1712
X-RAY TOPOGRAPHIC STUDIES ON LITHIUM NIOBATE AND LITHIUM TANTALATE SINGLE CRYSTALS. 11-1830
TATIST/ COMPARISON OF FERROELECTRICITY IN ISOMORPHIC LITHIUM NIOBATE AND LITHIUM TANTALATE USING AUTHOR'S S 03-0366
N ON CRYSTAL INHOMOGENEITY (THEORY AND TEST ON BARIUM SODIUM NIOBATE AND NEODYMIUM DOPED YAG LASER) /RATIO 13-2043
(MECHANISM OF) OPTICAL INDEX DAMAGE IN LITHIUM NIOBATE AND OTHER PYROELECTRIC INSULATORS. 11-1735
LTRASONIC WAVES IN A LAYERED STRUCTURE CONSISTING OF LITHIUM NIOBATE AND P-TYPE SILICON. /ION OF RAYLEIGH U 11-1747
/ELECTROOPTIC EFFECTS AND Q SWITCHING PERFORMANCE IN LITHIUM NIOBATE AND POTASSIUM DIDEUTERIUM PHOSPHATE P/ 11-1710
E SY/ PHASE EQUILIBRIA IN THE POTASSIUM NIOBATE- STRONTIUM DINIOBATE AND POTASSIUM NIOBATE- BARIUM DINIOBAT 13-2016
GROWTH OF SINGLE CRYSTALS OF LEAD (ZINCATE NIOBATE) AND PROPERTIES OF THIS COMPOSITION. 13-1990
DING ON THE PROPAGATION OF ACOUSTIC SURFACE WAVES ON LITHIUM NIOBATE AND QUARTZ. EFFECT OF MASS LOA 11-1757
OR SIGNAL PROCESSING PURPOSES. (TECHNIQUE APPLIED TO LITHIUM NIOBATE AND QUARTZ) /F SURFACE ELASTIC WAVES F 11-1738
OOLING: AN AID TO CZOCHRALSKI CRYSTAL GROWTH. (BARIUM SODIUM NIOBATE AND SODIUM LANTHANUM TANTALATE) /ASE C 13-1928
S CRYSTALS. (NIOBIUM-93 AND SODIUM-23 RESONANCE IN POTASSIUM NIOBATE AND SODIUM NIOBATE) /ONS IN PEROVSKITE 08-1283
EPENDENCE OF THE OPTICAL PROPERTIES OF FERROELECTRIC LITHIUM NIOBATE AND TANTALATE. TEMPERATURE D 11-1770
/TACKING FAULT MODEL FOR STOICHIOMETRY DEVIATIONS IN LITHIUM NIOBATE AND TANTALATE AND THE EFFECT ON THE C/ 11-1776
/TACKING FAULT MODEL FOR STOICHIOMETRY DEVIATIONS IN LITHIUM NIOBATE AND TANTALATE AND THE EFFECT ON THE C/ 11-1777
RTIES OF THE EQUATIONS OF PIEZOELECTRIC VIBRATION OF LITHIUM NIOBATE AND TANTALATE CRYSTALS. SOME PROPE 11-1740
/ANTIFERROELECTRICITY AND FERROELECTRICITY APPLIED TO SODIUM NIOBATE AND THE SYSTEM POTASSIUM NIOBATE- SOD/ 03-0218
SOLUTIONS OF P/ HIGH TEMPERATURE PHASE TRANSITIONS IN SODIUM NIOBATE AND THE USE OF TILTING SCHEMES IN THE 08-1336
STRUCTURAL DI- AND FERROELECTRIC PROPERTIES OF X POTASSIUM DINIOBATE AND (1-X) BARIUM TITANATE. 04-0623
RANSITION) ELECTRICAL PROPERTIES OF LITHIUM NIOBATE. (ANOMALY AT 263 DEG K IS STRUCTURAL T 11-1709
X-RAY DIFFUSE SCATTERING FROM SODIUM NIOBATE AS A FUNCTION OF TEMPERATURE. 08-1313
TH BY KYROPOULO/ DEVELOPMENT OF SINGLE CRYSTAL BARIUM SODIUM NIOBATE AS A NONLINEAR OPTICAL MATERIAL. (GROW 13-2021
IBLE HOLOGRAPHIC STORAGE. RH DOPED LITHIUM NIOBATE AS AN IMPROVED NEW MATERIAL FOR REVERS 11-1359
DIFFUSE SCATTERING IN SODIUM NIOBATE AT ROOM TEMPERATURE. 08-1304
N THE PARAELECTRIC PHASE. (POTASSIUM TANTALATE AND TANTALATE NIOBATE, BARIUM AND STRONTIUM TITANATES) /ES I 04-0690
ENSE BY ANOMALOUS SCATTERING AT AN ABSORPTION EDGE. (LITHIUM NIOBATE, BARIUM MANGANESE TETRAFLUORIDE) /TY S 11-1673
/E TYPE STRUCTURES (POTASSIUM (BARIUM OR STRONTIUM) TITANATE NIOBATE, (BARIUM OR STRONTIUM) BISMUTH TITANA/ 13-1962
OLOGRAPHIC PATTERN FIXING IN ELECTROOPTIC CRYSTALS. (LITHIUM NIOBATE, BARIUM SODIUM NIOBATE) H 11-1661
UM TANTALATE (FORCE CONSTANT/ VIBRATIONAL SPECTRA OF LITHIUM NIOBATE, BARIUM SODIUM NIOBATE AND BARIUM SODI 11-1808
DOMAINS OF PEROVSKITE TYPE OXIDES. (LEAD TITANATE, POTASSIUM NIOBATE, BARIUM TITANATE) /S ON FERROELECTRIC 04-0993
TUNGSTEN BRONZE FIELD IN THE PSEUDOTERNARY SYSTEM SODIUM NIOBATE BARIUM TITANATE BARIUM DINIOBATE. 13-1970
ELEVANCE TO SOFT PHONON MODES. (X-RAY / STRUCTURES OF SODIUM NIOBATE BETWEEN 480 AND 575 DEG C, AND THEIR R 08-1273
ACE WAVES BY NONLINEAR MIXING IN ANISOTROPIC MEDIA. (LITHIUM NIOBATE, BISMUTH(12) GERMANIUM OXIDE(20)) /URF 11-1778
/CS HANDBOOK. VOL 1: SURFACE WAVE VELOCITIES. (BARIUM SODIUM NIOBATE, BISMUTH(12) GERMANIUM OXIDE(20), LIT/ 01-0028
DAMAGE TO THE SURFACE OF LITHIUM NIOBATE BY LIGHT. 11-1866
THE DISTRIBUTION COEFFICIENTS OF LITHIUM OXIDE INTO LITHIUM NIOBATE BY PHASE MATCHING TEMPERATURES. /ON OF 11-1841
FRACTURE OF NONLINEAR CRYSTALS (KDP AND LITHIUM NIOBATE) BY RADIATION FROM A RUBY LASER. 11-1663
/SS TOWARDS THE GROWTH OF GOOD OPTICAL QUALITY BARIUM SODIUM NIOBATE (BY USING COMPOSITION WHERE LIQUIDUS / 13-2047
DISPLAY OF TWINNING IN LITHIUM NIOBATE BY X-RAY DIFFRACTION TOPOGRAPHY. 11-1845
ELASTIC CONSTANT OF ORTHORHOMBIC POTASSIUM NIOBATE BY X-RAY DIFFUSE SCATTERING. 08-1416
ELL PARAMETERS) RADIATION DAMAGE IN LITHIUM NIOBATE. (CAPACITANCE, OPTICAL SPECTRA, ESR, C 11-1705
AND THERMAL EXPANSION STUDY OF A SODIUM(0.88) LITHIUM(0.12) NIOBATE CERAMIC. X-RAY 13-1999
THE SYSTEM LEAD TITANATE- STRONTIUM COPPER NIOBATE. (CERAMIC SYNTHESIS) 05-1044
PROPERTIES OF FERROGLASS CERAMICS BASED ON LEAD BARIUM NIOBATE COMPOUNDS. 13-2006
KI) CRYSTAL GROWTH. BARIUM SODIUM NIOBATE: CONSTITUTIONAL STUDIES AND (CZOCHRALS 13-1904
/OPERTIES OF SINGLE DOMAIN CRYSTALS OF FERROELECTRIC LITHIUM NIOBATE. (CRACK-FREE, THROUGH ADDITION OF MAG/ 11-1774
BIREFRINGENCE INDUCED IN A LITHIUM NIOBATE CRYSTAL. 11-1767
FERMI RESONANCE OF POLARITON WITH BIPHONON IN AN LITHIUM NIOBATE CRYSTAL. 11-1761
NONLINEAR INTERACTION OF HYPERSONIC WAVES IN A LITHIUM NIOBATE CRYSTAL. 11-1802
PARAMETRIC LUMINESCENCE INTENSITY IN THE LITHIUM NIOBATE CRYSTAL. 11-1750
ORRELATIONS OF SECOND HARMONIC GENERATED LIGHT (FROM LITHIUM NIOBATE CRYSTAL, AGREEMENT WITH THEORY). /ON C 11-1701
RADIATION EXPOSURE OF A LITHIUM NIOBATE CRYSTAL AT HIGH TEMPERATURES. 11-1721
/RM AND ALTERABLE SHG PHASE MATCHING TEMPERATURES IN LITHIUM NIOBATE (CRYSTAL GROWTH BY ADDING MAGNESIUM O/ 11-1684
/H FREQUENCY ACOUSTIC HARMONICS IN A (SINGLE DOMAIN) LITHIUM NIOBATE CRYSTAL (NONLINEAR ELASTIC PROPERTIES/ 11-1707
) TRANSVERSE POLARITONS IN A LITHIUM NIOBATE CRYSTAL. (RAMAN SPECTRUM, PERMITTIVITY 11-1762
NERGY LEVELS OF NEODYMIUM(3+) AND ERBIUM(3+) IONS IN LITHIUM NIOBATE CRYSTALS. / LUMINESCENCE SPECTRA AND E 11-1713
LLINEAR INTERACTION OF LONGITUDINAL ELASTIC WAVES IN LITHIUM NIOBATE CRYSTALS. CO 11-1678
CZOCHRALSKI GROWTH OF BARIUM STRONTIUM NIOBATE CRYSTALS. 13-1913
DISORDER OF LITHIUM NIOBATE CRYSTALS. 11-1679
DOMAIN STRUCTURE OF LITHIUM METANIOBATE CRYSTALS. 11-1708
GROWTH OF UNCRACKED BARIUM SODIUM NIOBATE CRYSTALS. 13-1901
TIGATION OF A SYSTEM FOR BEAM DEFLECTION BY MEANS OF LITHIUM NIOBATE CRYSTALS. INVES 11-1768
GHT BEAM DEFLECTION CONTROLLED WITH A PRISM OF LEAD MAGNESIO NIOBATE CRYSTALS. LI 08-1280
TION (STOICHIOMETRY) DEPENDENCE OF POCKELS EFFECT IN LITHIUM NIOBATE CRYSTALS. MELT COMPOSI 11-1728

PHOTOELASTIC CONSTANTS OF LITHIUM NIOBATE CRYSTALS. 11-1745

SPACE CHARGE LIMITED CURRENTS IN SEMICONDUCTING POTASSIUM NIOBATE CRYSTALS. 08-1412

POLARITONS. PART-2: THE CASE OF CADMIUM SULFIDE AND LITHIUM NIOBATE CRYSTALS. /TANEOUS RAMAN SCATTERING BY 11-1787

STABILITY REGIONS FCR THE GROWTH OF BARIUM STRONTIUM NIOBATE CRYSTALS. 13-2056

SURFACE CHARGE LAYERS CN FERROELECTRIC POTASSIUM NIOBATE CRYSTALS. 08-1369

GROWTH OF BARIUM SODIUM NIOBATE CRYSTALS BY THE KYROPOULOS TECHNIQUE. 13-1909

(10 TO 3/ CALCULATION OF THE (ABSORPTION) SPECTRA OF LITHIUM NIOBATE CRYSTALS CONTAINING CHROMIUM(3+) IONS 11-1831

GROWTH TWINNING (UNDER DIRECT CURRENT) IN C-AXIS LITHIUM NIOBATE CRYSTALS (CZOCHRALSKI METHOD). 11-1672

DOUBL/ CATASTROPHIC SURFACE DAMAGE PRODUCED IN BARIUM SODIUM NIOBATE CRYSTALS DURING INTRACAVITY FREQUENCY 13-2052

PRECISION MACHINING OF LITHIUM NIOBATE CRYSTALS FOR ELECTROOPTIC ELEMENTS. 11-1742

W BIREFRINGENCE VARIATIONS) / GROWTH OF HIGH QUALITY LITHIUM NIOBATE CRYSTALS (FOR NONLINEAR OPTICS WITH LO 11-1688

N/ ANISOTROPIC CRITICAL X-RAY DIFFUSE SCATTERING FROM SODIUM NIOBATE CRYSTALS (NEAR THE 640 DEG C PHASE TRA 08-1311

PHASE MATCHING ANGLES AND TEMPERATURE OF LITHIUM METANIOBATE CRYSTALS OF DIFFERENT STOICHIOMETRIES. 11-1665

LIGHT SCATTERING BY ULTRASONIC WAVES IN LITHIUM NIOBATE CRYSTALS. (PHOTOELASTIC CONSTANT) 11-1754

WITH A LOW HALF WAVE VOLTAGE. CUTS OF LITHIUM NIOBATE CRYSTALS SUITABLE FOR LIGHT MODULATORS 11-1715

GENERATION AND PROPAGATION OF HYPERSONIC WAVES IN LITHIUM NIOBATE CRYSTALS. (80-950 DEG K, 200-2000 MHZ) 11-1753

NEW FERROELECTRIC: ANTIMONY ORTHONIOBATE. (CURIE TEMPERATURE) 36-3589

ND OPTICAL CONSTANTS) ROOM/ ELASTIC ANOMALY OF BARIUM SODIUM NIOBATE (ELASTIC, PIEZOELECTRIC, DIELECTRIC, A 13-2060

PIEZOELECTRIC LEAKY SURFACE WAVE IN LITHIUM NIOBATE (ELECTROMAGNETIC COUPLING). 11-1832

TION SCATTERING CF SOUND IN PIEZOELECTRICS. (THEORY, LITHIUM NIOBATE EXAMPLE) STIMULATED COMBINA 03-0431

TRUCTURE. (MIXED OXIDES ABO(3)) (BARIUM TITANATE AND LITHIUM NIOBATE FILMS ON PLATINUM OR SILICON) / TYPE S 04-0914

TOR. BARIUM SODIUM NIOBATE FOR A TUNABLE STIMULATED RAMAN OSCILLA 13-1917

HING. THINNING OF LITHIUM NIOBATE FOR ACOUSTOOPTIC DEFLECTORS BY ION ETC 11-1862

PARAMETRIC FLUORESCENCE IN BARIUM SODIUM NIOBATE FOR NINE CW PUMP FREQUENCIES. 13-1950

5-528.7 MICRONS). PARAMETRIC FLUORESCENCE IN BARIUM SODIUM NIOBATE FOR 10 CW PUMP WAVELENGTHS (RANGE 454. 13-1983

MECHANISM AND KINETICS OF LITHIUM METANIOBATE FORMATION BY A SOLID PHASE REACTION. 11-1840

CRYSTALLIZATION OF LEAD METANIOBATE FROM GLASS. 08-1275

CRYSTALLIZATION OF SOLID SOLUTIONS OF SODIUM NIOBATE FROM GLASS. 08-1386

NGITUDI/ MANDELSTAM BRILLOUIN SCATTERING OF LIGHT IN LITHIUM NIOBATE (FROM TRANSVERSE (MORE INTENSE) AND LO 11-1743

DEFLECTION. ELECTROOPTIC POTASSIUM TANTALATE NIOBATE GRATINGS FOR LIGHT BEAM MODULATION AND 13-1929

TO L/ NEW (UNIAXIAL) NONLINEAR OPTIC CRYSTAL BARIUM LITHIUM NIOBATE HAVING NO MICROTWINNING (AND RESISTANT 13-1958

STRUCTURE OF SODIUM NIOBATE HETEROTYPES. 08-1395

/DEPENDENCE OF THE DIELECTRIC PROPERTIES OF BARIUM STRONTIUM NIOBATE IN THE FREQUENCY RANGE OF 10 KILOHERT/ 13-1938

FORMATION PROCESSES OF LEAD MAGNESIO NIOBATE IN THE SOLID STATE REACTION. 08-1354

ELECTROOPTIC EFFECT IN LITHIUM NIOBATE IN THE (SUB) MILLIMETER RANGE. 11-1844

ETICAL AND E/ BISMUTH(12) GERMANIUM OXIDE(20) VERSUS LITHIUM NIOBATE IN THE SURFACE WAVE AMPLIFIER. A THEOR 11-1720

THERMAL EXPANSION OF LITHIUM NIOBATE IN THE TEMPERATURE RANGE OF 70-300 K. 11-1685

/E DEPENDENCE OF HYPERSONIC WAVE DAMPING IN RUBY AND LITHIUM NIOBATE IN THE THREE CENTIMETER RADIO FREQUEN/ 11-1718

/ THE TRANSITION CF THE (ORTHORHOMBIC) FERROELECTRIC (LEAD DINIOBATE INTO THE NONFERROELECTRIC (RHOMBOHEDR/ 08-1285

ROELECTRIC DOMAIN BOUNDARIES (DOPED AND UNDOPED TGS, LITHIUM NIOBATE, KDP, ABSTRACT ONLY). /CONTRAST AT FER 11-1671

HASIZING ELECTROOPTIC MODULATION, LITHIUM TANTALATE, LITHIUM NIOBATE, KDP, 113 REFS). /ICATIONS (REVIEW EMP 02-0056

/OF PIEZOELECTRIC CRYSTALS. (LEAD TITANATE, BARIUM STRONTIUM NIOBATE, LEAD NIOBIUM(4) OXIDE(11), SILVER GA/ 05-1066

LITHIUM NIOBATE LIGHT MODULATOR. 11-1782

OPERTIES (BIREFRINGENCE AND REFRACTIVE INDICES) OF POTASSIUM NIOBATE (LINEAR ELECTROOPTICS). OPTICAL PR 08-1468

(950 MHZ) ACOUSTIC WAVE ALONG THE (001) DIRECTION OF LITHIUM NIOBATE (LIQUID HELIUM TEMPERATURE). /EQUENCY 11-1835

TRATE, LITHIUM THALLIUM TARTRATE, LITHIUM TANTALATE, LITHIUM NIOBATE, LITHIUM HYDRAZINIUM SULFATE) /IUM TAR 02-0052

/CEPTIBILITIES. (BARIUM TITANATE, LITHIUM TANTALATE, LITHIUM NIOBATE, LITHIUM IODATE, KDP, ADP, AND OTHER / 02-0107

/CTROOPTIC MATERIALS, SINGLE CRYSTALS OF POTASSIUM TANTALATE NIOBATE, LITHIUM NIOBATE, AND BISMUTH TITANAT/ 02-0073

/ FOURTH HARMONICS OF 1.06 MICRON RADIATION IN BARIUM SODIUM NIOBATE, LITHIUM NIOBATE, AND LITHIUM IODATE/ 11-1779

/TE, POTASSIUM DIDEUTERIUM PHOSPHATE, POTASSIUM TANTALATE OR NIOBATE, LITHIUM NIOBATE OR TANTALATE, BARIUM/ 01-0021

/PTICAL POLARIZABILITY OF THE NIOBIUM- OXYGEN BOND. (LITHIUM NIOBATE, LITHIUM TANTALATE, POTASSIUM LITHIUM/ 11-1734

CRYSTALS. (KDP, ADP, RUBIDIUM DIHYDROGEN PHOSPHATE, LITHIUM NIOBATE, LITHIUM TANTALATE, QUARTZ) /NONLINEAR 11-1865

/ENTS OF ACOUSTIC SURFACE WAVE VELOCITY AND DELAY ON LITHIUM NIOBATE, LITHIUM TANTALATE, QUARTZ, AND TELLU/ 11-1821

LTS. PHASE COMPOSITION OF (LEAD, BARIUM) DINIOBATE. METHOD OF INVESTIGATION AND SOME RESU 13-1945

SECONDARY (ELECTRO) PHOTOELASTIC EFFECT IN LITHIUM NIOBATE (ON ELECTRON IRRADIATION). 11-1801

LTS FOR SAPPH/ EXPANSION, CRAZING AND EXFOLIATION OF LITHIUM NIOBATE ON ION BOMBARDMENT AND COMPARISON RESU 11-1800

NDENCE OF SECOND HARMONIC GENERATION COEFFICIENTS OF LITHIUM NIOBATE ON MELT COMPOSITION. DEPE 11-1769

, DIE/ OPTICAL AND FERROELECTRIC PROPERTIES OF BARIUM SODIUM NIOBATE. (OPTICAL ABSORPTION, REFRACTIVE INDEX 13-2024

/NATE TITANATE MODIFIED WITH SODIUM OR POTASSIUM ANTIMONATE, NIOBATE, OR BISMUTHATE, OR BY BISMUTH OR LANT/ 09-1539

/F NONLINEAR OPTICAL COEFFICIENTS CF POLAR CRYSTALS (LITHIUM NIOBATE OR TANTALATE, BARIUM TITANATE, ZINC O/ 02-0098

/ION) COEFFICIENTS CF POLAR (PYROELECTRIC) CRYSTALS (LITHIUM NIOBATE OR TANTALATE OR GALLATE, BARIUM TITAN/ 02-0095

CIES AND EIGENVECTORS OF POLARITONS WITH APPLICATION TO LEAD NIOBATE. PART-1: LONG OPTICAL PHONONS. /REQUEN 08-1284

/ AND EIGENVECTORS OF POLARITONS WITH APPLICATION TO LITHIUM NIOBATE. PART-2: POLARITONS OF SMALL WAVE VEC/ 11-1766

NG CONDITIONS ON THE STRUCTURE AND PROPERTIES OF LEAD BARIUM NIOBATE PIEZOCERAMICS. EFFECTS OF ROASTI 13-2046

FREQUENCY RANGE WITH LOW RAT/ LITHIUM TANTALATE AND LITHIUM NIOBATE PIEZOELECTRIC RESONATORS IN THE MEDIUM 11-1724

TEMPERATURE DEPENDENCE OF RESONANT FREQUENCIES OF LITHIUM NIOBATE PLATE RESONATORS. (COMMENTS) 11-1812

ITUDINAL ELECTROOPTIC EFFECT IN OBLIQUE CUT STRONTIUM BARIUM NIOBATE PLATES. LONG 08-1472

ITUDINAL ELECTROOPTIC EFFECT IN OBLIQUE CUT STRONTIUM BARIUM NIOBATE PLATES. LONG 13-2063

/TASSIUM TUNGSTEN BRONZE STRUCTURE (POTASSIUM SODIUM LITHIUM NIOBATE, POTASSIUM STRONTIUM LITHIUM NIOBATE / 13-1923

CURRENT IN FERROELECTRIC OXIDE CRYSTALS. (POTASSIUM LITHIUM NIOBATE, POTASSIUM STRONTIUM NIOBATE) /MULATED 13-1926

TPIC DETECTORS. CHARACTERISTICS OF POTASSIUM TANTALATE NIOBATE (POTASSIUM TANTALATE NIOBATE) PYROELEC 13-2030

35), CURIE TEMPERATUR/ RAMAN SPECTRUM OF POTASSIUM TANTALATE NIOBATE. (POTASSIUM TANTALATE(0.65) NIOBATE(0. 08-1391

ELECTROOPTIC LIGHT BEAM DEFLECTION WITH BARIUM STRONTIUM NIOBATE PRISM. 13-2036

OPTICALLY INDUCED PHYSICAL DAMAGE TO LITHIUM NIOBATE, PROUSTITE, AND LITHIUM IODATE. 08-1328

CHARACTERISTICS OF KTN (POTASSIUM TANTALATE NIOBATE) PYROELECTRIC DETECTORS. 13-2030

PHOTOMIXING AT 10.6 MICRONS WITH STRONTIUM BARIUM NIOBATE PYROELECTRIC DETECTORS. 09-1477

TIVE CAPACITANCES. LOW TEMPERATURE COEFFICIENT OF LITHIUM NIOBATE RESONATORS BY USE OF TEMPERATURE SENSI 11-1670

TIES OF (THE TUNGSTEN BRONZE SYSTEM) POTASSIUM SODIUM BARIUM NIOBATE (SECOND HARMONIC GENERATION). / PROPER 13-2028

UREMENT OF SPUTTER ETCHING RATES FOR MONOCRYSTALLINE LITHIUM NIOBATE, SILICON DIOXIDE, AND SILICON. MEAS 11-1714

LITHIUM NIOBATE SILICON SURFACE WAVE CONVOLUTER. 11-1788

FERROELECTRIC PROPERTIES OF STRONTIUM NIOBATE SINGLE CRYSTAL. 08-1404

LAPPING CHARACTERISTICS OF LITHIUM NIOBATE SINGLE CRYSTAL. 11-1783

PIEZOELECTRIC PROPERTIES OF POTASSIUM TANTALATE NIOBATE SINGLE CRYSTAL. 08-1271

PREPARATION OF POTASSIUM NIOBATE SINGLE CRYSTAL FOR OPTICAL APPLICATION 08-1332

ONZE). DIELECTRIC PROPERTY OF BARIUM TITANATE NIOBATE SINGLE CRYSTAL (TETRAGONAL TUNGSTEN BR 13-1969

/ECTRIC AND ELECTROOPTIC PROPERTIES OF BARIUM SODIUM YTTRIUM NIOBATE SINGLE CRYSTAL WITH TUNGSTEN BRONZE T/ 13-1971

E SCATTERING FROM POTASSIUM MANGANESE TRIFLUORIDE AND SODIUM NIOBATE SINGLE CRYSTALS. /RITICAL X-RAY DIFFUS 08-1312

FORMATION AND RELAXATION BEHAVIOR IN FERROELECTRIC POTASSIUM NIOBATE SINGLE CRYSTALS. DE 08-1420

ERMINATION OF ELASTIC AND PIEZOELECTRIC CONSTANTS OF LITHIUM NIOBATE SINGLE CRYSTALS. DET 11-1749

DOMAIN STRUCTURE IN LITHIUM NIOBATE SINGLE CRYSTALS. 11-1703

DOMAINS IN FERROELECTRIC POTASSIUM NIOBATE SINGLE CRYSTALS. 08-1315
ELECTRICAL CONDUCTIVITY OF LITHIUM NIOBATE SINGLE CRYSTALS. 11-1733
HEAT TREATMENT INDUCED POINT DEFECTS IN LITHIUM METANIOBATE SINGLE CRYSTALS. 11-1668
INTERFEROMETRIC STUDIES OF DOMAIN STRUCTURES IN POTASSIUM NIOBATE SINGLE CRYSTALS. 08-1318
INTERFEROMETRIC STUDIES OF DOMAIN STRUCTURE IN POTASSIUM NIOBATE SINGLE CRYSTALS. 08-1319
LAPPING CHARACTERISTICS OF LITHIUM NIOBATE SINGLE CRYSTALS. 11-1784
ITCHING OF 60 DEGREE DOMAIN WALLS IN FERROELECTRIC POTASSIUM NIOBATE SINGLE CRYSTALS. MECHANICAL SW 08-1421
OPTICAL STUDIES OF DOMAIN STRUCTURE OF POTASSIUM NIOBATE SINGLE CRYSTALS. 08-1320
SYNTHESIS AND STUDY OF LEAD SCANDIUM NIOBATE SINGLE CRYSTALS. 08-1325
SORPTION OF 9.4 GHZ HYPERSOUND IN ALUMINUM OXIDE AND LITHIUM NIOBATE SINGLE CRYSTALS. /EPENDENCES OF THE AB 11-1719
DIELECTRIC, PIEZOELECTRIC, AND ELASTIC PROPERTIES OF LITHIUM NIOBATE SINGLE CRYSTALS. /ERMAL DEPENDENCE OF 11-1693
ROPERTIES. FLUX GROWN POTASSIUM TANTALATE NIOBATE SINGLE CRYSTALS AND THEIR ELECTRICAL P 09-1533
ADDERS ON PSEUDOCUBIC (001) / DOMAIN STRUCTURES IN POTASSIUM NIOBATE SINGLE CRYSTALS ASSOCIATED WITH STEP L 08-1316
INTERNAL FRICTION OF LITHIUM NIOBATE SINGLE CRYSTALS AT 1-40 HZ. 11-1798
HOD. GROWTH OF BARIUM LITHIUM NIOBATE SINGLE CRYSTALS BY THE CZOCHRALSKI MET 13-1991
IELECTRIC/ PREPARATION AND STUDY OF BARIUM(X) STRONTIUM(1-X) NIOBATE SINGLE CRYSTALS (CZOCHRALSKI GROWTH, D 13-1936
NS. (BY PULLING FROM THE MELT, STRU/ GROWTH OF BARIUM SODIUM NIOBATE SINGLE CRYSTALS FOR OPTICAL APPLICATIO 13-2045
CLEAVAGE PLANES. DOMAINS IN POTASSIUM NIOBATE SINGLE CRYSTALS FROM PSEUDOCUBIC (001) 08-1317
NONLINEAR OPTICAL PROPERTIES OF POTASSIUM NIOBATE SINGLE CRYSTALS. (REFRACTIVE INDICES) 08-1454
NS. DIELECTRIC CONSTANT OF POTASSIUM NIOBATE SINGLE CRYSTALS UNDER BIASING CONDITIO 08-1381
ON. POINT DEFECTS IN LITHIUM NIOBATE SINGLE CRYSTALS USED IN LASER MODULATI 11-1737
SER RADIATIO/ INVESTIGATION OF CERTAIN PARAMETERS OF LITHIUM NIOBATE SINGLE CRYSTALS USING SCATTERING OF LA 11-1797
GROWTH OF LITHIUM NIOBATE SINGLE DOMAIN CRYSTALS. 11-1781
/M TITANATE, POTASSIUM TANTALATE, BARIUM TITANATE, POTASSIUM NIOBATE, SODIUM NITRITE, POTASSIUM DIDEUTERIU/ 02-0059
ECT OF ELECTRIC FIELD ON PHASE COMPOSITION OF (LEAD, BARIUM) NIOBATE SOLID SOLUTIONS. EFF 13-1943
DIELECTRIC PROPERTIES OF LEAD TITANATE- LEAD (FERRATE NIOBATE) SOLID SOLUTIONS AT HIGH PRESSURES. 05-1071
/RTIES OF LITHIUM CONTAINING PEROVSKITES. (STRONTIUM LITHIUM NIOBATE, STRONTIUM LITHIUM TANTALATE, STRONTI/ 13-1292
POTASSIUM TANTALATE NIOBATE, STRUCTURAL AND DYNAMICAL STUDIES. 13-2068
LAYER WAVE AMPLIFICATION IN A CADMIUM SULFIDE FLUID- LITHIUM NIOBATE STRUCTURE. 11-1667
RFACE WAVE CONVOLUTION USING A CADMIUM SULFIDE FLUID LITHIUM NIOBATE STRUCTURE. SU 11-1700
ULTRASONIC SURFACE WAVES BY THIN GOLD FILM LAYERS ON LITHIUM NIOBATE SUBSTRATES. /ACTION AND REFLECTION OF 11-1789
ULTRASONIC SURFACE WAVES BY THIN GOLD FILM LAYERS ON LITHIUM NIOBATE SUBSTRATES. /ACTION AND REFLECTION OF 11-1790
LOW POWER DENSITY GALLIUM ARSENIDE- LITHIUM NIOBATE SURFACE WAVE AMPLIFIER. 11-1825
NONLINEAR EFFECTS IN MICROWAVE ACOUSTIC LITHIUM NIOBATE SURFACE WAVE DELAY LINES. 11-1818
DECORATION OF ION BOMBARDED LITHIUM NIOBATE SURFACES. 11-1799
YSTALS FROM THE POTASSIUM NIOBATE- LITHIUM NIOBATE- BARIUM DINIOBATE SYSTEM. /PERTIES OF TUNGSTEN BRONZE CR 13-2003
AND PYROELECTRIC PROPERTIES OF THE LEAD DINIOBATE- BARIUM DINIOBATE SYSTEM. DIELECTRIC, PIEZOELECTRIC 08-1384
AND PYROELECTRIC PROPERTIES OF THE LEAD METANIOBATE- BARIUM NIOBATE SYSTEM. DIELECTRIC, PIEZOELECTRIC 08-1385
FERROELECTRICS IN THE LITHIUM POTASSIUM NIOBATE SYSTEM. 11-1660
AGONAL TRANSITION IN RELATION TO SC/ THE STRUCTURE OF SODIUM NIOBATE T(2) AT 600 DEG C, AND THE CUBIC- TETR 08-1337
(THEORY, APPLIED TO BARIUM STRONTIUM TITANATE AND POTASSIUM NIOBATE TANTALATE) /ROELECTRIC SOLID SOLUTION. 03-0260
NIOBIUM-93 NMR STUDY OF THE LITHIUM NIOBATE TANTALATE SOLID SOLUTION SYSTEM. 11-1794
/TEN OXYGEN BRONZE TYPE: QUADRATIC LEAD OR BISMUTH TANTALATE NIOBATE, TANTALATE TUNGSTATE, OR TUNGSTATE NI/ 13-1985
/ FREQUENCY ACOUSTIC SURFACE WAVE LOSS MECHANISMS ON LITHIUM NIOBATE (TEMPERATURE AND FREQUENCY DEPENDENCE/ 11-1817
/DRUPOLE RESONANCE INVESTIGATION OF (STOICHIOMETRIC) LITHIUM NIOBATE (TEMPERATURE DEPENDENCE OF COUPLING C/ 11-1811
HASE RELATIONS IN THE LEAD ZIRCONATE LEAD TITANATE- LEAD METANIOBATE TERNARY SYSTEM. /YSTAL STRUCTURE AND P 09-1596
SOLID SOLUTIONS IN A LEAD ZIRCONATE- LEAD TITANATE- LEAD METANIOBATE TERNARY SYSTEM. CURIE TEMPERATURES OF 09-1595
E ORTHORHOMBIC- TETRAGONAL PHASE TRANSITION IN BARIUM SODIUM NIOBATE. (THERMAL EXPANSION) /TRIC STUDY OF TH 11-1894
GES. CONTROL OF THE SUSCEPTIBILITY OF LITHIUM NIOBATE TO LASER INDUCED REFRACTIVE INDEX CHAN 11-1791
X-RAY STUDY OF HIGH TEMPERATURE PHASE TRANSITIONS OF SODIUM NIOBATE (TO 500C). 13-1965
OPTIMUM CUT FOR A LITHIUM NIOBATE TRANSVERSE LIGHT MODULATOR. 11-1755
VOLTAGE, T/ OPTIMUM CRYSTAL ORIENTATION FOR A BARIUM SODIUM NIOBATE TRANSVERSE LIGHT MODULATOR. (HALF WAVE 13-1973
SUPERLATTICE STRUCTURE OF STRONTIUM SODIUM LITHIUM NIOBATE. (TUNGSTEN BRONZE) TYPE FERROELECTRICS. 13-2002
. ELECTROOPTICAL PROPERTIES OF STRONTIUM POTASSIUM LITHIUM NIOBATE. (TUNGSTEN BRONZE TYPE FERROELECTRICS) 13-2051
POLYMORPHISM OF STRONTIUM METANIOBATE (TWO REVERSIBLE PHASES). 36-3535
ND PIEZOELECTRIC CONSTANTS OF STRONTIUM(4) POTASSIUM LITHIUM NIOBATE TYPE FERROELECTRIC CRYSTAL. ELASTIC A 13-1946
ELECTROOPTIC PROPERTIES OF STRONTIUM LITHIUM NIOBATE TYPE FERROELECTRICS (SINGLE CRYSTAL). 13-2062
TRIC CONSTANT ON MELT COMPOSITION (STOICHIOMETRY) IN LITHIUM NIOBATE (VERY LITTLE EFFECT). /FECT AND DIELEC 11-1842
/RROELECTRIC (PYROCHLORE TYPE) COMPOUND (AMMONO THIO CADMIUM NIOBATE, WITH 4 ELECTRIC PHASES, 2 FERROELECT/ 36-3529
EFFECT TUNABLE OPTICAL EMISSION FROM LITHIUM NIOBATE WITHOUT A RESONATOR. 11-1858
FERROELECTRIC PROPERTIES OF TETRAGONAL (BARIUM, LEAD) (NIOBATE, ZIRCONATE). (CURIE TEMPERATURES) 08-1293
/ DIELECTRIC AND FERROELECTRIC PROPERTIES OF THE BARIUM LEAD NIOBATE ZIRCONATE SYSTEM. (TETRAGONAL AND ORT/ 13-1925
DEFLECTORS AND LIGHT MODULATORS. (POTASSIUM TANTALATE(0.65) NIOBATE(0.35)) /ROVSKITE PARAELECTRICS IN BEAM 13-1924
/ OF POTASSIUM TANTALATE NIOBATE. (POTASSIUM TANTALATE(0.65) NIOBATE(0.35), CURIE TEMPERATURE EQUALS 10 DE/ 08-1391
FERROELECTRIC MATERIALS. (BARIUM TITANATE, LEAD FERRATE(0.5) NIOBATE(0.5)) / THE PROPERTIES OF A SERIES OF 04-0780
AL LEAD MAGNESATE(0.33) NIOBATE(0.67), AND LEAD FERRATE(0.5) NIOBATE(0.5)) /OLYCRYSTALLINE AND SINGLE CRYST 13-1375
IN (DOPED AND UNDOPED CERAMIC PEROVSKITE) LEAD FERRATE(0.5) NIOBATE(0.5). /URE COEFFICIENT OF RESISTIVITY) 13-2000
E SOLID SOL/ FERROELECTRIC PROPERTIES OF LEAD (STANNATE(0.5) NIOBATE(0.5)) (1-X) ZIRCONATE(X) PEROVSKITE TYP 09-1548
/SIS OF PEROVSKITE COMPLEX COMPOSITIONS: LEAD MANGANATE(0.5) NIOBATE(0.5) AND LEAD MANGANATE(0.33) NIOBATE(/ 08-1453
/NTANEOUS MAGNETOELECTRIC MAGNETIZATION OF LEAD FERRATE(0.5) NIOBATE(0.5) AND LEAD MANGANATE(0.5) NIOBATE(/ 13-2069
/NETS. (LEAD FERRATE(0.67) TUNGSTATE(0.33) AND LEAD FERRATE(0.5) NIOBATE(0.5), BISMUTH FERRATE AND ITS SOLID S/ 08-1368
EW PIEZOELECTRIC CERAMICS. (LEAD TITANATE- LEAD FERRATE(0.5) NIOBATE(0.5), LEAD TITANATE- LEAD ZIRCONATE) N 09-1544
NEODYMIUM: YTTRIUM ALUMINUM GARNET- BARIUM SODIUM NIOBATE 0.66 MILLIMICRON HARMONIC SOURCE. 13-1975
S OF HOT PRESSED CERAMICS IN THE SYSTEM LEAD (ZINCATE(0.333) NIOBATE(0.667)). ELECTROOPTIC EFFECT 13-1900
ON SYSTEM LEAD TITANATE- LEAD ZIRCONATE- LEAD (ZINCATE(0.33) NIOBATE(0.67)). /ROPERTIES OF THE SOLID SOLUTI 13-1951
TERING) AND PROPERTIES OF PEROVSKITE TYPE LEAD ZINCATE(0.33) NIOBATE(0.67). /SIS (SOLID STATE REACTION, SIN 08-1393
S: LEAD MANGANATE(0.5) NIOBATE(0.5) AND LEAD MANGANATE(0.33) NIOBATE(0.67). /PEROVSKITE COMPLEX COMPOSITION 08-1453
E EFFECT OF PRESSURE ON THE TRANSITION OF LEAD ZINCATE(0.33) NIOBATE(0.67). TH 08-1473
THE SYSTEM LEAD TITANATE- STRONTIUM CUPRATE(0.33) NIOBATE(0.67). 05-1045
/ECTRIC AND PIEZOELECTRIC PROPERTIES OF LEAD COBALTATE(0.33) NIOBATE(0.67)- TITANATE- ZIRCONATE SOLID SOLU/ 09-1549
(FERROELECTRIC- PARAE/ CRYSTAL GROWTH OF LEAD (CADMATE(0.33) NIOBATE(0.67) AND ITS DIELECTRIC PROPERTIES. 13-1959
/E. (POLYCRYSTALLINE AND SINGLE CRYSTAL LEAD MAGNESATE(0.33) NIOBATE(0.67), AND LEAD FERRATE(0.5) NIOBATE(/ 08-1375
CIENTS) ELECTROOPTICAL EFFECT IN LEAD ZINCATE(0.33) NIOBATE(0.67). (DISPERSION OF QUADRATIC COEFFI 13-1982
/TRUCTURAL AND DIELECTRIC PROPERTIES OF BARIUM ZINCATE(0.33) NIOBATE(0.67) (NOT FERROELECTRIC -160 TO +250/ 36-3609
DIELECTRIC AND OPTICAL PROPERTIES OF LEAD ZINCATE(0.33) NIOBATE(0.67) SINGLE CRYSTAL. 13-2065
OXIDE FLUX). FERROELECTRIC PROPERTIES OF LEAD ZINCATE(0.33) NIOBATE(0.67) (SINGLE CRYSTALS GROWN FROM LEAD 13-2066
TIES IN A LEAD TITANATE- LEAD ZIRCONATE- LEAD (ZINCATE(0.33) NIOBATE(0.67)) SYSTEM. /D PIEZOELECTRIC PROPER 09-1617
/XATION POLARIZATION OF A FERROELECTRIC LEAD MAGNESATE(0.33) NIOBATE(0.67) WITH A DIFFUSE PHASE TRANSITION. 13-1980
TERNARY SYSTEM LEAD TITANATE- LEAD ZIRCONATE- LEAD NIOBATE(0.67) ZINCATE(0.33). 09-1507
AESORPTION EDGE SPLITTING IN POTASSIUM TANTALATE(X) NIOBATE(1-X). 13-1987

D PROPERTIES OF FERROELECTRIC POTASSIUM LITHIUM TANTALATE(X) NIOBATE(1-X).   GROWTH AN   11-1711
OTASSIUM(X) SODIUM(1-X) TANTALATE AND POTASSIUM TANTALATE(1-X). / IN THE MIXED CRYSTAL SYSTEMS P 13-1933
TYPE) FERROELECTRIC (SYSTEM) POTASSIUM LITHIUM TANTALATE(X) NIOBATE(1-X). /CRYSTALS IN THE TUNGSTEN BRONZE 13-1947
OPERTIES OF FERROELECTRIC POTASSIUM LITHIUM TANTALATE(0-0.5) NIOBATE(1-0.5) SINGLE CRYSTALS. /ZOELECTRIC PR 12-1890
/ERTIES. (BARIUM TITANATE, LEAD ZIRCONATE TITANATE, LEAD METANIOBATE, 20-150 DEG C, PRESSURE TO 1200 KG/SQ/ 04-0985
BARIUM TITANATE ZIRCONATE (OR STANNATE), POTASSIUM TANTALATE NIOBATE, 228 REFS). /IUM) STRONTIUM TITANATE, 08-1281
(BACKWARD WAVE) ACOUSTIC PARAMETRIC OSCILLATIONS IN LITHIUM NIOBATE (60 DB GAIN).   11-1837
OF THERMAL EXPANSION OF GADOLINIUM SUBSTITUTED BARIUM SODIUM NIOBATES.   CAUSES 13-2048
OMPOSITION AND STRUCTURE OF SPUTTERED FILMS OF FERROELECTRIC NIOBATES.   C 13-1941
GROWTH AND NONLINEAR OPTICAL PROPERTIES OF POTASSIUM LITHIUM NIOBATES.   CRYSTAL 13-2015
ELECTRIC STUDIES OF THE CATION SUBSTITUTED MIXED CRYSTALS OF NIOBATES.   DI 11-1848
ARED ABSORPTION SPECTROPHOTOMETRIC STUDY CF MONOVALENT METAL NIOBATES.   INFR 02-0113
R THERMAL EXPANSIONS OF GADOLINIUM SUBSTITUTED BARIUM SODIUM NIOBATES.   LINEA 13-1996
MICROCHEMICAL ANALYSIS OF GADOLINIUM DOPED BARIUM SODIUM NIOBATES.   13-1953
DISPLACEMENT TYPE. (BORACITES, MOLYBDATES, TUNGSTEN BRONZES, NIOBATES)   NEW CLASSES OF FERROELECTRICS OF 02-0118
G ELECTRON MICROSCOPIC EXAMINATION OF SINTERED BARIUM SODIUM NIOBATES.   SCANNIN 13-1955
LITHIUM OXIDE- NIOBIUM PENTOXIDE SYSTEM. (POTASSIUM LITHIUM NIOBATES) /LIDUS EQUILIBRIA IN POTASSIUM OXIDE- 13-1963
STRUCTURE, AND PROPERTIES. (REVIEW CF WORK TC SEP 1/ LITHIUM NIOBATE- A NEW TYPE OF FERROELECTRIC: GROWTH, 11-1775
PREPARATION OF SODIUM AND BARIUM NIOBATES AND LEAD MOLYBDATE.   04-0657
LS OF THE 3D SE/ ELECTRICAL PROPERTIES AND STRUCTURES OF METANIOBATES AND METATANTALATES OF TRANSITION META 36-3527
THALLIUM CONTAINING PEROVSKITES. (TITANATES, HAFNATES, MIXED NIOBATES AND TANTALATES) / OF NEW CADMIUM AND 08-1458
DIELECTRIC PROPERTIES OF NIOBATES AND TUNGSTATES.   08-1322
RIC, PIEZOELECTRIC AND PYROELECTRIC PROPERTIES OF THE LEAD DINIOBATE- BARIUM DINIOBATE SYSTEM.   DIELECT 08-1384
C, PIEZOELECTRIC AND PYROELECTRIC PROPERTIES OF THE LEAD METANIOBATE- BARIUM NIOBATE SYSTEM.   DIELECTRI 08-1385
/RROELECTRIC TRANSITION TEMPERATURES CF TUNGSTEN BRONZE TYPE NIOBATES (CORRELATION OF CURIE TEMPERATURE WI/ 13-2049
STUDY OF DIVALENT BINARY AND TERNARY NIOBATES IN THE SOLID STATE.   08-1286
DIELECTRIC CONSTANTS OF THE SYSTEM SAMARIUM NIOBATE- LANTHANUM NIOBATE.   08-1268
/C PROPERTIES OF TUNGSTEN BRONZE CRYSTALS FROM THE POTASSIUM NIOBATE- LITHIUM NIOBATE- BARIUM DINIOBATE SY/ 13-2003
/HE (OXIDE) SYSTEM POTASSIA LITHIA NICBIA (POTASSIUM LITHIUM NIOBATES, PHASE RELATIONSHIPS, SOLID SOLUTION/ 13-2017
TRANSVERSE ACOUSTOELECTRIC EFFECT IN A LAYERED LITHIUM NIOBATE- SILICON STRUCTURE.   11-1748
ECTRICITY APPLIED TO SODIUM NIOBATE AND THE SYSTEM POTASSIUM NIOBATE- SODIUM NIOBATE. /CTRICITY AND FERROEL 03-0218
BATE- BARIUM DINIOBATE SY/ PHASE EQUILIBRIA IN THE POTASSIUM NIOBATE- STRONTIUM DINIOBATE AND POTASSIUM NIO 13-2016
IN PROPERTIES OF NEW COMPOUNDS WITH A PYROCHLORE STRUCTURE. (NIOBATES, TANTALATES)   SYNTHESIS AND CERTA 02-0047
/ESTIGATION OF FERROELECTRICS OF ILMENITE STRUCTURE. (ABO(3) NIOBATES, TANTALATES, AND MIXTURES WITH THEM) 08-1438
ROMATES, COBA/ FLUX GROWTH OF SOME COMPLEX OXIDE MATERIALS. (NIOBATES, TANTALATES, STANNATES, TITANATES, CH 02-0129
NZE STRUC/ NONLINEAR OPTICAL PROPERTIES OF POTASSIUM LITHIUM NIOBATES (TRANSPARENT, TETRAGONAL TUNGSTEN BRO 13-2026
VARIOUS FERROELECTRICS AND ANTIFERROELECTRICS. (ZIRCONATES, NIOBATES, TUNGSTATES) /BOVE THE CURIE POINT IN 02-0130
/IC, DIELECTRIC, AND NONLINEAR OPTICAL PROPERTIES OF THE ABC NIOBATES (WHERE A AND B ARE CALCIUM, STRONTIU/ 04-0912
RE.   SODIUM BARIUM RARE EARTH NIOBATES WITH THE TUNGSTEN BRONZE TYPE STRUCTU 13-1972
/ (LEAD ZIRCONATE TITANATE) FERROELECTRICS (WITH AND WITHOUT NIOBIA ADDITION) DUE TO INTERACTION OF DOMAIN/ 09-1584
TUNGSTEN BRONZE FIELD IN THE (OXIDE) SYSTEM POTASSIA LITHIA NIOBIA (POTASSIUM LITHIUM NIOBATES, PHASE REL/ 13-2017
DIELECTRIC PROPERTIES OF SINGLE CRYSTAL BARIUM TITANATE WITH NIOBIUM ADDITIONS.   OPTICAL AND 04-0875
S.   A SEMICONDUCTING BARIUM TITANATE CERAMIC DOPED WITH NIOBIUM AND MANGANESE IONS AND ITS APPLICATION 04-0734
IMPURITY EFFECTS IN SODIUM TUNGSTEN BRONZES: NIOBIUM AND TANTALUM DOPING.   13-2053
DIELECTRIC PROPERTIES OF BARIUM TITANATE CONTAINING NIOBIUM AND THE EFFECT OF ADDITIVES.   04-0575
IES OF LEAD ZIRCONATE TITANATE STANNATE. (CERAMIC DOPED WITH NIOBIUM AND TITANIUM)   MOSSBAUER STUD 09-1604
M, CADMIUM, YTTRIUM, GADOLINIUM, DYSPROSIUM, IRON, SELENIUM, NIOBIUM, COBALT, SCANDIUM, CR COPPER) /AGNESIU 10-1640
ONTIUM TITANATE SINGLE CRYSTALS (HYDROGEN REDUCED, CERIUM OR NIOBIUM DOPED). /PECTRA) OF SEMICONDUCTING STR 06-1198
INFLUENCE OF GRAIN GROWTH ON DIELECTRIC PROPERTIES OF NIOBIUM DOPED BARIUM TITANATE.   04-0763
PREPARATION OF SMALL GRAINED AND LARGE GRAINED CERAMICS FROM NIOBIUM DOPED BARIUM TITANATE.   04-0765
PARAMAGNETIC RESONANCE OF PLATINUM IONS IN NIOBIUM DOPED BARIUM TITANATE SINGLE CRYSTAL.   04-0977
FERROELECTRICITY AND ANTIFERROELECTRICITY IN PURE AND NIOBIUM DOPED LEAD ZIRCONATE.   07-1251
/N FERROELECTRIC AND ANTIFERROELECTRIC PHASES. (LANTHANUM OR NIOBIUM DOPED LEAD ZIRCONATE STANNATE TITANAT/ 09-1487
FREE CARRIER OPTICAL ABSORPTION IN NIOBIUM DOPED STRONTIUM TITANATE.   06-1214
DOPED) POTASSIUM/ ELECTRON TUNNELING AND BAND STRUCTURE OF (NIOBIUM DOPED) STRONTIUM TITANATE AND (CALCIUM 06-1222
/VELOCITY AT FERROELECTRIC-1, FERROELECTRIC-2 TRANSITION IN (NIOBIUM DOPED) ZIRCONATE(0.95) TITANATE(0.05). 09-1511
SUREMENT TECHNIQUE.   DIFFUSION COEFFICIENT OF NIOBIUM IN BARIUM TITANATE BY A DIELECTRIC MEA 04-0926
EFFECTS OF GRAIN GROWTH ON THE DISTRIBUTION OF NIOBIUM IN BARIUM TITANATE CERAMICS.   04-0595
BIC" PEROVSKITE STRUCTURE OF LEAD/ ORDERING OF MAGNESIUM AND NIOBIUM IN THE OCTAHEDRAL POSITIONS OF THE "CU 08-1379
0.47) (EXPERIMENT AND STRUCTURA/ SUBSTITUTION OF BISMUTH AND NIOBIUM IONS IN LEAD ZIRCONATE(0.53) TITANATE(09-1480
TANATE.   PRECIPITATION OF ZIRCONIA PHASE IN NIOBIUM MODIFIED CERAMICS OF LEAD ZIRCONATE TI 09-1540
SINTERING SCANDIUM AND NIOBIUM MODIFIED LEAD ZIRCONATE TITANATE.   09-1555
PROPERTIES OF S/ EFFECT OF THE SUBSTITUTION OF TANTALUM FOR NIOBIUM ON THE CRYSTALLOGRAPHIC AND DIELECTRIC 13-1921
BARIUM(3) BISMUTH(/ NEW FERROELECTRICS. (BARIUM(2) BISMUTH (NIOBIUM OR TANTALUM OR VANADIUM) OXIDE(6), AND 36-3619
CRYSTAL STRUCTURE OF BISMUTH TITANIUM NIOBIUM OXIDE.   13-2058
ZOELECTRIC CONSTANTS OF STIBOTANTALITE. (ANTIMONY (TANTALUM, NIOBIUM) OXIDE(4)) /TURE DEPENDENCE OF THE PIE 36-3601
GENATED QUADRATIC TUNGSTEN BRONZE TYPE. (STRONTIUM POTASSIUM NIOBIUM OXYFLUORIDE) /UORIDE PHASES OF THE OXY 13-2011
/LIKE LAYER TYPE FERROELECTRICS. (BISMUTH TUNGSTATE, BISMUTH NIOBIUM OXYFLUORIDE, BISMUTH TANTALUM OXYFLUO/ 13-1966
FERROELECTRICS IN THE POTASSIUM OXIDE- STRONTIUM OXIDE- NIOBIUM PENTOXIDE.   13-1896
IUM OXIDES.   BINARY AND PSEUDOBINARY PHASES BETWEEN NIOBIUM PENTOXIDE AND LEAD, BARIUM, AND STRONT 04-0590
/COEFFICIENT OF RESISTANCE) BEHAVIOR CF BARIUM TITANATE WITH NIOBIUM PENTOXIDE AND MANGANESE DIOXIDE ADDIT/ 04-0837
NATE(0.52) TITANATE(0.48) CERAMICS CAUSED BY THE ADDITION OF NIOBIUM PENTOXIDE (BELOW 0.2 PERCENT). / ZIRCO 09-1606
MPROVED ELECTRICAL PROPERTIES. (DIELECTRIC CON/ SYNTHESIS OF NIOBIUM PENTOXIDE DOPED BARIUM TITANATE WITH I 04-0840
/ATIONS IN THE TERNARY SYSTEM BARIUM OXIDE- STRONTIUM OXIDE- NIOBIUM PENTOXIDE. (OXIDES, ESPECIALLY THE NI/ 13-1920
GEN DEPLETED CERAMIC LEAD ZIRCONATE TITANATE (PLUS 1 PERCENT NIOBIUM PENTOXIDE) SURFACE LAYERS. /IES OF OXY 09-1576
ND MELT GROWTH OF CRYSTALS IN THE SODIUM OXIDE- BARIUM OXIDE NIOBIUM PENTOXIDE SYSTEM. /ND SOLIDUS) FIELD A 13-1949
PHASE EQUILIBRIA AND CRYSTAL GROWTH IN THE LITHIUM OXIDE- NIOBIUM PENTOXIDE SYSTEM.   11-1683
/ OF SUBSOLIDUS EQUILIBRIA IN POTASSIUM OXIDE- LITHIUM OXIDE- NIOBIUM PENTOXIDE SYSTEM. (POTASSIUM LITHIUM / 13-1963
/SKITE AND PYROCHLORE STRUCTURES. (BARIUM AND BISMUTH(0.5)) (NIOBIUM, TANTALUM, OR VANADIUM) (0.5) OXIDE, (/ 08-1462
/ BRONZES. (SODIUM(X) TUNGSTEN(1-X) Z(Y) OXIDE(3) WHERE Z IS NIOBIUM, TANTALUM, ZIRCONIUM, VANADIUM, OR CO/ 13-1944
/NTIUM LITHIUM NIOBATE TITANATE, POTASSIUM LANTHANUM LITHIUM NIOBIUM TITANATE, BARIUM LITHIUM TITANIUM NIO/ 13-1923
/POUNDS OF TUNGSTEN BRONZE TYPE A(6)B(10)O(30). (WHERE B IS (NIOBIUM, TITANIUM) OR (NIOBIUM, TUNGSTEN), CO/ 13-1964
TURE AND PRESSURE ENFORC/ RELEASE OF ELECTRIC ENERGY IN LEAD NIOBIUM ZIRCONATE TITANATE STANNATE BY TEMPERA 09-1488
/S IN A FERROELECTRIC CERAMIC (THEORY, APPLIED TO LEAD(0.99) NIOBIUM(0.02) ZIRCONATE(0.65) TITANATE(0.35)). 09-1537
/TERED LEAD(0.92) BISMUTH(0.07) LANTHANUM(0.01) (IRON(0.405) NIOBIUM(0.325) ZIRCONIUM(0.27)) OXYGEN(3) FER/ 37-3629
RAMAN SPECTRUM OF POTASSIUM TANTALUM(0.64) NIOBIUM(0.36) OXIDE.   13-1989
/O SHARP FERROELECTRIC PHASE TRANSITION SYSTEMS LEAD (B(0.5) NIOBIUM(0.5)) (1-X) B'(X) OXYGEN(3), WHERE B I/ 13-2044
OF SOLID SOLUTIONS IN THE SYSTEM BARIUM TITANATE- STRONTIUM NIOBIUM(0.8) OXIDE(3). /RROELECTRIC PROPERTIES 04-0707
RIC CRYSTAL STRUCTURE OF THE TUNGSTEN BRONZE TYP/ BISMUTH(3) NIOBIUM(17) OXIDE(47): POTENTIALLY FERROELECT 13-1978
NEW DESCRIPTION OF THE PYROCHLORE STRUCTURE (CADMIUM(2) NIOBIUM(2) OXIDE(6) SULFIDE).   36-3602

A NEW FERROELECTRIC SINGLE CRYSTAL STRONTIUM(2) NIOBIUM(2) OXIDE(7).   13-1997
CRYSTAL STRUCTURE AND PROPERTIES OF CALCIUM(2) NIOBIUM(2) OXIDE(7).   13-1911
ONE TEC/ GROWTH OF FERROELECTRIC SINGLE CRYSTAL STRONTIUM(2) NIOBIUM(2) OXIDE(7) BY MEANS OF THE FLOATING Z 13-2037
OELECTRIC DOMAINS.   STRUCTURE OF THE BRONZE SODIUM(13) NIOBIUM(35) OXIDE(94) AND THE GEOMETRY OF FERR 13-1931
/IC CRYSTALS. (LEAD TITANATE, BARIUM STRONTIUM NIOBATE, LEAD NIOBIUM(4) OXIDE(11), SILVER GALLIUM DISULFID/ 05-1066
NONLINEAR OPTICAL PROPERTIES OF FERROELECTRIC LEAD NIOBIUM(4) OXIDE(12).   13-2023
M TANTALATE, POTASS/ NONLINEAR OPTICAL POLARIZABILITY OF THE NIOBIUM- OXYGEN BOND. (LITHIUM NIOBATE, LITHIU 11-1734
/STRONTIUM OXIDE- NIOBIUM PENTOXIDE. (OXIDES, ESPECIALLY THE NIOBIUM-RICH REGION AND THE BINARY STRONTIUM / 13-1920
DIUM NIOBATE.   NIOBIUM-93 AND SODIUM-23 IN POLYCRYSTALLINE SO 08-1469
M / NMR STUDY OF PHASE TRANSITIONS IN PEROVSKITES CRYSTALS. (NIOBIUM-93 AND SODIUM-23 RESONANCE IN POTASSIU 08-1283
C LITHIUM NIOBATE.   NIOBIUM-93 NMR LINE WIDTHS IN NONSTOICHIOMETRI 11-1795
NTALATE SOLID SOLUTION SYSTEM.   NIOBIUM-93 NMR STUDY OF THE LITHIUM NIOBATE TA 11-1794
IGATION OF (STOICHIOMETRIC) LITHIUM NIOBATE (TEMPERATURE DE/ NIOBIUM-93 NUCLEAR QUADRUPOLE RESONANCE INVEST 11-1811
CRYSTAL STRUCTURE AND PHASE TRANSITION OF DIGLYCINE NITRATE.   16-2312
CRYSTAL STRUCTURE OF FERROELECTRIC GLYCINE SILVER NITRATE.   16-2297
CRYSTAL STRUCTURE OF FERROELECTRIC GLYCINE SILVER NITRATE.   16-2333
CRYSTAL STRUCTURE OF GLYCINE SILVER NITRATE.   16-2348
DETERMINATION OF THE ELASTIC CONSTANTS OF POTASSIUM NITRATE.   08-1399
ICAL CONDUCTIVITY STUDIES IN POTASSIUM NITRITE AND POTASSIUM NITRATE.   DIELECTRIC AND ELECTR 16-2340
DIES ON THE FERROELECTRIC PHASE TRANSITION OF GLYCINE SILVER NITRATE.   DIELECTRIC AND X-RAY STU 16-2347
AVIOR OF (FERROELECTRIC) FILMS OF VACUUM DEPOSITED POTASSIUM NITRATE.   DIELECTRIC BEH 16-2356
DIELECTRIC DISPERSION IN THE PARAELECTRIC PHASE OF POTASSIUM NITRATE.   16-2341
DIELECTRIC MEASUREMENTS OF AMMONIUM NITRATE.   16-2338
DIELECTRIC STUDIES OF PHASE 2 OF AMMONIUM NITRATE.   16-2339
NS ON THE FACES PARALLEL TO THE FERROELECTRIC-AXIS OF SODIUM NITRATE.   DOMAIN ETCH PATTER 16-2349
RONTIUM(2+) IONS ON THE ELECTRICAL CONDUCTIVITY OF POTASSIUM NITRATE.   EFFECT OF ADDING TITANIUM(+) AND ST 16-2319
ECT OF HYDROSTATIC PRESSURE ON FERROELECTRICITY OF POTASSIUM NITRATE.   EFF 16-2344
ELECTRICAL PROPERTIES OF RHOMBOHEDRAL POTASSIUM NITRATE.   16-2367
ARED SPECTRA AND FERROELECTRIC PHASE TRANSITION OF POTASSIUM NITRATE.   FAR INFR 16-2308
RS AROUND THE PHASE TRANSITIONS IN SODIUM NITRITE AND SODIUM NITRATE.   GRUNEISEN PARAMETE 16-2390
A, BETA TO GAMMA AND GAMMA TO ALPHA TRANSITIONS IN POTASSIUM NITRATE. /H IN THE ALPHA TO BETA, BETA TO ALPH 16-2259
OF NITROGEN DIOXIDE AND NITRITE ION IN IRRADIATED POTASSIUM NITRATE. /GNETIC RESONANCE STUDY OF PRODUCTION 16-2404
PARAMAGNETIC SPECIES IN IRRADIATED NITRATE.   16-2405
PHONON DISPERSION IN FERROELECTRIC POTASSIUM NITRATE.   16-2334
PRESSURE INDUCED FERROELECTRIC PHASE TRANSITION IN POTASSIUM NITRATE.   16-2378
YDROGEN PHOSPHATE, POTASSIUM FERROCYANIDE TRIHYDRATE, SODIUM NITRATE) /STICS OF THEIR GROWTH. (AMMONIUM DIH 16-2316
STRUCTURE OF THE DISORDERED PHASE OF POTASSIUM NITRATE.   16-2379
SWITCHING BEHAVIOR IN FERROELECTRIC POTASSIUM NITRATE.   16-2285
SWITCHING PROPERTIES OF THICK FILM FERROELECTRIC POTASSIUM NITRATE.   16-2282
TWO MODES OF ELECTRICAL CONDUCTIVITY IN POTASSIUM NITRATE.   16-2368
MEASUREMENTS ON THERMAL EXPANSION OF FERROELECTRIC POTASSIUM NITRATE.   X-RAY 16-2322
X-RAY STUDY OF PHASE TRANSITION IN SODIUM NITRATE.   16-2380
D STRUCTURE ABOVE THE FERROELECTRIC CURIE POINT IN POTASSIUM NITRATE.   X-RAY STUDY ON THE DISORDERE 16-2366
URE.   DIELECTRIC ANOMALIES OF SODIUM NITRATE ABOVE THE FERROELECTRIC CURIE TEMPERAT 16-2376
)   RHYTHMIC GROWTH OF NEW PHASE IN POTASSIUM NITRATE. (ALPHA TO BETA POLYMORPHIC TRANSITION 16-2263
ELECTRICAL PROPERTIES OF RUBIDIUM NITRATE AND CESIUM NITRATE.   16-2286
FERROELECTRIC- PARAELECT/ PHASE TRANSFORMATIONS IN POTASSIUM NITRATE AND POTASSIUM NITRITE. (DOES NOT SEEM 16-2369
/FRARED ABSORPTION) STUDIES ON FERROELECTRICITY IN POTASSIUM NITRATE AND RELATED COMPOUNDS. (POTASSIUM(1-X/ 16-2402
ANHARMONICITY IN SODIUM NITRATE AND SODIUM NITRITE.   16-2388
TANT AND SPONTANEOUS POLARIZATION OF FERROELECTRIC POTASSIUM NITRATE AND SODIUM NITRITE. /E DIELECTRIC CONS 16-2331
S.   INFRARED SPECTRA OF POTASSIUM NITRATE AND SODIUM NITRITE AT PHASE TRANSITION 16-2352
SHI/ NMR STUDY OF THE FERROELECTRIC TRANSITIONS IN DIGLYCINE NITRATE AND TRIS-SARCOSINE CALCIUM CHLORIDE. (16-2267
TIVITY ACCOMPANYING THE POLYMORPHIC TRANSITIONS IN POTASSIUM NITRATE AT 1 ATMOSPHERE. /IN ELECTRICAL CONDUC 16-2391
IONS OF CRYSTALLIZATION ON THE FORM OF CRYSTALS OF POTASSIUM NITRATE. (CRYSTAL GROWTH) /UENCE OF THE CONDIT 16-2332
ATR INFRARED SPECTRA OF UNIAXIAL NITRATE CRYSTALS.   16-2280
ELECTRICAL RESISTIVITIES OF SODIUM NITRITE AND POTASSIUM NITRATE CRYSTALS.   16-2258
/LIARITIES IN THE VIBRATIONAL SPECTRA OF CESIUM AND RUBIDIUM NITRATE CRYSTALS IN ITS LOW TEMPERATURE PHASE/ 16-2323
DIFFUSIVITY AND THERMAL CONDUCTIVITY OF CALCIUM DOPED SODIUM NITRATE CRYSTALS NEAR THE TRANSITION POINT. / 16-2283
/, NONDESTRUCTIVE READOUT MEMORY ELEMENT UTILIZING POTASSIUM NITRATE. (FUSED THICK FILM OR VACUUM DEPOSITE/ 16-2269
TRANSITION RATES OF POTASSIUM NITRATE HIGH PRESSURE POLYMORPHS.   16-2279
ALCOHOL)   METASTABLE POTASSIUM NITRATE(III) FROM SOLUTION. (GROWN IN WATER OR 16-2329
LATTICE DYNAMICS MODEL OF POTASSIUM NITRATE IN ITS FERROELECTRIC PHASE.   16-2377
LUMINESCENCE OF POTASSIUM NITRATE IRRADIATED WITH GAMMA-RAYS.   16-2335
THERMAL EXPANSION OF SOME FERROELECTRICS. (POTASSIUM NITRATE, KDP, SODIUM NITRITE)   16-2298
FERROELECTRICITY IN AMMONIUM(X) POTASSIUM(1-X) NITRATE MIXED CRYSTAL.   16-2401
PHISM.   RUBIDIUM NITRATE NEAR THE MELTING POINT AND ITS POLYMOR 16-2330
LE BRIDGMAN TECHNIQUE).   PHASE TRANSITIONS IN POTASSIUM NITRATE NITRITE MIXED CRYSTALS (GROWTH BY DOUB 16-2345
/ X-RAY CRITICAL SCATTERING IN TRIGLYCINE SULFATE AND SODIUM NITRATE. (ORDER DISORDER TRANSITION, FERROELEC 16-2400
/ CRYSTAL GROWTH AT POLYMORPHIC TRANSFORMATIONS IN POTASSIUM NITRATE, SILVER NITRATE, AND AMMONIUM NITRATE/ 16-2260
MELTING PROPERTIES OF POTASSIUM NITRATE UNDER HIGH PRESSURE.   16-2265
VACUUM ULTRAVIOLET ABSORPTION OF THE ALKALI NITRITES AND NITRATES.   16-2398
VIOLET ABSORPTION SPECT/ STUDIES OF PHASE TRANSFORMATIONS IN NITRATES AND NITRITES. PART-1: CHANGE IN ULTRA 16-2278
DIPOLE CORRELATIONS IN MOLTEN ALKALI METAL NITRATES AND SODIUM NITRITE.   16-2389
/PERATURES IN ALKALI NITRITES (SODIUM, POTASSIUM) AND ALKALI NITRATES. (LITHIUM, SODIUM, RUBIDIUM, CESIUM) 16-2399
ACOUSTIC NUCLEAR MAGNETIC RESONANCE IN FERROELECTRIC SODIUM NITRITE.   16-2321
TIC RESONANCE OF MANGANESE(2+) DOPED IN FERROELECTRIC SODIUM NITRITE. /RANSITIONS IN THE ELECTRON PARAMAGNE 16-2383
AN X-RAY STUDY OF THE PHASE TRANSITION IN SODIUM NITRITE.   16-2318
ANALYSIS OF LOW FREQUENCY ELECTROOPTIC RESPONSE IN SODIUM NITRITE.   16-2353
SION OF LARGE WAVE VECTOR POLARITONS IN FERROELECTRIC SODIUM NITRITE.   ANGULAR DISPER 16-2256
ANHARMONICITY IN SODIUM NITRATE AND SODIUM NITRITE.   16-2388
CRITICAL SCATTERING OF X-RAYS IN SODIUM NITRITE.   16-2270
DIELECTRIC RELAXATION IN FERROELECTRIC SILVER SODIUM DINITRITE.   16-2290
DIELECTRIC RELAXATION IN FERROELECTRIC SILVER SODIUM DINITRITE.   16-2292
IES ON THE FERROELECTRIC PHASE TRANSITION IN SILVER SODIUM DINITRITE.   DIELECTRIC STUD 16-2289
POLE CORRELATIONS IN MOLTEN ALKALI METAL NITRATES AND SODIUM NITRITE.   DI 16-2389
LATTICE RELAXATION TIME NEAR THE NEEL TEMPERATURE IN SODIUM NITRITE.   DIPOLE 16-2300
ONANCE OF MANGANESE(2+) IN THE FERROELECTRIC PHASE OF SODIUM NITRITE.   ELECTRON PARAMAGNETIC RES 16-2358
F ADDITIONAL DISORDER IN THE RHOMBOHEDRAL FORM OF POTASSIUM NITRITE.   EVIDENCE O 16-2371
FERROELECTRIC PHASE TRANSITION IN SILVER SODIUM DINITRITE   16-2266
FERROELECTRIC POLARIZATION SWITCHING IN SODIUM NITRITE.   16-2301

DIELECTRIC RELAXATION TIME IN FERROELECTRIC SILVER SODIUM DINITRITE.
CONDITIONS ON THE NMR RELAXATION RATE OF SODIUM-23 IN SODIUM NITRITE.
ABSORPTION IN GAMMA-RAY IRRADIATED SINGLE CRYSTAL OF SODIUM NITRITE.
ND RAMAN SCATTERING IN FERROELECTRIC AND PARAELECTRIC SODIUM NITRITE.
ON STUDIES OF FERROELECTRIC POLARIZATION REVERSALS IN SODIUM NITRITE.
NEUTRON DIFFRACTION STUDY OF SODIUM NITRITE.
OGEN-14 NUCLEAR QUADRUPOLE RESONANCE IN FERROELECTRIC SODIUM NITRITE.
NMR STUDY OF SODIUM-23 IN SILVER SODIUM DINITRITE.
NONLINEAR OPTICAL PROPERTIES OF FERROELECTRIC SODIUM NITRITE.
N FERROELECTRICS. (POTASSIUM FERROCYANIDE TRIHYDRATE, SODIUM NITRITE)
NUCLEAR QUADRUPOLE RELAXATION OF NITROGEN-14 IN SODIUM NITRITE.
NUCLEAR QUADRUPOLE RESONANCE OF NITROGEN IN SODIUM NITRITE.
ION AND CRITICAL DYNAMICS OF FERROELECTRIC CRYSTALS. (SODIUM NITRITE)
US BIAXIAL CRYSTALS: METHYL MESITYL OXIDE OXALATE AND SODIUM NITRITE.
NERATION DURING FERROELECTRIC POLARIZATION REVERSAL. (SODIUM NITRITE)
PHASE TRANSITION IN SODIUM NITRITE.
PHENOMENOLOGICAL THEORY OF PHASE TRANSITIONS IN SODIUM NITRITE.
VERSAL IN A WIDE RANGE OF IMPRESSING RATE OF FIELD IN SODIUM NITRITE.
OLARIZED INFRARED AND RAMAN SPECTRA OF SINGLE CRYSTAL CESIUM NITRITE.
RIC STUDY ON THE SPONTANEOUS POLARIZATION IN SILVER SODIUM DINITRITE.
N PHONON RELAXATION AND FERROELECTRIC TRANSITION. (IN SODIUM NITRITE)
QUASI STATIC SWITCHING CURRENT IN SODIUM NITRITE.
RAMAN SCATTERING FROM FERROELECTRIC SODIUM NITRITE.
DEPENDENCE OF B1 PHONON AND DIELECTRIC RELAXATION IN SODIUM NITRITE.
SECOND HARMONIC GENERATION IN FERROELECTRIC SODIUM NITRITE.
SPECIFIC HEAT ANOMALY IN FERROELECTRIC SILVER SODIUM DINITRITE.
SODIUM-23 NMR CENTRAL LINE IN FERROELECTRIC SILVER SODIUM DINITRITE.
LECTRIC RELAXATION TIME NEAR THE TRANSITION POINTS IN SODIUM NITRITE.
S POLARIZATION OF FERROELECTRIC POTASSIUM NITRATE AND SODIUM NITRITE.
CE OF NONLINEAR OPTICAL COEFFICIENTS IN FERROELECTRIC SODIUM NITRITE.
MPERATURE DEPENDENCE OF SECOND HARMONIC GENERATION IN SODIUM NITRITE.
DENCE OF THE ELECTRIC FIELD GRADIENT IN FERROELECTRIC SODIUM NITRITE.
MPERATURE DEPENDENCE OF THE FORCE CONSTANTS OF ION IN SODIUM NITRITE.
DEPENDENCE OF THE OPTICAL NONLINEAR SUSCEPTIBILITY OF SODIUM NITRITE.
THE SODIUM-23 NUCLEAR QUADRUPOLE COUPLING CONSTANT IN SODIUM NITRITE.
THERMAL EXPANSION IN FERROELECTRIC SILVER SODIUM DINITRITE.
SION OF SOME FERROELECTRICS. (POTASSIUM NITRATE, KDP, SODIUM NITRITE)
TWO PHONON ABSORPTION IN FERROELECTRIC SODIUM NITRITE.
RASONIC ATTENUATION NEAR THE TWO TRANSITION POINTS IN SODIUM NITRITE.
OF FERROELECTRIC DOMAINS IN THE PARAELECTRIC PHASE OF SODIUM NITRITE.
-RAY AND NEUTRON STUDY ON THE PHASE TRANSFORMATION OF SODIUM NITRITE.
TASSIUM SODIUM TARTRATE, TGS, AMMONIUM FLUOBERYLLATE, SODIUM NITRITE, ALUMS) /CS. (BARIUM TITANATE, KDP, PO 02-0134
PART-2: DIELECTRIC CONSTANTS AND DIELECTRIC LOSSES OF SODIUM NITRITE AND GLYCINE SULFATE. /FERROELECTRICS. 16-2354
DIELECTRIC AND ELECTRICAL CONDUCTIVITY STUDIES IN POTASSIUM NITRITE AND POTASSIUM NITRATE. 16-2340
ELECTRICAL RESISTIVITIES OF SODIUM NITRITE AND POTASSIUM NITRATE CRYSTALS. 16-2258
PHASE DIAGRAMS OF SODIUM NITRITE AND POTASSIUM NITRITE TO 40 KBAR. 16-2361
AL THERMAL ANALYSIS)   PROPERTIES OF SILVER SODIUM DINITRITE AND SILVER SODIUM NITRITE. (DIFFERENTI 16-2370
GRUNEISEN PARAMETERS AROUND THE PHASE TRANSITIONS IN SODIUM NITRITE AND SODIUM NITRATE. 16-2390
/-RAY DIFFRACTION TOPOGRAPHIC METHOD TO FERROELECTRIC SODIUM NITRITE AND TGS CRYSTALS. (DOMAIN BOUNDARIES / 16-2375
163 DEG C).   FERROELECTRIC PHASE TRANSITION IN SODIUM NITRITE AS STUDIED BY EPR (PHASE TRANSITION AT 16-2359
/ULATION OF THE VARIATION OF THE RAMAN FREQUENCIES IN SODIUM NITRITE AT ORDINARY TEMPERATURES AS A FUNCTIO/ 16-2363
INFRARED SPECTRA OF POTASSIUM NITRATE AND SODIUM NITRITE AT PHASE TRANSITIONS. 16-2352
BRILLOUIN SCATTERING IN SODIUM NITRITE AT ROOM TEMPERATURE. 16-2305
DIENT AND LONG RANGE ORDER PARAMETER OF FERROELECTRIC SODIUM NITRITE. (AT THE SODIUM-23 NUCLEUS) /FIELD GRA 16-2395
STRUCTURE OF SODIUM NITRITE AT 150, 185, AND 225 DEGREES C. 16-2327
X-RAY CRITICAL SCATTERING OF SODIUM NITRITE AT 160 DEG C AND AT 215 DEG C. 16-2343
DISORDERED STRUCTURE OF SODIUM NITRITE AT 185 DEG C. 16-2325
AN SPECTRA AND MODE FREQUENCY SHIFTS OF FERROELECTRIC SODIUM NITRITE AT 77 AND 294 DEG K.   RAM 16-2262
PERATURES.   MEASUREMENT OF THE ELASTIC CONSTANTS OF SODIUM NITRITE BY BRILLOUIN DIFFUSION AT ORDINARY TEM 16-2316
LONG RANGE ORDER IN SODIUM NITRITE BY OPTICAL SECOND HARMONIC GENERATION. 16-2314
MODULATION OF THE MILLIMETER WAVE ABSORPTION IN SODIUM NITRITE BY POLARIZATION REVERSAL. 16-2392
FERROELECTRIC HYSTERESIS LOOP OF SILVER SODIUM DINITRITE BY QUASI STATIC ELECTRIC FIELD. 16-2293
YSIS OF THE TEMPERATURE DEPENDENT PHONON STRUCTURE IN SODIUM NITRITE BY RAMAN SPECTROSCOPY.   ANAL 16-2299
ATURE DEPENDENCE)   STUDY OF PHASE TRANSITION OF SODIUM NITRITE BY SECOND HARMONIC GENERATION. (TEMPER 16-2403
TRIC RELAXATION MEASUREMENT ON FERROELECTRIC SILVER SODIUM DINITRITE BY STEP FUNCTION METHOD.   DIELEC 16-2291
SLOWING DOWN OF THE DIELECTRIC RELAXATION IN SILVER SODIUM DINITRITE BY THERMAL NOISE MEASUREMENTS. /TICAL 16-2382
NTIFERROELECTRIC PHASE)   LATTICE VIBRATION OF SODIUM NITRITE. (CALCULATIONS TO RELATE TO SPECIFIC A 16-2317
O-PHONON SCATTERING AND FERROELECTRIC TRANSITION IN A SODIUM NITRITE CRYSTAL.   TW 16-2277
UM.   INVESTIGATION OF FERROELECTRIC TRANSITION IN SODIUM NITRITE CRYSTAL BY OBSERVATION OF RAMAN SPECTR 16-2275
ELECTRICAL BREAKDOWN IN FERROELECTRIC SODIUM NITRITE CRYSTALS. 16-2288
THERMAL EXPANSION IN FERROELECTRIC SODIUM NITRITE CRYSTALS. 16-2351
RUM IN THE REGION OF THE ANTISYMMETR/ PHASE CHANGE IN SODIUM NITRITE CRYSTALS AND COMBINATION SCATTER SPECT 16-2276
GREE DOMAIN WALLS. TOPOGRAPHIC STUDY ON FERROELECTRIC SODIUM NITRITE CRYSTALS. PART-1: STRUCTURE OF 180 DE 16-2372
ATION REVERSAL.   TOPOGRAPHIC STUDY OF FERROELECTRIC SODIUM NITRITE CRYSTALS. PART-2; MECHANISM OF POLARIZAT 16-2374
RE DEPENDENCE OF FAR INFRARED REFLECTIVITY SPECTRA OF SODIUM NITRITE CRYSTALS. (0.85 DEG K, 30-200 DEG C) / 16-2373
PROPERTIES OF SILVER SODIUM DINITRITE AND SILVER SODIUM NITRITE. (DIFFERENTIAL THERMAL ANALYSIS) 16-2370
CT/ PHASE TRANSFORMATIONS IN POTASSIUM NITRATE AND POTASSIUM NITRITE. (DOES NOT SEEM FERROELECTRIC- PARAELE 16-2369
)   TEMPERATURE DEPENDENCE OF ABSORPTION OF SODIUM NITRITE IN THE FAR INFRARED. (350-1100 MICRONS 16-2387
N T/ NUCLEAR MAGNETIC RESONANCE STUDY ON SODIUM-23 IN SODIUM NITRITE IN THE VICINITY OF THE PHASE TRANSITIO 16-2396
GNETIC RESONANCE STUDY OF PRODUCTION OF NITROGEN DIOXIDE AND NITRITE ION IN IRRADIATED POTASSIUM NITRATE. / 16-2404
MAN TECHNIQUE).   PHASE TRANSITIONS IN POTASSIUM NITRATE NITRITE MIXED CRYSTALS (GROWTH BY DOUBLE BRIDG 16-2345
POINT.   SPECIFIC HEAT OF SODIUM NITRITE NEAR THE ANTIFERROELECTRIC TRANSITION 16-2303
CRYSTAL DYNAMICS OF SODIUM NITRITE. (NEUTRON SCATTERING) 16-2284
/EIR PHASE TRANSITIONS. PART-1: TGS. PART-2: SILVER SODIUM DINITRITE. PART-3: ROCHELLE SALT. PART-4: LITHI/ 02-0077
ASSIUM TANTALATE, BARIUM TITANATE, POTASSIUM NIOBATE, SODIUM NITRITE, POTASSIUM DIDEUTERIUM PHOSPHATE) /POT 02-0059
ACCURATE FORCE CONSTANTS FROM ISOTOPIC SUBSTITUTION IN THE NITRITE RADICAL IN SODIUM NITRITE. 16-2309
ELASTIC CONSTANTS AND ULTRASONIC ABSORPTION OF SODIUM NITRITE SINGLE CRYSTALS. 16-2357
/L DYNAMICS AND THE FERROELECTRIC PHASE TRANSITION OF SODIUM NITRITE (STUDIED BY COHERENT NEUTRON INELASTI/ 16-2364
/RIC SUBSTRATES. (TELLURIUM ON TRIGLYCINE SULFATE AND SODIUM NITRITE SURFACES) (DECORATION METHOD OF DOMAI/ 16-2394
/UM-23 IN THE PARAELECTRIC AND FERROELECTRIC PHASE OF SODIUM NITRITE. (THERMODYNAMICS OF THE PHASE TRANSIT/ 16-2264

HYDROSTATIC PRESSURE EFFECT ON THE 16-2294
INFLUENCE OF THE ELECTRICAL BOUNDARY 16-2350
INFRARED 16-2336
/TION RULES FOR INFRARED ABSORPTION A 16-2281
NEUTRON DIFFRACTI 16-2365
16-2326
NITR 16-2320
16-2324
16-2272
/E STUDIES OF THE PHASE TRANSITIONS I 16-2255
16-2254
16-2313
/LEAR QUADRUPOLE SPIN LATTICE RELAXAT 16-2268
OPTICAL ACTIVITY IN NON-ENANTIOMORPHO 16-2273
OPTICAL SECOND HARMONIC GE 16-2384
16-2397
03-0254
POLARIZATION RE 16-2381
P 16-2342
PYROELECT 16-2295
QUADRUPOLE SPI 16-2362
16-2261
08-1344
RELATIONSHIP OF THE TEMPERATURE 16-2257
16-2385
16-2296
SPLITTING OF THE 16-2307
/ATIC ELECTRIC SUSCEPTIBILITY AND DIE 16-2302
/E DIELECTRIC CONSTANT AND SPONTANEOU 16-2331
TEMPERATURE DEPENDEN 16-2274
TE 16-2315
TEMPERATURE DEPEN 16-2337
TE 16-2311
TEMPERATURE 16-2386
TEMPERATURE DEPENDENCE OF 16-2393
16-2346
THERMAL EXPAN 16-2298
16-2310
ULT 16-2304
X-RAY ANALYSIS 16-2271
X 16-2328

DIPOLES)   THE ORDER DISORDER TRANSITION OF SODIUM NITRITE. (TWO SUBLATTICE MODEL OF INTERACTING 16-2287
                    VACUUM ULTRAVIOLET ABSORPTION OF THE ALKALI NITRITES AND NITRATES.   16-2398
TION SPECT/ STUDIES OF PHASE TRANSFORMATIONS IN NITRATES AND NITRITES. PART-1: CHANGE IN ULTRAVIOLET ABSORP 16-2278
/(5-23 EV AT ROOM AND LIQUID NITROGEN TEMPERATURES IN ALKALI NITRITES (SODIUM, POTASSIUM) AND ALKALI NITRA/ 16-2399
POTASSIUM NI/ PARAMAGNETIC RESONANCE STUDY OF PRODUCTION OF NITROGEN DIOXIDE AND NITRITE ION IN IRRADIATED 16-2404
                    NUCLEAR QUADRUPOLE RESONANCE OF NITROGEN IN SODIUM NITRITE.   16-2313
            CHEMICAL CHANGES IN BARIUM TITANATE FIRED UNDER NITROGEN IN THE PRESENCE OF CARBON.   04-0872
FERROELECTRIC TRANSITION IN TRIGLYCINE SULFATE.   PULSED NITROGEN PROTON DOUBLE RESONANCE STUDY OF THE 27-2968
U/ VACUUM ULTRAVIOLET ABSORPTION (5-23 EV AT ROOM AND LIQUID NITROGEN TEMPERATURES IN ALKALI NITRITES (SODI 16-2399
EA.   STARK EFFECT ON THE NUCLEAR QUADRUPOLE RESONANCE OF NITROGEN-14 AND THE FERROELECTRICITY OF THIOUR 25-2909
/E RESONANCE STUDY OF THE NUCLEAR QUADRUPOLE INTERACTIONS OF NITROGEN-14 IN PARAELECTRIC TRIGLYCINE SULFAT/ 27-2969
                    NUCLEAR QUADRUPOLE RELAXATION OF NITROGEN-14 IN SODIUM NITRITE.   16-2254
RROELECTRIC SODIUM NITRITE.   NITROGEN-14 NUCLEAR QUADRUPOLE RESONANCE IN FE 16-2320
 OF LITHIUM HYDRAZINIUM SULFATE.   NITROGEN-14 NUCLEAR QUADRUPOLE RESONANCE STUDY 32-3345
C AMMONIUM SULFATE.   NITROGEN-14 QUADRUPOLE COUPLING IN PARAELECTRI 19-2749
                                                        NMR SEE NUCLEAR MAGNETIC RESONANCE
).   AUTOMATIC IMPEDANCE MEASUREMENTS USING THERMAL NOISE ANALYSIS (APPLICATION TO BARIUM TITANATE 04-0844
CTRIC CONDENSER UNDER A DC FIELD.   NOISE AND IMPEDANCE MEASUREMENTS OF A FERROELE 37-3645
ALS OF SEIGNETTE SALT.   CORRELATION BETWEEN BARKHAUSEN NOISE AND SURFACE MICRODAMAGES ON SINGLE CRYST 28-3208
            INFLUENCE OF AN ELECTRIC FIELD ON THE NOISE AND THE PYROELECTRIC EFFECT IN TGS. 27-3076
UM TITANATE.   HIGH L/F NOISE ANOMALY IN SEMICONDUCTING BARIUM STRONTI 04-0866
THERMODYNAMIC METHOD)   SPONTANEOUS POLARIZATION NOISE IN POLAR DIELECTRICS. (THEORY, EINSTEIN 03-0219
BASED ON FERROELECTRICS I/ POSSIBLE REALIZATION OF ULTRA LOW NOISE LEVEL PARAELECTRIC MICROWAVE AMPLIFIER (37-3664
DY OF THE CRITICAL BEHAVIOR OF TRIGLYCINE SULFATE BY THERMAL NOISE MEASUREMENTS.   STU 27-3147
 DIELECTRIC RELAXATION IN SILVER SODIUM DINITRITE BY THERMAL NOISE MEASUREMENTS. /TICAL SLOWING DOWN OF THE 16-2382
CHELLE SALT, BARIUM TITANATE)   NOISE MEASUREMENTS IN FERROELECTRICS. (TGS, RO 04-0698
ESITYL OXIDE OXALATE AND SODIUM NITRITE. OPTICAL ACTIVITY IN NON-ENANTIOMORPHOUS BIAXIAL CRYSTALS: METHYL M 16-2273
RYSTAL GROWTH/ SILVER DITHORIUM PHOSPHATE- A NEW CASE OF THE NON-HYDROGEN BONDED PHOSPHATE FERROELECTRIC (C 36-3610
ERIUM) PHOSPHATE CRYSTAL/ CONTRIBUTION TO THE PREPARATION ON NON-WEDGE SHAPED RUBIDIUM DIHYDROGEN OR DIDEUT 17-2637
NIOBATE.   MIXING OF NONCOLLINEAR ELASTIC SURFACE WAVES ON LITHIUM 11-1690
EPRESENTATION THEORY APPLIED, 18 REFS)   COLLINEAR AND NONCOLLINEAR FERRO- AND ANTIFERROELECTRICS. (R 03-0173
FECT IN CADMIUM SULFIDE AND STRONTIUM TITANATE.   NONCOLLINEAR PARAMETRIC 4 PHOTON RECIPROCAL EF 06-1122
NIOBATE.   NONCOLLINEAR PHASE MATCHING EFFECTS IN LITHIUM 11-1675
/LOGRAMS IN A FERROELECTRIC PHOTOCONDUCTOR DEVICE (ERASABLE, NONDESTRUCTIVE READOUT, BASED ON BISMUTH TITA/ 37-3646
G POTASSIUM NITRATE. (/ PROGRESS TOWARD A FAST, NONVOLATILE, NONDESTRUCTIVE READOUT MEMORY ELEMENT UTILIZIN 16-2269
TARTRATE MONOHYDRATE. (EFFECT ON LATTICE/ SOME PROPERTIES OF NONDEUTERATED AND DEUTERATED LITHIUM THALLIUM 29-3239
CN PROCESSES IN FERROELECTRIC ANTIMONY SULFOIO/ INFLUENCE OF NONEQUILIBRIUM CARRIER SCREENING ON POLARIZATI 15-2158
E OF THE FERROELECTRIC SEMICONDUCTOR ANTIMONY / INFLUENCE OF NONEQUILIBRIUM CARRIERS ON THE DOMAIN STRUCTUR 15-2104
N IN FERROELECTRIC SEMICONDUCTOR (COMPOUNDS) A/ INFLUENCE OF NONEQUILIBRIUM CARRIERS ON THE PHASE TRANSITIO 15-2145
ODIUM NITRATE. (ORDER DISORDER TRANSITION, FERROELECTRIC AND NONFERROELECTRIC) /IN TRIGLYCINE SULFATE AND S 16-2400
OF THE (ORTHORHOMBIC) FERROELECTRIC (LEAD DINIOBATE INTO THE NONFERROELECTRIC (RHOMBOHEDRIC) FORM. /SITION 08-1285
RROELECTRIC PLATES NEAR THE CURIE POI/ THEORY OF DOMAINLESS, NONHOMOGENEOUS, SPONTANEOUS POLARIZATION OF FE 03-0321
ICEUTERIUM(X) DISELENATE CRYSTALS.   ISOTOPIC NONISOMORPHISM OF RUBIDIUM TRIHYDROGEN (1-X) TR 22-2873
(SHS) OR SODIUM TRIDEUTERIUM S/ INFRARED SPECTRA OF ISOTOPIC NONISOMORPHISM OF SODIUM TRIHYDROGEN SELENITE 22-2836
RIUM(2+) AND TITANIUM(4+) IONS ARE REPLACED BY ISOVALENT AND NONISOVALENT CATIONS. /RIUM TITANATE, WHERE BA 04-0960
ON THE ELECTROOPTICAL PROPERTIES OF BARIUM TITAN/ EFFECT OF NONISOVALENT SUBSTITUTION OF TITANIUM(4+) IONS 04-0959
PHOSPHATE, RUBIDIUM DIHYDROGEN/ MEASUREMENT OF THE RELATIVE NONLINEAR COEFFICIENTS OF POTASSIUM DIHYDROGEN 08-1413
/ OF A TRANSVERSE INHOMOGENEITY OF THE REFRACTIVE INDEX OF A NONLINEAR CRYSTAL ON SECOND HARMONIC GENERATI/ 11-1687
OGE/ AUTOFOCUSING OF LASER RADIATION IN ACTIVE MATERIALS AND NONLINEAR CRYSTALS. (KDP, ADP, RUBIDIUM DIHYDR 11-1865
Y RADIATION FROM A RUBY LASER.   FRACTURE OF NONLINEAR CRYSTALS (KDP AND LITHIUM NIOBATE) B 11-1663
            EFFECT OF OPTICAL INHOMOGENEITIES ON PHASE MATCHING IN NONLINEAR CRYSTALS (LITHIUM NIOBATE). 11-1773
ON OF BARIUM (OR LEAD OR CALCIUM) STR/ MICROWAVE BEHAVIOR OF NONLINEAR DIELECTRICS (EXPERIMENT AND DISCUSSI 08-1281
C TUNING. (WITH ELECTRODES REDUCING POLARIZATION SWITCHING / NONLINEAR DIELECTRICS FOR HIGH POWER ELECTRONI 37-3660
NSITION IN ANTIMONY SULFCIODIDE CRYSTALS.   NONLINEAR EFFECTS AND THE LOW TEMPERATURE TRA 15-2247
THIUM NIOBATE)   NONLINEAR EFFECTS IN FERROCERAMICS. (PZT-8, LI 09-1546
M NIOBATE SURFACE WAVE DELAY LINES.   NONLINEAR EFFECTS IN MICROWAVE ACOUSTIC LITHIU 11-1818
AMPLITUDE HYPERSONIC WAVES IN FERROELECTRIC CRYSTALS.   NONLINEAR EFFECTS IN THE PROPAGATION OF FINITE 03-0498
STIC HARMONICS IN A (SINGLE DOMAIN) LITHIUM NIOBATE CRYSTAL (NONLINEAR ELASTIC PROPERTIES). /FREQUENCY ACOU 11-1707
D HARMONIC GENERATION IN STRONTIUM TITANATE SINGLE CRYSTALS (NONLINEAR ELASTIC PROPERTIES). ULTRASONIC THIR 06-1181
VSKITE STRUCTURE.   ELECTRICAL PROPERTIES OF NONLINEAR FERROELECTRIC CERAMICS WITH THE PERO 08-1461
 LITHIUM NIOBATE CRYSTAL.   NONLINEAR INTERACTION OF HYPERSONIC WAVES IN A 11-1802
ELDS AND SOUND IN LITHIUM NIOBATE.   NONLINEAR INTERACTION OF MICROWAVE ELECTRIC FI 11-1838
CS TO MICROWAVE SIGNAL PROCESSING.   APPLICATION OF NONLINEAR INTERACTIONS IN FERROELECTRIC CERAMI 37-3647
C CERAMICS. THIN FILM MICROWAVE FERROELECTRIC CAPACITORS AND NONLINEAR MICROWAVE PROPERTIES OF FERROELECTRI 37-3643
INE SOLID SOLUTIONS BASED ON BARIUM TITANATE/ DIELECTRIC AND NONLINEAR MICROWAVE PROPERTIES OF POLYCRYSTALL 04-1019
M NI/ POSSIBILITY OF SWITCHING AND STEERING SURFACE WAVES BY NONLINEAR MIXING IN ANISOTROPIC MEDIA. (LITHIU 11-1778
 HAVING NO MICROTWINNING (AND RESISTANT TO L/ NEW (UNIAXIAL) NONLINEAR OPTIC CRYSTAL BARIUM LITHIUM NIOBATE 13-1958
OF PHYSICAL PROPERTIES TO THE SYMMETRY OF POTASSIUM IODATE. (NONLINEAR OPTIC, REFRACTIVE INDICES) RELATION 08-1296
IONS / GROWTH OF HIGH QUALITY LITHIUM NIOBATE CRYSTALS (FOR NONLINEAR OPTICS WITH LOW BIREFRINGENCE VARIAT 11-1688
S OF DIELECTRICS AND FERROELECTRICS. (REVIEW, 195 REFERENCE/ NONLINEAR OPTICAL AND ELECTROOPTICAL PROPERTIE 02-0081
D SECOND HARMONIC GENERATION IN LITHIUM IODATE.   LARGE NONLINEAR OPTICAL COEFFICIENT AND PHASE MATCHE 08-1405
/(OPTICAL ABSORPTION, REFRACTIVE INDEX, DIELECTRIC CONSTANT, NONLINEAR OPTICAL COEFFICIENTS, AND LINEAR EL/ 13-2024
C SODIUM NITRITE.   TEMPERATURE DEPENDENCE OF NONLINEAR OPTICAL COEFFICIENTS IN FERROELECTRI 16-2274
LS (LITHIUM NIOBATE OR TANTALATE, BARIUM / RELATIVE SIGNS OF NONLINEAR OPTICAL COEFFICIENTS OF POLAR CRYSTA 02-0098
HYDROGEN PHOSPHATE.   ABSOLUTE SIGNS OF THE NONLINEAR OPTICAL COEFFICIENTS OF POTASSIUM DI 17-2597
RMONIC GENERATION AND OPTICAL PARAMETRIC OSCILLATION IN FER/ NONLINEAR OPTICAL DEVICES. (BASED IN SECOND HA 02-0097
LO/ DEVELOPMENT OF SINGLE CRYSTAL BARIUM SODIUM NIOBATE AS A NONLINEAR OPTICAL MATERIAL. (GROWTH BY KYROPOU 13-2021
                                                        NONLINEAR OPTICAL MATERIALS.   02-0055
ECTION, MATERIALS NEEDS, 118 REFS)   NONLINEAR OPTICAL MATERIALS (REVIEW, DATA COLL 02-0049
M- OXYGEN BOND. (LITHIUM NIOBATE, LITHIUM TANTALATE, POTASS/ NONLINEAR OPTICAL POLARIZABILITY OF THE NIOBIU 11-1734
/RIZATION OF LUMINOUS WAVES IN CUBIC ELECTROOPTIC CRYSTALS. (NONLINEAR OPTICAL PROPERTIES, ELECTROOPTIC CO/ 02-0044
                                                        NONLINEAR OPTICAL PROPERTIES OF CRYSTALS.   03-0379
LEAD NIOBIUM(4) OXIDE(12).   NONLINEAR OPTICAL PROPERTIES OF FERROELECTRIC 13-2023
SODIUM NITRITE.   NONLINEAR OPTICAL PROPERTIES OF FERROELECTRIC 16-2272
LEAD(5) GERMANIUM(3) OXIDE(11).   NONLINEAR OPTICAL PROPERTIES OF FERROELECTRIC 35-3506
LEAD TITANATE.   NONLINEAR OPTICAL PROPERTIES OF FERROELECTRIC 05-1080
LEAD TITANATE.   NONLINEAR OPTICAL PROPERTIES OF FERROELECTRIC 05-1079
CRYSTALS. (SECOND HARMONIC GENERATION, OPTICAL PARAMETRIC O/ NONLINEAR OPTICAL PROPERTIES OF FERROELECTRIC 03-0383
YBDATE AND TERBIUM MOLYBDATE.   NONLINEAR OPTICAL PROPERTIES OF GADOLINIUM MOL 30-3305
AND BISMUTH TITANATE CRYSTALS.   NONLINEAR OPTICAL PROPERTIES OF LEAD TITANATE 04-0941

TALS.   NONLINEAR OPTICAL PROPERTIES OF MOLECULAR CRYS 02-0078
LIDENE FLUORIDE FILMS.   PYROELECTRIC AND NONLINEAR OPTICAL PROPERTIES OF POLED POLYVINY 36-3595
IUM NIOBATE.   NONLINEAR OPTICAL PROPERTIES OF POTASSIUM LITH 13-2027
IUM NIOBATES (TRANSPARENT, TETRAGONAL TUNGSTEN BRONZE STRUC/   NONLINEAR OPTICAL PROPERTIES OF POTASSIUM LITH 13-2026
IUM NIOBATES.   CRYSTAL GROWTH AND NONLINEAR OPTICAL PROPERTIES OF POTASSIUM LITH 13-2015
ATE SINGLE CRYSTALS. (REFRACTIVE INDICES)   NONLINEAR OPTICAL PROPERTIES OF POTASSIUM NIOB 08-1454
/ COMPARATIVE STUDY OF THE CRYSTALLOGRAPHIC, DIELECTRIC, AND   NONLINEAR OPTICAL PROPERTIES OF THE ABC NIOBAT 04-0912
COEFFICIENTS OF POLAR (PYROELECTRIC) CRY/ ABSOLUTE SIGNS OF   NONLINEAR OPTICAL (SECOND HARMONIC GENERATION) 02-0095
SUSCEPTIBILITY IN LITHIUM NIOBATE.   NONLINEAR OPTICAL SPECTROSCOPY OF THIRD ORDER 11-1855
RMATE MONOHYDRATE.   NONLINEAR OPTICAL SUSCEPTIBILITY OF LITHIUM FO 26-2945
CONVOLUTION IN LITHIUM NIOBATE.   DETERMINATION OF A   NONLINEAR PARAMETER FOR ACOUSTIC SURFACE WAVE 11-1681
OPTICAL HARMONICS AND   NONLINEAR PHENOMENA. (THEORY, REVIEW)   02-0067
STRONTIUM BISMUTH TITANATES.   NONLINEAR PROPERTIES AND PHASE TRANSITIONS OF 13-1952
PHOSPHATE CRYSTAL.   NONLINEAR PROPERTIES OF A RUBIDIUM DIHYDROGEN 17-2508
OWAVE FREQUENCIES. (TRIGLYCINE SULFATE, POLYCRYSTALLINE BAR/   NONLINEAR PROPERTIES OF FERROELECTRICS AT MICR 04-0870
HOSPHATE. (DETERMINATION OF BETA IN THE P EXPANSION OF FREE/   NONLINEAR PROPERTIES OF POTASSIUM DIHYDROGEN P 17-2543
DIELECTRIC CONSTANT VERSUS ALTERNATING FIELD VOLTAGE, TEMPE/   NONLINEAR PROPERTIES OF TRIGLYCINE SELENATE. (27-3158
PHASE SYNCHRONIZATION IN   NONLINEAR RUBIDIUM DIHYDROGEN PHOSPHATE.   17-2473
N OF RADIATION AND FIRST HARMONIC, CALCULATIONS FOR LITHIUM/   NONLINEAR SADOVSKII EFFECT (THEORY, PROPAGATIO 03-0186
VALUES FOR QUARTZ, AMMONIUM DIHYDROGEN PHOSPHATE, A/ OPTICAL   NONLINEAR SUSCEPTIBILITIES: ACCURATE RELATIVE 17-2539
LITHIUM TANTAL/ DETERMINATION OF ABSOLUTE SIGNS OF MICROWAVE   NONLINEAR SUSCEPTIBILITIES. (BARIUM TITANATE, 02-0107
RAMETRIC SCATTERING, IDLER ABSORPTION, AND DISPERSION IN THE   NONLINEAR SUSCEPTIBILITY IN LITHIUM NIOBATE. / 11-1771
OGEN PHOSPHATE CRYSTAL.   CURIE-WEISS LAW FOR THE   NONLINEAR SUSCEPTIBILITY OF A POTASSIUM DIHYDR 17-2686
TEMPERATURE DEPENDENCE OF THE OPTICAL   NONLINEAR SUSCEPTIBILITY OF SODIUM NITRITE.   16-2386
ATION REGIME OF A FERROEL/ APPLICATIONS OF THE PIEZOELECTRIC   NONLINEARITIES IN THE TEMPERATURE AUTOSTABILIZ 17-2502
TE, AMMONIUM DIHYDROGEN/ RELATIVE MEASUREMENT OF THE OPTICAL   NONLINEARITIES OF POTASSIUM DIHYDROGEN PHOSPHA 11-1677
4 COMPOSITIONS EACH) STRONTIUM TITANATE SYSTEMS./ DIELECTRIC   NONLINEARITY AND LOSS IN THE BARIUM (OR LEAD, 09-1522
ROM POTASSIUM FLUO/ FERROELECTRIC SINGLE CRYSTALS WITH LARGE   NONLINEARITY. (BARIUM TITANATE HAFNATE GROWN F 08-1373
NEAR THE PHASE TRANSI/ ANOMALOUS BEHAVIOR OF THE DIELECTRIC   NONLINEARITY COEFFICIENT IN STRONTIUM TITANATE 06-1105
TALS.   MECHANICAL RELAXATION AND   NONLINEARITY IN STRONTIUM TITANATE SINGLE CRYS 06-1218
VSKITE LIKE FERROELECTRICS, U/ INVESTIGATION OF POLARIZATION   NONLINEARITY NEAR THE PHASE TRANSITION IN PERO 03-0249
IN A NONUNIFORM ELECTRIC FIELD.   ELECTRICAL   NONLINEARITY OF A FERROELECTRIC SEMICONDUCTOR 03-0500
S.   MICROWAVE DYNAMIC   NONLINEARITY OF BARIUM TITANATE SINGLE CRYSTAL 04-0749
RROELECTRIC PROPERTIES OF MATERIALS.   NONLINEARITY OF DIELECTRIC POLARIZATION AND FE 03-0185
AL PROPERTIES OF FERROELECTRICS. (BARIUM TITANATE CERAMICS)   NONLINEARITY OF ELECTROMECHANICAL AND MECHANIC 04-0986
TALS NEAR PHASE TRANSITION AT 110 DEGREES K.   DIELECTRIC   NONLINEARITY OF STRONTIUM TITANATE SINGLE CRYS 06-1106
ROELECTRIC TRIGLYCINE SULFATE.   NONLINEARITY OF THE DIELECTRIC CONSTANT OF FER 27-3070
ACOUSTIC CONVOLUTION AND CORRELATION AND THE ASSOCIATED   NONLINEARITY PARAMETERS IN LITHIUM NIOBATE.   11-1758
TEMPERATURE, POLARIZATI/ THEORY OF THE ELASTOOPTIC EFFECT IN   NONMETALLIC CRYSTALS. (STRAIN DEPENDENT CURIE 03-0513
PIEZOELECTRIC CERAMICS. (NONMETALLIC SOLIDS, VOL-3, REVIEW)   NONMETALLIC SOLIDS, VOL-3, REVIEW)   01-0015
GEN PHOSPHATE AND RUBIDIUM DIHYDROGEN PHOSPHATE IN POLAR AND   NONPOLAR PHASES. /RYSTALS OF POTASSIUM DIHYDRO 17-2457
. (150-500 DEG C, 20-50 KILOHERTZ, LISSAJOUS FI/ ENERGY LOSS   NONPOLAR STATE POLYCRYSTALLINE BARIUM TITANATE 04-0831
IONS, SPONTANEOUS) P/ POTASSIUM IODATE, A FERROELECTRIC WITH   NONREVERSIBLE, BUT TILTABLE (TO 3 STABLE POSIT 08-1346
IBRATOR MODEL. (FERROELECTRICS)   NONSTATIONARY POLARIZATION IN THE ANHARMONIC V 03-0345
CRYSTALS.   PREPARATION AND PHYSICAL PROPERTIES OF   NONSTOICHIOMETRIC ANTIMONY SULFOIODIDE SINGLE 15-2199
CALCULATIONS OF ELECTRONIC CONDUCTIVITY IN   NONSTOICHIOMETRIC BARIUM TITANATE.   04-0821
ELECTRON PARAMAGNETIC RESONANCE IN   NONSTOICHIOMETRIC BARIUM TITANATE.   04-0640
IGH TEMPERATURE THERMODYNAMIC AND X-RAY DIFFRACTION STUDY OF   NONSTOICHIOMETRIC BARIUM TITANATE.   H 04-0885
PARAMAGNETIC RESONANCE OF AN F CENTER IN   NONSTOICHIOMETRIC BARIUM TITANATE.   04-0976
GUIDE.   SEMICONDUCTING AND   NONSTOICHIOMETRIC BARIUM TITANATE. A REFERENCE 01-0022
H TEMPERATURE ELECTRICAL CONDUCTIVITY MEASURE/ TITANIUM RICH   NONSTOICHIOMETRIC BARIUM TITANATE. PART-1: HIG 04-0822
H TEMPERATURE THERMODYNAMIC AND X-RAY DIFFRAC/ TITANIUM RICH   NONSTOICHIOMETRIC BARIUM TITANATE. PART-3: HIG 04-0886
NIOBIUM-93 NMR LINE WIDTHS IN   NONSTOICHIOMETRIC LITHIUM NIOBATE.   11-1795
ONTIUM TI/ TOP SEEDED SOLUTION GROWTH OF OXIDE CRYSTALS FROM   NONSTOICHIOMETRIC MELTS. (BARIUM TITANATE, STR 04-0553
EPR STUDY OF THE CATALYTIC ACTIVITY OF   NONSTOICHIOMETRIC TUNGSTEN TRIOXIDE.   10-1628
/TRON MICROSCOPY IN THE STUDY OF EXTENDED DEFECTS RELATED TO   NONSTOICHIOMETRY (AND ANTIFERROELECTRIC DOMAI/ 36-3522
NIOBATE.   NONSTOICHIOMETRY AND CRYSTAL GROWTH OF LITHIUM 11-1691
                                                                         NONSTOICHIOMETRY IN BARIUM TITANATE.   04-0631
E TITANATE.   INTRINSIC   NONSTOICHIOMETRY IN SINGLE PHASE LEAD ZIRCONAT 09-1534
TANATE)   CATION   NONSTOICHIOMETRY IN TERNARY OXIDES. (BARIUM TI 04-0969
E SYSTEM.   INTRINSIC   NONSTOICHIOMETRY IN THE LEAD ZIRCONATE TITANAT 09-1535
E SYSTEM. (DEFECT STRUCTURE)   INTRINSIC AND EXTRINSIC   NONSTOICHIOMETRY IN THE LEAD ZIRCONATE TITANAT 09-1536
TE SINGLE CRYSTALS.   HETEROVALENT SUBSTITUTION AND OXYGEN   NONSTOICHIOMETRY OF BARIUM TITANATE AND HEMATI 04-0874
LECTRICAL NONLINEARITY OF A FERROELECTRIC SEMICONDUCTOR IN A   NONUNIFORM ELECTRIC FIELD.   E 03-0500
PARAELECTRIC SEMICONDUCTOR IN A   NONUNIFORM ELECTRIC FIELD.   03-0502
DEBYE SCREENING OF A   NONUNIFORM ELECTRIC FIELD IN A FERROELECTRIC. 03-0499
/ATION OF X-RAY ABSORPTION COEFFICIENTS AND THE THICKNESS OF   NONUNIFORM THIN SAMPLES (STRONTIUM TITANATE A/ 06-1179
MENT UTILIZING POTASSIUM NITRATE. (/ PROGRESS TOWARD A FAST,   NONVOLATILE, NONDESTRUCTIVE READOUT MEMORY ELE 16-2269
IN FERROELECTRIC CRYSTALS OF THE TRIGLYCINE SULFATE GROUP.   (NORMAL AND DEUTERATED) /OMORPHOUS SUBSTITUTION 27-3102
ERROELECTRIC STATE.   (THEORY)   PHASE TRANSITION FROM   NORMAL FERROELECTRIC TO MAGNETICALLY ORDERED F 03-0394
/TRICS (HAMILTONIAN IN TERMS OF LOCALIZED D-STRAIN AND SOFT-   NORMAL MODE COORDINATES AND TEMPERATURE INDEP/ 03-0432
POLARIZATION VECTOR, P(S), ARISES PERPENDICULAR TO THE WAVE   NORMAL OF THE SOFT WAVES. (THEORY) / WHICH THE 03-0141
OF P/   SYMMETRY OF   NORMAL OSCILLATIONS IN A FERROELECTRIC CRYSTAL 17-2623
PHATE SINGLE CRYSTAL (FROM SOLUTION) IN STATIC REGIME AND AT   NORMAL PH. /IC OF THE AMMONIUM DIHYDROGEN PHOS 17-2516
LS AT LOW TEMPERAT/ INFLUENCE OF THE PHASE TRANSITION ON THE   NORMAL VIBRATION MODES OF FERROELECTRIC CRYSTA 17-2456
E.   NORMAL VIBRATIONS OF STRONTIUM TITANATE LATTIC 06-1128
ROVSKITES (POTASSIUM TANTALATE, STRONT/ DETERMINATION OF THE   NORMAL VIBRATIONAL DISPLACEMENTS OF SEVERAL PE 06-1139
PE CRYSTALS.   NOVEL CUT OF POTASSIUM DIHYDROGEN PHOSPHATE TY 17-2530
TAGE LIGHT MODULATION.   NOVEL TYPE OF CUT FOR KDP CRYSTALS FOR LOW VOL 17-2531
(POTASSIUM DIHYDROGEN ARSENATE)   FERROELECTRIC MODE AND   NQR (NUCLEAR QUADRUPOLE RESONANCE) FREQUENCY. 17-2713
NS IN THE IRRA/ ELECTRON PARAMAGNETIC RESONANCE AND ELECTRON   NUCLEAR DOUBLE RESONANCE OBSERVATIONS OF DOMAI 18-2726
R IN X-IRRADIA/ ELECTRON PARAMAGNETIC RESONANCE AND ELECTRON   NUCLEAR DOUBLE RESONANCE OF THE ARSENATE CENTE 18-2727
/ DETECTION BY ELECTRON PARAMAGNETIC RESONANCE AND ELECTRON   NUCLEAR DOUBLE RESONANCE OF THE PROTON LATTICE 18-2724
IN FERROELECTRIC POTASSIUM DIHYDROGEN PHOSPHATE.   NUCLEAR DOUBLE RESONANCE STUDY OF POTASSIUM-39 17-2599
D SINGLE CRYSTALS OF ICE.   ENDOR (ELECTRON   NUCLEAR DOUBLE RESONANCE) STUDY OF X-IRRADIATE 34-3413
SEMICONDUCTING COMPOUNDS OF THE TYPE A(2)(V)B(3)(VI)AND A(/   NUCLEAR GAMMA RESONANCE SPECTROSCOPIC STUDY OF 15-2108
RANSITION IN AMMONIUM DIHYDROGEN PHOSPHATE SINGLE CRYSTAL BY   NUCLEAR MAGNETIC RESONANCE /ON OF THE PHASE T 17-2545
F ICE POLYMORPHS BY NEUTRON DIFFRACTION, PROTON AND DEUTERON   NUCLEAR MAGNETIC RESONANCE. /IN PARAELECTRIC A 24-2889
ND FERROELECTRIC POTASSIUM FERROCYANIDE TRIHYDRATE BY PULSED   NUCLEAR MAGNETIC RESONANCE AND DIELECTRIC RELA 18-2718
XATION STUDIES OF LONG RANGE CORRELATION IN FERROELECTRIC P/

F POTASSIUM FERROCYANIDE TRIHYDRATE CRYSTALS.
ROELECTRIC SILVER SODIUM DINITRI/ SPLITTING OF THE SODIUM-23 NUCLEAR MAGNETIC RESONANCE CENTRAL LINE IN FER 16-2307
DIUM NITRITE.   ACOUSTIC NUCLEAR MAGNETIC RESONANCE IN FERROELECTRIC SO 16-2321
OGEN PHOSPHATE / ABSENCE OF MOTIONAL NARROWING IN THE PROTON NUCLEAR MAGNETIC RESONANCE IN POTASSIUM DIHYDR 17-2594
ETERMINATION OF THE ORIENTATIONS OF PROTON PAIRS (IN THE WA/ NUCLEAR MAGNETIC RESONANCE IN ROCHELLE SALT, D 28-3203
IN FERROELECTRICS: LITHIUM OR SODIUM TRIHYDROGEN DISELENITE, NUCLEAR MAGNETIC RESONANCE INVESTIGATION. /SM 22-2837
IGLYCINE SULFATE.   NUCLEAR MAGNETIC RESONANCE INVESTIGATION OF TR 27-3059
ERROELECTRIC DICALCIUM STRONTIUM PROPIONATE.   NUCLEAR MAGNETIC RESONANCE INVESTIGATIONS OF F 26-2931
CE CRYSTALS. PART-1: PROTON MAGNETIC RESONANCE IN WATER ICE/ NUCLEAR MAGNETIC RESONANCE INVESTIGATIONS OF I 34-3471
HTHALOCYANINE.   NUCLEAR MAGNETIC RESONANCE INVESTIGATIONS OF P 36-3551
TOICHIOMETRIC LITHIUM NIOBATE.   NIOBIUM-93 NUCLEAR MAGNETIC RESONANCE LINE WIDTHS IN NONS 11-1795
ROELECTRIC CRYSTALS POTASSIUM FERROCYANIDE TRIHYDRATE BY THE NUCLEAR MAGNETIC RESONANCE METHOD. /OF THE FER 24-2898
   FERROELECTRIC SODIUM DEUTERO AMMONIUM SELENATE DIDEUTERATE. NUCLEAR MAGNETIC RESONANCE OF DEUTERIUM IN THE 22-2828
PERDEUTERO AMMONIUM FLUOBERYLLATE CRYSTAL.   NUCLEAR MAGNETIC RESONANCE OF DEUTERIUM IN THE 19-2744
THIUM AMMONIUM TARTRATE.   NUCLEAR MAGNETIC RESONANCE OF FERROELECTRIC LI 29-3240
THIUM AMMONIUM TARTRATE.   NUCLEAR MAGNETIC RESONANCE OF FERROELECTRIC LI 29-3241
   SODIUM AMMONIUM SELENATE DIHYDRATE CRYSTAL.   NUCLEAR MAGNETIC RESONANCE OF SODIUM-23 IN THE 22-2883
/ FERROCYANIDE AND ISOMORPHOUS CRYSTALS. (NEW MODEL BASED ON NUCLEAR MAGNETIC RESONANCE OF UNTWINNED SAMPL/ 24-2893
TS IN POTASSIUM DIHYDROGEN PHOSPHATE TYPE CRYSTALS.   NUCLEAR MAGNETIC RESONANCE PROTON SECOND MOMEN 17-2408
SODI/ INFLUENCE OF THE ELECTRICAL BOUNDARY CONDITIONS ON THE NUCLEAR MAGNETIC RESONANCE RELAXATION RATE OF 16-2350
UM DIHYDROGEN ARSENATE AND DEUTERO AMMONIUM DIDE/ ARSENIC-75 NUCLEAR MAGNETIC RESONANCE RESONANCE IN AMMONI 18-2736
/ ACOUSTIC DIPOLE AND QUADRUPOLE SATURATION OF THE LITHIUM-7 NUCLEAR MAGNETIC RESONANCE SIGNALS IN LITHIUM 11-1666
NT IN ORDER DISORDE/ NATURE OF THE TEMPERATURE DEPENDENCE OF NUCLEAR MAGNETIC RESONANCE SPECTRA SECOND MOME 03-0460
MOTION IN LITHIUM NI/ TEMPERATURE DEPENDENCE OF THE LITHIUM NUCLEAR MAGNETIC RESONANCE SPECTRUM AND ATOMIC 11-1722
ROGEN BONDED FERROELECTRICS. (DEUTERATED ROCHELLE SALT, POT/ NUCLEAR MAGNETIC RESONANCE STUDIES IN SOME HYD 18-2719
HYDRAZINIUM SULFATE.   DEUTERON NUCLEAR MAGNETIC RESONANCE STUDIES OF LITHIUM 32-3354
CTAHEDRA FERROELECTRICS.   NUCLEAR MAGNETIC RESONANCE STUDIES OF OXYGEN O 13-2059
OUS POLARIZATION IN ROCHELLE SALT.   NUCLEAR MAGNETIC RESONANCE STUDIES OF SPONTANE 28-3194
E TRANSITIONS IN FERROELECTRICS. (POTASSIUM FERROCYANIDE TR/ NUCLEAR MAGNETIC RESONANCE STUDIES OF THE PHAS 16-2255
E TRANSITION IN STANNOUS CHLORIDE DIHYDRATE. DIELECTRIC AND NUCLEAR MAGNETIC RESONANCE STUDIES OF THE PHAS 36-3582
DIFFUSION AND DEUTERIUM AND SODIUM-23 RELAXATION IN SODIUM / NUCLEAR MAGNETIC RESONANCE STUDY OF DEUTERIUM 22-2849
TRIC TRANSITION IN (MANGANESE DOPED) AMMONIUM SULFATE.   (NUCLEAR MAGNETIC RESONANCE STUDY OF) FERROELEC 19-2784
RIC LITHIUM HYDRAZINIUM SULFATE.   NUCLEAR MAGNETIC RESONANCE STUDY OF FERROELECT 32-3343
SITIONS IN PEROVSKITES CRYSTALS. (NIOBIUM-93 AND SODIUM-23 / NUCLEAR MAGNETIC RESONANCE STUDY OF PHASE TRAN 08-1283
SITIONS IN METALLIC SODIUM TUNGSTEN BRONZES.   TRANSIENT NUCLEAR MAGNETIC RESONANCE STUDY OF PHASE TRAN 13-1910
SITIONS IN ROCHELLE SALT.   NUCLEAR MAGNETIC RESONANCE STUDY OF PHASE TRAN 28-3195
IN SILVER SODIUM DINITRITE.   NUCLEAR MAGNETIC RESONANCE STUDY OF SODIUM-23 16-2324
LECTRIC TRANSITIONS IN DIGLYCINE NITRATE AND TRIS-SARCOSINE/ NUCLEAR MAGNETIC RESONANCE STUDY OF THE FERRO 16-2267
LECTRIC TRANSITIONS IN SODIUM TRIHYDROGEN AND TRIDEUTERIUM / NUCLEAR MAGNETIC RESONANCE STUDY OF THE FERROE 22-2832
LECTRIC TRANSITION IN DEUT/ BERYLLIUM-9 QUADRUPOLE PERTURBED NUCLEAR MAGNETIC RESONANCE STUDY OF THE FERROE 27-2965
M NIOBATE TANTALATE SOLID SOLUTION SYSTEM.   NIOBIUM-93 NUCLEAR MAGNETIC RESONANCE STUDY OF THE LITHIU 11-1794
TRANSITION IN STRONTIUM TITANATE.   NUCLEAR MAGNETIC RESONANCE STUDY OF THE PHASE 06-1237
URAL PHASE TRANSITION IN LANTHANUM ALUMINATE.   NUCLEAR MAGNETIC RESONANCE STUDY OF THE STRUCT 36-3533
IN SODIUM NITRITE IN THE VICINITY OF THE PHASE TRANSITION T/ NUCLEAR MAGNETIC RESONANCE STUDY ON SODIUM-23 16-2396
AND SPIN LATTICE RELAXATION) INVESTIGATIONS OF THE PHASE TR/ NUCLEAR MAGNETIC RESONANCE (WIDE LINE SPECTRA 20-2802
IN FERROELECTRIC ANTIMONY SULFOIODIDE.   NUCLEAR QUADRUPOLE AND ELECTROACOUSTIC ECHOES 15-2189
NITRITE.   TEMPERATURE DEPENDENCE OF THE SODIUM-23 NUCLEAR QUADRUPOLE COUPLING CONSTANT IN SODIUM 22-2393
TIMES IN FERROELECTRIC HYDROGEN CHLORIDE.   NUCLEAR QUADRUPOLE FREQUENCIES AND RELAXATION 31-3336
CTRIC DIAMMONIUM TRIHYDROGEN PERIODATE PROTON CROSSOVER REL/ NUCLEAR QUADRUPOLE INTERACTION IN ANTIFERROLE 36-3578
CTRIC DIAMMONIUM TRIHYDROGEN PERIODATE BY IODINE PROTON CRO/ NUCLEAR QUADRUPOLE INTERACTION IN ANTIFERROELE 36-3579
TYPE COMPOUNDS OF HAFNIUM-181 STUDIED BY THE PERTURBED ANGU/ NUCLEAR QUADRUPOLE INTERACTIONS IN PEROVSKITE 08-1276
IN PARAELECTRIC TRIGL/ PULSED DOUBLE RESONANCE STUDY OF THE NUCLEAR QUADRUPOLE INTERACTIONS OF NITROGEN-14 27-2969
N SODIUM NITRITE.   NUCLEAR QUADRUPOLE RELAXATION OF NITROGEN-14 I 16-2254
ODATE CRYSTAL STRUCTURE STUDIED BY INFRARED SPECTROSCOPY AND NUCLEAR QUADRUPOLE RESONANCE.   POTASSIUM I 08-1278
SSIUM DIHYDROGEN ARSENATE)   FERROELECTRIC MODE AND NQR (NUCLEAR QUADRUPOLE RESONANCE) FREQUENCY. (POTA 17-2713
HYDROGEN BROMIDE.   BROMINE NUCLEAR QUADRUPOLE RESONANCE IN FERROELECTRIC 31-3332
SODIUM NITRITE.   NITROGEN-14 NUCLEAR QUADRUPOLE RESONANCE IN FERROELECTRIC 16-2320
OF THE ANTIMONY SULFOIODIDE (AND BROMIDE) TYPE.   NUCLEAR QUADRUPOLE RESONANCE IN FERROELECTRICS 15-2216
(STOICHIOMETRIC) LITHIUM NIOBATE (TEMPERATURE DE/ NIOBIUM-93 NUCLEAR QUADRUPOLE RESONANCE INVESTIGATION OF 11-1811
SILVER PERIODATE.   NUCLEAR QUADRUPOLE RESONANCE OF IODINE-127 IN 08-1426
SOLID HYDROGEN IODIDE.   NUCLEAR QUADRUPOLE RESONANCE OF IODINE-127 IN 31-3335
ROELECTRIC COMPOUNDS. (LITHIUM AMMONIUM TARTRATE, LITHIUM T/ NUCLEAR QUADRUPOLE RESONANCE OF LITHIUM IN FER 02-0052
DIUM NITRITE.   NUCLEAR QUADRUPOLE RESONANCE OF NITROGEN IN SO 16-2313
D THE FERROELECTRICITY OF THIOUREA.   STARK EFFECT ON THE NUCLEAR QUADRUPOLE RESONANCE OF NITROGEN-14 AN 25-2909
TIONS, AND FERROELECTRICITY OF ALKALI IODATES.   NUCLEAR QUADRUPOLE RESONANCE, PHASE TRANSFORMA 08-1350
LLY DEUTERATED FERROELECTRIC CRYSTALS CESIUM DIH/ ARSENIC-75 NUCLEAR QUADRUPOLE RESONANCE SPECTRA OF PARTIA 18-2740
RROELECTRIC CRYSTAL CESIUM DIHYDROGEN A/ STARK EFFECT IN THE NUCLEAR QUADRUPOLE RESONANCE SPECTRA OF THE FE 18-2722
HYDRAZINIUM SULFATE.   NITROGEN-14 NUCLEAR QUADRUPOLE RESONANCE STUDY OF LITHIUM 32-3345
ANSITION IN AMMONIUM MONOCHLORO ACETATE, MONOCHLORO ACETIC / NUCLEAR QUADRUPOLE RESONANCE STUDY OF PHASE TR 36-3625
CRITICAL DYNAMICS OF FERROELECTRIC CRYSTALS. (SODIUM NITRI/ NUCLEAR QUADRUPOLE SPIN LATTICE RELAXATION AND 16-2268
DYNAMICAL PROPERTIES OF FERROELECTRIC CRYSTALS. (THEORY) NUCLEAR QUADRUPOLE SPIN LATTICE RELAXATION AND 03-0187
.   INVESTIGATIONS OF NUCLEAR RELAXATION OF TRIGLYCINE FLUOBERYLLATE 27-3104
OF THE STRUCTURE OF ANTIFERROELECTRIC AMMONIUM PERIODATE BY NUCLEAR RESONANCE METHODS.   STUDY 36-3580
IC TRIGLYCINE SULFATE.   NUCLEAR SPIN LATTICE RELAXATION IN FERROELECTR 27-2967
LECTRIC AMMONIUM SALTS. (AMMONIUM SULFATE, FLUOBERYLLATE, P/ NUCLEAR SPIN LATTICE RELAXATION IN SOME FERROE 19-2769
N THE PARAELECTRIC AND FERROELECTRIC PHASE OF SODIUM NITRIT/ NUCLEAR SPIN LATTICE RELAXATION OF SODIUM-23 I 16-2264
NTS IN SODIUM DEUTERIUM SELENATE.   NUCLEAR SPIN LATTICE RELAXATION TIME MEASUREME 22-2847
ER IN X-RAY IRRADIATED SINGLE CRYST/ OBSERVATION OF ELECTRON NUCLEAR TRIPLE RESONANCE FOR THE ARSENATE CENT 18-2729
IN BARIUM TITANATE AT FIELDS FROM 2 TO 450 KV/CM.   NUCLEATION AND GROWTH OF FERROELECTRIC DOMAINS 04-0979
NSITION IN BARIUM TITANATE.   FIELD INDUCED NUCLEATION AT THE (CUBIC TETRAGONAL) PHASE TRA 04-0753
F THE DYNAMICS OF THE BARIUM TITANATE TRANSITION IN TERMS OF NUCLEATION. (CONCEPTS)   INTERPRETATION O 04-0606
               THICKNESS DEPENDENCE OF THE NUCLEATION FIELD OF TRIGLYCINE SULFATE.   27-3036
FRONT, AND OBSERVATIONS IN BARIUM TITANATE.   NUCLEATION MODELS FOR A FIRST ORDER TRANSITION 04-0608
E SULFATE.   MOVEMENT OF DOMAIN WALLS AND THE NUCLEATION OF DOMAINS IN CRYSTALS OF TRIGLYCIN 27-3071
Y, WITH PHOTOS)   WALL MOTION AND NUCLEATION OF DOMAINS IN TGS CRYSTALS. (SUMMAR 27-3072
RIUM ON TRIGLYCINE SULFATE AND SODIUM NITRITE SUR/ THIN FILM NUCLEATION ON FERROELECTRIC SUBSTRATES. (TELLU 16-2394
- TETRAGONAL PHASE TRANSITIONS.   NUCLEATION PROCESS IN BARIUM TITANATE AT CUBIC 04-0754
LICATIONS TO PYROELECTRIC DETECTO/ LASER STUDY OF REVERSIBLE NUCLEATION SITES IN TRIGLYCINE SULFATE AND APP 27-3011
AXATION TIMES IN FERROELECTRIC/ TONE BURSTING OF QUADRUPOLAR NUCLEI. DETERMINATION OF THE SPIN LATTICE REL 11-1829
E T/ ACOUSTIC AND MAGNETIC EFFECTS INVOLVING THE FLUORINE-19 NUCLEI IN ANTIFERROMAGNETIC POTASSIUM MANGANES 33-3390

NUCLEAR MAGNETIC RESONANCE AND X-RAY STUDIES O 24-2894

/BARIUM TITANATE. (AND MICROSCOPIC MODEL OF LOCALLY ORDERED "NUCLEI", PHASE STRUCTURE ABOVE CURIE POINT, R/ 04-0647
PARAMETER OF FERROELECTRIC SODIUM NITRITE. (AT THE SODIUM-23 NUCLEUS) /FIELD GRADIENT AND LONG RANGE ORDER 16-2395
AGING PROCESSES IN BARIUM TITANATE (DECREASE IN NUMBER OF 180 DEG DOMAINS). 04-0911
E/ OXYGEN CONCENTRATION CELL MEASUREMENTS OF IONIC TRANSPORT NUMBERS IN PZT (LEAD ZIRCONATE TITANATE) FERRO 09-1512

O

LONGITUDINAL ELECTROOPTIC EFFECT IN OBLIQUE CUT STRONTIUM BARIUM NIOBATE PLATES. 08-1472
LONGITUDINAL ELECTROOPTIC EFFECT IN OBLIQUE CUT STRONTIUM BARIUM NIOBATE PLATES. 13-2063
LONGITUDINAL ELECTROOPTICAL EFFECT IN OBLIQUE SECTIONS OF LITHIUM NIOBATE. 11-1863
OF) FERROELECTRICITY. (ITS NATURE AND THE PREDICTION OF ITS OCCURANCE) DO WE REALLY UNDERSTAND (THE CAUSE 03-0510
TRAGONAL TRANSITION, AND PRO/ MONODOMAIN STRONTIUM TITANATE (OCCURANCE IN THIN (110) PLATES BELOW CUBIC- TE 06-1164
NUCLEAR MAGNETIC RESONANCE STUDIES OF OXYGEN OCTAHEDRA FERROELECTRICS. 13-2059
/D BIREFRINGENCE AND POLARIZATION) OF LEAD CONTAINING OXYGEN OCTAHEDRA FERROELECTRICS: CERAMIC LEAD BARIUM/ 13-2039
/ ABOVE FERROELECTRIC CRITICAL TEMPERATURE (ROTATIONS OF BO6 OCTAHEDRA IN ABO(3) PEROVSKITES, STRONTIUM TI/ 06-1205
ELECTROMECHANICAL PROPERTIES OF OXYGEN OCTAHEDRAL FERROELECTRIC CRYSTALS. 13-2061
STRUCTURE OF LEAD/ ORDERING OF MAGNESIUM AND NIOBIUM IN THE OCTAHEDRAL POSITIONS OF THE "CUBIC" PEROVSKITE 08-1379
IC MATERIALS. (HETEROTYPES, REVIEW) ROLE OF THE REGULAR OCTAHEDRON AND LARGER POLYHEDRA IN FERROELECTR 02-0094
ERATURE. (BARIUM(0-1./ MICROWAVE DIELECTRIC LOSSES IN OXYGEN OCTAHEDRON FERROELECTRICS ABOVE THE CURIE TEMP 04-0559
TE. OHMIC CONTACTS TO SEMICONDUCTING BARIUM TITANA 04-0764
ELECTROOPTICAL METHOD. (ANTIFERROELECTRIC) ANTIPOLARIZATION IN COPPER FORMATE CRYSTALS BY THE 26-2925
TE. COEXISTENCE OF THE ANTIPOLARIZED AND POLARIZED STATE IN BISMUTH TITANA 08-1422
FERROELECTRIC CERAMIC CYLINDERS. (HYDROPHONE ELEMENTS) OPEN CIRCUIT SENSITIVITY OF AXIALLY POLARIZED 37-3652
FERROELECTRIC CERAMIC CYLINDERS. (HYDROPHONE ELEMENTS) OPEN CIRCUIT SENSITIVITY OF RADIALLY POLARIZED 37-3651
FERROELECTRIC CERAMIC HOLLOW SPHERES. (HYDROPHONE ELEMENTS) OPEN CIRCUIT SENSITIVITY OF RADIALLY POLARIZED 37-3653
ASYMMETRIC MINOR LOOP AND TRANSIENT OPEN LOOP COMPUTATIONS FOR FERROELECTRICS. 03-0296
IELECTRIC. (THEO/ INPUT ADMITTANCE AND FIELDS RADIATED BY AN OPEN WAVEGUIDE FILLED WITH HIGH PERMITTIVITY D 03-0359
AMMONIUM DIHYDROGEN PHOSPHATE CRYSTALS OF LIGHT FROM A LASER OPERATING IN THE FREE GENERATION MODE. /E AND 17-2717
LEAD TITANATE CERAMIC RESONATORS OPERATING IN VHF BAND. 05-1069
ALS OF ANTIMONY SULFOIODIDE. OPPOSING FERROELECTRIC DOMAINS IN SINGLE CRYST 15-2252
EN PHOSPHATE AND POTAS/ LIGHT SCATTERING STUDIES OF THE SOFT OPTIC AND ACOUSTIC MODES OF POTASSIUM DIHYDROG 17-2497
MICROTWINNING (AND RESISTANT TO L/ NEW (UNIAXIAL) NONLINEAR OPTIC CRYSTAL BARIUM LITHIUM NIOBATE HAVING NO 13-1958
M TITANATE. PRESSURE DEPENDENCE OF ACOUSTIC MODE- SOFT OPTIC MODE INTERACTIONS IN FERROELECTRIC BARIU 04-0893
M TITANATE. ACOUSTIC SOFT OPTIC MODE INTERACTIONS IN FERROELECTRIC BARIU 04-0671
ACOUSTIC PHONON SOFT OPTIC PHONON INTERACTION IN BARIUM TITANATE. 04-0806
MANIUM(3) OXIDE(11). SOFT OPTIC PHONON MODE IN FERROELECTRIC LEAD(5) GER 35-3495
L PROPERTIES TO THE SYMMETRY OF POTASSIUM IODATE. (NONLINEAR OPTIC, REFRACTIVE INDICES) RELATION OF PHYSICA 08-1296
/WTH OF HIGH QUALITY LITHIUM NIOBATE CRYSTALS (FOR NONLINEAR OPTICS WITH LOW BIREFRINGENCE VARIATIONS) FRO/ 11-1688
RTIES OF BARIUM TITANATE. (RESISTIVITY, SEEBECK COEFFICIENT, OPTICAL ABSORPTION) SOME SEMICONDUCTING PROPE 04-0919
TRIC BORACITES. (IRON, NICKEL, OR / CRYSTAL FIELD THEORY AND OPTICAL ABSORPTION (CALCULATIONS) OF FERROELEC 14-2095
POLARIZATION FLUCTUATIONS AND THE OPTICAL ABSORPTION EDGE IN BARIUM TITANATE. 04-1028
OPTICAL ABSORPTION EDGE OF STRONTIUM TITANATE. 06-1187
AROUND THE 105 DEGREES K PHASE TRANSITION. OPTICAL ABSORPTION EDGE OF STRONTIUM TITANATE 06-1099
PHOTOEMISSION OF GLYCOL SULFATE. OPTICAL ABSORPTION, ELECTRIC CONDUCTIVITY AND 19-2780
TITANATE. FREE CARRIER OPTICAL ABSORPTION IN NIOBIUM DOPED STRONTIUM 06-1214
E CUBIC TO TETRAGONAL TRANSITION. DISTORTION ENHANCED OPTICAL ABSORPTION IN STRONTIUM TITANATE AT TH 06-1109
TES. CRYSTAL FIELD THEORY AND OPTICAL ABSORPTION OF COBALT AND NICKEL BORACI 14-2096
LE CRYSTALS AND (POTASSIUM SODIUM/ DIELECTRIC PROPERTIES AND OPTICAL ABSORPTION OF POTASSIUM TANTALATE SING 08-1272
IN STRONTIUM TITANATE AND LANTHA/ PARAMAGNETIC RESONANCE AND OPTICAL ABSORPTION OF TRANSITION ELEMENT IONS 06-1165
CKEL CHLORINE AND NICKEL IODINE BORACITES (AND NICKEL DOPED/ OPTICAL ABSORPTION SPECTRA OF NICKEL(2+) IN NI 14-2074
M TITANATE. OPTICAL ABSORPTION SPECTRA OF REDUCED STRONTIU 06-1245
G K) OF FERROELECTRIC IRON(2+) BORACITES. (CHLORIDE, BROMID/ OPTICAL ABSORPTION (625-25 MICRONS, 77-650 DE 14-2097
L CRYSTALS: METHYL MESITYL OXIDE OXALATE AND SODIUM NITRITE. OPTICAL ACTIVITY IN NON-ENANTIOMORPHOUS BIAXIA 14-2093
RONTIUM PROPIONATE. (TRANSITION AT 209 DEGREES K) OPTICAL ACTIVITY OF FERROELECTRIC DICALCIUM ST 27-3040
ANIUM(3) OXIDE(11) SINGLE CRYSTALS. OPTICAL ACTIVITY OF FERROELECTRIC LEAD(5) GERM 35-3503
S (PYROELECTRICITY, TGS, ROCHELLE SALT). OPTICAL ACTIVITY OF SOME FERROELECTRIC CRYSTAL 27-3017
ON RESONANCE, ABSORPTION AND LUMINESCENCE SPECTRA, INFRARED, OPTICAL, AND ACOUSTIC SCATTERING). /S (CYCLOTR 03-0427
YSTAL BARIUM TITANATE WITH NIOBIUM ADDITIONS. OPTICAL AND DIELECTRIC PROPERTIES OF SINGLE CR 04-0875
STALS IN THE TUNGSTEN BRONZE TYPE) FERR/ (X-RAY) STRUCTURAL (OPTICAL) AND DIELECTRIC STUDIES OF (SINGLE CRY 13-1947
RAMICS. RECENT IMPROVEMENTS IN THE OPTICAL AND ELECTROOPTIC PROPERTIES OF PLZT CE 09-1529
/RROELECTRIC TRANSITION IN AMMONIUM SULFATE, THE DIELECTRIC, OPTICAL, AND ELECTROOPTICAL PROPERTIES IN THE/ 19-2745
CTRICS AND FERROELECTRICS. (REVIEW, 195 REFERENCE/ NONLINEAR OPTICAL AND ELECTROOPTICAL PROPERTIES OF DIELE 02-0081
SODIUM NIOBATE. (OPTICAL ABSORPTION, REFRACTIVE INDEX, DIE/ OPTICAL AND FERROELECTRIC PROPERTIES OF BARIUM 13-2024
UNGSTEN BRONZE SYSTEM) POTASSIUM SODIUM BARIUM NIOBATE (SEC/ OPTICAL AND FERROELECTRIC PROPERTIES OF (THE T 13-2028
TRANSITION METAL DOPED LITHIUM NIOBATE. OPTICAL AND HOLOGRAPHIC STORAGE PROPERTIES OF 11-1796
/HARACTERISTICS OF BARIUM TITANATE PYROELECTRIC DETECTOR FOR OPTICAL AND INFRARED RADIATION (0.5 TO 2 MICR/ 04-0751
ECTRIC GUANIDINIUM ALUMINUM SULFATE HEXAHYDRATE CLEAVAGES A/ OPTICAL AND INTERFEROMETRIC STUDIES ON FERROEL 21-2816
) EUROPIUM SILICATE. (UNSTABLE FERROELECT/ CRYSTALLOGRAPHIC, OPTICAL, AND MAGNETIC PROPERTIES OF (MONOCLINIC 36-3620
RIC GADOLINIUM MOLYBDATE. ELECTRICAL, OPTICAL, AND MECHANICAL BEHAVIOR OF FERROELECT 30-3267
/CTRIC CONTROL OF ELASTIC COMPLIANCE, NICKEL IODINE BORACITE OPTICAL ANISOTROPY, BISMUTH TITANATE, POSSIBL/ 02-0060
CITE DUE TO QUADRUPOLE TRANSITIONS. OPTICAL ANISOTROPY OF CUBIC NICKEL IODINE BORA 14-2093
BORACITE. OPTICAL ANISOTROPY OF THE CUBIC COBALT IODINE 14-2094
OPTICAL ANOMALIES IN FERROELECTRIC CRYSTALS. 02-0082
PHATE CRYSTALS. EFFECT OF ANNEALING ON OPTICAL ANOMALIES OF POTASSIUM DIHYDROGEN PHOS 17-2427
PREPARATION OF POTASSIUM NIOBATE SINGLE CRYSTAL FOR OPTICAL APPLICATION. 08-1332
T, STRU/ GROWTH OF BARIUM SODIUM NIOBATE SINGLE CRYSTALS FOR OPTICAL APPLICATIONS. (BY PULLING FROM THE MEL 13-2045
IOBATE SINGLE CRYSTALS IN THE REG/ ANOMALOUS BEHAVIOR OF THE OPTICAL ASBSORPTION COEFFICIENT OF LEAD MAGNON 13-1974
HIUM TANTALATE CRYSTAL. OPTICAL BEAM STEERING USING A MULTICHANNEL LIT 12-1881
GADOLINIUM MOLYBDATE. AN EXPLANATION OF ANOMALOUS OPTICAL BEHAVIOR OF THE IMPROPER FERROELECTRIC 30-3292
THE CURIE TEMPERATURE OF A FERROELECTRIC SEMICON/ METASTABLE OPTICAL CHARGE EXCHANGE AND "FROZEN" SHIFT OF 15-2160
ACTIVE COMPENSATORS FOR FERROELECTRIC OPTICAL CIRCUITS. 37-3675
ARMONIC GENERATION IN LITHIUM IODATE. LARGE NONLINEAR OPTICAL COEFFICIENT AND PHASE MATCHED SECOND H 08-1405
ITRITE. TEMPERATURE DEPENDENCE OF NONLINEAR OPTICAL COEFFICIENTS IN FERROELECTRIC SODIUM N 16-2274
M NIOBATE OR TANTALATE, BARIUM / RELATIVE SIGNS OF NONLINEAR OPTICAL COEFFICIENTS OF POLAR CRYSTALS (LITHIU 02-0098
HOSPHATE. ABSOLUTE SIGNS OF THE NONLINEAR OPTICAL COEFFICIENTS OF POTASSIUM DIHYDROGEN P 17-2597
ON OF LEAD LANTHANATE ZIRCONATE CERAMICS FOR APPLICATIONS IN OPTICAL COMMUNICATIONS. EVALUATI 07-1254
CTROOPTIC MODUL/ (HIGH SPEED, SMALL APERTURE) MODULATORS FOR OPTICAL COMMUNICATIONS (REVIEW EMPHASIZING ELE 02-0056
UM MANGANESE TRI/ VIBRATIONAL SPECTRUM AND DISPERSION OF THE OPTICAL CONSTANTS OF ANTIFERROMAGNETIC POTASSI 33-3385
ODATE. ELECTRICAL AND OPTICAL CONSTANTS OF FERROELECTRIC POTASSIUM I 08-1427
RARED REGION. OPTICAL CONSTANTS OF ICE I OVER THE ENTIRE INF 34-3492

E NEAR INFRARED TO FAR INFRARED AND IN THE RADIO FREQUENCY / OPTICAL CONSTANTS OF TRIGLYCINE SELENATE IN TH 27-3007
/IUM SODIUM NIOBATE (ELASTIC, PIEZOELECTRIC, DIELECTRIC, AND OPTICAL CONSTANTS) ROOM TEMPERATURE TO THE CU/ 13-2060
CHANGES IN BARIUM TITANATE (EXPERIMENTS TO CHARACTERIZE THE OPTICAL DAMAGE). /LLY INDUCED REFRACTIVE INDEX 04-1004
NT IN FERROELECTRIC OXIDE CRYSTALS. (PCTASS/ TRAPPING SITES, OPTICAL DAMAGE, AND THERMALLY STIMULATED CURRE 13-1926
SECONDS WHEN ILLUMINATED UNDER APPLIED ELECTR/ (REVERSIBLE) OPTICAL DAMAGE IN KDP (WITH RELAXATION TIME OF 17-2703
OPTICAL DAMAGE IN KTN. 13-1979
NATURE OF INTERNAL ELECTRIC FIELD DURING OPTICAL DAMAGE PROCESS IN LITHIUM NIOBATE. 11-1861
NIUM MOLYBDATE. OPTICAL DEVICE ELEMENT OF FERROELECTRIC GADOLI 30-3295
ERATION AND OPTICAL PARAMETRIC OSCILLATION IN FER/ NONLINEAR OPTICAL DEVICES. (BASED IN SECOND HARMONIC GEN 02-0097
A RESONATOR. EFFECT TUNABLE OPTICAL EMISSION FROM LITHIUM NIOBATE WITHOUT 11-1858
TEMPERATURE CHARACTERISTICS OF FERROELECTRIC TRIGLYCINE SE/ (OPTICAL) ENERGY GAP (AND DIELECTRIC CONSTANT) 27-3021
IN PARAELECTRIC BARIUM TITANATE. OPTICAL EVIDENCE FOR POLARIZATION FLUCTUATIONS 04-0737
EXCITON MODEL OF SPIN AND OPTICAL EXCITATIONS OF AN ANTIFERRODIELECTRIC. 03-0416
N SPECTRA OF STRONTIUM TITANATE AND POTASSIU/ COUPLED MODES (OPTICAL, FERROELECTRIC- ACOUSTIC) IN THE PHONO 06-1157
ATOMIC DISPLACEMENTS IN PEROVSKITE STRONTIUM TITANATE (AT OPTICAL FREQUENCIES, CALCULATIONS) 06-1142
RINE, ZINC CHLORINE, CADMIUM CHLORINE, CADMIUM IODINE, IRO/ (OPTICAL) FREQUENCY DOUBLING IN (MAGNESIUM CHLO 14-2080
ROTATING WAVEPLATE OPTICAL FREQUENCY SHIFTING IN LITHIUM NIOBATE. 11-1689
DETERMINATION OF THE NATURE OF THE OPTICAL GAP OF STRONTIUM TITANATE. 06-1108
FROM URBACH TAIL BEHAVIOR) OPTICAL GAP OF STRONTIUM TITANATE. (DEVIATION 06-1111
, BARIUM TITANATE, AND POTASSIUM DI/ QUANTITATIVE STUDIES OF OPTICAL HARMONIC GENERATION IN CADMIUM SULFIDE 04-0852
ATE AND POTASSIUM TANTALATE. ELECTRIC FIELD INDUCED OPTICAL HARMONIC GENERATION IN STRONTIUM TITAN 06-1203
TITANATE. GENERATION OF SECOND OPTICAL HARMONIC IN THE CUBIC PHASE OF BARIUM 04-0538
EORY, REVIEW) OPTICAL HARMONICS AND NONLINEAR PHENOMENA. (TH 02-0067
EFFECTS OF DISPERSION AND FOCUSING ON THE PRODUCTION OF OPTICAL HARMONICS. (KDP) 17-2579
HER PYROELECTRIC INSULATORS. (MECHANISM OF) OPTICAL INDEX DAMAGE IN LITHIUM NIOBATE AND OT 11-1735
ONLINEAR CRYSTALS (LITHIUM NIOBATE). EFFECT OF OPTICAL INHOMOGENEITIES ON PHASE MATCHING IN N 11-1773
ERROELEC/ AN ANALYSIS OF MULTIPLE LIGHT SCATTERING BY RANDOM OPTICAL INHOMOGENEITIES WITH APPLICATIONS TO F 37-3634
IES OF POTASSIUM DIHYDROGEN PHOSPHATE. ELECTRICAL AND OPTICAL INVESTIGATION OF FERROELECTRIC PROPERT 17-2449
/ENSION OF THE POULET- LCUDON APPROXIMATION IN THE THEORY OF OPTICAL LATTICE VIBRATIONS. (QUARTZ, BARIUM T/ 03-0380
/MENT OF SINGLE CRYSTAL BARIUM SODIUM NIOBATE AS A NONLINEAR OPTICAL MATERIAL. (GROWTH BY KYROPOULOS AND C/ 13-2021
NONLINEAR OPTICAL MATERIALS. 02-0055
TERIALS NEEDS, 118 REFS) NONLINEAR OPTICAL MATERIALS (REVIEW, DATA COLLECTION, MA 02-0049
FERROELECTRIC LIGHT VALVE ARRAYS FOR OPTICAL MEMORIES. 37-3676
AISAL. FERROELECTRICS IN OPTICAL MEMORIES AND DISPLAYS: A CRITICAL APPR 02-0039
AISAL. FERROELECTRICS IN OPTICAL MEMORIES AND DISPLAYS: A CRITICAL APPR 37-3628
T. (KDP CRYSTAL FILTER) OPTICAL MIXING OF COHERENT AND INCOHERENT LIGH 17-2669
S BY TRIGLYCINE SULFATE CRYSTAL) OPTICAL MIXING. (OF PULSED RUBY LASER EMISSION 27-2958
TE BY THE MOSSBAUER EFFECT. TEMPERATURE DEPENDENT OPTICAL MODE IN ANTIFERROELECTRIC LEAD ZIRCONA 07-1260
ELASTIC CONSTANTS AND SOFT OPTICAL MODES IN GADOLINIUM MOLYBDATE. 30-3320
EORY, STRONTIUM TITANATE DATA) OPTICAL MODES OF THE PEROVSKITE STRUCTURE. (TH 03-0304
RE DEPENDENCE OF POLARITON DISPERSION IN LITHIUM TANTALATE. (OPTICAL MODULATION VIA TEMPERATURE) TEMPERATU 12-1873
FABRICATION OF A LITHIUM TANTALATE TEMPERATURE STABILIZED OPTICAL MODULATOR. 12-1872
PERFORMANCE OF AN 11 GHZ OPTICAL MODULATOR USING LITHIUM TANTALATE. 12-1888
ELATIVE VALUES FOR QUARTZ, AMMONIUM DIHYDROGEN PHOSPHATE, A/ OPTICAL NONLINEAR SUSCEPTIBILITIES: ACCURATE R 17-2539
RITE. TEMPERATURE DEPENDENCE OF THE OPTICAL NONLINEAR SUSCEPTIBILITY OF SODIUM NIT 16-2386
PHOSPHATE, AMMONIUM DIHYDROGEN/ RELATIVE MEASUREMENT OF THE OPTICAL NONLINEARITIES OF POTASSIUM DIHYDROGEN 11-1677
TE. OPTICAL PARAMETRIC OSCILLATION IN LITHIUM IODA 08-1338
C GENERATION (IN LITHIUM NIOBATE). SIMULTANEOUS OPTICAL PARAMETRIC OSCILLATION, SECOND HARMONI 11-1859
N CUBIC PEROVSKITE STRONTIUM T/ ELECTRIC FIELD DEPENDENCE OF OPTICAL PHONON FREQUENCIES (RAMAN SCATTERING I 06-1241
WAVEVECTOR DEPENDENT RELAXATION OF AN OPTICAL PHONON IN GADOLINIUM MOLYBDATE. 30-3320
ION IN BARIUM AND STRONTIUM TITANATES AND POTASSIUM/ COUPLED OPTICAL PHONON MODE THEORY OF INFRARED DISPERS 04-0547
ASSIGNMENTS OF OPTICAL PHONON MODES IN LITHIUM NIOBATE. 11-1694
/ TO THE LEAD TITANATE ZIRCONATE SYSTEM (MEASUREMENT OF SOFT OPTICAL PHONON MODES IN TETRAGONAL FERROELECT/ 09-1497
/S, TRANSVERSE ISING MODEL, SOFT MODE COUPLING TO TRANSVERSE OPTICAL PHONON). PART-3: (POTASSIUM DIHYDROGE/ 03-0384
OF POLARITONS WITH APPLICATION TO LEAD NIOBATE. PART-1: LONG OPTICAL PHONONS. /REQUENCIES AND EIGENVECTORS 08-1284
ONY SULFCIODIDE. (POLARIZED INFRARED AND RA/ LONG WAVELENGTH OPTICAL PHONONS AND PHASE TRANSITIONS IN ANTIM 15-2106
ORDER DISORDER AND DISPLACIVE FERROELECTRICS. (ANTIMONY SUL/ OPTICAL PHONONS AND PHASE TRANSITIONS IN SOME 15-2208
PITAXI/ EFFECT OF TEMPERATURE AND PROPAGATION DIRECTION UPON OPTICAL PHONONS IN CUBIC SODIUM CHLORIDE AND E 29-3243
/RING OF CONDUCTION ELECTRONS BY TRANSVERSE AND LONGITUDINAL OPTICAL PHONONS IN IONIC FERROELECTRIC SEMICO/ 03-0292
TEMPERATURE DEPENDENT OPTICAL PHONONS IN LEAD TITANATE. 05-1083
DIRECTIONAL DISPERSION AND ASSIGNMENT OF OPTICAL PHONONS IN LITHIUM NIOBATE. 11-1695
OVSKITES. (THEORY, ABO(3) TYPE) DIELECTRIC PROPERTIES AND OPTICAL PHONONS IN PARA- AND FERROELECTRIC PER 03-0413
ATE. INTERACTIONS BETWEEN PROTON TUNNELING AND OPTICAL PHONONS IN POTASSIUM DIHYDROGEN PHOSPH 17-2592
ATE. INTERACTION BETWEEN PROTON TUNNELING AND OPTICAL PHONONS IN POTASSIUM DIHYDROGEN PHOSPH 17-2593
BOND. (LITHIUM NIOBATE, LITHIUM TANTALATE, POTASS/ NONLINEAR OPTICAL POLARIZABILITY OF THE NIOBIUM- OXYGEN 11-1734
NIC GENERATION (IN LITHIUM NIOBATE). OPTICAL PROBING OF ACOUSTIC SURFACE WAVE HARMO 11-1752
E IN POTASSIUM DIHYDROGEN PHOSPHATE CRYSTALS BY POLARIZATION OPTICAL PROCEDURE. /RVATION OF DOMAIN STRUCTUR 17-2493
NGLE CRYSTAL NEAR ITS CURIE POINT. PART-1: PIEZOELECTRIC AND OPTICAL PROPERTIES. /R OF LITHIUM TANTALATE SI 12-1877
TE PHOTOCONDUCTIVITY, CALCIUM FLUORIDE PHOTOCONDUCTIVITY AND OPTICAL PROPERTIES) /SEARCH. (STRONTIUM TITANA 06-1140
IN NEW SOLID STATE MATERIALS (INCLUDING FERROELEC/ STUDY OF OPTICAL PROPERTIES AND COLLECTIVE OSCILLATIONS 02-0104
SOME EFFECTS OF MELT STOICHIOMETRY ON THE OPTICAL PROPERTIES BARIUM SODIUM NIOBATE. 13-1902
VE INDICES) OF POTASSIUM NIOBATE (LINEAR ELECTROOPTICS). OPTICAL PROPERTIES (BIREFRINGENCE AND REFRACTI 08-1468
VERSUS TEMPERATURE, ABSORPTION SPECTRA) OF S/ ELECTRICAL AND OPTICAL PROPERTIES (CONDUCTIVITY AND MOBILITY 06-1198
FRINGENCE OF (HOT PRESSED) FERROELECTRIC CERAMICS. (PZT) OPTICAL PROPERTIES (ELECTRICALLY VARIABLE BIRE 09-1554
/F LUMINOUS WAVES IN CUBIC ELECTROOPTIC CRYSTALS. (NONLINEAR OPTICAL PROPERTIES, ELECTROOPTIC COEFFICIENTS) 02-0044
THE INFRARED. OPTICAL PROPERTIES OF ANTIMONY SELENOIODIDE IN 15-2219
GLASSY AND CRYSTALLINE STATES. PRODUCTION AND SOME OPTICAL PROPERTIES OF ANTIMONY SULFOBROMIDE IN 15-2204
N PARAELECTRIC AND FERROELECTRIC/ ELECTRON STATE DENSITY AND OPTICAL PROPERTIES OF ANTIMONY SULFOIODIDE. (I 15-2204
SOME EFFECTS OF MELT STOICHIOMETRY ON THE OPTICAL PROPERTIES OF BARIUM SODIUM NIOBATE. 13-1903
D TITANATE THIN SECTIONS. ELECTRICAL AND OPTICAL PROPERTIES OF (BARIUM, STRONTIUM) MIXE 04-0603
PRODUCTION METHODS AND SOME OPTICAL PROPERTIES OF BISMUTH TELLUROIODIDE. 15-2141
NONLINEAR OPTICAL PROPERTIES OF CRYSTALS. 03-0379
E TRIHYDROGEN DISELENITE FAMILY. (ESPECI/ (FERROELECTRIC AND OPTICAL PROPERTIES OF) CRYSTALS OF THE ALKALIN 22-2848
/RYSTALLOGRAPHIC, PHYSICAL, THERMAL MAGNETIC, ELECTRONIC AND OPTICAL PROPERTIES OF EACH OF THE II-VI BINAR 01-0025
(SECOND HARMONIC GENERATION, OPTICAL PARAMETRIC O/ NONLINEAR OPTICAL PROPERTIES OF FERROELECTRIC CRYSTALS. 03-0383
/ERRED TEMPERATURE DEPENDENCES OF ELECTRICAL, MECHANICAL AND OPTICAL PROPERTIES OF FERROELECTRIC FERROELAS/ 30-3253
UM(4) OXIDE(12). NONLINEAR OPTICAL PROPERTIES OF FERROELECTRIC LEAD NIOBI 13-2023
ATE. NONLINEAR OPTICAL PROPERTIES OF FERROELECTRIC LEAD TITAN 05-1080
ATE. NONLINEAR OPTICAL PROPERTIES OF FERROELECTRIC LEAD TITAN 05-1079
RMANIUM(3) OXIDE(11). NONLINEAR OPTICAL PROPERTIES OF FERROELECTRIC LEAD(5) GE 35-3506

OBATE AND TANTALATE.   TEMPERATURE DEPENDENCE OF THE OPTICAL PROPERTIES OF FERROELECTRIC LITHIUM NI 11-1770
RITE.   NONLINEAR OPTICAL PROPERTIES OF FERROELECTRIC SODIUM NIT 16-2272
ONS FROM PHENOMENOLOGICAL THEORY. (AMMONIUM SULFATE, AMMONI/ OPTICAL PROPERTIES OF FERROELECTRICS. DIGRESSI 02-0040
  THEIR DEVICE APPLICATIONS.   OPTICAL PROPERTIES OF GADOLINIUM MOLYBDATE AND 30-3299
  THEIR DEVICE APPLICATIONS.   OPTICAL PROPERTIES OF GADOLINIUM MOLYBDATE AND 30-3300
  TEREIUM MOLYBDATE.   NONLINEAR OPTICAL PROPERTIES OF GADOLINIUM MOLYBDATE AND 30-3305
H TITANATE CRYSTALS.   NONLINEAR OPTICAL PROPERTIES OF LEAD TITANATE AND BISMUT 04-0941
TE(0.67) SINGLE CRYSTAL.   DIELECTRIC OPTICAL PROPERTIES OF LEAD ZINCATE(0.33) NIOBA 13-2065
DE(11) AND ITS ISOMORPHOUS COMPOUND LEAD / FERROELECTRIC AND OPTICAL PROPERTIES OF LEAD(5) GERMANIUM(3) OXI 35-3501
                                                NONLINEAR OPTICAL PROPERTIES OF MOLECULAR CRYSTALS.   02-0078
IDIUM DIHYDROGEN PHOSPHATE CRYSTALS.   ELECTROOPTICAL AND OPTICAL PROPERTIES OF PARTIALLY DEUTERATED RUB 17-2698
ORIDE FILMS.   PYROELECTRIC AND NONLINEAR OPTICAL PROPERTIES OF POLED POLYVINYLIDENE FLU 36-3595
E.   NONLINEAR OPTICAL PROPERTIES OF POTASSIUM LITHIUM NIOBAT 13-2027
ES (TRANSPARENT, TETRAGONAL TUNGSTEN BRONZE STRUC/ NONLINEAR OPTICAL PROPERTIES OF POTASSIUM LITHIUM NIOBAT 13-2026
ES.   CRYSTAL GROWTH AND NONLINEAR OPTICAL PROPERTIES OF POTASSIUM LITHIUM NIOBAT 13-2015
  CRYSTALS. (REFRACTIVE INDICES)   NONLINEAR OPTICAL PROPERTIES OF POTASSIUM NIOBATE SINGLE 08-1454
ATE.   OPTICAL PROPERTIES OF POTASSIUM TANTALATE NIOB 36-3593
/VE STUDY OF THE CRYSTALLOGRAPHIC, DIELECTRIC, AND NONLINEAR OPTICAL PROPERTIES OF THE ABC NIOBATES (WHERE/ 04-0912
H TELLURDIODIDE.   OPTICAL PROPERTIES OF THE SEMICONDUCTOR BISMUT 15-2169
                                                        OPTICAL PROPERTIES OF THIOUREA.   25-2913
C GLASS CERAMIC SYSTEMS.   DIELECTRIC AND OPTICAL PROPERTIES OF TRANSPARENT FERROELECTRI 04-0576
C / EFFECTS OF GRAIN SIZE AND POROSITY ON THE ELECTRICAL AND OPTICAL PROPERTIES OF TRANSPARENT FERROELECTRI 13-2004
ND BELOW TRANSITION TEMPERATURE. (NO COCHRAN TYPE / INFRARED OPTICAL PROPERTIES OF VANADIUM DIOXIDE ABOVE A 36-3526
ND STRUCTURE) OF SOME ABC(3) FERRO/ VISIBLE AND ULTRAVIOLET OPTICAL PROPERTIES (REFLECTANCE SPECTRA AND BA 08-1383
G COMPOSITION WHERE LIC/ PROGRESS TOWARDS THE GROWTH OF GOOD OPTICAL QUALITY BARIUM SODIUM NIOBATE (BY USIN 13-2047
ECIALLY ON LITHIUM NIOBATE AND LIT/ CHEMICAL COMPOSITION AND OPTICAL QUALITY OF FERROELECTRIC CRYSTALS, ESP 11-1780
YSTALS.   STOICHIOMETRY AND OPTICAL QUALITY OF LITHIUM TANTALATE SINGLE CR 12-1883
AMMONIUM TRIHYDROGEN PERIODATE CRYSTALS, PULSED MAGNETIC AND OPTICAL RESONANCE. /ELAXATION) EXPERIMENTS IN 36-3577
D(5) GERMANIUM(3) OXIDE(11).   FERROELECTRIC OPTICAL ROTATION DOMAINS IN SINGLE CRYSTAL LEA 36-3550
) GERMANIUM(3) OXIDE(11) SINGLE CRYSTAL.   SWITCHING OF OPTICAL ROTATORY POWER IN FERROELECTRIC LEAD(5 36-3570
              LONG RANGE ORDER IN SODIUM NITRITE BY OPTICAL SECOND HARMONIC GENERATION.   16-2314
NTS OF POLAR (PYROELECTRIC) CRY/ ABSOLUTE SIGNS OF NONLINEAR OPTICAL (SECOND HARMONIC GENERATION) COEFFICIE 02-0095
OELECTRIC POLARIZATION REVERSAL. (SODIUM NITRITE)   OPTICAL SECOND HARMONIC GENERATION DURING FERR 16-2384
SULFOIODIDE.   OPTICAL SECOND HARMONIC GENERATION IN ANTIMONY 15-2164
CTRIC CRYSTALS. (KDP TYPE)   OPTICAL SECOND HARMONIC GENERATION IN PIEZOELE 17-2582
LIDENE FLUORIDE FILMS (AFTER COOLING ST/ PYROELECTRICITY AND OPTICAL SECOND HARMONIC GENERATION IN POLYVINY 36-3530
IUM DIHYDROGEN PHOSPHATE (AND DEU/ TEMPERATURE DEPENDENCE OF OPTICAL (SECOND) HARMONIC GENERATION IN POTASS 17-2695
PERTIES,) FUNCTION AND APPLICATIONS OF GADOLINIUM MOLYBDATE (OPTICAL SHUTTER, COLOR MODULATOR).   (PRO 30-3298
          RADIATION DAMAGE IN LITHIUM NIOBATE. (CAPACITANCE, OPTICAL SPECTRA, ESR, CELL PARAMETERS)   11-1705
ANUM ALUMINATE.   OPTICAL SPECTRA OF CHROMIUM(3+) PAIRS IN LANTH 36-3616
LITY IN LITHIUM NIOBATE.   NONLINEAR OPTICAL SPECTROSCOPY OF THIRD ORDER SUSCEPTIBI 11-1855
          BISMUTH TITANATE FERROELECTRIC- PHOTOCONDUCTOR OPTICAL STORAGE MEDIUM FURTHER DEVELOPMENTS.   08-1371
WRITE)   BISMUTH TITANATE PHOTOCONDUCTOR OPTICAL STORAGE MEDIUM. (HIGH RESOLUTION READ 08-1445
ATE.   FERROELECTRIC PHOTOCONDUCTOR OPTICAL STORAGE MEDIUM UTILIZING BISMUTH TITAN 04-0769
UM NIOBATE SINGLE CRYSTALS.   OPTICAL STUDIES OF DOMAIN STRUCTURE OF POTASSI 08-1320
F ULTRASONIC SURFACE WAVES BY THIN GOLD FILM LAYERS ON LITH/ OPTICAL STUDIES OF REFRACTION AND REFLECTION O 11-1789
ROELECTRIC DICALCIUM STRONTIUM HEXAACETATE.   X-RAY AND OPTICAL STUDIES ON THE PHASE TRANSITION OF FER 36-3584
ATE.   OPTICAL STUDY OF DOMAIN WALLS IN BISMUTH TITAN 08-1301
PERATURE.   (PREPARATION AND ELECTRICAL AND OPTICAL) STUDY OF TUNGSTEN TRIOXIDE AT LOW TEM 10-1643
SSIUM DIHYDROGEN OR DEUTERIUM PHOSPHATE (INFRARED PHOSPHATE/ OPTICAL STUDY ON THE PHASE TRANSITIONS IN POTA 17-2646
      DIRECT OBSERVATION OF SINGLE DOMAIN STRONTIUM TITANATE (OPTICAL STUDY UNDER UNIAXIAL STRESS).   06-1112
HYDRATE.   NONLINEAR OPTICAL SUSCEPTIBILITY OF LITHIUM FORMATE MONO 26-2945
  BISMUTH(4) TITANIUM(3) OXIDE(12).   ELECTRICAL AND OPTICAL SWITCHING PROPERTIES OF SINGLE CRYSTAL 35-3517
M TUNGSTATE.   OPTICAL TRANSITION TO THE FERMI LEVEL IN SODIU 10-1634
IONS IN ANTIFERRODIELECTRIC ELECTRIC AND MAGNETIC MOMENTS OF OPTICAL TRANSITIONS. /CTRONIC AND SPIN EXCITAT 03-0418
           TUNGSTEN BRONZE TYPE FERROELECTRICS FOR OPTICAL USE (REVIEW, 22 REFS).   02-0054
HOLOGRAMS.   OPTICALLY ERASABLE AND REWRITABLE SOLID STATE 37-3638
ETHOD OF DETERMINING HALF WAVE VOLTA/ ELECTROOPTIC EFFECT IN OPTICALLY HETEROGENEOUS CRYSTALS (THEORY AND M 03-0461
UM NIOBATE AND LITHIUM TANTALATE.   INHOMOGENEITY OF THE OPTICALLY INDUCED INDEX OF REFRACTION IN LITHI 11-1664
IOBATE, PROUSTITE, AND LITHIUM IODATE.   OPTICALLY INDUCED PHYSICAL DAMAGE TO LITHIUM N 08-1328
BARIUM TITANATE (EXPERIMENTS TO CHARACTERIZE THE OPTICAL DA/ OPTICALLY INDUCED REFRACTIVE INDEX CHANGES IN 04-1004
BARIUM TITANATE.   OPTICALLY INDUCED REFRACTIVE INDEX CHANGES IN 04-1005
AND MEMORY APPLICATIONS.   A NEW METHOD OF OPTICALLY READING DOMAINS IN BISMUTH TITANATE 08-1298
S. (KDP)   OPTIMISATION OF ELECTROOPTIC CRYSTAL MODULATOR 17-2541
TRIGLYCINE FLUOBERYLLATE) AT PHASE TRANSITIONS OF THE SECOND ORDER. /SINGLE CRYSTAL TRIGLYCINE SULFATE AND 27-3137
ASE TRANSITION OF FERROELECTRIC POTASSIUM TANTALATE. (SECOND ORDER)   CHARACTER OF PH 08-1307
/SICAL ANHARMONIC OSCILLATOR. A UNIFIED OSCILLATOR MODEL FOR ORDER DISORDER AND DISPLACIVE FERROELECTRICS. 03-0405
(ANTIMONY SUL/ OPTICAL PHONONS AND PHASE TRANSITIONS IN SOME ORDER DISORDER AND DISPLACIVE FERROELECTRICS. 15-2208
ATIONS.   ENERGY SPECTRUM OF AN ORDER DISORDER FERROELECTRIC CONTAINING DISLOC 03-0485
IVE CORRECTION.   STATISTICAL PERTURBATION THEORY OF ORDER DISORDER FERROELECTRICS: FIRST PERTURBAT 03-0399
           DIELECTRIC DISPERSION IN ORDER DISORDER FERROELECTRICS. (THEORY)   03-0373
(FREQUENCY DEPENDENT SUSCEPTIBILITY).   COUPLED ORDER DISORDER PHONON SYSTEM WITH DAMPING: KDP 17-2667
SALT, SILVER PERIODATE, / FERROELECTRICS. (THEORY, DYNAMICAL ORDER DISORDER PROPERTIES, KDP, TGS, ROCHELLE 02-0050
       STATISTICAL CONSIDERATIONS OF THE ORDER DISORDER REACTION IN AMMONIUM PERIODATE. 36-3581
         INFLUENCE OF IMPURITIES ON THE ORDER DISORDER REACTION IN HEXAGONAL ICE I(H). 34-3473
PHOSPHATE TYPE FERROELECTRICS.   ORDER DISORDER THEORY OF POTASSIUM DIHYDROGEN 17-2685
S. (THEORY)   CRITICAL DYNAMICS OF THE ORDER DISORDER TRANSFORMATION IN FERROELECTRIC 03-0281
/TICAL SCATTERING IN TRIGLYCINE SULFATE AND SODIUM NITRATE. (ORDER DISORDER TRANSITION, FERROELECTRIC AND / 16-2400
MECHANISM)   ENTROPY OF AN ORDER DISORDER TRANSITION. (IN FERROELECTRICS, 03-0192
TRIGLYCINE SULFATE, HIGHLY ACCURATE TEMPERATURE CONTROL)   ORDER DISORDER TRANSITION IN FERROELECTRICS. (27-2982
N SELENATE.   MODELS FOR THE ORDER DISORDER TRANSITION IN SODIUM TRIHYDROGE 22-2862
TWO SUBLATTICE MODEL OF INTERACTING DIPOLES)   THE ORDER DISORDER TRANSITION OF SODIUM NITRITE. (16-2287
/OF (ANOMALOUS) DIELECTRIC RELAXATION IN SEVERAL CRYSTALS OF ORDER DISORDER TYPE. (COPPER FORMATE TETRAHYD/ 02-0087
F THE TEMPERATURE DEPENDENCE OF NMR SPECTRA SECOND MOMENT IN ORDER DISORDER TYPE FERROELECTRICS. NATURE O 03-0460
ROELECTRICS   DYNAMICS OF ORDER DISORDER TYPE FERROELECTRICS AND ANTIFER 03-0179
ON FLUCTUATI/ SPIN LATTICE RELAXATION BY QUASI SPIN WAVES IN ORDER DISORDER TYPE FERROELECTRICS: POLARIZATI 26-2928
TTICE RELAXATION TIME NEAR THE TRANSITION TEMPERATURE IN THE ORDER DISORDER TYPE FERROELECTRICS. (THEORY) / 03-0488
T RANGE INTER/ CRITICAL BEHAVIOR OF THE PSEUDOSPIN MODEL FOR ORDER DISORDER TYPE FERROELECTRICS WITH A SHOR 03-0423
PERMITTIVITY DISPERSION PECULIARITIES IN FERROELECTRICS WITH ORDER DISORDER TYPE PHASE TRANSITION. /ECTRIC 03-0412
           TEMPERATURE DEPENDENCE OF THE FIRST ORDER ELASTIC CONSTANTS OF STRONTIUM TITANATE. 06-1160

RAMAN SPECTROSCOPY OF PHONON INSTABILITY IN FIRST	ORDER FERROELECTRIC PHASE TRANSITIONS.	03-0163
THERMODYNAMIC CLASSIFICATION OF SECOND	ORDER FERROELECTRIC PHASE TRANSITIONS.	03-0303
CALORIMETRIC INVESTIGATIONS OF SECOND	ORDER FERROELECTRIC TRANSITIONS.	03-0435
FERROELECTRIC PHASE TRANSITION OF FIRST	ORDER IN BARIUM TITANATE.	04-1030
ONIC GENERATION. LONG RANGE	ORDER IN SODIUM NITRITE BY OPTICAL SECOND HARM	16-2314
ANISOTROPIC SUBSTRATES: GOLD ON LIT/ FUNDAMENTAL AND HIGHER	ORDER MODE SURFACE WAVE PROPAGATION ON LAYERED	11-1846
LOCYANINE IN THE FERROELECTRIC STATE. FIRST	ORDER NEGATIVE PHOTODIELECTRIC EFFECT IN PHTHA	36-3539
DIELECTRIC SUSCEPTIBILITY AND THE	ORDER OF FERROELECTRIC PHASE TRANSITIONS.	03-0223
N OF POTASSIUM DIHYDROGEN PHOSPHATE.	ORDER OF THE FERROELECTRIC PHASE TRANSFORMATIO	17-2555
EXPERIMENTS ON THE	ORDER OF THE TRANSITION IN KDP.	17-2603
SE TRANSITION OF POTASSIUM DIHYDROGEN PHOSPHATE. PART-1: THE	ORDER OF TRANSITION. /OF THE FERROELECTRIC PHA	17-2557
SED STRONTIUM TITANATE. (EPR OF IRON- VANADIUM PAIRS)	ORDER PARAMETER AND PHASE TRANSITIONS OF STRES	06-1170
E. (AT THE SODIUM-23/ ELECTRIC FIELD GRADIENT AND LONG RANGE	ORDER PARAMETER OF FERROELECTRIC SODIUM NITRIT	16-2395
'S THEORY OF FERROELECTRICS. A NEW TYPE OF SECOND	ORDER PHASE TRANSITION DERIVED FROM DEVONSHIRE	03-0226
TO THE FERROMAGNETIC AND FERROELECTRIC STA/ THEORY OF SECOND	ORDER PHASE TRANSITION FROM THE FERROMAGNETIC	03-0395
PHOSPHATE CRYSTALS. FIRST	ORDER PHASE TRANSITION IN POTASSIUM DEUTERIUM	17-2677
L MODEL OF A FERROELECTRIC. SECOND	ORDER PHASE TRANSITION IN THE THREE DIMENSIONA	03-0450
ON RELAXATION PROCESSES NEAR THE SECOND	ORDER PHASE TRANSITION POINT.	03-0385
INGLE DOMAIN POLARIZATION OF A FERROELECTRIC HAVING A SECOND	ORDER PHASE TRANSITION (THEORY). S	03-0202
THEORY OF SECOND	ORDER PHASE TRANSITIONS.	03-0364
TIN TELLURIDE, POTASSIUM TANTALATE) THEORY OF SECOND	ORDER PHASE TRANSITIONS AT LOW TEMPERATURES. (	03-0433
ALS. DOUBLE HYSTERESIS LOOPS AND FIRST	ORDER PHASE TRANSITIONS IN FERROELECTRIC CRYST	03-0456
ELF CONSISTENT POLARIZATION SCHEME FOR ANH/ FIRST AND SECOND	ORDER PHASE TRANSITIONS IN FERROELECTRICS: A S	03-0355
TIC RESONANCE IN ROCHELLE SALT. SECOND	ORDER QUADRUPOLE SHIFTS OF THE SODIUM-23 MAGNE	28-3186
NONLINEAR OPTICAL SPECTROSCOPY OF THIRD	ORDER SUSCEPTIBILITY IN LITHIUM NIOBATE.	11-1855
R CALCIUM TITANATE OR ZIRCONATE, TEST OF SHORT OR LONG RANGE	ORDER THEORIES). /AIN PEROVSKITES (STRONTIUM O	03-0414
OGICAL THEORY OF ANTIFERROELECTRIC TRANSITION. PART-2: FIRST	ORDER TRANSITION. PHENOMENOL	03-0404
RATION OF HARMONICS IN A FERROELECTRIC CRYSTAL WITH A SECOND	ORDER TRANSITION ABOVE THE CURIE POINT. / GENE	03-0331
RIUM TITANATE. NUCLEATION MODELS FOR A FIRST	ORDER TRANSITION FRONT, AND OBSERVATIONS IN BA	04-0608
HATE. DIELECTRIC EVIDENCE OF A FIRST	ORDER TRANSITION IN POTASSIUM DIHYDROGEN PHOSP	17-2602
ION AND VELOCITY OF ULTRASOUND IN FERROELECTRICS NEAR SECOND	ORDER TRANSITION POINT. /LE FORCES ON ATTENUAT	03-0253
THEORY OF ANOMALIES OF THERMODYNAMIC PARAMETERS NEAR SECOND	ORDER TRANSITION POINTS IN FERROELECTRICS. /AL	03-0362
SHIFT WITH APPLIED FIELD IN FERROELECTRICS POSSESSING FIRST	ORDER TRANSITION. (THEORY, CALCULATION) /POINT	03-0327
/IONS. (THERMODYNAMICS, THEORY, MAGNETISM, FERROELECTRICI,	ORDERED ALLOYS, SUPERFLUIDITY, SUPERCONDUCTIV/	03-0183
CTURES OF THE FORMS OF ICES. (ICES II, VIII AND IX ARE FULLY	ORDERED (ANTIFERROELECTRIC STRUCTURES)) STRU	34-3451
EFS). MAGNETICALLY	ORDERED FERROELECTRIC MATERIALS (REVIEW, 251 R	02-0117
PHASE TRANSITION FROM NORMAL FERROELECTRIC TO MAGNETICALLY	ORDERED FERROELECTRIC STATE. (THEORY)	03-0394
COUPLED ELECTROMAGNETIC SPIN WAVES IN MAGNETICALLY	ORDERED FERROELECTRICS.	03-0148
G ELECTRIC FIELD. BEHAVIOR OF MAGNETICALLY	ORDERED FERROELECTRICS IN A RAPIDLY OSCILLATIN	03-0147
POSSIBILITY OF EXCITATION OF SPIN WAVES IN MAGNETICALLY	ORDERED FERROELECTRICS. (THEORY)	03-0150
HYDRATE BY ANTIFERROMAGNETIC RESONANCE. STUDY OF THE	ORDERED MAGNETIC STATE OF COPPER FORMATE TETRA	26-2943
/CTRUM OF BARIUM TITANATE. (AND MICROSCOPIC MODEL OF LOCALLY	ORDERED "NUCLEI", PHASE STRUCTURE ABOVE CURIE/	04-0647
/TELLER EFFECTS IN ORIGIN OF FERROELECTRICITY, OCCURRENCE OF	ORDERED PHASE IN PEROVSKITE TYPE FERROELECTRI/	04-0964
RONS FROM POTASSIUM MANGANESE TRIFLUORIDE IN THE SHORT RANGE	ORDERED REGION. /RAMAGNETIC SCATTERING OF NEUT	33-3371
Y(2) OXIDE(4) (MAY BE FERROELECTRIC DUE TO ANTIFERROMAGNETIC	ORDERING). /N THE ANTIFERROMAGNET IRON ANTIMON	36-3563
ES OF RARE EARTH COMPOUNDS. (WITH MAGNETIC AND FERROELECTRIC	ORDERING) PREPARATION AND PROPERTI	04-0131
ONAL ICE. RELAXATIONAL PROTON	ORDERING AND GLASSY CRYSTALLINE STATE IN HEXAG	34-3436
EMICAL BOND. (COVALENCY, DIRECTIVITY CONDITIONS FOR ELECTRIC	ORDERING AND STABILITY) /OVSKITE FAMILY AND CH	03-0243
ENHANCED	ORDERING IN PLZT CERAMICS.	09-1570
SODIUM	ORDERING IN SODIUM TUNGSTEN BRONZES.	13-2012
AND FERROELECTRIC PHASE TRANSITION IN POTASSIUM FERROCYANID/	ORDERING OF DIPOLE MOMENTS OF WATER MOLECULES	24-2893
MICS. ENHANCED	ORDERING OF FERROELECTRIC DOMAINS IN PLZT CERA	09-1579
MICS. ENHANCED	ORDERING OF FERROELECTRIC DOMAINS IN PLZT CERA	09-1580
EDRAL POSITIONS OF THE "CUBIC" PEROVSKITE STRUCTURE OF LEAD/	ORDERING OF MAGNESIUM AND NIOBIUM IN THE OCTAH	08-1379
LIGHT SCATTERING BY POLARITONS ASSOCIATED WITH	ORDINARY PHOTONS IN LITHIUM NIOBATE.	11-1697
OBSERVATION OF	ORDINARY POLARITONS IN LITHIUM NIOBATE.	11-1854
BRILLOUIN EFFECT AT	ORDINARY TEMPERATURE IN AMMONIUM SULFATE.	19-2766
LASTIC CONSTANTS OF SODIUM NITRITE BY BRILLOUIN DIFFUSION AT	ORDINARY TEMPERATURES. MEASUREMENT OF THE E	16-2306
/THE VARIATION OF THE RAMAN FREQUENCIES IN SODIUM NITRITE AT	ORDINARY TEMPERATURES AS A FUNCTION OF CRYSTA/	16-2363
ECTRIC MATERIALS. (FINE POWDERS PREPARED BY DECOMPOSITION OF	ORGANIC COMPOUNDS) /ICAL PROPERTIES OF FERROEL	02-0093
STARK EFFECT AND FERROELECTRICITY IN THE	ORGANIC SEMICONDUCTOR COPPER PHTHALOCYANINE.	36-3532
LOW TEMPERATURE PHASE TRANSITION IN THE POLAR STATE OF THE	ORGANIC SEMICONDUCTOR PHTHALOCYANINE.	36-3623
TE TO DNA. (FERROELECTRICITY, BIOLOGICAL MECHANISMS, MEMORY,	ORGANICS) FROM BARIUM TITANA	02-0092
EXCLUSIVE, THEREFORE FERROELECTRIC, BUT NOT SUPERCONDUCTIVE	ORGANICS EXIST). /ROELECTRICITY (SEEM MUTUALLY	03-0377
UM NITRITE AT ORDINARY TEMPERATURES AS A FUNCTION OF CRYSTAL	ORIENTATION / OF THE RAMAN FREQUENCIES IN SODI	16-2363
ST AND BRIGHTNESS IN GADOLINIUM MOLYBDATE LIGHT VALVES.	ORIENTATION AND THICKNESS DEPENDENCE OF CONTRA	30-3291
ERSE LIGHT MODULATOR. (HALF WAVE VOLTAGE, T/ OPTIMUM CRYSTAL	ORIENTATION FOR A BARIUM SODIUM NIOBATE TRANSV	13-1973
E COMPOUNDS (PROTON MAGN/ FERROELECTRICS WITH AMMONIUM GROUP	ORIENTATION MOBILITY: METAL AMMONIUM (BX4) TYP	22-2826
ELECTRICS. (APPLIED TO ADP, TUNGSTEN TRIOXIDE, SODIUM NIOBA/	ORIENTATION OF COMPOSITION PLANES IN ANTIFERRO	08-1327
ES IN FERROELECTRICS. (FACTORS INFLUENCING) TYPE AND	ORIENTATION OF THE EQUILIBRIUM DOMAIN STRUCTUR	03-0246
/R MAGNETIC RESONANCE IN ROCHELLE SALT, DETERMINATION OF THE	ORIENTATIONS OF PROTON PAIRS (IN THE WATER MO/	28-3203
A MODEL OF KANZIG REGIONS TAKING INTO ACCOUNT DIFFERENT	ORIENTATIONS OF THE SPONTANEOUS POLARIZATION.	03-0439
TWO DIMENSIONAL MODEL. (ICE I) ATOMIC VIBRATIONS IN	ORIENTATIONALLY DISORDERED SYSTEMS. PART-1: A	34-3474
INYLIDENE FLUORIDE. PYROELECTRICITY AND PIEZOELECTRICITY IN	ORIENTED FILMS OF POLYVINYL FLUORIDE AND POLYV	36-3544
DIELECTRIC ABSORPTION IN	ORIENTED POLYVINYLIDENE FLUORIDE.	36-3572
/OPING PROBLEMS IN (TITANATE) FERROELECTRIC SEMICONDUCTORS.	(ORIGIN AND EFFECTS OF GRAIN BOUNDARY BLOCKING/	02-0074
SKITE) STRUCTURES WITH SPECIAL DIELECTRIC PROPERTIES.	ORIGIN OF DIPOLE CONFIGURATIONS IN SOME (PEROV	08-1390
RED PHASE IN PEROVSKITE TYPE/ ROLE OF JAHN TELLER EFFECTS IN	ORIGIN OF FERROELECTRICITY, OCCURRENCE OF ORDE	04-0964
MATERIALS AND THEIR INFLUENCE ON BEHAVIOR.	ORIGIN OF LATTICE DEFECTS IN (BARIUM) TITANATE	04-0784
ACTIVITY IN LEAD ZIRCONATE(X) TITANATE(1-X) CERAMICS NEAR /	ORIGIN OF THE MAXIMUM IN THE ELECTROMECHANICAL	09-1499
ACTIVITY IN LEAD ZIRCONATE(X) TITANATE(1-X) CERAMICS NEAR /	ORIGIN OF THE MAXIMUM IN THE ELECTROMECHANICAL	09-1500
ISOTOPE EFFECTS IN FERROELECTRIC POTASSIUM DIHYDROGEN PHOS/	ORIGIN OF THE SPONTANEOUS POLARIZATION AND THE	17-2433
ADOLINIUM MOLYBDATE. (SYMMETRY CHANGES)	ORIGIN OF THE STRUCTURAL PHASE TRANSITION IN G	30-3273
CTURE AND ITS INTERACTION WITH CRYSTAL DISLOCATIONS. (DOMAIN	ORIGIN, SHAPES, MOVEMENT, 34 REFS) /OMAIN STRU	03-0390
LY ACTIVE A-IONS IN ABO(3) STRUCTURES. (ANTIFERROELECTRICITY	ORIGINS) FERROELECTRICAL	08-1364
OVSKITES. (ABSTRACT ONLY) CONDITIONS FAVORING THE	ORIGINATION OF SPONTANEOUS POLARIZATION IN PER	08-1457
	ORTHO... SEE THE PREFIXED WORD.	
DEPENDENCE OF THE DIELECTRIC CONSTANTS OF TRIGLYCINE SULFATE	ORTHOGONAL TO THE FERROELECTRIC AXIS. /RATURE	27-2975
CHLORIDES AT TEMPERATURES ABOVE 80 DEGREES/ RAMAN SPECTRA OF	ORTHORHOMBIC AND CUBIC HYDROGEN AND DEUTERIUM	31-3330
1: TEMPERATURE DEPENDEN/ MOLECULAR AND LATTICE VIBRATIONS OF	ORTHORHOMBIC AND FERROELECTRIC THIOUREA. PART-	25-2920

2: CALCULATION OF THE I/ MOLECULAR AND LATTICE VIBRATIONS OF ORTHORHOMBIC AND FERROELECTRIC THIOUREA. PART- 25-2907
TANATE.   DIELECTRIC ABSORPTION IN THE ORTHORHOMBIC AND QUADRATIC PHASES OF BARIUM TI 04-0900
TANATE.   DIELECTRIC ABSORPTION IN THE ORTHORHOMBIC AND QUADRATIC PHASES OF BARIUM TI 04-0899
POLARIZATION PROCESSES OF BARIUM TITANATE SINGLE CRYSTALS IN ORTHORHOMBIC AND RHOMBOHEDRAL PHASES.   RE 04-0963
STANCE. (WHEN THE B ION IS FERROELECTRIC/ DIPOLE PATTERNS IN ORTHORHOMBIC AND TRIGONAL PHASES OF ABO(3) SUB 08-1363
RICAITE) ATOMIC DISPLACEMENTS IN FERROELECTRIC TRIGONAL AND ORTHORHOMBIC BORACITE STRUCTURES. (BORACITE, E 14-2075
OXIDE(20) AND BARI/ PIEZOELECTRIC SURFACE WAVES IN CUBIC AND ORTHORHOMBIC CRYSTALS. (BISMUTH(12) GERMANIUM 13-2041
TRIC, NOT FERROELECTRIC)   ORTHORHOMBIC DIGLYCINE SULFATE. (NOT PIEZOELEC 36-3624
TO THE NCNFERROE/ STRUCTURAL STUDY OF THE TRANSITION OF THE (ORTHORHOMBIC) FERROELECTRIC (LEAD DINIOBATE IN 08-1285
          CRYSTAL SIZE EFFECTS IN THE RAMAN SPECTRA OF ORTHORHOMBIC HYDROGEN CHLORIDE CRYSTALS.   31-3338
   CALCULATION OF TETRAGONA/ ELECTROSTATIC FIELD IN A SLIGHTLY ORTHORHOMBIC IONIC CRYSTAL. APPLICATION TO THE 04-0578
. (EFFECT OF SUBSTITUTIONS)   THE   ORTHORHOMBIC PHASE IN MODIFIED BARIUM TITANATE 04-0617
               THERMODYNAMIC THEORY OF THE CUBIC- ORTHORHOMBIC PHASE TRANSITION IN BORACITES.   03-0231
OF THE BARIUM LEAD NIOBATE ZIRCONATE SYSTEM. (TETRAGONAL AND ORTHORHOMBIC PHASES, ABSTRACT ONLY) /OPERTIES 13-1925
E SCATTERING.   ELASTIC CONSTANT OF ORTHORHOMBIC POTASSIUM NIOBATE BY X-RAY DIFFUS 12-1416
ARIUM SODIUM NIOBATE. (THERMAL EX/ DILATOMETRIC STUDY OF THE ORTHORHOMBIC- TETRAGONAL PHASE TRANSITION IN B 13-1894
GSTEN TRIOXIDE PHASE.   THERMAL EXPANSION AND THE ORTHORHOMBIC- TETRAGONAL TRANSITION OF THE TUN 10-1626
ELASTIC AND ELECTROELASTIC EQUATIONS OF STATE: ISOTROPIC AND ORTHOTROPIC (FERROELECTRIC CERAMIC) CASES. /R 03-0277
BEHAVIOR OF MAGNETICALLY ORDERED FERROELECTRICS IN A RAPIDLY OSCILLATING ELECTRIC FIELD.   03-0147
IC CRYSTALS. (SECOND HARMONIC GENERATION, OPTICAL PARAMETRIC OSCILLATION) /PTICAL PROPERTIES OF FERROELECTR 03-0383
/(BASED IN SECOND HARMONIC GENERATION AND OPTICAL PARAMETRIC OSCILLATION IN FERROELECTRIC CRYSTALS, REVIEW/ 02-0097
        PARAMETRIC LIGHT AMPLIFICATION AND OSCILLATION IN KDP WITH MODE LOCKED PUMP.   17-2468
                OPTICAL PARAMETRIC OSCILLATION IN LITHIUM IODATE.   08-1338
BISMUTH(12) GERMANIUM O/ LIGHT AND ELECTRIC FIELD DEPENDENT OSCILLATION OF SPACE CHARGE LIMITED CURRENT IN 35-3505
THIUM NIOBATE).   SIMULTANEOUS OPTICAL PARAMETRIC OSCILLATION, SECOND HARMONIC GENERATION (IN LI 11-1859
IN FERROELECTRIC ANTIMONY SULFOIODIDE. (PIEZOELECTRIC DOMAIN OSCILLATIONS)   ANOMALOUS ECHO 15-2211
         SYMMETRY OF NORMAL OSCILLATIONS IN A FERROELECTRIC CRYSTAL OF KDP 17-2623
        (BACKWARD WAVE) ACOUSTIC PARAMETRIC OSCILLATIONS IN LITHIUM NIOBATE (60 DB GAIN). 11-1837
LUDING FERROELEC/ STUDY OF OPTICAL PROPERTIES AND COLLECTIVE OSCILLATIONS IN NEW SOLID STATE MATERIALS (INC 02-0104
TANTALATE.   STUDY OF PIEZOELECTRIC OSCILLATIONS IN WIDEBAND PYROELECTRIC LITHIUM 12-1875
EQUENCY OF APPLIED FIELD CAUSED BY SHEARING DUE TO POLARIZA/ OSCILLATIONS OF FERROELECTRIC BODIES. (WITH FR 03-0197
EARIUM SODIUM NIOBATE FOR A TUNABLE STIMULATED RAMAN OSCILLATOR.   13-1917
ER DISORDER / DYNAMIC SUSCEPTIBILITY OF CLASSICAL ANHARMONIC OSCILLATOR. A UNIFIED OSCILLATOR MODEL FOR ORD 03-0405
ACOUSTIC WAVE GENERATION IN PIEZOELECTRICS COUPLED TO A GUNN OSCILLATOR. (LITHIUM NIOBATE) /EASUREMENTS OF 11-1731
ECTRIC PHASE TRANSITION IN TERMS OF A SINGLE MODE ANHARMONIC OSCILLATOR MODEL. /ND SPECIFIC HEAT IN FERROEL 03-0406
ITION IN PEROVSKITE LIKE FERROELECTRICS, USING AN ANHARMONIC OSCILLATOR MODEL. /EARITY NEAR THE PHASE TRANS 03-0249
        VISIBLE CW PARAMETRIC OSCILLATOR USING BARIUM SODIUM NIOBATE.   13-1984
SITY WITH TEMPERATURE IN THE FERROELECTRIC PHASE POTASSIUM / OSCILLATORY BEHAVIOR OF RAMAN SCATTERING INTEN 17-2662
FECT OF AUTOMATIC TEMPERATURE STABILIZATION IN FERROELECTRIC OSCILLATORY SYSTEMS.   EF 03-0294
N NON-ENANTIOMORPHOUS BIAXIAL CRYSTALS: METHYL MESITYL OXIDE OXALATE AND SODIUM NITRITE. OPTICAL ACTIVITY I 16-2273
APPARATUS, GROWTH OF ADP, KDP, HYDROIODIC ACID, AND AMMONIUM OXALATE MONOHYDRATE). /ROM AQUEOUS SOLUTIONS (17-2420
E DOPED BARIUM TITANATE.   OXIDATION OF CARBON MONOXIDE ON LANTHANUM OXID 04-0947
BARIUM TITANATE.   CATALYTIC OXIDATION OF CARBON MONOXIDE ON SEMICONDUCTIVE 04-0945
ELECTROCOLORATION IN STRONTIUM TITANATE: VACANCY DRIFT AND OXIDATION REDUCTION OF TRANSITION METALS.   06-1098
NATE- LANTHANUM TITANATE SYSTEM.   OXIDATION REDUCTION PHENOMENA IN THE LEAD TITA 09-1587
ON THE APPARENT CONTACT FERROELECTRIC EFFECT OF MANGANESE DIOXIDE.   A NOTE 36-3606
        CRYSTAL STRUCTURE OF BISMUTH TITANIUM NIOBIUM OXIDE.   13-2058
        CRYSTAL STRUCTURE OF BISMUTH TITANIUM OXIDE.   13-1935
TEMPERATURE) CHARACTERISTICS OF SOLID SOLUTIONS OF TUNGSTEN OXIDE.   DIELECTRIC (AND 10-1657
DIELECTRIC PROPERTIES OF BARIUM TITANATE DOPED WITH TUNGSTEN OXIDE.   04-0696
        EFFECT OF DOPING BARIUM TITANATE WITH MANGANESE DIOXIDE.   04-0562
RAMIC (LEAD STRONTIUM TITANATE ZIRCONATE DOPED WITH CHROMIUM OXIDE). / AND PIEZOELECTRIC PROPERTIES OF A CE 09-1591
YL STRETCHING VIBRATION IN STRONTIUM TITANATE AND TITANIUM DIOXIDE. /RE AND PHASE TRANSITIONS ON THE HYDROX 06-1209
NGLE CRYSTALS OF SOLID SOLUTIONS OF BARIUM TITANATE AND ZINC OXIDE.   ELECTROOPTICAL PROPERTIES OF SI 04-0953
OF POLYCRYSTALLINE STRONTIUM TITANATE CONTAINING MANGANESE DIOXIDE.   EPR INVESTIGATION 06-1162
        EVAPORATED FILMS OF TUNGSTEN OXIDE.   10-1647
DISPERSION IN BARIUM AND STRONTIUM TITANATES AND TITANIUM DIOXIDE.   FAR INFRARED DIELECTRIC 04-0975
ECTRICS IN THE POTASSIUM OXIDE- STRONTIUM OXIDE- NIOBIUM PENTOXIDE.   FERROEL 13-1896
FECTIVE DIELECTRIC CONSTANT (DUE TO ADDING CUPROUS OR FERRIC OXIDE). /ARIUM TITANATE CERAMIC WITH A HIGH EF 04-0581
PHASE TRANSITION IN BARIUM BISMUTH(5+) BISMUTH(3+) OXIDE.   13-1940
N OF A NEW FERROELECTRIC OXIDE, STRONTIUM OXIDE- TELLURIUM DIOXIDE.   PHASE TRANSITIO 36-3627
AD TITANATE CERAMICS. (CONTAINING 1 MOLE PERCENT MANGANESE DIOXIDE) /ION AND PIEZOELECTRIC PROPERTIES OF LE 05-1061
EPITAXY OF SPUTTERED FILMS OF FERROELECTRIC BISMUTH TITANIUM OXIDE.   PREPARATION AND 13-2038
TITANIUM OXIDE(20) AND THE SYSTEM BISMUTH OXIDE- TITANIUM DIOXIDE.   PROPERTIES OF BISMUTH(12) 13-1994
RAMAN SPECTRUM OF POTASSIUM TANTALUM(0.64) NIOBIUM(0.36) OXIDE.   13-1989
LITHIUM NIOBATE, LITHIUM TANTALATE, QUARTZ, AND TELLURIUM DIOXIDE. /TIC SURFACE WAVE VELOCITY AND DELAY ON 11-1821
(NO COCHRAN TYPE / INFRARED OPTICAL PROPERTIES OF VANADIUM DIOXIDE ABOVE AND BELOW TRANSITION TEMPERATURE. 36-3526
/TION BETWEEN LEAD MONOXIDE AND PZT. PART-4: EFFECT OF BORON OXIDE ADDITION ON THE PRODUCT OF PZT. PART-5:/ 09-1624
ANATE COMPACT   EFFECT OF OXIDE ADDITIONS ON THE SINTERING OF BARIUM TIT 04-0996
OR OF BARIUM TITANATE WITH NIOBIUM PENTOXIDE AND MANGANESE DIOXIDE ADDITIVES. /ICIENT OF RESISTANCE) BEHAVI 04-0837
NATE CERAMICS.   THE ROLE OF IRON OXIDE ADULTERANT IN A LEAD ZIRCONATE LEAD TITA 09-1592
(PHASE TRANSITIONS AND) THERMOCURRENTS IN TITANIUM DIOXIDE AND BARIUM TITANATE SINGLE CRYSTALS.   04-0697
ITH ISOTOPE 170.   FLUX GROWTH OF MAGNESIUM OXIDE AND LANTHANUM ALUMINATE CRYSTALS DOPED W 36-3558
BINARY AND PSEUDOBINARY PHASES BETWEEN NIOBIUM PENTOXIDE AND LEAD, BARIUM, AND STRONTIUM OXIDES. 04-0590
RYSTALS (LITHIUM NIOBATE OR TANTALATE, BARIUM TITANATE, ZINC OXIDE, AND LITHIUM GALLIUM OXIDE). /OF POLAR C 02-0098
ENDENCES OF THE ABSORPTION OF 9.4 GHZ HYPERSOUND IN ALUMINUM OXIDE AND LITHIUM NIOBATE SINGLE CRYSTALS. /EP 11-1719
OF RESISTANCE) BEHAVIOR OF BARIUM TITANATE WITH NIOBIUM PENTOXIDE AND MANGANESE DIOXIDE ADDITIVES. /ICIENT 04-0837
ACTION OF BISMUTH(4) TITANIUM(3) OXIDE(12) WITH VANADIUM PENTOXIDE AND MOLYBDENUM AND TUNGSTEN TRIOXIDE. /E 13-2029
NI/ PARAMAGNETIC RESONANCE STUDY OF PRODUCTION OF NITROGEN DIOXIDE AND NITRITE ION IN IRRADIATED POTASSIUM 16-2404
AGE PARTICLE SIZE / STUDIES ON THE SINTERING BETWEEN LEAD MONOXIDE AND PZT. PART-2: EFFECTS OF AVER 09-1624
ETCHING RATES FOR MONOCRYSTALLINE LITHIUM NIOBATE, SILICON DIOXIDE, AND SILICON.   MEASUREMENT OF SPUTTER 11-1714
SYSTEMS, METAL CONTACTS) SOME ELECTRONIC PROPERTIES OF ZINC OXIDE AND STRONTIUM TITANATE. (SURFACE BARRIER 06-1176
/T. PART-1: EFFECTS OF ADDITION OF GLASSES CONTAINING BARIUM OXIDE AND TITANIUM DIOXIDE (WITH SILICA AND A/ 04-0829
/CAPACITORS. PART-10: STRUCTURE OF SEMICONDUCTIVE TITANIUM DIOXIDE AND TITANIUM DIOXIDE- BARIUM TITANATE C/ 04-0774
/COMPONENTS OF THE CRYSTALLOGRAPHIC TRANSITION IN VANADIUM DIOXIDE. (ANTIFERROELECTRIC TEMPERATURE 340 DEG/ 36-3562
/Y OF THE X-RAY PHOTOELECTRON SPECTRUM OF TUNGSTEN- TUNGSTEN OXIDE AS A FUNCTION OF THICKNESS OF THE SURFA/ 10-1631
LANTHANUM DOPED BARIUM TITANATE CERAMICS WITH BISMUTH OXIDE AS A GRAIN BOUNDARY ELEMENT.   04-0799
/IUM AND BISMUTH(0.5)) (NIOBIUM, TANTALUM, OR VANADIUM)(0.5) OXIDE, (BARIUM OR LEAD) BISMUTH(0.67) (TUNGST/ 08-1462
ITANATE(0.48) CERAMICS CAUSED BY THE ADDITION OF NIOBIUM PENTOXIDE (BELOW 0.2 PERCENT). / ZIRCONATE(0.52) T 09-1606
GROWTH OF SINGLE CRYSTALS IN LEAD MONOXIDE- GERMANIUM DIOXIDE BINARY SYSTEM.   35-3514

SYNTHESIS OF OXIDE BRONZES. (REVIEW, 88 REFS)   02-0042
XIDE/ ELECTROCHEMICAL DEPOSITION AND DISSOLUTION OF TUNGSTEN OXIDE BRONZES. (SODIUM(X) TUNGSTEN(1-X) Z(Y) O 13-1944
THIUM NICBATE OR TANTALATE OR GALLATE, BARIUM TITANATE, ZINC OXIDE, CADMIUM SULFIDE. /ECTRIC) CRYSTALS (LI 02-0095
DEFECTS IN OXIDE CERAMICS. (STRONTIUM TITANATE)   06-1163
SISTANCE ON LANTHANUM(0.002) BARIUM(0.998) TITANATE- BISMUTH OXIDE COMPOSITE CERAMICS. /LOUS INCREASE OF RE 04-0801
PERTIES OF TERNARY BARIUM OXIDE- LANTHANUM OXIDE- TITANIUM OXIDE COMPOUNDS.   ELECTRICAL PRO 13-1912
/ OF MADELUNG ENERGY AND COVALENCY ON THE STRUCTURE OF AB(5) OXIDE COMPOUNDS. (BARIUM TITANATE, RUBIDIUM N/ 04-0761
BARIUM TITANATE, STRONTIUM TI/ TOP SEEDED SOLUTION GROWTH OF OXIDE CRYSTALS FROM NONSTOICHIOMETRIC MELTS. (04-0553
/L DAMAGE, AND THERMALLY STIMULATED CURRENT IN FERROELECTRIC OXIDE CRYSTALS. (POTASSIUM LITHIUM NIOBATE, 13-1926
/IDE(11) AND ITS ISOMORPHOUS COMPOUND LEAD GERMANIUM SILICON OXIDE. (DIELECTRIC CONSTANTS, SPONTANEOUS POL/ 35-3501
ELECTRICAL PROPERTIES OF SCANDIUM TRIOXIDE DOPED, RARE EARTH OXIDE DOPED, AND UNDOPED BARIUM TITANATE. /ND 04-0839
OXIDATION OF CARBON MONOXIDE ON LANTHANUM OXIDE DOPED BARIUM TITANATE.   04-0947
TRICAL PROPERTIES. (DIELECTRIC CON/ SYNTHESIS OF NIOBIUM PENTOXIDE DOPED BARIUM TITANATE WITH IMPROVED ELEC 04-0840
X-RAY STUDY OF THE SYSTEM BISMUTH OXIDE FERRIC OXIDE.   08-1376
/ELECTRIC FIELD EFFECT STUDIES AT LOW TEMPERATURES (WITH TIN OXIDE FILMS ON LEAD ZIRCONATE(0.65) TITANATE(/ 09-1505
TE (0.35) ZIRCO/ FERROELECTRIC FIELD EFFECT STUDIES. (ON TIN OXIDE FILMS ON LEAD(0.97) BISMUTH(0.02) TITANA 09-1506
ZINCATE(0.33) NIOBATE(0.67) (SINGLE CRYSTALS GROWN FROM LEAD OXIDE FLUX). FERROELECTRIC PROPERTIES OF LEAD 13-2066
MODE FOR THE ANTIMONY SULFOIODIDE- TIN DIOXIDE HETEROSTRUCTURE SWITCH.   15-2243
SWITCHING AND MEMORY EFFECT IN ANTIMONY SULFOIODIDE- TIN DIOXIDE HETEROSTRUCTURE THIN FILM. 15-2166
OF LANTHANUM DOPED BARIUM TITANATE.   EFFECT OF OXIDE IMPURITIES ON THE ELECTRICAL RESISTIVITY 04-0587
/CTION OF BARIUM TITANATE FROM BARIUM SULFATE AND TITANIUM DIOXIDE IN THE PRESENCE OF SODIUM CARBONATE AT / 04-0624
E/ DETERMINATION OF THE DISTRIBUTION COEFFICIENTS OF LITHIUM OXIDE INTO LITHIUM NIOBATE BY PHASE MATCHING T 11-1841
TES, TITANATES, CHROMATES, COBA/ FLUX GROWTH OF SOME COMPLEX OXIDE MATERIALS. (NIOBATES, TANTALATES, STANNA 02-0129
IELD AND MELT GROWTH OF CRYSTALS IN THE SODIUM OXIDE- BARIUM OXIDE NIOBIUM PENTOXIDE SYSTEM. /ND SOLIDUS) F 13-1949
OXIDATION OF CARBON MONOXIDE ON LANTHANUM OXIDE DOPED BARIUM TITANATE 04-0947
CATALYTIC OXIDATION OF CARBON MONOXIDE ON SEMICONDUCTIVE BARIUM TITANATE.   04-0945
ECTRIC MATERIALS. (BARIUM TITANATE, LEAD / EFFECT OF LITHIUM OXIDE ON THE PROPERTIES OF A SERIES OF FERROEL 04-0780
RIUM TITANATE SINGLE CRYSTALS.   EFFECT OF FERRIC OXIDE ON THE 120 DEGREE PHASE TRANSITION IN BA 04-0544
VITY IN NON-ENANTIOMORPHOUS BIAXIAL CRYSTALS: METHYL MESITYL OXIDE OXALATE AND SODIUM NITRITE. OPTICAL ACTI 16-2273
/E TERNARY SYSTEM BARIUM OXIDE- STRONTIUM OXIDE- NIOBIUM PENTOXIDE. (OXIDES, ESPECIALLY THE NIOBIUM-RICH R/ 13-1920
DOMAIN ALIGNMENT IN LEAD ZIRCONATE- LEAD TITANATE- LANTHANUM OXIDE (PLZT) CERAMIC. /EPENDENT FERROELECTRIC 09-1586
/EW LAYERED FERROELECTRICS. (LANTHANUM BISMUTH TITANIUM IRON OXIDE, PRASEODYMIUM BISMUTH TITANIUM IRON OXI/ 13-1993
FERROELECTRICS AND OXIDE SEMICONDUCTORS.   01-0027
ON BEAM WRITING OF FERROELECTRIC DOMAINS IN BISMUTH TITANIUM OXIDE SINGLE CRYSTALS.   ELECTR 08-1300
TIES OF THE THERMOCURRENTS IN BARIUM TITANATE AND TITANIUM DIOXIDE SINGLE CRYSTALS.   SOME PROPER 04-0622
UNGSTEN TRIOXIDE).   PREPARATION OF TUNGSTIC OXIDE SINGLE CRYSTALS (SUBLIMATION IN AIR OF T 10-1644
/RE OF POTASSIUM TUNGSTEN BRONZE. (POTASSIUM BARIUM TANTALUM OXIDE, SODIUM BARIUM TANTALUM OXIDE, POTASSIU/ 13-2050
E SOLID ELECTROLYTE FUE/ IONIC CONDUCTION IN PEROVSKITE TYPE OXIDE SOLID SOLUTION AND ITS APPLICATION TO TH 08-1444
ELECTROOPTICAL PROPERTIES OF BARIUM TITANATE- TANTALUM PENTOXIDE SOLID SOLUTION SINGLE CRYSTALS. 04-0955
STOICHIOMETRY OF DONOR DOPED BARIUM TITANATE AND TITANIUM DIOXIDE. (STOICHIOMETRIC COMPENSATION) OXYGEN 04-0664
PHASE TRANSITION OF A NEW FERROELECTRIC OXIDE, STRONTIUM OXIDE- TELLURIUM DIOXIDE.   36-3627
CERAMIC LEAD ZIRCONATE TITANATE (PLUS 1 PERCENT NIOBIUM PENTOXIDE) SURFACE LAYERS. /IES OF OXYGEN DEPLETED 09-1576
TH OF CRYSTALS IN THE SODIUM OXIDE- BARIUM OXIDE NIOBIUM PENTOXIDE SYSTEM. /ND SOLIDUS) FIELD AND MELT GROW 13-1949
ILIBRIA AND CRYSTAL GROWTH IN THE LITHIUM OXIDE- NIOBIUM PENTOXIDE SYSTEM.   PHASE EQU 11-1683
M LITHIUM NIOBATES, PHASE REL/ TUNGSTEN BRONZE FIELD IN THE (OXIDE) SYSTEM POTASSIA LITHIA NIOBIA (POTASSIU 13-2017
DUS EQUILIBRIA IN POTASSIUM OXIDE- LITHIUM OXIDE- NIOBIUM PENTOXIDE SYSTEM. (POTASSIUM LITHIUM NIOBATES) /LI 13-1963
TURES IN LITHIUM NIOBATE (CRYSTAL GROWTH BY ADDING MAGNESIUM OXIDE TO A CONGRUENT MELT) / MATCHING TEMPERA 11-1684
LITHIUM NIOBATE. (CRACK-FREE, THROUGH ADDITION OF MAGNESIUM OXIDE TO MELT) /MAIN CRYSTALS OF FERROELECTRIC 11-1774
ODIUM OR POTASSIUM) PHASES OF THE TETRAGONAL TUNGSTEN BRONZE OXIDE TYPE. / STRONTIUM, OR BARIUM, AND C IS S 04-0912
FERROELECTRICITY IN BISMUTH OXIDE TYPE LAYER STRUCTURE COMPOUNDS.   13-1932
/ ADDITION OF GLASSES CONTAINING BARIUM OXIDE AND TITANIUM DIOXIDE (WITH SILICA AND ALUMINA) ON PROPERTIES/ 04-0829
ACOUSTOOPTIC PROPERTY OF SINGLE CRYSTAL LEAD(5) GERMANIUM(3) OXIDE(11).   35-3509
CRYSTAL STRUCTURE OF FERROELECTRIC LEAD(5) GERMANIUM(3) OXIDE(11).   35-3504
ICAL ROTATION DOMAINS IN SINGLE CRYSTAL LEAD(5) GERMANIUM(3) OXIDE(11).   FERROELECTRIC OPT 36-3550
LITHIUM THALLIUM TARTRATE MONOHYDRATE, LEAD(5) GERMANIUM(3) OXIDE(11)) /CTRIC CRYSTALS. (BISMUTH TITANATE, 02-0061
FERROELECTRICITY IN LEAD(5) GERMANIUM(3) OXIDE(11).   35-3508
EAR OPTICAL PROPERTIES OF FERROELECTRIC LEAD(5) GERMANIUM(3) OXIDE(11).   NONLIN 35-3506
ATTERING FROM THE FERROELECTRIC MODE IN LEAD(5) GERMANIUM(3) OXIDE(11).   RAMAN SC 35-3498
SOFT OPTIC PHONON MODE IN FERROELECTRIC LEAD(5) GERMANIUM(3) OXIDE(11).   35-3495
OF 180 DEGREE DOMAINS IN FERROELECTRIC LEAD(5) GERMANIUM(3) OXIDE(11).   X-RAY TOPOGRAPHIC OBSERVATION 35-3516
/ERROELECTRIC AND OPTICAL PROPERTIES OF LEAD(5) GERMANIUM(3) OXIDE(11) AND ITS ISOMORPHOUS COMPOUND LEAD G/ 35-3501
LECTRIC PROPERTY AND ENANTIOMORPHISM OF LEAD(5) GERMANIUM(3) OXIDE(11) AND THE RELATED MATERIALS. FERROE 35-3499
FERROELECTRIC LEAD(5) GERMANIUM(3) OXIDE(11) CRYSTAL.   35-3500
CTRIC CONSTANTS AND REFRACTIVE INDICES) LEAD(5) GERMANIUM(3) OXIDE(11) CRYSTAL, A NEW FERROELECTRIC. (DIELE 35-3502
EZOELECTRIC PROPERTIES OF FERROELECTRIC LEAD(5) GERMANIUM(3) OXIDE(11) CRYSTALS.   ELASTIC AND PI 35-3520
/. (LEAD TITANATE, BARIUM STRONTIUM NIOBATE, LEAD NIOBIUM(4) OXIDE(11), SILVER GALLIUM DISULFIDE, SILICON / 05-1066
OPTICAL ROTATORY POWER IN FERROELECTRIC LEAD(5) GERMANIUM(3) OXIDE(11) SINGLE CRYSTAL.   SWITCHING OF 36-3570
CRYSTAL GROWTH AND PROPERTIES OF LEAD(5) GERMANIUM(3) OXIDE(11) SINGLE CRYSTALS.   35-3513
OPTICAL ACTIVITY OF FERROELECTRIC LEAD(5) GERMANIUM(3) OXIDE(11) SINGLE CRYSTALS.   35-3503
WITCHING PROPERTIES OF SINGLE CRYSTAL BISMUTH(4) TITANIUM(3) OXIDE(12).   ELECTRICAL AND OPTICAL S 35-3517
GROWTH OF LARGE SINGLE CRYSTAL BISMUTH(4) TITANIUM(3) OXIDE(12).   35-3507
ONLINEAR OPTICAL PROPERTIES OF FERROELECTRIC LEAD NIOBIUM(4) OXIDE(12).   N 13-2023
FERROELECTRIC PHOTOCONDUCTOR DEVICE. (BISMUTH(4) TITANIUM(3) OXIDE(12))   PHASE HOLOGRAMS IN A 13-1977
IUM(M+1) OXIDE(3) (M+2) (M=2,3,4). (DATA FOR THE TITANIUM(3) OXIDE(12), BARIUM TITANIUM(4) OXIDE(15), AND / 02-0066
RUBIDIUM IRON TRIFLUORIDE, BORACITES, BISMUTH(4) TITANIUM(3) OXIDE(12), HAFNIUM DIVANADITE /M ANTIMONIDE, 02-0036
FERROELECTRIC DIELECTRIC ANOMALY IN BISMUTH(4) TITANIUM(3) OXIDE(12) SINGLE CRYSTALS.   13-2009
/ERMOGRAPHIC STUDY OF THE REACTION OF BISMUTH(4) TITANIUM(3) OXIDE(12) WITH VANADIUM PENTOXIDE AND MOLYBDE/ 13-2029
/4). (DATA FOR THE TITANIUM(3) OXIDE(12), BARIUM TITANIUM(4) OXIDE(15), AND BARIUM(2) TITANIUM(5) OXYGEN(1/ 02-0066
F CADMIUM SULFIDE, LITHIUM NIOBATE AND BISMUTH(12) GERMANIUM OXIDE(20). /ACE WAVE CONVOLUTION ON CRYSTALS O 11-1692
CRYSTAL STRUCTURE OF PIEZOELECTRIC BISMUTH(12) GERMANIUM OXIDE(20).   35-3493
OELECTRIC, AND DIELECTRIC CONSTANTS OF BISMUTH(12) GERMANIUM OXIDE(20).   ELASTIC, PIEZ 35-3512
ION OF SPACE CHARGE LIMITED CURRENT IN BISMUTH(12) GERMANIUM OXIDE(20). / ELECTRIC FIELD DEPENDENT OSCILLAT 35-3505
ECTRIC SUBSTRATES (LITHIUM NIOBATE AND BISMUTH(12) GERMANIUM OXIDE(20)). / ELASTIC SURFACE WAVES ON PIEZOEL 11-1751
N ANISOTROPIC MEDIA. (LITHIUM NIOBATE, BISMUTH(12) GERMANIUM OXIDE(20)) /URFACE WAVES BY NONLINEAR MIXING I 11-1778
RAMAN SPECTRUM OF BISMUTH(12) GERMANIUM OXIDE(20).   35-3519
S IN CUBIC AND ORTHORHOMBIC CRYSTALS. (BISMUTH(12) GERMANIUM OXIDE(20) AND BARIUM SODIUM NIOBATE) /ACE WAVE 13-2041
/STIMULATED CURRENTS AND LUMINESCENCE IN BISMUTH(12) SILICON OXIDE(20) AND BISMUTH(12) GERMANIUM OXIDE(20). 36-3585
UM DIOXIDE.   PROPERTIES OF BISMUTH(12) TITANIUM OXIDE(20) AND THE SYSTEM BISMUTH OXIDE- TITANI 13-1994
ONIC ATTENUATION IN SINGLE CRYSTALS OF BISMUTH(12) GERMANIUM OXIDE(20) AT LOW TEMPERATURES.   ULTRAS 35-3510
OUSTIC SURFACE WAVE LOSS MECHANISMS ON BISMUTH(12) GERMANIUM OXIDE(20) AT MICROWAVE FREQUENCIES.   AC 35-3511

/E VELOCITIES. (BARIUM SODIUM NIOBATE, BISMUTH(12) GERMANIUM OXIDE(20), LITHIUM NIOBATE, LITHIUM TANTALATE/ 01-0028
NIC WAVE PULSES IN LITHIUM NIOBATE AND BISMUTH(12) GERMANIUM OXIDE(20) PIEZOELECTRIC CRYSTALS. /RSE HYPERSO 11-1815
E, PROUSTITE, CALCIUM PYRONIOBATE, AND BISMUTH(12) GERMANIUM OXIDE(20). (REVIEW) / NIOBATE, CUPROUS CHLORID 01-0021
E WAVE AMPLIFIER. A THEORETICAL AND E/ BISMUTH(12) GERMANIUM OXIDE(20) VERSUS LITHIUM NIOBATE IN THE SURFAC 11-1720
UTIONS IN THE SYSTEM BARIUM TITANATE- STRONTIUM NIOBIUM(0.8) OXIDE(3). /RROELECTRIC PROPERTIES OF SOLID SOL 04-0707
/LECTRICS OF THE FAMILY BISMUTH(4) BARIUM(M-2) TITANIUM(M+1) OXIDE(3) (M+2) (M=2,3,4). (DATA FOR THE TITANI/ 02-0066
/ON OF TUNGSTEN OXIDE BRONZES. (SODIUM(X) TUNGSTEN(1-X) Z(Y) OXIDE(3) WHERE Z IS NIOBIUM, TANTALUM, ZIRCON/ 13-1944
EXAMINATION OF SUBSTOICHIOMETRIC TUNGSTEN OXIDE(3-X) CRYSTALS BY ELECTRON MICROSCOPY. 10-1627
C CONSTANTS OF STIBOTANTALITE. (ANTIMONY (TANTALUM, NIOBIUM) OXIDE(4)) /TURE DEPENDENCE OF THE PIEZOELECTRI 36-3601
/NETOELECTRIC EFFECT IN THE ANTIFERROMAGNET IRON ANTIMONY(2) OXIDE(4) (MAY BE FERROELECTRIC DUE TO ANTIFER/ 36-3563
STRUCTURE OF THE TUNGSTEN BRONZE TYP/ BISMUTH(3) NIOBIUM(17) OXIDE(47): POTENTIALLY FERROELECTRIC CRYSTAL 13-1978
/TRICS. (BARIUM(2) BISMUTH (NIOBIUM OR TANTALUM OR VANADIUM) OXIDE(6), AND BARIUM(3) BISMUTH(2) (TUNGSTEN / 36-3619
SCRIPTION OF THE PYROCHLORE STRUCTURE (CADMIUM(2) NIOBIUM(2) OXIDE(6) SULFIDE). NEW DE 36-3602
A NEW FERROELECTRIC SINGLE CRYSTAL STRONTIUM(2) NIOBIUM(2) OXIDE(7). 13-1997
CRYSTAL STRUCTURE AND PROPERTIES OF CALCIUM(2) NIOBIUM(2) OXIDE(7). 13-1911
/WTH OF FERROELECTRIC SINGLE CRYSTAL STRONTIUM(2) NIOBIUM(2) OXIDE(7) BY MEANS OF THE FLOATING ZONE TECHNI/ 13-2037
/ND (BARIUM OR STRONTIUM) (3) COPPER (NIOBIUM OR TANTALUM) (2) OXIDE(9), AND VARIOUS TUNGSTEN SUBSTITUTED LE/ 36-3619
MAINS. STRUCTURE OF THE BRONZE SODIUM(13) NIOBIUM(35) OXIDE(94) AND THE GEOMETRY OF FERROELECTRIC DO 13-1931
ES BETWEEN NIOBIUM PENTOXIDE AND LEAD, BARIUM, AND STRONTIUM OXIDES. BINARY AND PSEUDOBINARY PHAS 04-0590
FERROELECTRICS WITH DIFFUSE PHASE TRANSITIONS. (MIXED LAYER OXIDES) /ELECTRIC AND STRUCTURAL PROPERTIES OF 13-2014
ELECTRICAL CONTACT TO N- AND P-TYPE FERROELECTRIC OXIDES. 03-0512
FERROELECTRIC CERAMIC OXIDES. 02-0038
TUNGSTEN BRONZE TYPE COMPLEX OXIDES. 13-1981
CHIOMETRY (AND ANTIFERROELECTRIC DOMAINS IN TRANSITION METAL OXIDES). / EXTENDED DEFECTS RELATED TO NONSTOI 36-3522
/ THIN FILM MATERIALS WITH PEROVSKITE TYPE STRUCTURE. (MIXED OXIDES ABO(3)) (BARIUM TITANATE AND LITHIUM N/ 04-0914
/ND SOLIDUS) FIELD AND MELT GROWTH OF CRYSTALS IN THE SODIUM OXIDE- BARIUM OXIDE NIOBIUM PENTOXIDE SYSTEM. 13-1949
CATION NONSTOICHIOMETRY IN TERNARY OXIDES. (BARIUM TITANATE) 04-0969
/STRUCTURE OF SEMICONDUCTIVE TITANIUM DIOXIDE AND TITANIUM DIOXIDE- BARIUM TITANATE COATINGS PRODUCED BY F/ 04-0774
TITANATE. IONIC CONDUCTION IN SINTERED OXIDES BASED ON CALCIUM TITANATE OR STRONTIUM 06-1141
/Y SYSTEM BARIUM OXIDE- STRONTIUM OXIDE- NIOBIUM PENTOXIDE. (OXIDES, ESPECIALLY THE NIOBIUM-RICH REGION AN/ 13-1920
GROWTH OF SINGLE CRYSTALS IN LEAD MONOXIDE- GERMANIUM DIOXIDE BINARY SYSTEM. 35-3514
UNDS. ELECTRICAL PROPERTIES OF TERNARY BARIUM OXIDE- LANTHANUM OXIDE- TITANIUM DIOXIDE COMPO 13-1912
CE OF PEROVSKITE PHASES IN THE SYSTEM LEAD OXIDE- TITANIUM DIOXIDE- LANTHANUM TRIOXIDE. RANGE OF EXISTEN 08-1349
/OSCOPIC STUDIES ON FERROELECTRIC DOMAINS OF PEROVSKITE TYPE OXIDES. (LEAD TITANATE, POTASSIUM NIOBATE, BA/ 04-0993
. (POTASSIUM LIT/ STUDY OF SUBSOLIDUS EQUILRIA IN POTASSIUM OXIDE- LITHIUM OXIDE- NIOBIUM PENTOXIDE SYSTEM 13-1963
PHASE EQUILIBRIA AND CRYSTAL GROWTH IN THE LITHIUM OXIDE- NIOBIUM PENTOXIDE SYSTEM. 11-1683
/DIELECTRIC CONSTANTS OF CUBIC IONIC COMPOUNDS. (HALIDES AND OXIDES, RELATION BETWEEN CURIE CONSTANT AND C/ 03-0188
ELECTRICAL CONDUCTION IN METAL OXIDES. (REVIEW, 319 REFS) 02-0109
FERROELECTRICS IN THE POTASSIUM OXIDE- STRONTIUM OXIDE- NIOBIUM PENTOXIDE. 13-1896
IDE/ PHASE EQUILIBRIA RELATIONS IN THE TERNARY SYSTEM BARIUM OXIDE- STRONTIUM OXIDE- NIOBIUM PENTOXIDE. (OX 13-1920
PHASE TRANSITION OF A NEW FERROELECTRIC OXIDE, STRONTIUM OXIDE- TELLURIUM DIOXIDE. 36-3627
IES OF BISMUTH(12) TITANIUM OXIDE(20) AND THE SYSTEM BISMUTH OXIDE- TITANIUM DIOXIDE. PROPERT 13-1994
RANGE OF EXISTENCE OF PEROVSKITE PHASES IN THE SYSTEM LEAD OXIDE- TITANIUM DIOXIDE- LANTHANUM TRIOXIDE. 08-1349
THE BISMUTH OXIDE- TUNGSTEN TRIOXIDE SYSTEM. 10-1656
CRYSTAL CHEMISTRY OF MIXED BISMUTH OXIDES WITH LAYER TYPE STRUCTURE. 13-2032
INTERATOMIC DISTANCES IN OXIDES WITH THE PEROVSKITE STRUCTURE. 08-1439
S AND SPONTANEOUS DEFORMATION IN THE FERROELECTRIC PHASES OF OXIDES WITH THE PEROVSKITE STRUCTURE. /ACEMENT 08-1377
OTASSIUM (BARIUM OR STRONTIUM) TITANATE NIOBATE, (B/ COMPLEX OXIDES WITH TUNGSTEN BRONZE TYPE STRUCTURES (P 13-1962
FERROELECTRICITY AND THE CHEMICAL BOND IN PEROVSKITE TYPE OXIDES. (43 REFS) 02-0102
OELECTRIC- ANTIFERROMAGNETIC LITHIUM IRON(0.5) TANTALUM(0.5) OXY FLUORIDE. MAGNETOELECTRIC EFFECT IN FERR 36-3568
LEAD MONOXIDE ON THE SINTERING. PART-10: PROPERTIES OF LEAD OXY HYDROXIDE POWDER. /FFECT OF MOTHER SALT OF 09-1624
OXY... SEE THE PREFIXED WORD.
RANSITION IN ANTIMONY SULFOIODIDE SINGLE CRYSTALS. EFFECT OF OXYGEN ADSORPTION ON THE FERROELECTRIC PHASE T 15-2187
ION OF ANTIMONY BY BISMUTH OR ARSENIC, SULFUR BY SELENIUM OR OXYGEN, AND IODINE BY BROMINE OR CHLORINE) /UT 15-2205
NGLE CRYSTAL BARIUM TITANATE FILMS GROWN FROM THE MELT IN AN OXYGEN ATMOSPHERE. SI 04-0612
/ ON SINTERING. PART-6: FORMATION OF PZT AND SINTERING UNDER OXYGEN ATMOSPHERE. PART-7: EFFECT OF CALCINAT/ 09-1624
TE, POTASS/ NONLINEAR OPTICAL POLARIZABILITY OF THE NIOBIUM- OXYGEN BOND. (LITHIUM NIOBATE, LITHIUM TANTALA 11-1734
TANTALATE NIOBATE, TA/ NEW PHASES OF THE TETRAGONAL TUNGSTEN OXYGEN BRONZE TYPE: QUADRATIC LEAD OR BISMUTH 13-1985
CTIVITY MEASUREMENTS ON PZT FERROELECTRICS. OXYGEN CONCENTRATION CELL AND ELECTRICAL CONDU 09-1495
C TRANSPORT NUMBERS IN PZT (LEAD ZIRCONATE TITANATE) FERROE/ OXYGEN CONCENTRATION CELL MEASUREMENTS OF IONI 09-1512
ERTIES OF TUNGSTEN TRIOXIDE CRYSTALS. EFFECT OF OXYGEN DEFICIENCY ON ELECTRICAL TRANSPORT PROP 10-1629
E (PLUS 1 PERCENT NIOBIUM PENTOXIDE) SURFACE / PROPERTIES OF OXYGEN DEPLETED CERAMIC LEAD ZIRCONATE TITANAT 09-1576
STRONTIUM TITANATE, BARIUM TITANATE) MEASUREMENT OF OXYGEN DIFFUSION CONSTANT IN SOME TITANATES. (04-0559
AL CONDUCTIVITY OF BARIUM TITANATE SINGLE CRYSTALS HEATED IN OXYGEN (EFFECT TIED TO GOLD ANODES). /ELECTRIC 04-0997
HEMATITE SINGLE CRYSTALS. HETEROVALENT SUBSTITUTION AND OXYGEN NONSTOICHIOMETRY OF BARIUM TITANATE AND 04-0874
NUCLEAR MAGNETIC RESONANCE STUDIES OF OXYGEN OCTAHEDRA FERROELECTRICS. 13-2059
/ INDUCED BIREFRINGENCE AND POLARIZATION) OF LEAD CONTAINING OXYGEN OCTAHEDRA FERROELECTRICS: CERAMIC LEAD/ 13-2039
ELECTROMECHANICAL PROPERTIES OF OXYGEN OCTAHEDRAL FERROELECTRIC CRYSTALS. 13-2061
IE TEMPERATURE. (BARIUM(0-1./ MICROWAVE DIELECTRIC LOSSES IN OXYGEN OCTAHEDRON FERROELECTRICS ABOVE THE CUR 04-0559
DE SINGLE CRYSTALS. EFFECT OF THE ADSORPTION OF OXYGEN ON THE PROPERTIES OF ANTIMONY SULFOIODI 15-2154
ONDUCTING BARIUM TITANATE. INFLUENCE OF OXYGEN PARTIAL PRESSURE ON PROPERTIES OF SEMIC 04-0999
HE PROPERTIES AND PROCESSING OF SINGLE CRYSTAL BA/ EFFECT OF OXYGEN STOICHIOMETRY AND HYDROXYL CONTENT ON T 13-1915
CRYSTALS. DETERMINATION OF THE DEVIATION FROM OXYGEN STOICHIOMETRY OF BARIUM TITANATE SINGLE 04-0736
ANATE AND TITANIUM DIOXIDE. (STOICHIOMETRIC COMPENSATION) OXYGEN STOICHIOMETRY OF DONOR DOPED BARIUM TIT 04-0664
OF STRONTIUM TITANATE. ANALYSIS OF THE IRON OXYGEN VACANCY CENTER IN THE TETRAGONAL PHASE 06-1231
OR FERROELECTR/ NEW BORACITES OF FORMULA LITHIUM(4) BORON(7) OXYGEN(12) CHLORIDE OR BROMIDE (NOT EXAMINED F 14-2089
12), BARIUM TITANIUM(4) OXIDE(15), AND BARIUM(2) TITANIUM(5) OXYGEN(18) COMPOUNDS) / THE TITANIUM(3) OXIDE(02-0066
ONTIUM TITANATE (ESR MEASUREMENTS, M/ CHARGE COMPENSATION BY OXYGEN(2-) VACANCIES IN CHROMIUM(3+) DOPED STR 06-1161
/INVESTIGATIONS OF THE SYSTEMS YTTRIUM MANGANESE(1-X) (B(X)) OXYGEN(3) (B=TRIVALENT IRON, CHROMIUM OR ALUM/ 36-3569
LANTHANUM(0.01) (IRON(0.405) NIOBIUM(0.325) ZIRCONIUM(0.27)) OXYGEN(3) FERROELECTRIC MEMORY FILMS. /(0.07) 37-3629
M, CHROMIUM, MANGANESE, IRO/ SUPERSTRUCTURE OF BISMUTH METAL OXYGEN(3) PEROVSKITES. (WHERE METAL IS SCANDIU 08-1451
HIGH PRESSURE SYNTHESIS OF LEAD METAL- OXYGEN(3) TYPE PEROVSKITE. 08-1331
/SE TRANSITION SYSTEMS LEAD (B(0.5) NIOBIUM(0.5)) (1-X) B'(X) OXYGEN(3), WHERE B IS IRON OR SCANDIUM, AND B/ 13-2044
RONTIUM POTAS/ FIRST FERROELECTRIC OXYFLUORIDE PHASES OF THE OXYGENATED QUADRATIC TUNGSTEN BRONZE TYPE. (ST 13-2011

/ FERROCYANIDE TRIHYDRATE SINGLE CRYSTALS, P-N JUNCTIONS AND P- AND N- RANGES IN SEMICONDUCTORS. (SILICON / 24-2904
/CTURE IN POTASSIUM FERROCYANIDE TRIHYDRATE SINGLE CRYSTALS, P-N JUNCTIONS AND P- AND N- RANGES IN SEMICON/ 24-2904
VES IN A LAYERED STRUCTURE CONSISTING OF LITHIUM NIOBATE AND P-TYPE SILICON. /ION OF RAYLEIGH ULTRASONIC WA 11-1747
TIONS OF STRESSED STRONTIUM TITANATE. (EPR OF IRON- VANADIUM PAIRS)   ORDER PARAMETER AND PHASE TRANSI 06-1170
                         OPTICAL SPECTRA OF CHROMIUM(3+) PAIRS IN LANTHANUM ALUMINATE.   36-3616
EVELS OF CHROMIUM(3+) IONS AND EXCHANGE COUPLED CHROMIUM(3+) PAIRS IN LANTHANUM ALUMINATE. /UM AND ENERGY L 36-3565
N ROCHELLE SALT, DETERMINATION OF THE ORIENTATIONS OF PROTON PAIRS (IN THE WATER MOLECULES). /C RESONANCE I 28-3203
ABO(3) TYPE)   DIELECTRIC PROPERTIES AND OPTICAL PHONONS IN PARA- AND FERROELECTRIC PEROVSKITES. (THEORY, 03-0413
RAMAN SPECTRUM OF POTASSIUM DIDEUTERIUM PHOSPHATE CRYSTAL IN PARA- AND FERROELECTRIC PHASE.   17-2647
D ABSORPTION SPECTRUM OF A SINGLE CRYSTAL OF THIOUREA IN THE PARA- AND FERROELECTRIC PHASES.   INFRARE 25-2910
       INFRARED ABSORPTION SPECTRUM OF THIOUREA IN THE PARA- AND FERROELECTRIC PHASES.   25-2908
IDIUM HYDROGEN SULFATE AND RUBIDIUM DEUTERIUM SULFATE IN THE PARA- AND FERROELECTRIC PHASES. /PECTRA OF RUB 19-2768
POTASSIUM DIHYDROGEN PHOSPHATE IN THE INFRARED REGION IN THE PARA- AND FERROELECTRIC PHASES. /NSMISSION OF 17-2459
        VIBRATION SPECTRA OF POTASSIUM DIHYDROGEN PHOSPHATE IN PARA- AND FERROELECTRIC PHASES.   17-2480
UORIDE CRYSTAL STRUCTURE (SPACE GROUP AND LAT/ FERROELECTRIC PARAELASTIC PARAMAGNETIC BARIUM COBALT TETRAFL 33-3386
Y STUDY OF THE FERROELECTRIC- FERROELASTIC AND PARAELECTRIC- PARAELASTIC PHASES OF GADOLINIUM MOLYBDATE. /A 30-3287
NITRATE AND POTASSIUM NITRITE. (DOES NOT SEEM FERROELECTRIC- PARAELASTIC, ABSTRACT) /MATIONS IN POTASSIUM 16-2369
              NITROGEN-14 QUADRUPOLE COUPLING IN PARAELECTRIC AMMONIUM SULFATE.   19-2749
              RAMAN SPECTRA OF ADP IN THE PARAELECTRIC AND ANTIFERROELECTRIC PHASES.   17-2431
IODIDE (PHONON ASSIGNMENTS).   (POLARIZED) RAMAN SPECTRUM OF (PARAELECTRIC AND) FERROELECTRIC ANTIMONY SULFO 15-2207
NITRIT/ NUCLEAR SPIN LATTICE RELAXATION OF SODIUM-23 IN THE PARAELECTRIC AND FERROELECTRIC PHASE OF SODIUM 16-2264
AYER IN BARIUM TITANATE (X-RAY DIFFRACTION, DC FIELD EFFECT, PARAELECTRIC AND FERROELECTRIC PHASES). /ACE L 04-0857
HOSPHATE ANION IN A RUBIDIUM DIHYDROGEN PHOSPHATE CRYSTAL IN PARAELECTRIC AND FERROELECTRIC PHASES. / THE P 17-2472
CYANIDE TRIHYDRATE BY PULSED NMR.   STUDY OF HYDROGEN-1 IN PARAELECTRIC AND FERROELECTRIC POTASSIUM FERRO 24-2889
DENSITY AND OPTICAL PROPERTIES OF ANTIMONY SULFOIODIDE. (IN PARAELECTRIC AND FERROELECTRIC STATES) / STATE 15-2204
          X-RAY STUDY ON THE LATTICE STRAINS OF (PARAELECTRIC AND) (IRON- IODINE- BORACITE).   14-2085
CRYSTALS OF THE POTASSIUM DIHYDROGEN PHOSPHATE TYPE IN THEIR PARAELECTRIC AND LOW TEMPERATURE PHASES. / OF 17-2515
            NOTES ON ABSORPTION BAND EDGE OF (PARAELECTRIC) ANTIMONY SULFOIODIDE.   15-2177
COMMENTS ON THE NEUTRON SPECTRUM OF FERROELECTRIC AND PARAELECTRIC BARIUM TITANATE.   04-0813
     OPTICAL EVIDENCE FOR POLARIZATION FLUCTUATIONS IN PARAELECTRIC BARIUM TITANATE.   04-0737
          ELECTRIC FIELD INDUCED RAMAN EFFECT IN PARAELECTRIC CRYSTALS. (POTASSIUM TANTALATE)   08-1326
T.   EFFECT OF IONIC MOTION ON THE REFRACTIVE INDEX OF PARAELECTRIC KDP IN THE VICINITY OF CURIE POIN 17-2548
     EFFECT OF PRESSURE ON THE DIELECTRIC PROPERTIES OF A PARAELECTRIC MATERIAL. (POTASSIUM TANTALATE)   08-1335
NTIUM TI/ PRESSURE DEPENDENCE OF THE DIELECTRIC CONSTANTS OF PARAELECTRIC MATERIALS. (BARIUM TITANATE, STRO 04-0856
ROELECTRICS I/ POSSIBLE REALIZATION OF ULTRA LOW NOISE LEVEL PARAELECTRIC MICROWAVE AMPLIFIER (BASED ON FER 37-3664
SOME REMARKS ABOUT ANTIFERROELECTRIC PHASE TRANSITIONS (TO PARAELECTRIC OR FERROELECTRIC, THEORY).   03-0171
ION OF DIELECTRIC PROPERTIES OF SOME FERRO-, ANTIFERRO- AND PARAELECTRIC PEROVSKITES.   EFFECT OF IRRADIAT 08-1474
STAL STRUCTURE OF TRIGLYCINE FLUOBERYLLATE IN THE FERRO- AND PARAELECTRIC PHASE.   CRY 27-3175
OELECTRIC ACTIVITY OF BARIUM TITANATE SINGLE CRYSTALS IN THE PARAELECTRIC PHASE.   PIEZ 04-0771
CRYSTALS OF THE POTASSIUM DIHYDROGEN PHOSPHATE GROUP IN THE PARAELECTRIC PHASE. /OPERTIES OF FERROELECTRIC 17-2696
SECOND HARMONIC GENERATION IN THE TRIGLYCINE CRYSTAL IN THE PARAELECTRIC PHASE.   27-3105
NDAMENTAL VIBRATIONS OF POTASSIUM DIHYDROGEN ARSENATE IN THE PARAELECTRIC PHASE. /TION. PART-2: INTERNAL FU 18-2738
ROGEN DICHLORO ACETATE. PART-1: THE CRYSTAL STRUCTURE OF THE PARAELECTRIC PHASE. /RANSITION OF AMMONIUM HYD 36-3567
RIC).   RAMAN SPECTRUM OF SODIUM TRIHYDROGEN DISELENITE (PARAELECTRIC PHASE, COMPARISON WITH FERROELECT 22-2864
/ SPECTRA OF INTERNAL VIBRATIONS OF THE PHOSPHATE ION IN THE PARAELECTRIC PHASE OF KDP CRYSTALS (INFRARED A 17-2479
          DIELECTRIC DISPERSION IN THE PARAELECTRIC PHASE OF POTASSIUM NITRATE.   16-2341
          X-RAY ANALYSIS OF FERROELECTRIC DOMAINS IN THE PARAELECTRIC PHASE OF SODIUM NITRITE.   16-2271
A/ ELECTROOPTIC PROPERTIES OF SOME ABO(3) PEROVSKITES IN THE PARAELECTRIC PHASE. (POTASSIUM TANTALATE AND T 04-0690
/OELECTRICS TOGETHER WITH EXAMINATION OF CHARACTERS OF THEIR PARAELECTRIC PHASE TRANSFORMATIONS. (AND SUGG/ 03-0138
IOBATE(0.67)) AND ITS DIELECTRIC PROPERTIES. (FERROELECTRIC- PARAELECTRIC PHASE TRANSITION) /ADMATE(0.33) N 13-1959
LATE.   NATURE OF THE FERROELECTRIC- PARAELECTRIC PHASE TRANSITION IN LITHIUM TANTA 12-1879
/ENTIAL DISTRIBUTION IN ANTIMONY SULFOIODIDE CRYSTAL (IN THE PARAELECTRIC PHASE, UNDER DC FIELD ALONG C-AX/ 15-2173
YSTALS.   OBSERVATION OF PHASE BOUNDARIES BETWEEN FERRO- AND PARAELECTRIC PHASES IN GADOLINIUM MOLYBDATE CR 30-3306
NEUTRON MEASUREMENTS ON THE HYDROGEN BOND POTENTIAL IN PARAELECTRIC POTASSIUM DIHYDROGEN PHOSPHATE.   17-2511
D POTA/ FOURIER TRANSFORM FAR INFRARED REFLECTION SPECTRA OF PARAELECTRIC POTASSIUM DIHYDROGEN PHOSPHATE AN 03-0356
EUTRON SCATTERING).   HYDROGEN DENSITY DISTRIBUTION IN PARAELECTRIC POTASSIUM DIHYDROGEN PHOSPHATE (N 17-2490
OLTAGE AMPERE CHARACTERISTICS OF ANTIMONY SULFOIODIDE IN THE PARAELECTRIC REGION. /ION ON THE FORM OF THE V 15-2119
     PYROELECTRIC BEHAVIOR OF TRIGLYCINE SULFATE IN THE PARAELECTRIC REGION.   27-3066
SIUM CHLORIDE.   PARAELECTRIC RESONANCE OF LITHIUM ION IN POTAS 31-3325
     EFFECT OF MECHANICAL STRESS ON DIELECTRIC PROPERTIES OF PARAELECTRIC ROCHELLE SALT.   28-3225
CTRIC FIELD.   PARAELECTRIC SEMICONDUCTOR IN A NONUNIFORM ELE 03-0502
NFRARED ABSORPTION AND RAMAN SCATTERING IN FERROELECTRIC AND PARAELECTRIC SODIUM NITRITE. /TION RULES FOR I 16-2281
          NEUTRON DIFFRACTION STUDY ON PARAELECTRIC SODIUM TRIHYDROGEN DISELENITE.   22-2846
     ELASTIC ENERGY AND THE FERROELECTRIC PARAELECTRIC TRANSITION.   03-0237
     EFFECT OF SEVERAL PARAMETERS ON THE FERROELECTRIC- PARAELECTRIC TRANSITION IN TRIGLYCINE SULFATE.   27-2981
     EFFECT OF GRAIN SIZE ON THE FERROELECTRIC PARAELECTRIC TRANSITION OF BARIUM TITANATE.   04-0917
SPECIFIC HEAT ANOMALY IN THE NEIGHBORHOOD OF A FERRO- PARAELECTRIC TRANSITION. (TGS)   27-2962
UDY OF THE NUCLEAR QUADRUPOLE INTERACTIONS OF NITROGEN-14 IN PARAELECTRIC TRIGLYCINE SULFATE. /RESONANCE ST 27-2969
ULATORS. (POTASSIUM TANTALATE(0.65) NIOBA/ USE OF PEROVSKITE PARAELECTRICS IN BEAM DEFLECTORS AND LIGHT MOD 13-1924
/EHENSIVE X-RAY STUDY OF THE FERROELECTRIC- FERROELASTIC AND PARAELECTRIC- PARAELASTIC PHASES OF GADOLINIU/ 30-3287
          CRYSTAL STUDY OF BARIUM TITANATE WITH PARALLEL INTERFERENCE FRINGES.   04-0705
          STUDY OF CRYSTALS OF BARIUM TITANATE HAVING PARALLEL INTERFERENCE FRINGES.   04-0897
/EPOLARIZATION OF LINEARLY POLARIZED LIGHT DUE TO HEATING OF PARALLEL PLANES OF POTASSIUM DIHYDROGEN PHOSP/ 17-2540
ITRATE.   DOMAIN ETCH PATTERNS ON THE FACES PARALLEL TO THE FERROELECTRIC-AXIS OF SODIUM N 16-2349
AL STRUCTURE (SPACE GROUP AND LAT/ FERROELECTRIC PARAELASTIC PARAMAGNETIC BARIUM COBALT TETRAFLUORIDE CRYST 33-3386
L STRUCTURE OF THE TRANSITION ME/ FERROELECTRIC FERROELASTIC PARAMAGNETIC BETA GADOLINIUM MOLYBDATE. CRYSTA 30-3289
LSKI SINGLE C/ (ROOM TEMPERATURE) FERROELECTRIC FERROELASTIC PARAMAGNETIC (BETA) TERBIUM MOLYBDATE (CZOCHRA 30-3290
YSTALS.   EPR STUDY OF PARAMAGNETIC ION COMPLEXES IN FERROELECTRIC CR 03-0477
. EPR STUDIES OF CHROMIUM(3+), COPPER(2+), AND MANGANESE(2+) PARAMAGNETIC ION COMPLEXES IN GLYCINE CRYSTALS 27-3135
ING IN FERROELECTRIC TRIGLYCINE SULFATE, PURE AND DOPED WITH PARAMAGNETIC IONS.   SWITCH 27-3020
/ING: MAGNETIC RESONANCE LINE SHAPES AND SPIN CORRELATION IN PARAMAGNETIC POTASSIUM MANGANESE TRIFLUORIDE,/ 33-3376
NS IN FERROELECTRIC KDP.   PARAMAGNETIC RELAXATION AND DIVALENT COPPER IO 17-2629
RE IN VANADIUM DOPED TRIGLYCINE SULFATE OBSERVED BY ELECTRON PARAMAGNETIC RESONANCE /OF THE DOMAIN STRUCTU 27-3013
ITION TEMPERATURES IN CESIUM LEAD TRICHLORIDE USING ELECTRON PARAMAGNETIC RESONANCE   DETERMINATION OF TRANS 33-3408
PHOTOCHROMIC IRON IN STRONTIUM TITANATE.   EVIDENCE FROM PARAMAGNETIC RESONANCE.   06-1171
/ON ON HIGH DIELECTRIC CONSTANT MATERIALS GAINED BY ELECTRON PARAMAGNETIC RESONANCE. (ABSTRACT ONLY, 11 RE/ 02-0101
UBLE RESONANCE OF THE PROTON LATTICE / DETECTION BY ELECTRON PARAMAGNETIC RESONANCE AND ELECTRON NUCLEAR DO 18-2724

UBLE RESONANCE OF THE ARSENATE CENTER IN X-IRRADIA/ ELECTRON PARAMAGNETIC RESONANCE AND ELECTRON NUCLEAR DO 18-2727
UBLE RESONANCE OBSERVATIONS OF DOMAINS IN THE IRRA/ ELECTRON PARAMAGNETIC RESONANCE AND ELECTRON NUCLEAR DO 18-2726
OF TRANSITION ELEMENT IONS IN STRONTIUM TITANATE AND LANTHA/ PARAMAGNETIC RESONANCE AND OPTICAL ABSORPTION 06-1165
   THE CURIE TEMPERATURE.   ELECTRON PARAMAGNETIC RESONANCE IN BARIUM TITANATE NEAR 04-0928
GLE CRYSTAL OF FERROELECTRIC LITHIUM TRIHYDROGEN D/ ELECTRON PARAMAGNETIC RESONANCE IN GAMMA-IRRADIATED SIN 22-2842
RIUM TITANATE.   ELECTRON PARAMAGNETIC RESONANCE IN NONSTOICHIOMETRIC BA 04-0640
+) AND MANGANESE(2+) IN LITHIUM NICBATE.   AN ELECTRON PARAMAGNETIC RESONANCE INVESTIGATION OF IRON(3 11-1725
M NIOBATE.   LCW FREQUENCY (510 MHZ) PARAMAGNETIC RESONANCE INVESTIGATION OF LITHIU 11-1793
YSTALLINE STRONTIUM TITANATE CONTAINING MANGANESE / ELECTRON PARAMAGNETIC RESONANCE INVESTIGATION OF POLYCR 06-1162
TERNAL ELECTRIC FIELD IN COPPER DOPED ROCHELLE SALT ELECTRON PARAMAGNETIC RESONANCE INVESTIGATION OF THE IN 28-3233
ER) DOPED FERROELECTRIC CRYSTALS. (TRIGLYCINE SULF/ ELECTRON PARAMAGNETIC RESONANCE INVESTIGATIONS OF (COPP 27-3132
NESE(2+) COMPLEXES IN FERROELECTRIC TRIS-SARCOSINE/ ELECTRON PARAMAGNETIC RESONANCE INVESTIGATIONS OF MANGA 36-3525
   MANGANESE TRIFLUORI/ SHIFT OF NEEL TEMPERATURE AND ELECTRON PARAMAGNETIC RESONANCE LINE WIDTH OF POTASSIUM 33-3377
FATE.   TEMPERATURE DEPENDENCE OF THE ELECTRON PARAMAGNETIC RESONANCE OF AMMONIUM CADMIUM SUL 19-2792
OICHIOMETRIC BARIUM TITANATE.   PARAMAGNETIC RESONANCE OF AN F CENTER IN NONST 04-0976
                                             ELECTRON PARAMAGNETIC RESONANCE OF CHROMIUM(3+) IN ADP. 17-2596
   ANTIFERROELECTRIC ADP CRYSTALS.   ELECTRON PARAMAGNETIC RESONANCE OF CHROMIUM(3+) IONS IN 17-2546
IUM TANTALATE.   ELECTRON PARAMAGNETIC RESONANCE OF COBALT(2+) IN POTASS 08-1342
LLE SALT.   ELECTRON PARAMAGNETIC RESONANCE OF COPPER IONS IN ROCHE 28-3176
   STRON/ INTERPRETATION OF ELECTRIC FIELD EFFECTS IN ELECTRON PARAMAGNETIC RESONANCE OF GADOLINIUM(3+) DOPED 06-1168
   LITHIUM NIOBATE.   ELECTRON PARAMAGNETIC RESONANCE OF GADOLINIUM(3+) DOPED 11-1833
THIUM NICBATE.   ELECTRON PARAMAGNETIC RESONANCE OF GADOLINIUM(3+) IN LI 11-1763
HELLE SALT.   ELECTRON PARAMAGNETIC RESONANCE OF GAMMA-IRRADIATED ROC 28-3180
UM NIOBATE.   ELECTRON PARAMAGNETIC RESONANCE OF IRON(3+) DOPED LITHI 11-1765
ITANATE AT LOW TEMPERATURES.   ELECTRON PARAMAGNETIC RESONANCE OF IRON(3+) IN BARIUM T 04-0927
ITANATE IN THE RHOMBOHEDRAL AND THE CUBIC PHASES. ELECTRON PARAMAGNETIC RESONANCE OF IRON(3+) IN BARIUM T 04-0929
/PHASE TRANSITIONS OF STRESSED STRONTIUM TITANATE. (ELECTRON PARAMAGNETIC RESONANCE OF IRON- VANADIUM PAIR/ 06-1170
I/ FORBIDDEN LINES IN THE HYPERFINE SPECTRUM OF THE ELECTRON PARAMAGNETIC RESONANCE OF MANGANESE(+2) IN BAR 04-0877
AMMONIUM CADMIUM SULFATE.   ELECTRON PARAMAGNETIC RESONANCE OF MANGANESE(2+) DOPED 19-2791
   ALLOWED AND FORBIDDEN HYPERFINE TRANSITIONS IN THE ELECTRON PARAMAGNETIC RESONANCE OF MANGANESE(2+) DOPED/ 16-2383
ASSIUM TANTALATE.   ELECTRON PARAMAGNETIC RESONANCE OF MANGANESE(2+) IN POT 08-1343
   FERROELECTRIC PHASE OF SODIUM NITRITE.   ELECTRON PARAMAGNETIC RESONANCE OF MANGANESE(2+) IN THE 16-2358
ONIUM TITANATE.   EFFECTIVE CHARGE OF TITANIUM(4+) FROM THE ELECTRON PARAMAGNETIC RESONANCE OF MANGANESE(4+) IN STR 04-1167
BIUM DOPED BARIUM TITANATE SINGLE CRYSTAL.   PARAMAGNETIC RESONANCE OF PLATINUM IONS IN NIO 04-0977
                                       THE ELECTRON PARAMAGNETIC RESONANCE OF POTASSIUM SELENATE. 22-2850
RITIES IN STRONTIUM TITANATE.   ELECTRON PARAMAGNETIC RESONANCE OF SOME RARE EARTH IMPU 06-1192
   IN TRIGLYCINE FLUCBERYLLATE.   ELECTRON PARAMAGNETIC RESONANCE OF THE CHROMIUM COMPLEX 27-3133
COMPLEX WITH SPIN S=2 IN TRIGLYCINE FLUOBERYLLATE / ELECTRON PARAMAGNETIC RESONANCE OF THE FOUR COPPER ION 27-3134
IONS IN STRONTIUM AND BARIUM TITANATES.   ELECTRON PARAMAGNETIC RESONANCE OF TRIVALENT GADOLINIUM 04-0920
/C PHASE TRANSITION IN SODIUM NITRITE AS STUDIED BY ELECTRON PARAMAGNETIC RESONANCE (PHASE TRANSITION AT 1/ 16-2359
/ DETERMINATION OF CHROMIUM VALENCY FROM STUDIES OF ELECTRON PARAMAGNETIC RESONANCE SPECTRA AND STATIC MAGN 04-0542
RIC CERAMIC CONTAINING TITANI/ PECULIARITIES OF THE ELECTRON PARAMAGNETIC RESONANCE SPECTRA IN A FERROELECT 04-1020
IONS/ INFLUENCE CF SPONTANEOUS POLARIZATION ON THE ELECTRON PARAMAGNETIC RESONANCE SPECTRA OF CHROMIUM(3+) 27-3136
DOPED FERROEL/ EFFECT OF APPLIED ELECTRIC FIELD ON ELECTRON PARAMAGNETIC RESONANCE SPECTRA OF (COPPER(2+)) 28-3185
AL IMPUR/ EFFECT OF TEMPERATURE AND PRESSURE ON THE ELECTRON PARAMAGNETIC RESONANCE SPECTRA OF SUBSTITUTION 06-1193
DOPED) LITHIUM NICBATE AND (COPPER(2+) DCPED) ZINC/ ELECTRON PARAMAGNETIC RESONANCE SPECTRUM IN (MANGANESE 11-1792
STRONTIUM TITANATE DUE TO NEAREST NE/ STRONG AXIAL ELECTRON PARAMAGNETIC RESONANCE SPECTRUM OF IRON(3+) IN 06-1144
5+) IN STRONTIUM TITANATE: AN EXAMPLE OF THE DYNAM/ ELECTRON PARAMAGNETIC RESONANCE SPECTRUM OF MOLYBDENUM(06-1125
, COPPER (2+), AND MANGANESE(2+) PARAMAGNETIC ION C/ ELECTRON PARAMAGNETIC RESONANCE STUDIES AND DYNAMICS IN 06-1232
D VANADYI- DOPED TRIGLYCINE SULFATE SINGLE CRYSTALS ELECTRON PARAMAGNETIC RESONANCE STUDIES OF CHROMIUM(3+) 27-3135
D PARAMETERS IN IRON(3+) DOPED LITHIUM NIOBATE. ELECTRON PARAMAGNETIC RESONANCE STUDIES OF CHROMIUM- AN 27-3161
/ELECTRICS: EVIDENCE FOR SLATER CONFIGURATIONS FROM ELECTRON PARAMAGNETIC RESONANCE STUDIES OF CRYSTAL FIEL 11-1839
   POTASSIUM DIHYDROGEN ARSENATE, POTASSIUM DIDEUTER/ ELECTRON PARAMAGNETIC RESONANCE STUDIES OF X-IRRADIATE/ 18-2728
R(2+) COMPLEXES IN CRYSTALS OF GUANIDINIUM ALUMINU/ ELECTRON PARAMAGNETIC RESONANCE STUDIES OF X-IRRADIATED 18-2725
ON COMPLEXES IN FERROELECTRIC CRYSTALS.   ELECTRON PARAMAGNETIC RESONANCE STUDY OF MAGNETIC COPPE 21-2820
NITROGEN DIOXIDE AND NITRITE ION IN IRRADIATED POTASSIUM NI/ PARAMAGNETIC RESONANCE STUDY OF PARAMAGNETIC I 03-0477
S IN GUANIDINIUM ALUMINUM SULFATE HEXAHYDRATE.   ELECTRON PARAMAGNETIC RESONANCE STUDY OF PRODUCTION OF 16-2404
ACTIVITY CF NONSTCICHIOMETRIC TUNGSTEN TRIOXIDE. ELECTRON PARAMAGNETIC RESONANCE STUDY OF SOME METAL ION 21-2817
FERROELECTRICS.   ELECTRON PARAMAGNETIC RESONANCE STUDY OF THE CATALYTIC 10-1628
SIUM MANGANESE TRIFLUORIDE IN THE SHORT RANGE ORDERED REGIO/ ELECTRON PARAMAGNETIC RESONANCE STUDY OF X-RAY DAMAGED 02-0079
E CRYSTAL OF FERROELECTRIC LITHIUM TRIHYDROGEN DISELENITE. PARAMAGNETIC SCATTERING OF NEUTRONS FROM POTAS 33-3371
                                                     PARAMAGNETIC SPECIES IN GAMMA-IRRADIATED SINGL 22-2843
                                                     PARAMAGNETIC SPECIES IN IRRADIATED NITRATE. 16-2405
IUM DIHYDR/ EFFECTS OF UNIAXIAL STRESS AND ELECTRIC FIELD ON PARAMAGNETIC SPECTRA OF CHROMIUM(3+) IN POTASS 17-2554
NU/ RELAXATION PHENOMENA AND MOSSBAUER SPECTRA OF MAGNETIZED PARAMAGNETIC SUBSTANCES. PART-1: IRON(X) ALUMI 19-2751
   PHOSPHATE AND POTASSIUM DIHYDROGEN PHOSPHATE EXCITED BY A / PARAMETRIC FLUORESCENCE IN AMMONIUM DIHYDROGEN 17-2486
TE FOR NINE CW PUMP FREQUENCIES.   PARAMETRIC FLUORESCENCE IN BARIUM SODIUM NIOBA 13-1950
TE FOR 10 CW PUMP WAVELENGTHS (RANGE 454.5-528.7 MICRONS). PARAMETRIC FLUORESCENCE IN BARIUM SODIUM NIOBA 13-1983
GHT BY THE ELECTROOPTIC EFFECT IN POTASSIUM DIHYDROGEN PHOS/ PARAMETRIC FREQUENCY CONVERSION OF COHERENT LI 17-2616
BLE ACROSS THE VISIBLE SPECTRUM.   EFFICIENT HIGH GRAIN PARAMETRIC GENERATION IN ADP CONTINUOUSLY TUNA 17-2712
IN KDP WITH MODE LOCKED PUMP.   PARAMETRIC LIGHT AMPLIFICATION AND OSCILLATION 17-2468
Y POLARITONS. (LITHIUM NIOBATE)   PARAMETRIC LUMINESCENCE AND LIGHT SCATTERING B 11-1746
UM NIOBATE CRYSTAL.   PARAMETRIC LUMINESCENCE INTENSITY IN THE LITHI 11-1750
FERROELECTRIC CRYSTALS. (SECOND HARMONIC GENERATION, OPTICAL PARAMETRIC OSCILLATION) /PTICAL PROPERTIES OF 03-0383
/L DEVICES. (BASED IN SECOND HARMONIC GENERATION AND OPTICAL PARAMETRIC OSCILLATION IN FERROELECTRIC CRYST/ 02-0097
                                              OPTICAL PARAMETRIC OSCILLATION IN LITHIUM IODATE. 08-1338
TION (IN LITHIUM NIOBATE).   SIMULTANEOUS OPTICAL PARAMETRIC OSCILLATION, SECOND HARMONIC GENERA 11-1859
DB GAIN).   (BACKWARD WAVE) ACOUSTIC PARAMETRIC OSCILLATIONS IN LITHIUM NIOBATE (60 11-1837
ATE.   VISIBLE CW PARAMETRIC OSCILLATOR USING BARIUM SODIUM NIOB 13-1984
ISPERSION IN THE NONLINEAR SUSCEPTIBILITY IN LI/ SPONTANEOUS PARAMETRIC SCATTERING, IDLER ABSORPTION, AND D 11-1771
CTIVE INDEX OF ADP AND KDP CRYSTALS IN THE INFRARED RANGE BY PARAMETRIC SCATTERING OF LIGHT. / OF THE REFRA 17-2553
TE (INDEX OF REFRACTION 1.2-7/ (CCNTINUOUS WAVE) SPONTANEOUS PARAMETRIC SCATTERING OF LIGHT IN LITHIUM IODA 08-1291
UM SULFIDE AND STRONTIUM TITANATE.   NONCOLLINEAR PARAMETRIC 4 PHOTON RECIPROCAL EFFECT IN CADMI 06-1122
CRYSTAL.   TRAPPING OF PARAMETRICALLY AMPLIFIED LIGHT WAVES IN A KDP 17-2428
AND REPOLARIZATI/ CONNECTION BETWEEN THE STARTING FIELD FOR PARKHAUSEN STEPS AND THE CRITICAL POLARIZATION 03-0318
TION SPECTRA OF BARIUM TITANATE AND STRONTIUM TIT/ EFFECT OF PARTICLE SIZE AND SHAPE ON THE INFRARED ABSORP 04-0823
/N LEAD MONOXIDE AND PZT. PART-1. PART-2: EFFECTS OF AVERAGE PARTICLE SIZE OF RAW POWDER ON SINTERING. PAR/ 09-1624
OTHERS, BY SELECTIVE DEPOSITION OF CHARGED POLYSTYRENE LATEX PARTICLES). /FACES OF IONIC CRYSTALS (TGS AND 27-2987
OMALOUS CRYSTALLOGRAPHIC PROPERTIES OF SMALL BARIUM TITANATE PARTICLES.   ON THE AN 04-0935
NITUDE OF SPONTANEOUS POLARIZATION IN THIN CRYSTALS AND FINE PARTICLES OF BARIUM TITANATE.   MAG 04-0607

TURE ON THE DIELECTRIC BEHAVIOR OF ULTRAFINE BARIUM TITANATE PARTICULATES. EFFECT OF SURFACE MOIS 04-0859
ECTRIC, AND GENERAL ISING MODELS. ZEROES OF THE PARTITION FUNCTION FOR THE HEISENBERG, FERROEL 03-0481
ROELECTRICITY (KDP MODEL, ANT/ DISTRIBUTION OF ZEROES OF THE PARTITION FUNCTION FOR THE SLATER MODEL OF FER 03-0308
OUIN SCATTERING STUDY CF ANTIMONY SULFOICDIDE USING A DOUBLE PASSED, STABILIZED SCANNING INTERFEROMETER. /L 15-2225
ASSING COHERENT LIGHT THROUGH CEUTERATED KDP HAVING A CHARGE PATTERN). /ON BY ELECTROOPTIC MODULATION (BY P 17-2507
IUM NICBATE, BARIUM SODIUM NIOBATE) HOLOGRAPHIC PATTERN FIXING IN ELECTROOPTIC CRYSTALS. (LITH 11-1661
 MCVING STRIPE PATTERN IN BARIUM TITANATE. 04-0643
DINIUM ALUMINUM SULFATE HEXAHYDRATE CLEAVAGES ANC THEIR ETCH PATTERNS. /TRIC STUDIES ON FERROELECTRIC GUANI 21-2816
EFFICIENT WHITE LIGHT REACING OF DOMAIN PATTERNS IN BISMUTH TITANATE. 08-1299
I/ ELECTRICAL CONTROL OF FIXATION AND ERASURE OF HOLOGRAPHIC PATTERNS IN FERROELECTRIC MATERIALS. (BARIUM T 04-0847
F ABO(3) SUBSTANCE. (WHEN THE B ICN IS FERROELECTRIC/ DIPOLE PATTERNS IN ORTHORHOMBIC AND TRIGONAL PHASES O 08-1363
ICS AND ANTIFERROELECTRICS. (BARIUM TITANATE, KDP, P/ DIPOLE PATTERNS IN THE STRUCTURES OF SOME FERROELECTR 02-0134
YSTALS. DIP ETCH PATTERNS ON THE FACES OF TRIGLYCINE SULFATE CR 27-3127
CTRIC-AXIS OF SODIUM NITRATE. DOMAIN ETCH PATTERNS ON THE FACES PARALLEL TO THE FERROELE 16-2349
F ANTIMONY INTO THE BARIUM TITANATE LATTICE. (STRUCTURE, AND PCT RESISTORS) INCORPORATION O 04-0934
ODATE. PECULIAR FERROELECTRIC BEHAVIOR OF POTASSIUM I 08-1347
DIAM/ CONJECTURED MECHANISM OF THE PHASE TRANSFORMATION OF A PECULIAR FERROELECTRIC- FERROELASTIC CRYSTAL: 19-2741
SODIUM AM/ INFERRED MECHANISM OF THE PHASE TRANSFORMATION OF PECULIAR FERROELECTRIC- FERROELASTIC CRYSTAL: 29-3238
ORDER TYPE PHASE TRANSIT/ DIELECTRIC PERMITTIVITY DISPERSION PECULIARITIES IN FERROELECTRICS WITH ORDER DIS 03-0412
SIUM AND RUBIDIUM NITRATE CRYSTALS IN ITS LOW TEMPERATURE P/ PECULIARITIES IN THE VIBRATIONAL SPECTRA OF CE 16-2323
TE SINGLE CRYSTALS WITH DOUBLE HYSTERESIS LOOP. PECULIARITIES OF POLARIZATION OF BARIUM TITANA 04-0709
F CADMIUM PYRONIOBATE. PECULIARITIES OF THE DIELECTRIC POLARIZATION O 36-3561
ONANCE SPECTRA IN A FERROELECTRIC CERAMIC CONTAINING TITANI/ PECULIARITIES OF THE ELECTRON PARAMAGNETIC RES 04-1020
INTERACTION OF DOMAIN BOUNDARIES AND POINT DEFECTS (TORSION PENDULUM AND RESONANCE TECHNIQUES, 20-400C). / 09-1584
 PENTA... SEE THE PREFIXED WORD.
 CRYSTAL PERFECTION OF AMMONIUM DIHYDROGEN PHOSPHATE. 17-2426
TIONS FROM SATELLITE SCATTERING DATA) LONG PERIOD STRUCTURE OF THIOUREA CRYSTAL. (CALCULA 25-2911
NUCLEAR QUADRUPOLE RESONANCE CF IODINE-127 IN SILVER PERIODATE. 08-1426
AL CONSIDERATICNS OF THE ORCER DISCRCER REACTION IN AMMONIUM PERIODATE. STATISTIC 36-3581
TTERING SPECTRA OF POTASSIUM DIHYDROGEN PHOSPHATE AND SILVER PERIODATE BELOW THE CRITICAL TEMPERATURE. /SCA 17-2687
ROELECTRIC AMMONIUM SALTS. (AMMCNIUM SULFATE, FLUOBERYLLATE, PERIODATE, BISULFATE) / RELAXATION IN SOME FER 19-2769
/OLE INTERACTICN IN ANTIFERROELECTRIC DIAMMONIUM TRIHYDROGEN PERIODATE BY IODINE PROTON CROSSOVER RELAXATI/ 36-3579
STUDY OF THE STRUCTURE OF ANTIFERROELECTRIC AMMONIUM PERIODATE BY NUCLEAR RESONANCE METHODS. 36-3580
GEL GROWTH OF SILVER PERIODATE CRYSTALS. 36-3523
/SING (CROSS RELAXATION) EXPERIMENTS IN AMMONIUM TRIHYDROGEN PERIODATE CRYSTALS, PULSED MAGNETIC AND OPTIC/ 36-3577
PCLE INTERACTION IN ANTIFERROELECTRIC DIAMMONIUM TRIHYDROGEN PERIODATE PROTON CROSSOVER RELAXATION. /QUADRU 36-3578
L ORDER DISORDER PROPERTIES, KDF, TGS, ROCHELLE SALT, SILVER PERIODATE, SODIUM TRIHYDROGEN SELENITE /AMICA 02-0050
REFLECTION OF ELASTIC SURFACE WAVE FROM PERIODIC ARRAYS ON LITHIUM NIOBATE. 11-1853
OF TRIGLYCINE SELENATE DEPENDING CN THE CONDITI/ DIELECTRIC PERMEABILITY AND DIELECTRIC LOSS ANGLE TANGENT 27-3125
S IN THE PHASE TRANSITION REGION. DIELECTRIC PERMEABILITY OF BARIUM TITANATE SINGLE CRYSTAL 04-0645
HASE TRANSITION. DIELECTRIC PERMEABILITY OF TRIGLYCINE SELENATE NEAR THE P 27-3166
IES (THEORY AND TABULATION) PERMISSIBLE DOMAIN WALLS IN FERROELECTRIC SPEC 03-0247
IN COMPLEX CERAMICS BASED ON BARIUM TITANATE. (CAPACITANCE, PERMITTIVITY) AGING 04-0724
IGLYCINE SULFATE SINGLE CRYSTALS (HEAT CAPACITY AND ELECTRIC PERMITTIVITY). /ND ELECTRICAL PROPERTIES OF TR 27-3141
RIC (BARIUM TITANATE TYPE) CERAMIC WITH CONTROLLED MICROWAVE PERMITTIVITY. /CHARACTERISTICS OF A FERROELECT 04-0904
SE POLARITONS IN A LITHIUM NIOBATE CRYSTAL. (RAMAN SPECTRUM, PERMITTIVITY) TRANSVER 11-1762
DIELECTRIC DISPERSION IN TGS CRYSTALS. (PERMITTIVITY AND LOSS FROM 10 GHZ TO 100 GHZ) 27-3094
ELECTRIC CRYSTALS (TRIGLYCINE SULFA/ BEHAVIOR OF THE COMPLEX PERMITTIVITY (AUDIO FREQUENCIES) OF SOME FERRO 02-0125
LCW LOSS (FERROELECTRIC CERAMICS) WITH GREAT DIELECTRIC PERMITTIVITY. (BARIUM TITANATE TYPE MATERIALS) 04-0779
STAL. (DIELECTRIC HEATING AND POLARIZATION REVERSAL, COMPLEX PERMITTIVITY CALCULATION) /F FERROELECTRIC CRY 03-0245
ED ON BARIUM TITANATE. ENERGY STORAGE IN HIGH PERMITTIVITY CERAMICS: SURVEY OF MATERIALS BAS 04-0733
ERY SLOWLY CHANGING FIELD. PERMITTIVITY CHANGE OF BARIUM TITANATE UNDER V 04-0868
CE AND FIELDS RADIATED BY AN OPEN WAVEGUIDE FILLED WITH HIGH PERMITTIVITY DIELECTRIC. (THEORY) /UT ADMITTAN 03-0359
E FERROELECTRICS. (SINGLE CRYSTALS AND CERAMICS) PERMITTIVITY DISPERSION OF BARIUM TITANATE TYP 04-0903
ELECTRICS WITH ORDER DISCRDER TYPE PHASE TRANSIT/ DIELECTRIC PERMITTIVITY DISPERSION PECULIARITIES IN FERRO 03-0412
STEP RADIA/ MEASUREMENT OF THE PYROELECTRIC COEFFICIENT AND PERMITTIVITY FROM THE PYROELECTRIC RESPONSE TO 03-0469
IDEUTERIUM(X) TR/ UNUSUAL EFFECT OF AN ELECTRIC FIELD ON THE PERMITTIVITY OF A NEW DELTA PHASE OF SODIUM TR 22-2875
D BY DISPERSIVE FOURIER TRANSF/ DETERMINATION OF THE COMPLEX PERMITTIVITY OF ADP AND KDP IN THE FAR INFRARE 17-2609
S. THICKNESS DEPENDENCE OF THE PERMITTIVITY OF BARIUM TITANATE SINGLE CRYSTAL 04-0637
S. INFLUENCE OF THE AMPLITUDE OF FREE VIBRATIONS ON THE PERMITTIVITY OF BARIUM TITANATE SINGLE CRYSTAL 04-0768
. EFFECT OF THERMAL NEUTRON IRRADIATION ON THE RELATIVE PERMITTIVITY OF BARIUM TITANATE SINGLE CRYSTAL 04-0781
S AT MICROWAVE FREQUENCIES. MEASUREMENT OF THE PERMITTIVITY OF BARIUM TITANATE SINGLE CRYSTAL 04-0759
/RIMENTAL CHECK OF THE THEORY OF MICROWAVE DISPERSION OF THE PERMITTIVITY OF BARIUM TITANATE TYPE FERROELE/ 04-0649
ITE TYPE STRUCTURE. NATURE OF THE HIGH PERMITTIVITY OF CERIUM ALUMINATES WITH PEROVSK 08-1355
MMONIUM DIHYDROGEN PHOSPHATE. DIELECTRIC PERMITTIVITY OF FINELY DIVIDED POTASSIUM AND A 17-2689
LOW TEMPERATURE DIELECTRIC CELL AND PERMITTIVITY OF HEXAGONAL ICE TO 2 K. 34-3432
TIME DEPENDENCE OF THE DIELECTRIC PERMITTIVITY OF HYDROXYL DOPED BARIUM TITANATE 04-0550
AND AMMONIUM DIHYDROGEN PHOSPHATE CRYSTALS AT MIC/ RELATIVE PERMITTIVITY OF POTASSIUM DIHYDROGEN PHOSPHATE 17-2714
CONTRIBUTION OF DOMAIN WALL MOTION TO THE PERMITTIVITY OF ROCHELLE SALT. 28-3198
WALLS. (ABSTRACT ONLY) LOW FREQUENCY DISPERSIONS OF THE PERMITTIVITY OF ROCHELLE SALT CAUSED BY DOMAIN 28-3199
 PERMITTIVITY OF ROCHELLE SALT DURING SWITCHING 28-3196
SULFATE IN A STRONG ELECTRIC FIELD. PERMITTIVITY OF SINGLE CRYSTALS OF TRIGLYCINE 27-2978
1000 DEG C. (BARIUM T/ TEMPERATURE DEPENDENCE OF THE COMPLEX PERMITTIVITY OF SOME TITANATES BETWEEN 20 AND 04-0820
 PERMITTIVITY OF STRONTIUM TITANATE. 06-1174
TEMPERATURE AND FIELD STRENGTH DEPENDENCE OF LARGE SIGNAL PERMITTIVITY OF TGS. (10 KHZ-1 MHZ) 27-3068
OMAIN WALL IN BARIUM TITANATE (THECRY, CALCULA/ SMALL SIGNAL PERMITTIVITY OF THE STATIONARY (100)-180 DEG D 03-0358
(NEAR THE CURIE TEMPERATURE) LARGE SIGNAL PERMITTIVITY OF TRIGLYCINE SULFATE AT 100 MHZ. 27-3029
EFFECT CF TEMPERATURE). (WEAK FIELD) DIELECTRIC (PERMITTIVITY) PROPERTIES OF BISMUTH TITANATE (36-3555
VE FIELD) OF BISMUTH TITANATE AT LCW T/ DIELECTRIC BEHAVIOR (PERMITTIVITY, SPONTANEOUS POLARIZATION, COERCI 08-1436
D LEAD TITANATES, EFFECT/ PREPARATICN AND PROPERTIES OF HIGH PERMITTIVITY THIN FILM DIELECTRIC. (BARIUM AN 04-0670
HIGH PRESSURE SYNTHESIS OF LEAD METAL- OXYGEN(3) TYPE PEROVSKITE. 08-1331
GROUP ANALYSIS OF LATTICE VIBRATIONS OF CUBIC PEROVSKITE ABO(3). 08-1321
AND BISMUTH(0.5)) (NIOBIUM, TANTALU/ NEW FERROELECTRICS WITH PEROVSKITE AND PYROCHLORE STRUCTURES. (BARIUM 08-1462
STUDY OF TH/ NEW FERROELECTRICS ANC SEIGNETTO MAGNETS OF THE PEROVSKITE AND PYROCHLORE TYPE. SYNTHESIS AND 02-0128
PENDENCE OF TRANSITION TEMPERATURES AND ELECTROSTRICTIONS IN PEROVSKITE BARIUM TITANATE. PRESSURE DE 04-0853
E(0.5) NIOBATE(0.5) AND LEAD MANGANATE(0.33) N/ SYNTHESIS OF PEROVSKITE COMPLEX COMPOSITIONS: LEAD MANGANAT 08-1453
/N OF INTERNAL ELECTRIC FIELDS AND THEIR GRADIENTS IN ABO(3) PEROVSKITE COMPOUNDS WITH DISTINCTIVE DIELECT/ 08-1455
SIUM TANTALA/ LIFETIME OF THE "FERROELECTRIC" PHONON IN SOME PEROVSKITE CRYSTALS (STRONTIUM TITANATE, POTAS 03-0227
Y, DIRECTIVITY CONDITIONS FOR ELECTRIC ORDERING AND STABILI/ PEROVSKITE FAMILY AND CHEMICAL BOND. (COVALENC 03-0243
/LCULATION OF THE ELECTROSTATIC INTERACTIONS IN FINE GRAINED PEROVSKITE FERROELECTRIC CERAMICS. (BARIUM TI/ 04-0642

MOVEMENT OF 180 DEGREE DOMAIN BOUNDARIES IN PEROVSKITE FERROELECTRICS. 08-1305
RADIATION AND ACOUSTIC EXCITATION ON ANOMALOUS LAYERS OF PEROVSKITE FERROELECTRICS. 08-1388
ORPHOTROPIC PHASE TRANSITION TETRAGONAL- RHOMBOHEDRAL IN THE PEROVSKITE FERROELECTRICS. /IC THEORY OF THE M 03-0172
S OF LOCALIZED D-STRAIN AND SOFT- NORMAL MODE COO/ THEORY OF PEROVSKITE FERROELECTRICS (HAMILTONIAN IN TERM 03-0432
OPIC DEBYE- WALLER FAC/ LOW FREQUENCY ZONE BOUNDARY MODES IN PEROVSKITE FERROELECTRICS INDICATED BY ANISOTR 08-1351
RE COEFFICIENT OF RESISTIVITY) IN (DOPED AND UNDOPED CERAMIC PEROVSKITE) LEAD FERRATE(0.5) NIOBATE(0.5). /U 13-2000
/N OF POLARIZATION NONLINEARITY NEAR THE PHASE TRANSITION IN PEROVSKITE LIKE FERROELECTRICS, USING AN ANHA/ 03-0249
SMUTH TUNGSTATE, BISMUTH NIOBIUM OXYFL/ CRYSTAL CHEMISTRY OF PEROVSKITE LIKE LAYER TYPE FERROELECTRICS. (BI 13-1966
BISMUTH TITANATE FERR/ MOSSBAUER STUDIES OF IRON-57 IN SOME PEROVSKITE LIKE LAYER TYPE FERROELECTRICS. (IN 13-2035
ND THE ANTIFERR/ HIGH TEMPERATURE X-RAY INVESTIGATION OF THE PEROVSKITE MODIFICATION OF CADMIUM TITANATE (A 08-1387
/BOUT THE CURIE POINT AND NATURE OF DIELECTRIC PROPERTIES IN PEROVSKITE MODIFICATION OF CADMIUM TITANATE. / 08-1424
D LIGHT MODULATORS. (POTASSIUM TANTALATE(0.65) NIOBA/ USE OF PEROVSKITE PARAELECTRICS IN BEAM DEFLECTORS AN 13-1924
TANIUM DIOXIDE- LANTHANUM TRIOXIDE. RANGE OF EXISTENCE OF PEROVSKITE PHASES IN THE SYSTEM LEAD OXIDE- TI 08-1349
IC G STRUCT/ NEUTRON DIFFRACTION STUDY OF (TETRAGONAL- CUBIC PEROVSKITE) STRONTIUM IRON TUNGSTATE . (MAGNET 10-1655
UENCIES, CALCULATIONS). ATOMIC DISPLACEMENTS IN PEROVSKITE STRONTIUM TITANATE (AT OPTICAL FREQ 06-1142
/CE OF OPTICAL PHONON FREQUENCIES (RAMAN SCATTERING IN CUBIC PEROVSKITE STRONTIUM TITANATE, 8-250K, 0.2-12/ 06-1241
ICAL PROPERTIES OF NONLINEAR FERROELECTRIC CERAMICS WITH THE PEROVSKITE STRUCTURE. ELECTR 08-1461
ROWTH AND STUDY OF NEW SINGLE CRYSTAL CADMIUM COMPOUNDS WITH PEROVSKITE STRUCTURE. G 08-1429
INTERATOMIC DISTANCES IN OXIDES WITH THE PEROVSKITE STRUCTURE. 08-1439
S DEFORMATION IN THE FERROELECTRIC PHASES OF OXIDES WITH THE PEROVSKITE STRUCTURE. /ACEMENTS AND SPONTANEOU 08-1377
ECT IN THE REPOLARIZATION PROCESS OF FERROELECTRICS WITH THE PEROVSKITE STRUCTURE. /VERSE PIEZOELECTRIC EFF 08-1374
C THEORY OF DOMAIN WALLS IN FERROELECTRIC MATERIALS WITH THE PEROVSKITE STRUCTURE. THERMODYNAMI 03-0190
D DATA, CLASSIFICATI/ ANTIFERROELECTRICITY IN COMPOUNDS WITH PEROVSKITE STRUCTURE. (EXAMINATION OF PUBLISHE 01-0018
/SIUM AND NIOBIUM IN THE OCTAHEDRAL POSITIONS OF THE "CUBIC" PEROVSKITE STRUCTURE OF LEAD MAGNESIUM NIOBAT/ 08-1379
/URE AND PROPERTIES OF SOME NEW FERROELECTRIC MATERIALS WITH PEROVSKITE STRUCTURE (PEROVSKITE COMPOUNDS CO/ 08-1460
ICITY TO LATTICE DYNAMI/ ACOUSTIC PROPERTIES OF MATERIALS OF PEROVSKITE STRUCTURE. (RELATION OF FERROELECTR 02-0043
ATE DATA) OPTICAL MODES OF THE PEROVSKITE STRUCTURE. (THEORY, STRONTIUM TITAN 03-0304
). FORBIDDEN ZONE WIDTHS IN POLYCRYSTALLINE DIELECTRICS WITH PEROVSKITE STRUCTURE (TITANATES AND ZIRCONATES 08-1288
/EFRACTIVE INDEX AND ELECTROOPTICAL PHENOMENA IN CRYSTALS OF PEROVSKITE STRUCTURE (WITH APPLICATION TO BAR/ 03-0518
M NIOBATE AND THE USE OF TILTING SCHEMES IN THE SOLUTIONS OF PEROVSKITE STRUCTURES. /E TRANSITIONS IN SODIU 08-1336
TITANATE. LATTICE DYNAMICS OF CUBIC PEROVSKITE STRUCTURES, IN PARTICULAR STRONTIUM 06-1185
PROPERTIES. ORIGIN OF DIPOLE CONFIGURATIONS IN SOME (PEROVSKITE) STRUCTURES WITH SPECIAL DIELECTRIC 08-1390
PROPOSED MECHANISM FOR THERMOPHOTOTROPIC BEHAVIOR IN PEROVSKITE STRUCTURED TITANATES. 03-0400
ITION AND PROPERTIES IN FERROELECTRIC SOLID SOLUTIONS OF THE PEROVSKITE TYPE. /THE RELATIONS BETWEEN COMPOS 08-1330
/ONTANEOUS POLARIZATION AND PHASE TRANSITIONS IN CRYSTALS OF PEROVSKITE TYPE AND OF ONES WITH TETRAHEDRAL / 03-0175
LATTICE DYNAMICS OF PHASE TRANSITIONS IN PEROVSKITE TYPE ANTIFERROELECTRICS. 08-1475
X-RAY AND DIELECTRIC STUDIES OF IRRADIATED PEROVSKITE TYPE COMPOUNDS. 08-1443
IUM(X) MANGANATE (PH/ PREPARATION AND PHYSICAL PROPERTIES OF PEROVSKITE TYPE COMPOUNDS, LANTHANUM(1-X) CALC 08-1356
ED BY THE PERTURBED ANGU/ NUCLEAR QUADRUPOLE INTERACTIONS IN PEROVSKITE TYPE COMPOUNDS OF HAFNIUM-181 STUDI 08-1276
/NTERNAL ELECTRIC FIELDS AND ELECTRIC FIELD GRADIENTS IN THE PEROVSKITE TYPE COMPOUNDS WITH SPECIAL DIELEC/ 08-1442
STRUCTURAL PHASE TRANSITIONS IN PEROVSKITE TYPE CRYSTALS. (EXTENDED ABSTRACT) 08-1450
/LATTICE DYNAMICAL THEORY OF STRUCTURAL PHASE TRANSITIONS IN PEROVSKITE TYPE CRYSTALS. (STRONTIUM TITANATE/ 03-0424
/ ORIGIN OF FERROELECTRICITY, OCCURRENCE OF ORDERED PHASE IN PEROVSKITE TYPE FERROELECTRICS. (BARIUM TITAN/ 04-0964
FS) POINT DEFECTS IN PEROVSKITE TYPE FERROELECTRICS. (REVIEW, 77 RE 02-0041
/NTHESIS (SOLID STATE REACTION, SINTERING) AND PROPERTIES OF PEROVSKITE TYPE LEAD ZINCATE(0.33) NIOBATE(0./ 08-1393
PPLICATION TO THE SOLID ELECTROLYTE FUE/ IONIC CONDUCTION IN PEROVSKITE TYPE OXIDE SOLID SOLUTION AND ITS A 08-1444
IU/ ELECTRON MICROSCOPIC STUDIES ON FERROELECTRIC DOMAINS IN PEROVSKITE TYPE OXIDES. (LEAD TITANATE, POTASS 08-0993
FERROELECTRICITY AND THE CHEMICAL BOND IN PEROVSKITE TYPE OXIDES. (43 REFS) 02-0102
/ESTIGATION OF SOME FERROELECTRICS AND ANTIFERROELECTRICS OF PEROVSKITE TYPE. (POLYCRYSTALLINE AND SINGLE / 08-1375
PEROVSKITE TYPE SEIGNETTO MAGNETS. 08-1459
BARDMENT. SPLIT INTERSTITIAL CONFIGURATION INDUCED IN PEROVSKITE TYPE SINGLE CRYSTALS BY NEUTRON BOM 08-1467
RTIES OF LEAD (STANNATE(0.5) NIOBATE(0.5)) (1-X) ZIRCONATE(X) PEROVSKITE TYPE SOLID SOLUTIONS. /ECTRIC PROPE 09-1548
URE TO 6/ MOSSBAUER STUDIES OF BISMUTH FERRATE LEAD TITANATE PEROVSKITE TYPE SOLID SOLUTIONS (ROOM TEMPERAT 08-1471
F THE CHEMICAL BONDS IN ABO(3) FERROELECTRIC CRYSTALS WITH A PEROVSKITE TYPE STRUCTURE. CHARACTER O 08-1341
NATURE OF THE HIGH PERMITTIVITY OF CERIUM ALUMINATES WITH PEROVSKITE TYPE STRUCTURE. 08-1355
)) (/ PREPARATION AND PROPERTIES OF THIN FILM MATERIALS WITH PEROVSKITE TYPE STRUCTURE. (MIXED OXIDES ABO(3 04-0914
/ICS OF POLARIZATION OF FERROELECTRIC SOLID SOLUTIONS OF THE PEROVSKITE TYPE WITHIN A WIDE RANGE OF TEMPER/ 08-1380
INFRARED AND RAMAN SPECTRA OF SEVERAL PEROVSKITE ZIRCONATES. 08-1411
(SOLID SOLUTIONS AND MIXED CATION COMPOUNDS, KANZIG REGIONS, PEROVSKITES) /LAXATION IN SOME FERROELECTRICS 03-0291
PROBLEM OF FERROELECTRICITY AND ANTIFERROELECTRICITY IN THE PEROVSKITES. CERTAIN ASPECTS OF THE 08-1456
L CHEMISTRY OF THE ABX(3) FAMILY WITH PARTICULAR EMPHASIS ON PEROVSKITES. CRYSTA 08-1403
DISPLACEMENT PHASE TRANSITIONS IN PEROVSKITES. 08-1274
TRIC PROPERTIES OF SOME FERRO-, ANTIFERRO- AND PARAELECTRIC PEROVSKITES. EFFECT OF IRRADIATION OF DIELEC 08-1474
ZIRCONATES, 2 SUBSTITUTED LEAD ZIRCONATE STANNATES, COMPLEX PEROVSKITES) /CS. (8 SUBSTITUTED LEAD TITANATE 08-1418
NAL ELECTRIC FIELD GRADIENTS IN FERROELECTRIC FERROMAGNETS. (PEROVSKITES) / THEORETICAL ESTIMATION OF INTER 08-1401
THE PROBLEM OF FERROELECTRICITY AND ANTIFERROELECTRICITY IN PEROVSKITES. SOME ASPECTS OF 03-0504
ENDENT SPECTRUM OF AN ANTIFERROELECTRIC LINEAR CHAIN MODEL. (PEROVSKITES) TEMPERATURE DEP 03-0468
THEORY OF STRUCTURAL PHASE TRANSITION IN PEROVSKITES. 03-0242
IONS FAVORING THE ORIGINATION OF SPONTANEOUS POLARIZATION IN PEROVSKITES. (ABSTRACT ONLY) CONDIT 08-1457
ATION. (REVIEW OF TECHNIQUE AND APPLICATION TO THIN FILMS OF PEROVSKITES AND III-V COMPOUNDS, 8 REFS). /POR 02-0112
STUDY OF THE DIELECTRIC PROPERTIES OF NEW CADMIUM-CONTAINING PEROVSKITES AND PYROCHLORES. /Y ANALYSIS, AND 08-1309
MUT/ (CRYSTALLOGRAPHIC AND DIELECTRIC) PROPERTIES OF BISMUTH PEROVSKITES. (AND THE FERROELECTRIC STATE, BIS 08-1289
ANALYSIS. SYNTHESIS OF THALLIUM CONTAINING PEROVSKITES AT HIGH PRESSURES AND THEIR X-RAY 08-1463
(NOT FERROELECTRIC/ SYNTHESIS AND X-RAY INVESTIGATION OF THE PEROVSKITES BARIUM OR STRONTIUM METAPLUMBATES 36-3607
/RDER IN CRYSTALS (CASE OF SILICON, QUARTZ AND FERROELECTRIC PEROVSKITES). (BARIUM TITANATE AND POTASSIUM / 04-0629
LATE) FAR INFRARED DIELECTRIC DISPERSION OF PEROVSKITES. (BARIUM TITANATE, POTASSIUM TANTA 04-0546
ANTIFERROMAGNETIC PROPERTIES OF SOME PEROVSKITES. (BISMUTH FERRATE) 08-1441
3 RESONANCE IN POTASSIUM / NMR STUDY OF PHASE TRANSITIONS IN PEROVSKITES CRYSTALS. (NIOBIUM-93 AND SODIUM-2 08-1283
ANALYSIS OF BAND SCHEME, STRONTIUM / ENERGY BAND CHANGES IN PEROVSKITES DUE TO LATTICE POLARIZATION. (LCAO 06-1101
ICATION) CRYSTALLOCHEMISTRY OF PEROVSKITES. (EXISTENCE CONDITIONS AND CLASSIF 08-1324
IUM TANTALATE AND TA/ ELECTROOPTIC PROPERTIES OF SOME ABO(3) PEROVSKITES IN THE PARAELECTRIC PHASE. (POTASS 04-0690
/ATION OF SPONTANEOUS POLARIZATION AND COERCIVE FORCE OF THE PEROVSKITES IN THE SYSTEM LEAD ZIRCONATE TITA/ 09-1597
PRESSURE STUDIES OF FERROELECTRIC PROPERTIES. (PEROVSKITES, KDP TYPE) 08-1435
/CTRIC PROPERTIES AND PHASE TRANSITIONS OF THE FERROELECTRIC PEROVSKITES: LEAD TITANATE AND BISMUTH TITAN/ 08-1434
IC PROPERTIES AND PHASE TRANSITIONS OF THE ANTIFERROELECTRIC PEROVSKITES LEAD ZIRCONATE AND LEAD HAFNATE. / 08-1433
DATE. PHASE TRANSITIONS IN COMPLEX PEROVSKITES OF THE TYPE BARIUM LANTHANON MOLYB 30-3261
STRUCTURE AND TRANSITIONS IN PEROVSKITES. (ORTHOFERRITES, SODIUM NIOBATE) 08-1396
/MINATION OF THE NORMAL VIBRATIONAL DISPLACEMENTS OF SEVERAL PEROVSKITES (POTASSIUM TANTALATE, STRONTIUM T/ 06-1139
T/ STRUCTURE AND DIELECTRIC PROPERTIES OF LITHIUM CONTAINING PEROVSKITES. (STRONTIUM LITHIUM NIOBATE, STRON 08-1292
/ENDED FINE STRUCTURE IN X-RAY ABSORPTION SPECTRA OF CERTAIN PEROVSKITES (STRONTIUM OR CALCIUM TITANATE OR/ 03-0414

C CRITICAL TEMPERATURE (ROTATIONS OF BO6 OCTAHEDRA IN ABO(3) PEROVSKITES, STRONTIUM TITANATE). /ERROELECTRI 06-1205
/EISEN PARAMETER OF THE SOFT FERROELECTRIC MODE IN THE CUBIC PEROVSKITES. (STRONTIUM TITANATE, POTASSIUM T/ 02-0114
STRUCTURE AND DIELECTRIC LOSSES OF FERROELECTRIC PEROVSKITES SYNTHESIZED AT HIGH PRESSURES. 08-1452
BDATE. (WHERE A IS ZIRCONIUM, HAFNIUM, CERIUM, THO/ THALLIUM PEROVSKITES. THALLIUM A(1-X) TUNGSTATE OR MOLY 10-1640
IC PROPERTIES AND OPTICAL PHONONS IN PARA- AND FERROELECTRIC PEROVSKITES. (THEORY, ABO(3) TYPE)   DIELECTR 03-0413
/CTURE AND PROPERTIES OF NEW CADMIUM AND THALLIUM CONTAINING PEROVSKITES. (TITANATES, HAFNATES, MIXED NIOB/ 08-1458
WITH ELASTIC ST/ THEORY OF A STRUCTURAL PHASE TRANSITION IN PEROVSKITE- TYPE CRYSTALS, PART-2, INTERACTION 03-0241
M, MANGANESE, IRO/. SUPERSTRUCTURE OF BISMUTH METAL OXYGEN(3) PEROVSKITES. (WHERE METAL IS SCANDIUM, CHROMIU 08-1451
STUDY OF THEIR STRUCT/ GROWTH OF SINGLE CRYSTALS OF VARIOUS PEROVSKITES WITH COMPLEX COMPOSITIONS, AND THE 08-1428
STRONTIUM TITANATE AND POTASSIUM T/ THERMAL CONDUCTIVITY IN PEROVSKITES (3 PHONON INTERACTION PROPOSED FOR 06-1095
ATOMIC STRUCTURE IN THE PEROVSKITIC FERRATE. 08-1398
/ATE, A MODEL IN WHICH THE POLARIZATION VECTOR, P(S), ARISES PERPENDICULAR TO THE WAVE NORMAL OF THE SOFT / 03-0141
NG TO ELASTIC SURFACE WAVES ON PIEZOELECTRIC SUBSTRATES (LI/ PERTURBATION THEORY FOR ELECTROMAGNETIC COUPLI 11-1751
CTRICS: FIRST PERTURBATIVE CORRECTION.   STATISTICAL PERTURBATION THEORY OF ORDER DISORDER FERROELE 03-0399
PERTURBATION THEORY OF ORDER DISORDER FERROELECTRICS: FIRST PERTURBATIVE CORRECTION.   STATISTICAL 03-0399
/ IN PEROVSKITE TYPE COMPOUNDS OF HAFNIUM-181 STUDIED BY THE PERTURBED ANGULAR CORRELATION TECHNIQUE. (LEA/ 08-1276
IELD GRADIENT IN ANTIFERROELECTRIC LEAD ZIRCONATE STUDIED BY PERTURBED ANGULAR CORRELATIONS. /AL ELECTRIC F 07-1255
44 WITHIN THE LATTICE OF FERROELECTRIC BARIUM TITANATE. PERTURBED DIRECTIONAL CORRELATION OF SCANDIUM- 04-0691
44 IN LEAD TITANATE. PERTURBED DIRECTIONAL CORRELATION OF SCANDIUM- 05-1055
44 IN THE FERROELECTRIC PHASES OF BARIUM TITANATE. PERTURBED DIRECTIONAL CORRELATION OF SCANDIUM- 04-0558
ITION IN DEUTERATED TRIGLYCINE FLUOB/ BERYLLIUM-9 QUADRUPOLE PERTURBED NMR STUDY OF THE FERROELECTRIC TRANS 27-2965
INGLE CRYSTAL (FROM SOLUTION) IN STATIC REGIME AND AT NORMAL PH. /IC OF THE AMMONIUM DIHYDROGEN PHOSPHATE S 17-2516
E OF TRIGLYCINE FLUOBERYLLATE IN THE FERRO- AND PARAELECTRIC PHASE.   CRYSTAL STRUCTUR 27-3175
S OF (MONOCLINIC) EUROPIUM SILICATE. (UNSTABLE FERROELECTRIC PHASE) /RAPHIC, OPTICAL AND MAGNETIC PROPERTIE 36-3620
STIC COMPLIANCE IS TEMPERATURE INDEPENDENT IN THE PROTOTYPIC PHASE. /HAT THE ELECTRIC SUSCEPTIBILITY OR ELA 03-0140
ICE DYNAMICS MODEL OF POTASSIUM NITRATE IN ITS FERROELECTRIC PHASE.   LATT 16-2377
TRITE. (CALCULATIONS TO RELATE TO SPECIFIC ANTIFERROELECTRIC PHASE)   LATTICE VIBRATION OF SODIUM NI 16-2317
IVITY OF BARIUM TITANATE SINGLE CRYSTALS IN THE PARAELECTRIC PHASE.   PIEZOELECTRIC ACT 04-0771
IUM DIDEUTERIUM PHOSPHATE CRYSTAL IN PARA- AND FERROELECTRIC PHASE.   RAMAN SPECTRUM OF POTASS 17-2647
THE POTASSIUM DIHYDROGEN PHOSPHATE GROUP IN THE PARAELECTRIC PHASE. /OPERTIES OF FERROELECTRIC CRYSTALS OF 17-2696
NIC GENERATION IN THE TRIGLYCINE CRYSTAL IN THE PARAELECTRIC PHASE.   SECOND HARMO 27-3105
RATIONS OF POTASSIUM DIHYDROGEN ARSENATE IN THE PARAELECTRIC PHASE. /TION. PART-2: INTERNAL FUNDAMENTAL VIB 18-2738
ORTHORHOMBIC- TETRAGONAL TRANSITION OF THE TUNGSTEN TRIOXIDE PHASE.   THERMAL EXPANSION AND THE 10-1626
STRUCTURE OF AMMONIUM SULFATE IN FERROELECTRIC PHASE AND TRANSITION. 19-2774
Y, APPLIED TO LEAD(0.99) NIOBI/ MEASUREMENT OF PIEZOELECTRIC PHASE ANGLES IN A FERROELECTRIC CERAMIC (THEOR 09-1537
DEL FOR HYDROGEN BONDED FERROELECTRICS.   RANDOM PHASE APPROXIMATION FOR THE BLINC-DE GENNES MO 03-0338
THEORETICAL DYNAMIC STUDY OF CUBIC PHASE BARIUM TITANATE. 03-0261
IC PHASES IN GADOLINIUM MOLYBDATE CRYSTALS.   OBSERVATION OF PHASE BOUNDARIES BETWEEN FERRO- AND PARAELECTR 30-3306
STALS NEAR CURIE POINT. PHASE BOUNDARIES IN BARIUM TITANATE SINGLE CRY 04-0577
AD ZIRCONATE(X) TITANATE(1-X) CERAMICS NEAR THE MORPHOTROPIC PHASE BOUNDARY. /CTROMECHANICAL ACTIVITY IN LE 09-1499
IN LEAD ZIRCONATE(X) TITANATE(1-X) CERAMICS NEAR THE MORPHIC PHASE BOUNDARY. /E ELECTROMECHANICAL ACTIVITY 09-1500
PLZT CERAMICS. PHASE BOUNDARY AND DEFECT STRUCTURE STUDIES IN 09-1590
AND FIELD DEPENDENCE OF THE ANTIFERROELECTRIC FERROELECTRIC PHASE BOUNDARY IN PLZT CERAMICS.   TEMPERATURE 09-1491
ON ON THE FERROELECTRIC PROPERTIES OF ANTIMONY SULFOIODIDE. (PHASE BOUNDARY SHAPE STRIATIONS IN DC FIELD) 15-2197
LUTIONS (THERMODYNAMIC EXISTENCE CONDITIONS FOR MORPHOTROPIC PHASE BOUNDARY, STOICHIOMETRY DEPENDENCE). /SO 09-1541
NIDE OR SULFIDE, ANT/ GROWING SINGLE CRYSTALS FROM THE VAPOR PHASE BY A DYNAMIC METHOD (ANTIMONY SESQUISELE 15-2162
F CADMIUM TITANATE (AND THE ANTIFERROELECTRIC- FERROELECTRIC PHASE CHANGE). / THE PEROVSKITE MODIFICATION O 08-1387
DIFFRACTION STUDY OF GADOLINIUM MOLYBDATE (CRYSTALLOGRAPHIC PHASE CHANGE AT CURIE TEMPERATURE).   X-RAY 30-3312
C AND MAGNETIC TRA/ HIGHLY SENSITIVE, SIMPLE CALORIMETER FOR PHASE CHANGE DETECTION (INCLUDING FERROELECTRI 28-3201
SPIN RESONANCE AND THE / INVESTIGATION OF THE FERROELECTRIC PHASE CHANGE IN BARIUM TITANATE USING ELECTRON 04-0827
MBINATION SCATTER SPECTRUM IN THE REGION OF THE ANTISYMMETR/ PHASE CHANGE IN SODIUM NITRITE CRYSTALS AND CO 16-2276
C MODEL. (THEORY) PHASE CHANGE IN THREE DIMENSIONAL FERROELECTRI 03-0451
ESR STUDY OF GAMMA-IRRADIATED POTASSIUM SELENATE (STRUCTURAL PHASE CHANGES). 22-2825
/CONDUCTORS OF THE A(V)B(VI)C(VII) TYPE WITH LOW TEMPERATURE PHASE CHANGES. (EFFECT OF PHOTOCONDUCTIVITY, / 15-2161
TITANATE FERROELECTRICS. (26 REFS) PHASE CHANGES IN TRIGLYCINE SULFATE AND BARIUM 04-0621
IN A FERROELECTRIC MATERIAL. (PZT) PHASE CHANGES INDUCED BY HYDROSTATIC PRESSURE 09-1521
THE DIELECTRIC PROPERTIES OF TRIGLYCINE SELENATE. (AGING AND PHASE CHANGES NEAR THE CURIE POINT) /FIELD ON 27-3131
SAL OF ANTIMONY SULFOIODIDE SINGLE CRYSTALS IN THE REGION OF PHASE COEXISTENCE.   POLARIZATION REVER 15-2296
AMAN SPECTRUM OF SODIUM TRIHYDROGEN DISELENITE (PARAELECTRIC PHASE, COMPARISON WITH FERROELECTRIC).   R 22-2864
METHOD OF INVESTIGATION AND SOME RESULTS. PHASE COMPOSITION OF (LEAD, BARIUM) DINIOBATE. 13-1945
LID SOLUTIONS.   EFFECT OF ELECTRIC FIELD ON PHASE COMPOSITION OF (LEAD, BARIUM) NIOBATE SO 13-1943
RIUM TITANATE.   ON THE PHASE COMPOSITION OF REDUCED AND REOXIDIZED BA 04-0541
DIELECTRIC STUDY OF THE LEAD ZIRCONATE LEAD HAFNATE (PHASE) DIAGRAM. 08-1340
ISMUTH SELENOIODIDE- ANTIMONY SELENOIODIDE SYSTEM ALLOYS. PHASE DIAGRAM AND ELECTRICAL CONDUCTIVITY OF B 15-2122
(ZINCATE NIOBATE)- BARIUM (ZINCATE NIOBATE)- LEAD TITANATE. (PHASE DIAGRAM, LATTICE CONSTANTS) /YSTEM LEAD 05-1070
/A IN WATER SYSTEMS. PART-1: MOLECULAR INTERPRETATION OF THE PHASE DIAGRAM OF ICE. PART-2: DIELECTRIC AND / 34-3485
X) PLUS 0.8 PERCENT TUNGSTEN TRIOXIDE. PHASE DIAGRAM OF LEAD ZIRCONATE(1-X) TITANATE(09-1581
ROGEN(1-X) PHOSPHATE SYSTEM VERSUS CONCENTRATION, TEMPERATU/ PHASE DIAGRAM OF SODIUM TRIDEUTERIUM(X) TRIHYD 17-2666
ERROELECTRIC IRON(1-X) SULFIDE. PHASE DIAGRAM OF THE SEMICONDUCTING UNIAXIAL F 36-3617
OWTH/ OF HEXAGONAL SINGLE CRYSTALS OF RARE EARTH MANGANATES (PHASE DIAGRAMS, CONVECTION EFFECTS). / FLUX GR 36-3536
/STATE, BISMUTH(0.5) (POTASSIUM OR SODIUM(0.5) TITANATES AND PHASE DIAGRAMS FOR BISMUTH(X) POTASSIUM(X) BA/ 08-1289
E AND POTASSIUM DIHYDROGEN ARSENATE TO 52 KBAR AND 400 DEG / PHASE DIAGRAMS OF AMMONIUM DIHYDROGEN PHOSPHAT 17-2478
NITRITE TO 40 KBAR. PHASE DIAGRAMS OF SODIUM NITRITE AND POTASSIUM 16-2361
HIUM OXIDE- NIOBIUM PENTOXIDE SYSTEM. PHASE EQUILIBRIA AND CRYSTAL GROWTH IN THE LIT 11-1683
NTHANUM TITANATE).   SYSTEM LANTHANA TITANIA PHASE EQUILIBRIA AND ELECTRICAL PROPERTIES (LA 36-3592
E SYSTEM.   HIGH TEMPERATURE PHASE EQUILIBRIA IN THE LEAD ZIRCONATE TITANAT 09-1577
ONTIUM DINIOBATE AND POTASSIUM NIOBATE- BARIUM DINIOBATE SY/ PHASE EQUILIBRIA IN THE POTASSIUM NIOBATE- STR 13-2016
- BARIUM GERMANATE. PHASE EQUILIBRIA IN THE SYSTEM BARIUM TITANATE 04-0710
EM BARIUM OXIDE- STRONTIUM OXIDE- NIOBIUM PENTOXIDE. (OXIDE/ PHASE EQUILIBRIA RELATIONS IN THE TERNARY SYST 13-1920
OME PROBLEMS AND RESULTS.   NEW TWO- AND THREE PHASE FERROELECTRIC FERROMAGNETIC MATERIALS. S 03-0360
ARIUM TIT/ ELECTRIC AND MAGNETIC MEASUREMENTS OF CERTAIN TWO PHASE FERROELECTRIC FERROMAGNETIC MIXTURES. (B 04-0565
TRICS- SILVER (BARIUM TITANATE AND BARIUM TITANATE- MAG/ TWO PHASE FERROELECTRIC SYSTEMS. PART-2: FERROELEC 04-0816
MECHANISM AND KINETICS OF SOLID PHASE FORMATION OF LEAD METANIOBATE. 13-2042
OBSERVATION OF PHASE FRONT MOTION IN KDP CRYSTAL. 17-2589
TOR DEVICE. (BISMUTH(4) TITANIUM(3) OXIDE(12)) PHASE HOLOGRAMS IN A FERROELECTRIC PHOTOCONDUC 13-1977
MULTIPLE STORAGE OF THICK PHASE HOLOGRAMS IN LITHIUM NIOBATE. 11-1827
SUBSTITUTIONS)   THE ORTHORHOMBIC PHASE IN MODIFIED BARIUM TITANATE. (EFFECT OF 04-0617
CONATE TITANATE.   PRECIPITATION OF ZIRCONIA PHASE IN NIOBIUM MODIFIED CERAMICS OF LEAD ZIR 09-1540
/FFECTS IN ORIGIN OF FERROELECTRICITY, OCCURRENCE OF ORDERED PHASE IN PEROVSKITE TYPE FERROELECTRICS. (BAR/ 04-0964
YMORPHIC TRANSITION)   RHYTHMIC GROWTH OF NEW PHASE IN POTASSIUM NITRATE. (ALPHA TO BETA POL 16-2263

ATE- LANTHANUM FERRATE.   ANTIFERROELECTRIC PHASE IN THE SYSTEM LEAD TITANATE- LEAD ZIRCON 13-1961
YSTERESIS NEAR THE TRANSITION POINT OF DIFFUSE FERROELECTRIC PHASE. (KAENZIG MODEL) /ATION AND DIELECTRIC H 03-0437
                       INTRINSIC NONSTOICHIOMETRY IN SINGLE PHASE LEAD ZIRCONATE TITANATE.   09-1534
THIUM IODATE.   LARGE NONLINEAR OPTICAL COEFFICIENT AND PHASE MATCHED SECOND HARMONIC GENERATION IN LI 08-1405
UM METANIOBATE CRYSTALS OF DIFFERENT STOICHIOMETRIES.   PHASE MATCHING ANGLES AND TEMPERATURE OF LITHI 11-1665
                                         NONCOLLINEAR PHASE MATCHING EFFECTS IN LITHIUM NIOBATE.   11-1675
NIOBATE).   EFFECT OF OPTICAL INHOMOGENEITIES ON PHASE MATCHING IN NONLINEAR CRYSTALS (LITHIUM 11-1773
BUTION COEFFICIENTS OF LITHIUM OXIDE INTO LITHIUM NIOBATE BY PHASE MATCHING TEMPERATURES. /ON OF THE DISTRI 11-1841
    (CRYSTAL GROWTH BY ADD/ SPATIALLY UNIFORM AND ALTERABLE SHG PHASE MATCHING TEMPERATURES IN LITHIUM NIOBATE 11-1841
                     SUPERSTRUCTURE IN THE FERROELECTRIC PHASE OF AMMONIUM ROCHELLE SALT.   28-3224
ONS IN THE CUBIC AND TETRAGONAL (FERROELECTRICALLY POLARIZED PHASE OF BARIUM TITANATE. /ANNELING OF LIGHT I 04-0688
                     DIELECTRIC ABSORPTION IN THE RHOMOHEDRAL PHASE OF BARIUM TITANATE.   04-0809
             GENERATION OF SECOND OPTICAL HARMONIC IN THE CUBIC PHASE OF BARIUM TITANATE.   04-0538
C CONSTANT)   BAND STRUCTURE OF THE CUBIC PHASE OF BARIUM TITANATE. (IMAGINARY DIELECTRI 04-0845
         UNIT CELL AND SPACE GROUP OF THE ANTIFERROELECTRIC PHASE OF COPPER FORMATE TETRAHYDRATE.   26-2927
N DIFFRACTION.   APPROXIMATE STRUCTURE FOR ANTIFERROELECTRIC PHASE OF COPPER FORMATE TETRAHYDRATE BY NEUTRO 26-2935
         DEUTERON MAGNETIC RESONANCE OF THE HIGH TEMPERATURE PHASE OF FERROELECTRIC THIOUREA.   25-2918
HARMONIC APPROXIMATI/ MOLECULAR MOTION IN THE FERROELECTRIC PHASE OF HYDROGEN CHLORIDE. (CALCULATED IN THE 31-3340
/NTERNAL VIBRATIONS OF THE PHOSPHATE ION IN THE PARAELECTRIC PHASE OF KDP CRYSTALS (INFRARED ABSORPTION AN/ 17-2479
         CRYSTAL GROWTH AND OBSERVATION OF THE FERROELECTRIC PHASE OF LEAD ZIRCONATE.   07-1264
                 DIELECTRIC DISPERSION IN THE PARAELECTRIC PHASE OF POTASSIUM NITRATE.   16-2341
                           STRUCTURE OF THE DISORDERED PHASE OF POTASSIUM NITRATE.   16-2379
PARAMAGNETIC RESONANCE OF MANGANESE(2+) IN THE FERROELECTRIC PHASE OF SODIUM NITRITE.   ELECTRON 16-2358
        X-RAY ANALYSIS OF FERROELECTRIC DOMAINS IN THE PARAELECTRIC PHASE OF SODIUM NITRITE.   16-2271
/LAXATION OF SODIUM-23 IN THE PARAELECTRIC AND FERROELECTRIC PHASE OF SODIUM NITRITE. (THERMODYNAMICS OF T/ 16-2264
/ECT OF AN ELECTRIC FIELD ON THE PERMITTIVITY OF A NEW DELTA PHASE OF SODIUM TRIDEUTERIUM(X) TRIHYDROGEN(1/ 22-2875
IC FIELDS.   DOMAIN STRUCTURE OF THE BETA PHASE OF SODIUM TRIHYDROGEN SELENATE IN ELECTR 22-2830
ANALYSIS OF THE IRON OXYGEN VACANCY CENTER IN THE TETRAGONAL PHASE OF STRONTIUM TITANATE.   06-1231
         THE IRON(3+)- VACANCY CENTER IN THE TETRAGONAL PHASE OF STRONTIUM TITANATE.   06-1234
/ULIAR DOMAIN STRUCTURE AND ITS UNUSUAL BEHAVIOR IN THE BETA PHASE OF THE FERROELECTRIC SODIUM TRIHYDROGEN/ 22-2874
                   CRYSTAL STRUCTURE OF THE IV PHASE OF THIOUREA. (SATELLITE SCATTERING)   25-2923
         FERROELECTRIC DOMAIN SWITCHING IN RHOMBOHEDRAL PHASE PLZT CERAMICS.   09-1563
/ SCATTERING INTENSITY WITH TEMPERATURE IN THE FERROELECTRIC PHASE POTASSIUM DIHYDROGEN PHOSPHATE. (DIFFRA/ 17-2662
/C PROPERTIES OF SOME ABO(3) PEROVSKITES IN THE PARAELECTRIC PHASE. (POTASSIUM TANTALATE AND TANTALATE NIO/ 04-0690
ISM AND KINETICS OF LITHIUM METANIOBATE FORMATION BY A SOLID PHASE REACTION.   MECHAN 11-1840
TITE, IRON(1-X) SULFIDE.   PHASE RELATIONS AND SUPERSTRUCTURES OF PYRRHO 36-3597
ANATE- LEAD METANIOBATE TERNARY SYSTE/ CRYSTAL STRUCTURE AND PHASE RELATIONS IN THE LEAD ZIRCONATE LEAD TIT 09-1596
    AND MOLYBDATES.   STRUCTURAL AND PHASE RELATIONSHIPS AMONG TRIVALENT TUNGSTATES 10-1648
    AND MOLYBDATES.   STRUCTURAL AND PHASE RELATIONSHIPS AMONG TRIVALENT TUNGSTATES 10-1649
) SYSTEM POTASSIA LITHIA NIOBIA (POTASSIUM LITHIUM NIOBATES, PHASE RELATIONSHIPS, SOLID SOLUTIONS). /(OXIDE 13-2017
(0.35) STRONTIUM(0.65/ (COMPACT, SURFACE WAVE) FERROELECTRIC PHASE SHIFTERS FOR VHF AND UHF. (BASED ON LEAD 37-3633
         EXAGGERATED GRAIN GROWTH IN LIQUID PHASE SINTERING OF BARIUM TITANATE.   04-0835
AMMA-IRRADIATED FERROELECTRIC MATERIALS. (ROCHELLE SALT, TR/ PHASE STABILIZATION (HINDERED TRANSITION) IN G 27-3165
/TANATE. (AND MICROSCOPIC MODEL OF LOCALLY ORDERED "NUCLEI", PHASE STRUCTURE ABOVE CURIE POINT, RAMAN SPEC/ 04-0647
EMS. (WHERE A IS STRONTIUM OR BARIUM, AND B IS HAFNIU/ X-RAY PHASE STUDY OF THE CADMIUM BORATE- ABO(3) SYST 36-3528
HYDROGEN PHOSPHATE.   PHASE SYNCHRONIZATION IN NONLINEAR RUBIDIUM DI 17-2473
ND DYNAMICS IN STRONTIUM TITANATE IN THE NEIGHBORHOOD OF THE PHASE TRANSFORMATION.   EPR STUDIES 06-1232
CTROOPTICAL EFFECT IN RUBIDIUM TRIHYDROGEN OR DEUTERIUM DIS/ PHASE TRANSFORMATION AND THERMOOPTICAL AND ELE 22-2840
/C STUDY OF GLASS TRANSITION OF THE AMORPHOUS ICE AND OF THE PHASE TRANSFORMATION BETWEEN THE CUBIC AND HE/ 34-3476
LUTIONS. (THEORY OF COMPOSITION DE/ PARTICULAR PROPERTIES OF PHASE TRANSFORMATION IN FERROELECTRIC SOLID SO 03-0250
HOSPHATE.   NEUTRON MEASUREMENTS ON THE FERROELECTRIC PHASE TRANSFORMATION IN POTASSIUM DIHYDROGEN P 17-2418
    (TERBIUM MOLYBDATE)   MECHANISM OF THE FERROELECTRIC PHASE TRANSFORMATION IN RARE EARTH MOLYBDATES. 30-3254
         NEUTRON SCATTERING STUDY OF THE FERROELECTRIC PHASE TRANSFORMATION IN TERBIUM MOLYBDATE.   30-3268
IC- FERROELASTIC CRYSTAL: DIAM/ CONJECTURED MECHANISM OF THE PHASE TRANSFORMATION OF A PECULIAR FERROELECTR 19-2741
RISULFATE, A MODEL IN WHICH / MECHANISM OF THE FERROELECTRIC PHASE TRANSFORMATION OF DIAMMONIUM DICADMIUM T 03-0141
- FERROELASTIC CRYSTAL: SODIUM AM/ INFERRED MECHANISM OF THE PHASE TRANSFORMATION OF PECULIAR FERROELECTRIC 29-3238
HOSPHATE.   ORDER OF THE FERROELECTRIC PHASE TRANSFORMATION OF POTASSIUM DIHYDROGEN P 17-2555
         X-RAY AND NEUTRON STUDY ON THE PHASE TRANSFORMATION OF SODIUM NITRITE.   16-2328
/RIC AND ANTIFERROELECTRIC CERAMICS. PART-1: DEFORMATION AND PHASE TRANSFORMATION. PART-2: ELECTRIC CONDUC/ 03-0169
ALKALI IODATES.   NUCLEAR QUADRUPOLE RESONANCE, PHASE TRANSFORMATIONS, AND FERROELECTRICITY OF 08-1350
KBAR (BY DTA).   PHASE TRANSFORMATIONS AND MELTING IN KDP TO 40 17-2634
/GETHER WITH EXAMINATION OF CHARACTERS OF THEIR PARAELECTRIC PHASE TRANSFORMATIONS. (AND SUGGESTIONS FOR D/ 03-0138
/VSKITE TYPE COMPOUNDS, LANTHANUM(1-X) CALCIUM(X) MANGANATE (PHASE TRANSFORMATIONS CONFIGURATIONAL ENERGY). 08-1356
         DYNAMICAL EFFECTS IN SOLID STATE PHASE TRANSFORMATIONS. (FERROELECTRICS)   03-0154
CHARACTERISTICS OF THE STRUCTURAL CHANGES INVOLVED IN PHASE TRANSFORMATIONS IN ANTIFERROELECTRICS.   03-0526
LEAD TITANATES. (/ (IN PILE, FAST NEUTRON) RADIATION INDUCED PHASE TRANSFORMATIONS IN (CERAMIC) BARIUM AND 04-0971
IN PRESENCE OF DOMAIN WALLS. (THEORY)   PHASE TRANSFORMATIONS IN FERROELECTRIC CRYSTAL 03-0346
         NEUTRON STUDIES OF DISPLACIVE STRUCTURAL PHASE TRANSFORMATIONS. (IN FERROELECTRICS)   03-0156
. PART-1: CHANGE IN ULTRAVIOLET ABSORPTION SPECT/ STUDIES OF PHASE TRANSFORMATIONS IN NITRATES AND NITRITES 16-2278
POTASSIUM NITRITE. (DOES NOT SEEM FERROELECTRIC- PARAELECT/ PHASE TRANSFORMATIONS IN POTASSIUM NITRATE AND 16-2369
         NEW DATA FROM AN X-RAY STUDY OF PHASE TRANSFORMATIONS IN SODIUM NIOBATE.   08-1361
         X-RAY DIFFRACTOMETRY OF LOW TEMPERATURE PHASE TRANSFORMATIONS IN STRONTIUM TITANATE.   06-1156
CS.   THEORY OF VIBRONIC PHASE TRANSFORMATIONS IN WIDE GAP FERROELECTRI 03-0349
FERROELECTRICS)   NEUTRON STUDIES OF DISPLACIVE STRUCTURAL PHASE TRANSFORMATIONS. (INELASTIC DIFFUSION IN 03-0155
CS.   PHASE TRANSFORMATIONS OF IMPROPER FERROELECTRI 03-0335
SAMARIUM MOLYBDATE (SOLID PHASE SYNTHESIS, N/ FORMATION AND PHASE TRANSFORMATIONS OF THE M MODIFICATION OF 30-3270
AD COBALTATE(0.5) TUNGSTATE(0.5) AND LEAD COBALTATE(0.5) TU/ PHASE TRANSISTION IN THE ANTIFERROELECTRET LE 13-1964
/ABSORPTION SPECTRA OF 23.8 KEV GAMMA-RAYS OF TIN-119 DURING PHASE TRANSISTIONS OF THE BISMUTH FERRATE STR/ 08-1400
SCATTERING FROM SODIUM NIOBATE CRYSTALS (NEAR THE 640 DEG C PHASE TRANSITION). /PIC CRITICAL X-RAY DIFFUSE 08-1311
EHAVIOR OF POTASSIUM MANGANESE TRIFLUORIDE NEAR A STRUCTURAL PHASE TRANSITION.   ANOMALOUS ELASTIC B 33-3392
ATION OF THE UNIVERSALITY HYPOTHESIS IN A LATTICE MODEL OF A PHASE TRANSITION.   APPARENT VIOL 03-0515
F STRONTIUM TITANATE PRODUCED BY THE 105 DEGREE K STRUCTURAL PHASE TRANSITION.   BIREFRINGENCE O 06-1116
TANNATE SOLID SOLUTIONS IN THE VICINITY OF THE FERROELECTRIC PHASE TRANSITION. /IN-119 IN BARIUM TITANATE S 04-0573
         CRITICAL BEHAVIOR OF STRONTIUM TITANATE NEAR THE 105 K PHASE TRANSITION.   06-1194
AND ITS DIELECTRIC PROPERTIES. (FERROELECTRIC- PARAELECTRIC PHASE TRANSITION) /ADMATE(0.33) NIOBATE(0.67)) 13-1959
         DIELECTRIC PERMEABILITY OF TRIGLYCINE SELENATE NEAR THE PHASE TRANSITION.   27-3166
ION PECULIARITIES IN FERROELECTRICS WITH ORDER DISORDER TYPE PHASE TRANSITION. /ECTRIC PERMITTIVITY DISPERS 03-0412
STIC PROPERTIES OF GADOLINIUM MOLYBDATE AT ITS FERROELECTRIC PHASE TRANSITION.   ELA 30-3283
         ELASTOOPTICAL EFFECT IN FERROELECTRIC WITH DIFFUSE PHASE TRANSITION.   03-0307
TROOPTICAL EFFECT IN FERROELECTRIC CRYSTALS HAVING A SMEARED PHASE TRANSITION.   ELEC 03-0174

F AMMONIUM BISULFATE IN THE VICINITY OF THE HIGH TEMPERATURE PHASE TRANSITION. / FERROELECTRIC PROPERTIES O 19-2787
E LEAD TITANATE ZIRCONATE COMPOSITIONS NEAR THE MORPHOTROPIC PHASE TRANSITION. /ROPERTIES OF POLYCRYSTALLIN 09-1557
ACE EFFECTS ON THE PYROELECTRIC BEHAVIOR OF TGS CLOSE TO THE PHASE TRANSITION.   INFLUENCE OF SURF 27-3064
                      MICROWAVE LOSSES IN STRONTIUM TITANATE ABOVE THE PHASE TRANSITION.   06-1199
SORPTION EDGE OF STRONTIUM TITANATE AROUND THE 105 DEGREES K PHASE TRANSITION.   OPTICAL AB 06-1099
                 POLARIZATION FLUCTUATIONS NEAR A FERROELECTRIC PHASE TRANSITION.   03-0368
RRANGEMENT OF THE DOMAIN STRUCTURE IN ROCHELLE SALT NEAR THE PHASE TRANSITION.   REA 28-3181
ROELECTRIC LEAD MAGNESATE(0.33) NIOBATE(0.67) WITH A DIFFUSE PHASE TRANSITION. /ATION POLARIZATION OF A FER 13-1980
ATTERING SPECTRUM OF FERROELECTRIC DEUTERATED KDP DURING THE PHASE TRANSITION. /TURE DEPENDENCE OF RAMAN SC 17-2626
            THEORY OF RAYLEIGH AND BRILLOUIN SCATTERING NEAR THE PHASE TRANSITION.   03-0334
OF ADP AND DEUTERATED ADP IN THE REGION OF ANTIFERROELECTRIC PHASE TRANSITION.   THERMAL EXPANSION 17-2445
POTASSIUM DIHYD/ STRUCTURAL SENSITIVITY OF THE FERROELECTRIC PHASE TRANSITION AND DIELECTRIC PROPERTIES OF 17-2615
ONY SULFOIODIDE CRYSTALS. (MICRO CINEMATOGRAPHY)   PHASE TRANSITION AND DOMAIN STRUCTURE IN ANTIM 15-2184
        ANOMALOUS ABSORPTION OF SOUND NEAR THE FERROELECTRIC PHASE TRANSITION (ANTIMONY SULFOIODIDE).   15-2221
/ FERROELECTRIC SEMICONDUCTORS AT TEMPERATURES NEAR A TYPE-1 PHASE TRANSITION. (ANTIMONY SULFOIODIDE, SELE/ 15-2114
/DUCTIVITY ANOMALY AS AN EVIDENCE OF THE LOCAL FERROELECTRIC PHASE TRANSITION AROUND 47 DEG K IN STRONTIUM/ 06-1249
TRIC NONLINEARITY COEFFICIENT IN STRONTIUM TITANATE NEAR THE PHASE TRANSITION AT 110 DEG K. / OF THE DIELEC 06-1105
TRIC NONLINEARITY OF STRONTIUM TITANATE SINGLE CRYSTALS NEAR PHASE TRANSITION AT 110 DEGREES K.   DIELEC 06-1106
ANTIFERROELECTRIC DOMAINS OF AMMONIUM DIHYDROGEN PHOSPHATE. (PHASE TRANSITION AT 148 DEG K) /BSERVATION OF 17-2643
HE NATURE OF DIELECTRIC POLARIZATION RELAXATION I/ CAUSES OF PHASE TRANSITION BROADENING (DIFFUSNESS) AND T 03-0291
PIEZOELECTRIC PROPERTIES OF A CERAM/ EFFECT OF FERROELECTRIC PHASE TRANSITION CONDITIONS ON DIELECTRIC AND 09-1591
ORY OF FERROELECTRICS.   A NEW TYPE OF SECOND ORDER PHASE TRANSITION DERIVED FROM DEVONSHIRE'S THE 03-0226
       THE BEHAVIORS OF THE MICRODOMAINS ON THE DIFFUSE PHASE TRANSITION FERROELECTRICS.   03-0283
MAGNETICALLY ORDERED FERROELECTRIC STATE. (THEORY)   PHASE TRANSITION FROM NORMAL FERROELECTRIC TO 03-0394
 FERROMAGNETIC AND FERROELECTRIC STA/ THEORY OF SECOND ORDER PHASE TRANSITION FROM THE FERROMAGNETIC TO THE 03-0395
ECTRIC DIPOLE STRUCTURE: CESIUM COPPER TRICHLORIDE.   A PHASE TRANSITION IN A COMPOUND WITH HELICAL EL 33-3388
GEN BONDS/ EFFECT OF VIOLATION OF ELECTRIC NEUTRALITY ON THE PHASE TRANSITION IN A FERROELECTRIC WITH HYDRO 03-0449
/ERVATION OF FERROELECTRIC DOMAIN STRUCTURES, MODULATED BY A PHASE TRANSITION, IN A SCANNING ELECTRON MICR/ 03-0474
ODEL (FERROELECTRIC SQUARE LATTICE).   PHASE TRANSITION IN A SIXTEEN VERTEX LATTICE M/ 03-0517
E SPECTRA AND SPIN LATTICE RELAXATION) INVESTIGATIONS OF THE PHASE TRANSITION IN ALUMS. /ESONANCE (WIDE LIN 20-2802
AND DEUTERATED AMMONIUM DICHLORO ACETA/ FERROELECTRICITY AND PHASE TRANSITION IN AMMONIUM DICHLORO ACETATE 36-3538
ATE.   A THEORY OF THE PHASE TRANSITION IN AMMONIUM DIHYDROGEN PHOSPH 03-0286
ATE SINGLE CRYSTAL BY NMR.   INVESTIGATION OF THE PHASE TRANSITION IN AMMONIUM DIHYDROGEN PHOSPH 17-2545
E, MONOCHLORO ACETIC / NUCLEAR QUADRUPOLE RESONANCE STUDY OF PHASE TRANSITION IN AMMONIUM MONOCHLORO ACETAT 36-3625
HEORY)   MECHANISM OF FERROELECTRIC PHASE TRANSITION IN AMMONIUM ROCHELLE SALT. (T 28-3223
               LATTICE MODES AND PHASE TRANSITION IN ANTIMONY SULFOIODIDE.   15-2110
        SOFT PHONON COUPLINGS IN THE PRESSURE INDUCED PHASE TRANSITION IN ANTIMONY SULFOIODIDE.   15-2240
ALS.   MOTION PICTURES OF THE PHASE TRANSITION IN ANTIMONY SULFOIODIDE CRYST 15-2120
E CRYSTALS.   FEATURES OF THE PHASE TRANSITION IN ANTIMONY SULFOIODIDE SINGL 15-2148
E CRYSTALS. EFFECT OF OXYGEN ADSORPTION ON THE FERROELECTRIC PHASE TRANSITION IN ANTIMONY SULFOIODIDE SINGL 15-2187
(3+) OXIDE.   PHASE TRANSITION IN BARIUM BISMUTH(5+) BISMUTH 13-1940
ERMAL EX/ DILATOMETRIC STUDY OF THE ORTHORHOMBIC- TETRAGONAL PHASE TRANSITION IN BARIUM SODIUM NIOBATE. (TH 13-1894
        FIELD INDUCED NUCLEATION AT THE (CUBIC TETRAGONAL) PHASE TRANSITION IN BARIUM TITANATE.   04-0753
        ULTRASONIC STUDIES OF FIELD INDUCED PHASE TRANSITION IN BARIUM TITANATE.   04-0549
STALS.   EFFECT OF FERRIC OXIDE ON THE 120 DEGREE PHASE TRANSITION IN BARIUM TITANATE SINGLE CRY 04-0544
         THERMODYNAMIC THEORY OF THE CUBIC- ORTHORHOMBIC PHASE TRANSITION IN BORACITES.   03-0231
ECT/ DIELECTRIC CONSTANT BEHAVIOR IN THE NEIGHBORHOOD OF THE PHASE TRANSITION IN CHARACTERISTICALLY FERROEL 03-0522
        X-RAY DIFFRACTION STUDY OF PHASE TRANSITION IN COPPER FORMATE.   26-2938
ECTRICS, PIEZOELECTRICS)   LOW TEMPERATURE DISPLACEMENT TYPE PHASE TRANSITION IN CRYSTALS. (THEORY, FERROEL 03-0320
G STRUCTURAL GROUPS. (SODIUM AM/ MECHANISMS OF FERROELECTRIC PHASE TRANSITION IN CRYSTALS WITH REORIENTATIN 22-2827
        ULTRASONIC ATTENUATION NEAR THE PHASE TRANSITION IN DICALCIUM LEAD PROPIONATE 26-2949
ROPIONATES SUBSTITUTED PARTIALLY BY ACETATE IONS.   PHASE TRANSITION IN DIVALENT METAL DICALCIUM P 26-2944
GADOLINIUM MOLYBDATE.   ANOMALOUS PHONON BEHAVIOR NEAR THE PHASE TRANSITION IN FERROELECTRIC FERROELECTRIC 30-3276
L AMMONIUM ALUMINUM SULFATE, AMMONIUM ALUM/ INVESTIGATION OF PHASE TRANSITION IN FERROELECTRIC ALUMS (METHY 20-2809
FOIODIDE CRYSTALS USING THE RAMAN SCATTERING SPECTRUM. (LAT/ PHASE TRANSITION IN FERROELECTRIC ANTIMONY SUL 15-2143
R (COMPOUNDS) A/ INFLUENCE OF NONEQUILIBRIUM CARRIERS ON THE PHASE TRANSITION IN FERROELECTRIC SEMICONDUCTO 15-2145
EPWISE) CHANGE IN PROBABILITY OF THE MOSSBAUER EFFECT AT THE PHASE TRANSITION IN FERROELECTRICS. ABRUPT (ST 03-0276
EN BONDS. (KDP,/ INFLUENCE OF POLAR STATES OF PROTONS ON THE PHASE TRANSITION IN FERROELECTRICS WITH HYDROG 17-2671
                     PHASE TRANSITION IN GADOLINIUM MOLYBDATE.   30-3316
        EFFECT OF HYDROSTATIC PRESSURE ON PHASE TRANSITION IN GADOLINIUM MOLYBDATE.   30-3321
LECTRIC AND EL/ PHENOMENOLOGICAL THEORY OF THE FERROELECTRIC PHASE TRANSITION IN GADOLINIUM MOLYBDATE. (DIE 30-3303
METRY CHANGES)   ORIGIN OF THE STRUCTURAL PHASE TRANSITION IN GADOLINIUM MOLYBDATE. (SYM 30-3273
                  STRUCTURE AND PHASE TRANSITION IN HYDROGEN CHLORIDE.   31-3337
                     PHASE TRANSITION IN ICE.   34-3442
POLARIZING ENERGY IN IT. (THEORY)   PHASE TRANSITION IN KDP AND THE ROLE OF THE DE 03-0229
ASHI'S DYNAMICAL THEORY).   PHASE TRANSITION IN KDP (MODIFICATION OF KOBAY 17-2488
ACT ONLY)   POLARIZATION (DOMAIN STRUCTURE) AND PHASE TRANSITION IN KDP SINGLE CRYSTAL. (ABSTR 17-2505
        NMR STUDY OF THE STRUCTURAL PHASE TRANSITION IN LANTHANUM ALUMINATE.   36-3533
                  INVESTIGATION OF PHASE TRANSITION IN LEAD CADMIUM TUNGSTATE.   10-1633
    THE ANOMALOUS BEHAVIOR OF PHONON NEAR THE 60 DEGREES C PHASE TRANSITION IN LEAD DICALCIUM PROPIONATE. 26-2942
        ATHERMAL PHASE TRANSITION IN LEAD ZIRCONATE.   07-1258
        NATURE OF THE FERROELECTRIC- PARAELECTRIC PHASE TRANSITION IN LITHIUM TANTALATE.   12-1879
CRYSTALS BY MODIFIED MOLECULAR FIELD APPROXIMATION. STUDY OF PHASE TRANSITION IN MIXED KDP- DEUTERATED KDP 17-2528
RICS, U/ INVESTIGATION OF POLARIZATION NONLINEARITY NEAR THE PHASE TRANSITION IN PEROVSKITE LIKE FERROELECT 03-0249
        THEORY OF STRUCTURAL PHASE TRANSITION IN PEROVSKITES.   03-0242
    PART-2, INTERACTION WITH ELASTIC ST/ THEORY OF A STRUCTURAL PHASE TRANSITION IN PEROVSKITE- TYPE CRYSTALS, 03-0241
ATE CRYSTALS.   FIRST ORDER PHASE TRANSITION IN POTASSIUM DEUTERIUM PHOSPH 17-2671
HATE TYPE FERROELECTRICS. (24 REFS)   MODELS OF THE PHASE TRANSITION IN POTASSIUM DIHYDROGEN PHOSP 03-0489
/RING OF DIPOLE MOMENTS OF WATER MOLECULES AND FERROELECTRIC PHASE TRANSITION IN POTASSIUM FERROCYANIDE AN/ 24-2893
HYDRATE.   MOSSBAUER STUDY OF THE FERROELECTRIC PHASE TRANSITION IN POTASSIUM FERROCYANIDE TRI 24-2901
ORIDE.   SUPERSTRUCTURE ARISING DURING THE PHASE TRANSITION IN POTASSIUM MANGANESE TRIFLU 33-3402
        PRESSURE INDUCED FERROELECTRIC PHASE TRANSITION IN POTASSIUM NITRATE.   16-2378
        THE STRUCTURE AND PHASE TRANSITION IN POTASSIUM SELENATE.   22-2872
ATE BY SCATTERING OF COLD NEUTRONS.   HIGH TEMPERATURE PHASE TRANSITION IN RUBIDIUM DIHYDROGEN PHOSPH 17-2489
ITE.   IMPROPER FERROELECTRIC PHASE TRANSITION IN RUBIDIUM TRIHYDROGEN SELEN 22-2834
        FERROELECTRIC PHASE TRANSITION IN SILVER SODIUM DINITRITE.   16-2266
        DIELECTRIC STUDIES ON THE FERROELECTRIC PHASE TRANSITION IN SILVER SODIUM DINITRITE.   16-2289
HYDRATE./ DIELECTRIC AND SOME OTHER STUDIES OF FERROELECTRIC PHASE TRANSITION IN SODIUM AMMONIUM SULFATE DI 19-2767
        DIFFUSE X-RAY STUDY OF 640 DEGREES C PHASE TRANSITION IN SODIUM NIOBATE.   08-1358
        X-RAY STUDY OF PHASE TRANSITION IN SODIUM NITRATE.   16-2380

```
                                                 PHASE TRANSITION IN SODIUM NITRITE.                    16-2397
                             AN X-RAY STUDY OF THE PHASE TRANSITION IN SODIUM NITRITE.                    16-2318
BY EPR (PHASE TRANSITION AT 163 DEG C).          FERROELECTRIC PHASE TRANSITION IN SODIUM NITRITE AS STUDIED 16-2359
E.                    DIELECTRIC AND NMR STUDIES OF THE PHASE TRANSITION IN STANNOUS CHLORIDE DIHYDRAT 36-3582
E                            AN X-RAY STUDY OF THE PHASE TRANSITION IN STRONTIUM FERRATE TANTALAT 08-1382
                  ABSORPTION EDGE AND STRUCTURAL PHASE TRANSITION IN STRONTIUM TITANATE.          06-1107
                     CUBIC TO TRIGONAL STRESS INDUCED PHASE TRANSITION IN STRONTIUM TITANATE.          06-1235
ERING STUDIES OF THE SOFT MODES NEAR THE CUBIC TO TETRAGONAL PHASE TRANSITION IN STRONTIUM TITANATE. /SCATT 06-1242
SERVATIONS FROM NEUTRON SCATTERING STUDIES OF THE STRUCTURAL PHASE TRANSITION IN STRONTIUM TITANATE. NEW OB 06-1210
                   NUCLEAR MAGNETIC RESONANCE STUDY OF THE PHASE TRANSITION IN STRONTIUM TITANATE.          06-1237
FLUORESCENCE STUDY OF CHROMIUM(3+) THROUGH A STRESS INDUCED PHASE TRANSITION IN STRONTIUM TITANATE. /RIZED 06-1115
              SOUND PROPAGATION NEAR THE STRUCTURAL PHASE TRANSITION IN STRONTIUM TITANATE.          06-1155
ACT)      SPECIFIC HEAT ANOMALY ASSOCIATED WITH THE 110 DEG K PHASE TRANSITION IN STRONTIUM TITANATE. (ABSTR 06-1226
ARMONIC CSCI/ FREE ENERGY AND SPECIFIC HEAT IN FERROELECTRIC PHASE TRANSITION IN TERMS OF A SINGLE MODE ANH 03-0406
NSTANT, X-RAY SCATTERING)     STUDIES ON THE MECHANISM OF THE PHASE TRANSITION IN TGS. (STATIC DIELECTRIC CO 27-3093
LIUM (DIHYDROGEN) PHOSPHATES.                    PHASE TRANSITION IN THALLIUM AND AMMONIUM THAL 17-2571
ANIC SEMICONDUCTOR PHTHALOCYANINE.     LOW TEMPERATURE PHASE TRANSITION IN THE POLAR STATE OF THE ORG 36-3623
L OF A FERROELECTRIC.             SECOND ORDER PHASE TRANSITION IN THE THREE DIMENSIONAL MODE 03-0450
                        NEW HIGH PRESSURE PHASE TRANSITION IN TRIGLYCINE SELENATE.          27-3090
        APPLICABILITY OF THE PIPPARD RELATIONS TO THE PHASE TRANSITION IN TRIGLYCINE SULFATE.           27-3157
                 INVESTIGATION OF THE PHASE TRANSITION IN TRIGLYCINE SULFATE.           27-3054
THERMAL CONTROL)    EXPERIMENTAL DETAILED STUDY OF THE PHASE TRANSITION IN TRIGLYCINE SULFATE. (CLOSE 27-3030
GH PRESSURE.                         PHASE TRANSITION IN TUNGSTEN TRIOXIDE UNDER HI 10-1658
ONTIUM TITANATE. (ABSTRACT)  RAMAN STUDIES OF STRESS INDUCED PHASE TRANSITION NEAR CURIE TEMPERATURE IN STR 06-1236
STRONTIUM OXIDE- TELLURIUM DIOXIDE.              PHASE TRANSITION OF A NEW FERROELECTRIC OXIDE,  36-3627
ACETATE. PART-1: THE CRYSTAL STR/ THE CRYSTAL STRUCTURE AND PHASE TRANSITION OF AMMONIUM HYDROGEN DICHLORO 36-3567
AND THE STRUCTURE OF THE FERROELECTRIC PHASE.         ON THE PHASE TRANSITION OF AMMONIUM HYDROGEN SULFATE 19-2777
      EFFECT OF HYDROSTATIC PRESSURE ON THE FERROELECTRIC PHASE TRANSITION OF AMMONIUM SULFATE.           16-2360
E CRY/ INFLUENCE OF ONE DIMENSIONAL MECHANICAL STRESS ON THE PHASE TRANSITION OF ANTIMONY SULFOIODIDE SINGL 15-2245
E.                   MAGNETISM AND THE PHASE TRANSITION OF COPPER FORMATE TETRAHYDRAT 26-2930
                   CRYSTAL STRUCTURE AND PHASE TRANSITION OF DIGLYCINE NITRATE.           16-2312
RONTIUM HEXAACETATE.      X-RAY AND OPTICAL STUDIES ON THE PHASE TRANSITION OF FERROELECTRIC DICALCIUM ST 36-3584
NTALATE. (SECOND ORDER)             CHARACTER OF PHASE TRANSITION OF FERROELECTRIC POTASSIUM TA 08-1307
EN BONDS KDP TYPE.  ROLE OF PROTON PHONON INTERACTION IN THE PHASE TRANSITION OF FERROELECTRICS WITH HYDROG 17-2674
NATE.                          FERROELECTRIC PHASE TRANSITION OF FIRST ORDER IN BARIUM TITA 04-1030
      ABNORMAL ABSORPTION OF ELASTIC WAVES NEAR THE PHASE TRANSITION OF GADOLINIUM MOLYBDATE.           30-3265
      DIELECTRIC AND X-RAY STUDIES ON THE FERROELECTRIC PHASE TRANSITION OF GLYCINE SILVER NITRATE.        16-2347
                 CRYSTAL STRUCTURE AND PHASE TRANSITION OF HYDROGEN CHLORIDE.           31-3334
HATE. PART-1:/ X-RAY DILATOMETRIC STUDY OF THE FERROELECTRIC PHASE TRANSITION OF POTASSIUM DIHYDROGEN PHOSP 17-2557
ICE.              DIELECTRIC DISPERSION AND PHASE TRANSITION OF POTASSIUM HYDROXIDE DOPED 34-3452
ORIDE AT 184 DEGREES K.            X-RAY SCATTERING AND THE PHASE TRANSITION OF POTASSIUM MANGANESE TRIFLU 33-3394
      FAR INFRARED SPECTRA AND FERROELECTRIC PHASE TRANSITION OF POTASSIUM NITRATE.           16-2308
ARMONIC GENERATION. (TEMPERATURE DEPENDENCE)     STUDY OF PHASE TRANSITION OF SODIUM NITRITE BY SECOND H 16-2403
COHERENT NEUTRON IN/ CRYSTAL DYNAMICS AND THE FERROELECTRIC PHASE TRANSITION OF SODIUM NITRITE (STUDIED BY 16-2364
5 DEG K. ANALYSIS OF THE ELASTIC ANOMALIES AT THE STRUCTURAL PHASE TRANSITION OF STRONTIUM TITANATE NEAR 10 06-1183
TTERING.              STUDY OF THE PHASE TRANSITION OF TGS BY SECOND HARMONIC SCA 27-2992
      ELECTROMECHANICAL COUPLING NEAR THE FERROELECTRIC PHASE TRANSITION OF TRIGLYCINE SULFATE.           27-3174
/LING EFFECTS ON THE DIELECTRIC PROPERTIES AND FERROELECTRIC PHASE TRANSITION ON LITHIUM THALLATE TARTRATE. 29-3251
ECTRUM OF THE FERROELECTRIC RUBID/ EFFECT OF TEMPERATURE AND PHASE TRANSITION ON THE HYDROXIDE VIBRATION SP 17-2470
OF FERROELECTRIC CRYSTALS AT LOW TEMPERAT/ INFLUENCE OF THE PHASE TRANSITION ON THE NORMAL VIBRATION MODES 17-2456
DUCTION BANDS.           EFFECT OF THE 110 DEGREE K PHASE TRANSITION ON THE STRONTIUM TITANATE CON 06-1158
TTENUATION OF SOUND WAVES IN DEUTERATED ROCHELLE SALT AT THE PHASE TRANSITION POINT.                    A 28-3204
TTENUATION OF SOUND WAVES IN DEUTERATED ROCHELLE SALT AT THE PHASE TRANSITION POINT.                    A 28-3205
EQUATION FOR QUADRUPOLE RESONANCE IN FERROELECTRICS NEAR THE PHASE TRANSITION POINT.            KINETIC 03-0309
      ON RELAXATION PROCCESSES NEAR THE SECOND ORDER PHASE TRANSITION POINT.                    03-0385
CR/ METHODS OF DETERMINING SPONTANEOUS POLARIZATION NEAR THE PHASE TRANSITION POINT OF TRIGLYCINE SELENATE 27-2994
ECTROOPT/ PHYSICAL PHENOMENA IN FERROELECTRICS WITH DIFFUSED PHASE TRANSITION. (POLARIZATION MECHANISMS, EL 02-0119
      DIELECTRIC DISPERSION OF IRRADIATED BARIUM TITANATE IN THE PHASE TRANSITION REGION.                    04-0652
CTRIC PERMEABILITY OF BARIUM TITANATE SINGLE CRYSTALS IN THE PHASE TRANSITION REGION.            DIELE 04-0645
      ON THE NATURE OF DIELECTRIC DISPERSION IN THE FERROELECTRIC PHASE TRANSITION REGION. /TERACTION OF DIPOLES 03-0417
THE ANOMALOUS ELECTRIC CONDUCTIVITY OF FERROELECTRICS IN THE PHASE TRANSITION REGION.            NATURE OF 03-0508
LARIZATION OF A RUBIDIUM DIHYDROGEN PHOSPHATE CRYSTAL IN THE PHASE TRANSITION REGION.            PO 17-2506
LARIZATION OF A RUBIDIUM DIHYDROGEN PHOSPHATE CRYSTAL IN THE PHASE TRANSITION REGION. (COERCIVE FIELD) PO 17-2504
      DIELECTRIC, ELASTIC AND MAGNETIC PROPERTY, SYMMETRY, AND PHASE TRANSITION REPRESENTATION. (18 REFS) 03-0316
0.5))(1-X) B'(X) OXYGEN(3), / DIFFUSE TO SHARP FERROELECTRIC PHASE TRANSITION SYSTEMS LEAD (B(0.5) NIOBIUM( 13-2044
/CT OF AN EXTERNAL ELECTRIC FIELD ON A DIFFUSE FERROELECTRIC PHASE TRANSITION, TALKING INTO ACCOUNT THE IN/ 03-0441
HE ABSORPTION OF ULTRASOUND IN ANTIMONY SULFOIODIDE NEAR THE PHASE TRANSITION TEMPERATURE. /LUMINATION IN T 15-2240
NGLE CRYSTALS OF FERROELECTRIC ANTIMONY SULFOIODIDE NEAR THE PHASE TRANSITION TEMPERATURE. /LECTION FROM SI 15-2151
CRYSTALS OF THE PO/ INFLUENCE OF HYDROSTATIC PRESSURE ON THE PHASE TRANSITION TEMPERATURE OF FERROELECTRIC 17-2495
STUDY ON SODIUM-23 IN SODIUM NITRITE IN THE VICINITY OF THE PHASE TRANSITION TEMPERATURES. /ETIC RESONANCE 16-2396
EAD ZIRCONATE TITANATE CERAMIC/ STABILIZATION EFFECTS AT THE PHASE TRANSITION TETRAGONAL- RHOMBOHEDRAL IN L 09-1611
HE PEROVSKITE FERR/ THERMODYNAMIC THEORY OF THE MORPHOTROPIC PHASE TRANSITION TETRAGONAL- RHOMBOHEDRAL IN T 03-0172
DOMAIN POLARIZATION OF A FERROELECTRIC HAVING A SECOND ORDER PHASE TRANSITION. (THEORY).            SINGLE 03-0202
OF POTASSIUM DIHYDROGEN PHOSPHATE NEAR THE HIGH TEMPERATURE PHASE TRANSITION. (THERMAL EXPANSION) /PERTIES 17-2627
ECTRIC (BARIUM, STRONTIUM) TITANATE SOLID SOLUTIONS NEAR THE PHASE TRANSITION, UNDER HIGH PRESSURES. /RROEL 04-0680
/ ENFORCED FERROELECTRICITY IN GLASS BONDED LEAD ZIRCONATE. (PHASE TRANSITION 200 DEG C BELOW THE CURIE PO/ 07-1252
ACTION IN THE KANZIG REGIONS MODEL FOR DIFFUSE FERROELECTRIC PHASE TRANSITIONS.            ACCOUNTING FOR INTER 03-0438
AL ABSORPTION EDGE IN TRIHYDROGEN SELENITE CRYSTALS AT THEIR PHASE TRANSITIONS.            BEHAVIOR OF THE FUNDAMENT 22-2839
      CRITICAL POINT INDICES OF FERROELECTRIC AND FERROMAGNETIC PHASE TRANSITIONS.                    03-0264
      DIELECTRIC SUSCEPTIBILITY AND THE ORDER OF FERROELECTRIC PHASE TRANSITIONS.                    03-0223
AIN STRUCTURE OF POLYCRYSTALLINE FERROELECTRICS WITH DIFFUSE PHASE TRANSITIONS.            DOM 03-0505
YNAMICS OF POTASSIUM DIHYDROGEN PHOSPHATE TYPE FERROELECTRIC PHASE TRANSITIONS.            D 17-2692
HYDROSTATIC PRESSURE ON THE NATURE OF DIFFUSE FERROELECTRIC PHASE TRANSITIONS.            EFFECT OF 03-0442
ND RELATIONS AND THEIR APPLICATION TO FERROELECTRIC (LAMBDA) PHASE TRANSITIONS. /ATION OF THE PIPPARD-GARLA 03-0299
INFRARED SPECTRA OF POTASSIUM NITRATE AND SODIUM NITRITE AT PHASE TRANSITIONS.                    16-2352
              INTERBAND MECHANISMS FOR FERROELECTRIC PHASE TRANSITIONS.                    02-0085
NUCLEATION PROCESS IN BARIUM TITANATE AT CUBIC- TETRAGONAL PHASE TRANSITIONS.                    04-0754
ON THE SYMMETRY OF THE SOFT MODES IN THE FERROELECTRIC PHASE TRANSITIONS.                    03-0357
CTROSCOPY OF PHONON INSTABILITY IN FIRST ORDER FERROELECTRIC PHASE TRANSITIONS.            RAMAN SPE 03-0163
ONATE TITANATE STANNATE BY TEMPERATURE AND PRESSURE ENFORCED PHASE TRANSITIONS. /NERGY IN LEAD NIOBIUM ZIRC 09-1488
```

STATISTICAL MECHANICS OF FERROELECTRIC PHASE TRANSITIONS.   03-0198
THEORY OF SECOND ORDER PHASE TRANSITIONS.   03-0364
THERMODYNAMIC CLASSIFICATION OF SECOND ORDER FERROELECTRIC PHASE TRANSITIONS.   03-0303
ODYNAMICS AND STATISTICAL MECHANICS OF DIFFUSE FERROELECTRIC PHASE TRANSITIONS.   THERM 03-0445
RIMENT. PART-1: INTRODUCTION AND MAGNETIC RES/ FERROELECTRIC PHASE TRANSITIONS: A REVIEW OF THEORY AND EXPE 03-0397
-5: FERROELECTRIC AND ANTIFERRO/ ULTRASONIC INVESTIGATION OF PHASE TRANSITIONS AND CRITICAL POINTS. SECTION 02-0070
/CTURAL DEFECTS PRODUCED BY GAMMA AND REACTOR IRRADIATION ON PHASE TRANSITIONS AND DIELECTRIC PROPERTIES O/ 17-2614
UM DIHYDROGEN PHOSPHATE TYPE CRYSTALS.   HIGH TEMPERATURE PHASE TRANSITIONS AND METASTABILITY IN POTASSI 17-2513
IMENTS. (REVIEW)   FERRO- AND ANTIFERROELECTRIC PHASE TRANSITIONS AND RELATED SCATTERING EXPER 02-0132
STRUCTURAL PHASE TRANSITIONS AND SOFT MODES.   03-0176
/LOGRAPHICAL CLASSIFICATION OF FERROELECTRICS, FERROELECTRIC PHASE TRANSITIONS AND SPECIAL FEATURES OF DOM/ 03-0480
URE OF (HIGH SENSITIVITY) BARIUM TITANATE.   DIFFUSIONLESS PHASE TRANSITIONS AND THE STABLE DOMAIN STRUCT 04-0888
IUM DIOXIDE AND BARIUM TITANATE SINGLE CRYSTALS.   (PHASE TRANSITIONS AND) THERMOCURRENTS IN TITAN 04-0697
UM TITANATE AND ((BARIUM(X) STRONT/ INFLUENCE OF PRESSURE ON PHASE TRANSITIONS AT LOW TEMPERATURES: STRONTI 04-0898
LLURIDE, POTASSIUM TANTALATE.   THEORY OF SECOND ORDER PHASE TRANSITIONS AT LOW TEMPERATURES. (TIN TE 03-0433
/DIELECTRIC ANOMALIES IN FERROELECTRIC MATERIAL WITH SEVERAL PHASE TRANSITIONS AT TEMPERATURES CLOSE TOGET/ 03-0363
(FAST ION SCATTERING) SHADOW EFFECT IN (DISPLACIVE) PHASE TRANSITIONS. (BARIUM TITANATE)   03-0533
/ (EXAMINATION OF) ANOMALIES OF DIELECTRIC PROPERTIES AT THE PHASE TRANSITIONS. (BASED ON LANDAU PHENOMENOL 19-2765
PHASE TRANSITIONS IN A MODEL FERROELECTRIC.   03-0257
PHASE TRANSITIONS IN A MODEL FERROELECTRIC.   03-0256
FLUORIDE CRYSTAL.   RAMAN SCATTERING AND PHASE TRANSITIONS IN A POTASSIUM MANGANESE TRI 33-3400
RCONATE AND FERROELECTRIC L/ MOSSBAUER EFFECT STUDIES OF THE PHASE TRANSITIONS IN ANTIFERROELECTRIC LEAD ZI 07-1253
LARIZED INFRARED AND RA/ LONG WAVELENGTH OPTICAL PHONONS AND PHASE TRANSITIONS IN ANTIMONY SULFOIODIDE. (PO 15-2106
DOMAIN INTERACTIONS DURING PHASE TRANSITIONS IN BARIUM TITANATE.   04-0651
SYMMETRY ASPECTS OF THE PHASE TRANSITIONS IN BORACITES.   14-2077
THEORY OF PHASE TRANSITIONS IN BORACITES.   03-0269
ISOTHERMAL PHASE TRANSITIONS IN CERAMIC LEAD ZIRCONATE.   08-1466
EXPERIMENTAL STUDIES OF STRUCTURAL PHASE TRANSITIONS IN CESIUM LEAD TRICHLORIDE.   33-3378
MICROSCOPIC OBSERVATIONS OF PHASE TRANSITIONS IN CESIUM LEAD TRICHLORIDE.   33-3383
(MECHANISM OF) STRUCTURAL PHASE TRANSITIONS IN CESIUM LEAD TRICHLORIDE.   33-3380
ELECTRICAL CONDUCTION AND (NONFERROELECTRIC) PHASE TRANSITIONS IN CESIUM LEAD TRICHLORIDE.   33-3365
(LOSS OF A CENTER OF SYMMETRY AT 194 DEGREES K) PHASE TRANSITIONS IN CESIUM LEAD TRICHLORIDE. 33-3368
E TYPE BARIUM LANTHANON MOLYBDATE.   PHASE TRANSITIONS IN COMPLEX PEROVSKITES OF TH 30-3261
SINGLE DOMAIN FERROELECTRICS BELOW CURIE POINT).   COMPLEX PHASE TRANSITIONS IN CRYSTALS (APPLICATION TO 03-0319
PE AND OF ONES / MICROTHEORY OF SPONTANEOUS POLARIZATION AND PHASE TRANSITIONS IN CRYSTALS OF PEROVSKITE TY 03-0175
UNGSTEN BRONZE STRUCTURE (POTASSIU/ FORMATION CONDITIONS AND PHASE TRANSITIONS IN CRYSTALS WITH POTASSIUM T 13-1923
ROELECTRIC MATERIALS.   NATURE OF PHASE TRANSITIONS IN FERROELECTRIC AND ANTIFER 03-0311
DOUBLE HYSTERESIS LOOPS AND FIRST ORDER PHASE TRANSITIONS IN FERROELECTRIC CRYSTALS.   03-0456
PHASE TRANSITIONS IN FERROELECTRICS.   01-0011
NSISTENT POLARIZATION SCHEME FOR ANH/ FIRST AND SECOND ORDER PHASE TRANSITIONS IN FERROELECTRICS: A SELF CO 03-0355
AGNE/ CHANGE OF COMPLETE SYMMETRY OF THE CRYSTALS DURING THE PHASE TRANSITIONS IN FERROELECTRICS AND FERROM 03-0527
M FERROCYANIDE TR/ NUCLEAR MAGNETIC RESONANCE STUDIES OF THE PHASE TRANSITIONS IN FERROELECTRICS. (POTASSIU 16-2255
PPLIED TO BARIUM T/ CORRELATION EFFECTS IN DISPLACEMENT TYPE PHASE TRANSITIONS IN FERROELECTRICS. (THEORY A 03-0497
ATE.   HIGH TEMPERATURE (TO 236 DEG C) PHASE TRANSITIONS IN (HIGH PURITY) LEAD ZIRCON 07-1266
ITANATE SINGLE CRYSTALS/ CUBIC TETRAGONAL (MARTENSITIC TYPE) PHASE TRANSITIONS IN HIGH RESISTIVITY BARIUM T 04-0887
RIUM PHOSPHATE. NEUTRON DIFFRACTION STUDIES OF FERROELECTRIC PHASE TRANSITIONS IN KDP AND POTASSIUM DIDEUTE 17-2661
STRUCTURAL PHASE TRANSITIONS IN LANGBEINITES.   36-3552
TATE(0.5), TITANAT/ ELECTRICAL MAGNETIC AND X-RAY STUDIES OF PHASE TRANSITIONS IN LEAD COBALTATE(0.5) TUNGS 13-1960
BRONZES.   TRANSIENT NMR STUDY OF PHASE TRANSITIONS IN METALLIC SODIUM TUNGSTEN 13-1910
ELECTRICS.   LATTICE DYNAMICS OF PHASE TRANSITIONS IN PEROVSKITE TYPE ANTIFERRO 08-1475
(ST/ SELF CONSISTENT LATTICE DYNAMICAL THEORY OF STRUCTURAL PHASE TRANSITIONS IN PEROVSKITE TYPE CRYSTALS. 03-0424
(EXTENDED ABSTRACT)   STRUCTURAL PHASE TRANSITIONS IN PEROVSKITE TYPE CRYSTALS. 08-1450
DISPLACEMENT PHASE TRANSITIONS IN PEROVSKITES.   01-1274
OBIUM-93 AND SODIUM-23 RESONANCE IN POTASSIUM / NMR STUDY OF PHASE TRANSITIONS IN PEROVSKITES CRYSTALS. (NI 08-1283
ACOUSTIC DETECTION OF FERROELECTRIC PHASE TRANSITIONS IN PLZT CERAMICS.   09-1545
EUTERIUM PHOSPHATE (INFRARED PHOSPHATE/ OPTICAL STUDY ON THE PHASE TRANSITIONS IN POTASSIUM DIHYDROGEN OR D 17-2646
UORIDE CRYSTALS.   RAMAN SCATTERING AND PHASE TRANSITIONS IN POTASSIUM MANGANESE TRIFL 33-3372
UORIDE.   NEUTRON SCATTERING STUDY OF THE LATTICE DYNAMICAL PHASE TRANSITIONS IN POTASSIUM MANGANESE TRIFL 33-3405
UORIDE: TEMPERATURE DISCONTINUITIE/ TWO PHONON PROCESSES AND PHASE TRANSITIONS IN POTASSIUM MANGANESE TRIFL 33-3363
MIXED CRYSTALS (GROWTH BY DOUBLE BRIDGMAN TECHNIQUE).   PHASE TRANSITIONS IN POTASSIUM NITRATE NITRITE 16-2345
/SODIUM) (0.5) BISMUTH(0.5) TITANATE AND THE HIGH TEMPERATURE PHASE TRANSITIONS IN POTASSIUM(0.5) BISMUTH(0/ 08-1362
MICROSCOPIC THEORY OF THE PHASE TRANSITIONS IN ROCHELLE SALT.   03-0482
NMR STUDY OF PHASE TRANSITIONS IN ROCHELLE SALT.   28-3195
ANATE, ANTIMONY SULFOIO/ DIELECTRIC AND X-RAY STUDIES OF THE PHASE TRANSITIONS IN ROCHELLE SALT, BARIUM TIT 02-0099
ANISOTROPIC DIFFUSE X-RAY SCATTERING AND PHASE TRANSITIONS IN SODIUM NIOBATE.   01-1357
E OF TILTING SCHEMES IN THE SOLUTIONS OF P/ HIGH TEMPERATURE PHASE TRANSITIONS IN SODIUM NIOBATE AND THE US 08-1336
PHENOMENOLOGICAL THEORY OF PHASE TRANSITIONS IN SODIUM NITRITE.   03-0254
NITRATE.   GRUNEISEN PARAMETERS AROUND THE PHASE TRANSITIONS IN SODIUM NITRITE AND SODIUM 16-2390
NITE.   ESR STUDY OF THE FERROELECTRIC PHASE TRANSITIONS IN SODIUM TRIHYDROGEN DISELE 22-2871
INGLE CRYSTAL X-RAY AND NEUTRON DIFFRA/ FERROELECTRICITY AND PHASE TRANSITIONS IN SOLID HYDROGEN HALIDES (S 31-3331
ISPLACIVE FERROELECTRICS. (ANTIMONY SUL/ OPTICAL PHONONS AND PHASE TRANSITIONS IN SOME ORDER DISORDER AND D 15-2208
TE.   RAMAN SCATTERING AND PHASE TRANSITIONS IN STRESSED STRONTIUM TITANA 06-1103
TURE OF POTASSIUM TUNGSTEN BRONZE. (POTASSIUM BARIUM TANTAL/ PHASE TRANSITIONS IN TANTALATES WITH THE STRUC 12-2050
TE SYSTEM.   STUDY OF PHASE TRANSITIONS IN THE LEAD ZIRCONATE TITANA 09-1620
TITANATE- BISMUTH/ MOSSBAUER EFFECT STUDIES OF FERROELECTRIC PHASE TRANSITIONS IN THE LEAD ZIRCONATE- LEAD 09-1498
TITANATE SYSTEM.   STUDY OF PHASE TRANSITIONS IN THE LEAD ZIRCONATE- LEAD 09-1621
STRONTIUM(1-X) BARIUM(X) TITANATE.   PHASE TRANSITIONS IN THE MIXED CRYSTAL SYSTEM 04-1016
TERING)   STUDIES ON THE MECHANISM OF THE PHASE TRANSITIONS IN THIOUREA. (SATTELITE SCAT 25-2922
EFFECT OF CONSTANT ELECTRICAL FIELD ON PHASE TRANSITIONS IN TRIGLYCINE FLUOBERYLLATE.   27-3167
SOME ADDITIONAL FEATURES OF FERROELECTRIC PHASE TRANSITIONS IN VIBRONIC MODEL.   03-0339
RIC AND STRUCTURAL PROPERTIES OF FERROELECTRICS WITH DIFFUSE PHASE TRANSITIONS. (MIXED LAYER OXIDES) /ELECT 13-2014
URE IN STRONTIUM TITA/ SOFT MODE STUDY OF THE STRESS INDUCED PHASE TRANSITIONS NEAR THE TRANSITION TEMPERAT 06-1196
TE CRYSTALS.   EFFECT OF HIGH PRESSURE ON PHASE TRANSITIONS OF ALKALI TRIHYDROGEN SELENI 22-2882
URE OF HYDROGEN BONDING AN/ LOW TEMPERATURE STUDIES. PART-4: PHASE TRANSITIONS OF AMMONIUM SULFATE: THE NAT 19-2783
F RARE EARTH MOLYBDATES.   PHASE TRANSITIONS OF ANHYDROUS MODIFICATIONS O 30-3272
KINETICS OF THE PHASE TRANSITIONS OF BARIUM TITANATE.   04-0650
BORACITE AT LOW TEMPERATURES. (MONOCLINIC A/ X-RAY STUDY ON PHASE TRANSITIONS OF FERROELECTRIC IRON IODINE 14-2086
/OINT DEFECTS ON DIELECTRIC PROPERTIES, DOMAIN DYNAMICS AND PHASE TRANSITIONS OF FERROELECTRICS. (GENERAL/ 03-0495
ISOSTRUCTURAL COMPOUNDS (EUROPIUM TERBIUM, DYSPROSIUM AND H/ PHASE TRANSITIONS OF GADOLINIUM MOLYBDATE AND 30-3271
(IMPORTANCE OF HYDROGEN BONDING IN THE FERROELECTRIC MECHAN/ PHASE TRANSITIONS OF HYDROGEN HALIDE CRYSTALS 31-3329

X-RAY STUDY OF HIGH TEMPERATURE PHASE TRANSITIONS OF SODIUM NIOBATE (TO 500C). 13-1965
PHASE TRANSITIONS OF SOLID HYDROGEN HALIDES. 31-3327
EL/ EFFECT OF THE CHANGE OF ELECTROSTATIC CONSTRAINTS ON THE PHASE TRANSITIONS OF SOME PHENOMENOLOGICAL MOD 03-0280
TE. (EPR OF IRON- VANADIUM PAIRS) CRDER PARAMETER AND PHASE TRANSITIONS OF STRESSED STRONTIUM TITANA 06-1170
ES. NONLINEAR PROPERTIES AND PHASE TRANSITIONS OF STRONTIUM BISMUTH TITANAT 13-1952
/AND TEMPERATURE DEPENDENCE OF THE DIELECTRIC PROPERTIES AND PHASE TRANSITIONS OF THE ANTIFERROELECTRIC PE/ 08-1433
/AND TEMPERATURE DEPENDENCE OF THE DIELECTRIC PROPERTIES AND PHASE TRANSITIONS OF THE FERROELECTRIC PEROVS/ 08-1434
E SALT. HIGH PRESSURE PHASE TRANSITIONS OF THE FIRST KIND IN ROCHELL 28-3214
CRYSTAL TRIGLYCINE SULFATE AND TRIGLYCINE FLUOBERYLLATE) AT PHASE TRANSITIONS OF THE SECOND ORDER. /SINGLE 27-3137
BARIUM TITANATE. INFLUENCE OF PHASE TRANSITIONS ON ELECTRICAL RESISTANCE OF 04-0599
IBRATION IN STRONTIUM TITANATE A/ EFFECTS OF TEMPERATURE AND PHASE TRANSITIONS ON THE HYDROXYL STRETCHING V 06-1209
/RUCTURE ANALYSIS OF FERROELECTRICS IN CONNECTION WITH THEIR PHASE TRANSITIONS. PART-1: TGS. PART-2: SILVE/ 02-0077
LATTICE DYNAMICS ABOVE STRUCTURAL PHASE TRANSITIONS: STRONTIUM TITANATE. 06-1124
IUM TANTALATE, B/ NEUTRON INELASTIC SCATTERING AT STRUCTURAL PHASE TRANSITIONS. (STRONTIUM TITANATE, POTASS 02-0059
C RARE EARTH MOLYBDATES. (REVIEW, FORMATION, CRYSTAL GROWTH, PHASE TRANSITIONS, STRUCTURE, 49 REFS) /LECTRI 02-0064
ELECTROGYRATION AND FERROELECTRIC PHASE TRANSITIONS (TGS). 27-2970
/ATIONSHIP BETWEEN THE SUPERCONDUCTING AND THE FERROELECTRIC PHASE TRANSITIONS (TIN TELLURIDE PURE AND ALL/ 36-3564
CTRIC, THEORY). SOME REMARKS ABOUT ANTIFERROELECTRIC PHASE TRANSITIONS (TO PARAELECTRIC OR FERROELE 03-0171
IUM DIHYDROGEN AND DIDEUT/ LASER LIGHT SCATTERING STUDIES OF PHASE TRANSITIONS. (TRIGLYCINE SULFATE, POTASS 17-2481
THEORY OF FERROELECTRIC PHASE TRANSITIONS WITH VIBRONIC MECHANISM. 03-0351
DIELECTRIC DISPERSION OF IRRADIATED BARIUM TITANATE NEAR THE PHASE TRASNITION. 04-0648
IBUTION IN ANTIMONY SULFOIODIDE CRYSTAL (IN THE PARAELECTRIC PHASE, UNDER DC FIELD ALONG C-AXIS). /AL DISTR 15-2173
DIELECTRIC STUDIES OF PHASE 2 OF AMMONIUM NITRATE. 16-2339
FERROELECTRIC DOMAINS OF BARIUM TITANATE IN LOW TEMPERATURE PHASES. ELECTRON MICROSCOPIC STUDY OF 04-1034
RON(3+) IN BARIUM TITANATE IN THE RHOMBOHEDRAL AND THE CUBIC PHASES. ELECTRON PARAMAGNETIC RESONANCE OF I 04-0929
A SINGLE CRYSTAL OF THIOUREA IN THE PARA- AND FERROELECTRIC PHASES. INFRARED ABSORPTION SPECTRUM OF 25-2910
SORPTION SPECTRUM OF THIOUREA IN THE PARA- AND FERROELECTRIC PHASES. INFRARED AB 25-2908
ND RUBIDIUM DEUTERIUM SULFATE IN THE PARA- AND FERROELECTRIC PHASES. /PECTRA OF RUBIDIUM HYDROGEN SULFATE A 19-2768
HATE AND RUBIDIUM DIHYDROGEN PHOSPHATE IN POLAR AND NONPOLAR PHASES. /RYSTALS OF POTASSIUM DIHYDROGEN PHOSP 17-2457
CESIUM AND RUBIDIUM NITRATE CRYSTALS IN ITS LOW TEMPERATURE PHASES. /ARITIES IN THE VIBRATIONAL SPECTRA OF 16-2323
FCLYMORPHISM OF STRONTIUM METANIOBATE (TWO REVERSIBLE PHASES). 36-3535
MAN SPECTRA OF ADP IN THE PARAELECTRIC AND ANTIFERROELECTRIC PHASES. RA 17-2431
GEN PHOSPHATE TYPE IN THEIR PARAELECTRIC AND LOW TEMPERATURE PHASES. / OF CRYSTALS OF THE POTASSIUM DIHYDRO 17-2515
ODULATED ABSORPTION IN STRONTIUM TITANATE WITHOUT COEXISTING PHASES. REINTERPRETATION OF WAVELENGTH M 06-1188
UM TITANATE SINGLE CRYSTALS IN ORTHORHOMBIC AND RHOMBOHEDRAL PHASES. REPOLARIZATION PROCESSES OF BARI 04-0963
DIFFRACTION, DC FIELD EFFECT, PARAELECTRIC AND FERROELECTRIC PHASES). /ACE LAYER IN BARIUM TITANATE (X-RAY 04-0857
SPHATE IN THE INFRARED REGION IN THE PARA- AND FERROELECTRIC PHASES. /NSMISSION OF POTASSIUM DIHYDROGEN PHO 17-2459
OF POTASSIUM DIHYDROGEN PHOSPHATE IN PARA- AND FERROELECTRIC PHASES. VIBRATION SPECTRA 17-2480
HYDROGEN PHOSPHATE CRYSTAL IN PARAELECTRIC AND FERROELECTRIC PHASES. / THE PHOSPHATE ANION IN A RUBIDIUM DI 17-2472
E BORACITE AT LOW TEMPERATURES. (MONOCLINIC AND RHOMBOHEDRAL PHASES) /ANSITIONS OF FERROELECTRIC IRON IODIN 14-2086
LEAD NIOBATE ZIRCONATE SYSTEM. (TETRAGONAL AND ORTHORHOMBIC PHASES, ABSTRACT ONLY) /OPERTIES OF THE BARIUM 13-1925
UBIC ELECTROOPTIC CRYSTALS. (NONLINEAR OPTICAL PROPERTIES, / PHASES AND POLARIZATION OF LUMINOUS WAVES IN C 02-0044
THE FERROELECTRIC PROPERTIES OF ABO(3) AND A(0.5)BO(3) TYPE PHASES AND THE IONIC MASS OF THE B IONS. /LATE 03-0203
IUM, AND STRONTIUM OXIDES. BINARY AND PSEUDOBINARY PHASES BETWEEN NIOBIUM PENTOXIDE AND LEAD, BAR 04-0590
ERVATION OF PHASE BOUNDARIES BETWEEN FERRO- AND PARAELECTRIC PHASES IN GADOLINIUM MOLYBDATE CRYSTALS. OBS 30-3306
IDE- LANTHANUM TRIOXIDE. RANGE OF EXISTENCE OF PEROVSKITE PHASES IN THE SYSTEM LEAD OXIDE- TITANIUM DIOX 08-1349
/OSTRICTIVE CONSTANTS IN FERROELECTRIC AND ANTIFERROELECTRIC PHASES. (LANTHANUM OR NIOBIUM DOPED LEAD ZIRC/ 09-1487
FERROELECTRIC/ DIPOLE PATTERNS IN ORTHORHOMBIC AND TRIGONAL PHASES OF ABO(3) SUBSTANCE. (WHEN THE B ION IS 08-1363
DIELECTRIC ABSORPTION IN THE ORTHORHOMBIC AND QUADRATIC PHASES OF BARIUM TITANATE. 04-0900
DIELECTRIC ABSORPTION IN THE ORTHORHOMBIC AND QUADRATIC PHASES OF BARIUM TITANATE. 04-0899
DIRECTIONAL CORRELATION OF SCANDIUM-44 IN THE FERROELECTRIC PHASES OF BARIUM TITANATE. PERTURBED 04-0558
HE FERROELECTRIC- FERROELASTIC AND PARAELECTRIC- PARAELASTIC PHASES OF GADOLINIUM MOLYBDATE. /AY STUDY OF T 30-3287
/PLACEMENTS AND SPONTANEOUS DEFORMATION IN THE FERROELECTRIC PHASES OF OXIDES WITH THE PEROVSKITE STRUCTUR/ 08-1377
POT/ CRYSTAL CHEMISTRY AND ELECTRICAL PROPERTIES OF TWO NEW PHASES OF QUADRATIC TUNGSTEN BRONZES STRONTIUM 13-2010
ONZE TYPE. (STRONTIUM POTAS/ FIRST FERROELECTRIC OXYFLUORIDE PHASES OF THE OXYGENATED QUADRATIC TUNGSTEN BR 13-2011
/ALCIUM, STRONTIUM, OR BARIUM, AND C IS SODIUM OR POTASSIUM) PHASES OF THE TETRAGONAL TUNGSTEN BRONZE OXID/ 04-0912
E TYPE: QUADRATIC LEAD OR BISMUTH TANTALATE NIOBATE, TA/ NEW PHASES OF THE TETRAGONAL TUNGSTEN OXYGEN BRONZ 13-1985
/POUNDS LEAD ZIRCONATE(X) TITANATE(1-X) FOR TWO RHOMBOHEDRAL PHASES. (X=0.9 AND 0.58) (CORRELATION OF CATI/ 09-1574
TYPE) COMPOUND (AMMONO THIO CADMIUM NIOBATE, WITH 4 ELECTRIC PHASES, 2 FERROELECTRIC). /ECTRIC (PYROCHLORE 36-3529
PARAMAGNETIC SUBSTANCES. PART-1: IRON(X) ALUMINU/ RELAXATION PHENOMENA AND MOSSBAUER SPECTRA OF MAGNETIZED 19-2751
THERMODYNAMIC THEORY OF DISPERSION PHENOMENA. (FERROELECTRICITY) 03-0420
HELLE SALT, TGS) SHORT LIVED FERROELECTRIC "AFTER EFFECT" PHENOMENA. (HYSTERESIS LOOP CONSTRICTIONS, ROC 27-3156
STATE) CRITICAL PHENOMENA IN ANTIFERROELECTRICS. (EQUATION OF 03-0343
CRITICAL PHENOMENA IN ANTIMONY SULFOIODIDE. 15-2231
MECHANISM FOR THERMOELECTRIC CURRENTS) TRANSPORT PHENOMENA IN ANTIMONY SULFOIODIDE. (CONDUCTION 15-2241
PHOTOCONDUCTIVITY AND PHOTOFERROELECTRIC PHENOMENA IN BARIUM TITANATE. 04-0574
/NOMENOLOGICAL THEORY OF REFRACTIVE INDEX AND ELECTROOPTICAL PHENOMENA IN CRYSTALS OF PEROVSKITE STRUCTURE/ 03-0518
WAVELENGTH DEPENDENCE OF HOLOGRAPHIC PHENOMENA IN DOPED LITHIUM NIOBATE. 11-1785
OF DISSIPATIVE EFFECTS WHICH ACCOMPANY DISACCOMODATION TYPE PHENOMENA IN FERRITES AND FERROELECTRICS. /URE 03-0447
ION OF FIELD NEAR SURFACE) MICROSCOPIC THEORY OF SURFACE PHENOMENA IN FERROELECTRIC CRYSTALS. (CALCULAT 03-0501
EORY) INSTABILITY PHENOMENA IN FERROELECTRIC SEMICONDUCTORS. (TH 03-0201
E TRANSITION. (POLARIZATION MECHANISMS, ELECTROOPT/ PHYSICAL PHENOMENA IN FERROELECTRICS WITH DIFFUSED PHAS 02-0119
DIFFUSION AND RELAXATION PHENOMENA IN ICE. 34-3470
ONTIUM TITANATE. (ABSTRACT) ULTRASONIC STUDIES OF CRITICAL PHENOMENA IN INSULATING AND SEMICONDUCTING STR 06-1118
ATE AND ZIRCONATE. (SOLID SOLUTIONS LEAD MAGNESATE/ PHYSICAL PHENOMENA IN PIEZOCERAMICS BASED ON LEAD TITAN 05-1063
DIELECTRIC PHENOMENA IN ROCHELLE SALT. 28-3182
EFFECT OF THE CUPRIC ION ON DELAYED PHENOMENA IN ROCHELLE SALT. 28-3221
ON THE EFFECTS OF X-RAY IRRADIATION ON TRANSIENT PHENOMENA IN ROCHELLE SALT. 28-3222
PHOTOCONDUCTIVITY AND TRAPPING PHENOMENA IN STRONTIUM TITANATE. 06-1091
POLAR DIELECTRICS. (POTASSIUM DIHYDROGEN PHOSPHAT/ TRANSIENT PHENOMENA IN THE DISCHARGE OF CONDENSERS WITH 17-2631
NATE SYSTEM. OXIDATION REDUCTION PHENOMENA IN THE LEAD TITANATE- LANTHANUM TITA 09-1587
INTERPRETATION OF THE PHASE DIAGRAM OF ICE. PART-/ MOLECULAR PHENOMENA IN WATER SYSTEMS. PART-1: MOLECULAR 34-3485
OPTICAL HARMONICS AND NONLINEAR PHENOMENA. (THEORY, REVIEW) 02-0067
ERROELASTICITY. PHENOMENOLOGICAL LATTICE DYNAMICAL THEORY OF F 03-0144
/SPECTRA AND BAND STRUCTURE) OF SOME ABC(3) FERROELECTRICS. (PHENOMENOLOGICAL MODEL OF ELECTROOPTICAL EFFE/ 08-1383
F ELECTROSTATIC CONSTRAINTS ON THE PHASE TRANSITIONS OF SOME PHENOMENOLOGICAL MODELS OF FERROELECTRICS. / O 03-0280
RACITE/ INTERACTIONS BETWEEN MAGNETIZATION AND POLARIZATION: PHENOMENOLOGICAL SYMMETRY CONSIDERATIONS ON BO 14-2072
MONI/ OPTICAL PROPERTIES OF FERROELECTRICS. DIGRESSIONS FROM PHENOMENOLOGICAL THEORY. (AMMONIUM SULFATE, AM 02-0040
ERRODIELECTRICS WITH AN ARBITRARY MAGNETIC SYMMET/ SPIN WAVE PHENOMENOLOGICAL THEORY IN FINITE SAMPLES OF F 03-0164
DYNAMIC PARAMETERS NEAR SECOND ORDER TRANSITION POINTS IN F/ PHENOMENOLOGICAL THEORY OF ANOMALIES OF THERMO 03-0362

RANSITION. PART-2: FIRST ORDER TRANSITION. PHENOMENOLOGICAL THEORY OF ANTIFERROELECTRIC T 03-0404
Y AND FERROELECTRICITY APPLIED TO SODIUM NIOBATE AND THE SY/ PHENOMENOLOGICAL THEORY OF ANTIFERROELECTRICIT 03-0218
ANICAL PROPERTIES OF IMPROPER FERROELECTRIC CRYSTALS. (AMMO/ PHENOMENOLOGICAL THEORY OF DIELECTRIC AND MECH 14-2084
S IN FERROELECTRIC MATERIAL WITH SEVERAL PHASE TRANSITIONS / PHENOMENOLOGICAL THEORY OF DIELECTRIC ANOMALIE 03-0363
FERROELECTRIC CRYSTALS. QUASI PHENOMENOLOGICAL THEORY OF ELASTIC ANOMALY IN 03-0374
NTIFERROELECTRIC CRYSTALS. PHENOMENOLOGICAL THEORY OF FERROELECTRIC AND A 03-0270
IZATION REVERSAL. PHENOMENOLOGICAL THEORY OF FERROELECTRIC POLAR 03-0282
AN INTRODUCTION TO THE PHENOMENOLOGICAL THEORY OF FERROELECTRICITY. 01-0013
FERROELECTRICS. (AT TEMPERATURES ABOVE THE CURIE POINT) PHENOMENOLOGICAL THEORY OF MICROWAVE LOSSES IN 03-0503
N SODIUM NITRITE. PHENOMENOLOGICAL THEORY OF PHASE TRANSITIONS I 03-0254
D ELECTROOPTICAL PHENOMENA IN CRYSTALS OF PEROVSKITE STRUCT/ PHENOMENOLOGICAL THEORY OF REFRACTIVE INDEX AN 03-0518
HASE TRANSITION IN GADOLINIUM MOLYBDATE. (DIELECTRIC AND EL/ PHENOMENOLOGICAL THEORY OF THE FERROELECTRIC P 30-3303
ION IN FERROELECTRICS. PHENOMENOLOGICAL THEORY OF ULTRASONIC ATTENUAT 03-0228
/CTRIC PROPERTIES AT THE PHASE TRANSITIONS. (BASED ON LANDAU PHENOMENOLOGICAL THEORY, TESTED ON AMMONIUM F/ 19-2765
THERMODYNAMICS AND THE PHENOMENOLOGY OF FERROELECTRICITY. 03-0457
YSTAL AND CERAMIC SYSTEMS. THERMODYNAMIC PHENOMENOLOGY OF FERROELECTRICITY IN SINGLE CR 03-0214
ITE. STUDY OF THE AGING PHENOMENON IN BARIUM TITANATE SINGLE CRYSTALS. 04-0545
RITE. TWO PHONON ABSORPTION IN FERROELECTRIC SODIUM NITR 16-2310
UM OF (PARAELECTRIC AND) FERROELECTRIC ANTIMONY SULFOIODIDE RELATIONSHIP OF THE TEMPERATURE DEPENDENCE OF B1 PHONON AND DIELECTRIC RELAXATION IN SODIUM NIT 16-2257
ERROELASTIC FERROELECTRIC GADOLINIUM MOLYBDATE. ANOMALOUS MICROWAVE (PHONON ASSIGNMENTS). (POLARIZED) RAMAN SPECTR 15-2207
AND POTASSIUM DIHYDROGEN ARSENATE. PROTON PHONON ATTENUATION IN X CUT LITHIUM NIOBATE. 11-1756
OTASSIUM DIHYDROGEN PHOSPHATE. (COLLECTIV/ EFFECT OF PROTON- PHONON BEHAVIOR NEAR THE PHASE TRANSITION IN F 30-3276
TRANSITION IN ANTIMONY SULFOIODIDE. SOFT PHONON COUPLING IN CESIUM DIHYDROGEN ARSENATE 18-2733
ITRATE. PHONON COUPLING ON THE FERROELECTRIC MODE IN P 17-2655
LITHIUM NIOBATE. DIRECTIONAL DISPERSION OF PHONON COUPLINGS IN THE PRESSURE INDUCED PHASE 15-2240
PEROVSKITE STRONTIUM T/ ELECTRIC FIELD DEPENDENCE OF OPTICAL PHONON DISPERSION IN FERROELECTRIC POTASSIUM N 16-2334
FERMI RESONANCE OF POLARITON WITH BI PHONON FREQUENCIES AND POLARITON SCATTERING IN 11-1696
WAVEVECTOR DEPENDENT RELAXATION OF AN OPTICAL PHONON FREQUENCIES (RAMAN SCATTERING IN CUBIC 06-1241
TITANATE, POTASSIUM TANTALA/ LIFETIME OF THE "FERROELECTRIC" PHONON IN AN LITHIUM NIOBATE CRYSTAL. 11-1761
LCULATION OF THE RESPONSE FUNCTION OF THE SOFT ZONE BOUNDARY PHONON IN GADOLINIUM MOLYBDATE. 30-3320
C PHASE TRANSITIONS. RAMAN SPECTROSCCPY OF PHONON IN SOME PEROVSKITE CRYSTALS (STRONTIUM 03-0227
EFFECT OF THE QUARTIC ANHARMONICITY ON THE SOFT PHONON IN STRONTIUM TITANATE. MICROSCOPIC CA 06-1213
CYCLOTRON RESONANCE, ABSORPTION AND LUM/ EFFECTS OF ELECTRON PHONON INSTABILITY IN FIRST ORDER FERROELECTRI 03-0163
OGEN BONDS. (KDP TYPE) PROTON PHONON INSTABILITY OF STRONTIUM TITANATE. 06-1211
UM TITANATE). PHOTON PHONON INTERACTION IN FERROELECTRIC CRYSTALS (03-0427
FERROELECTRICS WITH HYDROGEN BONDS KDP TYPE. ROLE OF PROTON PHONON INTERACTION IN FERROELECTRICS WITH HYDR 17-2672
NATE AND POTASSIUM T/ THERMAL CONDUCTIVITY IN PEROVSKITES (3 PHONON INTERACTION IN THE NEAR INFRARED. (BARI 04-0925
OSPHATE. PROTON PHONON INTERACTION IN THE PHASE TRANSITION OF 17-2674
OIODIDE (BY RAMAN SCATTERING). SOFT PHONON INTERACTION PROPOSED FOR STRONTIUM TITA 06-1095
N MAGNETIC MOMENT OF CHROMIUM- POTASSIUM ALUM (UNSTABLE SOFT PHONON INTERACTIONS IN POTASSIUM DIHYDROGEN PH 17-2653
(3) OXIDE(11). SOFT OPTIC PHONON MODE AND MODE COUPLING IN ANTIMONY SULF 15-2167
YDRATE (NEAR THE TRANSITION TEMPERATURE, NO ANOMALY BUT SOFT PHONON MODE, DISPLACIVE FERROELECTRICS). / MEA 20-2806
ARIUM AND STRONTIUM TITANATES AND POTASSIUM/ COUPLED OPTICAL PHONON MODE IN FERROELECTRIC LEAD(5) GERMANIUM 35-3495
C, AND THE CUBIC- TETRAGONAL TRANSITION IN RELATION TO SOFT PHONON MODE STILL POSSIBLE). /ERROCYANIDE TRIH 24-2891
ASSIGNMENTS OF PHONON MODE THEORY OF INFRARED DISPERSION IN B 04-0547
/LEAD TITANATE ZIRCONATE SYSTEM (MEASUREMENT OF SOFT OPTICAL PHONON MODES. / SODIUM NIOBATE T(2) AT 600 DEG 08-1337
SSIUM(X) SODIUM(1-X) TANTALATE AND POTASSIUM TANTALATE/ SOFT PHONON MODES IN LITHIUM NIOBATE. 11-1694
OBATE BETWEEN 480 AND 575 DEG C, AND THEIR RELEVANCE TO SOFT PHONON MODES IN TETRAGONAL FERROELECTRIC REGI/ 09-1497
IN LEAD DICALCIUM PROFICNATE. THE ANOMALOUS BEHAVIOR OF PHONON MODES IN THE MIXED CRYSTAL SYSTEMS POTA 13-1933
/VERSE ISING MODEL, SOFT MODE COUPLING TO TRANSVERSE OPTICAL PHONON MODES. (X-RAY DIFFRACTION) /F SODIUM NI 08-1273
N ALUMINUM PHOSPHATE AND CESIUM DIHYDROGEN ARSENATE. PHONON NEAR THE 60 DEGREES C PHASE TRANSITION 26-2942
SSIUM MANGANESE TRIFLUORIDE: TEMPERATURE DISCONTINUITIE/ TWO PHONON). PART-3: (POTASSIUM DIHYDROGEN ARSENA/ 03-0384
ON SYSTEM IN HYDROGEN BONDED FERROELECTRICS OF THE KDP TYPE PHONON PHONON AND PROTON PHONON INTERACTIONS I 18-2739
. (IN SODIUM NITRITE) QUADRUPOLE SPIN (PHONON PROCESSES AND PHASE TRANSITIONS IN POTA 33-3363
TITANATE. ACOUSTIC PHONON PROTON INTERACTION). /S OF THE PROTON I 17-2670
SIU/ COUPLED MODES (OPTICAL, FERROELECTRIC- ACOUSTIC) IN THE PHONON RELAXATION AND FERROELECTRIC TRANSITION 16-2302
IMITATION OF THE CRYSTAL SIZE. PHONON SOFT OPTIC PHONON INTERACTION IN BARIUM 04-0806
E RESONANT INTERFERENCE IN RAMAN SCATTERING FROM ONE AND TWO PHONON SPECTRA OF STRONTIUM TITANATE AND POTAS 06-1157
ECTROSCOPY. ANALYSIS OF THE TEMPERATURE DEPENDENT PHONON SPECTRUM OF FERROELECTRICS STUDIED BY L 03-0486
ENDENT SUSCEPTIBILITY). COUPLED ORDER DISORDER PHONON STATES OF BARIUM TITANATE. AUGER-LIK 04-0923
PHONON STRUCTURE IN SODIUM NITRITE BY RAMAN SP 16-2299
F TETRAGONAL POTASSIUM DIHYDROGEN PHOSPHATE. (T/ ANALYSIS OF PHONON SYSTEM WITH DAMPING: KDP (FREQUENCY DEP 17-2667
OF ANISOTROPIC RELAXATION ATTENUATION OF QUASI LONGITUDINAL PHONON TRANSPORT IN BARIUM TITANATE. 04-0830
ITONS WITH APPLICATION TO LEAD NIOBATE. PART-1: LONG OPTICAL PHONON TUNNELING MODES IN THE RAMAN SPECTRUM O 03-0396
IUM NIOBATE (FROM TRANSVERSE (MORE INTENSE) AND LONGITUDINAL PHONONS) /E SULFATE SINGLE CRYSTALS. (THEORY 27-3058
OIODIDE. (POLARIZED INFRARED AND RA/ LONG WAVELENGTH OPTICAL PHONONS. /REQUENCIES AND EIGENVECTORS OF POLAR 08-1284
SORDER AND DISPLACIVE FERROELECTRICS. (ANTIMONY SUL/ OPTICAL PHONONS). /ILLOUIN SCATTERING OF LIGHT IN LITH 11-1743
SCATTERING OF LIGHT BY ACOUSTIC PHONONS AND PHASE TRANSITIONS IN ANTIMONY SULF 15-2106
IC MATERIALS. PHONONS AND PHASE TRANSITIONS IN SOME ORDER DI 15-2208
/FFECT OF TEMPERATURE AND PROPAGATION DIRECTION UPON OPTICAL PHONONS AND POLARITONS IN LITHIUM TANTALATE. 12-1878
STALS. LIGHT SCATTERING BY ACOUSTIC PHONONS, ELECTRONS, AND PROTONS IN FERROELECTR 02-0116
/CONDUCTION ELECTRONS BY TRANSVERSE AND LONGITUDINAL OPTICAL PHONONS IN CUBIC SODIUM CHLORIDE AND EPITAXIA/ 29-3243
TEMPERATURE DEPENDENT OPTICAL PHONONS IN FERROELECTRIC AND PIEZOELECTRIC CRY 03-0471
DIRECTIONAL DISPERSION AND ASSIGNMENT OF OPTICAL PHONONS IN IONIC FERROELECTRIC SEMICONDUCTORS. 03-0292
TE. SCATTERING OF LIGHT ON ACOUSTIC PHONONS IN LEAD TITANATE. 05-1083
. (THEORY, ABO(3) TYPE) DIELECTRIC PROPERTIES AND OPTICAL PHONONS IN LITHIUM NIOBATE. 11-1695
INTERACTIONS BETWEEN PROTON TUNNELING AND OPTICAL PHONONS IN LITHIUM NIOBATE AND LITHIUM TANTALA 11-1744
INTERACTION BETWEEN PROTON TUNNELING AND OPTICAL PHONONS IN PARA- AND FERROELECTRIC PEROVSKITES 03-0413
DISPERSION AND DAMPING OF SOFT ZONE BONDARY PHONONS IN POTASSIUM DIHYDROGEN PHOSPHATE. 17-2592
A THEORY OF THE PHASE TRANSITION IN AMMONIUM DIHYDROGEN SOFT PHONONS IN POTASSIUM DIHYDROGEN PHOSPHATE. 17-2593
F THE NONLINEAR OPTICAL COEFFICIENTS OF POTASSIUM DIHYDROGEN PHONONS IN POTASSIUM MANGANESE TRIFLUORIDE. 33-3374
BASED ON THE MICHELSON INTERFEROMETER. (AMMONIUM DIHYDROGEN PHONONS IN POTASSIUM TANTALATE NIOBATE. 13-2040
ILLOUIN DIFFUSION IN A SINGLE CRYSTAL OF RUBIDIUM DIHYDROGEN PHOSPHATE. 03-0286
LLOUIN SCATTERING BY A SINGLE CRYSTAL OF RUBIDIUM DIHYDROGEN PHOSPHATE. ABSOLUTE SIGNS O 17-2597
CRYSTAL PERFECTION OF AMMONIUM DIHYDROGEN PHOSPHATE) AMPLITUDE MODULATORS 17-2715
PHOSPHATE. BR 17-2520
UM CHLORIDE, POTASSIUM SODIUM TARTRATE, POTASSIUM DIHYDROGEN PHOSPHATE. BRI 17-2519
-D...O ECNE. (INCLUDING FULLY DEUTERATED AMMONIUM DIHYDROGEN PHOSPHATE) / BROMIDE, POTASSIUM CHLORIDE, SODI 17-2707
PHOSPHATE) /TUDY OF SOME CRYSTALS CONTAINING O 03-0204
17-2426

THERMAL VARIATION OF REFRACTIVE INDEX IN AMMONIUM DIHYDROGEN PHOSPHATE AND TRIGLYCINE SULFATE CRYSTALS. 17-2474
HATE CRYSTAL IN PARAELECTRIC AND / VIBRATION SPECTRUM OF THE PHOSPHATE ANION IN A RUBIDIUM DIHYDROGEN PHOSP 17-2472
VANISHING OF FERROELECTRICITY IN POTASSIUM DIHYDROGEN PHOSPHATE AT HIGH PRESSURES. 17-2442
MOTIONAL NARROWING IN THE PROTON NMR IN POTASSIUM DIHYDROGEN PHOSPHATE AT HIGH TEMPERATURES. ABSENCE OF 17-2594
EPENDENCE OF THE DIELECTRIC CONSTANT OF POTASSIUM DIHYDROGEN PHOSPHATE AT LAMBDA EQUAL 2.14 MM. /PERATURE D 17-2501
ERATURE. ACOUSTIC PROPAGATION IN POTASSIUM DEUTERIUM PHOSPHATE AT THE FERROELECTRIC TRANSITION TEMP 17-2573
MEASUREMENT OF THE ELASTIC CONSTANTS OF POTASSIUM DIHYDROGEN PHOSPHATE BY BRILLOUIN DIFFUSION. 17-2454
R/ FERROELECTRIC (SOFT) MODE MOTION IN POTASSIUM DIDEUTERIUM PHOSPHATE . (BY NEUTRON TRIPLE AXIS SPECTROMET 17-2668
HIGH TEMPERATURE PHASE TRANSITION IN RUBIDIUM PHOSPHATE BY SCATTERING OF CCLD NEUTRONS. 17-2489
/ COUPLING ON THE FERROELECTRIC MODE IN POTASSIUM DIHYDROGEN PHOSPHATE. (COLLECTIVE PSEUDOSPIN MODEL OF KD/ 17-2655
/ATURE DEPENDENT FERROELECTRIC MODE IN POTASSIUM DIDEUTERIUM PHOSPHATE (COMPARED WITH KDP AND WITH NEUTRON/ 17-2705
W FOR THE NONLINEAR SUSCEPTIBILITY OF A POTASSIUM DIHYDROGEN PHOSPHATE CRYSTAL. CURIE-WEISS LA 17-2686
ECTRIC PROPERTIES OF A GAMMA-IRRADIATED POTASSIUM DIHYDROGEN PHOSPHATE CRYSTAL. DIEL 17-2467
AVE MODULATOR OF RUBY LASER LIGHT USING POTASSIUM DIHYDROGEN PHOSPHATE CRYSTAL. MICROW 17-2640
NONLINEAR PROPERTIES OF A RUBIDIUM DIHYDROGEN PHOSPHATE CRYSTAL. 17-2508
OF THE FERROELECTRIC MODIFICATION OF A POTASSIUM DIHYDROGEN PHOSPHATE CRYSTAL. RAMAN SPECTRA 17-2622
MAN SPECTRUM OF FERROELECTRIC MODE IN A POTASSIUM DIHYDROGEN PHOSPHATE CRYSTAL. RA 17-2663
NCE IN AN ISOLATED AND A SHORTCIRCUITED POTASSIUM DIHYDROGEN PHOSPHATE CRYSTAL. STRESS INDUCED BIREFRINGE 17-2452
LE OF THE SECOND HARMONIC GENERATION IN POTASSIUM DIHYDROGEN PHOSPHATE CRYSTAL. /ENCE OF INDEX MATCHING ANG 17-2544
/TROOPTIC CHARACTERISTICS OF DEUTERATED POTASSIUM DIHYDROGEN PHOSPHATE CRYSTAL AND ITS APPLICATION FOR A H/ 17-2499
HASE. RAMAN SPECTRUM OF POTASSIUM DIDEUTERIUM PHOSPHATE CRYSTAL IN PARA- AND FERROELECTRIC P 17-2647
ON. (COERCIVE FIELD) POLARIZATION OF A RUBIDIUM DIHYDROGEN PHOSPHATE CRYSTAL IN THE PHASE TRANSITION REGI 17-2504
ON. POLARIZATION OF A RUBIDIUM DIHYDROGEN PHOSPHATE CRYSTAL IN THE PHASE TRANSITION REGI 17-2506
ROCALORI/ SPONTANEOUS POLARIZATION OF A POTASSIUM DIHYDROGEN PHOSPHATE CRYSTAL NEAR THE CURIE POINT. (ELECT 17-2678
ELECTROOPTICAL SHUTTER BASED ON Z CUT POTASSIUM DIDEUTERIUM PHOSPHATE CRYSTALS. CHARACTERISTICS OF AN 17-2460
CRITICAL BEHAVIOR OF KDP AND RUBIDIUM DIHYDROGEN PHOSPHATE CRYSTALS. 17-2676
HT DUE TO HEATING OF PARALLEL PLANES OF POTASSIUM DIHYDROGEN PHOSPHATE CRYSTALS. /OF LINEARLY POLARIZED LIG 17-2540
ELECTRIC PROPERTIES OF GAMMA-IRRADIATED POTASSIUM DIHYDROGEN PHOSPHATE CRYSTALS. DI 17-2542
POLARIZED LIGHT BY ULTRASONIC WAVES IN POTASSIUM DIHYDROGEN PHOSPHATE CRYSTALS. DIFFRACTION OF 17-2407
DOUBLE DIELECTRIC HYSTERESIS LOOP OF POTASSIUM DIDEUTERIUM PHOSPHATE CRYSTALS. 17-2503
CT OF ANNEALING ON OPTICAL ANOMALIES OF POTASSIUM DIHYDROGEN PHOSPHATE CRYSTALS. EFFE 17-2427
EFFECT OF DISLOCATIONS ON THE GROWTH OF POTASSIUM DIHYDROGEN PHOSPHATE CRYSTALS. 17-2491
FFECT OF IMPURITIES ON THE STRUCTURE OF POTASSIUM DIHYDROGEN PHOSPHATE CRYSTALS. E 17-2606
CTRONIC EFFECT AND ELASTIC PROPERTIES OF RUBIDIUM DIHYDROGEN PHOSPHATE CRYSTALS. ELE 17-2617
TICAL PROPERTIES OF PARTIALLY DEUTERATED RUBIDIUM DIHYDROGEN PHOSPHATE CRYSTALS. ELECTROOPTICAL AND OP 17-2698
FAR INFRARED SPECTRA OF RUBIDIUM DIHYDROGEN PHOSPHATE CRYSTALS. 17-2638
FIRST ORDER PHASE TRANSITION IN POTASSIUM DEUTERIUM PHOSPHATE CRYSTALS. 17-2677
LANDAU-KHALATNIKOV ATTENUATION IN POTASSIUM DIDEUTERIUM PHOSPHATE CRYSTALS. 17-2656
LIGHT AND CURRENT PULSES FROM X-RAYED POTASSIUM DIHYDROGEN PHOSPHATE CRYSTALS. 17-2716
HE PRIMARY LINEAR ELECTROOPTIC EFFECT IN AMMONIUM DIHYDROGEN PHOSPHATE CRYSTALS. MEASUREMENT OF T 17-2654
THE (MECHANICAL) STRENGTH PROPERTIES OF POTASSIUM DIHYDROGEN PHOSPHATE CRYSTALS. STUDY OF 17-2648
C STUDY OF THE DISLOCATIONS PRODUCED IN POTASSIUM DIHYDROGEN PHOSPHATE CRYSTALS BY GROWTH FROM SOLUTION. /I 17-2492
CED/ OBSERVATION OF DOMAIN STRUCTURE IN POTASSIUM DIHYDROGEN PHOSPHATE CRYSTALS BY POLARIZATION OPTICAL PRO 17-2493
/ESTIGATION OF THE THERMAL EXPANSION OF POTASSIUM DIHYDROGEN PHOSPHATE CRYSTALS BY USING SEMICONDUCTOR FIL/ 17-2565
/SPECIFIC HEAT (BY A DYNAMIC METHOD) OF POTASSIUM DIHYDROGEN PHOSPHATE CRYSTALS FOR STUDY OF THE PYROELECT/ 17-2566
TION OPTIMUM CUT IN X DIHYDROGEN PHOSPHATE CRYSTALS FOR TRANSVERSE LIGHT MODULA/ 17-2494
/ROWTH LAYERS AND OTHER MACRODEFECTS IN AMMONIUM DIHYDROGEN PHOSPHATE CRYSTALS GROWN FROM AN AQUEOUS SOLU/ 17-2496
KINETICS OF GROWTH AND DISSOLUTION OF AMMONIUM DIHYDROGEN PHOSPHATE CRYSTALS IN SOLUTION. 17-2412
TATE IONS. GROWTH AND ETCHING OF POTASSIUM DIHYDROGEN PHOSPHATE CRYSTALS IN SOLUTIONS CONTAINING ACE 17-2567
/RANSITION AND DIELECTRIC PROPERTIES OF POTASSIUM DIHYDROGEN PHOSPHATE CRYSTALS IRRADIATED WITH FAST NEUTR/ 17-2615
/ION ON NON-WEDGE SHAPED RUBIDIUM DIHYDROGEN OR DIDEUTERIUM) PHOSPHATE CRYSTALS THEORY AND PRACTICAL CONSI/ 17-2637
ANSION OF FREE/ NONLINEAR PROPERTIES OF POTASSIUM DIHYDROGEN PHOSPHATE. (DETERMINATION OF BETA IN THE P EXP 17-2543
/ORPTION, NEUTRON INELASTIC SCATTERING, POTASSIUM DIHYDROGEN PHOSPHATE, DEUTERATED POTASSIUM DIHYDROGEN PH/ 03-0401
TEMPERATURE IN THE FERROELECTRIC PHASE POTASSIUM DIHYDROGEN PHOSPHATE. (DIFFRACTION BY DOMAIN WALLS) /WITH 17-2662
S POLARIZATION) ELECTRONIC STUDIES OF POTASSIUM DIHYDROGEN PHOSPHATE. (ELECTROCALORIC STUDIES, SPONTANEOU 17-2429
/ULATION OF A DOUBLE 45 DEGREES, Z CUT POTASSIUM DIDEUTERIUM PHOSPHATE. ELECTROOPTIC Q SWITCH BY DESK-TOP / 17-2702
/ DITHORIUM PHOSPHATE- A NEW CASE OF THE NON-HYDROGEN BONDED PHOSPHATE FERROELECTRIC (CRYSTAL GROWTH AND S/ 36-3610
EMPERATURE DEPENDENCE OF SOME PROPERTIES OF SODIUM DITHORIUM PHOSPHATE FERROELECTRIC CRYSTALS. T 36-3611
IAL ELECTROOPTIC MIXED CRYSTAL AMMONIUM POTASSIUM DIHYDROGEN PHOSPHATE FILM WAVEGUIDE. EPITAX 17-2633
AND POLISHING IN UNIAXIAL CRYSTALS OF THE TYPE X DIHYDROGEN PHOSPHATE FOR LIGHT MODULATORS. CUTTING 17-2559
A TUNNELING MODEL/ DEUTERON MOTIONS IN POTASSIUM DIDEUTERIUM PHOSPHATE (FROM POLARIZED REFLECTION SPECTRA, 17-2645
CRYSTALLIZATION OF RUBIDIUM DIDEUTERIUM PHOSPHATE FROM SOLUTIONS IN HEAVY WATER. 17-2583
UCTURE OF FERROELECTRIC CRYSTALS OF THE POTASSIUM DIHYDROGEN PHOSPHATE GROUP. STR 17-2607
ERTIES OF FERROELECTRIC CRYSTALS OF THE POTASSIUM DIHYDROGEN PHOSPHATE GROUP IN THE PARAELECTRIC PHASE. /OP 17-2696
GROWTH BEHAVIOR OF SMALL SEEDS OF AMMONIUM DIHYDROGEN PHOSPHATE IN A MODIFIED CIRCULATION APPARATUS. 17-2704
SPECTRA OF POLYMORPHIC MODIFICATIONS OF POTASSIUM DIHYDROGEN PHOSPHATE IN AN ELECTRIC FIELD. RAMAN 17-2425
VIBRATION SPECTRA OF POTASSIUM DIHYDROGEN PHOSPHATE IN PARA- AND FERROELECTRIC PHASES. 17-2480
AND FERROELECTRIC PHAS/ TRANSMISSION OF POTASSIUM DIHYDROGEN PHOSPHATE IN THE INFRARED REGION IN THE PARA- 17-2459
OSPHATE IN HE/ ENTHALPIES OF SOLUTION OF RUBIDIUM DIDEUTERIUM PHOSPHATE IN WATER AND RUBIDIUM DIDEUTERIUM PH 17-2585
/ THE PHASE TRANSITIONS IN POTASSIUM DIHYDROGEN OR DEUTERIUM PHOSPHATE. (INFRARED PHOSPHATE STRETCHING, WID/ 17-2646
/D THE ISOTOPE EFFECTS IN FERROELECTRIC POTASSIUM DIHYDROGEN PHOSPHATE. (INTERBAND INTERACTION AND DIPOLE / 17-2433
ALT. STRUCTURE OF DIHYDROGEN PHOSPHATE ION IN THE FERROELECTRIC POTASSIUM S 17-2624
CRYSTALS (INFRARED A/ SPECTRA OF INTERNAL VIBRATIONS OF THE PHOSPHATE ION IN THE PARAELECTRIC PHASE OF KDP / 17-2479
/IALS AND NONLINEAR CRYSTALS. (KDP, ADP, RUBIDIUM DIHYDROGEN PHOSPHATE, LITHIUM NIOBATE, LITHIUM TANTALATE/ 11-1865
LASER INDUCED DAMAGE IN X DIHYDROGEN PHOSPHATE MATERIALS. 17-2475
ISOTOPIC EFFECT IN POTASSIUM DIHYDROGEN(X) DIDEUTERIUM(1-X) PHOSPHATE MIXED CRYSTALS. / FIELD STUDY OF THE 17-2563
HYDROGEN PHOSPHATE- POTASSIUM DIHYDROGEN(1-X) DIDEUTERIUM(X) PHOSPHATE MIXED CRYSTALS. /IES OF POTASSIUM DI 17-2680
MODIFIED POTASSIUM DIHYDROGEN PHOSPHATE MODEL IN A STAGGERED FIELD. 17-2710
Y ON THERMAL EXPANSION OF FERROELECTRIC POTASSIUM DIHYDROGEN PHOSPHATE. (NEAR CURIE TEMPERATURE) X-RAY STUD 17-2556
SITION. (T/ THERMOELASTIC PROPERTIES OF POTASSIUM DIHYDROGEN PHOSPHATE NEAR THE HIGH TEMPERATURE PHASE TRAN 17-2627
EN DENSITY DISTRIBUTION IN PARAELECTRIC POTASSIUM DIHYDROGEN PHOSPHATE (NEUTRON SCATTERING). HYDROG 17-2490
/PTIC MODULATOR MATERIALS. (POTASSIUM OR AMMONIUM DIHYDROGEN PHOSPHATE OR ARSENATE, POTASSIUM DIDEUTERIUM / 01-0021
F THE FERROELECTRIC PHASE TRANSITION OF POTASSIUM DIHYDROGEN PHOSPHATE. PART-1: THE ORDER OF TRANSITION. /O 17-2557
ERVATION OF ANTIFERROELECTRIC DOMAINS OF AMMONIUM DIHYDROGEN PHOSPHATE. (PHASE TRANSITION AT 148 DEG K) /BS 17-2643
ING. LOSSLESS POTASSIUM DIDEUTERIUM PHOSPHATE POCKELS CELL FOR HIGH POWER Q SWITCH 17-2529
ING PERFORMANCE IN LITHIUM NIOBATE AND POTASSIUM DIDEUTERIUM PHOSPHATE POCKELS CELLS. (COMMENTS) / Q SWITCH 11-1710
/NANCE AND RELAXATION IN FERROELECTRIC POTASSIUM DIDEUTERIUM PHOSPHATE, POTASSIUM DIDEUTERIUM ARSENATE, AN/ 17-2438
RU/ DIELECTRIC ANOMALIES IN CRYSTALS OF POTASSIUM DIHYDROGEN PHOSPHATE POTASSIUM DIDEUTERIUM PHOSPHATE AND 17-2613
TEMPERATURE DEPENDENT RAMAN SPECTRA OF POTASSIUM DIHYDROGEN PHOSPHATE, POTASSIUM DIDEUTERIUM PHOSPHATE, P/ 17-2409
/SR STUDY OF RADIATION DAMAGE CENTERS IN AMMONIUM DIHYDROGEN PHOSPHATE POTASSIUM DIHYDROGEN ARSENATE MIXED/ 17-2547

/Y OF ELECTROOPTIC LASER MODULATORS WITH AMMONIUM DIHYDROGEN PHOSPHATE, POTASSIUM DIHYDROGEN PHOSPHATE AND/ 17-2675
/S AND CHARACTERISTICS OF THEIR GROWTH. (AMMONIUM DIHYDROGEN PHOSPHATE, POTASSIUM FERROCYANIDE TRIHYDRATE,/ 16-2316
/ESONANCE IN FERROELECTRICS. (MODEL FOR POTASSIUM DIHYDROGEN PHOSPHATE PRESENTED, TRIGLYCINE SULFATE AND B/ 03-0180
DEUTERATION ON THE ELASTIC PROPERTIES OF RUBIDIUM DIHYDROGEN PHOSPHATE (RDP).   EFFECT OF 17-2584
/CIFIC HEAT OF SOME KDP TYPE CRYSTALS. (POTASSIUM DIHYDROGEN PHOSPHATE, RUBIDIUM AND AMMONIUM ANALOGS SOME/ 17-2679
/THE RELATIVE NONLINEAR COEFFICIENTS OF AMMONIUM DIHYDROGEN PHOSPHATE, RUBIDIUM DIHYDROGEN PHOSPHATE, RUB/ 08-1413
INVESTIGATION OF THE PHASE TRANSITION IN AMMONIUM DIHYDROGEN PHOSPHATE SINGLE CRYSTAL BY NMR.   17-2545
ATIC REGIME AND / GROWING KINETIC OF THE AMMONIUM DIHYDROGEN PHOSPHATE SINGLE CRYSTAL (FROM SOLUTION) IN ST 17-2516
          GROWTH OF AMMONIUM DIHYDROGEN PHOSPHATE SINGLE CRYSTAL FROM SOLUTIONS.   17-2411
          HEAT CAPACITY OF RUBIDIUM DIHYDROGEN PHOSPHATE SINGLE CRYSTALS.   17-2415
          LIGHT ABSORPTION IN POTASSIUM DIHYDROGEN PHOSPHATE SINGLE CRYSTALS.   17-2421
SOLID SOLUTIONS IN POTASSIUM DIHYDROGEN(1-X) DIDEUTERIUM(X) PHOSPHATE SINGLE CRYSTALS. REGULAR BEHAVIOR OF 17-2458
CE OF THE MECHANICAL CHARACTERISTICS OF POTASSIUM DIHYDROGEN PHOSPHATE SINGLE CRYSTALS. /MPERATURE DEPENDEN 17-2649
          GROWTH OF POTASSIUM DIHYDROGEN PHOSPHATE SINGLE CRYSTALS IN GEL.   17-2455
/SELENATE OR FLUOBERYLLATE, POTASSIUM OR RUBIDIUM DIHYDROGEN PHOSPHATE, SODIUM, POTASSIUM(2), OR LITHIUM T/ 02-0089
S IN THE AMMONIUM DIHYDROGEN PHOSPHATE- POTASSIUM DIHYDROGEN PHOSPHATE SYSTEM. /ROPERTIES OF SINGLE CRYSTAL 17-2572
S FERROELECTRIC SYSTEMS. (DISCUSSED FOR POTASSIUM DIHYDROGEN PHOSPHATE SYSTEM) /E POINT) IN SOME ISOMORPHOU 17-2487
TU/ PHASE DIAGRAM OF SODIUM TRIDEUTERIUM(X) TRIHYDROGEN(1-X) PHOSPHATE SYSTEM VERSUS CONCENTRATION, TEMPERA 17-2666
/THE FERROELECTRIC TRANSITION REGION IN POTASSIUM DIHYDROGEN PHOSPHATE. (TEMPERATURE AND FIELD DEPENDENCE) 17-2576
/N OF LONGITUDINAL ULTRASONIC WAVES IN POTASSIUM DIDEUTERIUM PHOSPHATE (TEMPERATURE DEPENDENCE OF THE YOUN/ 17-2660
DES IN THE RAMAN SPECTRUM OF TETRAGONAL POTASSIUM DIHYDROGEN PHOSPHATE. (THEORY) /IS OF PHONON TUNNELING MO 03-0396
TC/ (EFFECT OF TEMPERATURE ON) QUASI SYMMETRICAL HARD IONS (PHOSPHATE, TITANATE, SULFATE, FLUOBERYLLATE, E 03-0330
FAR INFRARED DIELECTRIC MEASUREMENTS ON POTASSIUM DIHYDROGEN PHOSPHATE, TRIGLYCINE SULFATE, AND RUTILE.   17-2423
RATURE OF FERROELECTRIC CRYSTALS OF THE POTASSIUM DIHYDROGEN PHOSPHATE TYPE. /ON THE PHASE TRANSITION TEMPE 17-2495
N BONDS.   DYNAMICAL THEORY OF AMMONIUM DIHYDROGEN PHOSPHATE TYPE ANTIFERROELECTRICS WITH HYDROGE 03-0476
INE WIDTH TRANSITION FOR THE PROTONS IN POTASSIUM DIHYDROGEN PHOSPHATE TYPE CRYSTALS.   ABSENCE OF A L 17-2436
NS OF HYDROGEN BONDS. (CALCULATIONS FOR POTASSIUM DIHYDROGEN PHOSPHATE TYPE CRYSTALS)   COLLECTIVE MOTIO 17-2483
PHASE TRANSITIONS AND METASTABILITY IN POTASSIUM DIHYDROGEN PHOSPHATE TYPE CRYSTALS.   HIGH TEMPERATURE 17-2513
          NMR PROTON SECOND MOMENTS IN POTASSIUM DIHYDROGEN PHOSPHATE TYPE CRYSTALS.   17-2408
          NOVEL CUT OF POTASSIUM DIHYDROGEN PHOSPHATE TYPE CRYSTALS.   17-2530
THE LATTICE PARAMETERS AND STRUCTURE OF POTASSIUM DIHYDROGEN PHOSPHATE TYPE CRYSTALS.   PRESSURE EFFECTS ON 17-2588
N CADMIUM SULFIDE, BARIUM TITANATE, AND POTASSIUM DIHYDROGEN PHOSPHATE TYPE CRYSTALS. /ARMONIC GENERATION I 04-0852
THEORY OF DOMAIN WALL MOTION IN POTASSIUM DIHYDROGEN PHOSPHATE TYPE CRYSTALS.   17-2652
LECTRIC AND ANTIFERROELECTRIC STATES IN POTASSIUM DIHYDROGEN PHOSPHATE TYPE CRYSTALS AT HIGH PRESSURE. /ROE 17-2642
          OPTIMUM CUT OF POTASSIUM DIHYDROGEN PHOSPHATE TYPE CRYSTALS FOR LIGHT MODULATION.   17-2532
ONS AND THE DYNAMICAL SUSCEPTIBILITY OF POTASSIUM DIHYDROGEN PHOSPHATE TYPE FERROELECTRIC CRYSTALS. /ERACTI 17-2570
.   DYNAMICS OF POTASSIUM DIHYDROGEN PHOSPHATE TYPE FERROELECTRIC PHASE TRANSITIONS 17-2692
          ORDER DISORDER THEORY OF POTASSIUM DIHYDROGEN PHOSPHATE TYPE FERROELECTRICS.   17-2685
ERGY PARAMETERS AND PROTON TUNNELING IN POTASSIUM DIHYDROGEN PHOSPHATE TYPE FERROELECTRICS. (THEORY)   EN 03-0506
MODELS OF THE PHASE TRANSITION IN POTASSIUM DIHYDROGEN PHOSPHATE TYPE FERROELECTRICS. (24 REFS) 03-0489
EMPER/ RAMAN SPECTRA OF CRYSTALS OF THE POTASSIUM DIHYDROGEN PHOSPHATE TYPE IN THEIR PARAELECTRIC AND LOW T 17-2515
CONDITIONS FOR THE GROWTH OF POTASSIUM DIHYDROGEN PHOSPHATE WHISKERS.   17-2628
/URE INFRARED ABSORPTION SPECTRA OF AMMONIUM DIHYDROGEN ORTHOPHOSPHATE (4000-10000 PER CM) AND HARMONIC VI/ 17-2432
FERROELECTRIC DOMAINS OF POTASSIUM DIHYDROGEN PHOSPHATE. (93 REFS)   17-2447
VIBRATION SPECTRUM OF THE FERROELECTRIC RUBIDIUM DIHYDROGEN PHOSPHATES. /PHASE TRANSITION ON THE HYDROXIDE 17-2470
SE TRANSITION IN THALLIUM AND AMMONIUM THALLIUM (DIHYDROGEN) PHOSPHATES.   PHA 17-2571
TION AND STRUCTURE OF) RUBIDIUM AND CESIUM SEMIMETALLIC ORTHOPHOSPHATES.   (PREPARA 17-2598
ED PHOSPHATE FERROELECTRIC (CRYSTAL GROWTH/ SILVER DITHORIUM PHOSPHATE- A NEW CASE OF THE NON-HYDROGEN BOND 36-3610
POTASSIUM DIHYDROGEN PHOSPHATE)/ RELAXATION OF FERROELECTRIC PHOSPHATES (AMMONIUM DIHYDROGEN PHOSPHATE AND 17-2580
/ SCATTERING FROM FERROELECTRIC MODES IN THE KDP ISOMORPHOUS PHOSPHATES AND ARSENATES. (KDP, AMMONIUM DIHY/ 17-2639
/URES OF POTASSIUM, RUBIDIUM, CESIUM AND AMMONIUM DIHYDROGEN PHOSPHATES AND CESIUM AND AMMONIUM DIHYDROGEN/ 17-2577
DIHYDRO/ ANOMALIES IN THE INTERNAL STRESS FOR FERROELECTRIC PHOSPHATES NEAR THEIR CURIE POINTS. (POTASSIUM 17-2665
STRUCTURAL STUDIES OF THE SYSTEM POTASSIUM DIHYDROGEN PHOSPHATE- POTASSIUM DIDEUTERIUM PHOSPHATE.   17-2590
/IC PROPERTIES OF SINGLE CRYSTALS IN THE AMMONIUM DIHYDROGEN PHOSPHATE- POTASSIUM DIHYDROGEN PHOSPHATE SYS/ 17-2572
ER SYSTEM.   SOLUBILITY ISOBAR IN THE AMMONIUM DIHYDROGEN PHOSPHATE- POTASSIUM DIHYDROGEN PHOSPHATE- WAT 17-2658
ER SYSTE/ POLYTHERM OF SOLUBILITY OF THE AMMONIUM DIHYDROGEN PHOSPHATE- POTASSIUM DIHYDROGEN PHOSPHATE- WAT 17-2434
UM(/ SOME THERMODYNAMICAL PROPERTIES OF POTASSIUM DIHYDROGEN PHOSPHATE- POTASSIUM DIHYDROGEN(1-X) DIDEUTERI 17-2680
/R STUDY OF AMMONIUM IONS IN IRRADIATED AMMONIUM DIDEUTERIUM PHOSPHATE- 5 PERCENT POTASSIUM DIDEUTERIUM AR/ 17-2684
FERROELECTRIC TRANSITION IN RUBIDIUM DIHYDROGEN PHOSPHIDE USING RAMAN SCATTERING SPECTRA.   17-2620
L GROWTH OF ANTIMONY SULFOIODIDE, NICKEL CARBONATE AND BLACK PHOSPHORUS.   HIGH GAS PRESSURE CRYSTA 15-2137
UM TITANATE.   PHOTOCHROISM IN TRANSITION METAL DOPED STRONTI 06-1126
ATE, AND CALCIUM TI/ ERASE MODE RECORDING CHARACTERISTICS OF PHOTOCHROMIC CALCIUM FLUORIDE, STRONTIUM TITAN 06-1119
E OF IRON AND MOLYBDENUM IN BARIUM SODIUM NIOBATE.   PHOTOCHROMIC EFFECT AND ELECTRON SPIN RESONANC 13-1930
ENCE FROM PARAMAGNETIC RESONANCE.   PHOTOCHROMIC IRON IN STRONTIUM TITANATE. EVID 06-1171
TANATE PHOTOCONDUCTIVITY, CALCIUM FLUORIDE PHOTOCONDUCTIVIT/ PHOTOCHROMIC MATERIALS RESEARCH. (STRONTIUM TI 06-1140
ND BLEACHING).   PHOTOCONDUCTIVITY (TRANSIENTS) IN PHOTOCHROMIC STRONTIUM TITANATE (ON COLORING A 06-1240
ATERIALS.   STUDY OF THE DYNAMICS OF PHOTOCHROMIC SWITCHING IN STRONTIUM TITANATE M 06-1092
          PHOTOCONDUCTION OF POTASSIUM TANTALATE.   08-1408
S OF ANTIMONY SULFOIODIDE. (PHASE BOUNDARY SHAPE / EFFECT OF PHOTOCONDUCTION ON THE FERROELECTRIC PROPERTIE 15-2197
PHOTOVOLTAIC EFFECT IN FERROELECTRIC AND PHOTOCONDUCTIVE ANTIMONY SULFOIODIDE.   15-2174
MENA IN BARIUM TITANATE.   PHOTOCONDUCTIVITY AND PHOTOFERROELECTRIC PHENO 04-0574
RONTIUM TITANATE.   PHOTOCONDUCTIVITY AND TRAPPING PHENOMENA IN ST 06-1091
E LOCAL FERROELECTRIC PHASE TRANSITION AROUND 47 DEG K IN S/ PHOTOCONDUCTIVITY ANOMALY AS AN EVIDENCE OF TH 06-1249
/)C(VII) TYPE WITH LOW TEMPERATURE PHASE CHANGES. (EFFECT OF PHOTOCONDUCTIVITY, ANTIMONY SULFOIODIDE(X) BR/ 15-2161
CTIVIT/ PHOTOCHROMIC MATERIALS RESEARCH. (STRONTIUM TITANATE PHOTOCONDUCTIVITY, CALCIUM FLUORIDE PHOTOCONDU 06-1140
FACE LAYERS. (BARIUM TITANATE, ANTIMONY SULFOIODI/ EXTRINSIC PHOTOCONDUCTIVITY IN FERROELECTRICS DUE TO SUR 04-0681
DUCTOR ANTIMONY SULFOIODIDE.   INDUCED IMPURITY PHOTOCONDUCTIVITY IN THE FERROELECTRIC SEMICON 15-2159
GLE CRYSTALS.   PHOTOELECTRIC PROPERTIES (PHOTOCONDUCTIVITY) OF ANTIMONY SULFOIODIDE SIN 15-2134
TOR ANTIMONY SULFOIODIDE.   PHOTOCONDUCTIVITY OF FERROELECTRIC PHOTOCONDUC 15-2226
STRONTIUM TITANATE (ON COLORING AND BLEACHING).   PHOTOCONDUCTIVITY (TRANSIENTS) IN PHOTOCHROMIC 06-1240
ING ON THE FERROELECTRIC CURIE POINT (UNIAXIAL FERROELECTRIC PHOTOCONDUCTOR).   INFLUENCE OF SCREEN 03-0392
THIN FILM FERROELECTRIC PHOTOCONDUCTOR MEMORY DEVICE.   PHOTOCONDUCTOR AND ELECTRODE REQUIREMENTS FOR 37-3671
PHOTOCONDUCTIVITY OF FERROELECTRIC PHOTOCONDUCTOR ANTIMONY SULFOIODIDE.   15-2226
OXIDE(12))   PHASE HOLOGRAMS IN A FERROELECTRIC PHOTOCONDUCTOR DEVICE. (BISMUTH(4) TITANIUM(3) 13-1977
E READOUT, BASED ON/ STORAGE OF HOLOGRAMS IN A FERROELECTRIC PHOTOCONDUCTOR DEVICE (ERASABLE, NONDESTRUCTIV 37-3646
ES. (PLZT CERAMICS)   SCATTERING MODE FERROELECTRIC PHOTOCONDUCTOR IMAGE STORAGE AND DISPLAY DEVIC 09-1599
ES OPERATED IN A REFLECTION MO/ STRAIN BIASED FERROELECTRIC- PHOTOCONDUCTOR IMAGE STORAGE AND DISPLAY DEVIC 09-1564
ES. (PLZT)   STRAIN BIASED FERROELECTRIC PHOTOCONDUCTOR IMAGE STORAGE AND DISPLAY DEVIC 09-1565
EQUIVALENT CIRCUIT ANALYSIS OF A FERROELECTRIC PHOTOCONDUCTOR MEMORY DEVICE.   37-3644
FERROELECTRIC- PHOTOCONDUCTOR MEMORY DEVICE. (PLZT)   09-1568

THIN FILM FERROELECTRIC PHOTOCONDUCTOR MEMORY DEVICE. (PLZT)   09-1503
RATIONS RELATING TO SWITCHING AND REMANENCE IN FERROELECTRIC PHOTOCONDUCTOR MEMORY DEVICES. /ETICAL CONSIDE 03-0378
DEVELOPMENTS.   BISMUTH TITANATE FERROELECTRIC- PHOTOCONDUCTOR OPTICAL STORAGE MEDIUM FURTHER   08-1371
ESOLUTION READ WRITE)   BISMUTH TITANATE PHOTOCONDUCTOR OPTICAL STORAGE MEDIUM. (HIGH R   08-1445
G BISMUTH TITANATE.   FERROELECTRIC PHOTOCONDUCTOR OPTICAL STORAGE MEDIUM UTILIZIN 04-0769
         DESIGN AND PERFORMANCE OF A THIN FILM FERROELECTRIC PHOTOCONDUCTOR STORAGE DEVICE.   37-3632
                        PHOTOEFFECTS AT THE PHOTOCONDUCTOR- FERROELECTRIC BOUNDARY.   03-0519
SINGLE CRYSTALS.   PHOTODIELECTRIC EFFECT IN ANTIMONY SULFOIODIDE 15-2236
E FERROELECTRIC STATE.   FIRST ORDER NEGATIVE PHOTODIELECTRIC EFFECT IN PHTHALOCYANINE IN TH   36-3539
                           MECHANISM OF THE PHOTODOMAIN EFFECT IN ANTIMONY SULFOIODIDE.   15-2146
MONY SULFOIODIDE CRYSTALS.   PHOTODOMAIN EFFECT IN BARIUM TITANATE.   04-0682
MONY SULFOIODIDE CRYSTALS.   PHOTODOMAIN EFFECT IN NATURALLY POLARIZED ANTI 15-2133
RIC BOUNDARY.   PHOTODOMAIN EFFECT IN NATURALLY POLARIZED ANTI 15-2135
                                          PHOTOEFFECTS AT THE PHOTOCONDUCTOR- FERROELECT   03-0519
                                                             PHOTOEFFECTS IN POTASSIUM TANTALATE.   08-1415
     INFLUENCE OF CRITICAL POLARIZATION FLUCTUATIONS ON THE PHOTOELASTIC BEHAVIOR OF BARIUM TITANATE.   04-0628
SCATTERING BY ULTRASONIC WAVES IN LITHIUM NIOBATE CRYSTALS. (PHOTOELASTIC CONSTANT)   LIGHT   11-1754
TALS.   PHOTOELASTIC CONSTANTS OF LITHIUM NIOBATE CRYS 11-1745
CTRON IRRADIATION).   SECONDARY (ELECTRO) PHOTOELASTIC EFFECT IN LITHIUM NIOBATE (ON ELE 11-1801
ASSIUM AND AMMONIUM DIHYDROGEN PHOSPHATE.   PHOTOELASTIC EFFECT IN 45 DEGREE X-CUTS OF POT 17-2443
                           PIEZOELECTRIC AND PHOTOELASTIC PROPERTIES OF LITHIUM IODATE.   08-1465
                           PIEZOELECTRIC AND PHOTOELASTIC PROPERTIES OF LITHIUM IODATE.   08-1464
POLARIZATION IN ANTIMONY SULFOIODIDE.   EFFECT OF PHOTOELECTRET CHARGE ON VALUE OF FERROELECTRIC   15-2186
SOLUTIONS. (DIELECTRIC CONSTANTS, PYROELECTRI/ FERROELECTRIC PHOTOELECTRETS BASED ON A(V)B(VI)C(VII) SOLID 15-2157
          CHARACTERISTICS OF THE PHOTOELECTRIC EMF OF ANTIMONY SULFOIODIDE.   15-2129
          CHARACTERISTICS OF THE PHOTOELECTRIC EMF OF ANTIMONY SULFOIODIDE.   15-2153
DE.   ELECTRICAL AND PHOTOELECTRIC PROPERTIES OF ANTIMONY SULFOIODI   15-2107
DE CRYSTAL/ EFFECT OF COMPOSITION ON GROWTH, DIELECTRIC, AND PHOTOELECTRIC PROPERTIES OF ANTIMONY SULFOIODI 15-2118
E CRYSTALS.   PHOTOELECTRIC PROPERTIES OF BISMUTH SULFOIODID 15-2170
EN ANTIMONY SULFOIODIDE SINGLE CRYSTALS AND / ELECTRICAL AND PHOTOELECTRIC PROPERTIES OF THE CONTACTS BETWE 15-2251
F ANTIMONY SULFOIODIDE SINGLE CRYSTALS.   PHOTOELECTRIC PROPERTIES (PHOTOCONDUCTIVITY) O 15-2134
    BAND STRUCTURE OF STRONTIUM TITANATE AS MEASURED BY X-RAY PHOTOELECTRON SPECTROSCOPY (ESCA).   06-1100
XIDE AS A FUNCTION OF THICKNESS OF THE S/ STUDY OF THE X-RAY PHOTOELECTRON SPECTRUM OF TUNGSTEN- TUNGSTEN O 10-1631
ANTIMONY SULFOIODIDE.   PHOTOEMF'S IN THE FERROELECTRIC SEMICONDUCTOR 15-2132
IMONY SULFOIODIDE CRYSTALS)   SHORT CIRCUIT PHOTOEMF IN FERROELECTRIC SEMICONDUCTORS. (ANT 15-2105
                              SOME FEATURES OF THE PHOTOEMF SPECTRUM OF ANTIMONY SULFOIODIDE.   15-2136
     OBSERVATION OF THE SURFACES OF FERROELECTRIC CRYSTALS IN A PHOTOEMISSION ELECTRON MICROSCOPE.   37-3679
ARIUM TIT/ STUDY OF DOMAINS IN FERROELECTRIC CRYSTALS WITH A PHOTOEMISSION MICROSCOPE (TUNGSTEN TRIOXIDE, B 04-0811
URFACE OF TUNGSTEN TRIOXIDE CRYSTALS.   PHOTOEMISSION MICROSCOPIC OBSERVATION OF THE S 10-1645
URFACE OF BARIUM TITANATE CRYSTALS.   PHOTOEMISSION MICROSCOPIC OBSERVATION OF THE S 04-0810
          OPTICAL ABSORPTION, ELECTRIC CONDUCTIVITY AND PHOTOEMISSION OF GLYCOL SULFATE.   19-2780
IN IMAGING, ABSTRACT ONLY)   SOME RESULTS ABOUT PHOTOEMISSION ON FERROELECTRIC CRYSTALS. (DOMA 03-0386
              ANOMALOUS PROPERTIES IN PHOTOFERROELECTRIC ANTIMONY SULFOIODIDE.   15-2196
AND BARIUM TITANATE TYPE FERROELECTRICS.   PHOTOFERROELECTRIC EFFECTS IN A(V)B(VI)C(VII) 04-0552
MICONDUCTORS OF THE A(V)B(VI)C(VII) TYPE WITH LOW TEMPERATU/ PHOTOFERROELECTRIC EFFECTS IN FERROELECTRIC SE 15-2161
OF ANTIMONY SULFOIODIDE.   DILATOMETRIC INVESTIGATION OF PHOTOFERROELECTRIC EFFECTS IN SINGLE CRYSTALS 15-2156
E.   PHOTOCONDUCTIVITY AND PHOTOFERROELECTRIC PHENOMENA IN BARIUM TITANAT   04-0574
IUM NIOBATE PYROELECTRIC DETECTORS.   PHOTOMIXING AT 10.6 MICRONS WITH STRONTIUM BAR 09-1477
ED LIGHT (FROM LITHIUM NIOBATE CRYSTAL, AGR/ MEASUREMENTS OF PHOTON CORRELATIONS OF SECOND HARMONIC GENERAT 11-1701
. (BARIUM TITANATE).   PHOTON PHONON INTERACTION IN THE NEAR INFRARED 04-0925
D STRONTIUM TITANATE.   NONCOLLINEAR PARAMETRIC 4 PHOTON RECIPROCAL EFFECT IN CADMIUM SULFIDE AN 06-1122
/-8: ULTRASOUND PROPAGATION. PART-9: SPECIFIC HEAT. PART-10: PHOTON RENORMALIZATION AND THEORY OF SPONTANE/ 03-0397
                ANGULAR CORRELATION OF ANNIHILATION PHOTONS IN ICE SINGLE CRYSTALS.   34-3457
     LIGHT SCATTERING BY POLARITONS ASSOCIATED WITH ORDINARY PHOTONS IN LITHIUM NIOBATE.   11-1697
T VALVES IN DISPLAY AND MODULATOR APPLICATIONS. USE OF CUBIC PHOTOSENSITIVE CRYSTALS AS POCKELS EFFECT LIGH 37-3661
TRIGLYCINE/ INFLUENCE OF THE SPONTANEOUS POLARIZATION ON THE PHOTOSTIMULATED EXOELECTRON EMISSION YIELD OF 27-2961
CONDUCTIVE ANTIMONY SULFOIODIDE.   PHOTOVOLTAIC EFFECT IN FERROELECTRIC AND PHOTO 15-2174
               DEBYE TYPE ABSORPTION IN POLAR PHTHALOCYANINE.   36-3621
E TRANSITION IN THE POLAR STATE OF THE ORGANIC SEMICONDUCTOR PHTHALOCYANINE.   LOW TEMPERATURE PHAS 36-3623
                          NMR INVESTIGATIONS OF PHTHALOCYANINE.   36-3551
ECT AND FERROELECTRICITY IN THE ORGANIC SEMICONDUCTOR COPPER PHTHALOCYANINE.   STARK EFF   36-3532
ELECTRIC STATE.   ELECTRICAL PROPERTIES OF PHTHALOCYANINE AS A SEMICONDUCTOR IN THE FERRO   36-3622
         FIRST ORDER NEGATIVE PHOTODIELECTRIC EFFECT IN PHTHALOCYANINE IN THE FERROELECTRIC STATE.   36-3539
C STATE)   PYROELECTRIC EFFECT IN METAL-FREE PHTHALOCYANINE. (IN THE POLAR, OR FERROELECTRI 36-3540
/AN SCATTERING TECHNIQUES APPLIED TO PROBLEMS IN SOLID STATE PHYSICS (DOMAIN WALL THICKNESS IN GADOLINIUM / 30-3319
TION, REVIEW)   CHEMISTRY AND PHYSICS OF (ALKALI) METAL TANTALATES. (PREPARA   02-0048
YPE.   RADIATION PHYSICS OF FERROELECTRICS OF BARIUM TITANATE T   04-0972
                               CHEMICAL PHYSICS OF ICE.   01-0009
          REVIEW ON THE PROBLEMS OF THE PHYSICS OF ICE.   02-0072
68 MUNICH)   PHYSICS OF ICE. (PROC OF INT SYMP, 9-14 SEP 19 01-0026
                        MODERN ASPECTS OF PHYSICS OF SOLIDS. (FERROELECTRICITY)   01-0019
Y, MAGNETISM, FERROELECTRICITY, ORDERED ALLOYS, SUPERFLUIDI/ PHYSICS OF TRANSITIONS. (THERMODYNAMICS, THEOR 03-0183
M BARIUM TITANATE PREPARED BY RF SPUTTERING ON SILICON SUBS/ PHYSICAL AND ELECTRICAL PROPERTIES OF THIN FIL 04-0828
ICS OF A LEAD ZIRCONATE TITANATE FERROELECTRIC CERAMIC. SOME PHYSICAL AND ELECTRICAL SWITCHING CHARACTERIST 09-1601
AND LITHIUM IODATE.   OPTICALLY INDUCED PHYSICAL DAMAGE TO LITHIUM NIOBATE, PROUSTITE, 08-1328
USED PHASE TRANSITION. (POLARIZATION MECHANISMS, ELECTROOPT/ PHYSICAL PHENOMENA IN FERROELECTRICS WITH DIFF 02-0119
EAD TITANATE AND ZIRCONATE. (SOLID SOLUTIONS LEAD MAGNESATE/ PHYSICAL PHENOMENA IN PIEZOCERAMICS BASED ON L 05-1063
SPECIALLY METHYL AMMONIUM ALUMINUM SULFATE)   PHYSICAL PROPERITES OF FERROELECTRIC ALUMS. (E 20-2810
CRYSTALS. (LATTICE PA/ EFFECT OF SOME ADMIXED NICKEL ON SOME PHYSICAL PROPERTIES OF BARIUM TITANATE SINGLE 04-0909
.   EFFECT OF CRYSTALLINE GRAIN SIZE ON THE PHYSICAL PROPERTIES OF CERAMIC BARIUM TITANATE 04-0785
/ANSITIONS AND SPECIAL FEATURES OF DOMAIN STRUCTURE AND SOME PHYSICAL PROPERTIES OF FERROELECTRICS OF DIFF/ 03-0480
LECTRIC CONSTANTS, PIEZOELECTRIC / SINGLE CRYSTAL GROWTH AND PHYSICAL PROPERTIES OF LITHIUM TANTALATE. (DIE 12-1876
ONY SULFOIODIDE SINGLE CRYSTALS.   PREPARATION AND PHYSICAL PROPERTIES OF NONSTOICHIOMETRIC ANTIM 15-2199
DS, LANTHANUM(1-X) CALCIUM(X) MANGANATE (PH/ PREPARATION AND PHYSICAL PROPERTIES OF PEROVSKITE TYPE COMPOUN 08-1356
UM TITANATE WITH COBALT ADDED.   SOME CRYSTAL CHEMICAL AND PHYSICAL PROPERTIES OF SINGLE CRYSTALS OF BARI 04-0618
/GATION OF THE THERMAL CONDUCTIVITY IN CONNECTION WITH OTHER PHYSICAL PROPERTIES OF SOME FERROELECTRICS BE/ 04-0721
          INFLUENCE OF IMPURITIES ON THE GROWTH AND SOME PHYSICAL PROPERTIES OF TGS SINGLE CRYSTALS.   27-3084
UM IODATE. (NONLINEAR OPTIC, REFRACTIVE INDICES) RELATION OF PHYSICAL PROPERTIES TO THE SYMMETRY OF POTASSI 08-1296
/ING COMPOUNDS - DATA TABLES. (MECHANICAL, CRYSTALLOGRAPHIC, PHYSICAL, THERMAL MAGNETIC, ELECTRONIC AND OP/ 01-0025
/OF DIELECTRIC PROPERTIES OF BARIUM TITANATE CERAMICS ON THE PHYSICO- CHEMICAL NATURE OF THE SOLID SOLUTIO/ 04-0789

GENERATION OF FAR INFRARED RADIATION BY PICOSECOND LIGHT PULSES IN LITHIUM NIOBATE. 11-1857
ULFOIODIDE CRYSTALS. MOTION PICTURES OF THE PHASE TRANSITION IN ANTIMONY S 15-2120
SELF CONSISTENCY OF LANDAU'S MODEL IN THE TRANSITION FROM PIEZO TO FERROELECTRICITY (THEORY). 03-0507
TIONS ON THE STRUCTURE AND PROPERTIES OF LEAD BARIUM NIOBATE PIEZOCERAMICS. EFFECTS OF ROASTING CONDI 13-2046
NATE. (SOLID SOLUTIONS LEAD MAGNESATE/ PHYSICAL PHENOMENA IN PIEZOCERAMICS BASED ON LEAD TITANATE AND ZIRCO 05-1063
WITH COMPLEX ADDITIVES CONTAINING GERMANIUM. PIEZOCERAMICS BASED ON LEAD ZIRCONATE TITANATE 09-1510
SCREENING THE SPONTANEOUS INDUCTION. DEPENDENCE OF THE PIEZOEFFECT IN BARIUM TITANATE ON THE MEANS OF 04-1021
ITCHING IN POLYCRYSTALLINE FERROELECTRICS AT / ZERO SHIFT IN PIEZOELECTRIC ACCELEROMETERS. (POLARIZATION SW 28-3216
LE CRYSTALS IN THE PARAELECTRIC PHASE. PIEZOELECTRIC ACTIVITY OF BARIUM TITANATE SING 04-0771
TANATE(1-X) CERAMICS. POLARIZATION, ELECTROSTRICTION, AND PIEZOELECTRIC ACTIVITY OF LEAD ZIRCONATE(X) TI 09-1501
MUTH(12) GERMANIUM OXIDE(20). ELASTIC, PIEZOELECTRIC, AND DIELECTRIC CONSTANTS OF BIS 35-3512
HIUM TANTALATE AND L/ TEMPERATURE DEPENDENCE OF THE ELASTIC, PIEZOELECTRIC, AND DIELECTRIC CONSTANTS OF LIT 11-1823
D TITANATE ZIRCONATE CERAMICS AT LOW TEMPERATURES. PIEZOELECTRIC AND DIELECTRIC PROPERTIES OF LEA 09-1519
TITANATE. EFFECT OF GRAIN SIZE ON THE PIEZOELECTRIC AND ELASTIC PROPERTIES OF BARIUM 04-1011
UM NIOBATE SINGLE CRYSTAL/ THERMAL DEPENDENCE OF DIELECTRIC, PIEZOELECTRIC, AND ELASTIC PROPERTIES OF LITHI 11-1693
M TANTALATE CHARACTERISTICS. PIEZOELECTRIC AND ELASTIC PROPERTIES OF LITHIU 12-1893
VIEW, 991 REFS) FERROELECTRIC, PIEZOELECTRIC, AND ELECTROOPTIC MATERIALS. (RE 02-0058
ELECTRIC POTASSIUM IODATE. PIEZOELECTRIC AND ELECTROOPTIC MODULI OF FERRO 08-1431
. ACOUSTIC ATTENUATION AND AMPLIFICATION IN PIEZOELECTRIC AND FERROELECTRIC SEMICONDUCTORS 03-0275
THIUM TANTALATE SINGLE CRYSTAL NEAR ITS CURIE POINT. PART-1: PIEZOELECTRIC AND OPTICAL PROPERTIES. /R OF LI 12-1877
ITHIUM IODATE. PIEZOELECTRIC AND PHOTOELASTIC PROPERTIES OF L 08-1465
ITHIUM IODATE. PIEZOELECTRIC AND PHOTOELASTIC PROPERTIES OF L 08-1464
HE LEAD DINIOBATE- BARIUM DINIOBATE SYSTEM. DIELECTRIC, PIEZOELECTRIC AND PYROELECTRIC PROPERTIES OF T 08-1386
HE LEAD METANIOBATE- BARIUM NIOBATE SYSTEM. DIELECTRIC, PIEZOELECTRIC AND PYROELECTRIC PROPERTIES OF T 08-1385
ITANATE CERAMICS. (DIELECTRIC CONST/ EFFECTS OF ADDITIVES ON PIEZOELECTRIC AND RELATED PROPERTIES OF LEAD T 05-1084
IELDS ON MODIFIED LEAD ZIRCONATE- LEAD TITANATE CERAMICS FOR PIEZOELECTRIC APPLICATIONS. /F HIGH ELECTRIC F 09-1485
CRYSTAL STRUCTURE OF PIEZOELECTRIC BISMUTH(12) GERMANIUM OXIDE(20). 35-3493
/ICATION OF ULTRASOUND IN A TWO-LAYER SYSTEM CONSISTING OF A PIEZOELECTRIC CERAMIC (BARIUM TITANATE) WITH / 04-1037
REFS) PIEZOELECTRIC CERAMIC SUBSTANCES. (REVIEW, 23 02-0105
ISSIPATION ASSOCIATED WITH THE ELECTROMECHANICAL COUPLING IN PIEZOELECTRIC CERAMICS. MEASUREMENT OF THE D 03-0514
/ ELECTROPHYSICAL AND STRUCTURAL PARAMETERS OF FERROELECTRIC PIEZOELECTRIC CERAMICS AND METHODS OF THEIR C/ 03-0217
ROPERTIES. ADVANCES IN PRODUCTION OF (COMMERCIAL) PIEZOELECTRIC CERAMICS AND (TABLES OF) 02-0110
ERRATE(0.5) NIOBATE(0.5), LEAD TITANATE- LEAD ZIRCONATE/ NEW PIEZOELECTRIC CERAMICS. (LEAD TITANATE- LEAD F 09-1544
ATE) LEAD TITANATE PIEZOELECTRIC CERAMICS. (MATSUSHITA LEAD TITAN 09-1616
OL-3, REVIEW) PIEZOELECTRIC CERAMICS. (NONMETALLIC SOLIDS, V 01-0015
ATE MODIFIED WITH SODIUM OR POTASSIUM ANTIMONATE, NIOBATE, / PIEZOELECTRIC CERAMICS OF LEAD ZIRCONATE TITAN 09-1539
LDS. (SUE/ BEHAVIOR OF MODIFIED LEAD ZIRCONATE LEAD TITANATE PIEZOELECTRIC CERAMICS UNDER HIGH ELECTRIC FIE 09-1484
E CERAMIC EXHIBITED WHEN IT IS GIVEN AN IMPACT. PIEZOELECTRIC CERAMIC CHARACTERISTIC OF BARIUM TITANAT 04-0881
ATE. PIEZOELECTRIC COEFFICIENT OF GADOLINIUM MOLYBD 30-3317
STATE OF IRON- IODINE- BORACITE. (TEMPERATURE DEPENDENCE OF PIEZOELECTRIC CONSTANT / IN THE FERROELECTRIC 14-2083
ON OF THEIR ELECTRICAL PROPERTIES. (DIELECTRIC, ELASTIC, AND PIEZOELECTRIC CONSTANTS) /SURE AND INVESTIGATI 15-2138
ALS. (ROCHELLE SALT) MEASUREMENTS OF COMPLEX PIEZOELECTRIC CONSTANTS IN FERROELECTRIC CRYST 28-3213
GLE CRYSTALS. DETERMINATION OF ELASTIC AND PIEZOELECTRIC CONSTANTS OF LITHIUM NIOBATE SIN 11-1749
TIMONY (/ DOMAIN STRUCTURE AND TEMPERATURE DEPENDENCE OF THE PIEZOELECTRIC CONSTANTS OF STIBOTANTALITE. (AN 36-3601
IUM LITHIUM NIOBATE TYPE FERROELECTRIC CRYSTAL. ELASTIC AND PIEZOELECTRIC CONSTANTS OF STRONTIUM(4) POTASS 13-1946
/CAL PROPERTIES OF LITHIUM TANTALATE. (DIELECTRIC CONSTANTS, PIEZOELECTRIC CONSTANTS, SPONTANEOUS POLARIZA/ 12-1876
ABSOLUTE SIGNS OF SECOND HARMONIC GENERATION COEFFICIENTS OF PIEZOELECTRIC CRYSTALS. 02-0096
ELASTIC AND ELECTROMAGNETIC WAVES IN PIEZOELECTRIC CRYSTALS. 03-0371
ULSES IN LITHIUM NIOBATE AND BISMUTH(12) GERMANIUM OXIDE(20) PIEZOELECTRIC CRYSTALS. /RSE HYPERSONIC WAVE P 11-1815
LIGHT SCATTERING BY ACOUSTIC PHONONS IN FERROELECTRIC AND PIEZOELECTRIC CRYSTALS. 03-0477
OPTICAL SECOND HARMONIC GENERATION IN PIEZOELECTRIC CRYSTALS. (KDP TYPE) 17-2582
/BSOLUTE SIGNS OF SECOND HARMONIC GENERATION COEFFICIENTS OF PIEZOELECTRIC CRYSTALS. (LEAD TITANATE, BARIU 05-1066
OTASSIUM DIHYDROGEN ARSENATE AND CES/ DIELECTRIC RESPONSE IN PIEZOELECTRIC CRYSTALS. (RAMAN SCATTERING IN P 18-2723
Y. RAMAN SCATTERING IN PIEZOELECTRIC CRYSTALS WITH TETRAGONAL SYMMETR 03-0407
TS) ROOM/ ELASTIC ANOMALY OF BARIUM SODIUM NIOBATE (ELASTIC, PIEZOELECTRIC, DIELECTRIC, AND OPTICAL CONSTAN 13-2060
ANOMALOUS ECHO IN FERROELECTRIC ANTIMONY SULFOIODIDE. (PIEZOELECTRIC DOMAIN OSCILLATIONS) 15-2211
GAMMA-IRRADIATED ROCHELLE SALT CRYSTALS. PIEZOELECTRIC EFFECT AND INTERNAL FRICTION IN 28-3229
TITANATE CERAMIC. PIEZOELECTRIC EFFECT IN ALUMINUM DOPED BARIUM 04-0536
RYSTALS. PIEZOELECTRIC EFFECT IN LEAD TITANATE SINGLE C 05-1049
MS. DIRECT PIEZOELECTRIC EFFECT IN POLYVINYL CHLORIDE FIL 36-3543
CESS OF FERROELECTRICS WITH THE PEROVSKITE STRU/ THE INVERSE PIEZOELECTRIC EFFECT IN THE REPOLARIZATION PRO 08-1374
ANSVERSE ACOUSTIC WAVE FROM DOMAIN BOUNDARIES IN FERROELECT/ PIEZOELECTRIC EFFECT ON THE REFLECTION OF A TR 03-0312
S (AND THE THEORY OF PIEZOELECTRIC ENERGY CONVERSION). PIEZOELECTRIC EFFECTS IN FERROELECTRIC CERAMIC 03-0278
OBATE (ELECTROMAGNETIC COUPLING). PIEZOELECTRIC LEAKY SURFACE WAVE IN LITHIUM NI 11-1832
IOBATE. PROPAGATION AND AMPLIFICATION OF RAYLEIGH WAVES AND PIEZOELECTRIC LEAKY SURFACE WAVES IN LITHIUM N 11-1856
CRYSTAL STRUCTURE OF PIEZOELECTRIC LITHIUM FORMATE MONOHYDRATE. 26-2940
FIELD) DIFFRACTION SPREADING OF SURFACE (ACOUSTIC) WAVES ON (PIEZOELECTRIC) LITHIUM NIOBATE. (FAR 11-1850
/ITATION OF LONGITUDINAL AND TRANSVERSE HYPERSONIC PULSES IN PIEZOELECTRIC LITHIUM NIOBATE AND BISMUTH BIG/ 11-1814
E SULFATE. DYNAMIC DETERMINATION OF (ELASTIC AND PIEZOELECTRIC) MATERIAL CONSTANTS OF TRIGLYCIN 27-2984
NEW PIEZOELECTRIC MATERIALS. 02-0076
/ION OF SPONTANEOUS POLARIZATION BY ELASTIC INSTABILITIES IN PIEZOELECTRIC MATERIALS. (GADOLINIUM MOLYBDAT/ 30-3282
IC FIELDS. BARIUM TITANATE PIEZOELECTRIC MICROPROBE FOR STUDYING ULTRASON 04-0938
F THE INTERNAL FRICTION IN BARIUM TITANATE SINGLE CRYSTALS. (PIEZOELECTRIC MODULUS) AMPLITUDE DEPENDENCE O 04-0791
E AUTOSTABILIZATION REGIME OF A FERROEL/ APPLICATIONS OF THE PIEZOELECTRIC NONLINEARITIES IN THE TEMPERATUR 17-2502
ORTHORHOMBIC DIGLYCINE SULFATE. (NOT PIEZOELECTRIC, NOT FERROELECTRIC) 36-3624
ENERGY GENERATION FROM IMPACT ON PIEZOELECTRIC (OR FERROELECTRIC) MATERIALS. 03-0222
CTRIC LITHIUM TANTALATE. STUDY OF PIEZOELECTRIC OSCILLATIONS IN WIDEBAND PYROELE 12-1875
CERAMIC (THEORY, APPLIED TO LEAD(0.99) NIOBI/ MEASUREMENT OF PIEZOELECTRIC PHASE ANGLES IN A FERROELECTRIC 09-1537
FAR INFRARED SPECTRA OF PIEZOELECTRIC POLYVINYLIDENE FLUORIDE. 36-3590
PZT-5 UNDER PRESSURE, DIELECTRIC AND PIEZOELECTRIC PROPERTIES. 09-1496
EAD ZIRCONATE- LEAD (ZINCATE(0.33) NIOBAT/ FERROELECTRIC AND PIEZOELECTRIC PROPERTIES IN A LEAD TITANATE- L 09-1617
LEAD (ZINCATE NIOBATE)- BARIUM (ZINCATE NIO/ FERROELECTRIC AND PIEZOELECTRIC PROPERTIES IN THE TERNARY SYSTEM 05-1070
/FERROELECTRIC PHASE TRANSITION CONDITIONS ON DIELECTRIC AND PIEZOELECTRIC PROPERTIES OF A CERAMIC (LEAD S/ 09-1591
DE IN A CONSTANT ELECTRIC FIELD. DIELECTRIC AND PIEZOELECTRIC PROPERTIES OF ANTIMONY SULFOIODI 15-2249
NGLE CRYSTALS WITH LAMINAR DOMAIN S/ DIELECTRIC, ELASTIC AND PIEZOELECTRIC PROPERTIES OF BARIUM TITANATE SI 04-1010
E IN STRONG ELECTRIC FIELDS. ELASTIC AND PIEZOELECTRIC PROPERTIES OF CADMIUM PYRONIOBAT 13-1967
CITE. PIEZOELECTRIC PROPERTIES OF COBALT IODINE BORA 14-2100
(5) GERMANIUM(3) OXIDE(11) CRYSTALS. ELASTIC AND PIEZOELECTRIC PROPERTIES OF FERROELECTRIC LEAD 35-3520
SSIUM LITHIUM TANTALATE(0-0.5) NIOBATE(1-0.5) SINGLE CRYSTA/ PIEZOELECTRIC PROPERTIES OF FERROELECTRIC POTA 12-1890
SINGLE CRYSTAL. (FOUND NOT FERROELECTRIC) ELECTROOPTIC AND PIEZOELECTRIC PROPERTIES OF LANTHANUM TITANATE 36-3576

3) NIOBATE(0.67)- TITANATE- ZIRCONATE SOLID / DIELECTRIC AND PIEZOELECTRIC PROPERTIES OF LEAD COBALTATE(0.3 09-1549
MICS. (CONTAINING 1 MOLE PERCENT MANGANESE/ POLARIZATION AND PIEZOELECTRIC PROPERTIES OF LEAD TITANATE CERA 05-1061
MICS. PIEZOELECTRIC PROPERTIES OF LEAD TITANATE CERA 05-1062
A ZIRCOKIUM TO TITANIU/ LANTHANUM DEPENDENCE OF ELASTIC AND PIEZOELECTRIC PROPERTIES OF PLZT CERAMICS WITH 09-1573
ITANATE CERAMICS SUBJECTED TC UNIAXIAL TRANSVERSE COMPRESSI/ PIEZOELECTRIC PROPERTIES OF POLARIZED BARIUM T 04-0962
E NIOBATE SINGLE CRYSTAL. PIEZOELECTRIC PROPERTIES OF POTASSIUM TANTALAT 08-1271
ONATE- LEAD TITANATE CERAMIC MEMORY PLATE) A FERROELECTRIC PIEZOELECTRIC RANDOM ACCESS MEMORY. (LEAD ZIRC 09-1504
ENOIOIIDE CRYSTALS. PIEZOELECTRIC RESISTANCE EFFECT IN BISMUTH SEL 15-2128
CY RANGE WITH LOW RAT/ LITHIUM TANTALATE AND LITHIUM NIOBATE PIEZOELECTRIC RESONATORS IN THE MEDIUM FREQUEN 11-1724
RY ARRAY. PIEZOELECTRIC RESPONSE OF A FERROELECTRIC MEMO 37-3635
/RY FOR ELECTROMAGNETIC COUPLING TC ELASTIC SURFACE WAVES ON PIEZOELECTRIC SUBSTRATES (LITHIUM NIOBATE AND 11-1751
RHOMBIC CRYSTALS. (BISMUTH(12) GERMANIUM OXIDE(20) AND BARI/ PIEZOELECTRIC SURFACE WAVES IN CUBIC AND ORTHO 13-2041
OF A FERROELECTRIC (CLASS 6MM). PIEZOELECTRIC SURFACE WAVES ON A C-DOMAIN WALL 03-0421
. PIEZOELECTRIC SURFACE WAVE ON LITHIUM NIOBATE 11-1824
SHELLS. DYNAMIC PIEZOELECTRIC THEORY FOR FERROELECTRIC CERAMIC 03-0225
ISKERS IN LEAD ZIRCONATE TITANATE) NEW COMPOSITE CERAMIC PIEZOELECTRIC TRANSDUCER MATERIAL. (ALUMINA WH 09-1556
TANTALATE CRYSTALS. SOME PROPERTIES CF THE EQUATIONS OF PIEZOELECTRIC VIBRATION OF LITHIUM NIOBATE AND 11-1740
DE CRYSTAIS NEAR THE CURIE POINT (EFFECTS OF TEMPERATURE, F/ PIEZOELECTRIC VIBRATIONS IN ANTIMONY SULFOIODI 15-2121
D LITHIUM TANT/ SIMPLIFIED CHARACTERISTIC EQUATION FOR PLANE PIEZOELECTRIC VIBRATIONS OF LITHIUM NIOBATE AN 11-1739
TYPE PHASE TRANSITION IN CRYSTALS. (THEORY, FERROELECTRICS, PIEZOELECTRICS) LOW TEMPERATURE DISPLACEMENT 03-0320
/ ACCUSTOOPTICAL MEASUREMENTS OF ACOUSTIC WAVE GENERATION IN PIEZOELECTRICS COUPLED TO A GUNN OSCILLATOR. (11-1731
ADAPTIVE CONTROL VOLTAGE MO/ ANALOG MEMORY DEVICES EMPLOYING PIEZOELECTRIC- FERROELECTRIC INTERACTIONS FOR 37-3663
LE) STIMULATED COMBINATION SCATTERING OF SOUND IN PIEZOELECTRICS. (THEORY, LITHIUM NIOBATE EXAMP 03-0431
IC ANTIMCNY SULFCIODIDE. ELECTROSTRICTION, PIEZOELECTRICITY AND ELASTICITY IN FERROELECTR 15-2165
ONY SULFOIODIDE SINGLE CRYSTALS. PIEZOELECTRICITY AND ELECTROSTRICTION OF ANTIM 15-2227
YCINE SULFATE. PIEZOELECTRICITY AND ELECTROSTRICTION OF TRIGL 27-3118
L FLUORIDE AND POLYVINYLIDENE FLUCRIDE. PYROELECTRICITY AND PIEZOELECTRICITY IN ORIENTED FILMS OF POLYVINY 36-3544
LUORIDE FILMS. PIEZOELECTRICITY IN POLARIZED POLYVINYLIDENE F 36-3557
PIEZOELECTRICITY OF POLYVINYLIDENE FLUORIDE. 36-3573
IOBATE)- LEAD TITANATE. CRYSTAL STRUCTURE AND PIEZOELECTRICITY OF THE SYSTEM LEAD (ZINCATE N 13-2067
CTROSTRICTION CONSTANT OF POLYVINYLIDENE FLUORIDE. PIEZOELECTRICITY, PYROELECTRICITY, AND THE ELE 36-3598
N PHOSPHATE AND AMMONIUM DIHYDROGEN FHOS/ ELECTROOPTICAL AND PIEZOOPTICAL PROPERTIES OF POTASSIUM DIHYDROGE 17-2697
SULFOICDIDE (DARK CONDUCTIVITY NEAR TR/ INVESTIGATION OF THE PIEZORESISTANCE OF (POLYCRYSTALLINE) ANTIMONY 15-2250
EARIUM STRONTIUM TITANATE, DOPANT NCT NAMED) PIEZORESISTIVITY OF DOPED BARIUM TITANATE. (OR 04-0729
ANSFORMATIONS IN (CERAMIC) BARIUM AND LEAD TITANATES. (/ (IN PILE, FAST NEUTRON) RADIATION INDUCED PHASF TR 04-0971
PERTIES CF LEAD ZIRCONATE. EFFECT OF PILE NEUTRON IRRADIATION ON THE DIELECTRIC PRO 07-1259
S. DIELECTRIC MEASUREMENTS ON PILE-IRRADIATED BARIUM TITANATE SINGLE CRYSTAL 04-0910
RIGLYCINE SULFATE. APPLICABILITY OF THE PIPPARD RELATIONS TO THE PHASE TRANSITION IN T 27-3159
N TO FERROELECTRIC (LAMBDA) PHASE TRA/ GENERALIZATION OF THE PIPPARD-GARLAND RELATIONS AND THEIR APPLICATIO 03-0299
ITANATE SINGLE CRYSTALS. ETCH PITS CORRESPONDING TO DISLOCATIONS IN BARIUM T 04-1023
ETCH PITS IN BARIUM TITANATE SINGLE CRYSTALS. 04-0922
S) ANALYTICAL PROPERTIES OF A CLASS OF PLANAR FERROELECTRIC MODELS. (THEORY, KDP TYPE 03-0258
PLANAR GROWTH OF ICE FROM THE PURE MELT. 34-3444
STATIC SCALING HYPOTHESIS ANC ITS IMPLICATIONS FOR THE PLANAR SPIN MODEL OF STRONTIUM TITANATE. 06-1223
DIRECT SKY BLUE DYE. GROWTH RATE OF THE (010) PLANE OF ROCHELLE SALT IN THE PRESENCE OF PURE 28-3217
ATE AND LITHIUM TANT/ SIMPLIFIED CHARACTERISTIC EQUATION FOR PLANE PIEZOELECTRIC VIBRATIONS OF LITHIUM NIOB 11-1739
E CRYSTALS ASSOCIATED WITH STEP LADDERS ON PSEUDOCUBIC (001) PLANES. /STRUCTURES IN POTASSIUM NIOBATE SINGL 08-1316
SIUM NIOBATE SINGLE CRYSTALS FROM PSEUDOCUBIC (001) CLEAVAGE PLANES. DOMAINS IN POTAS 08-1317
TUNGSTEN TRIOXIDE, SODIUM NIOBA/ ORIENTATION OF COMPOSITION PLANES IN ANTIFERROELECTRICS. (APPLIED TO ADP, 08-1327
/TION OF LINEARLY POLARIZED LIGHT DUE TO HEATING OF PARALLEL PLANES OF POTASSIUM DIHYDROGEN PHOSPHATE CRYS/ 17-2540
LAYERED FERROELECTRIC CERAMICS CBTAINED BY PROJECTION WITH A PLASMA TORCH OR FRITTING. (BARIUM TITANATE) /I 04-0778
INFLUENCE OF THE SURFACE LAYER IN PLASTIC DEFORMATION OF ICE SINGLE CRYSTALS. 34-3459
MS OF HYDROGEN REORIENTATION. IMPURITY EFFECTS ON THE PLASTICITY OF ICE AND THEIR EXPLANATION IN TER 34-3449
ACCESS MEMORY. (LEAD ZIRCONATE- LEAD TITANATE CERAMIC MEMORY PLATE) A FERROELECTRIC PIEZOELECTRIC RANDOM 09-1504
ELASTIC SURFACE WAVE ON THE PZT CERAMIC PLATE. 09-1612
AN ELECTROPTIC SWITCH USING ROTATED Y PLATE OF LITHIUM NIOBATE. 11-1732
RATURE DEPENDENCE OF RESONANT FREQUENCIES OF LITHIUM NIOBATE PLATE RESONATORS. (COMMENTS) TEMPE 11-1812
ELECTROOPTIC EFFECT IN OBLIQUE CUT STRONTIUM BARIUM NIOBATE PLATES. LONGITUDINAL 08-1472
ELECTROOPTIC EFFECT IN OBLIQUE CUT STRONTIUM BARIUM NIOBATE PLATES. LONGITUDINAL 13-2063
UM TITANATE TYPE FERROELECTRIC CRYSTALS IN THE FORM OF THICK PLATES. MACROSCOPIC THEORY OF BARI 04-0770
ASTIC WAVES ON BARIUM TITANATE(0.95) ZIRCCNATE(0.05) CERAMIC PLATES. /AGATION CHARACTERISTICS OF SURFACE EL 09-1613
NNING ELECTRON MICROSCOPE. (TUNGSTEN TRIOXIDE SINGLE CRYSTAL PLATES) /NS OF FERROELECTRIC CRYSTALS WITH SCA 10-1646
NES WITH INTERDIGITAL TRANSDUCERS ON UNPOLARIZED PZT CERAMIC PLATES. SURFACE WAVE DELAY LI 09-1614
PRO/ MONODOMAIN STRONTIUM TITANATE (OCCURANCE IN THIN (110) PLATES BELOW CUBIC- TETRAGONAL TRANSITION, AND 06-1164
S, NONHOMOGENEOUS, SPONTANEOUS POLARIZATICN OF FERROELECTRIC PLATES NEAR THE CURIE POINT. /ORY OF DOMAINLES 03-0321
E CHARGE LIMITED CURRENTS IN HEXAGONAL BARIUM TITANATE(0.75) PLATINATE(0.25) CRYSTALS. SPAC 04-0936
TANATE- METAL. (METHOD OF STUDYING THE TYPE OF CONDUCTIVITY, PLATINUM ELECTRODES) /HE SYSTEM METAL- LEAD TI 05-1057
SINGLE CRYSTAL. PARAMAGNETIC RESONANCE OF PLATINUM IONS IN NIOBIUM DOPED BARIUM TITANATE 04-0977
EPITAXY OF PLATINUM ON BARIUM STRONTIUM NIOBATE. 13-1907
N OF THIN BARIUM TITANATE FILMS BY DC DIODE SPUTTERING (ONTO PLATINUM OR FUSED QUARTZ). PREPARATIO 04-0948
OXIDES AEO(3)) (BARIUM TITANATE ANC LITHIUM NIOBATE FILMS ON PLATINUM OR SILICON) / TYPE STRUCTURE. (MIXED 04-0914
RAY INVESTIGATION OF THE PEROVSKITES BARIUM OR STRONTIUM METAPLUMBATES (NOT FERROELECTRICS). /THESIS AND X- 36-3607
LONGITUDINAL DISPLAY MODE FOR CERAMIC ELECTROOPTIC DEVICES. (PLZT) A NEW 09-1523
A SEMICONDUCTOR- FERROELECTRIC MEMORY DEVICE. (PLZT) 09-1589
PTIC CERAMICS AS WAVELENGTH SELECTION DEVICES IN DYE LASERS (PLZT). ELECTROO 09-1618
FERROELECTRIC- PHOTOCONDUCTOR MEMORY DEVICE. (PLZT) 09-1568
RETENTION IN THE THIN FERROELECTRIC FILMS. (PLZT) 09-1593
OELECTRIC PHOTOCONDUCTOR IMAGE STORAGE AND DISPLAY DEVICES. (PLZT) STRAIN BIASED FERR 09-1565
STORAGE AND DISPLAY DEVICES OPERATED IN A REFLECTION MODE. (PLZT) /SED FERROELECTRIC- PHOTOCONDUCTOR IMAGE 09-1564
THIN FILM FERROELECTRIC PHOTOCONDUCTOR MEMORY DEVICE. (PLZT) 09-1503
PUTERS AND OTHER DEVICES. (LEAD ZIRCONATE- LEAD TITANATE AND PLZT) USE OF FERROELECTRIC CERAMICS IN COM 09-1532
OLD PRESSING AND LOW TEMPERATURE SINTERING OF ALKOXY-DERIVED PLZT AND PLZT. C 09-1493
ALIGNMENT IN LEAD ZIRCONATE- LEAD TITANATE- LANTHANUM OXIDE (PLZT) CERAMIC. /EPENDENT FERROELECTRIC DOMAIN 09-1586
ACOUSTIC DETECTION OF FERROELECTRIC PHASE TRANSITIONS IN PLZT CERAMICS. 09-1545
DEVICE APPLICATIONS OF PLZT CERAMICS. 09-1569
ELECTROELASTIC CONSTANT MEASUREMENTS FOR PLZT CERAMICS. 09-1598
ENHANCED ORDERING IN PLZT CERAMICS. 09-1570
ENHANCED ORDERING OF FERROELECTRIC DOMAINS IN PLZT CERAMICS. 09-1579
ENHANCED CRDERING CF FERROELECTRIC DOMAINS IN PLZT CERAMICS. 09-1580
FERROELECTRIC DOMAIN SWITCHING IN RHOMBOHEDRAL PHASE PLZT CERAMICS. 09-1563

OTHERMAL GRAIN GROWTH AND ELECTRICAL BEHAVIOR OF FULLY DENSE PLZT CERAMICS.   IS 09-1553
                        PHASE BOUNDARY AND DEFECT STRUCTURE STUDIES IN PLZT CERAMICS.   09-1590
    PYROELECTRIC PROPERTIES OF THE LANTHANUM DOPED FERROELECTRIC PLZT CERAMICS.   09-1558
T IMPROVEMENTS IN THE OPTICAL AND ELECTROCPTIC PROPERTIES OF PLZT CERAMICS.   RECEN   09-1529
OELECTRIC PHCTOCONDUCTOR IMAGE STORAGE AND CISPLAY DEVICES. (PLZT CERAMICS)   SCATTERING MODE FERR 09-1599
NCE OF THE ANTIFERROELECTRIC FERRCELECTRIC PHASE BOUNDARY IN PLZT CERAMICS.   TEMPERATURE AND FIELD DEPENDE 09-1491
                        FABRICATION OF ELECTROOPTIC PLZT CERAMICS BY ATMOSPHERE SINTERING.   09-1600
/HANUM DEFENDENCE OF ELASTIC AND PIEZOELECTRIC PROPERTIES OF PLZT CERAMICS WITH A ZIRCONIUM TO TITANIUM RA/ 09-1573
                              . DISPLAY APPLICATIONS OF PLZT FERROELECTRIC CERAMICS.   09-1561
              ELECTROOPTIC DEVICES USING STRAIN BIASED PLZT FERROELECTRIC CERAMICS.   09-1562
                    SOME FATIGUING EFFECTS IN 8/65/35 PLZT FINE GRAINED FERROELECTRIC CERAMIC.   08-1432
        LINEAR ELECTROOPTIC EFFECT IN FERROELECTRIC CERAMICS: PLZT 12-40-60. (USED FOR SENSING LOW VOLTAGES) 09-1610
              LOSSLESS POTASSIUM DIDEUTERIUM PHOSPHATE POCKELS CELL FOR HIGH POWER Q SWITCHING.   17-2529
FECTS AND Q SWITCHING PERFORMANCE IN LITHIUM NIOBATE AND KDP POCKELS CELLS.   TRANSIENT ELASTOOPTIC EF   11-1727
MANCE IN LITHIUM NICBATE AND POTASSIUM DIDEUTERIUM PHOSPHATE POCKELS CELLS. (COMMENTS) / Q SWITCHING PERFOR 11-1710
GEN PHOSPHA/ (FINITE APERTURE LIGHT SWITCH USING) TRANSVERSE POCKELS EFFECT COMMUTATORS. (POTASSIUM DIHYDRO 17-2509
E ABOVE ROOM TEMPERATURE.   POCKELS EFFECT IN AMMONIUM DIHYDROGEN PHOSPHAT 17-2518
                MELT COMPOSITION (STOICHIOMETRY) DEPENDENCE OF POCKELS EFFECT IN LITHIUM NIOBATE CRYSTALS.   11-1728
ULATOR APPLICATIONS. USE OF CUBIC PHOTOSENSITIVE CRYSTALS AS POCKELS EFFECT LIGHT VALVES IN DISPLAY AND MOD 37-3661
IMONY SULFOIODIDE FERROELECTRIC SEMICONDUCTOR NEAR ITS CURIE POINT.   ABSORPTION OF ULTRASOUND IN AN ANT 15-2115
CONDUCTIVITY OF A FERROELECTRIC SEMICONDUCTOR NEAR THE CUBIC POINT.   ANOMALIES IN THE ELECTRICAL 03-0268
ND WAVES IN DEUTERATED ROCHELLE SALT AT THE PHASE TRANSITION POINT.   ATTENUATION OF SOU 28-3204
ND WAVES IN DEUTERATED ROCHELLE SALT AT THE PHASE TRANSITION POINT.   ATTENUATION OF SOU 28-3205
TALS AT TEMPERATURES ABOVE AND BELOW THE FERROELECTRIC CURIE POINT. /F PROTONS IN THIN BARIUM TITANATE CRYS 04-0687
ALS (APPLICATION TO SINGLE DOMAIN FERROELECTRICS BELOW CURIE POINT).   COMPLEX PHASE TRANSITIONS IN CRYST 04-0319
LLIMETER WAVELENGTHS IN BARIUM TITANATE ABOVE THE TRANSITION POINT.   DISPERSION AT MI 04-0601
TRIGLYCINE SULFATE CRYSTALS GROWN BELOW AND ABOVE THE CURIE POINT.   DOMAIN STRUCTURE OF PURE AND DOPED 27-3130
OF TRIGLYCINE SELENATE CRYSTAL (DURING COOLING THROUGH CURIE POINT).   DOMAIN STRUCTURE 27-3043
ARGE INJECTION IN SOLIDS. (USING BARIUM TITANATE ABOVE CURIE POINT)   DOUBLE SPACE CH 04-0638
TRIGLYCINE SELENATE. (AGING AND PHASE CHANGES NEAR THE CURIE POINT) /FIELD ON THE DIELECTRIC PROPERTIES OF 27-3131
EFRACTIVE INDEX OF PARAELECTRIC KDP IN THE VICINITY OF CURIE POINT.   EFFECT OF IONIC MOTION ON THE R 17-2548
ECTRIC, AND CPTICAL CONSTANTS) ROOM TEMPERATURE TO THE CURIE POINT. / NIOBATE (ELASTIC, PIEZOELECTRIC, DIEL 13-2060
      ELASTIC CONSTANTS OF GADOLINIUM MOLYBDATE (NEAR THE CURIE POINT).   30-3274
STRUCTURE OF TRIGLYCINE SULFATE IN THE VICINITY OF THE CURIE POINT.   EQUILIBRIUM DOMAIN 27-3112
LEAD ZIRCONATE. (PHASE TRANSITION 200 DEG C BELOW THE CURIE POINT) /ORCED FERROELECTRICITY IN GLASS BONDED 07-1252
HE AESORPTION OF SOUND IN ROCHELLE SALT NEAR ITS UPPER CURIE POINT.   FREQUENCY DEPENDENCE OF T 28-3183
OF ULTRASOUND IN FERROELECTRICS NEAR SECOND ORDER TRANSITION POINT. /LE FORCES ON ATTENUATION AND VELOCITY 03-0253
RUPOLE RESONANCE IN FERROELECTRICS NEAR THE PHASE TRANSITION POINT.   KINETIC EQUATION FOR QUAD 03-0309
RELAXATION PROCESSES NEAR THE SECOND ORDER PHASE TRANSITION POINT.   ON 03-0385
ASE BOUNDARIES IN BARIUM TITANATE SINGLE CRYSTALS NEAR CURIE POINT.   PH 04-0577
E LOSSES IN FERROELECTRICS. (AT TEMPERATURES ABOVE THE CURIE POINT) PHENOMENOLOGICAL THEORY OF MICROWAV 03-0503
OLARIZATION FLUCTUATION OF TRIGLYCINE SULFATE NEAR THE CURIE POINT.   P 27-3077
CTRIC CRYSTAL WITH A SECOND ORDER TRANSITION ABOVE THE CURIE POINT. / GENERATION OF HARMONICS IN A FERROELE 03-0331
AND THERMAL PROPERTIES OF TRIGLYCINE SELENATE NEAR THE CURIE POINT.   RELATION BETWEEN ELECTRICAL 27-3168
HEAT OF SODIUM NITRITE NEAR THE ANTIFERROELECTRIC TRANSITION POINT.   SPECIFIC 16-2303
OF CALCIUM DOPED SODIUM NITRATE CRYSTALS NEAR THE TRANSITION POINT. / DIFFUSIVITY AND THERMAL CONDUCTIVITY 16-2283
NTANEOUS POLARIZATION OF FERROELECTRIC PLATES NEAR THE CURIE POINT. /ORY OF DOMAINLESS, NONHOMOGENEOUS, SPO 03-0321
          THERMAL CONDUCTIVITY OF TRIGLYCINE SULFATE NEAR THE CURIE POINT.   27-2983
          THERMAL CONDUCTIVITY OF TRIGLYCINE SULFATE NEAR THE CURIE POINT.   27-3016
                RUBIDIUM NITRATE NEAR THE MELTING POINT AND ITS POLYMORPHISM.   16-2330
EROVSKITE MODIFICATION OF CADMIUM TITANATE./ ABOUT THE CURIE POINT AND NATURE OF DIELECTRIC PROPERTIES IN P 08-1424
FERROELECTRIC MATERIALS (ROCHELLE SALT NEAR THE UPPER CURIE POINT AS EXAMPLE). /ENCE OF THE LOSS FACTOR IN 28-3220
HAVIOR OF DISPLACIVE TYPE FERROELECTRICS NEAR THE TRANSITION POINT. (BARIUM TITANATE, STRONTIUM TITANATE) / 04-0992
D ELECTROOPTICAL PROPERTIES IN THE NEIGHBORHOOD OF THE CURIE POINT (BIREFRINGENCE ANOMALIES). / OPTICAL, AN 24-2745
IC TGS AND DEUTERATED TGS (HYSTERESIS LOOPS JUST BELOW CURIE POINT CORRECTED FOR ELECTROCALORIC EFFECT). /R 27-2966
TITANATE.   POINT DEFECTS AND SINTERING OF LEAD ZIRCONATE 09-1478
YSTALS.   HEAT TREATMENT INDUCED POINT DEFECTS IN LITHIUM METANIOBATE SINGLE CR 11-1668
LS USED IN LASER MODULATION.   POINT DEFECTS IN LITHIUM NIOBATE SINGLE CRYSTA 11-1737
S. (REVIEW, 77 REFS)   POINT DEFECTS IN PEROVSKITE TYPE FERROELECTRIC 02-0041
TRICAL AND MECHANICAL PROPERTIES. (REVIEW, 56/ STRUCTURE AND POINT DEFECTS OF ICE: THEIR EFFECT ON THE ELEC 02-0071
DYNAMICS, AND PHASE TRANSITIONS OF FERROELECT/ INFLUENCE OF POINT DEFECTS ON DIELECTRIC PROPERTIES, DOMAIN 03-0495
EN BONDED FERROELECTRICS.   EFFECT OF POINT DEFECTS ON THE ENERGY SPECTRUM OF HYDROG 03-0486
CRYSTALLINE LEAD ZIRCONAT/ INFLUENCE OF THE CONCENTRATION OF POINT DEFECTS ON THE INTERNAL FRICTION IN POLY 09-1582
/IOBIA ADDITION) DUE TO INTERACTION OF DOMAIN BOUNDARIES AND POINT DEFECTS (TORSION PENDULUM AND RESONANCE/ 09-1584
ION TO LEAD TITANATE.   POINT DIPOLE ELECTROOPTIC MODEL BY AN APPLICAT 05-1073
/ VIBRATIONS IN ANTIMONY SULFOIODIDE CRYSTALS NEAR THE CURIE POINT (EFFECTS OF TEMPERATURE, FIELD STRENGTH/ 15-2121
N OF A POTASSIUM DIHYDROGEN PHOSPHATE CRYSTAL NEAR THE CURIE POINT. (ELECTROCALORIC EFFECT) /US POLARIZATIO 17-2678
MOVEMENT OF THE RADICAL IN TRIGLYCINE SULFATE NEAR THE CURIE POINT. (ESR OF GAMMA-IRRADIATED SPECIMENS) /E 27-3033
C RESPONSE OF TGS IN THE FERROELECTRIC REGION NEAR THE CURIE POINT (IMPROVED METHOD OF MEASUREMENT). /ECTRI 27-2979
              PYROCURRENTS HIGHER THAN THE CURIE POINT IN ANTIMONY SULFOIODIDE.   15-2131
                                          CRITICAL POINT IN FERROELECTRIC ANTIMONY SULFOIODIDE.   15-2237
DIELECTRIC RELAXATION PROCESSES NEAR THE CRITICAL POINT IN FERROELECTRICS.   03-0272
ALOUS / ULTRASONIC ATTENUATION NEAR THE SOFT MODE TRANSITION POINT IN POTASSIUM MANGANESE TRIFLUORIDE (ANOM 33-3373
DY ON THE DISORDERED STRUCTURE ABOVE THE FERROELECTRIC CURIE POINT IN POTASSIUM NITRATE.   X-RAY STU 16-2366
MS. (DISCUSSE/ UNIT CELL VOLUME EFFECTS (RELATION WITH CURIE POINT) IN SOME ISOMORPHOUS FERROELECTRIC SYSTE 17-2487
/ENGTH DISPERSION IN THE DIELECTRIC CONSTANT ABOVE THE CURIE POINT IN VARIOUS FERROELECTRICS AND ANTIFERRO/ 02-0124
IC PHASE TRANSITIONS.   CRITICAL POINT INDICES OF FERROELECTRIC AND FERROMAGNET 03-0243
/ POLARIZATION AND DIELECTRIC HYSTERESIS NEAR THE TRANSITION POINT OF DIFFUSE FERROELECTRIC PHASE. (KAENZI/ 03-0437
          RELAXATION ABSORPTION OF SOUND NEAR THE CURIE POINT OF FERROELECTRIC SEMICONDUCTORS.   03-0492
TERNAL FRICTION IN THE REGION OF THE ANTIFERROMAGNETIC CURIE POINT OF LEAD MAGNESATE(0.5) TUNGSTATE(0.5). / 13-2020
LARON H/ ELECTRICAL CONDUCTIVITY (SEEBECK COEFFICIENT, CURIE POINT) OF REDUCED BARIUM TITANATE CRYSTALS (PO 04-0918
          DOUBLE HYSTERESIS LOOPS AT THE UPPER CURIE POINT OF SODIUM TRIHYDROGEN SELENITE.   22-2831
TERMINING SPONTANEOUS POLARIZATION NEAR THE PHASE TRANSITION POINT OF TRIGLYCINE SELENATE CRYSTALS. / OF DE 27-2994
.   ANOMALIES NEAR FERROELECTRIC TRANSITIONS FROM THE POINT OF VIEW OF THE THEORY OF CRITICAL POINTS 03-0298
BEHAVIOR OF LITHIUM TANTALATE SINGLE CRYSTAL NEAR ITS CURIE POINT. PART-1: PIEZOELECTRIC AND OPTICAL PROP/ 12-1877
BEHAVIOR OF LITHIUM TANTALATE SINGLE CRYSTAL NEAR ITS CURIE POINT. PART-2, DIELECTRIC AND ULTRASONIC PROP/ 12-1885
DEL OF LOCALLY ORDERED "NUCLEI", PHASE STRUCTURE ABOVE CURIE POINT, RAMAN SPECTRUM OF STRONTIUM TITANATE) / 04-0647
CS POSSESSING FIRST ORDER TRANSITION. (THEORY, CALCUL/ CURIE POINT SHIFT WITH APPLIED FIELD IN FERROELECTRI 03-0327
E MAGNETIZATION IN THE TWO DIMENSIONAL ANTIFERROMAGNET/ ZERO POINT SPIN DEVIATION AND SPONTANEOUS SUBLATTIC 26-2929
COMPLEX DIELECTRIC CONSTANT OF FERROELECTRICS NEAR THE CURIE POINT. (THEORY) /MPERATURE DEPENDENCES OF THE 03-0361

) MICROWAVE DISPERSION OF FERROELECTRICS NEAR THE TRANSITION POINT (THEORY, MEASUREMENTS ON TGS). /E ON THE 03-0370
INFLUENCE OF SCREENING ON THE FERROELECTRIC CURIE POINT (UNIAXIAL FERROELECTRIC PHOTOCONDUCTOR). 03-0392
INE) ANTIMONY SULFOIODIDE (DARK CONDUCTIVITY NEAR TRANSITION POINT, 1-1000 ATM). /SISTANCE OF (POLYCRYSTALL 15-2250
ARIZATION AND MAGNETIZATION IN NICKEL IODIDE BORACITE (CURIE POINT 64 DEG K). /THE SPONTANEOUS ELECTRIC POL 14-2090
AL ULTRASONIC WAVES IN ROCHELLE SALT CRYSTALS NEAR THE CURIE POINTS. ABSORPTION OF LONGITUDIN 28-3211
TRANSITIONS FROM THE POINT OF VIEW OF THE THEORY OF CRITICAL POINTS. ANOMALIES NEAR FERROELECTRIC 03-0298
ERATURE RANGE 4.2-300 DEG K (LATTICE PARAMETERS VERSUS CURIE POINTS). /ATE LEAD TITANATE SYSTEM IN THE TEMP 01-1308
NATE, POTASSIUM TANTALATE AND BARIUM TITANATE NEAR THE CURIE POINTS. THERMAL CONDUCTIVITY OF STRONTIUM TITA 04-0989
RIC INSTABILITY (FERROELECTRIC TWINNING) TREATED AS CRITICAL POINTS IN CRYSTALS. (ANTIMONY SULFOIODIDE) /CT 15-2176
IES OF THERMODYNAMIC PARAMETERS NEAR SECOND ORDER TRANSITION POINTS IN FERROELECTRICS. /AL THEORY OF ANOMAL 03-0362
YSTALS BY HYDROSTATIC PRESSURE. DISPLACEMENT OF THE CURIE POINTS IN KDP AND LITHIUM THALLIUM TARTRATE CR 17-2523
PTIBILITY AND DIELECTRIC RELAXATION TIME NEAR THE TRANSITION POINTS IN SODIUM NITRITE. /ATIC ELECTRIC SUSCE 16-2302
ULTRASONIC ATTENUATION NEAR THE TWO TRANSITION POINTS IN SODIUM NITRITE. 16-2304
OMPOUNDS./ DETERMINATION OF THE LATTICE PARAMETERS AND CURIE POINTS OF SOLID SOLUTIONS OF A(V)B(VI)C(VII) C 15-2116
. EFFECT OF A TWO DIMENSIONAL PRESSURE ON THE CURIE POINTS OF TRIGLYCINE SULFATE AND ROCHELLE SALT 27-3082
/TERNAL STRESS FOR FERROELECTRIC PHOSPHATES NEAR THEIR CURIE POINTS. (POTASSIUM DIHYDROGEN PHOSPHATE, ALSO/ 17-2665
/ ULTRASONIC INVESTIGATION OF PHASE TRANSITIONS AND CRITICAL POINTS. SECTION-5: FERROELECTRIC AND ANTIFERRO 02-0070
LEAD COBALT TUNGSTATE- CADMIUM MANGANATE SYSTEM (BOTH CURIE POINTS STUDIED, VARIOUS COMPOSITIONS) / IN THE 36-3604
DISKS. MEASUREMENT OF POISSON'S RATIO IN POLED FERROELECTRIC CERAMIC 37-3658
ERROELECTRIC POTASSIUM IODATE CRYSTALS. POLAR AND LATTICE DYNAMIC CHARACTERISTICS OF F 08-1430
UM DIHYDROGEN PHOSPHATE AND RUBIDIUM DIHYDROGEN PHOSPHATE IN POLAR AND NONPOLAR PHASES. /RYSTALS OF POTASSI 17-2457
CLASSIFICATION OF ANISOTROPIC FERROELECTRIC AND ANTI- POLAR- AND POLAR- NEUTRAL CRYSTAL STRUCTURES. (03-0336
BARIUM / RELATIVE SIGNS OF NONLINEAR OPTICAL COEFFICIENTS OF POLAR CRYSTALS (LITHIUM NIOBATE OR TANTALATE, 02-0098
FERROELECTRICITY AND ANTIFERROELECTRICITY IN POLAR CRYSTALS. (THEORY, LEAD ZIRCONATE) 03-0372
HAT/ TRANSIENT PHENOMENA IN THE DISCHARGE OF CONDENSERS WITH POLAR DIELECTRICS. (POTASSIUM DIHYDROGEN PHOSP 17-2631
AMIC METHOD) SPONTANEOUS POLARIZATION NOISE IN POLAR DIELECTRICS. (THEORY, EINSTEIN THERMODYN 03-0219
DYNAMIC POLAR MODEL OF MONOCRYSTALLINE ICE I(H). 34-3424
PYROELECTRIC EFFECT IN METAL-FREE PHTHALOCYANINE. (IN THE POLAR, OR FERROELECTRIC STATE) 36-3650
DEBYE TYPE ABSORPTION IN POLAR PHTHALOCYANINE. 36-3621
/LINEAR OPTICAL (SECOND HARMONIC GENERATION) COEFFICIENTS OF POLAR (PYROELECTRIC) CRYSTALS (LITHIUM NIOBAT/ 02-0095
LOCYANINE. LOW TEMPERATURE PHASE TRANSITION IN THE POLAR STATE OF THE ORGANIC SEMICONDUCTOR PHTHA 36-3623
N IN FERROELECTRICS WITH HYDROGEN BONDS. (KDP,/ INFLUENCE OF POLAR STATES OF PROTONS ON THE PHASE TRANSITIO 17-2671
URES. (CLASSIFICATION OF ANISOTROPIC FERROELECTRIC AND ANTI/ POLAR- POLAR AND POLAR- NEUTRAL CRYSTAL STRUCT 03-0336
COUPLED POLARITIONS OF A1 SYMMETRY IN BARIUM TITANATE. 04-0619
MODE IN LITHIUM NIOBATE. TEMPERATURE STUDY OF THE POLARITON ASSOCIATED WITH THE 248 CM(-1) SOFT 11-1807
TICAL MODULATION VIA TEMPERATURE) TEMPERATURE DEPENDENCE OF POLARITON DISPERSION IN LITHIUM TANTALATE. (OP 12-1873
IC BARIUM TITANATE. (CALCULATIONS) POLARITON DISPERSION OF TETRAGONAL FERROELECTR 04-0842
TITANATE. POLARITON DISPERSION RELATION IN CUBIC BARIUM 04-0878
RAMAN SCATTERING LINE SHAPE OF THE SOFT E POLARITON MODE IN BARIUM TITANATE. 04-0803
UNGSTEN BRONZE CRYSTAL STRUCTURE. POLARITON RESULTS IN FERROELECTRICS WITH THE T 13-1918
DIRECTIONAL DISPERSION OF PHONON FREQUENCIES AND POLARITON SCATTERING IN LITHIUM NIOBATE. 11-1696
CRYSTAL. FERMI RESONANCE OF POLARITON WITH BIPHONON IN AN LITHIUM NIOBATE 11-1761
LITHIUM NIOBATE. LIGHT SCATTERING BY POLARITONS ASSOCIATED WITH ORDINARY PHOTONS IN 11-1697
UM NIOBATE. POLARITONS IN A BIAXIAL CRYSTAL OF BARIUM SODI 13-1992
N SPECTRUM, PERMITTIVITY) POLARITONS IN A LITHIUM NIOBATE CRYSTAL. (RAMA 11-1762
TRANSVERSE POLARITONS IN FERROELECTRIC SODIUM NITRITE. 16-2256
ANGULAR DISPERSION OF LARGE WAVE VECTOR POLARITONS IN LITHIUM IODATE. 08-1294
OBSERVATION OF ORDINARY POLARITONS IN LITHIUM NIOBATE. 11-1854
SCATTERING OF LIGHT BY ACOUSTIC PHONONS AND POLARITONS IN LITHIUM TANTALATE. 12-1878
RY OF THE RAMAN LINE SHAPE. POLARITONS IN TETRAGONAL BARIUM TITANATE: THEO 04-0556
PARAMETRIC LUMINESCENCE AND LIGHT SCATTERING BY POLARITONS. (LITHIUM NIOBATE) 11-1746
E AND LITHIUM NIOBATE CRYST/ SPONTANEOUS RAMAN SCATTERING BY POLARITONS. PART-2: THE CASE OF CADMIUM SULFID 11-1787
ART-1: LONG OPTICAL PH/ EIGENFREQUENCIES AND EIGENVECTORS OF POLARITONS WITH APPLICATION TO LEAD NIOBATE. P 08-1284
. PART-2: POLARITONS O/ EIGENFREQUENCIES AND EIGENVECTORS OF POLARITONS WITH APPLICATION TO LITHIUM NIOBATE 11-1766
BSORPTION EDGE. (LITHIUM NIOBATE, BA/ X-RAY DETERMINATION OF POLARITY SENSE BY ANOMALOUS SCATTERING AT AN A 11-1673
IN BARIUM AND STRONTIUM TITANATES. TEMPERATURE DEPENDENT POLARIZABILITIES AND COCHRAN MODE ASSIGNMENTS 04-0805
E. TEMPERATURE DEPENDENCE OF ELECTRONIC POLARIZABILITY OF STRONTIUM AND BARIUM TITANAT 04-0738
ITHIUM NIOBATE, LITHIUM TANTALATE, POTASS/ NONLINEAR OPTICAL POLARIZABILITY OF THE NIOBIUM- OXYGEN BOND. (L 11-1734
AKING INTO ACCOUNT DIFFERENT ORIENTATIONS OF THE SPONTANEOUS POLARIZATION. A MODEL OF KANZIG REGIONS T 03-0439
PARAMETERS OF FERROELECTRIC CERAMIC RESONATORS UPON STATE OF POLARIZATION. DEPENDENCE OF THE SMALL SIGNAL 37-3649
M DIHYDROGEN PHOSPHATE. (ELECTROCALORIC STUDIES, SPONTANEOUS POLARIZATION) ELECTRONIC STUDIES OF POTASSIU 17-2429
C CONSTANT VERSUS ALTERNATING FIELD VOLTAGE, TEMPERATURE AND POLARIZATION) /TRIGLYCINE SELENATE. (DIELECTRI 27-3158
ERROELECTRICITY. INDUCED POLARIZATION AND A ONE POLARIZATION MODEL OF F 03-0329
MICS. SPACE CHARGE POLARIZATION AND AGING OF BARIUM TITANATE CERA 04-0880
TE (FLUX GROWN SINGLE CRYSTALS). SPONTANEOUS POLARIZATION AND COERCIVE FIELD OF LEAD TITANA 05-1050
ITES IN THE SYSTEM LEAD ZI/ THERMAL VARIATION OF SPONTANEOUS POLARIZATION AND COERCIVE FORCE OF THE PEROVSK 09-1597
/RE" ICE I(H), SINGLE CRYSTALS. PART-2: MOLECULAR MODELS FOR POLARIZATION AND CONDUCTION. PART-3: EXTRINSI/ 34-3455
E TRANSITION POINT OF DIFFUSE FERROELECTRIC PHASE. (KAEN/ AC POLARIZATION AND DIELECTRIC HYSTERESIS NEAR TH 03-0437
ATERIALS. NONLINEARITY OF DIELECTRIC POLARIZATION AND FERROELECTRIC PROPERTIES OF M 03-0185
THEORY OF POLARIZATION AND HYSTERESIS OF FERROELECTRICS. 03-0494
E BORACITE (CURIE P/ COEXISTENCE OF THE SPONTANEOUS ELECTRIC POLARIZATION AND MAGNETIZATION IN NICKEL IODID 14-2090
OF PEROVSKITE TYPE AND OF ONES / MICROTHEORY OF SPONTANEOUS POLARIZATION AND PHASE TRANSITIONS IN CRYSTALS 03-0175
EAD TITANATE CERAMICS. (CONTAINING 1 MOLE PERCENT MANGANESE/ POLARIZATION AND PIEZOELECTRIC PROPERTIES OF L 05-1061
/EN THE STARTING FIELD FOR PARKHAUSEN STEPS AND THE CRITICAL POLARIZATION AND REPOLARIZATION FIELDS IN FER/ 03-0318
RIGLYCINE SULFATE CRYSTALS AND / EFFECT OF ULTRASOUND ON THE POLARIZATION AND REPOLARIZATION PROCESSES IN T 04-0924
NY SULFOIODIDE. RELATION BETWEEN THE ELECTRIC POLARIZATION AND THE ABSORPTION EDGE IN ANTIMO 15-2175
LECTRIC POTASSIUM DIHYDROGEN PHOS/ ORIGIN OF THE SPONTANEOUS POLARIZATION AND THE ISOTOPE EFFECTS IN FERROE 17-2433
D CERAMICS IN THE LEAD TITANATE ZIRCONATE/ DIELECTRIC SYSTEM POLARIZATION AND THERMAL EXPANSION OF POLARIZE 09-1603
UM TANTALATE USING AUTHOR'S STATISTICAL THEORY. (SPONTANEOUS POLARIZATION, BONDING) /HIUM NIOBATE AND LITHI 03-0366
ELECTRIC MATERIALS. (GADOLINIUM M/ PRODUCTION OF SPONTANEOUS POLARIZATION BY ELASTIC INSTABILITIES IN PIEZO 30-3282
NE SULFATE. (DIELECTRIC DISPLACEMENT AND COERCIVE FIELD) (POLARIZATION CHANGE) AFTER EFFECTS IN TRIGLYCI 27-2952
OF FERROELECTRIC TRIGLYCINE SULFATE AND SELENATE CRYSTALS. (POLARIZATION CHARGE) /HIN FILMS ON A SUBSTRATE 27-3046
INJECTION CURRENTS AND ACCUMULATION OF POLARIZATION CHARGE IN ANTIMONY SULFOIODIDE. 15-2188
ATE AT LOW T/ DIELECTRIC BEHAVIOR (PERMITTIVITY, SPONTANEOUS POLARIZATION, COERCIVE FIELD) OF BISMUTH TITAN 08-1436
FERROELECTRIC SEMICONDUCTOR INTERFACE. POLARIZATION DISTRIBUTION IN THE VICINITY OF A 03-0167
SITION IN KDP SINGLE CRYSTAL. (ABSTRACT ONLY) POLARIZATION (DOMAIN STRUCTURE) AND PHASE TRAN 17-2505
M DIHYDROGEN PHOSPHATE) MACROSCOPIC STIMULATED POLARIZATION ECHO IN FERROELECTRICS. (POTASSIU 17-2419
CRYSTAL POTASSIUM DIHYDROGEN PHOSPHATE. POLARIZATION ECHO IN THE FERROELECTRIC SINGLE 17-2564
DUE TO DIPOLE RELAXATION) LOW TEMPERATURE POLARIZATION EFFECTS IN ICE. (CURRENT REVERSAL 34-3418
LOW TEMPERATURES. ELECTRIC POLARIZATION EFFECTS IN PURE AND DOPED ICE AT 34-3412

ON OF ULTRASOUND IN ANTIMONY SULFOIODIDE CRYSTALS. — POLARIZATION EFFECTS IN THE ANOMALOUS ABSORPTI 15-2123
RIC ACTIVITY OF LEAD ZIRCONATE(X) TITANATE(1-X) CERAMICS. — POLARIZATION, ELECTROSTRICTION, AND PIEZOELECT 09-1501
) SEMICONDUCTOR F/ EFFECT OF FERROELECTRIC (BARIUM TITANATE) — POLARIZATION FIELDS ON (GERMANIUM OR TELLURIUM 04-0594
NEAR THE CURIE POINT. — POLARIZATION FLUCTUATION OF TRIGLYCINE SULFATE 27-3077
RPTION EDGE IN BARIUM TITANATE. — POLARIZATION FLUCTUATIONS AND THE OPTICAL ABSO 04-1028
AMMONIUM DIHYDROGEN PHOSPHATE. LONG WAVELENGTH — POLARIZATION FLUCTUATIONS IN ANTIFERROELECTRIC 17-2461
(ABOVE THE FERROELECTRIC CURIE TEMPERATURE, AS SH/ CRITICAL — (POLARIZATION) FLUCTUATIONS IN BARIUM TITANATE 04-1032
/ BY QUASI SPIN WAVES IN ORDER DISORDER TYPE FERROELECTRICS: — POLARIZATION FLUCTUATIONS IN DICALCIUM STRONT/ 26-2928
UM TITANATE. OPTICAL EVIDENCE FOR — POLARIZATION FLUCTUATIONS IN PARAELECTRIC BARI 04-0737
/ NONMETALLIC CRYSTALS. (STRAIN DEPENDENT CURIE TEMPERATURE, — POLARIZATION FLUCTUATIONS, LITHIUM NIOBATE, A/ 03-0513
PHASE TRANSITION. — POLARIZATION FLUCTUATIONS NEAR A FERROELECTRIC 03-0368
BEHAVIOR OF BARIUM TITANATE. INFLUENCE OF CRITICAL — POLARIZATION FLUCTUATIONS ON THE PHOTOELASTIC 04-0628
EFFECT OF PHOTOELECTRET CHARGE ON VALUE OF FERROELECTRIC — POLARIZATION IN ANTIMONY SULFOIODIDE. 15-2186
YSTALS. INSTABILITY OF FERROELECTRIC — POLARIZATION IN ANTIMONY SULFOIODIDE SINGLE CR 15-2152
RELATIONSHIP BETWEEN DIRECTIONS OF THE ELECTRIC AND MAGNETIC — POLARIZATION IN BORACITE AND OTHER CRYSTALS. 14-2087
IONIC DISPLACEMENT AND THE CURIE TEMPERATURE AND SPONTANEOUS — POLARIZATION IN DISPLACIVE FERROELECTRICS. /N 03-0453
A MORE EXACT DYNAMIC METHOD OF DETERMINING — POLARIZATION IN FERROELECTRICS. 03-0381
X-RAY DIFFRACTION STUDIES OF ELECTRON — POLARIZATION IN FERROELECTRICS. 03-0199
/HERMAL EXPANSION ANOMALIES AND THE DIRECTION OF SPONTANEOUS — POLARIZATION IN FERROELECTRICS. (ABSTRACT ONL/ 03-0350
Y OF THE ELECTRIC FIELD. DEPENDENCE OF EFFECTIVE — POLARIZATION IN FERROELECTRICS ON THE INTENSIT 03-0382
CONDUCTION ANOMALIES AND — POLARIZATION IN ICE AT LOW TEMPERATURES. 34-3428
OF THE LIMITATION OF THE ANGULAR FIELD BY THE COMMUTATOR OF — POLARIZATION IN KDP, IN A DIGITAL DEVIATOR. /Y 17-2524
/ PRIMARY AND SECONDARY PYROELECTRIC EFFECTS AND SPONTANEOUS — POLARIZATION IN LITHIUM TANTALATE (THERMAL EXP 12-1884
CONDITIONS FAVORING THE ORIGINATION OF SPONTANEOUS — POLARIZATION IN PEROVSKITES. (ABSTRACT ONLY) 08-1457
. MEASUREMENTS OF THE SPONTANEOUS — POLARIZATION IN POTASSIUM DIHYDROGEN PHOSPHATE 17-2430
NMR STUDIES OF SPONTANEOUS — POLARIZATION IN ROCHELLE SALT. 28-3194
PYROELECTRIC STUDY ON THE SPONTANEOUS — POLARIZATION IN SILVER SODIUM DINITRITE. 16-2295
(FERROELECTRICS) NONSTATIONARY — POLARIZATION IN THE ANHARMONIC VIBRATOR MODEL. 03-0345
ES OF BARIUM TITANATE. MAGNITUDE OF SPONTANEOUS — POLARIZATION IN THIN CRYSTALS AND FINE PARTICL 04-0607
MICONDUCTING SUBSTRATES. (TGS ON/ THERMODYNAMIC STABILITY OF — POLARIZATION IN THIN FERROELECTRIC FILMS ON SE 27-3163
BLOCKING OF SPONTANEOUS — POLARIZATION IN TRIGLYCINE SULFATE. 27-3008
. (WITH FREQUENCY OF APPLIED FIELD CAUSED BY SHEARING DUE TO — POLARIZATION INDUCED STRAIN) /OELECTRIC BODIES
TRONTIUM / ENERGY BAND CHANGES IN PEROVSKITES DUE TO LATTICE — POLARIZATION. (LCAO ANALYSIS OF BAND SCHEME, S 06-1101
/HENOMENA IN FERROELECTRICS WITH DIFFUSED PHASE TRANSITION. (— POLARIZATION MECHANISMS, ELECTROOPTICAL PROPE/ 02-0119
TWO — POLARIZATION MODEL THEORY OF FERROELECTRICITY. 03-0332
TRICS OF THE ANTIMONY SULFOIODIDE TYPE. — POLARIZATION NATURE IN SEMICONDUCTOR FERROELEC 15-2181
F TRIGLYCINE SELENATE CR/ METHODS OF DETERMINING SPONTANEOUS — POLARIZATION NEAR THE PHASE TRANSITION POINT O 27-2994
RY, EINSTEIN THERMODYNAMIC METHOD/ SPONTANEOUS — POLARIZATION NOISE IN POLAR DIELECTRICS. (THEO 03-0219
ITION IN PEROVSKITE LIKE FERROELECTRICS, U/ INVESTIGATION OF — POLARIZATION NONLINEARITY NEAR THE PHASE TRANS 03-0249
EFFECT OF THE SURFACE ON — POLARIZATION OF A DIELECTRIC (THEORY). 03-0271
D ORDER PHASE TRANSITION (THEORY). SINGLE DOMAIN — POLARIZATION OF A FERROELECTRIC HAVING A SECON 03-0202
(0.33) NIOBATE(0.67) WITH A DIFFUSE PHASE TRANSI/ RELAXATION — POLARIZATION OF A FERROELECTRIC LEAD MAGNESATE 13-1980
TE CRYSTAL NEAR THE CURIE POINT. (ELECTROCALORI/ SPONTANEOUS — POLARIZATION OF A POTASSIUM DIHYDROGEN PHOSPHA 17-2678
E CRYSTAL IN THE PHASE TRANSITION REGION. (COERCIVE FIELD) — POLARIZATION OF A RUBIDIUM DIHYDROGEN PHOSPHAT 17-2504
E CRYSTAL IN THE PHASE TRANSITION REGION. — POLARIZATION OF A RUBIDIUM DIHYDROGEN PHOSPHAT 17-2506
(TEMPERATURE DEPENDENCE OF) THE SPONTANEOUS — POLARIZATION OF AMMONIUM SULFATE. 19-2799
S. STABILIZATION OF SPONTANEOUS — POLARIZATION OF BARIUM TITANATE SINGLE CRYSTAL 04-0795
S WITH DOUBLE HYSTERESIS LOOP. PECULIARITIES OF — POLARIZATION OF BARIUM TITANATE SINGLE CRYSTAL 04-0709
PECULIARITIES OF THE DIELECTRIC — POLARIZATION OF CADMIUM PYRONIOBATE. 36-3561
NTIUM TITANATE). INERTIAL ELECTRICAL — POLARIZATION OF CRYSTALS (ALKALI HALIDES, STRO 06-1197
AND BENDING DEFORMATION IN THE ELECTRIC FIELD. (OBSERVATION/ — POLARIZATION OF FERROELECTRIC FILM BY BENDING 03-0195
CURIE POI/ THEORY OF DOMAINLESS, NONHOMOGENEOUS, SPONTANEOUS — POLARIZATION OF FERROELECTRIC PLATES NEAR THE 03-0321
/SSURE DEPENDENCE OF THE DIELECTRIC CONSTANT AND SPONTANEOUS — POLARIZATION OF FERROELECTRIC POTASSIUM NITRA/ 16-2331
OF THE PEROVSKITE TYPE WIT/ INVESTIGATION OF THE DYNAMICS OF — POLARIZATION OF FERROELECTRIC SOLID SOLUTIONS 08-1380
URE DEPENDENCE 150-165 / PYROELECTRIC EFFECT AND SPONTANEOUS — POLARIZATION OF GADOLINIUM MOLYBDATE (TEMPERAT 30-3314
/CURIE TEMPERATURE AND TEMPERATURE DEPENDENCE OF SPONTANEOUS — POLARIZATION OF HEXAGONAL (RHOMBOHEDRAL) FERR/ 03-0290
/TROOPTIC G COEFFICIENTS (RELATING INDUCED BIREFRINGENCE AND — POLARIZATION) OF LEAD CONTAINING OXYGEN OCTAH/ 13-2039
SOLUTIONS (THERMODYNAMIC EXISTENCE CONDITIONS FO/ PHASES AND — POLARIZATION OF LEAD TITANATE ZIRCONATE SOLID 09-1541
OOPTIC CRYSTALS. (NONLINEAR OPTICAL PROPERTIES, / PHASES AND — POLARIZATION OF LUMINOUS WAVES IN CUBIC ELECTR 02-0044
. SYMMETRY LIMITATIONS TO — POLARIZATION OF POLYCRYSTALLINE FERROELECTRICS 03-0434
TITANATE BY MEANS OF A ST/ OBSERVATION OF THE CHANGE IN THE — POLARIZATION OF SINGLE CRYSTAL FILMS OF BARIUM 04-0942
LFOIODIDE. INFLUENCE OF LIGHT ON SELF — POLARIZATION OF SINGLE CRYSTALS OF ANTIMONY SU 15-2228
TITANATE COMPOUNDS BETWEEN 4 AND 100 DEGREES K. — POLARIZATION OF SOLID BARIUM(X) STRONTIUM(1-X) 04-0719
L VIBRATIONS AND STRUCTURE OF THE HYDROGEN BONDS IN THE FER/ — POLARIZATION OF THE RAMAN BANDS OF THE HYDROXY 17-2471
/VERSIBLE, BUT TILTABLE (TO 3 STABLE POSITIONS, SPONTANEOUS) — POLARIZATION (ON APPLYING AN ELECTRIC FIELD). 08-1346
) IONS IN TRIGLYCINE FLUOBERYLLATE. INFLUENCE OF SPONTANEOUS — POLARIZATION ON THE EPR SPECTRA OF CHROMIUM(3+ 27-3136
N EMISSION YIELD OF TRIGLYCINE/ INFLUENCE OF THE SPONTANEOUS — POLARIZATION ON THE PHOTOSTIMULATED EXOELECTRO 27-2961
MAIN STRUCTURE IN POTASSIUM DIHYDROGEN PHOSPHATE CRYSTALS BY — POLARIZATION OPTICAL PROCEDURE. /RVATION OF DO 17-2493
T. PART-10: PHOTON RENORMALIZATION AND THEORY OF SPONTANEOUS — POLARIZATION. PART-11: CONCLUSIONS. /CIFIC HEA 03-0397
ERATIONS ON BORACITE/ INTERACTIONS BETWEEN MAGNETIZATION AND — POLARIZATION: PHENOMENOLOGICAL SYMMETRY CONSID 14-2072
LOW TEMPERATURE./ DIELECTRIC AND ELECTROOPTIC STUDIES OF THE — POLARIZATION PROCESS IN STRONTIUM TITANATE AT 06-1117
NY SULFOIO/ INFLUENCE OF NONEQUILIBRIUM CARRIER SCREENING ON — POLARIZATION PROCESSES IN FERROELECTRIC ANTIMO 15-2158
CITY OF IONIC CRYSTALS. (THEORY, NO REFS) — POLARIZATION, PYROELECTRICTY, AND FERROELECTRI 03-0143
SING FERROELECTRICS. (BARIUM TITANATE) — POLARIZATION REFLECTION TYPE LIGHT MODULATOR U 04-0714
THE FERROELECTRIC TRANSITION REGION OF POTASSIUM DIDEUTERIU/ — POLARIZATION RELAXATION AND SUSCEPTIBILITY IN 17-2575
/SITION BROADENING (DIFFUSNESS) AND THE NATURE OF DIELECTRIC — POLARIZATION RELAXATION IN SOME FERROELECTRIC/ 03-0291
SPHATE (TEMPERATURE DEPENDENCE OF THE YOUNGS MODULUS AND THE — POLARIZATION RELAXATION TIME). /IDEUTERIUM PHO 17-2660
ATION OF THE MILLIMETER WAVE ABSORPTION IN SODIUM NITRITE BY — POLARIZATION REVERSAL. MODUL 16-2392
PHENOMENOLOGICAL THEORY OF FERROELECTRIC — POLARIZATION REVERSAL. 03-0282
TIME DEPENDENCE OF MATERIAL CONSTANTS OF TGS AFTER — POLARIZATION REVERSAL. 27-2953
FERROELECTRIC SODIUM NITRITE CRYSTALS. PART-2: MECHANISM OF — POLARIZATION REVERSAL. TOPOGRAPHIC STUDY OF 16-2374
OMAGNETOELECTRIC MATERIAL THAT ALLOWS NO 180 DEGREE ELECTRIC — POLARIZATION REVERSAL. /FERROELECTRIC AND FERR 14-2098
IUM HYDROGEN SULFATE. — POLARIZATION REVERSAL CHARACTERISTICS OF AMMON 19-2788
/ILIZATION OF FERROELECTRIC CRYSTAL. (DIELECTRIC HEATING AND — POLARIZATION REVERSAL, COMPLEX PERMITTIVITY C/ 03-0245
SSING RATE OF FIELD IN SODIUM NITRITE. — POLARIZATION REVERSAL IN A WIDE RANGE OF IMPRE 16-2381
EFFECT OF PRESSURE (TO ABOUT 2KBAR) ON — POLARIZATION REVERSAL IN BARIUM TITANATE. 04-0718
TRIGLYCINE SULFATE) SUSCEPTIBILITY INCREASE DURING — POLARIZATION REVERSAL. (IN BARIUM TITANATE AND 04-0981
E CRYSTALS. THICKNESS DEPENDENCE OF — POLARIZATION REVERSAL. (IN BARIUM TITANATE SINGL 04-0949
S. STATISTICAL METHOD FOR INVESTIGATING (DYNAMICS OF) — POLARIZATION REVERSAL IN FERROELECTRIC CERAMIC 03-0493
ULATIONS OF SWITCHING CURRENT) THEORY OF — POLARIZATION REVERSAL IN FERROELECTRICS. (CALC 03-0422

C BISMUTH TITANATE.   DOMAIN STRUCTURE AND POLARIZATION REVERSAL IN FILMS OF FERROELECTRI 08-1470
ELECTRICS. (8 SUBSTITUTED LEAD TITANATE ZIRCONAT/ HIGH FIELD POLARIZATION REVERSAL IN POLYCRYSTALLINE FERRO 08-1418
                                                             POLARIZATION REVERSAL IN TGS SINGLE CRYSTALS.   27-2980
   MEANS OF A LASER BEAM. APPLIC/ IRREVERSIBLE AND SPONTANEOUS POLARIZATION REVERSAL IN TRIGLYCINE SULFATE BY 27-3010
                POSSIBILITY OF GENERATING INFRARED RADIATION BY POLARIZATION REVERSAL OF A FERROELECTRIC.   03-0162
SINGLE CRYSTALS IN THE REGION OF PHASE COEXISTENCE.   POLARIZATION REVERSAL OF ANTIMONY SULFOIODIDE   15-2246
AL.   THEORY OF POLARIZATION REVERSAL OF BARIUM TITANATE CRYST 03-0326
SODIUM NIOBATE.   ROLE OF HYDROGEN IN POLARIZATION REVERSAL OF FERROELECTRIC BARIUM   13-2025
S.   ON SURFACE STATE AND POLARIZATION REVERSAL OF FERROELECTRIC CRYSTAL   03-0323
       OPTICAL SECOND HARMONIC GENERATION DURING FERROELECTRIC POLARIZATION REVERSAL. (SODIUM NITRITE)   16-2384
RIUM TITANATE CRYSTALS (CALCULATIO/ HIGH FIELD (1.5-6 KV/CM) POLARIZATION REVERSALS IN LIQUID ELECTRODED BA 04-0994
          NEUTRON DIFFRACTION STUDIES OF FERROELECTRIC POLARIZATION REVERSALS IN SODIUM NITRITE.   16-2365
/RDER PHASE TRANSITIONS IN FERROELECTRICS: A SELF CONSISTENT POLARIZATION SCHEME FOR ANHARMONIC POTENTIALS/ 03-0355
/R OF FRCTONS THROUGH PURE ICE I(H) SINGLE CRYSTALS. PART-1: POLARIZATION SPECTRA OF ICE I(H). PART-2: MOL/ 34-3486
/OF PROTCNS THROUGH "PURE" ICE I(E) SINGLE CRYSTALS. SECT-1: POLARIZATION SPECTRA OF ICE I(H). SECT-2: MOL/ 34-3482
                          CHARGE AND POLARIZATION STORAGE IN ICE CRYSTALS.   34-3456
IES OF A DIGITAL LIGHT DEFLECTOR UTILIZING LITHIUM TANTALATE POLARIZATION SWITCHES. /AND DEFLECTION CAPACIT 11-1809
OELECTRICS AT / ZERC SHIFT IN PIEZOELECTRIC ACCELEROMETERS. (POLARIZATION SWITCHING IN POLYCRYSTALLINE FERR 28-3216
                            FERROELECTRIC POLARIZATION SWITCHING IN SODIUM NITRITE.   16-2301
/FOR HIGH POWER ELECTRONIC TUNING. (WITH ELECTRODES REDUCING POLARIZATION SWITCHING TIME, LEAD(0.35) STRON/ 37-3660
/(DIELECTRIC CONSTANTS, PIEZOELECTRIC CONSTANTS, SPONTANEOUS POLARIZATION, THERMAL EXPANSION COEFFICIENTS,/ 12-1876
/GERMANIUM SILICON CXIDE. (DIELECTRIC CONSTANTS, SPONTANEOUS POLARIZATION, THERMAL EXPANSION, INDICES OF R/ 35-3501
/CN OF DIAMMONIUM DICADMIUM TRISULFATE, A MODEL IN WHICH THE POLARIZATION VECTOR, P(S), ARISES PERPENDICUL/ 03-0141
/IC PROPERTIES OF GUANIDINIUM ALUMINUM SULFATE HEXAHYDRATE. (POLARIZATION VERSUS ELECTRIC FIELD, -80C TO R/ 21-2813
MIC FERROELECTRIC. (PZT)   METHOD OF MEASURING POLARIZATION VIBRATIONS OF A PREPOLARIZED CERA   09-1615
OPMENTS IN THEORY OF FERROELECTRIC PROPERTIES OF CRYSTALS, / POLARIZATION WAVES IN SOLIDS. (INCLUDING DEVEL 03-0306
      ACOUSTIC SURFACE WAVE GUIDE, AVAILING FERROELECTRIC POLARIZATIONS.   37-3680
DISTRIBUTION OF POTENTIAL ALONG TITANATE CERAMICS PREVIOUSLY POLARIZED.   04-0788
                PHOTODOMAIN EFFECT IN NATURALLY POLARIZED ANTIMONY SULFOIODIDE CRYSTALS.   15-2133
                PHOTODOMAIN EFFECT IN NATURALLY POLARIZED ANTIMONY SULFOIODIDE CRYSTALS.   15-2135
O UNIAXIAL TRANSVERSE COMPRESSI/ PIEZOELECTRIC PROPERTIES OF POLARIZED BARIUM TITANATE CERAMICS SUBJECTED T 04-0962
              SURFACE CONDUCTIVITY IN POLARIZED BARIUM TITANATE CRYSTALS.   04-0569
      METHOD OF MEASURING POLARIZATION VIBRATIONS OF A PREPOLARIZED CERAMIC FERROELECTRIC. (PZT)   09-1615
ATE/ DIELECTRIC SYSTEM POLARIZATION AND THERMAL EXPANSION OF POLARIZED CERAMICS IN THE LEAD TITANATE ZIRCON 09-1603
ATE, BARIUM / ULTRAHIGH FREQUENCY DIELECTRIC ANISOTROPY OF A POLARIZED FERROELECTRIC CERAMIC. (BARIUM TITAN 04-0767
DROPHONE ELEMENTS)   OPEN CIRCUIT SENSITIVITY OF RADIALLY POLARIZED FERROELECTRIC CERAMIC CYLINDERS. (HY 37-3651
DROPHONE ELEMENTS)   OPEN CIRCUIT SENSITIVITY OF AXIALLY POLARIZED FERROELECTRIC CERAMIC CYLINDERS. (HY 37-3652
. (HYDROPHONE ELEMENTS) OPEN CIRCUIT SENSITIVITY OF RADIALLY POLARIZED FERROELECTRIC CERAMIC HOLLOW SPHERES 37-3653
          VIBRATIONS OF COAXIALLY SEGMENTED, LONGITUDINALLY POLARIZED FERROELECTRIC CYLINDRICAL TUBES.   03-0375
                                              VIBRATIONS OF LONGITUDINALLY POLARIZED FERROELECTRIC TUBES.   03-0376
HROUGH A STRESS INDUCED PHASE TRANSITION IN STRONTIUM TITAN/ POLARIZED FLUORESCENCE STUDY OF CHROMIUM(3+) T 06-1115
/HEAT CAPACITY, ENTROPY, MAGNETIC MOMENT OF THE ELECTRICALLY POLARIZED FORM FROM 0.4 TO 4.2 DEGREES K WITH/ 30-3275
              EQUILIBRIUM STRUCTURE OF POLARIZED ICE.   34-3411
ICAL PHONONS AND PHASE TRANSITIONS IN ANTIMONY SULFOIODIDE. (POLARIZED INFRARED AND RAMAN SPECTRA) /GTH OPT 15-2106
   CRYSTAL CESIUM NITRITE.   POLARIZED INFRARED AND RAMAN SPECTRA OF SINGLE   16-2342
/ STRUCTURAL STUDIES OF HYDROGEN BONDED FERROELECTRICS USING POLARIZED IR RADIATION. PART-2: INTERNAL FUNDA 18-2738
UM DIHYDROGEN PHOSPHATE CRYSTALS.   DIFFRACTION OF POLARIZED LIGHT BY ULTRASONIC WAVES IN POTASSI   17-2407
NES OF POTASSIUM DIHYDROGEN PHOS/ DEPOLARIZATION OF LINEARLY POLARIZED LIGHT DUE TO HEATING OF PARALLEL PLA 17-2540
OF LIGHT IONS IN THE CUBIC AND TETRAGONAL (FERROELECTRICALLY POLARIZED PHASE OF BARIUM TITANATE. /ANNELING 04-0688
                    PIEZOELECTRICITY IN POLARIZED POLYVINYLIDENE FLUORIDE FILMS.   36-3557
) FERROELECTRIC ANTIMONY SULFOIODIDE (PHONON ASSIGNMENTS).   (POLARIZED) RAMAN SPECTRUM OF (PARAELECTRIC AND 15-2207
L/ DEUTERON MOTIONS IN POTASSIUM DIDEUTERIUM PHOSPHATE (FROM POLARIZED REFLECTION SPECTRA, A TUNNELING MODE 17-2645
          COEXISTENCE OF THE ANTIPOLARIZED AND POLARIZED STATE IN BISMUTH TITANATE.   08-1422
ARIUM TITANATE)   POLARIZED STATES OF FERROELECTRIC CERAMICS. (B   04-0666
                                                 POLARIZED TRIGLYCINE SULFATES.   27-3058
                                   ADP POLARIZER FOR LASERS.   17-2551
URE OF COPPER AND CHROMIUM IONS ON THE DIELECTR/ EFFECT OF A POLARIZING FIELD AND OF THE ADDITION OF A MIXT 27-3045
      COMPLEX DIELECTRIC CONSTANT OF LEAD ZIRCONATE UNDER POLARIZING FIELD, 370-520 DEG K.   07-1256
ERVATION OF 180 DEG DOMAINS IN BARIUM TITANATE BY MEANS OF A POLARIZING MICROSCOPE.   DIRECT OBS 04-0952
EFFICIENT, CURIE POINT) OF REDUCED BARIUM TITANATE CRYSTALS (POLARON HOPPING INTERPRETATION). / (SEEBECK CO 04-0918
TANATE.   POLARON MECHANISM OF CONDUCTIVITY IN BARIUM TI 04-0609
TION OF THE THEORY OF TRANSITIONS FROM LARGE TO NEARLY SMALL POLARONS. /TE BARRIERS INTERPRETED AS CONFIRMA 03-0233
CALCULATIONS FOR STRONTIUM TITANATE).   POLARONS IN ANISOTROPIC ENERGY BANDS (THEORY,   03-0305
S.   SMALL POLARONS IN CONDUCTING BARIUM TITANATE CRYSTAL 04-0613
          MEASUREMENT OF POISSON'S RATIO IN POLED FERROELECTRIC CERAMIC DISKS.   37-3658
   PYROELECTRIC AND NONLINEAR OPTICAL PROPERTIES OF POLED POLYVINYLIDENE FLUORIDE FILMS.   36-3595
MUTH TITANATE, (BY ANNEALING).   PREPARATION OF POLED, TWIN-FREE CRYSTALS OF FERROELECTRIC BIS 08-1352
POLYCRYSTALLINE LEAD TITANATE ZIRCONATE WITH THOROUGHNESS OF POLING. /ATION OF ELECTROELASTIC CONSTANTS OF 09-1489
IHYDROGEN PHOSPHATE FOR LIGHT MODULATORS.   CUTTING AND POLISHING IN UNIAXIAL CRYSTALS OF THE TYPE X D 17-2559
TO ADP AND KDP).   AN APPARATUS FOR POLISHING WATER SOLUBLE CRYSTALS (APPLICATION   17-2632
          ELECTRET EFFECT IN A MIXTURE OF POLY METHACRYLATE AND BARIUM TITANATE   04-0838
ELAXATION TIME IN FERROELECTRICS NEAR THE CURIE TEMPERATURE (POLY VERSUS MONODISPERSIVE BEHAVIOR). /ON OF R 03-0521
          ELECTROCHEMISTRY ON THE POLYCRYSTAL OF POTASSIUM DIHYDROGEN PHOSPHATE.   17-2536
/ FERROELECTRICS AND ANTIFERROELECTRICS OF PEROVSKITE TYPE. (POLYCRYSTALLINE AND SINGLE CRYSTAL LEAD MAGNE/ 08-1375
          HEAT CAPACTIY OF POLYCRYSTALLINE ANTIMONY SULFOIODIDE.   15-2238
NDUCTIVITY NEAR TR/ INVESTIGATION OF THE PIEZORESISTANCE OF (POLYCRYSTALLINE) ANTIMONY SULFOIODIDE (DARK CO 15-2250
          ANOMALOUS DECAY EFFECT IN POLYCRYSTALLINE BARIUM TITANATE.   04-0752
          CONDUCTIVITY INJECTION AND EXTRACTION IN POLYCRYSTALLINE BARIUM TITANATE.   04-0814
          DIELECTRIC BREAKDOWN OF POLYCRYSTALLINE BARIUM TITANATE.   04-1015
          ELECTROCALORIC EFFECT IN POLYCRYSTALLINE BARIUM TITANATE.   04-0766
      FLASH EVAPORATION OF FERROELECTRIC THIN FILMS OF POLYCRYSTALLINE BARIUM TITANATE.   04-0968
OF A CONSTANT ELECTRIC FIELD ON THE DIELECTRIC PROPERTIES OF POLYCRYSTALLINE BARIUM TITANATE.   INFLUENCE 04-1012
      MECHANISM OF MICROWAVE DIELECTRIC DISPERSION IN POLYCRYSTALLINE BARIUM TITANATE.   04-0745
ERROELECTRICS AT MICROWAVE FREQUENCIES. (TRIGLYCINE SULFATE, POLYCRYSTALLINE BARIUM TITANATE) /PERTIES OF F 04-0870
E DOMAIN REORIENTATION PRODUCED BY ABRASION AND ANNEALING OF POLYCRYSTALLINE BARIUM TITANATE.   SURFAC 04-0639
AGE CHARACTERISTICS, 150-37/ SPACE CHARGE LIMITED CURRENT IN POLYCRYSTALLINE BARIUM TITANATE. (CURRENT VOLT 04-0557
ONSTANT, CONDUCTIVITY) ELECTRICAL PROPERTIES OF HIGH PURITY POLYCRYSTALLINE BARIUM TITANATE. (DIELECTRIC C 04-0704
C, 20-50 KILOHERTZ, LISSAJOUS FI/ ENERGY LOSS NONPOLAR STATE POLYCRYSTALLINE BARIUM TITANATE. (150-500 DEG 04-0831
RUCTURE (TITANATES AND ZIRCONATES). FORBIDDEN ZONE WIDTHS IN POLYCRYSTALLINE DIELECTRICS WITH PEROVSKITE ST 08-1288
      SYMMETRY LIMITATIONS TO POLARIZATION OF POLYCRYSTALLINE FERROELECTRICS.   03-0434

/IN PIEZOELECTRIC ACCELEROMETERS. (POLARIZATION SWITCHING IN POLYCRYSTALLINE FERROELECTRICS AT VERY LOW FI/ 28-3216
E AND SAME WITH 1 PERCENT/ UHF DIELECTRIC PROPERTIES OF SOME POLYCRYSTALLINE FERROELECTRICS (BARIUM TITANAT 04-0818
                MACROSCOPIC ANALOG OF THE SPIN- ECHO EFFECT IN POLYCRYSTALLINE FERROELECTRICS. (THEORY) 03-0310
ASE TRANSITIONS.   DOMAIN STRUCTURE OF POLYCRYSTALLINE FERROELECTRICS WITH DIFFUSE PH 03-0505
  LEAD TITANATE ZIRCONAT/ HIGH FIELD POLARIZATION REVERSAL IN POLYCRYSTALLINE FERROELECTRICS. (8 SUBSTITUTED 08-1418
ITIONS N/ INFLUENCE OF DOMAIN STRUCTURE ON THE PROPERTIES OF POLYCRYSTALLINE LEAD TITANATE ZIRCONATE COMPOS 09-1557
HOROUGHNESS OF POL/ VIBRATION OF ELECTROELASTIC CONSTANTS OF POLYCRYSTALLINE LEAD TITANATE ZIRCONATE WITH T 09-1489
E CONCENTRATION OF POINT DEFECTS ON THE INTERNAL FRICTION IN POLYCRYSTALLINE LEAD ZIRCONATE TITANATE. /F TH 09-1582
                     NIOBIUM-93 AND SODIUM-23 IN POLYCRYSTALLINE SODIUM NIOBATE.   08-1469
M TITANATE/ DIELECTRIC AND NONLINEAR MICROWAVE PROPERTIES OF POLYCRYSTALLINE SOLID SOLUTIONS BASED ON BARIU 04-1019
, S/ CONCENTRATION DEPENDENCE OF THE CURIE-WEISS CONSTANT IN POLYCRYSTALLINE SOLID SOLUTIONS IN THE (BARIUM 04-0586
/THE TEMPERATURE DEPENDENCE OF THE VELOCITY OF ULTRASOUND IN POLYCRYSTALLINE SOLID SOLUTIONS OF BARIUM AND/ 04-0894
ND BISMUTH TITANATES.   STRUCTURE OF POLYCRYSTALLINE SOLID SOLUTIONS OF STRONTIUM A 06-1145
D TITANATE ZIRCONATE SYSTEM (MEASUREMENT C/ RAMAN SPECTRA OF POLYCRYSTALLINE SOLIDS: APPLICATION TO THE LEA 09-1497
                    INFRARED SPECTRA OF SINGLE AND POLYCRYSTALLINE STRONTIUM TITANATE.   06-1250
/CT) STUDIES OF FERROELECTRIC CERAMICS AT LOW TEMPERATURES. (POLYCRYSTALLINE STRONTIUM TITANATE AND BARIUM/ 04-0720
/ECTRICS BELOW 80 DEG K. (SINGLE CRYSTAL STRONTIUM TITANATE, POLYCRYSTALLINE, STRONTIUM TITANATE, AND SOLI/ 04-0721
MANGANESE DIOXIDE.   EPR INVESTIGATION OF POLYCRYSTALLINE STRONTIUM TITANATE CONTAINING 06-1162
YPES, REVIEW)   ROLE OF THE REGULAR OCTAHEDRON AND LARGER POLYHEDRA IN FERROELECTRIC MATERIALS. (HETEROT 02-0094
            TRANSITION RATES OF POTASSIUM NITRATE HIGH PRESSURE POLYMORPHS.   16-2279
DEUTERON NUCLEAR MAGNETIC RESONAN/ STRUCTURAL STUDIES OF ICE POLYMORPHS BY NEUTRON DIFFRACTION, PROTON AND 34-3466
GEN PHOSPHATE IN AN ELECTRIC FIELD.   RAMAN SPECTRA OF POLYMORPHIC MODIFICATIONS OF POTASSIUM DIHYDRO 17-2425
TE, SILVER NITRATE, AND AMM/ MORPHOLOGY OF CRYSTAL GROWTH AT POLYMORPHIC TRANSFORMATIONS IN POTASSIUM NITRA 16-2260
MIC GROWTH OF NEW PHASE IN POTASSIUM NITRATE. (ALPHA TO BETA POLYMORPHIC TRANSITION)   RHYTH 16-2263
T 1 ATM/ CHANGES IN ELECTRICAL CONDUCTIVITY ACCOMPANYING THE POLYMORPHIC TRANSITIONS IN POTASSIUM NITRATE A 16-2391
        RUBIDIUM NITRATE NEAR THE MELTING POINT AND ITS POLYMORPHISM.   16-2330
ERSIBLE PHASES).   POLYMORPHISM OF STRONTIUM METANIOBATE (TWO REV 36-3535
CRYSTALS (TGS AND OTHERS, BY SELECTIVE DEPOSITION OF CHARGED POLYSTYRENE LATEX PARTICLES). /FACES OF IONIC 27-2987
OGEN PHOSPHATE- POTASSIUM DIHYDROGEN PHOSPHATE- WATER SYSTE/ POLYTHERM OF SOLUBILITY OF THE AMMONIUM DIHYDR 17-2434
STALS POTASSIUM FERROCYANIDE TRIHYDRATE BY THE NMR METHOD.   POLYTYPE AND TWINNING OF THE FERROELECTRIC CRY 24-2898
            DIRECT PIEZOELECTRIC EFFECT IN POLYVINYL CHLORIDE FILMS.   36-3543
. PYROELECTRICITY AND PIEZOELECTRICITY IN ORIENTED FILMS OF POLYVINYL FLUORIDE AND POLYVINYLIDENE FLUORIDE 36-3544
        DIELECTRIC ABSORPTION IN ORIENTED POLYVINYLIDENE FLUORIDE.   36-3572
            FAR INFRARED SPECTRA OF PIEZOELECTRIC POLYVINYLIDENE FLUORIDE.   36-3590
    FERROELECTRIC FILMS AND THEIR DEVICE APPLICATIONS. (POLYVINYLIDENE FLUORIDE)   36-3556
                PIEZOELECTRICITY OF POLYVINYLIDENE FLUORIDE.   36-3573
ICITY, PYROELECTRICITY, AND THE ELECTROSTRICTION CONSTANT OF POLYVINYLIDENE FLUORIDE.   PIEZOELECTR 36-3598
PIEZOELECTRICITY IN ORIENTED FILMS OF POLYVINYL FLUORIDE AND POLYVINYLIDENE FLUORIDE.   PYROELECTRICITY AND 36-3544
ED DETECTION.   PYROELECTRIC PROPERTIES OF POLYVINYLIDENE FLUORIDE AND ITS USE FOR INFRAR 36-3560
            PIEZOELECTRICITY IN POLARIZED POLYVINYLIDENE FLUORIDE FILMS.   36-3557
    PYROELECTRIC AND NONLINEAR OPTICAL PROPERTIES OF POLED POLYVINYLIDENE FLUORIDE FILMS.   36-3595
T/ PYROELECTRICITY AND OPTICAL SECOND HARMONIC GENERATION IN POLYVINYLIDENE FLUORIDE FILMS (AFTER COOLING S 36-3530
IES OF TRANSPARENT FERROELECTRIC / EFFECTS OF GRAIN SIZE AND POROSITY ON THE ELECTRICAL AND OPTICAL PROPERT 13-2004
A FUNCTION OF TEMPERATURE FOR ROCHELLE SALT AND TRIGLYCINE/ POSITION OF THE FUNDAMENTAL ABSORPTION EDGE AS 27-3160
HE CRYSTAL STRUCTURE OF FERROELECTRIC ACID LITHIUM SELENITE/ POSITION OF THE LITHIUM ION.   REFINEMENT OF T 22-2859
RYSTAL STRUCTURE OF FERROELECTRIC LITHIUM HYDROGEN SELENITE: POSITION OF THE LITHIUM ION. /INEMENT OF THE C 22-2860
OF LEAD/ ORDERING OF MAGNESIUM AND NIOBIUM IN THE OCTAHEDRAL POSITIONS OF THE "CUBIC" PEROVSKITE STRUCTURE 08-1379
/FERROELECTRIC WITH NONREVERSIBLE, BUT TILTABLE (TO 3 STABLE POSITIONS, SPONTANEOUS) POLARIZATION (ON APPL/ 08-1346
/D SOLUTIONS LEAD MAGNESATE(0.5) TUNGSTATE(0.5)), EFFECTS OF POSITIVE OR NEGATIVE HIGH DC FIELDS SUMMARY O/ 05-1063
                SEE BOTH POSITIVE TEMPERATURE COEFFICIENT AND PTC.
AND PTCR.   SEE BOTH POSITIVE TEMPERATURE COEFFICIENT OF RESISTANCE
) BEHAVIOR OF BARIUM TITANATE WITH NIOBIUM PENTOXIDE / PTCR (POSITIVE TEMPERATURE COEFFICIENT OF RESISTANCE 04-0837
IN DOPED LEAD TITANATE.   POSITIVE TEMPERATURE COEFFICIENT OF RESISTANCE 05-1051
OF BARIUM TITANATE SINGLE CRYSTALS IN THE REGION OF THE CU/ POSITIVE TEMPERATURE COEFFICIENT OF RESISTANCE 04-1036
Y)   RESISTIVITY ANOMALY IN SEMICONDUCTING BARIUM TITANATE. (POSITIVE TEMPERATURE COEFFICIENT OF RESISTIVIT 04-0584
Y) IN (DOPED AND UNDOPED CERAMIC PEROVSKITE) LE/ PTC EFFECT (POSITIVE TEMPERATURE COEFFICIENT OF RESISTIVIT 13-2000
/RICAL CONDUCTIVITY MEASUREMENTS. (INTRINSIC BAND GAP, PTCR (POSITIVE TEMPERATURE COEFFICIENT OF RESISTIVI/ 04-0822
TY.   SEMICONDUCTING BARIUM AND STRONTIUM TITANATES WITH POSITIVE TEMPERATURE COEFFICIENTS OF RESISTIVI 04-0995
SALT.   INFLUENCE OF THE DOMAIN STRUCTURE ON POSITRON (TWO-PHONON) ANNIHILATION IN ROCHELLE 28-3191
TRY ASPECTS OF FERROELECTRICITY. (DESCRIPTIVE AND PREDICTIVE POSSIBILITIES)   (MACROSCOPIC) SYMME 03-0464
ZIG REGIONS IN FERROELECTRICS.   SOME POSSIBILITIES OF DETERMINING THE VOLUME OF KAN 03-0444
ATES, PHASE REL/ TUNGSTEN BRONZE FIELD IN THE (OXIDE) SYSTEM POTASSIA LITHIA NIOBATE (POTASSIA LITHIUM NIOB 13-2017
/ELECTRIC AND CERAMIC PROPERTIES OF (BISMUTH(0.5) (SODIUM OR POTASSIUM) (0.5) (X) LEAD(1-X) TITANATE SYSTEMS 13-2013
SOME CHEMICAL REACTIONS IN SILICA GELS. PART-3: FORMATION OF POTASSIUM ACID TARTRATE CRYSTALS.   29-3246
/PERATURE VARIATION OF THE MEAN MAGNETIC MOMENT OF CHROMIUM- POTASSIUM ALUM (UNSTABLE SOFT PHONON MODE, DI/ 20-2806
/ND LIQUID NITROGEN TEMPERATURES IN ALKALI NITRITES (SODIUM, POTASSIUM) AND ALKALI NITRATES. (LITHIUM, SOD/ 16-2399
        PHOTOELASTIC EFFECT IN 45 DEGREE X-CUTS OF POTASSIUM AND AMMONIUM DIHYDROGEN PHOSPHATE. 17-2443
    DIELECTRIC PERMITTIVITY OF FINELY DIVIDED POTASSIUM AND AMMONIUM DIHYDROGEN PHOSPHATE. 17-2689
/CERAMICS OF LEAD ZIRCONATE TITANATE MODIFIED WITH SODIUM OR POTASSIUM ANTIMONATE, NIOBATE, OR BISMUTHATE,/ 09-1539
TE, (B/ COMPLEX OXIDES WITH TUNGSTEN BRONZE TYPE STRUCTURES (POTASSIUM (BARIUM OR STRONTIUM) TITANATE NIOBA 13-1962
UTRON MEASUREMENT IN TWO SYSTEMS OF HYDROGEN BONDS. (KDP AND POTASSIUM BICARBONATE)   NE 13-2417
    PREPARATION AND DIELECTRIC PROPERTIES OF POTASSIUM BISMUTH ZINC NIOBATE.   13-2001
/DYNAMICS OF DIATOMIC CRYSTALS OF HIGH DIELECTRIC CONSTANT. (POTASSIUM BROMIDE, LEAD TELLURIDE, TIN TELLUR/ 36-3541
CHLORIDE,/ CRYSTALLIZATION POTENTIALS IN AQUEOUS SOLUTIONS. (POTASSIUM BROMIDE, POTASSIUM CHLORIDE, SODIUM 27-2707
PERTIES OF POTASSIUM DIHYDROGEN PH/ EFFECT OF REPLACEMENT OF POTASSIUM BY ALUMINUM ON THE FERROELECTRIC PRO 17-2560
    PARAELECTRIC RESONANCE OF LITHIUM ION IN POTASSIUM CHLORIDE.   31-3325
            DIELECTRIC PROPERTIES OF POTASSIUM CHLORIDE ICE.   34-3454
/RIC DIPOLE SYSTEMS (SPIN WAVES IN HYDROXYL DIPOLE SYSTEM IN POTASSIUM CHLORIDE LATTICE, FERROELECTRIC STA/ 03-0452
/THERMAL CONDUCTIONS IN THE LOW DIMENSIONAL ANTIFERROMAGNETS POTASSIUM COPPER TRIFLUORIDE AND COPPER FORMA/ 26-2939
TRIC TRANSITION TEMPERATURE.   ACOUSTIC PROPAGATION IN POTASSIUM DEUTERIUM PHOSPHATE AT THE FERROELEC 17-2573
    FIRST ORDER PHASE TRANSITION IN POTASSIUM DEUTERIUM PHOSPHATE CRYSTALS.   17-2677
ICNS IN IRRADIATED AMMONIUM DIDEUTERIUM PHOSPHATE- 5 PERCENT POTASSIUM DIDEUTERIUM ARSENATE. / OF AMMONIUM 17-2684
            FERROELECTRIC TRANSITION IN POTASSIUM DIDEUTERIUM PHOSPHATE.   17-2611
R INDUCED SELF FOCUSING IN AMMONIUM DIHYDROGEN PHOSPHATE AND POTASSIUM DIDEUTERIUM PHOSPHATE.   LASE 17-2476
    LIGHT SCATTERING FROM CRITICAL FLUCTUATIONS IN POTASSIUM DIDEUTERIUM PHOSPHATE.   17-2463
ACTION STUDIES OF FERROELECTRIC PHASE TRANSITIONS IN KDP AND POTASSIUM DIDEUTERIUM PHOSPHATE. NEUTRON DIFFR 17-2661
    PRESSURE DEPENDENCE OF THE FERROELECTRIC PROPERTIES OF POTASSIUM DIDEUTERIUM PHOSPHATE.   17-2641
OBSERVATION OF THE SOFT FERROELECTRIC AND ACOUSTIC MODES OF POTASSIUM DIDEUTERIUM PHOSPHATE. SIMULTANEOUS 17-2636
            THE CRYSTAL STRUCTURE OF MONOCLINIC POTASSIUM DIDEUTERIUM PHOSPHATE.   17-2591
ULTRASONIC STRESS WAVE INTERACTION WITH DOMAIN WALLS IN POTASSIUM DIDEUTERIUM PHOSPHATE.   17-2659

/ND SUSCEPTIBILITY IN THE FERROELECTRIC TRANSITION REGION OF POTASSIUM DIDEUTERIUM PHOSPHATE (ANALYSIS AND/ 17-2575
AMMONIU/ DETERMINATION OF THE EIGENVECTORS OF SOFT MODES IN POTASSIUM DIDEUTERIUM PHOSPHATE AND DEUTERATED 17-2700
/LE THEORETICAL EXPLANATION OF SHIFT OF CURIE TEMPERATURE IN POTASSIUM DIDEUTERIUM PHOSPHATE AND DEUTERO P/ 17-2601
TRIPLE AXIS SPECTROMETR/ FERROELECTRIC (SOFT) MODE MOTION IN POTASSIUM DIDEUTERIUM PHOSPHATE . (BY NEUTRON 17-2668
/ATTERING BY THE TEMPERATURE DEPENDENT FERROELECTRIC MODE IN POTASSIUM DIDEUTERIUM PHOSPHATE (COMPARED WIT/ 17-2705
A- AND FERROELECTRIC PHASE.   RAMAN SPECTRUM OF POTASSIUM DIDEUTERIUM PHOSPHATE CRYSTAL IN PAR 17-2647
CHARACTERISTICS OF AN ELECTROOPTICAL SHUTTER BASED ON Z CUT POTASSIUM DIDEUTERIUM PHOSPHATE CRYSTALS.   17-2460
                      DOUBLE DIELECTRIC HYSTERESIS LOOP OF POTASSIUM DIDEUTERIUM PHOSPHATE CRYSTALS.   17-2503
                      LANDAU-KHALATNIKOV ATTENUATION IN POTASSIUM DIDEUTERIUM PHOSPHATE CRYSTALS.   17-2656
Q SWITCH BY DESK-T/ SIMULATION OF A DOUBLE 45 DEGREES, Z CUT POTASSIUM DIDEUTERIUM PHOSPHATE. ELECTROOPTIC 17-2702
D REFLECTION SPECTRA, A TUNNELING MODEL/ DEUTERON MOTIONS IN POTASSIUM DIDEUTERIUM PHOSPHATE (FROM POLARIZE 17-2645
OR HIGH POWER Q SWITCHING.   LOSSLESS POTASSIUM DIDEUTERIUM PHOSPHATE POCKELS CELL F 17-2529
/ EFFECTS AND Q SWITCHING PERFORMANCE IN LITHIUM NIOBATE AND POTASSIUM DIDEUTERIUM PHOSPHATE POCKELS CELLS/ 11-1710
DEUTERON MAGNETIC RESONANCE AND RELAXATION IN FERROELECTRIC POTASSIUM DIDEUTERIUM PHOSPHATE, POTASSIUM DI/ 17-2438
E/ ANOMALOUS PROPAGATION OF LONGITUDINAL ULTRASONIC WAVES IN POTASSIUM DIDEUTERIUM PHOSPHATE (TEMPERATURE D 17-2660
/ATTERING STUDIES OF PHASE TRANSITIONS. (TRIGLYCINE SULFATE, POTASSIUM DIHYDROGEN AND DIDEUTERIUM SULFATE) 17-2481
LEAR DOUBLE RESONANCE OF THE ARSENATE CENTER IN X-IRRADIATED POTASSIUM DIHYDROGEN ARSENATE. /D ELECTRON NUC 18-2727
TRIC MODE AND NQR (NUCLEAR QUADRUPOLE RESONANCE) FREQUENCY. (POTASSIUM DIHYDROGEN ARSENATE)   FERROELEC 17-2713
ELAXATION STUDIES OF LONG RANGE CORRELATION IN FERROELECTRIC POTASSIUM DIHYDROGEN ARSENATE. /D DIELECTRIC R 18-2718
R THE ARSENATE CENTER IN X-RAY IRRADIATED SINGLE CRYSTALS OF POTASSIUM DIHYDROGEN ARSENATE. /E RESONANCE FO 18-2729
PROTON PHONON COUPLING IN CESIUM DIHYDROGEN ARSENATE AND POTASSIUM DIHYDROGEN ARSENATE.   18-2733
                      PROTONIC CONDUCTION IN POTASSIUM DIHYDROGEN ARSENATE.   18-2737
                      SEE BOTH KDA AND POTASSIUM DIHYDROGEN ARSENATE.
/ FOR SLATER CONFIGURATIONS FROM EPR STUDIES OF X-IRRADIATED POTASSIUM DIHYDROGEN ARSENATE AND AMMONIUM DI/ 18-2728
/IC RESPONSE IN PIEZOELECTRIC CRYSTALS. (RAMAN SCATTERING IN POTASSIUM DIHYDROGEN ARSENATE AND CESIUM DIHY/ 18-2723
/XCHANGE OF THE ARSENATE RADICAL IN IRRADIATED FERROELECTRIC POTASSIUM DIHYDROGEN ARSENATE AND POTASSIUM D/ 18-2734
/CE OBSERVATIONS OF DOMAINS IN THE IRRADIATED FERROELECTRICS POTASSIUM DIHYDROGEN ARSENATE AND POTASSIUM D/ 18-2726
DEUTERIUM ARSENATE.   THERMODYNAMIC PROPERTIES OF POTASSIUM DIHYDROGEN ARSENATE AND POTASSIUM DI 18-2730
/ SOFT MODE COUPLING TO TRANSVERSE OPTICAL PHONON). PART-3: (POTASSIUM DIHYDROGEN ARSENATE, CESIUM DIHYDRO/ 03-0384
/ED IR RADIATION. PART-2: INTERNAL FUNDAMENTAL VIBRATIONS OF POTASSIUM DIHYDROGEN ARSENATE IN THE PARAELEC/ 03-0384
/F RADIATION DAMAGE CENTERS IN AMMONIUM DIHYDROGEN PHOSPHATE POTASSIUM DIHYDROGEN ARSENATE MIXED CRYSTALS. 17-2547
/LEAR DOUBLE RESONANCE OF THE PROTON LATTICE COUPLED MODE IN POTASSIUM DIHYDROGEN ARSENATE, POTASSIUM DEUT/ 18-2724
TER/ ELECTRON PARAMAGNETIC RESONANCE STUDIES OF X-IRRADIATED POTASSIUM DIHYDROGEN ARSENATE, POTASSIUM DIDEU 18-2725
00 DEG / PHASE DIAGRAMS OF AMMONIUM DIHYDROGEN PHOSPHATE AND POTASSIUM DIHYDROGEN ARSENATE TO 52 KBAR AND 4 17-2478
NFRARED PHOSPHATE/ OPTICAL STUDY ON THE PHASE TRANSITIONS IN POTASSIUM DIHYDROGEN OR DEUTERIUM PHOSPHATE (I 17-2646
ABSOLUTE SIGNS OF THE NONLINEAR OPTICAL COEFFICIENTS OF POTASSIUM DIHYDROGEN PHOSPHATE.   17-2597
DIELECTRIC EVIDENCE OF A FIRST ORDER TRANSITION IN POTASSIUM DIHYDROGEN PHOSPHATE.   17-2602
DIELECTRIC PROPERTIES OF DEUTERATED POTASSIUM DIHYDROGEN PHOSPHATE.   17-2581
DEUTERONS) IN FERROELECTRIC CRYSTALS WITH HYDROGEN BONDING. (POTASSIUM DIHYDROGEN PHOSPHATE) /PROTONS (AND 17-2568
EFFECT OF DEUTERON SUBSTITUTION ON DOMAIN DYNAMICS IN POTASSIUM DIHYDROGEN PHOSPHATE.   17-2437
ND ELECTRIC FIELD ON PARAMAGNETIC SPECTRA OF CHROMIUM(3+) IN POTASSIUM DIHYDROGEN PHOSPHATE. /XIAL STRESS A 17-2554
CAL AND OPTICAL INVESTIGATION OF FERROELECTRIC PROPERTIES OF POTASSIUM DIHYDROGEN PHOSPHATE.   ELECTRI 17-2449
ELECTROCHEMISTRY ON THE POLYCRYSTAL OF POTASSIUM DIHYDROGEN PHOSPHATE.   17-2536
ELECTRON DELOCALIZATION WEIGHT OF THE HYDROGEN BOND IN POTASSIUM DIHYDROGEN PHOSPHATE.   17-2525
ELECTRON SPIN RESONANCE OF IRRADIATED AND DEUTERATED POTASSIUM DIHYDROGEN PHOSPHATE.   17-2533
AMPLITUDE MODULATION OF LASER GENERATED SECOND HARMONICS IN POTASSIUM DIHYDROGEN PHOSPHATE.   ELECTROOPTIC 17-2694
GATIONS OF X-RAY IRRADIATED SINGLE CRYSTALS OF FERROELECTRIC POTASSIUM DIHYDROGEN PHOSPHATE.   ESR INVESTI 17-2558
ENCE OF DISLOCATIONS AT DOMAIN TIPS IN FERROELECTRIC CRYSTAL POTASSIUM DIHYDROGEN PHOSPHATE.   EXIST 17-2450
EXPERIMENTAL PROOF FOR PROTON TUNNELING IN POTASSIUM DIHYDROGEN PHOSPHATE.   17-2510
LIGHT SWITCH USING) TRANSVERSE POCKELS EFFECT COMMUTATORS. (POTASSIUM DIHYDROGEN PHOSPHATE) /NITE APERTURE 17-2509
OR OF HYDROGEN BONDED FERROELECTRICS: TRIGLYCINE SULFATE AND POTASSIUM DIHYDROGEN PHOSPHATE. /QUENCY BEHAVI 17-2526
INDEX OF REFRACTION OF POTASSIUM DIHYDROGEN PHOSPHATE.   17-2484
INFLUENCE OF TUNNELING ON DIELECTRIC BEHAVIOR OF POTASSIUM DIHYDROGEN PHOSPHATE.   17-2527
INTERACTION BETWEEN PROTON TUNNELING AND OPTICAL PHONONS IN POTASSIUM DIHYDROGEN PHOSPHATE.   17-2593
INTERACTIONS BETWEEN PROTON TUNNELING AND OPTICAL PHONONS IN POTASSIUM DIHYDROGEN PHOSPHATE.   17-2592
INTERNAL STRESS AND THE FERROELECTRIC TRANSITION IN POTASSIUM DIHYDROGEN PHOSPHATE.   17-2424
LOW FREQUENCY HYDROGEN VIBRATIONS IN POTASSIUM DIHYDROGEN PHOSPHATE.   17-2699
LOW FREQUENCY HYDROGEN VIBRATIONS IN POTASSIUM DIHYDROGEN PHOSPHATE.   17-2608
TEMPERATURE THERMAL EXPANSION MEASUREMENTS ON SINGLE CRYSTAL POTASSIUM DIHYDROGEN PHOSPHATE.   LOW 17-2464
MACROSCOPIC STIMULATED POLARIZATION ECHO IN FERROELECTRICS. (POTASSIUM DIHYDROGEN PHOSPHATE)   17-2419
MEASUREMENT OF CHEMICAL SHIFT IN SOLID POTASSIUM DIHYDROGEN PHOSPHATE.   17-2688
MEASUREMENTS OF THE SPONTANEOUS POLARIZATION IN POTASSIUM DIHYDROGEN PHOSPHATE.   17-2430
N OF SINGLE CRYSTALS BY THE GRADUAL ADDITION OF A REACTANT. (POTASSIUM DIHYDROGEN PHOSPHATE) /TH IN SOLUTIO 17-2595
MODE INTERACTIONS IN POTASSIUM DIHYDROGEN PHOSPHATE.   17-2701
MEASUREMENTS ON THE HYDROGEN BOND POTENTIAL IN PARAELECTRIC POTASSIUM DIHYDROGEN PHOSPHATE.   NEUTRON 17-2511
ON MEASUREMENTS ON THE FERROELECTRIC PHASE TRANSFORMATION IN POTASSIUM DIHYDROGEN PHOSPHATE.   NEUTR 17-2418
LEAR DOUBLE RESONANCE STUDY OF POTASSIUM-39 IN FERROELECTRIC POTASSIUM DIHYDROGEN PHOSPHATE.   NUC 17-2599
LATIVE VALUES FOR QUARTZ, AMMONIUM DIHYDROGEN PHOSPHATE, AND POTASSIUM DIHYDROGEN PHOSPHATE. /: ACCURATE RE 17-2539
ORDER OF THE FERROELECTRIC PHASE TRANSFORMATION OF POTASSIUM DIHYDROGEN PHOSPHATE.   17-2555
Y CONVERSION OF COHERENT LIGHT BY THE ELECTROOPTIC EFFECT IN POTASSIUM DIHYDROGEN PHOSPHATE. /TRIC FREQUENC 17-2616
POLARIZATION ECHO IN THE FERROELECTRIC SINGLE CRYSTAL POTASSIUM DIHYDROGEN PHOSPHATE.   17-2564
PROTON PHONON INTERACTIONS IN POTASSIUM DIHYDROGEN PHOSPHATE.   17-2653
PULSE DISTORTIONS IN MISMATCHED SECOND HARMONIC GENERATION. (POTASSIUM DIHYDROGEN PHOSPHATE)   17-2690
RAMAN SCATTERING FROM FERROELECTRIC CRYSTALS OF POTASSIUM DIHYDROGEN PHOSPHATE.   17-2552
RAMAN SCATTERING IN POTASSIUM DIHYDROGEN PHOSPHATE.   17-2709
RAMAN SPECTRA OF TETRAGONAL POTASSIUM DIHYDROGEN PHOSPHATE.   17-2657
RANDOM MOTION OF DEUTERONS IN DEUTERATED POTASSIUM DIHYDROGEN PHOSPHATE.   17-2651
SEE BOTH KDP AND POTASSIUM DIHYDROGEN PHOSPHATE.
SPIN LATTICE RELAXATION BY THE FERROELECTRIC MODE IN POTASSIUM DIHYDROGEN PHOSPHATE.   17-2440
SYMMETRY OF ATOMIC VIBRATIONS IN POTASSIUM DIHYDROGEN PHOSPHATE.   17-2586
THERMAL HYSTERESIS OF THE FERROELECTRIC TRANSITION IN POTASSIUM DIHYDROGEN PHOSPHATE.   17-2682
/RESS FOR FERROELECTRIC PHOSPHATES NEAR THEIR CURIE POINTS. (POTASSIUM DIHYDROGEN PHOSPHATE, ALSO RUBIDIUM/ 17-2665
ROGEN/ RELATIVE MEASUREMENT OF THE OPTICAL NONLINEARITIES OF POTASSIUM DIHYDROGEN PHOSPHATE, AMMONIUM DIHYD 11-1677
/ENCIES) OF SOME FERROELECTRIC CRYSTALS (TRIGLYCINE SULFATE, POTASSIUM DIHYDROGEN PHOSPHATE, AMMONIUM SULF/ 02-0125
HYDROGEN PHOS/ ELECTROOPTICAL AND PIEZOOPTICAL PROPERTIES OF POTASSIUM DIHYDROGEN PHOSPHATE AND AMMONIUM DI 17-2697
HYDROGEN PHOSPHATE CRYSTALS AT MIC/ RELATIVE PERMITTIVITY OF POTASSIUM DIHYDROGEN PHOSPHATE AND AMMONIUM DI 17-2714
HYDROGEN PHOSPHATE CRYSTALS OF LIG/ THERMAL SELF FOCUSING IN POTASSIUM DIHYDROGEN PHOSPHATE AND AMMONIUM DI 17-2711
HYDROGEN PHOSPHATE/ ULTRAVIOLET REFLECTION AND ABSORPTION OF POTASSIUM DIHYDROGEN PHOSPHATE AND AMMONIUM DI 17-2422
/ATURE DEPENDENCE OF OPTICAL (SECOND) HARMONIC GENERATION IN POTASSIUM DIHYDROGEN PHOSPHATE (AND DEUTERATE/ 17-2695
/OOPTIC LASER MODULATORS WITH AMMONIUM DIHYDROGEN PHOSPHATE, POTASSIUM DIHYDROGEN PHOSPHATE AND PARTIALLY / 17-2675

HATE SINGLE CRYSTALS. REGULAR BEHAVIOR OF SOLID SOLUTIONS IN POTASSIUM DIHYDROGEN(1-X) DIDEUTERIUM(X) PHOSP 17-2458
STRUCTURAL DI- AND FERROELECTRIC PROPERTIES OF X POTASSIUM DINIOBATE AND (1-X) BARIUM TITANATE. 04-0623
DOMAIN STRUCTURE IN THE FERROELECTRIC POTASSIUM FERROCYANIDE. 24-2895
/TS OF WATER MOLECULES AND FERROELECTRIC PHASE TRANSITION IN POTASSIUM FERROCYANIDE AND ISOMORPHOUS CRYSTA/ 24-2893
E THE CURIE TEMP/ NEUTRON DIFFRACTION STUDY OF FERROELECTRIC POTASSIUM FERROCYANIDE TRIDEUTERO HYDRATE ABOV 24-2903
ANOMALOUS SPIN LATTICE RELAXATION BY QUASI SPIN WAVES IN POTASSIUM FERROCYANIDE TRIHYDRATE. 24-2899
ANOMALOUS SPECIFIC HEAT CF FERROELECTRIC THIOUREA AND POTASSIUM FERROCYANIDE TRIHYDRATE. 24-2902
NTERNAL FIELD IN GENERAL DIPOLE LATTICES. (THEORY APPLIED TO POTASSIUM FERROCYANIDE TRIHYDRATE) I 03-0220
RIC TRANSITIONS IN FERRIC AMMONIUM SULFATE DODECAHYDRATE AND POTASSIUM FERROCYANIDE TRIHYDRATE. /FERROELECT 19-2755
RIC TRANSITIONS IN FERRIC AMMONIUM SULFATE DODECAHYDRATE AND POTASSIUM FERROCYANIDE TRIHYDRATE. /FERROELECT 24-2892
MOSSBAUER STUDY OF THE FERROELECTRIC PHASE TRANSITION IN POTASSIUM FERROCYANIDE TRIHYDRATE. 24-2901
MOSSBAUER RESONANCE OF SINGLE CRYSTAL POTASSIUM FERROCYANIDE TRIHYDRATE. 24-2890
OUS COMPOUNDS.   RADIO SPECTROSCOPIC INVESTIGATION OF POTASSIUM FERROCYANIDE TRIHYDRATE AND ISOMORPH 24-2900
R.   STUDY OF HYDROGEN-1 IN PARAELECTRIC AND FERROELECTRIC POTASSIUM FERROCYANIDE TRIHYDRATE BY PULSED NM 24-2889
ETHOD.   POLYTYPE AND TWINNING OF THE FERROELECTRIC CRYSTALS POTASSIUM FERROCYANIDE TRIHYDRATE BY THE NMR M 24-2898
R HIGH HYDROSTATIC PRESSURE (/ FERROELECTRIC TRANSITION IN A POTASSIUM FERROCYANIDE TRIHYDRATE CRYSTAL UNDE 24-2896
NUCLEAR MAGNETIC RESONANCE AND X-RAY STUDIES OF POTASSIUM FERROCYANIDE TRIHYDRATE CRYSTALS. 24-2894
HIGH PRESSURE. (TO 5100 KG/SQ CM)   FERROELECTRIC BEHAVIOR OF POTASSIUM FERROCYANIDE TRIHYDRATE CRYSTALS AT 24-2897
ANSITION TEMPERATUR/ MOSSBAUER EFFECT STUDY OF FERROELECTRIC POTASSIUM FERROCYANIDE TRIHYDRATE (NEAR THE TR 24-2891
/ION OF THE DEW METHOD FOR REVEALING THE DOMAIN STRUCTURE IN POTASSIUM FERROCYANIDE TRIHYDRATE SINGLE CRYS/ 24-2904
/CTERISTICS OF THEIR GROWTH. (AMMONIUM DIHYDROGEN PHOSPHATE, POTASSIUM FERROCYANIDE TRIHYDRATE, SODIUM NIT/ 16-2316
/ONANCE STUDIES OF THE PHASE TRANSITIONS IN FERROELECTRICS. (POTASSIUM FERROCYANIDE TRIHYDRATE, SODIUM NIT/ 16-2255
E FINE STRUCTURE OF BARIUM TITANATE CRYSTALS FLUX GROWN FROM POTASSIUM FLUORIDE). INVESTIGATION OF TH 04-0708
RIUM(1-X) STRONTIUM(X) TITANATE SINGLE CRYSTALS GROWN FROM A POTASSIUM FLUORIDE FLUX. /RIC PROPERTIES OF BA 04-0635
ON THE CRYSTAL GROWTH OF BARIUM TITANATE SINGLE CRYSTALS IN POTASSIUM FLUORIDE FLUX.   DISCUSSION 04-1022
WITH LARGE NONLINEARITY. (BARIUM TITANATE HAFNATE GROWN FROM POTASSIUM FLUORIDE FLUX) /RIC SINGLE CRYSTALS 04-1373
OF BARIUM TITANATE SINGLE CRYSTALS GROWN FROM SOLUTION IN A POTASSIUM FLUORIDE MELT. /CTRICAL CONDUCTIVITY 04-1035
F (BARIUM(1-X) STRONTIUM(X)) TITANATE- MIXED CRYSTALS FROM A POTASSIUM FLUORIDE MELT.   DEFECT STRUCTURE O 04-0873
STRUCTURE OF BARIUM TITANATE CRYSTALS GROWN FROM SOLUTION IN POTASSIUM FLUORIDE MELT. /H CONDITIONS ON THE 04-0598
NOTE ON THE SPACE GROUP OF POTASSIUM HYDROGEN IODATE. 08-1370
DIELECTRIC DISPERSION AND PHASE TRANSITION OF POTASSIUM HYDROXIDE DOPED ICE. 34-3452
CRYSTALLIZATION OF POTASSIUM IODATE. 08-1367
ELECTRICAL AND OPTICAL CONSTANTS OF FERROELECTRIC POTASSIUM IODATE. 08-1427
PECULIAR FERROELECTRIC BEHAVIOR OF POTASSIUM IODATE. 08-1347
PIEZOELECTRIC AND ELECTROOPTIC MODULI OF FERROELECTRIC POTASSIUM IODATE. 08-1431
(DEPENDENCE OF DIELECTRIC CONSTANT ON THERMAL AND FIELD HI/ POTASSIUM IODATE - A DISORDERED FERROELECTRIC. 08-1348
RSIBLE, BUT TILTABLE (TO 3 STABLE POSITIONS, SPONTANEOUS) P/ POTASSIUM IODATE, A FERROELECTRIC WITH NONREVE 08-1346
INFRARED SPECTROSCOPY AND NUCLEAR QUADRUPOLE RESONANCE. POTASSIUM IODATE CRYSTAL STRUCTURE STUDIED BY 08-1278
POLAR AND LATTICE DYNAMIC CHARACTERISTICS OF FERROELECTRIC POTASSIUM IODATE CRYSTALS. 08-1430
INDICES) RELATION OF PHYSICAL PROPERTIES TO THE SYMMETRY OF POTASSIUM IODATE. (NONLINEAR OPTIC, REFRACTIVE 08-1296
IOR)   FERROELECTRICITY OF POTASSIUM IODATE. (UNUSUAL FERROELECTRIC BEHAV 08-1345
TYPE CRYSTALS GROWN IN MOLTEN SOLUTIONS OF THE PSEUDOSYSTEM POTASSIUM LITHIUM BARIUM NIOBATE. /STEN BRONZE 13-1948
RROELECTRIC TUNGSTEN BRONZE TYPE CRYSTAL STRUCTURES. PART-3: POTASSIUM LITHIUM NIOBATE. FE 13-1895
LT, STRUCTURE- PROPERTY CORRELATION WITH LITHIUM NIOBATE AND POTASSIUM LITHIUM NIOBATE) /ULLING FROM THE ME 13-2045
NONLINEAR OPTICAL PROPERTIES OF POTASSIUM LITHIUM NIOBATE. 13-2027
/ NIOBIUM- OXYGEN BOND. (LITHIUM NIOBATE, LITHIUM TANTALATE, POTASSIUM LITHIUM NIOBATE, BARIUM SODIUM NIOB/ 11-1734
/RMALLY STIMULATED CURRENT IN FERROELECTRIC OXIDE CRYSTALS. (POTASSIUM LITHIUM NIOBATE, POTASSIUM STRONTIU/ 13-1926
PE FERROELECTRICS). ELECTROOPTICAL PROPERTIES OF STRONTIUM POTASSIUM LITHIUM NIOBATE. (TUNGSTEN BRONZE TY 13-2051
RYSTAL. ELASTIC AND PIEZOELECTRIC CONSTANTS OF STRONTIUM(4) POTASSIUM LITHIUM NIOBATE TYPE FERROELECTRIC C 13-1946
CRYSTAL GROWTH AND NONLINEAR OPTICAL PROPERTIES OF POTASSIUM LITHIUM NIOBATES. 13-2015
/ BRONZE FIELD IN THE (OXIDE) SYSTEM POTASSIA LITHIA NIOBIA (POTASSIUM LITHIUM NIOBATES, PHASE RELATIONSHI/ 13-2017
GONAL TUNGSTEN BRONZE STRUC/ NONLINEAR OPTICAL PROPERTIES OF POTASSIUM LITHIUM NIOBATES (TRANSPARENT, TETRA 13-2026
GROWTH AND PROPERTIES OF FERROELECTRIC POTASSIUM LITHIUM TANTALATE(X) NIOBATE(1-X). 11-1711
CRYSTALS IN THE TUNGSTEN BRONZE TYPE) FERROELECTRIC (SYSTEM) POTASSIUM LITHIUM TANTALATE(X) NIOBATE(1-X). / 13-1947
.5) SINGLE CRYSTA/ PIEZOELECTRIC PROPERTIES OF FERROELECTRIC POTASSIUM LITHIUM TANTALATE(0-0.5) NIOBATE(1-0 12-1860
FFECTS INVOLVING THE FLUORINE-19 NUCLEI IN ANTIFERROMAGNETIC POTASSIUM MANGANESE TRIFLUORIDE. /D MAGNETIC E 33-3390
COVALENCY IN POTASSIUM MANGANESE TRIFLUORIDE. 33-3391
CRITICAL NEUTRON SCATTERING IN STRONTIUM TITANATE AND POTASSIUM MANGANESE TRIFLUORIDE. 06-1208
DISPERSION AND DAMPING OF SOFT ZONE BONDARY PHONONS IN POTASSIUM MANGANESE TRIFLUORIDE. 33-3374
FINE STRUCTURE OF EXCITON- MAGNON ABSORPTION OF LIGHT IN POTASSIUM MANGANESE TRIFLUORIDE. 33-3364
SCATTERING BY MAGNONS IN NICKEL DOPED RUBIDIUM MANGANATE AND POTASSIUM MANGANESE TRIFLUORIDE. /CE OF RAMAN 33-3395
INFRARED ACTIVE MODE BELOW THE STRUCTURAL TRANSITION IN POTASSIUM MANGANESE TRIFLUORIDE. 33-3362
LUMINESCENCE OF EUROPIUM(3+) ION IN ANTIFERROMAGNETIC POTASSIUM MANGANESE TRIFLUORIDE. 33-3381
MAGNETIC EXCITATIONS IN NICKEL DOPED POTASSIUM MANGANESE TRIFLUORIDE. 33-3407
MAGNETOELASTIC EFFECTS IN POTASSIUM MANGANESE TRIFLUORIDE. 33-3389
ATTERING STUDY OF THE LATTICE DYNAMICAL PHASE TRANSITIONS IN POTASSIUM MANGANESE TRIFLUORIDE.   NEUTRON SC 33-3405
OBSERVATION OF THE ANOMALY OF THE THERMAL CONDUCTIVITY OF POTASSIUM MANGANESE TRIFLUORIDE. 33-3382
RESONANCE BEHAVIOR IN CANTED ANTIFERROMAGNETIC POTASSIUM MANGANESE TRIFLUORIDE. 33-3403
SPECIFIC HEAT OF POTASSIUM MANGANESE TRIFLUORIDE. 33-3387
SUPERSTRUCTURE ARISING DURING THE PHASE TRANSITION IN POTASSIUM MANGANESE TRIFLUORIDE. 33-3402
THEORY OF SUPEREXCHANGE REACTION. PART-1: POTASSIUM MANGANESE TRIFLUORIDE. 03-0252
TWO MAGNON SCATTERING OF LIGHT IN ANTIFERROMAGNETIC POTASSIUM MANGANESE TRIFLUORIDE. 33-3401
/ND DISPERSION OF THE OPTICAL CONSTANTS OF ANTIFERROMAGNETIC POTASSIUM MANGANESE TRIFLUORIDE AND POTASSIUM/ 33-3385
/SCATTERING OF LIGHT IN MANGANESE FLUORIDE, COBALT FLUORIDE, POTASSIUM MANGANESE TRIFLUORIDE AND RUBIDIUM / 33-3399
BATE SIN/ ANISOTROPIC CRITICAL X-RAY DIFFUSE SCATTERING FROM POTASSIUM MANGANESE TRIFLUORIDE AND SODIUM NIO 08-1312
/TRASONIC ATTENUATION NEAR THE SOFT MODE TRANSITION POINT IN POTASSIUM MANGANESE TRIFLUORIDE (ANOMALOUS HE/ 33-3373
K.   X-RAY SCATTERING AND THE PHASE TRANSITION OF POTASSIUM MANGANESE TRIFLUORIDE AT 184 DEGREES 33-3394
RAMAN SCATTERING AND PHASE TRANSITIONS IN A POTASSIUM MANGANESE TRIFLUORIDE CRYSTAL. 33-3400
RAMAN SCATTERING AND PHASE TRANSITIONS IN POTASSIUM MANGANESE TRIFLUORIDE CRYSTALS. 33-3372
RHOOD OF / ATTENUATION AND DISPERSION OF ULTRASONIC WAVES IN POTASSIUM MANGANESE TRIFLUORIDE IN THE NEIGHBO 33-3370
ANGE ORDERED REGIO/ PARAMAGNETIC SCATTERING OF NEUTRONS FROM POTASSIUM MANGANESE TRIFLUORIDE IN THE SHORT R 33-3371
RAL PHASE TRANSITION.   ANOMALOUS ELASTIC BEHAVIOR OF POTASSIUM MANGANESE TRIFLUORIDE NEAR A STRUCTU 33-3392
ALT TRIFLUORIDE, AND RUBIDIU/ INFRARED ABSORPTION SPECTRA OF POTASSIUM MANGANESE TRIFLUORIDE, POTASSIUM COB 33-3384
/ RESONANCE LINE SHAPES AND SPIN CORRELATION IN PARAMAGNETIC POTASSIUM MANGANESE TRIFLUORIDE, RUBIDIUM MAG/ 33-3376
/CAL ANISOTROPIC FLUCTUATIONS AT THE 184 DEG K TRANSITION OF POTASSIUM MANGANESE TRIFLUORIDE SINGLE CRYSTA/ 33-3369
/RMINATION OF LARGE, PURE RUBIDIUM MANGANESE TRIFLUORIDE AND POTASSIUM MANGANESE TRIFLUORIDE SINGLE CRYSTA/ 33-3397
ISCONTINUITIE/ TWO PHONON PROCESSES AND PHASE TRANSITIONS IN POTASSIUM MANGANESE TRIFLUORIDE: TEMPERATURE D 33-3363
DOPING.   SHIFT OF NEEL TEMPERATURE AND EPR LINE WIDTH OF POTASSIUM MANGANESE TRIFLUORIDE WITH MAGNESIUM 33-3377
ANGANESE RE/ ANTIFERROELECTRICITY IN (PURE AND) COBALT DOPED POTASSIUM MANGANESE TRIFLUORIDE. (20 PERCENT M 33-3366
INFRARED AND RAMAN SPECTROSCOPY OF POTASSIUM METANIOBATE. 08-1303

TE, POTASSIUM MOLYBDATE, AND POTASSIUM TAN/ ENERGY BANDS FOR POTASSIUM NICKEL TRIFLUORIDE, STRONTIUM TITANA 02-0091
CT OF INTENSE LIGHT ON THE REFRACTIVE INDEX OF A CRYSTALLINE POTASSIUM NIOBATE.   EFFE 08-1410
APHIC PATTERNS IN FERROELECTRIC MATERIALS. (BARIUM TITANATE, POTASSIUM NIOBATE) /TION AND ERASURE OF HOLOGR 04-0847
                                            ENERGY BANDS OF POTASSIUM NIOBATE.   08-1372
TROOPTICAL APPLICATIONS. (LEAD ZIRCONATE TITANATE AND SODIUM POTASSIUM NIOBATE) /ELECTRIC CERAMICS FOR ELEC 09-1527
QUARTZ AND FERROELECTRIC PEROVSKITES). (BARIUM TITANATE AND POTASSIUM NIOBATE) /CRYSTALS (CASE OF SILICON, 04-0629
                      NEUTRON STUDY OF DIFFUSE SCATTERING IN CUBIC POTASSIUM NIOBATE.   08-1407
                      PREPARATION AND FABRICATION OF SODIUM POTASSIUM NIOBATE.   13-1937
ECTRIC MAGNETS. (LEAD AND BARIUM TITANATES, BISMUTH FERRATE, POTASSIUM NIOBATE) /FERROELECTRICS AND FERROEL 04-1038
RUCTURE DISORDER AND RAMAN SCATTERING IN BARIUM TITANATE AND POTASSIUM NIOBATE.   ST 04-0676
SEEDED FLUX) CRYSTAL GROWTH AND NEUTRON CHARACTERIZATION OF POTASSIUM NIOBATE.   (TOP 08-1353
PEROVSKITES CRYSTALS. (NIOBIUM-93 AND SODIUM-23 RESONANCE IN POTASSIUM NIOBATE AND SODIUM NIOBATE) /ONS IN 08-1283
OELECTRIC DOMAINS OF PEROVSKITE TYPE OXIDES. (LEAD TITANATE, POTASSIUM NIOBATE, BARIUM TITANATE) /S ON FERR 04-0993
              ELASTIC CONSTANT OF ORTHORHOMBIC POTASSIUM NIOBATE BY X-RAY DIFFUSE SCATTERING.   08-1416
              SPACE CHARGE LIMITED CURRENTS IN SEMICONDUCTING POTASSIUM NIOBATE CRYSTALS.   08-1412
              SURFACE CHARGE LAYERS ON FERROELECTRIC POTASSIUM NIOBATE CRYSTALS.   08-1369
OPTICAL PROPERTIES (BIREFRINGENCE AND REFRACTIVE INDICES) OF POTASSIUM NIOBATE (LINEAR ELECTROOPTICS).   08-1468
PPLICATION.   PREPARATION OF POTASSIUM NIOBATE SINGLE CRYSTAL FOR OPTICAL A 08-1332
DEFORMATION AND RELAXATION BEHAVIOR IN FERROELECTRIC POTASSIUM NIOBATE SINGLE CRYSTALS.   08-1420
                    DOMAINS IN FERROELECTRIC POTASSIUM NIOBATE SINGLE CRYSTALS.   08-1315
        INTERFEROMETRIC STUDIES OF DOMAIN STRUCTURES IN POTASSIUM NIOBATE SINGLE CRYSTALS.   08-1318
        INTERFEROMETRIC STUDIES OF DOMAIN STRUCTURE IN POTASSIUM NIOBATE SINGLE CRYSTALS.   08-1319
HANICAL SWITCHING OF 60 DEGREE DOMAIN WALLS IN FERROELECTRIC POTASSIUM NIOBATE SINGLE CRYSTALS.   MEC 08-1421
OPTICAL STUDIES OF DOMAIN STRUCTURE OF FERROELECTRIC POTASSIUM NIOBATE SINGLE CRYSTALS.   08-1320
ITH STEP LADDERS ON PSEUDOCUBIC (001) / DOMAIN STRUCTURES IN POTASSIUM NIOBATE SINGLE CRYSTALS ASSOCIATED W 08-1316
UBIC (001) CLEAVAGE PLANES.   DOMAINS IN POTASSIUM NIOBATE SINGLE CRYSTALS FROM PSEUDOC 08-1317
INDICES)   NONLINEAR OPTICAL PROPERTIES OF POTASSIUM NIOBATE SINGLE CRYSTALS. (REFRACTIVE 08-1454
G CONDITIONS.   DIELECTRIC CONSTANT OF POTASSIUM NIOBATE SINGLE CRYSTALS UNDER BIASIN 08-1381
          FERROELECTRICS IN THE LITHIUM POTASSIUM NIOBATE SYSTEM.   11-1660
SOLUTION. (THEORY, APPLIED TO BARIUM STRONTIUM TITANATE AND POTASSIUM NIOBATE TANTALATE) /ROELECTRIC SOLID 03-0260
/LECTROOPTIC PROPERTIES OF TUNGSTEN BRONZE CRYSTALS FROM THE POTASSIUM NIOBATE- LITHIUM NIOBATE- BARIUM DI/ 13-2003
ND FERROELECTRICITY APPLIED TO SODIUM NIOBATE AND THE SYSTEM POTASSIUM NIOBATE- SODIUM NIOBATE. /CTRICITY A 03-0218
ASSIUM NIOBATE- BARIUM DINIOBATE SY/ PHASE EQUILIBRIA IN THE POTASSIUM NIOBATE- STRONTIUM DINIOBATE AND POT 13-2016
OF THE OXYGENATED QUADRATIC TUNGSTEN BRONZE TYPE. (STRONTIUM POTASSIUM NIOBIUM OXYFLUORIDE) /UORIDE PHASES 13-2011
DETERMINATION OF THE ELASTIC CONSTANTS OF POTASSIUM NITRATE.   08-1399
ECTRIC BEHAVIOR OF (FERROELECTRIC) FILMS OF VACUUM DEPOSITED POTASSIUM NITRATE.   DIEL 16-2356
DIELECTRIC DISPERSION IN THE PARAELECTRIC PHASE OF POTASSIUM NITRATE.   16-2341
(+) AND STRONTIUM(2+) IONS ON THE ELECTRICAL CONDUCTIVITY OF POTASSIUM NITRATE.   EFFECT OF ADDING TITANIUM 16-2319
EFFECT OF HYDROSTATIC PRESSURE ON FERROELECTRICITY OF POTASSIUM NITRATE.   16-2344
ELECTRICAL PROPERTIES OF RHOMBOHEDRAL POTASSIUM NITRATE.   16-2367
FAR INFRARED SPECTRA AND FERROELECTRIC PHASE TRANSITION OF POTASSIUM NITRATE.   16-2308
TA TO ALPHA, BETA TO GAMMA AND GAMMA TO ALPHA TRANSITIONS IN POTASSIUM NITRATE. /H IN THE ALPHA TO BETA, BE 16-2259
PRODUCTION OF NITROGEN DIOXIDE AND NITRITE ION IN IRRADIATED POTASSIUM NITRATE. /GNETIC RESONANCE STUDY OF 16-2404
PHONON DISPERSION IN FERROELECTRIC POTASSIUM NITRATE.   16-2334
PRESSURE INDUCED FERROELECTRIC PHASE TRANSITION IN POTASSIUM NITRATE.   16-2378
STRUCTURE OF THE DISORDERED PHASE OF POTASSIUM NITRATE.   16-2379
SWITCHING BEHAVIOR IN FERROELECTRIC POTASSIUM NITRATE.   16-2285
SWITCHING PROPERTIES OF THICK FILM FERROELECTRIC POTASSIUM NITRATE.   16-2282
TWO MODES OF ELECTRICAL CONDUCTIVITY IN POTASSIUM NITRATE.   16-2368
X-RAY MEASUREMENTS ON THERMAL EXPANSION OF FERROELECTRIC POTASSIUM NITRATE.   16-2322
DISORDERED STRUCTURE ABOVE THE FERROELECTRIC CURIE POINT IN POTASSIUM NITRATE.   X-RAY STUDY ON THE 16-2366
TRANSITION)   RHYTHMIC GROWTH OF NEW PHASE IN POTASSIUM NITRATE. (ALPHA TO BETA POLYMORPHIC 16-2263
NOT SEEM FERROELECTRIC- PARAELECT/ PHASE TRANSFORMATIONS IN POTASSIUM NITRATE AND POTASSIUM NITRITE. (DOES 16-2369
/NTS AND INFRARED ABSORPTION) STUDIES ON FERROELECTRICITY IN POTASSIUM NITRATE AND RELATED COMPOUNDS. (POT/ 16-2402
CTRIC CONSTANT AND SPONTANEOUS POLARIZATION OF FERROELECTRIC POTASSIUM NITRATE AND SODIUM NITRITE. /E DIELE 16-2331
TRANSITIONS.   INFRARED SPECTRA OF POTASSIUM NITRATE AND SODIUM NITRITE AT PHASE 16-2352
CAL CONDUCTIVITY ACCOMPANYING THE POLYMORPHIC TRANSITIONS IN POTASSIUM NITRATE AT 1 ATMOSPHERE. /IN ELECTRI 16-2391
THE CONDITIONS OF CRYSTALLIZATION ON THE FORM OF CRYSTALS OF POTASSIUM NITRATE. (CRYSTAL GROWTH) /UENCE OF 16-2332
ELECTRICAL RESISTIVITIES OF SODIUM NITRITE AND POTASSIUM NITRATE CRYSTALS.   16-2258
/ONVOLATILE, NONDESTRUCTIVE READOUT MEMORY ELEMENT UTILIZING POTASSIUM NITRATE. (FUSED THICK FILM OR VACUU/ 16-2269
TRANSITION RATES OF POTASSIUM NITRATE HIGH PRESSURE POLYMORPHS.   16-2279
N WATER OR ALCOHOL)   METASTABLE POTASSIUM NITRATE(III) FROM SOLUTION. (GROWN I 16-2329
LATTICE DYNAMICS MODEL OF POTASSIUM NITRATE IN ITS FERROELECTRIC PHASE.   16-2377
LUMINESCENCE OF POTASSIUM NITRATE IRRADIATED WITH GAMMA-RAYS.   16-2335
THERMAL EXPANSION OF SOME FERROELECTRICS. (POTASSIUM NITRATE, KDP, SODIUM NITRITE)   16-2298
TH BY DOUBLE BRIDGMAN TECHNIQUE).   PHASE TRANSITIONS IN POTASSIUM NITRATE NITRITE MIXED CRYSTALS (GROW 16-2345
/PHOLOGY OF CRYSTAL GROWTH AT POLYMORPHIC TRANSFORMATIONS IN POTASSIUM NITRATE, SILVER NITRATE, AND AMMONI/ 16-2260
MELTING PROPERTIES OF POTASSIUM NITRATE UNDER HIGH PRESSURE.   16-2265
EVIDENCE OF ADDITIONAL DISORDER IN THE RHOMBOHEDRAL FORM OF POTASSIUM NITRITE.   16-2371
DIELECTRIC AND ELECTRICAL CONDUCTIVITY STUDIES IN POTASSIUM NITRITE AND POTASSIUM NITRATE.   16-2340
PHASE DIAGRAMS OF SODIUM NITRITE AND POTASSIUM NITRITE TO 40 KBAR.   16-2345
ARSENATE, POTASSI/ LINEAR ELECTROOPTIC MODULATOR MATERIALS. (POTASSIUM OR AMMONIUM DIHYDROGEN PHOSPHATE OR 01-0021
/SOLUTION GROWTH, TGS, TRIGLYCINE SELENATE OR FLUOBERYLLATE, POTASSIUM OR RUBIDIUM DIHYDROGEN PHOSPHATE, S/ 02-0089
/ HYDROGEN BONDED FERROELECTRICS. (DEUTERATED ROCHELLE SALT, POTASSIUM OR RUBIDIUM OR AMMONIUM DIHYDROGEN / 18-2719
/TERMINATION OF THE SYMMETRY OF THE FERROELECTRIC COMPOUNDS (POTASSIUM OR SODIUM) (0.5) BISMUTH(0.5) TITANA/ 08-1362
/TH PEROVSKITES. (AND THE FERROELECTRIC STATE, BISMUTH(0.5) (POTASSIUM OR SODIUM(0.5) TITANATES AND PHASE / 08-1289
IDE SYSTEM. (POTASSIUM LIT/ STUDY OF SUBSOLIDUS EQUILBRIA IN POTASSIUM OXIDE- LITHIUM OXIDE- NIOBIUM PENTOX 13-1963
OXIDE.   FERROELECTRICS IN THE POTASSIUM OXIDE- STRONTIUM OXIDE- NIOBIUM PENT 13-1896
/AND B ARE CALCIUM, STRONTIUM, OR BARIUM, AND C IS SODIUM OR POTASSIUM) PHASES OF THE TETRAGONAL TUNGSTEN / 04-0912
/TRIC TRANSITION IN KDP TYPE CRYSTALS (CURIE TEMPERATURES OF POTASSIUM, RUBIDIUM, CESIUM AND AMMONIUM DIHY/ 17-2577
/TE OR SELENATE OR FLUOBERYLLATE (METALS CHOSEN FROM SODIUM, POTASSIUM, RUBIDIUM, CESIUM, OR AMMONIUM, ABS/ 19-2743
STRUCTURE OF DIHYDROGEN PHOSPHATE ION IN THE FERROELECTRIC POTASSIUM SALT.   17-2624
CRYSTAL STRUCTURE OF POTASSIUM SELENATE.   22-2845
THE ELECTRON PARAMAGNETIC RESONANCE OF POTASSIUM SELENATE.   22-2850
THE STRUCTURE AND PHASE TRANSITION IN POTASSIUM SELENATE.   22-2872
ESR STUDY OF GAMMA-IRRADIATED POTASSIUM SELENATE (STRUCTURAL PHASE CHANGES).   22-2825
BELOW 93K).   DIELECTRIC AND THERMAL STUDY OF POTASSIUM SELENATE TRANSITIONS (FERROELECTRIC 22-2824
/ND FERROELECTRIC PROPERTIES OF (THE TUNGSTEN BRONZE SYSTEM) POTASSIUM SODIUM BARIUM NIOBATE (SECOND HARMO/ 13-2028
CONTINUOUS HOT PRESSING OF POTASSIUM SODIUM NIOBATE.   13-2007
/RROELECTRICS AND ANTIFERROELECTRICS. (BARIUM TITANATE, KDP, POTASSIUM SODIUM TARTRATE, TGS, AMMONIUM FLUO/ 02-0134

OF INFRARED DISPERSION IN BARIUM AND STRONTIUM TITANATES AND POTASSIUM TANTALATE. /ICAL PHONON MODE THEORY 04-0547
EFFECT OF PRESSURE ON THE STATIC DIELECTRIC CONSTANT OF POTASSIUM TANTALATE. 08-1269
RE ON THE DIELECTRIC PROPERTIES OF A PARAELECTRIC MATERIAL. (POTASSIUM TANTALATE) EFFECT OF PRESSU 08-1335
NDUCED OPTICAL HARMONIC GENERATION IN STRONTIUM TITANATE AND POTASSIUM TANTALATE. ELECTRIC FIELD I 06-1203
ECTRIC FIELD INDUCED RAMAN EFFECT IN PARAELECTRIC CRYSTALS. (POTASSIUM TANTALATE) EL 08-1326
ELECTRICAL PROPERTIES AND DEFECT STRUCTURE OF POTASSIUM TANTALATE. 08-1314
ELECTRON PARAMAGNETIC RESONANCE OF MANGANESE(2+) IN POTASSIUM TANTALATE. 08-1343
EPR OF COBALT(2+) IN POTASSIUM TANTALATE. 08-1342
ARED DIELECTRIC DISPERSION OF PEROVSKITES. (BARIUM TITANATE, POTASSIUM TANTALATE) FAR INFR 04-0546
INFRARED ACTIVE MODES IN POTASSIUM TANTALATE. 08-1366
ATTERING ANALYSIS OF THE LINEAR DISPLACEMENT CORRELATIONS IN POTASSIUM TANTALATE. NEUTRON SC 36-3546
PHOTOCONDUCTION OF POTASSIUM TANTALATE. 08-1408
PHOTOEFFECTS IN POTASSIUM TANTALATE. 08-1415
ORDER PHASE TRANSITIONS AT LOW TEMPERATURES. (TIN TELLURIDE, POTASSIUM TANTALATE) THEORY OF SECOND 03-0433
ES (3 PHONON INTERACTION PROPOSED FOR STRONTIUM TITANATE AND POTASSIUM TANTALATE). /NDUCTIVITY IN PEROVSK 06-1095
BAND CALCULATIONS ON RHENIUM TRIOXIDE, SODIUM TUNGSTATE, AND POTASSIUM TANTALATE. TIGHT BONDING 08-1334
RIC PROPERTIES OF SOLID SOLUTIONS BASED ON THE FERROELECTRIC POTASSIUM TANTALATE. /RAY ANALYSIS AND DIELECT 08-1310
/S OF TWO NEW PHASES OF QUADRATIC TUNGSTEN BRONZES STRONTIUM POTASSIUM TANTALATE AND BARIUM POTASSIUM TANT/ 13-2010
HE CURIE POINTS. THERMAL CONDUCTIVITY OF STRONTIUM TITANATE, POTASSIUM TANTALATE AND BARIUM TITANATE NEAR T 04-0989
/TIES OF SOME ABO(3) PEROVSKITES IN THE PARAELECTRIC PHASE. (POTASSIUM TANTALATE AND TANTALATE NIOBATE, BA/ 04-0690
/IC" PHONON IN SOME PEROVSKITE CRYSTALS (STRONTIUM TITANATE, POTASSIUM TANTALATE) AT LOW TEMPERATURES. (TH/ 03-0227
/ATTERING STUDY ON SOFT MODES. (REVIEW OF BROOKHAVEN WORK ON POTASSIUM TANTALATE, BARIUM AND STRONTIUM TIT/ 02-0115
/LECTRIC MODE IN THE CUBIC PEROVSKITES. (STRONTIUM TITANATE, POTASSIUM TANTALATE, BARIUM TITANATE, LEAD TI/ 02-0114
/ERING AT STRUCTURAL PHASE TRANSITIONS. (STRONTIUM TITANATE, POTASSIUM TANTALATE, BARIUM TITANATE, POTASSI/ 02-0059
/RATION VS NET IONIZED DONOR CONCENTRATION IN SINGLE CRYSTAL POTASSIUM TANTALATE. (CONDUCTIVE, KYROPOULOS / 08-1437
SINGLE CRYSTAL GROWTH OF STRONTIUM POTASSIUM TANTALATE FROM MOLTEN SALTS. 13-2034
C- ACOUSTIC) IN THE PHONON SPECTRA OF STRONTIUM TITANATE AND POTASSIUM TANTALATE (LATTICE DYNAMICS). /ECTRI 06-1157
E TEMPERATURE RANGE 4.2-300 DEG K (LATTICE PARAMETERS VERSU/ POTASSIUM TANTALATE LEAD TITANATE SYSTEM IN TH 08-1308
ABSORPTION EDGE SPLITTING IN POTASSIUM TANTALATE NIOBATE. 13-1988
GROWTH OF HIGH QUALITY CRYSTALS OF POTASSIUM TANTALATE NIOBATE. 13-2055
LONGITUDINAL ELECTROOPTIC EFFECT IN POTASSIUM TANTALATE NIOBATE 13-1942
ROELECTRICS INDICATED BY ANISOTROPIC DEBYE- WALLER FACTORS. (POTASSIUM TANTALATE NIOBATE) /N PEROVSKITE FER 08-1351
NEUTRON SCATTERING STUDY OF THE SOFT MODES IN CUBIC POTASSIUM TANTALATE NIOBATE. 13-2064
OPTICAL DAMAGE IN POTASSIUM TANTALATE NIOBATE 13-1979
OPTICAL PROPERTIES OF POTASSIUM TANTALATE NIOBATE. 36-3593
· SOFT MODES AND STRUCTURE OF FERROELECTRIC TETRAGONAL POTASSIUM TANTALATE NIOBATE. 13-1957
SOFT PHONONS IN POTASSIUM TANTALATE NIOBATE. 13-2040
BEAM MODULATION AND DEFLECTION. ELECTROOPTIC POTASSIUM TANTALATE NIOBATE GRATINGS FOR LIGHT 13-1929
/FERROELECTRICS. (ELECTROOPTIC MATERIALS, SINGLE CRYSTALS OF POTASSIUM TANTALATE NIOBATE, LITHIUM NIOBATE,/ 02-0073
TE NIOBATE) PYROELECTRIC DETECTORS. CHARACTERISTICS OF POTASSIUM TANTALATE NIOBATE (POTASSIUM TANTALA 13-2030
ATE(0.65) NIOBATE(0.35), CURIE TEMPERATUR/ RAMAN SPECTRUM OF POTASSIUM TANTALATE NIOBATE. (POTASSIUM TANTAL 08-1391
CTORS. CHARACTERISTICS OF KTN (POTASSIUM TANTALATE NIOBATE) PYROELECTRIC DETE 13-2030
PIEZOELECTRIC PROPERTIES OF POTASSIUM TANTALATE NIOBATE SINGLE CRYSTAL. 08-1271
D THEIR ELECTRICAL PROPERTIES. FLUX GROWN POTASSIUM TANTALATE NIOBATE SINGLE CRYSTALS AN 09-1533
NAMICAL STUDIES. POTASSIUM TANTALATE NIOBATE, STRUCTURAL AND DY 13-2068
STRONTIUM TITANATE, BARIUM TITANATE ZIRCONATE (OR STANNATE), POTASSIUM TANTALATE NIOBATE, 228 REFS). /IUM 08-1281
· ELECTRON TUNNELING INTO POTASSIUM TANTALATE SCHOTTKY BARRIER JUNCTIONS 08-1365
/E OF (NIOBIUM DOPED) STRONTIUM TITANATE AND (CALCIUM DOPED) POTASSIUM TANTALATE (SCHOTTKY BARRIERS WITH I/ 06-1222
CHARACTER OF PHASE TRANSITION OF FERROELECTRIC POTASSIUM TANTALATE. (SECOND ORDER) 08-1307
SIUM SODIUM/ DIELECTRIC PROPERTIES AND OPTICAL ABSORPTION OF POTASSIUM TANTALATE SINGLE CRYSTALS AND (POTAS 08-1272
/HE NORMAL VIBRATIONAL DISPLACEMENTS OF SEVERAL PEROVSKITES (POTASSIUM TANTALATE, STRONTIUM TITANATE, RUBI/ 06-1139
ABSORPTION EDGE SPLITTING IN POTASSIUM TANTALATE(X) NIOBATE(1-X). 13-1987
MIXED CRYSTAL SYSTEMS POTASSIUM(X) SODIUM(1-X) TANTALATE AND POTASSIUM TANTALATE(X) NIOBATE(1-X). / IN THE 13-1933
KITE PARAELECTRICS IN BEAM DEFLECTORS AND LIGHT MODULATORS. (POTASSIUM TANTALATE(0.65) NIOBATE(0.35)) /ROVS 13-1924
RAMAN SPECTRUM OF POTASSIUM TANTALUM(0.64) NIOBIUM(0.36) OXIDE. 13-1989
IED MECHANISM UNDERLYING URBACH'S RULE. (STRONTIUM TITANATE, POTASSIUM TITANATE) IS THERE A UNIF 06-1110
/ILLOUIN SCATTERING OF LIGHT IN MANGANESE FLUORIDE CRYSTALS (POTASSIUM TRIFLUORIDE, RUBIDIUM MANGANESE TRI/ 33-3398
MAGNETIC RESONANCE STUDY OF SODIUM TRIHYDROGEN SELENATE AND POTASSIUM TRIHYDROGEN SELENATE. PROTON 22-2876
ANOMALOUS THERMAL EXPANSION OF POTASSIUM TRIHYDROGEN SELENATE CRYSTALS. 22-2880
A NEUTRON DIFFRACTION STUDY OF POTASSIUM TRIHYDROGEN SELENITE. 22-2866
UM TRIDEUTERIUM SELENITE, AS DE/ THE HYDROGEN BOND SYSTEM IN POTASSIUM TRIHYDROGEN SELENITE, AND IN POTASSI 22-2854
YDROGEN SELENITE. CRYSTAL STRUCTURES OF POTASSIUM TRIHYDROGEN SELENITE AND SODIUM TRIH 22-2838
ANTAL/ PHASE TRANSITIONS IN TANTALATES WITH THE STRUCTURE OF POTASSIUM TUNGSTEN BRONZE. (POTASSIUM BARIUM T 13-2050
FORMATION CONDITIONS AND PHASE TRANSITIONS IN CRYSTALS WITH POTASSIUM TUNGSTEN BRONZE STRUCTURE (POTASSIU/ 13-1923
/ OR SODIUM(0.5) TITANATES AND PHASE DIAGRAMS FOR BISMUTH(X) POTASSIUM(X) BARIUM(1-2X) TITANATE, BISMUTH (0/ 08-1289
UM TANTALATE/ SOFT PHONON MODES IN THE MIXED CRYSTAL SYSTEMS POTASSIUM(X) SODIUM (1-X) TANTALATE AND POTASSI 13-1933
TEMPERATURE DEPENDENCE OF DIELECTRIC CONSTANT IN THE SYSTEM POTASSIUM YTTRIUM MOLYBDATE- POTASSIUM DYSPRO/ 30-3309
(0.5) TITANATE AND THE HIGH TEMPERATURE PHASE TRANSITIONS IN POTASSIUM(0.5) BISMUTH(0.5) TITANATE. /BISMUTH 08-1362
ERROELECTRICITY IN POTASSIUM NITRATE AND RELATED COMPOUNDS. (POTASSIUM(1-X) AMMONIUM(X) NITRATE) /DIES ON F 16-2402
FERROELECTRICITY IN AMMONIUM(X) POTASSIUM(1-X) NITRATE MIXED CRYSTAL. 16-2401
ROPERTIES AND SOFT MODES IN THE FERROELECTRIC MIXED CRYSTALS POTASSIUM(1-X) SODIUM(X) TANTALATE. /LECTRIC P 08-1306
/YLLATE, POTASSIUM OR RUBIDIUM DIHYDROGEN PHOSPHATE, SODIUM, POTASSIUM(2), OR LITHIUM THALLIUM TARTRATE, L/ 02-0089
HOT PRESSING OF POTASSIUM- SODIUM NIOBATE. 09-1542
OGEN PHOSPHATE. NUCLEAR DOUBLE RESONANCE STUDY OF POTASSIUM-39 IN FERROELECTRIC POTASSIUM DIHYDR 17-2599
OLARIZED. DISTRIBUTION OF POTENTIAL ALONG TITANATE CERAMICS PREVIOUSLY P 04-0788
ROGEN HALIDES. INTERMOLECULAR POTENTIAL AND FERROELECTRIC TRANSITIONS IN HYD 31-3339
EXPERIMENT ON LINEAR CHAIN OF ATOMS LYING IN DOUBLE MINIMUM POTENTIAL. (BARIUM TITANATE) COMPUTER 03-0287
TANATE. (PTCR DEVICES) POTENTIAL BARRIERS ON SEMICONDUCTING BARIUM TI 04-0851
CRYSTAL (IN THE PARAELECTRIC PHASE, UNDER DC FIELD ALONG C/ POTENTIAL DISTRIBUTION IN ANTIMONY SULFOIODIDE 15-2173
MICS AND CRYSTALS. INFLUENCE OF HUMIDITY ON THE POTENTIAL DISTRIBUTION IN BARIUM TITANATE CERA 04-0883
) STRONTIUM(1-X) TITANATE CERAMIC/ "VIBRATING WIRE" STUDY OF POTENTIAL DISTRIBUTION (STABILITY) IN BARIUM(X 04-0699
PYROELECTRIC AMMONIUM IODATE, A POTENTIAL FERROELASTIC: CRYSTAL STRUCTURE. 36-3574
THERMODIELECTRIC EFFECT AND FREEZING POTENTIAL IN GROWING ICE. 34-3417
PHOSPHATE. NEUTRON MEASUREMENTS ON THE HYDROGEN BOND POTENTIAL IN PARAELECTRIC POTASSIUM DIHYDROGEN 17-2511
RESEARCH STATUS AND DEVICE POTENTIAL OF FERROELECTRIC THIN FILMS. 37-3637
RIC CRYSTALS. (THEORY) THERMODYNAMIC POTENTIALS AND SECONDARY EFFECTS IN FERROELECT 03-0387
OMIDE, POTASSIUM CHLORIDE, SODIUM CHLORIDE,/ CRYSTALLIZATION POTENTIALS IN AQUEOUS SOLUTIONS. (POTASSIUM BR 17-2707
F PTC MATERIALS. DIFFUSION POTENTIALS IN BARIUM TITANATE AND THE THEORY O 03-0352
CTRICS: A SELF CONSISTENT POLARIZATION SCHEME FOR ANHARMONIC POTENTIALS. (THEORY) / TRANSITIONS IN FERROELE 03-0355
THE TUNGSTEN BRONZE TYP/ BISMUTH(3) NIOBIUM(17) OXIDE(47): POTENTIALLY FERROELECTRIC CRYSTAL STRUCTURE OF 13-1978

OPTICAL LATTICE VIBRATIONS. (QUARTZ, BARIU/ EXTENSION CF THE | POULET- LOUDON APPROXIMATION IN THE THEORY OF | 03-0380
/ZT. PART-1. PART-2: EFFECTS OF AVERAGE PARTICLE SIZE OF RAW | POWDER ON SINTERING. PART-3: SINTERING FROCES/ | 09-1624
E FERROELECTRICS. (LATTICE VIBRATIONS) | POWDER RAMAN SPECTRA: APPLICATION TO DISPLACIV | 03-0194
| POWDER RAMAN STUDY FOR BARIUM TITANATE. | 04-0610

VARIATIONS IN LITHIUM NICBATE AND LITHIUM TANTALATE BY RAMAN | POWDER SPECTROSCOPY. STOICHIOMETRY | 11-1813
ABSORPTICN SPECTRA OF BARIUM TITANATE AND STRONTIUM TITANATE | POWDERS. /ICLE SIZE AND SHAPE ON THE INFRARED | 04-0823
LCOHCL. PREPARATICN OF BARIUM TITANATE AND OTHER CERAMIC | POWDERS BY COPRECIPITATION OF CITRATES IN AN A | 04-0861
/TITANATE CERAMICS. PART-2: EFFECTS OF THE ADDITION OF GLASS | POWDERS CONTAINING BARIUM OXIDE AND TITANIUM / | 04-0829
/AND ELECTRICAL PROPERTIES OF FERROELECTRIC MATERIALS. (FINE | POWDERS PREPARED BY DECOMPOSITION OF ORGANIC / | 02-0093
RON FOR ZIRCONIUM AND TITANIUM, EFFECT CK ACOUSTIC RADIATION | POWER) /ON OF STRONTIUM OR CALCIUM FOR LEAD, I | 09-1484
, THERMAL EXPANSION, INDICES OF REFRACTICN, OPTICAL ROTATORY | POWER) /IC CONSTANTS, SPONTANEOUS POLARIZATION | 35-3501
INFRARED EMISSION IN LITHIUM NIOBATE. | POWER AND LINE WIDTH OF TUNABLE STIMULATED FAR | 11-1736
E SURFACE WAVE AMPLIFIER. LOW | POWER DENSITY GALLIUM ARSENIDE- LITHIUM NIOBAT | 11-1825
CING POLARIZATION SWITCHING / NONLINEAR DIELECTRICS FOR HIGH | POWER ELECTRONIC TUNING. (WITH ELECTRODES REDU | 37-3660
IDE(11) SINGLE CRYSTAL. SWITCHING OF OPTICAL ROTATORY | POWER IN FERROELECTRIC LEAD(5) GERMANIUM(3) OX | 36-3570
NG BARIUM TITANATE CRYSTAL) MINIMUM DETECTABLE | POWER OF A PYROELECTRIC THERMAL RECEIVER. (USI | 04-0632
SSLESS POTASSIUM DIDEUTERIUM PHCSFHATE POCKELS CELL FOR HIGH | POWER Q SWITCHING. LO | 17-2529
/RED FERROELECTRICS. (LANTHANUM BISMUTH TITANIUM IRON OXIDE, | PRASEODYMIUM BISMUTH TITANIUM IRON OXIDE, BIS/ | 13-1993
/OMALIES IN THE (1 KHZ) DIELECTRIC CONSTANT (5-300K) OF THE | PRASEODYMIUM(1-X) NEODYMIUM(X) (ORTHO) ALUMIN/ | 36-3600
IFIED CERAMICS OF LEAD ZIRCONATE TITANATE. | PRECIPITATION OF ZIRCONIA PHASE IN NIOBIUM MOD | 09-1540
ERS AND THE COEFFICIENTS OF THERMAL EXPANSION OF BISMUTH FE/ | PRECISION DETERMINATION OF THE LATTICE PARAMET | 08-1287
S FOR ELECTROOPTIC ELEMENTS. | PRECISION MACHINING OF LITHIUM NIOBATE CRYSTAL | 11-1742
TE TYPE RARE EARTH MOLYEDATES. (CRYSTAL STRUCTURE) | PRECISION PARAMETERS OF SOME LANTHANON MOLYBDA | 30-3262
EARTH MCLYEDATES: LANTHANUM MCLYEDATE. | PRECISION PARAMETERS OF THE FERROELECTRIC RARE | 30-3264
ERSTAND (THE CAUSE OF) FERROELECTRICITY. (ITS NATURE AND THE | PREDICTION OF ITS OCCURANCE) DO WE REALLY UND | 03-0510
OPIC) SYMMETRY ASPECTS OF FERRCELECTRICITY. (DESCRIPTIVE AND | PREDICTIVE POSSIBILITIES) (MACROSC | 03-0464
LENATE DIHYDRATE CRYSTALS. | PREPARATION AND ANALYSIS OF SODIUM AMMONIUM SE | 22-2863
IVED STRONTIUM ZIRCONATE AND STRONTIUM TITANATE (OF HIGH PU/ | PREPARATION AND CHARACTERIZATION OF ALKOXY-DER | 06-1217
A(2+) E(3+) (2)O(4) TYPE CALCIUM AND STRONTIUM SCANDATES. | PREPARATION AND CRYSTALLOGRAPHIC PROPERTIES OF | 36-3537
SIUM BISMUTH ZINC NIOBATE. | PREPARATION AND DIELECTRIC PROPERTIES OF POTAS | 13-2001
ION OF LARGE, PURE RUBIDIUM MANGANESE TRIFLUORIDE AND POTAS/ | PREPARATION AND DISLOCATION DENSITY DETERMINAT | 33-3397
OF TUNGSTEN TRIOXIDE AT LOW TEMPERATURE. | (PREPARATION AND ELECTRICAL AND OPTICAL) STUDY | 10-1643
CERAMIC MATERIALS AT MICROWAVE FREQUENCIES. (TITANATE ZIRCO/ | PREPARATION AND ELECTRICAL PROPERTIES OF SOME | 09-1602
FERROELECTRIC BISMUTH TITANIUM OXIDE. | PREPARATION AND EPITAXY OF SPUTTERED FILMS OF | 13-2038
M NIOBATE. | PREPARATION AND FABRICATION OF SODIUM TITANATE | 13-1937
S OF ILMENITE STRUCTURE. (ABC(3) NIOBATES, TANTALATES, AND / | PREPARATION AND INVESTIGATION OF FERROELECTRIC | 08-1438
CHIOMETRIC ANTIMONY SULFOIODIDE SINGLE CRYSTALS. | PREPARATION AND PHYSICAL PROPERTIES OF NONSTOI | 15-2199
ITE TYPE COMPCUNDS, LANTHANUM(1-X) CALCIUM(X) MANGANATE (PH/ | PREPARATION AND PHYSICAL PROPERTIES OF PEROVSK | 08-1356
UM) TITANATE SINGLE CRYSTALS. | PREPARATION AND PROPERTIES OF (BARIUM, STRONTI | 04-0560
DES. (FERROELECTRICITY INDETERMINITE) | PREPARATION AND PROPERTIES OF CESIUM LEAD HALI | 33-3359
CONTAINING FERROELECTRIC CRYSTALS (BARIUM OR LEAD TITANATE,/ | PREPARATION AND PROPERTIES OF GLASS- CERAMICS | 04-0783
Y THIN FILM DIELECTRICS. (BARIUM AND LEAD TITANATES, EFFECT/ | PREPARATION AND PROPERTIES OF HIGH PERMITTIVIT | 04-0670
ARIUM TITANATE SINGLE CRYSTALS. | PREPARATION AND PROPERTIES OF PURE AND DOPED B | 04-0668
UNDS. (WITH MAGNETIC AND FERROELECTRIC ORDERING) | PREPARATION AND PROPERTIES OF RARE EARTH COMPO | 02-0131
NATE FILMS. | PREPARATION AND PROPERTIES OF THIN BARIUM TITA | 04-0669
ALS WITH PEROVSKITE TYPE STRUCTURE. (MIXED OXIDES ABO(3)) (/ | PREPARATION AND PROPERTIES OF THIN FILM MATERI | 04-0914
RIUM TITANATE. | PREPARATION AND PROPERTIES OF THIN FILMS OF BA | 04-0843
LS (SYSTEM WIT/ FERROELECTRIC PROPERTIES OF BISMUTH FERRATE | (PREPARATION AND STRUCTURE) AND RELATED MATERIA | 08-1333
IUM SEMIMETALLIC CRTHOPHOSPHATES. | (PREPARATION AND STRUCTURE OF) RUBIDIUM AND CES | 17-2598
-X) NIOBATE SINGLE CRYSTALS (CZOCHRALSKI GROWTH, DIELECTRIC/ | PREPARATION AND STUDY OF BARIUM(X) STRONTIUM(1 | 13-1936
, STRONTIUM) TITANATE. | PREPARATION AND STUDY OF THIN FILMS OF (BARIUM | 04-1000
ES OF BARIUM TITANATE. INFLUENCE OF MICROSTRUCTURES | (PREPARATION CONDITIONS) ON DIELECTRIC PROPERTI | 04-0798
TIES OF FILAMENTARY CRYSTALS OF ADP. | PREPARATION (MORPHOLOGY) AND MECHANICAL PROPER | 17-2630
S. | PREPARATION OF AMMONIUM SULFATE SINGLE CRYSTAL | 19-2753
ROM THE MELT UNDER PRESSURE AND INVESTIGATION OF THEIR ELEC/ | PREPARATION OF ANTIMONY SULFOIODIDE TEXTURES F | 15-2138
IC POWDERS FY COPRECIPITATION OF CITRATES IN AN ALCOHOL. | PREPARATION OF BARIUM TITANATE AND OTHER CERAM | 04-0861
ACTIONS (INVOLVING TITANIUM ESTERS TC YIELD SM/ HYDROTHERMAL | PREPARATION OF BARIUM TITANATE BY TRANSPORT RE | 04-0625
ACTIONS. HYDROTHERMAL | PREPARATION OF BARIUM TITANATE BY TRANSPORT RE | 04-0626
TION. | PREPARATION OF BARIUM TITANATE FILM BY EVAPORA | 04-0743
IUM TITANATE BY VACUUM EVAPORATION. THIN FILMS: | PREPARATION OF FERROELECTRIC THIN FILMS OF BAR | 04-0834
NATE. | PREPARATION OF FLUORINE CONTAINING BARIUM TITA | 04-0984
ROTHERMAL SYNTHESIS. | PREPARATION OF NICKEL BORACITE CRYSTALS BY HYD | 14-2079
RROELECTRIC BISMUTH TITANATE, (BY ANNEALING). | PREPARATION OF POLED, TWIN-FREE CRYSTALS OF FE | 08-1352
L FOR OPTICAL APPLICATION. | PREPARATION OF POTASSIUM NIOBATE SINGLE CRYSTA | 08-1353
F TRIGLYCINE SULFATE SINGLE CRYSTALS. | PREPARATION OF PURE GLYCINE USED FOR GROWING O | 27-3085
ME INVESTIGATIONS OF THEIR QUALITY. (VOLUME DISTRIBUTION OF/ | PREPARATION OF PURE TGS SINGLE CRYSTALS AND SO | 27-3086
ION EXCHANGE) | PREPARATION OF PURE TRIGLYCINE SULFATE. (USING | 27-3087
FATE. | PREPARATION OF SINGLE CRYSTALS OF AMMONIUM SUL | 19-2758
CERAMICS FROM NIOBIUM DOPED BARIUM TITANATE. | PREPARATION OF SMALL GRAINED AND LARGE GRAINED | 04-0765
LEAD MOLYBDATE. | PREPARATION OF SODIUM AND BARIUM NIOBATES AND | 04-0657
) LEAD(0.12) CALCIUM(0.08) TITANATE. | PREPARATION OF THE TERNARY TITANATE BARIUM(0.8 | 36-3587
C DIODE SPUTTERING (ONTO PLATINUM CR FUSED QUARTZ). | PREPARATION OF THIN BARIUM TITANATE FILMS BY D | 04-0948
ICS. (HCT PRESSING) | PREPARATION OF TRANSPARENT FERROELECTRIC CERAM | 02-0103
(SUBLIMATION IN AIR OF TUNGSTEN TRIOXIDE). | PREPARATION OF TUNGSTIC OXIDE SINGLE CRYSTALS | 10-1644
ROGEN OR DIDEUTERIUM) PHCSPHATE CRYSTAL/ CONTRIBUTION TO THE | PREPARATION ON NON-WEDGE SHAPED RUBIDIUM DIHYD | 17-2637
CHEMISTRY AND PHYSICS OF (ALKALI) METAL TANTALATES. (PREPARATION, REVIEW) | 02-0048
EMICONDUCTOR WITH PZT. ATTEMPT TO | PREPARE HIGH DOPED BARIUM STRONTIUM TITANATE S | 04-0717
/TRICAL PROPERTIES OF FERROELECTRIC MATERIALS. (FINE POWDERS | PREPARED BY DECOMPOSITION OF ORGANIC COMPOUND/ | 02-0093
/ICAL AND ELECTRICAL PROPERTIES OF THIN FILM BARIUM TITANATE | PREPARED BY RF SPUTTERING ON SILICON SUBSTRAT/ | 04-0828
DE OR GALLIUM ARSENIDE. BARIUM TITANATE FILMS | PREPARED BY RF SPUTTERING ONTO INDIUM ANTIMONI | 04-0741
TIMONY OR BISMUTH, SULFUR, SELENIUM/ HYDROTHERMAL METHCD FOR | PREPARING A(V)B(VI)C(VII) COMPOUNDS. (USING AN | 15-2194
.333) NIOBATE(0.667)). ELECTROOPTIC EFFECTS OF HOT | PRESSED CERAMICS IN THE SYSTEM LEAD (ZINCATE(0 | 13-1900
/ION OF LEAD ZIRCCNATE- LEAD TITANATE MATERIALS. PART-2: HOT | PRESSED COMPOSITIONS AT NEAR THEORETICAL DENS/ | 09-1578
HANUM ZIRCONATE TITANATE SYSTEM. IMPROVED HOT | PRESSED ELECTROOPTIC CERAMICS IN THE LEAD LANT | 09-1528
ERAMICS. (IN RELATION TO DENSI/ ELECTRICAL PROPERTIES OF HOT | PRESSED FERROELECTRIC (BARIUM TITANATE TYPE) C | 04-0879
CAL APPLICATIONS. (LEAD ZIRCONATE TITANATE AND SODIUM P/ HOT | PRESSED FERROELECTRIC CERAMICS FOR ELECTROOPTI | 09-1527
CAL PROPERTIES (ELECTRICALLY VARIAELE BIREFRINGENCE) OF (HOT | PRESSED) FERROELECTRIC CERAMICS. (PZT) OPTI | 09-1554
CERAMICS FOR ELECTROOPTICAL APPLICATIONS. HOT | PRESSED FERROELECTRIC LEAD ZIRCONATE TITANATE | 09-1524
) FERROELECTRIC CERAMICS FOR ELECTROOPTIC APPLICATIONS. HOT | PRESSED (LEAD, LANTHANUM) (ZIRCONATE, TITANATE | 09-1526
INING BISMUTH. | HOT PRESSED LEAD ZIRCONATE TITANATE CERAMICS CONTA | 09-1525

IN SIZE, / RECENT DEVELOPMENTS IN ELECTROOPTIC CERAMICS (HOT PRESSED LEAD ZIRCONATE TITANATE, EFFECT OF GRA 02-0086
/ICS- NEW MATERIALS FOR INFORMATION STORAGE AND DISPLAY (HOT PRESSED LEAD ZIRCONATE TITANATE, TUTORIAL, 17/ 09-1550
    PREPARATION OF TRANSPARENT FERROELECTRIC CERAMICS. (HOT PRESSING) 02-0103
XY-DERIVED PLZT AND PLZT.   COLD PRESSING AND LOW TEMPERATURE SINTERING OF ALKO 09-1493
         CONTINUOUS HOT PRESSING OF POTASSIUM SODIUM NIOBATE. 13-2007
               HOT PRESSING OF POTASSIUM- SODIUM NIOBATE. 09-1542
IN KDP AND LITHIUM THALLIUM TARTRATE CRYSTALS BY HYDROSTATIC PRESSURE.   DISPLACEMENT OF THE CURIE POINTS 17-2523
M TITANATE ZIRCONATES) IN STRONG ELECTRIC FIELDS AND AT HIGH PRESSURE. /ES OF CERAMIC FERROELECTRICS (BARIU 14-2078
FERROELECTRIC TRANSITION IN COBALT- IODINE- BORACITE UNDER PRESSURE. 14-2078
E(1-X) SOLID SOLUTIONS DEPEND ON TEMPERATURE AND HYDROSTATIC PRESSURE. /S OF ANTIMONY SULFOIODIDE(X) BROMID 15-2149
  MELTING PROPERTIES OF POTASSIUM NITRATE UNDER HIGH PRESSURE. 16-2265
OSPHATE SYSTEM VERSUS CONCENTRATION, TEMPERATURE, FIELD, AND PRESSURE. /TRIDEUTERIUM(X) TRIHYDROGEN(1-X) PH 17-2666
  PHASE TRANSITION IN TUNGSTEN TRIOXIDE UNDER HIGH PRESSURE. 10-1658
AND DEUTERO POTASSIUM DIDEUTERIUM PHOSPHATE WITH HYDROSTATIC PRESSURE. /IN POTASSIUM DIDEUTERIUM PHOSPHATE 17-2601
  THE BEHAVIOR OF DIAMMONIUM DICADMIUM TRISULFATE UNDER PRESSURE. 19-2757
ATES IN POTASSIUM DIHYDROGEN PHOSPHATE TYPE CRYSTALS AT HIGH PRESSURE. /ROELECTRIC AND ANTIFERROELECTRIC ST 17-2642
/RATION OF ANTIMONY SULFOIODIDE TEXTURES FROM THE MELT UNDER PRESSURE AND INVESTIGATION OF THEIR ELECTRICA/ 15-2138
LECTRIC PROPERTIES AND PHASE TRANSITIONS OF THE FERROELECTR/ PRESSURE AND TEMPERATURE DEPENDENCE OF THE DIE 08-1434
LECTRIC PROPERTIES AND PHASE TRANSITIONS OF THE ANTIFERROEL/ PRESSURE AND TEMPERATURE DEPENDENCE OF THE DIE 08-1433
ER OF STRONTIUM TITANATE.   HIGH PRESSURE COMPRESSIBILITY AND GRUNEISEN PARAMET 06-1121
E, NICKEL CARBONATE AND BLACK PHOSPHORUS.   HIGH GAS PRESSURE CRYSTAL GROWTH OF ANTIMONY SULFOIODID 15-2137
IC MODE INTERACTIONS IN FERROELECTRIC BARIUM TITANATE. PRESSURE DEPENDENCE OF ACOUSTIC MODE- SOFT OPT 04-0893
TION TEMPERATURE OF CESIUM LEAD TRICHLORIDE. PRESSURE DEPENDENCE OF CUBIC TETRAGONAL TRANSI 33-3393
TTICE DEFECTS. (ICE)   INTERPRETATION OF THE PRESSURE DEPENDENCE OF PROPERTIES CAUSED BY LA 34-3434
COEFFICIENT OF ICE I(H) SINGLE CRYSTALS. PRESSURE DEPENDENCE OF THE COMPLEX DISPERSION 34-3479
AND SPONTANEOUS POLARIZATION OF FERROELECT/ TEMPERATURE AND PRESSURE DEPENDENCE OF THE DIELECTRIC CONSTANT 16-2331
S OF PARAELECTRIC MATERIALS. (BARIUM TITANATE, STRONTIUM TI/ PRESSURE DEPENDENCE OF THE DIELECTRIC CONSTANT 04-0856
OF STRONTIUM TITANATE NEAR THE TRANSITION TEMPERATURE. PRESSURE DEPENDENCE OF THE ELASTIC COMPLIANCE 06-1219
          STRONTIUM TITANATE. PRESSURE DEPENDENCE OF THE ELASTIC CONSTANT OF 06-1097
F STRONTIUM TITANATE. PRESSURE DEPENDENCE OF THE ELASTIC CONSTANTS O 06-1096
RTIES OF POTASSIUM DIDEUTERIUM PHOSPHATE. PRESSURE DEPENDENCE OF THE FERROELECTRIC PROPE 17-2641
AND ELECTROSTRICTIONS IN PEROVSKITE BARIUM TITANATE. PRESSURE DEPENDENCE OF TRANSITION TEMPERATURES 04-0853
IES.   PZT-5 UNDER PRESSURE, DIELECTRIC AND PIEZOELECTRIC PROPERT 09-1496
IME IN FERROELECTRIC SILVER SODIUM DINITRITE. HYDROSTATIC PRESSURE EFFECT ON THE DIELECTRIC RELAXATION T 16-2294
SODIUM TRIDEUTERIUM(X) TRIHYDROGEN(1-X) SELENITE CRYSTALS. (PRESSURE EFFECTS) /ITY OF A NEW DELTA PHASE OF 22-2875
DAMAGE ELECTRON SPIN RESONANCE SP/ DIELECTRIC PROPERTIES AND PRESSURE EFFECTS ON SELENITE (ION). RADIATION 22-2870
STRUCTURE OF POTASSIUM DIHYDROGEN PHOSPHATE TYPE CRYSTALS. PRESSURE EFFECTS ON THE LATTICE PARAMETERS AND 17-2588
LEAD NIOBIUM ZIRCONATE TITANATE STANNATE BY TEMPERATURE AND PRESSURE ENFORCED PHASE TRANSITIONS. /NERGY IN 09-1488
    DEUTERON ARRANGEMENTS IN THE HIGH PRESSURE FORMS OF ICE. 34-3438
    PHASE CHANGES INDUCED BY HYDROSTATIC PRESSURE IN A FERROELECTRIC MATERIAL. (PZT) 09-1521
N IN POTASSIUM NITRATE. PRESSURE INDUCED FERROELECTRIC PHASE TRANSITIO 16-2378
SULFOIODIDE.   SOFT PHONON COUPLINGS IN THE PRESSURE INDUCED PHASE TRANSITION IN ANTIMONY 15-2240
.   VAPOR PRESSURE ISOTOPE EFFECT OF ICE AND ITS ISOMERS 34-3491
WITH NEW RESULTS ON LEAD ZIRCONATE / EFFECTS OF HYDROSTATIC PRESSURE ON FERROELECTRIC PROPERTIES. (SUMMARY 07-1263
ATE.   EFFECT OF HYDROSTATIC PRESSURE ON FERROELECTRICITY OF POTASSIUM NITR 16-2344
YBDATE.   EFFECT OF HYDROSTATIC PRESSURE ON PHASE TRANSITION IN GADOLINIUM MOL 30-3321
RES: STRONTIUM TITANATE AND ((BARIUM(X) STRONT/ INFLUENCE OF PRESSURE ON PHASE TRANSITIONS AT LOW TEMPERATU 04-0898
ROGEN SELENITE CRYSTALS.   EFFECT OF HIGH PRESSURE ON PHASE TRANSITIONS OF ALKALI TRIHYD 22-2882
M TITANATE.   INFLUENCE OF OXYGEN PARTIAL PRESSURE ON PROPERTIES OF SEMICONDUCTING BARIU 04-0999
/DITION ON THE PRODUCT OF PZT. PART-5: EFFECT OF THE FORMING PRESSURE ON SINTERING. PART-6: FORMATION OF P/ 09-1624
FATE AND ROCHELLE SALT.   EFFECT OF A TWO DIMENSIONAL PRESSURE ON THE CURIE POINTS OF TRIGLYCINE SUL 27-3082
AELECTRIC MATERIAL. (POTASSIUM TANTALATE)   EFFECT OF PRESSURE ON THE DIELECTRIC PROPERTIES OF A PAR 08-1335
E SPECTRA OF SUBSTITUTIONAL IMPUR/ EFFECT OF TEMPERATURE AND PRESSURE ON THE ELECTRON PARAMAGNETIC RESONANC 06-1193
OF AMMONIUM SULFATE.   EFFECT OF HYDROSTATIC PRESSURE ON THE FERROELECTRIC PHASE TRANSITION 16-2360
C PHASE TRANSITIONS.   EFFECT OF HYDROSTATIC PRESSURE ON THE NATURE OF DIFFUSE FERROELECTRI 03-0442
F FERROELECTRIC CRYSTALS OF THE PO/ INFLUENCE OF HYDROSTATIC PRESSURE ON THE PHASE TRANSITION TEMPERATURE O 17-2495
POTASSIUM TANTALATE.   EFFECT OF PRESSURE ON THE STATIC DIELECTRIC CONSTANT OF 08-1269
X-RAY INVESTIGATION OF THE EFFECT OF HYDROSTATIC PRESSURE ON THE STRUCTURE OF BARIUM TITANATE. 04-0760
AN X-RAY STUDY OF THE EFFECT OF HYDROSTATIC PRESSURE ON THE STRUCTURE OF LEAD TITANATE. 05-1064
ULFATE.   EFFECT OF PRESSURE ON THE SWITCHING RATE OF TRIGLYCINE S 27-3014
3) NIOBATE(0.67).   THE EFFECT OF PRESSURE ON THE TRANSITION OF LEAD ZINCATE (0.3 04-1473
NIUM SULFATE.   EFFECT OF THE HYDROSTATIC PRESSURE ON THE TRANSITION TEMPERATURE IN AMMO 19-2796
INE SULFATE.   EFFECT OF HYDROSTATIC PRESSURE ON THE UNIPOLAR PROPERTIES OF TRIGLYC 27-3053
TE.   NEW HIGH PRESSURE PHASE TRANSITION IN TRIGLYCINE SELENA 27-3090
N ROCHELLE SALT.   HIGH PRESSURE PHASE TRANSITIONS OF THE FIRST KIND I 28-3214
TRANSITION RATES OF POTASSIUM NITRATE HIGH PRESSURE POLYMORPHS. 16-2279
OF ADDITIVES ON DENSIFICATION AND PROPERTIES OF SINTERED AN PRESSURE SINTERED BARIUM TITANATE.   INFLUENCE 04-1026
(PEROVSKITES, KDP TYPE) PRESSURE STUDIES OF FERROELECTRIC PROPERTIES. 08-1435
PE PEROVSKITE.   HIGH PRESSURE SYNTHESIS OF LEAD METAL- OXYGEN(3) TY 08-1331
ERING/ AND PROPERTIES OF PEROVSKITE TYPE LEAD ZINCATE(/ HIGH PRESSURE SYNTHESIS (SOLID STATE REACTION, SINT 01-1393
RSAL IN BARIUM TITANATE.   EFFECT OF PRESSURE (TO ABOUT 2KBAR) ON POLARIZATION REVE 04-0718
TE, LEAD ZIRCONATE TITANATE, LEAD METANIOBATE, 20-150 DEG C, PRESSURE (TO 1200 KG/SQ CM) /ES. (BARIUM TITANA 04-0985
.1) STRONTIUM(0.9-1) TITANATE SOLID SOLUTIONS BY HYDROSTATIC PRESSURE. (TO 1500 ATM) /ERATURE OF BARIUM(0-0 04-0722
HAVIOR OF POTASSIUM FERROCYANIDE TRIHYDRATE CRYSTALS AT HIGH PRESSURE. (TO 5100 KG/SQ CM) FERROELECTRIC BE 24-2897
/SIUM FERROCYANIDE TRIHYDRATE CRYSTAL UNDER HIGH HYDROSTATIC PRESSURE (TO 5600 KG/SQ CM, TEMPERATURE DEPEN/ 24-2896
TANATE SOLID SOLUTIONS NEAR THE PHASE TRANSITION, UNDER HIGH PRESSURES. /RROELECTRIC (BARIUM, STRONTIUM) TI 04-0680
EAD TITANATE- LEAD (FERRATE NIOBATE) SOLID SOLUTIONS AT HIGH PRESSURES.   DIELECTRIC PROPERTIES OF L 05-1071
TRIC LOSSES OF FERROELECTRIC PEROVSKITES SYNTHESIZED AT HIGH PRESSURES.   STRUCTURE AND DIELEC 08-1452
F FERROELECTRICITY IN POTASSIUM DIHYDROGEN PHOSPHATE AT HIGH PRESSURES.   VANISHING O 17-2442
SYNTHESIS OF THALLIUM CONTAINING PEROVSKITES AT HIGH PRESSURES AND THEIR X-RAY ANALYSIS. 08-1463
TRIC ANOMALIES IN SOLID METHANE AT LOW TEMPERATURES AND HIGH PRESSURES. (POSSIBLY FERROELECTRIC) DIELEC 36-3547
SPONTANEOUS POLARIZATION IN LITHIUM TANTALATE (THERMAL EXP/ PRESSURES. PRIMARY AND SECONDARY PYROELECTRIC EFFECTS AND 12-1884
DIHYDROGEN PHOSPHATE CRYSTALS.   MEASUREMENT OF THE PRIMARY LINEAR ELECTROOPTIC EFFECT IN AMMONIUM 17-2654
TROOPTIC LIGHT BEAM DEFLECTION WITH BARIUM STRONTIUM NIOBATE PRISM.   ELEC 13-2036
LIGHT BEAM DEFLECTION CONTROLLED WITH A PRISM OF LEAD MAGNESIO NIOBATE CRYSTALS. 08-1280
REACTIONS (INVOLVING TITANIUM ESTERS TO YIELD SMALL CRYSTAL PRISMS). /TION OF BARIUM TITANATE BY TRANSPORT 04-0625
SE TRANSITION IN FERROELECTRICS. ABRUPT (STEPWISE) CHANGE IN PROBABILITY OF THE MOSSBAUER EFFECT AT THE PHA 03-0276
NATE SOLID SOLUTIONS IN THE / CHANGE OF THE MOSSBAUER EFFECT PROBABILITY ON TIN-119 IN BARIUM TITANATE STAN 04-0573
PORATED BARIUM TITANATE FILMS.   ELECTRON PROBE X-RAY MICROANALYSIS ON IMPURITIES IN EVA 04-0742
RATION (IN LITHIUM NIOBATE)   OPTICAL PROBING OF ACOUSTIC SURFACE WAVE HARMONIC GENE 11-1752

R INTERACTIONS IN FERROELECTRIC CERAMICS TO MICROWAVE SIGNAL PROCESSING.   APPLICATION OF NONLINEA 37-3647
CTRICALLY CONTR/ ELECTROOPTIC CERAMIC (INFORMATION) STORAGE (PROCESSING) AND DISPLAY DEVICES. (BASED ON ELE 37-3650
/EN STOICHIOMETRY AND HYDROXYL CONTENT ON THE PROPERTIES AND PROCESSING OF SINGLE CRYSTAL BARIUM SODIUM NI/ 13-1915
H/ ELECTRON BEAM SENSING OF SURFACE ELASTIC WAVES FOR SIGNAL PROCESSING PURPOSES. (TECHNIQUE APPLIED TO LIT 11-1738
/XIDE AND PZT. PART-4: EFFECT OF BORON OXIDE ADDITION ON THE PRODUCT OF PZT. PART-5: EFFECT OF THE FORMING/ 09-1624
MONY SULFOBROMIDE IN GLASSY AND CRYSTALLINE STATES.   PRODUCTION AND SOME OPTICAL PROPERTIES OF ANTI 15-2142
TITANATE SINGLE CRYSTAL   PRODUCTION AND STUDY OF PURE AND DOPED BARIUM 04-0667
SEMICONDUCTING BARIUM TITANATE CERAMICS.   PRODUCTION AND STUDY OF THE INNER BARRIERS IN 04-0583
IN THIN (110) PLATES BELOW CUBIC- TETRAGONAL TRANSITION, AND PRODUCTION BY SHAPING AND STRESSING). (URANCE 06-1164
OF BISMUTH TELLUROIODIDE.   PRODUCTION METHODS AND SOME OPTICAL PROPERTIES 15-2141
ATE AND TITANIUM DIOXIDE IN THE PRESENCE OF SODIUM CARB/ THE PRODUCTION OF BARIUM TITANATE FROM BARIUM SULF 04-0624
ICS AND (TABLES OF) THEIR PROPERTIES.   ADVANCES IN PRODUCTION OF (COMMERCIAL) PIEZOELECTRIC CERAM 02-0110
                                                             PRODUCTION OF LARGE SEED CRYSTALS OF KDP. 17-2413
IN IRRADIATED POTASSIUM NI/ PARAMAGNETIC RESONANCE STUDY OF PRODUCTION OF NITROGEN DIOXIDE AND NITRITE ION 16-2404
EFFECTS OF DISPERSION AND FOCUSING ON THE PRODUCTION OF OPTICAL HARMONICS. (KDP) 17-2579
TIC INSTABILITIES IN PIEZOELECTRIC MATERIALS. (GADOLINIUM M/ PRODUCTION OF SPONTANEOUS POLARIZATION BY ELAS 30-3282
NEW CRYSTAL FOR COMMUNICATIONS SYSTEMS ENTERS PRODUCTION STAGE. (LITHIUM TANTALATE) 12-1867
HE BIT SIZE OF THE FE-PC MEMO/ EFFECT OF GAUSSIAN IRRADIANCE PROFILE ON THE SWITCHING CHARACTERISTICS AND T 37-3670
/CS OF THIN, MULTILAYERED FERROELECTRIC CERAMICS OBTAINED BY PROJECTION WITH A PLASMA TORCH OR FRITTING. (/ 04-0778
OGEN PHOSPHATE.   EXPERIMENTAL PROOF FOR PROTON TUNNELING IN POTASSIUM DIHYDR 17-2510
/OF SECOND AND THIRD HARMONICS BY A TRANSVERSE ACOUSTIC WAVE PROPAGATING ITSELF IN THE (001) DIRECTION OF / 11-1682
S AND PIEZOELECTRIC LEAKY SURFACE WAVES IN LITHIUM NIOBATE. PROPAGATION AND AMPLIFICATION OF RAYLEIGH WAVE 11-1856
ERROELEC/ EXPERIMENTAL METHODS FOR INVESTIGATING STRAIN WAVE PROPAGATION AND ASSOCIATED CHARGE RELEASE TO F 37-3630
WAVES ON BARIUM TITANATE(0.95) ZIRCONATE(0.05) CERAMIC PLA/ PROPAGATION CHARACTERISTICS OF SURFACE ELASTIC 09-1613
CUBIC SODIUM CHLORIDE AND EPITAXI/ EFFECT OF TEMPERATURE AND PROPAGATION DIRECTION UPON OPTICAL PHONONS IN 29-3243
T THE FERROELECTRIC TRANSITION TEMPERATURE.   ACOUSTIC PROPAGATION IN POTASSIUM DEUTERIUM PHOSPHATE A 17-2573
ON IN STRONTIUM TITANATE.   SOUND PROPAGATION NEAR THE STRUCTURAL PHASE TRANSITI 06-1155
STRONTIUM TITANATE.   THERMAL EXPANSIVITY AND ULTRASONIC PROPAGATION NEAR THE STRUCTURAL TRANSITION OF 06-1135
UM NIOBATE AND QUARTZ.   EFFECT OF MASS LOADING ON THE PROPAGATION OF ACOUSTIC SURFACE WAVES ON LITHI 11-1757
EEN DOMAINS. (IN KDP)   VELOCITY OF PROPAGATION OF DOMAIN WALLS: INTERACTIONS BETW 17-2453
LECTRIC CERAMIC/ EFFECTS OF THERMALLY DIFFUSED IMPURITIES ON PROPAGATION OF ELASTIC SURFACE WAVES ON FERROE 03-0487
ES IN FERROELECTRIC CRYSTALS.   NONLINEAR EFFECTS IN THE PROPAGATION OF FINITE AMPLITUDE HYPERSONIC WAV 03-0498
BATE CRYSTALS. (80-950 DEG K, 200-2000 MHZ)   GENERATION AND PROPAGATION OF HYPERSONIC WAVES IN LITHIUM NIO 11-1753
N POTASSIUM DIDEUTERIUM PHOSPHATE (TEMPERATURE DE/ ANOMALOUS PROPAGATION OF LONGITUDINAL ULTRASONIC WAVES I 17-2660
ALCULATIONS FOR LITHIUM/ NONLINEAR SADOVSKII EFFECT (THEORY, PROPAGATION OF RADIATION AND FIRST HARMONIC, C 03-0186
GOLD ON LIT/ FUNDAMENTAL AND HIGHER ORDER MODE SURFACE WAVE PROPAGATION ON LAYERED ANISOTROPIC SUBSTRATES: 11-1846
/: DIELECTRIC AND SWITCHING MEASUREMENTS. PART-8: ULTRASOUND PROPAGATION. PART-9: SPECIFIC HEAT. PART-10: / 03-0397
ARAMETERS AT THE DISPLACIVE TRANSITION IN STRONT/ ULTRASONIC PROPAGATION, STRESS EFFECTS, AND INTERACTION P 06-1127
METHYL AMMONIUM ALUMINUM SULFATE)   PHYSICAL PROPERITES OF FERROELECTRIC ALUMS. (ESPECIALLY 20-2810
AXATION TIME NEAR THE CRITICAL TEMPERATURE OF DICALCIUM LEAD PROPIONATE. /LY OF THE PROTON SPIN LATTICE REL 26-2947
OBSERVATION OF FERROELECTRIC DOMAINS OF DICALCIUM STRONTIUM PROPIONATE.   ELECTRON MIRROR MACROSCOPIC 26-2946
OELECTRIC DOMAINS OF BARIUM TITANATE AND DICALCIUM STRONTIUM PROPIONATE. /IRROR MICROSCOPIC ASPECTS OF FERR 04-0973
NMR INVESTIGATIONS OF FERROELECTRIC DICALCIUM STRONTIUM PROPIONATE. 26-2931
OELECTRICS: POLARIZATION FLUCTUATIONS IN DICALCIUM STRONTIUM PROPIONATE. /WAVES IN ORDER DISORDER TYPE FERR 26-2928
PERATURE DEPENDENCE OF SOUND VELOCITY IN DICALCIUM STRONTIUM PROPIONATE.   TEM 26-2934
NON NEAR THE 60 DEGREES C PHASE TRANSITION IN LEAD DICALCIUM PROPIONATE.   THE ANOMALOUS BEHAVIOR OF PHO 26-2942
TTENUATION NEAR THE CURIE TEMPERATURE IN DICALCIUM STRONTIUM PROPIONATE.   ULTRASONIC A 26-2948
ONIC ATTENUATION NEAR THE PHASE TRANSITION IN DICALCIUM LEAD PROPIONATE   ULTRAS 26-2949
OPTICAL ACTIVITY OF FERROELECTRIC DICALCIUM STRONTIUM PROPIONATE. (TRANSITION AT 209 DEGREES K) 27-3040
ONS.   PHASE TRANSITION IN DIVALENT METAL DICALCIUM PROPIONATES SUBSTITUTED PARTIALLY BY ACETATE I 26-2944
-2: THE T(1 RHO) MINIMUM.   PROTEIN SPIN RELAXATION IN HEXAGONAL ICE. PART 34-3488
/TRUCTURAL STUDIES OF ICE POLYMORPHS BY NEUTRON DIFFRACTION, PROTON AND DEUTERON NUCLEAR MAGNETIC RESONANC/ 34-3466
UDIES OF FERROELECTRIC SOLIDS BY MAGNETIC RESONANCE PART-18: PROTON AND DEUTERON RESONANCE OF THIOUREA.   ST 25-2919
.   PROTON- PROTON AND PROTON- LATTICE INTERACTIONS IN ICE 34-3430
ACTION IN ANTIFERROELECTRIC DIAMMONIUM TRIHYDROGEN PERIODATE PROTON CROSSOVER RELAXATION. /QUADRUPOLE INTER 36-3578
ANTIFERROELECTRIC DIAMMONIUM TRIHYDROGEN PERIODATE BY IODINE PROTON CROSSOVER RELAXATION. / INTERACTION IN 36-3579
TRIC TRANSITION IN TRIGLYCINE SULFATE.   PULSED NITROGEN PROTON DOUBLE RESONANCE STUDY OF THE FERROELEC 27-2968
                                                             PROTON DYNAMICS IN FERROELECTRIC COLEMANITE. 23-2887
ICS. PART-2: APPROXIMATED SOLUTIONS TO THE KINETIC EQUATION/ PROTON DYNAMICS IN HYDROGEN BONDED FERROELECTR 03-0205
/NTUM THEORY OF FERROELECTRICITY, PART-1: SPECTRAL ANALYSIS (PROTON DYNAMICS IN HYDROGEN BONDED FERROELECT/ 03-0402
ICS ABOVE THE CURIE TEMPERATURE.   PROTON DYNAMICS IN HYDROGEN BONDED FERROELECTR 03-0206
DIED BY INFRARED ABSORPTION.   PROTON DYNAMICS IN KDP TYPE FERROELECTRICS STU 17-2708
TRICS OF THE KDP TYPE (PHONON FRO/ COUPLED VIBRATIONS OF THE PROTON ION SYSTEM IN HYDROGEN BONDED FERROELEC 17-2670
/ETIC RESONANCE AND ELECTRON NUCLEAR DOUBLE RESONANCE OF THE PROTON LATTICE COUPLED MODE IN POTASSIUM DIHY/ 18-2724
SUSCEPTIBILITY OF POTASSIUM DIHYDROGEN PHOSPHATE TYPE FERRO/ PROTON LATTICE INTERACTIONS AND THE DYNAMICAL 17-2570
FERROELECTRIC CRYSTALS. (THEORY, KDP TYPE)   PROTON LATTICE INTERACTIONS IN HYDROGEN BONDED 03-0181
/ AMMONIUM ALUMINUM SULFATE AND A 36:64 RATIO MIXED CRYSTAL, PROTON MAGNETIC RESONANCE, DIELECTRIC CONSTAN/ 20-2809
: DEUTERON SPIN/ NMR INVESTIGATIONS OF ICE CRYSTALS. PART-1: PROTON MAGNETIC RESONANCE IN WATER ICE. PART-2 34-3471
YDROGEN SELENATE AND POTASSIUM TRIHYDROGEN SELENATE.   PROTON MAGNETIC RESONANCE STUDY OF SODIUM TRIH 22-2876
MMETRY STATES OF AMMONIUM IONS IN SOLIDS. (AMMONIUM SULFATE/ PROTON MAGNETIC RESONANCE STUDY OF THE SPIN SY 19-2800
/ ORIENTATION MOBILITY: METAL AMMONIUM (BX4) TYPE COMPOUNDS (PROTON MAGNETIC RESONANCE, 84 DEG K AND 120-3/ 22-2826
C AND ANTIFERROELECTRIC SOLIDS. PART-1: ENERGY LOSS NEUTRON/ PROTON MOTIONS IN HYDROGEN BONDED FERROELECTR 17-2687
T HIGH TEMPERATURES.   ABSENCE OF MOTIONAL NARROWING IN THE PROTON NMR IN POTASSIUM DIHYDROGEN PHOSPHATE A 17-2594
N HEXAGONAL ICE.   RELAXATIONAL PROTON ORDERING AND GLASSY CRYSTALLINE STATE I 34-3436
NANCE IN ROCHELLE SALT, DETERMINATION OF THE ORIENTATIONS OF PROTON PAIRS (IN THE WATER MOLECULES). /C RESO 28-3203
SENATE AND POTASSIUM DIHYDROGEN ARSENATE.   PROTON PHONON COUPLING IN CESIUM DIHYDROGEN AR 18-2733
TH HYDROGEN BONDS. (KDP TYPE)   PROTON PHONON INTERACTION IN FERROELECTRICS WI 17-2672
ION OF FERROELECTRICS WITH HYDROGEN BONDS KDP TYPE. ROLE OF PROTON PHONON INTERACTION IN THE PHASE TRANSIT 17-2674
TE AND CESIUM DIHYDROGEN ARSENATE.   PHONON PHONON AND PROTON PHONON INTERACTIONS IN ALUMINUM PHOSPHA 18-2739
OGEN PHOSPHATE.   PROTON PHONON INTERACTIONS IN POTASSIUM DIHYDR 17-2678
THE THREE DIMENSIONAL EIGHT VERTEX MODEL AND THE PROTON PROTON CORRELATION FUNCTIONS IN ICE. 34-3481
AND SPIN DEVIATION IN COPPER FORMATE TETRADEUTERATE.   PROTON RESONANCE STUDY OF SUBLATTICE ROTATIONS 26-2950
DRAZINIUM SULFATE.   PROTON ROTATING FRAME RELAXATION IN LITHIUM HY 32-3347
PHOSPHATE TYPE CRYSTALS.   NMR PROTON SECOND MOMENTS IN POTASSIUM DIHYDROGEN 17-2408
C SODIUM AMMONIUM TARTRATE.   PROTON SPIN LATTICE RELAXATION IN FERROELECTRI 29-3248
C COLEMANITE.   PROTON SPIN LATTICE RELAXATION IN FERROELECTRI 23-2885
C SODIUM TRIHYDROGEN SELENATE.   PROTON SPIN LATTICE RELAXATION IN FERROELECTRI 22-2852
E.   INTERPRETATION OF THE PROTON SPIN LATTICE RELAXATION IN HEXAGONAL IC 34-3475
ODIUM TRIHYDROGEN SELENATE.   PROTON SPIN LATTICE RELAXATION IN IRRADIATED S 22-2853
RITICAL TEMPERATURE OF DICALCIUM LEAD PROPIO/ ANOMALY OF THE PROTON SPIN LATTICE RELAXATION TIME NEAR THE C 26-2947

IUM DIHYDROGEN PHOSPHATE. INTERACTIONS BETWEEN	PROTON TUNNELING AND OPTICAL PHONONS IN POTASS 17-2592
IUM DIHYDROGEN PHOSPHATE. INTERACTION BETWEEN	PROTON TUNNELING AND OPTICAL PHONONS IN POTASS 17-2593
HATE. EXPERIMENTAL PROOF FOR	PROTON TUNNELING IN POTASSIUM DIHYDROGEN PHOSP 17-2510
HATE TYPE FERROELECTRICS. (THEORY) ENERGY PARAMETERS AND	PROTON TUNNELING IN POTASSIUM DIHYDROGEN PHOSP 03-0506
ALS WITH HYDROGEN BONDING. (POTASSIUM DIHYDROGE/ DYNAMICS OF	PROTONS (AND DEUTERONS) IN FERROELECTRIC CRYST 17-2568
PHONONS, ELECTRONS, AND	PROTONS IN FERROELECTRIC MATERIALS. 02-0116
CRYSTALS. ABSENCE OF A LINE WIDTH TRANSITION FOR THE	PROTONS IN POTASSIUM DIHYDROGEN PHOSPHATE TYPE 17-2436
ALT AND AMMONIUM ROCHELLE SALT BY SPIN LATTICE RELAXATION OF	PROTONS IN ROTATING FRAME. /ULES IN ROCHELLE S 28-3232
MPERATURES ABOVE AND BELOW THE FERROELECTRIC / CHANNELING OF	PROTONS IN THIN BARIUM TITANATE CRYSTALS AT TE 04-0687
ICS WITH HYDROGEN BONDS. (KDP,/ INFLUENCE OF POLAR STATES OF	PROTONS ON THE PHASE TRANSITION IN FERROELECTR 17-2671
ODE IN POTASSIUM DIHYDROGEN PHOSPHATE. (COLLECTIV/ EFFECT OF	PROTON- PHONON COUPLING ON THE FERROELECTRIC M 17-2655
S IN ICE.	PROTON- PROTON AND PROTON- LATTICE INTERACTION 34-3430
LS. PART-2: MOLECULAR MODELS FOR POLARIZATION A/ TRANSFER OF	PROTONS THROUGH "PURE" ICE I(H), SINGLE CRYSTA 34-3486
PART-1: POLARIZATION SPECTRA OF ICE I(H). PART/ TRANSFER OF	PROTONS THROUGH PURE ICE I(H) SINGLE CRYSTALS. 34-3486
/OUS SOLUTIONS AND THEIR INTERPRETATION. PART-8: TRANSFER OF	PROTONS THROUGH "PURE" ICE I(H) SINGLE CRYSTA/ 34-3482
/FERROELECTRICS. PART-1: DAMPINGS AND (FREQUENCY) SHIFTS (OF	PROTONIC COLLECTIVE MOTION, TUNNELING, CALCUL/ 03-0475
SENATE.	PROTONIC CONDUCTION IN POTASSIUM DIHYDROGEN AR 18-2737
RATURE REGION. PART-2: LOW TEMPERATURE REGION.	PROTONIC CONDUCTION OF ICE. PART-1: HIGH TEMPE 34-3414
	PROTONIC SEMICONDUCTORS. (ICE) 34-3461
LITY OR ELASTIC COMPLIANCE IS TEMPERATURE INDEPENDENT IN THE	PROTOTYPIC PHASE. /HAT THE ELECTRIC SUSCEPTIBI 03-0140
OPTICALLY INDUCED PHYSICAL DAMAGE TO LITHIUM NIOBATE,	PROUSTITE, AND LITHIUM IODATE. 08-1328
/BATE OR TANTALATE, BARIUM SODIUM NIOBATE, CUPROUS CHLORIDE,	PROUSTITE, CALCIUM PYRONIOBATE, AND BISMUTH(1/ 01-0021
AND LEAD, BARIUM, AND STRONTIUM OXIDES. BINARY AND	PSEUDOBINARY PHASES BETWEEN NIOBIUM PENTOXIDE 04-0590
ON MICROSCOPE STUDY OF REDUCED TUNGSTEN TRIOXIDE AND RELATED	PSEUDOBINARY SYSTEMS. /RUCTURE TYPE: AN ELECTR 10-1630
L BARIUM TITANATE STRUCTURE TYPES. ELECTROSTATIC FIELD IN A	PSEUDOCUBIC IONIC STRUCTURE. APPLICATION TO AL 04-1008
DOMAINS IN POTASSIUM NIOBATE SINGLE CRYSTALS FROM	PSEUDOCUBIC (001) CLEAVAGE PLANES. 08-1317
SIUM NIOBATE SINGLE CRYSTALS ASSOCIATED WITH STEP LADDERS ON	PSEUDOCUBIC (001) PLANES. /STRUCTURES IN POTAS 08-1316
ELECTRICS WITH A SHORT RANGE INTER/ CRITICAL BEHAVIOR OF THE	PSEUDOSPIN MODEL FOR ORDER DISORDER TYPE FERRO 03-0423
ELECTRIC MODE IN POTASSIUM DIHYDROGEN PHOSPHATE. (COLLECTIVE	PSEUDOSPIN MODEL OF KDP IS QUESTIONABLE) /ERRO 17-2655
/GSTEN BRONZE TYPE CRYSTALS GROWN IN MOLTEN SOLUTIONS OF THE	PSEUDOSYSTEM POTASSIUM LITHIUM BARIUM NIOBATE. 13-1948
ANATE BARIUM DINIOBATE. TUNGSTEN BRONZE FIELD IN THE	PSEUDOTERNARY SYSTEM SODIUM NIOBATE BARIUM TIT 13-1970
F RESISTIVITY) IN (DOPED AND UNDOPED CERAMIC PEROVSKITE) LE/	PTC EFFECT (POSITIVE TEMPERATURE COEFFICIENT O 13-2000
DIFFUSION POTENTIALS IN BARIUM TITANATE AND THE THEORY OF	PTC MATERIALS. 03-0352
A NEW	PTC THERMISTOR MATERIAL. (BARIUM TITANATE) 04-0931
. EFFECT OF HEAT TREATMENT ON THE	PTCR ANOMALY IN SEMICONDUCTING BARIUM TITANATE 04-0762
POTENTIAL BARRIERS ON SEMICONDUCTING BARIUM TITANATE. (PTCR DEVICES)	04-0851
/ ELECTRICAL CONDUCTIVITY MEASUREMENTS. (INTRINSIC BAND GAP,	PTCR (POSITIVE TEMPERATURE COEFFICIENT OF RES/ 04-0822
STANCE) BEHAVIOR OF BARIUM TITANATE WITH NIOBIUM PENTOXIDE /	PTCR (POSITIVE TEMPERATURE COEFFICIENT OF RESI 04-0837
TIUM EARI/ MODIFICATION TO THE CZOCHRALSKI METHOD OF CRYSTAL	PULLING (CONTROL OF CRYSTAL DIAMETER OF STRON 13-1914
/ODIUM NIOBATE SINGLE CRYSTALS FOR OPTICAL APPLICATIONS. (BY	PULLING FROM THE MELT, STRUCTURE- PROPERTY CO/ 13-2045
C GENERATION. (POTASSIUM DIHYDROGEN PHOSPHATE)	PULSE DISTORTIONS IN MISMATCHED SECOND HARMONI 17-2690
ECTRIC TRIGLYCINE FLUOBERYLLATE CRYSTALS DURING SWITCHING IN	PULSE ELECTRIC FIELD. IMPEDANCE OF FERROEL 27-3027
RS.	PULSE MEASUREMENT OF R(63) IN KDP. 17-2477
ERROELECTRICS IN MEMORY DEVICE/ UTILIZATION OF THE (CRITICAL	PULSE RESPONSE OF ZERO DEGREE CUT ADP MODULATO 17-2635
N FROM BARIUM TITANATE CRYSTALS SUBJECTED TO UNIAXIAL STRESS	PULSE WIDTH) PARTIAL SWITCHING PROPERTIES OF F 27-3144
E FERROELECTRIC SURFACES. (NEW MODEL REPRODUCES EXPERIMENTAL	PULSES. LIGHT EMISSIO 04-0672
Y SULFOIODIDE SINGLE CRYSTALS EXCITED WITH MICROSECOND LIGHT	PULSES) ON THE SWITCHING PROPERTIES OF TH 37-3641
TALS. GENERATOR OF BIPOLAR	PULSES. RELAXATION PROCESSES IN ANTIMON 15-2202
HATE CRYSTALS. LIGHT AND CURRENT	PULSES FOR INVESTIGATION OF FERROELECTRIC CRYS 37-3673
GENERATION OF FAR INFRARED RADIATION BY PICOSECOND LIGHT	PULSES FROM X-RAYED POTASSIUM DIHYDROGEN PHOSP 17-2716
/T EXCITATION OF LONGITUDINAL AND TRANSVERSE HYPERSONIC WAVE	PULSES IN LITHIUM NIOBATE. 11-1857
/PARATE EXCITATION OF LONGITUDINAL AND TRANSVERSE HYPERSONIC	PULSES IN LITHIUM NIOBATE AND BISMUTH(12) GER/ 11-1815
IUM N/ PYROELECTRIC VOLTAGE RESPONSE TO SHORT INFRARED LASER	PULSES IN PIEZOELECTRIC LITHIUM NIOBATE AND B/ 11-1814
UADRUPOLE INTERACTIONS OF NITROGEN-14 IN PARAELECTRIC TRIGL/	PULSES IN TRIGLYCINE SULFATE AND BARIUM STRONT 13-2018
ION) EXPERIMENTS IN AMMONIUM TRIHYDROGEN PERIODATE CRYSTALS,	PULSED DOUBLE RESONANCE STUDY OF THE NUCLEAR Q 27-2969
OF THE FERROELECTRIC TRANSITION IN TRIGLYCINE SULFATE.	PULSED MAGNETIC AND OPTICAL RESONANCE. /ELAXAT 36-3577
CTRIC AND FERROELECTRIC POTASSIUM FERROCYANIDE TRIHYDRATE BY	PULSED NITROGEN PROTON DOUBLE RESONANCE STUDY 27-2968
ATE CRYSTAL) OPTICAL MIXING.	PULSED NMR. STUDY OF HYDROGEN-1 IN PARAELE 24-2889
POTASSIUM DIHYDROGEN PHOSPHATE EXCITED BY A 2573 ANGSTROM CW	PULSED RUBY LASER EMISSIONS BY TRIGLYCINE SULF 27-2958
LIGHT AMPLIFICATION AND OSCILLATION IN KDP WITH MODE LOCKED	PUMP. /E IN AMMONIUM DIHYDROGEN PHOSPHATE AND 17-2486
PARAMETRIC FLUORESCENCE IN BARIUM SODIUM NIOBATE FOR NINE CW	PUMP. PARAMETRIC 17-2468
PARAMETRIC FLUORESCENCE IN BARIUM SODIUM NIOBATE FOR 10 CW	PUMP FREQUENCIES. 13-1950
/TING AND THE FERROELECTRIC PHASE TRANSITIONS (TIN TELLURIDE	PUMP WAVELENGTHS (RANGE 454.5-528.7 MICRONS). 13-1983
FLUORIDE. (20 PERCENT MANGANESE RE/ ANTIFERROELECTRICITY IN	(PURE AND) ALLOYED WITH SMALL AMOUNTS OF GERMAN/ 36-3564
	(PURE AND) COBALT DOPED POTASSIUM MANGANESE TRI 33-3366
PRODUCTION AND STUDY OF	PURE AND DOPED BARIUM TITANATE SINGLE CRYSTAL 04-0667
PREPARATION AND PROPERTIES OF	PURE AND DOPED BARIUM TITANATE SINGLE CRYSTALS 04-0668
ELECTRIC POLARIZATION EFFECTS IN	PURE AND DOPED ICE AT LOW TEMPERATURES. 34-3412
SPECIFIC HEAT OF	PURE AND DOPED ICE NEAR 120 DEGREES K. 34-3463
WN BELOW AND ABOVE THE CURIE POINT. DOMAIN STRUCTURE OF	PURE AND DOPED TRIGLYCINE SULFATE CRYSTALS GRO 27-3130
SWITCHING IN FERROELECTRIC TRIGLYCINE SULFATE,	PURE AND DOPED WITH PARAMAGNETIC IONS. 27-3020
FERROELECTRICITY AND ANTIFERROELECTRICITY IN	PURE AND NIOBIUM DOPED LEAD ZIRCONATE. 07-1251
RATE OF THE (010) PLANE OF ROCHELLE SALT IN THE PRESENCE OF	PURE DIRECT SKY BLUE DYE. GROWTH 28-3217
WET SYNTHETIC METHOD OF MAKING	PURE, EXACTLY STOICHIOMETRIC BARIUM TITANATE. 04-0790
LFATE SINGLE CRYSTALS. PREPARATION OF	PURE GLYCINE USED FOR GROWING OF TRIGLYCINE SU 27-3085
DIELECTRIC CONSTANT OF	PURE ICE I(H) SINGLE CRYSTALS. 34-3478
ATION SPECTRA OF ICE I(H). PART/ TRANSFER OF PROTONS THROUGH	PURE ICE I(H) SINGLE CRYSTALS. PART-1: POLARIZ 34-3486
ULAR MODELS FOR POLARIZATION A/ TRANSFER OF PROTONS THROUGH "PURE"	ICE I(H), SINGLE CRYSTALS. PART-2: MOLEC 34-3455
/ THEIR INTERPRETATION. PART-8: TRANSFER OF PROTONS THROUGH "PURE"	ICE I(H) SINGLE CRYSTALS. SECT-1: POLAR/ 34-3482
PLANAR GROWTH OF ICE FROM THE	PURE MELT. 34-3444
ONIUM DIHYDROGEN ARSENATE.	PURE QUADRUPOLE RESONANCE OF ARSENIC-75 IN AMM 18-2732
PREPARATION AND DISLOCATION DENSITY DETERMINATION OF LARGE,	PURE RUBIDIUM MANGANESE TRIFLUORIDE AND POTAS/ 33-3397
) SINGLE CRYSTALS. CONDUCTIVITY OF BARIUM TITANATE,	PURE SINGLE CRYSTALS AND DOPED (IRON, HYDROXYL 04-0695
NS OF THEIR QUALITY. (VOLUME DISTRIBUTION OF/ PREPARATION OF	PURE TGS SINGLE CRYSTALS AND SOME INVESTIGATIO 27-3086
PREPARATION OF	PURE TRIGLYCINE SULFATE. (USING ION EXCHANGE) 27-3087
/DERIVED STRONTIUM ZIRCONATE AND STRONTIUM TITANATE (OF HIGH	PURITY AND SURFACE ACTIVITY VIA HYDROLYTIC DE/ 06-1217
HIGH TEMPERATURE (TO 236 DEG C) PHASE TRANSITIONS IN (HIGH	PURITY) LEAD ZIRCONATE. 07-1266
CTRIC CONSTANT, CONDUCTIVITY) ELECTRICAL PROPERTIES OF HIGH	PURITY POLYCRYSTALLINE BARIUM TITANATE. (DIELE 04-0704
/BEAM SENSING OF SURFACE ELASTIC WAVES FOR SIGNAL PROCESSING	PURPOSES. (TECHNIQUE APPLIED TO LITHIUM NIOBA/ 11-1738
/SIS AND X-RAY DIFFRACTION STUDY OF SINGLE CRYSTALS OF A NEW	PYROCHLORE CONTAINING LITHIUM, STRONTIUM LANT/ 36-3613

IDE(6) SULFIDE). NEW DESCRIPTION OF THE PYROCHLORE STRUCTURE (CADMIUM(2) NIOBIUM(2) OX 36-3602
SYNTHESIS AND CERTAIN PROPERTIES OF NEW COMPOUNDS WITH A PYROCHLORE STRUCTURE. (NIOBATES, TANTALATES) 02-0047
)) (NIOBIUM, TANTALU/ NEW FERROELECTRICS WITH PEROVSKITE AND PYROCHLORE STRUCTURES. (BARIUM AND BISMUTH(0.5 08-1462
NIOBATE, WITH 4 ELECTRIC PHASES, 2 FERR/ NEW FERROELECTRIC (PYROCHLORE TYPE) COMPOUND (AMMONO THIO CADMIUM 36-3529
/ FERROELECTRICS AND SEIGNETTO MAGNETS OF THE PEROVSKITE AND PYROCHLORE TYPE. SYNTHESIS AND STUDY OF THE C/ 02-0128
/E METHOD FOR ESTIMATING THE PARAMETER X IN COMPOUNDS OF THE PYROCHLORE TYPE WITH GENERAL FORMULA A(2)B(2)/ 36-3599
LECTRIC PROPERTIES OF NEW CADMIUM-CONTAINING PEROVSKITES AND PYROCHLORES. /Y ANALYSIS, AND STUDY OF THE DIE 01-1309
TIMONY SULFOIODIDE. PYROCURRENTS HIGHER THAN THE CURIE POINT IN AN 15-2131
LFOIODIDE. LIGHT INDUCED PYROCURRENTS IN FERROSEMICONDUCTOR ANTIMONY SU 15-2130
OELASTIC: CRYSTAL STRUCTURE. PYROELECTRIC AMMONIUM IODATE, A POTENTIAL FERR 36-3567
OF POLED POLYVINYLIDENE FLUORIDE FILMS. PYROELECTRIC AND NONLINEAR OPTICAL PROPERTIES 36-3595
DOPED TRIGLYCINE SULFATE FOR PYROELECTRIC APPLICATIONS. 27-3057
IUM MOLYBDATE. PYROELECTRIC BEHAVIOR OF FERROELECTRIC GADOLIN 30-3279
E TRANSITION. INFLUENCE OF SURFACE EFFECTS ON THE PYROELECTRIC BEHAVIOR OF TGS CLOSE TO THE PHAS 27-3064
THE PARAELECTRIC REGION. PYROELECTRIC BEHAVIOR OF TRIGLYCINE SULFATE IN 27-3066
NT OF FERROELECTRICS. SIMULTANEOUS MEASUREMENT OF PYROELECTRIC COEFFICIENT AND DIELECTRIC CONSTA 03-0333
THE PYROELECTRIC RESPONSE TO STEP RADIA/ MEASUREMENT OF THE PYROELECTRIC COEFFICIENT AND PERMITTIVITY FROM 03-0469
CHNIQUE AND APPLICATION TO A NANOSECOND RESPONSE TIME DETEC/ PYROELECTRIC COEFFICIENT DIRECT MEASUREMENT TE 13-1919
IRCONATE TITANATE) MEASUREMENT OF THE PYROELECTRIC COEFFICIENT OF 65-35 PZT. (LEAD Z 09-1567
NEW METHOD OF MEASURING PYROELECTRIC COEFFICIENTS. (TGS EXAMPLE) 17-2517
/OPTICAL (SECOND HARMONIC GENERATION) COEFFICIENTS OF POLAR (PYROELECTRIC) CRYSTALS (LITHIUM NIOBATE OR TA/ 02-0095
/ ON A(V)B(VI)C(VII) SOLID SOLUTIONS. (DIELECTRIC CONSTANTS, PYROELECTRIC CURRENTS, ANTIMONY SULFOIODIDE (X/ 15-2157
M MOLYBDATE. PYROELECTRIC DETECTION PROPERTIES OF GADOLINIU 30-3324
RADIATION (0.5 TO 2 MIC/ CHARACTERISTICS OF BARIUM TITANATE PYROELECTRIC DETECTOR FOR OPTICAL AND INFRARED 04-0751
CHARACTERISTICS OF KIN (POTASSIUM TANTALATE NIOBATE) PYROELECTRIC DETECTORS. 13-2030
HIGH D*, FAST, LEAD ZIRCONATE TITANATE PYROELECTRIC DETECTORS. 09-1560
E NUCLEATION SITES IN TRIGLYCINE SULFATE AND APPLICATIONS TO PYROELECTRIC DETECTORS. /ER STUDY OF REVERSIBL 27-3011
PHOTOMIXING AT 10.6 MICRONS WITH STRONTIUM BARIUM NIOBATE PYROELECTRIC DETECTORS. 09-1477
AMPLE) THERMAL ANALYSIS OF PYROELECTRIC DETECTORS. (TRIGLYCINE SULFATE EX 27-2964
UNCTION SYSTEMS. (INFLUENCE OF THE ELECTRODE MATERIAL ON THE PYROELECTRIC EFFECT) /- FERROELECTRIC- METAL J 02-0127
DOLINIUM MOLYBDATE. PYROELECTRIC EFFECT AND DOMAIN STRUCTURE OF GA 30-3315
ON OF GADOLINIUM MOLYBDATE (TEMPERATURE DEPENDENCE 150-165 / PYROELECTRIC EFFECT AND SPONTANEOUS POLARIZATI 30-3314
INE SULFATE, A/ DETECTION OF ELECTROMAGNETIC RADIATION USING PYROELECTRIC EFFECT. (BARIUM TITANATE, TRIGLYC 04-0978
NO IODIDE SINGLE CRYSTALS. PYROELECTRIC EFFECT IN ANTIMONY SULFO AND SELE 15-2163
PYROELECTRIC EFFECT IN GADOLINIUM MOLYBDATE. 30-3323
NE. (IN THE POLAR, OR FERROELECTRIC STATE) PYROELECTRIC EFFECT IN METAL-FREE PHTHALOCYANI 36-3540
INFLUENCE OF AN ELECTRIC FIELD ON THE NOISE AND THE PYROELECTRIC EFFECT IN TGS. 27-3076
R/ TRANSFORMATION OF ELECTROMAGNETIC RADIATION INTO MOTION. (PYROELECTRIC EFFECT IN TGS, ETHYL DIAMINE TART 04-0854
E SULFATE AND BARIUM TITANATE. STUDY OF THE PYROELECTRIC EFFECT IN THIN FILMS OF TRIGLYCIN 04-0686
ION IN LITHIUM TANTALATE (THERMAL EXP/ PRIMARY AND SECONDARY PYROELECTRIC EFFECTS AND SPONTANEOUS POLARIZAT 12-1884
LEAD TITANATE PYROELECTRIC INFRARED DETECTOR. 05-1087
GLYCINE SULFATE, TRIGLYCINE FLUOROBERYLL/ IMPROVEMENT IN THE PYROELECTRIC INFRARED RADIATION DETECTOR. (TRI 02-0046
HANISM OF) OPTICAL INDEX DAMAGE IN LITHIUM NIOBATE AND OTHER PYROELECTRIC INSULATORS. (MEC 11-1735
CTURE STABILITY IN FERROELECTRICS. (ROCHELLE SALT, TRIGLYCI/ PYROELECTRIC INVESTIGATIONS OF THE DOMAIN STRU 02-0083
STUDY OF PIEZOELECTRIC OSCILLATIONS IN WIDEBAND PYROELECTRIC LITHIUM TANTALATE. 12-1875
STICS OF A SMALL TRIGLYCINE SULFATE DETECTOR OPERATED IN THE PYROELECTRIC MODE. PERFORMANCE CHARACTERI 27-3119
OF POTASSIUM DIHYDROGEN PHOSPHATE CRYSTALS FOR STUDY OF THE PYROELECTRIC PROPERTIES. /BY A DYNAMIC METHOD) 17-2566
RIDE AND ITS USE FOR INFRARED DETECTION. PYROELECTRIC PROPERTIES OF POLYVINYLIDENE FLUO 36-3560
/ EFFECT OF SAMPLE THICKNESS ON THE DOMAIN STRUCTURE AND THE PYROELECTRIC PROPERTIES OF SINGLE CRYSTALS OF/ 04-1039
EFFECTS OF ADDITIVES ON THE PYROELECTRIC PROPERTIES OF TGS. 27-3037
FERROELECTRIC PLZT CERAMICS. PYROELECTRIC PROPERTIES OF THE LANTHANUM DOPED 09-1558
BARIUM DINIOBATE SYSTEM. DIELECTRIC, PIEZOELECTRIC AND PYROELECTRIC PROPERTIES OF THE LEAD DINIOBATE- 08-1384
E- BARIUM NIOBATE SYSTEM. DIELECTRIC, PIEZOELECTRIC AND PYROELECTRIC PROPERTIES OF THE LEAD METANIOBAT 08-1385
RIC REGION NEAR THE CURIE POINT (IMPROVED METHOD OF MEASURE/ PYROELECTRIC RESPONSE OF TGS IN THE FERROELECT 27-2979
TION IN SILVER SODIUM DINITRITE. PYROELECTRIC STUDY ON THE SPONTANEOUS POLARIZA 16-2295
PYROELECTRIC THERMAL IMAGING DEVICES. (TGS) 27-3107
ITANATE CRYSTAL) MINIMUM DETECTABLE POWER OF A PYROELECTRIC THERMAL RECEIVER. (USING BARIUM T 04-0632
NFRARED SIGNALS IN TRIGLYCINE SULFATE AND STRONTIUM BARIUM / PYROELECTRIC VOLTAGE RESPONSE TO RECTANGULAR I 13-2019
D LASER PULSES IN TRIGLYCINE SULFATE AND BARIUM STRONTIUM N/ PYROELECTRIC VOLTAGE RESPONSE TO SHORT INFRARE 13-2018
OF INFRARED RADIATION IN TRIGLYCINE SULFATE AND STRONTIUM B/ PYROELECTRIC VOLTAGE RESPONSE TO STEP SIGNALS 13-2022
HEORY AND APPLICATION OF FERROELECTRIC RADIATION DETECTORS. (PYROELECTRICS) T 37-3669
NERATION IN POLYVINYLIDENE FLUORIDE FILMS (AFTER COOLING ST/ PYROELECTRICITY AND OPTICAL SECOND HARMONIC GE 36-3530
ED FILMS OF POLYVINYL FLUORIDE AND POLYVINYLIDENE FLUORIDE. PYROELECTRICITY AND PIEZOELECTRICITY IN ORIENT 36-3544
ELECTRICAL CONSTANTS OF TRIGLYCINE SULFATE, APPLICATIONS TO PYROELECTRICITY AND THE DETECTION OF FAR INFR/ 17-2514
TANT OF POLYVINYLIDENE FLUORIDE. PIEZOELECTRICITY, PYROELECTRICITY, AND THE ELECTROSTRICTION CONS 36-3598
OPTICAL ACTIVITY OF SOME FERROELECTRIC CRYSTALS (PYROELECTRICS, TGS, ROCHELLE SALT). 27-3017
CRYSTALS. (THEORY, NO REFS) POLARIZATION, PYROELECTRICTY, AND FERROELECTRICITY OF IONIC 03-0143
PECULIARITIES OF THE DIELECTRIC POLARIZATION OF CADMIUM PYRONIOBATE. 36-3561
/BARIUM SODIUM NIOBATE, CUPROUS CHLORIDE, PROUSTITE, CALCIUM PYRONIOBATE, AND BISMUTH(12) GERMANIUM OXIDE(/ 01-0021
ELASTIC AND PIEZOELECTRIC PROPERTIES OF CADMIUM PYRONIOBATE IN STRONG ELECTRIC FIELDS. 13-1967
PHASE RELATIONS AND SUPERSTRUCTURES OF PYRRHOTITE, IRON(1-X) SULFIDE. 36-3597
R - A SOLID STATE ANALOG MEMORY DEVICE. (USING, FOR EXAMPLE, PZT) AN ADAPTIVE FERROELECTRIC TRANSFORME 09-1566
PARE HIGH DOPED BARIUM STRONTIUM TITANATE SEMICONDUCTOR WITH PZT. ATTEMPT TO PRE 04-0717
FERROELECTRIC CERAMICS FOR INFORMATION STORAGE AND DISPLAY. (PZT) 09-1552
IZATION VIBRATIONS OF A PREPOLARIZED CERAMIC FERROELECTRIC. (PZT) METHOD OF MEASURING POLAR 09-1615
BLE BIREFRINGENCE) OF (HOT PRESSED) FERROELECTRIC CERAMICS. (PZT) OPTICAL PROPERTIES (ELECTRICALLY VARIA 09-1554
NDUCED BY HYDROSTATIC PRESSURE IN A FERROELECTRIC MATERIAL. (PZT) PHASE CHANGES I 09-1521
ELASTIC SURFACE WAVE ON THE PZT CERAMIC PLATE. 09-1612
AVE DELAY LINES WITH INTERDIGITAL TRANSDUCERS ON UNPOLARIZED PZT CERAMIC PLATES. SURFACE W 09-1614
NCENTRATION CELL AND ELECTRICAL CONDUCTIVITY MEASUREMENTS ON PZT FERROELECTRICS. OXYGEN CO 09-1495
MEASUREMENT OF THE PYROELECTRIC COEFFICIENT OF 65-35 PZT. (LEAD ZIRCONATE TITANATE) 09-1567
/NCENTRATION CELL MEASUREMENTS OF IONIC TRANSPORT NUMBERS IN PZT (LEAD ZIRCONATE TITANATE) FERROELECTRICS. 09-1512
LE SIZE / STUDIES ON THE SINTERING BETWEEN LEAD MONOXIDE AND PZT. PART-1. PART-2: EFFECTS OF AVERAGE PARTIC 09-1624
TRIC PROPERTIES. PZT-5 UNDER PRESSURE, DIELECTRIC AND PIEZOELEC 09-1496
RIC TRANSFORMERS WITH IMPROVED TEMPERATURE CHARACTERISTICS. (PZT-5H) ADAPTIVE FERROELECT 09-1583
NONLINEAR EFFECTS IN FERROCERAMICS. (PZT-8, LITHIUM NIOBATE) 09-1546

Q

DEGREES, Z CUT POTASSIUM DIDEUTERIUM PHOSPHATE. ELECTROOPTIC Q SWITCH BY DESK-TOP COMPUTER. /F A DOUBLE 45 17-2702
POTASSIUM DIDEUTERIUM PHOSPHATE POCKELS CELL FOR HIGH POWER Q SWITCHING.   LOSSLESS 17-2529
POTASSIUM DIDEUTERIUM P/ TRANSIENT ELECTROOPTIC EFFECTS AND Q SWITCHING PERFORMANCE IN LITHIUM NIOBATE AND 11-1710
KDP POCKELS CELLS.   TRANSIENT ELASTOOPTIC EFFECTS AND Q SWITCHING PERFORMANCE IN LITHIUM NIOBATE AND 11-1727
L EFFECT IN LEAD ZINCATE(0.33) NICEATE(0.67). (DISPERSION OF QUADRATIC COEFFICIENTS)   ELECTROOPTICA 13-1982
UM TITANATE (SINGLE) C/ INTERFEROMETRIC DETERMINATION OF THE QUADRATIC ELECTROOPTIC COEFFICIENTS IN STRONTI 06-1130
A/ NEW PHASES OF THE TETRAGONAL TUNGSTEN OXYGEN BRONZE TYPE: QUADRATIC LEAD OR BISMUTH TANTALATE NIOBATE, T 13-1985
DIELECTRIC ABSORPTION IN THE ORTHORHOMBIC AND QUADRATIC PHASES OF BARIUM TITANATE.   04-0900
DIELECTRIC ABSORPTION IN THE ORTHORHOMBIC AND QUADRATIC PHASES OF BARIUM TITANATE.   04-0899
ONTIUM(1-X) TITANATE (RECIPROCAL DIELECTRIC CONSTANT FOLLOWS QUADRATIC TEMPERATURE LAW). /D ((BARIUM(X) STR 04-0898
AS/ FIRST FERROELECTRIC OXYFLUORIDE PHASES OF THE OXYGENATED QUADRATIC TUNGSTEN BRONZE TYPE. (STRONTIUM POT 13-2011
/AL CHEMISTRY AND ELECTRICAL PROPERTIES OF TWO NEW PHASES OF QUADRATIC TUNGSTEN BRONZES STRONTIUM POTASSIU- 13-2010
DIHYDROGEN PHOSPHATE AND CESIUM/ RUBIDIUM-87 AND ARSENIC-75 QUADRUPOLAR COUPLING IN FERROELECTRIC RUBIDIUM 17-2439
LATTICE RELAXATION TIMES IN FERROELECTRIC/ TONE BURSTING OF QUADRUPOLAR NUCLEI. DETERMINATION OF THE SPIN 11-1829
ELECTRIC ANTIMONY SULFOIODIDE.   NUCLEAR QUADRUPOLE AND ELECTROACOUSTIC ECHOES IN FERRO 15-2189
ONIUM SULFATE.   DEUTERON QUADRUPOLE COUPLING CONSTANT IN DEUTERATED AMM 19-2761
.   TEMPERATURE DEPENDENCE OF THE SODIUM-23 NUCLEAR QUADRUPOLE COUPLING CONSTANT IN SODIUM NITRITE 16-2393
DEUTERIUM ARSENATE.   DEUTERON QUADRUPOLE COUPLING IN FERROELECTRIC CESIUM DI 18-2721
ULFATE.   NITROGEN-14 QUADRUPOLE COUPLING IN PARAELECTRIC AMMONIUM S 19-2749
FERROELECTRIC HYDROGEN CHLORIDE.   NUCLEAR QUADRUPOLE FREQUENCIES AND RELAXATION TIMES IN 31-3336
AMMONIUM TRIHYDROGEN PERIODATE PROTON CROSSOVER REL/ NUCLEAR QUADRUPOLE INTERACTION IN ANTIFERROELECTRIC DI 36-3578
AMMONIUM TRIHYDROGEN PERIODATE BY IODINE PROTON CRO/ NUCLEAR QUADRUPOLE INTERACTION IN ANTIFERROELECTRIC DI 36-3579
M TITANATE.   STATIC QUADRUPOLE INTERACTION OF SCANDIUM-44 IN BARIU 04-0692
IONS IN BARIUM AND LEAD TITANATE.   STATIC ELECTRIC QUADRUPOLE INTERACTION OF TANTALUM AND HAFNIUM 04-0937
POUNDS OF HAFNIUM-181 STUDIED BY THE PERTURBED ANGU/ NUCLEAR QUADRUPOLE INTERACTIONS IN PEROVSKITE TYPE COM 08-1276
ELECTRIC TRIGL/ PULSED DOUBLE RESONANCE STUDY OF THE NUCLEAR QUADRUPOLE INTERACTIONS OF NITROGEN-14 IN PARA 27-2969
CTRIC TRANSITION IN DEUTERATED TRIGLYCINE FLUOB/ BERYLLIUM-9 NUCLEAR QUADRUPOLE PERTURBED NMR STUDY OF THE FERROELE 27-2965
NITRITE.   NUCLEAR QUADRUPOLE RELAXATION OF NITROGEN-14 IN SODIUM 16-2254
YSTAL STRUCTURE STUDIED BY INFRARED SPECTROSCOPY AND NUCLEAR QUADRUPOLE RESONANCE.   POTASSIUM IODATE CR 08-1278
HYDROGEN ARSENATE)   FERROELECTRIC MODE AND NQR (NUCLEAR QUADRUPOLE RESONANCE) FREQUENCY. (POTASSIUM DI 17-2713
BROMIDE.   BROMINE NUCLEAR QUADRUPOLE RESONANCE IN FERROELECTRIC HYDROGEN 31-3332
ITRITE.   NITROGEN-14 NUCLEAR QUADRUPOLE RESONANCE IN FERROELECTRIC SODIUM N 16-2320
E PHASE TRANSITION POINT.   KINETIC EQUATION FOR QUADRUPOLE RESONANCE IN FERROELECTRICS NEAR TH 03-0309
ANTIMONY SULFOIODIDE (AND BROMIDE) TYPE.   NUCLEAR QUADRUPOLE RESONANCE IN FERROELECTRICS OF THE 15-2216
OMETRIC) LITHIUM NIOBATE (TEMPERATURE DE/ NIOBIUM-93 NUCLEAR QUADRUPOLE RESONANCE INVESTIGATION OF (STOICHI 11-1811
DIHYDROGEN ARSENATE.   PURE QUADRUPOLE RESONANCE OF ARSENIC-75 IN AMMONIUM 18-2732
ERIODATE.   NUCLEAR QUADRUPOLE RESONANCE OF IODINE-127 IN SILVER P 08-1426
DROGEN IODIDE.   NUCLEAR QUADRUPOLE RESONANCE OF IODINE-127 IN SOLID HY 31-3335
IC COMPOUNDS. (LITHIUM AMMONIUM TARTRATE, LITHIUM T/ NUCLEAR QUADRUPOLE RESONANCE OF LITHIUM IN FERROELECTR 02-0052
RITE.   NUCLEAR QUADRUPOLE RESONANCE OF NITROGEN IN SODIUM NIT 16-2313
RROELECTRICITY OF THIOUREA.   STARK EFFECT ON THE NUCLEAR QUADRUPOLE RESONANCE OF NITROGEN-14 AND THE FE 25-2909
ND FERROELECTRICITY OF ALKALI IODATES.   NUCLEAR QUADRUPOLE RESONANCE, PHASE TRANSFORMATIONS, A 08-1350
ERATED FERROELECTRIC CRYSTALS CESIUM DIH/ ARSENIC-75 NUCLEAR QUADRUPOLE RESONANCE SPECTRA OF PARTIALLY DEUT 18-2740
RIC CRYSTAL CESIUM DIHYDROGEN A/ STARK EFFECT IN THE NUCLEAR QUADRUPOLE RESONANCE SPECTRA OF THE FERROELECT 18-2722
IUM SULFATE.   NITROGEN-14 NUCLEAR QUADRUPOLE RESONANCE STUDY OF LITHIUM HYDRAZIN 32-3345
IN AMMONIUM MONOCHLORO ACETATE, MONOCHLORO ACETIC / NUCLEAR QUADRUPOLE RESONANCE STUDY OF PHASE TRANSITION 36-3625
NALS IN LITHIUM NIOBATE.   ACOUSTIC DIPOLE AND QUADRUPOLE SATURATION OF THE LITHIUM-7 NMR SIG 11-1666
SONANCE IN ROCHELLE SALT.   SECOND ORDER QUADRUPOLE SHIFTS OF THE SODIUM-23 MAGNETIC RE 28-3186
L DYNAMICS OF FERROELECTRIC CRYSTALS. (SODIUM NITRI/ NUCLEAR QUADRUPOLE SPIN LATTICE RELAXATION AND CRITICA 16-2268
AL PROPERTIES OF FERROELECTRIC CRYSTALS. (THEORY)   NUCLEAR QUADRUPOLE SPIN LATTICE RELAXATION AND DYNAMIC 03-0187
CTRIC TRANSITION. (IN SODIUM NITRITE)   QUADRUPOLE SPIN PHONON RELAXATION AND FERROELE 16-2362
OPTICAL ANISOTROPY OF CUBIC NICKEL IODINE BORACITE DUE TO QUADRUPOLE TRANSITIONS.   14-2093
ITION WHERE LIQ/ PROGRESS TOWARDS THE GROWTH OF GOOD OPTICAL QUALITY BARIUM SODIUM NIOBATE (BY USING COMPOS 13-2047
E.   GROWTH OF HIGH QUALITY CRYSTALS OF POTASSIUM TANTALATE NIOBAT 13-2055
MICS.   EFFECTS OF IMPURITIES ON THE MECHANICAL QUALITY FACTOR OF LEAD ZIRCONATE TITANATE CERA 09-1608
R OPTICS WITH LOW BIREFRINGENCE VARIATIONS) / GROWTH OF HIGH QUALITY LITHIUM NIOBATE CRYSTALS (FOR NONLINEA 11-1688
ON LITHIUM NIOBATE AND LIT/ CHEMICAL COMPOSITION AND OPTICAL QUALITY OF FERROELECTRIC CRYSTALS, ESPECIALLY 11-1780
STOICHIOMETRY AND OPTICAL QUALITY OF LITHIUM TANTALATE SINGLE CRYSTALS. 12-1883
ON THEIR DIELECTRIC PROPERTIES. (REVIEW)   INFLUENCE OF THE QUALITY OF TRIGLYCINE SULFATE SINGLE CRYSTALS 02-0088
OF PURE TGS SINGLE CRYSTALS AND SOME INVESTIGATIONS OF THEIR QUALITY. (VOLUME DISTRIBUTION OF DEFECTS) /ON 27-3086
ATE.   RESONANCE ABSORPTION OF GAMMA QUANTA IN BARIUM, STRONTIUM, AND CALCIUM STANN 36-3566
H HYDROGEN BONDS.   QUANTUM EFFECTS IN FERROELECTRIC MATERIALS WIT 03-0347
ICAL MODEL OF THE FERROELECTRIC KDP. PART-1. PART-2.   QUANTUM FIELD THEORETICAL TREATMENT OF A DYNAM 03-0279
ECTRAL ANALYSIS (PROTON DYNAMICS IN HYDROGEN BONDED FERROEL/ QUANTUM THEORY OF FERROELECTRICITY, PART-1: SP 03-0402
ERMODYNAMIC BEHAVIOR OF FERROELECTRIC SPINS.   QUANTUM THEORY OF FERROELECTRICITY. PART-2: TH 03-0403
LECTRIC DIPOLE SYSTEMS (SPIN WAVES IN HYDROXYL DIPOL/ ON THE QUANTUM THEORY OF THE FERROELECTRIC STATE IN E 03-0452
FERROMAGNETICS WITHOUT A SYMMETRY CENTER.   QUANTUM THEORY OF VIBRATIONS IN FERROELECTRIC 03-0166
BILITY OF STRONTIUM TITANATE.   EFFECT OF THE QUARTIC ANHARMONICITY ON THE SOFT PHONON INSTA 06-1211
UM DIHYDROGEN PHOSPHATE, LITHIUM NIOBATE, LITHIUM TANTALATE, QUARTZ) /NONLINEAR CRYSTALS. (KDP, ADP, RUBIDI 11-1865
PROPAGATION OF ACOUSTIC SURFACE WAVES IN LITHIUM NIOBATE AND QUARTZ.   EFFECT OF MASS LOADING ON THE 11-1757
OCESSING PURPOSES. (TECHNIQUE APPLIED TO LITHIUM NIOBATE AND QUARTZ) /F SURFACE ELASTIC WAVES FOR SIGNAL PR 11-1738
ITANATE FILMS BY DC DIODE SPUTTERING (ONTO PLATINUM OR FUSED QUARTZ).   PREPARATION OF THIN BARIUM T 04-0948
/AL NONLINEAR SUSCEPTIBILITIES: ACCURATE RELATIVE VALUES FOR QUARTZ, AMMONIUM DIHYDROGEN PHOSPHATE, AND PO/ 17-2539
TITANATE AND/ LINEAR DISORDER IN CRYSTALS (CASE OF SILICON, QUARTZ AND FERROELECTRIC PEROVSKITES). (BARIUM 04-0629
DDITION OF TWO CRYSTAL RESONATOR FREQUENCIES: APPLICATION TO QUARTZ AND LITHIUM TANTALATE. /MPENSATION BY A 11-1686
VE VELOCITY AND DELAY ON LITHIUM NIOBATE, LITHIUM TANTALATE, QUARTZ, AND TELLURIUM DIOXIDE. /TIC SURFACE WA 11-1821
APPROXIMATION IN THE THEORY OF OPTICAL LATTICE VIBRATIONS. (QUARTZ, BARIUM TITANATE) /F THE POULET- LOUDON 03-0380
DIELECTRIC SUSCEPTIBILITY OF QUASI ELASTIC SCATTERING IN LEAD GERMANATE.   35-3497
WAVES BY 180 DOMAIN BOUNDARIES/ REFLECTION AND REFRACTION OF QUASI FERROELECTRIC STRONTIUM TITANATE.   06-1133
CRYSTALS. (THEORY OF ANISOTROPIC RELAXATION ATTENUATION OF QUASI LONGITUDINAL AND QUASI TRANSVERSE SOUND 03-0314
LY IN FERROELECTRIC CRYSTALS.   QUASI LONGITUDINAL PHONONS) /E SULFATE SINGLE 27-3078
ING CHARACTERISTICS AND VARIOUS CONCEPTIONS OF THE TUNNELING QUASI PHENOMENOLOGICAL THEORY OF ELASTIC ANOMA 03-0374
P AND DEUTERATED KDP.   LASER RAMAN STUDY OF QUASI SPIN MODEL. / (KDP TYPE) PART-2: SCATTER 03-0478
LECTRICS: POLARIZATION FLUCTUATI/ SPIN LATTICE RELAXATION BY QUASI SPIN WAVE HYDROGEN TUNNELING MODES IN KD 17-2569
HYDRATE.   ANOMALOUS SPIN LATTICE RELAXATION BY QUASI SPIN WAVES IN ORDER DISORDER TYPE FERROE 26-2928
FERROELECTRIC HYSTERESIS LOOP OF SILVER SODIUM DINITRITE BY QUASI SPIN WAVES IN POTASSIUM FERROCYANIDE TRI 24-2899
QUASI STATIC ELECTRIC FIELD.   16-2293

TE.   QUASI STATIC SWITCHING CURRENT IN SODIUM NITRI 16-2261
ATE, SULFATE, FLUOBERYLLATE, ETC/ (EFFECT OF TEMPERATURE ON) QUASI SYMMETIRICAL HARD IONS (PHOSPHATE, TITAN 03-0330
ELECTRIC CRYSTAL WITH A SECOND ORDER TRANSITION ABOVE THE C/ QUASISTATIC GENERATION OF HARMONICS IN A FERRO 03-0331

R

DERS. (HYDROPHONE ELEMENTS)   OPEN CIRCUIT SENSITIVITY OF RADIALLY POLARIZED FERROELECTRIC CERAMIC CYLIN 37-3651
W SPHERES. (HYDROPHONE ELEMENTS) OPEN CIRCUIT SENSITIVITY OF RADIALLY POLARIZED FERROELECTRIC CERAMIC HOLLO 37-3653
A SOURCE OF ULTRASONIC VIBRATIONS RADIATING IN THE DIRECTIO/ RADIALLY VIBRATING FERROCERAMIC TRANSDUCER AS 37-3657
 PERMITIVITY DIELECTRIC. (THEO/ INPUT ADMITTANCE AND FIELDS RADIATED BY AN OPEN WAVEGUIDE FILLED WITH HIGH 03-0359
/ERROCERAMIC TRANSDUCER AS A SOURCE OF ULTRASONIC VIBRATIONS RADIATING IN THE DIRECTION OF THE TRANSDUCER / 37-3657
 CHANGE IN COOLED KDP TEMPERATURE DUE TO 337 MICRON INCIDENT RADIATION.   AN OBSERVATION OF THE   17-2711
LAYERS OF PEROVSKITE FERROELECTRICS.   RADIATION AND ACOUSTIC EXCITATION ON ANOMALOUS 08-1388
 LITHIUM/ NONLINEAR SADOVSKII EFFECT (THEORY, PROPAGATION OF RADIATION AND FIRST HARMONIC, CALCULATIONS FOR 03-0186
OF LITHIUM NIOBATE SINGLE CRYSTALS USING SCATTERING OF LASER RADIATION BY HYPERSONIC VIBRATIONS. /RAMETERS 11-1797
BATE.   ACOUSTIC RADIATION BY INTERDIGITAL GRIDS ON LITHIUM NIO 11-1699
M NIOBATE.   GENERATION OF FAR INFRARED RADIATION BY PICOSECOND LIGHT PULSES IN LITHIU 11-1857
LECTRIC.   POSSIBILITY OF GENERATING INFRARED RADIATION BY POLARIZATION REVERSAL OF A FERROE 03-0162
IN ROCHELLE SALT.   RADIATION DAMAGE AND THE FERROELECTRIC EFFECT 28-3207
N PHOSPHATE POTASSIUM DIHYDROGEN ARSENATE MIXE/ ESR STUDY OF RADIATION DAMAGE CENTERS IN AMMONIUM DIHYDROGE 17-2547
/ELECTRIC PROPERTIES AND PRESSURE EFFECTS ON SELENITE (ION). RADIATION DAMAGE ELECTRON SPIN RESONANCE SPEC/ 22-2870
NCE, OPTICAL SPECTRA, ESR, CELL PARAMETERS)   RADIATION DAMAGE IN LITHIUM NIOBATE. (CAPACITA 11-1705
CRYSTALS.   X-RAY TOPOGRAPHIC OBSERVATIONS OF RADIATION DAMAGE IN TRIGLYCINE SULFATE SINGLE 27-3106
LLE SALT UP TO HIGH DOSAGE (DESTR/ FERROELECTRIC BEHAVIOR OF RADIATION DAMAGED TRIGLYCINE SULFATE AND ROCHE 27-3098
YCINE FLUOROBERYLL/ IMPROVEMENT IN THE PYROELECTRIC INFRARED RADIATION DETECTOR. (TRIGLYCINE SULFATE, TRIGL 02-0046
      THEORY AND APPLICATION OF FERROELECTRIC RADIATION DETECTORS. (PYROELECTRICS)   37-3669
TE.   RADIATION EFFECTS IN SODIUM TRIHYDROGEN SELENA 22-2865
L AT HIGH TEMPERATURES.   RADIATION EXPOSURE OF A LITHIUM NIOBATE CRYSTA 11-1721
 FRACTURE OF NONLINEAR CRYSTALS (KDP AND LITHIUM NIOBATE) BY RADIATION FROM A RUBY LASER.   11-1663
YSTALS. (KDP, ADP, RUBIDIUM DIHYDROGE/ AUTOFOCUSING OF LASER RADIATION IN ACTIVE MATERIALS AND NONLINEAR CR 11-1865
/N OF THE SECOND, THIRD, AND FOURTH HARMONICS OF 1.06 MICRON RADIATION IN BARIUM SODIUM NIOBATE, LITHIUM N/ 11-1779
TRON DIFFRACTI/ STRUCTURAL EFFECTS OF (DAMAGE FROM) IONIZING RADIATION IN FERROELECTRIC ROCHELLE SALT. (NEU 28-3188
   COMBINATION OF THE FREQUENCIES OF COHERENT AND INCOHERENT RADIATION IN KDP CRYSTAL.   17-2500
                   VISUALIZATION OF ACOUSTIC RADIATION IN LITHIUM NIOBATE.   11-1864
B/ PYROELECTRIC VOLTAGE RESPONSE TO STEP SIGNALS OF INFRARED RADIATION IN TRIGLYCINE SULFATE AND STRONTIUM 13-2022
NATE TITANATE CERAMICS.   EFFECTS OF RADIATION INDUCED DAMAGE CENTERS IN LEAD ZIRCO 09-1520
LLE SALT AND TRIGLYCINE SULFATE.   NATURE OF RADIATION INDUCED DEFECTS IN CRYSTALS OF ROCHE 27-3159
RAMIC) BARIUM AND LEAD TITANATES. (/ (IN PILE, FAST NEUTRON) RADIATION INDUCED PHASE TRANSFORMATIONS IN (CE 04-0971
 TGS, ETHYL DIAMINE TARTR/ TRANSFORMATION OF ELECTROMAGNETIC RADIATION INTO MOTION. (PYROELECTRIC EFFECT IN 04-0854
/TUDIES OF HYDROGEN BONDED FERROELECTRICS USING POLARIZED IR RADIATION. PART-2: INTERNAL FUNDAMENTAL VIBRA/ 18-2738
TITANATE TYPE.   RADIATION PHYSICS OF FERROELECTRICS OF BARIUM 04-0972
OR LEAD, IRON FOR ZIRCONIUM AND TITANIUM, EFFECT ON ACOUSTIC RADIATION POWER/ /ON OF STRONTIUM OR CALCIUM F 09-1484
IENT AND PERMITTIVITY FROM THE PYROELECTRIC RESPONSE TO STEP RADIATION SIGNALS IN FERROELECTRICS. / COEFFIC 03-0469
ITANATE, TRIGLYCINE SULFATE, A/ DETECTION OF ELECTROMAGNETIC RADIATION USING PYROELECTRIC EFFECT. (BARIUM T 04-0978
RIUM TITANATE PYROELECTRIC DETECTOR FOR OPTICAL AND INFRARED RADIATION (0.5 TO 2 MICRONS). /TERISTICS OF BA 04-0751
CATIONS TO PYROELECTRICITY AND THE DETECTION OF FAR INFRARED RADIATIONS. /ANTS OF TRIGLYCINE SULFATE, APPLI 17-2514
CENTER IN LANTHANUM ALUMINATE.   RADIATIONLESS TRANSITIONS IN THE EUROPIUM(3+) 36-3531
                                                          A RADICAL IN FERROELECTRIC DGN CRYSTAL.   16-2253
                                                        THE RADICAL IN GAMMA-IRRADIATED KDP.   17-2406
/ON OF SLATER CONFIGURATIONS AND OF EXCHANGE OF THE ARSENATE RADICAL IN IRRADIATED FERROELECTRIC POTASSIUM/ 18-2734
TE FORCE CONSTANTS FROM ISOTOPIC SUBSTITUTION IN THE NITRITE RADICAL IN SODIUM NITRITE.   ACCURA 16-2309
OINT. (ESR OF GAMMA-IRRADIATED S/ DISPLACIVE MOVEMENT OF THE RADICAL IN TRIGLYCINE SULFATE NEAR THE CURIE P 27-3033
C ARSENATE SINGLE CRYSTALS.   ESR OF FREE RADICALS IN FERROELECTRIC AND ANTIFERROELECTRI 18-2731
 DAMPING IN RUBY AND LITHIUM NIOBATE IN THE THREE CENTIMETER RADIO FREQUENCY RANGE. /NCE OF HYPERSONIC WAVE 11-1718
INE SELENATE IN THE NEAR INFRARED TO FAR INFRARED AND IN THE RADIO FREQUENCY REGION. / CONSTANTS OF TRIGLYC 27-3007
FERROCYANIDE TRIHYDRATE AND ISOMORPHOUS COMPOUNDS.   RADIO SPECTROSCOPIC INVESTIGATION OF POTASSIUM 24-2900
/ATE AND RELATED COMPOUNDS. (GADOLINIUM MOLYBDATE, GROWTH BY RADIOFREQUENCY CZOCHRALSKI AND FROM TOP SEEDE/ 08-1323
ON SILICON.   CHARACTERISTICS OF RADIOFREQUENCY SPUTTERED BARIUM TITANATE FILMS 04-0930
FILMS.   FABRICATION OF RADIOFREQUENCY SPUTTERED BARIUM TITANATE THIN 04-0907
FILMS.   CHARACTERISTICS OF RADIOFREQUENCY SPUTTERED BARIUM TITANATE THIN 04-0906
/CTRICAL PROPERTIES OF THIN FILM BARIUM TITANATE PREPARED BY RADIOFREQUENCY SPUTTERING ON SILICON SUBSTRAT/ 04-0828
DE OR GALLIUM ARSENIDE.   BARIUM TITANATE FILMS PREPARED BY RADIOFREQUENCY SPUTTERING ONTO INDIUM ANTIMONI 04-0741
ETRAHYDRATE.   RAMAN AND INFRARED SPECTRA OF COPPER FORMATE T 26-2936
ES.   RAMAN AND INFRARED SPECTRA OF INORGANIC FORMAT 26-2937
C TRANSITION IN AMMONIUM SULFATE.   RAMAN AND INFRARED STUDIES OF THE FERROELECTRI 19-2793
UCTURE OF THE HYDROGEN BONDS IN THE FER/ POLARIZATION OF THE RAMAN BANDS OF THE HYDROXYL VIBRATIONS AND STR 17-2471
RONTIUM TITANATE.   TEMPERATURE DEPENDENCE OF THE RAMAN CROSS SECTIONS IN BARIUM TITANATE AND ST 04-0548
IUM TANTALATE)   ELECTRIC FIELD INDUCED RAMAN EFFECT IN PARAELECTRIC CRYSTALS. (POTASS 08-1326
IC PROPERTIES OF AMMONIUM DIHYDROGEN PHOSPHATE AND THE DAVAI RAMAN ELASTICITY THEORY.   ELAST 17-2410
Y TEMPERATURES AS A FUN/ CALCULATION OF THE VARIATION OF THE RAMAN FREQUENCIES IN SODIUM NITRITE AT ORDINAR 16-2363
                                                            RAMAN LIGHT SCATTERING IN ROCHELLE SALT.   28-3231
    POLARITONS IN TETRAGONAL BARIUM TITANATE: THEORY OF THE RAMAN LINE SHAPE.   04-0556
        BARIUM SODIUM NIOBATE FOR A TUNABLE STIMULATED RAMAN OSCILLATOR.   13-1917
METRY VARIATIONS IN LITHIUM NIOBATE AND LITHIUM TANTALATE BY RAMAN POWDER SPECTROSCOPY.   STOICHIO 11-1813
OF THE KDP GROUP.   SPECTRUM OF RAMAN SCATTERED LASER LIGHT IN FERROELECTRICS 17-2625
         DOMAIN WALL STRUCTURE IN GADOLINIUM MOLYBDATE BY RAMAN SCATTERING.   30-3318
FT PHONON MODE AND MODE COUPLING IN ANTIMONY SULFOIODIDE BY RAMAN SCATTERING).   SO 15-2167
PARAELECTRIC PHASE OF KDP CRYSTALS (INFRARED ABSORPTION AND RAMAN SCATTERING). /F THE PHOSPHATE ION IN THE 17-2479
TASSIUM MANGANESE TRIFLUORIDE CRYSTAL.   RAMAN SCATTERING AND PHASE TRANSITIONS IN A PO 33-3400
SSIUM MANGANESE TRIFLUORIDE CRYSTALS.   RAMAN SCATTERING AND PHASE TRANSITIONS IN POTA 33-3372
SSED STRONTIUM TITANATE.   RAMAN SCATTERING AND PHASE TRANSITIONS IN STRE 06-1103
                                                            RAMAN SCATTERING BY BARIUM STRONTIUM NIOBATE. 13-1899
LUOBERYLLATE CRYSTALS.   LOW FREQUENCY RAMAN SCATTERING BY FERROELECTRIC TRIGLYCINE F 27-3113
BIDIUM MANGANATE AND P/ IMPURITY CONCENTRATION DEPENDENCE OF RAMAN SCATTERING BY MAGNONS IN NICKEL DOPED RU 33-3395
SE OF CADMIUM SULFIDE AND LITHIUM NIOBATE CRYST/ SPONTANEOUS RAMAN SCATTERING BY POLARITONS. PART-2: THE CA 11-1787
FERROELECTRIC MODE IN POTASSIUM DIDEUTERIUM PHOSPHATE (COMP/ RAMAN SCATTERING BY THE TEMPERATURE DEPENDENT 11-1705
F POTASSIUM DIHYDROGEN PHOSPHATE.   RAMAN SCATTERING FROM FERROELECTRIC CRYSTALS O 17-2552
HE KDP ISOMORPHOUS PHOSPHATES AND ARSENATES. (KDP, AMMONIUM/ RAMAN SCATTERING FROM FERROELECTRIC MODES IN T 17-2639
RITE.   RAMAN SCATTERING FROM FERROELECTRIC SODIUM NIT 08-1344
S OF BARIUM TITANATE.   AUGER-LIKE RESONANT INTERFERENCE IN RAMAN SCATTERING FROM ONE AND TWO PHONON STATE 04-0923

N LEAD(5) GERMANIUM(3) OXIDE(11).
ULFATE CRYSTAL.
        LINEAR DISORDER AND TEMPERATURE DEPENDENCE OF
IUM NIOBATE.
        STRUCTURE DISORDER AND
T/ ELECTRIC FIELD DEPENDENCE OF OPTICAL PHONON FREQUENCIES
TRI/ MULTIPHONON SELECTION RULES FOR INFRARED ABSORPTION AND
R ANTIMONY SULFOIODIDE.

H TETRAGONAL SYMMETRY.
ATE AND CES/ DIELECTRIC RESPONSE IN PIEZOELECTRIC CRYSTALS.
HATE.
        EFFECT OF UNIAXIAL STRESS ON
EAD BARIUM TITANATE.
        THE FERROELECTRIC PHASE POTASSIUM / OSCILLATORY BEHAVIOR OF
RITON MODE IN BARIUM TITANATE.
        EFFECT OF CHARGE TRANSFER ON
/DUCTIVITY. PART-3: NEUTRON SCATTERING. PART-4: INFRARED AND
ROELECTRIC TRANSITION IN RUBIDIUM DIHYDROGEN PHOSPHIDE USING
/ATURE DEPENDENCE OF THE INTENSITY OF HYDROGEN BOND BANDS IN
/ON IN FERROELECTRIC ANTIMONY SULFOIODIDE CRYSTALS USING THE
TERATED KDP DURING THE PHASE TRAN/ TEMPERATURE DEPENDENCE OF
S IN SOLID STATE PHYSICS (DOMAIN WALL THICKNESS IN GADOLINI/
TRANSITIONS IN ANTIMONY SULFOIODIDE. (POLARIZED INFRARED AND
ROELECTRIC SODIUM NITRITE AT 77 AND 294 DEG K.
THIOUREA. PART-1: TEMPERATURE DEPENDENCE OF THE INFRARED AND
ELECTRICS. (LATTICE VIBRATIONS)   POWDER
/GITUDINAL MODES IN HYDROGEN AND DEUTERIUM HALIDE CRYSTALS.
        ISOTOPE EFFECT IN THE
NTIFERROELECTRIC PHASES.

YDROGEN PHOSPHATE TYPE IN THEIR PARAELECTRIC AND LOW TEMPER/
MMONIUM ALUMINUM (SULFATE OR SELENATE) DODECAH/ INFRARED AND
TRIGLYCINE SULFATE TYPE.   LOW FREQUENCY
RY)   FAR INFRARED DISPERSION AND
        CRYSTAL GROWTH AND TEMPERATURE DEPENDENT
EN AND DEUTERIUM CHLORIDES AT TEMPERATURES ABOVE 80 DEGREES/
E CRYSTALS.   CRYSTAL SIZE EFFECTS IN THE
CATION TO THE LEAD TITANATE ZIRCONATE SYSTEM (MEASUREMENT O/
POTASSIUM DIHYDROGEN PHOSPHATE IN AN ELECTRIC FIELD.
E, POTASSIUM DIDEUTERIUM PHOSPHATE, P/ TEMPERATURE DEPENDENT
.   INFRARED AND
.   POLARIZED INFRARED AND
EN PHOSPHATE.
N OF A POTASSIUM DIHYDROGEN PHOSPHATE CRYSTAL.
TEMPERATURE DEPENDENT PHONON STRUCTURE IN SODIUM NITRITE BY
ICS, CESIUM DIHYDROGEN ARSENATE, STRONTIUM TITANA/ SOFT MODE
RST ORDER FERROELECTRIC PHASE TRANSITIONS.
        INFRARED AND
CTRIC TRANSITION IN SODIUM NITRITE CRYSTAL BY OBSERVATION OF
IN KDP.   TEMPERATURE DEPENDENCE OF
ATE.
TION OF TEMPERATURE.   LOW FREQUENCY
        TEMPERATURE DEPENDENCE OF THE BRILLOUIN-
20).
SSIUM DIHYDROGEN PHOSPHATE CRYSTAL.
        TEMPERATURE DEPENDENCE OF THE
TATION WITH A MIXED COULOMB VALENCE DYNAMICAL/ LOW FREQUENCY
TRIC ANTIMONY SULFOIODIDE (PHONON ASSIGNMENTS). (POLARIZED)
ATE CRYSTAL IN PARA- AND FERROELECTRIC PHASE.
(POTASSIUM TANTALATE(0.65) NIOBATE(0.35), CURIE TEMPERATUR/
BIUM(0.36) OXIDE.
E (PARAELECTRIC PHASE, COMPARISON WITH FERROELECTRIC).
LOCALLY ORDERED "NUCLEI", PHASE STRUCTURE ABOVE CURIE POINT,
GEN PHOSPHATE. (T/ ANALYSIS OF PHONON TUNNELING MODES IN THE

        TRANSVERSE POLARITONS IN A LITHIUM NIOBATE CRYSTAL.
OF THIOUREA.   INFRARED AND
ON NEAR CURIE TEMPERATURE IN STRONTIUM TITANATE. (ABSTRACT)
R DAMPED SOFT MODES IN LEAD TITANATE.
        POWDER
IUM AND RUBIDIUM TUNGSTATES)
ING MODES IN KDP AND DEUTERATED KDP.   LASER
ROELECTRICS AS A FUNCTION OF TEMPERATURE USING INFRARED AND
TANATE CERAMIC MEMORY PLATE)   A FERROELECTRIC PIEZOELECTRIC
SIUM DIHYDROGEN PHOSPHATE.
NS TO FERROELEC/ AN ANALYSIS OF MULTIPLE LIGHT SCATTERING BY
NNES MODEL FOR HYDROGEN BONDED FERROELECTRICS.
/IDE TRIHYDRATE SINGLE CRYSTALS, P-N JUNCTIONS AND P- AND N-
/ON IN TUNGSTEN TRIOXIDE AND ITS USE IN DEDUCING MEAN DAMAGE
        BEHAVIOR OF MAGNETICALLY ORDERED FERROELECTRICS IN A
ELECTRIC ORDERING)
        PREPARATION AND PROPERTIES OF
        CZOCHRALSKI SYNTHESIS AND PROPERTIES OF
        ELECTRON PARAMAGNETIC RESONANCE OF SOME
TION IN (CERAMIC) STRONTIUM TITANATES CONTAINING (TRIVALENT)
/ABRICATION (BY FLUX GROWTH) OF HEXAGONAL SINGLE CRYSTALS OF
        PHASE TRANSITIONS OF ANHYDROUS MODIFICATIONS OF
PRECISION PARAMETERS OF SOME LANTHANON MOLYBDATE TYPE
PRECISION PARAMETERS OF THE FERROELECTRIC

RAMAN SCATTERING FROM THE FERROELECTRIC MODE I 35-3498
RAMAN SCATTERING IN A FERROELECTRIC AMMONIUM S 19-2786
RAMAN SCATTERING IN BARIUM TITANATE. 04-0675
RAMAN SCATTERING IN BARIUM TITANATE AND POTASS 04-0676
RAMAN SCATTERING IN CUBIC PEROVSKITE STRONTIUM 06-1241
RAMAN SCATTERING IN FERROELECTRIC AND PARAELEC 16-2281
RAMAN SCATTERING IN FERROELECTRIC SEMICONDUCTO 15-2111
RAMAN SCATTERING IN FERROELECTRICS. (REVIEW) 02-0120
RAMAN SCATTERING IN PIEZOELECTRIC CRYSTALS WIT 03-0407
RAMAN SCATTERING IN POTASSIUM DIHYDROGEN ARSEN 18-2723
RAMAN SCATTERING IN POTASSIUM DIHYDROGEN PHOSP 17-2709
RAMAN SCATTERING IN STRONTIUM TITANATE. 06-1102
RAMAN SCATTERING IN THE FERROELECTRIC SYSTEM L 04-0611
RAMAN SCATTERING INTENSITY WITH TEMPERATURE IN 17-2662
RAMAN SCATTERING LINE SHAPE OF THE SOFT E POLA 04-0803
RAMAN SCATTERING OF HYDROGEN HALIDES. 31-3328
RAMAN SCATTERING. PART-5: MOSSBAUER EFFECT. P/ 03-0397
RAMAN SCATTERING SPECTRA. FER 17-2620
RAMAN SCATTERING SPECTRA OF GYPSUM AND ROCHEL/ 28-3189
RAMAN SCATTERING SPECTRUM. (LATTICE VIBRATION/ 15-2143
RAMAN SCATTERING SPECTRUM OF FERROELECTRIC DEU 17-2626
RAMAN SCATTERING TECHNIQUES APPLIED TO PROBLEM 30-3319
RAMAN SPECTRA) /GTH OPTICAL PHONONS AND PHASE 15-2106
RAMAN SPECTRA AND MODE FREQUENCY SHIFTS OF FER 16-2262
RAMAN SPECTRA AND THE CALCULATED ENTROPY. /IC 25-2920
RAMAN SPECTRA: APPLICATION TO DISPLACIVE FERRO 03-0194
RAMAN SPECTRA, DIPOLAR COUPLING MODEL COMPUTA/ 31-3326
RAMAN SPECTRA OF A DEUTERATED KDP CRYSTAL. 17-2621
RAMAN SPECTRA OF ADP IN THE PARAELECTRIC AND A 17-2431
RAMAN SPECTRA OF BARIUM SODIUM NIOBATE. 13-1906
RAMAN SPECTRA OF CRYSTALS OF THE POTASSIUM DIH 17-2515
RAMAN SPECTRA OF FERROELECTRIC ALUMS: METHYL A 20-2803
RAMAN SPECTRA OF FERROELECTRIC CRYSTALS OF THE 27-3114
RAMAN SPECTRA OF FERROELECTRIC CRYSTALS. (THEO 03-0165
RAMAN SPECTRA OF LEAD TITANATE. 05-1074
RAMAN SPECTRA OF ORTHORHOMBIC AND CUBIC HYDROG 31-3330
RAMAN SPECTRA OF ORTHORHOMBIC HYDROGEN CHLORID 31-3338
RAMAN SPECTRA OF POLYCRYSTALLINE SOLIDS: APPLI 09-1497
RAMAN SPECTRA OF POLYMORPHIC MODIFICATIONS OF 17-2425
RAMAN SPECTRA OF POTASSIUM DIHYDROGEN PHOSPHAT 17-2409
RAMAN SPECTRA OF SEVERAL PEROVSKITE ZIRCONATES 08-1411
RAMAN SPECTRA OF SINGLE CRYSTAL CESIUM NITRITE 16-2342
RAMAN SPECTRA OF TETRAGONAL POTASSIUM DIHYDROG 17-2657
RAMAN SPECTRA OF THE FERROELECTRIC MODIFICATIO 17-2622
RAMAN SPECTROSCOPY. ANALYSIS OF THE 16-2299
RAMAN SPECTROSCOPY: COUPLED MODES (FERROELECTR 06-1207
RAMAN SPECTROSCOPY OF PHONON INSTABILITY IN FI 03-0163
RAMAN SPECTROSCOPY OF POTASSIUM METANIOBATE. 08-1303
RAMAN SPECTROSCOPY OF TUNGSTEN TRIOXIDE. 10-1635
RAMAN SPECTRUM. INVESTIGATION OF FERROELE 16-2275
RAMAN SPECTRUM AND SECOND HARMONIC GENERATION 17-2550
RAMAN SPECTRUM AND STRUCTURE OF TERBIUM MOLYBD 30-3296
RAMAN SPECTRUM OF A THIOUREA CRYSTAL AS A FUNC 25-2906
RAMAN SPECTRUM OF BARIUM TITANATE. 04-0746
RAMAN SPECTRUM OF BARIUM TITANATE. 04-0807
RAMAN SPECTRUM OF BISMUTH(12) GERMANIUM OXIDE (35-3519
RAMAN SPECTRUM OF FERROELECTRIC MODE IN A POTA 17-2663
RAMAN SPECTRUM OF GADOLINIUM MOLYBDATE. 30-3285
RAMAN SPECTRUM OF ICE I(H). TENTATIVE INTERPRE 34-3451
RAMAN SPECTRUM OF (PARAELECTRIC AND) FERROELEC 15-2207
RAMAN SPECTRUM OF POTASSIUM DIDEUTERIUM PHOSPH 17-2647
RAMAN SPECTRUM OF POTASSIUM TANTALATE NIOBATE. 08-1391
RAMAN SPECTRUM OF POTASSIUM TANTALUM(0.64) NIO 13-1989
RAMAN SPECTRUM OF SODIUM TRIHYDROGEN DISELENIT 22-2864
RAMAN SPECTRUM OF STRONTIUM TITANATE. /DEL OF 04-0647
RAMAN SPECTRUM OF TETRAGONAL POTASSIUM DIHYDRO 03-0396
RAMAN SPECTRUM OF TRIGLYCINE SELENATE. 27-2957
(RAMAN SPECTRUM, PERMITTIVITY) 11-1762
RAMAN STUDIES IN THE FERROELECTRIC TRANSITION 25-2916
RAMAN STUDIES OF STRESS INDUCED PHASE TRANSITI 06-1236
RAMAN STUDIES (TEMPERATURE DEPENDENCE) OF UNDE 05-1041
RAMAN STUDY FOR BARIUM TITANATE. 04-0610
RAMAN STUDY OF METALLIC TUNGSTEN BRONZES. (SOD 10-1654
RAMAN STUDY OF QUASI SPIN WAVE HYDROGEN TUNNEL 17-2569
RAMAN TECHNIQUES. /TE MATERIALS (INCLUDING FER 02-0104
RANDOM ACCESS MEMORY. (LEAD ZIRCONATE- LEAD TI 09-1504
RANDOM MOTION OF DEUTERONS IN DEUTERATED POTAS 17-2569
RANDOM OPTICAL INHOMOGENEITIES WITH APPLICATIO 37-3634
RANDOM PHASE APPROXIMATION FOR THE BLINC-DE GE 03-0338
RANGES. (SILICON INTEGRATED/ 10-1642
RANGES. (STUDIES ON BOMBARDMENT INDUCED DISOR/ 03-0147
RAPIDLY OSCILLATING ELECTRIC FIELD. 02-0131
RARE EARTH COMPOUNDS. (WITH MAGNETIC AND FERRO 35-3496
RARE EARTH DOPED BISMUTH GERMANATE. 06-1192
RARE EARTH IMPURITIES IN STRONTIUM TITANATE. 06-1143
RARE EARTH IONS. DIELECTRIC RELAXA 36-3536
RARE EARTH MANGANATES (PHASE DIAGRAMS, CONVEC/ 30-3272
RARE EARTH MOLYBDATES. 30-3262
RARE EARTH MOLYBDATES. (CRYSTAL STRUCTURE) 30-3264
RARE EARTH MOLYBDATES: LANTHANUM MOLYBDATE.

STAL GROWTH, PHASE TRANSITIONS, STRUCTURE, 49/ FERROELECTRIC RARE EARTH MOLYBDATES. (REVIEW, FORMATION, CRY 02-0064
MECHANISM OF THE FERROELECTRIC PHASE TRANSFORMATION IN RARE EARTH MOLYBDATES. (TERBIUM MOLYBDATE) 30-3254
YPE STRUCTURE. SODIUM BARIUM RARE EARTH NIOBATES WITH THE TUNGSTEN BRONZE T 13-1972
ITES ELECTRICAL TRANSPORT IN (SEMICONDUCTING AND MAGNETIC) RARE EARTH ORTHOCHROMITES, MANGANITES AND FERR 36-3608
MAGNETIC PROPERTIES OF HEAVY RARE EARTH ORTHOMANGANATES. 36-3545
/CTURE AND ELECTRICAL PROPERTIES OF SCANDIUM TRIOXIDE DOPED, RARE EARTH OXIDE DOPED, AND UNDOPED BARIUM TI/ 04-0839
CTRIC PROPERTIES OF HEXAGONAL ORTHOMANGANITES OF YTTRIUM AND RARE EARTHS. (YTTERBIUM, HOLMIUM, ERBIUM) /ELE 36-3542
S OF CRYSTALLIZATION OF ANTIMONY SULFOIODIDE (CRYSTAL GROWTH RATE, ANISOTROPY) KINETIC 15-2215
TEMS. (THEORY) DETERMINATION OF THE GROWTH RATE OF FERROELECTRIC CRYSTALS FROM BINARY SYS 03-0178
POLARIZATION REVERSAL IN A WIDE RANGE OF IMPRESSING RATE OF FIELD IN SODIUM NITRITE. 16-2381
OF THEIR GROWTH. (AMMONIUM DIHYDROGEN PHOSPHATE, POTASSIUM / RATE OF ROTATING CRYSTALS AND CHARACTERISTICS 16-2316
OF THE ELECTRICAL BOUNDARY CONDITIONS ON THE NMR RELAXATION RATE OF SODIUM-23 IN SODIUM NITRITE. INFLUENCE 16-2350
E PRESENCE OF PURE DIRECT SKY BLUE DYE. GROWTH RATE OF THE (010) PLANE OF ROCHELLE SALT IN TH 28-3217
EFFECT OF PRESSURE ON THE SWITCHING RATE OF TRIGLYCINE SULFATE. 27-3014
TEMPERATURE DEPENDENCE OF THE SWITCHING RATE OF TRIGLYCINE SULFATE. 27-3015
OPERTIES OF TGS SINGLE CRYSTALS. THE EFFECT OF GROWTH RATE ON THE DEFECT STRUCTURE AND DIELECTRIC PR 27-3067
RIUM STRONTIUM NIOBATE) VARIATION OF CRYSTAL GROWTH RATE WITH SUPERSATURATION IN FLUXED MELTS. (BA 13-1939
ICON DIOXIDE, AND SILICON. MEASUREMENT OF SPUTTER ETCHING RATES FOR MONOCRYSTALLINE LITHIUM NIOBATE, SIL 11-1714
ORPHS. TRANSITION RATES OF POTASSIUM NITRATE HIGH PRESSURE POLYM 16-2279
AR COMPOUNDS. STUDIES ON ANALYSIS OF ELECTROCERAMICS AND ITS RAW MATERIALS. PART-1. PART-2. /NATE AND SIMIL 05-1068
/ND PZT. PART-1. PART-2: EFFECTS OF AVERAGE PARTICLE SIZE OF RAW POWDER ON SINTERING. PART-3: SINTERING PR/ 09-1624
SE TRANSITION. THEORY OF RAYLEIGH AND BRILLOUIN SCATTERING NEAR THE PHA 03-0334
E. INTENSITY OF SOME BRILLOUIN AND RAYLEIGH LINES IN RUBIDIUM DIHYDROGEN PHOSPHAT 17-2521
SULFOIODIDE. SOFT MODE COUPLING AND CRITICAL RAYLEIGH SCATTERING IN FERROELECTRIC ANTIMONY 15-2232
RE CONSISTING OF LITHIUM NIOBATE AND P-TYP/ AMPLIFICATION OF RAYLEIGH ULTRASONIC WAVES IN A LAYERED STRUCTU 11-1747
WAVES IN LITHIUM NIOBATE. PROPAGATION AND AMPLIFICATION OF RAYLEIGH WAVES AND PIEZOELECTRIC LEAKY SURFACE 11-1856
ON THE ELASTIC PROPERTIES OF RUBIDIUM DIHYDROGEN PHOSPHATE (RDP). EFFECT OF DEUTERATION 17-2584
IN SOLUTION OF SINGLE CRYSTALS BY THE GRADUAL ADDITION OF A REACTANT. (POTASSIUM DIHYDROGEN PHOSPHATE) /TH 17-2595
MATION PROCESSES OF LEAD MAGNESIUM NIOBATE IN THE SOLID STATE REACTION. FOR 08-1354
D KINETICS OF LITHIUM METANIOBATE FORMATION BY A SOLID PHASE REACTION. MECHANISM AN 11-1840
/N SINTERING. PART-3: SINTERING PROCESS AND MECHANISM OF THE REACTION BETWEEN LEAD MONOXIDE AND PZT. PART-/ 09-1624
STATISTICAL CONSIDERATIONS OF THE ORDER DISORDER REACTION IN AMMONIUM PERIODATE. 36-3581
INFLUENCE OF IMPURITIES ON THE ORDER DISORDER REACTION IN HEXAGONAL ICE I(H). 34-3473
ITH VANADIUM PENTOXIDE AND MOLYB/ THERMOGRAPHIC STUDY OF THE REACTION OF BISMUTH(4) TITANIUM(3) OXIDE(12) W 13-2029
IDE. THEORY OF SUPEREXCHANGE REACTION. PART-1: POTASSIUM MANGANESE TRIFLUOR 03-0252
TE IN AN INDUSTRIAL ROTARY FURNACE. REACTION SEQUENCE IN FORMING BARIUM METATITANA 04-1003
ITE TYPE LEAD ZINCATE(/ HIGH PRESSURE SYNTHESIS (SOLID STATE REACTION, SINTERING) AND PROPERTIES OF PEROVSK 08-1393
FORMATION OF THIN LAYERS OF MIXED TITANATES BY SOLID SOLID REACTIONS. 13-2005
HYDROTHERMAL PREPARATION OF BARIUM TITANATE BY TRANSPORT REACTIONS. 04-0626
POTASSIUM ACID TARTRATE CRYSTALS. SOME CHEMICAL REACTIONS IN SILICA GELS. PART-3: FORMATION OF 29-3246
SM/ HYDROTHERMAL PREPARATION OF BARIUM TITANATE BY TRANSPORT REACTIONS (INVOLVING TITANIUM ESTERS TO YIELD 04-0625
IELECTRI/ EFFECT OF STRUCTURAL DEFECTS PRODUCED BY GAMMA AND REACTOR IRRADIATION ON PHASE TRANSITIONS AND D 17-2614
NATE PHOTOCONDUCTOR OPTICAL STORAGE MEDIUM. (HIGH RESOLUTION READ WRITE) BISMUTH TITA 08-1445
APPLICATIONS. A NEW METHOD OF OPTICALLY READING DOMAINS IN BISMUTH TITANATE AND MEMORY 08-1299
. EFFICIENT WHITE LIGHT READING OF DOMAIN PATTERNS IN BISMUTH TITANATE 08-1299
/RROELECTRIC PHOTOCONDUCTOR DEVICE (ERASABLE, NONDESTRUCTIVE READOUT, BASED ON BISMUTH TITANATE AND ZINC S/ 37-3646
RATE. (/ PROGRESS TOWARD A FAST, NONVOLATILE, NONDESTRUCTIVE READOUT MEMORY ELEMENT UTILIZING POTASSIUM NIT 16-2269
TOR OF THE GROWTH ST/ ELECTRICAL STRUCTURE OF THE SURFACE OF REAL CRYSTAL SUBSTRATES AS THE DETERMINING FAC 27-3039
OOPTIC MODULATION (BY PASSING COHERENT LIGHT THROUGH DEUTER/ REAL-TIME HOLOGRAPHIC RECONSTRUCTION BY ELECTR 17-2507
RED RESPONSE USING SURFACE WAVE TRANSDUCERS ON LEAD ZIRCONA/ REALIZATION OF A 40 MHZ COLOR TELEVISION INFRA 09-1509
RIC MICROWAVE AMPLIFIER (BASED ON FERROELECTRICS I/ POSSIBLE REALIZATION OF ULTRA LOW NOISE LEVEL PARAELECT 37-3664
LLE SALT NEAR THE PHASE TRANSITION. REARRANGEMENT OF THE DOMAIN STRUCTURE IN ROCHE 28-3181
MINIMUM DETECTABLE POWER OF A PYROELECTRIC THERMAL RECEIVER. (USING BARIUM TITANATE CRYSTAL) 04-0632
/STRONTIUM TITANATE AND ((BARIUM(X) STRONTIUM(1-X) TITANATE (RECIPROCAL DIELECTRIC CONSTANT FOLLOWS QUADRA/ 04-0898
RECIPROCAL DOMAINS IN LEAD TITANATE CRYSTALS. 05-1048
TIUM TITANATE. NONCOLLINEAR PARAMETRIC 4 PHOTON RECIPROCAL EFFECT IN CADMIUM SULFIDE AND STRON 06-1122
PASSING COHERENT LIGHT THROUGH DEUTER/ REAL-TIME HOLOGRAPHIC RECONSTRUCTION BY ELECTROOPTIC MODULATION (BY 17-2507
IUM FLUORIDE, STRONTIUM TITANATE, AND CALCIUM TI/ ERASE MODE RECORDING CHARACTERISTICS OF PHOTOCHROMIC CALC 06-1119
/ECTS (INTERNAL GENERATION OF STRONG FIELDS) DURING HOLOGRAM RECORDING IN CRYSTALS (IRON DOPED STRONTIUM T/ 06-1090
HOLOGRAPHIC RECORDING IN LITHIUM NIOBATE. 11-1662
AN APPROXIMATIVE ANALYSIS OF A FERROELECTRIC RECORDING SYSTEM. 37-3674
FATE AND STRONTIUM BARIUM / PYROELECTRIC VOLTAGE RESPONSE TO RECTANGULAR INFRARED SIGNALS IN TRIGLYCINE SUL 13-2019
XAHYDRATE. ZEEMAN EFFECT (STUDY) OF TWO FINE LINES IN (THE RED REGION OF) CHROMIUM GUANIDINIUM SULFATE HE 21-2814
CRYSTAL GUANIDINIUM ALUMINUM SULFATE HEXAHYDRATE AND ITS CH/ REDETERMINATION OF STRUCTURE OF FERROELECTRIC 21-2819
ON THE PHASE COMPOSITION OF REDUCED AND REOXIDIZED BARIUM TITANATE. 04-0541
LOWERING THE CURIE TEMPERATURE IN REDUCED BARIUM TITANATE. 04-0711
ANOMALY OF RESISTANCE IN REDUCED BARIUM TITANATE CRYSTAL. 04-0988
/ECTRICAL CONDUCTIVITY (SEEBECK COEFFICIENT, CURIE POINT) OF REDUCED BARIUM TITANATE CRYSTALS (POLARON HOP/ 04-0918
SEMICONDUCTING STRONTIUM TITANATE SINGLE CRYSTALS (HYDROGEN REDUCED, CERIUM OR NIOBIUM DOPED). /PECTRA) OF 06-1198
ABSORPTION OF ELASTIC WAVES IN REDUCED LITHIUM NIOBATE. 11-1816
OPTICAL ABSORPTION SPECTRA OF REDUCED STRONTIUM TITANATE. 06-1245
TRANSPORT PROPERTIES IN REDUCED STRONTIUM TITANATE. (ABSTRACT) 06-1152
. ELECTRONIC CONDUCTION IN SLIGHTLY REDUCED STRONTIUM TITANATE AT LOW TEMPERATURES 06-1151
HZ-10 MHZ, EFFE/ DIELECTRIC (CONSTANT AND LOSS) BEHAVIOR OF REDUCED STRONTIUM TITANATE SINGLE CRYSTALS (40 06-1148
/UM TRIOXIDE STRUCTURE TYPE: AN ELECTRON MICROSCOPE STUDY OF REDUCED TUNGSTEN TRIOXIDE AND RELATED PSEUDOB/ 10-1630
/LECTRICS FOR HIGH POWER ELECTRONIC TUNING. (WITH ELECTRODES REDUCING POLARIZATION SWITCHING TIME, LEAD (O./ 37-3660
STRONTIUM TITANATE SINGLE CRYSTALS (40 HZ-10 MHZ, EFFECTS OF REDUCTION METHODS). /OSS) BEHAVIOR OF REDUCED 06-1148
CHEMICAL PROCESSES DURING HYDROGEN REDUCTION OF BARIUM TITANATE CRYSTALS. 04-0540
OLORATION IN STRONTIUM TITANATE: VACANCY DRIFT AND OXIDATION REDUCTION OF TRANSITION METALS. ELECTROC 06-1098
HANUM TITANATE SYSTEM. OXIDATION REDUCTION PHENOMENA IN THE LEAD TITANATE- LANT 09-1587
FERROELECTRIC TRIGLYCINE SULFATE. REEXAMINATION OF THE THERMAL EXPANSION OF THE 27-3121
SULFATE. REFINEMENT OF CRYSTAL STRUCTURE OF TRIGLYCINE 27-3024
ECTRIC ACID LITHIUM SELENITE, POSITION OF THE LITHIUM ION. REFINEMENT OF THE CRYSTAL STRUCTURE OF FERROEL 22-2859
ECTRIC LITHIUM HYDROGEN SELENITE: POSITION OF THE LITHIUM I/ REFINEMENT OF THE CRYSTAL STRUCTURE OF FERROEL 22-2860
AMMONIUM TARTRATE MONOHYDRATE. X-RAY REFINEMENT OF THE CRYSTAL STRUCTURE OF LITHIUM 29-3244
OF AMMONIUM HYDROGEN SULFATE ABOVE THE FERROEL/ CONSTRAINED REFINEMENT TECHNIQUES APPLIED TO THE STRUCTURE 16-2355
ROGEN PHOSPHATE AND AMMONIUM DIHYDROGEN PHOSPHATE CRYSTALS. (REFLECTANCE, REFRACTIVE INDICES) /ASSIUM DIHYD 17-2422
E ABO(3) FERROE/ VISIBLE AND ULTRAVIOLET OPTICAL PROPERTIES (REFLECTANCE SPECTRA AND BAND STRUCTURE) OF SOM 08-1383
GEN PHOSPHATE AND AMMONIUM DIHYDROGEN PHOSPHATE/ ULTRAVIOLET REFLECTION AND ABSORPTION OF POTASSIUM DIHYDRO 17-2422
L AND QUASI TRANSVERSE SOUND WAVES BY 180 DOMAIN BOUNDARIES/ REFLECTION AND REFRACTION OF QUASI LONGITUDINA 03-0314

WAVES AT DOMAIN BOUNDARIES IN FERROELECTRICS. (THEORY) REFLECTION AND REFRACTION OF TRANSVERSE SOUND 03-0313
OTOCONDUCTOR IMAGE STORAGE AND DISPLAY DEVICES OPERATED IN A REFLECTION MODE. (PLZT) /SED FERROELECTRIC- PH 09-1564
DOMAIN BOUNDARIES IN FERROELECT/ PIEZOELECTRIC EFFECT ON THE REFLECTION OF A TRANSVERSE ACOUSTIC WAVE FROM 03-0312
IC ARRAYS ON LITHIUM NIOBATE. REFLECTION OF ELASTIC SURFACE WAVE FROM PERIOD 11-1853
 GOLD FILM LAYERS ON LITH/ OPTICAL STUDIES OF REFRACTION AND REFLECTION OF ULTRASONIC SURFACE WAVES BY THIN 11-1789
 GOLD FILM LAYERS ON LITHIUM NIOBATE SUBSTRA/ REFRACTION AND REFLECTION OF ULTRASONIC SURFACE WAVES BY THIN 11-1790
N MOTIONS IN POTASSIUM DIDEUTERIUM PHOSPHATE (FROM POLARIZED REFLECTION SPECTRA, A TUNNELING MODEL). /UTERO 17-2645
STALS. ANISOTROPY OF THE REFLECTION SPECTRA OF ANTIMONY SULFOIODIDE CRY 15-2125
ANATES IN THE 2-1000 MICRON/ TEMPERATURE STUDIES OF INFRARED REFLECTION SPECTRA OF BARIUM AND STRONTIUM TIT 04-0863
ONG WAVELENGTH INFRARED REGION. ABSORPTION AND REFLECTION SPECTRA OF BARIUM TITANATE IN THE L 04-0644
IHYDROGEN PHOSPHATE AND POTA/ FOURIER TRANSFORM FAR INFRARED REFLECTION SPECTRA OF PARAELECTRIC POTASSIUM D 03-0356
ALS OF GROUPS V, VI AND VII. (ANTIMONY AND BISMUTH SULFO HA/ REFLECTION SPECTRA OF SOME FERROELECTRIC CRYST 15-2210
CTRICS. (BARIUM TITANATE) POLARIZATION REFLECTION TYPE LIGHT MODULATOR USING FERROELE 04-0714
/ENCE OF THE DEBYE- WALLER FACTORS OF BARIUM TITANATE (BRAGG REFLECTIONS C-AXIS, 20-180 DEG C, CALCULATION/ 04-0892
STRONTIUM TITANATE FILMS. FAR INFRARED REFLECTIVITY AND TRANSMISSION MEASUREMENTS ON 06-1154
 FAR INFRARED REFLECTIVITY OF ANTIMONY SULFOIODIDE. 15-2209
 FAR INFRARED REFLECTIVITY OF BARIUM TITANATE. 04-0744
 REFLECTIVITY ON ANTIMONY SULFOIODIDE. 15-2244
 FAR INFRARED REFLECTIVITY SPECTRA OF ANTIMONY SULFOIODIDE. 15-2234
 FAR INFRARED REFLECTIVITY SPECTRA OF ANTIMONY SULFOIODIDE. 15-2235
E. FAR INFRARED REFLECTIVITY SPECTRA OF CESIUM LEAD TRICHLORID 33-3379
R CM, 83-300K). FAR INFRARED REFLECTIVITY SPECTRA OF KDP CRYSTAL (16-400 PE 17-2681
S. (0.85 DEG K, 30-2/ TEMPERATURE DEPENDENCE OF FAR INFRARED REFLECTIVITY SPECTRA OF SODIUM NITRITE CRYSTAL 16-2373
ELENITE CRYSTAL. FAR INFRARED REFLECTIVITY SPECTRA OF SODIUM TRIHYDROGEN DIS 22-2877
 K) OF THE LATTICE VIBRATIONS IN LITHIUM TANTALAT/ INFRARED (REFLECTIVITY) STUDY (20-10,000 PER CM, 300 DEG 12-1870
 USE OF KDP TYPE CRYSTALS AS ELECTROOPTICAL REFLECTORS. 17-2512
E WAVES BY THIN GOLD FILM LAYERS ON LITH/ OPTICAL STUDIES OF REFRACTION AND REFLECTION OF ULTRASONIC SURFAC 11-1789
E WAVES BY THIN GOLD FILM LAYERS ON LITHIUM NIOBATE SUBSTRA/ REFRACTION AND REFLECTION OF ULTRASONIC SURFAC 11-1790
ALATE. INHOMOGENEITY OF THE OPTICALLY INDUCED INDEX OF REFRACTION IN LITHIUM NIOBATE AND LITHIUM TANT 11-1664
 INDEX OF REFRACTION OF POTASSIUM DIHYDROGEN PHOSPHATE. 17-2484
NSVERSE SOUND WAVES BY 180 DOMAIN BOUNDARIES/ REFLECTION AND REFRACTION OF QUASI LONGITUDINAL AND QUASI TRA 03-0314
 BOUNDARIES IN FERROELECTRICS. (THEORY) REFLECTION AND REFRACTION OF TRANSVERSE SOUND WAVES AT DOMAIN 03-0313
NTS, SPONTANEOUS POLARIZATION, THERMAL EXPANSION, INDICES OF REFRACTION, OPTICAL ROTATORY POWER) /IC CONSTA 35-3501
S PARAMETRIC SCATTERING OF LIGHT IN LITHIUM IODATE (INDEX OF REFRACTION 1.2-7 MICRONS). /S WAVE) SPONTANEOU 08-1291
PONTANEOUS POLARIZATION, THERMAL EXPANSION COEFFICIENTS, AND REFRACTIVE INDEX) / PIEZOELECTRIC CONSTANTS, S 12-1876
IN CRYSTALS OF PEROVSKITE STRUCT/ PHENOMENOLOGICAL THEORY OF REFRACTIVE INDEX AND ELECTROOPTICAL PHENOMENA 03-0518
OL OF THE SUSCEPTIBILITY OF LITHIUM NIOBATE TO LASER INDUCED REFRACTIVE INDEX CHANGES. CONTR 11-1791
 OPTICALLY INDUCED REFRACTIVE INDEX CHANGES IN BARIUM TITANATE. 04-1005
XPERIMENTS TO CHARACTERIZE THE OPTICAL DA/ OPTICALLY INDUCED REFRACTIVE INDEX CHANGES IN BARIUM TITANATE (E 04-1004
/C PROPERTIES OF BARIUM SODIUM NIOBATE. (OPTICAL ABSORPTION, REFRACTIVE INDEX, DIELECTRIC CONSTANT, NONLIN/ 13-2024
ATE AND TRIGLYCINE SULFATE CRYSTALS. THERMAL VARIATION OF REFRACTIVE INDEX IN AMMONIUM DIHYDROGEN PHOSPH 17-2474
OBATE. EFFECT OF INTENSE LIGHT ON THE REFRACTIVE INDEX OF A CRYSTALLINE POTASSIUM NI 08-1410
OND HARMONIC/ INFLUENCE OF A TRANSVERSE INHOMOGENEITY OF THE REFRACTIVE INDEX OF A NONLINEAR CRYSTAL ON SEC 11-1687
E INFRARED RANGE BY PARAMETRIC SCATTERIN/ MEASUREMENT OF THE REFRACTIVE INDEX OF ADP AND KDP CRYSTALS IN TH 17-2553
CINITY OF CURIE POINT. EFFECT OF IONIC MOTION ON THE REFRACTIVE INDEX OF PARAELECTRIC KDP IN THE VI 17-2548
(11) CRYSTAL, A NEW FERROELECTRIC. (DIELECTRIC CONSTANTS AND REFRACTIVE INDICES) LEAD(5) GERMANIUM(3) OXIDE 35-3502
AR OPTICAL PROPERTIES OF POTASSIUM NIOBATE SINGLE CRYSTALS. (REFRACTIVE INDICES) NONLINE 08-1454
RTIES TO THE SYMMETRY OF POTASSIUM IODATE. (NONLINEAR OPTIC, REFRACTIVE INDICES) RELATION OF PHYSICAL PROPE 08-1296
TE AND AMMONIUM DIHYDROGEN PHOSPHATE CRYSTALS. (REFLECTANCE, REFRACTIVE INDICES) /ASSIUM DIHYDROGEN PHOSPHA 17-2422
EPENDENCE ON TEMPERATURE, WAVELENGTH, APPLIED FIELD STRENGT/ REFRACTIVE INDICES OF ANTIMONY SULFOIODIDE. (D 15-2206
 TEMPERATURE DEPENDENCE OF THE REFRACTIVE INDICES OF BARIUM SODIUM NIOBATE. 13-2008
AR ELECTROOPTICS. OPTICAL PROPERTIES (BIREFRINGENCE AND REFRACTIVE INDICES) OF POTASSIUM NIOBATE (LINE 08-1468
BATE HAVING NO MICROTWINNING (AND RESISTANT TO LASER INDUCED REFRACTIVE INHOMOGENEITIES). /RIUM LITHIUM NIO 13-1958
 SOME EXPERIMENTS ON THE REGELATION OF ICE. 34-3435
IHYDROGEN PHOSPHATE SINGLE CRYSTAL (FROM SOLUTION) IN STATIC REGIME AND AT NORMAL PH. /IC OF THE AMMONIUM D 17-2516
/LECTRIC NONLINEARITIES IN THE TEMPERATURE AUTOSTABILIZATION REGIME OF A FERROELECTRIC CRYSTAL. (TGS TANDE/ 17-2502
N SPECTRA OF BARIUM TITANATE IN THE LONG WAVELENGTH INFRARED REGION. ABSORPTION AND REFLECTIO 04-0644
ERSION OF IRRADIATED BARIUM TITANATE IN THE PHASE TRANSITION REGION. DIELECTRIC DISP 04-0652
Y OF BARIUM TITANATE SINGLE CRYSTALS IN THE PHASE TRANSITION REGION. DIELECTRIC PERMEABILIT 04-0645
 DIELECTRIC DISPERSION IN THE FERROELECTRIC PHASE TRANSITION REGION. /TERACTION OF DIPOLES ON THE NATURE OF 03-0417
TRANSFORMATION. PART-2: ELECTRIC CONDUCTION IN THE INDUCTION REGION. /AMICS. PART-1: DEFORMATION AND PHASE 03-0169
 CHARACTERISTICS OF ANTIMONY SULFOIODIDE IN THE PARAELECTRIC REGION. /ION ON THE FORM OF THE VOLTAGE AMPERE 15-2211
CTRIC CONDUCTIVITY OF FERROELECTRICS IN THE PHASE TRANSITION REGION. NATURE OF THE ANOMALOUS ELE 03-0508
 OPTICAL CONSTANTS OF ICE I OVER THE ENTIRE INFRARED REGION. 34-3492
THE NEAR INFRARED TO FAR INFRARED AND IN THE RADIO FREQUENCY REGION. / CONSTANTS OF TRIGLYCINE SELENATE IN 27-3007
M POTASSIUM MANGANESE TRIFLUORIDE IN THE SHORT RANGE ORDERED REGION. /RAMAGNETIC SCATTERING OF NEUTRONS FRO 33-3371
UBIDIUM DIHYDROGEN PHOSPHATE CRYSTAL IN THE PHASE TRANSITION REGION. POLARIZATION OF A R 17-2506
OELECTRIC BEHAVIOR OF TRIGLYCINE SULFATE IN THE PARAELECTRIC REGION. PYR 27-3066
ENT OF SOFT OPTICAL PHONON MODES IN TETRAGONAL FERROELECTRIC REGION). / TITANATE ZIRCONATE SYSTEM (MEASUREM 09-1497
ECTRA OF BARIUM AND STRONTIUM TITANATES IN THE 20-1000 MICRON REGION. /URE STUDIES OF INFRARED REFLECTION SP 04-0863
/DE- NIOBIUM PENTOXIDE. (OXIDES, ESPECIALLY THE NIOBIUM-RICH REGION AND THE BINARY STRONTIUM OXIDE- NIOBIU/ 13-1920
UBIDIUM DIHYDROGEN PHOSPHATE CRYSTAL IN THE PHASE TRANSITION REGION. (COERCIVE FIELD) POLARIZATION OF A R 17-2504
 MANGANESE(2+) ABSORPTION NEAR THE CRITICAL REGION IN AMMONIUM SULFATE. 19-2641
 CRITICAL REGION IN FERROELECTRIC TRIGLYCINE SULFATE. 27-2985
/GATION (15 MHZ SHEAR WAVES) OF THE FERROELECTRIC TRANSITION REGION IN POTASSIUM DIHYDROGEN PHOSPHATE. (TE/ 17-2576
/NSMISSION OF POTASSIUM DIHYDROGEN PHOSPHATE IN THE INFRARED REGION IN THE PARA- AND FERROELECTRIC PHASES. 17-2459
F MEASURE/ PYROELECTRIC RESPONSE OF TGS IN THE FERROELECTRIC REGION NEAR THE CURIE POINT (IMPROVED METHOD O 27-2979
TION COEFFICIENT OF LEAD MAGNONIOBATE SINGLE CRYSTALS IN THE REGION OF A BROAD FERROELECTRIC TRANSITION. /P 13-1974
 THERMAL EXPANSION OF ADP AND DEUTERATED ADP IN THE REGION OF ANTIFERROELECTRIC PHASE TRANSITION. 17-2445
DRATE. ZEEMAN EFFECT (STUDY) OF TWO FINE LINES IN (THE RED REGION OF) CHROMIUM GUANIDINIUM SULFATE HEXAHY 21-2814
TION REVERSAL OF ANTIMONY SULFOIODIDE SINGLE CRYSTALS IN THE REGION OF PHASE COEXISTENCE. POLARIZA 15-2246
/LAXATION AND SUSCEPTIBILITY IN THE FERROELECTRIC TRANSITION REGION OF POTASSIUM DIDEUTERIUM PHOSPHATE (AN/ 17-2525
 LEAD / ANOMALIES OF ELASTICITY AND INTERNAL FRICTION IN THE REGION OF THE ANTIFERROELECTRIC CURIE POINT OF 13-2020
/UM NITRITE CRYSTALS AND COMBINATION SCATTER SPECTRUM IN THE REGION OF THE ANTISYMMETRICAL VIBRATION OF TH/ 16-2276
IENT OF RESISTANCE OF BARIUM TITANATE SINGLE CRYSTALS IN THE REGION OF THE CURIE TEMPERATURE. /TURE COEFFIC 04-1036
AD STANNATE NIOBATE. SOLID SOLUTIONS IN ANTIFERROELECTRIC REGION OF THE SYSTEM LEAD HAFNATE TITANATE, LE 13-1956
 PROTONIC CONDUCTION OF ICE. PART-1: HIGH TEMPERATURE REGION. PART-2: LOW TEMPERATURE REGION. 34-3414
EL). HEAT CAPACITY IN THE FERROELECTRIC TRANSITION REGION (THEORY, RELATION TO KAENZIG DOMAIN MOD 03-0443
HASE TRANSITION, TALKING INTO ACCOUNT THE INTERACTING KANZIG REGIONS. /C FIELD ON A DIFFUSE FERROELECTRIC P 03-0441
BATE CRYSTALS. STABILITY REGIONS FOR THE GROWTH OF BARIUM STRONTIUM NIO 13-2056

YSTALS. (TOPOGRAPHY OF IMPERFECTIONS, DOMAIN STRUCTU/ GROWTH REGIONS IN FERROELECTRIC TRIGLYCINE SULFATE CR 27-2999
UTIONS IN DIELECTRICS, WITH SPECIAL EMPHASIS ON NEAR SURFACE REGIONS IN FERROELECTRICS. /TRIC FIELD DISTRIB 03-0182
SCME POSSIBILITIES OF DETERMINING THE VOLUME OF KANZIG REGIONS IN FERROELECTRICS. 03-0444
TRANSITIONS. ACCOUNTING FOR INTERACTION IN THE KANZIG REGIONS MODEL FOR DIFFUSE FERROELECTRIC PHASE 03-0438
E CRYSTALS. SPECIAL ELECTRICAL PROPERTIES OF REGIONS NEAR DOMAIN WALLS IN TRIGLYCINE SULFAT 27-2988
CRYSTAL LATTICES. ESTIMATION OF THE DIMENSIONS OF KANZIG REGIONS ON THE BASIS OF THE DYNAMIC THEORY OF 03-0446
LECTRICS (SOLID SOLUTIONS AND MIXED CATION COMPOUNDS, KANZIG REGIONS, PEROVSKITES) /LAXATION IN SOME FERROE 03-0291
IONS OF THE SPONTANEOUS POLARIZATION. A MODEL OF KANZIG REGIONS TAKING INTO ACCOUNT DIFFERENT ORIENTAT 03-0439
UM DIHYDROGEN(1-X) DIDEUTERIUM(X) PHOSPHATE SINGLE CRYSTALS. REGULAR BEHAVIOR OF SOLID SOLUTIONS IN POTASSI 17-2458
FERROELECTRIC TRANSFORMATIONS OF TENSORIAL PROPERTIES IN REGULAR FERROELECTRICS. 03-0139
ROELECTRIC MATERIALS. (HETEROTYPES, REVIEW) ROLE OF THE REGULAR OCTAHEDRON AND LARGER POLYHEDRA IN FER 02-0094
PTION IN STRONTIUM TITANATE WITHOUT COEXISTING PHASES. REINTERPRETATION OF WAVELENGTH MODULATED ABSOR 06-1188
TITANATE CERAMIC. REJUVENATION OF A BARIUM(0.72) STRONTIUM(0.28) 04-0817
LATTICE VIBRATION CF SODIUM NITRITE. (CALCULATIONS TO RELATE TO SPECIFIC ANTIFERROELECTRIC PHASE) 16-2317
OLYBDATES. STRUCTURAL AND PHASE RELATIONSHIPS AMONG TRIVALENT TUNGSTATES AND M 10-1648
OLYBDATES. STRUCTURAL AND PHASE RELATIONSHIPS AMONG TRIVALENT TUNGSTATES AND M 10-1649
EM POTASSIA LITHIA NIOBIA (POTASSIUM LITHIUM NIOBATES, PHASE RELATIONSHIPS, SOLID SOLUTIONS). /(OXIDE) SYST 13-2017
FERRO- AND ANTIFERROELECTRIC COUPLING AND DIELECTRIC RELAXATION. 03-0354
POLARIZATION EFFECTS IN ICE. (CURRENT REVERSAL DUE TO DIPOLE RELAXATION) LOW TEMPERATURE 34-3418
ROELECTRIC DIAMMONIUM TRIHYDROGEN PERIODATE PROTON CROSSOVER RELAXATION. /QUADRUPOLE INTERACTION IN ANTIFER 36-3578
DIAMMONIUM TRIHYDROGEN PERIODATE BY IODINE PROTON CROSSOVER RELAXATION. / INTERACTION IN ANTIFERROELECTRIC 36-3579
POINT OF FERROELECTRIC SEMICONDUCTORS. RELAXATION ABSORPTION OF SOUND NEAR THE CURIE 03-0492
CRYSTALS OF BARIUM TITANATE IN THE PRESENCE OF A STEADY ELE/ RELAXATION ABSORPTION OF ULTRASOUND IN SINGLE 04-0850
RIC CRYSTALS. (SODIUM NITRI/ NUCLEAR QUADRUPOLE SPIN LATTICE RELAXATION AND CRITICAL DYNAMICS OF FERROELEC 16-2268
ECTRIC KDP. PARAMAGNETIC RELAXATION AND DIVALENT COPPER IONS IN FERROEL 17-2629
ECTRIC CRYSTALS. (THEORY) NUCLEAR QUADRUPOLE SPIN LATTICE RELAXATION AND DYNAMICAL PROPERTIES OF FERROEL 03-0187
ID SOLUTIONS IN THE SYSTEM BARIUM TITANATE- STRONTIUM NIOBI/ RELAXATION AND FERROELECTRIC PROPERTIES OF SOL 04-0707
ODIUM NITRITE) QUADRUPOLE SPIN PHONON RELAXATION AND FERROELECTRIC TRANSITION. (IN S 16-2362
ATE SINGLE CRYSTALS. MECHANICAL RELAXATION AND NONLINEARITY IN STRONTIUM TITAN 06-1218
UM DIDEUTERIUM ARSENATE AND CESIUM / CESIUM-183 SPIN LATTICE RELAXATION AND RESONANCE IN FERROELECTRIC CESI 18-2720
TRIC TRANSITION REGION OF POTASSIUM DIDEUTERIU/ POLARIZATION RELAXATION AND SUSCEPTIBILITY IN THE FERROELEC 17-2575
YSTAL. DIPOLAR RELAXATION AT LOW TEMPERATURE OF ICE SINGLE CR 34-3447
/TRIGLYCINE SULFATE SINGLE CRYSTALS. (THEORY OF ANISOTROPIC RELAXATION ATTENUATION OF QUASI LONGITUDINAL/ 27-3078
NIOBATE SINGLE CRYSTALS. DEFORMATION AND RELAXATION BEHAVIOR IN FERROELECTRIC POTASSIUM 08-1420
RIC SODIUM TRIHYDROGEN DISELENITE AND ITS THERMAL EXPANSION. RELAXATION BETWEEN THE STRUCTURE OF FERROELECT 22-2851
CE SINGLE CRYSTALS. DIELECTRIC RELAXATION, BULK AND SURFACE CONDUCTIVITY OF I 34-3459
ER TYPE FERROELECTRICS: POLARIZATION FLUCTUATI/ SPIN LATTICE RELAXATION BY QUASI SPIN WAVES IN ORDER DISORD 26-2928
RROCYANIDE TRIHYDRATE. ANOMALOUS SPIN LATTICE RELAXATION BY QUASI SPIN WAVES IN POTASSIUM FE 24-2899
IUM DIHYDROGEN PHOSPHATE. SPIN LATTICE RELAXATION BY THE FERROELECTRIC MODE IN SODIUM 17-2440
N PERIODATE CRYSTALS, PULSED MAGNETIC/ LEVEL CROSSING (CROSS RELAXATION) EXPERIMENTS IN AMMONIUM TRIHYDROGE 36-3577
NTAINING (TRIVALENT) RARE EARTH IONS. DIELECTRIC RELAXATION IN (CERAMIC) STRONTIUM TITANATES CO 06-1143
GNETIC RESONANCE IN WATER ICE. PART-2: DEUTERON SPIN LATTICE RELAXATION IN DEUTERATED ICE. /RT-1: PROTON MA 34-3471
STUDY OF CONDUCTIVITY AND DIPOLAR RELAXATION IN DOPED ICE SINGLE CRYSTALS. 34-3458
ERATED ISOMORPHS. (THEORY) DIELECTRIC RESONANCE AND RELAXATION IN ELECTRICALLY BIASED KDP AND DEUT 03-0467
PROTON SPIN LATTICE RELAXATION IN FERROELECTRIC COLEMANITE. 23-2885
M HYDROGEN SULFATE. DEUTERON SPIN LATTICE RELAXATION IN FERROELECTRIC DEUTERATED AMMONIU 19-2795
IUM ALUMINUM SULFATE DODECAHYDRATE. (COMPLEX DIE/ DIELECTRIC RELAXATION IN FERROELECTRIC MASD, METHYL AMMON 20-2804
IUM PHOSPHATE, POTASSIUM DI/ DEUTERON MAGNETIC RESONANCE AND RELAXATION IN FERROELECTRIC POTASSIUM DIDEUTER 17-2438
TRITE. DIELECTRIC RELAXATION IN FERROELECTRIC SILVER SODIUM DINI 16-2290
TRITE. DIELECTRIC RELAXATION IN FERROELECTRIC SILVER SODIUM DINI 16-2292
RTRATE. PROTON SPIN LATTICE RELAXATION IN FERROELECTRIC SODIUM AMMONIUM TA 29-3248
SELENATE. PROTON SPIN LATTICE RELAXATION IN FERROELECTRIC SODIUM TRIHYDROGEN 22-2852
. NUCLEAR SPIN LATTICE RELAXATION IN FERROELECTRIC TRIGLYCINE SULFATE 27-2967
REPANCY). NOTE ON THE DIELECTRIC RELAXATION IN FERROELECTRICS (THEORETICAL DISC 03-0391
INTERPRETATION OF THE PROTON SPIN LATTICE RELAXATION IN HEXAGONAL ICE. 34-3475
STRONG COLLISION LIMIT OF SPIN LATTICE RELAXATION IN HEXAGONAL ICE. 34-3480
HO) MINIMUM. PROTEIN SPIN RELAXATION IN HEXAGONAL ICE. PART-2: THE T(1 R 34-3488
LENATE. PROTON SPIN LATTICE RELAXATION IN IRRADIATED SODIUM TRIHYDROGEN SE 22-2853
PROTON ROTATING FRAME RELAXATION IN LITHIUM HYDRAZINIUM SULFATE. 32-3337
NE SULFATE. DIELECTRIC RELAXATION IN MULTIDOMAIN CRYSTALS OF TRIGLYCI 27-3002
DIELECTRIC RELAXATION IN MULTIDOMAIN TGS SINGLE CRYSTALS. 27-3001
ER / CRITICAL SLOWING DOWN PROCESS OF (ANOMALOUS) DIELECTRIC RELAXATION IN SEVERAL CRYSTALS OF ORDER DISORD 02-0087
AL NOI/ STUDY OF THE CRITICAL SLOWING DOWN OF THE DIELECTRIC RELAXATION IN SILVER SODIUM DINITRITE BY THERM 16-2382
IP OF THE TEMPERATURE DEPENDENCE OF B1 PHONON AND DIELECTRIC RELAXATION IN SODIUM NITRITE. RELATIONSH 16-2257
NMR STUDY OF DEUTERIUM DIFFUSION AND SODIUM-23 RELAXATION IN SODIUM TRIDEUTERIUM SELENATE. 22-2849
S. (AMMONIUM SULFATE, FLUOBERYLLATE, P/ NUCLEAR SPIN LATTICE RELAXATION IN SOME FERROELECTRIC AMMONIUM SALT 19-2769
/NING (DIFFUSNESS) AND THE NATURE OF DIELECTRIC POLARIZATION RELAXATION IN SOME FERROELECTRICS (SOLID SOLU/ 03-0291
/LEAR MAGNETIC RESONANCE (WIDE LINE SPECTRA AND SPIN LATTICE RELAXATION) INVESTIGATIONS OF THE PHASE TRANS/ 20-2802
SODIUM DINITRITE BY STEP FUNCTION METHOD. DIELECTRIC RELAXATION MEASUREMENT ON FERROELECTRIC SILVER 16-2291
ELECTRIC TRIGLYCINE SULFATE. ULTRASONIC RELAXATION NEAR THE CURIE TEMPERATURE OF FERRO 27-3097
MOLYBDATE. WAVEVECTOR DEPENDENT RELAXATION OF AN OPTICAL PHONON IN GADOLINIUM 30-3320
EFFECTIVE FIELD, AND ACTIVATION ENERGY. APPLICATION TO LEAD/ RELAXATION OF ANTIFERROELECTRIC DOMAIN WALLS, 07-1257
TE. (TEMPERATURE DEPENDENCES OF THE COMPLEX DIELECTRIC CONS/ RELAXATION OF DOMAIN WALLS IN TRIGLYCINE SULFA 27-3103
UM DIHYDROGEN PHOSPHATE AND POTASSIUM DIHYDROGEN PHOSPHATE/ RELAXATION OF FERROELECTRIC PHOSPHATES (AMMONI 17-2580
N LITHIUM HYDRAZINIUM SULFATE. DIPOLAR RELAXATION OF LITHIUM-7 BY HINDERED ROTATORS I 32-3350
DIELECTRIC RELAXATION OF MULTIDOMAIN TRIGLYCINE SULFATE. 27-3050
NUCLEAR QUADRUPOLE RELAXATION OF NITROGEN-14 IN SODIUM NITRITE. 16-2254
IN ROCHELLE SALT AND AMMONIUM ROCHELLE SALT BY SPIN LATTICE RELAXATION OF PROTONS IN ROTATING FRAME. /ULES 28-3232
D FERROELECTRIC PHASE OF SODIUM NITRIT/ NUCLEAR SPIN LATTICE RELAXATION OF SODIUM-23 IN THE PARAELECTRIC AN 16-2264
INVESTIGATIONS OF NUCLEAR RELAXATION OF TRIGLYCINE FLUOBERYLLATE. 27-3104
MAGNETIZED PARAMAGNETIC SUBSTANCES. PART-1: IRON(X) ALUMINU/ RELAXATION PHENOMENA AND MOSSBAUER SPECTRA OF 19-2751
DIFFUSION AND RELAXATION PHENOMENA IN ICE. 34-3470
D MAGNESATE(0.33) NIOBATE(0.67) WITH A DIFFUSE PHASE TRANSI/ RELAXATION POLARIZATION OF A FERROELECTRIC LEA 13-1980
INGLE CRYSTALS EXCITED WITH MICROSECOND LIGHT PULSES. RELAXATION PROCESSES IN ANTIMONY SULFOIODIDE S 15-2202
N FERROELECTRICS. DIELECTRIC RELAXATION PROCESSES NEAR THE CRITICAL POINT I 03-0272
SE TRANSITION POINT. ON RELAXATION PROCESSES NEAR THE SECOND ORDER PHA 03-0385
INGLE CRYSTALS. RELAXATION PROCESSES OF ANTIMONY SULFOIODIDE S 15-2190
. INFLUENCE OF THE ELECTRICAL BOUNDARY CONDITIONS ON THE NMR RELAXATION RATE OF SODIUM-23 IN SODIUM NITRITE 16-2350
SOLUTIONS AND THEIR INTERPRETATION. PART-8: TRAN/ DIELECTRIC RELAXATION SPECTRA OF WATER, ICE, AND AQUEOUS 34-3482
N FERROELECTRIC POTASSIUM DIHYDROGEN ARS/ NMR AND DIELECTRIC RELAXATION STUDIES OF LONG RANGE CORRELATION I 18-2718

RATURE DEPENDENCE OF THE YOUNGS MODULUS AND THE POLARIZATION RELAXATION TIME). /IDEUTERIUM PHOSPHATE '(TEMPE 17-2660
/CRYSTAL DEFECTS UPON FERROELECTRIC PROPERTIES. (DISPERSION, RELAXATION TIME AND STRENGTH, HYSTERESIS, 26 / 03-0496
DINITRITE.   HYDROSTATIC PRESSURE EFFECT ON THE DIELECTRIC RELAXATION TIME IN FERROELECTRIC SILVER SODIUM 16-2294
IE TEMPERATURE (POLY VERSUS MONODISPERSIVE / DISTRIBUTION OF RELAXATION TIME IN FERROELECTRICS NEAR THE CUR 03-0521
UM SELENATE.   NUCLEAR SPIN LATTICE RELAXATION TIME MEASUREMENTS IN SODIUM DEUTERI 22-2847
OF DICALCIUM LEAD PROPIO/ ANOMALY OF THE PROTON SPIN LATTICE RELAXATION TIME NEAR THE CRITICAL TEMPERATURE 26-2947
ODIUM NITRITE.   DIPOLE LATTICE RELAXATION TIME NEAR THE NEEL TEMPERATURE IN S 16-2300
SODIUM NITRIT/ STATIC ELECTRIC SUSCEPTIBILITY AND DIELECTRIC RELAXATION TIME NEAR THE TRANSITION POINTS IN 16-2302
E IN THE ORDER DISORDER TYPE FE/ ANOMALY OF THE SPIN LATTICE RELAXATION TIME NEAR THE TRANSITION TEMPERATUR 03-0488
DIELECTRIC CONSTANT AND DIELECTRIC RELAXATION TIME OF ROCHELLE SALT. 28-3179
DER APPLIED ELECTR/ (REVERSIBLE) OPTICAL DAMAGE IN KDP (WITH RELAXATION TIME OF SECONDS WHEN ILLUMINATED UN 17-2703
ORIDE.   NUCLEAR QUADRUPOLE FREQUENCIES AND RELAXATION TIMES IN FERROELECTRIC HYDROGEN CHL 31-3336
/G OF QUADRUPOLAR NUCLEI. DETERMINATION OF THE SPIN LATTICE RELAXATION TIMES IN FERROELECTRIC LITHIUM TAN/ 11-1829
DISTRIBUTION OF RELAXATION TIMES IN SOME FERROELECTRICS. 03-0288
LLINE STATE IN HEXAGONAL ICE.   RELAXATIONAL PROTON ORDERING AND GLASSY CRYSTA 34-3436
CONATE TITANATE STANNATE BY TEMPERATURE AND PRESSURE ENFORC/ RELEASE OF ELECTRIC ENERGY IN LEAD NIOBIUM ZIR 09-1488
INVESTIGATING STRAIN WAVE PROPAGATION AND ASSOCIATED CHARGE RELEASE TO FERROELECTRIC MATERIALS. /THODS FOR 37-3630
/URES OF SODIUM NIOBATE BETWEEN 480 AND 575 DEG C, AND THEIR RELEVANCE TO SOFT PHONON MODES. (X-RAY DIFFRA/ 08-1273
R/ SOME THEORETICAL CONSIDERATIONS RELATING TO SWITCHING AND REMANENCE IN FERROELECTRIC PHOTOCONDUCTOR MEMO 03-0378
/RASOUND PROPAGATION. PART-9: SPECIFIC HEAT. PART-10: PHOTON RENORMALIZATION AND THEORY OF SPONTANEOUS POL/ 03-0397
/CHANISMS OF FERROELECTRIC PHASE TRANSITION IN CRYSTALS WITH REORIENTATING STRUCTURAL GROUPS. (SODIUM AMMO/ 22-2827
PLASTICITY OF ICE AND THEIR EXPLANATION IN TERMS OF HYDROGEN REORIENTATION.   IMPURITY EFFECTS ON THE 34-3449
OF AMMONIUM SULFATE: THE NATURE OF HYDROGEN BONDING AND THE REORIENTATION OF THE AMMONIUM IONS. /ANSITIONS 19-2783
NG OF POLYCRYSTALLINE BARIUM TITANATE.   SURFACE DOMAIN REORIENTATION PRODUCED BY ABRASION AND ANNEALI 04-0639
ON THE PHASE COMPOSITION OF REDUCED AND REOXIDIZED BARIUM TITANATE. 04-0541
RROELECTRIC PROPERTIES OF POTASSIUM DIHYDROGEN PH/ EFFECT OF REPLACEMENT OF POTASSIUM BY ALUMINUM ON THE FE 17-2560
ROELECTRICS (BARIUM TITANATE AND SAME WITH 1 PERCENT SILICON REPLACING TITANIUM). /SOME POLYCRYSTALLINE FER 04-0818
CRYSTALS DEPENDENT ON GROWTH CONDITIONS.   MAXIMUM REPOLARIZATION CURRENT FOR TRIGLYCINE SULFATE 27-3154
FIELD FOR PARKHAUSEN STEPS AND THE CRITICAL POLARIZATION AND REPOLARIZATION FIELDS IN FERROELECTRICS /TING 03-0318
ALS.   EFFECT OF EXTERNAL ACTIONS ON THE REPOLARIZATION OF BARIUM TITANATE SINGLE CRYST 04-0957
SINGLE CRYSTALS.   EFFECT OF ILLUMINATION ON REPOLARIZATION PROCESS IN ANTIMONY SULFOIODIDE 15-2220
THE PEROVSKITE STRU/ THE INVERSE PIEZOELECTRIC EFFECT IN THE REPOLARIZATION PROCESS OF FERROELECTRICS WITH 08-1374
CRYSTALS AND / EFFECT OF ULTRASOUND ON THE POLARIZATION AND REPOLARIZATION PROCESSES IN TRIGLYCINE SULFATE 04-0924
NGLE CRYSTALS IN ORTHORHOMBIC AND RHOMBOHEDRAL PHASES. REPOLARIZATION PROCESSES OF BARIUM TITANATE SI 04-0963
COLLINEAR AND NONCOLLINEAR FERRO- AND ANTIFERROELECTRICS. (REPRESENTATION THEORY APPLIED, 18 REFS) 03-0173
LASTIC AND MAGNETIC PROPERTY, SYMMETRY, AND PHASE TRANSITION REPRESENTATION. (18 REFS)   DIELECTRIC, E 03-0316
ITCHING PROPERTIES OF THE FERROELECTRIC SURFACES. (NEW MODEL REPRODUCES EXPERIMENTAL PULSES)   ON THE SW 37-3641
LATTICE STATISTICS OF HYDROGEN BONDED CRYSTALS, PART-1, THE RESIDUAL ENTROPY OF ICE, PART-2, HYDROGEN BON/ 03-0388
E.   RESIDUAL ENTROPY OF SODIUM TRIHYDROGEN SELENAT 22-2835
FATE AND TRIGLYCINE SELENATE CRYSTALS ON CHANGES IN SPECIFIC RESISTANCE. /OWTH CONDITIONS OF TRIGLYCINE SUL 27-3069
EMICONDUCTING BARIUM TITANATE. (DIELECTRIC CONSTANT, CONTACT RESISTANCE)   S 04-0730
SEE BOTH POSITIVE TEMPERATURE COEFFICIENT OF RESISTANCE AND PTCR.
IOBIUM PENTOXIDE / PTCR (POSITIVE TEMPERATURE COEFFICIENT OF RESISTANCE) BEHAVIOR OF BARIUM TITANATE WITH N 04-0837
TALS.   PIEZOELECTRIC RESISTANCE EFFECT IN BISMUTH SELENOIODIDE CRYS 15-2218
POSITIVE TEMPERATURE COEFFICIENT OF RESISTANCE IN DOPED LEAD TITANATE. 05-1051
ANOMALY OF RESISTANCE IN REDUCED BARIUM TITANATE CRYSTAL. 04-0988
INFLUENCE OF PHASE TRANSITIONS ON ELECTRICAL RESISTANCE OF BARIUM TITANATE. 04-0599
IN THE REGION OF THE CU/ POSITIVE TEMPERATURE COEFFICIENT OF RESISTANCE OF BARIUM TITANATE SINGLE CRYSTALS 04-1036
ERAMICS.   EFFECT OF EXCESS TITANIA ON THE ELECTRIC RESISTANCE OF SEMICONDUCTIVE BARIUM TITANATE C 04-0966
ITANATE- BISMUTH OXIDE COMPO/ TWO STEP ANOMALOUS INCREASE OF RESISTANCE ON LANTHANUM(0.002) BARIUM(0.998) T 04-0801
/CRYSTAL BARIUM LITHIUM NIOBATE HAVING NO MICROTWINNING (AND RESISTANT TO LASER INDUCED REFRACTIVE INHOMOG/ 13-1958
NITRATE CRYSTALS.   ELECTRICAL RESISTIVITIES OF SODIUM NITRITE AND POTASSIUM 16-2258
PROPERTIES OF LEAD TITANATE CERAMICS. (DIELECTRIC CONSTANT, RESISTIVITY) /VES ON PIEZOELECTRIC AND RELATED 05-1084
TRONTIUM TITANATES WITH POSITIVE TEMPERATURE COEFFICIENTS OF RESISTIVITY.   SEMICONDUCTING BARIUM AND S 04-0995
RESISTIVITY AND SINTERING OF BARIUM TITANATE. 04-0819
ITANATE. (POSITIVE TEMPERATURE COEFFICIENT OF RESISTIVITY) RESISTIVITY ANOMALY IN SEMICONDUCTING BARIUM T 04-0584
ITANATE CERAMICS.   EFFECT OF GRAIN SIZE ON THE RESISTIVITY ANOMALY IN SEMICONDUCTIVE BARIUM T 04-0735
/BIC TETRAGONAL (MARTENSITIC TYPE) PHASE TRANSITIONS IN HIGH RESISTIVITY BARIUM TITANATE SINGLE CRYSTALS. / 04-0887
BARIUM ZIRCONIUM TITANATE. (TEMPERATURE DEPENDENCE) RESISTIVITY BEHAVIOR OF SEMICONDUCTING YTTRIUM 13-1995
/TRINSIC BAND GAP, PTCR (POSITIVE TEMPERATURE COEFFICIENT OF RESISTIVITY) DEVICES) PART-2: ANALYSIS OF DEF/ 04-0822
OVSKITE) LE/ PTC EFFECT (POSITIVE TEMPERATURE COEFFICIENT OF RESISTIVITY) IN (DOPED AND UNDOPED CERAMIC PER 13-2000
.   EFFECT OF OXIDE IMPURITIES ON THE RESISTIVITY OF LANTHANUM DOPED BARIUM TITANATE 04-0587
S CONTAINING IMPURITIES.   ELECTRICAL RESISTIVITY OF LEAD ZIRCONATE TITANATE CERAMIC 09-1607
- BISMUTH TRIOXIDE COMPOSITE/ TWO-STEP ANOMALOUS INCREASE OF RESISTIVITY ON LANTHANUM DOPED BARIUM TITANATE 04-0800
RPTION) SOME SEMICONDUCTING PROPERTIES OF BARIUM TITANATE. (RESISTIVITY, SEEBECK COEFFICIENT, OPTICAL ABSO 04-0919
(FLUX) GROWTH AND FERROELECTRIC PROPERTIES OF HIGH RESISTIVITY SINGLE CRYSTALS OF LEAD TITANATE. 05-1072
TIMONY INTO THE BARIUM TITANATE LATTICE. (STRUCTURE, AND PCT RESISTORS)   INCORPORATION OF AN 04-0934
RAGE DIPOLE MOMENT IN KDP TYPE FERROELECTRICS.   RESOLUTION OF A DISCREPANCY CONCERNING THE AVE 17-2618
ISMUTH TITANATE PHOTOCONDUCTOR OPTICAL STORAGE MEDIUM. (HIGH RESOLUTION READ WRITE)   B 08-1445
M DOPED TRIGLYCINE SULFATE OBSERVED BY ELECTRON PARAMAGNETIC RESONANCE /OF THE DOMAIN STRUCTURE IN VANADIU 27-3013
TURES IN CESIUM LEAD TRICHLORIDE USING ELECTRON PARAMAGNETIC RESONANCE DETERMINATION OF TRANSITION TEMPERA 33-3408
TRIC TRIGLYCINE SULFATE (SEVERAL KHZ TO 23 KMHZ, DOMAIN WALL RESONANCE). DIELECTRIC DISPERSION IN FERROELEC 27-3032
NIUM DIHYDROGEN PHOSPHATE SINGLE CRYSTAL BY NUCLEAR MAGNETIC RESONANCE /ON OF THE PHASE TRANSITION IN AMMO 17-2545
TRIHYDROGEN PERIODATE CRYSTALS, PULSED MAGNETIC AND OPTICAL RESONANCE. /ELAXATION) EXPERIMENTS IN AMMONIUM 36-3577
OMIC IRON IN STRONTIUM TITANATE. EVIDENCE FROM PARAMAGNETIC RESONANCE.   PHOTOCHR 06-1171
TURE STUDIED BY INFRARED SPECTROSCOPY AND NUCLEAR QUADRUPOLE RESONANCE.   POTASSIUM IODATE CRYSTAL STRUC 08-1278
BY NEUTRON DIFFRACTION, PROTON AND DEUTERON NUCLEAR MAGNETIC RESONANCE. /UCTURAL STUDIES OF ICE POLYMORPHS 34-3446
POTASSIUM FERROCYANIDE TRIHYDRATE BY PULSED NUCLEAR MAGNETIC RESONANCE. /IN PARAELECTRIC AND FERROELECTRIC 24-2889
IC STATE OF COPPER FORMATE TETRAHYDRATE BY ANTIFERROMAGNETIC RESONANCE.   STUDY OF THE ORDERED MAGNET 26-2943
/RON PHONON INTERACTION IN FERROELECTRIC CRYSTALS (CYCLOTRON RESONANCE, ABSORPTION AND LUMINESCENCE SPECTR/ 03-0427
, STRONTIUM, AND CALCIUM STANNATE.   RESONANCE ABSORPTION OF GAMMA QUANTA IN BARIUM 36-3566
-RAYS OF TIN-119 DURING PHASE TRANSITIONS OF THE/ CHANGE OF RESONANCE ABSORPTION SPECTRA OF 23.8 KEV GAMMA 08-1400
IELECTRIC CONSTANT MATERIALS GAINED BY ELECTRON PARAMAGNETIC RESONANCE. (ABSTRACT ONLY, 11 REFS) /ON HIGH D 02-0101
LONG RANGE CORRELATION IN FERROELECTRIC P/ NUCLEAR MAGNETIC RESONANCE AND DIELECTRIC RELAXATION STUDIES OF 18-2718
E OF THE PROTON LATTICE / DETECTION BY ELECTRON PARAMAGNETIC RESONANCE AND ELECTRON NUCLEAR DOUBLE RESONANC 18-2724
E OF THE ARSENATE CENTER IN X-IRRADIA/ ELECTRON PARAMAGNETIC RESONANCE AND ELECTRON NUCLEAR DOUBLE RESONANC 18-2727
E OBSERVATIONS OF DOMAINS IN THE IRRA/ ELECTRON PARAMAGNETIC RESONANCE AND ELECTRON NUCLEAR DOUBLE RESONANC 18-2726
ELEMENT IONS IN STRONTIUM TITANATE AND LANTHA/ PARAMAGNETIC RESONANCE AND OPTICAL ABSORPTION OF TRANSITION 06-1165
D KDP AND DEUTERATED ISOMORPHS. (THEORY)   DIELECTRIC RESONANCE AND RELAXATION IN ELECTRICALLY BIASE 03-0467
SSIUM DIDEUTERIUM PHOSPHATE, POTASSIUM DI/ DEUTERON MAGNETIC RESONANCE AND RELAXATION IN FERROELECTRIC POTA 17-2438

ELECTRIC PHASE CHANGE IN BARIUM TITANATE USING ELECTRON SPIN RESONANCE AND THE MOSSBAUER EFFECT. /THE FERRO 04-0827
ANTIFERROELECTRICITY.   MAGNETIC RESONANCE AND THE PROBLEM OF THE DEFINITION OF 03-0266
CYANIDE TRIHYDRATE CRYSTALS.   NUCLEAR MAGNETIC RESONANCE AND X-RAY STUDIES OF POTASSIUM FERRO 24-2894
ICE)   ELECTRIC RESONANCE: APPLICATION TO THE HYDROGEN BOND. (34-3477
POTASSIUM MANGANESE TRIFLUORIDE.   RESONANCE BEHAVIOR IN CANTED ANTIFERROMAGNETIC 33-3403
ITHIUM NIOBATE. (CAPACITANCE, OPTICAL SPECTRA, ELECTRON SPIN RESONANCE CELL PARAMETERS) /ATION DAMAGE IN L 11-1705
SODIUM DINITRI/ SPLITTING OF THE SODIUM-23 NUCLEAR MAGNETIC RESONANCE CENTRAL LINE IN FERROELECTRIC SILVER 16-2307
NUM SULFATE AND A 36:64 RATIO MIXED CRYSTAL, PROTON MAGNETIC RESONANCE, DIELECTRIC CONSTANT). /MONIUM ALUMI 20-2809
RRODIELECTRIC SINGLE CRYSTALS HAVING DOMAIN STRUCTURE.   RESONANCE EFFECTS IN MAGNETICALLY EQUIAXIAL FE 03-0509
Y IN STRONTIUM TITANATE.   NEW ELECTRON SPIN RESONANCE EVIDENCE FOR THE 65 DEGREES K ANOMAL 06-1230
ADIATED SINGLE CRYST/ OBSERVATION OF ELECTRON NUCLEAR TRIPLE RESONANCE FOR THE ARSENATE CENTER IN X-RAY IRR 18-2729
OF THERMAL EXPANSION OF ROCHELLE SALT BY USING THE SHIFT OF RESONANCE FREQUENCY IN CAVITY.   MEASUREMENT 28-3202
SENATE)   FERROELECTRIC MODE AND NQR (NUCLEAR QUADRUPOLE RESONANCE) FREQUENCY. (POTASSIUM DIHYDROGEN AR 17-2713
ONS AND OF EXCHANGE OF THE ARSENATE RADICAL I/ ELECTRON SPIN RESONANCE IDENTIFICATION OF SLATER CONFIGURATI 18-2734
DEUTERO AMMONIUM DIDEUTERIUM ARSENATE.   ARSENIC-75 NMR RESONANCE IN AMMONIUM DIHYDROGEN ARSENATE AND 18-2736
MPERATURE.   ELECTRON PARAMAGNETIC RESONANCE IN BARIUM TITANATE NEAR THE CURIE TE 04-0928
ARSENATE AND CESIUM / CESIUM-183 SPIN LATTICE RELAXATION AND RESONANCE IN FERROELECTRIC CESIUM DIDEUTERIUM 18-2720
BROMINE NUCLEAR QUADRUPOLE RESONANCE IN FERROELECTRIC HYDROGEN BROMIDE. 31-3332
ACOUSTIC NUCLEAR MAGNETIC RESONANCE IN FERROELECTRIC SODIUM NITRITE. 16-2321
NITROGEN-14 NUCLEAR QUADRUPOLE RESONANCE IN FERROELECTRIC SODIUM NITRITE. 16-2320
IUM DIHYDROGEN PHOSPHATE PRESENTED, TRIGLYCINE SUL/ MAGNETIC RESONANCE IN FERROELECTRICS. (MODEL FOR POTASS 03-0180
NSITION POINT.   KINETIC EQUATION FOR QUADRUPOLE RESONANCE IN FERROELECTRICS NEAR THE PHASE TRA 03-0309
LFOIODIDE (AND BROMIDE) TYPE.   NUCLEAR QUADRUPOLE RESONANCE IN FERROELECTRICS OF THE ANTIMONY SU 15-2216
F FERROELECTRIC LITHIUM TRIHYDROGEN D/ ELECTRON PARAMAGNETIC RESONANCE IN GAMMA-IRRADIATED SINGLE CRYSTAL O 22-2842
ELECTRON SPIN RESONANCE IN GAMMA-IRRADIATED TGS. 27-3035
.   ELECTRON PARAMAGNETIC RESONANCE IN NONSTOICHIOMETRIC BARIUM TITANATE 04-0640
/BSENCE OF MOTIONAL NARROWING IN THE PROTON NUCLEAR MAGNETIC RESONANCE IN POTASSIUM DIHYDROGEN PHOSPHATE A/ 17-2594
/NSITIONS IN PEROVSKITES CRYSTALS. (NIOBIUM-93 AND SODIUM-23 RESONANCE IN POTASSIUM NIOBATE AND SODIUM NIO/ 08-1283
SECOND ORDER QUADRUPOLE SHIFTS OF THE SODIUM-23 MAGNETIC RESONANCE IN ROCHELLE SALT. 28-3186
HE ORIENTATIONS OF PROTON PAIRS (IN THE WA/ NUCLEAR MAGNETIC RESONANCE IN ROCHELLE SALT, DETERMINATION OF T 28-3203
ADOLINIUM(3+) OR IRC/ ELECTRIC FIELD EFFECT OF ELECTRON SPIN RESONANCE IN STRONTIUM TITANATE. (DOPED WITH G 06-1229
NMR INVESTIGATIONS OF ICE CRYSTALS. PART-1: PROTON MAGNETIC RESONANCE IN WATER ICE. PART-2: DEUTERON SPIN/ 34-3471
: LITHIUM OR SODIUM TRIHYDROGEN DISELENITE, NUCLEAR MAGNETIC RESONANCE INVESTIGATION. /SM IN FERROELECTRICS 22-2837
ESE(2+) IN LITHIUM NIOBATE.   AN ELECTRON PARAMAGNETIC RESONANCE INVESTIGATION OF IRON(3+) AND MANGAN 11-1725
LOW FREQUENCY (510 MHZ) PARAMAGNETIC RESONANCE INVESTIGATION OF LITHIUM NIOBATE. 11-1793
ONTIUM TITANATE CONTAINING MANGANESE / ELECTRON PARAMAGNETIC RESONANCE INVESTIGATION OF POLYCRYSTALLINE STR 06-1162
THIUM NIOBATE (TEMPERATURE DE/ NIOBIUM-93 NUCLEAR QUADRUPOLE RESONANCE INVESTIGATION OF (STOICHIOMETRIC) LI 11-1811
IC FIELD IN COPPER DOPED ROCHELLE SALT ELECTRON PARAMAGNETIC RESONANCE INVESTIGATION OF THE INTERNAL ELECTR 28-3233
NUCLEAR MAGNETIC RESONANCE INVESTIGATION OF TRIGLYCINE SULFATE. 27-3059
ROELECTRIC CRYSTALS. (TRIGLYCINE SULF/ ELECTRON PARAMAGNETIC RESONANCE INVESTIGATIONS OF (COPPER) DOPED FER 27-3132
LCIUM STRONTIUM PROPIONATE.   NUCLEAR MAGNETIC RESONANCE INVESTIGATIONS OF FERROELECTRIC DICA 26-2931
-1: PROTON MAGNETIC RESONANCE IN WATER ICE/ NUCLEAR MAGNETIC RESONANCE INVESTIGATIONS OF ICE CRYSTALS. PART 34-3471
LEXES IN FERROELECTRIC TRIS-SARCOSINE/ ELECTRON PARAMAGNETIC RESONANCE INVESTIGATIONS OF MANGANESE(2+) COMP 36-3525
NUCLEAR MAGNETIC RESONANCE INVESTIGATIONS OF PHTHALOCYANINE. 36-3551
INGLE CRYSTALS OF FERROELECTRIC POTASSIUM DIH/ ELECTRON SPIN RESONANCE INVESTIGATIONS OF X-RAY IRRADIATED S 17-2558
PARAMAGNETIC POTASSIUM MANGANE/ EXCHANGE NARROWING: MAGNETIC RESONANCE LINE SHAPES AND SPIN CORRELATION IN 33-3376
NETS.   FERROMAGNETIC RESONANCE LINE WIDTH IN FERROELECTRIC FERROMAG 03-0207
IFLUORI/ SHIFT OF NEEL TEMPERATURE AND ELECTRON PARAMAGNETIC RESONANCE LINE WIDTH OF POTASSIUM MANGANESE TR 33-3377
HIUM NIOBATE.   NIOBIUM-93 NUCLEAR MAGNETIC RESONANCE LINE WIDTHS IN NONSTOICHIOMETRIC LIT 11-1795
ROMAGNET COPPER FORMATE TE/ ANOMALOUS SHIFT OF ELECTRON SPIN RESONANCE LINES IN THE TWO DIMENSIONAL ANTIFER 26-2941
CIES IN CHROMIUM(3+) DOPED STRONTIUM TITANATE (ELECTRON SPIN RESONANCE MEASUREMENTS, MODEL PROPOSED). /ACAN 06-1161
LS POTASSIUM FERROCYANIDE TRIHYDRATE BY THE NUCLEAR MAGNETIC RESONANCE METHOD. /OF THE FERROELECTRIC CRYSTA 24-2898
STRUCTURE OF ANTIFERROELECTRIC AMMONIUM PERIODATE BY NUCLEAR RESONANCE METHODS.   STUDY OF THE 36-3580
ACOUSTIC OR ELECTROMAGNETIC WAVE IN A FERROELECTRIC ANTIFE/ RESONANCE OF A LOW FREQUENCY SPIN WAVE WITH AN 03-0284
TEMPERATURE DEPENDENCE OF THE ELECTRON PARAMAGNETIC RESONANCE OF AMMONIUM CADMIUM SULFATE. 19-2792
BARIUM TITANATE.   PARAMAGNETIC RESONANCE OF AN F CENTER IN NONSTOICHIOMETRIC 04-0976
ARSENATE.   PURE QUADRUPOLE RESONANCE OF ARSENIC-75 IN AMMONIUM DIHYDROGEN 18-2732
ELECTRON PARAMAGNETIC RESONANCE OF CHROMIUM(3+) IN ADP. 17-2596
RONTIUM TITANATE SINGLE CRYSTALS.   ELECTRON SPIN RESONANCE OF CHROMIUM(3+) IN CHROMIUM DOPED ST 06-1147
CTRIC ADP CRYSTALS.   ELECTRON PARAMAGNETIC RESONANCE OF CHROMIUM(3+) IONS IN ANTIFERROELE 17-2546
.   ELECTRON PARAMAGNETIC RESONANCE OF COBALT(2+) IN POTASSIUM TANTALATE 08-1342
ELECTRON PARAMAGNETIC RESONANCE OF COPPER IONS IN ROCHELLE SALT. 28-3176
TARTRATE.   ELECTRON SPIN RESONANCE OF COPPER(2+) DOPED SODIUM AMMONIUM 29-3252
DIUM DEUTERO AMMONIUM SELENATE DIDEUTERATE. NUCLEAR MAGNETIC RESONANCE OF DEUTERIUM IN THE FERROELECTRIC SO 22-2828
IUM FLUOBERYLLATE CRYSTAL.   NUCLEAR MAGNETIC RESONANCE OF DEUTERIUM IN THE PERDEUTERO AMMON 19-2744
RTRATE.   NUCLEAR MAGNETIC RESONANCE OF FERROELECTRIC LITHIUM AMMONIUM TA 29-3240
RTRATE.   NUCLEAR MAGNETIC RESONANCE OF FERROELECTRIC LITHIUM AMMONIUM TA 29-3241
D ANTIFERROELECTRIC ARSENATE SINGLE CRYSTALS. ELECTRON SPIN RESONANCE OF FREE RADICALS IN FERROELECTRIC AN 18-2731
ATE.   ELECTRON PARAMAGNETIC RESONANCE OF GADOLINIUM(3+) DOPED LITHIUM NIOB 11-1833
/RETATION OF ELECTRIC FIELD EFFECTS IN ELECTRON PARAMAGNETIC RESONANCE OF GADOLINIUM(3+) DOPED STRONTIUM T/ 06-1168
CERAMIC SEMICONDUCTOR.   TWO SETS OF ELECTRON SPIN RESONANCE OF GADOLINIUM(3+) IN BARIUM TITANATE 04-0990
ELECTRON PARAMAGNETIC RESONANCE OF GADOLINIUM(3+) IN LITHIUM NIOBATE 11-1763
MONIUM HYDROGEN SULFATE.   ELECTRON SPIN RESONANCE OF GAMMA DEFECTS IN FERROELECTRIC AM 19-2748
.   ELECTRON SPIN RESONANCE OF GAMMA-IRRADIATED AMMONIUM SULFATE 19-2790
ELECTRON SPIN RESONANCE OF GAMMA-IRRADIATED ROCHELLE SALT. 28-3180
L IN TRIGLYCINE SULFATE NEAR THE CURIE POINT. (ELECTRON SPIN RESONANCE OF GAMMA-IRRADIATED SPECIMENS) /DICA 27-3033
ERNAL MOTION OF HYDROGEN AND THE ACTIVATION E/ ELECTRON SPIN RESONANCE OF GAMMA-IRRADIATED TGS. PART-2: INT 27-3034
ATE.   ELECTRON SPIN RESONANCE OF GAMMA-IRRADIATED TRIGLYCINE SELEN 27-3139
NUCLEAR QUADRUPOLE RESONANCE OF IODINE-127 IN SILVER PERIODATE. 08-1426
DE.   NUCLEAR QUADRUPOLE RESONANCE OF IODINE-127 IN SOLID HYDROGEN IODI 31-3335
IUM NIOBATE.   PHOTOCHROMIC EFFECT AND ELECTRON SPIN RESONANCE OF IRON AND MOLYBDENUM IN BARIUM SOD 13-1930
ELECTRON PARAMAGNETIC RESONANCE OF IRON(3+) DOPED LITHIUM NIOBATE. 11-1765
W TEMPERATURES.   ELECTRON PARAMAGNETIC RESONANCE OF IRON(3+) IN BARIUM TITANATE AT LO 04-0927
E RHOMBOHEDRAL AND THE CUBIC PHASES. ELECTRON PARAMAGNETIC RESONANCE OF IRON(3+) IN BARIUM TITANATE IN TH 04-0929
TIONS OF STRESSED STRONTIUM TITANATE. (ELECTRON PARAMAGNETIC RESONANCE OF IRON- VANADIUM PAIRS) /ASE TRANSI 06-1170
UM DIHYDROGEN PHOSPHATE.   ELECTRON SPIN RESONANCE OF IRRADIATED AND DEUTERATED POTASSI 17-2533
S. (LITHIUM AMMONIUM TARTRATE, LITHIUM T/ NUCLEAR QUADRUPOLE RESONANCE OF LITHIUM IN FERROELECTRIC COMPOUND 02-0052
PARAELECTRIC RESONANCE OF LITHIUM ION IN POTASSIUM CHLORIDE 31-3325
/INES IN THE HYPERFINE SPECTRUM OF THE ELECTRON PARAMAGNETIC RESONANCE OF MANGANESE(+2) IN BARIUM TITANATE. 04-0877
IUM SULFATE.   ELECTRON PARAMAGNETIC RESONANCE OF MANGANESE(2+) DOPED AMMONIUM CADM 19-2791

/ORBIDDEN HYPERFINE TRANSITIONS IN THE ELECTRON PARAMAGNETIC RESONANCE OF MANGANESE(2+) DOPED IN FERROELEC/ 16-2383
ATE.   ELECTRON SPIN RESONANCE OF MANGANESE(2+) IN BARIUM TITANATE. 04-0747
C PHASE OF SODIUM NITRITE.   ELECTRON PARAMAGNETIC RESONANCE OF MANGANESE(2+) IN POTASSIUM TANTAL 08-1343
TE.   EFFECTIVE CHARGE OF TITANIUM(4+) FROM THE PARAMAGNETIC RESONANCE OF MANGANESE(2+) IN THE FERROELECTRI 16-2358
                                                                        RESONANCE OF MANGANESE(2+) IN STRONTIUM TITANA 16-1167
                                         NUCLEAR QUADRUPOLE RESONANCE OF NITROGEN IN SODIUM NITRITE. 16-2313
ITY OF THIOUREA.   STARK EFFECT ON THE NUCLEAR QUADRUPOLE RESONANCE OF NITROGEN-14 AND THE FERROELECTRIC 25-2909
RIUM TITANATE SINGLE CRYSTAL.   PARAMAGNETIC RESONANCE OF PLATINUM IONS IN NIOBIUM DOPED BA 04-0977
HIUM NIOBATE CRYSTAL.   FERMI RESONANCE OF POLARITON WITH BIPHONON IN AN LIT 11-1761
                       THE ELECTRON PARAMAGNETIC RESONANCE OF POTASSIUM SELENATE. 22-2850
NIDE TRIHYDRATE.   MOSSBAUER RESONANCE OF SINGLE CRYSTAL POTASSIUM FERROCYA 24-2890
SELENATE DIHYDRATE CRYSTAL.   NUCLEAR MAGNETIC RESONANCE OF SODIUM-23 IN THE SODIUM AMMONIUM 22-2883
ONTIUM TITANATE.   ELECTRON PARAMAGNETIC RESONANCE OF SOME RARE EARTH IMPURITIES IN STR 06-1192
                                      ANOMALOUS RESONANCE OF STRONTIUM TITANATE. 06-1173
E FLUOBERYLLATE.   ELECTRON PARAMAGNETIC RESONANCE OF THE CHROMIUM COMPLEX IN TRIGLYCIN 27-3133
SPIN S=2 IN TRIGLYCINE FLUOBERYLLATE / ELECTRON PARAMAGNETIC RESONANCE OF THE FOUR COPPER ION COMPLEX WITH 27-3134
ROELECTRIC THIOUREA.   DEUTERON MAGNETIC RESONANCE OF THE HIGH TEMPERATURE PHASE OF FER 25-2918
NTIUM AND BARIUM TITANATES.   ELECTRON PARAMAGNETIC RESONANCE OF TRIVALENT GADOLINIUM IONS IN STRO 04-0920
D ISOMORPHOUS CRYSTALS. (NEW MODEL BASED ON NUCLEAR MAGNETIC RESONANCE OF UNTWINNED SAMPLES) /RROCYANIDE AN 24-2893
CE OF THIOUREA. STUDIES OF FERROELECTRIC SOLIDS BY MAGNETIC RESONANCE PART-18: PROTON AND DEUTERON RESONAN 25-2919
/OF THEORY AND EXPERIMENT. PART-1: INTRODUCTION AND MAGNETIC RESONANCE. PART-2: THERMAL CONDUCTIVITY. PART/ 03-0397
CTRICITY OF ALKALI IODATES.   NUCLEAR QUADRUPOLE RESONANCE PHASE TRANSFORMATIONS, AND FERROELE 08-1350
SITION IN SODIUM NITRITE AS STUDIED BY ELECTRON PARAMAGNETIC RESONANCE (PHASE TRANSITION AT 163 DEG C). /AN 16-2359
IHYDROGEN PHOSPHATE TYPE CRYSTALS.   NUCLEAR MAGNETIC RESONANCE PROTON SECOND MOMENTS IN POTASSIUM D 17-2408
/ THE ELECTRICAL BOUNDARY CONDITIONS ON THE NUCLEAR MAGNETIC RESONANCE RELAXATION RATE OF SODIUM-23 IN SOD/ 16-2350
ENATE AND DEUTERO AMMONIUM DIDE/ ARSENIC-75 NUCLEAR MAGNETIC RESONANCE RESONANCE IN AMMONIUM DIHYDROGEN ARS 18-2736
AND QUADRUPOLE SATURATION OF THE LITHIUM-7 NUCLEAR MAGNETIC RESONANCE SIGNALS IN LITHIUM NIOBATE. / DIPOLE 11-1666
/N OF CHROMIUM VALENCY FROM STUDIES OF ELECTRON PARAMAGNETIC RESONANCE SPECTRA AND STATIC MAGNETIC SUSCEPT/ 04-0542
ONTAINING TITANI/ PECULIARITIES OF THE ELECTRON PARAMAGNETIC RESONANCE SPECTRA IN A FERROELECTRIC CERAMIC C 04-1020
/E EFFECTS ON SELENITE (ICN). RADIATION DAMAGE ELECTRON SPIN RESONANCE SPECTRA IN FERROELECTRIC SODIUM TRI/ 22-2870
UM(3+) IN LANTHANUM ALUMINATE.   ELECTRON SPIN RESONANCE SPECTRA OF CHROMIUM(3+) AND GADOLINI 36-3583
/CE OF SPONTANEOUS POLARIZATION ON THE ELECTRON PARAMAGNETIC RESONANCE SPECTRA OF CHROMIUM(3+) IONS IN TRI/ 27-3136
L/ EFFECT OF APPLIED ELECTRIC FIELD ON ELECTRON PARAMAGNETIC RESONANCE SPECTRA OF (COPPER(2+)) DOPED FERROE 28-3185
UM NIOBATE.   FORBIDDEN HYPERFINE LINES IN ELECTRON SPIN RESONANCE SPECTRA OF MANGANESE(2+) DOPED LITHI 11-1805
OELECTRIC CRYSTALS CESIUM DIH/ ARSENIC-75 NUCLEAR QUADRUPOLE RESONANCE SPECTRA OF PARTIALLY DEUTERATED FERR 18-2740
/CT OF TEMPERATURE AND PRESSURE ON THE ELECTRON PARAMAGNETIC RESONANCE SPECTRA OF SUBSTITUTIONAL IMPURITIE/ 06-1193
CESIUM DIHYDROGEN A/ STARK EFFECT IN THE NUCLEAR QUADRUPOLE RESONANCE SPECTRA OF THE FERROELECTRIC CRYSTAL 18-2722
DE/ NATURE OF THE TEMPERATURE DEPENDENCE OF NUCLEAR MAGNETIC RESONANCE SPECTRA SECOND MOMENT IN ORDER DISOR 03-0460
SOME FERROELECTRICS AND FERROELECTRIC FERROM/ ELECTRON SPIN RESONANCE (SPECTRA, TEMPERATURE DEPENDENCE) IN 08-1368
G COMPOUNDS OF THE TYPE A(2) (V) B(3) (VI) AND A(/ NUCLEAR GAMMA RESONANCE SPECTROSCOPIC STUDY OF SEMICONDUCTIN 15-2108
M NI/ TEMPERATURE DEPENDENCE OF THE LITHIUM NUCLEAR MAGNETIC RESONANCE SPECTRUM AND ATOMIC MOTION IN LITHIU 11-1722
RIDE.   ANTIFERROMAGNETIC RESONANCE SPECTRUM IN BARIUM MANGANESE TRIFLUO 33-3396
M NIOBATE AND (COPPER(2+) DOPED) ZINC/ ELECTRON PARAMAGNETIC RESONANCE SPECTRUM IN (MANGANESE DOPED) LITHIU 11-1792
TANATE DUE TO NEAREST NE/ STRONG AXIAL ELECTRON PARAMAGNETIC RESONANCE SPECTRUM OF IRON(3+) IN STRONTIUM TI 06-1144
IUM TITANATE: AN EXAMPLE OF THE DYNAM/ ELECTRON PARAMAGNETIC RESONANCE SPECTRUM OF MOLYBDENUM(5+) IN STRONT 06-1125
TANATE IN THE NEIGHBORHOOD OF THE PHA/ ELECTRON PARAMAGNETIC RESONANCE STUDIES AND DYNAMICS IN STRONTIUM TI 06-1232
OELECTRICS. (DEUTERATED ROCHELLE SALT, POT/ NUCLEAR MAGNETIC RESONANCE STUDIES IN SOME HYDROGEN BONDED FERR 18-2719
AND MANGANESE(2+) PARAMAGNETIC ION C/ ELECTRON PARAMAGNETIC RESONANCE STUDIES OF CHROMIUM(3+), COPPER(2+), 27-3135
IODATE.   ELECTRON SPIN RESONANCE STUDIES OF CHROMIUM(3+) IN LITHIUM N 11-1804
PED TRIGLYCINE SULFATE SINGLE CRYSTALS ELECTRON PARAMAGNETIC RESONANCE STUDIES OF CHROMIUM- AND VANADYL- DO 27-3161
IN IRON(3+) DOPED LITHIUM NIOBATE.   ELECTRON PARAMAGNETIC RESONANCE STUDIES OF CRYSTAL FIELD PARAMETERS 11-1839
IN MANGANESE(2+) DOPED LITHIUM NIOBATE.   ELECTRON PARAMAGNETIC RESONANCE STUDIES OF CRYSTAL FIELD PARAMETERS 11-1803
TE.   DEUTERON NUCLEAR MAGNETIC RESONANCE STUDIES OF LITHIUM HYDRAZINIUM SULFA 32-3354
CTRICS.   NUCLEAR MAGNETIC RESONANCE STUDIES OF OXYGEN OCTAHEDRA FERROELE 13-2059
IN ROCHELLE SALT.   NUCLEAR MAGNETIC RESONANCE STUDIES OF SPONTANEOUS POLARIZATION 28-3194
TANNOUS CHLORIDE DIHYDRATE. DIELECTRIC AND NUCLEAR MAGNETIC RESONANCE STUDIES OF THE PHASE TRANSITION IN S 36-3582
FERROELECTRICS. (POTASSIUM FERROCYANIDE TR/ NUCLEAR MAGNETIC RESONANCE STUDIES OF THE PHASE TRANSITIONS IN 16-2255
/IDENCE FOR SLATER CONFIGURATIONS FROM ELECTRON PARAMAGNETIC RESONANCE STUDIES OF X-IRRADIATED POTASSIUM D/ 18-2727
HYDROGEN ARSENATE, POTASSIUM DIDEUTER/ ELECTRON PARAMAGNETIC RESONANCE STUDIES OF X-IRRADIATED POTASSIUM DI 18-2725
                                       ELECTRON SPIN RESONANCE STUDY IN STRONTIUM TITANATE. 06-1228
                                       ELECTRON SPIN RESONANCE STUDY OF ADP- KDA MIXED CRYSTALS. 17-2683
AMMONIUM DIDEUTERIUM PHOSPHATE- 5 PERCENT PO/ ELECTRON SPIN RESONANCE STUDY OF AMMONIUM IONS IN IRRADIATED 17-2684
TERIUM AND SODIUM-23 RELAXATION IN SODIUM / NUCLEAR MAGNETIC RESONANCE STUDY OF DEUTERIUM DIFFUSION AND DEU 22-2849
ZINIUM SULFATE.   NUCLEAR MAGNETIC RESONANCE STUDY OF FERROELECTRIC LITHIUM HYDRA 32-3343
N (MANGANESE DOPED) AMMONIUM SULFATE.   (NUCLEAR MAGNETIC RESONANCE STUDY OF) FERROELECTRIC TRANSITION I 19-2784
HYDROGEN SELENITE AT ROOM TEMPERATURE.   ELECTRON SPIN RESONANCE STUDY OF GAMMA-IRRADIATED CESIUM TRI 22-2833
SELENATE (STRUCTURAL PHASE CHANGES).   ELECTRON SPIN RESONANCE STUDY OF GAMMA-IRRADIATED POTASSIUM 22-2825
.   NITROGEN-14 NUCLEAR QUADRUPOLE RESONANCE STUDY OF LITHIUM HYDRAZINIUM SULFATE 32-3345
IN FERROELECTRIC CRYSTALS.   ELECTRON PARAMAGNETIC RESONANCE STUDY OF MAGNETIC COPPER(2+) COMPLEX 21-2820
M MONOCHLORO ACETATE, MONOCHLORO ACETIC / NUCLEAR QUADRUPOLE RESONANCE STUDY OF PARAMAGNETIC ION COMPLEXES 03-0477
IC SODIUM TUNGSTEN BRONZES.   TRANSIENT NUCLEAR MAGNETIC RESONANCE STUDY OF PHASE TRANSITION IN AMMONIU 36-3625
KITES CRYSTALS. (NIOBIUM-93 AND SODIUM-23 / NUCLEAR MAGNETIC RESONANCE STUDY OF PHASE TRANSITIONS IN METALL 13-1910
LE SALT.   NUCLEAR MAGNETIC RESONANCE STUDY OF PHASE TRANSITIONS IN PEROVS 08-1283
IC POTASSIUM DIHYDROGEN PHOSPHATE.   NUCLEAR DOUBLE RESONANCE STUDY OF PHASE TRANSITIONS IN ROCHEL 28-3195
IDE AND NITRITE ION IN IRRADIATED POTASSIUM NI/ PARAMAGNETIC RESONANCE STUDY OF POTASSIUM-39 IN FERROELECTR 17-2599
AMMONIUM DIHYDROGEN PHOSPHATE POTASSIUM DIHY/ ELECTRON SPIN RESONANCE STUDY OF PRODUCTION OF NITROGEN DIOX 16-2404
AND POTASSIUM TRIHYDROGEN SELENATE.   PROTON MAGNETIC RESONANCE STUDY OF RADIATION DAMAGE CENTERS IN 17-2547
DINITRITE.   NUCLEAR MAGNETIC RESONANCE STUDY OF SODIUM TRIHYDROGEN SELENATE 22-2876
D...O BOND. (INCLUDING FULLY DEUTERATED A/ DEUTERON MAGNETIC RESONANCE STUDY OF SODIUM-23 IN SILVER SODIUM 16-2324
IUM ALUMINUM SULFATE HEXAHYDRATE.   ELECTRON PARAMAGNETIC RESONANCE STUDY OF SOME CRYSTALS CONTAINING O- 03-0204
IN DEVIATION IN COPPER FORMATE TETRADEUTERATE.   PROTON RESONANCE STUDY OF SOME METAL IONS IN GUANIDIN 21-2817
ONSTOICHIOMETRIC TUNGSTEN TRIOXIDE.   ELECTRON PARAMAGNETIC RESONANCE STUDY OF SUBLATTICE ROTATIONS AND SP 26-2950
NSITIONS IN SODIUM TRIHYDROGEN DISELENITE.   ELECTRON SPIN RESONANCE STUDY OF THE CATALYTIC ACTIVITY OF N 10-1628
NS IN DIGLYCINE NITRATE AND TRIS-SARCOSINE/ NUCLEAR MAGNETIC RESONANCE STUDY OF THE FERROELECTRIC PHASE TRA 22-2871
N IN DEUT/ BERYLLIUM-9 QUADRUPOLE PERTURBED NUCLEAR MAGNETIC RESONANCE STUDY OF THE FERROELECTRIC TRANSITIO 16-2267
N IN TRIGLYCINE SULFATE.   PULSED NITROGEN PROTON DOUBLE RESONANCE STUDY OF THE FERROELECTRIC TRANSITIO 27-2965
NS IN SODIUM TRIHYDROGEN AND TRIDEUTERIUM / NUCLEAR MAGNETIC RESONANCE STUDY OF THE FERROELECTRIC TRANSITIO 27-2966
TE SOLID SOLUTION SYSTEM.   NIOBIUM-93 NUCLEAR MAGNETIC RESONANCE STUDY OF THE FERROELECTRIC TRANSITIO 22-2832
RACTIONS OF NITROGEN-14 IN PARAELECTRIC TRIGL/ PULSED DOUBLE RESONANCE STUDY OF THE LITHIUM NIOBATE TANTALA 11-1794
                                                                        RESONANCE STUDY OF THE NUCLEAR QUADRUPOLE INTE 27-2969

ONTIUM TITANATE.   NUCLEAR MAGNETIC RESONANCE STUDY OF THE PHASE TRANSITION IN STR 06-1237
   AMMONIUM IONS IN SOLIDS. (AMMONIUM SULFATE/ PROTON MAGNETIC RESONANCE STUDY OF THE SPIN SYMMETRY STATES OF 19-2800
TION IN LANTHANUM ALUMINATE.   NUCLEAR MAGNETIC RESONANCE STUDY OF THE STRUCTURAL PHASE TRANSI 36-3533
LS OF ICE.   ENDOR (ELECTRON NUCLEAR DOUBLE RESONANCE) STUDY OF X-IRRADIATED SINGLE CRYSTA 34-3413
S.   ELECTRON PARAMAGNETIC RESONANCE STUDY OF X-RAY DAMAGED FERROELECTRIC 02-0079
   IN THE VICINITY OF THE PHASE TRANSITION T/ NUCLEAR MAGNETIC RESONANCE STUDY ON SODIUM-23 IN SODIUM NITRITE 16-2396
OF DOMAIN BOUNDARIES AND POINT DEFECTS (TORSION PENDULUM AND RESONANCE TECHNIQUES, 20-400C). / INTERACTION 09-1584
RELAXATION) INVESTIGATIONS OF THE PHASE TR/ NUCLEAR MAGNETIC RESONANCE (WIDE LINE SPECTRA AND SPIN LATTICE 20-2802
/ILITY: METAL AMMONIUM (BX4) TYPE COMPOUNDS (PROTON MAGNETIC RESONANCE, 84 DEG K AND 120-300 DEG K, EXAMPL/ 22-2826
              LOW FREQUENCY MAGNETOELECTRIC RESONANCES IN BARIUM COBALT TETRAFLUORIDE.   33-3409
RESONATORS. (COMMENTS)   TEMPERATURE DEPENDENCE OF RESONANT FREQUENCIES OF LITHIUM NIOBATE PLATE 11-1812
EXPANDER BARS.   ZERO TEMPERATURE COEFFICIENT OF RESONANT FREQUENCY IN LITHIUM TANTALATE LENGTH 11-1810
   ONE AND TWO PHONON STATES OF BARIUM TITANATE.   AUGER-LIKE RESONANT INTERFERENCE IN RAMAN SCATTERING FROM 04-0923
FECT TUNABLE OPTICAL EMISSION FROM LITHIUM NIOBATE WITHOUT A RESONATOR.   EF 11-1858
/E RANGE TEMPERATURE COMPENSATION BY ADDITION OF TWO CRYSTAL RESONATOR FREQUENCIES: APPLICATION TO QUARTZ / 11-1686
ACTERISTICS OF ROCHELLE TYPE CERAMICS. (BARI/ UHF (WAVEGUIDE RESONATOR) METHOD OF STUDYING TEMPERATURE CHAR 04-0905
SINGLE CRYSTALS OF FERROELECTRIC SUBSTANCES BY THE COMPOSITE RESONATOR METHOD. (TGS) /INTERNAL FRICTION IN 27-3080
/EMPERATURE DEPENDENCE OF AN X CUT LITHIUM TANTALATE CRYSTAL RESONATOR VIBRATING IN THICKNESS SHEAR MODE O/ 12-1868
ACITANCES.   LOW TEMPERATURE COEFFICIENT OF LITHIUM NIOBATE RESONATORS BY USE OF TEMPERATURE SENSITIVE CAP 11-1670
   DEPENDENCE OF RESONANT FREQUENCIES OF LITHIUM NIOBATE PLATE RESONATORS. (COMMENTS)   TEMPERATURE 11-1812
LOW RAT/ LITHIUM TANTALATE AND LITHIUM NIOBATE PIEZOELECTRIC RESONATORS IN THE MEDIUM FREQUENCY RANGE WITH 11-1724
ALATE.   MEDIUM FREQUENCY RESONATORS OF LITHIUM NIOBATE AND LITHIUM TANT 11-1851
              LEAD TITANATE CERAMIC RESONATORS OPERATING IN VHF BAND.   05-1069
ENCE OF THE SMALL SIGNAL PARAMETERS OF FERROELECTRIC CERAMIC RESONATORS UPON STATE OF POLARIZATION.   DEPEND 37-3649
D TITANATE)   ELECTRON SPIN RESOURCE STUDIES AT DEFECTS IN FLUORITES. (LEA 05-1086
TES.   DIELECTRIC RESPONSE FUNCTION OF COPPER FORMATE TETRAHYDRA 26-2933
ONON IN STRONTIUM TITANATE.   MICROSCOPIC CALCULATION OF THE RESPONSE FUNCTION OF THE SOFT ZONE BOUNDARY PH 06-1213
   SPONTANEOUS BIREFRINGENCE AND ELECTROOPTIC RESPONSE IN GADOLINIUM MOLYBDATE.   30-3307
TTERING IN POTASSIUM DIHYDROGEN ARSENATE AND CES/ DIELECTRIC RESPONSE IN PIEZOELECTRIC CRYSTALS. (RAMAN SCA 18-2723
         ANALYSIS OF LOW FREQUENCY ELECTROOPTIC RESPONSE IN SODIUM NITRITE.   16-2353
              PIEZOELECTRIC RESPONSE OF A FERROELECTRIC MEMORY ARRAY.   37-3635
ONS)   DIELECTRIC RESPONSE OF CRYSTALS. (FERROELECTRIC FLUCTUATI 03-0210
/THE PHASE DIAGRAM OF ICE. PART-2: DIELECTRIC AND MECHANICAL RESPONSE OF ICE I(H) SINGLE CRYSTALS AND ITS / 34-3485
NTERPRETATION.   DIELECTRIC AND MECHANICAL RESPONSE OF ICE I(H) SINGLE CRYSTALS AND ITS I 34-3483
AR THE CURIE POINT (IMPROVED METHOD OF MEASURE/ PYROELECTRIC RESPONSE OF TGS IN THE FERROELECTRIC REGION NE 27-2979
              PULSE RESPONSE OF ZERO DEGREE CUT ADP MODULATORS.   17-2553
/IRECT MEASUREMENT TECHNIQUE AND APPLICATION TO A NANOSECOND RESPONSE TIME DETECTOR. (STRONTIUM BARIUM NIO/ 13-1919
IGLYCINE SULFATE AND STRONTIUM BARIUM / PYROELECTRIC VOLTAGE RESPONSE TO RECTANGULAR INFRARED SIGNALS IN TR 13-2019
GLYCINE SULFATE AND BARIUM STRONTIUM N/ PYROELECTRIC VOLTAGE RESPONSE TO SHORT INFRARED LASER PULSES IN TRI 13-2018
/ELECTRIC COEFFICIENT AND PERMITTIVITY FROM THE PYROELECTRIC RESPONSE TO STEP RADIATION SIGNALS IN FERROEL/ 03-0469
   IN TRIGLYCINE SULFATE AND STRONTIUM B/ PYROELECTRIC VOLTAGE RESPONSE TO STEP SIGNALS OF INFRARED RADIATION 13-2022
D ZIRCONA/ REALIZATION OF A 40 MHZ COLOR TELEVISION INFRARED RESPONSE USING SURFACE WAVE TRANSDUCERS ON LEA 09-1509
ZT)   RETENTION IN THE THIN FERROELECTRIC FILMS. (PL 09-1593
S.   SURFACE STRUCTURES AND CORRESPONDING LIGHT FIGURES REVEALED BY WATER ETCHED ROCHELLE SALT CRYSTAL 28-3234
RROCYANIDE TRIHYDRATE SIN/ APPLICATION OF THE DEW METHOD FOR REVEALING THE DOMAIN STRUCTURE IN POTASSIUM FE 24-2904
   BY SELECTIVE CRYSTALLIZATION OF ANTHRAQUINONE (SUBLIMED ON/ REVEALING THE DOMAIN STRUCTURE OF TGS CRYSTALS 27-3041
N GUANIDINIUM ALUMINUM SULFATE HEXAHYDRATE (AND MECHANISM OF REVERSAL).   FERROELECTRIC MOMENT I 21-2818
MILLIMETER WAVE ABSORPTION IN SODIUM NITRITE BY POLARIZATION REVERSAL.   MODULATION OF THE 16-2392
   PHENOMENOLOGICAL THEORY OF FERROELECTRIC POLARIZATION REVERSAL.   03-0282
E DEPENDENCE OF MATERIAL CONSTANTS OF TGS AFTER POLARIZATION REVERSAL.   TIM 27-2953
C SODIUM NITRITE CRYSTALS. PART-2: MECHANISM OF POLARIZATION REVERSAL.   TOPOGRAPHIC STUDY OF FERROELECTRI 16-2374
RIC MATERIAL THAT ALLOWS NO 180 DEGREE ELECTRIC POLARIZATION REVERSAL. /FERROELECTRIC AND FERROMAGNETOELECT 14-2098
YSTAL BISMUTH TITANATE.   SELF REVERSAL AND WAITING TIME EFFECTS IN SINGLE CR 08-1448
SULFATE.   POLARIZATION REVERSAL CHARACTERISTICS OF AMMONIUM HYDROGEN 19-2788
   FERROELECTRIC CRYSTAL. (DIELECTRIC HEATING AND POLARIZATION REVERSAL, COMPLEX PERMITTIVITY CALCULATION) /F 03-0245
   LOW TEMPERATURE POLARIZATION EFFECTS IN ICE. (CURRENT REVERSAL DUE TO DIPOLE RELAXATION) 34-3418
FIELD IN SODIUM NITRITE.   POLARIZATION REVERSAL IN A WIDE RANGE OF IMPRESSING RATE OF 16-2381
   EFFECT OF PRESSURE (TO ABOUT 2KBAR) ON POLARIZATION REVERSAL IN BARIUM TITANATE.   04-0718
ULFATE)   SUSCEPTIBILITY INCREASE DURING POLARIZATION REVERSAL. (IN BARIUM TITANATE AND TRIGLYCINE S 04-0981
         THICKNESS DEPENDENCE OF POLARIZATION REVERSAL IN BARIUM TITANATE SINGLE CRYSTALS. 04-0949
TISTICAL METHOD FOR INVESTIGATING (DYNAMICS OF) POLARIZATION REVERSAL IN FERROELECTRIC CERAMICS.   STA 03-0493
WITCHING CURRENT)   THEORY OF POLARIZATION REVERSAL IN FERROELECTRICS. (CALCULATIONS OF S 03-0422
ANATE.   DOMAIN STRUCTURE AND POLARIZATION REVERSAL IN FILMS OF FERROELECTRIC BISMUTH TIT 08-1470
              FERROELECTRIC DOMAIN REVERSAL IN LITHIUM METATANTALATE.   12-1869
   SUBSTITUTED LEAD TITANATE ZIRCONAT/ HIGH FIELD POLARIZATION REVERSAL IN POLYCRYSTALLINE FERROELECTRICS. (8 08-1418
              POLARIZATION REVERSAL IN TGS SINGLE CRYSTALS.   27-2980
ASER BEAM. APPLIC/ IRREVERSIBLE AND SPONTANEOUS POLARIZATION REVERSAL IN TRIGLYCINE SULFATE BY MEANS OF A L 27-3010
ER BEAM.   REVERSIBLE DOMAIN REVERSAL IN TRIGLYCINE SULFATE CAUSED BY A LAS 27-3012
OR SODIUM TRIHYDROGEN DISELENITE, NUCLEAR MAGNETIC RESONANC/ REVERSAL MECHANISM IN FERROELECTRICS: LITHIUM 22-2837
   POSSIBILITY OF GENERATING INFRARED RADIATION BY POLARIZATION REVERSAL OF A FERROELECTRIC.   03-0162
LS IN THE REGION OF PHASE COEXISTENCE.   POLARIZATION REVERSAL OF ANTIMONY SULFOIODIDE SINGLE CRYSTA 15-2246
              THEORY OF POLARIZATION REVERSAL OF BARIUM TITANATE CRYSTAL.   03-0326
E.   ROLE OF HYDROGEN IN POLARIZATION REVERSAL OF FERROELECTRIC BARIUM SODIUM NIOBAT 13-2025
   ON SURFACE STATE AND POLARIZATION REVERSAL OF FERROELECTRIC CRYSTALS.   03-0323
RROELECTRIC CERAMICS.   UHF REVERSAL PROPERTIES OF BARIUM TITANATE TYPE FE 04-1014
SECOND HARMONIC GENERATION DURING FERROELECTRIC POLARIZATION REVERSAL. (SODIUM NITRITE)   OPTICAL 16-2384
CRYSTALS (CALCULATIO/ HIGH FIELD (1.5-6 KV/CM) POLARIZATION REVERSALS IN LIQUID ELECTRODED BARIUM TITANATE 04-0994
   NEUTRON DIFFRACTION STUDIES OF FERROELECTRIC POLARIZATION REVERSALS IN SODIUM NITRITE.   16-2365
TE CAUSED BY A LASER BEAM.   REVERSIBLE DOMAIN REVERSAL IN TRIGLYCINE SULFA 27-3012
   RH DOPED LITHIUM NIOBATE AS AN IMPROVED NEW MATERIAL FOR REVERSIBLE HOLOGRAPHIC STORAGE.   08-1359
ATE AND APPLICATIONS TO PYROELECTRIC DETECTO/ LASER STUDY OF REVERSIBLE NUCLEATION SITES IN TRIGLYCINE SULF 27-3011
TION TIME OF SECONDS WHEN ILLUMINATED UNDER APPLIED ELECTR/ (REVERSIBLE) OPTICAL DAMAGE IN KDP (WITH RELAXA 17-2703
   POLYMORPHISM OF STRONTIUM METANIOBATE (TWO REVERSIBLE PHASES).   36-3535
   OPTICALLY ERASABLE AND REWRITABLE SOLID STATE HOLOGRAMS.   37-3638
              RF SEE RADIOFREQUENCY
TERIAL FOR REVERSIBLE HOLOGRAPHIC STORAGE.   RH DOPED LITHIUM NIOBATE AS AN IMPROVED NEW MA 08-1359
IUM TANTALATE.   TIGHT BONDING BAND CALCULATIONS ON RHENIUM TRIOXIDE, SODIUM TUNGSTATE, AND POTASS 08-1334
ICRO/ CS (CRYSTALLOGRAPHIC SHEAR)   FAMILIES DERIVED FROM THE RHENIUM TRIOXIDE STRUCTURE TYPE: AN ELECTRON M 10-1630
D BARIUM TITANATE.   CORRELATION BETWEEN RHEOLOGICAL AND ELECTRICAL BEHAVIOR IN SINTERE 04-1027
PARAMAGNETIC RESONANCE OF IRON(3+) IN BARIUM TITANATE IN THE RHOMBOHEDRAL AND THE CUBIC PHASES.   ELECTRON 04-0929

ERATURE DEPENDENCE OF SPONTANEOUS POLARIZATION OF HEXAGONAL (RHOMBOHEDRAL) FERROELECTRICS. /RATURE AND TEMP 03-0290
C/ STABILIZATION EFFECTS AT THE PHASE TRANSITION TETRAGONAL- RHOMBOHEDRAL IN LEAD ZIRCONATE TITANATE CERAMI 09-1611
/MIC THEORY OF THE MORPHOTROPIC PHASE TRANSITION TETRAGONAL- RHOMBOHEDRAL IN THE PEROVSKITE FERROELECTRICS. 03-0172
FERROELECTRIC DOMAIN SWITCHING IN RHOMBOHEDRAL PHASE PLZT CERAMICS. 09-1563
ESSES OF BARIUM TITANATE SINGLE CRYSTALS IN ORTHORHOMBIC AND RHOMBOHEDRAL PHASES. REPOLARIZATION PROC 04-0963
IC IRON IODINE BORACITE AT LOW TEMPERATURES. (MONOCLINIC AND RHOMBOHEDRAL PHASES) /ANSITIONS OF FERROELECTR 14-2086
/OELECTRIC COMPOUNDS LEAD ZIRCONATE(X) TITANATE(1-X) FOR TWO RHOMBOHEDRAL PHASES. (X=0.9 AND 0.58) (CORREL/ 09-1574
ELECTRICAL PROPERTIES OF RHOMBOHEDRAL POTASSIUM NITRATE. 16-2367
IC) FERROELECTRIC (LEAD LINIOBATE INTO THE NONFERROELECTRIC (RHOMBOHEDRIC) FORM. /SITION OF THE (ORTHORHOMB 08-1285
EVIDENCE OF ADDITIONAL DISORDER IN THE RHOMBOHEDRRAL FORM OF POTASSIUM NITRITE. 16-2371
DIELECTRIC ABSORPTION IN THE RHOMOHEDRAL PHASE OF BARIUM TITANATE. 04-0809
ATE. (ALPHA TO BETA POLYMORPHIC TRANSITION) RHYTHMIC GROWTH OF NEW PHASE IN POTASSIUM NITR 16-2263
LATTICE DYNAMICS OF A RIGID ION MODEL OF GADOLINIUM MOLYBDATE. 30-3281
DYNAMICAL THEORY OF RIGID LATTICES OF ANTIFERROELECTRICS. 03-0344
RTIES OF LEAD BARIUM NIOBATE PIEZOCERAMICS. EFFECTS OF ROASTING CONDITIONS ON THE STRUCTURE AND PROPE 13-2046
ANALYSIS OF DOUBLE HYSTERESIS LOOPS. (ROCHELLE SALT) 28-3212
RSIVE FIELD AT VERY LOW FREQUENCIES AND JERKY WALL MOTION IN ROCHELLE SALT. ANOMALOUS DISPERSION OF COE 28-3177
CONTRIBUTION OF DOMAIN WALL MOTION TO THE PERMITTIVITY OF ROCHELLE SALT. 28-3198
DIELECTRIC CONSTANT AND DIELECTRIC RELAXATION TIME OF ROCHELLE SALT. 28-3179
DIELECTRIC PHENOMENA IN ROCHELLE SALT. 28-3182
DYNAMICS OF FERROELECTRIC ROCHELLE SALT. 28-3237
DYNAMICS OF FERROELECTRIC ROCHELLE SALT. 28-3236
IONAL PRESSURE ON THE CURIE POINTS OF TRIGLYCINE SULFATE AND ROCHELLE SALT. EFFECT OF A TWO DIMENS 27-3082
RIC FIELD ON EPR SPECTRA OF (COPPER(2+)) DOPED FERROELECTRIC ROCHELLE SALT. EFFECT OF APPLIED ELECT 28-3185
F MECHANICAL STRESS ON DIELECTRIC PROPERTIES OF PARAELECTRIC ROCHELLE SALT. EFFECT O 28-3225
EFFECT OF THE CUPRIC ION ON DELAYED PHENOMENA IN ROCHELLE SALT. 28-3221
ACOUSTIC WAVES IN FERROELECTRICS. (THEORY, BARIUM TITANATE, ROCHELLE SALT) ELECTROMAGNETIC AND 03-0455
ELECTRON PARAMAGNETIC RESONANCE OF COPPER IONS IN ROCHELLE SALT. 28-3176
ELECTRON PARAMAGNETIC RESONANCE OF GAMMA-IRRADIATED ROCHELLE SALT. 28-3180
OPPER) LCPED FERROELECTRIC CRYSTALS. (TRIGLYCINE SULFATE AND ROCHELLE SALT) /RESONANCE INVESTIGATIONS OF (C 27-3132
INVESTIGATION OF THE INTERNAL ELECTRIC FIELD IN COPPER DOPED ROCHELLE SALT. EPR 28-3233
FAR INFRARED ABSORPTION SPECTRA OF ROCHELLE SALT. 28-3210
FERROELECTRICITY IN ROCHELLE SALT. 28-3218
C MEASUREMENTS OF THE CHANGES OF WATER OF CRYSTALLIZATION OF ROCHELLE SALT. GRAVIMETRI 28-3235
HIGH PRESSURE PHASE TRANSITIONS OF THE FIRST KIND IN ROCHELLE SALT. 28-3214
N (INCLUDING FERROELECTRIC AND MAGNETIC TRANSITIONS, AMPLE ROCHELLE SALT) /ETER FOR PHASE CHANGE DETECTIO 28-3201
ION ON BARKHAUSEN EFFECT IN FERROELECTRIC MATERIALS (TGS AND ROCHELLE SALT). INFLUENCE OF GAMMA-IRRADIAT 27-3123
HE DOMAIN STRUCTURE ON POSITRON (TWO-PHONON) ANNIHILATION IN ROCHELLE SALT. INFLUENCE OF T 28-3191
INTERNAL FRICTION AS A FUNCTION OF TEMPERATURE FOR ROCHELLE SALT. 28-3227
COMPLEX PIEZOELECTRIC CONSTANTS IN FERROELECTRIC CRYSTALS. (ROCHELLE SALT) MEASUREMENTS OF 28-3213
MICROSCOPIC THEORY OF THE PHASE TRANSITIONS IN ROCHELLE SALT. 03-0482
NMR STUDIES OF SPONTANEOUS POLARIZATION IN ROCHELLE SALT. 28-3194
NMR STUDY OF PHASE TRANSITIONS IN ROCHELLE SALT. 28-3195
N THE EFFECTS OF X-RAY IRRADIATION ON TRANSIENT PHENOMENA IN ROCHELLE SALT. O 28-3222
TIVITY OF SOME FERROELECTRIC CRYSTALS (PYROELECTRICITY, TGS, ROCHELLE SALT). OPTICAL AC 27-3017
H SEVERAL PHASE TRANSITIONS AT TEMPERATURES CLOSE TOGETHER. (ROCHELLE SALT) / IN FERROELECTRIC MATERIAL WIT 03-0363
RADIATION DAMAGE AND THE FERROELECTRIC EFFECT IN ROCHELLE SALT. 28-3207
RAMAN LIGHT SCATTERING IN ROCHELLE SALT. 28-3231
DER QUADRUPOLE SHIFTS OF THE SODIUM-23 MAGNETIC RESONANCE IN ROCHELLE SALT. SECOND OR 28-3186
RIUM TITANATE AND TITANATE ZIRCONATE, TRIGLYCINE SULFATE AND ROCHELLE SALT) /LECTRICS OF VARIOUS TYPES. (BA 02-0108
SUPERSTRUCTURE IN THE FERROELECTRIC PHASE OF AMMONIUM ROCHELLE SALT. 28-3224
THERMAL CONDUCTIVITY OF ROCHELLE SALT. 28-3226
ULTRASONIC ANOMALIES IN ROCHELLE SALT. 28-3197
YCRYSTALLINE FERROELECTRICS AT VERY LOW FIELDS AND STRESSES, ROCHELLE SALT) /(POLARIZATION SWITCHING IN POL 28-3216
IN LATTICE RELAX/ STUDY OF SLOW MOTION OF WATER MOLECULES IN ROCHELLE SALT AND AMMONIUM ROCHELLE SALT BY SP 28-3232
/NESCENCE STUDIES OF THE GAMMA-RAY IRRADIATED FERROELECTRICS ROCHELLE SALT AND GUANIDINE ALUMINUM SULFATE / 21-2812
SEE BOTH ROCHELLE SALT AND SEIGNETTE SALT.
MICROWAVE BEHAVIOR OF ROCHELLE SALT AND TGS (18 REFS). 27-3088
OF INTERNAL FRICTION IN SINGLE CRYSTALS OF FERROELECTRICS. (ROCHELLE SALT AND TRIGLYCINE SULFATE) /ENDENCE 27-3122
NATURE OF RADIATION INDUCED DEFECTS IN CRYSTALS OF ROCHELLE SALT AND TRIGLYCINE SULFATE. 27-3159
NEW MEASUREMENTS OF DIVERSE THERMODYNAMICAL COEFFICIENTS OF ROCHELLE SALT AND TRIGLYCINE SULFATE. 27-2995
/ON AND COPPER AND IRON IMPURITY) ABSORPTION BANDS OF DEFECT ROCHELLE SALT AND TRIGLYCINE SULFATE CRYSTALS. 27-3108
DISTINCTIVE FEATURES OF BARKHAUSEN EFFECT IN ROCHELLE SALT AND TRIGLYCINE SULFATE CRYSTALS. 27-3110
/UNDAMENTAL ABSORPTION EDGE AS A FUNCTION OF TEMPERATURE FOR ROCHELLE SALT AND TRIGLYCINE SULFATE CRYSTALS. 27-3160
RESULTS OF EXPERIMENTAL STUDY OF GAMMA-RADIATION EFFECTS IN ROCHELLE SALT AND TRIGLYCINE SULFATE. (ON ELE/ 27-3101
LD METHOD. CRITICAL PROPERTIES OF TGS AND ROCHELLE SALT AS DETERMINED BY AN ELECTRIC FIE 27-3056
DIFFRACTION STUDIES OF FERROELECTRICS. (BARIUM TITANATE AND ROCHELLE SALT AS EXAMPLES) NEUTRON 04-0679
ATTENUATION OF SOUND WAVES IN DEUTERATED ROCHELLE SALT AT THE PHASE TRANSITION POINT. 28-3204
ATTENUATION OF SOUND WAVES IN DEUTERATED ROCHELLE SALT AT THE PHASE TRANSITION POINT. 28-3205
/ SULFATE, POTASSIUM DIHYDROGEN PHOSPHATE, AMMONIUM SULFATE, ROCHELLE SALT) AT THE TRANSITION TEMPERATURES. 02-0125
RATURE. MEASUREMENT OF THE COMPLEX DIELECTRIC CONSTANT OF ROCHELLE SALT AT 10 GHZ AS A FUNCTION OF TEMPE 28-3184
NOISE MEASUREMENTS IN FERROELECTRICS. (TGS, ROCHELLE SALT, BARIUM TITANATE) 04-0698
IO/ DIELECTRIC AND X-RAY STUDIES OF THE PHASE TRANSITIONS IN ROCHELLE SALT, BARIUM TITANATE, ANTIMONY SULFO 02-0099
FREQUENCY IN CAVITY. MEASUREMENT OF THERMAL EXPANSION OF ROCHELLE SALT BY USING THE SHIFT OF RESONANCE 28-3202
NENCE FOR FERROELEC/ SPECIFIC HEAT ANOMALY IN FERROELECTRIC ROCHELLE SALT (CALCULATION OF TEMPERATURE DEPE 28-3187
T ONLY) LOW FREQUENCY DISPERSIONS OF THE PERMITTIVITY OF ROCHELLE SALT CAUSED BY DOMAIN WALLS. (ABSTRAC 28-3199
DOMAIN WALLS CAUGHT IN (FINE LINE) "SUDARES" IN ROCHELLE SALT CRYSTAL. 28-3215
ELECTRONIC EFFECTS (OF X-RAY AND GAMMA-RAY IRRADIATION) IN ROCHELLE SALT CRYSTALS. 28-3192
LOW FREQUENCY DIELECTRIC DISPERSION IN ROCHELLE SALT CRYSTALS. 28-3228
EZOELECTRIC EFFECT AND INTERNAL FRICTION IN GAMMA-IRRADIATED ROCHELLE SALT CRYSTALS. PI 28-3229
SPONTANEOUS ELECTROOPTICAL EFFECT IN ROCHELLE SALT CRYSTALS. 28-3209
RES AND CORRESPONDING LIGHT FIGURES REVEALED BY WATER ETCHED ROCHELLE SALT CRYSTALS. SURFACE STRUCTU 28-3234
YDROGEN BOND BANDS IN RAMAN SCATTERING SPECTRA OF GYPSUM AND ROCHELLE SALT CRYSTALS. /OF THE INTENSITY OF H 28-3189
ABSORPTION OF LONGITUDINAL ULTRASONIC WAVES IN ROCHELLE SALT CRYSTALS NEAR THE CURIE POINTS. 28-3211
NS OF PROTON PAIRS (IN THE WA/ NUCLEAR MAGNETIC RESONANCE IN ROCHELLE SALT, DETERMINATION OF THE ORIENTATIO 28-3203
PERMITTIVITY OF ROCHELLE SALT DURING SWITCHING. 28-3196
YS. COERCIVE FIELDS OF ROCHELLE SALT HEAVILY IRRADIATED WITH GAMMA-RA 28-3178
CRYSTAL GROWTH) CRYSTALLIZATION OF ROCHELLE SALT IN ELECTRIC FIELDS (KINETICS OF 28-3200
INFLUENCE OF CRYSTAL DEFECTS ON THE DIELECTRIC BEHAVIOR OF ROCHELLE SALT IN MICROWAVE RANGE. 28-3219
KY BLUE DYE. GROWTH RATE OF THE (010) PLANE OF ROCHELLE SALT IN THE PRESENCE OF PURE DIRECT S 28-3217

FREQUENCY DEPENDENCE OF THE ABSORPTION OF SOUND IN ROCHELLE SALT NEAR ITS UPPER CURIE POINT. 28-3183
REARRANGEMENT OF THE DOMAIN STRUCTURE IN ROCHELLE SALT NEAR THE PHASE TRANSITION. 28-3181
/Y DEPENDENCE OF THE LOSS FACTOR IN FERROELECTRIC MATERIALS (ROCHELLE SALT NEAR THE UPPER CURIE POINT AS E/ 28-3220
/FFECTS OF (DAMAGE FROM) IONIZING RADIATION IN FERROELECTRIC ROCHELLE SALT. (NEUTRON DIFFRACTION MEASUREME/ 28-3188
INFLUENCE OF THE CRYSTALLIZATION TEMPERATURE OF ROCHELLE SALT ON ITS FERROELECTRIC PROPERTIES. 28-3193
/IONS. PART-1: TGS. PART-2: SILVER SODIUM DINITRITE. PART-3: ROCHELLE SALT. PART-4: LITHIUM AMMONIUM TARTR/ 02-0077
/STUDIES IN SOME HYDROGEN BONDED FERROELECTRICS. (DEUTERATED ROCHELLE SALT, POTASSIUM OR RUBIDIUM OR AMMON/ 18-2719
/CS. (THEORY, DYNAMICAL ORDER DISORDER PROPERTIES, KDP, TGS, ROCHELLE SALT, SILVER PERIODATE, SODIUM TRIHY/ 02-0050
TRIC FIELD ON DOMAIN STRUCTURE FORMATION IN SEIGNETTES SALT (ROCHELLE SALT). (TEMPERATURE ANHYSTERESIS). /C 28-3190
IC "AFTER EFFECT" PHENOMENA. (HYSTERESIS LOOP CONSTRICTIONS, ROCHELLE SALT, TGS)   SHORT LIVED FERROELECTR 27-3156
MECHANISM OF FERROELECTRIC PHASE TRANSITION IN AMMONIUM ROCHELLE SALT. (THEORY) 28-3223
THE ROLE OF LORENTZ FACTORS IN ROCHELLE SALT. (THEORY) 03-0289
ED TRANSITION) IN GAMMA-IRRADIATED FERROELECTRIC MATERIALS. (ROCHELLE SALT, TRIGLYCINE SULFATE) /ON (HINDER 27-3165
/TIONS OF THE DOMAIN STRUCTURE STABILITY IN FERROELECTRICS. (ROCHELLE SALT, TRIGLYCINE SULFATE, SELENATE, / 02-0083
/ECTRIC BEHAVIOR OF RADIATION DAMAGED TRIGLYCINE SULFATE AND ROCHELLE SALT UP TO HIGH DOSAGE (DESTRUCTION / 27-3098
/ESONATOR) METHOD OF STUDYING TEMPERATURE CHARACTERISTICS OF ROCHELLE TYPE CERAMICS. (BARIUM CALCIUM TITAN/ 04-0905
I NITRITES (SODIU/ VACUUM ULTRAVIOLET ABSORPTION (5-23 EV AT ROOM AND LIQUID NITROGEN TEMPERATURES IN ALKAL 16-2399
BRILLOUIN SCATTERING IN SODIUM NITRITE AT ROOM TEMPERATURE. 16-2305
TE HEXAHYDRATE. (POLARIZATION VERSUS ELECTRIC FIELD, -80C TO ROOM TEMPERATURE) / GUANIDINIUM ALUMINUM SULFA 21-2813
DIFFUSE SCATTERING IN SODIUM NIOBATE AT ROOM TEMPERATURE. 08-1304
ESR STUDY OF GAMMA-IRRADIATED CESIUM TRIHYDROGEN SELENITE AT ROOM TEMPERATURE. 22-2833
EUTRON SCATTERING ANALYSIS OF CESIUM TRIHYDROGEN SELENITE AT ROOM TEMPERATURE.   N 22-2869
POCKELS EFFECT IN AMMONIUM DIHYDROGEN PHOSPHATE ABOVE ROOM TEMPERATURE. 17-2518
OELECTRIC SODIUM TRIDEUTERO SELENITE. ROOM TEMPERATURE CRYSTAL STRUCTURE OF THE FERR 22-2861
ARAMAGNETIC (BETA) TERBIUM MOLYBDATE (CZOCHRALSKI SINGLE C/ (ROOM TEMPERATURE) FERROELECTRIC FERROELASTIC P 30-3290
(ELASTIC, PIEZOELECTRIC, DIELECTRIC, AND OPTICAL CONSTANTS) (ROOM TEMPERATURE TO THE CURIE POINT. / NIOBATE 13-2060
SMITH FERRATE LEAD TITANATE PEROVSKITE TYPE SOLID SOLUTIONS (ROOM TEMPERATURE TO 600 DEG C). /STUDIES OF BI 08-1471
ION SEQUENCE IN FORMING BARIUM METATITANATE IN AN INDUSTRIAL ROTARY FURNACE.   REACT 04-1003
AN ELECTROOPTIC SWITCH USING ROTATED Y PLATE OF LITHIUM NIOBATE. 11-1732
GROWTH. (AMMONIUM DIHYDROGEN PHOSPHATE, POTASSIUM / RATE OF ROTATING CRYSTALS AND CHARACTERISTICS OF THEIR 16-2316
ONIUM ROCHELLE SALT BY SPIN LATTICE RELAXATION OF PROTONS IN ROTATING FRAME. /ULES IN ROCHELLE SALT AND AMM 28-3232
PROTON ROTATING FRAME RELAXATION IN LITHIUM HYDRAZINI 32-3347
IN LITHIUM NIOBATE. ROTATING WAVEPLATE OPTICAL FREQUENCY SHIFTING 11-1689
MANIUM(3) OXIDE(11). FERROELECTRIC OPTICAL ROTATION DOMAINS IN SINGLE CRYSTAL LEAD(5) GER 36-3550
TETRADEUTERATE.   PROTON RESONANCE STUDY OF SUBLATTICE ROTATIONS AND SPIN DEVIATION IN COPPER FORMATE 26-2950
/TURAL TRANSITIONS ABOVE FERROELECTRIC CRITICAL TEMPERATURE (ROTATIONS OF BO6 OCTAHEDRA IN ABO(3) PEROVSKI/ 06-1205
C TRANSITION IN AMMONIUM SULFATE STUDIED BY CCLD NEUTRONS (/ ROTATIONAL MOTIONS IN SOLIDS: THE FERROELECTRI 19-2752
DIPOLAR RELAXATION OF LITHIUM-7 BY HINDERED ROTATORS IN LITHIUM HYDRAZINIUM SULFATE. 32-3350
ARIZATION, THERMAL EXPANSION, INDICES OF REFRACTION, OPTICAL ROTATORY POWER) /IC CONSTANTS, SPONTANEOUS POL 35-3501
IUM(3) OXIDE(11) SINGLE CRYSTAL.   SWITCHING OF OPTICAL ROTATORY POWER IN FERROELECTRIC LEAD(5) GERMAN 36-3550
CRYSTAL FIELD ENERGY LEVEL SPLITTINGS OF TITANIUM(3+) ION IN RUBIDIUM ALUMINUM DISULFATE DODECAHYDRATE. /F 19-2754
/R THEIR CURIE POINTS. (POTASSIUM DIHYDROGEN PHOSPHATE, ALSO RUBIDIUM ANALOG AND POTASSIUM DIDEUTERIUM PHO/ 17-2665
/OF SOME KDP TYPE CRYSTALS. (POTASSIUM DIHYDROGEN PHOSPHATE, RUBIDIUM AND AMMONIUM ANALOGS SOME DEUTERATE) 17-2679
MICS.   ELECTRICAL PROPERTIES OF RUBIDIUM AND CESIUM DOPED BARIUM TITANATE CERA 04-0860
ES.   (PREPARATION AND STRUCTURE OF) RUBIDIUM AND CESIUM SEMIMETALLIC ORTHOPHOSPHAT 17-2598
RROELECTRIC / CHANGES IN THE INFRARED SPECTRA OF AMMONIUM OR RUBIDIUM BISULFATE, AND AMMONIUM SULFATE AT FE 19-2772
S (SODIUM, POTASSIUM) AND ALKALI NITRATES. (LITHIUM, SODIUM, RUBIDIUM, CESIUM) /PERATURES IN ALKALI NITRITE 16-2399
/TION IN KDP TYPE CRYSTALS (CURIE TEMPERATURES OF POTASSIUM, RUBIDIUM, CESIUM AND AMMONIUM DIHYDROGEN PHOS/ 17-2577
/ATE OR FLUOBERYLLATE (METALS CHOSEN FROM SODIUM, POTASSIUM, RUBIDIUM, CESIUM, OR AMMONIUM, ABSTRACT ONLY). 19-2743
IUM MANGANESE TRIFLUORIDE, POTASSIUM COBALT TRIFLUORIDE, AND RUBIDIUM COBALT TRIFLUORIDE. /PECTRA OF POTASS 33-3384
IN HEAVY WATER.   CRYSTALLIZATION OF RUBIDIUM DIDEUTERIUM PHOSPHATE FROM SOLUTIONS 17-2583
ERATED FERROELECTRIC CRYSTALS CESIUM DIHYDROGEN ARSENATE AND RUBIDIUM DIHYDROGEN ARSENATE. / PARTIALLY DEUT 18-2740
/ASSIUM DIHYDROGEN ARSENATE, POTASSIUM DIDEUTERIUM ARSENATE, RUBIDIUM DIHYDROGEN ARSENATE, RUBIDIUM DIDEUT/ 18-2725
CRYSTAL/ CONTRIBUTION TO THE PREPARATION ON NON-WEDGE SHAPED RUBIDIUM DIHYDROGEN OR DIDEUTERIUM) PHOSPHATE 17-2637
BRILLOUIN DIFFUSION IN A SINGLE CRYSTAL OF RUBIDIUM DIHYDROGEN PHOSPHATE. 17-2520
BRILLOUIN SCATTERING BY A SINGLE CRYSTAL OF RUBIDIUM DIHYDROGEN PHOSPHATE. 17-2519
INTENSITY OF SOME BRILLOUIN AND RAYLEIGH LINES IN RUBIDIUM DIHYDROGEN PHOSPHATE. 17-2521
PHASE SYNCHRONIZATION IN NONLINEAR RUBIDIUM DIHYDROGEN PHOSPHATE. 17-2473
/IUM-87 AND ARSENIC-75 QUADRUPOLAR COUPLING IN FERROELECTRIC RUBIDIUM DIHYDROGEN PHOSPHATE AND CESIUM DIHY/ 17-2439
/RUCTURE OF THE HYDROGEN BONDS IN THE FERROELECTRIC CRYSTALS RUBIDIUM DIHYDROGEN PHOSPHATE AND RUBIDIUM DI/ 17-2471
/UM DIHYDROGEN PHOSPHATE POTASSIUM DIDEUTERIUM PHOSPHATE AND RUBIDIUM DIHYDROGEN PHOSPHATE AT HIGH TEMPERA/ 17-2613
COLD NEUTRONS.   HIGH TEMPERATURE PHASE TRANSITION IN RUBIDIUM DIHYDROGEN PHOSPHATE BY SCATTERING OF 17-2489
NONLINEAR PROPERTIES OF A RUBIDIUM DIHYDROGEN PHOSPHATE CRYSTAL. 17-2508
LECTRIC AND / VIBRATION SPECTRUM OF THE PHOSPHATE ANION IN A RUBIDIUM DIHYDROGEN PHOSPHATE CRYSTAL IN PARAE 17-2472
HASE TRANSITION REGION. (COERCIVE FIELD)   POLARIZATION OF A RUBIDIUM DIHYDROGEN PHOSPHATE CRYSTAL IN THE P 17-2504
HASE TRANSITION REGION.   POLARIZATION OF A RUBIDIUM DIHYDROGEN PHOSPHATE CRYSTAL IN THE P 17-2506
CRITICAL BEHAVIOR OF KDP AND RUBIDIUM DIHYDROGEN PHOSPHATE CRYSTALS. 17-2676
ELECTRONIC EFFECT AND ELASTIC PROPERTIES OF RUBIDIUM DIHYDROGEN PHOSPHATE CRYSTALS. 17-2617
LECTROOPTICAL AND OPTICAL PROPERTIES OF PARTIALLY DEUTERATED RUBIDIUM DIHYDROGEN PHOSPHATE CRYSTALS.   E 17-2698
FAR INFRARED SPECTRA OF RUBIDIUM DIHYDROGEN PHOSPHATE CRYSTALS. 17-2638
/ECTRA OF MONOCRYSTALS OF POTASSIUM DIHYDROGEN PHOSPHATE AND RUBIDIUM DIHYDROGEN PHOSPHATE IN POLAR AND NO/ 17-2457
IDIUM DIDEUTERIUM PHOSPHATE IN HE/ ENTHALPIES OF SOLUTION OF RUBIDIUM DIHYDROGEN PHOSPHATE IN WATER AND RUB 17-2585
/TION IN ACTIVE MATERIALS AND NONLINEAR CRYSTALS. (KDP, ADP, RUBIDIUM DIHYDROGEN PHOSPHATE, LITHIUM NIOBAT/ 11-1865
EFFECT OF DEUTERATION ON THE ELASTIC PROPERTIES OF RUBIDIUM DIHYDROGEN PHOSPHATE (RDP). 17-2584
/E NONLINEAR COEFFICIENTS OF POTASSIUM DIHYDROGEN PHOSPHATE, RUBIDIUM DIHYDROGEN PHOSPHATE, RUBIDIUM DIHYD/ 08-1413
HEAT CAPACITY OF RUBIDIUM DIHYDROGEN PHOSPHATE SINGLE CRYSTALS. 17-2415
/TH, TGS, TRIGLYCINE SELENATE OR FLUOBERYLLATE, POTASSIUM OR RUBIDIUM DIHYDROGEN PHOSPHATE, SODIUM, POTASS/ 02-0089
ION ON THE HYDROXIDE VIBRATION SPECTRUM OF THE FERROELECTRIC RUBIDIUM DIHYDROGEN PHOSPHATES. /PHASE TRANSIT 17-2470
TERING SPECTRA.   FERROELECTRIC TRANSITION IN RUBIDIUM DIHYDROGEN PHOSPHIDE USING RAMAN SCAT 17-2620
CRYSTAL STRUCTURE OF FERROELECTRIC RUBIDIUM HYDROGEN SULFATE. (ABSTRACT) 19-2746
UM SULFATE IN THE PARA- AND FERROELECTR/ INFRARED SPECTRA OF RUBIDIUM HYDROGEN SULFATE AND RUBIDIUM DEUTERI 19-2768
TRAFLUORIDE.   FERROELASTIC EFFECT IN RUBIDIUM IRON TETRAFLUORIDE AND CESIUM IRON TE 33-3358
CTRIC IRON(2+) BORACITES. (CHLORIDE, BROMIDE AND IODIDE, AND RUBIDIUM IRON TRIFLUORIDE /DEG K) OF FERROELE 14-2097
/INIUM MOLYBDATE, KDP, ORTHOFERRITES, DYSPROSIUM ANTIMONIDE, RUBIDIUM IRON TRIFLUORIDE, BORACITES, BISMUTH/ 02-0036
/ORRELATION IN PARAMAGNETIC POTASSIUM MANGANESE TRIFLUORIDE, RUBIDIUM MANGANESE TRIFLUORIDE, AND MANGANESE 33-3376
/N DEPENDENCE OF RAMAN SCATTERING BY MAGNONS IN NICKEL DOPED RUBIDIUM MANGANATE AND POTASSIUM MANGANESE TR/ 33-3395
ERATURE.   THERMAL EXPANSIVITY OF RUBIDIUM MANGANESE FLUORIDE NEAR THE NEEL TEMP 33-3375
/RATION AND DISLOCATION DENSITY DETERMINATION OF LARGE, PURE RUBIDIUM MANGANESE TRIFLUORIDE AND POTASSIUM / 33-3397
/VERAL PEROVSKITES (POTASSIUM TANTALATE, STRONTIUM TITANATE, RUBIDIUM MANGANESE TRIFLUORIDE) BY INELASTIC / 06-1139
UORIDE, COBALT FLUORIDE, POTASSIUM MANGANESE TRIFLUORIDE AND RUBIDIUM MANGANESE TRIFLUORIDE CRYSTALS. /E FL 33-3399

/IGHT IN MANGANESE FLUORIDE CRYSTALS (POTASSIUM TRIFLUORIDE, RUBIDIUM MANGANESE TRIFLUORIDE, MANGANESE DIF/ 33-3398
ON THE STRUCTURE OF AB(5) OXIDE COMPOUNDS. (BARIUM TITANATE, RUBIDIUM NIOBATE) /ELUNG ENERGY AND COVALENCY 04-0761
                              ELECTRICAL PROPERTIES OF RUBIDIUM NITRATE AND CESIUM NITRATE.   16-2286
RE P/ PECULIARITIES IN THE VIBRATIONAL SPECTRA OF CESIUM AND RUBIDIUM NITRATE CRYSTALS IN ITS LOW TEMPERATU 16-2323
S POLYMORPHISM.   RUBIDIUM NITRATE NEAR THE MELTING POINT AND IT   16-2330
/DED FERROELECTRICS. (DEUTERATED ROCHELLE SALT, POTASSIUM OR RUBIDIUM OR AMMONIUM DIHYDROGEN ARSENATES, RU/ 18-2719
. (ANSWER-NO)   IS SODIUM RUBIDIUM TARTRATE TETRAHYDRATE A FERROELECTRIC 29-3245
/ANSFORMATION AND THERMOOPTICAL AND ELECTROOPTICAL EFFECT IN RUBIDIUM TRIHYDROGEN OR DEUTERIUM DISELENITE / 22-2840
                             CRYSTAL STRUCTURE OF RUBIDIUM TRIHYDROGEN SELENITE.   22-2881
       IMPROPER FERROELECTRIC PHASE TRANSITION IN RUBIDIUM TRIHYDROGEN SELENITE.   22-2834
LENATE CRYSTALS.   ISOTOPIC NONISOMORPHISM OF RUBIDIUM TRIHYDROGEN(1-X) TRIDEUTERIUM(X) DISE   22-2873
      RAMAN STUDY OF METALLIC TUNGSTEN BRONZES. (SODIUM AND RUBIDIUM TUNGSTATES)   10-1654
G IN FERROELECTRIC RUBIDIUM DIHYDROGEN PHOSPHATE AND CESIUM/ RUBIDIUM-87 AND ARSENIC-75 QUADRUPOLAR COUPLIN 17-2439
ER RAD/ TEMPERATURE DEPENDENCE OF HYPERSONIC WAVE DAMPING IN RUBY AND LITHIUM NIOBATE IN THE THREE CENTIMET 11-1718
INEAR CRYSTALS (KDP AND LITHIUM NIOBATE) BY RADIATION FROM A RUBY LASER.   FRACTURE OF NONL   11-1663
               INTRACAVITY SECOND HARMONIC GENERATION WITH RUBY LASER BY KDP.   17-2604
STAL)   OPTICAL MIXING. (OF PULSED RUBY LASER EMISSIONS BY TRIGLYCINE SULFATE CRY   27-2958
OSPHATE CRYSTAL.   MICROWAVE MODULATOR OF RUBY LASER LIGHT USING POTASSIUM DIHYDROGEN PH   17-2640
                        CIRCLE THEOREM FOR ICE RULE FERROELECTRIC MODELS.   03-0200
      IS THERE A UNIFIED MECHANISM UNDERLYING URBACH'S RULE. (STRONTIUM TITANATE, POTASSIUM TITANATE)   06-1110
PLICATIONS TO LITHIUM NIOBATE.   GROUP THEORETICAL SELECTION RULES FOR INELASTIC NEUTRON SCATTERING WITH AP 11-1704
RING IN FERROELECTRIC AND PARAELECTRI/ MULTIPHONON SELECTION RULES FOR INFRARED ABSORPTION AND RAMAN SCATTE 16-2281
S ON POTASSIUM DIHYDROGEN PHOSPHATE, TRIGLYCINE SULFATE, AND RUTILE.   FAR INFRARED DIELECTRIC MEASUREMENT 17-2423

S

TION AND FIRST HARMONIC, CALCULATIONS FOR LITHIUM/ NONLINEAR SADOVSKII EFFECT (THEORY, PROPAGATION OF RADIA 03-0186
TRIC EFFECT IN TGS, ETHYL DIAMINE TARTRATE, BARIUM TITANATE, SALICYLIDENE- 4-BROMO ANILINE. /ION. (PYROELEC 04-0854
                 TENSILE AND FLEXURE PROPERTIES OF SALINE ICE.   34-3422
              ANALYSIS OF DOUBLE HYSTERESIS LOOPS. (ROCHELLE SALT)   28-3212
LD AT VERY LOW FREQUENCIES AND JERKY WALL MOTION IN ROCHELLE SALT.   ANOMALOUS DISPERSION OF COERSIVE FIE 28-3177
BUTION OF DOMAIN WALL MOTION TO THE PERMITTIVITY OF ROCHELLE SALT.   CONTRI   28-3198
ISE AND SURFACE MICRODAMAGES ON SINGLE CRYSTALS OF SEIGNETTE SALT.   CORRELATION BETWEEN BARKHAUSEN NO   28-3208
ELECTRIC CONSTANT AND DIELECTRIC RELAXATION TIME OF ROCHELLE SALT.   DI   28-3179
                    DIELECTRIC PHENOMENA IN ROCHELLE SALT.   28-3182
                    DYNAMICS OF FERROELECTRIC ROCHELLE SALT.   28-3237
                    DYNAMICS OF FERROELECTRIC ROCHELLE SALT.   28-3236
SSURE ON THE CURIE POINTS OF TRIGLYCINE SULFATE AND ROCHELLE SALT.   EFFECT OF A TWO DIMENSIONAL PRE   27-3082
ON EPR SPECTRA OF (COPPER(2+)) DOPED FERROELECTRIC ROCHELLE SALT.   EFFECT OF APPLIED ELECTRIC FIELD   28-3185
CAL STRESS ON DIELECTRIC PROPERTIES OF PARAELECTRIC ROCHELLE SALT.   EFFECT OF MECHANI   28-3225
      EFFECT OF THE CUPRIC ION ON DELAYED PHENOMENA IN ROCHELLE SALT.   28-3221
WAVES IN FERROELECTRICS. (THEORY, BARIUM TITANATE, ROCHELLE SALT)   ELECTROMAGNETIC AND ACOUSTIC   03-0455
PED FERROELECTRIC CRYSTALS. (TRIGLYCINE SULFATE AND ROCHELLE SALT) /RESONANCE INVESTIGATIONS OF (COPPER) DO 27-3132
ELECTRON PARAMAGNETIC RESONANCE OF COPPER IONS IN ROCHELLE SALT.   28-3176
ELECTRON PARAMAGNETIC RESONANCE OF GAMMA-IRRADIATED ROCHELLE SALT.   28-3180
TION OF THE INTERNAL ELECTRIC FIELD IN COPPER DOPED ROCHELLE SALT.   EPR INVESTIGA   28-3233
         FAR INFRARED ABSORPTION SPECTRA OF ROCHELLE SALT.   28-3210
                FAST AFTEREFFECT OF SEIGNETTE SALT.   28-3206
                    FERROELECTRICITY IN ROCHELLE SALT.   28-3218
MENTS OF THE CHANGES OF WATER OF CRYSTALLIZATION OF ROCHELLE SALT.   GRAVIMETRIC MEASURE   28-3235
IGH PRESSURE PHASE TRANSITIONS OF THE FIRST KIND IN ROCHELLE SALT.   H   28-3214
ING FERROELECTRIC AND MAGNETIC TRANSITIONS, EXAMPLE ROCHELLE SALT) /ETER FOR PHASE CHANGE DETECTION (INCLUD 28-3201
RKHAUSEN EFFECT IN FERROELECTRIC MATERIALS (TGS AND ROCHELLE SALT).   INFLUENCE OF GAMMA-IRRADIATION ON BA 27-3123
STRUCTURE ON POSITRON (TWO-PHONON) ANNIHILATION IN ROCHELLE SALT.   INFLUENCE OF THE DOMAIN   28-3191
INTERNAL FRICTION AS A FUNCTION OF TEMPERATURE FOR ROCHELLE SALT.   28-3227
PIEZOELECTRIC CONSTANTS IN FERROELECTRIC CRYSTALS. (ROCHELLE SALT)   MEASUREMENTS OF COMPLEX   28-3213
         MICROSCOPIC THEORY OF THE PHASE TRANSITIONS IN ROCHELLE SALT.   03-0482
                NMR STUDIES OF SPONTANEOUS POLARIZATION IN ROCHELLE SALT.   28-3194
                NMR STUDY OF PHASE TRANSITIONS IN ROCHELLE SALT.   28-3195
ECTS OF X-RAY IRRADIATION ON TRANSIENT PHENOMENA IN ROCHELLE SALT.   ON THE EFF   28-3222
SOME FERROELECTRIC CRYSTALS (PYROELECTRICITY, TGS, ROCHELLE SALT).   OPTICAL ACTIVITY OF   27-3017
PHASE TRANSITIONS AT TEMPERATURES CLOSE TOGETHER. (ROCHELLE SALT) / IN FERROELECTRIC MATERIAL WITH SEVERAL 03-0363
        RADIATION DAMAGE AND THE FERROELECTRIC EFFECT IN ROCHELLE SALT.   28-3207
                RAMAN LIGHT SCATTERING IN ROCHELLE SALT.   28-3231
UPOLE SHIFTS OF THE SODIUM-23 MAGNETIC RESONANCE IN ROCHELLE SALT.   SECOND ORDER QUADR   28-3186
E OF DIHYDROGEN PHOSPHATE ION IN THE FERROELECTRIC POTASSIUM SALT.   STRUCTUR   17-2624
NATE AND TITANATE ZIRCONATE, TRIGLYCINE SULFATE AND ROCHELLE SALT) /LECTRICS OF VARIOUS TYPES. (BARIUM TITA 02-0108
PERSTRUCTURE IN THE FERROELECTRIC PHASE OF AMMONIUM ROCHELLE SALT.   SU 28-3224
                THERMAL CONDUCTIVITY OF ROCHELLE SALT.   28-3226
                ULTRASONIC ANOMALIES IN ROCHELLE SALT.   28-3197
INE FERROELECTRICS AT VERY LOW FIELDS AND STRESSES, ROCHELLE SALT) / (POLARIZATION SWITCHING IN POLYCRYSTALL 28-3216
E RELAX/ STUDY OF SLOW MOTION OF WATER MOLECULES IN ROCHELLE SALT AND AMMONIUM ROCHELLE SALT BY SPIN LATTIC 28-3232
/STUDIES OF THE GAMMA-RAY IRRADIATED FERROELECTRICS ROCHELLE SALT AND GUANIDINE ALUMINUM SULFATE HEXAHYDRA/ 21-2812
                       SEE BOTH ROCHELLE SALT AND SEIGNETTE SALT.
                MICROWAVE BEHAVIOR OF ROCHELLE SALT AND TGS (18 REFS).   27-3088
NAL FRICTION IN SINGLE CRYSTALS OF FERROELECTRICS. (ROCHELLE SALT AND TRIGLYCINE SULFATE) /ENDENCE OF INTER 27-3122
NATURE OF RADIATION INDUCED DEFECTS IN CRYSTALS OF ROCHELLE SALT AND TRIGLYCINE SULFATE.   27-3159
UREMENTS OF DIVERSE THERMODYNAMICAL COEFFICIENTS OF ROCHELLE SALT AND TRIGLYCINE SULFATE.   NEW MEAS   27-2995
      DISTINCTIVE FEATURES OF BARKHAUSEN EFFECT IN ROCHELLE SALT AND TRIGLYCINE SULFATE CRYSTALS.   27-3110
OPPER AND IRON IMPURITY) ABSORPTION BANDS OF DEFECT ROCHELLE SALT AND TRIGLYCINE SULFATE CRYSTALS. /N AND C 27-3108
AL ABSORPTION EDGE AS A FUNCTION OF TEMPERATURE FOR ROCHELLE SALT AND TRIGLYCINE SULFATE CRYSTALS. /NDAMENT 27-3160
/F EXPERIMENTAL STUDY OF GAMMA-RADIATION EFFECTS IN ROCHELLE SALT AND TRIGLYCINE SULFATE. (ON ELECTRONIC P/ 27-3101
.   CRITICAL PROPERTIES OF TGS AND ROCHELLE SALT AS DETERMINED BY AN ELECTRIC FIELD METHOD 27-3056
ION STUDIES OF FERROELECTRICS. (BARIUM TITANATE AND ROCHELLE SALT AS EXAMPLES)   NEUTRON DIFFRACT   04-0679
        ATTENUATION OF SOUND WAVES IN DEUTERATED ROCHELLE SALT AT THE PHASE TRANSITION POINT.   28-3204
        ATTENUATION OF SOUND WAVES IN DEUTERATED ROCHELLE SALT AT THE PHASE TRANSITION POINT.   28-3205
, POTASSIUM DIHYDROGEN PHOSPHATE, AMMONIUM SULFATE, ROCHELLE SALT) AT THE TRANSITION TEMPERATURES. /SULFATE 02-0125
MEASUREMENT OF THE COMPLEX DIELECTRIC CONSTANT OF ROCHELLE SALT AT 10 GHZ AS A FUNCTION OF TEMPERATURE.   28-3184
     NOISE MEASUREMENTS IN FERROELECTRICS. (TGS, ROCHELLE SALT, BARIUM TITANATE)   04-0698
/TRIC AND X-RAY STUDIES OF THE PHASE TRANSITIONS IN ROCHELLE SALT, BARIUM TITANATE, ANTIMONY SULFOIODIDE, / 02-0099

IN CAVITY.   MEASUREMENT OF THERMAL EXPANSION OF ROCHELLE SALT BY USING THE SHIFT OF RESONANCE FREQUENCY 28-3202
R FERROELEC/ SPECIFIC HEAT ANOMALY IN FERROELECTRIC ROCHELLE SALT (CALCULATION OF TEMPERATURE DEPENDENCE FO 28-3187
LOW FREQUENCY DISPERSIONS OF THE PERMITTIVITY OF ROCHELLE SALT CAUSED BY DOMAIN WALLS. (ABSTRACT ONLY) 28-3199
DOMAIN WALLS CAUGHT IN (FINE LINE) "SUDARES" IN ROCHELLE SALT CRYSTAL. 28-3215
NIC EFFECTS (OF X-RAY AND GAMMA-RAY IRRADIATION) IN ROCHELLE SALT CRYSTALS.   ELECTRO 28-3192
ICW FREQUENCY DIELECTRIC DISPERSION IN ROCHELLE SALT CRYSTALS. 28-3228
IC EFFECT AND INTERNAL FRICTION IN GAMMA-IRRADIATED ROCHELLE SALT CRYSTALS.   PIEZOELECTR 28-3229
SPONTANEOUS ELECTROOPTICAL EFFECT IN ROCHELLE SALT CRYSTALS. 28-3209
ORRESPONDING LIGHT FIGURES REVEALED BY WATER ETCHED ROCHELLE SALT CRYSTALS.   SURFACE STRUCTURES AND C 28-3234
OND BANDS IN RAMAN SCATTERING SPECTRA OF GYPSUM AND ROCHELLE SALT CRYSTALS. /OF THE INTENSITY OF HYDROGEN B 28-3189
ABSORPTION OF LONGITUDINAL ULTRASONIC WAVES IN ROCHELLE SALT CRYSTALS NEAR THE CURIE POINTS. 28-3211
TON PAIRS (IN THE WA/ NUCLEAR MAGNETIC RESONANCE IN ROCHELLE SALT, DETERMINATION OF THE ORIENTATIONS OF PRO 28-3203
PERMITTIVITY OF ROCHELLE SALT DURING SWITCHING. 28-3196
EFFECT OF A DC FIELD ON DOMAIN STRUCTURE IN SEIGNETTE SALT. (EFFECT OF AGING AND COPPER(2+) DOPING) 28-3230
COERCIVE FIELDS OF ROCHELLE SALT HEAVILY IRRADIATED WITH GAMMA-RAYS. 28-3178
ROWTH)   CRYSTALLIZATION OF ROCHELLE SALT IN ELECTRIC FIELDS (KINETICS OF CRYSTAL G 28-3200
CE OF CRYSTAL DEFECTS ON THE DIELECTRIC BEHAVIOR OF ROCHELLE SALT IN MICROWAVE RANGE.   INFLUEN 28-3219
YE.   GROWTH RATE OF THE (010) PLANE OF ROCHELLE SALT IN THE PRESENCE OF PURE DIRECT SKY BLUE D 28-3217
FREQUENCY DEPENDENCE CF THE ABSORPTION OF SOUND IN ROCHELLE SALT NEAR ITS UPPER CURIE POINT. 28-3183
REARRANGEMENT OF THE DOMAIN STRUCTURE IN ROCHELLE SALT NEAR THE PHASE TRANSITION. 28-3181
ENCE OF THE LOSS FACTOR IN FERROELECTRIC MATERIALS (ROCHELLE SALT NEAR THE UPPER CURIE POINT AS EXAMPLE). / 28-3220
F (DAMAGE FROM) IONIZING RADIATION IN FERROELECTRIC ROCHELLE SALT. (NEUTRON DIFFRACTION MEASUREMENTS) /TS O 28-3188
/NOCHEMISTRY OF THE SINTERING BODY. PART-9: EFFECT OF MOTHER SALT OF LEAD MONOXIDE ON THE SINTERING. PART-/ 09-1624
INFLUENCE OF THE CRYSTALLIZATION TEMPERATURE OF ROCHELLE SALT ON ITS FERROELECTRIC PROPERTIES. 28-3193
/T-1: TGS. PART-2: SILVER SODIUM DINITRITE. PART-3: ROCHELLE SALT. PART-4: LITHIUM AMMONIUM TARTRATE HYDRA/ 02-0077
/N SOME HYDROGEN BONDED FERROELECTRICS. (DEUTERATED ROCHELLE SALT, POTASSIUM OR RUBIDIUM OR AMMONIUM DIHYD/ 18-2719
/ ELECTRIC FIELD ON DOMAIN STRUCTURE FORMATION IN SEIGNETTES SALT (ROCHELLE SALT). (TEMPERATURE ANHYSTERES/ 28-3190
/RY, DYNAMICAL ORDER DISORDER PROPERTIES, KDP, TGS, ROCHELLE SALT, SILVER PERIODATE, SODIUM TRIHYDROGEN SE/ 02-0050
EFFECT" PHENOMENA. (HYSTERESIS LOOP CONSTRICTIONS, ROCHELLE SALT, TGS)   SHORT LIVED FERROELECTRIC "AFTER 27-3156
ANISM OF FERROELECTRIC PHASE TRANSITION IN AMMONIUM ROCHELLE SALT. (THEORY)   MECH 28-3223
THE ROLE OF LORENTZ FACTORS IN ROCHELLE SALT. (THEORY) 03-0289
TION) IN GAMMA-IRRADIATED FERROELECTRIC MATERIALS. (ROCHELLE SALT, TRIGLYCINE SULFATE) /ON (HINDERED TRANSI 27-3165
/THE DOMAIN STRUCTURE STABILITY IN FERROELECTRICS. (ROCHELLE SALT, TRIGLYCINE SULFATE, FLUOBERYL/ 02-0083
/HAVIOR OF RADIATION DAMAGED TRIGLYCINE SULFATE AND ROCHELLE SALT UP TO HIGH DOSAGE (DESTRUCTION OR MASKIN/ 27-3098
CRYSTAL GROWTH OF STRONTIUM POTASSIUM TANTALATE FROM MOLTEN SALTS.   SINGLE 13-2034
/LEAR SPIN LATTICE RELAXATION IN SOME FERROELECTRIC AMMONIUM SALTS. (AMMONIUM SULFATE, FLUOBERYLLATE, PERI/ 19-2769
/ORMATION AND PHASE TRANSFORMATIONS OF THE M MODIFICATION OF SAMARIUM MOLYBDATE (SOLID PHASE SYNTHESIS, NE/ 30-3270
DIELECTRIC CONSTANTS OF THE SYSTEM SAMARIUM NIOBATE- LANTHANUM NIOBATE. 08-1268
FERROELECTRIC HYSTERESIS TRACER FEATURING COMPENSATION AND SAMPLE GROUNDING. 37-3668
EAR THE CURIE POINT (EFFECTS OF TEMPERATURE, FIELD STRENGTH, SAMPLE ILLUMINATION). / SULFOIODIDE CRYSTALS N 15-2121
ECTRICS NEAR THE/ (EFFECT OF TEMPERATURE GRADIENT WITHIN THE SAMPLE ON THE) MICROWAVE DISPERSION OF FERROEL 03-0370
HE PYROELECTRIC PROPERTIES O/ INVESTIGATION OF THE EFFECT OF SAMPLE THICKNESS ON THE DOMAIN STRUCTURE AND T 04-1039
TRIGLYCINE SULFATE PARAMETERS.   EFFECT OF SAMPLE TREATMENT AND ELECTRODE NATURE ON SOME 27-3148
C ISOMORPHOUS CRYSTALS. (NEW MODEL BASED ON NMR OF UNTWINNED SAMPLES) /NSITION IN POTASSIUM FERROCYANIDE AN 24-2893
MAGNETIC SYMMET/ SPIN WAVE PHENOMENOLOGICAL THEORY IN FINITE SAMPLES OF FERRODIELECTRICS WITH AN ARBITRARY 03-0164
ABSORPTION COEFFICIENTS AND THE THICKNESS CF NONUNIFORM THIN SAMPLES (STRONTIUM TITANATE AS EXAMPLE). /RAY 06-1179
ITHIUM NIOBATE ON ION BOMBARDMENT AND COMPARISON RESULTS FOR SAPPHIRE. /NSION, CRAZING AND EXFOLIATION OF L 11-1800
THE CRYSTAL STRUCTURE OF TRIS-SARCOSINE CALCIUM CHLORIDE. 36-3524
/THE FERROELECTRIC TRANSITIONS IN DIGLYCINE NITRATE AND TRIS-SARCOSINE CALCIUM CHLORIDE. (SHIFT OF CURIE T/ 16-2267
/STIGATIONS OF MANGANESE(2+) COMPLEXES IN FERROELECTRIC TRIS-SARCOSINE CALCIUM CHLORIDE (TSCC) SINGLE CRYS/ 36-3525
CRYSTAL STRUCTURE OF THE IV PHASE OF THIOUREA. (SATELLITE SCATTERING) 25-2923
ONG PERIOD STRUCTURE OF THIOUREA CRYSTAL. (CALCULATIONS FROM SATELLITE SCATTERING DATA)   L 25-2911
LATION OF THIOUREA.   SATELLITE X-RAY SCATTERING AND STRUCTURAL MODU 25-2921
DIES ON THE MECHANISM CF THE PHASE TRANSITIONS IN THIOUREA. (SATELLITE SCATTERING)   STU 25-2922
HIUM NIOBATE.   ACOUSTIC DIPOLE AND QUADRUPOLE SATURATION OF THE LITHIUM-7 NMR SIGNALS IN LIT 11-1666
CADMIUM SELENIDE LAYERS ON BARIUM TITANATE- STRONTIUM TITAN/ SATURATION OF VOLT- AMPERE CHARACTERISTICS OF 04-0592
E PLANAR SPIN MODEL OF STRONTIUM TITANATE.   STATIC SCALING HYPOTHESIS AND ITS IMPLICATIONS FOR TH 06-1223
C PROPERTIES OF A(2+) B(3+) (2) O(4) TYPE CALCIUM AND STRONTIUM SCANDATES.   PREPARATION AND CRYSTALLOGRAPHI 36-3537
(0.5) NIOBIUM(0.5)) (1-X) B'(X) OXYGEN(3), WHERE B IS IRON OR SCANDIUM, AND B' IS ZIRCONIUM OR HAFNIUM. / (B 13-2044
ITANATE.   SINTERING SCANDIUM AND NIOBIUM MODIFIED LEAD ZIRCONATE T 09-1555
/URE OF BISMUTH METAL OXYGEN(3) PEROVSKITES. (WHERE METAL IS SCANDIUM, CHROMIUM, MANGANESE, IRON, COBALT, / 08-1451
0.5).   SOLUTION KINETICS OF LUTETIUM AND SCANDIUM IONS IN LEAD TITANATE(0.5) ZIRCONATE(09-1538
SYNTHESIS AND STUDY OF LEAD SCANDIUM NIOBATE SINGLE CRYSTALS. 08-1325
UM, GADOLINIUM, DYSPROSIUM, IRON, SELENIUM, NIOBIUM, COBALT, SCANDIUM, OR COPPER) /AGNESIUM, CADMIUM, YTTRI 10-1640
D, AND UNDOPED / MICROSTRUCTURE AND ELECTRICAL PROPERTIES OF SCANDIUM TRIOXIDE DOPED, RARE EARTH OXIDE DOPE 04-0399
STATIC QUADRUPOLE INTERACTION OF SCANDIUM-44 IN BARIUM TITANATE. 04-0692
PERTURBED DIRECTIONAL CORRELATION OF SCANDIUM-44 IN LEAD TITANATE. 05-1055
IUM TITANATE.   PERTURBED DIRECTIONAL CORRELATION OF SCANDIUM-44 IN THE FERROELECTRIC PHASES OF BAR 04-0558
C BARIUM TITANATE.   PERTURBED DIRECTIONAL CORRELATION OF SCANDIUM-44 WITHIN THE LATTICE OF FERROELECTRI 04-0691
A SELF SCANNED FERROELECTRIC IMAGE SENSOR. 37-3667
DIRECT OBSERVATION OF BARIUM TITANATE CRYSTALS BY SCANNING ELECTRON MICROSCOPE. 04-0808
ION OF FERROELECTRIC DOMAINS IN TRIGLYCINE SULFATE USING THE SCANNING ELECTRON MICROSCOPE.   DIRECT OBSERVAT 27-3051
RIC DOMAIN STRUCTURES, MODULATED BY A PHASE TRANSITION, IN A SCANNING ELECTRON MICROSCOPE. /N OF FERROELECT 03-0474
STUDY OF THE SURFACE OF ICE WITH A SCANNING ELECTRON MICROSCOPE. 34-3420
SURFACE STUDIES OF BARIUM TITANATE CERAMICS USING THE SCANNING ELECTRON MICROSCOPE. 04-0567
CTRIC DOMAINS IN TGS.   SCANNING ELECTRON MICROSCOPE STUDY OF FERROELE 27-3052
DE SINGLE C/ STUDY OF DOMAINS OF FERROELECTRIC CRYSTALS WITH SCANNING ELECTRON MICROSCOPE. (TUNGSTEN TRIOXI 10-1646
INTERED BARIUM SODIUM NIOBATES.   SCANNING ELECTRON MICROSCOPIC EXAMINATION OF S 13-1955
DY OF ANTIMONY SULFOIODIDE USING A DOUBLE PASSED, STABILIZED SCANNING INTERFEROMETER. /LOUIN SCATTERING STU 15-2225
ETR/ PHASE CHANGE IN SODIUM NITRITE CRYSTALS AND COMBINATION SCATTER SPECTRUM IN THE REGION OF THE ANTISYMM 16-2276
KDP GROUP.   SPECTRUM OF RAMAN SCATTERED LASER LIGHT IN FERROELECTRICS OF THE 17-2625
. (20 PERCENT MANGANESE REPLACED BY COBALT, COHERENT NEUTRON SCATTERING) /D POTASSIUM MANGANESE TRIFLUORIDE 33-3366
ASSIUM MANGANESE TRIFLUORIDE SINGLE CRYSTALS, (X-RAY DIFFUSE SCATTERING). / THE 184 DEG K TRANSITION OF POT 33-3369
VE THE FERROELECTRIC CURIE TEMPERATURE, AS SHOWN) BY NEUTRON SCATTERING. /CTUATIONS IN BARIUM TITANATE (ABO 04-1032
ION OF SODIUM NITRITE (STUDIED BY COHERENT NEUTRON INELASTIC SCATTERING). / THE FERROELECTRIC PHASE TRANSIT 16-2364
CRYSTAL DYNAMICS OF SODIUM NITRITE. (NEUTRON SCATTERING) 16-2284
CRYSTAL STRUCTURE OF THE IV PHASE OF THIOUREA. (SATELLITE SCATTERING) 25-2923
TANATE, RUBIDIUM MANGANESE TRIFLUORIDE) BY INELASTIC NEUTRON SCATTERING. /POTASSIUM TANTALATE, STRONTIUM TI 06-1139
DOMAIN WALL STRUCTURE IN GADOLINIUM MOLYBDATE BY RAMAN SCATTERING. 30-3318
ON AND LUMINESCENCE SPECTRA, INFRARED, OPTICAL, AND ACOUSTIC SCATTERING). /S (CYCLOTRON RESONANCE, ABSORPTI 03-0427

CONSTANT OF ORTHORHOMBIC POTASSIUM NIOBATE BY X-RAY DIFFUSE SCATTERING.   ELASTIC 08-1416
TION IN PARAELECTRIC POTASSIUM DIHYDROGEN PHOSPHATE (NEUTRON SCATTERING).   HYDROGEN DENSITY DISTRIBU 17-2490
               ICE I: LATTICE DYNAMICS AND INCOHERENT NEUTRON SCATTERING.   34-3465
RIC TRANSITION IN AMMONIUM SULFATE STUDIED BY COLD NEUTRONS (SCATTERING). /OTIONS IN SOLIDS: THE FERROELECT 19-2752
SOFT FERROELECTRIC MODES IN LEAD TITANATE (NEUTRON INELASTIC SCATTERING).   05-1077
NON MODE AND MODE COUPLING IN ANTIMONY SULFOIODIDE (BY RAMAN SCATTERING).   SOFT PHO 15-2167
LECTRIC PHASE OF KDP CRYSTALS (INFRARED ABSORPTION AND RAMAN SCATTERING). /F THE PHOSPHATE ION IN THE PARAE 17-2479
E MECHANISM OF THE PHASE TRANSITIONS IN THIOUREA. (SATELLITE SCATTERING)   STUDIES ON TH 25-2922
  PHASE TRANSITION IN TGS. (STATIC DIELECTRIC CONSTANT, X-RAY SCATTERING)   STUDIES ON THE MECHANISM OF THE 27-3093
        STUDY OF THE PHASE TRANSITION OF TGS BY SECOND HARMONIC SCATTERING.   27-2992
ITION IN TIN TELLURIDE- GERMANIUM TELLURIDE ALLOYS. (NEUTRON SCATTERING)   TRANS 36-3586
NITE AT ROOM TEMPERATURE.   NEUTRON SCATTERING ANALYSIS OF CESIUM TRIHYDROGEN SELE 22-2869
  CORRELATIONS IN POTASSIUM TANTALATE.   NEUTRON SCATTERING ANALYSIS OF THE LINEAR DISPLACEMENT 36-3546
AND DISPLAY DEVICES. (BASED ON ELECTRICALLY CONTROLLED LIGHT SCATTERING AND BIREFRINGENCE, REVIEW) /SSING) 37-3650
ODIUM NITRITE CRYSTAL.   TWO-PHONON SCATTERING AND FERROELECTRIC TRANSITION IN A S 16-2277
M MANGANESE TRIFLUORIDE CRYSTAL.   RAMAN SCATTERING AND PHASE TRANSITIONS IN A POTASSIU 33-3400
MANGANESE TRIFLUORIDE CRYSTALS.   RAMAN SCATTERING AND PHASE TRANSITIONS IN POTASSIUM 33-3372
BATE.   ANISOTROPIC DIFFUSE X-RAY SCATTERING AND PHASE TRANSITIONS IN SODIUM NIO 08-1357
TRONTIUM TITANATE.   RAMAN SCATTERING AND PHASE TRANSITIONS IN STRESSED S 06-1103
EA.   SATELLITE X-RAY SCATTERING AND STRUCTURAL MODULATION OF THIOUR 25-2921
UM MANGANESE TRIFLUORIDE AT 184 DEGREES K.   X-RAY SCATTERING AND THE PHASE TRANSITION OF POTASSI 33-3394
Y IN LITHIUM HYDRAZINIUM SULFATE.   ANOMALOUS NEUTRON SCATTERING AND THE QUESTION OF FERROELECTRICIT 32-3341
BATE, BA/ X-RAY DETERMINATION OF POLARITY SENSE BY ANOMALOUS SCATTERING AT AN ABSORPTION EDGE. (LITHIUM NIO 11-1673
TRONTIUM TITANATE, POTASSIUM TANTALATE, B/ NEUTRON INELASTIC SCATTERING AT STRUCTURAL PHASE TRANSITIONS. (S 02-0059
(THEORY, APPLIED TO BARIUM STRONTIUM TITANATE AND P/ NEUTRON SCATTERING BY A FERROELECTRIC SOLID SOLUTION. 03-0260
YDROGEN PHOSPHATE.   BRILLOUIN SCATTERING BY A SINGLE CRYSTAL OF RUBIDIUM DIH 17-2519
C AND PIEZOELECTRIC CRYSTALS.   LIGHT SCATTERING BY ACOUSTIC PHONONS IN FERROELECTRI 03-0471
                                                        RAMAN SCATTERING BY BARIUM STRONTIUM NIOBATE.   13-1899
EMPERATURE.   CRITICAL X-RAY INCOHERENT SCATTERING BY BARIUM TITANATE NEAR THE CURIE T 04-0535
INE SULFATE.   SECOND HARMONIC LIGHT SCATTERING BY DOMAINS IN FERROELECTRIC TRIGLYC 27-2991
YLLATE CRYSTALS.   LOW FREQUENCY RAMAN SCATTERING BY FERROELECTRIC TRIGLYCINE FLUOBER 27-3113
T LOW TEMPERATURES. PART-1: GENE/ THEORY OF COHERENT NEUTRON SCATTERING BY HYDROGEN BONDED FERROELECTRICS A 03-0478
  MANGANATE AND P/ IMPURITY CONCENTRATION DEPENDENCE OF RAMAN SCATTERING BY MAGNONS IN NICKEL DOPED RUBIDIUM 33-3395
ARY PHOTONS IN LITHIUM NIOBATE.   LIGHT SCATTERING BY POLARITONS ASSOCIATED WITH ORDIN 11-1697
              PARAMETRIC LUMINESCENCE AND LIGHT SCATTERING BY POLARITONS. (LITHIUM NIOBATE)   11-1746
CADMIUM SULFIDE AND LITHIUM NIOBATE CRYST/ SPONTANEOUS RAMAN SCATTERING BY POLARITONS. PART-2: THE CASE OF 11-1787
ITH APPLICATIONS TO FERROELEC/ AN ANALYSIS OF MULTIPLE LIGHT SCATTERING BY RANDOM OPTICAL INHOMOGENEITIES W 37-3634
LECTRIC MODE IN POTASSIUM DIDEUTERIUM PHOSPHATE (COMP/ RAMAN SCATTERING BY THE TEMPERATURE DEPENDENT FERROE 17-2705
ATE CRYSTALS. (PHOTOELASTIC CONSTANT)   RAMAN SCATTERING BY ULTRASONIC WAVES IN LITHIUM NIOB 11-1754
LECTRICS.   INELASTIC SCATTERING CROSS SECTION OF NEUTRONS IN FERROE 03-0297
  STRUCTURE OF THIOUREA CRYSTAL. (CALCULATIONS FROM SATELLITE SCATTERING DATA)   LONG PERIOD 25-2911
  FERRO- AND ANTIFERROELECTRIC PHASE TRANSITIONS AND RELATED SCATTERING EXPERIMENTS. (REVIEW)   02-0132
SIUM DIDEUTERIUM PHOSPHATE.   LIGHT SCATTERING FROM CRITICAL FLUCTUATIONS IN POTAS 17-2463
STIC SCATTERING)   FERROELECTRIC CRITICAL SCATTERING FROM DEUTERATED KDP. (NEUTRON INELA 17-2610
SSIUM DIHYDROGEN PHOSPHATE.   RAMAN SCATTERING FROM FERROELECTRIC CRYSTALS OF POTA 17-2552
  ISOMORPHOUS PHOSPHATES AND ARSENATES. (KDP, AMMONIUM/ RAMAN SCATTERING FROM FERROELECTRIC MODES IN THE KDP 17-2639
                                                        RAMAN SCATTERING FROM FERROELECTRIC SODIUM NITRITE. 08-1344
                                             INCOHERENT NEUTRON SCATTERING FROM HYDROGEN BOND IN KDP AND ADP. 17-2416
D FROM FIRST PRINCIPLES, APPLIED TO DISPL/ THEORY OF NEUTRON SCATTERING FROM LATTICE VIBRATIONS (REDEVELOP 03-0484
ARIUM TITANATE.   AUGER-LIKE RESONANT INTERFERENCE IN RAMAN SCATTERING FROM ONE AND TWO PHONON STATES OF B 04-0923
E AND SODIUM NIOBATE SIN/ ANISOTROPIC CRITICAL X-RAY DIFFUSE SCATTERING FROM POTASSIUM MANGANESE TRIFLUORID 08-1312
(LATTICE DYNAMICS)   INELASTIC NEUTRON SCATTERING FROM SINGLE DOMAIN BARIUM TITANATE. 04-0946
F TEMPERATURE.   X-RAY DIFFUSE SCATTERING FROM SODIUM NIOBATE AS A FUNCTION O 08-1313
THE 640 DEG C PHASE TRAN/ ANISOTROPIC CRITICAL X-RAY DIFFUSE SCATTERING FROM SODIUM NIOBATE CRYSTALS (NEAR 08-1311
                                     CRITICAL NEUTRON SCATTERING FROM STRONTIUM TITANATE.   06-1195
(5) GERMANIUM(3) OXIDE(11).   RAMAN SCATTERING FROM THE FERROELECTRIC MODE IN LEAD 35-3498
N THE NONLINEAR SUSCEPTIBILITY IN LI/ SPONTANEOUS PARAMETRIC SCATTERING, IDLER ABSORPTION, AND DISPERSION I 11-1771
  CRYSTAL.   RAMAN SCATTERING IN A FERROELECTRIC AMMONIUM SULFATE 19-2786
        LINEAR DISORDER AND TEMPERATURE DEPENDENCE OF RAMAN SCATTERING IN BARIUM TITANATE.   04-0675
OBATE.   STRUCTURE DISORDER AND RAMAN SCATTERING IN BARIUM TITANATE AND POTASSIUM NI 04-0676
/CTRIC FIELD DEPENDENCE OF OPTICAL PHONON FREQUENCIES (RAMAN SCATTERING IN CUBIC PEROVSKITE STRONTIUM TITA/ 06-1241
                          NEUTRON STUDY OF DIFFUSE SCATTERING IN CUBIC POTASSIUM NIOBATE.   08-1410
/LTIPHONON SELECTION RULES FOR INFRARED ABSORPTION AND RAMAN SCATTERING IN FERROELECTRIC AND PARAELECTRIC / 16-2281
DE.   SOFT MODE COUPLING AND CRITICAL RAYLEIGH SCATTERING IN FERROELECTRIC ANTIMONY SULFOIODI 15-2232
MONY SULFOIODIDE.   RAMAN SCATTERING IN FERROELECTRIC SEMICONDUCTOR ANTI 15-2111
                                    X-RAY CRITICAL SCATTERING IN FERROELECTRIC TRIGLYCINE SULFATE 27-2998
                                                        RAMAN SCATTERING IN FERROELECTRICS. (REVIEW)   02-0120
                                    QUASI ELASTIC SCATTERING IN LEAD GERMANATE.   35-3497
                                           BRILLOUIN SCATTERING IN LITHIUM NIOBATE.   11-1786
  DIRECTIONAL DISPERSION OF PHONON FREQUENCIES AND POLARITON SCATTERING IN LITHIUM NIOBATE.   11-1696
AGONAL SYMMETRY.   RAMAN SCATTERING IN PIEZOELECTRIC CRYSTALS WITH TETR 03-0407
D CES/ DIELECTRIC RESPONSE IN PIEZOELECTRIC CRYSTALS. (RAMAN SCATTERING IN POTASSIUM DIHYDROGEN ARSENATE AN 18-2723
                                                        RAMAN SCATTERING IN POTASSIUM DIHYDROGEN PHOSPHATE. 17-2709
                                   RAMAN LIGHT SCATTERING IN ROCHELLE SALT.   28-3231
RE.   DIFFUSE SCATTERING IN SODIUM NIOBATE AT ROOM TEMPERATU 08-1304
RE.   BRILLOUIN SCATTERING IN SODIUM NITRITE AT ROOM TEMPERATU 16-2305
                  EFFECT OF UNIAXIAL STRESS ON RAMAN SCATTERING IN STRONTIUM TITANATE.   06-1102
MANGANESE TRIFLUORIDE.   CRITICAL NEUTRON SCATTERING IN STRONTIUM TITANATE AND POTASSIUM 06-1208
LS IN THE TEMPERATURE RANGE 5 DEG K-300 DEG K.   BRILLOUIN SCATTERING IN STRONTIUM TITANATE SINGLE CRYSTA 06-1149
RIUM TITANATE.   RAMAN SCATTERING IN THE FERROELECTRIC SYSTEM LEAD BA 04-0611
        EFFECTS OF DOMAIN SHAPES ON SECOND HARMONIC SCATTERING IN TRIGLYCINE SULFATE.   27-2989
TRATE. (ORDER DISORDER TRANSITION, FERROELEC/ X-RAY CRITICAL SCATTERING IN TRIGLYCINE SULFATE AND SODIUM NI 16-2400
ERROELECTRIC PHASE POTASSIUM / OSCILLATORY BEHAVIOR OF RAMAN SCATTERING INTENSITY WITH TEMPERATURE IN THE F 17-2662
MODE IN BARIUM TITANATE.   RAMAN SCATTERING LINE SHAPE OF THE SOFT E POLARITON 04-0803
MAGE STORAGE AND DISPLAY DEVICES. (PLZT CERAMICS)   SCATTERING MODE FERROELECTRIC PHOTOCONDUCTOR I 09-1599
                THEORY OF RAYLEIGH AND BRILLOUIN SCATTERING NEAR THE PHASE TRANSITION.   03-0334
ERATURE PHASE TRANSITION IN RUBIDIUM DIHYDROGEN PHOSPHATE BY SCATTERING OF COLD NEUTRONS.   HIGH TEMP 17-2489
SE AND LONGITUDINAL OPTICAL PHONONS IN IONIC FERROELECTRIC / SCATTERING OF CONDUCTION ELECTRONS BY TRANSVER 03-0292
                EFFECT OF CHARGE TRANSFER ON RAMAN SCATTERING OF HYDROGEN HALIDES.   31-3328
/CERTAIN PARAMETERS OF LITHIUM NIOBATE SINGLE CRYSTALS USING SCATTERING OF LASER RADIATION BY HYPERSONIC V/ 11-1797

OF ADP AND KDP CRYSTALS IN THE INFRARED RANGE BY PARAMETRIC SCATTERING OF LIGHT. / OF THE REFRACTIVE INDEX 17-2553
LARITONS IN LITHIUM TANTALATE. SCATTERING OF LIGHT BY ACOUSTIC PHONONS AND PO 12-1878
SIUM MANGANESE TRIFLUORIDE. TWO MAGNON SCATTERING OF LIGHT IN ANTIFERROMAGNETIC POTAS 33-3401
F REFRACTION 1.2-7/ (CONTINUOUS WAVE) SPONTANEOUS PARAMETRIC SCATTERING OF LIGHT IN LITHIUM IODATE (INDEX O 08-1291
RANSVERSE (MORE INTENSE) AND LONGITUDI/ MANDELSTAM BRILLOUIN SCATTERING OF LIGHT IN LITHIUM NIOBATE (FROM T 11-1743
ALT FLUORIDE, POTASSIUM MANGANESE TRIF/ MANDELSTAM BRILLOUIN SCATTERING OF LIGHT IN MANGANESE FLUORIDE, COB 33-3399
TALS (POTASSIUM TRIFLUORIDE, RUBIDIUM / MANDELSTAM BRILLOUIN SCATTERING OF LIGHT IN MANGANESE FLUORIDE CRYS 33-3398
HIUM NIOBATE AND LITHIUM TANTALATE. SCATTERING OF LIGHT ON ACOUSTIC PHONONS IN LIT 11-1744
E TRIFLUORIDE IN THE SHORT RANGE ORDERED REGIO/ PARAMAGNETIC SCATTERING OF NEUTRONS FROM POTASSIUM MANGANES 33-3371
AT 215 DEG C. X-RAY CRITICAL SCATTERING OF SODIUM NITRITE AT 160 DEG C AND 16-2343
, LITHIUM NIOBATE EXAMPLE) STIMULATED COMBINATION SCATTERING OF SOUND IN PIEZOELECTRICS. (THEORY 03-0431
CRITICAL SCATTERING OF X-RAYS IN SODIUM NITRITE. 16-2270
LATTICE DYNAMICS BY NEUTRON SCATTERING. PARTS 1 AND 2. (REVIEW) 02-0051
/IC RESONANCE. PART-2: THERMAL CONDUCTIVITY. PART-3: NEUTRON SCATTERING. PART-4: INFRARED AND RAMAN SCATTE/ 03-0397
/DED FERROELECTRICS. (INFRARED ABSORPTION, NEUTRON INELASTIC SCATTERING, POTASSIUM DIHYDROGEN PHOSPHATE, D/ 03-0401
RACT ONLY, DISCUSSION) GENERALIZED SUSCEPTIBILITY AND THE SCATTERING PROPERTIES OF FERROELECTRICS. (ABST 03-0209
RY) DYNAMICAL (DIELECTRIC) AND SCATTERING PROPERTIES OF FERROELECTRICS. (THEO 03-0208
E TRANSITIONS. (BARIUM TITANATE) (FAST ION SCATTERING) SHADOW EFFECT IN (DISPLACIVE) PHAS 04-0533
TRIC TRANSITION IN RUBIDIUM DIHYDROGEN PHOSPHIDE USING RAMAN SCATTERING SPECTRA. FERROELEC 17-2620
/DEPENDENCE OF THE INTENSITY OF HYDROGEN BOND BANDS IN RAMAN SCATTERING SPECTRA OF GYPSUM AND ROCHELLE SAL/ 28-3189
/ERROELECTRIC SOLIDS. PART-1: ENERGY LOSS NEUTRON INCOHERENT SCATTERING SPECTRA OF POTASSIUM DIHYDROGEN PH/ 17-2687
/FERROELECTRIC ANTIMONY SULFOIODIDE CRYSTALS USING THE RAMAN SCATTERING SPECTRUM. (LATTICE VIBRATIONS, BON/ 15-2143
D KDP DURING THE PHASE TRAN/ TEMPERATURE DEPENDENCE OF RAMAN SCATTERING SPECTRUM OF FERROELECTRIC DEUTERATE 17-2626
LYCINE SULFATE, POTASSIUM DIHYDROGEN AND DIDEUT/ LASER LIGHT SCATTERING STUDIES OF PHASE TRANSITIONS. (TRIG 17-2481
CUBIC TO TETRAGONAL PHASE TRANSITION IN STRONTIUM TIT/ LIGHT SCATTERING STUDIES OF THE SOFT MODES NEAR THE 06-1242
TIC MODES OF POTASSIUM DIHYDROGEN PHOSPHATE AND POTAS/ LIGHT SCATTERING STUDIES OF THE SOFT OPTIC AND ACOUS 17-2497
NSITION IN STRONTIUM TITANATE. NEW OBSERVATIONS FROM NEUTRON SCATTERING STUDIES OF THE STRUCTURAL PHASE TRA 06-1210
A DOUBLE PASSED, STABIL/ (ELASTIC CONSTANTS FROM) BRILLOUIN SCATTERING STUDY OF ANTIMONY SULFOIODIDE USING 15-2225
TITANATE. NEUTRON SCATTERING STUDY OF SOFT MODES IN CUBIC BARIUM 04-0712
ANSFORMATION IN TERBIUM MOLYBDATE. NEUTRON SCATTERING STUDY OF THE FERROELECTRIC PHASE TR 30-3268
ON IN KDP. LIGHT SCATTERING STUDY OF THE FERROELECTRIC TRANSITI 17-2462
RONTIUM TITANATE: HARMONIC MODELS. NEUTRON INELASTIC SCATTERING STUDY OF THE LATTICE DYNAMICS OF ST 06-1225
E TRANSITIONS IN POTASSIUM MANGANESE TRIFLUORIDE. NEUTRON SCATTERING STUDY OF THE LATTICE DYNAMICAL PHAS 33-3405
TASSIUM TANTALATE NIOBATE. NEUTRON SCATTERING STUDY OF THE SOFT MODES IN CUBIC PO 13-2064
OKHAVEN WORK ON POTASSIUM TANTALATE, BARI/ NEUTRON INELASTIC SCATTERING STUDY ON SOFT MODES. (REVIEW OF BRO 02-0115
OLID STATE PHYSICS (DOMAIN WALL THICKNESS IN GADOLINI/ RAMAN SCATTERING TECHNIQUES APPLIED TO PROBLEMS IN S 30-3319
E. GROUP THEORETICAL SELECTION RULES FOR INELASTIC NEUTRON SCATTERING WITH APPLICATIONS TO LITHIUM NIOBAT 11-1704
RANSITIONS IN FERROELECTRICS: A SELF CONSISTENT POLARIZATION SCHEME FOR ANHARMONIC POTENTIALS. (THEORY) / T 03-0355
OVSKITES DUE TO LATTICE POLARIZATION. (LCAO ANALYSIS OF BAND SCHEME, STRONTIUM TITANATE) /ND CHANGES IN PER 06-1101
/ PHASE TRANSITIONS IN SODIUM NIOBATE AND THE USE OF TILTING SCHEMES IN THE SOLUTIONS OF PEROVSKITE STRUCT/ 08-1335
ELECTRON TUNNELING INTO POTASSIUM TANTALATE SCHOTTKY BARRIER JUNCTIONS. 08-1365
STRONTIUM TITANATE AND (CALCIUM DOPED) POTASSIUM TANTALATE (SCHOTTKY BARRIERS WITH INDIUM). /IOBIUM DOPED) 06-1222
FERROELECTRIC. DEBYE SCREENING OF A NONUNIFORM ELECTRIC FIELD IN A 03-0499
ECTRIC ANTIMONY SULFOIO/ INFLUENCE OF NONEQUILIBRIUM CARRIER SCREENING ON POLARIZATION PROCESSES IN FERROEL 15-2158
IAXIAL FERROELECTRIC PHOTOCONDUCTOR). INFLUENCE OF SCREENING ON THE FERROELECTRIC CURIE POINT (UN 03-0392
NDENCE OF THE PIEZOEFFECT IN BARIUM TITANATE ON THE MEANS OF SCREENING THE SPONTANEOUS INDUCTION. DEPE 04-1021
TRIHYDROGEN SELENITE (SHS) OR SODIUM TRIDEUTERIUM SELENITE (SDS). /RA OF ISOTOPIC NONISOMORPHISM OF SODIUM 22-2836
N BRONZE STRUCTURE. THE SEARCH FOR NEW FERROELECTRICS WITH THE TUNGSTE 13-1897
DE SINGLE CRYSTALS. GENERATION OF THE SECOND ACOUSTIC HARMONIC IN ANTIMONY SULFOIODI 15-2224
DE SINGLE CRYSTALS. GENERATION OF THE SECOND ACOUSTIC HARMONIC IN ANTIMONY SULFOIODI 15-2223
USTIC WAVE PROPAGATING ITSELF IN THE (001) DI/ GENERATION OF SECOND AND THIRD HARMONICS BY A TRANSVERSE ACO 11-1682
GENERATION OF THE SECOND HARMONIC AND FERROELECTRICITY. (THEORY) 02-0063
S) GENERATION OF THE SECOND HARMONIC BY TRIGLYCINE SULFATE. (84 REF 27-2990
NIOBATE CRYSTAL, AGR/ MEASUREMENTS OF PHOTON CORRELATIONS OF SECOND HARMONIC GENERATED LIGHT (FROM LITHIUM 11-1701
LONG RANGE ORDER IN SODIUM NITRITE BY OPTICAL SECOND HARMONIC GENERATION. 16-2314
THE TUNGSTEN BRONZE SYSTEM) POTASSIUM SODIUM BARIUM NIOBATE (SECOND HARMONIC GENERATION). / PROPERTIES OF (13-2028
RIC OSCILLATION IN FER/ NONLINEAR OPTICAL DEVICES. (BASED IN SECOND HARMONIC GENERATION AND OPTICAL PARAMET 02-0097
IN THE FREE AND CLAMPED VALUES OF THE DIELE/ MEASUREMENTS OF SECOND HARMONIC GENERATION AND THE VARIATIONS 13-1998
HIUM NIOBATE ON MELT COMPOSITION. DEPENDENCE OF SECOND HARMONIC GENERATION COEFFICIENTS OF LIT 11-1769
ZOELECTRIC CRYSTALS. ABSOLUTE SIGNS OF SECOND HARMONIC GENERATION COEFFICIENTS OF PIE 02-0096
ZOELECTRIC CRYSTA/ FURTHER MEASUREMENTS OF ABSOLUTE SIGNS OF SECOND HARMONIC GENERATION COEFFICIENTS OF PIE 05-1066
LAR (PYROELECTRIC) CRY/ ABSOLUTE SIGNS OF NONLINEAR OPTICAL (SECOND HARMONIC GENERATION) COEFFICIENTS OF PO 02-0095
C POLARIZATION REVERSAL. (SODIUM NITRITE) OPTICAL SECOND HARMONIC GENERATION DURING FERROELECTRI 16-2384
FLUORESCEIN DYE LASERS. EFFICIENT SECOND HARMONIC GENERATION IN ADP WITH TWO NEW 17-2538
DIDE. OPTICAL SECOND HARMONIC GENERATION IN ANTIMONY SULFOIO 15-2164
DIUM NITRITE. SECOND HARMONIC GENERATION IN FERROELECTRIC SO 16-2385
TEMPERATURE DEPENDENCE OF RAMAN SPECTRUM AND SECOND HARMONIC GENERATION IN KDP. 17-2550
LARGE NONLINEAR OPTICAL COEFFICIENT AND PHASE MATCHED SECOND HARMONIC GENERATION IN LITHIUM IODATE. 08-1405
OPTIMUM SECOND HARMONIC GENERATION IN LITHIUM NIOBATE. 11-1764
). SIMULTANEOUS OPTICAL PARAMETRIC OSCILLATION, SECOND HARMONIC GENERATION (IN LITHIUM NIOBATE 11-1859
GADOLINIUM MOLYBDATE AND TERBIUM MOLYBDATE. SECOND HARMONIC GENERATION IN MOLYBDATES (BETA 30-3286
YSTALS. (KDP TYPE) OPTICAL SECOND HARMONIC GENERATION IN PIEZOELECTRIC CR 17-2582
LUORIDE FILMS (AFTER COOLING ST/ PYROELECTRICITY AND OPTICAL SECOND HARMONIC GENERATION IN POLYVINYLIDENE F 36-3530
OGEN / TEMPERATURE DEPENDENCE OF INDEX MATCHING ANGLE OF THE SECOND HARMONIC GENERATION IN POTASSIUM DIHYDR 17-2544
ROGEN PHOSPHATE (AND DEU/ TEMPERATURE DEPENDENCE OF OPTICAL (SECOND) HARMONIC GENERATION IN POTASSIUM DIHYD 17-2695
TEMPERATURE DEPENDENCE OF SECOND HARMONIC GENERATION IN SODIUM NITRITE. 16-2315
ERROELECTRIC STRONTIUM TITANATE. SECOND HARMONIC GENERATION IN STRESS INDUCED F 06-1131
RYSTAL IN THE PARAELECTRIC PHASE. SECOND HARMONIC GENERATION IN THE TRIGLYCINE C 27-3105
TE CRYSTALS. SECOND HARMONIC GENERATION IN TRIGLYCINE SULFA 27-3138
E, IRON CHLORINE IRON IODINE AND NICKEL CHLORINE) BORACITES (SECOND HARMONIC GENERATION INDEX MATCHING). /N 14-2080
/MOGENEITY OF THE REFRACTIVE INDEX OF A NONLINEAR CRYSTAL ON SECOND HARMONIC GENERATION. (LITHIUM NIOBATE) 11-1687
CRYSTAL BY THE INDEX MATCHING METHOD. SECOND HARMONIC GENERATION OF LIGHT IN THE KDP 17-2693
NEITY (THEORY AND TEST ON BARIUM SODIUM NIOBA/ DEPENDENCE OF SECOND HARMONIC GENERATION ON CRYSTAL INHOMOGE 13-2043
O/ NONLINEAR OPTICAL PROPERTIES OF FERROELECTRIC CRYSTALS. (SECOND HARMONIC GENERATION, OPTICAL PARAMETRIC 03-0383
GEN PHOSPHATE) PULSE DISTORTIONS IN MISMATCHED SECOND HARMONIC GENERATION. (POTASSIUM DIHYDRO 17-2690
DENCE) STUDY OF PHASE TRANSITION OF SODIUM NITRITE BY SECOND HARMONIC GENERATION. (TEMPERATURE DEPEN 16-2403
KDP. INTRACAVITY SECOND HARMONIC GENERATION WITH RUBY LASER BY 17-2604
FERROELECTRIC TRIGLYCINE SULFATE. SECOND HARMONIC LIGHT SCATTERING BY DOMAINS IN 27-2991
THEORETICAL STUDY OF THE GENERATION OF THE ACOUSTIC SECOND HARMONIC. (LITHIUM NIOBATE) 11-1772
STUDY OF THE PHASE TRANSITION OF TGS BY SECOND HARMONIC SCATTERING. 27-2992

TE.   EFFECTS CF DOMAIN SHAPES ON SECOND HARMONIC SCATTERING IN TRIGLYCINE SULFA 27-2989
OR ELECTROMETER. (WITH A TGS SINGLE CRYSTAL)   A NEW SECOND HARMONIC TYPE FERROELECTRIC MODULATOR F 27-2951
HATE.   ELECTROOPTIC AMPLITUDE MODULATION OF LASER GENERATED SECOND HARMONICS IN POTASSIUM DIHYDROGEN PHOSP 17-2694
TY. A EUROPEAN VIEW. AN AMERICAN VIEW.   REPORT ON THE SECOND INTERNATIONAL MEETING ON FERROELECTRICI 01-0008
/ARS OF FERROELECTRICITY: A REVIEW OF THE PROCEEDINGS OF THE SECOND INTERNATIONAL MEETING ON FERROELECTRIC/ 01-0010
TRICS.   NATURE OF THE TEMPERATURE DEPENDENCE OF NMR SPECTRA SECOND MOMENT IN ORDER DISORDER TYPE FERROELEC 03-0460
TE TYPE CRYSTALS.   NMR PROTON SECOND MOMENTS IN POTASSIUM DIHYDROGEN PHOSPHA 17-2408
BARIUM TITANATE.   GENERATION OF SECOND OPTICAL HARMONIC IN THE CUBIC PHASE OF 04-0538
TE AND TRIGLYCINE FLUOBERYLLATE) AT PHASE TRANSITIONS OF THE SECOND ORDER. /SINGLE CRYSTAL TRIGLYCINE SULFA 27-3137
R OF PHASE TRANSITION CF FERROELECTRIC POTASSIUM TANTALATE. (SECOND ORDER)   CHARACTE 08-1307
THERMODYNAMIC CLASSIFICATION OF SECOND ORDER FERROELECTRIC PHASE TRANSITIONS. 03-0303
CALORIMETRIC INVESTIGATIONS OF SECOND ORDER FERROELECTRIC TRANSITIONS. 03-0435
ONSHIRE'S THEORY OF FERROELECTRICS.   A NEW TYPE OF SECOND ORDER PHASE TRANSITION DERIVED FROM DEV 03-0226
GNETIC TC THE FERROMAGNETIC AND FERROELECTRIC STA/ THEORY OF SECOND ORDER PHASE TRANSITION FROM THE FERROMA 03-0395
ENSIONAL MODEL OF A FERROELECTRIC.   SECOND ORDER PHASE TRANSITION IN THE THREE DIM 03-0450
ON RELAXATION PROCESSES NEAR THE SECOND ORDER PHASE TRANSITION POINT. 03-0385
SINGLE DOMAIN POLARIZATION OF A FERROELECTRIC HAVING A SECOND ORDER PHASE TRANSITION (THEORY). 03-0202
THEORY OF SECOND ORDER PHASE TRANSITIONS. 03-0364
URES. (TIN TELLURIDE, POTASSIUM TANTALATE)   THEORY OF SECOND ORDER PHASE TRANSITIONS AT LOW TEMPERAT 03-0433
CS: A SELF CONSISTENT POLARIZATION SCHEME FOR ANH/ FIRST AND SECOND ORDER PHASE TRANSITIONS IN FERROELECTRI 03-0355
3 MAGNETIC RESONANCE IN ROCHELLE SALT.   SECOND ORDER QUADRUPOLE SHIFTS OF THE SODIUM-2 28-3186
/C GENERATION OF HARMONICS IN A FERROELECTRIC CRYSTAL WITH A SECOND ORDER TRANSITION ABOVE THE CURIE POINT. 03-0331
TTENUATION AND VELOCITY OF ULTRASOUND IN FERROELECTRICS NEAR SECOND ORDER TRANSITION POINT. /LE FORCES ON A 03-0253
/OGICAL THEORY OF ANOMALIES OF THERMODYNAMIC PARAMETERS NEAR SECOND ORDER TRANSITION POINTS IN FERROELECTR/ 03-0362
CRON RADI/ OBSERVATION OF THE SIMULTANEOUS GENERATION OF THE SECOND, THIRD, AND FOURTH HARMONICS OF 1.06 MI 11-1779
(REVERSIBLE) OPTICAL DAMAGE IN KCP (WITH RELAXATION TIME OF SECONDS WHEN ILLUMINATED UNDER APPLIED ELECTR/ 17-2703
THEORY)   THERMODYNAMIC POTENTIALS AND SECONDARY EFFECTS IN FERROELECTRIC CRYSTALS. (03-0387
HIUM NIOBATE (ON ELECTRON IRRADIATION).   SECONDARY (ELECTRO) PHOTOELASTIC EFFECT IN LIT 11-1801
POLARIZATION IN LITHIUM TANTALATE (THERMAL EXP/ PRIMARY AND SECONDARY PYROELECTRIC EFFECTS AND SPONTANEOUS 15-1884
MICONDUCTOR ANTIMONY SU/ TEMPERATURE DEPENDENCE OF THE CROSS SECTION OF CAPTURE CENTERS IN FERROELECTRIC SE 15-2147
INELASTIC SCATTERING CROSS SECTION OF NEUTRONS IN FERROELECTRICS. 03-0297
PTICAL PROPERTIES OF (BARIUM, STRONTIUM) MIXED TITANATE THIN SECTIONS.   ELECTRICAL AND O 04-0603
NATE.   TEMPERATURE DEPENDENCE OF THE RAMAN CROSS SECTIONS IN BARIUM TITANATE AND STRONTIUM TITA 04-0548
LONGITUDINAL ELECTROOPTICAL EFFECT IN OBLIQUE SECTIONS OF LITHIUM NIOBATE. 11-1863
ARIUM TITANATE CRYSTALS (POLARON H/ ELECTRICAL CONDUCTIVITY (SEEBECK COEFFICIENT, CURIE POINT) OF REDUCED B 04-0918
SEMICONDUCTING PROPERTIES OF BARIUM TITANATE. (RESISTIVITY, SEEBECK COEFFICIENT, OPTICAL ABSORPTION)   SOME 04-0919
.2-300 DEG / ELECTRONIC TRANSPORT (CONDUCTIVITY AND HALL AND SEEBECK COEFFICIENTS) IN STRONTIUM TITANATE (4 06-1129
NTACT SYSTEMS.   SEEBECK EFFECT IN METAL LEAD TITANATE METAL CO 05-1056
PRODUCTION OF LARGE SEED CRYSTALS OF KDP. 17-2413
DIFIED CIRCULATION APPARATUS.   GROWTH BEHAVIOR OF SMALL SEEDS OF AMMONIUM DIHYDROGEN PHOSPHATE IN A MO 17-2704
TERIZATION OF POTASSIUM NIOBATE.   (TOP SEEDED FLUX) CRYSTAL GROWTH AND NEUTRON CHARAC 08-1353
(GADOLINIUM MOLYBDATE, GROWTH BY RF CZOCHRALSKI AND FROM TOP SEEDED SOLUTION) /NATE AND RELATED COMPOUNDS. 08-1323
NONSTOICHIOMETRIC MELTS. (BARIUM TITANATE, STRONTIUM TI/ TOP SEEDED SOLUTION GROWTH OF OXIDE CRYSTALS FROM 04-0553
RIC TUBES.   VIBRATIONS OF COAXIALLY SEGMENTED, LONGITUDINALLY POLARIZED FERROELECT 03-0376
KHAUSEN NOISE AND SURFACE MICRODAMAGES ON SINGLE CRYSTALS OF SEIGNETTE SALT.   CORRELATION BETWEEN BAR 28-3208
FAST AFTEREFFECT OF SEIGNETTE SALT. 28-3206
SEE BOTH ROCHELLE SALT AND SEIGNETTE SALT.
) DOPING)   EFFECT OF A DC FIELD ON DOMAIN STRUCTURE IN SEIGNETTE SALT. (EFFECT OF AGING AND COPPER(2+ 28-3230
A/ EFFECT OF ELECTRIC FIELD ON DOMAIN STRUCTURE FORMATION IN SEIGNETTES SALT (ROCHELLE SALT). (TEMPERATURE 28-3190
PEROVSKITE TYPE SEIGNETTO MAGNETS. 08-1459
LORE TYPE. SYNTHESIS AND STUDY OF TH/ NEW FERROELECTRICS AND SEIGNETTO MAGNETS OF THE PEROVSKITE AND PYROCH 02-0128
FERRATE(0.67) TUNGSTATE(0.33)- LEAD TITANATE SYSTEM. SEIGNETTO MAGNETIC SOLID SOLUTIONS OF THE LEAD 05-1065
ELECTROOPTIC CERAMICS AS WAVELENGTH SELECTION DEVICES IN DYE LASERS (PLZT). 09-1618
GUIDE LINES FOR THE SELECTION OF ACOUSTOOPTIC MATERIALS. 03-0425
NG WITH APPLICATIONS TO LITHIUM NIOBATE.   GROUP THEORETICAL SELECTION RULES FOR INELASTIC NEUTRON SCATTERI 11-1704
MAN SCATTERING IN FERROELECTRIC AND PARAELECTRI/ MULTIPHONON SELECTION RULES FOR INFRARED ABSORPTION AND RA 16-2281
DOMAIN STRUCTURE OF ANTIMONY SULFOIODIDE CRYSTALS (FROM SELECTIVE CRYSTALLIZATION AND ETCHING). 15-2183
BLIMED CN/ REVEALING THE DOMAIN STRUCTURE OF TGS CRYSTALS BY SELECTIVE CRYSTALLIZATION OF ANTHRAQUINONE (SU 27-3001
/RTIES CF THE SURFACES OF IONIC CRYSTALS (TGS AND OTHERS, BY SELECTIVE DEPOSITION OF CHARGED POLYSTYRENE L/ 27-2987
LOCATION STRUCTURES IN TRIGLYCINE SULFATE.   APPLICATION OF SELECTIVE ETCHING TO THE STUDY OF TWIN AND DIS 27-3042
CRYSTAL STRUCTURE OF POTASSIUM SELENATE. 22-2845
ELECTRON SPIN RESONANCE OF GAMMA-IRRADIATED TRIGLYCINE SELENATE. 27-3139
DELS FOR THE ORDER DISORDER TRANSITION IN SODIUM TRIHYDROGEN SELENATE.   MO 22-2862
NEW HIGH PRESSURE PHASE TRANSITION IN TRIGLYCINE SELENATE. 27-3090
ND DEUTERIUM AND SODIUM-23 RELAXATION IN SODIUM TRIDEUTERIUM SELENATE.   NMR STUDY OF DEUTERIUM DIFFUSION A 22-2849
OELECTRIC TRANSITIONS IN SODIUM TRIHYDROGEN AND TRIDEUTERIUM SELENATE. /AGNETIC RESONANCE STUDY OF THE FERR 22-2832
PIN LATTICE RELAXATION TIME MEASUREMENTS IN SODIUM TRIHYDROGEN SELENATE.   NUCLEAR S 22-2847
ANT) TEMPERATURE CHARACTERISTICS CF FERROELECTRIC TRIGLYCINE SELENATE. /L) ENERGY GAP (AND DIELECTRIC CONST 27-3021
PROPERTIES OF MIXED CRYSTALS OF TRIGLYCINE SULFATE AND SELENATE. 27-3022
TON SPIN LATTICE RELAXATION IN IRRADIATED SODIUM TRIHYDROGEN SELENATE.   PRO 22-2853
SPIN LATTICE RELAXATION IN FERROELECTRIC SODIUM TRIHYDROGEN SELENATE.   PROTON 22-2852
RADIATION EFFECTS IN SODIUM TRIHYDROGEN SELENATE. 22-2865
RAMAN SPECTRUM OF TRIGLYCINE SELENATE. 27-2957
RESIDUAL ENTROPY OF SODIUM TRIHYDROGEN SELENATE. 22-2835
THE ELECTRON PARAMAGNETIC RESONANCE OF POTASSIUM SELENATE. 22-2850
OR IN THE BETA PHASE OF THE FERROELECTRIC SODIUM TRIHYDROGEN SELENATE. /IN STRUCTURE AND ITS UNUSUAL BEHAVI 22-2874
THE STRUCTURE AND PHASE TRANSITION IN POTASSIUM SELENATE. 22-2872
/CT OF A DC FIELD ON THE DIELECTRIC PROPERTIES OF TRIGLYCINE SELENATE. (AGING AND PHASE CHANGES NEAR THE C/ 27-3131
PROTON MAGNETIC RESONANCE STUDY OF SODIUM TRIHYDROGEN SELENATE AND POTASSIUM TRIHYDROGEN SELENATE. 22-2876
INVESTIGATION OF FERROELECTRICITY IN SODIUM AMMONIUM SELENATE AND SODIUM AMMONIUM SULFATE. 19-2742
RTIES OF HYDROGEN BONDED FERROELECTRICS. (SODIUM TRIHYDROGEN SELENATE AND SODIUM TRIDEUTERIUM SELENATE) /PE 22-2855
POINT).   DOMAIN STRUCTURE OF TRIGLYCINE SELENATE CRYSTAL (DURING COOLING THROUGH CURIE 27-3043
ANOMALOUS THERMAL EXPANSION OF POTASSIUM TRIHYDROGEN SELENATE CRYSTALS. 22-2880
ONISOMORPHISM OF RUBIDIUM TRIHYDROGEN(1-X) TRIDEUTERIUM(X) DISELENATE CRYSTALS.   ISOTOPIC N 22-2873
S POLARIZATION NEAR THE PHASE TRANSITION POINT OF TRIGLYCINE SELENATE CRYSTALS. / OF DETERMINING SPONTANEOU 27-2994
/E OF GROWTH CONDITIONS CF TRIGLYCINE SULFATE AND TRIGLYCINE SELENATE CRYSTALS ON CHANGES IN SPECIFIC RESI/ 27-3069
FILMS CN A SUBSTRATE OF FERROELECTRIC TRIGLYCINE SULFATE AND SELENATE CRYSTALS. (POLARIZATION CHARGE) /HIN 27-3046
/ERMEABILITY AND DIELECTRIC LOSS ANGLE TANGENT OF TRIGLYCINE SELENATE DEPENDING ON THE CONDITIONS OF GROWI/ 27-3125
CE OF DEUTERIUM IN THE FERROELECTRIC SODIUM DEUTERO AMMONIUM SELENATE DIDEUTERATE. NUCLEAR MAGNETIC RESONAN 22-2828
ING FIELD VOLTAGE, TEMPE/ NONLINEAR PROPERTIES OF TRIGLYCINE SELENATE. (DIELECTRIC CONSTANT VERSUS ALTERNAT 27-3158

ROMECHANICAL PROPERTIES CF THE FERROELECTRIC SODIUM AMMONIUM SELENATE DIHYDRATE.   ELECT 22-2884
SONANCE, 84 DEG K AND 120-300 DEG K, EXAMPLE SODIUM AMMONIUM SELENATE DIHYDRATE). /UNDS (PROTON MAGNETIC RE 22-2826
R ELECTROOPTICAL EFFECT IN THE FERROELECTRIC SODIUM AMMONIUM SELENATE DIHYDRATE.   LINEA 22-2829
STALS WITH REORIENTATING STRUCTURAL GROUPS. (SODIUM AMMONIUM SELENATE DIHYDRATE) /C PHASE TRANSITION IN CRY 22-2827
CLEAR MAGNETIC RESONANCE OF SODIUM-23 IN THE SODIUM AMMONIUM SELENATE DIHYDRATE CRYSTAL.   NU 22-2883
             PREPARATION AND ANALYSIS CF SODIUM AMMONIUM SELENATE DIHYDRATE CRYSTALS.   22-2863
/F FERROELECTRIC ALUMS: METHYL AMMONIUM ALUMINUM (SULFATE OR SELENATE) DODECAHYDRATE. (AND THE DEUTERATED / 20-2803
/LITY IN FERROELECTRICS. (ROCHELLE SALT, TRIGLYCINE SULFATE, SELENATE, FLUOBERYLLATE AND BARIUM TITANATE, / 02-0083
/ATE SINGLE CRYSTALS WITH ISOMORPHOUS COMPOUNDS. (TRIGLYCINE SELENATE, FLUOBERYLLATE, AND THE DEUTERATED S/ 27-2972
      DOMAIN STRUCTURE OF THE BETA·PHASE CF SODIUM TRIHYDROGEN SELENATE IN ELECTRIC FIELDS.   22-2830
AND IN THE RADIO FREQUENCY / OPTICAL CONSTANTS OF TRIGLYCINE SELENATE IN THE NEAR INFRARED TO FAR INFRARED 27-3007
TION BETWEEN ELECTRICAL AND THERMAL PROPERTIES OF TRIGLYCINE SELENATE NEAR THE CURIE POINT.   RELA 27-3168
           DIELECTRIC PERMEABILITY OF TRIGLYCINE SELENATE NEAR THE PHASE TRANSITION.   27-3166
/ROELECTRICS WITH GENERAL FORMULA METAL-1 METAL-2 SULFATE OR SELENATE OR FLUOBERYLLATE (METALS CHOSEN FROM/ 19-2743
/OF FERROELECTRIC CRYSTALS (SOLUTICN GROWTH, TGS, TRIGLYCINE SELENATE OR FLUOBERYLLATE, POTASSIUM OR RUBID/ 02-0089
           ESR STUCY CF GAMMA-IRRADIATED POTASSIUM SELENATE (STRUCTURAL PHASE CHANGES).   22-2825
.   DIELECTRIC AND THERMAL STUDY OF POTASSIUM SELENATE TRANSITIONS (FERROELECTRIC BELOW 93K) 22-2824
SULFATE AND ITS VARIATION UNDER X-I/ BAND GAP OF TRIGLYCINE SELENATE, TRIGLYCINE FLUOBERYLLATE, TRIGLYCINE 27-3124
ION IN POTASSIUM DIHYDROGEN PHOSPHATE AND SODIUM TRIHYDROGEN SELENATE TYPE CRYSTALS. ULTRASLOW HYDROGEN MOT 17-2441
MMETRY STATES OF AMMONIUM IONS IN SOLIDS. (AMMONIUM SULFATE, SELENATE, VANADATE AND 15 OTHERS) /THE SPIN SY 19-2800
, NONDESTRUCTIVE READOUT, BASED ON BISMUTH TITANATE AND ZINC SELENIDE). /IC PHOTOCONDUCTOR DEVICE (ERASABLE 37-3646
LECTRIC AMPLIFICATION OF SURFACE WAVE STRUCTURE OF A CADMIUM SELENIDE FILM ON LITHIUM NIOBATE.   ACOUSTOE 11-1723
TITAN/ SATURATION OF VOLT- AMPERE CHARACTERISTICS OF CADMIUM SELENIDE LAYERS ON BARIUM TITANATE- STRONTIUM 04-0592
/LS FROM THE VAPOR PHASE BY A DYNAMIC METHOD (ANTIMONY SESQUISELENIDE OR SULFIDE, ANTIMONY SULFO OR SELENO/ 15-2162
/(2)(V)E(3)(VI) AND A(V)B(VI)C(VII). (ANTIMONY TRISULFIDE, TRISELENIDE, SULFOIODIDE, SELENOIODIDE, TELLUROI/ 15-2108
       A NEUTRON DIFFRACTION STUDY OF POTASSIUM TRIHYDROGEN SELENITE.   22-2866
T AND CONFIGURATIONAL ENTROPY CHANGE IN SODIUM TRIHYDROGEN DISELENITE.   ANOMALOUS SPECIFIC HEA 22-2856
         ANOMALOUS SPECIFIC HEAT CF SODIUM TRIHYDROGEN DISELENITE.   22-2823
NUM SULFATE DODECAHYDRATE, COLEMANITE, SODIUM TRIDEUTERIUM DISELENITE) /ROMIUM ALUMS, METHYL AMMONIUM ALUMI 02-0087
       CRYSTAL STRUCTURE OF RUBIDIUM TRIHYDROGEN DISELENITE.   22-2881
SPIN RESONANCE SPECTRA IN FERROELECTRIC SODIUM TRIHYDROGEN DISELENITE. /E (ION). RADIATION DAMAGE ELECTRON 22-2870
TERESIS LOOPS AT THE UPPER CURIE POINT CF SODIUM TRIHYDROGEN DISELENITE.   DOUBLE HYS 22-2831
DIATED SINGLE CRYSTAL OF FERROELECTRIC LITHIUM TRIHYDROGEN DISELENITE. /ARAMAGNETIC RESONANCE IN GAMMA-IRRA 22-2842
THE FERROELECTRIC PHASE TRANSITIONS IN SODIUM TRIHYDROGEN DISELENITE.   ESR STUDY OF 22-2871
DP, TGS, ROCHELLE SALT, SILVER PERIODATE, SODIUM TRIHYDROGEN SELENITE) /AMICAL ORDER DISORDER PROPERTIES, K 02-0050
            HEAT OF FORMATICN OF LITHIUM TRIHYDROGEN DISELENITE.   22-2867
STUDIES. PART-58: CRYSTAL STRUCTURE OF AMMONIUM TRIHYDROGEN SELENITE.   HYDROGEN BOND 22-2879
ROPER FERROELECTRIC PHASE TRANSITION IN RUBIDIUM TRIHYDROGEN SELENITE.   IMP 22-2834
UTRON DIFFRACTION STUDY ON PARAELECTRIC SODIUM TRIHYDROGEN DISELENITE.   NE 22-2846
DIATED SINGLE CRYSTAL OF FERROELECTRIC LITHIUM TRIHYDROGEN DISELENITE. PARAMAGNETIC SPECIES IN GAMMA-IRRA 22-2843
URE CRYSTAL STRUCTURE OF THE FERRCELECTRIC SODIUM TRIDEUTERO SELENITE.   ROOM TEMPERAT 22-2861
NSION OF HYDROGEN BONDED FERROELECTRICS. LITHIUM TRIHYDROGEN SELENITE.   THERMAL EXPA 22-2857
TE, AS DE/ THE HYDROGEN BOND SYSTEM IN POTASSIUM TRIHYDROGEN SELENITE, AND IN POTASSIUM TRIDEUTERIUM SELENI 22-2854
ERMAL EXPANSION IN SODIUM TRIHYDROGEN(1-X) TRIDEUTERIUM(X) DISELENITE AND ISOTOPE EFFECT.   TH 22-2858
BETWEEN THE STRUCTURE OF FERROELECTRIC SODIUM TRIHYDROGEN DISELENITE AND ITS THERMAL EXPANSION. RELAXATION 22-2851
       CRYSTAL STRUCTURES OF POTASSIUM TRIHYDROGEN SELENITE AND SODIUM TRIHYDROGEN SELENITE.   22-2838
     ANOMALOUS SPECIFIC HEAT OF SODIUM TRIHYDROGEN SELENITE ASSOCIATED COMBINATORIAL PROBLEM.   22-2822
     ESR STUDY OF GAMMA-IRRADIATED CESIUM TRIHYDROGEN SELENITE AT ROOM TEMPERATURE.   22-2833
     NEUTRON SCATTERING ANALYSIS OF CESIUM TRIHYDROGEN SELENITE AT ROOM TEMPERATURE.   22-2869
FAR INFRARED REFLECTIVITY SPECTRA OF SODIUM TRIHYDROGEN DISELENITE CRYSTAL.   22-2877
N DIFFRACTION STUDY OF THE FERRCELECTRIC LITHIUM TRIHYDROGEN SELENITE. (CRYSTAL STRUCTURE) /RT-54: A NEUTRO 22-2878
OF HIGH PRESSURE ON PHASE TRANSITICNS CF ALKALI TRIHYDROGEN SELENITE CRYSTALS.   EFFECT 22-2882
ELECTROOPTICAL EFFECT IN RUBIDIUM TRIHYDROGEN OR DEUTERIUM DISELENITE CRYSTALS. /ION AND THERMOOPTICAL AND 22-2840
BEHAVIOR OF THE FUNDAMENTAL ABSORPTION EDGE IN TRIHYDROGEN SELENITE CRYSTALS AT THEIR PHASE TRANSITIONS.   22-2839
A NEW DELTA PHASE OF SODIUM TRIDEUTERIUM(X) TRIHYDROGEN(1-X) SELENITE CRYSTALS. (PRESSURE EFFECTS) /ITY OF 22-2875
       COERCIVE FIELD IN LITHIUM TRIHYDROGEN DISELENITE. (EXPERIMENTAL)   22-2844
/TICAL PROPERTIES OF) CRYSTALS OF THE ALKALINE TRIHYDROGEN DISELENITE FAMILY. (ESPECIALLY THE SODIUM MEMBE/ 22-2848
RESONANCE SP/ DIELECTRIC PROPERTIES AND PRESSURE EFFECTS ON SELENITE (ION). RADIATION DAMAGE ELECTRON SPIN 22-2870
/ECHANISM IN FERROELECTRICS: LITHIUM OR SODIUM TRIHYDROGEN DISELENITE. NUCLEAR MAGNETIC RESONANCE INVESTIG/ 22-2837
FERROELECTRIC).   RAMAN SPECTRUM OF SODIUM TRIHYDROGEN DISELENITE (PARAELECTRIC PHASE, COMPARISON WITH 22-2864
EMENT OF THE CRYSTAL STRUCTURE OF FERROELECTRIC ACID LITHIUM SELENITE, POSITION OF THE LITHIUM ION.   REFIN 22-2859
T OF THE CRYSTAL STRUCTURE OF FERROELECTRIC LITHIUM HYDROGEN SELENITE: POSITION OF THE LITHIUM ION. /INEMEN 22-2860
/ED SPECTRA OF ISOTOPIC NONISOMORPHISM OF SODIUM TRIHYDROGEN SELENITE (SHS) OR SODIUM TRIDEUTERIUM SELENIT/ 22-2836
NIUM SULFATE DIHYDRATE, SODIUM TRIHYDROGEN OR TRIDEUTERIUM DISELENITE. (SUMMARY ONLY) /UDIES OF SODIUM AMMO 19-2779
          THERMAL STABILITY OF LITHIUM TRIHYDROGEN SELENITES   22-2868
   THERMOOPTIC AND ELECTROOPTIC PROPERTIES OF TRIHYDROGEN SELENITES.   22-2841
/ MAGNESIUM, CADMIUM, YTTRIUM, GADOLINIUM, DYSPROSIUM, IRON, SELENIUM, NIOBIUM, COBALT, SCANDIUM, OR COPPE/ 10-1640
/L SUBSTITUTION OF ANTIMONY BY BISMUTH OR ARSENIC, SULFUR BY SELENIUM OR OXYGEN, AND IODINE BY BROMINE OR / 15-2205
/)B(VI)C(VII) COMPOUNDS. (USING ANTIMONY CR BISMUTH, SULFUR, SELENIUM OR TELLURIUM, AND CHLORINE, BROMINE / 15-2194
/DE SINGLE CRYSTALS, V IS ANTIMONY OR BISMUTH, VI IS SULFUR, SELENIUM OR TELLURIUM, VII IS IODINE, BROMINE/ 15-2214
UNDER HYDROTHERMAL CONDITIONS. (ANTIMONY AND BISMUTH SULFO, SELENO, AND TELLUROIODIDES AND BROMIDES) /VII) 15-2213
ETHOD (ANTIMONY SESQUISELENIDE OR SULFIDE, ANTIMONY SULFO OR SELENO IODIDE). /HE VAPOR PHASE BY A DYNAMIC M 15-2162
        PYROELECTRIC EFFECT IN ANTIMONY SULFO AND SELENO IODIDE SINGLE CRYSTALS.   15-2163
/DS. ANTIMONY(X) BISMUTH(1-X) SULFOIODIDE, ANTIMONY(X) SELENO(1-X) IODIDE, ANTIMONY(X) ARSENIC(1-X) / 15-2166
ECTRIC PROPERTIES, AND BAND STRUCTURE CF CRYSTALLINE BISMUTH SELENOIODIDE.   CONDUCTIVITY, THERMOEL 15-2126
/URES NEAR A TYPE-1 PHASE TRANSITION. (ANTIMONY SULFOIODIDE, SELENOIODIDE, BISMUTH SULFOIODIDE, ANTIMONY S/ 15-2114
       PIEZOELECTRIC RESISTANCE EFFECT IN ANTIMONY SELENOIODIDE CRYSTALS.   15-2128
           OPTICAL PROPERTIES OF ANTIMONY SELENOIODIDE IN THE INFRARED.   15-2219
ND ELECTRICAL CONDUCTIVITY CF BISMUTH SELENOIODIDE- ANTIMONY SELENOIODIDE SYSTEM ALLOYS.   PHASE DIAGRAM A 15-2122
/(VI)C(VII). (ANTIMONY TRISULFIDE, TRISELENIDE, SULFOIODIDE, SELENOIODIDE, TELLUROIODIDE, TRITELLURIDE, TR/ 15-2108
OYS.   PHASE DIAGRAM AND ELECTRICAL CONDUCTIVITY OF BISMUTH SELENOIODIDE- ANTIMONY SELENOIODIDE SYSTEM ALL 15-2122
SITION FROM PIEZO TO FERROELECTRICITY (THEORY).   SELF CONSISTENCY OF LANDAU'S MODEL IN THE TRAN 03-0507
N.   ENERGY BAND STRUCTURE OF STRONTIUM TITANATE FROM A SELF CONSISTENT FIELD TIGHT BINDING CALCULATIO 06-1221
RUCTURAL PHASE TRANSITIONS IN PEROVSKITE TYPE CRYSTALS. (ST/ SELF CONSISTENT LATTICE DYNAMICAL THEORY OF ST 03-0424
/RST AND SECOND ORDER PHASE TRANSITIONS IN FERROELECTRICS: A SELF CONSISTENT POLARIZATION SCHEME FOR ANHAR/ 03-0355
AND POTASSIUM DIDEUTERIUM PHOSPHATE.   LASER INDUCED SELF FOCUSING IN AMMONIUM DIHYDROGEN PHOSPHATE 17-2476
E AND AMMONIUM DIHYDROGEN PHOSPHATE CRYSTALS OF LIG/ THERMAL SELF FOCUSING IN POTASSIUM DIHYDROGEN PHOSPHAT 17-2717
ON OF LITHIUM TANTALATE.   SELF INDUCED THERMAL EFFECTS ON LIGHT EXTINCTI 12-1889
NY SULFOIODIDE.   INFLUENCE OF LIGHT ON SELF POLARIZATION OF SINGLE CRYSTALS OF ANTIMO 15-2228
LE CRYSTAL BISMUTH TITANATE.   SELF REVERSAL AND WAITING TIME EFFECTS IN SING 08-1448

A SELF SCANNED FERROELECTRIC IMAGE SENSOR. 37-3667
F BARIUM TITANATE CRYSTALS. SEM OBSERVATION OF THE FERROELECTRIC SURFACE O 04-0812
NEW FERROELECTRIC: SEMICARBAZIDE HYDROCHLORIDE. 36-3603
HROMITES, MANGANITES AND FERRITES ELECTRICAL TRANSPORT IN (SEMICONDUCTING AND MAGNETIC) RARE EARTH ORTHOC 36-3608
TANATE. A REFERENCE GUIDE. SEMICONDUCTING AND NONSTOICHIOMETRIC BARIUM TI 01-0022
WITH POSITIVE TEMPERATURE COEFFICIENTS OF RESISTIVITY. SEMICONDUCTING BARIUM AND STRONTIUM TITANATES 04-0995
HIGH L/F NOISE ANOMALY IN SEMICONDUCTING BARIUM STRONTIUM TITANATE. 04-0866
EFFECT OF HEAT TREATMENT ON THE PTCR ANOMALY IN SEMICONDUCTING BARIUM TITANATE. 04-0762
INFLUENCE OF OXYGEN PARTIAL PRESSURE ON PROPERTIES OF SEMICONDUCTING BARIUM TITANATE. 04-0999
INVERSION OF ELECTRICAL PROPERTIES OF SEMICONDUCTING BARIUM TITANATE. 04-0572
OHMIC CONTACTS TO SEMICONDUCTING BARIUM TITANATE. 04-0764
PROPERTIES OF BLACK SEMICONDUCTING BARIUM TITANATE. 04-0702
SOME ASPECTS OF SEMICONDUCTING BARIUM TITANATE. 04-0758
THERMOSTABILIZING PROPERTIES OF SEMICONDUCTING BARIUM TITANATE. 04-0570
ITH NIOBIUM AND MANGANESE IONS AND ITS APPLICATIONS. A SEMICONDUCTING BARIUM TITANATE CERAMIC DOPED W 04-0734
PRODUCTION AND STUDY OF THE INNER BARRIERS IN SEMICONDUCTING BARIUM TITANATE CERAMICS. 04-0583
NSTANT, CONTACT RESISTANCE) SEMICONDUCTING BARIUM TITANATE. (DIELECTRIC CO 04-0730
ERATURE COEFFICIENT OF RESISTIVITY) RESISTIVITY ANOMALY IN SEMICONDUCTING BARIUM TITANATE. (POSITIVE TEMP 04-0584
POTENTIAL BARRIERS ON SEMICONDUCTING BARIUM TITANATE. (PTCR DEVICES) 04-0851
NICAL, CRYSTALLOGRAPHIC, PHYSICAL, THERMAL MAGNETIC, / II-VI SEMICONDUCTING COMPOUNDS - DATA TABLES. (MECHA 01-0025
II). (HIGH GRADE SINGLE CRY/ HYDROTHERMAL CRYSTALLIZATION OF SEMICONDUCTING COMPOUNDS OF GROUP A(V)B(VI)C(V 15-2214
3)(VI)AND A(/ NUCLEAR GAMMA RESONANCE SPECTROSCOPIC STUDY OF SEMICONDUCTING COMPOUNDS OF THE TYPE A(2)(V)B(15-2108
INVESTIGATIONS OF SURFACES OF FERROELECTRICS (TGS) WITH SEMICONDUCTING ELEMENTS. 27-3089
OME FEATURES OF TRAPPING LEVELS IN ANTIMONY SULFOIODIDE TYPE SEMICONDUCTING FERROELECTRICS. S 15-2182
IS/ SYNTHESIS AND CHARACTERIZATION OF THIN FERROELECTRIC AND SEMICONDUCTING FILMS. (THE MAJOR PORTION OF TH 02-0057
FORMATION OF BARIUM TITANATE CERAMIC SEMICONDUCTING LAYER BY DIFFUSION. 04-0684
SPACE CHARGE LIMITED CURRENTS IN SEMICONDUCTING POTASSIUM NIOBATE CRYSTALS. 08-1412
(RESISTIVITY, SEEBECK COEFFICIENT, OPTICAL ABSORPTION) SOME SEMICONDUCTING PROPERTIES OF BARIUM TITANATE. 04-0919
NATE. EFFECT OF SILICA ADDITIONS ON THE SEMICONDUCTING PROPERTIES OF DOPED BARIUM TITA 04-0660
EROVALENT CATIONIC S/ ELECTROCONDUCTIVITY AND HALL EFFECT IN SEMICONDUCTING SOLID SOLUTIONS OF ISO- AND HET 04-0943
ULTRASONIC STUDIES OF CRITICAL PHENOMENA IN INSULATING AND SEMICONDUCTING STRONTIUM TITANATE. (ABSTRACT) 06-1118
/ITY AND MOBILITY VERSUS TEMPERATURE, ABSORPTION SPECTRA) OF SEMICONDUCTING STRONTIUM TITANATE SINGLE CRYS/ 06-1198
MIC STABILITY OF POLARIZATION IN THIN FERROELECTRIC FILMS ON SEMICONDUCTING SUBSTRATES. (TGS ON SILICON) /A 27-3163
) SULFIDE. PHASE DIAGRAM OF THE SEMICONDUCTING UNIAXIAL FERROELECTRIC IRON(1-X 36-3617
) SULFIDE. PART-1. PART-2. MODEL FOR THE SEMICONDUCTING UNIAXIAL FERROELECTRIC IRON(1-X 36-3615
TE. (TEMPERATURE DEPENDENCE) RESISTIVITY BEHAVIOR OF SEMICONDUCTING YTTRIUM BARIUM ZIRCONIUM TITANA 04-1995
CATALYTIC OXIDATION OF CARBON MONOXIDE ON SEMICONDUCTIVE BARIUM TITANATE. 04-0945
EFFECT OF GRAIN SIZE ON THE RESISTIVITY ANOMALY IN SEMICONDUCTIVE BARIUM TITANATE CERAMICS. 04-0735
EFFECT OF EXCESS TITANIA ON THE ELECTRIC RESISTANCE OF SEMICONDUCTIVE BARIUM TITANATE CERAMICS. 04-0966
INSULATION LAYERS ON THE GRAIN BOUNDARIES OF BARIUM TITANATE SEMICONDUCTIVE CERAMICS. /CHANISM FOR FORMING 04-0867
THERMOELECTRIC EFFECT IN SEMICONDUCTIVE LEAD ZIRCONATE. 07-1267
/E SPRAYED BARIUM TITANATE CAPACITORS. PART-10: STRUCTURE OF SEMICONDUCTIVE TITANIUM DIOXIDE AND TITANIUM / 04-0774
SEMICONDUCTIVITY OF THIOUREA. 25-2924
RAMIC (BARIUM TITANATE) WITH LARGE DIELECTRIC CONSTANT AND A SEMICONDUCTOR. /NSISTING OF A PIEZOELECTRIC CE 04-1037
SPIN RESONANCE OF GADOLINIUM(3+) IN BARIUM TITANATE CERAMIC SEMICONDUCTOR. TWO SETS OF ELECTRON 04-0990
EORY, TITANATES DISCU/ FIELD EFFECT AT THE CONTACT BETWEEN A SEMICONDUCTOR AND A C-DOMAIN FERROELECTRIC (TH 03-0511
BAND STRUCTURE OF THE FERROELECTRIC SEMICONDUCTOR ANTIMONY SULFOIODIDE. 15-2179
INDUCED IMPURITY PHOTOCONDUCTIVITY IN THE FERROELECTRIC SEMICONDUCTOR ANTIMONY SULFOIODIDE. 15-2159
IBRIUM CARRIERS ON THE DOMAIN STRUCTURE OF THE FERROELECTRIC SEMICONDUCTOR ANTIMONY SULFOIODIDE. / NONEQUIL 15-2104
LUMINESCENCE OF SINGLE CRYSTALS OF THE FERROELECTRIC SEMICONDUCTOR ANTIMONY SULFOIODIDE. 15-2229
PHOTOEMF'S IN THE FERROELECTRIC SEMICONDUCTOR ANTIMONY SULFOIODIDE. 15-2132
RAMAN SCATTERING IN FERROELECTRIC SEMICONDUCTOR ANTIMONY SULFOIODIDE. 15-2111
NCE OF THE CROSS SECTION OF CAPTURE CENTERS IN FERROELECTRIC SEMICONDUCTOR ANTIMONY SULFOIODIDE. /E DEPENDE 15-2147
/ "FROZEN" SHIFT OF THE CURIE TEMPERATURE OF A FERROELECTRIC SEMICONDUCTOR. (ANTIMONY SULFOIODIDE BROMIDE) 15-2160
ELECTRIC FIELD. AGING OF THE FERROELECTRIC SEMICONDUCTOR ANTIMONY SULFOIODIDE IN A STATIC 15-2113
BAND STRUCTURE OF THE FERROELECTRIC SEMICONDUCTOR ANTIMONY TRISULFIDE. 36-3575
OPTICAL PROPERTIES OF THE SEMICONDUCTOR BISMUTH TELLUROIODIDE. 15-2169
/OPERTIES OF CRYSTALLITES AND INTERCRYSTALLITE LAYERS IN THE SEMICONDUCTOR CERAMIC BARIUM TITANATE. (CERIU/ 04-0568
QUILIBRIUM CARRIERS ON THE PHASE TRANSITION IN FERROELECTRIC SEMICONDUCTOR (COMPOUNDS) A(V)B(VI)C(VII). /NE 15-2141
STARK EFFECT AND FERROELECTRICITY IN THE ORGANIC SEMICONDUCTOR COPPER PHTHALOCYANINE. 36-3532
THEORY OF ELECTRIC INDUCTION WAVES IN FERROELECTRIC SEMICONDUCTOR CRYSTALS. 03-0466
ORS ESPECIALLY ON TRIGLYCINE SULFATE. AFTER EFFECTS IN SEMICONDUCTOR ELECTRODES EVAPORATED ON INSULAT 27-3170
MAIN STRUCTURE AND LOCAL STATES IN ANTIMONY SULFOIODIDE TYPE SEMICONDUCTOR FERROELECTRICS. DO 15-2127
ULFOIODIDE TYPE. POLARIZATION NATURE IN SEMICONDUCTOR FERROELECTRICS OF THE ANTIMONY S 15-2181
XPANSION OF POTASSIUM DIHYDROGEN PHOSPHATE CRYSTALS BY USING SEMICONDUCTOR FILMS. /IGATION OF THE THERMAL E 17-2565
/M TITANATE) POLARIZATION FIELDS ON (GERMANIUM OR TELLURIUM) SEMICONDUCTOR FILMS. (AND USE OF THESE FILMS / 04-0594
ELECTRICAL NONLINEARITY OF A FERROELECTRIC SEMICONDUCTOR IN A NONUNIFORM ELECTRIC FIELD. 03-0500
PARAELECTRIC SEMICONDUCTOR IN A NONUNIFORM ELECTRIC FIELD. 03-0502
ELECTRICAL PROPERTIES OF PHTHALOCYANINE AS A SEMICONDUCTOR IN THE FERROELECTRIC STATE. 36-3622
POLARIZATION DISTRIBUTION IN THE VICINITY OF A FERROELECTRIC SEMICONDUCTOR INTERFACE. 03-0167
PTION OF ULTRASOUND IN AN ANTIMONY SULFOIODIDE FERROELECTRIC SEMICONDUCTOR NEAR ITS CURIE POINT. ABSOR 15-2115
ANOMALIES IN THE ELECTRICAL CONDUCTIVITY OF A FERROELECTRIC SEMICONDUCTOR NEAR THE CUBIC POINT. 03-0268
CONVOLUTION OF SURFACE WAVES IN A STRUCTURE OF SEMICONDUCTOR ON LITHIUM NIOBATE. 11-1847
MPERATURE PHASE TRANSITION IN THE POLAR STATE OF THE ORGANIC SEMICONDUCTOR PHTHALOCYANINE. LOW TE 36-3623
STANNATE) SOLID SOLUTIONS DOPED BY TUNGSTEN TRIOXIDE. SEMICONDUCTOR PROPERTIES OF BARIUM (TITANATE, 04-0965
ATTEMPT TO PREPARE HIGH DOPED BARIUM STRONTIUM TITANATE SEMICONDUCTOR WITH PZT. 04-0717
NUATION AND AMPLIFICATION IN PIEZOELECTRIC AND FERROELECTRIC SEMICONDUCTORS. ACOUSTIC ATTE 03-0275
FERROELECTRICS AND OXIDE SEMICONDUCTORS. 01-0027
ON ABSORPTION OF SOUND NEAR THE CURIE POINT OF FERROELECTRIC SEMICONDUCTORS. RELAXATI 03-0492
ERSE AND LONGITUDINAL OPTICAL PHONONS IN IONIC FERROELECTRIC SEMICONDUCTORS. /ONDUCTION ELECTRONS BY TRANSV 03-0292
THERMODYNAMIC THEORY OF FERROELECTRIC SEMICONDUCTORS. 03-0409
THERMODYNAMIC THEORY OF FERROELECTRIC SEMICONDUCTORS. 03-0410
) SHORT CIRCUIT PHOTOEMF IN FERROELECTRIC SEMICONDUCTORS. (ANTIMONY SULFOIODIDE CRYSTALS 15-2105
57 REFS) FERROELECTRIC SEMICONDUCTORS. (ANTIMONY SULFOIODIDE, REVIEW, 02-0084
H/ ABSORPTION OF ULTRASOUND IN A(V)B(VI)C(VII) FERROELECTRIC SEMICONDUCTORS AT TEMPERATURES NEAR A TYPE-1 P 15-2114
LZT) A SEMICONDUCTOR- FERROELECTRIC MEMORY DEVICE. (P 09-1589
PROTONIC SEMICONDUCTORS. (ICE) 34-3461
H LOW TEMPERATU/ PHOTOFERROELECTRIC EFFECTS IN FERROELECTRIC SEMICONDUCTORS OF THE A(V)B(VI)C(VII) TYPE WIT 15-2161
II) UNDER HYDROTHERMAL CONDITIONS. (ANTI/ CRYSTALLIZATION OF SEMICONDUCTORS OF THE COMPOSITION A(V)B(VI)C(V 15-2213
OUNDARY BLOCKIN/ DOPING PROBLEMS IN (TITANATE) FERROELECTRIC SEMICONDUCTORS. (ORIGIN AND EFFECTS OF GRAIN B 02-0074

/RATE SINGLE CRYSTALS, P-N JUNCTICNS AND P- AND N- RANGES IN SEMICONDUCTORS. (SILICON INTEGRATED CIRCUITS) 24-2904
INSTABILITY PHENOMENA IN FERROELECTRIC SEMICONDUCTORS. (THEORY) 03-0201
(PREPARATION AND STRUCTURE OF) RUBIDIUM AND CESIUM SEMIMETALLIC ORTHOPHOSPHATES. 17-2598
EDGE. (LITHIUM NIOBATE, BA/ X-RAY DETERMINATION OF POLARITY SENSE BY ANOMALOUS SCATTERING AT AN ABSORPTION 11-1673
C EFFECT IN FERROELECTRIC CERAMICS: PLZT 12-40-60. (USED FOR SENSING LOW VOLTAGES) LINEAR ELECTROOPTI 09-1610
OCESSING PURPOSES. (TECHNIQUE APPLIED TO LITH/ ELECTRON BEAM SENSING OF SURFACE ELASTIC WAVES FOR SIGNAL PR 11-1738
FFICIENT OF LITHIUM NIOBATE RESONATORS BY USE OF TEMPERATURE SENSITIVE CAPACITANCES. LOW TEMPERATURE COE 11-1670
DETECTION (INCLUDING FERROELECTRIC AND MAGNETIC TRA/ SENSITIVE, SIMPLE CALORIMETER FOR PHASE CHANGE 28-3201
S PHASE TRANSITIONS AND THE STABLE DOMAIN STRUCTURE OF (HIGH SENSITIVITY) BARIUM TITANATE. DIFFUSIONLES 04-0888
CERAMIC CYLINDERS. (HYDROPHONE ELEMENTS) OPEN CIRCUIT SENSITIVITY OF AXIALLY POLARIZED FERROELECTRIC 37-3652
C CERAMIC CYLINDERS. (HYDROPHONE ELEMENTS) OPEN CIRCUIT SENSITIVITY OF RADIALLY POLARIZED FERROELECTRI 37-3651
C CERAMIC HOLLOW SPHERES. (HYDROPHONE ELEMENTS) OPEN CIRCUIT SENSITIVITY OF RADIALLY POLARIZED FERROELECTRI 37-3653
ION AND DIELECTRIC PROPERTIES OF POTASSIUM DIHYD/ STRUCTURAL SENSITIVITY OF THE FERROELECTRIC PHASE TRANSIT 17-2615
A SELF SCANNED FERROELECTRIC IMAGE SENSOR. 37-3667
ERSE HYPERSONIC PULSES IN PIEZOELECTRIC LITHIUM NIOBATE AND/ SEPARATE EXCITATION OF LONGITUDINAL AND TRANSV 11-1814
L PROPERTIES OF FLASH EVAPORATED BARIUM TITANATE THIN FILMS (SEPARATED FROM SUBSTRATE). /URE) AND ELECTRICA 04-0605
INDUSTRIAL ROTARY FURNACE. REACTION SEQUENCE IN FORMING BARIUM METATITANATE IN AN 04-1003
TANIOBATES AND METATANTALATES OF TRANSITION METALS OF THE 3D SERIES. /RICAL PROPERTIES AND STRUCTURES OF ME 36-3527
ANATE, LEAD / EFFECT OF LITHIUM OXIDE ON THE PROPERTIES OF A SERIES OF FERROELECTRIC MATERIALS. (BARIUM TIT 04-0780
SESQUI... SEE THE PREFIXED WORD.
3+) IN BARIUM TITANATE CERAMIC SEMICONDUCTOR. TWO SETS OF ELECTRON SPIN RESONANCE OF GADOLINIUM(04-0990
S. (BARIUM TITANATE) (FAST ION SCATTERING) SHADOW EFFECT IN (DISPLACIVE) PHASE TRANSITION 04-0533
/UDYING SMALL DEFORMATIONS OF A CRYSTAL LATTICE BASED ON THE SHADOW EFFECT. (NEUTRON DIFFRACTION, BARIUM T/ 04-0533
TONS IN TETRAGONAL BARIUM TITANATE: THEORY OF THE RAMAN LINE SHAPE. POLARI 04-0556
URE, CURIE CONSTANT) OF FERROELECTRIC CRYSTALS. (5 COMPOUND/ SHAPE AND COMPOSITION OF FERROELECTRIC CRYSTAL 02-0089
S (SOLUTION GRO/ EFFECT OF CRYSTALLIZATION CONDITIONS ON THE SHAPE DEPENDENCE OF PROPERTIES (CURIE TEMPERAT 27-3120
ELECTRICS. THEORY OF X-RAY DIFFRACTION LINE SHAPE EFFECTS DUE TO DOMAIN FORMATION IN FERRO 03-0236
IS LOOP. INFLUENCE OF THE SHAPE OF DOMAINS ON THE FERROELECTRIC HYSTERES 03-0300
FERROELECTRICS. TEMPERATURE DEPENDENCE OF THE SHAPE OF THE DOMAIN WALL IN FERROMAGNETICS AND 03-0191
ITANATE. RAMAN SCATTERING LINE SHAPE OF THE SOFT E POLARITON MODE IN BARIUM T 04-0803
RIUM TITANATE AND STRONTIUM TIT/ EFFECT OF PARTICLE SIZE AND SHAPE ON THE INFRARED ABSORPTION SPECTRA OF BA 04-0823
ELECTRIC PROPERTIES OF ANTIMONY SULFOIODIDE. (PHASE BOUNDARY SHAPE STRIATIONS IN DC FIELD) /ON ON THE FERRO 15-2197
TASSIUM MANGANE/ EXCHANGE NARROWING: MAGNETIC RESONANCE LINE SHAPES AND SPIN CORRELATION IN PARAMAGNETIC PO 33-3376
D ITS INTERACTION WITH CRYSTAL DISLOCATIONS. (DOMAIN ORIGIN, SHAPES, MOVEMENT, 34 REFS) /OMAIN STRUCTURE AN 03-0390
CINE SULFATE. EFFECTS OF DOMAIN SHAPES ON SECOND HARMONIC SCATTERING IN TRIGLY 27-2989
SPHATE CRYSTAL/ CONTRIBUTION TO THE PREPARATION ON NON-WEDGE SHAPED RUBIDIUM DIHYDROGEN OR DIDEUTERIUM) PHO 17-2637
PLATES BELOW CUBIC- TETRAGONAL TRANSITION, AND PRODUCTION BY SHAPING AND STRESSING). /URANCE IN THIN (110) 06-1164
NTIUM TITANATE. SHARED HOLES TRAPPED BY CHARGE DEFECTS IN STRO 06-1123
EAD (B(0.5) NIOBIUM(0.5))(1-X) B'(X) OXYGEN(3), / DIFFUSE TO SHARP FERROELECTRIC PHASE TRANSITION SYSTEMS L 13-2044
EXPERIMENTAL CRITERION OF FERROELECTRICITY. (SHARP OR DIFFUSE TRANSITIONS) 03-0184
XIDE STRUCTURE TYPE: AN ELECTRON MICRO/ CS (CRYSTALLOGRAPHIC SHEAR) FAMILIES DERIVED FROM THE RHENIUM TRIO 10-1630
T LITHIUM TANTALATE CRYSTAL RESONATOR VIBRATING IN THICKNESS SHEAR MODE OF MOTION. /E DEPENDENCE OF AN X CU 12-1868
DE AND A CONTRIBUTION TO THE UNDERSTANDING OF THE SO-CALLED "SHEAR STRUCTURES". /RUCTURE OF TUNGSTEN TRIOXI 10-1653
TRIOXIDE. FORMATION OF SHEAR STRUCTURES IN SUBSTOICHIOMETRIC TUNGSTEN 10-1659
EGION IN POTASSIUM DIHYDRO/ ULTRASONIC INVESTIGATION (15 MHZ SHEAR WAVES) OF THE FERROELECTRIC TRANSITION R 17-2576
OELECTRIC BODIES. (WITH FREQUENCY OF APPLIED FIELD CAUSED BY SHEARING DUE TO POLARIZATION INDUCED STRAIN) / 03-0197
IUM TITANATE. STATIC SHELL MODEL FOR BARIUM AND LEAD TITANATES. 04-0826
SHELL MODEL OF INTERIONIC INTERACTIONS FOR BAR 04-0716
DYNAMIC SHELL THEORY FOR FERROELECTRIC CERAMICS. 03-0224
DYNAMIC PIEZOELECTRIC THEORY FOR FERROELECTRIC CERAMIC SHELLS. 03-0225
BATE (CRYSTAL GROWTH BY ADD/ SPATIALLY UNIFORM AND ALTERABLE SHG PHASE MATCHING TEMPERATURES IN LITHIUM NIO 11-1684
CURIE SHIFT IN BARIUM TITANATE BY HETEROSUBSTITUTION 04-0727
ZATION SWITCHING IN POLYCRYSTALLINE FERROELECTRICS AT / ZERO CAN THE ISOTOPIC SHIFT IN KDP BE EXPLAINED BY TUNNELING ONLY. 17-2562
MEASUREMENT OF CHEMICAL SHIFT IN PIEZOELECTRIC ACCELEROMETERS. (POLARI 28-3216
S IN DIGLYCINE NITRATE AND TRIS-SARCOSINE CALCIUM CHLORIDE. SHIFT IN SOLID POTASSIUM DIHYDROGEN PHOSPHATE. 17-2638
ERIUM PHOSPHATE AND DEU/ POSSIBLE THEORETICAL EXPLANATION OF SHIFT OF CURIE TEMPERATURE) /ECTRIC TRANSITION 16-2267
FERROMAGNET COPPER FORMATE TETRAHYDRATE. ANOMALOUS SHIFT OF CURIE TEMPERATURE IN POTASSIUM DIDEUT 17-2601
F POTASSIUM MANGANESE TRIFLUORIDE WITH MAGNESIUM DOPING. SHIFT OF ESR LINES IN THE TWO DIMENSIONAL ANTI 26-2941
ASUREMENT OF THERMAL EXPANSION OF ROCHELLE SALT BY USING THE SHIFT OF NEEL TEMPERATURE AND EPR LINE WIDTH O 33-3377
RIC SEMICON/ METASTABLE OPTICAL CHARGE EXCHANGE AND "FROZEN" SHIFT OF RESONANCE FREQUENCY IN CAVITY. ME 28-3202
SESSING FIRST ORDER TRANSITION. (THEORY, CALCUL/ CURIE POINT SHIFT OF THE CURIE TEMPERATURE OF A FERROELECT 15-2160
IUM TITANATE) IONIC CHARACTERS FROM MOSSBAUER ISOMER SHIFT WITH APPLIED FIELD IN FERROELECTRICS POS 03-0327
ND 294 DEG K. RAMAN SPECTRA AND MODE FREQUENCY SHIFTS. (DEGREE OF COVALENCY IN IRON DOPED BAR 04-0563
/GEN BONDED FERROELECTRICS. PART-1: DAMPINGS AND (FREQUENCY) SHIFTS OF FERROELECTRIC SODIUM NITRITE AT 77 A 16-2262
ROCHELLE SALT. SECOND ORDER QUADRUPOLE SHIFTS (OF PROTONIC COLLECTIVE MOTION, TUNNEL/ 03-0475
/HOMBOHEDRAL PHASES. (X=0.9 AND 0.58) (CORRELATION OF CATION SHIFTS OF THE SODIUM-23 MAGNETIC RESONANCE IN 28-3186
STRONTIUM(0.65/ (COMPACT, SURFACE WAVE) FERROELECTRIC PHASE SHIFTS WITH BISMUTH FERRATE, LITHIUM NIOBATE / 09-1574
FERROELECTRIC DOMAIN SHIFTERS FOR VHF AND UHF. (BASED ON LEAD(0.35) 37-3633
ROTATING WAVEPLATE OPTICAL FREQUENCY SHIFTING DEVICES. (GADOLINIUM MOLYBDATE) 30-3280
UM(0-0.1) STRONTIUM(0.9-1) TITANATE SOLID SOLUTIONS BY HYDR/ SHIFTING IN LITHIUM NIOBATE. 11-1689
SHIFTING OF THE TRANSITION TEMPERATURE OF BARI 04-0722
BARIUM TITANATE AND LEAD ZIRCONATE TITANATE. TWO WAVE SHOCK INDUCED TRANSITION IN BARIUM TITANATE. 04-0589
N IN FERROELECTRIC MATERIALS. SHOCK STRUCTURES IN THE FERROELECTRIC CERAMICS 04-0916
TITANATE. SHOCK WAVE INDUCED ELECTROMECHANICAL CONVERSIO 03-0491
NDUCTORS. (ANTIMONY SULFOIODIDE CRYSTALS) SHOCK WAVE STUDIES OF SURFACE LAYERS IN BARIUM 04-0588
ATE AND BARIUM STRONTIUM N/ PYROELECTRIC VOLTAGE RESPONSE TO SHORT CIRCUIT PHOTOEMF IN FERROELECTRIC SEMICO 15-2105
MENA. (HYSTERESIS LOOP CONSTRICTIONS, ROCHELLE SALT, TGS) SHORT INFRARED LASER PULSES IN TRIGLYCINE SULF 13-2018
VSKITES (STRONTIUM OR CALCIUM TITANATE OR ZIRCONATE, TEST OF SHORT LIVED FERROELECTRIC "AFTER EFFECT" PHENO 27-3156
EUDOSPIN MODEL FOR ORDER DISORDER TYPE FERROELECTRICS WITH A SHORT OR LONG RANGE ORDER THEORIES) /AIN PERO 03-0414
RING OF NEUTRONS FROM POTASSIUM MANGANESE TRIFLUORIDE IN THE SHORT RANGE INTERACTION. (THEORY) /R OF THE PS 03-0423
CRYSTAL. STRESS INDUCED BIREFRINGENCE IN AN ISOLATED AND A SHORT RANGE ORDERED REGION. /RAMAGNETIC SCATTE 33-3371
WEIGHT OF SHORTCIRCUITED POTASSIUM DIHYDROGEN PHOSPHATE 17-2452
FERROELECTRIC SWITCHING AND THE SHOTTKY BARRIERS IN FERROELECTRICS. 03-0459
A OF ISOTOPIC NONISOMORPHISM OF SODIUM TRIHYDROGEN SELENITE (SHS) OR SODIUM TRIDEUTERIUM SELENITE (SDS). /R 22-2836
ROGEN PHOSPHATE CRYSTAL AND ITS APPLICATION FOR A HIGH SPEED SHUTTER. /ISTICS OF DEUTERATED POTASSIUM DIHYD 17-2499
HOSPHATE CRYSTALS. CHARACTERISTICS OF AN ELECTROOPTICAL SHUTTER BASED ON Z CUT POTASSIUM DIDEUTERIUM P 17-2460
) FUNCTION AND APPLICATIONS OF GADOLINIUM MOLYBDATE (OPTICAL SHUTTER, COLOR MODULATOR). (PROPERTIES, 30-3298
THE) HYSTERESIS LOOP OF FERROELECTRICS. SIDEWAYS DOMAIN WALL MOTION AND (MECHANISM OF 03-0301
FERROELECTRIC SWITCHING AND THE SIEVERT INTEGRAL. 03-0239
(BARIUM TITANATE, LITHIUM TANTAL/ DETERMINATION OF ABSOLUTE SIGNS OF MICROWAVE NONLINEAR SUSCEPTIBILITIES. 02-0107

AR CRYSTALS (LITHIUM NIOBATE OR TANTALATE, BARIUM / RELATIVE SIGNS OF NONLINEAR OPTICAL COEFFICIENTS OF POL 02-0098
NERATION) COEFFICIENTS OF POLAR (PYROELECTRIC) CRY/ ABSOLUTE SIGNS OF NONLINEAR OPTICAL (SECOND HARMONIC GE 02-0095
TS OF PIEZOELECTRIC CRYSTALS.   ABSOLUTE SIGNS OF SECOND HARMONIC GENERATION COEFFICIEN 02-0096
TS OF PIEZOELECTRIC CRYSTA/ FURTHER MEASUREMENTS OF ABSOLUTE SIGNS OF SECOND HARMONIC GENERATION COEFFICIEN 05-1066
HIUM TANTALATE.   SIGNS OF THE ELECTROOPTIC COEFFICIENTS FOR LIT 12-1880
POTASSIUM DIHYDROGEN PHOSPHATE.   ABSOLUTE SIGNS OF THE NONLINEAR OPTICAL COEFFICIENTS OF 17-2597
ONATORS UPON STATE OF POLARIZATION. DEPENDENCE OF THE SMALL SIGNAL PARAMETERS OF FERROELECTRIC CERAMIC RES 37-3649
TEMPERATURE AND FIELD STRENGTH DEPENDENCE OF LARGE SIGNAL PERMITTIVITY OF TGS. (10 KHZ-1 MHZ) 27-3068
0 DEG DOMAIN WALL IN BARIUM TITANATE (THEORY, CALCULA/ SMALL SIGNAL PERMITTIVITY OF THE STATIONARY (100)-18 03-0358
00 MHZ. (NEAR THE CURIE TEMPERATURE)   LARGE SIGNAL PERMITTIVITY OF TRIGLYCINE SULFATE AT 1 27-3029
ONLINEAR INTERACTIONS IN FERROELECTRIC CERAMICS TO MICROWAVE SIGNAL PROCESSING.   APPLICATION OF N 37-3647
TO LITH/ ELECTRON BEAM SENSING OF SURFACE ELASTIC WAVES FOR SIGNAL PROCESSING PURPOSES. (TECHNIQUE APPLIED 11-1738
ERMITTIVITY FROM THE PYROELECTRIC RESPONSE TO STEP RADIATION SIGNALS IN FERROELECTRICS. / COEFFICIENT AND P 03-0469
OUSTIC DIPOLE AND QUADRUPOLE SATURATION OF THE LITHIUM-7 NMR SIGNALS IN LITHIUM NIOBATE.   AC 11-1666
RIUM / PYROELECTRIC VOLTAGE RESPONSE TO RECTANGULAR INFRARED SIGNALS IN TRIGLYCINE SULFATE AND STRONTIUM BA 13-2019
LFATE AND STRONTIUM B/ PYROELECTRIC VOLTAGE RESPONSE TO STEP SIGNALS OF INFRARED RADIATION IN TRIGLYCINE SU 13-2022
AD TITANATE, FROM SYSTEMS BARIA OR LITHIA, TITANIA, ALUMINA, SILICA). /FERROELECTRIC CRYSTALS (BARIUM OR LE 04-0783
IES OF DOPED BARIUM TITANATE.   EFFECT OF SILICA ADDITIONS ON THE SEMICONDUCTING PROPERT 04-0660
/ GLASSES CONTAINING BARIUM OXIDE AND TITANIUM DIOXIDE (WITH SILICA AND ALUMINA) ON PROPERTIES OF BARIUM T/ 04-0829
ID TARTRATE CRYSTALS.   SOME CHEMICAL REACTIONS IN SILICA GELS. PART-3: FORMATION OF POTASSIUM AC 29-3246
ERAMICS.   EFFECTS OF SILICA ON MANGANESE DOPED STRONTIUM TITANATE C 06-1172
OPTICAL ANISOTROPY, BISMUTH TITANATE, POSSIBLE FERROELECTRIC SILICATE) /COMPLIANCE, NICKEL IODINE BORACITE 02-0060
IC, OPTICAL AND MAGNETIC PROPERTIES OF (MONOCLINIC) EUROPIUM SILICATE. (UNSTABLE FERROELECTRIC PHASE) /RAPH 36-3620
A LAYERED STRUCTURE CONSISTING OF LITHIUM NIOBATE AND P-TYPE SILICON. /ION OF RAYLEIGH ULTRASONIC WAVES IN 11-1747
CHARACTERISTICS OF RF SPUTTERED BARIUM TITANATE FILMS ON SILICON.   04-0930
)) (BARIUM TITANATE AND LITHIUM NIOBATE FILMS ON PLATINUM OR SILICON) / TYPE STRUCTURE. (MIXED OXIDES ABO(3 04-0914
IN FERROELECTRIC FILMS ON SEMICONDUCTING SUBSTRATES. (TGS ON SILICON) /AMIC STABILITY OF POLARIZATION IN TH 27-3163
IOBATE, LEAD NIOBIUM(4) OXIDE(11), SILVER GALLIUM DISULFIDE, SILICON CARBIDE) /TITANATE, BARIUM STRONTIUM N 05-1066
F SPUTTER ETCHING RATES FOR MONOCRYSTALLINE LITHIUM NIOBATE, SILICON DIOXIDE, AND SILICON.   MEASUREMENT O 11-1714
TALS, P-N JUNCTIONS AND P- AND N- RANGES IN SEMICONDUCTORS. (SILICON INTEGRATED CIRCUITS) /RATE SINGLE CRYS 24-2904
/UM(3) OXIDE(11) AND ITS ISOMORPHOUS COMPOUND LEAD GERMANIUM SILICON OXIDE. (DIELECTRIC CONSTANTS, SPONTAN/ 35-3501
/ERMALLY STIMULATED CURRENTS AND LUMINESCENCE IN BISMUTH(12) SILICON OXIDE(20) AND BISMUTH(12) GERMANIUM O/ 36-3585
. (BARIUM TITANATE AND/ LINEAR DISORDER IN CRYSTALS (CASE OF SILICON, QUARTZ AND FERROELECTRIC PEROVSKITES) 04-0629
LINE FERROELECTRICS (BARIUM TITANATE AND SAME WITH 1 PERCENT SILICON REPLACING TITANIUM). /SOME POLYCRYSTAL 04-0818
NSVERSE ACOUSTOELECTRIC EFFECT IN A LAYERED LITHIUM NIOBATE- SILICON STRUCTURE.   TRA 11-1748
ES OF THIN FILM BARIUM TITANATE PREPARED BY RF SPUTTERING ON SILICON SUBSTRATES. /L AND ELECTRICAL PROPERTI 04-0828
LITHIUM NIOBATE SILICON SURFACE WAVE CONVOLUTER.   11-1788
MONY SULFOIODIDE SINGLE CRYSTALS AND SOME METALS. (ANTIMONY, SILVER, ALUMINUM) /F THE CONTACTS BETWEEN ANTI 15-2251
AG/ TWO PHASE FERROELECTRIC SYSTEMS. PART-2: FERROELECTRICS- SILVER (BARIUM TITANATE AND BARIUM TITANATE- M 04-0816
NON-HYDROGEN BONDED PHOSPHATE FERROELECTRIC (CRYSTAL GROWTH/ SILVER DITHORIUM PHOSPHATE- A NEW CASE OF THE 36-3610
TANATE, BARIUM STRONTIUM NIOBATE, LEAD NIOBIUM(4) OXIDE(11), SILVER GALLIUM DISULFIDE, SILICON CARBIDE) /TI 05-1066
CRYSTAL STRUCTURE OF FERROELECTRIC GLYCINE SILVER NITRATE.   16-2297
CRYSTAL STRUCTURE OF FERROELECTRIC GLYCINE SILVER NITRATE.   16-2333
CRYSTAL STRUCTURE OF GLYCINE SILVER NITRATE.   16-2348
RAY STUDIES ON THE FERROELECTRIC PHASE TRANSITION OF GLYCINE SILVER NITRATE.   DIELECTRIC AND X- 16-2347
/GROWTH AT POLYMORPHIC TRANSFORMATIONS IN POTASSIUM NITRATE, SILVER NITRATE, AND AMMONIUM NITRATE SINGLE C/ 16-2260
NUCLEAR QUADRUPOLE RESONANCE OF IODINE-127 IN SILVER PERIODATE.   08-1426
/NT SCATTERING SPECTRA OF POTASSIUM DIHYDROGEN PHOSPHATE AND SILVER PERIODATE BELOW THE CRITICAL TEMPERATU/ 17-2687
GEL GROWTH OF SILVER PERIODATE CRYSTALS.   36-3523
/NAMICAL ORDER DISORDER PROPERTIES, KDP, TGS, ROCHELLE SALT, SILVER PERIODATE, SODIUM TRIHYDROGEN SELENITE) 02-0050
DIELECTRIC RELAXATION IN FERROELECTRIC SILVER SODIUM DINITRITE.   16-2290
DIELECTRIC RELAXATION IN FERROELECTRIC SILVER SODIUM DINITRITE.   16-2292
DIELECTRIC STUDIES ON THE FERROELECTRIC PHASE TRANSITION IN SILVER SODIUM DINITRITE.   16-2289
FERROELECTRIC PHASE TRANSITION IN SILVER SODIUM DINITRITE   16-2266
RE EFFECT ON THE DIELECTRIC RELAXATION TIME IN FERROELECTRIC SILVER SODIUM DINITRITE.   HYDROSTATIC PRESSU 16-2294
NMR STUDY OF SODIUM-23 IN SILVER SODIUM DINITRITE.   16-2324
PYROELECTRIC STUDY ON THE SPONTANEOUS POLARIZATION IN SILVER SODIUM DINITRITE.   16-2295
SPECIFIC HEAT ANOMALY IN FERROELECTRIC SILVER SODIUM DINITRITE.   16-2296
SPLITTING OF THE SODIUM-23 NMR CENTRAL LINE IN FERROELECTRIC SILVER SODIUM DINITRITE.   16-2307
THERMAL EXPANSION IN FERROELECTRIC SILVER SODIUM DINITRITE.   16-2346
ITE. (DIFFERENTIAL THERMAL ANALYSIS)   PROPERTIES OF SILVER SODIUM DINITRITE AND SILVER SODIUM NITR 16-2370
IC FIELD.   FERROELECTRIC HYSTERESIS LOOP OF SILVER SODIUM DINITRITE BY QUASI STATIC ELECTR 16-2293
D.   DIELECTRIC RELAXATION MEASUREMENT ON FERROELECTRIC SILVER SODIUM DINITRITE BY STEP FUNCTION METHO 16-2291
/F THE CRITICAL SLOWING DOWN OF THE DIELECTRIC RELAXATION IN SILVER SODIUM DINITRITE BY THERMAL NOISE MEAS/ 16-2382
/NNECTION WITH THEIR PHASE TRANSITIONS. PART-1: TGS. PART-2: SILVER SODIUM DINITRITE. PART-3: ROCHELLE SAL/ 02-0077
(INCLUDING FERROELECTRIC AND MAGNETIC TRA/ HIGHLY SENSITIVE, SIMPLE CALORIMETER FOR PHASE CHANGE DETECTION 28-3201
AVIOR.   A SIMPLE MECHANICAL MODEL WITH FERROELECTRIC BEH 03-0151
IEZOELECTRIC VIBRATIONS OF LITHIUM NIOBATE AND LITHIUM TANT/ SIMPLIFIED CHARACTERISTIC EQUATION FOR PLANE P 11-1739
SIUM DIDEUTERIUM PHOSPHATE. ELECTROOPTIC Q SWITCH BY DESK-T/ SIMULATION OF A DOUBLE 45 DEGREES, Z CUT POTAS 17-2702
AND FOURTH HARMONICS OF 1.06 MICRON RADI/ OBSERVATION OF THE SIMULTANEOUS GENERATION OF THE SECOND, THIRD, 11-1779
ICIENT AND DIELECTRIC CONSTANT OF FERROELECTRICS. SIMULTANEOUS MEASUREMENT OF PYROELECTRIC COEFF 03-0333
TRIC AND ACOUSTIC MODES OF POTASSIUM DIDEUTERIUM PHOSPHATE. SIMULTANEOUS OBSERVATION OF THE SOFT FERROELEC 17-2636
ECOND HARMONIC GENERATION (IN LITHIUM NIOBATE). SIMULTANEOUS OPTICAL PARAMETRIC OSCILLATION, S 11-1859
S IN WHICH FERROMAGNETIC AND FERROELECTRIC PROPERTIES APPEAR SIMULTANEOUSLY.   SYMMETRY OF CRYSTAL 03-0465
INFRARED SPECTRA OF SINGLE AND POLYCRYSTALLINE STRONTIUM TITANATE. 06-1250
C TYPE FERROELECTRIC MODULATOR FOR ELECTROMETER. (WITH A TGS SINGLE CRYSTAL)   A NEW SECOND HARMONI 27-2951
C AND OPTICAL PROPERTIES OF LEAD ZINCATE(0.33) NIOBATE(0.67) SINGLE CRYSTAL.   DIELECTRI 13-2065
DIPOLAR RELAXATION AT LOW TEMPERATURE OF ICE SINGLE CRYSTAL.   34-3447
IRRADIATION ON THE RELATIVE PERMITTIVITY OF BARIUM TITANATE SINGLE CRYSTAL.   EFFECT OF THERMAL NEUTRON 04-0781
ELECTROOPTIC PROPERTIES OF FERROELECTRIC LEAD GERMANATE SINGLE CRYSTAL.   35-3518
PROPERTIES OF STRONTIUM LITHIUM NIOBATE TYPE FERROELECTRICS (SINGLE CRYSTAL).   ELECTROOPTIC 13-2062
FERROELECTRIC PROPERTIES OF STRONTIUM NIOBATE SINGLE CRYSTAL.   08-1404
CTRIC FERROELECTRIC CONDENSER UNDER CONTINUOUS HEATING. (TGS SINGLE CRYSTAL) /EDANCE MEASUREMENT OF A DIELE 27-3028
E QUADRATIC ELECTROOPTIC COEFFICIENTS IN STRONTIUM TITANATE (SINGLE) CRYSTAL. /ROMETRIC DETERMINATION OF TH 06-1130
LAPPING CHARACTERISTICS OF LITHIUM NIOBATE SINGLE CRYSTAL.   11-1783
RESONANCE OF PLATINUM IONS IN NIOBIUM DOPED BARIUM TITANATE SINGLE CRYSTAL.   PARAMAGNETIC 04-0977
PIEZOELECTRIC PROPERTIES OF POTASSIUM TANTALATE SINGLE CRYSTAL.   08-1271
PRODUCTION AND STUDY OF PURE AND DOPED BARIUM TITANATE SINGLE CRYSTAL   04-0667
TATORY POWER IN FERROELECTRIC LEAD(5) GERMANIUM(3) OXIDE(11) SINGLE CRYSTAL.   SWITCHING OF OPTICAL RO 36-3570
X-RAY STUDY OF THE SURFACE LAYER ON BARIUM TITANATE SINGLE CRYSTAL.   04-0858

POLARIZATICN (DOMAIN STRUCTURE) AND PHASE TRANSITION IN KDP SINGLE CRYSTAL. (ABSTRACT ONLY)   17-2505
              THERMODYNAMIC PHENOMENOLOGY OF FERROELECTRICITY IN SINGLE CRYSTAL AND CERAMIC SYSTEMS.   03-0214
TRY AND HYDROXYL CONTENT ON THE PROPERTIES AND PROCESSING OF SINGLE CRYSTAL BARIUM SODIUM NIOBATE. /ICHIOME   13-1915
INEAR OPTICAL MATERIAL. (GROWTH BY KYROPOULO/ DEVELOFMENT OF SINGLE CRYSTAL BARIUM SODIUM NIOBATE AS A NONL   13-2021
                       BINARY LIGHT BEAM DEFLECTION IN SINGLE CRYSTAL BARIUM TITANATE.   04-0653
                       BINARY LIGHT BEAM DETECTION IN SINGLE CRYSTAL BARIUM TITANATE.   04-0654
M THE MELT IN AN OXYGEN ATMOSPHERE.   SINGLE CRYSTAL BARIUM TITANATE FILMS GROWN FRO   04-0612
DITIONS.   OPTICAL AND DIELECTRIC PROPERTIES OF SINGLE CRYSTAL BARIUM TITANATE WITH NIOBIUM AD   04-0875
ELECTRIC NATURE).   DIELECTRIC HYSTERESIS IN (FLUX GROWN) SINGLE CRYSTAL BISMUTH FERRATE (CONFIRMS FERRO   08-1449
              SELF REVERSAL AND WAITING TIME EFFECTS IN SINGLE CRYSTAL BISMUTH TITANATE.   08-1448
MPOUNDS. (GADOLINIUM MOLYBDATE, GR/ GROWTH AND EVALUATION OF SINGLE CRYSTAL BISMUTH TITANATE AND RELATED CO   08-1323
).   GROWTH OF LARGE SINGLE CRYSTAL BISMUTH(4) TITANIUM(3) OXIDE(12   35-3507
).   ELECTRICAL AND OPTICAL SWITCHING PROPERTIES OF SINGLE CRYSTAL BISMUTH(4) TITANIUM(3) OXIDE(12   35-3517
ION OF THE PHASE TRANSITION IN AMMONIUM DIHYDROGEN PHOSPHATE SINGLE CRYSTAL BY NMR.   INVESTIGAT   17-2545
TE STRUCTURE.   GROWTH AND STUDY OF NEW SINGLE CRYSTAL CADMIUM COMPOUNDS WITH PEROVSKI   08-1429
PORTION CF THIS REPCRT IS CCNCERNED WITH ICN IMPLANTATION OF SINGLE CRYSTAL CADMIUM SULFIDE) /. (THE MAJOR   02-0057
             POLARIZED INFRARED AND RAMAN SPECTRA OF SINGLE CRYSTAL CESIUM NITRITE.   16-2342
                 LATTICE DISTORTIONS IN A STRONTIUM TITANATE SINGLE CRYSTAL ELECTRET.   06-1238
          MICRCSTRUCTURE OF DCMAINS AND DOMAIN WALLS IN SINGLE CRYSTAL FILMS OF BARIUM TITANATE.   04-1031
NS OF A ST/ OBSERVATION OF THE CHANGE IN THE POLARIZATION OF SINGLE CRYSTAL FILMS OF BARIUM TITANATE BY MEA   04-0942
RON MICRCSC/ MCTION OF DOMAIN BOUNDARIES DURING SWITCHING IN SINGLE CRYSTAL FILMS OF BARIUM TITANATE (ELECT   04-0940
               PREPARATICN OF POTASSIUM NIOBATE SINGLE CRYSTAL FOR OPTICAL APPLICATION.   08-1332
CTROOPTIC AND PIEZOELECTRIC PROPERTIES OF LANTHANUM TITANATE SINGLE CRYSTAL. (FOUND NOT FERROELECTRIC) ELE   36-3576
E AND / GROWING KINETIC OF THE AMMONIUM DIHYDROGEN PHOSPHATE SINGLE CRYSTAL (FROM SOLUTION) IN STATIC REGIM   17-2516
                GROWTH OF AMMCNIUM DIHYDROGEN PHOSPHATE SINGLE CRYSTAL FROM SOLUTIONS.   17-2411
NATE)   A FLOATING ZONE SINGLE CRYSTAL GROWING APPARATUS. (BARIUM TITA   04-0998
OF LITHIUM TANTALATE. (DIELECTRIC CONSTANTS, PIEZOELECTRIC / SINGLE CRYSTAL GROWTH AND PHYSICAL PROPERTIES   12-1876
        CRYSTALLIZATION BEHAVIOR OF ANTIMONY SULFOIODIDE. (SINGLE CRYSTAL GROWTH, CHEMICAL TRANSPORT)   15-2201
N THE DIELECTRIC CONSTAN/ EFFECT CF TRIGLYCINE FLUOBERYLLATE SINGLE CRYSTAL GROWTH CONDITIONS ON A CHANGE I   27-3151
N HALIDES.   STRESS FREE SINGLE CRYSTAL GROWTH OF FERROELECTRIC HYDROGE   31-3333
ES. (SODIUM TUNGSTATE)   KYROPOULOS SINGLE CRYSTAL GROWTH OF SODIUM TUNGSTEN BRONZ   13-2054
ANTALATE FROM MOLTEN SALTS.   SINGLE CRYSTAL GROWTH OF STRONTIUM POTASSIUM T   13-2034
/ANTIFERROELECTRICS OF PEROVSKITE TYPE. (POLYCRYSTALLINE AND SINGLE CRYSTAL LEAD MAGNESATE(0.33) NIOBATE(0/   08-1375
           GROWTH AND DIELECTRIC PROPERTIES OF SINGLE CRYSTAL LEAD ZIRCONATE   07-1265
             ACOUSTOOPTIC PROPERTY OF SINGLE CRYSTAL LEAD(5) GERMANIUM(3) OXIDE(11).   35-3509
       FERROELECTRIC OPTICAL ROTATION DOMAINS IN SINGLE CRYSTAL LEAD(5) GERMANIUM(3) OXIDE(11).   36-3550
TEMPERATURES. (BISMUTH FERRATE AND SYSTEMS BASED ON IT, AND SINGLE CRYSTAL LITHIUM NIOBATE) /TH HIGH CURIE   08-1378
IEZOELECTRIC AND OPTICAL PRCF/ BEHAVICR OF LITHIUM TANTALATE SINGLE CRYSTAL NEAR ITS CURIE POINT. PART-1: P   12-1877
IELECTRIC AND ULTRASONIC PROP/ BEHAVIOR OF LITHIUM TANTALATE SINGLE CRYSTAL NEAR ITS CURIE POINT, PART-2, D   12-1885
ROGEN DISELENITE.   PARAMAGNETIC SPECIES IN GAMMA-IRRADIATED SINGLE CRYSTAL OF FERROELECTRIC LITHIUM TRIHYD   22-2843
ROGEN D/ ELECTRON PARAMAGNETIC RESCNANCE IN GAMMA-IRRADIATED SINGLE CRYSTAL OF FERROELECTRIC LITHIUM TRIHYD   22-2842
E.   BRILLOUIN SCATTERING BY A SINGLE CRYSTAL OF RUBIDIUM DIHYDROGEN PHOSPHAT   17-2519
E.   BRILLCUIN DIFFUSION IN A SINGLE CRYSTAL OF RUBIDIUM DIHYDROGEN PHOSPHAT   17-2520
        INFRARED ABSORPTION IN GAMMA-RAY IRRADIATED SINGLE CRYSTAL OF SODIUM NITRITE.   16-2336
RROELECTRIC PHASES.   INFRARED ABSORPTION SPECTRUM OF A SINGLE CRYSTAL OF THIOUREA IN THE PARA- AND FE   25-2910
YSTALS WITH SCANNING ELECTRCN MICFCSCOPE. (TUNGSTEN TRIOXIDE SINGLE CRYSTAL PLATES) /NS OF FERROELECTRIC CR   10-1646
    LOW TEMPERATURE THERMAL EXPANSION MEASUREMENTS ON SINGLE CRYSTAL POTASSIUM DIHYDROGEN PHOSPHATE.   17-2464
         POLARIZATION ECHO IN THE FERROELECTRIC SINGLE CRYSTAL POTASSIUM DIHYDROGEN PHOSPHATE.   17-2564
TE.   MOSSBAUER RESONANCE OF SINGLE CRYSTAL POTASSIUM FERROCYANIDE TRIHYDRA   24-2890
CALCIUM CONCENTRATION VS NET IONIZED DCNOR CONCENTRATION IN SINGLE CRYSTAL POTASSIUM TANTALATE. (CONDUCTI/   08-1437
          APPARATUS FOR GROWTH OF SINGLE CRYSTAL, SINGLE DOMAIN LITHIUM NIOBATE.   11-1726
OSITION, LATTICE AND CURIE TEMPERATURE ME/ LITHIUM TANTALATE SINGLE CRYSTAL STOICHIOMETRY (VERSUS MELT COMP   12-1871
/PHYSICAL FROPERTIES OF SCME FERRCELECTRICS BELOW 80 DEG K. (SINGLE CRYSTAL STRONTIUM TITANATE, POLYCRYSTA/   04-0721
).   A NEW FERROELECTRIC SINGLE CRYSTAL STRONTIUM(2) NIOBIUM(2) OXIDE(7   13-1997
) BY MEANS OF THE FLOATING ZONE TEC/ GROWTH OF FERROELECTRIC SINGLE CRYSTAL STRONTIUM(2) NIOBIUM(2) OXIDE(7   13-2037
CIELECTRIC PROPERTY OF BARIUM TITANATE NIOBATE SINGLE CRYSTAL (TETRAGONAL TUNGSTEN BRONZE).   13-1969
ODS OF GROWTH)   SINGLE CRYSTAL TITANATES AND ZIRCONATES. (METH   02-0045
NE FLUOBER/ ANOMALY OF THERMAL PROPERTIES OF FERROELECTRICS (SINGLE CRYSTAL TRIGLYCINE SULFATE AND TRIGLYCI   27-3137
/ND ELECTROOPTIC PROPERTIES OF BARIUM SODIUM YTTRIUM NIOBATE SINGLE CRYSTAL WITH TUNGSTEN BRONZE TYPE STRU/   13-1971
/LECTRICITY AND PHASE TRANSITIONS IN SCLID HYDROGEN HALIDES (SINGLE CRYSTAL X-RAY AND NEUTRON DIFFRACTION).   31-3331
        ANGULAR CORRELATION OF ANNIHILATION PHOTONS IN ICE SINGLE CRYSTALS.   34-3457
RING FROM POTASSIUM MANGANESE TRIFLUORIDE AND SODIUM NIOBATE SINGLE CRYSTALS. /RITICAL X-RAY DIFFUSE SCATTE   08-1312
F HETEROVALENT FERRIC IONS IN THE LATTICE OF BARIUM TITANATE SINGLE CRYSTALS.   COMPENSATION O   04-0539
STAL GRCWTH AND PROPERTIES OF LEAC(5) GERMANIUM(3) OXIDE(11) SINGLE CRYSTALS.   CRY   35-3513
N AND RELAXATION BEHAVIOR IN FERROELECTRIC POTASSIUM NIOBATE SINGLE CRYSTALS.   DEFORMATIO   08-1420
ON OF ELASTIC AND PIEZOELECTRIC CCNSTANTS OF LITHIUM NIOBATE SINGLE CRYSTALS.   DETERMINATI   11-1749
F THE DEVIATION FROM OXYGEN STOICHIOMETRY OF BARIUM TITANATE SINGLE CRYSTALS.   DETERMINATION O   04-0736
               DIELECTRIC CONSTANT OF PURE ICE I(H) SINGLE CRYSTALS.   34-3478
      DIELECTRIC MEASUREMENTS ON PILE-IRRADIATED BARIUM TITANATE SINGLE CRYSTALS.   04-0910
     DIELECTRIC RELAXATION, BULK AND SURFACE CONDUCTIVITY OF ICE SINGLE CRYSTALS.   34-3469
            DIELECTRIC RELAXATION IN MULTIDOMAIN TGS SINGLE CRYSTALS.   27-3001
       DILATOMETRIC INVESTIGATION OF ANTIMONY SULFOIODIDE SINGLE CRYSTALS.   15-2155
               DISLOCATIONS IN BARIUM TITANATE SINGLE CRYSTALS.   04-1007
              DOMAIN STRUCTURE IN LITHIUM NIOBATE SINGLE CRYSTALS.   11-1703
            DOMAINS IN FERRCELECTRIC POTASSIUM NIOBATE SINGLE CRYSTALS.   08-1315
EFFECT OF ADDING NICKEL ON THE PROPERTIES OF BARIUM TITANATE SINGLE CRYSTALS.   04-0793
OF EXTERNAL ACTIONS ON THE REPOLARIZATION OF BARIUM TITANATE SINGLE CRYSTALS.   EFFECT   04-0957
OXIDE ON THE 120 DEGREE PHASE TRANSITION IN BARIUM TITANATE SINGLE CRYSTALS.   EFFECT OF FERRIC   04-0544
GROWTH CONDITIONS ON CIELECTRIC LCSSES IN TRIGLYCINE SULFATE SINGLE CRYSTALS.   EFFECT OF   27-3152
IUMINATICN ON REPOLARIZATION PROCESS IN ANTIMONY SULFOIODIDE SINGLE CRYSTALS.   EFFECT OF IL   15-2220
PRESSION ON THE UHF DIELECTRIC FROPERTIES OF BARIUM TITANATE SINGLE CRYSTALS.   EFFECT OF LINEAR COM   04-0896
4+) IONS IN THE ELECTROOPTICAL FRCFERTIES OF BARIUM TITANATE SINGLE CRYSTALS. /NT SUBSTITUTION OF TITANIUM(04-0959
N THE FERROELECTRIC PHASE TRANSITION IN ANTIMONY SULFOIODIDE SINGLE CRYSTALS. EFFECT OF OXYGEN ADSORPTION O   15-2187
SORPTION OF OXYGEN CN THE PROPERTIES CF ANTIMONY SULFOIODIDE SINGLE CRYSTALS.   EFFECT OF THE AD   15-2154
ERSE COMPRESSION OF CIELECTRIC FROPERTIES OF BARIUM TITANATE SINGLE CRYSTALS.   EFFECT OF TRANSV   04-0956
IASTIC CCNSTANTS AND ULTRASONIC ABSORPTION OF SODIUM NITRITE SINGLE CRYSTALS.   E   16-2357
EIASTIC FRCFERTIES OF YTTRIUM DOPED STRONTIUM TITANATE SINGLE CRYSTALS.   06-1180
              ELECTRICAL CONDUCTIVITY CF LITHIUM NIOBATE SINGLE CRYSTALS.   11-1733
              ELECTRICAL CONDUCTIVITY CF AMMONIUM SULFATE SINGLE CRYSTALS.   19-2781
          ELECTROMECHANICAL FRCFERTIES OF STRONTIUM TITANATE SINGLE CRYSTALS.   06-1206

M WRITING OF FERROELECTRIC DOMAINS IN BISMUTH TITANIUM OXIDE SINGLE CRYSTALS. ELECTRON BEA 08-1300
ERTIES OF BARIUM TITANATE- TANTALUM PENTOXIDE SOLID SOLUTION SINGLE CRYSTALS. ELECTROOPTICAL PROP 04-0955
ELECTROOPTICAL PROPERTIES OF LITHIUM IODATE SINGLE CRYSTALS. 08-1423
EXES IN FERROELECTRIC TRIS-SARCOSINE CALCIUM CHLORIDE (TSCC) SINGLE CRYSTALS. /TIONS OF MANGANESE(2+) COMPL 36-3525
R STUDIES OF CHROMIUM- AND VANADYL- DOPED TRIGLYCINE SULFATE SINGLE CRYSTALS. EP 27-3161
ESR OF CHROMIUM(3+) IN CHROMIUM DOPED STRONTIUM TITANATE SINGLE CRYSTALS. 06-1147
REE RADICALS IN FERROELECTRIC AND ANTIFERROELECTRIC ARSENATE SINGLE CRYSTALS. ESR OF F 18-2731
ETCH PITS CORRESPONDING TO DISLOCATIONS IN BARIUM TITANATE SINGLE CRYSTALS. 04-1023
ETCH PITS IN BARIUM TITANATE SINGLE CRYSTALS. 04-0922
EVOLUTION OF FERROELECTRIC DOMAINS IN TGS SINGLE CRYSTALS. 27-3003
FEATURES OF THE PHASE TRANSITION IN ANTIMONY SULFOIODIDE SINGLE CRYSTALS. 15-2148
CTRIC DIELECTRIC ANOMALY IN BISMUTH(4) TITANIUM(3) OXIDE(12) SINGLE CRYSTALS. FERROELE 13-2009
FERROELECTRIC PROPERTIES OF LEAD TITANATE SINGLE CRYSTALS. 05-1046
ENCE OF ANOMALOUS ULTRASOUND DAMPING IN ANTIMONY SULFOIODIDE SINGLE CRYSTALS. FREQUENCY DEPEND 15-2139
TION OF THE SECOND ACOUSTIC HARMONIC IN ANTIMONY SULFOIODIDE SINGLE CRYSTALS. GENERA 15-2224
TION OF THE SECOND ACOUSTIC HARMONIC IN ANTIMONY SULFOIODIDE SINGLE CRYSTALS. GENERA 15-2223
GROWTH OF TRIGLYCINE SULFATE SINGLE CRYSTALS. 27-3083
HEAT CAPACITY OF RUBIDIUM DIHYDROGEN PHOSPHATE SINGLE CRYSTALS. 17-2415
HEAT TREATMENT INDUCED POINT DEFECTS IN LITHIUM METANIOBATE SINGLE CRYSTALS. 11-1668
AND OXYGEN NONSTOICHIOMETRY OF BARIUM TITANATE AND HEMATITE SINGLE CRYSTALS. HETEROVALENT SUBSTITUTION 04-0874
INFLUENCE OF DEFECTS ON SOME DIELECTRIC PROPERTIES OF TGS SINGLE CRYSTALS. 27-3063
IMPURITIES ON THE GROWTH AND SOME PHYSICAL PROPERTIES OF TGS SINGLE CRYSTALS. INFLUENCE OF 27-3084
NICAL STRESS ON THE PHASE TRANSITION OF ANTIMONY SULFOIODIDE SINGLE CRYSTALS. /NCE OF ONE DIMENSIONAL MECHA 15-2245
DE OF FREE VIBRATIONS ON THE PERMITTIVITY OF BARIUM TITANATE SINGLE CRYSTALS. INFLUENCE OF THE AMPLITU 04-0768
UENCE OF THE SURFACE LAYER ON THE PLASTIC DEFORMATION OF ICE SINGLE CRYSTALS. INFL 34-3459
LECTRICAL PROPERTIES AND DOMAIN STRUCTURE OF BARIUM TITANATE SINGLE CRYSTALS. /AL TENSILE STRESSES ON THE E 04-0961
BILITY OF FERROELECTRIC POLARIZATION IN ANTIMONY SULFOIODIDE SINGLE CRYSTALS. INSTA 15-2152
FEROMETRIC STUDIES OF DOMAIN STRUCTURES IN POTASSIUM NIOBATE SINGLE CRYSTALS. INTER 08-1318
RFEROMETRIC STUDIES OF DOMAIN STRUCTURE IN POTASSIUM NIOBATE SINGLE CRYSTALS. INTE 08-1319
INTERNAL FRICTION IN BARIUM TITANATE SINGLE CRYSTALS. 04-0794
STUDIES OF ELECTRIC CONDUCTION IN IRON DOPED BARIUM TITANATE SINGLE CRYSTALS. KERR EFFECT 04-0703
LAPPING CHARACTERISTICS OF LITHIUM NIOBATE SINGLE CRYSTALS. 11-1784
LATTICE DYNAMICAL THEORY OF SWITCHING IN BARIUM TITANATE SINGLE CRYSTALS. 03-0398
LIGHT ABSORPTION IN POTASSIUM DIHYDROGEN PHOSPHATE SINGLE CRYSTALS. 17-2421
MECHANICAL PROPERTIES OF ICE SINGLE CRYSTALS. 34-3441
MECHANICAL RELAXATION AND NONLINEARITY IN STRONTIUM TITANATE SINGLE CRYSTALS. 06-1218
OF 60 DEGREE DOMAIN WALLS IN FERROELECTRIC POTASSIUM NIOBATE SINGLE CRYSTALS. MECHANICAL SWITCHING 08-1421
MICROWAVE DYNAMIC NONLINEARITY OF BARIUM TITANATE SINGLE CRYSTALS. 04-0749
S IN POTASSIUM NITRATE, SILVER NITRATE, AND AMMONIUM NITRATE SINGLE CRYSTALS. /T POLYMORPHIC TRANSFORMATION 16-2260
ELECTRIC PROPERTIES OF FERRIC AMMONIUM SULFATE DODECAHYDRATE SINGLE CRYSTALS. MOSSBAUER STUDY ON FERRO 19-2770
CAL ACTIVITY OF FERROELECTRIC LEAD(5) GERMANIUM(3) OXIDE(11) SINGLE CRYSTALS. OPTI 35-3503
OPTICAL STUDIES OF DOMAIN STRUCTURE OF POTASSIUM NIOBATE SINGLE CRYSTALS. 08-1320
AND) THERMOCURRENTS IN TITANIUM DIOXIDE AND BARIUM TITANATE SINGLE CRYSTALS. (PHASE TRANSITIONS 04-0697
PHOTODIELECTRIC EFFECT IN ANTIMONY SULFOIODIDE SINGLE CRYSTALS. 15-2236
CTRIC PROPERTIES (PHOTOCONDUCTIVITY) OF ANTIMONY SULFOIODIDE SINGLE CRYSTALS. PHOTOELE 15-2134
PIEZOELECTRIC EFFECT IN LEAD TITANATE SINGLE CRYSTALS. 05-1049
ROELECTRIC POTASSIUM LITHIUM TANTALATE(0-0.5) NIOBATE(1-0.5) SINGLE CRYSTALS. /ZOELECTRIC PROPERTIES OF FER 12-1890
IEZOELECTRICITY AND ELECTROSTRICTION OF ANTIMONY SULFOIODIDE SINGLE CRYSTALS. P 15-2227
POLARIZATION REVERSAL IN TGS SINGLE CRYSTALS. 27-2980
UM MANGANESE TRIFLUORIDE AND POTASSIUM MANGANESE TRIFLUORIDE SINGLE CRYSTALS. /NATION OF LARGE, PURE RUBIDI 33-3397
HYSICAL PROPERTIES OF NONSTOICHIOMETRIC ANTIMONY SULFOIODIDE SINGLE CRYSTALS. PREPARATION AND P 15-2199
PREPARATION AND PROPERTIES OF (BARIUM, STRONTIUM) TITANATE SINGLE CRYSTALS. 04-0560
PREPARATION AND PROPERTIES OF PURE AND DOPED BARIUM TITANATE SINGLE CRYSTALS. 04-0668
PREPARATION OF AMMONIUM SULFATE SINGLE CRYSTALS. 19-2753
ATION OF PURE GLYCINE USED FOR GROWING OF TRIGLYCINE SULFATE SINGLE CRYSTALS. PREPAR 27-3085
DEPENDENCE OF THE COMPLEX DISPERSION COEFFICIENT OF ICE I(H) SINGLE CRYSTALS. PRESSURE 34-3479
PROBLEM OF THE SURFACE LAYER IN BARIUM TITANATE SINGLE CRYSTALS. 04-1013
PROPERTIES OF DEUTERATED TRIGLYCINE FLUOBERYLLATE SINGLE CRYSTALS. 27-2973
PYROELECTRIC EFFECT IN ANTIMONY SULFO AND SELENO IODIDE SINGLE CRYSTALS. 15-2163
UTIONS IN POTASSIUM DIHYDROGEN(1-X) DIDEUTERIUM(X) PHOSPHATE SINGLE CRYSTALS. REGULAR BEHAVIOR OF SOLID SOL 17-2458
RELAXATION PROCESSES OF ANTIMONY SULFOIODIDE SINGLE CRYSTALS. 15-2190
KS ABOUT SWITCHING IN BARIUM TITANATE AND TRIGLYCINE SULFATE SINGLE CRYSTALS. REMAR 04-0915
ROELASTIC PARAMAGNETIC (BETA) TERBIUM MOLYBDATE (CZOCHRALSKI SINGLE CRYSTALS). /PERATURE) FERROELECTRIC FER 30-3290
F THE THERMOCURRENTS IN BARIUM TITANATE AND TITANIUM DIOXIDE SINGLE CRYSTALS. SOME PROPERTIES O 04-0622
POLARIZATION AND COERCIVE FIELD OF LEAD TITANATE (FLUX GROWN SINGLE CRYSTALS). SPONTANEOUS 05-1050
STABILIZATION OF SPONTANEOUS POLARIZATION OF BARIUM TITANATE SINGLE CRYSTALS. 04-0795
STOICHIOMETRY AND OPTICAL QUALITY OF LITHIUM TANTALATE SINGLE CRYSTALS. 12-1883
STUDY OF CONDUCTIVITY AND DIPOLAR RELAXATION IN DOPED ICE SINGLE CRYSTALS. 34-3458
STUDY OF THE AGING PHENOMENON IN BARIUM TITANATE SINGLE CRYSTALS. 04-0545
SYNTHESIS AND STUDY OF LEAD SCANDIUM NIOBATE SINGLE CRYSTALS. 08-1325
MPERATURE DEPENDENCE OF INTERNAL FRICTION OF BARIUM TITANATE SINGLE CRYSTALS. TE 04-0796
DEPENDENCE OF THE SWITCHING PROCESS IN ANTIMONY SULFOIODIDE SINGLE CRYSTALS. TEMPERATURE 15-2191
MECHANICAL CHARACTERISTICS OF POTASSIUM DIHYDROGEN PHOSPHATE SINGLE CRYSTALS. /MPERATURE DEPENDENCE OF THE 17-2649
OF 9.4 GHZ HYPERSOUND IN ALUMINUM OXIDE AND LITHIUM NIOBATE SINGLE CRYSTALS. /EPENDENCES OF THE ABSORPTION 11-1719
HE HALL AND DRIFT MOBILITIES OF HOLES IN ANTIMONY TRISULFIDE SINGLE CRYSTALS. TEMPERATURE DEPENDENCES OF T 15-2193
ATE ON THE DEFECT STRUCTURE AND DIELECTRIC PROPERTIES OF TGS SINGLE CRYSTALS. THE EFFECT OF GROWTH R 27-3067
IC, PIEZOELECTRIC, AND ELASTIC PROPERTIES OF LITHIUM TANTALATE SINGLE CRYSTALS. /ERMAL DEPENDENCE OF DIELECTR 11-1693
THERMALLY STIMULATED CONDUCTIVITY IN STRONTIUM TITANATE SINGLE CRYSTALS. 06-1227
KNESS DEPENDENCE OF POLARIZATION REVERSAL IN BARIUM TITANATE SINGLE CRYSTALS. THIC 04-0949
THICKNESS DEPENDENCE OF THE PERMITTIVITY OF BARIUM TITANATE SINGLE CRYSTALS. 04-0637
VIBRATIONAL SPECTRA OF IODIC ACID SINGLE CRYSTALS. 36-3548
CTION TOPOGRAPHIC STUDIES OF THE DEFORMATION BEHAVIOR OF ICE SINGLE CRYSTALS. X-RAY DIFFRA 34-3426
ON TOPOGRAPHY STUDY OF DEFECTS (DISLOCATIONS) IN KDP AND ADP SINGLE CRYSTALS. X-RAY DIFFRACTI 17-2578
SERVATION OF 180 DEG DOMAINS IN FERROELECTRIC LEAD GERMANATE SINGLE CRYSTALS. X-RAY TOPOGRAPHIC OB 35-3515
APHIC OBSERVATIONS OF RADIATION DAMAGE IN TRIGLYCINE SULFATE SINGLE CRYSTALS. X-RAY TOPOGR 27-3106
TOPOGRAPHIC STUDIES ON LITHIUM NIOBATE AND LITHIUM TANTALATE SINGLE CRYSTALS. X-RAY 11-1830
ITTIVITY DISPERSION OF BARIUM TITANATE TYPE FERROELECTRICS. (SINGLE CRYSTALS AND CERAMICS) PERM 04-0903
GLE CRYSTALS. CONDUCTIVITY OF BARIUM TITANATE, PURE SINGLE CRYSTALS AND DOPED (IRON, HYDROXYL) SIN 04-0695
DIELECTRIC AND MECHANICAL RESPONSE OF ICE I(H) SINGLE CRYSTALS AND ITS INTERPRETATION. 34-3483
ICE. PART-2: DIELECTRIC AND MECHANICAL RESPONSE OF ICE I(H) SINGLE CRYSTALS AND ITS INTERPRETATION. /AM OF 34-3485

/IC PROPERTIES AND OPTICAL ABSORPTION CF POTASSIUM TANTALATE SINGLE CRYSTALS AND (POTASSIUM SODIUM TANTALA/ 08-1272
IR QUALITY. (VOLUME DISTRIBUTION OF/ PREPARATION OF PURE TGS SINGLE CRYSTALS AND SOME INVESTIGATIONS OF THE 27-3086
/RIC PROPERTIES OF THE CONTACTS BETWEEN ANTIMONY SULFOIODIDE SINGLE CRYSTALS AND SOME METALS. (ANTIMONY, S/ 15-2251
S.   FLUX GROWN KTN SINGLE CRYSTALS AND THEIR ELECTRICAL PROPERTIE 09-1533
TYPE) PHASE TRANSITIONS IN HIGH RESISTIVITY BARIUM TITANATE SINGLE CRYSTALS. (AND TWINNING TENDENCY) /ITIC 04-0887
/CTROOPTIC AND FERROELECTRIC PROPERTIES CF LITHIUM TANTALATE SINGLE CRYSTALS AS A FUNCTION OF MELT COMPOSI/ 12-1874
N PSEUDOCUBIC (001) / DOMAIN STRUCTURES IN POTASSIUM NIOBATE SINGLE CRYSTALS ASSOCIATED WITH STEP LADDERS O 08-1316
K).   DIELECTRIC PROPERTIES OF STRONTIUM TITANATE (ANNEALED SINGLE CRYSTALS) AT LOW TEMPERATURE (5-300 DEG 06-1201
MEASUREMENT OF THE PERMITTIVITY OF BARIUM TITANATE SINGLE CRYSTALS AT MICROWAVE FREQUENCIES.   04-0759
INTERNAL FRICTION OF LITHIUM NIOBATE SINGLE CRYSTALS AT 1-40 HZ.   11-1798
E FLUCTUATION ABSORPTION OF ULTRASOUND IN TRIGLYCINE SULFATE SINGLE CRYSTALS (AT 10, 30, 50 MHZ). /CE OF TH 27-3079
SPLIT INTERSTITIAL CONFIGURATION INDUCED IN PEROVSKITE TYPE SINGLE CRYSTALS BY NEUTRON BOMBARDMENT.   08-1467
GROWTH OF BARIUM LITHIUM NIOBATE SINGLE CRYSTALS BY THE CZOCHRALSKI METHOD.   13-1991
EACTANT. (POTASSIUM DIHYDRO/ METHOD CF GROWTH IN SOLUTION OF SINGLE CRYSTALS BY THE GRADUAL ADDITION OF A R 17-2595
C/ PREPARATION AND STUDY CF BARIUM(X) STRONTIUM(1-X) NIOBATE SINGLE CRYSTALS (CZOCHRALSKI GROWTH, DIELECTRI 13-1936
DIELECTRIC PROPERTIES OF BARIUM TITANATE SINGLE CRYSTALS DOPED BY HYDROXIDE GROUPS.   04-0634
PULSES.   RELAXATION PROCESSES IN ANTIMONY SULFOIODIDE SINGLE CRYSTALS EXCITED WITH MICROSECOND LIGHT 15-2202
PULLING FROM THE MELT, STRU/ GROWTH OF BARIUM SODIUM NIOBATE SINGLE CRYSTALS FOR OPTICAL APPLICATIONS. (BY 13-2045
SOLI/ EFFECT OF EXTERNAL ACTIONS ON THE DOMAIN STRUCTURE OF SINGLE CRYSTALS FORMED BY BARIUM TITANATE BASE 04-0954
E PLANES.   DOMAINS IN POTASSIUM NIOBATE SINGLE CRYSTALS FROM PSEUDOCUBIC (001) CLEAVAG 08-1317
MIC METHOD (ANTIMONY SESQUISELENIDE OR SULFIDE, ANT/ GROWING SINGLE CRYSTALS FROM THE VAPOR PHASE BY A DYNA 15-2162
/ DIELECTRIC PROPERTIES OF BARIUM(1-X) STRONTIUM(X) TITANATE SINGLE CRYSTALS GROWN FROM A POTASSIUM FLUORID 04-0635
ERROELECTRIC PROPERTIES OF LEAD ZINCATE(0.33) NIOBATE(0.67) (SINGLE CRYSTALS GROWN FROM LEAD OXIDE FLUX). F 13-2066
/NISOTROPY OF THE ELECTRICAL CONDUCTIVITY OF BARIUM TITANATE SINGLE CRYSTALS GROWN FROM SOLUTION IN A POTA/ 04-1035
RESONANCE EFFECTS IN MAGNETICALLY EQUIAXIAL FERRODIELECTRIC SINGLE CRYSTALS HAVING DOMAIN STRUCTURE.   03-0509
RMI/ THERMAL AND ELECTRICAL PROPERTIES OF TRIGLYCINE SULFATE SINGLE CRYSTALS (HEAT CAPACITY AND ELECTRIC PE 27-3141
/EPENDENCE OF THE ELECTRICAL CONDUCTIVITY OF BARIUM TITANATE SINGLE CRYSTALS HEATED IN OXYGEN (EFFECT TIED/ 04-0997
/E, ABSORPTION SPECTRA) OF SEMICONDUCTING STRONTIUM TITANATE SINGLE CRYSTALS (HYDROGEN REDUCED, CERIUM OR / 06-1198
DIELECTRIC CONSTANT OF TRIGLYCINE SULFATE SINGLE CRYSTALS IN A STRONG ELECTRIC FIELD.   27-3150
TEMPERATURE AUTOSTABILIZATION EFFECT OF TRIGLYCINE SULFATE SINGLE CRYSTALS IN AN AC ELECTRIC FIELD.   27-3005
GROWTH OF POTASSIUM DIHYDROGEN PHOSPHATE SINGLE CRYSTALS IN GEL.   17-2455
OXIDE BINARY SYSTEM.   GROWTH OF SINGLE CRYSTALS IN LEAD MONOXIDE- GERMANIUM DI 35-3514
AL PHASES.   REPOLARIZATION PROCESSES OF BARIUM TITANATE SINGLE CRYSTALS IN ORTHORHOMBIC AND RHOMBOHEDR 04-0963
DISCUSSION ON THE CRYSTAL GROWTH OF BARIUM TITANATE SINGLE CRYSTALS IN POTASSIUM FLUORIDE FLUX.   04-1022
SPHATE- POTASSIUM DIHYDROGEN PHOSP/ DIELECTRIC PROPERTIES OF SINGLE CRYSTALS IN THE AMMONIUM DIHYDROGEN PHO 17-2572
PIEZOELECTRIC ACTIVITY OF BARIUM TITANATE SINGLE CRYSTALS IN THE PARAELECTRIC PHASE.   04-0771
.   DIELECTRIC PERMEABILITY OF BARIUM TITANATE SINGLE CRYSTALS IN THE PHASE TRANSITION REGION 04-0645
/OF THE OPTICAL ABSORPTION COEFFICIENT OF LEAD MAGNONIOBATE SINGLE CRYSTALS IN THE REGION OF A BROAD FERR/ 13-1974
ENCE.   POLARIZATION REVERSAL CF ANTIMONY SULFOIODIDE SINGLE CRYSTALS IN THE REGION OF PHASE COEXIST 15-2246
/VE TEMPERATURE COEFFICIENT OF RESISTANCE OF BARIUM TITANATE SINGLE CRYSTALS IN THE REGION OF THE CURIE TE/ 15-2201
/ DEPENDENCE OF THE ELASTIC COMPLIANCE OF STRONTIUM TITANATE SINGLE CRYSTALS IN THE TEMPERATURE RANGE 20 T/ 04-1036
K-300 DEG K.   BRILLOUIN SCATTERING IN STRONTIUM TITANATE SINGLE CRYSTALS IN THE TEMPERATURE RANGE 5 DEG 06-1220
ERR/ (X-RAY) STRUCTURAL (OPTICAL) AND DIELECTRIC STUDIES OF (SINGLE CRYSTALS IN THE TUNGSTEN BRONZE TYPE) F 06-1149
DMIXED NICKEL ON SOME PHYSICAL PROPERTIES OF BARIUM TITANATE SINGLE CRYSTALS. (LATTICE PARAMETERS) / SOME A 13-1947
PHASE BOUNDARIES IN BARIUM TITANATE SINGLE CRYSTALS NEAR CURIE POINT.   04-0909
EGREES K.   DIELECTRIC NONLINEARITY OF STRONTIUM TITANATE SINGLE CRYSTALS NEAR PHASE TRANSITION AT 110 D 04-0577
. ULTRASONIC THIRD HARMONIC GENERATION IN STRONTIUM TITANATE SINGLE CRYSTALS (NONLINEAR ELASTIC PROPERTIES) 06-1181
LITHIUM, STRONTIU/ SYNTHESIS AND X-RAY DIFFRACTION STUDY OF SINGLE CRYSTALS OF A NEW PYROCHLORE CONTAINING 36-3613
PREPARATION OF SINGLE CRYSTALS OF AMMONIUM SULFATE.   19-2758
ANISOTROPY OF ANOMALOUS ABSORPTION OF ULTRASOUND IN SINGLE CRYSTALS OF ANTIMONY SULFOIODIDE.   15-2222
DILATOMETRIC INVESTIGATION OF PHOTOFERROELECTRIC EFFECTS IN SINGLE CRYSTALS OF ANTIMONY SULFOIODIDE.   15-2156
INFLUENCE OF LIGHT ON SELF POLARIZATION IN SINGLE CRYSTALS OF ANTIMONY SULFOIODIDE.   15-2228
OPPOSING FERROELECTRIC DOMAINS IN SINGLE CRYSTALS OF ANTIMONY SULFOIODIDE.   15-2252
SOLUTIONS.   DIELECTRIC AND ELASTIC PROPERTIES OF SINGLE CRYSTALS OF BARIUM LEAD TITANATE SOLID 04-0597
DYNAMIC FATIGUE OF SINGLE CRYSTALS OF BARIUM TITANATE.   04-0792
S ON THE DOMAIN STRUCTURE AND THE PYROELECTRIC PROPERTIES OF SINGLE CRYSTALS OF BARIUM TITANATE. / THICKNES 04-1039
ENCE OF A STEADY ELE/ RELAXATION ABSORPTION OF ULTRASOUND IN SINGLE CRYSTALS OF BARIUM TITANATE IN THE PRES 04-0850
ADDED.   SOME CRYSTAL CHEMICAL AND PHYSICAL PROPERTIES OF SINGLE CRYSTALS OF BARIUM TITANATE WITH COBALT 04-0618
ATE SOLID SOLUTIONS.   ELECTROOPTICAL EFFECT IN SINGLE CRYSTALS OF BARIUM TITANATE- ZINC TITAN 04-0958
(20) AT LOW TEMPERATURES.   ULTRASONIC ATTENUATION IN SINGLE CRYSTALS OF BISMUTH(12) GERMANIUM OXIDE 35-3510
OIODIDE NEAR THE PHASE TRANSITION TE/ ELECTROREFLECTION FROM SINGLE CRYSTALS OF FERROELECTRIC ANTIMONY SULF 15-2151
YDROGEN PHOSPHATE.   ESR INVESTIGATIONS OF X-RAY IRRADIATED SINGLE CRYSTALS OF FERROELECTRIC POTASSIUM DIH 17-2558
THE COMPOSITE RESONATO/ MEASUREMENT OF INTERNAL FRICTION IN SINGLE CRYSTALS OF FERROELECTRIC SUBSTANCES BY 27-3080
ALT AND TRIGLY/ AMPLITUDE DEPENDENCE OF INTERNAL FRICTION IN SINGLE CRYSTALS OF FERROELECTRICS. (ROCHELLE S 27-3122
ROELECTRIC AND FERROELASTIC SUBSTA/ GROWTH AND MORPHOLOGY OF SINGLE CRYSTALS OF GADOLINIUM MOLYBDATE, A FER 30-3259
OR (ELECTRON NUCLEAR DOUBLE RESONANCE) STUDY OF X-IRRADIATED SINGLE CRYSTALS OF ICE.   END 34-3413
HARDNESS ANISOTROPY OF SINGLE CRYSTALS OF ICE I(H).   34-3460
LUX) GROWTH AND FERROELECTRIC PROPERTIES OF HIGH RESISTIVITY SINGLE CRYSTALS OF LEAD TITANATE.   (F 05-1072
S PROPERTIES.   GROWTH OF SINGLE CRYSTALS OF LEAD ZINCATE NIOBATE AND IT 36-3594
PROPERTIES OF THIS COMPOSITION.   GROWTH OF SINGLE CRYSTALS OF LEAD (ZINCATE NIOBATE) AND 13-1990
/RIPLE RESONANCE FOR THE ARSENATE CENTER IN X-RAY IRRADIATED SINGLE CRYSTALS OF POTASSIUM DIHYDROGEN ARSEN/ 18-2729
/RECENT ADVANCES IN FERROELECTRICS. (ELECTROOPTIC MATERIALS, SINGLE CRYSTALS OF POTASSIUM TANTALATE NIOBAT/ 02-0073
E DIAGRAMS, CONVE/ FABRICATION (BY FLUX GROWTH) OF HEXAGONAL SINGLE CRYSTALS OF RARE EARTH MANGANATES (PHAS 36-3536
ELATION BETWEEN BARKHAUSEN NOISE AND SURFACE MICRODAMAGES ON SINGLE CRYSTALS OF SEIGNETTE SALT.   CORR 28-3208
IUM TITANATE, WHERE BARIUM(2+) AND TITANIUM(4+) ION/ GROWING SINGLE CRYSTALS OF SOLID SOLUTION BASED ON BAR 04-0960
ITANATE AND ZINC OXIDE.   ELECTROOPTICAL PROPERTIES OF SINGLE CRYSTALS OF SOLID SOLUTIONS OF BARIUM T 04-0953
CTOR ANTIMONY SULFOIODIDE.   LUMINESCENCE OF SINGLE CRYSTALS OF THE FERROELECTRIC SEMICONDU 15-2229
UM) TITANATE. (SYNTHESIS AND LOW FREQUENCY DIELECTRIC PROPE/ SINGLE CRYSTALS OF THE SYSTEM (BARIUM, STRONTI 04-0561
ONG ELECTRIC FIELD.   PERMITTIVITY OF SINGLE CRYSTALS OF TRIGLYCINE SULFATE IN A STR 27-2978
/N ELECTRICAL BOUNDARY CONDITIONS ON THE THERMAL CAPACITY OF SINGLE CRYSTALS OF TRIGLYCINE SULFATE (NEAR T/ 27-3140
MPLEX COMPOSITIONS, AND THE STUDY OF THEIR STRUCT/ GROWTH OF SINGLE CRYSTALS OF VARIOUS PEROVSKITES WITH CO 08-1428
THE ANGLE/ EFFECT OF GROWTH CONDITIONS OF TRIGLYCINE SULFATE SINGLE CRYSTALS ON A CHANGE IN THE TANGENT OF 27-3153
. (REVIEW)   INFLUENCE OF THE QUALITY OF TRIGLYCINE SULFATE SINGLE CRYSTALS ON THEIR DIELECTRIC PROPERTIES 02-0088
DEPENDENCE OF DIELECTRIC DISPERSION IN BARIUM TITANATE SINGLE CRYSTALS ON THEIR DOMAIN STRUCTURE.   04-0895
DEPENDENCE OF COERCIVE FIELD OF BARIUM TITANATE SINGLE CRYSTALS ON THEIR THICKNESS.   04-0693
/G THE DOMAIN STRUCTURE IN POTASSIUM FERROCYANIDE TRIHYDRATE SINGLE CRYSTALS, P-N JUNCTIONS AND P- AND N- / 24-2904
OF ICE I(H). PART/ TRANSFER OF PROTONS THROUGH PURE ICE I(H) SINGLE CRYSTALS. PART-1: POLARIZATION SPECTRA 34-3486
POLARIZATION A/ TRANSFER OF PROTONS THROUGH "PURE" ICE I(H). SINGLE CRYSTALS. PART-2: MOLECULAR MODELS FOR 34-3455
ITUDE DEPENDENCE OF THE INTERNAL FRICTION IN BARIUM TITANATE SINGLE CRYSTALS. (PIEZOELECTRIC MODULUS)   AMPL 04-0791
NONLINEAR OPTICAL PROPERTIES OF POTASSIUM NIOBATE SINGLE CRYSTALS. (REFRACTIVE INDICES)   08-1454

/TATION. PART-8: TRANSFER OF PROTCNS THROUGH "PURE" ICE I(H) SINGLE CRYSTALS. SECT-1: POLARIZATION SPECTRA/ 34-3482
ES.   BARIUM TITANATE SINGLE CRYSTALS SUBJECTED TO MECHANICAL IMPULS 04-1006
N TRIOXIDE).   PREPARATION OF TUNGSTIC OXIDE SINGLE CRYSTALS (SUBLIMATION IN AIR OF TUNGSTE 10-1644
ATION / ANISCTROPY CF SOUND ABSORPTION IN TRIGLYCINE SULFATE SINGLE CRYSTALS. (THEORY OF ANISOTROPIC RELAX 27-3078
           CIELECTRIC CONSTANT OF POTASSIUM NIOBATE SINGLE CRYSTALS UNDER BIASING CONDITIONS.   08-1381
                    POINT DEFECTS IN LITHIUM NIOBATE SINGLE CRYSTALS USED IN LASER MODULATION.   11-1737
ATIO/ INVESTIGATION OF CERTAIN PARAMETERS OF LITHIUM NIOBATE SINGLE CRYSTALS USING SCATTERING OF LASER RADI 11-1797
/ICONDUCTING COMPOUNDS OF GROUP A(V)B(VI)C(VII). (HIGH GRADE SINGLE CRYSTALS, V IS ANTIMONY OR BISMUTH, VI/ 15-2214
           PECULIARITIES OF POLARIZATION OF BARIUM TITANATE SINGLE CRYSTALS WITH DOUBLE HYSTERESIS LOOP.   04-0709
/TION OF SOLID SOLUTIONS OF FERROELECTRIC TRIGLYCINE SULFATE SINGLE CRYSTALS WITH ISOMORPHOUS COMPOUNDS. (/ 27-2972
/IC, ELASTIC ANC PIEZOELECTRIC PRCFERTIES OF BARIUM TITANATE SINGLE CRYSTALS WITH LAMINAR DOMAIN STRUCTURE. 04-1010
UM TITANATE HAFNATE GROWN FROM POTASSIUM FLUO/ FERROELECTRIC SINGLE CRYSTALS WITH LARGE NONLINEARITY. (BARI 08-1373
THE 184 DEG K TRANSITION OF POTASSIUM MANGANESE TRIFLUORIDE SINGLE CRYSTALS, (X-RAY DIFFUSE SCATTERING). / 33-3369
/ (CONSTANT ANC LOSS) PEHAVIOR CF REDUCED STRONTIUM TITANATE SINGLE CRYSTALS (40 HZ-10 MHZ, EFFECTS OF RED/ 06-1148
                          ELECTRICAL CONDUCTIVITY OF SINGLE CRYSTALLINE BARIUM TITANATE.   04-0663
PROPERTIES OF THE IRON TRANSITICN GROUP IN THE LATTICE OF SINGLE CRYSTALLINE LITHIUM NIOBATE.   11-1669
CS)   INELASTIC NEUTRON SCATTERING FROM SINGLE DOMAIN BARIUM TITANATE. (LATTICE DYNAMI 04-0946
                          DEPOLING OF SINGLE DOMAIN BISMUTH TITANATE.   08-1446
                     GROWTH OF LITHIUM NIOBATE SINGLE DOMAIN CRYSTALS.   11-1781
M NIOBATE. (CRACK-FREE, THROUGH AD/ GROWTH AND PROPERTIES OF SINGLE DOMAIN CRYSTALS OF FERROELECTRIC LITHIU 11-1774
     MECHANICAL MODEL OF LONG WAVELENGTH VIBRATIONS IN A (SINGLE DOMAIN) FERROELECTRIC TYPE CRYSTAL.   03-0342
).   CCMPLEX PHASE TRANSITIONS IN CRYSTALS (APPLICATION TO SINGLE DOMAIN FERROELECTRICS BELOW CURIE POINT 03-0319
/ GENERATION OF SUPERHIGH FREQUENCY ACOUSTIC HARMONICS IN A (SINGLE DOMAIN) LITHIUM NIOBATE CRYSTAL (NONLIN 11-1707
                DIELECTRIC PROPERTIES OF SINGLE DOMAIN MELT GROWN BARIUM TITANATE.   04-0616
HAVING A SECOND ORDER PHASE TRANSITION (THEORY).   SINGLE DOMAIN POLARIZATION OF A FERROELECTRIC   03-0202
Y UNDER UNIAXIAL STRESS).   DIRECT OBSERVATION OF SINGLE DOMAIN STRONTIUM TITANATE (OPTICAL STUD 06-1112
R LAYER DIELECTRICS.   SINGLE FIRE PROCESS FOR BARIUM TITANATE BARRIE 04-1029
PECIFIC HEAT IN FERROELECTRIC PHASE TRANSITION IN TERMS OF A SINGLE MODE ANHARMONIC OSCILLATOR MODEL. /ND S 03-0406
                     INTRINSIC NONSTOICHIOMETRY IN SINGLE PHASE LEAD ZIRCONATE TITANATE.   09-1534
TIES CF FERROELECTRI/ INFLUENCE OF FIRING OF BARIUM TITANATE SINTERS ON ELECTRICAL AND TECHNOLOGICAL PROPER 04-1002
ADDITIONS ON BARIUM TITANATE CERAMICS. CRYSTAL STRUCTURE AND SINTERABILITY.   EFFECT OF CLAY AND DOLOMITE 04-0739
    INFLUENCE OF ADDITIVES ON DENSIFICATION AND PROPERTIES OF SINTERED AN PRESSURE SINTERED BARIUM TITANATE. 04-1026
          SCANNING ELECTRON MICRCSCCPIC EXAMINATION OF SINTERED BARIUM SODIUM NIOBATES.   13-1955
CORRELATION BETWEEN RHEOLOGICAL AND ELECTRICAL BEHAVIOR IN SINTERED BARIUM TITANATE.   04-1027
N SIZE.   VARIATION IN CERTAIN ELECTRICAL PROPERTIES OF SINTERED BARIUM TITANATE AS A FUNCTION OF GRAI 04-0627
TRONTIUM TITANATE.   IONIC CONDUCTION IN SINTERED OXIDES BASED ON CALCIUM TITANATE OR S 06-1141
FABRICATION OF ELECTROOPTIC PLZT CERAMICS BY ATMOSPHERE SINTERING.   09-1600
ZIRCONATE TITANATE CERAMICS.   SINTERING AND FERROELECTRIC PROPERTIES OF LEAD 09-1481
EAD ZINCATE(/ HIGH PRESSURE SYNTHESIS (SOLID STATE REACTION, SINTERING) AND PROPERTIES OF PEROVSKITE TYPE L 08-1393
1. PART-2: EFFECTS OF AVERAGE PARTICLE SIZE / STUDIES ON THE SINTERING BETWEEN LEAD MONOXIDE AND PZT. PART- 09-1624
          COLD PRESSING AND LOW TEMPERATURE SINTERING OF ALKOXY-DERIVED PLZT AND PLZT.   09-1493
          EXAGGERATED GRAIN GROWTH IN LIQUID PHASE SINTERING OF BARIUM TITANATE.   04-0835
                RESISTIVITY AND SINTERING OF BARIUM TITANATE.   04-0819
          EFFECT OF OXIDE ADDITIONS ON THE SINTERING OF BARIUM TITANATE COMPACT   04-0996
                     POINT DEFECTS AND SINTERING OF LEAD ZIRCONATE TITANATE.   09-1478
          EFFECTS OF CALCINING ON SINTERING OF LEAD ZIRCONATE TITANATE CERAMICS. 09-1494
IRCONATE TITANATE.   SINTERING SCANDIUM AND NIOBIUM MODIFIED LEAD Z 09-1555
O PYROELECTRIC DETECTO/ LASER STUDY OF REVERSIBLE NUCLEATION SITES IN TRIGLYCINE SULFATE AND APPLICATIONS T 27-3011
D CURRENT IN FERROELECTRIC OXIDE CRYSTALS. (POTASS/ TRAPPING SITES, OPTICAL DAMAGE, AND THERMALLY STIMULATE 13-1926
UARE LATTICE).   PHASE TRANSITION IN A SIXTEEN VERTEX LATTICE MODEL (FERROELECTRIC SQ 03-0517
CTRUM OF FERROELECTRICS STUDIED BY LIMITATION OF THE CRYSTAL SIZE.   PHONON SPE 03-0196
ROPERTIES OF SINTERED BARIUM TITANATE AS A FUNCTION OF GRAIN SIZE.   VARIATION IN CERTAIN ELECTRICAL P 04-0627
L PROPERTIES OF TRANSPARENT FERROELECTRIC / EFFECTS OF GRAIN SIZE AND POROSITY ON THE ELECTRICAL AND OPTICA 13-2004
TRA OF BARIUM TITANATE AND STRONTIUM TIT/ EFFECT OF PARTICLE SIZE AND SHAPE ON THE INFRARED ABSORPTION SPEC 04-0823
HARACTERISTICS OF BARIUM TITANATE.   GRAIN SIZE DEPENDENCE OF DIELECTRIC AND STRUCTURAL C 04-0582
RAMICS (HOT PRESSED LEAD ZIRCONATE TITANATE, EFFECT OF GRAIN SIZE, DOPANTS, REVIEW). /TS IN ELECTROOPTIC CE 02-0086
TE)   SIZE EFFECTS IN FERROELECTRICS. (BARIUM TITANA 04-0656
BIC HYDROGEN CHLORIDE CRYSTALS.   CRYSTAL SIZE EFFECTS IN THE RAMAN SPECTRA OF ORTHORHOM 31-3338
/NOXIDE AND PZT. PART-1. PART-2: EFFECTS OF AVERAGE PARTICLE SIZE OF RAW POWDER ON SINTERING. PART-3: SINT/ 09-1624
ADIANCE PROFILE ON THE SWITCHING CHARACTERISTICS AND THE PIT SIZE OF THE FE-PC MEMORY DEVICE. /GAUSSIAN IRR 37-3670
ION OF BARIUM TITANATE.   EFFECT OF GRAIN SIZE ON THE FERROELECTRIC PARAELECTRIC TRANSIT 04-0917
IUM TITANATE.   EFFECT OF CRYSTALLINE GRAIN SIZE ON THE PHYSICAL PROPERTIES OF CERAMIC BAR 04-0785
ES OF BARIUM TITANATE.   EFFECT OF GRAIN SIZE ON THE PIEZOELECTRIC AND ELASTIC PROPERTI 04-1011
IVE BARIUM TITANATE CERAMICS.   EFFECT OF GRAIN SIZE ON THE RESISTIVITY ANOMALY IN SEMICONDUCT 04-0735
(010) PLANE OF ROCHELLE SALT IN THE PRESENCE OF PURE DIRECT SKY BLUE DYE.   GROWTH RATE OF THE 28-3217
RSENATE RADICAL IN IRRADIATED FERROEL/ ESR IDENTIFICATION OF SLATER CONFIGURATIONS AND OF EXCHANGE OF THE A 18-2734
RADIATED POTAS/ HYDROGEN BONDED FERROELECTRICS: EVIDENCE FCR SLATER CONFIGURATIONS FROM EPR STUDIES OF X-IR 18-2728
                LYNAMIC MODEL OF THE SLATER KDP FERROELECTRIC.   03-0285
NT/ DISTRIBUTION OF ZEROES OF THE PARTITION FUNCTION FOR THE SLATER MODEL OF FERROELECTRICITY (KDP MODEL, A 03-0308
ON TO THE CALCULATION OF TETRAGCNA/ ELECTROSTATIC FIELD IN A SLIGHTLY ORTHORHOMBIC IONIC CRYSTAL. APPLICATI 04-0578
          ELECTRONIC CONDUCTION IN SLIGHTLY REDUCED STRONTIUM TITANATE AT LOW TEM 06-1151
          GROWTH OF TRIGLYCINE SULFATE FROM SLIGHTLY SUPERSATURATED SOLUTIONS.   27-3096
GLYCINE SULFATE).   CRITICAL SLOWING DOWN AT FERROELECTRIC TRANSITIONS (TRI 27-3155
ILVER SODIUM DINITRITE BY THERMAL NOI/ STUDY OF THE CRITICAL SLOWING DOWN OF THE DIELECTRIC RELAXATION IN S 16-2382
RELAXATION IN SEVERAL CRYSTALS OF ORDER DISORDER / CRITICAL SLOWING DOWN PROCESS OF (ANOMALOUS) DIELECTRIC 02-0087
ELECTROOPTICAL EFFECT IN FERROELECTRIC CRYSTALS HAVING A SMEARED PHASE TRANSITION.   03-0174
TE DURING SWITCHING (CALCULATIONS COMPARED WITH DA/ (STEP OR SMOOTH) DOMAIN WALL STRUCTURE IN BARIUM TITANA 04-0677
OTASSIUM(X) SODIUM(0.5-X) TITANATE, BISMUTH(X) (POTASSIUM OR SODIUM)(X) LEAD (1-2X) ZIRCONATES). /TH(0.5) P 08-1289
/F THE SYMMETRY OF THE FERROELECTRIC COMPOUNDS (POTASSIUM OR SODIUM)(0.5) BISMUTH(0.5) TITANATE AND THE HI/ 08-1362
ULFATE.   INVESTIGATION OF FERROELECTRICITY IN SODIUM AMMONIUM SELENATE AND SODIUM AMMONIUM S 19-2742
     ELECTROMECHANICAL PRCPERTIES OF THE FERROELECTRIC SODIUM AMMONIUM SELENATE DIHYDRATE.   22-2884
OTON MAGNETIC RESONANCE, 84 DEG K AND 120-300 DEG K, EXAMPLE SODIUM AMMONIUM SELENATE DIHYDRATE). /UNDS (PR 22-2826
     LINEAR ELECTROOPTICAL EFFECT IN THE FERROELECTRIC SODIUM AMMONIUM SELENATE DIHYDRATE.   22-2829
RANSITION IN CRYSTALS WITH REORIENTATING STRUCTURAL GROUPS. (SODIUM AMMONIUM SELENATE DIHYDRATE) /C PHASE T 22-2827
NUCLEAR MAGNETIC RESCNANCE CF SODIUM-23 IN THE SODIUM AMMONIUM SELENATE DIHYDRATE CRYSTAL.   22-2883
          PREPARATION AND ANALYSIS OF SODIUM AMMONIUM SELENATE DIHYDRATE CRYSTALS.   22-2863
                DOMAIN STRUCTURE OF FERROELECTRIC SODIUM AMMONIUM SULFATE DIHYDRATE.   19-2778
/AND SOME OTHER STUDIES OF FERROELECTRIC PHASE TRANSITION IN SODIUM AMMONIUM SULFATE DIHYDRATE. (CRYSTALS) 19-2767
YDROGEN CR TRIDEUTERIUM D/ THERMAL AND SOME OTHER STUDIES OF SODIUM AMMONIUM SULFATE DIHYDRATE, SODIUM TRIH 19-2779
          ESR OF COPPER(2+) DOPED SODIUM AMMONIUM TARTRATE.   29-3252

NSFORMATION OF PECULIAR FERROELECTRIC- FERROELASTIC CRYSTAL: SODIUM AMMONIUM TARTRATE. /SM OF THE PHASE TRA 29-3238
PROTCN SPIN LATTICE RELAXATION IN FERROELECTRIC SODIUM AMMONIUM TARTRATE. 29-3248
PREPARATION OF SODIUM AND BARIUM NIOBATES AND LEAD MOLYBDATE. 04-0657
RAMAN STUDY OF METALLIC TUNGSTEN BRONZES. (SODIUM AND RUBIDIUM TUNGSTATES) 10-1654
/ECTRIC PROPERTIES OF (THE TUNGSTEN BRONZE SYSTEM) POTASSIUM SODIUM BARIUM NIOBATE (SECOND HARMONIC GENERA/ 13-2028
GSTEN BRONZE TYPE STRUCTURE. SODIUM BARIUM RARE EARTH NIOBATES WITH THE TUN 13-1972
/OTASSIUM TUNGSTEN BRONZE. (POTASSIUM BARIUM TANTALUM OXIDE, SODIUM BARIUM TANTALUM OXIDE, POTASSIUM STRON/ 13-2050
FROM BARIUM SULFATE AND TITANIUM DIOXIDE IN THE PRESENCE OF SODIUM CARBONATE AT HIGH TEMPERATURE. /ITANATE 04-0624
THE VIBRATIONAL SPECTRUM AND DIELECTRIC BEHAVIOR OF SODIUM CHLORATE. (NOT FERROELECTRIC) 36-3549
/URE AND PROPAGATION DIRECTION UPON OPTICAL PHONONS IN CUBIC SODIUM CHLORIDE AND EPITAXIAL FERROELECTRIC S/ 29-3243
/ AQUEOUS SOLUTIONS. (POTASSIUM BROMIDE, POTASSIUM CHLORIDE, SODIUM CHLORIDE, POTASSIUM SODIUM TARTRATE, P/ 17-2707
N TUNGSTEN BRONZES, HYDROGEN(X) TUNGSTATE. (ANALOGOUS TO THE SODIUM COMPOUNDS) / SUSCEPTIBILITY) OF HYDROGE 13-1934
NUCLEAR SPIN LATTICE RELAXATION TIME MEASUREMENTS IN SODIUM DEUTERIUM SELENATE. 22-2847
NUCLEAR MAGNETIC RESONANCE OF DEUTERIUM IN THE FERROELECTRIC SODIUM DEUTERO AMMONIUM SELENATE DIDEUTERATE. 22-2828
DIELECTRIC RELAXATION IN FERROELECTRIC SILVER SODIUM DINITRITE. 16-2290
DIELECTRIC RELAXATION IN FERROELECTRIC SILVER SODIUM DINITRITE. 16-2292
TRIC STUDIES ON THE FERROELECTRIC PHASE TRANSITION IN SILVER SODIUM DINITRITE. DIELEC 16-2289
FERROELECTRIC PHASE TRANSITION IN SILVER SODIUM DINITRITE 16-2266
CT ON THE DIELECTRIC RELAXATION TIME IN FERROELECTRIC SILVER SODIUM DINITRITE. HYDROSTATIC PRESSURE EFFE 16-2294
NMR STUDY OF SODIUM-23 IN SILVER SODIUM DINITRITE. 16-2324
PYROELECTRIC STUDY ON THE SPONTANEOUS POLARIZATION IN SILVER SODIUM DINITRITE. 16-2295
SPECIFIC HEAT ANOMALY IN FERROELECTRIC SILVER SODIUM DINITRITE. 16-2296
NG OF THE SODIUM-23 NMR CENTRAL LINE IN FERROELECTRIC SILVER SODIUM DINITRITE. SPLITTI 16-2307
THERMAL EXPANSION IN FERROELECTRIC SILVER SODIUM DINITRITE. 16-2346
IFFERENTIAL THERMAL ANALYSIS) PROPERTIES OF SILVER SODIUM DINITRITE AND SILVER SODIUM NITRITE. (D 16-2370
D. FERROELECTRIC HYSTERESIS LOOP OF SILVER SODIUM DINITRITE BY QUASI STATIC ELECTRIC FIEL 16-2293
DIELECTRIC RELAXATION MEASUREMENT ON FERROELECTRIC SILVER SODIUM DINITRITE BY STEP FUNCTION METHOD. 16-2291
/RITICAL SLOWING DOWN OF THE DIELECTRIC RELAXATION IN SILVER SODIUM DINITRITE BY THERMAL NOISE MEASUREMENT/ 16-2382
/N WITH THEIR PHASE TRANSITIONS. PART-1: TGS. PART-2: SILVER SODIUM DINITRITE. PART-3: ROCHELLE SALT. PART/ 02-0077
ALS. TEMPERATURE DEPENDENCE OF SOME PROPERTIES OF SODIUM DITHORIUM PHOSPHATE FERROELECTRIC CRYST 36-3611
/RYSTALS WITH POTASSIUM TUNGSTEN BRONZE STRUCTURE (POTASSIUM SODIUM LITHIUM NIOBATE, POTASSIUM STRONTIUM L/ 13-1923
FERROELECTRICS. SUPERLATTICE STRUCTURE OF STRONTIUM SODIUM LITHIUM NIOBATE (TUNGSTEN BRONZE) TYPE 13-2002
/ (WHERE A IS ZIRCONIUM, HAFNIUM, CERIUM, THORIUM, TITANIUM, SODIUM, MAGNESIUM, CADMIUM, YTTRIUM, GADOLINI/ 10-1640
THE ALKALINE TRIHYDROGEN DISELENITE FAMILY. (ESPECIALLY THE SODIUM MEMBER, AND EFFECTS OF DEUTERATION) /OF 22-2848
NISOTROPIC DIFFUSE X-RAY SCATTERING AND PHASE TRANSITIONS IN SODIUM NIOBATE. A 08-1357
CONTINUOUS HOT PRESSING OF POTASSIUM SODIUM NIOBATE. 13-2007
CRYSTAL GROWTH OF SODIUM NIOBATE. 08-1409
DIFFUSE X-RAY STUDY OF 640 DEGREES C PHASE TRANSITION IN SODIUM NIOBATE. 08-1358
NT ON THE PROPERTIES AND PROCESSING OF SINGLE CRYSTAL BARIUM SODIUM NIOBATE. /ICHIOMETRY AND HYDROXYL CONTE 13-1915
S OF CHANGES IN MELT COMPOSITION ON CRYSTAL GROWTH OF BARIUM SODIUM NIOBATE. EFFECT 13-1908
FAR INFRARED DIELECTRIC DISPERSION IN BARIUM SODIUM NIOBATE. 13-1986
RN FIXING IN ELECTROOPTIC CRYSTALS. (LITHIUM NIOBATE, BARIUM SODIUM NIOBATE) HOLOGRAPHIC PATTE 11-1661
HOLOGRAPHIC STORAGE IN DOPED BARIUM SODIUM NIOBATE. 13-1898
HOT PRESSING OF POTASSIUM- SODIUM NIOBATE. 09-1542
LOW FREQUENCY VIBRATIONAL MODES IN BARIUM SODIUM NIOBATE. 13-1905
DIELECTRIC CONSTANTS AND ELECTROOPTIC COEFFICIENTS IN BARIUM SODIUM NIOBATE. /EE AND CLAMPED VALUES OF THE 13-1998
NEW DATA FROM AN X-RAY STUDY OF PHASE TRANSFORMATIONS IN SODIUM NIOBATE. 08-1361
NIOBIUM-93 AND SODIUM-23 IN POLYCRYSTALLINE SODIUM NIOBATE. 08-1469
(NIOBIUM-93 AND SODIUM-23 RESONANCE IN POTASSIUM NIOBATE AND SODIUM NIOBATE) /ONS IN PEROVSKITES CRYSTALS. 08-1283
S IN ANTIFERROELECTRICS. (APPLIED TO ADP, TUNGSTEN TRIOXIDE, SODIUM NIOBATE) /ENTATION OF COMPOSITION PLANE 08-1327
AND ELECTRON SPIN RESONANCE OF IRON AND MOLYBDENUM IN BARIUM SODIUM NIOBATE. PHOTOCHROMIC EFFECT 13-1930
HOMBIC CRYSTALS. (BISMUTH(12) GERMANIUM OXIDE(20) AND BARIUM SODIUM NIOBATE) /ACE WAVES IN CUBIC AND ORTHOR 13-2041
POLARITONS IN A BIAXIAL CRYSTAL OF BARIUM SODIUM NIOBATE. 13-1992
RAMAN SPECTRA OF BARIUM SODIUM NIOBATE. 13-1906
OF HYDROGEN IN POLARIZATION REVERSAL OF FERROELECTRIC BARIUM SODIUM NIOBATE. ROLE 13-2025
FECTS OF MELT STOICHIOMETRY ON THE OPTICAL PROPERTIES OF BARIUM SODIUM NIOBATE. SOME EF 13-1902
TS OF MELT STOICHIOMETRY ON THE OPTICAL PROPERTIES OF BARIUM SODIUM NIOBATE. SOME EFFEC 13-1903
STRUCTURE AND TRANSITIONS IN PEROVSKITES. (ORTHOFERRITES, SODIUM NIOBATE) 08-1396
TEMPERATURE DEPENDENCE OF THE REFRACTIVE INDICES OF BARIUM SODIUM NIOBATE. 13-2008
TRANSFORMATIONS IN SODIUM NIOBATE. 08-1397
VISIBLE CW PARAMETRIC OSCILLATOR USING BARIUM SODIUM NIOBATE. 13-1984
THE CRYSTALLOGRAPHIC AND DIELECTRIC PROPERTIES OF STRONTIUM SODIUM NIOBATE AND BARIUM SODIUM NIOBATE. / ON 13-1921
RCE CONSTANT/ VIBRATIONAL SPECTRA OF LITHIUM NIOBATE, BARIUM SODIUM NIOBATE AND BARIUM SODIUM TANTALATE (FO 11-1808
/ERATION ON CRYSTAL INHOMOGENEITY (THEORY AND TEST ON BARIUM SODIUM NIOBATE AND NEODYMIUM DOPED YAG LASER). 13-2043
/BASE COOLING: AN AID TO CZOCHRALSKI CRYSTAL GROWTH. (BARIUM SODIUM NIOBATE AND SODIUM LANTHANUM TANTALATE) 13-1928
/ORY OF ANTIFERROELECTRICITY AND FERROELECTRICITY APPLIED TO SODIUM NIOBATE AND THE SYSTEM POTASSIUM NIOBA/ 03-0218
IN THE SOLUTIONS OF P/ HIGH TEMPERATURE PHASE TRANSITIONS IN SODIUM NIOBATE AND THE USE OF TILTING SCHEMES 08-1336
X-RAY DIFFUSE SCATTERING FROM SODIUM NIOBATE AS A FUNCTION OF TEMPERATURE. 08-1313
. (GROWTH BY KYROPOULO/ DEVELOPMENT OF SINGLE CRYSTAL BARIUM SODIUM NIOBATE AS A NONLINEAR OPTICAL MATERIAL 13-2021
DIFFUSE SCATTERING IN SODIUM NIOBATE AT ROOM TEMPERATURE. 08-1304
IOBATE, LITHIUM TANTALATE, POTASSIUM LITHIUM NIOBATE, BARIUM SODIUM NIOBATE, BARIUM STRONTIUM NIOBATE) /M N 11-1734
E. TUNGSTEN BRONZE FIELD IN THE PSEUDOTERNARY SYSTEM SODIUM NIOBATE BARIUM TITANATE BARIUM DINIOBAT 13-1970
THEIR RELEVANCE TO SOFT PHONON MODES. (X-RAY / STRUCTURES OF SODIUM NIOBATE BETWEEN 480 AND 575 DEG C, AND 08-1273
/ACOUSTICS HANDBOOK. VOL 1: SURFACE WAVE VELOCITIES. (BARIUM SODIUM NIOBATE, BISMUTH(12) GERMANIUM OXIDE(2/ 01-0028
/ PROGRESS TOWARDS THE GROWTH OF GOOD OPTICAL QUALITY BARIUM SODIUM NIOBATE. (BY USING COMPOSITION WHERE LIQ 13-2047
OCHRALSKI) CRYSTAL GROWTH. BARIUM SODIUM NIOBATE: CONSTITUTIONAL STUDIES AND (CZ 13-1904
GROWTH OF UNCRACKED BARIUM SODIUM NIOBATE CRYSTALS. 13-1901
NIQUE. GROWTH OF BARIUM SODIUM NIOBATE CRYSTALS BY THE KYROPOULOS TECH 13-1909
QUENCY DOUBL/ CATASTROPHIC SURFACE DAMAGE PRODUCED IN BARIUM SODIUM NIOBATE CRYSTALS DURING INTRACAVITY FRE 13-2052
ASE TRAN/ ANISOTROPIC CRITICAL X-RAY DIFFUSE SCATTERING FROM SODIUM NIOBATE CRYSTALS (NEAR THE 640 DEG C PH 08-1311
/ TANTALATE OR NIOBATE, LITHIUM NIOBATE OR TANTALATE, BARIUM SODIUM NIOBATE, CUPROUS CHLORIDE, PROUSTITE, / 01-0021
TRIC, AND OPTICAL CONSTANTS) ROOM/ ELASTIC ANOMALY OF BARIUM SODIUM NIOBATE (ELASTIC, PIEZOELECTRIC, DIELEC 13-2060
OSCILLATOR. BARIUM SODIUM NIOBATE FOR A TUNABLE STIMULATED RAMAN 13-1917
PARAMETRIC FLUORESCENCE IN BARIUM SODIUM NIOBATE FOR NINE CW PUMP FREQUENCIES. 13-1950
GE 454.5-528.7 MICRONS). PARAMETRIC FLUORESCENCE IN BARIUM SODIUM NIOBATE FOR 10 CW PUMP WAVELENGTHS (RAN 13-1983
CRYSTALLIZATION OF SOLID SOLUTIONS OF SODIUM NIOBATE FROM GLASS. 08-1386
STRUCTURE OF SODIUM NIOBATE HETEROTYPES. 08-1395
/RD, AND FOURTH HARMONICS OF 1.06 MICRON RADIATION IN BARIUM SODIUM NIOBATE, LITHIUM NIOBATE, AND LITHIUM / 11-1779
E INDEX, CIE/ OPTICAL AND FERROELECTRIC PROPERTIES OF BARIUM SODIUM NIOBATE. (OPTICAL ABSORPTION, REFRACTIV 13-2024
DIFFUSE SCATTERING FROM POTASSIUM MANGANESE TRIFLUORIDE AND SODIUM NIOBATE SINGLE CRYSTALS. /RITICAL X-RAY 08-1312

LICATIONS. (PY PULLING FROM THE MELT, STRU/ GROWTH OF BARIUM SODIUM NIOBATE SINGLE CRYSTALS FOR OPTICAL APP 13-2045
C- TETRAGONAL TRANSITION IN RELATION TO SO/ THE STRUCTURE OF SODIUM NIOBATE T(2) AT 600 DEG C, AND THE CUBI 08-1337
Y OF THE ORTHORHOMBIC- TETRAGONAL PHASE TRANSITION IN BARIUM SODIUM NIOBATE. (THERMAL EXPANSION) /TRIC STUD 13-1894
X-RAY STUDY OF HIGH TEMPERATURE PHASE TRANSITIONS OF SODIUM NIOBATE (TO 500C). 13-1965
LF WAVE VOLTAGE, T/ OPTIMUM CRYSTAL ORIENTATION FOR A BARIUM SODIUM NIOBATE TRANSVERSE LIGHT MODULATOR. (HA 13-1973
E. NEODYMIUM: YTTRIUM ALUMINUM GARNET- BARIUM SODIUM NIOBATE 0.66 MILLIMICRON HARMONIC SOURC 13-1975
CAUSES OF THERMAL EXPANSION OF GADOLINIUM SUBSTITUTED BARIUM SODIUM NIOBATES. 13-2048
LINEAR THERMAL EXPANSIONS OF GADOLINIUM SUBSTITUTED BARIUM SODIUM NIOBATES. 13-1996
MICROCHEMICAL ANALYSIS OF GADOLINIUM DOPED BARIUM SODIUM NIOBATES. 13-1953
SCANNING ELECTRON MICROSCOPIC EXAMINATION OF SINTERED BARIUM SODIUM NIOBATES. 13-1955
PATTERNS ON THE FACES PARALLEL TO THE FERROELECTRIC-AXIS OF SODIUM NITRATE. DOMAIN ETCH 16-2349
IUM DIHYDROGEN PHOSPHATE, POTASSIUM FERROCYANIDE TRIHYDRATE, SODIUM NITRATE) /STICS OF THEIR GROWTH. (AMMON 16-2316
X-RAY STUDY OF PHASE TRANSITION IN SODIUM NITRATE. 16-2380
EMPERATURE. DIELECTRIC ANOMALIES OF SODIUM NITRATE ABOVE THE FERROELECTRIC CURIE T 16-2376
ANHARMONICITY IN SODIUM NITRATE AND SODIUM NITRITE. 16-2388
/ERMAL DIFFUSIVITY AND THERMAL CONDUCTIVITY OF CALCIUM DOPED SODIUM NITRATE CRYSTALS NEAR THE TRANSITION P/ 16-2283
RROELEC/ X-RAY CRITICAL SCATTERING IN TRIGLYCINE SULFATE AND SODIUM NITRATE. (ORDER DISORDER TRANSITION, FE 16-2400
NSTANTS FROM ISOTOPIC SUBSTITUTION IN THE NITRITE RADICAL IN SODIUM NITRITE. ACCURATE FORCE CO 16-2309
ACOUSTIC NUCLEAR MAGNETIC RESONANCE IN FERROELECTRIC SODIUM NITRITE. 16-2321
RAMAGNETIC RESONANCE OF MANGANESE(2+) DOPED IN FERROELECTRIC SODIUM NITRITE. /RANSITIONS IN THE ELECTRON PA 16-2383
AN X-RAY STUDY OF THE PHASE TRANSITION IN SODIUM NITRITE. 16-2318
ANALYSIS OF LOW FREQUENCY ELECTROOPTIC RESPONSE IN SODIUM NITRITE. 16-2353
DISPERSION OF LARGE WAVE VECTOR POLARITONS IN FERROELECTRIC SODIUM NITRITE. ANGULAR 16-2256
CRITICAL SCATTERING OF X-RAYS IN SODIUM NITRITE. 16-2270
DIPOLE CORRELATIONS IN MOLTEN ALKALI METAL NITRATES AND SODIUM NITRITE. 16-2389
DIPOLE LATTICE RELAXATION TIME NEAR THE NEEL TEMPERATURE IN SODIUM NITRITE. 16-2300
TIC RESONANCE OF MANGANESE(2+) IN THE FERROELECTRIC PHASE OF SODIUM NITRITE. ELECTRON PARAMAGNE 16-2358
FERROELECTRIC POLARIZATION SWITCHING IN SODIUM NITRITE. 16-2301
UNDARY CONDITIONS ON THE NMR RELAXATION RATE OF SODIUM-23 IN SODIUM NITRITE. INFLUENCE OF THE ELECTRICAL BO 16-2350
NFRARED ABSORPTION IN GAMMA-RAY IRRADIATED SINGLE CRYSTAL OF SODIUM NITRITE. I 16-2336
PTION AND RAMAN SCATTERING IN FERROELECTRIC AND PARAELECTRIC SODIUM NITRITE. /TION RULES FOR INFRARED ABSOR 16-2281
FFRACTION STUDIES OF FERROELECTRIC POLARIZATION REVERSALS IN SODIUM NITRITE. NEUTRON DI 16-2365
NEUTRON DIFFRACTION STUDY OF SODIUM NITRITE. 16-2326
NITROGEN-14 NUCLEAR QUADRUPOLE RESONANCE IN FERROELECTRIC SODIUM NITRITE. 16-2320
NONLINEAR OPTICAL PROPERTIES OF FERROELECTRIC SODIUM NITRITE. 16-2272
TIONS IN FERROELECTRICS. (POTASSIUM FERROCYANIDE TRIHYDRATE, SODIUM NITRITE) /E STUDIES OF THE PHASE TRANSI 16-2255
NUCLEAR QUADRUPOLE RELAXATION OF NITROGEN-14 IN SODIUM NITRITE. 16-2254
NUCLEAR QUADRUPOLE RESONANCE OF NITROGEN IN SODIUM NITRITE. 16-2313
RELAXATION AND CRITICAL DYNAMICS OF FERROELECTRIC CRYSTALS. (SODIUM NITRITE) /LEAR QUADRUPOLE SPIN LATTICE 16-2268
OMORPHOUS BIAXIAL CRYSTALS: METHYL MESITYL OXIDE OXALATE AND SODIUM NITRITE. OPTICAL ACTIVITY IN NON-ENANTI 16-2273
ONIC GENERATION DURING FERROELECTRIC POLARIZATION REVERSAL. (SODIUM NITRITE) OPTICAL SECOND HARM 16-2384
PHASE TRANSITION IN SODIUM NITRITE. 16-2397
PHENOMENOLOGICAL THEORY OF PHASE TRANSITIONS IN SODIUM NITRITE. 03-0254
TION REVERSAL IN A WIDE RANGE OF IMPRESSING RATE OF FIELD IN SODIUM NITRITE. POLARIZA 16-2381
OLE SPIN PHONON RELAXATION AND FERROELECTRIC TRANSITION. (IN SODIUM NITRITE) QUADRUP 16-2362
QUASI STATIC SWITCHING CURRENT IN SODIUM NITRITE. 16-2261
RAMAN SCATTERING FROM FERROELECTRIC SODIUM NITRITE. 08-1344
ERATURE DEPENDENCE OF B1 PHONON AND DIELECTRIC RELAXATION IN SODIUM NITRITE. RELATIONSHIP OF THE TEMP 16-2257
SECOND HARMONIC GENERATION IN FERROELECTRIC SODIUM NITRITE. 16-2385
AND DIELECTRIC RELAXATION TIME NEAR THE TRANSITION POINTS IN SODIUM NITRITE. /ATIC ELECTRIC SUSCEPTIBILITY 16-2302
NTANEOUS POLARIZATION OF FERROELECTRIC POTASSIUM NITRATE AND SODIUM NITRITE. /E DIELECTRIC CONSTANT AND SPO 16-2331
EPENDENCE OF NONLINEAR OPTICAL COEFFICIENTS IN FERROELECTRIC SODIUM NITRITE. TEMPERATURE D 16-2274
TEMPERATURE DEPENDENCE OF SECOND HARMONIC GENERATION IN SODIUM NITRITE. 16-2315
TEMPERATURE DEPENDENCE OF THE FORCE CONSTANTS OF ION IN SODIUM NITRITE. 16-2311
E DEPENDENCE OF THE ELECTRIC FIELD GRADIENT IN FERROELECTRIC SODIUM NITRITE. TEMPERATUR 16-2337
NCE OF THE SODIUM-23 NUCLEAR QUADRUPOLE COUPLING CONSTANT IN SODIUM NITRITE. TEMPERATURE DEPENDE 16-2393
RATURE DEPENDENCE OF THE OPTICAL NONLINEAR SUSCEPTIBILITY OF SODIUM NITRITE. TEMPE 16-2386
L EXPANSION OF SOME FERROELECTRICS. (POTASSIUM NITRATE, KDP, SODIUM NITRITE) THERMA 16-2298
TWO PHONON ABSORPTION IN FERROELECTRIC SODIUM NITRITE. 16-2310
ULTRASONIC ATTENUATION NEAR THE TWO TRANSITION POINTS IN SODIUM NITRITE. 16-2304
ALYSIS OF FERROELECTRIC DOMAINS IN THE PARAELECTRIC PHASE OF SODIUM NITRITE. X-RAY AN 16-2271
X-RAY AND NEUTRON STUDY ON THE PHASE TRANSFORMATION OF SODIUM NITRITE. 16-2328
TRICS. PART-2: DIELECTRIC CONSTANTS AND DIELECTRIC LOSSES OF SODIUM NITRITE AND GLYCINE SULFATE. /FERROELEC 16-2354
ELECTRICAL RESISTIVITIES OF SODIUM NITRITE AND POTASSIUM NITRATE CRYSTALS. 16-2258
R. PHASE DIAGRAMS OF SODIUM NITRITE AND POTASSIUM NITRATE TO 40 KBA 16-2361
GRUNEISEN PARAMETERS AROUND THE PHASE TRANSITIONS IN SODIUM NITRITE AND SODIUM NITRATE. 16-2390
/F THE X-RAY DIFFRACTION TOPOGRAPHIC METHOD TO FERROELECTRIC SODIUM NITRITE AND TGS CRYSTALS. (DOMAIN BOUN/ 16-2375
TION AT 163 DEG C). FERROELECTRIC PHASE TRANSITION IN SODIUM NITRITE AS STUDIED BY EPR (PHASE TRANSI 16-2359
UN/ CALCULATION OF THE VARIATION OF THE RAMAN FREQUENCIES IN SODIUM NITRITE AT ORDINARY TEMPERATURES AS A F 16-2363
INFRARED SPECTRA OF POTASSIUM NITRATE AND SODIUM NITRITE AT PHASE TRANSITIONS. 16-2352
BRILLOUIN SCATTERING IN SODIUM NITRITE AT ROOM TEMPERATURE. 16-2305
ELD GRADIENT AND LONG RANGE ORDER PARAMETER OF FERROELECTRIC SODIUM NITRITE. (AT THE SODIUM-23 NUCLEUS) /FI 16-2395
STRUCTURE OF SODIUM NITRITE AT 150, 185, AND 225 DEGREES C. 16-2327
X-RAY CRITICAL SCATTERING OF SODIUM NITRITE AT 160 DEG C AND AT 215 DEG C. 16-2343
DISORDERED STRUCTURE OF SODIUM NITRITE AT 185 DEG C. 16-2325
RAMAN SPECTRA AND MODE FREQUENCY SHIFTS OF FERROELECTRIC SODIUM NITRITE AT 77 AND 294 DEG K. 16-2262
ARY TEMPERATURES. MEASUREMENT OF THE ELASTIC CONSTANTS OF SODIUM NITRITE BY BRILLOUIN DIFFUSION AT ORDIN 16-2306
RATION. LONG RANGE ORDER IN SODIUM NITRITE BY OPTICAL SECOND HARMONIC GENE 16-2314
MODULATION OF THE MILLIMETER WAVE ABSORPTION IN SODIUM NITRITE BY POLARIZATION REVERSAL. 16-2392
ANALYSIS OF THE TEMPERATURE DEPENDENT PHONON STRUCTURE IN SODIUM NITRITE BY RAMAN SPECTROSCOPY. 16-2299
(TEMPERATURE DEPENDENCE) STUDY OF PHASE TRANSITION OF SODIUM NITRITE BY SECOND HARMONIC GENERATION. 16-2403
CIFIC ANTIFERROELECTRIC PHASE) LATTICE VIBRATION OF SODIUM NITRITE. (CALCULATIONS TO RELATE TO SPE 16-2317
TWO-PHONON SCATTERING AND FERROELECTRIC TRANSITION IN A SODIUM NITRITE CRYSTAL. 16-2277
SPECTRUM. INVESTIGATION OF FERROELECTRIC TRANSITION IN SODIUM NITRITE CRYSTAL BY OBSERVATION OF RAMAN 16-2275
ELECTRICAL BREAKDOWN IN FERROELECTRIC SODIUM NITRITE CRYSTALS. 16-2288
THERMAL EXPANSION IN SODIUM NITRITE CRYSTALS. 16-2351
R SPECTRUM IN THE REGION OF THE ANTISYMMETR/ PHASE CHANGE IN SODIUM NITRITE CRYSTALS AND COMBINATION SCATTE 16-2276
180 DEGREE DOMAIN WALLS. TOPOGRAPHIC STUDY ON FERROELECTRIC SODIUM NITRITE CRYSTALS. PART-1: STRUCTURE OF 16-2372
POLARIZATION REVERSAL. TOPOGRAPHIC STUDY OF FERROELECTRIC SODIUM NITRITE CRYSTALS. PART-2: MECHANISM OF 16-2374
/PERATURE DEPENDENCE OF FAR INFRARED REFLECTIVITY SPECTRA OF SODIUM NITRITE CRYSTALS. (0.85 DEG K, 30-200 / 16-2373

MICRONS)   TEMPERATURE DEPENDENCE OF ABSORPTION OF SODIUM NITRITE IN THE FAR INFRARED. (350-1100   16-2387
ANSITION T/ NUCLEAR MAGNETIC RESONANCE STUDY ON SODIUM-23 IN SODIUM NITRITE IN THE VICINITY OF THE PHASE TR   16-2396
SITION POINT.   SPECIFIC HEAT OF SODIUM NITRITE NEAR THE ANTIFERROELECTRIC TRAN   16-2303
                           CRYSTAL DYNAMICS OF SODIUM NITRITE. (NEUTRON SCATTERING)   16-2284
/E, POTASSIUM TANTALATE, BARIUM TITANATE, POTASSIUM NIOBATE, SODIUM NITRITE, POTASSIUM DIDEUTERIUM PHOSPHA/   02-0059
ELASTIC CONSTANTS AND ULTRASONIC ABSORPTION OF SODIUM NITRITE SINGLE CRYSTALS.   16-2357
/ CRYSTAL DYNAMICS AND THE FERROELECTRIC PHASE TRANSITION OF SODIUM NITRITE (STUDIED BY COHERENT NEUTRON IN   16-2364
/ROELECTRIC SUBSTRATES. (TELLURIUM ON TRIGLYCINE SULFATE AND SODIUM NITRITE SURFACES) (DECORATION METHOD O/   16-2394
/OF SODIUM-23 IN THE PARAELECTRIC AND FERROELECTRIC PHASE OF SODIUM NITRITE. (THERMODYNAMICS OF THE PHASE /   16-2264
ACTING DIPOLES)   THE ORDER DISORDER TRANSITION OF SODIUM NITRITE. (TWO SUBLATTICE MODEL OF INTER   16-2287
E SYSTE/ DIELECTRIC AND CERAMIC PROPERTIES OF (BISMUTH(0.5) (SODIUM OR POTASSIUM)(0.5) (X) LEAD(1-X) TITANAT   13-2013
/OELECTRIC CERAMICS OF LEAD ZIRCONATE TITANATE MODIFIED WITH SODIUM OR POTASSIUM ANTIMONATE, NIOBATE, OR B/   09-1539
/ (WHERE A AND B ARE CALCIUM, STRONTIUM, OR BARIUM, AND C IS SODIUM OR POTASSIUM) PHASES OF THE TETRAGONAL/   04-0912
                                          SODIUM ORDERING IN SODIUM TUNGSTEN BRONZES.   13-2012
/UIDUS AND SOLIDUS) FIELD AND MELT GROWTH OF CRYSTALS IN THE SODIUM OXIDE- BARIUM OXIDE NIOBIUM PENTOXIDE /   13-1949
/T ROOM AND LIQUID NITROGEN TEMPERATURES IN ALKALI NITRITES (SODIUM, POTASSIUM) AND ALKALI NITRATES. (LITH/   16-2399
OR ELECTROOPTICAL APPLICATIONS. (LEAD ZIRCONATE TITANATE AND SODIUM POTASSIUM NIOBATE) /ELECTRIC CERAMICS F   09-1527
                      PREPARATION AND FABRICATION OF SODIUM POTASSIUM NIOBATE.   13-1937
/-2 SULFATE OR SELENATE CR FLUOBERYLLATE (METALS CHOSEN FROM SODIUM, POTASSIUM, RUBIDIUM, CESIUM, OR AMMON/   19-2743
/ FLUOBERYLLATE, POTASSIUM OR RUBIDIUM DIHYDROGEN PHOSPHATE, SODIUM, POTASSIUM(2), OR LITHIUM THALLIUM TAR/   02-0089
LECTRIC. (ANSWER-NO)   IS SODIUM RUBIDIUM TARTRATE TETRAHYDRATE A FERROE   29-3245
            INTERNAL ELECTRIC FIELDS IN CRYSTALS OF SODIUM TANTALATE AND CADMIUM TITANATE.   08-1389
/PTION OF POTASSIUM TANTALATE SINGLE CRYSTALS AND (POTASSIUM SODIUM TANTALATE MIXED CRYSTALS (100 HZ-10 MH/   08-1272
/CS AND ANTIFERROELECTRICS. (BARIUM TITANATE, KDP, POTASSIUM SODIUM TARTRATE, TGS, AMMONIUM FLUOBERYLLATE,/   02-0134
METHYL AMMONIUM ALUMINUM SULFATE DODECAHYDRATE, COLEMANITE, SODIUM TRIDEUTERIUM DISELENITE) /ROMIUM ALUMS,   02-0087
EUTERIUM DIFFUSION AND DEUTERIUM AND SODIUM-23 RELAXATION IN SODIUM TRIDEUTERIUM SELENATE.   NMR STUDY OF D 02-0087
ATE SYSTEM VERSUS CONCENTRATION, TEMPERATU/ PHASE DIAGRAM OF SODIUM TRIDEUTERIUM(X) TRIHYDROGEN(1-X) PHOSPH   17-2666
/ ELECTRIC FIELD ON THE PERMITTIVITY OF A NEW DELTA PHASE OF SODIUM TRIDEUTERIUM(X) TRIHYDROGEN(1-X) SELEN/   22-2875
ROOM TEMPERATURE CRYSTAL STRUCTURE OF THE FERROELECTRIC SODIUM TRIDEUTERIUM SELENITE.   22-2861
/AGNETIC RESONANCE STUDY OF THE FERROELECTRIC TRANSITIONS IN SODIUM TRIHYDROGEN AND TRIDEUTERIUM SELENATE.   22-2832
NOMALOUS SPECIFIC HEAT AND CONFIGURATIONAL ENTROPY CHANGE IN SODIUM TRIHYDROGEN DISELENITE.   A   22-2856
TION DAMAGE ELECTRON SPIN RESONANCE SPECTRA IN FERROELECTRIC SODIUM TRIHYDROGEN DISELENITE. /E (ION). RADIA   22-2870
ESR STUDY OF THE FERROELECTRIC PHASE TRANSITIONS IN SODIUM TRIHYDROGEN DISELENITE.   22-2871
            NEUTRON DIFFRACTION STUDY ON PARAELECTRIC SODIUM TRIHYDROGEN DISELENITE.   22-2846
EXPANSION. RELAXATION BETWEEN THE STRUCTURE OF FERROELECTRIC SODIUM TRIHYDROGEN DISELENITE AND ITS THERMAL   22-2851
          FAR INFRARED REFLECTIVITY SPECTRA OF SODIUM TRIHYDROGEN DISELENITE CRYSTAL.   22-2877
C RESONANC/ REVERSAL MECHANISM IN FERROELECTRICS: LITHIUM OR SODIUM TRIHYDROGEN DISELENITE, NUCLEAR MAGNETI   22-2837
ASE, COMPARISON WITH FERROELECTRIC).   RAMAN SPECTRUM OF SODIUM TRIHYDROGEN DISELENITE (PARAELECTRIC PH   22-2864
            MODELS FOR THE ORDER DISORDER TRANSITION IN SODIUM TRIHYDROGEN SELENATE.   22-2862
          PROTON SPIN LATTICE RELAXATION IN IRRADIATED SODIUM TRIHYDROGEN SELENATE.   22-2853
        PROTON SPIN LATTICE RELAXATION IN FERROELECTRIC SODIUM TRIHYDROGEN SELENATE.   22-2852
                          RADIATION EFFECTS IN SODIUM TRIHYDROGEN SELENATE.   22-2865
                        RESIDUAL ENTROPY OF SODIUM TRIHYDROGEN SELENATE.   22-2835
ITS UNUSUAL BEHAVIOR IN THE BETA PHASE OF THE FERROELECTRIC SODIUM TRIHYDROGEN SELENATE. /IN STRUCTURE AND   22-2874
YDROGEN SELENATE.   PROTON MAGNETIC RESONANCE STUDY OF SODIUM TRIHYDROGEN SELENATE AND POTASSIUM TRIH   22-2876
E/ DIELECTRIC PROPERTIES OF HYDROGEN BONDED FERROELECTRICS. (SODIUM TRIHYDROGEN SELENATE AND SODIUM TRIDEUT   22-2855
.   DOMAIN STRUCTURE OF THE BETA PHASE OF SODIUM TRIHYDROGEN SELENATE IN ELECTRIC FIELDS   22-2830
RASLOW HYDROGEN MOTION IN POTASSIUM DIHYDROGEN PHOSPHATE AND SODIUM TRIHYDROGEN SELENATE TYPE CRYSTALS. ULT   17-2441
                    ANOMALOUS SPECIFIC HEAT OF SODIUM TRIHYDROGEN SELENITE.   22-2823
       CRYSTAL STRUCTURES OF POTASSIUM TRIHYDROGEN SELENITE AND SODIUM TRIHYDROGEN SELENITE.   22-2838
        DOUBLE HYSTERESIS LOOPS AT THE UPPER CURIE POINT OF SODIUM TRIHYDROGEN SELENITE.   22-2831
ORDER PROPERTIES, KDP, TGS, ROCHELLE SALT, SILVER PERIODATE, SODIUM TRIHYDROGEN SELENITE) /AMICAL ORDER DIS   02-0050
TORIAL PROBLEM.   ANOMALOUS SPECIFIC HEAT OF SODIUM TRIHYDROGEN SELENITE ASSOCIATED COMBINA   22-2822
IDEUTERIUM S/ INFRARED SPECTRA OF ISOTOPIC NONISOMORPHISM OF SODIUM TRIHYDROGEN SELENITE (SHS) OR SODIUM TR   22-2836
NITE AND ISOTOPE EFFECT.   THERMAL EXPANSION IN SODIUM TRIHYDROGEN(1-X) TRIDEUTERIUM(X) DISELE   22-2858
              OPTICAL TRANSITION TO THE FERMI LEVEL IN SODIUM TUNGSTATE.   10-1634
  TIGHT BONDING BAND CALCULATIONS ON RHENIUM TRIOXIDE, SODIUM TUNGSTATE, AND POTASSIUM TANTALATE.   08-1334
       TRANSIENT NMR STUDY OF PHASE TRANSITIONS IN METALLIC SODIUM TUNGSTEN BRONZES.   13-1910
DOPING.   IMPURITY EFFECTS IN SODIUM TUNGSTEN BRONZES: NIOBIUM AND TANTALUM   13-2053
          KYROPOULOS SINGLE CRYSTAL GROWTH OF SODIUM TUNGSTEN BRONZES. (SODIUM TUNGSTATE)   13-2054
           THERMAL CONDUCTIVITY OF CUBIC SODIUM TUNGSTEN BRONZES. (SODIUM TUNGSTATES)   13-2057
OFT MODES IN THE FERROELECTRIC MIXED CRYSTALS POTASSIUM(1-X) SODIUM(X) TANTALATE. /LECTRIC PROPERTIES AND S   08-1306
/ICAL DEPOSITION AND DISSOLUTION OF TUNGSTEN OXIDE BRONZES. (SODIUM(X) TUNGSTEN(1-X) Z(Y) OXIDE(3) WHERE Z/   13-1944
GSTEN BRON/ DIELECTRIC AND ELECTROOPTIC PROPERTIES OF BARIUM SODIUM YTTRIUM NIOBATE SINGLE CRYSTAL WITH TUN   13-1971
/S. (AND THE FERROELECTRIC STATE, BISMUTH(0.5) (POTASSIUM OR SODIUM(0.5) TITANATES AND PHASE DIAGRAMS FOR /   08-1289
/TASSIUM(X) BARIUM(1-2X) TITANATE, BISMUTH(0.5) POTASSIUM(X) SODIUM(0.5-X) TITANATE, BISMUTH(X) (POTASSIUM/   08-1289
            X-RAY AND THERMAL EXPANSION STUDY OF A SODIUM(0.88) LITHIUM(0.12) NIOBATE CERAMIC.   13-1999
SOFT PHONON MODES IN THE MIXED CRYSTAL SYSTEMS POTASSIUM(X) SODIUM(1-X) TANTALATE AND POTASSIUM TANTALATE/   13-1933
TRY OF FERROELECTRIC DOMAINS.   STRUCTURE OF THE BRONZE SODIUM(13) NIOBIUM(35) OXIDE(94) AND THE GEOME   13-1931
                         NIOBIUM-23 AND SODIUM-23 IN POLYCRYSTALLINE SODIUM NIOBATE.   08-1469
                                   NMR STUDY OF SODIUM-23 IN SILVER SODIUM DINITRITE.   16-2324
ELECTRICAL BOUNDARY CONDITIONS ON THE NMR RELAXATION RATE OF SODIUM-23 IN SODIUM NITRITE. INFLUENCE OF THE   16-2350
THE PHASE TRANSITION T/ NUCLEAR MAGNETIC RESONANCE STUDY ON SODIUM-23 IN SODIUM NITRITE IN THE VICINITY OF   16-2396
C PHASE OF SODIUM NITRIT/ NUCLEAR SPIN LATTICE RELAXATION OF SODIUM-23 IN THE PARAELECTRIC AND FERROELECTRI   16-2264
DRATE CRYSTAL.   NUCLEAR MAGNETIC RESONANCE OF SODIUM-23 IN THE SODIUM AMMONIUM SELENATE DIHY   22-2883
              SECOND ORDER QUADRUPOLE SHIFTS OF THE SODIUM-23 MAGNETIC RESONANCE IN ROCHELLE SALT.   28-3186
LVER SODIUM DINITRITE.   SPLITTING OF THE SODIUM-23 NMR CENTRAL LINE IN FERROELECTRIC SI   16-2307
IN SODIUM NITRITE.   TEMPERATURE DEPENDENCE OF THE SODIUM-23 NUCLEAR QUADRUPOLE COUPLING CONSTANT   16-2393
NGE ORDER PARAMETER OF FERROELECTRIC SODIUM NITRITE. (AT THE SODIUM-23 NUCLEUS) /FIELD GRADIENT AND LONG RA   16-2395
LENATE.   NMR STUDY OF DEUTERIUM DIFFUSION AND DEUTERIUM AND SODIUM-23 RELAXATION IN SODIUM TRIDEUTERIUM SE   22-2849
/ PHASE TRANSITIONS IN PEROVSKITES CRYSTALS. (NIOBIUM-93 AND SODIUM-23 RESONANCE IN POTASSIUM NIOBATE AND /   08-1283
            RAMAN SCATTERING LINE SHAPE OF THE SOFT E POLARITON MODE IN BARIUM TITANATE.   04-0803
SIUM DIDEUTERIUM PHOSPHATE. SIMULTANEOUS OBSERVATION OF THE SOFT FERROELECTRIC AND ACOUSTIC MODES OF POTAS   17-2636
ES. (STRONTIUM TITANATE, POT/ THE GRUNEISEN PARAMETER OF THE SOFT FERROELECTRIC MODE IN THE CUBIC PEROVSKIT   02-0114
TRON INELASTIC SCATTERING).   SOFT FERROELECTRIC MODES IN LEAD TITANATE (NEU   05-1077
    VIBRONIC EFFECT IN THE TEMPERATURE DEPENDENCE OF THE SOFT FREQUENCY OF BARIUM TITANATE.   04-0662
IN ANTIMONY SULFOIODIDE.   SOFT LATTICE VIBRATIONS AND VIBRATION COUPLING   15-2233
M DIHYDROGEN PHOSPHATE AND POTASSIUM DIDEUTERIUM PHOSPHATE. (SOFT MADE THEORY) /RA OF PARAELECTRIC POTASSIU   03-0356
(THEORY, KDP)   SOFT MODE BEHAVIOR IN IMPROPER FERROELECTRICS.   03-0230
ERING IN FERROELECTRIC ANTIMONY SULFOIODIDE.   SOFT MODE COUPLING AND CRITICAL RAYLEIGH SCATT   15-2232

/EL.) PART-2: (DYNAMICAL PROPERTIES, TRANSVERSE ISING MODEL, SOFT MODE COUPLING TO TRANSVERSE OPTICAL PHON/ 03-0384
E.   NEW TYPE OF FAR INFRARED SOFT MODE IN FERROELECTRIC GADOLINIUM MOLYBDAT 30-3310
           FERROELECTRIC DOMAINS AND THE CONDENSED SOFT MODE IN GADOLINIUM MOLYBDATE.   30-3260
                       NEW TYPE OF FERROELECTRIC SOFT MODE IN GADOLINIUM MOLYBDATE.   30-3311
RATURE STUDY OF THE POLARITON ASSOCIATED WITH THE 248 CM(-1) SOFT MODE IN LITHIUM NIOBATE.   TEMPE 11-1807
DEG K TRANSITION.   TEMPERATURE DEPENDENCE OF THE SOFT MODE IN STRONTIUM TITANATE ABOVE THE 105   06-1178
                          ZONE BOUNDARY SOFT MODE IN TERBIUM MOLYBDATE.   30-3269
TE.   APPARENT SOFT MODE LINE WIDTH DIVERGENCE IN LEAD TITANA 05-1078
SPHATE . (BY NEUTRON TRIPLE AXIS SPECTROMETR/ FERROELECTRIC (SOFT) MODE MOTION IN POTASSIUM DIDEUTERIUM PHO 17-2668
ERROELECTRICS, CESIUM DIHYDROGEN ARSENATE, STRONTIUM TITANA/ SOFT MODE RAMAN SPECTROSCOPY: COUPLED MODES (F 06-1207
ANSITIONS NEAR THE TRANSITION TEMPERATURE IN STRONTIUM TITA/ SOFT MODE STUDY OF THE STRESS INDUCED PHASE TR 06-1196
ATE(3-X) AND CALCIUM(Y) STRONTIUM(1-Y) TITANATE(3-X).   SOFT MODE SUPERCONDUCTIVITY IN STRONTIUM TITAN 06-1093
ELECTRIC PROPERTIES OF AN ANHARMONIC CRYSTAL. (EXTENSION OF "SOFT MODE" THEORY)   DI 03-0212
ESE TRIFLUORIDE (ANOMALOUS / ULTRASONIC ATTENUATION NEAR THE SOFT MODE TRANSITION POINT IN POTASSIUM MANGAN 33-3373
                        STRUCTURAL PHASE TRANSITIONS AND SOFT MODES.   03-0176
AGONAL POTASSIUM TANTALATE NIOBATE.   SOFT MODES AND STRUCTURE OF FERROELECTRIC TETR 13-1957
              NEUTRON SCATTERING STUDY OF SOFT MODES IN CUBIC BARIUM TITANATE.   04-0712
E.   NEUTRON SCATTERING STUDY OF THE SOFT MODES IN CUBIC POTASSIUM TANTALATE NIOBAT 13-2064
         RAMAN STUDIES (TEMPERATURE DEPENDENCE) OF UNDER DAMPED SOFT MODES IN LEAD TITANATE.   05-1041
AND DEUTERATED AMMONIU/ DETERMINATION OF THE EIGENVECTORS OF SOFT MODES IN POTASSIUM DIDEUTERIUM PHOSPHATE 17-2700
 POTASSIUM(1-X) SODIUM(X) TANTALA/ DIELECTRIC PROPERTIES AND SOFT MODES IN THE FERROELECTRIC MIXED CRYSTALS 08-1306
ONS.   ON THE SYMMETRY OF THE SOFT MODES IN THE FERROELECTRIC PHASE TRANSITI 03-0357
TRANSITION IN STRONTIUM TIT/ LIGHT SCATTERING STUDIES OF THE SOFT MODES NEAR THE CUBIC TO TETRAGONAL PHASE 06-1242
SSIUM TANTALATE, BARI/ NEUTRON INELASTIC SCATTERING STUDY ON SOFT MODES. (REVIEW OF BROOKHAVEN WORK ON POTA 02-0115
YDROGEN PHOSPHATE AND POTAS/ LIGHT SCATTERING STUDIES OF THE SOFT OPTIC AND ACOUSTIC MODES OF POTASSIUM DIH 17-2497
BARIUM TITANATE.   PRESSURE DEPENDENCE OF ACOUSTIC MODE- SOFT OPTIC MODE INTERACTIONS IN FERROELECTRIC 04-0893
BARIUM TITANATE.   ACOUSTIC SOFT OPTIC MODE INTERACTIONS IN FERROELECTRIC 04-0671
TE.   ACOUSTIC PHONON SOFT OPTIC PHONON INTERACTION IN BARIUM TITANA 04-0806
) GERMANIUM(3) OXIDE(11) .   SOFT OPTIC PHONON MODE IN FERROELECTRIC LEAD(5 35-3495
                  ELASTIC CONSTANTS AND SOFT OPTICAL MODES IN GADOLINIUM MOLYBDATE.   30-3284
/ATION TO THE LEAD TITANATE ZIRCONATE SYSTEM (MEASUREMENT OF SOFT OPTICAL PHONON MODES IN TETRAGONAL FERRO/ 09-1497
PHASE TRANSITION IN ANTIMONY SULFOIODIDE.   SOFT PHONON COUPLINGS IN THE PRESSURE INDUCED 15-2240
          EFFECT OF THE QUARTIC ANHARMONICITY ON THE SOFT PHONON INSTABILITY OF STRONTIUM TITANATE. 06-1211
SULFOIODIDE (BY RAMAN SCATTERING).   SOFT PHONON MODE AND MODE COUPLING IN ANTIMONY . 15-2167
/ MEAN MAGNETIC MOMENT OF CHROMIUM- POTASSIUM ALUM (UNSTABLE SOFT PHONON MODE, DISPLACIVE FERROELECTRICS) . 20-2806
TRIHYDRATE (NEAR THE TRANSITION TEMPERATURE, NO ANOMALY BUT SOFT PHONON MODE STILL POSSIBLE) . /ERROCYANIDE 24-2891
0 DEG C, AND THE CUBIC- TETRAGONAL TRANSITION IN RELATION TO SOFT PHONON MODES. / SODIUM NIOBATE T(2) AT 60 08-1337
POTASSIUM(X) SODIUM(1-X) TANTALATE AND POTASSIUM TANTALATE/ SOFT PHONON MODES IN THE MIXED CRYSTAL SYSTEMS 13-1933
UM NIOBATE BETWEEN 480 AND 575 DEG C, AND THEIR RELEVANCE TO SOFT PHONON MODES. (X-RAY DIFFRACTION) /F SODI 08-1273
                          SOFT PHONONS IN POTASSIUM TANTALATE NIOBATE.   13-2040
VECTOR, P(S), ARISES PERPENDICULAR TO THE WAVE NORMAL OF THE SOFT WAVES. (THEORY) / WHICH THE POLARIZATION 03-0141
SE TRIFLUORIDE.   DISPERSION AND DAMPING OF SOFT ZONE BOUNDARY PHONONS IN POTASSIUM MANGANE 33-3374
E.   MICROSCOPIC CALCULATION OF THE RESPONSE FUNCTION OF THE SOFT ZONE BOUNDARY PHONON IN STRONTIUM TITANAT 06-1213
/ROELECTRICS (HAMILTONIAN IN TERMS OF LOCALIZED D-STRAIN AND SOFT- NORMAL MODE COORDINATES AND TEMPERATURE/ 03-0432
NDS BETWEEN 4 AND 100 DEGREES K.   POLARIZATION OF SOLID BARIUM(X) STRONTIUM(1-X) TITANATE COMPOU 04-0719
ION OF DOMAIN STRUCTURE IN LOW TEMPERATURE FERROELECTRICS BY SOLID DEW TECHNIQUE.   OBSERVAT 03-0341
EFS)   ANISOTROPIC, SOLID DIELECTRICS AND FERROELECTRICITY. (106 R 02-0065
                          PHASE TRANSITIONS OF SOLID HYDROGEN HALIDES.   31-3327
ND NEUTRON DIFFRA/ FERROELECTRICITY AND PHASE TRANSITIONS IN SOLID HYDROGEN HALIDES (SINGLE CRYSTAL X-RAY A 31-3331
            NUCLEAR QUADRUPOLE RESONANCE OF IODINE-127 IN SOLID HYDROGEN IODIDE.   31-3335
SSURES. (POSSIBLY FERROELECTRIC)   DIELECTRIC ANOMALIES IN SOLID METHANE AT LOW TEMPERATURES AND HIGH PRE 36-3547
                  MECHANISM AND KINETICS OF SOLID PHASE FORMATION OF LEAD METANIOBATE.   13-2042
MECHANISM AND KINETICS OF LITHIUM METANIOBATE FORMATION BY A SOLID PHASE REACTION.   11-1840
TRANSFORMATIONS OF THE M MODIFICATION OF SAMARIUM MOLYBDATE (SOLID PHASE SYNTHESIS, NEW MODIFICATION). /SE 30-3270
        MEASUREMENT OF CHEMICAL SHIFT IN SOLID POTASSIUM DIHYDROGEN PHOSPHATE.   17-2688
      FORMATION OF THIN LAYERS OF MIXED TITANATES BY SOLID SOLID REACTIONS.   13-2005
STRUCTURE OF SINGLE CRYSTALS FORMED BY BARIUM TITANATE BASE SOLID SOLUTION. /XTERNAL ACTIONS ON THE DOMAIN 04-0954
D ELECTROLYTE FUE/ IONIC CONDUCTION IN PEROVSKITE TYPE OXIDE SOLID SOLUTION AND ITS APPLICATION TO THE SOLI 08-1444
BARIUM(2+) AND TITANIUM(4+) ION/ GROWING SINGLE CRYSTALS OF SOLID SOLUTION BASED ON BARIUM TITANATE, WHERE 04-0960
UCTURE.   STUDIES ON THE (STRONTIUM, CALCIUM) TITANATE SOLID SOLUTION CERAMIC WITH BOUNDARY LAYER STR 06-1246
S OF LEAD COBALTATE(0.33) NIOBATE(0.67)- TITANATE- ZIRCONATE SOLID SOLUTION CERAMICS. /ZOELECTRIC PROPERTIE 09-1549
IUM TITANATE CERAMICS ON THE PHYSICO- CHEMICAL NATURE OF THE SOLID SOLUTION FORMED FROM TWO TITANATES. /BAR 04-0789
TROOPTICAL PROPERTIES OF BARIUM TITANATE- TANTALUM PENTOXIDE SOLID SOLUTION SINGLE CRYSTALS.   ELEC 04-0955
NIOBIUM-93 NMR STUDY OF THE LITHIUM NIOBATE TANTALATE SOLID SOLUTION SYSTEM.   11-1794
ONATE- LEAD (ZINCATE(0.33/ ELECTROPHYSICAL PROPERTIES OF THE SOLID SOLUTION SYSTEM LEAD TITANATE- LEAD ZIRC 13-1951
ONTIUM TITANATE AND P/ NEUTRON SCATTERING BY A FERROELECTRIC SOLID SOLUTION. (THEORY, APPLIED TO BARIUM STR 03-0260
            BISMUTH TITANATE SOLID SOLUTIONS.   08-1277
CRYSTAL CHEMICAL PROBLEMS OF (BARIUM, STRONTIUM) TITANATE SOLID SOLUTIONS.   04-0633
LASTIC PROPERTIES OF SINGLE CRYSTALS OF BARIUM LEAD TITANATE SOLID SOLUTIONS.   DIELECTRIC AND E 04-0597
INE STRONTIUM TITANATE AND BARIUM(X) STRONTIUM(1-X) TITANATE SOLID SOLUTIONS) / TEMPERATURES. (POLYCRYSTALL 04-0720
DIELECTRIC PROPERTIES OF ANTIMONY SULFOIODIDE- SULFOBROMIDE SOLID SOLUTIONS.   15-2242
LECTRIC FIELD ON PHASE COMPOSITION OF (LEAD, BARIUM) NIOBATE SOLID SOLUTIONS.   EFFECT OF E 13-1943
EFFECT IN SINGLE CRYSTALS OF BARIUM TITANATE- ZINC TITANATE SOLID SOLUTIONS.   ELECTROOPTICAL 04-0958
TANNATE(0.5) NIOBATE(0.5)) (1-X) ZIRCONATE(X) PEROVSKITE TYPE SOLID SOLUTIONS. /ECTRIC PROPERTIES OF LEAD (S 09-1548
FERROELECTRIC SOLID SOLUTIONS.   02-0069
HYDROTHERMAL CRYSTALLIZATION OF LEAD TITANATE ZIRCONATE SOLID SOLUTIONS.   09-1482
E(0.5) AND LEAD COBALTATE(0.5) TUNGSTATE(0.5)- LEAD TITANATE SOLID SOLUTIONS. /LEAD COBALTATE(0.5) TUNGSTAT 13-1968
HIA NIOBIA (POTASSIUM LITHIUM NIOBATES, PHASE RELATIONSHIPS, SOLID SOLUTIONS) . /(OXIDE) SYSTEM POTASSIA LIT 13-2017
/ DIELECTRIC POLARIZATION RELAXATION IN SOME FERROELECTRICS (SOLID SOLUTIONS AND MIXED CATION COMPOUNDS, K/ 03-0291
/CAL CONDUCTIVITY AND RELATED PROPERTIES OF 60 COMPOUNDS, 58 SOLID SOLUTIONS, AND 10 TECHNICAL FERROCERAMI/ 01-0014
ELECTRIC PROPERTIES OF LEAD TITANATE- LEAD (FERRATE NIOBATE) SOLID SOLUTIONS AT HIGH PRESSURES.   DI 05-1071
STRONTIUM TITANATE, POLYCRYSTALLINE, STRONTIUM TITANATE, AND SOLID SOLUTIONS BARIUM STRONTIUM TITANATE) /L 04-0721
SOLID SOLUTIONS BASED ON BARIUM TITANATE.   04-0786
FORMATION OF FLUORINE CONTAINING SOLID SOLUTIONS BASED ON BARIUM TITANATE.   04-0932
/CTRIC AND NONLINEAR MICROWAVE PROPERTIES OF POLYCRYSTALLINE SOLID SOLUTIONS BASED ON BARIUM TITANATE. (BA/ 04-1019
ASSIUM TANTALAT/ X-RAY ANALYSIS AND DIELECTRIC PROPERTIES OF SOLID SOLUTIONS BASED ON THE FERROELECTRIC POT 08-1310
/TION TEMPERATURE OF BARIUM (0-0.1) STRONTIUM (0.9-1) TITANATE SOLID SOLUTIONS BY HYDROSTATIC PRESSURE. (TO / 04-0722
/ELECTRIC PROPERTIES OF ANTIMONY SULFOIODIDE(X) BROMIDE(1-X) SOLID SOLUTIONS DEPEND ON TEMPERATURE AND HYD/ 15-2149
ECTRI/ FERROELECTRIC PHOTOELECTRETS BASED ON A(V)B(VI)C(VII) SOLID SOLUTIONS. (DIELECTRIC CONSTANTS, PYROEL 15-2157
SEMICONDUCTOR PROPERTIES OF BARIUM (TITANATE, STANNATE) SOLID SOLUTIONS DOPED BY TUNGSTEN TRIOXIDE.   04-0965

/TRIC TRANSITION IN THE LEAD (ZIRCONATE, STANNATE, TITANATE) SOLID SOLUTIONS. (ELECTROSTRICTION CONSTANTS) 09-1486
NATE- LEAD METANIOBATE TERNARY SYSTEM. CURIE TEMPERATURES OF SOLID SOLUTIONS IN A LEAD ZIRCONATE- LEAD TITA 09-1595
THE SYSTEM LEAD HAFNATE TITANATE, LEAD STANNATE NIOBATE. SOLID SOLUTIONS IN ANTIFERROELECTRIC REGION OF 13-1956
ICEUTERIUM(X) PHOSPHATE SINGLE CRYSTALS. REGULAR BEHAVIOR OF SOLID SOLUTIONS IN POTASSIUM DIHYDROGEN(1-X) D 17-2458
IUM FLUCEERYLLATE SYSTEM.   FERROELECTRIC PROPERTIES OF SOLID SOLUTIONS IN THE AMMONIUM SULFATE- AMMON 19-2764
IUM FLUCEERYLLATE SYSTEM.   FERROELECTRIC PROPERTIES OF SOLID SOLUTIONS IN THE AMMONIUM SULFATE- AMMON 19-2763
/N DEFENCENCE OF THE CURIE-WEISS CONSTANT IN POLYCRYSTALLINE SOLID SOLUTIONS IN THE (BARIUM, STRONTIUM) TI 04-0586
/TURE AND MAGNETIC PROPERTIES OF FERROELECTRIC FERROMAGNETIC SOLID SOLUTIONS IN THE LEAD COBALT TUNGSTATE-/ 36-3604
- AMMCNIUM FLUCEERYLLATE.   FERROELECTRIC PROPERTIES OF SOLID SOLUTIONS IN THE SYSTEM AMMONIUM SULFATE 19-2762
STRONTIUM NIOBI/ RELAXATION AND FERROELECTRIC PROPERTIES OF SOLID SOLUTIONS IN THE SYSTEM BARIUM TITANATE- 04-0707
/R EFFECT PROBABILITY ON TIN-119 IN BARIUM TITANATE STANNATE SOLID SOLUTIONS IN THE VICINITY OF THE FERROE/ 04-0573
/ENA IN PIEZOCERAMICS BASED ON LEAD TITANATE AND ZIRCONATE. (SOLID SOLUTIONS LEAD MAGNESATE(0.5) TUNGSTATE/ 05-1063
/IC PROPERTIES OF FERROELECTRIC (BARIUM, STRONTIUM) TITANATE SOLID SOLUTIONS NEAR THE PHASE TRANSITION, UN/ 04-0680
DETERMINATION OF THE LATTICE PARAMETERS AND CURIE POINTS OF SOLID SOLUTIONS OF A(V)B(VI)C(VII) COMPOUNDS./ 15-2116
/DEPENDENCE OF THE VELOCITY OF ULTRASOUND IN POLYCRYSTALLINE SOLID SOLUTIONS OF BARIUM AND STRONTIUM TITAN/ 04-0894
E.   MOBILITY DETERMINATIONS FROM WEIGHT MEASUREMENTS IN SOLID SOLUTIONS OF (BARIUM, STRONTIUM) TITANAT 04-0689
IDE.   ELECTROOPTICAL PROPERTIES OF SINGLE CRYSTALS OF SOLID SOLUTIONS OF BARIUM TITANATE AND ZINC OX 04-0953
ALS.   X-RAY IRRADIATION EFFECT ON CIELECTRIC PROPERTIES OF SOLID SOLUTIONS OF BARIUM TITANATE TYPE MATERI 04-0782
TITANATE.   DIELECTRIC AGING IN TETRAGONAL SOLID SOLUTIONS OF CALCIUM TITANATE IN BARIUM 08-1295
LFATE SINGLE CRYSTALS WITH I/ GROWTH AND CHARACTERIZATION OF SOLID SOLUTIONS OF FERROELECTRIC TRIGLYCINE SU 27-2972
NIC S/ ELECTROCONDUCTIVITY AND HALL EFFECT IN SEMICONDUCTING SOLID SOLUTIONS OF ISO- AND HETEROVALENT CATIO 04-0943
THERMODYNAMIC THECRY OF SOLID SOLUTIONS OF ISOMORPHOUS FERROELECTRICS. 03-0302
MUTH AND MANGANESE CONTAINING ADDITIVES ON THE PROPERTIES OF SOLID SOLUTIONS OF LEAD ZIRCONATE TITANATE. /S 09-1588
CRYSTALLIZATION OF SOLID SOLUTIONS OF SODIUM NIOBATE FROM GLASS. 08-1386
ATES.   STRUCTURE OF POLYCRYSTALLINE SOLID SOLUTIONS OF STRONTIUM AND BISMUTH TITAN 06-1145
STATE(0.33)- LEAD TITANATE SYSTEM.   SEIGNETTO MAGNETIC SOLID SOLUTIONS OF THE LEAD FERRATE(0.67) TUNG 05-1065
ELATIONS BETWEEN COMPOSITION AND PROPERTIES IN FERROELECTRIC SOLID SOLUTIONS OF THE PEROVSKITE TYPE. /THE R 08-1330
/ESTIGATION OF THE DYNAMICS OF POLARIZATION OF FERROELECTRIC SOLID SOLUTIONS OF THE PEROVSKITE TYPE WITHIN/ 08-1380
/ONCENTRATION DEPENDENCE OF SOME FERROELECTRIC PROPERTIES OF SOLID SOLUTIONS OF TRIGLYCINE SULFATE WITH IS/ 27-2971
DIELECTRIC (AND TEMPERATURE) CHARACTERISTICS OF SOLID SOLUTIONS OF TUNGSTEN OXIDE. 10-1657
THERMODYNAMICS OF FERROELECTRIC SOLID SOLUTIONS. PART-1. PART-2. 03-0524
/ER STUDIES OF BISMUTH FERRATE LEAD TITANATE PEROVSKITE TYPE SOLID SOLUTIONS (ROOM TEMPERATURE TO 600 DEG / 08-1471
THERMODYNAMICS OF FERROELECTRIC SOLID SOLUTIONS. (THEORY) 03-0523
/TICULAR PROPERTIES OF PHASE TRANSFORMATION IN FERROELECTRIC SOLID SOLUTIONS. (THEORY OF COMPOSITION DEPEN/ 03-0250
TIONS FO/ DIELECTRIC POLARIZATION OF LEAD TITANATE ZIRCONATE SOLID SOLUTIONS (THERMODYNAMIC EXISTENCE CONDI 09-1541
URE AND PROPERTIES OF FERROELECTRIC LEAD TITANATE (GERMANATE SOLID SOLUTIONS TO 7 OR 8 MO1). /ON THE STRUCT 05-1043
ON FLUCTUATIONS.   DIELECTRIC CONSTANT OF FERROELECTRIC SOLID SOLUTIONS WITH ALLOWANCE FOR CONCENTRATI 03-0440
/33) LEAD FERRATE(0.5) NIOBATE(0.5), BISMUTH FERRATE AND ITS SOLID SOLUTIONS WITH BARIUM TITANATE, AND LAN/ 08-1368
EFFECT OF ELASTIC STRESSES ON PARAMETERS OF SOLID SOLUTIONS WITH FERROELECTRIC PROPERTIES. 03-0265
/PROPERTIES IN PEROVSKITE MODIFICATION OF CADMIUM TITANATE. (SOLID SOLUTIONS WITH STRONTIUM TITANATE LITHI/ 08-1424
STUDY OF DIVALENT BINARY AND TERNARY NIOBATES IN THE SOLID STATE. 08-1286
EXAMPLE, PZT)   AN ADAPTIVE FERROELECTRIC TRANSFORMER - A SOLID STATE ANALOG MEMORY DEVICE. (USING, FOR 09-1566
FERROELECTRICS AND THEIR APPLICATION IN SOLID STATE DEVICES AS ADAPTIVE CONTROL. 37-3665
OPTICALLY ERASABLE AND REWRITABLE SOLID STATE HOLOGRAMS. 37-3638
/DY OF OPTICAL PROPERTIES AND COLLECTIVE OSCILLATIONS IN NEW SOLID STATE MATERIALS (INCLUDING FERROELECTRI/ 02-0104
RICS)   DYNAMICAL EFFECTS IN SOLID STATE PHASE TRANSFORMATIONS. (FERROELECT 03-0154
GADOLINI/ RAMAN SCATTERING TECHNIQUES APPLIED TO PROBLEMS IN SOLID STATE PHYSICS (DOMAIN WALL THICKNESS IN 30-3319
FORMATION PROCESSES OF LEAD MAGNESIO NIOBATE IN THE SOLID STATE REACTION. 08-1354
S OF PEROVSKITE TYPE LEAD ZINCATE(/ HIGH PRESSURE SYNTHESIS (SOLID STATE REACTION, SINTERING) AND PROPERTIE 08-1393
SCOPE (BARRIER LAYERS IN / IMAGING ELECTRIC MICRO FIELDS (OF SOLID SURFACES) IN THE EMISSION ELECTRON MICRO 04-0913
OF THE PHASE TRANSITION IN CHARACTERISTICALLY FERROELECTRIC SOLIDS. /CONSTANT BEHAVIOR IN THE NEIGHBORHOOD 03-0522
/NANCE STUDY OF THE SPIN SYMMETRY STATES OF AMMONIUM IONS IN SOLIDS. (AMMONIUM SULFATE, SELENATE, VANADATE/ 19-2800
NATE SYSTEM (MEASUREMENT O/ RAMAN SPECTRA OF POLYCRYSTALLINE SOLIDS: APPLICATION TO THE LEAD TITANATE ZIRCO 09-1497
ND DEUTERON RESONANCE OF THIOUREA.   STUDIES OF FERROELECTRIC SOLIDS BY MAGNETIC RESONANCE PART-18: PROTON A 25-2963
MODERN ASPECTS OF PHYSICS OF SOLIDS. (FERROELECTRICITY) 01-0019
ERROELECTRIC PROPERTIES OF CRYSTALS, / POLARIZATION WAVES IN SOLIDS. (INCLUDING DEVELOPMENTS IN THEORY OF F 03-0306
DETECTION OF ATOM TUNNELING IN SOLIDS. (KDP) 17-2612
/IONS IN HYDROGEN BONDED FERROELECTRIC AND ANTIFERROELECTRIC SOLIDS. PART-1: ENERGY LOSS NEUTRON INCOHEREN/ 17-2687
UM SULFATE STUDIED BY COLD NEUTRONS / ROTATIONAL MOTIONS IN SOLIDS: THE FERROELECTRIC TRANSITION IN AMMONI 19-2752
NT)   DOUBLE SPACE CHARGE INJECTION IN SOLIDS. (USING BARIUM TITANATE ABOVE CURIE POI 04-0638
PIEZOELECTRIC CERAMICS. (NONMETALLIC SOLIDS, VOL-3, REVIEW) 01-0015
TE.   SOLIDIFICATION STUDY OF BARIUM STRONTIUM NIOBA 13-1954
RIUM SODIUM NIOBATE (BY USING COMPOSITION WHERE LIQUIDUS AND SOLIDUS COINCIDE). /OF GOOD OPTICAL QUALITY BA 13-2047
OXIDE- NICBIUM PENTOXIDE SYSTEM. (POTASSIUM LIT/ STUDY OF SUBSOLIDUS EQUILBRIA IN POTASSIUM OXIDE- LITHIUM 13-1963
THE SODIUM OXIDE-/ (EXTENT OF) TUNGSTEN BRONZE (LIQUIDUS AND SOLIDUS) FIELD AND MELT GROWTH OF CRYSTALS IN 13-1949
ALS ADP AND KDP.   SOLUBILITY IN THE GROWTH OF ELECTROOPTIC CRYST 17-2446
HOSPHATE- POTASSIUM DIHYDROGEN PHOSPHATE- WATER SYSTEM.   SOLUBILITY ISOBAR IN THE AMMONIUM DIHYDROGEN P 17-2658
) TITANATE IN BARIUM TITANATE.   SOLUBILITY LIMIT OF LANTHANUM(0.5) LITHIUM(0.5 04-0889
ATE.   SOLUBILITY OF ALUMINUM IN LEAD ZIRCONATE TITAN 09-1479
E- POTASSIUM DIHYDROGEN PHOSPHATE- WATER SYSTE/ POLYTHERM OF SOLUBILITY OF THE AMMONIUM DIHYDROGEN PHOSPHAT 17-2434
AN APPARATUS FOR POLISHING WATER SOLUBLE CRYSTALS (APPLICATION TO ADP AND KDP). 17-2632
TURE OF SINGLE CRYSTALS FORMED BY BARIUM TITANATE BASE SOLID SOLUTION. /XTERNAL ACTIONS ON THE DOMAIN STRUC 04-0954
NIUM MOLYBDATE, GROWTH BY RF CZOCHRALSKI AND FROM TOP SEEDED SOLUTION) /NATE AND RELATED COMPOUNDS. (GADOLI 08-1323
OTASSIUM DIHYDROGEN PHOSPHATE CRYSTAL GROWN FROM AN AQUEOUS SOLUTION. / LAYERS AND OTHER MACRODEFECTS IN P 17-2496
AND DISSOLUTION OF AMMONIUM DIHYDROGEN PHOSPHATE CRYSTALS IN SOLUTION.   KINETICS OF GROWTH 17-2412
ED IN POTASSIUM DIHYDROGEN PHOSPHATE CRYSTALS BY GROWTH FROM SOLUTION. /IC STUDY OF THE DISLOCATIONS PRODUC 17-2492
TROLYTE FUE/ IONIC CONDUCTION IN PEROVSKITE TYPE OXIDE SOLID SOLUTION AND ITS APPLICATION TO THE SOLID ELEC 08-1444
M(2+) AND TITANIUM(4+) ION/ GROWING SINGLE CRYSTALS OF SOLID SOLUTION BASED ON BARIUM TITANATE, WHERE BARIU 04-0960
.   STUDIES ON THE (STRONTIUM, CALCIUM) TITANATE SOLID SOLUTION CERAMIC WITH BOUNDARY LAYER STRUCTURE 06-1246
EAD COBALTATE(0.33) NIOBATE(0.67)- TITANATE- ZIRCONATE SOLID SOLUTION CERAMICS. /ZOELECTRIC PROPERTIES OF L 09-1549
TANATE CERAMICS ON THE PHYSICO- CHEMICAL NATURE OF THE SOLID SOLUTION FORMED FROM TWO TITANATES. /BARIUM TI 04-0789
METASTABLE POTASSIUM NITRATE(III) FROM SOLUTION. (GROWN IN WATER OR ALCOHOL) 16-2329
GROWTH OF IODIC ACID AND IODATE CRYSTALS BY A SOLUTION GROWTH METHOD. 08-1392
CHIOMETRIC MELTS. (BARIUM TITANATE, STRONTIUM TI/ TOP SEEDED SOLUTION GROWTH OF OXIDE CRYSTALS FROM NONSTOI 04-0553
/ONS ON THE SHAPE AND COMPOSITION CF FERROELECTRIC CRYSTALS (SOLUTION GROWTH, TGS, TRIGLYCINE SELENATE OR / 02-0089
L CONDUCTIVITY OF BARIUM TITANATE SINGLE CRYSTALS GROWN FROM SOLUTION IN A POTASSIUM FLUORIDE MELT. /CTRICA 04-1035
IONS ON THE STRUCTURE OF BARIUM TITANATE CRYSTALS GROWN FROM SOLUTION IN POTASSIUM FLUORIDE MELT. /H CONDIT 04-0598
IC OF THE AMMONIUM DIHYDROGEN PHOSPHATE SINGLE CRYSTAL (FROM SOLUTION) IN STATIC REGIME AND AT NORMAL PH. / 17-2516
S IN LEAD TITANATE(0.5) ZIRCONATE(0.5).   SOLUTION KINETICS OF LUTETIUM AND SCANDIUM ION 09-1538
SOLUTION OF AN ELASTIC FERROELECTRIC MODEL. 03-0211

TRICS IN ARBITRARY EXTERNAL ELECTRIC FIELD.   EXACT SOLUTION OF MODEL OF TWO DIMENSIONAL FERROELEC 03-0520
ATER AND RUBIDIUM DIDEUTERIUM PHOSPHATE IN HE/ ENTHALPIES OF SOLUTION OF RUBIDIUM DIHYDROGEN PHOSPHATE IN W 17-2585
ITION OF A REACTANT. (POTASSIUM DIHYDRO/ METHOD OF GROWTH IN SOLUTION OF SINGLE CRYSTALS BY THE GRADUAL ADD 17-2595
ICAL PROPERTIES OF BARIUM TITANATE- TANTALUM PENTOXIDE SOLID SOLUTION SINGLE CRYSTALS.   ELECTROOPT 04-0955
NIOBIUM-93 NMR STUDY OF THE LITHIUM NIOBATE TANTALATE SOLID SOLUTION SYSTEM.   11-1794
LEAD (ZINCATE(0.33) ELECTROPHYSICAL PROPERTIES OF THE SOLID SOLUTION SYSTEM LEAD TITANATE- LEAD ZIRCONATE- 13-1951
TITANATE AND P/ NEUTRON SCATTERING BY A FERROELECTRIC SOLID SOLUTION. (THEORY, APPLIED TO BARIUM STRONTIU 03-0260
                                   BISMUTH TITANATE SOLID SOLUTIONS.   08-1277
STAL CHEMICAL PROBLEMS OF (BARIUM, STRONTIUM) TITANATE SOLID SOLUTIONS.   CRY 04-0633
PROPERTIES OF SINGLE CRYSTALS OF BARIUM LEAD TITANATE SOLID SOLUTIONS.   DIELECTRIC AND ELASTIC 04-0597
RONTIUM TITANATE AND BARIUM(X) STRONTIUM(1-X) TITANATE SOLID SOLUTIONS) / TEMPERATURES. (POLYCRYSTALLINE ST 04-0720
CTRIC PROPERTIES OF ANTIMONY SULFOIODIDE- SULFOBROMIDE SOLID SOLUTIONS.   DIELE 15-2242
C FIELD ON PHASE COMPOSITION OF (LEAD, BARIUM) NIOBATE SOLID SOLUTIONS.   EFFECT OF ELECTRI 13-1943
T IN SINGLE CRYSTALS OF BARIUM TITANATE- ZINC TITANATE SOLID SOLUTIONS.   ELECTROOPTICAL EFFEC 04-0958
E(0.5) NIOBATE(0.5))(1-X) ZIRCONATE(X) PEROVSKITE TYPE SOLID SOLUTIONS. /ECTRIC PROPERTIES OF LEAD (STANNAT 04-0548
                                  FERROELECTRIC SOLID SOLUTIONS.   02-0069
GROWTH OF AMMONIUM DIHYDROGEN PHOSPHATE SINGLE CRYSTAL FROM SOLUTIONS.   17-2411
GROWTH OF TRIGLYCINE SULFATE FROM SLIGHTLY SUPERSATURATED SOLUTIONS.   27-3096
YDROTHERMAL CRYSTALLIZATION OF LEAD TITANATE ZIRCONATE SOLID SOLUTIONS.   H 09-1482
AND LEAD COBALTATE(0.5) TUNGSTATE(0.5)- LEAD TITANATE SOLID SOLUTIONS. /LEAD COBALTATE(0.5) TUNGSTATE(0.5) 13-1968
OBIA (POTASSIUM LITHIUM NIOBATES, PHASE RELATIONSHIPS) SOLID SOLUTIONS). /(OXIDE) SYSTEM POTASSIA LITHIA NI 13-2017
/CTRIC POLARIZATION RELAXATION IN SOME FERROELECTRICS (SOLID SOLUTIONS AND MIXED CATION COMPOUNDS, KANZIG / 03-0291
AN/ DIELECTRIC RELAXATION SPECTRA OF WATER, ICE, AND AQUEOUS SOLUTIONS AND THEIR INTERPRETATION. PART-8: TR 34-3482
/NDUCTIVITY AND RELATED PROPERTIES OF 60 COMPOUNDS, 58 SOLID SOLUTIONS, AND 10 TECHNICAL FERROCERAMICS, CU/ 01-0014
OIODIC ACID,/ GROWTH OF ELECTROOPTICAL CRYSTALS FROM AQUEOUS SOLUTIONS (APPARATUS, GROWTH OF ADP, KDP, HYDR 17-2420
IC PROPERTIES OF LEAD TITANATE- LEAD (FERRATE NIOBATE) SOLID SOLUTIONS AT HIGH PRESSURES.   DIELECTR 05-1071
IUM TITANATE, POLYCRYSTALLINE, STRONTIUM TITANATE, AND SOLID SOLUTIONS BARIUM STRONTIUM TITANATE) /L STRONT 04-0721
                FORMATION OF FLUORINE CONTAINING SOLID SOLUTIONS BASED ON BARIUM TITANATE.   04-0932
                                            SOLID SOLUTIONS BASED ON BARIUM TITANATE.   04-0786
/AND NONLINEAR MICROWAVE PROPERTIES OF POLYCRYSTALLINE SOLID SOLUTIONS BASED ON BARIUM TITANATE. (BARIUM S/ 04-1019
TANTALAT/ X-RAY ANALYSIS AND DIELECTRIC PROPERTIES OF SOLID SOLUTIONS BASED ON THE FERROELECTRIC POTASSIUM 08-1310
/EMPERATURE OF BARIUM(0-0.1) STRONTIUM(0.9-1) TITANATE SOLID SOLUTIONS BY HYDROSTATIC PRESSURE. (TO 1500 A/ 04-0722
TH AND ETCHING OF POTASSIUM DIHYDROGEN PHOSPHATE CRYSTALS IN SOLUTIONS CONTAINING ACETATE IONS.   GROW 17-2567
/IC PROPERTIES OF ANTIMONY SULFOIODIDE(X) BROMIDE(1-X) SOLID SOLUTIONS DEPEND ON TEMPERATURE AND HYDROSTAT/ 15-2149
FERROELECTRIC PHOTOELECTRETS BASED ON A(V)B(VI)C(VII) SOLID SOLUTIONS. (DIELECTRIC CONSTANTS, PYROELECTRI/ 15-2157
EMICONDUCTOR PROPERTIES OF BARIUM (TITANATE, STANNATE) SOLID SOLUTIONS DOPED BY TUNGSTEN TRIOXIDE.   S 04-0965
TRANSITION IN THE LEAD (ZIRCONATE, STANNATE, TITANATE) SOLID SOLUTIONS. (ELECTROSTRICTION CONSTANTS) /TRIC 09-1486
LEAD METANIOBATE TERNARY SYSTEM. CURIE TEMPERATURES OF SOLID SOLUTIONS IN A LEAD ZIRCONATE- LEAD TITANATE- 09-1595
YSTEM LEAD HAFNATE TITANATE, LEAD STANNATE NIOBATE.   SOLID SOLUTIONS IN ANTIFERROELECTRIC REGION OF THE S 13-1956
CRYSTALLIZATION OF RUBIDIUM DIDEUTERIUM PHOSPHATE FROM SOLUTIONS IN HEAVY WATER.   17-2583
RIUM(X) PHOSPHATE SINGLE CRYSTALS. REGULAR BEHAVIOR OF SOLID SOLUTIONS IN POTASSIUM DIHYDROGEN(1-X) DIDEUTE 17-2458
UOBERYLLATE SYSTEM.   FERROELECTRIC PROPERTIES OF SOLID SOLUTIONS IN THE AMMONIUM SULFATE- AMMONIUM FL 19-2764
UOBERYLLATE SYSTEM.   FERROELECTRIC PROPERTIES OF SOLID SOLUTIONS IN THE AMMONIUM SULFATE- AMMONIUM FL 19-2763
/NDENCE OF THE CURIE-WEISS CONSTANT IN POLYCRYSTALLINE SOLID SOLUTIONS IN THE (BARIUM, STRONTIUM) TITANATE/ 04-0586
/ND MAGNETIC PROPERTIES OF FERROELECTRIC FERROMAGNETIC SOLID SOLUTIONS IN THE LEAD COBALT TUNGSTATE- CADMI 36-3604
NIUM FLUOBERYLLATE.   FERROELECTRIC PROPERTIES OF SOLID SOLUTIONS IN THE SYSTEM AMMONIUM SULFATE- AMMO 19-2762
TIUM NIOBI/ RELAXATION AND FERROELECTRIC PROPERTIES OF SOLID SOLUTIONS IN THE SYSTEM BARIUM TITANATE- STRON 04-0707
/CT PROBABILITY ON TIN-119 IN BARIUM TITANATE STANNATE SOLID SOLUTIONS IN THE VICINITY OF THE FERROELECTRI/ 04-0573
/ PIEZOCERAMICS BASED ON LEAD TITANATE AND ZIRCONATE. (SOLID SOLUTIONS LEAD MAGNESATE(0.5) TUNGSTATE(0.5))/ 05-1063
/PERTIES OF FERROELECTRIC (BARIUM, STRONTIUM) TITANATE SOLID SOLUTIONS NEAR THE PHASE TRANSITION, UNDER HI/ 04-0680
/INATION OF THE LATTICE PARAMETERS AND CURIE POINTS OF SOLID SOLUTIONS OF A(V)B(VI)C(VII) COMPOUNDS. ANTIM/ 15-2116
DENCE OF THE VELOCITY OF ULTRASOUND IN POLYCRYSTALLINE SOLID SOLUTIONS OF BARIUM AND STRONTIUM TITANATES. / 04-0894
MOBILITY DETERMINATIONS FROM WEIGHT MEASUREMENTS IN SOLID SOLUTIONS OF (BARIUM, STRONTIUM) TITANATE.   04-0689
ELECTROOPTICAL PROPERTIES OF SINGLE CRYSTALS OF SOLID SOLUTIONS OF BARIUM TITANATE AND ZINC OXIDE.   04-0953
X-RAY IRRADIATION EFFECT ON DIELECTRIC PROPERTIES OF SOLID SOLUTIONS OF BARIUM TITANATE TYPE MATERIALS.   04-0782
TE.   DIELECTRIC AGING IN TETRAGONAL SOLID SOLUTIONS OF CALCIUM TITANATE IN BARIUM TITANA 04-0721
SINGLE CRYSTALS WITH I/ GROWTH AND CHARACTERIZATION OF SOLID SOLUTIONS OF FERROELECTRIC TRIGLYCINE SULFATE 27-2972
ELECTROCONDUCTIVITY AND HALL EFFECT IN SEMICONDUCTING SOLID SOLUTIONS OF ISO- AND HETEROVALENT CATIONIC S/ 04-0943
THERMODYNAMIC THEORY OF SOLID SOLUTIONS OF ISOMORPHOUS FERROELECTRICS.   03-0302
ND MANGANESE CONTAINING ADDITIVES ON THE PROPERTIES OF SOLID SOLUTIONS OF LEAD ZIRCONATE TITANATE. /SMUTH A 09-1588
IONS IN SODIUM NIOBATE AND THE USE OF TILTING SCHEMES IN THE SOLUTIONS OF PEROVSKITE STRUCTURES. /E TRANSIT 08-1336
CRYSTALLIZATION OF POLYCRYSTALLINE SOLID SOLUTIONS OF SODIUM NIOBATE FROM GLASS.   08-1336
STRUCTURE OF POLYCRYSTALLINE SOLID SOLUTIONS OF STRONTIUM AND BISMUTH TITANATES.   06-1145
0.33)- LEAD TITANATE SYSTEM.   SEIGNETTO MAGNETIC SOLID SOLUTIONS OF THE LEAD FERRATE(0.67) TUNGSTATE(05-1065
NS BETWEEN COMPOSITION AND PROPERTIES IN FERROELECTRIC SOLID SOLUTIONS OF THE PEROVSKITE TYPE. /THE RELATIO 08-1330
/TION OF THE DYNAMICS OF POLARIZATION OF FERROELECTRIC SOLID SOLUTIONS OF THE PEROVSKITE TYPE WITHIN A WID/ 08-1380
/ (TETRAGONAL) TUNGSTEN BRONZE TYPE CRYSTALS GROWN IN MOLTEN SOLUTIONS OF THE PSEUDOSYSTEM POTASSIUM LITHIU 13-1946
/RATION DEPENDENCE OF SOME FERROELECTRIC PROPERTIES OF SOLID SOLUTIONS OF TRIGLYCINE SULFATE WITH ISOMORPH/ 27-2971
DIELECTRIC (AND TEMPERATURE) CHARACTERISTICS OF SOLID SOLUTIONS OF TUNGSTEN OXIDE.   10-1657
THERMODYNAMICS OF FERROELECTRIC SOLID SOLUTIONS. PART-1. PART-2.   03-0524
IDE, SODIUM CHLORIDE,/ CRYSTALLIZATION POTENTIALS IN AQUEOUS SOLUTIONS. (POTASSIUM BROMIDE, POTASSIUM CHLOR 17-2707
UDIES OF BISMUTH FERRATE LEAD TITANATE PEROVSKITE TYPE SOLID SOLUTIONS (ROOM TEMPERATURE TO 600 DEG C). /ST 08-1471
THERMODYNAMICS OF FERROELECTRIC SOLID SOLUTIONS. (THEORY)   03-0523
/R PROPERTIES OF PHASE TRANSFORMATION IN FERROELECTRIC SOLID SOLUTIONS. (THEORY OF COMPOSITION DEPENDENCE / 03-0520
FO/ DIELECTRIC POLARIZATION OF LEAD TITANATE ZIRCONATE SOLID SOLUTIONS (THERMODYNAMIC EXISTENCE CONDITIONS 09-1541
MICS IN HYDROGEN BONDED FERROELECTRICS. PART-2: APPROXIMATED SOLUTIONS TO THE KINETIC EQUATIONS. /OTON DYNA 03-0205
D PROPERTIES OF FERROELECTRIC LEAD TITANATE (GERMANATE SOLID SOLUTIONS TO 7 OR 8 MO). /ON THE STRUCTURE AN 05-1043
CTUATIONS.   DIELECTRIC CONSTANT OF FERROELECTRIC SOLID SOLUTIONS WITH ALLOWANCE FOR CONCENTRATION FLU 03-0440
/AD FERRATE(0.5) NIOBATE(0.5), BISMUTH FERRATE AND ITS SOLID SOLUTIONS WITH BARIUM TITANATE, AND LANTHANUM/ 08-1368
EFFECT OF ELASTIC STRESSES ON PARAMETERS OF SOLID SOLUTIONS WITH FERROELECTRIC PROPERTIES.   03-0265
/TIES IN PEROVSKITE MODIFICATION OF CADMIUM TITANATE. (SOLID SOLUTIONS WITH STRONTIUM TITANATE LITHIUM TAN/ 08-1424
IRRADIATION PRODUCED SOLVATED ELECTRONS IN ICE.   34-3423
(STRONTIUM TITANATE, LEAD(X) STRONTIUM(1-X) TITANATE, LEAD/ SOLVENT ZONE GROWTH OF FERROELECTRIC CRYSTALS. 05-1042
CRYSTALS. (THEORY OF ANISOTROPIC RELAXATION / ANISOTROPY OF SOUND ABSORPTION IN TRIGLYCINE SULFATE SINGLE 27-3078
DISPLACIVE TYPE FERROELECTRICS. (THEORY, BAR/ HIGH FREQUENCY SOUND ATTENUATION AND THERMAL CONDUCTIVITY IN 03-0160
NONLINEAR INTERACTION OF MICROWAVE ELECTRIC FIELDS AND SOUND IN LITHIUM NIOBATE.   11-1838
ATE EXAMPLE)   STIMULATED COMBINATION SCATTERING OF SOUND IN PIEZOELECTRICS. (THEORY, LITHIUM NIOB 03-0431
INT.   FREQUENCY DEPENDENCE OF THE ABSORPTION OF SOUND IN ROCHELLE SALT NEAR ITS UPPER CURIE PO 28-3183
MICONDUCTORS.   RELAXATION ABSORPTION OF SOUND NEAR THE CURIE POINT OF FERROELECTRIC SE 03-0492
(ANTIMONY SULFOIODIDE).   ANOMALOUS ABSORPTION OF SOUND NEAR THE FERROELECTRIC PHASE TRANSITION 15-2221

ANSITION IN STRONTIUM TITANATE. SOUND PROPAGATION NEAR THE STRUCTURAL PHASE TR 06-1155
IC-2 TRANSITION IN (NIOBIUM DOPED) ZIRCONATE(0/ VARIATION OF SOUND VELOCITY AT FERROELECTRIC-1, FERROELECTR 09-1511
TE.   TEMPERATURE DEPENDENCE OF SOUND VELOCITY IN DICALCIUM STRONTIUM PROPIONA 26-2934
RICS. (THEORY)   REFLECTION AND REFRACTION OF TRANSVERSE SOUND WAVES AT DOMAIN BOUNDARIES IN FERROELECT 03-0313
/N AND REFRACTION OF QUASI LONGITUDINAL AND QUASI TRANSVERSE SOUND WAVES BY 180 DOMAIN BOUNDARIES IN FERRO/ 03-0314
 PHASE TRANSITION POINT.   ATTENUATION OF SOUND WAVES IN DEUTERATED ROCHELLE SALT AT THE 28-3204
 PHASE TRANSITION POINT.   ATTENUATION OF SOUND WAVES IN DEUTERATED ROCHELLE SALT AT THE 28-3205
INUM GARNET- BARIUM SODIUM NIOBATE 0.66 MILLIMICRON HARMONIC SOURCE.   NEODYMIUM: YTTRIUM ALUM 13-1975
HE DIRECTIO/ RADIALLY VIBRATING FERROCERAMIC TRANSDUCER AS A SOURCE OF ULTRASONIC VIBRATIONS RADIATING IN T 37-3657
AL BEHAVIOR OF ICE AT LOW TEMPERATURES, FERROELECTRICITY AND SPACE CHARGE.   ELECTRIC 34-3448
 CERAMICS CAUSED BY THE ADDITION OF IMPURITIES. SPACE CHARGE EFFECT IN LEAD ZIRCONATE TITANATE 09-1609
M TITANATE ABOVE CURIE POINT)   DOUBLE SPACE CHARGE INJECTION IN SOLIDS. (USING BARIU 04-0638
RMANIUM O/ LIGHT AND ELECTRIC FIELD DEPENDENT OSCILLATION OF SPACE CHARGE LIMITED CURRENT IN BISMUTH(12) GE 35-3505
E BARIUM TITANATE. (CURRENT VOLTAGE CHARACTERISTICS, 150-37/ SPACE CHARGE LIMITED CURRENT IN POLYCRYSTALLIN 04-0557
                                       SPACE CHARGE LIMITED CURRENTS IN A METAL FERRO 03-0348
S.   THEORY OF SPACE CHARGE LIMITED CURRENTS IN FERROELECTRIC 03-0251
IUM TITANATE(0.75) PLATINATE(0.25) CRYSTALS.   SPACE CHARGE LIMITED CURRENTS IN HEXAGONAL BAR 04-0936
G POTASSIUM NIOBATE CRYSTALS.   SPACE CHARGE LIMITED CURRENTS IN SEMICONDUCTIN 08-1412
TITANATE CERAMICS.   SPACE CHARGE POLARIZATION AND AGING OF BARIUM 04-0880
TERO HYDRATE ABOVE THE CURIE TEMPERATURE (LATTICE CONSTANTS, SPACE GROUP). /C POTASSIUM FERROCYANIDE TRIDEU 24-2903
 PARAMAGNETIC BARIUM COBALT TETRAFLUORIDE CRYSTAL STRUCTURE (SPACE GROUP AND LATTICE CONSTANTS). /RAELASTIC 33-3386
SODIUM NIOBATE AND BARIUM SODIUM TANTALATE (FORCE CONSTANTS, SPACE GROUP ASSIGNMENT). /IUM NIOBATE, BARIUM 11-1808
               NOTE ON THE SPACE GROUP OF POTASSIUM HYDROGEN IODATE. 08-1370
COPPER FORMATE TETRAHYDRATE.   UNIT CELL AND SPACE GROUP OF THE ANTIFERROELECTRIC PHASE OF 26-2927
HING TEMPERATURES IN LITHIUM NIOBATE (CRYSTAL GROWTH BY ADD/ SPATIALLY UNIFORM AND ALTERABLE SHG PHASE MATC 11-1684
 DIGLYCINE SULFATE AN INTERESTING NEW DIELECTRIC CRYSTAL SPECIES.   27-2993
FERROELECTRIC LITHIUM TRIHYDROGEN DISELENITE.   PARAMAGNETIC SPECIES IN GAMMA-IRRADIATED SINGLE CRYSTAL OF 22-2843
                   PARAMAGNETIC SPECIES IN IRRADIATED NITRATE. 16-2405
STALS (PREVIOUS THEORY EXTENDED TO MAGNETIC CRYSTA/ POSSIBLE SPECIES OF FERROMAGNETIC, AND FERROELASTIC CRY 03-0142
             PERMISSIBLE DOMAIN WALLS IN FERROELECTRIC SPECIES (THEORY AND TABULATION) 03-0247
TICE VIBRATION OF SODIUM NITRITE. (CALCULATIONS TO RELATE TO SPECIFIC ANTIFERROELECTRIC PHASE)   LAT 16-2317
E OF THE VELOCITY OF ULTRASOUND IN POLYCRYSTALLINE SOLID SO/ SPECIFIC FEATURES OF THE TEMPERATURE DEPENDENC 04-0894
GE IN SODIUM TRIHYDROGEN DISELENITE.   ANOMALOUS SPECIFIC HEAT AND CONFIGURATIONAL ENTROPY CHAN 22-2856
ETHYL AMMONIUM ALUM ABOVE THE CRITICAL TEMPERATURE. SPECIFIC HEAT AND SUSCEPTIBILITY IN CHROMIUM M 20-2805
DEG K PHASE TRANSITION IN STRONTIUM TITANATE. (ABSTRACT) SPECIFIC HEAT ANOMALY ASSOCIATED WITH THE 110 06-1226
E SALT (CALCULATION OF TEMPERATURE DEPENDENCE FOR FERROELEC/ SPECIFIC HEAT ANOMALY IN FERROELECTRIC ROCHELL 28-3187
SODIUM DINITRITE.   SPECIFIC HEAT ANOMALY IN FERROELECTRIC SILVER 16-2296
FERRO- PARAELECTRIC TRANSITION. (TGS)   SPECIFIC HEAT ANOMALY IN THE NEIGHBORHOOD OF A 27-2962
IODINE BORACITE.   SPECIFIC HEAT ANOMALY OF FERROELECTRIC COBALT 14-2101
UM DIHYDROGEN PHOSPHATE CRYSTALS FOR STUDY OF THE PY/ USE OF SPECIFIC HEAT (BY A DYNAMIC METHOD) OF POTASSI 17-2656
OF FERROELECTRIC CERAMICS AT LOW TE/ DIELECTRIC AND THERMAL (SPECIFIC HEAT, ELECTROCALORIC EFFECT) STUDIES 04-0720
N IN TERMS OF A SINGLE MODE ANHARMONIC OSCI/ FREE ENERGY AND SPECIFIC HEAT IN FERROELECTRIC PHASE TRANSITIO 03-0406
        EFFECTS OF DEUTERATION ON THE SPECIFIC HEAT OF ADP CRYSTALS. 17-2414
THE TEMPERATURE RANGE FROM -70 TO +14 DEG C.   SPECIFIC HEAT OF AMMONIUM HYDROGEN SULFATE IN 19-2789
                    SPECIFIC HEAT OF BARIUM TITANATE. 04-0715
TASSIUM FERROCYANIDE TRIHYDRATE.   ANOMALOUS SPECIFIC HEAT OF FERROELECTRIC THIOUREA AND PO 24-2902
               SPECIFIC HEAT OF ICE I(H). 34-3464
          MEASUREMENT OF THE SPECIFIC HEAT OF KDP BY A DYNAMIC METHOD. 17-2619
DE.   SPECIFIC HEAT OF POTASSIUM MANGANESE TRIFLUORI 33-3387
EGREES K.   SPECIFIC HEAT OF PURE AND DOPED ICE NEAR 120 D 34-3463
ERROELECTRIC TRANSITION POINT.   SPECIFIC HEAT OF SODIUM NITRITE NEAR THE ANTIF 16-2303
               ANOMALOUS SPECIFIC HEAT OF SODIUM TRIHYDROGEN SELENITE. 22-2823
SSOCIATED COMBINATORIAL PROBLEM.   ANOMALOUS SPECIFIC HEAT OF SODIUM TRIHYDROGEN SELENITE A 22-2822
SSIUM DIHYDROGEN PHOSPHATE, RUBIDIUM AND AMMONIUM ANALOGS S/ SPECIFIC HEAT OF SOME KDP TYPE CRYSTALS. (POTA 17-2679
                  ANOMALOUS SPECIFIC HEAT OF STRONTIUM TITANATE. 06-1159
TRUCTURAL TRANSITION.   SPECIFIC HEAT OF STRONTIUM TITANATE NEAR THE S 06-1132
               FERROELECTRIC SPECIFIC HEAT OF TRIGLYCINE SULFATE. 27-3146
/CHING MEASUREMENTS. PART-8: ULTRASOUND PROPAGATION. PART-9: SPECIFIC HEAT. PART-10: PHOTON RENORMALIZATIO/ 03-0397
 CONDUCTIVITY OF CALCIUM DOPED SODIUM NITRATE CRYSTALS NEAR/ SPECIFIC HEAT, THERMAL DIFFUSIVITY AND THERMAL 16-2283
 ELECTROCALORIC EFFECT IN TRIGLYCINE SULFATE.   SPECIFIC HEAT UNDER APPLIED ELECTRIC FIELD AND 27-2960
YCINE SULFATE AND TRIGLYCINE SELENATE CRYSTALS ON CHANGES IN SPECIFIC RESISTANCE. /OWTH CONDITIONS OF TRIGL 27-3069
TION IN RUBIDIUM DIHYDROGEN PHOSPHIDE USING RAMAN SCATTERING SPECTRA.   FERROELECTRIC TRANSI 17-2620
TIONS IN ANTIMONY SULFOIODIDE. (POLARIZED INFRARED AND RAMAN SPECTRA) /GTH OPTICAL PHONONS AND PHASE TRANSI 15-2106
N POTASSIUM DIDEUTERIUM PHOSPHATE (FROM POLARIZED REFLECTION SPECTRA, A TUNNELING MODEL). /UTERON MOTIONS I 17-2645
ROE/ VISIBLE AND ULTRAVIOLET OPTICAL PROPERTIES (REFLECTANCE SPECTRA AND BAND STRUCTURE) OF SOME ABO(3) FER 08-1383
ERBIUM(3+) IONS IN LITHIUM NIO/ ABSORPTION AND LUMINESCENCE SPECTRA AND ENERGY LEVELS OF NEODYMIUM(3+) AND 11-1713
POTASSIUM NITRATE.   FAR INFRARED SPECTRA AND FERROELECTRIC PHASE TRANSITION OF 16-2308
TRIC SODIUM NITRITE AT 77 AND 294 DEG K.   RAMAN SPECTRA AND MODE FREQUENCY SHIFTS OF FERROELEC 16-2262
TIONS OF THE PHASE TR/ NUCLEAR MAGNETIC RESONANCE (WIDE LINE SPECTRA AND SPIN LATTICE RELAXATION) INVESTIGA 20-2802
CHROM/ DETERMINATION OF CHROMIUM VALENCY FROM STUDIES OF EPR SPECTRA AND STATIC MAGNETIC SUSCEPTIBILITY OF 04-0542
EA. PART-1: TEMPERATURE DEPENDENCE OF THE INFRARED AND RAMAN SPECTRA AND THE CALCULATED ENTROPY. /IC THIOUR 25-2920
UM TITANATE.   ON THE INTERPRETATION OF INFRARED SPECTRA AND THE LATTICE DYNAMICS OF CUBIC BARI 04-0659
ICS. (LATTICE VIBRATIONS)   POWDER RAMAN SPECTRA: APPLICATION TO DISPLACIVE FERROELECTR 03-0194
INAL MODES IN HYDROGEN AND DEUTERIUM HALIDE CRYSTALS. (RAMAN SPECTRA, DIPOLAR COUPLING MODEL COMPUTATION) / 31-3326
 RADIATION DAMAGE IN LITHIUM NIOBATE. (CAPACITANCE, OPTICAL SPECTRA, ESR, CELL PARAMETERS) 11-1705
            FAR INFRARED ABSORPTION SPECTRA FOR AMMONIUM DIHYDROGEN PHOSPHATE. 17-2587
TITANI/ PECULIARITIES OF THE ELECTRON PARAMAGNETIC RESONANCE SPECTRA IN A FERROELECTRIC CERAMIC CONTAINING 04-1020
/ON SELENITE (ION). RADIATION DAMAGE ELECTRON SPIN RESONANCE SPECTRA IN FERROELECTRIC SODIUM TRIHYDROGEN D/ 22-2870
/ CRYSTALS (CYCLOTRON RESONANCE, ABSORPTION AND LUMINESCENCE SPECTRA, INFRARED, OPTICAL, AND ACOUSTIC SCAT/ 03-0427
         ISOTOPE EFFECT IN THE RAMAN SPECTRA OF A DEUTERATED KDP CRYSTAL. 17-2621
ROELECTRIC PHASES.   RAMAN SPECTRA OF ADP IN THE PARAELECTRIC AND ANTIFER 17-2431
(4000-10000 PER CM) AND/ LOW TEMPERATURE INFRARED ABSORPTION SPECTRA OF AMMONIUM DIHYDROGEN ORTHOPHOSPHATE 17-2432
AMMONIUM SULFATE AT FERROELECTRIC / CHANGES IN THE INFRARED SPECTRA OF AMMONIUM OR RUBIDIUM BISULFATE, AND 19-2772
            FAR INFRARED REFLECTIVITY SPECTRA OF ANTIMONY SULFOIODIDE. 15-2234
            FAR INFRARED REFLECTIVITY SPECTRA OF ANTIMONY SULFOIODIDE. 15-2235
        ANISOTROPY OF THE REFLECTION SPECTRA OF ANTIMONY SULFOIODIDE CRYSTALS. 15-2125
HE 2-1000 MICRON/ TEMPERATURE STUDIES OF INFRARED REFLECTION SPECTRA OF BARIUM AND STRONTIUM TITANATES IN T 04-0863
               RAMAN SPECTRA OF BARIUM SODIUM NIOBATE. 13-1906
/FFECT OF PARTICLE SIZE AND SHAPE ON THE INFRARED ABSORPTION SPECTRA OF BARIUM TITANATE AND STRONTIUM TITA/ 04-0823
GTH INFRARED REGION.   ABSORPTION AND REFLECTION SPECTRA OF BARIUM TITANATE IN THE LONG WAVELEN 04-0644

TELLUROIODIDE CRYSTALS.	ABSORPTION SPECTRA OF BISMUTH TELLUROBROMIDE AND BISMUTH	15-2217
AICIUM TITANATE/ EXTENDED FINE STRUCTURE IN X-RAY ABSORPTION	SPECTRA OF CERTAIN PEROVSKITES (STRONTIUM OR C	03-0414
S IN ITS LOW TEMPERATURE P/ PECULIARITIES IN THE VIBRATIONAL	SPECTRA OF CESIUM AND RUBIDIUM NITRATE CRYSTAL	16-2323
FAR INFRARED REFLECTIVITY	SPECTRA OF CESIUM LEAD TRICHLORIDE.	33-3379
LANTHANUM ALUMINATE. ELECTRON SPIN RESONANCE	SPECTRA OF CHROMIUM(3+) AND GADOLINIUM(3+) IN	36-3583
/FECTS OF UNIAXIAL STRESS AND ELECTRIC FIELD ON PARAMAGNETIC	SPECTRA OF CHROMIUM(3+) IN POTASSIUM DIHYDROG/	17-2554
OBERYLLATE. INFLUENCE OF SPONTANEOUS POLARIZATION ON THE EPR	SPECTRA OF CHROMIUM(3+) IONS IN TRIGLYCINE FLU	27-3136
MINATE. OPTICAL	SPECTRA OF CHROMIUM(3+) PAIRS IN LANTHANUM ALU	36-3616
RAMAN AND INFRARED	SPECTRA OF COPPER FORMATE TETRAHYDRATE.	26-2936
CHELLE SALT. EFFECT OF APPLIED ELECTRIC FIELD ON EPR	SPECTRA OF (COPPER(2+)) DOPED FERROELECTRIC RO	28-3185
N PHOSPHATE TYPE IN THEIR PARAELECTRIC AND LOW TEMPER/ RAMAN	SPECTRA OF CRYSTALS OF THE POTASSIUM DIHYDROGE	17-2515
M ALUMINUM (SULFATE OR SELENATE) DODECAH/ INFRARED AND RAMAN	SPECTRA OF FERROELECTRIC ALUMS: METHYL AMMONIU	20-2803
YCINE SULFATE TYPE. LOW FREQUENCY RAMAN	SPECTRA OF FERROELECTRIC CRYSTALS OF THE TRIGL	27-3114
FAR INFRARED DISPERSION AND RAMAN	SPECTRA OF FERROELECTRIC CRYSTALS. (THEORY)	03-0165
REA. LATTICE VIBRATIONAL	SPECTRA OF FIVE CRYSTAL MODIFICATIONS OF THIOU	25-2905
. INFRARED	SPECTRA OF FLUORINE CONTAINING BARIUM TITANATE	04-0983
/OF THE INTENSITY OF HYDROGEN BOND BANDS IN RAMAN SCATTERING	SPECTRA OF GYPSUM AND ROCHELLE SALT CRYSTALS.	28-3189
VIBRATION	SPECTRA OF HYDROGEN BONDS. (KDP)	17-2482
/THROUGH PURE ICE I(H) SINGLE CRYSTALS. PART-1: POLARIZATION	SPECTRA OF ICE I(H). PART-2: MOLECULAR MODELS/	34-3486
RAMAN AND INFRARED	SPECTRA OF INORGANIC FORMATES.	26-2937
E ION IN THE PARAELECTRIC PHASE OF KDP CRYSTALS (INFRARED A/	SPECTRA OF INTERNAL VIBRATIONS OF THE PHOSPHAT	17-2479
VIBRATIONAL	SPECTRA OF IODIC ACID SINGLE CRYSTALS.	36-3548
RIHYDROGEN SELENITE (SHS) OR SODIUM TRIDEUTERIUM S/ INFRARED	SPECTRA OF ISOTOPIC NONISOMORPHISM OF SODIUM T	22-2836
D 300 DEG K) FAR INFRARED	SPECTRA OF KDP AND ADP. (20-550 PER CM, 200 AN	17-2549
). FAR INFRARED REFLECTIVITY	SPECTRA OF KDP CRYSTAL (16-400 PER CM, 83-300K	17-2681
CRYSTAL GROWTH AND TEMPERATURE DEPENDENT RAMAN	SPECTRA OF LEAD TITANATE.	05-1074
ATE AND BARIUM SODIUM TANTALATE (FORCE CONSTANT/ VIBRATIONAL	SPECTRA OF LITHIUM NIOBATE, BARIUM SODIUM NIOB	11-1808
CHROMIUM(3+) IONS (10 TO 3/ CALCULATION OF THE (ABSORPTION)	SPECTRA OF LITHIUM NIOBATE CRYSTALS CONTAINING	11-1831
PART-1: IRON(X) ALUMINU/ RELAXATION PHENOMENA AND MOSSBAUER	SPECTRA OF MAGNETIZED PARAMAGNETIC SUBSTANCES.	19-2751
. FORBIDDEN HYPERFINE LINES IN ESR	SPECTRA OF MANGANESE(2+) DOPED LITHIUM NIOBATE	11-1805
N PHOSPHATE AND RUBIDIUM / LOW FREQUENCY INFRARED ABSORPTION	SPECTRA OF MONOCRYSTALS OF POTASSIUM DIHYDROGE	17-2457
ICKEL IODINE BORACITES (AND NICKEL DOPED/ OPTICAL ABSORPTION	SPECTRA OF NICKEL(2+) IN NICKEL CHLORINE AND N	14-2074
DEUTERIUM CHLORIDES AT TEMPERATURES ABOVE 80 DEGREES/ RAMAN	SPECTRA OF ORTHORHOMBIC AND CUBIC HYDROGEN AND	31-3330
TALS. CRYSTAL SIZE EFFECTS IN THE RAMAN	SPECTRA OF ORTHORHOMBIC HYDROGEN CHLORIDE CRYS	31-3338
HOSPHATE AND POTA/ FOURIER TRANSFORM FAR INFRARED REFLECTION	SPECTRA OF PARAELECTRIC POTASSIUM DIHYDROGEN P	03-0356
CRYSTALS CESIUM DIH/ ARSENIC-75 NUCLEAR QUADRUPOLE RESONANCE	SPECTRA OF PARTIALLY DEUTERATED FERROELECTRIC	18-2740
DE. FAR INFRARED	SPECTRA OF PIEZOELECTRIC POLYVINYLIDENE FLUORI	36-3590
TO THE LEAD TITANATE ZIRCONATE SYSTEM (MEASUREMENT O/ RAMAN	SPECTRA OF POLYCRYSTALLINE SOLIDS: APPLICATION	09-1497
IUM DIHYDROGEN PHOSPHATE IN AN ELECTRIC FIELD. RAMAN	SPECTRA OF POLYMORPHIC MODIFICATIONS OF POTASS	17-2425
/C SOLIDS. PART-1: ENERGY LOSS NEUTRON INCOHERENT SCATTERING	SPECTRA OF POTASSIUM DIHYDROGEN PHOSPHATE AND/	17-2687
ARA- AND FERROELECTRIC PHASES. VIBRATION	SPECTRA OF POTASSIUM DIHYDROGEN PHOSPHATE IN P	17-2480
ASSIUM DIDEUTERIUM PHOSPHATE, P/ TEMPERATURE DEPENDENT RAMAN	SPECTRA OF POTASSIUM DIHYDROGEN PHOSPHATE, POT	17-2409
TASSIUM COBALT TRIFLUORIDE, AND RUBIDIU/ INFRARED ABSORPTION	SPECTRA OF POTASSIUM MANGANESE TRIFLUORIDE, PO	33-3384
E AT PHASE TRANSITIONS. INFRARED	SPECTRA OF POTASSIUM NITRATE AND SODIUM NITRIT	16-2352
OPTICAL ABSORPTION	SPECTRA OF REDUCED STRONTIUM TITANATE.	06-1245
FAR INFRARED ABSORPTION	SPECTRA OF ROCHELLE SALT.	28-3210
ALS. FAR INFRARED	SPECTRA OF RUBIDIUM DIHYDROGEN PHOSPHATE CRYST	17-2638
IUM DEUTERIUM SULFATE IN THE PARA- AND FERROELECTR/ INFRARED	SPECTRA OF RUBIDIUM HYDROGEN SULFATE AND RUBID	19-2768
/S (CONDUCTIVITY AND MOBILITY VERSUS TEMPERATURE, ABSORPTION	SPECTRA) OF SEMICONDUCTING STRONTIUM TITANATE/	06-1198
INFRARED AND RAMAN	SPECTRA OF SEVERAL PEROVSKITE ZIRCONATES.	08-1411
M TITANATE. INFRARED	SPECTRA OF SINGLE AND POLYCRYSTALLINE STRONTIU	06-1250
POLARIZED INFRARED AND RAMAN	SPECTRA OF SINGLE CRYSTAL CESIUM NITRITE.	16-2342
K, 30-2/ TEMPERATURE DEPENDENCE OF FAR INFRARED REFLECTIVITY	SPECTRA OF SODIUM NITRITE CRYSTALS. (0.85 DEG	16-2373
AL. FAR INFRARED REFLECTIVITY	SPECTRA OF SODIUM TRIHYDROGEN DISELENITE CRYST	22-2877
PS V, VI AND VII. (ANTIMONY AND BISMUTH SULFO HA/ INFRARED	SPECTRA OF SOME FERROELECTRIC CRYSTALS OF GROU	15-2210
ATES. VIBRATIONAL	SPECTRA OF STRONTIUM, BARIUM AND CALCIUM TITAN	04-0864
/PLED MODES (OPTICAL, FERROELECTRIC- ACOUSTIC) IN THE PHONON	SPECTRA OF STRONTIUM TITANATE AND POTASSIUM T/	06-1157
/ERATURE AND PRESSURE ON THE ELECTRON PARAMAGNETIC RESONANCE	SPECTRA OF SUBSTITUTIONAL IMPURITIES IN CUBIC/	06-1193
SPHATE. RAMAN	SPECTRA OF TETRAGONAL POTASSIUM DIHYDROGEN PHO	17-2657
HYDROGEN A/ STARK EFFECT IN THE NUCLEAR QUADRUPOLE RESONANCE	SPECTRA OF THE FERROELECTRIC CRYSTAL CESIUM DI	18-2722
POTASSIUM DIHYDROGEN PHOSPHATE CRYSTAL. RAMAN	SPECTRA OF THE FERROELECTRIC MODIFICATION OF A	17-2652
AND CALCIUM. (75 REFS) VIBRATIONAL	SPECTRA OF THE TITANATES OF STRONTIUM, BARIUM,	04-0865
ATR INFRARED	SPECTRA OF UNIAXIAL NITRATE CRYSTALS.	16-2280
ND THEIR INTERPRETATION. PART-8: TRAN/ DIELECTRIC RELAXATION	SPECTRA OF WATER, ICE, AND AQUEOUS SOLUTIONS A	34-3482
NG PHASE TRANSITIONS OF THE/ CHANGE OF RESONANCE ABSORPTION	SPECTRA OF 23.8 KEV GAMMA-RAYS OF TIN-119 DURI	08-1400
/ATES AND NITRITES. PART-1: CHANGE IN ULTRAVIOLET ABSORPTION	SPECTRA ON MELTING. PART-2: CHANGE IN ULTRAVI/	16-2278
ERROELECTRICS. NATURE OF THE TEMPERATURE DEPENDENCE OF NMR	SPECTRA SECOND MOMENT IN ORDER DISORDER TYPE F	03-0460
ELECTRICS AND FERROELECTRIC FERROM/ ELECTRON SPIN RESONANCE	(SPECTRA, TEMPERATURE DEPENDENCE) IN SOME FERRO	08-1368
INUM SULFATE HEXAHYDRAT/ LOW TEMPERATURE INFRARED ABSORPTION	SPECTRA (5200-4200 PER CM) OF GUANIDINIUM ALUM	21-2811
BONDED FERROEL/ QUANTUM THEORY OF FERROELECTRICITY, PART-1:	SPECTRAL ANALYSIS (PROTON DYNAMICS IN HYDROGEN	03-0402
IN POTASSIUM DIDEUTERIUM PHOSPHATE . (BY NEUTRON TRIPLE AXIS	SPECTROMETRY) /RROELECTRIC (SOFT) MODE MOTION	17-2668
IOBATES. INFRARED ABSORPTION	SPECTROPHOTOMETRIC STUDY OF MONOVALENT METAL N	02-0113
CYANIDE TRIHYDRATE AND ISOMORPHOUS COMPOUNDS. RADIO	SPECTROSCOPIC INVESTIGATION OF POTASSIUM FERRO	24-2900
CTRICS. (LITHIUM NIOBATE AND GADOLINIUM MOLYBDATE) LASER AND	SPECTROSCOPIC PROPERTIES OF ACTIVATED FERROELE	11-1741
D GADOLINIUM MOLYBDATE CRYSTALS. CERTAIN GROWTH, LASING AND	SPECTROSCOPIC PROPERTIES OF NEODYMIUM(3+) DOPE	30-3255
AMMONIUM ALUMS.	SPECTROSCOPIC STUDIES OF FERROELECTRIC METHYL	20-2808
S OF THE TYPE A(2) (V) B(3) (VI) AND A(/ NUCLEAR GAMMA RESONANCE	SPECTROSCOPIC STUDY OF SEMICONDUCTING COMPOUND	15-2108
RATURE DEPENDENT PHONON STRUCTURE IN SODIUM NITRITE BY RAMAN	SPECTROSCOPY. ANALYSIS OF THE TEMPE	16-2299
AND KDP IN THE FAR INFRARED BY DISPERSIVE FOURIER TRANSFORM	SPECTROSCOPY. /THE COMPLEX PERMITTIVITY OF ADP	17-2609
TH LEAD ZIRCONATE AND WITH LEAD ZIRCONATE TITANATE MOSSBAUER	SPECTROSCOPY. /ND RELATED MATERIALS (SYSTEM WI	08-1333
OELECTRICITY AS CAPACITOR DIELECTRICS, A STUDY IN DIELECTRIC	SPECTROSCOPY. /RIC CONSTANT MATERIALS AND FERR	37-3678
ONS IN LITHIUM NIOBATE AND LITHIUM TANTALATE BY RAMAN POWDER	SPECTROSCOPY. STOICHIOMETRY VARIATI	11-1813
POTASSIUM IODATE CRYSTAL STRUCTURE STUDIED BY INFRARED	SPECTROSCOPY AND NUCLEAR QUADRUPOLE RESONANCE.	08-1278
ESIUM DIHYDROGEN ARSENATE, STRONTIUM TITANA/ SOFT MODE RAMAN	SPECTROSCOPY: COUPLED MODES (FERROELECTRICS, C	06-1207
URE OF STRONTIUM TITANATE AS MEASURED BY X-RAY PHOTOELECTRON	SPECTROSCOPY (ESCA). BAND STRUCT	06-1100
DER FERROELECTRIC PHASE TRANSITIONS. RAMAN	SPECTROSCOPY OF PHONON INSTABILITY IN FIRST OR	03-0163
INFRARED AND RAMAN	SPECTROSCOPY OF POTASSIUM METANIOBATE.	08-1303
LITHIUM NIOBATE. NONLINEAR OPTICAL	SPECTROSCOPY OF THIRD ORDER SUSCEPTIBILITY IN	11-1855
RAMAN	SPECTROSCOPY OF TUNGSTEN TRIOXIDE.	10-1635
IC GENERATION IN ADP CONTINUOUSLY TUNABLE ACROSS THE VISIBLE	SPECTRUM. EFFICIENT HIGH GRAIN PARAMETR	17-2712

TRANSITION IN SODIUM NITRITE CRYSTAL BY OBSERVATION OF RAMAN SPECTRUM.   INVESTIGATION OF FERROELECTRIC 16-2275
TEMPERATURE DEPENDENCE OF THE LITHIUM NMR SPECTRUM AND ATOMIC MOTION IN LITHIUM NIOBATE. 11-1722
ORATE. (NOT FERROELECTRIC)   THE VIBRATIONAL SPECTRUM AND DIELECTRIC BEHAVIOR OF SODIUM CHL 36-3549
TS OF ANTIFERROMAGNETIC POTASSIUM MANGANESE TRI/ VIBRATIONAL SPECTRUM AND DISPERSION OF THE OPTICAL CONSTAN 33-3385
S AND EXCHANGE COUPLED CHROMIUM(3+) PAIRS IN LANTHANUM ALUM/ SPECTRUM AND ENERGY LEVELS OF CHROMIUM(3+) ION 36-3565
ESE TRIFLUORIDE: TEMPERATURE DISCONTINUITIES IN THE INFRARED SPECTRUM AND GROUP THEORETICAL ANALYSIS. /NGAN 33-3363
.   TEMPERATURE DEPENDENCE OF RAMAN SPECTRUM AND SECOND HARMONIC GENERATION IN KDP 17-2550
RAMAN SPECTRUM AND STRUCTURE OF TERBIUM MOLYBDATE. 30-3296
ANTIFERROMAGNETIC RESONANCE SPECTRUM IN BARIUM MANGANESE TRIFLUORIDE. 33-3396
AND (COPPER(2+) DOPED) ZINC/ ELECTRON PARAMAGNETIC RESONANCE SPECTRUM IN (MANGANESE DOPED) LITHIUM NIOBATE 11-1792
/E CHANGE IN SODIUM NITRITE CRYSTALS AND COMBINATION SCATTER SPECTRUM IN THE REGION OF THE ANTISYMMETRICAL/ 16-2276
RIC ANTIMONY SULFOIODIDE CRYSTALS USING THE RAMAN SCATTERING SPECTRUM. (LATTICE VIBRATIONS, BONDING) /ELECT 15-2143
ONLY)   EFFECT OF FREE CARRIERS ON THE INFRARED SPECTRUM OF A FERROELECTRIC. (THEORY, ABSTRACT 03-0408
E PARA- AND FERROELECTRIC PHASES.   INFRARED ABSORPTION SPECTRUM OF A SINGLE CRYSTAL OF THIOUREA IN TH 25-2910
F TEMPERATURE.   LOW FREQUENCY RAMAN SPECTRUM OF A THIOUREA CRYSTAL AS A FUNCTION O 25-2906
MODEL. (PEROVSKITES)   TEMPERATURE DEPENDENT SPECTRUM OF AN ANTIFERROELECTRIC LINEAR CHAIN 03-0468
NTAINING DISLOCATIONS.   ENERGY SPECTRUM OF AN ORDER DISORDER FERROELECTRIC CO 03-0485
SOME FEATURES OF THE PHOTOEMF SPECTRUM OF ANTIMONY SULFOIODIDE.   • 15-2136
RAMAN SPECTRUM OF BARIUM TITANATE. 04-0746
TEMPERATURE DEPENDENCE OF THE BRILLOUIN- RAMAN SPECTRUM OF BARIUM TITANATE. 04-0807
MODEL OF LOCALLY ORDERED "NUCLEI", PHASE STRUCTU/ DIELECTRIC SPECTRUM OF BARIUM TITANATE. (AND MICROSCOPIC 04-0647
RAMAN SPECTRUM OF BISMUTH(12) GERMANIUM OXIDE(20). 35-3519
FREQUENCY SPECTRUM OF FERROCERAMIC TRANSDUCERS. 37-3656
IUM TITANATE.   COMMENTS ON THE NEUTRON SPECTRUM OF FERROELECTRIC AND PARAELECTRIC BAR 04-0813
G THE PHASE TRAN/ TEMPERATURE DEPENDENCE OF RAMAN SCATTERING SPECTRUM OF FERROELECTRIC DEUTERATED KDP DURIN 17-2626
DIHYDROGEN PHOSPHATE CRYSTAL.   RAMAN SPECTRUM OF FERROELECTRIC MODE IN A POTASSIUM 17-2663
ON OF THE CRYSTAL SIZE.   PHONON SPECTRUM OF FERROELECTRICS STUDIED BY LIMITATI 03-0196
TEMPERATURE DEPENDENCE OF THE RAMAN SPECTRUM OF GADOLINIUM MOLYBDATE. 30-3285
EFFECT OF POINT DEFECTS ON THE ENERGY SPECTRUM OF HYDROGEN BONDED FERROELECTRICS. 03-0486
NFRARED ABSORPTION, NEUTRON INELASTIC SCATTERING,/ FREQUENCY SPECTRUM OF HYDROGEN BONDED FERROELECTRICS. (I 03-0401
M(-1).   INFRARED SPECTRUM OF ICE I(H) IN THE RANGE 4000 TO 15 C 34-3490
WITH A MIXED COULOMB VALENCE DYNAMICAL/ LOW FREQUENCY RAMAN SPECTRUM OF ICE I(H). TENTATIVE INTERPRETATION 34-3425
TO NEAREST NE/ STRONG AXIAL ELECTRON PARAMAGNETIC RESONANCE SPECTRUM OF IRON(3+) IN STRONTIUM TITANATE DUE 06-1144
TE: AN EXAMPLE OF THE DYNAM/ ELECTRON PARAMAGNETIC RESONANCE SPECTRUM OF MOLYBDENUM(5+) IN STRONTIUM TITANA 06-1125
NTIMONY SULFOIODIDE (PHONON ASSIGNMENTS). (POLARIZED) RAMAN SPECTRUM OF (PARAELECTRIC AND) FERROELECTRIC A 15-2207
YSTAL IN PARA- AND FERROELECTRIC PHASE.   RAMAN SPECTRUM OF POTASSIUM DIDEUTERIUM PHOSPHATE CR 17-2647
SSIUM TANTALATE(0.65) NIOBATE(0.35), CURIE TEMPERATUR/ RAMAN SPECTRUM OF POTASSIUM TANTALATE NIOBATE. (POTA 08-1391
.36) OXIDE.   RAMAN SPECTRUM OF POTASSIUM TANTALUM(0.64) NIOBIUM(0 13-1989
ROELECTRICS OF THE KDP GROUP.   SPECTRUM OF RAMAN SCATTERED LASER LIGHT IN FER 17-2625
AELECTRIC PHASE, COMPARISON WITH FERROELECTRIC).   RAMAN SPECTRUM OF SODIUM TRIHYDROGEN DISELENITE (PAR 22-2864
OSPHATE. (T/ ANALYSIS OF PHONON TUNNELING MODES IN THE RAMAN SPECTRUM OF TETRAGONAL POTASSIUM DIHYDROGEN PH 03-0396
E OF MANGANESE(+2) IN BARI/ FORBIDDEN LINES IN THE HYPERFINE SPECTRUM OF THE ELECTRON PARAMAGNETIC RESONANC 04-0877
/TEMPERATURE AND PHASE TRANSITION ON THE HYDROXIDE VIBRATION SPECTRUM OF THE FERROELECTRIC RUBIDIUM DIHYDR/ 17-2470
TETRAHYDRATE AND COPPER DEUTERO FORMATE TETRADEUT/ VIBRATION SPECTRUM OF THE FORMATE ION IN COPPER FORMATE 26-2926
DIHYDROGEN PHOSPHATE CRYSTAL IN PARAELECTRIC AND / VIBRATION SPECTRUM OF THE PHOSPHATE ANION IN A RUBIDIUM 17-2472
FAR INFRARED TRANSMISSION SPECTRUM OF THIOUREA. 25-2917
CTRIC PHASES.   INFRARED ABSORPTION SPECTRUM OF THIOUREA IN THE PARA- AND FERROELE 25-2908
RAMAN SPECTRUM OF TRIGLYCINE SELENATE. 27-2957
TION OF THICKNESS OF THE S/ STUDY OF THE X-RAY PHOTOELECTRON SPECTRUM OF TUNGSTEN- TUNGSTEN OXIDE AS A FUNC 10-1631
TRANSVERSE POLARITONS IN A LITHIUM NIOBATE CRYSTAL. (RAMAN SPECTRUM, PERMITTIVITY) 11-1762
DIHYDROGEN PHOSPHATE CRYSTAL AND ITS APPLICATION FOR A HIGH SPEED SHUTTER. /ISTICS OF DEUTERATED POTASSIUM 17-2499
COMMUNICATIONS (REVIEW EMPHASIZING ELECTROOPTIC MODUL/ (HIGH SPEED, SMALL APERTURE) MODULATORS FOR OPTICAL 02-0056
NSITIVITY OF RADIALLY POLARIZED FERROELECTRIC CERAMIC HOLLOW SPHERES. (HYDROPHONE ELEMENTS) OPEN CIRCUIT SE 37-3653
ELECTRIC.   EXCITON MODEL OF SPIN AND OPTICAL EXCITATIONS OF AN ANTIFERRODI 03-0416
GANE/ EXCHANGE NARROWING: MAGNETIC RESONANCE LINE SHAPES AND SPIN CORRELATION IN PARAMAGNETIC POTASSIUM MAN 33-3376
ETIZATION IN THE TWO DIMENSIONAL ANTIFERROMAGNET/ ZERO POINT SPIN DEVIATION AND SPONTANEOUS SUBLATTICE MAGN 26-2929
E.   PROTON RESONANCE STUDY OF SUBLATTICE ROTATIONS AND SPIN DEVIATION IN COPPER FORMATE TETRADEUTERAT 26-2950
IC AND MAGNETIC MOMENTS OF OPTICAL TRANSITIO/ ELECTRONIC AND SPIN EXCITATIONS IN ANTIFERRODIELECTRIC ELECTR 03-0418
ODECAHYDRATE.   MOSSBAUER STUDY OF ELECTRONIC SPIN FLIP PROCESSES IN AMMONIUM IRON SULFATE D 19-2771
OF FERROELECTRIC CRYSTALS. (SODIUM NITRI/ NUCLEAR QUADRUPOLE SPIN LATTICE RELAXATION AND CRITICAL DYNAMICS 16-2268
ES OF FERROELECTRIC CRYSTALS. (THEORY)   NUCLEAR QUADRUPOLE SPIN LATTICE RELAXATION AND DYNAMICAL PROPERTI 03-0187
ELECTRIC CESIUM DIDEUTERIUM ARSENATE AND CESIUM / CESIUM-183 SPIN LATTICE RELAXATION AND RESONANCE IN FERRO 18-2720
POTASSIUM FERROCYANIDE TRIHYDRATE.   ANOMALOUS SPIN LATTICE RELAXATION BY QUASI SPIN WAVES IN 24-2899
ORDER DISORDER TYPE FERROELECTRICS: POLARIZATION FLUCTUATI/ SPIN LATTICE RELAXATION BY QUASI SPIN WAVES IN 26-2928
ODE IN POTASSIUM DIHYDROGEN PHOSPHATE.   SPIN LATTICE RELAXATION BY THE FERROELECTRIC M 17-2640
-1: PROTON MAGNETIC RESONANCE IN WATER ICE. PART-2: DEUTERON SPIN LATTICE RELAXATION IN DEUTERATED ICE. /RT 34-3471
ANITE.   PROTON SPIN LATTICE RELAXATION IN FERROELECTRIC COLEM 23-2885
RATED AMMONIUM HYDROGEN SULFATE.   DEUTERON SPIN LATTICE RELAXATION IN FERROELECTRIC DEUTE 19-2795
M AMMONIUM TARTRATE.   PROTON SPIN LATTICE RELAXATION IN FERROELECTRIC SODIU 29-3248
M TRIHYDROGEN SELENATE.   PROTON SPIN LATTICE RELAXATION IN FERROELECTRIC SODIU 22-2852
YCINE SULFATE.   NUCLEAR SPIN LATTICE RELAXATION IN FERROELECTRIC TRIGL 27-2967
INTERPRETATION OF THE PROTON SPIN LATTICE RELAXATION IN HEXAGONAL ICE. 34-3475
STRONG COLLISION LIMIT OF SPIN LATTICE RELAXATION IN HEXAGONAL ICE. 34-3480
RIHYDROGEN SELENATE.   PROTON SPIN LATTICE RELAXATION IN IRRADIATED SODIUM T 22-2853
AMMONIUM SALTS. (AMMONIUM SULFATE, FLUOBERYLLATE, P/ NUCLEAR SPIN LATTICE RELAXATION IN SOME FERROELECTRIC 19-2769
PHASE TR/ NUCLEAR MAGNETIC RESONANCE (WIDE LINE SPECTRA AND SPIN LATTICE RELAXATION) INVESTIGATIONS OF THE 20-2802
/ER MOLECULES IN ROCHELLE SALT AND AMMONIUM ROCHELLE SALT BY SPIN LATTICE RELAXATION OF PROTONS IN ROTATIN/ 28-3232
RAELECTRIC AND FERROELECTRIC PHASE OF SODIUM NITRIT/ NUCLEAR SPIN LATTICE RELAXATION OF SODIUM-23 IN THE PA 16-2264
ODIUM DEUTERIUM SELENATE.   NUCLEAR SPIN LATTICE RELAXATION TIME MEASUREMENTS IN S 22-2847
TEMPERATURE OF DICALCIUM LEAD PROPIO/ ANOMALY OF THE PROTON SPIN LATTICE RELAXATION TIME NEAR THE CRITICAL 26-2947
ON TEMPERATURE IN THE ORDER DISORDER TYPE FE/ ANOMALY OF THE SPIN LATTICE RELAXATION TIME NEAR THE TRANSITI 03-0488
/ TONE BURSTING OF QUADRUPOLAR NUCLEI. DETERMINATION OF THE SPIN LATTICE RELAXATION TIMES IN FERROELECTRIC 11-1829
ARACTERISTICS AND VARIOUS CONCEPTIONS OF THE TUNNELING QUASI SPIN MODEL. / (KDP TYPE) PART-2: SCATTERING CH 03-0478
TATIC SCALING HYPOTHESIS AND ITS IMPLICATIONS FOR THE PLANAR SPIN MODEL OF STRONTIUM TITANATE.   S 06-1223
ITION. (IN SODIUM NITRITE)   QUADRUPOLE SPIN PHONON RELAXATION AND FERROELECTRIC TRANS 16-2362
T(1 RHO) MINIMUM.   PROTON SPIN RELAXATION IN HEXAGONAL ICE. PART-2: THE 34-3488
FERROELECTRIC PHASE CHANGE IN BARIUM TITANATE USING ELECTRON SPIN RESONANCE AND THE MOSSBAUER EFFECT. /THE 04-0827
IN LITHIUM NIOBATE. (CAPACITANCE, OPTICAL SPECTRA, ELECTRON SPIN RESONANCE CELL PARAMETERS) /ATION DAMAGE 11-1705
NOMALY IN STRONTIUM TITANATE.   NEW ELECTRON SPIN RESONANCE EVIDENCE FOR THE 65 DEGREES K A 06-1230
URATIONS AND OF EXCHANGE OF THE ARSENATE RADICAL I/ ELECTRON SPIN RESONANCE IDENTIFICATION OF SLATER CONFIG 18-2734

ELECTRON SPIN RESONANCE IN GAMMA-IRRADIATED TGS. 27-3035
ITH GADOLINIUM(3+) OR IRC/ ELECTRIC FIELD EFFECT OF ELECTRON SPIN RESONANCE IN STRONTIUM TITANATE. (DOPED W 06-1229
TED SINGLE CRYSTALS OF FERROELECTRIC POTASSIUM DIH/ ELECTRON SPIN RESONANCE INVESTIGATIONS OF X-RAY IRRADIA 17-2558
TIFERROMAGNET COPPER FORMATE TE/ ANOMALOUS SHIFT OF ELECTRON SPIN RESONANCE LINES IN THE TWO DIMENSIONAL AN 26-2941
/ACANCIES IN CHROMIUM(3+) DOPED STRONTIUM TITANATE (ELECTRON SPIN RESONANCE MEASUREMENTS, MODEL PROPOSED). 06-1161
ED STRONTIUM TITANATE SINGLE CRYSTALS. ELECTRON SPIN RESONANCE OF CHROMIUM(3+) IN CHROMIUM DOP 06-1147
NIUM TARTRATE. ELECTRON SPIN RESONANCE OF COPPER(2+) DOPED SODIUM AMMO 29-3252
IC AND ANTIFERROELECTRIC ARSENATE SINGLE CRYSTALS. ELECTRON SPIN RESONANCE OF FREE RADICALS IN FERROELECTR 18-2731
ANATE CERAMIC SEMICONDUCTOR.   TWO SETS OF ELECTRON SPIN RESONANCE OF GADOLINIUM(3+) IN BARIUM TIT 04-0990
IC AMMONIUM HYDROGEN SULFATE. ELECTRON SPIN RESONANCE OF GAMMA DEFECTS IN FERROELECTR 19-2748
LFATE. ELECTRON SPIN RESONANCE OF GAMMA-IRRADIATED AMMONIUM SU 19-2790
/DICAL IN TRIGLYCINE SULFATE NEAR THE CURIE POINT. (ELECTRON SPIN RESONANCE OF GAMMA-IRRADIATED SPECIMENS) 27-3033
: INTERNAL MOTION OF HYDROGEN AND THE ACTIVATION E/ ELECTRON SPIN RESONANCE OF GAMMA-IRRADIATED TGS. PART-2 27-3034
SELENATE. ELECTRON SPIN RESONANCE OF GAMMA-IRRADIATED TRIGLYCINE 27-3139
M SODIUM NIOBATE.   PHOTOCHROMIC EFFECT AND ELECTRON SPIN RESONANCE OF IRON AND MOLYBDENUM IN BARIU 13-1930
TASSIUM DIHYDROGEN PHOSPHATE. ELECTRON SPIN RESONANCE OF IRRADIATED AND DEUTERATED PO 17-2533
NATE. ELECTRON SPIN RESONANCE OF MANGANESE(2+) IN BARIUM TITA 04-0747
/ESSURE EFFECTS ON SELENITE (ION). RADIATION DAMAGE ELECTRON SPIN RESONANCE SPECTRA IN FERROELECTRIC SODIU/ 22-2870
OLINIUM(3+) IN LANTHANUM ALUMINATE. ELECTRON SPIN RESONANCE SPECTRA OF CHROMIUM(3+) AND GAD 36-3583
LITHIUM NIOBATE.   FORBIDDEN HYPERFINE LINES IN ELECTRON SPIN RESONANCE SPECTRA OF MANGANESE(2+) DOPED 11-1805
E) IN SOME FERROELECTRICS AND FERROELECTRIC FERROM/ ELECTRON SPIN RESONANCE (SPECTRA, TEMPERATURE DEPENDENC 08-1368
IUM NIOBATE. ELECTRON SPIN RESONANCE STUDIES OF CHROMIUM(3+) IN LITH 11-1804
TERS IN MANGANESE(2+) DOPED LITHIUM NIOBATE. ELECTRON SPIN RESONANCE STUDIES OF CRYSTAL FIELD PARAME 11-1803
ELECTRON SPIN RESONANCE STUDY IN STRONTIUM TITANATE. 06-1228
S. ELECTRON SPIN RESONANCE STUDY OF ADP- KDA MIXED CRYSTAL 17-2683
IATED AMMONIUM DIDEUTERIUM PHOSPHATE- 5 PERCENT PO/ ELECTRON SPIN RESONANCE STUDY OF AMMONIUM IONS IN IRRAD 17-2684
M TRIHYDROGEN SELENITE AT ROOM TEMPERATURE. ELECTRON SPIN RESONANCE STUDY OF GAMMA-IRRADIATED CESIU 22-2833
SIUM SELENATE (STRUCTURAL PHASE CHANGES). ELECTRON SPIN RESONANCE STUDY OF GAMMA-IRRADIATED POTAS 22-2825
RS IN AMMONIUM DIHYDROGEN PHOSPHATE POTASSIUM DIHY/ ELECTRON SPIN RESONANCE STUDY OF RADIATION DAMAGE CENTE 17-2547
E TRANSITIONS IN SODIUM TRIHYDROGEN DISELENITE. ELECTRON SPIN RESONANCE STUDY OF THE FERROELECTRIC PHAS 22-2871
(LEAD TITANATE) ELECTRON SPIN RESOURCE STUDIES AT DEFECTS IN FLUORITES. 05-1086
AL.   EPR OF THE FOUR COPPER ION COMPLEX WITH SPIN S=2 IN TRIGLYCINE FLUOBERYLLATE MONOCRYST 27-3134
S. (AMMONIUM SULFATE/ PROTON MAGNETIC RESONANCE STUDY OF SPIN SYMMETRY STATES OF AMMONIUM IONS IN SOLID 19-2800
DEUTERATED KDP.   LASER RAMAN STUDY OF QUASI SPIN WAVE HYDROGEN TUNNELING MODES IN KDP AND 17-2569
MPLES OF FERRODIELECTRICS WITH AN ARBITRARY MAGNETIC SYMMET/ SPIN WAVE PHENOMENOLOGICAL THEORY IN FINITE SA 03-0164
WAVE IN A FERROELECTRIC ANTIFE/ RESONANCE OF A LOW FREQUENCY SPIN WAVE WITH AN ACOUSTIC OR ELECTROMAGNETIC 03-0284
N EXTERNAL FIELDS. (THEORY)   INSTABILITY OF SPIN WAVES IN FERROELECTRIC ANTIFERROMAGNETS I 03-0149
INTERACTION OF LIGHT WITH SPIN WAVES IN FERROELECTRICS. (THEORY) 03-0158
/EORY OF THE FERROELECTRIC STATE IN ELECTRIC DIPOLE SYSTEMS (SPIN WAVES IN HYDROXYL DIPOLE SYSTEM IN POTAS/ 03-0452
ICS.   COUPLED ELECTROMAGNETIC SPIN WAVES IN MAGNETICALLY ORDERED FERROELECTR 03-0148
ICS. (THEORY)   POSSIBILITY OF EXCITATION OF SPIN WAVES IN MAGNETICALLY ORDERED FERROELECTR 03-0150
ELECTRICITY. PART-2: THERMODYNAMIC BEHAVIOR OF FERROELECTRIC SPINS.   QUANTUM THEORY OF FERRO 03-0403
TRICS. (THEORY)   MACROSCOPIC ANALOG OF THE SPIN- ECHO EFFECT IN POLYCRYSTALLINE FERROELEC 03-0310
DYNAMICAL BEHAVIOR OF FERROELECTRIC SPINS. (KDP TYPE CRYSTALS) 17-2600
ROVSKITE TYPE SINGLE CRYSTALS BY NEUTRON BOMBARDMENT. SPLIT INTERSTITIAL CONFIGURATION INDUCED IN PE 08-1467
X).   ABSORPTION EDGE SPLITTING IN POTASSIUM TANTALATE NIOBATE. 13-1988
ABSORPTION EDGE SPLITTING IN POTASSIUM TANTALATE(X) NIOBATE(1- 13-1987
FERROELECTRIC SILVER SODIUM DINITRITE. SPLITTING OF THE SODIUM-23 NMR CENTRAL LINE IN 16-2307
MINUM DISULFATE D/ CALCULATION OF CRYSTAL FIELD ENERGY LEVEL SPLITTINGS OF TITANIUM(3+) ION IN RUBIDIUM ALU 19-2754
PONSE IN GADOLINIUM MOLYBDATE. SPONTANEOUS BIREFRINGENCE AND ELECTROOPTIC RES 30-3307
CTRIC GADOLINIUM MOLYBDATE.   INDUCED AND SPONTANEOUS BIREFRINGENCE IN IMPROPER FERROELE 30-3278
HASES OF / RELATIONSHIP BETWEEN THE ATOMIC DISPLACEMENTS AND SPONTANEOUS DEFORMATION IN THE FERROELECTRIC P 08-1377
/TRIC PIEZOELECTRIC CERAMICS AND METHODS OF THEIR CHANGING. (SPONTANEOUS DEFORMATION RELATED TO THEIR CHAR/ 03-0217
IUM TITANATE. SPONTANEOUS DESTRUCTION OF FERROCERAMICS. (BAR 04-0787
ATION IN NICKEL IODIDE BORACITE (CURIE P/ COEXISTENCE OF THE SPONTANEOUS ELECTRIC POLARIZATION AND MAGNETIZ 14-2090
SALT CRYSTALS. SPONTANEOUS ELECTROOPTICAL EFFECT IN ROCHELLE 28-3209
PIEZOEFFECT IN BARIUM TITANATE ON THE MEANS OF SCREENING THE SPONTANEOUS INDUCTION.   DEPENDENCE OF THE 04-1021
OPTICAL PROPERTIES OF FERROELECTRIC (IRON IODINE) BORACITE. (SPONTANEOUS LATTICE STRAINS)   ELECTRO 14-2082
EAD FERRATE(0.5) NIOBATE(0.5) AND LEAD M/ SUPPRESSION OF THE SPONTANEOUS MAGNETOELECTRIC MAGNETIZATION OF L 13-2069
PTION, AND DISPERSION IN THE NONLINEAR SUSCEPTIBILITY IN LI/ SPONTANEOUS PARAMETRIC SCATTERING, IDLER ABSOR 11-1771
LITHIUM IODATE (INDEX OF REFRACTION 1.2-7/ (CONTINUOUS WAVE) SPONTANEOUS PARAMETRIC SCATTERING OF LIGHT IN 08-1291
IG REGIONS TAKING INTO ACCOUNT DIFFERENT ORIENTATIONS OF THE SPONTANEOUS POLARIZATION.   A MODEL OF KANZ 03-0439
OF POTASSIUM DIHYDROGEN PHOSPHATE. (ELECTROCALORIC STUDIES) SPONTANEOUS POLARIZATION)   ELECTRONIC STUDIES 17-2429
LEAD TITANATE (FLUX GROWN SINGLE CRYSTALS). SPONTANEOUS POLARIZATION AND COERCIVE FIELD OF 05-1050
THE PEROVSKITES IN THE SYSTEM LEAD ZI/ THERMAL VARIATION OF SPONTANEOUS POLARIZATION AND COERCIVE FORCE OF 09-1597
IN CRYSTALS OF PEROVSKITE TYPE AND OF ONES / MICROTHEORY OF SPONTANEOUS POLARIZATION AND PHASE TRANSITIONS 03-0175
TS IN FERROELECTRIC POTASSIUM DIHYDROGEN PHOS/ ORIGIN OF THE SPONTANEOUS POLARIZATION AND THE ISOTOPE EFFEC 17-2433
TE AND LITHIUM TANTALATE USING AUTHOR'S STATISTICAL THEORY. (SPONTANEOUS POLARIZATION, BONDING) /HIUM NIOBA 03-0366
IES IN PIEZOELECTRIC MATERIALS. (GADOLINIUM M/ PRODUCTION OF SPONTANEOUS POLARIZATION BY ELASTIC INSTABILIT 30-3282
ISMUTH TITANATE AT LOW T/ DIELECTRIC BEHAVIOR (PERMITTIVITY, SPONTANEOUS POLARIZATION, COERCIVE FIELD) OF B 08-1436
/IP BETWEEN IONIC DISPLACEMENT AND THE CURIE TEMPERATURE AND SPONTANEOUS POLARIZATION IN DISPLACIVE FERROE/ 03-0453
ESTRACT ON/ THERMAL EXPANSION ANOMALIES AND THE DIRECTION OF SPONTANEOUS POLARIZATION IN FERROELECTRICS. (A 03-0350
(THERMAL EXP/ PRIMARY AND SECONDARY PYROELECTRIC EFFECTS AND SPONTANEOUS POLARIZATION IN LITHIUM TANTALATE 12-1884
RACT ONLY)   CONDITIONS FAVORING THE ORIGINATION OF SPONTANEOUS POLARIZATION IN PEROVSKITES. (ABST 08-1457
EN PHOSPHATE.   MEASUREMENTS OF THE SPONTANEOUS POLARIZATION IN POTASSIUM DIHYDROG 17-2430
NMR STUDIES OF SPONTANEOUS POLARIZATION IN ROCHELLE SALT. 28-3194
TRITE.   PYROELECTRIC STUDY ON THE SPONTANEOUS POLARIZATION IN SILVER SODIUM DINI 16-2295
FINE PARTICLES OF BARIUM TITANATE.   MAGNITUDE OF SPONTANEOUS POLARIZATION IN THIN CRYSTALS AND 04-0607
.   BLOCKING OF SPONTANEOUS POLARIZATION IN TRIGLYCINE SULFATE 27-3008
TION POINT OF TRIGLYCINE SELENATE CR/ METHODS OF DETERMINING SPONTANEOUS POLARIZATION NEAR THE PHASE TRANSI 27-2994
TRICS. (THEORY, EINSTEIN THERMODYNAMIC METHOD) SPONTANEOUS POLARIZATION NOISE IN POLAR DIELEC 03-0219
OGEN PHOSPHATE CRYSTAL NEAR THE CURIE POINT. (ELECTROCALORI/ SPONTANEOUS POLARIZATION OF A POTASSIUM DIHYDR 17-2678
(TEMPERATURE DEPENDENCE OF) THE SPONTANEOUS POLARIZATION OF AMMONIUM SULFATE. 19-2799
NGLE CRYSTALS.   STABILIZATION OF SPONTANEOUS POLARIZATION OF BARIUM TITANATE SI 04-0795
ES NEAR THE CURIE POI/ THEORY OF DOMAINLESS, NONHOMOGENEOUS, SPONTANEOUS POLARIZATION OF FERROELECTRIC PLAT 03-0321
/TURE AND PRESSURE DEPENDENCE OF THE DIELECTRIC CONSTANT AND SPONTANEOUS POLARIZATION OF FERROELECTRIC POT/ 16-2331
TE (TEMPERATURE DEPENDENCE 150-165 / PYROELECTRIC EFFECT AND SPONTANEOUS POLARIZATION OF GADOLINIUM MOLYBDA 30-3314
/TION OF THE CURIE TEMPERATURE AND TEMPERATURE DEPENDENCE OF SPONTANEOUS POLARIZATION OF HEXAGONAL (RHOMBO/ 03-0290
/IC WITH NONREVERSIBLE, BUT TILTABLE (TO 3 STABLE POSITIONS, SPONTANEOUS) POLARIZATION (ON APPLYING AN ELE/ 08-1346
CHROMIUM(3+) IONS IN TRIGLYCINE FLUOBERYLLATE. INFLUENCE OF SPONTANEOUS POLARIZATION ON THE EPR SPECTRA OF 27-3136

D EXOELECTRON EMISSION YIELD OF TRIGLYCINE/ INFLUENCE OF THE SPONTANEOUS POLARIZATION ON THE PHOTOSTIMULATE 27-2961

/PECIFIC HEAT. PART-10: PHOTON RENORMALIZATION AND THEORY OF SPONTANEOUS POLARIZATION. PART-11: CONCLUSION/ 03-0397

E SULFATE BY MEANS OF A LASER BEAM. APPLIC/ IRREVERSIBLE AND SPONTANEOUS POLARIZATION REVERSAL IN TRIGLYCIN 27-3010

/ TANTALATE. (DIELECTRIC CONSTANTS, PIEZOELECTRIC CONSTANTS, SPONTANEOUS POLARIZATION, THERMAL EXPANSION C/ 12-1876

/MPOUND LEAD GERMANIUM SILICON OXIDE. (DIELECTRIC CONSTANTS, SPONTANEOUS POLARIZATION, THERMAL EXPANSION, 35-3501

RT-2: THE CASE OF CADMIUM SULFIDE AND LITHIUM NIOBATE CRYST/ SPONTANEOUS RAMAN SCATTERING BY POLARITONS. PA 11-1787

DETERMINATION OF THE STATE PARAMETERS AND FORMULATION OF SPONTANEOUS STRAIN FOR FERROELASTICS (THEORY). 03-0137

O DIMENSIONAL ANTIFERROMAGNET/ ZERO POINT SPIN DEVIATION AND SPONTANEOUS SUBLATTICE MAGNETIZATION IN THE TW 26-2929

RIC. FLAME SPRAYED BARIUM TITANATE AS A CAPACITOR DIELECT 04-0773

TRUCTURE OF SEMICONDUCTIVE TITANIUM DIOXID/ STUDIES OF FLAME SPRAYED BARIUM TITANATE CAPACITORS. PART-10: S 04-0774

OF THICK FILM BARIUM TITANATE DIELECTRICS PRODUCED BY FLAME SPRAYING (CAPACITORS). ELECTRICAL PROPERTIES 04-0772

/ITANIUM DIOXIDE- BARIUM TITANATE COATINGS PRODUCED BY FLAME SPRAYING. PART-11: ELECTRICAL CHARACTERISTICS/ 04-0774

OELECTRIC) LITHIUM NIOBATE. (FAR FIELD) DIFFRACTION SPREADING OF SURFACE (ACOUSTIC) WAVES ON (PIEZ 11-1850

DIFFRACTION SPREADING OF SURFACE WAVES ON LITHIUM NIOBATE. 11-1849

IUM NIOBATE, SILICON DIOXIDE, AND SILICON. MEASUREMENT OF SPUTTER ETCHING RATES FOR MONOCRYSTALLINE LITH 11-1714

CHARACTERISTICS OF RF SPUTTERED BARIUM TITANATE FILMS ON SILICON. 04-0930

CHARACTERISTICS OF RF SPUTTERED BARIUM TITANATE THIN FILMS. 04-0906

FABRICATION OF RF SPUTTERED BARIUM TITANATE THIN FILMS. 04-0907

IUM OXIDE. PREPARATION AND EPITAXY OF SPUTTERED FILMS OF FERROELECTRIC BISMUTH TITAN 13-2038

COMPOSITION AND STRUCTURE OF SPUTTERED FILMS OF FERROELECTRIC NIOBATES. 13-1941

FERROELECTRICITY IN SPUTTERED LEAD TITANATE ZIRCONATE FILMS. 09-1559

.01) (IRON(0.405) NIOBIUM(0.325) ZIRCONIUM(0/ PERFORMANCE OF SPUTTERED LEAD(0.92) BISMUTH(0.07) LANTHANUM(0 37-3629

AND PROPERTIES OF BARIUM TITANATE FILMS PRODUCED BY CATHODE SPUTTERING. STRUCTURE 04-0832

RICAL PROPERTIES OF THIN FILM BARIUM TITANATE PREPARED BY RF SPUTTERING ON SILICON SUBSTRATES. /L AND ELECT 04-0828

RSENIDE. BARIUM TITANATE FILMS PREPARED BY RF SPUTTERING ONTO INDIUM ANTIMONIDE OR GALLIUM A 04-0741

PREPARATION OF THIN BARIUM TITANATE FILMS BY DC DIODE SPUTTERING (ONTO PLATINUM OR FUSED QUARTZ). 04-0948

CRYSTAL STRUCTURE OF BARIUM TITANATE PRODUCING SQUARE INTERFERENCE FRINGES. 04-0706

GENERALIZED FERROELECTRIC MODEL ON A SQUARE LATTICE. 03-0168

TRANSITION IN A SIXTEEN VERTEX LATTICE MODEL (FERROELECTRIC SQUARE LATTICE). PHASE 03-0517

(COVALENCY, DIRECTIVITY CONDITIONS FOR ELECTRIC ORDERING AND STABILITY) /OVSKITE FAMILY AND CHEMICAL BOND. 03-0243

E CERAMIC/ "VIBRATING WIRE" STUDY OF POTENTIAL DISTRIBUTION (STABILITY) IN BARIUM(X) STRONTIUM(1-X) TITANAT 04-0699

RIGLYCI/ PYROELECTRIC INVESTIGATIONS OF THE DOMAIN STRUCTURE STABILITY IN FERROELECTRICS. (ROCHELLE SALT, T 02-0083

THERMAL STABILITY OF LITHIUM HYDROGEN SELENITES. 22-2868

C FILMS ON SEMICONDUCTING SUBSTRATES. (TGS ON/ THERMODYNAMIC STABILITY OF POLARIZATION IN THIN FERROELECTRI 27-3163

CINE SULFATE) THERMODYNAMIC STABILITY OF THIN FERROELECTRIC FILMS. (TRIGLY 27-2959

ONTIUM NIOBATE CRYSTALS. STABILITY REGIONS FOR THE GROWTH OF BARIUM STR 13-2056

TETRAGONAL- RHOMBOHEDRAL IN LEAD ZIRCONATE TITANATE CERAMIC/ STABILIZATION EFFECTS AT THE PHASE TRANSITION 09-1611

RRADIATED FERROELECTRIC MATERIALS. (ROCHELLE SALT, TR/ PHASE STABILIZATION (HINDERED TRANSITION) IN GAMMA-I 27-3165

TEMS. EFFECT OF AUTOMATIC TEMPERATURE STABILIZATION IN FERROELECTRIC OSCILLATORY SYS 03-0294

ARIUM TITANATE SINGLE CRYSTALS. STABILIZATION OF SPONTANEOUS POLARIZATION OF B 04-0795

TOR DIELECTRICS. STABILIZED BARIUM TITANATE CERAMICS FOR CAPACI 04-0825

FABRICATION OF A LITHIUM TANTALATE TEMPERATURE STABILIZED OPTICAL MODULATOR. 12-1872

TTERING STUDY OF ANTIMONY SULFOIODIDE USING A DOUBLE PASSED, STABILIZED SCANNING INTERFEROMETER. /LOUIN SCA 15-2225

BARIUM TITANATE. DIFFUSIONLESS PHASE TRANSITIONS AND THE STABLE DOMAIN STRUCTURE OF (HIGH SENSITIVITY) 04-0888

T/ METHOD FOR ACHIEVING HIGH DIELECTRIC CONSTANT TEMPERATURE STABLE FERROELECTRICS. (BARIUM TITANATE, STRON 04-0551

/ATE, A FERROELECTRIC WITH NONREVERSIBLE, BUT TILTABLE (TO 3 STABLE POSITIONS, SPONTANEOUS) POLARIZATION (/ 08-1346

ONS IN LITHIUM NIOBATE AND TANTALATE AND THE EFFECT ON THE / STACKING FAULT MODEL FOR STOICHIOMETRY DEVIATI 11-1776

ONS IN LITHIUM NIOBATE AND TANTALATE AND THE EFFECT ON THE / STACKING FAULT MODEL FOR STOICHIOMETRY DEVIATI 11-1777

L CRYSTAL SUBSTRATES AS THE DETERMINING FACTOR OF THE GROWTH STAGE IN EPITAXY. (TRIGLYCINE SULFATE) /OF REA 27-3039

NEW CRYSTAL FOR COMMUNICATIONS SYSTEMS ENTERS PRODUCTION STAGE. (LITHIUM TANTALATE) 12-1867

MODIFIED POTASSIUM DIHYDROGEN PHOSPHATE MODEL IN A STAGGERED FIELD. 17-2710

GEN PHOSPHATES AND CESIUM AND AMMONIUM DIHYDROGEN ARSENATES, STANDARD METHOD PROPOSED). /D AMMONIUM DIHYDRO 17-2577

MPERATURE STABLE FERROELECTRICS. (BARIUM TITANATE, STRONTIUM STANNATE) /HIEVING HIGH DIELECTRIC CONSTANT TE 04-0551

ABSORPTION OF GAMMA QUANTA IN BARIUM, STRONTIUM, AND CALCIUM STANNATE. RESONANCE 36-3566

RICS- SILVER (BARIUM TITANATE AND BARIUM TITANATE- MAGNESIUM STANNATE). /ECTRIC SYSTEMS. PART-2: FERROELECT 04-0816

BARIUM TITANATE. (BARIUM STRONTIUM TITANATE, BARIUM TITANATE STANNATE, BARIUM TITANATE ZIRCONATE /ASED ON 04-1019

/LEASE OF ELECTRIC ENERGY IN LEAD NIOBIUM ZIRCONATE TITANATE STANNATE BY TEMPERATURE AND PRESSURE ENFORCED/ 09-1488

NIUM) MOSSBAUER STUDIES OF LEAD ZIRCONATE TITANATE STANNATE. (CERAMIC DOPED WITH NIOBIUM AND TITA 09-1604

TRIC). DIELECTRIC PROPERTIES OF DIAMMONIUM HEXABROMO STANNATE (DIELECTRIC TRANSITIONS NOT FERROELEC 36-3612

RROELECTRIC REGION OF THE SYSTEM LEAD HAFNATE TITANATE, LEAD STANNATE NIOBATE. SOLID SOLUTIONS IN ANTIFE 13-1956

/ CALCIUM) STRONTIUM TITANATE, BARIUM TITANATE ZIRCONATE (OR STANNATE), POTASSIUM TANTALATE NIOBATE, 228 N/ 08-1281

IOXIDE. SEMICONDUCTOR PROPERTIES OF BARIUM (TITANATE, STANNATE) SOLID SOLUTIONS DOPED BY TUNGSTEN TR 04-0965

/ MOSSBAUER EFFECT PROBABILITY ON TIN-119 IN BARIUM TITANATE STANNATE SOLID SOLUTIONS IN THE VICINITY OF T/ 04-0573

OELECTRIC PHASES. (LANTHANUM OR NIOBIUM DOPED LEAD ZIRCONATE STANNATE TITANATE) /FERROELECTRIC AND ANTIFERR 09-1487

/OELECTRIC- FERROELECTRIC TRANSITION IN THE LEAD (ZIRCONATE, STANNATE, TITANATE) SOLID SOLUTIONS. (ELECTRO/ 09-1486

9 DURING PHASE TRANSISTIONS OF THE BISMUTH FERRATE STRONTIUM STANNATE(0.33) MANGANATE(0.67) SYSTEM. /TIN-11 08-1400

PEROVSKITE TYPE SOLID SOL/ FERROELECTRIC PROPERTIES OF LEAD (STANNATE(0.5) NIOBATE(0.5)) (1-X) ZIRCONATE(X) 09-1548

TUTED LEAD TITANATE ZIRCONATES, 2 SUBSTITUTED LEAD ZIRCONATE STANNATES, COMPLEX PEROVSKITES) /CS. (8 SUBSTI 08-1418

OWTH OF SOME COMPLEX OXIDE MATERIALS. (NIOBATES, TANTALATES, STANNATES, TITANATES, CHROMATES, COBALTATES) / 02-0129

DIELECTRIC AND NMR STUDIES OF THE PHASE TRANSITION IN STANNOUS CHLORIDE DIHYDRATE. 36-3582

IC SEMICONDUCTOR COPPER PHTHALOCYANINE. STARK EFFECT AND FERROELECTRICITY IN THE ORGAN 36-3532

CE SPECTRA OF THE FERROELECTRIC CRYSTAL CESIUM DIHYDROGEN A/ STARK EFFECT IN THE NUCLEAR QUADRUPOLE RESONAN 18-2722

CE OF NITROGEN-14 AND THE FERROELECTRICITY OF THIOUREA. STARK EFFECT ON THE NUCLEAR QUADRUPOLE RESONAN 25-2909

ATE. EFFECT OF PRESSURE ON THE STATIC DIELECTRIC CONSTANT OF POTASSIUM TANTAL 08-1269

STUDIES ON THE MECHANISM OF THE PHASE TRANSITION IN TGS. (STATIC DIELECTRIC CONSTANT, X-RAY SCATTERING) 27-3093

TALS. (CRYSTALLOGRAPHIC VIEWPOINT) MICROSCOPIC THEORY OF STATIC DOMAIN STRUCTURES OF FERROELECTRIC CRYS 03-0340

OF THE FERROELECTRIC SEMICONDUCTOR ANTIMONY SULFOIODIDE IN A STATIC ELECTRIC FIELD. AGING 15-2113

ELECTRIC HYSTERESIS LOOP OF SILVER SODIUM DINITRITE BY QUASI STATIC ELECTRIC FIELD. FERRO 16-2293

ALUM AND HAFNIUM IONS IN BARIUM AND LEAD TITANATE. STATIC ELECTRIC QUADRUPOLE INTERACTION OF TANT 04-0937

RELAXATION TIME NEAR THE TRANSITION POINTS IN SODIUM NITRIT/ STATIC ELECTRIC SUSCEPTIBILITY AND DIELECTRIC 16-2302

/INATION OF CHROMIUM VALENCY FROM STUDIES OF EPR SPECTRA AND STATIC MAGNETIC SUSCEPTIBILITY OF CHROMIUM DO/ 04-0542

PPLICATION TO CALCULATING THE WALL ENERGY AND T/ THEORETICAL STATIC MODEL FOR BARIUM AND LEAD TITANATES - A 03-0189

N BARIUM TITANATE. STATIC QUADRUPOLE INTERACTION OF SCANDIUM-44 I 04-0692

ONIUM DIHYDROGEN PHOSPHATE SINGLE CRYSTAL (FROM SOLUTION) IN STATIC REGIME AND AT NORMAL PH. /IC OF THE AMM 17-2516

FOR THE PLANAR SPIN MODEL OF STRONTIUM TITANATE. STATIC SCALING HYPOTHESIS AND ITS IMPLICATIONS 06-1223

ES. STATIC SHELL MODEL FOR BARIUM AND LEAD TITANAT 04-0826

QUASI STATIC SWITCHING CURRENT IN SODIUM NITRITE. 16-2261

RECENT PROBLEMS IN DOMAIN STATICS AND DYNAMICS. (56 REFS) 03-0248

TITANATE (THEORY, CALCULA/ SMALL SIGNAL PERMITTIVITY OF THE STATIONARY (100)-180 DEG DOMAIN WALL IN BARIUM 03-0358

DOMAIN STATISTICS AND FERROELECTRIC TRANSIENTS. 03-0216

	IMPURITY STATISTICS IN ICE.	34-3472
, THE RESIDUAL ENTROPY OF ICE, PART-2, HYDROGEN BON/	LATTICE STATISTICS OF HYDROGEN BONDED CRYSTALS, PART-1	03-0388
ECTRICITY. (THEORY OF ENERGY STATES INVOLVED)	STATISTICAL AND ELECTRONIC PROBLEMS IN FERROEL	03-0325
ER REACTION IN AMMONIUM PERIODATE.	STATISTICAL CONSIDERATIONS OF THE ORDER DISORD	36-3581
ONE DIMENSIONAL KCP MODEL IN	STATISTICAL MECHANICS.	03-0389
PHASE TRANSITIONS.	STATISTICAL MECHANICS OF DIFFUSE FERROELECTRIC	03-0445
THERMODYNAMICS AND	STATISTICAL MECHANICS OF FERROELECTRIC PHASE T	03-0198
RANSITIONS.	STATISTICAL METHOD FOR INVESTIGATING (DYNAMICS	03-0493
CF) POLARIZATION REVERSAL IN FERROELECTRIC CERAMICS.	STATISTICAL PERTURBATION THEORY OF ORDER DISOR	03-0399
DER FERROELECTRICS: FIRST PERTURBATIVE CORRECTION.	STATISTICAL THEORY FOR DISPLACEMENT FERROELECT	03-0369
RICS.	STATISTICAL THEORY FOR FERROELECTRICITY IN TRI	03-0263
GLYCINE SULFATE.	STATISTICAL THEORY. (SPONTANEOUS POLARIZATION/	03-0366
/ORPHIC LITHIUM NIOBATE AND LITHIUM TANTALATE USING AUTHOR'S	STEADY ELECTRIC FIELD. /BSORPTION OF ULTRASOUN	04-0850
D IN SINGLE CRYSTALS OF BARIUM TITANATE IN THE PRESENCE OF A	STEERING CUT FOR ACOUSTIC SURFACE WAVES ON LIT	11-1819
HIUM NIOBATE. NEW HIGH FREQUENCY HIGH COUPLING LOW BEAM	STEERING SURFACE WAVES BY NONLINEAR MIXING IN	11-1778
ANISOTROPIC MEDIA. (LITHIUM NI/ POSSIBILITY OF SWITCHING AND	STEERING USING A MULTICHANNEL LITHIUM TANTALAT	12-1881
E CRYSTAL. OPTICAL BEAM	STEP ANOMALOUS INCREASE OF RESISTANCE ON LANTH	04-0801
ANUM(0.002) BARIUM(0.998) TITANATE- BISMUTH OXIDE COMPO/ TWO	STEP FUNCTION METHOD. DIELECTRIC RELAXA	16-2291
TION MEASUREMENT ON FERROELECTRIC SILVER SODIUM DINITRITE BY	STEP GROWTH MODEL.	34-3410
LOW ENTROPY FORM OF ICE I(H) OBTAINED FROM A LINEAR	STEP LADDERS ON PSEUDOCUBIC (001) PLANES. /STR	08-1316
UCTURES IN POTASSIUM NIOBATE SINGLE CRYSTALS ASSOCIATED WITH	(STEP OR SMOOTH) DOMAIN WALL STRUCTURE IN BAPIU	04-0677
M TITANATE DURING SWITCHING (CALCULATIONS COMPARED WITH DA/	STEP RADIATION SIGNALS IN FERROELECTRICS. / CO	03-0469
EFFICIENT AND PERMITTIVITY FROM THE PYROELECTRIC RESPONSE TO	STEP SIGNALS OF INFRARED RADIATION IN TRIGLYCI	13-2022
NE SULFATE AND STRONTIUM B/ PYROELECTRIC VOLTAGE RESPONSE TO	STEPS AND THE CRITICAL POLARIZATION AND REPOLA	03-0318
RIZATI/ CONNECTION BETWEEN THE STARTING FIELD FOR PARKHAUSEN	(STEPWISE) CHANGE IN PROBABILITY OF THE MOSSBAU	03-0276
ER EFFECT AT THE PHASE TRANSITION IN FERROELECTRICS. ABRUPT	STIBOTANTALITE.	36-3559
FERROELECTRIC BEHAVIOR OF THE MINERAL	STIBOTANTALITE. (ANTIMONY (TANTALUM, NIOBIUM)/	36-3601
/ND TEMPERATURE DEPENDENCE OF THE PIEZOELECTRIC CONSTANTS OF	STIMULATED COMBINATION SCATTERING OF SOUND IN	03-0431
PIEZOELECTRICS. (THEORY, LITHIUM NIOBATE EXAMPLE)	STIMULATED CONDUCTIVITY IN STRONTIUM TITANATE	06-1227
SINGLE CRYSTALS. THERMALLY	STIMULATED CURRENT IN FERROELECTRIC OXIDE CRYS	13-1926
TALS. (POTASS/ TRAPPING SITES, OPTICAL DAMAGE, AND THERMALLY	STIMULATED CURRENTS AND LUMINESCENCE IN BISMUT	36-3585
H(12) SILICON OXIDE(20) AND BISMUTH(12) GERMANIUM/	STIMULATED FAR INFRARED EMISSION IN LITHIUM NI	11-1736
OBATE. POWER AND LINE WIDTH OF TUNABLE	STIMULATED POLARIZATION ECHO IN FERROELECTRICS	17-2419
. (POTASSIUM DIHYDROGEN PHOSPHATE) MACROSCOPIC	STIMULATED RAMAN OSCILLATOR.	13-1917
BARIUM SODIUM NIOBATE FOR A TUNABLE	STOCKBARGER METHOD) CRYSTAL GROWTH IN THE	15-2230
ANTIMONY SULFOIODIDE- ANTIMONY TRISULFIDE SYSTEM. (BRIDGMAN-	STOICHIOMETRIC BARIUM TITANATE.	04-0790
WET SYNTHETIC METHOD OF MAKING PURE, EXACTLY	(STOICHIOMETRIC COMPENSATION) OXYGEN STOICHI	04-0664
OMETRY OF DONOR DOPED BARIUM TITANATE AND TITANIUM DIOXIDE.	(STOICHIOMETRIC) LITHIUM NIOBATE (TEMPERATURE D	11-1811
E/ NIOBIUM-93 NUCLEAR QUADRUPOLE RESONANCE INVESTIGATION OF	EXAMINATION OF SUBSTOICHIOMETRIC TUNGSTEN OXIDE(3-X) CRYSTALS BY	10-1627
ELECTRON MICROSCOPY.	FORMATION OF SHEAR STRUCTURES IN SUBSTOICHIOMETRIC TUNGSTEN TRIOXIDE.	10-1659
AND TEMPERATURE OF LITHIUM METANIOBATE CRYSTALS OF DIFFERENT	STOICHIOMETRIES. PHASE MATCHING ANGLES	11-1665
ERTIES AND PROCESSING OF SINGLE CRYSTAL BA/ EFFECT OF OXYGEN	STOICHIOMETRY AND HYDROXYL CONTENT ON THE PROP	13-1915
ANTALATE SINGLE CRYSTALS.	STOICHIOMETRY AND OPTICAL QUALITY OF LITHIUM T	12-1883
YNAMIC EXISTENCE CONDITIONS FOR MORPHOTROPIC PHASE BOUNDARY,	STOICHIOMETRY DEPENDENCE). /SOLUTIONS (THERMOD	09-1541
LITHIUM NIOBATE CRYSTALS. MELT COMPOSITION	(STOICHIOMETRY) DEPENDENCE OF POCKELS EFFECT IN	11-1728
D TANTALATE AND THE EFFECT ON THE / STACKING FAULT MODEL FOR	STOICHIOMETRY DEVIATIONS IN LITHIUM NIOBATE AN	11-1776
D TANTALATE AND THE EFFECT ON THE / STACKING FAULT MODEL FOR	STOICHIOMETRY DEVIATIONS IN LITHIUM NIOBATE AN	11-1777
HARGE MECHANISM IN FERROELECTRICS. (LEAD ZIRCONATE TITANATE,	STOICHIOMETRY EFFECTS) C	03-0259
/ROOPTIC EFFECT AND DIELECTRIC CONSTANT ON MELT COMPOSITION	(STOICHIOMETRY) IN LITHIUM NIOBATE (VERY LITTL/	11-1842
LS. DETERMINATION OF THE DEVIATION FROM OXYGEN	STOICHIOMETRY OF BARIUM TITANATE SINGLE CRYSTA	04-0736
ND TITANIUM DIOXIDE. (STOICHIOMETRIC COMPENSATION) OXYGEN	STOICHIOMETRY OF DONOR DOPED BARIUM TITANATE A	04-0664
SODIUM NIOBATE. SOME EFFECTS OF MELT	STOICHIOMETRY ON THE OPTICAL PROPERTIES BARIUM	13-1902
IUM SODIUM NIOBATE. SOME EFFECTS OF MELT	STOICHIOMETRY ON THE OPTICAL PROPERTIES OF BAR	13-1903
D LITHIUM TANTALATE BY RAMAN POWDER SPECTROSCOPY.	STOICHIOMETRY VARIATIONS IN LITHIUM NIOBATE AN	11-1813
E AND CURIE TEMPERATURE ME/ LITHIUM TANTALATE SINGLE CRYSTAL	STOICHIOMETRY (VERSUS MELT COMPOSITION, LATTIC	12-1871
OBATE AS AN IMPROVED NEW MATERIAL FOR REVERSIBLE HOLOGRAPHIC	STORAGE. RH DOPED LITHIUM NI	08-1359
ECTION MO/ STRAIN BIASED FERROELECTRIC- PHOTOCONDUCTOR IMAGE	STORAGE AND DISPLAY DEVICES OPERATED IN A REFL	09-1564
STRAIN BIASED FERROELECTRIC PHOTOCONDUCTOR IMAGE	STORAGE AND DISPLAY DEVICES. (PLZT)	09-1565
SCATTERING MODE FERROELECTRIC PHOTOCONDUCTOR IMAGE	STORAGE AND DISPLAY DEVICES. (PLZT CERAMICS)	09-1599
ERROELECTRIC CERAMICS. (LEAD ZIRCONATE- LEAD TITANATE) IMAGE	STORAGE AND DISPLAY DEVICES USING FINE GRAIN F	09-1572
E TITA/ ELECTROOPTIC CERAMICS- NEW MATERIALS FOR INFORMATION	STORAGE AND DISPLAY (HOT PRESSED LEAD ZIRCONAT	09-1550
FERROELECTRIC CERAMICS FOR INFORMATION	STORAGE AND DISPLAY. (PZT)	09-1552
AND PERFORMANCE OF A THIN FILM FERROELECTRIC PHOTOCONDUCTOR	STORAGE DEVICE. DESIGN	37-3632
D BARIUM TITANATE CRYSTALS. HOLOGRAPHIC	STORAGE, ELECTRICAL FIXING AND ERASING IN DOPE	04-0848
ANATE, BARIUM TITANATE)	STORAGE IN CERAMIC DIELECTRICS. (STRONTIUM TIT	04-0602
HOLOGRAPHIC	STORAGE IN DOPED BARIUM SODIUM NIOBATE.	13-1898
ATERIALS BASED UPON BARIUM TITANATE. ENERGY	STORAGE IN FERROELECTRIC CERAMICS: SURVEY OF M	02-0075
OF MATERIALS BASED ON BARIUM TITANATE. ENERGY	STORAGE IN HIGH PERMITTIVITY CERAMICS: SURVEY	04-0733
CHARGE AND POLARIZATION	STORAGE IN ICE CRYSTALS.	34-3456
COUPLED WAVE ANALYSIS OF HOLOGRAPHIC	STORAGE IN LITHIUM NIOBATE.	11-1826
BISMUTH TITANATE FERROELECTRIC- PHOTOCONDUCTOR OPTICAL	STORAGE MEDIUM FURTHER DEVELOPMENTS.	08-1371
BISMUTH TITANATE PHOTOCONDUCTOR OPTICAL	STORAGE MEDIUM. (HIGH RESOLUTION READ WRITE)	08-1445
FERROELECTRIC PHOTOCONDUCTOR OPTICAL	STORAGE MEDIUM UTILIZING BISMUTH TITANATE.	04-0769
ONDUCTOR DEVICE (ERASABLE, NONDESTRUCTIVE READOUT, BASED ON/	STORAGE OF HOLOGRAMS IN A FERROELECTRIC PHOTOC	37-3646
OBATE. MULTIPLE	STORAGE OF THICK PHASE HOLOGRAMS IN LITHIUM NI	11-1827
ED ON ELECTRICALLY CONTR/ ELECTROOPTIC CERAMIC (INFORMATION)	STORAGE (PROCESSING) AND DISPLAY DEVICES. (BAS	37-3650
ITHIUM NIOBATE. OPTICAL AND HOLOGRAPHIC	STORAGE PROPERTIES OF TRANSITION METAL DOPED L	11-1796
APPLIED FIELD CAUSED BY SHEARING DUE TO POLARIZATION INDUCED	STRAIN) /OELECTRIC BODIES. (WITH FREQUENCY OF	03-0197
USED BY ILLUMINATION IN DC ELECTRIC FIELD.	STRAIN ALONG C-AXIS OF ANTIMONY SULFOIODIDE CA	15-2239
GE STORAGE AND DISPLAY DEVICES. (PLZT)	STRAIN BIASED FERROELECTRIC PHOTOCONDUCTOR IMA	09-1565
AGE STORAGE AND DISPLAY DEVICES OPERATED IN A REFLECTION MO/	STRAIN BIASED FERROELECTRIC- PHOTOCONDUCTOR IM	09-1564
ELECTROOPTIC DEVICES USING	STRAIN BIASED PLZT FERROELECTRIC CERAMICS.	09-1562
/ THEORY OF THE ELASTOOPTIC EFFECT IN NONMETALLIC CRYSTALS.	(STRAIN DEPENDENT CURIE TEMPERATURE, POLARIZATI	03-0513
ATION OF THE STATE PARAMETERS AND FORMULATION OF SPONTANEOUS	STRAIN FOR FERROELASTICS (THEORY). DETERMIN	03-0137
INE- BORACITE. (TEMPERATURE DEPENDENCE OF PI/ LATENT LATTICE	STRAIN IN THE FERROELECTRIC STATE OF IRON- IOD	14-2083
/PEROVSKITE- TYPE CRYSTALS, PART-2, INTERACTION WITH ELASTIC	STRAIN (STRONTIUM TITANATE AND LANTHANUM ALUM/	03-0241
RELEASE TO FERROELEC/ EXPERIMENTAL METHODS FOR INVESTIGATING	STRAIN WAVE PROPAGATION AND ASSOCIATED CHARGE	37-3630
TIUM TITANATE. INTERACTION OF ELASTIC	STRAIN WITH THE STRUCTURAL TRANSITION OF STRON	06-1216
F FERROELECTRIC (IRON IODINE) BORACITE. (SPONTANEOUS LATTICE	STRAINS) ELECTROOPTICAL PROPERTIES O	14-2082
ZIRCONATE TITANATE CERAMICS WHEN SUBJECTED TO LARGE TENSILE	STRAINS. /ROELASTIC BEHAVIOR OF LEAD LANTHANUM	09-1571

EMICONDUCTOR FILMS. (AND USE OF THESE FILMS TO STUDY SURFACE STRAINS IN THE FERROELECTRIC CRYSTAL) /RIUM) S 04-0594
LATTICE STRAINS OF BORACITE CRYSTALS. 14-2081
ORACITE). X-RAY STUDY ON THE LATTICE STRAINS OF (PARAELECTRIC AND) (IRON- IODINE- B 14-2085
E SULFATE CRYSTALS ON TEMPERATURE AND APPLIED ELECTRIC FIELD STRENGTH. /F THE DOMAIN STRUCTURE OF TRIGLYCIN 27-3129
ODIDE. (DEPENDENCE ON TEMPERATURE, WAVELENGTH, APPLIED FIELD STRENGTH) /FRACTIVE INDICES OF ANTIMONY SULFOI 15-2206
TY OF TGS. (10 KHZ-1 MHZ) TEMPERATURE AND FIELD STRENGTH DEPENDENCE OF LARGE SIGNAL PERMITTIVI 27-3068
N FERROELECTRIC PROPERTIES. (DISPERSION, RELAXATION TIME AND STRENGTH, HYSTERESIS, 26 REFS) /AL DEFECTS UPO 03-0496
OSPHATE CRYSTALS. STUDY OF THE (MECHANICAL) STRENGTH PROPERTIES OF POTASSIUM DIHYDROGEN PH 17-2648
CRYSTALS NEAR THE CURIE POINT (EFFECTS OF TEMPERATURE, FIELD STRENGTH, SAMPLE ILLUMINATION). / SULFOIODIDE 15-2121
NGLE DOMAIN STRONTIUM TITANATE (OPTICAL STUDY UNDER UNIAXIAL STRESS). DIRECT OBSERVATION OF SI 06-1112
TRA OF CHROMIUM(3+) IN POTASSIUM DIHYDR/ EFFECTS OF UNIAXIAL STRESS AND ELECTRIC FIELD ON PARAMAGNETIC SPEC 17-2554
NCE IN CHROMIUM(3+) DOPED STRONTIUM TITANATE. ANALYSIS OF STRESS AND TEMPERATURE DEPENDENCE OF FLUORESCE 06-1215
ASSIUM DIHYDROGEN PHOSPHATE. INTERNAL STRESS AND THE FERROELECTRIC TRANSITION IN POT 17-2424
THE DISPLACIVE TRANSITION IN STRONT/ ULTRASONIC PROPAGATION, STRESS EFFECTS, AND INTERACTION PARAMETERS AT 06-1127
CURIE POINTS. (POTASSIUM DIHYDRO/ ANOMALIES IN THE INTERNAL STRESS FOR FERROELECTRIC PHOSPHATES NEAR THEIR 17-2665
TRIC HYDROGEN HALIDES. STRESS FREE SINGLE CRYSTAL GROWTH OF FERROELEC 31-3333
D A SHORTCIRCUITED POTASSIUM DIHYDROGEN PHOSPHATE CRYSTAL. STRESS INDUCED BIREFRINGENCE IN AN ISOLATED AN 17-2452
E. SECOND HARMONIC GENERATION IN STRESS INDUCED FERROELECTRIC STRONTIUM TITANAT 06-1131
ITANATE. STRESS INDUCED FERROELECTRICITY IN STRONTIUM T 06-1104
ITAN/ POLARIZED FLUORESCENCE STUDY OF CHROMIUM(3+) THROUGH A STRESS INDUCED PHASE TRANSITION IN STRONTIUM T 06-1115
ITANATE. CUBIC TO TRIGONAL STRESS INDUCED PHASE TRANSITION IN STRONTIUM T 06-1235
PERATURE IN STRONTIUM TITANATE. (ABSTRACT) RAMAN STUDIES OF STRESS INDUCED PHASE TRANSITION NEAR CURIE TEM 06-1236
SITION TEMPERATURE IN STRONTIUM TITA/ SOFT MODE STUDY OF THE STRESS INDUCED PHASE TRANSITIONS NEAR THE TRAN 06-1196
C ROCHELLE SALT. EFFECT OF MECHANICAL STRESS ON DIELECTRIC PROPERTIES OF PARAELECTRI 28-3225
TE. EFFECT OF UNIAXIAL STRESS ON RAMAN SCATTERING IN STRONTIUM TITANA 06-1102
FOIODIDE SINGLE CRY/ INFLUENCE OF ONE DIMENSIONAL MECHANICAL STRESS ON THE PHASE TRANSITION OF ANTIMONY SUL 15-2245
RATURE OF STRONTIUM TITANATE. EFFECT OF STRESS ON THE SUPERCONDUCTING TRANSITION TEMPE 06-1182
EMISSION FROM BARIUM TITANATE CRYSTALS SUBJECTED TO UNIAXIAL STRESS PULSES. LIGHT 04-0672
OTASSIUM DIDEUTERIUM PHOSPHATE. ULTRASONIC STRESS WAVE INTERACTION WITH DOMAIN WALLS IN P 17-2659
DOMAIN STRUCTURE OF STRONTIUM TITANATE UNDER UNIAXIAL STRESSES. 06-1113
FERENCE IN TEMPERATURES FOR MAXIMUM DIELECTRIC CON/ INTERNAL STRESSES IN BARIUM TITANATE CERAMICS. (AND DIF 04-0849
FERROELECTRIC PROPERTIES. EFFECT OF ELASTIC STRESSES ON PARAMETERS OF SOLID SOLUTIONS WITH 03-0265
IN STRUCTURE OF BARIUM TITANA/ INFLUENCE OF UNIAXIAL TENSILE STRESSES ON THE ELECTRICAL PROPERTIES AND DOMA 04-0961
ING IN POLYCRYSTALLINE FERROELECTRICS AT VERY LOW FIELDS AND STRESSES, ROCHELLE SALT) /(POLARIZATION SWITCH 28-3216
RAMAN SCATTERING AND PHASE TRANSITIONS IN STRESSED STRONTIUM TITANATE. 06-1103
ADIUM PAIRS) ORDER PARAMETER AND PHASE TRANSITIONS OF STRESSED STRONTIUM TITANATE. (EPR OF IRON- VAN 06-1170
CUBIC- TETRAGONAL TRANSITION, AND PRODUCTION BY SHAPING AND STRESSING). /URANCE IN THIN (110) PLATES BELOW 06-1164
/ GENERATION IN POLYVINYLIDENE FLUORIDE FILMS (AFTER COOLING STRETCHED FILM IN ELECTRIC FIELD, THERMOELECT/ 36-3530
/FFECTS OF TEMPERATURE AND PHASE TRANSITIONS ON THE HYDROXYL STRETCHING VIBRATION IN STRONTIUM TITANATE AN/ 06-1209
ASSIUM DIHYDROGEN OR DEUTERIUM PHOSPHATE (INFRARED PHOSPHATE STRETCHING, WIDE TEMPERATURE RANGE). /S IN POT 17-2646
IC PROPERTIES OF ANTIMONY SULFOIODIDE. (PHASE BOUNDARY SHAPE STRIATIONS IN DC FIELD) /ON ON THE FERROELECTR 15-2197
MOVING STRIPE PATTERN IN BARIUM TITANATE. AN EFFECT O 04-0643
F HEAT TREATMENT ON TUNGSTEN TRIOXIDE CRYSTAL. PART-2: GREEN STRIPES. 10-1639
/ CENTIMETER AND DECIMETER WAVELENGTHS EMPLOYING A SYMMETRIC STRIPLINE. (BARIUM TITANATE AND BARIUM CALCIU/ 04-0855
/ON OF SINGLE CRYSTAL FILMS OF BARIUM TITANATE BY MEANS OF A STROBOSCOPIC TRANSMISSION ELECTRON MICROSCOPE. 04-0942
PECTRUM OF IRON(3+) IN STRONTIUM TITANATE DUE TO NEAREST NE/ STRONG AXIAL ELECTRON PARAMAGNETIC RESONANCE S 06-1144
ION IN HEXAGONAL ICE. STRONG COLLISION LIMIT OF SPIN LATTICE RELAXAT 34-3480
ELECTRIC CONSTANT OF TRIGLYCINE SULFATE SINGLE CRYSTALS IN A STRONG ELECTRIC FIELD. DI 27-3150
CHANGE IN THE TANGENT OF THE ANGLE OF DIELECTRIC LOSSES IN A STRONG ELECTRIC FIELD. / SINGLE CRYSTALS ON A 27-3153
IC CONSTANT AND TANGENT OF THE ANGLE OF DIELECTRIC LOSS IN A STRONG ELECTRIC FIELD. /CHANGE IN THE DIELECTR 27-3151
PERMITTIVITY OF SINGLE CRYSTALS OF TRIGLYCINE SULFATE IN A STRONG ELECTRIC FIELD. 27-2978
ASTIC AND PIEZOELECTRIC PROPERTIES OF CADMIUM PYRONIOBATE IN STRONG ELECTRIC FIELDS. EL 13-1967
ES OF CERAMIC FERROELECTRICS (BARIUM TITANATE ZIRCONATES) IN STRONG ELECTRIC FIELDS AND AT HIGH PRESSURE. / 09-1605
/RROELECTRICS, SUBJECTED BEFOREHAND TO AN ELECTRIC FIELD, IN STRONG ELECTRIC FIELDS. (ANISOTROPY DEVELOPED/ 04-0661
/ ELECTRON THERMAL DIFFUSION EFFECTS (INTERNAL GENERATION OF STRONG FIELDS) DURING HOLOGRAM RECORDING IN CR 06-1090
/3) BISMUTH(2) (TUNGSTEN OR MOLYBDENUM) OXIDE(6), (BARIUM OR STRONTIUM) (2) COPPER TUNGSTEN OXIDE(6), AND (/ 36-3619
/M OR STRONTIUM) (2) COPPER TUNGSTEN OXIDE(6), AND (BARIUM OR STRONTIUM) (3) COPPER (NIOBIUM OR TANTALUM) (2)/ 36-3619
OR ANALYZING LEAD ZIRCONATE TITANATES CONTAINING BISMUTH AND STRONTIUM ADDITIVES. CHROMATOGRAPHIC METHOD F 09-1625
TEMPERATURE DEPENDENCE OF ELECTRONIC POLARIZABILITY OF STRONTIUM AND BARIUM TITANATES. 04-0738
CTRON PARAMAGNETIC RESONANCE OF TRIVALENT GADOLINIUM IONS IN STRONTIUM AND BARIUM TITANATES. ELE 04-0920
STRUCTURE OF POLYCRYSTALLINE SOLID SOLUTIONS OF STRONTIUM AND BISMUTH TITANATES. 06-1145
RESONANCE ABSORPTION OF GAMMA QUANTA IN BARIUM, STRONTIUM, AND CALCIUM STANNATE. 36-3566
VIBRATIONAL SPECTRA OF STRONTIUM, BARIUM AND CALCIUM TITANATES. 04-0864
VIBRATIONAL SPECTRA OF THE TITANATES OF STRONTIUM, BARIUM, AND CALCIUM. (75 REFS) 04-0865
FERROELECTRICITY MEASUREMENTS ON STRONTIUM BARIUM NIOBATE. 13-1922
I METHOD OF CRYSTAL PULLING. (CONTROL OF CRYSTAL DIAMETER OF STRONTIUM BARIUM NIOBATE) /N TO THE CZOCHRALSK 13-1914
QUE AND APPLICATION TO A NANOSECOND RESPONSE TIME DETECTOR. (STRONTIUM BARIUM NIOBATE) / MEASUREMENT TECHNI 13-1919
SE TO RECTANGULAR INFRARED SIGNALS IN TRIGLYCINE SULFATE AND STRONTIUM BARIUM NIOBATE. /TRIC VOLTAGE RESPON 13-2019
STEP SIGNALS OF INFRARED RADIATION IN TRIGLYCINE SULFATE AND STRONTIUM BARIUM NIOBATE. /OLTAGE RESPONSE TO 13-2022
LONGITUDINAL ELECTROOPTIC EFFECT IN OBLIQUE CUT STRONTIUM. BARIUM NIOBATE PLATES. 08-1472
LONGITUDINAL ELECTROOPTIC EFFECT IN OBLIQUE CUT STRONTIUM BARIUM NIOBATE PLATES. 13-2063
S. PHOTOMIXING AT 10.6 MICRONS WITH STRONTIUM BARIUM NIOBATE PYROELECTRIC DETECTOR 09-1477
NONLINEAR PROPERTIES AND PHASE TRANSITIONS OF STRONTIUM BISMUTH TITANATES. 13-1952
RAMIC DIELECTRICS. STUDIES ON THE (STRONTIUM, CALCIUM) TITANATE BOUNDARY LAYER CE 06-1247
RAMIC WITH BOUNDARY LAYER STRUCTURE. STUDIES ON THE (STRONTIUM, CALCIUM) TITANATE SOLID SOLUTION CE 06-1246
THE SYSTEM LEAD TITANATE- STRONTIUM COPPER NIOBATE. (CERAMIC SYNTHESIS) 05-1044
THE SYSTEM LEAD TITANATE- STRONTIUM CUPRATE(0.33) NIOBATE(0.67). 05-1045
IUM DINIOBATE SY/ PHASE EQUILIBRIA IN THE POTASSIUM NIOBATE- STRONTIUM DINIOBATE AND POTASSIUM NIOBATE- BAR 13-2016
AN X-RAY STUDY OF THE PHASE TRANSITION IN STRONTIUM FERRATE TANTALATE 08-1382
EFFECT OF IRON(4+) IN THE SYSTEM STRONTIUM FERRATE TITANATE. 13-1927
TION TECHNIQUE. (LEAD HAFNATE, TIN HAFNATE, CALCIUM HAFNATE, STRONTIUM HAFNATE) / PERTURBED ANGULAR CORRELA 08-1276
INE SULFATE, HYDROGEN AMMONIUM MONOCHLORO ACETATE, DICALCIUM STRONTIUM HEXAACETATE. /IDE, THIOUREA, TRIGLYC 02-0099
L STUDIES ON THE PHASE TRANSITION OF FERROELECTRIC DICALCIUM STRONTIUM HEXAACETATE. X-RAY AND OPTICA 36-3584
NEUTRON DIFFRACTION STUDY OF (TETRAGONAL- CUBIC PEROVSKITE) STRONTIUM IRON TUNGSTATE . (MAGNETIC G STRUCT/ 10-1655
Y OF SINGLE CRYSTALS OF A NEW PYROCHLORE CONTAINING LITHIUM, STRONTIUM LANTHANUM TANTALUM OXYFLUORIDE. /TUD 36-3613
S APPLICATION TO THE SOLID ELECTROLYTE FUEL CELL. ((CALCIUM, STRONTIUM) LANTHANUM TITANATE- ALUMINATES) /IT 08-1444
/D DIELECTRIC PROPERTIES OF LITHIUM CONTAINING PEROVSKITES. (STRONTIUM LITHIUM NIOBATE, STRONTIUM LITHIUM / 08-1292
/ONZE STRUCTURE (POTASSIUM SODIUM LITHIUM NIOBATE, POTASSIUM STRONTIUM LITHIUM NIOBATE TITANATE, POTASSIUM/ 13-1923
(SINGLE CRYSTAL). ELECTROOPTIC PROPERTIES OF STRONTIUM LITHIUM NIOBATE TYPE FERROELECTRICS 13-2062
FOLYMORPHISM OF STRONTIUM METANIOBATE (TWO REVERSIBLE PHASES). 36-3535

/THESIS AND X-RAY INVESTIGATION OF THE PEROVSKITES BARIUM OR STRONTIUM METAPLUMBATES (NOT FERROELECTRICS).   36-3607
ELECTRICAL AND OPTICAL PROPERTIES OF (BARIUM, STRONTIUM) MIXED TITANATE THIN SECTIONS.   04-0603
EPITAXY OF PLATINUM ON BARIUM STRONTIUM NIOBATE.   13-1907
SULFATE, TRIGLYCINE FLUOROBERYLLATE, LITHIUM SULFATE, BARIUM STRONTIUM NIOBATE) /ION DETECTOR. (TRIGLYCINE   02-0046
TE, POTASSIUM LITHIUM NIOBATE, BARIUM SODIUM NIOBATE, BARIUM STRONTIUM NIOBATE) /M NIOBATE, LITHIUM TANTALA   11-1734
SHORT INFRARED LASER PULSES IN TRIGLYCINE SULFATE AND BARIUM STRONTIUM NIOBATE. /CTRIC VOLTAGE RESPONSE TO   13-2018
RAMAN SCATTERING BY BARIUM STRONTIUM NIOBATE.   13-1899
SOLIDIFICATION STUDY OF BARIUM STRONTIUM NIOBATE.   13-1954
ECTRIC OXIDE CRYSTALS. (POTASSIUM LITHIUM NIOBATE, POTASSIUM STRONTIUM NIOBATE) /MULATED CURRENT IN FERROEL   13-1926
AL GROWTH RATE WITH SUPERSATURATION IN FLUXED MELTS. (BARIUM STRONTIUM NIOBATE)   VARIATION OF CRYST   13-1939
CZOCHRALSKI GROWTH OF BARIUM STRONTIUM NIOBATE CRYSTALS.   13-1913
STABILITY REGIONS FOR THE GROWTH OF BARIUM STRONTIUM NIOBATE CRYSTALS.   13-2056
/MPERATURE DEPENDENCE OF THE DIELECTRIC PROPERTIES OF BARIUM STRONTIUM NIOBATE IN THE FREQUENCY RANGE OF 1/   13-1938
/FFICIENTS OF PIEZOELECTRIC CRYSTALS. (LEAD TITANATE, BARIUM STRONTIUM NIOBATE, LEAD NIOBIUM(4) OXIDE(11),/   05-1066
ELECTROOPTIC LIGHT BEAM DEFLECTION WITH BARIUM STRONTIUM NIOBATE PRISM.   13-2036
FERROELECTRIC PROPERTIES OF STRONTIUM NIOBATE SINGLE CRYSTAL.   08-1404
PROPERTIES OF SOLID SOLUTIONS IN THE SYSTEM BARIUM TITANATE- STRONTIUM NIOBIUM(0.8) OXIDE(3). /RROELECTRIC   04-0707
/SE STUDY OF THE CADMIUM BORATE- ABO(3) SYSTEMS. (WHERE A IS STRONTIUM OR BARIUM, AND B IS HAFNIUM, TITANI/   36-3528
/ PROPERTIES OF THE ABC NIOBATES (WHERE A AND B ARE CALCIUM, STRONTIUM, OR BARIUM, AND C IS SODIUM OR POTA/   04-0912
/CTRIC CERAMICS UNDER HIGH ELECTRIC FIELDS. (SUBSTITUTION OF STRONTIUM OR CALCIUM FOR LEAD, IRON FOR ZIRCO/   09-1484
/RUCTURE IN X-RAY ABSORPTION SPECTRA OF CERTAIN PEROVSKITES (STRONTIUM OR CALCIUM TITANATE OR ZIRCONATE, T/   03-0414
INARY PHASES BETWEEN NIOBIUM PENTOXIDE AND LEAD, BARIUM, AND STRONTIUM OXIDES.   BINARY AND PSEUDOB   04-0590
FERROELECTRICS IN THE POTASSIUM OXIDE- STRONTIUM OXIDE- NIOBIUM PENTOXIDE.   13-1896
/SE EQUILIBRIA RELATIONS IN THE TERNARY SYSTEM BARIUM OXIDE- STRONTIUM OXIDE- NIOBIUM PENTOXIDE. (OXIDES, /   13-1920
PHASE TRANSITION OF A NEW FERROELECTRIC OXIDE, STRONTIUM OXIDE- TELLURIUM DIOXIDE.   36-3627
BRONZE TYPE FERROELECTRICS).   ELECTROOPTICAL PROPERTIES OF STRONTIUM POTASSIUM LITHIUM NIOBATE. (TUNGSTEN   13-2051
DE PHASES OF THE OXYGENATED QUADRATIC TUNGSTEN BRONZE TYPE. (STRONTIUM POTASSIUM NIOBIUM OXYFLUORIDE) /UORI   13-2011
/ PROPERTIES OF TWO NEW PHASES OF QUADRATIC TUNGSTEN BRONZES STRONTIUM POTASSIUM TANTALATE AND BARIUM POTA/   13-2010
S.   SINGLE CRYSTAL GROWTH OF STRONTIUM POTASSIUM TANTALATE FROM MOLTEN SALT   13-2034
ACROSCOPIC OBSERVATION OF FERROELECTRIC DOMAINS OF DICALCIUM STRONTIUM PROPIONATE.   ELECTRON MIRROR M   26-2946
TS OF FERROELECTRIC DOMAINS OF BARIUM TITANATE AND DICALCIUM STRONTIUM PROPIONATE. /IRROR MICROSCOPIC ASPEC   04-0973
NMR INVESTIGATIONS OF FERROELECTRIC DICALCIUM STRONTIUM PROPIONATE.   26-2931
TYPE FERROELECTRICS: POLARIZATION FLUCTUATIONS IN DICALCIUM STRONTIUM PROPIONATE. /WAVES IN ORDER DISORDER   26-2928
TEMPERATURE DEPENDENCE OF SOUND VELOCITY IN DICALCIUM STRONTIUM PROPIONATE.   26-2934
TRASONIC ATTENUATION NEAR THE CURIE TEMPERATURE IN DICALCIUM STRONTIUM PROPIONATE.   UL   26-2948
ES K)   OPTICAL ACTIVITY OF FERROELECTRIC DICALCIUM STRONTIUM PROPIONATE. (TRANSITION AT 209 DEGRE   27-3040
ALLOGRAPHIC PROPERTIES OF A(2+)B(3+)(2)O(4) TYPE CALCIUM AND STRONTIUM SCANDATES.   PREPARATION AND CRYST   36-3537
NZE) TYPE FERROELECTRICS.   SUPERLATTICE STRUCTURE OF STRONTIUM SODIUM LITHIUM NIOBATE (TUNGSTEN BRO   13-2002
/IOBIUM ON THE CRYSTALLOGRAPHIC AND DIELECTRIC PROPERTIES OF STRONTIUM SODIUM NIOBATE AND BARIUM SODIUM NI/   13-1921
ONSTANT TEMPERATURE STABLE FERROELECTRICS. (BARIUM TITANATE, STRONTIUM STANNATE /HIEVING HIGH DIELECTRIC C   04-0551
/OF TIN-119 DURING PHASE TRANSITIONS OF THE BISMUTH FERRATE STRONTIUM STANNATE(0.33) MANGANATE(0.67) SYST/   08-1400
/IUM TANTALUM OXIDE, SODIUM BARIUM TANTALUM OXIDE, POTASSIUM STRONTIUM TANTALUM OXIDE, SODIUM STRONTIUM TA/   13-2050
NEW FERROELECTRIC COMPOUND STRONTIUM TELLURATE.   36-3626
ABSORPTION EDGE AND STRUCTURAL PHASE TRANSITION IN STRONTIUM TITANATE.   06-1107
TEMPERATURE DEPENDENCE OF FLUORESCENCE IN CHROMIUM(3+) DOPED STRONTIUM TITANATE.   ANALYSIS OF STRESS AND   06-1215
OF THE IRON OXYGEN VACANCY CENTER IN THE TETRAGONAL PHASE OF STRONTIUM TITANATE.   ANALYSIS   06-1231
ANOMALOUS FERROELECTRIC HYSTERESIS LOOPS. (LEAD STRONTIUM TITANATE)   05-1067
ANOMALOUS RESONANCE OF STRONTIUM TITANATE.   06-1173
ANOMALOUS SPECIFIC HEAT OF STRONTIUM TITANATE.   06-1159
ANOMALOUS ULTRASONIC ATTENUATION AT THE 105K TRANSITION IN STRONTIUM TITANATE.   06-1189
CERAMIC DIELECTRICS. (BARIUM TITANATE, STRONTIUM TITANATE)   04-0683
CRITICAL NEUTRON SCATTERING FROM STRONTIUM TITANATE.   06-1195
CUBIC TO TRIGONAL STRESS INDUCED PHASE TRANSITION IN STRONTIUM TITANATE.   06-1235
DEFECT STUDIES IN STRONTIUM TITANATE.   06-1244
DEFECTS IN OXIDE CERAMICS. (STRONTIUM TITANATE)   06-1163
DETERMINATION OF THE NATURE OF THE OPTICAL GAP OF STRONTIUM TITANATE.   06-1108
ARY LAYER STRUCTURE FOR HIGH FREQUENCY APPLICATION. (BARIUM, STRONTIUM TITANATE) /CTRIC CERAMICS WITH BOUND   04-0833
UCLEI", PHASE STRUCTURE ABOVE CURIE POINT, RAMAN SPECTRUM OF STRONTIUM TITANATE) /DEL OF LOCALLY ORDERED "N   04-0647
DIELECTRIC SUSCEPTIBILITY OF QUASI FERROELECTRIC STRONTIUM TITANATE.   06-1133
MPERATURE (ROTATIONS OF BO6 OCTAHEDRA IN ABO(3) PEROVSKITES, STRONTIUM TITANATE). /ERROELECTRIC CRITICAL TE   06-1205
FERROELECTRICS NEAR THE TRANSITION POINT. (BARIUM TITANATE, STRONTIUM TITANATE) /HAVIOR OF DISPLACIVE TYPE   04-0992
UM TARTRATE, AND MIXED CRYSTALS, BARIUM TITANATE, AND BARIUM STRONTIUM TITANATE) / TARTRATE, LITHIUM AMMONI   02-0089
F DOMAIN STRUCTURE ON THE FLUORESCENCE OF CHROMIUM(3+) DOPED STRONTIUM TITANATE.   EFFECT O   06-1114
T OF STRESS ON THE SUPERCONDUCTING TRANSITION TEMPERATURE OF STRONTIUM TITANATE.   EFFEC   06-1182
ETIC RESONANCE SPECTRA OF SUBSTITUTIONAL IMPURITIES IN CUBIC STRONTIUM TITANATE. / ON THE ELECTRON PARAMAGN   06-1193
THE QUARTIC ANHARMONICITY ON THE SOFT PHONON INSTABILITY OF STRONTIUM TITANATE.   EFFECT OF   06-1211
EFFECT OF UNIAXIAL STRESS ON RAMAN SCATTERING IN STRONTIUM TITANATE.   06-1102
EFFECTS OF LONGITUDINAL MAGNETIC FIELD ON ULTRASOUND IN STRONTIUM TITANATE.   06-1243
NIUM(4+) FROM THE PARAMAGNETIC RESONANCE OF MANGANESE(4+) IN STRONTIUM TITANATE.   EFFECTIVE CHARGE OF TITA   06-1167
ELASTIC CONSTANT ANOMALIES IN STRONTIUM TITANATE.   06-1177
ELECTRIC PROPERTIES OF HYDROGEN DOPED STRONTIUM TITANATE.   06-1088
TRON PARAMAGNETIC RESONANCE OF SOME RARE EARTH IMPURITIES IN STRONTIUM TITANATE.   ELEC   06-1192
ELECTRON SPIN RESONANCE STUDY IN STRONTIUM TITANATE.   06-1228
NG FIELDS) DURING HOLOGRAM RECORDING IN CRYSTALS (IRON DOPED STRONTIUM TITANATE). /ERNAL GENERATION OF STRO   06-1090
DUE TO LATTICE POLARIZATION. (LCAO ANALYSIS OF BAND SCHEME, STRONTIUM TITANATE) /ND CHANGES IN PEROVSKITES   06-1101
FAR INFRARED FERROELECTRIC VIBRATION MODE IN STRONTIUM TITANATE.   06-1094
FORCE CONSTANTS IN BARIUM TITANATE AND STRONTIUM TITANATE.   04-0748
FREE CARRIER OPTICAL ABSORPTION IN NIOBIUM DOPED STRONTIUM TITANATE.   06-1214
FUNDAMENTAL ABSORPTION EDGE OF STRONTIUM TITANATE.   06-1186
HIGH L/F NOISE ANOMALY IN SEMICONDUCTING BARIUM STRONTIUM TITANATE.   04-0866
HIGH PRESSURE COMPRESSIBILITY AND GRUNEISEN PARAMETER OF STRONTIUM TITANATE.   06-1121
NERTIAL ELECTRICAL POLARIZATION OF CRYSTALS (ALKALI HALIDES, STRONTIUM TITANATE).   I   06-1197
INFRARED SPECTRA OF SINGLE AND POLYCRYSTALLINE STRONTIUM TITANATE.   06-1250
ERACTION OF ELASTIC STRAIN WITH THE STRUCTURAL TRANSITION OF STRONTIUM TITANATE.   INT   06-1216
ION OF ELECTRIC FIELD EFFECTS IN EPR OF GADOLINIUM(3+) DOPED STRONTIUM TITANATE.   INTERPRETAT   06-1168
C CONDUCTION IN SINTERED OXIDES BASED ON CALCIUM TITANATE OR STRONTIUM TITANATE.   IONI   06-1141
LATTICE DYNAMICS ABOVE STRUCTURAL PHASE TRANSITIONS: STRONTIUM TITANATE.   06-1124
TTICE DYNAMICS OF CUBIC PEROVSKITE STRUCTURES, IN PARTICULAR STRONTIUM TITANATE.   LA   06-1185
LATTICE DYNAMICS OF STRONTIUM TITANATE.   06-1224
SOFT MODES NEAR THE CUBIC TO TETRAGONAL PHASE TRANSITION IN STRONTIUM TITANATE. /SCATTERING STUDIES OF THE   06-1242

PROPERTIES OF FERROELECTRICS AT SUBMILLIMETER WAVELENGTHS. (STRONTIUM TITANATE) MEASUREMENT OF DIELECTRIC 06-1089
MEASUREMENT OF ELECTRON ENERGY LOSS FUNCTION IN BULK STRONTIUM TITANATE. 06-1138
OF THE RESPONSE FUNCTION OF THE SOFT ZONE BOUNDARY PHONON IN STRONTIUM TITANATE. MICROSCOPIC CALCULATION 06-1213
MICROWAVE ABSORPTION IN CUBIC STRONTIUM TITANATE. 06-1212
IONS FROM WEIGHT MEASUREMENTS IN SOLID SOLUTIONS OF (BARIUM, STRONTIUM) TITANATE. MOBILITY DETERMINAT 04-0689
MONODOMAIN STRONTIUM TITANATE. 06-1169
MOSSBAUER STUDY OF COLOR CENTERS IN STRONTIUM TITANATE. 06-1153
NEW ESR EVIDENCE FOR THE 65 DEGREES K ANOMALY IN STRONTIUM TITANATE. 06-1230
RON SCATTERING STUDIES OF THE STRUCTURAL PHASE TRANSITION IN STRONTIUM TITANATE. NEW OBSERVATIONS FROM NEUT 06-1210
PARAMETRIC 4 PHOTON RECIPROCAL EFFECT IN CADMIUM SULFIDE AND STRONTIUM TITANATE. NONCOLLINEAR 06-1122
NUCLEAR MAGNETIC RESONANCE STUDY OF THE PHASE TRANSITION IN STRONTIUM TITANATE. 06-1237
OPTICAL ABSORPTION EDGE OF STRONTIUM TITANATE. 06-1187
OPTICAL ABSORPTION SPECTRA OF REDUCED STRONTIUM TITANATE. 06-1245
FERROELECTRIC CERAMIC CONTAINING TITANIUM. (BARIUM TITANATE, STRONTIUM TITANATE) /C RESONANCE SPECTRA IN A 04-1020
PERMITTIVITY OF STRONTIUM TITANATE. 06-1174
PHOTOCHROISM IN TRANSITION METAL DOPED STRONTIUM TITANATE. 06-1126
PHOTOCONDUCTIVITY AND TRAPPING PHENOMENA IN STRONTIUM TITANATE. 06-1091
THE LOCAL FERROELECTRIC PHASE TRANSITION AROUND 47 DEG K IN STRONTIUM TITANATE. /ANOMALY AS AN EVIDENCE OF 06-1249
OF CHROMIUM(3+) THROUGH A STRESS INDUCED PHASE TRANSITION IN STRONTIUM TITANATE. /RIZED FLUORESCENCE STUDY 06-1115
LARONS IN ANISOTROPIC ENERGY BANDS (THEORY, CALCULATIONS FOR STRONTIUM TITANATE). PO 03-0305
PREPARATION AND STUDY OF THIN FILMS OF (BARIUM, STRONTIUM) TITANATE. 04-1000
PRESSURE DEPENDENCE OF THE ELASTIC CONSTANTS OF STRONTIUM TITANATE. 06-1096
PRESSURE DEPENDENCE OF THE ELASTIC CONSTANT OF STRONTIUM TITANATE. 06-1097
CTRIC CONSTANTS OF PARAELECTRIC MATERIALS. (BARIUM TITANATE, STRONTIUM TITANATE) /E DEPENDENCE OF THE DIELE 04-0856
RAMAN SCATTERING AND PHASE TRANSITIONS IN STRESSED STRONTIUM TITANATE. 06-1113
SECOND HARMONIC GENERATION IN STRESS INDUCED FERROELECTRIC STRONTIUM TITANATE. 06-1131
SHARED HOLES TRAPPED BY CHARGE DEFECTS IN STRONTIUM TITANATE. 06-1123
: COUPLED MODES (FERROELECTRICS, CESIUM DIHYDROGEN ARSENATE, STRONTIUM TITANATE) /T MODE RAMAN SPECTROSCOPY 06-1207
INDUCED PHASE TRANSITIONS NEAR THE TRANSITION TEMPERATURE IN STRONTIUM TITANATE. /MODE STUDY OF THE STRESS 06-1196
SOUND PROPAGATION NEAR THE STRUCTURAL PHASE TRANSITION IN STRONTIUM TITANATE. 06-1155
HYPOTHESIS AND ITS IMPLICATIONS FOR THE PLANAR SPIN MODEL OF STRONTIUM TITANATE. STATIC SCALING 06-1223
STRESS INDUCED FERROELECTRICITY IN STRONTIUM TITANATE. 06-1104
SURFACE BARRIER ENERGIES ON STRONTIUM TITANATE. 06-1175
ND) ELECTRIC FIELD DEPENDENCE OF A HYDROGEN IMPURITY MODE IN STRONTIUM TITANATE. (TEMPERATURE A 06-1146
TEMPERATURE DEPENDENCE OF THE DIELECTRIC CONSTANT OF STRONTIUM TITANATE. 06-1134
MPERATURE DEPENDENCE OF THE FIRST ORDER ELASTIC CONSTANTS OF STRONTIUM TITANATE. TE 06-1160
EPENDENCE OF THE RAMAN CROSS SECTIONS IN BARIUM TITANATE AND STRONTIUM TITANATE. TEMPERATURE D 04-0548
THE IRON(3+)- VACANCY CENTER IN THE TETRAGONAL PHASE OF STRONTIUM TITANATE. 06-1234
AND ULTRASONIC PROPAGATION NEAR THE STRUCTURAL TRANSITION OF STRONTIUM TITANATE. THERMAL EXPANSIVITY 06-1135
TIME EFFECTS IN THE HYSTERESIS LOOP OF (LEAD, STRONTIUM) TITANATE. 05-1052
TIME EFFECTS IN THE HYSTERESIS LOOP OF LEAD STRONTIUM TITANATE. 05-1053
ULTRASONIC DISPERSION IN STRONTIUM TITANATE. 06-1239
, AND INTERACTION PARAMETERS AT THE DISPLACIVE TRANSITION IN STRONTIUM TITANATE. /OPAGATION, STRESS EFFECTS 06-1127
ULTRASONIC VELOCITY ABOVE TANTALUM IN STRONTIUM TITANATE. 06-1137
Y DIFFRACTOMETRY OF LOW TEMPERATURE PHASE TRANSFORMATIONS IN STRONTIUM TITANATE. X-RA 06-1156
URE. CRITICAL ASYMMETRY IN LOCAL FLUCTUATIONS IN STRONTIUM TITANATE ABOVE THE CRITICAL TEMPERAT 06-1166
URE. CRITICAL DYNAMICS OF STRONTIUM TITANATE ABOVE THE CRITICAL TEMPERAT 06-1204
URE. FLUCTUATIONS AND CORRELATIONS IN STRONTIUM TITANATE ABOVE THE CRITICAL TEMPERAT 06-1233
MICROWAVE LOSSES IN STRONTIUM TITANATE ABOVE THE PHASE TRANSITION. 06-1199
ION. TEMPERATURE DEPENDENCE OF THE SOFT MODE IN STRONTIUM TITANATE ABOVE THE 105 DEG K TRANSIT 06-1178
OF STRESS INDUCED PHASE TRANSITION NEAR CURIE TEMPERATURE IN STRONTIUM TITANATE. (ABSTRACT) RAMAN STUDIES 06-1236
AT ANOMALY ASSOCIATED WITH THE 110 DEG K PHASE TRANSITION IN STRONTIUM TITANATE. (ABSTRACT) SPECIFIC HE 06-1226
TRANSPORT PROPERTIES IN REDUCED STRONTIUM TITANATE. (ABSTRACT) 06-1152
UDIES OF CRITICAL PHENOMENA IN INSULATING AND SEMICONDUCTING STRONTIUM TITANATE. (ABSTRACT) ULTRASONIC ST 06-1118
/ECTRON PARAMAGNETIC RESONANCE SPECTRUM OF MOLYBDENUM(5+) IN STRONTIUM TITANATE: AN EXAMPLE OF THE DYNAMIC/ 06-1125
/ERROELECTRIC CERAMICS AT LOW TEMPERATURES. (POLYCRYSTALLINE STRONTIUM TITANATE AND BARIUM(X) STRONTIUM(1-/ 04-0720
/UENCE OF PRESSURE ON PHASE TRANSITIONS AT LOW TEMPERATURES: STRONTIUM TITANATE AND ((BARIUM(X) STRONTIUM(/ 04-0898
UM/ ELECTRON TUNNELING AND BAND STRUCTURE OF (NIOBIUM DOPED) STRONTIUM TITANATE AND (CALCIUM DOPED) POTASSI 06-1222
/RECORDING CHARACTERISTICS OF PHOTOCHROMIC CALCIUM FLUORIDE, STRONTIUM TITANATE, AND CALCIUM TITANATE CRYS/ 06-1119
ITE- TYPE CRYSTALS, PART-2, INTERACTION WITH ELASTIC STRAIN (STRONTIUM TITANATE AND LANTHANUM ALUMINATE). / 03-0241
SONANCE AND OPTICAL ABSORPTION OF TRANSITION ELEMENT IONS IN STRONTIUM TITANATE AND LANTHANUM ALUMINATE. / 06-1165
SOME TITANATES BETWEEN 20 AND 1000 DEG C. (BARIUM TITANATE, STRONTIUM TITANATE AND OTHER MATERIALS) /TY OF 04-0820
FLUORIDE. CRITICAL NEUTRON SCATTERING IN STRONTIUM TITANATE AND POTASSIUM MANGANESE TRI 06-1208
/ A FERROELECTRIC SOLID SOLUTION. (THEORY, APPLIED TO BARIUM STRONTIUM TITANATE AND POTASSIUM NIOBATE TANT/ 03-0260
NDUCTIVITY IN PEROVSKITES (3 PHONON INTERACTION PROPOSED FOR STRONTIUM TITANATE AND POTASSIUM TANTALATE)./ 06-1095
ELECTRIC FIELD INDUCED OPTICAL HARMONIC GENERATION IN STRONTIUM TITANATE AND POTASSIUM TANTALATE. 06-1203
/(OPTICAL, FERROELECTRIC- ACOUSTIC) IN THE PHONON SPECTRA OF STRONTIUM TITANATE AND POTASSIUM TANTALATE (L/ 06-1157
ND PHASE TRANSITIONS ON THE HYDROXYL STRETCHING VIBRATION IN STRONTIUM TITANATE AND TITANIUM DIOXIDE. /RE A 06-1209
AT LOW TEMPERATURE (5-300 DEG K). DIELECTRIC PROPERTIES OF STRONTIUM TITANATE (ANNEALED SINGLE CRYSTALS) 06-1201
ASE TRANSITION. OPTICAL ABSORPTION EDGE OF STRONTIUM TITANATE AROUND THE 105 DEGREES K PH 06-1099
N COEFFICIENTS AND THE THICKNESS OF NONUNIFORM THIN SAMPLES (STRONTIUM TITANATE AS EXAMPLE). /RAY ABSORPTIO 06-1179
LECTRON SPECTROSCOPY (ESCA). BAND STRUCTURE OF STRONTIUM TITANATE AS MEASURED BY X-RAY PHOTOE 06-1100
/RIC AND ELECTROOPTIC STUDIES OF THE POLARIZATION PROCESS IN STRONTIUM TITANATE AT THE CUBIC TO TETRAGONAL 06-1117
DIELECTRIC PROPERTIES OF STRONTIUM TITANATE AT LOW TEMPERATURE. (BIREF/ 06-1200
ELECTRONIC CONDUCTION IN SLIGHTLY REDUCED STRONTIUM TITANATE AT LOW TEMPERATURES. 06-1151
FIELD INDUCED SWITCHING OF TETRAGONAL DOMAINS IN STRONTIUM TITANATE AT LOW TEMPERATURES. 06-1184
RATURE DEPENDENCE, ANISOTROPY) DIELECTRIC PROPERTIES OF STRONTIUM TITANATE AT LOW TEMPERATURES. (TEMPE 06-1202
LCULATIONS). ATOMIC DISPLACEMENTS IN PEROVSKITE STRONTIUM TITANATE (AT OPTICAL FREQUENCIES, CA 06-1142
TRANSITION. DISTORTION ENHANCED OPTICAL ABSORPTION IN STRONTIUM TITANATE AT THE CUBIC TO TETRAGONAL 06-1109
ION. ULTRASONIC PROPERTIES OF STRONTIUM TITANATE AT THE 105 DEGREE K TRANSIT 06-1191
ENERGY STORAGE IN CERAMIC DIELECTRICS. (STRONTIUM TITANATE, BARIUM TITANATE) 04-0602
MEASUREMENT OF OXYGEN DIFFUSION CONSTANT IN SOME TITANATES. (STRONTIUM TITANATE, BARIUM TITANATE) 04-1033
/YSTALLINE SOLID SOLUTIONS BASED ON BARIUM TITANATE. (BARIUM STRONTIUM TITANATE, BARIUM TITANATE STANNATE,/ 04-1019
/S (EXPERIMENT AND DISCUSSION OF BARIUM (OR LEAD OR CALCIUM) STRONTIUM TITANATE, BARIUM TITANATE ZIRCONATE/ 08-1281
FIRMATION OF THE THEORY OF TRAN/ TUNNELING RESULTS IN INDIUM STRONTIUM TITANATE BARRIERS INTERPRETED AS CON 03-0233
ECTRICS. STUDIES ON THE BARIUM STRONTIUM TITANATE BOUNDARY LAYER CERAMIC DIEL 04-1025
STUDIES ON STRONTIUM TITANATE BOUNDARY LAYER DIELECTRICS. 06-1248
EFFECTS OF SILICA ON MANGANESE DOPED STRONTIUM TITANATE CERAMICS. 06-1172
EFFECT OF THE 110 DEGREE K PHASE TRANSITION ON THE STRONTIUM TITANATE CONDUCTION BANDS. 06-1158
E. EPR INVESTIGATION OF POLYCRYSTALLINE STRONTIUM TITANATE CONTAINING MANGANESE DIOXID 06-1162

OPTICAL MODES OF THE PEROVSKITE STRUCTURE. (THEORY, STRONTIUM TITANATE DATA) 03-0304
L BEHAVIOR/ OPTICAL GAP OF STRONTIUM TITANATE. (DEVIATION FROM URBACH TAI 06-1111
ROGEN PHOSPHATE PRESENTED, TRIGLYCINE SULFATE AND BARIUM AND STRONTIUM TITANATE DISCUSSED) /POTASSIUM DIHYD 03-0180
PIEZORESISTIVITY OF DOPED BARIUM TITANATE. (OR BARIUM STRONTIUM TITANATE, DOPANT NOT NAMED) 04-0729
OR IRO/ ELECTRIC FIELD EFFECT OF ELECTRON SPIN RESONANCE IN STRONTIUM TITANATE. (DOPED WITH GADOLINIUM(3+) 06-1229
/IAL ELECTRON PARAMAGNETIC RESONANCE SPECTRUM OF IRON(3+) IN STRONTIUM TITANATE DUE TO NEAREST NEIGHBOR CH/ 06-1144
RS) ORDER PARAMETER AND PHASE TRANSITIONS OF STRESSED STRONTIUM TITANATE. (EPR OF IRON- VANADIUM PAI 06-1170
/ COMPENSATION BY OXYGEN(2-) VACANCIES IN CHROMIUM(3+) DOPED STRONTIUM TITANATE (ESR MEASUREMENTS, MODEL P/ 06-1161
C RESONANCE. PHOTOCHROMIC IRON IN STRONTIUM TITANATE. EVIDENCE FROM PARAMAGNETI 06-1171
FAR INFRARED REFLECTIVITY AND TRANSMISSION MEASUREMENTS ON STRONTIUM TITANATE FILMS. 06-1154
C TIGHT BINDING CALCULATION. ENERGY BAND STRUCTURE OF STRONTIUM TITANATE FROM A SELF CONSISTENT FIEL 06-1221
IDE CRYSTALS FROM NONSTOICHIOMETRIC MELTS. (BARIUM TITANATE, STRONTIUM TITANATE, GERMANATES) / GROWTH OF OX 04-0553
THREE APPLICATION AREAS FOR STRONTIUM TITANATE GLASS CERAMICS. 06-1150
EUTRON INELASTIC SCATTERING STUDY OF THE LATTICE DYNAMICS OF STRONTIUM TITANATE: HARMONIC MODELS. N 06-1225
PHASE TRANSFORMATION. EPR STUDIES AND DYNAMICS IN STRONTIUM TITANATE IN THE NEIGHBORHOOD OF THE 06-1232
F STRUCTURAL PHASE TRANSITIONS IN PEROVSKITE TYPE CRYSTALS. (STRONTIUM TITANATE, LANTHANUM ALUMINATE) /RY O 03-0424
NORMAL VIBRATIONS OF STRONTIUM TITANATE LATTICE. 06-1128
ANATE, LEAD/ SOLVENT ZONE GROWTH OF FERROELECTRIC CRYSTALS. (STRONTIUM TITANATE, LEAD(X) STRONTIUM(1-X) TIT 05-1042
/ITE MODIFICATION OF CADMIUM TITANATE. (SOLID SOLUTIONS WITH STRONTIUM TITANATE LITHIUM TANTALATE, LITHIUM/ 08-1424
STUDY OF THE DYNAMICS OF PHOTOCHROMIC SWITCHING IN STRONTIUM TITANATE MATERIALS. 06-1092
D ANHARMONIC EFFECTS IN CRYSTALS. (AND CALCULATIONS BASED ON STRONTIUM TITANATE MODEL) /FERROELECTRICITY AN 03-0213
/LOUS BEHAVIOR OF THE DIELECTRIC NONLINEARITY COEFFICIENT IN STRONTIUM TITANATE NEAR THE PHASE TRANSITION / 06-1105
ION. SPECIFIC HEAT OF STRONTIUM TITANATE NEAR THE STRUCTURAL TRANSIT 06-1132
ION. THERMAL EXPANSIVITY OF STRONTIUM TITANATE NEAR THE STRUCTURAL TRANSIT 06-1136
TURE. PRESSURE DEPENDENCE OF THE ELASTIC COMPLIANCE OF STRONTIUM TITANATE NEAR THE TRANSITION TEMPERA 06-1219
TION. CRITICAL BEHAVIOR OF STRONTIUM TITANATE NEAR THE 105 K PHASE TRANSI 06-1194
THE ELASTIC ANOMALIES AT THE STRUCTURAL PHASE TRANSITION OF STRONTIUM TITANATE NEAR 105 DEG K. ANALYSIS OF 06-1183
/ WITH TUNGSTEN BRONZE TYPE STRUCTURES (POTASSIUM (BARIUM OR STRONTIUM) TITANATE NIOBATE, (BARIUM OR STRON/ 13-1962
ATES BELOW CUBIC- TETRAGONAL TRANSITION, AND PRO/ MONODOMAIN STRONTIUM TITANATE (OCCURANCE IN THIN (110) PL 06-1164
. PHOTOCONDUCTIVITY (TRANSIENTS) IN PHOTOCHROMIC STRONTIUM TITANATE (ON COLORING AND BLEACHING) 06-1240
AL STRESS). DIRECT OBSERVATION OF SINGLE DOMAIN STRONTIUM TITANATE (OPTICAL STUDY UNDER UNIAXI 06-1112
FLUORIDE PHOTOCONDUCTIVIT/ PHOTOCHROMIC MATERIALS RESEARCH. (STRONTIUM TITANATE PHOTOCONDUCTIVITY, CALCIUM 06-1140
/TIES OF SOME FERROELECTRICS BELOW 80 DEG K. (SINGLE CRYSTAL STRONTIUM TITANATE, POLYCRYSTALLINE, STRONTIU/ 04-0721
OTASSIUM TAN/ ENERGY BANDS FOR POTASSIUM NICKEL TRIFLUORIDE, STRONTIUM TITANATE, POTASSIUM MOLYBDATE, AND P 02-0091
RIUM TITANATE NEAR THE CURIE POINTS. THERMAL CONDUCTIVITY IN STRONTIUM TITANATE, POTASSIUM TANTALATE AND BA 04-0989
/ OF THE "FERROELECTRIC" PHONON IN SOME PEROVSKITE CRYSTALS (STRONTIUM TITANATE, POTASSIUM TANTALATE) AT L/ 03-0227
/TRON INELASTIC SCATTERING AT STRUCTURAL PHASE TRANSITIONS. (STRONTIUM TITANATE, POTASSIUM TANTALATE, BARI/ 02-0059
/R OF THE SOFT FERROELECTRIC MODE IN THE CUBIC PEROVSKITES. (STRONTIUM TITANATE, POTASSIUM TANTALATE, BARI/ 02-0114
IS THERE A UNIFIED MECHANISM UNDERLYING URBACH'S RULE. (STRONTIUM TITANATE, POTASSIUM TITANATE) 06-1110
PE ON THE INFRARED ABSORPTION SPECTRA OF BARIUM TITANATE AND STRONTIUM TITANATE POWDERS. /ICLE SIZE AND SHA 04-0823
K STRUCTURAL PHASE TRANSITION. BIREFRINGENCE OF STRONTIUM TITANATE PRODUCED BY THE 105 DEGREE 06-1116
/ DISPLACEMENTS OF SEVERAL PEROVSKITES (POTASSIUM TANTALATE, STRONTIUM TITANATE, RUBIDIUM MANGANESE TRIFLU/ 06-1139
ATTEMPT TO PREPARE HIGH DOPED BARIUM STRONTIUM TITANATE SEMICONDUCTOR WITH PZT. 04-0717
DETERMINATION OF THE QUADRATIC ELECTROOPTIC COEFFICIENTS IN STRONTIUM TITANATE (SINGLE) CRYSTAL. /ROMETRIC 06-1130
LATTICE DISTORTIONS IN A STRONTIUM TITANATE SINGLE CRYSTAL ELECTRET. 06-1238
ELASTIC PROPERTIES OF YTTRIUM DOPED STRONTIUM TITANATE SINGLE CRYSTALS. 06-1180
ELECTROMECHANICAL PROPERTIES OF STRONTIUM TITANATE SINGLE CRYSTALS. 06-1206
ESR OF CHROMIUM(3+) IN CHROMIUM DOPED STRONTIUM TITANATE SINGLE CRYSTALS. 06-1147
MECHANICAL RELAXATION AND NONLINEARITY IN STRONTIUM TITANATE SINGLE CRYSTALS. 06-1218
PREPARATION AND PROPERTIES OF (BARIUM, STRONTIUM) TITANATE SINGLE CRYSTALS. 04-0560
THERMALLY STIMULATED CONDUCTIVITY IN STRONTIUM TITANATE SINGLE CRYSTALS. 06-1227
/Y VERSUS TEMPERATURE, ABSORPTION SPECTRA) OF SEMICONDUCTING STRONTIUM TITANATE SINGLE CRYSTALS (HYDROGEN / 06-1198
ERATURE RANGE 5 DEG K-300 DEG K. BRILLOUIN SCATTERING IN STRONTIUM TITANATE SINGLE CRYSTALS IN THE TEMP 06-1149
ERATURE/ TEMPERATURE DEPENDENCE OF THE ELASTIC COMPLIANCE OF STRONTIUM TITANATE SINGLE CRYSTALS IN THE TEMP 06-1220
TRANSITION AT 110 DEGREES K. DIELECTRIC NONLINEARITY OF STRONTIUM TITANATE SINGLE CRYSTALS NEAR PHASE 06-1106
ELASTIC PROPERTIES). ULTRASONIC THIRD HARMONIC GENERATION IN STRONTIUM TITANATE SINGLE CRYSTALS (NONLINEAR 06-1181
HZ, EFFE/ DIELECTRIC (CONSTANT AND LOSS) BEHAVIOR OF REDUCED STRONTIUM TITANATE SINGLE CRYSTALS (40 HZ-10 M 06-1148
CRYSTAL CHEMICAL PROBLEMS OF (BARIUM, STRONTIUM) TITANATE SOLID SOLUTIONS. 04-0633
HASE TRANSI/ DIELECTRIC PROPERTIES OF FERROELECTRIC (BARIUM, STRONTIUM) TITANATE SOLID SOLUTIONS NEAR THE P 04-0680
/RACTERISTICS OF CADMIUM SELENIDE LAYERS ON BARIUM TITANATE- STRONTIUM TITANATE SUBSTRATES AT SUPERSONIC D/ 04-0592
METAL CONTACTS) SOME ELECTRONIC PROPERTIES OF ZINC OXIDE AND STRONTIUM TITANATE. (SURFACE BARRIER SYSTEMS, 06-1176
NCY DIELECTRIC PROPE/ SINGLE CRYSTALS OF THE SYSTEM (BARIUM, STRONTIUM) TITANATE. (SYNTHESIS AND LOW FREQUE 04-0561
CONSTANT IN POLYCRYSTALLINE SOLID SOLUTIONS IN THE (BARIUM, STRONTIUM) TITANATE SYSTEM. /F THE CURIE-WEISS 04-0586
/ARITY AND LOSS IN THE BARIUM (OR LEAD, 4 COMPOSITIONS EACH) STRONTIUM TITANATE SYSTEMS. (FOR VOLTAGE DEPE/ 09-1522
DOMAIN STRUCTURE OF STRONTIUM TITANATE UNDER UNIAXIAL STRESSES. 06-1113
N REDUCTION OF TRANSITION METALS. ELECTROCOLORATION IN STRONTIUM TITANATE: VACANCY DRIFT AND OXIDATIO 06-1098
REINTERPRETATION OF WAVELENGTH MODULATED ABSORPTION IN STRONTIUM TITANATE WITHOUT COEXISTING PHASES. 06-1188
/ DIELECTRIC AND PIEZOELECTRIC PROPERTIES OF A CERAMIC (LEAD STRONTIUM TITANATE ZIRCONATE DOPED WITH CHROM/ 09-1591
LOW TEMPERATURE ELASTIC MODULI OF STRONTIUM TITANATE (10 TO 130 DEG K). 06-1190
UM(1-Y) TITANATE(3-X). SOFT MODE SUPERCONDUCTIVITY IN STRONTIUM TITANATE(3-X) AND CALCIUM(Y) STRONTI 06-1093
RANSPORT (CONDUCTIVITY AND HALL AND SEEBECK COEFFICIENTS) IN STRONTIUM TITANATE (4.2-300 DEG K) /ECTRONIC T 06-1129
CAL PHONON FREQUENCIES (RAMAN SCATTERING IN CUBIC PEROVSKITE STRONTIUM TITANATE, 8-250K, 0.2-12 KV/CM). /TI 06-1241
HASE. (POTASSIUM TANTALATE AND TANTALATE NIOBATE, BARIUM AND STRONTIUM TITANATES) /ES IN THE PARAELECTRIC P 04-0690
ULTRASOUND IN POLYCRYSTALLINE SOLID SOLUTIONS OF BARIUM AND STRONTIUM TITANATES. /DENCE OF THE VELOCITY OF 04-0894
POLARIZABILITIES AND COCHRAN MODE ASSIGNMENTS IN BARIUM AND STRONTIUM TITANATES. TEMPERATURE DEPENDENT 04-0805
ICAL PHONON MODE THEORY OF INFRARED DISPERSION IN BARIUM AND STRONTIUM TITANATES AND POTASSIUM TANTALATE. / 04-0547
FAR INFRARED DIELECTRIC DISPERSION IN BARIUM AND STRONTIUM TITANATES AND TITANIUM DIOXIDE. 04-0975
E EARTH IONS. DIELECTRIC RELAXATION IN (CERAMIC) STRONTIUM TITANATES CONTAINING (TRIVALENT) RAR 06-1143
/RATURE STUDIES OF INFRARED REFLECTION SPECTRA OF BARIUM AND STRONTIUM TITANATES IN THE 2-1000 MICRON REGI/ 04-0863
REVIEW OF BROOKHAVEN WORK ON POTASSIUM TANTALATE, BARIUM AND STRONTIUM TITANATES, KDP, 22 REFS) /T MODES. (02-0115
IC CERAMICS. DISPERSION IN STRONTIUM TITANATE- LANTHANUM FERRATE DIELECTR 06-1120
COEFFICIENTS OF RESISTIVITY. SEMICONDUCTING BARIUM AND STRONTIUM TITANATES WITH POSITIVE TEMPERATURE 04-0995
OM A POTASSIUM FLUORID/ DIELECTRIC PROPERTIES OF BARIUM(1-X) STRONTIUM(X) TITANATE SINGLE CRYSTALS GROWN FR 04-0635
POTASSIUM FLUORIDE MELT. DEFECT STRUCTURE OF (BARIUM(1-X) STRONTIUM(X)) TITANATE- MIXED CRYSTALS FROM A 04-0873
ATURES. ELECTRICAL CONDUCTIVITY OF STRONTIUM ZIRCONATE AND HAFNATE AT HIGH TEMPER 08-1417
HIGH PU/ PREPARATION AND CHARACTERIZATION OF ALKOXY-DERIVED STRONTIUM ZIRCONATE AND STRONTIUM TITANATE (OF 06-1217
REJUVENATION OF A BARIUM(0.72) STRONTIUM(0.28) TITANATE CERAMIC. 04-0817
E, LEAD(0.35) STRONTIUM(0.65) TITANATE, BARIUM(0.65 OR 0.75) STRONTIUM(0.35 OR 0.25) TITANATES) /TCHING TIM 37-3660
LECTRIC PHASE SHIFTERS FOR VHF AND UHF. (BASED ON LEAD(0.35) STRONTIUM(0.65) TITANATE) /URFACE WAVE) FERROE 37-3633
/ELECTRODES REDUCING POLARIZATION SWITCHING TIME, LEAD(0.35) STRONTIUM(0.65) TITANATE, BARIUM(0.65 OR 0.75/ 37-3660

YDR/ SHIFTING OF THE TRANSITION TEMPERATURE OF BARIUM(0-0.1) STRONTIUM(0.9-1) TITANATE SOLID SOLUTIONS BY H 04-0722
N FERROELECTRICS ABOVE THE CURIE TEMPERATURE. (BARIUM(0-1.0) STRONTIUM(1.0-0) TITANATE) /N OXYGEN OCTAHEDRO 04-0559
PHASE TRANSITIONS IN THE MIXED CRYSTAL SYSTEM STRONTIUM(1-X) BARIUM(X) TITANATE. 04-1016
ES. (STRONTIUM LITHIUM NIOBATE, STRONTIUM LITHIUM TANTALATE, STRONTIUM(1-X) LEAD(X) LITHIUM NIOBATE) /VSKIT 08-1292
ALSKI GROWTH, DIELECTRIC/ PREPARATION AND STUDY OF BARIUM(X) STRONTIUM(1-X) NIOBATE SINGLE CRYSTALS (CZOCHR 13-1936
RE" STUDY OF POTENTIAL DISTRIBUTION (STABILITY) IN BARIUM(X) STRONTIUM(1-X) TITANATE CERAMICS. /IBRATING WI 04-0699
D 100 DEGREES K. POLARIZATION OF SOLID BARIUM(X) STRONTIUM(1-X) TITANATE COMPOUNDS BETWEEN 4 AN 04-0719
/WTH OF FERROELECTRIC CRYSTALS. (STRONTIUM TITANATE, LEAD(X) STRONTIUM(1-X) TITANATE, LEAD(X) TITANIUM(1-X/ 05-1042
/ONS AT LOW TEMPERATURES: STRONTIUM TITANATE AND ((BARIUM(X) STRONTIUM(1-X) TITANATE (RECIPROCAL DIELECTRI/ 04-0898
PERATURES. (POLYCRYSTALLINE STRONTIUM TITANATE AND BARIUM(X) STRONTIUM(1-X) TITANATE SOLID SOLUTIONS) / TEM 04-0720
SUPERCONDUCTIVITY IN STRONTIUM TITANATE(3-X) AND CALCIUM(Y) STRONTIUM(1-Y) TITANATE(3-X). SOFT MODE 06-1093
ITY OF POTASSIUM NITRATE. EFFECT OF ADDING TITANIUM(+) AND STRONTIUM(2+) IONS ON THE ELECTRICAL CONDUCTIV 16-2319
A NEW FERROELECTRIC SINGLE CRYSTAL STRONTIUM(2) NIOBIUM(2) OXIDE(7). 13-1997
HE FLOATING ZONE TEC/ GROWTH OF FERROELECTRIC SINGLE CRYSTAL STRONTIUM(2) NIOBIUM(2) OXIDE(7) BY MEANS OF T 13-2037
RROELECTRIC CRYSTAL. ELASTIC AND PIEZOELECTRIC CONSTANTS OF STRONTIUM(4) POTASSIUM LITHIUM NIOBATE TYPE FE 13-1946
VARIATION OF THE DECAY CONSTANT OF STRONTIUM-85 IN BARIUM TITANATE. 04-0876
GSTEN TRIOXIDE DUE TO NEUTRON IRRADIATION. STRUCTURAL AND DIELECTRIC MODIFICATIONS OF TUN 10-1651
ZINCATE(0.33) NICBATE(0.67) (NCT FERROELECTRIC -160 TO +25/ STRUCTURAL AND DIELECTRIC PROPERTIES OF BARIUM 36-3609
PCTASSIUM TANTALATE NICBATE, STRUCTURAL AND DYNAMICAL STUDIES. 13-2068
LENT TUNGSTATES AND MOLYBDATES. STRUCTURAL AND PHASE RELATIONSHIPS AMONG TRIVA 10-1648
LENT TUNGSTATES AND MOLYBDATES. STRUCTURAL AND PHASE RELATIONSHIPS AMONG TRIVA 10-1649
ELASTICITY. (SYMMETRY RELATIONS) STRUCTURAL BASIS OF FERROELECTRICITY AND FERRO 03-0135
SMUTH TITANATE FAMILY. STRUCTURAL BASIS OF FERROELECTRICITY IN THE BI 08-1406
ATIONS IN ANTIFERROELECTRICS. CHARACTERISTICS OF THE STRUCTURAL CHANGES INVOLVED IN PHASE TRANSFORM 03-0526
GRAIN SIZE DEPENDENCE OF DIELECTRIC AND STRUCTURAL CHARACTERISTICS OF BARIUM TITANATE. 04-0582
THEORY OF THE MOBILITY OF STRUCTURAL DEFECTS IN ICE. 03-0244
OR IRRADIATION ON PHASE TRANSITIONS AND DIELECTRI/ EFFECT OF STRUCTURAL DEFECTS PRODUCED BY GAMMA AND REACT 17-2614
X POTASSIUM DINIOBATE AND (1-X) BARIUM TITANATE. STRUCTURAL DI- AND FERROELECTRIC PROPERTIES OF 04-0623
IONS IN LEAD ZIRCONATE(0.53) TITANATE(0.47) (EXPERIMENT AND STRUCTURAL DISCUSSION). /F BISMUTH AND NIOBIUM 09-1480
ADIATION IN FERROELECTRIC ROCHELLE SALT. (NEUTRON DIFFRACTI/ STRUCTURAL EFFECTS OF (DAMAGE FROM) IONIZING R 28-3188
F FERROELECTRICS AND FERROELECTRIC MAGNETS. (LEAD AND BARIU/ STRUCTURAL ELECTRON MICRODIFFRACTION STUDIES O 04-1038
/RROELECTRIC PHASE TRANSITION IN CRYSTALS WITH REORIENTATING STRUCTURAL GROUPS. (SODIUM AMMONIUM SELENATE / 22-2827
NG IN TRIGLYCINE SULFATE. PART-2: X-IRRADIATION TREATMENT. STRUCTURAL INHIBITION OF FERROELECTRIC SWITCHI 17-2469
NG IN TRIGLYCINE SULFATE. PART-1: ADDITIVES. STRUCTURAL INHIBITION OF FERROELECTRIC SWITCHI 27-3038
SOME STRUCTURAL MECHANISMS IN FERROELECTRICITY. 03-0411
FERROELECTRIC AND FERROMAGNETIC SUBSTANCES. (LEAD TITANATE) STRUCTURAL MICROELECTRON DIFFRACTION STUDY OF 05-1082
SATELLITE X-RAY SCATTERING AND STRUCTURAL MODULATION OF THIOUREA. 25-2921
(SINGLE CRYSTALS IN THE TUNGSTEN BRONZE TYPE) FERR/ (X-RAY) STRUCTURAL (OPTICAL) AND DIELECTRIC STUDIES OF 13-1947
ECTRIC CERAMI/ RELATION BETWEEN THE MAIN ELECTROPHYSICAL AND STRUCTURAL PARAMETERS OF FERROELECTRIC PIEZOEL 03-0217
ESR STUDY OF GAMMA-IRRADIATED POTASSIUM SELENATE (STRUCTURAL PHASE CHANGES). 22-2825
CTRICS) NEUTRON STUDIES OF DISPLACIVE STRUCTURAL PHASE TRANSFORMATIONS. (IN FERROELE 03-0156
IFFUSION IN FERROELECTRICS) NEUTRON STUDIES OF DISPLACIVE STRUCTURAL PHASE TRANSFORMATIONS. (INELASTIC D 03-0155
S ELASTIC BEHAVIOR OF POTASSIUM MANGANESE TRIFLUORIDE NEAR A STRUCTURAL PHASE TRANSITION. ANOMALOU 33-3392
FRINGENCE OF STRONTIUM TITANATE PRODUCED BY THE 105 DEGREE K STRUCTURAL PHASE TRANSITION. BIRE 06-1116
BDATE. (SYMMETRY CHANGES) ORIGIN OF THE STRUCTURAL PHASE TRANSITION IN GADOLINIUM MOLY 30-3273
NATE. NMR STUDY OF THE STRUCTURAL PHASE TRANSITION IN LANTHANUM ALUMI 36-3533
THEORY OF STRUCTURAL PHASE TRANSITION IN PEROVSKITES. 03-0242
E CRYSTALS, PART-2, INTERACTION WITH ELASTIC ST/ THEORY OF A STRUCTURAL PHASE TRANSITION IN PEROVSKITE- TYP 03-0241
ATE. ABSORPTION EDGE AND STRUCTURAL PHASE TRANSITION IN STRONTIUM TITAN 06-1107
ATE. NEW OBSERVATIONS FROM NEUTRON SCATTERING STUDIES OF THE STRUCTURAL PHASE TRANSITION IN STRONTIUM TITAN 06-1210
ATE. SOUND PROPAGATION NEAR THE STRUCTURAL PHASE TRANSITION IN STRONTIUM TITAN 06-1155
ATE NEAR 105 DEG K. ANALYSIS OF THE ELASTIC ANOMALIES AT THE STRUCTURAL PHASE TRANSITION OF STRONTIUM TITAN 06-1183
STRUCTURAL PHASE TRANSITIONS AND SOFT MODES. 03-0176
ICHLORIDE. EXPERIMENTAL STUDIES OF STRUCTURAL PHASE TRANSITIONS IN CESIUM LEAD TR 33-3378
ICHLORIDE. (MECHANISM OF) STRUCTURAL PHASE TRANSITIONS IN CESIUM LEAD TR 33-3380
STRUCTURAL PHASE TRANSITIONS IN LANGBEINITES. 36-3552
E CRYSTALS. (ST/ SELF CONSISTENT LATTICE DYNAMICAL THEORY OF STRUCTURAL PHASE TRANSITIONS IN PEROVSKITE TYP 03-0424
E CRYSTALS. (EXTENDED ABSTRACT) STRUCTURAL PHASE TRANSITIONS IN PEROVSKITE TYP 08-1450
ATE, POTASSIUM TANTALATE, B/ NEUTRON INELASTIC SCATTERING AT STRUCTURAL PHASE TRANSITIONS. (STRONTIUM TITAN 02-0059
TE. LATTICE DYNAMICS ABOVE STRUCTURAL PHASE TRANSITIONS: STRONTIUM TITANA 06-1124
IFFUSE PHASE TRANSITIONS. (MIXED LAYER OXIDE/ DIELECTRIC AND STRUCTURAL PROPERTIES OF FERROELECTRICS WITH D 13-2014
ASE TRANSITION AND DIELECTRIC PROPERTIES OF POTASSIUM DIHYD/ STRUCTURAL SENSITIVITY OF THE FERROELECTRIC PH 17-2615
ELECTRIC MAGNETS. (ELECTRON MICRODIFFRACTION, / (METHODS OF) STRUCTURAL STUDIES OF FERROELECTRICS AND FERRO 03-0490
CTRICS USING POLARIZED IR RADIATION. PART-2: INTERNAL FUNDA/ STRUCTURAL STUDIES OF HYDROGEN BONDED FERROELE 18-2738
N DIFFRACTION, PROTON AND DEUTERON NUCLEAR MAGNETIC RESONAN/ STRUCTURAL STUDIES OF ICE POLYMORPHS BY NEUTRO 34-3466
YDROGEN PHOSPHATE- POTASSIUM DIDEUTERIUM PHOSPHATE. STRUCTURAL STUDIES OF THE SYSTEM POTASSIUM DIH 17-2590
ALS. STRUCTURAL STUDY OF GADOLINIUM MOLYBDATE CRYST 30-3302
HORHOMBIC) FERROELECTRIC (LEAD DINIOBATE INTO THE NONFERROE/ STRUCTURAL STUDY OF THE TRANSITION OF THE (ORT 08-1285
N POTASSIUM MANGANESE TRIFLUORIDE IN THE NEIGHBORHOOD OF THE STRUCTURAL TRANSITION. / OF ULTRASONIC WAVES I 33-3370
ICAL PROPERTIES OF LITHIUM NIOBATE. (ANOMALY AT 263 DEG K IS STRUCTURAL TRANSITION) ELECTR 11-1709
SPECIFIC HEAT OF STRONTIUM TITANATE NEAR THE STRUCTURAL TRANSITION. 06-1132
THERMAL EXPANSIVITY OF STRONTIUM TITANATE NEAR THE STRUCTURAL TRANSITION. 06-1136
RIFLUORIDE. INFRARED ACTIVE MODE BELOW THE STRUCTURAL TRANSITION IN POTASSIUM MANGANESE T 33-3362
THERMAL EXPANSIVITY AND ULTRASONIC PROPAGATION NEAR THE STRUCTURAL TRANSITION OF STRONTIUM TITANATE. 06-1135
INTERACTION OF ELASTIC STRAIN WITH THE STRUCTURAL TRANSITION OF STRONTIUM TITANATE. 06-1216
TICAL TEMPERATURE (ROTATIONS OF BO6 OCTAHEDRA I/ DYNAMICS AT STRUCTURAL TRANSITIONS ABOVE FERROELECTRIC CRI 06-1205
IN CRYSTALS OF PEROVSKITE TYPE AND OF ONES WITH TETRAHEDRAL STRUCTURAL UNITS. /ATION AND PHASE TRANSITIONS 03-0175
GING PROCESS IN TRIGLYCINE SULFATE. (HYSTERESIS LAPS, DOMAIN STRUCTURE) A 27-3128
TRIGLYCINE SULFATE. (TANDEL OR CAPACITANCE PROPERTIES DOMAIN STRUCTURE) AGING PROCESSES IN 27-3081
ONDS IN ABO(3) FERROELECTRIC CRYSTALS WITH A PEROVSKITE TYPE STRUCTURE. CHARACTER OF THE CHEMICAL B 08-1341
CRYSTAL CHEMISTRY OF MIXED BISMUTH OXIDES WITH LAYER TYPE STRUCTURE. 13-2032
ISPERSION IN BARIUM TITANATE SINGLE CRYSTALS ON THEIR DOMAIN STRUCTURE. DEPENDENCE OF DIELECTRIC D 04-0895
IUM YTTRIUM NIOBATE SINGLE CRYSTAL WITH TUNGSTEN BRONZE TYPE STRUCTURE. /CTROOPTIC PROPERTIES OF BARIUM SOD 13-1971
RTIES OF BARIUM TITANATE SINGLE CRYSTALS WITH LAMINAR DOMAIN STRUCTURE. /C, ELASTIC AND PIEZOELECTRIC PROPE 04-1010
TIES OF NONLINEAR FERROELECTRIC CERAMICS WITH THE PEROVSKITE STRUCTURE. ELECTRICAL PROPER 08-1461
TUDY OF NEW SINGLE CRYSTAL CADMIUM COMPOUNDS WITH PEROVSKITE STRUCTURE. GROWTH AND S 08-1429
TH OF LEAD TITANATE CRYSTALS AND EXAMINATION OF THEIR DOMAIN STRUCTURE. GROW 05-1047
YCINE SULFATE CRYSTALS. (TOPOGRAPHY OF IMPERFECTIONS, DOMAIN STRUCTURE) /WTH REGIONS IN FERROELECTRIC TRIGL 27-2999
OF THE FERROELECTRIC LITHIUM TRIHYDROGEN SELENITE. (CRYSTAL STRUCTURE) /RT-54: A NEUTRON DIFFRACTION STUDY 22-2878
INTERATOMIC DISTANCES IN OXIDES WITH THE PEROVSKITE STRUCTURE. 08-1439

STOICHIOMETRY IN THE LEAD ZIRCONATE TITANATE SYSTEM. (DEFECT STRUCTURE)
VE AMPLIFICATION IN A CADMIUM SULFIDE FLUID- LITHIUM NIOBATE STRUCTURE.
HIGH PERMITTIVITY OF CERIUM ALUMINATES WITH PEROVSKITE TYPE STRUCTURE.
N RESULTS IN FERROELECTRICS WITH THE TUNGSTEN BRONZE CRYSTAL STRUCTURE.
OME LANTHANON MOLYBDATE TYPE RARE EARTH MOLYBDATES. (CRYSTAL STRUCTURE)
OELECTRIC AMMONIUM IODATE, A POTENTIAL FERROELASTIC: CRYSTAL STRUCTURE.
ON IN THE FERROELECTRIC PHASES OF OXIDES WITH THE PEROVSKITE STRUCTURE.
ALLY EQUIAXIAL FERRODIELECTRIC SINGLE CRYSTALS HAVING DOMAIN STRUCTURE.
IUM BARIUM RARE EARTH NIOBATES WITH THE TUNGSTEN BRONZE TYPE STRUCTURE.
CALCIUM) TITANATE SOLID SOLUTION CERAMIC WITH BOUNDARY LAYER STRUCTURE.
VE CONVOLUTION USING A CADMIUM SULFIDE FLUID LITHIUM NIOBATE STRUCTURE.
REPOLARIZATION PROCESS OF FERROELECTRICS WITH THE PEROVSKITE STRUCTURE.
THE SEARCH FOR NEW FERROELECTRICS WITH THE TUNGSTEN BRONZE STRUCTURE.
DOMAIN WALLS IN FERROELECTRIC MATERIALS WITH THE PEROVSKITE STRUCTURE.
FFICIENT OF RESISTIVITY) DEVICES) PART-2: ANALYSIS OF DEFECT STRUCTURE.
ACOUSTOELECTRIC EFFECT IN A LAYERED LITHIUM NIOBATE- SILICON STRUCTURE.
PREPARATION AND INVESTIGATION OF FERROELECTRICS OF ILMENITE STRUCTURE. (ABO(3) NIOBATES, TANTALATES, AND /
/. (AND MICROSCOPIC MODEL OF LOCALLY ORDERED "NUCLEI", PHASE STRUCTURE ABOVE CURIE POINT, RAMAN SPECTRUM O/
IN POTASSIUM NITRATE.   X-RAY STUDY ON THE DISORDERED STRUCTURE ABOVE THE FERROELECTRIC CURIE POINT
/L- CUBIC PEROVSKITE) STRONTIUM IRON TUNGSTATE . (MAGNETIC G STRUCTURE ANALOGOUS TO ANTIFERROELECTRIC STRU/
TION WITH THEIR PHASE TRANSITIONS. PART-1: TGS. PA/ ACCURATE STRUCTURE ANALYSIS OF FERROELECTRICS IN CONNEC
DIDE.   CRYSTAL STRUCTURE AND CONDUCTIVITY OF ANTIMONY SULFOIO
CONTAINING PEROVSKITES. (STRONTIUM LITHIUM NIOBATE, STRONT/ STRUCTURE AND DIELECTRIC LOSSES OF FERROELECTR
GLE CRYSTALS.   THE EFFECT OF GROWTH RATE ON THE DEFECT STRUCTURE AND DIELECTRIC PROPERTIES OF LITHIUM
TRIC SUBSTANCES. (BARIUM TITANATE)   CORRELATION BETWEEN STRUCTURE AND DIELECTRIC PROPERTIES OF TGS SIN
LANTHANUM DOPED LEAD TITANATE ZIRCONATE CERAMICS. STRUCTURE AND ELECTRIC PROPERTIES OF FERROELEC
LANTHANUM DOPED LEAD TITANATE ZIRCONATE CERAMICS. STRUCTURE AND ELECTROMECHANICAL PROPERTIES OF
ZIRCONATE.   DIPOLE STRUCTURE AND ELECTROMECHANICAL PROPERTIES OF
OPHYSICAL CLASSIFICATION OF FERROELECTRICS. (BASED ON DOMAIN STRUCTURE AND INTERNAL ELECTRIC FIELDS IN LEAD
LOCATIONS. (DOMAIN ORIGIN, SHAPES, MOV/ FERROELECTRIC DOMAIN STRUCTURE AND ITS APPLICATIONS)   CRYSTALL
PHASE OF THE FERROELECTRIC SODIUM TRIH/ THE PECULIAR DOMAIN STRUCTURE AND ITS INTERACTION WITH CRYSTAL DIS
DIDE TYPE SEMICONDUCTOR FERROELECTRICS.   DOMAIN STRUCTURE AND ITS UNUSUAL BEHAVIOR IN THE BETA
TRIC FERROMAGNETIC SOLID SOLUTIONS IN THE LEAD COBALT TUNGS/ STRUCTURE AND LOCAL STATES IN ANTIMONY SULFOIO
INCORPORATION OF ANTIMONY INTO THE BARIUM TITANATE LATTICE. STRUCTURE AND MAGNETIC PROPERTIES OF FERROELEC
ONATE LEAD TITANATE- LEAD METANIOBATE TERNARY SYSTE/ CRYSTAL (STRUCTURE, AND PCT RESISTORS)
ORIDE.   STRUCTURE AND PHASE RELATIONS IN THE LEAD ZIRC
CRYSTAL. (ABSTRACT ONLY)   POLARIZATION (DOMAIN STRUCTURE AND PHASE TRANSITION IN HYDROGEN CHL
LENATE.   THE STRUCTURE) AND PHASE TRANSITION IN KDP SINGLE
ROGEN DICHLORO ACETATE. PART-1: THE CRYSTAL STR/ THE CRYSTAL STRUCTURE AND PHASE TRANSITION IN POTASSIUM SE
TRATE.   CRYSTAL STRUCTURE AND PHASE TRANSITION OF AMMONIUM HYD
ORIDE.   CRYSTAL STRUCTURE AND PHASE TRANSITION OF DIGLYCINE NI
EAD (ZINCATE NIOBATE)- LEAD TITANATE.   CRYSTAL STRUCTURE AND PHASE TRANSITION OF HYDROGEN CHL
CT ON THE ELECTRICAL AND MECHANICAL PROPERTIES. (REVIEW, 56/ STRUCTURE AND PIEZOELECTRICITY OF THE SYSTEM L
F FERROELECTRIC BISMUTH TITANATE.   DOMAIN STRUCTURE AND POINT DEFECTS OF ICE: THEIR EFFE
LMS PRODUCED BY CATHODE SPUTTERING.   STRUCTURE AND POLARIZATION REVERSAL IN FILMS O
(2) OXIDE(7).   CRYSTAL STRUCTURE AND PROPERTIES OF BARIUM TITANATE FI
TITANATE (GERMANATE SOLID SOLU/ EFFECT OF GERMANIUM ON THE STRUCTURE AND PROPERTIES OF CALCIUM (2) NIOBIUM
E PIEZOCERAMICS.   EFFECTS OF ROASTING CONDITIONS ON THE STRUCTURE AND PROPERTIES OF FERROELECTRIC LEAD
ALLIUM CONTAINING PEROVSKITES. (TITANATES, HAFNATES/ CRYSTAL STRUCTURE AND PROPERTIES OF LEAD BARIUM NIOBAT
TRIC MATERIALS WITH PEROVSKITE STRUCTURE (PEROVSKITE COMPOU/ STRUCTURE AND PROPERTIES OF NEW CADMIUM AND TH
/ITE AND PYROCHLORE TYPE. SYNTHESIS AND STUDY OF THE CRYSTAL STRUCTURE AND PROPERTIES OF SOME NEW FERROELEC
SEP 1/ LITHIUM NIOBATE- A NEW TYPE OF FERROELECTRIC: GROWTH, STRUCTURE AND PROPERTIES. (REVIEW OF KARPOV I/
/ERROELECTRIC PROPERTIES OF BISMUTH FERRATE (PREPARATION AND STRUCTURE, AND PROPERTIES. (REVIEW OF WORK TO
AND DOLOMITE ADDITIONS ON BARIUM TITANATE CERAMICS. CRYSTAL STRUCTURE) AND RELATED MATERIALS (SYSTEM WITH/
/ROELECTRIC PHASE TRANSITIONS AND SPECIAL FEATURES OF DOMAIN STRUCTURE AND SINTERABILITY.   EFFECT OF CLAY
EZOELECTRIC CONSTANTS OF STIBOTANTALITE. (ANTIMONY (/ DOMAIN STRUCTURE AND SOME PHYSICAL PROPERTIES OF FER/
                    AMMONIUM HYDROGEN SULFATE, ITS STRUCTURE AND TEMPERATURE DEPENDENCE OF THE PI
/VESTIGATION OF THE EFFECT OF SAMPLE THICKNESS ON THE DOMAIN STRUCTURE AND THE FERROELECTRIC TRANSITION.
HOFERRITES, SODIUM NIOBATE)   STRUCTURE AND THE PYROELECTRIC PROPERTIES OF /
STRUCTURE TYPES. ELECTROSTATIC FIELD IN A PSEUDOCUBIC IONIC STRUCTURE AND TRANSITIONS IN PEROVSKITES. (ORT
IDE).   NEW DESCRIPTION OF THE PYROCHLORE STRUCTURE. APPLICATION TO ALL BARIUM TITANATE
PHASE TRANSITION IN A COMPOUND WITH HELICAL ELECTRIC DIPOLE STRUCTURE: CESIUM COPPER TRICHLORIDE.   A
            FERROELECTRICITY IN BISMUTH OXIDE TYPE LAYER STRUCTURE COMPOUNDS.
TYP/ AMPLIFICATION OF RAYLEIGH ULTRASONIC WAVES IN A LAYERED STRUCTURE CONSISTING OF LITHIUM NIOBATE AND P-
IUM TITANATE AND POTASSIUM NIOBATE.   STRUCTURE DISORDER AND RAMAN SCATTERING IN BAR
ROELECTRICS. BASED ON NEW TREATMENT AND INTERPRETATION OF X/ STRUCTURE DISORDER OF BARIUM TITANATE TYPE FER
SSIFICATI/ ANTIFERROELECTRICITY IN COMPOUNDS WITH PEROVSKITE STRUCTURE. (EXAMINATION OF PUBLISHED DATA, CL/
UM LITHIUM NIOBATES (TRANSPARENT, TETRAGONAL TUNGSTEN BRONZE STRUCTURE, FERROELECTRIC). /PERTIES OF POTASSI
R FORMATE TETRAHYDRATE BY NEUTRON DIFFRACTION.   APPROXIMATE STRUCTURE FOR ANTIFERROELECTRIC PHASE OF COPPE
IUM, STRONTIUM TITA/ DIELECTRIC CERAMICS WITH BOUNDARY LAYER STRUCTURE FOR HIGH FREQUENCY APPLICATION. (BAR
TUDIES OF IRRADIATION FIELD TREATED TRIGLYCINE SULFATE: NEW STRUCTURE FORM.   CRYSTALLOGRAPHIC S
LE SALT). (TEMPERATURE A/ EFFECT OF ELECTRIC FIELD ON DOMAIN STRUCTURE FORMATION IN SEIGNETTES SALT (ROCHEL
-HYDROGEN BONDED PHOSPHATE FERROELECTRIC (CRYSTAL GROWTH AND STRUCTURE HYSTERESIS LOOP RESULTS) . /F THE NON
ICRO CINEMATOGRAPHY)   PHASE TRANSITION AND DOMAIN STRUCTURE IN ANTIMONY SULFOIODIDE CRYSTALS. (M
(CALCULATIONS COMPARED WITH DA/ (STEP OR SMOOTH) DOMAIN WALL STRUCTURE IN BARIUM TITANATE DURING SWITCHING
TTERING.   DOMAIN WALL STRUCTURE IN GADOLINIUM MOLYBDATE BY RAMAN SCA
                                               DOMAIN STRUCTURE IN LITHIUM NIOBATE SINGLE CRYSTALS.
SOLID DEW TECHNIQUE.   OBSERVATION OF DOMAIN STRUCTURE IN LOW TEMPERATURE FERROELECTRICS BY
YSTALS BY POLARIZATION OPTICAL PROCED/ OBSERVATION OF DOMAIN STRUCTURE IN POTASSIUM DIHYDROGEN PHOSPHATE CR
SIN/ APPLICATION OF THE DEW METHOD FOR REVEALING THE DOMAIN STRUCTURE IN POTASSIUM FERROCYANIDE TRIHYDRATE
.   INTERFEROMETRIC STUDIES OF DOMAIN STRUCTURE IN POTASSIUM NIOBATE SINGLE CRYSTALS
SITION.   REARRANGEMENT OF THE DOMAIN STRUCTURE IN ROCHELLE SALT NEAR THE PHASE TRAN
AND COPPER(2+) DOPING)   EFFECT OF A DC FIELD ON DOMAIN STRUCTURE IN SEIGNETTE SALT. (EFFECT OF AGING
OPY.   ANALYSIS OF THE TEMPERATURE DEPENDENT PHONON STRUCTURE IN SODIUM NITRITE BY RAMAN SPECTROSC
CYANIDE.   DOMAIN STRUCTURE IN THE FERROELECTRIC POTASSIUM FERRO
                                        ATOMIC STRUCTURE IN THE PEROVSKITIC FERRATE.
DISLOCATIONS, SURFACE LAYER AND X-RAY IRRADIATION ON DOMAIN STRUCTURE IN TRIGLYCINE SULFATE. INFLUENCE OF
OBSERVED BY EPR.   CHANGES OF THE DOMAIN STRUCTURE IN VANADIUM DOPED TRIGLYCINE SULFATE
IN PEROVSKITES (STRONTIUM OR CALCIUM TITANATE/ EXTENDED FINE STRUCTURE IN X-RAY ABSORPTION SPECTRA OF CERTA

INTRINSIC AND EXTRINSIC NON 09-1536
LAYER WA 11-1667
NATURE OF THE 08-1355
POLARITO 13-1918
PRECISION PARAMETERS OF S 30-3262
PYR 36-3574
/ACEMENTS AND SPONTANEOUS DEFORMATI 08-1377
RESONANCE EFFECTS IN MAGNETIC 03-0509
SOD 13-1972
STUDIES ON THE (STRONTIUM, 06-1246
SURFACE WA 11-1700
/VERSE PIEZOELECTRIC EFFECT IN THE 08-1374
13-1897
THERMODYNAMIC THEORY OF 03-0190
/AP, PTCR (POSITIVE TEMPERATURE COE 04-0822
TRANSVERSE 11-1748
08-1438
04-0647
16-2366
10-1655
02-0077
15-2112
08-1452
08-1292
27-3067
04-0728
09-1502
09-1531
07-1261
03-0463
03-0390
22-2874
15-2127
36-3604
04-0934
09-1596
31-3337
17-2505
28-2872
36-3567
16-2312
31-3334
13-2067
02-0071
08-1470
04-0832
13-1911
05-1043
13-2046
08-1458
08-1460
02-0128
11-1775
08-1333
04-0739
03-0480
36-3601
19-2773
04-1039
08-1396
04-1008
36-3602
33-3388
13-1932
11-1747
04-0676
04-0630
01-0018
13-2026
26-2935
08-0833
27-2996
28-3190
36-3610
15-2184
04-0677
30-3318
11-1703
03-0341
17-2493
24-2904
08-1319
28-3181
28-3230
16-2299
24-2895
08-1398
27-3004
27-3013
03-0414

LEAD TITANATE WITH A-SITE AND B-SITE VACANCIES.   X-RAY STRUCTURE INVESTIGATION OF LANTHANUM MODIFIED 05-1059
ERTIES OF FLASH EVAPORATED BARIUM TITANATE THIN FILM/ GROWTH STRUCTURE (MICROSTRUCTURE) AND ELECTRICAL PROP 04-0605
/ AND PROPERTIES OF THIN FILM MATERIALS WITH PEROVSKITE TYPE STRUCTURE. (MIXED OXIDES ABO(3)) (BARIUM TITA/ 04-0914
IS AND CERTAIN PROPERTIES OF NEW COMPOUNDS WITH A PYROCHLORE STRUCTURE. (NIOBATES, TANTALATES)   SYNTHES 02-0047
M NIOBATE.   ACOUSTOELECTRIC AMPLIFICATION OF SURFACE WAVE STRUCTURE OF A CADMIUM SELENIDE FILM ON LITHIU 11-1723
OF A WALL MOTION IN BARIUM TITANATE)   STRUCTURE OF A MOVING DOMAIN WALL. (MECHANISM 04-0678
NG AGING.   INVESTIGATION OF THE DOMAIN STRUCTURE OF A TRIGLYCINE SULFATE CRYSTAL DURI 17-2561
TANATE, R/ INFLUENCE OF MADELUNG ENERGY AND COVALENCY ON THE STRUCTURE OF AB(5) OXIDE COMPOUNDS. (BARIUM TI 04-0761
  TEMPERATURE.   STRUCTURE OF AMMONIUM FLUOBERYLLATE AT AMBIENT 19-2776
HE FERROEL/ CONSTRAINED REFINEMENT TECHNIQUES APPLIED TO THE STRUCTURE OF AMMONIUM HYDROGEN SULFATE ABOVE T 16-2355
HE FERROELEC/ X-RAY DIFFRACTION DETERMINATION OF THE CRYSTAL STRUCTURE OF AMMONIUM HYDROGEN SULFATE ABOVE T 19-2775
                         ELECTRON DIFFRACTION STUDY OF THE STRUCTURE OF AMMONIUM SULFATE.   19-2797
  PHASE AND TRANSITION.   STRUCTURE OF AMMONIUM SULFATE IN FERROELECTRIC 19-2774
                  HYDROGEN BOND STUDIES. PART-58: CRYSTAL STRUCTURE OF AMMONIUM TRIHYDROGEN SELENITE.   22-2879
ATE BY NUCLEAR RESONANCE METHODS.   STUDY OF THE STRUCTURE OF ANTIFERROELECTRIC AMMONIUM PERIOD 36-3580
OM SELECTIVE CRYSTALLIZATION AND ETCHING).   DOMAIN STRUCTURE OF ANTIMONY SULFOIODIDE CRYSTALS (FR 15-2183
THOD)   EFFECTS OF ILLUMINATION ON THE DOMAIN STRUCTURE OF ANTIMONY SULFOIODIDE. (ETCHING ME 15-2124
Y INVESTIGATION OF THE EFFECT OF HYDROSTATIC PRESSURE ON THE STRUCTURE OF BARIUM TITANATE.   X-RA 04-0760
WN FROM POTASSIUM FLUORIDE).   INVESTIGATION OF THE FINE STRUCTURE OF BARIUM TITANATE CRYSTALS FLUX GRO 04-0708
OM SOLUTI/ EFFECT OF IMPURITIES AND GROWTH CONDITIONS ON THE STRUCTURE OF BARIUM TITANATE CRYSTALS GROWN FR 04-0598
                         GROWTH AND STRUCTURE OF BARIUM TITANATE EVAPORATED FILMS. 04-0604
INTERFERENCE FRINGES.   CRYSTAL STRUCTURE OF BARIUM TITANATE PRODUCING SQUARE 04-0706
/AL TENSILE STRESSES ON THE ELECTRICAL PROPERTIES AND DOMAIN STRUCTURE OF BARIUM TITANATE SINGLE CRYSTALS. 04-0961
TE- MIXED CRYSTALS FROM A POTASSIUM FLUORIDE MELT.   DEFECT STRUCTURE OF (BARIUM(1-X) STRONTIUM(X)) TITANA 04-0873
                                                CRYSTAL STRUCTURE OF BISMUTH TITANIUM NIOBIUM OXIDE. 13-2058
                                                CRYSTAL STRUCTURE OF BISMUTH TITANIUM OXIDE.   13-1935
                      DOUBLETS CONCERNING THE STRUCTURE OF BORACITES.   14-2088
ECHANICAL PROPERTIES. (BARIUM TITANATE,/ INFLUENCE OF DOMAIN STRUCTURE OF CERAMIC FERROELECTRICS ON THEIR M 04-0985
            CONDUCTIVITY, THERMOELECTRIC PROPERTIES, AND BAND STRUCTURE OF CRYSTALLINE BISMUTH SELENOIODIDE. 15-2126
ATE.   BAND STRUCTURE OF CUBIC AND TETRAGONAL BARIUM TITAN 04-0846
  CONCENTRATION DETERMINATION.   STRUCTURE OF DEUTERATED KDP CRYSTAL: DEUTERIUM 17-2535
ERROELECTRIC POTASSIUM SALT.   STRUCTURE OF DIHYDROGEN PHOSPHATE ION IN THE F 17-2624
HT IN POTASSIUM MANGANESE TRIFLUORIDE.   FINE STRUCTURE OF EXCITON- MAGNON ABSORPTION OF LIG 33-3364
TE, POSITION OF THE LITHIUM ION.   REFINEMENT OF THE CRYSTAL STRUCTURE OF FERROELECTRIC ACID LITHIUM SELENI 22-2859
  ALUMINUM SULFATE HEXAHYDRATE AND ITS CH/ REDETERMINATION OF STRUCTURE OF FERROELECTRIC CRYSTAL GUANIDINIUM 21-2819
ASSIUM DIHYDROGEN PHOSPHATE GROUP.   STRUCTURE OF FERROELECTRIC CRYSTALS OF THE POT 17-2607
E.   THE CRYSTAL STRUCTURE OF FERROELECTRIC GADOLINIUM MOLYBDAT 30-3288
ATE.   CRYSTAL STRUCTURE OF FERROELECTRIC GLYCINE SILVER NITR 16-2297
ATE.   CRYSTAL STRUCTURE OF FERROELECTRIC GLYCINE SILVER NITR 16-2333
) OXIDE(11).   CRYSTAL STRUCTURE OF FERROELECTRIC LEAD(5) GERMANIUM(4 35-3504
  SULFATE.   CRYSTAL STRUCTURE OF FERROELECTRIC LITHIUM HYDRAZINIUM 32-3356
LENITE: POSITION OF THE LITHIUM I/ REFINEMENT OF THE CRYSTAL STRUCTURE OF FERROELECTRIC LITHIUM HYDROGEN SE 22-2860
ULFATE. (ABSTRACT)   CRYSTAL STRUCTURE OF FERROELECTRIC RUBIDIUM HYDROGEN S 19-2746
FATE DIHYDRATE.   DOMAIN STRUCTURE OF FERROELECTRIC SODIUM AMMONIUM SUL 19-2778
DISELENITE AND ITS THERMAL EXPANSION. RELAXATION BETWEEN THE STRUCTURE OF FERROELECTRIC SODIUM TRIHYDROGEN 22-2851
M TANTALATE NIOBATE.   SOFT MODES AND STRUCTURE OF FERROELECTRIC TETRAGONAL POTASSIU 13-1957
         INFLUENCE OF COMPENSATING CHARGES ON THE C-DOMAIN STRUCTURE OF FERROELECTRICS.   03-0458
                                                    DOMAIN STRUCTURE OF GADOLINIUM MOLYBDATE.   30-3301
                          PYROELECTRIC EFFECT AND DOMAIN STRUCTURE OF GADOLINIUM MOLYBDATE.   30-3315
                                                    DOMAIN STRUCTURE OF GAMMA-IRRADIATED TGS CRYSTALS. 27-3164
                                                    CRYSTAL STRUCTURE OF GLYCINE SILVER NITRATE.   16-2348
E.   DIFFUSIONLESS PHASE TRANSITIONS AND THE STABLE DOMAIN STRUCTURE OF (HIGH SENSITIVITY) BARIUM TITANAT 04-0888
IOBIUM IN THE OCTAHEDRAL POSITIONS OF THE "CUBIC" PEROVSKITE STRUCTURE OF LEAD MAGNESIUM NIOBATE. /UM AND N 08-1379
AN X-RAY STUDY OF THE EFFECT OF HYDROSTATIC PRESSURE ON THE STRUCTURE OF LEAD TITANATE.   05-1064
RATE.   X-RAY REFINEMENT OF THE CRYSTAL STRUCTURE OF LITHIUM AMMONIUM TARTRATE MONOHYD 29-3244
                                                    CRYSTAL STRUCTURE OF LITHIUM HYDRAZINIUM SULFATE. 32-3342
                                 ABSOLUTE STRUCTURE OF LITHIUM IODATE CRYSTALS.   08-1290
                                                    DOMAIN STRUCTURE OF LITHIUM METANIOBATE CRYSTALS. 11-1708
PHOSPHATE.   THE CRYSTAL STRUCTURE OF MONOCLINIC POTASSIUM DIDEUTERIUM 17-2591
E AND (CALCIUM DOPED) POTASSIUM/ ELECTRON TUNNELING AND BAND STRUCTURE OF (NIOBIUM DOPED) STRONTIUM TITANAT 06-1222
UM OXIDE(20).   CRYSTAL STRUCTURE OF PIEZOELECTRIC BISMUTH(12) GERMANI 35-3493
OHYDRATE.   CRYSTAL STRUCTURE OF PIEZOELECTRIC LITHIUM FORMATE MON 26-2940
                          EQUILIBRIUM STRUCTURE OF POLARIZED ICE.   34-3411
TH DIFFUSE PHASE TRANSITIONS.   DOMAIN STRUCTURE OF POLYCRYSTALLINE FERROELECTRICS WI 03-0505
F STRONTIUM AND BISMUTH TITANATES.   STRUCTURE OF POLYCRYSTALLINE SOLID SOLUTIONS O 06-1145
YSTALS.   EFFECT OF IMPURITIES ON THE STRUCTURE OF POTASSIUM DIHYDROGEN PHOSPHATE CR 17-2606
PE CRYSTALS.   PRESSURE EFFECTS ON THE LATTICE PARAMETERS AND STRUCTURE OF POTASSIUM DIHYDROGEN PHOSPHATE TY 17-2588
              OPTICAL STUDIES OF DOMAIN STRUCTURE OF POTASSIUM NIOBATE SINGLE CRYSTALS 08-1320
                                                    CRYSTAL STRUCTURE OF POTASSIUM SELENATE.   22-2845
                      ELECTRICAL PROPERTIES AND DEFECT STRUCTURE OF POTASSIUM TANTALATE.   08-1314
SIUM BARIUM TANTAL/ PHASE TRANSITIONS IN TANTALATES WITH THE STRUCTURE OF POTASSIUM TUNGSTEN BRONZE. (POTAS 13-2050
CRYSTALS GROWN BELOW AND ABOVE THE CURIE POINT.   DOMAIN STRUCTURE OF PURE AND DOPED TRIGLYCINE SULFATE 27-3130
ORTHOPHOSPHATES.   (PREPARATION AND STRUCTURE OF) RUBIDIUM AND CESIUM SEMIMETALLIC 17-2598
                                                    CRYSTAL STRUCTURE OF RUBIDIUM TRIHYDROGEN SELENITE. 22-2881
/UDIES OF FLAME SPRAYED BARIUM TITANATE CAPACITORS. PART-10: STRUCTURE OF SEMICONDUCTIVE TITANIUM DIOXIDE / 04-0704
                CONVOLUTION OF SURFACE WAVES IN A STRUCTURE OF SEMICONDUCTOR ON LITHIUM NIOBATE. 11-1847
TITANATE BASE SOLI/ EFFECT OF EXTERNAL ACTIONS ON THE DOMAIN STRUCTURE OF SINGLE CRYSTALS FORMED BY BARIUM 04-0954
                                                           STRUCTURE OF SODIUM NIOBATE HETEROTYPES.   08-1395
AND THE CUBIC- TETRAGONAL TRANSITION IN RELATION TO SO/ THE STRUCTURE OF SODIUM NIOBATE T(2) AT 600 DEG C, 08-1337
25 DEGREES C.   STRUCTURE OF SODIUM NITRITE AT 150, 185, AND 2 16-2327
                                          DISORDERED STRUCTURE OF SODIUM NITRITE AT 185 DEG C.   16-2325
/LTRAVIOLET OPTICAL PROPERTIES (REFLECTANCE SPECTRA AND BAND STRUCTURE) OF SOME ABO(3) FERROELECTRICS. (PH/ 08-1383
NIOBATES.   COMPOSITION AND STRUCTURE OF SPUTTERED FILMS OF FERROELECTRIC 13-1941
(TUNGSTEN BRONZE) TYPE FERROELECTRICS.   SUPERLATTICE STRUCTURE OF STRONTIUM SODIUM LITHIUM NIOBATE 13-2002
X-RAY PHOTOELECTRON SPECTROSCOPY (ESCA).   BAND STRUCTURE OF STRONTIUM TITANATE AS MEASURED BY 06-1100
NSISTENT FIELD TIGHT BINDING CALCULATION.   ENERGY BAND STRUCTURE OF STRONTIUM TITANATE FROM A SELF CO 06-1221
STRESSES.   DOMAIN STRUCTURE OF STRONTIUM TITANATE UNDER UNIAXIAL 06-1113
DEVELOPMENT OF FERROELECTRIC CHARACTER IN AN APP/ MONOCLINIC STRUCTURE OF SYNTHETIC CALCIUM CHLORAPATITE. (36-3591
                            RAMAN SPECTRUM AND STRUCTURE OF TERBIUM MOLYBDATE.   30-3296
                                                           STRUCTURE OF TETRAGONAL TIN TUNGSTEN BRONZES. 13-2031

LIZATION OF ANTHRAQUINONE (SUBLIMED ON/ REVEALING THE DOMAIN STRUCTURE OF TGS CRYSTALS BY SELECTIVE CRYSTAL 27-3041
GEN SELENATE IN ELECTRIC FIELDS.   DOMAIN STRUCTURE OF THE BETA PHASE OF SODIUM TRIHYDRO 22-2830
OXIDE(94) AND THE GEOMETRY OF FERROELECTRIC DOMAINS.   STRUCTURE OF THE BRONZE SODIUM(13) NIOBIUM(35) 13-1931
E. (IMAGINARY DIELECTRIC CONSTANT)   BAND STRUCTURE OF THE CUBIC PHASE OF BARIUM TITANAT 04-0845
NITRATE.   STRUCTURE OF THE DISORDERED PHASE OF POTASSIUM 16-2379
ON THE PHASE TRANSITION OF AMMONIUM HYDROGEN SULFATE AND THE STRUCTURE OF THE FERROELECTRIC PHASE.   19-2777
NTIMONY / INFLUENCE OF NONEQUILIBRIUM CARRIERS ON THE DOMAIN STRUCTURE OF THE FERROELECTRIC SEMICONDUCTOR A 15-2104
NTIMONY SULFOIODIDE.   BAND STRUCTURE OF THE FERROELECTRIC SEMICONDUCTOR A 15-2179
NTIMONY TRISULFIDE.   BAND STRUCTURE OF THE FERROELECTRIC SEMICONDUCTOR A 36-3575
RO SELENITE.   ROOM TEMPERATURE CRYSTAL STRUCTURE OF THE FERROELECTRIC SODIUM TRIDEUTE 22-2861
/ARIZATION OF THE RAMAN BANDS OF THE HYDROXYL VIBRATIONS AND STRUCTURE OF THE HYDROGEN BONDS IN THE FERROE/ 17-2471
ITE SCATTERING)   CRYSTAL STRUCTURE OF THE IV PHASE OF THIOUREA. (SATELL 25-2923
RATES AS THE DETERMINING FACTOR OF THE GROWTH ST/ ELECTRICAL STRUCTURE OF THE SURFACE OF REAL CRYSTAL SUBST 27-3039
/ERROELASTIC PARAMAGNETIC BETA GADOLINIUM MOLYBDATE. CRYSTAL STRUCTURE OF THE TRANSITION METAL MOLYBDATES 30-3289
3) NIOBIUM(17) OXIDE(47): POTENTIALLY FERROELECTRIC CRYSTAL STRUCTURE OF THE TUNGSTEN BRONZE TYPE. /SMUTH(13-1978
               X-RAY TOPOGRAPHY OF THE DEFECT STRUCTURE OF THIOUREA.   25-2914
ROM SATELLITE SCATTERING DATA)   LONG PERIOD STRUCTURE OF THIOUREA CRYSTAL. (CALCULATIONS F 25-2911
ERRO- AND PARAELECTRIC PHASE.   CRYSTAL STRUCTURE OF TRIGLYCINE FLUOBERYLLATE IN THE F 27-3175
NG COOLING THROUGH CURIE POINT).   DOMAIN STRUCTURE OF TRIGLYCINE SELENATE CRYSTAL (DURI 27-3043
             REFINEMENT OF CRYSTAL STRUCTURE OF TRIGLYCINE SULFATE.   27-3024
WITH DOMAIN STRUCTURES)   DEFECT STRUCTURE OF TRIGLYCINE SULFATE. (CORRELATION 27-3044
MPERATURE AND APPLIED ELECTRIC FIE/ DEPENDENCE OF THE DOMAIN STRUCTURE OF TRIGLYCINE SULFATE CRYSTALS ON TE 27-3129
Y OF THE CURIE POINT.   EQUILIBRIUM DOMAIN STRUCTURE OF TRIGLYCINE SULFATE IN THE VICINIT 27-3112
                             THE CRYSTAL STRUCTURE OF TRIS-SARCOSINE CALCIUM CHLORIDE. 36-3524
ION TO THE UNDERSTANDING OF THE SO-CALL/ SOME ASPECTS OF THE STRUCTURE OF TUNGSTEN TRIOXIDE AND A CONTRIBUT 10-1653
HIC STUDY ON FERROELECTRIC SODIUM NITRITE CRYSTALS. PART-1: STRUCTURE OF 180 DEGREE DOMAIN WALLS. TOPOGRAP 16-2372
TITANATE.   STRUCTURE OF 90 DEGREE DOMAIN WALLS IN BARIUM 04-0641
N IN ROCHELLE SALT.   INFLUENCE OF THE DOMAIN STRUCTURE ON POSITRON (TWO-PHONON) ANNIHILATIO 28-3191
DOPED STRONTIUM TITANATE.   EFFECT OF DOMAIN STRUCTURE ON THE FLUORESCENCE OF CHROMIUM(3+) 06-1114
LEAD TITANATE ZIRCONATE COMPOSITIONS N/ INFLUENCE OF DOMAIN STRUCTURE ON THE PROPERTIES OF POLYCRYSTALLINE 09-1557
LYCINE SULFA/ INFLUENCE OF THE ELECTRIC FIELD AND THE DOMAIN STRUCTURE ON THE ULTRASONIC ABSORPTION IN TRIG 27-2955
/HASE TRANSITIONS IN CRYSTALS WITH POTASSIUM TUNGSTEN BRONZE STRUCTURE (POTASSIUM SODIUM LITHIUM NIOBATE, (41-1923
TTICE DYNAMI/ ACOUSTIC PROPERTIES OF MATERIALS OF PEROVSKITE STRUCTURE. (RELATION OF FERROELECTRICITY TO LA 02-0043
/ARAELASTIC PARAMAGNETIC BARIUM COBALT TETRAFLUORIDE CRYSTAL STRUCTURE (SPACE GROUP AND LATTICE CONSTANTS). 33-3386
LE SALT, TRIGLYCI/ PYROELECTRIC INVESTIGATIONS OF THE DOMAIN STRUCTURE STABILITY IN FERROELECTRICS. (ROCHEL 02-0083
                   PHASE BOUNDARY AND DEFECT STRUCTURE STUDIES IN PLZT CERAMICS.   09-1590
NUCLEAR QUADRUPOLE RESONANCE.   POTASSIUM IODATE CRYSTAL STRUCTURE STUDIED BY INFRARED SPECTROSCOPY AND 08-1278
            OPTICAL MODES OF THE PEROVSKITE STRUCTURE. (THEORY, STRONTIUM TITANATE DATA) 03-0304
N ZONE WIDTHS IN POLYCRYSTALLINE DIELECTRICS WITH PEROVSKITE STRUCTURE (TITANATES AND ZIRCONATES). FORBIDDE 08-1288
/OGRAPHIC SHEAR) FAMILIES DERIVED FROM THE RHENIUM TRIOXIDE STRUCTURE TYPE: AN ELECTRON MICROSCOPE STUDY / 10-1630
/NDEX AND ELECTROOPTICAL PHENOMENA IN CRYSTALS OF PEROVSKITE STRUCTURE (WITH APPLICATION TO BARIUM TITANAT/ 03-0518
ATES. (REVIEW, FORMATION, CRYSTAL GROWTH, PHASE TRANSITIONS, STRUCTURE, 49 REFS) /LECTRIC RARE EARTH MOLYBD 02-0064
OF PUBLISHED DATA, CLASSIFICATION, LISTING AND DISCUSSION OF STRUCTURES) /ROVSKITE STRUCTURE. (EXAMINATION 01-0018
NITRITE AND TGS CRYSTALS. (DOMAIN BOUNDARIES AND METASTABLE STRUCTURES) /IC METHOD TO FERROELECTRIC SODIUM 16-2375
CT STRUCTURE OF TRIGLYCINE SULFATE. (CORRELATION WITH DOMAIN STRUCTURES)   DEFE 27-3044
ND THE USE OF TILTING SCHEMES IN THE SOLUTIONS OF PEROVSKITE STRUCTURES. /E TRANSITIONS IN SODIUM NIOBATE A 08-1336
A CONTRIBUTION TO THE UNDERSTANDING OF THE SO-CALLED "SHEAR STRUCTURES". /RUCTURE OF TUNGSTEN TRIOXIDE AND 10-1653
EALED BY WATER ETCHED ROCHELLE SALT CRYSTALS.   SURFACE STRUCTURES AND CORRESPONDING LIGHT FIGURES REV 28-3234
EROVSKITES WITH COMPLEX COMPOSITIONS, AND THE STUDY OF THEIR STRUCTURES AND DIELECTRIC PROPERTIES. /RIOUS P 08-1428
             FERROELECTRICITY (IN BARIUM TITANATE TYPE STRUCTURES) AND LATTICE DYNAMICS.   04-0950
                FERROELECTRICALLY ACTIVE A-IONS IN ABO(3) STRUCTURES. (ANTIFERROELECTRICITY ORIGINS) 08-1364
, TANTALU/ NEW FERROELECTRICS WITH PEROVSKITE AND PYROCHLORE STRUCTURES. (BARIUM AND BISMUTH(0.5)) (NIOBIUM 08-1462
ACEMENTS IN FERROELECTRIC TRIGONAL AND ORTHORHOMBIC BORACITE STRUCTURES. (BORACITE, ERICAITE) ATOMIC DISPL 14-2075
ROELECTRIC AND ANTI/ POLAR- POLAR AND POLAR- NEUTRAL CRYSTAL STRUCTURES. (CLASSIFICATION OF ANISOTROPIC FER 03-0336
INFLUENCING) TYPE AND ORIENTATION OF THE EQUILIBRIUM DOMAIN STRUCTURES IN FERROELECTRICS. (FACTORS 03-0246
             CONTRIBUTION TO THE THEORY OF DOMAIN STRUCTURES IN MAGNETS AND FERROELECTRICS. 03-0430
          LATTICE DYNAMICS OF CUBIC PEROVSKITE STRUCTURES, IN PARTICULAR STRONTIUM TITANATE. 06-1185
S ASSOCIATED WITH STEP LADDERS ON PSEUDOCUBIC (001) / DOMAIN STRUCTURES IN POTASSIUM NIOBATE SINGLE CRYSTAL 08-1316
S.   INTERFEROMETRIC STUDIES OF DOMAIN STRUCTURES IN POTASSIUM NIOBATE SINGLE CRYSTAL 08-1318
IDE.   FORMATION OF SHEAR STRUCTURES IN SUBSTOICHIOMETRIC TUNGSTEN TRIOX 10-1659
M TITANATE AND LEAD ZIRCONATE TITANATE.   TWO WAVE SHOCK STRUCTURES IN THE FERROELECTRIC CERAMICS BARIU 04-0916
ON OF SELECTIVE ETCHING TO THE STUDY OF TWIN AND DISLOCATION STRUCTURES IN TRIGLYCINE SULFATE.   APPLICATI 27-3042
N A SCANNING ELECTRON M/ OBSERVATION OF FERROELECTRIC DOMAIN STRUCTURES, MODULATED BY A PHASE TRANSITION, I 03-0474
LOGRAPHIC VIEWPOINT)   MICROSCOPIC THEORY OF STATIC DOMAIN STRUCTURES OF FERROELECTRIC CRYSTALS. (CRYSTAL 03-0340
4 REFS)   NEUTRON DIFFRACTION INVESTIGATIONS OF THE ATOMIC STRUCTURES OF INORGANIC MATERIALS. (REVIEW, 24 02-0133
OF TRANSITION METALS OF THE 3D SE/ ELECTRICAL PROPERTIES AND STRUCTURES OF METANIOBATES AND METATANTALATES 36-3527
ND SODIUM TRIHYDROGEN SELENITE.   CRYSTAL STRUCTURES OF POTASSIUM TRIHYDROGEN SELENITE A 22-2838
75 DEG C, AND THEIR RELEVANCE TO SOFT PHONON MODES. (X-RAY / STRUCTURES OF SODIUM NIOBATE BETWEEN 480 AND 5 08-1273
OELECTRICS. (BARIUM TITANATE, KDP, P/ DIPOLE PATTERNS IN THE STRUCTURES OF SOME FERROELECTRICS AND ANTIFERR 02-0134
ZIRCONATE(X) TITANATE(1-X) FOR TWO RHOMBOHEDRAL PHA/ ATOMIC STRUCTURES OF THE FERROELECTRIC COMPOUNDS LEAD 09-1574
I AND IX ARE FULLY ORDERED (ANTIFERROELECTRIC STRUCTURES)) STRUCTURES OF THE FORMS OF ICES. (ICES II, VII 34-3451
E.   LEED STUDY OF SURFACE STRUCTURES ON THE (001) FACE OF BARIUM TITANAT 04-0529
        FERROELECTRIC TUNGSTEN BRONZE TYPE CRYSTAL STRUCTURES. PART-3: POTASSIUM LITHIUM NIOBATE. 13-1895
TANATE NIOBATE, (B/ COMPLEX OXIDES WITH TUNGSTEN BRONZE TYPE STRUCTURES (POTASSIUM (BARIUM OR STRONTIUM) TI 13-1962
/YSTALS FOR OPTICAL APPLICATIONS. (BY PULLING FROM THE MELT, STRUCTURE- PROPERTY CORRELATION WITH LITHIUM / 13-2045
     ORIGIN OF DIPOLE CONFIGURATIONS IN SOME (PEROVSKITE) STRUCTURES WITH SPECIAL DIELECTRIC PROPERTIES. 08-1390
                                                             SUB... SEE THE PREFIXED WORD.
L ANTIFERROMAGNET/ ZERO POINT SPIN DEVIATION AND SPONTANEOUS SUBLATTICE MAGNETIZATION IN THE TWO DIMENSIONA 26-2929
        THE ORDER DISORDER TRANSITION OF SODIUM NITRITE. (TWO SUBLATTICE MODEL OF INTERACTING DIPOLES) 16-2287
PER FORMATE TETRADEUTERATE.   PROTON RESONANCE STUDY OF SUBLATTICE ROTATIONS AND SPIN DEVIATION IN COP 26-2950
     PREPARATION OF TUNGSTIC OXIDE SINGLE CRYSTALS (SUBLIMATION IN AIR OF TUNGSTEN TRIOXIDE). 10-1644
TGS CRYSTALS BY SELECTIVE CRYSTALLIZATION OF ANTHRAQUINONE (SUBLIMED ONTO TGS). /G THE DOMAIN STRUCTURE OF 27-3041
LS OF GADOLINIUM MOLYBDATE, A FERROELECTRIC AND FERROELASTIC SUBSTANCE. /TH AND MORPHOLOGY OF SINGLE CRYSTA 30-3259
/POLE PATTERNS IN ORTHORHOMBIC AND TRIGONAL PHASES OF ABO(3) SUBSTANCE. (WHEN THE B ION IS FERROELECTRICAL/ 08-1363
ACOUSTOELECTRIC INTERACTION IN FERROELECTRIC SUBSTANCES.   03-0353
ES OF SOLID SOLUTIONS OF TRIGLYCINE SULFATE WITH ISOMORPHOUS SUBSTANCES. /CE OF SOME FERROELECTRIC PROPERTI 27-2971
IN TABELLEN, NEUE SERIE. VOL 3: FERRO- AND ANTIFERROELECTRIC SUBSTANCES.   LANDOLT-BORNSTE 01-0002
N BETWEEN STRUCTURE AND ELECTRIC PROPERTIES OF FERROELECTRIC SUBSTANCES. (BARIUM TITANATE)   CORRELATIO 04-0728
ATURE DEPENDENCE OF THE DIELECTRIC CONSTANT OF FERROELECTRIC SUBSTANCES. (BARIUM TITANATE CERAMICS) TEMPER 04-0591
/NT OF INTERNAL FRICTION IN SINGLE CRYSTALS OF FERROELECTRIC SUBSTANCES BY THE COMPOSITE RESONATOR METHOD./ 27-3080

LECTRON DIFFRACTION STUDY OF FERROELECTRIC AND FERROMAGNETIC SUBSTANCES. (LEAD TITANATE)   STRUCTURAL MICROE 05-1082
/ PHENOMENA AND MOSSBAUER SPECTRA OF MAGNETIZED PARAMAGNETIC SUBSTANCES. PART-1: IRON(X) ALUMINUM(1-X) AMM/ 19-2751
                                            PIEZOELECTRIC CERAMIC SUBSTANCES. (REVIEW, 23 REFS)   02-0105
                      CAUSES OF THERMAL EXPANSION OF GADOLINIUM SUBSTITUTED BARIUM SODIUM NIOBATES.   13-2048
                       LINEAR THERMAL EXPANSIONS OF GADOLINIUM SUBSTITUTED BARIUM SODIUM NIOBATES.   13-1996
/POLARIZATION REVERSAL IN POLYCRYSTALLINE FERROELECTRICS. (8 SUBSTITUTED LEAD TITANATE ZIRCONATES, 2 SUBST/   08-1418
PPER (NIOBIUM OR TANTALUM)(2) OXIDE(9), AND VARIOUS TUNGSTEN SUBSTITUTED LEAD TUNGSTATES) /STRONTIUM) (3) CO 36-3619
                  DIELECTRIC STUDIES OF THE CATION SUBSTITUTED MIXED CRYSTALS OF NIOBATES.   11-1848
         PHASE TRANSITION IN DIVALENT METAL CICALCIUM PROPIONATES SUBSTITUTED PARTIALLY BY ACETATE IONS.   26-2944
C CONSTANTS OF THE ALKALINE EARTH FLUORIDES BY THE METHOD OF SUBSTITUTION.   LOW FREQUENCY DIELECTRI   33-3361
RIUM TITANATE AND HEMATITE SINGLE CRYSTALS.   HETEROVALENT SUBSTITUTION AND OXYGEN NONSTOICHIOMETRY OF BA   04-0874
TRIGLYCINE SULFA/ UNUSUAL EFFECT OF ISOTOPIC AND ISOMORPHOUS SUBSTITUTION IN FERROELECTRIC CRYSTALS OF THE   27-3102
NITRITE.   ACCURATE FORCE CONSTANTS FROM ISOTOPIC SUBSTITUTION IN THE NITRITE RADICAL IN SODIUM   16-2309
/) COMPOUNDS OF ANTIMONY SULFOICDIDE TYPE. (PARTIAL CR TOTAL SUBSTITUTION OF ANTIMONY BY BISMUTH OR ARSENI/   15-2205
AD ZIRCONATE(0.53) TITANATE(0.47) (EXPERIMENT AND STRUCTURA/ SUBSTITUTION OF BISMUTH AND NIOBIUM IONS IN LE   09-1480
/ITANATE PIEZOELECTRIC CERAMICS UNDER HIGH ELECTRIC FIELDS. (SUBSTITUTION OF STRONTIUM OR CALCIUM FOR LEAD/   09-1484
YSTALLOGRAPHIC AND DIELECTRIC PROPERTIES OF S/ EFFECT OF THE SUBSTITUTION OF TANTALUM FOR NIOBIUM ON THE CR   13-1921
ROOPTICAL PROPERTIES OF BARIUM TITAN/ EFFECT OF NONISCVALENT SUBSTITUTION OF TITANIUM(4+) IONS ON THE ELECT   04-0959
IHYDROGEN PHOSPHATE.   EFFECT OF DEUTERON SUBSTITUTION ON DOMAIN DYNAMICS IN POTASSIUM D   17-2437
E ORTHORHOMBIC PHASE IN MODIFIED BARIUM TITANATE. (EFFECT OF SUBSTITUTIONS)   TH   04-0617
CONDUCTING SOLID SOLUTIONS OF ISO- AND HETEROVALENT CATIONIC SUBSTITUTIONS IN BARIUM TITANATE. /ECT IN SEMI 04-0943
/ PRESSURE ON THE ELECTRON PARAMAGNETIC RESONANCE SPECTRA OF SUBSTITUTIONAL IMPURITIES IN CUBIC STRONTIUM / 06-1193
FLASH EVAPORATED BARIUM TITANATE THIN FILMS (SEPARATED FROM SUBSTRATE). /URE) AND ELECTRICAL PROPERTIES OF   04-0605
AND SELENATE CRYS/ FIELD EFFECT IN TELLURIUM THIN FILMS ON A SUBSTRATE OF FERROELECTRIC TRIGLYCINE SULFATE   27-3046
IC SURFACE WAVES BY THIN GOLD FILM LAYERS ON LITHIUM NIOBATE SUBSTRATES. /ACTION AND REFLECTION OF ULTRASON 11-1789
IN FILM BARIUM TITANATE PREPARED BY RF SPUTTERING ON SILICON SUBSTRATES. /L AND ELECTRICAL PROPERTIES OF TH 04-0828
IC SURFACE WAVES BY THIN GOLD FILM LAYERS ON LITHIUM NIOBATE SUBSTRATES. /ACTION AND REFLECTION OF ULTRASON 11-1790
FERROELECTRIC FIELD EFFECT IN TELLURIUM FILMS ON TRIGLYCINE SUBSTRATES.   TEMPERATURE DEPENDENCE OF THE   27-3018
OWTH ST/ ELECTRICAL STRUCTURE OF THE SURFACE OF REAL CRYSTAL SUBSTRATES AS THE DETERMINING FACTOR OF THE GR 27-3039
/MIUM SELENIDE LAYERS ON BARIUM TITANATE- STRONTIUM TITANATE SUBSTRATES AT SUPERSONIC DRIFT VELOCITY OF CU/ 04-0592
R ORDER MODE SURFACE WAVE PROPAGATION ON LAYERED ANISOTROPIC SUBSTRATES: GOLD ON LITHIUM NIOBATE. /ND HIGHE 11-1846
/MAGNETIC COUPLING TO ELASTIC SURFACE WAVES ON PIEZOELECTRIC SUBSTRATES (LITHIUM NIOBATE AND BISMUTH(12) G/ 11-1751
ND SODIUM NITRITE SUR/ THIN FILM NUCLEATION ON FERROELECTRIC SUBSTRATES (TELLURIUM ON TRIGLYCINE SULFATE A   16-2394
F POLARIZATION IN THIN FERROELECTRIC FILMS ON SEMICONDUCTING SUBSTRATES. (TGS ON SILICON) /AMIC STABILITY O 27-3163
            DOMAIN WALLS CAUGHT IN (FINE LINE) "SUDARES" IN ROCHELLE SALT CRYSTAL.   28-3215
DUCER FOR USE IN HEAT TRANSFER AND FLOW STUDIES. (TRIGLYCINE SULFATE)   A NEW FERROELECTRIC TRANS   27-3031
ON FROM FERROELECTRIC CRYSTALS. (BARIUM TITANATE, TRIGLYCINE SULFATE   ACOUSTIC EMISSI   04-0593
ELECTRODES EVAPORATED ON INSULATORS ESPECIALLY ON TRIGLYCINE SULFATE.   AFTER EFFECTS IN SEMICONDUCTOR   27-3170
CHARGES APPLIED TO TELLURIUM FILMS ON UHV CLEAVED TRIGLYCINE SULFATE /THE FIELD EFFECT AT VERY LARGE GATE   27-3171
LE CRYSTALS CF FERROELECTRICS. (ROCHELLE SALT AND TRIGLYCINE SULFATE) /ENDENCE OF INTERNAL FRICTION IN SING 27-3122
        ANALYSIS OF HYSTERESIS LOOPS IN LITHIUM HYDRAZINIUM SULFATE.   32-3351
AND THE QUESTION OF FERROELECTRICITY IN LITHIUM HYDRAZINIUM SULFATE.   ANOMALOUS NEUTRON SCATTERING   32-3341
THE PIPPARD RELATIONS TO THE PHASE TRANSITION IN TRIGLYCINE SULFATE.   APPLICABILITY OF   27-3157
O THE STUDY OF TWIN AND DISLOCATION STRUCTURES IN TRIGLYCINE SULFATE. APPLICATION OF SELECTIVE ETCHING T   27-3042
          BLOCKING OF SPONTANEOUS POLARIZATION IN TRIGLYCINE SULFATE.   27-3008
          BRILLOUIN EFFECT AT ORDINARY TEMPERATURE IN AMMONIUM SULFATE.   19-2766
AR FERROELECTRIC- FERROELASTIC CRYSTAL: DIAMMONIUM DICADMIUM SULFATE. /THE PHASE TRANSFORMATION OF A PECULI 19-2741
RES DUE TO THE FERROELECTRIC DOMAIN BOUNDARIES OF TRIGLYCINE SULFATE.   CORROSION FIGU   27-3142
                    CRITICAL FLUCTUATIONS IN TRIGLYCINE SULFATE.   27-2974
                    CRITICAL REGION IN FERROELECTRIC TRIGLYCINE SULFATE.   27-2985
ITICAL SLOWING DOWN AT FERROELECTRIC TRANSITIONS (TRIGLYCINE SULFATE).   CR   27-3155
        CRYSTAL STRUCTURE OF FERROELECTRIC LITHIUM HYDRAZINIUM SULFATE.   32-3356
        CRYSTAL STRUCTURE OF LITHIUM HYDRAZINIUM SULFATE.   32-3342
          DAMAGE PRODUCED BY X-RAYS ON TRIGLYCINE SULFATE.   27-3073
          DEUTERON NMR STUDIES OF LITHIUM HYDRAZINIUM SULFATE.   32-3354
DEUTERON QUADRUPOLE COUPLING CONSTANT IN DEUTERATED AMMONIUM SULFATE.   19-2761
ICE RELAXATION IN FERROELECTRIC DEUTERATED AMMONIUM HYDROGEN SULFATE.   DEUTERON SPIN LATT   19-2795
          DIELECTRIC BEHAVIOR OF GAMMA-RAY IRRADIATED TRIGLYCINE SULFATE.   27-3149
                    DIELECTRIC CONSTANT OF TRIGLYCINE SULFATE.   27-3049
                    DIELECTRIC DISPERSION OF TRIGLYCINE SULFATE   17-2574
                    DIELECTRIC DISPERSION OF TRIGLYCINE SULFATE   27-3060
          DIELECTRIC PROPERTIES OF LITHIUM HYDRAZINIUM SULFATE.   32-3344
          DIELECTRIC PROPERTIES OF LITHIUM HYDRAZINIUM SULFATE.   32-3352
          DIELECTRIC PROPERTIES OF LITHIUM HYDRAZINIUM SULFATE.   32-3353
DIELECTRIC RELAXATION IN MULTIDOMAIN CRYSTALS OF TRIGLYCINE SULFATE.   27-3002
          DIELECTRIC RELAXATION CF MULTIDOMAIN TRIGLYCINE SULFATE.   27-3050
ION OF LITHIUM-7 BY HINDERED ROTATORS IN LITHIUM HYDRAZINIUM SULFATE.   DIPOLAR RELAXAT   32-3350
(ELASTIC AND PIEZOELECTRIC) MATERIAL CONSTANTS OF TRIGLYCINE SULFATE.   DYNAMIC DETEPMINATION OF   27-2984
                    DYNAMICS OF AMMONIUM IONS IN AMMONIUM SULFATE.   19-2794
AND CHROMIUM IONS ON THE DIELECTRIC PROPERTIES OF TRIGLYCINE SULFATE. /THE ADDITION OF A MIXTURE OF COPPER   27-3045
ALLIC ELECTRODES ON THE TRANSITION TEMPERATURE OF TRIGLYCINE SULFATE.   EFFECT OF EVAPORATED MET   27-3100
C PRESSURE ON THE FERROELECTRIC PHASE TRANSITION OF AMMONIUM SULFATE.   EFFECT OF HYDROSTATI   16-2360
YDROSTATIC PRESSURE ON THE UNIPOLAR PROPERTIES OF TRIGLYCINE SULFATE.   EFFECT OF H   27-3053
    EFFECT OF PRESSURE ON THE SWITCHING RATE IN TRIGLYCINE SULFATE.   27-3014
ON THE FERROELECTRIC- PARAELECTRIC TRANSITION IN TRIGLYCINE SULFATE.   EFFECT OF SEVERAL PARAMETERS   27-2981
EAR THE TRANSITION POINT (THEORY, MEASUREMENTS ON TRIGLYCINE SULFATE . /WAVE DISPERSION OF FERROELECTRICS N 03-0370
CROSTATIC PRESSURE ON THE TRANSITION TEMPERATURE IN AMMONIUM SULFATE.   EFFECT OF THE HY   19-2796
- AND GAMMA-RADIATION ON THE SWITCHING PROCESS IN TRIGLYCINE SULFATE.   EFFECT OF X   27-3019
TS OF ADDITIVES ON THE PYROELECTRIC PROPERTIES OF TRIGLYCINE SULFATE   EFFEC   27-3037
OF DOMAIN SHAPES ON SECOND HARMONIC SCATTERING IN TRIGLYCINE SULFATE.   EFFECTS   27-2989
TERMINING FACTOR OF THE GROWTH STAGE IN EPITAXY. (TRIGLYCINE SULFATE) /OF REAL CRYSTAL SUBSTRATES AS THE DE 27-3039
CTROGYRATION AND FERROELECTRIC PHASE TRANSITIONS (TRIGLYCINE SULFATE .   ELE   27-2970
UPLING NEAR THE FERROELECTRIC PHASE TRANSITION OF TRIGLYCINE SULFATE.   ELECTROMECHANICAL CO   27-3174
        ELECTRON DIFFRACTION STUDY OF THE STRUCTURE OF AMMONIUM SULFATE.   19-2797
RAMAGNETIC RESONANCE OF MANGANESE(2+) DOPED AMMONIUM CADMIUM SULFATE.   ELECTRON PA   19-2791
ESONANCE OF GAMMA DEFECTS IN FERROELECTRIC AMMONIUM HYDROGEN SULFATE.   ELECTRON SPIN R   19-2748
              ESR IN GAMMA-IRRADIATED TRIGLYCINE SULFATE   27-3035
              ESR OF GAMMA-IRRADIATED AMMONIUM SULFATE.   19-2790
RAMICROSCOPIC STUDIES OF FERROELECTRIC DOMAINS IN TRIGLYCINE SULFATE.   ETCHING AND ULT   27-3143
          FERROELECTRIC SPECIFIC HEAT OF TRIGLYCINE SULFATE.   27-3146

FERROELECTRIC TRANSITION OF AMMONIUM SULFATE. 19-2798
FERROELECTRIC-LIKE BEHAVIOR OF LITHIUM HYDRAZINIUM SULFATE. 32-3348
TY OF TELLURIUM AND LEAD TELLURIDE ON UHV CLEAVED TRIGLYCINE SULFATE. FIELD EFFECT AND HALL MOBILI 27-3172
IGH INTERNAL BIAS FIELDS IN L-ALANINE SUBSTITUTED TRIGLYCINE SULFATE H 27-2977
HYDROGEN BONDING IN LITHIUM HYDRAZINIUM SULFATE. 32-3346
FIELD ON THE NOISE AND THE PYROELECTRIC EFFECT IN TRIGLYCINE SULFATE INFLUENCE OF AN ELECTRIC 27-3076
AYER AND X-RAY IRRADIATION ON DOMAIN STRUCTURE IN TRIGLYCINE SULFATE. INFLUENCE OF DISLOCATIONS, SURFACE L 27-3004
DOMAIN STRUCTURE ON THE ULTRASONIC ABSORPTION IN TRIGLYCINE SULFATE. /LUENCE OF THE ELECTRIC FIELD AND THE 27-2955
OELECTRICITY IN SODIUM AMMONIUM SELENATE AND SODIUM AMMONIUM SULFATE. INVESTIGATION OF FERR 27-2742
INVESTIGATION OF THE PHASE TRANSITION IN TRIGLYCINE SULFATE. 27-3054
ANGANESE(2+) ABSORPTION NEAR THE CRITICAL REGION IN AMMONIUM SULFATE. M 19-2785
IC SUBSTANCES BY THE COMPOSITE RESONATOR METHOD. (TRIGLYCINE /ON IN SINGLE CRYSTALS OF FERROELECTR 27-3080
ONSTANTS AND DIELECTRIC LOSSES OF SODIUM NITRITE AND GLYCINE SULFATE. /FERROELECTRICS. PART-2: DIELECTRIC C 16-2354
TRIC MEASUREMENTS OF GAMMA-IRRADIATED FERROELECTRIC AMMONIUM SULFATE. MICROWAVE DIELEC 19-2750
MICROWAVE DIELECTRIC STUDY OF LITHIUM HYDRAZINIUM SULFATE. 32-3357
ALLS AND THE NUCLEATION OF DOMAINS IN CRYSTALS OF TRIGLYCINE SULFATE. MOVEMENT OF DOMAIN W 27-3071
INDUCED DEFECTS IN CRYSTALS OF ROCHELLE SALT AND TRIGLYCINE SULFATE. NATURE OF RADIATION 27-3159
NEUTRON DIFFRACTION STUDY OF TRIGLYCINE SULFATE. 27-3099
NEW FERROELECTRIC LANGBEINITE THALLIUM CADMIUM SULFATE. 36-3534
THERMODYNAMICAL COEFFICIENTS OF ROCHELLE SALT AND TRIGLYCINE SULFATE. NEW MEASUREMENTS OF DIVERSE 27-2995
14 NUCLEAR QUADRUPOLE RESONANCE STUDY OF LITHIUM HYDRAZINIUM SULFATE. NITROGEN- 32-3345
NITROGEN-14 QUADRUPOLE COUPLING IN PARAELECTRIC AMMONIUM SULFATE. 19-2749
Y OF) FERROELECTRIC TRANSITION IN (MANGANESE DOPED) AMMONIUM SULFATE. (NMR STUD 19-2784
ARITY OF THE DIELECTRIC CONSTANT OF FERROELECTRIC AMMONIUM SULFATE. NONLINE 27-3070
NUCLEAR MAGNETIC RESONANCE INVESTIGATION OF TRIGLYCINE SULFATE. 27-3059
AGNETIC RESONANCE STUDY OF FERROELECTRIC LITHIUM HYDRAZINIUM SULFATE. NUCLEAR M 32-3343
ATE, LITHIUM TANTALATE, LITHIUM NIOBATE, LITHIUM HYDRAZINIUM SULFATE) /IUM TARTRATE, LITHIUM THALLIUM TARTR 02-0052
NUCLEAR SPIN LATTICE RELAXATION IN FERROELECTRIC TRIGLYCINE SULFATE. 27-2967
MONIUM SALTS. (AMMONIUM SULFATE, FLUOBERYLLATE, PERIODATE, BISULFATE) / RELAXATION IN SOME FERROELECTRIC AM 19-2769
BSORPTION, ELECTRIC CONDUCTIVITY AND PHOTOEMISSION OF GLYCOL SULFATE. OPTICAL A 19-2780
RADIATED FERROELECTRIC MATERIALS. (ROCHELLE SALT, TRIGLYCINE SULFATE) /ON (HINDERED TRANSITION) IN GAMMA-IR 27-3165
OF FERROELECTRIC ALUMS. (ESPECIALLY METHYL AMMONIUM ALUMINUM SULFATE) PHYSICAL PROPERITES 20-2810
PIEZOELECTRICITY AND ELECTROSTRICTION OF TRIGLYCINE SULFATE. 27-3118
POLARIZATION REVERSAL CHARACTERISTICS OF AMMONIUM HYDROGEN SULFATE. 19-2788
PREPARATION OF SINGLE CRYSTALS OF AMMONIUM SULFATE. 19-2758
PROPERTIES OF FERROELECTRIC CADMIUM AMMONIUM SULFATE. 19-2756
PROTON ROTATING FRAME RELAXATION IN LITHIUM HYDRAZINIUM SULFATE. 32-3347
UPOLE INTERACTIONS OF NITROGEN-14 IN PARAELECTRIC TRIGLYCINE SULFATE. /RESONANCE STUDY OF THE NUCLEAR QUADR 27-2969
ESONANCE STUDY OF THE FERROELECTRIC TRANSITION IN TRIGLYCINE SULFATE. PULSED NITROGEN PROTON DOUBLE R 27-2968
PYROELECTRIC THERMAL IMAGING DEVICES. (TRIGLYCINE SULFATE 27-3107
INFRARED STUDIES OF THE FERROELECTRIC TRANSITION IN AMMONIUM SULFATE. RAMAN AND 19-2793
ION OF THE THERMAL EXPANSION OF THE FERROELECTRIC TRIGLYCINE SULFATE. REEXAMINAT 27-3121
REFINEMENT OF CRYSTAL STRUCTURE OF TRIGLYCINE SULFATE. 27-3024
TRON MICROSCOPE STUDY OF FERROELECTRIC DOMAINS IN TRIGLYCINE SULFATE SCANNING ELEC 27-3052
ONIC LIGHT SCATTERING BY DOMAINS IN FERROELECTRIC TRIGLYCINE SULFATE. SECOND HARM 27-2991
A. (HYSTERESIS LOOP CONSTRICTIONS, ROCHELLE SALT, TRIGLYCINE SULFATE /ERROELECTRIC "AFTER EFFECT" PHENOMEN 27-3156
EIGHBORHOOD OF A FERRO- PARAELECTRIC TRANSITION. (TRIGLYCINE SULFATE SPECIFIC HEAT ANOMALY IN THE N 27-2962
PLIED ELECTRIC FIELD AND ELECTROCALORIC EFFECT IN TRIGLYCINE SULFATE. SPECIFIC HEAT UNDER AP 27-2960
STATISTICAL THEORY FOR FERROELECTRICITY IN TRIGLYCINE SULFATE. 03-0263
STUDY OF SWITCHING PARAMETERS IN TRIGLYCINE SULFATE. 27-3095
SUPERSONIC DOMAIN WALL MOTION IN TRIGLYCINE SULFATE. 27-2963
NG POLARIZATION REVERSAL. (IN BARIUM TITANATE AND TRIGLYCINE SULFATE) SUSCEPTIBILITY INCREASE DURI 04-0981
E OF THE ELECTRON PARAMAGNETIC RESONANCE OF AMMONIUM CADMIUM SULFATE. TEMPERATURE DEPENDENC 19-2792
TURE DEPENDENCE OF) THE SPONTANEOUS POLARIZATION OF AMMONIUM SULFATE. (TEMPERA 19-2799
TEMPERATURE DEPENDENCE OF THE SWITCHING RATE OF TRIGLYCINE SULFATE. 27-3015
LD ON THE ULTRASONIC ATTENUATION IN FERROELECTRIC TRIGLYCINE SULFATE. THE INFLUENCE OF THE ELECTRIC FIE 27-2956
THERMAL EXPANSION OF TRIGLYCINE SULFATE. 27-3000
MODYNAMIC STABILITY OF THIN FERROELECTRIC FILMS. (TRIGLYCINE SULFATE) THER 27-2959
THERMODYNAMIC THEORY OF IRRADIATED TRIGLYCINE SULFATE. 03-0152
THICKNESS DEPENDENCE OF THE NUCLEATION FIELD OF TRIGLYCINE SULFATE. 27-3036
ATION NEAR THE CURIE TEMPERATURE OF FERROELECTRIC TRIGLYCINE SULFATE. ULTRASONIC RELAX 27-3097
IES OF FERROELECTRICS IN MEMORY DEVICES. (EXAMPLE TRIGLYCINE SULFATE) /LSE WIDTH) PARTIAL SWITCHING PROPERT 27-3144
X-RAY CRITICAL SCATTERING IN FERROELECTRIC TRIGLYCINE SULFATE. 27-2998
X-RAY DAMAGE IN TRIGLYCINE SULFATE. 27-3074
IC STUDY OF DISLOCATIONS AND FERROELECTRIC DOMAINS IN GLYCOL SULFATE. X-RAY TOPOGRAPH 19-2747
-RAY TOPOGRAPHIC STUDY OF DOMAIN WALL MOVEMENT IN TRIGLYCINE SULFATE X 27-3026
/RROELECTRIC PHASE TRANSFORMATION OF DIAMMONIUM DICADMIUM TRISULFATE, A MODEL IN WHICH THE POLARIZATION VE/ 03-0141
ENT TECHNIQUES APPLIED TO THE STRUCTURE OF AMMONIUM HYDROGEN SULFATE ABOVE THE FERROELECTRIC TRANSITION. /M 16-2355
DETERMINATION OF THE CRYSTAL STRUCTURE OF AMMONIUM HYDROGEN SULFATE ABOVE THE FERROELECTRIC TRANSITION. /N 19-2775
CRYSTAL STRUCTURE OF FERROELECTRIC RUBIDIUM HYDROGEN SULFATE. (ABSTRACT) 19-2746
TIME DEPENDENCE OF MATERIAL CONSTANTS OF TRIGLYCINE SULFATE AFTER POLARIZATION REVERSAL. 27-2953
/TRANSITION IN FERROELECTRIC ALUMS (METHYL AMMONIUM ALUMINUM SULFATE, AMMONIUM ALUMINUM SULFATE AND A 36:6/ 20-2809
/CTRICS. DIGRESSIONS FROM PHENOMENOLOGICAL THEORY. (AMMONIUM SULFATE, AMMONIUM FLUOBERYLLATE, GUANIDINIUM / 02-0040
/BARIUM TITANATE, KDP, POTASSIUM SODIUM TARTRATE, TRIGLYCINE SULFATE, AMMONIUM FLUOBERYLLATE, SODIUM NITRI/ 02-0134
SPECIES. DIGLYCINE SULFATE AN INTERESTING NEW DIELECTRIC CRYSTAL 27-2993
RE INFRARED STUDIES. PART-8: FERROELECTRIC AMMONIUM HYDROGEN SULFATE AND AMMONIUM DEUTERIUM SULFATE. /ERATU 19-2782
/ CHANGES IN THE INFRARED SPECTRA OF AMMONIUM OR RUBIDIUM BISULFATE, AND AMMONIUM SULFATE AT FERROELECTRIC 19-2772
TO/ LASER STUDY OF REVERSIBLE NUCLEATION SITES IN TRIGLYCINE SULFATE, AND APPLICATIONS TO PYROELECTRIC DETEC 27-3011
/EL FOR POTASSIUM DIHYDROGEN PHOSPHATE PRESENTED, TRIGLYCINE SULFATE AND BARIUM AND STRONTIUM TITANATE DIS/ 03-0180
OLTAGE RESPONSE TO SHORT INFRARED LASER PULSES IN TRIGLYCINE SULFATE AND BARIUM STRONTIUM NIOBATE. /CTRIC V 13-2018
EXPERIMENTAL AND THEORETICAL STUDY OF DEFECTS IN TRIGLYCINE SULFATE AND BARIUM TITANATE. 04-0620
STUDY OF THE PYROELECTRIC EFFECT IN THIN FILMS OF TRIGLYCINE SULFATE AND BARIUM TITANATE. 04-0686
6 REFS) PHASE CHANGES IN TRIGLYCINE SULFATE AND BARIUM TITANATE FERROELECTRICS. (2 04-0621
NS). DIELECTRIC DISPERSION OF TRIGLYCINE SULFATE AND DEUTERATED KDP (THEORY, CALCULATIO 17-2534
UST BELOW CUR/ CRITICAL BEHAVIOR OF FERROELECTRIC TRIGLYCINE SULFATE AND DEUTERATED TGS (HYSTERESIS LOOPS J 27-2966
/F TRIGLYCINE SELENATE, TRIGLYCINE FLUOBERYLLATE, TRIGLYCINE SULFATE AND ITS VARIATION UNDER X-IRRADIATION. 27-3124
/AL PROPERTIES OF THE SURFACES OF IONIC CRYSTALS (TRIGLYCINE SULFATE AND OTHERS, BY SELECTIVE DEPOSITION O/ 27-2987
UENCY BEHAVIOR OF HYDROGEN BONDED FERROELECTRICS: TRIGLYCINE SULFATE AND POTASSIUM DIHYDROGEN PHOSPHATE. /Q 17-2526
A TWO DIMENSIONAL PRESSURE ON THE CURIE POINTS OF TRIGLYCINE SULFATE AND ROCHELLE SALT. EFFECT OF 27-3082
ATIONS OF (COPPER) DOPED FERROELECTRIC CRYSTALS. (TRIGLYCINE SULFATE AND ROCHELLE SALT) /RESONANCE INVESTIG 27-3132

ON BARKHAUSEN EFFECT IN FERROELECTRIC MATERIALS (TRIGLYCINE SULFATE AND ROCHELLE SALT). /GAMMA-IRRADIATION 27-3123
S TYPES. (BARIUM TITANATE AND TITANATE ZIRCONATE, TRIGLYCINE SULFATE AND ROCHELLE SALT) /LECTRICS OF VARIOU 02-0108
ELECTRIC FIELD METHOD.   CRITICAL PROPERTIES OF TRIGLYCINE SULFATE AND ROCHELLE SALT AS DETERMINED BY AN 27-3056
ESTR/ FERROELECTRIC BEHAVIOR OF RADIATION DAMAGED TRIGLYCINE SULFATE AND ROCHELLE SALT UP TO HIGH DOSAGE (D 27-3098
PARA- AND FERROELECTR/ INFRARED SPECTRA OF RUBIDIUM HYDROGEN SULFATE AND RUBIDIUM DEUTERIUM SULFATE IN THE 19-2768
C MEASUREMENTS ON POTASSIUM DIHYDROGEN PHOSPHATE, TRIGLYCINE SULFATE, AND RUTILE.   FAR INFRARED DIELECTRI 17-2423
                   PROPERTIES OF MIXED CRYSTALS OF TRIGLYCINE SULFATE AND SELENATE.   27-3022
/URIUM THIN FILMS ON A SUBSTRATE OF FERROELECTRIC TRIGLYCINE SULFATE AND SELENATE CRYSTALS. (POLARIZATION / 27-3046
ANSITION, FERROELEC/ X-RAY CRITICAL SCATTERING IN TRIGLYCINE SULFATE AND SODIUM NITRATE. (ORDER DISORDER TR 16-2400
/ATION ON FERROELECTRIC SUBSTRATES. (TELLURIUM ON TRIGLYCINE SULFATE AND SODIUM NITRITE SURFACES) (DECOPAT/ 16-2394
LTAGE RESPONSE TO RECTANGULAR INFRARED SIGNALS IN TRIGLYCINE SULFATE AND STRONTIUM BARIUM NIOBATE. /TRIC VO 13-2019
RESPONSE TO STEP SIGNALS OF INFRARED RADIATION IN TRIGLYCINE SULFATE AND STRONTIUM BARIUM NIOBATE. /OLTAGE 13-2022
PHASE.   ON THE PHASE TRANSITION OF AMMONIUM HYDROGEN SULFATE AND THE STRUCTURE OF THE FERROELECTRIC 19-2777
F SODIUM CARB/ THE PRODUCTION OF BARIUM TITANATE FROM BARIUM SULFATE AND TITANIUM DIOXIDE IN THE PRESENCE O 04-0624
/MAL PROPERTIES OF FERROELECTRICS (SINGLE CRYSTAL TRIGLYCINE SULFATE AND TRIGLYCINE FLUOBERYLLATE) AT PHAS/ 27-3137
ANGES IN SPECI/ INFLUENCE OF GROWTH CONDITIONS OF TRIGLYCINE SULFATE AND TRIGLYCINE SELENATE CRYSTALS ON CH 27-3069
HE DETECTION OF FAR INFR/ ELECTRICAL CONSTANTS OF TRIGLYCINE SULFATE, APPLICATIONS TO PYROELECTRICITY AND T 17-2514
THERMAL HYSTERESIS OF THE DIELECTRIC CONSTANT OF TRIGLYCINE SULFATE AS A FUNCTION OF THE CONDITIONS OF GR/ 27-3126
ARED SPECTRA OF AMMONIUM OR RUBIDIUM BISULFATE, AND AMMONIUM SULFATE AT FERROELECTRIC TRANSITIONS. /HE INFR 19-2772
                   DIELECTRIC DISPERSION OF TRIGLYCINE SULFATE AT LOW FREQUENCIES.   27-3023
          MEASUREMENT OF THE DIELECTRIC CONSTANT OF TRIGLYCINE SULFATE AT MICROWAVE FREQUENCIES.   27-2954
E)   LARGE SIGNAL PERMITTIVITY OF TRIGLYCINE SULFATE AT 100 MHZ. (NEAR THE CURIE TEMPERATUR 27-3029
TION USING PYROELECTRIC EFFECT. (BARIUM TITANATE, TRIGLYCINE SULFATE, AT 5.6 AND 10.5 GHZ) /OMAGNETIC RADIA 04-0978
/TRIC STUDY OF CRITICAL BEHAVIOR OF FERROELECTRIC TRIGLYCINE SULFATE BY A DIGITAL TECHNIQUE. (MINUTE CHANG/ 27-3116
/ERSIBLE AND SPONTANEOUS POLARIZATION REVERSAL IN TRIGLYCINE SULFATE BY MEANS OF A LASER BEAM. APPLICATION/ 27-3010
          STUDY OF THE PHASE TRANSITION OF TRIGLYCINE SULFATE BY SECOND HARMONIC SCATTERING.   27-2992
          STUDY OF THE CRITICAL BEHAVIOR OF TRIGLYCINE SULFATE BY THERMAL NOISE MEASUREMENTS.   27-3147
          REVERSIBLE DOMAIN REVERSAL IN TRIGLYCINE SULFATE CAUSED BY A LASER BEAM.   27-3012
          EFFECT OF DEUTERATION IN AMMONIUM BISULFATE (CHANGE IN TRANSITION TEMPERATURES).   19-2760
               THIN FILMS ON TRIGLYCINE SULFATE CLEAVED IN ULTRAHIGH VACUUM.   27-3173
IMENTAL DETAILED STUDY OF THE PHASE TRANSITION IN TRIGLYCINE SULFATE. (CLOSE THERMAL CONTROL)   EXPER 27-3030
F SURFACE EFFECTS ON THE PYROELECTRIC BEHAVIOR OF TRIGLYCINE SULFATE CLOSE TO THE PHASE TRANSITION. /ENCE O 27-3064
               DEFECT STRUCTURE OF TRIGLYCINE SULFATE. (CORRELATION WITH DOMAIN STRUCTURES)   27-3044
PTICAL MIXING. (OF PULSED RUBY LASER EMISSIONS BY TRIGLYCINE SULFATE CRYSTAL)   O 27-2958
          RAMAN SCATTERING IN A FERROELECTRIC AMMONIUM SULFATE CRYSTAL.   19-2786
     INVESTIGATION OF THE DOMAIN STRUCTURE OF A TRIGLYCINE SULFATE CRYSTAL DURING AGING.   17-2561
     BARKHAUSEN JUMPS AND SWITCHING CURRENT IN TRIGLYCINE SULFATE CRYSTALS.   27-3109
               DIP ETCH PATTERNS ON THE FACES OF TRIGLYCINE SULFATE CRYSTALS.   27-3127
EATURES OF BARKHAUSEN EFFECT IN ROCHELLE SALT AND TRIGLYCINE SULFATE CRYSTALS.   DISTINCTIVE F 27-3110
               DOMAIN STRUCTURE OF GAMMA-IRRADIATED TRIGLYCINE SULFATE CRYSTALS.   27-3164
MAIN WALL CAUGHT IN DISLOCATIONS IN FERROELECTRIC TRIGLYCINE SULFATE CRYSTALS.   DO 27-3091
     FREQUENCY DEPENDENCE OF THE COERCIVE FIELD OF TRIGLYCINE SULFATE CRYSTALS.   27-3062
               INDUCED GROWTH ANISOTROPY IN TRIGLYCINE SULFATE CRYSTALS.   27-3162
THE PHOTOSTIMULATED EXOELECTRON EMISSION YIELD OF TRIGLYCINE SULFATE CRYSTALS. /PONTANEOUS POLARIZATION ON 27-2961
          INVESTIGATION OF BARKHAUSEN EFFECT IN TRIGLYCINE SULFATE CRYSTALS.   27-3111
ITY) ABSORPTION BANDS OF DEFECT ROCHELLE SALT AND TRIGLYCINE SULFATE CRYSTALS. /N AND COPPER AND IRON IMPUR 27-3108
S A FUNCTION OF TEMPERATURE FOR ROCHELLE SALT AND TRIGLYCINE SULFATE CRYSTALS. /NDAMENTAL ABSORPTION EDGE A 27-3160
          SECOND HARMONIC GENERATION IN TRIGLYCINE SULFATE CRYSTALS.   27-3138
TRICAL PROPERTIES OF REGIONS NEAR DOMAIN WALLS IN TRIGLYCINE SULFATE CRYSTALS.   SPECIAL ELEC 27-2988
ACTIVE INDEX IN AMMONIUM DIHYDROGEN PHOSPHATE AND TRIGLYCINE SULFATE CRYSTALS.   THERMAL VARIATION OF REFR 17-2474
/THE POLARIZATION AND REPOLARIZATION PROCESSES IN TRIGLYCINE SULFATE CRYSTALS AND BARIUM TITANATE CERAMIC. 04-0924
OF ANTHRAQUINO/ REVEALING THE DOMAIN STRUCTURE OF TRIGLYCINE SULFATE CRYSTALS BY SELECTIVE CRYSTALLIZATION 27-3041
S.   MAXIMUM REPOLARIZATION CURRENT FOR TRIGLYCINE SULFATE CRYSTALS DEPENDENT ON GROWTH CONDITION 27-3154
/APHIC METHOD TO FERROELECTRIC SODIUM NITRITE AND TRIGLYCINE SULFATE CRYSTALS. (DOMAIN BOUNDARIES AND META/ 16-2375
          LIGHT EMISSION FROM TRIGLYCINE SULFATE (CRYSTALS DURING HEATING)   27-3117
AND GAMMA-RAYS.   DIELECTRIC LOSSES OF TRIGLYCINE SULFATE CRYSTALS EXPOSED TO SMALL DOSES OF X- 27-2976
IE POINT.   DOMAIN STRUCTURE OF PURE AND DOPED TRIGLYCINE SULFATE CRYSTALS GROWN BELOW AND ABOVE THE CUR 27-3130
OND/ X-RAY TOPOGRAPHIC STUDY OF GROWTH DEFECTS IN TRIGLYCINE SULFATE CRYSTALS IN RELATION TO THEIR GROWTH C 27-3129
ECTRIC FIE/ DEPENDENCE OF THE DOMAIN STRUCTURE OF TRIGLYCINE SULFATE CRYSTALS ON TEMPERATURE AND APPLIED EL 27-3129
10 GHZ TO 100 GHZ)   DIELECTRIC DISPERSION IN TRIGLYCINE SULFATE CRYSTALS. (PERMITTIVITY AND LOSS FROM 27-3094
ILMS.   THERMAL EXPANSION OF TRIGLYCINE SULFATE CRYSTALS STUDIED WITH TELLURIUM THIN F 27-3047
     WALL MOTION AND NUCLEATION OF DOMAINS IN TRIGLYCINE SULFATE CRYSTALS. (SUMMARY, WITH PHOTOS) 27-3072
, DOMAIN STRUCTU/ GROWTH REGIONS IN FERROELECTRIC TRIGLYCINE SULFATE CRYSTALS. (TOPOGRAPHY OF IMPERFECTIONS 27-2999
          VARIATION OF THE UNIPOLARITY OF TRIGLYCINE SULFATE CRYSTALS WITH TIME.   27-3115
MODE.   PERFORMANCE CHARACTERISTICS OF A SMALL TRIGLYCINE SULFATE DETECTOR OPERATED IN THE PYROELECTRIC 27-3119
FIELD)   (POLARIZATION CHANGE) AFTER EFFECTS IN TRIGLYCINE SULFATE. (DIELECTRIC DISPLACEMENT AND COERCIVE 27-2952
DOMAIN STRUCTURE OF FERROELECTRIC SODIUM AMMONIUM SULFATE DIHYDRATE.   19-2778
STUDIES OF FERROELECTRIC PHASE TRANSITION IN SODIUM AMMONIUM SULFATE DIHYDRATE. (CRYSTALS) /AND SOME OTHER 19-2767
UTERIUM D/ THERMAL AND SOME OTHER STUDIES OF SODIUM AMMONIUM SULFATE DIHYDRATE, SODIUM TRIHYDROGEN OR TRIDE 19-2779
LEVEL SPLITTINGS OF TITANIUM(3+) ION IN RUBIDIUM ALUMINUM DISULFATE DODECAHYDRATE. /F CRYSTAL FIELD ENERGY 19-2754
UER STUDY OF ELECTRONIC SPIN FLIP PROCESSES IN AMMONIUM IRON SULFATE DODECAHYDRATE.   MOSSBA 19-2771
AMAGNETIC SUBSTANCES. PART-1: IRON(X) ALUMINUM(1-X) AMMONIUM SULFATE DODECAHYDRATE. /CTRA OF MAGNETIZED PAR 19-2751
/T STUDY OF THE FERROELECTRIC TRANSITIONS IN FERRIC AMMONIUM SULFATE DODECAHYDRATE AND POTASSIUM FERROCYAN/ 19-2755
/T STUDY OF THE FERROELECTRIC TRANSITIONS IN FERRIC AMMONIUM SULFATE DODECAHYDRATE AND POTASSIUM FERROCYAN/ 24-2892
/ AMMONIUM IRON AND CHROMIUM ALUMS, METHYL AMMONIUM ALUMINUM SULFATE DODECAHYDRATE, COLEMANITE, SODIUM TRI/ 02-0087
/ RELAXATION IN FERROELECTRIC MASD, METHYL AMMONIUM ALUMINUM SULFATE DODECAHYDRATE. (COMPLEX DIELECTRIC CO/ 20-2804
SSBAUER STUDY ON FERROELECTRIC PROPERTIES OF FERRIC AMMONIUM SULFATE DODECAHYDRATE SINGLE CRYSTALS.   MO 19-2770
               ELASTIC ANOMALY IN AMMONIUM SULFATE. (ELECTROSTRICTIVE CONSTANTS)   19-2759
/C RADIATION INTO MOTION. (PYROELECTRIC EFFECT IN TRIGLYCINE SULFATE ETHYL DIAMINE TARTRATE, BARIUM TITAN/ 04-0854
W METHOD OF MEASURING PYROELECTRIC COEFFICIENTS. (TRIGLYCINE SULFATE EXAMPLE)   NE 17-2517
     THERMAL ANALYSIS OF PYROELECTRIC DETECTORS. (TRIGLYCINE SULFATE EXAMPLE)   27-2964
/TURE ON) QUASI SYMMETRICAL HARD IONS (PHOSPHATE, TITANATE, SULFATE, FLUOBERYLLATE, ETC) OF SOME MONOCLIN/ 03-0330
/ RELAXATION IN SOME FERROELECTRIC AMMONIUM SALTS. (AMMONIUM SULFATE, FLUOBERYLLATE, PERIODATE, BISULFATE) 19-2769
               DOPED TRIGLYCINE SULFATE FOR PYROELECTRIC APPLICATIONS.   27-3057
          GROWTH OF TRIGLYCINE SULFATE FROM SLIGHTLY SUPERSATURATED SOLUTIONS 27-3096
ATIONS.   MATERIAL RESEARCH OF FERROELECTRIC TRIGLYCINE SULFATE FROM THE VIEWPOINT OF PRACTICAL APPLIC 27-3065
/AL PROPERTIES OF IMPROPER FERROELECTRIC CRYSTALS. (AMMONIUM SULFATE, GADOLINIUM MOLYBDATE, IRON IODINE BO/ 14-2084
ELASTIC PROPERTIES OF FERROELECTRIC CRYSTALS THE TRIGLYCINE SULFATE GROUP. /TERATION EFFECT ON THERMAL AND 27-2986
OUS SUBSTITUTION IN FERROELECTRIC CRYSTALS OF THE TRIGLYCINE SULFATE GROUP. (NORMAL AND DEUTERATED) /OMORPH 27-3102
C RESONANCE STUDY OF SOME METAL IONS IN GUANIDINIUM ALUMINUM SULFATE HEXAHYDRATE.   ELECTRON PARAMAGNETI 21-2817

TIC COPPER(2+) COMPLEXES IN CRYSTALS OF GUANIDINIUM ALUMINUM SULFATE HEXAHYDRATE.   EPR STUDY OF MAGNE 21-2820
                          SEE BOTH GASH AND GUANIDINIUM ALUMINUM SULFATE HEXAHYDRATE.
RADIATED FERROELECTRICS ROCHELLE SALT AND GUANIDINE ALUMINUM SULFATE HEXAHYDRATE. /DIES OF THE GAMMA-RAY IR 21-2812
   DILUTICN EFFECT OF ABSORPTION LINES IN GUANIDINIUM CHROMIUM SULFATE HEXAHYDRATE.   ZEEMAN EFFECT AND 21-2821
F TWO FINE LINES IN (THE RED REGION OF) CHROMIUM GUANIDINIUM SULFATE HEXAHYDRATE.   ZEEMAN EFFECT (STUDY) O 21-2814
/SORPTICN SPECTRA (5200-4200 PER CM) OF GUANIDINIUM ALUMINUM SULFATE HEXAHYDRATE AND FREQUENCIES OF THE FU/ 21-2811
/ OF STRUCTURE OF FERROELECTRIC CRYSTAL GUANIDINIUM ALUMINUM SULFATE HEXAHYDRATE AND ITS CHROMIUM ISOMORPH. 21-2819
).   FERROELECTRIC MOMENT IN GUANIDINIUM ALUMINUM SULFATE HEXAHYDRATE (AND MECHANISM OF REVERSAL 21-2818
          MAGNETIC SUSCEPTIBILITY OF GUANIDINIUM VANADIUM SULFATE HEXAHYDRATE AT LOW TEMPERATURES.   21-2815
/TERFEROMETRIC STUDIES ON FERROELECTRIC GUANIDINIUM ALUMINUM SULFATE HEXAHYDRATE CLEAVAGES AND THEIR ETCH / 21-2816
/TS AND THE FERROELECTRIC PROPERTIES OF GUANIDINIUM ALUMINUM SULFATE HEXAHYDRATE. (POLARIZATION VERSUS ELE/ 21-2813
        ORDER DISORDER TRANSITION IN FERRCELECTRICS. (TRIGLYCINE SULFATE, HIGHLY ACCURATE TEMPERATURE CONTROL) 27-2982
/BARIUM TITANATE, ANTIMONY SULFICCIDE, TBICUREA, TRIGLYCINE SULFATE, HYDROGEN AMMONIUM MONOCHLORO ACETATE/ 02-0099
                          AGING PROCESS IN TRIGLYCINE SULFATE. (HYSTERESIS LAPS, DOMAIN STRUCTURE)   27-3128
        PERMITTIVITY OF SINGLE CRYSTALS OF TRIGLYCINE SULFATE IN A STRONG ELECTRIC FIELD.   27-2978
                    STRUCTURE OF AMMONIUM SULFATE IN FERROELECTRIC PHASE AND TRANSITION.   19-2774
ENCENCE OF DIELECTRIC PROPERTIES OF FERROELECTRIC TRIGLYCINE SULFATE IN THE FAR INFRARED. TEMPERATURE DEP 27-3061
REQUENC/ DISPERSION OF THE DIELECTRIC CONSTANT OF TRIGLYCINE SULFATE IN THE FAR INFRARED (TEMPERATURE AND F 27-3009
URIE POINT (IMPROVED ME/ PYROELECTRIC RESPONSE OF TRIGLYCINE SULFATE IN THE FERROELECTRIC REGION NEAR THE C 27-2979
              PYROELECTRIC BEHAVIOR OF TRIGLYCINE SULFATE IN THE PARAELECTRIC REGION.   27-3066
14 DEG C.   SPECIFIC HEAT OF AMMONIUM HYDROGEN SULFATE IN THE TEMPERATURE RANGE FROM -70 TO + 19-2789
          EQUILIBRIUM DOMAIN STRUCTURE OF TRIGLYCINE SULFATE IN THE VICINITY OF THE CURIE POINT.   27-3112
/INVESTIGATION OF THE FERROELECTRIC PROPERTIES OF AMMONIUM BISULFATE IN THE VICINITY OF THE HIGH TEMPERATU/ 27-2787
RANSITION.   AMMONIUM HYDROGEN SULFATE, ITS STRUCTURE AND THE FERROELECTRIC T 19-2773
/RROELECTRIC DOMAIN BOUNDARIES (DOPED AND UNDOPED TRIGLYCINE SULFATE LITHIUM NIOBATE, KDP, ABSTRACT ONLY). 11-1671
TELY)   ELECTROMECHANICAL PROPERTIES OF TRIGLYCINE SULFATE. (MEASUREMENTS VERIFY THEORY APPROXIMA 27-3145
UATION OF STATE FOR THE COOPERATIVE TRANSITION OF TRIGLYCINE SULFATE NEAR CURIE TEMPERATURE.   EQ 27-3006
         POLARIZATION FLUCTUATION OF TRIGLYCINE SULFATE NEAR THE CURIE POINT.   27-3077
           THERMAL CONDUCTIVITY OF TRIGLYCINE SULFATE NEAR THE CURIE POINT.   ST 27-2983
           THERMAL CONDUCTIVITY OF TRIGLYCINE SULFATE NEAR THE CURIE POINT.   27-3016
RADIATED S/ DISPLACIVE MOVEMENT OF THE RADICAL IN TRIGLYCINE SULFATE NEAR THE CURIE POINT. (ESR OF GAMMA-IR 27-3033
ONS ON THE THERMAL CAPACITY OF SINGLE CRYSTALS OF TRIGLYCINE SULFATE (NEAR THE FERROELECTRIC TRANSITION). / 27-3140
TALLOGRAPHIC STUDIES OF IRRADIATION FIELD TREATED TRIGLYCINE SULFATE: NEW STRUCTURE FORM.   CRYS 27-2996
)   ORTHORHOMBIC DIGLYCINE SULFATE. (NOT PIEZOELECTRIC, NOT FERROELECTRIC 36-3624
CHANGES CF THE DOMAIN STRUCTURE IN VANADIUM DOPED TRIGLYCINE SULFATE OBSERVED BY EPR.   27-3013
Y OF GAMMA-RADIATION EFFECTS IN ROCHELLE SALT AND TRIGLYCINE SULFATE. (ON ELECTRONIC PROPERTIES.) /TAL STUD 27-3101
E.   DEPENDENCE CF THE COERCIVE FIELD OF TRIGLYCINE SULFATE ON FREQUENCY, AMPLITUDE AND TEMPERATUR 27-3092
ERROELECTRIC FILMS ON SEMICONDUCTING SUBSTRATES. (TRIGLYCINE SULFATE ON SILICON) /OF POLARIZATION IN THIN F 27-3163
/N SPECTRA OF FERROELECTRIC ALUMS: METHYL AMMONIUM ALUMINUM (SULFATE OR SELENATE) DODECAHYDRATE. (AND THE / 20-2803
/TION OF FERROELECTRICS WITH GENERAL FORMULA METAL-1 METAL-2 SULFATE OR SELENATE OR FLUOBERYLLATE (METALS / 19-2743
/RATURE DEPENDENCE OF THE DIELECTRIC CONSTANTS OF TRIGLYCINE SULFATE ORTHOGONAL TO THE FERROELECTRIC AXIS. / 27-2975
 OF SAMPLE TREATMENT AND ELECTRODE NATURE ON SOME TRIGLYCINE SULFATE PARAMETERS.   EFFECT 27-3148
RUCTURAL INHIBITION OF FERROELECTRIC SWITCHING IN TRIGLYCINE SULFATE.PART-1: ADDITIVES.   ST 27-3038
ND THE ACTIVATION ENERGY/ ESR OF GAMMA-IRRADIATED TRIGLYCINE SULFATE PART-2: INTERNAL MOTION OF HYDROGEN A 27-3034
/CONNECTION WITH THEIR PHASE TRANSITIONS. PART-1: TRIGLYCINE SULFATE PART-2: SILVER SODIUM DINITRITE. PAR/ 02-0077
RUCTURAL INHIBITION OF FERROELECTRIC SWITCHING IN TRIGLYCINE SULFATE PART-2: X-IRRADIATION TREATMENT. ST 17-2469
TIES OF FERROELECTRICS AT MICRCWAVE FREQUENCIES. (TRIGLYCINE SULFATE, POLYCRYSTALLINE BARIUM TITANATE) /PER 04-0870
/ LIGHT SCATTERING STUDIES OF PHASE TRANSITIONS. (TRIGLYCINE SULFATE, POTASSIUM DIHYDROGEN AND DIDEUTERIUM/ 17-2481
/DIO FREQUENCIES) OF SOME FERROELECTRIC CRYSTALS (TRIGLYCINE SULFATE, POTASSIUM DIHYDROGEN PHOSPHATE, AMMO/ 02-0125
              SWITCHING IN FERROELECTRIC TRIGLYCINE SULFATE, PURE AND DOPED WITH PARAMAGNETIC IONS 27-3020
OF FERROELECTRIC CRYSTALS. (5 COMPCUNDS MEASURED, TRIGLYCINE SULFATE RESULTS GIVEN) /TURE, CURIE CONSTANT) 27-3120
 OF SOME FERROELECTRIC CRYSTALS (PYROELECTRICITY, TRIGLYCINE SULFATE ROCHELLE SALT).   OPTICAL ACTIVITY 27-3017
            NOISE MEASUREMENTS IN FERROELECTRICS. (TRIGLYCINE SULFATE ROCHELLE SALT, BARIUM TITANATE) 04-0698
/HEORY, DYNAMICAL ORDER DISORDER PROPERTIES, KDP, TRIGLYCINE SULFATE ROCHELLE SALT, SILVER PERIODATE, SOD/ 02-0050
/URE STABILITY IN FERROELECTRICS. (ROCHELLE SALT, TRIGLYCINE SULFATE, SELENATE, FLUOBERYLLATE AND BARIUM T/ 02-0083
E SPIN SYMMETRY STATES OF AMMONIUM IONS IN SOLIDS. (AMMONIUM SULFATE, SELENATE, VANADATE AND 15 OTHERS) /TH 19-2800
ESONANCE). DIELECTRIC DISPERSION IN FERROELECTRIC TRIGLYCINE SULFATE (SEVERAL KHZ TO 23 KMHZ, DOMAIN WALL R 27-3032
FERROELECTRIC MODULATOR FOR ELECTROMETER. (WITH A TRIGLYCINE SULFATE SINGLE CRYSTAL) /SECOND HARMONIC TYPE 27-2951
ERROELECTRIC CONDENSER UNDER CONTINUOUS HEATING. (TRIGLYCINE SULFATE SINGLE CRYSTAL) /ENT OF A DIELECTRIC F 27-3028
         DIELECTRIC RELAXATION IN MULTIDOMAIN TRIGLYCINE SULFATE SINGLE CRYSTALS.   27-3001
FECT OF GROWTH CONDITIONS ON DIELECTRIC LOSSES IN TRIGLYCINE SULFATE SINGLE CRYSTALS.   EF 27-3152
       ELECTRICAL CONDUCTIVITY OF AMMONIUM SULFATE SINGLE CRYSTALS.   19-2781
EPR STUDIES OF CHROMIUM- ANC VANADYL- DOPED TRIGLYCINE SULFATE SINGLE CRYSTALS.   27-3161
   EVOLUTICN OF FERROELECTRIC DOMAINS IN TRIGLYCINE SULFATE SINGLE CRYSTALS.   27-3003
              GROWTH OF TRIGLYCINE SULFATE SINGLE CRYSTALS.   27-3083
UENCE OF DEFECTS ON SOME DIELECTRIC PROPERTIES OF TRIGLYCINE SULFATE SINGLE CRYSTALS.   INFL 27-3063
IES ON THE GROWTH AND SOME PHYSICAL PROPERTIES OF TRIGLYCINE SULFATE SINGLE CRYSTALS. INFLUENCE OF IMPURIT 27-3084
              POLARIZATION REVERSAL IN TRIGLYCINE SULFATE SINGLE CRYSTALS.   27-2980
              PREPARATION OF AMMONIUM SULFATE SINGLE CRYSTALS.   19-2753
    PREPARATION OF PURE GLYCINE USED FOR GROWING OF TRIGLYCINE SULFATE SINGLE CRYSTALS.   27-3085
    REMARKS ABOUT SWITCHING IN BARIUM TITANATE AND TRIGLYCINE SULFATE SINGLE CRYSTALS.   04-0915
THE DEFECT STRUCTURE AND DIELECTRIC PROPERTIES OF TRIGLYCINE SULFATE SINGLE CRYSTALS. /T OF GROWTH RATE ON 27-3067
Y TOPOGRAPHIC OBSERVATIONS OF RADIATION DAMAGE IN TRIGLYCINE SULFATE SINGLE CRYSTALS.   X-RA 27-3106
S OF THEIR QUALITY. (VOLUME / PREPARATICN OF PURE TRIGLYCINE SULFATE SINGLE CRYSTALS AND SOME INVESTIGATION 27-3086
CE OF THE FLUCTUATION ABSORPTION OF ULTRASOUND IN TRIGLYCINE SULFATE SINGLE CRYSTALS (AT 10, 30, 50 MHZ). / 27-3079
CTRIC PERMI/ THERMAL AND ELECTRICAL PRCPERTIES OF TRIGLYCINE SULFATE SINGLE CRYSTALS (HEAT CAPACITY AND ELE 27-3141
IELD.   DIELECTRIC CONSTANT OF TRIGLYCINE SULFATE SINGLE CRYSTALS IN A STRONG ELECTRIC F 27-3150
D.   TEMPERATURE AUTOSTABILIZATION EFFECT OF TRIGLYCINE SULFATE SINGLE CRYSTALS IN AN AC ELECTRIC FIEL 27-3005
GENT OF THE ANGLE/ EFFECT OF GROWTH CCNDITIONS OF TRIGLYCINE SULFATE SINGLE CRYSTALS ON A CHANGE IN THE TAN 27-3153
OPERTIES. (REVIEW)   INFLUENCE OF THE QUALITY OF TRIGLYCINE SULFATE SINGLE CRYSTALS ON THEIR DIELECTRIC PR 02-0088
IC RELAXATION / ANISOTROPY OF SOUND ABSORPTION IN TRIGLYCINE SULFATE SINGLE CRYSTALS. (THEORY OF ANISOTROP 27-3078
/ACTERIZATION OF SOLID SOLUTIONS OF FERROELECTRIC TRIGLYCINE SULFATE SINGLE CRYSTALS WITH ISOMORPHOUS COMP/ 27-2972
/DIES ON THE MECHANISM OF THE PHASE TRANSITION IN TRIGLYCINE SULFATE (STATIC DIELECTRIC CONSTANT, X-RAY S/ 27-3093
/MOTIONS IN SOLIDS: THE FERROELECTRIC TRANSITION IN AMMONIUM SULFATE STUDIED BY COLD NEUTRONS (SCATTERING). 19-2752
STABILIZATION REGIME OF A FERROELECTRIC CRYSTAL. (TRIGLYCINE SULFATE TANDEL) /ITIES IN THE TEMPERATURE AUTO 17-2502
AIN STRUCTURE)   AGING PROCESSES IN TRIGLYCINE SULFATE. (TANDEL OR CAPACITANCE PROPERTIES DOM 27-3081
EX DIELECTRIC CONS/ RELAXATION OF DOMAIN WALLS IN TRIGLYCINE SULFATE. (TEMPERATURE DEPENDENCES OF THE COMPL 27-3103
PTICAL PROPERTIES IN T/ FERROELECTRIC TRANSITION IN AMMONIUM SULFATE, THE DIELECTRIC, OPTICAL, AND ELECTROO 19-2745
/ TEMPERATURE STUDIES. PART-4: FHASE TRANSITIONS OF AMMONIUM SULFATE: THE NATURE OF HYDROGEN BONDING AND T/ 19-2783
KNESS ANC ENERGY DE/ FERROELECTRIC DOMAIN WALL IN TRIGLYCINE SULFATE. (THEORY AND CALCULATIONS OF WALL THIC 27-2997

/N THE PYROELECTRIC INFRARED RADIATION DETECTOR. (TRIGLYCINE SULFATE, TRIGLYCINE FLUOROBERYLLATE, LITHIUM / 02-0046
/TION OF FERROELECTRIC CRYSTALS (SOLUTION GROWTH, TRIGLYCINE SULFATE TRIGLYCINE SELENATE OR FLUOBERYLLATE/ 02-0089
CY RAMAN SPECTRA OF FERROELECTRIC CRYSTALS OF THE TRIGLYCINE SULFATE TYPE.   LOW FREQUEN 27-3114
                        THE BEHAVIOR OF DIAMMONIUM DICADMIUM TRISULFATE UNDER PRESSURE.   19-2757
                      PREPARATION OF PURE TRIGLYCINE SULFATE. (USING ION EXCHANGE)   27-3087
. DIRECT OBSERVATION OF FERROELECTRIC DOMAINS IN TRIGLYCINE SULFATE USING THE SCANNING ELECTRON MICROSCOPE 27-3051
ME FERROELECTRIC PROPERTIES OF SOLID SOLUTIONS OF TRIGLYCINE SULFATE WITH ISOMORPHOUS SUBSTANCES. /CE OF SO 27-2971
INVESTIGATIONS OF SURFACES OF FERROELECTRICS (TRIGLYCINE SULFATE WITH SEMICONDUCTING ELEMENTS.   27-3089
RENGTH DEPENDENCE OF LARGE SIGNAL PERMITTIVITY OF TRIGLYCINE SULFATE (10 KHZ-1 MHZ) /PERATURE AND FIELD ST 27-3068
MICROWAVE BEHAVIOR OF ROCHELLE SALT AND TRIGLYCINE SULFATE (18 REFS)   27-3088
GENERATION OF THE SECOND HARMONIC BY TRIGLYCINE SULFATE. (84 REFS)   27-2990
                                        POLARIZED TRIGLYCINE SULFATES.   27-3058
LECTRIC PROPERTIES OF SOLID SOLUTIONS IN THE SYSTEM AMMONIUM SULFATE- AMMONIUM FLUOBERYLLATE.   FERROE 19-2762
FERROELECTRIC PROPERTIES OF SOLID SOLUTIONS IN THE AMMONIUM SULFATE- AMMONIUM FLUOBERYLLATE SYSTEM.   19-2764
FERROELECTRIC PROPERTIES OF SOLID SOLUTIONS IN THE AMMONIUM SULFATE- AMMONIUM FLUOBERYLLATE SYSTEM.   19-2763
R TANTALATE OR GALLATE, BARIUM TITANATE, ZINC OXIDE, CADMIUM SULFIDE). /ECTRIC) CRYSTALS (LITHIUM NIOBATE O 02-0095
            ANOMALIES OF THERMAL PROPERTIES OF TRIGLYCINE SULFIDE.   27-3048
AND STRUCTURE OF THE FERROELECTRIC SEMICONDUCTOR ANTIMONY TRISULFIDE.   B 36-3575
OF THE PYROCHLORE STRUCTURE (CADMIUM(2) NIOBIUM(2) OXIDE(6) SULFIDE).   NEW DESCRIPTION 36-3602
AGRAM OF THE SEMICONDUCTING UNIAXIAL FERROELECTRIC IRON(1-X) SULFIDE.   PHASE DI 36-3617
HASE RELATIONS AND SUPERSTRUCTURES OF PYRRHOTITE, IRON(1-X) SULFIDE.   P 36-3597
IS CONCERNED WITH ION IMPLANTATION OF SINGLE CRYSTAL CADMIUM SULFIDE) /. (THE MAJOR PORTION OF THIS REPORT 02-0057
RAMAN SCATTERING BY POLARITONS. PART-2: THE CASE OF CADMIUM SULFIDE AND LITHIUM NIOBATE CRYSTALS. /TANEOUS 11-1787
ONCOLLINEAR PARAMETRIC 4 PHOTON RECIPROCAL EFFECT IN CADMIUM SULFIDE AND STRONTIUM TITANATE.   N 06-1122
VAPOR PHASE BY A DYNAMIC METHOD (ANTIMONY SESQUISELENIDE OR SULFIDE, ANTIMONY SULFO OR SELENO IODIDE). /HE 15-2162
/NTITATIVE STUDIES OF OPTICAL HARMONIC GENERATION IN CADMIUM SULFIDE, BARIUM TITANATE, AND POTASSIUM DIHYD/ 04-0852
SURFACE WAVE CONVOLUTION USING A CADMIUM SULFIDE FLUID LITHIUM NIOBATE STRUCTURE.   11-1700
LAYER WAVE AMPLIFICATION IN A CADMIUM SULFIDE FLUID- LITHIUM NIOBATE STRUCTURE.   11-1667
ANT. (POTASSIUM BROMIDE, LEAD TELLURIDE, TIN TELLURIDE, LEAD SULFIDE, GERMANIUM TELLURIDE) /IELECTRIC CONST 36-3541
NI/ ACOUSTIC SURFACE WAVE CONVOLUTION ON CRYSTALS OF CADMIUM SULFIDE, LITHIUM NIOBATE AND BISMUTH(12) GERMA 11-1692
CCEL FOR THE SEMICONDUCTING UNIAXIAL FERROELECTRIC IRON(1-X) SULFIDE. PART-1. PART-2.   M 36-3615
RONTIUM NIOBATE, LEAD NIOBIUM(4) OXIDE(11), SILVER GALLIUM DISULFIDE, SILICON CARBIDE) /TITANATE, BARIUM ST 05-1066
CES OF THE HALL AND DRIFT MOBILITIES OF HOLES IN ANTIMONY TRISULFIDE SINGLE CRYSTALS. TEMPERATURE DEPENDEN 15-2193
CRYSTAL GROWTH IN THE ANTIMONY SULFOIODIDE- ANTIMONY TRISULFIDE SYSTEM. (BRIDGMAN- STOCKBARGER METHOD) 15-2230
4)   FERROELECTRICITY IN THE IRON DEFICIENT FERROUS SULFIDE SYSTEM. (EXISTS TO DEFICIENCIES OF 0.0 36-3614
/F THE TYPE A(2)(V)B(3)(VI) AND A(V)B(VI)C(VII). (ANTIMONY TRISULFIDE, TRISELENIDE, SULFOIODIDE, SELENOIODI/ 15-2108
PYROELECTRIC EFFECT IN ANTIMONY SULFO AND SELENO IODIDE SINGLE CRYSTALS.   15-2163
TRIC CRYSTALS OF GROUPS V, VI AND VII. (ANTIMONY AND BISMUTH SULFO HALIDES) /TION SPECTRA OF SOME FERROELEC 15-2210
DYNAMIC METHOD (ANTIMONY SESQUISELENIDE OR SULFIDE, ANTIMONY SULFO OR SELENO IODIDE). /HE VAPOR PHASE BY A 15-2162
/C(VII) UNDER HYDROTHERMAL CONDITIONS. (ANTIMONY AND BISMUTH SULFO, SELENO, AND TELLUROIODIDES AND BROMIDE/ 15-2213
ONY SULFOIODIDE, SELENOIODIDE, BISMUTH SULFOIODIDE, ANTIMONY SULFO SELENOIODIDE, CURIE TEMPERATURES) /ANTIM 15-2114
/) COMPOUNDS. ANTIMONY(X) BISMUTH(1-X) SULFOIODIDE, ANTIMONY SULFO(X) SELENO(1-X) IODIDE, ANTIMONY(X) ARSE/ 15-2116
D CHEMICAL SYNTHESIS UNDER HYDROTHERMAL CONDITIONS. (BISMUTH SULFOBROMIDE,   CRYSTAL GROWTH AN 15-2218
PRODUCTION AND SOME OPTICAL PROPERTIES OF ANTIMONY SULFOBROMIDE IN GLASSY AND CRYSTALLINE STATES. 15-2142
DIELECTRIC PROPERTIES OF ANTIMONY SULFOIODIDE- SULFOBROMIDE SOLID SOLUTIONS.   15-2242
LOUS ABSORPTION OF ULTRASOUND IN SINGLE CRYSTALS OF ANTIMONY SULFOIODIDE.   ANISOTROPY OF ANOMA 15-2222
N OF SOUND NEAR THE FERROELECTRIC PHASE TRANSITION (ANTIMONY SULFOIODIDE).   ANOMALOUS ABSORPTIO 15-2221
ANOMALOUS PROPERTIES IN PHOTOFERROELECTRIC ANTIMONY SULFOIODIDE.   15-2196
BAND STRUCTURE OF THE FERROELECTRIC SEMICONDUCTOR ANTIMONY SULFOIODIDE.   15-2179
BCSON ECHOES. (ANTIMONY SULFOIODIDE)   15-2144
CHARACTERISTICS OF THE PHOTOELECTRIC EMF OF ANTIMONY SULFOIODIDE.   15-2129
CHARACTERISTICS OF THE PHOTOELECTRIC EMF OF ANTIMONY SULFOIODIDE.   15-2153
CRITICAL PHENOMENA IN ANTIMONY SULFOIODIDE.   15-2231
CRITICAL POINT IN FERROELECTRIC ANTIMONY SULFOIODIDE.   15-2237
CRYSTAL FIELD GRADIENTS IN ANTIMONY SULFOIODIDE.   15-2185
CRYSTAL STRUCTURE AND CONDUCTIVITY OF ANTIMONY SULFOIODIDE.   15-2112
RE DEPENDENCE OF LATTICE CONSTANTS IN FERROELECTRIC ANTIMONY SULFOIODIDE.   DETERMINATION OF THE TEMPERATU 15-2140
OF PHOTOFERROELECTRIC EFFECTS IN SINGLE CRYSTALS OF ANTIMONY SULFOIODIDE.   DILATOMETRIC INVESTIGATION 15-2156
ET CHARGE ON VALUE OF FERROELECTRIC POLARIZATION IN ANTIMONY SULFOIODIDE.   EFFECT OF PHOTOELECTR 15-2186
ELECTRICAL AND PHOTOELECTRIC PROPERTIES OF ANTIMONY SULFOIODIDE.   15-2107
ELECTROOPTIC EFFECTS OF THE ABSORPTION EDGE OF ANTIMONY SULFOIODIDE.   15-2198
ELECTROOPTICAL AND ELECTROMECHANICAL EFFECT IN ANTIMONY SULFOIODIDE.   15-2171
N, PIEZOELECTRICITY AND ELASTICITY IN FERROELECTRIC ANTIMONY SULFOIODIDE.   ELECTROSTRICTIO 15-2165
OELECTRICS DUE TO SURFACE LAYERS. (BARIUM TITANATE, ANTIMONY SULFOIODIDE) /RINSIC PHOTOCONDUCTIVITY IN FERR 04-0681
FAR INFRARED REFLECTIVITY OF ANTIMONY SULFOIODIDE.   15-2209
FAR INFRARED REFLECTIVITY SPECTRA OF ANTIMONY SULFOIODIDE.   15-2234
FAR INFRARED REFLECTIVITY SPECTRA OF ANTIMONY SULFOIODIDE.   15-2235
FERROELECTRIC BARKHAUSEN EFFECT IN ANTIMONY SULFOIODIDE.   15-2172
FERROELECTRIC TRANSITION OF ANTIMONY SULFOIODIDE.   15-2195
TWINNING) TREATED AS CRITICAL POINTS IN CRYSTALS. (ANTIMONY SULFOIODIDE) /CTRIC INSTABILITY (FERROELECTRIC 15-2176
FIELD EMISSION FROM ANTIMONY SULFOIODIDE.   15-2203
HEAT CAPACTIY OF POLYCRYSTALLINE ANTIMONY SULFOIODIDE.   15-2238
HOTOCONDUCTIVITY IN THE FERROELECTRIC SEMICONDUCTOR ANTIMONY SULFOIODIDE.   INDUCED IMPURITY P 15-2159
OF LIGHT ON SELF POLARIZATION OF SINGLE CRYSTALS OF ANTIMONY SULFOIODIDE.   INFLUENCE 15-2228
DOMAIN STRUCTURE OF THE FERROELECTRIC SEMICONDUCTOR ANTIMONY SULFOIODIDE. / NONEQUILIBRIUM CARRIERS ON THE 15-2104
CREENING ON POLARIZATION PROCESSES IN FERROELECTRIC ANTIMONY SULFOIODIDE. /ENCE OF NONEQUILIBRIUM CARRIER S 15-2158
CURRENTS AND ACCUMULATION OF POLARIZATION CHARGE IN ANTIMONY SULFOIODIDE.   INJECTION 15-2188
WAVES WITH MOVING DOMAIN WALLS IN FERROELECTRICS. (ANTIMONY SULFOIODIDE)   INTERACTION OF ULTRASONIC 15-2192
LATTICE MODES AND PHASE TRANSITION IN ANTIMONY SULFOIODIDE.   15-2110
LIGHT INDUCED PYROCURRENTS IN FERROSEMICONDUCTOR ANTIMONY SULFOIODIDE.   15-2130
SINGLE CRYSTALS OF THE FERROELECTRIC SEMICONDUCTOR ANTIMONY SULFOIODIDE.   LUMINESCENCE OF 15-2229
MECHANISM OF THE PHOTODOMAIN EFFECT IN ANTIMONY SULFOIODIDE.   15-2146
NOTES ON ABSORPTION BAND EDGE OF (PARAELECTRIC) ANTIMONY SULFOIODIDE.   15-2177
DRUPOLE AND ELECTROACOUSTIC ECHOES IN FERROELECTRIC ANTIMONY SULFOIODIDE.   NUCLEAR QUA 15-2189
PPOSING FERROELECTRIC DOMAINS IN SINGLE CRYSTALS OF ANTIMONY SULFOIODIDE.   O 15-2252
OPTICAL SECOND HARMONIC GENERATION IN ANTIMONY SULFOIODIDE.   15-2164
PHOTOCONDUCTIVITY OF FERROELECTRIC PHOTOCONDUCTOR ANTIMONY SULFOIODIDE.   15-2226
PHOTOEMF'S IN THE FERROELECTRIC SEMICONDUCTOR ANTIMONY SULFOIODIDE.   15-2132
VOLTAIC EFFECT IN FERROELECTRIC AND PHOTOCONDUCTIVE ANTIMONY SULFOIODIDE.   PHOTO 15-2174
PYROCURRENTS HIGHER THAN THE CURIE POINT IN ANTIMONY SULFOIODIDE.   15-2131

```
                    RAMAN SCATTERING IN FERROELECTRIC SEMICONDUCTOR ANTIMONY SULFOIODIDE.                                                    15-2111
                                                        REFLECTIVITY ON ANTIMONY SULFOIODIDE.                                                15-2244
HE ELECTRIC POLARIZATION AND THE ABSORPTION EDGE IN ANTIMONY SULFOIODIDE.                          RELATION BETWEEN T 15-2175
     SOFT LATTICE VIBRATIONS AND VIBRATICN COUPLING IN ANTIMONY SULFOIODIDE.                                                                15-2233
G AND CRITICAL RAYLEIGH SCATTERING IN FERROELECTRIC ANTIMONY SULFOIODIDE.                          SOFT MODE COUPLIN 15-2232
UPLINGS IN THE PRESSURE INDUCED PHASE TRANSITION IN ANTIMONY SULFOIODIDE.                          SOFT PHONON CO 15-2240
          SOME FEATURES OF THE PHCTOEMF SFECTRUM OF ANTIMONY SULFOIODIDE.                                                                    15-2136
N OF CAPTURE CENTERS IN FERROELECTRIC SEMICONDUCTOR ANTIMONY SULFOIODIDE. /E DEPENDENCE OF THE CROSS SECTIO 15-2147
                           X-RAY CRYSTALLOGRAPHIC STUDIES OF ANTIMONY SULFOIODIDE.                                                           15-2109
CLEAR QUACRUPOLE RESONANCE IN FERRCELECTRICS OF THE ANTIMONY SULFOIODIDE (AND BROMIDE) TYPE.                    NU 15-2216
/ICNS OF A(V)B(VI)C(VII) COMPOUNDS. ANTIMONY(X) BISMUTH(1-X) SULFOIODIDE, ANTIMONY SULFO(X) SELENO(1-X) IO/ 15-2116
9.8 GHZ, 0-50 C).            DIELECTRIC PROPERTIES OF ANTIMONY SULFOIODIDE AT MICROWAVE FREQUENCIES (3.3 AND 15-2171
URIE TEMPERATURE OF A FERROELECTRIC SEMICONDUCTOR. (ANTIMONY SULFOIODIDE BROMIDE) / "FROZEN" SHIFT OF THE C 15-2160
          SOFT PHONON MODE AND MCDE COUPLING IN ANTIMONY SULFOIODIDE (BY RAMAN SCATTERING).                                                 15-2167
RIC FIELD.         STRAIN ALONG C-AXIS OF ANTIMONY SULFOIODIDE CAUSED BY ILLUMINATION IN DC ELECT 15-2239
LECTRIC CURRENTS)         TRANSPORT PHENOMENA IN ANTIMONY SULFOIODIDE. (CONDUCTION MECHANISM FOR THERMOE 15-2241
          KINETICS CF CRYSTALLIZATION CF ANTIMONY SULFOIODIDE (CRYSTAL GROWTH RATE, ANISOTROPY) 15-2215
, UNDER DC FIELD ALONG C/ POTENTIAL DISTRIBUTION IN ANTIMONY SULFOIODIDE CRYSTAL (IN THE PARAELECTRIC PHASE 15-2173
      ANISOTROPY OF THE REFLECTION SPECTRA OF ANTIMONY SULFOIODIDE CRYSTALS.                                                                 15-2125
               DIELECTRIC DISPERSION IN ANTIMONY SULFOIODIDE CRYSTALS.                                                                       15-2117
GROWTH, DIELECTRIC, AND PHOTOELECTRIC PROPERTIES CF ANTIMONY SULFOIODIDE CRYSTALS. /FECT OF COMPOSITION ON 15-2118
T TRANSITIONS AND ABSORPTION IN THE MID-INFRAREC IN ANTIMONY SULFOIODIDE CRYSTALS.                  INDIREC 15-2150
     MOTION PICTURES OF THE PHASE TRANSITION IN ANTIMONY SULFOIODIDE CRYSTALS.                                                               15-2120
NEAR EFFECTS AND THE LOW TEMPERATURE TRANSITION IN ANTIMONY SULFOIODIDE CRYSTALS.                   NONLI 15-2247
     PHOTODOMAIN EFFECT IN NATURALLY POLARIZED ANTIMONY SULFOIODIDE CRYSTALS.                                                                15-2133
     PHOTODOMAIN EFFECT IN NATURALLY PCLARIZED ANTIMONY SULFOIODIDE CRYSTALS.                                                                15-2135
          PHOTOELECTRIC PROPERTIES OF BISMUTH SULFOIODIDE CRYSTALS.                                                                          15-2170
FFECTS IN THE ANCMALOUS ABSORPTION OF ULTRASOUND IN ANTIMONY SULFOIODIDE CRYSTALS.           POLARIZATION E 15-2123
CIRCUIT PHOTOEMF IN FERROELECTRIC SEMICONDUCTORS. (ANTIMONY SULFOIODIDE CRYSTALS)                 SHORT 15-2105
ZATION ANC ETCHING).      DCMAIN STRUCTURE OF ANTIMONY SULFOIODIDE CRYSTALS (FROM SELECTIVE CRYSTALLI 15-2183
     PHASE TRANSITION AND DOMAIN STRUCTURE IN ANTIMONY SULFOIODIDE CRYSTALS. (MICRO CINEMATOGRAPHY) 15-2184
ECTS OF TEMPERATURE, F/ FIEZOELECTRIC VIBRATIONS IN ANTIMONY SULFOIODIDE CRYSTALS NEAR THE CURIE POINT (EFF 15-2121
OL OF NEEDLE MORPHOLOGY.        GRCWTH OF LARGE ANTIMONY SULFOIODIDE CRYSTALS (TO 1 CM DIAMETER): CONTR 15-2200
G SPECTRUM. (LAT/ PHASE TRANSITION IN FERROELECTRIC ANTIMONY SULFOIODIDE CRYSTALS USING THE RAMAN SCATTERIN 15-2143
/GATION CF THE PIEZCRESISTANCE OF (POLYCRYSTALLINE) ANTIMONY SULFOIODIDE (DARK CONDUCTIVITY NEAR TRANSITIO/ 15-2250
ENGTH, APPLIED FIELD STRENGT/ REFRACTIVE INDICES OF ANTIMONY SULFOIODIDE. (DEPENDENCE ON TEMPERATURE, WAVEL 15-2206
EFFECTS OF ILLUMINATION ON THE DOMAIN STRUCTURE OF ANTIMONY SULFOIODIDE. (ETCHING METHOD) 15-2124
TS CURIE POINT.       ABSORPTION OF ULTRASOUND IN AN ANTIMONY SULFOIODIDE FERROELECTRIC SEMICONDUCTOR NEAR I 15-2115
               CRYSTALLIZATION OF BISMUTH SULFOIODIDE. (HYDROTHERMAL SYNTHESIS) 15-2212
     CIELECTRIC AND PIEZOELECTRIC PROPERTIES OF ANTIMONY SULFOIODIDE IN A CONSTANT ELECTRIC FIELD. 15-2249
     AGING OF THE FERRCELECTRIC SEMICONDUCTOR ANTIMONY SULFOIODIDE IN A STATIC ELECTRIC FIELD. 15-2113
C/ ELECTRON STATE DENSITY AND OPTICAL PROPERTIES OF ANTIMONY SULFOIODIDE. (IN PARAELECTRIC AND FERROELECTRI 15-2204
N THE FORM OF THE VCLTAGE AMPERE CHARACTERISTICS OF ANTIMONY SULFOIODIDE IN THE PARAELECTRIC REGION. /ION O 15-2119
SOME ORDER DISORDER AND CISPLACIVE FERROELECTRICS. (ANTIMONY SULFOIODIDE, KDP) /S AND PHASE TRANSITIONS IN 15-2208
/ROREFLECTION FROM SINGLE CRYSTALS OF FERROELECTRIC ANTIMONY SULFOIODIDE NEAR THE PHASE TRANSITION TEMPERA/ 15-2151
/ND ILLUMINATION ON THE ABSORPTION OF ULTRASOUND IN ANTIMONY SULFOIODIDE NEAR THE PHASE TRANSITION TEMPERA/ 15-2248
ORUS.       HIGH GAS PRESSURE CRYSTAL GROWTH OF ANTIMONY SULFOIODIDE, NICKEL CARBONATE AND BLACK PHOSPH 15-2137
/PHOTOCONDUCTION ON THE FERROELECTRIC PROPERTIES CF ANTIMONY SULFOIODIDE. (PHASE BOUNDARY SHAPE STRIATIONS) 15-2197
RAMAN SPECTRUM OF (PARAELECTRIC AND) FERROELECTRIC ANTIMONY SULFOIODIDE (PHONON ASSIGNMENTS). (POLARIZED) 15-2207
S)       ANOMALOUS ECHO IN FERROELECTRIC ANTIMONY SULFOIODIDE. (PIEZOELECTRIC DOMAIN OSCILLATION 15-2141
/AVELENGTH OPTICAL PHONONS AND PHASE TRANSITIONS IN ANTIMONY SULFOIODIDE. (POLARIZED INFRARED AND RAMAN SP/ 15-2106
          FERROELECTRIC SEMICONDUCTORS. (ANTIMONY SULFOIODIDE, REVIEW, 57 REFS) 02-0084
/S AT TEMPERATURES NEAR A TYPE-1 PHASE TRANSITION. ANTIMONY SULFOIODIDE, SELENOIODIDE, BISMUTH SULFOIODID/ 15-2114
/(VI) AND A(V)B(VI)C(VII). (ANTIMONY TRISULFIDE, TRISELENIDE, ANTIMONY SULFOIODIDE, SELENOIODIDE, TELLUROIODIDE, TRI/ 15-2108
TRANSPORT)       CRYSTALLIZATICN BEHAVIOR OF ANTIMONY SULFOIODIDE. (SINGLE CRYSTAL GROWTH, CHEMICAL 15-2201
          DILATOMETRIC INVESTIGATION OF ANTIMONY SULFOIODIDE SINGLE CRYSTALS. 15-2155
EFFECT OF ILLUMINATION ON REPOLARIZATION PROCESS IN ANTIMONY SULFOIODIDE SINGLE CRYSTALS. 15-2220
ADSORPTION ON THE FERROELECTRIC PHASE TRANSITION IN ANTIMONY SULFOIODIDE SINGLE CRYSTALS. EFFECT OF OXYGEN 15-2187
CT OF THE ADSORPTION OF OXYGEN ON THE PROPERTIES OF ANTIMONY SULFOIODIDE SINGLE CRYSTALS.              EFFE 15-2154
     FEATURES OF THE PHASE TRANSITION IN ANTIMONY SULFOIODIDE SINGLE CRYSTALS. 15-2148
UENCY CEPENDENCE OF ANOMALOUS ULTRASOUND DAMPING IN ANTIMONY SULFOIODIDE SINGLE CRYSTALS.           FREQ 15-2139
     GENERATION OF THE SECOND ACCUSTIC HARMONIC IN ANTIMONY SULFOIODIDE SINGLE CRYSTALS. 15-2224
     GENERATION OF THE SECOND ACOUSTIC HARMONIC IN ANTIMONY SULFOIODIDE SINGLE CRYSTALS. 15-2223
SIONAL MECHANICAL STRESS ON THE PHASE TRANSITION OF ANTIMONY SULFOIODIDE SINGLE CRYSTALS. /NCE OF ONE DIMEN 15-2245
     INSTABILITY OF FERROELECTRIC POLARIZATION IN ANTIMONY SULFOIODIDE SINGLE CRYSTALS. 15-2152
          PHOTODIELECTRIC EFFECT IN ANTIMONY SULFOIODIDE SINGLE CRYSTALS. 15-2236
     PHOTOELECTRIC PROPERTIES (PHOTCCONDUCTIVITY) OF ANTIMONY SULFOIODIDE SINGLE CRYSTALS. 15-2134
          PIEZOELECTRICITY AND ELECTROSTRICTION OF ANTIMONY SULFOIODIDE SINGLE CRYSTALS. 15-2227
RATION AND PHYSICAL PROPERTIES OF NONSTOICHIOMETRIC ANTIMONY SULFOIODIDE SINGLE CRYSTALS.           PREPA 15-2199
          RELAXATION FROCESSES OF ANTIMONY SULFOIODIDE SINGLE CRYSTALS. 15-2190
TEMPERATURE DEPENCENCE OF THE SWITCHING FROCESS IN ANTIMONY SULFOIODIDE SINGLE CRYSTALS. 15-2191
/D PHOTOELECTRIC PROPERTIES OF THE CONTACTS BETWEEN ANTIMONY SULFOIODIDE SINGLE CRYSTALS AND SOME METALS. / 15-2251
SECONC LIGHT PULSES.      RELAXATION PROCESSES IN ANTIMONY SULFOIODIDE SINGLE CRYSTALS EXCITED WITH MICRO 15-2202
HASE COEXISTENCE.      POLARIZATICN REVERSAL OF ANTIMONY SULFOIODIDE SINGLE CRYSTALS IN THE REGION OF P 15-2146
URE AND INVESTIGATION OF THEIR ELEC/ PREPARATION OF ANTIMONY SULFOIODIDE TEXTURES FROM THE MELT UNDER PRESS 15-2138
/ASE TRANSITIONS IN ROCHELLE SALT, BARIUM TITANATE, ANTIMONY SULFOIODIDE, THIOUREA, TRIGLYCINE SULFATE, HY/ 02-0099
ATION NATURE IN SEMICONLUCTOR FERRCELECTRICS CF THE ANTIMONY SULFOIODIDE TYPE.                  POLARIZ 15-2181
          MOSSBAUER STUDIES OF ANTIMONY SULFOIODIDE TYPE CRYSTALS. 15-2180
ON/ NEW FERROELECTRIC A(V)B(VII)C(VII) COMPCUNDS OF ANTIMONY SULFOIODIDE TYPE. (PARTIAL OR TOTAL SUBSTITUTI 15-2205
.          SOME FEATURES CF TRAFPING LEVELS IN ANTIMONY SULFOIODIDE TYPE SEMICONDUCTING FERROELECTRICS 15-2182
     DOMAIN STRUCTURE AND LOCAL STATES IN ANTIMONY SULFOIODIDE TYPE SEMICONDUCTOR FERROELECTRICS. 15-2127
/STIC CONSTANTS FROM) BRILLOUIN SCATTERING STUDY OF ANTIMONY SULFOIODIDE USING A DOUBLE PASSED, STABILIZED/ 15-2225
IONS. (CIELECTRIC CONSTANTS, PYROELECTRIC CURRENTS, ANTIMONY SULFOIODIDE(X) BROMIDE(1-X)) /VII) SOLID SOLUT 15-2157
RATURE PHASE CHANGES. (EFFECT OF PHOTOCONDUCTIVITY, ANTIMONY SULFOIODIDE(X) BROMIDE(1-X)) /E WITH LOW TEMPE 15-2161
PEND ON TEMPERATU/ HOW THE DIELECTRIC PROPERTIES OF ANTIMONY SULFOIODIDE(X) BROMIDE(1-X) SOLID SOLUTIONS DE 15-2149
GMAN- STCCKEARGER METHOD)       CRYSTAL GROWTH IN THE ANTIMONY SULFOIODIDE- ANTIMONY TRISULFIDE SYSTEM. (BRID 15-2230
          DIELECTRIC FROPERTIES OF ANTIMONY SULFOIODIDE- SULFOBROMIDE SOLID SOLUTIONS. 15-2242
H.       MODE FOR THE ANTIMONY SULFOIODIDE- TIN DIOXIDE HETEROSTRUCTURE SWITC 15-2243
FILM.       SWITCHING ANC MEMORY EFFECT IN ANTIMONY SULFOIODIDE- TIN DIOXIDE HETEROSTRUCTURE THIN 15-2166
/AL OR TOTAL SUBSTITUTION OF ANTIMONY BY BISMUTH OR ARSENIC, SULFUR BY SELENIUM OR OXYGEN, AND IODINE BY B/ 15-2205
```

/RING A(V)B(VI)C(VII) COMPOUNDS. (USING ANTIMONY OR BISMUTH, SULFUR, SELENIUM OR TELLURIUM, AND CHLORINE, / 15-2194
/HIGH GRADE SINGLE CRYSTALS, V IS ANTIMONY OR BISMUTH, VI IS SULFUR, SELENIUM OR TELLURIUM, VII IS IODINE,/ 15-2214
ANSITIONS (TIN TELLUR/ EXPERIMENTAL RELATIONSHIP BETWEEN THE SUPERCONDUCTING AND THE FERROELECTRIC PHASE TR 36-3564
NTIUM TITANATE.   EFFECT OF STRESS ON THE SUPERCONDUCTING TRANSITION TEMPERATURE OF STRO 06-1182
Y (SEEM MUTUALLY EXCLUSIVE, THEREFORE FERROELECTRIC, BUT NOT SUPERCONDUCTIVE ORGANICS EXIST). /ROELECTRICIT 03-0377
MAGNETISM, FERROELECTRICITY, ORDERED ALLOYS, SUPERFLUIDITY, SUPERCONDUCTIVITY /. (THERMODYNAMICS, THEORY, 03-0183
ND CALCIUM(Y) STRONTIUM(1-Y) TITANATE(3-X).   SOFT MODE SUPERCONDUCTIVITY IN STRONTIUM TITANATE(3-X) A 06-1093
M MUTUALLY EXCLUSIVE, THEREFORE FERROELECTRIC, BUT NOT MAG/ SUPERCONDUCTIVITY VERSUS FERROELECTRICITY (SEE 03-0377
ANESE TRIFLUORIDE.   THEORY OF SUPEREXCHANGE REACTION. PART-1: POTASSIUM MANG 03-0252
NAMICS, THEORY, MAGNETISM, FERROELECTRICITY, ORDERED ALLOYS, SUPERFLUIDITY, SUPERCONDUCTIVITY) /. (THERMODY 03-0183
HIUM NIOBATE (TUNGSTEN BRONZE) TYPE FERROELECTRICS. SUPERLATTICE STRUCTURE OF STRONTIUM SODIUM LIT 13-2002
GROWTH OF TRIGLYCINE SULFATE FROM SLIGHTLY SUPERSATURATED SOLUTIONS.   27-3096
TIUM NIOBATE)   VARIATION OF CRYSTAL GROWTH RATE WITH SUPERSATURATION IN FLUXED MELTS. (BARIUM STRON 13-1939
LFATE.   SUPERSONIC DOMAIN WALL MOTION IN TRIGLYCINE SU 27-2963
/LAYERS ON BARIUM TITANATE- STRONTIUM TITANATE SUBSTRATES AT SUPERSONIC DRIFT VELOCITY OF CURRENT CARRIERS. 04-0592
TION IN POTASSIUM MANGANESE TRIFLUORIDE. SUPERSTRUCTURE ARISING DURING THE PHASE TRANSI 33-3402
MMONIUM ROCHELLE SALT.   SUPERSTRUCTURE IN THE FERROELECTRIC PHASE OF A 28-3224
VSKITES. (WHERE METAL IS SCANDIUM, CHROMIUM, MANGANESE, IRO/ SUPERSTRUCTURE OF BISMUTH METAL OXYGEN(3) PERO 08-1451
IDE.   PHASE RELATIONS AND SUPERSTRUCTURES OF PYRRHOTITE, IRON(1-X) SULF 36-3597
MAGNETIZATION OF LEAD FERRATE(0.5) NIOBATE(0.5) AND LEAD M/ SUPPRESSION OF THE SPONTANEOUS MAGNETOELECTRIC 13-2069
ALATE.   TEMPERATURE DEPENDENCE OF SURFACE ACOUSTIC WAVE VELOCITY IN LITHIUM TANT 12-1886
THIUM NIOBATE.   (FAR FIELD) DIFFRACTION SPREADING OF SURFACE (ACOUSTIC) WAVES ON (PIEZOELECTRIC) LI 11-1850
/ONTIUM ZIRCONATE AND STRONTIUM TITANATE (OF HIGH PURITY AND SURFACE ACTIVITY VIA HYDROLYTIC DECOMPOSITION/ 06-1217
/G. PART-5: MOSSBAUER EFFECT. PART-6: ELECTRICAL CONDUCTIVE, SURFACE, AND ELECTROOPTIC PROPERTIES. PART-7:/ 03-0397
ANATE.   CHARACTERIZATION OF SURFACE AND GRAIN BOUNDARY LAYER OF BARIUM TIT 04-0701
                                          SURFACE BARRIER ENERGIES ON STRONTIUM TITANATE 06-1175
ED BARIUM TITANATE- BISMUTH TRIOXIDE COMPOSITE CERAMICS WITH SURFACE BARRIER LAYER. /IVITY ON LANTHANUM DOP 04-0800
ELECTRONIC PROPERTIES OF ZINC OXIDE AND STRONTIUM TITANATE. (SURFACE BARRIER SYSTEMS, METAL CONTACTS) SOME 06-1176
.   SURFACE BREAKDOWN IN VACUUM ON BARIUM TITANATE 04-0596
UM NIOBATE CRYSTALS.   SURFACE CHARGE LAYERS ON FERROELECTRIC POTASSI 08-1369
ATE CRYSTALS.   SURFACE CONDUCTIVITY IN POLARIZED BARIUM TITAN 04-0569
DIELECTRIC RELAXATION, BULK AND SURFACE CONDUCTIVITY OF ICE SINGLE CRYSTALS.   34-3469
TE CRYSTALS DURING INTRACAVITY FREQUENCY DOUBL/ CATASTROPHIC SURFACE DAMAGE PRODUCED IN BARIUM SODIUM NIOBA 13-2052
ION AND ANNEALING OF POLYCRYSTALLINE BARIUM TITANATE. SURFACE DOMAIN REORIENTATION PRODUCED BY ABRAS 04-0639
ATE.   SURFACE EFFECT IN FERROELECTRIC YTTRIUM MANGAN 36-3588
F TGS CLOSE TO THE PHASE TRANSITION.   INFLUENCE OF SURFACE EFFECTS ON THE PYROELECTRIC BEHAVIOR O 27-3064
RPOSES. (TECHNIQUE APPLIED TO LITH/ ELECTRON BEAM SENSING OF SURFACE ELASTIC WAVES FOR SIGNAL PROCESSING PU 11-1738
ZIRCONATE(0.05) CERAMIC PLA/ PROPAGATION CHARACTERISTICS OF SURFACE ELASTIC WAVES ON BARIUM TITANATE(0.95) 09-1613
NDARIES IN A FERROELECTRIC.   SURFACE ELASTOPOLARIZATION WAVES AT DOMAIN BOU 03-0315
STRUCTURE IN TRIGLYCINE SULFATE. INFLUENCE OF DISLOCATIONS, SURFACE LAYER AND X-RAY IRRADIATION ON DOMAIN 27-3004
LS.   PROBLEM OF THE SURFACE LAYER IN BARIUM TITANATE SINGLE CRYSTA 04-1013
CTION, DC FIELD EFFECT, PARAELECTRIC AND FERROELECTRIC PHAS/ SURFACE LAYER IN BARIUM TITANATE (X-RAY DIFFRA 04-0857
L.   X-RAY STUDY OF THE SURFACE LAYER ON BARIUM TITANATE SINGLE CRYSTA 04-0858
E SINGLE CRYSTALS.   INFLUENCE OF THE SURFACE LAYER ON THE PLASTIC DEFORMATION OF IC 34-3459
C LEAD ZIRCONATE TITANATE (PLUS 1 PERCENT NIOBIUM PENTOXIDE) SURFACE LAYERS. /IES OF OXYGEN DEPLETED CERAMI 09-1576
FOIOCI/ EXTRINSIC PHOTOCONDUCTIVITY IN FERROELECTRICS DUE TO SURFACE LAYERS. (BARIUM TITANATE, ANTIMONY SUL 04-0681
SHOCK WAVE STUDIES OF SURFACE LAYERS IN BARIUM TITANATE.   04-0588
GNETTE SALT.   CORRELATION BETWEEN BARKHAUSEN NOISE AND SURFACE MICRODAMAGES ON SINGLE CRYSTALS OF SEI 28-3208
TITANATE NEAR THE CURIE TEMPERATURE.   EMISSION OF LIGHT BY SURFACE MICROPLASMA OF CERAMIC (BARIUM, LEAD) 04-0982
ULTRAFINE BARIUM TITANATE PARTICULATES.   EFFECT OF SURFACE MOISTURE ON THE DIELECTRIC BEHAVIOR OF 04-0859
/E TEMPERATURE DEPENDENCE OF LEED INTENSITIES FROM THE (001) SURFACE OF BARIUM TITANATE BETWEEN 20 DEGREES/ 04-0531
PHOTOEMISSION MICROSCOPIC OBSERVATION OF THE SURFACE OF BARIUM TITANATE CRYSTALS.   04-0810
SEM OBSERVATION OF THE FERROELECTRIC SURFACE OF BARIUM TITANATE CRYSTALS.   04-0812
COPE.   STUDY OF THE SURFACE OF ICE WITH A SCANNING ELECTRON MICROS 34-3420
HOUT CRATERS.   GROWTH OF CRATERS ON ION ETCHED SURFACE OF LITHIUM NIOBATE AND ION ETCHING WIT 11-1860
DAMAGE TO THE SURFACE OF LITHIUM NIOBATE BY LIGHT.   11-1866
RMINING FACTOR OF THE GROWTH ST/ ELECTRICAL STRUCTURE OF THE SURFACE OF REAL CRYSTAL SUBSTRATES AS THE DETE 27-3039
PHOTOEMISSION MICROSCOPIC OBSERVATION OF THE SURFACE OF TUNGSTEN TRIOXIDE CRYSTALS.   10-1645
Y).   EFFECT OF THE SURFACE ON POLARIZATION OF A DIELECTRIC (THEOR 03-0271
F TUNGSTEN- TUNGSTEN OXIDE AS A FUNCTION OF THICKNESS OF THE SURFACE OXIDE LAYER. /PHOTOELECTRON SPECTRUM O 10-1631
CALCULATION OF FIELD NEAR SURFACE)   MICROSCOPIC THEORY OF SURFACE PHENOMENA IN FERROELECTRIC CRYSTALS. (03-0501
DISTRIBUTIONS IN DIELECTRICS, WITH SPECIAL EMPHASIS ON NEAR SURFACE REGIONS IN FERROELECTRICS. /TRIC FIELD 03-0182
ROELECTRIC CRYSTALS.   ON SURFACE STATE AND POLARIZATION REVERSAL OF FER 03-0323
/RIUM) SEMICONDUCTOR FILMS. (AND USE OF THESE FILMS TO STUDY SURFACE STRAINS IN THE FERROELECTRIC CRYSTAL) 04-0594
URES REVEALED BY WATER ETCHED ROCHELLE SALT CRYSTALS. SURFACE STRUCTURES AND CORRESPONDING LIGHT FIG 28-3234
TITANATE.   LEED STUDY OF SURFACE STRUCTURES ON THE (001) FACE OF BARIUM 04-0529
ING THE SCANNING ELECTRON MICROSCOPE. SURFACE STUDIES OF BARIUM TITANATE CERAMICS US 04-0567
CONDUCTION. SECT-3: EXTRINSIC VERSUS INTRINSIC POLARIZATION, SURFACE VERSUS VOLUME CONDUCTION. /ZATION AND 34-3482
/ONDUCTION. PART-3: EXTRINSIC VERSUS INTRINSIC POLARIZATION, SURFACE VERSUS VOLUME CONDUCTION (INCLUDES EL/ 34-3455
LOW POWER DENSITY GALLIUM ARSENIDE- LITHIUM NIOBATE SURFACE WAVE AMPLIFIER.   11-1825
/SMUTH(12) GERMANIUM OXIDE(20) VERSUS LITHIUM NIOBATE IN THE SURFACE WAVE AMPLIFIER. A THEORETICAL AND EXP/ 11-1720
LITHIUM NIOBATE SILICON SURFACE WAVE CONVOLUTER.   11-1788
DETERMINATION OF A NONLINEAR PARAMETER FOR ACOUSTIC SURFACE WAVE CONVOLUTION IN LITHIUM NIOBATE.   11-1681
M SULFIDE, LITHIUM NIOBATE AND BISMUTH(12) GERMANI/ ACOUSTIC SURFACE WAVE CONVOLUTION ON CRYSTALS OF CADMIU 11-1692
DE FLUID LITHIUM NIOBATE STRUCTURE. SURFACE WAVE CONVOLUTION USING A CADMIUM SULFI 11-1700
NONLINEAR EFFECTS IN MICROWAVE ACOUSTIC LITHIUM NIOBATE SURFACE WAVE DELAY LINES.   11-1818
NSDUCERS ON UNPOLARIZED PZT CERAMIC PLATES. SURFACE WAVE DELAY LINES WITH INTERDIGITAL TRA 09-1614
SURFACE WAVE DIFFRACTION IN LITHIUM NIOBATE.   11-1698
VHF AND UHF. (BASED ON LEAD(0.35) STRONTIUM(0.65/ (COMPACT, SURFACE WAVE) FERROELECTRIC PHASE SHIFTERS FOR 37-3633
IOBATE.   REFLECTION OF ELASTIC SURFACE WAVE FROM PERIODIC ARRAYS ON LITHIUM N 11-1853
ARIZATIONS.   ACOUSTIC SURFACE WAVE GUIDE. AVAILING FERROELECTRIC POL 37-3680
IOBATE)   OPTICAL PROBING OF ACOUSTIC SURFACE WAVE HARMONIC GENERATION (IN LITHIUM N 11-1752
IC COUPLING).   PIEZOELECTRIC LEAKY SURFACE WAVE IN LITHIUM NIOBATE (ELECTROMAGNET 11-1832
RMANIUM OXIDE(20) AT MICROWAVE FREQUENCIES.   ACOUSTIC SURFACE WAVE LOSS MECHANISMS ON BISMUTH(12) GE 35-3511
E (TEMPERATURE AND FREQUENCY D/ MICROWAVE FREQUENCY ACOUSTIC SURFACE WAVE LOSS MECHANISMS ON LITHIUM NIOBAT 11-1817
ELASTIC SURFACE WAVE ON THE PZT CERAMIC PLATE.   09-1612
C SUBSTRATES: GOLD ON LIT/ FUNDAMENTAL AND HIGHER ORDER MODE SURFACE WAVE PROPAGATION ON LAYERED ANISOTROPI 11-1846
ILM ON LITHIUM NIOBATE.   ACOUSTOELECTRIC AMPLIFICATION OF SURFACE WAVE STRUCTURE OF A CADMIUM SELENIDE F 11-1723
UM TANTALATE.   SURFACE WAVE TEMPERATURE COEFFICIENTS ON LITHI 11-1852
/ZATION OF A 40 MHZ COLOR TELEVISION INFRARED RESPONSE USING SURFACE WAVE TRANSDUCERS ON LEAD ZIRCONATE TI/ 09-1509

MEASURED ELECTRICAL CHARACTERISTICS OF INTERDIGITAL SURFACE WAVE TRANSDUCERS ON LITHIUM NIOBATE.   11-1702
E, BISMUTH(12) GERMANI/ MICROWAVE ACOUSTICS HANDBOOK. VOL 1: SURFACE WAVE VELOCITIES. (BARIUM SODIUM NIOBAT 01-0028
BATE, LITHIUM TANTALAT/ TEMPERATURE COEFFICIENTS OF ACOUSTIC SURFACE WAVE VELOCITY AND DELAY ON LITHIUM NIO 11-1821
IC MEDIA. (LITHIUM NI/ POSSIBILITY OF SWITCHING AND STEERING SURFACE WAVES BY NONLINEAR MIXING IN ANISOTROP 11-1778
/ OPTICAL STUDIES OF REFRACTION AND REFLECTION OF ULTRASONIC SURFACE WAVES BY THIN GOLD FILM LAYERS ON LITH 11-1789
IUM NIOBATE SUBSTRA/ REFRACTION AND REFLECTION OF ULTRASONIC SURFACE WAVES BY THIN GOLD FILM LAYERS ON LITH 11-1790
ON LITHIUM NIOBATE.   CONVOLUTION OF SURFACE WAVES IN A STRUCTURE OF SEMICONDUCTOR 11-1847
LS. (BISMUTH(12) GERMANIUM OXIDE(20) AND BARI/ PIEZOELECTRIC SURFACE WAVES IN CUBIC AND ORTHORHOMBIC CRYSTA 13-2041
 AND AMPLIFICATION OF RAYLEIGH WAVES AND PIEZOELECTRIC LEAKY SURFACE WAVES IN LITHIUM NIOBATE. PROPAGATION 11-1856
CTRIC (CLASS 6MM).   PIEZOELECTRIC SURFACE WAVES ON A C-DOMAIN WALL OF A FERROELE 03-0421
/ OF THERMALLY DIFFUSED IMPURITIES ON PROPAGATION OF ELASTIC SURFACE WAVES ON FERROELECTRIC CERAMICS. (IMP/ 03-0487
                   DIFFRACTION SPREADING OF SURFACE WAVES ON LITHIUM NIOBATE.   11-1849
              MIXING OF NONCOLLINEAR ELASTIC SURFACE WAVES ON LITHIUM NIOBATE.   11-1690
       NEW HIGH COUPLING LOW DIFFRACTION CUT FOR ACOUSTIC SURFACE WAVES ON LITHIUM NIOBATE.   11-1820
H FREQUENCY HIGH COUPLING LOW BEAM STEERING CUT FOR ACOUSTIC SURFACE WAVES ON LITHIUM NIOBATE.   NEW HIG 11-1819
                          PIEZOELECTRIC SURFACE WAVES ON LITHIUM NIOBATE.   11-1824
      EFFECT OF MASS LOADING ON THE PROPAGATION OF ACOUSTIC SURFACE WAVES ON LITHIUM NIOBATE AND QUARTZ. 11-1757
PERTURBATION THEORY FOR ELECTROMAGNETIC COUPLING TO ELASTIC SURFACE WAVES ON PIEZOELECTRIC SUBSTRATES (LI/ 11-1751
           DECORATION OF ION BOMBARDED LITHIUM NIOBATE SURFACES.   11-1799
/TRATES. (TELLURIUM ON TRIGLYCINE SULFATE AND SODIUM NITRITE SURFACES) (DECORATION METHOD OF DOMAIN STUDY) 16-2394
(BARRIER LAYERS IN / IMAGING ELECTRIC MICRO FIELDS (OF SOLID SURFACES) IN THE EMISSION ELECTRON MICROSCOPE 04-0913
ULSES)   ON THE SWITCHING PROPERTIES OF THE FERROELECTRIC SURFACES. (NEW MODEL REPRODUCES EXPERIMENTAL P 37-3641
MISSION ELECTRON MICROSCOPE.   OBSERVATION OF THE SURFACES OF FERROELECTRIC CRYSTALS IN A PHOTOE 37-3679
UCTING ELEMENTS.   INVESTIGATIONS OF SURFACES OF FERROELECTRICS (TGS) WITH SEMICOND 27-3089
 SELECTIVE DEPOSITION OF CHARG/ ELECTRICAL PROPERTIES OF THE SURFACES OF IONIC CRYSTALS (TGS AND OTHERS, BY 27-2987
 QUARTZ, AMMONIUM DIHYDROGEN PHOSPHATE, A/ OPTICAL NONLINEAR SUSCEPTIBILITIES: ACCURATE RELATIVE VALUES FOR 17-2539
NTAL/ DETERMINATION OF ABSOLUTE SIGNS OF MICROWAVE NONLINEAR SUSCEPTIBILITIES. (BARIUM TITANATE, LITHIUM TA 02-0107
ISORDER PHONON SYSTEM WITH DAMPING: KDP (FREQUENCY DEPENDENT SUSCEPTIBILITY).   COUPLED ORDER D 17-2667
NEAR THE TRANSITION POINTS IN SODIUM NITRIT/ STATIC DIELECTRIC SUSCEPTIBILITY AND DIELECTRIC RELAXATION TIME 16-2302
PHASE TRANSITIONS.   DIELECTRIC SUSCEPTIBILITY AND THE ORDER OF FERROELECTRIC 03-0223
F FERROELECTRICS. (ABSTRACT ONLY, DISCUSSION)   GENERALIZED SUSCEPTIBILITY AND THE SCATTERING PROPERTIES O 03-0209
M ABOVE THE CRITICAL TEMPERATURE.   SPECIFIC HEAT AND SUSCEPTIBILITY IN CHROMIUM METHYL AMMONIUM ALU 20-2805
        NONLINEAR OPTICAL SPECTROSCOPY OF THIRD ORDER SUSCEPTIBILITY IN LITHIUM NIOBATE.   11-1855
CATTERING, IDLER ABSORPTION, AND DISPERSION IN THE NONLINEAR SUSCEPTIBILITY IN LITHIUM NIOBATE. /RAMETRIC S 11-1771
 REGION OF POTASSIUM DIDEUTERIU/ POLARIZATION RELAXATION AND SUSCEPTIBILITY IN THE FERROELECTRIC TRANSITION 17-2575
VERSAL. (IN BARIUM TITANATE AND TRIGLYCINE SULFATE)   SUSCEPTIBILITY INCREASE DURING POLARIZATION RE 04-0981
HATE CRYSTAL.   CURIE-WEISS LAW FOR THE NONLINEAR SUSCEPTIBILITY OF A POTASSIUM DIHYDROGEN PHOSP 17-2686
                           ELECTRICAL SUSCEPTIBILITY OF ADP (THEORY).   17-2686
/IUM VALENCY FROM STUDIES OF EPR SPECTRA AND STATIC MAGNETIC SUSCEPTIBILITY OF CHROMIUM DOPED BARIUM METAT/ 04-0542
TOR. A UNIFIED OSCILLATOR MODEL FOR ORDER DISORDER / DYNAMIC SUSCEPTIBILITY OF CLASSICAL ANHARMONIC OSCILLA 03-0405
HEXAHYDRATE AT LOW TEMPERATURES.   MAGNETIC SUSCEPTIBILITY OF GUANIDINIUM VANADIUM SULFATE 21-2815
HYDROGEN(X) T/ ELECTRONIC PROPERTIES (CONDUCTIVITY, MAGNETIC SUSCEPTIBILITY) OF HYDROGEN TUNGSTEN BRONZES, 13-1934
             NONLINEAR OPTICAL SUSCEPTIBILITY OF LITHIUM FORMATE MONOHYDRATE. 26-2945
UCED REFRACTIVE INDEX CHANGES.   CONTROL OF THE SUSCEPTIBILITY OF LITHIUM NIOBATE TO LASER IND 11-1791
TE TYPE FERRO/ PROTON LATTICE INTERACTIONS AND THE DYNAMICAL SUSCEPTIBILITY OF POTASSIUM DIHYDROGEN PHOSPHA 17-2570
M TITANATE.   DIELECTRIC SUSCEPTIBILITY OF QUASI FERROELECTRIC STRONTIU 06-1133
     TEMPERATURE DEPENDENCE OF THE OPTICAL NONLINEAR SUSCEPTIBILITY OF SODIUM NITRITE.   16-2386
/ OF FERROELECTRICS AND FERROELASTICS SUCH THAT THE ELECTRIC SUSCEPTIBILITY OR ELASTIC COMPLIANCE IS TEMPE/ 03-0140
DE FOR THE ANTIMONY SULFOIODIDE- TIN DIOXIDE HETEROSTRUCTURE SWITCH.   MO 15-2243
GREES, Z CUT POTASSIUM DIDEUTERIUM PHOSPHATE. ELECTROOPTIC Q SWITCH BY DESK-TOP COMPUTER. /F A DOUBLE 45 DE 17-2702
E.   AN ELECTROOPTIC SWITCH USING ROTATED Y PLATE OF LITHIUM NIOBAT 11-1732
ATORS. (POTASSIUM DIHYDROGEN PHOSPHA/ (FINITE APERTURE LIGHT SWITCH USING) TRANSVERSE POCKELS EFFECT COMMUT 17-2509
TAL LIGHT DEFLECTOR UTILIZING LITHIUM TANTALATE POLARIZATION SWITCHES. /AND DEFLECTION CAPACITIES OF A DIGI 11-1809
          DOMAIN WALL MOTION IN FERROELECTRIC SWITCHING.   03-0273
            FERROELECTRIC DOMAIN SWITCHING.   37-3642
OTASSIUM DIDEUTERIUM PHOSPHATE POCKELS CELL FOR HIGH POWER Q SWITCHING.   LOSSLESS P 17-2529
        PERMITTIVITY OF ROCHELLE SALT DURING SWITCHING.   28-3196
ODIDE- TIN DIOXIDE HETEROSTRUCTURE THIN FILM.   SWITCHING AND MEMORY EFFECT IN ANTIMONY SULFOI 15-2166
CONDUCTOR MEMOR/ SOME THEORETICAL CONSIDERATIONS RELATING TO SWITCHING AND REMANENCE IN FERROELECTRIC PHOTO 03-0378
EAR MIXING IN ANISOTROPIC MEDIA. (LITHIUM NI/ POSSIBILITY OF SWITCHING AND STEERING SURFACE WAVES BY NONLIN 11-1778
            FERROELECTRIC SWITCHING AND THE SIEVERT INTEGRAL.   03-0239
          YTTRIUM MANGANATE SWITCHING BEHAVIOR.   36-3571
NITRATE.   SWITCHING BEHAVIOR IN FERROELECTRIC POTASSIUM 16-2285
P OR SMOOTH) DOMAIN WALL STRUCTURE IN BARIUM TITANATE DURING SWITCHING (CALCULATIONS COMPARED WITH DATA). / 04-0677
THE FE-PC MEMO/ EFFECT OF GAUSSIAN IRRADIANCE PROFILE ON THE SWITCHING CHARACTERISTICS AND THE BIT SIZE OF 37-3670
TITANATE FERROELECTRIC CERAMIC. SOME PHYSICAL AND ELECTRICAL SWITCHING CHARACTERISTICS OF A LEAD ZIRCONATE 09-1601
OF POLARIZATION REVERSAL IN FERROELECTRICS. (CALCULATIONS OF SWITCHING CURRENT)   THEORY 03-0422
           QUASI STATIC SWITCHING CURRENT IN SODIUM NITRITE.   16-2261
      BARKHAUSEN JUMPS AND SWITCHING CURRENT IN TGS CRYSTALS.   27-3109
TALS.   ANTIPARALLEL SWITCHING IN ALPHA DOMAIN BARIUM TITANATE CRYS 04-0655
       EFFECT OF "A" DOMAINS UPON "C" DOMAIN SWITCHING IN BARIUM TITANATE.   04-0615
LFATE SINGLE CRYSTALS.   REMARKS ABOUT SWITCHING IN BARIUM TITANATE AND TRIGLYCINE SU 04-0915
       LATTICE DYNAMICAL THEORY OF SWITCHING IN BARIUM TITANATE SINGLE CRYSTALS.   03-0398
PURE AND DOPED WITH PARAMAGNETIC IONS.   SWITCHING IN FERROELECTRIC TRIGLYCINE SULFATE, 27-3020
/ ZERO SHIFT IN PIEZOELECTRIC ACCELEROMETERS. (POLARIZATION SWITCHING IN POLYCRYSTALLINE FERROELECTRICS AT 28-3216
CE OF FERROELECTRIC TRIGLYCINE FLUOBERYLLATE CRYSTALS DURING SWITCHING IN PULSE ELECTRIC FIELD.   IMPEDAN 27-3027
            FERROELECTRIC DOMAIN SWITCHING IN RHOMBOHEDRAL PHASE PLZT CERAMICS. 09-1563
TANATE (ELECTRON MICROSC/ MOTION OF DOMAIN BOUNDARIES DURING SWITCHING IN SINGLE CRYSTAL FILMS OF BARIUM TI 04-0940
               FERROELECTRIC POLARIZATION SWITCHING IN SODIUM NITRITE.   16-2301
    STUDY OF THE DYNAMICS OF PHOTOCHROMIC SWITCHING IN STRONTIUM TITANATE MATERIALS.   06-1092
IVES.   STRUCTURAL INHIBITION OF FERROELECTRIC SWITCHING IN TRIGLYCINE SULFATE. PART-1: ADDIT 27-3038
ADIATION TREATMENT.   STRUCTURAL INHIBITION OF FERROELECTRIC SWITCHING IN TRIGLYCINE SULFATE. PART-2: X-IRR 17-2469
/URFACE, AND ELECTROOPTIC PROPERTIES. PART-7: DIELECTRIC AND SWITCHING MEASUREMENTS. PART-8: ULTRASOUND PR/ 03-0397
AMICS WITH INDIUM ELECTRODES.   IMPROVED AGING AND SWITCHING OF LEAD ZIRCONATE- LEAD TITANATE CER 09-1516
AMICS WITH INDIUM ELECTRODES.   IMPROVED AGING AND SWITCHING OF LEAD ZIRCONATE- LEAD TITANATE CER 09-1517
ECTRIC LEAD(5) GERMANIUM(3) OXIDE(11) SINGLE CRYSTAL.   SWITCHING OF OPTICAL ROTATORY POWER IN FERROEL 36-3570
ITANATE AT LOW TEMPERATURES.   FIELD INDUCED SWITCHING OF TETRAGONAL DOMAINS IN STRONTIUM T 06-1184
ECTRIC POTASSIUM NIOBATE SINGLE CRYSTALS.   MECHANICAL SWITCHING OF 60 DEGREE DOMAIN WALLS IN FERROEL 08-1421
            STUDY OF SWITCHING PARAMETERS IN TRIGLYCINE SULFATE. 27-3095
   KINETICS OF DOMAIN WALL MOTION IN FERROELECTRIC SWITCHING. PART-1: GENERAL FORMULATION.   03-0274

DP POCKELS CELLS.　　　TRANSIENT ELASTOOPTIC EFFECTS AND Q SWITCHING PERFORMANCE IN LITHIUM NIOBATE AND K 11-1727
OTASSIUM DIDEUTERIUM P/ TRANSIENT ELECTROOPTIC EFFECTS AND Q SWITCHING PERFORMANCE IN LITHIUM NIOBATE AND P 11-1710
LE CRYSTALS.　　　TEMPERATURE DEPENDENCE OF THE SWITCHING PROCESS IN ANTIMONY SULFOIODIDE SING 15-2191
EFFECT OF X- AND GAMMA-RADIATION ON THE SWITCHING PROCESS IN TRIGLYCINE SULFATE. 27-3019
FAMILY BISMUTH(4) BARIUM(M-2) TITANIUM(M+1) OXIDE(3) (M+2) (/ SWITCHING PROPERTIES IN FERROELECTRICS OF THE 02-0066
OLEMANITE.　　　SWITCHING PROPERTIES OF BIASED FERROELECTRIC C 23-2888
RY DEVICE/ UTILIZATION OF THE (CRITICAL PULSE WIDTH) PARTIAL SWITCHING PROPERTIES OF FERROELECTRICS IN MEMO 27-3144
(4) TITANIUM(3) OXIDE(12).　　　ELECTRICAL AND OPTICAL SWITCHING PROPERTIES OF SINGLE CRYSTAL BISMUTH 35-3517
ACES. (NEW MODEL REPRODUCES EXPERIMENTAL PULSES)　　ON THE SWITCHING PROPERTIES OF THE FERROELECTRIC SURF 37-3641
IC POTASSIUM NITRATE.　　　SWITCHING PROPERTIES OF THICK FILM FERROELECTR 16-2282
EFFECT OF PRESSURE ON THE SWITCHING RATE OF TRIGLYCINE SULFATE. 27-3014
TEMPERATURE DEPENDENCE OF THE SWITCHING RATE OF TRIGLYCINE SULFATE. 27-3015
/R ELECTRONIC TUNING. (WITH ELECTRODES REDUCING POLARIZATION SWITCHING TIME, LEAD(0.35) STRONTIUM(0.65) TI/ 37-3660
ULFATE, FLUOBERYLLATE, ETC/ (EFFECT OF TEMPERATURE ON) QUASI SYMMETIRICAL HARD IONS (PHOSPHATE, TITANATE, S 03-0330
/TERIALS AT CENTIMETER AND DECIMETER WAVELENGTHS EMPLOYING A SYMMETRIC STRIPLINE. (BARIUM TITANATE AND BAR/ 04-0855
RAMAN SCATTERING IN PIEZOELECTRIC CRYSTALS WITH TETRAGONAL SYMMETRY. 03-0407
INITE SAMPLES OF FERRODIELECTRICS WITH AN ARBITRARY MAGNETIC SYMMETRY. /N WAVE PHENOMENOLOGICAL THEORY IN F 03-0164
IC ICE.　　　SYMMETRY ANALYSIS AND ELECTRONIC STATES IN CUB 34-3462
(18 REFS)　　　DIELECTRIC, ELASTIC AND MAGNETIC PROPERTY, SYMMETRY, AND PHASE TRANSITION REPRESENTATION. 03-0316
TIVE AND PREDICTIVE POSSIBILITIES)　　　(MACROSCOPIC) SYMMETRY ASPECTS OF FERROELECTRICITY. (DESCRIP 03-0464
ORACITES.　　　SYMMETRY ASPECTS OF THE PHASE TRANSITIONS IN A 14-2077
TRANSITIONS IN CESIUM LEAD TRICHLORIDE. (LOSS OF A CENTER OF SYMMETRY AT 194 DEGREES K)　　　PHASE 33-3368
EORY OF VIBRATIONS IN FERROELECTRIC FERROMAGNETICS WITHOUT A SYMMETRY CENTER.　　　QUANTUM TH 03-0166
OF THE STRUCTURAL PHASE TRANSITION IN GADOLINIUM MOLYBDATE. (SYMMETRY CHANGES)　　　ORIGIN 30-3273
ONS BETWEEN MAGNETIZATION AND POLARIZATION: PHENOMENOLOGICAL SYMMETRY CONSIDERATIONS ON BORACITES. /TERACTI 14-2072
ATION TO BARIUM TITANAT/ FERROELECTRIC AND ANTIFERROELECTRIC SYMMETRY GROUPS. (AND THEIR COMPARISON, APPLIC 04-0933
COUPLED POLARITONS OF A1 SYMMETRY IN BARIUM TITANATE. 04-0619
YSTALLINE FERROELECTRICS.　　　SYMMETRY LIMITATIONS TO POLARIZATION OF POLYCR 03-0434
YDROGEN PHOSPHATE.　　　SYMMETRY OF ATOMIC VIBRATIONS IN POTASSIUM DIH 17-2586
D FERROELECTRIC PROPERTIES APPEAR SIMULTANEOUSLY.　　　SYMMETRY OF CRYSTALS IN WHICH FERROMAGNETIC AN 03-0465
OME COMMENTS)　　　(MACROSCOPIC AND MICROSCOPIC) SYMMETRY OF FERROELECTRIC ENERGY FUNCTIONS. (S 03-0193
TRIC CRYSTAL OF KDP.　　　SYMMETRY OF NORMAL OSCILLATIONS IN A FERROELEC 17-2623
, REFRACTIVE INDICES) RELATION OF PHYSICAL PROPERTIES TO THE SYMMETRY OF POTASSIUM IODATE. (NONLINEAR OPTIC 08-1296
SITIONS IN FERROELECTRICS AND FERROMAGNE/ CHANGE OF COMPLETE SYMMETRY OF THE CRYSTALS DURING THE PHASE TRAN 03-0527
SIUM OR SODIUM) (0.5) BISMUTH(0.5/ X-RAY DETERMINATION OF THE SYMMETRY OF THE FERROELECTRIC COMPOUNDS (POTAS 08-1362
ON THE SYMMETRY OF THE HYDROGEN BONDS IN ICE-VII. 34-3445
C PHASE TRANSITIONS.　　　ON THE SYMMETRY OF THE SOFT MODES IN THE FERROELECTRI 03-0357
STRUCTURAL BASIS OF FERROELECTRICITY AND FERROELASTICITY. (SYMMETRY RELATIONS) 03-0135
MMONIUM SULFATE/ PROTON MAGNETIC RESONANCE STUDY OF THE SPIN SYMMETRY STATES OF AMMONIUM IONS IN SOLIDS. (A 19-2800
FERROELECTRICITY AND SYMMETRY. (41 REFS) 03-0528
SYMPOSIUM ON APPLICATIONS OF FERROELECTRICS. 01-0005
OTORS RES LABS, WARREN, MICHIGAN, PROC. FERROELECTRICITY. SYMPOSIUM ON FERROELECTRICITY, 1966, GENERAL M 01-0034
EN PHOSPHATE.　　　PHASE SYNCHRONIZATION IN NONLINEAR RUBIDIUM DIHYDROG 17-2473
RTAIN NEW LAYERED FERROELECTRICS. (LANTHANUM BISMUTH TITANI/ SYNTHESES, X-RAY, AND DIELECTRIC STUDIES OF CE 13-1993
CRYSTALLIZATION OF BISMUTH SULFOIODIDE. (HYDROTHERMAL SYNTHESIS) 15-2212
PREPARATION OF NICKEL BORACITE CRYSTALS BY HYDROTHERMAL SYNTHESIS. 14-2079
THE SYSTEM LEAD TITANATE- STRONTIUM COPPER NIOBATE. (CERAMIC SYNTHESIS) 05-1044
NDS WITH A PYROCHLORE STRUCTURE. (NIOBATES, TANTALATES) SYNTHESIS AND CERTAIN PROPERTIES OF NEW COMPOU 02-0047
ECTRIC AND SEMICONDUCTING FILMS. (THE MAJOR PORTION OF THIS/ SYNTHESIS AND CHARACTERIZATION OF THIN FERROEL 02-0057
/INGLE CRYSTALS OF THE SYSTEM (BARIUM, STRONTIUM) TITANATE. (SYNTHESIS AND LOW FREQUENCY DIELECTRIC PROPER/ 04-0561
ISMUTH GERMANATE.　　　CZOCHRALSKI SYNTHESIS AND PROPERTIES OF RARE EARTH DOPED B 35-3496
INGLE CRYSTALS.　　　SYNTHESIS AND STUDY OF LEAD SCANDIUM NIOBATE S 08-1325
/ND SEIGNETTO MAGNETS OF THE PEROVSKITE AND PYROCHLORE TYPE. SYNTHESIS AND STUDY OF THE CRYSTAL STRUCTURE / 02-0128
E CRYSTALS OF A NEW PYROCHLORE CONTAINING LITHIUM, STRONTIU/ SYNTHESIS AND X-RAY DIFFRACTION STUDY OF SINGL 36-3613
SKITES BARIUM OR STRONTIUM METAPLUMBATES (NOT FERROELECTRIC/ SYNTHESIS AND X-RAY INVESTIGATION OF THE PEROV 36-3607
ONS OF THE M MODIFICATION OF SAMARIUM MOLYBDATE (SOLID PHASE SYNTHESIS, NEW MODIFICATION). /SE TRANSFORMATI 30-3270
KITE.　　　HIGH PRESSURE SYNTHESIS OF LEAD METAL- OXYGEN(3) TYPE PEROVS 08-1331
TANATE WITH IMPROVED ELECTRICAL PROPERTIES. (DIELECTRIC CON/ SYNTHESIS OF NIOBIUM PENTOXIDE DOPED BARIUM TI 04-0840
SYNTHESIS OF OXIDE BRONZES. (REVIEW, 88 REFS) 02-0042
LEAD MANGANATE(0.5) NIOBATE(0.5) AND LEAD MANGANATE(0.33) N/ SYNTHESIS OF PEROVSKITE COMPLEX COMPOSITIONS: 08-1453
T HIGH PRESSURES AND THEIR X-RAY ANALYSIS.　　　SYNTHESIS OF THALLIUM CONTAINING PEROVSKITES A 08-1463
D PROPERTIES OF PEROVSKITE TYPE ZINCATE(/ HIGH PRESSURE SYNTHESIS (SOLID STATE REACTION, SINTERING) AN 08-1393
UTH SULFOBROMIDE)　　　CRYSTAL GROWTH AND CHEMICAL SYNTHESIS UNDER HYDROTHERMAL CONDITIONS. (BISM 15-2218
ELECTRIC PROPERTIES OF NEW CADMIUM-CONTAINING PEROVSKITES A/ SYNTHESIS, X-RAY ANALYSIS, AND STUDY OF THE DI 08-1309
STRUCTURE AND DIELECTRIC LOSSES OF FERROELECTRIC PEROVSKITES SYNTHESIZED AT HIGH PRESSURES. 08-1452
F FERROELECTRIC CHARACTER IN AN APP/ MONOCLINIC STRUCTURE OF SYNTHETIC CALCIUM CHLORAPATITE. (DEVELOPMENT O 36-3591
HIOMETRIC BARIUM TITANATE.　　　WET SYNTHETIC METHOD OF MAKING PURE, EXACTLY STOIC 04-0790

T

ERMISSIBLE DOMAIN WALLS IN FERROELECTRIC SPECIES (THEORY AND TABULATION)　　　P 03-0247
OPTICAL GAP OF STRONTIUM TITANATE. (DEVIATION FROM URBACH TAIL BEHAVIOR) 06-1111
/ELECTRIC FIELD ON A DIFFUSE FERROELECTRIC PHASE TRANSITION, TALKING INTO ACCOUNT THE INTERACTING KANZIG R/ 03-0441
RE AUTOSTABILIZATION REGIME OF A FERROELECTRIC CRYSTAL. (TGS TANDEL) /CTRIC NONLINEARITIES IN THE TEMPERATU 17-2502
ABILIZATION EFFECT IN FERROELECTRICS. (CAPACITOR PROPERTIES, TANDEL)　　　TEMPERATURE AUTOST 37-3655
URE)　　　AGING PROCESSES IN TRIGLYCINE SULFATE. (TANDEL OR CAPACITANCE PROPERTIES DOMAIN STRUCT 27-3081
/ROWTH CONDITIONS ON A CHANGE IN THE DIELECTRIC CONSTANT AND TANGENT OF THE ANGLE OF DIELECTRIC LOSS IN A / 27-3151
/NS OF TRIGLYCINE SULFATE SINGLE CRYSTALS ON A CHANGE IN THE TANGENT OF THE ANGLE OF DIELECTRIC LOSSES IN / 27-3153
E CONDITI/ DIELECTRIC PERMEABILITY AND DIELECTRIC LOSS ANGLE TANGENT OF TRIGLYCINE SELENATE DEPENDING ON TH 27-3125
AN X-RAY STUDY OF THE PHASE TRANSITION IN STRONTIUM FERRATE TANTALATE 08-1382
ION SHIFTS WITH BISMUTH FERRATE, LITHIUM NIOBATE AND LITHIUM TANTALATE) /=0.9 AND 0.58) (CORRELATION OF CAT 09-1574
ELECTRIC CRYSTALS, ESPECIALLY ON LITHIUM NIOBATE AND LITHIUM TANTALATE. /ITION AND OPTICAL QUALITY OF FERRO 11-1780
CONGRUENT MELTING COMPOSITION OF LITHIUM METATANTALATE. 12-1882
D DISPERSION IN BARIUM AND STRONTIUM TITANATES AND POTASSIUM TANTALATE. /ICAL PHONON MODE THEORY OF INFRARE 04-0547
CRYSTAL GROWTH. (BARIUM SODIUM NIOBATE AND SODIUM LANTHANUM TANTALATE) /ASE COOLING: AN AID TO CZOCHRALSKI 13-1928
CURIE TEMPERATURE OF FERROELECTRIC LITHIUM TANTALATE. 12-1887
IN THE FERROELECTRIC MIXED CRYSTALS POTASSIUM(1-X) SODIUM(X) TANTALATE. /LECTRIC PROPERTIES AND SOFT MODES 08-1306
DIELECTRIC PROPERTIES OF A PARAELECTRIC MATERIAL. (POTASSIUM TANTALATE)　　　EFFECT OF PRESSURE ON THE 08-1335
T OF PRESSURE ON THE STATIC DIELECTRIC CONSTANT OF POTASSIUM TANTALATE.　　　EFFEC 08-1269
ICAL HARMONIC GENERATION IN STRONTIUM TITANATE AND POTASSIUM TANTALATE.　　　ELECTRIC FIELD INDUCED OPT 06-1203

LD INDUCED RAMAN EFFECT IN PARAELECTRIC CRYSTALS. (POTASSIUM TANTALATE) ELECTRIC FIE 08-1326,
ELECTRICAL PROPERTIES AND DEFECT STRUCTURE OF POTASSIUM TANTALATE. 08-1314
LECTRON PARAMAGNETIC RESONANCE OF MANGANESE(2+) IN POTASSIUM TANTALATE. E 08-1343
RIDE, STRONTIUM TITANATE, POTASSIUM MOLYBDATE, AND POTASSIUM TANTALATE. /BANDS FOR POTASSIUM NICKEL TRIFLUO 02-0091
EPR OF COBALT(2+) IN POTASSIUM TANTALATE. 08-1342
CTRIC DISPERSION OF PEROVSKITES. (BARIUM TITANATE, POTASSIUM TANTALATE) FAR INFRARED DIELE 04-0546
FERROELECTRIC DOMAIN REVERSAL IN LITHIUM METATANTALATE. 12-1869
INFRARED ACTIVE MODES IN POTASSIUM TANTALATE. 08-1366
Y INDUCED INDEX OF REFRACTION IN LITHIUM NIOBATE AND LITHIUM TANTALATE. INHOMOGENEITY OF THE OPTICALL 11-1664
INTERACTION OF TWO HYPERSOUND WAVES IN LITHIUM TANTALATE. 12-1891
MEDIUM FREQUENCY RESONATORS OF LITHIUM NIOBATE AND LITHIUM TANTALATE. 11-1851
THE FERROELECTRIC- PARAELECTRIC PHASE TRANSITION IN LITHIUM TANTALATE. NATURE OF 12-1879
NALYSIS OF THE LINEAR DISPLACEMENT CORRELATIONS IN POTASSIUM TANTALATE. NEUTRON SCATTERING A 36-3546
, APPLIED TO BARIUM STRONTIUM TITANATE AND POTASSIUM NIOBATE TANTALATE) /ROELECTRIC SOLID SOLUTION. (THEORY 03-0260
FOR COMMUNICATIONS SYSTEMS ENTERS PRODUCTION STAGE. (LITHIUM TANTALATE) NEW CRYSTAL 12-1867
PERFORMANCE OF AN 11 GHZ OPTICAL MODULATOR USING LITHIUM TANTALATE. 12-1888
PHOTOCONDUCTION OF POTASSIUM TANTALATE. 08-1408
PHOTOEFFECTS IN POTASSIUM TANTALATE. 08-1415
ERING OF LIGHT BY ACOUSTIC PHONONS AND POLARITONS IN LITHIUM TANTALATE. SCATT 12-1878
OF LIGHT ON ACOUSTIC PHONONS IN LITHIUM NIOBATE AND LITHIUM TANTALATE. SCATTERING 11-1744
SELF INDUCED THERMAL EFFECTS ON LIGHT EXTINCTION IN LITHIUM TANTALATE. 12-1889
SIGNS OF THE ELECTROOPTIC COEFFICIENTS FOR LITHIUM TANTALATE. 12-1880
PIEZOELECTRIC OSCILLATIONS IN WIDEBAND PYROELECTRIC LITHIUM TANTALATE. STUDY OF 12-1875
SURFACE WAVE TEMPERATURE COEFFICIENTS ON LITHIUM TANTALATE. 11-1852
TURE DEPENDENCE OF SURFACE ACOUSTIC WAVE VELOCITY IN LITHIUM TANTALATE. TEMPERA 12-1886
THE OPTICAL PROPERTIES OF FERROELECTRIC LITHIUM NIOBATE AND TANTALATE. TEMPERATURE DEPENDENCE OF 11-1770
E TRANSITIONS AT LOW TEMPERATURES. (TIN TELLURIDE, POTASSIUM TANTALATE) THEORY OF SECOND ORDER PHAS 03-0433
ON INTERACTION PROPOSED FOR STRONTIUM TITANATE AND POTASSIUM TANTALATE). /NDUCTIVITY IN PEROVSKITES (3 PHON 06-1095
LATIONS ON RHENIUM TRIOXIDE, SODIUM TUNGSTATE, AND POTASSIUM TANTALATE. TIGHT BONDING BAND CALCU 08-1334
TAL RESONATOR FREQUENCIES: APPLICATION TO QUARTZ AND LITHIUM TANTALATE. /MPENSATION BY ADDITION OF TWO CRYS 11-1686
TIES OF SOLID SOLUTIONS BASED ON THE FERROELECTRIC POTASSIUM TANTALATE. /RAY ANALYSIS AND DIELECTRIC PROPER 08-1310
NEW PHASES OF QUADRATIC TUNGSTEN BRONZES STRONTIUM POTASSIUM TANTALATE AND BARIUM POTASSIUM TANTALATE. /WO 13-2010
OINTS. THERMAL CONDUCTIVITY OF STRONTIUM TITANATE, POTASSIUM TANTALATE AND BARIUM TITANATE NEAR THE CURIE P 04-0989
INTERNAL ELECTRIC FIELDS IN CRYSTALS OF SODIUM TANTALATE AND CADMIUM TITANATE. 08-1389
ELASTIC, PIEZOELECTRIC, AND DIELECTRIC CONSTANTS OF LITHIUM TANTALATE AND LITHIUM NIOBATE. /ENDENCE OF THE 11-1823
F THE SPIN LATTICE RELAXATION TIMES IN FERROELECTRIC LITHIUM TANTALATE AND LITHIUM NIOBATE. /ETERMINATION O 11-1829
SONATORS IN THE MEDIUM FREQUENCY RANGE WITH LOW RAT/ LITHIUM TANTALATE AND LITHIUM NIOBATE PIEZOELECTRIC RE 11-1724
/, BISMUTH(12) GERMANIUM OXIDE(20), LITHIUM NIOBATE, LITHIUM TANTALATE, AND OTHER NONFERROELECTRIC SUBSTAN/ 01-0028
/MODES IN THE MIXED CRYSTAL SYSTEMS POTASSIUM(X) SODIUM(1-X) TANTALATE AND POTASSIUM TANTALATE(X) NIOBATE(/ 13-1933
/ME AEO(3) PEROVSKITES IN THE PARAELECTRIC PHASE. (POTASSIUM TANTALATE AND TANTALATE NIOBATE, BARIUM AND S/ 04-0690
/T MODEL FOR STOICHIOMETRY DEVIATIONS IN LITHIUM NIOBATE AND TANTALATE AND THE EFFECT ON THE CURIE TEMPERA/ 11-1770
/T MODEL FOR STOICHIOMETRY DEVIATIONS IN LITHIUM NIOBATE AND TANTALATE AND THE EFFECT ON THE CURIE TEMPERA/ 11-1777
N IN SOME PEROVSKITE CRYSTALS (STRONTIUM TITANATE, POTASSIUM TANTALATE) AT LOW TEMPERATURES. (THEORY) /HONO 03-0227
/TUDY ON SOFT MODES. (REVIEW OF BROCKHAVEN WORK ON POTASSIUM TANTALATE, BARIUM AND STRONTIUM TITANATES, KD/ 02-0115
/DE IN THE CUBIC PEROVSKITES. (STRONTIUM TITANATE, POTASSIUM TANTALATE, BARIUM TITANATE, LEAD TITANATE, LE/ 02-0114
/TRUCTURAL PHASE TRANSITIONS. (STRONTIUM TITANATE, POTASSIUM TANTALATE, BARIUM TITANATE, POTASSIUM NIOBATE/ 02-0059
/ OPTICAL COEFFICIENTS OF POLAR CRYSTALS (LITHIUM NIOBATE OR TANTALATE, BARIUM TITANATE, ZINC OXIDE, AND L/ 02-0098
STOICHIOMETRY VARIATIONS IN LITHIUM NIOBATE AND LITHIUM TANTALATE BY RAMAN POWDER SPECTROSCOPY. 11-1813
PIEZOELECTRIC AND ELASTIC PROPERTIES OF LITHIUM TANTALATE CHARACTERISTICS. 12-1893
NET IONIZED DONOR CONCENTRATION IN SINGLE CRYSTAL POTASSIUM TANTALATE. (CONDUCTIVE, KYROPOULOS GROWN) / VS 08-1437
OPTICAL BEAM STEERING USING A MULTICHANNEL LITHIUM TANTALATE CRYSTAL. 12-1881
NESS SHEAR MODE / TEMPERATURE DEPENDENCE OF AN X CUT LITHIUM TANTALATE CRYSTAL RESONATOR VIBRATING IN THICK 12-1868
LANE PIEZOELECTRIC PROPERTIES OF LITHIUM NIOBATE AND LITHIUM TANTALATE CRYSTALS. /ACTERISTIC EQUATION FOR P 11-1739
EQUATIONS OF PIEZOELECTRIC VIBRATION OF LITHIUM NIOBATE AND TANTALATE CRYSTALS. SOME PROPERTIES OF THE 11-1740
,000 PER CM, 300 DEG K) OF THE LATTICE VIBRATIONS IN LITHIUM TANTALATE (DIELECTRIC CONSTANTS). /TUDY (20-10 12-1870
C / SINGLE CRYSTAL GROWTH AND PHYSICAL PROPERTIES OF LITHIUM TANTALATE. (DIELECTRIC CONSTANTS, PIEZOELECTRI 12-1876
ANOMALOUS (LONGITUDINAL WAVE) ATTENUATION IN LITHIUM TANTALATE (EFFECT OF VACANCIES). 12-1892
ONS FOR IONIC MOTION IN DISPLACEMENT FERROELECTRICS (LITHIUM TANTALATE EXAMPLE). INTERCELL CORRECTI 03-0367
GROWTH AND PROPERTIES OF LITHIUM NIOBATE AND LITHIUM TANTALATE FILMS. 11-1712
/OF LITHIUM NIOBATE, BARIUM SODIUM NIOBATE AND BARIUM SODIUM TANTALATE (FORCE CONSTANTS, SPACE GROUP ASSIG/ 11-1808
SINGLE CRYSTAL GROWTH OF STRONTIUM POTASSIUM TANTALATE FROM MOLTEN SALTS. 13-2034
C) IN THE PHONON SPECTRA OF STRONTIUM TITANATE AND POTASSIUM TANTALATE (LATTICE DYNAMICS). /ECTRIC- ACOUSTI 06-1117
URE RANGE 4.2-300 DEG K (LATTICE PARAMETERS VERSU/ POTASSIUM TANTALATE LEAD TITANATE SYSTEM IN THE TEMPERAT 08-1308
/RENT FERROELECTRIC CERAMICS. ((LEAD, LANTHANUM) (ZIRCONATE, TANTALATE), (LEAD, TUNGSTEN) (ZIRCONATE TANTA/ 13-2004
ERO TEMPERATURE COEFFICIENT OF RESONANT FREQUENCY IN LITHIUM TANTALATE LENGTH EXPANDER BARS. Z 11-1810
M TITANATE. (SOLID SOLUTIONS WITH STRONTIUM TITANATE LITHIUM TANTALATE, LITHIUM NIOBATE) /ICATION OF CADMIU 08-1424
CATIONS (REVIEW EMPHASIZING ELECTROOPTIC MODULATION, LITHIUM TANTALATE, LITHIUM NIOBATE, KDP, 113 REFS). /I 02-0056
/THIUM AMMONIUM TARTRATE, LITHIUM THALLIUM TARTRATE, LITHIUM TANTALATE, LITHIUM NIOBATE, LITHIUM HYDRAZINI/ 02-0052
/OWAVE NONLINEAR SUSCEPTIBILITIES. (BARIUM TITANATE, LITHIUM TANTALATE, LITHIUM NIOBATE, LITHIUM IODATE, K/ 02-0107
ABSORPTION EDGE SPLITTING IN POTASSIUM TANTALATE NIOBATE. 13-1988
GROWTH OF HIGH QUALITY CRYSTALS OF POTASSIUM TANTALATE NIOBATE 13-2055
LONGITUDINAL ELECTROOPTIC EFFECT IN POTASSIUM TANTALATE NIOBATE 13-1942
S INDICATED BY ANISOTROPIC DEBYE- WALLER FACTORS. (POTASSIUM TANTALATE NIOBATE) /N PEROVSKITE FERROELECTRIC 08-1351
EUTRON SCATTERING STUDY OF THE SOFT MODES IN CUBIC POTASSIUM TANTALATE NIOBATE N 13-2064
OPTICAL DAMAGE IN POTASSIUM TANTALATE NIOBATE 13-1979
OPTICAL PROPERTIES OF POTASSIUM TANTALATE NIOBATE. 36-3593
FT MODES AND STRUCTURE OF FERROELECTRIC TETRAGONAL POTASSIUM TANTALATE NIOBATE. SO 13-1957
SOFT PHONONS IN POTASSIUM TANTALATE NIOBATE. 13-2040
LATION AND DEFLECTION. ELECTROOPTIC POTASSIUM TANTALATE NIOBATE GRATINGS FOR LIGHT BEAM MODU 13-1929
/RICS. (ELECTROOPTIC MATERIALS, SINGLE CRYSTALS OF POTASSIUM TANTALATE NIOBATE, LITHIUM NIOBATE, AND BISMU/ 02-0073
) PYROELECTRIC DETECTORS. CHARACTERISTICS OF POTASSIUM TANTALATE NIOBATE (POTASSIUM TANTALATE NIOBATE 13-2030
NIOBATE(0.35), CURIE TEMPERATUR/ RAMAN SPECTRUM OF POTASSIUM TANTALATE NIOBATE. (POTASSIUM TANTALATE(0.65) 08-1391
CHARACTERISTICS OF KTN (POTASSIUM TANTALATE NIOBATE) PYROELECTRIC DETECTORS. 13-2030
PIEZOELECTRIC PROPERTIES OF POTASSIUM TANTALATE NIOBATE SINGLE CRYSTAL. 08-1271
ECTRICAL PROPERTIES. FLUX GROWN POTASSIUM TANTALATE NIOBATE SINGLE CRYSTALS AND THEIR EL 09-1533
UDIES. POTASSIUM TANTALATE NIOBATE, STRUCTURAL AND DYNAMICAL ST 13-2068
/ONAL TUNGSTEN OXYGEN BRONZE TYPE: QUADRATIC LEAD OR BISMUTH TANTALATE NIOBATE, TANTALATE TUNGSTATE, OR TU/ 13-1985
TITANATE, BARIUM TITANATE ZIRCONATE (OR STANNATE), POTASSIUM TANTALATE NIOBATE, 228 REFS). /IUM) STRONTIUM 08-1281
) TEMPERATURE DEPENDENCE OF POLARITON DISPERSION IN LITHIUM TANTALATE. (OPTICAL MODULATION VIA TEMPERATURE 12-1873
/CIENTS OF POLAR (PYROELECTRIC) CRYSTALS (LITHIUM NIOBATE OR TANTALATE OR GALLATE, BARIUM TITANATE, ZINC O/ 02-0095

/ATE OR ARSENATE, POTASSIUM DIDEUTERIUM PHOSPHATE, POTASSIUM TANTALATE OR NIOBATE, LITHIUM NIOBATE OR TANT/ 01-0021
ON CAPACITIES OF A DIGITAL LIGHT DEFLECTOR UTILIZING LITHIUM TANTALATE POLARIZATION SWITCHES. /AND DEFLECTI 11-1809
/LITY OF THE NIOBIUM- OXYGEN BOND. (LITHIUM NIOBATE, LITHIUM TANTALATE, POTASSIUM LITHIUM NIOBATE, BARIUM / 11-1734
ADP, RUBIDIUM DIHYDROGEN PHOSPHATE, LITHIUM NIOBATE, LITHIUM TANTALATE, QUARTZ) /NONLINEAR CRYSTALS. (KDP, 11-1865
SURFACE WAVE VELOCITY AND DELAY ON LITHIUM NIOBATE, LITHIUM TANTALATE, QUARTZ, AND TELLURIUM DIOXIDE. /TIC 11-1821
ELECTRON TUNNELING INTO POTASSIUM TANTALATE SCHOTTKY BARRIER JUNCTIONS. 08-1365
BIUM DOPED) STRONTIUM TITANATE AND (CALCIUM DOPED) POTASSIUM TANTALATE (SCHOTTKY BARRIERS WITH INDIUM). /IO 06-1222
CHARACTER OF PHASE TRANSITION OF FERROELECTRIC POTASSIUM TANTALATE. (SECOND ORDER) 08-1307
PART-1: PIEZOELECTRIC AND OPTICAL PROP/ BEHAVIOR OF LITHIUM TANTALATE SINGLE CRYSTAL NEAR ITS CURIE POINT, 12-1877
PART-2, DIELECTRIC AND ULTRASONIC PROP/ BEHAVIOR OF LITHIUM TANTALATE SINGLE CRYSTAL NEAR ITS CURIE POINT, 12-1885
MELT COMPOSITION, LATTICE AND CURIE TEMPERATURE ME/ LITHIUM TANTALATE SINGLE CRYSTAL STOICHIOMETRY (VERSUS 12-1871
STOICHIOMETRY AND OPTICAL QUALITY OF LITHIUM TANTALATE SINGLE CRYSTALS. 12-1883
X-RAY TOPOGRAPHIC STUDIES ON LITHIUM NIOBATE AND LITHIUM TANTALATE SINGLE CRYSTALS. 11-1830
M/ DIELECTRIC PROPERTIES AND OPTICAL ABSORPTION OF POTASSIUM TANTALATE SINGLE CRYSTALS AND (POTASSIUM SODIU 08-1272
T COMP/ ELECTROOPTIC AND FERROELECTRIC PROPERTIES OF LITHIUM TANTALATE SINGLE CRYSTALS AS A FUNCTION OF MEL 12-1874
NIOBIUM-93 NMR STUDY OF THE LITHIUM NIOBATE TANTALATE SOLID SOLUTION SYSTEM. 11-1794
/VIBRATIONAL DISPLACEMENTS OF SEVERAL PEROVSKITES (POTASSIUM TANTALATE, STRONTIUM TITANATE, RUBIDIUM MANGA- 06-1139
/ PEROVSKITES. (STRONTIUM LITHIUM NIOBATE, STRONTIUM LITHIUM TANTALATE, STRONTIUM(1-X) LEAD(X) LITHIUM NIO/ 08-1292
ATOR. FABRICATION OF A LITHIUM TANTALATE TEMPERATURE STABILIZED OPTICAL MODUL 12-1872
PYROELECTRIC EFFECTS AND SPONTANEOUS POLARIZATION IN LITHIUM TANTALATE (THERMAL EXPANSION). /AND SECONDARY 12-1884
/ FERROELECTRICITY IN ISOMORPHIC LITHIUM NIOBATE AND LITHIUM TANTALATE USING AUTHOR'S STATISTICAL THEORY. / 03-0366
ABSORPTION EDGE SPLITTING IN POTASSIUM TANTALATE(X) NIOBATE(1-X). 13-1987
GROWTH AND PROPERTIES OF FERROELECTRIC POTASSIUM LITHIUM TANTALATE(X) NIOBATE(1-X). 11-1711
TAL SYSTEMS POTASSIUM(X) SODIUM(1-X) TANTALATE AND POTASSIUM TANTALATE(X) NIOBATE(1-X). / IN THE MIXED CRYS 13-1933
NGSTEN BRONZE TYPE) FERROELECTRIC (SYSTEM) POTASSIUM LITHIUM TANTALATE(X) NIOBATE(1-X). /CRYSTALS IN THE TU 13-1947
LECTRICS IN BEAM DEFLECTORS AND LIGHT MODULATORS. (POTASSIUM TANTALATE(0.65) NIOBATE(0.35)) /ROVSKITE PARAE 13-1924
R/ RAMAN SPECTRUM OF POTASSIUM TANTALATE NIOBATE. (POTASSIUM TANTALATE(0.65) NIOBATE(0.35), CURIE TEMPERATU 08-1391
PIEZOELECTRIC PROPERTIES OF FERROELECTRIC POTASSIUM LITHIUM TANTALATE(0-0.5) NIOBATE(1-0.5) SINGLE CRYSTA/ 12-1890
AINING PEROVSKITES. (TITANATES, HAFNATES, MIXED NIOBATES AND TANTALATES) / OF NEW CADMIUM AND THALLIUM CONT 08-1458
IES OF NEW COMPOUNDS WITH A PYROCHLORE STRUCTURE. (NIOBATES, TANTALATES) SYNTHESIS AND CERTAIN PROPERT 02-0047
N OF FERROELECTRICS OF ILMENITE STRUCTURE. (ABO(3) NIOBATES, TANTALATES, AND MIXTURES WITH THEM) /ESTIGATIO 08-1438
/LECTRICAL PROPERTIES AND STRUCTURES OF METANIOBATES AND METATANTALATES OF TRANSITION METALS OF THE 3D SER/ 36-3527
CHEMISTRY AND PHYSICS OF (ALKALI) METAL TANTALATES. (PREPARATION, REVIEW) 02-0048
OBA/ FLUX GROWTH OF SOME COMPLEX OXIDE MATERIALS. (NIOBATES, TANTALATES, STANNATES, TITANATES, CHROMATES, C 02-0129
GSTEN BRONZE. (POTASSIUM BARIUM TANTAL/ PHASE TRANSITIONS IN TANTALATES WITH THE STRUCTURE OF POTASSIUM TUN 13-2050
/N OXIDE(6), AND (BARIUM OR STRONTIUM) (3) COPPER (NIOBIUM OR TANTALUM) (2) OXIDE(9), AND VARIOUS TUNGSTEN S/ 36-3619
ITANATE. STATIC ELECTRIC QUADRUPLE INTERACTION OF TANTALUM AND HAFNIUM IONS IN BARIUM AND LEAD T 04-0937
IMPURITY EFFECTS IN SODIUM TUNGSTEN BRONZES: NIOBIUM AND TANTALUM DOPING. 13-2053
NC DIELECTRIC PROPERTIES OF S/ EFFECT OF THE SUBSTITUTION OF TANTALUM FOR NIOBIUM ON THE CRYSTALLOGRAPHIC A 13-1921
ULTRASONIC VELOCITY ABOVE TANTALUM IN STRONTIUM TITANATE. 06-1110
OF THE PIEZOELECTRIC CONSTANTS OF STIBOTANTALITE. (ANTIMONY (TANTALUM, NIOBIUM) OXIDE(4)) /TURE DEPENDENCE 36-3601
/ PYROCHLORE STRUCTURES. (BARIUM AND BISMUTH(0.5)) (NIOBIUM, TANTALUM, OR VANADIUM) (0.5) OXIDE, (BARIUM OR/ 08-1462
BISMUTH(/ NEW FERROELECTRICS. (BARIUM(2) BISMUTH (NIOBIUM OR TANTALUM OR VANADIUM) OXIDE(6), AND BARIUM(3) 36-3619
/E STRUCTURE OF POTASSIUM TUNGSTEN BRONZE. (POTASSIUM BARIUM TANTALUM OXIDE, SODIUM BARIUM TANTALUM OXIDE,/ 13-2050
CS. (BISMUTH TUNGSTATE, BISMUTH NIOBIUM OXYFLUORIDE, BISMUTH TANTALUM OXYFLUORIDE) /LAYER TYPE FERROELECTRI 13-1966
OF A NEW PYROCHLORE CONTAINING LITHIUM, STRONTIUM LANTHANUM TANTALUM OXYFLUORIDE. /TUDY OF SINGLE CRYSTALS 36-3613
ALS. ELECTROOPTICAL PROPERTIES OF BARIUM TITANATE- TANTALUM PENTOXIDE SOLID SOLUTION SINGLE CRYST 04-0955
. (SODIUM(X) TUNGSTEN(1-X) Z(Y) OXIDE(3) WHERE Z IS NIOBIUM, TANTALUM, ZIRCONIUM, VANADIUM, OR COBALT) /ZES 13-1944
EFFECT IN FERROELECTRIC- ANTIFERROMAGNETIC LITHIUM IRON(0.5) TANTALUM(0.5) OXY FLUORIDE. MAGNETOELECTRIC 36-3568
RAMAN SPECTRUM OF POTASSIUM TANTALUM(0.64) NIOBIUM(0.36) OXIDE. 13-1989
IN CUBIC SODIUM CHLORIDE AND EPITAXIAL FERROELECTRIC SODIUM TARTRATE. /TION DIRECTION UPON OPTICAL PHONONS 29-3243
RTIES AND FERROELECTRIC PHASE TRANSITION ON LITHIUM THALLATE TARTRATE. /ING EFFECTS ON THE DIELECTRIC PROPE 29-3251
ESR OF COPPER(2+) DOPED SODIUM AMMONIUM TARTRATE. 29-3252
ECULIAR FERROELECTRIC- FERROELASTIC CRYSTAL: SODIUM AMMONIUM TARTRATE. /SM OF THE PHASE TRANSFORMATION OF P 29-3238
NUCLEAR MAGNETIC RESONANCE OF FERROELECTRIC LITHIUM AMMONIUM TARTRATE. 29-3240
NUCLEAR MAGNETIC RESONANCE OF FERROELECTRIC LITHIUM AMMONIUM TARTRATE. 29-3241
TON SPIN LATTICE RELAXATION IN FERROELECTRIC SODIUM AMMONIUM TARTRATE. PRO 29-3248
SOME PROPERTIES OF THE FERROELECTRIC LITHIUM THALLIUM TARTRATE. 29-3242
/ION INTO MOTION. (PYROELECTRIC EFFECT IN TGS, ETHYL DIAMINE TARTRATE, BARIUM TITANATE, SALICYLIDENE- 4-BR/ 04-0854
EACTIONS IN SILICA GELS. PART-3: FORMATION OF POTASSIUM ACID TARTRATE CRYSTALS. SOME CHEMICAL R 29-3246
DISPLACEMENT OF THE CURIE POINTS IN KDP AND LITHIUM THALLIUM TARTRATE CRYSTALS BY HYDROSTATIC PRESSURE. 17-2523
M DINITRITE. PART-3: ROCHELLE SALT. PART-4: LITHIUM AMMONIUM TARTRATE HYDRATE. /: TGS. PART-2: SILVER SODIU 02-0077
OF THE ELASTIC COMPLIANCE OF FERROELECTRIC LITHIUM THALLIUM TARTRATE HYDRATE (LTT) BY AN ELECTRIC FIELD. / 29-3249
/DROGEN PHOSPHATE, SODIUM, POTASSIUM(2), OR LITHIUM THALLIUM TARTRATE, LITHIUM AMMONIUM TARTRATE, AND MIXE/ 02-0089
/CE OF LITHIUM IN FERROELECTRIC COMPOUNDS. (LITHIUM AMMONIUM TARTRATE, LITHIUM THALLIUM TARTRATE, LITHIUM / 02-0052
CRYSTAL DATA FOR LITHIUM THALLIUM TARTRATE MONOHYDRATE. 29-3247
-RAY REFINEMENT OF THE CRYSTAL STRUCTURE OF LITHIUM AMMONIUM TARTRATE MONOHYDRATE. X 29-3244
/PROPERTIES OF NONDEUTERATED AND DEUTERATED LITHIUM THALLIUM TARTRATE MONOHYDRATE. (EFFECT ON LATTICE AND / 29-3239
/AND CONDUCTION IN FERROELECTRIC CRYSTALS. (LITHIUM THALLIUM TARTRATE MONOHYDRATE, ELECTRIC CONTROL OF ELA/ 02-0060
/FERROELECTRIC CRYSTALS. (BISMUTH TITANATE, LITHIUM THALLIUM TARTRATE MONOHYDRATE, LEAD(5) GERMANIUM(3) OX/ 02-0050
OMIDE, POTASSIUM CHLORIDE, SODIUM CHLORIDE, POTASSIUM SODIUM TARTRATE, POTASSIUM DIHYDROGEN PHOSPHATE / BR 17-2707
-NO) IS SODIUM RUBIDIUM TARTRATE TETRAHYDRATE A FERROELECTRIC. (ANSWER 29-3245
/ANTIFERROELECTRICS. (BARIUM TITANATE, KDP, POTASSIUM SODIUM TARTRATE, TGS, AMMONIUM FLUOBERYLLATE, SODIUM/ 02-0134
FIELD CONTROL OF THE ELASTIC COMPLIANCE IN LITHIUM THALLIUM TARTRATE (TUNING AND DELAY DEVICES). ELECTRIC 29-3250
LECTRIC TRANSITION. CHANGE IN THE DECAY CONSTANT OF TECHNETIUM-99 IN BARIUM TITANATE BY THE FERROE 04-0871
UENCE OF FIRING OF BARIUM TITANATE SINTERS ON ELECTRICAL AND TECHNOLOGICAL PROPERTIES OF FERROELECTRICS. /L 04-1002
E TRANSDUCERS ON LEAD ZIRCONA/ REALIZATION OF A 40 MHZ COLOR TELEVISION INFRARED RESPONSE USING SURFACE WAV 09-1509
UM(5+) IN STRONTIUM TITANATE: AN EXAMPLE OF THE DYNAMIC JAHN TELLER EFFECT. /RESONANCE SPECTRUM OF MOLYBDEN 06-1125
OCCURRENCE OF ORDERED PHASE IN PEROVSKITE TYPE/ ROLE OF JAHN TELLER EFFECTS IN ORIGIN OF FERROELECTRICITY, 04-0964
NEW FERROELECTRIC COMPOUND STRONTIUM TELLURATE. 36-3626
TRANSITION IN TIN TELLURIDE- GERMANIUM TELLURIDE ALLOYS. (NEUTRON SCATTERING) 36-3586
FIELD EFFECT AND HALL MOBILITY OF TELLURIUM AND LEAD TELLURIDE ON UHV CLEAVED TRIGLYCINE SULFATE. 27-3172
OF SECOND ORDER PHASE TRANSITIONS AT LOW TEMPERATURES. (TIN TELLURIDE, POTASSIUM TANTALATE) THEORY 03-0433
/UPERCONDUCTING AND THE FERROELECTRIC PHASE TRANSITIONS (TIN TELLURIDE PURE AND ALLOYED WITH SMALL AMOUNTS/ 36-3564
/STALS OF HIGH DIELECTRIC CONSTANT. (POTASSIUM BROMIDE, LEAD TELLURIDE, TIN TELLURIDE, LEAD SULFIDE, GERMA/ 36-3541
/, TRISELENIDE, SULFOIODIDE, SELENOIODIDE, TELLUROIODIDE, TRITELLURIDE, TRIIODIDE, TRIBROMIDE, INDIUM ANTI/ 15-2108
ONIC CRYSTALS WITH HIGH DIELECTRIC CONSTANTS. (LEAD AND TIN TELLURIDES) LATTICE DYNAMICS OF I 36-3553
N SCATTERING) TRANSITION IN TIN TELLURIDE- GERMANIUM TELLURIDE ALLOYS. (NEUTRO 36-3586
) COMPOUNDS. (USING ANTIMONY OR BISMUTH, SULFUR, SELENIUM OR TELLURIUM, AND CHLORINE, BROMINE OR IODINE) /I 15-2194
IGLYCINE SULFATE. FIELD EFFECT AND HALL MOBILITY OF TELLURIUM AND LEAD TELLURIDE ON UHV CLEAVED TR 27-3172

SE TRANSITION OF A NEW FERROELECTRIC OXIDE, STRONTIUM OXIDE- TELLURIUM DIOXIDE.   PHA 36-3627
AND DELAY ON LITHIUM NIOBATE, LITHIUM TANTALATE, QUARTZ, AND TELLURIUM DIOXIDE. /TIC SURFACE WAVE VELOCITY 11-1821
  TEMPERATURE DEPENDENCE OF THE FERROELECTRIC FIELD EFFECT IN TELLURIUM FILMS ON TRIGLYCINE SUBSTRATES.   27-3018
TS IN THE FIELD EFFECT AT VERY LARGE GATE CHARGES APPLIED TO TELLURIUM FILMS ON UHV CLEAVED TGS. /TER EFFEC 27-3171
RITE SUR/ THIN FILM NUCLEATION ON FERROELECTRIC SUBSTRATES. (TELLURIUM ON TRIGLYCINE SULFATE AND SODIUM NIT 16-2394
/TRIC (BARIUM TITANATE) POLARIZATION FIELDS ON (GERMANIUM OR TELLURIUM) SEMICONDUCTOR FILMS. (AND USE OF T/ 04-0594
HERMAL EXPANSION OF TRIGLYCINE SULFATE CRYSTALS STUDIED WITH TELLURIUM THIN FILMS.   T 27-3047
ECTRIC TRIGLYCINE SULFATE AND SELENATE CRYS/ FIELD EFFECT IN TELLURIUM THIN FILMS ON A SUBSTRATE OF FERROEL 27-3046
/YSTALS, V IS ANTIMONY OR BISMUTH, VI IS SULFUR, SELENIUM OR TELLURIUM, VII IS IODINE, BROMINE OR CHLORINE) 15-2214
ALS.   ABSORPTION SPECTRA OF BISMUTH TELLUROBROMIDE AND BISMUTH TELLUROIODIDE CRYST 15-2217
                OPTICAL PROPERTIES OF THE SEMICONDUCTOR BISMUTH TELLUROIODIDE.   15-2169
  PRODUCTION METHODS AND SOME OPTICAL PROPERTIES OF BISMUTH TELLUROIODIDE.   15-2141
    AESORPTION SPECTRA OF BISMUTH TELLUROBROMIDE AND BISMUTH TELLUROIODIDE CRYSTALS.   15-2217
/NTIMONY TRISULFIDE, TRISELENIDE, SULFOICIDE, SELENOIODIDE, TELLUROIODIDE, TRITELLURIDE, TRIIODIDE, TRIBR/ 15-2108
THERMAL CONDITIONS. (ANTIMONY AND BISMUTH SULFO, SELENO, AND TELLUROIODIDES AND BROMIDES) /VII) UNDER HYDRO 15-2213
OTASSIUM DEUTERIUM PHOSPHATE AT THE FERROELECTRIC TRANSITION TEMPERATURE.   ACOUSTIC PROPAGATION IN P 17-2573
                                                                                                          16-2305
    BRILLOUIN SCATTERING IN SODIUM NITRITE AT ROOM TEMPERATURE.   16-2305
  LOCAL FLUCTUATIONS IN STRONTIUM TITANATE ABOVE THE CRITICAL TEMPERATURE.   CRITICAL ASYMMETRY IN 06-1166
    CRITICAL DYNAMICS OF STRONTIUM TITANATE ABOVE THE CRITICAL TEMPERATURE.   06-1204
-RAY INCOHERENT SCATTERING BY BARIUM TITANATE NEAR THE CURIE TEMPERATURE.   CRITICAL X 04-0535
XAHYDRATE. (POLARIZATION VERSUS ELECTRIC FIELD, -80C TO ROOM TEMPERATURE) / GUANIDINIUM ALUMINUM SULFATE HE 21-2813
CIVE FIELD OF TRIGLYCINE SULFATE ON FREQUENCY, AMPLITUDE AND TEMPERATURE.   DEPENDENCE OF THE COER 27-3092
IC ANOMALIES OF SODIUM NITRATE ABOVE THE FERROELECTRIC CURIE TEMPERATURE.   DIELECTR 16-2376
                                                                                                          08-1304
    DIFFUSE SCATTERING IN SODIUM NIOBATE AT ROOM TEMPERATURE.   ELECT 04-0928
RON PARAMAGNETIC RESONANCE IN BARIUM TITANATE NEAR THE CURIE TEMPERATURE.   04-0928
FERROELECTRIC ANTIMONY SULFOIODIDE NEAR THE PHASE TRANSITION TEMPERATURE. /LECTION FROM SINGLE CRYSTALS OF 15-2151
ICROPLASMA OF CERAMIC (BARIUM, LEAD) TITANATE NEAR THE CURIE TEMPERATURE.   EMISSION OF LIGHT BY SURFACE M 04-0982
  THE COOPERATIVE TRANSITION OF TRIGLYCINE SULFATE NEAR CURIE TEMPERATURE.   EQUATION OF STATE FOR 27-3006
TUDY OF GAMMA-IRRADIATED CESIUM TRIHYDROGEN SELENITE AT ROOM TEMPERATURE.   ESR S 22-2833
NS AND CORRELATIONS IN STRONTIUM TITANATE ABOVE THE CRITICAL TEMPERATURE.   FLUCTUATIO 06-1233
  ALONG THE (001) DIRECTION OF LITHIUM NIOBATE (LIQUID HELIUM TEMPERATURE). /EQUENCY (950 MHZ) ACOUSTIC WAVE 11-1835
RMITTIVITY OF TRIGLYCINE SULFATE AT 100 MHZ. (NEAR THE CURIE TEMPERATURE)   LARGE SIGNAL PE 27-3029
QUENCY RAMAN SPECTRUM OF A THIOUREA CRYSTAL AS A FUNCTION OF TEMPERATURE.   LOW FRE 25-2906
LECTRIC CONSTANT OF ROCHELLE SALT AT 10 GHZ AS A FUNCTION OF TEMPERATURE.   MEASUREMENT OF THE COMPLEX DIE 28-3184
N SCATTERING ANALYSIS OF CESIUM TRIHYDROGEN SELENITE AT ROOM TEMPERATURE.   NEUTRO 22-2869
      NEW FERROELECTRIC: ANTIMONY ORTHONIOBATE. (CURIE TEMPERATURE)   36-3589
NITRATE AND TRIS-SARCOSINE CALCIUM CHLORIDE. (SHIFT OF CURIE TEMPERATURE) /ECTRIC TRANSITIONS IN DIGLYCINE 16-2267
  STRONTIUM OXIDE- NIOBIUM PENTOXIDE SYSTEM, DTA, X-RAY CURIE TEMPERATURE). /BIUM-RICH REGION AND THE BINARY 13-1920
  POCKELS EFFECT IN AMMONIUM DIHYDROGEN PHOSPHATE ABOVE ROOM TEMPERATURE.   17-2518
ND ELECTRICAL AND OPTICAL) STUDY OF TUNGSTEN TRIOXIDE AT LOW TEMPERATURE.   (PREPARATION A 10-1643
  ELASTIC COMPLIANCE OF STRONTIUM TITANATE NEAR THE TRANSITION TEMPERATURE.   PRESSURE DEPENDENCE OF THE 06-1219
N DYNAMICS IN HYDROGEN BONDED FERROELECTRICS ABOVE THE CURIE TEMPERATURE.   PROTO 03-0206
DIHYDROGEN PHOSPHATE AND SILVER PERIODATE BELOW THE CRITICAL TEMPERATURE. /SCATTERING SPECTRA OF POTASSIUM 17-2687
TIBILITY IN CHROMIUM METHYL AMMONIUM ALUM ABOVE THE CRITICAL TEMPERATURE.   SPECIFIC HEAT AND SUSCEP 20-2805
IN LITHIUM NIOBATE AND TANTALATE AND THE EFFECT ON THE CURIE TEMPERATURE. /EL FOR STOICHIOMETRY DEVIATIONS 11-1776
IN LITHIUM NIOBATE AND TANTALATE AND THE EFFECT ON THE CURIE TEMPERATURE. /EL FOR STOICHIOMETRY DEVIATIONS 11-1777
      STRUCTURE OF AMMONIUM FLUOBERYLLATE AT AMBIENT TEMPERATURE.   19-2778
TITANIUM DIOXIDE IN THE PRESENCE OF SODIUM CARBONATE AT HIGH TEMPERATURE. /ITANATE FROM BARIUM SULFATE AND 04-0624
MAL EXPANSIVITY OF RUBIDIUM MANGANESE FLUORIDE NEAR THE NEEL TEMPERATURE.   THER 33-3375
SUREMENT OF THE ELECTROCALORIC EFFECT IN KDP (NEAR THE CURIE TEMPERATURE).   ULTRASONIC MEA 17-2664
RIC (PERMITTIVITY) PROPERTIES OF BISMUTH TITANATE (EFFECT OF TEMPERATURE).   (WEAK FIELD) DIELECT 36-3555
GADOLINIUM MOLYBDATE (CRYSTALLOGRAPHIC PHASE CHANGE AT CURIE TEMPERATURE).   X-RAY DIFFRACTION STUDY OF 30-3312
-RAY DIFFUSE SCATTERING FROM SODIUM NIOBATE AS A FUNCTION OF TEMPERATURE.   X 08-1313
OF FERROELECTRIC POTASSIUM DIHYDROGEN PHOSPHATE. (NEAR CURIE TEMPERATURE) X-RAY STUDY ON THERMAL EXPANSION 17-2556
IUM MANGANESE TRIFLUORIDE (ANOMALOUS HEAT CAPACITY, CRITICAL TEMPERATURE ABOUT 186-65 DEG K). /NT IN POTASS 33-3373
/AL AND OPTICAL PROPERTIES (CONDUCTIVITY AND MOBILITY VERSUS TEMPERATURE, ABSORPTION SPECTRA) OF SEMICONDU/ 06-1198
/E OF THE DOMAIN STRUCTURE OF TRIGLYCINE SULFATE CRYSTALS ON TEMPERATURE AND APPLIED ELECTRIC FIELD STRENG/ 27-3129
A HYDROGEN IMPURITY MODE IN STRONTIUM TITANATE.   (TEMPERATURE AND) ELECTRIC FIELD DEPENDENCE OF 06-1146
NGANESE TRIFLUORIDE WITH MAGNESIUM DOPING.   SHIFT OF NEEL TEMPERATURE AND EPR LINE WIDTH OF POTASSIUM MA 33-3377
ECTRIC TRANSITION REGION IN POTASSIUM DIHYDROGEN PHOSPHATE. (TEMPERATURE AND FIELD DEPENDENCE) /THE FERROEL 17-2576
RROELECTRIC FERROELECTRIC PHASE BOUNDARY IN PLZT CERAMICS.   TEMPERATURE AND FIELD DEPENDENCE OF THE ANTIFE 09-1491
ARGE SIGNAL PERMITTIVITY OF TGS. (10 KHZ-1 MHZ)   TEMPERATURE AND FIELD STRENGTH DEPENDENCE OF L 27-3068
RSION OF THE DIELECTRIC CONSTANT OF TGS IN THE FAR INFRARED (TEMPERATURE AND FREQUENCY DEPENDENCE).   DISPE 27-3009
SS FACTOR IN FERROELECTRIC MATERIALS (ROCHELLE SALT NEAR TH/ TEMPERATURE AND FREQUENCY DEPENDENCE OF THE LO 28-3220
/Y ACOUSTIC SURFACE WAVE LOSS MECHANISMS ON LITHIUM NIOBATE (TEMPERATURE AND FREQUENCY DEPENDENCE, 0.5-5 G/ 11-1817
LECTRICAL PROPERTIES OF LITHIUM METANIOBATE.   TEMPERATURE AND FREQUENCY DEPENDENCES OF THE E 11-1717
TIMONY SULFOIODIDE(X) BROMIDE(1-X) SOLID SOLUTIONS DEPEND ON TEMPERATURE AND HYDROSTATIC PRESSURE. /S OF AN 15-2149
IDE VIBRATION SPECTRUM OF THE FERROELECTRIC RUBID/ EFFECT OF TEMPERATURE AND PHASE TRANSITION ON THE HYDROX 17-2470
XYL STRETCHING VIBRATION IN STRONTIUM TITANATE A/ EFFECTS OF TEMPERATURE AND PHASE TRANSITIONS ON THE HYDRO 06-1209
NATE. (DIELECTRIC CONSTANT VERSUS ALTERNATING FIELD VOLTAGE, TEMPERATURE AND POLARIZATION) /TRIGLYCINE SELE 27-3158
LECTRIC CONSTANT AND SPONTANEOUS POLARIZATION OF FERROELECT/ TEMPERATURE AND PRESSURE DEPENDENCE OF THE DIE 16-2331
/CTRIC ENERGY IN LEAD NIOBIUM ZIRCONATE TITANATE STANNATE BY TEMPERATURE AND PRESSURE ENFORCED PHASE TRANS/ 09-1488
AGNETIC RESONANCE SPECTRA OF SUBSTITUTIONAL IMPUR/ EFFECT OF TEMPERATURE AND PRESSURE ON THE ELECTRON PARAM 06-1193
ICAL PHONONS IN CUBIC SODIUM CHLORIDE AND EPITAXI/ EFFECT OF TEMPERATURE AND PROPAGATION DIRECTION UPON OPT 29-3243
SPLAC/ RELATIONSHIP BETWEEN IONIC DISPLACEMENT AND THE CURIE TEMPERATURE AND SPONTANEOUS POLARIZATION IN DI 03-0453
TANEOUS PO/ X-RAY DIFFRACTOMETRIC DETERMINATION OF THE CURIE TEMPERATURE AND TEMPERATURE DEPENDENCE OF SPON 03-0290
AIN STRUCTURE FORMATION IN SEIGNETTES SALT (ROCHELLE SALT). (TEMPERATURE ANHYSTERESIS). /CTRIC FIELD ON DOM 28-3190
/CTUATIONS IN BARIUM TITANATE (ABOVE THE FERROELECTRIC CURIE TEMPERATURE, AS SHOWN) BY NEUTRON SCATTERING. 04-1032
LECTRICS.   ON THE FREQUENCY DEPENDENCE OF THE TEMPERATURE AUTOSTABILIZATION EFFECT IN FERROE 03-0479
LECTRICS. (CAPACITOR PROPERTIES, TANDEL)   TEMPERATURE AUTOSTABILIZATION EFFECT IN FERROE 37-3655
CINE SULFATE SINGLE CRYSTALS IN AN AC ELECTRIC FIELD.   TEMPERATURE AUTOSTABILIZATION EFFECT OF TRIGLY 27-3005
  CRYSTAL. (DIELECTRIC HEATING AND POLARIZ/ INTERPRETATION OF TEMPERATURE AUTOSTABILIZATION OF FERROELECTRIC 03-0245
OEL/ APPLICATIONS OF THE PIEZOELECTRIC NONLINEARITIES IN THE TEMPERATURE AUTOSTABILIZATION REGIME OF A FERR 17-2502
/ LOSSES IN OXYGEN OCTAHEDRON FERROELECTRICS ABOVE THE CURIE TEMPERATURE. (BARIUM (0-1.0) STRONTIUM (1.0-0) / 04-0559
IES OF THE POLARIZATION PROCESS IN STRONTIUM TITANATE AT LOW TEMPERATURE. (BIREFRINGENCE, 20-275 DEG K) /UD 06-1117
RIGLYCINE SE/ (OPTICAL) ENERGY GAP (AND DIELECTRIC CONSTANT) TEMPERATURE CHARACTERISTICS OF FERROELECTRIC T 27-3021
ERAMICS. (BARI/ UHF (WAVEGUIDE RESONATOR) METHOD OF STUDYING TEMPERATURE CHARACTERISTICS OF ROCHELLE TYPE C 04-0905
S OF TUNGSTEN OXIDE.   DIELECTRIC (AND TEMPERATURE) CHARACTERISTICS OF SOLID SOLUTION 10-1657
/POWDERS CONTAINING BARIUM OXIDE AND TITANIUM DIOXIDE ON THE TEMPERATURE CHARACTERISTICS OF THE DIELECTRIC/ 04-0829
  ADAPTIVE FERROELECTRIC TRANSFORMERS WITH IMPROVED TEMPERATURE CHARACTERISTICS. (PZT-5H) 09-1583

                              SEE BOTH POSITIVE TEMPERATURE COEFFICIENT AND PTC.
ONATORS BY USE OF TEMPERATURE SENSITIVE CAPACITANCES.   LOW TEMPERATURE COEFFICIENT OF LITHIUM NIOBATE RES 11-1670
.   SEE BOTH POSITIVE TEMPERATURE COEFFICIENT OF RESISTANCE AND PTCR
R OF BARIUM TITANATE WITH NIOBIUM PENTOXIDE / PTCR (POSITIVE TEMPERATURE COEFFICIENT OF RESISTANCE) BEHAVIO 04-0837
  LEAD TITANATE.   POSITIVE TEMPERATURE COEFFICIENT OF RESISTANCE IN DOPED 05-1051
M TITANATE SINGLE CRYSTALS IN THE REGION OF THE CU/ POSITIVE TEMPERATURE COEFFICIENT OF RESISTANCE OF BARIU 04-1036
STIVITY ANOMALY IN SEMICONDUCTING BARIUM TITANATE. (POSITIVE TEMPERATURE COEFFICIENT OF RESISTIVITY)   RESI 04-0584
PED AND UNDOPED CERAMIC PEROVSKITE LE/ PTC EFFECT (POSITIVE TEMPERATURE COEFFICIENT OF RESISTIVITY) IN (DO 13-2000
IN LITHIUM TANTALATE LENGTH EXPANDER BARS.   ZERO TEMPERATURE COEFFICIENT OF RESONANT FREQUENCY 11-1810
AVE VELOCITY AND DELAY ON LITHIUM NIOBATE, LITHIUM TANTALAT/ TEMPERATURE COEFFICIENTS OF ACOUSTIC SURFACE W 11-1821
/DIUM FREQUENCY RANGE WITH LOW RATIOS OF CAPACITANCE AND LOW TEMPERATURE COEFFICIENTS OF FREQUENCY (DEVICE/ 11-1724
  SEMICONDUCTING BARIUM AND STRONTIUM TITANATES WITH POSITIVE TEMPERATURE COEFFICIENTS OF RESISTIVITY.   04-0995
                              SURFACE WAVE TEMPERATURE COEFFICIENTS ON LITHIUM TANTALATE. 11-1852
YSTAL RESONATOR FREQUENCIES: APPLICATION TO QUAR/ WIDE RANGE TEMPERATURE COMPENSATION BY ADDITION OF TWO CR 11-1686
TION IN FERROELECTRICS. (TRIGLYCINE SULFATE, HIGHLY ACCURATE TEMPERATURE CONTROL)   ORDER DISORDER TRANSI 27-2982
TRIC SODIUM TRIDEUTERO SELENITE.   ROOM TEMPERATURE CRYSTAL STRUCTURE OF THE FERROELEC 22-2861
CRYSTALS. (5 COMPOUND/ SHAPE DEPENDENCE OF PROPERTIES (CURIE TEMPERATURE, CURIE CONSTANT) OF FERROELECTRIC 27-3120
C PROPERTIES OF A (SMALL DOSE) GAMMA-IRRADIATED KDP CRYSTAL (TEMPERATURE DEPENDENCE) .   DIELECTRI 17-2466
TE SINGLE CRYSTALS AS A FUNCTION OF MELT COMPOSITION. (CURIE TEMPERATURE DEPENDENCE) /ES OF LITHIUM TANTALA 12-1874
HAVIOR OF SEMICONDUCTING YTTRIUM BARIUM ZIRCONIUM TITANATE. (TEMPERATURE DEPENDENCE)   RESISTIVITY BE 13-1995
TRANSITION OF SODIUM NITRITE BY SECOND HARMONIC GENERATION. (TEMPERATURE DEPENDENCE)   STUDY OF PHASE 16-2403
CTRIC PROPERTIES OF STRONTIUM TITANATE AT LOW TEMPERATURES. (TEMPERATURE DEPENDENCE, ANISOTROPY)   DIELE 06-1202
/HEAT ANOMALY IN FERROELECTRIC ROCHELLE SALT (CALCULATION OF TEMPERATURE DEPENDENCE FOR FERROELECTRIC CONT/ 28-3187
AND FERROELECTRIC FERROM/ ELECTRON SPIN RESONANCE (SPECTRA, TEMPERATURE DEPENDENCE) IN SOME FERROELECTRICS 08-1368
NITRITE IN THE FAR INFRARED. (350-1100 MICRONS)   TEMPERATURE DEPENDENCE OF ABSORPTION OF SODIUM 16-2387
TALATE CRYSTAL RESONATOR VIBRATING IN THICKNESS SHEAR MODE / TEMPERATURE DEPENDENCE OF AN X CUT LITHIUM TAN 12-1868
TRIC RELAXATION IN SODIUM NITRITE.   RELATIONSHIP OF THE TEMPERATURE DEPENDENCE OF B1 PHONON AND DIELEC 16-2257
BARIUM TITANATE.   TEMPERATURE DEPENDENCE OF CARRIER MOBILITY IN 04-0614
/ESONANCE INVESTIGATION OF (STOICHIOMETRIC) LITHIUM NIOBATE (TEMPERATURE DEPENDENCE OF COUPLING CONSTANT 2/ 11-1811
/ CRYSTAL UNDER HIGH HYDROSTATIC PRESSURE (TO 5600 KG/SQ CM, TEMPERATURE DEPENDENCE OF DIELECTRIC CONSTANT/ 24-2896
IN THE SYSTEM POTASSIUM YTTRIUM MOLYBDATE- POTASSIUM DYSPRO/ TEMPERATURE DEPENDENCE OF DIELECTRIC CONSTANT 30-3309
OF CUBIC IONIC COMPOUNDS. (HALIDES AND OXIDES, RELATION BE/ TEMPERATURE DEPENDENCE OF DIELECTRIC CONSTANTS 03-0188
S OF FERROELECTRIC TRIGLYCINE SULFATE IN THE FAR INFRARED. TEMPERATURE DEPENDENCE OF DIELECTRIC PROPERTIE 27-3061
ILITY OF STRONTIUM AND BARIUM TITANATE.   TEMPERATURE DEPENDENCE OF ELECTRONIC POLARIZAB 04-0738
IVITY SPECTRA OF SODIUM NITRITE CRYSTALS. (0.85 DEG K, 30-2/ TEMPERATURE DEPENDENCE OF FAR INFRARED REFLECT 16-2373
MIUM(3+) DOPED STRONTIUM TITANATE.   ANALYSIS OF STRESS AND TEMPERATURE DEPENDENCE OF FLUORESCENCE IN CHRO 06-1215
ING IN RUBY AND LITHIUM NIOBATE IN THE THREE CENTIMETER RAD/ TEMPERATURE DEPENDENCE OF HYPERSONIC WAVE DAMP 11-1718
OF THE SECOND HARMONIC GENERATION IN POTASSIUM DIHYDROGEN / TEMPERATURE DEPENDENCE OF INDEX MATCHING ANGLE 17-2544
BARIUM TITANATE SINGLE CRYSTALS.   TEMPERATURE DEPENDENCE OF INTERNAL FRICTION OF 04-0796
FERROELECTRIC ANTIMONY SULFOIODIDE.   DETERMINATION OF THE TEMPERATURE DEPENDENCE OF LATTICE CONSTANTS IN 15-2140
M THE (001) SURFACE OF BARIUM TITANATE BETWEEN/ STUDY OF THE TEMPERATURE DEPENDENCE OF LEED INTENSITIES FRO 04-0531
OMENT IN ORDER DISORDER TYPE FERROELECTRICS.   NATURE OF THE TEMPERATURE DEPENDENCE OF NMR SPECTRA SECOND M 03-0460
EFFICIENTS IN FERROELECTRIC SODIUM NITRITE.   TEMPERATURE DEPENDENCE OF NONLINEAR OPTICAL CO 16-2274
MONIC GENERATION IN POTASSIUM DIHYDROGEN PHOSPHATE (AND DEU/ TEMPERATURE DEPENDENCE OF OPTICAL (SECOND) HAR 17-2695
/RAIN IN THE FERROELECTRIC STATE OF IRON- IODINE- BORACITE. (TEMPERATURE DEPENDENCE OF PIEZOELECTRIC CONST/ 14-2083
IN LITHIUM TANTALATE. (OPTICAL MODULATION VIA TEMPERATURE) TEMPERATURE DEPENDENCE OF POLARITON DISPERSION 12-1873
BARIUM TITANATE.   LINEAR DISORDER AND TEMPERATURE DEPENDENCE OF RAMAN SCATTERING IN 04-0675
CTRUM OF FERROELECTRIC DEUTERATED KDP DURING THE PHASE TRAN/ TEMPERATURE DEPENDENCE OF RAMAN SCATTERING SPE 17-2626
ECOND HARMONIC GENERATION IN KDP.   TEMPERATURE DEPENDENCE OF RAMAN SPECTRUM AND S 17-2550
OF LITHIUM NIOBATE PLATE RESONATORS. (COMMENTS)   TEMPERATURE DEPENDENCE OF RESONANT FREQUENCIES 11-1812
RATION IN SODIUM NITRITE.   TEMPERATURE DEPENDENCE OF SECOND HARMONIC GENE 16-2315
ODIUM DITHORIUM PHOSPHATE FERROELECTRIC CRYSTALS.   TEMPERATURE DEPENDENCE OF SOME PROPERTIES OF S 36-3611
CALCIUM STRONTIUM PROPIONATE.   TEMPERATURE DEPENDENCE OF SOUND VELOCITY IN DI 26-2934
E VELOCITY IN LITHIUM TANTALATE.   TEMPERATURE DEPENDENCE OF SURFACE ACOUSTIC WAV 12-1886
  SPECTRUM OF BARIUM TITANATE.   TEMPERATURE DEPENDENCE OF THE BRILLOUIN- RAMAN 04-0807
VITY OF SOME TITANATES BETWEEN 20 AND 1000 DEG C. (BARIUM T/ TEMPERATURE DEPENDENCE OF THE COMPLEX PERMITTI 04-0820
  CAPTURE CENTERS IN FERROELECTRIC SEMICONDUCTOR ANTIMONY SU/ TEMPERATURE DEPENDENCE OF THE CROSS SECTION OF 15-2147
CTORS OF BARIUM TITANATE (BRAGG REFLECTIONS C-AXIS, 20-180 / TEMPERATURE DEPENDENCE OF THE DEBYE- WALLER FA 04-0892
ANT OF STRONTIUM TITANATE.   TEMPERATURE DEPENDENCE OF THE DIELECTRIC CONST 06-1134
ANTS OF TRIGLYCINE SULFATE ORTHOGONAL TO THE FERROELECTRIC / TEMPERATURE DEPENDENCE OF THE DIELECTRIC CONST 27-2975
ANT OF POTASSIUM DIHYDROGEN PHOSPHATE AT LAMBDA EQUAL 2.14 / TEMPERATURE DEPENDENCE OF THE DIELECTRIC CONST 17-2501
ANT OF FERROELECTRIC SUBSTANCES. (BARIUM TITANATE CERAMICS) TEMPERATURE DEPENDENCE OF THE DIELECTRIC CONST 04-0591
RTIES AND PHASE TRANSITIONS OF THE FERROELECR/ PRESSURE AND TEMPERATURE DEPENDENCE OF THE DIELECTRIC PROPE 08-1434
RTIES OF BARIUM STRONTIUM NIOBATE IN THE FREQUENCY RANGE OF/ TEMPERATURE DEPENDENCE OF THE DIELECTRIC PROPE 13-1938
RTIES AND PHASE TRANSITIONS OF THE ANTIFERROEL/ PRESSURE AND TEMPERATURE DEPENDENCE OF THE DIELECTRIC PROPE 08-1433
CE OF STRONTIUM TITANATE SINGLE CRYSTALS IN THE TEMPERATURE/ TEMPERATURE DEPENDENCE OF THE ELASTIC COMPLIAN 06-1220
ECTRIC, AND DIELECTRIC CONSTANTS OF LITHIUM TANTALATE AND L/ TEMPERATURE DEPENDENCE OF THE ELASTIC, PIEZOEL 11-1823
RADIENT IN FERROELECTRIC SODIUM NITRITE.   TEMPERATURE DEPENDENCE OF THE ELECTRIC FIELD G 16-2337
CTIVITY OF FERROELECTRICS.   MEASUREMENT OF THE TEMPERATURE DEPENDENCE OF THE ELECTRICAL CONDU 03-0525
NETIC RESONANCE OF AMMONIUM CADMIUM SULFATE.   TEMPERATURE DEPENDENCE OF THE ELECTRON PARAMAG 19-2792
ELD EFFECT IN TELLURIUM FILMS ON TRIGLYCINE SUBSTRATES.   TEMPERATURE DEPENDENCE OF THE FERROELECTRIC FI 27-3010
TIC CONSTANTS OF STRONTIUM TITANATE.   TEMPERATURE DEPENDENCE OF THE FIRST ORDER ELAS 06-1160
OF ION IN SODIUM NITRITE.   TEMPERATURE DEPENDENCE OF THE FORCE CONSTANTS 16-2311
/RATIONS OF ORTHORHOMBIC AND FERROELECTRIC THIOUREA. PART-1: TEMPERATURE DEPENDENCE OF THE INFRARED AND RA/ 25-2920
ROGEN BOND BANDS IN RAMAN SCATTERING SPECTRA OF GYPSUM AND / TEMPERATURE DEPENDENCE OF THE INTENSITY OF HYD 28-3189
TRUM AND ATOMIC MOTION IN LITHIUM NIOBATE.   TEMPERATURE DEPENDENCE OF THE LITHIUM NMR SPEC 11-1722
CTERISTICS OF POTASSIUM DIHYDROGEN PHOSPHATE SINGLE CRYSTAL/ TEMPERATURE DEPENDENCE OF THE MECHANICAL CHARA 17-2649
R SUSCEPTIBILITY OF SODIUM NITRITE.   TEMPERATURE DEPENDENCE OF THE OPTICAL NONLINEA 16-2386
ES OF FERROELECTRIC LITHIUM NIOBATE AND TANTALATE.   TEMPERATURE DEPENDENCE OF THE OPTICAL PROPERTI 11-1770
NSTANTS OF STIBOTANTALITE. (ANTIMONY (/ DOMAIN STRUCTURE AND TEMPERATURE DEPENDENCE OF THE PIEZOELECTRIC CO 36-3601
IONS IN BARIUM TITANATE AND STRONTIUM TITANATE.   TEMPERATURE DEPENDENCE OF THE RAMAN CROSS SECT 04-0548
F GADOLINIUM MOLYBDATE.   TEMPERATURE DEPENDENCE OF THE RAMAN SPECTRUM O 30-3285
ES OF BARIUM SODIUM NIOBATE.   TEMPERATURE DEPENDENCE OF THE REFRACTIVE INDIC 13-2008
AIN WALL IN FERROMAGNETICS AND FERROELECTRICS.   TEMPERATURE DEPENDENCE OF THE SHAPE OF THE DOM 03-0191
R QUADRUPOLE COUPLING CONSTANT IN SODIUM NITRITE.   TEMPERATURE DEPENDENCE OF THE SODIUM-23 NUCLEA 16-2393
F BARIUM TITANATE.   VIBRONIC EFFECT IN THE TEMPERATURE DEPENDENCE OF THE SOFT FREQUENCY O 04-0662
ONTIUM TITANATE ABOVE THE 105 DEG K TRANSITION.   TEMPERATURE DEPENDENCE OF THE SOFT MODE IN STR 06-1178
ARIZATION OF AMMONIUM SULFATE.   (TEMPERATURE DEPENDENCE OF) THE SPONTANEOUS POL 19-2799
S IN ANTIMONY SULFOIODIDE SINGLE CRYSTALS.   TEMPERATURE DEPENDENCE OF THE SWITCHING PROCES 15-2191
F TRIGLYCINE SULFATE.   TEMPERATURE DEPENDENCE OF THE SWITCHING RATE O 27-3015

ASOUND IN POLYCRYSTALLINE SOLID SC/ SPECIFIC FEATURES OF THE TEMPERATURE DEPENDENCE OF THE VELOCITY OF ULTR 04-0894
/UDINAL ULTRASONIC WAVES IN POTASSIUM DIDEUTERIUM PHOSPHATE (TEMPERATURE DEPENDENCE OF THE YOUNGS MODULUS / 17-2660
ODES IN LEAD TITANATE.   RAMAN STUDIES (TEMPERATURE DEPENDENCE) OF UNDER DAMPED SOFT M 05-1041
EFFECT AND SPONTANEOUS POLARIZATION OF GADOLINIUM MOLYBDATE (TEMPERATURE DEPENDENCE 150-165 C). /OELECTRIC 30-3314
CAL AND OPTICAL PROPERTIES OF FERROELECTRIC FERROE/ INFERRED TEMPERATURE DEPENDENCES OF ELECTRICAL, MECHANI 30-3253
.4 GHZ HYPERSOUND IN ALUMINUM OXIDE AND LITHIUM NIOBATE SIN/ TEMPERATURE DEPENDENCES OF THE ABSORPTION OF 9 11-1719
RIC CONS/ RELAXATION OF DOMAIN WALLS IN TRIGLYCINE SULFATE. (TEMPERATURE DEPENDENCES OF THE COMPLEX DIELECT 27-3103
RIC CONSTANT OF FERROELECTRICS NEAR THE CURIE/ FREQUENCY AND TEMPERATURE DEPENDENCES OF THE COMPLEX DIELECT 03-0361
MOBILITIES OF HOLES IN ANTIMONY TRISULFIDE SINGLE CRYSTALS. TEMPERATURE DEPENDENCES OF THE HALL AND DRIFT 15-2193
   FERROELECTRIC CERAMIC.   TIME AND TEMPERATURE DEPENDENCES OF THE PARAMETERS OF A 09-1514
GNMENT IN LEAD ZIRCONATE- LEAD TITANATE- LANTHANUM OXIDE (P/ TEMPERATURE DEPENDENT FERROELECTRIC DOMAIN ALI 09-1586
TASSIUM DIDEUTERIUM PHOSPHATE (CCMF/ RAMAN SCATTERING BY THE TEMPERATURE DEPENDENT FERROELECTRIC MODE IN PO 17-2705
OELECTRIC LEAD ZIRCONATE BY THE MOSSBAUER EFFECT.   TEMPERATURE DEPENDENT OPTICAL MODE IN ANTIFERR 07-1260
TITANATE.   TEMPERATURE DEPENDENT OPTICAL PHONONS IN LEAD 05-1083
UM NITRITE BY RAMAN SPECTROSCOPY.   ANALYSIS OF THE TEMPERATURE DEPENDENT PHONON STRUCTURE IN SODI 16-2299
HRAN MODE ASSIGNMENTS IN BARIUM AND STRONTIUM TITANATES.   TEMPERATURE DEPENDENT POLARIZABILITIES AND COC 04-0805
TANATE.   CRYSTAL GROWTH AND TEMPERATURE DEPENDENT RAMAN SPECTRA OF LEAD TI 05-1074
UM DIHYDROGEN PHOSPHATE, POTASSIUM DIDEUTERIUM PHOSPHATE, P/ TEMPERATURE DEPENDENT RAMAN SPECTRA OF POTASSI 17-2409
ELECTRIC LINEAR CHAIN MODEL. (PEROVSKITES)   TEMPERATURE DEPENDENT SPECTRUM OF AN ANTIFERRO 03-0468
F HEXAGONAL ICE TO 2 K.   LOW TEMPERATURE DIELECTRIC CELL AND PERMITTIVITY O 34-3432
/S AND PHASE TRANSITIONS IN POTASSIUM MANGANESE TRIFLUORIDE: TEMPERATURE DISCONTINUITIES IN THE INFRARED S/ 33-3363
 IN CRYSTALS. (THEORY, FERROELECTRICS, PIEZOELECTRICS) LOW TEMPERATURE DISPLACEMENT TYPE PHASE TRANSITION 03-0320
ON.   AN OBSERVATION OF THE CHANGE IN COOLED KDP TEMPERATURE DUE TO 337 MICRON INCIDENT RADIATI 17-2711
TE (10 TO 130 DEG K).   LOW TEMPERATURE ELASTIC MODULI OF STRONTIUM TITANA 06-1190
ON THE ABSORPTION OF ULTRASOUND IN ANTIMONY SULF/ EFFECTS OF TEMPERATURE, ELECTRIC FIELD, AND ILLUMINATION 15-2248
/TANIUM RICH NONSTOICHIOMETRIC BARIUM TITANATE. PART-1: HIGH TEMPERATURE ELECTRICAL CONDUCTIVITY MEASUREME/ 04-0822
ATE NIOBATE. (POTASSIUM TANTALATE(0.65) NIOBATE(0.35), CURIE TEMPERATURE EQUALS 10 DEG C, ABSTRACT) /TANTAL 08-1391
RIUM TITANATE.   TEMPERATURE FACTORS OF TETRAGONAL AND CUBIC BA 04-0532
GNETIC (BETA) TERBIUM MOLYBDATE (CZOCHRALSKI SINGLE C/ (ROOM TEMPERATURE) FERROELECTRIC FERROELASTIC PARAMA 30-3290
QUE.   OBSERVATION OF DOMAIN STRUCTURE IN LOW TEMPERATURE FERROELECTRICS BY SOLID DEW TECHNI 03-0341
M(X) TRIHYDROGEN(1-X) PHOSPHATE SYSTEM VERSUS CONCENTRATION, TEMPERATURE, FIELD, AND PRESSURE. /TRIDEUTERIU 17-2666
/IMONY SULFOIODIDE CRYSTALS NEAR THE CURIE POINT (EFFECTS OF TEMPERATURE, FIELD STRENGTH, SAMPLE ILLUMINAT/ 15-2121
              INTERNAL FRICTION AS A FUNCTION OF TEMPERATURE FOR ROCHELLE SALT.   28-3227
/OSITION OF THE FUNDAMENTAL ABSORPTION EDGE AS A FUNCTION OF TEMPERATURE FOR ROCHELLE SALT AND TRIGLYCINE / 27-3160
MICROWAVE DISPERSION OF FERROELECTRICS NEAR THE/ (EFFECT OF TEMPERATURE GRADIENT WITHIN THE SAMPLE ON THE) 03-0370
       ON THE INFLUENCE OF DEFECTS ON THE TEMPERATURE HYSTERESIS OF BARIUM TITANATE.   04-0987
            BRILLOUIN EFFECT AT ORDINARY TEMPERATURE IN AMMONIUM SULFATE.   19-2766
EFFECT OF THE HYDROSTATIC PRESSURE ON THE TRANSITION TEMPERATURE IN AMMONIUM SULFATE.   19-2796
           ULTRASONIC ATTENUATION NEAR THE CURIE TEMPERATURE IN DICALCIUM STRONTIUM PROPIONATE. 26-2948
/ OBSERVATION OF A VERY LOW FREQUENCY MODE AT THE TRANSITION TEMPERATURE IN KDP AND DEUTERATED KDP (DOPED / 17-2465
AND DEU/ POSSIBLE THEORETICAL EXPLANATION OF SHIFT OF CURIE TEMPERATURE IN POTASSIUM DIDEUTERIUM PHOSPHATE 17-2601
            LOWERING THE CURIE TEMPERATURE IN REDUCED BARIUM TITANATE.   04-0711
DIPOLE LATTICE RELAXATION TIME NEAR THE NEEL TEMPERATURE IN SODIUM NITRITE.   16-2300
OF THE STRESS INDUCED PHASE TRANSITIONS NEAR THE TRANSITION TEMPERATURE IN STRONTIUM TITANATE. /MODE STUDY 06-1196
RAMAN STUDIES OF STRESS INDUCED PHASE TRANSITION NEAR CURIE TEMPERATURE IN STRONTIUM TITANATE. (ABSTRACT) 06-1236
UM / OSCILLATORY BEHAVIOR OF RAMAN SCATTERING INTENSITY WITH TEMPERATURE IN THE FERROELECTRIC PHASE POTASSI 17-2662
/ALY OF THE SPIN LATTICE RELAXATION TIME NEAR THE TRANSITION TEMPERATURE IN THE ORDER DISORDER TYPE FERROE/ 03-0488
/H THAT THE ELECTRIC SUSCEPTIBILITY OR ELASTIC COMPLIANCE IS TEMPERATURE INDEPENDENT IN THE PROTOTYPIC PHA/ 03-0140
OF LOCALIZED D-STRAIN AND SOFT- NORMAL MODE COORDINATES AND TEMPERATURE INDEPENDENT PARAMETERS). /IN TERMS 03-0432
ONIUM DIHYDROGEN ORTHOPHOSPHATE (4000-10000 PER CM) AND/ LOW TEMPERATURE INFRARED ABSORPTION SPECTRA OF AMM 17-2432
4200 PER CM) OF GUANIDINIUM ALUMINUM SULFATE HEXAHYDRAT/ LOW TEMPERATURE INFRARED ABSORPTION SPECTRA (5200- 21-2811
CTRIC AMMONIUM HYDROGEN SULFATE AND AMMONIUM DEUTERIUM / LOW TEMPERATURE INFRARED STUDIES. PART-8: FERROELE 19-2782
MODULATORS WITH AMMONIUM DIHYDROGEN PHOSPHATE, POTASSIUM DI/ TEMPERATURE INSTABILITY OF ELECTROOPTIC LASER 17-2675
/C POTASSIUM FERROCYANIDE TRIDEUTERO HYDRATE ABOVE THE CURIE TEMPERATURE (LATTICE CONSTANTS, SPACE GROUP). 24-2903
) TITANATE (RECIPROCAL DIELECTRIC CONSTANT FOLLOWS QUADRATIC TEMPERATURE LAW). /D ((BARIUM(X) STRONTIUM(1-X 04-0898
AL STOICHIOMETRY (VERSUS MELT COMPOSITION, LATTICE AND CURIE TEMPERATURE MEASUREMENTS). /ALATE SINGLE CRYST 12-1851
/TRIC POTASSIUM FERROCYANIDE TRIHYDRATE (NEAR THE TRANSITION TEMPERATURE, NO ANOMALY BUT SOFT PHONON MODE / 24-2891
/L PROPERTIES OF VANADIUM DIOXIDE ABOVE AND BELOW TRANSITION TEMPERATURE. (NO COCHRAN TYPE FERROELECTRIC M/ 36-3526
/BLE OPTICAL CHARGE EXCHANGE AND "FROZEN" SHIFT OF THE CURIE TEMPERATURE OF A FERROELECTRIC SEMICONDUCTOR. 15-2160
TITANATE SOLID SOLUTIONS BY HYDR/ SHIFTING OF THE TRANSITION TEMPERATURE OF BARIUM(0-0.1) STRONTIUM(0.9-1) 04-0722
PRESSURE DEPENDENCE OF CUBIC TETRAGONAL TRANSITION TEMPERATURE OF CESIUM LEAD TRICHLORIDE.   33-3393
OF THE PROTON SPIN LATTICE RELAXATION TIME NEAR THE CRITICAL TEMPERATURE OF DICALCIUM LEAD PROPIONATE. /LY 26-2947
C/ INFLUENCE OF HYDROSTATIC PRESSURE ON THE PHASE TRANSITION TEMPERATURE OF FERROELECTRIC CRYSTALS OF THE P 17-2495
.   CURIE TEMPERATURE OF FERROELECTRIC LITHIUM TANTALATE 12-1887
E.   ULTRASONIC RELAXATION NEAR THE CURIE TEMPERATURE OF FERROELECTRIC TRIGLYCINE SULFAT 27-3097
                DIPOLAR RELAXATION AT LOW TEMPERATURE OF ICE SINGLE CRYSTAL.   34-3447
DIFFERENT STOICHIOMETRIES.   PHASE MATCHING ANGLES AND TEMPERATURE OF LITHIUM METANIOBATE CRYSTALS OF 11-1665
RIC PROPERTIES.   INFLUENCE OF THE CRYSTALLIZATION TEMPERATURE OF ROCHELLE SALT ON ITS FERROELECT 28-3193
     EFFECT OF STRESS ON THE SUPERCONDUCTING TRANSITION TEMPERATURE OF STRONTIUM TITANATE.   06-1182
  EFFECT OF EVAPORATED METALLIC ELECTRODES ON THE TRANSITION TEMPERATURE OF TRIGLYCINE SULFATE.   27-3100
PHOSPHATE, TITANATE, SULFATE, FLUOBERYLLATE, ETC/ (EFFECT OF TEMPERATURE ON) QUASI SYMMETIRICAL HARD IONS / 03-0330
/LECTRIC SEMICONDUCTORS OF THE A(V)B(VI)C(VII) TYPE WITH LOW TEMPERATURE PHASE CHANGES. (EFFECT OF PHOTOCO/ 15-2161
NATE TITANATE SYSTEM.   HIGH TEMPERATURE PHASE EQUILIBRIA IN THE LEAD ZIRCO 09-1577
           DEUTERON MAGNETIC RESONANCE OF THE HIGH TEMPERATURE PHASE OF FERROELECTRIC THIOUREA. 25-2918
TITANATE.   X-RAY DIFFRACTOMETRY OF LOW TEMPERATURE PHASE TRANSFORMATIONS IN STRONTIUM 06-1156
PROPERTIES OF AMMONIUM BISULFATE IN THE VICINITY OF THE HIGH TEMPERATURE PHASE TRANSITION. / FERROELECTRIC 19-2787
ROGEN PHOSPHATE BY SCATTERING OF COLD NEUTRONS.   HIGH TEMPERATURE PHASE TRANSITION IN RUBIDIUM DIHYD 17-2489
E OF THE ORGANIC SEMICONDUCTOR PHTHALOCYANINE.   LOW TEMPERATURE PHASE TRANSITION IN THE POLAR STAT 36-3623
/ PROPERTIES OF POTASSIUM DIHYDROGEN PHOSPHATE NEAR THE HIGH TEMPERATURE PHASE TRANSITION. (THERMAL EXPANS/ 17-2627
Y IN POTASSIUM DIHYDROGEN PHOSPHATE TYPE CRYSTALS.   HIGH TEMPERATURE PHASE TRANSITIONS AND METASTABILIT 17-2513
/OTASSIUM OR SODIUM)(0.5) BISMUTH(0.5) TITANATE AND THE HIGH TEMPERATURE PHASE TRANSITIONS IN POTASSIUM(0./ 08-1362
E AND THE USE OF TILTING SCHEMES IN THE SOLUTIONS OF P/ HIGH TEMPERATURE PHASE TRANSITIONS IN SODIUM NIOBAT 08-1336
E (TO 500C).   X-RAY STUDY OF HIGH TEMPERATURE PHASE TRANSITIONS OF SODIUM NIOBAT 13-1965
PIC STUDY OF FERROELECTRIC DOMAINS OF BARIUM TITANATE IN LOW TEMPERATURE PHASES.   ELECTRON MICROSCO 04-1034
L SPECTRA OF CESIUM AND RUBIDIUM NITRATE CRYSTALS IN ITS LOW TEMPERATURE PHASES. /ARITIES IN THE VIBRATIONA 16-2323
SIUM DIHYDROGEN PHOSPHATE TYPE IN THEIR PARAELECTRIC AND LOW TEMPERATURE PHASES. / OF CRYSTALS OF THE POTAS 17-2515
ENT REVERSAL DUE TO DIPOLE RELAXATION)   LOW TEMPERATURE POLARIZATION EFFECTS IN ICE. (CURR 34-3418
/TIC EFFECT IN NONMETALLIC CRYSTALS. (STRAIN DEPENDENT CURIE TEMPERATURE, POLARIZATION FLUCTUATIONS, LITHI/ 03-0513
/IBUTION OF RELAXATION TIME IN FERROELECTRICS NEAR THE CURIE TEMPERATURE (POLY VERSUS MONODISPERSIVE BEHAV/ 03-0521
OR DEUTERIUM PHOSPHATE (INFRARED PHOSPHATE STRETCHING, WIDE TEMPERATURE RANGE). /S IN POTASSIUM DIHYDROGEN 17-2646

SPECIFIC HEAT OF AMMONIUM HYDROGEN SULFATE IN THE TEMPERATURE RANGE FROM -70 TO +14 DEG C. 19-2789
THERMAL EXPANSION OF LITHIUM NIOBATE IN THE TEMPERATURE RANGE OF 70-300 K. 11-1685
FTERS VERSU/ POTASSIUM TANTALATE LEAD TITANATE SYSTEM IN THE TEMPERATURE RANGE 4.2-300 DEG K (LATTICE PARAM 08-1308
OUIN SCATTERING IN STRONTIUM TITANATE SINGLE CRYSTALS IN THE TEMPERATURE RANGE 5 DEG K-300 DEG K. BRILL 06-1149
GION. PROTONIC CONDUCTION OF ICE. PART-1: HIGH TEMPERATURE REGION. PART-2: LOW TEMPERATURE RE 34-3414
/MICS AT STRUCTURAL TRANSITIONS ABOVE FERROELECTRIC CRITICAL TEMPERATURE (ROTATIONS OF BO6 OCTAHEDRA IN AB/ 06-1205
ND PLZT. COLD PRESSING AND LOW TEMPERATURE SINTERING OF ALKOXY-DERIVED PLZT A 09-1493
ILLATORY SYSTEMS. EFFECT OF AUTOMATIC TEMPERATURE STABILIZATION IN FERROELECTRIC OSC 03-0294
FABRICATION OF A LITHIUM TANTALATE TEMPERATURE STABILIZED OPTICAL MODULATOR. 12-1872
ANATE, STRONT/ METHOD FOR ACHIEVING HIGH DIELECTRIC CONSTANT TEMPERATURE STABLE FERROELECTRICS. (BARIUM TIT 04-0551
CTRA OF BARIUM AND STRONTIUM TITANATES IN THE 2-1000 MICRON/ TEMPERATURE STUDIES OF INFRARED REFLECTION SPE 04-0863
OF AMMONIUM SULFATE: THE NATURE OF HYDROGEN BONDING AN/ LOW TEMPERATURE STUDIES. PART-4: PHASE TRANSITIONS 19-2783
WITH THE 248 CM(-1) SOFT MODE IN LITHIUM NIOBATE. TEMPERATURE STUDY OF THE POLARITON ASSOCIATED 11-1807
SINGLE CRYSTAL POTASSIUM DIHYDROGEN PHOSPHATE. LOW TEMPERATURE THERMAL EXPANSION MEASUREMENTS ON 17-2464
/TANIUM RICH NONSTOICHIOMETRIC BARIUM TITANATE. PART-3: HIGH TEMPERATURE THERMODYNAMIC AND X-RAY DIFFRACTI 04-0886
N STUDY OF NONSTOICHIOMETRIC BARIUM TITANATE. HIGH TEMPERATURE THERMODYNAMIC AND X-RAY DIFFRACTIO 04-0885
STIC, PIEZOELECTRIC, DIELECTRIC, AND OPTICAL CONSTANTS) ROOM TEMPERATURE TO THE CURIE POINT. / NIOBATE (ELA 12-2060
N (HIGH PURITY) LEAD ZIRCONATE. HIGH TEMPERATURE (TO 236 DEG C) PHASE TRANSITIONS I 07-1266
FERRATE LEAD TITANATE PEROVSKITE TYPE SOLID SOLUTIONS (ROOM TEMPERATURE TO 600 DEG C). /STUDIES OF BISMUTH 08-1471
CRYSTALS. NONLINEAR EFFECTS AND THE LOW TEMPERATURE TRANSITION IN ANTIMONY SULFOIODIDE 15-2247
/STATE MATERIALS (INCLUDING FERROELECTRICS) AS A FUNCTION OF TEMPERATURE TRANSITION OF INFRARED REFLECTION SPE 02-0104
ENT OF CHROMIUM- POTASS/ POSSIBLE EXPLANATION OF THE UNUSUAL TEMPERATURE VARIATION OF THE MEAN MAGNETIC MOM 20-2806
/ REFRACTIVE INDICES OF ANTIMONY SULFOIODIDE. (DEPENDENCE ON TEMPERATURE, WAVELENGTH, APPLIED FIELD STRENGT 15-2206
TURES OF TUNGSTEN BRONZE TYPE NIOBATES (CORRELATION OF CURIE TEMPERATURE WITH CHEMICAL COMPOSITION). /MPERA 13-2049
ITE MODIFICATION OF CADMIUM TITANATE (AND THE ANTIFERR/ HIGH TEMPERATURE X-RAY INVESTIGATION OF THE PEROVSK 08-1387
LOGRAPHIC TRANSITION IN VANADIUM DIOXIDE. (ANTIFERROELECTRIC TEMPERATURE 340 DEG K) /PONENTS OF THE CRYSTAL 36-3562
TIES OF STRONTIUM TITANATE (ANNEALED SINGLE CRYSTALS) AT LOW TEMPERATURE (5-300 DEG K). DIELECTRIC PROPER 06-1201
OUS POLARIZATION, COERCIVE FIELD) OF BISMUTH TITANATE AT LOW TEMPERATURE (5-350K). /(PERMITTIVITY, SPONTANE 08-1436
IN THE PROTON NMR IN POTASSIUM DIHYDROGEN PHOSPHATE AT LOW TEMPERATURES.) ABSENCE OF MOTIONAL NARROWING 17-2594
OELECTRICITY AND ANTIFERROELECTRIC MATERIALS. (TABLE OF NEEL TEMPERATURES) ANTIFERR 01-0020
HOSPHATE, AMMONIUM SULFATE, ROCHELLE SALT) AT THE TRANSITION TEMPERATURES. /SULFATE, POTASSIUM DIHYDROGEN P 02-0125
CONDUCTION ANOMALIES AND POLARIZATION IN ICE AT LOW TEMPERATURES. 34-3428
DEFECT LEAD TITANATES WITH DIVERSE CURIE TEMPERATURES. 05-1076
ENTS OF LITHIUM OXIDE INTO LITHIUM NIOBATE BY PHASE MATCHING TEMPERATURES. /ON OF THE DISTRIBUTION COEFFICI 11-1841
EUTERIUM PHOSPHATE AND RUBIDIUM DIHYDROGEN PHOSPHATE AT HIGH TEMPERATURES. /YDROGEN PHOSPHATE POTASSIUM DID 17-2613
DIELECTRIC BEHAVIOR OF CUBIC AND HEXAGONAL ICES AT LOW TEMPERATURES. 34-3431
DIELECTRIC PROPERTIES OF STRONTIUM TITANATE AT LOW TEMPERATURES. 06-1200
T OF DEUTERATION IN AMMONIUM BISULFATE (CHANGE IN TRANSITION TEMPERATURES). EFFEC 19-2760
ELASTIC ANOMALIES OF ICE AT LOW TEMPERATURES. 34-3439
ELECTRIC POLARIZATION EFFECTS IN PURE AND DOPED ICE AT LOW TEMPERATURES. 34-3412
ICAL CONDUCTIVITY OF STRONTIUM ZIRCONATE AND HAFNATE AT HIGH TEMPERATURES. ELECTR 08-1417
PARAMAGNETIC RESONANCE OF IRON(3+) IN BARIUM TITANATE AT LOW TEMPERATURES. ELECTRON 04-0927
NIC CONDUCTION IN SLIGHTLY REDUCED STRONTIUM TITANATE AT LOW TEMPERATURES. ELECTRO 06-1151
PHOSPHATE AND AMMONIUM DIHYDROGEN PHOSPHATE CRYSTALS AT HIGH TEMPERATURES. /ERTIES OF POTASSIUM DIHYDROGEN 17-2697
ES OF TETRAGONAL (BARIUM, LEAD) (NIOBATE, ZIRCONATE). (CURIE TEMPERATURES) FERROELECTRIC PROPERTI 08-1293
SWITCHING OF TETRAGONAL DOMAINS IN STRONTIUM TITANATE AT LOW TEMPERATURES. FIELD INDUCED 06-1184
PTIBILITY OF GUANIDINIUM VANADIUM SULFATE HEXAHYDRATE AT LOW TEMPERATURES. MAGNETIC SUSCE 21-2815
IELECTRIC CONSTANTS FOR A FEW FERROELECTRIC MATERIALS AT LOW TEMPERATURES. MEASUREMENT OF D 02-0035
NSTANTS OF SODIUM NITRITE BY BRILLOUIN DIFFUSION AT ORDINARY TEMPERATURES. MEASUREMENT OF THE ELASTIC CO 16-2350
23 IN SODIUM NITRITE IN THE VICINITY OF THE PHASE TRANSITION TEMPERATURES. /ETIC RESONANCE STUDY ON SODIUM- 16-2396
EM IN POTASSIUM CHLORIDE LATTICE, FERROELECTRIC STATE AT LOW TEMPERATURES). / WAVES IN HYDROXYL DIPOLE SYST 03-0452
ECTRIC PROPERTIES OF LEAD TITANATE ZIRCONATE CERAMICS AT LOW TEMPERATURES. PIEZOELECTRIC AND DIEL 09-1519
RADIATION EXPOSURE OF A LITHIUM NIOBATE CRYSTAL AT HIGH TEMPERATURES. 11-1721
R (NIOBIUM, TUNGSTEN), COMPILATION, LATTICE CONSTANTS, CURIE TEMPERATURES) /HERE B IS (NIOBIUM, TITANIUM) O 13-1964
IN SINGLE CRYSTALS OF BISMUTH(12) GERMANIUM OXIDE(20) AT LOW TEMPERATURES. ULTRASONIC ATTENUATION 35-3510
/ CHANNELING OF PROTONS IN THIN BARIUM TITANATE CRYSTALS AT TEMPERATURES ABOVE AND BELOW THE FERROELECTRIC 04-0687
ENOLOGICAL THEORY OF MICROWAVE LOSSES IN FERROELECTRICS. (AT TEMPERATURES ABOVE THE CURIE POINT) PHENOM 03-0503
F ORTHORHOMBIC AND CUBIC HYDROGEN AND DEUTERIUM CHLORIDES AT TEMPERATURES ABOVE 80 DEGREES K. /AN SPECTRA O 31-3330
/LID SOLUTIONS OF THE PEROVSKITE TYPE WITHIN A WIDE RANGE OF TEMPERATURES AND ELECTRIC FIELD INTENSITIES A/ 08-1380
TE BARIUM TITANATE. PRESSURE DEPENDENCE OF TRANSITION TEMPERATURES AND ELECTROSTRICTIONS IN PEROVSKI 04-0853
ROELECTRIC) DIELECTRIC ANOMALIES IN SOLID METHANE AT LOW TEMPERATURES AND HIGH PRESSURES. (POSSIBLY FER 36-3547
/TION OF THE RAMAN FREQUENCIES IN SODIUM NITRITE AT ORDINARY TEMPERATURES AS A FUNCTION OF CRYSTAL ORIENTA/ 16-2363
/ SOME FERROELECTRICS AND ANTIFERROELECTRICS WITH HIGH CURIE TEMPERATURES. (BISMUTH FERRATE AND SYSTEMS BA/ 08-1378
IN FERROELECTRIC MATERIAL WITH SEVERAL PHASE TRANSITIONS AT TEMPERATURES CLOSE TOGETHER. (ROCHELLE SALT) / 03-0363
E. ELECTRICAL BEHAVIOR OF ICE AT LOW TEMPERATURES, FERROELECTRICITY AND SPACE CHARG 34-3448
/AL STRESSES IN BARIUM TITANATE CERAMICS. (AND DIFFERENCE IN TEMPERATURES FOR MAXIMUM DIELECTRIC CONSTANT / 04-0849
/ULTRAVIOLET ABSORPTION (5-23 EV AT ROOM AND LIQUID NITROGEN TEMPERATURES IN ALKALI NITRITES (SODIUM, POTA/ 16-2399
EPR. DETERMINATION OF TRANSITION TEMPERATURES IN CESIUM LEAD TRICHLORIDE USING 33-3408
ROELECTRICS (WITH AND WITHOUT / INTERNAL FRICTION (AND CURIE TEMPERATURES) IN (LEAD ZIRCONATE TITANATE) FER 09-1584
H BY ADD/ SPATIALLY UNIFORM AND ALTERABLE SHG PHASE MATCHING TEMPERATURES IN LITHIUM NIOBATE (CRYSTAL GROWT 11-1684
THE NORMAL VIBRATION MODES OF FERROELECTRIC CRYSTALS AT LOW TEMPERATURES. (KDP TYPE CRYSTALS) /ANSITION ON 17-2456
/SE TRANSITIONS OF FERROELECTRIC IRON IODINE BORACITE AT LOW TEMPERATURES. (MONOCLINIC AND RHOMBOHEDRAL PH/ 14-2086
/TRASOUND IN A(V) B(VI) C(VII) FERROELECTRIC SEMICONDUCTORS AT TEMPERATURES NEAR A TYPE-1 PHASE TRANSITION. / 15-2114
/OF THE FERROELECTRIC TRANSITION IN KDP TYPE CRYSTALS (CURIE TEMPERATURES OF POTASSIUM, RUBIDIUM, CESIUM A/ 17-2577
ONATE- LEAD TITANATE- LEAD METANIOBATE TERNARY SYSTEM. CURIE TEMPERATURES OF SOLID SOLUTIONS IN A LEAD ZIRC 09-1595
/CAL FACTORS FOR CALCULATION OF THE FERROELECTRIC TRANSITION TEMPERATURES OF TUNGSTEN BRONZE TYPE NIOBATES/ 13-2049
S, 58 SOLID SOLUTIONS, AND 10 TECHNICAL FERROCERAMICS, CURIE TEMPERATURES OF 290 COMPOUNDS TABULATED) /OUND 01-0014
/NEUTRON SCATTERING BY HYDROGEN BONDED FERROELECTRICS AT LOW TEMPERATURES. PART-1: GENERAL EXPRESSION FOR / 03-0478
/TROCALORIC EFFECT) STUDIES OF FERROELECTRIC CERAMICS AT LOW TEMPERATURES. (POLYCRYSTALLINE STRONTIUM TITA/ 04-0720
X) STRONT/ INFLUENCE OF PRESSURE ON PHASE TRANSITIONS AT LOW TEMPERATURES: STRONTIUM TITANATE AND ((BARIUM(04-0898
OPY) DIELECTRIC PROPERTIES OF STRONTIUM TITANATE AT LOW TEMPERATURES. (TEMPERATURE DEPENDENCE, ANISOTR 06-1202
TE CRYSTALS (STRONTIUM TITANATE, POTASSIUM TANTALATE) AT LOW TEMPERATURES. (THEORY) /HONON IN SOME PEROVSK 03-0227
ATE) THEORY OF SECOND ORDER PHASE TRANSITIONS AT LOW TEMPERATURES. (TIN TELLURIDE, POTASSIUM TANTAL 03-0433
CONATE(0.65) TITA/ FERROELECTRIC FIELD EFFECT STUDIES AT LOW TEMPERATURES (WITH TIN OXIDE FILMS ON LEAD ZIR 09-1505
TENSILE AND FLEXURE PROPERTIES OF SALINE ICE. 34-3422
ANTHANUM ZIRCONATE TITANATE CERAMICS WHEN SUBJECTED TO LARGE TENSILE STRAINS. /ROELASTIC BEHAVIOR OF LEAD L 09-1571
AND DOMAIN STRUCTURE OF BARIUM TITANA/ INFLUENCE OF UNIAXIAL TENSILE STRESSES ON THE ELECTRICAL PROPERTIES 04-0961
. FERROELECTRIC TRANSFORMATIONS OF TENSORIAL PROPERTIES IN REGULAR FERROELECTRICS 03-0139
VALENCE DYNAMICAL/ LOW FREQUENCY RAMAN SPECTRUM OF ICE I(H). TENTATIVE INTERPRETATION WITH A MIXED COULOMB 34-3425
F GADOLINIUM MOLYBDATE AND ISOSTRUCTURAL COMPOUNDS (EUROPIUM TERBIUM, DYSPROSIUM AND HOLMIUM MOLYBDATES). / 30-3271

ERROELECTRIC PHASE TRANSFORMATION IN RARE EARTH MOLYBDATES. (TERBIUM MOLYBDATE)   MECHANISM OF THE F 30-3254
CATTERING STUDY OF THE FERROELECTRIC PHASE TRANSFORMATION IN TERBIUM MOLYBDATE.   NEUTRON S 30-3268
   NONLINEAR OPTICAL PROPERTIES OF GADOLINIUM MOLYBDATE AND TERBIUM MOLYBDATE.   30-3305
   RAMAN SPECTRUM AND STRUCTURE OF TERBIUM MOLYBDATE.   30-3296
ONIC GENERATION IN MOLYBDATES (BETA GADOLINIUM MOLYBDATE AND TERBIUM MOLYBDATE.   SECOND HARM 30-3286
   ZONE BOUNDARY SOFT MODE IN TERBIUM MOLYBDATE.   30-3269
/TEMPERATURE) FERROELECTRIC FERROELASTIC PARAMAGNETIC (BETA) TERBIUM MOLYBDATE (CZOCHRALSKI SINGLE CRYSTAL/ 30-3290
/ERGY AND SPECIFIC HEAT IN FERROELECTRIC PHASE TRANSITION IN TERMS OF A SINGLE MODE ANHARMONIC OSCILLATOR / 03-0406
TY EFFECTS ON THE PLASTICITY OF ICE AND THEIR EXPLANATION IN TERMS OF HYDROGEN REORIENTATION.   IMPURI 34-3449
ODE CCO/ THEORY OF PEROVSKITE FERROELECTRICS (HAMILTONIAN IN TERMS OF LOCALIZED D-STRAIN AND SOFT- NORMAL M 03-0432
ETATION OF THE DYNAMICS CF THE BARIUM TITANATE TRANSITION IN TERMS OF NUCLEATION. (CONCEPTS)   INTERPR 04-0606
M DIOXIDE COMPOUNDS.   ELECTRICAL PROPERTIES OF TERNARY BARIUM OXIDE- LANTHANUM OXIDE- TITANIU 13-1912
   STUDY OF DIVALENT BINARY AND TERNARY NIOBATES IN THE SOLID STATE.   08-1286
   CATION NONSTOICHIOMETRY IN TERNARY OXIDES. (BARIUM TITANATE)   04-0969
ATIONS IN THE LEAD ZIRCONATE LEAD TITANATE- LEAD METANIOBATE TERNARY SYSTEM. /YSTAL STRUCTURE AND PHASE REL 09-1596
LUTIONS IN A LEAD ZIRCONATE- LEAD TITANATE- LEAD METANIOBATE TERNARY SYSTEM. CURIE TEMPERATURES OF SOLID SO 09-1595
NIOBIUM PENTOXIDE. (OXIDE/ PHASE EQUILIBRIA RELATIONS IN THE TERNARY SYSTEM BARIUM OXIDE- STRONTIUM OXIDE- 13-1920
LEAD NIOBATE(0.67) ZINCATE(0.33).   TERNARY SYSTEM LEAD TITANATE- LEAD ZIRCONATE- 09-1507
(ZINCATE NIO/ DIELECTRIC AND PIEZOELECTRIC PROPERTIES IN THE TERNARY SYSTEM LEAD (ZINCATE NIOBATE)- BARIUM 05-1070
M(0.08) TITANATE.   PREPARATION OF THE TERNARY TITANATE BARIUM(0.8) LEAD(0.12) CALCIU 36-3587
AIN PEROVSKITES (STRONTIUM OR CALCIUM TITANATE OR ZIRCONATE, TEST OF SHORT OR LONG RANGE ORDER THEORIES). / 03-0414
/ND HARMONIC GENERATION ON CRYSTAL INHOMOGENEITY (THEORY AND TEST ON BARIUM SODIUM NIOBATE AND NEODYMIUM D/ 13-2043
/HASE TRANSITIONS. (BASED ON LANDAU PHENOMENOLOGICAL THEORY, TESTED ON AMMONIUM FLUOBERYLLATE, ABSTRACT ON/ 19-2765
N APPLICATION TO LEAD TITANATE.   TESTING A POINT DIPOLE ELECTROOPTIC MODEL BY A 05-1073
   TETRA... SEE THE PREFIXED WORD.
   TEMPERATURE FACTORS OF TETRAGONAL AND CUBIC BARIUM TITANATE.   04-0532
/IC PROPERTIES OF THE BARIUM LEAD NIOBATE ZIRCONATE SYSTEM. (TETRAGONAL AND ORTHORHOMBIC PHASES, ABSTRACT / 13-1925
. (CURIE TEMPERATURES)   FERROELECTRIC PROPERTIES OF TETRAGONAL (BARIUM, LEAD) (NIOBATE, ZIRCONATE) 08-1293
   BAND STRUCTURE OF CUBIC AND TETRAGONAL BARIUM TITANATE.   04-0846
   VIBRATIONAL AMPLITUDES IN FERROELECTRIC TETRAGONAL BARIUM TITANATE.   04-0713
   X-RAY AND NEUTRON DIFFRACTION STUDY OF TETRAGONAL BARIUM TITANATE.   04-0740
N LINE SHAPE.   POLARITONS IN TETRAGONAL BARIUM TITANATE: THEORY OF THE RAMA 04-0556
/THORHOMBIC IONIC CRYSTAL. APPLICATION TO THE CALCULATION OF TETRAGONAL BIREFRINGENCE IN BARIUM AND LEAD T/ 04-0578
.   THERMAL HYSTERESIS IN THE TETRAGONAL CUBIC TRANSITION OF BARIUM TITANATE 04-0921
W TEMPERATURES.   FIELD INDUCED SWITCHING OF TETRAGONAL DOMAINS IN STRONTIUM TITANATE AT LO 06-1184
   AGING IN TETRAGONAL FERROELECTRIC BARIUM TITANATE.   04-0579
CULATIONS)   POLARITON DISPERSION OF TETRAGONAL FERROELECTRIC BARIUM TITANATE. (CAL 04-0842
IRCONATE SYSTEM (MEASUREMENT OF SOFT OPTICAL PHONON MODES IN TETRAGONAL FERROELECTRIC REGION). / TITANATE Z 09-1497
OF BARIUM TITANAT/ CHANNELING OF LIGHT IONS IN THE CUBIC AND TETRAGONAL (FERROELECTRICALLY POLARIZED PHASE 04-0688
S IN HIGH RESISTIVITY BARIUM TITANATE SINGLE CRYSTALS/ CUBIC TETRAGONAL (MARTENSITIC TYPE) PHASE TRANSITION 04-0887
   ANALYSIS OF THE IRON OXYGEN VACANCY CENTER IN THE TETRAGONAL PHASE OF STRONTIUM TITANATE.   06-1231
   THE IRON(3+)- VACANCY CENTER IN THE TETRAGONAL PHASE OF STRONTIUM TITANATE.   06-1234
IOBATE. (THERMAL EX/ DILATOMETRIC STUDY OF THE ORTHORHOMBIC- TETRAGONAL PHASE TRANSITION IN BARIUM SODIUM N 13-1894
E.   FIELD INDUCED NUCLEATION AT THE (CUBIC TETRAGONAL) PHASE TRANSITION IN BARIUM TITANAT 04-0753
/IGHT SCATTERING STUDIES OF THE SOFT MODES NEAR THE CUBIC TO TETRAGONAL PHASE TRANSITION IN STRONTIUM TITA/ 06-1242
   NUCLEATION PROCESS IN BARIUM TITANATE AT CUBIC- TETRAGONAL PHASE TRANSITIONS.   04-0754
   RAMAN SPECTRA OF TETRAGONAL POTASSIUM DIHYDROGEN PHOSPHATE.   17-2657
ANALYSIS OF PHONON TUNNELING MODES IN THE RAMAN SPECTRUM OF TETRAGONAL POTASSIUM DIHYDROGEN PHOSPHATE. (T/ 03-0396
   SOFT MODES AND STRUCTURE OF FERROELECTRIC TETRAGONAL POTASSIUM TANTALATE NIOBATE.   13-1957
IN BARIUM TITANATE.   DIELECTRIC AGING IN TETRAGONAL SOLID SOLUTIONS OF CALCIUM TITANATE 08-1295
   RAMAN SCATTERING IN PIEZOELECTRIC CRYSTALS WITH TETRAGONAL SYMMETRY.   03-0407
   STRUCTURE OF TETRAGONAL TIN TUNGSTEN BRONZES.   13-2031
CED OPTICAL ABSORPTION IN STRONTIUM TITANATE AT THE CUBIC TO TETRAGONAL TRANSITION.   DISTORTION ENHAN 06-1109
/NTIUM TITANATE (OCCURANCE IN THIN (110) PLATES BELOW CUBIC- TETRAGONAL TRANSITION, AND PRODUCTION BY SHAP/ 06-1164
/RUCTURE OF SODIUM NIOBATE T(2) AT 600 DEG C, AND THE CUBIC- TETRAGONAL TRANSITION IN RELATION TO SOFT PHO/ 13-1337
PHASE.   THERMAL EXPANSION AND THE ORTHORHOMBIC- TETRAGONAL TRANSITION OF THE TUNGSTEN TRIOXIDE 10-1626
AD TRICHLORIDE.   PRESSURE DEPENDENCE OF CUBIC TETRAGONAL TRANSITION TEMPERATURE OF CESIUM LE 33-3393
ELECTRIC PROPERTY OF BARIUM TITANATE NIOBATE SINGLE CRYSTAL (TETRAGONAL TUNGSTEN BRONZE).   DI 13-1969
TIUM, OR BARIUM, AND C IS SODIUM OR POTASSIUM) PHASES OF THE TETRAGONAL TUNGSTEN BRONZE OXIDE TYPE. / STRON 04-0912
/ICAL PROPERTIES OF POTASSIUM LITHIUM NIOBATES (TRANSPARENT, TETRAGONAL TUNGSTEN BRONZE STRUCTURE, FERROEL/ 13-2026
N IN MOLTEN SOLUTIONS OF THE PSEUDOSYSTEM POTASSIUM LITHIU/ (TETRAGONAL) TUNGSTEN BRONZE TYPE CRYSTALS GROW 13-1948
TIC LEAD OR BISMUTH TANTALATE NIOBATE, TA/ NEW PHASES OF THE TETRAGONAL TUNGSTEN OXYGEN BRONZE TYPE: QUADRA 13-1985
UNGSTATE . (MAGNETIC G STRUCT/ NEUTRON DIFFRACTION STUDY OF (TETRAGONAL- CUBIC PEROVSKITE) STRONTIUM IRON T 10-1655
ANATE CERAMIC/ STABILIZATION EFFECTS AT THE PHASE TRANSITION TETRAGONAL- RHOMBOHEDRAL IN LEAD ZIRCONATE TIT 09-1611
R/ THERMODYNAMIC THEORY OF THE MORPHOTROPIC PHASE TRANSITION TETRAGONAL- RHOMBOHEDRAL IN THE PEROVSKITE FER 03-0172
TRANSITIONS IN CRYSTALS OF PEROVSKITE TYPE AND OF ONES WITH TETRAHEDRAL STRUCTURAL UNITS. /ATION AND PHASE 03-0175
   DIELECTRIC PROPERTIES AND DOMAIN TEXTURE OF BARIUM TITANATE CRYSTALS.   04-0537
STIGATION OF THEIR ELEC/ PREPARATION OF ANTIMONY SULFOIODIDE TEXTURES FROM THE MELT UNDER PRESSURE AND INVE 15-2138
   TGS SEE TRIGLYCINE SULFATE
RIC PROPERTIES AND FERROELECTRIC PHASE TRANSITION ON LITHIUM THALLATE TARTRATE. /ING EFFECTS ON THE DIELECT 29-3251
OSPHATES.   PHASE TRANSITION IN THALLIUM AND AMMONIUM THALLIUM (DIHYDROGEN) PH 17-2571
   NEW FERROELECTRIC LANGBEINITE THALLIUM CADMIUM SULFATE.   36-3534
RES AND THEIR X-RAY ANALYSIS.   SYNTHESIS OF THALLIUM CONTAINING PEROVSKITES AT HIGH PRESSU 08-1463
AFNATES/ CRYSTAL STRUCTURE AND PROPERTIES OF NEW CADMIUM AND THALLIUM CONTAINING PEROVSKITES. (TITANATES, H 08-1458
   NEW FERROELECTRIC: THE DOUBLE AMMONIUM THALLIUM MONOARSENATE OR ATLAS.   18-2735
E OR MOLYBDATE. (WHERE A IS ZIRCONIUM, HAFNIUM, CERIUM, THO/ THALLIUM PEROVSKITES. THALLIUM A(1-X) TUNGSTAT 10-1640
   SOME PROPERTIES OF THE FERROELECTRIC LITHIUM THALLIUM TARTRATE.   29-3242
SURE.   DISPLACEMENT OF THE CURIE POINTS IN KDP AND LITHIUM THALLIUM TARTRATE CRYSTALS BY HYDROSTATIC PRES 17-2523
/ CONTROL OF THE ELASTIC COMPLIANCE OF FERROELECTRIC LITHIUM THALLIUM TARTRATE HYDRATE (LTT) BY AN ELECTRIC 29-3249
/DIUM DIHYDROGEN PHOSPHATE, SODIUM, POTASSIUM(2), OR LITHIUM THALLIUM TARTRATE, LITHIUM AMMONIUM TARTRATE,/ 02-0089
/ERROELECTRIC COMPOUNDS. (LITHIUM AMMONIUM TARTRATE, LITHIUM THALLIUM TARTRATE, LITHIUM TANTALATE, LITHIUM/ 02-0052
   CRYSTAL DATA FOR LITHIUM THALLIUM TARTRATE MONOHYDRATE.   29-3247
ICE/ SOME PROPERTIES OF NONDEUTERATED AND DEUTERATED LITHIUM THALLIUM TARTRATE MONOHYDRATE. (EFFECT ON LATT 29-3239
/CTRICITY AND CONDUCTION IN FERROELECTRIC CRYSTALS. (LITHIUM THALLIUM TARTRATE MONOHYDRATE, ELECTRIC CONTR/ 02-0060
/CTION IN FERROELECTRIC CRYSTALS. (BISMUTH TITANATE, LITHIUM THALLIUM TARTRATE MONOHYDRATE, LEAD(5) GERMAN/ 02-0061
ELECTRIC FIELD CONTROL CF THE ELASTIC COMPLIANCE IN LITHIUM THALLIUM TARTRATE (TUNING AND DELAY DEVICES). 29-3250
   CIRCLE THEOREM FOR ICE RULE FERROELECTRIC MODELS.   03-0200
MPERATURE DISCONTINUITIES IN THE INFRARED SPECTRUM AND GROUP THEORETICAL ANALYSIS. /NGANESE TRIFLUORIDE: TE 33-3363
(20) VERSUS LITHIUM NIOBATE IN THE SURFACE WAVE AMPLIFIER. A THEORETICAL AND EXPERIMENTAL COMPARISON. /XIDE 11-1720

90 DEGREE TYPE DOMAIN WALL OF BARIUM TITANATE CRYSTAL. (THEORY, CALCULATIONS)   03-0324
POLARONS IN ANISOTROPIC ENERGY BANDS (THEORY, CALCULATIONS FOR STRONTIUM TITANATE).   03-0305
DP, TGS, ROCHELLE SALT, SILVER PERIODATE, / FERROELECTRICS. (THEORY, DYNAMICAL ORDER DISORDER PROPERTIES, K   02-0050
SPONTANEOUS POLARIZATION NOISE IN POLAR DIELECTRICS. (THEORY, EINSTEIN THERMODYNAMIC METHOD)   03-0219
/ECIES OF FERROMAGNETIC, AND FERROELASTIC CRYSTALS (PREVIOUS THEORY EXTENDED TO MAGNETIC CRYSTALS AND TO I/   03-0142
TEMPERATURE DISPLACEMENT TYPE PHASE TRANSITION IN CRYSTALS. (THEORY, FERROELECTRICS, PIEZOELECTRICS)   LOW   03-0320
STATISTICAL THEORY FOR DISPLACEMENT FERROELECTRICS.   03-0369
SURFACE WAVES ON PIEZOELECTRIC SUBSTRATES (LI/ PERTURBATION THEORY FOR ELECTROMAGNETIC COUPLING TO ELASTIC   11-1751
DYNAMIC PIEZOELECTRIC THEORY FOR FERROELECTRIC CERAMIC SHELLS.   03-0225
DYNAMIC SHELL THEORY FOR FERROELECTRIC CERAMICS.   03-0224
ATE.   STATISTICAL THEORY FOR FERROELECTRICITY IN TRIGLYCINE SULF   03-0263
AGNESIUM NIOBATE.   A MICROSCOPIC THEORY FOR THE DIELECTRIC PROPERTIES OF LEAD M   08-1440
ITH AN ARBITRARY MAGNETIC SYMMET/ SPIN WAVE PHENOMENOLOGICAL THEORY IN FINITE SAMPLES OF FERRODIELECTRICS W   03-0164
SOFT MODE BEHAVIOR IN IMPROPER FERROELECTRICS. (THEORY, KDP)   03-0230
ICE INTERACTIONS IN HYDROGEN BONDED FERROELECTRIC CRYSTALS. (THEORY, KDP TYPE)   PROTON LATT   03-0181
TICAL PROPERTIES OF A CLASS OF PLANAR FERROELECTRIC MODELS. (THEORY, KDP TYPES)   ANALY   03-0258
ERROELECTRICITY AND ANTIFERROELECTRICITY IN POLAR CRYSTALS. (THEORY, LEAD ZIRCONATE)   F   03-0372
IMULATED COMBINATION SCATTERING OF SOUND IN PIEZOELECTRICS. (THEORY, LITHIUM NIOBATE EXAMPLE)   ST   03-0431
LLOYS, SUPERFLUIDI/ PHYSICS OF TRANSITIONS. (THERMODYNAMICS, THEORY, MAGNETISM, FERROELECTRICITY, ORDERED A   03-0183
WAVE DISPERSION OF FERROELECTRICS NEAR THE TRANSITION POINT (THEORY, MEASUREMENTS ON TGS). /E ON THE) MICRO   03-0370
ON, PYROELECTRICTY, AND FERROELECTRICITY OF IONIC CRYSTALS. (THEORY, NO REFS)   POLARIZATI   03-0143
OVSKITE- TYPE CRYSTALS, PART-2, INTERACTION WITH ELASTIC ST/ THEORY OF A STRUCTURAL PHASE TRANSITION IN PER   03-0241
NTIFERROELECTRICS WITH HYDROGEN BONDS.   DYNAMICAL THEORY OF AMMONIUM DIHYDROGEN PHOSPHATE TYPE A   03-0476
/F SOUND ABSORPTION IN TRIGLYCINE SULFATE SINGLE CRYSTALS. (THEORY OF ANISOTROPIC RELAXATION ATTENUATION /   27-3078
S NEAR SECOND ORDER TRANSITION POINTS IN F/ PHENOMENOLOGICAL THEORY OF ANOMALIES OF THERMODYNAMIC PARAMETER   03-0362
: FIRST ORDER TRANSITION.   PHENOMENOLOGICAL THEORY OF ANTIFERROELECTRIC TRANSITION. PART-2   03-0404
ICITY APPLIED TO SODIUM NIOBATE AND THE SY/ PHENOMENOLOGICAL THEORY OF ANTIFERROELECTRICITY AND FERROELECTR   03-0218
THEORY OF BARIUM TITANATE.   03-0483
RYSTALS IN THE FORM OF THICK PLATES.   MACROSCOPIC THEORY OF BARIUM TITANATE TYPE FERROELECTRIC C   04-0770
THERMODYNAMICAL THEORY OF BARIUM TITANATE TYPE FERROELECTRICS.   03-0322
GEN BONDED FERROELECTRICS AT LOW TEMPERATURES. PART-1: GENE/ THEORY OF COHERENT NEUTRON SCATTERING BY HYDRO   03-0478
/ OF PHASE TRANSFORMATION IN FERROELECTRIC SOLID SOLUTIONS. (THEORY OF COMPOSITION DEPENDENCE OF PHASE TRA/   03-0250
NEAR FERROELECTRIC TRANSITIONS FROM THE POINT OF VIEW OF THE THEORY OF CRITICAL POINTS.   ANOMALIES   03-0298
THE DIMENSIONS OF KANZIG REGIONS ON THE BASIS OF THE DYNAMIC THEORY OF CRYSTAL LATTICES.   ESTIMATION OF   03-0446
D FERROMAGNETIC PROPERTIES.   THERMODYNAMIC THEORY OF CRYSTALS POSSESSING FERROELECTRIC AN   03-0472
OF IMPROPER FERROELECTRIC CRYSTALS. (AMMO/ PHENOMENOLOGICAL THEORY OF DIELECTRIC AND MECHANICAL PROPERTIES   14-2084
C MATERIAL WITH SEVERAL PHASE TRANSITIONS / PHENOMENOLOGICAL THEORY OF DIELECTRIC ANOMALIES IN FERROELECTRI   03-0363
ITY)   THERMODYNAMIC THEORY OF DISPERSION PHENOMENA. (FERROELECTRIC   03-0420
ROELECTRICS.   CONTRIBUTION TO THE THEORY OF DOMAIN STRUCTURES IN MAGNETS AND FER   03-0430
DROGEN PHOSPHATE TYPE CRYSTALS.   THEORY OF DOMAIN WALL MOTION IN POTASSIUM DIHY   17-2652
ALS WITH THE PEROVSKITE STRUCTURE.   THERMODYNAMIC THEORY OF DOMAIN WALLS IN FERROELECTRIC MATERI   03-0190
OUS POLARIZATION OF FERROELECTRIC PLATES NEAR THE CURIE POI/ THEORY OF DOMAINLESS, NONHOMOGENEOUS, SPONTANE   03-0321
STALS.   QUASI PHENOMENOLOGICAL THEORY OF ELASTIC ANOMALY IN FERROELECTRIC CRY   03-0374
CTRIC SEMICONDUCTOR CRYSTALS.   THEORY OF ELECTRIC INDUCTION WAVES IN FERROELE   03-0466
ITH APPLICATION TO BARIUM TITANATE.   THEORY OF ELECTROSTRICTION IN CUBIC CRYSTALS W   03-0267
STATISTICAL AND ELECTRONIC PROBLEMS IN FERROELECTRICITY. (THEORY OF ENERGY STATES INVOLVED)   03-0325
PHENOMENOLOGICAL LATTICE DYNAMICAL THEORY OF FERROELASTICITY.   03-0144
CRYSTALS.   PHENOMENOLOGICAL THEORY OF FERROELECTRIC AND ANTIFERROELECTRIC   03-0270
VIBRONIC MECHANISM.   THEORY OF FERROELECTRIC PHASE TRANSITIONS WITH   03-0351
PHENOMENOLOGICAL THEORY OF FERROELECTRIC POLARIZATION REVERSAL.   03-0282
, / POLARIZATION WAVES IN SOLIDS. (INCLUDING DEVELOPMENTS IN THEORY OF FERROELECTRIC PROPERTIES OF CRYSTALS   03-0306
THERMODYNAMIC THEORY OF FERROELECTRIC SEMICONDUCTORS.   03-0409
THERMODYNAMIC THEORY OF FERROELECTRIC SEMICONDUCTORS.   03-0410
THE ANHARMONIC LATTICE PROBLEM. (WITH AN APPLICATION TO THE THEORY OF FERROELECTRIC TRANSITIONS) /ROACH TO   03-0221
E OF SECOND ORDER PHASE TRANSITION DERIVED FROM DEVONSHIRE'S THEORY OF FERROELECTRICS.   A NEW TYP   03-0226
MACROSO MACROSCOPIC THEORY OF FERROELECTRICS.   03-0295
AN INTRODUCTION TO THE PHENOMENOLOGICAL THEORY OF FERROELECTRICITY.   01-0013
INTRODUCTION TO THE THEORY OF FERROELECTRICITY.   03-0470
TWO POLARIZATION MODEL THEORY OF FERROELECTRICITY.   03-0332
CTS IN CRYSTALS. (AND CALCU/ (MICROSCOPIC AND THERMODYNAMIC) THEORY OF FERROELECTRICITY AND ANHARMONIC EFFE   03-0213
NALYSIS (PROTON DYNAMICS IN HYDROGEN BONDED FERROEL/ QUANTUM THEORY OF FERROELECTRICITY, PART-1: SPECTRAL A   03-0402
MIC BEHAVIOR OF FERROELECTRIC SPINS.   QUANTUM THEORY OF FERROELECTRICITY. PART-2: THERMODYNA   03-0403
THERMODYNAMIC THEORY OF GADOLINIUM MOLYBDATE.   03-0232
-1: (FORMALISM OF LOCAL MODE MODEL.) PART-2: (DYNAMICAL PRO/ THEORY OF HYDROGEN BONDED FERROELECTRICS. PART   03-0384
-1: DAMPINGS AND (FREQUENCY) SHIFTS (OF PROTONIC / DYNAMICAL THEORY OF HYDROGEN BONDED FERROELECTRICS. PART   03-0475
RONTIUM TITANATES AND POTASSIUM/ COUPLED OPTICAL PHONON MODE THEORY OF INFRARED DISPERSION IN BARIUM AND ST   04-0547
THERMODYNAMIC THEORY OF IRRADIATED TRIGLYCINE SULFATE.   03-0152
RICS IN THE FREQUENCY RANGE OF DOUBLE ELECTRONIC EXC/ ON THE THEORY OF LIGHT ABSORPTION BY ANTIFERRODIELECT   03-0494
VITY OF BARIUM TITANATE TYPE FERR/ EXPERIMENTAL CHECK OF THE THEORY OF MICROWAVE DISPERSION OF THE PERMITTI   04-0649
(AT TEMPERATURES ABOVE THE CURIE POINT)   PHENOMENOLOGICAL THEORY OF MICROWAVE LOSSES IN FERROELECTRICS.   03-0503
ATIONS (REDEVELOPED FROM FIRST PRINCIPLES, APPLIED TO DISPL/ THEORY OF NEUTRON SCATTERING FROM LATTICE VIBR   04-0484
BARIU/ EXTENSION OF THE POULET- LUDON APPROXIMATION IN THE THEORY OF OPTICAL LATTICE VIBRATIONS. (QUARTZ,   03-0380
PERTURBATIVE CORRECTION.   STATISTICAL PERTURBATION THEORY OF ORDER DISORDER FERROELECTRICS: FIRST   03-0399
AN IN TERMS OF LOCALIZED D-STRAIN AND SOFT- NORMAL MODE COO/ THEORY OF PEROVSKITE FERROELECTRICS (HAMILTONI   03-0432
THEORY OF PHASE TRANSITIONS IN BORACITES.   03-0269
PHENOMENOLOGICAL THEORY OF PHASE TRANSITIONS IN SODIUM NITRITE.   03-0254
THEORY OF PIEZOELECTRIC ENERGY CONVERSION).   03-0278
PIEZOELECTRIC EFFECTS IN FERROELECTRIC CERAMICS (AND THE THEORY OF POLARIZATION AND HYSTERESIS OF FERRO   03-0494
ELECTRICS.   THEORY OF POLARIZATION REVERSAL IN FERROELECTR   03-0422
ICS. (CALCULATIONS OF SWITCHING CURRENT)   THEORY OF POLARIZATION REVERSAL OF BARIUM TITA   03-0326
NATE CRYSTAL.   ORDER DISORDER THEORY OF POTASSIUM DIHYDROGEN PHOSPHATE TYPE   17-2685
FERROELECTRICS.   THEORY OF PTC MATERIALS.   03-0352
DIFFUSION POTENTIALS IN BARIUM TITANATE AND THE THEORY OF RAYLEIGH AND BRILLOUIN SCATTERING NE   03-0334
AR THE PHASE TRANSITION.   THEORY OF REFRACTIVE INDEX AND ELECTROOPTICAL   03-0518
PHENOMENA IN CRYSTALS OF PEROVSKITE STRUCT/ PHENOMENOLOGICAL THEORY OF RIGID LATTICES OF ANTIFERROELECTRICS   03-0344
.   DYNAMICAL THEORY OF SECOND ORDER PHASE TRANSITION FROM T   03-0395
HE FERROMAGNETIC TO THE FERROMAGNETIC AND FERROELECTRIC STA/ THEORY OF SECOND ORDER PHASE TRANSITIONS.   03-0364
W TEMPERATURES. (TIN TELLURIDE, POTASSIUM TANTALATE)   THEORY OF SECOND ORDER PHASE TRANSITIONS AT LO   03-0433
ELECTRICS.   THERMODYNAMIC THEORY OF SOLID SOLUTIONS OF ISOMORPHOUS FERRO   03-0302
ROELECTRICS.   THEORY OF SPACE CHARGE LIMITED CURRENTS IN FER   03-0251

CTRIC CRYSTALS. (CRYSTALLOGRAPHIC VIEWPOINT)   MICROSCOPIC THEORY OF STATIC DOMAIN STRUCTURES OF FERROELE 03-0340
SKITES.   THEORY OF STRUCTURAL PHASE TRANSITION IN PEROV 03-0242
VSKITE TYPE CRYSTALS. (ST/ SELF CONSISTENT LATTICE DYNAMICAL THEORY OF STRUCTURAL PHASE TRANSITIONS IN PERO 03-0424
SSIUM MANGANESE TRIFLUORIDE.   THEORY OF SUPEREXCHANGE REACTION. PART-1: POTA 03-0252
RYSTALS. (CALCULATION OF FIELD NEAR SURFACE)   MICROSCOPIC THEORY OF SURFACE PHENOMENA IN FERROELECTRIC C 03-0501
CRYSTALS.   LATTICE DYNAMICAL THEORY OF SWITCHING IN BARIUM TITANATE SINGLE 03-0398
TION IN EORACITES.   THERMODYNAMIC THEORY OF THE CUBIC- ORTHORHOMBIC PHASE TRANSI 03-0231
ITANATE.   THEORY OF THE DIELECTRIC CONSTANTS OF BARIUM T 04-0776
C CRYSTALS. (STRAIN DEPENDENT CURIE TEMPERATURE, POLARIZATI/ THEORY OF THE ELASTOOPTIC EFFECT IN NONMETALLI 03-0513
N GADOLINIUM MOLYEDATE. (DIELECTRIC AND EL/ PHENOMENOLOGICAL THEORY OF THE FERROELECTRIC PHASE TRANSITION I 30-3303
DIPOLE SYSTEMS (SPIN WAVES IN HYDROXYL DIPOL/ ON THE QUANTUM THEORY OF THE FERROELECTRIC STATE IN ELECTRIC 03-0452
N ICE.   THEORY OF THE MOBILITY OF STRUCTURAL DEFECTS I 03-0244
TRAGONAL- RHOMBOHEDRAL IN THE PEROVSKITE FERR/ THERMODYNAMIC THEORY OF THE MORPHOTROPIC PHASE TRANSITION TE 03-0172
YDROGEN PHOSPHATE.   A THEORY OF THE PHASE TRANSITION IN AMMONIUM DIH 03-0286
LT.   MICROSCOPIC THEORY OF THE PHASE TRANSITIONS IN ROCHELLE SA 03-0482
              POLARITONS IN TETRAGONAL BARIUM TITANATE: THEORY OF THE RAMAN LINE SHAPE.   04-0556
/ONTIUM TITANATE BARRIERS INTERPRETED AS CONFIRMATION OF THE THEORY OF TRANSITIONS FROM LARGE TO NEARLY SM/ 03-0233
RICS.   PHENOMENOLOGICAL THEORY OF ULTRASONIC ATTENUATION IN FERROELECT 03-0228
NETICS WITHOUT A SYMMETRY CENTER.   QUANTUM THEORY OF VIBRATIONS IN FERROELECTRIC FERROMAG 03-0166
DE GAP FERROELECTRICS.   THEORY OF VIBRONIC PHASE TRANSFORMATIONS IN WI 03-0349
CTRICS.   SUPPLEMENT TC A PREVIOUS THEORY OF WEAK FERROELASTICS AND WEAK FERROELE 03-0146
DUE TO DOMAIN FORMATION IN FERROELECTRICS.   THEORY OF X-RAY DIFFRACTION LINE SHAPE EFFECTS 03-0236
/ION IMAGES OF COHERENT DOMAIN ECUNDARIES. PART-1: DYNAMICAL THEORY. PART-2: OBSERVATIONS. (FERROELECTRIC / 03-0255
MONIC. CALCULATIONS FOR LITHIUM/ NONLINEAR SADOVSKII EFFECT (THEORY, PROPAGATION OF RADIATION AND FIRST HAR 03-0186
HEAT CAPACITY IN THE FERROELECTRIC TRANSITION REGION (THEORY, RELATION TO KAENZIG DOMAIN MODEL).   03-0443
OPTICAL HARMONICS AND NONLINEAR PHENOMENA. (THEORY, REVIEW) 02-0067
IUM NIOBATE AND LITHIUM TANTALATE USING AUTHOR'S STATISTICAL THEORY. (SPONTANEOUS POLARIZATION, BONDING) /H 03-0366
OPTICAL MODES OF THE PEROVSKITE STRUCTURE. (THEORY, STRONTIUM TITANATE DATA)   03-0304
/AT THE PHASE TRANSITIONS. (BASED CN LANDAU PHENOMENOLOGICAL THEORY, TESTED ON AMMONIUM FLUOBERYLLATE, ABS/ 19-2765
ONTACT BETWEEN A SEMICONDUCTOR AND A C-DOMAIN FERROELECTRIC (THEORY, TITANATES DISCUSSED). /EFFECT AT THE C 03-0511
THE PHASE TRANSITION IN A FERROELECTRIC WITH HYDROGEN BONDS (THEORY, WU MODEL). /OF ELECTRIC NEUTRALITY ON 03-0449
ER SODIUM DINITRITE AND SILVER SODIUM NITRITE. (DIFFERENTIAL THERMAL ANALYSIS)   PROPERTIES OF SILV 16-2370
RIGLYCINE SULFATE EXAMPLE)   THERMAL ANALYSIS OF PYROELECTRIC DETECTORS. (T 27-2964
C CRYSTALS THE TRIGLYCINE SULFATE GRO/ DEUTERATION EFFECT ON THERMAL AND ELASTIC PROPERTIES OF FERROELECTRI 27-2986
E SULFATE SINGLE CRYSTALS (HEAT CAPACITY AND ELECTRIC PERMI/ THERMAL AND ELECTRICAL PROPERTIES OF TRIGLYCIN 27-3141
ORDERED FERROELECTRIC. (DEPENDENCE OF DIELECTRIC CONSTANT ON THERMAL AND FIELD HISTORY) /IUM IODATE - A DIS 08-1348
IUM SULFATE DIHYDRATE, SODIUM TRIHYDROGEN OR TRIDEUTERIUM D/ THERMAL AND SOME OTHER STUDIES OF SODIUM AMMON 19-2779
INE SULFAT/ EFFECTS ON ELECTRICAL BOUNDARY CONDITIONS ON THE THERMAL CAPACITY OF SINGLE CRYSTALS OF TRIGLYC 27-3140
RIUM TITANATE TYPE) CERAMIC WITH CONTROLLED MICROWAVE PERMI/ THERMAL CHARACTERISTICS OF A FERROELECTRIC (BA 04-0904
RROMAGNETIC COPPER FORMATE TETRAHYDRATE.   THERMAL CONDUCTION IN A TWO DIMENSIONAL ANTIFE 26-2932
IFERROMAGNETS POTASSIUM COPPER TRIFLUOR/ MEASUREMENTS OF THE THERMAL CONDUCTIONS IN THE LOW DIMENSIONAL ANT 26-2939
PHYSICAL PROPERTIES OF SOME FERROELECT/ INVESTIGATION OF THE THERMAL CONDUCTIVITY IN CONNECTION WITH OTHER 04-0721
LECTRICS. (THEORY, BAR/ HIGH FREQUENCY SOUND ATTENUATION AND THERMAL CONDUCTIVITY IN DISPLACIVE TYPE FERROE 03-0160
INTERACTION PROPOSED FOR STRONTIUM TITANATE AND POTASSIUM T/ THERMAL CONDUCTIVITY IN PEROVSKITES (3 PHONON 06-1095
BRONZES. (SODIUM TUNGSTATES)   THERMAL CONDUCTIVITY OF CUBIC SODIUM TUNGSTEN 13-2057
UM ALUM.   THERMAL CONDUCTIVITY OF METHYL AMMONIUM ALUMIN 20-2807
IFLUORIDE.   OBSERVATION OF THE ANOMALY OF THE THERMAL CONDUCTIVITY OF POTASSIUM MANGANESE TR 33-3382
                                                            THERMAL CONDUCTIVITY OF ROCHELLE SALT.   28-3226
TASSIUM TANTALATE AND BARIUM TITANATE NEAR THE CURIE POINTS. THERMAL CONDUCTIVITY OF STRONTIUM TITANATE, PO 04-0989
R THE CURIE POINT.   THERMAL CONDUCTIVITY OF TRIGLYCINE SULFATE NEA 27-2983
R THE CURIE POINT.   THERMAL CONDUCTIVITY OF TRIGLYCINE SULFATE NEA 27-3016
/IMENT. PART-1: INTRODUCTION AND MAGNETIC RESONANCE. PART-2: THERMAL CONDUCTIVITY. PART-3: NEUTRON SCATTER/ 03-0397
STUDY OF THE PHASE TRANSITION IN TRIGLYCINE SULFATE. (CLOSE THERMAL CONTROL)   EXPERIMENTAL DETAILED 27-3030
C, AND ELASTIC PROPERTIES OF LITHIUM NIOBATE SINGLE CRYSTAL/ THERMAL DEPENDENCE OF DIELECTRIC, PIEZOELECTRI 11-1693
OF STRONG FIELDS) DURING HOLOGRAM RECORDING IN CR/ ELECTRON THERMAL DIFFUSION EFFECTS (INTERNAL GENERATION 06-1090
F CALCIUM DOPED SODIUM NITRATE CRYSTALS NEAR/ SPECIFIC HEAT, THERMAL DIFFUSIVITY AND THERMAL CONDUCTIVITY O 16-2283
TANTALATE.   SELF INDUCED THERMAL EFFECTS ON LIGHT EXTINCTION OF LITHIUM 12-1889
MBIC- TETRAGONAL PHASE TRANSITION IN BARIUM SODIUM NIOBATE. (THERMAL EXPANSION) /TRIC STUDY OF THE ORTHORHO 13-1894
C EFFECTS AND SPONTANEOUS POLARIZATION IN LITHIUM TANTALATE (THERMAL EXPANSION). /AND SECONDARY PYROELECTRI 12-1884
CTURE OF FERROELECTRIC SODIUM TRIHYDROGEN DISELENITE AND ITS THERMAL EXPANSION. RELAXATION BETWEEN THE STRU 22-2851
ROGEN PHOSPHATE NEAR THE HIGH TEMPERATURE PHASE TRANSITION. (THERMAL EXPANSION) /PERTIES OF POTASSIUM DIHYD 17-2627
ONAL TRANSITION OF THE TUNGSTEN TRIOXIDE PHASE.   THERMAL EXPANSION AND THE ORTHORHOMBIC- TETRAG 10-1626
OF SPONTANEOUS POLARIZATION IN FERROELECTRICS. (ABSTRACT ON/ THERMAL EXPANSION ANOMALIES AND THE DIRECTION 03-0350
/NSTANTS, PIEZOELECTRIC CONSTANTS, SPONTANEOUS POLARIZATION, THERMAL EXPANSION COEFFICIENTS, AND REFRACTIV/ 12-1876
UM DINITRITE.   THERMAL EXPANSION IN FERROELECTRIC SILVER SODI 16-2346
                                                            THERMAL EXPANSION IN SODIUM NITRITE CRYSTALS. 16-2351
RIDEUTERIUM(X) DISELENITE AND ISOTOPE EFFECT.   THERMAL EXPANSION IN SODIUM TRIHYDROGEN(1-X) T 22-2858
/CON OXIDE. (DIELECTRIC CONSTANTS, SPONTANEOUS POLARIZATION, THERMAL EXPANSION, INDICES OF REFRACTION, OPT/ 35-3501
AL POTASSIUM DIHYDROGEN PHOSPHATE.   LOW TEMPERATURE THERMAL EXPANSION MEASUREMENTS ON SINGLE CRYST 17-2464
THE REGION OF ANTIFERROELECTRIC PHASE TRANSITION.   THERMAL EXPANSION OF ADP AND DEUTERATED ADP IN 17-2445
HATE.   X-RAY DETERMINATION OF THE THERMAL EXPANSION OF AMMONIUM DIHYDROGEN PHOSP 17-2485
ERMINATION OF THE LATTICE PARAMETERS AND THE COEFFICIENTS OF THERMAL EXPANSION OF BISMUTH FERRATE. /ION DET 08-1287
MOLYBDATE.   X-RAY STUDY OF THERMAL EXPANSION OF FERROELECTRIC GADOLINIUM 30-3294
IHYDROGEN PHOSPHATE. (NEAR CURIE TEMPERATURE) X-RAY STUDY ON THERMAL EXPANSION OF FERROELECTRIC POTASSIUM D 17-2556
ITRATE.   X-RAY MEASUREMENTS ON THERMAL EXPANSION OF FERROELECTRIC POTASSIUM N 16-2322
RIUM SODIUM NIOBATES.   CAUSES OF THERMAL EXPANSION OF GADOLINIUM SUBSTITUTED BA 13-2048
TRICS. LITHIUM TRIHYDROGEN SELENITE.   THERMAL EXPANSION OF HYDROGEN BONDED FERROELEC 22-2857
MPERATURE RANGE OF 70-300 K.   THERMAL EXPANSION OF LITHIUM NIOBATE IN THE TE 11-1685
LEAD TITANATE ZIRCONATE/ DIELECTRIC SYSTEM POLARIZATION AND THERMAL EXPANSION OF POLARIZED CERAMICS IN THE 09-1603
PHATE CRYSTALS BY USING SEMICONDUCTOR / INVESTIGATION OF THE THERMAL EXPANSION OF POTASSIUM DIHYDROGEN PHOS 17-2565
ENATE CRYSTALS.   ANOMALOUS THERMAL EXPANSION OF POTASSIUM TRIHYDROGEN SEL 22-2880
E SHIFT CF RESONANCE FREQUENCY IN CAVITY.   MEASUREMENT OF THERMAL EXPANSION OF ROCHELLE SALT BY USING TH 28-3202
NTIFERROELECTRICS WITH HIGH CURIE/ DIELECTRIC PROPERTIES AND THERMAL EXPANSION OF SOME FERROELECTRICS AND A 08-1378
ASSIUM NITRATE, KDP, SODIUM NITRITE)   THERMAL EXPANSION OF SOME FERROELECTRICS. (POT 16-2298
INE SULFATE.   REEXAMINATION CF THE THERMAL EXPANSION OF THE FERROELECTRIC TRIGLYC 27-3121
CRYSTALS IN THE FERROELECTRIC TRANSITION RANGE.   THERMAL EXPANSION OF TRIGLYCINE FLUOBERYLLATE 27-3169
                                                            THERMAL EXPANSION OF TRIGLYCINE SULFATE.   27-3000
LS STUDIED WITH TELLURIUM THIN FILMS.   THERMAL EXPANSION OF TRIGLYCINE SULFATE CRYSTA 27-3047
IUM(0.12) NIOBATE CERAMIC.   X-RAY AND THERMAL EXPANSION STUDY OF A SODIUM(0.88) LITH 13-1999
ARIUM SODIUM NIOBATES.   LINEAR THERMAL EXPANSIONS OF GADOLINIUM SUBSTITUTED B 13-1996

NEAR THE STRUCTURAL TRANSITION OF STRONTIUM TITANATE.   THERMAL EXPANSIVITY AND ULTRASONIC PROPAGATION 06-1135
RIDE NEAR THE NEEL TEMPERATURE.   THERMAL EXPANSIVITY OF RUBIDIUM MANGANESE FLUO 33-3375
THE STRUCTURAL TRANSITION.   THERMAL EXPANSIVITY OF STRONTIUM TITANATE NEAR 06-1136
/NELASTIC COHERENT SCATTERING OF SLOW NEUTRONS AND EFFECTIVE THERMAL FACTORS. (KDP TYPE) PART-2: SCATTERIN/ 03-0478
NSTION OF BARIUM TITANATE.   THERMAL HYSTERESIS IN THE TETRAGONAL CUBIC TRA 04-0921
OF TRIGLYCINE SULFATE AS A FUNCTION OF THE CONDITIONS OF GR/ THERMAL HYSTERESIS OF THE DIELECTRIC CONSTANT 27-3126
TION IN POTASSIUM DIHYDROGEN PHOSPHATE.   THERMAL HYSTERESIS OF THE FERROELECTRIC TRANSI 17-2682
                                    PYROELECTRIC THERMAL IMAGING DEVICES. (TGS) 27-3107
/NDS - DATA TABLES. (MECHANICAL, CRYSTALLOGRAPHIC, PHYSICAL, THERMAL MAGNETIC, ELECTRONIC AND OPTICAL PROP/ 01-0025
RMITTIVITY OF BARIUM TITANATE SINGLE CRYSTAL.   EFFECT OF THERMAL NEUTRON IRRADIATION ON THE RELATIVE PE 04-0781
TITANATE).   AUTOMATIC IMPEDANCE MEASUREMENTS USING THERMAL NOISE ANALYSIS (APPLICATION TO BARIUM 04-0844
N OF THE DIELECTRIC RELAXATION IN SILVER SODIUM DINITRITE BY THERMAL NOISE MEASUREMENTS. /TICAL SLOWING DOW 16-2382
STUDY OF THE CRITICAL BEHAVIOR OF TRIGLYCINE SULFATE BY THERMAL NOISE MEASUREMENTS. 27-3147
RYSTAL TRIGLYCINE SULFATE AND TRIGLYCINE FLUOBER/ ANOMALY OF THERMAL PROPERTIES OF FERROELECTRICS (SINGLE C 27-3137
THE CURIE POINT.   RELATION BETWEEN ELECTRICAL AND THERMAL PROPERTIES OF TRIGLYCINE SELENATE NEAR 27-3168
                            ANOMALIES OF THERMAL PROPERTIES OF TRIGLYCINE SULFIDE. 27-3048
AL)   MINIMUM DETECTABLE POWER OF A PYROELECTRIC THERMAL RECEIVER. (USING BARIUM TITANATE CRYST 04-0632
PHOSPHATE AND AMMONIUM DIHYDROGEN PHOSPHATE CRYSTALS OF LIG/ THERMAL SELF FOCUSING IN POTASSIUM DIHYDROGEN 17-2717
STUDIES OF FERROELECTRIC CERAMICS AT LOW TE/ DIELECTRIC AND THERMAL (SPECIFIC HEAT, ELECTROCALORIC EFFECT) 04-0720
S   THERMAL STABILITY OF LITHIUM HYDROGEN SELENITE 22-2868
NATE.   THERMAL STUDIES OF CALCIUM(X) BARIUM(1-X) TITA 04-0694
S (FERROELECTRIC BELOW 93K).   DIELECTRIC AND THERMAL STUDY OF POTASSIUM SELENATE TRANSITION 22-2824
ART-2: CHANGE IN ULTRAVIOLET ABSORPTION SPECTRA ACCOMPANYING THERMAL TRANSFORMATIONS IN THE CRYSTALS. /G. P 16-2278
IUM DIHYDROGEN PHOSPHATE AND TRIGLYCINE SULFATE CRYSTALS.   THERMAL VARIATION OF REFRACTIVE INDEX IN AMMON 17-2474
AND COERCIVE FORCE OF THE PEROVSKITES IN THE SYSTEM LEAD ZI/ THERMAL VARIATION OF SPONTANEOUS POLARIZATION 09-1597
F ELASTIC SURFACE WAVES ON FERROELECTRIC CERAMIC/ EFFECTS OF THERMALLY DIFFUSED IMPURITIES ON PROPAGATION O 03-0487
   THERMALLY FIXED HOLOGRAMS IN LITHIUM NIOBATE. 11-1828
TITANATE SINGLE CRYSTALS.   THERMALLY STIMULATED CONDUCTIVITY IN STRONTIUM 06-1227
OXIDE CRYSTALS. (POTASS/ TRAPPING SITES, OPTICAL DAMAGE, AND THERMALLY STIMULATED CURRENT IN FERROELECTRIC 13-1926
IN BISMUTH(12) SILICON OXIDE(20) AND BISMUTH(12) GERMANIUM/ THERMALLY STIMULATED CURRENTS AND LUMINESCENCE 36-3585
   A NEW PTC THERMISTOR MATERIAL. (BARIUM TITANATE) 04-0931
DIOXIDE SINGLE CRYSTALS.   SOME PROPERTIES OF THE THERMOCURRENTS IN BARIUM TITANATE AND TITANIUM 04-0622
TITANATE SINGLE CRYSTALS.   (PHASE TRANSITIONS AND) THERMOCURRENTS IN TITANIUM DIOXIDE AND BARIUM 04-0697
IN GROWING ICE.   THERMODIELECTRIC EFFECT AND FREEZING POTENTIAL 34-3417
/NONSTOICHIOMETRIC BARIUM TITANATE. PART-3: HIGH TEMPERATURE THERMODYNAMIC AND X-RAY DIFFRACTION MEASUREME/ 04-0886
ONSTOICHIOMETRIC BARIUM TITANATE.   HIGH TEMPERATURE THERMODYNAMIC AND X-RAY DIFFRACTION STUDY OF N 04-0885
   QUANTUM THEORY OF FERROELECTRICITY. PART-2: THERMODYNAMIC BEHAVIOR OF FERROELECTRIC SPINS. 03-0403
NERGY DEPENDENCE OF 180 DEG DOMAIN WALL IN BARIUM TITANATE. (THERMODYNAMIC CALCULATIONS)   ANGULAR E 04-0658
ERROELECTRIC PHASE TRANSITIONS.   THERMODYNAMIC CLASSIFICATION OF SECOND ORDER F 03-0303
/IC POLARIZATION OF LEAD TITANATE ZIRCONATE SOLID SOLUTIONS (THERMODYNAMIC EXISTENCE CONDITIONS FOR MORPHO/ 09-1541
S POLARIZATION NOISE IN POLAR DIELECTRICS. (THEORY, EINSTEIN THERMODYNAMIC METHOD)   SPONTANEOU 03-0219
NSITION POINTS IN F/ PHENOMENOLOGICAL THEORY OF ANOMALIES OF THERMODYNAMIC PARAMETERS NEAR SECOND ORDER TRA 03-0362
Y IN SINGLE CRYSTAL AND CERAMIC SYSTEMS.   THERMODYNAMIC PHENOMENOLOGY OF FERROELECTRICIT 03-0214
IN FERROELECTRIC CRYSTALS. (THEORY)   THERMODYNAMIC POTENTIALS AND SECONDARY EFFECTS 03-0387
EN ARSENATE AND POTASSIUM DIDEUTERIUM ARSENATE.   THERMODYNAMIC PROPERTIES OF POTASSIUM DIHYDROG 18-2730
N FERROELECTRIC FILMS ON SEMICONDUCTING SUBSTRATES. (TGS ON/ THERMODYNAMIC STABILITY OF POLARIZATION IN THI 27-3163
FILMS. (TRIGLYCINE SULFATE)   THERMODYNAMIC STABILITY OF THIN FERROELECTRIC 27-2959
RROELECTRIC AND FERROMAGNETIC PROPERTIES.   THERMODYNAMIC THEORY OF CRYSTALS POSSESSING FE 03-0472
(FERROELECTRICITY)   THERMODYNAMIC THEORY OF DISPERSION PHENOMENA. 03-0420
LECTRIC MATERIALS WITH THE PEROVSKITE STRUCTURE.   THERMODYNAMIC THEORY OF DOMAIN WALLS IN FERROE 03-0190
UCTORS.   THERMODYNAMIC THEORY OF FERROELECTRIC SEMICOND 03-0409
UCTORS.   THERMODYNAMIC THEORY OF FERROELECTRIC SEMICOND 03-0410
ANHARMONIC EFFECTS IN CRYSTALS. (AND CALCU/ (MICROSCOPIC AND THERMODYNAMIC) THEORY OF FERROELECTRICITY AND 03-0213
   THERMODYNAMIC THEORY OF GADOLINIUM MOLYBDATE. 03-0232
SULFATE.   THERMODYNAMIC THEORY OF IRRADIATED TRIGLYCINE 03-0152
MORPHOUS FERROELECTRICS.   THERMODYNAMIC THEORY OF SOLID SOLUTIONS OF ISO 03-0302
C PHASE TRANSITION IN BORACITES.   THERMODYNAMIC THEORY OF THE CUBIC- ORTHORHOMBI 03-0231
TRANSITION TETRAGONAL- RHOMBOHEDRAL IN THE PEROVSKITE FERR/ THERMODYNAMIC THEORY OF THE MORPHOTROPIC PHASE 03-0172
FFUSE FERROELECTRIC PHASE TRANSITIONS.   THERMODYNAMICS AND STATISTICAL MECHANICS OF DI 03-0445
LECTRICITY.   THERMODYNAMICS AND THE PHENOMENOLOGY OF FERROE 03-0457
EL.   THERMODYNAMICS OF AN ELASTIC FERROELECTRIC MOD 03-0240
S. (THEORY)   THERMODYNAMICS OF FERROELECTRIC SOLID SOLUTION 03-0523
S. PART-1. PART-2.   THERMODYNAMICS OF FERROELECTRIC SOLID SOLUTION 03-0524
TITANATE.   THERMODYNAMICS OF FERROELECTRICITY IN BISMUTH 08-1297
TITANATE.   THERMODYNAMICS OF FERROELECTRICITY IN BISMUTH 08-1419
LIED TO FERROELECTRIC TRANSITIONS/ DIELECTRIC DISPERSION AND THERMODYNAMICS OF IRREVERSIBLE PROCESSES. (APP 03-0415
THE PARAELECTRIC AND FERROELECTRIC PHASE OF SODIUM NITRITE. (THERMODYNAMICS OF THE PHASE TRANSITION) /3 IN 16-2264
ICITY, ORDERED ALLOYS, SUPERFLUIDI/ PHYSICS OF TRANSITIONS. (THERMODYNAMICS, THEORY, MAGNETISM, FERPOELECTR 03-0183
AND TRIGLYCINE SULFATE.   NEW MEASUREMENTS OF DIVERSE THERMODYNAMICAL COEFFICIENTS OF ROCHELLE SALT 27-2995
OGEN PHOSPHATE- POTASSIUM DIHYDROGEN(1-X) DIDEUTERIUM(/ SOME THERMODYNAMICAL PROPERTIES OF POTASSIUM DIHYDR 17-2620
FERROELECTRICS.   THERMODYNAMICAL THEORY OF BARIUM TITANATE TYPE 03-0322
EN PHOSPHATE NEAR THE HIGH TEMPERATURE PHASE TRANSITION. (T/ THERMOELASTIC PROPERTIES OF POTASSIUM DIHYDROG 17-2627
ORIDE FILMS. (AFTER COOLING STRETCHED FILM IN ELECTRIC FIELD, THERMOELECTRETS). /ATION IN POLYVINYLIDENE FLU 36-3530
PHENOMENA IN ANTIMONY SULFOIODIDE. (CONDUCTION MECHANISM FOR THERMOELECTRIC CURRENTS)   TRANSPORT 15-2241
   THERMOELECTRIC EFFECT IN ICE. 34-3446
IRCONATE.   THERMOELECTRIC EFFECT IN SEMICONDUCTIVE LEAD Z 07-1267
AD TITANATE- METAL. (METHOD OF STUDYING THE TYPE OF CONDUCT/ THERMOELECTRIC EFFECTS IN THE SYSTEM METAL- LE 05-1057
OF CRYSTALLINE BISMUTH SELENOIODIDE.   CONDUCTIVITY, THERMOELECTRIC PROPERTIES, AND BAND STRUCTURE 15-2126
(4) TITANIUM(3) OXIDE(12) WITH VANADIUM PENTOXIDE AND MOLYB/ THERMOGRAPHIC STUDY OF THE REACTION OF BISMUTH 13-2029
   X-RAY AND THERMOLUMINESCENCE OF BARIUM TITANATE. 04-0750
RADIATED FERROELECTRICS ROCHELLE SALT AND GUANIDINE ALUMINU/ THERMOLUMINESCENCE STUDIES OF THE GAMMA-RAY IR 21-2812
HYDROGEN SELENITES.   THERMOOPTIC AND ELECTROOPTIC PROPERTIES OF TRI 22-2841
IDIUM TRIHYDROGEN OR DEUTERIUM DIS/ PHASE TRANSFORMATION AND THERMOOPTICAL AND ELECTROOPTICAL EFFECT IN RUB 22-2840
TURED TITANATES.   PROPOSED MECHANISM FOR THERMOPHOTOTROPIC BEHAVIOR IN PEROVSKITE STRUC 03-0400
BARIUM TITANATE.   THERMOSTABILIZING PROPERTIES OF SEMICONDUCTING 04-0570
D BY FLAME SPRAYING (CAPACITORS).   ELECTRICAL PROPERTIES OF THICK FILM BARIUM TITANATE DIELECTRICS PRODUCE 04-0772
   SWITCHING PROPERTIES OF THICK FILM FERROELECTRIC POTASSIUM NITRATE. 16-2282
E READOUT MEMORY ELEMENT UTILIZING POTASSIUM NITRATE. (FUSED THICK FILM OR VACUUM DEPOSITED THIN FILM) /TIV 16-2269
   MULTIPLE STORAGE OF THICK PHASE HOLOGRAMS IN LITHIUM NIOBATE. 11-1827
F BARIUM TITANATE TYPE FERROELECTRIC CRYSTALS IN THE FORM OF THICK PLATES.   MACROSCOPIC THEORY O 04-0770

F COERCIVE FIELD OF BARIUM TITANATE SINGLE CRYSTALS ON THEIR THICKNESS.   DEPENDENCE O 04-0693
WALL IN TRIGLYCINE SULFATE. (THEORY AND CALCULATIONS OF WALL THICKNESS AND ENERGY DENSITY) /LECTRIC DOMAIN 27-2997
FLASH EVAPORATED FERROELECTRIC BARIUM TITANATE THIN FILMS. (THICKNESS DFPENDENCE) ELECTRICAL PROPERTIES OF 04-0967
S IN GADOLINIUM MOLYBDATE LIGHT VALVES.   ORIENTATION AND THICKNESS DEPENDENCE OF CONTRAST AND BRIGHTNES 30-3291
IN BARIUM TITANATE SINGLE CRYSTALS.   THICKNESS DEPENDENCE OF POLARIZATION REVERSAL 04-0949
F TRIGLYCINE SULFATE.   THICKNESS DEPENDENCE OF THE NUCLEATION FIELD O 27-3036
RIUM TITANATE SINGLE CRYSTALS.   THICKNESS DEPENDENCE OF THE PERMITTIVITY OF BA 04-0637
QUES APPLIED TO PROBLEMS IN SOLID STATE PHYSICS (DOMAIN WALL THICKNESS IN GADOLINIUM MOLYBDATE). /NG TECHNI 30-3319
FERROELASTIC CRYSTALS. (BARIUM TITANATE, GADOLINIUM MOLYBD/ THICKNESS OF DOMAIN WALLS IN FERROELECTRIC AND 04-0777
M TI/ DETERMINATICN OF X-RAY ABSCRPTION COEFFICIENTS AND THE THICKNESS OF NONUNIFORM THIN SAMPLES (STRONTIU 06-1179
ECTRON SPECTRUM CF TUNGSTEN- TUNGSTEN OXIDE AS A FUNCTION OF THICKNESS OF THE SURFACE OXIDE LAYER. /PHOTOEL 10-1631
ELECTRIC PROPERTIES O/ INVESTIGATION OF THE EFFECT OF SAMPLE THICKNESS ON THE DOMAIN STRUCTURE AND THE PYRO 04-1039
OF AN X CUT LITHIUM TANTALATE CRYSTAL RESCNATOR VIBRATING IN THICKNESS SHEAR MODE OF MOTION. /E DEPENDENCE 12-1868
ABOVE AND BELOW THE FERROELECTRIC / CHANNELING OF PROTONS IN THIN BARIUM TITANATE CRYSTALS AT TEMPERATURES 04-0687
PREPARATICN AND PROPERTIES OF THIN BARIUM TITANATE FILMS.   04-0669
ING (ONTO PLATINUM OR FUSED QUARTZ).   PREPARATION OF THIN BARIUM TITANATE FILMS BY DC DIODE SPUTTER 04-0948
ANATE.   MAGNITUDE OF SPONTANEOUS POLARIZATION IN THIN CRYSTALS AND FINE PARTICLES OF BARIUM TIT 04-0607
THE MAJOR PORTION OF THIS/ SYNTHESIS AND CHARACTERIZATION OF THIN FERROELECTRIC AND SEMICONDUCTING FILMS. (02-0057
DEPOLARIZATION FIELD IN THIN FERROELECTRIC FILMS.   37-3659
STRATES. (TGS ON/ THERMODYNAMIC STABILITY OF POLARIZATION IN THIN FERROELECTRIC FILMS ON SEMICONDUCTING SUB 27-3163
RETENTION IN THE THIN FERROELECTRIC FILMS. (PLZT)   09-1593
THERMODYNAMIC STABILITY OF THIN FERROELECTRIC FILMS. (TRIGLYCINE SULFATE) 27-2959
MANUFACTURING METHCD AND FROPERTIES OF FERROELECTRIC THIN FILM.   37-3681
ING POTASSIUM NITRATE. (FUSED THICK FILM OR VACUUM DEPOSITED THIN FILM) /TIVE READOUT MEMORY ELEMENT UTILIZ 16-2269
EFFECT IN ANTIMONY SULFOIODIDE- TIN DIOXIDE HETEROSTRUCTURE THIN FILM.   SWITCHING AND MEMORY 15-2166
ERING ON SILICON SUBS/ PHYSICAL AND ELECTRICAL PROPERTIES OF THIN FILM BARIUM TITANATE PREPARED BY RF SPUTT 04-0828
TES, EFFECT/ PREPARATION AND PROPERTIES OF HIGH PERMITTIVITY THIN FILM DIELECTRICS. (BARIUM AND LEAD TITANA 04-0670
FORMATION PROCESS OF BARIUM TITANATE THIN FILM EVAPORATED IN VACUO.   04-0862
DEVICE. (PLZT)   THIN FILM FERROELECTRIC PHOTOCONDUCTOR MEMORY 09-1503
DEVICE.   PHOTOCONDUCTOR AND ELECTRODE REQUIREMENTS FOR THIN FILM FERROELECTRIC PHOTOCONDUCTOR MEMORY 37-3631
DEVICE.   DESIGN AND PERFORMANCE OF A THIN FILM FERROELECTRIC PHOTOCONDUCTOR STORAGE 37-3632
TURE. (MIXED OXIDES ABO(3)) (/ PREPARATION AND PROPERTIES OF THIN FILM MATERIALS WITH PEROVSKITE TYPE STRUC 04-0914
ND NONLINEAR MICROWAVE PROPERTIES OF FERROELECTRIC CERAMICS. THIN FILM MICROWAVE FERROELECTRIC CAPACITORS A 37-3643
ES. (TELLURIUM ON TRIGLYCINE SULFATE AND SODIUM NITRITE SUR/ THIN FILM NUCLEATION ON FERROELECTRIC SUBSTRAT 16-2394
DOWN TO 3 MIC/ DIELECTRIC CONSTANT AND LATTICE DYNAMICS OF A THIN FILM OF A FERROELECTRIC (BARIUM TITANATE 04-0600
CHARACTERISTICS OF RF SPUTTERED BARIUM TITANATE THIN FILMS.   04-0906
FABRICATION OF RF SPUTTERED BARIUM TITANATE THIN FILMS.   04-0907
RESEARCH STATUS AND DEVICE POTENTIAL OF FERROELECTRIC THIN FILMS.   37-3637
ANSION OF TRIGLYCINE SULFATE CRYSTALS STUDIED WITH TELLURIUM THIN FILMS.   THERMAL EXP 27-3047
PREPARATION AND STUDY OF THIN FILMS OF (BARIUM, STRONTIUM) TITANATE.   04-1000
PREPARATION AND PROPERTIES OF THIN FILMS OF BARIUM TITANATE.   04-0843
/ FLASH EVAPORATION. (REVIEW OF TECHNIQUE AND APPLICATION TO THIN FILMS OF PEROVSKITES AND III-V COMPOUNDS, 02-0112
FLASH EVAPORATION OF FERROELECTRIC THIN FILMS OF POLYCRYSTALLINE BARIUM TITANATE. 04-0968
TANATE.   STUDY OF THE PYROELECTRIC EFFECT IN THIN FILMS OF TRIGLYCINE SULFATE AND BARIUM TI 04-0686
GLYCINE SULFATE AND SELENATE CRYS/ FIELD EFFECT IN TELLURIUM THIN FILMS ON A SUBSTRATE OF FERROELECTRIC TRI 27-3046
THIN FILMS ON TGS CLEAVED IN ULTRAHIGH VACUUM. 27-3173
FILMS OF BARIUM TITANATE BY VACUUM EVAPORATION.   THIN FILMS: PREPARATION OF FERROELECTRIC THIN 04-0834
ND ELECTRICAL PROPERTIES OF FLASH EVAPORATED BARIUM TITANATE THIN FILMS (SEPARATED FROM SUBSTRATE). /URE) A 04-0605
PROPERTIES OF FLASH EVAPORATED FERROELECTRIC BARIUM TITANATE THIN FILMS. (THICKNESS DEPENDENCE) ELECTRICAL 04-0967
/OF REFRACTION AND REFLECTION OF ULTRASONIC SURFACE WAVES BY THIN GOLD FILM LAYERS ON LITHIUM NIOBATE SUBS/ 11-1789
RA/ REFRACTION AND REFLECTION OF ULTRASONIC SURFACE WAVES BY THIN GOLD FILM LAYERS ON LITHIUM NIOBATE SUBST 11-1790
REACTIONS.   FORMATION OF THIN LAYERS OF MIXED TITANATES BY SOLID SOLID 13-2005
INED BY PROJECTION WITH A PLASMA TORCH O/ CHARACTERISTICS OF THIN, MULTILAYERED FERROELECTRIC CERAMICS OBTA 04-0778
/RAY ABSORPTION COEFFICIENTS AND THE THICKNESS OF NONUNIFORM THIN SAMPLES (STRONTIUM TITANATE AS EXAMPLE). 06-1179
AND OPTICAL PROPERTIES OF (BARIUM, STRONTIUM) MIXED TITANATE THIN SECTIONS.   ELECTRICAL 04-0603
SITION, AND PRO/ MONODOMAIN STRONTIUM TITANATE (OCCURANCE IN THIN (110) PLATES BELOW CUBIC- TETRAGONAL TRAN 06-1164
EFLECTORS BY ION ETCHING.   THINNING OF LITHIUM NIOBATE FOR ACOUSTOOPTIC D 11-1862
2 FERR/ NEW FERROELECTRIC (PYROCHLORE TYPE) COMPOUND (AMMONO THIO CADMIUM NIOBATE, WITH 4 ELECTRIC PHASES, 36-3529
TIC RESONANCE OF THE HIGH TEMPERATURE PHASE OF FERROELECTRIC THIOUREA.   DEUTERON MAGNE 25-2918
DIELECTRIC DISPERSION OF UREA AND THIOUREA.   25-2912
FAR INFRARED TRANSMISSION SPECTRUM OF THIOUREA.   25-2917
NFRARED AND RAMAN STUDIES IN THE FERROELECTRIC TRANSITION OF THIOUREA.   I 25-2916
LATTICE VIBRATIONAL SPECTRA OF FIVE CRYSTAL MODIFICATIONS OF THIOUREA.   25-2905
OPTICAL PROPERTIES OF THIOUREA.   25-2913
SATELLITE X-RAY SCATTERING AND STRUCTURAL MODULATION OF THIOUREA.   25-2921
SEMICONDUCTIVITY OF THIOUREA.   25-2924
DRUPOLE RESONANCE OF NITROGEN-14 AND THE FERROELECTRICITY OF THIOUREA.   STARK EFFECT ON THE NUCLEAR QUA 25-2909
MAGNETIC RESONANCE PART-18: PROTON AND DEUTERON RESONANCE OF THIOUREA.   STUDIES OF FERROELECTRIC SOLIDS BY 25-2919
VIBRATIONAL LINE WIDTHS OF THIOUREA.   25-2915
X-RAY TOPOGRAPHY OF THE DEFECT STRUCTURE OF THIOUREA.   25-2914
ANOMALOUS SPECIFIC HEAT OF FERROELECTRIC THIOUREA AND POTASSIUM FERROCYANIDE TRIHYDRATE 24-2902
LOW FREQUENCY RAMAN SPECTRUM OF A THIOUREA CRYSTAL AS A FUNCTION OF TEMPERATURE. 25-2906
SCATTERING DATA)   LONG PERIOD STRUCTURE OF THIOUREA CRYSTAL. (CALCULATIONS FROM SATELLITE 25-2911
INFRARED ABSORPTION SPECTRUM OF A SINGLE CRYSTAL OF THIOUREA IN THE PARA- AND FERROELECTRIC PHASES 25-2910
INFRARED ABSCRPTION SPECTRUM OF THIOUREA IN THE PARA- AND FERROELECTRIC PHASES 25-2908
/AR AND LATTICE VIBRATIONS CF ORTHORHOMBIC AND FERROELECTRIC THIOUREA. PART-1: TEMPERATURE DEPENDENCE OF T/ 25-2920
/AR AND LATTICE VIBRATIONS CF OFTHORHOMBIC AND FERROELECTRIC THIOUREA. PART-2: CALCULATION OF THE INTERMOL/ 25-2907
CRYSTAL STRUCTURE OF THE IV PHASE OF THIOUREA. (SATELLITE SCATTERING)   25-2923
STUDIES ON THE MECHANISM CF THE PHASE TRANSITIONS IN THIOUREA. (SATELLITE SCATTERING)   25-2922
/NS IN ROCHELLE SALT, BARIUM TITANATE, ANTIMONY SULFOIODIDE, THIOUREA, TRIGLYCINE SULFATE, HYDROGEN AMMONI/ 02-0099
I/ OBSERVATION OF THE SIMULTANEOUS GENERATICN OF THE SECOND, THIRD, AND FOURTH HARMONICS OF 1.06 MICRON RAD 11-1779
E SINGLE CRYSTALS (NONLINEAR ELASTIC PROPERTIES). ULTRASONIC THIRD HARMONIC GENERATION IN STRONTIUM TITANAT 06-1181
Y (950 MHZ) ACOUSTIC WAVE ALONG THE (001)/ GENERATION OF THE THIRD HARMONIC OF A LONGITUDINAL HIGH FREQUENC 11-1835
PROPAGATING ITSELF IN THE (001) DI/ GENERATION OF SECOND AND THIRD HARMONICS BY A TRANSVERSE ACOUSTIC WAVE 11-1682
NONLINEAR CPTICAL SPECTROSCOPY OF THIRD ORDER SUSCEPTIBILITY IN LITHIUM NIOBATE. 11-1855
TEMPERATURE DEPENDENCE OF SOME PROPERTIES OF SODIUM DITHORIUM PHOSPHATE FERROELECTRIC CRYSTALS. 36-3611
GEN BONDED PHOSPHATE FERROELECTRIC (CRYSTAL GROWTH/ SILVER DITHORIUM PHOSPHATE- A NEW CASE OF THE NON-HYDRO 36-3610
/STATE CR MOLYBDATE. (WHERE A IS ZIRCONIUM, HAFNIUM, CERIUM, THORIUM, TITANIUM, SODIUM, MAGNESIUM, CADMIUM/ 10-1640
IC CONSTANTS OF POLYCRYSTALLINE LEAD TITANATE ZIRCONATE WITH THOROUGHNESS OF POLING. /ATION OF ELECTROELAST 09-1489
STRUCTURE OF STRONTIUM TITANATE FROM A SELF CONSISTENT FIELD TIGHT BINDING CALCULATION.   ENERGY BAND 06-1221

OXIDE, SODIUM TUNGSTATE, AND POTASSIUM TANTALATE.
P/ POTASSIUM IODATE, A FERROELECTRIC WITH NONREVERSIBLE, BUT
/PERATURE PHASE TRANSITIONS IN SODIUM NIOBATE AND THE USE OF
NDENCE OF THE YOUNGS MODULUS AND THE POLARIZATION RELAXATION
ATION OF THE UNIPOLARITY OF TRIGLYCINE SULFATE CRYSTALS WITH
FECTS UPON FERROELECTRIC PROPERTIES. (DISPERSION, RELAXATION
TERS OF A FERROELECTRIC CERAMIC.
FTER POLARIZATION REVERSAL.
OF HYDROXYL DOPED BARIUM TITANATE.
OF BARIUM TITANATE SINGLE CRYSTALS HEATED IN OXYGEN (EFFEC/
ASUREMENT TECHNIQUE AND APPLICATION TO A NANOSECOND RESPONSE
E.   SELF REVERSAL AND WAITING
STRONTIUM) TITANATE.
RONTIUM TITANATE.
HYDROSTATIC PRESSURE EFFECT ON THE DIELECTRIC RELAXATION
URE (POLY VERSUS MONODISPERSIVE / DISTRIBUTION OF RELAXATION
/IC TUNING. (WITH ELECTRODES REDUCING POLARIZATION SWITCHING
.   NUCLEAR SPIN LATTICE RELAXATION
M LEAD PROPIO/ ANOMALY OF THE PROTON SPIN LATTICE RELAXATION
TE.   DIPOLE LATTICE RELAXATION
IT/ STATIC ELECTRIC SUSCEPTIBILITY AND DIELECTRIC RELAXATION
DER DISORDER TYPE FE/ ANOMALY OF THE SPIN LATTICE RELAXATION
DIELECTRIC CONSTANT AND DIELECTRIC RELAXATION
ELECTR/ (REVERSIBLE) OPTICAL DAMAGE IN KDP (WITH RELAXATION
NUCLEAR QUADRUPOLE FREQUENCIES AND RELAXATION
/POLAR NUCLEI. DETERMINATION OF THE SPIN LATTICE RELAXATION
DISTRIBUTION OF RELAXATION
IDE RANGE OF TEMPERATURES AND ELECTRIC FIELD INTENSITIES AND
ERE A IS STRONTIUM OR BARIUM, AND B IS HAFNIUM, TITANIUM, OR
MODE FOR THE ANTIMONY SULFOIODIDE-
SWITCHING AND MEMORY EFFECT IN ANTIMONY SULFOIODIDE-
/THE PERTURBED ANGULAR CORRELATION TECHNIQUE. (LEAD HAFNATE,
/ERROELECTRIC FIELD EFFECT STUDIES AT LOW TEMPERATURES (WITH
TANATE (0.35) ZIRCO/ FERROELECTRIC FIELD EFFECT STUDIES (ON
/GH DIELECTRIC CONSTANT. (POTASSIUM BROMIDE, LEAD TELLURIDE,
EORY OF SECOND ORDER PHASE TRANSITIONS AT LOW TEMPERATURES.
/HE SUPERCONDUCTING AND THE FERROELECTRIC PHASE TRANSITIONS
OF IONIC CRYSTALS WITH HIGH DIELECTRIC CONSTANTS. (LEAD AND
UTRON SCATTERING)
A POLARIZED FERROELECTRIC CERAMIC. (BARIUM TITANATE, BARIUM
STRUCTURE OF TETRAGONAL
/E OF RESONANCE ABSORPTION SPECTRA OF 23.8 KEV GAMMA-RAYS OF
TIONS IN THE / CHANGE OF THE MOSSBAUER EFFECT PROBABILITY ON
OGEN PHOSPHATE.   EXISTENCE OF DISLOCATIONS AT DOMAIN
A FLOATING ZONE SINGLE CRYSTAL GROWING APPARATUS. (BARIUM
A NEW PTC THERMISTOR MATERIAL. (BARIUM
ABSORPTION EDGE AND STRUCTURAL PHASE TRANSITION IN STRONTIUM
ACOUSTIC PHONON SOFT OPTIC PHONON INTERACTION IN BARIUM
COUSTIC SOFT OPTIC MODE INTERACTIONS IN FERROELECTRIC BARIUM
AGING IN TETRAGONAL FERROELECTRIC BARIUM
THE EFFECT OF HYDROSTATIC PRESSURE ON THE STRUCTURE OF LEAD
E DEPENDENCE OF FLUORESCENCE IN CHROMIUM(3+) DOPED STRONTIUM
N OXYGEN VACANCY CENTER IN THE TETRAGONAL PHASE OF STRONTIUM
ANOMALOUS DECAY EFFECT IN POLYCRYSTALLINE BARIUM
ANOMALOUS FERROELECTRIC HYSTERESIS LOOPS. (LEAD STRONTIUM
ANOMALOUS RESONANCE OF STRONTIUM
ANOMALOUS SPECIFIC HEAT OF STRONTIUM
S ULTRASONIC ATTENUATION AT THE 105K TRANSITION IN STRONTIUM
APPARENT SOFT MODE LINE WIDTH DIVERGENCE IN LEAD
IN RAMAN SCATTERING FROM ONE AND TWO PHONON STATES OF BARIUM
UREMENTS USING THERMAL NOISE ANALYSIS (APPLICATION TO BARIUM
BAND STRUCTURE OF CUBIC AND TETRAGONAL BARIUM
BINARY LIGHT BEAM DEFLECTION IN SINGLE CRYSTAL BARIUM
EINARY LIGHT BEAM DETECTION IN SINGLE CRYSTAL BARIUM
CALCULATION OF (DOMAIN) WALL ENERGY IN BARIUM
S IN FINE GRAINED PEROVSKITE FERROELECTRIC CERAMICS. (BARIUM
TIONS OF ELECTRONIC CONDUCTIVITY IN NONSTOICHIOMETRIC BARIUM
ALYTIC OXIDATION OF CARBON MONOXIDE ON SEMICONDUCTIVE BARIUM
CATION NONSTOICHIOMETRY IN TERNARY OXIDES. (BARIUM
CERAMIC ACOUSTIC DETECTORS. (BARIUM
AND TETRAGONAL (FERROELECTRICALLY POLARIZED PHASE OF BARIUM
CHARACTERISTICS OF DEFECT LEAD
AINED BY PROJECTION WITH A PLASMA TORCH OR FRITTING. (BARIUM
ARACTERIZATION OF SURFACE AND GRAIN BOUNDARY LAYER IN BARIUM
XISTENCE OF THE ANTIPOLARIZED AND POLARIZED STATE IN BISMUTH
HE NEUTRON SPECTRUM OF FERROELECTRIC AND PARAELECTRIC BARIUM
IFTERS FOR VHF AND UHF. (BASED ON LEAD(0.35) STRONTIUM(0.65)
AR CHAIN OF ATOMS LYING IN DOUBLE MINIMUM POTENTIAL. (BARIUM
DUCTIVITY INJECTION AND EXTRACTION IN POLYCRYSTALLINE BARIUM
TWEEN RHEOLOGICAL AND ELECTRICAL BEHAVIOR IN SINTERED BARIUM
AND ELECTRIC PROPERTIES OF FERROELECTRIC SUBSTANCES. (BARIUM
ASE TRANSITIONS IN FERROELECTRICS. (THEORY APPLIED TO BARIUM
COUPLED POLARITONS OF A1 SYMMETRY IN BARIUM
CRITICAL FIELD IN CERAMIC BARIUM
CRITICAL NEUTRON SCATTERING FROM STRONTIUM
YSTAL GROWTH AND TEMPERATURE DEPENDENT RAMAN SPECTRA OF LEAD
PIEZOELECTRICITY OF THE SYSTEM LEAD (ZINCATE NIOBATE)- LEAD
BIC TO TRIGONAL STRESS INDUCED PHASE TRANSITION IN STRONTIUM
CURRENT EXCITATION IN GAMMA-IRRADIATED BARIUM
DEFECT STUDIES IN STRONTIUM

TIGHT BONDING BAND CALCULATIONS ON RHENIUM TRI 08-1334
TILTABLE (TO 3 STABLE POSITIONS, SPONTANEOUS) 08-1346
TILTING SCHEMES IN THE SOLUTIONS OF PEROVSKIT/ 08-1336
TIME). /IDEUTERIUM PHOSPHATE (TEMPERATURE DEPE 17-2660
TIME.   VARI 27-3115
TIME AND STRENGTH, HYSTERESIS, 26 REFS) /AL DE 03-0496
TIME AND TEMPERATURE DEPENDENCES OF THE PARAME 09-1514
TIME DEPENDENCE OF MATERIAL CONSTANTS OF TGS A 27-2953
TIME DEPENDENCE OF THE DIELECTRIC PERMITTIVITY 04-0550
TIME DEPENDENCE OF THE ELECTRICAL CONDUCTIVITY 04-0997
TIME DETECTOR. (STRONTIUM BARIUM NIOBATE) / ME 13-1919
TIME EFFECTS IN SINGLE CRYSTAL BISMUTH TITANAT 08-1448
TIME EFFECTS IN THE HYSTERESIS LOOP OF (LEAD, 05-1052
TIME EFFECTS IN THE HYSTERESIS LOOP OF LEAD ST 05-1053
TIME IN FERROELECTRIC SILVER SODIUM DINITRITE. 16-2294
TIME IN FERROELECTRICS NEAR THE CURIE TEMPERAT 03-0521
TIME, LEAD(0.35) STRONTIUM(0.65) TITANATE, BA/ 37-3660
TIME MEASUREMENTS IN SODIUM DEUTERIUM SELENATE 22-2847
TIME NEAR THE CRITICAL TEMPERATURE OF DICALCIU 26-2947
TIME NEAR THE NEEL TEMPERATURE IN SODIUM NITRI 16-2300
TIME NEAR THE TRANSITION POINTS IN SODIUM NITR 16-2302
TIME NEAR THE TRANSITION TEMPERATURE IN THE OR 03-0488
TIME OF ROCHELLE SALT. 28-3179
TIME OF SECONDS WHEN ILLUMINATED UNDER APPLIED 17-2703
TIMES IN FERROELECTRIC HYDROGEN CHLORIDE. 31-3336
TIMES IN FERROELECTRIC LITHIUM TANTALATE AND / 11-1829
TIMES IN SOME FERROELECTRICS. 03-0288
TIMES OF APPLICATION. /OVSKITE TYPE WITHIN A W 08-1380
TIN) / THE CADMIUM BORATE- ABO(3) SYSTEMS. (WH 36-3528
TIN DIOXIDE HETEROSTRUCTURE SWITCH. 15-2243
TIN DIOXIDE HETEROSTRUCTURE THIN FILM. 15-2166
TIN HAFNATE, CALCIUM HAFNATE, STRONTIUM HAFNA/ 08-1276
TIN OXIDE FILMS ON LEAD ZIRCONATE(0.65) TITAN/ 09-1505
TIN OXIDE FILMS ON LEAD(0.97) BISMUTH(0.02) TI 09-1506
TIN TELLURIDE, LEAD SULFIDE, GERMANIUM TELLUR/ 36-3541
TIN TELLURIDE, POTASSIUM TANTALATE) TH 03-0433
TIN TELLURIDE PURE AND ALLOYED WITH SMALL AMO/ 36-3564
TIN TELLURIDES)   LATTICE DYNAMICS 36-3553
TIN TELLURIDE- GERMANIUM TELLURIDE ALLOYS. (NE 36-3586
TIN TITANATE, LEAD BARIUM DINIOBATE) /TROPY OF 04-0767
TIN TUNGSTEN BRONZES. 13-2031
TIN-119 DURING PHASE TRANSITIONS OF THE BISM/ 08-1400
TIN-119 IN BARIUM TITANATE STANNATE SOLID SOLU 04-0573
TIPS IN FERROELECTRIC CRYSTAL POTASSIUM DIHYDR 17-2450
TITANATE) 04-0998
TITANATE) 04-0931
TITANATE. 06-1107
TITANATE. 04-0806
TITANATE. A 04-0671
TITANATE. 04-0579
TITANATE. AN X-RAY STUDY OF 05-1064
TITANATE. ANALYSIS OF STRESS AND TEMPERATUR 06-1215
TITANATE. ANALYSIS OF THE IRO 06-1231
TITANATE. 04-0752
TITANATE) 05-1067
TITANATE. 06-1173
TITANATE. 06-1159
TITANATE. ANOMALOU 06-1189
TITANATE. 05-1078
TITANATE. AUGER-LIKE RESONANT INTERFERENCE 04-0923
TITANATE). AUTOMATIC IMPEDANCE MEAS 04-0844
TITANATE. 04-0846
TITANATE. 04-0653
TITANATE. 04-0654
TITANATE. 04-0673
TITANATE) /ON OF THE ELECTROSTATIC INTERACTION 04-0642
TITANATE. CALCULA 04-0821
TITANATE. CAT 04-0945
TITANATE) 04-0969
TITANATE) 01-0003
TITANATE. /ANNELING OF LIGHT IONS IN THE CUBIC 04-0688
TITANATE. 05-1075
TITANATE) /ILAYERED FERROELECTRIC CERAMICS OBT 04-0778
TITANATE. CH 04-0701
TITANATE. COE 08-1422
TITANATE. COMMENTS ON T 04-0813
TITANATE) /URFACE WAVE) FERROELECTRIC PHASE SH 37-3633
TITANATE. COMPUTER EXPERIMENT ON LINE 03-0287
TITANATE. CON 04-0814
TITANATE. 04-1027
TITANATE. CORRELATION BETWEEN STRUCTURE 04-0728
TITANATE) /ION EFFECTS IN DISPLACEMENT TYPE PH 03-0497
TITANATE. 04-0619
TITANATE. 04-0951
TITANATE. 06-1195
TITANATE. CR 05-1074
TITANATE. CRYSTAL STRUCTURE AND 13-2067
TITANATE. CU 06-1235
TITANATE. 04-0884
TITANATE. 06-1244

DEFECTS IN CXIDE CERAMICS. (STRONTIUM TITANATE) 06-1163
DEPOLING OF SINGLE DOMAIN BISMUTH TITANATE. 08-1446
STATIC MAGNETIC SUSCEPTIBILITY OF CHROMIUM DOPED BARIUM METATITANATE. /NCY FROM STUDIES OF EPR SPECTRA AND 04-0542
DETERMINATION OF THE NATURE OF THE OPTICAL GAP OF STRONTIUM TITANATE. 06-1108
BSORPTICN IN THE ORTHORHOMBIC AND QUADRATIC PHASES OF BARIUM TITANATE. DIELECTRIC A 04-0900
BSORPTION IN THE ORTHORHOMBIC AND QUADRATIC PHASES OF BARIUM TITANATE. DIELECTRIC A 04-0899
DIELECTRIC AESORPTICN IN THE RHOMCHEDRAL PHASE OF BARIUM TITANATE. 04-0809
DIELECTRIC BREAKDOWN IN BARIUM TITANATE. 04-0580
DIELECTRIC BREAKDCWN OF POLYCRYSTALLINE BARIUM TITANATE. 04-1015
STRUCTURE FOR HIGH FREQUENCY APPLICATICN. (BARIUM, STRONTIUM TITANATE) /CTRIC CERAMICS WITH BOUNDARY LAYER 04-0833
DIELECTRIC DISPERSICN IN BARIUM TITANATE. 04-0646
DIELECTRIC PROPERTIES OF SINGLE DOMAIN MELT GROWN BARIUM TITANATE. 04-0616
DIELECTRIC SUSCEPTIBILITY OF QUASI FERROELECTRIC STRONTIUM TITANATE. 06-1133
AND THE STABLE DOMAIN STRUCTURE CF (HIGH SENSITIVITY) BARIUM TITANATE. DIFFUSIONLESS PHASE TRANSITIONS 04-0888
DISPERSION IN FERROELECTRICS. (THEORY, BARIUM TITANATE) 03-0454
DISPERSION OF THE ELECTROOPTIC EFFECT IN BARIUM TITANATE. 04-0756
DISTRIBUTION OF VACANCIES IN LANTHANA DOPED LEAD TITANATE. 05-1058
DOMAIN INTERACTIONS DURING PHASE TRANSITIONS IN BARIUM TITANATE. 04-0651
AND POLARIZATION REVERSAL IN FILMS OF FERROELECTRIC BISMUTH TITANATE. DOMAIN STRUCTURE 08-1470
DYNAMIC FATIGUE CF SINGLE CRYSTALS OF BARIUM TITANATE. 04-0792
(ROTATICNS OF BO6 OCTAHEDRA IN ABC(3) PEROVSKITES, STRONTIUM TITANATE). /ERROELECTRIC CRITICAL TEMPERATURE 06-1205
EFFECT OF "A" DOMAINS UPCN "C" DOMAIN SWITCHING IN BARIUM TITANATE. 04-0615
TIVES CN THE PROPERTIES OF SOLID SCLUTICNS OF LEAD ZIRCONATE TITANATE. /SMUTH AND MANGANESE CONTAINING ADDI 09-1588
LINE GRAIN SIZE ON THE PHYSICAL PROPERTIES OF CERAMIC BARIUM TITANATE. EFFECT OF CRYSTAL 04-0785
TRUCTURE ON THE FLUORESCENCE OF CHROMIUM(3+) DOPED STRONTIUM TITANATE. EFFECT OF DOMAIN S 06-1114
SIZE ON THE FERROELECTRIC PARAELECTRIC TRANSITICN OF BARIUM TITANATE. EFFECT OF GRAIN 04-0917
N SIZE CN THE PIEZCELECTRIC AND ELASTIC PROPERTIES OF BARIUM TITANATE. EFFECT OF GRAI 04-1011
HEAT TREATMENT CN THE PTCR ANOMALY IN SEMICONDUCTING BARIUM TITANATE. EFFECT OF 04-0762
EFFECT OF IRON(4+) IN THE SYSTEM STRONTIUM FERRATE TITANATE. 13-1927
TIES CN THE ELECTRICAL RESISTIVITY OF LANTHANUM DOPED BARIUM TITANATE. EFFECT OF OXIDE IMPURI 04-0587
PRESSURE (TO ABOUT 2KBAR) ON POLARIZATION REVERSAL IN BARIUM TITANATE. EFFECT OF 04-0718
A ADDITIONS ON THE SEMICONDUCTING PROPERTIES OF DOPED BARIUM TITANATE. EFFECT OF SILIC 04-0660
S ON THE SUPERCONDUCTING TRANSITICN TEMPERATURE OF STRONTIUM TITANATE. EFFECT OF STRES 06-1182
ANCE SPECTRA OF SUBSTITUTIONAL IMPURITIES IN CUBIC STRONTIUM TITANATE. / ON THE ELECTRON PARAMAGNETIC RESON 06-1193
IC ANHARMONICITY ON THE SOFT PHONON INSTABILITY OF STRONTIUM TITANATE. EFFECT OF THE QUART 06-1211
EFFECT OF UNIAXIAL STRESS ON RAMAN SCATTERING IN STRONTIUM TITANATE. 06-1102
OF ELASTIC CONSTANTS ON DISLOCATIONS IN FERROELECTRIC BARIUM TITANATE. EFFECTS OF ANISOTROPY 04-0674
TANTALATE), LEAD (LANTHANATE, NIOBATE)- LEAD ZIRCONATE- LEAD TITANATE) /LATE), (LEAD, TUNGSTEN) (ZIRCONATE 13-2004
TS OF LCNGITUCINAL MAGNETIC FIELD CN ULTRASOUND IN STRONTIUM TITANATE. EFFEC 06-1243
ROM THE PARAMAGNETIC RESONANCE OF MANGANESE(4+) IN STRONTIUM TITANATE. EFFECTIVE CHARGE OF TITANIUM(4+) F 06-1167
EFFICIENT WHITE LIGHT READING OF DOMAIN PATTERNS IN BISMUTH TITANATE. 08-1299
ELASTIC CONSTANT ANOMALIES IN STRONTIUM TITANATE. 06-1177
ELECTRET EFFECT IN A MIXTURE OF POLY METHACRYLATE AND BARIUM TITANATE 04-0838
TAIN TWO PHASE FERROELECTRIC FERRCMAGNETIC MIXTUFES. (BARIUM TITANATE) /IC AND MAGNETIC MEASUREMENTS OF CER 04-0565
ELECTRIC PROPERTIES OF HYDROGEN DOPED STRONTIUM TITANATE. 06-1088
ELECTRICAL CONDUCTIVITY OF SINGLE CRYSTALLINE BARIUM TITANATE. 04-0663
-90 VOL PERCENT, FERROELECTRIC, CCMPARED WITH CERAMIC BARIUM TITANATE) /ERIALS CRYSTALLIZED FROM GLASS. (60 04-0725
ELECTROCALORIC EFFECT IN PCLYCRYSTALLINE BARIUM TITANATE. 04-0766
NS OF ISC- AND HETEROVALENT CATIONIC SUBSTITUTIONS IN BARIUM TITANATE. /ECT IN SEMICONDUCTING SOLID SOLUTIO 04-0943
EORY. PART-2: OBSERVATIONS. (FERRCELECTRIC DCMAINS IN BARIUM TITANATE) /IN BOUNDARIES. PART-1: DYNAMICAL TH 03-0255
IC OBSERVATION OF FERROELECTRIC DCMAINS. (GOLD COATED BARIUM TITANATE) ELECTRON MIRROR MICROSCOP 04-0974
ELECTRON PARAMAGNETIC RESONANCE IN NONSTOICHIOMETRIC BARIUM TITANATE. 04-0640
AGNETIC RESONANCE OF SOME RARE EARTH IMPURITIES IN STRONTIUM TITANATE. ELECTRON PARAM 06-1192
ELECTRON SPIN RESONANCE OF MANGANESE(2+) IN BARIUM TITANATE. 04-0747
ELECTRON SPIN RESCNANCE STUDY IN STRONTIUM TITANATE. 06-1228
LECTRON SPIN RESOURCE STUDIES AT DEFECTS IN FLUORITES. (LEAD TITANATE) E 05-1086
DURING HCLOGRAM RECORDING IN CRYSTALS (IRON DOPED STRONTIUM TITANATE). /ERNAL GENERATION OF STRONG FIELDS) 06-1090
CTROOPTIC EFFECTS IN FERROELECTRIC CERAMICS. (LEAD ZIRCONATE TITANATE) ELE 09-1551
YGEN OCTAHEDRA FERROELECTRICS: CERAMIC LEAD BARIUM ZIRCONATE TITANATE. /POLARIZATION) OF LEAD CONTAINING OX 13-2039
PHASES. (LANTHANUM OR NIOBIUM DOPED LEAD ZIRCONATE STANNATE TITANATE) /FERROELECTRIC AND ANTIFERROELECTRIC 09-1487
TTICE POLARIZATION. (LCAO ANALYSIS OF BAND SCHEME, STRONTIUM TITANATE) /ND CHANGES IN PEROVSKITES DUE TO LA 06-1101
ERROELECTRIC CERAMICS: SURVEY OF MATERIALS BASED UPON BARIUM TITANATE. ENERGY STORAGE IN F 02-0075
H PERMITTIVITY CERAMICS: SURVEY OF MATERIALS BASED ON BARIUM TITANATE. ENERGY STORAGE IN HIG 04-0733
EXAGGERATED GRAIN GRCWTH IN LIQUID PHASE SINTERING OF BARIUM TITANATE. 04-0835
HEORETICAL STUDY OF DEFECTS IN TRIGLYCINE SULFATE AND BARIUM TITANATE. EXPERIMENTAL AND T 04-0620
IN THE THEORY OF OPTICAL LATTICE VIBRATIONS. (QUARTZ, BARIUM TITANATE) /F THE POULET- LOUDON APPROXIMATION 03-0380
FAR INFRARED FERROELECTRIC VIBRATICN MODE IN STRONTIUM TITANATE. 06-1094
FAR INFRARED REFLECTIVITY OF BARIUM TITANATE. 04-0744
NG) SHADOW EFFECT IN (DISPLACIVE) PHASE TRANSITIONS. (BARIUM TITANATE) (FAST ION SCATTERI 04-0533
LITY OF ELECTROOPTIC DEVICES UTILIZING FERROELECTRIC BARIUM TITANATE. FEASIBI 08-1447
YMMETRY GROUPS. (AND THEIR CCMPARISON, APPLICATION TC BARIUM TITANATE) /RROELECTRIC AND ANTIFERROELECTRIC S 04-0933
CS AND CONVERSION OF MECHANOELECTRIC ENERGY. (LEAD ZIRCONATE TITANATE) FERROELECTRIC CERAMI 09-1483
ERROELECTRIC CERAMICS AND ENERGY CONVERSION. (LEAD ZIRCONATE TITANATE) F 09-1508
FERROELECTRIC FAMILY OF BARIUM TITANATE. 04-1018
FERROELECTRIC MEMORY DEVICE PARAMETER STUDY. (LEAD ZIRCONATE TITANATE) 09-1594
FERROELECTRIC PHASE TRANSITION OF FIRST ORDER IN BARIUM TITANATE. 04-1030
TRIC PHOTOCONDUCTOR OPTICAL STORAGE MEDIUM UTILIZING BISMUTH TITANATE. FERROELEC 04-0769
BANDS OF FERROELECTRICS IN FAR INFRARED. (APPLIED TO BARIUM TITANATE) FIELD DEPENDENCE OF ABSORPTION 04-0665
LEATION AT THE (CUBIC TETRAGONAL) PHASE TRANSITION IN BARIUM TITANATE. FIELD INDUCED NUC 04-0753
RATION OF FERROELECTRIC THIN FILMS CF PCLYCRYSTALLINE BARIUM TITANATE. FLASH EVAPO 04-0968
CTRIC PROPERTIES OF HIGH RESISTIVITY SINGLE CRYSTALS OF LEAD TITANATE. (FLUX) GROWTH AND FERROELE 05-1072
E ELECTRON PARAMAGNETIC RESONANCE OF MANGANESE(+2) IN BARIUM TITANATE. /NES IN THE HYPERFINE SPECTRUM OF TH 04-0877
ATION CF FLUORINE CCNTAINING SOLID SOLUTIONS BASED ON BARIUM TITANATE. FORM 04-0932
FREE CARRIER OPTICAL ABSORPTION IN NIOBIUM DOPED STRONTIUM TITANATE. 06-1214
FUNDAMENTAL AESORPTION EDGE OF STRONTIUM TITANATE. 06-1186
TION OF SECOND OPTICAL HARMONIC IN THE CUBIC PHASE OF BARIUM TITANATE. GENERA 04-0538
DENCE OF DIELECTRIC AND STRUCTURAL CHARACTERISTICS OF BARIUM TITANATE. GRAIN SIZE DEPEN 04-0582
DUCTIVITY IN DISPLACIVE TYPE FERROELECTRICS. (THEORY, BARIUM TITANATE) /Y SOUND ATTENUATION AND THERMAL CON 03-0160
HIGH L/F NOISE ANCMALY IN SEMICONDUCTING BARIUM STRONTIUM TITANATE. 04-0866
RESSURE COMPRESSIEILITY AND GRUNEISEN PARAMETER OF STRONTIUM TITANATE. HIGH P 06-1121

AMIC AND X-RAY DIFFRACTION STUDY OF NONSTOICHIOMETRIC BARIUM TITANATE.	HIGH TEMPERATURE THERMODYN	04-0885
HYDROXYL AS DEFECT IN BARIUM TITANATE.		04-0636
ING FINE GRAIN FERROELECTRIC CERAMICS. (LEAD ZIRCONATE- LEAD TITANATE)	IMAGE STORAGE AND DISPLAY DEVICES US	09-1572
ECTRICAL POLARIZATION OF CRYSTALS (ALKALI HALIDES, STRONTIUM TITANATE).	INERTIAL EL	06-1197
FIELD ON THE DIELECTRIC PROPERTIES OF POLYCRYSTALLINE BARIUM TITANATE.	INFLUENCE OF A CONSTANT ELECTRIC	04-1012
ATION AND PROPERTIES OF SINTERED AN PRESSURE SINTERED BARIUM TITANATE.	INFLUENCE OF ADDITIVES ON DENSIFIC	04-1026
RIZATION FLUCTUATIONS ON THE PHOTOELASTIC BEHAVIOR OF BARIUM TITANATE.	INFLUENCE OF CRITICAL POLA	04-0628
RAIN GROWTH ON DIELECTRIC PROPERTIES OF NIOBIUM DOPED BARIUM TITANATE.	INFLUENCE OF G	04-0763
(PREPARATION CONDITIONS) ON DIELECTRIC PROPERTIES OF BARIUM TITANATE.	INFLUENCE OF MICROSTRUCTURES	04-0798
YGEN PARTIAL PRESSURE ON PROPERTIES OF SEMICONDUCTING BARIUM TITANATE.	INFLUENCE OF OX	04-0999
ENCE OF PHASE TRANSITIONS ON ELECTRICAL RESISTANCE OF BARIUM TITANATE.	INFLU	04-0599
S ON THE INTERNAL FRICTION IN POLYCRYSTALLINE LEAD ZIRCONATE TITANATE.	/F THE CONCENTRATION OF POINT DEFECT	09-1582
INFRARED SPECTRA OF FLUORINE CONTAINING BARIUM TITANATE.		04-0983
INFRARED SPECTRA OF SINGLE AND POLYCRYSTALLINE STRONTIUM TITANATE.		06-1250
TRANSPARENT CERAMIC FERROELECTRICS. (LEAD LANTHANUM ZIRCONUM TITANATE)	/CRIPTION OF HOLOGRAPHIC RESULTS IN	09-1575
INTERACTION BETWEEN THE 90 DEG AND 180 DEG DOMAINS IN BARIUM TITANATE.		04-0585
F ELASTIC STRAIN WITH THE STRUCTURAL TRANSITION OF STRONTIUM TITANATE.	INTERACTION O	06-1216
ELECTRIC FIELDS IN CRYSTALS OF SODIUM TANTALATE AND CADMIUM TITANATE.	INTERNAL	08-1389
CTRIC FIELD EFFECTS IN EPR OF GADOLINIUM(3+) DOPED STRONTIUM TITANATE.	INTERPRETATION OF ELE	06-1168
INTRINSIC NONSTOICHIOMETRY IN SINGLE PHASE LEAD ZIRCONATE TITANATE.		09-1534
INVERSION OF ELECTRICAL PROPERTIES OF SEMICONDUCTING BARIUM TITANATE.		04-0572
AND THE PYROELECTRIC PROPERTIES OF SINGLE CRYSTALS OF BARIUM TITANATE.	/ THICKNESS ON THE DOMAIN STRUCTURE	04-1039
UER ISOMER SHIFTS. (DEGREE OF COVALENCY IN IRON DOPED BARIUM TITANATE)	IONIC CHARACTERS FROM MOSSBA	04-0563
KINETICS OF THE PHASE TRANSITIONS OF BARIUM TITANATE.		04-0650
TTICE DYNAMICS ABOVE STRUCTURAL PHASE TRANSITIONS: STRONTIUM TITANATE.	LA	06-1124
MICS OF CUBIC PEROVSKITE STRUCTURES, IN PARTICULAR STRONTIUM TITANATE.	LATTICE DYNA	06-1185
LATTICE DYNAMICS OF STRONTIUM TITANATE.		06-1224
LEED STUDY OF SURFACE STRUCTURES ON THE (001) FACE OF BARIUM TITANATE.		04-0529
S NEAR THE CUBIC TO TETRAGONAL PHASE TRANSITION IN BARIUM TITANATE.	/SCATTERING STUDIES OF THE SOFT MODE	06-1242
DER AND TEMPERATURE DEPENDENCE OF RAMAN SCATTERING IN BARIUM TITANATE.	LINEAR DISOR	04-0675
LOWERING THE CURIE TEMPERATURE IN REDUCED BARIUM TITANATE.		04-0711
S POLARIZATION IN THIN CRYSTALS AND FINE PARTICLES OF BARIUM TITANATE.	MAGNITUDE OF SPONTANEOU	04-0607
S OF FERROELECTRICS AT SUBMILLIMETER WAVELENGTHS. (STRONTIUM TITANATE)	MEASUREMENT OF DIELECTRIC PROPERTIE	06-1089
ASUREMENT OF ELECTRON ENERGY LOSS FUNCTION IN BULK STRONTIUM TITANATE.	ME	06-1138
F THE PYROELECTRIC COEFFICIENT OF 65-35 PZT. (LEAD ZIRCONATE TITANATE)	MEASUREMENT O	09-1567
MECHANISM OF ELECTROCONDUCTIVITY IN BARIUM TITANATE.		04-0908
OF MICROWAVE DIELECTRIC DISPERSION IN POLYCRYSTALLINE BARIUM TITANATE.	MECHANISM	04-0745
PONSE FUNCTION OF THE SOFT ZONE BOUNDARY PHONON IN STRONTIUM TITANATE.	MICROSCOPIC CALCULATION OF THE RES	06-1213
M TRIOXIDE DOPED, RARE EARTH OXIDE DOPED, AND UNDOPED BARIUM TITANATE.	/ND ELECTRICAL PROPERTIES OF SCANDIU	04-0839
F DOMAINS AND DOMAIN WALLS IN SINGLE CRYSTAL FILMS OF BARIUM TITANATE.	MICROSTRUCTURE O	04-1031
MICROWAVE ABSORPTION IN CUBIC STRONTIUM TITANATE.		04-1212
ABOVE THE CURIE TEMPERATURE. (BARIUM(0-1.0) STRONTIUM(1.0-0) TITANATE)	/N OXYGEN OCTAHEDRON FERROELECTRICS	04-0559
EIGHT MEASUREMENTS IN SOLID SOLUTIONS OF (BARIUM, STRONTIUM) TITANATE.	MOBILITY DETERMINATIONS FROM W	04-0689
MONODOMAIN STRONTIUM TITANATE.		06-1169
MOSSBAUER EFFECT FOR IRON-57 IN FERROELECTRIC LEAD TITANATE.		05-1040
MOSSBAUER EFFECT OF COBALT-57 IN BARIUM TITANATE.		04-0723
MOSSBAUER STUDY OF COLOR CENTERS IN STRONTIUM TITANATE.		06-1153
MOVING STRIPE PATTERN IN BARIUM TITANATE.		04-0643
NEUTRON SCATTERING STUDY OF SOFT MODES IN CUBIC BARIUM TITANATE.		04-0712
RIC TRANSDUCER MATERIAL. (ALUMINA WHISKERS IN LEAD ZIRCONATE TITANATE)	NEW COMPOSITE CERAMIC PIEZOELECT	09-1556
NEW ESR EVIDENCE FOR THE 65 DEGREES K ANOMALY IN STRONTIUM TITANATE.		06-1230
RING STUDIES OF THE STRUCTURAL PHASE TRANSITION IN BARIUM TITANATE.	NEW OBSERVATIONS FROM NEUTRON SCATTE	06-1210
MEASUREMENTS IN FERROELECTRICS. (TGS, ROCHELLE SALT, BARIUM TITANATE)	NOISE	04-0698
4 PHOTON RECIPROCAL EFFECT IN CADMIUM SULFIDE AND STRONTIUM TITANATE.	NONCOLLINEAR PARAMETRIC	06-1122
NONLINEAR OPTICAL PROPERTIES OF FERROELECTRIC LEAD TITANATE.		05-1080
NONLINEAR OPTICAL PROPERTIES OF FERROELECTRIC LEAD TITANATE.		05-1079
AVE FREQUENCIES. (TRIGLYCINE SULFATE, POLYCRYSTALLINE BARIUM TITANATE)	/PERTIES OF FERROELECTRICS AT MICROW	04-0870
NONSTOICHIOMETRY IN BARIUM TITANATE.		04-0631
AGNETIC RESONANCE STUDY OF THE PHASE TRANSITION IN STRONTIUM TITANATE.	NUCLEAR M	06-1237
R A FIRST ORDER TRANSITION FRONT, AND OBSERVATIONS IN BARIUM TITANATE.	NUCLEATION MODELS FO	04-0608
OHMIC CONTACTS TO SEMICONDUCTING BARIUM TITANATE.		04-0764
INFLUENCE OF DEFECTS ON THE TEMPERATURE HYSTERESIS OF BARIUM TITANATE.	ON THE	04-0987
OF INFRARED SPECTRA AND THE LATTICE DYNAMICS OF CUBIC BARIUM TITANATE.	ON THE INTERPRETATION	04-0659
ON THE LATTICE VIBRATION OF CUBIC BARIUM TITANATE.		04-0775
ON THE PHASE COMPOSITION OF REDUCED AND REOXIDIZED BARIUM TITANATE.		04-0541
OPTICAL ABSORPTION EDGE OF STRONTIUM TITANATE.		06-1187
OPTICAL ABSORPTION SPECTRA OF REDUCED STRONTIUM TITANATE.		06-1245
VIDENCE FOR POLARIZATION FLUCTUATIONS IN PARAELECTRIC BARIUM TITANATE.	OPTICAL E	04-0737
OPTICAL STUDY OF DOMAIN WALLS IN BISMUTH TITANATE.		08-1301
OPTICALLY INDUCED REFRACTIVE INDEX CHANGES IN BARIUM TITANATE.		04-1005
OXIDATION OF CARBON MONOXIDE ON LANTHANUM OXIDE DOPED BARIUM TITANATE.		04-0947
AGNETIC RESONANCE OF AN F CENTER IN NONSTOICHIOMETRIC BARIUM TITANATE.	PARAM	04-0976
PERMITTIVITY OF STRONTIUM TITANATE.		06-1174
ON OF SCANDIUM-44 WITHIN THE LATTICE OF FERROELECTRIC BARIUM TITANATE.	PERTURBED DIRECTIONAL CORRELATI	04-0691
PERTURBED DIRECTIONAL CORRELATION OF SCANDIUM-44 IN LEAD TITANATE.		05-1055
ELATION OF SCANDIUM-44 IN THE FERROELECTRIC PHASES OF LEAD TITANATE.	PERTURBED DIRECTIONAL CORR	04-0558
SITIONS IN THE MIXED CRYSTAL SYSTEM STRONTIUM(1-X) BARIUM(X) TITANATE.	PHASE TRAN	04-1016
CRYSTALS OF PEROVSKITE STRUCTURE (WITH APPLICATION TO BARIUM TITANATE).	/X AND ELECTROOPTICAL PHENOMENA IN	03-0518
PHONON TRANSPORT IN BARIUM TITANATE.		04-0830
PHOTOCHROISM IN TRANSITION METAL DOPED STRONTIUM TITANATE.		06-1126
PHOTOCONDUCTIVITY AND PHOTOFERROELECTRIC PHENOMENA IN BARIUM TITANATE.		04-0574
PHOTOCONDUCTIVITY AND TRAPPING PHENOMENA IN STRONTIUM TITANATE.		06-1091
FERROELECTRIC PHASE TRANSITION AROUND 47 DEG K IN STRONTIUM TITANATE.	/ANOMALY AS AN EVIDENCE OF THE LOCAL	06-1249
PHOTODOMAIN EFFECT IN BARIUM TITANATE.		04-0682
PHOTON PHONON INTERACTION IN THE NEAR INFRARED. (BARIUM TITANATE)		04-0925
POINT DEFECTS AND SINTERING OF LEAD ZIRCONATE TITANATE.		09-1478
POLARITON DISPERSION RELATION IN CUBIC BARIUM TITANATE.		04-0878
ATION FLUCTUATIONS AND THE OPTICAL ABSORPTION EDGE IN BARIUM TITANATE.	POLARIZ	04-1028
EFLECTION TYPE LIGHT MODULATOR USING FERROELECTRICS. (BARIUM TITANATE)	POLARIZATION R	04-0714
M(3+) THROUGH A STRESS INDUCED PHASE TRANSITION IN STRONTIUM TITANATE.	/RIZED FLUORESCENCE STUDY OF CHROMIU	06-1115

POLARIZED STATES OF FERROELECTRIC CERAMICS. (BARIUM TITANATE)   04-0666
POLARON MECHANISM OF CONDUCTIVITY IN BARIUM TITANATE.   04-0609
ANISOTROPIC ENERGY BANDS (THEORY, CALCULATIONS FOR STRONTIUM TITANATE).   POLARONS IN   03-0305
POSITIVE TEMPERATURE COEFFICIENT OF RESISTANCE IN DOPED LEAD TITANATE.   05-1051
POWDER RAMAN STUDY FOR BARIUM TITANATE.   04-0610
IRCONIA PHASE IN NIOBIUM MODIFIED CERAMICS OF LEAD ZIRCONATE TITANATE.   PRECIPITATION OF Z   09-1540
PREPARATION AND PROPERTIES OF THIN FILMS OF BARIUM TITANATE.   04-0843
PREPARATION AND STUDY OF THIN FILMS OF (BARIUM, STRONTIUM) TITANATE.   04-1000
PREPARATION OF FLUORINE CONTAINING BARIUM TITANATE.   04-0984
GRAINED AND LARGE GRAINED CERAMICS FROM NIOBIUM DOPED BARIUM TITANATE.   PREPARATION OF SMALL   04-0765
C MODE- SOFT OPTIC MODE INTERACTIONS IN FERROELECTRIC BARIUM TITANATE.   PRESSURE DEPENDENCE OF ACOUSTI   04-0893
PRESSURE DEPENDENCE OF THE ELASTIC CONSTANTS OF STRONTIUM TITANATE.   06-1096
PRESSURE DEPENDENCE OF THE ELASTIC CONSTANT OF STRONTIUM TITANATE.   06-1097
TION TEMPERATURES AND ELECTROSTRICTIONS IN PEROVSKITE BARIUM TITANATE.   PRESSURE DEPENDENCE OF TRANSI   04-0853
PROPERTIES OF BLACK SEMICONDUCTING BARIUM TITANATE.   04-0702
RAMAN SCATTERING AND PHASE TRANSITIONS IN STRESSED STRONTIUM TITANATE.   06-1103
RAMAN SCATTERING IN THE FERROELECTRIC SYSTEM LEAD BARIUM TITANATE.   04-0611
SCATTERING LINE SHAPE OF THE SOFT E POLARITON MODE IN BARIUM TITANATE.   RAMAN   04-0803
RAMAN SPECTRUM OF BARIUM TITANATE.   04-0746
(TEMPERATURE DEPENDENCE) OF UNDER DAMPED SOFT MODES IN LEAD TITANATE.   RAMAN STUDIES   05-1041
ED RESPONSE USING SURFACE WAVE TRANSDUCERS ON LEAD ZIRCONATE TITANATE. /OF A 40 MHZ COLOR TELEVISION INFRAR   09-1509
N BETWEEN ELECTRICAL PROPERTIES AND MICROSTRUCTURE OF BARIUM TITANATE.   RELATIO   04-0980
RESISTIVITY AND SINTERING OF BARIUM TITANATE.   04-0819
OF ORDERED PHASE IN PEROVSKITE TYPE FERROELECTRICS. (BARIUM TITANATE) /GIN OF FERROELECTRICITY, OCCURRENCE   04-0964
ARMONIC GENERATION IN STRESS INDUCED FERROELECTRIC STRONTIUM TITANATE.   SECOND H   06-1131
REVERSAL AND WAITING TIME EFFECTS IN SINGLE CRYSTAL BISMUTH TITANATE.   SELF   08-1448
SHARED HOLES TRAPPED BY CHARGE DEFECTS IN STRONTIUM TITANATE.   06-1123
SHELL MODEL OF INTERIONIC INTERACTIONS FOR BARIUM TITANATE.   04-0716
SHOCK INDUCED TRANSITION IN BARIUM TITANATE.   04-0589
SHOCK WAVE STUDIES OF SURFACE LAYERS IN BARIUM TITANATE.   04-0588
SINTERING SCANDIUM AND NIOBIUM MODIFIED LEAD ZIRCONATE TITANATE.   09-1555
SIZE EFFECTS IN FERROELECTRICS. (BARIUM TITANATE)   04-0656
MODES (FERROELECTRICS, CESIUM DIHYDROGEN ARSENATE, STRONTIUM TITANATE) /T MODE RAMAN SPECTROSCOPY: COUPLED   06-1207
ASE TRANSITIONS NEAR THE TRANSITION TEMPERATURE IN STRONTIUM TITANATE. /MODE STUDY OF THE STRESS INDUCED PH   06-1196
SOLID SOLUTIONS BASED ON BARIUM TITANATE.   04-0786
SOLUBILITY OF ALUMINUM IN LEAD ZIRCONATE TITANATE.   09-1479
SOME ASPECTS OF SEMICONDUCTING BARIUM TITANATE.   04-0758
OF POTASSIUM TANTALATE NIOBATE, LITHIUM NIOBATE, AND BISMUTH TITANATE) /ROOPTIC MATERIALS, SINGLE CRYSTALS   02-0073
ROPAGATION NEAR THE STRUCTURAL PHASE TRANSITION IN STRONTIUM TITANATE.   SOUND P   06-1155
SPECIFIC HEAT OF BARIUM TITANATE.   06-1071 5
SPONTANEOUS DESTRUCTION OF FERROCERAMICS. (BARIUM TITANATE)   04-0787
STATE OF CHROMIUM ADDED TO BARIUM METATITANATE.   04-0543
INTERACTION OF TANTALUM AND HAFNIUM IONS IN BARIUM AND LEAD TITANATE.   STATIC ELECTRIC QUADRUPOLE   04-0937
STATIC QUADRUPOLE INTERACTION OF SCANDIUM-44 IN BARIUM TITANATE.   04-0692
AND ITS IMPLICATIONS FOR THE PLANAR SPIN MODEL OF STRONTIUM TITANATE.   STATIC SCALING HYPOTHESIS   06-1223
STRESS INDUCED FERROELECTRICITY IN STRONTIUM TITANATE.   06-1104
LECTRIC PROPERTIES OF X POTASSIUM NIOBATE AND (1-X) BARIUM TITANATE.   STRUCTURAL DI- AND FERROE   04-0623
N STUDY OF FERROELECTRIC AND FERROMAGNETIC SUBSTANCES. (LEAD TITANATE) STRUCTURAL MICROELECTRON DIFFRACTIO   05-1082
A MOVING DOMAIN WALL. (MECHANISM OF A WALL MOTION IN BARIUM TITANATE)   STRUCTURE OF   04-0678
STRUCTURE OF 90 DEGREE DOMAIN WALLS IN BARIUM TITANATE.   04-0641
S WITH A PHOTOEMISSION MICROSCOPE (TUNGSTEN TRIOXIDE, BARIUM TITANATE). /F DOMAINS IN FERROELECTRIC CRYSTAL   04-0811
ECTRIC EFFECT IN THIN FILMS OF TRIGLYCINE SULFATE AND BARIUM TITANATE.   STUDY OF THE PYROEL   04-0686
ICE BASED ON THE SHADOW EFFECT. (NEUTRON DIFFRACTION, BARIUM TITANATE) /MALL DEFORMATIONS OF A CRYSTAL LATT   04-0534
SURFACE BARRIER ENERGIES ON STRONTIUM TITANATE.   06-1175
SURFACE BREAKDOWN IN VACUUM ON BARIUM TITANATE.   04-0596
PRODUCED BY ABRASION AND ANNEALING OF POLYCRYSTALLINE BARIUM TITANATE.   SURFACE DOMAIN REORIENTATION   04-0639
ITANIA PHASE EQUILIBRIA AND ELECTRICAL PROPERTIES (LANTHANUM TITANATE).   SYSTEM LANTHANA T   36-3592
IC FIELD DEPENDENCE OF A HYDROGEN IMPURITY MODE IN STRONTIUM TITANATE.   (TEMPERATURE AND) ELECTR   06-1146
TEMPERATURE DEPENDENCE OF CARRIER MOBILITY IN BARIUM TITANATE.   04-0614
ENDENCE OF ELECTRONIC POLARIZABILITY OF STRONTIUM AND BARIUM TITANATE.   TEMPERATURE DEP   04-0738
RATURE DEPENDENCE OF THE BRILLOUIN- RAMAN SPECTRUM OF BARIUM TITANATE.   TEMPE   04-0807
MPERATURE DEPENDENCE OF THE DIELECTRIC CONSTANT OF STRONTIUM TITANATE.   TE   06-1134
DEPENDENCE OF THE FIRST ORDER ELASTIC CONSTANTS OF STRONTIUM TITANATE.   TEMPERATURE   06-1160
TEMPERATURE DEPENDENT OPTICAL PHONONS IN LEAD TITANATE.   05-1083
TEMPERATURE FACTORS OF TETRAGONAL AND CUBIC BARIUM TITANATE.   04-0532
A POINT DIPOLE ELECTROOPTIC MODEL BY AN APPLICATION TO LEAD TITANATE.   TESTING   05-1073
RON(3+) - VACANCY CENTER IN THE TETRAGONAL PHASE OF STRONTIUM TITANATE.   THE I   06-1234
THEORETICAL DYNAMIC STUDY OF CUBIC PHASE BARIUM TITANATE.   03-0261
THEORY OF BARIUM TITANATE.   03-0483
LECTROSTRICTION IN CUBIC CRYSTALS WITH APPLICATION TO BARIUM TITANATE.   THEORY OF E   03-0267
THEORY OF THE DIELECTRIC CONSTANTS OF BARIUM TITANATE.   04-0776
ONIC PROPAGATION NEAR THE STRUCTURAL TRANSITION OF STRONTIUM TITANATE.   THERMAL EXPANSIVITY AND ULTRAS   06-1135
RMAL HYSTERESIS IN THE TETRAGONAL CUBIC TRANSITION OF BARIUM TITANATE.   THE   04-0921
THERMAL STUDIES OF CALCIUM(X) BARIUM(1-X) TITANATE.   04-0694
THERMODYNAMICS OF FERROELECTRICITY IN BISMUTH TITANATE.   08-1297
THERMODYNAMICS OF FERROELECTRICITY IN BISMUTH TITANATE.   08-1419
THERMOSTABILIZING PROPERTIES OF SEMICONDUCTING BARIUM TITANATE.   04-0570
ENCE OF THE DIELECTRIC PERMITTIVITY OF HYDROXYL DOPED BARIUM TITANATE.   TIME DEPEND   04-0550
TIME EFFECTS IN THE HYSTERESIS LOOP OF (LEAD, STRONTIUM) TITANATE.   05-1052
TIME EFFECTS IN THE HYSTERESIS LOOP OF LEAD STRONTIUM TITANATE.   05-1053
DEVELOPED BY SANDIA LABORATORIES. (LEAD LANTHANUM ZIRCONATE TITANATE)   TRANSPARENT CERAMIC   09-1476
ULTRASONIC DISPERSION IN STRONTIUM TITANATE.   06-1239
RACTION PARAMETERS AT THE DISPLACIVE TRANSITION IN STRONTIUM TITANATE. /OPAGATION, STRESS EFFECTS, AND INTE   06-1127
TRASONIC STUDIES OF FIELD INDUCED PHASE TRANSITION IN BARIUM TITANATE.   UL   04-0549
ULTRASONIC VELOCITY ABOVE TANTALUM IN STRONTIUM TITANATE.   06-1137
VACUUM EVAPORATION OF BARIUM TITANATE.   04-1001
VARIATION OF THE DECAY CONSTANT OF STRONTIUM-85 IN BARIUM TITANATE.   04-0876
VIBRATIONAL AMPLITUDES IN FERROELECTRIC TETRAGONAL BARIUM TITANATE.   04-0726
N THE TEMPERATURE DEPENDENCE OF THE SOFT FREQUENCY OF BARIUM TITANATE.   VIBRONIC EFFECT I   04-0662
NTHETIC METHOD OF MAKING PURE, EXACTLY STOICHIOMETRIC BARIUM TITANATE.   WET SY   04-0790

X-RAY AND NEUTRON DIFFRACTION STUDY OF TETRAGONAL BARIUM TITANATE. 04-0713
X-RAY AND NEUTRON DIFFRACTION STUDY OF TETRAGONAL BARIUM TITANATE. 04-0740
X-RAY AND THERMOLUMINESCENCE OF BARIUM TITANATE. 04-0750
OMETRY OF LOW TEMPERATURE PHASE TRANSFORMATIONS IN STRONTIUM TITANATE. X-RAY DIFFRACT 06-1156
X-RAY INVESTIGATION OF IRRADIATED LEAD TITANATE. 05-1081
HE EFFECT OF HYDROSTATIC PRESSURE ON THE STRUCTURE OF BARIUM TITANATE. X-RAY INVESTIGATION OF T 04-0760
SEMICONDUCTING AND NONSTOICHIOMETRIC BARIUM TITANATE. A REFERENCE GUIDE. 01-0022
DOUBLE SPACE CHARGE INJECTION IN SOLIDS. (USING BARIUM TITANATE ABOVE CURIE POINT) 04-0638
CRITICAL ASYMMETRY IN LOCAL FLUCTUATIONS IN STRONTIUM TITANATE ABOVE THE CRITICAL TEMPERATURE. 06-1166
CRITICAL DYNAMICS OF STRONTIUM TITANATE ABOVE THE CRITICAL TEMPERATURE. 06-1204
FLUCTUATIONS AND CORRELATIONS IN STRONTIUM TITANATE ABOVE THE CRITICAL TEMPERATURE. 06-1233
ATURE, AS SH/ CRITICAL (POLARIZATION) FLUCTUATIONS IN BARIUM TITANATE (ABOVE THE FERROELECTRIC CURIE TEMPER 04-1032
MICROWAVE LOSSES IN STRONTIUM TITANATE ABOVE THE PHASE TRANSITION. 06-1199
DISPERSION AT MILLIMETER WAVELENGTHS IN BARIUM TITANATE ABOVE THE TRANSITION POINT. 04-0601
TEMPERATURE DEPENDENCE OF THE SOFT MODE IN STRONTIUM TITANATE ABOVE THE 105 DEG K TRANSITION. 06-1178
INDUCED PHASE TRANSITION NEAR CURIE TEMPERATURE IN STRONTIUM TITANATE. (ABSTRACT) RAMAN STUDIES OF STRESS 06-1236
ASSOCIATED WITH THE 110 DEG K PHASE TRANSITION IN STRONTIUM TITANATE. (ABSTRACT) SPECIFIC HEAT ANOMALY 06-1226
TRANSPORT PROPERTIES IN REDUCED STRONTIUM TITANATE. (ABSTRACT) 06-1152
RITICAL PHENOMENA IN INSULATING AND SEMICONDUCTING STRONTIUM TITANATE. (ABSTRACT) ULTRASONIC STUDIES OF C 06-1118
/AMAGNETIC RESONANCE SPECTRUM OF MOLYBDENUM(5+) IN STRONTIUM TITANATE: AN EXAMPLE OF THE DYNAMIC JAHN TELL/ 06-1125
CIMETER WAVELENGTHS EMPLOYING A SYMMETRIC STRIPLINE. (BARIUM TITANATE AND BARIUM CALCIUM TITANATE) / AND DE 04-0855
TRATE, LITHIUM AMMONIUM TARTRATE, AND MIXED CRYSTALS, BARIUM TITANATE, AND BARIUM STRONTIUM TITANATE) / TAR 02-0089
/RROELECTRIC SYSTEMS. PART-2: FERROELECTRICS- SILVER (BARIUM TITANATE AND BARIUM TITANATE- MAGNESIUM STANN/ 04-0816
/RESSURE ON PHASE TRANSITIONS AT LOW TEMPERATURES: STRONTIUM TITANATE AND ((BARIUM(X) STRONTIUM(1-X) TITAN/ 04-0898
/IC CERAMICS AT LOW TEMPERATURES. (POLYCRYSTALLINE STRONTIUM TITANATE AND BARIUM(X) STRONTIUM(1-X) TITANAT/ 04-0720
/D, IN STRONG ELECTRIC FIELDS. (ANISOTROPY DEVELOPED, BARIUM TITANATE AND BARIUM(0.95) CALCIUM(0.05) TITAN/ 04-0661
ND PHASE TRANSITIONS OF THE FERROELECTRIC PEROVSKITES: LEAD TITANATE AND BISMUTH TITANATE. /C PROPERTIES A 04-1434
NONLINEAR OPTICAL PROPERTIES OF LEAD TITANATE AND BISMUTH TITANATE CRYSTALS. 04-0941
/N TUNNELING AND BAND STRUCTURE OF (NIOBIUM DOPED) STRONTIUM TITANATE AND (CALCIUM DOPED) POTASSIUM TANTAL/ 06-1222
CHARACTERISTICS OF PHOTOCHROMIC CALCIUM FLUORIDE, STRONTIUM TITANATE, AND CALCIUM TITANATE CRYSTALS. /DING 06-1119
RS) RELATIONSHIP BETWEEN THE BREAKDOWN MECHANISM IN BARIUM TITANATE AND COLORATION. (INJECTED COLOR CENTE 04-0944
IRROR MICROSCOPIC ASPECTS OF FERROELECTRIC DOMAINS OF BARIUM TITANATE AND DICALCIUM STRONTIUM PROPIONATE. / 04-0973
EROVALENT SUBSTITUTION AND OXYGEN NONSTOICHIOMETRY OF BARIUM TITANATE AND HEMATITE SINGLE CRYSTALS. HET 04-0874
D OPTICAL ABSORPTION OF TRANSITION ELEMENT IONS IN STRONTIUM TITANATE AND LANTHANUM ALUMINATE. /ESONANCE AN 06-1165
CRYSTALS, PART-2, INTERACTION WITH ELASTIC STRAIN (STRONTIUM TITANATE AND LANTHANUM ALUMINATE). /ITE- TYPE 03-0241
TE(0.5), BISMUTH FERRATE AND ITS SOLID SOLUTIONS WITH BARIUM TITANATE, AND LANTHANUM FERRATE) /E(0.5) NIOBA 08-1368
O WAVE SHOCK STRUCTURES IN THE FERROELECTRIC CERAMICS BARIUM TITANATE AND LEAD ZIRCONATE TITANATE. TW 04-0916
TRIC CERAMICS. (REVIEW, 31 REFERENCES) AGING OF BARIUM TITANATE AND LEAD ZIRCONATE TITANATE FERROELEC 02-0062
/TH PEROVSKITE TYPE STRUCTURE. (MIXED OXIDES ABO(3)) (BARIUM TITANATE AND LITHIUM NIOBATE FILMS ON PLATINU/ 04-0914
A NEW METHOD OF OPTICALLY READING DOMAINS IN BISMUTH TITANATE AND MEMORY APPLICATIONS. 08-1298
DERED "NUCLEI", PHASE STRUCTU/ DIELECTRIC SPECTRUM OF BARIUM TITANATE. (AND MICROSCOPIC MODEL OF LOCALLY OR 04-0647
ITATION OF CITRATES IN AN ALCOHOL. PREPARATION OF BARIUM TITANATE AND OTHER CERAMIC POWDERS BY COPRECIP 04-0861
SEE BOTH LEAD LANTHANUM ZIRCONATE TITANATE AND PLZT.
RAMICS IN COMPUTERS AND OTHER DEVICES. (LEAD ZIRCONATE- LEAD TITANATE AND PLZT) USE OF FERROELECTRIC CE 09-1532
/S OF OPTICAL HARMONIC GENERATION IN CADMIUM SULFIDE, BARIUM TITANATE, AND POTASSIUM DIHYDROGEN PHOSPHATE / 04-0852
CRITICAL NEUTRON SCATTERING IN STRONTIUM TITANATE AND POTASSIUM MANGANESE TRIFLUORIDE. 06-1208
E OF SILICON, QUARTZ AND FERROELECTRIC PEROVSKITES). (BARIUM TITANATE AND POTASSIUM NIOBATE) /CRYSTALS (CAS 04-0629
STRUCTURE DISORDER AND RAMAN SCATTERING IN BARIUM TITANATE AND POTASSIUM NIOBATE. 04-0676
LECTRIC SOLID SOLUTION. (THEORY, APPLIED TO BARIUM STRONTIUM TITANATE AND POTASSIUM NIOBATE TANTALATE) /ROE 03-0260
CTRIC FIELD INDUCED OPTICAL HARMONIC GENERATION IN STRONTIUM TITANATE AND POTASSIUM TANTALATE. ELE 06-1203
IN PEROVSKITES (3 PHONON INTERACTION PROPOSED FOR STRONTIUM TITANATE AND POTASSIUM TANTALATE). /NDUCTIVITY 06-1095
/FERROELECTRIC- ACOUSTIC) IN THE PHONON SPECTRA OF STRONTIUM TITANATE AND POTASSIUM TANTALATE (LATTICE DYN/ 06-1157
SEE BOTH LEAD ZIRCONATE TITANATE AND PZT.
LYBDATE, GR/ GROWTH AND EVALUATION OF SINGLE CRYSTAL BISMUTH TITANATE AND RELATED COMPOUNDS. (GADOLINIUM MO 08-1323
NEUTRON DIFFRACTION STUDIES OF FERROELECTRICS. (BARIUM TITANATE AND ROCHELLE SALT AS EXAMPLES) 04-0679
/C PROPERTIES OF SOME POLYCRYSTALLINE FERROELECTRICS (BARIUM TITANATE AND SAME WITH 1 PERCENT SILICON REPL/ 04-0818
/CT CHELATOMETRIC DETERMINATION OF LEAD AND TITANIUM IN LEAD TITANATE AND SIMILAR COMPOUNDS. STUDIES ON AN/ 05-1068
IC CERAMICS FOR ELECTROOPTICAL APPLICATIONS. (LEAD ZIRCONATE TITANATE AND SODIUM POTASSIUM NIOBATE) /ELECTR 09-1527
FORCE CONSTANTS IN BARIUM TITANATE AND STRONTIUM TITANATE. 04-0748
TEMPERATURE DEPENDENCE OF THE RAMAN CROSS SECTIONS IN BARIUM TITANATE AND STRONTIUM TITANATE. 04-0548
SIZE AND SHAPE ON THE INFRARED ABSORPTION SPECTRA OF BARIUM TITANATE AND STRONTIUM TITANATE POWDERS. /ICLE 04-0823
/RAY INVESTIGATION OF THE PEROVSKITE MODIFICATION OF CADMIUM TITANATE (AND THE ANTIFERROELECTRIC- FERROELE/ 08-1387
/OELECTRIC COMPOUNDS (POTASSIUM OR SODIUM)(0.5) BISMUTH(0.5) TITANATE AND THE HIGH TEMPERATURE PHASE TRANS/ 08-1362
DIFFUSION POTENTIALS IN BARIUM TITANATE AND THE THEORY OF PTC MATERIALS. 03-0352
/TRIC DISPERSION IN FERROELECTRICS OF VARIOUS TYPES. (BARIUM TITANATE AND TITANATE ZIRCONATE, TRIGLYCINE S/ 02-0108
RANSITIONS ON THE HYDROXYL STRETCHING VIBRATION IN STRONTIUM TITANATE AND TITANIUM DIOXIDE. /RE AND PHASE T 06-1209
SOME PROPERTIES OF THE THERMOCURRENTS IN BARIUM TITANATE AND TITANIUM DIOXIDE SINGLE CRYSTALS. 04-0622
COMPENSATION) OXYGEN STOICHIOMETRY OF DONOR DOPED BARIUM TITANATE AND TITANIUM DIOXIDE. (STOICHIOMETRIC 04-0664
EPTIBILITY INCREASE DURING POLARIZATION REVERSAL. (IN BARIUM TITANATE AND TRIGLYCINE SULFATE) SUSC 04-0981
S. REMARKS ABOUT SWITCHING IN BARIUM TITANATE AND TRIGLYCINE SULFATE SINGLE CRYSTAL 04-0915
L PROPERTIES OF SINGLE CRYSTALS OF SOLID SOLUTIONS OF BARIUM TITANATE AND ZINC OXIDE. ELECTROOPTICA 04-0953
R DEVICE (ERASABLE, NONDESTRUCTIVE READOUT, BASED ON BISMUTH TITANATE AND ZINC SELENIDE) . /IC PHOTOCONDUCTO 37-3646
MAGNESATE/ PHYSICAL PHENOMENA IN PIEZOCERAMICS BASED ON LEAD TITANATE AND ZIRCONATE. (SOLID SOLUTIONS LEAD 05-1063
PERATURE (5-300 DEG K). DIELECTRIC PROPERTIES OF STRONTIUM TITANATE (ANNEALED SINGLE CRYSTALS) AT LOW TEM 06-1201
ONDUCTIVITY IN FERROELECTRICS DUE TO SURFACE LAYERS. (BARIUM TITANATE, ANTIMONY SULFOIODIDE) /RINSIC PHOTOC 04-0681
/Y STUDIES OF THE PHASE TRANSITIONS IN ROCHELLE SALT, BARIUM TITANATE, ANTIMONY SULFOIODIDE, THIOUREA, TRI/ 02-0099
TION. OPTICAL ABSORPTION EDGE OF STRONTIUM TITANATE AROUND THE 105 DEGREES K PHASE TRANSI 06-1099
FLAME SPRAYED BARIUM TITANATE AS A CAPACITOR DIELECTRIC. 04-0773
ARIATION IN CERTAIN ELECTRICAL PROPERTIES OF SINTERED BARIUM TITANATE AS A FUNCTION OF GRAIN SIZE. V 04-0627
ENTS AND THE THICKNESS OF NONUNIFORM THIN SAMPLES (STRONTIUM TITANATE AS EXAMPLE). /RAY ABSORPTION COEFFICI 06-1179
ECTROSCOPY (ESCA). BAND STRUCTURE OF STRONTIUM TITANATE AS MEASURED BY X-RAY PHOTOELECTRON SP 06-1100
ULTRASONIC INTERFEROMETER WITH LEAD ZIRCONATE TITANATE AS TRANSDUCER. 09-1585
S. NUCLEATION PROCESS IN BARIUM TITANATE AT CUBIC- TETRAGONAL PHASE TRANSITION 04-0754
NUCLEATION AND GROWTH OF FERROELECTRIC DOMAINS IN BARIUM TITANATE AT FIELDS FROM 2 TO 450 KV/CM. 04-0979
/ECTROOPTIC STUDIES OF THE POLARIZATION PROCESS IN STRONTIUM TITANATE AT LOW TEMPERATURE. (BIREFRINGENCE, / 06-1117
TIVITY, SPONTANEOUS POLARIZATION, COERCIVE FIELD) OF BISMUTH TITANATE AT LOW TEMPERATURE (5-350K). /(PERMIT 08-1436
DIELECTRIC PROPERTIES OF STRONTIUM TITANATE AT LOW TEMPERATURES. 06-1200
ELECTRON PARAMAGNETIC RESONANCE OF IRON(3+) IN BARIUM TITANATE AT LOW TEMPERATURES. 04-0927
ELECTRONIC CONDUCTION IN SLIGHTLY REDUCED STRONTIUM TITANATE AT LOW TEMPERATURES. 06-1151
FIELD INDUCED SWITCHING OF TETRAGONAL DOMAINS IN STRONTIUM TITANATE AT LOW TEMPERATURES. 06-1184

ENDENCE, ANISOTROPY)   DIELECTRIC PROPERTIES OF STRONTIUM TITANATE AT LOW TEMPERATURES. (TEMPERATURE DEP 06-1202
).   ATOMIC DISPLACEMENTS IN PEROVSKITE STRONTIUM TITANATE (AT OPTICAL FREQUENCIES, CALCULATIONS 06-1142
.   DISTORTION ENHANCED OPTICAL ABSORPTION IN STRONTIUM TITANATE AT THE CUBIC TO TETRAGONAL TRANSITION 06-1109
                          ULTRASONIC PROPERTIES OF STRONTIUM TITANATE AT THE 105 DEGREE K TRANSITION.   06-1191
NCE / CHANGE IN THE DECAY CONSTANT OF ZIRCONIUM-89 IN BARIUM TITANATE (BARIUM CHLORIDE FLUX GROWTH IN PRESE 04-0685
ONZE FIELD IN THE PSEUDOTERNARY SYSTEM SODIUM NIOBATE BARIUM TITANATE BARIUM DINIOBATE.   TUNGSTEN BR 13-1970
/IC GENERATION COEFFICIENTS OF PIEZOELECTRIC CRYSTALS. (LEAD TITANATE, BARIUM STRONTIUM NIOBATE, LEAD NIOB/ 05-1066
/OPERTIES OF POLYCRYSTALLINE SOLID SOLUTIONS BASED ON BARIUM TITANATE. (BARIUM STRONTIUM TITANATE, BARIUM / 04-1019
/IC ANISOTROPY OF A POLARIZED FERROELECTRIC CERAMIC. (BARIUM TITANATE, BARIUM TIN TITANATE, LEAD BARIUM DI/ 04-0767
              ENERGY STORAGE IN CERAMIC DIELECTRICS. (STRONTIUM TITANATE, BARIUM TITANATE)   04-0602
T OF OXYGEN DIFFUSION CONSTANT IN SOME TITANATES. (STRONTIUM TITANATE, BARIUM TITANATE)   MEASUREMEN 04-1033
/ENT AND DISCUSSION OF BARIUM (OR LEAD OR CALCIUM) STRONTIUM TITANATE, BARIUM TITANATE ZIRCONATE (OR STANN/ 08-1281
/ING POLARIZATION SWITCHING TIME, LEAD(0.35) STRONTIUM(0.65) TITANATE, BARIUM(0.65 OR 0.75) STRONTIUM(0.35/ 37-3660
TITANATE.   PREPARATION OF THE TERNARY TITANATE BARIUM(0.8) LEAD(0.12) CALCIUM(0.08)   36-3587
                     SINGLE FIRE PROCESS FOR BARIUM TITANATE BARRIER LAYER DIELECTRICS.   04-1029
OF THE THEORY OF TRAN/ TUNNELING RESULTS ON INDIUM STRONTIUM TITANATE BARRIERS INTERPRETED AS CONFIRMATION 03-0233
ON THE DOMAIN STRUCTURE OF SINGLE CRYSTALS FORMED BY BARIUM TITANATE BASE SOLID SOLUTION. /XTERNAL ACTIONS 04-0954
/NDENCE OF LEED INTENSITIES FROM THE (001) SURFACE OF BARIUM TITANATE BETWEEN 20 DEGREES C AND 650 DEGREES/ 04-0531
         DISPERSION OF ELECTROOPTIC EFFECT IN BARIUM TITANATE. (BIREFRINGENCE, PRINCIPAL INDICES)   04-0755
/AND PHASE DIAGRAMS FOR BISMUTH(X) POTASSIUM(X) BARIUM(1-2X) TITANATE, BISMUTH(0.5) POTASSIUM(X) SODIUM(0./ 04-1289
              STUDIES ON THE BARIUM STRONTIUM TITANATE BOUNDARY LAYER CERAMIC DIELECTRICS.   04-1025
         STUDIES ON THE (STRONTIUM, CALCIUM) TITANATE BOUNDARY LAYER CERAMIC DIELECTRICS.   06-1247
                STUDIES ON STRONTIUM TITANATE BOUNDARY LAYER DIELECTRICS.   06-1248
/MPERATURE DEPENDENCE OF THE DEBYE- WALLER FACTORS OF BARIUM TITANATE (BRAGG REFLECTIONS C-AXIS, 20-180 DE/ 04-0892
.   DIFFUSION COEFFICIENT OF NIOBIUM IN BARIUM TITANATE BY A DIELECTRIC MEASUREMENT TECHNIQUE 04-0926
RATION OF POLED, TWIN-FREE CRYSTALS OF FERROELECTRIC BISMUTH TITANATE, (BY ANNEALING).   PPEPA 08-1352
ON.   BARIUM TITANATE BY CONVERGENT BEAM ELECTRON DIFFRACTI 04-0991
              CURIE SHIFT IN BARIUM TITANATE BY HETEROSUBSTITUTION.   04-0727
    DIRECT OBSERVATION OF 180 DEG DOMAINS IN BARIUM TITANATE BY MEANS OF A POLARIZING MICROSCOPE.   04-0952
/HANGE IN THE POLARIZATION OF SINGLE CRYSTAL FILMS OF BARIUM TITANATE BY MEANS OF A STROBOSCOPIC TRANSMISS/ 04-0942
   CHANGE IN THE DECAY CONSTANT OF TECHNETIUM-99 IN BARIUM TITANATE BY THE FERROELECTRIC TRANSITION.   04-0871
              HYDROTHERMAL PREPARATION OF BARIUM TITANATE BY TRANSPORT REACTIONS.   04-0626
ANIUM ESTERS TO YIELD SM/ HYDROTHERMAL PREPARATION OF BARIUM TITANATE BY TRANSPORT REACTIONS (INVOLVING TIT 04-0625
HIN FILMS: PREPARATION OF FERROELECTRIC THIN FILMS OF BARIUM TITANATE BY VACUUM EVAPORATION.   T 04-0834
       POLARITON DISPERSION OF TETRAGONAL FERROELECTRIC BARIUM TITANATE. (CALCULATIONS)   04-0842
/MAIN WALL CONTRIBUTION TO THE DIELECTRIC CONSTANT OF BARIUM TITANATE. (CALCULATIONS, SUMMARY ONLY, DISCUS/ 04-0804
               AGING IN COMPLEX CERAMICS BASED ON BARIUM TITANATE. (CAPACITANCE, PERMITTIVITY)   04-0724
ICONDUCTIVE TITANIUM DIOXID/ STUDIES OF FLAME SPRAYED BARIUM TITANATE CAPACITORS. PART-10: STRUCTURE OF SEM 04-0774
               DOUBLE INJECTION IN BARIUM TITANATE CERAMIC.   04-0554
RIZATION PROCESSES IN TRIGLYCINE SULFATE CRYSTALS AND BARIUM TITANATE CERAMIC. /THE POLARIZATION AND REPOLA 04-0994
    EFFECTS OF CALCINING ON THE FIRING OF LEAD ZIRCONATE TITANATE CERAMIC   09-1622
         FERROELECTRIC LOSSES IN (PURE) BARIUM TITANATE CERAMIC.   04-0555
         PIEZOELECTRIC EFFECT IN ALUMINUM DOPED BARIUM TITANATE CERAMIC.   04-0536
         REJUVENATION OF A BARIUM(0.72) STRONTIUM(0.28) TITANATE CERAMIC.   04-0817
E CHARACTERISTICS OF ROCHELLE TYPE CERAMICS. (BARIUM CALCIUM TITANATE CERAMIC) /THOD OF STUDYING TEMPERATUR 04-0905
                          BARIUM TITANATE CERAMIC: A NEW ELECTRON EMITTER.   04-0802
ESE IONS AND ITS APPLICATIONS.   A SEMICONDUCTING BARIUM TITANATE CERAMIC DOPED WITH NIOBIUM AND MANGAN 04-0734
IMPACT.   PIEZOELECTRIC CHARACTERISTIC OF BARIUM TITANATE CERAMIC EXHIBITED WHEN IT IS GIVEN AN 04-0881
IC PIEZOELECTRIC RANDOM ACCESS MEMORY. (LEAD ZIRCONATE- LEAD TITANATE CERAMIC MEMORY PLATE)   A FERROELECTR 04-1504
AND.   LEAD TITANATE CERAMIC RESONATORS OPERATING IN VHF B 05-1069
SION.   FORMATION OF BARIUM TITANATE CERAMIC SEMICONDUCTING LAYER BY DIFFU 04-0684
SETS OF ELECTRON SPIN RESONANCE OF GADOLINIUM(3+) IN BARIUM TITANATE CERAMIC SEMICONDUCTOR.   TWO 04-0990
/ BOUNDARY BARRIER LAYERS IN (N TYPE, ANTIMONY DOPED) BARIUM TITANATE CERAMIC WITH A HIGH EFFECTIVE DIELEC/ 04-0581
              DC BIAS EFFECTS IN BARIUM TITANATE CERAMICS.   04-0566
              DC BIAS EFFECTS IN LEAD ZIRCONATE TITANATE CERAMICS.   09-1490
TITANIA ON THE ELECTRIC RESISTANCE OF SEMICONDUCTIVE BARIUM TITANATE CERAMICS.   EFFECT OF EXCESS 04-0966
AIN SIZE ON THE RESISTIVITY ANOMALY IN SEMICONDUCTIVE BARIUM TITANATE CERAMICS.   EFFECT OF GR 04-0735
F INTERNAL BOUNDARIES ON THE DIELECTRIC PROPERTIES OF BARIUM TITANATE CERAMICS.   EFFECT O 04-0891
EUTRON IRRADIATION ON THE FERROELECTRIC PROPERTIES OF BARIUM TITANATE CERAMICS.   EFFECT OF N 04-0732
    EFFECTS OF CALCINING ON SINTERING OF LEAD ZIRCONATE TITANATE CERAMICS.   09-1494
CTS OF GRAIN GROWTH ON THE DISTRIBUTION OF NIOBIUM IN BARIUM TITANATE CERAMICS.   EFFE 04-0595
MPURITIES ON THE MECHANICAL QUALITY FACTOR OF LEAD ZIRCONATE TITANATE CERAMICS.   EFFECTS OF I 09-1608
FFECTS OF RADIATION INDUCED DAMAGE CENTERS IN LEAD ZIRCONATE TITANATE CERAMICS.   E 09-1520
    EFFECTS OF SILICA ON MANGANESE DOPED STRONTIUM TITANATE CERAMICS.   06-1172
    ELECTRICAL PROPERTIES OF RUBIDIUM AND CESIUM DOPED BARIUM TITANATE CERAMICS.   04-0860
    ELECTROMECHANICAL PROPERTIES AND APPLICATION OF LEAD TITANATE CERAMICS.   05-1085
LECTROSTRICTIVELY GENERATED AND MODULATED FRACTURE IN BARIUM TITANATE CERAMICS.   E 04-0700
EMISSION ELECTRON MICROSCOPE (BARRIER LAYERS IN DOPED BARIUM TITANATE CERAMICS). /F SOLID SURFACES) IN THE 04-0913
FLUENCE OF VIBRATIONS ON THE DIELECTRIC PROPERTIES OF BARIUM TITANATE CERAMICS.   IN 04-0890
              INFLUENCE OF WATER VAPOR ON BARIUM TITANATE CERAMICS.   04-0882
HANICAL AND MECHANICAL PROPERTIES OF FERROELECTRICS. (BARIUM TITANATE CERAMICS)   NONLINEARITY OF ELECTROMEC 04-0986
              PIEZOELECTRIC PROPERTIES OF LEAD TITANATE CERAMICS.   05-1062
ION AND STUDY OF THE INNER BARRIERS IN SEMICONDUCTING BARIUM TITANATE CERAMICS.   PRODUCT 04-0583
    SINTERING AND FERROELECTRIC PROPERTIES OF LEAD ZIRCONATE TITANATE CERAMICS.   09-1481
         SPACE CHARGE POLARIZATION AND AGING OF BARIUM TITANATE CERAMICS.   04-0880
PHASE TRANSITION TETRAGONAL- RHOMBOHEDRAL IN LEAD ZIRCONATE TITANATE CERAMICS. /ABILIZATION EFFECTS AT THE 09-1611
THE DIELECTRIC CONSTANT OF FERROELECTRIC SUBSTANCES. (BARIUM TITANATE CERAMICS)   TEMPERATURE DEPENDENCE OF 04-0591
THE ROLE OF IRON OXIDE ADULTERANT IN A LEAD ZIRCONATE LEAD TITANATE CERAMICS.   09-1592
TENTIAL DISTRIBUTION (STABILITY) IN BARIUM(1-X) STRONTIUM TITANATE CERAMICS. /IBRATING WIRE" STUDY OF PO 04-0699
         X-RAY ANALYSIS OF MODIFIED LEAD ZIRCONATE- LEAD TITANATE CERAMICS.   09-1619
NFLUENCE OF HUMIDITY ON THE POTENTIAL DISTRIBUTION IN BARIUM TITANATE CERAMICS AND CRYSTALS.   I 04-0883
URES FOR MAXIMUM DIELECTRIC CON/ INTERNAL STRESSES IN BARIUM TITANATE CERAMICS. (AND DIFFERENCE IN TEMPERAT 04-0849
ROELECTRIC PROPERTIES OF (INSULATING, BOUNDARY LAYER) BARIUM TITANATE CERAMICS AT GRAIN BOUNDARY.   FER 04-1024
ADMIUM TR/ CHANGES IN THE FERROELECTRIC PROPERTIES OF BARIUM TITANATE CERAMICS CAUSED BY IRRADIATION WITH C 04-0731
PURITIES.   SPACE CHARGE EFFECT IN LEAD ZIRCONATE TITANATE CERAMICS CAUSED BY THE ADDITION OF IM 09-1609
         HOT PRESSED LEAD ZIRCONATE TITANATE CERAMICS CONTAINING BISMUTH.   09-1525
    ELECTRICAL RESISTIVITY OF LEAD ZIRCONATE TITANATE CERAMICS CONTAINING IMPURITIES.   09-1607
ANESE.   ELECTROMECHANICAL PROPERTIES OF LEAD TITANATE CERAMICS CONTAINING LANTANUM AND MANG 05-1060
MANGANESE/ POLARIZATION AND PIEZOELECTRIC PROPERTIES OF LEAD TITANATE CERAMICS. (CONTAINING 1 MOLE PERCENT 05-1061
RABILITY.   EFFECT OF CLAY AND DOLOMITE ADDITIONS ON BARIUM TITANATE CERAMICS. CRYSTAL STRUCTURE AND SINTE 04-0739

/F ADDITIVES ON PIEZOELECTRIC AND RELATED PROPERTIES OF LEAD TITANATE CERAMICS. (DIELECTRIC CONSTANT, RESI/ 05-1084
STABILIZED BARIUM TITANATE CERAMICS FOR CAPACITOR DIELECTRICS. 04-0825
ONS. HOT PRESSED FERROELECTRIC LEAD ZIRCONATE TITANATE CERAMICS FOR ELECTROOPTICAL APPLICATI 09-1524
/TS OF HIGH ELECTRIC FIELDS ON MODIFIED LEAD ZIRCONATE- LEAD TITANATE CERAMICS FOR PIEZOELECTRIC APPLICATI/ 09-1485
URE OF THE SC/ DEPENDENCE OF DIELECTRIC PROPERTIES OF BARIUM TITANATE CERAMICS ON THE PHYSICO- CHEMICAL NAT 04-0789
DISTRIBUTION OF POTENTIAL ALONG BARIUM TITANATE CERAMICS PREVIOUSLY POLARIZED. 04-0788
INTERNAL FRICTION OF MODIFIED LEAD ZIRCONATE- LEAD TITANATE CERAMICS. (PROPRIETARY COMPOSITIONS) 09-1623
ERSE COMPRESSI/ PIEZOELECTRIC PROPERTIES OF POLARIZED BARIUM TITANATE CERAMICS SUBJECTED TO UNIAXIAL TRANSV 04-0962
MICROSCOPE. SURFACE STUDIES OF BARIUM TITANATE CERAMICS USING THE SCANNING ELECTRON 04-0567
ILE STRAI/ FERROELASTIC BEHAVIOR OF LEAD LANTHANUM ZIRCONATE TITANATE CERAMICS WHEN SUBJECTED TO LARGE TENS 09-1571
N BOUNDARY ELEMENT. LANTHANUM DOPED BARIUM TITANATE CERAMICS WITH BISMUTH OXIDE AS A GRAI 04-0799
T. PART-1: EFFECTS OF ADDITION OF GLASSES CONTAINING/ BARIUM TITANATE CERAMICS WITH HIGH DIELECTRIC CONSTAN 04-0829
IMPROVED AGING AND SWITCHING OF LEAD ZIRCONATE- LEAD TITANATE CERAMICS WITH INDIUM ELECTRODES. 09-1516
IMPROVED AGING AND SWITCHING OF LEAD ZIRCONATE- LEAD TITANATE CERAMICS WITH INDIUM ELECTRODES. 09-1517
INTERCRYSTALLITE LAYERS IN THE SEMICONDUCTOR CERAMIC BARIUM TITANATE. (CERIUM DOPED) / OF CRYSTALLITES AND 04-0568
EFFECT OF OXIDE ADDITIONS ON THE SINTERING OF BARIUM TITANATE COMPACT 04-0996
/GETIC) ELECTRON AND (COBALT-60) GAMMA-IRRADIATION ON BARIUM TITANATE, (COMPARED TO STUDY DAMAGE MECHANISM) 04-0970
TEMPERATURE FOR FERROELECTRICS AND ANTIFERROELECTRICS BARIUM TITANATE, COMPLEX LAYER OXIDES) /NT AND CURIE 03-0188
. POLARIZATION OF SOLID BARIUM(X) STRONTIUM(1-X) TITANATE COMPOUNDS BETWEEN 4 AND 100 DEGREES K 04-0719
IONIC CONTRIBUTION TO INJECTION AND CONDUCTION IN BARIUM TITANATE (CONDENSER CHARGE- DISCHARGE METHOD). 04-0571
EFFECT OF THE 110 DEGREE K PHASE TRANSITION ON THE STRONTIUM TITANATE CONDUCTION BANDS. 06-1158
EPR INVESTIGATION OF POLYCRYSTALLINE STRONTIUM TITANATE CONTAINING MANGANESE DIOXIDE. 06-1162
ADDITIVES. DIELECTRIC PROPERTIES OF BARIUM TITANATE CONTAINING NIOBIUM AND THE EFFECT OF 04-0575
ANOMALY OF RESISTANCE IN REDUCED BARIUM TITANATE CRYSTAL. 04-0988
ABLE POWER OF A PYROELECTRIC THERMAL RECEIVER. (USING BARIUM TITANATE CRYSTAL) MINIMUM DETECT 04-0632
THEORY OF POLARIZATION REVERSAL OF BARIUM TITANATE CRYSTAL. 03-0326
90 DEGREE TYPE DOMAIN WALL OF BARIUM TITANATE CRYSTAL. (THEORY, CALCULATIONS) 03-0324
ANTIPARALLEL SWITCHING IN ALPHA DOMAIN BARIUM TITANATE CRYSTALS. 04-0655
CHEMICAL PROCESSES DURING HYDROGEN REDUCTION OF BARIUM TITANATE CRYSTALS. 04-0540
DIELECTRIC PROPERTIES AND DOMAIN TEXTURE OF BARIUM TITANATE CRYSTALS. 04-0537
APHIC STORAGE, ELECTRICAL FIXING AND ERASING IN DOPED BARIUM TITANATE CRYSTALS. HOLOGR 04-0848
LEED STUDIES ON THE (001) FACE OF BARIUM TITANATE CRYSTALS. 04-0530
OTOEMISSION MICROSCOPIC OBSERVATION OF THE SURFACE OF BARIUM TITANATE CRYSTALS. PH 04-0810
RECIPROCAL DOMAINS IN LEAD TITANATE CRYSTALS. 05-1048
SEM OBSERVATION OF THE FERROELECTRIC SURFACE OF BARIUM TITANATE CRYSTALS. 04-0812
SMALL POLARONS IN CONDUCTING BARIUM TITANATE CRYSTALS. 04-0613
SURFACE CONDUCTIVITY IN POLARIZED BARIUM TITANATE CRYSTALS. 04-0569
AIN STRUCTURE. GROWTH OF LEAD TITANATE CRYSTALS AND EXAMINATION OF THEIR DOM 05-1047
LOW THE FERROELECTRIC / CHANNELING OF PROTONS IN THIN BARIUM TITANATE CRYSTALS AT TEMPERATURES ABOVE AND BE 04-0687
OPE. DIRECT OBSERVATION OF BARIUM TITANATE CRYSTALS BY SCANNING ELECTRON MICROSC 04-0808
-6 KV/CM) POLARIZATION REVERSALS IN LIQUID ELECTRODED BARIUM TITANATE CRYSTALS (CALCULATIONS). / FIELD (1.5 04-0994
UORIDE). INVESTIGATION OF THE FINE STRUCTURE OF BARIUM TITANATE CRYSTALS FLUX GROWN FROM POTASSIUM FL 04-0708
/IMPURITIES AND GROWTH CONDITIONS ON THE STRUCTURE OF BARIUM TITANATE CRYSTALS GROWN FROM SOLUTION IN POTA/ 04-0598
/TIVITY (SEEBECK COEFFICIENT, CURIE POINT) OF REDUCED BARIUM TITANATE CRYSTALS (POLARON HOPPING INTERPRETA/ 04-0918
PULSES. LIGHT EMISSION FROM BARIUM TITANATE CRYSTALS SUBJECTED TO UNIAXIAL STRESS 04-0672
0-37/ SPACE CHARGE LIMITED CURRENT IN POLYCRYSTALLINE BARIUM TITANATE. (CURRENT VOLTAGE CHARACTERISTICS, 15 04-0557
PTICAL MODES OF THE PEROVSKITE STRUCTURE. (THEORY, STRONTIUM TITANATE DATA) O 03-0304
S). AGING PROCESSES IN BARIUM TITANATE (DECREASE IN NUMBER OF 180 DEG DOMAIN 04-0911
) OPTICAL GAP OF STRONTIUM TITANATE. (DEVIATION FROM URBACH TAIL BEHAVIOR 06-1111
ELECTRICAL PROPERTIES OF HIGH PURITY POLYCRYSTALLINE BARIUM TITANATE. (DIELECTRIC CONSTANT, CONDUCTIVITY) 04-0704
ANCE). SEMICONDUCTING BARIUM TITANATE. (DIELECTRIC CONSTANT, CONTACT RESIST 04-0730
G (CAPACITORS). ELECTRICAL PROPERTIES OF THICK FILM BARIUM TITANATE DIELECTRICS PRODUCED BY FLAME SPRAYIN 04-0772
PHATE PRESENTED, TRIGLYCINE SULFATE AND BARIUM AND STRONTIUM TITANATE DISCUSSED) /POTASSIUM DIHYDROGEN PHOS 03-0180
/ECTRIC FIELD EFFECT OF ELECTRON SPIN RESONANCE IN STRONTIUM TITANATE. (DOPED WITH GADOLINIUM(3+) OR IRON(/ 06-1229
DIELECTRIC PROPERTIES OF BARIUM TITANATE DOPED WITH TUNGSTEN OXIDE. 04-0696
D LATTICE DYNAMICS OF A THIN FILM OF A FERROELECTRIC (BARIUM TITANATE DOWN TO 3 MICRONS). /TRIC CONSTANT AN 04-0600
/ON PARAMAGNETIC RESONANCE SPECTRUM OF IRON(3+) IN STRONTIUM TITANATE DUE TO NEAREST NEIGHBOR CHARGE COMPE/ 06-1144
ED WITH DA/ (STEP OR SMOOTH) DOMAIN WALL STRUCTURE IN BARIUM TITANATE DURING SWITCHING (CALCULATIONS COMPAR 04-0677
/PMENTS IN ELECTROOPTIC CERAMICS (HOT PRESSED LEAD ZIRCONATE TITANATE, EFFECT OF GRAIN SIZE, DOPANTS, REVI/ 02-0086
THE ORTHORHOMBIC PHASE IN MODIFIED BARIUM TITANATE. (EFFECT OF SUBSTITUTIONS) 04-0617
(WEAK FIELD) DIELECTRIC (PERMITTIVITY) PROPERTIES OF BISMUTH TITANATE (EFFECT OF TEMPERATURE). 36-3555
FLUORINE CONTAINING BARIUM TITANATE. (ELECTRICAL PROPERTIES) 04-0869
OUNDARIES DURING SWITCHING IN SINGLE CRYSTAL FILMS OF BARIUM TITANATE (ELECTRON MICROSCOPY). /N OF DOMAIN B 04-0940
H AND PROPERTIES OF EPITAXIAL FILMS OF FERROELECTRIC BISMUTH TITANATE. (ELECTROOPTICS) GROWT 08-1329
ORDER PARAMETER AND PHASE TRANSITIONS OF STRESSED STRONTIUM TITANATE. (EPR OF IRON- VANADIUM PAIRS) 06-1170
TION BY OXYGEN(2-) VACANCIES IN CHROMIUM(3+) DOPED STRONTIUM TITANATE (ESR MEASUREMENTS, MODEL PROPOSED). / 06-1161
GROWTH AND STRUCTURE OF BARIUM TITANATE EVAPORATED FILMS. 04-0604
E. PHOTOCHROMIC IRON IN STRONTIUM TITANATE. EVIDENCE FROM PARAMAGNETIC RESONANC 06-1171
CAL DA/ OPTICALLY INDUCED REFRACTIVE INDEX CHANGES IN BARIUM TITANATE (EXPERIMENTS TO CHARACTERIZE THE OPTI 04-1004
STRUCTURAL BASIS OF FERROELECTRICITY IN THE BISMUTH TITANATE FAMILY. 18-1406
SOME PEROVSKITE LIKE LAYER TYPE FERROELECTRICS. (IN BISMUTH TITANATE FERRATES) /AUER STUDIES OF IRON-57 IN 13-2035
AND ELECTRICAL SWITCHING CHARACTERISTICS OF A LEAD ZIRCONATE TITANATE FERROELECTRIC CERAMIC. SOME PHYSICAL 09-1601
BARIUM ZIRCONATE LEAD TITANATE FERROELECTRIC CERAMICS. 09-1492
TIC APPLICATIONS. HOT PRESSED (LEAD, LANTHANUM) (ZIRCONATE, TITANATE) FERROELECTRIC CERAMICS FOR ELECTROOP 09-1526
N AND EFFECTS OF GRAIN BOUNDARY BLOCKIN/ DOPING PROBLEMS IN (TITANATE) FERROELECTRIC SEMICONDUCTORS. (ORIGI 02-0074
ASUREMENTS OF IONIC TRANSPORT NUMBERS IN PZT (LEAD ZIRCONATE TITANATE) FERROELECTRICS. /NCENTRATION CELL ME 09-1512
STORAGE MEDIUM FURTHER DEVELOPMENTS. BISMUTH TITANATE FERROELECTRIC- PHOTOCONDUCTOR OPTICAL 08-1371
/TERNAL FRICTION (AND CURIE TEMPERATURES) IN (LEAD ZIRCONATE TITANATE) FERROELECTRICS (WITH AND WITHOUT NI/ 09-1584
PHASE CHANGES IN TRIGLYCINE SULFATE AND BARIUM TITANATE FERROELECTRICS. (26 REFS) 04-0621
PREPARATION OF BARIUM TITANATE FILM BY EVAPORATION. 04-0743
PROBE X-RAY MICROANALYSIS ON IMPURITIES IN EVAPORATED BARIUM TITANATE FILMS. ELECTRON 04-0742
ARED REFLECTIVITY AND TRANSMISSION MEASUREMENTS ON STRONTIUM TITANATE FILMS. FAR INFR 06-1154
EXPERIMENTAL STUDY OF EVAPORATED BARIUM TITANATE FILMS. 04-0939
PREPARATION AND PROPERTIES OF THIN BARIUM TITANATE FILMS. 04-0669
ATINUM OR FUSED QUARTZ). PREPARATION OF THIN BARIUM TITANATE FILMS BY DC DIODE SPUTTERING (ONTO PL 04-0948
N ATMOSPHERE. SINGLE CRYSTAL BARIUM TITANATE FILMS GROWN FROM THE MELT IN AN OXYGE 04-0612
CHARACTERISTICS OF RF SPUTTERED BARIUM TITANATE FILMS ON SILICON. 04-0930
INDIUM ANTIMONIDE OR GALLIUM ARSENIDE. BARIUM TITANATE FILMS PREPARED BY RF SPUTTERING ONTO. 04-0741
STRUCTURE AND PROPERTIES OF BARIUM TITANATE FILMS PRODUCED BY CATHODE SPUTTERING. 04-0832
OF CARBON. CHEMICAL CHANGES IN BARIUM TITANATE FIRED UNDER NITROGEN IN THE PRESENCE 04-0872

SPONTANEOUS POLARIZATION AND COERCIVE FIELD OF LEAD TITANATE (FLUX GROWN SINGLE CRYSTALS). 05-1050
NDING CALCULATION. ENERGY BAND STRUCTURE OF STRONTIUM TITANATE FROM A SELF CONSISTENT FIELD TIGHT BI 06-1221
IDE IN THE PRESENCE OF SODIUM CARB/ THE PRODUCTION OF BARIUM TITANATE FROM BARIUM SULFATE AND TITANIUM DIOX 04-0624
/ CERAMICS CONTAINING FERROELECTRIC CRYSTALS (BARIUM OR LEAD TITANATE, FROM SYSTEMS BARIA OR LITHIA, TITAN/ 04-0783
IN WALLS IN FERROELECTRIC AND FERROELASTIC CRYSTALS. (BARIUM TITANATE, GADOLINIUM MOLYBDATE) /KNESS OF DOMA 04-0777
/ANIUM ON THE STRUCTURE AND PROPERTIES OF FERROELECTRIC LEAD TITANATE (GERMANATE SOLID SOLUTIONS TO 7 OR 8/ 04-1043
THREE APPLICATION AREAS FOR STRONTIUM TITANATE GLASS CERAMICS. 06-1150
APPLICATION OF DIELECTRIC MIXTURE FORMULAS TO (LEAD TITANATE) GLASS- CERAMIC SYSTEMS. 05-1054
/ROELECTRIC SINGLE CRYSTALS WITH LARGE NONLINEARITY. (BARIUM TITANATE HAFNIUM GROWN FROM POTASSIUM FLUORID/ 08-1373
LASTIC SCATTERING STUDY OF THE LATTICE DYNAMICS OF STRONTIUM TITANATE: HARMONIC MODELS. NEUTRON INE 06-1225
STUDY OF CRYSTALS OF BARIUM TITANATE HAVING PARALLEL INTERFERENCE FRINGES. 04-0897
BAND STRUCTURE OF THE CUBIC PHASE OF BARIUM TITANATE. (IMAGINARY DIELECTRIC CONSTANT) 04-0845
REACTION SEQUENCE IN FORMING BARIUM METATITANATE IN AN INDUSTRIAL ROTARY FURNACE. 04-1003
DIELECTRIC AGING IN TETRAGONAL SOLID SOLUTIONS OF CALCIUM TITANATE IN BARIUM TITANATE. 08-1295
SOLUBILITY LIMIT OF LANTHANUM(0.5) LITHIUM(0.5) TITANATE IN BARIUM TITANATE. 04-0889
LECTRON MICROSCOPIC STUDY OF FERROELECTRIC DOMAINS OF BARIUM TITANATE IN LOW TEMPERATURE PHASES. E 04-1034
FREQUENCIES (MICROWAVES). FERROELECTRIC PROPERTIES OF BARIUM TITANATE IN THE CENTIMETER RANGE OF SUPERHIGH 04-0901
N. ABSORPTION AND REFLECTION SPECTRA OF BARIUM TITANATE IN THE LONG WAVELENGTH INFRARED REGIO 04-0644
SFORMATION. EPR STUDIES AND DYNAMICS IN STRONTIUM TITANATE IN THE NEIGHBORHOOD OF THE PHASE TRAN 06-1232
DIELECTRIC DISPERSION OF IRRADIATED BARIUM TITANATE IN THE PHASE TRANSITION REGION. 04-0652
/ATION ABSORPTION OF ULTRASOUND IN SINGLE CRYSTALS OF BARIUM TITANATE IN THE PRESENCE OF A STEADY ELECTRIC/ 04-0850
SES. ELECTRON PARAMAGNETIC RESONANCE OF IRON(3+) IN BARIUM TITANATE IN THE RHOMBOHEDRAL AND THE CUBIC PHA 04-0929
/URES OF SOME FERROELECTRICS AND ANTIFERROELECTRICS. (BARIUM TITANATE, KDP, POTASSIUM SODIUM TARTRATE, TGS/ 02-0134
AL PHASE TRANSITIONS IN PEROVSKITE TYPE CRYSTALS. (BARIUM TITANATE, LANTHANUM ALUMINATE) /RY OF STRUCTUR 03-0424
NORMAL VIBRATIONS OF STRONTIUM TITANATE LATTICE. 06-1128
INELASTIC NEUTRON SCATTERING FROM SINGLE DOMAIN BARIUM TITANATE. (LATTICE DYNAMICS) 04-0946
MOSSBAUER STUDIES OF LOWER TRANSITIONS IN BARIUM TITANATE. (LATTICE INSTABILITIES) 04-0564
S) INCORPORATION OF ANTIMONY INTO THE BARIUM TITANATE LATTICE. (STRUCTURE, AND PCT RESISTOR 04-0934
E PROPERTIES OF A SERIES OF FERROELECTRIC MATERIALS. (BARIUM TITANATE, LEAD FERRATE(0.5) NIOBATE(0.5)) / TH 04-0780
TIONS IN ANTIFERROELECTRIC REGION OF THE SYSTEM LEAD HAFNATE TITANATE, LEAD STANNATE NIOBATE. SOLID SOLU 13-1956
D/ SOLVENT ZONE GROWTH OF FERROELECTRIC CRYSTALS. (STRONTIUM TITANATE, LEAD(X) STRONTIUM(1-X) TITANATE, LEA 05-1042
FERROELECTRIC DISPLAYS. (BISMUTH TITANATE, LEAD ZIRCONATE TITANATE) 09-1515
NATURE OF AGING IN FERROELECTRIC CERAMICS. (BARIUM TITANATE, LEAD ZIRCONATE TITANATE) 04-0757
/AMIC FERROELECTRICS ON THEIR MECHANICAL PROPERTIES. (BARIUM TITANATE, LEAD ZIRCONATE TITANATE, LEAD METAN/ 04-0985
/LUTE SIGNS OF MICROWAVE NONLINEAR SUSCEPTIBILITIES. (BARIUM TITANATE, LITHIUM TANTALATE, LITHIUM NIOBATE,/ 02-0107
/CTRICITY AND CONDUCTION IN FERROELECTRIC CRYSTALS. (BISMUTH TITANATE, LITHIUM THALLIUM TARTRATE MONOHYDRA/ 02-0061
STUDY OF THE DYNAMICS OF PHOTOCHROMIC SWITCHING IN STRONTIUM TITANATE MATERIALS. 06-1092
VIOR. ORIGIN OF LATTICE DEFECTS IN (BARIUM) TITANATE MATERIALS AND THEIR INFLUENCE ON BEHA 04-0784
ITIONS AT NEAR THEORET/ VAPORIZATION OF LEAD ZIRCONATE- LEAD TITANATE MATERIALS. PART-2: HOT PRESSED COMPOS 09-1578
SEEBECK EFFECT IN METAL LEAD TITANATE METAL CONTACT SYSTEMS. 05-1056
ERCIVE FORCE OF THE PEROVSKITES IN THE SYSTEM LEAD ZIRCONATE TITANATE METANIOBATE. /OUS POLARIZATION AND CO 09-1597
IC EFFECTS IN CRYSTALS. (AND CALCULATIONS BASED ON STRONTIUM TITANATE MODEL) /FERROELECTRICITY AND ANHARMON 03-0213
TION) FERROELECTRIC TRANSFORMATION IN BARIUM TITANATE. (MODEL GIVES NEARLY COMPLETE DESCRIP 04-0836
IMONATE, NIOBATE, / PIEZOELECTRIC CERAMICS OF LEAD ZIRCONATE TITANATE MODIFIED WITH SODIUM OR POTASSIUM ANT 09-1539
ATERIALS (SYSTEM WITH LEAD ZIRCONATE AND WITH LEAD ZIRCONATE TITANATE MOSSBAUER SPECTROSCOPY. /ND RELATED M 08-1333
CRITICAL X-RAY INCOHERENT SCATTERING BY BARIUM TITANATE NEAR THE CURIE TEMPERATURE. 04-0535
ELECTRON PARAMAGNETIC RESONANCE IN BARIUM TITANATE NEAR THE CURIE TEMPERATURE. 04-0928
ON OF LIGHT BY SURFACE MICROPLASMA OF CERAMIC (BARIUM, LEAD) TITANATE NEAR THE CURIE TEMPERATURE. EMISSI 04-0982
/IOR OF THE DIELECTRIC NONLINEARITY COEFFICIENT IN STRONTIUM TITANATE NEAR THE PHASE TRANSITION AT 110 DEG/ 06-1105
DIELECTRIC DISPERSION OF IRRADIATED BARIUM TITANATE NEAR THE PHASE TRASNITION. 04-0648
SPECIFIC HEAT OF STRONTIUM TITANATE NEAR THE STRUCTURAL TRANSITION. 06-1132
THERMAL EXPANSIVITY OF STRONTIUM TITANATE NEAR THE STRUCTURAL TRANSITION. 06-1136
PRESSURE DEPENDENCE OF THE ELASTIC COMPLIANCE OF STRONTIUM TITANATE NEAR THE TRANSITION TEMPERATURE. 06-1219
CRITICAL BEHAVIOR OF STRONTIUM TITANATE NEAR THE 105 K PHASE TRANSITION. 06-1194
IC ANOMALIES AT THE STRUCTURAL PHASE TRANSITION OF STRONTIUM TITANATE NEAR 105 DEG K. ANALYSIS OF THE ELAST 06-1183
SOFT FERROELECTRIC MODES IN LEAD TITANATE (NEUTRON INELASTIC SCATTERING) 05-1077
/TEN BRONZE TYPE STRUCTURES (POTASSIUM (BARIUM OR STRONTIUM) TITANATE NIOBATE, (BARIUM OR STRONTIUM) BISMU/ 13-1962
NGSTEN BRONZE). DIELECTRIC PROPERTY OF BARIUM TITANATE NIOBATE SINGLE CRYSTAL (TETRAGONAL TU 13-1969
CUBIC- TETRAGONAL TRANSITION, AND PRO/ MONODOMAIN STRONTIUM TITANATE (OCCURANCE IN THIN (110) PLATES BELOW 06-1164
/IZATION OF ALKOXY-DERIVED STRONTIUM ZIRCONATE AND STRONTIUM TITANATE (OF HIGH PURITY AND SURFACE ACTIVITY/ 06-1217
PHOTOCONDUCTIVITY (TRANSIENTS) IN PHOTOCHROMIC STRONTIUM TITANATE (ON COLORING AND BLEACHING). 06-1240
EOUS INDUCTION. DEPENDENCE OF THE PIEZOEFFECT IN BARIUM TITANATE ON THE MEANS OF SCREENING THE SPONTAN 04-1021
. DIRECT OBSERVATION OF SINGLE DOMAIN STRONTIUM TITANATE (OPTICAL STUDY UNDER UNIAXIAL STRESS) 06-1112
T NOT NAMED) PIEZORESISTIVITY OF DOPED BARIUM TITANATE. (OR BARIUM STRONTIUM TITANATE, DOPAN 04-0729
IONIC CONDUCTION IN SINTERED OXIDES BASED ON CALCIUM TITANATE OR STRONTIUM TITANATE. 06-1141
/RPTION SPECTRA OF CERTAIN PEROVSKITES (STRONTIUM OR CALCIUM TITANATE OR ZIRCONATE, TEST OF SHORT OR LONG / 03-0414
CONDUCTIVITY MEASURE/ TITANIUM RICH NONSTOICHIOMETRIC BARIUM TITANATE. PART-1: HIGH TEMPERATURE ELECTRICAL 04-0822
IC AND X-RAY DIFFRAC/ TITANIUM RICH NONSTOICHIOMETRIC BARIUM TITANATE. PART-3: HIGH TEMPERATURE THERMODYNAM 04-0886
ON THE ANOMALOUS CRYSTALLOGRAPHIC PROPERTIES OF SMALL BARIUM TITANATE PARTICLES. 04-0935
FACE MOISTURE ON THE DIELECTRIC BEHAVIOR OF ULTRAFINE BARIUM TITANATE PARTICULATES. EFFECT OF SUR 04-0859
TEMPERATURE TO 6/ MOSSBAUER STUDIES OF BISMUTH FERRATE LEAD TITANATE PEROVSKITE TYPE SOLID SOLUTIONS (ROOM 08-1471
YSTEM LEAD (ZINCATE NIOBATE)- BARIUM (ZINCATE NIOBATE)- LEAD TITANATE. (PHASE DIAGRAM, LATTICE CONSTANTS) 05-1070
HOTOCONDUCTIVIT/ PHOTOCHROMIC MATERIALS RESEARCH. (STRONTIUM TITANATE PHOTOCONDUCTIVITY, CALCIUM FLUORIDE P 06-1140
. (HIGH RESOLUTION READ WRITE) BISMUTH TITANATE PHOTOCONDUCTOR OPTICAL STORAGE MEDIUM 08-1445
EAD TITANATE) LEAD TITANATE PIEZOELECTRIC CERAMICS. (MATSUSHITA L 09-1616
CTRIC FIELDS. (SUB/ BEHAVIOR OF MODIFIED LEAD ZIRCONATE LEAD TITANATE PIEZOELECTRIC CERAMICS UNDER HIGH ELE 09-1484
ULTRASONIC FIELDS. BARIUM TITANATE PIEZOELECTRIC MICROPROBE FOR STUDYING 04-0938
RFACE / PROPERTIES OF OXYGEN DEPLETED CERAMIC LEAD ZIRCONATE TITANATE (PLUS 1 PERCENT NIOBIUM PENTOXIDE) SU 09-1576
TELLURIUM) SEMICONDUCTOR F/ EFFECT OF FERROELECTRIC (BARIUM TITANATE) POLARIZATION FIELDS ON (GERMANIUM OR 04-0594
/ME FERROELECTRICS BELOW 80 DEG K. (SINGLE CRYSTAL STRONTIUM TITANATE, POLYCRYSTALLINE, STRONTIUM TITANATE/ 04-0721
RESISTIVITY) RESISITIVITY ANOMALY IN SEMICONDUCTING BARIUM TITANATE. (POSITIVE TEMPERATURE COEFFICIENT OF 04-0584
MPLIANCE, NICKEL IODINE BORACITE OPTICAL ANISOTROPY, BISMUTH TITANATE, POSSIBLE FERROELECTRIC SILICATE) /CO 02-0060
/SODIUM LITHIUM NIOBATE, POTASSIUM STRONTIUM LITHIUM NIOBATE, POTASSIUM LANTHANUM LITHIUM NIOBIUM/ 13-1923
AN/ ENERGY BANDS FOR POTASSIUM NICKEL TRIFLUORIDE, STRONTIUM TITANATE, POTASSIUM MOLYBDATE, AND POTASSIUM T 02-0091
OF HOLOGRAPHIC PATTERNS IN FERROELECTRIC MATERIALS. (BARIUM TITANATE, POTASSIUM NIOBATE) /TION AND ERASURE 04-0847
/S ON FERROELECTRIC DOMAINS OF PEROVSKITE TYPE OXIDES. (LEAD TITANATE, POTASSIUM NIOBATE, BARIUM TITANATE) 04-0993
FAR INFRARED DIELECTRIC DISPERSION OF PEROVSKITES. (BARIUM TITANATE, POTASSIUM TANTALATE) 04-0546
ATE NEAR THE CURIE POINTS. THERMAL CONDUCTIVITY OF STRONTIUM TITANATE, POTASSIUM TANTALATE AND BARIUM TITAN 04-0989
/ERROELECTRIC" PHONON IN SOME PEROVSKITE CRYSTALS (STRONTIUM TITANATE, POTASSIUM TANTALATE) AT LOW TEMPERA/ 03-0227
/OFT FERROELECTRIC MODE IN THE CUBIC PEROVSKITES. (STRONTIUM TITANATE, POTASSIUM TANTALATE, BARIUM TITANAT/ 02-0114

/STIC SCATTERING AT STRUCTURAL PHASE TRANSITIONS. (STRONTIUM TITANATE, POTASSIUM TANTALATE, BARIUM TITANAT/ 02-0059
ERE A UNIFIED MECHANISM UNDERLYING URBACH'S RULE. (STRONTIUM TITANATE, POTASSIUM TITANATE)   IS TH 06-1110
SUBS/ PHYSICAL AND ELECTRICAL PROPERTIES OF THIN FILM BARIUM TITANATE PREPARED BY RF SPUTTERING ON SILICON 04-0828
AL PHASE TRANSITION.   BIREFRINGENCE OF STRONTIUM TITANATE PRODUCED BY THE 105 DEGREE K STRUCTUR 06-1116
.   CRYSTAL STRUCTURE OF BARIUM TITANATE PRODUCING SQUARE INTERFERENCE FRINGES 04-0706
          POTENTIAL BARRIERS ON SEMICONDUCTING BARIUM TITANATE. (PTCR DEVICES)   04-0851
, HYDROXYL) SINGLE CRYSTALS.   CONDUCTIVITY OF BARIUM TITANATE, PURE SINGLE CRYSTALS AND DOPED (IRON 04-0695
INFRARED RADIATION (0.5 TO 2 MIC/ CHARACTERISTICS OF BARIUM TITANATE PYROELECTRIC DETECTOR FOR OPTICAL AND 04-0751
          HIGH D*, FAST, LEAD ZIRCONATE TITANATE PYROELECTRIC DETECTORS.   09-1560
          LEAD TITANATE PYROELECTRIC INFRARED DETECTOR.   05-1087
PTICAL ABSORPTION) SOME SEMICONDUCTING PROPERTIES OF BARIUM TITANATE. (RESISTIVITY, SEEBECK COEFFICIENT, O 04-0919
GNETIC AND ACOUSTIC WAVES IN FERROELECTRICS. (THEORY, BARIUM TITANATE, ROCHELLE SALT)   ELECTROMA 03-0455
/ENTS OF SEVERAL PEROVSKITES (POTASSIUM TANTALATE, STRONTIUM TITANATE, RUBIDIUM MANGANESE TRIFLUORIDE) BY / 06-1139
COVALENCY ON THE STRUCTURE OF AB(5) OXIDE COMPOUNDS. (BARIUM TITANATE, RUBIDIUM NIOBATE) /ELUNG ENERGY AND 04-0761
(PYROELECTRIC EFFECT IN TGS, ETHYL DIAMINE TARTRATE, BARIUM TITANATE, SALICYLIDENE- 4-BROMO ANILINE. /ION. 04-0854
FORMING INSULATION LAYERS ON THE GRAIN BOUNDARIES OF BARIUM TITANATE SEMICONDUCTIVE CERAMICS. /CHANISM FOR 04-0867
          ATTEMPT TO PREPARE HIGH DOPED BARIUM STRONTIUM TITANATE SEMICONDUCTOR WITH PZT.   04-0717
L NEUTRON IRRADIATION ON THE RELATIVE PERMITTIVITY OF BARIUM TITANATE SINGLE CRYSTAL.   EFFECT OF THERMA 04-0781
TION OF THE QUADRATIC ELECTROOPTIC COEFFICIENTS IN STRONTIUM TITANATE (SINGLE) CRYSTAL. /ROMETRIC DETERMINA 06-1130
AMAGNETIC RESONANCE OF PLATINUM IONS IN NIOBIUM DOPED BARIUM TITANATE SINGLE CRYSTAL.   PAR 04-0977
          PRODUCTION AND STUDY OF PURE AND DOPED BARIUM TITANATE SINGLE CRYSTAL   04-0667
          X-RAY STUDY OF THE SURFACE LAYER ON BARIUM TITANATE SINGLE CRYSTAL.   04-0858
          LATTICE DISTORTIONS IN A STRONTIUM TITANATE SINGLE CRYSTAL ELECTRET.   06-1238
RIC) ELECTROOPTIC AND PIEZOELECTRIC PROPERTIES OF LANTHANUM TITANATE SINGLE CRYSTAL. (FOUND NOT FERROELECT 36-3576
NSATION OF HETEROVALENT FERRIC IONS IN THE LATTICE OF BARIUM TITANATE SINGLE CRYSTALS.   COMPE 04-0539
INATION OF THE DEVIATION FROM OXYGEN STOICHIOMETRY OF BARIUM TITANATE SINGLE CRYSTALS.   DETERM 04-0736
          DIELECTRIC MEASUREMENTS ON PILE-IRRADIATED BARIUM TITANATE SINGLE CRYSTALS.   04-0910
          DISLOCATIONS IN BARIUM TITANATE SINGLE CRYSTALS.   04-1007
          EFFECT OF ADDING NICKEL ON THE PROPERTIES OF BARIUM TITANATE SINGLE CRYSTALS.   04-0793
          EFFECT OF EXTERNAL ACTIONS ON THE REPOLARIZATION OF BARIUM TITANATE SINGLE CRYSTALS.   04-0957
OF FERRIC OXIDE ON THE 120 DEGREE PHASE TRANSITION IN BARIUM TITANATE SINGLE CRYSTALS.   EFFECT 04-0544
INEAR COMPRESSION ON THE UHF DIELECTRIC PROPERTIES OF BARIUM TITANATE SINGLE CRYSTALS.   EFFECT OF L 04-0896
TITANIUM(4+) IONS ON THE ELECTROOPTICAL PROPERTIES OF BARIUM TITANATE SINGLE CRYSTALS. /NT SUBSTITUTION OF 04-0959
OF TRANSVERSE COMPRESSION OF DIELECTRIC PROPERTIES OF BARIUM TITANATE SINGLE CRYSTALS.   EFFECT 04-0956
          ELASTIC PROPERTIES OF YTTRIUM DOPED STRONTIUM TITANATE SINGLE CRYSTALS.   06-1180
          ELECTROMECHANICAL PROPERTIES OF STRONTIUM TITANATE SINGLE CRYSTALS.   06-1206
          ESR OF CHROMIUM(3+) IN CHROMIUM DOPED STRONTIUM TITANATE SINGLE CRYSTALS.   06-1147
          ETCH PITS CORRESPONDING TO DISLOCATIONS IN BARIUM TITANATE SINGLE CRYSTALS.   04-1023
          ETCH PITS IN BARIUM TITANATE SINGLE CRYSTALS.   04-0922
          FERROELECTRIC PROPERTIES OF LEAD TITANATE SINGLE CRYSTALS.   05-1046
E AMPLITUDE OF FREE VIBRATIONS ON THE PERMITTIVITY OF BARIUM TITANATE SINGLE CRYSTALS. INFLUENCE OF TH 04-0768
ON THE ELECTRICAL PROPERTIES AND DOMAIN STRUCTURE OF BARIUM TITANATE SINGLE CRYSTALS. /AL TENSILE STRESSES 04-0961
          INTERNAL FRICTION IN BARIUM TITANATE SINGLE CRYSTALS.   04-0794
R EFFECT STUDIES OF ELECTRIC CONDUCTION IN IRON DOPED BARIUM TITANATE SINGLE CRYSTALS.   KER 04-0703
          LATTICE DYNAMICAL THEORY OF SWITCHING IN BARIUM TITANATE SINGLE CRYSTALS.   03-0398
          MECHANICAL RELAXATION AND NONLINEARITY IN STRONTIUM TITANATE SINGLE CRYSTALS.   06-1218
          MICROWAVE DYNAMIC NONLINEARITY OF BARIUM TITANATE SINGLE CRYSTALS.   04-0749
ANSITIONS AND) THERMOCURRENTS IN TITANIUM DIOXIDE AND BARIUM TITANATE SINGLE CRYSTALS.   (PHASE TR 04-0697
          PIEZOELECTRIC EFFECT IN LEAD TITANATE SINGLE CRYSTALS.   05-1049
          PREPARATION AND PROPERTIES OF (BARIUM, STRONTIUM) TITANATE SINGLE CRYSTALS.   04-0560
          PREPARATION AND PROPERTIES OF PURE AND DOPED BARIUM TITANATE SINGLE CRYSTALS.   04-0668
          PROBLEM OF THE SURFACE LAYER IN BARIUM TITANATE SINGLE CRYSTALS.   04-1013
          STABILIZATION OF SPONTANEOUS POLARIZATION OF BARIUM TITANATE SINGLE CRYSTALS.   04-0795
          STUDY OF THE AGING PHENOMENON IN BARIUM TITANATE SINGLE CRYSTALS.   04-0545
          TEMPERATURE DEPENDENCE OF INTERNAL FRICTION OF BARIUM TITANATE SINGLE CRYSTALS.   04-0796
          THERMALLY STIMULATED CONDUCTIVITY IN STRONTIUM TITANATE SINGLE CRYSTALS.   06-1227
          THICKNESS DEPENDENCE OF THE PERMITTIVITY OF BARIUM TITANATE SINGLE CRYSTALS.   04-0637
          THICKNESS DEPENDENCE OF POLARIZATION REVERSAL IN BARIUM TITANATE SINGLE CRYSTALS.   04-0949
/TENSITIC TYPE) PHASE TRANSITIONS IN HIGH RESISTIVITY BARIUM TITANATE SINGLE CRYSTALS. (AND TWINNING TENDE 04-0887
IES.   MEASUREMENT OF THE PERMITTIVITY OF BARIUM TITANATE SINGLE CRYSTALS AT MICROWAVE FREQUENC 04-0759
OUPS.   DIELECTRIC PROPERTIES OF BARIUM TITANATE SINGLE CRYSTALS DOPED BY HYDROXIDE GR 04-0634
M FLUORIC/ DIELECTRIC PROPERTIES OF BARIUM(1-X) STRONTIUM(X) TITANATE SINGLE CRYSTALS GROWN FROM A POTASSIU 04-0895
N A POT/ ANISOTROPY OF THE ELECTRICAL CONDUCTIVITY OF BARIUM TITANATE SINGLE CRYSTALS GROWN FROM SOLUTION I 04-1035
EC/ TIME DEPENDENCE OF THE ELECTRICAL CONDUCTIVITY OF BARIUM TITANATE SINGLE CRYSTALS HEATED IN OXYGEN (EFF 04-0997
/EMPERATURE, ABSORPTION SPECTRA) OF SEMICONDUCTING STRONTIUM TITANATE SINGLE CRYSTALS (HYDROGEN REDUCED, C/ 06-1198
HOMBOHEDRAL PHASES.   REPOLARIZATION PROCESSES OF BARIUM TITANATE SINGLE CRYSTALS IN ORTHORHOMBIC AND R 04-0963
FLUX.   DISCUSSION ON THE CRYSTAL GROWTH OF BARIUM TITANATE SINGLE CRYSTALS IN POTASSIUM FLUORIDE 04-1022
HASE.   PIEZOELECTRIC ACTIVITY OF BARIUM TITANATE SINGLE CRYSTALS IN THE PARAELECTRIC P 04-0771
ON REGION.   DIELECTRIC PERMEABILITY OF BARIUM TITANATE SINGLE CRYSTALS IN THE PHASE TRANSITI 04-0645
CU/ POSITIVE TEMPERATURE COEFFICIENT OF RESISTANCE OF BARIUM TITANATE SINGLE CRYSTALS IN THE REGION OF THE 04-1036
NGE 5 DEG K-300 DEG K.   BRILLOUIN SCATTERING IN STRONTIUM TITANATE SINGLE CRYSTALS IN THE TEMPERATURE RA 06-1149
/MPERATURE DEPENDENCE OF THE ELASTIC COMPLIANCE OF STRONTIUM TITANATE SINGLE CRYSTALS IN THE TEMPERATURE R/ 06-1220
/F SOME ADMIXED NICKEL ON SOME PHYSICAL PROPERTIES OF BARIUM TITANATE SINGLE CRYSTALS. (LATTICE PARAMETERS) 04-0909
          PHASE BOUNDARIES IN BARIUM TITANATE SINGLE CRYSTALS NEAR CURIE POINT.   04-0577
AT 110 DEGREES K.   DIELECTRIC NONLINEARITY OF STRONTIUM TITANATE SINGLE CRYSTALS NEAR PHASE TRANSITION 06-1106
OPERTIES). ULTRASONIC THIRD HARMONIC GENERATION IN STRONTIUM TITANATE SINGLE CRYSTALS (NONLINEAR ELASTIC PR 06-1181
TURE.   DEPENDENCE OF DIELECTRIC DISPERSION IN BARIUM TITANATE SINGLE CRYSTALS ON THEIR DOMAIN STRUC 04-0895
          DEPENDENCE OF COERCIVE FIELD OF BARIUM TITANATE SINGLE CRYSTALS ON THEIR THICKNESS 04-0693
US) AMPLITUDE DEPENDENCE OF THE INTERNAL FRICTION IN BARIUM TITANATE SINGLE CRYSTALS. (PIEZOELECTRIC MODUL 04-0791
AL IMPULSES.   BARIUM TITANATE SINGLE CRYSTALS SUBJECTED TO MECHANIC 04-1006
S LOOP.   PECULIARITIES OF POLARIZATION OF BARIUM TITANATE SINGLE CRYSTALS WITH DOUBLE HYSTERESI 04-0709
/ DIELECTRIC, ELASTIC AND PIEZOELECTRIC PROPERTIES OF BARIUM TITANATE SINGLE CRYSTALS WITH LAMINAR DOMAIN S 04-1010
/IELECTRIC (CONSTANT AND LOSS) BEHAVIOR OF REDUCED STRONTIUM TITANATE SINGLE CRYSTALS (40 HZ-10 MHZ, EFFEC/ 06-1148
AL PROPERTIES OF FERROELECTRI/ INFLUENCE OF FIRING OF BARIUM TITANATE SINTERS ON ELECTRICAL AND TECHNOLOGIC 04-1002
LAYER STRUCTURE.   STUDIES ON THE (STRONTIUM, CALCIUM) TITANATE SOLID SOLUTION CERAMIC WITH BOUNDARY 06-1246
          BISMUTH TITANATE SOLID SOLUTIONS.   08-1277
          CRYSTAL CHEMICAL PROBLEMS OF (BARIUM, STRONTIUM) TITANATE SOLID SOLUTIONS.   04-0633
RIC AND ELASTIC PROPERTIES OF SINGLE CRYSTALS OF BARIUM LEAD TITANATE SOLID SOLUTIONS.   DIELECT 04-0597
ROOPTICAL EFFECT IN SINGLE CRYSTALS OF BARIUM TITANATE- ZINC TITANATE SOLID SOLUTIONS.   ELECT 04-0958
TUNGSTATE(0.5) AND LEAD COBALTATE(0.5) TUNGSTATE(0.5)- LEAD TITANATE SOLID SOLUTIONS. /LEAD COBALTATE(0.5) 13-1968

/HE TRANSITION TEMPERATURE OF BARIUM(0-0.1) STRONTIUM(0.9-1) TITANATE SOLID SOLUTIONS BY HYDROSTATIC PRESS/ 04-0722
/ FERROELECTRIC TRANSITION IN THE LEAD (ZIRCONATE, STANNATE, TITANATE) SOLID SOLUTIONS. (ELECTROSTRICTION / 09-1486
/ DIELECTRIC PROPERTIES OF FERROELECTRIC (BARIUM, STRONTIUM) TITANATE SOLID SOLUTIONS NEAR THE PHASE TRANSI 04-0680
/DIELECTRIC PROPERTIES IN PEROVSKITE MODIFICATION OF CADMIUM TITANATE. (SOLID SOLUTIONS WITH STRONTIUM TIT/ 08-1424
ENFORC/ RELEASE OF ELECTRIC ENERGY IN LEAD NIOBIUM ZIRCONATE TITANATE STANNATE BY TEMPERATURE AND PRESSURE 09-1488
AND TITANIUM)   MOSSBAUER STUDIES OF LEAD ZIRCONATE TITANATE STANNATE. (CERAMIC DOPED WITH NIOBIUM 09-1604
UNGSTEN TRIOXIDE.   SEMICONDUCTOR PROPERTIES OF BARIUM (TITANATE, STANNATE) SOLID SOLUTIONS DOPED BY T 04-0965
/GE OF THE MOSSBAUER EFFECT PROBABILITY CN TIN-119 IN BARIUM TITANATE STANNATE SOLID SOLUTIONS IN THE VICI/ 04-0573
CHARGE MECHANISM IN FERRCELECTRICS. (LEAD ZIRCONATE TITANATE, STOICHIOMETRY EFFECTS) 03-0259
ELECTRIC CONSTANT TEMPERATURE STABLE FERROELECTRICS. (BARIUM TITANATE, STRONTIUM STANNATE) /HIEVING HIGH DI 04-0551
CERAMIC DIELECTRICS. (BARIUM TITANATE, STRONTIUM TITANATE) 04-0683
ACIVE TYPE FERROELECTRICS NEAR THE TRANSITION POINT. (BARIUM TITANATE, STRONTIUM TITANATE) /HAVIOR OF DISPL 04-0992
CTRA IN A FERROELECTRIC CERAMIC CONTAINING TITANIUM. (BARIUM TITANATE, STRONTIUM TITANATE) /C RESONANCE SPE 04-1020
THE DIELECTRIC CONSTANTS OF PARAELECTRIC MATERIALS. (BARIUM TITANATE, STRONTIUM TITANATE) /E DEPENDENCE OF 04-0856
/TIVITY OF SOME TITANATES BETWEEN 20 AND 1000 DEG C. (BARIUM TITANATE, STRONTIUM TITANATE AND OTHER MATERI/ 04-0820
OWTH OF OXIDE CRYSTALS FROM NONSTOICHIOMETRIC MELTS. (BARIUM TITANATE, STRONTIUM TITANATE, GERMANATES) / GR 04-0553
IN A PSEUDOCUEIC IONIC STRUCTURE. APPLICATION TO ALL BARIUM TITANATE STRUCTURE TYPES.   ELECTROSTATIC FIELD 04-1008
/CS OF CADMIUM SELENIDE LAYERS ON BARIUM TITANATE- STRONTIUM TITANATE SUBSTRATES AT SUPERSONIC DRIFT VELOC/ 04-0592
/CF TEMPERATURE ON) QUASI SYMMETIRICAL HARD IONS (PHOSPHATE, TITANATE, SULFATE, FLUOBERYLLATE, ETC) OF SOM/ 03-0330
SALT, TRIGLYCINE SULFATE, SELENATE, FLUOBERYLLATE AND BARIUM TITANATE, SUMMARY ONLY) /ELECTRICS. (ROCHELLE 02-0083
ACTS) SOME ELECTRONIC PROPERTIES CF ZINC OXIDE AND STRONTIUM TITANATE. (SURFACE BARRIER SYSTEMS, METAL CONT 06-1176
RIC PROPE/ SINGLE CRYSTALS OF THE SYSTEM (BARIUM, STRONTIUM) TITANATE. (SYNTHESIS AND LOW FREQUENCY DIELECT 04-0561
N POLYCRYSTALLINE SOLID SOLUTIONS IN THE (BARIUM, STRONTIUM) TITANATE SYSTEM. /F THE CURIE-WEISS CONSTANT I 04-0586
HIGH TEMPERATURE PHASE EQUILIBRIA IN THE LEAD ZIRCONATE TITANATE SYSTEM. 09-1577
RESSED ELECTROOPTIC CERAMICS IN THE LEAD LANTHANUM ZIRCONATE TITANATE SYSTEM.   IMPROVED HOT P 09-1528
INTRINSIC NONSTOICHIOMETRY IN THE LEAD ZIRCONATE TITANATE SYSTEM. 09-1535
D ELECTROOPTIC PROPERTIES OF THE (LEAD, LANTHANUM) (HAFNATE, TITANATE) SYSTEM.   MATERIAL AN 08-1302
XIDATION REDUCTION PHENOMENA IN THE LEAD TITANATE- LANTHANUM TITANATE SYSTEM.   O 09-1587
ID SOLUTIONS OF THE LEAD FERRATE(0.67) TUNGSTATE(0.33)- LEAD TITANATE SYSTEM.   SEIGNETTO MAGNETIC SOL 05-1065
STUDY OF PHASE TRANSITIONS IN THE LEAD ZIRCONATE TITANATE SYSTEM. 09-1620
STUDY OF PHASE TRANSITIONS IN THE LEAD ZIRCONATE- LEAD TITANATE SYSTEM. 09-1621
TRINSIC AND EXTRINSIC NONSTOICHIOMETRY IN THE LEAD ZIRCONATE TITANATE SYSTEM. (DEFECT STRUCTURE)   IN 09-1536
00 DEG K (LATTICE PARAMETERS VERSU/ POTASSIUM TANTALATE LEAD TITANATE SYSTEM IN THE TEMPERATURE RANGE 4.2-3 08-1308
IES OF (BISMUTH(0.5) (SODIUM OR POTASSIUM) (0.5) (X) LEAD(1-X) TITANATE SYSTEMS. /LECTRIC AND CERAMIC PROPERT 13-2013
/LOSS IN THE BARIUM (LEAD, 4 COMPOSITIONS EACH) STRONTIUM TITANATE SYSTEMS. (FOR VOLTAGE DEPENDENT CAPA/ 09-1522
ISTIVITY BEHAVIOR OF SEMICONDUCTING YTTRIUM BARIUM ZIRCONIUM TITANATE. (TEMPERATURE DEPENDENCE)   RES 13-1995
ACCUSTIC EMISSION FROM FERROELECTRIC CRYSTALS. (BARIUM TITANATE, TGS) 04-0593
TIVITY OF THE STATIONARY (100)-180 DEG DOMAIN WALL IN BARIUM TITANATE (THEORY, CALCULATIONS). /IGNAL PERMIT 03-0358
POLARITONS IN TETRAGONAL BARIUM TITANATE: THEORY OF THE RAMAN LINE SHAPE. 04-0556
ANGULAR ENERGY DEPENDENCE OF 180 DEG DOMAIN WALL IN BARIUM TITANATE. (THERMODYNAMIC CALCULATIONS) 04-0658
FORMATION PROCESS OF BARIUM TITANATE THIN FILM EVAPORATED IN VACUO. 04-0862
CHARACTERISTICS OF RF SPUTTERED BARIUM TITANATE THIN FILMS. 04-0906
FABRICATION OF RF SPUTTERED BARIUM TITANATE THIN FILMS. 04-0907
/CTURE) AND ELECTRICAL PROPERTIES OF FLASH EVAPORATED BARIUM TITANATE THIN FILMS (SEPARATED FROM SUBSTRATE/ 04-0605
ECTRICAL PROPERTIES OF FLASH EVAPORATED FERROELECTRIC BARIUM TITANATE THIN FILMS. (THICKNESS DEPENDENCE) EL 04-0967
ECTRICAL AND OPTICAL PROPERTIES OF (BARIUM, STRONTIUM) MIXED TITANATE THIN SECTIONS.   EL 04-0603
MECHANISMS, MEMORY, ORGANICS)   FROM BARIUM TITANATE TO DNA. (FERROELECTRICITY, BIOLOGICAL 02-0092
FRACTURE TOUGHNESS CF COMMERCIAL BARIUM TITANATE TRANSDUCER MATERIAL. 04-0841
CNCEPTS)   INTERPRETATION OF THE DYNAMICS OF THE BARIUM TITANATE TRANSITION IN TERMS OF NUCLEATION. (C 04-0606
/LECTROMAGNETIC RADIATION USING PYROELECTRIC EFFECT. (BARIUM TITANATE, TRIGLYCINE SULFATE, AT 5.6 AND 10.5/ 04-0978
INFORMATION STORAGE AND DISPLAY (HOT PRESSED LEAD ZIRCONATE TITANATE, TUTORIAL, 17 REFS). /W MATERIALS FOR 09-1550
CALCULATION OF ELECTRIC FIELDS IN IONIC CRYSTALS. (BARIUM TITANATE TYPE) 04-0824
RFECTIONS ON THE DIELECTRIC PROPERTIES OF MATERIALS. (BARIUM TITANATE TYPE)   EFFECT OF CHARGED LATTICE IMPE 04-1017
EW TRANSPARENT FERROELECTRIC CERAMICS. (LEAD ZIRCONATE- LEAD TITANATE TYPE)   PROPERTIES OF N 09-1518
RADIATION PHYSICS OF FERROELECTRICS OF BARIUM TITANATE TYPE. 04-0972
VE PERMI/ THERMAL CHARACTERISTICS CF A FERROELECTRIC (BARIUM TITANATE TYPE) CERAMIC WITH CONTROLLED MICROWA 04-0904
/ ELECTRICAL PROPERTIES OF HOT PRESSED FERROELECTRIC (BARIUM TITANATE TYPE) CERAMICS. (IN RELATION TO DENSI 04-0879
UHF REVERSAL PROPERTIES OF BARIUM TITANATE TYPE FERROELECTRIC CERAMICS. 04-1014
RM OF THICK PLATES.   MACROSCOPIC THEORY OF BARIUM TITANATE TYPE FERROELECTRIC CRYSTALS IN THE FO 04-0770
THEORY OF MICROWAVE DISPERSION CF THE PERMITTIVITY OF BARIUM TITANATE TYPE FERROELECTRICS. /L CHECK OF THE 04-0649
MICROWAVE DISPERSION MECHANISM IN BARIUM TITANATE TYPE FERROELECTRICS. 04-0902
PHOTOFERROELECTRIC EFFECTS IN A(V)B(VI)C(VII) AND BARIUM TITANATE TYPE FERROELECTRICS. 04-0552
THERMODYNAMICAL THEORY OF BARIUM TITANATE TYPE FERROELECTRICS. 03-0322
ATMENT AND INTERPRETATION OF X/ STRUCTURE DISORDER OF BARIUM TITANATE TYPE FERROELECTRICS. BASED ON NEW TRE 04-0630
DIELECTRIC DISPERSION IN BARIUM TITANATE TYPE FERROELECTRICS IN THE METER BAND 04-1009
AND CERAMICS)   PERMITTIVITY DISPERSION OF BARIUM TITANATE TYPE FERROELECTRICS. (SINGLE CRYSTALS 04-0903
ECTRIC CERAMICS) WITH GREAT DIELECTRIC PERMITTIVITY. (BARIUM TITANATE TYPE MATERIALS)   LOW LOSS (FERROEL 04-0779
EFFECT ON DIELECTRIC PROPERTIES OF SOLID SOLUTIONS OF BARIUM TITANATE TYPE MATERIALS.   X-RAY IRRADIATION 04-0782
FERROELECTRICITY (IN BARIUM TITANATE TYPE STRUCTURES) AND LATTICE DYNAMICS 04-0950
DOMAIN STRUCTURE OF STRONTIUM TITANATE UNDER UNIAXIAL STRESSES. 06-1113
PERMITTIVITY CHANGE OF BARIUM TITANATE UNDER VERY SLOWLY CHANGING FIELD. 04-0868
/ INVESTIGATION OF THE FERROELECTRIC PHASE CHANGE IN BARIUM TITANATE USING ELECTRON SPIN RESONANCE AND THE 04-0827
N OF TRANSITION METALS.   ELECTROCOLORATION IN STRONTIUM TITANATE: VACANCY DRIFT AND OXIDATION REDUCTIO 06-1098
N/ GROWING SINGLE CRYSTALS OF SOLID SOLUTION BASED ON BARIUM TITANATE, WHERE BARIUM(2+) AND TITANIUM(4+) IO 04-0960
X-RAY STRUCTURE INVESTIGATION OF LANTHANUM MODIFIED LEAD TITANATE WITH A-SITE AND B-SITE VACANCIES. 05-1059
HEMICAL AND PHYSICAL PROPERTIES OF SINGLE CRYSTALS OF BARIUM TITANATE WITH COBALT ADDED.   SOME CRYSTAL C 04-0618
MANIUM.   PIEZOCERAMICS BASED ON LEAD ZIRCONATE TITANATE WITH COMPLEX ADDITIVES CONTAINING GER 09-1510
DEVELOPMENT OF BARIUM TITANATE WITH CONSISTENT DIELECTRIC PROPERTIES 04-0797
(DIELECTRIC CON/ SYNTHESIS OF NIOBIUM PENTOXIDE DOPED BARIUM TITANATE WITH IMPROVED ELECTRICAL PROPERTIES. 04-0840
/-LAYER SYSTEM CONSISTING OF A PIEZOELECTRIC CERAMIC (BARIUM TITANATE) WITH LARGE DIELECTRIC CONSTANT AND / 04-1037
EFFECT OF DOPING BARIUM TITANATE WITH MANGANESE DIOXIDE. 04-0562
OPTICAL AND DIELECTRIC PROPERTIES OF SINGLE CRYSTAL BARIUM TITANATE WITH NIOBIUM ADDITIONS. 04-0875
/E TEMPERATURE COEFFICIENT OF RESISTANCE) BEHAVIOR OF BARIUM TITANATE WITH NIOBIUM PENTOXIDE AND MANGANESE/ 04-0837
CRYSTAL STUDY OF BARIUM TITANATE WITH PARALLEL INTERFERENCE FRINGES. 04-0705
TERPRETATION OF WAVELENGTH MODULATED ABSORPTION IN STRONTIUM TITANATE WITHOUT COEXISTING PHASES.   REIN 06-1188
PHASE DIAGRAM OF LEAD ZIRCONATE(1-X) TITANATE(X) PLUS 0.8 PERCENT TUNGSTEN TRIOXIDE 09-1581
PARAELECTRIC AND FERROELECTRIC PHAS/ SURFACE LAYER IN BARIUM TITANATE (X-RAY DIFFRACTION, DC FIELD EFFECT, 04-0857
/NTS OF POLAR CRYSTALS (LITHIUM NIOBATE OR TANTALATE, BARIUM TITANATE, ZINC OXIDE, AND LITHIUM GALLIUM OXI/ 02-0098
C) CRYSTALS (LITHIUM NIOBATE OR TANTALATE OR GALLATE, BARIUM TITANATE, ZINC OXIDE, CADMIUM SULFIDE). /ECTRI 02-0095
S IN ANTIFERROELECTRIC LEAD ZIRCONATE AND FERROELECTRIC LEAD TITANATE ZIRCONATE. /S OF THE PHASE TRANSITION 07-1253

IBUTION OF A-SITE AND B-SITE VACANCIES IN (LEAD, LANTHANUM) (TITANATE, ZIRCONATE) CERAMICS.   DISTR 09-1530
URE AND ELECTROMECHANICAL PROPERTIES OF LANTHANUM DOPED LEAD TITANATE ZIRCONATE CERAMICS.   STRUCT 09-1502
URE AND ELECTROMECHANICAL PROPERTIES OF LANTHANUM DOPED LEAD TITANATE ZIRCONATE CERAMICS.   STRUCT 09-1531
S.   PIEZOELECTRIC AND DIELECTRIC PROPERTIES OF LEAD TITANATE ZIRCONATE CERAMICS AT LOW TEMPERATURE 09-1519
/ DOMAIN STRUCTURE ON THE PROPERTIES OF POLYCRYSTALLINE LEAD TITANATE ZIRCONATE COMPOSITIONS NEAR THE MORP/ 09-1557
/C AND PIEZOELECTRIC PROPERTIES OF A CERAMIC (LEAD STRONTIUM TITANATE ZIRCONATE DOPED WITH CHROMIUM OXIDE). 09-1591
FERROELECTRICITY IN SPUTTERED LEAD TITANATE ZIRCONATE FILMS.   09-1559
PERTIES OF SOME CERAMIC MATERIALS AT MICROWAVE FREQUENCIES. (TITANATE ZIRCONATE GARNET MIXTURES) /RICAL PRO 09-1602
/ION AND THERMAL EXPANSION OF POLARIZED CERAMICS IN THE LEAD TITANATE ZIRCONATE MAGNESATE(0.5) TUNGSTATE(0/ 09-1603
HYDROTHERMAL CRYSTALLIZATION OF LEAD TITANATE ZIRCONATE SOLID SOLUTIONS.   09-1482
MIC EXISTENCE CONDITIONS FO/ DIELECTRIC POLARIZATION OF LEAD TITANATE ZIRCONATE SOLID SOLUTIONS (THERMODYNA 09-1541
OF PHASE TRANSITIONS IN LEAD COBALTATE(0.5) TUNGSTATE(0.5), TITANATE, ZIRCONATE SYSTEM. /AND X-RAY STUDIES 13-1960
/ SPECTRA OF POLYCRYSTALLINE SOLIDS: APPLICATION TO THE LEAD TITANATE ZIRCONATE SYSTEM (MEASUREMENT OF SOF/ 09-1497
STUDY OF DOMAIN WALL MOBILITY AND MECHANICAL DAMPING IN LEAD TITANATE ZIRCONATE TRANSDUCER CERAMICS. /SION 09-1547
/BRATION OF ELECTROELASTIC CONSTANTS OF POLYCRYSTALLINE LEAD TITANATE ZIRCONATE WITH THOROUGHNESS OF POLIN/ 09-1489
/CTROMECHANICAL PROPERTIES OF CERAMIC FERROELECTRICS (BARIUM TITANATE ZIRCONATES) IN STRONG ELECTRIC FIELD/ 09-1605
/RSAL IN POLYCRYSTALLINE FERROELECTRICS. (8 SUBSTITUTED LEAD ZIRCO, 2 SUBSTITUTED LEAD ZIRCO/ 08-1418
ERROELECTRIC-2 TRANSITION IN (NIOBIUM DOPED) ZIRCONATE(0.95) TITANATE(0.05). /ELOCITY AT FERROELECTRIC-1, F 09-1511
(THEORY, APPLIED TO LEAD(0.99) NIOBIUM(0.02) ZIRCONATE(0.65) TITANATE(0.35)). / IN A FERROELECTRIC CERAMIC 09-1537
W TEMPERATURES (WITH TIN OXIDE FILMS ON LEAD ZIRCONATE(0.65) TITANATE(0.35) CERAMIC. /EFFECT STUDIES AT LO 09-1505
/CT STUDIES. (ON TIN OXIDE FILMS ON LEAD(0.97) BISMUTH(0.02) TITANATE (0.35) ZIRCONATE(0.65), SUMMARY ONLY) 09-1506
/ITUTION OF BISMUTH AND NIOBIUM IONS IN LEAD ZIRCONATE(0.53) TITANATE(0.47) (EXPERIMENT AND STRUCTURAL DIS/ 09-1480
/OF THE ELECTROMECHANICAL PROPERTIES OF LEAD ZIRCONATE(0.52) TITANATE(0.48) CERAMICS CAUSED BY THE ADDITIO/ 09-1606
SOLUTION KINETICS OF LUTETIUM AND SCANDIUM IONS IN LEAD TITANATE(0.5) ZIRCONATE(0.5).   09-1538
SPACE CHARGE LIMITED CURRENTS IN HEXAGONAL BARIUM TITANATE(0.75) PLATINATE(0.25) CRYSTALS.   04-0936
/PAGATION CHARACTERISTICS OF SURFACE ELASTIC WAVES ON BARIUM TITANATE(0.95) ZIRCONATE(0.05) CERAMIC PLATES. 09-1613
ROSTRICTION, AND PIEZOELECTRIC ACTIVITY OF LEAD ZIRCONATE(X) TITANATE(1-X) CERAMICS.   POLARIZATION, ELECT 09-1501
/IMUM IN THE ELECTROMECHANICAL ACTIVITY IN LEAD ZIRCONATE(X) TITANATE(1-X) CERAMICS NEAR THE MORPHIC PHASE/ 09-1500
/IMUM IN THE ELECTROMECHANICAL ACTIVITY IN LEAD ZIRCONATE(X) TITANATE(1-X) CERAMICS NEAR THE MORPHOTROPIC / 09-1499
/STRUCTURES OF THE FERROELECTRIC COMPOUNDS LEAD ZIRCONATE(X) TITANATE(1-X) FOR TWO RHOMBOHEDRAL PHASES. (X/ 09-1574
LOW TEMPERATURE ELASTIC MODULI OF STRONTIUM TITANATE (10 TO 130 DEG K).   06-1190
SAJOUS FI/ ENERGY LOSS NONPOLAR STATE POLYCRYSTALLINE BARIUM TITANATE. (150-500 DEG C, 20-50 KILOHERTZ, LIS 04-0831
TANATE(3-X).   SOFT MODE SUPERCONDUCTIVITY IN STRONTIUM TITANATE(3-X) AND CALCIUM(Y) STRONTIUM(1-Y) TI 06-1093
CONDUCTIVITY AND HALL AND SEEBECK COEFFICIENTS) IN STRONTIUM TITANATE (4.2-300 DEG K) /ECTRONIC TRANSPORT (06-1129
FREQUENCIES (RAMAN SCATTERING IN CUBIC PEROVSKITE STRONTIUM TITANATE, 8-250K, 0.2-12 KV/CM). /TICAL PHONON 06-1241
YSICO- CHEMICAL NATURE OF THE SOLID SOLUTION FORMED FROM TWO TITANATES. /BARIUM TITANATE CERAMICS ON THE PH 04-0789
SONANCE OF TRIVALENT GADOLINIUM IONS IN STRONTIUM AND BARIUM TITANATES.   ELECTRON PARAMAGNETIC RE 04-0920
ASSIUM TANTALATE AND TANTALATE NIOBATE, BARIUM AND STRONTIUM TITANATES) /ES IN THE PARAELECTRIC PHASE. (POT 04-0690
E CALCULATION OF TETRAGONAL BIREFRINGENCE IN BARIUM AND LEAD TITANATES. /C IONIC CRYSTAL. APPLICATION TO TH 04-0578
HYDROTHERMAL CRYSTALLIZATION AND SOME PROPERTIES OF BISMUTH TITANATES   08-1279
0.65) TITANATE, BARIUM(0.65 OR 0.75) STRONTIUM(0.35 OR 0.25) TITANATES) /TCHING TIME, LEAD(0.35) STRONTIUM(37-3660
LINEAR PROPERTIES AND PHASE TRANSITIONS OF STRONTIUM BISMUTH TITANATES.   NON 13-1952
NISM FOR THERMOPHOTOTROPIC BEHAVIOR IN PEROVSKITE STRUCTURED TITANATES.   PROPOSED MECHA 03-0400
D IN POLYCRYSTALLINE SOLID SOLUTIONS OF BARIUM AND STRONTIUM TITANATES. /DENCE OF THE VELOCITY OF ULTRASOUN 04-0894
STATIC SHELL MODEL FOR BARIUM AND LEAD TITANATES.   04-0826
OF POLYCRYSTALLINE SOLID SOLUTIONS OF STRONTIUM AND BISMUTH TITANATES.   STRUCTURE 06-1145
ILITIES AND COCHRAN MODE ASSIGNMENTS IN BARIUM AND STRONTIUM TITANATES.   TEMPERATURE DEPENDENT POLARIZAB 04-0805
VIBRATION FREQUENCIES OF TITANATES.   08-1394
VIBRATIONAL SPECTRA OF STRONTIUM, BARIUM AND CALCIUM TITANATES.   04-0864
L ENERGY AND T/ THEORETICAL STATIC MODEL FOR BARIUM AND LEAD TITANATES - APPLICATION TO CALCULATING THE WAL 03-0189
DIELECTRIC PROPERTIES OF TITANATE- ALUMINA CERAMICS.   08-1414
SOLID ELECTROLYTE FUEL CELL. ((CALCIUM, STRONTIUM) LANTHANUM TITANATE- ALUMINATES) /ITS APPLICATION TO THE 08-1444
/FERROELECTRIC STATE, BISMUTH(0.5) (POTASSIUM OR SODIUM(0.5) TITANATES AND PHASE DIAGRAMS FOR BISMUTH(X) P/ 08-1289
N MODE THEORY OF INFRARED DISPERSION IN BARIUM AND STRONTIUM TITANATES AND POTASSIUM TANTALATE. /ICAL PHONO 04-0547
FAR INFRARED DIELECTRIC DISPERSION IN BARIUM AND STRONTIUM TITANATES AND TITANIUM DIOXIDE.   04-0975
HS IN POLYCRYSTALLINE DIELECTRICS WITH PEROVSKITE STRUCTURE (TITANATES AND ZIRCONATES). FORBIDDEN ZONE WIDT 08-1288
SINGLE CRYSTAL TITANATES AND ZIRCONATES. (METHODS OF GROWTH) 02-0045
PHASE EQUILIBRIA IN THE SYSTEM BARIUM TITANATE- BARIUM GERMANATE.   04-0710
DEMONSTRATION OF AGING EFFECTS IN BARIUM TITANATE- BARIUM ZIRCONATE.   09-1543
/ TEMPERATURE DEPENDENCE OF THE COMPLEX PERMITTIVITY OF SOME TITANATES BETWEEN 20 AND 1000 DEG C. (BARIUM T 04-0820
FERROELECTRIC PHASE TRANSITIONS IN THE LEAD ZIRCONATE- LEAD TITANATE- BISMUTH FERRATE. / EFFECT STUDIES OF 09-1498
/ FERROELECTRICS AND FERROELECTRIC MAGNETS. (LEAD AND BARIUM TITANATES, BISMUTH FERRATE, POTASSIUM NIOBATE) 04-1038
OUS INCREASE OF RESISTANCE ON LANTHANUM(0.002) BARIUM(0.998) TITANATE- BISMUTH OXIDE COMPOSITE CERAMICS. /L 04-0801
/ANOMALOUS INCREASE OF RESISTIVITY ON LANTHANUM DOPED BARIUM TITANATE- BISMUTH TRIOXIDE COMPOSITE CERAMICS/ 04-0800
FORMATION OF THIN LAYERS OF MIXED TITANATES BY SOLID SOLID REACTIONS.   13-2005
E COMPLEX OXIDE MATERIALS. (NIOBATES, TANTALATES, STANNATES, TITANATES, CHROMATES, COBALTATES) /OWTH OF SOM 02-0129
ITIVES.   CHROMATOGRAPHIC METHOD FOR ANALYZING LEAD ZIRCONATE TITANATES CONTAINING BISMUTH AND STRONTIUM ADD 09-1625
NS.   DIELECTRIC RELAXATION IN (CERAMIC) STRONTIUM TITANATES CONTAINING (TRIVALENT) RARE EARTH IO 06-1143
ETWEEN A SEMICONDUCTOR AND A C-DOMAIN FERROELECTRIC (THEORY, TITANATES DISCUSSED). /EFFECT AT THE CONTACT B 03-0511
/F HIGH PERMITTIVITY THIN FILM DIELECTRICS. (BARIUM AND STRONTIUM TITANATES, EFFECT OF FERROELECTRIC CHARACTERI/ 04-0670
/ERTIES OF NEW CADMIUM AND THALLIUM CONTAINING PEROVSKITES. (TITANATES, HAFNATES, MIXED NIOBATES AND TANTA/ 08-1458
UDIES OF INFRARED REFLECTION SPECTRA OF BARIUM AND STRONTIUM TITANATES IN THE 2-1000 MICRON REGION. /URE ST 04-0863
BROOKHAVEN WORK ON POTASSIUM TANTALATE, BARIUM AND STRONTIUM TITANATES, KDP, 22 REFS) /T MODES. (REVIEW OF 02-0115
S.   DISPERSION IN STRONTIUM TITANATE- LANTHANUM FERRATE DIELECTRIC CERAMIC 06-1120
NDENT FERROELECTRIC DOMAIN ALIGNMENT IN LEAD ZIRCONATE- LEAD TITANATE- LANTHANUM OXIDE (PLZT) CERAMIC. /EPE 09-1586
OXIDATION REDUCTION PHENOMENA IN THE LEAD TITANATE- LANTHANUM TITANATE SYSTEM.   09-1587
ELECTROMECHANICAL PROPERTIES IN THE BINARY SYSTEM LEAD TITANATE- LEAD CADMATE(0.5) TUNGSTATE(0.5). 10-1638
NS AT HIGH PRESSURES.   DIELECTRIC PROPERTIES OF LEAD TITANATE- LEAD (FERRATE NIOBATE) SOLID SOLUTIO 05-1071
TITANATE- LEAD ZIRCONATE) NEW PIEZOELECTRIC CERAMICS. (LEAD TITANATE- LEAD FERRATE(0.5) NIOBATE(0.5), LEAD 09-1544
TAL STRUCTURE AND PHASE RELATIONS IN THE LEAD ZIRCONATE LEAD TITANATE- LEAD METANIOBATE TERNARY SYSTEM. /YS 09-1596
IE TEMPERATURES OF SOLID SOLUTIONS IN A LEAD ZIRCONATE- LEAD TITANATE- LEAD METANIOBATE TERNARY SYSTEM. CUR 09-1595
ANTIFERROELECTRIC PHASE IN THE SYSTEM LEAD TITANATE- LEAD ZIRCONATE- LANTHANUM FERRATE. 13-1961
INCATE(0.33).   TERNARY SYSTEM LEAD TITANATE- LEAD ZIRCONATE- LEAD NIOBATE(0.67) Z 09-1507
NIOBAT/ FERROELECTRIC AND PIEZOELECTRIC PROPERTIES IN A LEAD TITANATE- LEAD ZIRCONATE- LEAD (ZINCATE(0.33) 09-1617
/LECTROPHYSICAL PROPERTIES OF THE SOLID SOLUTION SYSTEM LEAD TITANATE- LEAD ZIRCONATE- LEAD (ZINCATE(0.33)/ 13-1951
. PART-2: FERROELECTRICS- SILVER (BARIUM TITANATE AND BARIUM TITANATE- MAGNESIUM STANNATE). /ECTRIC SYSTEMS 04-0816
TWO-PHASE FERROELECTRIC SYSTEM. PART-1: BARIUM TITANATE- METAL.   04-0815
OF CONDUCT/ THERMOELECTRIC EFFECTS IN THE SYSTEM METAL- LEAD TITANATE- METAL. (METHOD OF STUDYING THE TYPE 05-1057
RIDE MELT.   DEFECT STRUCTURE OF (BARIUM(1-X) STRONTIUM(X)) TITANATE- MIXED CRYSTALS FROM A POTASSIUM FLUO 04-0873
75 REFS)   VIBRATIONAL SPECTRA OF THE TITANATES OF STRONTIUM, BARIUM, AND CALCIUM. (04-0865

YNTHESIS)   THE SYSTEM LEAD TITANATE- STRONTIUM COPPER NIOBATE. (CERAMIC S 05-1044
).   THE SYSTEM LEAD TITANATE- STRONTIUM CUPRATE(0.33) NIOBATE(0.67 05-1045
OELECTRIC PROPERTIES OF SOLID SOLUTIONS IN THE SYSTEM BARIUM TITANATE- STRONTIUM NIOBIUM(0.8) OXIDE(3). /RR 04-0707
E)   MEASUREMENT OF OXYGEN DIFFUSION CONSTANT IN SOME TITANATES. (STRONTIUM TITANATE, BARIUM TITANAT 04-1033
/AMPERE CHARACTERISTICS OF CADMIUM SELENIDE LAYERS ON BARIUM TITANATE- STRONTIUM TITANATE SUBSTRATES AT SU/ 04-0592
N INDUCED PHASE TRANSFORMATIONS IN (CERAMIC) BARIUM AND LEAD TITANATES. (SUMMARY ONLY) /T NEUTRON) RADIATIO 04-0971
NGLE CRYSTALS.   ELECTROOPTICAL PROPERTIES OF BARIUM TITANATE- TANTALUM PENTOXIDE SOLID SOLUTION SI 04-0955
                                    DEFECT LEAD TITANATES WITH DIVERSE CURIE TEMPERATURES.   05-1076
TS OF RESISTIVITY.   SEMICONDUCTING BARIUM AND STRONTIUM TITANATES WITH POSITIVE TEMPERATURE COEFFICIEN 04-0995
         ELECTROOPTICAL EFFECT IN SINGLE CRYSTALS OF BARIUM TITANATE- ZINC TITANATE SOLID SOLUTIONS.   04-0958
ZOELECTRIC PROPERTIES OF LEAD COBALTATE(0.33) NIOBATE(0.67)- TITANATE- ZIRCONATE SOLID SOLUTION CERAMICS. / 09-1549
TALS (BARIUM OR LEAD TITANATE, FROM SYSTEMS BARIA OR LITHIA, TITANIA, ALUMINA, SILICA). /FERROELECTRIC CRYS 04-0783
UCTIVE BARIUM TITANATE CERAMICS.   EFFECT OF EXCESS TITANIA ON THE ELECTRIC RESISTANCE OF SEMICOND 04-0966
TIES (LANTHANUM TITANATE).   SYSTEM LANTHANA TITANIA PHASE EQUILIBRIA AND ELECTRICAL PROPER 36-3592
ZIRCONATE TITANATE STANNATE. (CERAMIC DOPED WITH NIOBIUM AND TITANIUM)   MOSSBAUER STUDIES OF LEAD 09-1604
S (BARIUM TITANATE AND SAME WITH 1 PERCENT SILICON REPLACING TITANIUM). /SOME POLYCRYSTALLINE FERROELECTRIC 04-0818
TRICAL CONDUCTIVITY OF POTASSIUM NITRATE.   EFFECT OF ADDING TITANIUM(+) AND STRONTIUM(2+) IONS ON THE ELEC 16-2319
/TIC RESONANCE SPECTRA IN A FERROELECTRIC CERAMIC CONTAINING TITANIUM. (BARIUM TITANATE, STRONTIUM TITANAT/ 04-1020
THE HYDROXYL STRETCHING VIBRATION IN STRONTIUM TITANATE AND TITANIUM DIOXIDE. /RE AND PHASE TRANSITIONS ON 06-1209
DIELECTRIC DISPERSION IN BARIUM AND STRONTIUM TITANATES AND TITANIUM DIOXIDE.   FAR INFRARED 04-0975
YSTALS.   (PHASE TRANSITIONS AND) THERMOCURRENTS IN TITANIUM DIOXIDE AND BARIUM TITANATE SINGLE CR 04-0697
/M TITANATE CAPACITORS. PART-10: STRUCTURE OF SEMICONDUCTIVE TITANIUM DIOXIDE AND TITANIUM DIOXIDE- BARIUM/ 04-0774
CTRICAL PROPERTIES OF TERNARY BARIUM OXIDE- LANTHANUM OXIDE- TITANIUM DIOXIDE COMPOUNDS.   ELE 13-1912
B/ THE PRODUCTION OF BARIUM TITANATE FROM BARIUM SULFATE AND TITANIUM DIOXIDE IN THE PRESENCE OF SODIUM CAR 04-0624
SOME PROPERTIES OF THE THERMOCURRENTS IN BARIUM TITANATE AND TITANIUM DIOXIDE SINGLE CRYSTALS.   04-0622
)   OXYGEN STOICHIOMETRY OF DONOR DOPED BARIUM TITANATE AND TITANIUM DIOXIDE. (STOICHIOMETRIC COMPENSATION 04-0664
/ EFFECTS OF ADDITION OF GLASSES CONTAINING BARIUM OXIDE AND TITANIUM DIOXIDE (WITH SILICA AND ALUMINA) ON/ 04-0829
OF EXISTENCE OF PEROVSKITE PHASES IN THE SYSTEM LEAD OXIDE- TITANIUM DIOXIDE- LANTHANUM TRIOXIDE.   RANGE 08-1349
/ON OF STRONTIUM OR CALCIUM FOR LEAD, IRON FOR ZIRCONIUM AND TITANIUM, EFFECT ON ACOUSTIC RADIATION POWER) 09-1484
/RATION OF BARIUM TITANATE BY TRANSPORT REACTIONS (INVOLVING TITANIUM ESTERS TO YIELD SMALL CRYSTAL PRISMS/ 04-0625
OELECTRIC EFFECT IN FERROELECTRIC- ANTIFERROMAGNETIC BISMUTH TITANIUM FERRATE.   MAGNET 08-1360
S. STUDIES O/ DIRECT CHELATOMETRIC DETERMINATION OF LEAD AND TITANIUM IN LEAD TITANATE AND SIMILAR COMPOUND 05-1068
/S OF CERTAIN NEW LAYERED FERROELECTRICS. (LANTHANUM BISMUTH TITANIUM IRON OXIDE, PRASEODYMIUM BISMUTH TIT/ 13-1993
/TIES IN FERROELECTRICS OF THE FAMILY BISMUTH(4) BARIUM(M-2) TITANIUM(M+1) OXIDE(3)(M+2) (M=2,3,4). (DATA / 02-0066
POTASSIUM LANTHANUM LITHIUM NIOBIUM TITANATE, BARIUM LITHIUM TITANIUM NIOBATE). /LITHIUM NIOBATE TITANATE, 13-1923
/TE NIOBATE, STRONTIUM TITANATE, POTASSIUM LANTHANUM TITANIUM NIOBATE, POTASSIUM STRONTIUM NIOBATE/ 13-1962
         CRYSTAL STRUCTURE OF BISMUTH TITANIUM NIOBIUM OXIDE.   13-2058
/ TUNGSTEN BRONZE TYPE A(6)B(10)O(30). (WHERE B IS (NIOBIUM, TITANIUM) OR (NIOBIUM, TUNGSTEN), COMPILATION/ 13-1964
SYSTEMS. (WHERE A IS STRONTIUM OR BARIUM, AND B IS HAFNIUM, TITANIUM, OR TIN) / THE CADMIUM BORATE- ABO(3) 36-3528
         CRYSTAL STRUCTURE OF BISMUTH TITANIUM OXIDE.   13-1935
TION AND EPITAXY OF SPUTTERED FILMS OF FERROELECTRIC BISMUTH TITANIUM OXIDE SINGLE CRYSTALS.   PREPARA 13-2038
ELECTRON BEAM WRITING OF FERROELECTRIC DOMAINS IN BISMUTH TITANIUM OXIDE SINGLE CRYSTALS.   08-1300
E- TITANIUM DIOXIDE.   PROPERTIES OF BISMUTH(12) TITANIUM OXIDE(20) AND THE SYSTEM BISMUTH OXID 13-1994
IEZOELECTRIC PROPERTIES OF PLZT CERAMICS WITH A ZIRCONIUM TO TITANIUM RATIO OF 65 TO 35. / OF ELASTIC AND P 09-1573
E. PART-1: HIGH TEMPERATURE ELECTRICAL CONDUCTIVITY MEASURE/ TITANIUM RICH NONSTOICHIOMETRIC BARIUM TITANAT 04-0822
E. PART-3: HIGH TEMPERATURE THERMODYNAMIC AND X-RAY DIFFRAC/ TITANIUM RICH NONSTOICHIOMETRIC BARIUM TITANAT 04-0886
/MOLYBDATE. (WHERE A IS ZIRCONIUM, HAFNIUM, CERIUM, THORIUM, TITANIUM, SODIUM, MAGNESIUM, CADMIUM, YTTRIUM/ 10-1640
STRONTIUM TITANATE, LEAD(X) STRONTIUM(1-X) TITANATE, LEAD(X) TITANIUM(1-X) TITANATE) /OELECTRIC CRYSTALS. (05-1042
E D/ CALCULATION OF CRYSTAL FIELD ENERGY LEVEL SPLITTINGS OF TITANIUM(3+) ION IN RUBIDIUM ALUMINUM DISULFAT 19-2754
ND OPTICAL SWITCHING PROPERTIES OF SINGLE CRYSTAL BISMUTH(4) TITANIUM(3) OXIDE(12).   ELECTRICAL A 35-3517
         GROWTH OF LARGE SINGLE CRYSTAL BISMUTH(4) TITANIUM(3) OXIDE(12).   35-3507
OGRAMS IN A FERROELECTRIC PHOTOCONDUCTOR DEVICE. (BISMUTH(4) TITANIUM(3) OXIDE(12))   PHASE HOL 13-1977
/M(M-2) TITANIUM(M+1) OXIDE(3)(M+2) (M=2,3,4). (DATA FOR THE TITANIUM(3) OXIDE(12), BARIUM TITANIUM(4) OXI/ 02-0066
ANTIMONIDE, RUBIDIUM IRON TRIFLUORIDE, BORACITES, BISMUTH(4) TITANIUM(3) OXIDE(12), HAFNIUM DIVANADIDE /M 02-0036
         FERROELECTRIC DIELECTRIC ANOMALY IN BISMUTH(4) TITANIUM(3) OXIDE(12) SINGLE CRYSTALS.   13-2009
AND MOLYB/ THERMOGRAPHIC STUDY OF THE REACTION OF BISMUTH(4) TITANIUM(3) OXIDE(12) WITH VANADIUM PENTOXIDE 13-2029
F MANGANESE(4+) IN STRONTIUM TITANATE.   EFFECTIVE CHARGE OF TITANIUM(4+) FROM THE PARAMAGNETIC RESONANCE O 06-1167
/LID SOLUTION BASED ON BARIUM TITANATE, WHERE BARIUM(2+) AND TITANIUM(4+) IONS ARE REPLACED BY ISOVALENT A/ 04-0960
TIES OF BARIUM TITAN/ EFFECT OF NONISOVALENT SUBSTITUTION OF TITANIUM(4+) IONS ON THE ELECTROOPTICAL PROPER 04-0959
/M+2) (M=2,3,4). (DATA FOR THE TITANIUM(3) OXIDE(12), BARIUM TITANIUM(4) OXIDE(15), AND BARIUM(2) TITANIUM/ 02-0066
UM(3) OXIDE(12), BARIUM TITANIUM(4) OXIDE(15), AND BARIUM(2) TITANIUM(5) OXYGEN(18) COMPOUNDS) / THE TITANI 02-0066
ATION OF THE SPIN LATTICE RELAXATION TIMES IN FERROELECTRIC/ TONE BURSTING OF QUADRUPOLAR NUCLEI. DETERMIN 11-1829
ARACTERIZATION OF POTASSIUM NIOBATE.   (TOP SEEDED FLUX) CRYSTAL GROWTH AND NEUTRON CH 08-1353
DS. (GADOLINIUM MOLYBDATE, GROWTH BY RF CZOCHRALSKI AND FROM TOP SEEDED SOLUTION /NATE AND RELATED COMPOUN 08-1323
ROM NONSTOICHIOMETRIC MELTS. (BARIUM TITANATE, STRONTIUM TI/ TOP SEEDED SOLUTION GROWTH OF OXIDE CRYSTALS F 04-0553
RITE AND TGS CRYSTALS./ APPLICATION OF THE X-RAY DIFFRACTION TOPOGRAPHIC METHOD TO FERROELECTRIC SODIUM NIT 16-2375
FERROELECTRIC LEAD GERMANATE SINGLE CRYSTALS.   X-RAY TOPOGRAPHIC OBSERVATION OF 180 DEG DOMAINS IN 35-3515
IN FERROELECTRIC LEAD(5) GERMANIUM(3) OXIDE(11).   X-RAY TOPOGRAPHIC OBSERVATION OF 180 DEGREE DOMAINS 35-3516
N TRIGLYCINE SULFATE SINGLE CRYSTALS.   X-RAY TOPOGRAPHIC OBSERVATIONS OF RADIATION DAMAGE I 27-3106
R OF ICE SINGLE CRYSTALS.   X-RAY DIFFRACTION TOPOGRAPHIC STUDIES OF THE DEFORMATION BEHAVIO 34-3426
HIUM TANTALATE SINGLE CRYSTALS.   X-RAY TOPOGRAPHIC STUDIES ON LITHIUM NIOBATE AND LIT 11-1830
CTRIC DOMAINS IN GLYCOL SULFATE.   X-RAY TOPOGRAPHIC STUDY OF DISLOCATIONS AND FERROELE 19-2747
GS.   X-RAY TOPOGRAPHIC STUDY OF DOMAIN WALL MOVEMENT IN T 27-3026
ITE CRYSTALS. PART-2: MECHANISM OF POLARIZATION REVERSAL.   TOPOGRAPHIC STUDY OF FERROELECTRIC SODIUM NITR 16-2374
INE SULFATE CRYSTALS IN RELATION TO THEIR GROWTH COND/ X-RAY TOPOGRAPHIC STUDY OF GROWTH DEFECTS IN TRIGLYC 27-3025
IN POTASSIUM DIHYDROGEN PHOSPHATE CRYSTALS BY GROWTH/ X-RAY TOPOGRAPHIC STUDY OF THE DISLOCATIONS PRODUCED 17-2492
ITE CRYSTALS.   PART-1: STRUCTURE OF 180 DEGREE DOMAIN WALLS. TOPOGRAPHIC STUDY ON FERROELECTRIC SODIUM NITR 16-2372
ECTRIC DOMAIN BOUNDARIES (DOPED AND UNDOPED TGS, LITH/ X-RAY TOPOGRAPHICAL STUDY OF THE CONTRAST AT FERROEL 11-1671
DISPLAY OF TWINNING IN LITHIUM NIOBATE BY X-RAY DIFFRACTION TOPOGRAPHY.   11-1845
         X-RAY TOPOGRAPHY OF FERROELECTRIC CRYSTALS.   02-0121
/OWTH REGIONS IN FERROELECTRIC TRIGLYCINE SULFATE CRYSTALS. (TOPOGRAPHY OF IMPERFECTIONS, DOMAIN STRUCTURE) 27-2999
.   X-RAY TOPOGRAPHY OF THE DEFECT STRUCTURE OF THIOUREA 25-2914
KDP AND ADP SINGLE CRYSTALS.   X-RAY DIFFRACTION TOPOGRAPHY STUDY OF DEFECTS (DISLOCATIONS) IN 17-2578
FERROELECTRIC CERAMICS OBTAINED BY PROJECTION WITH A PLASMA TORCH OR FRITTING. (BARIUM TITANATE) /ILAYERED 04-0778
/ DUE TO INTERACTION OF DOMAIN BOUNDARIES AND POINT DEFECTS (TORSION PENDULUM AND RESONANCE TECHNIQUES, 20/ 09-1584
/)C(VII) COMPOUNDS OF ANTIMONY SULFOIODIDE TYPE. (PARTIAL OR TOTAL SUBSTITUTION OF ANTIMONY BY BISMUTH OR / 15-2205
UCER MATERIAL.   FRACTURE TOUGHNESS OF COMMERCIAL BARIUM TITANATE TRANSD 04-0841
DING.   FERROELECTRIC HYSTERESIS TRACER FEATURING COMPENSATION AND SAMPLE GROUN 37-3668
         NEW TYPE OF LOOP TRACER FOR FERROELECTRICS.   37-3666
DESCRIBED)   VARIOUS HYSTERESIS LOOP TRACERS FOR A WIDE FREQUENCY RANGE. (CIRCUITRY 37-3640

ULTRASONIC INTERFEROMETER WITH LEAD ZIRCONATE TITANATE AS TRANSDUCER. 09-1585
S RADIATING IN THE DIRECTIO/ RADIALLY VIBRATING FERROCERAMIC TRANSDUCER AS A SOURCE OF ULTRASONIC VIBRATION 37-3657
L MOBILITY AND MECHANICAL DAMPING IN LEAD TITANATE ZIRCONATE TRANSDUCER CERAMICS. /SION STUDY OF DOMAIN WAL 09-1547
TUDIES. (TRIGLYCINE SULFATE)   A NEW FERROELECTRIC TRANSDUCER FOR USE IN HEAT TRANSFER AND FLOW S 27-3031
          FRACTURE TOUGHNESS OF COMMERCIAL BARIUM TITANATE TRANSDUCER MATERIAL. 04-0841
ZIRCONATE TITANATE)   NEW COMPOSITE CERAMIC PIEZOELECTRIC TRANSDUCER MATERIAL. (ALUMINA WHISKERS IN LEAD 09-1556
TIC TRANSIENTS GENERATED BY ANTIFERROELECTRIC- FERROELECTRIC TRANSDUCERS.   ACOUS 37-3672
            FREQUENCY SPECTRUM OF FERROCERAMIC TRANSDUCERS. 37-3656
40 MHZ COLOR TELEVISION INFRARED RESPONSE USING SURFACE WAVE TRANSDUCERS ON LEAD ZIRCONATE TITANATE. /OF A 09-1509
URED ELECTRICAL CHARACTERISTICS OF INTERDIGITAL SURFACE WAVE TRANSDUCERS ON LITHIUM NIOBATE.   MEAS 11-1702
        SURFACE WAVE DELAY LINES WITH INTERDIGITAL TRANSDUCERS ON UNPOLARIZED PZT CERAMIC PLATES. 09-1614
)   A NEW FERROELECTRIC TRANSDUCER FOR USE IN HEAT TRANSFER AND FLOW STUDIES. (TRIGLYCINE SULFATE) 27-3031
INGLE CRYSTALS. PART-2: MOLECULAR MODELS FOR POLARIZATION A/ TRANSFER OF PROTONS THROUGH "PURE" ICE I(H), S 34-3455
LE CRYSTALS. PART-1: POLARIZATION SPECTRA OF ICE I(H). PART/ TRANSFER OF PROTONS THROUGH PURE ICE I(H) SING 34-3486
/CE, AND AQUEOUS SOLUTIONS AND THEIR INTERPRETATION. PART-8/ TRANSFER OF PROTONS THROUGH "PURE" ICE I(H) S/ 34-3482
ES.   EFFECT OF CHARGE TRANSFER ON RAMAN SCATTERING OF HYDROGEN HALID 31-3328
ARAELECTRIC POTASSIUM DIHYDROGEN PHOSPHATE AND POTA/ FOURIER TRANSFORM FAR INFRARED REFLECTION SPECTRA OF P 03-0356
ITY OF ADP AND KDP IN THE FAR INFRARED BY DISPERSIVE FOURIER TRANSFORM SPECTROSCOPY. /THE COMPLEX PERMITTIV 17-2609
AMICS IN STRONTIUM TITANATE IN THE NEIGHBORHOOD OF THE PHASE TRANSFORMATION.   EPR STUDIES AND DYN 06-1232
TICAL EFFECT IN RUBIDIUM TRIHYDROGEN OR DEUTERIUM DIS/ PHASE TRANSFORMATION AND THERMOOPTICAL AND ELECTROOP 22-2840
/Y OF GLASS TRANSITION OF THE AMORPHOUS ICE AND OF THE PHASE TRANSFORMATION BETWEEN THE CUBIC AND HEXAGONA/ 34-3476
S NEARLY COMPLETE DESCRIPTION)   FERROELECTRIC TRANSFORMATION IN BARIUM TITANATE. (MODEL GIVE 04-0836
S. (THEORY OF COMPOSITION DE/ PARTICULAR PROPERTIES OF PHASE TRANSFORMATION IN FERROELECTRIC SOLID SOLUTION 03-0250
        CRITICAL DYNAMICS OF THE ORDER DISORDER TRANSFORMATION IN FERROELECTRICS. (THEORY) 03-0281
TE.   NEUTRON MEASUREMENTS ON THE FERROELECTRIC PHASE TRANSFORMATION IN POTASSIUM DIHYDROGEN PHOSPHA 17-2418
IUM MOLYBDATE)   MECHANISM OF THE FERROELECTRIC PHASE TRANSFORMATION IN RARE EARTH MOLYBDATES. (TERB 30-3254
    NEUTRON SCATTERING STUDY OF THE FERROELECTRIC PHASE TRANSFORMATION IN TERBIUM MOLYBDATE. 30-3268
RROELASTIC CRYSTAL: DIAM/ CONJECTURED MECHANISM OF THE PHASE TRANSFORMATION OF A PECULIAR FERROELECTRIC- FE 19-2741
ATE, A MODEL IN WHICH / MECHANISM OF THE FERROELECTRIC PHASE TRANSFORMATION OF DIAMMONIUM DICADMIUM TRISULF 03-0141
TO MOTION. (PYROELECTRIC EFFECT IN TGS, ETHYL DIAMINE TARTR/ TRANSFORMATION OF ELECTROMAGNETIC RADIATION IN 04-0854
OELASTIC CRYSTAL: SODIUM AM/ INFERRED MECHANISM OF THE PHASE TRANSFORMATION OF PECULIAR FERROELECTRIC- FERR 29-3238
TE.   ORDER OF THE FERROELECTRIC PHASE TRANSFORMATION OF POTASSIUM DIHYDROGEN PHOSPHA 17-2555
    X-RAY AND NEUTRON STUDY ON THE PHASE TRANSFORMATION OF SODIUM NITRITE. 16-2328
/D ANTIFERROELECTRIC CERAMICS. PART-1: DEFORMATION AND PHASE TRANSFORMATION. PART-2: ELECTRIC CONDUCTION I/ 03-0169
I IODATES.   NUCLEAR QUADRUPOLE RESONANCE, PHASE TRANSFORMATIONS, AND FERROELECTRICITY OF ALKAL 08-1350
(BY DTA).   PHASE TRANSFORMATIONS AND MELTING IN KDP TO 40 KBAR 17-2634
/ WITH EXAMINATION OF CHARACTERS OF THEIR PARAELECTRIC PHASE TRANSFORMATIONS. (AND SUGGESTIONS FOR DISCOVE/ 03-0138
E TYPE COMPOUNDS, LANTHANUM(1-X) CALCIUM(X) MANGANATE (PHASE TRANSFORMATIONS CONFIGURATIONAL ENERGY). /SKIT 08-1356
        DYNAMICAL EFFECTS IN SOLID STATE TRANSFORMATIONS. (FERROELECTRICS) 03-0154
CHARACTERISTICS OF THE STRUCTURAL CHANGES INVOLVED IN PHASE TRANSFORMATIONS IN ANTIFERROELECTRICS. 03-0526
ITANATES. (/ (IN PILE, FAST NEUTRON) RADIATION INDUCED PHASE TRANSFORMATIONS IN (CERAMIC) BARIUM AND LEAD T 04-0971
ESENCE OF DOMAIN WALLS. (THEORY)   PHASE TRANSFORMATIONS IN FERROELECTRIC CRYSTAL IN PR 03-0346
    NEUTRON STUDIES OF DISPLACIVE STRUCTURAL PHASE TRANSFORMATIONS. (IN FERROELECTRICS) 03-0156
-1: CHANGE IN ULTRAVIOLET ABSORPTION SPECT/ STUDIES OF PHASE TRANSFORMATIONS IN NITRATES AND NITRITES. PART 16-2278
SIUM NITRITE. (DOES NOT SEEM FERROELECTRIC- PARAELECT/ PHASE TRANSFORMATIONS IN POTASSIUM NITRATE AND POTAS 16-2369
ITRATE, AND AMM/ MORPHOLOGY OF CRYSTAL GROWTH AT POLYMORPHIC TRANSFORMATIONS IN POTASSIUM NITRATE, SILVER N 16-2260
        TRANSFORMATIONS IN SODIUM NIOBATE. 08-1397
    NEW DATA FROM AN X-RAY STUDY OF PHASE TRANSFORMATIONS IN SODIUM NIOBATE. 08-1361
    X-RAY DIFFRACTOMETRY OF LOW TEMPERATURE PHASE TRANSFORMATIONS IN STRONTIUM TITANATE. 06-1156
    THEORY OF VIBRONIC PHASE TRANSFORMATIONS IN WIDE GAP FERROELECTRICS. 03-0349
ELECTRICS)   NEUTRON STUDIES OF DISPLACIVE STRUCTURAL PHASE TRANSFORMATIONS. (INELASTIC DIFFUSION IN FERRO 03-0155
        PHASE TRANSFORMATIONS OF IMPROPER FERROELECTRICS. 03-0335
ULAR FERROELECTRICS.   FERROELECTRIC TRANSFORMATIONS OF TENSORIAL PROPERTIES IN REG 03-0139
IUM MOLYBDATE (SOLID PHASE SYNTHESIS, N/ FORMATION AND PHASE TRANSFORMATIONS OF THE M MODIFICATION OF SAMAR 30-3270
CE. (USING, FOR EXAMPLE, PZT)   AN ADAPTIVE FERROELECTRIC TRANSFORMER - A SOLID STATE ANALOG MEMORY DEVI 09-1566
ERISTICS. (PZT-5H)   ADAPTIVE FERROELECTRIC TRANSFORMERS WITH IMPROVED TEMPERATURE CHARACT 09-1565
PERFORMANCE IN LITHIUM NIOBATE AND KDP POCKELS CELLS. TRANSIENT ELASTOOPTIC EFFECTS AND Q SWITCHING 11-1727
PERFORMANCE IN LITHIUM NIOBATE AND POTASSIUM DIDEUTERIUM P/ TRANSIENT ELECTROOPTIC EFFECTS AND Q SWITCHING 11-1710
TALLIC SODIUM TUNGSTEN BRONZES.   TRANSIENT NMR STUDY OF PHASE TRANSITIONS IN ME 13-1910
TRICS.   ASYMMETRIC MINOR LOOP AND TRANSIENT OPEN LOOP COMPUTATIONS FOR FERROELEC 03-0296
    ON THE EFFECTS OF X-RAY IRRADIATION ON TRANSIENT PHENOMENA IN ROCHELLE SALT. 28-3222
SERS WITH POLAR DIELECTRICS. (POTASSIUM DIHYDROGEN PHOSPHAT/ TRANSIENT PHENOMENA IN THE DISCHARGE OF CONDEN 17-2631
    DOMAIN STATISTICS AND FERROELECTRIC TRANSIENTS. 03-0216
ROELECTRIC TRANSDUCERS.   ACOUSTIC TRANSIENTS GENERATED BY ANTIFERROELECTRIC- FER 37-3672
(ON COLORING AND BLEACHING)   PHOTOCHROMIC (TRANSIENTS) IN PHOTOCHROMIC STRONTIUM TITANATE 06-1240
ALTATE(0.5) TUNGSTATE(0.5) AND LEAD COBALTATE(0.5) TU/ PHASE TRANSISITION IN THE ANTIFERROELECTRET LEAD COB 13-1968
/TION SPECTRA OF 23.8 KEV GAMMA-RAYS OF TIN-119 DURING PHASE TRANSISITIONS OF THE BISMUTH FERRATE STRONTIUM/ 08-1400
MONIUM HYDROGEN SULFATE, ITS STRUCTURE AND THE FERROELECTRIC TRANSITION.   AM 19-2773
ERING FROM SODIUM NIOBATE CRYSTALS (NEAR THE 640 DEG C PHASE TRANSITION). /PIC CRITICAL X-RAY DIFFUSE SCATT 08-1311
OBATE SINGLE CRYSTALS IN THE REGION OF A BROAD FERROELECTRIC TRANSITION. /PTION COEFFICIENT OF LEAD MAGNONI 13-1974
R OF POTASSIUM MANGANESE TRIFLUORIDE NEAR A STRUCTURAL PHASE TRANSITION.   ANOMALOUS ELASTIC BEHAVIO 33-3392
OF THE UNIVERSALITY HYPOTHESIS IN A LATTICE MODEL OF A PHASE TRANSITION.   APPARENT VIOLATION 03-0515
MANGANESE TRIFLUORIDE IN THE NEIGHBORHOOD OF THE STRUCTURAL TRANSITION. / OF ULTRASONIC WAVES IN POTASSIUM 33-3370
NTIUM TITANATE PRODUCED BY THE 105 DEGREE K STRUCTURAL PHASE TRANSITION.   BIREFRINGENCE OF STRO 06-1116
ANT OF TECHNETIUM-99 IN BARIUM TITANATE BY THE FERROELECTRIC TRANSITION.   CHANGE IN THE DECAY CONST 04-0871
E SOLID SOLUTIONS IN THE VICINITY OF THE FERROELECTRIC PHASE TRANSITION. /IN-119 IN BARIUM TITANATE STANNAT 04-0573
RUCTURE OF AMMONIUM HYDROGEN SULFATE ABOVE THE FERROELECTRIC TRANSITION. /MENT TECHNIQUES APPLIED TO THE ST 16-2355
CRITICAL BEHAVIOR OF STRONTIUM TITANATE NEAR THE 105 K PHASE TRANSITION. 06-1194
TS DIELECTRIC PROPERTIES. (FERROELECTRIC- PARAELECTRIC PHASE TRANSITION) /ADMATE(0.33) NIOBATE(0.67)) AND I 13-1959
IELECTRIC PERMEABILITY OF TRIGLYCINE SELENATE NEAR THE PHASE TRANSITION.   D 27-3166
CULIARITIES IN FERROELECTRICS WITH ORDER DISORDER TYPE PHASE TRANSITION. /ECTRIC PERMITTIVITY DISPERSION PE 03-0412
ABSORPTION IN STRONTIUM TITANATE AT THE CUBIC TO TETRAGONAL TRANSITION.   DISTORTION ENHANCED OPTICAL 06-1109
INGLE CRYSTALS OF TRIGLYCINE SULFATE (NEAR THE FERROELECTRIC TRANSITION). /ONS ON THE THERMAL CAPACITY OF S 27-3140
    ELASTIC ENERGY AND THE FERROELECTRIC PARAELECTRIC TRANSITION. 03-0237
ROPERTIES OF GADOLINIUM MOLYBDATE AT ITS FERROELECTRIC PHASE TRANSITION.   ELASTIC P 30-3283
    ELASTOOPTICAL EFFECT IN FERROELECTRIC WITH DIFFUSE PHASE TRANSITION. 03-0307
TIES OF LITHIUM NIOBATE. (ANOMALY AT 263 DEG K IS STRUCTURAL TRANSITION)   ELECTRICAL PROPER 11-1709
ICAL EFFECT IN FERROELECTRIC CRYSTALS HAVING A SMEARED PHASE TRANSITION.   ELECTROOPT 03-0174
NIUM BISULFATE IN THE VICINITY OF THE HIGH TEMPERATURE PHASE TRANSITION. / FERROELECTRIC PROPERTIES OF AMMO 19-2787
TITANATE ZIRCONATE COMPOSITIONS NEAR THE MORPHOTROPIC PHASE TRANSITION. /ROPERTIES OF POLYCRYSTALLINE LEAD 09-1557

FECTS ON THE PYROELECTRIC BEHAVIOR OF TGS CLOSE TO THE PHASE TRANSITION.   INFLUENCE OF SURFACE EF 27-3064
MICROWAVE LOSSES IN STRONTIUM TITANATE ABOVE THE PHASE TRANSITION.   06-1199
ECTRIC PHASE OF SODIUM NITRITE. (THERMODYNAMICS OF THE PHASE TRANSITION) /3 IN THE PARAELECTRIC AND FERROEL 16-2264
ON EDGE OF STRONTIUM TITANATE AROUND THE 105 DEGREES K PHASE TRANSITION.   OPTICAL ABSORPTI 06-1099
SOLID SOLUTIONS. (THEORY OF COMPOSITION DEPENDENCE OF PHASE TRANSITION) /E TRANSFORMATION IN FERROELECTRIC 03-0250
POLARIZATION FLUCTUATIONS NEAR A FERROELECTRIC PHASE TRANSITION.   03-0368
MENT OF THE DOMAIN STRUCTURE IN ROCHELLE SALT NEAR THE PHASE TRANSITION.   REARRANGE 28-3181
TRIC LEAD MAGNESATE(0.33) NIOBATE(0.67) WITH A DIFFUSE PHASE TRANSITION. /ATION POLARIZATION OF A FERROELEC 13-1980
F NEW PHASE IN POTASSIUM NITRATE. (ALPHA TO BETA POLYMORPHIC TRANSITION) RHYTHMIC GROWTH O 16-2263
ITHIUM THALLIUM TARTRATE MONOHYDRATE. (EFFECT ON LATTICE AND TRANSITION) /OF NONDEUTERATED AND DEUTERATED L 29-3239
SPECIFIC HEAT OF STRONTIUM TITANATE NEAR THE STRUCTURAL TRANSITION.   06-1132
STRUCTURE OF AMMONIUM SULFATE IN FERROELECTRIC PHASE AND TRANSITION.   19-2774
NG SPECTRUM OF FERROELECTRIC DEUTERATED KDP DURING THE PHASE TRANSITION. /TURE DEPENDENCE OF RAMAN SCATTERI 17-2626
E OF THE SOFT MODE IN STRONTIUM TITANATE ABOVE THE 105 DEG K TRANSITION.   TEMPERATURE DEPENDENC 06-1178
THEORY OF RAYLEIGH AND BRILLOUIN SCATTERING NEAR THE PHASE TRANSITION.   03-0334
AND DEUTERATED ADP IN THE REGION OF ANTIFERROELECTRIC PHASE TRANSITION.   THERMAL EXPANSION OF ADP 17-2445
HERMAL EXPANSIVITY OF STRONTIUM TITANATE NEAR THE STRUCTURAL TRANSITION.   T 06-1136
RASONIC PROPERTIES OF STRONTIUM TITANATE AT THE 105 DEGREE K TRANSITION.   ULT 06-1191
RUCTURE OF AMMONIUM HYDROGEN SULFATE ABOVE THE FERROELECTRIC TRANSITION. /N DETERMINATION OF THE CRYSTAL ST 19-2775
OF HARMONICS IN A FERROELECTRIC CRYSTAL WITH A SECOND ORDER PHASE TRANSITION ABOVE THE CURIE POINT. / GENERATION 03-0331
IUM DIHYD/ STRUCTURAL SENSITIVITY OF THE FERROELECTRIC PHASE TRANSITION AND DIELECTRIC PROPERTIES OF POTASS 17-2615
LFOIODIDE CRYSTALS. (MICRO CINEMATOGRAPHY) PHASE TRANSITION AND DOMAIN STRUCTURE IN ANTIMONY SU 15-2184
/ATE (OCCURANCE IN THIN (110) PLATES BELOW CUBIC- TETRAGONAL TRANSITION, AND PRODUCTION BY SHAPING AND STR/ 06-1164
ANOMALOUS ABSORPTION OF SOUND NEAR THE FERROELECTRIC PHASE TRANSITION (ANTIMONY SULFOIODIDE). 15-2221
/ELECTRIC SEMICONDUCTORS AT TEMPERATURES NEAR A TYPE-1 PHASE TRANSITION. (ANTIMONY SULFOIODIDE, SELENOIODI/ 15-2114
/ITY ANOMALY AS AN EVIDENCE OF THE LOCAL FERROELECTRIC PHASE TRANSITION AROUND 47 DEG K IN STRONTIUM TITAN/ 06-1249
ONLINEARITY COEFFICIENT IN STRONTIUM TITANATE NEAR THE PHASE TRANSITION AT 110 DEG K. / OF THE DIELECTRIC N 06-1105
ONLINEARITY OF STRONTIUM TITANATE SINGLE CRYSTALS NEAR THE PHASE TRANSITION AT 110 DEGREES K.   DIELECTRIC N 06-1106
RROELECTRIC DOMAINS OF AMMONIUM DIHYDROGEN PHOSPHATE. (PHASE TRANSITION AT 148 DEG K) /BSERVATION OF ANTIFE 17-2643
L ACTIVITY OF FERROELECTRIC DICALCIUM STRONTIUM PROPIONATE. (TRANSITION AT 209 DEGREES K) OPTICA 27-3040
URE OF DIELECTRIC POLARIZATION RELAXATION I/ CAUSES OF PHASE TRANSITION BROADENING (DIFFUSNESS) AND THE NAT 03-0291
LECTRIC PROPERTIES OF A CERAM/ EFFECT OF FERROELECTRIC PHASE TRANSITION CONDITIONS ON DIELECTRIC AND PIEZOE 09-1591
FERROELECTRICS. A NEW TYPE OF SECOND ORDER PHASE TRANSITION DERIVED FROM DEVONSHIRE'S THEORY OF 03-0226
AND LANTHA/ PARAMAGNETIC RESONANCE AND OPTICAL ABSORPTION OF TRANSITION ELEMENT IONS IN STRONTIUM TITANATE 06-1165
/G IN TRIGLYCINE SULFATE AND SODIUM NITRATE. (ORDER DISORDER TRANSITION, FERROELECTRIC AND NONFERROELECTRI/ 16-2400
THE BEHAVIORS OF THE MICRODOMAINS ON THE DIFFUSE PHASE TRANSITION FERROELECTRICS. 03-0283
OGEN PHOSPHATE TYPE CRYSTALS. ABSENCE OF A LINE WIDTH TRANSITION FOR THE PROTONS IN POTASSIUM DIHYDR 17-2436
ICALLY ORDERED FERROELECTRIC STATE. (THEORY) PHASE TRANSITION FROM NORMAL FERROELECTRIC TO MAGNET 03-0394
ORY). SELF CONSISTENCY OF LANDAU'S MODEL IN THE PHASE TRANSITION FROM PIEZO TO FERROELECTRICITY (THE 03-0507
MAGNETIC AND FERROELECTRIC STA/ THEORY OF SECOND ORDER PHASE TRANSITION FROM THE FERROMAGNETIC TO THE FERRO 03-0395
ITANATE. NUCLEATION MODELS FOR A FIRST ORDER TRANSITION FRONT, AND OBSERVATIONS IN BARIUM T 04-0608
TALLINE LITHIUM NIOBATE. PROPERTIES OF THE IRON TRANSITION GROUP IN THE LATTICE OF SINGLE CRYS 11-1669
DIPOLE STRUCTURE: CESIUM COPPER TRICHLORIDE. A PHASE TRANSITION IN A COMPOUND WITH HELICAL ELECTRIC 33-3388
NDS/ EFFECT OF VIOLATION OF ELECTRIC NEUTRALITY ON THE PHASE TRANSITION IN A FERROELECTRIC WITH HYDROGEN BO 03-0449
ATE CRYSTAL UNDER HIGH HYDROSTATIC PRESSURE (/ FERROELECTRIC TRANSITION IN A POTASSIUM FERROCYANIDE TRIHYDR 24-2896
/ON OF FERROELECTRIC DOMAIN STRUCTURES, MODULATED BY A PHASE TRANSITION, IN A SCANNING ELECTRON MICROSCOPE. 03-0474
FERROELECTRIC SQUARE LATTICE). PHASE TRANSITION IN A SIXTEEN VERTEX LATTICE MODEL (03-0517
TWO-PHONON SCATTERING AND FERROELECTRIC TRANSITION IN A SODIUM NITRITE CRYSTAL. 16-2277
TRA AND SPIN LATTICE RELAXATION INVESTIGATIONS OF THE PHASE TRANSITION IN ALUMS. /ESONANCE (WIDE LINE SPEC 20-2802
UTERATED AMMONIUM DICHLORO ACETA/ FERROELECTRICITY AND PHASE TRANSITION IN AMMONIUM DICHLORO ACETATE AND DE 36-3538
A THEORY OF THE PHASE TRANSITION IN AMMONIUM DIHYDROGEN PHOSPHATE. 03-0286
NGLE CRYSTAL BY NMR. INVESTIGATION OF THE PHASE TRANSITION IN AMMONIUM DIHYDROGEN PHOSPHATE SI 17-2545
OCHLORO ACETIC / NUCLEAR QUADRUPOLE RESONANCE STUDY OF PHASE TRANSITION IN AMMONIUM MONOCHLORO ACETATE, MON 36-3625
MECHANISM OF FERROELECTRIC PHASE TRANSITION IN AMMONIUM ROCHELLE SALT. (THEORY) 28-3223
RAMAN AND INFRARED STUDIES OF THE FERROELECTRIC TRANSITION IN AMMONIUM SULFATE. 19-2793
NEUTRONS (/ ROTATIONAL MOTIONS IN SOLIDS: THE FERROELECTRIC TRANSITION IN AMMONIUM SULFATE STUDIED BY COLD 19-2752
, OPTICAL, AND ELECTROOPTICAL PROPERTIES IN T/ FERROELECTRIC TRANSITION IN AMMONIUM SULFATE, THE DIELECTRIC 19-2745
LATTICE MODES AND PHASE TRANSITION IN ANTIMONY SULFOIODIDE. 15-2110
SOFT PHONON COUPLINGS IN THE PRESSURE INDUCED PHASE TRANSITION IN ANTIMONY SULFOIODIDE. 15-2240
NONLINEAR EFFECTS AND THE LOW TEMPERATURE TRANSITION IN ANTIMONY SULFOIODIDE CRYSTALS. 15-2247
MOTION PICTURES OF THE PHASE TRANSITION IN ANTIMONY SULFOIODIDE CRYSTALS. 15-2120
TALS. FEATURES OF THE PHASE TRANSITION IN ANTIMONY SULFOIODIDE SINGLE CRYS 15-2148
TALS. EFFECT OF OXYGEN ADSORPTION ON THE FERROELECTRIC PHASE TRANSITION IN ANTIMONY SULFOIODIDE SINGLE CRYS 15-2187
XIDE. PHASE TRANSITION IN BARIUM BISMUTH(5+) BISMUTH(3+) O 13-1940
EX/ DILATOMETRIC STUDY OF THE ORTHORHOMBIC- TETRAGONAL PHASE TRANSITION IN BARIUM SODIUM NIOBATE. (THERMAL 13-1894
FIELD INDUCED NUCLEATION AT THE (CUBIC TETRAGONAL) PHASE TRANSITION IN BARIUM TITANATE. 04-0753
SHOCK INDUCED PHASE TRANSITION IN BARIUM TITANATE. 04-0589
ULTRASONIC STUDIES OF FIELD INDUCED PHASE TRANSITION IN BARIUM TITANATE. 04-0549
EFFECT OF FERRIC OXIDE ON THE 120 DEGREE PHASE TRANSITION IN BARIUM TITANATE SINGLE CRYSTALS. 04-0544
THERMODYNAMIC THEORY OF THE CUBIC- ORTHORHOMBIC PHASE TRANSITION IN BORACITES. 03-0231
/ELECTRIC CONSTANT BEHAVIOR IN THE NEIGHBORHOOD OF THE PHASE TRANSITION IN CHARACTERISTICALLY FERROELECTRI/ 03-0522
FERROELECTRIC TRANSITION IN COBALT IODINE BORACITE. 14-2099
RESSURE. FERROELECTRIC TRANSITION IN COBALT- IODINE- BORACITE UNDER P 14-2078
X-RAY DIFFRACTION STUDY OF PHASE TRANSITION IN COPPER FORMATE. 26-2938
S, PIEZOELECTRICS) LOW TEMPERATURE DISPLACEMENT TYPE PHASE TRANSITION IN CRYSTALS. (THEORY, FERROELECTRIC 03-0320
CTURAL GROUPS. (SODIUM AM/ MECHANISMS OF FERROELECTRIC PHASE TRANSITION IN CRYSTALS WITH REORIENTATING STRU 22-2937
/LLIUM-9 QUADRUPOLE PERTURBED NMR STUDY OF THE FERROELECTRIC PHASE TRANSITION IN DEUTERATED TRIGLYCINE FLUOBERYL/ 27-2965
ULTRASONIC ATTENUATION NEAR THE PHASE TRANSITION IN DICALCIUM LEAD PROPIONATE 26-2949
ATES SUBSTITUTED PARTIALLY BY ACETATE IONS. PHASE TRANSITION IN DIVALENT METAL DICALCIUM PROPION 26-2944
INIUM MOLYBDATE. ANOMALOUS PHONON BEHAVIOR NEAR THE PHASE TRANSITION IN FERROELASTIC FERROELECTRIC GADOL 30-3276
NIUM ALUMINUM SULFATE, AMMONIUM ALUM/ INVESTIGATION OF PHASE TRANSITION IN FERROELECTRIC ALUMS (METHYL AMMO 20-2809
DE CRYSTALS USING THE RAMAN SCATTERING SPECTRUM. (LAT/ PHASE TRANSITION IN FERROELECTRIC ANTIMONY SULFOIODI 15-2143
POUNDS) A/ INFLUENCE OF NONEQUILIBRIUM CARRIERS ON THE PHASE TRANSITION IN FERROELECTRIC SEMICONDUCTOR (COM 15-2145
) CHANGE IN PROBABILITY OF THE MOSSBAUER EFFECT AT THE PHASE TRANSITION IN FERROELECTRICS. ABRUPT (STEPWISE 03-0276
ENTROPY OF AN ORDER DISORDER TRANSITION. (IN FERROELECTRICS, MECHANISM) 03-0192
ATE, HIGHLY ACCURATE TEMPERATURE CONTROL) ORDER DISORDER TRANSITION IN FERROELECTRICS. (TRIGLYCINE SULF 27-2982
DS. (KDP,/ INFLUENCE OF POLAR STATES OF PROTONS ON THE PHASE TRANSITION IN FERROELECTRICS WITH HYDROGEN BON 17-2671
EFFECT OF HYDROSTATIC PRESSURE ON PHASE TRANSITION IN GADOLINIUM MOLYBDATE. 30-3321
PHASE TRANSITION IN GADOLINIUM MOLYBDATE. 30-3316
C AND EL/ PHENOMENOLOGICAL THEORY OF THE FERROELECTRIC PHASE TRANSITION IN GADOLINIUM MOLYBDATE. (DIELECTRI 30-3303

CHANGES)   ORIGIN OF THE STRUCTURAL PHASE TRANSITION IN GADOLINIUM MOLYBDATE. (SYMMETRY   30-3273
MATERIALS. (ROCHELLE SALT, TR/ PHASE STABILIZATION (HINDERED TRANSITION) IN GAMMA-IRRADIATED FERROELECTRIC   27-3165
                                            STRUCTURE AND PHASE TRANSITION IN HYDROGEN CHLORIDE.   31-3337
                                                           PHASE TRANSITION IN ICE.   34-3442
PHOSPHATE.   MOSSBAUER STUDY OF ANTIFERROELECTRIC TRANSITION IN IRON-57(3X) AMMONIUM DIHYDROGEN   17-2644
                             EXPERIMENTS ON THE ORDER OF THE TRANSITION IN KDP.   17-2603
                     LIGHT SCATTERING STUDY OF THE FERROELECTRIC TRANSITION IN KDP.   17-2462
ZING ENERGY IN IT. (THEORY)   PHASE TRANSITION IN KDP AND THE ROLE OF THE DEPOLARI   03-0229
DYNAMICAL THEORY).   PHASE TRANSITION IN KDP (MODIFICATION OF KOBAYASHI'S   17-2488
LY)   POLARIZATION (DOMAIN STRUCTURE) AND PHASE TRANSITION IN KDP SINGLE CRYSTAL. (ABSTRACT ON   17-2505
TURES OF POTASSIUM, RUBIDIUM/ DTA STUDY OF THE FERROELECTRIC TRANSITION IN KDP TYPE CRYSTALS (CURIE TEMPERA   17-2577
                         NMR STUDY OF THE STRUCTURAL PHASE TRANSITION IN LANTHANUM ALUMINATE.   36-3533
                             INVESTIGATION OF PHASE TRANSITION IN LEAD CADMIUM TUNGSTATE.   10-1633
THE ANOMALOUS BEHAVIOR OF PHONON NEAR THE 60 DEGREES C PHASE TRANSITION IN LEAD DICALCIUM PROPIONATE.   26-2942
                                    ATHERMAL PHASE TRANSITION IN LEAD ZIRCONATE.   07-1258
        NATURE OF THE FERROELECTRIC- PARAELECTRIC PHASE TRANSITION IN LITHIUM TANTALATE.   12-1879
TE.   (NMR STUDY OF) FERROELECTRIC TRANSITION IN (MANGANESE DOPED) AMMONIUM SULFA   19-2784
LS BY MODIFIED MOLECULAR FIELD APPROXIMATION. STUDY OF PHASE TRANSITION IN MIXED KDP- DEUTERATED KDP CRYSTA   17-2528
/ATION OF SOUND VELOCITY AT FERROELECTRIC-1, FERROELECTRIC-2 TRANSITION IN (NIOBIUM DOPED) ZIRCONATE(0.95)/   09-1511
U/ INVESTIGATION OF POLARIZATION NONLINEARITY NEAR THE PHASE TRANSITION IN PEROVSKITE LIKE FERROELECTRICS,   03-0249
                              THEORY OF STRUCTURAL PHASE TRANSITION IN PEROVSKITES.   03-0242
2, INTERACTION WITH ELASTIC ST/ THEORY OF A STRUCTURAL PHASE TRANSITION IN PEROVSKITE- TYPE CRYSTALS, PART-   03-0241
YSTALS.   FIRST ORDER PHASE TRANSITION IN POTASSIUM DEUTERIUM PHOSPHATE CR   17-2677
                         FERROELECTRIC TRANSITION IN POTASSIUM DIDEUTERIUM PHOSPHATE.   17-2611
        INTERNAL STRESS AND THE FERROELECTRIC TRANSITION IN POTASSIUM DIHYDROGEN PHOSPHATE.   17-2424
        THERMAL HYSTERESIS OF THE FERROELECTRIC TRANSITION IN POTASSIUM DIHYDROGEN PHOSPHATE.   17-2682
        DIELECTRIC EVIDENCE OF A FIRST ORDER TRANSITION IN POTASSIUM DIHYDROGEN PHOSPHATE.   17-2602
YPE FERROELECTRICS. (24 REFS)   MODELS OF THE PHASE TRANSITION IN POTASSIUM DIHYDROGEN PHOSPHATE T   03-0489
/F DIPOLE MOMENTS OF WATER MOLECULES AND FERROELECTRIC PHASE TRANSITION IN POTASSIUM FERROCYANIDE AND `ISOM   24-2893
E.   MOSSBAUER STUDY OF THE FERROELECTRIC PHASE TRANSITION IN POTASSIUM FERROCYANIDE TRIHYDRAT   24-2901
     INFRARED ACTIVE MODE BELOW THE STRUCTURAL TRANSITION IN POTASSIUM MANGANESE TRIFLUORIDE.   33-3362
     SUPERSTRUCTURE ARISING DURING THE PHASE TRANSITION IN POTASSIUM MANGANESE TRIFLUORIDE.   33-3402
     PRESSURE INDUCED FERROELECTRIC PHASE TRANSITION IN POTASSIUM NITRATE.   16-2378
                        THE STRUCTURE AND PHASE TRANSITION IN POTASSIUM SELENATE.   22-2872
SODIUM NIOBATE T(2) AT 600 DEG C, AND THE CUBIC- TETRAGONAL TRANSITION IN RELATION TO SOFT PHONON MODES. /   08-1337
SCATTERING OF COLD NEUTRONS.   HIGH TEMPERATURE PHASE TRANSITION IN RUBIDIUM DIHYDROGEN PHOSPHATE BY   17-2489
ING RAMAN SCATTERING SPECTRA.   FERROELECTRIC PHASE TRANSITION IN RUBIDIUM DIHYDROGEN PHOSPHIDE US   17-2620
     IMPROPER FERROELECTRIC PHASE TRANSITION IN RUBIDIUM TRIHYDROGEN SELENITE.   22-2834
DIELECTRIC STUDIES ON THE FERROELECTRIC PHASE TRANSITION IN SILVER SODIUM DINITRITE.   16-2289
                    FERROELECTRIC PHASE TRANSITION IN SILVER SODIUM DINITRITE   16-2266
E./ DIELECTRIC AND SOME OTHER STUDIES OF FERROELECTRIC PHASE TRANSITION IN SODIUM AMMONIUM SULFATE DIHYDRAT   19-2767
     DIFFUSE X-RAY STUDY OF 640 DEGREES C PHASE TRANSITION IN SODIUM NIOBATE.   08-1358
                     X-RAY STUDY OF PHASE TRANSITION IN SODIUM NITRATE.   16-2380
            AN X-RAY STUDY OF THE PHASE TRANSITION IN SODIUM NITRITE.   16-2318
                              PHASE TRANSITION IN SODIUM NITRITE.   16-2397
     QUADRUPOLE SPIN PHONON RELAXATION AND FERROELECTRIC TRANSITION. (IN SODIUM NITRITE)   16-2362
(PHASE TRANSITION AT 163 DEG C).   FERROELECTRIC PHASE TRANSITION IN SODIUM NITRITE AS STUDIED BY EPR   16-2359
ATION OF RAMAN SPECTRUM.   INVESTIGATION OF FERROELECTRIC TRANSITION IN SODIUM NITRITE CRYSTAL BY OBSERV   16-2275
        MODELS FOR THE ORDER DISORDER PHASE TRANSITION IN SODIUM TRIHYDROGEN SELENATE.   22-2842
DIELECTRIC AND NMR STUDIES OF THE PHASE TRANSITION IN STANNOUS CHLORIDE DIHYDRATE.   36-3582
     AN X-RAY STUDY OF THE PHASE TRANSITION IN STRONTIUM FERRATE TANTALATE   08-1382
     ABSORPTION EDGE AND STRUCTURAL PHASE TRANSITION IN STRONTIUM TITANATE.   06-1107
ANOMALOUS ULTRASONIC ATTENUATION AT THE 105K TRANSITION IN STRONTIUM TITANATE.   06-1189
            CUBIC TO TRIGONAL STRESS INDUCED PHASE TRANSITION IN STRONTIUM TITANATE.   06-1235
STUDIES OF THE SOFT MODES NEAR THE CUBIC TO TETRAGONAL PHASE TRANSITION IN STRONTIUM TITANATE. /SCATTERING   06-1242
IONS FROM NEUTRON SCATTERING STUDIES OF THE STRUCTURAL PHASE TRANSITION IN STRONTIUM TITANATE. NEW OBSERVAT   06-1210
            NUCLEAR MAGNETIC RESONANCE STUDY OF THE PHASE TRANSITION IN STRONTIUM TITANATE.   06-1237
ESCENCE STUDY OF CHROMIUM(3+) THROUGH A STRESS INDUCED PHASE TRANSITION IN STRONTIUM TITANATE. /RIZED FLUOR   06-1115
            SOUND PROPAGATION NEAR THE STRUCTURAL PHASE TRANSITION IN STRONTIUM TITANATE.   06-1155
STRESS EFFECTS, AND INTERACTION PARAMETERS AT THE DISPLACIVE TRANSITION IN STRONTIUM TITANATE. /OPAGATION,   06-1127
SPECIFIC HEAT ANOMALY ASSOCIATED WITH THE 110 DEG K PHASE TRANSITION IN STRONTIUM TITANATE. (ABSTRACT)   06-1226
C OSCI/ FREE ENERGY AND SPECIFIC HEAT IN FERROELECTRIC PHASE TRANSITION IN TERMS OF A SINGLE MODE ANHARMONI   03-0406
     INTERPRETATION OF THE DYNAMICS OF THE BARIUM TITANATE TRANSITION IN TERMS OF NUCLEATION. (CONCEPTS)   04-0606
, X-RAY SCATTERING)   STUDIES ON THE MECHANISM OF THE PHASE TRANSITION IN TGS. (STATIC DIELECTRIC CONSTANT   27-3093
DIHYDROGEN) PHOSPHATES.   PHASE TRANSITION IN THALLIUM AND AMMONIUM THALLIUM (17-2571
ITANATE) SOLID SOLUTIONS. / ANTIFERROELECTRIC- FERROELECTRIC TRANSITION IN THE LEAD (ZIRCONATE, STANNATE, T   09-1486
EMICONDUCTOR PHTHALOCYANINE.   LOW TEMPERATURE PHASE TRANSITION IN THE POLAR STATE OF THE ORGANIC S   36-3623
FERROELECTRIC.   SECOND ORDER PHASE TRANSITION IN THE THREE DIMENSIONAL MODEL OF A   03-0450
DE ALLOYS. (NEUTRON SCATTERING)   TRANSITION IN TIN TELLURIDE- GERMANIUM TELLURI   36-3596
                    NEW HIGH PRESSURE PHASE TRANSITION IN TRIGLYCINE SELENATE.   27-3090
     APPLICABILITY OF THE PIPPARD RELATIONS TO THE PHASE TRANSITION IN TRIGLYCINE SULFATE.   27-3157
ECT OF SEVERAL PARAMETERS ON THE FERROELECTRIC- PARAELECTRIC TRANSITION IN TRIGLYCINE SULFATE.   EFF   27-2981
            INVESTIGATION OF THE PHASE TRANSITION IN TRIGLYCINE SULFATE.   27-3054
NITROGEN PROTON DOUBLE RESONANCE STUDY OF THE FERROELECTRIC TRANSITION IN TRIGLYCINE SULFATE.   PULSED   27-2968
AL CONTROL)   EXPERIMENTAL DETAILED STUDY OF THE PHASE TRANSITION IN TRIGLYCINE SULFATE. (CLOSE THERM   27-3030
SSURE.   PHASE TRANSITION IN TUNGSTEN TRIOXIDE UNDER HIGH PRE   10-1658
TRIC TEMPERATURE 340/ TWO COMPONENTS OF THE CRYSTALLOGRAPHIC TRANSITION IN VANADIUM DIOXIDE. (ANTIFERROELEC   36-3562
LLATE, ETC) OF SOME MONOCLINIC CRYSTALS IN THE FERROELECTRIC TRANSITION MECHANISM. /NATE, SULFATE, FLUOBERY   03-0330
     OPTICAL AND HOLOGRAPHIC STORAGE PROPERTIES OF TRANSITION METAL DOPED LITHIUM NIOBATE.   11-1796
                    PHOTOCHROISM IN TRANSITION METAL DOPED STRONTIUM TITANATE.   06-1126
/AGNETIC BETA GADOLINIUM MOLYBDATE. CRYSTAL STRUCTURE OF THE TRANSITION METAL MOLYBDATES AND TUNGSTATES. P/   30-3289
ELATED TO NONSTOICHIOMETRY (AND ANTIFERROELECTRIC DOMAINS IN TRANSITION METAL OXIDES). / EXTENDED DEFECTS R   36-3522
STRONTIUM TITANATE: VACANCY DRIFT AND OXIDATION REDUCTION OF TRANSITION METALS.   ELECTROCOLORATION IN   06-1098
PERTIES AND STRUCTURES OF METANIOBATES AND METATANTALATES OF TRANSITION METALS OF THE 3D SERIES. /RICAL PRO   06-1236
TITANATE. (ABSTRACT)   RAMAN STUDIES OF STRESS INDUCED PHASE TRANSITION NEAR CURIE TEMPERATURE IN STRONTIUM   06-1236
TIUM OXIDE- TELLURIUM DIOXIDE.   PHASE TRANSITION OF A NEW FERROELECTRIC OXIDE, STRON   36-3627
TE. PART-1: THE CRYSTAL STR/ THE CRYSTAL STRUCTURE AND PHASE TRANSITION OF AMMONIUM HYDROGEN DICHLORO ACETA   36-3567
E STRUCTURE OF THE FERROELECTRIC PHASE.   ON THE PHASE TRANSITION OF AMMONIUM HYDROGEN SULFATE AND TH   19-2711
EFFECT OF HYDROSTATIC PRESSURE ON THE FERROELECTRIC PHASE TRANSITION OF AMMONIUM SULFATE.   16-2360
                    FERROELECTRIC TRANSITION OF AMMONIUM SULFATE.   19-2798

FERROELECTRIC TRANSITION OF ANTIMONY SULFOIODIDE. 15-2195
INFLUENCE OF ONE DIMENSIONAL MECHANICAL STRESS ON THE PHASE TRANSITION OF ANTIMONY SULFOIODIDE SINGLE CRY/ 15-2245
EFFECT OF GRAIN SIZE ON THE FERROELECTRIC PARAELECTRIC TRANSITION OF BARIUM TITANATE. 04-0917
THERMAL HYSTERESIS IN THE TETRAGONAL CUBIC TRANSITION OF BARIUM TITANATE. 04-0921
MAGNETISM AND THE PHASE TRANSITION OF COPPER FORMATE TETRAHYDRATE. 26-2930
CRYSTAL STRUCTURE AND PHASE TRANSITION OF DIGLYCINE NITRATE. 16-2312
M HEXAACETATE. X-RAY AND OPTICAL STUDIES ON THE PHASE TRANSITION OF FERROELECTRIC DICALCIUM STRONTIU 36-3584
E. (SECOND ORDER) CHARACTER OF PHASE TRANSITION OF FERROELECTRIC POTASSIUM TANTALAT 08-1307
DS KDP TYPE. ROLE OF PROTON PHONON INTERACTION IN THE PHASE TRANSITION OF FERROELECTRICS WITH HYDROGEN BON 17-2674
FERROELECTRIC PHASE TRANSITION OF FIRST ORDER IN BARIUM TITANATE. 04-1030
ABNORMAL ABSORPTION OF ELASTIC WAVES NEAR THE PHASE TRANSITION OF GADOLINIUM MOLYBDATE. 30-3265
DIELECTRIC AND X-RAY STUDIES ON THE FERROELECTRIC PHASE TRANSITION OF GLYCINE SILVER NITRATE. 16-2347
CRYSTAL STRUCTURE AND PHASE TRANSITION OF HYDROGEN CHLORIDE. 31-3334
THE EFFECT OF PRESSURE ON THE TRANSITION OF LEAD ZINCATE(0.33) NIOBATE(0.67) 08-1473
PART-1:/ X-RAY DILATOMETRIC STUDY OF THE FERROELECTRIC PHASE TRANSITION OF POTASSIUM DIHYDROGEN PHOSPHATE. 17-2557
DIELECTRIC DISPERSION AND PHASE TRANSITION OF POTASSIUM HYDROXIDE DOPED ICE. 34-3452
SINGLE C/ CRITICAL ANISOTROPIC FLUCTUATIONS AT THE 184 DEG K TRANSITION OF POTASSIUM MANGANESE TRIFLUORIDE 33-3369
AT 184 DEGREES K. X-RAY SCATTERING AND THE PHASE TRANSITION OF POTASSIUM MANGANESE TRIFLUORIDE 33-3394
FAR INFRARED SPECTRA AND FERROELECTRIC PHASE TRANSITION OF POTASSIUM NITRATE. 16-2308
C GENERATION. (TEMPERATURE DEPENDENCE) STUDY OF PHASE TRANSITION OF SODIUM NITRITE BY SECOND HARMONI 16-2403
ENT NEUTRON IN/ CRYSTAL DYNAMICS AND THE FERROELECTRIC PHASE TRANSITION OF SODIUM NITRITE (STUDIED BY COHER 16-2364
MODEL OF INTERACTING DIPOLES) THE ORDER DISORDER TRANSITION OF SODIUM NITRITE. (TWO SUBLATTICE 16-2287
INTERACTION OF ELASTIC STRAIN WITH THE STRUCTURAL TRANSITION OF STRONTIUM TITANATE. 06-1216
L EXPANSIVITY AND ULTRASONIC PROPAGATION NEAR THE STRUCTURAL TRANSITION OF STRONTIUM TITANATE. THERMA 06-1135
K. ANALYSIS OF THE ELASTIC ANOMALIES AT THE STRUCTURAL PHASE TRANSITION OF STRONTIUM TITANATE NEAR 105 DEG 06-1183
G. STUDY OF THE PHASE TRANSITION OF TGS BY SECOND HARMONIC SCATTERIN 27-2992
SE TRANSFORMATION BETWEEN THE C/ CALORIMETRIC STUDY OF GLASS TRANSITION OF THE AMORPHOUS ICE AND OF THE PHA 34-3476
(LEAD DINIOBATE INTO THE NONFERROE/ STRUCTURAL STUDY OF THE TRANSITION OF THE (ORTHORHOMBIC) FERROELECTRIC 08-1285
THERMAL EXPANSION AND THE ORTHORHOMBIC- TETRAGONAL TRANSITION OF THE TUNGSTEN TRIOXIDE PHASE. 10-1626
INFRARED AND RAMAN STUDIES IN THE FERROELECTRIC TRANSITION OF THIOUREA. 25-2916
ELECTROMECHANICAL COUPLING NEAR THE FERROELECTRIC PHASE TRANSITION OF TRIGLYCINE SULFATE. 27-3174
MPERATURE. EQUATION OF STATE FOR THE COOPERATIVE TRANSITION OF TRIGLYCINE SULFATE NEAR CURIE TE 27-3006
EFFECTS ON THE DIELECTRIC PROPERTIES AND FERROELECTRIC PHASE TRANSITION ON LITHIUM THALLATE TARTRATE. /ING 29-3251
OF THE FERROELECTRIC RUBID/ EFFECT OF TEMPERATURE AND PHASE TRANSITION ON THE HYDROXIDE VIBRATION SPECTRUM 17-2470
RROELECTRIC CRYSTALS AT LOW TEMPERAT/ INFLUENCE OF THE PHASE TRANSITION ON THE NORMAL VIBRATION MODES OF FE 17-2456
N BANDS. EFFECT OF THE 110 DEGREE K TRANSITION ON THE STRONTIUM TITANATE CONDUCTIO 06-1158
PHENOMENOLOGICAL THEORY OF ANTIFERROELECTRIC TRANSITION. PART-2: FIRST ORDER TRANSITION. 03-0404
TION OF SOUND WAVES IN DEUTERATED ROCHELLE SALT AT THE PHASE TRANSITION POINT. ATTENUA 28-3204
TION OF SOUND WAVES IN DEUTERATED ROCHELLE SALT AT THE PHASE TRANSITION POINT. ATTENUA 28-3205
RSION AT MILLIMETER WAVELENGTHS IN BARIUM TITANATE ABOVE THE TRANSITION POINT. DISPE 04-0601
D VELOCITY OF ULTRASOUND IN FERROELECTRICS NEAR SECOND ORDER TRANSITION POINT. /LE FORCES ON ATTENUATION AN 03-0253
ON FOR QUADRUPOLE RESONANCE IN FERROELECTRICS NEAR THE PHASE TRANSITION POINT. KINETIC EQUATI 03-0309
ON RELAXATION PROCESSES NEAR THE SECOND ORDER PHASE TRANSITION POINT. 03-0385
SPECIFIC HEAT OF SODIUM NITRITE NEAR THE ANTIFERROELECTRIC TRANSITION POINT. 16-2303
NDUCTIVITY OF CALCIUM DOPED SODIUM NITRATE CRYSTALS NEAR THE TRANSITION POINT. / DIFFUSIVITY AND THERMAL CO 16-2283
/NAMICAL BEHAVIOR OF DISPLACIVE TYPE FERROELECTRICS NEAR THE TRANSITION POINT. (BARIUM TITANATE, STRONTIUM/ 04-0992
ORIDE (ANOMALOUS / ULTRASONIC ATTENUATION NEAR THE SOFT MODE TRANSITION POINT IN POTASSIUM MANGANESE TRIFLU 33-3373
E. (KAEN/ AC POLARIZATION AND DIELECTRIC HYSTERESIS NEAR THE TRANSITION POINT OF DIFFUSE FERROELECTRIC PHAS 03-0437
/HODS OF DETERMINING SPONTANEOUS POLARIZATION NEAR THE PHASE TRANSITION POINT OF TRIGLYCINE SELENATE CRYST/ 27-2994
/PLE ON THE) MICROWAVE DISPERSION OF FERROELECTRICS NEAR THE TRANSITION POINT (THEORY, MEASUREMENTS ON TGS/ 03-0370
OLYCRYSTALLINE) ANTIMONY SULFOIODIDE (DARK CONDUCTIVITY NEAR TRANSITION POINT, 1-1000 ATM) /SISTANCE OF (P 15-2250
Y OF ANOMALIES OF THERMODYNAMIC PARAMETERS NEAR SECOND ORDER TRANSITION POINTS IN FERROELECTRICS. /AL THEOR 03-0362
CTRIC SUSCEPTIBILITY AND DIELECTRIC RELAXATION TIME NEAR THE TRANSITION POINTS IN SODIUM NITRITE. /ATIC ELE 16-2302
ULTRASONIC ATTENUATION NEAR THE TWO TRANSITION POINTS IN SODIUM NITRITE. 16-2304
PT/ PHYSICAL PHENOMENA IN FERROELECTRICS WITH DIFFUSED PHASE TRANSITION. (POLARIZATION MECHANISMS, ELECTROO 02-0119
ON OF TRIGLYCINE FLUOBERYLLATE CRYSTALS IN THE FERROELECTRIC TRANSITION RANGE. THERMAL EXPANSI 27-3169
SSURE POLYMORPHS. TRANSITION RATES OF POTASSIUM NITRATE HIGH PRE 16-2279
ECTRIC DISPERSION OF IRRADIATED BARIUM TITANATE IN THE PHASE TRANSITION REGION. DIEL 04-0652
PERMEABILITY OF BARIUM TITANATE SINGLE CRYSTALS IN THE PHASE TRANSITION REGION. DIELECTRIC 04-0645
E NATURE OF DIELECTRIC DISPERSION IN THE FERROELECTRIC PHASE TRANSITION REGION. /TERACTION OF DIPOLES ON TH 03-0417
OMALOUS ELECTRIC CONDUCTIVITY OF FERROELECTRICS IN THE PHASE TRANSITION REGION. NATURE OF THE AN 03-0508
TION OF A RUBIDIUM DIHYDROGEN PHOSPHATE CRYSTAL IN THE PHASE TRANSITION REGION. POLARIZA 17-2506
TION OF A RUBIDIUM DIHYDROGEN PHOSPHATE CRYSTAL IN THE PHASE TRANSITION REGION. (COERCIVE FIELD) POLARIZA 17-2504
/NIC INVESTIGATION (15 MHZ SHEAR WAVES) OF THE FERROELECTRIC TRANSITION REGION IN POTASSIUM DIHYDROGEN PHO/ 17-2576
/RIZATION RELAXATION AND SUSCEPTIBILITY IN THE FERROELECTRIC TRANSITION REGION OF POTASSIUM DIDEUTERIUM PH/ 17-2575
DOMAIN MODEL). HEAT CAPACITY IN THE FERROELECTRIC TRANSITION REGION (THEORY, RELATION TO KAENZIG 03-0443
ELECTRIC, ELASTIC AND MAGNETIC PROPERTY, SYMMETRY, AND PHASE TRANSITION REPRESENTATION. (18 REFS) DI 03-0316
1-X) B'(X) OXYGEN(3), / DIFFUSE TO SHARP FERROELECTRIC PHASE TRANSITION SYSTEMS LEAD (B(0.5) NIOBIUM(0.5))(13-2044
/AN EXTERNAL ELECTRIC FIELD ON A DIFFUSE FERROELECTRIC PHASE TRANSITION, TALKING INTO ACCOUNT THE INTERACT/ 03-0441
GATION IN POTASSIUM DEUTERIUM PHOSPHATE AT THE FERROELECTRIC TRANSITION TEMPERATURE. ACOUSTIC PROPA 17-2573
ORPTION OF ULTRASOUND IN ANTIMONY SULFOIODIDE NEAR THE PHASE TRANSITION TEMPERATURE. /LUMINATION ON THE ABS 15-2248
RYSTALS OF FERROELECTRIC ANTIMONY SULFOIODIDE NEAR THE PHASE TRANSITION TEMPERATURE. /LECTION FROM SINGLE C 15-2151
NCE OF THE ELASTIC COMPLIANCE OF STRONTIUM TITANATE NEAR THE TRANSITION TEMPERATURE. PRESSURE DEPENDE 06-1219
EFFECT OF THE HYDROSTATIC PRESSURE ON THE TRANSITION TEMPERATURE IN AMMONIUM SULFATE. 19-2796
/AUER EFFECT OBSERVATION OF A VERY LOW FREQUENCY MODE AT THE TRANSITION TEMPERATURE IN KDP AND DEUTERATED / 17-2465
/MODE STUDY OF THE STRESS INDUCED PHASE TRANSITIONS NEAR THE TRANSITION TEMPERATURE IN STRONTIUM TITANATE. 06-1196
YPE FE/ ANOMALY OF THE SPIN LATTICE RELAXATION TIME NEAR THE TRANSITION TEMPERATURE IN THE ORDER DISORDER T 03-0488
/F FERROELECTRIC POTASSIUM FERROCYANIDE TRIHYDRATE (NEAR THE TRANSITION TEMPERATURE, NO ANOMALY BUT SOFT P/ 24-2891
/ARED OPTICAL PROPERTIES OF VANADIUM DIOXIDE ABOVE AND BELOW TRANSITION TEMPERATURE. (NO COCHRAN TYPE FERR/ 36-3526
IUM(0.9-1) TITANATE SOLID SOLUTIONS BY HYDR/ SHIFTING OF THE TRANSITION TEMPERATURE OF BARIUM(0-0.1) STRONT 04-0722
IDE. PRESSURE DEPENDENCE OF CUBIC TETRAGONAL TRANSITION TEMPERATURE OF CESIUM LEAD TRICHLOR 33-3393
LS OF THE PO/ INFLUENCE OF HYDROSTATIC PRESSURE ON THE PHASE TRANSITION TEMPERATURE OF FERROELECTRIC CRYSTA 17-2495
EFFECT OF STRESS ON THE SUPERCONDUCTING TRANSITION TEMPERATURE OF STRONTIUM TITANATE. 06-1182
EFFECT OF EVAPORATED METALLIC ELECTRODES ON THE TRANSITION TEMPERATURE OF TRIGLYCINE SULFATE. 27-3100
IHYDROGEN PHOSPHATE, AMMONIUM SULFATE, ROCHELLE SALT) AT THE TRANSITION TEMPERATURES. /SULFATE, POTASSIUM D 02-0125
EFFECT OF DEUTERATION IN AMMONIUM BISULFATE (CHANGE IN TRANSITION TEMPERATURES). 19-2760
ON SODIUM-23 IN SODIUM NITRITE IN THE VICINITY OF THE PHASE TRANSITION TEMPERATURES. /ETIC RESONANCE STUDY 16-2396
IN PEROVSKITE BARIUM TITANATE. PRESSURE DEPENDENCE OF TRANSITION TEMPERATURES AND ELECTROSTRICTIONS 04-0853
RIDE USING EPR. DETERMINATION OF TRANSITION TEMPERATURES IN CESIUM LEAD TRICHLO 33-3408
E NI/ EMPIRICAL FACTORS FOR CALCULATION OF THE FERROELECTRIC TRANSITION TEMPERATURES OF TUNGSTEN BRONZE TYP 13-2049

RCONATE TITANATE CERAMIC/ STABILIZATION EFFECTS AT THE PHASE TRANSITION TETRAGONAL- RHOMBOHEDRAL IN LEAD ZI 09-1611
OVSKITE FERR/ THERMODYNAMIC THEORY OF THE MORPHOTROPIC PHASE TRANSITION TETRAGONAL- RHOMBOHEDRAL IN THE PER 03-0172
IC HEAT ANOMALY IN THE NEIGHBORHOOD OF A FERRO- PARAELECTRIC TRANSITION. (TGS) SPECIF 27-2962
POLARIZATION OF A FERROELECTRIC HAVING A SECOND ORDER PHASE TRANSITION (THEORY). SINGLE DOMAIN 03-0202
WITH APPLIED FIELD IN FERROELECTRICS POSSESSING FIRST ORDER TRANSITION. (THEORY, CALCULATION) /POINT SHIFT 03-0327
TASSIUM DIHYDROGEN PHOSPHATE NEAR THE HIGH TEMPERATURE PHASE TRANSITION. (THERMAL EXPANSION) /PERTIES OF PO 17-2627
ATE. OPTICAL TRANSITION TO THE FERMI LEVEL IN SODIUM TUNGST 10-1634
(BARIUM, STRONTIUM) TITANATE SOLID SOLUTIONS NEAR THE PHASE TRANSITION, UNDER HIGH PRESSURES. /RROELECTRIC 04-0680
RCED FERROELECTRICITY IN GLASS BONDED LEAD ZIRCONATE. (PHASE TRANSITION 200 DEG C BELOW THE CURIE POINT) /O 07-1252
IN THE KANZIG REGIONS MODEL FOR DIFFUSE FERROELECTRIC PHASE TRANSITIONS. ACCOUNTING FOR INTERACTION 03-0438
ORPTION EDGE IN TRIHYDROGEN SELENITE CRYSTALS AT THEIR PHASE TRANSITIONS. BEHAVIOR OF THE FUNDAMENTAL ABS 22-2839
CALORIMETRIC INVESTIGATIONS OF SECOND ORDER FERROELECTRIC TRANSITIONS. 03-0435
OR RUBIDIUM BISULFATE, AND AMMONIUM SULFATE AT FERROELECTRIC TRANSITIONS. /HE INFRARED SPECTRA OF AMMONIUM 19-2772
TICAL POINT INDICES OF FERROELECTRIC AND FERROMAGNETIC PHASE TRANSITIONS. CRI 03-0264
ELECTRIC SUSCEPTIBILITY AND THE ORDER OF FERROELECTRIC PHASE TRANSITIONS. DI 03-0223
RUCTURE OF POLYCRYSTALLINE FERROELECTRICS WITH DIFFUSE PHASE TRANSITIONS. DOMAIN ST 03-0505
S OF POTASSIUM DIHYDROGEN PHOSPHATE TYPE FERROELECTRIC PHASE TRANSITIONS. DYNAMIC 17-2692
STATIC PRESSURE ON THE NATURE OF DIFFUSE FERROELECTRIC PHASE TRANSITIONS. EFFECT OF HYDRO 03-0442
ANTIFERROELECTRIC ELECTRIC AND MAGNETIC MOMENTS OF OPTICAL TRANSITIONS. /CTRONIC AND SPIN EXCITATIONS IN 03-0418
XPERIMENTAL CRITERION OF FERROELECTRICITY. (SHARP OR DIFFUSE TRANSITIONS) E 03-0184
ATIONS AND THEIR APPLICATION TO FERROELECTRIC (LAMBDA) PHASE TRANSITIONS. /ATION OF THE PIPPARD-GARLAND REL 03-0299
RED SPECTRA OF POTASSIUM NITRATE AND SODIUM NITRITE AT PHASE TRANSITIONS. INFRA 16-2352
INTERBAND MECHANISMS FOR FERROELECTRIC PHASE TRANSITIONS. 02-0085
EATION PROCESS IN BARIUM TITANATE AT CUBIC- TETRAGONAL PHASE TRANSITIONS. NUCL 04-0754
ON THE SYMMETRY OF THE SOFT MODES IN THE FERROELECTRIC PHASE TRANSITIONS. 03-0357
ANISOTROPY OF CUBIC NICKEL IODINE BORACITE DUE TO QUADRUPOLE TRANSITIONS. OPTICAL 14-2093
OPY OF PHONON INSTABILITY IN FIRST ORDER FERROELECTRIC PHASE TRANSITIONS. RAMAN SPECTROSC 03-0163
TITANATE STANNATE BY TEMPERATURE AND PRESSURE ENFORCED PHASE TRANSITIONS. /NERGY IN LEAD NIOBIUM ZIRCONATE 09-1488
STATISTICAL MECHANICS OF FERROELECTRIC PHASE TRANSITIONS. 03-0198
THEORY OF SECOND ORDER PHASE TRANSITIONS. 03-0364
MODYNAMIC CLASSIFICATION OF SECOND ORDER FERROELECTRIC PHASE TRANSITIONS. THER 03-0303
ICS AND STATISTICAL MECHANICS OF DIFFUSE FERROELECTRIC PHASE TRANSITIONS. THERMODYNAM 03-0445
PROBLEM. (WITH AN APPLICATION TO THE THEORY OF FERROELECTRIC TRANSITIONS) /ROACH TO THE ANHARMONIC LATTICE 03-0221
. PART-1: INTRODUCTION AND MAGNETIC RES/ FERROELECTRIC PHASE TRANSITIONS: A REVIEW OF THEORY AND EXPERIMENT 03-0397
RATURE (ROTATIONS OF BO6 OCTAHEDRA I/ DYNAMICS AT STRUCTURAL TRANSITIONS ABOVE FERROELECTRIC CRITICAL TEMPE 06-1205
IN ANTIMONY SULFOIODIDE CRYSTALS. INDIRECT TRANSITIONS AND ABSORPTION IN THE MID-INFRARED 15-2150
RROELECTRIC AND ANTIFERRO/ ULTRASONIC INVESTIGATION OF PHASE TRANSITIONS AND CRITICAL POINTS. SECTION-5: FE 02-0070
/ DEFECTS PRODUCED BY GAMMA AND REACTOR IRRADIATION ON PHASE TRANSITIONS AND DIELECTRIC PROPERTIES OF KDP / 17-2614
YDROGEN PHOSPHATE TYPE CRYSTALS. HIGH TEMPERATURE PHASE TRANSITIONS AND METASTABILITY IN POTASSIUM DIH 17-2513
. (REVIEW) FERRO- AND ANTIFERROELECTRIC PHASE TRANSITIONS AND RELATED SCATTERING EXPERIMENTS 02-0132
STRUCTURAL PHASE TRANSITIONS AND SOFT MODES. 03-0176
/HICAL CLASSIFICATION OF FERROELECTRICS, FERROELECTRIC PHASE TRANSITIONS AND SPECIAL FEATURES OF DOMAIN ST/ 03-0480
(HIGH SENSITIVITY) BARIUM TITANATE. DIFFUSIONLESS PHASE TRANSITIONS AND THE STABLE DOMAIN STRUCTURE OF 04-0888
OXIDE AND BARIUM TITANATE SINGLE CRYSTALS. (PHASE TRANSITIONS AND) THERMOCURRENTS IN TITANIUM DI 04-0697
ANATE AND ((BARIUM(X) STRONT/ INFLUENCE OF PRESSURE ON PHASE TRANSITIONS AT LOW TEMPERATURES: STRONTIUM TIT 04-0898
E, POTASSIUM TANTALATE) THEORY OF SECOND ORDER PHASE TRANSITIONS AT LOW TEMPERATURES. (TIN TELLURID 03-0433
/TRIC ANOMALIES IN FERROELECTRIC MATERIAL WITH SEVERAL PHASE TRANSITIONS AT TEMPERATURES CLOSE TOGETHER. (/ 03-0363
(FAST ION SCATTERING) SHADOW EFFECT IN (DISPLACIVE) PHASE TRANSITIONS. (BARIUM TITANATE) 04-0533
/INATION OF) ANOMALIES OF DIELECTRIC PROPERTIES AT THE PHASE TRANSITIONS. (BASED ON LANDAU PHENOMENOLOGICA/ 19-2765
PHASE CHANGE DETECTION (INCLUDING FERROELECTRIC AND MAGNETIC TRANSITIONS, EXAMPLE ROCHELLE SALT) /ETER FOR 28-3201
DIELECTRIC AND THERMAL STUDY OF POTASSIUM SELENATE TRANSITIONS (FERROELECTRIC BELOW 93K). 22-2824
/ANATE BARRIERS INTERPRETED AS CONFIRMATION OF THE THEORY OF TRANSITIONS FROM LARGE TO NEARLY SMALL POLARO/ 03-0233
RY OF CRITICAL POINTS. ANOMALIES NEAR FERROELECTRIC TRANSITIONS FROM THE POINT OF VIEW OF THE THEO 03-0298
PHASE TRANSITIONS IN A MODEL FERROELECTRIC. 03-0257
PHASE TRANSITIONS IN A MODEL FERROELECTRIC. 03-0256
DE CRYSTAL. RAMAN SCATTERING AND PHASE TRANSITIONS IN A POTASSIUM MANGANESE TRIFLUORI 33-3400
ENATE STUDIED BY INFRARED ABSORPTION. ANTIFERROELECTRIC TRANSITIONS IN ADP AND AMMONIUM DIHYDROGEN ARS 17-2706
E AND FERROELECTRIC L/ MOSSBAUER EFFECT STUDIES OF THE PHASE TRANSITIONS IN ANTIFERROELECTRIC LEAD ZIRCONAT 07-1253
D INFRARED AND RA/ LONG WAVELENGTH OPTICAL PHONONS AND PHASE TRANSITIONS IN ANTIMONY SULFOIODIDE. (POLARIZE 15-2106
DOMAIN INTERACTIONS DURING PHASE TRANSITIONS IN BARIUM TITANATE. 04-0651
BILITIES) MOSSBAUER STUDIES OF LOWER TRANSITIONS IN BARIUM TITANATE. (LATTICE INSTA 04-0564
SYMMETRY ASPECTS OF THE PHASE TRANSITIONS IN BORACITES. 14-2077
THEORY OF PHASE TRANSITIONS IN BORACITES. 03-0269
ISOTHERMAL PHASE TRANSITIONS IN CERAMIC LEAD ZIRCONATE. 08-1466
ELECTRICAL CONDUCTION AND (NONFERROELECTRIC) PHASE TRANSITIONS IN CESIUM LEAD TRICHLORIDE. 33-3365
EXPERIMENTAL STUDIES OF STRUCTURAL PHASE TRANSITIONS IN CESIUM LEAD TRICHLORIDE. 33-3378
(MECHANISM OF) STRUCTURAL PHASE TRANSITIONS IN CESIUM LEAD TRICHLORIDE. 33-3380
MICROSCOPIC OBSERVATIONS OF PHASE TRANSITIONS IN CESIUM LEAD TRICHLORIDE. 33-3383
OF A CENTER OF SYMMETRY AT 194 DEGREES K) PHASE TRANSITIONS IN CESIUM LEAD TRICHLORIDE. (LOSS 33-3368
BARIUM LANTHANON MOLYBDATE. PHASE TRANSITIONS IN COMPLEX PEROVSKITES OF THE TYPE 30-3261
DOMAIN FERROELECTRICS BELOW CURIE POINT). COMPLEX PHASE TRANSITIONS IN CRYSTALS (APPLICATION TO SINGLE 03-0319
OF ONES / MICROTHEORY OF SPONTANEOUS POLARIZATION AND PHASE TRANSITIONS IN CRYSTALS OF PEROVSKITE TYPE AND 03-0175
N BRONZE STRUCTURE (POTASSIU/ FORMATION CONDITIONS AND PHASE TRANSITIONS IN CRYSTALS WITH POTASSIUM TUNGSTE 13-1923
OSINE CALCIUM CHLORIDE. (SHI/ NMR STUDY OF THE FERROELECTRIC TRANSITIONS IN DIGLYCINE NITRATE AND TRIS-SARC 16-2267
YDRATE AND POTA/ MOSSBAUER EFFECT STUDY OF THE FERROELECTRIC TRANSITIONS IN FERRIC AMMONIUM SULFATE DODECAH 19-2755
YDRATE AND POTA/ MOSSBAUER EFFECT STUDY OF THE FERROELECTRIC TRANSITIONS IN FERRIC AMMONIUM SULFATE DODECAH 24-2892
TRIC MATERIALS. NATURE OF PHASE TRANSITIONS IN FERROELECTRIC AND ANTIFERROELEC 03-0311
DOUBLE HYSTERESIS LOOPS AND FIRST ORDER PHASE TRANSITIONS IN FERROELECTRIC CRYSTALS. 03-0456
PHASE TRANSITIONS IN FERROELECTRICS. 01-0011
NT POLARIZATION SCHEME FOR ANH/ FIRST AND SECOND ORDER PHASE TRANSITIONS IN FERROELECTRICS: A SELF CONSISTE 03-0355
/HANGE OF COMPLETE SYMMETRY OF THE CRYSTALS DURING THE PHASE TRANSITIONS IN FERROELECTRICS AND FERROMAGNET/ 03-0527
OCYANIDE TR/ NUCLEAR MAGNETIC RESONANCE STUDIES OF THE PHASE TRANSITIONS IN FERROELECTRICS. (POTASSIUM FERR 16-2255
TO BARIUM T/ CORRELATION EFFECTS IN DISPLACEMENT TYPE PHASE TRANSITIONS IN FERROELECTRICS. (THEORY APPLIED 03-0497
HIGH TEMPERATURE (TO 236 DEG C) PHASE TRANSITIONS IN (HIGH PURITY) LEAD ZIRCONATE. 07-1266
E SINGLE CRYSTALS/ CUBIC TETRAGONAL (MARTENSITIC TYPE) PHASE TRANSITIONS IN HIGH RESISTIVITY BARIUM TITANAT 04-0887
INTERMOLECULAR POTENTIAL AND FERROELECTRIC PHASE TRANSITIONS IN HYDROGEN HALIDES. 33-3339
HOSPHATE. NEUTRON DIFFRACTION STUDIES OF FERROELECTRIC PHASE TRANSITIONS IN KDP AND POTASSIUM DIDEUTERIUM P 17-2661
STRUCTURAL PHASE TRANSITIONS IN LANGBEINITES. 36-3552
.5), TITANAT/ ELECTRICAL MAGNETIC AND X-RAY STUDIES OF PHASE TRANSITIONS IN LEAD COBALTATE(0.5) TUNGSTATE(0 13-1960
S. TRANSIENT NMR STUDY OF PHASE TRANSITIONS IN METALLIC SODIUM TUNGSTEN BRONZE 13-1910

ICS.   LATTICE DYNAMICS OF PHASE TRANSITIONS IN PEROVSKITE TYPE ANTIFERROELECTR 08-1475
NDED ABSTRACT)   STRUCTURAL PHASE TRANSITIONS IN PEROVSKITE TYPE CRYSTALS. (EXTE 08-1450
/ELF CONSISTENT LATTICE DYNAMICAL THEORY OF STRUCTURAL PHASE TRANSITIONS IN PEROVSKITE TYPE CRYSTALS. (STR/ 03-0424
                                            DISPLACEMENT PHASE TRANSITIONS IN PEROVSKITES.   08-1274
93 AND SODIUM-23 RESONANCE IN POTASSIUM / NMR STUDY OF PHASE TRANSITIONS IN PEROVSKITES CRYSTALS. (NIOBIUM- 08-1283
DIUM NIOBATE)   STRUCTURE AND TRANSITIONS IN PEROVSKITES. (ORTHOFERRITES, SO 08-1396
         ACCUSTIC DETECTION OF FERROELECTRIC PHASE TRANSITIONS IN PLZT CERAMICS.   09-1545
UM PHOSPHATE (INFRARED PHOSPHATE/ OPTICAL STUDY ON THE PHASE TRANSITIONS IN POTASSIUM DIHYDROGEN OR DEUTERI 17-2646
CRYSTALS.   RAMAN SCATTERING AND PHASE TRANSITIONS IN POTASSIUM MANGANESE TRIFLUORIDE 33-3372
.   NEUTRON SCATTERING STUDY OF THE LATTICE DYNAMICAL PHASE TRANSITIONS IN POTASSIUM MANGANESE TRIFLUORIDE 33-3405
: TEMPERATURE DISCONTINUITIE/ TWO PHONON PROCESSES AND PHASE TRANSITIONS IN POTASSIUM MANGANESE TRIFLUORIDE 33-3363
PHA TO BETA, BETA TO ALPHA, BETA TO GAMMA AND GAMMA TO ALPHA TRANSITIONS IN POTASSIUM NITRATE. /H IN THE AL 16-2259
/GES IN ELECTRICAL CONDUCTIVITY ACCOMPANYING THE POLYMORPHIC TRANSITIONS IN POTASSIUM NITRATE AT 1 ATMOSPH/ 16-2391
CRYSTALS (GROWTH BY DOUBLE BRIDGMAN TECHNIQUE).   PHASE TRANSITIONS IN POTASSIUM NITRATE NITRITE MIXED 16-2345
/) (0.5) BISMUTH(0.5) TITANATE AND THE HIGH TEMPERATURE PHASE TRANSITIONS IN POTASSIUM(0.5) BISMUTH(0.5) TI/ 08-1362
                   MICROSCOPIC THEORY OF THE PHASE TRANSITIONS IN ROCHELLE SALT.   03-0482
                      NMR STUDY OF PHASE TRANSITIONS IN ROCHELLE SALT.   28-3195
ANTIMONY SULFOIO/ DIELECTRIC AND X-RAY STUDIES OF THE PHASE TRANSITIONS IN ROCHELLE SALT, BARIUM TITANATE, 02-0099
       ANISOTROPIC DIFFUSE X-RAY SCATTERING AND PHASE TRANSITIONS IN SODIUM NIOBATE.   08-1357
ILTING SCHEMES IN THE SOLUTIONS OF P/ HIGH TEMPERATURE PHASE TRANSITIONS IN SODIUM NIOBATE AND THE USE OF T 08-1336
              PHENOMENOLOGICAL THEORY OF PHASE TRANSITIONS IN SODIUM NITRITE.   03-0254
TE.   GRUNEISEN PARAMETERS AROUND THE PHASE TRANSITIONS IN SODIUM NITRITE AND SODIUM NITRA 16-2390
RIUM / NUCLEAR MAGNETIC RESONANCE STUDY OF THE FERROELECTRIC TRANSITIONS IN SODIUM TRIHYDROGEN AND TRIDEUTE 22-2832
       ESR STUDY OF THE FERROELECTRIC PHASE TRANSITIONS IN SODIUM TRIHYDROGEN DISELENITE.   22-2871
CRYSTAL X-RAY AND NEUTRON DIFFRA/ FERROELECTRICITY AND PHASE TRANSITIONS IN SOLID HYDROGEN HALIDES (SINGLE 31-3331
IVE FERROELECTRICS. (ANTIMONY SUL/ OPTICAL PHONONS AND PHASE TRANSITIONS IN SOME ORDER DISORDER AND DISPLAC 15-2208
            RAMAN SCATTERING AND PHASE TRANSITIONS IN STRESSED STRONTIUM TITANATE.   06-1103
F POTASSIUM TUNGSTEN BRONZE. (POTASSIUM BARIUM TANTAL/ PHASE TRANSITIONS IN TANTALATES WITH THE STRUCTURE O 13-2050
ANCE OF MANGANESE(2+) DOPED/ ALLOWED AND FORBIDDEN HYPERFINE TRANSITIONS IN THE ELECTRON PARAMAGNETIC RESON 16-2383
HANUM ALUMINATE.   RADIATIONLESS TRANSITIONS IN THE EUROPIUM(3+) CENTER IN LANT 36-3531
TEM.   STUDY OF PHASE TRANSITIONS IN THE LEAD ZIRCONATE TITANATE SYS 09-1620
TE- BISMUTH/ MOSSBAUER EFFECT STUDIES OF FERROELECTRIC PHASE TRANSITIONS IN THE LEAD ZIRCONATE- LEAD TITANA 09-1498
TE SYSTEM.   STUDY OF PHASE TRANSITIONS IN THE LEAD ZIRCONATE- LEAD TITANA 09-1621
IUM(1-X) BARIUM(X) TITANATE.   PHASE TRANSITIONS IN THE MIXED CRYSTAL SYSTEM STRONT 04-1016
)   STUDIES ON THE MECHANISM OF THE PHASE TRANSITIONS IN THIOUREA. (SATTELITE SCATTERING 25-2922
        EFFECT OF CONSTANT ELECTRICAL FIELD ON PHASE TRANSITIONS IN TRIGLYCINE FLUOBERYLLATE.   27-3167
SOME ADDITIONAL FEATURES OF FERROELECTRIC PHASE TRANSITIONS IN VIBRONIC MODEL.   03-0339
D STRUCTURAL PROPERTIES OF FERROELECTRICS WITH DIFFUSE PHASE TRANSITIONS. (MIXED LAYER OXIDES) /ELECTRIC AN 13-2014
STRONTIUM TITA/ SOFT MODE STUDY OF THE STRESS INDUCED PHASE TRANSITIONS NEAR THE TRANSITION TEMPERATURE IN 06-1196
TRIC PROPERTIES OF DIAMMONIUM HEXABROMO STANNATE (DIELECTRIC TRANSITIONS NOT FERROELECTRIC). DIELEC 36-3612
STALS.   EFFECT OF HIGH PRESSURE ON PHASE TRANSITIONS OF ALKALI TRIHYDROGEN SELENITE CRY 22-2882
HYDROGEN BONDING AN/ LOW TEMPERATURE STUDIES. PART-4: PHASE TRANSITIONS OF AMMONIUM SULFATE: THE NATURE OF 19-2783
EARTH MOLYBDATES.   PHASE TRANSITIONS OF ANHYDROUS MODIFICATIONS OF RARE 30-3272
          KINETICS OF THE PHASE TRANSITIONS OF BARIUM TITANATE.   04-0650
ITE AT LOW TEMPERATURES. (MONOCLINIC A/ X-RAY STUDY ON PHASE TRANSITIONS OF FERROELECTRIC IRON IODINE BORAC 14-2086
/EFFECTS ON DIELECTRIC PROPERTIES, DOMAIN DYNAMICS, AND PHASE TRANSITIONS OF FERROELECTRICS. (GENERAL DISCU/ 03-0495
UCTURAL COMPOUNDS (EUROPIUM TERBIUM, DYSPROSIUM AND H/ PHASE TRANSITIONS OF GADOLINIUM MOLYBDATE AND ISOSTR 30-3271
TANCE OF HYDROGEN BONDING IN THE FERROELECTRIC MECHAN/ PHASE TRANSITIONS OF HYDROGEN HALIDE CRYSTALS (IMPOR 31-3329
      X-RAY STUDY OF HIGH TEMPERATURE PHASE TRANSITIONS OF SODIUM NIOBATE (TO 500C).   13-1965
                         PHASE TRANSITIONS OF SOLID HYDROGEN HALIDES.   31-3327
/ECT OF THE CHANGE OF ELECTROSTATIC CONSTRAINTS ON THE PHASE TRANSITIONS OF SOME PHENOMENOLOGICAL MODELS O/ 03-0280
PR OF IRON- VANADIUM PAIRS)   ORDER PARAMETER AND PHASE TRANSITIONS OF STRESSED STRONTIUM TITANATE. (E 06-1170
              NONLINEAR PROPERTIES AND PHASE TRANSITIONS OF STRONTIUM BISMUTH TITANATES.   13-1952
/MPERATURE DEPENDENCE OF THE DIELECTRIC PROPERTIES AND PHASE TRANSITIONS OF THE ANTIFERROELECTRIC PEROVSKI/ 08-1433
/MPERATURE DEPENDENCE OF THE DIELECTRIC PROPERTIES AND PHASE TRANSITIONS OF THE FERROELECTRIC PEROVSKITES:/ 08-1434
              HIGH PRESSURE PHASE TRANSITIONS OF THE FIRST KIND IN ROCHELLE SALT. 28-3214
AL TRIGLYCINE SULFATE AND TRIGLYCINE FLUOBERYLLATE) AT PHASE TRANSITIONS OF THE SECOND ORDER. /SINGLE CRYST 27-3137
TITANATE.   INFLUENCE OF PHASE TRANSITIONS ON ELECTRICAL RESISTANCE OF BARIUM 04-0599
ON IN STRONTIUM TITANATE A/ EFFECTS OF TEMPERATURE AND PHASE TRANSITIONS ON THE HYDROXYL STRETCHING VIBRATI 06-1209
/E ANALYSIS OF FERROELECTRICS IN CONNECTION WITH THEIR PHASE TRANSITIONS. PART-1: TGS. PART-2: SILVER SODI/ 02-0077
        LATTICE DYNAMICS ABOVE STRUCTURAL PHASE TRANSITIONS: STRONTIUM TITANATE.   06-1124
NTALATE, B/ NEUTRON INELASTIC SCATTERING AT STRUCTURAL PHASE TRANSITIONS. (STRONTIUM TITANATE, POTASSIUM TA 02-0059
EARTH MOLYBDATES. (REVIEW, FORMATION, CRYSTAL GROWTH, PHASE TRANSITIONS, STRUCTURE, 49 REFS) /LECTRIC RARE 02-0064
       ELECTROGYRATION AND FERROELECTRIC PHASE TRANSITIONS (TGS).   27-2970
M, FERROELECTRICITY, ORDERED ALLOYS, SUPERFLUIDI/ PHYSICS OF TRANSITIONS. (THERMODYNAMICS, THEORY, MAGNETIS 03-0183
/HIP BETWEEN THE SUPERCONDUCTING AND THE FERROELECTRIC PHASE TRANSITIONS (TIN TELLURIDE PURE AND ALLOYED W/ 36-3564
THEORY).   SOME REMARKS ABOUT ANTIFERROELECTRIC PHASE TRANSITIONS (TO PARAELECTRIC OR FERROELECTRIC, 03-0171
          CRITICAL SLOWING DOWN AT FERROELECTRIC TRANSITIONS (TRIGLYCINE SULFATE).   27-3155
HYDROGEN AND DIDEUT/ LASER LIGHT SCATTERING STUDIES OF PHASE TRANSITIONS. (TRIGLYCINE SULFATE, POTASSIUM DI 17-2481
ITY (FERROELECTRIC TWINNING) TREATED AS CRITI/ FERROELECTRIC TRANSITIONS WITH A GENUINE DIELECTRIC INSTABIL 15-2176
     THEORY OF FERROELECTRIC PHASE TRANSITIONS WITH VIBRONIC MECHANISM.   03-0351
YNAMICS OF IRREVERSIBLE PROCESSES. (APPLIED TO FERROELECTRIC TRANSITIONS, 31 REFS) / DISPERSION AND THERMOD 03-0415
CRYSTAL FILMS OF BARIUM TITANATE BY MEANS OF A STROBOSCOPIC TRANSMISSION ELECTRON MICROSCOPE. /N OF SINGLE 04-0942
IES. PART-1: DYNAMICAL THEORY. PART-2: / ELECTRON MICROSCOPE TRANSMISSION IMAGES OF COHERENT DOMAIN BOUNDAR 03-0255
E FILMS.   FAR INFRARED REFLECTIVITY AND TRANSMISSION MEASUREMENTS ON STRONTIUM TITANAT 06-1154
IN THE INFRARED REGION IN THE PARA- AND FERROELECTRIC PHAS/ TRANSMISSION OF POTASSIUM DIHYDROGEN PHOSPHATE 17-2459
              FAR INFRARED TRANSMISSION SPECTRUM OF THIOUREA.   25-2917
BARIUM TITANATE CERAMICS CAUSED BY IRRADIATION WITH CADMIUM TRANSMITTED NEUTRONS. /OELECTRIC PROPERTIES OF 04-0731
TORIES. (LEAD LANTHANUM ZIRCONATE TITANATE)   TRANSPARENT CERAMIC DEVELOPED BY SANDIA LABORA 09-1476
HANUM ZIRCONUM TITANA/ INSCRIPTION OF HOLOGRAPHIC RESULTS IN TRANSPARENT CERAMIC FERROELECTRICS. (LEAD LANT 09-1575
ING)   PREPARATION OF TRANSPARENT FERROELECTRIC CERAMICS. (HOT PRESS 02-0103
/ZE AND POROSITY ON THE ELECTRICAL AND OPTICAL PROPERTIES OF TRANSPARENT FERROELECTRIC CERAMICS. ((LEAD, L/ 13-2004
ONATE- LEAD TITANATE TYPE)   PROPERTIES OF NEW TRANSPARENT FERROELECTRIC CERAMICS. (LEAD ZIRC 09-1518
S.   DIELECTRIC AND OPTICAL PROPERTIES OF TRANSPARENT FERROELECTRIC GLASS CERAMIC SYSTEM 04-0576
NONLINEAR OPTICAL PROPERTIES OF POTASSIUM LITHIUM NIOBATES (TRANSPARENT, TETRAGONAL TUNGSTEN BRONZE STRUC/ 13-2026
OR OF ANTIMONY SULFOIODIDE. (SINGLE CRYSTAL GROWTH, CHEMICAL TRANSPORT)   CRYSTALLIZATION BEHAVI 15-2201
OEFFICIENTS) IN STRONTIUM TITANATE (4.2-300 DEG / ELECTRONIC TRANSPORT (CONDUCTIVITY AND HALL AND SEEBECK C 06-1129
              PHONON TRANSPORT IN BARIUM TITANATE.   04-0830
E EARTH ORTHOCHROMITES, MANGANITES AND FERRITES   ELECTRICAL TRANSPORT IN (SEMICONDUCTING AND MAGNETIC) RAR 36-3608
ATE) FERROE/ OXYGEN CONCENTRATION CELL MEASUREMENTS OF IONIC TRANSPORT NUMBERS IN PZT (LEAD ZIRCONATE TITAN 09-1512

CONDUCTION MECHANISM FOR THERMOELECTRIC CURRENTS)
NATE. (ABSTRACT)
TALS.   EFFECT OF OXYGEN DEFICIENCY ON ELECTRICAL
HYDROTHERMAL PREPARATION OF BARIUM TITANATE BY
TO YIELD SM/ HYDROTHERMAL PREPARATION OF BARIUM TITANATE BY
S IN FERROELECT/ PIEZOELECTRIC EFFECT ON THE REFLECTION OF A
THE (001) LI/ GENERATION OF SECOND AND THIRD HARMONICS BY A
LITHIUM NIOBATE- SILICON STRUCTURE.
IONIC FERROELECTRIC / SCATTERING OF CONDUCTION ELECTRONS BY
OF POLARIZED BARIUM TITANATE CERAMICS SUBJECTED TO UNIAXIAL
S OF BARIUM TITANATE SINGLE CRYSTALS.   EFFECT OF
LITHIUM NIOBATE AND/ SEPARATE EXCITATION OF LONGITUDINAL AND
IOBATE AND BISMU/ INDEPENDENT EXCITATION OF LONGITUDINAL AND
EX OF A NONLINEAR CRYSTAL ON SECOND HARMONIC/ INFLUENCE OF A
/MALISM OF LOCAL MODE MODEL.) PART-2: (DYNAMICAL PROPERTIES,
OPTIMUM CUT IN X DIHYDROGEN PHOSPHATE CRYSTALS FOR
OPTIMUM CUT FOR A LITHIUM NIOBATE
, T/ OPTIMUM CRYSTAL ORIENTATION FOR A BARIUM SODIUM NIOBATE
/STAM BRILLOUIN SCATTERING OF LIGHT IN LITHIUM NIOBATE (FROM
IUM DIHYDROGEN PHOSPHA/ (FINITE APERTURE LIGHT SWITCH USING)
STAL. (RAMAN SPECTRUM, PERMITTIVITY)
FERROELECTRICS. (THEORY)   REFLECTION AND REFRACTION OF
S/ REFLECTION AND REFRACTION OF QUASI LONGITUDINAL AND QUASI
E.   SHARED HOLES
EMICONDUCTING FERROELECTRICS.   SOME FEATURES OF
ES IN A KDP CRYSTAL.
                         PHOTOCONDUCTIVITY AND
STIMULATED CURRENT IN FERROELECTRIC OXIDE CRYSTALS. (POTASS/
TRIC DISPERSION OF IRRADIATED BARIUM TITANATE NEAR THE PHASE

                         SECOND HARMONIC GENERATION IN THE
RBED NMR STUDY OF THE FERROELECTRIC TRANSITION IN DEUTERATED
                         CRITICAL BEHAVIOR OF
EFFECT OF CONSTANT ELECTRICAL FIELD ON PHASE TRANSITIONS IN
                         EPR OF THE CHROMIUM COMPLEX IN
EOUS POLARIZATION ON THE EPR SPECTRA OF CHROMIUM(3+) IONS IN
LOW FREQUENCY RAMAN SCATTERING BY FERROELECTRIC
HING IN PULSE ELECTRIC FIELD.   IMPEDANCE OF FERROELECTRIC
ELECTRIC TRANSITION RANGE.   THERMAL EXPANSION OF
AELECTRIC PHASE.   CRYSTAL STRUCTURE OF
EPR OF THE FOUR COPPER ION COMPLEX WITH SPIN S=2 IN
CONDITIONS ON A CHANGE IN THE DIELECTRIC CONSTAN/ EFFECT OF
                         PROPERTIES OF DEUTERATED
ELECTRON SPIN RESONANCE OF GAMMA-IRRADIATED
                         NEW HIGH PRESSURE PHASE TRANSITION IN
CTRIC CONSTANT) TEMPERATURE CHARACTERISTICS OF FERROELECTRIC
                         RAMAN SPECTRUM OF
NEAR T/ EFFECT OF A DC FIELD ON THE DIELECTRIC PROPERTIES OF
ROUGH CURIE POINT).   DOMAIN STRUCTURE OF
SPONTANEOUS POLARIZATION NEAR THE PHASE TRANSITION POINT OF
/IELECTRIC PERMEABILITY AND DIELECTRIC LOSS ANGLE TANGENT OF
US ALTERNATING FIELD VOLTAGE, TEMPE/ NONLINEAR PROPERTIES OF
R INFRARED AND IN THE RADIO FREQUENCY / OPTICAL CONSTANTS OF
RELATION BETWEEN ELECTRICAL AND THERMAL PROPERTIES OF
                         DIELECTRIC PERMEABILITY OF
/OMPOSITION OF FERROELECTRIC CRYSTALS (SOLUTION GROWTH, TGS,
TRIGLYCINE SULFATE AND ITS VARIATION UNDER X-I/ BAND GAP OF
ENCE OF THE FERROELECTRIC FIELD EFFECT IN TELLURIUM FILMS ON
CTRIC TRANSDUCER FOR USE IN HEAT TRANSFER AND FLOW STUDIES. (TRIGLYCINE SULFATE)
STIC EMISSION FROM FERROELECTRIC CRYSTALS. (BARIUM TITANATE,
ICONDUCTOR ELECTRODES EVAPORATED ON INSULATORS ESPECIALLY ON
LARGE GATE CHARGES APPLIED TO TELLURIUM FILMS ON UHV CLEAVED
ION IN SINGLE CRYSTALS OF FERROELECTRICS. (ROCHELLE SALT AND
CABILITY OF THE PIPPARD RELATIONS TO THE PHASE TRANSITION IN
E ETCHING TO THE STUDY OF TWIN AND DISLOCATION STRUCTURES IN
                         BLOCKING OF SPONTANEOUS POLARIZATION IN
ROSION FIGURES DUE TO THE FERROELECTRIC DOMAIN BOUNDARIES OF
                         CRITICAL FLUCTUATIONS IN
                         CRITICAL REGION IN FERROELECTRIC
CRITICAL SLOWING DOWN AT FERROELECTRIC TRANSITIONS (TRIGLYCINE SULFATE).
                         DAMAGE PRODUCED BY X-RAYS ON
DIELECTRIC BEHAVIOR OF GAMMA-RAY IRRADIATED
                         DIELECTRIC CONSTANT OF
                         DIELECTRIC DISPERSION OF
                         DIELECTRIC DISPERSION OF
DIELECTRIC RELAXATION IN MULTIDOMAIN CRYSTALS OF
                         DIELECTRIC RELAXATION OF MULTIDOMAIN
INATION OF (ELASTIC AND PIEZOELECTRIC) MATERIAL CONSTANTS OF
OF COPPER AND CHROMIUM IONS ON THE DIELECTRIC PROPERTIES OF
PORATED METALLIC ELECTRODES ON THE TRANSITION TEMPERATURE OF
EFFECT OF HYDROSTATIC PRESSURE ON THE UNIPOLAR PROPERTIES OF
                         EFFECT OF PRESSURE ON THE SWITCHING RATE OF
PARAMETERS ON THE FERROELECTRIC- PARAELECTRIC TRANSITION IN
ELECTRICS NEAR THE TRANSITION POINT (THEORY, MEASUREMENTS ON
EFFECT OF X- AND GAMMA-RADIATION ON THE SWITCHING PROCESS IN
EFFECTS OF ADDITIVES ON THE PYROELECTRIC PROPERTIES OF
EFFECTS OF DOMAIN SHAPES ON SECOND HARMONIC SCATTERING IN
S AS THE DETERMINING FACTOR OF THE GROWTH STAGE IN EPITAXY. (TRIGLYCINE SULFATE) /OF REAL CRYSTAL SUBSTRATE
ELECTROGYRATION AND FERROELECTRIC PHASE TRANSITIONS (TRIGLYCINE SULFATE .

TRANSPORT PHENOMENA IN ANTIMONY SULFOIODIDE. (15-2241
TRANSPORT PROPERTIES IN REDUCED STRONTIUM TITA   06-1152
TRANSPORT PROPERTIES OF TUNGSTEN TRIOXIDE CRYS   10-1629
TRANSPORT REACTIONS.   04-0626
TRANSPORT REACTIONS (INVOLVING TITANIUM ESTERS   04-0625
TRANSVERSE ACOUSTIC WAVE FROM DOMAIN BOUNDARIE   03-0312
TRANSVERSE ACOUSTIC WAVE PROPAGATING ITSELF IN   11-1682
TRANSVERSE ACOUSTOELECTRIC EFFECT IN A LAYERED   11-1748
TRANSVERSE AND LONGITUDINAL OPTICAL PHONONS IN   03-0292
TRANSVERSE COMPRESSION. /ZOELECTRIC PROPERTIES   04-0962
TRANSVERSE COMPRESSION OF DIELECTRIC PROPERTIE   04-0956
TRANSVERSE HYPERSONIC PULSES IN PIEZOELECTRIC   11-1814
TRANSVERSE HYPERSONIC WAVE PULSES IN LITHIUM N   11-1815
TRANSVERSE INHOMOGENEITY OF THE REFRACTIVE IND   11-1687
TRANSVERSE ISING MODEL, SOFT MODE COUPLING TO/   03-0384
TRANSVERSE LIGHT MODULATION   17-2494
TRANSVERSE LIGHT MODULATION   11-1755
TRANSVERSE LIGHT MODULATOR. (HALF WAVE VOLTAGE   13-1973
TRANSVERSE (MORE INTENSE) AND LONGITUDINAL PH/   11-1743
TRANSVERSE POCKELS EFFECT COMMUTATORS. (POTASS   17-2509
TRANSVERSE POLARITONS IN A LITHIUM NIOBATE CRY   11-1762
TRANSVERSE SOUND WAVES AT DOMAIN BOUNDARIES IN   03-0313
TRANSVERSE SOUND WAVES BY 180 DOMAIN BOUNDARIE   03-0314
TRAPPED BY CHARGE DEFECTS IN STRONTIUM TITANAT   06-1123
TRAPPING LEVELS IN ANTIMONY SULFOIODIDE TYPE S   15-2182
TRAPPING OF PARAMETRICALLY AMPLIFIED LIGHT WAV   17-2428
TRAPPING PHENOMENA IN STRONTIUM TITANATE.   06-1091
TRAPPING SITES, OPTICAL DAMAGE, AND THERMALLY   13-1926
TRASNITION.   DIELEC   04-0648
TRI... SEE THE PREFIXED WORD.
TRIGLYCINE CRYSTAL IN THE PARAELECTRIC PHASE.   27-3105
TRIGLYCINE FLUOBERYLLATE. /-9 QUADRUPOLE PERTU   27-2965
TRIGLYCINE FLUOBERYLLATE.   27-3075
TRIGLYCINE FLUOBERYLLATE.   27-3167
TRIGLYCINE FLUOBERYLLATE.   27-3133
TRIGLYCINE FLUOBERYLLATE. INFLUENCE OF SPONTAN   27-3136
TRIGLYCINE FLUOBERYLLATE.   27-3104
TRIGLYCINE FLUOBERYLLATE CRYSTALS.   27-3113
TRIGLYCINE FLUOBERYLLATE CRYSTALS DURING SWITC   27-3027
TRIGLYCINE FLUOBERYLLATE CRYSTALS IN THE FERRO   27-3169
TRIGLYCINE FLUOBERYLLATE IN THE FERRO- AND PAR   27-3175
TRIGLYCINE FLUOBERYLLATE MONOCRYSTAL.   27-3134
TRIGLYCINE FLUOBERYLLATE SINGLE CRYSTAL GROWTH   27-3151
TRIGLYCINE FLUOBERYLLATE SINGLE CRYSTALS.   27-2973
TRIGLYCINE SELENATE.   27-3139
TRIGLYCINE SELENATE.   27-3090
TRIGLYCINE SELENATE. /L) ENERGY GAP (AND DIELE   27-3021
TRIGLYCINE SELENATE.   27-2957
TRIGLYCINE SELENATE. (AGING AND PHASE CHANGES   27-3131
TRIGLYCINE SELENATE CRYSTAL (DURING COOLING TH   27-3043
TRIGLYCINE SELENATE CRYSTALS. / OF DETERMINING   27-2994
TRIGLYCINE SELENATE DEPENDING ON THE CONDITIO/   27-3125
TRIGLYCINE SELENATE. (DIELECTRIC CONSTANT VERS   27-3158
TRIGLYCINE SELENATE IN THE NEAR INFRARED TO FA   27-3007
TRIGLYCINE SELENATE NEAR THE CURIE POINT.   27-3168
TRIGLYCINE SELENATE NEAR THE PHASE TRANSITION.   27-3166
TRIGLYCINE SELENATE OR FLUOBERYLLATE, POTASSI/   02-0089
TRIGLYCINE SELENATE, TRIGLYCINE FLUOBERYLLATE,   27-3124
TRIGLYCINE SUBSTRATES.   TEMPERATURE DEPEND   27-3018
TRIGLYCINE SULFATE)   A NEW FERROELE   27-3031
TRIGLYCINE SULFATE   ACOU   04-0593
TRIGLYCINE SULFATE.   AFTER EFFECTS IN SEM   27-3170
TRIGLYCINE SULFATE /THE FIELD EFFECT AT VERY   27-3171
TRIGLYCINE SULFATE) /ENDENCE OF INTERNAL FRICT   27-3122
TRIGLYCINE SULFATE.   APPLI   27-3157
TRIGLYCINE SULFATE.   APPLICATION OF SELECTIV   27-3042
TRIGLYCINE SULFATE.   27-3008
TRIGLYCINE SULFATE.   COR   27-3142
TRIGLYCINE SULFATE.   27-2974
TRIGLYCINE SULFATE.   27-2985
TRIGLYCINE SULFATE).   27-3155
TRIGLYCINE SULFATE.   27-3073
TRIGLYCINE SULFATE.   27-3149
TRIGLYCINE SULFATE.   27-3049
TRIGLYCINE SULFATE   17-2574
TRIGLYCINE SULFATE   27-3060
TRIGLYCINE SULFATE.   27-3002
TRIGLYCINE SULFATE.   27-3050
TRIGLYCINE SULFATE.   DYNAMIC DETERM   27-2984
TRIGLYCINE SULFATE. /THE ADDITION OF A MIXTURE   27-3045
TRIGLYCINE SULFATE.   EFFECT OF EVA   27-3100
TRIGLYCINE SULFATE.   27-3053
TRIGLYCINE SULFATE.   27-3014
TRIGLYCINE SULFATE.   EFFECT OF SEVERAL   27-2981
TRIGLYCINE SULFATE . /WAVE DISPERSION OF FERRO   03-0370
TRIGLYCINE SULFATE.   27-3019
TRIGLYCINE SULFATE   27-3037
TRIGLYCINE SULFATE.   27-2989
TRIGLYCINE SULFATE) /OF REAL CRYSTAL SUBSTRATE   27-3039
TRIGLYCINE SULFATE .   27-2970

CHANICAL COUPLING NEAR THE FERROELECTRIC PHASE TRANSITION OF TRIGLYCINE SULFATE.   ELECTROME 27-3174
ESR IN GAMMA-IRRADIATED TRIGLYCINE SULFATE   27-3035
ING AND ULTRAMICROSCOPIC STUDIES OF FERROELECTRIC DOMAINS IN TRIGLYCINE SULFATE.   ETCH 27-3143
FERROELECTRIC SPECIFIC HEAT OF TRIGLYCINE SULFATE.   27-3146
HALL MOBILITY OF TELLURIUM AND LEAD TELLURIDE ON UHV CLEAVED TRIGLYCINE SULFATE.   FIELD EFFECT AND 27-3172
HIGH INTERNAL BIAS FIELDS IN L-ALANINE SUBSTITUTED TRIGLYCINE SULFATE   27-2977
N ELECTRIC FIELD ON THE NOISE AND THE PYROELECTRIC EFFECT IN TRIGLYCINE SULFATE   INFLUENCE OF A 27-3076
, SURFACE LAYER AND X-RAY IRRADIATION ON DOMAIN STRUCTURE IN TRIGLYCINE SULFATE. INFLUENCE OF DISLOCATIONS 27-3004
ELL AND THE DOMAIN STRUCTURE ON THE ULTRASONIC ABSORPTION IN TRIGLYCINE SULFATE. /LUENCE OF THE ELECTRIC FI 27-2955
INVESTIGATION OF THE PHASE TRANSITION IN TRIGLYCINE SULFATE.   27-3054
FERROELECTRIC SUBSTANCES BY THE COMPOSITE RESONATOR METHOD. (TRIGLYCINE SULFATE /ON IN SINGLE CRYSTALS OF 27-3080
OF DOMAIN WALLS AND THE NUCLEATION OF DOMAINS IN CRYSTALS OF TRIGLYCINE SULFATE.   MOVEMENT 27-3071
F RADIATION INDUCED DEFECTS IN CRYSTALS OF ROCHELLE SALT AND TRIGLYCINE SULFATE.   NATURE O 27-3159
NEUTRON DIFFRACTION STUDY OF TRIGLYCINE SULFATE.   27-3099
OF DIVERSE THERMODYNAMICAL COEFFICIENTS OF ROCHELLE SALT AND TRIGLYCINE SULFATE.   NEW MEASUREMENTS 27-2995
NONLINEARITY OF THE DIELECTRIC CONSTANT OF FERROELECTRIC TRIGLYCINE SULFATE.   27-3070
NUCLEAR MAGNETIC RESONANCE INVESTIGATION OF TRIGLYCINE SULFATE.   27-3059
NUCLEAR SPIN LATTICE RELAXATION IN FERROELECTRIC TRIGLYCINE SULFATE.   27-2967
IN GAMMA-IRRADIATED FERROELECTRIC MATERIALS. (ROCHELLE SALT, TRIGLYCINE SULFATE) /ON (HINDERED TRANSITION) 27-3165
PIEZOELECTRICITY AND ELECTROSTRICTION OF TRIGLYCINE SULFATE.   27-3118
CLEAR QUADRUPOLE INTERACTIONS OF NITROGEN-14 IN PARAELECTRIC TRIGLYCINE SULFATE. /RESONANCE STUDY OF THE NU 27-2969
ON DOUBLE RESONANCE STUDY OF THE FERROELECTRIC TRANSITION IN TRIGLYCINE SULFATE.   PULSED NITROGEN PROT 27-2968
PYROELECTRIC THERMAL IMAGING DEVICES. (TRIGLYCINE SULFATE   27-3107
REEXAMINATION OF THE THERMAL EXPANSION OF THE FERROELECTRIC TRIGLYCINE SULFATE.   27-3121
REFINEMENT OF CRYSTAL STRUCTURE OF TRIGLYCINE SULFATE.   27-3024
ANNING ELECTRON MICROSCOPE STUDY OF FERROELECTRIC DOMAINS IN TRIGLYCINE SULFATE   SC 27-3052
SECOND HARMONIC LIGHT SCATTERING BY DOMAINS IN FERROELECTRIC TRIGLYCINE SULFATE   27-2991
T" PHENOMENA. (HYSTERESIS LOOP CONSTRICTIONS, ROCHELLE SALT, TRIGLYCINE SULFATE /ERROELECTRIC "AFTER EFFEC 27-3156
LY IN THE NEIGHBORHOOD OF A FERRO- PARAELECTRIC TRANSITION. (TRIGLYCINE SULFATE   SPECIFIC HEAT ANOMA 27-2962
AT UNDER APPLIED ELECTRIC FIELD AND ELECTROCALORIC EFFECT IN TRIGLYCINE SULFATE.   SPECIFIC HE 27-2960
STATISTICAL THEORY FOR FERROELECTRICITY IN TRIGLYCINE SULFATE.   03-0263
STUDY OF SWITCHING PARAMETERS IN TRIGLYCINE SULFATE.   27-3095
SUPERSONIC DOMAIN WALL MOTION IN TRIGLYCINE SULFATE.   27-2963
CREASE DURING POLARIZATION REVERSAL. (IN BARIUM TITANATE AND TRIGLYCINE SULFATE) SUSCEPTIBILITY IN 04-0981
TEMPERATURE DEPENDENCE OF THE SWITCHING RATE OF TRIGLYCINE SULFATE.   27-3015
LECTRIC FIELD ON THE ULTRASONIC ATTENUATION IN FERROELECTRIC TRIGLYCINE SULFATE.   THE INFLUENCE OF THE E 27-2956
THERMAL EXPANSION OF TRIGLYCINE SULFATE.   27-3000
THERMODYNAMIC STABILITY OF THIN FERROELECTRIC FILMS. (TRIGLYCINE SULFATE)   27-2959
THERMODYNAMIC THEORY OF IRRADIATED TRIGLYCINE SULFATE.   03-0152
THICKNESS DEPENDENCE OF THE NUCLEATION FIELD OF TRIGLYCINE SULFATE.   27-3036
SONIC RELAXATION NEAR THE CURIE TEMPERATURE OF FERROELECTRIC TRIGLYCINE SULFATE.   ULTRA 27-3097
ING PROPERTIES OF FERROELECTRICS IN MEMORY DEVICES. (EXAMPLE TRIGLYCINE SULFATE) /LSE WIDTH) PARTIAL SWITCH 27-3144
X-RAY CRITICAL SCATTERING IN FERROELECTRIC TRIGLYCINE SULFATE.   27-2998
X-RAY DAMAGE IN TRIGLYCINE SULFATE.   27-3074
X-RAY TOPOGRAPHIC STUDY OF DOMAIN WALL MOVEMENT IN TRIGLYCINE SULFATE   27-3026
.   TIME DEPENDENCE OF MATERIAL CONSTANTS OF TRIGLYCINE SULFATE AFTER POLARIZATION REVERSAL 27-2953
/LECTRICS. (BARIUM TITANATE, KDP, POTASSIUM SODIUM TARTRATE, TRIGLYCINE SULFATE AMMONIUM FLUOBERYLLATE, S/ 02-0134
CTRIC DETECTO/ LASER STUDY OF REVERSIBLE NUCLEATION SITES IN TRIGLYCINE SULFATE AND APPLICATIONS TO PYROELE 27-3011
/TRICS. (MODEL FOR POTASSIUM DIHYDROGEN PHOSPHATE PRESENTED, TRIGLYCINE SULFATE AND BARIUM AND STRONTIUM T/ 03-0180
/ELECTRIC VOLTAGE RESPONSE TO SHORT INFRARED LASER PULSES IN TRIGLYCINE SULFATE AND BARIUM STRONTIUM NIOBA/ 13-2018
EXPERIMENTAL AND THEORETICAL STUDY OF DEFECTS IN TRIGLYCINE SULFATE AND BARIUM TITANATE.   04-0620
STUDY OF THE PYROELECTRIC EFFECT IN THIN FILMS OF TRIGLYCINE SULFATE AND BARIUM TITANATE.   04-0686
ECTRICS. (26 REFS)   PHASE CHANGES IN TRIGLYCINE SULFATE AND BARIUM TITANATE FERROEL 04-0621
CALCULATIONS).   DIELECTRIC DISPERSION OF TRIGLYCINE SULFATE AND DEUTERATED KDP (THEORY, 17-2534
SIS LOOPS JUST BELOW CUR/ CRITICAL BEHAVIOR OF FERROELECTRIC TRIGLYCINE SULFATE AND DEUTERATED TGS (HYSTERE 27-2966
PO/ ELECTRICAL PROPERTIES OF THE SURFACES OF IONIC CRYSTALS (TRIGLYCINE SULFATE AND OTHERS, BY SELECTIVE DE 27-2997
/ HIGH FREQUENCY BEHAVIOR OF HYDROGEN BONDED FERROELECTRICS: TRIGLYCINE SULFATE AND POTASSIUM DIHYDROGEN PH 17-2526
EFFECT OF A TWO DIMENSIONAL PRESSURE ON THE CURIE POINTS OF TRIGLYCINE SULFATE AND ROCHELLE SALT.   27-3082
CE INVESTIGATIONS OF (COPPER) DOPED FERROELECTRIC CRYSTALS. (TRIGLYCINE SULFATE AND ROCHELLE SALT) /RESONAN 27-3132
IRRADIATION ON BARKHAUSEN EFFECT IN FERROELECTRIC MATERIALS (TRIGLYCINE SULFATE AND ROCHELLE SALT). /GAMMA- 27-3123
S OF VARIOUS TYPES. (BARIUM TITANATE AND TITANATE ZIRCONATE, TRIGLYCINE SULFATE AND ROCHELLE SALT) /LECTRIC 02-0108
INED BY AN ELECTRIC FIELD METHOD.   CRITICAL PROPERTIES OF TRIGLYCINE SULFATE AND ROCHELLE SALT AS DETERM 27-3056
H DOSAGE (DESTR/ FERROELECTRIC BEHAVIOR OF RADIATION DAMAGED TRIGLYCINE SULFATE AND ROCHELLE SALT UP TO HIG 27-3098
D DIELECTRIC MEASUREMENTS ON POTASSIUM DIHYDROGEN PHOSPHATE, TRIGLYCINE SULFATE, AND RUTILE.   FAR INFRARE 17-2423
PROPERTIES OF MIXED CRYSTALS OF TRIGLYCINE SULFATE AND SELENATE.   27-3022
/ECT IN TELLURIUM THIN FILMS ON A SUBSTRATE OF FERROELECTRIC TRIGLYCINE SULFATE AND SELENATE CRYSTALS. (PO/ 27-3046
DISORDER TRANSITION, FERROELEC/ X-RAY CRITICAL SCATTERING IN TRIGLYCINE SULFATE AND SODIUM NITRATE. (ORDER 16-2400
/ FILM NUCLEATION ON FERROELECTRIC SUBSTRATES. (TELLURIUM ON TRIGLYCINE SULFATE AND SODIUM NITRITE SURFACE/ 16-2394
/LECTRIC VOLTAGE RESPONSE TO RECTANGULAR INFRARED SIGNALS IN TRIGLYCINE SULFATE AND STRONTIUM BARIUM NIOBA/ 13-2019
/C VOLTAGE RESPONSE TO STEP SIGNALS OF INFRARED RADIATION IN TRIGLYCINE SULFATE AND STRONTIUM BARIUM NIOBA/ 13-2022
/ALY OF THERMAL PROPERTIES OF FERROELECTRICS (SINGLE CRYSTAL TRIGLYCINE SULFATE AND TRIGLYCINE FLUOBERYLLA/ 27-3137
STALS ON CHANGES IN SPECI/ INFLUENCE OF GROWTH CONDITIONS OF TRIGLYCINE SULFATE AND TRIGLYCINE SELENATE CRY 27-3069
ICITY AND THE DETECTION OF FAR INFR/ ELECTRICAL CONSTANTS OF TRIGLYCINE SULFATE, APPLICATIONS TO PYROELECTR 17-2514
IONS OF GR/ THERMAL HYSTERESIS OF THE DIELECTRIC CONSTANT OF TRIGLYCINE SULFATE AS A FUNCTION OF THE CONDIT 27-3126
DIELECTRIC DISPERSION OF TRIGLYCINE SULFATE AT LOW FREQUENCIES.   27-3023
MEASUREMENT OF THE DIELECTRIC CONSTANT OF TRIGLYCINE SULFATE AT MICROWAVE FREQUENCIES. 27-2954
TEMPERATURE)   LARGE SIGNAL PERMITTIVITY OF TRIGLYCINE SULFATE AT 100 MHZ. (NEAR THE CURIE 27-3029
NETIC RADIATION USING PYROELECTRIC EFFECT. (BARIUM TITANATE, TRIGLYCINE SULFATE, AT 5.6 AND 10.5 GHZ) /OMAG 04-0978
NUTE/ DIELECTRIC STUDY OF CRITICAL BEHAVIOR OF FERROELECTRIC TRIGLYCINE SULFATE BY A DIGITAL TECHNIQUE. (MI 27-3116
PPLIC/ IRREVERSIBLE AND SPONTANEOUS POLARIZATION REVERSAL IN TRIGLYCINE SULFATE BY MEANS OF A LASER BEAM. A 27-3010
NG.   STUDY OF THE PHASE TRANSITION OF TRIGLYCINE SULFATE BY SECOND HARMONIC SCATTERI 27-2992
TS.   STUDY OF THE CRITICAL BEHAVIOR OF TRIGLYCINE SULFATE BY THERMAL NOISE MEASUREMEN 27-3147
REVERSIBLE DOMAIN REVERSAL IN TRIGLYCINE SULFATE CAUSED BY A LASER BEAM.   27-3012
.   THIN FILMS ON TRIGLYCINE SULFATE CLEAVED IN ULTRAHIGH VACUUM 27-3173
EXPERIMENTAL DETAILED STUDY OF THE PHASE TRANSITION IN TRIGLYCINE SULFATE. (CLOSE THERMAL CONTROL) 27-3030
/NFLUENCE OF SURFACE EFFECTS ON THE PYROELECTRIC BEHAVIOR OF TRIGLYCINE SULFATE CLOSE TO THE PHASE TRANSIT/ 27-3064
TRUCTURES)   DEFECT STRUCTURE OF TRIGLYCINE SULFATE. (CORRELATION WITH DOMAIN S 27-3044
OPTICAL MIXING. (OF PULSED RUBY LASER EMISSIONS BY TRIGLYCINE SULFATE CRYSTAL) 27-2958
INVESTIGATION OF THE DOMAIN STRUCTURE OF A TRIGLYCINE SULFATE CRYSTAL DURING AGING.   17-2561
BARKHAUSEN JUMPS AND SWITCHING CURRENT IN TRIGLYCINE SULFATE CRYSTALS.   27-3109

LIP ETCH PATTERNS ON THE FACES OF TRIGLYCINE SULFATE CRYSTALS. 27-3127
STINCTIVE FEATURES OF BARKHAUSEN EFFECT IN ROCHELLE SALT AND TRIGLYCINE SULFATE CRYSTALS. DI 27-3110
DOMAIN STRUCTURE OF GAMMA-IRRADIATED TRIGLYCINE SULFATE CRYSTALS. 27-3164
DOMAIN WALL CAUGHT IN DISLOCATIONS IN FERROELECTRIC TRIGLYCINE SULFATE CRYSTALS. 27-3091
FREQUENCY DEPENDENCE OF THE COERCIVE FIELD OF TRIGLYCINE SULFATE CRYSTALS. 27-3062
INDUCED GROWTH ANISOTROPY IN TRIGLYCINE SULFATE CRYSTALS. 27-3162
IZATION ON THE PHOTOSTIMULATED EXCELECTRON EMISSION YIELD OF TRIGLYCINE SULFATE CRYSTALS. /PONTANEOUS POLAR 27-2961
INVESTIGATION OF BARKHAUSEN EFFECT IN TRIGLYCINE SULFATE CRYSTALS. 27-3111
IRON IMPURITY) ABSORPTION BANDS OF DEFECT ROCHELLE SALT AND TRIGLYCINE SULFATE CRYSTALS. /N AND COPPER AND 27-3108
TION EDGE AS A FUNCTION OF TEMPERATURE FOR ROCHELLE SALT AND TRIGLYCINE SULFATE CRYSTALS. /NDAMENTAL ABSORP 27-3160
SECOND HARMONIC GENERATION IN TRIGLYCINE SULFATE CRYSTALS. 27-3138
PECIAL ELECTRICAL PROPERTIES OF REGIONS NEAR DOMAIN WALLS IN TRIGLYCINE SULFATE CRYSTALS. S 27-2988
ION OF REFRACTIVE INDEX IN AMMONIUM DIHYDROGEN PHOSPHATE AND TRIGLYCINE SULFATE CRYSTALS. THERMAL VARIAT 17-2474
/RASCUND ON THE POLARIZATION AND REPOLARIZATION PROCESSES IN TRIGLYCINE SULFATE CRYSTALS AND BARIUM TITANA/ 04-0924
ALLIZATION OF ANTHRAQUINO/ REVEALING THE DOMAIN STRUCTURE OF TRIGLYCINE SULFATE CRYSTALS BY SELECTIVE CRYST 27-3041
H CONDITIONS. MAXIMUM REPOLARIZATION CURRENT FOR TRIGLYCINE SULFATE CRYSTALS DEPENDENT ON GROWT 27-3154
/TION TOPOGRAPHIC METHOD TO FERROELECTRIC SODIUM NITRITE AND TRIGLYCINE SULFATE CRYSTALS. (DOMAIN BOUNDARI/ 16-2375
LIGHT EMISSION FROM TRIGLYCINE SULFATE (CRYSTALS DURING HEATING) 27-3117
OSES OF X- AND GAMMA-RAYS. DIELECTRIC LOSSES OF TRIGLYCINE SULFATE CRYSTALS EXPOSED TO SMALL D 27-2976
OVE THE CURIE POINT. DOMAIN STRUCTURE OF PURE AND DOPED TRIGLYCINE SULFATE CRYSTALS GROWN BELOW AND AB 27-3130
IR GROWTH COND/ X-RAY TOPOGRAPHIC STUDY OF GROWTH DEFECTS IN TRIGLYCINE SULFATE CRYSTALS IN RELATION TO THE 27-3025
APPLIED ELECTRIC FIE/ DEPENDENCE OF THE DOMAIN STRUCTURE OF TRIGLYCINE SULFATE CRYSTALS ON TEMPERATURE AND 27-3129
LOSS FROM 10 GHZ TO 100 GHZ) DIELECTRIC DISPERSION IN TRIGLYCINE SULFATE CRYSTALS. (PERMITTIVITY AND 27-3094
RIUM THIN FILMS. THERMAL EXPANSION OF TRIGLYCINE SULFATE CRYSTALS STUDIED WITH TELLU 27-3047
OTOS) WALL MOTION AND NUCLEATION OF DOMAINS IN TRIGLYCINE SULFATE CRYSTALS. (SUMMARY, WITH PH 27-3072
PERFECTIONS, DOMAIN STRUCTU/ GROWTH REGIONS IN FERROELECTRIC TRIGLYCINE SULFATE CRYSTALS. (TOPOGRAPHY OF IM 27-2999
VARIATION OF THE UNIPOLARITY OF TRIGLYCINE SULFATE CRYSTALS WITH TIME. 27-3115
ROELECTRIC MODE. PERFORMANCE CHARACTERISTICS OF A SMALL TRIGLYCINE SULFATE DETECTOR OPERATED IN THE PY 27-3119
NE COERCIVE FIELD) (POLARIZATION CHANGE) AFTER EFFECTS IN TRIGLYCINE SULFATE. (DIELECTRIC DISPLACEMENT A 27-2952
/CTROMAGNETIC RADIATION INTO MOTION. (PYROELECTRIC EFFECT IN TRIGLYCINE SULFATE ETHYL DIAMINE TARTRATE, B/ 04-0854
NEW METHOD OF MEASURING PYROELECTRIC COEFFICIENTS. (TRIGLYCINE SULFATE EXAMPLE) 17-2517
THERMAL ANALYSIS OF PYROELECTRIC DETECTORS. (TRIGLYCINE SULFATE EXAMPLE) 27-2964
NS. DOPED TRIGLYCINE SULFATE FOR PYROELECTRIC APPLICATIO 27-3057
D SOLUTIONS. GROWTH OF TRIGLYCINE SULFATE FROM SLIGHTLY SUPERSATURATE 27-3096
ICAL APPLICATIONS. MATERIAL RESEARCH OF FERROELECTRIC TRIGLYCINE SULFATE FROM THE VIEWPOINT OF PRACT 27-3065
THERMAL AND ELASTIC PROPERTIES OF FERROELECTRIC CRYSTALS THE TRIGLYCINE SULFATE GROUP. /TERATION EFFECT ON 27-2986
/D ISOMORPHOUS SUBSTITUTION IN FERROELECTRIC CRYSTALS OF THE TRIGLYCINE SULFATE GROUP. (NORMAL AND DEUTERA/ 27-3102
E CONTROL) ORDER DISORDER TRANSITION IN FERROELECTRICS. (TRIGLYCINE SULFATE, HIGHLY ACCURATE TEMPERATUR 27-2982
/ELLE SALT, BARIUM TITANATE, ANTIMONY SULFOIODIDE, THIOUREA, TRIGLYCINE SULFATE, HYDROGEN AMMONIUM MONOCHL/ 02-0099
TRUCTURE) AGING PROCESS IN TRIGLYCINE SULFATE. (HYSTERESIS LAPS, DOMAIN S 27-3128
PERMITTIVITY OF SINGLE CRYSTALS OF TRIGLYCINE SULFATE IN A STRONG ELECTRIC FIELD. 27-2978
ERATURE DEPENDENCE OF DIELECTRIC PROPERTIES OF FERROELECTRIC TRIGLYCINE SULFATE IN THE FAR INFRARED. TEMP 27-3061
ATURE AND FREQUENC/ DISPERSION OF THE DIELECTRIC CONSTANT OF TRIGLYCINE SULFATE IN THE FAR INFRARED (TEMPER 27-3009
NEAR THE CURIE POINT (IMPROVED ME/ PYROELECTRIC RESPONSE OF TRIGLYCINE SULFATE IN THE FERROELECTRIC REGION 27-2979
PYROELECTRIC BEHAVIOR OF TRIGLYCINE SULFATE IN THE PARAELECTRIC REGION. 27-3066
E POINT. EQUILIBRIUM DOMAIN STRUCTURE OF TRIGLYCINE SULFATE IN THE VICINITY OF THE CURI 27-3112
/TRAST AT FERROELECTRIC DOMAIN BOUNDARIES (DOPED AND UNDOPED TRIGLYCINE SULFATE LITHIUM NIOBATE, KDP, ABS/ 11-1671
Y APPROXIMATELY) ELECTROMECHANICAL PROPERTIES OF TRIGLYCINE SULFATE. (MEASUREMENTS VERIFY THEOR 27-3145
EQUATION OF STATE FOR THE COOPERATIVE TRANSITION OF TRIGLYCINE SULFATE NEAR CURIE TEMPERATURE. 27-3006
POLARIZATION FLUCTUATION OF TRIGLYCINE SULFATE NEAR THE CURIE POINT. 27-3077
THERMAL CONDUCTIVITY OF TRIGLYCINE SULFATE NEAR THE CURIE POINT. 27-2983
THERMAL CONDUCTIVITY OF TRIGLYCINE SULFATE NEAR THE CURIE POINT. 27-3016
OF GAMMA-IRRADIATED S/ DISPLACIVE MOVEMENT OF THE RADICAL IN TRIGLYCINE SULFATE NEAR THE CURIE POINT. (ESR 27-3033
/RY CONDITIONS ON THE THERMAL CAPACITY OF SINGLE CRYSTALS OF TRIGLYCINE SULFATE (NEAR THE FERROELECTRIC TR/ 27-3140
CRYSTALLOGRAPHIC STUDIES OF IRRADIATION FIELD TREATED TRIGLYCINE SULFATE: NEW STRUCTURE FORM. 27-2996
CHANGES OF THE DOMAIN STRUCTURE IN VANADIUM DOPED TRIGLYCINE SULFATE OBSERVED BY EPR. 27-3013
/ENTAL STUDY OF GAMMA-RADIATION EFFECTS IN ROCHELLE SALT AND TRIGLYCINE SULFATE. (ON ELECTRONIC PROPERTIES/ 27-3101
TEMPERATURE. DEPENDENCE OF THE COERCIVE FIELD OF TRIGLYCINE SULFATE ON FREQUENCY, AMPLITUDE AND 27-3092
N IN THIN FERROELECTRIC FILMS ON SEMICONDUCTING SUBSTRATES. (TRIGLYCINE SULFATE ON SILICON) /OF POLARIZATIO 27-3163
TRIC / TEMPERATURE DEPENDENCE OF THE DIELECTRIC CONSTANTS OF TRIGLYCINE SULFATE ORTHOGONAL TO THE FERROELEC 27-2975
EFFECT OF SAMPLE TREATMENT AND ELECTRODE NATURE ON SOME TRIGLYCINE SULFATE PARAMETERS. 27-3148
STRUCTURAL INHIBITION OF FERROELECTRIC SWITCHING IN TRIGLYCINE SULFATE. PART-1: ADDITIVES. 27-3038
HYDROGEN AND THE ACTIVATION ENERGY/ ESR OF GAMMA-IRRADIATED TRIGLYCINE SULFATE PART-2: INTERNAL MOTION OF 27-3034
/ECTRICS IN CONNECTION WITH THEIR PHASE TRANSITIONS. PART-1: TRIGLYCINE SULFATE PART-2: SILVER SODIUM DIN/ 02-0077
TMENT. STRUCTURAL INHIBITION OF FERROELECTRIC SWITCHING IN TRIGLYCINE SULFATE. PART-2: X-IRRADIATION TREA 17-2469
/EAR PROPERTIES OF FERROELECTRICS AT MICROWAVE FREQUENCIES. (TRIGLYCINE SULFATE, POLYCRYSTALLINE BARIUM TI/ 04-0870
IDEUT/ LASER LIGHT SCATTERING STUDIES OF PHASE TRANSITIONS. (TRIGLYCINE SULFATE, POTASSIUM DIHYDROGEN AND D 17-2481
/TTIVITY (AUDIO FREQUENCIES) OF SOME FERROELECTRIC CRYSTALS (TRIGLYCINE SULFATE, POTASSIUM DIHYDROGEN PHOS/ 02-0125
GNETIC IONS. SWITCHING IN FERROELECTRIC TRIGLYCINE SULFATE, PURE AND DOPED WITH PARAMA 27-3020
CONSTANT) OF FERROELECTRIC CRYSTALS. (5 COMPOUNDS MEASURED, TRIGLYCINE SULFATE RESULTS GIVEN) /TURE, CURIE 27-3120
AL ACTIVITY OF SOME FERROELECTRIC CRYSTALS (PYROELECTRICITY, TRIGLYCINE SULFATE ROCHELLE SALT). OPTIC 27-3017
NATE) NOISE MEASUREMENTS IN FERROELECTRICS. (TRIGLYCINE SULFATE ROCHELLE SALT, BARIUM TITA 04-0698
/ECTRICS. (THEORY, DYNAMICAL ORDER DISORDER PROPERTIES, KDP, TRIGLYCINE SULFATE ROCHELLE SALT, SILVER PER/ 02-0050
/MAIN STRUCTURE STABILITY IN FERROELECTRICS. (ROCHELLE SALT, TRIGLYCINE SULFATE, SELENATE, FLUOBERYLLATE A/ 02-0083
MAIN WALL RESONANCE). DIELECTRIC DISPERSION IN FERROELECTRIC TRIGLYCINE SULFATE (SEVERAL KHZ TO 23 KMHZ, DO 27-3032
MONIC TYPE FERROELECTRIC MODULATOR FOR ELECTROMETER. (WITH A TRIGLYCINE SULFATE SINGLE CRYSTAL) /SECOND HAR 27-2951
IELECTRIC FERROELECTRIC CONDENSER UNDER CONTINUOUS HEATING. (TRIGLYCINE SULFATE SINGLE CRYSTAL) /ENT OF A D 27-3028
DIELECTRIC RELAXATION IN MULTIDOMAIN TRIGLYCINE SULFATE SINGLE CRYSTALS. 27-3001
EFFECT OF GROWTH CONDITIONS ON DIELECTRIC LOSSES IN TRIGLYCINE SULFATE SINGLE CRYSTALS. 27-3152
EPR STUDIES OF CHROMIUM- AND VANADYL- DOPED TRIGLYCINE SULFATE SINGLE CRYSTALS. 27-3161
EVOLUTION OF FERROELECTRIC DOMAINS IN TRIGLYCINE SULFATE SINGLE CRYSTALS. 27-3003
GROWTH OF TRIGLYCINE SULFATE SINGLE CRYSTALS. 27-3083
INFLUENCE OF DEFECTS ON SOME DIELECTRIC PROPERTIES OF TRIGLYCINE SULFATE SINGLE CRYSTALS. 27-3063
OF IMPURITIES ON THE GROWTH AND SOME PHYSICAL PROPERTIES OF TRIGLYCINE SULFATE SINGLE CRYSTALS. INFLUENCE 27-3084
POLARIZATION REVERSAL IN TRIGLYCINE SULFATE SINGLE CRYSTALS. 27-2980
PREPARATION OF PURE GLYCINE USED FOR GROWING OF TRIGLYCINE SULFATE SINGLE CRYSTALS. 27-3085
REMARKS ABOUT SWITCHING IN BARIUM TITANATE AND TRIGLYCINE SULFATE SINGLE CRYSTALS. 04-0915
TH RATE ON THE DEFECT STRUCTURE AND DIELECTRIC PROPERTIES OF TRIGLYCINE SULFATE SINGLE CRYSTALS. /T OF GROW 27-3067
X-RAY TOPOGRAPHIC OBSERVATIONS OF RADIATION DAMAGE IN TRIGLYCINE SULFATE SINGLE CRYSTALS. 27-3106
VESTIGATIONS OF THEIR QUALITY. (VOLUME / PREPARATION OF PURE TRIGLYCINE SULFATE SINGLE CRYSTALS AND SOME IN 27-3086

/Y DEPENDENCE OF THE FLUCTUATION ABSORPTION OF ULTRASOUND IN TRIGLYCINE SULFATE SINGLE CRYSTALS (AT 10, 30/ 27-3079
ITY AND ELECTRIC PERMI/ THERMAL AND ELECTRICAL PROPERTIES OF TRIGLYCINE SULFATE SINGLE CRYSTALS (HEAT CAPAC 27-3141
  ELECTRIC FIELD.   DIELECTRIC CONSTANT OF TRIGLYCINE SULFATE SINGLE CRYSTALS IN A STRONG 27-3150
ECTRIC FIELD.   TEMPERATURE AUTOSTABILIZATION EFFECT OF TRIGLYCINE SULFATE SINGLE CRYSTALS IN AN AC EL 27-3005
  IN THE TANGENT OF THE ANGLE/ EFFECT OF GROWTH CONDITIONS OF TRIGLYCINE SULFATE SINGLE CRYSTALS ON A CHANGE 27-3153
ELECTRIC PROPERTIES. (REVIEW)   INFLUENCE OF THE QUALITY OF TRIGLYCINE SULFATE SINGLE CRYSTALS ON THEIR DI 02-0088
F ANISOTROPIC RELAXATION / ANISOTROPY OF SOUND ABSORPTION IN TRIGLYCINE SULFATE SINGLE CRYSTALS.   (THEORY O 27-3078
/TH AND CHARACTERIZATION OF SOLID SOLUTIONS OF FERROELECTRIC TRIGLYCINE SULFATE SINGLE CRYSTALS WITH ISOMO/ 27-2972
T, X-RA/ STUDIES ON THE MECHANISM OF THE PHASE TRANSITION IN TRIGLYCINE SULFATE   (STATIC DIELECTRIC CONSTAN 27-3093
RATURE AUTOSTABILIZATION REGIME OF A FERROELECTRIC CRYSTAL. (TRIGLYCINE SULFATE TANDEL) /ITIES IN THE TEMPE 17-2502
PERTIES DOMAIN STRUCTURE)   AGING PROCESSES IN TRIGLYCINE SULFATE. (TANDEL OR CAPACITANCE PRO 27-3081
F THE COMPLEX DIELECTRIC CONS/ RELAXATION OF DOMAIN WALLS IN TRIGLYCINE SULFATE. (TEMPERATURE DEPENDENCES O 27-3103
F WALL THICKNESS AND ENERGY DE/ FERROELECTRIC DOMAIN WALL IN TRIGLYCINE SULFATE. (THEORY AND CALCULATIONS O 27-2997
/PROVEMENT IN THE PYROELECTRIC INFRARED RADIATION DETECTOR. (TRIGLYCINE SULFATE, TRIGLYCINE FLUOROBERYLLAT/ 02-0046
/AND COMPOSITION OF FERROELECTRIC CRYSTALS (SOLUTION GROWTH, TRIGLYCINE SULFATE TRIGLYCINE SELENATE OR FL/ 02-0089
LOW FREQUENCY RAMAN SPECTRA OF FERROELECTRIC CRYSTALS OF THE TRIGLYCINE SULFATE TYPE.   27-3114
                    PREPARATION OF PURE TRIGLYCINE SULFATE. (USING ION EXCHANGE)   27-3087
MICROSCOPE.   DIRECT OBSERVATION OF FERROELECTRIC DOMAINS IN TRIGLYCINE SULFATE USING THE SCANNING ELECTRON 27-3051
/ENCE OF SOME FERROELECTRIC PROPERTIES OF SOLID SOLUTIONS OF TRIGLYCINE SULFATE WITH ISOMORPHOUS SUBSTANCE/ 27-2971
TS.   INVESTIGATIONS OF SURFACES OF FERROELECTRICS (TRIGLYCINE SULFATE WITH SEMICONDUCTING ELEMEN 27-3089
ND FIELD STRENGTH DEPENDENCE OF LARGE SIGNAL PERMITTIVITY OF TRIGLYCINE SULFATE (10 KHZ-1 MHZ) /PERATURE A 27-3068
            MICROWAVE BEHAVIOR OF ROCHELLE SALT AND TRIGLYCINE SULFATE (18 REFS).   27-3088
            GENERATION OF THE SECOND HARMONIC BY TRIGLYCINE SULFATE. (84 REFS)   27-2990
                POLARIZED TRIGLYCINE SULFATES.   27-3058
            ANOMALIES OF THERMAL PROPERTIES OF TRIGLYCINE SULFIDE.   27-3048
(BORACITE, ERICAITE) ATOMIC DISPLACEMENTS IN FERROELECTRIC TRIGONAL AND ORTHORHOMBIC BORACITE STRUCTURES. 14-2075
C AND FERROMAGNETOELECTRIC MATERIAL THAT ALLOWS NO 180 DEGR/ TRIGONAL BORACITES: A NEW TYPE OF FERROELECTRI 14-2098
B ION IS FERROELECTRIC/ DIPOLE PATTERNS IN ORTHORHOMBIC AND TRIGONAL PHASES OF ABO(3) SUBSTANCE. (WHEN THE 08-1363
RONTIUM TITANATE.   CUBIC TO TRIGONAL STRESS INDUCED PHASE TRANSITION IN ST 06-1235
                A NEW FORM OF TUNGSTEN TRIOXIDE.   10-1652
                ELECTRICAL CONDUCTIVITY OF TUNGSTEN TRIOXIDE.   10-1637
TUDY OF THE CATALYTIC ACTIVITY OF NONSTOICHIOMETRIC TUNGSTEN TRIOXIDE.   EPR S 10-1628
  FORMATION OF SHEAR STRUCTURES IN SUBSTOICHIOMETRIC TUNGSTEN TRIOXIDE.   10-1659
AND ANTIFERROELECTRICITY OF THE AO3 TYPE CRYSTAL. (TUNGSTEN TRIOXIDE)   ON FERROELECTRICITY 10-1641
OF LEAD ZIRCONATE(1-X) TITANATE(X) PLUS 0.8 PERCENT TUNGSTEN TRIOXIDE.   PHASE DIAGRAM 09-1581
NGSTIC OXIDE SINGLE CRYSTALS (SUBLIMATION IN AIR OF TUNGSTEN TRIOXIDE).   PREPARATION OF TU 10-1644
                RAMAN SPECTROSCOPY OF TUNGSTEN TRIOXIDE.   10-1635
PHASES IN THE SYSTEM LEAD OXIDE- TITANIUM DIOXIDE- LANTHANUM TRIOXIDE.   RANGE OF EXISTENCE OF PEROVSKITE 08-1349
ARIUM (TITANATE, STANNATE) SOLID SOLUTIONS DOPED BY TUNGSTEN TRIOXIDE.   SEMICONDUCTOR PROPERTIES OF B 04-0965
XIDE(12) WITH VANADIUM PENTOXIDE AND MOLYBDENUM AND TUNGSTEN TRIOXIDE. /EACTION OF BISMUTH(4) TITANIUM(3) O 13-2029
NG OF THE SO-CALL/ SOME ASPECTS OF THE STRUCTURE OF TUNGSTEN TRIOXIDE AND A CONTRIBUTION TO THE UNDERSTANDI 10-1653
ANGES. (STUDI/ BOMBARDMENT INDUCED AMORPHIZATION IN TUNGSTEN TRIOXIDE AND ITS USE IN DEDUCING MEAN DAMAGE R 10-1642
(PREPARATION AND ELECTRICAL AND OPTICAL) STUDY OF TUNGSTEN TRIOXIDE AT LOW TEMPERATURE.   10-1643
OELECTRIC CRYSTALS WITH A PHOTOEMISSION MICROSCOPE (TUNGSTEN TRIOXIDE, BARIUM TITANATE). /F DOMAINS IN FERR 04-0811
/ OF RESISTIVITY ON LANTHANUM DOPED BARIUM TITANATE- BISMUTH TRIOXIDE COMPOSITE CERAMICS WITH SURFACE BARR/ 04-0800
          AN EFFECT OF HEAT TREATMENT ON TUNGSTEN TRIOXIDE CRYSTAL. PART-2: GREEN STRIPES.   10-1639
EN DEFICIENCY ON ELECTRICAL TRANSPORT PROPERTIES OF TUNGSTEN TRIOXIDE CRYSTALS.   EFFECT OF OXYG 10-1629
OEMISSION MICROSCOPIC OBSERVATION OF THE SURFACE OF TUNGSTEN TRIOXIDE CRYSTALS.   PHOT 10-1645
DOPED / MICROSTRUCTURE AND ELECTRICAL PROPERTIES OF SCANDIUM TRIOXIDE DOPED, RARE EARTH OXIDE DOPED, AND UN 04-0839
        STRUCTURAL AND DIELECTRIC MODIFICATIONS OF TUNGSTEN TRIOXIDE DUE TO NEUTRON IRRADIATION.   10-1651
            THE COLOR PROBLEM OF TUNGSTEN TRIOXIDE ELECTRICAL CONDUCTIVITY.   10-1636
AND THE ORTHORHOMBIC- TETRAGONAL TRANSITION OF THE TUNGSTEN TRIOXIDE PHASE.   THERMAL EXPANSION 10-1626
ECTRIC CRYSTALS WITH SCANNING ELECTRON MICROSCOPE. (TUNGSTEN TRIOXIDE SINGLE CRYSTAL PLATES) /NS OF FERROEL 10-1646
TION PLANES IN ANTIFERROELECTRICS. (APPLIED TO ADP, TUNGSTEN TRIOXIDE, SODIUM NIOBATE) /ENTATION OF COMPOSI 08-1327
ALATE.   TIGHT BONDING BAND CALCULATIONS ON RHENIUM TRIOXIDE, SODIUM TUNGSTATE, AND POTASSIUM TANT 08-1334
/(CRYSTALLOGRAPHIC SHEAR) FAMILIES DERIVED FROM THE RHENIUM TRIOXIDE STRUCTURE TYPE: AN ELECTRON MICROSCO/ 10-1630
            THE BISMUTH OXIDE- TUNGSTEN TRIOXIDE SYSTEM.   10-1656
            PHASE TRANSITION IN TUNGSTEN TRIOXIDE UNDER HIGH PRESSURE.   10-1658
            THE COLOR PROBLEM OF TUNGSTEN TRIOXIDE. X-RAY STUDIES.   10-1632
MODE MOTION IN POTASSIUM DIDEUTERIUM PHOSPHATE . (BY NEUTRON TRIPLE AXIS SPECTROMETRY) /RROELECTRIC (SOFT) 17-2668
RAY IRRADIATED SINGLE CRYST/ OBSERVATION OF ELECTRON NUCLEAR TRIPLE RESONANCE FOR THE ARSENATE CENTER IN X- 18-2729
IUM TITANATES.   ELECTRON PARAMAGNETIC RESONANCE OF TRIVALENT GADOLINIUM IONS IN STRONTIUM AND BAR 04-0920
TRIC RELAXATION IN (CERAMIC) STRONTIUM TITANATES CONTAINING (TRIVALENT) RARE EARTH IONS.   DIELEC 06-1143
        STRUCTURAL AND PHASE RELATIONSHIPS AMONG TRIVALENT TUNGSTATES AND MOLYBDATES.   10-1648
        STRUCTURAL AND PHASE RELATIONSHIPS AMONG TRIVALENT TUNGSTATES AND MOLYBDATES.   10-1649
COMPLEXES IN FERROELECTRIC TRIS-SARCOSINE CALCIUM CHLORIDE (TSCC) SINGLE CRYSTALS. /TIONS OF MANGANESE(2+) 36-3525
            MICROSTRUCTURE OF TSTS-19 TYPE FERROELECTRIC CERAMICS.   36-3554
COAXIALLY SEGMENTED, LONGITUDINALLY POLARIZED FERROELECTRIC TUBES.   VIBRATIONS OF 03-0376
ATIONS OF LONGITUDINALLY POLARIZED FERROELECTRIC CYLINDRICAL TUBES.   VIBR 03-0375
FICIENT HIGH GRAIN PARAMETRIC GENERATION IN ADP CONTINUOUSLY TUNABLE ACROSS THE VISIBLE SPECTRUM.   EF 17-2712
WITHOUT A RESONATOR.   EFFECT TUNABLE OPTICAL EMISSION FROM LITHIUM NIOBATE 11-1858
THIUM NIOBATE.   POWER AND LINE WIDTH OF TUNABLE STIMULATED FAR INFRARED EMISSION IN LI 11-1736
            BARIUM SODIUM NIOBATE FOR A TUNABLE STIMULATED RAMAN OSCILLATOR.   13-1917
GROWTH, CRYSTALLOGRAPHY AND DIELECTRIC PROPERTIES OF BISMUTH TUNGSTATE.   10-1650
            INVESTIGATION OF PHASE TRANSITION IN LEAD CADMIUM TUNGSTATE.   10-1633
OS SINGLE CRYSTAL GROWTH OF SODIUM TUNGSTEN BRONZES. (SODIUM TUNGSTATE)   KYROPOUL 13-2054
            OPTICAL TRANSITION TO THE FERMI LEVEL IN SODIUM TUNGSTATE.   10-1634
/C SUSCEPTIBILITY) OF HYDROGEN TUNGSTEN BRONZES, HYDROGEN(X) TUNGSTATE. (ANALOGOUS TO THE SODIUM COMPOUNDS) 13-1934
TIGHT BONDING BAND CALCULATIONS ON RHENIUM TRIOXIDE, SODIUM TUNGSTATE, AND POTASSIUM TANTALATE.   08-1334
/STRY OF PEROVSKITE LIKE LAYER TYPE FERROELECTRICS. (BISMUTH TUNGSTATE, BISMUTH NIOBIUM OXYFLUORIDE, BISMU/ 13-1966
/TION STUDY OF (TETRAGONAL- CUBIC PEROVSKITE) STRONTIUM IRON TUNGSTATE . (MAGNETIC G STRUCTURE ANALOGOUS T/ 10-1655
HAFNIUM, CERIUM, THO/ THALLIUM PEROVSKITES. THALLIUM A(1-X) TUNGSTATE OR MOLYBDATE. (WHERE A IS ZIRCONIUM, 10-1640
TYPE: QUADRATIC LEAD OR BISMUTH TANTALATE NIOBATE, TANTALATE TUNGSTATE, OR TUNGSTATE NIOBATE. /YGEN BRONZE 13-1985
SEIGNETTO MAGNETIC SOLID SOLUTIONS OF THE LEAD FERRATE(0.67) TUNGSTATE(0.33)- LEAD TITANATE SYSTEM.   05-1065
/ECTRICS AND FERROELECTRIC FERROMAGNETS. (LEAD FERRATE(0.67) TUNGSTATE(0.33) LEAD FERRATE(0.5) NIOBATE(0.5/ 08-1368
OF THE ANTIFERROELECTRIC CURIE POINT OF LEAD MAGNESATE(0.5) TUNGSTATE(0.5). /TERNAL FRICTION IN THE REGION 13-2020
RIZED CERAMICS IN THE LEAD TITANATE ZIRCONATE MAGNESATE(0.5) TUNGSTATE(0.5). /AND THERMAL EXPANSION OF POLA 09-1603
ERTIES IN THE BINARY SYSTEM LEAD TITANATE- LEAD CADMATE(0.5) TUNGSTATE(0.5).   ELECTROMECHANICAL PROP 10-1638
/ LEAD COBALTATE(0.5) TUNGSTATE(0.5) AND LEAD COBALTATE(0.5) TUNGSTATE(0.5)- LEAD TITANATE SOLID SOLUTIONS. 13-1968
/E TRANSISITION IN THE ANTIFERROELECTRET LEAD COBALTATE(0.5) TUNGSTATE(0.5) AND LEAD COBALTATE(0.5) TUNGST/ 13-1968

/ITANATE AND ZIRCONATE. (SOLID SOLUTIONS LEAD MAGNESATE(0.5) TUNGSTATE(0.5)), EFFECTS OF POSITIVE OR NEGAT/ 05-1063
ND X-RAY STUDIES OF PHASE TRANSITIONS IN LEAD COBALTATE(0.5) TUNGSTATE(0.5), TITANATE, ZIRCONATE SYSTEM. /A 13-1960
                    DIELECTRIC PROPERTIES OF NIOBATES AND TUNGSTATES.   08-1322
ERROELECTRICS AND ANTIFERROELECTRICS. (ZIRCONATES, NIOBATES, TUNGSTATES) /BOVE THE CURIE POINT IN VARIOUS F 02-0124
TANTALUM) (2) OXIDE(9), AND VARIOUS TUNGSTEN SUBSTITUTED LEAD TUNGSTATES) /STRONTIUM) (3) COPPER (NIOBIUM OR 36-3619
MAN STUDY OF METALLIC TUNGSTEN BRONZES. (SODIUM AND RUBIDIUM TUNGSTATES)   RA 10-1654
ERMAL CONDUCTIVITY OF CUBIC SODIUM TUNGSTEN BRONZES. (SODIUM TUNGSTATES)   TH 13-2057
            STRUCTURAL AND PHASE RELATIONSHIPS AMONG TRIVALENT TUNGSTATES AND MOLYBDATES.   10-1648
            STRUCTURAL AND PHASE RELATIONSHIPS AMONG TRIVALENT TUNGSTATES AND MOLYBDATES.   10-1649
/ROELECTRIC FERROMAGNETIC SOLID SOLUTIONS IN THE LEAD COBALT TUNGSTATE- CADMIUM MANGANATE SYSTEM (BOTH CUR/ 36-3604
TE. CRYSTAL STRUCTURE OF THE TRANSITION METAL MOLYBDATES AND TUNGSTATES. PART-6. /C BETA GADOLINIUM MOLYBDA 30-3289
OPERTY OF BARIUM TITANATE NIOBATE SINGLE CRYSTAL (TETRAGONAL TUNGSTEN BRONZE).   DIELECTRIC PR 13-1969
            POLARITON RESULTS IN FERROELECTRICS WITH THE TUNGSTEN BRONZE CRYSTAL STRUCTURE.   13-1918
OBATE- LITHIU/ CRYSTAL GROWTH AND ELECTROOPTIC PROPERTIES OF TUNGSTEN BRONZE CRYSTALS FROM THE POTASSIUM NI 13-2003
FS)   PROPERTIES OF TUNGSTEN BRONZE FERROELECTRICS. (REVIEW, 33 RE 02-0053
TASSIA LITHIA NIOBIA (POTASSIUM LITHIUM NIOBATES, PHASE REL/ TUNGSTEN BRONZE FIELD IN THE (OXIDE) SYSTEM PO 13-2017
TEM SODIUM NIOBATE BARIUM TITANATE BARIUM DINIOBATE.   TUNGSTEN BRONZE FIELD IN THE PSEUDOTERNARY SYS 13-1970
ND MELT GROWTH OF CRYSTALS IN THE SODIUM OXIDE-/ (EXTENT OF) TUNGSTEN BRONZE (LIQUIDUS AND SOLIDUS) FIELD A 13-1949
RIUM, AND C IS SODIUM OR POTASSIUM) PHASES OF THE TETRAGONAL TUNGSTEN BRONZE OXIDE TYPE. / STRONTIUM, OR BA 04-0912
/E TRANSITIONS IN TANTALATES WITH THE STRUCTURE OF POTASSIUM TUNGSTEN BRONZE. (POTASSIUM BARIUM TANTALUM O/ 13-2050
            THE SEARCH FOR NEW FERROELECTRICS WITH THE TUNGSTEN BRONZE STRUCTURE.   13-1897
RTIES OF POTASSIUM LITHIUM NIOBATES (TRANSPARENT, TETRAGONAL TUNGSTEN BRONZE STRUCTURE, FERROELECTRIC). /PE 13-2026
/CONDITIONS AND PHASE TRANSITIONS IN CRYSTALS WITH POTASSIUM TUNGSTEN BRONZE STRUCTURE (POTASSIUM SODIUM L/ 13-1923
M NIOBATE (SEC/ OPTICAL AND FERROELECTRIC PROPERTIES OF (THE TUNGSTEN BRONZE SYSTEM) POTASSIUM SODIUM BARIU 13-2028
IDE(47): POTENTIALLY FERROELECTRIC CRYSTAL STRUCTURE OF THE TUNGSTEN BRONZE TYPE. /SMUTH(3) NIOBIUM(17) OX 13-1978
IS (NIOBIUM, TITANIUM) OR (NIOBIUM, TUNGS/ SOME COMPOUNDS OF TUNGSTEN BRONZE TYPE A(6)B(10)O(30). (WHERE B 13-1964
            TUNGSTEN BRONZE TYPE COMPLEX OXIDES.   13-1981
3: POTASSIUM LITHIUM NIOBATE.   FERROELECTRIC TUNGSTEN BRONZE TYPE CRYSTAL STRUCTURES. PART- 13-1895
SOLUTIONS OF THE PSEUDOSYSTEM POTASSIUM LITHIU/ (TETRAGONAL) TUNGSTEN BRONZE TYPE CRYSTALS GROWN IN MOLTEN 13-1948
/(OPTICAL) AND DIELECTRIC STUDIES OF (SINGLE CRYSTALS IN THE TUNGSTEN BRONZE TYPE) FERROELECTRIC (SYSTEM) / 13-1947
OOPTICAL PROPERTIES OF STRONTIUM POTASSIUM LITHIUM NIOBATE. (TUNGSTEN BRONZE TYPE FERROELECTRICS).   ELECTR 13-2051
SUPERLATTICE STRUCTURE OF STRONTIUM SODIUM LITHIUM NIOBATE (TUNGSTEN BRONZE) TYPE FERROELECTRICS.   13-2002
L USE (REVIEW, 22 REFS).   TUNGSTEN BRONZE TYPE FERROELECTRICS FOR OPTICA 02-0054
/CALCULATION OF THE FERROELECTRIC TRANSITION TEMPERATURES OF TUNGSTEN BRONZE TYPE NIOBATES (CORRELATION OF/ 13-2049
/ERROELECTRIC OXYFLUORIDE PHASES OF THE OXYGENATED QUADRATIC TUNGSTEN BRONZE TYPE. (STRONTIUM POTASSIUM NI/ 13-2011
PERTIES OF BARIUM SODIUM YTTRIUM NIOBATE SINGLE CRYSTAL WITH TUNGSTEN BRONZE TYPE STRUCTURE. /CTROOPTIC PRO 13-1971
            SODIUM BARIUM RARE EARTH NIOBATES WITH THE TUNGSTEN BRONZE TYPE STRUCTURE.   13-1972
RIUM OR STRONTIUM) TITANATE NIOBATE, (B/ COMPLEX OXIDES WITH TUNGSTEN BRONZE TYPE STRUCTURES (POTASSIUM (BA 13-1962
            SODIUM ORDERING IN SODIUM TUNGSTEN BRONZES.   13-2012
            STRUCTURE OF TETRAGONAL TIN TUNGSTEN BRONZES.   13-2031
TRANSIENT NMR STUDY OF PHASE TRANSITIONS IN METALLIC SODIUM TUNGSTEN BRONZES.   13-1910
/PERTIES (CONDUCTIVITY, MAGNETIC SUSCEPTIBILITY) OF HYDROGEN TUNGSTEN BRONZES, HYDROGEN(X) TUNGSTATE. (ANA/ 13-1934
FERROELECTRICS OF DISPLACEMENT TYPE. (BORACITES, MOLYBDATES, TUNGSTEN BRONZES, NIOBATES)   NEW CLASSES OF 02-0118
            IMPURITY EFFECTS IN SODIUM TUNGSTEN BRONZES: NIOBIUM AND TANTALUM DOPING. 13-2053
TES)   RAMAN STUDY OF METALLIC TUNGSTEN BRONZES. (SODIUM AND RUBIDIUM TUNGSTA 10-1654
            KYROPOULOS SINGLE CRYSTAL GROWTH OF SODIUM TUNGSTEN BRONZES. (SODIUM TUNGSTATE) 13-2054
            THERMAL CONDUCTIVITY OF CUBIC SODIUM TUNGSTEN BRONZES. (SODIUM TUNGSTATES) 13-2057
/RY AND ELECTRICAL PROPERTIES OF TWO NEW PHASES OF QUADRATIC TUNGSTEN BRONZES STRONTIUM POTASSIUM TANTALAT/ 13-2010
UM, OR VANADIUM) (0.5) OXIDE, (BARIUM OR LEAD) BISMUTH(0.67) (TUNGSTEN OR MOLYBDENUM) (0.33) OXIDE /, TANTAL 08-1462
/R TANTALUM OR VANADIUM) OXIDE(6), AND BARIUM(3) BISMUTH(2) (TUNGSTEN OR MOLYBDENUM) OXIDE(6), (BARIUM OR / 36-3619
TRIC (AND TEMPERATURE) CHARACTERISTICS OF SOLID SOLUTIONS OF TUNGSTEN OXIDE.   DIELEC 10-1657
DIELECTRIC PROPERTIES OF BARIUM TITANATE DOPED WITH TUNGSTEN OXIDE.   04-0696
            EVAPORATED FILMS OF TUNGSTEN OXIDE.   10-1647
HE S/ STUDY OF THE X-RAY PHOTOELECTRON SPECTRUM OF TUNGSTEN- TUNGSTEN OXIDE AS A FUNCTION OF THICKNESS OF T 10-1631
X) Z(Y) OXIDE/ ELECTROCHEMICAL DEPOSITION AND DISSOLUTION OF TUNGSTEN OXIDE BRONZES. (SODIUM(X) TUNGSTEN(1- 13-1944
SCOPY.   EXAMINATION OF SUBSTOICHIOMETRIC TUNGSTEN OXIDE(3-X) CRYSTALS BY ELECTRON MICRO 10-1627
BISMUTH TANTALATE NIOBATE, TA/ NEW PHASES OF THE TETRAGONAL TUNGSTEN OXYGEN BRONZE TYPE: QUADRATIC LEAD OR 13-1985
            A NEW FORM OF TUNGSTEN TRIOXIDE.   10-1652
            ELECTRICAL CONDUCTIVITY OF TUNGSTEN TRIOXIDE.   10-1637
EPR STUDY OF THE CATALYTIC ACTIVITY OF NONSTOICHIOMETRIC TUNGSTEN TRIOXIDE.   10-1628
FORMATION OF SHEAR STRUCTURES IN SUBSTOICHIOMETRIC TUNGSTEN TRIOXIDE.   10-1659
ECTRICITY AND ANTIFERROELECTRICITY OF THE AO3 TYPE CRYSTAL. (TUNGSTEN TRIOXIDE)   ON FERROEL 10-1641
DIAGRAM OF LEAD ZIRCONATE(1-X) TITANATE(X) PLUS 0.8 PERCENT TUNGSTEN TRIOXIDE.   PHASE 09-1581
ION OF TUNGSTIC OXIDE SINGLE CRYSTALS (SUBLIMATION IN AIR OF TUNGSTEN TRIOXIDE).   PREPARAT 10-1644
            RAMAN SPECTROSCOPY OF TUNGSTEN TRIOXIDE.   10-1635
TIES OF BARIUM (TITANATE, STANNATE) SOLID SOLUTIONS DOPED BY TUNGSTEN TRIOXIDE.   SEMICONDUCTOR PROPER 04-0965
NIUM(3) OXIDE(12) WITH VANADIUM PENTOXIDE AND MOLYBDENUM AND TUNGSTEN TRIOXIDE. /EACTION OF BISMUTH(4) TITA 13-2029
DERSTANDING OF THE SO-CALL/ SOME ASPECTS OF THE STRUCTURE OF TUNGSTEN TRIOXIDE AND A CONTRIBUTION TO THE UN 10-1653
DAMAGE RANGES. (STUDI/ BOMBARDMENT INDUCED AMORPHIZATION IN TUNGSTEN TRIOXIDE AND ITS USE IN DEDUCING MEAN 10-1642
/IDE STRUCTURE TYPE: AN ELECTRON MICROSCOPE STUDY OF REDUCED TUNGSTEN TRIOXIDE AND RELATED PSEUDOBINARY SY/ 10-1630
            (PREPARATION AND ELECTRICAL AND OPTICAL) STUDY OF TUNGSTEN TRIOXIDE AT LOW TEMPERATURE. 10-1643
S IN FERROELECTRIC CRYSTALS WITH A PHOTOEMISSION MICROSCOPE (TUNGSTEN TRIOXIDE, BARIUM TITANATE). /F DOMAIN 04-0811
ES.   AN EFFECT OF HEAT TREATMENT ON TUNGSTEN TRIOXIDE CRYSTAL. PART-2: GREEN STRIP 10-1639
T OF OXYGEN DEFICIENCY ON ELECTRICAL TRANSPORT PROPERTIES OF TUNGSTEN TRIOXIDE CRYSTALS.   EFFEC 10-1629
PHOTOEMISSION MICROSCOPIC OBSERVATION OF THE SURFACE OF TUNGSTEN TRIOXIDE CRYSTALS.   10-1645
STRUCTURAL AND DIELECTRIC MODIFICATIONS OF TUNGSTEN TRIOXIDE DUE TO NEUTRON IRRADIATION. 10-1651
            THE COLOR PROBLEM OF TUNGSTEN TRIOXIDE ELECTRICAL CONDUCTIVITY. 10-1636
EXPANSION AND THE ORTHORHOMBIC- TETRAGONAL TRANSITION OF THE TUNGSTEN TRIOXIDE PHASE.   THERMAL 10-1626
F FERROELECTRIC CRYSTALS WITH SCANNING ELECTRON MICROSCOPE. (TUNGSTEN TRIOXIDE SINGLE CRYSTAL PLATES) /NS O 10-1646
F COMPOSITION PLANES IN ANTIFERROELECTRICS. (APPLIED TO ADP, TUNGSTEN TRIOXIDE, SODIUM NIOBATE) /ENTATION O 08-1327
            THE BISMUTH OXIDE- TUNGSTEN TRIOXIDE SYSTEM.   10-1656
            PHASE TRANSITION IN TUNGSTEN TRIOXIDE UNDER HIGH PRESSURE. 10-1658
            THE COLOR PROBLEM OF TUNGSTEN TRIOXIDE. X-RAY STUDIES. 10-1632
/CERAMICS. ((LEAD, LANTHANUM) (ZIRCONATE, TANTALATE), (LEAD, TUNGSTEN) (ZIRCONATE TANTALATE), LEAD (LANTHA/ 13-2004
/ITION AND DISSOLUTION OF TUNGSTEN OXIDE BRONZES. (SODIUM(X) TUNGSTEN(1-X) Z(Y) OXIDE(3) WHERE Z IS NIOBIU/ 13-1944
KNESS OF THE S/ STUDY OF THE X-RAY PHOTOELECTRON SPECTRUM OF TUNGSTEN- TUNGSTEN OXIDE AS A FUNCTION OF THIC 10-1631
AIR OF TUNGSTEN TRIOXIDE).   PREPARATION OF TUNGSTIC OXIDE SINGLE CRYSTALS (SUBLIMATION IN 10-1644
TROL OF THE ELASTIC COMPLIANCE IN LITHIUM THALLIUM TARTRATE (TUNING AND DELAY DEVICES).   ELECTRIC FIELD CON 29-3250
SWITCHING / NONLINEAR DIELECTRICS FOR HIGH POWER ELECTRONIC TUNING. (WITH ELECTRODES REDUCING POLARIZATION 37-3660
) STRONTIUM TITANATE AND (CALCIUM DOPED) POTASSIUM/ ELECTRON TUNNELING AND BAND STRUCTURE OF (NIOBIUM DOPED 06-1222

YDROGEN PHOSPHATE.   INTERACTIONS BETWEEN PROTON TUNNELING AND OPTICAL PHONONS IN POTASSIUM DIH 17-2592
YDROGEN PHOSPHATE.   INTERACTION BETWEEN PROTON TUNNELING AND OPTICAL PHONONS IN POTASSIUM DIH 17-2593
PINGS AND (FREQUENCY) SHIFTS (OF PROTONIC COLLECTIVE MOTION, TUNNELING, CALCULATIONS, INTERPRETATION). /DAM 03-0475
                    EXPERIMENTAL PROOF FOR PROTON TUNNELING IN POTASSIUM DIHYDROGEN PHOSPHATE.   17-2510
PE FERROELECTRICS. (THEORY)   ENERGY PARAMETERS AND PROTON TUNNELING IN POTASSIUM DIHYDROGEN PHOSPHATE TY 03-0506
                            DETECTION OF ATOM TUNNELING IN SOLIDS. (KDP)   17-2612
RRIER JUNCTIONS.   ELECTRON TUNNELING INTO POTASSIUM TANTALATE SCHOTTKY BA 08-1365
        LASER RAMAN STUDY OF QUASI SPIN WAVE HYDROGEN TUNNELING MODES IN KDP AND DEUTERATED KDP.   17-2569
GONAL POTASSIUM DIHYDROGEN PHOSPHATE. (T/ ANALYSIS OF PHONON TUNNELING MODES IN THE RAMAN SPECTRUM OF TETRA 03-0396
  DIDEUTERIUM PHOSPHATE (FROM POLARIZED REFLECTION SPECTRA, A TUNNELING MODEL). /UTERON MOTIONS IN POTASSIUM 17-2645
DIHYDROGEN PHOSPHATE.   INFLUENCE OF TUNNELING ON DIELECTRIC BEHAVIOR OF POTASSIUM 17-2527
        CAN THE ISOTOPIC SHIFT IN KDP BE EXPLAINED BY TUNNELING ONLY.   17-2562
2: SCATTERING CHARACTERISTICS AND VARIOUS CONCEPTIONS OF THE TUNNELING QUASI SPIN MODEL. / (KDP TYPE) PART- 03-0478
BARRIERS INTERPRETED AS CONFIRMATION OF THE THEORY OF TRAN/ TUNNELING RESULTS ON INDIUM STRONTIUM TITANATE 03-0233
SULFATE. APPLICATION OF SELECTIVE ETCHING TO THE STUDY OF TWIN AND DISLOCATION STRUCTURES IN TRIGLYCINE 27-3042
TANATE, (BY ANNEALING).   PREPARATION OF POLED, TWIN-FREE CRYSTALS OF FERROELECTRIC BISMUTH TI 08-1352
ON TOPOGRAPHY.   DISPLAY OF TWINNING IN LITHIUM NIOBATE BY X-RAY DIFFRACTI 11-1845
UM FERROCYANIDE TRIHYDRATE BY THE NMR METHOD. POLYTYPE AND TWINNING OF THE FERROELECTRIC CRYSTALS POTASSI 24-2898
NS IN HIGH RESISTIVITY BARIUM TITANATE SINGLE CRYSTALS. (AND TWINNING TENDENCY) /ITIC TYPE) PHASE TRANSITIO 04-0887
/ITIONS WITH A GENUINE DIELECTRIC INSTABILITY (FERROELECTRIC TWINNING) TREATED AS CRITICAL POINTS IN CRYST/ 15-2176
IUM NIOBATE CRYSTALS (CZOCHRALSKI METHOD).   GROWTH TWINNING (UNDER DIRECT CURRENT) IN C-AXIS LITH 11-1672
IC MATERIALS. SOME PROBLEMS AND RESULTS.   NEW TWO- AND THREE PHASE FERROELECTRIC FERROMAGNET 03-0360
  CERAMIC (B/ ABSORPTION AND AMPLIFICATION OF ULTRASOUND IN A TWO-LAYER SYSTEM CONSISTING OF A PIEZOELECTRIC 04-1037
TITANATE- METAL.   TWO-PHASE FERROELECTRIC SYSTEM. PART-1: BARIUM 04-0815
        INFLUENCE OF THE DOMAIN STRUCTURE ON POSITRON (TWO-PHONON) ANNIHILATION IN ROCHELLE SALT.   28-3191
TION IN A SODIUM NITRITE CRYSTAL.   TWO-PHONON SCATTERING AND FERROELECTRIC TRANSI 16-2277
LANTHANUM DOPED BARIUM TITANATE- BISMUTH TRIOXIDE COMPOSITE/ TWO-STEP ANOMALOUS INCREASE OF RESISTIVITY ON 04-0800
/) C (VII) FERROELECTRIC SEMICONDUCTORS AT TEMPERATURES NEAR A TYPE-1 PHASE TRANSITION. (ANTIMONY SULFOIODID/ 15-2114

                                                    U

                                                    UHF SEE ULTRAHIGH FREQUENCY
                                                    UHV SEE ULTRAHIGH VACUUM
FECT AT VERY LARGE GATE CHARGES APPLIED TO TELLURIUM FILMS ON UHV CLEAVED TGS. /TER EFFECTS IN THE FIELD EF 27-3171
D EFFECT AND HALL MOBILITY OF TELLURIUM AND LEAD TELLURIDE ON UHV CLEAVED TRIGLYCINE SULFATE.   FIEL 27-3172
MPLIFIER (BASED ON FERROELECTRICS I/ POSSIBLE REALIZATION OF ULTRA LOW NOISE LEVEL PARAELECTRIC MICROWAVE A 37-3664
    EFFECT OF SURFACE MOISTURE ON THE DIELECTRIC BEHAVIOR OF ULTRAFINE BARIUM TITANATE PARTICULATES.   04-0859
/ACT, SURFACE WAVE) FERROELECTRIC PHASE SHIFTERS FOR VHF AND ULTRAHIGH FREQUENCY (BASED ON LEAD(0.35) STR/ 37-3633
POLARIZED FERROELECTRIC CERAMIC. (BARIUM TITANATE, BARIUM / ULTRAHIGH FREQUENCY DIELECTRIC ANISOTROPY OF A 04-0767
ARIUM TITANATE SINGLE C/ EFFECT OF LINEAR COMPRESSION ON THE ULTRAHIGH FREQUENCY DIELECTRIC PROPERTIES OF B 04-0896
OME POLYCRYSTALLINE FERROELECTRICS (BARIUM TITANATE AND SAM/ ULTRAHIGH FREQUENCY DIELECTRIC PROPERTIES OF S 04-0818
IUM TITANATE TYPE FERROELECTRIC CERAMICS.   ULTRAHIGH FREQUENCY REVERSAL PROPERTIES OF BAR 04-1014
OD OF STUDYING TEMPERATURE CHARACTERISTICS OF ROCHELLE TYPE/ ULTRAHIGH FREQUENCY (WAVEGUIDE RESONATOR) METH 04-0905
                    THIN FILMS ON TGS CLEAVED IN ULTRAHIGH VACUUM.   27-3173
INS IN TRIGLYCINE SULFATE.   ETCHING AND ULTRAMICROSCOPIC STUDIES OF FERROELECTRIC DOMA 27-3143
GEN PHOSPHATE AND SODIUM TRIHYDROGEN SELENATE TYPE CRYSTALS. ULTRASLOW HYDROGEN MOTION IN POTASSIUM DIHYDRO 17-2441
LUENCE OF THE ELECTRIC FIELD AND THE DOMAIN STRUCTURE ON THE ULTRASONIC ABSORPTION IN TRIGLYCINE SULFATE. / 27-2955
CRYSTALS.   ELASTIC CONSTANTS AND ULTRASONIC ABSORPTION OF SODIUM NITRITE SINGLE 16-2357
                                                    ULTRASONIC ANOMALIES IN ROCHELLE SALT.   28-3197
YPE FERROELECTRICS.   ULTRASONIC ATTENUATION ANOMALY IN DISPLACIVE T 03-0393
IN STRONTIUM TITANATE.   ANOMALOUS ULTRASONIC ATTENUATION AT THE 105K TRANSITION 06-1189
CINE SULFATE. THE INFLUENCE OF THE ELECTRIC FIELD ON THE ULTRASONIC ATTENUATION IN FERROELECTRIC TRIGLY 27-2956
            PHENOMENOLOGICAL THEORY OF ULTRASONIC ATTENUATION IN FERROELECTRICS.   27-3055
ISMUTH(12) GERMANIUM OXIDE(20) AT LOW TEMPERATURES. ULTRASONIC ATTENUATION IN SINGLE CRYSTALS OF B 35-3510
ICS.(TGS, ABSTRACT ONLY)   ANISOTROPY OF ULTRASONIC ATTENUATION IN UNIAXIAL FERROELECTR 27-3055
URE IN DICALCIUM STRONTIUM PROPIONATE.   ULTRASONIC ATTENUATION NEAR THE CURIE TEMPERAT 26-2948
ON IN DICALCIUM LEAD PROPIONATE   ULTRASONIC ATTENUATION NEAR THE PHASE TRANSITI 26-2949
SITION POINT IN POTASSIUM MANGANESE TRIFLUORIDE (ANOMALOUS / ULTRASONIC ATTENUATION NEAR THE SOFT MODE TRAN 33-3373
  POINTS IN SODIUM NITRITE.   ULTRASONIC ATTENUATION NEAR THE TWO TRANSITION 16-2304
                                                    ULTRASONIC DISPERSION IN STRONTIUM TITANATE. 06-1239
        BARIUM TITANATE PIEZOELECTRIC MICROPROBE FOR STUDYING ULTRASONIC FIELDS.   04-0938
TITANATE AS TRANSDUCER.   ULTRASONIC INTERFEROMETER WITH LEAD ZIRCONATE 09-1585
AND CRITICAL POINTS. SECTION-5: FERROELECTRIC AND ANTIFERRO/ ULTRASONIC INVESTIGATION OF PHASE TRANSITIONS 02-0070
OF THE FERROELECTRIC TRANSITION REGION IN POTASSIUM DIHYDRO/ ULTRASONIC INVESTIGATION (15 MHZ SHEAR WAVES) 17-2576
FFECT IN KDP (NEAR THE CURIE TEMPERATURE)   ULTRASONIC MEASUREMENT OF THE ELECTROCALORIC E 17-2664
NSITION OF STRONTIUM TITANATE. THERMAL EXPANSIVITY AND ULTRASONIC PROPAGATION NEAR THE STRUCTURAL TRA 06-1135
TERACTION PARAMETERS AT THE DISPLACIVE TRANSITION IN STRONT/ ULTRASONIC PROPAGATION, STRESS EFFECTS, AND IN 06-1127
SINGLE CRYSTAL NEAR ITS CURIE POINT, PART-2, DIELECTRIC AND ULTRASONIC PROPERTIES. /R OF LITHIUM TANTALATE 12-1885
THE 105 DEGREE K TRANSITION.   ULTRASONIC PROPERTIES OF STRONTIUM TITANATE AT 06-1191
RE OF FERROELECTRIC TRIGLYCINE SULFATE.   ULTRASONIC RELAXATION NEAR THE CURIE TEMPERATU 27-3097
WALLS IN POTASSIUM DIDEUTERIUM PHOSPHATE.   ULTRASONIC STRESS WAVE INTERACTION WITH DOMAIN 17-2659
SULATING AND SEMICONDUCTING STRONTIUM TITANATE. (ABSTRACT) ULTRASONIC STUDIES OF CRITICAL PHENOMENA IN IN 06-1118
SITION IN BARIUM TITANATE.   ULTRASONIC STUDIES OF FIELD INDUCED PHASE TRAN 04-0549
ERS ON LITH/ OPTICAL STUDIES OF REFRACTION AND REFLECTION OF ULTRASONIC SURFACE WAVES BY THIN GOLD FILM LAY 11-1789
ERS ON LITHIUM NIOBATE SUBSTRA/ REFRACTION AND REFLECTION OF ULTRASONIC SURFACE WAVES BY THIN GOLD FILM LAY 11-1790
IUM TITANATE SINGLE CRYSTALS (NONLINEAR ELASTIC PROPERTIES). ULTRASONIC THIRD HARMONIC GENERATION IN STRONT 06-1181
M TITANATE.   ULTRASONIC VELOCITY ABOVE TANTALUM IN STRONTIU 06-1137
O/ RADIALLY VIBRATING FERROCERAMIC TRANSDUCER AS A SOURCE OF ULTRASONIC VIBRATIONS RADIATING IN THE DIRECTI 37-3657
TING OF LITHIUM NIOBATE AND P-TYP/ AMPLIFICATION OF RAYLEIGH ULTRASONIC WAVES IN A LAYERED STRUCTURE CONSIS 11-1747
(PHOTOELASTIC CONSTANT)   LIGHT SCATTERING BY ULTRASONIC WAVES IN LITHIUM NIOBATE CRYSTALS.   11-1754
PHATE (TEMPERATURE DE/ ANOMALOUS PROPAGATION OF LONGITUDINAL ULTRASONIC WAVES IN POTASSIUM DIDEUTERIUM PHOS 17-2660
HATE CRYSTALS.   DIFFRACTION OF POLARIZED LIGHT BY ULTRASONIC WAVES IN POTASSIUM DIHYDROGEN PHOSP 17-2407
ORIDE IN THE NEIGHBORHOOD OF / ATTENUATION AND DISPERSION OF ULTRASONIC WAVES IN POTASSIUM MANGANESE TRIFLU 33-3370
R THE CURIE POINTS.   ABSORPTION OF LONGITUDINAL ULTRASONIC WAVES IN ROCHELLE SALT CRYSTALS NEA 28-3211
ERROELECTRICS. (ANTIMONY SULFOIODIDE)   INTERACTION OF ULTRASONIC WAVES WITH MOVING DOMAIN WALLS IN F 15-2192
  BARIUM MANGANESE FLUORIDE, A NEW CRYSTAL FOR MICROWAVE ULTRASONICS.   33-3406
GLE CRYSTALS.   FREQUENCY DEPENDENCE OF ANOMALOUS ULTRASOUND DAMPING IN ANTIMONY SULFOIODIDE SIN 15-2139
  A PIEZOELECTRIC CERAMIC (B/ ABSORPTION AND AMPLIFICATION OF ULTRASOUND IN A TWO-LAYER SYSTEM CONSISTING OF 04-1037
MICONDUCTORS AT TEMPERATURES NEAR A TYPE-1 PH/ ABSORPTION OF ULTRASOUND IN A(V)B(VI)C(VII) FERROELECTRIC SE 15-2114
CTRIC SEMICONDUCTOR NEAR ITS CURIE POINT.   ABSORPTION OF ULTRASOUND IN AN ANTIMONY SULFOIODIDE FERROELE 15-2115

POLARIZATION EFFECTS IN THE ANOMALOUS ABSORPTION OF ULTRASOUND IN ANTIMONY SULFOIODIDE CRYSTALS.   15-2123
/TURE, ELECTRIC FIELE, AND ILLUMINATICN ON THE ABSORPTION OF ULTRASOUND IN ANTIMONY SULFOIODIDE NEAR THE P/   15-2248
/G RANGE DIPOLE DIPOLE FORCES ON ATTENUATION AND VELOCITY OF ULTRASOUND IN FERROELECTRICS NEAR SECOND ORDE/   03-0253
/C FEATURES OF THE TEMPERATURE DEPENDENCE OF THE VELOCITY OF ULTRASOUND IN POLYCRYSTALLINE SOLID SOLUTIONS/   04-0894
OIODIDE.   ANISCTROPY CF ANOMALOUS ABSORPTION OF ULTRASOUND IN SINGLE CRYSTALS OF ANTIMONY SULF   15-2222
TE IN THE PRESENCE OF A STEADY ELE/ RELAXATION ABSORPTION OF ULTRASOUND IN SINGLE CRYSTALS OF BARIUM TITANA   04-0850
          EFFECTS OF LONGITUCINAL MAGNETIC FIELD ON ULTRASOUND IN STRONTIUM TITANATE.   06-1243
LS (A/ FREQUENCY DEPENDENCE OF THE FLUCTUATION ABSORPTION OF ULTRASOUND IN.TRIGLYCINE SULFATE SINGLE CRYSTA   27-3079
ION PRCCESSES IN TRIGLYCINE SULFATE CRYSTALS AND / EFFECT OF ULTRASOUND ON THE POLARIZATION AND REPOLARIZAT   04-0924
/IES. PART-7: DIELECTRIC AND SWITCHING MEASUREMENTS. PART-8: ULTRASOUND PROPAGATION. PART-9: SPECIFIC HEAT/   03-0397
ANC NITRATES.   VACUUM ULTRAVIOLET ABSORPTION OF THE ALKALI NITRITES   16-2398
/TRANSFCRMATIONS IN NITRATES ANC NITRITES. PART-1: CHANGE IN ULTRAVIOLET ABSORPTION SPECTRA ON MELTING. /A   16-2278
QUID NITROGEN TEMPERATURES IN ALKALI NITRITES (SODIU/ VACUUM ULTRAVIOLET ABSORPTION (5-23 EV AT ROOM AND LI   16-2399
ECTRA AND BAND STRUCTURE) OF SOME AEO(3) FERROE/ VISIBLE AND ULTRAVIOLET OPTICAL PROPERTIES (REFLECTANCE SP   08-1383
SIUM DIHYDROGEN PHOSPHATE AND AMMONIUM DIHYDROGEN PHOSPHATE/ ULTRAVIOLET REFLECTION AND ABSORPTION OF POTAS   17-2422
          GROWTH OF UNCRACKED BARIUM SODIUM NIOBATE CRYSTALS.   13-1901
POTASSIUM TITANATE)   IS THERE A UNIFIED MECHANISM UNDERLYING URBACH'S RULE. (STRONTIUM TITANATE,   06-1110
TS NATURE ANC THE PREDICTION OF ITS OCCURANCE)   DO WE REALLY UNDERSTAND (THE CAUSE OF) FERROELECTRICITY. (I   03-0510
ICE RESEARCH.   MOLECULAR UNDERSTANDING OF ELECTROCHEMICAL FROCESSES BY   34-3484
/HE STRUCTURE OF TUNGSTEN TRIOXIDE AND A CONTRIBUTION TO THE UNDERSTANDING OF THE SO-CALLED "SHEAR STRUCTU/   10-1653
TIES OF SCANDIUM TRIOXIDE DOPED, RARE EARTH OXIDE DOPED, AND UNDOPED BARIUM TITANATE. /ND ELECTRICAL PROPER   04-0839
/ITIVE TEMPERATURE COEFFICIENT OF RESISTIVITY) IN (DOPED AND UNDOPED CERAMIC PEROVSKITE) LEAD FERRATE(0.5)/   13-2000
/ THE CONTRAST AT FERROELECTRIC DCMAIN BOUNDARIES (DOPED AND UNDOPED TGS, LITHIUM NIOBATE, KDP, ABSTRACT O/   11-1671
RROELECTRICS AND FERROELASTICS HAVING THE INDEX OF FAINTNESS UNEQUAL TO THE CELL MULTIPLICITY. /MPLES OF FE   03-0145
SPHATE FOR LIGHT MODULATORS.   CUTTING AND POLISHING IN UNIAXIAL CRYSTALS OF THE TYPE X DIHYDROGEN PHO   17-2559
          PHASE DIAGRAM OF THE SEMICONDUCTING UNIAXIAL FERROELECTRIC IRON(1-X) SULFIDE.   36-3617
-1. PART-2.   MODEL FOR THE SEMICONDUCTING UNIAXIAL FERROELECTRIC IRON(1-X) SULFIDE. PART   36-3615
INFLUENCE OF SCREENING ON THE FERROELECTRIC CURIE POINT (UNIAXIAL FERROELECTRIC PHOTOCONDUCTOR).   03-0392
          ANISOTROPY OF ULTRASONIC ATTENUATION IN UNIAXIAL FERROELECTRICS.(TGS, ABSTRACT ONLY)   27-3055
          ATR INFRARED SPECTRA OF UNIAXIAL NITRATE CRYSTALS.   16-2280
UM NICBATE HAVING NO MICROTWINNING (AND RESISTANT TO L/ NEW (UNIAXIAL) NONLINEAR OPTIC CRYSTAL BARIUM LITHI   13-1958
ION OF SINGLE DOMAIN STRONTIUM TITANATE (OPTICAL STUDY UNDER UNIAXIAL STRESS).   DIRECT OBSERVAT   06-1112
ETIC SPECTRA OF CHROMIUM(3+) IN POTASSIUM DIHYDR/ EFFECTS OF UNIAXIAL STRESS AND ELECTRIC FIELD ON PARAMAGN   17-2554
UM TITANATE.   EFFECT OF UNIAXIAL STRESS ON RAMAN SCATTERING IN STRONTI   06-1102
LIGHT EMISSION FROM BARIUM TITANATE CRYSTALS SUBJECTED TO UNIAXIAL STRESS PULSES.   04-0672
          DOMAIN STRUCTURE OF STRONTIUM TITANATE UNDER UNIAXIAL STRESSES.   06-1113
OPERTIES AND DCMAIN STRUCTURE OF BARIUM TITANA/ INFLUENCE OF UNIAXIAL TENSILE STRESSES ON THE ELECTRICAL PR   04-0961
ROPERTIES OF POLARIZED BARIUM TITANATE CERAMICS SUBJECTED TO UNIAXIAL TRANSVERSE COMPRESSION. /ZOELECTRIC P   04-0962
TRONTIUM TITANATE, POTASSIUM TITANATE)   IS THERE A UNIFIED MECHANISM UNDERLYING URBACH'S RULE. (S   06-1110
/YNAMIC SUSCEPTIBILITY OF CLASSICAL ANHARMONIC OSCILLATOR. A UNIFIED CSCILLATOR MODEL FOR ORDER DISORDER A/   03-0405
RATURES IN LITHIUM NIOBATE (CRYSTAL GROWTH BY ADD/ SPATIALLY UNIFORM AND ALTERABLE SHG PHASE MATCHING TEMPE   11-1684
EFFECT OF HYDROSTATIC PRESSURE ON THE UNIPOLAR PROPERTIES OF TRIGLYCINE SULFATE.   27-3053
H TIME.   VARIATION OF THE UNIPOLARITY OF TRIGLYCINE SULFATE CRYSTALS WIT   27-3115
A PHASE TRANSITION.   APPARENT VIOLATION OF THE UNIVERSALITY HYPOTHESIS IN A LATTICE MODEL OF   03-0515
SURFACE WAVE CELAY LINES WITH INTERDIGITAL TRANSDUCERS ON UNPOLARIZED PZT CERAMIC PLATES.   09-1614
AND MAGNETIC PROPERTIES OF (MONOCLINIC) EUROPIUM SILICATE. (UNSTABLE FERROELECTRIC PHASE) /RAPHIC, OPTICAL   36-3620
/ON OF THE MEAN MAGNETIC MOMENT OF CHROMIUM- POTASSIUM ALUM (UNSTABLE SOFT PHONON MODE, DISPLACIVE FERROEL/   20-2806
CYANIDE AND ISOMORPHOUS CRYSTALS. (NEW MODEL BASED ON NMR O/ UNTWINNED SAMPLES) /NSITION IN POTASSIUM FERRO   24-2893
ENCE OF THE ABSORPTION OF SCUNC IN ROCHELLE SALT NEAR ITS UPPER CURIE POINT.   FREQUENCY DEP   28-3183
SS FACTOR IN FERROELECTRIC MATERIALS (ROCHELLE SALT NEAR THE UPPER CURIE POINT AS EXAMPLE). /ENCE OF THE LO   28-3220
TE.   DOUBLE HYSTERESIS LOOPS AT THE UPPER CURIE POINT OF SODIUM TRIHYDROGEN SELENI   22-2831
TITANATE)   IS THERE A UNIFIED MECHANISM UNDERLYING URBACH'S RULE. (STRONTIUM TITANATE, POTASSIUM   06-1110
OPTICAL GAP OF STRONTIUM TITANATE. (DEVIATION FROM URBACH TAIL BEHAVIOR)   06-1111
          DIELECTRIC DISPERSION OF UREA AND THIOUREA.   25-2912
IAL SWITCHING PROPERTIES OF FERROELECTRICS IN MEMORY DEVICE/ UTILIZATION OF THE (CRITICAL PULSE WIDTH) PART   27-3144

V

NGITUDINAL WAVE) ATTENUATION IN LITHIUM TANTALATE (EFFECT OF VACANCIES).   ANOMALOUS (LO   12-1892
N OF LANTHANUM MODIFIED LEAD TITANATE WITH A-SITE AND B-SITE VACANCIES.   X-RAY STRUCTURE INVESTIGATIO   05-1059
NATE (ESR MEASUREMENTS, M/ CHARGE COMPENSATION BY OXYGEN(2-) VACANCIES IN CHROMIUM(3+) DOPED STRONTIUM TITA   06-1161
          DISTRIBUTION OF VACANCIES IN LANTHANA DOPED LEAD TITANATE.   05-1058
ONATE) CERAMICS.   DISTRIBUTION OF A-SITE AND B-SITE VACANCIES IN (LEAD, LANTHANUM) (TITANATE, ZIRC   09-1530
NTIUM TITANATE.   ANALYSIS OF THE IRON OXYGEN VACANCY CENTER IN THE TETRAGONAL PHASE OF STRO   06-1231
NTIUM TITANATE.   THE IRON(3+)- VACANCY CENTER IN THE TETRAGONAL PHASE OF STRO   06-1234
ITION METALS.   ELECTROCOLORATION IN SIRONTIUM TITANATE. VACANCY DRIFT AND OXIDATION REDUCTION OF TRANS   06-1098
TRANSFORMATIONS. (AND SUGGESTIONS FOR DISCOVERING MEMBERS OF VACANT CLASSES) / OF THEIR PARAELECTRIC PHASE   03-0138
FORMATION PROCESS OF BARIUM TITANATE THIN FILM EVAPORATED IN VACUO.   04-0862
          THIN FILMS CN TGS CLEAVED IN ULTRAHIGH VACUUM.   27-3173
DIELECTRIC BEHAVIOR OF (FERROELECTRIC) FILMS OF VACUUM DEPOSITED POTASSIUM NITRATE.   16-2356
RY ELEMENT UTILIZING POTASSIUM NITRATE. (FUSED THICK FILM OR VACUUM DEPOSITED THIN FILM) /TIVE READOUT MEMO   16-2269
REPARATICN OF FERROELECTRIC THIN FILMS OF BARIUM TITANATE BY VACUUM EVAPORATION.   THIN FILMS: P   04-0834
          VACUUM EVAPORATION OF BARIUM TITANATE.   04-1001
          SURFACE BREAKDOWN IN VACUUM ON BARIUM TITANATE.   04-0596
TRITES AND NITRATES.   VACUUM ULTRAVIOLET ABSORPTION OF THE ALKALI NI   16-2398
AND LIQUID NITROGEN TEMPERATURES IN ALKALI NITRITES (SODIU/ VACUUM ULTRAVIOLET ABSORPTION (5-23 EV AT ROOM   16-2399
M OF ICE I(H). TENTATIVE INTERPRETATION WITH A MIXED COULOMB VALENCE DYNAMICAL MODEL. /QUENCY RAMAN SPECTRU   34-3425
MAGNETIC SUSCEPTIBILITY OF CHROM/ DETERMINATION OF CHROMIUM VALENCY FROM STUDIES OF EPR SPECTRA AND STATIC   04-0542
Y SULFOIODILE.   EFFECT CF PHOTOELECTRET CHARGE ON VALUE OF FERROELECTRIC POLARIZATION IN ANTIMON   15-2186
TE, A/ OPTICAL NONLINEAR SUSCEPTIBILITIES: ACCURATE RELATIVE VALUES FOR QUARTZ, AMMONIUM DIHYDROGEN PHOSPHA   17-2539
/MONIC GENERATION AND THE VARIATICNS IN THE FREE AND CLAMPED VALUES OF THE DIELECTRIC CONSTANTS AND ELECTR/   13-1998
          FERROELECTRIC LIGHT VALVE ARRAYS FOR OPTICAL MEMORIES.   37-3676
NCE OF CCNTRAST AND BRIGHTNESS IN GADOLINIUM MOLYBDATE LIGHT VALVES.   ORIENTATION AND THICKNESS DEPENDE   30-3291
USE OF CUEIC PHOTCSENSITIVE CRYSTALS AS FCCKELS EFFECT LIGHT VALVES IN DISPLAY AND MODULATOR APPLICATIONS.   37-3661
TES OF AMMONIUM IONS IN SOLIDS. (AMMONIUM SULFATE, SELENATE, VANADATE AND 15 OTHERS) /THE SPIN SYMMETRY STA   19-2800
RIDE, BORACITES, BISMUTH(4) TITANIUM(3) OXIDE(12), HAFNIUM DIVANADIDE) /M ANTIMONIDE, RUBIDIUM IRON TRIFLUO   02-0036
/TRUCTURES. (BARIUM AND BISMUTH(0.5)) (NIOBIUM, TANTALUM, OR VANADIUM)(0.5) OXIDE, (BARIUM OR LEAD) BISMUT/   08-1462
MPERATURE. (NO COCHRAN TYPE / INFRARED OPTICAL PROPERTIES OF VANADIUM DIOXIDE ABOVE AND BELOW TRANSITION TE   36-3526
RE 340/ TWO COMPONENTS OF THE CRYSTALLOGRAPHIC TRANSITION IN VANADIUM DIOXIDE. (ANTIFERROELECTRIC TEMPERATU   36-3562
EPR.   CHANGES CF THE DOMAIN STRUCTURE IN VANADIUM DOPED TRIGLYCINE SULFATE OBSERVED BY   27-3013

(1-X) Z(Y) OXIDE(3) WHERE Z IS NIOBIUM, TANTALUM, ZIRCONIUM, VANADIUM, OR COBALT) /ZES. (SODIUM(X) TUNGSTEN 13-1944
/ FERROELECTRICS. (BARIUM(2) BISMUTH (NIOBIUM OR TANTALUM OR VANADIUM OXIDE(6), AND BARIUM(3) BISMUTH(2) / 36-3619
SE TRANSITIONS OF STRESSED STRONTIUM TITANATE. (EPR OF IRON- VANADIUM PAIRS)   ORDER PARAMETER AND PHA 06-1170
/DY OF THE REACTION OF BISMUTH(4) TITANIUM(3) OXIDE(12) WITH VANADIUM PENTOXIDE AND MOLYBDENUM AND TUNGSTE/ 13-2029
ES.   MAGNETIC SUSCEPTIBILITY OF GUANIDINIUM VANADIUM SULFATE HEXAHYDRATE AT LOW TEMPERATUR 21-2815
ALS.   EPR STUDIES OF CHROMIUM- AND VANADYL- DOPED TRIGLYCINE SULFATE SINGLE CRYST 27-3161
YDROGEN PHOSPHATE AT HIGH PRESSURES.   VANISHING OF FERROELECTRICITY IN POTASSIUM DIH 17-2442
ECTRIC STATES IN POTASSIUM DIHYDROGEN PHOSPHATE TYPE CRYSTA/ VANISHING OF THE FERROELECTRIC AND ANTIFERROEL 17-2642
              INFLUENCE OF WATER VAPOR ON BARIUM TITANATE CERAMICS.   04-0882
UISELENIDE OR SULFIDE, ANT/ GROWING SINGLE CRYSTALS FROM THE VAPOR PHASE BY A DYNAMIC METHOD (ANTIMONY SESQ 15-2162
SOMERS.   VAPOR PRESSURE ISOTOPE EFFECT OF ICE AND ITS I 34-3491
MATERIALS. PART-2: HOT PRESSED COMPOSITIONS AT NEAR THEORET/ VAPORIZATION OF LEAD ZIRCONATE- LEAD TITANATE 09-1578
ELECTRIC CERAMICS. (PZT)   OPTICAL PROPERTIES (ELECTRICALLY VARIABLE BIREFRINGENCE) OF (HOT PRESSED) FERRO 09-1554
/IUM DICADMIUM TRISULFATE, A MODEL IN WHICH THE POLARIZATION VECTOR, P(S), ARISES PERPENDICULAR TO THE WAV/ 03-0141
ITE.   ANGULAR DISPERSION OF LARGE WAVE VECTOR POLARITONS IN FERROELECTRIC SODIUM NITR 16-2256
ICATION TO LITHIUM NIOBATE. PART-2: POLARITONS OF SMALL WAVE VECTORS. /EIGENVECTORS OF POLARITONS WITH APPL 11-1766
) GERMANI/ MICROWAVE ACOUSTICS HANDBOOK. VOL 1: SURFACE WAVE VELOCITIES. (BARIUM SODIUM NIOBATE, BISMUTH (12 01-0028
CE OF CRITICAL ELECTRIC FIELDS.   DISPLACEMENTS AND VELOCITIES OF THE DOMAIN WALLS IN KDP. EXISTEN 17-2448
              ULTRASONIC VELOCITY ABOVE TANTALUM IN STRONTIUM TITANATE. 06-1137
TANTALAT/ TEMPERATURE COEFFICIENTS OF ACOUSTIC SURFACE WAVE VELOCITY AND DELAY ON LITHIUM NIOBATE, LITHIUM 11-1821
RANSITION IN (NIOBIUM DOPED) ZIRCONATE(0/ VARIATION OF SOUND VELOCITY AT FERROELECTRIC-1, FERROELECTRIC-2 T 09-1511
      TEMPERATURE DEPENDENCE OF SOUND VELOCITY IN DICALCIUM STRONTIUM PROPIONATE.   26-2934
      TEMPERATURE DEPENDENCE OF SURFACE ACOUSTIC WAVE VELOCITY IN LITHIUM TANTALATE.   12-1886
TITANATE- STRONTIUM TITANATE SUBSTRATES AT SUPERSONIC DRIFT VELOCITY OF CURRENT CARRIERS. /AYERS ON BARIUM 04-0592
ACTIONS BETWEEN DOMAINS. (IN KDP)   VELOCITY OF PROPAGATION OF DOMAIN WALLS: INTER 17-2453
/UENCE OF LONG RANGE DIPOLE DIPOLE FORCES ON ATTENUATION AND VELOCITY OF ULTRASOUND IN FERROELECTRIC NEAR/ 03-0253
D SO/ SPECIFIC FEATURES OF THE TEMPERATURE DEPENDENCE CF THE VELOCITY OF ULTRASOUND IN POLYCRYSTALLINE SOLI 04-0894
ROMECHANICAL PROPERTIES CF TRIGLYCINE SULFATE. (MEASUREMENTS VERIFY THEORY APPROXIMATELY)   ELECT 27-3145
TICE).   PHASE TRANSITION IN A SIXTEEN VERTEX LATTICE MODEL (FERROELECTRIC SQUARE LAT 03-0517
FUNCTIONS IN ICE.   THE THREE DIMENSIONAL EIGHT VERTEX MODEL AND THE PROTON PROTON CORRELATIO 34-3481
65/ (COMPACT, SURFACE WAVE) FERROELECTRIC PHASE SHIFTERS FOR VHF AND UHF. (BASED ON LEAD(0.35) STRONTIUM (0. 37-3633
      LEAD TITANATE CERAMIC RESONATORS OPERATING IN VHF BAND.   05-1069
OF ULTRASONIC VIBRATIONS RADIATING IN THE DIRECTIO/ RADIALLY VIBRATING FERROCERAMIC TRANSDUCER AS A SOURCE 37-3657
E DEPENDENCE OF AN X CUT LITHIUM TANTALATE CRYSTAL RESONATOR VIBRATING IN THICKNESS SHEAR MODE OF MOTION. / 12-1868
N (STABILITY) IN BARIUM(X) STRONTIUM(1-X) TITANATE CERAMIC/ "VIBRATING WIRE" STUDY OF POTENTIAL DISTRIBUTIO 04-0699
      SOFT LATTICE VIBRATIONS AND VIBRATION COUPLING IN ANTIMONY SULFOIODIDE. 15-2233
KDP FERROELECTRIC (CALCULATIONS).   INFRARED VIBRATION FREQUENCIES OF THE HYDROGEN BOND IN 17-2650
M DIHYDROGEN ORTHOPHOSPHATE (4000-10000 PER CM) AND HARMONIC VIBRATION FREQUENCIES OF THE HYDROGEN BONDS. / 17-2432
      VIBRATION FREQUENCIES OF TITANATES.   08-1394
/EMPERATURE AND PHASE TRANSITIONS ON THE HYDROXYL STRETCHING VIBRATION IN STRONTIUM TITANATE AND TITANIUM / 06-1209
FAR INFRARED FERROELECTRIC VIBRATION MODE IN STRONTIUM TITANATE.   06-1094
OW TEMPERA1/ INFLUENCE OF THE PHASE TRANSITION ON THE NORMAL VIBRATION MODES OF FERROELECTRIC CRYSTALS AT L 17-2456
ON THE LATTICE VIBRATION OF CUBIC BARIUM TITANATE.   04-0775
RYSTALLINE LEAD TITANATE ZIRCONATE WITH THOROUGHNESS OF POL/ VIBRATION OF ELECTROELASTIC CONSTANTS OF POLYC 09-1489
STALS.   SOME PROPERTIES OF THE EQUATIONS OF PIEZOELECTRIC VIBRATION OF LITHIUM NIOBATE AND TANTALATE CRY 11-1740
RELATE TO SPECIFIC ANTIFERROELECTRIC PHASE)   LATTICE VIBRATION OF SODIUM NITRITE. (CALCULATIONS TO 16-2317
NATION SCATTER SPECTRUM IN THE REGION OF THE ANTISYMMETRICAL VIBRATION OF THE NITRITE ANION. /ALS AND COMBI 16-2276
      VIBRATION SPECTRA OF HYDROGEN BONDS. (KDP) 17-2482
PHATE IN PARA- AND FERROELECTRIC PHASES.   VIBRATION SPECTRA OF POTASSIUM DIHYDROGEN PHOS 17-2480
EFFECT CF TEMPERATURE AND PHASE TRANSITION ON THE HYDROXIDE VIBRATION SPECTRUM OF THE FERROELECTRIC RUBID/ 17-2470
R FORMATE TETRAHYDRATE AND COPPER DEUTERO FORMATE TETRADEUT/ VIBRATION SPECTRUM OF THE FORMATE ION IN COPPE 26-2926
RUBIDIUM DIHYDROGEN PHOSPHATE CRYSTAL IN PARAELECTRIC AND / VIBRATION SPECTRUM OF THE PHOSPHATE ANION IN A 17-2472
ANTIFERROELECTRIC MODE IN LEAD ZIRCONATE. (LATTICE VIBRATIONS)   07-1262
E CRYSTALS USING SCATTERING OF LASER RADIATION BY HYPERSONIC VIBRATIONS. /RAMETERS OF LITHIUM NIOBATE SINGL 11-1797
SPECTRA: APPLICATION TO DISPLACIVE FERROELECTRICS. (LATTICE VIBRATIONS)   POWDER RAMAN 03-0194
TYPE FERROELECTRICS (THEORY).   ATTENUATION OF CRITICAL VIBRATIONS AND DIELECTRIC LOSSES IN DISPLACIVE 03-0159
IN THE FER/ POLARIZATION OF THE RAMAN BANDS OF THE HYDROXYL VIBRATIONS AND STRUCTURE OF THE HYDROGEN BONDS 17-2471
SULFOIODIDE.   SOFT LATTICE VIBRATIONS AND VIBRATION COUPLING IN ANTIMONY 15-2233
ODIDE CRYSTALS USING THE RAMAN SCATTERING SPECTRUM. (LATTICE VIBRATIONS, BONDING) /ELECTRIC ANTIMONY SULFOI 15-2143
TYPE CRYSTAL.   MECHANICAL MODEL OF LONG WAVELENGTH VIBRATIONS IN A (SINGLE DOMAIN) FERROELECTRIC 03-0342
AR THE CURIE POINT (EFFECTS CF TEMPERATURE, F/ PIEZOELECTRIC VIBRATIONS IN ANTIMONY SULFOIODIDE CRYSTALS NE 15-2121
HOUT A SYMMETRY CENTER.   QUANTUM THEORY OF VIBRATIONS IN FERROELECTRIC FERROMAGNETICS WIT 03-0166
/CTIVITY) STUDY (20-10,000 PER CM, 300 DEG K) OF THE LATTICE VIBRATIONS IN LITHIUM TANTALATE (DIELECTRIC C/ 12-1870
MS. PART-1: A TWO DIMENSIONAL MODEL. (ICE I)   ATOMIC VIBRATIONS IN ORIENTATIONALLY DISORDERED SYSTE 34-3474
SYMMETRY OF ATOMIC VIBRATIONS IN POTASSIUM DIHYDROGEN PHOSPHATE. 17-2586
LOW FREQUENCY HYDROGEN VIBRATIONS IN POTASSIUM DIHYDROGEN PHOSPHATE. 17-2699
LOW FREQUENCY HYDROGEN VIBRATIONS IN POTASSIUM DIHYDROGEN PHOSPHATE. 17-2608
TRIC. (PZT)   METHOD OF MEASURING POLARIZATION VIBRATIONS OF A PREPOLARIZED CERAMIC FERROELEC 09-1615
LLY POLARIZED FERROELECTRIC TUBES.   VIBRATIONS OF COAXIALLY SEGMENTED, LONGITUDINA 03-0376
GROUP ANALYSIS OF LATTICE VIBRATIONS OF CUBIC PEROVSKITE ABO(3).   08-1321
/ SIMPLIFIED CHARACTERISTIC EQUATION FOR PLANE PIEZOELECTRIC VIBRATIONS OF LITHIUM NIOBATE AND LITHIUM TANT 11-1739
ECTRIC CYLINDRICAL TUBES.   VIBRATIONS OF LONGITUDINALLY POLARIZED FERROEL 03-0375
HIOUREA. PART-1: TEMPERATURE DEPENDEN/ MOLECULAR AND LATTICE VIBRATIONS OF ORTHORHOMBIC AND FERROELECTRIC 25-2909
HIOUREA. PART-2: CALCULATION OF THE I/ MOLECULAR AND LATTICE VIBRATIONS OF ORTHORHOMBIC AND FERROELECTRIC T 25-2907
/ USING POLARIZED IR RADIATION. PART-2: INTERNAL FUNDAMENTAL VIBRATIONS OF POTASSIUM DIHYDROGEN ARSENATE I/ 18-2738
NORMAL VIBRATIONS OF STRONTIUM TITANATE LATTICE.   06-1128
CTRIC PHASE OF KDP CRYSTALS (INFRARED A/ SPECTRA OF INTERNAL VIBRATIONS OF THE PHOSPHATE ION IN THE PARAELE 17-2479
N BONDED FERROELECTRICS OF THE KDP TYPE (PHONON PRO/ COUPLED VIBRATIONS OF THE PROTON ION SYSTEM IN HYDROGE 17-2670
MINUM SULFATE HEXAHYDRATE AND FREQUENCIES OF THE FUNDAMENTAL VIBRATIONS OF THE WATER MOLECULES. /DINIUM ALU 21-2811
IUM TITANATE CERAMICS.   INFLUENCE OF VIBRATIONS ON THE DIELECTRIC PROPERTIES OF BAR 04-0890
ATE SINGLE CRYSTALS.   INFLUENCE OF THE AMPLITUDE OF FREE VIBRATIONS ON THE PERMITTIVITY OF BARIUM TITAN 04-0768
OULET- LCUDON APPROXIMATION IN THE THEORY OF OPTICAL LATTICE VIBRATIONS. (QUARTZ, BARIUM TITANATE) /F THE P 03-0380
/VIBRATING FERROCERAMIC TRANSDUCER AS A SOURCE OF ULTRASONIC VIBRATIONS RADIATING IN THE DIRECTION OF THE / 37-3657
APPLIED TO DISPL/ THEORY OF NEUTRON SCATTERING FROM LATTICE VIBRATIONS (REDEVELOPED FROM FIRST PRINCIPLES, 03-0484
CNAL BARIUM TITANATE.   VIBRATIONAL AMPLITUDES IN FERROELECTRIC TETRAG 04-0726
PROBLEM. (WITH AN APPLICATION TO THE THEORY OF FERROELECTR/ VIBRATIONAL APPROACH TO THE ANHARMONIC LATTICE 03-0221
ES (POTASSIUM TANTALATE, STRONT/ DETERMINATION OF THE NORMAL VIBRATIONAL DISPLACEMENTS OF SEVERAL PEROVSKIT 06-1139
      VIBRATIONAL LINE WIDTHS OF THIOUREA.   25-2915
LOW FREQUENCY VIBRATIONAL MODES IN BARIUM SODIUM NIOBATE.   13-1905
RATE CRYSTALS IN ITS LOW TEMPERATURE P/ PECULIARITIES IN THE VIBRATIONAL SPECTRA OF CESIUM AND RUBIDIUM NIT 16-2323

ONS OF THIOUREA. LATTICE VIBRATIONAL SPECTRA OF FIVE CRYSTAL MODIFICATI 25-2905
ALS. VIBRATIONAL SPECTRA OF IODIC ACID SINGLE CRYST 36-3548
SODIUM NIOBATE AND BARIUM SODIUM TANTALATE (FORCE CONSTANT/ VIBRATIONAL SPECTRA OF LITHIUM NIOBATE, BARIUM 11-1808
ALCIUM TITANATES. VIBRATIONAL SPECTRA OF STRONTIUM, BARIUM AND C 04-0864
IUM, BARIUM, AND CALCIUM. (75 REFS) VIBRATIONAL SPECTRA OF THE TITANATES OF STRONT 04-0865
F SODIUM CHLORATE. (NOT FERROELECTRIC) THE VIBRATIONAL SPECTRUM AND DIELECTRIC BEHAVIOR O 36-3549
ICAL CONSTANTS OF ANTIFERROMAGNETIC POTASSIUM MANGANESE TRI/ VIBRATIONAL SPECTRUM AND DISPERSION OF THE OPT 33-3385
NONSTATIONARY POLARIZATION IN THE ANHARMONIC VIBRATOR MODEL. (FERROELECTRICS) 03-0345
OF THE SOFT FREQUENCY OF BARIUM TITANATE. VIBRONIC EFFECT IN THE TEMPERATURE DEPENDENCE 04-0662
THEORY OF FERROELECTRIC PHASE TRANSITIONS WITH VIBRONIC MECHANISM. 03-0351
ME ADDITIONAL FEATURES OF FERROELECTRIC PHASE TRANSITIONS IN VIBRONIC MODEL. SO 03-0339
ROELECTRICS. THEORY OF VIBRONIC PHASE TRANSFORMATIONS IN WIDE GAP FER 03-0349
SECOND INTERNATIONAL MEETING ON FERROELECTRICITY. A EUROPEAN VIEW. AN AMERICAN VIEW. REPORT ON THE 01-0008
ANOMALIES NEAR FERROELECTRIC TRANSITIONS FROM THE POINT OF VIEW OF THE THEORY OF CRITICAL POINTS. 03-0298
MAIN STRUCTURES OF FERROELECTRIC CRYSTALS. (CRYSTALLOGRAPHIC VIEWPOINT) MICROSCOPIC THEORY OF STATIC DO 03-0340
TERIAL RESEARCH OF FERROELECTRIC TRIGLYCINE SULFATE FROM THE VIEWPOINT OF PRACTICAL APPLICATIONS. MA 27-3065
TRANSITION IN A FERROELECTRIC WITH HYDROGEN BONDS/ EFFECT OF VIOLATION OF ELECTRIC NEUTRALITY ON THE PHASE 03-0449
LATTICE MODEL OF A PHASE TRANSITION. APPARENT VIOLATION OF THE UNIVERSALITY HYPOTHESIS IN A 03-0515
FLECTANCE SPECTRA AND BAND STRUCTURE) OF SOME ABO(3) FERROE/ VISIBLE AND ULTRAVIOLET OPTICAL PROPERTIES (RE 08-1383
SODIUM NIOBATE. VISIBLE CW PARAMETRIC OSCILLATOR USING BARIUM 13-1984
PARAMETRIC GENERATION IN ADP CONTINUOUSLY TUNABLE ACROSS THE VISIBLE SPECTRUM. EFFICIENT HIGH GRAIN 17-2712
NIOBATE. VISUALIZATION OF ACOUSTIC RADIATION IN LITHIUM 11-1864
DE LAYERS ON BARIUM TITANATE- STRONTIUM TITAN/ SATURATION OF VOLT- AMPERE CHARACTERISTICS OF CADMIUM SELENI 04-0592
CRYSTALS SUITABLE FOR LIGHT MODULATORS WITH A LOW HALF WAVE VOLTAGE. CUTS OF LITHIUM NIOBATE 11-1715
GENEOUS CRYSTALS (THEORY AND METHOD OF DETERMINING HALF WAVE VOLTAGE). /TROOPTIC EFFECT IN OPTICALLY HETERO 03-0461
FOIODIDE IN TH/ INFLUENCE OF ILLUMINATION ON THE FORM OF THE VOLTAGE AMPERE CHARACTERISTICS OF ANTIMONY SUL 15-2119
LIMITED CURRENT IN POLYCRYSTALLINE BARIUM TITANATE. (CURRENT VOLTAGE CHARACTERISTICS, 150-375 DEG C) /ARGE 04-0557
FERROELECTRIC CERAMIC LIGHT GATES OPERATED IN A VOLTAGE CONTROLLED MODE. 37-3654
LEAD, 4 COMPOSITIONS EACH) STRONTIUM TITANATE SYSTEMS. (FOR VOLTAGE DEPENDENT CAPACITORS) / THE BARIUM (OR 09-1522
NOVEL TYPE OF CUT FOR KDP CRYSTALS FOR LOW VOLTAGE LIGHT MODULATION. 17-2531
EZOELECTRIC- FERROELECTRIC INTERACTIONS FOR ADAPTIVE CONTROL VOLTAGE MODULES. / MEMORY DEVICES EMPLOYING PI 37-3663
IS IN TRIGLYCINE SULFATE AND STRONTIUM BARIUM / PYROELECTRIC VOLTAGE RESPONSE TO RECTANGULAR INFRARED SIGNA 13-2019
S IN TRIGLYCINE SULFATE AND BARIUM STRONTIUM N/ PYROELECTRIC VOLTAGE RESPONSE TO SHORT INFRARED LASER PULSE 13-2018
ADIATION IN TRIGLYCINE SULFATE AND STRONTIUM B/ PYROELECTRIC VOLTAGE RESPONSE TO STEP SIGNALS OF INFRARED R 13-2022
CINE SELENATE. (DIELECTRIC CONSTANT VERSUS ALTERNATING FIELD VOLTAGE, TEMPERATURE AND POLARIZATION) /TRIGLY 27-3158
BARIUM SODIUM NIOBATE TRANSVERSE LIGHT MODULATOR. (HALF WAVE VOLTAGE, THEORY) /M CRYSTAL ORIENTATION FOR A 13-1973
FERROELECTRIC CERAMICS: PLZT 12-40-60. (USED FOR SENSING LOW VOLTAGES) LINEAR ELECTROOPTIC EFFECT IN 09-1610
A FERROELECTRIC MATERIAL. (THEORY) VOLTAGES AND CURRENTS DURING DEPOLARIZATION OF 03-0177
INTERACTION OF VOLUME ACOUSTIC MODES IN LITHIUM NIOBATE. 11-1806
T-3: EXTRINSIC VERSUS INTRINSIC POLARIZATION, SURFACE VERSUS VOLUME CONDUCTION. /ZATION AND CONDUCTION. SEC 34-3482
/-3: EXTRINSIC VERSUS INTRINSIC POLARIZATION, SURFACE VERSUS VOLUME CONDUCTION (INCLUDES ELECTRODE EFFECTS/ 34-3455
S SINGLE CRYSTALS AND SOME INVESTIGATIONS OF THEIR QUALITY. (VOLUME DISTRIBUTION OF DEFECTS) /ON OF PURE TG 27-3086
SOME ISOMORPHOUS FERROELECTRIC SYSTEMS. (DISCUSSE/ UNIT CELL VOLUME EFFECTS (RELATION WITH CURIE POINT) IN 17-2487
SOME POSSIBILITIES OF DETERMINING THE VOLUME OF KANZIG REGIONS IN FERROELECTRICS. 03-0444

W

TITANATE. SELF REVERSAL AND WAITING TIME EFFECTS IN SINGLE CRYSTAL BISMUTH 08-1448
RIGLYCINE SULFATE CRYSTALS. DOMAIN WALL CAUGHT IN DISLOCATIONS IN FERROELECTRIC T 27-3091
F BARIUM TITANATE. (CALCULATIONS, SUMMARY ON/ 180 DEG DOMAIN WALL CONTRIBUTION TO THE DIELECTRIC CONSTANT O 04-0804
R BARIUM AND LEAD TITANATES - APPLICATION TO CALCULATING THE WALL ENERGY AND THE DIELECTRIC CONSTANTS. / FO 03-0189
CALCULATION OF (DOMAIN) WALL ENERGY IN BARIUM TITANATE. 04-0673
/ SIGNAL PERMITTIVITY OF THE STATIONARY (100)-180 DEG DOMAIN WALL IN BARIUM TITANATE (THEORY, CALCULATIONS/ 03-0358
ATIONS) ANGULAR ENERGY DEPENDENCE OF 180 DEG DOMAIN WALL IN BARIUM TITANATE. (THERMODYNAMIC CALCUL 04-0658
TEMPERATURE DEPENDENCE OF THE SHAPE OF THE DOMAIN WALL IN FERROMAGNETICS AND FERROELECTRICS. 03-0191
ATIONS OF WALL THICKNESS AND ENERGY DE/ FERROELECTRIC DOMAIN WALL IN TRIGLYCINE SULFATE. (THEORY AND CALCUL 27-2997
TANATE) STRUCTURE OF A MOVING DOMAIN WALL. (MECHANISM OF A WALL MOTION IN BARIUM TI 04-0678
ITANATE ZIRCONATE TRANSDUCER CER/ DISPERSION STUDY OF DOMAIN WALL MOBILITY AND MECHANICAL DAMPING IN LEAD T 09-1547
LOOP OF FERROELECTRICS. SIDEWAYS DOMAIN WALL MOTION AND (MECHANISM OF THE) HYSTERESIS 03-0301
RYSTALS. (SUMMARY, WITH PHOTOS) DOMAIN WALL MOTION AND NUCLEATION OF DOMAINS IN TGS C 27-3072
DOMAIN WALL MOTION IN FERROELECTRIC SWITCHING. 03-0273
: GENERAL FORMULATION. KINETICS OF DOMAIN WALL MOTION IN FERROELECTRIC SWITCHING. PART-1 03-0274
CONTROL AND APPLICATION OF DOMAIN WALL MOTION IN GADOLINIUM MOLYBDATE. 30-3257
CONTROL AND APPLICATION OF DOMAIN WALL MOTION IN GADOLINIUM MOLYBDATE. 30-3256
TYPE CRYSTALS. THEORY OF DOMAIN WALL MOTION IN POTASSIUM DIHYDROGEN PHOSPHATE 17-2652
SPERSION OF COERSIVE FIELD AT VERY LOW FREQUENCIES AND JERKY WALL MOTION IN ROCHELLE SALT. ANOMALOUS DI 28-3177
SUPERSONIC DOMAIN WALL MOTION IN TRIGLYCINE SULFATE. 27-2963
LT. CONTRIBUTION OF DOMAIN WALL MOTION TO THE PERMITTIVITY OF ROCHELLE SA 28-3198
X-RAY TOPOGRAPHIC STUDY OF DOMAIN WALL MOVEMENT IN TGS. 27-3026
PIEZOELECTRIC SURFACE WAVES ON A C-DOMAIN WALL OF A FERROELECTRIC (CLASS 6MM). 03-0421
ULATIONS) 90 DEGREE TYPE DOMAIN WALL OF BARIUM TITANATE CRYSTAL. (THEORY, CALC 03-0324
OELECTRIC TRIGLYCINE SULFATE (SEVERAL KHZ TO 23 KMHZ, DOMAIN WALL RESONANCE). DIELECTRIC DISPERSION IN FERR 27-3032
HING (CALCULATIONS COMPARED WITH DA/ (STEP OR SMOOTH) DOMAIN WALL STRUCTURE IN BARIUM TITANATE DURING SWITC 04-0677
N SCATTERING. DOMAIN WALL STRUCTURE IN GADOLINIUM MOLYBDATE BY RAMA 30-3318
ECHNIQUES APPLIED TO PROBLEMS IN SOLID STATE PHYSICS (DOMAIN WALL THICKNESS IN GADOLINIUM MOLYBDATE). /NG T 30-3319
PHASE POTASSIUM DIHYDROGEN PHOSPHATE. (DIFFRACTION BY DOMAIN WALLS) /WITH TEMPERATURE IN THE FERROELECTRIC 17-2662
UM NITRITE CRYSTALS. PART-1: STRUCTURE OF 180 DEGREE DOMAIN WALLS. TOPOGRAPHIC STUDY ON FERROELECTRIC SODI 16-2372
RSIONS OF THE PERMITTIVITY OF ROCHELLE SALT CAUSED BY DOMAIN WALLS. (ABSTRACT ONLY) LOW FREQUENCY DISPE 28-3199
S OF TRIGLYCINE SULFATE. MOVEMENT OF DOMAIN WALLS AND THE NUCLEATION OF DOMAINS IN CRYSTAL 27-3071
LLE SALT CRYSTAL. DOMAIN WALLS CAUGHT IN (FINE LINE) "SUDARES" IN ROCHE 28-3215
APPLICATION TO LEAD/ RELAXATION OF ANTIFERROELECTRIC DOMAIN WALLS, EFFECTIVE FIELD, AND ACTIVATION ENERGY. 07-1257
STRUCTURE OF 90 DEGREE DOMAIN WALLS IN BARIUM TITANATE. 04-0641
OPTICAL STUDY OF DOMAIN WALLS IN BISMUTH TITANATE. 08-1301
LS. (BARIUM TITANATE, GADOLINIUM MOLYBD/ THICKNESS OF DOMAIN WALLS IN FERROELECTRIC AND FERROELASTIC CRYSTA 04-0777
VSKITE STRUCTURE. THERMODYNAMIC THEORY OF DOMAIN WALLS IN FERROELECTRIC MATERIALS WITH THE PERO 03-0190
E CRYSTALS. MECHANICAL SWITCHING OF 60 DEGREE DOMAIN WALLS IN FERROELECTRIC POTASSIUM NIOBATE SINGL 08-1421
ULATION) PERMISSIBLE DOMAIN WALLS IN FERROELECTRIC SPECIES (THEORY AND TAB 03-0247
) INTERACTION OF ULTRASONIC WAVES WITH MOVING DOMAIN WALLS IN FERROELECTRICS. (ANTIMONY SULFOIODIDE 15-2192
OBSERVATION OF DIFFRACTED LIGHT FROM DOMAIN WALLS IN GADOLINIUM MOLYBDATE. 30-3322
IELDS. DISPLACEMENTS AND VELOCITIES OF THE DOMAIN WALLS IN KDP. EXISTENCE OF CRITICAL ELECTRIC F 17-2448

	ULTRASONIC STRESS WAVE INTERACTION WITH DOMAIN	WALLS IN POTASSIUM DIDEUTERIUM PHOSPHATE.	17-2659
TE.	MICROSTRUCTURE OF DOMAINS AND DOMAIN	WALLS IN SINGLE CRYSTAL FILMS OF BARIUM TITANA	04-1031
	SPECIAL ELECTRICAL PROPERTIES OF REGIONS NEAR DOMAIN	WALLS IN TRIGLYCINE SULFATE CRYSTALS.	27-2988
NDENCES OF THE COMPLEX DIELECTRIC CONS/ RELAXATION OF DOMAIN	WALLS IN TRIGLYCINE SULFATE. (TEMPERATURE DEPE	27-3103
VELOCITY OF PROPAGATION OF DOMAIN	WALLS: INTERACTIONS BETWEEN DOMAINS. (IN KDP)	17-2453
LOCAL ELECTRON STATES IN DOMAIN	WALLS OF FERROELECTRICS.	03-0161
ANSFORMATIONS IN FERROELECTRIC CRYSTAL IN PRESENCE OF DOMAIN	WALLS. (THEORY) PHASE TR	03-0346
CTIONS C-AXIS, 20-180 / TEMPERATURE DEPENDENCE OF THE DEBYE-	WALLER FACTORS OF BARIUM TITANATE (BRAGG REFLE	04-0892
/N PEROVSKITE FERROELECTRICS INDICATED BY ANISOTROPIC DEBYE-	WALLER FACTORS. (POTASSIUM TANTALATE NIOBATE)	08-1351
ON OF RUBIDIUM DIDEUTERIUM PHOSPHATE FROM SOLUTIONS IN HEAVY	WATER. CRYSTALLIZATI	17-2583
/ ENTHALPIES OF SOLUTION OF RUBIDIUM DIHYDROGEN PHOSPHATE IN	WATER AND RUBIDIUM DIDEUTERIUM PHOSPHATE IN HE	17-2585
RFACE STRUCTURES AND CORRESPONDING LIGHT FIGURES REVEALED BY	WATER ETCHED ROCHELLE SALT CRYSTALS. SU	28-3234
TERPRETATION. PART-8: TRAN/ DIELECTRIC RELAXATION SPECTRA OF	WATER, ICE, AND AQUEOUS SOLUTIONS AND THEIR IN	34-3482
/TIONS OF ICE CRYSTALS. PART-1: PROTON MAGNETIC RESONANCE IN	WATER ICE. PART-2: DEUTERON SPIN LATTICE RELA/	34-3471
HYDRATE AND FREQUENCIES OF THE FUNDAMENTAL VIBRATIONS OF THE	WATER MOLECULES. /DINIUM ALUMINUM SULFATE HEXA	21-2811
T, DETERMINATION OF THE ORIENTATIONS OF PROTON PAIRS (IN THE	WATER MOLECULES). /C RESONANCE IN ROCHELLE SAL	28-3203
TION IN POTASSIUM FERROCYANID/ ORDERING OF DIPOLE MOMENTS OF	WATER MOLECULES AND FERROELECTRIC PHASE TRANSI	24-2893
ROCHELLE SALT BY SPIN LATTICE RELAX/ STUDY OF SLOW MOTION OF	WATER MOLECULES IN ROCHELLE SALT AND AMMONIUM	28-3232
GRAVIMETRIC MEASUREMENTS OF THE CHANGES OF	WATER OF CRYSTALLIZATION OF ROCHELLE SALT.	28-3235
METASTABLE POTASSIUM NITRATE (III) FROM SOLUTION. (GROWN IN	WATER OR ALCOHOL)	16-2329
KDP). AN APPARATUS FOR POLISHING	WATER SOLUBLE CRYSTALS (APPLICATION TO ADP AND	17-2632
MONIUM DIHYDROGEN PHOSPHATE- POTASSIUM DIHYDROGEN PHOSPHATE-	WATER SYSTEM. /LYTHERM OF SOLUBILITY OF THE AM	17-2434
MONIUM DIHYDROGEN PHOSPHATE- POTASSIUM DIHYDROGEN PHOSPHATE-	WATER SYSTEM. SOLUBILITY ISOBAR IN THE AM	17-2658
N OF THE PHASE DIAGRAM OF ICE. PART-/ MOLECULAR PHENOMENA IN	WATER SYSTEMS. PART-1: MOLECULAR INTERPRETATIO	34-3485
INFLUENCE OF	WATER VAPOR ON BARIUM TITANATE CERAMICS.	04-0882
ION REVERSAL. MODULATION OF THE MILLIMETER	WAVE ABSORPTION IN SODIUM NITRITE BY POLARIZAT	16-2392
IUM NIOBATE (60 DB GAIN). (BACKWARD)	WAVE ACOUSTIC PARAMETRIC OSCILLATIONS IN LITH	11-1837
/ARMONIC OF A LONGITUDINAL HIGH FREQUENCY (950 MHZ) ACOUSTIC	WAVE ALONG THE (001) DIRECTION OF LITHIUM NIO/	11-1835
LITHIUM NIOBATE STRUCTURE. LAYER	WAVE AMPLIFICATION IN A CADMIUM SULFIDE FLUID-	11-1667
LOW POWER DENSITY GALLIUM ARSENIDE- LITHIUM NIOBATE SURFACE	WAVE AMPLIFIER.	11-1825
/) GERMANIUM OXIDE (20) VERSUS LITHIUM NIOBATE IN THE SURFACE	WAVE AMPLIFIER. A THEORETICAL AND EXPERIMENTA/	11-1720
M NIOBATE. COUPLED	WAVE ANALYSIS OF HOLOGRAPHIC STORAGE IN LITHIU	11-1826
RECTI/ COLLINEAR INTERACTION BETWEEN A LONGITUDINAL ACOUSTIC	WAVE AND TWO LUMINOUS WAVES ALONG THE (100) DI	11-1680
OF VACANCIES). ANOMALOUS (LONGITUDINAL	WAVE) ATTENUATION IN LITHIUM TANTALATE (EFFECT	12-1892
LITHIUM NIOBATE SILICON SURFACE	WAVE CONVOLUTION.	11-1788
DETERMINATION OF A NONLINEAR PARAMETER FOR ACOUSTIC SURFACE	WAVE CONVOLUTION IN LITHIUM NIOBATE.	11-1681
E, LITHIUM NIOBATE AND BISMUTH (12) GERMANI/ ACOUSTIC SURFACE	WAVE CONVOLUTION ON CRYSTALS OF CADMIUM SULFID	11-1692
LITHIUM NIOBATE STRUCTURE. SURFACE	WAVE CONVOLUTION USING A CADMIUM SULFIDE FLUID	11-1700
E THREE CENTIMETER RAD/ TEMPERATURE DEPENDENCE OF HYPERSONIC	WAVE DAMPING IN RUBY AND LITHIUM NIOBATE IN TH	11-1718
LINEAR EFFECTS IN MICROWAVE ACOUSTIC LITHIUM NIOBATE SURFACE	WAVE DELAY LINES. NON	11-1818
ON UNPOLARIZED PZT CERAMIC PLATES. SURFACE	WAVE DELAY LINES WITH INTERDIGITAL TRANSDUCERS	09-1614
SURFACE	WAVE DIFFRACTION IN LITHIUM NIOBATE.	11-1698
UHF. (BASED ON LEAD (0.35) STRONTIUM (0.65/ (COMPACT, SURFACE	WAVE) FERROELECTRIC PHASE SHIFTERS FOR VHF AND	37-3633
/OELECTRIC EFFECT ON THE REFLECTION OF A TRANSVERSE ACOUSTIC	WAVE FROM DOMAIN BOUNDARIES IN FERROELECTRICS.	03-0312
REFLECTION OF ELASTIC SURFACE	WAVE FROM PERIODIC ARRAYS ON LITHIUM NIOBATE.	11-1853
GUNN OSCILLATOR. (/ ACOUSTOOPTICAL MEASUREMENTS OF ACOUSTIC	WAVE GENERATION IN PIEZOELECTRICS COUPLED TO A	11-1731
NS. ACOUSTIC SURFACE	WAVE GUIDE, AVAILING FERROELECTRIC POLARIZATIO	37-3680
OPTICAL PROBING OF ACOUSTIC SURFACE	WAVE HARMONIC GENERATION (IN LITHIUM NIOBATE)	11-1752
RATED KDP. LASER RAMAN STUDY OF QUASI SPIN	WAVE HYDROGEN TUNNELING MODES IN KDP AND DEUTE	17-2569
ING). DIFFRACTION OF A CONTINUOUS LIGHT	WAVE IN KDP.	17-2444
ERROELECTRIC MATERIALS. PIEZOELECTRIC LEAKY SURFACE	WAVE IN LITHIUM NIOBATE (ELECTROMAGNETIC COUPL	11-1832
M DIDEUTERIUM PHOSPHATE. SHOCK	WAVE INDUCED ELECTROMECHANICAL CONVERSION IN F	03-0491
OXIDE (20) AT MICROWAVE FREQUENCIES. ULTRASONIC STRESS	WAVE INTERACTION WITH DOMAIN WALLS IN POTASSIU	17-2659
RATURE AND FREQUENCY D/ MICROWAVE FREQUENCY ACOUSTIC SURFACE	WAVE LOSS MECHANISMS ON BISMUTH (12) GERMANIUM	35-3511
H THE POLARIZATION VECTOR, P(S), ARISES PERPENDICULAR TO THE	WAVE LOSS MECHANISMS ON LITHIUM NIOBATE (TEMPE	11-1817
ELASTIC SURFACE	WAVE NORMAL OF THE SOFT WAVES. (THEORY) / WHIC	03-0141
OF FERRODIELECTRICS WITH AN ARBITRARY MAGNETIC SYMMET/ SPIN	WAVE ON THE PZT CERAMIC PLATE.	09-1612
/TION OF SECOND AND THIRD HARMONICS BY A TRANSVERSE ACOUSTIC	WAVE PHENOMENOLOGICAL THEORY IN FINITE SAMPLES	03-0164
TO FERROELEC/ EXPERIMENTAL METHODS FOR INVESTIGATING STRAIN	WAVE PROPAGATING ITSELF IN THE (001) DIRECTIO/	11-1682
ATES: GOLD ON LIT/ FUNDAMENTAL AND HIGHER ORDER MODE SURFACE	WAVE PROPAGATION AND ASSOCIATED CHARGE RELEASE	37-3630
/ENDENT EXCITATION OF LONGITUDINAL AND TRANSVERSE HYPERSONIC	WAVE PROPAGATION ON LAYERED ANISOTROPIC SUBSTR	11-1846
AMICS BARIUM TITANATE AND LEAD ZIRCONATE TITANATE. TWO	WAVE PULSES IN LITHIUM NIOBATE AND BISMUTH (12/	11-1815
HT IN LITHIUM IODATE (INDEX OF REFRACTION 1.2-7/ (CONTINUOUS	WAVE) SHOCK STRUCTURES IN THE FERROELECTRIC CER	04-0916
ITHIUM NIOBATE. ACOUSTOELECTRIC AMPLIFICATION OF SURFACE	WAVE SPONTANEOUS PARAMETRIC SCATTERING OF LIG	08-1291
ATE. SHOCK	WAVE STRUCTURE OF A CADMIUM SELENIDE FILM ON L	11-1723
LATE. SURFACE	WAVE STUDIES OF SURFACE LAYERS IN BARIUM TITAN	04-0588
OF A 40 MHZ COLOR TELEVISION INFRARED RESPONSE USING SURFACE	WAVE TEMPERATURE COEFFICIENTS ON LITHIUM TANTA	11-1852
MEASURED ELECTRICAL CHARACTERISTICS OF INTERDIGITAL SURFACE	WAVE TRANSDUCERS ON LEAD ZIRCONATE TITANATE. /	09-1509
NITRITE. ANGULAR DISPERSION OF LARGE	WAVE TRANSDUCERS ON LITHIUM NIOBATE.	11-1702
APPLICATION TO LITHIUM NIOBATE. PART-2: POLARITONS OF SMALL	WAVE VECTOR POLARITONS IN FERROELECTRIC SODIUM	16-2256
TH (12) GERMANI/ MICROWAVE ACOUSTICS HANDBOOK. VOL 1: SURFACE	WAVE VECTORS. /EIGENVECTORS OF POLARITONS WITH	11-1766
THIUM TANTALAT/ TEMPERATURE COEFFICIENTS OF ACOUSTIC SURFACE	WAVE VELOCITIES. (BARIUM SODIUM NIOBATE, BISMU	01-0028
TEMPERATURE DEPENDENCE OF SURFACE ACOUSTIC	WAVE VELOCITY AND DELAY ON LITHIUM NIOBATE, LI	11-1821
OBATE CRYSTALS SUITABLE FOR LIGHT MODULATORS WITH A LOW HALF	WAVE VELOCITY IN LITHIUM TANTALATE.	12-1886
ETEROGENEOUS CRYSTALS (THEORY AND METHOD OF DETERMINING HALF	WAVE VOLTAGE. CUTS OF LITHIUM NI	11-1715
OR A BARIUM SODIUM NIOBATE TRANSVERSE LIGHT MODULATOR. (HALF	WAVE VOLTAGE). /TROOPTIC EFFECT IN OPTICALLY H	03-0461
IN A FERROELECTRIC ANTIFE/ RESONANCE OF A LOW FREQUENCY SPIN	WAVE VOLTAGE). /M CRYSTAL ORIENTATION F	13-1973
/CTION BETWEEN A LONGITUDINAL ACOUSTIC WAVE AND TWO LUMINOUS	WAVE WITH AN ACOUSTIC OR ELECTROMAGNETIC WAVE	03-0284
LITHIUM NIOBATE. PROPAGATION AND AMPLIFICATION OF RAYLEIGH	WAVES ALONG THE (100) DIRECTION OF LITHIUM NI/	11-1680
/ENTAL STUDY OF LIGHT DIFFRACTION BY HIGH FREQUENCY ACOUSTIC	WAVES AND PIEZOELECTRIC LEAKY SURFACE WAVES IN	11-1856
SURFACE ELASTOPOLARIZATION	WAVES. APPLICATION OF ACOUSTIC WAVES TO HARMO/	11-1834
(THEORY) REFLECTION AND REFRACTION OF TRANSVERSE SOUND	WAVES AT DOMAIN BOUNDARIES IN A FERROELECTRIC.	03-0315
. (LITHIUM NI/ POSSIBILITY OF SWITCHING AND STEERING SURFACE	WAVES AT DOMAIN BOUNDARIES IN FERROELECTRICS.	03-0313
/ STUDIES OF REFRACTION AND REFLECTION OF ULTRASONIC SURFACE	WAVES BY NONLINEAR MIXING IN ANISOTROPIC MEDIA	11-1778
ATE SUBSTRA/ REFRACTION AND REFLECTION OF ULTRASONIC SURFACE	WAVES BY THIN GOLD FILM LAYERS ON LITHIUM NIO/	11-1789
/REFRACTION OF QUASI LONGITUDINAL AND QUASI TRANSVERSE SOUND	WAVES BY THIN GOLD FILM LAYERS ON LITHIUM NIOB	11-1790
UE APPLIED TO LITH/ ELECTRON BEAM SENSING OF SURFACE ELASTIC	WAVES BY 180 DOMAIN BOUNDARIES IN FERROELECTR/	03-0314
INTERACTION OF LIGHT WITH MAGNETOACOUSTIC	WAVES FOR SIGNAL PROCESSING PURPOSES. (TECHNIQ	11-1738
TRAPPING OF PARAMETRICALLY AMPLIFIED LIGHT	WAVES IN A FERRODIELECTRIC (THEORY).	03-0157
	WAVES IN A KDP CRYSTAL.	17-2428

HIUM NICBATE AND P-TYP/ AMPLIFICATION OF RAYLEIGH ULTRASONIC WAVES IN A LAYERED STRUCTURE CONSISTING OF LIT 11-1747
NONLINEAR INTERACTION OF HYPERSONIC WAVES IN A LITHIUM NIOBATE CRYSTAL. 11-1802
UM NIOBATE. CONVOLUTION OF SURFACE WAVES IN A STRUCTURE OF SEMICONDUCTOR ON LITHI 11-1847
MUTH(12) GERMANIUM OXIDE(20) AND EARI/ PIEZOELECTRIC SURFACE WAVES IN CUBIC AND ORTHORHOMBIC CRYSTALS. (BIS 13-2041
AR OPTICAI PROPERTIES, / PHASES AND POLARIZATION OF LUMINOUS WAVES IN CUBIC ELECTROOPTIC CRYSTALS. (NONLINE 02-0044
TRANSITICN POINT. ATTENUATION OF SOUND WAVES IN DEUTERATED ROCHELLE SALT AT THE PHASE 28-3204
TRANSITICN POINT. ATTENUATION OF SOUND WAVES IN DEUTERATED ROCHELLE SALT AT THE PHASE 28-3205
ERNAL FIELDS. (THEORY) `INSTABILITY OF SPIN WAVES IN FERROELECTRIC ANTIFERROMAGNETS IN EXT 03-0149
AR EFFECTS IN THE PROPAGATION OF FINITE AMPLITUDE HYPERSONIC WAVES IN FERROELECTRIC CRYSTALS. NONLINE 03-0498
THEORY OF ELECTRIC INDUCTION WAVES IN FERROELECTRIC SEMICONDUCTOR CRYSTALS. 03-0466
DISPERSION OF ELASTIC WAVES IN FERROELECTRICS. (LITHIUM NIOBATE) 11-1706
INTERACTION OF LIGHT WITH SPIN WAVES IN FERROELECTRICS. (THEORY) 03-0158
ATE, ROCHELLE SALT) ELECTROMAGNETIC AND ACOUSTIC WAVES IN FERROELECTRICS. (THEORY, BARIUM TITAN 03-0455
/OF THE FERROELECTRIC STATE IN ELECTRIC DIPOLE SYSTEMS (SPIN WAVES IN HYDROXYL DIPOLE SYSTEM IN POTASSIUM / 03-0452
INTERACTIONS BETWEEN HIGH FREQUENCY ACOUSTICAL WAVES IN LITHIUM NIOBATE. 11-1836
COLLINEAR INTERACTION OF LONGITUDINAL ELASTIC WAVES IN LITHIUM NIOBATE CRYSTALS. 11-1678
IC CONSTANT) LIGHT SCATTERING BY ULTRASONIC WAVES IN LITHIUM NIOBATE CRYSTALS. (PHOTOELAST 11-1754
K, 200-2000 MHZ) GENERATICN AND PROPAGATION OF HYPERSONIC WAVES IN LITHIUM NIOBATE CRYSTALS. (80-950 DEG 11-1753
INTERACTION OF TWO HYPERSOUND WAVES IN LITHIUM TANTALATE. 12-1891
COUPLED ELECTROMAGNETIC SPIN WAVES IN MAGNETICALLY ORDERED FERROELECTRICS. 03-0148
(THEORY) POSSIBILITY OF EXCITATION OF SPIN WAVES IN MAGNETICALLY ORDERED FERROELECTRICS. 03-0150
OLARIZATION FLUCTUATI/ SPIN LATTICE RELAXATION BY QUASI SPIN WAVES IN ORDER DISORDER TYPE FERROELECTRICS: P 26-2928
ELASTIC AND ELECTROMAGNETIC WAVES IN PIEZOELECTRIC CRYSTALS. 03-0371
ERATURE DE/ ANOMALOUS PROPAGATION OF LONGITUDINAL ULTRASONIC WAVES IN POTASSIUM DIDEUTERIUM PHOSPHATE (TEMP 17-2660
LS. DIFFRACTION CF POLARIZED LIGHT BY ULTRASONIC WAVES IN POTASSIUM DIHYDROGEN PHOSPHATE CRYSTA 17-2407
ANOMALOUS SPIN LATTICE RELAXATION BY QUASI SPIN WAVES IN POTASSIUM FERROCYANIDE TRIHYDRATE. 24-2899
E NEIGHBCRHOOD OF / ATTENUATION AND CISPERSION OF ULTRASONIC WAVES IN POTASSIUM MANGANESE TRIFLUORIDE IN TH 33-3370
ABSORPTION OF ELASTIC WAVES IN REDUCED LITHIUM NIOBATE. 11-1816
POINTS. ABSORPTION OF LONGITUDINAL ULTRASONIC WAVES IN ROCHELLE SALT CRYSTALS NEAR THE CURIE 28-3211
EORY OF FERROELECTRIC PROPERTIES OF CRYSTALS, / POLARIZATION WAVES IN SOLIDS. (INCLUDING DEVELOPMENTS IN TH 03-0306
MOLYEDATE. ABNORMAL ABSORFTION OF ELASTIC WAVES NEAR THE PHASE TRANSITION OF GADOLINIUM 30-3265
IN POTASSIUM DIHYDRO/ ULTRASONIC INVESTIGATION (15 MHZ SHEAR WAVES) OF THE FERROELECTRIC TRANSITION REGION 17-2576
LASS 6MM). PIEZOELECTRIC SURFACE WAVES ON A C-DOMAIN WALL OF A FERROELECTRIC (C 03-0421
CERAMIC PLA/ PROPAGATION CHARACTERISTICS OF SURFACE ELASTIC WAVES ON BARIUM TITANATE(0.95) ZIRCONATE(0.05) 09-1613
/MALLY CIFFUSED IMPURITIES CN PROPAGATION OF ELASTIC SURFACE WAVES ON FERROELECTRIC CERAMICS. (IMPROVEMENT) 03-0487
DIFFRACTION SPREADING OF SURFACE WAVES ON LITHIUM NIOBATE. 11-1849
MIXING OF NONCOLLINEAR ELASTIC SURFACE WAVES ON LITHIUM NIOBATE. 11-1690
NEW HIGH COUPLING LOW DIFFRACTION CUT FOR ACOUSTIC SURFACE WAVES ON LITHIUM NIOBATE. 11-1820
NCY HIGH COUPLING LOW BEAM STEERING CUT FCR ACOUSTIC SURFACE WAVES ON LITHIUM NIOBATE. NEW HIGH FREQUE 11-1819
PIEZOELECTRIC SURFACE WAVES ON LITHIUM NIOBATE. 11-1824
FFECT OF MASS LOADING ON THE PROPAGATION OF ACOUSTIC SURFACE WAVES ON LITHIUM NIOBATE AND QUARTZ. E 11-1757
(FAR FIELD) DIFFRACTICN SPREADING OF SURFACE (ACOUSTIC) WAVES ON (PIEZOELECTRIC) LITHIUM NIOBATE. 11-1850
/TION THEORY FOR ELECTROMAGNETIC CCUPLING TO ELASTIC SURFACE WAVES ON PIEZOELECTRIC SUBSTRATES (LITHIUM NI/ 11-1751
R, P(S), ARISES PERPENDICULAR TO THE WAVE NORMAL OF THE SOFT WAVES. (THEORY) / WHICH THE POLARIZATION VECTO 03-0141
CS. (ANTIMONY SULFOIODIDE) INTERACTION OF ULTRASONIC WAVES WITH MOVING DOMAIN WALLS IN FERROELECTRI 15-2192
C MIXED CRYSTAL AMMONIUM POTASSIUM DIHYDROGEN PHOSPHATE FILM WAVEGUIDE. EPITAXIAL ELECTROOPTI 17-2633
TRIC. (THEO/ INPUT ADMITTANCE AND FIELDS RADIATED BY AN OPEN WAVEGUIDE FILLED WITH HIGH PERMITTIVITY DIELEC 03-0359
ATURE CHARACTERISTICS OF ROCHELLE TYPE CERAMICS. (BARI/ UHF (WAVEGUIDE RESONATOR) METHOD OF STUDYING TEMPER 04-0905
INDICES CF ANTIMONY SULFOIODIDE. (DEPENDENCE ON TEMPERATURE, WAVELENGTH, APPLIED FIELD STRENGTH) /FRACTIVE 15-2206
IN DOPED LITHIUM NIOBATE. WAVELENGTH DEPENDENCE OF HOLOGRAPHIC PHENOMENA 11-1785
NT ABOVE THE CURIE POINT IN VARIOUS FERROELECTRI/ MILLIMETER WAVELENGTH DISPERSION IN THE DIELECTRIC CONSTA 02-0124
RPTION AND REFLECTION SPECTRA OF EARIUM TITANATE IN THE LONG WAVELENGTH INFRARED REGION. ABSO 04-0644
ITANATE WITHOUT COEXISTING PHASES. REINTERPRETATION OF WAVELENGTH MODULATED ABSORPTION IN STRONTIUM T 06-1188
NS IN ANTIMONY SULFOICDIDE. (POLARIZED INFRARED AND RA/ LONG WAVELENGTH OPTICAL PHONONS AND PHASE TRANSITIO 15-2106
RROELECTRIC AMMONIUM DIHYDROGEN PHOSPHATE. LONG WAVELENGTH POLARIZATION FLUCTUATIONS IN ANTIFE 15-2461
BIREFRINGENCE OF CERTAIN CRYSTALS IN THE MILLIMETER WAVELENGTH RANGE. (LITHIUM NIOBATE) 11-1730
ZT). ELECTROOPTIC CERAMICS AS WAVELENGTH SELECTION DEVICES IN DYE LASERS (PL 09-1618
ROELECTRIC TYPE CRYSTAL. MECHANICAL MODEL OF LONG WAVELENGTH VIBRATIONS IN A (SINGLE DOMAIN) FER 03-0342
/TIES OF FERROELECTRIC MATERIALS AT CENTIMETER AND DECIMETER WAVELENGTHS EMPLOYING A SYMMETRIC STRIPLINE. / 04-0855
ITION PCINT. DISPERSION AT MILLIMETER WAVELENGTHS IN BARIUM TITANATE ABOVE THE TRANS 04-0601
AMETRIC FLUORESCENCE IN BARIUM SODIUM NIOBATE FOR 10 CW PUMP WAVELENGTHS (RANGE 454.5-528.7 MICRONS). PAR 13-1943
OF DIELECTRIC PROPERTIES OF FERRCELECTRICS AT SUBMILLIMETER WAVELENGTHS. (STRONTIUM TITANATE) MEASUREMENT 06-1089
M NIOBATE. ROTATING WAVEPLATE OPTICAL FREQUENCY SHIFTING IN LITHIU 11-1689
PHONON IN GADOLINIUM MOLYBDATE. WAVEVECTOR DEPENDENT RELAXATION OF AN OPTICAL 30-3320
SUPPLEMENT TO A PREVIOUS THEORY OF WEAK FERROELASTICS AND WEAK FERROELECTRICS. 03-0146
ES OF BISMUTH TITANATE (EFFECT OF TEMPERATURE). (WEAK FIELD) DIELECTRIC (PERMITTIVITY) PROPERTI 36-3555
IUM, STRONTIUM) TITANATE. MOBILITY DETERMINATIONS FROM WEIGHT MEASUREMENTS IN SOLID SOLUTIONS OF (BAR 04-0689
WEIGHT OF SHOTTKY BARRIERS IN FERROELECTRICS. 03-0459
ROGEN PHOSPHATE. ELECTRON DELOCALIZATION WEIGHT OF THE HYDROGEN BOND IN POTASSIUM DIHYD 17-2525
TOICHIOMETRIC BARIUM TITANATE. WET SYNTHETIC METHOD OF MAKING PURE, EXACTLY S 04-0790
CONDITIONS FOR THE GROWTH OF POTASSIUM DIHYDROGEN PHOSPHATE WHISKERS. 17-2628
OMPOSITE CERAMIC PIEZOELECTRIC TRANSDUCER MATERIAL. (ALUMINA WHISKERS IN LEAD ZIRCONATE TITANATE. NEW C 09-1556
UTH TITANATE. EFFICIENT WHITE LIGHT READING OF DOMAIN PATTERNS IN BISM 08-1299
STUDY OF PIEZOELECTRIC OSCILLATIONS IN WIDEBAND PYROELECTRIC LITHIUM TANTALATE. 12-1875
APPARENT SOFT MODE LINE WIDTH DIVERGENCE IN LEAD TITANATE. 05-1078
FERROMAGNETIC RESONANCE LINE WIDTH IN FERROELECTRIC FERROMAGNETS. 03-0207
MAGNESIUM DOPING. SHIFT OF NEEL TEMPERATURE AND EPR LINE WIDTH OF POTASSIUM MANGANESE TRIFLUORIDE WITH 33-3377
ION IN LITHIUM NIOBATE. POWER AND LINE WIDTH OF TUNABLE STIMULATED FAR INFRARED EMISS 11-1736
ECTRICS IN MEMORY DEVICE/ UTILIZATION OF THE (CRITICAL PULSE WIDTH) PARTIAL SWITCHING PROPERTIES OF FERROEL 27-3144
DIHYDROGEN PHOSPHATE TYPE CRYSTALS. ABSENCE OF A LINE WIDTH TRANSITION FOR THE PROTONS IN POTASSIUM 17-2436
NIOBIUM-93 NMR LINE WIDTHS IN NONSTOICHIOMETRIC LITHIUM NIOBATE. 11-1795
OVSKITE STRUCTURE (TITANATES AND ZIRCONATES). FORBIDDEN ZONE WIDTHS IN POLYCRYSTALLINE DIELECTRICS WITH PER 08-1288
VIBRATIONAL LINE WIDTHS OF THIOUREA. 25-2915
TY) IN BARIUM(X) STRONTIUM(1-X) TITANATE CERAMIC/ "VIBRATING WIRE" STUDY OF POTENTIAL DISTRIBUTION (STABILI 04-0699
/ASTIC SCATTERING STUDY ON SOFT MODES (REVIEW OF BROOKHAVEN WORK ON POTASSIUM TANTALATE, BARIUM AND STRON/ 02-0115
PHOTOCONDUCTOR OPTICAL STCRAGE MECIUM. (HIGH RESOLUTION READ WRITE) BISMUTH TITANATE 08-1445
TANIUM OXIDE SINGLE CRYSTALS. ELECTRON BEAM WRITING OF FERROELECTRIC DOMAINS IN BISMUTH TI 08-1300
F FERROELECTRICITY (KDP MODEL, ANTIFERROELECTRICITY F MODEL, WU KDP MODEL, CALCULATIONS). /E SLATER MODEL O 03-0308
E TRANSITION IN A FERROELECTRIC WITH HYDROGEN BONDS (THEORY, WU MODEL). /OF ELECTRIC NEUTRALITY ON THE PHAS 03-0449

X

S IN TRIGLYCINE SULFATE.   EFFECT OF X- AND GAMMA-RADIATION ON THE SWITCHING PROCES 27-3019
SES OF TRIGLYCINE SULFATE CRYSTALS EXPOSED TO SMALL DOSES OF X- AND GAMMA-RAYS.   DIELECTRIC LOS 27-2976
OSPHATE.   PHOTOELASTIC EFFECT IN 45 DEGREE X-CUTS OF POTASSIUM AND AMMONIUM DIHYDROGEN PH 17-2443
ELECTRON NUCLEAR DOUBLE RESONANCE OF THE ARSENATE CENTER IN X-IRRADIATED POTASSIUM DIHYDROGEN ARSENATE. /D 18-2727
/ICS: EVIDENCE FOR SLATER CONFIGURATIONS FROM EPR STUDIES OF X-IRRADIATED POTASSIUM DIHYDROGEN ARSENATE AN/ 18-2728
TASSIUM DIDEUTER/ ELECTRON PARAMAGNETIC RESONANCE STUDIES OF X-IRRADIATED POTASSIUM DIHYDROGEN ARSENATE, PO 18-2725
ENDOR (ELECTRON NUCLEAR DOUBLE RESONANCE) STUDY OF X-IRRADIATED SINGLE CRYSTALS OF ICE.   34-3413
NE FLUCEERYLLATE, TRIGLYCINE SULFATE AND ITS VARIATION UNDER X-IRRADIATION. / TRIGLYCINE SELENATE, TRIGLYCI 27-3124
ON OF FERROELECTRIC SWITCHING IN TRIGLYCINE SULFATE. PART-2: X-IRRADIATION TREATMENT.   STRUCTURAL INHIBITI 17-2469
RPTION BANDS OF DEFECT ROCHELLE SALT AND/ IONIZATION OF THE (X-RADIATION AND COPPER AND IRON IMPURITY) ABSO 27-3108
S OF NONUNIFORM THIN SAMPLES (STRONTIUM TI/ DETERMINATION OF X-RAY ABSORPTION COEFFICIENTS AND THE THICKNES 06-1179
S (STRONTIUM OR CALCIUM TITANATE/ EXTENDED FINE STRUCTURE IN X-RAY ABSORPTION SPECTRA OF CERTAIN PEROVSKITE 03-0414
THALLIUM CONTAINING PEROVSKITES AT HIGH PRESSURES AND THEIR X-RAY ANALYSIS.   SYNTHESIS OF 08-1463
LID SOLUTIONS BASED ON THE FERROELECTRIC POTASSIUM TANTALAT/ X-RAY ANALYSIS AND DIELECTRIC PROPERTIES OF SO 18-1310
OPERTIES OF NEW CADMIUM-CONTAINING PEROVSKITES A/ SYNTHESIS, X-RAY ANALYSIS, AND STUDY OF THE DIELECTRIC PR 08-1309
PARAELECTRIC PHASE OF SODIUM NITRITE.   X-RAY ANALYSIS OF FERROELECTRIC DOMAINS IN THE 16-2271
D TITANATE CERAMICS.   X-RAY ANALYSIS OF MODIFIED LEAD ZIRCONATE- LEA 27-3124
AYERED FERROELECTRICS. (LANTHANUM BISMUTH TITANI/ SYNTHESES, X-RAY, AND DIELECTRIC STUDIES OF CERTAIN NEW L 13-1993
OVSKITE TYPE COMPOUNDS.   X-RAY AND DIELECTRIC STUDIES OF IRRADIATED PER 08-1443
MS YTTRIUM MANGANESE(1-X) (B(X)) OXYGEN(3) (B=TRIVALENT IRO/ X-RAY AND ELECTRIC INVESTIGATIONS OF THE SYSTE 36-3569
ALT CRYSTALS.   ELECTRONIC EFFECTS (OF X-RAY AND GAMMA-RAY IRRADIATION) IN ROCHELLE S 28-3192
PHASE TRANSITIONS IN SOLID HYDROGEN HALIDES (SINGLE CRYSTAL X-RAY AND NEUTRON DIFFRACTION). /ECTRICITY AND 31-3331
OMAGNETIC)   FERROELECTRIC BISMUTH FERRATE X-RAY AND NEUTRON DIFFRACTION STUDY. (ANTIFERR 08-1402
NAL BARIUM TITANATE.   X-RAY AND NEUTRON DIFFRACTION STUDY OF TETRAGO 04-0713
NAL BARIUM TITANATE.   X-RAY AND NEUTRON DIFFRACTION STUDY OF TETRAGO 04-0740
ATION OF SODIUM NITRITE.   X-RAY AND NEUTRON STUDY ON THE PHASE TRANSFORM 16-2328
ICN OF FERROELECTRIC DICALCIUM STRONTIUM HEXAACETATE. X-RAY AND OPTICAL STUDIES ON THE PHASE TRANSIT 36-3584
0.88) LITHIUM(0.12) NIOBATE CERAMIC.   X-RAY AND THERMAL EXPANSION STUDY OF A SODIUM(13-1999
E.   X-RAY AND THERMOLUMINESCENCE OF BARIUM TITANAT 04-0750
GLYCINE SULFATE.   X-RAY CRITICAL SCATTERING IN FERROELECTRIC TRI 27-2998
E AND SODIUM NITRATE. (ORDER DISORDER TRANSITION, FERROELEC/ X-RAY CRITICAL SCATTERING IN TRIGLYCINE SULFAT 16-2400
160 DEG C AND AT 215 DEG C.   X-RAY CRITICAL SCATTERING OF SODIUM NITRITE AT 16-2343
FOIODIDE.   X-RAY CRYSTALLOGRAPHIC STUDIES OF ANTIMONY SUL 15-2109
D THE BINARY STRONTIUM OXIDE- NIOBIUM PENTOXIDE SYSTEM, DTA, X-RAY CURIE TEMPERATURE). /BIUM-RICH REGION AN 13-1920
EPR STUDY OF X-RAY DAMAGE IN TRIGLYCINE SULFATE.   27-3074
FERROELECTRICS. BASED ON NEW TREATMENT AND INTERPRETATION OF X-RAY DAMAGED FERROELECTRICS.   02-0079
LOUS SCATTERING AT AN ABSORPTION EDGE. (LITHIUM NIOBATE, BA/ X-RAY DATA) /DISORDER OF BARIUM TITANATE TYPE 04-0630
ROELECTRIC COMPOUNDS (POTASSIUM OR SODIUM) (0.5) BISMUTH(0.5/ X-RAY DETERMINATION OF POLARITY SENSE BY ANOMA 11-1673
F AMMONIUM DIHYDROGEN PHOSPHATE.   X-RAY DETERMINATION OF THE SYMMETRY OF THE FER 08-1362
BARKHAUSEN EFFECT IN FERROELECTRICS UNDER CONTINUOUS X-RAY DETERMINATION OF THE THERMAL EXPANSION O 17-2485
80 AND 575 DEG C, AND THEIR RELEVANCE TO SOFT PHONON MODES. (X-RAY DIFFRACTION.   03-0337
IC AND FERROELECTRIC PHAS/ SURFACE LAYER IN BARIUM TITANATE (X-RAY DIFFRACTION) /F SODIUM NIOBATE BETWEEN 4 08-1273
STRUCTURE OF AMMONIUM HYDROGEN SULFATE ABOVE THE FERROELEC/ X-RAY DIFFRACTION, DC FIELD EFFECT, PARAELECTR 04-0857
MAIN FORMATION IN FERROELECTRICS.   THEORY OF X-RAY DIFFRACTION DETERMINATION OF THE CRYSTAL 19-2775
BARIUM TITANATE. PART-3: HIGH TEMPERATURE THERMODYNAMIC AND X-RAY DIFFRACTION LINE SHAPE EFFECTS DUE TO DO 03-0236
TION IN FERROELECTRICS.   X-RAY DIFFRACTION MEASUREMENTS. /TOICHIOMETRIC 04-0886
E (CRYSTALLOGRAPHIC PHASE CHANGE AT CURIE TEMPERATURE). X-RAY DIFFRACTION STUDIES OF ELECTRON POLARIZA 03-0199
ARIUM TITANATE.   HIGH TEMPERATURE THERMODYNAMIC AND X-RAY DIFFRACTION STUDY OF GADOLINIUM MOLYBDAT 30-3312
COPPER FORMATE.   X-RAY DIFFRACTION STUDY OF NONSTOICHIOMETRIC B 04-0885
A NEW PYROCHLORE CONTAINING LITHIUM, STRONTIU/ SYNTHESIS AND X-RAY DIFFRACTION STUDY OF PHASE TRANSITION IN 26-2938
LECTRIC SODIUM NITRITE AND TGS CRYSTALS./ APPLICATION OF THE X-RAY DIFFRACTION STUDY OF SINGLE CRYSTALS OF 36-3613
EFORMATION BEHAVIOR OF ICE SINGLE CRYSTALS.   X-RAY DIFFRACTION TOPOGRAPHIC METHOD TO FERROE 12-2375
DISPLAY OF TWINNING IN LITHIUM NIOBATE BY X-RAY DIFFRACTION TOPOGRAPHIC STUDIES OF THE D 34-3426
CRYSTALS.   X-RAY DIFFRACTION TOPOGRAPHY.   11-1845
(DISLOCATIONS) IN KDP AND ADP SINGLE CRYSTALS.   X-RAY DIFFRACTION TOPOGRAPHY OF FERROELECTRIC 02-0121
IE TEMPERATURE AND TEMPERATURE DEPENDENCE OF SPONTANEOUS PO/ X-RAY DIFFRACTION TOPOGRAPHY STUDY OF DEFECTS 17-2578
TRANSFORMATIONS IN STRONTIUM TITANATE.   X-RAY DIFFRACTOMETRIC DETERMINATION OF THE CUR 03-0290
NSITION OF POTASSIUM MANGANESE TRIFLUORIDE SINGLE CRYSTALS, (X-RAY DIFFRACTOMETRY OF LOW TEMPERATURE PHASE 06-1150
ELASTIC CONSTANT OF ORTHORHOMBIC POTASSIUM NIOBATE BY X-RAY DIFFUSE SCATTERING). / THE 184 DEG K TRA 33-3369
ESE TRIFLUORIDE AND SODIUM NIOBATE SIN/ ANISOTROPIC CRITICAL X-RAY DIFFUSE SCATTERING.   08-1416
S A FUNCTION OF TEMPERATURE.   X-RAY DIFFUSE SCATTERING FROM POTASSIUM MANGAN 08-1312
RYSTALS (NEAR THE 640 DEG C PHASE TRAN/ ANISOTROPIC CRITICAL X-RAY DIFFUSE SCATTERING FROM SODIUM NIOBATE A 08-1313
PHASE TRANSITION OF POTASSIUM DIHYDROGEN PHOSPHATE. PART-1:/ X-RAY DIFFUSE SCATTERING FROM SODIUM NIOBATE C 08-1311
NEAR THE CURIE TEMPERATURE.   CRITICAL X-RAY DILATOMETRIC STUDY OF THE FERROELECTRIC 17-2557
E.   X-RAY INCOHERENT SCATTERING BY BARIUM TITANATE 04-0535
IC PRESSURE ON THE STRUCTURE OF BARIUM TITANATE.   X-RAY INVESTIGATION OF IRRADIATED LEAD TITANAT 05-1081
TION OF CADMIUM TITANATE (AND THE ANTIFERR/ HIGH TEMPERATURE X-RAY INVESTIGATION OF THE EFFECT OF HYDROSTAT 04-0760
OR STRONTIUM METAPLUMBATES (NOT FERROELECTRIC/ SYNTHESIS AND X-RAY INVESTIGATION OF THE PEROVSKITE MODIFICA 08-1387
RIC POTASSIUM DIHYDROGEN PHOSPHATE.   ESR INVESTIGATIONS OF X-RAY INVESTIGATION OF THE PEROVSKITES BARIUM 36-3607
/LECTRON NUCLEAR TRIPLE RESONANCE FOR THE ARSENATE CENTER IN X-RAY IRRADIATED SINGLE CRYSTALS OF FERROELECT 17-2558
IES OF SOLID SOLUTIONS OF BARIUM TITANATE TYPE MATERIALS. X-RAY IRRADIATED SINGLE CRYSTALS OF POTASSIUM/ 18-2729
YCINE SULFATE.   INFLUENCE OF DISLOCATIONS, SURFACE LAYER AND X-RAY IRRADIATION EFFECT ON DIELECTRIC PROPERT 04-0782
CHELLE SALT.   ON THE EFFECTS OF X-RAY IRRADIATION ON DOMAIN STRUCTURE IN TRIGL 27-3004
ROELECTRIC POTASSIUM NITRATE.   X-RAY IRRADIATION ON TRANSIENT PHENOMENA IN RO 28-3222
D BARIUM TITANATE FILMS.   ELECTRON PROBE X-RAY MEASUREMENTS ON THERMAL EXPANSION OF FER 16-2322
) SYSTEMS. (WHERE A IS STRONTIUM OR BARIUM, AND B IS HAFNIU/ X-RAY MICROANALYSIS ON IMPURITIES IN EVAPORATE 04-0742
BAND STRUCTURE OF STRONTIUM TITANATE AS MEASURED BY X-RAY PHASE STUDY OF THE CADMIUM BORATE- ABO(3 36-3528
STEN OXIDE AS A FUNCTION OF THICKNESS OF THE S/ STUDY OF THE X-RAY PHOTOELECTRON SPECTROSCOPY (ESCA).   06-1100
ITHIUM AMMONIUM TARTRATE MONOHYDRATE.   X-RAY PHOTOELECTRON SPECTRUM OF TUNGSTEN- TUNG 10-1631
OF THE PHASE TRANSITION IN TGS. (STATIC DIELECTRIC CONSTANT, X-RAY REFINEMENT OF THE CRYSTAL STRUCTURE OF L 29-3244
UM NIOBATE.   ANISOTROPIC DIFFUSE X-RAY SCATTERING) STUDIES ON THE MECHANISM 27-3093
THIOUREA.   SATELLITE X-RAY SCATTERING AND PHASE TRANSITIONS IN SODI 21-1357
OTASSIUM MANGANESE TRIFLUORIDE AT 184 DEGREES K.   X-RAY SCATTERING AND STRUCTURAL MODULATION OF 25-2921
DIES OF (SINGLE CRYSTALS IN THE TUNGSTEN BRONZE TYPE) FERR/ X-RAY SCATTERING AND THE PHASE TRANSITION OF P 33-3394
IFIED LEAD TITANATE WITH A-SITE AND B-SITE VACANCIES. (X-RAY) STRUCTURAL (OPTICAL) AND DIELECTRIC STU 13-1947
THE COLOR PROBLEM OF TUNGSTEN TRIOXIDE. X-RAY STRUCTURE INVESTIGATION OF LANTHANUM MOD 05-1059
X-RAY STUDIES.   10-1632

ALTATE(0.5) TUNGSTATE(0.5), TITANAT/ ELECTRICAL MAGNETIC AND X-RAY STUDIES OF PHASE TRANSITIONS IN LEAD COB 13-1960
RATE CRYSTALS.   NUCLEAR MAGNETIC RESONANCE AND X-RAY STUDIES OF POTASSIUM FERROCYANIDE TRIHYD 24-2894
ELLE SALT, BARIUM TITANATE, ANTIMONY SULFOIO/ DIELECTRIC AND X-RAY STUDIES OF THE PHASE TRANSITIONS IN ROCH 02-0099
ITION OF GLYCINE SILVER NITRATE.   DIELECTRIC AND X-RAY STUDIES ON THE FERROELECTRIC PHASE TRANS 16-2347
ONS OF SODIUM NIOPATE (TO 500C).   X-RAY STUDY OF HIGH TEMPERATURE PHASE TRANSITI 13-1965
  NIOBATE.   NEW DATA FROM AN X-RAY STUDY OF PHASE TRANSFORMATIONS IN SODIUM 08-1361
ATE.   X-RAY STUDY OF PHASE TRANSITION IN SODIUM NITR 16-2380
URE ON THE STRUCTURE OF LEAD TITANATE.   AN X-RAY STUDY OF THE EFFECT OF HYDROSTATIC PRESS 05-1064
  AND PARAELECTRIC- PARAELASTIC PHASES OF GADO/ COMPREHENSIVE X-RAY STUDY OF THE FERROELECTRIC- FERROELASTIC 30-3287
NITRITE.   AN X-RAY STUDY OF THE PHASE TRANSITION IN SODIUM 16-2318
UM FERRATE TANTALATE   AN X-RAY STUDY OF THE PHASE TRANSITION IN STRONTI 08-1382
ANATE SINGLE CRYSTAL.   X-RAY STUDY OF THE SURFACE LAYER ON BARIUM TIT 04-0858
  OXIDE.   X-RAY STUDY OF THE SYSTEM BISMUTH OXIDE FERRIC 08-1376
RIC GADOLINIUM MOLYBDATE.   X-RAY STUDY OF THERMAL EXPANSION OF FERROELECT 30-3294
IN SODIUM NIOBATE.   DIFFUSE X-RAY STUDY OF 640 DEGREES C PHASE TRANSITION 08-1358
RIC IRON IODINE BORACITE AT LOW TEMPERATURES. (MONOCLINIC A/ X-RAY STUDY ON PHASE TRANSITIONS OF FERROELECT 14-2086
THE FERROELECTRIC CURIE POINT IN POTASSIUM NITRATE.   X-RAY STUDY ON THE DISORDERED STRUCTURE ABOVE 14-2366
CTRIC AND) (IRON- IODINE- BORACITE).   X-RAY STUDY ON THE LATTICE STRAINS OF (PARALE 14-2085
RIC POTASSIUM DIHYDROGEN PHOSPHATE. (NEAR CURIE TEMPERATURE)   X-RAY STUDY ON THERMAL EXPANSION OF FERROELECT 17-2556
NS IN FERROELECTRIC LEAD GERMANATE SINGLE CRYSTALS.   X-RAY TOPOGRAPHIC OBSERVATION OF 180 DEG DOMAI 35-3515
MAINS IN FERROELECTRIC LEAD(5) GERMANIUM(3) OXIDE(11).   X-RAY TOPOGRAPHIC OBSERVATION OF 180 DEGREE DO 35-3516
MAGE IN TRIGLYCINE SULFATE SINGLE CRYSTALS.   X-RAY TOPOGRAPHIC OBSERVATIONS OF RADIATION DA 27-3106
ND LITHIUM TANTALATE SINGLE CRYSTALS.   X-RAY TOPOGRAPHIC STUDIES ON LITHIUM NIOBATE A 11-1830
RROELECTRIC DOMAINS IN GLYCOL SULFATE.   X-RAY TOPOGRAPHIC STUDY OF DISLOCATIONS AND FE 19-2747
T IN TGS.   X-RAY TOPOGRAPHIC STUDY OF DOMAIN WALL MOVEMEN 27-3026
RIGLYCINE SULFATE CRYSTALS IN RELATION TO THEIR GROWTH COND/ X-RAY TOPOGRAPHIC STUDY OF GROWTH DEFECTS IN T 27-3025
ODUCED IN POTASSIUM DIHYDROGEN PHOSPHATE CRYSTALS BY GROWTH/ X-RAY TOPOGRAPHIC STUDY OF THE DISLOCATIONS PR 17-2492
ERROELECTRIC DOMAIN BOUNDARIES (DOPED AND UNDOPED TGS, LITH/ X-RAY TOPOGRAPHICAL STUDY OF THE CONTRAST AT F 11-1671
IOUREA.   X-RAY TOPOGRAPHY OF THE DEFECT STRUCTURE OF TH 25-2914
                               CRITICAL SCATTERING OF X-RAYS IN SODIUM NITRITE.   16-2270
                               DAMAGE PRODUCED BY X-RAYS ON TRIGLYCINE SULFATE.   27-3073
S.   LIGHT AND CURRENT PULSES FROM X-RAYED POTASSIUM DIHYDROGEN PHOSPHATE CRYSTAL 17-2716

Y

        LIGHT MODULATORS USING 45 DEGREE X CUT AND 45 DEGREE Y CUT ADP CRYSTALS.   17-2605
                        AN ELECTROPTIC SWITCH USING ROTATED Y PLATE OF LITHIUM NIOBATE.   11-1732
THEORY AND TEST ON BARIUM SODIUM NIOBATE AND NEODYMIUM DOPED YAG LASER. /RATION ON CRYSTAL INHOMOGENEITY (13-2043
M NIOBATE AND KDP.   NEODYMIUM YAG LASER IRRADIATION INDUCED DAMAGE TO LITHIU 11-1674
OUS POLARIZATION ON THE PHOTOSTIMULATED EXOELECTRON EMISSION YIELD OF TRIGLYCINE SULFATE CRYSTALS. /PONTANE 27-2961
ITANATE BY TRANSPORT REACTIONS (INVOLVING TITANIUM ESTERS TO YIELD SMALL CRYSTAL PRISMS). /TION OF BARIUM T 04-0625
S AND COMBINATION SCATTER SPECTRUM IN THE REGION OF THE ANTISYMMETRICAL VIBRATION OF THE NITRITE ANION. /AL 16-2276
/ASSIUM DIDEUTERIUM PHOSPHATE (TEMPERATURE DEPENDENCE OF THE YOUNGS MODULUS AND THE POLARIZATION RELAXATIO/ 17-2660
ES OF HEXAGONAL ORTHOMANGANITES OF YTTRIUM AND RARE EARTHS. (YTTERBIUM, HOLMIUM, ERBIUM) /ELECTRIC PROPERTI 36-3542
ETAL IS SCANDIUM, CHROMIUM, MANGANESE, IRON, COBALT, NICKEL, YTTRIUM) /ETAL OXYGEN(3) PEROVSKITES. (WHERE M 08-1451
  0.66 MILLIMICRON HARMONIC SOURCE.   NEODYMIUM: YTTRIUM ALUMINUM GARNET- BARIUM SODIUM NIOBATE 13-1975
ER/ FERROELECTRIC PROPERTIES OF HEXAGONAL ORTHOMANGANITES OF YTTRIUM AND RARE EARTHS. (YTTERBIUM, HOLMIUM, 36-3542
E DEPENDENCE)   RESISTIVITY BEHAVIOR OF SEMICONDUCTING YTTRIUM BARIUM ZIRCONIUM TITANATE. (TEMPERATUR 13-1995
LS.   ELASTIC PROPERTIES OF YTTRIUM DOPED STRONTIUM TITANATE SINGLE CRYSTA 06-1180
/IUM, CERIUM, THORIUM, TITANIUM, SODIUM, MAGNESIUM, CADMIUM, YTTRIUM, GADOLINIUM, DYSPROSIUM, IRON, SELENI/ 10-1640
                   SURFACE EFFECT IN FERROELECTRIC YTTRIUM MANGANATE.   36-3588
                                                             YTTRIUM MANGANATE SWITCHING BEHAVIOR.   36-3571
VALENT IRC/ X-RAY AND ELECTRIC INVESTIGATIONS OF THE SYSTEMS YTTRIUM MANGANESE(1-X) (B(X)) OXYGEN(3) (B=TRI 36-3569
/E DEPENDENCE OF DIELECTRIC CONSTANT IN THE SYSTEM POTASSIUM YTTRIUM MOLYBDATE- POTASSIUM DYSPROSIUM MOLYB/ 30-3309
RON/ DIELECTRIC AND ELECTROOPTIC PROPERTIES OF BARIUM SODIUM YTTRIUM NIOBATE SINGLE CRYSTAL WITH TUNGSTEN B 13-1971

Z

N LINES IN GUANIDINIUM CHROMIUM SULFATE HEXAHYDRATE.   ZEEMAN EFFECT AND DILUTION EFFECT OF ABSORPTIO 21-2821
E RED REGION OF) CHROMIUM GUANIDINIUM SULFATE HEXAHYDRATE.   ZEEMAN EFFECT (STUDY) OF TWO FINE LINES IN (TH 21-2814
                               PULSE RESPONSE OF ZERO DEGREE CUT ADP MODULATORS.   17-2635
ATTICE MAGNETIZATION IN THE TWO DIMENSIONAL ANTIFERROMAGNET/ ZERO POINT SPIN DEVIATION AND SPONTANEOUS SUBL 26-2929
OLARIZATION SWITCHING IN POLYCRYSTALLINE FERROELECTRICS AT / ZERO SHIFT IN PIEZOELECTRIC ACCELEROMETERS. (P 28-3216
ENCY IN LITHIUM TANTALATE LENGTH EXPANDER BARS.   ZERO TEMPERATURE COEFFICIENT OF RESONANT FREQU 11-1810
NBERG, FERROELECTRIC, AND GENERAL ISING MODELS.   ZEROES OF THE PARTITION FUNCTION FOR THE HEISE 03-0481
R MODEL OF FERROELECTRICITY (KDP MODEL, ANT/ DISTRIBUTION OF ZEROES OF THE PARTITION FUNCTION FOR THE SLATE 03-0308
IN (MANGANESE DOPED) LITHIUM NIOBATE AND (COPPER(2+) DOPED) ZINC BORACITE. /ARAMAGNETIC RESONANCE SPECTRUM 11-1792
E, IRC/ (OPTICAL) FREQUENCY DOUBLING IN (MAGNESIUM CHLORINE, ZINC CHLORINE, CADMIUM CHLORINE, CADMIUM IODIN 14-2080
  PREPARATION AND DIELECTRIC PROPERTIES OF POTASSIUM BISMUTH ZINC NIOBATE.   13-2001
OF SINGLE CRYSTALS OF SOLID SOLUTIONS OF BARIUM TITANATE AND ZINC OXIDE.   ELECTROOPTICAL PROPERTIES 04-0953
LAR CRYSTALS (LITHIUM NIOBATE OR TANTALATE, BARIUM TITANATE, ZINC OXIDE, AND LITHIUM GALLIUM OXIDE). /OF PO 02-0098
RRIER SYSTEMS, METAL CONTACTS) SOME ELECTRONIC PROPERTIES OF ZINC OXIDE AND STRONTIUM TITANATE. (SURFACE BA 06-1176
S (LITHIUM NIOBATE OR TANTALATE OR GALLATE, BARIUM TITANATE, ZINC OXIDE, CADMIUM SULFIDE). /ECTRIC) CRYSTAL 02-0095
SABLE, NONDESTRUCTIVE READOUT, BASED ON BISMUTH TITANATE AND ZINC SELENIDE). /IC PHOTOCONDUCTOR DEVICE (ERA 37-3646
ELECTROOPTICAL EFFECT IN SINGLE CRYSTALS OF BARIUM TITANATE- ZINC TITANATE SOLID SOLUTIONS.   04-0958
/IC AND PIEZOELECTRIC PROPERTIES IN THE TERNARY SYSTEM LEAD (ZINCATE NIOBATE)- BARIUM (ZINCATE NIOBATE)- L/ 05-1070
  CRYSTAL STRUCTURE AND PIEZOELECTRICITY OF THE SYSTEM LEAD (ZINCATE NIOBATE)- LEAD TITANATE.   13-2067
            GROWTH OF SINGLE CRYSTALS OF LEAD (ZINCATE NIOBATE AND ITS PROPERTIES.   36-3594
ITION.   GROWTH OF SINGLE CRYSTALS OF LEAD (ZINCATE NIOBATE) AND PROPERTIES OF THIS COMPOS 13-1990
ARY SYSTEM LEAD TITANATE- LEAD ZIRCONATE- LEAD NIOBATE(0.67) ZINCATE(0.33).   TERN 09-1507
E SOLID SOLUTION SYSTEM LEAD TITANATE- LEAD ZIRCONATE- LEAD (ZINCATE(0.33) NIOBATE(0.67)). /ROPERTIES OF TH 13-1951
REACTION, SINTERING) AND PROPERTIES OF PEROVSKITE TYPE LEAD (ZINCATE(0.33) NIOBATE(0.67). /SIS (SOLID STATE 11-1393
  THE EFFECT OF PRESSURE ON THE TRANSITION OF LEAD (ZINCATE(0.33) NIOBATE(0.67).   08-1473
ADRATIC COEFFICIENTS)   ELECTROOPTICAL EFFECT IN LEAD (ZINCATE(0.33) NIOBATE(0.67). (DISPERSION OF QU 13-1982
-160 TO +25/ STRUCTURAL AND DIELECTRIC PROPERTIES OF BARIUM (ZINCATE(0.33) NIOBATE(0.67) (NOT FERROELECTRIC 36-3609
        DIELECTRIC AND OPTICAL PROPERTIES OF LEAD (ZINCATE(0.33) NIOBATE(0.67) SINGLE CRYSTAL.   13-2065
ROWN FROM LEAD OXIDE FLUX). FERROELECTRIC PROPERTIES OF LEAD (ZINCATE(0.33) NIOBATE(0.67) (SINGLE CRYSTALS G 13-2066
LECTRIC PROPERTIES IN A LEAD TITANATE- LEAD ZIRCONATE- LEAD (ZINCATE(0.33) NIOBATE(0.67) SYSTEM. /D PIEZOE 09-1617
TROOPTIC EFFECTS OF HOT PRESSED CERAMICS IN THE SYSTEM LEAD (ZINCATE(0.333) NIOBATE(0.667)).   ELEC 13-1900
          ATHERMAL PHASE TRANSITION IN LEAD ZIRCONATE.   07-1258

AL GROWTH AND OBSERVATION OF THE FERROELECTRIC PHASE OF LEAD ZIRCONATE. CRYST 07-1264
DEMONSTRATION OF AGING EFFECTS IN BARIUM TITANATE- BARIUM ZIRCONATE. 09-1543
TRONTIUM TITANATE, BARIUM TITANATE STANNATE, BARIUM TITANATE ZIRCONATE) /ASED ON BARIUM TITANATE. (BARIUM S 04-1019
DIPOLE STRUCTURE AND INTERNAL ELECTRIC FIELDS IN LEAD ZIRCONATE. 07-1261
ILE NEUTRON IRRADIATION ON THE DIELECTRIC PROPERTIES OF LEAD ZIRCONATE. EFFECT OF P 07-1259
TY AND ANTIFERROELECTRICITY IN POLAR CRYSTALS. (THEORY, LEAD ZIRCONATE) FERROELECTRICI 03-0372
CITY AND ANTIFERROELECTRICITY IN PURE AND NIOBIUM DOPED LEAD ZIRCONATE. FERROELECTRI 07-1251
GROWTH AND DIELECTRIC PROPERTIES OF SINGLE CRYSTAL LEAD ZIRCONATE 07-1265
ATURE (TO 236 DEG C) PHASE TRANSITIONS IN (HIGH PURITY) LEAD ZIRCONATE. HIGH TEMPER 07-1266
ISOTHERMAL PHASE TRANSITIONS IN CERAMIC LEAD ZIRCONATE. 08-1466
ITANATE- LEAD FERRATE(0.5) NIOBATE(0.5), LEAD TITANATE- LEAD ZIRCONATE) NEW PIEZOELECTRIC CERAMICS. (LEAD T 09-1544
EFFECTIVE FIELD, AND ACTIVATION ENERGY. APPLICATION TO LEAD ZIRCONATE. /OF ANTIFERROELECTRIC DOMAIN WALLS, 07-1257
THERMOELECTRIC EFFECT IN SEMICONDUCTIVE LEAD ZIRCONATE. 07-1267
/ STUDIES OF THE PHASE TRANSITIONS IN ANTIFERROELECTRIC LEAD ZIRCONATE AND FERROELECTRIC LEAD TITANATE ZIR/ 07-1253
ELECTRICAL CONDUCTIVITY OF STRONTIUM ZIRCONATE AND HAFNATE AT HIGH TEMPERATURES. 08-1417
PHASE TRANSITIONS OF THE ANTIFERROELECTRIC PEROVSKITES LEAD ZIRCONATE AND LEAD HAFNATE. /IC PROPERTIES AND 08-1433
FERROELECTRIC PROPERTIES. (SUMMARY WITH NEW RESULTS ON LEAD ZIRCONATE AND OTHER ANTIFERROELECTRICS) /RE ON 07-1263
/REPARATION AND CHARACTERIZATION OF ALKOXY-DERIVED STRONTIUM ZIRCONATE AND STRONTIUM TITANATE (OF HIGH PUR/ 06-1217
/TION AND STRUCTURE) AND RELATED MATERIALS (SYSTEM WITH LEAD ZIRCONATE AND WITH LEAD ZIRCONATE TITANATE MO/ 08-1333
TEMPERATURE DEPENDENT OPTICAL MODE IN ANTIFERROELECTRIC LEAD ZIRCONATE BY THE MOSSBAUER EFFECT. 07-1260
A-SITE AND B-SITE VACANCIES IN (LEAD, LANTHANUM) (TITANATE, ZIRCONATE) CERAMICS. DISTRIBUTION OF 09-1530
LECTROMECHANICAL PROPERTIES OF LANTHANUM DOPED LEAD TITANATE ZIRCONATE CERAMICS. STRUCTURE AND E 09-1502
LECTROMECHANICAL PROPERTIES OF LANTHANUM DOPED LEAD TITANATE ZIRCONATE CERAMICS. STRUCTURE AND E 09-1531
PIEZOELECTRIC AND DIELECTRIC PROPERTIES OF LEAD TITANATE ZIRCONATE CERAMICS AT LOW TEMPERATURES. 09-1519
COMMUNICATIONS. EVALUATION OF LEAD LANTHANATE ZIRCONATE CERAMICS FOR APPLICATIONS IN OPTICAL 07-1254
/TRUCTURE ON THE PROPERTIES OF POLYCRYSTALLINE LEAD TITANATE ZIRCONATE COMPOSITIONS NEAR THE MORPHOTROPIC / 09-1557
ROELECTRIC PROPERTIES OF TETRAGONAL (BARIUM, LEAD) (NIOBATE, ZIRCONATE). (CURIE TEMPERATURES) FER 08-1293
EZOELECTRIC PROPERTIES OF A CERAMIC (LEAD STRONTIUM TITANATE ZIRCONATE DOPED WITH CHROMIUM OXIDE). / AND PI 09-1591
FERROELECTRICITY IN SPUTTERED LEAD TITANATE ZIRCONATE FILMS. 09-1559
F SOME CERAMIC MATERIALS AT MICROWAVE FREQUENCIES. (TITANATE ZIRCONATE GARNET MIXTURES) /RICAL PROPERTIES O 09-1602
ANTIFERROELECTRIC MODE IN LEAD ZIRCONATE. (LATTICE VIBRATIONS) 07-1262
E, POTASSIUM TANTALATE, BARIUM TITANATE, LEAD TITANATE, LEAD ZIRCONATE, LEAD HAFNATE) /. (STRONTIUM TITANAT 02-0114
DIELECTRIC STUDY OF THE LEAD ZIRCONATE LEAD HAFNATE (PHASE) DIAGRAM. 08-1340
THE ROLE OF IRON OXIDE ADULTERANT IN A LEAD ZIRCONATE LEAD TITANATE CERAMICS. 09-1592
BARIUM ZIRCONATE LEAD TITANATE FERROELECTRIC CERAMICS 09-1492
UNDER HIGH ELECTRIC FIELDS. (SUB/ BEHAVIOR OF MODIFIED LEAD ZIRCONATE LEAD TITANATE PIEZOELECTRIC CERAMICS 09-1484
ARY SYSTE/ CRYSTAL STRUCTURE AND PHASE RELATIONS IN THE LEAD ZIRCONATE LEAD TITANATE- LEAD METANIOBATE TERN 09-1596
THERMAL EXPANSION OF POLARIZED CERAMICS IN THE LEAD TITANATE ZIRCONATE MAGNESATE(0.5) TUNGSTATE(0.5). /AND 09-1603
/UM (OR LEAD OR CALCIUM) STRONTIUM TITANATE, BARIUM TITANATE ZIRCONATE (OR STANNATE), POTASSIUM TANTALATE / 08-1281
HE CUR/ FIELD ENFORCED FERROELECTRICITY IN GLASS BONDED LEAD ZIRCONATE. (PHASE TRANSITION 200 DEG C BELOW T 07-1252
PROPERTIES OF LEAD COBALTATE(0.33) NIOBATE(0.67)- TITANATE- ZIRCONATE SOLID SOLUTION CERAMICS. /ZOELECTRIC 09-1549
HYDROTHERMAL CRYSTALLIZATION OF LEAD TITANATE ZIRCONATE SOLID SOLUTIONS. 09-1482
/SICAL PHENOMENA IN PIEZOCERAMICS BASED ON LEAD TITANATE AND ZIRCONATE. (SOLID SOLUTIONS LEAD MAGNESATE(0./ 05-1063
ENCE CONDITIONS FO/ DIELECTRIC POLARIZATION OF LEAD TITANATE ZIRCONATE SOLID SOLUTIONS (THERMODYNAMIC EXIST 09-1541
D ANTIFERROELECTRIC PHASES. (LANTHANUM OR NIOBIUM DOPED LEAD ZIRCONATE STANNATE TITANATE) /FERROELECTRIC AN 09-1487
. / ANTIFERROELECTRIC- FERROELECTRIC TRANSITION IN THE LEAD (ZIRCONATE, STANNATE, TITANATE) SOLID SOLUTIONS 09-1486
(8 SUBSTITUTED LEAD TITANATE ZIRCONATES, 2 SUBSTITUTED LEAD ZIRCONATE STANNATES, COMPLEX PEROVSKITES) /CS. 08-1418
/ INTERNAL ELECTRIC FIELD GRADIENT IN ANTIFERROELECTRIC LEAD ZIRCONATE STUDIED BY PERTURBED ANGULAR CORREL/ 07-1255
TRANSITIONS IN LEAD COBALTATE(0.5) TUNGSTATE(0.5), TITANATE, ZIRCONATE SYSTEM. /AND X-RAY STUDIES OF PHASE 13-1960
/OF POLYCRYSTALLINE SOLIDS: APPLICATION TO THE LEAD TITANATE ZIRCONATE SYSTEM (MEASUREMENT OF SOFT OPTICAL/ 09-1497
/RIC AND FERROELECTRIC PROPERTIES OF THE BARIUM LEAD NIOBATE ZIRCONATE SYSTEM. (TETRAGONAL AND ORTHORHOMBI/ 13-1925
BAUER MEASUREMENTS OF BISMUTH FERRATE- BISMUTH FERRATE- LEAD ZIRCONATE SYSTEMS. MOSS 08-1282
/ OF TRANSPARENT FERROELECTRIC CERAMICS. ((LEAD, LANTHANUM) (ZIRCONATE, TANTALATE), (LEAD, TUNGSTEN) (ZIRC/ 13-2004
/RA OF CERTAIN PEROVSKITES (STRONTIUM OR CALCIUM TITANATE OR ZIRCONATE, TEST OF SHORT OR LONG RANGE ORDER / 03-0414
INING ADDITIVES ON THE PROPERTIES OF SOLID SOLUTIONS OF LEAD ZIRCONATE TITANATE. /SMUTH AND MANGANESE CONTA 09-1588
ELECTROOPTIC EFFECTS IN FERROELECTRIC CERAMICS. (LEAD ZIRCONATE TITANATE) 09-1551
TAINING OXYGEN OCTAHEDRA FERROELECTRICS: CERAMIC LEAD BARIUM ZIRCONATE TITANATE. /POLARIZATION) OF LEAD CON 13-2039
RIC CERAMICS AND CONVERSION OF MECHANOELECTRIC ENERGY. (LEAD ZIRCONATE TITANATE) FERROELECT 09-1483
FERROELECTRIC CERAMICS AND ENERGY CONVERSION. (LEAD ZIRCONATE TITANATE) 09-1508
FERROELECTRIC DISPLAYS. (BISMUTH TITANATE, LEAD ZIRCONATE TITANATE) 09-1515
FERROELECTRIC MEMORY DEVICE PARAMETER STUDY. (LEAD ZIRCONATE TITANATE) 09-1594
INT DEFECTS ON THE INTERNAL FRICTION IN POLYCRYSTALLINE LEAD ZIRCONATE TITANATE. /F THE CONCENTRATION OF PO 09-1582
INTRINSIC NONSTOICHIOMETRY IN SINGLE PHASE LEAD ZIRCONATE TITANATE. 09-1534
SUREMENT OF THE PYROELECTRIC COEFFICIENT OF 65-35 PZT. (LEAD ZIRCONATE TITANATE) MEA 09-1567
E OF AGING IN FERROELECTRIC CERAMICS. (BARIUM TITANATE, LEAD ZIRCONATE TITANATE) NATUR 04-0757
PIEZOELECTRIC TRANSDUCER MATERIAL. (ALUMINA WHISKERS IN LEAD ZIRCONATE TITANATE) NEW COMPOSITE CERAMIC 09-1556
POINT DEFECTS AND SINTERING OF LEAD ZIRCONATE TITANATE. 09-1478
ATION OF ZIRCONIA PHASE IN NIOBIUM MODIFIED CERAMICS OF LEAD ZIRCONATE TITANATE. PRECIPIT 09-1540
ION INFRARED RESPONSE USING SURFACE WAVE TRANSDUCERS ON LEAD ZIRCONATE TITANATE. /OF A 40 MHZ COLOR TELEVIS 09-1509
SINTERING SCANDIUM AND NIOBIUM MODIFIED LEAD ZIRCONATE TITANATE. 09-1555
SOLUBILITY OF ALUMINUM IN LEAD ZIRCONATE TITANATE. 09-1479
NT CERAMIC DEVELOPED BY SANDIA LABORATORIES. (LEAD LANTHANUM ZIRCONATE TITANATE) TRANSPARE 09-1476
TURES IN THE FERROELECTRIC CERAMICS BARIUM TITANATE AND LEAD ZIRCONATE TITANATE. TWO WAVE SHOCK STRUC 04-0916
SEE BOTH LEAD LANTHANUM ZIRCONATE TITANATE AND PLZT.
SEE BOTH LEAD ZIRCONATE TITANATE AND PZT.
/RROELECTRIC CERAMICS FOR ELECTROOPTICAL APPLICATIONS. (LEAD ZIRCONATE TITANATE AND SODIUM POTASSIUM NIOBA/ 09-1527
ULTRASONIC INTERFEROMETER WITH LEAD ZIRCONATE TITANATE AS TRANSDUCER. 09-1585
EFFECTS OF CALCINING ON THE FIRING OF LEAD ZIRCONATE TITANATE CERAMIC 09-1622
DC BIAS EFFECTS IN LEAD ZIRCONATE TITANATE CERAMICS. 09-1490
EFFECTS OF CALCINING ON SINTERING OF LEAD ZIRCONATE TITANATE CERAMICS. 09-1494
FECTS OF IMPURITIES ON THE MECHANICAL QUALITY FACTOR OF LEAD ZIRCONATE TITANATE CERAMICS. EF 09-1608
EFFECTS OF RADIATION INDUCED DAMAGE CENTERS IN LEAD ZIRCONATE TITANATE CERAMICS. 09-1520
SINTERING AND FERROELECTRIC PROPERTIES OF LEAD ZIRCONATE TITANATE CERAMICS. 09-1481
CTS AT THE PHASE TRANSITION TETRAGONAL- RHOMBOHEDRAL IN LEAD ZIRCONATE TITANATE CERAMICS. /ABILIZATION EFFE 09-1611
TION OF IMPURITIES. SPACE CHARGE EFFECT IN LEAD ZIRCONATE TITANATE CERAMICS CAUSED BY THE ADDI 09-1609
. HOT PRESSED LEAD ZIRCONATE TITANATE CERAMICS CONTAINING BISMUTH 09-1525
IES. ELECTRICAL RESISTIVITY OF LEAD ZIRCONATE TITANATE CERAMICS CONTAINING IMPURIT 09-1607
APPLICATIONS. HOT PRESSED FERROELECTRIC LEAD ZIRCONATE TITANATE CERAMICS FOR ELECTROOPTICAL 09-1524
LARGE TENSILE STRAI/ FERROELASTIC BEHAVIOR OF LEAD LANTHANUM ZIRCONATE TITANATE CERAMICS WHEN SUBJECTED TO 09-1571
/ENT DEVELOPMENTS IN ELECTROOPTIC CERAMICS (HOT PRESSED LEAD ZIRCONATE TITANATE, EFFECT OF GRAIN SIZE, DOP/ 02-0086

PHYSICAL AND ELECTRICAL SWITCHING CHARACTERISTICS OF A LEAD ZIRCONATE TITANATE FERROELECTRIC CERAMIC. SOME 09-1601
R ELECTROOPTIC APPLICATIONS. HOT PRESSED (LEAD, LANTHANUM) (ZIRCONATE, TITANATE) FERROELECTRIC CERAMICS FO 09-1526
VIEW, 31 REFERENCES)   AGING OF BARIUM TITANATE AND LEAD ZIRCONATE TITANATE FERROELECTRIC CERAMICS. (RE 02-0062
ON CELL MEASUREMENTS OF IONIC TRANSPORT NUMBERS IN PZT (LEAD ZIRCONATE TITANATE) FERROELECTRICS. /NCENTRATI 09-1512
ITHOUT / INTERNAL FRICTION (AND CURIE TEMPERATURES) IN (LEAD ZIRCONATE TITANATE) FERROELECTRICS (WITH AND W 09-1584
/RICS ON THEIR MECHANICAL PROPERTIES. (BARIUM TITANATE, LEAD ZIRCONATE TITANATE, LEAD METANIOBATE, 20-150 / 04-0985
ION AND COERCIVE FORCE OF THE PEROVSKITES IN THE SYSTEM LEAD ZIRCONATE TITANATE METANIOBATE. /OUS POLARIZAT 09-1597
ASSIUM ANTIMONATE, NIOBATE, / PIEZOELECTRIC CERAMICS OF LEAD ZIRCONATE TITANATE MODIFIED WITH SODIUM OR POT 09-1539
TOXIDE) SURFACE / PROPERTIES OF OXYGEN DEPLETED CERAMIC LEAD ZIRCONATE TITANATE (PLUS 1 PERCENT NIOBIUM PEN 09-1576
                                          HIGH D*, FAST, LEAD ZIRCONATE TITANATE PYROELECTRIC DETECTORS.   09-1560
PRESSURE ENFORC/ RELEASE OF ELECTRIC ENERGY IN LEAD NIOBIUM ZIRCONATE TITANATE STANNATE BY TEMPERATURE AND 09-1488
TH NIOBIUM AND TITANIUM)   MOSSBAUER STUDIES OF LEAD ZIRCONATE TITANATE STANNATE. (CERAMIC DOPED WI 09-1604
           CHARGE MECHANISM IN FERROELECTRICS. (LEAD ZIRCONATE TITANATE, STOICHIOMETRY EFFECTS)   03-0259
              HIGH TEMPERATURE PHASE EQUILIBRIA IN THE LEAD ZIRCONATE TITANATE SYSTEM.   09-1577
OVED HOT PRESSED ELECTROOPTIC CERAMICS IN THE LEAD LANTHANUM ZIRCONATE TITANATE SYSTEM.   IMPR 09-1528
            INTRINSIC NONSTOICHIOMETRY IN THE LEAD ZIRCONATE TITANATE SYSTEM.   09-1535
                STUDY OF PHASE TRANSITIONS IN THE LEAD ZIRCONATE TITANATE SYSTEM.   09-1620
     INTRINSIC AND EXTRINSIC NONSTOICHIOMETRY IN THE LEAD ZIRCONATE TITANATE SYSTEM. (DEFECT STRUCTURE) 09-1536
ERIALS FOR INFORMATION STORAGE AND DISPLAY (HOT PRESSED LEAD ZIRCONATE TITANATE, TUTORIAL, 17 REFS). /W MAT 09-1550
AINING GERMANIUM.   PIEZOCERAMICS BASED ON LEAD ZIRCONATE TITANATE WITH COMPLEX ADDITIVES CONT 09-1510
ONTIUM ADDITIVES. CHROMATOGRAPHIC METHOD FOR ANALYZING LEAD ZIRCONATE TITANATES CONTAINING BISMUTH AND STR 09-1625
DOMAIN WALL MOBILITY AND MECHANICAL DAMPING IN LEAD TITANATE ZIRCONATE TRANSDUCER CERAMICS. /SION STUDY OF 09-1547
/ROELECTRICS OF VARIOUS TYPES. (BARIUM TITANATE AND TITANATE ZIRCONATE, TRIGLYCINE SULFATE AND ROCHELLE SA/ 02-0108
K.   COMPLEX DIELECTRIC CONSTANT OF LEAD ZIRCONATE UNDER POLARIZING FIELD, 370-520 DEG 07-1256
/ECTRIC PROPERTIES OF LEAD (STANNATE(0.5) NIOBATE(0.5)) (1-X) ZIRCONATE WITH THOROUGHNESS OF POLING. /ATION 09-1489
ZATION, ELECTROSTRICTION, AND PIEZOELECTRIC ACTIVITY OF LEAD ZIRCONATE(X) PEROVSKITE TYPE SOLID SOLUTIONS. 09-1548
/IN OF THE MAXIMUM IN THE ELECTROMECHANICAL ACTIVITY IN LEAD ZIRCONATE(X) TITANATE(1-X) CERAMICS.   POLARI 09-1501
/IN OF THE MAXIMUM IN THE ELECTROMECHANICAL ACTIVITY IN LEAD ZIRCONATE(X) TITANATE(1-X) CERAMICS NEAR THE / 09-1499
L PHA/ ATOMIC STRUCTURES OF THE FERROELECTRIC COMPOUNDS LEAD ZIRCONATE(X) TITANATE(1-X) CERAMICS NEAR THE / 09-1500
CTERISTICS OF SURFACE ELASTIC WAVES ON BARIUM TITANATE(0.95) ZIRCONATE(X) TITANATE(1-X) FOR TWO RHOMBOHEDRA 09-1574
KINETICS OF LUTETIUM AND SCANDIUM IONS IN LEAD TITANATE(0.5) ZIRCONATE(0.05) CERAMIC PLATES. /AGATION CHARA 09-1613
/ DETERIORATION OF THE ELECTROMECHANICAL PROPERTIES OF LEAD ZIRCONATE(0.5).   SOLUTION 09-1538
STRUCTURA/ SUBSTITUTION OF BISMUTH AND NIOBIUM IONS IN LEAD ZIRCONATE(0.52) TITANATE(0.48) CERAMICS CAUSED 09-1606
TIN OXIDE FILMS ON LEAD(0.97) BISMUTH(0.02) TITANATE (0.35) ZIRCONATE(0.53) TITANATE(0.47) (EXPERIMENT AND 09-1480
LECTRIC CERAMIC (THEORY, APPLIED TO LEAD(0.99) NIOBIUM(0.02) ZIRCONATE(0.65) TITANATE(0.35). / IN A FERROE 09-1506
CT STUDIES AT LOW TEMPERATURES (WITH TIN OXIDE FILMS ON LEAD ZIRCONATE(0.65) TITANATE(0.35) CERAMIC). /EFFE 09-1537
RROELECTRIC-1, FERROELECTRIC-2 TRANSITION IN (NIOBIUM DOPED) ZIRCONATE(0.95) TITANATE(0.05). /ELOCITY AT FE 09-1505
NGSTEN TRIOXIDE.   PHASE DIAGRAM OF LEAD ZIRCONATE(1-X) TITANATE(X) PLUS 0.8 PERCENT TU 09-1581
X) TITANATE, BISMUTH(X) (POTASSIUM OR SODIUM)(X) LEAD (1-2X) ZIRCONATES). /TH(0.5) POTASSIUM(X) SODIUM(0.5- 08-1289
TALLINE DIELECTRICS WITH PEROVSKITE STRUCTURE (TITANATES AND ZIRCONATES). FORBIDDEN ZONE WIDTHS IN POLYCRYS 08-1288
            INFRARED AND RAMAN SPECTRA OF SEVERAL PEROVSKITE ZIRCONATES.   08-1411
/NICAL PROPERTIES OF CERAMIC FERROELECTRICS (BARIUM TITANATE ZIRCONATES) IN STRONG ELECTRIC FIELDS AND AT / 09-1605
     ANTIFERROELECTRIC PHASE IN THE SYSTEM LEAD TITANATE- LEAD ZIRCONATE- LANTHANUM FERRATE.   13-1961
              TERNARY SYSTEM LEAD TITANATE- LEAD ZIRCONATE- LEAD NIOBATE(0.67) ZINCATE(0.33).   09-1507
TEN) (ZIRCONATE TANTALATE), LEAD (LANTHANATE, NIOBATE)- LEAD ZIRCONATE- LEAD TITANATE) /LATE), (LEAD, TUNGS 13-2004
SPLAY DEVICES USING FINE GRAIN FERROELECTRIC CERAMICS. (LEAD ZIRCONATE- LEAD TITANATE) IMAGE STORAGE AND DI 09-1572
FERROELECTRIC CERAMICS IN COMPUTERS AND OTHER DEVICES. (LEAD ZIRCONATE- LEAD TITANATE AND PLZT)   USE OF 09-1532
A FERROELECTRIC PIEZOELECTRIC RANDOM ACCESS MEMORY. (LEAD ZIRCONATE- LEAD TITANATE CERAMIC MEMORY PLATE) 09-1504
              X-RAY ANALYSIS OF MODIFIED LEAD ZIRCONATE- LEAD TITANATE CERAMICS.   09-1619
CTRIC APPL/ EFFECTS OF HIGH ELECTRIC FIELDS ON MODIFIED LEAD ZIRCONATE- LEAD TITANATE CERAMICS FOR PIEZOELE 09-1485
Y COMPOSITIONS)   INTERNAL FRICTION OF MODIFIED LEAD ZIRCONATE- LEAD TITANATE CERAMICS. (PROPRIETAR 09-1623
ELECTRODES.   IMPROVED AGING AND SWITCHING OF LEAD ZIRCONATE- LEAD TITANATE CERAMICS WITH INDIUM 09-1516
ELECTRODES.   IMPROVED AGING AND SWITCHING OF LEAD ZIRCONATE- LEAD TITANATE CERAMICS WITH INDIUM 09-1517
T PRESSED COMPOSITIONS AT NEAR THEORET/ VAPORIZATION OF LEAD ZIRCONATE- LEAD TITANATE MATERIALS. PART-2: HO 09-1578
            STUDY OF PHASE TRANSITIONS IN THE LEAD ZIRCONATE- LEAD TITANATE SYSTEM.   09-1621
PROPERTIES OF NEW TRANSPARENT FERROELECTRIC CERAMICS. (LEAD ZIRCONATE- LEAD TITANATE TYPE)   09-1518
FFECT STUDIES OF FERROELECTRIC PHASE TRANSITIONS IN THE LEAD ZIRCONATE- LEAD TITANATE- BISMUTH FERRATE. / E 09-1498
/EMPERATURE DEPENDENT FERROELECTRIC DOMAIN ALIGNMENT IN LEAD ZIRCONATE- LEAD TITANATE- LANTHANUM OXIDE (PL/ 09-1586
NARY SYSTEM. CURIE TEMPERATURES OF SOLID SOLUTIONS IN A LEAD ZIRCONATE- LEAD TITANATE- LEAD METANIOBATE TER 09-1595
/PROPERTIES OF THE SOLID SOLUTION SYSTEM LEAD TITANATE- LEAD ZIRCONATE- LEAD (ZINCATE(0.33) NIOBATE(0.67)). 13-1591
/CTRIC AND PIEZOELECTRIC PROPERTIES IN A LEAD TITANATE- LEAD ZIRCONATE- LEAD (ZINCATE(0.33) NIOBATE(0.67)/ 09-1617
              SINGLE CRYSTAL TITANATES AND ZIRCONATES. (METHODS OF GROWTH)   02-0045
RIE POINT IN VARIOUS FERROELECTRICS AND ANTIFERROELECTRICS. (ZIRCONATES, NIOBATES, TUNGSTATES) /BOVE THE CU 02-0124
/OLYCRYSTALLINE FERROELECTRICS. (8 SUBSTITUTED LEAD TITANATE ZIRCONATES, 2 SUBSTITUTED LEAD ZIRCONATE STAN/ 08-1418
LEAD ZIRCONATE TITANATE.   PRECIPITATION OF ZIRCONIA PHASE IN NIOBIUM MODIFIED CERAMICS OF 09-1540
/S. (SUBSTITUTION OF STRONTIUM OR CALCIUM FOR LEAD, IRON FOR ZIRCONIUM AND TITANIUM, EFFECT ON ACOUSTIC RA/ 09-1484
/SKITES. THALLIUM A(1-X) TUNGSTATE OR MOLYBDATE. (WHERE A IS ZIRCONIUM, HAFNIUM, CERIUM, THORIUM, TITANIUM, 10-1640
1-X) B'(X) OXYGEN(3), WHERE B IS IRON OR SCANDIUM, AND B' IS ZIRCONIUM OR HAFNIUM. / (B(0.5) NIOBIUM(0.5)) (13-2044
          RESISTIVITY BEHAVIOR OF SEMICONDUCTING YTTRIUM BARIUM ZIRCONIUM TITANATE. (TEMPERATURE DEPENDENCE) 13-1995
ELASTIC AND PIEZOELECTRIC PROPERTIES OF PLZT CERAMICS WITH A ZIRCONIUM TO TITANIUM RATIO OF 65 TO 35. / OF 09-1573
X) TUNGSTEN(1-X) Z(Y) OXIDE(3) WHERE Z IS NIOBIUM, TANTALUM, ZIRCONIUM, VANADIUM, OR COBALT) /ZES. (SODIUM(13-1944
/) BISMUTH(0.07) LANTHANUM(0.01) (IRON(0.405) NIOBIUM(0.325) ZIRCONIUM(0.27). OXYGEN(3) FERROELECTRIC MEMO/ 37-3629
DE FLUX GROWTH IN PRESENCE / CHANGE IN THE DECAY CONSTANT OF ZIRCONIUM-89 IN BARIUM TITANATE (BARIUM CHLORI 04-0685
SULTS IN TRANSPARENT CERAMIC FERROELECTRICS. (LEAD LANTHANUM ZIRCONUM TITANATE) /CRIPTION OF HOLOGRAPHIC RE 09-1575
IFLUORIDE.   DISPERSION AND DAMPING OF SOFT ZONE BONDARY PHONONS IN POTASSIUM MANGANESE TR 33-3374
CS INDICATED BY ANISOTROPIC DEBYE- WALLER FAC/ LOW FREQUENCY ZONE BOUNDARY MODES IN PEROVSKITE FERROELECTRI 08-1351
MICROSCOPIC CALCULATION OF THE RESPONSE FUNCTION OF THE SOFT ZONE BOUNDARY PHONON IN STRONTIUM TITANATE. 06-1213
                                           ZONE BOUNDARY SOFT MODE IN TERBIUM MOLYBDATE. 30-3269
IUM TITANATE, LEAD(X) STRONTIUM(1-X) TITANATE, LEAD/ SOLVENT ZONE GROWTH OF FERROELECTRIC CRYSTALS. (STRONT 05-1042
TITANATE)   A FLOATING ZONE SINGLE CRYSTAL GROWING APPARATUS. (BARIUM 04-0998
AL STRONTIUM(2) NIOBIUM(2) OXIDE(7) BY MEANS OF THE FLOATING ZONE TECHNIQUE. /OF FERROELECTRIC SINGLE CRYST 13-2037
H PEROVSKITE STRUCTURE (TITANATES AND ZIRCONATES). FORBIDDEN ZONE WIDTHS IN POLYCRYSTALLINE DIELECTRICS WIT 08-1288

NS IN LITHIUM TANTALAT/ INFRARED (REFLECTIVITY) STUDY (20-10,000 PER CM, 300 DEG K) OF THE LATTICE VIBRATIO 12-1870
/GITUDINAL HIGH FREQUENCY (950 MHZ) ACOUSTIC WAVE ALONG THE (001) DIRECTION OF LITHIUM NIOBATE (LIQUID HEL/ 11-1835
                        LEED STUDY OF SURFACE STRUCTURES ON THE (001) FACE OF BARIUM TITANATE.   04-0529
/OF THE TEMPERATURE DEPENDENCE OF LEED INTENSITIES FROM THE (001) SURFACE OF BARIUM TITANATE BETWEEN 20 DE/ 04-0531
 PURE DIRECT SKY BLUE DYE.   GROWTH RATE OF THE (010) PLANE OF ROCHELLE SALT IN THE PRESENCE OF 28-3217
ES OF ANTIMONY SULFOIODIDE AT MICROWAVE FREQUENCIES (3.3 AND 9.8 GHZ, 0-50 C).   DIELECTRIC PROPERTI 15-2171

AUTHOR-TITLE INDEX

	04-0717	ATTEMPT TO PREPARE HIGH DOPED BARIUM STRONTIUM TITANATE SEMICONDUCTOR WITH PZT.
ABDULLAEVA KHM	16-2318	AN X-RAY STUDY OF THE PHASE TRANSITION IN SODIUM NITRITE.
ABDULSABIROV RYU	28-3176	ELECTRON PARAMAGNETIC RESONANCE OF COPPER IONS IN ROCHELLE SALT.
ABE R	02-0035	MEASUREMENT OF DIELECTRIC CONSTANTS FOR A FEW FERROELECTRIC MATERIALS AT LOW TEMPERATU
	16-2253	A RADICAL IN FERROELECTRIC DGN CRYSTAL.
	17-2406	THE RADICAL IN GAMMA-IRRADIATED KDP.
	17-2683	ESR STUDY OF ADP- KDA MIXED CRYSTALS.
	17-2684	ESR STUDY OF AMMONIUM IONS IN IRRADIATED AMMONIUM DIDEUTERIUM PHOSPHATE- 5 PERCENT POT
	17-2691	EFFECT OF GAMMA-IRRADIATION ON THE DIELECTRIC CONSTANTS OF KDP.
	19-2790	ESR OF GAMMA-IRRADIATED AMMONIUM SULFATE.
	27-3033	DISPLACIVE MOVEMENT OF THE RADICAL IN TRIGLYCINE SULFATE NEAR THE CURIE POINT. (ESR OF
	27-3034	ESR OF GAMMA-IRRADIATED TGS. PART-2: INTERNAL MOTION OF HYDROGEN AND THE ACTIVATION EN
	27-3035	ESR IN GAMMA-IRRADIATED TGS.
	27-3070	NONLINEARITY OF THE DIELECTRIC CONSTANT OF FERROELECTRIC TRIGLYCINE SULFATE.
	27-3139	ELECTRON SPIN RESONANCE OF GAMMA-IRRADIATED TRIGLYCINE SELENATE.
	28-3177	ANOMALOUS DISPERSION OF COERSIVE FIELD AT VERY LOW FREQUENCIES AND JERKY WALL MOTION I
	28-3178	COERCIVE FIELDS OF ROCHELLE SALT HEAVILY IRRADIATED WITH GAMMA-RAYS.
	28-3179	DIELECTRIC CONSTANT AND DIELECTRIC RELAXATION TIME OF ROCHELLE SALT.
	28-3180	ELECTRON PARAMAGNETIC RESONANCE OF GAMMA-IRRADIATED ROCHELLE SALT.
	28-3202	MEASUREMENT OF THERMAL EXPANSION OF ROCHELLE SALT BY USING THE SHIFT OF RESONANCE FREQ
	28-3203	NUCLEAR MAGNETIC RESONANCE IN ROCHELLE SALT, DETERMINATION OF THE ORIENTATIONS OF PROT
	29-3252	ESR OF COPPER(2+) DOPED SODIUM AMMONIUM TARTRATE.
ABE S	16-2254	NUCLEAR QUADRUPOLE RELAXATION OF NITROGEN-14 IN SODIUM NITRITE.
ABE Y	03-0308	DISTRIBUTION OF ZEROES OF THE PARTITION FUNCTION FOR THE SLATER MODEL OF FERROELECTRIC
	16-2254	NUCLEAR QUADRUPOLE RELAXATION OF NITROGEN-14 IN SODIUM NITRITE.
ABE Z	27-2951	A NEW SECOND HARMONIC TYPE FERROELECTRIC MODULATOR FOR ELECTROMETER. (WITH A TGS SINGL
ABEL WR	08-1269	EFFECT OF PRESSURE ON THE STATIC DIELECTRIC CONSTANT OF POTASSIUM TANTALATE.
ABELL JS	13-1894	DILATOMETRIC STUDY OF THE ORTHORHOMBIC- TETRAGONAL PHASE TRANSITION IN BARIUM SODIUM N
ABERDAM D	04-0529	LEED STUDY OF SURFACE STRUCTURES ON THE (001) FACE OF BARIUM TITANATE.
	04-0530	LEED STUDIES ON THE (001) FACE OF BARIUM TITANATE CRYSTALS.
	04-0531	STUDY OF THE TEMPERATURE DEPENDENCE OF LEED INTENSITIES FROM THE (001) SURFACE OF BARI
	04-0532	TEMPERATURE FACTORS OF TETRAGONAL AND CUBIC BARIUM TITANATE.
ABOLINSH Y	28-3181	REARRANGEMENT OF THE DOMAIN STRUCTURE IN ROCHELLE SALT NEAR THE PHASE TRANSITION.
ABRAHAM DB	22-2822	ANOMALOUS SPECIFIC HEAT OF SODIUM TRIHYDROGEN SELENITE ASSOCIATED COMBINATORIAL PROBLE
	22-2823	ANOMALOUS SPECIFIC HEAT OF SODIUM TRIHYDROGEN SELENITE.
	34-3427	ANALYTIC PROPERTIES OF THE FREE ENERGY FOR THE "ICE" MODELS (FERROELECTRICS).
ABRAHAMS SC	02-0036	COUPLED FERROELASTIC CRYSTALS. (GADOLINIUM MOLYBDATE, KDP, ORTHOFERRITES, DYSPROSIUM A
	02-0037	FERROELASTICITY.
	03-0135	STRUCTURAL BASIS OF FERROELECTRICITY AND FERROELASTICITY. (SYMMETRY RELATIONS)
	03-0316	DIELECTRIC, ELASTIC AND MAGNETIC PROPERTY, SYMMETRY, AND PHASE TRANSITION REPRESENTATI
	08-1270	FERROELASTIC EFFECT IN LANTHANUM ORTHOFERRITE.
	11-1673	X-RAY DETERMINATION OF POLARITY SENSE BY ANOMALOUS SCATTERING AT AN ABSORPTION EDGE. (
	13-1895	FERROELECTRIC TUNGSTEN BRONZE TYPE CRYSTAL STRUCTURES. PART-3: POTASSIUM LITHIUM NIOBA
	14-2102	FERROELECTRIC AND FERROELASTIC PROPERTIES OF MAGNESIUM CHLORINE BORACITE.
	30-3289	FERROELECTRIC FERROELASTIC PARAMAGNETIC BETA GADOLINIUM MOLYBDATE. CRYSTAL STRUCTURE O
	30-3290	(ROOM TEMPERATURE) FERROELECTRIC FERROELASTIC PARAMAGNETIC (BETA) TERBIUM MOLYBDATE (C
	33-3358	FERROELECTRIC EFFECT IN RUBIDIUM IRON TETRAFLUORIDE AND CESIUM IRON TETRAFLUORIDE.
	33-3386	FERROELECTRIC PARAELASTIC PARAMAGNETIC BARIUM COBALT TETRAFLUORIDE CRYSTAL STRUCTURE (
	35-3493	CRYSTAL STRUCTURE OF PIEZOELECTRIC BISMUTH(12) GERMANIUM OXIDE(20).
	36-3574	PYROELECTRIC AMMONIUM IODATE, A POTENTIAL FERROELASTIC: CRYSTAL STRUCTURE.
ABRAMOVICH TE	11-1762	TRANSVERSE POLARITONS IN A LITHIUM NIOBATE CRYSTAL. (RAMAN SPECTRUM, PERMITTIVITY)
ABRAMS RL	09-1477	PHOTOMIXING AT 10.6 MICRONS WITH STRONTIUM BARIUM NIOBATE PYROELECTRIC DETECTORS.
	12-1875	STUDY OF PIEZOELECTRIC OSCILLATIONS IN WIDEBAND PYROELECTRIC LITHIUM TANTALATE.
ACKERMANN RJ	10-1626	THERMAL EXPANSION AND THE ORTHORHOMBIC- TETRAGONAL TRANSITION OF THE TUNGSTEN TRIOXIDE
ADACHI M	08-1271	PIEZOELECTRIC PROPERTIES OF POTASSIUM TANTALATE NIOBATE SINGLE CRYSTAL.
ADAMS LH	16-2279	TRANSITION RATES OF POTASSIUM NITRATE HIGH PRESSURE POLYMORPHS.
ADLER R	09-1509	REALIZATION OF A 40 MHZ COLOR TELEVISION INFRARED RESPONSE USING SURFACE WAVE TRANSDUC
ADRIAENSSENS GJ	17-2404	NMR PROTON SECOND MOMENTS IN POTASSIUM DIHYDROGEN PHOSPHATE TYPE CRYSTALS.
	18-2718	NMR AND DIELECTRIC RELAXATION STUDIES OF LONG RANGE CORRELATION IN FERROELECTRIC POTAS
ADRIANOVA II	17-2407	DIFFRACTION OF POLARIZED LIGHT BY ULTRASONIC WAVES IN POTASSIUM DIHYDROGEN PHOSPHATE C
	36-3521	ANISOTROPY OF THE ELECTROOPTICAL EFFECT IN CRYSTALS OF LEAD MAGNONIOBATE.
AEGERTER S	06-1088	ELECTRIC PROPERTIES OF HYDROGEN DOPED STRONTIUM TITANATE.
AGAEV FA	13-1923	FORMATION CONDITIONS AND PHASE TRANSITIONS IN CRYSTALS WITH POTASSIUM TUNGSTEN BRONZE
	13-2050	PHASE TRANSITIONS IN TANTALATES WITH THE STRUCTURE OF POTASSIUM TUNGSTEN BRONZE. (POTA
	36-3569	X-RAY AND ELECTRIC INVESTIGATIONS OF THE SYSTEMS YTTRIUM MANGANESE(1-X) (B(X)) OXYGEN(
AGARONOV BS	15-2104	INFLUENCE OF NONEQUILIBRIUM CARRIERS ON THE DOMAIN STRUCTURE OF THE FERROELECTRIC SEMI
	15-2105	SHORT CIRCUIT PHOTOEMF IN FERROELECTRIC SEMICONDUCTORS. (ANTIMONY SULFOIODIDE CRYSTALS
AGATHA H	25-2920	MOLECULAR AND LATTICE VIBRATIONS OF ORTHORHOMBIC AND FERROELECTRIC THIOUREA. PART-1: T
AGRANOVICH VM	04-0533	(FAST ION SCATTERING) SHADOW EFFECT IN (DISPLACIVE) PHASE TRANSITIONS. (BARIUM TITANAT
AGRAWAL DK	15-2106	LONG WAVELENGTH OPTICAL PHONONS AND PHASE TRANSITIONS IN ANTIMONY SULFOIODIDE.
	15-2207	(POLARIZED) RAMAN SPECTRUM OF (PARAELECTRIC AND) FERROELECTRIC ANTIMONY SULFOIODIDE (P
	15-2208	OPTICAL PHONONS AND PHASE TRANSITIONS IN SOME ORDER DISORDER AND DISPLACIVE FERROELECT
	17-2409	TEMPERATURE DEPENDENT RAMAN SPECTRA OF POTASSIUM DIHYDROGEN PHOSPHATE, POTASSIUM DIDEU
AGRAWAL MD	08-1272	DIELECTRIC PROPERTIES AND OPTICAL ABSORPTION OF POTASSIUM TANTALATE SINGLE CRYSTALS AN
AHTEE M	08-1273	STRUCTURES OF SODIUM NIOBATE BETWEEN 480 AND 575 DEG C, AND THEIR RELEVANCE TO SOFT PH

AHTEE M	08-1336	HIGH TEMPERATURE PHASE TRANSITIONS IN SODIUM NIOBATE AND THE USE OF TILTING SCHEMES IN
AIKI K	22-2824	DIELECTRIC AND THERMAL STUDY OF POTASSIUM SELENATE TRANSITIONS (FERROELECTRIC BELOW 93
	22-2825	ESR STUDY OF GAMMA-IRRADIATED POTASSIUM SELENATE (STRUCTURAL PHASE CHANGES).
AIKINS J	18-2718	NMR AND DIELECTRIC RELAXATION STUDIES OF LONG RANGE CORRELATION IN FERROELECTRIC POTAS
AINGER FW	02-0038	FERROELECTRIC CERAMIC OXIDES.
	11-1660	FERROELECTRICS IN THE LITHIUM POTASSIUM NIOBATE SYSTEM.
	13-1896	FERROELECTRICS IN THE POTASSIUM OXIDE- STRONTIUM OXIDE- NIOBIUM PENTOXIDE.
	13-1897	THE SEARCH FOR NEW FERROELECTRICS WITH THE TUNGSTEN BRONZE STRUCTURE.
	13-1926	TRAPPING SITES, OPTICAL DAMAGE, AND THERMALLY STIMULATED CURRENT IN FERROELECTRIC OXID
	33-3359	PREPARATION AND PROPERTIES OF CESIUM LEAD HALIDES. (FERROELECTRICITY INDETERMINITE)
AITKIN RB	09-1480	SUBSTITUTION OF BISMUTH AND NIOBIUM IONS IN LEAD ZIRCONATE(0.53) TITANATE(0.47) (EXPER
AIZU K	03-0136	CONSIDERATIONS OF PARTIALLY FERROELASTIC AND PARTIALLY ANTIFERROELASTIC CRYSTALS AND P
	03-0137	DETERMINATION OF THE STATE PARAMETERS AND FORMULATION OF SPONTANEOUS STRAIN FOR FERROE
	03-0138	DETERMINATION OF KIND (TYPE, CLASS) FOR ACTUAL FERROELECTRICS TOGETHER WITH EXAMINATIO
	03-0139	FERROELECTRIC TRANSFORMATIONS OF TENSORIAL PROPERTIES IN REGULAR FERROELECTRICS.
	03-0140	GENERAL CONSIDERATION OF FERROELECTRICS AND FERROELASTICS SUCH THAT THE ELECTRIC SUSCE
	03-0141	MECHANISM OF THE FERROELECTRIC PHASE TRANSFORMATION OF DIAMMONIUM DICADMIUM TRISULFATE
	03-0142	POSSIBLE SPECIES OF FERROMAGNETIC, AND FERROELASTIC CRYSTALS (PREVIOUS THEORY EXTENDED
	03-0143	POLARIZATION, PYROELECTRICTY, AND FERROELECTRICITY OF IONIC CRYSTALS. (THEORY, NO REFS
	03-0144	PHENOMENOLOGICAL LATTICE DYNAMICAL THEORY OF FERROELASTICITY.
	03-0145	PRESENTATION AND DISCUSSION OF EXAMPLES OF FERROELECTRICS AND FERROELASTICS HAVING THE
	03-0146	SUPPLEMENT TO A PREVIOUS THEORY OF WEAK FERROELASTICS AND WEAK FERROELECTRICS.
	19-2741	CONJECTURED MECHANISM OF THE PHASE TRANSFORMATION OF A PECULIAR FERROELECTRIC- FERROEL
	29-3238	INFERRED MECHANISM OF THE PHASE TRANSFORMATION OF PECULIAR FERROELECTRIC- FERROELASTIC
	30-3253	INFERRED TEMPERATURE DEPENDENCES OF ELECTRICAL, MECHANICAL AND OPTICAL PROPERTIES OF F
AKHIEZER IA	03-0147	BEHAVIOR OF MAGNETICALLY ORDERED FERROELECTRICS IN A RAPIDLY OSCILLATING ELECTRIC FIEL
	03-0148	COUPLED ELECTROMAGNETIC SPIN WAVES IN MAGNETICALLY ORDERED FERROELECTRICS.
	03-0149	INSTABILITY OF SPIN WAVES IN FERROELECTRIC ANTIFERROMAGNETS IN EXTERNAL FIELDS. (THEOR
	03-0150	POSSIBILITY OF EXCITATION OF SPIN WAVES IN MAGNETICALLY ORDERED FERROELECTRICS. (THEOR
AKHMETOVA BG	04-0534	STUDYING SMALL DEFORMATIONS OF A CRYSTAL LATTICE BASED ON THE SHADOW EFFECT. (NEUTRON
AL ALI NS	27-3062	FREQUENCY DEPENDENCE OF THE COERCIVE FIELD OF TRIGLYCINE SULFATE CRYSTALS.
ALBERS J	03-0151	A SIMPLE MECHANICAL MODEL WITH FERROELECTRIC BEHAVIOR.
	14-2100	PIEZOELECTRIC PROPERTIES OF COBALT IODINE BORACITE.
	14-2100	PIEZOELECTRIC PROPERTIES OF COBALT IODINE BORACITE.
	27-2952	(POLARIZATION CHANGE) AFTER EFFECTS IN TRIGLYCINE SULFATE. (DIELECTRIC DISPLACEMENT AN
	27-2953	TIME DEPENDENCE OF MATERIAL CONSTANTS OF TGS AFTER POLARIZATION REVERSAL.
	27-3016	THERMAL CONDUCTIVITY OF TRIGLYCINE SULFATE NEAR THE CURIE POINT.
	28-3226	THERMAL CONDUCTIVITY OF ROCHELLE SALT.
ALEAKSEYEVA VG	15-2107	ELECTRICAL AND PHOTOELECTRIC PROPERTIES OF ANTIMONY SULFOIODIDE.
ALEKSANDROV AYU	15-2108	NUCLEAR GAMMA RESONANCE SPECTROSCOPIC STUDY OF SEMICONDUCTING COMPOUNDS OF THE TYPE A(
ALEKSANDROV KS	08-1274	DISPLACEMENT PHASE TRANSITIONS IN PEROVSKITES.
	16-2255	NUCLEAR MAGNETIC RESONANCE STUDIES OF THE PHASE TRANSITIONS IN FERROELECTRICS. (POTASS
	17-2410	ELASTIC PROPERTIES OF AMMONIUM DIHYDROGEN PHOSPHATE AND THE LAVAL RAMAN ELASTICITY THE
	19-2742	INVESTIGATION OF FERROELECTRICITY IN SODIUM AMMONIUM SELENATE AND SODIUM AMMONIUM SULF
	19-2743	INVESTIGATION OF FERROELECTRICS WITH GENERAL FORMULA METAL-1 METAL-2 SULFATE OR SELENA
	22-2826	FERROELECTRICS WITH AMMONIUM GROUP ORIENTATION MOBILITY: METAL AMMONIUM (BX4) TYPE COM
	33-3360	ELASTIC PROPERTIES OF CESIUM LEAD TRICHLORIDE.
	33-3402	SUPERSTRUCTURE ARISING DURING THE PHASE TRANSITION IN POTASSIUM MANGANESE TRIFLUORIDE.
ALEKSANDROVA IA	09-1591	EFFECT OF FERROELECTRIC PHASE TRANSITION CONDITIONS ON DIELECTRIC AND PIEZOELECTRIC PR
ALEKSANDROVA IP	19-2742	INVESTIGATION OF FERROELECTRICITY IN SODIUM AMMONIUM SELENATE AND SODIUM AMMONIUM SULF
	19-2744	NUCLEAR MAGNETIC RESONANCE OF DEUTERIUM IN THE PERDEUTERO AMMONIUM FLUOBERYLLATE CRYST
	22-2826	FERROELECTRICS WITH AMMONIUM GROUP ORIENTATION MOBILITY: METAL AMMONIUM (BX4) TYPE COM
	22-2827	MECHANISMS OF FERROELECTRIC PHASE TRANSITION IN CRYSTALS WITH REORIENTATING STRUCTURAL
	22-2828	NUCLEAR MAGNETIC RESONANCE OF DEUTERIUM IN THE FERROELECTRIC SODIUM DEUTERO AMMONIUM S
	22-2883	NUCLEAR MAGNETIC RESONANCE OF SODIUM-23 IN THE SODIUM AMMONIUM SELENATE DIHYDRATE CRYS
ALEKSANDROVA MV	17-2413	PRODUCTION OF LARGE SEED CRYSTALS OF KDP.
ALEMANY C	03-0152	THERMODYNAMIC THEORY OF IRRADIATED TRIGLYCINE SULFATE.
	22-2857	THERMAL EXPANSION OF HYDROGEN BONDED FERROELECTRICS. LITHIUM TRIHYDROGEN SELENITE.
	27-3073	DAMAGE PRODUCED BY X-RAYS ON TRIGLYCINE SULFATE.
	27-3074	X-RAY DAMAGE IN TRIGLYCINE SULFATE.
ALESHECHKIN VN	06-1089	MEASUREMENT OF DIELECTRIC PROPERTIES OF FERROELECTRICS AT SUBMILLIMETER WAVELENGTHS. (
ALEXANDROPOULOS NG	04-0535	CRITICAL X-RAY INCOHERENT SCATTERING BY BARIUM TITANATE NEAR THE CURIE TEMPERATURE.
ALEXANDROVA LM	08-1461	ELECTRICAL PROPERTIES OF NONLINEAR FERROELECTRIC CERAMICS WITH THE PEROVSKITE STRUCTUR
ALEXANDRU HV	17-2411	GROWTH OF AMMONIUM DIHYDROGEN PHOSPHATE SINGLE CRYSTAL FROM SOLUTIONS.
	17-2412	KINETICS OF GROWTH AND DISSOLUTION OF AMMONIUM DIHYDROGEN PHOSPHATE CRYSTALS IN SOLUTI
	17-2516	GROWING KINETIC OF THE AMMONIUM DIHYDROGEN PHOSPHATE SINGLE CRYSTAL (FROM SOLUTION) IN
ALIKHANOV RA	27-3054	INVESTIGATION OF THE PHASE TRANSITION IN TRIGLYCINE SULFATE.
ALLEN GR	22-2862	MODELS FOR THE ORDER DISORDER TRANSITION IN SODIUM TRIHYDROGEN SELENATE.
ALLEN RE	04-0725	ELECTRICAL PROPERTIES OF MICROCRYSTALLINE FERROELECTRIC MATERIALS CRYSTALLIZED FROM GL
ALLEN RR	06-1237	NUCLEAR MAGNETIC RESONANCE STUDY OF THE PHASE TRANSITION IN STRONTIUM TITANATE.
ALLPRESS JG	10-1627	EXAMINATION OF SUBSTOICHIOMETRIC TUNGSTEN OXIDE(3-X) CRYSTALS BY ELECTRON MICROSCOPY.
ALQUIE AM	10-1628	EPR STUDY OF THE CATALYTIC ACTIVITY OF NONSTOICHIOMETRIC TUNGSTEN TRIOXIDE.
ALSDORF R	18-2737	PROTONIC CONDUCTION IN POTASSIUM DIHYDROGEN ARSENATE.
ALSHIN BI	13-2069	SUPPRESSION OF THE SPONTANEOUS MAGNETOELECTRIC MAGNETIZATION OF LEAD FERRATE(0.5) NIOB
	14-2070	MAGNETOELECTRIC PROPERTIES OF NICKEL IODINE BORACITE.
	33-3409	LOW FREQUENCY MAGNETOELECTRIC RESONANCES IN BARIUM COBALT TETRAFLUORIDE.
AMBROVIC P	27-3048	ANOMALIES OF THERMAL PROPERTIES OF TRIGLYCINE SULFIDE.
AMELINCKX S	03-0255	ELECTRON MICROSCOPE TRANSMISSION IMAGES OF COHERENT DOMAIN BOUNDARIES. PART-1: DYNAMIC
	36-3522	USE OF ELECTRON MICROSCOPY IN THE STUDY OF EXTENDED DEFECTS RELATED TO NONSTOICHIOMETR
AMIN M	04-0536	PIEZOELECTRIC EFFECT IN ALUMINUM DOPED BARIUM TITANATE CERAMIC.
	17-2414	EFFECTS OF DEUTERATION ON THE SPECIFIC HEAT OF ADP CRYSTALS.
	17-2415	HEAT CAPACITY OF RUBIDIUM DIHYDROGEN PHOSPHATE SINGLE CRYSTALS.
AMMANN EO	11-1859	SIMULTANEOUS OPTICAL PARAMETRIC OSCILLATION, SECOND HARMONIC GENERATION (IN LITHIUM NI
AMODEI JJ	06-1090	ELECTRON THERMAL DIFFUSION EFFECTS (INTERNAL GENERATION OF STRONG FIELDS) DURING HOLOG
	06-1091	PHOTOCONDUCTIVITY AND TRAPPING PHENOMENA IN STRONTIUM TITANATE.
	06-1092	STUDY OF THE DYNAMICS OF PHOTOCHROMIC SWITCHING IN STRONTIUM TITANATE MATERIALS.
	11-1661	HOLOGRAPHIC PATTERN FIXING IN ELECTROOPTIC CRYSTALS. (LITHIUM NIOBATE, BARIUM SODIUM N
	11-1662	HOLOGRAPHIC RECORDING IN LITHIUM NIOBATE.
	11-1796	OPTICAL AND HOLOGRAPHIC STORAGE PROPERTIES OF TRANSITION METAL DOPED LITHIUM NIOBATE.
	11-1826	COUPLED WAVE ANALYSIS OF HOLOGRAPHIC STORAGE IN LITHIUM NIOBATE.

AMODEI JJ	11-1827	MULTIPLE STORAGE OF THICK PHASE HOLOGRAMS IN LITHIUM NIOBATE.
	11-1828	THERMALLY FIXED HOLOGRAMS IN LITHIUM NIOBATE.
	13-1898	HOLOGRAPHIC STORAGE IN DOPED BARIUM SODIUM NIOBATE.
AMOROS JL	16-2298	THERMAL EXPANSION OF SOME FERROELECTRICS. (POTASSIUM NITRATE, KDP, SODIUM NITRITE)
	16-2343	X-RAY CRITICAL SCATTERING OF SODIUM NITRITE AT 160 DEG C AND AT 215 DEG C.
AMOV IG	24-2904	APPLICATION OF THE DEW METHOD FOR REVEALING THE DOMAIN STRUCTURE IN POTASSIUM FERROCYA
	27-3148	EFFECT OF SAMPLE TREATMENT AND ELECTRODE NATURE ON SOME TRIGLYCINE SULFATE PARAMETERS.
AMZALLAG E	11-1807	TEMPERATURE STUDY OF THE POLARITON ASSOCIATED WITH THE 248 CM(-1) SOFT MODE IN LITHIUM
	12-1873	TEMPERATURE DEPENDENCE OF POLARITON DISPERSION IN LITHIUM TANTALATE. (OPTICAL MODULATI
	13-1899	RAMAN SCATTERING BY BARIUM STRONTIUM NIOBATE.
	13-1922	FERROELECTRICITY MEASUREMENTS ON STRONTIUM BARIUM NIOBATE.
ANANEVA AA	01-0003	CERAMIC ACOUSTIC DETECTORS. (BARIUM TITANATE)
	09-1617	FERROELECTRIC AND PIEZOELECTRIC PROPERTIES IN A LEAD TITANATE- LEAD ZIRCONATE- LEAD (Z
	13-1951	ELECTROPHYSICAL PROPERTIES OF THE SOLID SOLUTION SYSTEM LEAD TITANATE- LEAD ZIRCONATE-
ANANIN OB	11-1663	FRACTURE OF NONLINEAR CRYSTALS (KDP AND LITHIUM NIOBATE) BY RADIATION FROM A RUBY LASE
ANDA E	16-2256	ANGULAR DISPERSION OF LARGE WAVE VECTOR POLARITONS IN FERROELECTRIC SODIUM NITRITE.
ANDEEN C	33-3361	LOW FREQUENCY DIELECTRIC CONSTANTS OF THE ALKALINE EARTH FLUORIDES BY THE METHOD OF SU
ANDERSON A	19-2793	RAMAN AND INFRARED STUDIES OF THE FERROELECTRIC TRANSITION IN AMMONIUM SULFATE.
	31-3338	CRYSTAL SIZE EFFECTS IN THE RAMAN SPECTRA OF ORTHORHOMBIC HYDROGEN CHLORIDE CRYSTALS.
ANDERSON HU	04-0631	NONSTOICHIOMETRY IN BARIUM TITANATE.
ANDERSON JC	16-2394	THIN FILM NUCLEATION ON FERROELECTRIC SUBSTRATES. (TELLURIUM ON TRIGLYCINE SULFATE AND
ANDERSON LK	02-0039	FERROELECTRICS IN OPTICAL MEMORIES AND DISPLAYS: A CRITICAL APPRAISAL.
	09-1564	STRAIN BIASED FERROELECTRIC- PHOTOCONDUCTOR IMAGE STORAGE AND DISPLAY DEVICES OPERATED
	37-3628	FERROELECTRICS IN OPTICAL MEMORIES AND DISPLAYS: A CRITICAL APPRAISAL.
ANDERSON MR	32-3341	ANOMALOUS NEUTRON SCATTERING AND THE QUESTION OF FERROELECTRICITY IN LITHIUM HYDRAZINI
ANDERSON RC	08-1275	CRYSTALLIZATION OF LEAD METANIOBATE FROM GLASS.
ANDERSON RL	27-2954	MEASUREMENT OF THE DIELECTRIC CONSTANT OF TRIGLYCINE SULFATE AT MICROWAVE FREQUENCIES.
ANDERSON TT	11-1721	RADIATION EXPOSURE OF A LITHIUM NIOBATE CRYSTAL AT HIGH TEMPERATURES.
	11-1801	SECONDARY (ELECTRO) PHOTOELASTIC EFFECT IN LITHIUM NIOBATE (ON ELECTRON IRRADIATION).
ANDO R	16-2258	ELECTRICAL RESISTIVITIES OF SODIUM NITRITE AND POTASSIUM NITRATE CRYSTALS.
ANDONOV P	04-0537	DIELECTRIC PROPERTIES AND DOMAIN TEXTURE OF BARIUM TITANATE CRYSTALS.
ANDRADE PR	04-0619	COUPLED POLARITONS OF A1 SYMMETRY IN BARIUM TITANATE.
	08-1276	NUCLEAR QUADRUPOLE INTERACTIONS IN PEROVSKITE TYPE COMPOUNDS OF HAFNIUM-181 STUDIED BY
	16-2257	RELATIONSHIP OF THE TEMPERATURE DEPENDENCE OF B1 PHONON AND DIELECTRIC RELAXATION IN S
ANDRE D	17-2598	(PREPARATION AND STRUCTURE OF) RUBIDIUM AND CESIUM SEMIMETALLIC ORTHOPHOSPHATES.
ANDREEVA GT	11-1792	ELECTRON PARAMAGNETIC RESONANCE SPECTRUM IN (MANGANESE DOPED) LITHIUM NIOBATE AND (COP
	14-2095	CRYSTAL FIELD THEORY AND OPTICAL ABSORPTION (CALCULATIONS) OF FERROELECTRIC BORACITES.
	14-2097	OPTICAL ABSORPTION (625-25 MICRONS, 77-650 DEG K) OF FERROELECTRIC IRON(2+) BORACITES
ANDRUSHKO AI	03-0153	AN APPARATUS FOR INVESTIGATING THE ELECTRICAL PROPERTIES OF FERROELECTRICS.
ANGERT NB	11-1664	INHOMOGENEITY OF THE OPTICALLY INDUCED INDEX OF REFRACTION IN LITHIUM NIOBATE AND LITH
	11-1665	PHASE MATCHING ANGLES AND TEMPERATURE OF LITHIUM METANIOBATE CRYSTALS OF DIFFERENT STO
ANISTRATOV AT	02-0040	OPTICAL PROPERTIES OF FERROELECTRICS. DIGRESSIONS FROM PHENOMENOLOGICAL THEORY. (AMMON
	19-2745	FERROELECTRIC TRANSITION IN AMMONIUM SULFATE, THE DIELECTRIC, OPTICAL, AND ELECTROOPTI
	22-2829	LINEAR ELECTROOPTICAL EFFECT IN THE FERROELECTRIC SODIUM AMMONIUM SELENATE DIHYDRATE.
ANKUDINOV MA	27-3164	DOMAIN STRUCTURE OF GAMMA-IRRADIATED TGS CRYSTALS.
ANNAGIEV MKH	16-2318	AN X-RAY STUDY OF THE PHASE TRANSITION IN SODIUM NITRITE.
ANNIS AD	27-3037	EFFECTS OF ADDITIVES ON THE PYROELECTRIC PROPERTIES OF TGS.
	27-3038	STRUCTURAL INHIBITION OF FERROELECTRIC SWITCHING IN TRIGLYCINE SULFATE. PART-1: ADDITI
ANSELL GS	04-0579	AGING IN TETRAGONAL FERROELECTRIC BARIUM TITANATE.
	08-1295	DIELECTRIC AGING IN TETRAGONAL SOLID SOLUTIONS OF CALCIUM TITANATE IN BARIUM TITANATE.
ANTOKOLSKII GL	11-1666	ACOUSTIC DIPOLE AND QUADRUPOLE SATURATION OF THE LITHIUM-7 NMR SIGNALS IN LITHIUM NIOB
ANTONIAK CE	03-0434	SYMMETRY LIMITATIONS TO POLARIZATION OF POLYCRYSTALLINE FERROELECTRICS.
ANTONINI M	17-2416	INCOHERENT NEUTRON SCATTERING FROM HYDROGEN BOND IN KDP AND ADP.
ANTOSHIN MK	03-0474	OBSERVATION OF FERROELECTRIC DOMAIN STRUCTURES, MODULATED BY A PHASE TRANSITION, IN A
AOKI M	13-2036	ELECTROOPTIC LIGHT BEAM DEFLECTION WITH BARIUM STRONTIUM NIOBATE PRISM.
APKARYANTS PA	26-2925	ANTIPOLARIZATION IN COPPER FORMATE CRYSTALS BY THE ELECTROOPTICAL METHOD. (ANTIFERROEL
APPEL J	06-1093	SOFT MODE SUPERCONDUCTIVITY IN STRONTIUM TITANATE(3-X) AND CALCIUM(Y) STRONTIUM(1-Y) T
ARABIDZE AA	04-0538	GENERATION OF SECOND OPTICAL HARMONIC IN THE CUBIC PHASE OF BARIUM TITANATE.
ARAGHI MN	11-1667	LAYER WAVE AMPLIFICATION IN A CADMIUM SULFIDE FLUID- LITHIUM NIOBATE STRUCTURE.
	11-1700	SURFACE WAVE CONVOLUTION USING A CADMIUM SULFIDE FLUID LITHIUM NIOBATE STRUCTURE.
AREND H	02-0041	POINT DEFECTS IN PEROVSKITE TYPE FERROELECTRICS. (REVIEW, 77 REFS)
	04-0539	COMPENSATION OF HETEROVALENT FERRIC IONS IN THE LATTICE OF BARIUM TITANATE SINGLE CRYS
	04-0540	CHEMICAL PROCESSES DURING HYDROGEN REDUCTION OF BARIUM TITANATE CRYSTALS.
	04-0541	ON THE PHASE COMPOSITION OF REDUCED AND REOXIDIZED BARIUM TITANATE.
	04-0618	SOME CRYSTAL CHEMICAL AND PHYSICAL PROPERTIES OF SINGLE CRYSTALS OF BARIUM TITANATE WI
	04-0637	THICKNESS DEPENDENCE OF THE PERMITTIVITY OF BARIUM TITANATE SINGLE CRYSTALS.
	36-3523	GEL GROWTH OF SILVER PERIODATE CRYSTALS.
ARIMA H	05-1070	DIELECTRIC AND PIEZOELECTRIC PROPERTIES IN THE TERNARY SYSTEM LEAD (ZINCATE NIOBATE)-
	13-1900	ELECTROOPTIC EFFECTS OF HOT PRESSED CERAMICS IN THE SYSTEM LEAD (ZINCATE(0.333) NIOBAT
ARIYA SM	04-0542	DETERMINATION OF CHROMIUM VALENCY FROM STUDIES OF EPR SPECTRA AND STATIC MAGNETIC SUSC
	04-0543	STATE OF CHROMIUM ADDED TO BARIUM METATITANATE.
ARMSTRONG RA	08-1277	BISMUTH TITANATE SOLID SOLUTIONS.
ARNDT R	15-2109	X-RAY CRYSTALLOGRAPHIC STUDIES OF ANTIMONY SULFOIODIDE.
ARNOLD D	34-3415	ELECTRICAL CONDUCTION IN ICE.
ARNOLD GL	13-1915	EFFECT OF OXYGEN STOICHIOMETRY AND HYDROXYL CONTENT ON THE PROPERTIES AND PROCESSING O
ARNOLD RT	04-0544	EFFECT OF FERRIC OXIDE ON THE 120 DEGREE PHASE TRANSITION IN BARIUM TITANATE SINGLE CR
	06-1160	TEMPERATURE DEPENDENCE OF THE FIRST ORDER ELASTIC CONSTANTS OF STRONTIUM TITANATE.
	06-1180	ELASTIC PROPERTIES OF YTTRIUM DOPED STRONTIUM TITANATE SINGLE CRYSTALS.
	06-1181	ULTRASONIC THIRD HARMONIC GENERATION IN STRONTIUM TITANATE SINGLE CRYSTALS (NONLINEAR
ARONOFF AD	03-0182	ELECTRIC FIELD DISTRIBUTIONS IN DIELECTRICS, WITH SPECIAL EMPHASIS ON NEAR SURFACE REG
ARSENEV PA	11-1668	HEAT TREATMENT INDUCED POINT DEFECTS IN LITHIUM METANIOBATE SINGLE CRYSTALS.
	11-1669	PROPERTIES OF THE IRON TRANSITION GROUP IN THE LATTICE OF SINGLE CRYSTALLINE LITHIUM N
ARSIC-ESKINJA M	17-2417	NEUTRON MEASUREMENT IN TWO SYSTEMS OF HYDROGEN BONDS. (KDP AND POTASSIUM BICARBONATE)
	17-2418	NEUTRON MEASUREMENTS ON THE FERROELECTRIC PHASE TRANSFORMATION IN POTASSIUM DIHYDROGEN
ARTHUR G	04-0587	EFFECT OF OXIDE IMPURITIES ON THE ELECTRICAL RESISTIVITY OF LANTHANUM DOPED BARIUM TIT
ARTOBOLEVSKAYA ES	03-0525	MEASUREMENT OF THE TEMPERATURE DEPENDENCE OF THE ELECTRICAL CONDUCTIVITY OF FERROELECT
	15-2251	ELECTRICAL AND PHOTOELECTRIC PROPERTIES OF THE CONTACTS BETWEEN ANTIMONY SULFOIODIDE S
	15-2252	OPPOSING FERROELECTRIC DOMAINS IN SINGLE CRYSTALS OF ANTIMONY SULFOIODIDE.
ASADOV YUG	16-2259	MORPHOLOGY OF CRYSTAL GROWTH IN THE ALPHA TO BETA, BETA TO ALPHA, BETA TO GAMMA AND GA
	16-2260	MORPHOLOGY OF CRYSTAL GROWTH AT POLYMORPHIC TRANSFORMATIONS IN POTASSIUM NITRATE, SILV

PAGE 617

ASADOV YUG   16-2263   RHYTHMIC GROWTH OF NEW PHASE IN POTASSIUM NITRATE. (ALPHA TO BETA POLYMORPHIC TRANSITI
ASADULLIN YY   17-2419   MACROSCOPIC STIMULATED POLARIZATION ECHO IN FERROELECTRICS. (POTASSIUM DIHYDROGEN PHOS
ASAO Y   16-2258   ELECTRICAL RESISTIVITIES OF SODIUM NITRITE AND POTASSIUM NITRATE CRYSTALS.
                    16-2261   QUASI STATIC SWITCHING CURRENT IN SODIUM NITRITE.
                    16-2301   FERROELECTRIC POLARIZATION SWITCHING IN SODIUM NITRITE.
ASAWA CK   16-2262   RAMAN SPECTRA AND MODE FREQUENCY SHIFTS OF FERROELECTRIC SODIUM NITRITE AT 77 AND 294
ASCHER E   14-2071   DIELECTRIC PROPERTIES OF BORACITES AND EVIDENCE FOR FERROELECTRICITY.
                    14-2072   INTERACTIONS BETWEEN MAGNETIZATION AND POLARIZATION: PHENOMENOLOGICAL SYMMETRY CONSIDE
                    30-3292   AN EXPLANATION OF ANOMALOUS OPTICAL BEHAVIOR OF THE IMPROPER FERROELECTRIC GADOLINIUM
ASCOLI A   17-2420   GROWTH OF ELECTROOPTICAL CRYSTALS FROM AQUEOUS SOLUTIONS (APPARATUS, GROWTH OF ADP, KD
ASHIDA S   30-3297   FERROELECTRIC FERROELASTIC CRYSTAL GADOLINIUM MOLYBDATE.
                    30-3308   FERROELECTRIC AND FERROELASTIC CRYSTAL, GADOLINIUM MOLYBDATE.
ASHIDA T   11-1670   LOW TEMPERATURE COEFFICIENT OF LITHIUM NIOBATE RESONATORS BY USE OF TEMPERATURE SENSIT
                    12-1868   TEMPERATURE DEPENDENCE OF AN X CUT LITHIUM TANTALATE CRYSTAL RESONATOR VIBRATING IN TH
                    12-1885   BEHAVIOR OF LITHIUM TANTALATE SINGLE CRYSTAL NEAR ITS CURIE POINT, PART-2, DIELECTRIC
                    36-3524   THE CRYSTAL STRUCTURE OF TRIS-SARCOSINE CALCIUM CHLORIDE.
ASHKIN M   06-1118   ULTRASONIC STUDIES OF CRITICAL PHENOMENA IN INSULATING AND SEMICONDUCTING STRONTIUM TI
ASHMORE JP   19-2746   CRYSTAL STRUCTURE OF FERROELECTRIC RUBIDIUM HYDROGEN SULFATE. (ABSTRACT)
ASHWORTH T   20-2801   HEAT CAPACITY OF METHYL AMMONIUM CHROMIUM ALUM.
ASKOCHENSKII AA   22-2830   DOMAIN STRUCTURE OF THE BETA PHASE OF SODIUM TRIHYDROGEN SELENATE IN ELECTRIC FIELDS.
                    22-2874   THE PECULIAR DOMAIN STRUCTURE AND ITS UNUSUAL BEHAVIOR IN THE BETA PHASE OF THE FERROE
ASTROV DN   13-2069   SUPPRESSION OF THE SPONTANEOUS MAGNETOELECTRIC MAGNETIZATION OF LEAD FERRATE(0.5) NIOB
                    14-2070   MAGNETOELECTRIC PROPERTIES OF NICKEL IODINE BORACITE.
                    33-3409   LOW FREQUENCY MAGNETOELECTRIC RESONANCES IN BARIUM COBALT TETRAFLUORIDE.
ATKIN RB   09-1478   POINT DEFECTS AND SINTERING OF LEAD ZIRCONATE TITANATE.
                    09-1479   SOLUBILITY OF ALUMINUM IN LEAD ZIRCONATE TITANATE.
                    09-1481   SINTERING AND FERROELECTRIC PROPERTIES OF LEAD ZIRCONATE TITANATE CERAMICS.
                    37-3629   PERFORMANCE OF SPUTTERED LEAD(0.92) BISMUTH(0.07) LANTHANUM(0.01) (IRON(0.405) NIOBIUM
AUBRY A   17-2457   LOW FREQUENCY INFRARED ABSORPTION SPECTRA OF MONOCRYSTALS OF POTASSIUM DIHYDROGEN PHOS
AUBRY S   03-0506   ENERGY PARAMETERS AND PROTON TUNNELING IN POTASSIUM DIHYDROGEN PHOSPHATE TYPE FERROELE
AUSTIN AE   02-0131   PREPARATION AND PROPERTIES OF RARE EARTH COMPOUNDS. (WITH MAGNETIC AND FERROELECTRIC O
                    36-3545   MAGNETIC PROPERTIES OF HEAVY RARE EARTH ORTHOMANGANATES.
AUTHIER A   11-1671   X-RAY TOPOGRAPHICAL STUDY OF THE CONTRAST AT FERROELECTRIC DOMAIN BOUNDARIES (DOPED AN
                    19-2747   X-RAY TOPOGRAPHIC STUDY OF DISLOCATIONS AND FERROELECTRIC DOMAINS IN GLYCOL SULFATE.
                    27-3025   X-RAY TOPOGRAPHIC STUDY OF GROWTH DEFECTS IN TRIGLYCINE SULFATE CRYSTALS IN RELATION T
                    27-3026   X-RAY TOPOGRAPHIC STUDY OF DOMAIN WALL MOVEMENT IN TGS.
AUVERT G   34-3410   LOW ENTROPY FORM OF ICE I(H) OBTAINED FROM A LINEAR STEP GROWTH MODEL.
AVERBACH VS   17-2421   LIGHT ABSORPTION IN POTASSIUM DIHYDROGEN PHOSPHATE SINGLE CRYSTALS.
AVERYANOVA LN   02-0047   SYNTHESIS AND CERTAIN PROPERTIES OF NEW COMPOUNDS WITH A PYROCHLORE STRUCTURE. (NIOBAT
                    36-3528   X-RAY PHASE STUDY OF THE CADMIUM BORATE- ABO(3) SYSTEMS. (WHERE A IS STRONTIUM OR BARI
AVOGADRO A   16-2264   NUCLEAR SPIN LATTICE RELAXATION OF SODIUM-23 IN THE PARAELECTRIC AND FERROELECTRIC PHA
                    24-2889   STUDY OF HYDROGEN-1 IN PARAELECTRIC AND FERROELECTRIC POTASSIUM FERROCYANIDE TRIHYDRAT
AWAZU K   13-2036   ELECTROOPTIC LIGHT BEAM DEFLECTION WITH BARIUM STRONTIUM NIOBATE PRISM.
AXE JD   03-0154   DYNAMICAL EFFECTS IN SOLID STATE PHASE TRANSFORMATIONS. (FERROELECTRICS)
                    03-0155   NEUTRON STUDIES OF DISPLACIVE STRUCTURAL PHASE TRANSFORMATIONS. (INELASTIC DIFFUSION I
                    03-0156   NEUTRON STUDIES OF DISPLACIVE STRUCTURAL PHASE TRANSFORMATIONS. (IN FERROELECTRICS)
                    04-0712   NEUTRON SCATTERING STUDY OF SOFT MODES IN CUBIC BARIUM TITANATE.
                    04-0946   INELASTIC NEUTRON SCATTERING FROM SINGLE DOMAIN BARIUM TITANATE. (LATTICE DYNAMICS)
                    05-1077   SOFT FERROELECTRIC MODES IN LEAD TITANATE (NEUTRON INELASTIC SCATTERING).
                    06-1139   DETERMINATION OF THE NORMAL VIBRATIONAL DISPLACEMENTS OF SEVERAL PEROVSKITES (POTASSIU
                    06-1208   CRITICAL NEUTRON SCATTERING IN STRONTIUM TITANATE AND POTASSIUM MANGANESE TRIFLUORIDE.
                    06-1210   NEW OBSERVATIONS FROM NEUTRON SCATTERING STUDIES OF THE STRUCTURAL PHASE TRANSITION IN
                    08-1407   NEUTRON STUDY OF DIFFUSE SCATTERING IN CUBIC POTASSIUM NIOBATE.
                    30-3254   MECHANISM OF THE FERROELECTRIC PHASE TRANSFORMATION IN RARE EARTH MOLYBDATES. (TERBIUM
                    30-3268   NEUTRON SCATTERING STUDY OF THE FERROELECTRIC PHASE TRANSFORMATION IN TERBIUM MOLYBDAT
                    30-3269   ZONE BOUNDARY SOFT MODE IN TERBIUM MOLYBDATE.
                    33-3374   DISPERSION AND DAMPING OF SOFT ZONE BONDARY PHONONS IN POTASSIUM MANGANESE TRIFLUORIDE
AYAKAWA T   09-1540   PRECIPITATION OF ZIRCONIA PHASE IN NIOBIUM MODIFIED CERAMICS OF LEAD ZIRCONATE TITANAT
AZARBAYEJANI GH   11-1672   GROWTH TWINNING (UNDER DIRECT CURRENT) IN C-AXIS LITHIUM NIOBATE CRYSTALS (CZOCHRALSKI
AZUMI T   04-0974   ELECTRON MIRROR MICROSCOPIC OBSERVATION OF FERROELECTRIC DOMAINS. (GOLD COATED BARIUM
BABA PC   04-0545   STUDY OF THE AGING PHENOMENON IN BARIUM TITANATE SINGLE CRYSTALS.
BABB SE   16-2265   MELTING PROPERTIES OF POTASSIUM NITRATE UNDER HIGH PRESSURE.
BABCOCK RV   34-3411   EQUILIBRIUM STRUCTURE OF POLARIZED ICE.
BACCARA N   03-0183   PHYSICS OF TRANSITIONS. (THERMODYNAMICS, THEORY, MAGNETISM, FERROELECTRICITY, ORDERED
BADDUR A   17-2677   FIRST ORDER PHASE TRANSITION IN POTASSIUM DEUTERIUM PHOSPHATE CRYSTALS.
                    17-2678   SPONTANEOUS POLARIZATION OF A POTASSIUM DIHYDROGEN PHOSPHATE CRYSTAL NEAR THE CURIE PO
                    17-2680   SOME THERMODYNAMICAL PROPERTIES OF POTASSIUM DIHYDROGEN PHOSPHATE- POTASSIUM DIHYDROGE
BADER VI   08-1292   STRUCTURE AND DIELECTRIC PROPERTIES OF LITHIUM CONTAINING PEROVSKITES. (STRONTIUM LITH
                    36-3613   SYNTHESIS AND X-RAY DIFFRACTION STUDY OF SINGLE CRYSTALS OF A NEW PYROCHLORE CONTAININ
BAE P   28-3182   DIELECTRIC PHENOMENA IN ROCHELLE SALT.
BAGDASAROV KHS   30-3255   CERTAIN GROWTH, LASING AND SPECTROSCOPIC PROPERTIES OF NEODYMIUM(3+) DOPED GADOLINIUM
BAICAN R   19-2750   MICROWAVE DIELECTRIC MEASUREMENTS OF GAMMA-IRRADIATED FERROELECTRIC AMMONIUM SULFATE.
BAISA DF   08-1278   POTASSIUM IODATE CRYSTAL STRUCTURE STUDIED BY INFRARED SPECTROSCOPY AND NUCLEAR QUADRU
BAJAK II   27-2955   INFLUENCE OF THE ELECTRIC FIELD AND THE DOMAIN STRUCTURE ON THE ULTRASONIC ABSORPTION
                    27-2956   THE INFLUENCE OF THE ELECTRIC FIELD ON THE ULTRASONIC ATTENUATION IN FERROELECTRIC TRI
BAKAI AS   03-0157   INTERACTION OF LIGHT WITH MAGNETOACOUSTIC WAVES IN A FERRODIELECTRIC (THEORY).
                    03-0158   INTERACTION OF LIGHT WITH SPIN WAVES IN FERROELECTRICS. (THEORY)
BAKER WE   03-0222   ENERGY GENERATION FROM IMPACT ON PIEZOELECTRIC (OR FERROELECTRIC) MATERIALS.
BAKUSHINSKII AB   07-1261   DIPOLE STRUCTURE AND INTERNAL ELECTRIC FIELDS IN LEAD ZIRCONATE.
BALAGUROV BYA   03-0159   ATTENUATION OF CRITICAL VIBRATIONS AND DIELECTRIC LOSSES IN DISPLACIVE TYPE FERROELECT
                    03-0160   HIGH FREQUENCY SOUND ATTENUATION AND THERMAL CONDUCTIVITY IN DISPLACIVE TYPE FERROELEC
BALAKHONOV NF   03-0164   SPIN WAVE PHENOMENOLOGICAL THEORY IN FINITE SAMPLES OF FERRODIELECTRICS WITH AN ARBITR
BALASUBRAMANIAN K   27-2957   RAMAN SPECTRUM OF TRIGLYCINE SELENATE.
BALDINI G   17-2422   ULTRAVIOLET REFLECTION AND ABSORPTION OF POTASSIUM DIHYDROGEN PHOSPHATE AND AMMONIUM D
BALKANSKI M   03-0163   RAMAN SPECTROSCOPY OF PHONON INSTABILITY IN FIRST ORDER FERROELECTRIC PHASE TRANSITION
                    15-2110   LATTICE MODES AND PHASE TRANSITION IN ANTIMONY SULFOIODIDE.
                    15-2111   RAMAN SCATTERING IN FERROELECTRIC SEMICONDUCTOR ANTIMONY SULFOIODIDE.
                    15-2177   NOTES ON ABSORPTION BAND EDGE OF (PARAELECTRIC) ANTIMONY SULFOIODIDE.
                    15-2198   ELECTROOPTIC EFFECTS OF THE ABSORPTION EDGE OF ANTIMONY SULFOIODIDE.
                    15-2240   SOFT PHONON COUPLINGS IN THE PRESSURE INDUCED PHASE TRANSITION IN ANTIMONY SULFOIODIDE

BALKANSKI M	16-2378	PRESSURE INDUCED FERROELECTRIC PHASE TRANSITION IN POTASSIUM NITRATE.
BALKAREI YUI	03-0161	LOCAL ELECTRON STATES IN DOMAIN WALLS OF FERROELECTRICS.
	03-0162	POSSIBILITY OF GENERATING INFRARED RADIATION BY POLARIZATION REVERSAL OF A FERROELECTR
BALLANTYNE JM	04-0546	FAR INFRARED DIELECTRIC DISPERSION OF PEROVSKITES. (BARIUM TITANATE, POTASSIUM TANTALA
BALLARD SS	11-1685	THERMAL EXPANSION OF LITHIUM NIOBATE IN THE TEMPERATURE RANGE OF 70-300 K.
	17-2464	LOW TEMPERATURE THERMAL EXPANSION MEASUREMENTS ON SINGLE CRYSTAL POTASSIUM DIHYDROGEN
BALLMAN AA	12-1869	FERROELECTRIC DOMAIN REVERSAL IN LITHIUM METATANTALATE.
	12-1870	INFRARED (REFLECTIVITY) STUDY (20-10,000 PER CM, 300 DEG K) OF THE LATTICE VIBRATIONS
	13-1901	GROWTH OF UNCRACKED BARIUM SODIUM NIOBATE CRYSTALS.
	13-1902	SOME EFFECTS OF MELT STOICHIOMETRY ON THE OPTICAL PROPERTIES BARIUM SODIUM NIOBATE.
	13-1903	SOME EFFECTS OF MELT STOICHIOMETRY ON THE OPTICAL PROPERTIES OF BARIUM SODIUM NIOBATE.
	35-3506	NONLINEAR OPTICAL PROPERTIES OF FERROELECTRIC LEAD(5) GERMANIUM(3) OXIDE(11).
BALTES HP	33-3362	INFRARED ACTIVE MODE BELOW THE STRUCTURAL TRANSITION IN POTASSIUM MANGANESE TRIFLUORID
	33-3363	TWO PHONON PROCESSES AND PHASE TRANSITIONS IN POTASSIUM MANGANESE TRIFLUORIDE: TEMPERA
BALTRUNAS DI	15-2108	NUCLEAR GAMMA RESONANCE SPECTROSCOPIC STUDY OF SEMICONDUCTING COMPOUNDS OF THE TYPE A(
PAN I	10-1647	EVAPORATED FILMS OF TUNGSTEN OXIDE.
BAN T	21-2821	ZEEMAN EFFECT AND DILUTION EFFECT OF ABSORPTION LINES IN GUANIDINIUM CHROMIUM SULFATE
BANDO S	36-3524	THE CRYSTAL STRUCTURE OF TRIS-SARCOSINE CALCIUM CHLORIDE.
BANDY A	25-2905	LATTICE VIBRATIONAL SPECTRA OF FIVE CRYSTAL MODIFICATIONS OF THIOUREA.
BANKS E	02-0042	SYNTHESIS OF OXIDE BRONZES. (REVIEW, 88 REFS)
BARABAN VS	04-1036	POSITIVE TEMPERATURE COEFFICIENT OF RESISTANCE OF BARIUM TITANATE SINGLE CRYSTALS IN T
BARABANOVA LA	04-0832	STRUCTURE AND PROPERTIES OF BARIUM TITANATE FILMS PRODUCED BY CATHODE SPUTTERING.
BARABASH AI	08-1278	POTASSIUM IODATE CRYSTAL STRUCTURE STUDIED BY INFRARED SPECTROSCOPY AND NUCLEAR QUADRU
BARANOV AI	04-0924	EFFECT OF ULTRASOUND ON THE POLARIZATION AND REPOLARIZATION PROCESSES IN TRIGLYCINE SU
	17-2666	PHASE DIAGRAM OF SODIUM TRIDEUTERIUM(X) TRIHYDROGEN(1-X) PHOSPHATE SYSTEM VERSUS CONCE
	22-2875	UNUSUAL EFFECT OF AN ELECTRIC FIELD ON THE PERMITTIVITY OF A NEW DELTA PHASE OF SODIUM
	22-2882	EFFECT OF HIGH PRESSURE ON PHASE TRANSITIONS OF ALKALI TRIHYDROGEN SELENITE CRYSTALS.
	30-3321	EFFECT OF HYDROSTATIC PRESSURE ON PHASE TRANSITION IN GADOLINIUM MOLYBDATE.
BARANOV EA	11-1669	PROPERTIES OF THE IRON TRANSITION GROUP IN THE LATTICE OF SINGLE CRYSTALLINE LITHIUM N
BARANSKII KN	11-1814	SEPARATE EXCITATION OF LONGITUDINAL AND TRANSVERSE HYPERSONIC PULSES IN PIEZOELECTRIC
	11-1815	INDEPENDENT EXCITATION OF LONGITUDINAL AND TRANSVERSE HYPERSONIC WAVE PULSES IN LITHIU
	28-3183	FREQUENCY DEPENDENCE OF THE ABSORPTION OF SOUND IN ROCHELLE SALT NEAR ITS UPPER CURIE
BARBE M	15-2112	CRYSTAL STRUCTURE AND CONDUCTIVITY OF ANTIMONY SULFOIODIDE.
BARBOSA GA	04-0548	TEMPERATURE DEPENDENCE OF THE RAMAN CROSS SECTIONS IN BARIUM TITANATE AND STRONTIUM TI
BARBUR I	19-2748	ELECTRON SPIN RESONANCE OF GAMMA DEFECTS IN FERROELECTRIC AMMONIUM HYDROGEN SULFATE.
	19-2750	MICROWAVE DIELECTRIC MEASUREMENTS OF GAMMA-IRRADIATED FERROELECTRIC AMMONIUM SULFATE.
BARCHUK LF	04-0984	PREPARATION OF FLUORINE CONTAINING BARIUM TITANATE.
BARENTZEN H	25-2920	MOLECULAR AND LATTICE VIBRATIONS OF ORTHORHOMBIC AND FERROELECTRIC THIOUREA. PART-1: T
BARHAM D	04-0624	THE PRODUCTION OF BARIUM TITANATE FROM BARIUM SULFATE AND TITANIUM DIOXIDE IN THE PRES
BARIYAKHTAR VG	03-0166	QUANTUM THEORY OF VIBRATIONS IN FERROELECTRIC FERROMAGNETICS WITHOUT A SYMMETRY CENTER
BARKER AS	03-0165	FAR INFRARED DISPERSION AND RAMAN SPECTRA OF FERROELECTRIC CRYSTALS. (THEORY)
	04-0547	COUPLED OPTICAL PHONON MODE THEORY OF INFRARED DISPERSION IN BARIUM AND STRONTIUM TITA
	06-1094	FAR INFRARED FERROELECTRIC VIBRATION MODE IN STRONTIUM TITANATE.
	12-1870	INFRARED (REFLECTIVITY) STUDY (20-10,000 PER CM, 300 DEG K) OF THE LATTICE VIBRATIONS
	17-2423	FAR INFRARED DIELECTRIC MEASUREMENTS ON POTASSIUM DIHYDROGEN PHOSPHATE, TRIGLYCINE SUL
	36-3526	INFRARED OPTICAL PROPERTIES OF VANADIUM DIOXIDE ABOVE AND BELOW TRANSITION TEMPERATURE
BARKLEY JR	30-3256	CONTROL AND APPLICATION OF DOMAIN WALL MOTION IN GADOLINIUM MOLYBDATE.
	30-3257	CONTROL AND APPLICATION OF DOMAIN WALL MOTION IN GADOLINIUM MOLYBDATE.
	30-3318	DOMAIN WALL STRUCTURE IN GADOLINIUM MOLYBDATE BY RAMAN SCATTERING.
BARKOVSKII LM	02-0044	PHASES AND POLARIZATION OF LUMINOUS WAVES IN CUBIC ELECTROOPTIC CRYSTALS. (NONLINEAR O
BARNEA Z	04-0713	X-RAY AND NEUTRON DIFFRACTION STUDY OF TETRAGONAL BARIUM TITANATE.
	04-0740	X-RAY AND NEUTRON DIFFRACTION STUDY OF TETRAGONAL BARIUM TITANATE.
BARNETT JD	03-0236	THEORY OF X-RAY DIFFRACTION LINE SHAPE EFFECTS DUE TO DOMAIN FORMATION IN FERROELECTRI
BARNOSKI MK	16-2262	RAMAN SPECTRA AND MODE FREQUENCY SHIFTS OF FERROELECTRIC SODIUM NITRITE AT 77 AND 294
BARNS RL	08-1270	FERROELASTIC EFFECT IN LANTHANUM ORTHOFERRITE.
	11-1673	X-RAY DETERMINATION OF POLARITY SENSE BY ANOMALOUS SCATTERING AT AN ABSORPTION EDGE. (
	12-1871	LITHIUM TANTALATE SINGLE CRYSTAL STOICHIOMETRY (VERSUS MELT COMPOSITION, LATTICE AND C
	14-2102	FERROELECTRIC AND FERROELASTIC PROPERTIES OF MAGNESIUM CHLORINE BORACITE.
BARRACLOUGH KG	13-1894	DILATOMETRIC STUDY OF THE ORTHORHOMBIC- TETRAGONAL PHASE TRANSITION IN BARIUM SODIUM N
	13-1904	BARIUM SODIUM NIOBATE: CONSTITUTIONAL STUDIES AND (CZOCHRALSKI) CRYSTAL GROWTH.
BARRETT HH	02-0043	ACOUSTIC PROPERTIES OF MATERIALS OF PEROVSKITE STRUCTURE. (RELATION OF FERROELECTRICIT
	06-1095	THERMAL CONDUCTIVITY IN PEROVSKITES (3 PHONON INTERACTION PROPOSED FOR STRONTIUM TITAN
BARSUKOVA ML	04-0941	NONLINEAR OPTICAL PROPERTIES OF LEAD TITANATE AND BISMUTH TITANATE CRYSTALS.
	08-1279	HYDROTHERMAL CRYSTALLIZATION AND SOME PROPERTIES OF BISMUTH TITANATES
	09-1482	HYDROTHERMAL CRYSTALLIZATION OF LEAD TITANATE ZIRCONATE SOLID SOLUTIONS.
BARTIS FJ	17-2424	INTERNAL STRESS AND THE FERROELECTRIC TRANSITION IN POTASSIUM DIHYDROGEN PHOSPHATE.
BARTLETT RW	11-1737	POINT DEFECTS IN LITHIUM NIOBATE SINGLE CRYSTALS USED IN LASER MODULATION.
BARTUCH H	36-3525	EPR INVESTIGATIONS OF MANGANESE(2+) COMPLEXES IN FERROELECTRIC TRIS-SARCOSINE CALCIUM
BARTZBEDNARCZYK D	04-0549	ULTRASONIC STUDIES OF FIELD INDUCED PHASE TRANSITION IN BARIUM TITANATE.
BASS M	11-1674	NEODYMIUM YAG LASER IRRADIATION INDUCED DAMAGE TO LITHIUM NIOBATE AND KDP.
	27-2958	OPTICAL MIXING. (OF PULSED RUBY LASER EMISSIONS BY TRIGLYCINE SULFATE CRYSTAL)
BATES HE	11-1675	NONCOLLINEAR PHASE MATCHING EFFECTS IN LITHIUM NIOBATE.
BATRA IP	03-0167	POLARIZATION DISTRIBUTION IN THE VICINITY OF A FERROELECTRIC SEMICONDUCTOR INTERFACE.
	27-2959	THERMODYNAMIC STABILITY OF THIN FERROELECTRIC FILMS. (TRIGLYCINE SULFATE)
	27-3163	THERMODYNAMIC STABILITY OF POLARIZATION IN THIN FERROELECTRIC FILMS ON SEMICONDUCTING
	37-3644	EQUIVALENT CIRCUIT ANALYSIS OF A FERROELECTRIC PHOTOCONDUCTOR MEMORY DEVICE.
BATSANOV SS	17-2425	RAMAN SPECTRA OF POLYMORPHIC MODIFICATIONS OF POTASSIUM DIHYDROGEN PHOSPHATE IN AN ELE
BATYTEVA IA	17-2421	LIGHT ABSORPTION IN POTASSIUM DIHYDROGEN PHOSPHATE SINGLE CRYSTALS.
BAUER F	09-1483	FERROELECTRIC CERAMICS AND CONVERSION OF MECHANOELECTRIC ENERGY. (LEAD ZIRCONATE TITAN
	09-1521	PHASE CHANGES INDUCED BY HYDROSTATIC PRESSURE IN A FERROELECTRIC MATERIAL. (PZT)
BAUMBER P	28-3184	MEASUREMENT OF THE COMPLEX DIELECTRIC CONSTANT OF ROCHELLE SALT AT 10 GHZ AS A FUNCTIO
BAUMBERGER C	27-2960	SPECIFIC HEAT UNDER APPLIED ELECTRIC FIELD AND ELECTROCALORIC EFFECT IN TRIGLYCINE SUL
	27-3077	POLARIZATION FLUCTUATION OF TRIGLYCINE SULFATE NEAR THE CURIE POINT.
BAXTER RJ	03-0168	GENERALIZED FERROELECTRIC MODEL ON A SQUARE LATTICE.
BAZUEV GV	36-3527	ELECTRICAL PROPERTIES AND STRUCTURES OF METANIOBATES AND METATANTALATES OF TRANSITION
BEADLE CW	37-3630	EXPERIMENTAL METHODS FOR INVESTIGATING STRAIN WAVE PROPAGATION AND ASSOCIATED CHARGE R
BEALS MD	02-0045	SINGLE CRYSTAL TITANATES AND ZIRCONATES. (METHODS OF GROWTH)
BEATTIE AG	06-1096	PRESSURE DEPENDENCE OF THE ELASTIC CONSTANTS OF STRONTIUM TITANATE.
	06-1097	PRESSURE DEPENDENCE OF THE ELASTIC CONSTANT OF STRONTIUM TITANATE.
BECKER WJ	14-2073	NEUTRON AND ELECTRON DIFFRACTION STUDY IN NICKEL IODINE BORACITE.

PAGE 619

BECHARCZYK J	04-0549	ULTRASONIC STUDIES OF FIELD INDUCED PHASE TRANSITION IN BARIUM TITANATE.
BEERMAN HP	02-0046	IMPROVEMENT IN THE PYROELECTRIC INFRARED RADIATION DETECTOR. (TRIGLYCINE SULFATE, TRIG
BEIGE H	04-0550	TIME DEPENDENCE OF THE DIELECTRIC PERMITTIVITY OF HYDROXYL DOPED BARIUM TITANATE.
BELDING JH	09-1484	BEHAVIOR OF MODIFIED LEAD ZIRCONATE LEAD TITANATE PIEZOELECTRIC CERAMICS UNDER HIGH EL
	09-1485	EFFECTS OF HIGH ELECTRIC FIELDS ON MODIFIED LEAD ZIRCONATE- LEAD TITANATE CERAMICS FOR
BELDING WA	04-0551	METHOD FOR ACHIEVING HIGH DIELECTRIC CONSTANT TEMPERATURE STABLE FERROELECTRICS. (BARI
BELJUSTIN AW	17-2496	GROWTH LAYERS AND OTHER MACRODEFECTS IN POTASSIUM DIHYDROGEN PHOSPHATE CRYSTALS GROWN
BELL RO	06-1199	MICROWAVE LOSSES IN STRONTIUM TITANATE ABOVE THE PHASE TRANSITION.
BELOTSKII DP	15-2122	PHASE DIAGRAM AND ELECTRICAL CONDUCTIVITY OF BISMUTH SELENOIODIDE- ANTIMONY SELENOIODI
BELOUS LP	13-1968	PHASE TRANSISION IN THE ANTIFERROELECTRET LEAD COBALTATE(0.5) TUNGSTATE(0.5) AND LEA
BELOV NV	03-0465	SYMMETRY OF CRYSTALS IN WHICH FERROMAGNETIC AND FERROELECTRIC PROPERTIES APPEAR SIMULT
BELOV VF	15-2180	MOSSBAUER STUDIES OF ANTIMONY SULFOIODIDE TYPE CRYSTALS.
BELOVA LA	05-1051	POSITIVE TEMPERATURE COEFFICIENT OF RESISTANCE IN DOPED LEAD TITANATE.
BELRUSS V	04-0553	TOP SEEDED SOLUTION GROWTH OF OXIDE CRYSTALS FROM NONSTOICHIOMETRIC MELTS. (BARIUM TIT
BELT RF	17-2426	CRYSTAL PERFECTION OF AMMONIUM DIHYDROGEN PHOSPHATE.
BELUGINA NV	27-3164	DOMAIN STRUCTURE OF GAMMA-IRRADIATED TGS CRYSTALS.
BELYAEV AD	15-2113	AGING OF THE FERROELECTRIC SEMICONDUCTOR ANTIMONY SULFOIODIDE IN A STATIC ELECTRIC PIE
	15-2114	ABSORPTION OF ULTRASOUND IN A(V)B(VI)C(VII) FERROELECTRIC SEMICONDUCTORS AT TEMPERATUR
	15-2115	ABSORPTION OF ULTRASOUND IN AN ANTIMONY SULFOIODIDE FERROELECTRIC SEMICONDUCTOR NEAR I
	15-2116	DETERMINATION OF THE LATTICE PARAMETERS AND CURIE POINTS OF SOLID SOLUTIONS OF A(V)B(V
	15-2119	INFLUENCE OF ILLUMINATION ON THE FORM OF THE VOLTAGE AMPERE CHARACTERISTICS OF ANTIMON
	15-2121	PIEZOELECTRIC VIBRATIONS IN ANTIMONY SULFOIODIDE CRYSTALS NEAR THE CURIE POINT (EFFECT
	15-2123	POLARIZATION EFFECTS IN THE ANOMALOUS ABSORPTION OF ULTRASOUND IN ANTIMONY SULFOIODIDE
BELYAEV IN	02-0047	SYNTHESIS AND CERTAIN PROPERTIES OF NEW COMPOUNDS WITH A PYROCHLORE STRUCTURE. (NIOBAT
	36-3528	X-RAY PHASE STUDY OF THE CADMIUM BORATE- ABO(3) SYSTEMS. (WHERE A IS STRONTIUM OR BARI
BELYAEV IM	04-0552	PHOTOFERROELECTRIC EFFECTS IN A(V)B(VI)C(VII) AND BARIUM TITANATE TYPE FERROELECTRICS.
	15-2108	NUCLEAR GAMMA RESONANCE SPECTROSCOPIC STUDY OF SEMICONDUCTING COMPOUNDS OF THE TYPE A(
	15-2118	EFFECT OF COMPOSITION ON GROWTH, DIELECTRIC, AND PHOTOELECTRIC PROPERTIES OF ANTIMONY
	15-2120	MOTION PICTURES OF THE PHASE TRANSITION IN ANTIMONY SULFOIODIDE CRYSTALS.
	15-2145	INFLUENCE OF NONEQUILIBRIUM CARRIERS ON THE PHASE TRANSITION IN FERROELECTRIC SEMICOND
	16-2266	FERROELECTRIC PHASE TRANSITION IN SILVER SODIUM DINITRITE
	27-2961	INFLUENCE OF THE SPONTANEOUS POLARIZATION ON THE PHOTOSTIMULATED EXOELECTRON EMISSION
BELYAEV YUN	17-2428	TRAPPING OF PARAMETRICALLY AMPLIFIED LIGHT WAVES IN A KDP CRYSTAL.
BELYAEVA AI	33-3364	FINE STRUCTURE OF EXCITON- MAGNON ABSORPTION OF LIGHT IN POTASSIUM MANGANESE TRIFLUORI
BELYATSKAS RP	15-2117	DIELECTRIC DISPERSION IN ANTIMONY SULFOIODIDE CRYSTALS.
BELYUSTIN AV	17-2427	EFFECT OF ANNEALING ON OPTICAL ANOMALIES OF POTASSIUM DIHYDROGEN PHOSPHATE CRYSTALS.
	17-2628	CONDITIONS FOR THE GROWTH CF POTASSIUM DIHYDROGEN PHOSPHATE WHISKERS.
	28-3217	GROWTH RATE OF THE (010) PLANE OF ROCHELLE SALT IN THE PRESENCE OF PURE DIRECT SKY BLU
BENDER VP	15-2124	EFFECTS OF ILLUMINATION ON THE DOMAIN STRUCTURE OF ANTIMONY SULFOIODIDE. (ETCHING METH
BENDIK OG	03-0170	MODEL OF THE FERROELECTRIC MODE.
BENDRIKOVA GG	27-2961	INFLUENCE OF THE SPONTANEOUS POLARIZATION ON THE PHOTOSTIMULATED EXOELECTRON EMISSION
BENEPE JW	17-2429	ELECTRONIC STUDIES OF POTASSIUM DIHYDROGEN PHOSPHATE. (ELECTROCALORIC STUDIES, SPONTAN
	17-2430	MEASUREMENTS OF THE SPONTANEOUS POLARIZATION IN POTASSIUM DIHYDROGEN PHOSPHATE.
BENGUIGUI L	03-0169	EFFECT OF ELECTRIC FIELD ON FERROELECTRIC AND ANTIFERROELECTRIC CERAMICS. PART-1: DEFO
	03-0171	SCME REMARKS ABOUT ANTIFERROELECTRIC PHASE TRANSITIONS (TO PARAELECTRIC OR FERROELECTR
	03-0172	THERMODYNAMIC THEORY OF THE MORPHOTROPIC PHASE TRANSITION TETRAGONAL- RHOMBOHEDRAL IN
	04-0554	DOUBLE INJECTION IN BARIUM TITANATE CERAMIC.
	04-0555	FERROELECTRIC LOSSES IN (PURE) BARIUM TITANATE CERAMIC.
	04-0557	SPACE CHARGE LIMITED CURRENT IN POLYCRYSTALLINE BARIUM TITANATE. (CURRENT VOLTAGE CHAR
	07-1251	FERROELECTRICITY AND ANTIFERROELECTRICITY IN PURE AND NIOBIUM DOPED LEAD ZIRCONATE.
	09-1486	ANTIFERROELECTRIC- FERROELECTRIC TRANSITION IN THE LEAD (ZIRCONATE, STANNATE, TITANATE
	09-1487	ELECTROSTRICTIVE CONSTANTS IN FERROELECTRIC AND ANTIFERROELECTRIC PHASES. (LANTHANUM O
BENNACEUR R	15-2198	ELECTROOPTIC EFFECTS OF THE ABSORPTION EDGE OF ANTIMONY SULFOIODIDE.
BENOIT JP	17-2431	RAMAN SPECTRA OF ADP IN THE PARAELECTRIC AND ANTIFERROELECTRIC PHASES.
	25-2906	LOW FREQUENCY RAMAN SPECTRUM OF A THIOUREA CRYSTAL AS A FUNCTION OF TEMPERATURE.
BENSON HJ	04-0556	POLARITONS IN TETRAGONAL BARIUM TITANATE: THEORY OF THE RAMAN LINE SHAPE.
BENTON RK	04-0866	HIGH L/F NOISE ANOMALY IN SEMICONDUCTING BARIUM STRONTIUM TITANATE.
BENYON A	36-3564	EXPERIMENTAL RELATIONSHIP BETWEEN THE SUPERCONDUCTING AND THE FERROELECTRIC PHASE TRAN
BERAK JM	10-1629	EFFECT OF OXYGEN DEFICIENCY ON ELECTRICAL TRANSPORT PROPERTIES OF TUNGSTEN TRIOXIDE CR
BERCHA DM	15-2125	ANISOTRCPY OF THE REFLECTION SPECTRA OF ANTIMONY SULFOIODIDE CRYSTALS.
	15-2126	CONDUCTIVITY, THERMOELECTRIC PROPERTIES, AND BAND STRUCTURE OF CRYSTALLINE BISMUTH SEL
	15-2127	DOMAIN STRUCTURE AND LOCAL STATES IN ANTIMONY SULFOIODIDE TYPE SEMICONDUCTOR FERROELEC
	15-2128	PIEZOELECTRIC RESISTANCE EFFECT IN BISMUTH SELENOIODIDE CRYSTALS.
BERDOV GI	06-1162	EPR INVESTIGATION OF POLYCRYSTALLINE STRONTIUM TITANATE CONTAINING MANGANESE DIOXIDE.
BEREZHNOI AA	03-0174	ELECTROOPTICAL EFFECT IN FERROELECTRIC CRYSTALS HAVING A SMEARED PHASE TRANSITION.
	08-1280	LIGHT BEAM DEFLECTION CONTROLLED WITH A PRISM OF LEAD MAGNESIO NIOBATE CRYSTALS.
	36-3521	ANISCTRCPY OF THE ELECTROOPTICAL EFFECT IN CRYSTALS OF LEAD MAGNONIOBATE.
BEREZHNOI BM	17-2698	ELECTROOPTICAL AND OPTICAL PROPERTIES OF PARTIALLY DEUTERATED RUBIDIUM DIHYDROGEN PHOS
BERGER J	26-2926	VIBRATICN SPECTRUM OF THE FORMATE ION IN COPPER FORMATE TETRAHYDRATE AND COPPER DEUTER
BERGMAN AG	17-2434	POLYTHERM OF SOLUBILITY OF THE AMMONIUM DIHYDROGEN PHOSPHATE- POTASSIUM DIHYDROGEN PHO
BERGMAN JG	02-0049	NONLINEAR OPTICAL MATERIALS (REVIEW, DATA COLLECTION, MATERIALS NEEDS, 118 REFS)
	08-1465	PIEZOELECTRIC AND PHOTOELASTIC PROPERTIES OF LITHIUM IODATE.
	36-3530	FYRCELECTRICITY AND OPTICAL SECOND HARMONIC GENERATION IN POLYVINYLIDENE FLUORIDE FILM
	36-3560	PYROELECTRIC PROPERTIES OF POLYVINYLIDENE FLUORIDE AND ITS USE FOR INFRARED DETECTION.
	36-3595	PYROELECTRIC AND NONLINEAR OPTICAL PROPERTIES OF POLED POLYVINYLIDENE FLUORIDE FILMS.
BERGMANN H	02-0048	CHEMISTRY AND PHYSICS OF (ALKALI) METAL TANTALATES. (PREPARATION, REVIEW)
BERLINCOURT D	09-1488	RELEASE OF ELECTRIC ENERGY IN LEAD NIOBIUM ZIRCONATE TITANATE STANNATE BY TEMPERATURE
	09-1489	VIBRATION OF ELECTROELASTIC CONSTANTS OF POLYCRYSTALLINE LEAD TITANATE ZIRCONATE WITH
BERLINGER W	06-1164	MONODCMAIN STRONTIUM TITANATE (OCCURANCE IN THIN (110) PLATES BELOW CUBIC- TETRAGONAL
	06-1166	CRITICAL ASYMMETRY IN LOCAL FLUCTUATIONS IN STRONTIUM TITANATE ABOVE THE CRITICAL TEMP
	06-1169	MONODOMAIN STRONTIUM TITANATE.
	06-1170	CRDEF PARAMETER AND PHASE TRANSITIONS OF STRESSED STRONTIUM TITANATE. (EPR OF IRON- VA
	06-1171	PHOTOCHROMIC IRON IN STRONTIUM TITANATE. EVIDENCE FROM PARAMAGNETIC RESONANCE.
	06-1231	ANALYSIS OF THE IRON OXYGEN VACANCY CENTER IN THE TETRAGONAL PHASE OF STRONTIUM TITANA
	06-1232	EPR STUDIES AND DYNAMICS IN STRONTIUM TITANATE IN THE NEIGHBORHOOD OF THE PHASE TRANSF
	06-1233	FLUCTUATIONS AND CORRELATIONS IN STRONTIUM TITANATE ABOVE THE CRITICAL TEMPERATURE.
	06-1234	THE IRON(3+)- VACANCY CENTER IN THE TETRAGONAL PHASE OF STRONTIUM TITANATE.
BERNARD C	36-3529	NEW FERROELECTRIC (PYROCHLORE TYPE) COMPOUND (AMMONO THIO CADMIUM NIOBATE, WITH 4 ELEC
BERNARD J	03-0491	SHOCK WAVE INDUCED ELECTROMECHANICAL CONVERSION IN FERROELECTRIC MATERIALS.
BERNARD M	03-0177	VOLTAGES AND CURRENTS DURING DEPOLARIZATION OF A FERROELECTRIC MATERIAL. (THEORY)

BERNARD M   27-2962   SPECIFIC HEAT ANOMALY IN THE NEIGHBORHOOD OF A FERRO- PARAELECTRIC TRANSITION. (TGS)
BERNARD MP   17-2432   LOW TEMPERATURE INFRARED ABSORPTION SPECTRA OF AMMONIUM DIHYDROGEN ORTHOPHOSPHATE (400
                     21-2811   LOW TEMPERATURE INFRARED ABSORPTION SPECTRA (5200-4200 PER CM) OF GUANIDINIUM ALUMINUM
BERNDES G   28-3212   ANALYSIS OF DOUBLE HYSTERESIS LOOPS. (ROCHELLE SALT)
BERNSTEIN JL   08-1270   FERROELASTIC EFFECT IN LANTHANUM ORTHOFERRITE.
                     13-1895   FERROELECTRIC TUNGSTEN BRONZE TYPE CRYSTAL STRUCTURES. PART-3: POTASSIUM LITHIUM NIOBA
                     30-3289   FERROELECTRIC FERROELASTIC PARAMAGNETIC BETA GADOLINIUM MOLYBDATE. CRYSTAL STRUCTURE O
                     33-3358   FERROELASTIC EFFECT IN RUBIDIUM IRON TETRAFLUORIDE AND CESIUM IRON TETRAFLUORIDE.
                     33-3386   FERROELECTRIC PARAELASTIC PARAMAGNETIC BARIUM COBALT TETRAFLUORIDE CRYSTAL STRUCTURE (
                     35-3493   CRYSTAL STRUCTURE OF PIEZOELECTRIC BISMUTH(12) GERMANIUM OXIDE(20).
                     36-3574   PYROELECTRIC AMMONIUM IODATE, A POTENTIAL FERROELASTIC: CRYSTAL STRUCTURE.
BERRE B   03-0176   STRUCTURAL PHASE TRANSITIONS AND SOFT MODES.
                     06-1127   ULTRASONIC PROPAGATION, STRESS EFFECTS, AND INTERACTION PARAMETERS AT THE DISPLACIVE T
BERSUKER IB   03-0175   MICROTHEORY OF SPONTANEOUS POLARIZATION AND PHASE TRANSITIONS IN CRYSTALS OF PEROVSKIT
                     17-2433   ORIGIN OF THE SPONTANEOUS POLARIZATION AND THE ISOTOPE EFFECTS IN FERROELECTRIC POTASS
BERTAIN LF   04-0558   PERTURBED DIRECTIONAL CORRELATION OF SCANDIUM-44 IN THE FERROELECTRIC PHASES OF BARIUM
BERTAUT EF   03-0173   COLLINEAR AND NONCOLLINEAR FERRO- AND ANTIFERROELECTRICS. (REPRESENTATION THEORY APPLI
                     36-3542   FERROELECTRIC PROPERTIES OF HEXAGONAL ORTHOMANGANITES OF YTTRIUM AND RARE EARTHS. (YTT
BERTHOLD JW   11-1676   THREE LAYER BROAD BAND ANTIREFLECTION COATINGS FOR LITHIUM NIOBATE.
BESPALOV VI   17-2421   LIGHT ABSORPTION IN POTASSIUM DIHYDROGEN PHOSPHATE SINGLE CRYSTALS.
BESSONOVA EN   09-1582   INFLUENCE OF THE CONCENTRATION OF POINT DEFECTS ON THE INTERNAL FRICTION IN POLYCRYSTA
BESWICK JA   11-1660   FERROELECTRICS IN THE LITHIUM POTASSIUM NIOBATE SYSTEM.
                     13-1896   FERROELECTRICS IN THE POTASSIUM OXIDE- STRONTIUM OXIDE- NIOBIUM PENTOXIDE.
BETHE K   04-0559   MICROWAVE DIELECTRIC LOSSES IN OXYGEN OCTAHEDRON FERROELECTRICS ABOVE THE CURIE TEMPER
                     04-0560   PREPARATION AND PROPERTIES OF (BARIUM, STRONTIUM) TITANATE SINGLE CRYSTALS.
                     04-0561   SINGLE CRYSTALS OF THE SYSTEM (BARIUM, STRONTIUM) TITANATE. (SYNTHESIS AND LOW FREQUEN
                     08-1281   MICROWAVE BEHAVIOR OF NONLINEAR DIELECTRICS (EXPERIMENT AND DISCUSSION OF BARIUM (OR L
BEZDETNYI NM   15-2104   INFLUENCE OF NONEQUILIBRIUM CARRIERS ON THE DOMAIN STRUCTURE OF THE FERROELECTRIC SEMI
                     15-2105   SHORT CIRCUIT PHOTOEMF IN FERROELECTRIC SEMICONDUCTORS. (ANTIMONY SULFOIODIDE CRYSTALS
                     15-2129   CHARACTERISTICS OF THE PHOTOELECTRIC EMF OF ANTIMONY SULFOIODIDE.
                     15-2130   LIGHT INDUCED PYROCURRENTS IN FERROSEMICONDUCTOR ANTIMONY SULFOIODIDE.
                     15-2131   PYROCURRENTS HIGHER THAN THE CURIE POINT IN ANTIMONY SULFOIODIDE.
                     15-2132   PHOTOEMF'S IN THE FERROELECTRIC SEMICONDUCTOR ANTIMONY SULFOIODIDE.
                     15-2133   PHOTODOMAIN EFFECT IN NATURALLY POLARIZED ANTIMONY SULFOIODIDE CRYSTALS.
                     15-2134   PHOTOELECTRIC PROPERTIES (PHOTOCONDUCTIVITY) OF ANTIMONY SULFOIODIDE SINGLE CRYSTALS.
                     15-2135   PHOTODOMAIN EFFECT IN NATURALLY POLARIZED ANTIMONY SULFOIODIDE CRYSTALS.
                     15-2136   SOME FEATURES OF THE PHOTOEMF SPECTRUM OF ANTIMONY SULFOIODIDE.
                     15-2228   INFLUENCE OF LIGHT ON SELF POLARIZATION OF SINGLE CRYSTALS OF ANTIMONY SULFOIODIDE.
BEZNOSIKOV BV   33-3360   ELASTIC PROPERTIES OF CESIUM LEAD TRICHLORIDE.
                     33-3364   FINE STRUCTURE OF EXCITON- MAGNON ABSORPTION OF LIGHT IN POTASSIUM MANGANESE TRIFLUORI
                     33-3401   TWO MAGNON SCATTERING OF LIGHT IN ANTIFERROMAGNETIC POTASSIUM MANGANESE TRIFLUORIDE.
BEZNOSIKOVA NV   22-2863   PREPARATION AND ANALYSIS OF SODIUM AMMONIUM SELENATE DIHYDRATE CRYSTALS.
BHALLA AS   30-3258   DISLOCATION ETCHANTS FOR BETA GADOLINIUM MOLYBDATE.
BHIDE VG   04-0562   EFFECT OF DOPING BARIUM TITANATE WITH MANGANESE DIOXIDE.
                     04-0563   IONIC CHARACTERS FROM MOSSBAUER ISOMER SHIFTS. (DEGREE OF COVALENCY IN IRON DOPED BARI
                     04-0564   MOSSBAUER STUDIES OF LOWER TRANSITIONS IN BARIUM TITANATE. (LATTICE INSTABILITIES)
                     05-1040   MOSSBAUER EFFECT FOR IRON-57 IN FERROELECTRIC LEAD TITANATE.
BIAZZO MR   12-1872   FABRICATION OF A LITHIUM TANTALATE TEMPERATURE STABILIZED OPTICAL MODULATOR.
BICHARD VM   17-2435   1.06 MICRON ABSORPTION COEFFICIENTS OF DEUTERATED KDP WITH 70-100% DEUTERATION.
                     35-3494   ELECTROOPTIC MEASUREMENTS ON LITHIUM GERMANATE AND LEAD GERMANATE. (LITHIUM GERMANATE
BICKLEY WP   11-1660   FERROELECTRICS IN THE LITHIUM POTASSIUM NIOBATE SYSTEM.
BIEDENKAPP D   16-2393   TEMPERATURE DEPENDENCE OF THE SODIUM-23 NUCLEAR QUADRUPOLE COUPLING CONSTANT IN SODIUM
BIEDRZYCKI K   04-0982   EMISSION OF LIGHT BY SURFACE MICROPLASMA OF CERAMIC (BARIUM, LEAD) TITANATE NEAR THE C
BIELIS IYA   28-3181   REARRANGEMENT OF THE DOMAIN STRUCTURE IN ROCHELLE SALT NEAR THE PHASE TRANSITION.
BIELSKA-LEWANDOWSKA H 03-0360 NEW TWO- AND THREE PHASE FERROELECTRIC FERROMAGNETIC MATERIALS. SOME PROBLEMS AND RESU
                     04-0565   ELECTRIC AND MAGNETIC MEASUREMENTS OF CERTAIN TWO PHASE FERROELECTRIC FERROMAGNETIC MI
BIERSTEDT PE   30-3262   PRECISION PARAMETERS OF SOME LANTHANON MOLYBDATE TYPE RARE EARTH MOLYBDATES. (CRYSTAL
                     30-3264   PRECISION PARAMETERS OF THE FERROELECTRIC RARE EARTH MOLYBDATES: LANTHANUM MOLYBDATE.
BIGGERS JV   04-0566   DC BIAS EFFECTS IN BARIUM TITANATE CERAMICS.
                     04-0567   SURFACE STUDIES OF BARIUM TITANATE CERAMICS USING THE SCANNING ELECTRON MICROSCOPE.
                     09-1490   DC BIAS EFFECTS IN LEAD ZIRCONATE TITANATE CERAMICS.
                     09-1491   TEMPERATURE AND FIELD DEPENDENCE OF THE ANTIFERROELECTRIC FERROELECTRIC PHASE BOUNDARY
                     09-1590   PHASE BOUNDARY AND DEFECT STRUCTURE STUDIES IN PLZT CERAMICS.
BILYK DI   04-0984   PREPARATION OF FLUORINE CONTAINING BARIUM TITANATE.
BINBREK OS   19-2544   RAMAN AND INFRARED STUDIES OF THE FERROELECTRIC TRANSITION IN AMMONIUM SULFATE.
BINGGELI B   27-2963   SUPERSONIC DOMAIN WALL MOTION IN TRIGLYCINE SULFATE.
BIRAN A   08-1282   MOSSBAUER MEASUREMENTS OF BISMUTH FERRATE- BISMUTH FERRATE- LEAD ZIRCONATE SYSTEMS.
                     19-2770   MOSSBAUER STUDY ON FERROELECTRIC PROPERTIES OF FERRIC AMMONIUM SULFATE DODECAHYDRATE S
BIRD MJ   26-2927   UNIT CELL AND SPACE GROUP OF THE ANTIFERROELECTRIC PHASE OF COPPER FORMATE TETRAHYDRAT
BIRMAN BI   03-0178   DETERMINATION OF THE GROWTH RATE OF FERROELECTRIC CRYSTALS FROM BINARY SYSTEMS. (THEOR
BISHOP PB   34-3412   ELECTRIC POLARIZATION EFFECTS IN PURE AND DOPED ICE AT LOW TEMPERATURES.
BISMUTH G   04-0847   ELECTRICAL CONTROL OF FIXATION AND ERASURE OF HOLOGRAPHIC PATTERNS IN FERROELECTRIC MA
                     04-0848   HOLOGRAPHIC STORAGE, ELECTRICAL FIXING AND ERASING IN DOPED BARIUM TITANATE CRYSTALS.
                     09-1575   INSCRIPTION OF HOLOGRAPHIC RESULTS IN TRANSPARENT CERAMIC FERROELECTRICS. (LEAD LANTHA
                     11-1767   BIREFRINGENCE INDUCED IN A LITHIUM NIOBATE CRYSTAL.
BISWAS AB   08-1293   FERROELECTRIC PROPERTIES OF TETRAGONAL (BARIUM, LEAD) (NIOBATE, ZIRCONATE). (CURIE TEM
                     13-1925   STUDIES ON THE DIELECTRIC AND FERROELECTRIC PROPERTIES OF THE BARIUM LEAD NIOBATE ZIRC
BJORKHOLM JE   11-1677   RELATIVE MEASUREMENT OF THE OPTICAL NONLINEARITIES OF POTASSIUM DIHYDROGEN PHOSPHATE.
BJORKSTAM JL   17-2408   NMR PROTON SECOND MOMENTS IN POTASSIUM DIHYDROGEN PHOSPHATE TYPE CRYSTALS.
                     17-2436   ABSENCE OF A LINE WIDTH TRANSITION FOR THE PROTONS IN POTASSIUM DIHYDROGEN PHOSPHATE T
                     17-2437   EFFECT OF DEUTERON SUBSTITUTION ON DOMAIN DYNAMICS IN POTASSIUM DIHYDROGEN PHOSPHATE.
                     17-2581   DIELECTRIC PROPERTIES OF DEUTERATED POTASSIUM DIHYDROGEN PHOSPHATE.
                     18-2718   NMR AND DIELECTRIC RELAXATION STUDIES OF LONG RANGE CORRELATION IN FERROELECTRIC POTAS
                     18-2719   NMR STUDIES IN SOME HYDROGEN BONDED FERROELECTRICS. (DEUTERATED ROCHELLE SALT, POTASSI
BLACKBURN H   27-2964   THERMAL ANALYSIS OF PYROELECTRIC DETECTORS. (TRIGLYCINE SULFATE EXAMPLE)
BLANC J   06-1098   ELECTROCOLORATION IN STRONTIUM TITANATE: VACANCY DRIFT AND OXIDATION REDUCTION OF TRAN
BLANCKENEAGEN PV   34-3467   LATTICE DYNAMICS OF ICE.
BLANK H   03-0255   ELECTRON MICROSCOPE TRANSMISSION IMAGES OF COHERENT DOMAIN BOUNDARIES. PART-1: DYNAMIC
BLASSE G   36-3531   RADIATIONLESS TRANSITIONS IN THE EUROPIUM(3+) CENTER IN LANTHANUM ALUMINATE.
BLAZEY KW   06-1099   OPTICAL ABSORPTION EDGE OF STRONTIUM TITANATE AROUND THE 105 DEGREES K PHASE TRANSITIO

BLECKMANN P   25-2907   MOLECULAR AND LATTICE VIBRATIONS OF ORTHORHOMBIC AND FERROELECTRIC THIOUREA. PART-2: C
                   25-2920   MOLECULAR AND LATTICE VIBRATIONS OF ORTHORHOMBIC AND FERROELECTRIC THIOUREA. PART-1: T
BLINC R   02-0050   FERROELECTRICS. (THEORY, DYNAMICAL ORDER DISORDER PROPERTIES, KDP, TGS, ROCHELLE SALT,
                   03-0179   DYNAMICS OF ORDER DISORDER TYPE FERROELECTRICS AND ANTIFERROELECTRICS
                   03-0180   MAGNETIC RESONANCE IN FERROELECTRICS. (MODEL FOR POTASSIUM DIHYDROGEN PHOSPHATE PRESEN
                   03-0181   PROTON LATTICE INTERACTIONS IN HYDROGEN BONDED FERROELECTRIC CRYSTALS. (THEORY, KDP TY
                   03-0357   ON THE SYMMETRY OF THE SOFT MODES IN THE FERROELECTRIC PHASE TRANSITIONS.
                   16-2267   NMR STUDY OF THE FERROELECTRIC TRANSITIONS IN DIGLYCINE NITRATE AND TRIS-SARCOSINE CAL
                   16-2320   NITROGEN-14 NUCLEAR QUADRUPOLE RESONANCE IN FERROELECTRIC SODIUM NITRITE.
                   17-2438   DEUTERON MAGNETIC RESONANCE AND RELAXATION IN FERROELECTRIC POTASSIUM DIDEUTERIUM PHOS
                   17-2439   RUBIDIUM-87 AND ARSENIC-75 QUADRUPOLAR COUPLING IN FERROELECTRIC RUBIDIUM DIHYDROGEN P
                   17-2440   SPIN LATTICE RELAXATION BY THE FERROELECTRIC MODE IN POTASSIUM DIHYDROGEN PHOSPHATE.
                   17-2441   ULTRASLOW HYDROGEN MOTION IN POTASSIUM DIHYDROGEN PHOSPHATE AND SODIUM TRIHYDROGEN SEL
                   17-2442   VANISHING OF FERROELECTRICITY IN POTASSIUM DIHYDROGEN PHOSPHATE AT HIGH PRESSURES.
                   17-2685   ORIER DISORDER THEORY OF POTASSIUM DIHYDROGEN PHOSPHATE TYPE FERROELECTRICS.
                   18-2720   CESIUM-183 SPIN LATTICE RELAXATION AND RESONANCE IN FERROELECTRIC CESIUM DIDEUTERIUM A
                   18-2721   DEUTERON QUADRUPOLE COUPLING IN FERROELECTRIC CESIUM DIDEUTERIUM ARSENATE.
                   19-2749   NITROGEN-14 QUADRUPOLE COUPLING IN PARAELECTRIC AMMONIUM SULFATE.
                   19-2769   NUCLEAR SPIN LATTICE RELAXATION IN SOME FERROELECTRIC AMMONIUM SALTS. (AMMONIUM SULFAT
                   22-2831   DOUBLE HYSTERESIS LOOPS AT THE UPPER CURIE POINT OF SODIUM TRIHYDROGEN SELENITE.
                   22-2832   NUCLEAR MAGNETIC RESONANCE STUDY OF THE FERROELECTRIC TRANSITIONS IN SODIUM TRIHYDROGE
                   23-2885   PROTON SPIN LATTICE RELAXATION IN FERROELECTRIC COLEMANITE.
                   26-2928   SPIN LATTICE RELAXATION BY QUASI SPIN WAVES IN ORDER DISORDER TYPE FERROELECTRICS: POL
                   27-2965   BERYLLIUM-9 QUADRUPOLE PERTURBED NMR STUDY OF THE FERROELECTRIC TRANSITION IN DEUTERAT
                   27-2966   CRITICAL BEHAVIOR OF FERROELECTRIC TGS AND DEUTERATED TGS (HYSTERESIS LOOPS JUST BELOW
                   27-2967   NUCLEAR SPIN LATTICE RELAXATION IN FERROELECTRIC TRIGLYCINE SULFATE.
                   27-2968   PULSED NITROGEN PROTON DOUBLE RESONANCE STUDY OF THE FERROELECTRIC TRANSITION IN TRIGL
                   27-2969   PULSED DOUBLE RESONANCE STUDY OF THE NUCLEAR QUADRUPOLE INTERACTIONS OF NITROGEN-14 IN
                   27-3056   CRITICAL PROPERTIES OF TGS AND ROCHELLE SALT AS DETERMINED BY AN ELECTRIC FIELD METHOD
                   28-3185   EFFECT OF APPLIED ELECTRIC FIELD ON EPR SPECTRA OF (COPPER(2+)) DOPED FERROELECTRIC RO
                   28-3186   SECOND ORDER QUADRUPOLE SHIFTS OF THE SODIUM-23 MAGNETIC RESONANCE IN ROCHELLE SALT.
                   28-3187   SPECIFIC HEAT ANOMALY IN FERROELECTRIC ROCHELLE SALT (CALCULATION OF TEMPERATURE DEPEN
                   28-3236   DYNAMICS OF FERROELECTRIC ROCHELLE SALT.
                   28-3237   DYNAMICS OF FERROELECTRIC ROCHELLE SALT.
BLINOV LM   36-3532   STARK EFFECT AND FERROELECTRICITY IN THE ORGANIC SEMICONDUCTOR COPPER PHTHALOCYANINE.
BLISTANOV AA   17-2648   STUDY OF THE (MECHANICAL) STRENGTH PROPERTIES OF POTASSIUM DIHYDROGEN PHOSPHATE CRYSTA
BLOEMBERGEN N   17-2695   TEMPERATURE DEPENDENCE OF OPTICAL (SECOND) HARMONIC GENERATION IN POTASSIUM DIHYDROGEN
BLOKH OG   17-2443   PHOTOELASTIC EFFECT IN 45 DEGREE X-CUTS OF POTASSIUM AND AMMONIUM DIHYDROGEN PHOSPHATE
                   27-2970   ELECTROGYRATION AND FERROELECTRIC PHASE TRANSITIONS (TGS).
BLOOMFIELD PE   03-0182   ELECTRIC FIELD DISTRIBUTIONS IN DIELECTRICS, WITH SPECIAL EMPHASIS ON NEAR SURFACE REG
PLUM W   28-3184   MEASUREMENT OF THE COMPLEX DIELECTRIC CONSTANT OF ROCHELLE SALT AT 10 GHZ AS A FUNCTIO
BLUMENSTOCK D   31-3325   PARAELECTRIC RESONANCE OF LITHIUM ION IN POTASSIUM CHLORIDE.
BLUMENTHAL RN   04-0822   TITANIUM RICH NONSTOICHIOMETRIC BARIUM TITANATE. PART-1: HIGH TEMPERATURE ELECTRICAL C
                   04-0885   HIGH TEMPERATURE THERMODYNAMIC AND X-RAY DIFFRACTION STUDY OF NONSTOICHIOMETRIC BARIUM
                   04-0886   TITANIUM RICH NONSTOICHIOMETRIC BARIUM TITANATE. PART-3: HIGH TEMPERATURE THERMODYNAMI
BOARD R   06-1100   BAND STRUCTURE OF STRONTIUM TITANATE AS MEASURED BY X-RAY PHOTOELECTRON SPECTROSCOPY (
BOARDMAN SM   37-3663   ANALOG MEMORY DEVICES EMPLOYING PIEZOELECTRIC- FERROELECTRIC INTERACTIONS FOR ADAPTIVE
BOBB LC   13-1905   LOW FREQUENCY VIBRATIONAL MODES IN BARIUM SODIUM NIOBATE.
                   13-1906   RAMAN SPECTRA OF BARIUM SODIUM NIOBATE.
BOCHKAREVA OB   13-2029   THERMOGRAPHIC STUDY OF THE REACTION OF BISMUTH(4) TITANIUM(3) OXIDE(12) WITH VANADIUM
BOCHKOV EG   36-3601   DOMAIN STRUCTURE AND TEMPERATURE DEPENDENCE OF THE PIEZOELECTRIC CONSTANTS OF STIBOTAN
BODI A   19-2750   MICROWAVE DIELECTRIC MEASUREMENTS OF GAMMA-IRRADIATED FERROELECTRIC AMMONIUM SULFATE.
BOERSCH H   17-2444   DIFFRACTION OF A CONTINUOUS LIGHT WAVE IN KDP.
BOFFI S   02-0051   LATTICE DYNAMICS BY NEUTRON SCATTERING. PARTS 1 AND 2. (REVIEW)
BOGACHEK EN   03-0342   MECHANICAL MODEL OF LONG WAVELENGTH VIBRATIONS IN A (SINGLE DOMAIN) FERROELECTRIC TYPE
BOGATIN AS   04-0568   ELECTRICAL PROPERTIES OF CRYSTALLITES AND INTERCRYSTALLITE LAYERS IN THE SEMICONDUCTOR
                   04-0570   THERMOSTABILIZING PROPERTIES OF SEMICONDUCTING BARIUM TITANATE.
BOGATKO VV   04-0569   SURFACE CONDUCTIVITY IN POLARIZED BARIUM TITANATE CRYSTALS.
BOGDANOV SV   03-0184   EXPERIMENTAL CRITERION OF FERROELECTRICITY. (SHARP OR DIFFUSE TRANSITIONS)
                   03-0185   NONLINEARITY OF DIELECTRIC POLARIZATION AND FERROELECTRIC PROPERTIES OF MATERIALS.
                   04-0864   VIBRATIONAL SPECTRA OF STRONTIUM, BARIUM AND CALCIUM TITANATES.
                   06-1145   STRUCTURE OF POLYCRYSTALLINE SOLID SOLUTIONS OF STRONTIUM AND BISMUTH TITANATES.
BOGDANOV VL   11-1678   COLLINEAR INTERACTION OF LONGITUDINAL ELASTIC WAVES IN LITHIUM NIOBATE CRYSTALS.
BOGOMOLOV AA   15-2220   EFFECT OF ILLUMINATION ON REPOLARIZATION PROCESS IN ANTIMONY SULFOIODIDE SINGLE CRYSTA
BOGOMOLOVA GA   30-3255   CERTAIN GROWTH, LASING AND SPECTROSCOPIC PROPERTIES OF NEODYMIUM(3+) DOPED GADOLINIUM
BOGUSLAVSKII AA   18-2722   STARK EFFECT IN THE NUCLEAR QUADRUPOLE RESONANCE SPECTRA OF THE FERROELECTRIC CRYSTAL
BOHM J   13-1907   EPITAXY OF PLATINUM ON BARIUM STRONTIUM NIOBATE.
                   30-3259   GROWTH AND MORPHOLOGY OF SINGLE CRYSTALS OF GADOLINIUM MOLYBDATE, A FERROELECTRIC AND
                   30-3301   DOMAIN STRUCTURE OF GADOLINIUM MOLYBDATE.
BOIKO AA   17-2445   THERMAL EXPANSION OF ADP AND DEUTERATED ADP IN THE REGION OF ANTIFERROELECTRIC PHASE T
BOIS GV   04-0572   INVERSION OF ELECTRICAL PROPERTIES OF SEMICONDUCTING BARIUM TITANATE.
BOISSIERE C   04-0571   IONIC CONTRIBUTION TO INJECTION AND CONDUCTION IN BARIUM TITANATE (CONDENSER CHARGE- D
BOKOV VA   01-0029   FERROELECTRICS AND ANTIFERROELECTRICS.
                   04-0573   CHANGE OF THE MOSSBAUER EFFECT PROBABILITY ON TIN-119 IN BARIUM TITANATE STANNATE SOLI
BOKSHA SS   15-2137   HIGH GAS PRESSURE CRYSTAL GROWTH OF ANTIMONY SULFOIODIDE, NICKEL CARBONATE AND BLACK P
                   15-2138   PREPARATION OF ANTIMONY SULFOIODIDE TEXTURES FROM THE MELT UNDER PRESSURE AND INVESTIG
BOKUT BV   03-0186   NONLINEAR SADOVSKII EFFECT (THEORY, PROPAGATION OF RADIATION AND FIRST HARMONIC, CALCU
BOLDRINI P   36-3603   NEW FERROELECTRIC: SEMICARBAZIDE HYDROCHLORIDE.
BOLDYREV VV   16-2319   EFFECT OF ADDING TITANIUM(+) AND STRONTIUM(2+) IONS ON THE ELECTRICAL CONDUCTIVITY OF
BOLK TR   04-0574   PHOTOCONDUCTIVITY AND PHOTOFERROELECTRIC PHENOMENA IN BARIUM TITANATE.
BOLLMANN W   11-1679   DISORDER OF LITHIUM NIOBATE CRYSTALS.
                   14-2103   OBSERVATIONS OF FERROELECTRIC DOMAINS IN BORACITES.
BOLOGNESI GP   17-2446   SOLUBILITY IN THE GROWTH OF ELECTROOPTIC CRYSTALS ADP AND KDP.
BOLTON HC   03-0279   QUANTUM FIELD THEORETICAL TREATMENT OF A DYNAMICAL MODEL OF THE FERROELECTRIC KDP. PAR
BONARDEL J   17-2447   FERROELECTRIC DOMAINS OF POTASSIUM DIHYDROGEN PHOSPHATE. (93 REFS)
BONDARENKO VS   04-0656   SIZE EFFECTS IN FERROELECTRICS. (BARIUM TITANATE)
                   04-0832   STRUCTURE AND PROPERTIES OF BARIUM TITANATE FILMS PRODUCED BY CATHODE SPUTTERING.
                   11-1693   THERMAL DEPENDENCE OF DIELECTRIC, PIEZOELECTRIC, AND ELASTIC PROPERTIES OF LITHIUM NIO
BONERA G   03-0187   NUCLEAR QUADRUPOLE SPIN LATTICE RELAXATION AND DYNAMICAL PROPERTIES OF FERROELECTRIC C
                   08-1283   NMR STUDY OF PHASE TRANSITIONS IN PEROVSKITES CRYSTALS. (NIOBIUM-93 AND SODIUM-23 RESO

BONERA G	13-1910	TRANSIENT NMR STUDY OF PHASE TRANSITIONS IN METALLIC SODIUM TUNGSTEN BRONZES.
	16-2268	NUCLEAR QUADRUPOLE SPIN LATTICE RELAXATION AND CRITICAL DYNAMICS OF FERROELECTRIC CRYS
BONIORT JY	04-0657	PREPARATION OF SODIUM AND BARIUM NIOBATES AND LEAD MOLYBDATE.
BONNER WA	13-1908	EFFECTS OF CHANGES IN MELT COMPOSITION ON CRYSTAL GROWTH OF BARIUM SODIUM NIOBATE.
	13-1909	GROWTH CF BARIUM SODIUM NIOBATE CRYSTALS BY THE KYROPOULOS TECHNIQUE.
	13-2023	NONLINEAR OPTICAL PROPERTIES OF FERROELECTRIC LEAD NIOBIUM(4) OXIDE(12).
	13-2045	GROWTH OF BARIUM SODIUM NIOBATE SINGLE CRYSTALS FOR OPTICAL APPLICATIONS. (BY PULLING
	26-2945	NONLINEAR OPTICAL SUSCEPTIBILITY OF LITHIUM FORMATE MONOHYDRATE.
BONSACK JP	04-0575	DIELECTRIC PROPERTIES OF BARIUM TITANATE CONTAINING NIOBIUM AND THE EFFECT OF ADDITIVE
BORETS AN	15-2141	PRODUCTION METHODS AND SOME OPTICAL PROPERTIES OF BISMUTH TELLUROIODIDE.
	15-2150	INDIRECT TRANSITIONS AND ABSORPTION IN THE MID-INFRARED IN ANTIMONY SULFOIODIDE CRYSTA
	15-2217	ABSORPTION SPECTRA OF BISMUTH TELLUROBROMIDE AND BISMUTH TELLUROIODIDE CRYSTALS.
BORISENKO AI	04-1003	REACTION SEQUENCE IN FORMING BARIUM METATITANATE IN AN INDUSTRIAL ROTARY FURNACE.
BORN RC	16-2269	PROGRESS TOWARD A FAST, NONVOLATILE, NONDESTRUCTIVE READOUT MEMORY ELEMENT UTILIZING P
BORNAREL J	17-2448	DISPLACEMENTS AND VELOCITIES OF THE DOMAIN WALLS IN KDP. EXISTENCE OF CRITICAL ELECTRI
	17-2450	EXISTENCE OF DISLOCATIONS AT DOMAIN TIPS IN FERROELECTRIC CRYSTAL POTASSIUM DIHYDROGEN
	17-2451	INTERDOMAIN AND DOMAIN EFFECT INTERACTIONS IN KDP (FERROELECTRIC DOMAINS).
	17-2452	STRESS INDUCED BIREFRINGENCE IN AN ISOLATED AND A SHORTCIRCUITED POTASSIUM DIHYDROGEN
	17-2453	VELOCITY OF PROPAGATION OF DOMAIN WALLS: INTERACTIONS BETWEEN DOMAINS. (IN KDP)
BORNAREL P	17-2449	ELECTRICAL AND OPTICAL INVESTIGATION OF FERROELECTRIC PROPERTIES OF POTASSIUM DIHYDROG
BORODIN VZ	04-0577	PHASE BOUNDARIES IN BARIUM TITANATE SINGLE CRYSTALS NEAR CURIE POINT.
BORODINA VA	04-0577	PHASE BOUNDARIES IN BARIUM TITANATE SINGLE CRYSTALS NEAR CURIE POINT.
BORRELLI NF	04-0576	DIELECTRIC AND OPTICAL PROPERTIES OF TRANSPARENT FERROELECTRIC GLASS CERAMIC SYSTEMS.
BORSA F	03-0187	NUCLEAR QUADRUPOLE SPIN LATTICE RELAXATION AND DYNAMICAL PROPERTIES OF FERROELECTRIC C
	08-1283	NMR STUDY OF PHASE TRANSITIONS IN PEROVSKITES CRYSTALS. (NIOBIUM-93 AND SODIUM-23 RESO
	13-1910	TRANSIENT NMR STUDY OF PHASE TRANSITIONS IN METALLIC SODIUM TUNGSTEN BRONZES.
	16-2268	NUCLEAR QUADRUPOLE SPIN LATTICE RELAXATION AND CRITICAL DYNAMICS OF FERROELECTRIC CRYS
	36-3533	NMR STUDY OF THE STRUCTURAL PHASE TRANSITION IN LANTHANUM ALUMINATE.
BORSTEL G	08-1284	EIGENFREQUENCIES AND EIGENVECTORS OF POLARITONS WITH APPLICATION TO LEAD NIOBATE. PART
	11-1694	ASSIGNMENTS OF OPTICAL PHONON MODES IN LITHIUM NIOBATE.
	11-1695	DIRECTIONAL DISPERSION AND ASSIGNMENT OF OPTICAL PHONONS IN LITHIUM NIOBATE.
	11-1697	LIGHT SCATTERING BY POLARITONS ASSOCIATED WITH ORDINARY PHOTONS IN LITHIUM NIOBATE.
	11-1766	EIGENFREQUENCIES AND EIGENVECTORS OF POLARITONS WITH APPLICATION TO LITHIUM NIOBATE. P
BOSMAN AJ	03-0188	TEMPERATURE DEPENDENCE OF DIELECTRIC CONSTANTS OF CUBIC IONIC COMPOUNDS. (HALIDES AND
BOUCHET G	04-0529	LEED STUDY OF SURFACE STRUCTURES ON THE (001) FACE OF BARIUM TITANATE.
BOUILLOT J	03-0189	THEORETICAL STATIC MODEL FOR BARIUM AND LEAD TITANATES - APPLICATION TO CALCULATING TH
	03-0261	THEORETICAL DYNAMIC STUDY OF CUBIC PHASE BARIUM TITANATE.
	04-0578	ELECTROSTATIC FIELD IN A SLIGHTLY ORTHORHOMBIC IONIC CRYSTAL. APPLICATION TO THE CALCU
	04-0826	STATIC SHELL MODEL FOR BARIUM AND LEAD TITANATES.
	04-1008	ELECTROSTATIC FIELD IN A PSEUDOCUBIC IONIC STRUCTURE. APPLICATION TO ALL BARIUM TITANA
	27-3040	OPTICAL ACTIVITY OF FERROELECTRIC DICALCIUM STRONTIUM PROPIONATE. (TRANSITION AT 209 D
BOUTIN H	28-3188	STRUCTURAL EFFECTS OF (DAMAGE FROM) IONIZING RADIATION IN FERROELECTRIC ROCHELLE SALT.
	32-3356	CRYSTAL STRUCTURE OF FERROELECTRIC LITHIUM HYDRAZINIUM SULFATE.
BOX HC	34-3413	ENDOR (ELECTRON NUCLEAR DOUBLE RESONANCE) STUDY OF X-IRRADIATED SINGLE CRYSTALS OF ICE
BOYD GD	11-1734	NONLINEAR OPTICAL POLARIZABILITY OF THE NIOBIUM- OXYGEN BOND. (LITHIUM NIOBATE, LITHIU
	11-1773	EFFECT OF OPTICAL INHOMOGENEITIES ON PHASE MATCHING IN NONLINEAR CRYSTALS (LITHIUM NIO
BOYER L	17-2454	MEASUREMENT OF THE ELASTIC CONSTANTS OF POTASSIUM DIHYDROGEN PHOSPHATE BY BRILLOUIN DI
	30-3260	FERROELECTRIC DOMAINS AND THE CONDENSED SOFT MODE IN GADOLINIUM MOLYBDATE.
	30-3281	LATTICE DYNAMICS OF A RIGID ION MODEL OF GADOLINIUM MOLYBDATE.
	30-3282	PRODUCTION OF SPONTANEOUS POLARIZATION BY ELASTIC INSTABILITIES IN PIEZOELECTRIC MATER
BOZHEVOLNOV EA	08-1428	GROWTH CF SINGLE CRYSTALS OF VARIOUS PEROVSKITES WITH COMPLEX COMPOSITIONS, AND THE ST
BRAAY PJ	14-2088	DOUBLETS CONCERNING THE STRUCTURE OF BORACITES.
BRADLER J	27-3106	X-RAY TOPOGRAPHIC OBSERVATIONS OF RADIATION DAMAGE IN TRIGLYCINE SULFATE SINGLE CRYSTA
BRADLEY FN	04-0582	GRAIN SIZE DEPENDENCE OF DIELECTRIC AND STRUCTURAL CHARACTERISTICS OF BARIUM TITANATE.
	13-1912	ELECTRICAL PROPERTIES OF TERNARY BARIUM OXIDE- LANTHANUM OXIDE- TITANIUM DIOXIDE COMPO
BRADT RC	04-0579	AGING IN TETRAGONAL FERROELECTRIC BARIUM TITANATE.
	08-1295	DIELECTRIC AGING IN TETRAGONAL SOLID SOLUTIONS OF CALCIUM TITANATE IN BARIUM TITANATE.
BRAHMECHA BG	04-0584	RESISTIVITY ANOMALY IN SEMICONDUCTING BARIUM TITANATE. (POSITIVE TEMPERATURE COEFFICIE
BRANDLE CD	30-3261	PHASE TRANSITIONS IN COMPLEX PEROVSKITES OF THE TYPE BARIUM LANTHANON MOLYBDATE.
BRANDON JK	02-0094	ROLE OF THE REGULAR OCTAHEDRON AND LARGER POLYHEDRA IN FERROELECTRIC MATERIALS. (HETER
	13-1911	CRYSTAL STRUCTURE AND PROPERTIES OF CALCIUM(2) NIOBIUM(2) OXIDE(7).
BRANSKI W	04-0815	TWO-PHASE FERROELECTRIC SYSTEM. PART-1: BARIUM TITANATE- METAL.
	04-0816	TWO PHASE FERROELECTRIC SYSTEMS. PART-2: FERROELECTRICS- SILVER (BARIUM TITANATE AND B
BRANWOOD A	04-0580	DIELECTRIC BREAKDOWN IN BARIUM TITANATE.
BRASLAU N	17-2669	OPTICAL MIXING OF COHERENT AND INCOHERENT LIGHT. (KDP CRYSTAL FILTER)
BRATSCHUN WR	09-1492	BARIUM ZIRCONATE LEAD TITANATE FERROELECTRIC CERAMICS.
BRATTON RJ	13-1995	RESISTIVITY BEHAVIOR OF SEMICONDUCTING YTTRIUM BARIUM ZIRCONIUM TITANATE. (TEMPERATURE
BRAUER H	04-0581	GRAIN BOUNDARY BARRIER LAYERS IN (N TYPE, ANTIMONY DOPED) BARIUM TITANATE CERAMIC WITH
	04-0583	PRODUCTION AND STUDY OF THE INNER BARRIERS IN SEMICONDUCTING BARIUM TITANATE CERAMICS.
BRAVO C	22-2857	THERMAL EXPANSION OF HYDROGEN BONDED FERROELECTRICS. LITHIUM TRIHYDROGEN SELENITE.
BRAZDZIUNAS PP	15-2139	FREQUENCY DEPENDENCE OF ANOMALOUS ULTRASOUND DAMPING IN ANTIMONY SULFOIODIDE SINGLE CR
BREBNER JL	06-1151	ELECTRONIC CONDUCTION IN SLIGHTLY REDUCED STRONTIUM TITANATE AT LOW TEMPERATURES.
BREHAT F	17-2456	INFLUENCE OF THE PHASE TRANSITION ON THE NORMAL VIBRATION MODES OF FERROELECTRIC CRYST
	17-2457	LOW FREQUENCY INFRARED ABSORPTION SPECTRA OF MONOCRYSTALS OF POTASSIUM DIHYDROGEN PHOS
	17-2459	TRANSMISSION OF POTASSIUM DIHYDROGEN PHOSPHATE IN THE INFRARED REGION IN THE PARA- AND
	25-2908	INFRARED ABSORPTION SPECTRUM OF THIOUREA IN THE PARA- AND FERROELECTRIC PHASES.
	27-3007	OPTICAL CONSTANTS OF TRIGLYCINE SELENATE IN THE NEAR INFRARED TO FAR INFRARED AND IN T
BRENMAN M	19-2769	NUCLEAR SPIN LATTICE RELAXATION IN SOME FERROELECTRIC AMMONIUM SALTS. (AMMONIUM SULFAT
	23-2885	PROTON SPIN LATTICE RELAXATION IN FERROELECTRIC COLEMANITE.
BREWS JR	06-1150	ENERGY BAND CHANGES IN PEROVSKITES DUE TO LATTICE POLARIZATION. (LCAO ANALYSIS OF BAND
BREZINA B	04-0585	INTERACTION BETWEEN THE 90 DEG AND 180 DEG DOMAINS IN BARIUM TITANATE.
	17-2455	GROWTH OF POTASSIUM DIHYDROGEN PHOSPHATE SINGLE CRYSTALS IN GEL.
	17-2458	REGULAR BEHAVIOR OF SOLID SOLUTIONS IN POTASSIUM DIHYDROGEN(1-X) DIDEUTERIUM(X) PHOSPH
	17-2527	INFLUENCE OF TUNNELING ON DIELECTRIC BEHAVIOR OF POTASSIUM DIHYDROGEN PHOSPHATE.
	19-2756	PROPERTIES OF FERROELECTRIC CADMIUM AMMONIUM SULFATE.
	27-2971	CONCENTRATION DEPENDENCE OF SOME FERROELECTRIC PROPERTIES OF SOLID SOLUTIONS OF TRIGLY
	27-2972	GROWTH AND CHARACTERIZATION OF SOLID SOLUTIONS OF FERROELECTRIC TRIGLYCINE SULFATE SIN
	27-2973	PROPERTIES OF DEUTERATED TRIGLYCINE FLUOBERYLLATE SINGLE CRYSTALS.
	29-3239	SOME PROPERTIES OF NONDEUTERATED AND DEUTERATED LITHIUM THALLIUM TARTRATE MONOHYDRATE.
	36-3534	NEW FERROELECTRIC LANGBEINITE THALLIUM CADMIUM SULFATE.

BRICE JC	13-1913	CZOCHRALSKI GROWTH OF BARIUM STRONTIUM NIOBATE CRYSTALS.
	13-1914	MODIFICATION TO THE CZOCHRALSKI METHOD OF CRYSTAL PULLING. (CONTROL OF CRYSTAL DIAMETE
BRICKLEY WP	13-1897	THE SEARCH FOR NEW FERROELECTRICS WITH THE TUNGSTEN BRONZE STRUCTURE.
BRIDENEAUGH PM	11-1683	PHASE EQUILIBRIA AND CRYSTAL GROWTH IN THE LITHIUM OXIDE- NIOBIUM PENTOXIDE SYSTEM.
	11-1684	SPATIALLY UNIFORM AND ALTERABLE SHG PHASE MATCHING TEMPERATURES IN LITHIUM NIOBATE (CR
	11-1691	NONSTOICHIOMETRY AND CRYSTAL GROWTH OF LITHIUM NIOBATE.
	11-1716	CONTROL OF LASER DAMAGE IN LITHIUM NIOBATE.
	11-1769	DEPENDENCE OF SECOND HARMONIC GENERATION COEFFICIENTS OF LITHIUM NIOBATE ON MELT COMPO
	11-1773	EFFECT OF OPTICAL INHOMOGENEITIES ON PHASE MATCHING IN NONLINEAR CRYSTALS (LITHIUM NIO
	11-1793	LOW FREQUENCY (510 MHZ) PARAMAGNETIC RESONANCE INVESTIGATION OF LITHIUM NIOBATE.
	11-1842	DEPENDENCE OF LINEAR ELECTROOPTIC EFFECT AND DIELECTRIC CONSTANT ON MELT COMPOSITION (
	13-1998	MEASUREMENTS OF SECOND HARMONIC GENERATION AND THE VARIATIONS IN THE FREE AND CLAMPED
BRIDOUX E	11-1680	COLLINEAR INTERACTION BETWEEN A LONGITUDINAL ACOUSTIC WAVE AND TWO LUMINOUS WAVES ALON
	11-1681	DETERMINATION OF A NONLINEAR PARAMETER FOR ACOUSTIC SURFACE WAVE CONVOLUTION IN LITHIU
	11-1682	GENERATION OF SECOND AND THIRD HARMONICS BY A TRANSVERSE ACOUSTIC WAVE PROPAGATING ITS
	11-1834	EXPERIMENTAL STUDY OF LIGHT DIFFRACTION BY HIGH FREQUENCY ACOUSTIC WAVES. APPLICATION
	11-1835	GENERATION OF THE THIRD HARMONIC OF A LONGITUDINAL HIGH FREQUENCY (950 MHZ) ACOUSTIC W
	11-1836	INTERACTIONS BETWEEN HIGH FREQUENCY ACOUSTICAL WAVES IN LITHIUM NIOBATE.
BRIEBWASSER S	02-0123	FERROELECTRIC MATERIALS. (REVIEW, 31 REFS)
BRIL A	36-3531	RADIATIONLESS TRANSITIONS IN THE EUROPIUM(3+) CENTER IN LANTHANUM ALUMINATE.
BRITH M	25-2915	VIBRATIONAL LINE WIDTHS OF THIOUREA.
BRIXNER LH	30-3256	CONTROL AND APPLICATION OF DOMAIN WALL MOTION IN GADOLINIUM MOLYBDATE.
	30-3257	CONTROL AND APPLICATION OF DOMAIN WALL MOTION IN GADOLINIUM MOLYBDATE.
	30-3262	PRECISION PARAMETERS OF SOME LANTHANON MOLYBDATE TYPE RARE EARTH MOLYBDATES. (CRYSTAL
	30-3263	PI- GMO: ANOTHER MODIFICATION OF GADOLINIUM MOLYBDATE.
	30-3264	PRECISION PARAMETERS OF THE FERROELECTRIC RARE EARTH MOLYBDATES: LANTHANUM MOLYBDATE.
BROBERG TW	17-2461	LONG WAVELENGTH POLARIZATION FLUCTUATIONS IN ANTIFERROELECTRIC AMMONIUM DIHYDROGEN PHO
	17-2655	EFFECT OF PROTON- PHONON COUPLING ON THE FERROELECTRIC MODE IN POTASSIUM DIHYDROGEN PH
	17-2657	RAMAN SPECTRA OF TETRAGONAL POTASSIUM DIHYDROGEN PHOSPHATE.
BRODALE GE	30-3275	MAGNETOTHERMODYNAMICS OF ANTIFERROMAGNETIC, FERROELECTRIC BETA GADOLINIUM MOLYBDATE. P
BRODY EM	17-2462	LIGHT SCATTERING STUDY OF THE FERROELECTRIC TRANSITION IN KDP.
	17-2463	LIGHT SCATTERING FROM CRITICAL FLUCTUATIONS IN POTASSIUM DIDEUTERIUM PHOSPHATE.
	17-2497	LIGHT SCATTERING STUDIES OF THE SOFT OPTIC AND ACOUSTIC MODES OF POTASSIUM DIHYDROGEN
BRODY PS	04-0588	SHOCK WAVE STUDIES OF SURFACE LAYERS IN BARIUM TITANATE.
	04-0589	SHOCK INDUCED TRANSITION IN BARIUM TITANATE.
BROG KC	02-0131	PREPARATION AND PROPERTIES OF RARE EARTH COMPOUNDS. (WITH MAGNETIC AND FERROELECTRIC O
	36-3545	MAGNETIC PROPERTIES OF HEAVY RARE EARTH ORTHOMANGANATES.
BROK AYA	04-0586	CONCENTRATION DEPENDENCE OF THE CURIE-WEISS CONSTANT IN POLYCRYSTALLINE SOLID SOLUTION
BROPHY JJ	27-2974	CRITICAL FLUCTUATIONS IN TRIGLYCINE SULFATE.
BROSOWSKI G	27-2975	TEMPERATURE DEPENDENCE OF THE DIELECTRIC CONSTANTS OF TRIGLYCINE SULFATE ORTHOGONAL TO
BROVEEV SF	17-2460	CHARACTERISTICS OF AN ELECTROOPTICAL SHUTTER BASED ON Z CUT POTASSIUM DIDEUTERIUM PHOS
BROWDER JS	11-1685	THERMAL EXPANSION OF LITHIUM NIOBATE IN THE TEMPERATURE RANGE OF 70-300 K.
	17-2464	LOW TEMPERATURE THERMAL EXPANSION MEASUREMENTS ON SINGLE CRYSTAL POTASSIUM DIHYDROGEN
BROWER WS	33-3368	PHASE TRANSITIONS IN CESIUM LEAD TRICHLORIDE. (LOSS OF A CENTER OF SYMMETRY AT 194 DEG
	33-3408	DETERMINATION OF TRANSITION TEMPERATURES IN CESIUM LEAD TRICHLORIDE USING EPR.
BROWN DJ	04-0587	EFFECT OF OXIDE IMPURITIES ON THE ELECTRICAL RESISTIVITY OF LANTHANUM DOPED BARIUM TIT
BROWN H	12-1869	FERROELECTRIC DOMAIN REVERSAL IN LITHIUM METATANTALATE.
	13-1902	SOME EFFECTS OF MELT STOICHIOMETRY ON THE OPTICAL PROPERTIES BARIUM SODIUM NIOBATE.
	13-1903	SOME EFFECTS OF MELT STOICHIOMETRY ON THE OPTICAL PROPERTIES OF BARIUM SODIUM NIOBATE.
BROWN ID	32-3341	ANOMALOUS NEUTRON SCATTERING AND THE QUESTION OF FERROELECTRICITY IN LITHIUM HYDRAZINI
	32-3342	CRYSTAL STRUCTURE OF LITHIUM HYDRAZINIUM SULFATE.
BROWN KR	08-1384	DIELECTRIC, PIEZOELECTRIC AND PYROELECTRIC PROPERTIES OF THE LEAD DINIOBATE- BARIUM DI
	08-1385	DIELECTRIC, PIEZOELECTRIC AND PYROELECTRIC PROPERTIES OF THE LEAD METANIOBATE- BARIUM
BROWN LM	02-0093	QUANTITATIVE ANALYSIS OF MICROSTRUCTURE AND ELECTRICAL PROPERTIES OF FERROELECTRIC MAT
	04-0839	MICROSTRUCTURE AND ELECTRICAL PROPERTIES OF SCANDIUM TRIOXIDE DOPED, RARE EARTH OXIDE
	04-0840	SYNTHESIS OF NIOBIUM PENTOXIDE DOPED BARIUM TITANATE WITH IMPROVED ELECTRICAL PROPERTI
	09-1493	COLD PRESSING AND LOW TEMPERATURE SINTERING OF ALKOXY-DERIVED PLZT AND PLZT.
BROWN RM	13-1915	EFFECT OF OXYGEN STOICHIOMETRY AND HYDROXYL CONTENT ON THE PROPERTIES AND PROCESSING O
BRUCKNER W	19-2751	RELAXATION PHENOMENA AND MOSSBAUER SPECTRA OF MAGNETIZED PARAMAGNETIC SUBSTANCES. PART
BRUECKMANN H	04-0591	TEMPERATURE DEPENDENCE OF THE DIELECTRIC CONSTANT OF FERROELECTRIC SUBSTANCES. (BARIUM
BRUHL HG	15-2140	DETERMINATION OF THE TEMPERATURE DEPENDENCE OF LATTICE CONSTANTS IN FERROELECTRIC ANTI
BRUINING J	06-1146	(TEMPERATURE AND) ELECTRIC FIELD DEPENDENCE OF A HYDROGEN IMPURITY MODE IN STRONTIUM T
BRULEBOIS D	15-2112	CRYSTAL STRUCTURE AND CONDUCTIVITY OF ANTIMONY SULFOIODIDE.
BRUNNER J	11-1686	WIDE RANGE TEMPERATURE COMPENSATION BY ADDITION OF TWO CRYSTAL RESONATOR FREQUENCIES:
BRUNSTEIN M	17-2465	MOSSBAUER EFFECT OBSERVATION OF A VERY LOW FREQUENCY MODE AT THE TRANSITION TEMPERATUR
BRUSSET H	04-0590	BINARY AND PSEUDOBINARY PHASES BETWEEN NIOBIUM PENTOXIDE AND LEAD, BARIUM, AND STRONTI
	08-1285	STRUCTURAL STUDY OF THE TRANSITION OF THE (ORTHORHOMBIC) FERROELECTRIC (LEAD DINIOBATE
	08-1286	STUDY OF DIVALENT BINARY AND TERNARY NIOBATES IN THE SOLID STATE.
	36-3535	POLYMORPHISM OF STRONTIUM METANIOBATE (TWO REVERSIBLE PHASES).
BRUTON TM	13-1916	LOW FREQUENCY LINEAR ELECTROOPTIC EFFECT IN LEAD DITANATLATE.
BRYUZGIN AR	04-0592	SATURATION OF VOLT- AMPERE CHARACTERISTICS OF CADMIUM SELENIDE LAYERS ON BARIUM TITANA
BUCCI JD	08-1287	PRECISION DETERMINATION OF THE LATTICE PARAMETERS AND THE COEFFICIENTS OF THERMAL EXPA
BUCHMAN F	04-0593	ACOUSTIC EMISSION FROM FERROELECTRIC CRYSTALS. (BARIUM TITANATE, TGS)
	04-0594	EFFECT OF FERROELECTRIC (BARIUM TITANATE) POLARIZATION FIELDS ON (GERMANIUM OR TELLURI
BUCKNER DA	09-1494	EFFECTS OF CALCINING ON SINTERING OF LEAD ZIRCONATE TITANATE CERAMICS.
	09-1622	EFFECTS OF CALCINING ON THE FIRING OF LEAD ZIRCONATE TITANATE CERAMIC
BUDIM NI	08-1288	FORBIDDEN ZONE WIDTHS IN POLYCRYSTALLINE DIELECTRICS WITH PEROVSKITE STRUCTURE (TITANA
BUDIN JP	04-0912	COMPARATIVE STUDY OF THE CRYSTALLOGRAPHIC, DIELECTRIC, AND NONLINEAR OPTICAL PROPERTIE
BUDREAU AJ	11-1817	MICROWAVE FREQUENCY ACOUSTIC SURFACE WAVE LOSS MECHANISMS ON LITHIUM NIOBATE (TEMPERAT
	35-3511	ACOUSTIC SURFACE WAVE LOSS MECHANISMS ON BISMUTH(12) GERMANIUM OXIDE(20) AT MICROWAVE
BUDZINSKI EE	34-3413	ENDOR (ELECTRON NUCLEAR DOUBLE RESONANCE) STUDY OF X-IRRADIATED SINGLE CRYSTALS OF ICE
BUESSEM WB	04-0595	EFFECTS OF GRAIN GROWTH ON THE DISTRIBUTION OF NIOBIUM IN BARIUM TITANATE CERAMICS.
BUGAEV SP	04-0596	SURFACE BREAKDOWN IN VACUUM ON BARIUM TITANATE.
BUHRER CF	08-1289	(CRYSTALLOGRAPHIC AND DIELECTRIC) PROPERTIES OF BISMUTH PEROVSKITES. (AND THE FERROELE
BUISSON G	36-3536	FABRICATION (BY FLUX GROWTH) OF HEXAGONAL SINGLE CRYSTALS OF RARE EARTH MANGANATES (PH
	36-3542	FERROELECTRIC PROPERTIES OF HEXAGONAL ORTHOMANGANITES OF YTTRIUM AND RARE EARTHS. (YTT
BULAEVSKII LN	03-0190	THERMODYNAMIC THEORY OF DOMAIN WALLS IN FERROELECTRIC MATERIALS WITH THE PEROVSKITE ST
	03-0191	TEMPERATURE DEPENDENCE OF THE SHAPE OF THE DOMAIN WALL IN FERROMAGNETICS AND FERROELEC
BULLEMER B	01-0026	PHYSICS OF ICE. (PROC OF INT SYMP, 9-14 SEP 1968 MUNICH)
	34-3410	LOW ENTROPY FORM OF ICE I(H) OBTAINED FROM A LINEAR STEP GROWTH MODEL.

BULLEMER B	34-3414	PROTONIC CONDUCTION OF ICE. PART-1: HIGH TEMPERATURE REGION. PART-2: LOW TEMPERATURE R
BUNGET I	04-0599	INFLUENCE OF PHASE TRANSITIONS ON ELECTRICAL RESISTANCE OF BARIUM TITANATE.
BUNINA IK	04-0597	DIELECTRIC AND ELASTIC PROPERTIES OF SINGLE CRYSTALS OF BARIUM LEAD TITANATE SOLID SOL
	04-0598	EFFECT OF IMPURITIES AND GROWTH CONDITIONS ON THE STRUCTURE OF BARIUM TITANATE CRYSTAL
BUNTING JG	20-2801	HEAT CAPACITY OF METHYL AMMONIUM CHROMIUM ALUM.
BURDANINA NA	17-2466	DIELECTRIC PROPERTIES OF A (SMALL DOSE) GAMMA-IRRADIATED KDP CRYSTAL (TEMPERATURE DEPE
	17-2467	DIELECTRIC PROPERTIES OF A GAMMA-IRRADIATED POTASSIUM DIHYDROGEN PHOSPHATE CRYSTAL.
	17-2542	DIELECTRIC PROPERTIES OF GAMMA-IRRADIATED POTASSIUM DIHYDROGEN PHOSPHATE CRYSTALS.
	27-2976	DIELECTRIC LOSSES OF TRIGLYCINE SULFATE CRYSTALS EXPOSED TO SMALL DOSES OF X- AND GAMM
BURDINA KP	08-1463	SYNTHESIS OF THALLIUM CONTAINING PEROVSKITES AT HIGH PRESSURES AND THEIR X-RAY ANALYSI
BURFOOT JC	03-0192	ENTROPY OF AN ORDER DISORDER TRANSITION. (IN FERROELECTRICS, MECHANISM)
	03-0193	(MACROSCOPIC AND MICROSCOPIC) SYMMETRY OF FERROELECTRIC ENERGY FUNCTIONS. (SOME COMMEN
	04-0603	ELECTRICAL AND OPTICAL PROPERTIES OF (BARIUM, STRONTIUM) MIXED TITANATE THIN SECTIONS.
	04-0604	GROWTH AND STRUCTURE OF BARIUM TITANATE EVAPORATED FILMS.
	04-0605	GROWTH STRUCTURE (MICROSTRUCTURE) AND ELECTRICAL PROPERTIES OF FLASH EVAPORATED BARIUM
	04-0606	INTERPRETATION OF THE DYNAMICS OF THE BARIUM TITANATE TRANSITION IN TERMS OF NUCLEATIO
	04-0608	NUCLEATION MODELS FOR A FIRST ORDER TRANSITION FRONT, AND OBSERVATIONS IN BARIUM TITAN
	04-0887	CUBIC TETRAGONAL (MARTENSITIC TYPE) PHASE TRANSITIONS IN HIGH RESISTIVITY BARIUM TITAN
	04-0888	DIFFUSIONLESS PHASE TRANSITIONS AND THE STABLE DOMAIN STRUCTURE OF (HIGH SENSITIVITY)
	04-0967	ELECTRICAL PROPERTIES OF FLASH EVAPORATED FERROELECTRIC BARIUM TITANATE THIN FILMS. (T
	04-0968	FLASH EVAPORATION OF FERROELECTRIC THIN FILMS OF POLYCRYSTALLINE BARIUM TITANATE.
	09-1496	PZT-5 UNDER PRESSURE, DIELECTRIC AND PIEZOELECTRIC PROPERTIES.
	13-1926	TRAPPING SITES, OPTICAL DAMAGE, AND THERMALLY STIMULATED CURRENT IN FERROELECTRIC OXID
	17-2609	DETERMINATION OF THE COMPLEX PERMITTIVITY OF ADP AND KDP IN THE FAR INFRARED BY DISPER
BURGAR M	27-2966	CRITICAL BEHAVIOR OF FERROELECTRIC TGS AND DEUTERATED TGS (HYSTERESIS LOOPS JUST BELOW
	27-3056	CRITICAL PROPERTIES OF TGS AND ROCHELLE SALT AS DETERMINED BY AN ELECTRIC FIELD METHOD
BURKE BE	11-1825	LOW POWER DENSITY GALLIUM ARSENIDE- LITHIUM NIOBATE SURFACE WAVE AMPLIFIER.
BURKE WJ	06-1102	EFFECT OF UNIAXIAL STRESS ON RAMAN SCATTERING IN STRONTIUM TITANATE.
	06-1103	RAMAN SCATTERING AND PHASE TRANSITIONS IN STRESSED STRONTIUM TITANATE.
	06-1104	STRESS INDUCED FERROELECTRICITY IN STRONTIUM TITANATE.
	06-1186	FUNDAMENTAL ABSORPTION EDGE OF STRONTIUM TITANATE.
	06-1187	OPTICAL ABSORPTION EDGE OF STRONTIUM TITANATE.
	06-1188	REINTERPRETATION OF WAVELENGTH MODULATED ABSORPTION IN STRONTIUM TITANATE WITHOUT COEX
BURN I	04-0602	ENERGY STORAGE IN CERAMIC DIELECTRICS. (STRONTIUM TITANATE, BARIUM TITANATE)
	07-1252	FIELD ENFORCED FERROELECTRICITY IN GLASS BONDED LEAD ZIRCONATE. (PHASE TRANSITION 200
BURNEIKA K	17-2468	PARAMETRIC LIGHT AMPLIFICATION AND OSCILLATION IN KDP WITH MODE LOCKED PUMP.
BURNS G	02-0052	NUCLEAR QUADRUPOLE RESONANCE OF LITHIUM IN FERROELECTRIC COMPOUNDS. (LITHIUM AMMONIUM
	02-0053	PROPERTIES OF TUNGSTEN BRONZE FERROELECTRICS. (REVIEW, 33 REFS)
	02-0054	TUNGSTEN BRONZE TYPE FERROELECTRICS FOR OPTICAL USE (REVIEW, 22 REFS).
	03-0194	POWDER RAMAN SPECTRA: APPLICATION TO DISPLACIVE FERROELECTRICS. (LATTICE VIBRATIONS)
	04-0610	POWDER RAMAN STUDY FOR BARIUM TITANATE.
	04-0611	RAMAN SCATTERING IN THE FERROELECTRIC SYSTEM LEAD BARIUM TITANATE.
	05-1041	RAMAN STUDIES (TEMPERATURE DEPENDENCE) OF UNDER DAMPED SOFT MODES IN LEAD TITANATE.
	05-1074	CRYSTAL GROWTH AND TEMPERATURE DEPENDENT RAMAN SPECTRA OF LEAD TITANATE.
	07-1264	CRYSTAL GROWTH AND OBSERVATION OF THE FERROELECTRIC PHASE OF LEAD ZIRCONATE.
	07-1265	GROWTH AND DIELECTRIC PROPERTIES OF SINGLE CRYSTAL LEAD ZIRCONATE
	09-1497	RAMAN SPECTRA OF POLYCRYSTALLINE SOLIDS: APPLICATION TO THE LEAD TITANATE ZIRCONATE SY
	11-1813	STOICHIOMETRY VARIATIONS IN LITHIUM NIOBATE AND LITHIUM TANTALATE BY RAMAN POWDER SPEC
	13-1917	BARIUM SODIUM NIOBATE FOR A TUNABLE STIMULATED RAMAN OSCILLATOR.
	13-1918	POLARITON RESULTS IN FERROELECTRICS WITH THE TUNGSTEN BRONZE CRYSTAL STRUCTURE.
	13-1948	(TETRAGONAL) TUNGSTEN BRONZE TYPE CRYSTALS GROWN IN MOLTEN SOLUTIONS OF THE PSEUDOSYST
	13-1949	(EXTENT OF) TUNGSTEN BRONZE (LIQUIDUS AND SOLIDUS) FIELD AND MELT GROWTH OF CRYSTALS I
	13-2003	CRYSTAL GROWTH AND ELECTROOPTIC PROPERTIES OF TUNGSTEN BRONZE CRYSTALS FROM THE POTASS
	13-2015	CRYSTAL GROWTH AND NONLINEAR OPTICAL PROPERTIES OF POTASSIUM LITHIUM NIOBATES.
	13-2016	PHASE EQUILIBRIA IN THE POTASSIUM NIOBATE- STRONTIUM DINIOBATE AND POTASSIUM NIOBATE-
	13-2017	TUNGSTEN BRONZE FIELD IN THE (OXIDE) SYSTEM POTASSIA LITHIA NIOBIA (POTASSIUM LITHIUM
	13-2026	NONLINEAR OPTICAL PROPERTIES OF POTASSIUM LITHIUM NIOBATES (TRANSPARENT, TETRAGONAL TU
	13-2027	NONLINEAR OPTICAL PROPERTIES OF POTASSIUM LITHIUM NIOBATE.
	13-2028	OPTICAL AND FERROELECTRIC PROPERTIES OF (THE TUNGSTEN BRONZE SYSTEM) POTASSIUM SODIUM
	17-2487	UNIT CELL VOLUME EFFECTS (RELATION WITH CURIE POINT) IN SOME ISOMORPHOUS FERROELECTRIC
	35-3495	SOFT OPTIC PHONON MODE IN FERROELECTRIC LEAD(5) GERMANIUM(3) OXIDE(11).
BURSIAN EV	03-0195	POLARIZATION OF FERROELECTRIC FILM BY BENDING AND BENDING DEFORMATION IN THE ELECTRIC
	03-0196	PHONON SPECTRUM OF FERROELECTRICS STUDIED BY LIMITATION OF THE CRYSTAL SIZE.
	04-0600	DIELECTRIC CONSTANT AND LATTICE DYNAMICS OF A THIN FILM OF A FERROELECTRIC (BARIUM TIT
	04-0601	DISPERSION AT MILLIMETER WAVELENGTHS IN BARIUM TITANATE ABOVE THE TRANSITION POINT.
	04-0607	MAGNITUDE OF SPONTANEOUS POLARIZATION IN THIN CRYSTALS AND FINE PARTICLES OF BARIUM TI
	04-0609	POLARON MECHANISM OF CONDUCTIVITY IN BARIUM TITANATE.
	04-0612	SINGLE CRYSTAL BARIUM TITANATE FILMS GROWN FROM THE MELT IN AN OXYGEN ATMOSPHERE.
	04-0613	SMALL POLARONS IN CONDUCTING BARIUM TITANATE CRYSTALS.
	04-0614	TEMPERATURE DEPENDENCE OF CARRIER MOBILITY IN BARIUM TITANATE.
	04-0944	RELATIONSHIP BETWEEN THE BREAKDOWN MECHANISM IN BARIUM TITANATE AND COLORATION. (INJEC
BURSILL LA	10-1630	CS (CRYSTALLOGRAPHIC SHEAR) FAMILIES DERIVED FROM THE RHENIUM TRIOXIDE STRUCTURE TYPE
BURT JG	03-0259	CHARGE MECHANISM IN FERROELECTRICS. (LEAD ZIRCONATE TITANATE, STOICHIOMETRY EFFECTS)
	09-1495	OXYGEN CONCENTRATION CELL AND ELECTRICAL CONDUCTIVITY MEASUREMENTS ON PZT FERROELECTRI
	09-1512	OXYGEN CONCENTRATION CELL MEASUREMENTS OF IONIC TRANSPORT NUMBERS IN PZT (LEAD ZIRCONA
BUSMUNDRUD O	33-3365	ELECTRICAL CONDUCTION AND (NONFERROELECTRIC) PHASE TRANSITIONS IN CESIUM LEAD TRICHLOR
BUTLER SR	09-1553	ISOTHERMAL GRAIN GROWTH AND ELECTRICAL BEHAVIOR OF FULLY DENSE PLZT CERAMICS.
BUTSYCHENKO VA	15-2191	TEMPERATURE DEPENDENCE OF THE SWITCHING PROCESS IN ANTIMONY SULFOIODIDE SINGLE CRYSTAL
BUTYAGIN OF	11-1665	PHASE MATCHING ANGLES AND TEMPERATURE OF LITHIUM METANIOBATE CRYSTALS OF DIFFERENT STO
	11-1687	INFLUENCE OF A TRANSVERSE INHOMOGENEITY OF THE REFRACTIVE INDEX OF A NONLINEAR CRYSTAL
BUYERS WJL	17-2610	FERROELECTRIC CRITICAL SCATTERING FROM DEUTERATED KDP. (NEUTRON INELASTIC SCATTERING)
	17-2611	FERROELECTRIC TRANSITION IN POTASSIUM DIDEUTERIUM PHOSPHATE.
	33-3366	ANTIFERROELECTRICITY IN (PURE AND) COBALT DOPED POTASSIUM MANGANESE TRIFLUORIDE. (20 P
	33-3407	MAGNETIC EXCITATIONS IN NICKEL DOPED POTASSIUM MANGANESE TRIFLUORIDE.
	36-3586	TRANSITION IN TIN TELLURIDE- GERMANIUM TELLURIDE ALLOYS. (NEUTRON SCATTERING)
BUZIN IM	06-1105	ANOMALOUS BEHAVIOR OF THE DIELECTRIC NONLINEARITY COEFFICIENT IN STRONTIUM TITANATE NE
	06-1106	DIELECTRIC NONLINEARITY OF STRONTIUM TITANATE SINGLE CRYSTALS NEAR PHASE TRANSITION AT
BUZIN VN	27-3054	INVESTIGATION OF THE PHASE TRANSITION IN TRIGLYCINE SULFATE.
BYE KL	17-2469	STRUCTURAL INHIBITION OF FERROELECTRIC SWITCHING IN TRIGLYCINE SULFATE. PART-2: X-IRRA
	27-2977	HIGH INTERNAL BIAS FIELDS IN L-ALANINE SUBSTITUTED TGS.

BYE KL   27-3037 EFFECTS OF ADDITIVES ON THE PYROELECTRIC PROPERTIES OF TGS.
                      27-3038 STRUCTURAL INHIBITION OF FERROELECTRIC SWITCHING IN TRIGLYCINE SULFATE. PART-1: ADDITI
                      36-3624 ORTHORHOMBIC DIGLYCINE SULFATE. (NOT PIEZOELECTRIC, NOT FERROELECTRIC)
BYER RL   11-1688 GROWTH OF HIGH QUALITY LITHIUM NIOBATE CRYSTALS (FOR NONLINEAR OPTICS WITH LOW BIREFRI
                      13-1919 PYROELECTRIC COEFFICIENT DIRECT MEASUREMENT TECHNIQUE AND APPLICATION TO A NANOSECOND
BYKOVSKII YUA   11-1663 FRACTURE OF NONLINEAR CRYSTALS (KDP AND LITHIUM NIOBATE) BY RADIATION FROM A RUBY LASE
CAGLIOTI G   02-0051 LATTICE DYNAMICS BY NEUTRON SCATTERING. PARTS 1 AND 2. (REVIEW)
CALLABY DR   03-0197 OSCILLATIONS OF FERROELECTRIC BODIES. (WITH FREQUENCY OF APPLIED FIELD CAUSED BY SHEAR
                      04-0615 EFFECT OF "A" DOMAINS UPON "C" DOMAIN SWITCHING IN BARIUM TITANATE.
CAMERON JA   34-3416 MOSSBAUER EFFECT OF FERROUS IONS IN CUBIC ICE.
CAMLIBEL I   04-0616 DIELECTRIC PROPERTIES OF SINGLE DOMAIN MELT GROWN BARIUM TITANATE.
CAMOBRECO SJ   33-3397 PREPARATION AND DISLOCATION DENSITY DETERMINATION OF LARGE, PURE RUBIDIUM MANGANESE TR
CAMP PR   34-3415 ELECTRICAL CONDUCTION IN ICE.
CAMP WJ   03-0198 STATISTICAL MECHANICS OF FERROELECTRIC PHASE TRANSITIONS.
CAMPBELL ID   08-1290 ABSOLUTE STRUCTURE OF LITHIUM IODATE CRYSTALS.
CAMPBELL JJ   11-1846 FUNDAMENTAL AND HIGHER ORDER MODE SURFACE WAVE PROPAGATION ON LAYERED ANISOTROPIC SUBS
CAMPBELL JP   11-1689 ROTATING WAVEPLATE OPTICAL FREQUENCY SHIFTING IN LITHIUM NIOBATE.
CAMPILLO AJ   08-1291 (CONTINUOUS WAVE) SPONTANEOUS PARAMETRIC SCATTERING OF LIGHT IN LITHIUM IODATE (INDEX
CANNER JP   07-1253 MOSSBAUER EFFECT STUDIES OF THE PHASE TRANSITIONS IN ANTIFERROELECTRIC LEAD ZIRCONATE
                      09-1498 MOSSBAUER EFFECT STUDIES OF FERROELECTRIC PHASE TRANSITIONS IN THE LEAD ZIRCONATE- LEA
CANUT ML   03-0199 X-RAY DIFFRACTION STUDIES OF ELECTRON POLARIZATION IN FERROELECTRICS.
                      16-2270 CRITICAL SCATTERING OF X-RAYS IN SODIUM NITRITE.
                      16-2271 X-RAY ANALYSIS OF FERROELECTRIC DOMAINS IN THE PARAELECTRIC PHASE OF SODIUM NITRITE.
                      16-2298 THERMAL EXPANSION OF SOME FERROELECTRICS. (POTASSIUM NITRATE, KDP, SODIUM NITRITE)
                      16-2343 X-RAY CRITICAL SCATTERING OF SODIUM NITRITE AT 160 DEG C AND AT 215 DEG C.
CAPIZZI M   06-1107 ABSORPTION EDGE AND STRUCTURAL PHASE TRANSITION IN STRONTIUM TITANATE.
                      06-1108 DETERMINATION OF THE NATURE OF THE OPTICAL GAP OF STRONTIUM TITANATE.
                      06-1109 DISTORTION ENHANCED OPTICAL ABSORPTION IN STRONTIUM TITANATE AT THE CUBIC TO TETRAGONA
                      06-1110 IS THERE A UNIFIED MECHANISM UNDERLYING URBACH'S RULE. (STRONTIUM TITANATE, POTASSIUM
                      06-1111 OPTICAL GAP OF STRONTIUM TITANATE. (DEVIATION FROM URBACH TAIL BEHAVIOR)
                      06-1164 MONODOMAIN STRONTIUM TITANATE (OCCURANCE IN THIN (110) PLATES BELOW CUBIC- TETRAGONAL
                      06-1169 MONODOMAIN STRONTIUM TITANATE.
CAPRON JP   17-2689 DIELECTRIC PERMITTIVITY OF FINELY DIVIDED POTASSIUM AND AMMONIUM DIHYDROGEN PHOSPHATE.
CARL K   09-1499 ORIGIN OF THE MAXIMUM IN THE ELECTROMECHANICAL ACTIVITY IN LEAD ZIRCONATE(X) TITANATE(
                      09-1500 ORIGIN OF THE MAXIMUM IN THE ELECTROMECHANICAL ACTIVITY IN LEAD ZIRCONATE(X) TITANATE(
                      09-1501 POLARIZATION, ELECTROSTRICTION, AND PIEZOELECTRIC ACTIVITY OF LEAD ZIRCONATE(X) TITANA
                      09-1502 STRUCTURE AND ELECTROMECHANICAL PROPERTIES OF LANTHANUM DOPED LEAD TITANATE ZIRCONATE
CARLIN RL   21-2815 MAGNETIC SUSCEPTIBILITY OF GUANIDINIUM VANADIUM SULFATE HEXAHYDRATE AT LOW TEMPERATURE
                      21-2817 ELECTRON PARAMAGNETIC RESONANCE STUDY OF SOME METAL IONS IN GUANIDINIUM ALUMINUM SULFA
CARLSON EH   33-3367 ELECTRIC FIELD GRADIENTS IN IONIC CHLORIDES: CESIUM LEAD TRICHLORIDE.
CARLSON RE   31-3326 LONGITUDINAL MODES IN HYDROGEN AND DEUTERIUM HALIDE CRYSTALS. (RAMAN SPECTRA, DIPOLAR
CARLSON TA   10-1511 STUDY OF THE X-RAY PHOTOELECTRON SPECTRUM OF TUNGSTEN- TUNGSTEN OXIDE AS A FUNCTION OF
CARLSON WG   04-0999 INFLUENCE OF OXYGEN PARTIAL PRESSURE ON PROPERTIES OF SEMICONDUCTING BARIUM TITANATE.
CARNEVALE A   11-1793 LOW FREQUENCY (510 MHZ) PARAMAGNETIC RESONANCE INVESTIGATION OF LITHIUM NIOBATE.
                      11-1794 NIOBIUM-93 NMR STUDY OF THE LITHIUM NIOBATE TANTALATE SOLID SOLUTION SYSTEM.
                      11-1795 NIOBIUM-93 NMR LINE WIDTHS IN NONSTOICHIOMETRIC LITHIUM NIOBATE.
CAROME EF   37-3672 ACOUSTIC TRANSIENTS GENERATED BY ANTIFERROELECTRIC- FERROELECTRIC TRANSDUCERS.
CARR PH   11-1690 MIXING OF NONCOLLINEAR ELASTIC SURFACE WAVES ON LITHIUM NIOBATE.
                      11-1817 MICROWAVE FREQUENCY ACOUSTIC SURFACE WAVE LOSS MECHANISMS ON LITHIUM NIOBATE (TEMPERAT
CARRUTHERS JR   02-0055 NONLINEAR OPTICAL MATERIALS.
                      11-1684 SPATIALLY UNIFORM AND ALTERABLE SHG PHASE MATCHING TEMPERATURES IN LITHIUM NIOBATE (CF
                      11-1691 NONSTOICHIOMETRY AND CRYSTAL GROWTH OF LITHIUM NIOBATE.
                      11-1794 NIOBIUM-93 NMR STUDY OF THE LITHIUM NIOBATE TANTALATE SOLID SOLUTION SYSTEM.
                      11-1811 NIOBIUM-93 NUCLEAR QUADRUPOLE RESONANCE INVESTIGATION OF (STOICHIOMETRIC) LITHIUM NIOB
                      12-1871 LITHIUM TANTALATE SINGLE CRYSTAL STOICHIOMETRY (VERSUS MELT COMPOSITION, LATTICE AND C
                      13-1901 GROWTH OF UNCRACKED BARIUM SODIUM NIOBATE CRYSTALS.
                      13-1908 EFFECTS OF CHANGES IN MELT COMPOSITION-ON CRYSTAL GROWTH OF BARIUM SODIUM NIOBATE.
                      13-1920 PHASE EQUILIBRIA RELATIONS IN THE TERNARY SYSTEM BARIUM OXIDE- STRONTIUM OXIDE- NIOBIU
CARTER JR   36-3537 PREPARATION AND CRYSTALLOGRAPHIC PROPERTIES OF A(2+)B(3+)(2)O(4) TYPE CALCIUM AND STRO
CARTHEY L   24-2890 MOSSBAUER RESONANCE OF SINGLE CRYSTAL POTASSIUM FERROCYANIDE TRIHYDRATE.
CASABELLA PA   28-3195 NMR STUDY OF PHASE TRANSITIONS IN ROCHELLE SALT.
CASSETTARI M   34-3417 THERMODIELECTRIC EFFECT AND FREEZING POTENTIAL IN GROWING ICE.
CASTELLANO JA   37-3631 MESOMORPHIC MATERIALS FOR ELECTROOPTICAL APPLICATION.
CASTELLIZ LM   04-0617 THE ORTHORHOMBIC PHASE IN MODIFIED BARIUM TITANATE. (EFFECT OF SUBSTITUTIONS)
CASTNER TG   26-2943 STUDY OF THE ORDERED MAGNETIC STATE OF COPPER FORMATE TETRAHYDRATE BY ANTIFERROMAGNETI
CAUFOVA P   04-0618 SOME CRYSTAL CHEMICAL AND PHYSICAL PROPERTIES OF SINGLE CRYSTALS OF BARIUM TITANATE WI
CAVELIUS E   16-2264 NUCLEAR SPIN LATTICE RELAXATION OF SODIUM-23 IN THE PARAELECTRIC AND FERROELECTRIC PHA
                      24-2889 STUDY OF HYDROGEN-1 IN PARAELECTRIC AND FERROELECTRIC POTASSIUM FERROCYANIDE TRIHYDRAT
CERISIER P   16-2367 ELECTRICAL PROPERTIES OF RHOMBOHEDRAL POTASSIUM NITRATE.
CESANI FA   22-2833 ESR STUDY OF GAMMA-IRRADIATED CESIUM TRIHYDROGEN SELENITE AT ROOM TEMPERATURE.
CESSAC GL   25-2905 LATTICE VIBRATIONAL SPECTRA OF FIVE CRYSTAL MODIFICATIONS OF THIOUREA.
CEVC P   22-2870 DIELECTRIC PROPERTIES AND PRESSURE EFFECTS ON SELENITE (ION). RADIATION DAMAGE ELECTRO
                      22-2871 ESR STUDY OF THE FERROELECTRIC PHASE TRANSITIONS IN SODIUM TRIHYDROGEN DISELENITE.
CHABIN M   27-2980 POLARIZATION REVERSAL IN TGS SINGLE CRYSTALS.
CHAKRAVORTY D   06-1117 DIELECTRIC AND ELECTROOPTIC STUDIES OF THE POLARIZATION PROCESS IN STRONTIUM TITANATE
CHALUPA B   17-2567 GROWTH AND ETCHING OF POTASSIUM DIHYDROGEN PHOSPHATE CRYSTALS IN SOLUTIONS CONTAINING
CHAMBERLAIN J   17-2609 DETERMINATION OF THE COMPLEX PERMITTIVITY OF ADP AND KDP IN THE FAR INFRARED BY DISPER
                      34-3418 LOW TEMPERATURE POLARIZATION EFFECTS IN ICE. (CURRENT REVERSAL DUE TO DIPOLE RELAXATIO
CHAMBERS J   11-1692 ACOUSTIC SURFACE WAVE CONVOLUTION ON CRYSTALS OF CADMIUM SULFIDE, LITHIUM NIOBATE AND
CHAMINADE JP   13-1921 EFFECT OF THE SUBSTITUTION OF TANTALUM FOR NIOBIUM ON THE CRYSTALLOGRAPHIC AND DIELECT
                      13-2010 CRYSTAL CHEMISTRY AND ELECTRICAL PROPERTIES OF TWO NEW PHASES OF QUADRATIC TUNGSTEN BR
CHAN VT   08-1292 STRUCTURE AND DIELECTRIC PROPERTIES OF LITHIUM CONTAINING PEROVSKITES. (STRONTIUM LITH
                      13-1923 FORMATION CONDITIONS AND PHASE TRANSITIONS IN CRYSTALS WITH POTASSIUM TUNGSTEN BRONZE
CHANEY PE   16-2265 MELTING PROPERTIES OF POTASSIUM NITRATE UNDER HIGH PRESSURE.
CHANG KS   03-0200 CIRCLE THEOREM FOR ICE RULE FERROELECTRIC MODELS.
CHANG TS   06-1112 DIRECT OBSERVATION OF SINGLE DOMAIN STRONTIUM TITANATE (OPTICAL STUDY UNDER UNIAXIAL S
                      06-1113 DOMAIN STRUCTURE OF STRONTIUM TITANATE UNDER UNIAXIAL STRESSES.
                      06-1114 EFFECT OF DOMAIN STRUCTURE ON THE FLUORESCENCE OF CHROMIUM(3+) DOPED STRONTIUM TITANAT
                      06-1115 POLARIZED FLUORESCENCE STUDY OF CHROMIUM(3+) THROUGH A STRESS INDUCED PHASE TRANSITION
                      11-1807 TEMPERATURE STUDY OF THE POLARITON ASSOCIATED WITH THE 248 CM(-1) SOFT MODE IN LITHIUM

CHANG TS	12-1873	TEMPERATURE DEPENDENCE OF POLARITON DISPERSION IN LITHIUM TANTALATE. (OPTICAL MODULATI
	13-1899	RAMAN SCATTERING BY BARIUM STRONTIUM NIOBATE.
	13-1922	FERROELECTRICITY MEASUREMENTS ON STRONTIUM BARIUM NIOBATE.
CHANG TT	33-3368	PHASE TRANSITIONS IN CESIUM LEAD TRICHLORIDE. (LOSS OF A CENTER OF SYMMETRY AT 194 DEG
	33-3408	DETERMINATION OF TRANSITION TEMPERATURES IN CESIUM LEAD TRICHLORIDE USING EPR.
CHANTURIYA GF	04-0538	GENERATION OF SECOND OPTICAL HARMONIC IN THE CUBIC PHASE OF BARIUM TITANATE.
CHANUSSOT G	04-0571	IONIC CONTRIBUTION TO INJECTION AND CONDUCTION IN BARIUM TITANATE (CONDENSER CHARGE- D
	04-0620	EXPERIMENTAL AND THEORETICAL STUDY OF DEFECTS IN TRIGLYCINE SULFATE AND BARIUM TITANAT
	04-0621	PHASE CHANGES IN TRIGLYCINE SULFATE AND BARIUM TITANATE FERROELECTRICS. (26 REFS)
	04-0622	SOME PROPERTIES OF THE THERMOCURRENTS IN BARIUM TITANATE AND TITANIUM DIOXIDE SINGLE C
	04-0697	(PHASE TRANSITIONS AND) THERMOCURRENTS IN TITANIUM DIOXIDE AND BARIUM TITANATE SINGLE
	27-2979	PYROELECTRIC RESPONSE OF TGS IN THE FERROELECTRIC REGION NEAR THE CURIE POINT (IMPROVE
	27-3064	INFLUENCE OF SURFACE EFFECTS ON THE PYROELECTRIC BEHAVIOR OF TGS CLOSE TO THE PHASE TR
	27-3066	PYROELECTRIC BEHAVIOR OF TRIGLYCINE SULFATE IN THE PARAELECTRIC REGION.
CHAPELLE JP	16-2305	BRILLOUIN SCATTERING IN SODIUM NITRITE AT ROOM TEMPERATURE.
	17-2520	BRILLOUIN DIFFUSION IN A SINGLE CRYSTAL OF RUBIDIUM DIHYDROGEN PHOSPHATE.
	17-2521	INTENSITY OF SOME BRILLOUIN AND RAYLEIGH LINES IN RUBIDIUM DIHYDROGEN PHOSPHATE.
	25-2906	LOW FREQUENCY RAMAN SPECTRUM OF A THIOUREA CRYSTAL AS A FUNCTION OF TEMPERATURE.
	27-3145	ELECTROMECHANICAL PROPERTIES OF TRIGLYCINE SULFATE. (MEASUREMENTS VERIFY THEORY APPROX
CHAPMAN DW	09-1503	THIN FILM FERROELECTRIC PHOTOCONDUCTOR MEMORY DEVICE. (PLZT)
	37-3632	DESIGN AND PERFORMANCE OF A THIN FILM FERROELECTRIC PHOTOCONDUCTOR STORAGE DEVICE.
CHASE AB	13-1979	OPTICAL DAMAGE IN KTN.
CHAVES A	04-0548	TEMPERATURE DEPENDENCE OF THE RAMAN CROSS SECTIONS IN BARIUM TITANATE AND STRONTIUM TI
	04-0619	COUPLED POLARITONS OF A1 SYMMETRY IN BARIUM TITANATE.
CHECHERNIKOV VI	08-1368	ELECTRON SPIN RESONANCE (SPECTRA, TEMPERATURE DEPENDENCE) IN SOME FERROELECTRICS AND F
CHEKIN VV	04-0573	CHANGE OF THE MOSSBAUER EFFECT PROBABILITY ON TIN-119 IN BARIUM TITANATE STANNATE SOLI
CHEN CL	11-1778	POSSIBILITY OF SWITCHING AND STEERING SURFACE WAVES BY NONLINEAR MIXING IN ANISOTROPIC
CHEN FS	02-0056	(HIGH SPEED, SMALL APERTURE) MODULATORS FOR OPTICAL COMMUNICATIONS (REVIEW EMPHASIZING
	07-1254	EVALUATION OF LEAD LANTHANATE ZIRCONATE CERAMICS FOR APPLICATIONS IN OPTICAL COMMUNICA
	13-1924	USE OF PEROVSKITE PARAELECTRICS IN BEAM DEFLECTORS AND LIGHT MODULATORS. (POTASSIUM TA
CHENSKII EV	03-0162	POSSIBILITY OF GENERATING INFRARED RADIATION BY POLARIZATION REVERSAL OF A FERROELECTR
	03-0201	INSTABILITY PHENOMENA IN FERROELECTRIC SEMICONDUCTORS. (THEORY)
	03-0202	SINGLE DOMAIN POLARIZATION OF A FERROELECTRIC HAVING A SECOND ORDER PHASE TRANSITION (
	03-0348	SPACE CHARGE LIMITED CURRENTS IN A METAL FERROELECTRIC METAL SYSTEM (THEORY).
	15-2152	INSTABILITY OF FERROELECTRIC POLARIZATION IN ANTIMONY SULFOIODIDE SINGLE CRYSTALS.
	15-2252	CPDCSING FERROELECTRIC DOMAINS IN SINGLE CRYSTALS OF ANTIMONY SULFOIODIDE.
CHEPUR DV	15-2116	DETERMINATION OF THE LATTICE PARAMETERS AND CURIE POINTS OF SOLID SOLUTIONS OF A(V)B(V
	15-2119	INFLUENCE OF ILLUMINATION ON THE FORM OF THE VOLTAGE AMPERE CHARACTERISTICS OF ANTIMON
	15-2125	ANISOTROPY OF THE REFLECTION SPECTRA OF ANTIMONY SULFOIODIDE CRYSTALS.
	15-2127	DOMAIN STRUCTURE AND LOCAL STATES IN ANTIMONY SULFOIODIDE TYPE SEMICONDUCTOR FERROELEC
	15-2128	PIEZOELECTRIC RESISTANCE EFFECT IN BISMUTH SELENOIODIDE CRYSTALS.
	15-2141	PRODUCTION METHODS AND SOME OPTICAL PROPERTIES OF BISMUTH TELLUROIODIDE.
	15-2142	PRODUCTION AND SOME OPTICAL PROPERTIES OF ANTIMONY SULFOBROMIDE IN GLASSY AND CRYSTALL
	15-2154	EFFECT OF THE ADSORPTION OF OXYGEN ON THE PROPERTIES OF ANTIMONY SULFOIODIDE SINGLE CR
	15-2186	EFFECT OF PHOTOELECTRET CHARGE ON VALUE OF FERROELECTRIC POLARIZATION IN ANTIMONY SULF
	15-2199	PREPARATION AND PHYSICAL PROPERTIES OF NONSTOICHIOMETRIC ANTIMONY SULFOIODIDE SINGLE C
	15-2217	ABSORPTION SPECTRA OF BISMUTH TELLUROBROMIDE AND BISMUTH TELLUROIODIDE CRYSTALS.
CHERN MJ	16-2272	NONLINEAR OPTICAL PROPERTIES OF FERROELECTRIC SODIUM NITRITE.
	16-2273	OPTICAL ACTIVITY IN NON-ENANTIOMORPHOUS BIAXIAL CRYSTALS: METHYL MESITYL OXIDE OXALATE
	16-2274	TEMPERATURE DEPENDENCE OF NONLINEAR OPTICAL COEFFICIENTS IN FERROELECTRIC SODIUM NITRI
CHERNOW F	02-0057	SYNTHESIS AND CHARACTERIZATION OF THIN FERROELECTRIC AND SEMICONDUCTING FILMS. (THE MA
CHERNYSHEV KR	04-1009	DIELECTRIC DISPERSION IN BARIUM TITANATE TYPE FERROELECTRICS IN THE METER BAND.
CHIBA T	03-0204	DEUTERON MAGNETIC RESONANCE STUDY OF SOME CRYSTALS CONTAINING O-D...O BOND. (INCLUDING
CHIHARA H	36-3538	FERROELECTRICITY AND PHASE TRANSITION IN AMMONIUM DICHLORO ACETATE AND DEUTERATED AMMO
	36-3625	NUCLEAR QUADRUPOLE RESONANCE STUDY OF PHASE TRANSITION IN AMMONIUM MONOCHLORO ACETATE,
CHILVERS JO	04-0624	THE PRODUCTION OF BARIUM TITANATE FROM BARIUM SULFATE AND TITANIUM DIOXIDE IN THE PRES
CHINCHOLKAR VS	03-0203	AN ATTEMPT TO CORRELATE THE FERROELECTRIC PROPERTIES OF ABO(3) AND A(0.5)BO(3) TYPE PH
	04-0623	STRUCTURAL DI- AND FERROELECTRIC PROPERTIES OF X POTASSIUM DINIOBATE AND (1-X) BARIUM
	04-0889	SOLUBILITY LIMIT OF LANTHANUM(0.5) LITHIUM(0.5) TITANATE IN BARIUM TITANATE.
	08-1293	FERROELECTRIC PROPERTIES OF TETRAGONAL (BARIUM, LEAD) (NIOBATE, ZIRCONATE). (CURIE TEM
	13-1925	STUDIES ON THE DIELECTRIC AND FERROELECTRIC PROPERTIES OF THE BARIUM LEAD NIOBATE ZIRC
CHISLER EV	15-2143	PHASE TRANSITION IN FERROELECTRIC ANTIMONY SULFOIODIDE CRYSTALS USING THE RAMAN SCATTE
	16-2275	INVESTIGATION OF FERROELECTRIC TRANSITION IN SODIUM NITRITE CRYSTAL BY OBSERVATION OF
	16-2276	PHASE CHANGE IN SODIUM NITRITE CRYSTALS AND COMBINATION SCATTER SPECTRUM IN THE REGION
	16-2277	TWO-PHONON SCATTERING AND FERROELECTRIC TRANSITION IN A SODIUM NITRITE CRYSTAL.
	17-2470	EFFECT OF TEMPERATURE AND PHASE TRANSITION ON THE HYDROXIDE VIBRATION SPECTRUM OF THE
	17-2471	POLARIZATION OF THE RAMAN BANDS OF THE HYDROXYL VIBRATIONS AND STRUCTURE OF THE HYDROG
	17-2472	VIBRATION SPECTRUM OF THE PHOSPHATE ANION IN A RUBIDIUM DIHYDROGEN PHOSPHATE CRYSTAL I
	17-2647	RAMAN SPECTRUM OF POTASSIUM DIDEUTERIUM PHOSPHATE CRYSTAL IN PARA- AND FERROELECTRIC P
	28-3189	TEMPERATURE DEPENDENCE OF THE INTENSITY OF HYDROGEN BOND BANDS IN RAMAN SCATTERING SPE
CHISTYAKOV EA	36-3539	FIRST ORDER NEGATIVE PHOTODIELECTRIC EFFECT IN PHTHALOCYANINE IN THE FERROELECTRIC STA
	36-3540	PYROELECTRIC EFFECT IN METAL-FREE PHTHALOCYANINE. (IN THE POLAR, OR FERROELECTRIC STAT
	36-3621	DEBYE TYPE ABSORPTION IN POLAR PHTHALOCYANINE.
	36-3622	ELECTRICAL PROPERTIES OF PHTHALOCYANINE AS A SEMICONDUCTOR IN THE FERROELECTRIC STATE.
	36-3623	LOW TEMPERATURE PHASE TRANSITION IN THE POLAR STATE OF THE ORGANIC SEMICONDUCTOR PHTHA
CHIZHIKOV SI	30-3265	ABNORMAL ABSORPTION OF ELASTIC WAVES NEAR THE PHASE TRANSITION OF GADOLINIUM MOLYBDATE
CHKALOVA VV	11-1693	THERMAL DEPENDENCE OF DIELECTRIC, PIEZOELECTRIC, AND ELASTIC PROPERTIES OF LITHIUM NIO
CHMEL IS	13-1923	FORMATION CONDITIONS AND PHASE TRANSITIONS IN CRYSTALS WITH POTASSIUM TUNGSTEN BRONZE
CHMELA P	17-2473	PHASE SYNCHRONIZATION IN NONLINEAR RUBIDIUM DIHYDROGEN PHOSPHATE.
CHOBOT MA	09-1595	CURIE TEMPERATURES OF SOLID SOLUTIONS IN A LEAD ZIRCONATE- LEAD TITANATE- LEAD METANIO
	09-1596	CRYSTAL STRUCTURE AND PHASE RELATIONS IN THE LEAD ZIRCONATE LEAD TITANATE- LEAD METANI
	09-1597	THERMAL VARIATION OF SPONTANEOUS POLARIZATION AND COERCIVE FORCE OF THE PEROVSKITES CR
CHOCK DP	03-0205	PROTON DYNAMICS IN HYDROGEN BONDED FERROELECTRICS. PART-2: APPROXIMATED SOLUTIONS TO T
	03-0206	PROTON DYNAMICS IN HYDROGEN BONDED FERROELECTRICS ABOVE THE CURIE TEMPERATURE.
CHODOROW M	17-2703	(REVERSIBLE) OPTICAL DAMAGE IN KDP (WITH RELAXATION TIME OF SECONDS WHEN ILLUMINATED U
CHORVATOVA Z	17-2474	THERMAL VARIATION OF REFRACTIVE INDEX IN AMMONIUM DIHYDROGEN PHOSPHATE AND TRIGLYCINE
CHOSSON A	34-3425	LOW FREQUENCY RAMAN SPECTRUM OF ICE I(H). TENTATIVE INTERPRETATION WITH A MIXED COULOM
CHRISTENSEN AN	04-0625	HYDROTHERMAL PREPARATION OF BARIUM TITANATE BY TRANSPORT REACTIONS (INVOLVING TITANIUM
	04-0626	HYDROTHERMAL PREPARATION OF BARIUM TITANATE BY TRANSPORT REACTIONS.
CHRISTMAS TM	17-2475	LASER INDUCED DAMAGE IN X DIHYDROGEN PHOSPHATE MATERIALS.

PAGE 627

CHRISTMAS TM   17-2476   LASER INDUCED SELF FOCUSING IN AMMONIUM DIHYDROGEN PHOSPHATE AND POTASSIUM DIDEUTERIUM
                        17-2477   PULSE MEASUREMENT OF R(63) IN KDP.
CHUNG TK   28-3211   ABSORPTION OF LONGITUDINAL ULTRASONIC WAVES IN ROCHELLE SALT CRYSTALS NEAR THE CURIE P
CHUPIS IE   03-0166   QUANTUM THEORY OF VIBRATIONS IN FERROELECTRIC FERROMAGNETICS WITHOUT A SYMMETRY CENTER
                        03-0207   FERROMAGNETIC RESONANCE LINE WIDTH IN FERROELECTRIC FERROMAGNETS.
CHUPRAKOV VF   06-1105   ANOMALOUS BEHAVIOR OF THE DIELECTRIC NONLINEARITY COEFFICIENT IN STRONTIUM TITANATE NE
CHURAGULCV BR   27-3090   NEW HIGH PRESSURE PHASE TRANSITION IN TRIGLYCINE SELENATE.
                        28-3214   HIGH PRESSURE PHASE TRANSITIONS OF THE FIRST KIND IN ROCHELLE SALT.
CHYLA A   28-3190   EFFECT OF ELECTRIC FIELD ON DOMAIN STRUCTURE FORMATION IN SEIGNETTES SALT (ROCHELLE SA
CIARRAPICO JO   04-0627   VARIATICN IN CERTAIN ELECTRICAL PROPERTIES OF SINTERED BARIUM TITANATE AS A FUNCTION O
CLARK CC   33-3359   PREPARATION AND PROPERTIES OF CESIUM LEAD HALIDES. (FERROELECTRICITY INDETERMINITE)
CLARK JB   17-2478   PHASE DIAGRAMS OF AMMONIUM DIHYDROGEN PHOSPHATE AND POTASSIUM DIHYDROGEN ARSENATE TO 5
CLARK JR   14-2075   ATOMIC DISPLACEMENTS IN FERROELECTRIC TRIGONAL AND ORTHORHOMBIC BORACITE STRUCTURES. (
CLARKE R   13-1896   FERROELECTRICS IN THE POTASSIUM OXIDE- STRONTIUM OXIDE- NIOBIUM PENTOXIDE.
                        13-1926   TRAPPING SITES, OPTICAL DAMAGE, AND THERMALLY STIMULATED CURRENT IN FERROELECTRIC OXID
CLAUDEL J   27-3007   OPTICAL CONSTANTS OF TRIGLYCINE SELENATE IN THE NEAR INFRARED TO FAR INFRARED ANC IN T
                        27-3009   DISPERSION OF THE DIELECTRIC CONSTANT OF TGS IN THE FAR INFRARED (TEMPERATURE AND FREQ
CLAUS R   08-1294   POLARITONS IN LITHIUM IODATE.
                        11-1694   ASSIGNMENTS OF OPTICAL PHONON MODES IN LITHIUM NIOBATE.
                        11-1695   DIRECTIONAL DISPERSION AND ASSIGNMENT OF OPTICAL PHONONS IN LITHIUM NIOBATE.
                        11-1696   DIRECTIONAL DISPERSION OF PHONCN FREQUENCIES AND POLARITON SCATTERING IN LITHIUM NIOBA
                        11-1697   LIGHT SCATTERING BY POLARITONS ASSOCIATED WITH ORDINARY PHOTONS IN LITHIUM NIOBATE.
                        11-1854   CBSERVATION OF ORDINARY POLARITONS IN LITHIUM NIOBATE.
CLAUSER MJ   24-2891   MOSSBAUER EFFECT STUDY OF FERROELECTRIC POTASSIUM FERROCYANIDE TRIHYDRATE (NEAR THE TR
CLAWSON AR   23-2888   SWITCHING PROPERTIES OF BIASED FERROELECTRIC COLEMANITE.
CLEAVER B   16-2278   STUDIES OF PHASE TRANSFORMATIONS IN NITRATES AND NITRITES. PART-1: CHANGE IN ULTRAVIOL
CLEVENGER TR   13-1927   EFFECT OF IRON(4+) IN THE SYSTEM STRONTIUM FERRATE TITANATE.
COCHRAN W   03-0208   DYNAMICAL (DIELECTRIC) AND SCATTERING PROPERTIES OF FERROELECTRICS. (THEORY)
                        03-0209   GENERALIZED SUSCEPTIBILITY AND THE SCATTERING PROPERTIES OF FERROELECTRICS. (ABSTRACT
                        13-2064   NEUTRON SCATTERING STUDY OF THE SOFT MODES IN CUBIC POTASSIUM TANTALATE NIOBATE.
                        17-2610   FERROELECTRIC CRITICAL SCATTERING FROM DEUTERATED KDP. (NEUTRON INELASTIC SCATTERING)
                        17-2611   FERROELECTRIC TRANSITION IN POTASSIUM DIDEUTERIUM PHOSPHATE.
                        17-2700   DETERMINATION OF THE EIGENVECTCRS OF SOFT MODES IN POTASSIUM DIDEUTERIUM PHOSPHATE AND
                        36-3541   LATTICE DYNAMICS OF DIATOMIC CRYSTALS OF HIGH DIELECTRIC CONSTANT. (POTASSIUM BROMIDE,
COCKAYNE B   13-1894   DILATOMETRIC STUDY OF THE ORTHORHOMBIC- TETRAGONAL PHASE TRANSITION IN BARIUM SODIUM N
                        13-1904   BARIUM SODIUM NIOBATE: CONSTITUTIONAL STUDIES AND (CZOCHRALSKI) CRYSTAL GROWTH.
                        13-1928   CRUCIBLE BASE COOLING: AN AID TO CZOCHRALSKI CRYSTAL GROWTH. (BARIUM SODIUM NIOBATE AN
                        13-2047   PROGRESS TOWARDS THE GROWTH OF GOOD OPTICAL QUALITY BARIUM SODIUM NIOBATE (BY USING CO
COEURE PH   36-3542   FERROELECTRIC PROPERTIES OF HEXAGONAL ORTHOMANGANITES OF YTTRIUM AND RARE EARTHS. (YTT
COHEN A   08-1295   DIELECTRIC AGING IN TETRAGONAL SOLID SOLUTIONS OF CALCIUM TITANATE IN BARIUM TITANATE.
COHEN J   36-3543   DIRECT PIEZOELECTRIC EFFECT IN POLYVINYL CHLORIDE FILMS.
                        36-3544   PYROELECTRICITY AND PIEZOELECTRICITY IN ORIENTED FILMS OF POLYVINYL FLUORIDE AND POLYV
COHEN MG   04-0628   INFLUENCE OF CRITICAL POLARIZATION FLUCTUATIONS ON THE PHOTOELASTIC BEHAVIOR OF BARIUM
                        13-1929   ELECTROCPTIC POTASSIUM TANTALATE NIOBATE GRATINGS FOR LIGHT BEAM MODULATION AND DEFLEC
COHEN MI   33-3368   PHASE TRANSITIONS IN CESIUM LEAD TRICHLORIDE. (LOSS OF A CENTER OF SYMMETRY AT 194 DEG
                        33-3408   DETERMINATION OF TRANSITION TEMPERATURES IN CESIUM LEAD TRICHLORIDE USING EPR.
COHN M   37-3633   (COMPACT, SURFACE WAVE) FERROELECTRIC PHASE SHIFTERS FOR VHF AND UHF. (BASED ON LEAD(O
COIGNAC JP   17-2479   SPECTRA OF INTERNAL VIBRATIONS OF THE PHOSPHATE ION IN THE PARAELECTRIC PHASE OF KDP C
                        17-2480   VIBRATION SPECTRA OF POTASSIUM DIHYDROGEN PHOSPHATE IN PARA- AND FERROELECTRIC PHASES.
COLE RH   34-3419   DIELECTRIC PROPERTIES OF ICE I.
COLLINGS EW   36-3545   MAGNETIC PROPERTIES OF HEAVY RARE EARTH ORTHOMANGANATES.
COLOT JL   25-2909   STARK EFFECT ON THE NUCLEAR QUADRUPOLE RESONANCE OF NITROGEN-14 AND THE FERROELECTRICI
COMES R   04-0629   LINEAR DISORDER IN CRYSTALS (CASE OF SILICON, QUARTZ AND FERROELECTRIC PEROVSKITES). (
                        04-0630   STRUCTURE DISORDER OF BARIUM TITANATE TYPE FERROELECTRICS. BASED ON NEW TREATMENT AND
                        08-1311   ANISOTROPIC CRITICAL X-RAY DIFFUSE SCATTERING FROM SODIUM NIOBATE CRYSTALS (NEAR THE 6
                        08-1313   X-RAY DIFFUSE SCATTERING FROM SODIUM NIOBATE AS A FUNCTION OF TEMPERATURE.
                        33-3369   CRITICAL ANISOTROPIC FLUCTUATIONS AT THE 184 DEG K TRANSITION OF POTASSIUM MANGANESE T
                        36-3546   NEUTRON SCATTERING ANALYSIS OF THE LINEAR DISPLACEMENT CORRELATIONS IN POTASSIUM TANTA
CONDRATE RA   08-1411   INFRARED AND RAMAN SPECTRA OF SEVERAL PEROVSKITE ZIRCONATES.
CONGER GJ   04-0631   NONSTOICHIOMETRY IN BARIUM TITANATE.
CONNOLLY TF   01-0004   FERROELECTRICITY AND FERROELECTRIC MATERIALS: A COMPREHENSIVE BIBLIOGRAPHY.
CONSTANTINE JG   13-1930   PHOTOCHROMIC EFFECT AND ELECTRON SPIN RESONANCE OF IRON AND MOLYBDENUM IN BARIUM SODIU
CONSTANTINESCU O   18-2731   ESR CF FREE RADICALS IN FERROELECTRIC AND ANTIFERROELECTRIC ARSENATE SINGLE CRYSTALS.
                        18-2734   ESR IDENTIFICATION OF SLATER CONFIGURATIONS AND OF EXCHANGE OF THE ARSENATE RADICAL IN
CONWAY ED   01-0028   MICROWAVE ACOUSTICS HANDBOOK. VOL 1: SURFACE WAVE VELOCITIES. (BARIUM SODIUM NIOBATE,
                        11-1819   NEW HIGH FREQUENCY HIGH COUPLING LOW BEAM STEERING CUT FOR ACOUSTIC SURFACE WAVES ON L
COOK AR   09-1560   HIGH D*, FAST, LEAD ZIRCONATE TITANATE PYROELECTRIC DETECTORS.
COOK RL   07-1258   ATHERMAL PHASE TRANSITION IN LEAD ZIRCONATE.
COOK WR   01-0015   PIEZOELECTRIC CERAMICS. (NONMETALLIC SOLIDS, VOL-3, REVIEW)
                        02-0058   FERROELECTRIC, PIEZOELECTRIC, AND ELECTROOPTIC MATERIALS. (REVIEW, 991 REFS)
                        28-3207   RADIATION DAMAGE AND THE FERROELECTRIC EFFECT IN ROCHELLE SALT.
COOMBS GJ   03-0210   DIELECTRIC RESPONSE OF CRYSTALS. (FERROELECTRIC FLUCTUATIONS)
                        18-2723   DIELECTRIC RESPONSE IN PIEZOELECTRIC CRYSTALS. (RAMAN SCATTERING IN POTASSIUM DIHYDROG
COOPER J   04-0632   MINIMUM DETECTABLE POWER OF A PYROELECTRIC THERMAL RECEIVER. (USING BARIUM TITANATE CR
COPLAN LA   03-0211   SOLUTION OF AN ELASTIC FERROELECTRIC MODEL.
COQUET E   03-0386   SOME RESULTS ABOUT PHOTOEMISSION ON FERROELECTRIC CRYSTALS. (DOMAIN IMAGING, ABSTRACT
                        11-1772   THEORETICAL STUDY OF THE GENERATION OF THE ACOUSTIC SECOND HARMONIC. (LITHIUM NIOBATE)
COQUIN GA   37-3634   AN ANALYSIS OF MULTIPLE LIGHT SCATTERING BY RANDOM OPTICAL INHOMOGENEITIES WITH APPLIC
CORONEL G   27-2983   THERMAL CONDUCTIVITY OF TRIGLYCINE SULFATE NEAR THE CURIE POINT.
CORROCHANO F   27-2981   EFFECT OF SEVERAL PARAMETERS ON THE FERROELECTRIC- PARAELECTRIC TRANSITION IN TRIGLYCI
                        27-2982   ORDER DISORDER TRANSITION IN FERROELECTRICS. (TRIGLYCINE SULFATE, HIGHLY ACCURATE TEMP
                        27-3030   EXPERIMENTAL DETAILED STUDY OF THE PHASE TRANSITION IN TRIGLYCINE SULFATE. (CLOSE THER
COSENTINC LS   09-1601   SOME PHYSICAL AND ELECTRICAL SWITCHING CHARACTERISTICS OF A LEAD ZIRCONATE TITANATE FE
COSIER RS   36-3624   ORTHORHOMBIC DIGLYCINE SULFATE. (NOT PIEZOELECTRIC, NOT FERROELECTRIC)
COSTANTINO MS   36-3547   DIELECTRIC ANOMALIES IN SOLID METHANE AT LOW TEMPERATURES AND HIGH PRESSURES. (POSSIBL
COT L   32-3355   CRYSTALLCGRAPHIC STUDY OF SEVERAL HYDRAZINIUM ORTHO FLUOBERYLLATES.
COTTINI M   17-2422   ULTRAVIOLET REFLECTION AND ABSCRPTION OF POTASSIUM DIHYDROGEN PHOSPHATE AND AMMONIUM D
COUFOVA P   04-0539   COMPENSATION OF HETEROVALENT FERRIC IONS IN THE LATTICE OF BARIUM TITANATE SINGLE CRYS
                        04-0540   CHEMICAL PROCESSES DURING HYDROGEN REDUCTION OF BARIUM TITANATE CRYSTALS.
                        04-0633   CRYSTAL CHEMICAL PROBLEMS OF (BARIUM, STRONTIUM) TITANATE SOLID SOLUTIONS.

CCUFOVA P

04-0634 DIELECTRIC PROPERTIES OF BARIUM TITANATE SINGLE CRYSTALS DOPED BY HYDROXIDE GROUPS.
04-0635 DIELECTRIC PROPERTIES OF BARIUM(1-X) STRONTIUM(X) TITANATE SINGLE CRYSTALS GROWN FROM
04-0636 HYDROXYL AS DEFECT IN BARIUM TITANATE.
04-0637 THICKNESS DEPENDENCE OF THE PERMITTIVITY OF BARIUM TITANATE SINGLE CRYSTALS.
04-0736 DETERMINATION OF THE DEVIATION FROM OXYGEN STOICHIOMETRY OF BARIUM TITANATE SINGLE CRY
04-0873 DEFECT STRUCTURE OF (BARIUM(1-X) STRONTIUM(X)) TITANATE- MIXED CRYSTALS FROM A POTASSI
04-0875 OPTICAL AND DIELECTRIC PROPERTIES OF SINGLE CRYSTAL BARIUM TITANATE WITH NIOBIUM ADDIT

COUND VM   12-1880 SIGNS CF THE ELECTROOPTIC COEFFICIENTS FOR LITHIUM TANTALATE.
COURDILIE JM   33-3370 ATTENUATION AND DISPERSION OF ULTRASONIC WAVES IN POTASSIUM MANGANESE TRIFLUORIDE IN T
COURTENS E   06-1116 BIREFRINGENCE OF STRONTIUM TITANATE PRODUCED BY THE 105 DEGREE K STRUCTURAL PHASE TRAN
COUSON B   04-0695 CONDUCTIVITY OF BARIUM TITANATE, PURE SINGLE CRYSTALS AND DOPED (IRON, HYDROXYL) SINGL
COUTURE L   21-2814 ZEEMAN EFFECT (STUDY OF TWO FINE LINES IN (THE RED REGION OF) CHROMIUM GUANIDINIUM SU
                        36-3548 VIBRATIONAL SPECTRA OF IODIC ACID SINGLE CRYSTALS.
COWLEY RA   02-0059 NEUTRON INELASTIC SCATTERING AT STRUCTURAL PHASE TRANSITIONS. (STRONTIUM TITANATE, POT
                        03-0210 DIELECTRIC RESPONSE OF CRYSTALS. (FERROELECTRIC FLUCTUATIONS)
                        03-0212 DIELECTRIC PROPERTIES OF AN ANHARMONIC CRYSTAL. (EXTENSION OF "SOFT MODE" THEORY)
                        03-0213 (MICROSCOPIC AND THERMODYNAMIC) THEORY OF FERROELECTRICITY AND ANHARMONIC EFFECTS IN C
                        06-1224 LATTICE DYNAMICS OF STRONTIUM TITANATE.
                        16-2284 CRYSTAL DYNAMICS OF SODIUM NITRITE. (NEUTRON SCATTERING)
                        16-2364 CRYSTAL DYNAMICS AND THE FERROELECTRIC PHASE TRANSITION OF SODIUM NITRITE (STUDIED BY
                        17-2610 FERROELECTRIC CRITICAL SCATTERING FROM DEUTERATED KDP. (NEUTRON INELASTIC SCATTERING)
                        17-2611 FERROELECTRIC TRANSITION IN POTASSIUM DIDEUTERIUM PHOSPHATE.
                        18-2723 DIELECTRIC RESPONSE IN PIEZOELECTRIC CRYSTALS. (RAMAN SCATTERING IN POTASSIUM DIHYDROG
                        33-3366 ANTIFERROELECTRICITY IN (PURE AND) COBALT DOPED POTASSIUM MANGANESE TRIFLUORIDE. (20 P
                        33-3407 MAGNETIC EXCITATIONS IN NICKEL DOPED POTASSIUM MANGANESE TRIFLUORIDE.
                        36-3586 TRANSITION IN TIN TELLURIDE- GERMANIUM TELLURIDE ALLOYS. (NEUTRON SCATTERING)
COX GA   04-0638 DOUBLE SPACE CHARGE INJECTION IN SOLIDS. (USING BARIUM TITANATE ABOVE CURIE POINT)
CRABB JC   11-1698 SURFACE WAVE DIFFRACTION IN LITHIUM NIOBATE.
CPAIG DC   13-1931 STRUCTURE OF THE BRONZE SODIUM(13) NIOBIUM(35) OXIDE(94) AND THE GEOMETRY OF FERROELEC
CRANE GR   08-1296 RELATICN OF PHYSICAL PROPERTIES TO THE SYMMETRY OF POTASSIUM IODATE. (NONLINEAR OPTIC,
                        08-1465 PIEZOELECTRIC AND PHOTOELASTIC PROPERTIES OF LITHIUM IODATE.
                        36-3530 PYROELECTRICITY AND OPTICAL SECOND HARMONIC GENERATION IN POLYVINYLIDENE FLUORIDE FILM
                        36-3595 PYROELECTRIC AND NONLINEAR OPTICAL PROPERTIES OF POLED POLYVINYLIDENE FLUORIDE FILMS.
CRAWFORD JC   09-1504 A FERROELECTRIC PIEZOELECTRIC RANDOM ACCESS MEMORY. (LEAD ZIRCONATE- LEAD TITANATE CER
                        09-1505 FERROELECTRIC FIELD EFFECT STUDIES AT LOW TEMPERATURES (WITH TIN OXIDE FILMS ON LEAD Z
                        09-1506 FERROELECTRIC FIELD EFFECT STUDIES. (ON TIN OXIDE FILMS ON LEAD(0.97) BISMUTH(0.02) TI
                        37-3635 PIEZOELECTRIC RESPONSE OF A FERROELECTRIC MEMORY ARRAY.
CRAWFORD RK   36-3547 DIELECTRIC ANOMALIES IN SOLID METHANE AT LOW TEMPERATURES AND HIGH PRESSURES. (POSSIBL
CREEL RB   11-1829 TONE BURSTING OF QUADRUPOLAR NUCLEI. DETERMINATION OF THE SPIN LATTICE RELAXATION TIM
CRIPPA ML   13-1910 TRANSIENT NMR STUDY OF PHASE TRANSITIONS IN METALLIC SODIUM TUNGSTEN BRONZES.
                        36-3533 NMR STUDY OF THE STRUCTURAL PHASE TRANSITION IN LANTHANUM ALUMINATE.
CROSS JD   34-3420 STUDY OF THE SURFACE OF ICE WITH A SCANNING ELECTRON MICROSCOPE.
CROSS LE   01-0005 SYMPOSIUM ON APPLICATIONS OF FERROELECTRICS.
                        02-0060 FERROELECTRICITY AND CONDUCTION IN FERROELECTRIC CRYSTALS. (LITHIUM THALLIUM TARTRATE
                        02-0061 FERROELECTRICITY AND CONDUCTION IN FERROELECTRIC CRYSTALS. (BISMUTH TITANATE, LITHIUM
                        03-0214 THERMODYNAMIC PHENOMENOLOGY OF FERROELECTRICITY IN SINGLE CRYSTAL AND CERAMIC SYSTEMS.
                        03-0456 DOUBLE HYSTERESIS LOOPS AND FIRST ORDER PHASE TRANSITIONS IN FERROELECTRIC CRYSTALS.
                        04-0891 EFFECT OF INTERNAL BOUNDARIES ON THE DIELECTRIC PROPERTIES OF BARIUM TITANATE CERAMICS
                        06-1117 DIELECTRIC AND ELECTROOPTIC STUDIES OF THE POLARIZATION PROCESS IN STRONTIUM TITANATE
                        06-1143 DIELECTRIC RELAXATION IN (CERAMIC) STRONTIUM TITANATES CONTAINING (TRIVALENT) RARE EAR
                        06-1201 DIELECTRIC PROPERTIES OF STRONTIUM TITANATE (ANNEALED SINGLE CRYSTALS) AT LOW TEMPERAT
                        08-1297 THERMODYNAMICS OF FERROELECTRICITY IN BISMUTH TITANATE.
                        08-1419 THERMODYNAMICS OF FERROELECTRICITY IN BISMUTH TITANATE.
                        08-1436 DIELECTRIC BEHAVIOR (PERMITTIVITY, SPONTANEOUS POLARIZATION, COERCIVE FIELD) OF BISMUT
                        08-1440 A MICROSCOPIC THEORY FOR THE DIELECTRIC PROPERTIES OF LEAD MAGNESIUM NIOBATE.
                        13-1932 FERROELECTRICITY IN BISMUTH OXIDE TYPE LAYER STRUCTURE COMPOUNDS.
                        14-2093 CPTICAL ANISOTROPY OF CUBIC NICKEL IODINE BORACITE DUE TO QUADRUPOLE TRANSITIONS.
                        14-2094 OPTICAL ANISOTROPY OF THE CUBIC COBALT IODINE BORACITE.
                        29-3242 SOME PROPERTIES OF THE FERROELECTRIC LITHIUM THALLIUM TARTRATE.
                        29-3249 CONTROL OF THE ELASTIC COMPLIANCE OF FERROELECTRIC LITHIUM THALLIUM TARTRATE HYDRATE (
                        29-3250 ELECTRIC FIELD CONTROL CF THE ELASTIC COMPLIANCE IN LITHIUM THALLIUM TARTRATE (TUNING
                        29-3251 ELECTROMECHANICAL COUPLING EFFECTS ON THE DIELECTRIC PROPERTIES AND FERROELECTRIC PHAS
                        30-3266 FERROELECTRICITY IN GADOLINIUM MOLYBDATE.
                        36-3550 FERROELECTRIC OPTICAL ROTATION DOMAINS IN SINGLE CRYSTAL LEAD(5) GERMANIUM(3) OXIDE(11
                        36-3555 (WEAK FIELD) DIELECTRIC (PERMITTIVITY) PROPERTIES OF BISMUTH TITANATE (EFFECT OF TEMPE
CRUICKSHANK DWJ   22-2845 CRYSTAL STRUCTURE OF POTASSIUM SELENATE.
CSILLAG A   04-0689 MOBILITY DETERMINATIONS FROM WEIGHT MEASUREMENTS IN SOLID SOLUTIONS OF (BARIUM, STRONT
CUMMINS HZ   17-2481 LASER LIGHT SCATTERING STUDIES OF PHASE TRANSITIONS. (TRIGLYCINE SULFATE, POTASSIUM DI
                        17-2497 LIGHT SCATTERING STUDIES OF THE SOFT OPTIC AND ACOUSTIC MODES OF POTASSIUM DIHYDROGEN
                        17-2636 SIMULTANEOUS OBSERVATION OF THE SOFT FERROELECTRIC AND ACOUSTIC MODES OF POTASSIUM DID
                        17-2709 RAMAN SCATTERING IN POTASSIUM DIHYDROGEN PHOSPHATE.
CUMMINS SE   08-1298 A NEW METHOD OF OPTICALLY READING DOMAINS IN BISMUTH TITANATE AND MEMORY APPLICATIONS.
                        08-1299 EFFICIENT WHITE LIGHT READING OF DOMAIN PATTERNS IN BISMUTH TITANATE.
                        08-1300 ELECTRON BEAM WRITING OF FERROELECTRIC DOMAINS IN BISMUTH TITANIUM OXIDE SINGLE CRYSTA
                        08-1301 OPTICAL STUDY OF DOMAIN WALLS IN BISMUTH TITANATE.
                        30-3267 ELECTRICAL, OPTICAL, AND MECHANICAL BEHAVIOR OF FERROELECTRIC GADOLINIUM MOLYBDATE.
CUTCHEN JT   08-1302 MATERIAL AND ELECTROOPTIC PROPERTIES OF THE (LEAD, LANTHANUM) (HAFNATE, TITANATE) SYST
CUTHBERT JD   32-3343 NUCLEAR MAGNETIC RESONANCE STUDY OF FERROELECTRIC LITHIUM HYDRAZINIUM SULFATE.
CUTTER IA   04-0639 SURFACE DOMAIN REORIENTATION PRODUCED BY ABRASION AND ANNEALING OF POLYCRYSTALLINE BAR
CYRLIN VP   03-0164 SPIN WAVE PHENOMENOLOGICAL THEORY IN FINITE SAMPLES OF FERRODIELECTRICS WITH AN ARBITR
DA COSTA RCT   03-0215 HARMONICS GENERATION IN FERROELECTRIC MEDIA.
DABIR A   22-2879 HYDROGEN BOND STUDIES. PART-58: CRYSTAL STRUCTURE OF AMMONIUM TRIHYDROGEN SELENITE.
DABU R   17-2595 METHOD CF GROWTH IN SOLUTION OF SINGLE CRYSTALS BY THE GRADUAL ADDITION OF A REACTANT.
DADOUREK K   27-2984 DYNAMIC DETERMINATION OF (ELASTIC AND PIEZOELECTRIC) MATERIAL CONSTANTS OF TRIGLYCINE
DAGONNIER R   03-0205 PROTON DYNAMICS IN HYDROGEN BONDED FERROELECTRICS. PART-2: APPROXIMATED SOLUTIONS TO T
DAHL J   13-1905 LOW FREQUENCY VIBRATIONAL MODES IN BARIUM SODIUM NIOBATE.
DAHLBORG U   19-2752 ROTATIONAL MOTIONS IN SOLIDS: THE FERROELECTRIC TRANSITION IN AMMONIUM SULFATE STUDIED
DALAL NS   18-2724 DETECTION BY ELECTRON PARAMAGNETIC RESONANCE AND ELECTRON NUCLEAR DOUBLE RESONANCE OF
                        18-2725 ELECTRON PARAMAGNETIC RESONANCE STUDIES OF X-IRRADIATED POTASSIUM DIHYDROGEN ARSENATE,
                        18-2726 ELECTRON PARAMAGNETIC RESONANCE AND ELECTRON NUCLEAR DOUBLE RESONANCE OBSERVATIONS OF

DALAL NS	18-2727	ELECTRON PARAMAGNETIC RESONANCE AND ELECTRON NUCLEAR DOUBLE RESONANCE OF THE ARSENATE
	18-2728	HYDROGEN BONDED FERROELECTRICS: EVIDENCE FOR SLATER CONFIGURATIONS FROM EPR STUDIES OF
	18-2729	OBSERVATION OF ELECTRON NUCLEAR TRIPLE RESONANCE FOR THE ARSENATE CENTER IN X-RAY IRRA
DALLY JW	37-3630	EXPERIMENTAL METHODS FOR INVESTIGATING STRAIN WAVE PROPAGATION AND ASSOCIATED CHARGE R
DALTON NW	03-0216	DOMAIN STATISTICS AND FERROELECTRIC TRANSIENTS.
DAMJANOVIC A	13-1944	ELECTROCHEMICAL DEPOSITION AND DISSOLUTION OF TUNGSTEN OXIDE BRONZES. (SODIUM(X) TUNGS
DANIEL MR	11-1699	ACOUSTIC RADIATION BY INTERDIGITAL GRIDS ON LITHIUM NIOBATE.
	11-1702	MEASURED ELECTRICAL CHARACTERISTICS OF INTERDIGITAL SURFACE WAVE TRANSDUCERS ON LITHIU
DANIELEWICZ-FERCHMIN I	03-0331	QUASISTATIC GENERATION OF HARMONICS IN A FERROELECTRIC CRYSTAL WITH A SECOND ORDER TRA
DANIELS WB	36-3547	DIELECTRIC ANOMALIES IN SOLID METHANE AT LOW TEMPERATURES AND HIGH PRESSURES. (POSSIBL
DANILYCHEVA MN	19-2789	SPECIFIC HEAT OF AMMONIUM HYDROGEN SULFATE IN THE TEMPERATURE RANGE FROM -70 TO +14 DE
DANILYUK YUL	04-0640	ELECTRON PARAMAGNETIC RESONANCE IN NONSTOICHIOMETRIC BARIUM TITANATE.
DANSAS P	34-3448	ELECTRICAL BEHAVIOR OF ICE AT LOW TEMPERATURES, FERROELECTRICITY AND SPACE CHARGE.
DANTL G	34-3421	ELASTIC MODULI OF ICE.
DANTSIGER AYA	03-0217	RELATION BETWEEN THE MAIN ELECTROPHYSICAL AND STRUCTURAL PARAMETERS OF FERROELECTRIC P
	08-1325	SYNTHESIS AND STUDY OF LEAD SCANDIUM NIOBATE SINGLE CRYSTALS.
	09-1507	TERNARY SYSTEM LEAD TITANATE- LEAD ZIRCONATE- LEAD NIOBATE(0.67) ZINCATE(0.33).
DAO NQ	08-1303	INFRARED AND RAMAN SPECTROSCOPY OF POTASSIUM METANIOBATE.
DARINSKII BM	04-0641	STRUCTURE OF 90 DEGREE DOMAIN WALLS IN BARIUM TITANATE.
	08-1305	MOVEMENT OF 180 DEGREE DOMAIN BOUNDARIES IN PEROVSKITE FERROELECTRICS.
	17-2542	DIELECTRIC PROPERTIES OF GAMMA-IRRADIATED POTASSIUM DIHYDROGEN PHOSPHATE CRYSTALS.
DARLINGTON AN	02-0094	ROLE OF THE REGULAR OCTAHEDRON AND LARGER POLYHEDRA IN FERROELECTRIC MATERIALS. (HETER
DARLINGTON CNW	03-0218	PHENOMENOLOGICAL THEORY OF ANTIFERROELECTRICITY AND FERROELECTRICITY APPLIED TO SODIUM
	08-1304	DIFFUSE SCATTERING IN SODIUM NIOBATE AT ROOM TEMPERATURE.
	09-1590	PHASE BOUNDARY AND DEFECT STRUCTURE STUDIES IN PLZT CERAMICS.
DAS P	11-1667	LAYER WAVE AMPLIFICATION IN A CADMIUM SULFIDE FLUID- LITHIUM NIOBATE STRUCTURE.
	11-1700	SURFACE WAVE CONVOLUTION USING A CADMIUM SULFIDE FLUID LITHIUM NIOBATE STRUCTURE.
DATE M	26-2941	ANOMALOUS SHIFT OF ESR LINES IN THE TWO DIMENSIONAL ANTIFERROMAGNET COPPER FORMATE TET
DAVID J	03-0491	SHOCK WAVE INDUCED ELECTROMECHANICAL CONVERSION IN FERROELECTRIC MATERIALS.
	09-1508	FERROELECTRIC CERAMICS AND ENERGY CONVERSION. (LEAD ZIRCONATE TITANATE)
DAVID TG	13-2040	SOFT PHONONS IN POTASSIUM TANTALATE NIOBATE.
DAVIDSON DW	34-3431	DIELECTRIC BEHAVIOR OF CUBIC AND HEXAGONAL ICES AT LOW TEMPERATURES.
DAVIDSON F	11-1701	MEASUREMENTS OF PHOTON CORRELATIONS OF SECOND HARMONIC GENERATED LIGHT (FROM LITHIUM N
DAVIES PH	12-1880	SIGNS OF THE ELECTROOPTIC COEFFICIENTS FOR LITHIUM TANTALATE.
	17-2435	1.06 MICRON ABSORPTION COEFFICIENTS OF DEUTERATED KDP WITH 70-100% DEUTERATION.
	35-3494	ELECTROOPTIC MEASUREMENTS ON LITHIUM GERMANATE AND LEAD GERMANATE. (LITHIUM GERMANATE
DAVIS BL	16-2279	TRANSITION RATES OF POTASSIUM NITRATE HIGH PRESSURE POLYMORPHS.
DAVIS KL	11-1778	POSSIBILITY OF SWITCHING AND STEERING SURFACE WAVES BY NONLINEAR MIXING IN ANISOTROPIC
DAVIS L	03-0219	SPONTANEOUS POLARIZATION NOISE IN POLAR DIELECTRICS. (THEORY, EINSTEIN THERMODYNAMIC M
DAVIS LW	04-0803	RAMAN SCATTERING LINE SHAPE OF THE SOFT E POLARITON MODE IN BARIUM TITANATE.
DAVIS TG	08-1306	DIELECTRIC PROPERTIES AND SOFT MODES IN THE FERROELECTRIC MIXED CRYSTALS POTASSIUM(1-X
	13-1933	SOFT PHONON MODES IN THE MIXED CRYSTAL SYSTEMS POTASSIUM(X) SODIUM(1-X) TANTALATE AND
DAVISSON JW	23-2886	BIAS FIELDS IN FERROELECTRIC COLEMANITE.
DAVYDOV IN	03-0147	BEHAVIOR OF MAGNETICALLY ORDERED FERROELECTRICS IN A RAPIDLY OSCILLATING ELECTRIC FIEL
	03-0148	COUPLED ELECTROMAGNETIC SPIN WAVES IN MAGNETICALLY ORDERED FERROELECTRICS.
	03-0149	INSTABILITY OF SPIN WAVES IN FERROELECTRIC ANTIFERROMAGNETS IN EXTERNAL FIELDS. (THEOR
DAVYDOV VYU	17-2470	EFFECT OF TEMPERATURE AND PHASE TRANSITION ON THE HYDROXIDE VIBRATION SPECTRUM OF THE
	17-2471	POLARIZATION OF THE RAMAN BANDS OF THE HYDROXYL VIBRATIONS AND STRUCTURE OF THE HYDROG
	17-2472	VIBRATION SPECTRUM OF THE PHOSPHATE ANION IN A RUBIDIUM DIHYDROGEN PHOSPHATE CRYSTAL I
DAVYDOVA TS	22-2881	CRYSTAL STRUCTURE OF RUBIDIUM TRIHYDROGEN SELENITE.
DAWSON P	36-3549	THE VIBRATIONAL SPECTRUM AND DIELECTRIC BEHAVIOR OF SODIUM CHLORATE. (NOT FERROELECTRI
DAWSON RD	13-1939	VARIATION OF CRYSTAL GROWTH RATE WITH SUPERSATURATION IN FLUXED MELTS. (BARIUM STRONTI
DE BOCK A	11-1756	MICROWAVE PHONON ATTUNUATION IN X CUT LITHIUM NIOBATE.
DE GENNES PG	17-2482	VIBRATION SPECTRA OF HYDROGEN BONDS. (KDP)
	17-2483	COLLECTIVE MOTIONS OF HYDROGEN BONDS. (CALCULATIONS FOR POTASSIUM DIHYDROGEN PHOSPHATE
DE JONG M	37-3636	APPLICATION OF FERROELECTRIC MATERIALS IN ELECTRICAL FILTERS.
DE KLERK J	11-1702	MEASURED ELECTRICAL CHARACTERISTICS OF INTERDIGITAL SURFACE WAVE TRANSDUCERS ON LITHIU
DE LAUNAY J	02-0062	AGING OF BARIUM TITANATE AND LEAD ZIRCONATE TITANATE FERROELECTRIC CERAMICS. (REVIEW,
	04-0642	CALCULATION OF THE ELECTROSTATIC INTERACTIONS IN FINE GRAINED PEROVSKITE FERROELECTRIC
DE PAOLIS MV	08-1437	CALCIUM CONCENTRATION VS NET IONIZED DONOR CONCENTRATION IN SINGLE CRYSTAL POTASSIUM T
DE POORTER JA	36-3531	RADIATIONLESS TRANSITIONS IN THE EUROPIUM(3+) CENTER IN LANTHANUM ALUMINATE.
DE WETLE FW	03-0220	INTERNAL FIELD IN GENERAL DIPOLE LATTICES. (THEORY APPLIED TO POTASSIUM FERROCYANIDE T
DEAN P	34-3474	ATOMIC VIBRATIONS IN ORIENTATIONALLY DISORDERED SYSTEMS. PART-1: A TWO DIMENSIONAL MOD
DEGUCHI K	04-0643	MOVING STRIPE PATTERN IN BARIUM TITANATE.
	27-2985	CRITICAL REGION IN FERROELECTRIC TRIGLYCINE SULFATE.
DEIS DW	06-1118	ULTRASONIC STUDIES OF CRITICAL PHENOMENA IN INSULATING AND SEMICONDUCTING STRONTIUM TI
DELAHAIGUE A	25-2910	INFRARED ABSORPTION SPECTRUM OF A SINGLE CRYSTAL OF THIOUREA IN THE PARA- AND FERROELE
DELANNOY M	11-1680	COLLINEAR INTERACTION BETWEEN A LONGITUDINAL ACOUSTIC WAVE AND TWO LUMINOUS WAVES ALON
	11-1835	GENERATION OF THE THIRD HARMONIC OF A LONGITUDINAL HIGH FREQUENCY (950 MHZ) ACOUSTIC W
DELAVIGNETTE P	03-0255	ELECTRON MICROSCOPE TRANSMISSION IMAGES OF COHERENT DOMAIN BOUNDARIES. PART-1: DYNAMIC
DEMARS GA	04-0920	ELECTRON PARAMAGNETIC RESONANCE OF TRIVALENT GADOLINIUM IONS IN STRONTIUM AND BARIUM T
	06-1192	ELECTRON PARAMAGNETIC RESONANCE OF SOME RARE EARTH IMPURITIES IN STRONTIUM TITANATE.
DEMESHINA AI	04-0644	ABSORPTION AND REFLECTION SPECTRA OF BARIUM TITANATE IN THE LONG WAVELENGTH INFRARED R
	04-0863	TEMPERATURE STUDIES OF INFRARED REFLECTION SPECTRA OF BARIUM AND STRONTIUM TITANATES I
	04-0864	VIBRATIONAL SPECTRA OF STRONTIUM, BARIUM AND CALCIUM TITANATES.
DEMIDENKO NI	04-1020	PECULIARITIES OF THE ELECTRON PARAMAGNETIC RESONANCE SPECTRA IN A FERROELECTRIC CERAMI
DEMIDOVICH NP	27-3154	MAXIMUM REPOLARIZATION CURRENT FOR TRIGLYCINE SULFATE CRYSTALS DEPENDENT ON GROWTH CON
DEMUROV DG	02-0128	NEW FERROELECTRICS AND SEIGNETTO MAGNETS OF THE PEROVSKITE AND PYROCHLORE TYPE. SYNTHE
	08-1307	CHARACTER OF PHASE TRANSITION OF FERROELECTRIC POTASSIUM TANTALATE. (SECOND ORDER)
	08-1308	POTASSIUM TANTALATE LEAD TITANATE SYSTEM IN THE TEMPERATURE RANGE 4.2-300 DEG K (LATTI
	08-1309	SYNTHESIS, X-RAY ANALYSIS, AND STUDY OF THE DIELECTRIC PROPERTIES OF NEW CADMIUM-CONTA
	08-1310	X-RAY ANALYSIS AND DIELECTRIC PROPERTIES OF SOLID SOLUTIONS BASED ON THE FERROELECTRIC
DEMYANENKO VP	04-0983	INFRARED SPECTRA OF FLUORINE CONTAINING BARIUM TITANATE.
	08-1278	POTASSIUM IODATE CRYSTAL STRUCTURE STUDIED BY INFRARED SPECTROSCOPY AND NUCLEAR QUADRU
DEMYANOV VV	04-0645	DIELECTRIC PERMEABILITY OF BARIUM TITANATE SINGLE CRYSTALS IN THE PHASE TRANSITION REG
	04-0646	DIELECTRIC DISPERSION IN BARIUM TITANATE.
	04-0647	DIELECTRIC SPECTRUM OF BARIUM TITANATE. (AND MICROSCOPIC MODEL OF LOCALLY ORDERED "NUC
	04-0648	DIELECTRIC DISPERSION OF IRRADIATED BARIUM TITANATE NEAR THE PHASE TRASNITION.
	04-0649	EXPERIMENTAL CHECK OF THE THEORY OF MICROWAVE DISPERSION OF THE PERMITTIVITY OF BARIUM
	04-0650	KINETICS OF THE PHASE TRANSITIONS OF BARIUM TITANATE.

DENIAU M   25-2906 LOW FREQUENCY RAMAN SPECTRUM OF A THIOUREA CRYSTAL AS A FUNCTION OF TEMPERATURE.
DENISOVA VM   09-1514 TIME AND TEMPERATURE DEPENDENCES OF THE PARAMETERS OF A FERROELECTRIC CERAMIC.
DENIZ KU   33-3371 PARAMAGNETIC SCATTERING OF NEUTRONS FROM POTASSIUM MANGANESE TRIFLUORIDE IN THE SHORT
DENNIS JH   17-2484 INDEX OF REFRACTION OF POTASSIUM DIHYDROGEN PHOSPHATE.
DENNIS MD   04-0651 DOMAIN INTERACTIONS DURING PHASE TRANSITIONS IN BARIUM TITANATE.
DENOYER F   08-1311 ANISOTROPIC CRITICAL X-RAY DIFFUSE SCATTERING FROM SODIUM NIOBATE CRYSTALS (NEAR THE 6
                    08-1312 ANISOTROPIC CRITICAL X-RAY DIFFUSE SCATTERING FROM POTASSIUM MANGANESE TRIFLUORIDE AND
                    08-1313 X-RAY DIFFUSE SCATTERING FROM SODIUM NIOBATE AS A FUNCTION OF TEMPERATURE.
                    33-3369 CRITICAL ANISOTROPIC FLUCTUATIONS AT THE 184 DEG K TRANSITION OF POTASSIUM MANGANESE T
DENYANOV VV   04-0652 DIELECTRIC DISPERSION OF IRRADIATED BARIUM TITANATE IN THE PHASE TRANSITION REGION.
DEPUIS T   02-0113 INFRARED ABSORPTION SPECTROPHOTOMETRIC STUDY OF MONOVALENT METAL NIOBATES.
DEPUTY GO   08-1314 ELECTRICAL PROPERTIES AND DEFECT STRUCTURE OF POTASSIUM TANTALATE.
DERBYSHIRE SW   04-0980 RELATION BETWEEN ELECTRICAL PROPERTIES AND MICROSTRUCTURE OF BARIUM TITANATE.
DERI M   01-0006 FERROELECTRIC CERAMICS.
DERIGHETTI B   36-3533 NMR STUDY OF THE STRUCTURAL PHASE TRANSITION IN LANTHANUM ALUMINATE.
DEROYON JP   11-1682 GENERATION OF SECOND AND THIRD HARMONICS BY A TRANSVERSE ACOUSTIC WAVE PROPAGATING ITS
                    11-1834 EXPERIMENTAL STUDY OF LIGHT DIFFRACTION BY HIGH FREQUENCY ACOUSTIC WAVES. APPLICATION
DESAI CC   21-2816 OPTICAL AND INTERFEROMETRIC STUDIES ON FERROELECTRIC GUANIDINIUM ALUMINUM SULFATE HEXA
DESCHAMPS L   33-3369 CRITICAL ANISOTROPIC FLUCTUATIONS AT THE 184 DEG K TRANSITION OF POTASSIUM MANGANESE T
DESHMUKH KG   08-1315 DOMAINS IN FERROELECTRIC POTASSIUM NIOBATE SINGLE CRYSTALS.
                    08-1316 DOMAIN STRUCTURES IN POTASSIUM NIOBATE SINGLE CRYSTALS ASSOCIATED WITH STEP LADDERS ON
                    08-1317 DOMAINS IN POTASSIUM NIOBATE SINGLE CRYSTALS FROM PSEUDOCUBIC (001) CLEAVAGE PLANES.
                    08-1318 INTERFEROMETRIC STUDIES OF DOMAIN STRUCTURES IN POTASSIUM NIOBATE SINGLE CRYSTALS.
                    08-1319 INTERFEROMETRIC STUDIES OF DOMAIN STRUCTURE IN POTASSIUM NIOBATE SINGLE CRYSTALS.
                    08-1320 OPTICAL STUDIES OF DOMAIN STRUCTURE OF POTASSIUM NIOBATE SINGLE CRYSTALS.
                    08-1369 SURFACE CHARGE LAYERS ON FERROELECTRIC POTASSIUM NIOBATE CRYSTALS.
                    11-1703 DOMAIN STRUCTURE IN LITHIUM NIOBATE SINGLE CRYSTALS.
DESHPANDE VT   17-2485 X-RAY DETERMINATION OF THE THERMAL EXPANSION OF AMMONIUM DIHYDROGEN PHOSPHATE.
DESLATTES RD   03-0414 EXTENDED FINE STRUCTURE IN X-RAY ABSORPTION SPECTRA OF CERTAIN PEROVSKITES (STRONTIUM
DESVIGNES JM   17-2637 CONTRIBUTION TO THE PREPARATION ON NON-WEDGE SHAPED RUBIDIUM DIHYDROGEN OR DIDEUTERIUM
                    19-2753 PREPARATION OF AMMONIUM SULFATE SINGLE CRYSTALS.
DEUTSCH T   06-1193 EFFECT OF TEMPERATURE AND PRESSURE ON THE ELECTRON PARAMAGNETIC RESONANCE SPECTRA OF S
DEVANARAYANAN S   03-0350 THERMAL EXPANSION ANOMALIES AND THE DIRECTION OF SPONTANEOUS POLARIZATION IN FERROELEC
                    22-2851 RELAXATION BETWEEN THE STRUCTURE OF FERROELECTRIC SODIUM TRIHYDROGEN DISELENITE AND IT
DEVIN A   11-1772 THEORETICAL STUDY OF THE GENERATION OF THE ACOUSTIC SECOND HARMONIC. (LITHIUM NIOBATE)
DEVINE SD   11-1704 GROUP THEORETICAL SELECTION RULES FOR INELASTIC NEUTRON SCATTERING WITH APPLICATIONS T
                    16-2281 MULTIPHONON SELECTION RULES FOR INFRARED ABSORPTION AND RAMAN SCATTERING IN FERROELECT
DEVLIKANEVA RO   09-1507 TERNARY SYSTEM LEAD TITANATE- LEAD ZIRCONATE- LEAD NIOBATE(0.67) ZINCATE(0.33).
DEVLIN JP   16-2280 ATR INFRARED SPECTRA OF UNIAXIAL NITRATE CRYSTALS.
DEVRIES AJ   09-1509 REALIZATION OF A 40 MHZ COLOR TELEVISION INFRARED RESPONSE USING SURFACE WAVE TRANSDUC
DEYDA H   28-3164 MEASUREMENT OF THE COMPLEX DIELECTRIC CONSTANT OF ROCHELLE SALT AT 10 GHZ AS A FUNCTIO
DHALL BS   16-2269 PROGRESS TOWARD A FAST, NONVOLATILE, NONDESTRUCTIVE READOUT MEMORY ELEMENT UTILIZING P
                    16-2282 SWITCHING PROPERTIES OF THICK FILM FERROELECTRIC POTASSIUM NITRATE.
DI BENEDETTO B   05-1042 SOLVENT ZONE GROWTH OF FERROELECTRIC CRYSTALS. (STRONTIUM TITANATE, LEAD(X) STRONTIUM(
DI MARCELLO FV   04-0825 STABILIZED BARIUM TITANATE CERAMICS FOR CAPACITOR DIELECTRICS.
DIAMOND B   04-0594 EFFECT OF FERROELECTRIC (BARIUM TITANATE) POLARIZATION FIELDS ON (GERMANIUM OR TELLURI
DIAS JF   09-1509 REALIZATION OF A 40 MHZ COLOR TELEVISION INFRARED RESPONSE USING SURFACE WAVE TRANSDUC
DICKENS PG   13-1934 ELECTRONIC PROPERTIES (CONDUCTIVITY, MAGNETIC SUSCEPTIBILITY) OF HYDROGEN TUNGSTEN BRO
DICKINSON JR   18-2725 ELECTRON PARAMAGNETIC RESONANCE STUDIES OF X-IRRADIATED POTASSIUM DIHYDROGEN ARSENATE,
DICKINSON SK   35-3496 CZOCHRALSKI SYNTHESIS AND PROPERTIES OF RARE EARTH DOPED BISMUTH GERMANATE.
DIDKOVSKAYA OS   04-0780 EFFECT OF LITHIUM OXIDE ON THE PROPERTIES OF A SERIES OF FERROELECTRIC MATERIALS. (BAR
                    05-1043 EFFECT OF GERMANIUM ON THE STRUCTURE AND PROPERTIES OF FERROELECTRIC LEAD TITANATE (GE
                    05-1071 DIELECTRIC PROPERTIES OF LEAD TITANATE- LEAD (FERRATE NIOBATE) SOLID SOLUTIONS AT HIGH
                    09-1510 PIEZOCERAMICS BASED ON LEAD ZIRCONATE TITANATE WITH COMPLEX ADDITIVES CONTAINING GERMA
                    09-1544 NEW PIEZOELECTRIC CERAMICS. (LEAD TITANATE- LEAD FERRATE(0.5) NIOBATE(0.5), LEAD TITAN
                    09-1588 EFFECT OF BISMUTH AND MANGANESE CONTAINING ADDITIVES ON THE PROPERTIES OF SOLID SOLUTI
DIDOMENICO M   03-0513 THEORY OF THE ELASTOOPTIC EFFECT IN NONMETALLIC CRYSTALS. (STRAIN DEPENDENT CURIE TEMP
                    04-0616 DIELECTRIC PROPERTIES OF SINGLE DOMAIN MELT GROWN BARIUM TITANATE.
                    04-0628 INFLUENCE OF CRITICAL POLARIZATION FLUCTUATIONS ON THE PHOTOELASTIC BEHAVIOR OF BARIUM
DIEFENDORF RJ   04-0828 PHYSICAL AND ELECTRICAL PROPERTIES OF THIN FILM BARIUM TITANATE PREPARED BY RF SPUTTER
DIETZ RE   33-3395 IMPURITY CONCENTRATION DEPENDENCE OF RAMAN SCATTERING BY MAGNONS IN NICKEL DOPED RUBID
DIEULESAINT E   11-1681 DETERMINATION OF A NONLINEAR PARAMETER FOR ACOUSTIC SURFACE WAVE CONVOLUTION IN LITHIU
DIKANT J   16-2283 SPECIFIC HEAT, THERMAL DIFFUSIVITY AND THERMAL CONDUCTIVITY OF CALCIUM DOPED SODIUM NI
                    27-3048 ANOMALIES OF THERMAL PROPERTIES OF TRIGLYCINE SULFIDE.
DIMANE M   15-2112 CRYSTAL STRUCTURE AND CONDUCTIVITY OF ANTIMONY SULFOIODIDE.
DIMAROVA EN   17-2627 THERMOELASTIC PROPERTIES OF POTASSIUM DIHYDROGEN PHOSPHATE NEAR THE HIGH TEMPERATURE P
                    27-2986 DEUTERATION EFFECT ON THERMAL AND ELASTIC PROPERTIES OF FERROELECTRIC LEAD CRYSTALS THE TRI
DIONNE GF   19-2754 CALCULATION OF CRYSTAL FIELD ENERGY LEVEL SPLITTINGS OF TITANIUM(3+) ION IN RUBIDIUM A
DISCHLER B   11-1725 AN EPR INVESTIGATION OF IRON(3+) AND MANGANESE(2+) IN LITHIUM NIOBATE.
DISTLER GI   27-2987 ELECTRICAL PROPERTIES OF THE SURFACES OF IONIC CRYSTALS (TGS AND OTHERS), BY SELECTIVE
                    27-2988 SPECIAL ELECTRICAL PROPERTIES OF REGIONS NEAR DOMAIN WALLS IN TRIGLYCINE SULFATE CRYST
                    27-3039 ELECTRICAL STRUCTURE OF THE SURFACE OF REAL CRYSTAL SUBSTRATES AS THE DETERMINING FACT
                    27-3041 REVEALING THE DOMAIN STRUCTURE OF TGS CRYSTALS BY SELECTIVE CRYSTALLIZATION OF ANTHRAQ
                    27-3044 DEFECT STRUCTURE OF TRIGLYCINE SULFATE. (CORRELATION WITH DOMAIN STRUCTURES)
DITZENBERGER JA   12-1870 INFRARED (REFLECTIVITY) STUDY (20-10,000 PER CM, 300 DEG K) OF THE LATTICE VIBRATIONS
DIVINO MC   17-2632 AN APPARATUS FOR POLISHING WATER SOLUBLE CRYSTALS (APPLICATION TO ADP AND KDP).
DOBIASOVA L   04-0875 OPTICAL AND DIELECTRIC PROPERTIES OF SINGLE CRYSTAL BARIUM TITANATE WITH NIOBIUM ADDIT
DOI K   08-1404 FERROELECTRIC PROPERTIES OF STRONTIUM NIOBATE SINGLE CRYSTAL.
                    13-1997 A NEW FERROELECTRIC SINGLE CRYSTAL STRONTIUM(2) NIOBIUM(2) OXIDE(7).
                    13-2000 PTC EFFECT (POSITIVE TEMPERATURE COEFFICIENT OF RESISTIVITY) IN (DOPED AND UNDOPED CER
                    13-2067 CRYSTAL STRUCTURE AND PIEZOELECTRICITY OF THE SYSTEM LEAD (ZINCATE NIOBATE)- LEAD TITA
                    35-3508 FERROELECTRICITY IN LEAD(5) GERMANIUM(3) OXIDE(11).
                    36-3576 ELECTROOPTIC AND PIEZOELECTRIC PROPERTIES OF LANTHANUM TITANATE SINGLE CRYSTAL. (FOUND
DOLGAYA ZHA   13-2046 EFFECTS OF ROASTING CONDITIONS ON THE STRUCTURE AND PROPERTIES OF LEAD BARIUM NIOBATE
DOLINO G   02-0063 GENERATION OF THE SECOND HARMONIC AND FERROELECTRICITY. (THEORY)
                    27-2989 EFFECTS OF DOMAIN SHAPES ON SECOND HARMONIC SCATTERING IN TRIGLYCINE SULFATE.
                    27-2990 GENERATION OF THE SECOND HARMONIC BY TRIGLYCINE SULFATE. (84 REFS)
                    27-2991 SECOND HARMONIC LIGHT SCATTERING BY DOMAINS IN FERROELECTRIC TRIGLYCINE SULFATE.
                    27-2992 STUDY OF THE PHASE TRANSITION OF TGS BY SECOND HARMONIC SCATTERING.
DOLLING G   16-2284 CRYSTAL DYNAMICS OF SODIUM NITRITE. (NEUTRON SCATTERING)

DOLLING G	16-2364	CRYSTAL DYNAMICS AND THE FERROELECTRIC PHASE TRANSITION OF SODIUM NITRITE (STUDIED BY
	36-3586	TRANSITION IN TIN TELLURIDE- GERMANIUM TELLURIDE ALLOYS. (NEUTRON SCATTERING)
DOLLOFF RT	06-1217	PREPARATION AND CHARACTERIZATION OF ALKOXY-DERIVED STRONTIUM ZIRCONATE AND STRONTIUM T
DOMINGUEZ E	27-2993	DIGLYCINE SULFATE AN INTERESTING NEW DIELECTRIC CRYSTAL SPECIES.
DONIACH S	03-0221	VIBRATIONAL APPROACH TO THE ANHARMONIC LATTICE PROBLEM. (WITH AN APPLICATION TO THE TH
DONTSOVA LI	04-0911	AGING PROCESSES IN BARIUM TITANATE (DECREASE IN NUMBER OF 180 DEG DOMAINS).
DORAN DG	09-1511	VARIATION OF SOUND VELOCITY AT FERROELECTRIC-1, FERROELECTRIC-2 TRANSITION IN (NIOBIUM
DORK RA	16-2285	SWITCHING BEHAVIOR IN FERROELECTRIC POTASSIUM NITRATE.
	16-2356	DIELECTRIC BEHAVIOR OF (FERROELECTRIC) FILMS OF VACUUM DEPOSITED POTASSIUM NITRATE.
	37-3668	FERROELECTRIC HYSTERESIS TRACER FEATURING COMPENSATION AND SAMPLE GROUNDING.
DORMANN E	14-2074	OPTICAL ABSORPTION SPECTRA OF NICKEL(2+) IN NICKEL CHLORINE AND NICKEL IODINE BORACITE
DORNER B	30-3254	MECHANISM OF THE FERROELECTRIC PHASE TRANSFORMATION IN RARE EARTH MOLYBDATES. (TERBIUM
	30-3268	NEUTRON SCATTERING STUDY OF THE FERROELECTRIC PHASE TRANSFORMATION IN TERBIUM MOLYBDAT
	30-3269	ZONE BOUNDARY SOFT MODE IN TERBIUM MOLYBDATE.
DORRIAN JF	08-1406	STRUCTURAL BASIS OF FERROELECTRICITY IN THE BISMUTH TITANATE FAMILY.
	13-1935	CRYSTAL STRUCTURE OF BISMUTH TITANIUM OXIDE.
DOUGHERTY JP	36-3550	FERROELECTRIC OPTICAL ROTATION DOMAINS IN SINGLE CRYSTAL LEAD(5) GERMANIUM(3) OXIDE(11
DOVCHENKO GV	17-2584	EFFECT OF DEUTERATION ON THE ELASTIC PROPERTIES OF RUBIDIUM DIHYDROGEN PHOSPHATE (RDP)
DOVE RC	03-0222	ENERGY GENERATION FROM IMPACT ON PIEZOELECTRIC (OR FERROELECTRIC) MATERIALS.
DOWELL ME	11-1705	RADIATION DAMAGE IN LITHIUM NIOBATE. (CAPACITANCE, OPTICAL SPECTRA, ESR, CELL PARAMETE
DOWLEY MW	17-2486	PARAMETRIC FLUORESCENCE IN AMMONIUM DIHYDROGEN PHOSPHATE AND POTASSIUM DIHYDROGEN PHOS
DOWTY E	14-2075	ATOMIC DISPLACEMENTS IN FERROELECTRIC TRIGONAL AND ORTHORHOMBIC BORACITE STRUCTURES. (
DRAEGERT DA	03-0223	DIELECTRIC SUSCEPTIBILITY AND THE ORDER OF FERROELECTRIC PHASE TRANSITIONS.
	13-2024	OPTICAL AND FERROELECTRIC PROPERTIES OF BARIUM SODIUM NIOBATE. (OPTICAL ABSORPTION, RE
DRAKE MD	04-0653	BINARY LIGHT BEAM DEFLECTION IN SINGLE CRYSTAL BARIUM TITANATE.
	04-0654	BINARY LIGHT BEAM DETECTION IN SINGLE CRYSTAL BARIUM TITANATE.
DRASKO G	19-2758	PREPARATION OF SINGLE CRYSTALS OF AMMONIUM SULFATE.
DRICKAMER HG	08-1435	PRESSURE STUDIES OF FERROELECTRIC PROPERTIES. (PEROVSKITES, KDP TYPE)
DRIKER GYA	08-1400	CHANGE OF RESONANCE ABSORPTION SPECTRA OF 23.8 KEV GAMMA-RAYS OF TIN-119 DURING PHASE
DROBYSHEV LA	02-0064	FERROELECTRIC RARE EARTH MOLYBDATES. (REVIEW, FORMATION, CRYSTAL GROWTH, PHASE TRANSIT
	30-3270	FORMATION AND PHASE TRANSFORMATIONS OF THE M MODIFICATION OF SAMARIUM MOLYBDATE (SOLID
	30-3271	PHASE TRANSITIONS OF GADOLINIUM MOLYBDATE AND ISOSTRUCTURAL COMPOUNDS (EUROPIUM TERBIU
	30-3272	PHASE TRANSITIONS OF ANHYDROUS MODIFICATIONS OF RARE EARTH MOLYBDATES.
DROSI M	17-2446	SOLUBILITY IN THE GROWTH OF ELECTROOPTIC CRYSTALS ADP AND KDP.
DROZDOWSKI M	17-2566	USE OF SPECIFIC HEAT (BY A DYNAMIC METHOD) OF POTASSIUM DIHYDROGEN PHOSPHATE CRYSTALS
DRUMHELLER DS	03-0224	DYNAMIC SHELL THEORY FOR FERROELECTRIC CERAMICS.
	03-0225	DYNAMIC PIEZOELECTRIC THEORY FOR FERROELECTRIC CERAMIC SHELLS.
DRUMHELLER JE	32-3352	DIELECTRIC PROPERTIES OF LITHIUM HYDRAZINIUM SULFATE.
	32-3353	DIELECTRIC PROPERTIES OF LITHIUM HYDRAZINIUM SULFATE.
	32-3357	MICROWAVE DIELECTRIC STUDY OF LITHIUM HYDRAZINIUM SULFATE.
DRUZHININ VV	14-2095	CRYSTAL FIELD THEORY AND OPTICAL ABSORPTION (CALCULATIONS) OF FERROELECTRIC BORACITES.
	14-2096	CRYSTAL FIELD THEORY AND OPTICAL ABSORPTION OF COBALT AND NICKEL BORACITES.
	14-2097	OPTICAL ABSORPTION (625-25 MICRONS, 77-650 DEG K) OF FERROELECTRIC IRON(2+) BORACITES
DUCROS P	04-0529	LEED STUDY OF SURFACE STRUCTURES ON THE (001) FACE OF BARIUM TITANATE.
	04-0530	LEED STUDIES ON THE (001) FACE OF BARIUM TITANATE CRYSTALS.
DUDA VM	04-0655	ANTIPARALLEL SWITCHING IN ALPHA DOMAIN BARIUM TITANATE CRYSTALS.
	04-0952	DIRECT OBSERVATION OF 180 DEG DOMAINS IN BARIUM TITANATE BY MEANS OF A POLARIZING MICR
	04-0963	REPOLARIZATION PROCESSES OF BARIUM TITANATE SINGLE CRYSTALS IN ORTHORHOMBIC AND RHOMBO
DUDAREV VYA	05-1081	X-RAY INVESTIGATION OF IRRADIATED LEAD TITANATE.
	08-1443	X-RAY AND DIELECTRIC STUDIES OF IRRADIATED PEROVSKITE TYPE COMPOUNDS.
DUDEK J	05-1056	SEEBECK EFFECT IN METAL LEAD TITANATE METAL CONTACT SYSTEMS.
	05-1057	THERMOELECTRIC EFFECTS IN THE SYSTEM METAL- LEAD TITANATE- METAL. (METHOD OF STUDYING
	27-2994	METHODS OF DETERMINING SPONTANEOUS POLARIZATION NEAR THE PHASE TRANSITION POINT OF TRI
DUDKEVICH VP	04-0656	SIZE EFFECTS IN FERROELECTRICS. (BARIUM TITANATE)
	04-0832	STRUCTURE AND PROPERTIES OF BARIUM TITANATE FILMS PRODUCED BY CATHODE SPUTTERING.
DUDKINA SI	08-1325	SYNTHESIS AND STUDY OF LEAD SCANDIUM NIOBATE SINGLE CRYSTALS.
	09-1507	TERNARY SYSTEM LEAD TITANATE- LEAD ZIRCONATE- LEAD NIOBATE(0.67) ZINCATE(0.33).
DUDNIK EF	04-0655	ANTIPARALLEL SWITCHING IN ALPHA DOMAIN BARIUM TITANATE CRYSTALS.
	04-0952	DIRECT OBSERVATION OF 180 DEG DOMAINS IN BARIUM TITANATE BY MEANS OF A POLARIZING MICR
	04-0954	EFFECT OF EXTERNAL ACTIONS ON THE DOMAIN STRUCTURE OF SINGLE CRYSTALS FORMED BY BARIUM
	04-0957	EFFECT OF EXTERNAL ACTIONS ON THE REPOLARIZATION OF BARIUM TITANATE SINGLE CRYSTALS.
	04-0960	GROWING SINGLE CRYSTALS OF SOLID SOLUTION BASED ON BARIUM TITANATE, WHERE BARIUM(2+) A
	04-0963	REPOLARIZATION PROCESSES OF BARIUM TITANATE SINGLE CRYSTALS IN ORTHORHOMBIC AND RHOMBO
	13-1936	PREPARATION AND STUDY OF BARIUM(X) STRONTIUM(1-X) NIOBATE SINGLE CRYSTALS (CZOCHRALSKI
DUDREVA E	36-3551	NMR INVESTIGATIONS OF PHTHALOCYANINE.
DUFOUR JP	17-2617	ELECTRONIC EFFECT AND ELASTIC PROPERTIES OF RUBIDIUM DIHYDROGEN PHOSPHATE CRYSTALS.
DUGAUTIER C	33-3395	IMPURITY CONCENTRATION DEPENDENCE OF RAMAN SCATTERING BY MAGNONS IN NICKEL DOPED RUBID
DUKEK G	03-0226	A NEW TYPE OF SECOND ORDER PHASE TRANSITION DERIVED FROM DEVONSHIRE'S THEORY OF FERROE
DUMAS J	04-0657	PREPARATION OF SODIUM AND BARIUM NIOBATES AND LEAD MOLYBDATE.
	33-3370	ATTENUATION AND DISPERSION OF ULTRASONIC WAVES IN POTASSIUM MANGANESE TRIFLUORIDE IN T
DUNCAN RC	06-1119	ERASE MODE RECORDING CHARACTERISTICS OF PHOTOCHROMIC CALCIUM FLUORIDE, STRONTIUM TITAN
DUNGAN RH	13-1956	SOLID SOLUTIONS IN ANTIFERROELECTRIC REGION OF THE SYSTEM LEAD HAFNATE TITANATE, LEAD
DUNN D	06-1110	IS THERE A UNIFIED MECHANISM UNDERLYING URBACH'S RULE. (STRONTIUM TITANATE, POTASSIUM
DUNN JW	27-3036	THICKNESS DEPENDENCE OF THE NUCLEATION FIELD OF TRIGLYCINE SULFATE.
DUNNE TG	17-2487	UNIT CELL VOLUME EFFECTS (RELATION WITH CURIE POINT) IN SOME ISOMORPHOUS FERROELECTRIC
DUPAS A	26-2929	ZERO POINT SPIN DEVIATION AND SPONTANEOUS SUBLATTICE MAGNETIZATION IN THE TWO DIMENSIO
DURGE VV	04-0564	MOSSBAUER STUDIES OF LOWER TRANSITIONS IN BARIUM TITANATE. (LATTICE INSTABILITIES)
DURPAIRE JP	27-2960	SPECIFIC HEAT UNDER APPLIED ELECTRIC FIELD AND ELECTROCALORIC EFFECT IN TRIGLYCINE SUL
DVEGUBSKIJ NS	16-2332	INFLUENCE OF THE CONDITIONS OF CRYSTALLIZATION ON THE FORM OF CRYSTALS OF POTASSIUM NI
DVORAK V	01-0007	INTERNATIONAL MEETING ON FERROELECTRICITY, PRAGUE, PROC.
	01-0008	REPORT ON THE SECOND INTERNATIONAL MEETING ON FERROELECTRICITY. A EUROPEAN VIEW. AN AM
	03-0227	LIFETIME OF THE "FERROELECTRIC" PHONON IN SOME PEROVSKITE CRYSTALS (STRONTIUM TITANATE
	03-0228	PHENOMENOLOGICAL THEORY OF ULTRASONIC ATTENUATION IN FERROELECTRICS.
	03-0229	PHASE TRANSITION IN KDP AND THE ROLE OF THE DEPOLARIZING ENERGY IN IT. (THEORY)
	03-0230	SOFT MODE BEHAVIOR IN IMPROPER FERROELECTRICS. (THEORY, KDP)
	03-0231	THERMODYNAMIC THEORY OF THE CUBIC- ORTHORHOMBIC PHASE TRANSITION IN BORACITES.
	03-0232	THERMODYNAMIC THEORY OF GADOLINIUM MOLYBDATE.
	04-0658	ANGULAR ENERGY DEPENDENCE OF 180 DEG DOMAIN WALL IN BARIUM TITANATE. (THERMODYNAMIC CA
	04-0659	ON THE INTERPRETATION OF INFRARED SPECTRA AND THE LATTICE DYNAMICS OF CUBIC BARIUM TIT
	08-1321	GROUP ANALYSIS OF LATTICE VIBRATIONS OF CUBIC PEROVSKITE ABO(3).

PAGE 632

DVORAK V	14-2076	BORACITES - AN EXAMPLE OF IMPROPER FERROELECTRICS.
	14-2077	SYMMETRY ASPECTS OF THE PHASE TRANSITIONS IN BORACITES.
	17-2488	PHASE TRANSITION IN KDP (MODIFICATION OF KOBAYASHI'S DYNAMICAL THEORY).
	22-2834	IMPROPER FERROELECTRIC PHASE TRANSITION IN RUBIDIUM TRIHYDROGEN SELENITE.
	27-3005	TEMPERATURE AUTOSTABILIZATION EFFECT OF TRIGLYCINE SULFATE SINGLE CRYSTALS IN AN AC EL
	30-3273	ORIGIN OF THE STRUCTURAL PHASE TRANSITION IN GADOLINIUM MOLYBDATE. (SYMMETRY CHANGES)
	30-3311	NEW TYPE OF FERROELECTRIC SOFT MODE IN GADOLINIUM MOLYBDATE.
	36-3552	STRUCTURAL PHASE TRANSITIONS IN LANGBEINITES.
	36-3596	MOSSBAUER EFFECT AND FERROELECTRIC PROPERTIES OF IONIC CRYSTALS.
DWORAKOWSKI J	28-3191	INFLUENCE OF THE DOMAIN STRUCTURE ON POSITRON (TWO-PHONON) ANNIHILATION IN ROCHELLE SA
DYKINS JE	34-3422	TENSILE AND FLEXURE PROPERTIES OF SALINE ICE.
DZHMUKHADZE DF	02-0128	NEW FERROELECTRICS AND SEIGNETTO MAGNETS OF THE PEROVSKITE AND PYROCHLORE TYPE. SYNTHE
	05-1044	THE SYSTEM LEAD TITANATE- STRONTIUM COPPER NIOBATE. (CERAMIC SYNTHESIS)
	05-1045	THE SYSTEM LEAD TITANATE- STRONTIUM CUPRATE(0.33) NIOBATE(0.67).
DZIEDZIC JM	11-1684	SPATIALLY UNIFORM AND ALTERABLE SHG PHASE MATCHING TEMPERATURES IN LITHIUM NIOBATE (CR
	13-1998	MEASUREMENTS OF SECOND HARMONIC GENERATION AND THE VARIATIONS IN THE FREE AND CLAMPED
EAGLES DM	03-0233	TUNNELING RESULTS ON INDIUM STRONTIUM TITANATE BARRIERS INTERPRETED AS CONFIRMATION OF
EASTMAN C	04-0660	EFFECT OF SILICA ADDITIONS ON THE SEMICONDUCTING PROPERTIES OF DOPED BARIUM TITANATE.
EDAHIRO T	06-1120	DISPERSION IN STRONTIUM TITANATE- LANTHANUM FERRATE DIELECTRIC CERAMICS.
	06-1172	EFFECTS OF SILICA ON MANGANESE DOPED STRONTIUM TITANATE CERAMICS.
EDELMAN S	36-3543	DIRECT PIEZOELECTRIC EFFECT IN POLYVINYL CHLORIDE FILMS.
	36-3544	PYROELECTRICITY AND PIEZOELECTRICITY IN ORIENTED FILMS OF POLYVINYL FLUORIDE AND POLYV
EDMONDS HD	13-2015	CRYSTAL GROWTH AND NONLINEAR OPTICAL PROPERTIES OF POTASSIUM LITHIUM NIOBATES.
	13-2026	NONLINEAR OPTICAL PROPERTIES OF POTASSIUM LITHIUM NIOBATES (TRANSPARENT, TETRAGONAL TU
	13-2027	NONLINEAR OPTICAL PROPERTIES OF POTASSIUM LITHIUM NIOBATE.
EDWARDS DA	13-1937	PREPARATION AND FABRICATION OF SODIUM POTASSIUM NIOBATE.
EDWARDS DF	17-2655	EFFECT OF PROTON- PHONON COUPLING ON THE FERROELECTRIC MODE IN POTASSIUM DIHYDROGEN PH
	17-2657	RAMAN SPECTRA OF TETRAGONAL POTASSIUM DIHYDROGEN PHOSPHATE.
EDWARDS LR	06-1121	HIGH PRESSURE COMPRESSIBILITY AND GRUNEISEN PARAMETER OF STRONTIUM TITANATE.
EFRON U	17-2489	HIGH TEMPERATURE PHASE TRANSITION IN RUBIDIUM DIHYDROGEN PHOSPHATE BY SCATTERING OF CO
EGAMI H	02-0076	NEW PIEZOELECTRIC MATERIALS.
EGERTON L	09-1542	HOT PRESSING OF POTASSIUM- SODIUM NIOBATE.
EGOROVA LS	13-2006	PROPERTIES OF FERROGLASS CERAMICS BASED ON LEAD BARIUM NIOBATE COMPOUNDS.
EHRENBERG L	27-2968	PULSED NITROGEN PROTON DOUBLE RESONANCE STUDY OF THE FERROELECTRIC TRANSITION IN TRIGL
	27-2969	PULSED DOUBLE RESONANCE STUDY OF THE NUCLEAR QUADRUPOLE INTERACTIONS OF NITROGEN-14 IN
EHSES KH	27-2995	NEW MEASUREMENTS OF DIVERSE THERMODYNAMICAL COEFFICIENTS OF ROCHELLE SALT AND TRIGLYCI
EIBEN K	34-3423	IRRADIATION PRODUCED SOLVATED ELECTRONS IN ICE.
EICHLER H	06-1122	NONCOLLINEAR PARAMETRIC 4 PHOTON RECIPROCAL EFFECT IN CADMIUM SULFIDE AND STRONTIUM TI
	17-2444	DIFFRACTION OF A CONTINUOUS LIGHT WAVE IN KDP.
EIKENBERG AF	37-3633	(COMPACT, SURFACE WAVE) FERROELECTRIC PHASE SHIFTERS FOR VHF AND UHF. (BASED ON LEAD(0
EIMER M	27-3142	CORROSION FIGURES DUE TO THE FERROELECTRIC DOMAIN BOUNDARIES OF TRIGLYCINE SULFATE.
EIRIKSSON VR	17-2590	STRUCTURAL STUDIES OF THE SYSTEM POTASSIUM DIHYDROGEN PHOSPHATE- POTASSIUM DIDEUTERIUM
EISNER IYA	28-3192	ELECTRONIC EFFECTS (OF X-RAY AND GAMMA-RAY IRRADIATION) IN ROCHELLE SALT CRYSTALS.
EISNER J	28-3193	INFLUENCE OF THE CRYSTALLIZATION TEMPERATURE OF ROCHELLE SALT ON ITS FERROELECTRIC PRO
ELCOMBE MM	36-3553	LATTICE DYNAMICS OF IONIC CRYSTALS WITH HIGH DIELECTRIC CONSTANTS. (LEAD AND TIN TELL
ELDRUP M	34-3457	ANGULAR CORRELATION OF ANNIHILATION PHOTONS IN ICE SINGLE CRYSTALS.
ELGARD AM	04-0661	INVESTIGATION OF ANISOTROPY IN THE DIELECTRIC PROPERTIES OF FERROELECTRICS, SUBJECTED
	04-0985	INFLUENCE OF DOMAIN STRUCTURE OF CERAMIC FERROELECTRICS ON THEIR MECHANICAL PROPERTIES
	09-1605	ELECTROMECHANICAL PROPERTIES OF CERAMIC FERROELECTRICS (BARIUM TITANATE ZIRCONATES) IN
ELINSON MI	15-2151	ELECTROREFLECTION FROM SINGLE CRYSTALS OF FERROELECTRIC ANTIMONY SULFOIODIDE NEAR THE
	15-2152	INSTABILITY OF FERROELECTRIC POLARIZATION IN ANTIMONY SULFOIODIDE SINGLE CRYSTALS.
ELLIOTT JC	36-3591	MONOCLINIC STRUCTURE OF SYNTHETIC CALCIUM CHLORAPATITE. (DEVELOPMENT OF FERROELECTRIC
ELLIOTT RJ	03-0423	CRITICAL BEHAVIOR OF THE PSEUDOSPIN MODEL FOR ORDER DISORDER TYPE FERROELECTRICS WITH
ELNAGGAR S	04-0802	BARIUM TITANATE CERAMIC: A NEW ELECTRON EMITTER.
ELSAFFAR ZM	25-2918	DEUTERON MAGNETIC RESONANCE OF THE HIGH TEMPERATURE PHASE OF FERROELECTRIC THIOUREA.
	25-2919	STUDIES OF FERROELECTRIC SOLIDS BY MAGNETIC RESONANCE PART-18: PROTON AND DEUTERON RES
	29-3240	NUCLEAR MAGNETIC RESONANCE OF FERROELECTRIC LITHIUM AMMONIUM TARTRATE.
	29-3241	NUCLEAR MAGNETIC RESONANCE OF FERROELECTRIC LITHIUM AMMONIUM TARTRATE.
ELSEN E	13-1938	TEMPERATURE DEPENDENCE OF THE DIELECTRIC PROPERTIES OF BARIUM STRONTIUM NIOBATE IN THE
ELWELL D	13-1939	VARIATION OF CRYSTAL GROWTH RATE WITH SUPERSATURATION IN FLUXED MELTS. (BARIUM STRONTI
ELYARD CA	04-0660	EFFECT OF SILICA ADDITIONS ON THE SEMICONDUCTING PROPERTIES OF DOPED BARIUM TITANATE.
EMMENEGGER FP	08-1322	DIELECTRIC PROPERTIES OF NIOBATES AND TUNGSTATES.
EMRICK RM	16-2331	TEMPERATURE AND PRESSURE DEPENDENCE OF THE DIELECTRIC CONSTANT AND SPONTANEOUS POLARIZ
EMTAGE PR	11-1699	ACOUSTIC RADIATION BY INTERDIGITAL GRIDS ON LITHIUM NIOBATE.
ENDEN NM	04-0542	DETERMINATION OF CHROMIUM VALENCY FROM STUDIES OF EPR SPECTRA AND STATIC MAGNETIC SUSC
ENGELEN JJ	17-2494	OPTIMUM CUT IN X DIHYDROGEN PHOSPHATE CRYSTALS FOR TRANSVERSE LIGHT MODULATION
ENGELHARDT H	01-0026	PHYSICS OF ICE. (PROC OF INT SYMP, 9-14 SEP 1968 MUNICH)
	34-3414	PROTONIC CONDUCTION OF ICE. PART-1: HIGH TEMPERATURE REGION. PART-2: LOW TEMPERATURE R
	34-3463	SPECIFIC HEAT OF PURE AND DOPED ICE NEAR 120 DEGREES K.
ENGELS W	04-0591	TEMPERATURE DEPENDENCE OF THE DIELECTRIC CONSTANT OF FERROELECTRIC SUBSTANCES. (BARIUM
ENGLMAN R	03-0234	ANHARMONIC MODE MIXING IN IONIC FERROELECTRICS. (LATTICE DYNAMICS)
	04-0662	VIBRONIC EFFECT IN THE TEMPERATURE DEPENDENCE OF THE SOFT FREQUENCY OF BARIUM TITANATE
ENOMOTO Y	14-2084	PHENOMENOLOGICAL THEORY OF DIELECTRIC AND MECHANICAL PROPERTIES OF IMPROPER FERROELECT
	17-2555	ORDER OF THE FERROELECTRIC PHASE TRANSFORMATION OF POTASSIUM DIHYDROGEN PHOSPHATE.
	17-2556	X-RAY STUDY ON THERMAL EXPANSION OF FERROELECTRIC POTASSIUM DIHYDROGEN PHOSPHATE. (NEA
	17-2557	X-RAY DILATOMETRIC STUDY OF THE FERROELECTRIC PHASE TRANSITION OF POTASSIUM DIHYDROGEN
ENSIGN TC	06-1123	SHARED HOLES TRAPPED BY CHARGE DEFECTS IN STRONTIUM TITANATE.
ENZ CP	06-1124	LATTICE DYNAMICS ABOVE STRUCTURAL PHASE TRANSITIONS: STRONTIUM TITANATE.
EPSTEIN DJ	08-1323	GROWTH AND EVALUATION OF SINGLE CRYSTAL BISMUTH TITANATE AND RELATED COMPOUNDS. (GADOL
	30-3274	ELASTIC CONSTANTS OF GADOLINIUM MOLYBDATE (NEAR THE CURIE POINT).
EPSTEIN M	11-1738	ELECTRON BEAM SENSING OF SURFACE ELASTIC WAVES FOR SIGNAL PROCESSING PURPOSES. (TECHNI
EREMENKO VV	33-3364	FINE STRUCTURE OF EXCITON- MAGNON ABSORPTION OF LIGHT IN POTASSIUM MANGANESE TRIFLUORI
	33-3372	RAMAN SCATTERING AND PHASE TRANSITIONS IN POTASSIUM MANGANESE TRIFLUORIDE CRYSTALS.
	33-3400	RAMAN SCATTERING AND PHASE TRANSITIONS IN A POTASSIUM MANGANESE TRIFLUORIDE CRYSTAL.
ERMACHENKOV NS	37-3639	FERROELECTRIC MICROTHERMOSTATS.
ERMILIN KK	11-1706	DISPERSION OF ELASTIC WAVES IN FERROELECTRICS. (LITHIUM NIOBATE)
	11-1707	GENERATION OF SUPERHIGH FREQUENCY ACOUSTIC HARMONICS IN A (SINGLE DOMAIN) LITHIUM NIOB
EROR NG	04-0663	ELECTRICAL CONDUCTIVITY OF SINGLE CRYSTALLINE BARIUM TITANATE.
	04-0664	OXYGEN STOICHIOMETRY OF DONOR DOPED BARIUM TITANATE AND TITANIUM DIOXIDE. (STOICHIOMET
ESOP B	05-1056	SEEBECK EFFECT IN METAL LEAD TITANATE METAL CONTACT SYSTEMS.

EVANS GA	08-1413	MEASUREMENT OF THE RELATIVE NONLINEAR COEFFICIENTS OF POTASSIUM DIHYDROGEN PHOSPHATE,
EVDOKIMCV VB	03-0235	MAGNETIC PROPERTIES OF FERRODIELECTRIC CATALYSTS.
EVENSON WE	03-0236	THEORY OF X-RAY DIFFRACTION LINE SHAPE EFFECTS DUE TO DOMAIN FORMATION IN FERROELECTRI
EVLANOVA NF	11-1708	DOMAIN STRUCTURE OF LITHIUM METANIOBATE CRYSTALS.
EVSEEVA RYA	16-2352	INFRARED SPECTRA OF POTASSIUM NITRATE AND SODIUM NITRITE AT PHASE TRANSITIONS.
EYRAUD I	02-0065	ANISOTROPIC, SOLID DIELECTRICS AND FERROELECTRICITY. (106 REFS)
	03-0237	ELASTIC ENERGY AND THE FERROELECTRIC PARAELECTRIC TRANSITION.
	03-0491	SHOCK WAVE INLUCED ELECTROMECHANICAL CONVERSION IN FERROELECTRIC MATERIALS.
	04-0917	EFFECT OF GRAIN SIZE CN THE FERROELECTRIC PARAELECTRIC TRANSITION OF BARIUM TITANATE.
	09-1521	PHASE CHANGES INDUCED BY HYDROSTATIC PRESSURE IN A FERROELECTRIC MATERIAL. (PZT)
	09-1581	PHASE CIAGRAM OF LEAD ZIRCONATE(1-X) TITANATE(X) PLUS 0.8 PERCENT TUNGSTEN TRIOXIDE.
	09-1615	METHOD OF MEASURING POLARIZATION VIBRATIONS OF A PREPOLARIZED CERAMIC FERROELECTRIC. (
	27-2962	SPECIFIC HEAT ANOMALY IN THE NEIGHBORHOOD OF A FERRO- PARAELECTRIC TRANSITION. (TGS)
EZHOV VM	02-0047	SYNTHESIS AND CERTAIN PROPERTIES OF NEW COMPOUNDS WITH A PYROCHLORE STRUCTURE. (NIOBAT
EZIS A	09-1512	OXYGEN CONCENTRATION CELL MEASUREMENTS OF IONIC TRANSPORT NUMBERS IN PZT (LEAD ZIRCONA
FADEEVA NV	10-1655	NEUTRON DIFFRACTION STUDY OF (TETRAGONAL- CUBIC PEROVSKITE) STRONTIUM IRON TUNGSTATE .
FAIRALL CW	18-2730	THERMODYNAMIC PROPERTIES OF POTASSIUM DIHYDROGEN ARSENATE AND POTASSIUM DIDEUTERIUM AR
FALK G	03-0226	A NEW TYPE OF SECOND CRDER PHASE TRANSITION DERIVED FROM DEVONSHIRE'S THEORY OF FERROE
	03-0238	CRITICAL STATES OF FERROELECTRICS. (THEORY)
FAN HY	08-1391	RAMAN SPECTRUM OF POTASSIUM TANTALATE NICBATE. (POTASSIUM TANTALATE(0.65) NIOBATE(0.35
	13-1987	ABSORPTION EDGE SPLITTING IN POTASSIUM TANTALATE(X) NIOBATE(1-X).
	13-1988	ABSORPTION EDGE SPLITTING IN POTASSIUM TANTALATE NIOBATE.
	13-1989	RAMAN SPECTRUM OF POTASSIUM TANTALUM(0.64) NIOBIUM(0.36) OXIDE.
FANG PH	02-0066	SWITCHING PROPERTIES IN FERROELECTRICS OF THE FAMILY BISMUTH(4) BARIUM(M-2) TITANIUM(M
	03-0239	FERROELECTRIC SWITCHING AND THE SIEVERT INTEGRAL.
FARACH HA	22-2833	ESR STUDY OF GAMMA-IRRADIATED CESIUM TRIHYDROGEN SELENITE AT ROOM TEMPERATURE.
FARAG MS	10-1632	THE COLCR PROBLEM CF TUNGSTEN TRIOXIDE. X-RAY STUDIES.
FARRELL EF	34-3485	MOLECULAR PHENOMENA IN WATER SYSTEMS. PART-1: MOLECULAR INTERPRETATION OF THE PHASE DI
FATUZZO E	02-0066	SWITCHING PROPERTIES IN FERROELECTRICS OF THE FAMILY BISMUTH(4) BARIUM(M-2) TITANIUM(M
	03-0197	OSCILLATIONS OF FERROELECTRIC BODIES. (WITH FREQUENCY OF APPLIED FIELD CAUSED BY SHEAR
	04-0665	FIELD DEPENDENCE OF ABSORPTION BANDS OF FERROELECTRICS IN FAR INFRARED. (APPLIED TO BA
	27-2963	SUPERSONIC DOMAIN WALL MOTION IN TRIGLYCINE SULFATE.
FAUGHMAN BW	06-1125	ELECTRON PARAMAGNETIC RESONANCE SPECTRUM OF MOLYBDENUM(5+) IN STRONTIUM TITANATE: AN E
	06-1126	PHOTOCHROISM IN TRANSITION METAL DOPED STRONTIUM TITANATE.
FAURE P	03-0240	THERMODYNAMICS OF AN ELASTIC FERROELECTRIC MODEL.
	34-3424	DYNAMIC POLAR MODEL OF MONOCRYSTALLINE ICE I(H).
	34-3425	LOW FREQUENCY RAMAN SPECTRUM OF ICE I(H). TENTATIVE INTERPRETATION WITH A MIXED COULOM
FEDAK VV	15-2199	PREPARATION AND PHYSICAL PROPERTIES OF NONSTOICHIOMETRIC ANTIMONY SULFOIODIDE SINGLE C
FEDCHENKO ED	04-0666	POLARIZED STATES OF FERROELECTRIC CERAMICS. (BARIUM TITANATE)
	36-3554	MICROSTRUCTURE OF TSTS-19 TYPE FERROELECTRIC CERAMICS.
FEDER J	03-0241	THEORY OF A STRUCTURAL PHASE TRANSITION IN PEROVSKITE- TYPE CRYSTALS, PART-2, INTERACT
	03-0242	THEORY OF STRUCTURAL PHASE TRANSITION IN PEROVSKITES.
	08-1450	STRUCTURAL FHASE TRANSITIONS IN PEROVSKITE TYPE CRYSTALS. (EXTENDED ABSTRACT)
	33-3365	ELECTRICAL CONDUCTION AND (NONFERROELECTRIC) PHASE TRANSITIONS IN CESIUM LEAD TRICHLOR
FEDOROV FI	02-0044	PHASES AND POLARIZATION OF LUMINOUS WAVES IN CUBIC ELECTROOPTIC CRYSTALS. (NONLINEAR O
FEDOSOV VN	04-0641	STRUCTURE OF 90 DEGREE DOMAIN WALLS IN BARIUM TITANATE.
	08-1305	MOVEMENT OF 180 DEGREE DOMAIN BOUNDARIES IN PEROVSKITE FERROELECTRICS.
FEDOTOV II	09-1513	ELASTIC CONSTANTS OF FERROCERAMICS.
FEDULOV SA	08-1424	ABOUT THE CURIE POINT AND NATURE OF DIELECTRIC PROPERTIES IN PEROVSKITE MODIFICATION O
	08-1438	PREPARATION AND INVESTIGATION OF FERROELECTRICS OF ILMENITE STRUCTURE. (ABO(3) NIOBATE
	11-1840	MECHANISM AND KINETICS OF LITHIUM METANIOBATE FORMATION BY A SOLID PHASE REACTION.
	12-1887	CURIE TEMPERATURE OF FERROELECTRIC LITHIUM TANTALATE.
	13-2042	MECHANISM AND KINETICS OF SOLID PHASE FORMATION OF LEAD METANIOBATE.
FEIGELSON RS	11-1688	GROWTH OF HIGH QUALITY LITHIUM NIOBATE CRYSTALS (FOR NONLINEAR OPTICS WITH LOW BIREFRI
	36-3537	FREPARATION AND CRYSTALLOGRAPHIC PROPERTIES OF A(2+)B(3+)(2)O(4) TYPE CALCIUM AND STRO
FELCHER G	17-2490	HYDROGEN DENSITY DISTRIBUTION IN PARAELECTRIC POTASSIUM DIHYDROGEN PHOSPHATE (NEUTRON
FELDMAN NB	09-1514	TIME AND TEMPERATURE DEPENDENCES OF THE PARAMETERS OF A FERROELECTRIC CERAMIC.
	09-1557	INFLUENCE OF DOMAIN STRUCTURE ON THE PROPERTIES OF POLYCRYSTALLINE LEAD TITANATE ZIRCO
	09-1591	EFFECT OF FERROELECTRIC PHASE TRANSITION CONDITIONS ON DIELECTRIC AND PIEZOELECTRIC PR
FELSER H	10-1652	A NEW FORM OF TUNGSTEN TRIOXIDE.
FELTZ A	04-0667	FROCUCTION AND STUDY OF PURE AND DOPED BARIUM TITANATE SINGLE CRYSTAL
	04-0668	PREPARATION AND STUDY OF PURE AND DOPED BARIUM TITANATE SINGLE CRYSTALS.
FENRICK HW	22-2833	ESR STUDY OF GAMMA-IRRADIATED CESIUM TRIHYDROGEN SELENITE AT ROOM TEMPERATURE.
FEOKTISTOVA NN	36-3589	NEW FERROELECTRIC: ANTIMONY ORTHONIOBATE. (CURIE TEMPERATURE)
FERIN VD	04-0953	ELECTROOPTICAL PROPERTIES OF SINGLE CRYSTALS OF SOLID SOLUTIONS OF BARIUM TITANATE ANU
FERMOR JH	11-1709	ELECTRICAL PROPERTIES OF LITHIUM NIOBATE. (ANOMALY AT 263 DEG K IS STRUCTURAL TRANSITI
	16-2286	ELECTRICAL PROPERTIES OF RUBIDIUM NITRATE AND CESIUM NITRATE.
FERONOV AD	05-1047	GROWTH OF LEAD TITANATE CRYSTALS AND EXAMINATION OF THEIR DOMAIN STRUCTURE.
FERY H	06-1122	NONCOLLINEAR PARAMETRIC 4 PHOTON RECIPROCAL EFFECT IN CADMIUM SULFIDE AND STRONTIUM TI
	17-2444	DIFFRACTION OF A CONTINUOUS LIGHT WAVE IN KDP.
FESENKO EG	03-0217	RELATION BETWEEN THE MAIN ELECTROPHYSICAL AND STRUCTURAL PARAMETERS OF FERROELECTRIC P
	03-0243	PEROVSKITE FAMILY AND CHEMICAL BOND. (COVALENCY, DIRECTIVITY CONDITIONS FOR ELECTRIC O
	04-0656	SIZE EFFECTS IN FERROELECTRICS. (BARIUM TITANATE)
	04-0832	STRUCTURE AND PROPERTIES OF BARIUM TITANATE FILMS PRODUCED BY CATHODE SPUTTERING.
	05-1046	FERROELECTRIC PROPERTIES OF LEAD TITANATE SINGLE CRYSTALS.
	05-1047	GROWTH OF LEAD TITANATE CRYSTALS AND EXAMINATION OF THEIR DOMAIN STRUCTURE.
	05-1048	RECIPROCAL DOMAINS IN LEAD TITANATE CRYSTALS.
	05-1049	PIEZOELECTRIC EFFECT IN LEAD TITANATE SINGLE CRYSTALS.
	05-1050	SPONTANEOUS POLARIZATION AND COERCIVE FIELD OF LEAD TITANATE (FLUX GROWN SINGLE CRYSTA
	08-1324	CRYSTALLOCHEMISTRY OF PEROVSKITES. (EXISTENCE CONDITIONS AND CLASSIFICATION)
	08-1325	SYNTHESIS AND STUDY OF LEAD SCANDIUM NIOBATE SINGLE CRYSTALS.
	08-1377	RELATIONSHIP BETWEEN THE ATOMIC DISPLACEMENTS AND SPONTANEOUS DEFORMATION IN THE FERRO
	08-1382	AN X-RAY STUDY OF THE PHASE TRANSITION IN STRONTIUM FERRATE TANTALATE
	08-1439	INTERATOMIC DISTANCES IN OXIDES WITH THE PEROVSKITE STRUCTURE.
	09-1507	TERNARY SYSTEM LEAD TITANATE- LEAD ZIRCONATE- LEAD NIOBATE(0.67) ZINCATE(0.33).
	10-1633	INVESTIGATION OF PHASE TRANSITION IN LEAD CADMIUM TUNGSTATE.
	13-1940	PHASE TRANSITION IN BARIUM BISMUTH(5+) BISMUTH(3+) OXIDE.
	36-3607	SYNTHESIS AND X-RAY INVESTIGATION OF THE PEROVSKITES BARIUM OR STRONTIUM METAPLUMBATES
FETIVEAU M	04-0917	EFFECT OF GRAIN SIZE ON THE FERROELECTRIC PARAELECTRIC TRANSITION OF BARIUM TITANATE.
FETIVEAU Y	03-0177	VOLTAGES AND CURRENTS DURING DEPOLARIZATION OF A FERROELECTRIC MATERIAL. (THEORY)

FETIVEAU Y	03-0491	SHOCK WAVE INDUCED ELECTROMECHANICAL CONVERSION IN FERROELECTRIC MATERIALS.
	04-0779	LOW LOSS (FERROELECTRIC CERAMICS) WITH GREAT DIELECTRIC PERMITTIVITY. (BARIUM TITANATE
	09-1483	FERROELECTRIC CERAMICS AND CONVERSION OF MECHANOELECTRIC ENERGY. (LEAD ZIRCONATE TITAN
	09-1508	FERROELECTRIC CERAMICS AND ENERGY CONVERSION. (LEAD ZIRCONATE TITANATE)
	09-1521	PHASE CHANGES INDUCED BY HYDROSTATIC PRESSURE IN A FERROELECTRIC MATERIAL. (PZT)
	09-1615	METHOD OF MEASURING POLARIZATION VIBRATIONS OF A PREPOLARIZED CERAMIC FERROELECTRIC. (
FEUERSANGER AE	04-0669	PREPARATION AND PROPERTIES OF THIN BARIUM TITANATE FILMS.
	04-0670	PREPARATION AND PROPERTIES OF HIGH PERMITTIVITY THIN FILM DIELECTRICS. (BARIUM AND LEA
FIALA J	03-0479	ON THE FREQUENCY DEPENDENCE OF THE TEMPERATURE AUTOSTABILIZATION EFFECT IN FERROELECTR
FILIMONOV AA	08-1423	ELECTROOPTICAL PROPERTIES OF LITHIUM IODATE SINGLE CRYSTALS.
	17-2686	CURIE-WEISS LAW FOR THE NONLINEAR SUSCEPTIBILITY OF A POTASSIUM DIHYDROGEN PHOSPHATE C
	27-3105	SECOND HARMONIC GENERATION IN THE TRIGLYCINE CRYSTAL IN THE PARAELECTRIC PHASE.
FILIPEV VS	08-1324	CRYSTALLOCHEMISTRY OF PEROVSKITES. (EXISTENCE CONDITIONS AND CLASSIFICATION)
	10-1633	INVESTIGATION OF PHASE TRANSITION IN LEAD CADMIUM TUNGSTATE.
FINCH ED	34-3466	STRUCTURAL STUDIES OF ICE POLYMORPHS BY NEUTRON DIFFRACTION, PROTON AND DEUTERON NUCLE
FINK T	10-1651	STRUCTURAL AND DIELECTRIC MODIFICATIONS OF TUNGSTEN TRIOXIDE DUE TO NEUTRON IRRADIATIO
FIRESTONE S	04-0907	FABRICATION OF RF SPUTTERED BARIUM TITANATE THIN FILMS.
FIRSOVA MM	11-1717	TEMPERATURE AND FREQUENCY DEPENDENCES OF THE ELECTRICAL PROPERTIES OF LITHIUM METANIOB
FISCHER SF	03-0244	THEORY OF THE MOBILITY OF STRUCTURAL DEFECTS IN ICE.
FISHER ME	03-0481	ZEROES OF THE PARTITION FUNCTION FOR THE HEISENBERG, FERROELECTRIC, AND GENERAL ISING
FISHER RA	30-3275	MAGNETOTHERMODYNAMICS OF ANTIFERROMAGNETIC, FERROELECTRIC BETA GADOLINIUM MOLYBDATE. P
FISHMAN YUM	17-2491	EFFECT OF DISLOCATIONS ON THE GROWTH OF POTASSIUM DIHYDROGEN PHOSPHATE CRYSTALS.
	17-2492	X-RAY TOPOGRAPHIC STUDY OF THE DISLOCATIONS PRODUCED IN POTASSIUM DIHYDROGEN PHOSPHATE
	17-2578	X-RAY DIFFRACTION TOPOGRAPHY STUDY OF DEFECTS (DISLOCATIONS) IN KDP AND ADP SINGLE CRY
FITZGERALD ME	28-3194	NMR STUDIES OF SPONTANEOUS POLARIZATION IN ROCHELLE SALT.
	28-3195	NMR STUDY OF PHASE TRANSITIONS IN ROCHELLE SALT.
FLEROV IN	33-3387	SPECIFIC HEAT OF POTASSIUM MANGANESE TRIFLUORIDE.
FLEROVA SA	04-0672	LIGHT EMISSION FROM BARIUM TITANATE CRYSTALS SUBJECTED TO UNIAXIAL STRESS PULSES.
	04-0951	CRITICAL FIELD IN CERAMIC BARIUM TITANATE.
	04-0954	EFFECT OF EXTERNAL ACTIONS ON THE DOMAIN STRUCTURE OF SINGLE CRYSTALS FORMED BY BARIUM
	04-0956	EFFECT OF TRANSVERSE COMPRESSION OF DIELECTRIC PROPERTIES OF BARIUM TITANATE SINGLE CR
	04-0957	EFFECT OF EXTERNAL ACTIONS ON THE REPOLARIZATION OF BARIUM TITANATE SINGLE CRYSTALS.
	04-0961	INFLUENCE OF UNIAXIAL TENSILE STRESSES ON THE ELECTRICAL PROPERTIES AND DOMAIN STRUCTU
	04-0962	PIEZOELECTRIC PROPERTIES OF POLARIZED BARIUM TITANATE CERAMICS SUBJECTED TO UNIAXIAL T
FLETCHER NH	01-0009	CHEMICAL PHYSICS OF ICE.
	34-3418	LOW TEMPERATURE POLARIZATION EFFECTS IN ICE. (CURRENT REVERSAL DUE TO DIPOLE RELAXATIO
FLETCHER SR	27-2996	CRYSTALLOGRAPHIC STUDIES OF IRRADIATION FIELD TREATED TRIGLYCINE SULFATE: NEW STRUCTU
FLEURY PA	01-0010	FIFTY YEARS OF FERROELECTRICITY: A REVIEW OF THE PROCEEDINGS OF THE SECOND INTERNATION
	04-0671	ACOUSTIC SOFT OPTIC MODE INTERACTIONS IN FERROELECTRIC BARIUM TITANATE.
	04-0806	ACOUSTIC PHONON SOFT OPTIC PHONON INTERACTION IN BARIUM TITANATE.
	04-0807	TEMPERATURE DEPENDENCE OF THE BRILLOUIN- RAMAN SPECTRUM OF BARIUM TITANATE.
	06-1241	ELECTRIC FIELD DEPENDENCE OF OPTICAL PHONON FREQUENCIES (RAMAN SCATTERING IN CUBIC PER
	08-1326	ELECTRIC FIELD INDUCED RAMAN EFFECT IN PARAELECTRIC CRYSTALS. (POTASSIUM TANTALATE)
	30-3276	ANOMALOUS PHONON BEHAVIOR NEAR THE PHASE TRANSITION IN FERROELASTIC FERROELECTRIC GADO
FLICK C	29-3240	NUCLEAR MAGNETIC RESONANCE OF FERROELECTRIC LITHIUM AMMONIUM TARTRATE.
FLUGGE S	16-2287	THE ORDER DISORDER TRANSITION OF SODIUM NITRITE. (TWO SUBLATTICE MODEL OF INTERACTING
FOK J	16-2288	ELECTRICAL BREAKDOWN IN FERROELECTRIC SODIUM NITRITE CRYSTALS.
FOKINA GO	11-1693	THERMAL DEPENDENCE OF DIELECTRIC PIEZOELECTRIC, AND ELASTIC PROPERTIES OF LITHIUM NIO
FOLDMAN AM	03-0310	MACROSCOPIC ANALOG OF THE SPIN- ECHO EFFECT IN POLYCRYSTALLINE FERROELECTRICS. (THEORY
FOLWEILER RC	04-0553	TOP SEEDED SOLUTION GROWTH OF OXIDE CRYSTALS FROM NONSTOICHIOMETRIC MELTS. (BARIUM TIT
FOMICHEV NN	17-2493	OBSERVATION OF DOMAIN STRUCTURE IN POTASSIUM DIHYDROGEN PHOSPHATE CRYSTALS BY POLARIZA
FOMICHEV OI	04-0708	INVESTIGATION OF THE FINE STRUCTURE OF BARIUM TITANATE CRYSTALS FLUX GROWN FROM POTASS
FOMICHEV VV	36-3554	MICROSTRUCTURE OF TSTS-19 TYPE FERROELECTRIC CERAMICS.
FOMIN VI	33-3372	RAMAN SCATTERING AND PHASE TRANSITIONS IN POTASSIUM MANGANESE TRIFLUORIDE CRYSTALS.
	33-3398	MANDELSTAM BRILLOUIN SCATTERING OF LIGHT IN MANGANESE FLUORIDE CRYSTALS (POTASSIUM TRI
	33-3399	MANDELSTAM BRILLOUIN SCATTERING OF LIGHT IN MANGANESE FLUORIDE, COBALT FLUORIDE, POTAS
	33-3400	RAMAN SCATTERING AND PHASE TRANSITIONS IN A POTASSIUM MANGANESE TRIFLUORIDE CRYSTAL.
	33-3401	TWO MAGNON SCATTERING OF LIGHT IN ANTIFERROMAGNETIC POTASSIUM MANGANESE TRIFLUORIDE.
FONGER WH	37-3646	STORAGE OF HOLOGRAMS IN A FERROELECTRIC PHOTOCONDUCTOR DEVICE (ERASABLE, NONDESTRUCTIV
FONTAINE G	04-0673	CALCULATION OF (DOMAIN) WALL ENERGY IN BARIUM TITANATE.
	04-0674	EFFECTS OF ANISOTROPY OF ELASTIC CONSTANTS ON DISLOCATIONS IN FERROELECTRIC BARIUM TIT
FONTANA MP	04-0675	LINEAR DISORDER AND TEMPERATURE DEPENDENCE OF RAMAN SCATTERING IN BARIUM TITANATE.
	04-0676	STRUCTURE DISORDER AND RAMAN SCATTERING IN BARIUM TITANATE AND POTASSIUM NIOBATE.
FONTANELLA J	33-3361	LOW FREQUENCY DIELECTRIC CONSTANTS OF THE ALKALINE EARTH FLUORIDES BY THE METHOD OF SU
FORKER M	07-1255	THE INTERNAL ELECTRIC FIELD GRADIENT IN ANTIFERROELECTRIC LEAD ZIRCONATE STUDIED BY PE
	08-1276	NUCLEAR QUADRUPOLE INTERACTIONS IN PEROVSKITE TYPE COMPOUNDS OF HAFNIUM-181 STUDIED BY
FORMIGONI NP	13-2038	PREPARATION AND EPITAXY OF SPUTTERED FILMS OF FERROELECTRIC BISMUTH TITANIUM OXIDE.
FORSCH K	27-2995	NEW MEASUREMENTS OF DIVERSE THERMODYNAMICAL COEFFICIENTS OF ROCHELLE SALT AND TRIGLYCI
FOSSHEIM K	03-0176	STRUCTURAL PHASE TRANSITIONS AND SOFT MODES.
	06-1127	ULTRASONIC PROPAGATION, STRESS EFFECTS, AND INTERACTION PARAMETERS AT THE DISPLACIVE T
FOSTER NF	13-1941	COMPOSITION AND STRUCTURE OF SPUTTERED FILMS OF FERROELECTRIC NIOBATES.
FOUASSIER C	14-2079	PREPARATION OF NICKEL BORACITE CRYSTALS BY HYDROTHERMAL SYNTHESIS.
	14-2089	NEW BORACITES OF FORMULA LITHIUM(4) BORON(7) OXYGEN(12) CHLORIDE OR BROMIDE (NOT EXAMI
FOUNTAIN WD	08-1328	OPTICALLY INDUCED PHYSICAL DAMAGE TO LITHIUM NIOBATE, PROUSTITE, AND LITHIUM IODATE.
	11-1710	TRANSIENT ELECTROOPTIC EFFECTS AND Q SWITCHING PERFORMANCE IN LITHIUM NIOBATE AND POTA
FOUSEK J	01-0008	REPORT ON THE SECOND INTERNATIONAL MEETING ON FERROELECTRICITY. A EUROPEAN VIEW. AN AM
	03-0245	INTERPRETATION OF TEMPERATURE AUTOSTABILIZATION OF FERROELECTRIC CRYSTAL. (DIELECTRIC
	03-0246	(FACTORS INFLUENCING) TYPE AND ORIENTATION OF THE EQUILIBRIUM DOMAIN STRUCTURES IN FER
	03-0247	PERMISSIBLE DOMAIN WALLS IN FERROELECTRIC SPECIES (THEORY AND TABULATION)
	03-0248	RECENT PROBLEMS IN DOMAIN STATICS AND DYNAMICS. (56 REFS)
	03-0358	SMALL SIGNAL PERMITTIVITY OF THE STATIONARY (100)-180 DEG DOMAIN WALL IN BARIUM TITANA
	04-0585	INTERACTION BETWEEN THE 90 DEG AND 180 DEG DOMAINS IN BARIUM TITANATE.
	04-0804	180 DEG DOMAIN WALL CONTRIBUTION TO THE DIELECTRIC CONSTANT OF BARIUM TITANATE. (CALCU
	08-1327	ORIENTATION OF COMPOSITION PLANES IN ANTIFERROELECTRICS. (APPLIED TO ADP, TUNGSTEN TRI
	14-2078	FERROELECTRIC TRANSITION IN COBALT- IODINE- BORACITE UNDER PRESSURE.
	14-2099	FERROELECTRIC TRANSITION IN COBALT IODINE BORACITE.
	17-2452	STRESS INDUCED BIREFRINGENCE IN AN ISOLATED AND A SHORTCIRCUITED POTASSIUM DIHYDROGEN
	19-2756	PROPERTIES OF FERROELECTRIC CADMIUM AMMONIUM SULFATE.
	27-2955	INFLUENCE OF THE ELECTRIC FIELD AND THE DOMAIN STRUCTURE ON THE ULTRASONIC ABSORPTION
	27-2997	FERROELECTRIC DOMAIN WALL IN TRIGLYCINE SULFATE. (THEORY AND CALCULATIONS OF WALL THIC

FOUSEK J	27-3062	FREQUENCY DEPENDENCE OF THE COERCIVE FIELD OF TRIGLYCINE SULFATE CRYSTALS.
	29-3242	SOME PROPERTIES OF THE FERROELECTRIC LITHIUM THALLIUM TARTRATE.
	30-3277	ELECTROOPTICAL PROPERTIES OF FERROELECTRIC GADOLINIUM MOLYBDATE.
	30-3278	INDUCED AND SPONTANEOUS BIREFRINGENCE IN IMPROPER FERROELECTRIC GADOLINIUM MOLYBDATE.
FOUSKOVA A	01-0007	INTERNATIONAL MEETING ON FERROELECTRICITY, PRAGUE, PROC.
	04-0677	(STEP OR SMOOTH) DOMAIN WALL STRUCTURE IN BARIUM TITANATE DURING SWITCHING (CALCULATIO
	04-0678	STRUCTURE OF A MOVING DOMAIN WALL. (MECHANISM OF A WALL MOTION IN BARIUM TITANATE)
	17-2449	ELECTRICAL AND OPTICAL INVESTIGATION OF FERROELECTRIC PROPERTIES OF POTASSIUM DIHYDROG
	17-2458	REGULAR BEHAVIOR OF SOLID SOLUTIONS IN POTASSIUM DIHYDROGEN(1-X) DIDEUTERIUM(X) PHOSPH
	27-3027	IMPEDANCE OF FERROELECTRIC TRIGLYCINE FLUOBERYLLATE CRYSTALS DURING SWITCHING IN PULSE
	28-3196	PERMITTIVITY OF ROCHELLE SALT DURING SWITCHING.
	36-3555	(WEAK FIELD) DIELECTRIC (PERMITTIVITY) PROPERTIES OF BISMUTH TITANATE (EFFECT OF TEMPE
FOX AJ	13-1942	LONGITUDINAL ELECTROOPTIC EFFECT IN KTN.
FRAITOVA D	06-1128	NORMAL VIBRATIONS OF STRONTIUM TITANATE LATTICE.
FRANCOIS GE	11-1755	OPTIMUM CUT FOR A LITHIUM NIOBATE TRANSVERSE LIGHT MODULATOR.
	17-2494	OPTIMUM CUT IN X DIHYDROGEN PHOSPHATE CRYSTALS FOR TRANSVERSE LIGHT MODULATION
FRANCOMBE MH	08-1329	GROWTH AND PROPERTIES OF EPITAXIAL FILMS OF FERROELECTRIC BISMUTH TITANATE. (ELECTROOP
	08-1470	DOMAIN STRUCTURE AND POLARIZATION REVERSAL IN FILMS OF FERROELECTRIC BISMUTH TITANATE.
	09-1515	FERROELECTRIC DISPLAYS. (BISMUTH TITANATE, LEAD ZIRCONATE TITANATE)
	13-2038	PREPARATION AND EPITAXY OF SPUTTERED FILMS OF FERROELECTRIC BISMUTH TITANIUM OXIDE.
	36-3556	FERROELECTRIC FILMS AND THEIR DEVICE APPLICATIONS. (POLYVINYLIDENE FLUORIDE)
	37-3637	RESEARCH STATUS AND DEVICE POTENTIAL OF FERROELECTRIC THIN FILMS.
FRANKEN PA	02-0067	OPTICAL HARMONICS AND NONLINEAR PHENOMENA. (THEORY, REVIEW)
	27-2958	OPTICAL MIXING. (OF PULSED RUBY LASER EMISSIONS BY TRIGLYCINE SULFATE CRYSTAL)
FRANKUS F	13-2044	DIFFUSE TO SHARP FERROELECTRIC PHASE TRANSITION SYSTEMS LEAD (B(0.5) NIOBIUM(0.5))(1-X
FRASER DB	09-1516	IMPROVED AGING AND SWITCHING OF LEAD ZIRCONATE- LEAD TITANATE CERAMICS WITH INDIUM ELE
	09-1517	IMPROVED AGING AND SWITCHING OF LEAD ZIRCONATE- LEAD TITANATE CERAMICS WITH INDIUM ELE
	09-1518	PROPERTIES OF NEW TRANSPARENT FERROELECTRIC CERAMICS. (LEAD ZIRCONATE- LEAD TITANATE T
	09-1572	IMAGE STORAGE AND DISPLAY DEVICES USING FINE GRAIN FERROELECTRIC CERAMICS. (LEAD ZIRCO
FRASLAVSKAYA IP	04-0960	GROWING SINGLE CRYSTALS OF SOLID SOLUTION BASED ON BARIUM TITANATE, WHERE BARIUM(2+) A
FRAZER BC	04-0679	NEUTRON DIFFRACTION STUDIES OF FERROELECTRICS. (BARIUM TITANATE AND ROCHELLE SALT AS E
	16-2325	DISORDERED STRUCTURE OF SODIUM NITRITE AT 185 DEG C.
	16-2328	X-RAY AND NEUTRON STUDY ON THE PHASE TRANSFORMATION OF SODIUM NITRITE.
	17-2668	FERROELECTRIC (SOFT) MODE MOTION IN POTASSIUM DIDEUTERIUM PHOSPHATE. (BY NEUTRON TRIP
	28-3188	STRUCTURAL EFFECTS OF (DAMAGE FROM) IONIZING RADIATION IN FERROELECTRIC ROCHELLE SALT.
FRECH R	16-2280	ATR INFRARED SPECTRA OF UNIAXIAL NITRATE CRYSTALS.
FREDERIKSE HPR	06-1129	ELECTRONIC TRANSPORT (CONDUCTIVITY AND HALL AND SEEBECK COEFFICIENTS) IN STRONTIUM TIT
FREDLEIN RA	13-1944	ELECTROCHEMICAL DEPOSITION AND DISSOLUTION OF TUNGSTEN OXIDE BRONZES. (SODIUM(X) TUNGS
FREIDMAN GI	17-2428	TRAPPING OF PARAMETRICALLY AMPLIFIED LIGHT WAVES IN A KDP CRYSTAL.
FREIMANIS VA	13-1943	EFFECT OF ELECTRIC FIELD ON PHASE COMPOSITION OF (LEAD, BARIUM) NIOBATE SOLID SOLUTION
	13-1945	PHASE COMPOSITION OF (LEAD, BARIUM) DINIOBATE. METHOD OF INVESTIGATION AND SOME RESULT
FRENOIS CH	15-2144	BOSCN ECHOES. (ANTIMONY SULFOIODIDE)
FRENZEL C	06-1219	PRESSURE DEPENDENCE OF THE ELASTIC COMPLIANCE OF STRONTIUM TITANATE NEAR THE TRANSITIO
	14-2078	FERROELECTRIC TRANSITION IN COBALT- IODINE- BORACITE UNDER PRESSURE.
	17-2495	INFLUENCE OF HYDROSTATIC PRESSURE ON THE PHASE TRANSITION TEMPERATURE OF FERROELECTRIC
	19-2757	THE BEHAVIOR OF DIAMMONIUM CADMIUM TRISULFATE UNDER PRESSURE.
FREUND HG	34-3413	ENDOR (ELECTRON NUCLEAR DOUBLE RESONANCE) STUDY OF X-IRRADIATED SINGLE CRYSTALS OF ICE
FRIDKIN VM	03-0251	THEORY OF SPACE CHARGE LIMITED CURRENTS IN FERROELECTRICS.
	04-0552	PHOTOFERROELECTRIC EFFECTS IN A(V)B(VI)C(VII) AND BARIUM TITANATE TYPE FERROELECTRICS.
	04-0681	EXTRINSIC PHOTOCONDUCTIVITY IN FERROELECTRICS DUE TO SURFACE LAYERS. (BARIUM TITANATE,
	04-0682	PHOTODOMAIN EFFECT IN BARIUM TITANATE.
	15-2124	EFFECTS OF ILLUMINATION ON THE DOMAIN STRUCTURE OF ANTIMONY SULFOIODIDE. (ETCHING METH
	15-2143	PHASE TRANSITION IN FERROELECTRIC ANTIMONY SULFOIODIDE CRYSTALS USING THE RAMAN SCATTE
	15-2145	INFLUENCE OF NONEQUILIBRIUM CARRIERS ON THE PHASE TRANSITION IN FERROELECTRIC SEMICOND
	15-2146	MECHANISM OF THE PHOTODOMAIN EFFECT IN ANTIMONY SULFOIODIDE.
	15-2147	TEMPERATURE DEPENDENCE OF THE CROSS SECTION OF CAPTURE CENTERS IN FERROELECTRIC SEMICO
	15-2157	FERROELECTRIC PHOTOELECTRETS BASED ON A(V)B(VI)C(VII) SOLID SOLUTIONS. (DIELECTRIC CON
	15-2158	INFLUENCE OF NONEQUILIBRIUM CARRIER SCREENING ON POLARIZATION PROCESSES IN FERROELECTR
	15-2159	INDUCED IMPURITY PHOTOCONDUCTIVITY IN THE FERROELECTRIC SEMICONDUCTOR ANTIMONY SULFOIO
	15-2160	METASTABLE OPTICAL CHARGE EXCHANGE AND "FROZEN" SHIFT OF THE CURIE TEMPERATURE OF A FE
	15-2161	PHOTOFERROELECTRIC EFFECTS IN FERROELECTRIC SEMICONDUCTORS OF THE A(V)B(VI)C(VII) TYPE
	16-2266	FERROELECTRIC PHASE TRANSITION IN SILVER SODIUM DINITRITE
FRIDMAN SS	17-2427	EFFECT OF ANNEALING ON OPTICAL ANOMALIES OF POTASSIUM DIHYDROGEN PHOSPHATE CRYSTALS.
FRIEDMANN SS	17-2496	GROWTH LAYERS AND OTHER MACRODEFECTS IN POTASSIUM DIHYDROGEN PHOSPHATE CRYSTALS GROWN
FRIEDRICH HB	31-3326	LONGITUDINAL MODES IN HYDROGEN AND DEUTERIUM HALIDE CRYSTALS. (RAMAN SPECTRA, DIPOLAR
FRITSBERG VYA	01-0011	PHASE TRANSITIONS IN FERROELECTRICS.
	03-0249	INVESTIGATION OF POLARIZATION NONLINEARITY NEAR THE PHASE TRANSITION IN PEROVSKITE LIK
	04-0680	DIELECTRIC PROPERTIES OF FERROELECTRIC (BARIUM, STRONTIUM) TITANATE SOLID SOLUTIONS NE
	04-0894	SPECIFIC FEATURES OF THE TEMPERATURE DEPENDENCE OF THE VELOCITY OF ULTRASOUND IN POLYC
	08-1330	INVESTIGATING THE RELATIONS BETWEEN COMPOSITION AND PROPERTIES IN FERROELECTRIC SOLID
FRITZ IJ	17-2497	LIGHT SCATTERING STUDIES OF THE SOFT OPTIC AND ACOUSTIC MODES OF POTASSIUM DIHYDROGEN
	17-2636	SIMULTANEOUS OBSERVATION OF THE SOFT FERROELECTRIC AND ACOUSTIC MODES OF POTASSIUM DID
FRITZBERG WJ	03-0250	PARTICULAR PROPERTIES OF PHASE TRANSFORMATION IN FERROELECTRIC SOLID SOLUTIONS. (THEOR
FROHLICH H	02-0068	DIELECTRIC INSTABILITIES. (THEORETICAL RELATIONSHIP TO FERROELECTRICITY)
FROLKINA IT	30-3270	FORMATION AND PHASE TRANSFORMATIONS OF THE M MODIFICATION OF SAMARIUM MOLYBDATE (SOLID
	30-3271	PHASE TRANSITIONS OF GADOLINIUM MOLYBDATE AND ISOSTRUCTURAL COMPOUNDS (EUROPIUM TERBIU
FROVA A	06-1107	ABSORPTION EDGE AND STRUCTURAL PHASE TRANSITION IN STRONTIUM TITANATE.
	06-1108	DETERMINATION OF THE NATURE OF THE OPTICAL GAP OF STRONTIUM TITANATE.
	06-1109	DISTORTION ENHANCED OPTICAL ABSORPTION IN STRONTIUM TITANATE AT THE CUBIC TO TETRAGONA
	06-1110	IS THERE A UNIFIED MECHANISM UNDERLYING URBACH'S RULE. (STRONTIUM TITANATE, POTASSIUM
	06-1111	OPTICAL GAP OF STRONTIUM TITANATE. (DEVIATION FROM URBACH TAIL BEHAVIOR)
FRUMAR M	15-2170	PHOTOELECTRIC PROPERTIES OF BISMUTH SULFOIODIDE CRYSTALS.
FUCHIKAMI N	03-0252	THEORY OF SUPEREXCHANGE REACTION. PART-1: POTASSIUM MANGANESE TRIFLUORIDE.
	22-2835	RESIDUAL ENTROPY OF SODIUM TRIHYDROGEN SELENATE.
FUJIEAYASHI K	19-2759	ELASTIC ANOMALY IN AMMONIUM SULFATE. (ELECTROSTRICTIVE CONSTANTS)
FUJII Y	04-1016	PHASE TRANSITIONS IN THE MIXED CRYSTAL SYSTEM STRONTIUM(1-X) BARIUM(X) TITANATE.
	06-1130	INTERFEROMETRIC DETERMINATION OF THE QUADRATIC ELECTROOPTIC COEFFICIENTS IN STRONTIUM
	06-1131	SECOND HARMONIC GENERATION IN STRESS INDUCED FERROELECTRIC STRONTIUM TITANATE.
	06-1203	ELECTRIC FIELD INDUCED OPTICAL HARMONIC GENERATION IN STRONTIUM TITANATE AND POTASSIUM
	16-2400	X-RAY CRITICAL SCATTERING IN TRIGLYCINE SULFATE AND SODIUM NITRATE. (ORDER DISORDER TR

FUJII Y	27-2998	X-RAY CRITICAL SCATTERING IN FERROELECTRIC TRIGLYCINE SULFATE.
	31-3327	PHASE TRANSITIONS OF SOLID HYDROGEN HALIDES.
	33-3394	X-RAY SCATTERING AND THE PHASE TRANSITION OF POTASSIUM MANGANESE TRIFLUORIDE AT 184 DE
FUJIMORI Y	33-3373	ULTRASONIC ATTENUATION NEAR THE SOFT MODE TRANSITION POINT IN POTASSIUM MANGANESE TRIF
FUJINO Y	11-1841	DETERMINATION OF THE DISTRIBUTION COEFFICIENTS OF LITHIUM OXIDE INTO LITHIUM NIOBATE B
	12-1874	ELECTROOPTIC AND FERROELECTRIC PROPERTIES OF LITHIUM TANTALATE SINGLE CRYSTALS AS A FU
	12-1889	SELF INDUCED THERMAL EFFECTS ON LIGHT EXTINCTION OF LITHIUM TANTALATE.
	13-2043	DEPENDENCE OF SECOND HARMONIC GENERATION ON CRYSTAL INHOMOGENEITY (THEORY AND TEST ON
FUJITA N	17-2546	EPR OF CHROMIUM(3+) IONS IN ANTIFERROELECTRIC ADP CRYSTALS.
	17-2596	ELECTRON PARAMAGNETIC RESONANCE OF CHROMIUM(3+) IN ADP.
FUJITA T	04-0790	WET SYNTHETIC METHOD OF MAKING PURE, EXACTLY STOICHIOMETRIC BARIUM TITANATE.
	08-1331	HIGH PRESSURE SYNTHESIS OF LEAD METAL- OXYGEN(3) TYPE PEROVSKITE.
FUJIWARA S	04-0683	CERAMIC DIELECTRICS. (BARIUM TITANATE, STRONTIUM TITANATE).
FUJIWARA T	17-2498	NOTE ON THE LATTICE DYNAMICAL ASPECT OF FERROELECTRIC MODES OF KDP. (CALCULATIONS)
FUKADA E	36-3557	PIEZOELECTRICITY IN POLARIZED POLYVINYLIDENE FLUORIDE FILMS.
FUKAMI T	04-0684	FORMATION OF BARIUM TITANATE CERAMIC SEMICONDUCTING LAYER BY DIFFUSION.
FUKUDA A	34-3426	X-RAY DIFFRACTION TOPOGRAPHIC STUDIES OF THE DEFORMATION BEHAVIOR OF ICE SINGLE CRYSTA
FUKUDA T	02-0069	FERROELECTRIC SOLID SOLUTIONS.
	08-1332	PREPARATION OF POTASSIUM NIOBATE SINGLE CRYSTAL FOR OPTICAL APPLICATION.
	08-1454	NONLINEAR OPTICAL PROPERTIES OF POTASSIUM NIOBATE SINGLE CRYSTALS. (REFRACTIVE INDICES
	11-1711	GROWTH AND PROPERTIES OF FERROELECTRIC POTASSIUM LITHIUM TANTALATE(X) NIOBATE(1-X).
	13-1947	(X-RAY) STRUCTURAL (OPTICAL) AND DIELECTRIC STUDIES OF (SINGLE CRYSTALS IN THE TUNGSTE
FUKUI S	16-2365	NEUTRON DIFFRACTION STUDIES OF FERROELECTRIC POLARIZATION REVERSALS IN SODIUM NITRITE.
FUKUMOTO A	11-1812	TEMPERATURE DEPENDENCE OF RESONANT FREQUENCIES OF LITHIUM NIOBATE PLATE RESONATORS. (C
	13-1946	ELASTIC AND PIEZOELECTRIC CONSTANTS OF STRONTIUM(4) POTASSIUM LITHIUM NIOBATE TYPE FER
FUKUNAGA O	08-1331	HIGH PRESSURE SYNTHESIS OF LEAD METAL- OXYGEN(3) TYPE PEROVSKITE.
FUKUNISHI S	11-1712	GROWTH AND PROPERTIES OF LITHIUM NIOBATE AND LITHIUM TANTALATE FILMS.
FUKUTOMI K	32-3344	DIELECTRIC PROPERTIES OF LITHIUM HYDRAZINIUM SULFATE.
FULRATH RM	09-1478	POINT DEFECTS AND SINTERING OF LEAD ZIRCONATE TITANATE.
	09-1479	SOLUBILITY OF ALUMINUM IN LEAD ZIRCONATE TITANATE.
	09-1480	SUBSTITUTION OF BISMUTH AND NIOBIUM IONS IN LEAD ZIRCONATE(0.53) TITANATE(0.47) (EXPER
	09-1534	INTRINSIC NONSTOICHIOMETRY IN SINGLE PHASE LEAD ZIRCONATE TITANATE.
	09-1535	INTRINSIC NONSTOICHIOMETRY IN THE LEAD ZIRCONATE TITANATE SYSTEM.
	09-1538	SOLUTION KINETICS OF LUTETIUM AND SCANDIUM IONS IN LEAD TITANATE(0.5) ZIRCONATE(0.5).
	09-1577	HIGH TEMPERATURE PHASE EQUILIBRIA IN THE LEAD ZIRCONATE TITANATE SYSTEM.
FURUHATA Y	27-2951	A NEW SECOND HARMONIC TYPE FERROELECTRIC MODULATOR FOR ELECTROMETER. (WITH A TGS SINGL
	27-2999	GROWTH REGIONS IN FERROELECTRIC TRIGLYCINE SULFATE CRYSTALS. (TOPOGRAPHY OF IMPERFECTI
	30-3293	APPLICATION OF ELECTRON MIRROR MICROSCOPY TO DIRECT OBSERVATION OF MOVING FERROELECTRI
FURUICHI J	01-0024	A NOTE ON THE CLASSIFICATION OF FERROELECTRICS. (BY CURIE-WEISS CONSTANT)
	02-0099	DIELECTRIC AND X-RAY STUDIES OF THE PHASE TRANSITIONS IN ROCHELLE SALT, BARIUM TITANAT
FURUKAWA K	17-2554	EFFECTS OF UNIAXIAL STRESS AND ELECTRIC FIELD ON PARAMAGNETIC SPECTRA OF CHROMIUM(3+)
FURUKAWA M	26-2930	MAGNETISM AND THE PHASE TRANSITION OF COPPER FORMATE TETRAHYDRATE.
	28-3201	HIGHLY SENSITIVE, SIMPLE CALORIMETER FOR PHASE CHANGE DETECTION (INCLUDING FERROELECTR
	33-3373	ULTRASONIC ATTENUATION NEAR THE SOFT MODE TRANSITION POINT IN POTASSIUM MANGANESE TRIF
FURUYA N	17-2499	ELECTROOPTIC CHARACTERISTICS OF DEUTERATED POTASSIUM DIHYDROGEN PHOSPHATE CRYSTAL AND
FUSHIMI S	35-3499	FERROELECTRIC PROPERTY AND ENANTIOMORPHISM OF LEAD(5) GERMANIUM(3) OXIDE(11) AND THE R
FUTAMA H	25-2911	LONG PERIOD STRUCTURE OF THIOUREA CRYSTAL. (CALCULATIONS FROM SATELLITE SCATTERING DAT
GABRICHIDZE ZA	19-2786	RAMAN SCATTERING IN A FERROELECTRIC AMMONIUM SULFATE CRYSTAL.
	28-3231	RAMAN LIGHT SCATTERING IN ROCHELLE SALT.
GABRIELYAN VT	11-1713	ABSORPTION AND LUMINESCENCE SPECTRA AND ENERGY LEVELS OF NEODYMIUM(3+) AND ERBIUM(3+)
GABUDA SP	16-2337	TEMPERATURE DEPENDENCE OF THE ELECTRIC FIELD GRADIENT IN FERROELECTRIC SODIUM NITRITE.
GAGNEUX S	04-0685	CHANGE IN THE DECAY CONSTANT OF ZIRCONIUM-89 IN BARIUM TITANATE (BARIUM CHLORIDE FLUX
	04-0876	VARIATION OF THE DECAY CONSTANT OF STRONTIUM-85 IN BARIUM TITANATE.
GAIDIDEI YUB	03-0419	ON THE THEORY OF LIGHT ABSORPTION BY ANTIFERRODIELECTRICS IN THE FREQUENCY RANGE OF DO
GAIDUCHENYA VF	17-2648	STUDY OF THE (MECHANICAL) STRENGTH PROPERTIES OF POTASSIUM DIHYDROGEN PHOSPHATE CRYSTA
GAILLARD J	18-2731	ESR OF FREE RADICALS IN FERROELECTRIC AND ANTIFERROELECTRIC ARSENATE SINGLE CRYSTALS.
	18-2734	ESR IDENTIFICATION OF SLATER CONFIGURATIONS AND OF EXCHANGE OF THE ARSENATE RADICAL IN
GAINER AV	17-2500	COMBINATION OF THE FREQUENCIES OF COHERENT AND INCOHERENT RADIATION IN KDP CRYSTAL.
GALANOV EK	22-2836	INFRARED SPECTRA OF ISOTOPIC NONISOMORPHISM OF SODIUM TRIHYDROGEN SELENITE (SHS) OR SO
GALLAGHER PK	04-0825	STABILIZED BARIUM TITANATE CERAMICS FOR CAPACITOR DIELECTRICS.
GALLANOV EK	03-0330	(EFFECT OF TEMPERATURE ON) QUASI SYMMETRICAL HARD IONS (PHOSPHATE, TITANATE, SULFATE,
GALUSHKINA RA	17-2434	POLYTHERM OF SOLUBILITY OF THE AMMONIUM DIHYDROGEN PHOSPHATE- POTASSIUM DIHYDROGEN PHO
GAMYNIN EV	15-2237	CRITICAL POINT IN FERROELECTRIC ANTIMONY SULFOIODIDE.
GANESAN S	27-3000	THERMAL EXPANSION OF TRIGLYCINE SULFATE.
GANGULY BN	30-3279	PYROELECTRIC BEHAVIOR OF FERROELECTRIC GADOLINIUM MOLYBDATE.
	30-3323	PYROELECTRIC EFFECT IN GADOLINIUM MOLYBDATE.
	30-3324	PYROELECTRIC DETECTION PROPERTIES OF GADOLINIUM MOLYBDATE.
GARBUZ NG	04-0656	SIZE EFFECTS IN FERROELECTRICS. (BARIUM TITANATE)
GARD R	17-2638	FAR INFRARED SPECTRA OF RUBIDIUM DIHYDROGEN PHOSPHATE CRYSTALS.
GARLAND CW	02-0070	ULTRASONIC INVESTIGATION OF PHASE TRANSITIONS AND CRITICAL POINTS. SECTION-5: FERROELE
	17-2576	ULTRASONIC INVESTIGATION (15 MHZ SHEAR WAVES) OF THE FERROELECTRIC TRANSITION REGION I
GARNIER PR	06-1132	SPECIFIC HEAT OF STRONTIUM TITANATE NEAR THE STRUCTURAL TRANSITION.
GARRETSON RM	04-0998	A FLOATING ZONE SINGLE CRYSTAL GROWING APPARATUS. (BARIUM TITANATE)
GARSHKA EP	03-0353	ACOUSTOELECTRIC INTERACTION IN FERROELECTRIC SUBSTANCES.
GARTON G	36-3558	FLUX GROWTH OF MAGNESIUM OXIDE AND LANTHANUM ALUMINATE CRYSTALS DOPED WITH ISOTOPE 170
GARVIN HL	11-1714	MEASUREMENT OF SPUTTER ETCHING RATES FOR MONOCRYSTALLINE LITHIUM NIOBATE, SILICON DIOX
GAUBERT C	04-0530	LEED STUDIES ON THE (001) FACE OF BARIUM TITANATE CRYSTALS.
	04-0531	STUDY OF THE TEMPERATURE DEPENDENCE OF LEED INTENSITIES FROM THE (001) SURFACE OF BARI
GAULT JD	34-3465	ICE I: LATTICE DYNAMICS AND INCOHERENT NEUTRON SCATTERING.
GAUSS KE	17-2501	TEMPERATURE DEPENDENCE OF THE DIELECTRIC CONSTANT OF POTASSIUM DIHYDROGEN PHOSPHATE AT
GAVARS PE	13-1945	PHASE COMPOSITION OF (LEAD, BARIUM) DINIOBATE. METHOD OF INVESTIGATION AND SOME RESULT
GAVIN AP	11-1721	RADIATION EXPOSURE OF A LITHIUM NIOBATE CRYSTAL AT HIGH TEMPERATURES.
GAVRIELIDES AT	31-3340	MOLECULAR MOTION IN THE FERROELECTRIC PHASE OF HYDROGEN CHLORIDE. (CALCULATED IN THE H
GAVRILISHINA AI	04-0956	EFFECT OF TRANSVERSE COMPRESSION OF DIELECTRIC PROPERTIES OF BARIUM TITANATE SINGLE CR
	04-0962	PIEZOELECTRIC PROPERTIES OF POLARIZED BARIUM TITANATE CERAMICS SUBJECTED TO UNIAXIAL T
GAVRILOVA ND	02-0083	PYROELECTRIC INVESTIGATIONS OF THE DOMAIN STRUCTURE STABILITY IN FERROELECTRICS. (ROCH
	04-0686	STUDY OF THE PYROELECTRIC EFFECT IN THIN FILMS OF TRIGLYCINE SULFATE AND BARIUM TITANA
	27-3115	VARIATION OF THE UNIPOLARITY OF TRIGLYCINE SULFATE CRYSTALS WITH TIME.
	36-3559	FERROELECTRIC BEHAVIOR OF THE MINERAL STIBOTANTALITE.
	36-3601	DOMAIN STRUCTURE AND TEMPERATURE DEPENDENCE OF THE PIEZOELECTRIC CONSTANTS OF STIBOTAN

GAVRILOVA-PODOLSKAYA GV   22-2837   REVERSAL MECHANISM IN FERROELECTRICS: LITHIUM OR SODIUM TRIHYDROGEN DISELENITE, NUCLEA
GAVRILYACHENKO VG   05-1046   FERROELECTRIC PROPERTIES OF LEAD TITANATE SINGLE CRYSTALS.
                          05-1047   GROWTH OF LEAD TITANATE CRYSTALS AND EXAMINATION OF THEIR DOMAIN STRUCTURE.
                          05-1048   RECIPROCAL DOMAINS IN LEAD TITANATE CRYSTALS.
                          05-1049   PIEZOELECTRIC EFFECT IN LEAD TITANATE SINGLE CRYSTALS.
                          05-1050   SPONTANEOUS POLARIZATION AND COERCIVE FIELD OF LEAD TITANATE (FLUX GROWN SINGLE CRYSTA
GAYDOS W   10-1651   STRUCTURAL AND DIELECTRIC MODIFICATIONS OF TUNGSTEN TRIOXIDE DUE TO NEUTRON IRRADIATIO
GAYLORD TK   37-3638   OPTICALLY ERASABLE AND REWRITABLE SOLID STATE HOLOGRAMS.
GEGUZINA GA   08-1439   INTERATOMIC DISTANCES IN OXIDES WITH THE PEROVSKITE STRUCTURE.
GEGUZINA SYA   03-0253   INFLUENCE OF LONG RANGE DIPOLE DIPOLE FORCES ON ATTENUATION AND VELOCITY OF ULTRASOUND
GEMMELL DS   04-0687   CHANNELING OF PROTONS IN THIN BARIUM TITANATE CRYSTALS AT TEMPERATURES ABOVE AND BELOW
                          04-0688   CHANNELING OF LIGHT IONS IN THE CUBIC AND TETRAGONAL (FERROELECTRICALLY POLARIZED PHAS
GENE VV   03-0307   ELASTOOPTICAL EFFECT IN FERROELECTRIC WITH DIFFUSE PHASE TRANSITION.
                          15-2148   FEATURES OF THE PHASE TRANSITION IN ANTIMONY SULFOIODIDE SINGLE CRYSTALS.
                          15-2190   RELAXATION PROCESSES OF ANTIMONY SULFOIODIDE SINGLE CRYSTALS.
GERBAUX X   27-3009   DISPERSION OF THE DIELECTRIC CONSTANT OF TGS IN THE FAR INFRARED (TEMPERATURE AND FREQ
GERLICH D   17-2513   HIGH TEMPERATURE PHASE TRANSITIONS AND METASTABILITY IN POTASSIUM DIHYDROGEN PHOSPHATE
GERNAND M   11-1679   DISORDER OF LITHIUM NIOBATE CRYSTALS.
GERSON R   07-1253   MOSSBAUER EFFECT STUDIES OF THE PHASE TRANSITIONS IN ANTIFERROELECTRIC LEAD ZIRCONATE
                          08-1333   FERROELECTRIC PROPERTIES OF BISMUTH FERRATE (PREPARATION AND STRUCTURE) AND RELATED MA
                          08-1398   ATOMIC STRUCTURE IN THE PEROVSKITIC FERRATE.
                          08-1402   FERROELECTRIC BISMUTH FERRATE X-RAY AND NEUTRON DIFFRACTION STUDY. (ANTIFERROMAGNETIC)
                          08-1449   DIELECTRIC HYSTERESIS IN (FLUX GROWN) SINGLE CRYSTAL BISMUTH FERRATE (CONFIRMS FERROEL
                          08-1471   MOSSEAUER STUDIES OF BISMUTH FERRATE LEAD TITANATE PEROVSKITE TYPE SOLID SOLUTIONS (RO
                          09-1519   PIEZOELECTRIC AND DIELECTRIC PROPERTIES OF LEAD TITANATE ZIRCONATE CERAMICS AT LOW TEM
                          09-1574   ATOMIC STRUCTURES OF THE FERROELECTRIC COMPOUNDS LEAD ZIRCONATE(X) TITANATE(1-X) FOR T
GERSTEIN BC   08-1334   TIGHT BONDING BAND CALCULATIONS ON RHENIUM TRIOXIDE, SODIUM TUNGSTATE, AND POTASSIUM T
GERTHSEN P   04-0689   MOBILITY DETERMINATIONS FROM WEIGHT MEASUREMENTS IN SOLID SOLUTIONS OF (BARIUM, STRONT
GERZANICH EI   15-2142   PRODUCTION AND SOME OPTICAL PROPERTIES OF ANTIMONY SULFOBROMIDE IN GLASSY AND CRYSTALL
                          15-2149   HOW THE DIELECTRIC PROPERTIES OF ANTIMONY SULFOIODIDE(X) BROMIDE(1-X) SOLID SOLUTIONS
                          15-2150   INDIRECT TRANSITIONS AND ABSORPTION IN THE MID-INFRARED IN ANTIMONY SULFOIODIDE CRYSTA
GESI K   03-0254   PHENOMENOLOGICAL THEORY OF PHASE TRANSITIONS IN SODIUM NITRITE.
                          16-2289   DIELECTRIC STUDIES ON THE FERROELECTRIC PHASE TRANSITION IN SILVER SODIUM DINITRITE.
                          16-2290   DIELECTRIC RELAXATION IN FERROELECTRIC SILVER SODIUM DINITRITE.
                          16-2291   DIELECTRIC RELAXATION MEASUREMENT ON FERROELECTRIC SILVER SODIUM DINITRITE BY STEP FUN
                          16-2292   DIELECTRIC RELAXATION IN FERROELECTRIC SILVER SODIUM DINITRITE.
                          16-2293   FERROELECTRIC HYSTERESIS LOOP OF SILVER SODIUM DINITRITE BY QUASI STATIC ELECTRIC FIEL
                          16-2294   HYDROSTATIC PRESSURE EFFECT ON THE DIELECTRIC RELAXATION TIME IN FERROELECTRIC SILVER
                          16-2295   PYROELECTRIC STUDY ON THE SPONTANEOUS POLARIZATION IN SILVER SODIUM DINITRITE.
                          16-2296   SPECIFIC HEAT ANOMALY IN FERROELECTRIC SILVER SODIUM DINITRITE.
                          16-2346   THERMAL EXPANSION IN FERROELECTRIC SILVER SODIUM DINITRITE.
                          16-2376   DIELECTRIC ANOMALIES OF SODIUM NITRATE ABOVE THE FERROELECTRIC CURIE TEMPERATURE.
                          33-3374   DISPERSION AND DAMPING OF SOFT ZONE BONDARY PHONONS IN POTASSIUM MANGANESE TRIFLUORIDE
GEUSIC JE   04-0690   ELECTROOPTIC PROPERTIES OF SOME ABO(3) PEROVSKITES IN THE PARAELECTRIC PHASE. (POTASSI
                          13-1924   USE OF PEROVSKITE PARAELECTRICS IN BEAM DEFLECTORS AND LIGHT MODULATORS. (POTASSIUM TA
                          13-2024   OPTICAL AND FERROELECTRIC PROPERTIES OF BARIUM SODIUM NIOBATE. (OPTICAL ABSORPTION, RE
                          30-3280   FERROELECTRIC DOMAIN SHIFTING DEVICES. (GADOLINIUM MOLYBDATE)
GEVERS R   03-0255   ELECTRON MICROSCOPE TRANSMISSION IMAGES OF COHERENT DOMAIN BOUNDARIES. PART-1: DYNAMIC
GHARBI A   16-2306   MEASUREMENT OF THE ELASTIC CONSTANTS OF SODIUM NITRITE BY BRILLOUIN DIFFUSION AT ORDIN
GIALLORENZI TG   11-1771   SPONTANEOUS PARAMETRIC SCATTERING, IDLER ABSORPTION, AND DISPERSION IN THE NONLINEAR S
GIAUQUE WF   30-3275   MAGNETOTHERMODYNAMICS OF ANTIFERROMAGNETIC, FERROELECTRIC BETA GADOLINIUM MOLYBDATE. P
GIBBON DL   08-1379   ORDERING OF MAGNESIUM AND NIOBIUM IN THE OCTAHEDRAL POSITIONS OF THE "CUBIC" PEROVSKIT
GIESS EA   13-1948   (TETRAGONAL) TUNGSTEN BRONZE TYPE CRYSTALS GROWN IN MOLTEN SOLUTIONS OF THE PSEUDOSYST
                          13-1949   (EXTENT OF) TUNGSTEN BRONZE (LIQUIDUS AND SOLIDUS) FIELD AND MELT GROWTH OF CRYSTALS I
                          13-2003   CRYSTAL GROWTH AND ELECTROOPTIC PROPERTIES OF TUNGSTEN BRONZE CRYSTALS FROM THE POTASS
                          13-2015   CRYSTAL GROWTH AND NONLINEAR OPTICAL PPOPERTIES OF POTASSIUM LITHIUM NIOBATES.
                          13-2016   PHASE EQUILIBRIA IN THE POTASSIUM NIOBATE- STRONTIUM DINIOBATE AND POTASSIUM NIOBATE-
                          13-2017   TUNGSTEN BRONZE FIELD IN THE (OXIDE) SYSTEM POTASSIA LITHIA NIOBIA (POTASSIUM LITHIUM
GIFEISMAN SHN   02-0127   FUNDAMENTAL PROPERTIES OF METAL- FERROELECTRIC- METAL JUNCTION SYSTEMS. (INFLUENCE OF
GILLETTA F   27-2980   POLARIZATION REVERSAL IN TGS SINGLE CRYSTALS.
                          27-3001   DIELECTRIC RELAXATION IN MULTIDOMAIN TGS SINGLE CRYSTALS.
                          27-3002   DIELECTRIC RELAXATION IN MULTIDOMAIN CRYSTALS OF TRIGLYCINE SULFATE.
                          27-3003   EVOLUTION OF FERROELECTRIC DOMAINS IN TGS SINGLE CRYSTALS.
                          27-3004   INFLUENCE OF DISLOCATIONS, SURFACE LAYER AND X-RAY IRRADIATION ON DOMAIN STRUCTURE IN
                          27-3050   DIELECTRIC RELAXATION OF MULTIDOMAIN TRIGLYCINE SULFATE.
                          27-3143   ETCHING AND ULTRAMICROSCOPIC STUDIES OF FERROELECTRIC DOMAINS IN TRIGLYCINE SULFATE.
GILLIER-PANDRAUD H   04-0590   BINARY AND PSEUDOBINARY PHASES BETWEEN NIOBIUM PENTOXIDE AND LEAD, BARIUM, AND STRONTI
                          08-1285   STRUCTURAL STUDY OF THE TRANSITION OF THE (ORTHORHOMBIC) FERROELECTRIC (LEAD DINIOBATE
GILLILAND JW   21-2812   THERMOLUMINESCENCE STUDIES OF THE GAMMA-RAY IRRADIATED FERROELECTRICS ROCHELLE SALT AN
GILLIS NS   03-0256   PHASE TRANSITIONS IN A MODEL FERROELECTRIC.
                          03-0257   PHASE TRANSITIONS IN A MODEL FERROELECTRIC.
                          08-1335   EFFECT OF PRESSURE ON THE DIELECTRIC PROPERTIES OF A PARAELECTRIC MATERIAL. (POTASSIUM
GINDULINA VZ   06-1162   EPR INVESTIGATION OF POLYCRYSTALLINE STRONTIUM TITANATE CONTAINING MANGANESE DIOXIDE.
GINTEL J   06-1133   DIELECTRIC SUSCEPTIBILITY OF QUASI FERROELECTRIC STRONTIUM TITANATE.
                          06-1134   TEMPERATURE DEPENDENCE OF THE DIELECTRIC CONSTANT OF STRONTIUM TITANATE.
GINZBURG VL   03-0191   TEMPERATURE DEPENDENCE OF THE SHAPE OF THE DOMAIN WALL IN FERROMAGNETICS AND FERROELEC
GIRSHBERG YAG   03-0196   PHONON SPECTRUM OF FERROELECTRICS STUDIED BY LIMITATION OF THE CRYSTAL SIZE.
                          04-0600   DIELECTRIC CONSTANT AND LATTICE DYNAMICS OF A THIN FILM OF A FERROELECTRIC (BARIUM TIT
                          04-0601   DISPERSION AT MILLIMETER WAVELENGTHS IN BARIUM TITANATE ABOVE THE TRANSITION POINT.
                          04-0609   POLARON MECHANISM OF CONDUCTIVITY IN BARIUM TITANATE.
                          04-0613   SMALL POLARONS IN CONDUCTING BARIUM TITANATE CRYSTALS.
                          04-0614   TEMPERATURE DEPENDENCE OF CARRIER MOBILITY IN BARIUM TITANATE.
GIRSHOVICHUS MN   15-2223   GENERATION OF THE SECOND ACOUSTIC HARMONIC IN ANTIMONY SULFOIODIDE SINGLE CRYSTALS.
                          15-2224   GENERATION OF THE SECOND ACOUSTIC HARMONIC IN ANTIMONY SULFOIODIDE SINGLE CRYSTALS.
GISIN BV   11-1715   CUTS OF LITHIUM NIOBATE CRYSTALS SUITABLE FOR LIGHT MODULATORS WITH A LOW HALF WAVE VO
GIULIANI A   10-1634   OPTICAL TRANSITION TO THE FERMI LEVEL IN SODIUM TUNGSTATE.
GLADKII VV   17-2503   DOUBLE DIELECTRIC HYSTERESIS LOOP OF POTASSIUM DIDEUTERIUM PHOSPHATE CRYSTALS.
                          17-2504   POLARIZATION OF A RUBIDIUM DIHYDROGEN PHOSPHATE CRYSTAL IN THE PHASE TRANSITION REGION
                          17-2505   POLARIZATION (DOMAIN STRUCTURE) AND PHASE TRANSITION IN KDP SINGLE CRYSTAL. (ABSTRACT
                          17-2506   POLARIZATION OF A RUBIDIUM DIHYDROGEN PHOSPHATE CRYSTAL IN THE PHASE TRANSITION REGION

PAGE 638

GLADKOVSKAYA AA   17-2434   POLYTHERM OF SOLUBILITY OF THE AMMONIUM DIHYDROGEN PHOSPHATE- POTASSIUM DIHYDROGEN PHO
GLANC A   17-2502   APPLICATIONS OF THE PIEZOELECTRIC NONLINEARITIES IN THE TEMPERATURE AUTOSTABILIZATION
                     27-3005   TEMPERATURE AUTOSTABILIZATION EFFECT OF TRIGLYCINE SULFATE SINGLE CRYSTALS IN AN AC EL
GLASBEEK M   06-1147   ESR OF CHROMIUM(3+) IN CHROMIUM DOPED STRONTIUM TITANATE SINGLE CRYSTALS.
GLASS AM   05-1072   (FLUX) GROWTH AND FERROELECTRIC PROPERTIES OF HIGH RESISTIVITY SINGLE CRYSTALS OF LEAD
                     09-1477   PHOTOMIXING AT 10.6 MICRONS WITH STRONTIUM BARIUM NIOBATE PYROELECTRIC DETECTORS.
                     11-1716   CONTROL OF LASER DAMAGE IN LITHIUM NIOBATE.
                     11-1791   CONTROL OF THE SUSCEPTIBILITY OF LITHIUM NIOBATE TO LASER INDUCED REFRACTIVE INDEX CHA
                     11-1793   LOW FREQUENCY (510 MHZ) PARAMAGNETIC RESONANCE INVESTIGATION OF LITHIUM NIOBATE.
                     12-1875   STUDY OF PIEZOELECTRIC OSCILLATIONS IN WIDEBAND PYROELECTRIC LITHIUM TANTALATE.
                     30-3290   (ROOM TEMPERATURE) FERROELECTRIC FERROELASTIC PARAMAGNETIC (BETA) TERBIUM MOLYBDATE (C
                     36-3560   PYROELECTRIC PROPERTIES OF POLYVINYLIDENE FLUORIDE AND ITS USE FOR INFRARED DETECTION.
GLASS JC   04-0691   PERTURBED DIRECTIONAL CORRELATION OF SCANDIUM-44 WITHIN THE LATTICE OF FERROELECTRIC B
                     04-0692   STATIC QUADRUPOLE INTERACTION OF SCANDIUM-44 IN BARIUM TITANATE.
                     05-1055   PERTURBED DIRECTIONAL CORRELATION OF SCANDIUM-44 IN LEAD TITANATE.
GLASSER ML   03-0258   ANALYTICAL PROPERTIES OF A CLASS OF PLANAR FERROELECTRIC MODELS. (THEORY, KDP TYPES)
                     34-3427   ANALYTIC PROPERTIES OF THE FREE ENERGY FOR THE "ICE" MODELS (FERROELECTRICS).
GLAZER AM   08-1273   STRUCTURES OF SODIUM NIOBATE BETWEEN 480 AND 575 DEG C, AND THEIR RELEVANCE TO SOFT PH
                     08-1336   HIGH TEMPERATURE PHASE TRANSITIONS IN SODIUM NIOBATE AND THE USE OF TILTING SCHEMES IN
                     08-1337   THE STRUCTURE OF SODIUM NIOBATE T(2) AT 600 DEG C, AND THE CUBIC- TETRAGONAL TRANSITIO
GLEASON TG   24-2892   MOSSBAUER EFFECT STUDY OF THE FERROELECTRIC TRANSITIONS IN FERRIC AMMONIUM SULFATE DOD
GLEASON TJ   13-1950   PARAMETRIC FLUORESCENCE IN BARIUM SODIUM NIOBATE FOR NINE CW PUMP FREQUENCIES.
                     13-1983   PARAMETRIC FLUORESCENCE IN BARIUM SODIUM NIOBATE FOR 10 CW PUMP WAVELENGTHS (RANGE 454
                     19-2755   MOSSBAUER EFFECT STUDY OF THE FERROELECTRIC TRANSITIONS IN FERRIC AMMONIUM SULFATE DOD
GLEN JW   02-0071   STRUCTURE AND POINT DEFECTS OF ICE: THEIR EFFECT ON THE ELECTRICAL AND MECHANICAL PROP
                     34-3412   ELECTRIC POLARIZATION EFFECTS IN PURE AND DOPED ICE AT LOW TEMPERATURES.
                     34-3449   IMPURITY EFFECTS ON THE PLASTICITY OF ICE AND THEIR EXPLANATION IN TERMS OF HYDROGEN R
GLOCKMANN HP   34-3428   CONDUCTION ANOMALIES AND POLARIZATION IN ICE AT LOW TEMPERATURES.
GLOGAR P   01-0007   INTERNATIONAL MEETING ON FERROELECTRICITY, PRAGUE, PROC.
                     03-0260   NEUTRON SCATTERING BY A FERROELECTRIC SOLID SOLUTION. (THEORY, APPLIED TO BARIUM STRON
                     04-0693   DEPENDENCE OF COERCIVE FIELD OF BARIUM TITANATE SINGLE CRYSTALS ON THEIR THICKNESS.
GLOGAROVA M   17-2452   STRESS INDUCED BIREFRINGENCE IN AN ISOLATED AND A SHORTCIRCUITED POTASSIUM DIHYDROGEN
                     19-2756   PROPERTIES OF FERROELECTRIC CADMIUM AMMONIUM SULFATE.
                     19-2757   THE BEHAVIOR OF DIAMMONIUM DICADMIUM TRISULFATE UNDER PRESSURE.
                     30-3304   STUDY OF FERROELECTRIC DOMAINS IN GADOLINIUM MOLYBDATE.
                     36-3534   NEW FERROELECTRIC LANGBEINITE THALLIUM CADMIUM SULFATE.
GLOWER DD   03-0259   CHARGE MECHANISM IN FERROELECTRICS. (LEAD ZIRCONATE TITANATE, STOICHIOMETRY EFFECTS)
                     04-0694   THERMAL STUDIES OF CALCIUM(X) BARIUM(1-X) TITANATE.
                     09-1520   EFFECTS OF RADIATION INDUCED DAMAGE CENTERS IN LEAD ZIRCONATE TITANATE CERAMICS.
                     09-1576   PROPERTIES OF OXYGEN DEPLETED CERAMIC LEAD ZIRCONATE TITANATE (PLUS 1 PERCENT NIOBIUM
                     37-3667   A SELF SCANNED FERROELECTRIC IMAGE SENSOR.
                     37-3669   THEORY AND APPLICATION OF FERROELECTRIC RADIATION DETECTORS. (PYROELECTRICS)
GLUSHKOVA TM   11-1717   TEMPERATURE AND FREQUENCY DEPENDENCES OF THE ELECTRICAL PROPERTIES OF LITHIUM METANIOB
GNININVI L   03-0261   THEORETICAL DYNAMIC STUDY OF CUBIC PHASE BARIUM TITANATE.
GOBRAN NK   28-3197   ULTRASONIC ANOMALIES IN ROCHELLE SALT.
GOBUSH W   34-3429   CALCULATION OF THE DIELECTRIC CORRELATION FACTOR OF CUBIC ICE.
GODARD J   19-2758   PREPARATION OF SINGLE CRYSTALS OF AMMONIUM SULFATE.
GODEFROY G   01-0012   EUROPEAN MEETING ON FERROELECTRICITY, 2ND PROC, DIJON, 1971.
                     04-0697   (PHASE TRANSITIONS AND) THERMOCURRENTS IN TITANIUM DIOXIDE AND BARIUM TITANATE SINGLE
                     04-0699   "VIBRATING WIRE" STUDY OF POTENTIAL DISTRIBUTION (STABILITY) IN BARIUM(X) STRONTIUM(1-
                     04-0788   DISTRIBUTION OF POTENTIAL ALONG TITANATE CERAMICS PREVIOUSLY POLARIZED.
                     19-2780   OPTICAL ABSORPTION, ELECTRIC CONDUCTIVITY AND PHOTOEMISSION OF GLYCOL SULFATE.
GODEFROY L   01-0012   EUROPEAN MEETING ON FERROELECTRICITY, 2ND PROC, DIJON, 1971.
                     04-0622   SOME PROPERTIES OF THE THERMOCURRENTS IN BARIUM TITANATE AND TITANIUM DIOXIDE SINGLE C
                     04-0695   CONDUCTIVITY OF BARIUM TITANATE, PURE SINGLE CRYSTALS AND DOPED (IRON, HYDROXYL) SINGL
                     04-0698   NOISE MEASUREMENTS IN FERROELECTRICS. (TGS, ROCHELLE SALT, BARIUM TITANATE)
                     04-0826   STATIC SHELL MODEL FOR BARIUM AND LEAD TITANATES.
                     04-0844   AUTOMATIC IMPEDANCE MEASUREMENTS USING THERMAL NOISE ANALYSIS (APPLICATION TO BARIUM T
                     04-0849   INTERNAL STRESSES IN BARIUM TITANATE CERAMICS. (AND DIFFERENCE IN TEMPERATURES FOR MAX
                     04-0915   REMARKS ABOUT SWITCHING IN BARIUM TITANATE AND TRIGLYCINE SULFATE SINGLE CRYSTALS.
                     27-2960   SPECIFIC HEAT UNDER APPLIED ELECTRIC FIELD AND ELECTROCALORIC EFFECT IN TRIGLYCINE SUL
                     27-3076   INFLUENCE OF AN ELECTRIC FIELD ON THE NOISE AND THE PYROELECTRIC EFFECT IN TGS.
                     27-3077   POLARIZATION FLUCTUATION OF TRIGLYCINE SULFATE NEAR THE CURIE POINT.
GODYCKI E   04-0696   DIELECTRIC PROPERTIES OF BARIUM TITANATE DOPED WITH TUNGSTEN OXIDE.
GOETTELMAN R   09-1511   VARIATION OF SOUND VELOCITY AT FERROELECTRIC-1, FERROELECTRIC-2 TRANSITION IN (NIOBIUM
GOETZ GG   17-2507   REAL-TIME HOLOGRAPHIC RECONSTRUCTION BY ELECTROOPTIC MODULATION (BY PASSING COHERENT L
GOKHBERG LS   13-1982   ELECTROOPTICAL EFFECT IN LEAD ZINCATE(0.33) NIOBATE(0.67). (DISPERSION OF QUADRATIC CO
GOLDBERG LS   08-1338   OPTICAL PARAMETRIC OSCILLATION IN LITHIUM IODATE.
GOLDING B   06-1135   THERMAL EXPANSIVITY AND ULTRASONIC PROPAGATION NEAR THE STRUCTURAL TRANSITION OF STRON
                     06-1136   THERMAL EXPANSIVITY OF STRONTIUM TITANATE NEAR THE STRUCTURAL TRANSITION.
                     06-1137   ULTRASONIC VELOCITY ABOVE TANTALUM IN STRONTIUM TITANATE.
                     33-3375   THERMAL EXPANSIVITY OF RUBIDIUM MANGANESE FLUORIDE NEAR THE NEEL TEMPERATURE.
GOLIK LL   15-2151   ELECTROREFLECTION FROM SINGLE CRYSTALS OF FERROELECTRIC ANTIMONY SULFOIODIDE NEAR THE
                     15-2152   INSTABILITY OF FERROELECTRIC POLARIZATION IN ANTIMONY SULFOIODIDE SINGLE CRYSTALS.
GOLOVCHENKO LS   18-2740   ARSENIC-75 NUCLEAR QUADRUPOLE RESONANCE SPECTRA OF PARTIALLY DEUTERATED FERROELECTRIC
GOLOVEI MP   17-2508   NONLINEAR PROPERTIES OF A RUBIDIUM DIHYDROGEN PHOSPHATE CRYSTAL.
GOLOVKO YUI   08-1325   SYNTHESIS AND STUDY OF LEAD SCANDIUM NIOBATE SINGLE CRYSTALS.
GOLOVNIN VA   17-2445   THERMAL EXPANSION OF ADP AND DEUTERATED ADP IN THE REGION OF ANTIFERROELECTRIC PHASE T
                     36-3618   A NEW MAGNETOFERROELECTRIC: CADMIUM IRON NIOBATE.
GOLOVSHCHIKOVA GI   36-3561   PECULIARITIES OF THE DIELECTRIC POLARIZATION OF CADMIUM PYRONIOBATE.
GOLOVYANKO AA   04-0953   ELECTROOPTICAL PROPERTIES OF SINGLE CRYSTALS OF SOLID SOLUTIONS OF BARIUM TITANATE AND
                     04-0955   ELECTROOPTICAL PROPERTIES OF BARIUM TITANATE- TANTALUM PENTOXIDE SOLID SOLUTION SINGLE
                     04-0958   ELECTROOPTICAL EFFECT IN SINGLE CRYSTALS OF BARIUM TITANATE- ZINC TITANATE SOLID SOLUT
                     04-0959   EFFECT OF NONISOVALENT SUBSTITUTION OF TITANIUM(4+) IONS ON THE ELECTROOPTICAL PROPERT
GOLTSOV YUI   02-0047   SYNTHESIS AND CERTAIN PROPERTIES OF NEW COMPOUNDS WITH A PYROCHLORE STRUCTURE. (NIOBAT
                     05-1051   POSITIVE TEMPERATURE COEFFICIENT OF RESISTANCE IN DOPED LEAD TITANATE.
                     13-1940   PHASE TRANSITION IN BARIUM BISMUTH(5+) BISMUTH(3+) OXIDE.
GOLUB AM   04-0932   FORMATION OF FLUORINE CONTAINING SOLID SOLUTIONS BASED ON BARIUM TITANATE.
GONCHAROV KV   03-0498   NONLINEAR EFFECTS IN THE PROPAGATION OF FINITE AMPLITUDE HYPERSONIC WAVES IN FERROELEC
                     12-1891   INTERACTION OF TWO HYPERSOUND WAVES IN LITHIUM TANTALATE.

FERROELECTRICS--AUTHOR-TITLE INDEX

GONCHARUK IN	16-2276	PHASE CHANGE IN SODIUM NITRITE CRYSTALS AND COMBINATION SCATTER SPECTRUM IN THE REGION
	16-2277	TWO-PHONON SCATTERING AND FERROELECTRIC TRANSITION IN A SODIUM NITRITE CRYSTAL.
GONNARD P	09-1521	PHASE CHANGES INDUCED BY HYDROSTATIC PRESSURE IN A FERROELECTRIC MATERIAL. (PZT)
	09-1581	PHASE DIAGRAM OF LEAD ZIRCONATE(1-X) TITANATE(X) PLUS 0.8 PERCENT TUNGSTEN TRIOXIDE.
	09-1615	METHOD OF MEASURING POLARIZATION VIBRATIONS OF A PREPOLARIZED CERAMIC FERROELECTRIC. (
GONSIOR A	16-2338	DIELECTRIC MEASUREMENTS OF AMMONIUM NITRATE.
GONZALO JA	03-0262	FERROELECTRIC FREE ENERGY EXPANSION COEFFICIENTS FROM DOUBLE HYSTERESIS LOOPS.
	03-0263	STATISTICAL THEORY FOR FERROELECTRICITY IN TRIGLYCINE SULFATE.
	04-0921	THERMAL HYSTERESIS IN THE TETRAGONAL CUBIC TRANSITION OF BARIUM TITANATE.
	16-2326	NEUTRON DIFFRACTION STUDY OF SODIUM NITRITE.
	27-2983	THERMAL CONDUCTIVITY OF TRIGLYCINE SULFATE NEAR THE CURIE POINT.
	27-3006	EQUATION OF STATE FOR THE COOPERATIVE TRANSITION OF TRIGLYCINE SULFATE NEAR CURIE TEMP
	27-3075	CRITICAL BEHAVIOR OF TRIGLYCINE FLUOBERYLLATE.
	27-3098	FERROELECTRIC BEHAVIOR OF RADIATION DAMAGED TRIGLYCINE SULFATE AND ROCHELLE SALT UP TO
	27-3146	FERROELECTRIC SPECIFIC HEAT OF TRIGLYCINE SULFATE.
GOODENOUGH JB	36-3562	TWO COMPONENTS OF THE CRYSTALLOGRAPHIC TRANSITION IN VANADIUM DIOXIDE. (ANTIFERROELECT
GOODMAN G	04-0700	ELECTROSTRICTIVELY GENERATED AND MODULATED FRACTURE IN BARIUM TITANATE CERAMICS.
GOPAL ESR	03-0264	CRITICAL POINT INDICES OF FERROELECTRIC AND FERROMAGNETIC PHASE TRANSITIONS.
GORAK YAA	15-2141	PRODUCTION METHODS AND SOME OPTICAL PROPERTIES OF BISMUTH TELLUROIODIDE.
	15-2154	EFFECT OF THE ADSORPTION OF OXYGEN ON THE PROPERTIES OF ANTIMONY SULFOIODIDE SINGLE CR
GORBATOV GZ	15-2129	CHARACTERISTICS OF THE PHOTOELECTRIC EMF OF ANTIMONY SULFOIODIDE.
	15-2133	PHOTODOMAIN EFFECT IN NATURALLY POLARIZED ANTIMONY SULFOIODIDE CRYSTALS.
	15-2135	PHOTODOMAIN EFFECT IN NATURALLY POLARIZED ANTIMONY SULFOIODIDE CRYSTALS.
	15-2136	SOME FEATURES OF THE PHOTOEMF SPECTRUM OF ANTIMONY SULFOIODIDE.
	15-2153	CHARACTERISTICS OF THE PHOTOELECTRIC EMF OF ANTIMONY SULFOIODIDE.
	15-2228	INFLUENCE OF LIGHT ON SELF POLARIZATION OF SINGLE CRYSTALS OF ANTIMONY SULFOIODIDE.
GORBATYI LV	22-2838	CRYSTAL STRUCTURES OF POTASSIUM TRIHYDROGEN SELENITE AND SODIUM TRIHYDROGEN SELENITE.
GORBATYUK VA	04-0932	FORMATION OF FLUORINE CONTAINING SOLID SOLUTIONS BASED ON BARIUM TITANATE.
GORBOKON NV	17-2627	THERMOELASTIC PROPERTIES OF POTASSIUM DIHYDROGEN PHOSPHATE NEAR THE HIGH TEMPERATURE P
	27-2986	DEUTERATION EFFECT ON THERMAL AND ELASTIC PROPERTIES OF FERROELECTRIC CRYSTALS THE TRI
GORDEYEVA NV	22-2873	ISOTOPIC NONISOMORPHISM OF RUBIDIUM TRIHYDROGEN(1-X) TRIDEUTERIUM(X) DISELENATE CRYSTA
GORDON EI	13-1929	ELECTROOPTIC POTASSIUM TANTALATE NIOBATE GRATINGS FOR LIGHT BEAM MODULATION AND DEFLEC
GORELIK LF	13-1951	ELECTROPHYSICAL PROPERTIES OF THE SOLID SOLUTION SYSTEM LEAD TITANATE- LEAD ZIRCONATE-
GORELOV MI	04-0577	PHASE BOUNDARIES IN BARIUM TITANATE SINGLE CRYSTALS NEAR CURIE POINT.
GORNOSTAEV VF	27-3109	BARKHAUSEN JUMPS AND SWITCHING CURRENT IN TGS CRYSTALS.
GORNOSTANSKY S	34-3480	STRONG COLLISION LIMIT OF SPIN LATTICE RELAXATION IN HEXAGONAL ICE.
GORODETSKY G	36-3563	MAGNETOELECTRIC EFFECT IN THE ANTIFERROMAGNET IRON ANTIMONY(2) OXIDE(4) (MAY BE FERROE
GOSAR P	34-3430	PROTON- PROTON AND PROTON- LATTICE INTERACTIONS IN ICE.
GOSWAMI AK	04-0701	CHARACTERIZATION OF SURFACE AND GRAIN BOUNDARY LAYER OF BARIUM TITANATE.
	04-0702	PROPERTIES OF BLACK SEMICONDUCTING BARIUM TITANATE.
GOTO Y	04-0703	KERR EFFECT STUDIES OF ELECTRIC CONDUCTION IN IRON DOPED BARIUM TITANATE SINGLE CRYSTA
GOTTLIEB K	25-2920	MOLECULAR AND LATTICE VIBRATIONS OF ORTHORHOMBIC AND FERROELECTRIC THIOUREA. PART-1: T
GOUGH SR	34-3431	DIELECTRIC BEHAVIOR OF CUBIC AND HEXAGONAL ICES AT LOW TEMPERATURES.
	34-3432	LOW TEMPERATURE DIELECTRIC CELL AND PERMITTIVITY OF HEXAGONAL ICE TO 2 K.
GOULPEAU L	07-1256	COMPLEX DIELECTRIC CONSTANT OF LEAD ZIRCONATE UNDER POLARIZING FIELD, 370-520 DEG K.
	07-1257	RELAXATION OF ANTIFERROELECTRIC DOMAIN WALLS, EFFECTIVE FIELD, AND ACTIVATION ENERGY.
	08-1339	COMPLEX DIELECTRIC CONSTANT OF LEAD HAFNATE, 320-520 DEG K.
GOUTLE R	08-1340	DIELECTRIC STUDY OF THE LEAD ZIRCONATE LEAD HAFNATE (PHASE) DIAGRAM.
	03-0237	ELASTIC ENERGY AND THE FERROELECTRIC PARAELECTRIC TRANSITION.
GOUZERH J	17-2509	(FINITE APERTURE LIGHT SWITCH USING) TRANSVERSE POCKELS EFFECT COMMUTATORS. (POTASSIUM
	17-2524	STUDY OF THE LIMITATION OF THE ANGULAR FIELD BY THE COMMUTATOR OF POLARIZATION IN KDP,
GOVINDARAJAN K	03-0264	CRITICAL POINT INDICES OF FERROELECTRIC AND FERROMAGNETIC PHASE TRANSITIONS.
GOYAL PS	33-3371	PARAMAGNETIC SCATTERING OF NEUTRONS FROM POTASSIUM MANGANESE TRIFLUORIDE IN THE SHORT
GRABMAIER J	13-1938	TEMPERATURE DEPENDENCE OF THE DIELECTRIC PROPERTIES OF BARIUM STRONTIUM NIOBATE IN THE
GRAF P	13-1938	TEMPERATURE DEPENDENCE OF THE DIELECTRIC PROPERTIES OF BARIUM STRONTIUM NIOBATE IN THE
GRAHAM HC	04-0704	ELECTRICAL PROPERTIES OF HIGH PURITY POLYCRYSTALLINE BARIUM TITANATE. (DIELECTRIC CONS
GRANDE S	20-2802	NUCLEAR MAGNETIC RESONANCE (WIDE LINE SPECTRA AND SPIN LATTICE RELAXATION) INVESTIGATI
	26-2931	NMR INVESTIGATIONS OF FERROELECTRIC DICALCIUM STRONTIUM PROPIONATE.
	36-3551	NMR INVESTIGATIONS OF PHTHALOCYANINE.
GRANDITS DM	13-2054	KYROPOULOS SINGLE CRYSTAL GROWTH OF SODIUM TUNGSTEN BRONZES. (SODIUM TUNGSTATE)
GRANDJEAN D	27-3007	OPTICAL CONSTANTS OF TRIGLYCINE SELENATE IN THE NEAR INFRARED TO FAR INFRARED AND IN T
	27-3009	DISPERSION OF THE DIELECTRIC CONSTANT OF TGS IN THE FAR INFRARED (TEMPERATURE AND FREQ
GRANGE G	03-0177	VOLTAGES AND CURRENTS DURING DEPOLARIZATION OF A FERROELECTRIC MATERIAL. (THEORY)
	09-1581	PHASE DIAGRAM OF LEAD ZIRCONATE(1-X) TITANATE(X) PLUS 0.8 PERCENT TUNGSTEN TRIOXIDE.
GRANICHER H	02-0072	REVIEW ON THE PROBLEMS OF THE PHYSICS OF ICE.
	03-0266	MAGNETIC RESONANCE AND THE PROBLEM OF THE DEFINITION OF ANTIFERROELECTRICITY.
	04-0737	OPTICAL EVIDENCE FOR POLARIZATION FLUCTUATIONS IN PARAELECTRIC BARIUM TITANATE.
	04-0805	TEMPERATURE DEPENDENT POLARIZABILITIES AND COCHRAN MODE ASSIGNMENTS IN BARIUM AND STRO
	04-0856	PRESSURE DEPENDENCE OF THE DIELECTRIC CONSTANTS OF PARAELECTRIC MATERIALS. (BARIUM TIT
	06-1164	MONODOMAIN STRONTIUM TITANATE (OCCURANCE IN THIN (110) PLATES BELOW CUBIC- TETRAGONAL
	06-1169	MONODOMAIN STRONTIUM TITANATE.
	08-1347	PECULIAR FERROELECTRIC BEHAVIOR OF POTASSIUM IODATE.
	34-3433	EVALUATION OF DIELECTRIC DISPERSION DATA. (ICE)
	34-3434	INTERPRETATION OF THE PRESSURE DEPENDENCE OF PROPERTIES CAUSED BY LATTICE DEFECTS. (IC
	34-3471	NMR INVESTIGATIONS OF ICE CRYSTALS. PART-1: PROTON MAGNETIC RESONANCE IN WATER ICE. PA
	34-3479	PRESSURE DEPENDENCE OF THE COMPLEX DISPERSION COEFFICIENT OF ICE I(H) SINGLE CRYSTALS.
	36-3577	LEVEL CROSSING (CROSS RELAXATION) EXPERIMENTS IN AMMONIUM TRIHYDROGEN PERIODATE CRYSTA
	36-3578	NUCLEAR QUADRUPOLE INTERACTION IN ANTIFERROELECTRIC DIAMMONIUM TRIHYDROGEN PERIODATE P
	36-3579	NUCLEAR QUADRUPOLE INTERACTION IN ANTIFERROELECTRIC DIAMMONIUM TRIHYDROGEN PERIODATE B
	36-3581	STATISTICAL CONSIDERATIONS OF THE ORDER DISORDER REACTION IN AMMONIUM PERIODATE.
GRANOVSKII VG	03-0265	EFFECT OF ELASTIC STRESSES ON PARAMETERS OF SOLID SOLUTIONS WITH FERROELECTRIC PROPERT
	08-1341	CHARACTER OF THE CHEMICAL BONDS IN ABO(3) FERROELECTRIC CRYSTALS WITH A PEROVSKITE TYP
GRANT EH	25-2912	DIELECTRIC DISPERSION OF UREA AND THIOUREA.
GRASSIE ADC	36-3564	EXPERIMENTAL RELATIONSHIP BETWEEN THE SUPERCONDUCTING AND THE FERROELECTRIC PHASE TRAN
GRASSO M	11-1691	NONSTOICHIOMETRY AND CRYSTAL GROWTH OF LITHIUM NIOBATE.
	13-1920	PHASE EQUILIBRIA RELATIONS IN THE TERNARY SYSTEM BARIUM OXIDE- STRONTIUM OXIDE- NIOBIU
GRATE T	11-1721	RADIATION EXPOSURE OF A LITHIUM NIOBATE CRYSTAL AT HIGH TEMPERATURES.
GREGORIUS P	27-3088	MICROWAVE BEHAVIOR OF ROCHELLE SALT AND TGS (18 REFS).
GREKOV AA	04-0552	PHOTOFERROELECTRIC EFFECTS IN A(V)B(VI)C(VII) AND BARIUM TITANATE TYPE FERROELECTRICS.
	04-0574	PHOTOCONDUCTIVITY AND PHOTOFERROELECTRIC PHENOMENA IN BARIUM TITANATE.

GREKOV AA	04-0681	EXTRINSIC PHOTOCONDUCTIVITY IN FERROELECTRICS DUE TO SURFACE LAYERS. (BARIUM TITANATE,
	04-0682	PHOTODOMAIN EFFECT IN BARIUM TITANATE.
	15-2145	INFLUENCE OF NONEQUILIBRIUM CARRIERS ON THE PHASE TRANSITION IN FERROELECTRIC SEMICOND
	15-2147	TEMPERATURE DEPENDENCE OF THE CROSS SECTION OF CAPTURE CENTERS IN FERROELECTRIC SEMICO
	15-2155	DILATOMETRIC INVESTIGATION OF ANTIMONY SULFOIODIDE SINGLE CRYSTALS.
	15-2156	DILATOMETRIC INVESTIGATION OF PHOTOFERROELECTRIC EFFECTS IN SINGLE CRYSTALS OF ANTIMON
	15-2157	FERROELECTRIC PHOTOELECTRETS BASED ON A(V)B(VI)C(VII) SOLID SOLUTIONS. (DIELECTRIC CON
	15-2158	INFLUENCE OF NONEQUILIBRIUM CARRIER SCREENING ON POLARIZATION PROCESSES IN FERROELECTR
	15-2159	INDUCED IMPURITY PHOTOCONDUCTIVITY IN THE FERROELECTRIC SEMICONDUCTOR ANTIMONY SULFOIO
	15-2160	METASTABLE OPTICAL CHARGE EXCHANGE AND "FROZEN" SHIFT OF THE CURIE TEMPERATURE OF A FE
	15-2161	PHOTOFERROELECTRIC EFFECTS IN FERROELECTRIC SEMICONDUCTORS OF THE A(V)B(VI)C(VII) TYPE
GRENOUILLEAU P	04-0705	CRYSTAL STUDY OF BARIUM TITANATE WITH PARALLEL INTERFERENCE FRINGES.
	04-0706	CRYSTAL STRUCTURE OF BARIUM TITANATE PRODUCING SQUARE INTERFERENCE FRINGES.
GRESS EA	02-0053	PROPERTIES OF TUNGSTEN BRONZE FERROELECTRICS. (REVIEW, 33 REFS)
GREZNEV YUS	28-3176	ELECTRON PARAMAGNETIC RESONANCE OF COPPER IONS IN ROCHELLE SALT.
GRIB BN	17-2512	USE OF KDP TYPE CRYSTALS AS ELECTROOPTICAL REFLECTORS.
GRIFFITHS CR	05-1052	TIME EFFECTS IN THE HYSTERESIS LOOP OF (LEAD, STRONTIUM) TITANATE.
	05-1053	TIME EFFECTS IN THE HYSTERESIS LOOP OF LEAD STRONTIUM TITANATE.
	09-1522	DIELECTRIC NONLINEARITY AND LOSS IN THE BARIUM (OR LEAD, 4 COMPOSITIONS EACH) STRONTIU
GRIGAS B	15-2162	GROWING SINGLE CRYSTALS FROM THE VAPOR PHASE BY A DYNAMIC METHOD (ANTIMONY SESQUISELEN
	15-2163	PYROELECTRIC EFFECT IN ANTIMONY SULFO AND SELENO IODIDE SINGLE CRYSTALS.
GRIGAS IP	15-2117	DIELECTRIC DISPERSION IN ANTIMONY SULFOIODIDE CRYSTALS.
GRIGOREV MA	11-1718	TEMPERATURE DEPENDENCE OF HYPERSONIC WAVE DAMPING IN RUBY AND LITHIUM NIOBATE IN THE T
	11-1719	TEMPERATURE DEPENDENCES OF THE ABSORPTION OF 9.4 GHZ HYPERSOUND IN ALUMINUM OXIDE AND
GRIGOREVA EA	05-1047	GROWTH OF LEAD TITANATE CRYSTALS AND EXAMINATION OF THEIR DOMAIN STRUCTURE.
	08-1325	SYNTHESIS AND STUDY OF LEAD SCANDIUM NIOBATE SINGLE CRYSTALS.
GRILLI E	17-2422	ULTRAVIOLET REFLECTION AND ABSORPTION OF POTASSIUM DIHYDROGEN PHOSPHATE AND AMMONIUM D
GRIMM H	17-2418	NEUTRON MEASUREMENTS ON THE FERROELECTRIC PHASE TRANSFORMATION IN POTASSIUM DIHYDROGEN
	17-2510	EXPERIMENTAL PROOF FOR PROTON TUNNELING IN POTASSIUM DIHYDROGEN PHOSPHATE.
	17-2511	NEUTRON MEASUREMENTS ON THE HYDROGEN BOND POTENTIAL IN PARAELECTRIC POTASSIUM DIHYDROG
GRINBERG J	17-2465	MOSSBAUER EFFECT OBSERVATION OF A VERY LOW FREQUENCY MODE AT THE TRANSITION TEMPERATUR
GRINDLAY J	01-0013	AN INTRODUCTION TO THE PHENOMENOLOGICAL THEORY OF FERROELECTRICITY.
	03-0267	THEORY OF ELECTROSTRICTION IN CUBIC CRYSTALS WITH APPLICATION TO BARIUM TITANATE.
	03-0280	EFFECT OF THE CHANGE OF ELECTROSTATIC CONSTRAINTS ON THE PHASE TRANSITIONS OF SOME PHE
GRINVALD GZH	04-0894	SPECIFIC FEATURES OF THE TEMPERATURE DEPENDENCE OF THE VELOCITY OF ULTRASOUND IN POLYC
GROMASHEVSKII VL	15-2113	AGING OF THE FERROELECTRIC SEMICONDUCTOR ANTIMONY SULFOIODIDE IN A STATIC ELECTRIC FIE
	15-2114	ABSORPTION OF ULTRASOUND IN A(V)B(VI)C(VII) FERROELECTRIC SEMICONDUCTORS AT TEMPERATUR
	15-2115	ABSORPTION OF ULTRASOUND IN AN ANTIMONY SULFOIODIDE FERROELECTRIC SEMICONDUCTOR NEAR I
	15-2119	INFLUENCE OF ILLUMINATION ON THE FORM OF THE VOLTAGE AMPERE CHARACTERISTICS OF ANTIMON
	15-2121	PIEZOELECTRIC VIBRATIONS IN ANTIMONY SULFOIODIDE CRYSTALS NEAR THE CURIE POINT (EFFECT
	15-2123	POLARIZATION EFFECTS IN THE ANOMALOUS ABSORPTION OF ULTRASOUND IN ANTIMONY SULFOIODIDE
GROMOV AK	13-1936	PREPARATION AND GROWTH OF BARIUM(X) STRONTIUM(1-X) NIOBATE SINGLE CRYSTALS (CZOCHRALSKI
GROSESCU R	27-3104	INVESTIGATIONS OF NUCLEAR RELAXATION OF TRIGLYCINE FLUOBERYLLATE.
GROSHIK II	15-2242	DIELECTRIC PROPERTIES OF ANTIMONY SULFOIODIDE- SULFOBROMIDE SOLID SOLUTIONS.
GROSKREUTZ HE	24-2890	MOSSBAUER RESONANCE OF SINGLE CRYSTAL POTASSIUM FERROCYANIDE TRIHYDRATE.
GROSSMAN DG	05-1054	APPLICATION OF DIELECTRIC MIXTURE FORMULAS TO (LEAD TITANATE) GLASS- CERAMIC SYSTEMS.
GROZNOV IN	04-0707	RELAXATION AND FERROELECTRIC PROPERTIES OF SOLID SOLUTIONS IN THE SYSTEM BARIUM TITANA
GRUGULI U	34-3435	SOME EXPERIMENTS ON THE REGELATION OF ICE.
GRUNBERG J	17-2513	HIGH TEMPERATURE PHASE TRANSITIONS AND METASTABILITY IN POTASSIUM DIHYDROGEN PHOSPHATE
GRZHEGORZHEVSKII OG	15-2148	FEATURES OF THE PHASE TRANSITION IN ANTIMONY SULFOIODIDE SINGLE CRYSTALS.
GUBANOV AI	03-0268	ANOMALIES IN THE ELECTRICAL CONDUCTIVITY OF A FERROELECTRIC SEMICONDUCTOR NEAR THE CUB
GUBKIN AN	13-1952	NONLINEAR PROPERTIES AND PHASE TRANSITIONS OF STRONTIUM BISMUTH TITANATES.
GUENOK EP	04-0598	EFFECT OF IMPURITIES AND GROWTH CONDITIONS ON THE STRUCTURE OF BARIUM TITANATE CRYSTAL
	04-0708	INVESTIGATION OF THE FINE STRUCTURE OF BARIUM TITANATE CRYSTALS FLUX GROWN FROM POTASS
	04-0709	PECULIARITIES OF POLARIZATION OF BARIUM TITANATE SINGLE CRYSTALS WITH DOUBLE HYSTERESI
GUFAN YUM	03-0269	THEORY OF PHASE TRANSITIONS IN BORACITES.
	14-2070	MAGNETOELECTRIC PROPERTIES OF NICKEL IODINE BORACITE.
GUGGENHEIM HJ	33-3395	IMPURITY CONCENTRATION DEPENDENCE OF RAMAN SCATTERING BY MAGNONS IN NICKEL DOPED RUBID
	33-3406	BARIUM MANGANESE FLUORIDE, A NEW CRYSTAL FOR MICROWAVE ULTRASONICS.
	36-3526	INFRARED OPTICAL PROPERTIES OF VANADIUM DIOXIDE ABOVE AND BELOW TRANSITION TEMPERATURE
GUHA JP	04-0710	PHASE EQUILIBRIA IN THE SYSTEM BARIUM TITANATE- BARIUM GERMANATE.
GUHA S	16-2297	CRYSTAL STRUCTURE OF FERROELECTRIC GLYCINE SILVER NITRATE.
GUILLIEN R	28-3182	DIELECTRIC PHENOMENA IN ROCHELLE SALT.
GUINET P	36-3542	FERROELECTRIC PROPERTIES OF HEXAGONAL ORTHOMANGANITES OF YTTRIUM AND RARE EARTHS. (YTT
GUINIER A	04-0629	LINEAR DISORDER IN CRYSTALS (CASE OF SILICON, QUARTZ AND FERROELECTRIC PEROVSKITES). (
	04-0630	STRUCTURE DISORDER OF BARIUM TITANATE TYPE FERROELECTRICS. BASED ON NEW TREATMENT AND
GULISH OK	27-3053	EFFECT OF HYDROSTATIC PRESSURE ON THE UNIPOLAR PROPERTIES OF TRIGLYCINE SULFATE.
GULLEY JE	33-3376	EXCHANGE NARROWING: MAGNETIC RESONANCE LINE SHAPES AND SPIN CORRELATION IN PARAMAGNETI
GUNTERSDORFER M	04-0729	PIEZORESISTIVITY OF DOPED BARIUM TITANATE. (OR BARIUM STRONTIUM TITANATE, DOPANT NOT N
GUPTA KP	13-1953	MICROCHEMICAL ANALYSIS OF GADOLINIUM DOPED BARIUM SODIUM NIOBATES.
	13-1954	SOLIDIFICATION STUDY OF BARIUM STRONTIUM NIOBATE.
	13-1955	SCANNING ELECTRON MICROSCOPIC EXAMINATION OF SINTERED BARIUM SODIUM NIOBATES.
GUPTA LC	33-3377	SHIFT OF NEEL TEMPERATURE AND EPR LINE WIDTH OF POTASSIUM MANGANESE TRIFLUORIDE WITH M
	18-2732	PURE QUADRUPOLE RESONANCE OF ARSENIC-75 IN AMMONIUM DIHYDROGEN ARSENATE.
GUREVICH VM	01-0014	ELECTRIC CONDUCTIVITY OF FERROELECTRICS. (ELECTRICAL CONDUCTIVITY AND RELATED PROPERTI
GURK P	28-3198	CONTRIBUTION OF DOMAIN WALL MOTION TO THE PERMITTIVITY OF ROCHELLE SALT.
	28-3199	LOW FREQUENCY DISPERSIONS OF THE PERMITTIVITY OF ROCHELLE SALT CAUSED BY DOMAIN WALLS.
GURO GM	03-0511	FIELD EFFECT AT THE CONTACT BETWEEN A SEMICONDUCTOR AND A C-DOMAIN FERROELECTRIC (THEO
	04-1021	DEPENDENCE OF THE PIEZOEFFECT IN BARIUM TITANATE ON THE MEANS OF SCREENING THE SPONTAN
GUSEINOV NG	03-0270	PHENOMENOLOGICAL THEORY OF FERROELECTRIC AND ANTIFERROELECTRIC CRYSTALS.
GUSKOV VP	37-3639	FERROELECTRIC MICROTHERMOSTATS.
GUSTINETTI A	10-1634	OPTICAL TRANSITION TO THE FERMI LEVEL IN SODIUM TUNGSTATE.
GUTIERREZ M	16-2298	THERMAL EXPANSION OF SOME FERROELECTRICS. (POTASSIUM NITRATE, KDP, SODIUM NITRITE)
GUTNER OS	17-2656	LANDAU-KHALATNIKOV ATTENUATION IN POTASSIUM DIDEUTERIUM PHOSPHATE CRYSTALS.
GUYON P	17-2449	ELECTRICAL AND OPTICAL INVESTIGATION OF FERROELECTRIC PROPERTIES OF POTASSIUM DIHYDROG
HAAS MR	05-1055	PERTURBED DIRECTIONAL CORRELATION OF SCANDIUM-44 IN LEAD TITANATE.
HABUDA SP	16-2255	NUCLEAR MAGNETIC RESONANCE STUDIES OF THE PHASE TRANSITIONS IN FERROELECTRICS. (POTASS
	24-2893	ORDERING OF DIPOLE MOMENTS OF WATER MOLECULES AND FERROELECTRIC PHASE TRANSITION IN PO
HADNI A	17-2457	LOW FREQUENCY INFRARED ABSORPTION SPECTRA OF MONOCRYSTALS OF POTASSIUM DIHYDROGEN PHOS
	17-2459	TRANSMISSION OF POTASSIUM DIHYDROGEN PHOSPHATE IN THE INFRARED REGION IN THE PARA- AND

HADNI A 17-2514 ELECTRICAL CONSTANTS OF TRIGLYCINE SULFATE, APPLICATIONS TO PYROELECTRICITY AND THE DE
17-2587 FAR INFRARED ABSORPTION SPECTRA FOR AMMONIUM DIHYDROGEN PHOSPHATE.
25-2908 INFRARED ABSORPTION SPECTRUM OF THIOUREA IN THE PARA- AND FERROELECTRIC PHASES.
27-3007 OPTICAL CONSTANTS OF TRIGLYCINE SELENATE IN THE NEAR INFRARED TO FAR INFRARED AND IN T
27-3008 BLOCKING OF SPONTANEOUS POLARIZATION IN TRIGLYCINE SULFATE.
27-3009 DISPERSION OF THE DIELECTRIC CONSTANT OF TGS IN THE FAR INFRARED (TEMPERATURE AND FREQ
27-3010 IRREVERSIBLE AND SPONTANEOUS POLARIZATION REVERSAL IN TRIGLYCINE SULFATE BY MEANS OF A
27-3011 LASER STUDY OF REVERSIBLE NUCLEATION SITES IN TRIGLYCINE SULFATE AND APPLICATIONS TO P
27-3012 REVERSIBLE DOMAIN REVERSAL IN TRIGLYCINE SULFATE CAUSED BY A LASER BEAM.
28-3210 FAR INFRARED ABSORPTION SPECTRA OF ROCHELLE SALT.
HAERTLING GH 08-1302 MATERIAL AND ELECTROOPTIC PROPERTIES OF THE (LEAD, LANTHANUM) (HAFNATE, TITANATE) SYST
09-1523 A NEW LONGITUDINAL DISPLAY MODE FOR CERAMIC ELECTROOPTIC DEVICES. (PLZT)
09-1524 HOT PRESSED FERROELECTRIC LEAD ZIRCONATE TITANATE CERAMICS FOR ELECTROOPTICAL APPLICAT
09-1525 HOT PRESSED LEAD ZIRCONATE TITANATE CERAMICS CONTAINING BISMUTH.
09-1526 HOT PRESSED (LEAD, LANTHANUM) (ZIRCONATE, TITANATE) FERROELECTRIC CERAMICS FOR ELECTRO
09-1527 HOT PRESSED FERROELECTRIC CERAMICS FOR ELECTROOPTICAL APPLICATIONS. (LEAD ZIRCONATE TI
09-1528 IMPROVED HOT PRESSED ELECTROOPTIC CERAMICS IN THE LEAD LANTHANUM ZIRCONATE TITANATE SY
09-1529 RECENT IMPROVEMENTS IN THE OPTICAL AND ELECTROOPTIC PROPERTIES OF PLZT CERAMICS.
09-1554 OPTICAL PROPERTIES (ELECTRICALLY VARIABLE BIREFRINGENCE) OF (HOT PRESSED) FERROELECTRI
HAFELE HG 15-2164 OPTICAL SECOND HARMONIC GENERATION IN ANTIMONY SULFOIODIDE.
HAGENLOCKER AH 04-0669 PREPARATION AND PROPERTIES OF THIN BARIUM TITANATE FILMS.
HAGENMULLER P 04-0912 COMPARATIVE STUDY OF THE CRYSTALLOGRAPHIC, DIELECTRIC, AND NONLINEAR OPTICAL PROPERTIE
13-1921 EFFECT OF THE SUBSTITUTION OF TANTALUM FOR NIOBIUM ON THE CRYSTALLOGRAPHIC AND DIELECT
13-2011 FIRST FERROELECTRIC OXYFLUORIDE PHASES OF THE OXYGENATED QUADRATIC TUNGSTEN BRONZE TYP
14-2089 NEW BORACITES OF FORMULA LITHIUM(4) BORON(7) OXYGEN(12) CHLORIDE OR BROMIDE (NOT EXAMI
HAGON PJ 11-1720 BISMUTH(12) GERMANIUM OXIDE(20) VERSUS LITHIUM NIOBATE IN THE SURFACE WAVE AMPLIFIER.
HAHN EL 17-2599 NUCLEAR DOUBLE RESONANCE STUDY OF POTASSIUM-39 IN FERROELECTRIC POTASSIUM DIHYDROGEN P
HAHNE E 34-3435 SOME EXPERIMENTS ON THE REGELATION OF ICE.
HAIDA O 34-3436 RELAXATIONAL PROTON ORDERING AND GLASSY CRYSTALLINE STATE IN HEXAGONAL ICE.
HAJDUKOVIC G 16-2267 NMR STUDY OF THE FERROELECTRIC TRANSITIONS IN DIGLYCINE NITRATE AND TRIS-SARCOSINE CAL
HAJICEK P 27-2984 DYNAMIC DETERMINATION OF (ELASTIC AND PIEZOELECTRIC) MATERIAL CONSTANTS OF TRIGLYCINE
HALL CA 13-1956 SOLID SOLUTIONS IN ANTIFERROELECTRIC REGION OF THE SYSTEM LEAD HAFNATE TITANATE, LEAD
HALPERIN A 17-2716 LIGHT AND CURRENT PULSES FROM X-RAYED POTASSIUM DIHYDROGEN PHOSPHATE CRYSTALS.
HALSTEAD TK 11-1722 TEMPERATURE DEPENDENCE OF THE LITHIUM NMR SPECTRUM AND ATOMIC MOTION IN LITHIUM NIOBAT
HALTENORTH H 34-3437 DIFFUSION OF HYDROGEN FLUORIDE IN ICE.
HALVERSON SL 11-1721 RADIATION EXPOSURE OF A LITHIUM NIOBATE CRYSTAL AT HIGH TEMPERATURES.
HAM NS 25-2917 FAR INFRARED TRANSMISSION SPECTRUM OF THIOUREA.
HAMAKAWA Y 15-2166 SWITCHING AND MEMORY EFFECT IN ANTIMONY SULFOIODIDE- TIN DIOXIDE HETEROSTRUCTURE THIN
15-2165 ELECTROSTRICTION, PIEZOELECTRICITY AND ELASTICITY IN FERROELECTRIC ANTIMONY SULFOIODID
HAMANO K 16-2349 DOMAIN ETCH PATTERNS ON THE FACES PARALLEL TO THE FERROELECTRIC-AXIS OF SODIUM NITRATE
HAMAZAKI K 33-3382 OBSERVATION OF THE ANOMALY OF THE THERMAL CONDUCTIVITY OF POTASSIUM MANGANESE TRIFLUOR
HAMILTON WC 34-3438 DEUTERON ARRANGEMENTS IN THE HIGH PRESSURE FORMS OF ICE.
HAMMER H 17-2515 RAMAN SPECTRA OF CRYSTALS OF THE POTASSIUM DIHYDROGEN PHOSPHATE TYPE IN THEIR PARAELEC
HAMMER JM 06-1138 MEASUREMENT OF ELECTRON ENERGY LOSS FUNCTION IN BULK STRONTIUM TITANATE.
HAMMESFAHR A 07-1255 THE INTERNAL ELECTRIC FIELD GRADIENT IN ANTIFERROELECTRIC LEAD ZIRCONATE STUDIED BY PE
HANADATE H 16-2403 STUDY OF PHASE TRANSITION OF SODIUM NITRITE BY SECOND HARMONIC GENERATION. (TEMPERATUR
HANAFI Z 10-1632 THE COLOR PROBLEM OF TUNGSTEN TRIOXIDE. X-RAY STUDIES.
10-1636 THE COLOR PROBLEM OF TUNGSTEN TRIOXIDE ELECTRICAL CONDUCTIVITY.
HANAMURA E 31-3328 EFFECT OF CHARGE TRANSFER ON RAMAN SCATTERING OF HYDROGEN HALIDES.
31-3329 PHASE TRANSITIONS OF HYDROGEN HALIDE CRYSTALS (IMPORTANCE OF HYDROGEN BONDING IN THE F
HANANI M 28-3200 CRYSTALLIZATION OF ROCHELLE SALT IN ELECTRIC FIELDS (KINETICS OF CRYSTAL GROWTH)
HANDEREK J 04-0838 ELECTRET EFFECT IN A MIXTURE OF POLY METHACRYLATE AND BARIUM TITANATE
05-1056 SEEBECK EFFECT IN METAL LEAD TITANATE METAL CONTACT SYSTEMS.
05-1057 THERMOELECTRIC EFFECTS IN THE SYSTEM METAL- LEAD TITANATE- METAL. (METHOD OF STUDYING
HANEBREKKE H 11-1723 ACOUSTOELECTRIC AMPLIFICATION OF SURFACE WAVE STRUCTURE OF A CADMIUM SELENIDE FILM ON
HANG KW 07-1258 ATHERMAL PHASE TRANSITION IN LEAD ZIRCONATE.
36-3609 STRUCTURAL AND DIELECTRIC PROPERTIES OF BARIUM ZINCATE(0.33) NIOBATE(0.67) (NOT FERROE
HANGEA N 17-2411 GROWTH OF AMMONIUM DIHYDROGEN PHOSPHATE SINGLE CRYSTAL FROM SOLUTIONS.
17-2516 GROWING KINETIC OF THE AMMONIUM DIHYDROGEN PHOSPHATE SINGLE CRYSTAL (FROM SOLUTION) IN
HANN BF 36-3558 FLUX GROWTH OF MAGNESIUM OXIDE AND LANTHANUM ALUMINATE CRYSTALS DOPED WITH ISOTOPE 170
HANNON DM 08-1342 EPR OF COBALT(2+) IN POTASSIUM TANTALATE.
08-1343 ELECTRON PARAMAGNETIC RESONANCE OF MANGANESE(2+) IN POTASSIUM TANTALATE.
10-1635 RAMAN SPECTROSCOPY OF TUNGSTEN TRIOXIDE.
HANNON JJ 11-1724 LITHIUM TANTALATE AND LITHIUM NIOBATE PIEZOELECTRIC RESONATORS IN THE MEDIUM FREQUENCY
11-1851 MEDIUM FREQUENCY RESONATORS OF LITHIUM NIOBATE AND LITHIUM TANTALATE.
HANSCOMB JR 16-2288 ELECTRICAL BREAKDOWN IN FERROELECTRIC SODIUM NITRITE CRYSTALS.
HAPP H 16-2309 ACCURATE FORCE CONSTANTS FROM ISOTOPIC SUBSTITUTION IN THE NITRITE RADICAL IN SODIUM N
16-2385 SECOND HARMONIC GENERATION IN FERROELECTRIC SODIUM NITRITE.
16-2386 TEMPERATURE DEPENDENCE OF THE OPTICAL NONLINEAR SUSCEPTIBILITY OF SODIUM NITRITE.
16-2387 TEMPERATURE DEPENDENCE OF ABSORPTION OF SODIUM NITRITE IN THE FAR INFRARED. (350-1100
17-2501 TEMPERATURE DEPENDENCE OF THE DIELECTRIC CONSTANT OF POTASSIUM DIHYDROGEN PHOSPHATE AT
HARA H 04-0945 CATALYTIC OXIDATION OF CARBON MONOXIDE ON SEMICONDUCTIVE BARIUM TITANATE.
04-0947 OXIDATION OF CARBON MONOXIDE ON LANTHANUM OXIDE DOPED BARIUM TITANATE.
14-2082 ELECTROOPTICAL PROPERTIES OF FERROELECTRIC (IRON IODINE) BORACITE. (SPONTANEOUS LATTIC
HARADA J 04-0712 NEUTRON SCATTERING STUDY OF SOFT MODES IN CUBIC BARIUM TITANATE.
04-0713 X-RAY AND NEUTRON DIFFRACTION STUDY OF TETRAGONAL BARIUM TITANATE.
04-0740 X-RAY AND NEUTRON DIFFRACTION STUDY OF TETRAGONAL BARIUM TITANATE.
04-0946 INELASTIC NEUTRON SCATTERING FROM SINGLE DOMAIN BARIUM TITANATE. (LATTICE DYNAMICS)
05-1077 SOFT FERROELECTRIC MODES IN LEAD TITANATE (NEUTRON INELASTIC SCATTERING).
06-1139 DETERMINATION OF THE NORMAL VIBRATIONAL DISPLACEMENTS OF SEVERAL PEROVSKITES (POTASSIU
HARADOME M 10-1647 EVAPORATED FILMS OF TUNGSTEN OXIDE.
HARAGUCHI T 13-1962 COMPLEX OXIDES WITH TUNGSTEN BRONZE TYPE STRUCTURES (POTASSIUM (BARIUM OR STRONTIUM) T
13-1964 SOME COMPOUNDS OF TUNGSTEN BRONZE TYPE A(6)B(10)O(30). (WHERE B IS (NIOBIUM, TITANIUM)
HARBEKE G 15-2167 SOFT PHONON MODE AND MODE COUPLING IN ANTIMONY SULFOIODIDE (BY RAMAN SCATTERING).
15-2231 CRITICAL PHENOMENA IN ANTIMONY SULFOIODIDE.
15-2232 SOFT MODE COUPLING AND CRITICAL RAYLEIGH SCATTERING IN FERROELECTRIC ANTIMONY SULFOIOD
15-2233 SOFT LATTICE VIBRATIONS AND VIBRATION COUPLING IN ANTIMONY SULFOIODIDE.
HARBRINK B 03-0271 EFFECT OF THE SURFACE ON POLARIZATION OF A DIELECTRIC (THEORY).
HARDTL KH 04-0689 MOBILITY DETERMINATIONS FROM WEIGHT MEASUREMENTS IN SOLID SOLUTIONS OF (BARIUM, STRONT
04-0711 LOWERING THE CURIE TEMPERATURE IN REDUCED BARIUM TITANATE.

PAGE 642

HARDTL KH   05-1058   DISTRIBUTION OF VACANCIES IN LANTHANA DOPED LEAD TITANATE.
                         09-1499   ORIGIN OF THE MAXIMUM IN THE ELECTROMECHANICAL ACTIVITY IN LEAD ZIRCONATE(X) TITANATE(
                         09-1500   ORIGIN OF THE MAXIMUM IN THE ELECTROMECHANICAL ACTIVITY IN LEAD ZIRCONATE(X) TITANATE(
                         09-1501   POLARIZATION, ELECTROSTRICTION, AND PIEZOELECTRIC ACTIVITY OF LEAD ZIRCONATE(X) TITANA
                         09-1502   STRUCTURE AND ELECTROMECHANICAL PROPERTIES OF LANTHANUM DOPED LEAD TITANATE ZIRCONATE
                         09-1530   DISTRIBUTION OF A-SITE AND B-SITE VACANCIES IN (LEAD, LANTHANUM) (TITANATE, ZIRCONATE)
                         09-1531   STRUCTURE AND ELECTROMECHANICAL PROPERTIES OF LANTHANUM DOPED LEAD TITANATE ZIRCONATE
HARDWIG CM   08-1344   RAMAN SCATTERING FROM FERROELECTRIC SODIUM NITRITE.
HARDY JR   30-3260   FERROELECTRIC DOMAINS AND THE CONDENSED SOFT MODE IN GADOLINIUM MOLYBDATE.
                         30-3281   LATTICE DYNAMICS OF A RIGID ION MODEL OF GADOLINIUM MOLYBDATE.
                         30-3282   PRODUCTION OF SPONTANEOUS POLARIZATION BY ELASTIC INSTABILITIES IN PIEZOELECTRIC MATER
                         30-3285   TEMPERATURE DEPENDENCE OF THE RAMAN SPECTRUM OF GADOLINIUM MOLYBDATE.
                         30-3323   PYROELECTRIC EFFECT IN GADOLINIUM MOLYBDATE.
HARIDASAN TM   16-2334   PHONON DISPERSION IN FERROELECTRIC POTASSIUM NITRATE.
HARNIK E   17-2660   ANOMALOUS PROPAGATION OF LONGITUDINAL ULTRASONIC WAVES IN POTASSIUM DIDEUTERIUM PHOSPH
                         17-2664   ULTRASONIC MEASUREMENT OF THE ELECTROCALORIC EFFECT IN KDP (NEAR THE CURIE TEMPERATURE
HARRIS IR   13-1894   DILATOMETRIC STUDY OF THE ORTHORHOMBIC- TETRAGONAL PHASE TRANSITION IN BARIUM SODIUM N
                         13-1904   BARIUM SODIUM NIOBATE: CONSTITUTIONAL STUDIES AND (CZOCHRALSKI) CRYSTAL GROWTH.
HARRIS LB   17-2518   POCKELS EFFECT IN AMMONIUM DIHYDROGEN PHOSPHATE ABOVE ROOM TEMPERATURE.
HARRISON WB   09-1532   USE OF FERROELECTRIC CERAMICS IN COMPUTERS AND OTHER DEVICES. (LEAD ZIRCONATE- LEAD TI
HARSELL WB   09-1592   THE ROLE OF IRON OXIDE ADULTERANT IN A LEAD ZIRCONATE LEAD TITANATE CERAMICS.
HARTING CM   29-3243   EFFECT OF TEMPERATURE AND PROPAGATION DIRECTION UPON OPTICAL PHONONS IN CUBIC SODIUM C
HARTLEY NP   17-2517   NEW METHOD OF MEASURING PYROELECTRIC COEFFICIENTS. (TGS EXAMPLE)
HARTMAN P   10-1653   SOME ASPECTS OF THE STRUCTURE OF TUNGSTEN TRIOXIDE AND A CONTRIBUTION TO THE UNDERSTAN
HARTMANN E   27-3013   CHANGES OF THE DOMAIN STRUCTURE IN VANADIUM DOPED TRIGLYCINE SULFATE OBSERVED BY EPR.
HARTWICK TS   13-1979   OPTICAL DAMAGE IN KTN.
HARTWIG CM   16-2299   ANALYSIS OF THE TEMPERATURE DEPENDENT PHONON STRUCTURE IN SODIUM NITRITE BY RAMAN SPEC
HARUMI F   09-1624   STUDIES ON THE SINTERING BETWEEN LEAD MONOXIDE AND PZT. PART-1. PART-2: EFFECTS OF AVE
HASEGAWA T   04-0714   POLARIZATION REFLECTION TYPE LIGHT MODULATOR USING FERROELECTRICS. (BARIUM TITANATE)
HASHI T   17-2688   MEASUREMENT OF CHEMICAL SHIFT IN SOLID POTASSIUM DIHYDROGEN PHOSPHATE.
HASHKOZEV ZM   03-0471   LIGHT SCATTERING BY ACOUSTIC PHONONS IN FERROELECTRIC AND PIEZOELECTRIC CRYSTALS.
HASS M   11-1771   SPONTANEOUS PARAMETRIC SCATTERING, IDLER ABSORPTION, AND DISPERSION IN THE NONLINEAR S
HASTINGS RN   32-3345   NITROGEN-14 NUCLEAR QUADRUPOLE RESONANCE STUDY OF LITHIUM HYDRAZINIUM SULFATE.
HATTA I   03-0272   DIELECTRIC RELAXATION PROCESSES NEAR THE CRITICAL POINT IN FERROELECTRICS.
                         03-0521   DISTRIBUTION OF RELAXATION TIME IN FERROELECTRICS NEAR THE CURIE TEMPERATURE (POLY VER
                         04-0715   SPECIFIC HEAT OF BARIUM TITANATE.
                         16-2261   QUASI STATIC SWITCHING CURRENT IN SODIUM NITRITE.
                         16-2300   DIPOLE LATTICE RELAXATION TIME NEAR THE NEEL TEMPERATURE IN SODIUM NITRITE.
                         16-2301   FERROELECTRIC POLARIZATION SWITCHING IN SODIUM NITRITE.
                         16-2302   STATIC ELECTRIC SUSCEPTIBILITY AND DIELECTRIC RELAXATION TIME NEAR THE TRANSITION POIN
                         16-2303   SPECIFIC HEAT OF SODIUM NITRITE NEAR THE ANTIFERROELECTRIC TRANSITION POINT.
                         16-2304   ULTRASONIC ATTENUATION NEAR THE TWO TRANSITION POINTS IN SODIUM NITRITE.
HAURET G   16-2305   BRILLOUIN SCATTERING IN SODIUM NITRITE AT ROOM TEMPERATURE.
                         16-2306   MEASUREMENT OF THE ELASTIC CONSTANTS OF SODIUM NITRITE BY BRILLOUIN DIFFUSION AT ORDIN
                         17-2519   BRILLOUIN SCATTERING BY A SINGLE CRYSTAL OF RUBIDIUM DIHYDROGEN PHOSPHATE.
                         17-2520   BRILLOUIN DIFFUSION IN A SINGLE CRYSTAL OF RUBIDIUM DIHYDROGEN PHOSPHATE.
                         17-2521   INTENSITY OF SOME BRILLOUIN AND RAYLEIGH LINES IN RUBIDIUM DIHYDROGEN PHOSPHATE.
                         17-2522   STUDY OF THE BRILLOUIN EFFECT IN ADP (ALL ELASTIC CONSTANTS).
                         19-2766   BRILLOUIN EFFECT AT ORDINARY TEMPERATURE IN AMMONIUM SULFATE.
HAUSSUEHL S   08-1405   LARGE NONLINEAR OPTICAL COEFFICIENT AND PHASE MATCHED SECOND HARMONIC GENERATION IN LI
HAVINGA EE   03-0188   TEMPERATURE DEPENDENCE OF DIELECTRIC CONSTANTS OF CUBIC IONIC COMPOUNDS. (HALIDES AND
                         04-0716   SHELL MODEL OF INTERIONIC POTENTIALS FOR BARIUM TITANATE.
HAVRANKOVA M   17-2455   GROWTH OF POTASSIUM DIHYDROGEN PHOSPHATE SINGLE CRYSTALS IN GEL.
HAYAKAWA S   04-0734   A SEMICONDUCTING BARIUM TITANATE CERAMIC DOPED WITH NIOBIUM AND MANGANESE IONS AND ITS
                         04-0747   ELECTRON SPIN RESONANCE OF MANGANESE(2+) IN BARIUM TITANATE.
                         04-0872   CHEMICAL CHANGES IN BARIUM TITANATE FIRED UNDER NITROGEN IN THE PRESENCE OF CARBON.
                         04-0931   A NEW PTC THERMISTOR MATERIAL. (BARIUM TITANATE)
                         08-1393   HIGH PRESSURE SYNTHESIS (SOLID STATE REACTION, SINTERING) AND PROPERTIES OF PEROVSKITE
HAYAMA M   26-2950   PROTON RESONANCE STUDY OF SUBLATTICE ROTATIONS AND SPIN DEVIATION IN COPPER FORMATE TE
HAYASHI H   26-2932   THERMAL CONDUCTION IN A TWO DIMENSIONAL ANTIFERROMAGNETIC COPPER FORMATE TETRAHYDRATE.
                         26-2939   MEASUREMENTS OF THE THERMAL CONDUCTIONS IN THE LOW DIMENSIONAL ANTIFERROMAGNETS POTASS
                         33-3382   OBSERVATION OF THE ANOMALY OF THE THERMAL CONDUCTIVITY OF POTASSIUM MANGANESE TRIFLUOR
HAYASHI J   04-0717   ATTEMPT TO PREPARE HIGH DOPED BARIUM STRONTIUM TITANATE SEMICONDUCTOR WITH PZT.
HAYASHI M   03-0273   DOMAIN WALL MOTION IN FERROELECTRIC SWITCHING.
                         03-0274   KINETICS OF DOMAIN WALL MOTION IN FERROELECTRIC SWITCHING. PART-1: GENERAL FORMULATION
                         04-0718   EFFECT OF PRESSURE (TC ABOUT 2KBAR) ON POLARIZATION REVERSAL IN BARIUM TITANATE.
                         27-3014   EFFECT OF PRESSURE ON THE SWITCHING RATE OF TRIGLYCINE SULFATE.
                         27-3015   TEMPERATURE DEPENDENCE OF THE SWITCHING RATE OF TRIGLYCINE SULFATE.
                         27-3082   EFFECT OF A TWO DIMENSIONAL PRESSURE ON THE CURIE POINTS OF TRIGLYCINE SULFATE AND ROC
HAYASHI T   05-1087   LEAD TITANATE PYROELECTRIC INFRARED DETECTOR.
HEAPS JD   09-1558   PYROELECTRIC PROPERTIES OF THE LANTHANUM DOPED FERROELECTRIC PLZT CERAMICS.
HEASTIE R   31-3330   RAMAN SPECTRA OF ORTHORHOMBIC AND CUBIC HYDROGEN AND DEUTERIUM CHLORIDES AT TEMPERATUR
HEBENGARTH E   04-0719   POLARIZATION OF SOLID BARIUM(X) STRONTIUM(1-X) TITANATE COMPOUNDS BETWEEN 4 AND 100 DE
HEBER J   36-3565   SPECTRUM AND ENERGY LEVELS OF CHROMIUM(3+) IONS AND EXCHANGE COUPLED CHROMIUM(3+) PAIR
HEGDE MS   05-1040   MOSSBAUER EFFECT FOR IRON-57 IN FERROELECTRIC LEAD TITANATE.
HEGENEARTH E   04-0720   DIELECTRIC AND THERMAL (SPECIFIC HEAT, ELECTROCALORIC EFFECT) STUDIES OF FERROELECTRIC
                         04-0721   INVESTIGATION OF THE THERMAL CONDUCTIVITY IN CONNECTION WITH OTHER PHYSICAL PROPERTIES
                         04-0722   SHIFTING OF THE TRANSITION TEMPERATURE OF BARIUM(0-0.1) STRONTIUM(0.9-1) TITANATE SOLI
                         04-0898   INFLUENCE OF PRESSURE ON PHASE TRANSITIONS AT LOW TEMPERATURES: STRONTIUM TITANATE AND
                         06-1206   ELECTROMECHANICAL PROPERTIES OF STRONTIUM TITANATE SINGLE CRYSTALS.
                         06-1218   MECHANICAL RELAXATION AND NONLINEARITY IN STRONTIUM TITANATE SINGLE CRYSTALS.
                         06-1219   PRESSURE DEPENDENCE OF THE ELASTIC COMPLIANCE OF STRONTIUM TITANATE NEAR THE TRANSITIO
                         06-1220   TEMPERATURE DEPENDENCE OF THE ELASTIC COMPLIANCE OF STRONTIUM TITANATE SINGLE CRYSTALS
                         14-2078   FERROELECTRIC TRANSITION IN COBALT- IODINE- BORACITE UNDER PRESSURE.
                         17-2495   INFLUENCE OF HYDROSTATIC PRESSURE ON THE PHASE TRANSITION TEMPERATURE OF FERROELECTRIC
                         17-2523   DISPLACEMENT OF THE CURIE POINTS IN KDP AND LITHIUM THALLIUM TARTRATE CRYSTALS BY HYDR
                         17-2619   MEASUREMENT OF THE SPECIFIC HEAT OF KDP BY A DYNAMIC METHOD.
                         19-2757   THE BEHAVIOR OF DIAMMONIUM DICADMIUM TRISULFATE UNDER PRESSURE.
HEIMAN N   04-0723   MOSSBAUER EFFECT OF COBALT-57 IN BARIUM TITANATE.
HELG U   08-1345   FERROELECTRICITY OF POTASSIUM IODATE. (UNUSUAL FERROELECTRIC BEHAVIOR)

HELG U	08-1346	POTASSIUM IODATE, A FERROELECTRIC WITH NONREVERSIBLE, BUT TILTABLE (TO 3 STABLE POSITI
	08-1347	PECULIAR FERROELECTRIC BEHAVIOR OF POTASSIUM IODATE.
	08-1348	POTASSIUM IODATE - A DISORDERED FERROELECTRIC. (DEPENDENCE OF DIELECTRIC CONSTANT ON T
HELLWEGE KH	36-3565	SPECTRUM AND ENERGY LEVELS OF CHROMIUM(3+) IONS AND EXCHANGE COUPLED CHROMIUM(3+) PAIR
HELMREICH D	34-3439	ELASTIC ANOMALIES OF ICE AT LOW TEMPERATURES.
	34-3440	MOLECULAR FORCES OF HEAVY AND LIGHT ICE.
	34-3473	INFLUENCE OF IMPURITIES ON THE ORDER DISORDER REACTION IN HEXAGONAL ICE I(H).
HELWIG J	27-3016	THERMAL CONDUCTIVITY OF TRIGLYCINE SULFATE NEAR THE CURIE POINT.
	28-3226	THERMAL CONDUCTIVITY CF ROCHELLE SALT.
HENNINGS D	05-1058	DISTRIBUTION OF VACANCIES IN LANTHANA DOPED LEAD TITANATE.
	05-1059	X-RAY STRUCTURE INVESTIGATION OF LANTHANUM MODIFIED LEAD TITANATE WITH A-SITE AND B-SI
	08-1349	RANGE OF EXISTENCE OF PEROVSKITE PHASES IN THE SYSTEM LEAD OXIDE- TITANIUM DIOXIDE- LA
	09-1530	DISTRIBUTION OF A-SITE AND B-SITE VACANCIES IN (LEAD, LANTHANUM) (TITANATE, ZIRCONATE)
HEPNER G	17-2509	(FINITE APERTURE LIGHT SWITCH USING) TRANSVERSE POCKELS EFFECT COMMUTATORS. (POTASSIUM
	17-2524	STUDY OF THE LIMITATION OF THE ANGULAR FIELD BY THE COMMUTATOR OF POLARIZATION IN KDP,
HERAK R	22-2866	A NEUTRON DIFFRACTION STUDY OF POTASSIUM TRIHYDROGEN SELENITE.
HERBERT JM	02-0073	SOME RECENT ADVANCES IN FERROELECTRICS. (ELECTROOPTIC MATERIALS, SINGLE CRYSTALS OF PO
	04-0724	AGING IN COMPLEX CERAMICS BASED ON BARIUM TITANATE. (CAPACITANCE, PERMITTIVITY)
HERCZOG A	04-0725	ELECTRICAL PROPERTIES OF MICROCRYSTALLINE FERROELECTRIC MATERIALS CRYSTALLIZED FROM GL
HERLACH F	08-1350	NUCLEAR QUADRUPOLE RESONANCE, PHASE TRANSFORMATIONS, AND FERROELECTRICITY OF ALKALI IO
HERMANN F	06-1122	NONCOLLINEAR PARAMETRIC 4 PHOTON RECIPROCAL EFFECT IN CADMIUM SULFIDE AND STRONTIUM TI
HERMELBRACHT K	27-3017	OPTICAL ACTIVITY OF SOME FERROELECTRIC CRYSTALS (PYROELECTRICITY, TGS, ROCHELLE SALT).
HERMOSIN A	09-1575	INSCRIPTION OF HOLOGRAPHIC RESULTS IN TRANSPARENT CERAMIC FERROELECTRICS. (LEAD LANTHA
HERRICK V	30-3274	ELASTIC CONSTANTS OF GADOLINIUM MOLYBDATE (NEAR THE CURIE POINT).
HERRINGTON JB	11-1725	AN EPR INVESTIGATION OF IRON(3+) AND MANGANESE(2+) IN LITHIUM NIOBATE.
HERVOUET C	03-0275	ACOUSTIC ATTENUATION AND AMPLIFICATION IN PIEZOELECTRIC AND FERROELECTRIC SEMICONDUCTO
HERZOG P	04-0937	STATIC ELECTRIC QUADRUPOLE INTERACTION OF TANTALUM AND HAFNIUM IONS IN BARIUM AND LEAD
HESTER DL	09-1520	EFFECTS OF RADIATION INDUCED DAMAGE CENTERS IN LEAD ZIRCONATE TITANATE CERAMICS.
HETZLER U	27-3018	TEMPERATURE DEPENDENCE OF THE FERROELECTRIC FIELD EFFECT IN TELLURIUM FILMS ON TRIGLYC
HEWAT AW	04-0726	VIBRATIONAL AMPLITUDES IN FERROELECTRIC TETRAGONAL BARIUM TITANATE.
	08-1351	LOW FREQUENCY ZONE BOUNDARY MODES IN PEROVSKITE FERROELECTRICS INDICATED BY ANISOTROPI
	13-1957	SOFT MODES AND STRUCTURE OF FERROELECTRIC TETRAGONAL POTASSIUM TANTALATE NIOBATE.
	13-2068	POTASSIUM TANTALATE NIOBATE, STRUCTURAL AND DYNAMICAL STUDIES.
HEYMAN PM	06-1140	PHOTOCHROMIC MATERIALS RESEARCH. (STRONTIUM TITANATE PHOTOCONDUCTIVITY, CALCIUM FLUORI
HEYNS AM	19-2783	LOW TEMPERATURE STUDIES. PART-4: PHASE TRANSITIONS OF AMMONIUM SULFATE: THE NATURE OF
HEYWANG W	02-0074	DOPING PROBLEMS IN (TITANATE) FERROELECTRIC SEMICONDUCTORS. (ORIGIN AND EFFECTS OF GRA
	04-0727	CURIE SHIFT IN BARIUM TITANATE BY HETEROSUBSTITUTION.
	04-0728	CORRELATION BETWEEN STRUCTURE AND ELECTRIC PROPERTIES OF FERROELECTRIC SUBSTANCES. (BA
	04-0729	PIEZORESISTIVITY OF DOPED BARIUM TITANATE. (OR BARIUM STRONTIUM TITANATE, DOPANT NOT N
	04-0730	SEMICONDUCTING BARIUM TITANATE. (DIELECTRIC CONSTANT, CONTACT RESISTANCE)
HIDAKA T	17-2525	ELECTRON DELOCALIZATION WEIGHT OF THE HYDROGEN BOND IN POTASSIUM DIHYDROGEN PHOSPHATE.
HIEN PZ	03-0276	ABRUPT (STEPWISE) CHANGE IN PROBABILITY OF THE MOSSBAUER EFFECT AT THE PHASE TRANSITIO
	08-1400	CHANGE OF RESONANCE ABSORPTION SPECTRA OF 23.8 KEV GAMMA-RAYS OF TIN-119 DURING PHASE
	36-3566	RESONANCE ABSORPTION OF GAMMA QUANTA IN BARIUM, STRONTIUM, AND CALCIUM STANNATE.
HIGASHI A	34-3426	X-RAY DIFFRACTION TOPOGRAPHIC STUDIES OF THE DEFORMATION BEHAVIOR OF ICE SINGLE CRYSTA
	34-3441	MECHANICAL PROPERTIES OF ICE SINGLE CRYSTALS.
	34-3442	PHASE TRANSITION IN ICE.
HIGASHI N	09-1533	FLUX GROWN KTN SINGLE CRYSTALS AND THEIR ELECTRICAL PROPERTIES.
HIKITA T	16-2307	SPLITTING OF THE SODIUM-23 NMR CENTRAL LINE IN FERROELECTRIC SILVER SODIUM DINITRITE.
	16-2324	NMR STUDY OF SODIUM-23 IN SILVER SODIUM DINITRITE.
HILBERG RP	11-1727	TRANSIENT ELASTOOPTIC EFFECTS AND Q SWITCHING PERFORMANCE IN LITHIUM NIOBATE AND KDP P
	17-2529	LOSSLESS POTASSIUM DIDEUTERIUM PHOSPHATE POCKELS CELL FOR HIGH POWER Q SWITCHING.
HILCZER B	04-0731	CHANGES IN THE FERROELECTRIC PROPERTIES OF BARIUM TITANATE CERAMICS CAUSED BY IRRADIAT
	04-0732	EFFECT OF NEUTRON IRRADIATION ON THE FERROELECTRIC PROPERTIES OF BARIUM TITANATE CERAM
	07-1259	EFFECT OF PILE NEUTRON IRRADIATION ON THE DIELECTRIC PROPERTIES OF LEAD ZIRCONATE.
	21-2813	DEFECTS AND THE FERROELECTRIC PROPERTIES OF GUANIDINIUM ALUMINUM SULFATE HEXAHYDRATE.
	27-3019	EFFECT OF X- AND GAMMA-RADIATION ON THE SWITCHING PROCESS IN TRIGLYCINE SULFATE.
	27-3020	SWITCHING IN FERROELECTRIC TRIGLYCINE SULFATE, PURE AND DOPED WITH PARAMAGNETIC IONS.
HILL AE	27-2958	OPTICAL MIXING. (OF PULSED RUBY LASER EMISSIONS BY TRIGLYCINE SULFATE CRYSTAL)
HILL BH	08-1300	ELECTRON BEAM WRITING OF FERROELECTRIC DOMAINS IN BISMUTH TITANIUM OXIDE SINGLE CRYSTA
HILL GJ	02-0075	ENERGY STORAGE IN FERROELECTRIC CERAMICS: SURVEY OF MATERIALS BASED UPON BARIUM TITANA
	04-0733	ENERGY STORAGE IN HIGH PERMITTIVITY CERAMICS: SURVEY OF MATERIALS BASED ON BARIUM TITA
HILL JC	16-2308	FAR INFRARED SPECTRA AND FERROELECTRIC PHASE TRANSITION OF POTASSIUM NITRATE.
HILL OF	13-1913	CZOCHRALSKI GROWTH OF BARIUM STRONTIUM NIOBATE CRYSTALS.
	13-1914	MODIFICATION TO THE CZOCHRALSKI METHOD OF CRYSTAL PULLING. (CONTROL OF CRYSTAL DIAMETE
HILL RM	17-2526	HIGH FREQUENCY BEHAVIOR OF HYDROGEN BONDED FERROELECTRICS: TRIGLYCINE SULFATE AND POTA
HILTON RM	11-1726	APPARATUS FOR GROWTH OF SINGLE CRYSTAL, SINGLE DOMAIN LITHIUM NIOBATE.
	35-3496	CZOCHRALSKI SYNTHESIS AND PROPERTIES OF RARE EARTH DOPED BISMUTH GERMANATE.
HINAZUMI H	02-0077	ACCURATE STRUCTURE ANALYSIS OF FERROELECTRICS IN CONNECTION WITH THEIR PHASE TRANSITIO
	29-3244	X-RAY REFINEMENT OF THE CRYSTAL STRUCTURE OF LITHIUM AMMONIUM TARTRATE MONOHYDRATE.
HINENO M	17-2711	AN OBSERVATION OF THE CHANGE IN COOLED KDP TEMPERATURE DUE TO 337 MICRON INCIDENT RADI
HINTERMANN A	03-0240	THERMODYNAMICS OF AN ELASTIC FERROELECTRIC MODEL.
HIPOLOTO O	34-3443	A STUDY OF BROUT'S MODEL FOR FERROELECTRICS. INVESTIGATIONS ON HEXAGONAL ICE.
HIRABAYASHI H	24-2894	NUCLEAR MAGNETIC RESONANCE AND X-RAY STUDIES OF POTASSIUM FERROCYANIDE TRIHYDRATE CRYS
HIRAKAWA K	26-2930	MAGNETISM AND THE PHASE TRANSITION OF COPPER FORMATE TETRAHYDRATE.
	26-2932	THERMAL CONDUCTION IN A TWO DIMENSIONAL ANTIFERROMAGNETIC COPPER FORMATE TETRAHYDRATE.
	26-2939	MEASUREMENTS OF THE THERMAL CONDUCTIONS IN THE LOW DIMENSIONAL ANTIFERROMAGNETS POTASS
	28-3201	HIGHLY SENSITIVE, SIMPLE CALORIMETER FOR PHASE CHANGE DETECTION (INCLUDING FERROELECTR
	33-3373	ULTRASONIC ATTENUATION NEAR THE SOFT MODE TRANSITION POINT IN POTASSIUM MANGANESE TRIF
	33-3382	OBSERVATION OF THE ANOMALY OF THE THERMAL CONDUCTIVITY OF POTASSIUM MANGANESE TRIFLUOR
HIRANO H	11-1711	GROWTH AND PROPERTIES OF FERROELECTRIC POTASSIUM LITHIUM TANTALATE(X) NIOBATE(1-X).
	11-1728	MELT COMPOSITION (STOICHIOMETRY) DEPENDENCE OF POCKELS EFFECT IN LITHIUM NIOBATE CRYST
	13-1958	NEW (UNIAXIAL) NONLINEAR OPTIC CRYSTAL BARIUM LITHIUM NIOBATE HAVING NO MICROTWINNING
HIRANO M	33-3381	LUMINESCENCE OF EUROPIUM(3+) ION IN ANTIFERROMAGNETIC POTASSIUM MANGANESE TRIFLUORIDE.
HIROSE N	04-0734	A SEMICONDUCTING BARIUM TITANATE CERAMIC DOPED WITH NIOBIUM AND MANGANESE IONS AND ITS
	04-0735	EFFECT OF GRAIN SIZE ON THE RESISTIVITY ANOMALY IN SEMICONDUCTIVE BARIUM TITANATE CERA
HIROSE T	10-1637	ELECTRICAL CONDUCTIVITY OF TUNGSTEN TRIOXIDE.
HIROTSU S	33-3378	EXPERIMENTAL STUDIES OF STRUCTURAL PHASE TRANSITIONS IN CESIUM LEAD TRICHLORIDE.
	33-3379	FAR INFRARED REFLECTIVITY SPECTRA OF CESIUM LEAD TRICHLORIDE.
	33-3380	(MECHANISM OF) STRUCTURAL PHASE TRANSITIONS IN CESIUM LEAD TRICHLORIDE.

HISANO K	35-3497	QUASI ELASTIC SCATTERING IN LEAD GERMANATE.
	35-3498	RAMAN SCATTERING FROM THE FERROELECTRIC MODE IN LEAD(5) GERMANIUM(3) OXIDE(11).
HITTERMAN RL	24-2903	NEUTRON DIFFRACTION STUDY OF FERROELECTRIC POTASSIUM FERROCYANIDE TRIDEUTERO HYDRATE A
HLASIVCOVA N	04-0736	DETERMINATION OF THE DEVIATION FROM OXYGEN STOICHIOMETRY OF BARIUM TITANATE SINGLE CRY
	04-0873	DEFECT STRUCTURE OF (BARIUM(1-X) STRONTIUM(X)) TITANATE- MIXED CRYSTALS FROM A POTASSI
HOAHINO S	31-3334	CRYSTAL STRUCTURE AND PHASE TRANSITION OF HYDROGEN CHLORIDE.
HOBBS PV	34-3444	PLANAR GROWTH OF ICE FROM THE PURE MELT.
HOCHLI UT	30-3283	ELASTIC PROPERTIES OF GADOLINIUM MOLYBDATE AT ITS FERROELECTRIC PHASE TRANSITION.
	30-3284	ELASTIC CONSTANTS AND SOFT OPTICAL MODES IN GADOLINIUM MOLYBDATE.
HODGKINS CE	04-0696	DIELECTRIC PROPERTIES OF BARIUM TITANATE DOPED WITH TUNGSTEN OXIDE.
HOENEISEN B	06-1173	ANOMALOUS RESONANCE OF STRONTIUM TITANATE.
HOEVE CAJ	34-3429	CALCULATION OF THE DIELECTRIC CORRELATION FACTOR OF CUBIC ICE.
HOFACKER GL	03-0244	THEORY OF THE MOBILITY OF STRUCTURAL DEFECTS IN ICE.
HOFMANN R	04-0737	OPTICAL EVIDENCE FOR POLARIZATION FLUCTUATIONS IN PARAELECTRIC BARIUM TITANATE.
	04-0738	TEMPERATURE DEPENDENCE OF ELECTRONIC POLARIZABILITY OF STRONTIUM AND BARIUM TITANATE.
HOGAN EM	30-3256	CONTROL AND APPLICATION OF DOMAIN WALL MOTION IN GADOLINIUM MOLYBDATE.
	30-3257	CONTROL AND APPLICATION OF DOMAIN WALL MOTION IN GADOLINIUM MOLYBDATE.
HOLAH GD	16-2309	ACCURATE FORCE CONSTANTS FROM ISOTOPIC SUBSTITUTION IN THE NITRITE RADICAL IN SODIUM N
	16-2310	TWO PHONON ABSORPTION IN FERROELECTRIC SODIUM NITRITE.
	16-2311	TEMPERATURE DEPENDENCE OF THE FORCE CONSTANTS OF ION IN SODIUM NITRITE.
HOLAKOVSKY J	17-2527	INFLUENCE OF TUNNELING ON DIELECTRIC BEHAVIOR OF POTASSIUM DIHYDROGEN PHOSPHATE.
	17-2528	STUDY OF PHASE TRANSITION IN MIXED KDP- DEUTERATED KDP CRYSTALS BY MODIFIED MOLECULAR
HOLDEN BJ	30-3285	TEMPERATURE DEPENDENCE OF THE RAMAN SPECTRUM OF GADOLINIUM MOLYBDATE.
HOLDEN TM	33-3407	MAGNETIC EXCITATIONS IN NICKEL DOPED POTASSIUM MANGANESE TRIFLUORIDE.
HOLLAND MG	06-1095	THERMAL CONDUCTIVITY IN PEROVSKITES (3 PHONON INTERACTION PROPOSED FOR STRONTIUM TITAN
	11-1812	TEMPERATURE DEPENDENCE OF RESONANT FREQUENCIES OF LITHIUM NIOBATE PLATE RESONATORS. (C
	12-1886	TEMPERATURE DEPENDENCE OF SURFACE ACOUSTIC WAVE VELOCITY IN LITHIUM TANTALATE.
HOLLAND R	03-0277	COMPILATION OF CONVERSION FORMULAS INTERRELATING THE LINEAR ELASTIC AND ELECTROELASTIC
	03-0278	PIEZOELECTRIC EFFECTS IN FERROELECTRIC CERAMICS (AND THE THEORY OF PIEZOELECTRIC ENERG
	09-1537	MEASUREMENT OF PIEZOELECTRIC PHASE ANGLES IN A FERROELECTRIC CERAMIC (THEORY, APPLIED
	09-1551	ELECTROOPTIC EFFECTS IN FERROELECTRIC CERAMICS. (LEAD ZIRCONATE TITANATE)
HOLMAN RL	09-1480	SUBSTITUTION OF BISMUTH AND NIOBIUM IONS IN LEAD ZIRCONATE(0.53) TITANATE(0.47) (EXPER
	09-1534	INTRINSIC NONSTOICHIOMETRY IN SINGLE PHASE LEAD ZIRCONATE TITANATE.
	09-1535	INTRINSIC NONSTOICHIOMETRY IN THE LEAD ZIRCONATE TITANATE SYSTEM.
	09-1536	INTRINSIC AND EXTRINSIC NONSTOICHIOMETRY IN THE LEAD ZIRCONATE TITANATE SYSTEM. (DEFEC
	09-1538	SOLUTION KINETICS OF LUTETIUM AND SCANDIUM IONS IN LEAD TITANATE(0.5) ZIRCONATE(0.5).
HOLZAPFEL WB	34-3445	ON THE SYMMETRY OF THE HYDROGEN BONDS IN ICE-VII.
HOLZRICHTER JF	06-1112	DIRECT OBSERVATION OF SINGLE DOMAIN STRONTIUM TITANATE (OPTICAL STUDY UNDER UNIAXIAL S
	06-1115	POLARIZED FLUORESCENCE STUDY OF CHROMIUM(3+) THROUGH A STRESS INDUCED PHASE TRANSITION
HONE D	33-3376	EXCHANGE NARROWING: MAGNETIC RESONANCE LINE SHAPES AND SPIN CORRELATION IN PARAMAGNETI
HONEYMAN WN	27-3021	(OPTICAL) ENERGY GAP (AND DIELECTRIC CONSTANT) TEMPERATURE CHARACTERISTICS OF FERROELE
	27-3022	PROPERTIES OF MIXED CRYSTALS OF TRIGLYCINE SULFATE AND SELENATE.
HONIG JM	06-1100	BAND STRUCTURE OF STRONTIUM TITANATE AS MEASURED BY X-RAY PHOTOELECTRON SPECTROSCOPY (
HONJO G	04-0993	ELECTRON MICROSCOPIC STUDIES ON FERROELECTRIC DOMAINS OF PEROVSKITE TYPE OXIDES. (LEAD
	04-1034	ELECTRON MICROSCOPIC STUDY OF FERROELECTRIC DOMAINS OF BARIUM TITANATE IN LOW TEMPERAT
	08-1357	ANISOTROPIC DIFFUSE X-RAY SCATTERING AND PHASE TRANSITIONS IN SODIUM NIOBATE.
	08-1358	DIFFUSE X-RAY STUDY OF 640 DEGREES C PHASE TRANSITION IN SODIUM NIOBATE.
	13-1965	X-RAY STUDY OF HIGH TEMPERATURE PHASE TRANSITIONS OF SODIUM NIOBATE (TO 500C).
HOOK WR	11-1727	TRANSIENT ELASTOOPTIC EFFECTS AND Q SWITCHING PERFORMANCE IN LITHIUM NIOBATE AND KDP P
	17-2529	LOSSLESS POTASSIUM DIDEUTERIUM PHOSPHATE POCKELS CELL FOR HIGH POWER Q SWITCHING.
HOOKABE K	17-2530	NOVEL CUT OF POTASSIUM DIHYDROGEN PHOSPHATE TYPE CRYSTALS.
	17-2531	NOVEL TYPE OF CUT FOR KDP CRYSTALS FOR LOW VOLTAGE LIGHT MODULATION.
	17-2532	OPTIMUM CUT OF POTASSIUM DIHYDROGEN PHOSPHATE TYPE CRYSTALS FOR LIGHT MODULATION.
HOPFIELD JJ	04-0547	COUPLED OPTICAL PHONON MODE THEORY OF INFRARED DISPERSION IN BARIUM AND STRONTIUM TITA
HOPKINS MM	08-1352	PREPARATION OF POLED, TWIN-FREE CRYSTALS OF FERROELECTRIC BISMUTH TITANATE, (BY ANNEAL
HORAK J	15-2168	CHALCOGENIDE BROMIDES OF ANTIMONY AND BISMUTH.
	15-2169	OPTICAL PROPERTIES OF THE SEMICONDUCTOR BISMUTH TELLUROIODIDE.
	15-2170	PHOTOELECTRIC PROPERTIES OF BISMUTH SULFOIODIDE CRYSTALS.
HORDVIK A	11-1729	LUMINESCENCE FROM LITHIUM NIOBATE.
HORIE T	10-1639	AN EFFECT OF HEAT TREATMENT ON TUNGSTEN TRIOXIDE CRYSTAL. PART-2: GREEN STRIPES.
HORIOKA M	28-3202	MEASUREMENT OF THERMAL EXPANSION OF ROCHELLE SALT BY USING THE SHIFT OF RESONANCE FREQ
HORNUNG EW	30-3283	MAGNETOTHERMODYNAMICS OF ANTIFERROMAGNETIC, FERROELECTRIC BETA GADOLINIUM MOLYBDATE. P
HOSEMANN R	16-2271	X-RAY ANALYSIS OF FERROELECTRIC DOMAINS IN THE PARAELECTRIC PHASE OF SODIUM NITRITE.
HOSHINO M	04-0739	EFFECT OF CLAY AND DOLOMITE ADDITIONS ON BARIUM TITANATE CERAMICS. CRYSTAL STRUCTURE A
HOSHINO S	04-0857	SURFACE LAYER IN BARIUM TITANATE (X-RAY DIFFRACTION, DC FIELD EFFECT, PARAELECTRIC AND
	16-2312	CRYSTAL STRUCTURE AND PHASE TRANSITION OF DIGLYCINE NITRATE.
	16-2397	PHASE TRANSITION IN SODIUM NITRITE.
	27-3121	REEXAMINATION OF THE THERMAL EXPANSION OF THE FERROELECTRIC TRIGLYCINE SULFATE.
	31-3327	PHASE TRANSITIONS OF SOLID HYDROGEN HALIDES.
	31-3331	FERROELECTRICITY AND PHASE TRANSITIONS IN SOLID HYDROGEN HALIDES (SINGLE CRYSTAL X-RAY
	31-3333	STRESS FREE SINGLE CRYSTAL GROWTH OF FERROELECTRIC HYDROGEN HALIDES.
	31-3337	STRUCTURE AND PHASE TRANSITION IN HYDROGEN CHLORIDE.
HOSLER WR	06-1129	ELECTRONIC TRANSPORT (CONDUCTIVITY AND HALL AND SEEBECK COEFFICIENTS) IN STRONTIUM TIT
HOSOKAI M	37-3681	MANUFACTURING METHOD AND PROPERTIES OF FERROELECTRIC THIN FILM.
HOSOYA M	02-0099	DIELECTRIC AND X-RAY STUDIES OF THE PHASE TRANSITIONS IN ROCHELLE SALT, BARIUM TITANAT
	15-2171	DIELECTRIC PROPERTIES OF ANTIMONY SULFOIODIDE AT MICROWAVE FREQUENCIES (3.3 AND 9.8 GH
	25-2922	STUDIES ON THE MECHANISM OF THE PHASE TRANSITIONS IN THIOUREA. (SATELLITE SCATTERING)
HOTTMANN H	15-2201	CRYSTALLIZATION BEHAVIOR OF ANTIMONY SULFOIODIDE. (SINGLE CRYSTAL GROWTH, CHEMICAL TRA
HOUSTON GD	03-0279	QUANTUM FIELD THEORETICAL TREATMENT OF A DYNAMICAL MODEL OF THE FERROELECTRIC KDP. PAR
HOWARTH LE	04-0975	FAR INFRARED DIELECTRIC DISPERSION IN BARIUM AND STRONTIUM TITANATES AND TITANIUM DIOX
HOWELL FL	32-3346	HYDROGEN BONDING IN LITHIUM HYDRAZINIUM SULFATE.
	32-3353	DIELECTRIC PROPERTIES OF LITHIUM HYDRAZINIUM SULFATE.
	32-3354	DEUTERON NMR STUDIES OF LITHIUM HYDRAZINIUM SULFATE.
	32-3357	MICROWAVE DIELECTRIC STUDY OF LITHIUM HYDRAZINIUM SULFATE.
HRDLICKA J	27-3068	TEMPERATURE AND FIELD STRENGTH DEPENDENCE OF LARGE SIGNAL PERMITTIVITY OF TGS. (10 KHZ
	27-3086	PREPARATION OF PURE TGS SINGLE CRYSTALS AND SOME INVESTIGATIONS OF THEIR QUALITY. (VOL
	37-3640	VARIOUS HYSTERESIS LOOP TRACERS FOR A WIDE FREQUENCY RANGE. (CIRCUITRY DESCRIBED)
HRUSKA K	17-2537	DEPENDENCE OF THE ELASTIC COEFFICIENTS OF ADP ON THE ELECTRIC FIELD.
HUANG CC	03-0280	EFFECT OF THE CHANGE OF ELECTROSTATIC CONSTRAINTS ON THE PHASE TRANSITIONS OF SOME PHE
HUBER DL	03-0281	CRITICAL DYNAMICS OF THE ORDER DISORDER TRANSFORMATION IN FERROELECTRICS. (THEORY)

HUBER P	04-0685	CHANGE IN THE DECAY CONSTANT OF ZIRCONIUM-89 IN BARIUM TITANATE (BARIUM CHLORIDE FLUX
	04-0876	VARIATION OF THE DECAY CONSTANT OF STRONTIUM-85 IN BARIUM TITANATE.
HUBMANN M	34-3479	PRESSURE DEPENDENCE OF THE COMPLEX DISPERSION COEFFICIENT OF ICE I(H) SINGLE CRYSTALS.
HUGHES WE	17-2533	ELECTRON SPIN RESONANCE OF IRRADIATED AND DEUTERATED POTASSIUM DIHYDROGEN PHOSPHATE.
HUKUDA K	17-2546	EPR OF CHROMIUM(3+) IONS IN ANTIFERROELECTRIC ADP CRYSTALS.
	17-2554	EFFECTS OF UNIAXIAL STRESS AND ELECTRIC FIELD ON PARAMAGNETIC SPECTRA OF CHROMIUM(3+)
	17-2596	ELECTRON PARAMAGNETIC RESONANCE OF CHROMIUM(3+) IN ADP.
	22-2824	DIELECTRIC AND THERMAL STUDY OF POTASSIUM SELENATE TRANSITIONS (FERROELECTRIC BELOW 93
	22-2850	THE ELECTRON PARAMAGNETIC RESONANCE OF POTASSIUM SELENATE.
HULM JK	06-1118	ULTRASONIC STUDIES OF CRITICAL PHENOMENA IN INSULATING AND SEMICONDUCTING STRONTIUM TI
HULME KF	12-1880	SIGNS OF THE ELECTROOPTIC COEFFICIENTS FOR LITHIUM TANTALATE.
	17-2435	1.06 MICRON ABSORPTION COEFFICIENTS OF DEUTERATED KDP WITH 70-100% DEUTERATION.
	35-3494	ELECTROOPTIC MEASUREMENTS ON LITHIUM GERMANATE AND LEAD GERMANATE. (LITHIUM GERMANATE
HUMMEL FA	06-1143	DIELECTRIC RELAXATION IN (CERAMIC) STRONTIUM TITANATES CONTAINING (TRIVALENT) RARE EAR
HURD JD	04-0580	DIELECTRIC BREAKDOWN IN BARIUM TITANATE.
HURDITCH RJ	13-1934	ELECTRONIC PROPERTIES (CONDUCTIVITY, MAGNETIC SUSCEPTIBILITY) OF HYDROGEN TUNGSTEN BRO
HURST JJ	08-1353	(TOP SEEDED FLUX) CRYSTAL GROWTH AND NEUTRON CHARACTERIZATION OF POTASSIUM NIOBATE.
HUSIMI K	03-0282	PHENOMENOLOGICAL THEORY OF FERROELECTRIC POLARIZATION REVERSAL.
HUSSON E	08-1303	INFRARED AND RAMAN SPECTROSCOPY OF POTASSIUM METANIOBATE.
HYDE BG	10-1630	CS (CRYSTALLOGRAPHIC SHEAR) FAMILIES DERIVED FROM THE RHENIUM TRIOXIDE STRUCTURE TYPE
IBEAS JG	37-3641	ON THE SWITCHING PROPERTIES OF THE FERROELECTRIC SURFACES. (NEW MODEL REPRODUCES EXPER
IBRAIMOV NS	04-0666	POLARIZED STATES OF FERROELECTRIC CERAMICS. (BARIUM TITANATE)
ICHIKAWA M	02-0099	DIELECTRIC AND X-RAY STUDIES OF THE PHASE TRANSITIONS IN ROCHELLE SALT, BARIUM TITANAT
	36-3567	THE CRYSTAL STRUCTURE AND PHASE TRANSITION OF AMMONIUM HYDROGEN DICHLORO ACETATE. PART
ICHIKI SK	17-2526	HIGH FREQUENCY BEHAVIOR OF HYDROGEN BONDED FERROELECTRICS: TRIGLYCINE SULFATE AND POTA
ICHINOSE N	02-0076	NEW PIEZOELECTRIC MATERIALS.
	03-0283	THE BEHAVIORS OF THE MICRODOMAINS ON THE DIFFUSE PHASE TRANSITION FERROELECTRICS.
	10-1638	ELECTROMECHANICAL PROPERTIES IN THE BINARY SYSTEM LEAD TITANATE- LEAD CADMATE(0.5) TUN
	13-1959	CRYSTAL GROWTH OF LEAD (CADMATE(0.33) NIOBATE(0.67)) AND ITS DIELECTRIC PROPERTIES. (F
	13-1960	ELECTRICAL MAGNETIC AND X-RAY STUDIES OF PHASE TRANSITIONS IN LEAD COBALTATE(0.5) TUNG
IDA I	11-1742	PRECISION MACHINING OF LITHIUM NIOBATE CRYSTALS FOR ELECTROOPTIC ELEMENTS.
	11-1783	LAPPING CHARACTERISTICS OF LITHIUM NIOBATE SINGLE CRYSTAL.
	11-1784	LAPPING CHARACTERISTICS OF LITHIUM NIOBATE SINGLE CRYSTALS.
IDA M	27-3023	DIELECTRIC DISPERSION OF TRIGLYCINE SULFATE AT LOW FREQUENCIES.
IFIRI S	17-2605	LIGHT MODULATORS USING 45 DEGREE X CUT AND 45 DEGREE Y CUT ADP CRYSTALS.
IGNATAVICIUS M	17-2468	PARAMETRIC LIGHT AMPLIFICATION AND OSCILLATION IN KDP WITH MODE LOCKED PUMP.
IHARA H	17-2589	OBSERVATION OF PHASE FRONT MOTION IN KDP CRYSTAL.
IIDA H	08-1354	FORMATION PROCESSES OF LEAD MAGNESIO NIOBATE IN THE SOLID STATE REACTION.
IIDA S	04-0741	BARIUM TITANATE FILMS PREPARED BY RF SPUTTERING ONTO INDIUM ANTIMONIDE OR GALLIUM ARSE
	16-2382	STUDY OF THE CRITICAL SLOWING DOWN OF THE DIELECTRIC RELAXATION IN SILVER SODIUM DINIT
	27-3147	STUDY OF THE CRITICAL BEHAVIOR OF TRIGLYCINE SULFATE BY THERMAL NOISE MEASUREMENTS.
IIJIMA Y	04-0742	ELECTRON PROBE X-RAY MICROANALYSIS ON IMPURITIES IN EVAPORATED BARIUM TITANATE FILMS.
	04-0743	PREPARATION OF BARIUM TITANATE FILM BY EVAPORATION.
IIO K	16-2403	STUDY OF PHASE TRANSITION OF SODIUM NITRITE BY SECOND HARMONIC GENERATION. (TEMPERATUR
IKEDA R	16-2313	NUCLEAR QUADRUPOLE RESONANCE OF NITROGEN IN SODIUM NITRITE.
IKEDA T	08-1376	X-RAY STUDY OF THE SYSTEM BISMUTH OXIDE FERRIC OXIDE.
	09-1539	PIEZOELECTRIC CERAMICS OF LEAD ZIRCONATE TITANATE MODIFIED WITH SODIUM OR POTASSIUM AN
	09-1540	PRECIPITATION OF ZIRCONIA PHASE IN NIOBIUM MODIFIED CERAMICS OF LEAD ZIRCONATE TITANAT
	13-1961	ANTIFERROELECTRIC PHASE IN THE SYSTEM LEAD TITANATE- LEAD ZIRCONATE- LANTHANUM FERRATE
	13-1962	COMPLEX OXIDES WITH TUNGSTEN BRONZE TYPE STRUCTURES (POTASSIUM (BARIUM OR STRONTIUM) T
	13-1963	STUDY CF SUBSOLIDUS EQUILBRIA IN POTASSIUM OXIDE- LITHIUM OXIDE- NIOBIUM PENTOXIDE SYS
	13-1964	SOME COMPOUNDS OF TUNGSTEN BRONZE TYPE A(6)B(10)O(30). (WHERE B IS (NIOBIUM, TITANIUM)
	19-2759	ELASTIC ANOMALY IN AMMONIUM SULFATE. (ELECTROSTRICTIVE CONSTANTS)
IKEGAMI S	04-0744	FAR INFRARED REFLECTIVITY OF BARIUM TITANATE.
	04-0745	MECHANISM OF MICROWAVE DIELECTRIC DISPERSION IN POLYCRYSTALLINE BARIUM TITANATE.
	04-0746	RAMAN SPECTRUM OF BARIUM TITANATE.
	04-1015	DIELECTRIC BREAKDOWN OF POLYCRYSTALLINE BARIUM TITANATE.
	05-1060	ELECTROMECHANICAL PROPERTIES OF LEAD TITANATE CERAMICS CONTAINING LANTANUM AND MANGANE
	05-1061	POLARIZATION AND PIEZOELECTRIC PROPERTIES OF LEAD TITANATE CERAMICS. (CONTAINING 1 MOL
	05-1062	PIEZOELECTRIC PROPERTIES OF LEAD TITANATE CERAMICS.
	05-1085	ELECTROMECHANICAL PROPERTIES AND APPLICATION OF LEAD TITANATE CERAMICS.
	09-1616	LEAD TITANATE PIEZOELECTRIC CERAMICS. (MATSUSHITA LEAD TITANATE)
IKUSHIMA A	04-0715	SPECIFIC HEAT OF BARIUM TITANATE.
	16-2300	DIPOLE LATTICE RELAXATION TIME NEAR THE NEEL TEMPERATURE IN SODIUM NITRITE.
	16-2303	SPECIFIC HEAT OF SODIUM NITRITE NEAR THE ANTIFERROELECTRIC TRANSITION POINT.
IKUSHIMA H	04-0747	ELECTRON SPIN RESONANCE OF MANGANESE(2+) IN BARIUM TITANATE.
ILINSKII YUA	11-1687	INFLUENCE OF A TRANSVERSE INHOMOGENEITY OF THE REFRACTIVE INDEX OF A NONLINEAR CRYSTAL
IMAI K	15-2172	FERROELECTRIC BARKHAUSEN EFFECT IN ANTIMONY SULFOIODIDE.
IMAOKA K	33-3383	MICROSCOPIC OBSERVATIONS OF PHASE TRANSITIONS IN CESIUM LEAD TRICHLORIDE.
IMBUSCH GF	06-1112	DIRECT OBSERVATION OF SINGLE DOMAIN STRONTIUM TITANATE (OPTICAL STUDY UNDER UNIAXIAL S
	06-1114	EFFECT OF DOMAIN STRUCTURE ON THE FLUORESCENCE OF CHROMIUM(3+) DOPED STRONTIUM TITANAT
	06-1115	POLARIZED FLUORESCENCE STUDY OF CHROMIUM(3+) THROUGH A STRESS INDUCED PHASE TRANSITION
IMOTO F	08-1354	FORMATION PROCESSES OF LEAD MAGNESIO NIOBATE IN THE SOLID STATE REACTION.
IMRY Y	17-2612	DETECTION OF ATOM TUNNELING IN SOLIDS. (KDP)
INABA A	36-3538	FERROELECTRICITY AND PHASE TRANSITION IN AMMONIUM DICHLORO ACETATE AND DEUTERATED AMMO
INGEBRIGTSEN KA	11-1723	ACOUSTOELECTRIC AMPLIFICATION OF SURFACE WAVE STRUCTURE OF A CADMIUM SELENIDE FILM ON
INGLE SG	08-1315	DOMAINS IN FERROELECTRIC POTASSIUM NIOBATE SINGLE CRYSTALS.
	08-1316	DOMAIN STRUCTURES IN POTASSIUM NIOBATE SINGLE CRYSTALS ASSOCIATED WITH STEP LADDERS ON
	08-1317	DOMAINS IN POTASSIUM NIOBATE SINGLE CRYSTALS FROM PSEUDOCUBIC (001) CLEAVAGE PLANES.
	08-1318	INTERFEROMETRIC STUDIES OF DOMAIN STRUCTURES IN POTASSIUM NIOBATE SINGLE CRYSTALS.
	08-1319	INTERFEROMETRIC STUDIES OF DOMAIN STRUCTURE IN POTASSIUM NIOBATE SINGLE CRYSTALS.
	08-1320	OPTICAL STUDIES OF DOMAIN STRUCTURE OF POTASSIUM NIOBATE SINGLE CRYSTALS.
	08-1381	DIELECTRIC CONSTANT OF POTASSIUM NIOBATE SINGLE CRYSTALS UNDER BIASING CONDITIONS.
INOKUMA T	37-3681	MANUFACTURING METHOD AND PROPERTIES OF FERROELECTRIC THIN FILM.
INOUE K	16-2314	LONG RANGE ORDER IN SODIUM NITRITE BY OPTICAL SECOND HARMONIC GENERATION.
	16-2315	TEMPERATURE DEPENDENCE OF SECOND HARMONIC GENERATION IN SODIUM NITRITE.
INUISHI Y	11-1861	NATURE OF INTERNAL ELECTRIC FIELD DURING OPTICAL DAMAGE PROCESS IN LITHIUM NIOBATE.
	17-2550	TEMPERATURE DEPENDENCE OF RAMAN SPECTRUM AND SECOND HARMONIC GENERATION IN KDP.
INYUSHKIN GV	16-2316	RATE OF ROTATING CRYSTALS AND CHARACTERISTICS OF THEIR GROWTH. (AMMONIUM DIHYDROGEN PH
IOFFE IV	03-0284	RESONANCE OF A LOW FREQUENCY SPIN WAVE WITH AN ACOUSTIC OR ELECTROMAGNETIC WAVE IN A F

IOFFE VA 08-1355 NATURE OF THE HIGH PERMITTIVITY OF CERIUM ALUMINATES WITH PEROVSKITE TYPE STRUCTURE.
IORDANOVA M 04-0789 DEPENDENCE OF DIELECTRIC PROPERTIES OF BARIUM TITANATE CERAMICS ON THE PHYSICO- CHEMIC
IRIE K 15-2173 POTENTIAL DISTRIBUTION IN ANTIMONY SULFOIODIDE CRYSTAL (IN THE PARAELECTRIC PHASE, UND
15-2174 PHOTOVOLTAIC EFFECT IN FERROELECTRIC AND PHOTOCONDUCTIVE ANTIMONY SULFOIODIDE.
IRISOVA NA 11-1730 BIREFRINGENCE OF CERTAIN CRYSTALS IN THE MILLIMETER WAVELENGTH RANGE. (LITHIUM NIOBATE
IRSLINGER C 15-2164 OPTICAL SECOND HARMONIC GENERATION IN ANTIMONY SULFOIODIDE.
IRVINE RD 03-0285 DYNAMIC MODEL OF THE SLATER KDP FERROELECTRIC.
ISARD JO 05-1054 APPLICATION OF DIELECTRIC MIXTURE FORMULAS TO (LEAD TITANATE) GLASS- CERAMIC SYSTEMS.
ISERENTANT CM 08-1356 PREPARATION AND PHYSICAL PROPERTIES OF PEROVSKITE TYPE COMPOUNDS, LANTHANUM(1-X) CALCI
ISHERWOOD BJ 17-2535 STRUCTURE OF DEUTERATED KDP CRYSTAL: DEUTERIUM CONCENTRATION DETERMINATION.
ISHIEASHI Y 03-0286 A THEORY OF THE PHASE TRANSITION IN AMMONIUM DIHYDROGEN PHOSPHATE.
03-0287 COMPUTER EXPERIMENT ON LINEAR CHAIN OF ATOMS LYING IN DOUBLE MINIMUM POTENTIAL. (BARIU
03-0288 DISTRIBUTION OF RELAXATION TIMES IN SOME FERROELECTRICS.
03-0289 THE ROLE OF LORENTZ FACTORS IN ROCHELLE SALT. (THEORY)
03-0482 MICROSCOPIC THEORY OF THE PHASE TRANSITIONS IN ROCHELLE SALT.
04-0748 FORCE CONSTANTS IN BARIUM TITANATE AND STRONTIUM TITANATE.
04-0775 ON THE LATTICE VIBRATION OF CUBIC BARIUM TITANATE.
04-0868 PERMITTIVITY CHANGE OF BARIUM TITANATE UNDER VERY SLOWLY CHANGING FIELD.
04-0981 SUSCEPTIBILITY INCREASE DURING POLARIZATION REVERSAL. (IN BARIUM TITANATE AND TRIGLYCI
10-1641 ON FERROELECTRICITY AND ANTIFERROELECTRICITY OF THE AO3 TYPE CRYSTAL. (TUNGSTEN TRIOXI
16-2317 LATTICE VIBRATION OF SODIUM NITRITE. (CALCULATIONS TO RELATE TO SPECIFIC ANTIFERROELEC
16-2344 EFFECT OF HYDROSTATIC PRESSURE ON FERROELECTRICITY OF POTASSIUM NITRATE.
16-2345 PHASE TRANSITIONS IN POTASSIUM NITRATE NITRITE MIXED CRYSTALS (GROWTH BY DOUBLE BRIDGM
16-2357 ELASTIC CONSTANTS AND ULTRASONIC ABSORPTION OF SODIUM NITRITE SINGLE CRYSTALS.
17-2534 DIELECTRIC DISPERSION OF TGS AND DEUTERATED KDP (THEORY, CALCULATIONS).
19-2796 EFFECT OF THE HYDROSTATIC PRESSURE ON THE TRANSITION TEMPERATURE IN AMMONIUM SULFATE.
26-2933 DIELECTRIC RESPONSE FUNCTION OF COPPER FORMATE TETRAHYDRATES.
26-2934 TEMPERATURE DEPENDENCE OF SOUND VELOCITY IN DICALCIUM STRONTIUM PROPIONATE.
27-3116 DIELECTRIC STUDY OF CRITICAL BEHAVIOR OF FERROELECTRIC TRIGLYCINE SULFATE BY A DIGITAL
33-3383 MICROSCOPIC OBSERVATIONS OF PHASE TRANSITIONS IN CESIUM LEAD TRICHLORIDE.
33-3393 PRESSURE DEPENDENCE OF CUBIC TETRAGONAL TRANSITION TEMPERATURE OF CESIUM LEAD TRICHLOR
36-3612 DIELECTRIC PROPERTIES OF DIAMMONIUM HEXABROMO STANNATE (DIELECTRIC TRANSITIONS NOT FER
37-3642 FERROELECTRIC DOMAIN SWITCHING.
ISHIDA A 08-1359 RH DOPED LITHIUM NIOBATE AS AN IMPROVED NEW MATERIAL FOR REVERSIBLE HOLOGRAPHIC STORAG
11-1731 ACOUSTOOPTICAL MEASUREMENTS OF ACOUSTIC WAVE GENERATION IN PIEZOELECTRICS COUPLED TO A
ISHIDA K 02-0077 ACCURATE STRUCTURE ANALYSIS OF FERROELECTRICS IN CONNECTION WITH THEIR PHASE TRANSITIO
03-0391 NOTE ON THE DIELECTRIC RELAXATION IN FERROELECTRICS (THEORETICAL DISCREPANCY).
08-1357 ANISOTROPIC DIFFUSE X-RAY SCATTERING AND PHASE TRANSITIONS IN SODIUM NIOBATE.
08-1358 DIFFUSE X-RAY STUDY OF 640 DEGREES C PHASE TRANSITION IN SODIUM NIOBATE.
13-1965 X-RAY STUDY OF HIGH TEMPERATURE PHASE TRANSITIONS OF SODIUM NIOBATE (TO 500C).
27-3093 STUDIES ON THE MECHANISM OF THE PHASE TRANSITION IN TGS. (STATIC DIELECTRIC CONSTANT,
ISHIDATE T 16-2314 LONG RANGE ORDER IN SODIUM NITRITE BY OPTICAL SECOND HARMONIC GENERATION.
ISHIGURO T 16-2304 ULTRASONIC ATTENUATION NEAR THE TWO TRANSITION POINTS IN SODIUM NITRITE.
ISHIHARA K 17-2536 ELECTROCHEMISTRY ON THE POLYCRYSTAL OF POTASSIUM DIHYDROGEN PHOSPHATE.
ISHIKAWA K 04-0734 A SEMICONDUCTING BARIUM TITANATE CERAMIC DOPED WITH NIOBIUM AND MANGANESE IONS AND ITS
15-2175 RELATION BETWEEN THE ELECTRIC POLARIZATION AND THE ABSORPTION EDGE IN ANTIMONY SULFOIO
15-2241 TRANSPORT PHENOMENA IN ANTIMONY SULFOIODIDE. (CONDUCTION MECHANISM FOR THERMOELECTRIC
ISMAILZADE IG 08-1361 NEW DATA FROM AN X-RAY STUDY OF PHASE TRANSFORMATIONS IN SODIUM NIOBATE.
13-1923 FORMATION CONDITIONS AND PHASE TRANSITIONS IN CRYSTALS WITH POTASSIUM TUNGSTEN BRONZE
13-1993 SYNTHESES, X-RAY, AND DIELECTRIC STUDIES OF CERTAIN NEW LAYERED FERROELECTRICS. (LANTH
13-2050 PHASE TRANSITIONS IN TANTALATES WITH THE STRUCTURE OF POTASSIUM TUNGSTEN BRONZE. (POTA
16-2318 AN X-RAY STUDY OF THE PHASE TRANSITION IN SODIUM NITRITE.
ISMAILZADE IH 03-0290 X-RAY DIFFRACTOMETRIC DETERMINATION OF THE CURIE TEMPERATURE AND TEMPERATURE DEPENDENC
08-1360 MAGNETOELECTRIC EFFECT IN FERROELECTRIC- ANTIFERROMAGNETIC BISMUTH TITANIUM FERRATE.
13-1966 CRYSTAL CHEMISTRY OF PEROVSKITE LIKE LAYER TYPE FERROELECTRICS. (BISMUTH TUNGSTATE, BI
13-2035 MOSSBAUER STUDIES OF IRON-57 IN SOME PEROVSKITE LIKE LAYER TYPE FERROELECTRICS. (IN BI
36-3568 MAGNETOELECTRIC EFFECT IN FERROELECTRIC- ANTIFERROMAGNETIC LITHIUM IRON(0.5) TANTALUM(
36-3569 X-RAY AND ELECTRIC INVESTIGATIONS OF THE SYSTEMS YTTRIUM MANGANESE(1-X) (B(X)) OXYGEN(
ISUPOV VA 01-0029 FERROELECTRICS AND ANTIFERROELECTRICS.
02-0118 NEW CLASSES OF FERROELECTRICS OF DISPLACEMENT TYPE. (BORACITES, MOLYBDATES, TUNGSTEN B
03-0291 CAUSES OF PHASE TRANSITION BROADENING (DIFFUSNESS) AND THE NATURE OF DIELECTRIC POLARI
03-0360 NEW TWO- AND THREE PHASE FERROELECTRIC FERROMAGNETIC MATERIALS. SOME PROBLEMS AND RESU
04-0767 ULTRAHIGH FREQUENCY DIELECTRIC ANISOTROPY OF A POLARIZED FERROELECTRIC CERAMIC. (BARIU
05-1063 PHYSICAL PHENOMENA IN PIEZOCERAMICS BASED ON BARIUM TITANATE AND ZIRCONATE. (SOLID SOLUT
09-1541 DIELECTRIC POLARIZATION OF LEAD TITANATE ZIRCONATE SOLID SOLUTIONS (THERMODYNAMIC EXIS
09-1603 DIELECTRIC SYSTEM POLARIZATION AND THERMAL EXPANSION OF POLARIZED CERAMICS IN THE LEAD
13-1923 FORMATION CONDITIONS AND PHASE TRANSITIONS IN CRYSTALS WITH POTASSIUM TUNGSTEN BRONZE
13-1967 ELASTIC AND PIEZOELECTRIC PROPERTIES OF CADMIUM PYRONIOBATE IN STRONG ELECTRIC FIELDS.
13-1968 PHASE TRANSISITION IN THE ANTIFERROELECTRET LEAD COBALTATE(0.5) TUNGSTATE(0.5) AND LEA
13-1980 RELAXATION POLARIZATION OF A FERROELECTRIC LEAD MAGNESATE(0.33) NIOBATE(0.67) WITH A D
13-2050 PHASE TRANSITIONS IN TANTALATES WITH THE STRUCTURE OF POTASSIUM TUNGSTEN BRONZE. (POTA
36-3561 PECULIARITIES OF THE DIELECTRIC POLARIZATION OF CADMIUM PYRONIOBATE.
ISUPOVA LA 17-2425 RAMAN SPECTRA OF POLYMORPHIC MODIFICATIONS OF POTASSIUM DIHYDROGEN PHOSPHATE IN AN ELE
ITAKURA M 11-1732 AN ELECTROOPTIC SWITCH USING ROTATED Y PLATE OF LITHIUM NIOBATE.
ITOH K 15-2239 STRAIN ALONG C-AXIS OF ANTIMONY SULFOIODIDE CAUSED BY ILLUMINATION IN DC ELECTRIC FIEL
17-2536 ELECTROCHEMISTRY ON THE POLYCRYSTAL OF POTASSIUM DIHYDROGEN PHOSPHATE.
27-3024 REFINEMENT OF CRYSTAL STRUCTURE OF TRIGLYCINE SULFATE.
27-3093 STUDIES ON THE MECHANISM OF THE PHASE TRANSITION IN TGS. (STATIC DIELECTRIC CONSTANT,
ITOH Y 13-1969 DIELECTRIC PROPERTY OF BARIUM TITANATE NIOBATE SINGLE CRYSTAL (TETRAGONAL TUNGSTEN BRO
13-1970 TUNGSTEN BRONZE FIELD IN THE PSEUDOTERNARY SYSTEM SODIUM NIOBATE BARIUM TITANATE BARIU
35-3515 X-RAY TOPOGRAPHIC OBSERVATION OF 180 DEG DOMAINS IN FERROELECTRIC LEAD GERMANATE SINGL
35-3516 X-RAY TOPOGRAPHIC OBSERVATION OF 180 DEGREE DOMAINS IN FERROELECTRIC LEAD(5) GERMANIUM
ITSKOVSKII MA 03-0292 SCATTERING OF CONDUCTION ELECTRONS BY TRANSVERSE AND LONGITUDINAL OPTICAL PHONONS IN I
03-0508 NATURE OF THE ANOMALOUS ELECTRIC CONDUCTIVITY OF FERROELECTRICS IN THE PHASE TRANSITIO
ITTU M 17-2555 METHOD OF GROWTH IN SOLUTION OF SINGLE CRYSTALS BY THE GRADUAL ADDITION OF A REACTANT.
IVANCHIK II 03-0293 CONCERNING THE CRITERION OF FERROELECTRICITY IN THE MICROSCOPIC THEORY.
03-0295 MACROSO MACROSCOPIC THEORY OF FERROELECTRICS.
03-0511 FIELD EFFECT AT THE CONTACT BETWEEN A SEMICONDUCTOR AND A C-DOMAIN FERROELECTRIC (THEO
04-1021 DEPENDENCE OF THE PIEZOEFFECT IN BARIUM TITANATE ON THE MEANS OF SCREENING THE SPONTAN
IVANII GM 04-0750 X-RAY AND THERMOLUMINESCENCE OF BARIUM TITANATE.

IVANOV IV 03-0294 EFFECT OF AUTOMATIC TEMPERATURE STABILIZATION IN FERROELECTRIC OSCILLATORY SYSTEMS.
04-0749 MICROWAVE DYNAMIC NONLINEARITY OF BARIUM TITANATE SINGLE CRYSTALS.
04-1019 DIELECTRIC AND NONLINEAR MICROWAVE PROPERTIES OF POLYCRYSTALLINE SOLID SOLUTIONS BASED
06-1105 ANOMALOUS BEHAVIOR OF THE DIELECTRIC NONLINEARITY COEFFICIENT IN STRONTIUM TITANATE NE
06-1106 DIELECTRIC NONLINEARITY OF STRONTIUM TITANATE SINGLE CRYSTALS NEAR PHASE TRANSITION AT
37-3639 FERROELECTRIC MICROTHERMOSTATS.
37-3643 THIN FILM MICROWAVE FERROELECTRIC CAPACITORS AND NONLINEAR MICROWAVE PROPERTIES OF FER

IVANOV NR 03-0330 (EFFECT OF TEMPERATURE ON) QUASI SYMMETIRICAL HARD IONS (PHOSPHATE, TITANATE, SULFATE,
17-2666 PHASE DIAGRAM OF SODIUM TRIDEUTERIUM(X) TRIHYDROGEN(1-X) PHOSPHATE SYSTEM VERSUS CONCE
22-2836 INFRARED SPECTRA OF ISOTOPIC NONISOMORPHISM OF SODIUM TRIHYDROGEN SELENITE (SHS) OR SO
22-2839 BEHAVIOR OF THE FUNDAMENTAL ABSORPTION EDGE IN TRIHYDROGEN SELENITE CRYSTALS AT THEIR
22-2840 PHASE TRANSFORMATION AND THERMOOPTICAL AND ELECTROOPTICAL EFFECT IN RUBIDIUM TRIHYDROG
22-2841 THERMOOPTIC AND ELECTROOPTIC PROPERTIES OF TRIHYDROGEN SELENITES.
22-2873 ISOTOPIC NONISOMORPHISM OF RUBIDIUM TRIHYDROGEN(1-X) TRIDEUTERIUM(X) DISELENATE CRYSTA
22-2875 UNUSUAL EFFECT OF AN ELECTRIC FIELD ON THE PERMITTIVITY OF A NEW DELTA PHASE OF SODIUM

IVANOV VV 15-2220 EFFECT OF ILLUMINATION ON REPOLARIZATION PROCESS IN ANTIMONY SULFOIODIDE SINGLE CRYSTA
IVANOVA TSV 04-0789 DEPENDENCE OF DIELECTRIC PROPERTIES OF BARIUM TITANATE CERAMICS ON THE PHYSICO- CHEMIC
IVANOVA VV 02-0128 NEW FERROELECTRICS AND SEIGNETTO MAGNETS OF THE PEROVSKITE AND PYROCHLORE TYPE. SYNTHE
08-1362 X-RAY DETERMINATION OF THE SYMMETRY OF THE FERROELECTRIC COMPOUNDS (POTASSIUM OR SODIU
08-1459 PEROVSKITE TYPE SEIGNETTO MAGNETS.
10-1655 NEUTRON DIFFRACTION STUDY OF (TETRAGONAL- CUBIC PEROVSKITE) STRONTIUM IRON TUNGSTATE .
36-3618 A NEW MAGNETOFERROELECTRIC: CADMIUM IRON NIOBATE.
IVANOVSHITS AK 15-2238 HEAT CAPACTIY OF POLYCRYSTALLINE ANTIMONY SULFOIODIDE.
IVLEVA LI 11-1733 ELECTRICAL CONDUCTIVITY OF LITHIUM NIOBATE SINGLE CRYSTALS.
IWAHARA H 06-1141 IONIC CONDUCTION IN SINTERED OXIDES BASED ON CALCIUM TITANATE OR STRONTIUM TITANATE.
08-1444 IONIC CONDUCTION IN PEROVSKITE TYPE OXIDE SOLID SOLUTION AND ITS APPLICATION TO THE SO
IWAI T 10-1639 AN EFFECT OF HEAT TREATMENT ON TUNGSTEN TRIOXIDE CRYSTAL. PART-2: GREEN STRIPES.
IWASAKI H 04-0751 CHARACTERISTICS OF BARIUM TITANATE PYROELECTRIC DETECTOR FOR OPTICAL AND INFRARED RADI
12-1876 SINGLE CRYSTAL GROWTH AND PHYSICAL PROPERTIES OF LITHIUM TANTALATE. (DIELECTRIC CONSTA
12-1877 BEHAVIOR OF LITHIUM TANTALATE SINGLE CRYSTAL NEAR ITS CURIE POINT. PART-1: PIEZOELECTR
12-1882 CONGRUENT MELTING COMPOSITION OF LITHIUM METATANTALATE.
12-1883 STOICHIOMETRY AND OPTICAL QUALITY OF LITHIUM TANTALATE SINGLE CRYSTALS.
12-1893 PIEZOELECTRIC AND ELASTIC PROPERTIES OF LITHIUM TANTALATE CHARACTERISTICS.
13-1969 DIELECTRIC PROPERTY OF BARIUM TITANATE NIOBATE SINGLE CRYSTAL (TETRAGONAL TUNGSTEN BRO
13-1970 TUNGSTEN BRONZE FIELD IN THE PSEUDOTERNARY SYSTEM SODIUM NIOBATE BARIUM TITANATE BARIU
13-1971 DIELECTRIC AND ELECTROOPTIC PROPERTIES OF BARIUM SODIUM YTTRIUM NIOBATE SINGLE CRYSTAL
13-1972 SODIUM BARIUM RARE EARTH NIOBATES WITH THE TUNGSTEN BRONZE TYPE STRUCTURE.
13-2060 ELASTIC ANOMALY OF BARIUM SODIUM NIOBATE (ELASTIC, PIEZOELECTRIC, DIELECTRIC, AND OPTI
15-2236 PHOTODIELECTRIC EFFECT IN ANTIMONY SULFOIODIDE SINGLE CRYSTALS.
22-2842 ELECTRON PARAMAGNETIC RESONANCE IN GAMMA-IRRADIATED SINGLE CRYSTAL OF FERROELECTRIC LI
22-2843 PARAMAGNETIC SPECIES IN GAMMA-IRRADIATED SINGLE CRYSTAL OF FERROELECTRIC LITHIUM TRIHY
35-3499 FERROELECTRIC PROPERTY AND ENANTIOMORPHISM OF LEAD(5) GERMANIUM(3) OXIDE(11) AND THE R
35-3500 FERROELECTRIC LEAD(5) GERMANIUM(3) OXIDE(11) CRYSTAL.
35-3501 FERROELECTRIC AND OPTICAL PROPERTIES OF LEAD(5) GERMANIUM(3) OXIDE(11) AND ITS ISOMORP
35-3502 LEAD(5) GERMANIUM(3) OXIDE(11) CRYSTAL, A NEW FERROELECTRIC. (DIELECTRIC CONSTANTS AND
35-3503 OPTICAL ACTIVITY OF FERROELECTRIC LEAD(5) GERMANIUM(3) OXIDE(11) SINGLE CRYSTALS.
35-3513 CRYSTAL GROWTH AND PROPERTIES OF LEAD(5) GERMANIUM(3) OXIDE(11) SINGLE CRYSTALS.
35-3514 GROWTH OF SINGLE CRYSTALS IN LEAD MONOXIDE- GERMANIUM DIOXIDE BINARY SYSTEM.
35-3515 X-RAY TOPOGRAPHIC OBSERVATION OF 180 DEG DOMAINS IN FERROELECTRIC LEAD GERMANATE SINGL
35-3516 X-RAY TOPOGRAPHIC OBSERVATION OF 180 DEGREE DOMAINS IN FERROELECTRIC LEAD(5) GERMANIUM
35-3518 ELECTROOPTIC PROPERTIES OF FERROELECTRIC LEAD GERMANATE SINGLE CRYSTAL.
35-3520 ELASTIC AND PIEZOELECTRIC PROPERTIES OF FERROELECTRIC LEAD(5) GERMANIUM(3) OXIDE(11) C
36-3570 SWITCHING OF OPTICAL ROTATORY POWER IN FERROELECTRIC LEAD(5) GERMANIUM(3) OXIDE(11) SI
36-3626 NEW FERROELECTRIC COMPOUND STRONTIUM TELLURATE.
36-3627 PHASE TRANSITION OF A NEW FERROELECTRIC OXIDE, STRONTIUM OXIDE- TELLURIUM DIOXIDE.
IWATA Y 16-2365 NEUTRON DIFFRACTION STUDIES OF FERROELECTRIC POLARIZATION REVERSALS IN SODIUM NITRITE.
17-2661 NEUTRON DIFFRACTION STUDIES OF FERROELECTRIC PHASE TRANSITIONS IN KDP AND POTASSIUM DI
IZRAEL A 27-3025 X-RAY TOPOGRAPHIC STUDY OF GROWTH DEFECTS IN TRIGLYCINE SULFATE CRYSTALS IN RELATION T
27-3026 X-RAY TOPOGRAPHIC STUDY OF DOMAIN WALL MOVEMENT IN TGS.
IZRAILENKO AN 17-2698 ELECTROOPTICAL AND OPTICAL PROPERTIES OF PARTIALLY DEUTERATED RUBIDIUM DIHYDROGEN PHOS
26-2925 ANTIPOLARIZATION IN COPPER FORMATE CRYSTALS BY THE ELECTROOPTICAL METHOD. (ANTIFERROEL
JABRAILOVA GA 16-2260 MORPHOLOGY OF CRYSTAL GROWTH AT POLYMORPHIC TRANSFORMATIONS IN POTASSIUM NITRATE, SILV
JACCARD C 34-3446 THERMOELECTRIC EFFECT IN ICE.
JACKOWIAK I 27-3131 EFFECT OF A DC FIELD ON THE DIELECTRIC PROPERTIES OF TRIGLYCINE SELENATE. (AGING AND P
JACOBS JT 03-0216 DOMAIN STATISTICS AND FERROELECTRIC TRANSIENTS.
03-0296 ASYMMETRIC MINOR LOOP AND TRANSIENT OPEN LOOP COMPUTATIONS FOR FERROELECTRICS.
13-1976 A NEW CLASS OF FERROELECTRIC CERAMICS OF THE TYPE AB(2) (XO(4)) (3).
37-3644 EQUIVALENT CIRCUIT ANALYSIS OF A FERROELECTRIC PHOTOCONDUCTOR MEMORY DEVICE.
37-3659 DEPOLARIZATION FIELD IN THIN FERROELECTRIC FILMS.
JACOBSMEYER VP 17-2558 ESR INVESTIGATIONS OF X-RAY IRRADIATED SINGLE CRYSTALS OF FERROELECTRIC POTASSIUM DIHY
JAEGER RE 09-1542 HOT PRESSING OF POTASSIUM- SODIUM NIOBATE.
JAFFE B 01-0015 PIEZOELECTRIC CERAMICS. (NONMETALLIC SOLIDS, VOL-3, REVIEW)
09-1488 RELEASE OF ELECTRIC ENERGY IN LEAD NIOBIUM ZIRCONATE TITANATE STANNATE BY TEMPERATURE
JAFFE H 01-0015 PIEZOELECTRIC CERAMICS. (NONMETALLIC SOLIDS, VOL-3, REVIEW)
09-1488 RELEASE OF ELECTRIC ENERGY IN LEAD NIOBIUM ZIRCONATE TITANATE STANNATE BY TEMPERATURE
JAIN AP 07-1260 TEMPERATURE DEPENDENT OPTICAL MODE IN ANTIFERROELECTRIC LEAD ZIRCONATE BY THE MOSSBAUE
JAISWAL VK 03-0297 INELASTIC SCATTERING CROSS SECTION OF NEUTRONS IN FERROELECTRICS.
06-1142 ATOMIC DISPLACEMENTS IN PEROVSKITE STRONTIUM TITANATE (AT OPTICAL FREQUENCIES, CALCULA
JAMES JA 17-2535 STRUCTURE OF DEUTERATED KDP CRYSTAL: DEUTERIUM CONCENTRATION DETERMINATION.
JAMES LW 33-3390 ACOUSTIC AND MAGNETIC EFFECTS INVOLVING THE FLUORINE-19 NUCLEI IN ANTIFERROMAGNETIC PO
JAMES WF 08-1402 FERROELECTRIC BISMUTH FERRATE X-RAY AND NEUTRON DIFFRACTION STUDY. (ANTIFERROMAGNETIC)
JAMES WJ 07-1253 MOSSBAUER EFFECT STUDIES OF THE PHASE TRANSITIONS IN ANTIFERROELECTRIC LEAD ZIRCONATE
08-1287 PRECISION DETERMINATION OF THE LATTICE PARAMETERS AND THE COEFFICIENTS OF THERMAL EXPA
08-1333 FERROELECTRIC PROPERTIES OF BISMUTH FERRATE (PREPARATION AND STRUCTURE) AND RELATED MA
08-1398 ATOMIC STRUCTURE IN THE PEROVSKITIC FERRATE.
08-1449 DIELECTRIC HYSTERESIS IN (FLUX GROWN) SINGLE CRYSTAL BISMUTH FERRATE (CONFIRMS FERROEL
08-1471 MOSSBAUER STUDIES OF BISMUTH FERRATE LEAD TITANATE PEROVSKITE TYPE SOLID SOLUTIONS (RO
09-1574 ATOMIC STRUCTURES OF THE FERROELECTRIC COMPOUNDS LEAD ZIRCONATE(X) TITANATE(1-X) FOR T
JAMIESON PB 13-1895 FERROELECTRIC TUNGSTEN BRONZE TYPE CRYSTAL STRUCTURES. PART-3: POTASSIUM LITHIUM NIOBA
35-3493 CRYSTAL STRUCTURE OF PIEZOELECTRIC BISMUTH(12) GERMANIUM OXIDE(20) .

JAMSIK-VILFAN M   16-2267   NMR STUDY OF THE FERROELECTRIC TRANSITIONS IN DIGLYCINE NITRATE AND TRIS-SARCOSINE CAL
                   17-2438   DEUTERON MAGNETIC RESONANCE AND RELAXATION IN FERROELECTRIC POTASSIUM DIDEUTERIUM PHOS
                   18-2721   DEUTERON QUADRUPOLE COUPLING IN FERROELECTRIC CESIUM DIDEUTERIUM ARSENATE.
JANIK L   17-2537   DEPENDENCE OF THE ELASTIC COEFFICIENTS OF ADP ON THE ELECTRIC FIELD.
JANNIN M   27-3028   IMPEDANCE MEASUREMENT OF A DIELECTRIC FERROELECTRIC CONDENSER UNDER CONTINUOUS HEATING
                   37-3645   NOISE AND IMPEDANCE MEASUREMENTS OF A FERROELECTRIC CONDENSER UNDER A DC FIELD.
JANOUSEK V   04-0634   DIELECTRIC PROPERTIES OF BARIUM TITANATE SINGLE CRYSTALS DOPED BY HYDROXIDE GROUPS.
                   04-0635   DIELECTRIC PROPERTIES OF BARIUM(1-X) STRONTIUM(X) TITANATE SINGLE CRYSTALS GROWN FROM
                   04-0873   DEFECT STRUCTURE OF (BARIUM(1-X) STRONTIUM(X)) TITANATE- MIXED CRYSTALS FROM A POTASSI
                   27-3005   TEMPERATURE AUTOSTABILIZATION EFFECT OF TRIGLYCINE SULFATE SINGLE CRYSTALS IN AN AC EL
                   27-3027   IMPEDANCE OF FERROELECTRIC TRIGLYCINE FLUOBERYLLATE CRYSTALS DURING SWITCHING IN PULSE
                   28-3196   PERMITTIVITY OF ROCHELLE SALT DURING SWITCHING.
                   29-3239   SOME PROPERTIES OF NONDEUTERATED AND DEUTERATED LITHIUM THALLIUM TARTRATE MONOHYDRATE.
JANOVEC V   03-0298   ANOMALIES NEAR FERROELECTRIC TRANSITIONS FROM THE POINT OF VIEW OF THE THEORY OF CRITI
                   03-0299   GENERALIZATION OF THE PIPPARD-GARLAND RELATIONS AND THEIR APPLICATION TO FERROELECTRIC
                   03-0302   THERMODYNAMIC THEORY OF SOLID SOLUTIONS OF ISOMORPHOUS FERROELECTRICS.
                   03-0303   THERMODYNAMIC CLASSIFICATION OF SECOND ORDER FERROELECTRIC PHASE TRANSITIONS.
                   04-0658   ANGULAR ENERGY DEPENDENCE OF 180 DEG DOMAIN WALL IN BARIUM TITANATE. (THERMODYNAMIC CA
                   04-0659   ON THE INTERPRETATION OF INFRARED SPECTRA AND THE LATTICE DYNAMICS OF CUBIC BARIUM TIT
                   04-0693   DEPENDENCE OF COERCIVE FIELD OF BARIUM TITANATE SINGLE CRYSTALS ON THEIR THICKNESS.
                   08-1327   ORIENTATION OF COMPOSITION PLANES IN ANTIFERROELECTRICS. (APPLIED TO ADP, TUNGSTEN TRI
                   15-2176   FERROELECTRIC TRANSITIONS WITH A GENUINE DIELECTRIC INSTABILITY (FERROELECTRIC TWINNIN
                   27-3005   TEMPERATURE AUTOSTABILIZATION EFFECT OF TRIGLYCINE SULFATE SINGLE CRYSTALS IN AN AC EL
                   36-3596   MOSSBAUER EFFECT AND FERROELECTRIC PROPERTIES OF IONIC CRYSTALS.
JANSKIEWICZ A   04-0549   ULTRASONIC STUDIES OF FIELD INDUCED PHASE TRANSITION IN BARIUM TITANATE.
JANTA J   02-0088   INFLUENCE OF THE QUALITY OF TRIGLYCINE SULFATE SINGLE CRYSTALS ON THEIR DIELECTRIC PRO
                   03-0300   INFLUENCE OF THE SHAPE OF DOMAINS ON THE FERROELECTRIC HYSTERESIS LOOP.
                   03-0301   SIDEWAYS DOMAIN WALL MOTION AND (MECHANISM OF THE) HYSTERESIS LOOP OF FERROELECTRICS.
                   27-3029   LARGE SIGNAL PERMITTIVITY OF TRIGLYCINE SULFATE AT 100 MHZ. (NEAR THE CURIE TEMPERATUR
                   27-3064   INFLUENCE OF SURFACE EFFECTS ON THE PYROELECTRIC BEHAVIOR OF TGS CLOSE TO THE PHASE TR
                   27-3106   X-RAY TOPOGRAPHIC OBSERVATIONS OF RADIATION DAMAGE IN TRIGLYCINE SULFATE SINGLE CRYSTA
                   37-3655   TEMPERATURE AUTOSTABILIZATION EFFECT IN FERROELECTRICS. (CAPACITOR PROPERTIES, TANDEL)
JASKIEWICZ A   04-0752   ANOMALOUS DECAY EFFECT IN BARIUM TITANATE.
                   04-0753   FIELD INDUCED NUCLEATION AT THE (CUBIC TETRAGONAL) PHASE TRANSITION IN BARIUM TITANATE
                   04-0754   NUCLEATION PROCESS IN BARIUM TITANATE AT CUBIC- TETRAGONAL PHASE TRANSITIONS.
                   08-1363   DIPOLE PATTERNS IN ORTHORHOMBIC AND TRIGONAL PHASES OF ABO(3) SUBSTANCE. (WHEN THE B I
                   08-1364   FERROELECTRICALLY ACTIVE A-IONS IN ABO(3) STRUCTURES. (ANTIFERROELECTRICITY ORIGINS)
JEFFERY JW   25-2913   OPTICAL PROPERTIES OF THIOUREA.
JEGGO CR   11-1734   NONLINEAR OPTICAL POLARIZABILITY OF THE NIOBIUM- OXYGEN BOND. (LITHIUM NIOBATE, LITHIU
                   30-3286   SECOND HARMONIC GENERATION IN MOLYBDATES (BETA GADOLINIUM MOLYBDATE AND TERBIUM MOLYBD
JEITSCHKO W   30-3287   COMPREHENSIVE X-RAY STUDY OF THE FERROELECTRIC- FERROELASTIC AND PARAELECTRIC- PARAELA
                   30-3288   THE CRYSTAL STRUCTURE OF FERROELECTRIC GADOLINIUM MOLYBDATE.
JENEVEAU A   34-3447   DIPOLAR RELAXATION AT LOW TEMPERATURE OF ICE SINGLE CRYSTAL.
                   34-3448   ELECTRICAL BEHAVIOR OF ICE AT LOW TEMPERATURES, FERROELECTRICITY AND SPACE CHARGE.
JENNINGS DA   17-2538   EFFICIENT SECOND HARMONIC GENERATION IN ADP WITH TWO NEW FLUORESCEIN DYE LASERS.
JERPHAGNON J   02-0078   NONLINEAR OPTICAL PROPERTIES OF MOLECULAR CRYSTALS.
                   17-2539   OPTICAL NONLINEAR SUSCEPTIBILITIES: ACCURATE RELATIVE VALUES FOR QUARTZ, AMMONIUM DIHY
JIMENEZ B   02-0079   EPR STUDY OF X-RAY DAMAGED FERROELECTRICS.
                   22-2844   COERCIVE FIELD IN LITHIUM TRIHYDROGEN DISELENITE. (EXPERIMENTAL)
                   27-2981   EFFECT OF SEVERAL PARAMETERS ON THE FERROELECTRIC- PARAELECTRIC TRANSITION IN TRIGLYCI
                   27-2982   ORDER DISORDER TRANSITION IN FERROELECTRICS. (TRIGLYCINE SULFATE, HIGHLY ACCURATE TEMP
                   27-2993   DIGLYCINE SULFATE AN INTERESTING NEW DIELECTRIC CRYSTAL SPECIES.
                   27-3030   EXPERIMENTAL DETAILED STUDY OF THE PHASE TRANSITION IN TRIGLYCINE SULFATE. (CLOSE THER
                   36-3571   YTTRIUM MANGANATE SWITCHING BEHAVIOR.
JOFFRIN J   15-2144   BOSON ECHOES. (ANTIMONY SULFOIODIDE)
JOHN V   30-3302   STRUCTURAL STUDY OF GADOLINIUM MOLYBDATE CRYSTALS.
JOHNSON AR   16-2353   ANALYSIS OF LOW FREQUENCY ELECTROOPTIC RESPONSE IN SODIUM NITRITE.
JOHNSON BC   11-1736   POWER AND LINE WIDTH OF TUNABLE STIMULATED FAR INFRARED EMISSION IN LITHIUM NIOBATE.
                   11-1858   EFFECT TUNABLE OPTICAL EMISSION FROM LITHIUM NIOBATE WITHOUT A RESONATOR.
                   12-1873   TEMPERATURE DEPENDENCE OF POLARITON DISPERSION IN LITHIUM TANTALATE. (OPTICAL MODULATI
JOHNSON DC   13-1973   OPTIMUM CRYSTAL ORIENTATION FOR A BARIUM SODIUM NIOBATE TRANSVERSE LIGHT MODULATOR. (H
JOHNSON DW   06-1143   DIELECTRIC RELAXATION IN (CERAMIC) STRONTIUM TITANATES CONTAINING (TRIVALENT) RARE EAR
JOHNSON KW   08-1365   ELECTRON TUNNELING INTO POTASSIUM TANTALATE SCHOTTKY BARRIER JUNCTIONS.
JOHNSTON AR   04-0755   DISPERSION OF ELECTROOPTIC EFFECT IN BARIUM TITANATE. (BIREFRINGENCE, PRINCIPAL INDICE
                   04-0756   DISPERSION OF THE ELECTROOPTIC EFFECT IN BARIUM TITANATE.
JOHNSTON WD   11-1735   (MECHANISM OF) OPTICAL INDEX DAMAGE IN LITHIUM NIOBATE AND OTHER PYROELECTRIC INSULATO
JOLLS KR   27-3031   A NEW FERROELECTRIC TRANSDUCER FOR USE IN HEAT TRANSFER AND FLOW STUDIES. (TRIGLYCINE
JONA F   28-3188   STRUCTURAL EFFECTS OF (DAMAGE FROM) IONIZING RADIATION IN FERROELECTRIC ROCHELLE SALT.
JONES CK   06-1118   ULTRASONIC STUDIES OF CRITICAL PHENOMENA IN INSULATING AND SEMICONDUCTING STRONTIUM TI
JONES ED   17-2436   ABSENCE OF A LINE WIDTH TRANSITION FOR THE PROTONS IN POTASSIUM DIHYDROGEN PHOSPHATE T
JONES GR   35-3494   ELECTROOPTIC MEASUREMENTS ON LITHIUM GERMANATE AND LEAD GERMANATE. (LITHIUM GERMANATE
JONES SJ   34-3449   IMPURITY EFFECTS ON THE PLASTICITY OF ICE AND THEIR EXPLANATION IN TERMS OF HYDROGEN R
JONES WR   11-1789   OPTICAL STUDIES OF REFRACTION AND REFLECTION OF ULTRASONIC SURFACE WAVES BY THIN GOLD
                   11-1846   FUNDAMENTAL AND HIGHER ORDER MODE SURFACE WAVE PROPAGATION ON LAYERED ANISOTROPIC SUBS
JONGEJANS A   03-0359   INPUT ADMITTANCE AND FIELDS RADIATED BY AN OPEN WAVEGUIDE FILLED WITH HIGH PERMITTIVIT
JONKER GH   04-0757   NATURE OF AGING IN FERROELECTRIC CERAMICS. (BARIUM TITANATE, LEAD ZIRCONATE TITANATE)
                   04-0758   SOME ASPECTS OF SEMICONDUCTING BARIUM TITANATE.
                   09-1543   DEMONSTRATION OF AGING EFFECTS IN BARIUM TITANATE- BARIUM ZIRCONATE.
JORGENSEN PJ   11-1737   POINT DEFECTS IN LITHIUM NIOBATE SINGLE CRYSTALS USED IN LASER MODULATION.
JOSEPH RI   03-0304   OPTICAL MODES OF THE PEROVSKITE STRUCTURE. (THEORY, STRONTIUM TITANATE DATA)
                   04-0653   BINARY LIGHT BEAM DEFLECTION IN SINGLE CRYSTAL BARIUM TITANATE.
                   04-0654   BINARY LIGHT BEAM DETECTION IN SINGLE CRYSTAL BARIUM TITANATE.
                   08-1366   INFRARED ACTIVE MODES IN POTASSIUM TANTALATE.
JOSHI SG   11-1738   ELECTRON BEAM SENSING OF SURFACE ELASTIC WAVES FOR SIGNAL PROCESSING PURPOSES. (TECHNI
JOUBERT JC   14-2079   PREPARATION OF NICKEL BORACITE CRYSTALS BY HYDROTHERMAL SYNTHESIS.
JOUVE P   25-2910   INFRARED ABSORPTION SPECTRUM OF A SINGLE CRYSTAL OF THIOUREA IN THE PARA- AND FERROELE
JUNG B   17-2540   DEPOLARIZATION OF LINEARLY POLARIZED LIGHT DUE TO HEATING OF PARALLEL PLANES OF POTASS
KAATZE U   04-0759   MEASUREMENT OF THE PERMITTIVITY OF BARIUM TITANATE SINGLE CRYSTALS AT MICROWAVE FREQUE
KABALKINA SA   04-0760   X-RAY INVESTIGATION OF THE EFFECT OF HYDROSTATIC PRESSURE ON THE STRUCTURE OF BARIUM T
                   05-1064   AN X-RAY STUDY OF THE EFFECT OF HYDROSTATIC PRESSURE ON THE STRUCTURE OF LEAD TITANATE

KABANOV AA	16-2319	EFFECT OF ADDING TITANIUM(+) AND STRONTIUM(2+) IONS ON THE ELECTRICAL CONDUCTIVITY OF
KABELKA V	17-2468	PARAMETRIC LIGHT AMPLIFICATION AND OSCILLATION IN KDP WITH MODE LOCKED PUMP.
KACHALOV NP	15-2237	CRITICAL POINT IN FERROELECTRIC ANTIMONY SULFOIODIDE.
KACHI S	04-0703	KERR EFFECT STUDIES OF ELECTRIC CONDUCTION IN IRON DOPED BARIUM TITANATE SINGLE CRYSTA
KACSON L	34-3416	MOSSBAUER EFFECT OF FERROUS IONS IN CUBIC ICE.
KACZMAREK F	27-3032	DIELECTRIC DISPERSION IN FERROELECTRIC TRIGLYCINE SULFATE (SEVERAL KHZ TO 23 KMHZ, DOM
KADABA PK	16-2320	NITROGEN-14 NUCLEAR QUADRUPOLE RESONANCE IN FERROELECTRIC SODIUM NITRITE.
	31-3332	BROMINE NUCLEAR QUADRUPOLE RESONANCE IN FERROELECTRIC HYDROGEN BROMIDE.
	31-3335	NUCLEAR QUADRUPOLE RESONANCE OF IODINE-127 IN SOLID HYDROGEN IODIDE.
KAFALAS JA	04-0761	INFLUENCE OF MADELUNG ENERGY AND COVALENCY ON THE STRUCTURE OF AB(5) OXIDE COMPOUNDS.
KAGAN YUA	36-3587	PREPARATION OF THE TERNARY TITANATE BARIUM(0.8) LEAD(0.12) CALCIUM(0.08) TITANATE.
KAHANE A	03-0306	POLARIZATION WAVES IN SOLIDS. (INCLUDING DEVELOPMENTS IN THEORY OF FERROELECTRIC PROPE
	34-3410	LOW ENTROPY FORM OF ICE I(H) OBTAINED FROM A LINEAR STEP GROWTH MODEL.
	34-3424	DYNAMIC POLAR MODEL OF MONOCRYSTALLINE ICE I(H).
	34-3450	EXPERIMENTAL AND THEORETICAL STUDIES ON THE DC CONDUCTIVITY OF ICE.
KAHN AH	03-0305	POLARONS IN ANISOTROPIC ENERGY BANDS (THEORY, CALCULATIONS FOR STRONTIUM TITANATE).
KAHN M	04-0595	EFFECTS OF GRAIN GROWTH ON THE DISTRIBUTION OF NIOBIUM IN BARIUM TITANATE CERAMICS.
	04-0762	EFFECT OF HEAT TREATMENT ON THE PTCR ANOMALY IN SEMICONDUCTING BARIUM TITANATE.
	04-0763	INFLUENCE OF GRAIN GROWTH ON DIELECTRIC PROPERTIES OF NIOBIUM DOPED BARIUM TITANATE.
	04-0764	OHMIC CONTACTS TO SEMICONDUCTING BARIUM TITANATE.
	04-0765	PREPARATION OF SMALL GRAINED AND LARGE GRAINED CERAMICS FROM NIOBIUM DOPED BARIUM TITA
KAKUDO M	36-3524	THE CRYSTAL STRUCTURE OF TRIS-SARCOSINE CALCIUM CHLORIDE.
KAKUTANI H	36-3572	DIELECTRIC ABSORPTION IN ORIENTED POLYVINYLIDENE FLUORIDE.
KALINKINA IN	17-2508	NONLINEAR PROPERTIES OF A RUBIDIUM DIHYDROGEN PHOSPHATE CRYSTAL.
KALISKI S	11-1739	SIMPLIFIED CHARACTERISTIC EQUATION FOR PLANE PIEZOELECTRIC VIBRATIONS OF LITHIUM NIOBA
	11-1740	SOME PROPERTIES OF THE EQUATIONS OF PIEZOELECTRIC VIBRATION OF LITHIUM NIOBATE AND TAN
KALITINA MP	08-1423	ELECTROOPTICAL PROPERTIES OF LITHIUM IODATE SINGLE CRYSTALS.
KALMAN A	22-2845	CRYSTAL STRUCTURE OF POTASSIUM SELENATE.
KALNAJS J	04-0553	TOP SEEDED SOLUTION GROWTH OF OXIDE CRYSTALS FROM NONSTOICHIOMETRIC MELTS. (BARIUM TIT
KALNINS A	03-0224	DYNAMIC SHELL THEORY FOR FERROELECTRIC CERAMICS.
KALYMNIOS D	17-2541	OPTIMISATION OF ELECTROOPTIC CRYSTAL MODULATORS. (KDP)
KAMAEV VE	27-3110	DISTINCTIVE FEATURES OF BARKHAUSEN EFFECT IN ROCHELLE SALT AND TRIGLYCINE SULFATE CRYS
	27-3111	INVESTIGATION OF BARKHAUSEN EFFECT IN TRIGLYCINE SULFATE CRYSTALS.
	27-3123	INFLUENCE OF GAMMA-IRRADIATION ON BARKHAUSEN EFFECT IN FERROELECTRIC MATERIALS (TGS AN
KAMB B	34-3438	DEUTERON ARRANGEMENTS IN THE HIGH PRESSURE FORMS OF ICE.
	34-3451	STRUCTURES OF THE FORMS OF ICES. (ICES II, VIII AND IX ARE FULLY ORDERED (ANTIFERROELE
KAMEYAMA H	26-2934	TEMPERATURE DEPENDENCE OF SOUND VELOCITY IN DICALCIUM STRONTIUM PROPIONATE.
KAMIMURA H	15-2177	NOTES ON ABSORPTION BAND EDGE OF (PARAELECTRIC) ANTIMONY SULFOIODIDE.
KAMINOW IP	01-0016	LINEAR ELECTROOPTICAL MATERIALS. (REVIEW, 138 REFS)
KAMINSKII AA	11-1713	ABSORPTION AND LUMINESCENCE SPECTRA AND ENERGY LEVELS OF NEODYMIUM(3+) AND ERBIUM(3+)
	11-1741	LASER AND SPECTROSCOPIC PROPERTIES OF ACTIVATED FERROELECTRICS. (LITHIUM NIOBATE AND G
	30-3255	CERTAIN GROWTH, LASING AND SPECTROSCOPIC PROPERTIES OF NEODYMIUM(3+) DOPED GADOLINIUM
KAMIURA Y	17-2544	TEMPERATURE DEPENDENCE OF INDEX MATCHING ANGLE OF THE SECOND HARMONIC GENERATION IN PO
	17-2548	EFFECT OF IONIC MOTION ON THE REFRACTIVE INDEX OF PARAELECTRIC KDP IN THE VICINITY OF
	17-2550	TEMPERATURE DEPENDENCE OF RAMAN SPECTRUM AND SECOND HARMONIC GENERATION IN KDP.
KAMYSHEVA LN	17-2466	DIELECTRIC PROPERTIES OF A (SMALL DOSE) GAMMA-IRRADIATED KDP CRYSTAL (TEMPERATURE DEPE
	17-2467	DIELECTRIC PROPERTIES OF A GAMMA-IRRADIATED POTASSIUM DIHYDROGEN PHOSPHATE CRYSTAL.
	17-2542	DIELECTRIC PROPERTIES OF GAMMA-IRRADIATED POTASSIUM DIHYDROGEN PHOSPHATE CRYSTALS.
	17-2543	NONLINEAR PROPERTIES OF POTASSIUM DIHYDROGEN PHOSPHATE. (DETERMINATION OF BETA IN THE
	27-2976	DIELECTRIC LOSSES OF TRIGLYCINE SULFATE CRYSTALS EXPOSED TO SMALL DOSES OF X- AND GAMM
KAMZINA LS	03-0307	ELASTOOPTICAL EFFECT IN FERROELECTRIC WITH DIFFUSE PHASE TRANSITION.
	13-1974	ANOMALOUS BEHAVIOR OF THE OPTICAL ABSORPTION COEFFICIENT OF LEAD MAGNONIOBATE SINGLE
	36-3521	ANISOTROPY OF THE ELECTROOPTICAL EFFECT IN CRYSTALS OF LEAD MAGNONIOBATE.
KANAMARA F	08-1393	HIGH PRESSURE SYNTHESIS (SOLID STATE REACTION, SINTERING) AND PROPERTIES OF PEROVSKITE
KANASHIRO T	16-2321	ACOUSTIC NUCLEAR MAGNETIC RESONANCE IN FERROELECTRIC SODIUM NITRITE.
KANATANI K	08-1472	LONGITUDINAL ELECTROOPTIC EFFECT IN OBLIQUE CUT STRONTIUM BARIUM NIOBATE PLATES.
	13-2063	LONGITUDINAL ELECTROOPTIC EFFECT IN OBLIQUE CUT STRONTIUM BARIUM NIOBATE PLATES.
KANNO K	17-2682	THERMAL HYSTERESIS OF THE FERROELECTRIC TRANSITION IN POTASSIUM DIHYDROGEN PHOSPHATE.
KANTOLA M	16-2322	X-RAY MEASUREMENTS ON THERMAL EXPANSION OF FERROELECTRIC POTASSIUM NITRATE.
KAPISHEV AG	10-1655	NEUTRON DIFFRACTION STUDY OF (TETRAGONAL- CUBIC PEROVSKITE) STRONTIUM IRON TUNGSTATE .
KAPLAN SF	22-2846	NEUTRON DIFFRACTION STUDY ON PARAELECTRIC SODIUM TRIHYDROGEN DISELENITE.
KAPYSHEV AG	02-0128	NEW FERROELECTRICS AND SEIGNETO MAGNETS OF THE PEROVSKITE AND PYROCHLORE TYPE. SYNTHE
	08-1362	X-RAY DETERMINATION OF THE SYMMETRY OF THE FERROELECTRIC COMPOUNDS (POTASSIUM OR SODIU
	08-1458	CRYSTAL STRUCTURE AND PROPERTIES OF NEW CADMIUM AND THALLIUM CONTAINING PEROVSKITES. (
	08-1460	STRUCTURE AND PROPERTIES OF SOME NEW FERROELECTRIC MATERIALS WITH PEROVSKITE STRUCTURE
	10-1640	THALLIUM PEROVSKITES. THALLIUM A(1-X) TUNGSTATE OR MOLYBDATE. (WHERE A IS ZIRCONIUM, H
	36-3619	NEW FERROELECTRICS. (BARIUM(2) BISMUTH (NIOBIUM OR TANTALUM OR VANADIUM) OXIDE(6), AND
KARAMYAN AA	33-3384	INFRARED ABSORPTION SPECTRA OF POTASSIUM MANGANESE TRIFLUORIDE, POTASSIUM COBALT TRIFL
	33-3385	VIBRATIONAL SPECTRUM AND DISPERSION OF THE OPTICAL CONSTANTS OF ANTIFERROMAGNETIC POTA
KARCHEVSKII AI	04-0766	ELECTROCALORIC EFFECT IN POLYCRYSTALLINE BARIUM TITANATE.
KARGOPOLOVA NP	04-0767	ULTRAHIGH FREQUENCY DIELECTRIC ANISOTROPY OF A POLARIZED FERROELECTRIC CERAMIC. (BARIU
KARIAN HG	08-1334	TIGHT BONDING BAND CALCULATIONS ON RHENIUM TRIOXIDE, SODIUM TUNGSTATE, AND POTASSIUM T
KARPOV SV	16-2323	PECULIARITIES IN THE VIBRATIONAL SPECTRA OF CESIUM AND RUBIDIUM NITRATE CRYSTALS IN IT
KARR MA	13-1975	NEODYMIUM: YTTRIUM ALUMINUM GARNET- BARIUM SODIUM NIOBATE 0.66 MILLIMICRON HARMONIC SO
KARYAKINA NF	36-3559	FERROELECTRIC BEHAVIOR OF THE MINERAL STIBOTANTALITE.
	36-3601	DOMAIN STRUCTURE AND TEMPERATURE DEPENDENCE OF THE PIEZOELECTRIC CONSTANTS OF STIBOTAN
KASAHARA M	16-2307	SPLITTING OF THE SODIUM-23 NMR CENTRAL LINE IN FERROELECTRIC SILVER SODIUM DINITRITE.
	16-2324	NMR STUDY OF SODIUM-23 IN SILVER SODIUM DINITRITE.
	19-2760	EFFECT OF DEUTERATION IN AMMONIUM BISULFATE (CHANGE IN TRANSITION TEMPERATURES).
KASAI T	11-1742	PRECISION MACHINING OF LITHIUM NIOBATE CRYSTALS FOR ELECTROOPTIC ELEMENTS.
KASATKIN AP	08-1367	CRYSTALLIZATION OF POTASSIUM IODATE.
KASHCHENKO EP	04-0768	INFLUENCE OF THE AMPLITUDE OF FREE VIBRATIONS ON THE PERMITTIVITY OF BARIUM TITANATE S
	04-0791	AMPLITUDE DEPENDENCE OF THE INTERNAL FRICTION IN BARIUM TITANATE SINGLE CRYSTALS. (PIE
	04-0794	INTERNAL FRICTION IN BARIUM TITANATE SINGLE CRYSTALS.
	04-0796	TEMPERATURE DEPENDENCE OF INTERNAL FRICTION OF BARIUM TITANATE SINGLE CRYSTALS.
KASHLINSKII AI	08-1368	ELECTRON SPIN RESONANCE (SPECTRA, TEMPERATURE DEPENDENCE) IN SOME FERROELECTRICS AND F
	36-3618	A NEW MAGNETOFERROELECTRIC: CADMIUM IRON NIOBATE.
KASHTANOVA AM	13-1952	NONLINEAR PROPERTIES AND PHASE TRANSITIONS OF STRONTIUM BISMUTH TITANATES.
KASPRZAK J	28-3230	EFFECT OF A DC FIELD ON DOMAIN STRUCTURE IN SEIGNETTE SALT. (EFFECT OF AGING AND COPPE
KASS M	34-3469	DIELECTRIC RELAXATION, BULK AND SURFACE CONDUCTIVITY OF ICE SINGLE CRYSTALS.

KATAOKA S	04-0741	BARIUM TITANATE FILMS PREPARED BY RF SPUTTERING ONTO INDIUM ANTIMONIDE OR GALLIUM ARSE
KATIYAR RS	04-0619	COUPLED POLARITIONS OF A1 SYMMETRY IN BARIUM TITANATE.
	16-2257	RELATIONSHIP OF THE TEMPERATURE DEPENDENCE OF B1 PHONON AND DIELECTRIC RELAXATION IN S
	17-2639	RAMAN SCATTERING FROM FERROELECTRIC MODES IN THE KDP ISOMORPHOUS PHOSPHATES AND ARSENA
	17-2705	RAMAN SCATTERING BY THE TEMPERATURE DEPENDENT FERROELECTRIC MODE IN POTASSIUM DIDEUTER
	18-2723	DIELECTRIC RESPONSE IN PIEZOELECTRIC CRYSTALS. (RAMAN SCATTERING IN POTASSIUM DIHYDROG
	18-2733	PROTON PHONON COUPLING IN CESIUM DIHYDROGEN ARSENATE AND POTASSIUM DIHYDROGEN ARSENATE
KATO M	04-0790	WET SYNTHETIC METHOD OF MAKING PURE, EXACTLY STOICHIOMETRIC BARIUM TITANATE.
KATO R	16-2398	VACUUM ULTRAVIOLET ABSORPTION OF THE ALKALI NITRITES AND NITRATES.
	16-2399	VACUUM ULTRAVIOLET ABSORPTION (5-23 EV AT ROOM AND LIQUID NITROGEN TEMPERATURES IN ALK
KATO T	27-3033	DISPLACIVE MOVEMENT OF THE RADICAL IN TRIGLYCINE SULFATE NEAR THE CURIE POINT. (ESR OF
	27-3034	ESR OF GAMMA-IRRADIATED TGS. PART-2: INTERNAL MOTION OF HYDROGEN AND THE ACTIVATION EN
	27-3035	ESR IN GAMMA-IRRADIATED TGS.
	28-3203	NUCLEAR MAGNETIC RESONANCE IN ROCHELLE SALT, DETERMINATION OF THE ORIENTATIONS OF PROT
KATO Y	27-2951	A NEW SECOND HARMONIC TYPE FERROELECTRIC MODULATOR FOR ELECTROMETER. (WITH A TGS SINGL
KATPATAL AG	08-1369	SURFACE CHARGE LAYERS ON FERROELECTRIC POTASSIUM NIOBATE CRYSTALS.
KATSURA S	03-0308	DISTRIBUTION OF ZEROES OF THE PAPTITION FUNCTION FOR THE SLATER MODEL OF FERROELECTRIC
KATSURI SR	17-2545	INVESTIGATION OF THE PHASE TRANSITION IN AMMONIUM DIHYDROGEN PHOSPHATE SINGLE CRYSTAL
KATZ L	02-0094	ROLE OF THE REGULAR OCTAHEDRON AND LARGER POLYHEDRA IN FERROELECTRIC MATERIALS. (HETER
KAUAMORI A	22-2852	PROTON SPIN LATTICE RELAXATION IN FERROELECTRIC SODIUM TRIHYDROGEN SELENATE.
KAUSHANSKII DA	04-0884	CURRENT EXCITATION IN GAMMA-IRRADIATED BARIUM TITANATE.
KAVERIN LD	11-1798	INTERNAL FRICTION OF LITHIUM NIOBATE SINGLE CRYSTALS AT 1-40 HZ.
KAWAEATA A	08-1271	PIEZOELECTRIC PROPERTIES OF POTASSIUM TANTALATE NIOBATE SINGLE CRYSTAL.
	09-1612	ELASTIC SURFACE WAVE ON THE PZT CERAMIC PLATE.
	09-1614	SURFACE WAVE DELAY LINES WITH INTERDIGITAL TRANSDUCERS ON UNPOLARIZED PZT CERAMIC PLAT
	27-3149	DIELECTRIC BEHAVIOR OF GAMMA-RAY IRRADIATED TRIGLYCINE SULFATE.
KAWABE K	01-0017	FERROELECTRICS.
	17-2544	TEMPERATURE DEPENDENCE OF INDEX MATCHING ANGLE OF THE SECOND HARMONIC GENERATION IN PO
	17-2548	EFFECT OF IONIC MOTION ON THE REFRACTIVE INDEX OF PARAELECTRIC KDP IN THE VICINITY OF
	17-2550	TEMPERATURE DEPENDENCE OF RAMAN SPECTRUM AND SECOND HARMONIC GENERATION IN KDP.
KAWADA S	27-3023	DIELECTRIC DISPERSION OF TRIGLYCINE SULFATE AT LOW FREQUENCIES.
	34-3452	DIELECTRIC DISPERSION AND PHASE TRANSITION OF POTASSIUM HYDROXIDE DOPED ICE.
KAWAI H	36-3573	PIEZOELECTRICITY OF POLYVINYLIDENE FLUORIDE.
KAWAKUBO T	24-2902	ANOMALOUS SPECIFIC HEAT OF FERROELECTRIC THIOUREA AND POTASSIUM FERROCYANIDE TRIHYDRAT
KAWAMURA A	16-2253	A RADICAL IN FERROELECTRIC DGN CRYSTAL.
KAWAMURA T	17-2549	FAR INFRARED SPECTRA OF KDP AND ADP. (20-550 PER CM, 200 AND 300 DEG K)
	17-2711	AN OBSERVATION OF THE CHANGE IN COOLED KDP TEMPERATURE DUE TO 337 MICRON INCIDENT RADI
KAWANA A	11-1712	GROWTH AND PROPERTIES OF LITHIUM NIOBATE AND LITHIUM TANTALATE FILMS.
KAWANO I	10-1637	ELECTRICAL CONDUCTIVITY OF TUNGSTEN TRIOXIDE.
KAWANO T	17-2546	EPR OF CHROMIUM(3+) IONS IN ANTIFERROELECTRIC ADP CRYSTALS.
	17-2547	ESR STUDY OF RADIATION DAMAGE CENTERS IN AMMONIUM DIHYDROGEN PHOSPHATE POTASSIUM DIHYD
	17-2596	ELECTRON PARAMAGNETIC RESONANCE OF CHROMIUM(3+) IN ADP.
KAY HF	27-3036	THICKNESS DEPENDENCE OF THE NUCLEATION FIELD OF TRIGLYCINE SULFATE.
KAY MI	13-1935	CRYSTAL STRUCTURE OF BISMUTH TITANIUM OXIDE.
	16-2325	DISORDERED STRUCTURE OF SODIUM NITRITE AT 185 DEG C.
	16-2326	NEUTRON DIFFRACTION STUDY OF SODIUM NITRITE.
	16-2327	STRUCTURE OF SODIUM NITRITE AT 150, 185, AND 225 DEGREES C.
	16-2328	X-RAY AND NEUTRON STUDY ON THE PHASE TRANSFORMATION OF SODIUM NITRITE.
	22-2846	NEUTRON DIFFRACTION STUDY ON PARAELECTRIC SODIUM TRIHYDROGEN DISELENITE.
	26-2935	APPROXIMATE STRUCTURE FOR ANTIFERROELECTRIC PHASE OF COPPER FORMATE TETRAHYDRATE BY NE
KAY SM	17-2705	RAMAN SCATTERING BY THE TEMPERATURE DEPENDENT FERROELECTRIC MODE IN POTASSIUM DIDEUTER
KAZAKEVICH LI	04-0573	CHANGE OF THE MOSSBAUER EFFECT PROBABILITY ON TIN-119 IN BARIUM TITANATE STANNATE SOLI
KAZAKOV AL	03-0284	RESONANCE OF A LOW FREQUENCY SPIN WAVE WITH AN ACOUSTIC OR ELECTROMAGNETIC WAVE IN A F
KAZAMI A	03-0323	ON SURFACE STATE AND POLARIZATION REVERSAL OF FERROELECTRIC CRYSTALS.
	03-0326	THEORY CF POLARIZATION REVERSAL OF BARIUM TITANATE CRYSTAL.
KEEFE S	25-2912	DIELECTRIC DISPERSION OF UREA AND THIOUREA.
KEESTER KL	13-1976	A NEW CLASS OF FERROELECTRIC CERAMICS OF THE TYPE AB(2)(XO(4))(3).
KEJST J	27-2955	INFLUENCE OF THE ELECTRIC FIELD AND THE DOMAIN STRUCTURE ON THE ULTRASONIC ABSORPTION
KELL RC	02-0080	MODERN APPLICATIONS OF FERROELECTRICITY.
KELLER G	14-2080	(OPTICAL) FREQUENCY DOUBLING IN (MAGNESIUM CHLORINE, ZINC CHLORINE, CADMIUM CHLORINE,
KELLY EJ	06-1221	ENERGY BAND STRUCTURE OF STRONTIUM TITANATE FROM A SELF CONSISTENT FIELD TIGHT BINDING
KELLY R	10-1642	BOMBARDMENT INDUCED AMORPHIZATION IN TUNGSTEN TRIOXIDE AND ITS USE IN DEDUCING MEAN DA
KEMPER G	08-1370	NOTE ON THE SPACE GROUP OF POTASSIUM HYDROGEN IODATE.
KENEMAN SA	04-0769	FERROELECTRIC PHOTOCONDUCTOR OPTICAL STORAGE MEDIUM UTILIZING BISMUTH TITANATE.
	08-1371	BISMUTH TITANATE FERROELECTRIC- PHOTOCONDUCTOR OPTICAL STORAGE MEDIUM FURTHER DEVELOPM
	08-1445	BISMUTH TITANATE PHOTOCONDUCTOR OPTICAL STORAGE MEDIUM. (HIGH RESOLUTION READ WRITE)
	13-1977	PHASE HOLOGRAMS IN A FERROELECTRIC PHOTOCONDUCTOR DEVICE. (BISMUTH(4) TITANIUM(3) OXID
	37-3646	STORAGE OF HOLOGRAMS IN A FERROELECTRIC PHOTOCONDUCTOR DEVICE (ERASABLE, NONDESTRUCTIV
	37-3675	ACTIVE COMPENSATORS FOR FERROELECTRIC OPTICAL CIRCUITS.
KENNEDY SW	16-2329	METASTABLE POTASSIUM NITRATE(III) FROM SOLUTION. (GROWN IN WATER OR ALCOHOL)
	16-2330	RUBIDIUM NITRATE NEAR THE MELTING POINT AND ITS POLYMORPHISM.
KENNEMAN SA	08-1446	DEPOLING OF SINGLE DOMAIN BISMUTH TITANATE.
KENNEY JF	08-1372	ENERGY BANDS OF POTASSIUM NIOBATE.
KEONG JT	16-2331	TEMPERATURE AND PRESSURE DEPENDENCE OF THE DIELECTRIC CONSTANT AND SPONTANEOUS POLARIZ
KERN R	15-2178	ELECTROOPTICAL AND ELECTROMECHANICAL EFFECT IN ANTIMONY SULFOIODIDE.
KESSEL AR	03-0309	KINETIC EQUATION FOR QUADRUPOLE RESONANCE IN FERROELECTRICS NEAR THE PHASE TRANSITION
	03-0310	MACROSCOPIC ANALOG OF THE SPIN- ECHO EFFECT IN POLYCRYSTALLINE FERROELECTRICS. (THEORY
	03-0311	NATURE OF PHASE TRANSITIONS IN FERROELECTRIC AND ANTIFERROELECTRIC MATERIALS.
KESSENIKH GG	03-0312	PIEZOELECTRIC EFFECT ON THE REFLECTION OF A TRANSVERSE ACOUSTIC WAVE FROM DOMAIN BOUND
	03-0313	REFLECTION AND REFRACTION OF TRANSVERSE SOUND WAVES AT DOMAIN BOUNDARIES IN FERROELECT
	03-0314	REFLECTION AND REFRACTION OF QUASI LONGITUDINAL AND QUASI TRANSVERSE SOUND WAVES BY 18
	03-0315	SURFACE ELASTOPOLARIZATION WAVES AT DOMAIN BOUNDARIES IN A FERROELECTRIC.
	28-3204	ATTENUATION OF SOUND WAVES IN DEUTERATED ROCHELLE SALT AT THE PHASE TRANSITION POINT.
	28-3205	ATTENUATION OF SOUND WAVES IN DEUTERATED ROCHELLE SALT AT THE PHASE TRANSITION POINT.
KESTIGIAN M	12-1892	ANOMALOUS (LONGITUDINAL WAVE) ATTENUATION IN LITHIUM TANTALATE (EFFECT OF VACANCIES).
KESZTHELYI L	34-3416	MOSSBAUER EFFECT OF FERROUS IONS IN CUBIC ICE.
KEVE ET	03-0135	STRUCTURAL BASIS OF FERROELECTRICITY AND FERROELASTICITY. (SYMMETRY RELATIONS)
	03-0136	DIELECTRIC, ELASTIC AND MAGNETIC PROPERTY, SYMMETRY, AND PHASE TRANSITION REPRESENTATI
	10-1649	STRUCTURAL AND PHASE RELATIONSHIPS AMONG TRIVALENT TUNGSTATES AND MOLYBDATES.
	11-1673	X-RAY DETERMINATION OF POLARITY SENSE BY ANOMALOUS SCATTERING AT AN ABSORPTION EDGE. (

KEVE ET   13-1978   BISMUTH(3) NIOBIUM(17) OXIDE(47): POTENTIALLY FERROELECTRIC CRYSTAL STRUCTURE OF THE
                      17-2469   STRUCTURAL INHIBITION OF FERROELECTRIC SWITCHING IN TRIGLYCINE SULFATE. PART-2: X-IRRA
                      27-2977   HIGH INTERNAL BIAS FIELDS IN L-ALANINE SUBSTITUTED TGS.
                      27-2996   CRYSTALLOGRAPHIC STUDIES OF IRRADIATION FIELD TREATED TRIGLYCINE SULFATE: NEW STRUCTU
                      27-3037   EFFECTS OF ADDITIVES ON THE PYROELECTRIC PROPERTIES OF TGS.
                      27-3038   STRUCTURAL INHIBITION OF FERROELECTRIC SWITCHING IN TRIGLYCINE SULFATE. PART-1: ADDITI
                      27-3058   POLARIZED TRIGLYCINE SULFATES.
                      30-3289   FERROELECTRIC FERROELASTIC PARAMAGNETIC BETA GADOLINIUM MOLYBDATE. CRYSTAL STRUCTURE O
                      30-3290   (ROOM TEMPERATURE) FERROELECTRIC FERROELASTIC PARAMAGNETIC (BETA) TERBIUM MOLYBDATE (C
                      33-3386   FERROELECTRIC PARAELASTIC PARAMAGNETIC BARIUM COBALT TETRAFLUORIDE CRYSTAL STRUCTURE (
                      36-3574   PYROELECTRIC AMMONIUM IODATE, A POTENTIAL FERROELASTIC: CRYSTAL STRUCTURE.
KHAMSKIJ EV   16-2332   INFLUENCE OF THE CONDITIONS OF CRYSTALLIZATION ON THE FORM OF CRYSTALS OF POTASSIUM NI
KHAN AA   17-2485   X-RAY DETERMINATION OF THE THERMAL EXPANSION OF AMMONIUM DIHYDROGEN PHOSPHATE.
KHANNA RK   16-2342   POLARIZED INFRARED AND RAMAN SPECTRA OF SINGLE CRYSTAL CESIUM NITRITE.
KHARASH EV   09-1591   EFFECT OF FERROELECTRIC PHASE TRANSITION CONDITIONS ON DIELECTRIC AND PIEZOELECTRIC PR
KHARCHENKO LT   33-3398   MANDELSTAM BRILLOUIN SCATTERING OF LIGHT IN MANGANESE FLUORIDE CRYSTALS (POTASSIUM TRI
KHARCHENKO VA   04-0970   EFFECTS OF (MCNCENERGETIC) ELECTRON AND (COBALT-60) GAMMA-IRRADIATION ON BARIUM TITANA
KHARITONOV EV   04-0640   ELECTRON PARAMAGNETIC RESONANCE IN NONSTOICHIOMETRIC BARIUM TITANATE.
KHARITONOV VD   03-0485   ENERGY SPECTRUM OF AN ORDER DISORDER FERROELECTRIC CONTAINING DISLOCATIONS.
                      03-0486   EFFECT OF POINT DEFECTS ON THE ENERGY SPECTRUM OF HYDROGEN BONDED FERROELECTRICS.
KHARITONOV YUN   03-0318   CONNECTION BETWEEN THE STARTING FIELD FOR PARKHAUSEN STEPS AND THE CRITICAL POLARIZATI
                      03-0337   BARKHAUSEN EFFECT IN FERROELECTRICS UNDER CONTINUOUS X-RAY DIFFRACTION.
KHARKYANEN VN   03-0317   ADMIXTURE OF THE 4T1G STATE TO THE 6A1G(6S) STATE OF THE MANGANESE(2+) ION IN A CUBIC
KHASABOV AG   15-2179   BAND STRUCTURE OF THE FERROELECTRIC SEMICONDUCTOR ANTIMONY SULFOIODIDE.
                      15-2204   ELECTRON STATE DENSITY AND OPTICAL PROPERTIES OF ANTIMONY SULFOIODIDE. (IN PARAELECTRI
                      36-3575   BAND STRUCTURE OF THE FERROELECTRIC SEMICONDUCTOR ANTIMONY TRISULFIDE.
KHASHKHCZHEV ZM   11-1743   MANDELSTAM BRILLOUIN SCATTERING OF LIGHT IN LITHIUM NIOBATE (FROM TRANSVERSE (MORE INT
                      11-1744   SCATTERING OF LIGHT ON ACOUSTIC PHONONS IN LITHIUM NIOBATE AND LITHIUM TANTALATE.
                      12-1878   SCATTERING OF LIGHT BY ACOUSTIC PHONONS AND POLARITONS IN LITHIUM TANTALATE.
KHEIKER DM   22-2838   CRYSTAL STRUCTURES OF POTASSIUM TRIHYDROGEN SELENITE AND SODIUM TRIHYDROGEN SELENITE.
KHELIFA B   25-2910   INFRARED ABSORPTION SPECTRUM OF A SINGLE CRYSTAL OF THIOUREA IN THE PARA- AND FERROELE
KHILLA MA   10-1632   THE COLOR PROBLEM OF TUNGSTEN TRIOXIDE. X-RAY STUDIES.
                      10-1636   THE COLOR PROBLEM OF TUNGSTEN TRIOXIDE ELECTRICAL CONDUCTIVITY.
KHIMICH TA   15-2180   MOSSBAUER STUDIES OF ANTIMONY SULFOIODIDE TYPE CRYSTALS.
KHLYUSTOV VG   33-3387   SPECIFIC HEAT OF POTASSIUM MANGANESE TRIFLUORIDE.
KHMELNITSKII DE   03-0319   COMPLEX PHASE TRANSITIONS IN CRYSTALS (APPLICATION TO SINGLE DOMAIN FERROELECTRICS BEL
                      03-0320   LOW TEMPERATURE DISPLACEMENT TYPE PHASE TRANSITION IN CRYSTALS. (THEORY, FERROELECTRIC
KHODAKOV AL   08-1373   FERROELECTRIC SINGLE CRYSTALS WITH LARGE NONLINEARITY. (BARIUM TITANATE HAFNATE GROWN
KHOKHLOV RV   03-0431   STIMULATED COMBINATION SCATTERING OF SOUND IN PIEZOELECTRICS. (THEORY, LITHIUM NIOBATE
KHOLODENKO LP   03-0321   THEORY OF DOMAINLESS, NONHOMOGENEOUS, SPONTANEOUS POLARIZATION OF FERROELECTRIC PLATES
                      03-0322   THERMODYNAMICAL THEORY OF BARIUM TITANATE TYPE FERROELECTRICS.
                      04-0770   MACROSCOPIC THEORY OF BARIUM TITANATE TYPE FERROELECTRIC CRYSTALS IN THE FORM OF THICK
KHOTCHENKOV AG   04-0771   PIEZOELECTRIC ACTIVITY OF BARIUM TITANATE SINGLE CRYSTALS IN THE PARAELECTRIC PHASE.
                      08-1374   THE INVERSE PIEZOELECTRIC EFFECT IN THE REPOLARIZATION PROCESS OF FERROELECTRICS WITH
KHUCHUA NP   08-1375   MICROWAVE INVESTIGATION OF SOME FERROELECTRICS AND ANTIFERROELECTRICS OF PEROVSKITE TY
                      08-1378   DIELECTRIC PROPERTIES AND THERMAL EXPANSION OF SOME FERROELECTRICS AND ANTIFERROELECTR
KIEBURG H   17-2551   ADP POLARIZER FOR LASERS.
KIELICH S   02-0081   NONLINEAR OPTICAL AND ELECTROOPTICAL PROPERTIES OF DIELECTRICS AND FERROELECTRICS. (RE
KIHLBORG L   04-0541   ON THE PHASE COMPOSITION OF REDUCED AND REOXIDIZED BARIUM TITANATE.
KIKINESHI AA   15-2127   DOMAIN STRUCTURE AND LOCAL STATES IN ANTIMONY SULFOIODIDE TYPE SEMICONDUCTOR FERROELEC
                      15-2181   POLARIZATION NATURE IN SEMICONDUCTOR FERROELECTRICS OF THE ANTIMONY SULFOIODIDE TYPE.
                      15-2182   SOME FEATURES OF TRAPPING LEVELS IN ANTIMONY SULFOIODIDE TYPE SEMICONDUCTING FERROELEC
KIKUCHI H   14-2081   LATTICE STRAINS OF BORACITE CRYSTALS.
KIKUCHI K   04-1025   STUDIES ON THE BARIUM STRONTIUM TITANATE BOUNDARY LAYER CERAMIC DIELECTRICS.
KILARSKA J   27-3046   FIELD EFFECT IN TELLURIUM THIN FILMS ON A SUBSTRATE OF FERROELECTRIC TRIGLYCINE SULFAT
KIM JJ   17-2552   RAMAN SCATTERING FROM FERROELECTRIC CRYSTALS OF POTASSIUM DIHYDROGEN PHOSPHATE.
KIM YM   11-1803   ELECTRON SPIN RESONANCE STUDIES OF CRYSTAL FIELD PARAMETERS IN MANGANESE(2+) DOPED LIT
                      11-1804   ELECTRON SPIN RESONANCE STUDIES OF CHROMIUM(3+) IN LITHIUM NIOBATE.
                      11-1805   FORBIDDEN HYPERFINE LINES IN ESR SPECTRA OF MANGANESE(2+) DOPED LITHIUM NIOBATE.
                      11-1839   EPR STUDIES OF CRYSTAL FIELD PARAMETERS IN IRON(3+) DOPED LITHIUM NIOBATE.
KIMENEZ B   03-0152   THERMODYNAMIC THEORY OF IRRADIATED TRIGLYCINE SULFATE.
KIMURA M   08-1404   FERROELECTRIC PROPERTIES OF STRONTIUM NIOBATE SINGLE CRYSTAL.
                      13-1997   A NEW FERROELECTRIC SINGLE CRYSTAL STRONTIUM(2) NIOBIUM(2) OXIDE(7).
                      13-2037   GROWTH OF FERROELECTRIC SINGLE CRYSTAL STRONTIUM(2) NIOBIUM(2) OXIDE(7) BY MEANS OF TH
                      36-3576   ELECTROOPTIC AND PIEZOELECTRIC PROPERTIES OF LANTHANUM TITANATE SINGLE CRYSTAL. (FOUND
KIMURA S   04-0772   ELECTRICAL PROPERTIES OF THICK FILM BARIUM TITANATE DIELECTRICS PRODUCED BY FLAME SPRA
                      04-0773   FLAME SPRAYED BARIUM TITANATE AS A CAPACITOR DIELECTRIC.
                      04-0774   STUDIES OF FLAME SPRAYED BARIUM TITANATE CAPACITORS. PART-10: STRUCTURE OF SEMICONDUCT
KIMURA T   17-2640   MICROWAVE MODULATOR OF RUBY LASER LIGHT USING POTASSIUM DIHYDROGEN PHOSPHATE CRYSTAL.
KINASE W   03-0323   ON SURFACE STATE AND POLARIZATION REVERSAL OF FERROELECTRIC CRYSTALS.
                      03-0324   90 DEGREE TYPE DOMAIN WALL OF BARIUM TITANATE CRYSTAL. (THEORY, CALCULATIONS)
                      03-0325   STATISTICAL AND ELECTRONIC PROBLEMS IN FERROELECTRICITY. (THEORY OF ENERGY STATES INVO
                      03-0326   THEORY OF POLARIZATION REVERSAL OF BARIUM TITANATE CRYSTAL.
                      04-0775   ON THE LATTICE VIBRATION OF CUBIC BARIUM TITANATE.
                      04-0776   THEORY OF THE DIELECTRIC CONSTANTS OF BARIUM TITANATE.
                      10-1641   ON FERROELECTRICITY AND ANTIFERROELECTRICITY OF THE AO3 TYPE CRYSTAL. (TUNGSTEN TRIOXI
KIND R   08-1426   NUCLEAR QUADRUPOLE RESONANCE OF IODINE-127 IN SILVER PERIODATE.
                      36-3577   LEVEL CROSSING (CROSS RELAXATION) EXPERIMENTS IN AMMONIUM TRIHYDROGEN PERIODATE CRYSTA
                      36-3578   NUCLEAR QUADRUPOLE INTERACTION IN ANTIFERROELECTRIC DIAMMONIUM TRIHYDROGEN PERIODATE P
                      36-3579   NUCLEAR QUADRUPOLE INTERACTION IN ANTIFERROELECTRIC DIAMMONIUM TRIHYDROGEN PERIODATE B
                      36-3580   STUDY OF THE STRUCTURE OF ANTIFERROELECTRIC AMMONIUM PERIODATE BY NUCLEAR RESONANCE ME
                      36-3581   STATISTICAL CONSIDERATIONS ON THE ORDER DISORDER REACTION IN AMMONIUM PERIODATE.
KING SR   13-1979   OPTICAL DAMAGE IN KTN.
KINGSTON RH   17-2484   INDEX OF REFRACTION OF POTASSIUM DIHYDROGEN PHOSPHATE.
KINOSHITA K   27-3040   OPTICAL ACTIVITY OF FERROELECTRIC DICALCIUM STRONTIUM PROPIONATE. (TRANSITION AT 209 D
KINSPEL RR   22-2847   NUCLEAR SPIN LATTICE RELAXATION TIME MEASUREMENTS IN SODIUM DEUTERIUM SELENATE.
KIRICHENKO NA   36-3532   STARK EFFECT AND FERROELECTRICITY IN THE ORGANIC SEMICONDUCTOR COPPER PHTHALOCYANINE.
KIRIKOV VA   22-2830   DOMAIN STRUCTURE OF THE BETA PHASE OF SODIUM TRIHYDROGEN SELENATE IN ELECTRIC FIELDS.
                      22-2874   THE PECULIAR DOMAIN STRUCTURE AND ITS UNUSUAL BEHAVIOR IN THE BETA PHASE OF THE FERROE
KIRILLOV VV   13-1980   RELAXATION POLARIZATION OF A FERROELECTRIC LEAD MAGNESATE(0.33) NIOBATE(0.67) WITH A D

KIRIPICHNIKOVA LF	17-2666	PHASE DIAGRAM OF SODIUM TRIDEUTERIUM(X) TRIHYDROGEN(1-X) PHOSPHATE SYSTEM VERSUS CONCE
KIRIYAMA H	24-2894	NUCLEAR MAGNETIC RESONANCE AND X-RAY STUDIES OF POTASSIUM FERROCYANIDE TRIHYDRATE CRYS
	36-3582	DIELECTRIC AND NMR STUDIES OF THE PHASE TRANSITION IN STANNOUS CHLORIDE DIHYDRATE.
KIRIYAMA R	24-2894	NUCLEAR MAGNETIC RESONANCE AND X-RAY STUDIES OF POTASSIUM FERROCYANIDE TRIHYDRATE CRYS
	36-3582	DIELECTRIC AND NMR STUDIES OF THE PHASE TRANSITION IN STANNOUS CHLORIDE DIHYDRATE.
KIRKPATRICK ES	06-1144	STRONG AXIAL ELECTRON PARAMAGNETIC RESONANCE SPECTRUM OF IRON(3+) IN STRONTIUM TITANAT
KIRO D	36-3583	ELECTRON SPIN RESONANCE SPECTRA OF CHROMIUM(3+) AND GADOLINIUM(3+) IN LANTHANUM ALUMIN
KIROV KI	27-3148	EFFECT OF SAMPLE TREATMENT AND ELECTRODE NATURE ON SOME TRIGLYCINE SULFATE PARAMETERS.
KIRPICHNIKOVA LF	22-2848	(FERROELECTRIC AND OPTICAL PROPERTIES OF) CRYSTALS OF THE ALKALINE TRIHYDROGEN DISELEN
	22-2873	ISOTOPIC NONISOMORPHISM OF RUBIDIUM TRIHYDROGEN(1-X) TRIDEUTERIUM(X) DISELENATE CRYSTA
KISAKA S	04-0744	FAR INFRARED REFLECTIVITY OF BARIUM TITANATE.
KISELEV DF	11-1717	TEMPERATURE AND FREQUENCY DEPENDENCES OF THE ELECTRICAL PROPERTIES OF LITHIUM METANIOB
KISELEVA KV	06-1145	STRUCTURE OF POLYCRYSTALLINE SOLID SOLUTIONS OF STRONTIUM AND BISMUTH TITANATES.
KISLOVSKII LD	03-0330	(EFFECT OF TEMPERATURE ON) QUASI SYMMETRICAL HARD IONS (PHOSPHATE, TITANATE, SULFATE,
KISZENICK W	34-3415	ELECTRICAL CONDUCTION IN ICE.
KISZKOWSKI P	03-0327	CURIE POINT SHIFT WITH APPLIED FIELD IN FERROELECTRICS POSSESSING FIRST ORDER TRANSITI
	03-0328	ELASTIC GIBBS FUNCTION FOR A FERROELECTRIC CRYSTAL.
	03-0329	INDUCED POLARIZATION AND A ONE POLARIZATION MODEL OF FERROELECTRICITY.
	03-0331	QUASISTATIC GENERATION OF HARMONICS IN A FERROELECTRIC CRYSTAL WITH A SECOND ORDER TRA
	03-0332	TWO POLARIZATION MODEL THEORY OF FERROELECTRICITY.
KITAHIRO I	13-2051	ELECTROOPTICAL PROPERTIES OF STRONTIUM POTASSIUM LITHIUM NIOBATE. (TUNGSTEN BRONZE TYP
KITAO A	05-1068	DIRECT CHELATOMETRIC DETERMINATION OF LEAD AND TITANIUM IN LEAD TITANATE AND SIMILAR C
KITTEL C	04-0777	THICKNESS OF DOMAIN WALLS IN FERROELECTRIC AND FERROELASTIC CRYSTALS. (BARIUM TITANATE
KIYOHASHI K	13-1963	STUDY OF SUBSOLIDUS EQUILBRIA IN POTASSIUM OXIDE- LITHIUM OXIDE- NIOBIUM PENTOXIDE SYS
KIZHAEV SA	11-1792	ELECTRON PARAMAGNETIC RESONANCE SPECTRUM IN (MANGANESE DOPED) LITHIUM NIOBATE AND (COP
KJEKSHUS A	11-1709	ELECTRICAL PROPERTIES OF LITHIUM NIOBATE. (ANOMALY AT 263 DEG K IS STRUCTURAL TRANSITI
	16-2286	ELECTRICAL PROPERTIES OF RUBIDIUM NITRATE AND CESIUM NITRATE.
KLADKEVICH MD	03-0333	SIMULTANEOUS MEASUREMENT OF PYROELECTRIC COEFFICIENT AND DIELECTRIC CONSTANT OF FERROE
KLAPPER H	25-2914	X-RAY TOPOGRAPHY OF THE DEFECT STRUCTURE OF THIOUREA.
KLAZAR J	15-2168	CHALCOGENIDE BROMIDES OF ANTIMONY AND BISMUTH.
KLEBBA J	11-1701	MEASUREMENTS OF PHOTON CORRELATIONS OF SECOND HARMONIC GENERATED LIGHT (FROM LITHIUM N
KLEIMANN H	04-0778	CHARACTERISTICS OF THIN, MULTILAYERED FERROELECTRIC CERAMICS OBTAINED BY PROJECTION WI
	04-0779	LOW LOSS (FERROELECTRIC CERAMICS) WITH GREAT DIELECTRIC PERMITTIVITY. (BARIUM TITANATE
KLEIN G	28-3206	FAST AFTEREFFECT OF SEIGNETTE SALT.
KLEIN PH	13-1930	PHOTOCHROMIC EFFECT AND ELECTRON SPIN RESONANCE OF IRON AND MOLYBDENUM IN BARIUM SODIU
KLEIN R	03-0334	THEORY OF RAYLEIGH AND BRILLOUIN SCATTERING NEAR THE PHASE TRANSITION.
KLEINBERG R	26-2935	APPROXIMATE STRUCTURE FOR ANTIFERROELECTRIC PHASE OF COPPER FORMATE TETRAHYDRATE BY NE
KLEINMAN AD	04-0852	QUANTITATIVE STUDIES OF OPTICAL HARMONIC GENERATION IN CADMIUM SULFIDE, BARIUM TITANAT
KLEINMAN DA	04-0975	FAR INFRARED DIELECTRIC DISPERSION IN BARIUM AND STRONTIUM TITANATES AND TITANIUM DIOX
KLEMENT'EV FM	04-0818	UHF DIELECTRIC PROPERTIES CF SCME POLYCRYSTALLINE FERROELECTRICS (BARIUM TITANATE AND
KLIMOV VV	04-0780	EFFECT OF LITHIUM OXIDE ON THE PROPERTIES OF A SERIES OF FERROELECTRIC MATERIALS. (BAR
	05-1043	EFFECT OF GERMANIUM ON THE STRUCTURE AND PROPERTIES OF FERROELECTRIC LEAD TITANATE (GE
	05-1071	DIELECTRIC PROPERTIES OF LEAD TITANATE- LEAD (FERRATE NIOBATE) SOLID SOLUTIONS AT HIGH
	09-1510	PIEZOCERAMICS BASED ON LEAD ZIRCONATE TITANATE WITH COMPLEX ADDITIVES CONTAINING GERMA
	09-1544	NEW PIEZOELECTRIC CERAMICS. (LEAD TITANATE- LEAD FERRATE(0.5) NIOBATE(0.5), LEAD TITAN
	09-1588	EFFECT OF BISMUTH AND MANGANESE CONTAINING ADDITIVES ON THE PROPERTIES OF SOLID SOLUTI
KLINE D	08-1469	NIOBIUM-93 AND SODIUM-23 IN POLYCRYSTALLINE SODIUM NIOBATE.
KLINGER J	34-3437	DIFFUSION OF HYDROGEN FLUORIDE IN ICE.
KLIWER JK	04-0692	STATIC QUADRUPOLE INTERACTION OF SCANDIUM-44 IN BARIUM TITANATE.
KLIYA MO	15-2120	MOTION PICTURES OF THE PHASE TRANSITION IN ANTIMONY SULFOIODIDE CRYSTALS.
	15-2183	DOMAIN STRUCTURE OF ANTIMONY SULFOIODIDE CRYSTALS (FROM SELECTIVE CRYSTALLIZATION AND
	15-2184	PHASE TRANSITION AND DOMAIN STRUCTURE IN ANTIMONY SULFOIODIDE CRYSTALS. (MICRO CINEMAT
KLOOTWIJK B	06-1146	(TEMPERATURE AND) ELECTRIC FIELD DEPENDENCE OF A HYDROGEN IMPURITY MODE IN STRONTIUM T
KLUDZIN VV	11-1745	PHOTOELASTIC CONSTANTS OF LITHIUM NIOBATE CRYSTALS.
KLUEV VP	11-1753	GENERATION AND PROPAGATION OF HYPERSONIC WAVES IN LITHIUM NIOBATE CRYSTALS. (80-950 DE
KLUKUHN AFW	06-1146	(TEMPERATURE AND) ELECTRIC FIELD DEPENDENCE OF A HYDROGEN IMPURITY MODE IN STRONTIUM T
KLYSHKO DN	11-1746	PARAMETRIC LUMINESCENCE AND LIGHT SCATTERING BY POLARITONS. (LITHIUM NIOBATE)
	17-2553	MEASUREMENT OF THE REFRACTIVE INDEX OF ADP AND KDP CRYSTALS IN THE INFRARED RANGE BY P
KMETZ AR	30-3291	ORIENTATION AND THICKNESS DEPENDENCE OF CONTRAST AND BRIGHTNESS IN GADOLINIUM MOLYBDAT
KMITA AM	11-1747	AMPLIFICATION OF RAYLEIGH ULTRASONIC WAVES IN A LAYERED STRUCTURE CONSISTING OF LITHIU
	11-1748	TRANSVERSE ACOUSTOELECTRIC EFFECT IN A LAYERED LITHIUM NIOBATE- SILICON STRUCTURE.
KNEUBUHL FK	33-3362	INFRARED ACTIVE MODE BELOW THE STRUCTURAL TRANSITION IN POTASSIUM MANGANESE TRIFLUORID
	33-3363	TWO PHONON PROCESSES AND PHASE TRANSITIONS IN POTASSIUM MANGANESE TRIFLUORIDE: TEMPERA
KNISPEL RR	19-2761	DEUTERON QUADRUPOLE COUPLING CONSTANT IN DEUTERATED AMMONIUM SULFATE.
	22-2849	NMR STUDY OF DEUTERIUM DIFFUSION AND DEUTERIUM AND SODIUM-23 RELAXATION IN SODIUM TRID
	32-3347	PROTON ROTATING FRAME RELAXATION IN LITHIUM HYDRAZINIUM SULFATE.
KNOLL DB	34-3482	DIELECTRIC RELAXATION SPECTRA OF WATER, ICE, AND AQUEOUS SOLUTIONS AND THEIR INTERPRET
	34-3486	TRANSFER OF PROTONS THROUGH PURE ICE I(H) SINGLE CRYSTALS. PART-1: POLARIZATION SPECTR
KNOWLES HE	13-2012	SODIUM ORDERING IN SODIUM TUNGSTEN BRONZES.
KNYAZEV AS	06-1250	INFRARED SPECTRA OF SINGLE AND POLYCRYSTALLINE STRONTIUM TITANATE.
KOBAYASHI J	02-0082	OPTICAL ANOMALIES IN FERROELECTRIC CRYSTALS.
	03-0335	PHASE TRANSFORMATIONS OF IMPROPER FERROELECTRICS.
	04-0973	ELECTRON MIRROP MICROSCOPIC ASPECTS OF FERROELECTRIC DOMAINS OF BARIUM TITANATE AND DI
	04-0974	ELECTRON MIRROR MICROSCOPIC OBSERVATION OF FERROELECTRIC DOMAINS. (GOLD COATED BARIUM
	14-2081	LATTICE STRAINS OF BORACITE CRYSTALS.
	14-2082	ELECTROOPTICAL PROPERTIES OF FERROELECTRIC (IRON IODINE) BORACITE. (SPONTANEOUS LATTIC
	14-2083	LATENT LATTICE STRAIN IN THE FERROELECTRIC STATE OF IRON- IODINE- BORACITE. (TEMPERATU
	14-2084	PHENOMENOLOGICAL THEORY OF DIELECTRIC AND MECHANICAL PROPERTIES OF IMPROPER FERROELECT
	14-2085	X-RAY STUDY ON THE LATTICE STRAINS OF (PARAELECTRIC AND) (IRON- IODINE- BORACITE).
	14-2086	X-RAY STUDY ON PHASE TRANSITIONS OF FERROELECTRIC IRON IODINE BORACITE AT LOW TEMPERAT
	17-2555	ORDER OF THE FERROELECTRIC PHASE TRANSFORMATION OF POTASSIUM DIHYDROGEN PHOSPHATE.
	17-2556	X-RAY STUDY ON THERMAL EXPANSION OF FERROELECTRIC POTASSIUM DIHYDROGEN PHOSPHATE. (NEA
	17-2557	X-RAY DILATOMETRIC STUDY OF THE FERROELECTRIC PHASE TRANSITION OF POTASSIUM DIHYDROGEN
	26-2946	ELECTRON MIRROR MACROSCOPIC OBSERVATION OF FERROELECTRIC DOMAINS OF DICALCIUM STRONTIU
	27-3040	OPTICAL ACTIVITY OF FERROELECTRIC DICALCIUM STRONTIUM PROPIONATE. (TRANSITION AT 209 D
	30-3292	AN EXPLANATION OF ANOMALOUS OPTICAL BEHAVIOR OF THE IMPROPER FERROELECTRIC GADOLINIUM
	30-3293	APPLICATION OF ELECTRON MIRROR MICROSCOPY TO DIRECT OBSERVATION OF MOVING FERROELECTRI
	30-3294	X-RAY STUDY OF THERMAL EXPANSION OF FERROELECTRIC GADOLINIUM MOLYBDATE.
	36-3584	X-RAY AND OPTICAL STUDIES ON THE PHASE TRANSITION OF FERROELECTRIC DICALCIUM STRONTIUM
KOBAYASHI S	04-0781	EFFECT OF THERMAL NEUTRON IRRADIATION ON THE RELATIVE PERMITTIVITY OF BARIUM TITANATE

KOBAYASHI S   04-0782   X-RAY IRRADIATION EFFECT ON DIELECTRIC PROPERTIES OF SOLID SOLUTICNS OF BARIUM TITANAT
                   05-1085   ELECTROMECHANICAL PROPERTIES AND APPLICATION OF LEAD TITANATE CERAMICS.
KOBAYASHI T   17-2554   EFFECTS OF UNIAXIAL STRESS AND ELECTRIC FIELD ON PARAMAGNETIC SPECTRA OF CHROMIUM(3+)
                   22-2824   DIELECTRIC AND THERMAL STUDY OF POTASSIUM SELENATE TRANSITIONS (FERROELECTRIC BELOW 93
                   22-2850   THE ELECTRON PARAMAGNETIC RESONANCE OF POTASSIUM SELENATE.
KOBEL HR   04-0876   VARIATION OF THE DECAY CONSTANT OF STRONTIUM-85 IN BARIUM TITANATE.
KOBETS MI   30-3309   TEMPERATURE DEPENDENCE OF DIELECTRIC CONSTANT IN THE SYSTEM POTASSIUM YTTRIUM MOLYBDAT
KOBLOVA MM   11-1768   INVESTIGATION OF A SYSTEM FOR BEAM DEFLECTION BY MEANS OF LITHIUM NIOBATE CRYSTALS.
KOBZAREVA SA   27-2988   SPECIAL ELECTRICAL PROPERTIES CF REGIONS NEAR DOMAIN WALLS IN TRIGLYCINE SULFATE CRYST
                   27-3039   ELECTRICAL STRUCTURE OF THE SURFACE OF REAL CRYSTAL SUBSTRATES AS THE DETERMINING FACT
                   27-3041   REVEALING THE DOMAIN STRUCTURE OF TGS CRYSTALS BY SELECTIVE CRYSTALLIZATION OF ANTHRAQ
KOCHERIL GP   17-2558   ESR INVESTIGATIONS OF X-RAY IRRADIATED SINGLE CRYSTALS OF FERROELECTRIC POTASSIUM DIHY
KOCKEL A   10-1650   GROWTH, CRYSTALLOGRAPHY AND DIELECTRIC PROPERTIES OF BISMUTH TUNGSTATE.
KOEHLER TR   03-0257   PHASE TRANSITIONS IN A MODEL FERROELECTRIC.
KOERBER GG   11-1824   PIEZOELECTRIC SURFACE WAVES ON LITHIUM NIOBATE.
KOGA H   22-2824   DIELECTRIC AND THERMAL STUDY OF POTASSIUM SELENATE TRANSITIONS (FERROELECTRIC BELOW 93
KOGA K   06-1247   STUDIES ON THE (STRONTIUM, CALCIUM) TITANATE BOUNDARY LAYER CERAMIC DIELECTRICS.
KOGA M   30-3295   OPTICAL DEVICE ELEMENT OF FERROELECTRIC GADOLINIUM MOLYBDATE.
KOGAN OI   27-3103   RELAXATION OF DOMAIN WALLS IN TRIGLYCINE SULFATE. (TEMPERATURE DEPENDENCES OF THE COMP
KOIDE S   11-1711   GROWTH AND PROPERTIES OF FERROELECTRIC POTASSIUM LITHIUM TANTALATE(X) NIOBATE(1-X).
                   12-1890   PIEZOELECTRIC PROPERTIES OF FERROELECTRIC POTASSIUM LITHIUM TANTALATE(0-0.5) NIOBATE(1
                   13-1958   NEW (UNIAXIAL) NONLINEAR OPTIC CRYSTAL BARIUM LITHIUM NIOBATE HAVING NO MICROTWINNING
                   13-1981   TUNGSTEN BRONZE TYPE COMPLEX OXIDES.
KOIKOV SN   15-2185   CRYSTAL FIELD GRADIENTS IN ANTIMONY SULFOIODIDE.
KOIRANSKAYA EYU   04-0824   CALCULATION OF ELECTRIC FIELDS IN IONIC CRYSTALS. (BARIUM TITANATE TYPE)
KOIZUMI H   08-1376   X-RAY STUDY OF THE SYSTEM BISMUTH OXIDE FERRIC OXIDE.
                   32-3348   FERROELECTRIC-LIKE BEHAVIOR OF LITHIUM HYDRAZINIUM SULFATE.
                   35-3501   FERROELECTRIC AND OPTICAL PROPERTIES OF LEAD(5) GERMANIUM(3) OXIDE(11) AND ITS ISOMORP
                   35-3504   CRYSTAL STRUCTURE OF FERROELECTRIC LEAD(5) GERMANIUM(3) OXIDE(11).
KOIZUMI M   08-1393   HIGH PRESSURE SYNTHESIS (SOLID STATE REACTION, SINTERING) AND PROPERTIES OF PEROVSKITE
KOJIE-PRCDIC B   36-3610   SILVER DITHORIUM PHOSPHATE- A NEW CASE OF THE NON-HYDROGEN BONDED PHOSPHATE FERROELECT
KOJIMA F   13-2001   PREPARATION AND DIELECTRIC PROPERTIES OF POTASSIUM BISMUTH ZINC NIOBATE.
KOJIMA H   17-2559   CUTTING AND POLISHING IN UNIAXIAL CRYSTALS OF THE TYPE X DIHYDROGEN PHOSPHATE FOR LIGH
KOJIMA S   16-2254   NUCLEAR QUADRUPOLE RELAXATION OF NITROGEN-14 IN SODIUM NITRITE.
KOKEVA VP   04-0538   GENERATION OF SECOND OPTICAL HARMONIC IN THE CUBIC PHASE OF BARIUM TITANATE.
KOKUBO T   04-0783   PREPARATION AND PROPERTIES OF GLASS- CERAMICS CONTAINING FERROELECTRIC CRYSTALS (BARIU
KOLAR D   04-0710   PHASE EQUILIBRIA IN THE SYSTEM BARIUM TITANATE- BARIUM GERMANATE.
KOLBENEVA GI   17-2560   EFFECT OF REPLACEMENT OF POTASSIUM BY ALUMINUM ON THE FERROELECTRIC PROPERTIES OF POTA
                   17-2572   DIELECTRIC PROPERTIES OF SINGLE CRYSTALS IN THE AMMONIUM DIHYDROGEN PHOSPHATE- POTASSI
                   19-2763   FERROELECTRIC PROPERTIES OF SOLID SOLUTIONS IN THE AMMONIUM SULFATE- AMMONIUM FLUOBERY
                   19-2764   FERROELECTRIC PROPERTIES OF SOLID SOLUTIONS IN THE AMMONIUM SULFATE- AMMONIUM FLUOBERY
KOLESNICHENKO KA   04-0965   SEMICONDUCTOR PROPERTIES OF BARIUM (TITANATE, STANNATE) SOLID SOLUTIONS DOPED BY TUNGS
KOLESOVA RV   08-1377   RELATICNSHIP BETWEEN THE ATOMIC DISPLACEMENTS AND SPONTANEOUS DEFORMATION IN THE FERRO
KOLLER A   04-0784   ORIGIN OF LATTICE DEFECTS IN (BARIUM) TITANATE MATERIALS AND THEIR INFLUENCE ON BEHAVI
KOLODZIEJ HB   28-3191   INFLUENCE OF THE DOMAIN STRUCTURE ON POSITRON (TWO-PHONON) ANNIHILATION IN ROCHELLE SA
KOMAROV VD   04-0785   EFFECT CF CRYSTALLINE GRAIN SIZE ON THE PHYSICAL PROPERTIES OF CERAMIC BARIUM TITANATE
                   04-0786   SOLID SOLUTIONS BASED ON BARIUM TITANATE.
                   04-0787   SPONTANEOUS DESTRUCTION OF FERROCERAMICS. (BARIUM TITANATE)
                   04-1009   DIELECTRIC DISPERSION IN BARIUM TITANATE TYPE FERROELECTRICS IN THE METER BAND.
                   04-1011   EFFECT OF GRAIN SIZE ON THE PIEZOELECTRIC AND ELASTIC PROPERTIES OF BARIUM TITANATE.
                   04-1012   INFLUENCE OF A CONSTANT ELECTRIC FIELD ON THE DIELECTRIC PROPERTIES OF POLYCRYSTALLINE
                   04-1014   UHF REVERSAL PROPERTIES OF BARIUM TITANATE TYPE FERROELECTRIC CERAMICS.
KOMINIAK GJ   33-3406   BARIUM MANGANESE FLUORIDE, A NEW CRYSTAL FOR MICROWAVE ULTRASONICS.
KOMLYAKOVA NS   03-0318   CONNECTION BETWEEN THE STARTING FIELD FOR PARKHAUSEN STEPS AND THE CRITICAL POLARIZATI
                   03-0337   BARKHAUSEN EFFECT IN FERROELECTRICS UNDER CONTINUOUS X-RAY DIFFRACTION.
                   27-3123   INFLUENCE OF GAMMA-IRRADIATION ON BARKHAUSEN EFFECT IN FERROELECTRIC MATERIALS (TGS AN
KONAK C   30-3277   ELECTROOPTICAL PROPERTIES OF FERROELECTRIC GADOLINIUM MOLYBDATE.
                   30-3278   INDUCED AND SPONTANEOUS BIREFRINGENCE IN IMPROPER FERROELECTRIC GADOLINIUM MOLYBDATE.
KONDO T   30-3306   OBSERVATION OF PHASE BOUNDARIES BETWEEN FERRO- AND PARAELECTRIC PHASES IN GADOLINIUM M
                   30-3307   SPONTANEOUS BIREFRINGENCE AND ELECTROOPTIC RESPONSE IN GADOLINIUM MOLYBDATE.
KONGAROVA NI   09-1514   TIME AND TEMPERATURE DEPENDENCES OF THE PARAMETERS OF A FERROELECTRIC CERAMIC.
KONINGSTEIN JA   30-3296   RAMAN SPECTRUM AND STRUCTURE OF TERBIUM MOLYBDATE.
KONKOVA TS   17-2585   ENTHALPIES OF SOLUTION OF RUBIDIUM DIHYDROGEN PHOSPHATE IN WATER AND RUBIDIUM DIDEUTER
KONSIN PI   02-0085   INTERBAND MECHANISMS FOR FERROELECTRIC PHASE TRANSITIONS.
                   03-0339   SOME ADDITIONAL FEATURES OF FERROELECTRIC PHASE TRANSITIONS IN VIBRONIC MODEL.
                   03-0349   THEORY OF VIBRONIC PHASE TRANSFORMATIONS IN WIDE GAP FERROELECTRICS.
                   03-0351   THEORY OF FERROELECTRIC PHASE TRANSITIONS WITH VIBRONIC MECHANISM.
KONSTANTINOVA B   27-3130   DOMAIN STRUCTURE OF PURE AND DOPED TRIGLYCINE SULFATE CRYSTALS GROWN BELOW AND ABOVE T
KONSTANTINOVA VP   17-2561   INVESTIGATION OF THE DOMAIN STRUCTURE OF A TRIGLYCINE SULFATE CRYSTAL DURING AGING.
                   27-3041   REVEALING THE DOMAIN STRUCTURE OF TGS CRYSTALS BY SELECTIVE CRYSTALLIZATION OF ANTHRAQ
                   27-3042   APPLICATION OF SELECTIVE ETCHING TO THE STUDY OF TWIN AND DISLOCATION STRUCTURES IN TR
                   27-3043   DOMAIN STRUCTURE OF TRIGLYCINE SELENATE CRYSTAL (DURING COOLING THROUGH CURIE POINT)
                   27-3044   DEFECT STRUCTURE OF TRIGLYCINE SULFATE. (CORRELATION WITH DOMAIN STRUCTURES)
KONWENT H   03-0338   RANDOM PHASE APPROXIMATION FOR THE BLINC-DE GENNES MODEL FOR HYDROGEN BONDED FERROELEC
                   08-1363   DIPOLE PATTERNS IN ORTHORHOMBIC AND TRIGONAL PHASES OF ABO(3) SUBSTANCE. (WHEN THE B I
                   08-1364   FERROELECTRICALLY ACTIVE A-IONS IN ABO(3) STRUCTURES. (ANTIFERROELECTRICITY ORIGINS)
KOPERLES BM   15-2142   PRODUCTION AND SOME OPTICAL PROPERTIES OF ANTIMONY SULFOBROMIDE IN GLASSY AND CRYSTALL
                   15-2242   DIELECTRIC PROPERTIES OF ANTIMONY SULFOIODIDE- SULFOBROMIDE SOLID SOLUTIONS.
KOPINETS IF   15-2154   EFFECT OF THE ADSORPTION OF OXYGEN ON THE PROPERTIES OF ANTIMONY SULFOIODIDE SINGLE CR
                   15-2186   EFFECT OF PHOTOELECTRET CHARGE ON VALUE OF FERROELECTRIC POLARIZATION IN ANTIMONY SULF
                   15-2187   EFFECT OF OXYGEN ADSORPTION ON THE FERROELECTRIC PHASE TRANSITION IN ANTIMONY SULFOIOD
KOPSIK VA   02-0083   PYROELECTRIC INVESTIGATIONS OF THE DOMAIN STRUCTURE STABILITY IN FERROELECTRICS. (ROCH
                   17-2679   SPECIFIC HEAT OF SOME KDP TYPE CRYSTALS. (POTASSIUM DIHYDROGEN PHOSPHATE, RUBIDIUM AND
KOPSKY V   17-2562   CAN THE ISOTOPIC SHIFT IN KDP BE EXPLAINED BY TUNNELING ONLY.
                   17-2563   MOLECULAR FIELD STUDY OF THE ISOTOPIC EFFECT IN POTASSIUM DIHYDROGEN(X) DIDEUTERIUM(1-
KOPTSIK VA   03-0336   POLAR- POLAR AND POLAR- NEUTRAL CRYSTAL STRUCTURES. (CLASSIFICATION OF ANISOTROPIC FER
                   03-0340   MICROSCOPIC THEORY OF STATIC DOMAIN STRUCTURES OF FERROELECTRIC CRYSTALS. (CRYSTALLOGR
                   03-0341   OBSERVATION OF DOMAIN STRUCTURE IN LOW TEMPERATURE FERROELECTRICS BY SOLID DEW TECHNIQ
                   17-2676   CRITICAL BEHAVIOR OF KDP AND RUBIDIUM DIHYDROGEN PHOSPHATE CRYSTALS.
                   17-2678   SPONTANEOUS POLARIZATION OF A POTASSIUM DIHYDROGEN PHOSPHATE CRYSTAL NEAR THE CURIE PO

KOPTSIK VA	17-2680	SOME THERMODYNAMICAL PROPERTIES OF POTASSIUM DIHYDROGEN PHOSPHATE- POTASSIUM DIHYDROGE
	19-2787	EXPERIMENTAL INVESTIGATION OF THE FERROELECTRIC PROPERTIES OF AMMONIUM BISULFATE IN TH
	24-2895	DOMAIN STRUCTURE IN THE FERROELECTRIC POTASSIUM FERROCYANIDE.
	24-2898	POLYTYPE AND TWINNING OF THE FERROELECTRIC CRYSTALS POTASSIUM FERROCYANIDE TRIHYDRATE
	28-3211	ABSORPTION OF LONGITUDINAL ULTRASONIC WAVES IN ROCHELLE SALT CRYSTALS NEAR THE CURIE P
	36-3559	FERROELECTRIC BEHAVIOR OF THE MINERAL STIBOTANTALITE.
	36-3601	DOMAIN STRUCTURE AND TEMPERATURE DEPENDENCE OF THE PIEZOELECTRIC CONSTANTS OF STIBOTAN
KOPVILLEM UKH	17-2419	MACROSCOPIC STIMULATED POLARIZATION ECHO IN FERROELECTRICS. (POTASSIUM DIHYDROGEN PHOS
	17-2564	POLARIZATION ECHO IN THE FERROELECTRIC SINGLE CRYSTAL POTASSIUM DIHYDROGEN PHOSPHATE.
KOPYLOV YUL	13-1936	PREPARATION AND STUDY OF BARIUM(X) STRONTIUM(1-X) NIOBATE SINGLE CRYSTALS (CZOCHRALSKI
KORABLIN LN	15-2180	MOSSBAUER STUDIES OF ANTIMONY SULFOIODIDE TYPE CRYSTALS.
KORCHEMKIN MA	03-0309	KINETIC EQUATION FOR QUADRUPOLE RESONANCE IN FERROELECTRICS NEAR THE PHASE TRANSITION
	03-0311	NATURE OF PHASE TRANSITIONS IN FERROELECTRIC AND ANTIFERROELECTRIC MATERIALS.
KORINA RV	27-3124	BAND GAP OF TRIGLYCINE SELENATE, TRIGLYCINE FLUOBERYLLATE, TRIGLYCINE SULFATE AND ITS
KOROBOV AI	06-1106	DIELECTRIC NONLINEARITY OF STRONTIUM TITANATE SINGLE CRYSTALS NEAR PHASE TRANSITION AT
KOROLEV YUG	17-2407	DIFFRACTION OF POLARIZED LIGHT BY ULTRASONIC WAVES IN POTASSIUM DIHYDROGEN PHOSPHATE C
KOROLYUK AP	11-1749	DETERMINATION OF ELASTIC AND PIEZOELECTRIC CONSTANTS OF LITHIUM NIOBATE SINGLE CRYSTAL
KOROTKOV PA	17-2512	USE OF KDP TYPE CRYSTALS AS ELECTROOPTICAL REFLECTORS.
KORZHUEV MA	17-2676	CRITICAL BEHAVIOR OF KDP AND RUBIDIUM DIHYDROGEN PHOSPHATE CRYSTALS.
	17-2678	SPONTANEOUS POLARIZATION OF A POTASSIUM DIHYDROGEN PHOSPHATE CRYSTAL NEAR THE CURIE PO
KOSEK F	15-2170	PHOTOELECTRIC PROPERTIES OF BISMUTH SULFOIODIDE CRYSTALS.
KOSEVICH AM	03-0342	MECHANICAL MODEL OF LONG WAVELENGTH VIBRATIONS IN A (SINGLE DOMAIN) FERROELECTRIC TYPE
KOSMAN MS	15-2188	INJECTION CURRENTS AND ACCUMULATION OF POLARIZATION CHARGE IN ANTIMONY SULFOIODIDE.
KOSONOCKY WF	37-3676	FERROELECTRIC LIGHT VALVE ARRAYS FOR OPTICAL MEMORIES.
KOSONOGOV NA	04-0552	PHOTOFERROELECTRIC EFFECTS IN A(V) B(VI) C(VII) AND BARIUM TITANATE TYPE FERROELECTRICS.
	04-0574	PHOTOCONDUCTIVITY AND PHOTOFERROELECTRIC PHENOMENA IN BARIUM TITANATE.
	04-0682	PHOTODOMAIN EFFECT IN BARIUM TITANATE.
KOSOUROV GI	17-2508	NONLINEAR PROPERTIES OF A RUBIDIUM DIHYDROGEN PHOSPHATE CRYSTAL.
KOSTRUB VI	04-0796	TEMPERATURE DEPENDENCE OF INTERNAL FRICTION OF BARIUM TITANATE SINGLE CRYSTALS.
KOSYRBASOVA MG	16-2370	PROPERTIES OF SILVER SODIUM DINITRITE AND SILVER SODIUM NITRITE. (DIFFERENTIAL THERMAL
KOTELYANSKII IM	11-1747	AMPLIFICATION OF RAYLEIGH ULTRASONIC WAVES IN A LAYERED STRUCTURE CONSISTING OF LITHIU
KOUKAL V	04-0788	DISTRIBUTION OF POTENTIAL ALONG TITANATE CERAMICS PREVIOUSLY POLARIZED.
KOVACH DSH	15-2141	PRODUCTION METHODS AND SOME OPTICAL PROPERTIES OF BISMUTH TELLUROIODIDE.
	15-2150	INDIRECT TRANSITIONS AND ABSORPTION IN THE MID-INFRARED IN ANTIMONY SULFOIODIDE CRYSTA
	15-2217	ABSORPTION SPECTRA OF BISMUTH TELLUROBROMIDE AND BISMUTH TELLUROIODIDE CRYSTALS.
KOVALENKO AN	27-2976	DIELECTRIC LOSSES OF TRIGLYCINE SULFATE CRYSTALS EXPOSED TO SMALL DOSES OF X- AND GAMM
KOVALEV OV	14-2087	RELATIONSHIP BETWEEN DIRECTIONS OF THE ELECTRIC AND MAGNETIC POLARIZATION IN BORACITE
KOVTONYUK NF	04-0569	SURFACE CONDUCTIVITY IN POLARIZED BARIUM TITANATE CRYSTALS.
KOWALCHIK M	15-2200	GROWTH OF LARGE ANTIMONY SULFOIODIDE CRYSTALS (TO 1 CM DIAMETER): CONTROL OF NEEDLE MO
KOYANO N	16-2365	NEUTRON DIFFRACTION STUDIES OF FERROELECTRIC POLARIZATION REVERSALS IN SODIUM NITRITE.
	17-2661	NEUTRON DIFFRACTION STUDIES OF FERROELECTRIC PHASE TRANSITIONS IN KDP AND POTASSIUM DI
KOZAKOVA M	15-2168	CHALCOGENIDE BROMIDES OF ANTIMONY AND BISMUTH.
KOZLOV GV	11-1730	BIREFRINGENCE OF CERTAIN CRYSTALS IN THE MILLIMETER WAVELENGTH RANGE. (LITHIUM NIOBATE
	11-1844	ELECTROOPTIC EFFECT IN LITHIUM NIOBATE IN THE (SUB) MILLIMETER RANGE.
KOZLOVA LA	36-3618	A NEW MAGNETOFERROELECTRIC: CADMIUM IRON NIOBATE.
KOZLOVSKII LV	04-0789	DEPENDENCE OF DIELECTRIC PROPERTIES OF BARIUM TITANATE CERAMICS ON THE PHYSICO- CHEMIC
KOZLOVSKII VKH	03-0343	CRITICAL PHENOMENA IN ANTIFERROELECTRICS. (EQUATION OF STATE)
	03-0344	DYNAMICAL THEORY OF RIGID LATTICES OF ANTIFERROELECTRICS.
	03-0345	NONSTATIONARY POLARIZATION IN THE ANHARMONIC VIBRATOR MODEL. (FERROELECTRICS)
	03-0346	PHASE TRANSFORMATIONS IN FERROELECTRIC CRYSTAL IN PRESENCE OF DOMAIN WALLS. (THEORY)
	03-0347	QUANTUM EFFECTS IN FERROELECTRIC MATERIALS WITH HYDROGEN BONDS.
KOZYREVA MS	06-1198	ELECTRICAL AND OPTICAL PROPERTIES (CONDUCTIVITY AND MOBILITY VERSUS TEMPERATURE, ABSOR
	08-1288	FORBIDDEN ZONE WIDTHS IN POLYCRYSTALLINE DIELECTRICS WITH PEROVSKITE STRUCTURE (TITANA
KRAINIK NN	01-0018	ANTIFERROELECTRICITY IN COMPOUNDS WITH PEROVSKITE STRUCTURE. (EXAMINATION OF PUBLISHED
	01-0029	FERROELECTRICS AND ANTIFERROELECTRICS.
	03-0307	ELASTOOPTICAL EFFECT IN FERROELECTRIC WITH DIFFUSE PHASE TRANSITION.
	08-1292	STRUCTURE AND DIELECTRIC PROPERTIES OF LITHIUM CONTAINING PEROVSKITES. (STRONTIUM LITH
	08-1378	DIELECTRIC PROPERTIES AND THERMAL EXPANSION OF SOME FERROELECTRICS AND ANTIFERROELECTR
	13-1923	FORMATION CONDITIONS AND PHASE TRANSITIONS IN CRYSTALS WITH POTASSIUM TUNGSTEN BRONZE
	13-1974	ANOMALOUS BEHAVIOR OF THE OPTICAL ABSORPTION COEFFICIENT OF LEAD MAGNONIOBATE SINGLE
	13-2050	PHASE TRANSITIONS IN TANTALATES WITH THE STRUCTURE OF POTASSIUM TUNGSTEN BRONZE. (POTA
	15-2185	CRYSTAL FIELD GRADIENTS IN ANTIMONY SULFOIODIDE.
	15-2189	NUCLEAR QUADRUPOLE AND ELECTROACOUSTIC ECHOES IN FERROELECTRIC ANTIMONY SULFOIODIDE.
	15-2211	ANOMALOUS ECHO IN FERROELECTRIC ANTIMONY SULFOIODIDE. (PIEZOELECTRIC DOMAIN OSCILLATIO
	15-2216	NUCLEAR QUADRUPOLE RESONANCE IN FERROELECTRICS OF THE ANTIMONY SULFOIODIDE (AND BROMID
	36-3521	ANISOTROPY OF THE ELECTROOPTICAL EFFECT IN CRYSTALS OF LEAD MAGNONIOBATE.
KRAJEWSKI T	17-2565	INVESTIGATION OF THE THERMAL EXPANSION OF POTASSIUM DIHYDROGEN PHOSPHATE CRYSTALS BY U
	17-2566	USE OF SPECIFIC HEAT (BY A DYNAMIC METHOD) OF POTASSIUM DIHYDROGEN PHOSPHATE CRYSTALS
	27-3045	EFFECT OF A POLARIZING FIELD AND OF THE ADDITION OF A MIXTURE OF COPPER AND CHROMIUM I
	27-3046	FIELD EFFECT IN TELLURIUM THIN FILMS ON A SUBSTRATE OF FERROELECTRIC TRIGLYCINE SULFAT
	27-3047	THERMAL EXPANSION OF TRIGLYCINE SULFATE CRYSTALS STUDIED WITH TELLURIUM THIN FILMS.
KRAKOWSKI RA	09-1495	OXYGEN CONCENTRATION CELL AND ELECTRICAL CONDUCTIVITY MEASUREMENTS ON PZT FERROELECTRI
	09-1512	OXYGEN CONCENTRATION CELL MEASUREMENTS OF IONIC TRANSPORT NUMBERS IN PZT (LEAD ZIRCONA
KRANIK NN	01-0030	FERROELECTRICS AND ANTIFERROELECTRICS.
KRAPIVIN VF	03-0348	SPACE CHARGE LIMITED CURRENTS IN A METAL FERROELECTRIC METAL SYSTEM (THEORY).
KRASILNIKOV VA	11-1706	DISPERSION OF ELASTIC WAVES IN FERROELECTRICS. (LITHIUM NIOBATE)
	11-1707	GENERATION OF SUPERHIGH FREQUENCY ACOUSTIC HARMONICS IN A (SINGLE DOMAIN) LITHIUM NIOB
KRASNIKOVA AYA	24-2895	DOMAIN STRUCTURE IN THE FERROELECTRIC POTASSIUM FERROCYANIDE.
	24-2896	FERROELECTRIC TRANSITION IN A POTASSIUM FERROCYANIDE TRIHYDRATE CRYSTAL UNDER HIGH HYD
	24-2897	FERROELECTRIC BEHAVIOR OF POTASSIUM FERROCYANIDE TRIHYDRATE CRYSTALS AT HIGH PRESSURE.
	24-2898	POLYTYPE AND TWINNING OF THE FERROELECTRIC CRYSTALS POTASSIUM FERROCYANIDE TRIHYDRATE
KRASOTKIN IS	04-1002	INFLUENCE OF FIRING OF BARIUM TITANATE SINTERS ON ELECTRICAL AND TECHNOLOGICAL PROPERT
KRAUNIK NN	13-1982	ELECTROOPTICAL EFFECT IN LEAD ZINCATE(0.33) NIOBATE(0.67). (DISPERSION OF QUADRATIC CO
KRAUSE HB	08-1379	ORDERING OF MAGNESIUM AND NIOBIUM IN THE OCTAHEDRAL POSITIONS OF THE "CUBIC" PEROVSKIT
KRAUSE JT	09-1545	ACOUSTIC DETECTION OF FERROELECTRIC PHASE TRANSITIONS IN PLZT CERAMICS.
	09-1546	NONLINEAR EFFECTS IN FERROCERAMICS. (PZT-8, LITHIUM NIOBATE)
KRAUT EA	11-1757	EFFECT OF MASS LOADING ON THE PROPAGATION OF ACOUSTIC SURFACE WAVES ON LITHIUM NIOBATE
	37-3647	APPLICATION OF NONLINEAR INTERACTIONS IN FERROELECTRIC CERAMICS TO MICROWAVE SIGNAL PR
KRAUZMAN M	36-3548	VIBRATIONAL SPECTRA OF IODIC ACID SINGLE CRYSTALS.
KRAVCHENKO VB	13-1936	PREPARATION AND STUDY OF BARIUM(X) STRONTIUM(1-X) NIOBATE SINGLE CRYSTALS (CZOCHRALSKI

KRAYUKHINA EK   06-1198   ELECTRICAL AND OPTICAL PROPERTIES (CONDUCTIVITY AND MOBILITY VERSUS TEMPERATURE, ABSOR
KREHER K   02-0084   FERROELECTRIC SEMICONDUCTORS. (ANTIMONY SULFOIODIDE, REVIEW, 57 REFS)
                     03-0251   THEORY OF SPACE CHARGE LIMITED CURRENTS IN FERROELECTRICS.
KREMENCHUGSKII LS   02-0127   FUNDAMENTAL PROPERTIES OF METAL- FERROELECTRIC- METAL JUNCTION SYSTEMS. (INFLUENCE OF
                     03-0333   SIMULTANEOUS MEASUREMENT OF PYROELECTRIC COEFFICIENT AND DIELECTRIC CONSTANT OF FERROE
KREMNEV VV   04-0596   SURFACE BREAKDOWN IN VACUUM ON BARIUM TITANATE.
KRINDACH DP   11-1750   PARAMETRIC LUMINESCENCE INTENSITY IN THE LITHIUM NIOBATE CRYSTAL.
KRISHNAN RS   03-0350   THERMAL EXPANSION ANOMALIES AND THE DIRECTION OF SPONTANEOUS POLARIZATION IN FERROELEC
                     16-2333   CRYSTAL STRUCTURE OF FERROELECTRIC GLYCINE SILVER NITRATE.
                     16-2334   PHONON DISPERSION IN FERROELECTRIC POTASSIUM NITRATE.
                     20-2803   INFRARED AND RAMAN SPECTRA OF FERROELECTRIC ALUMS: METHYL AMMONIUM ALUMINUM (SULFATE O
                     22-2851   RELAXATION BETWEEN THE STRUCTURE OF FERROELECTRIC SODIUM TRIHYDROGEN DISELENITE AND IT
                     26-2936   RAMAN AND INFRARED SPECTRA OF COPPER FORMATE TETRAHYDRATE.
                     26-2937   RAMAN AND INFRARED SPECTRA OF INORGANIC FORMATES.
                     27-2957   RAMAN SPECTRUM OF TRIGLYCINE SELENATE.
                     29-3245   IS SODIUM RUBIDIUM TARTRATE TETRAHYDRATE A FERROELECTRIC. (ANSWER-NO)
KRISTOFFEL NN   02-0085   INTERBAND MECHANISMS FOR FERROELECTRIC PHASE TRANSITIONS.
                     03-0339   SOME ADDITIONAL FEATURES OF FERROELECTRIC PHASE TRANSITIONS IN VIBRONIC MODEL.
                     03-0349   THEORY OF VIBRONIC PHASE TRANSFORMATIONS IN WIDE GAP FERROELECTRICS.
                     03-0351   THEORY OF FERROELECTRIC PHASE TRANSITIONS WITH VIBRONIC MECHANISM.
KRIVOGLAZ MA   03-0253   INFLUENCE OF LONG RANGE DIPOLE DIPOLE FORCES ON ATTENUATION AND VELOCITY OF ULTRASOUND
KRIVOSHCHEKOV GV   17-2500   COMBINATION OF THE FREQUENCIES OF COHERENT AND INCOHERENT RADIATION IN KDP CRYSTAL.
KRIVSHICH VV   15-2114   ABSORPTION OF ULTRASOUND IN A(V)B(VI)C(VII) FERROELECTRIC SEMICONDUCTORS AT TEMPERATUR
KRIZ HM   14-2088   DOUBLETS CONCERNING THE STRUCTURE OF BORACITES.
KROESE CJ   33-3388   A PHASE TRANSITION IN A COMPOUND WITH HELICAL ELECTRIC DIPOLE STRUCTURE: CESIUM COPPER
KROL LM   11-1750   PARAMETRIC LUMINESCENCE INTENSITY IN THE LITHIUM NIOBATE CRYSTAL.
KROLEVETS NM   15-2229   LUMINESCENCE OF SINGLE CRYSTALS OF THE FERROELECTRIC SEMICONDUCTOR ANTIMONY SULFOIODID
KROSI M   17-2420   GROWTH OF ELECTROOPTICAL CRYSTALS FROM AQUEOUS SOLUTIONS (APPARATUS, GROWTH OF ADP, KD
KRUCHAN YAYA   01-0011   PHASE TRANSITIONS IN FERROELECTRICS.
                     13-1943   EFFECT OF ELECTRIC FIELD ON PHASE COMPOSITION OF (LEAD, BARIUM) NIOBATE SOLID SOLUTION
                     13-1945   PHASE COMPOSITION OF (LEAD, BARIUM) DINIOBATE. METHOD OF INVESTIGATION AND SOME RESULT
KRUEGER HHA   09-1488   RELEASE OF ELECTRIC ENERGY IN LEAD NIOBIUM ZIRCONATE STANNATE BY TEMPERATURE
                     09-1547   DISPERSION STUDY OF DOMAIN WALL MOBILITY AND MECHANICAL DAMPING IN LEAD TITANATE ZIRCO
                     28-3207   RADIATION DAMAGE AND THE FERROELECTRIC EFFECT IN ROCHELLE SALT.
KRUGER JS   13-1950   PARAMETRIC FLUORESCENCE IN BARIUM SODIUM NIOBATE FOR NINE CW PUMP FREQUENCIES.
                     13-1983   PARAMETRIC FLUORESCENCE IN BARIUM SODIUM NIOBATE FOR 10 CW PUMP WAVELENGTHS (RANGE 454
KRUGLIK AI   33-3402   SUPERSTRUCTURE ARISING DURING THE PHASE TRANSITION IN POTASSIUM MANGANESE TRIFLUORIDE.
KRUGLOV SV   17-2500   COMBINATION OF THE FREQUENCIES OF COHERENT AND INCOHERENT RADIATION IN KDP CRYSTAL.
KRUMIN AE   08-1380   INVESTIGATION OF THE DYNAMICS OF POLARIZATION OF FERROELECTRIC SOLID SOLUTIONS OF THE
KRUPNAYA VP   22-2826   FERROELECTRICS WITH AMMONIUM GROUP ORIENTATION MOBILITY: METAL AMMONIUM (BX4) TYPE COM
KRUPNYI AI   33-3360   ELASTIC PROPERTIES OF CESIUM LEAD TRICHLORIDE.
KRYLOV EI   36-3527   ELECTRICAL PROPERTIES AND STRUCTURES OF METANIOBATES AND METATANTALATES OF TRANSITION
KRYSINKA M   27-3129   DEPENDENCE OF THE DOMAIN STRUCTURE OF TRIGLYCINE SULFATE CRYSTALS ON TEMPERATURE AND A
KUBAREV YUG   03-0460   NATURE OF THE TEMPERATURE DEPENDENCE OF NMR SPECTRA SECOND MOMENT IN ORDER DISORDER TY
                     24-2899   ANOMALOUS SPIN LATTICE RELAXATION BY QUASI SPIN WAVES IN POTASSIUM FERROCYANIDE TRIHYD
KUBICAR I   27-3048   ANOMALIES OF THERMAL PROPERTIES OF TRIGLYCINE SULFIDE.
KUBO M   16-2313   NUCLEAR QUADRUPOLE RESONANCE OF NITROGEN IN SODIUM NITRITE.
KUBO T   04-0790   WET SYNTHETIC METHOD OF MAKING PURE, EXACTLY STOICHIOMETRIC BARIUM TITANATE.
KUBOTA Y   04-0881   PIEZOELECTRIC CHARACTERISTIC OF BARIUM TITANATE CERAMIC EXHIBITED WHEN IT IS GIVEN AN
KUCHAR F   09-1548   FERROELECTRIC PROPERTIES OF LEAD (STANNATE(0.5) NIOBATE(0.5))(1-X) ZIRCONATE(X) PEROVS
                     13-2044   DIFFUSE TO SHARP FERROELECTRIC PHASE TRANSITION SYSTEMS LEAD (B(0.5) NIOBIUM(0.5))(1-X
KUCHEROV IYA   04-1037   ABSORPTION AND AMPLIFICATION OF ULTRASOUND IN A TWO-LAYER SYSTEM CONSISTING OF A PIEZO
KUDO T   09-1549   DIELECTRIC AND PIEZOELECTRIC PROPERTIES OF LEAD COBALTATE(0.33) NIOBATE(0.67)- TITANAT
KUDRYAVTSEVA AP   11-1665   PHASE MATCHING ANGLES AND TEMPERATURE OF LITHIUM METANIOBATE CRYSTALS OF DIFFERENT STO
KUDZIN AYU   04-0597   DIELECTRIC AND ELASTIC PROPERTIES OF SINGLE CRYSTALS OF BARIUM LEAD TITANATE SOLID SOL
                     04-0598   EFFECT OF IMPURITIES AND GROWTH CONDITIONS ON THE STRUCTURE OF BARIUM TITANATE CRYSTAL
                     04-0708   INVESTIGATION OF THE FINE STRUCTURE OF BARIUM TITANATE CRYSTALS FLUX GROWN FROM POTASS
                     04-0709   PECULIARITIES OF POLARIZATION OF BARIUM TITANATE SINGLE CRYSTALS WITH DOUBLE HYSTERESI
                     04-0768   INFLUENCE OF THE AMPLITUDE OF FREE VIBRATIONS ON THE PERMITTIVITY OF BARIUM TITANATE S
                     04-0791   AMPLITUDE DEPENDENCE OF THE INTERNAL FRICTION IN BARIUM TITANATE SINGLE CRYSTALS. (PIE
                     04-0792   DYNAMIC FATIGUE OF SINGLE CRYSTALS OF BARIUM TITANATE.
                     04-0793   EFFECT OF ADDING NICKEL ON THE PROPERTIES OF BARIUM TITANATE SINGLE CRYSTALS.
                     04-0794   INTERNAL FRICTION IN BARIUM TITANATE SINGLE CRYSTALS.
                     04-0795   STABILIZATION OF SPONTANEOUS POLARIZATION OF BARIUM TITANATE SINGLE CRYSTALS.
                     04-0796   TEMPERATURE DEPENDENCE OF INTERNAL FRICTION OF BARIUM TITANATE SINGLE CRYSTALS.
                     04-1035   ANISOTROPY OF THE ELECTRICAL CONDUCTIVITY OF BARIUM TITANATE SINGLE CRYSTALS GROWN FRO
                     04-1036   POSITIVE TEMPERATURE COEFFICIENT OF RESISTANCE OF BARIUM TITANATE SINGLE CRYSTALS IN T
                     15-2190   RELAXATION PROCESSES OF ANTIMONY SULFOIODIDE SINGLE CRYSTALS.
                     15-2191   TEMPERATURE DEPENDENCE OF THE SWITCHING PROCESS IN ANTIMONY SULFOIODIDE SINGLE CRYSTAL
                     15-2245   INFLUENCE OF ONE DIMENSIONAL MECHANICAL STRESS ON THE PHASE TRANSITION OF ANTIMONY SUL
                     15-2246   POLARIZATION REVERSAL OF ANTIMONY SULFOIODIDE SINGLE CRYSTALS IN THE REGION OF PHASE C
KUKUSHKIN LS   03-0508   NATURE OF THE ANOMALOUS ELECTRIC CONDUCTIVITY OF FERROELECTRICS IN THE PHASE TRANSITIO
KULEK J   07-1259   EFFECT OF PILE NEUTRON IRRADIATION ON THE DIELECTRIC PROPERTIES OF LEAD ZIRCONATE.
KULKARNI AK   04-0797   DEVELOPMENT OF BARIUM TITANATE WITH CONSISTENT DIELECTRIC PROPERTIES.
                     04-0798   INFLUENCE OF MICROSTRUCTURES (PREPARATION CONDITIONS) ON DIELECTRIC PROPERTIES OF BARI
KULKARNI RH   08-1381   DIELECTRIC CONSTANT OF POTASSIUM NIOBATE SINGLE CRYSTALS UNDER BIASING CONDITIONS.
KULWICKI BM   03-0352   DIFFUSION POTENTIALS IN BARIUM TITANATE AND THE THEORY OF PTC MATERIALS.
KULYUPIN YUA   16-2335   LUMINESCENCE OF POTASSIUM NITRATE IRRADIATED WITH GAMMA-RAYS.
KUMADA A   14-2082   ELECTROOPTICAL PROPERTIES OF FERROELECTRIC (IRON IODINE) BORACITE. (SPONTANEOUS LATTIC
                     30-3297   FERROELECTRIC FERROELASTIC CRYSTAL GADOLINIUM MOLYBDATE.
                     30-3298   (PROPERTIES,) FUNCTION AND APPLICATIONS OF GADOLINIUM MOLYBDATE (OPTICAL SHUTTER, COLO
                     30-3299   OPTICAL PROPERTIES OF GADOLINIUM MOLYBDATE AND THEIR DEVICE APPLICATIONS.
                     30-3300   OPTICAL PROPERTIES OF GADOLINIUM MOLYBDATE AND THEIR DEVICE APPLICATIONS.
                     30-3306   OBSERVATION OF PHASE BOUNDARIES BETWEEN FERRO- AND PARAELECTRIC PHASES IN GADOLINIUM M
                     30-3307   SPONTANEOUS BIREFRINGENCE AND ELECTROOPTIC RESPONSE IN GADOLINIUM MOLYBDATE.
                     37-3648   ELECTROOPTIC EFFECT IN FERROELECTRICS.
KUNG R   13-1915   EFFECT OF OXYGEN STOICHIOMETRY AND HYDROXYL CONTENT ON THE PROPERTIES AND PROCESSING O
KUNIGELIS VF   03-0353   ACOUSTOELECTRIC INTERACTION IN FERROELECTRIC SUBSTANCES.
                     15-2139   FREQUENCY DEPENDENCE OF ANOMALOUS ULTRASOUND DAMPING IN ANTIMONY SULFOIODIDE SINGLE CR
                     15-2192   INTERACTION OF ULTRASONIC WAVES WITH MOVING DOMAIN WALLS IN FERROELECTRICS. (ANTIMONY

KUNIGELIS VF   15-2221 ANOMALOUS ABSORPTION OF SOUND NEAR THE FERROELECTRIC PHASE TRANSITION (ANTIMONY SULFOI
                   15-2222 ANISOTROPY OF ANOMALOUS ABSORPTION OF ULTRASOUND IN SINGLE CRYSTALS OF ANTIMONY SULFOI
                   15-2223 GENERATION OF THE SECOND ACOUSTIC HARMONIC IN ANTIMONY SULFOIODIDE SINGLE CRYSTALS.
                   15-2224 GENERATION OF THE SECOND ACOUSTIC HARMONIC IN ANTIMONY SULFOIODIDE SINGLE CRYSTALS.
KUNII TL   33-3391 COVALENCY IN POTASSIUM MANGANESE TRIFLUORIDE.
KUNZLER JV   08-1276 NUCLEAR QUADRUPOLE INTERACTIONS IN PEROVSKITE TYPE COMPOUNDS OF HAFNIUM-181 STUDIED BY
KUPRIYANOV MF   08-1324 CRYSTALLOCHEMISTRY OF PEROVSKITES. (EXISTENCE CONDITIONS AND CLASSIFICATION)
                   08-1382 AN X-RAY STUDY OF THE PHASE TRANSITION IN STRONTIUM FERRATE TANTALATE
                   09-1507 TERNARY SYSTEM LEAD TITANATE- LEAD ZIRCONATE- LEAD NIOBATE(0.67) ZINCATE(0.33).
KURASAWA Y   10-1641 ON FERROELECTRICITY AND ANTIFERROELECTRICITY OF THE AO3 TYPE CRYSTAL. (TUNGSTEN TRIOXI
KURBATOV LV   03-0164 SPIN WAVE PHENOMENOLOGICAL THEORY IN FINITE SAMPLES OF FERRODIELECTRICS WITH AN ARBITR
KURIHARA K   13-1960 ELECTRICAL MAGNETIC AND X-RAY STUDIES OF PHASE TRANSITIONS IN LEAD COBALTATE(0.5) TUNG
KURODA N   22-2852 PROTON SPIN LATTICE RELAXATION IN FERROELECTRIC SODIUM TRIHYDROGEN SELENATE.
KUROKI H   04-0868 PERMITTIVITY CHANGE OF BARIUM TITANATE UNDER VERY SLOWLY CHANGING FIELD.
                   04-0981 SUSCEPTIBILITY INCREASE DURING POLARIZATION REVERSAL. (IN BARIUM TITANATE AND TRIGLYCI
KURSTEN HD   30-3259 GROWTH AND MORPHOLOGY OF SINGLE CRYSTALS OF GADOLINIUM MOLYBDATE, A FERROELECTRIC AND
                   30-3301 DOMAIN STRUCTURE OF GADOLINIUM MOLYBDATE.
KURTZ SK   01-0005 SYMPOSIUM ON APPLICATIONS OF FERROELECTRICS.
                   02-0049 NONLINEAR OPTICAL MATERIALS (REVIEW, DATA COLLECTION, MATERIALS NEEDS, 118 REFS)
                   04-0690 ELECTROOPTIC PROPERTIES OF SOME ABO(3) PEROVSKITES IN THE PARAELECTRIC PHASE. (POTASSI
                   08-1383 VISIBLE AND ULTRAVIOLET OPTICAL PROPERTIES (REFLECTANCE SPECTRA AND BAND STRUCTURE) OF
                   13-1902 SOME EFFECTS OF MELT STOICHIOMETRY ON THE OPTICAL PROPERTIES BARIUM SODIUM NIOBATE.
                   13-1903 SOME EFFECTS OF MELT STOICHIOMETRY ON THE OPTICAL PROPERTIES OF BARIUM SODIUM NIOBATE.
                   13-1924 USE OF PEROVSKITE PARAELECTRICS IN BEAM DEFLECTORS AND LIGHT MODULATORS. (POTASSIUM TA
                   17-2539 OPTICAL NONLINEAR SUSCEPTIBILITIES: ACCURATE RELATIVE VALUES FOR QUARTZ, AMMONIUM DIHY
KURZ PF   29-3246 SOME CHEMICAL REACTIONS IN SILICA GELS. PART-3: FORMATION OF POTASSIUM ACID TARTRATE C
KUSCHKE R   04-0583 PRODUCTION AND STUDY OF THE INNER BARRIERS IN SEMICONDUCTING BARIUM TITANATE CERAMICS.
KUSHNIR VR   11-1665 PHASE MATCHING ANGLES AND TEMPERATURE OF LITHIUM METANIOBATE CRYSTALS OF DIFFERENT STO
KUSZ J   28-3208 CORRELATION BETWEEN BARKHAUSEN NOISE AND SURFACE MICRODAMAGES ON SINGLE CRYSTALS OF SE
KUTNYI IV   27-2970 ELECTROGYRATION AND FERROELECTRIC PHASE TRANSITIONS.
KUWABARA M   04-0799 LANTHANUM DOPED BARIUM TITANATE CERAMICS WITH BISMUTH OXIDE AS A GRAIN BOUNDARY ELEMEN
                   04-0800 TWO-STEP ANOMALOUS INCREASE OF RESISTIVITY ON LANTHANUM DOPED BARIUM TITANATE- BISMUTH
                   04-0801 TWO STEP ANOMALOUS INCREASE OF RESISTANCE ON LANTHANUM(0.002) BARIUM(0.998) TITANATE-
KUZMIN EN   04-0666 POLARIZED STATES OF FERROELECTRIC CERAMICS. (BARIUM TITANATE)
KUZMIN II   04-0970 EFFECTS OF (MONOENERGETIC) ELECTRON AND (COBALT-60) GAMMA-IRRADIATION ON BARIUM TITANA
                   04-0971 (IN PILE, FAST NEUTRON) RADIATION INDUCED PHASE TRANSFORMATIONS IN (CERAMIC) BARIUM AN
                   04-0972 RADIATION PHYSICS OF FERROELECTRICS OF BARIUM TITANATE TYPE.
                   05-1081 X-RAY INVESTIGATION OF IRRADIATED LEAD TITANATE.
                   08-1443 X-RAY AND DIELECTRIC STUDIES OF IRRADIATED PEROVSKITE TYPE COMPOUNDS.
                   08-1474 EFFECT OF IRRADIATION OF DIELECTRIC PROPERTIES OF SOME FERRO-, ANTIFERRO- AND PARAELE
KUZMIN RN   36-3618 A NEW MAGNETOFERROELECTRIC: CADMIUM IRON NIOBATE.
KUZMINOV YUS   11-1733 ELECTRICAL CONDUCTIVITY OF LITHIUM NIOBATE SINGLE CRYSTALS.
                   13-2014 DIELECTRIC AND STRUCTURAL PROPERTIES OF FERROELECTRICS WITH DIFFUSE PHASE TRANSITIONS.
KUZNECOVA MN   03-0235 MAGNETIC PROPERTIES OF FERRODIELECTRIC CATALYSTS.
KUZNETSOV GF   13-1936 PREPARATION AND STUDY OF BARIUM(X) STRONTIUM(1-X) NIOBATE SINGLE CRYSTALS (CZOCHRALSKI
KUZNETSOV VA   04-0941 NONLINEAR OPTICAL PROPERTIES OF LEAD TITANATE AND BISMUTH TITANATE CRYSTALS.
                   08-1279 HYDROTHERMAL CRYSTALLIZATION AND SOME PROPERTIES OF BISMUTH TITANATES
                   09-1482 HYDROTHERMAL CRYSTALLIZATION OF LEAD TITANATE ZIRCONATE SOLID SOLUTIONS.
KVAJIC G   34-3457 ANGULAR CORRELATION OF ANNIHILATION PHOTONS IN ICE SINGLE CRYSTALS.
KVAPIL J   17-2567 GROWTH AND ETCHING OF POTASSIUM DIHYDROGEN PHOSPHATE CRYSTALS IN SOLUTIONS CONTAINING
                   17-2567 GROWTH AND ETCHING OF POTASSIUM DIHYDROGEN PHOSPHATE CRYSTALS IN SOLUTIONS CONTAINING
                   30-3302 STRUCTURAL STUDY OF GADOLINIUM MOLYBDATE CRYSTALS.
KWUN SI   16-2336 INFRARED ABSORPTION IN GAMMA-RAY IRRADIATED SINGLE CRYSTAL OF SODIUM NITRITE.
KYI UA   08-1286 STUDY OF DIVALENT BINARY AND TERNARY NIOBATES IN THE SOLID STATE.
LAGENDIJK A   06-1147 ESR OF CHROMIUM(3+) IN CHROMIUM DOPED STRONTIUM TITANATE SINGLE CRYSTALS.
LAGUTINA ZP   27-3167 EFFECT OF CONSTANT ELECTRICAL FIELD ON PHASE TRANSITIONS IN TRIGLYCINE FLUOBERYLLATE.
LAHAJNAR G   16-2267 NMR STUDY OF THE FERROELECTRIC TRANSITIONS IN DIGLYCINE NITRATE AND TRIS-SARCOSINE CAL
                   26-2928 SPIN LATTICE RELAXATION BY QUASI SPIN WAVES IN ORDER DISORDER TYPE FERROELECTRICS: POL
                   27-2967 NUCLEAR SPIN LATTICE RELAXATION IN FERROELECTRIC TRIGLYCINE SULFATE.
LAJZEROWICZ J   02-0063 GENERATION OF THE SECOND HARMONIC AND FERROELECTRICITY. (THEORY)
                   03-0354 FERRO- AND ANTIFERROELECTRIC COUPLING AND DIELECTRIC RELAXATION.
                   17-2448 DISPLACEMENTS AND VELOCITIES OF THE DOMAIN WALLS IN KDP. EXISTENCE OF CRITICAL ELECTRI
                   17-2449 ELECTRICAL AND OPTICAL INVESTIGATION OF FERROELECTRIC PROPERTIES OF POTASSIUM DIHYDROG
                   17-2451 INTERDOMAIN AND DOMAIN EFFECT INTERACTIONS IN KDP (FERROELECTRIC DOMAINS).
                   17-2453 VELOCITY OF PROPAGATION OF DOMAIN WALLS: INTERACTIONS BETWEEN DOMAINS. (IN KDP)
                   27-2991 SECOND HARMONIC LIGHT SCATTERING BY DOMAINS IN FERROELECTRIC TRIGLYCINE SULFATE.
                   27-2992 STUDY OF THE PHASE TRANSITION OF TGS BY SECOND HARMONIC SCATTERING.
LAKIN KM   11-1720 BISMUTH(12) GERMANIUM OXIDE(20) VERSUS LITHIUM NIOBATE IN THE SURFACE WAVE AMPLIFIER.
                   11-1751 PERTURBATION THEORY FOR ELECTROMAGNETIC COUPLING TO ELASTIC SURFACE WAVES ON PIEZOELEC
LAL HB   06-1148 DIELECTRIC (CONSTANT AND LOSS) BEHAVIOR OF REDUCED STRONTIUM TITANATE SINGLE CRYSTALS
LAM NQ   10-1642 BOMBARDMENT INDUCED AMORPHIZATION IN TUNGSTEN TRIOXIDE AND ITS USE IN DEDUCING MEAN DA
LAMACCHIA JT   04-1004 OPTICALLY INDUCED REFRACTIVE INDEX CHANGES IN BARIUM TITANATE (EXPERIMENTS TO CHARACTE
                   04-1005 OPTICALLY INDUCED REFRACTIVE INDEX CHANGES IN BARIUM TITANATE.
LAMBERT M   04-0537 DIELECTRIC PROPERTIES AND DOMAIN TEXTURE OF BARIUM TITANATE CRYSTALS.
                   04-0629 LINEAR DISORDER IN CRYSTALS (CASE OF SILICON, QUARTZ AND FERROELECTRIC PEROVSKITES). (
                   04-0630 STRUCTURE DISORDER OF BARIUM TITANATE TYPE FERROELECTRICS. BASED ON NEW TREATMENT AND
                   04-0675 LINEAR DISORDER AND TEMPERATURE DEPENDENCE OF RAMAN SCATTERING IN BARIUM TITANATE.
                   04-0676 STRUCTURE DISORDER AND RAMAN SCATTERING IN BARIUM TITANATE AND POTASSIUM NIOBATE.
                   04-0910 DIELECTRIC MEASUREMENTS ON PILE-IRRADIATED BARIUM TITANATE SINGLE CRYSTALS.
                   08-1311 ANISOTROPIC CRITICAL X-RAY DIFFUSE SCATTERING FROM SODIUM NIOBATE CRYSTALS (NEAR THE 6
                   08-1312 ANISOTROPIC CRITICAL X-RAY DIFFUSE SCATTERING FROM POTASSIUM MANGANESE TRIFLUORIDE AND
                   08-1313 X-RAY DIFFUSE SCATTERING FROM SODIUM NIOBATE AS A FUNCTION OF TEMPERATURE.
                   33-3369 CRITICAL ANISOTROPIC FLUCTUATIONS AT THE 184 DEG K TRANSITION OF POTASSIUM MANGANESE T
LAMBERT VL   06-1238 LATTICE DISTORTIONS IN A STRONTIUM TITANATE SINGLE CRYSTAL ELECTRET.
LAMOTTE B   17-2568 DYNAMICS OF PROTONS (AND DEUTERONS) IN FERROELECTRIC CRYSTALS WITH HYDROGEN BONDING. (
                   18-2731 ESR OF FREE RADICALS IN FERROELECTRIC AND ANTIFERROELECTRIC ARSENATE SINGLE CRYSTALS.
                   18-2734 ESR IDENTIFICATION OF SLATER CONFIGURATIONS AND OF EXCHANGE OF THE ARSENATE RADICAL IN
LAMPRECHT G   03-0380 EXTENSION OF THE POULET- LOUDON APPROXIMATION IN THE THEORY OF OPTICAL LATTICE VIBRATI
LAMY C   10-1628 EPR STUDY OF THE CATALYTIC ACTIVITY OF NONSTOICHIOMETRIC TUNGSTEN TRIOXIDE.
LAN B   18-2737 PROTONIC CONDUCTION IN POTASSIUM DIHYDROGEN ARSENATE.

LAND CE	02-0086	RECENT DEVELOPMENTS IN ELECTROOPTIC CERAMICS (HOT PRESSED LEAD ZIRCONATE TITANATE, EFF
	09-1526	HOT PRESSED (LEAD, LANTHANUM) (ZIRCONATE, TITANATE) FERROELECTRIC CERAMICS FOR ELECTRO
	09-1529	RECENT IMPROVEMENTS IN THE OPTICAL AND ELECTROOPTIC PROPERTIES OF PLZT CERAMICS.
	09-1550	ELECTROOPTIC CERAMICS- NEW MATERIALS FOP INFORMATION STORAGE AND DISPLAY (HOT PRESSED
	09-1551	ELECTROOPTIC EFFECTS IN FERROELECTRIC CERAMICS. (LEAD ZIRCONATE TITANATE)
	09-1552	FERROELECTRIC CERAMICS FOR INFORMATION STORAGE AND DISPLAY. (PZT)
	09-1554	OPTICAL PROPERTIES (ELECTRICALLY VARIABLE BIREFRINGENCE) OF (HOT PRESSED) FERROELECTRI
	09-1599	SCATTERING MODE FERROELECTRIC PHOTOCONDUCTOR IMAGE STORAGE AND DISPLAY DEVICES. (PLZT
	37-3649	DEPENDENCE OF THE SMALL SIGNAL PARAMETERS OF FERROELECTRIC CERAMIC RESONATORS UPON STA
	37-3650	ELECTROOPTIC CERAMIC (INFORMATION) STORAGE (PROCESSING) AND DISPLAY DEVICES. (BASED ON
LAND DN	27-3021	(OPTICAL) ENERGY GAP (AND DIELECTRIC CONSTANT) TEMPERATURE CHARACTERISTICS OF FERROELE
LANDSBERG EG	15-2107	ELECTRICAL AND PHOTOELECTRIC PROPERTIES OF ANTIMONY SULFOIODIDE.
LANDURE Y	03-0491	SHOCK WAVE INDUCED ELECTROMECHANICAL CONVERSION IN FERROELECTRIC MATERIALS.
	09-1508	FERROELECTRIC CERAMICS AND ENERGY CONVERSION. (LEAD ZIRCONATE TITANATE)
LANE R	08-1384	DIELECTRIC, PIEZOELECTRIC AND PYROELECTRIC PROPERTIES OF THE LEAD DINIOBATE- BARIUM DI
	08-1385	DIELECTRIC, PIEZOELECTRIC AND PYROELECTRIC PROPERTIES OF THE LEAD METANIOBATE- BARIUM
LANGBEIN H	04-0667	PRODUCTION AND STUDY OF PURE AND DOPED BARIUM TITANATE SINGLE CRYSTAL
	04-0668	PREPARATION AND PROPERTIES OF PURE AND DOPED BARIUM TITANATE SINGLE CRYSTALS.
LANGMAN RA	09-1553	ISOTHERMAL GRAIN GROWTH AND ELECTRICAL BEHAVIOR OF FULLY DENSE PLZT CERAMICS.
LAO BY	03-0355	FIRST AND SECOND ORDER PHASE TRANSITIONS IN FERROELECTRICS: A SELF CONSISTENT POLARIZA
LAPLACA SJ	34-3438	DEUTERON ARRANGEMENTS IN THE HIGH PRESSURE FORMS OF ICE.
LAPSHIN VF	15-2122	PHASE DIAGRAM AND ELECTRICAL CONDUCTIVITY OF BISMUTH SELENOIODIDE- ANTIMONY SELENOIODI
LARSEN FK	22-2854	THE HYDROGEN BOND SYSTEM IN POTASSIUM TRIHYDROGEN SELENITE, AND IN POTASSIUM TRIDEUTER
LARSSON KE	19-2752	ROTATIONAL MOTIONS IN SOLIDS: THE FERROELECTRIC TRANSITION IN AMMONIUM SULFATE STUDIED
LASSABATERE L	32-3349	DETERMINATION OF FERROELECTRIC PROPERTIES OF LITHIUM HYDRAZINIUM FLUOBERYLLATE.
LAUBEREAU A	06-1149	BRILLOUIN SCATTERING IN STRONTIUM TITANATE SINGLE CRYSTALS IN THE TEMPERATURE RANGE 5
LAUER RB	36-3585	THERMALLY STIMULATED CURRENTS AND LUMINESCENCE IN BISMUTH(12) SILICON OXIDE(20) AND BI
LAUGHMAN L	04-0803	RAMAN SCATTERING LINE SHAPE OF THE SOFT E POLARITON MODE IN BARIUM TITANATE.
LAUGINIE P	27-3002	DIELECTRIC RELAXATION IN MULTIDOMAIN CRYSTALS OF TRIGLYCINE SULFATE.
	27-3004	INFLUENCE OF DISLOCATIONS, SURFACE LAYER AND X-RAY IRRADIATION ON DOMAIN STRUCTURE IN
	27-3049	DIELECTRIC CONSTANT OF TRIGLYCINE SULFATE.
	27-3050	DIELECTRIC RELAXATION OF MULTIDOMAIN TRIGLYCINE SULFATE.
LAULICHT I	25-2915	VIBRATIONAL LINE WIDTHS OF THIOUREA.
LAURENCE C	11-1701	MEASUREMENTS OF PHOTON CORRELATIONS OF SECOND HARMONIC GENERATED LIGHT (FROM LITHIUM N
	13-1984	VISIBLE CW PARAMETRIC OSCILLATOR USING BARIUM SODIUM NIOBATE.
LAURENT M	15-2112	CRYSTAL STRUCTURE AND CONDUCTIVITY OF ANTIMONY SULFOIODIDE.
LAUSCH W	04-0802	BARIUM TITANATE CERAMIC: A NEW ELECTRON EMITTER.
LAVERKO EN	03-0505	DOMAIN STRUCTURE OF POLYCRYSTALLINE FERROELECTRICS WITH DIFFUSE PHASE TRANSITIONS.
LAVRENCIC B	03-0356	FOURIER TRANSFORM FAR INFRARED REFLECTION SPECTRA OF PARAELECTRIC POTASSIUM DIHYDROGEN
	03-0357	ON THE SYMMETRY OF THE SOFT MODES IN THE FERROELECTRIC PHASE TRANSITIONS.
	17-2569	LASER RAMAN STUDY OF QUASI SPIN WAVE HYDROGEN TUNNELING MODES IN KDP AND DEUTERATED KD
	17-2570	PROTON LATTICE INTERACTIONS AND THE DYNAMICAL SUSCEPTIBILITY OF POTASSIUM DIHYDROGEN P
	22-2853	PROTON SPIN LATTICE RELAXATION IN IRRADIATED SODIUM TRIHYDROGEN SELENATE.
LAWLESS WN	01-0008	REPORT ON THE SECOND INTERNATIONAL MEETING ON FERROELECTRICITY. A EUROPEAN VIEW. AN AM
	03-0358	SMALL SIGNAL PERMITTIVITY OF THE STATIONARY (100)-180 DEG DOMAIN WALL IN BARIUM TITANA
	04-0804	180 DEG DOMAIN WALL CONTRIBUTION TO THE DIELECTRIC CONSTANT OF BARIUM TITANATE. (CALCU
	04-0805	TEMPERATURE DEPENDENT POLARIZABILITIES AND COCHRAN MODE ASSIGNMENTS IN BARIUM AND STRO
	06-1150	THREE APPLICATION AREAS FOR STRONTIUM TITANATE GLASS CERAMICS.
LAYTON MM	04-0576	DIELECTRIC AND OPTICAL PROPERTIES OF TRANSPARENT FERROELECTRIC GLASS CERAMIC SYSTEMS.
	04-0725	ELECTRICAL PROPERTIES OF MICROCRYSTALLINE FERROELECTRIC MATERIALS CRYSTALLIZED FROM GL
	08-1386	CRYSTALLIZATION OF SOLID SOLUTIONS OF SODIUM NIOBATE FROM GLASS.
LAZAY PD	04-0671	ACOUSTIC SOFT OPTIC MODE INTERACTIONS IN FERROELECTRIC BARIUM TITANATE.
	04-0806	ACOUSTIC PHONON SOFT OPTIC PHONON INTERACTION IN BARIUM TITANATE.
	04-0807	TEMPERATURE DEPENDENCE OF THE BRILLOUIN- RAMAN SPECTRUM OF BARIUM TITANATE.
LAZKO LA	27-2970	ELECTROGYRATION AND FERROELECTRIC PHASE TRANSITIONS (TGS).
LE BIHAN R	04-0808	DIRECT OBSERVATION OF BARIUM TITANATE CRYSTALS BY SCANNING ELECTRON MICROSCOPE.
	04-0810	PHOTOEMISSION MICROSCOPIC OBSERVATION OF THE SURFACE OF BARIUM TITANATE CRYSTALS.
	04-0811	STUDY OF DOMAINS IN FERROELECTRIC CRYSTALS WITH A PHOTOEMISSION MICROSCOPE (TUNGSTEN T
	04-0812	SEM OBSERVATION OF THE FERROELECTRIC SURFACE OF BARIUM TITANATE CRYSTALS.
	10-1643	(PREPARATION AND ELECTRICAL AND OPTICAL) STUDY OF TUNGSTEN TRIOXIDE AT LOW TEMPERATURE
	10-1644	PREPARATION OF TUNGSTIC OXIDE SINGLE CRYSTALS (SUBLIMATION IN AIR OF TUNGSTEN TRIOXIDE
	10-1645	PHOTOEMISSION MICROSCOPIC OBSERVATION OF THE SURFACE OF TUNGSTEN TRIOXIDE CRYSTALS.
	10-1646	STUDY OF DOMAINS OF FERROELECTRIC CRYSTALS WITH SCANNING ELECTRON MICROSCOPE. (TUNGSTE
	27-3051	DIRECT OBSERVATION OF FERROELECTRIC DOMAINS IN TRIGLYCINE SULFATE USING THE SCANNING E
	27-3052	SCANNING ELECTRON MICROSCOPE STUDY OF FERROELECTRIC DOMAINS IN TGS.
LE BLANC CL	37-3651	OPEN CIRCUIT SENSITIVITY OF RADIALLY POLARIZED FERROELECTRIC CERAMIC CYLINDERS. (HYDRO
	37-3652	OPEN CIRCUIT SENSITIVITY OF AXIALLY POLARIZED FERROELECTRIC CERAMIC CYLINDERS. (HYDROP
	37-3653	OPEN CIRCUIT SENSITIVITY OF RADIALLY POLARIZED FERROELECTRIC CERAMIC HOLLOW SPHERES. (
LE DONCHE L	17-2571	PHASE TRANSITION IN THALLIUM AND AMMONIUM THALLIUM (DIHYDROGEN) PHOSPHATES.
	18-2735	NEW FERROELECTRIC: THE DOUBLE AMMONIUM THALLIUM MONOARSENATE OR ATLAS.
LE FAOU C	03-0422	THEORY OF POLARIZATION REVERSAL IN FERROELECTRICS. (CALCULATIONS OF SWITCHING CURRENT)
LE MONTAGNER S	04-0809	DIELECTRIC ABSORPTION IN THE RHOMBOHEDRAL PHASE OF BARIUM TITANATE.
	08-1340	DIELECTRIC STUDY OF THE LEAD ZIRCONATE LEAD HAFNATE (PHASE) DIAGRAM.
	17-2571	PHASE TRANSITION IN THALLIUM AND AMMONIUM THALLIUM (DIHYDROGEN) PHOSPHATES.
	17-2631	TRANSIENT PHENOMENA IN THE DISCHARGE OF CONDENSERS WITH POLAR DIELECTRICS. (POTASSIUM
	18-2735	NEW FERROELECTRIC: THE DOUBLE AMMONIUM THALLIUM MONOARSENATE OR ATLAS.
	36-3529	NEW FERROELECTRIC (PYROCHLORE TYPE) COMPOUND (AMMONO THIO CADMIUM NIOBATE, WITH 4 ELEC
LE PARMENTIER L	13-1985	NEW PHASES OF THE TETRAGONAL TUNGSTEN OXYGEN BRONZE TYPE: QUADRATIC LEAD OR BISMUTH TA
LE TRAON A	04-0809	DIELECTRIC ABSORPTION IN THE RHOMBOHEDRAL PHASE OF BARIUM TITANATE.
LE TRAON F	04-0809	DIELECTRIC ABSORPTION IN THE RHOMBOHEDRAL PHASE OF BARIUM TITANATE.
LEAN EGH	11-1752	OPTICAL PROBING OF ACOUSTIC SURFACE WAVE HARMONIC GENERATION (IN LITHIUM NIOBATE)
LEAVITT RP	04-0588	SHOCK WAVE STUDIES OF SURFACE LAYERS IN BARIUM TITANATE.
LEBEDEV NN	15-2131	PYROCURRENTS HIGHER THAN THE CURIE POINT IN ANTIMONY SULFOIODIDE.
LEBEDEV VM	03-0504	SOME ASPECTS OF THE PROBLEM OF FERROELECTRICITY AND ANTIFERROELECTRICITY IN PEROVSKITE
	08-1387	HIGH TEMPERATURE X-RAY INVESTIGATION OF THE PEROVSKITE MODIFICATION OF CADMIUM TITANAT
	08-1424	ABOUT THE CURIE POINT AND NATURE OF DIELECTRIC PROPERTIES IN PEROVSKITE MODIFICATION O
	08-1456	CERTAIN ASPECTS OF THE PROBLEM OF FERROELECTRICITY AND ANTIFERROELECTRICITY IN THE PER
	08-1458	CRYSTAL STRUCTURE AND PROPERTIES OF NEW CADMIUM AND THALLIUM CONTAINING PEROVSKITES. (
	36-3619	NEW FERROELECTRICS. (BARIUM(2) BISMUTH (NIOBIUM OR TANTALUM OR VANADIUM) OXIDE(6), AND
LEBEDEVA NN	15-2104	INFLUENCE OF NONEQUILIBRIUM CARRIERS ON THE DOMAIN STRUCTURE OF THE FERROELECTRIC SEMI

LEBEDEVA NN   15-2105   SHORT CIRCUIT PHOTOEMF IN FERROELECTRIC SEMICONDUCTORS. (ANTIMONY SULFOIODIDE CRYSTALS
                   15-2129   CHARACTERISTICS OF THE PHOTOELECTRIC EMF OF ANTIMONY SULFOIODIDE.
                   15-2130   LIGHT INDUCED PYROCURRENTS IN FERROSEMICONDUCTOR ANTIMONY SULFOIODIDE.
                   15-2132   PHOTOEMF'S IN THE FERROELECTRIC SEMICONDUCTOR ANTIMONY SULFOIODIDE.
                   15-2133   PHOTODOMAIN EFFECT IN NATURALLY POLARIZED ANTIMONY SULFOIODIDE CRYSTALS.
                   15-2134   PHOTOELECTRIC PROPERTIES (PHOTOCONDUCTIVITY) OF ANTIMONY SULFOIODIDE SINGLE CRYSTALS.
                   15-2135   PHOTODOMAIN EFFECT IN NATURALLY POLARIZED ANTIMONY SULFOIODIDE CRYSTALS.
                   15-2136   SOME FEATURES OF THE PHOTOEMF SPECTRUM OF ANTIMONY SULFOIODIDE.
                   15-2153   CHARACTERISTICS OF THE PHOTOELECTRIC EMF OF ANTIMONY SULFOIODIDE.
                   15-2228   INFLUENCE OF LIGHT ON SELF POLARIZATION OF SINGLE CRYSTALS OF ANTIMONY SULFOIODIDE.
LEE C   06-1151   ELECTRONIC CONDUCTION IN SLIGHTLY REDUCED STRONTIUM TITANATE AT LOW TEMPERATURES.
                   06-1152   TRANSPORT PROPERTIES IN REDUCED STRONTIUM TITANATE. (ABSTRACT)
LEE DC   09-1555   SINTERING SCANDIUM AND NIOBIUM MODIFIED LEAD ZIRCONATE TITANATE.
LEE MK   27-3022   PROPERTIES OF MIXED CRYSTALS OF TRIGLYCINE SULFATE AND SELENATE.
LEE MN   16-2336   INFRARED ABSORPTION IN GAMMA-RAY IRRADIATED SINGLE CRYSTAL OF SODIUM NITRITE.
LEE S   17-2558   ESR INVESTIGATIONS OF X-RAY IRRADIATED SINGLE CRYSTALS OF FERROELECTRIC POTASSIUM DIHY
LEFEUVRE S   03-0359   INPUT ADMITTANCE AND FIELDS RADIATED BY AN OPEN WAVEGUIDE FILLED WITH HIGH PERMITTIVIT
LEFKOWITZ I   02-0094   ROLE OF THE REGULAR OCTAHEDRON AND LARGER POLYHEDRA IN FERROELECTRIC MATERIALS. (HETER
                   03-0182   ELECTRIC FIELD DISTRIBUTIONS IN DIELECTRICS, WITH SPECIAL EMPHASIS ON NEAR SURFACE REG
                   04-0813   COMMENTS ON THE NEUTRON SPECTRUM OF FERROELECTRIC AND PARAELECTRIC BARIUM TITANATE.
                   08-1388   RADIATION AND ACOUSTIC EXCITATION ON ANOMALOUS LAYERS OF PEROVSKITE FERROELECTRICS.
                   10-1651   STRUCTURAL AND DIELECTRIC MODIFICATIONS OF TUNGSTEN TRIOXIDE DUE TO NEUTRON IRRADIATIO
                   11-1705   RADIATION DAMAGE IN LITHIUM NIOBATE. (CAPACITANCE, OPTICAL SPECTRA, ESR, CELL PARAMETE
                   13-1905   LOW FREQUENCY VIBRATIONAL MODES IN BARIUM SODIUM NIOBATE.
                   13-1906   RAMAN SPECTRA OF BARIUM SODIUM NIOBATE.
                   36-3586   TRANSITION IN TIN TELLURIDE- GERMANIUM TELLURIDE ALLOYS. (NEUTRON SCATTERING)
LEHENY RF   10-1654   RAMAN STUDY OF METALLIC TUNGSTEN BRONZES. (SODIUM AND RUBIDIUM TUNGSTATES)
LEHMANN MS   22-2854   THE HYDROGEN BOND SYSTEM IN POTASSIUM TRIHYDROGEN SELENITE, AND IN POTASSIUM TRIDEUTER
LEHMPFUHL G   04-0991   BARIUM TITANATE BY CONVERGENT BEAM ELECTRON DIFFRACTION.
LEHOVEC K   04-0814   CONDUCTIVITY INJECTION AND EXTRACTION IN POLYCRYSTALLINE BARIUM TITANATE.
LEIBLER K   03-0360   NEW TWO- AND THREE PHASE FERROELECTRIC FERROMAGNETIC MATERIALS. SOME PROBLEMS AND RESU
                   04-0815   TWO-PHASE FERROELECTRIC SYSTEM. PART-1: BARIUM TITANATE- METAL.
LEIBOVICH TA   22-2863   PREPARATION AND ANALYSIS OF SODIUM AMMONIUM SELENATE DIHYDRATE CRYSTALS.
IFIKINA BB   04-0943   ELECTROCONDUCTIVITY AND HALL EFFECT IN SEMICONDUCTING SOLID SOLUTIONS OF ISO- AND HETE
LEILBLER K   04-0816   TWO PHASE FERROELECTRIC SYSTEMS. PART-2: FERROELECTRICS- SILVER (BARIUM TITANATE AND B
LEMANOV VV   11-1678   COLLINEAR INTERACTION OF LONGITUDINAL ELASTIC WAVES IN LITHIUM NIOBATE CRYSTALS.
                   11-1743   MANDELSTAM BRILLOUIN SCATTERING OF LIGHT IN LITHIUM NIOBATE (FROM TRANSVERSE (MORE INT
                   11-1744   SCATTERING OF LIGHT ON ACOUSTIC PHONONS IN LITHIUM NIOBATE AND LITHIUM TANTALATE.
                   11-1753   GENERATION AND PROPAGATION OF HYPERSONIC WAVES IN LITHIUM NIOBATE CRYSTALS. (80-950 DE
                   11-1754   LIGHT SCATTERING BY ULTRASONIC WAVES IN LITHIUM NIOBATE CRYSTALS. (PHOTOELASTIC CONSTA
                   11-1816   ABSORPTION OF ELASTIC WAVES IN REDUCED LITHIUM NIOBATE.
                   12-1878   SCATTERING OF LIGHT BY ACOUSTIC PHONONS AND POLARITONS IN LITHIUM TANTALATE.
                   17-2656   LANDAU-KHALATNIKOV ATTENUATION IN POTASSIUM DIDEUTERIUM PHOSPHATE CRYSTALS.
LEMONTAGNER S   07-1256   COMPLEX DIELECTRIC CONSTANT OF LEAD ZIRCONATE UNDER POLARIZING FIELD, 370-520 DEG K.
LENZO PV   35-3505   LIGHT AND ELECTRIC FIELD DEPENDENT OSCILLATION OF SPACE CHARGE LIMITED CURRENT IN BISM
LEONIDOVA GG   27-3053   EFFECT OF HYDROSTATIC PRESSURE ON THE UNIPOLAR PROPERTIES OF TRIGLYCINE SULFATE.
                   27-3054   INVESTIGATION OF THE PHASE TRANSITION IN TRIGLYCINE SULFATE.
                   27-3090   NEW HIGH PRESSURE PHASE TRANSITION IN TRIGLYCINE SELENATE.
                   28-3214   HIGH PRESSURE PHASE TRANSITIONS OF THE FIRST KIND IN ROCHELLE SALT.
LEONOV AI   08-1355   NATURE OF THE HIGH PERMITTIVITY OF CERIUM ALUMINATES WITH PEROVSKITE TYPE STRUCTURE.
LEONOVICI MR   04-0599   INFLUENCE OF PHASE TRANSITIONS ON ELECTRICAL RESISTANCE OF BARIUM TITANATE.
LESTER WW   09-1556   NEW COMPOSITE CERAMIC PIEZOELECTRIC TRANSDUCER MATERIAL. (ALUMINA WHISKERS IN LEAD ZIR
LEUENBERGER H   04-0685   CHANGE IN THE DECAY CONSTANT OF ZIRCONIUM-89 IN BARIUM TITANATE (BARIUM CHLORIDE FLUX
                   04-0876   VARIATION OF THE DECAY CONSTANT OF STRONTIUM-85 IN BARIUM TITANATE.
LEUTLOFF S   36-3565   SPECTRUM AND ENERGY LEVELS OF CHROMIUM(3+) IONS AND EXCHANGE COUPLED CHROMIUM(3+) PAIR
LEVANYUK AP   03-0361   FREQUENCY AND TEMPERATURE DEPENDENCES OF THE COMPLEX DIELECTRIC CONSTANT OF FERROELECT
                   03-0362   PHENOMENOLOGICAL THEORY OF ANOMALIES OF THERMODYNAMIC PARAMETERS NEAR SECOND ORDER TRA
                   03-0363   PHENOMENOLOGICAL THEORY OF DIELECTRIC ANOMALIES IN FERROELECTRIC MATERIAL WITH SEVERAL
                   03-0364   THEORY OF SECOND ORDER PHASE TRANSITIONS.
                   19-2765   (EXAMINATION OF) ANOMALIES OF DIELECTRIC PROPERTIES AT THE PHASE TRANSITIONS. (BASED O
                   27-3055   ANISOTROPY OF ULTRASONIC ATTENUATION IN UNIAXIAL FERROELECTRICS. (TGS, ABSTRACT ONLY)
                   30-3303   PHENOMENOLOGICAL THEORY OF THE FERROELECTRIC PHASE TRANSITION IN GADOLINIUM MOLYBDATE.
LEVASSEUR A   14-2079   PREPARATION OF NICKEL BORACITE CRYSTALS BY HYDROTHERMAL SYNTHESIS.
                   14-2089   NEW BORACITES OF FORMULA LITHIUM(4) BORON(7) OXYGEN(12) CHLORIDE OR BROMIDE (NOT EXAMI
LEVCHUK EA   11-1865   AUTOFOCUSING OF LASER RADIATION IN ACTIVE MATERIALS AND NONLINEAR CRYSTALS. (KDP, ADP,
                   11-1866   DAMAGE TO THE SURFACE OF LITHIUM NIOBATE BY LIGHT.
                   17-2717   THERMAL SELF FOCUSING IN POTASSIUM DIHYDROGEN PHOSPHATE AND AMMONIUM DIHYDROGEN PHOSPH
LEVEDEV VM   08-1460   STRUCTURE AND PROPERTIES OF SOME NEW FERROELECTRIC MATERIALS WITH PEROVSKITE STRUCTURE
LEVELUT A   15-2144   BOSON ECHOES. (ANTIMONY SULFOIODIDE)
LEVIALDI A   04-0627   VARIATION IN CERTAIN ELECTRICAL PROPERTIES OF SINTERED BARIUM TITANATE AS A FUNCTION O
                   04-0817   REJUVENATION OF A BARIUM(0.72) STRONTIUM(0.28) TITANATE CERAMIC.
LEVIN S   17-2513   HIGH TEMPERATURE PHASE TRANSITIONS AND METASTABILITY IN POTASSIUM DIHYDROGEN PHOSPHATE
                   17-2706   ANTIFERROELECTRIC TRANSITIONS IN ADP AND AMMONIUM DIHYDROGEN ARSENATE STUDIED BY INFRA
                   17-2708   PROTON DYNAMICS IN KDP TYPE FERROELECTRICS STUDIED BY INFRARED ABSORPTION.
LEVINA ME   16-2360   EFFECT OF HYDROSTATIC PRESSURE ON THE FERROELECTRIC PHASE TRANSITION OF AMMONIUM SULFA
                   17-2560   EFFECT OF REPLACEMENT OF POTASSIUM BY ALUMINUM ON THE FERROELECTRIC PROPERTIES OF POTA
                   17-2572   DIELECTRIC PROPERTIES OF SINGLE CRYSTALS IN THE AMMONIUM DIHYDROGEN PHOSPHATE- POTASSI
                   19-2762   FERROELECTRIC PROPERTIES OF SOLID SOLUTIONS IN THE SYSTEM AMMONIUM SULFATE- AMMONIUM F
                   19-2763   FERROELECTRIC PROPERTIES OF SOLID SOLUTIONS IN THE AMMONIUM SULFATE- AMMONIUM FLUOBERY
                   19-2764   FERROELECTRIC PROPERTIES OF SOLID SOLUTIONS IN THE AMMONIUM SULFATE- AMMONIUM FLUOBERY
LEVINA SS   05-1065   SEIGNETTO MAGNETIC SOLID SOLUTIONS OF THE LEAD FERRATE(0.67) TUNGSTATE(0.33)- LEAD TIT
LEVINSTEIN HJ   13-2025   ROLE OF HYDROGEN IN POLARIZATION REVERSAL OF FERROELECTRIC BARIUM SODIUM NIOBATE.
LEVITSKII RR   03-0476   DYNAMICAL THEORY OF AMMONIUM DIHYDROGEN PHOSPHATE TYPE ANTIFERROELECTRICS WITH HYDROGE
                   17-2670   COUPLED VIBRATIONS OF THE PROTON ION SYSTEM IN HYDROGEN BONDED FERROELECTRICS OF THE K
                   17-2674   ROLE OF PROTON PHONON INTERACTION IN THE PHASE TRANSITION OF FERROELECTRICS WITH HYDRO
LEVKINA AP   04-0709   PECULIARITIES OF POLARIZATION OF BARIUM TITANATE SINGLE CRYSTALS WITH DOUBLE HYSTERESI
LEVSTEK I   17-2569   LASER RAMAN STUDY OF QUASI SPIN WAVE HYDROGEN TUNNELING MODES IN KDP AND DEUTERATED KD
                   17-2570   PROTON LATTICE INTERACTIONS AND THE DYNAMICAL SUSCEPTIBILITY OF POTASSIUM DIHYDROGEN P
LEVSTIK A   22-2831   DOUBLE HYSTERESIS LOOPS AT THE UPPER CURIE POINT OF SODIUM TRIHYDROGEN SELENITE.
                   22-2855   DIELECTRIC PROPERTIES OF HYDROGEN BONDED FERROELECTRICS. (SODIUM TRIHYDROGEN SELENATE

LEVSTIK A	22-2870	DIELECTRIC PROPERTIES AND PRESSURE EFFECTS ON SELENITE (ION). RADIATION DAMAGE ELECTRO
	27-2966	CRITICAL BEHAVIOR OF FERROELECTRIC TGS AND DEUTERATED TGS (HYSTERESIS LOOPS JUST BELOW
	27-3056	CRITICAL PROPERTIES OF TGS AND ROCHELLE SALT AS DETERMINED BY AN ELECTRIC FIELD METHOD
LEWIS FA	35-3507	GROWTH OF LARGE SINGLE CRYSTAL BISMUTH(4) TITANIUM(3) OXIDE(12).
LEWIS JE	25-2916	INFRARED AND RAMAN STUDIES IN THE FERROELECTRIC TRANSITION OF THIOUREA.
LEY JM	17-2475	LASER INDUCED DAMAGE IN X DIHYDROGEN PHOSPHATE MATERIALS.
LEZGINTSEVA TN	09-1557	INFLUENCE OF DOMAIN STRUCTURE ON THE PROPERTIES OF POLYCRYSTALLINE LEAD TITANATE ZIRCO
LEZHEPEKCV AP	05-1071	DIELECTRIC PROPERTIES OF LEAD TITANATE- LEAD (FERRATE NIOBATE) SOLID SOLUTIONS AT HIGH
LI L	11-1713	ABSORPTION AND LUMINESCENCE SPECTRA AND ENERGY LEVELS OF NEODYMIUM(3+) AND ERBIUM(3+)
LIBBY WF	06-1088	ELECTRIC PROPERTIES OF HYDROGEN DOPED STRONTIUM TITANATE.
LIBERMAN ZA	04-0818	UHF DIELECTRIC PROPERTIES OF SOME POLYCRYSTALLINE FERROELECTRICS (BARIUM TITANATE AND
LIBRECHT FM	11-1755	OPTIMUM CUT FOR A LITHIUM NIOBATE TRANSVERSE LIGHT MODULATOR.
	17-2494	OPTIMUM CUT IN X DIHYDROGEN PHOSPHATE CRYSTALS FOR TRANSVERSE LIGHT MODULATION
LICIS MS	30-3262	PRECISION PARAMETERS OF SOME LANTHANON MOLYBDATE TYPE RARE EARTH MOLYBDATES. (CRYSTAL
	30-3264	PRECISION PARAMETERS OF THE FERROELECTRIC RARE EARTH MOLYBDATES: LANTHANUM MOLYBDATE.
LIDER VV	09-1482	HYDROTHERMAL CRYSTALLIZATION OF LEAD TITANATE ZIRCONATE SOLID SOLUTIONS.
LIEB EH	03-0365	TWO DIMENSIONAL FERROELECTRIC MODELS.
	22-2822	ANOMALOUS SPECIFIC HEAT OF SODIUM TRIHYDROGEN SELENITE ASSOCIATED COMBINATORIAL PROBLE
	22-2823	ANOMALOUS SPECIFIC HEAT OF SODIUM TRIHYDROGEN SELENITE.
	34-3427	ANALYTIC PROPERTIES OF THE FREE ENERGY FOR THE "ICE" MODELS (FERROELECTRICS).
	34-3453	ICE, FERRO- AND ANTIFERROELECTRICS.
LIEBERTZ J	10-1650	GROWTH, CRYSTALLOGRAPHY AND DIELECTRIC PROPERTIES OF BISMUTH TUNGSTATE.
LIEKENS W	11-1756	MICROWAVE PHONON ATTUNUATION IN X CUT LITHIUM NIOBATE.
LIGASOVA VD	19-2787	EXPERIMENTAL INVESTIGATION OF THE FERROELECTRIC PROPERTIES OF AMMONIUM BISULFATE IN TH
LILGA KT	34-3413	ENDOR (ELECTRON NUCLEAR DOUBLE RESONANCE) STUDY OF X-IRRADIATED SINGLE CRYSTALS OF ICE
LIM KO	03-0373	DIELECTRIC DISPERSION IN ORDER DISORDER FERROELECTRICS. (THEORY)
LIM TC	09-1546	NONLINEAR EFFECTS IN FERROCERAMICS. (PZT-8, LITHIUM NIOBATE)
	11-1757	EFFECT OF MASS LOADING ON THE PROPAGATION OF ACOUSTIC SURFACE WAVES ON LITHIUM NIOBATE
	37-3647	APPLICATION OF NONLINEAR INTERACTIONS IN FERROELECTRIC CERAMICS TO MICROWAVE SIGNAL PR
LIMAR TF	36-3587	PREPARATION OF THE TERNARY TITANATE BARIUM(0.8) LEAD(0.12) CALCIUM(0.08) TITANATE.
LIMINGA R	22-2878	HYDROGEN BOND STUDIES. PART-54: A NEUTRON DIFFRACTION STUDY OF THE FERROELECTRIC LITHI
	22-2879	HYDROGEN BOND STUDIES. PART-58: CRYSTAL STRUCTURE OF AMMONIUM TRIHYDROGEN SELENITE.
LIMOU P	08-1340	DIELECTRIC STUDY OF THE LEAD ZIRCONATE LEAD HAFNATE (PHASE) DIAGRAM.
LIN CC	19-2793	RAMAN AND INFRARED STUDIES OF THE FERROELECTRIC TRANSITION IN AMMONIUM SULFATE.
LINES ME	03-0366	COMPARISON OF FERROELECTRICITY IN ISOMORPHIC LITHIUM NIOBATE AND LITHIUM TANTALATE USI
	03-0367	INTERCELL CORRECTIONS FOR IONIC MOTION IN DISPLACEMENT FERROELECTRICS (LITHIUM TANTALA
	03-0368	POLARIZATION FLUCTUATIONS NEAR A FERROELECTRIC PHASE TRANSITION.
	03-0369	STATISTICAL THEORY FOR DISPLACEMENT FERROELECTRICS.
	11-1776	STACKING FAULT MODEL FOR STOICHIOMETRY DEVIATIONS IN LITHIUM NIOBATE AND TANTALATE AND
	11-1777	STACKING FAULT MODEL FOR STOICHIOMETRY DEVIATIONS IN LITHIUM NIOBATE AND TANTALATE AND
	12-1879	NATURE OF THE FERROELECTRIC- PARAELECTRIC PHASE TRANSITION IN LITHIUM TANTALATE.
LINGAFELTER EC	21-2818	FERROELECTRIC MOMENT IN GUANIDINIUM ALUMINUM SULFATE HEXAHYDRATE (AND MECHANISM OF REV
	21-2819	REDETERMINATION OF STRUCTURE OF FERROELECTRIC CRYSTAL GUANIDINIUM ALUMINUM SULFATE HEX
LINGSCHEIT JN	04-0819	RESISTIVITY AND SINTERING OF BARIUM TITANATE.
LINZ A	04-0553	TOP SEEDED SOLUTION GROWTH OF OXIDE CRYSTALS FROM NONSTOICHIOMETRIC MELTS. (BARIUM TIT
	04-0946	INELASTIC NEUTRON SCATTERING FROM SINGLE DOMAIN BARIUM TITANATE. (LATTICE DYNAMICS)
	04-1032	CRITICAL (POLARIZATION) FLUCTUATIONS IN BARIUM TITANATE (ABOVE THE FERROELECTRIC CURIE
	08-1353	(TOP SEEDED FLUX) CRYSTAL GROWTH AND NEUTRON CHARACTERIZATION OF POTASSIUM NIOBATE.
	33-3374	DISPERSION AND DAMPING OF SOFT ZONE BONDARY PHONONS IN POTASSIUM MANGANESE TRIFLUORIDE
	33-3405	NEUTRON SCATTERING STUDY OF THE LATTICE DYNAMICAL PHASE TRANSITIONS IN POTASSIUM MANGA
LIPAEVA GA	04-0820	TEMPERATURE DEPENDENCE OF THE COMPLEX PERMITTIVITY OF SOME TITANATES BETWEEN 20 AND 10
LIPPINCOTT ER	16-2342	POLARIZED INFRARED AND RAMAN SPECTRA OF SINGLE CRYSTAL CESIUM NITRITE.
	25-2905	LATTICE VIBRATIONAL SPECTRA OF FIVE CRYSTAL MODIFICATIONS OF THIOUREA.
LIPPOLD B	26-2931	NMR INVESTIGATIONS OF FERROELECTRIC DICALCIUM STRONTIUM PROPIONATE.
LIPSKIS KK	15-2193	TEMPERATURE DEPENDENCES OF THE HALL AND DRIFT MOBILITIES OF HOLES IN ANTIMONY TRISULFI
LIPSON HG	35-3496	CZOCHRALSKI SYNTHESIS AND PROPERTIES OF RARE EARTH DOPED BISMUTH GERMANATE.
LISSALDE FC	36-3588	SURFACE EFFECT IN FERROELECTRIC YTTRIUM MANGANATE.
LITHER G	17-2574	DIELECTRIC DISPERSION OF TGS.
LITOV E	17-2573	ACOUSTIC PROPAGATION IN POTASSIUM DEUTERIUM PHOSPHATE AT THE FERROELECTRIC TRANSITION
	17-2575	POLARIZATION RELAXATION AND SUSCEPTIBILITY IN THE FERROELECTRIC TRANSITION REGION OF P
	17-2576	ULTRASONIC INVESTIGATION (15 MHZ SHEAR WAVES) OF THE FERROELECTRIC TRANSITION REGION I
LITOVITZ TA	27-3097	ULTRASONIC RELAXATION NEAR THE CURIE TEMPERATURE OF FERROELECTRIC TRIGLYCINE SULFATE.
LITVIN BN	15-2194	HYDROTHERMAL METHOD FOR PREPARING A(V)B(VI)C(VII) COMPOUNDS. (USING ANTIMONY OR BISMUT
	15-2214	HYDROTHERMAL CRYSTALLIZATION OF SEMICONDUCTING COMPOUNDS OF GROUP A(V)B(VI)C(VII). (HI
	15-2215	KINETICS OF CRYSTALLIZATION OF ANTIMONY SULFOIODIDE (CRYSTAL GROWTH RATE, ANISOTROPY)
LITVINOV VI	17-2671	INFLUENCE OF POLAR STATES OF PROTONS ON THE PHASE TRANSITION IN FERROELECTRICS WITH HY
LIU NLH	06-1214	FREE CARRIER OPTICAL ABSORPTION IN NIOBIUM DOPED STRONTIUM TITANATE.
LIU ST	09-1558	PYROELECTRIC PROPERTIES OF THE LANTHANUM DOPED FERROELECTRIC PLZT CERAMICS.
LIVINGSTON R	16-2404	PARAMAGNETIC RESONANCE STUDY OF PRODUCTION OF NITROGEN DIOXIDE AND NITRITE ION IN IRRA
LJAKHOVICKJA VA	15-2138	PREPARATION OF ANTIMONY SULFOIODIDE TEXTURES FROM THE MELT UNDER PRESSURE AND INVESTIG
LLOYD P	11-1724	LITHIUM TANTALATE AND LITHIUM NIOBATE PIEZOELECTRIC RESONATORS IN THE MEDIUM FREQUENCY
	11-1851	MEDIUM FREQUENCY RESONATORS OF LITHIUM NIOBATE AND LITHIUM TANTALATE.
LOBACHEV AN	04-0941	NONLINEAR OPTICAL PROPERTIES OF LEAD TITANATE AND BISMUTH TITANATE CRYSTALS.
	08-1279	HYDROTHERMAL CRYSTALLIZATION AND SOME PROPERTIES OF BISMUTH TITANATES.
	09-1482	HYDROTHERMAL CRYSTALLIZATION OF LEAD TITANATE ZIRCONATE SOLID SOLUTIONS.
	15-2180	MOSSBAUER STUDIES OF ANTIMONY SULFOIODIDE TYPE CRYSTALS.
	15-2213	CRYSTALLIZATION OF SEMICONDUCTORS OF THE COMPOSITION A(V)B(VI)C(VII) UNDER HYDROTHERMA
	15-2214	HYDROTHERMAL CRYSTALLIZATION OF SEMICONDUCTING COMPOUNDS OF GROUP A(V)B(VI)C(VII). (HI
	36-3589	NEW FERROELECTRIC: ANTIMONY ORTHONIOBATE. (CURIE TEMPERATURE)
LOBO R	03-0215	HARMONICS GENERATION IN FERROELECTRIC MEDIA.
	34-3443	A STUDY OF BROUT'S MODEL FOR FERROELECTRICS. INVESTIGATIONS ON HEXAGONAL ICE.
LOCK PJ	27-3057	DOPED TRIGLYCINE SULFATE FOR PYROELECTRIC APPLICATIONS.
	27-3058	POLARIZED TRIGLYCINE SULFATES.
LOESCHE A	27-3059	NUCLEAR MAGNETIC RESONANCE INVESTIGATION OF TRIGLYCINE SULFATE.
LOGAN KW	34-3465	ICE I: LATTICE DYNAMICS AND INCOHERENT NEUTRON SCATTERING.
LOIACONO GM	17-2577	DTA STUDY OF THE FERROELECTRIC TRANSITION IN KDP TYPE CRYSTALS (CURIE TEMPERATURES OF
LOKTEV VM	03-0419	ON THE THEORY OF LIGHT ABSORPTION BY ANTIFERRODIELECTRICS IN THE FREQUENCY RANGE OF DO
LOMER TR	26-2927	UNIT CELL AND SPACE GROUP OF THE ANTIFERROELECTRIC PHASE OF COPPER FORMATE TETRAHYDRAT
LOMOVA LG	28-3209	SPONTANEOUS ELECTROOPTICAL EFFECT IN ROCHELLE SALT CRYSTALS.
LONG SA	04-0821	CALCULATIONS OF ELECTRONIC CONDUCTIVITY IN NONSTOICHIOMETRIC BARIUM TITANATE.

FERROELECTRICS--AUTHOR-TITLE INDEX

LONG SA	04-0822	TITANIUM RICH NONSTOICHIOMETRIC BARIUM TITANATE. PART-1: HIGH TEMPERATURE ELECTRICAL C
LONGINI RL	34-3411	EQUILIBRIUM STRUCTURE OF POLARIZED ICE.
	34-3472	IMPURITY STATISTICS IN ICE.
LOOMIS TC	08-1437	CALCIUM CONCENTRATION VS NET IONIZED DONOR CONCENTRATION IN SINGLE CRYSTAL POTASSIUM T
LOPEZ V	37-3641	ON THE SWITCHING PROPERTIES OF THE FERROELECTRIC SURFACES. (NEW MODEL REPRODUCES EXPER
LOPEZ-ALONSO JR	03-0263	STATISTICAL THEORY FOR FERROELECTRICITY IN TRIGLYCINE SULFATE.
LOSHMANOV AA	02-0133	NEUTRON DIFFRACTION INVESTIGATIONS OF THE ATOMIC STRUCTURES OF INORGANIC MATERIALS. (R
LOVGA IV	15-2128	PIEZOELECTRIC RESISTANCE EFFECT IN BISMUTH SELENOIODIDE CRYSTALS.
LOW W	36-3583	ELECTRON SPIN RESONANCE SPECTRA OF CHROMIUM(3+) AND GADOLINIUM(3+) IN LANTHANUM ALUMIN
LUBIMOV VN	08-1442	CALCULATIONS OF THE INTERNAL ELECTRIC FIELDS AND ELECTRIC FIELD GRADIENTS IN THE PEROV
LUCAS A	17-2650	INFRARED VIBRATION FREQUENCIES OF THE HYDROGEN BOND IN KDP FERROELECTRIC (CALCULATIONS
LUCAS J	36-3529	NEW FERROELECTRIC (PYROCHLORE TYPE) COMPOUND (AMMONO THIO CADMIUM NIOBATE, WITH 4 ELEC
	36-3602	NEW DESCRIPTION OF THE PYROCHLORE STRUCTURE (CADMIUM(2) NIOBIUM(2) OXIDE(6) SULFIDE).
LUDLOW JH	27-3107	PYROELECTRIC THERMAL IMAGING DEVICES. (TGS)
LUISKUTTY CT	06-1153	MOSSBAUER STUDY OF COLOR CENTERS IN STRONTIUM TITANATE.
LUKASZEWICZ K	02-0094	ROLE OF THE REGULAR OCTAHEDRON AND LARGER POLYHEDRA IN FERROELECTRIC MATERIALS. (HETER
LUKE TE	08-1298	A NEW METHOD OF OPTICALLY READING DOMAINS IN BISMUTH TITANATE AND MEMORY APPLICATIONS.
	08-1299	EFFICIENT WHITE LIGHT READING OF DOMAIN PATTERNS IN BISMUTH TITANATE.
LUNDIN AG	16-2255	NUCLEAR MAGNETIC RESONANCE STUDIES OF THE PHASE TRANSITIONS IN FERROELECTRICS. (POTASS
	16-2337	TEMPERATURE DEPENDENCE OF THE ELECTRIC FIELD GRADIENT IN FERROELECTRIC SODIUM NITRITE.
	24-2893	ORDERING OF DIPOLE MOMENTS OF WATER MOLECULES AND FERROELECTRIC PHASE TRANSITION IN PO
	24-2900	RADIO SPECTROSCOPIC INVESTIGATION OF POTASSIUM FERROCYANIDE TRIHYDRATE AND ISOMORPHOUS
LUONGO JP	36-3590	FAR INFRARED SPECTRA OF PIEZOELECTRIC POLYVINYLIDENE FLUORIDE.
LUPFER DA	09-1559	FERROELECTRICITY IN SPUTTERED LEAD TITANATE ZIRCONATE FILMS.
LURIO A	06-1154	FAR INFRARED REFLECTIVITY AND TRANSMISSION MEASUREMENTS ON STRONTIUM TITANATE FILMS.
	13-1986	FAR INFRARED DIELECTRIC DISPERSION IN BARIUM SODIUM NIOBATE.
LUSPIN Y	19-2766	BRILLOUIN EFFECT AT ORDINARY TEMPERATURE IN AMMONIUM SULFATE.
LUSTIV-SHUMSKII LF	17-2443	PHOTOELASTIC EFFECT IN 45 DEGREE X-CUTS OF POTASSIUM AND AMMONIUM DIHYDROGEN PHOSPHATE
	17-2697	ELECTROOPTICAL AND PIEZOOPTICAL PROPERTIES OF POTASSIUM DIHYDROGEN PHOSPHATE AND AMMON
LUTHER G	03-0370	(EFFECT OF TEMPERATURE GRADIENT WITHIN THE SAMPLE ON THE) MICROWAVE DISPERSION OF FERR
	27-2975	TEMPERATURE DEPENDENCE OF THE DIELECTRIC CONSTANTS OF TRIGLYCINE SULFATE ORTHOGONAL TO
	27-3060	DIELECTRIC DISPERSION OF TGS.
	27-3061	TEMPERATURE DEPENDENCE OF DIELECTRIC PROPERTIES OF FERROELECTRIC TRIGLYCINE SULFATE IN
	27-3088	MICROWAVE BEHAVIOR OF ROCHELLE SALT AND TGS (18 REFS).
	28-3206	FAST AFTEREFFECT OF SEIGNETTE SALT.
LUTHER-DAVIES B	12-1880	SIGNS OF THE ELECTROOPTIC COEFFICIENTS FOR LITHIUM TANTALATE.
LUTHI B	06-1155	SOUND PROPAGATION NEAR THE STRUCTURAL PHASE TRANSITION IN STRONTIUM TITANATE.
LUTSAU VG	17-2578	X-RAY DIFFRACTION TOPOGRAPHY STUDY OF DEFECTS (DISLOCATIONS) IN KDP AND ADP SINGLE CRY
LUUKKALAA M	11-1758	ACOUSTIC CONVOLUTION AND CORRELATION AND THE ASSOCIATED NONLINEARITY PARAMETERS IN LIT
LUXON JT	04-0823	EFFECT OF PARTICLE SIZE AND SHAPE ON THE INFRARED ABSORPTION SPECTRA OF BARIUM TITANAT
LVITSKII RR	17-2671	INFLUENCE OF POLAR STATES OF PROTONS ON THE PHASE TRANSITION IN FERROELECTRICS WITH HY
LYAKHOVITSKAYA VA	15-2108	NUCLEAR GAMMA RESONANCE SPECTROSCOPIC STUDY OF SEMICONDUCTING COMPOUNDS OF THE TYPE A(
	15-2118	EFFECT OF COMPOSITION ON GROWTH, DIELECTRIC, AND PHOTOELECTRIC PROPERTIES OF ANTIMONY
	15-2120	MOTION PICTURES OF THE PHASE TRANSITION IN ANTIMONY SULFOIODIDE CRYSTALS.
	15-2158	INFLUENCE OF NONEQUILIBRIUM CARRIER SCREENING ON POLARIZATION PROCESSES IN FERROELECTR
	15-2183	DOMAIN STRUCTURE OF ANTIMONY SULFOIODIDE CRYSTALS (FROM SELECTIVE CRYSTALLIZATION AND
	15-2184	PHASE TRANSITION AND DOMAIN STRUCTURE IN ANTIMONY SULFOIODIDE CRYSTALS. (MICRO CINEMAT
	15-2230	CRYSTAL GROWTH IN THE ANTIMONY SULFOIODIDE- ANTIMONY TRISULFIDE SYSTEM. (BRIDGMAN- STO
	15-2238	HEAT CAPACTIY OF POLYCRYSTALLINE ANTIMONY SULFOIODIDE.
	15-2248	EFFECTS OF TEMPERATURE, ELECTRIC FIELD, AND ILLUMINATION ON THE ABSORPTION OF ULTRASOU
LYKHIN VM	16-2319	EFFECT OF ADDING TITANIUM(+) AND STRONTIUM(2+) IONS ON THE ELECTRICAL CONDUCTIVITY OF
LYNBIMOV VN	08-1455	CALCULATION OF INTERNAL ELECTRIC FIELDS AND THEIR GRADIENTS IN ABO(3) PEROVSKITE COMPO
LYNCH RW	06-1121	HIGH PRESSURE COMPRESSIBILITY AND GRUNEISEN PARAMETER OF STRONTIUM TITANATE.
LYTLE FW	06-1156	X-RAY DIFFRACTOMETRY OF LOW TEMPERATURE PHASE TRANSFORMATIONS IN STRONTIUM TITANATE.
LYUBIMOV VN	03-0315	SURFACE ELASTOPOLARIZATION WAVES AT DOMAIN BOUNDARIES IN A FERROELECTRIC.
	03-0371	ELASTIC AND ELECTROMAGNETIC WAVES IN PIEZOELECTRIC CRYSTALS.
	03-0372	FERROELECTRICITY AND ANTIFERROELECTRICITY IN POLAR CRYSTALS. (THEORY, LEAD ZIRCONATE)
	03-0526	CHARACTERISTICS OF THE STRUCTURAL CHANGES INVOLVED IN PHASE TRANSFORMATIONS IN ANTIFER
	04-0824	CALCULATION OF ELECTRIC FIELDS IN IONIC CRYSTALS. (BARIUM TITANATE TYPE)
	07-1261	DIPOLE STRUCTURE AND INTERNAL ELECTRIC FIELDS IN LEAD ZIRCONATE.
	08-1389	INTERNAL ELECTRIC FIELDS IN CRYSTALS OF SODIUM TANTALATE AND CADMIUM TITANATE.
	08-1390	ORIGIN OF DIPOLE CONFIGURATIONS IN SOME (PEROVSKITE) STRUCTURES WITH SPECIAL DIELECTRI
	08-1459	PEROVSKITE TYPE SEIGNETTO MAGNETS.
LYUBUTIN IS	15-2108	NUCLEAR GAMMA RESONANCE SPECTROSCOPIC STUDY OF SEMICONDUCTING COMPOUNDS OF THE TYPE A(
MAARTENSE I	33-3399	MAGNETOELASTIC EFFECTS IN POTASSIUM MANGANESE TRIFLUORIDE.
MAASKANT WJA	33-3388	A PHASE TRANSITION IN A COMPOUND WITH HELICAL ELECTRIC DIPOLE STRUCTURE: CESIUM COPPER
MACBETH JW	09-1559	FERROELECTRICITY IN SPUTTERED LEAD TITANATE ZIRCONATE FILMS.
MACCHESNEY JB	04-0825	STABILIZED BARIUM TITANATE CERAMICS FOR CAPACITOR DIELECTRICS.
	36-3592	SYSTEM LANTHANA TITANIA PHASE EQUILIBRIA AND ELECTRICAL PROPERTIES (LANTHANUM TITANATE
MACHET R	03-0189	THEORETICAL STATIC MODEL FOR BARIUM AND LEAD TITANATES - APPLICATION TO CALCULATING TH
	04-0578	ELECTROSTATIC FIELD IN A SLIGHTLY ORTHORHOMBIC IONIC CRYSTAL. APPLICATION TO THE CALCU
	04-0826	STATIC SHELL MODEL FOR BARIUM AND LEAD TITANATES.
	04-1008	ELECTROSTATIC FIELD IN A PSEUDOCUBIC IONIC STRUCTURE. APPLICATION TO ALL BARIUM TITANA
MACK D	08-1385	DIELECTRIC, PIEZOELECTRIC AND PYROELECTRIC PROPERTIES OF THE LEAD METANIOBATE- BARIUM
MACK DL	08-1384	DIELECTRIC, PIEZOELECTRIC AND PYROELECTRIC PROPERTIES OF THE LEAD DINIOBATE- BARIUM DI
MACKIE PE	36-3591	MONOCLINIC STRUCTURE OF SYNTHETIC CALCIUM CHLORAPATITE. (DEVELOPMENT OF FERROELECTRIC
MACKOWIAK M	27-3134	EPR OF THE FOUR SPIN ION COMPLEX WITH SPIN S=2 IN TRIGLYCINE FLUOBERYLLATE MONOCRYST
MAEDA M	29-3252	ESR OF COPPER(2+) DOPED SODIUM AMMONIUM TARTRATE.
MAEKAWA S	04-0928	ELECTRON PARAMAGNETIC RESONANCE IN BARIUM TITANATE NEAR THE CURIE TEMPERATURE.
MAENO N	34-3454	DIELECTRIC PROPERTIES OF POTASSIUM CHLORIDE ICE.
MAGUIRE HG	04-0827	INVESTIGATION OF THE FERROELECTRIC PHASE CHANGE IN BARIUM TITANATE USING ELECTRON SPIN
MAHE R	04-0590	BINARY AND PSEUDOBINARY PHASES BETWEEN NIOBIUM PENTOXIDE AND LEAD, BARIUM, AND STRONTI
	08-1285	STRUCTURAL STUDY OF THE TRANSITION OF THE (ORTHORHOMBIC) FERROELECTRIC (LEAD DINIOBATE
	08-1286	STUDY OF DIVALENT BINARY AND TERNARY NIOBATES IN THE SOLID STATE.
MAHER GH	04-0828	PHYSICAL AND ELECTRICAL PROPERTIES OF THIN FILM BARIUM TITANATE PREPARED BY RF SPUTTER
MAHLER RJ	09-1560	HIGH D*, FAST, LEAD ZIRCONATE TITANATE PYROELECTRIC DETECTORS.
	33-3390	ACOUSTIC AND MAGNETIC EFFECTS INVOLVING THE FLUORINE-19 NUCLEI IN ANTIFERROMAGNETIC PO
MAIDIQUE MA	34-3455	TRANSFER OF PROTONS THROUGH "PURE" ICE I(H), SINGLE CRYSTALS. PART-2: MOLECULAR MODELS
	34-3482	DIELECTRIC RELAXATION SPECTRA OF WATER, ICE, AND AQUEOUS SOLUTIONS AND THEIR INTERPRET
MAINES JD	11-1698	SURFACE WAVE DIFFRACTION IN LITHIUM NIOBATE.

PAGE 661

MAJER J   11-1760   THE MEASUREMENT OF HYPERSONIC ATTENUATION IN LITHIUM NIOBATE.
MAJER M   11-1759   CRYSTAL FIELD IN CHROMIUM(3+) DOPED LITHIUM NIOBATE.
MAKAROV KV   03-0195   POLARIZATION OF FERROELECTRIC FILM BY BENDING AND BENDING DEFORMATION IN THE ELECTRIC
                  04-0600   DIELECTRIC CONSTANT AND LATTICE DYNAMICS OF A THIN FILM OF A FERROELECTRIC (BARIUM TIT
MAKER PD   17-2579   EFFECTS OF DISPERSION AND FOCUSING ON THE PRODUCTION OF OPTICAL HARMONICS. (KDP)
MAKI T   04-0829   BARIUM TITANATE CERAMICS WITH HIGH DIELECTRIC CONSTANT. PART-1: EFFECTS OF ADDITION OF
MAKITA Y   02-0087   CRITICAL SLOWING DOWN PROCESS OF (ANOMALOUS) DIELECTRIC RELAXATION IN SEVERAL CRYSTALS
                  16-2346   THERMAL EXPANSION IN FERROELECTRIC SILVER SODIUM DINITRITE.
                  19-2767   DIELECTRIC AND SOME OTHER STUDIES OF FERROELECTRIC PHASE TRANSITION IN SODIUM AMMONIUM
                  19-2778   DOMAIN STRUCTURE OF FERROELECTRIC SODIUM AMMONIUM SULFATE DIHYDRATE.
                  19-2779   THERMAL AND SOME OTHER STUDIES OF SODIUM AMMONIUM SULFATE DIHYDRATE, SODIUM TRIHYDROGE
                  20-2804   DIELECTRIC RELAXATION IN FERROELECTRIC MASD, METHYL AMMONIUM ALUMINUM SULFATE DODECAHY
                  22-2856   ANOMALOUS SPECIFIC HEAT AND CONFIGURATIONAL ENTROPY CHANGE IN SODIUM TRIHYDROGEN DISEL
                  22-2858   THERMAL EXPANSION IN SODIUM TRIHYDROGEN(1-X) TRIDEUTERIUM(X) DISELENITE AND ISOTOPE EF
                  26-2938   X-RAY DIFFRACTION STUDY OF PHASE TRANSITION IN COPPER FORMATE.
MAKOSZ JJ   16-2338   DIELECTRIC MEASUREMENTS OF AMMONIUM NITRATE.
                  16-2339   DIELECTRIC STUDIES OF PHASE 2 OF AMMONIUM NITRATE.
MALDONADO JR   09-1516   IMPROVED AGING AND SWITCHING OF LEAD ZIRCONATE- LEAD TITANATE CERAMICS WITH INDIUM ELE
                  09-1517   IMPROVED AGING AND SWITCHING OF LEAD ZIRCONATE- LEAD TITANATE CERAMICS WITH INDIUM ELE
                  09-1561   DISPLAY APPLICATIONS OF PLZT FERROELECTRIC CERAMICS.
                  09-1562   ELECTROOPTIC DEVICES USING STRAIN BIASED PLZT FERROELECTRIC CERAMICS.
                  09-1563   FERROELECTRIC DOMAIN SWITCHING IN RHOMBOHEDRAL PHASE PLZT CERAMICS.
                  09-1564   STRAIN BIASED FERROELECTRIC- PHOTOCONDUCTOR IMAGE STORAGE AND DISPLAY DEVICES OPERATED
                  09-1565   STRAIN BIASED FERROELECTRIC PHOTOCONDUCTOR IMAGE STORAGE AND DISPLAY DEVICES. (PLZT)
                  09-1572   IMAGE STORAGE AND DISPLAY DEVICES USING FINE GRAIN FERROELECTRIC CERAMICS. (LEAD ZIRCO
                  37-3654   FERROELECTRIC CERAMIC LIGHT GATES OPERATED IN A VOLTAGE CONTROLLED MODE.
MALDUTIS EK   11-1865   AUTOFOCUSING OF LASER RADIATION IN ACTIVE MATERIALS AND NONLINEAR CRYSTALS. (KDP, ADP,
                  17-2717   THERMAL SELF FOCUSING IN POTASSIUM DIHYDROGEN PHOSPHATE AND AMMONIUM DIHYDROGEN PHOSPH
MALECOT G   03-0237   ELASTIC ENERGY AND THE FERROELECTRIC PARAELECTRIC TRANSITION.
MALEK Z   02-0088   INFLUENCE OF THE QUALITY OF TRIGLYCINE SULFATE SINGLE CRYSTALS ON THEIR DIELECTRIC PRO
                  27-2979   PYROELECTRIC RESPONSE OF TGS IN THE FERROELECTRIC REGION NEAR THE CURIE POINT (IMPROVE
                  27-3062   FREQUENCY DEPENDENCE OF THE COERCIVE FIELD OF TRIGLYCINE SULFATE CRYSTALS.
                  27-3063   INFLUENCE OF DEFECTS ON SOME DIELECTRIC PROPERTIES OF TGS SINGLE CRYSTALS.
                  27-3064   INFLUENCE OF SURFACE EFFECTS ON THE PYROELECTRIC BEHAVIOR OF TGS CLOSE TO THE PHASE TR
                  27-3065   MATERIAL RESEARCH OF FERROELECTRIC TRIGLYCINE SULFATE FROM THE VIEWPOINT OF PRACTICAL
                  27-3066   PYROELECTRIC BEHAVIOR OF TRIGLYCINE SULFATE IN THE PARAELECTRIC REGION.
                  27-3067   THE EFFECT OF GROWTH RATE ON THE DEFECT STRUCTURE AND DIELECTRIC PROPERTIES OF TGS SIN
                  27-3068   TEMPERATURE AND FIELD STRENGTH DEPENDENCE OF LARGE SIGNAL PERMITTIVITY OF TGS. (10 KHZ
                  27-3086   PREPARATION OF PURE TGS SINGLE CRYSTALS AND SOME INVESTIGATIONS OF THEIR QUALITY. (VOL
                  37-3655   TEMPERATURE AUTOSTABILIZATION EFFECT IN FERROELECTRICS. (CAPACITOR PROPERTIES, TANDEL)
MALGRANGE C   30-3304   STUDY OF FERROELECTRIC DOMAINS IN GADOLINIUM MOLYBDATE.
MALI M   18-2720   CESIUM-183 SPIN LATTICE RELAXATION AND RESONANCE IN FERROELECTRIC CESIUM DIDEUTERIUM A
                  18-2721   DEUTERON QUADRUPOLE COUPLING IN FERROELECTRIC CESIUM DIDEUTERIUM ARSENATE.
                  19-2749   NITROGEN-14 QUADRUPOLE COUPLING IN PARAELECTRIC AMMONIUM SULFATE.
                  27-2968   PULSED NITROGEN PROTON DOUBLE RESONANCE STUDY OF THE FERROELECTRIC TRANSITION IN TRIGL
                  27-2969   PULSED DOUBLE RESONANCE STUDY OF THE NUCLEAR QUADRUPOLE INTERACTIONS OF NITROGEN-14 IN
                  28-3186   SECOND ORDER QUADRUPOLE SHIFTS OF THE SODIUM-23 MAGNETIC RESONANCE IN ROCHELLE SALT.
MALINEAU M   28-3210   FAR INFRARED ABSORPTION SPECTRA OF ROCHELLE SALT.
MALININA VG   15-2185   CRYSTAL FIELD GRADIENTS IN ANTIMONY SULFOIODIDE.
MALITSKAYA MA   15-2157   FERROELECTRIC PHOTOELECTRETS BASED ON A(V)B(VI)C(VII) SOLID SOLUTIONS. (DIELECTRIC CON
                  15-2161   PHOTOFERROELECTRIC EFFECTS IN FERROELECTRIC SEMICONDUCTORS OF THE A(V)B(VI)C(VII) TYPE
MALNEV AF   03-0333   SIMULTANEOUS MEASUREMENT OF PYROELECTRIC COEFFICIENT AND DIELECTRIC CONSTANT OF FERROE
MANABE K   05-1075   CHARACTERISTICS OF DEFECT LEAD TITANATE.
MANDEVILLE GD   12-1888   PERFORMANCE OF AN 11 GHZ OPTICAL MODULATOR USING LITHIUM TANTALATE.
MANI KK   07-1262   ANTIFERROELECTRIC MODE IN LEAD ZIRCONATE. (LATTICE VIBRATIONS)
MANLIEF SK   08-1391   RAMAN SPECTRUM OF POTASSIUM TANTALATE NIOBATE. (POTASSIUM TANTALATE(0.65) NIOBATE(0.35
                  13-1987   ABSORPTION EDGE SPLITTING IN POTASSIUM TANTALATE(X) NIOBATE(1-X).
                  13-1988   ABSORPTION EDGE SPLITTING IN POTASSIUM TANTALATE NIOBATE.
                  13-1989   RAMAN SPECTRUM OF POTASSIUM TANTALUM(0.64) NIOBIUM(0.36) OXIDE.
                  36-3593   OPTICAL PROPERTIES OF POTASSIUM TANTALATE NIOBATE.
MANOJLOVIC-MUR LJ 22-2866   A NEUTRON DIFFRACTION STUDY OF POTASSIUM TRIHYDROGEN SELENITE.
MANSINGH A   03-0373   DIELECTRIC DISPERSION IN ORDER DISORDER FERROELECTRICS. (THEORY)
                  16-2340   DIELECTRIC AND ELECTRICAL CONDUCTIVITY STUDIES IN POTASSIUM NITRITE AND POTASSIUM NITR
                  16-2341   DIELECTRIC DISPERSION IN THE PARAELECTRIC PHASE OF POTASSIUM NITRATE.
                  16-2369   PHASE TRANSFORMATIONS IN POTASSIUM NITRATE AND POTASSIUM NITRITE. (DOES NOT SEEM FERRO
MANTE AJH   04-0830   PHONON TRANSPORT IN BARIUM TITANATE.
                  06-1157   COUPLED MODES (OPTICAL, FERROELECTRIC- ACOUSTIC) IN THE PHONON SPECTRA OF STRONTIUM TI
MARCHENKO EI   08-1423   ELECTROOPTICAL PROPERTIES OF LITHIUM IODATE SINGLE CRYSTALS.
MARECEK V   02-0089   EFFECT OF CRYSTALLIZATION CONDITIONS ON THE SHAPE AND COMPOSITION OF FERROELECTRIC CRY
                  17-2637   CONTRIBUTION TO THE PREPARATION ON NON-WEDGE SHAPED RUBIDIUM DIHYDROGEN OR DIDEUTERIUM
                  19-2753   PREPARATION OF AMMONIUM SULFATE SINGLE CRYSTALS.
                  29-3239   SOME PROPERTIES OF NONDEUTERATED AND DEUTERATED LITHIUM THALLIUM TARTRATE MONOHYDRATE.
MARENNIKOV SI   17-2500   COMBINATION OF THE FREQUENCIES OF COHERENT AND INCOHERENT RADIATION IN KDP CRYSTAL.
MARGOLIN AM   04-0656   SIZE EFFECTS IN FERROELECTRICS. (BARIUM TITANATE)
                  04-0832   STRUCTURE AND PROPERTIES OF BARIUM TITANATE FILMS PRODUCED BY CATHODE SPUTTERING.
MARGOLIN LN   27-2978   PERMITTIVITY OF SINGLE CRYSTALS OF TRIGLYCINE SULFATE IN A STRONG ELECTRIC FIELD.
                  27-3069   INFLUENCE OF GROWTH CONDITIONS OF TRIGLYCINE SULFATE AND TRIGLYCINE SELENATE CRYSTALS
                  27-3125   DIELECTRIC PERMEABILITY AND DIELECTRIC LOSS ANGLE TANGENT OF TRIGLYCINE SELENATE DEPEN
                  27-3126   THERMAL HYSTERESIS OF THE DIELECTRIC CONSTANT OF TRIGLYCINE SULFATE AS A FUNCTION OF T
                  27-3150   DIELECTRIC CONSTANT OF TRIGLYCINE SULFATE SINGLE CRYSTALS IN A STRONG ELECTRIC FIELD.
                  27-3151   EFFECT OF TRIGLYCINE FLUOBERYLLATE SINGLE CRYSTAL GROWTH CONDITIONS ON A CHANGE IN THE
                  27-3152   EFFECT OF GROWTH CONDITIONS ON DIELECTRIC LOSSES IN TRIGLYCINE SULFATE SINGLE CRYSTALS
                  27-3153   EFFECT OF GROWTH CONDITIONS OF TRIGLYCINE SULFATE SINGLE CRYSTALS ON A CHANGE IN THE T
MARKOV M   27-3113   LOW FREQUENCY RAMAN SCATTERING BY FERROELECTRIC TRIGLYCINE FLUOBERYLLATE CRYSTALS.
                  27-3114   LOW FREQUENCY RAMAN SPECTRA OF FERROELECTRIC CRYSTALS OF THE TRIGLYCINE SULFATE TYPE.
MARKOVSKII VYU   17-2648   STUDY OF THE (MECHANICAL) STRENGTH PROPERTIES OF POTASSIUM DIHYDROGEN PHOSPHATE CRYSTA
                  17-2649   TEMPERATURE DEPENDENCE OF THE MECHANICAL CHARACTERISTICS OF POTASSIUM DIHYDROGEN PHOSP
MARKS GW   03-0434   SYMMETRY LIMITATIONS TO POLARIZATION OF POLYCRYSTALLINE FERROELECTRICS.
                  04-0831   ENERGY LOSS NONPOLAR STATE POLYCRYSTALLINE BARIUM TITANATE. (150-500 DEG C, 20-50 KILO
MARSH A   33-3359   PREPARATION AND PROPERTIES OF CESIUM LEAD HALIDES. (FERROELECTRICITY INDETERMINITE)

MARTIN DH	17-2580	RELAXATION OF FERROELECTRIC PHOSPHATES (AMMONIUM DIHYDROGEN PHOSPHATE AND POTASSIUM DI
MARTIN G	04-0719	POLARIZATION OF SOLID BARIUM(X) STRONTIUM(1-X) TITANATE COMPOUNDS BETWEEN 4 AND 100 DE
MARTIN GE	03-0375	VIBRATIONS OF LONGITUDINALLY POLARIZED FERROELECTRIC CYLINDRICAL TUBES.
	03-0376	VIBRATIONS OF COAXIALLY SEGMENTED, LONGITUDINALLY POLARIZED FERROELECTRIC TUBES.
MARTIN JJ	13-2057	THERMAL CONDUCTIVITY OF CUBIC SODIUM TUNGSTEN BRONZES. (SODIUM TUNGSTATES)
MARTIN-BRUNETIERE F	21-2814	ZEEMAN EFFECT (STUDY) OF TWO FINE LINES IN (THE RED REGION OF) CHROMIUM GUANIDINIUM SU
MARTIRENA HT	09-1496	PZT-5 UNDER PRESSURE, DIELECTRIC AND PIEZOELECTRIC PROPERTIES.
MARTYNENKO MA	05-1047	GROWTH OF LEAD TITANATE CRYSTALS AND EXAMINATION OF THEIR DOMAIN STRUCTURE.
	05-1050	SPONTANEOUS POLARIZATION AND COERCIVE FIELD OF LEAD TITANATE (FLUX GROWN SINGLE CRYSTA
MARTYNOV VG	19-2745	FERROELECTRIC TRANSITION IN AMMONIUM SULFATE, THE DIELECTRIC, OPTICAL, AND ELECTROOPTI
MARTYNOVA SV	04-0895	DEPENDENCE OF DIELECTRIC DISPERSION IN BARIUM TITANATE SINGLE CRYSTALS ON THEIR DOMAIN
	04-0896	EFFECT OF LINEAR COMPRESSION ON THE UHF DIELECTRIC PROPERTIES OF BARIUM TITANATE SINGL
MARTYNOVA VF	13-2042	MECHANISM AND KINETICS OF SOLID PHASE FORMATION OF LEAD METANIOBATE.
MARUSIC M	22-2871	ESR STUDY OF THE FERROELECTRIC PHASE TRANSITIONS IN SODIUM TRIHYDROGEN DISELENITE.
MARUTAKE M	03-0374	QUASI PHENOMENOLOGICAL THEORY OF ELASTIC ANOMALY IN FERROELECTRIC CRYSTALS.
MARUYAMA N	31-3331	FERROELECTRICITY AND PHASE TRANSITIONS IN SOLID HYDROGEN HALIDES (SINGLE CRYSTAL X-RAY
	31-3333	STRESS FREE SINGLE CRYSTAL GROWTH OF FERROELECTRIC HYDROGEN HALIDES.
MASAKI H	02-0035	MEASUREMENT OF DIELECTRIC CONSTANTS FOR A FEW FERROELECTRIC MATERIALS AT LOW TEMPERATU
MASCARENHAS S	34-3456	CHARGE AND POLARIZATION STORAGE IN ICE CRYSTALS.
MASE Y	17-2643	MICROSCOPIC OBSERVATION OF ANTIFERROELECTRIC DOMAINS OF AMMONIUM DIHYDROGEN PHOSPHATE.
MASON IM	11-1692	ACOUSTIC SURFACE WAVE CONVOLUTION ON CRYSTALS OF CADMIUM SULFIDE, LITHIUM NIOBATE AND
MASON WP	02-0090	FIFTY YEARS OF FERROELECTRICITY (REVIEW).
MASSEY GA	08-1328	OPTICALLY INDUCED PHYSICAL DAMAGE TO LITHIUM NIOBATE, PROUSTITE, AND LITHIUM IODATE.
	17-2712	EFFICIENT HIGH GRAIN PARAMETRIC GENERATION IN ADP CONTINUOUSLY TUNABLE ACROSS THE VISI
MASSON S	04-0834	THIN FILMS: PREPARATION OF FERROELECTRIC THIN FILMS OF BARIUM TITANATE BY VACUUM EVAPO
MASSOT M	15-2111	RAMAN SCATTERING IN FERROELECTRIC SEMICONDUCTOR ANTIMONY SULFOIODIDE.
	15-2195	FERROELECTRIC TRANSITION OF ANTIMONY SULFOIODIDE.
	15-2240	SOFT PHONON COUPLINGS IN THE PRESSURE INDUCED PHASE TRANSITION IN ANTIMONY SULFOIODIDE
MASTNER J	02-0088	INFLUENCE OF THE QUALITY OF TRIGLYCINE SULFATE SINGLE CRYSTALS ON THEIR DIELECTRIC PRO
	27-3068	TEMPERATURE AND FIELD STRENGTH DEPENDENCE OF LARGE SIGNAL PERMITTIVITY OF TGS. (10 KHZ
	37-3640	VARIOUS HYSTERESIS LOOP TRACERS FOR A WIDE FREQUENCY RANGE. (CIRCUITRY DESCRIBED)
MASUDA Y	13-2013	DIELECTRIC AND CERAMIC PROPERTIES OF (BISMUTH(0.5) (SODIUM OR POTASSIUM)(0.5)(X) LEAD(
MASUNO K	04-0833	DIELECTRIC CERAMICS WITH BOUNDARY LAYER STRUCTURE FOR HIGH FREQUENCY APPLICATION. (BAR
MATHIESON AMCL	08-1290	ABSOLUTE STRUCTURE OF LITHIUM IODATE CRYSTALS.
MATHIEU JP	36-3548	VIBRATIONAL SPECTRA OF IODIC ACID SINGLE CRYSTALS.
MATOSSI F	08-1394	VIBRATION FREQUENCIES OF TITANATES.
MATSAKOV LYA	11-1749	DETERMINATION OF ELASTIC AND PIEZOELECTRIC CONSTANTS OF LITHIUM NIOBATE SINGLE CRYSTAL
MATSUBARA T	03-0521	DISTRIBUTION OF RELAXATION TIME IN FERROELECTRICS NEAR THE CURIE TEMPERATURE (POLY VER
MATSUDA T	27-3070	NONLINEARITY OF THE DIELECTRIC CONSTANT OF FERROELECTRIC TRIGLYCINE SULFATE.
MATSUI N	17-2536	ELECTROCHEMISTRY ON THE POLYCRYSTAL OF POTASSIUM DIHYDROGEN PHOSPHATE.
MATSUMOTO M	05-1087	LEAD TITANATE PYROELECTRIC INFRARED DETECTOR.
MATSUMURA S	08-1392	GROWTH OF IODIC ACID AND IODATE CRYSTALS BY A SOLUTION GROWTH METHOD.
MATSUO T	34-3436	RELAXATIONAL PROTON ORDERING AND GLASSY CRYSTALLINE STATE IN HEXAGONAL ICE.
MATSUO U	04-0837	PTCR (POSITIVE TEMPERATURE COEFFICIENT OF RESISTANCE) BEHAVIOR OF BARIUM TITANATE WITH
MATSUO Y	04-0734	A SEMICONDUCTING BARIUM TITANATE CERAMIC DOPED WITH NIOBIUM AND MANGANESE IONS AND ITS
	04-0835	EXAGGERATED GRAIN GROWTH IN LIQUID PHASE SINTERING OF BARIUM TITANATE.
	04-0931	A NEW PTC THERMISTOR MATERIAL. (BARIUM TITANATE)
	08-1393	HIGH PRESSURE SYNTHESIS (SOLID STATE REACTION, SINTERING) AND PROPERTIES OF PEROVSKITE
	09-1587	OXIDATION REDUCTION PHENOMENA IN THE LEAD TITANATE- LANTHANUM TITANATE SYSTEM.
	13-1990	GROWTH OF SINGLE CRYSTALS OF LEAD (ZINCATE NIOBATE) AND PROPERTIES OF THIS COMPOSITION
	17-2530	NOVEL CUT OF POTASSIUM DIHYDROGEN PHOSPHATE TYPE CRYSTALS.
	17-2531	NOVEL TYPE OF CUT FOR KDP CRYSTALS FOR LOW VOLTAGE LIGHT MODULATION.
	17-2532	OPTIMUM CUT OF POTASSIUM DIHYDROGEN PHOSPHATE TYPE CRYSTALS FOR LIGHT MODULATION.
	36-3594	GROWTH OF SINGLE CRYSTALS OF LEAD ZINCATE NIOBATE AND ITS PROPERTIES.
MATSUOKA O	33-3391	COVALENCY IN POTASSIUM MANGANESE TRIFLUORIDE.
MATSUOKA T	04-0734	A SEMICONDUCTING BARIUM TITANATE CERAMIC DOPED WITH NIOBIUM AND MANGANESE IONS AND ITS
	04-0837	PTCR (POSITIVE TEMPERATURE COEFFICIENT OF RESISTANCE) BEHAVIOR OF BARIUM TITANATE WITH
MATSURA K	04-0775	ON THE LATTICE VIBRATION OF CUBIC BARIUM TITANATE.
MATSUSHITA S	12-1889	SELF INDUCED THERMAL EFFECTS ON LIGHT EXTINCTION OF LITHIUM TANTALATE.
	36-3576	ELECTROOPTIC AND PIEZOELECTRIC PROPERTIES OF LANTHANUM TITANATE SINGLE CRYSTAL. (FOUND
MATTES BL	04-0836	FERROELECTRIC TRANSFORMATION IN BARIUM TITANATE. (MODEL GIVES NEARLY COMPLETE DESCRIPT
MATTHEISS LF	02-0091	ENERGY BANDS FOR POTASSIUM NICKEL TRIFLUORIDE, STRONTIUM TITANATE, POTASSIUM MOLYBDATE
	06-1158	EFFECT OF THE 110 DEGREE K PHASE TRANSITION ON THE STRONTIUM TITANATE CONDUCTION BANDS
MATTHES H	13-1991	GROWTH OF BARIUM LITHIUM NIOBATE SINGLE CRYSTALS BY THE CZOCHRALSKI METHOD.
MATTHIAS BT	02-0092	FROM BARIUM TITANATE TO DNA. (FERROELECTRICITY, BIOLOGICAL MECHANISMS, MEMORY, ORGANIC
	03-0377	SUPERCONDUCTIVITY VERSUS FERROELECTRICITY (SEEM MUTUALLY EXCLUSIVE, THEREFORE FERROELE
MATUMURA O	04-0878	POLARITON DISPERSION RELATION IN CUBIC BARIUM TITANATE.
MAURER E	03-0152	THERMODYNAMIC THEORY OF IRRADIATED TRIGLYCINE SULFATE.
	27-2981	EFFECT OF SEVERAL PARAMETERS ON THE FERROELECTRIC- PARAELECTRIC TRANSITION IN TRIGLYCI
	27-2982	ORDER DISORDER TRANSITION IN FERROELECTRICS. (TRIGLYCINE SULFATE, HIGHLY ACCURATE TEMP
	27-3030	EXPERIMENTAL DETAILED STUDY OF THE PHASE TRANSITION IN TRIGLYCINE SULFATE. (CLOSE THER
MAURIN M	32-3355	CRYSTALLOGRAPHIC STUDY OF SEVERAL HYDRAZINIUM ORTHO FLUOBERYLLATES.
MAUSSION M	04-0808	DIRECT OBSERVATION OF BARIUM TITANATE CRYSTALS BY SCANNING ELECTRON MICROSCOPE.
	04-0812	SEM OBSERVATION OF THE FERROELECTRIC SURFACE OF BARIUM TITANATE CRYSTALS.
	27-3051	DIRECT OBSERVATION OF FERROELECTRIC DOMAINS IN TRIGLYCINE SULFATE USING THE SCANNING E
	27-3052	SCANNING ELECTRON MICROSCOPE STUDY OF FERROELECTRIC DOMAINS IN TGS.
MAVRIN BN	11-1761	FERMI RESONANCE OF POLARITON WITH BIPHONON IN AN LITHIUM NIOBATE CRYSTAL.
	11-1762	TRANSVERSE POLARITONS IN A LITHIUM NIOBATE CRYSTAL. (RAMAN SPECTRUM, PERMITTIVITY)
	13-1992	POLARITONS IN A BIAXIAL CRYSTAL OF BARIUM SODIUM NIOBATE.
MAY LF	03-0435	CALORIMETRIC INVESTIGATIONS OF SECOND ORDER FERROELECTRIC TRANSITIONS.
MAYER RJ	17-2581	DIELECTRIC PROPERTIES OF DEUTERATED POTASSIUM DIHYDROGEN PHOSPHATE.
MAZAK E	37-3656	FREQUENCY SPECTRUM OF FERROCERAMIC TRANSDUCERS.
	37-3657	RADIALLY VIBRATING FERROCERAMIC TRANSDUCER AS A SOURCE OF ULTRASONIC VIBRATIONS RADIAT
MAZDIYASNI KS	02-0093	QUANTITATIVE ANALYSIS OF MICROSTRUCTURE AND ELECTRICAL PROPERTIES OF FERROELECTRIC MAT
	04-0704	ELECTRICAL PROPERTIES OF HIGH PURITY POLYCRYSTALLINE BARIUM TITANATE. (DIELECTRIC CONS
	04-0839	MICROSTRUCTURE AND ELECTRICAL PROPERTIES OF SCANDIUM TRIOXIDE DOPED, RARE EARTH OXIDE
	04-0840	SYNTHESIS OF NIOBIUM PENTOXIDE DOPED BARIUM TITANATE WITH IMPROVED ELECTRICAL PROPERTI
	06-1217	PREPARATION AND CHARACTERIZATION OF ALKOXY-DERIVED STRONTIUM ZIRCONATE AND STRONTIUM T
	09-1493	COLD PRESSING AND LOW TEMPERATURE SINTERING OF ALKOXY-DERIVED PLZT AND PLZT.
MAZUR K	04-0838	ELECTRET EFFECT IN A MIXTURE OF POLY METHACRYLATE AND BARIUM TITANATE

MCCAMPFELL CB	09-1523	A NEW LONGITUDINAL DISPLAY MODE FOR CERAMIC ELECTROOPTIC DEVICES. (PLZT)
MCCARTHY GJ	29-3247	CRYSTAL DATA FOR LITHIUM THALLIUM TARTRATE MONOHYDRATE.
MCCORMIC WD	06-1159	ANOMALOUS SPECIFIC HEAT OF STRONTIUM TITANATE.
MCCUSKER JH	09-1566	AN ADAPTIVE FERROELECTRIC TRANSFORMER - A SOLID STATE ANALOG MEMORY DEVICE. (USING, FO
	09-1583	ADAPTIVE FERROELECTRIC TRANSFORMERS WITH IMPROVED TEMPERATURE CHARACTERISTICS. (PZT-5H
	37-3663	ANALOG MEMORY DEVICES EMPLOYING PIEZOELECTRIC- FERROELECTRIC INTERACTIONS FOR ADAPTIVE
MCDONALD PF	11-1763	EPR OF GADOLINIUM(3+) IN LITHIUM NIOBATE.
	11-1833	ELECTRON PARAMAGNETIC RESONANCE OF GADOLINIUM(3+) DOPED LITHIUM NIOBATE.
MCDOWELL CA	18-2724	DETECTION BY ELECTRON PARAMAGNETIC RESONANCE AND ELECTRON NUCLEAR DOUBLE RESONANCE OF
	18-2725	ELECTRON PARAMAGNETIC RESONANCE STUDIES OF X-IRRADIATED POTASSIUM DIHYDROGEN ARSENATE,
	18-2726	ELECTRON PARAMAGNETIC RESONANCE AND ELECTRON NUCLEAR DOUBLE RESONANCE OBSERVATIONS OF
	18-2727	ELECTRON PARAMAGNETIC RESONANCE AND ELECTRON NUCLEAR DOUBLE RESONANCE OF THE ARSENATE
	18-2728	HYDROGEN BONDED FERROELECTRICS: EVIDENCE FOR SLATER CONFIGURATIONS FROM EPR STUDIES OF
	18-2729	OBSERVATION OF ELECTRON NUCLEAR TRIPLE RESONANCE FOR THE ARSENATE CENTER IN X-RAY IRRA
MCELEARNEY JN	21-2815	MAGNETIC SUSCEPTIBILITY OF GUANIDINIUM VANADIUM SULFATE HEXAHYDRATE AT LOW TEMPERATURE
MCFEE JH	36-3530	PYROELECTRICITY AND OPTICAL SECOND HARMONIC GENERATION IN POLYVINYLIDENE FLUORIDE FILM
	36-3560	PYROELECTRIC PROPERTIES OF POLYVINYLIDENE FLUORIDE AND ITS USE FOR INFRARED DETECTION.
	36-3595	PYROELECTRIC AND NONLINEAR OPTICAL PROPERTIES OF POLED POLYVINYLIDENE FLUORIDE FILMS.
MCGEE TD	04-0819	RESISTIVITY AND SINTERING OF BARIUM TITANATE.
MCGEOCH MW	11-1764	OPTIMUM SECOND HARMONIC GENERATION IN LITHIUM NIOBATE.
MCGEVNA VG	09-1567	MEASUREMENT OF THE PYROELECTRIC COEFFICIENT OF 65-35 PZT. (LEAD ZIRCONATE TITANATE)
MCGUIRE GE	10-1631	STUDY OF THE X-RAY PHOTOELECTRON SPECTRUM OF TUNGSTEN- TUNGSTEN OXIDE AS A FUNCTION OF
MCKENZIE DR	25-2917	FAR INFRARED TRANSMISSION SPECTRUM OF THIOUREA.
MCKINNEY KR	04-0841	FRACTURE TOUGHNESS OF COMMERCIAL BARIUM TITANATE TRANSDUCER MATERIAL.
MCLAREN MG	09-1484	BEHAVIOR OF MODIFIED LEAD ZIRCONATE LEAD TITANATE PIEZOELECTRIC CERAMICS UNDER HIGH EL
MCLAY DB	16-2369	PHASE TRANSFORMATIONS IN POTASSIUM NITRATE AND POTASSIUM NITRITE. (DOES NOT SEEM FERRO
MCMAHON GW	37-3658	MEASUREMENT OF POISSON'S RATIO IN POLED FERROELECTRIC CERAMIC DISKS.
MCNEILLY JH	09-1567	MEASUREMENT OF THE PYROELECTRIC COEFFICIENT OF 65-35 PZT. (LEAD ZIRCONATE TITANATE)
MCPHERSON R	04-0639	SURFACE DOMAIN REORIENTATION PRODUCED BY ABRASION AND ANNEALING OF POLYCRYSTALLINE BAR
MEAD CA	06-1173	ANOMALOUS RESONANCE OF STRONTIUM TITANATE.
	06-1175	SURFACE BARRIER ENERGIES ON STRONTIUM TITANATE.
MEDVED AV	11-1747	AMPLIFICATION OF RAYLEIGH ULTRASONIC WAVES IN A LAYERED STRUCTURE CONSISTING OF LITHIU
	11-1748	TRANSVERSE ACOUSTOELECTRIC EFFECT IN A LAYERED LITHIUM NIOBATE- SILICON STRUCTURE.
MEEKS EL	06-1160	TEMPERATURE DEPENDENCE OF THE FIRST ORDER ELASTIC CONSTANTS OF STRONTIUM TITANATE.
MEGAW HD	02-0094	ROLE OF THE REGULAR OCTAHEDRON AND LARGER POLYHEDRA IN FERROELECTRIC MATERIALS. (HETER
	08-1273	STRUCTURES OF SODIUM NIOBATE BETWEEN 480 AND 575 DEG C, AND THEIR RELEVANCE TO SOFT PH
	08-1336	HIGH TEMPERATURE PHASE TRANSITIONS IN SODIUM NIOBATE AND THE USE OF TILTING SCHEMES IN
	08-1337	THE STRUCTURE OF SODIUM NIOBATE T(2) AT 600 DEG C, AND THE CUBIC- TETRAGONAL TRANSITIO
	08-1395	STRUCTURE OF SODIUM NIOBATE HETEROTYPES.
	08-1396	STRUCTURE AND TRANSITIONS IN PEROVSKITES. (ORTHOFERRITES, SODIUM NIOBATE)
	08-1397	TRANSFORMATIONS IN SODIUM NIOBATE.
	13-1911	CRYSTAL STRUCTURE AND PROPERTIES OF CALCIUM(2) NIOBIUM(2) OXIDE(7).
MEHRAN F	11-1765	ELECTRON PARAMAGNETIC RESONANCE OF IRON(3+) DOPED LITHIUM NIOBATE.
MEHTA RR	03-0378	SOME THEORETICAL CONSIDERATIONS RELATING TO SWITCHING AND REMANENCE IN FERROELECTRIC P
	09-1568	FERROELECTRIC- PHOTOCONDUCTOR MEMORY DEVICE. (PLZT)
	37-3632	DESIGN AND PERFORMANCE OF A THIN FILM FERROELECTRIC PHOTOCONDUCTOR STORAGE DEVICE.
	37-3659	DEPOLARIZATION FIELD IN THIN FERROELECTRIC FILMS.
	37-3671	PHOTOCONDUCTOR AND ELECTRODE REQUIREMENTS FOR THIN FILM FERROELECTRIC PHOTOCONDUCTOR M
MEIER W	25-2907	MOLECULAR AND LATTICE VIBRATIONS OF ORTHORHOMBIC AND FERROELECTRIC THIOUREA. PART-2: C
	25-2920	MOLECULAR AND LATTICE VIBRATIONS OF ORTHORHOMBIC AND FERROELECTRIC THIOUREA. PART-1: T
MEIERLING HD	06-1161	CHARGE COMPENSATION BY OXYGEN(2-) VACANCIES IN CHROMIUM(3+) DOPED STRONTIUM TITANATE (
MEIJER PH	20-2805	SPECIFIC HEAT AND SUSCEPTIBILITY IN CHROMIUM METHYL AMMONIUM ALUM ABOVE THE CRITICAL T
MEISNER LB	03-0379	NONLINEAR OPTICAL PROPERTIES OF CRYSTALS.
	08-1423	ELECTROOPTICAL PROPERTIES OF LITHIUM IODATE SINGLE CRYSTALS.
MEITZLER AH	09-1563	FERROELECTRIC DOMAIN SWITCHING IN RHOMBOHEDRAL PHASE PLZT CERAMICS.
	09-1565	STRAIN BIASED FERROELECTRIC PHOTOCONDUCTOR IMAGE STORAGE AND DISPLAY DEVICES. (PLZT)
	09-1569	DEVICE APPLICATIONS OF PLZT CERAMICS.
	09-1570	ENHANCED ORDERING IN PLZT CERAMICS.
	09-1571	FERROELASTIC BEHAVIOR OF LEAD LANTHANUM ZIRCONATE TITANATE CERAMICS WHEN SUBJECTED TO
	09-1572	IMAGE STORAGE AND DISPLAY DEVICES USING FINE GRAIN FERROELECTRIC CERAMICS. (LEAD ZIRCO
	09-1579	ENHANCED ORDERING OF FERROELECTRIC DOMAINS IN PLZT CERAMICS.
	09-1580	ENHANCED ORDERING OF FERROELECTRIC DOMAINS IN PLZT CERAMICS.
	37-3654	FERROELECTRIC CERAMIC LIGHT GATES OPERATED IN A VOLTAGE CONTROLLED MODE.
MELAMED VSH	04-0942	OBSERVATION OF THE CHANGE IN THE POLARIZATION OF SINGLE CRYSTAL FILMS OF BARIUM TITANA
MELCHER RL	33-3392	ANOMALOUS ELASTIC BEHAVIOR OF POTASSIUM MANGANESE TRIFLUORIDE NEAR A STRUCTURAL PHASE
MELESHINA VA	27-3053	EFFECT OF HYDROSTATIC PRESSURE ON THE UNIPOLAR PROPERTIES OF TRIGLYCINE SULFATE.
	27-3071	MOVEMENT OF DOMAIN WALLS AND THE NUCLEATION OF DOMAINS IN CRYSTALS OF TRIGLYCINE SULFA
	27-3072	WALL MOTION AND NUCLEATION OF DOMAINS IN TGS CRYSTALS. (SUMMARY, WITH PHOTOS)
	27-3141	THERMAL AND ELECTRICAL PROPERTIES OF TRIGLYCINE SULFATE SINGLE CRYSTALS (HEAT CAPACITY
	27-3164	DOMAIN STRUCTURE OF GAMMA-IRRADIATED TGS CRYSTALS.
	30-3255	CERTAIN GROWTH, LASING AND SPECTROSCOPIC PROPERTIES OF NEODYMIUM(3+) DOPED GADOLINIUM
	30-3265	ABNORMAL ABSORPTION OF ELASTIC WAVES NEAR THE PHASE TRANSITION OF GADOLINIUM MOLYBDATE
MELIK-SHANAZAROVA TA	08-1360	MAGNETOELECTRIC EFFECT IN FERROELECTRIC- ANTIFERROMAGNETIC BISMUTH TITANIUM FERRATE.
	36-3568	MAGNETOELECTRIC EFFECT IN FERROELECTRIC- ANTIFERROMAGNETIC LITHIUM IRON(0.5) TANTALUM(
MELNIKOVA SV	22-2829	LINEAR ELECTROOPTICAL EFFECT IN THE FERROELECTRIC SODIUM AMMONIUM SELENATE DIHYDRATE.
MELTZLER AH	09-1573	LANTHANUM DEPENDENCE OF ELASTIC AND PIEZOELECTRIC PROPERTIES OF PLZT CERAMICS WITH A Z
MELVEGER AJ	16-2342	POLARIZED INFRARED AND RAMAN SPECTRA OF SINGLE CRYSTAL CESIUM NITRITE.
MENDIOLA J	03-0152	THERMODYNAMIC THEORY OF IRRADIATED TRIGLYCINE SULFATE.
	16-2270	CRITICAL SCATTERING OF X-RAYS IN SODIUM NITRITE.
	16-2343	X-RAY CRITICAL SCATTERING OF SODIUM NITRITE AT 160 DEG C AND AT 215 DEG C.
	22-2857	THERMAL EXPANSION OF HYDROGEN BONDED FERROELECTRICS. LITHIUM TRIHYDROGEN SELENITE.
	27-2993	DIGLYCINE SULFATE AN INTERESTING NEW DIELECTRIC CRYSTAL SPECIES.
	27-3073	DAMAGE PRODUCED BY X-RAYS ON TRIGLYCINE SULFATE.
	27-3074	X-RAY DAMAGE IN TRIGLYCINE SULFATE.
MENEZES CA	08-1293	FERROELECTRIC PROPERTIES OF TETRAGONAL (BARIUM, LEAD) (NIOBATE, ZIRCONATE). (CURIE TEM
	13-1925	STUDIES ON THE DIELECTRIC AND FERROELECTRIC PROPERTIES OF THE BARIUM LEAD NIOBATE ZIRC
MERCADO A	27-3075	CRITICAL BEHAVIOR OF TRIGLYCINE FLUOBERYLLATE.
MERCHANT S	21-2815	MAGNETIC SUSCEPTIBILITY OF GUANIDINIUM VANADIUM SULFATE HEXAHYDRATE AT LOW TEMPERATURE
MERCOUROFF W	01-0019	MODERN ASPECTS OF PHYSICS OF SOLIDS. (FERROELECTRICITY)
MERIAKRI VV	06-1089	MEASUREMENT OF DIELECTRIC PROPERTIES OF FERROELECTRICS AT SUBMILLIMETER WAVELENGTHS. (

MERTEN L   03-0380 EXTENSION OF THE POULET- LOUDON APPROXIMATION IN THE THEORY OF OPTICAL LATTICE VIBRATI
                        04-0842 POLARITON DISPERSION OF TETRAGONAL FERROELECTRIC BARIUM TITANATE. (CALCULATIONS)
                        08-1284 EIGENFREQUENCIES AND EIGENVECTORS OF POLARITONS WITH APPLICATION TO LEAD NIOBATE. PART
                        11-1697 LIGHT SCATTERING BY PCLARITONS ASSOCIATED WITH ORDINARY PHOTONS IN LITHIUM NIOBATE.
                        11-1766 EIGENFREQUENCIES AND EIGENVECTORS OF POLARITONS WITH APPLICATION TO LITHIUM NIOBATE. P
MESHCHERYAKOV NA   06-1162 EPR INVESTIGATION OF POLYCRYSTALLINE STRONTIUM TITANATE CONTAINING MANGANESE DIOXIDE.
MESNARD MG   04-0843 PREPARATION ANC PROPERTIES OF THIN FILMS OF BARIUM TITANATE.
                        04-0845 BANC STRUCTURE OF THE CUBIC PHASE OF BARIUM TITANATE. (IMAGINARY DIELECTRIC CONSTANT)
METRAT G   04-0843 PREPARATION AND PROPERTIES OF THIN FILMS OF BARIUM TITANATE.
MEYENN KV   16-2287 THE ORDER DISORDER TRANSITION CF SODIUM NITRITE. (TWO SUBLATTICE MODEL OF INTERACTING
MEYER RA   12-1881 OPTICAL BEAM STEERING USING A MULTICHANNEL LITHIUM TANTALATE CRYSTAL.
MICHALEC R   17-2567 GROWTH AND ETCHING OF PCTASSIUM DIHYDROGEN PHOSPHATE CRYSTALS IN SOLUTIONS CONTAINING
MICHARD F   08-1399 CETERMINATION OF THE ELASTIC CCNSTANTS OF POTASSIUM NITRATE.
MICHEL C   08-1398 ATOMIC STRUCTURE IN THE PEROVSKITIC FERRATE.
                        08-1402 FERROELECTRIC BISMUTH FERRATE X-RAY AND NEUTRON DIFFRACTION STUDY. (ANTIFERROMAGNETIC)
                        09-1574 ATOMIC STRUCTURES OF THE FERROELECTRIC COMPOUNDS LEAD ZIRCONATE(X) TITANATE(1-X) FOR T
MICHEL-CALENDINI F   04-0845 BANC STRUCTURE OF THE CUBIC PHASE OF BARIUM TITANATE. (IMAGINARY DIELECTRIC CONSTANT)
                        04-0846 BAND STRUCTURE OF CUBIC AND TETRAGONAL BARIUM TITANATE.
MICHERON F   04-0622 SOME PROPERTIES OF THE THERMOCURRENTS IN BARIUM TITANATE AND TITANIUM DIOXIDE SINGLE C
                        04-0844 AUTOMATIC IMPEDANCE MEASUREMENTS USING THERMAL NOISE ANALYSIS (APPLICATION TO BARIUM T
                        04-0847 ELECTRICAL CONTROL OF FIXATION AND ERASURE OF HOLOGRAPHIC PATTERNS IN FERROELECTRIC MA
                        04-0848 HOLOGRAPHIC STORAGE, ELECTRICAL FIXING AND ERASING IN DCPED BARIUM TITANATE CRYSTALS.
                        04-0849 INTERNAL STRESSES IN BARIUM TITANATE CERAMICS. (AND DIFFERENCE IN TEMPERATURES FOR MAX
                        09-1575 INSCRIPTION OF HOLOGRAPHIC RESULTS IN TRANSPARENT CERAMIC FERROELECTRICS. (LEAD LANTHA
                        11-1767 BIREFRINGENCE INDUCED IN A LITHIUM NIOBATE CRYSTAL.
                        27-3076 INFLUENCE OF AN ELECTRIC FIELD ON THE NOISE AND THE PYROELECTRIC EFFECT IN TGS.
                        27-3077 POLARIZATION FLUCTUATION OF TRIGLYCINE SULFATE NEAR THE CURIE POINT.
MICHIELS L   11-1756 MICROWAVE PHONON ATTUNUATION IN X CUT LITHIUM NIOBATE.
MIDORIKAWA M   16-2344 EFFECT OF HYDROSTATIC PRESSURE ON FERROELECTRICITY OF POTASSIUM NITRATE.
                        16-2345 PHASE TRANSITIONS IN POTASSIUM NITRATE NITRITE MIXED CRYSTALS (GROWTH BY DOUBLE BRIDGM
                        17-2643 MICROSCOPIC OBSERVATION OF ANTIFERROELECTRIC DOMAINS OF AMMONIUM DIHYDROGEN PHOSPHATE.
                        33-3383 MICROSCOFIC OBSERVATIONS OF PHASE TRANSITIONS IN CESIUM LEAD TRICHLORIDE.
                        33-3393 PRESSURE DEPENDENCE OF CUBIC TETRAGONAL TRANSITION TEMPERATURE OF CESIUM LEAD TRICHLOR
MIELKE Z   19-2768 INFRAREL SPECTRA OF RUBIDIUM HYDROGEN SULFATE AND RUBIDIUM DEUTERIUM SULFATE IN THE PA
MIFUNE H   04-0734 A SEMICONDUCTING BARIUM TITANATE CERAMIC DOPED WITH NIOBIUM AND MANGANESE IONS AND ITS
MIHAILOV MP   03-0381 A MORE EXACT DYNAMIC METHOD OF DETERMINING POLARIZATION IN FERROELECTRICS.
                        03-0382 DEPENDENCE OF EFFECTIVE PCLARIZATION IN FERROELECTRICS ON THE INTENSITY OF THE ELECTRI
                        15-2196 ANOMALOUS PROPERTIES IN PHOTOFERROELECTRIC ANTIMONY SULFOIODIDE.
MIIKE H   26-2932 THERMAL CCNDUCTION IN A TWO DIMENSIONAL ANTIFERROMAGNETIC CCPPER FORMATE TETRAHYDRATE.
                        26-2939 MEASUREMENTS OF THE THERMAL CONDUCTIONS IN THE LOW DIMENSIONAL ANTIFERROMAGNETS POTASS
MIKAELYAN AL   11-1768 INVESTIGATION OF A SYSTEM FOR BEAM DEFLECTION BY MEANS OF LITHIUM NIOBATE CRYSTALS.
MIKALEVICIUS M   15-2162 GROWING SINGLE CRYSTALS FROM THE VAPOR PHASE BY A DYNAMIC METHOD (ANTIMONY SESQUISELEN
MIKALKYAVICHYUS MP   15-2193 TEMPERATURE DEPENDENCES OF THE HALL AND DRIFT MOBILITIES OF HOLES IN ANTIMONY TRISULFI
MIKAMI M   16-2313 NUCLEAR QUADRUPOLE RESONANCE OF NITROGEN IN SODIUM NITRITE.
MIKAMI O   08-1359 RH DOPED LITHIUM NIOBATE AS AN IMPROVED NEW MATERIAL FOR REVERSIBLE HOLOGRAPHIC STORAG
MIKHAILOVA NA   04-0572 INVERSION OF ELECTRICAL PROPERTIES OF SEMICONDUCTING BARIUM TITANATE.
MIKHAJLOV GM   16-2255 NUCLEAR MAGNETIC RESONANCE STUDIES OF THE PHASE TRANSITIONS IN FERROELECTRICS. (POTASS
MIKHALEVSKII AN   03-0499 CEBYE SCREENING OF A NONUNIFORM ELECTRIC FIELD IN A FERROELECTRIC.
                        03-0500 ELECTRICAL NONLINEARITY OF A FERROELECTRIC SEMICONDUCTOR IN A NONUNIFORM ELECTRIC FIEL
                        03-0502 PARAELECTRIC SEMICONDUCTOR IN A NONUNIFORM ELECTRIC FIELD.
MIKHELSON LM   04-0850 RELAXATION ABSORPTION OF ULTRASOUND IN SINGLE CRYSTALS OF BARIUM TITANATE IN THE PRESE
MIKI H   16-2346 THERMAL EXPANSION IN FERROELECTRIC SILVER SODIUM DINITRITE.
                        19-2779 THERMAL AND SOME OTHER STUDIES OF SODIUM AMMONIUM SULFATE DIHYDRATE, SODIUM TRIHYDROGE
                        22-2856 ANOMALOUS SPECIFIC HEAT AND CONFIGURATIONAL ENTROPY CHANGE IN SODIUM TRIHYDROGEN DISEL
                        22-2858 THERMAL EXPANSION IN SODIUM TRIHYDROGEN(1-X) TRIDEUTERIUM(X) DISELENITE AND ISOTOPE EF
MIKKELSON RC   04-0687 CHANNELING OF PROTONS IN THIN BARIUM TITANATE CRYSTALS AT TEMPERATURES ABOVE AND BELOW
                        04-0688 CHANNELING OF LIGHT IONS IN THE CUBIC AND TETRAGONAL (FERROELECTRICALLY POLARIZED PHAS
MIKOSHIBA N   16-2304 ULTRASONIC ATTENUATION NEAR THE TWO TRANSITION POINTS IN SODIUM NITRITE.
MIKULANNINETS SV   15-2154 EFFECT OF THE ADSORPTION OF OXYGEN ON THE PROPERTIES OF ANTIMONY SULFOIODIDE SINGLE CR
MILEK JT   01-0020 ANTIFERROELECTRICITY AND ANTIFERROELECTRIC MATERIALS. (TABLE OF NEEL TEMPERATURES)
                        01-0021 LINEAR ELECTROOPTIC MODULATOR MATERIALS. (POTASSIUM OR AMMONIUM DIHYDROGEN PHOSPHATE O
                        01-0022 SEMICONDUCTING AND NONSTOICHIOMETRIC BARIUM TITANATE. A REFERENCE GUIDE.
MILLER A   04-0769 FERROELECTRIC PHOTOCONDUCTOR OPTICAL STORAGE MEDIUM UTILIZING BISMUTH TITANATE.
                        08-1352 PREPARATION OF POLED, TWIN-FREE CRYSTALS OF FERROELECTRIC BISMUTH TITANATE, (BY ANNEAL
                        08-1371 BISMUTH TITANATE FERROELECTRIC- PHOTOCONDUCTOR OPTICAL STORAGE MEDIUM FURTHER DEVELOPM
                        08-1445 BISMUTH TITANATE PHOTOCONDUCTOR OPTICAL STORAGE MEDIUM. (HIGH RESOLUTION READ WRITE)
                        08-1446 CEPOLING OF SINGLE DOMAIN BISMUTH TITANATE.
                        08-1447 FEASIBILITY OF ELECTROOPTIC DEVICES UTILIZING FERROELECTRIC BISMUTH TITANATE.
                        13-1977 PHASE HCLOGRAMS IN A FERROELECTRIC PHOTOCONDUCTOR DEVICE. (BISMUTH(4) TITANIUM(3) OXID
                        35-3507 GROWTH OF LARGE SINGLE CRYSTAL BISMUTH(4) TITANIUM(3) OXIDE(12).
                        37-3646 STORAGE OF HOLOGRAMS IN A FERROELECTRIC PHOTOCONDUCTOR DEVICE (ERASAELE, NONDESTRUCTIV
MILLER CA   04-0851 POTENTIAL BARRIERS ON SEMICONDUCTING BARIUM TITANATE. (PTCR DEVICES)
MILLER DW   09-1576 PROPERTIES OF OXYGEN DEPLETED CERAMIC LEAD ZIRCONATE TITANATE (PLUS 1 PERCENT NIOBIUM
                        37-3669 THECRY AND APPLICATION OF FERROELECTRIC RADIATION DETECTORS. (PYROELECTRICS)
MILLER GR   06-1163 DEFECTS IN OXIDE CERAMICS. (STRONTIUM TITANATE)
                        06-1244 DEFECT STUDIES IN STRONTIUM TITANATE.
                        06-1245 OPTICAL ABSORPTION SPECTRA OF REDUCED STRONTIUM TITANATE.
MILLER RC   02-0095 ABSOLUTE SIGNS OF NONLINEAR OPTICAL (SECOND HARMONIC GENERATION) COEFFICIENTS OF POLAR
                        02-0096 ABSOLUTE SIGNS OF SECOND HARMONIC GENERATION COEFFICIENTS OF PIEZOELECTRIC CRYSTALS.
                        02-0097 NONLINEAR OPTICAL CEVICES. (BASED IN SECOND HARMONIC GENERATION AND OPTICAL PARAMETRIC
                        02-0098 RELATIVE SIGNS OF NONLINEAR OPTICAL COEFFICIENTS OF POLAR CRYSTALS (LITHIUM NIOBATE OR
                        03-0383 NONLINEAR OPTICAL PROPERTIES OF FERROELECTRIC CRYSTALS. (SECOND HARMONIC GENERATION, O
                        04-0852 QUANTITATIVE STUDIES OF OPTICAL HARMONIC GENERATION IN CADMIUM SULFIDE, BARIUM TITANAT
                        04-0975 FAR INFRARED DIELECTRIC DISPERSION IN BARIUM AND STRONTIUM TITANATES AND TITANIUM DIOX
                        05-1066 FURTHER MEASUREMENTS OF ABSOLUTE SIGNS OF SECOND HARMONIC GENERATION COEFFICIENTS OF P
                        11-1769 CEPENDENCE OF SECOND HARMONIC GENERATION COEFFICIENTS OF LITHIUM NIOBATE ON MELT COMPO
                        11-1770 TEMPERATURE DEPENDENCE OF THE OPTICAL PROPERTIES OF FERROELECTRIC LITHIUM NIOBATE AND
                        17-2582 CPTICAL SECOND HARMONIC GENERATION IN PIEZOELECTRIC CRYSTALS. (KDP TYPE)
                        30-3305 NONLINEAR OPTICAL PROPERTIES OF GADOLINIUM MOLYBDATE AND TERBIUM MOLYBDATE.
                        35-3506 NONLINEAR OPTICAL PROPERTIES OF FERROELECTRIC LEAD(5) GERMANIUM(3) OXIDE(11).

MILLER SR	19-2769	NUCLEAR SPIN LATTICE RELAXATION IN SOME FERROELECTRIC AMMONIUM SALTS. (AMMONIUM SULFAT
	23-2885	PROTON SPIN LATTICE RELAXATION IN FERROELECTRIC COLEMANITE.
MILLS DL	04-0556	POLARITONS IN TETRAGONAL BARIUM TITANATE: THEORY OF THE RAMAN LINE SHAPE.
MILOSEVIC-KVAJIC M	34-3457	ANGULAR CORRELATION OF ANNIHILATION PHOTONS IN ICE SINGLE CRYSTALS.
MINAEVA KA	13-2020	ANOMALIES OF ELASTICITY AND INTERNAL FRICTION IN THE REGION OF THE ANTIFERROELECTRIC C
	19-2788	POLARIZATION REVERSAL CHARACTERISTICS OF AMMONIUM HYDROGEN SULFATE.
	27-3055	ANISOTROPY OF ULTRASONIC ATTENUATION IN UNIAXIAL FERROELECTRICS. (TGS, ABSTRACT ONLY)
	27-3078	ANISOTRCPY OF SOUND ABSORPTION IN TRIGLYCINE SULFATE SINGLE CRYSTALS. (THEORY OF ANIS
	27-3079	FREQUENCY DEPENDENCE OF THE FLUCTUATION ABSORPTION OF ULTRASOUND IN TRIGLYCINE SULFATE
	27-3080	MEASUREMENT OF INTERNAL FRICTION IN SINGLE CRYSTALS OF FERROELECTRIC SUBSTANCES BY THE
	28-3211	ABSCRPTION OF LONGITUDINAL ULTRASONIC WAVES IN ROCHELLE SALT CRYSTALS NEAR THE CURIE P
MING-CHCNG I	13-1900	ELECTROOPTIC EFFECTS OF HCT PRESSED CERAMICS IN THE SYSTEM LEAD (ZINCATE(0.333) NIOBAT
MINKIEWICZ VJ	33-3394	X-RAY SCATTERING AND THE PHASE TRANSITION OF POTASSIUM MANGANESE TRIFLUORIDE AT 184 DE
	33-3405	NEUTRON SCATTERING STUDY OF THE LATTICE DYNAMICAL PHASE TRANSITIONS IN POTASSIUM MANGA
MINN SS	04-0705	CRYSTAL STUDY OF BARIUM TITANATE WITH PARALLEL INTERFERENCE FRINGES.
	04-0706	CRYSTAL STRUCTURE OF BARIUM TITANATE PRODUCING SQUARE INTERFERENCE FRINGES.
	04-0834	THIN FILMS: PREPARATION OF FERROELECTRIC THIN FILMS OF BARIUM TITANATE BY VACUUM EVAPO
MINOMURA S	04-0853	PRESSURE DEPENDENCE OF TRANSITION TEMPERATURES AND ELECTROSTRICTIONS IN PEROVSKITE BAR
	10-1658	PHASE TRANSITION IN TUNGSTEN TRIOXIDE UNDER HIGH PRESSURE.
MIRISHLI FA	13-1993	SYNTHESES, X-RAY, AND DIELECTRIC STUDIES OF CERTAIN NEW LAYERED FERROELECTRICS. (LANTH
	13-2035	MOSSBAUER STUDIES OF IRON-57 IN SOME PEROVSKITE LIKE LAYER TYPE FERROELECTRICS. (IN BI
MIRONENKO IG	37-3677	SOME PROPERTIES AND APPLICATION OF FERROELECTRICS AT MICROWAVES.
MISELYUK EG	04-0592	SATURATION OF VOLT- AMPERE CHARACTERISTICS CF CADMIUM SELENIDE LAYERS ON BARIUM TITANA
	04-1037	ABSORPTION AND AMPLIFICATION OF ULTRASOUND IN A TWO-LAYER SYSTEM CONSISTING OF A PIEZO
	15-2115	ABSORPTION OF ULTRASOUND IN AN ANTIMONY SULFOIODIDE FERROELECTRIC SEMICONDUCTOR NEAR I
	15-2116	DETERMINATION OF THE LATTICE PARAMETERS AND CURIE POINTS OF SOLID SOLUTIONS OF A(V)B(V
MISHCHENKO AV	17-2583	CRYSTALLIZATION OF RUBIDIUM DIDEUTERIUM PHOSPHATE FROM SOLUTIONS IN HEAVY WATER.
	17-2584	EFFECT CF DEUTERATION ON THE ELASTIC PROPERTIES OF RUBIDIUM DIHYDROGEN PHOSPHATE (RDP)
	17-2585	ENTHALPIES OF SOLUTION OF RUBIDIUM DIHYDROGEN PHOSPHATE IN WATER AND RUBIDIUM DIDEUTER
	17-2698	ELECTROOPTICAL AND OPTICAL PROPERTIES OF PARTIALLY DEUTERATED RUBIDIUM DIHYDROGEN PHOS
MITANI S	16-2347	DIELECTRIC AND X-RAY STUDIES ON THE FERROELECTRIC PHASE TRANSITION OF GLYCINE SILVER N
	16-2365	NEUTRON DIFFRACTION STUDIES OF FERROELECTRIC POLARIZATION REVERSALS IN SODIUM NITRITE.
MITROFANOV KP	08-1400	CHANGE OF RESONANCE ABSORPTION SPECTRA OF 23.8 KEV GAMMA-RAYS OF TIN-119 DURING PHASE
	08-1401	MOSSBAUER STUDY AND THEORETICAL ESTIMATION OF INTERNAL ELECTRIC FIELD GRADIENTS IN FER
MITSUI T	01-0024	A NOTE ON THE CLASSIFICATION OF FERROELECTRICS. (BY CURIE-WEISS CONSTANT)
	02-0077	ACCURATE STRUCTURE ANALYSIS OF FERROELECTRICS IN CONNECTION WITH THEIR PHASE TRANSITIO
	02-0099	DIELECTRIC AND X-RAY STUDIES OF THE PHASE TRANSITIONS IN ROCHELLE SALT, BARIUM TITANAT
	02-0100	FUTURE PROBLEMS ON FERROELECTRICS.
	27-3024	REFINEMENT OF CRYSTAL STRUCTURE OF TRIGLYCINE SULFATE.
	27-3093	STUDIES ON THE MECHANISM OF THE PHASE TRANSITION IN TGS. (STATIC DIELECTRIC CONSTANT,
	29-3244	X-RAY REFINEMENT OF THE CRYSTAL STRUCTURE OF LITHIUM AMMONIUM TARTRATE MONOHYDRATE.
MITSUICHI A	04-0744	FAR INFRARED REFLECTIVITY OF BARIUM TITANATE.
	17-2549	FAR INFRARED SPECTRA CF KDP AND ADP. (20-550 PER CM, 200 AND 300 DEG K)
MIURA S	10-1647	EVAPORATED FILMS OF TUNGSTEN OXIDE.
MIWA K	09-1624	STUDIES ON THE SINTERING BETWEEN LEAD MONOXIDE AND PZT. PART-1. PART-2: EFFECTS OF AVE
MIYAMOTO M	04-0734	A SEMICONDUCTING BARIUM TITANATE CERAMIC DOPED WITH NIOBIUM AND MANGANESE IONS AND ITS
MIYASHITA T	14-2090	COEXISTENCE OF THE SPCNTANEOUS ELECTRIC POLARIZATION AND MAGNETIZATION IN NICKEL IODID
	14-2091	CRYSTAL GROWTH OF NICKEL IODINE BORACITE.
MIYAZAWA S	08-1359	RH DOPED LITHIUM NIOBATE AS AN IMPROVED NEW MATERIAL FOR REVERSIBLE HOLOGRAPHIC STORAG
	11-1830	X-RAY TOPOGRAPHIC STUDIES ON LITHIUM NIOBATE AND LITHIUM TANTALATE SINGLE CRYSTALS.
	12-1876	SINGLE CRYSTAL GROWTH AND PHYSICAL PROPERTIES OF LITHIUM TANTALATE. (DIELECTRIC CONSTA
	12-1882	CONGRUENT MELTING COMPOSITION OF LITHIUM METATANTALATE.
	12-1883	STOICHIOMETRY AND OPTICAL QUALITY OF LITHIUM TANTALATE SINGLE CRYSTALS.
	13-1969	DIELECTRIC PROPERTY OF BARIUM TITANATE NIOBATE SINGLE CRYSTAL (TETRAGONAL TUNGSTEN BRO
	13-1971	DIELECTRIC AND ELECTROOPTIC PROPERTIES OF BARIUM SODIUM YTTRIUM NIOBATE SINGLE CRYSTAL
	35-3501	FERROELECTRIC AND OPTICAL PROPERTIES OF LEAD(5) GERMANIUM(3) OXIDE(11) AND ITS ISOMORP
	35-3513	CRYSTAL GROWTH AND PROPERTIES OF LEAD(5) GERMANIUM(3) OXIDE(11) SINGLE CRYSTALS.
	35-3514	GROWTH OF SINGLE CRYSTALS IN LEAD MONOXIDE- GERMANIUM DIOXIDE BINARY SYSTEM.
MIYAZAWA T	13-1999	X-RAY AND THERMAL EXPANSION STUDY OF A SODIUM(0.88) LITHIUM(0.12) NIOBATE CERAMIC.
MIZUNO O	28-3178	COERCIVE FIELDS OF ROCHELLE SALT HEAVILY IRRADIATED WITH GAMMA-RAYS.
	28-3203	NUCLEAR MAGNETIC RESONANCE IN ROCHELLE SALT, DETERMINATION OF THE ORIENTATIONS OF PROT
MIZUTANI I	14-2082	ELECTROOPTICAL PROPERTIES OF FERROELECTRIC (IRON IODINE) BORACITE. (SPONTANEOUS LATTIC
	14-2083	LATENT LATTICE STRAIN IN THE FERROELECTRIC STATE OF IRON- IODINE- BORACITE. (TEMPERATU
	14-2085	X-RAY STUDY ON THE LATTICE STRAINS OF (PARAELECTRIC AND) (IRON- IODINE- BORACITE) .
	17-2556	X-RAY STUDY ON THERMAL EXPANSION OF FERROELECTRIC POTASSIUM DIHYDROGEN PHOSPHATE. (NEA
MLAVSKY AI	05-1042	SOLVENT ZONE GROWTH OF FERROELECTRIC CRYSTALS. (STRONTIUM TITANATE, LEAD(X) STRONTIUM(
MLIKE H	33-3382	OBSERVATION OF THE ANOMALY OF THE THERMAL CONDUCTIVITY OF POTASSIUM MANGANESE TRIFLUOR
MNATSAKANYAN AV	17-2665	ANOMALIES IN THE INTERNAL STRESS FOR FERROELECTRIC PHOSPHATES NEAR THEIR CURIE POINTS.
MOCH P	33-3395	IMPURITY CONCENTRATION DEPENDENCE OF RAMAN SCATTERING BY MAGNONS IN NICKEL DOPED RUBID
MOCKEL P	04-0854	TRANSFORMATION OF ELECTROMAGNETIC RADIATION INTO MOTION. (PYROELECTRIC EFFECT IN TGS,
MOGENSEN O	34-3457	ANGULAR CORRELATION CF ANNIHILATION PHOTONS IN ICE SINGLE CRYSTALS.
MOHAN PV	16-2308	FAR INFRARED SPECTRA AND FERROELECTRIC PHASE TRANSITION OF POTASSIUM NITRATE.
MOHANA RAO JK	16-2333	CRYSTAL STRUCTURE OF FERROELECTRIC GLYCINE SILVER NITRATE.
	16-2348	CRYSTAL STRUCTURE OF GLYCINE SILVER NITRATE.
	22-2859	REFINEMENT OF THE CRYSTAL STRUCTURE OF FERROELECTRIC ACID LITHIUM SELENITE, POSITION O
	22-2860	REFINEMENT OF THE CRYSTAL STRUCTURE OF FERROELECTRIC LITHIUM HYDROGEN SELENITE: POSITI
	22-2861	ROCM TEMPERATURE CRYSTAL STRUCTURE OF THE FERROELECTRIC SODIUM TRIDEUTERO SELENITE.
	26-2940	CRYSTAL STRUCTURE OF PIEZCELECTRIC LITHIUM FORMATE MONOHYDRATE.
MOIYA AM	11-1863	LONGITUDINAL ELECTROOPTICAL EFFECT IN OBLIQUE SECTIONS OF LITHIUM NIOBATE.
MOK YW	11-1763	EPR OF GADOLINIUM(3+) IN LITHIUM NIOBATE.
	11-1833	ELECTRON PARAMAGNETIC RESONANCE OF GADOLINIUM(3+) DOPED LITHIUM NIOBATE.
MOKIEVSKII VA	17-2413	PRODUCTION OF LARGE SEED CRYSTALS OF KDP.
MOLCHANOV VI	04-0855	MEASUREMENT OF DIELECTRIC PROPERTIES OF FERROELECTRIC MATERIALS AT CENTIMETER AND DECI
MOLCHANCVA RA	04-0786	SOLID SOLUTIONS BASED ON BARIUM TITANATE.
	04-0787	SPONTANECUS DESTRUCTION OF FERROCERAMICS. (BARIUM TITANATE)
MOLL NJ	11-1788	LITHIUM NIOBATE SILICON SURFACE WAVE CONVOLUTER.
MOLNAR B	23-2886	BIAS FIELDS IN FERROELECTRIC COLEMANITE.
MOLOZOVA AP	13-2029	THERMOGRAPHIC STUDY CF THE REACTION OF BISMUTH(4) TITANIUM(3) OXIDE(12) WITH VANADIUM
MONAENKOVA AS	17-2585	ENTHALPIES OF SOLUTION OF RUBIDIUM DIHYDROGEN PHOSPHATE IN WATER AND RUBIDIUM DIDEUTER
MONTANO PA	08-1282	MOSSBAUER MEASUREMENTS OF BISMUTH FERRATE- BISMUTH FERRATE- LEAD ZIRCONATE SYSTEMS.

MONTANO PA	19-2770	MOSSBAUER STUDY ON FERROELECTRIC PROPERTIES OF FERRIC AMMONIUM SULFATE DODECAHYDRATE S
	24-2901	MOSSBAUER STUDY OF THE FERROELECTRIC PHASE TRANSITION IN POTASSIUM FERROCYANIDE TRIHYD
MONTGOMERY DJ	04-0823	EFFECT OF PARTICLE SIZE AND SHAPE ON THE INFRARED ABSORPTION SPECTRA OF BARIUM TITANAT
MONTGOMERY GP	11-1771	SPONTANEOUS PARAMETRIC SCATTERING, IDLER ABSORPTION, AND DISPERSION IN THE NONLINEAR S
MONTGOMERY H	17-2586	SYMMETRY OF ATOMIC VIBRATIONS IN POTASSIUM DIHYDROGEN PHOSPHATE.
MOON RL	09-1577	HIGH TEMPERATURE PHASE EQUILIBRIA IN THE LEAD ZIRCONATE TITANATE SYSTEM.
MOORE MA	03-0384	THEORY OF HYDROGEN BONDED FERROELECTRICS. PART-1: (FORMALISM OF LOCAL MODE MODEL.) PAR
MOORTHY VK	04-0797	DEVELOPMENT OF BARIUM TITANATE WITH CONSISTENT DIELECTRIC PROPERTIES.
	04-0798	INFLUENCE OF MICROSTRUCTURES (PREPARATION CONDITIONS) ON DIELECTRIC PROPERTIES OF BARI
MORAN PR	17-2545	INVESTIGATION OF THE PHASE TRANSITION IN AMMONIUM DIHYDROGEN PHOSPHATE SINGLE CRYSTAL
MORAN TJ	06-1155	SOUND PROPAGATION NEAR THE STRUCTURAL PHASE TRANSITION IN STRONTIUM TITANATE.
MORATIS CJ	13-1995	RESISTIVITY BEHAVIOR OF SEMICONDUCTING YTTRIUM BARIUM ZIRCONIUM TITANATE. (TEMPERATURE
MORAVEC F	02-0088	INFLUENCE OF THE QUALITY OF TRIGLYCINE SULFATE SINGLE CRYSTALS ON THEIR DIELECTRIC PRO
	27-3063	INFLUENCE OF DEFECTS ON SOME DIELECTRIC PROPERTIES OF TGS SINGLE CRYSTALS.
	27-3081	AGING PROCESSES IN TRIGLYCINE SULFATE. (TANDEL OR CAPACITANCE PROPERTIES DOMAIN STRUCT
	27-3083	GROWTH OF TRIGLYCINE SULFATE SINGLE CRYSTALS.
	27-3084	INFLUENCE OF IMPURITIES ON THE GROWTH AND SOME PHYSICAL PROPERTIES OF TGS SINGLE CRYST
	27-3085	PREPARATION OF PURE GLYCINE USED FOR GROWING OF TRIGLYCINE SULFATE SINGLE CRYSTALS.
	27-3086	PREPARATION OF PURE TGS SINGLE CRYSTALS AND SOME INVESTIGATIONS OF THEIR QUALITY. (VOL
	27-3087	PREPARATION OF PURE TRIGLYCINE SULFATE. (USING ION EXCHANGE)
	27-3096	GROWTH OF TRIGLYCINE SULFATE FROM SLIGHTLY SUPERSATURATED SOLUTIONS.
MORDANOV BP	09-1507	TERNARY SYSTEM LEAD TITANATE- LEAD ZIRCONATE- LEAD NIOBATE(0.67) ZINCATE(0.33).
MOREAU JM	08-1398	ATOMIC STRUCTURE IN THE PEROVSKITIC FERRATE.
	08-1402	FERROELECTRIC BISMUTH FERRATE X-RAY AND NEUTRON DIFFRACTION STUDY. (ANTIFERROMAGNETIC)
MOREL RJ	06-1147	ESR OF CHROMIUM(3+) IN CHROMIUM DOPED STRONTIUM TITANATE SINGLE CRYSTALS.
MORENO M	04-0856	PRESSURE DEPENDENCE OF THE DIELECTRIC CONSTANTS OF PARAELECTRIC MATERIALS. (BARIUM TIT
MORI H	03-0385	ON RELAXATION PROCESSES NEAR THE SECOND ORDER PHASE TRANSITION POINT.
MORI K	27-3082	EFFECT OF A TWO DIMENSIONAL PRESSURE ON THE CURIE POINTS OF TRIGLYCINE SULFATE AND ROC
MORI T	15-2197	EFFECT OF PHOTOCONDUCTION ON THE FERROELECTRIC PROPERTIES OF ANTIMONY SULFOIODIDE. (PH
MORIAMEZ C	11-1682	GENERATION OF SECOND AND THIRD HARMONICS BY A TRANSVERSE ACOUSTIC WAVE PROPAGATING ITS
MORIAMEZ M	11-1680	COLLINEAR INTERACTION BETWEEN A LONGITUDINAL ACOUSTIC WAVE AND TWO LUMINOUS WAVES ALON
	11-1681	DETERMINATION OF A NONLINEAR PARAMETER FOR ACOUSTIC SURFACE WAVE CONVOLUTION IN LITHIU
	11-1834	EXPERIMENTAL STUDY OF LIGHT DIFFRACTION BY HIGH FREQUENCY ACOUSTIC WAVES. APPLICATION
	11-1835	GENERATION OF THE THIRD HARMONIC OF A LONGITUDINAL HIGH FREQUENCY (950 MHZ) ACOUSTIC W
	11-1836	INTERACTIONS BETWEEN HIGH FREQUENCY ACOUSTICAL WAVES IN LITHIUM NIOBATE.
MORICE M	04-0555	FERROELECTRIC LOSSES IN (PURE) BARIUM TITANATE CERAMIC.
MORIMOTO K	29-3248	PROTON SPIN LATTICE RELAXATION IN FERROELECTRIC SODIUM AMMONIUM TARTRATE.
MORIMOTO N	36-3597	PHASE RELATIONS AND SUPERSTRUCTURES OF PYRRHOTITE, IRON(1-X) SULFIDE.
MORIMOTO S	16-2349	DOMAIN ETCH PATTERNS ON THE FACES PARALLEL TO THE FERROELECTRIC-AXIS OF SODIUM NITRATE
MORIMOTO Y	26-2941	ANOMALOUS SHIFT OF ESR LINES IN THE TWO DIMENSIONAL ANTIFERROMAGNET COPPER FORMATE TET
MORLON B	03-0386	SOME RESULTS ABOUT PHOTOEMISSION ON FERROELECTRIC CRYSTALS. (DOMAIN IMAGING, ABSTRACT
	11-1772	THEORETICAL STUDY OF THE GENERATION OF THE ACOUSTIC SECOND HARMONIC. (LITHIUM NIOBATE)
	19-2780	OPTICAL ABSORPTION, ELECTRIC CONDUCTIVITY AND PHOTOEMISSION OF GLYCOL SULFATE.
MORLOT G	17-2587	FAR INFRARED ABSORPTION SPECTRA FOR AMMONIUM DIHYDROGEN PHOSPHATE.
MOROSIN B	17-2588	PRESSURE EFFECTS ON THE LATTICE PARAMETERS AND STRUCTURE OF POTASSIUM DIHYDROGEN PHOSP
	22-2846	NEUTRON DIFFRACTION STUDY ON PARAELECTRIC SODIUM TRIHYDROGEN DISELENITE.
MOROZOV NA	04-0749	MICROWAVE DYNAMIC NONLINEARITY OF BARIUM TITANATE SINGLE CRYSTALS.
	04-1019	DIELECTRIC AND NONLINEAR MICROWAVE PROPERTIES OF POLYCRYSTALLINE SOLID SOLUTIONS BASED
	37-3643	THIN FILM MICROWAVE FERROELECTRIC CAPACITORS AND NONLINEAR MICROWAVE PROPERTIES OF FER
MORRISON AD	13-1994	PROPERTIES OF BISMUTH(12) TITANIUM OXIDE(20) AND THE SYSTEM BISMUTH OXIDE- TITANIUM DI
	35-3507	GROWTH OF LARGE SINGLE CRYSTAL BISMUTH(4) TITANIUM(3) OXIDE(12).
MORUP S	19-2771	MOSSBAUER STUDY OF ELECTRONIC SPIN FLIP PROCESSES IN AMMONIUM IRON SULFATE DODECAHYDRA
MOSKVICH YUN	24-2899	ANOMALOUS SPIN LATTICE RELAXATION BY QUASI SPIN WAVES IN POTASSIUM FERROCYANIDE TRIHYD
MOTEGI H	02-0099	DIELECTRIC AND X-RAY STUDIES OF THE PHASE TRANSITIONS IN ROCHELLE SALT, BARIUM TITANAT
	04-0857	SURFACE LAYER IN BARIUM TITANATE (X-RAY DIFFRACTION, DC FIELD EFFECT, PARAELECTRIC AND
	04-0858	X-RAY STUDY OF THE SURFACE LAYER ON BARIUM TITANATE SINGLE CRYSTAL.
	31-3331	FERROELECTRICITY AND PHASE TRANSITIONS IN SOLID HYDROGEN HALIDES (SINGLE CRYSTAL X-RAY
	31-3334	CRYSTAL STRUCTURE AND PHASE TRANSITION OF HYDROGEN CHLORIDE.
MOTOKI T	11-1782	LITHIUM NIOBATE LIGHT MODULATOR.
MOULTON WG	17-2533	ELECTRON SPIN RESONANCE OF IRRADIATED AND DEUTERATED POTASSIUM DIHYDROGEN PHOSPHATE.
MOUNIER S	34-3458	STUDY OF CONDUCTIVITY AND DIPOLAR RELAXATION IN DOPED ICE SINGLE CRYSTALS.
MOUNTVALA AJ	04-0859	EFFECT OF SURFACE MOISTURE ON THE DIELECTRIC BEHAVIOR OF ULTRAFINE BARIUM TITANATE PAR
	04-0860	ELECTRICAL PROPERTIES OF RUBIDIUM AND CESIUM DOPED BARIUM TITANATE CERAMICS.
MOUREY JF	16-2378	PRESSURE INDUCED FERROELECTRIC PHASE TRANSITION IN POTASSIUM NITRATE.
MUELLER KA	02-0101	INFORMATION ON HIGH DIELECTRIC CONSTANT MATERIALS GAINED BY ELECTRON PARAMAGNETIC RESO
	06-1144	STRONG AXIAL ELECTRON PARAMAGNETIC RESONANCE SPECTRUM OF IRON(3+) IN STRONTIUM TITANAT
	06-1164	MONODOMAIN STRONTIUM TITANATE (OCCURANCE IN THIN (110) PLATES BELOW CUBIC- TETRAGONAL
	06-1165	PARAMAGNETIC RESONANCE AND OPTICAL ABSORPTION OF TRANSITION ELEMENT IONS IN STRONTIUM
	08-1450	STRUCTURAL PHASE TRANSITIONS IN PEROVSKITE TYPE CRYSTALS. (EXTENDED ABSTRACT)
MUELLER MH	24-2903	NEUTRON DIFFRACTION STUDY OF FERROELECTRIC POTASSIUM FERROCYANIDE TRIDEUTERO HYDRATE A
MUESER HE	03-0370	(EFFECT OF TEMPERATURE GRADIENT WITHIN THE SAMPLE ON THE) MICROWAVE DISPERSION OF FERR
	27-3088	MICROWAVE BEHAVIOR OF ROCHELLE SALT AND TGS (18 REFS).
	27-3117	LIGHT EMISSION FROM TGS. (CRYSTALS DURING HEATING)
	27-3156	SHORT LIVED FERROELECTRIC "AFTER EFFECT" PHENOMENA. (HYSTERESIS LOOP CONSTRICTIONS, RO
MUGURUMA J	34-3459	INFLUENCE OF THE SURFACE LAYER ON THE PLASTIC DEFORMATION OF ICE SINGLE CRYSTALS.
MUHMOUD SA	28-3197	ULTRASONIC ANOMALIES IN ROCHELLE SALT.
MUIR KW	22-2866	A NEUTRON DIFFRACTION STUDY OF POTASSIUM TRIHYDROGEN SELENITE.
MUKHERJEE JL	13-1996	LINEAR THERMAL EXPANSIONS OF GADOLINIUM SUBSTITUTED BARIUM SODIUM NIOBATES.
	13-2048	CAUSES OF THERMAL EXPANSION OF GADOLINIUM SUBSTITUTED BARIUM SODIUM NIOBATE.
MUKHTAROVA II	17-2614	EFFECT OF STRUCTURAL DEFECTS PRODUCED BY GAMMA AND REACTOR IRRADIATION ON PHASE TRANSI
MULDAWER L	13-1905	LOW FREQUENCY VIBRATIONAL MODES IN BARIUM SODIUM NIOBATE.
	13-1906	RAMAN SPECTRA OF BARIUM SODIUM NIOBATE.
MULDER BJ	04-0861	PREPARATION OF BARIUM TITANATE AND OTHER CERAMIC POWDERS BY COPRECIPITATION OF CITRATE
MULLER D	16-2264	NUCLEAR SPIN LATTICE RELAXATION OF SODIUM-23 IN THE PARAELECTRIC AND FERROELECTRIC PHA
	16-2350	INFLUENCE OF THE ELECTRICAL BOUNDARY CONDITIONS ON THE NMR RELAXATION RATE OF SODIUM-2
	24-2889	STUDY OF HYDROGEN-1 IN PARAELECTRIC AND FERROELECTRIC POTASSIUM FERROCYANIDE TRIHYDRAT
MULLER F	30-3283	ELASTIC PROPERTIES OF GADOLINIUM MOLYBDATE AT ITS FERROELECTRIC PHASE TRANSITION.
MULLER J	14-2079	PREPARATION OF NICKEL BORACITE CRYSTALS BY HYDROTHERMAL SYNTHESIS.
MULLER KA	06-1166	CRITICAL ASYMMETRY IN LOCAL FLUCTUATIONS IN STRONTIUM TITANATE ABOVE THE CRITICAL TEMP
	06-1167	EFFECTIVE CHARGE OF TITANIUM(4+) FROM THE PARAMAGNETIC RESONANCE OF MANGANESE(4+) IN S

MULLER KA	06-1168	INTERPRETATION OF ELECTRIC FIELD EFFECTS IN EPR OF GADOLINIUM(3+) DOPED STRONTIUM TITA
	06-1169	MONODOMAIN STRONTIUM TITANATE.
	06-1170	ORDER PARAMETER AND PHASE TRANSITIONS OF STRESSED STRONTIUM TITANATE. (EPR OF IRON- VA
	06-1171	PHOTOCHROMIC IRON IN STRONTIUM TITANATE. EVIDENCE FROM PARAMAGNETIC RESONANCE.
	06-1231	ANALYSIS OF THE IRON OXYGEN VACANCY CENTER IN THE TETRAGONAL PHASE OF STRONTIUM TITANA
	06-1232	EPR STUDIES AND DYNAMICS IN STRONTIUM TITANATE IN THE NEIGHBORHOOD OF THE PHASE TRANSF
	06-1233	FLUCTUATIONS AND CORRELATIONS IN STRONTIUM TITANATE ABOVE THE CRITICAL TEMPERATURE.
	06-1234	THE IRON(3+)- VACANCY CENTER IN THE TETRAGONAL PHASE OF STRONTIUM TITANATE.
MULLER O	08-1403	CRYSTAL CHEMISTRY OF THE ABX(3) FAMILY WITH PARTICULAR EMPHASIS ON PEROVSKITES.
MULTANI MS	04-0562	EFFECT CF DOPING BARIUM TITANATE WITH MANGANESE DIOXIDE.
	04-0563	IONIC CHARACTERS FROM MOSSBAUER ISOMER SHIFTS. (DEGREE OF COVALENCY IN IRON DOPED BARI
MURAHTA M	05-1068	DIRECT CHELATOMETRIC DETERMINATION OF LEAD AND TITANIUM IN LEAD TITANATE AND SIMILAR C
MURAI T	13-2036	ELECTROOPTIC LIGHT BEAM DEFLECTION WITH BARIUM STRONTIUM NIOBATE PRISM.
MURAKAMI T	04-0833	DIELECTRIC CERAMICS WITH BOUNDARY LAYER STRUCTURE FOR HIGH FREQUENCY APPLICATION. (BAR
	04-1024	FERROELECTRIC PROPERTIES OF (INSULATING, BOUNDARY LAYER) BARIUM TITANATE CERAMICS AT G
	14-2090	COEXISTENCE OF THE SPONTANEOUS ELECTRIC POLARIZATION AND MAGNETIZATION IN NICKEL IODID
MURAYAMA Y	04-0862	FORMATION PROCESS OF BARIUM TITANATE THIN FILM EVAPORATED IN VACUO.
MURDOCH FJ	05-1067	ANOMALOUS FERROELECTRIC HYSTERESIS LOOPS. (LEAD STRONTIUM TITANATE)
	37-3660	NONLINEAR DIELECTRICS FOR HIGH POWER ELECTRONIC TUNING. (WITH ELECTRODES REDUCING POLA
MURUYAMA N	16-2351	THERMAL EXPANSION IN SODIUM NITRITE CRYSTALS.
MURZIN VN	04-0644	ABSORPTION AND REFLECTION SPECTRA OF BARIUM TITANATE IN THE LONG WAVELENGTH INFRARED R
	04-0863	TEMPERATURE STUDIES CF INFRARED REFLECTION SPECTRA OF BARIUM AND STRONTIUM TITANATES I
	04-0864	VIBRATIONAL SPECTRA OF STRONTIUM, BARIUM AND CALCIUM TITANATES.
	04-0865	VIBRATIONAL SPECTRA OF THE TITANATES OF STRONTIUM, BARIUM, AND CALCIUM. (75 REFS)
MUSER HE	01-0023	EUROPEAN MEETING ON FERROELECTRICITY, PROC.
	03-0387	THERMODYNAMIC POTENTIALS AND SECONDARY EFFECTS IN FERROELECTRIC CRYSTALS. (THEORY)
	27-2975	TEMPERATURE DEPENDENCE OF THE DIELECTRIC CONSTANTS OF TRIGLYCINE SULFATE ORTHOGONAL TO
	27-2995	NEW MEASUREMENTS OF DIVERSE THERMODYNAMICAL COEFFICIENTS OF ROCHELLE SALT AND TRIGLYCI
	27-3089	INVESTIGATIONS OF SURFACES OF FERROELECTRICS (TGS) WITH SEMICONDUCTING ELEMENTS.
	28-3212	ANALYSIS OF DOUBLE HYSTERESIS LOOPS. (ROCHELLE SALT)
	28-3213	MEASUREMENTS OF COMPLEX PIEZOELECTRIC CONSTANTS IN FERROELECTRIC CRYSTALS. (ROCHELLE S
MUTO K	13-2036	ELECTROOPTIC LIGHT BEAM DEFLECTION WITH BARIUM STRONTIUM NIOBATE PRISM.
MUZALEVSKII AA	03-0175	MICROTHEORY OF SPONTANEOUS POLARIZATION AND PHASE TRANSITIONS IN CRYSTALS OF PEROVSKIT
	17-2433	ORIGIN OF THE SPONTANEOUS POLARIZATION AND THE ISOTOPE EFFECTS IN FERROELECTRIC POTASS
MUZIKAR C	36-3596	MOSSBAUER EFFECT AND FERROELECTRIC PROPERTIES OF IONIC CRYSTALS.
MYASNIKOVA TP	16-2352	INFRARED SPECTRA OF POTASSIUM NITRATE AND SODIUM NITRITE AT PHASE TRANSITIONS.
	19-2772	CHANGES IN THE INFRARED SPECTRA OF AMMONIUM OR RUBIDIUM BISULFATE, AND AMMONIUM SULFAT
MYKOLAJEWYCZ R	34-3483	DIELECTRIC AND MECHANICAL RESPONSE OF ICE I(H) SINGLE CRYSTALS AND ITS INTERPRETATION.
MYLNIKOVA IE	03-0307	ELASTOOPTICAL EFFECT IN FERROELECTRIC WITH DIFFUSE PHASE TRANSITION.
	08-1292	STRUCTURE AND DIELECTRIC PROPERTIES OF LITHIUM CONTAINING PEROVSKITES. (STRONTIUM LITH
	08-1378	DIELECTRIC PROPERTIES AND THERMAL EXPANSION OF SOME FERROELECTRICS AND ANTIFERROELECTR
	13-1923	FORMATION CONDITIONS AND PHASE TRANSITIONS IN CRYSTALS WITH POTASSIUM TUNGSTEN BRONZE
	13-1982	ELECTROOPTIC EFFECT IN LEAD ZINCATE(0.33) NIOBATE(0.67). (DISPERSION OF QUADRATIC CO
	13-2050	PHASE TRANSITIONS IN TANTALATES WITH THE STRUCTURE OF POTASSIUM TUNGSTEN BRONZE. (POTA
	15-2189	NUCLEAR QUADRUPOLE AND ELECTROACOUSTIC ECHOES IN FERROELECTRIC ANTIMONY SULFOIODIDE.
	15-2216	NUCLEAR QUADRUPOLE RESONANCE IN FERROELECTRICS OF THE ANTIMONY SULFOIODIDE (AND BROMID
	36-3561	PECULIARITIES OF THE DIELECTRIC POLARIZATION OF CADMIUM PYRONIOBATE.
	36-3613	SYNTHESIS AND X-RAY DIFFRACTION STUDY OF SINGLE CRYSTALS OF A NEW PYROCHLORE CONTAININ
MYLOV VP	16-2360	EFFECT OF HYDROSTATIC PRESSURE ON THE FERROELECTRIC PHASE TRANSITION OF AMMONIUM SULFA
	27-3090	NEW HIGH PRESSURE PHASE TRANSITION IN TRIGLYCINE SELENATE.
	28-3214	HIGH PRESSURE PHASE TRANSITIONS OF THE FIRST KIND IN ROCHELLE SALT.
	30-3321	EFFECT OF HYDROSTATIC PRESSURE ON PHASE TRANSITION IN GADOLINIUM MOLYBDATE.
MYTTON RJ	04-0866	HIGH L/F NOISE ANOMALY IN SEMICONDUCTING BARIUM STRONTIUM TITANATE.
NAGAI K	11-1860	GROWTH OF CRATERS ON ION ETCHED SURFACE OF LITHIUM NIOBATE AND ION ETCHING WITHOUT CRA
	11-1862	THINNING OF LITHIUM NIOBATE FOR ACOUSTOOPTIC DEFLECTORS BY ION ETCHING.
NAGAI T	27-3093	STUDIES ON THE MECHANISM OF THE PHASE TRANSITION IN TGS. (STATIC DIELECTRIC CONSTANT,
NAGANO K	04-0734	A SEMICONDUCTING BARIUM TITANATE CERAMIC DOPED WITH NIOBIUM AND MANGANESE IONS AND ITS
NAGASAKI H	04-0853	PRESSURE DEPENDENCE OF TRANSITION TEMPERATURES AND ELECTROSTRICTIONS IN PEROVSKITE BAR
NAGASE K	04-0872	CHEMICAL CHANGES IN BARIUM TITANATE FIRED UNDER NITROGEN IN THE PRESENCE OF CARBON.
NAGATA K	13-2004	EFFECTS OF GRAIN SIZE AND POROSITY ON THE ELECTRICAL AND OPTICAL PROPERTIES OF TRANSPA
NAGATA T	05-1060	ELECTROMECHANICAL PROPERTIES OF LEAD TITANATE CERAMICS CONTAINING LANTANUM AND MANGANE
	05-1062	PIEZOELECTRIC PROPERTIES OF LEAD TITANATE CERAMICS.
	05-1069	LEAD TITANATE CERAMIC RESONATORS OPERATING IN VHF BAND.
NAGIBAROV VP	11-1797	INVESTIGATION OF CERTAIN PARAMETERS OF LITHIUM NIOBATE SINGLE CRYSTALS USING SCATTERIN
NAGLE JF	03-0388	LATTICE STATISTICS OF HYDROGEN BONDED CRYSTALS, PART-1, THE RESIDUAL ENTROPY OF ICE, P
	03-0389	ONE DIMENSIONAL KDP MODEL IN STATISTICAL MECHANICS.
	22-2862	MODELS FOR THE ORDER DISORDER TRANSITION IN SODIUM TRIHYDROGEN SELENATE.
NAGY G	34-3416	MOSSBAUER EFFECT OF FERROUS IONS IN CUBIC ICE.
NAITO F	09-1549	DIELECTRIC AND PIEZOELECTRIC PROPERTIES OF LEAD COBALTATE(0.33) NIOBATE(0.67)- TITANAT
NAITO M	17-2406	THE RADICAL IN GAMMA-IRRADIATED KDP.
NAKADA O	14-2082	ELECTROOPTICAL PROPERTIES OF FERROELECTRIC (IRON IODINE) BORACITE. (SPONTANEOUS LATTIC
NAKAGAWA T	08-1331	HIGH PRESSURE SYNTHESIS OF LEAD METAL- OXYGEN(3) TYPE PEROVSKITE.
	24-2902	ANOMALOUS SPECIFIC HEAT OF FERROELECTRIC THIOUREA AND POTASSIUM FERROCYANIDE TRIHYDRAT
NAKAHARA M	04-0867	MECHANISM FOR FORMING INSULATION LAYERS ON THE GRAIN BOUNDARIES OF BARIUM TITANATE SEM
	06-1120	DISPERSION IN STRONTIUM TITANATE- LANTHANUM FERRATE DIELECTRIC CERAMICS.
	06-1172	EFFECTS OF SILICA ON MANGANESE DOPED STRONTIUM TITANATE CERAMICS.
NAKAJIMA Y	05-1069	LEAD TITANATE CERAMIC RESONATORS OPERATING IN VHF BAND.
NAKAMURA D	16-2313	NUCLEAR QUADRUPOLE RESONANCE OF NITROGEN IN SODIUM NITRITE.
NAKAMURA E	01-0024	A NOTE ON THE CLASSIFICATION OF FERROELECTRICS. (BY CURIE-WEISS CONSTANT)
	02-0099	DIELECTRIC AND X-RAY STUDIES OF THE PHASE TRANSITIONS IN ROCHELLE SALT, BARIUM TITANAT
	03-0391	NOTE ON THE DIELECTRIC RELAXATION IN FERROELECTRICS (THEORETICAL DISCREPANCY).
	04-0643	MOVING STRIPE PATTERN IN BARIUM TITANATE.
	15-2171	DIELECTRIC PROPERTIES OF ANTIMONY SULFOIODIDE AT MICROWAVE FREQUENCIES (3.3 AND 9.8 GH
	16-2354	MEASUREMENT OF MICROWAVE DIELECTRIC CONSTANTS OF FERROELECTRICS. PART-2: DIELECTRIC CO
	27-2985	CRITICAL REGION IN FERROELECTRIC TRIGLYCINE SULFATE.
	27-3093	STUDIES ON THE MECHANISM OF THE PHASE TRANSITION IN TGS. (STATIC DIELECTRIC CONSTANT,
NAKAMURA H	27-3091	DOMAIN WALL CAUGHT IN DISLOCATIONS IN FERROELECTRIC TRIGLYCINE SULFATE CRYSTALS.
NAKAMURA K	36-3598	PIEZOELECTRICITY, PYROELECTRICITY, AND THE ELECTROSTRICTION CONSTANT OF POLYVINYLIDENE
NAKAMURA N	25-2923	CRYSTAL STRUCTURE OF THE IV PHASE OF THIOUREA. (SATELLITE SCATTERING)
	36-3538	FERROELECTRICITY AND PHASE TRANSITION IN AMMONIUM DICHLORO ACETATE AND DEUTERATED AMMO

NAKAMURA N	36-3625	NUCLEAR QUADRUPOLE RESONANCE STUDY OF PHASE TRANSITION IN AMMONIUM MONOCHLORO ACETATE,
NAKAMURA T	03-0390	FERROELECTRIC DOMAIN STRUCTURE AND ITS INTERACTION WITH CRYSTAL DISLOCATIONS. (DOMAIN
	04-0803	RAMAN SCATTERING LINE SHAPE OF THE SOFT E POLARITON MODE IN BARIUM TITANATE.
	04-0868	PERMITTIVITY CHANGE OF BARIUM TITANATE UNDER VERY SLOWLY CHANGING FIELD.
	04-0981	SUSCEPTIBILITY INCREASE DURING POLARIZATION REVERSAL. (IN BARIUM TITANATE AND TRIGLYCI
	15-2234	FAR INFRARED REFLECTIVITY SPECTRA OF ANTIMONY SULFOIODIDE.
	15-2235	FAR INFRARED REFLECTIVITY SPECTRA OF ANTIMONY SULFOIODIDE.
	16-2353	ANALYSIS OF LOW FREQUENCY ELECTROOPTIC RESPONSE IN SODIUM NITRITE.
	16-2373	TEMPERATURE DEPENDENCE OF FAR INFRARED REFLECTIVITY SPECTRA OF SODIUM NITRITE CRYSTALS
	17-2589	OBSERVATION OF PHASE FRONT MOTION IN KDP CRYSTAL.
	17-2681	FAR INFRARED REFLECTIVITY SPECTRA OF KDP CRYSTAL (16-400 PER CM, 83-300K).
	22-2877	FAR INFRARED REFLECTIVITY SPECTRA OF SODIUM TRIHYDROGEN DISELENITE CRYSTAL.
	27-3091	DOMAIN WALL CAUGHT IN DISLOCATIONS IN FERROELECTRIC TRIGLYCINE SULFATE CRYSTALS.
	28-3215	DOMAIN WALLS CAUGHT IN (FINE LINE) "SUDARES" IN ROCHELLE SALT CRYSTAL.
	30-3294	X-RAY STUDY OF THERMAL EXPANSION OF FERROELECTRIC GADOLINIUM MOLYBDATE.
	30-3306	OBSERVATION OF PHASE BOUNDARIES BETWEEN FERRO- AND PARAELECTRIC PHASES IN GADOLINIUM M
	30-3307	SPONTANEOUS BIREFRINGENCE AND ELECTROOPTIC RESPONSE IN GADOLINIUM MOLYBDATE.
NAKAO K	15-2198	ELECTROOPTIC EFFECTS OF THE ABSORPTION EDGE OF ANTIMONY SULFOIODIDE.
NAKATANI N	27-3092	DEPENDENCE OF THE COERCIVE FIELD OF TRIGLYCINE SULFATE ON FREQUENCY, AMPLITUDE AND TEM
NAKAZAWA H	36-3597	PHASE RELATIONS AND SUPERSTRUCTURES OF PYRRHOTITE, IRON(1-X) SULFIDE.
NAKHODNOVA AP	36-3587	PREPARATION OF THE TERNARY TITANATE BARIUM(0.8) LEAD(0.12) CALCIUM(0.08) TITANATE.
NAKONECHNYI YUS	15-2115	ABSORPTION OF ULTRASOUND IN AN ANTIMONY SULFOIODIDE FERROELECTRIC SEMICONDUCTOR NEAR I
	15-2199	PREPARATION AND PHYSICAL PROPERTIES OF NONSTOICHIOMETRIC ANTIMONY SULFOIODIDE SINGLE C
NANAMATSU S	08-1404	FERROELECTRIC PROPERTIES OF STRONTIUM NIOBATE SINGLE CRYSTAL.
	13-1997	A NEW FERROELECTRIC SINGLE CRYSTAL STRONTIUM(2) NIOBIUM(2) OXIDE(7).
	13-2037	GROWTH OF FERROELECTRIC SINGLE CRYSTAL STRONTIUM(2) NIOBIUM(2) OXIDE(7) BY MEANS OF TH
	13-2067	CRYSTAL STRUCTURE AND PIEZOELECTRICITY OF THE SYSTEM LEAD (ZINCATE NIOBATE)- LEAD TITA
	35-3508	FERROELECTRICITY IN LEAD(5) GERMANIUM(3) OXIDE(11).
	36-3576	ELECTROOPTIC AND PIEZOELECTRIC PROPERTIES OF LANTHANUM TITANATE SINGLE CRYSTAL. (FOUND
NAPIJALO M	36-3611	TEMPERATURE DEPENDENCE OF SOME PROPERTIES OF SODIUM DITHORIUM PHOSPHATE FERROELECTRIC
NARAYANAN PS	03-0350	THERMAL EXPANSION ANOMALIES AND THE DIRECTION OF SPONTANEOUS POLARIZATION IN FERROELEC
	20-2803	INFRARED AND RAMAN SPECTRA OF FERROELECTRIC ALUMS: METHYL AMMONIUM ALUMINUM (SULFATE O
	20-2808	SPECTROSCOPIC STUDIES OF FERROELECTRIC METHYL AMMONIUM ALUMS.
	22-2851	RELAXATION BETWEEN THE STRUCTURE OF FERROELECTRIC SODIUM TRIHYDROGEN DISELENITE AND IT
	22-2880	ANOMALOUS THERMAL EXPANSION OF POTASSIUM TRIHYDROGEN SELENATE CRYSTALS.
	29-3245	IS SODIUM RUBIDIUM TARTRATE TETRAHYDRATE A FERROELECTRIC. (ANSWER-NO)
NASH FR	11-1684	SPATIALLY UNIFORM AND ALTERABLE SHG PHASE MATCHING TEMPERATURES IN LITHIUM NIOBATE (CR
	11-1773	EFFECT OF OPTICAL INHOMOGENEITIES ON PHASE MATCHING IN NONLINEAR CRYSTALS (LITHIUM NIO
	11-1842	DEPENDENCE OF LINEAR ELECTROOPTIC EFFECT AND DIELECTRIC CONSTANT ON MELT COMPOSITION (
	13-1998	MEASUREMENTS OF SECOND HARMONIC GENERATION AND THE VARIATIONS IN THE FREE AND CLAMPED
NASIROV VI	16-2259	MORPHOLOGY OF CRYSTAL GROWTH IN THE ALPHA TO BETA, BETA TO ALPHA, BETA TO GAMMA AND GA
	16-2260	MORPHOLOGY OF CRYSTAL GROWTH AT POLYMORPHIC TRANSFORMATIONS IN POTASSIUM NITRATE, SILV
	16-2263	RHYTHMIC GROWTH OF NEW PHASE IN POTASSIUM NITRATE. (ALPHA TO BETA POLYMORPHIC TRANSITI
NASSAU K	10-1648	STRUCTURAL AND PHASE RELATIONSHIPS AMONG TRIVALENT TUNGSTATES AND MOLYBDATES.
	10-1649	STRUCTURAL AND PHASE RELATIONSHIPS AMONG TRIVALENT TUNGSTATES AND MOLYBDATES.
	11-1774	GROWTH AND PROPERTIES OF SINGLE DOMAIN CRYSTALS OF FERROELECTRIC LITHIUM NIOBATE. (CRA
	11-1775	LITHIUM NIOBATE- A NEW TYPE OF FERROELECTRIC: GROWTH, STRUCTURE, AND PROPERTIES. (REVI
	11-1776	STACKING FAULT MODEL FOR STOICHIOMETRY DEVIATIONS IN LITHIUM NIOBATE AND TANTALATE AND
	11-1777	STACKING FAULT MODEL FOR STOICHIOMETRY DEVIATIONS IN LITHIUM NIOBATE AND TANTALATE AND
	14-2092	MODIFIED TECHNIQUE FOR THE GROWTH OF BORACITE CRYSTALS.
	15-2200	GROWTH OF LARGE ANTIMONY SULFOIODIDE CRYSTALS (TO 1 CM DIAMETER): CONTROL OF NEEDLE MO
	30-3290	(ROOM TEMPERATURE) FERROELECTRIC FERROELASTIC PARAMAGNETIC (BETA) TERBIUM MOLYBDATE (C
	30-3305	NONLINEAR OPTICAL PROPERTIES OF GADOLINIUM MOLYBDATE AND TERBIUM MOLYBDATE.
NATH G	08-1405	LARGE NONLINEAR OPTICAL COEFFICIENT AND PHASE MATCHED SECOND HARMONIC GENERATION IN LI
NATTERMANN T	03-0392	INFLUENCE OF SCREENING ON THE FERROELECTRIC CURIE POINT (UNIAXIAL FERROELECTRIC PHOTOC
	03-0393	ULTRASONIC ATTENUATION ANOMALY IN DISPLACIVE TYPE FERROELECTRICS.
NAWROCIK W	27-3045	EFFECT OF A POLARIZING FIELD AND OF THE ADDITION OF A MIXTURE OF COPPER AND CHROMIUM I
NAYANOV VI	11-1718	TEMPERATURE DEPENDENCE OF HYPERSONIC WAVE DAMPING IN RUBY AND LITHIUM NIOBATE IN THE T
	11-1719	TEMPERATURE DEPENDENCES OF THE ABSORPTION OF 9.4 GHZ HYPERSOUND IN ALUMINUM OXIDE AND
NEDLIN GM	03-0394	PHASE TRANSITION FROM NORMAL FERROELECTRIC TO MAGNETICALLY ORDERED FERROELECTRIC STATE
	03-0395	THEORY OF SECOND ORDER PHASE TRANSITION FROM THE FERROMAGNETIC TO THE FERROMAGNETIC AN
NEELS H	15-2201	CRYSTALLIZATION BEHAVIOR OF ANTIMONY SULFOIODIDE. (SINGLE CRYSTAL GROWTH, CHEMICAL TRA
NEGRAN TJ	11-1791	CONTROL OF THE SUSCEPTIBILITY OF LITHIUM NIOBATE TO LASER INDUCED REFRACTIVE INDEX CHA
NEKRASOV MI	04-0932	FORMATION OF FLUORINE CONTAINING SOLID SOLUTIONS BASED ON BARIUM TITANATE.
	04-0983	INFRARED SPECTRA OF FLUORINE CONTAINING BARIUM TITANATE.
NEKRASOV MM	04-0592	SATURATION OF VOLT- AMPERE CHARACTERISTICS OF CADMIUM SELENIDE LAYERS ON BARIUM TITANA
	04-0869	FLUORINE CONTAINING BARIUM TITANATE. (ELECTRICAL PROPERTIES)
	04-0870	NONLINEAR PROPERTIES OF FERROELECTRICS AT MICROWAVE FREQUENCIES. (TRIGLYCINE SULFATE,
	27-3094	DIELECTRIC DISPERSION IN TGS CRYSTALS. (PERMITTIVITY AND LOSS FROM 10 GHZ TO 100 GHZ)
NELMES RJ	16-2355	CONSTRAINED REFINEMENT TECHNIQUES APPLIED TO THE STRUCTURE OF AMMONIUM HYDROGEN SULFAT
	17-2590	STRUCTURAL STUDIES OF THE SYSTEM POTASSIUM DIHYDROGEN PHOSPHATE- POTASSIUM DIDEUTERIUM
	17-2591	THE CRYSTAL STRUCTURE OF MONOCLINIC POTASSIUM DIDEUTERIUM PHOSPHATE.
	19-2773	AMMONIUM HYDROGEN SULFATE, ITS STRUCTURE AND THE FERROELECTRIC TRANSITION.
	19-2774	STRUCTURE OF AMMONIUM SULFATE IN FERROELECTRIC PHASE AND TRANSITION.
	19-2775	X-RAY DIFFRACTION DETERMINATION OF THE CRYSTAL STRUCTURE OF AMMONIUM HYDROGEN SULFATE
NELSON CW	02-0102	FERROELECTRICITY AND THE CHEMICAL BOND IN PEROVSKITE TYPE OXIDES. (43 REFS)
NELSON TJ	30-3280	FERROELECTRIC DOMAIN SHIFTING DEVICES. (GADOLINIUM MOLYBDATE)
NELSON WE	13-2021	DEVELOPMENT OF SINGLE CRYSTAL BARIUM SODIUM NIOBATE AS A NONLINEAR OPTICAL MATERIAL. (
NEPOMNYASHCHAYA VN	08-1423	ELECTROOPTICAL PROPERTIES OF LITHIUM IODATE SINGLE CRYSTALS.
NESTERENKO PS	15-2202	RELAXATION PROCESSES IN ANTIMONY SULFOIODIDE SINGLE CRYSTALS EXCITED WITH MICROSECOND
NESTERENKO VI	36-3569	X-RAY AND ELECTRIC INVESTIGATIONS OF THE SYSTEMS YTTRIUM MANGANESE(1-X) (B(X)) OXYGEN(
NESTEROVA NN	13-1974	ANOMALOUS BEHAVIOR OF THE OPTICAL ABSORPTION COEFFICIENT OF LEAD MAGNONIOBATE SINGLE
	14-2095	CRYSTAL FIELD THEORY AND OPTICAL ABSORPTION (CALCULATIONS) OF FERROELECTRIC BORACITES.
	14-2097	OPTICAL ABSORPTION (625-25 MICRONS, 77-650 DEG K) OF FERROELECTRIC IRON(2+) BORACITES
NETESOVA NP	27-3053	EFFECT OF HYDROSTATIC PRESSURE ON THE UNIPOLAR PROPERTIES OF TRIGLYCINE SULFATE.
NETTLETON RE	03-0396	ANALYSIS OF PHONON TUNNELING MODES IN THE RAMAN SPECTRUM OF TETRAGONAL POTASSIUM DIHYD
	03-0397	FERROELECTRIC PHASE TRANSITIONS: A REVIEW OF THEORY AND EXPERIMENT. PART-1: INTRODUCTI
	03-0398	LATTICE DYNAMICAL THEORY OF SWITCHING IN BARIUM TITANATE SINGLE CRYSTALS.
	03-0399	STATISTICAL PERTURBATION THEORY OF ORDER DISORDER FERROELECTRICS: FIRST PERTURBATIVE C
	17-2592	INTERACTIONS BETWEEN PROTON TUNNELING AND OPTICAL PHONONS IN POTASSIUM DIHYDROGEN PHOS

NETTLETON RE	17-2593	INTERACTION BETWEEN PROTON TUNNELING AND OPTICAL PHONONS IN POTASSIUM DIHYDROGEN PHOSP
NEUBERGER M	01-0025	II-VI SEMICONDUCTING COMPOUNDS - DATA TABLES. (MECHANICAL, CRYSTALLOGRAPHIC, PHYSICAL,
NEUDER SM	03-0400	PROPOSED MECHANISM FOR THERMOPHOTOTROPIC BEHAVIOR IN PEROVSKITE STRUCTURED TITANATES.
NEUMANN H	15-2140	DETERMINATION OF THE TEMPERATURE DEPENDENCE OF LATTICE CONSTANTS IN FERROELECTRIC ANTI
	15-2203	FIELD EMISSION FROM ANTIMONY SULFOIODIDE.
NEUMANN KH	28-3225	EFFECT OF MECHANICAL STRESS ON DIELECTRIC PROPERTIES OF PARAELECTRIC ROCHELLE SALT.
NEVILLE RC	06-1173	ANOMALOUS RESONANCE OF STRONTIUM TITANATE.
	06-1174	PERMITTIVITY OF STRONTIUM TITANATE.
	06-1175	SURFACE BARRIER ENERGIES ON STRONTIUM TITANATE.
	06-1176	SOME ELECTRONIC PROPERTIES OF ZINC OXIDE AND STRONTIUM TITANATE. (SURFACE BARRIER SYST
NEVOT L	27-3095	STUDY OF SWITCHING PARAMETERS IN TRIGLYCINE SULFATE.
NEWHOUSE VL	11-1778	POSSIBILITY OF SWITCHING AND STEERING SURFACE WAVES BY NONLINEAR MIXING IN ANISOTROPIC
NEWKIRK HW	10-1650	GROWTH, CRYSTALLOGRAPHY AND DIELECTRIC PROPERTIES OF BISMUTH TUNGSTATE.
NEWNHAM RE	08-1277	BISMUTH TITANATE SOLID SOLUTIONS.
	08-1406	STRUCTURAL BASIS OF FERROELECTRICITY IN THE BISMUTH TITANATE FAMILY.
	13-1935	CRYSTAL STRUCTURE OF BISMUTH TITANIUM OXIDE.
	13-2058	CRYSTAL STRUCTURE OF BISMUTH TITANIUM NIOBIUM OXIDE.
NEZAMAEVA NF	11-1840	MECHANISM AND KINETICS OF LITHIUM METANIOBATE FORMATION BY A SOLID PHASE REACTION.
NG WK	11-1779	OBSERVATION OF THE SIMULTANEOUS GENERATION OF THE SECOND, THIRD, AND FOURTH HARMONICS
NICHOLSON JY	17-2594	ABSENCE OF MOTIONAL NARROWING IN THE PROTON NMR IN POTASSIUM DIHYDROGEN PHOSPHATE AT H
	18-2736	ARSENIC-75 NMR RESONANCE IN AMMONIUM DIHYDROGEN ARSENATE AND DEUTERO AMMONIUM DIDEUTER
NICOLAS J	09-1575	INSCRIPTION OF HOLOGRAPHIC RESULTS IN TRANSPARENT CERAMIC FERROELECTRICS. (LEAD LANTHA
NICOLAU IF	17-2595	METHOD OF GROWTH IN SOLUTION OF SINGLE CRYSTALS BY THE GRADUAL ADDITION OF A REACTANT.
NICOLAU P	04-0599	INFLUENCE OF PHASE TRANSITIONS ON ELECTRICAL RESISTANCE OF BARIUM TITANATE.
NIGGLI A	15-2109	X-RAY CRYSTALLOGRAPHIC STUDIES OF ANTIMONY SULFOIODIDE.
NIIMORI K	17-2546	EPR OF CHROMIUM(3+) IONS IN ANTIFERROELECTRIC ADP CRYSTALS.
	17-2596	ELECTRON PARAMAGNETIC RESONANCE OF CHROMIUM(3+) IN ADP.
NIIMURA N	31-3334	CRYSTAL STRUCTURE AND PHASE TRANSITION OF HYDROGEN CHLORIDE.
	31-3337	STRUCTURE AND PHASE TRANSITION IN HYDROGEN CHLORIDE.
NIINO M	10-1637	ELECTRICAL CONDUCTIVITY OF TUNGSTEN TRIOXIDE.
NIIZEKI N	04-1007	DISLOCATIONS IN BARIUM TITANATE SINGLE CRYSTALS.
	08-1376	X-RAY STUDY OF THE SYSTEM BISMUTH OXIDE FERRIC OXIDE.
	11-1780	CHEMICAL COMPOSITION AND OPTICAL QUALITY OF FERROELECTRIC CRYSTALS, ESPECIALLY ON LITH
	11-1810	ZERO TEMPERATURE COEFFICIENT OF RESONANT FREQUENCY IN LITHIUM TANTALATE LENGTH EXPANDE
	11-1830	X-RAY TOPOGRAPHIC STUDIES ON LITHIUM NIOBATE AND LITHIUM TANTALATE SINGLE CRYSTALS.
	12-1868	TEMPERATURE DEPENDENCE OF AN X CUT LITHIUM TANTALATE CRYSTAL RESONATOR VIBRATING IN TH
	12-1876	SINGLE CRYSTAL GROWTH AND PHYSICAL PROPERTIES OF LITHIUM TANTALATE. (DIELECTRIC CONSTA
	12-1877	BEHAVIOR OF LITHIUM TANTALATE SINGLE CRYSTAL NEAR ITS CURIE POINT. PART-1: PIEZOELECTR
	12-1893	PIEZOELECTRIC AND ELASTIC PROPERTIES OF LITHIUM TANTALATE CHARACTERISTICS.
	13-2060	ELASTIC ANOMALY OF BARIUM SODIUM NIOBATE (ELASTIC, PIEZOELECTRIC, DIELECTRIC, AND OPTI
	24-2894	NUCLEAR MAGNETIC RESONANCE AND X-RAY STUDIES OF POTASSIUM FERROCYANIDE TRIHYDRATE CRYS
	32-3348	FERROELECTRIC-LIKE BEHAVIOR OF LITHIUM HYDRAZINIUM SULFATE.
	35-3501	FERROELECTRIC AND OPTICAL PROPERTIES OF LEAD(5) GERMANIUM(3) OXIDE(11) AND ITS ISOMORP
	35-3502	LEAD(5) GERMANIUM(3) OXIDE(11) CRYSTAL, A NEW FERROELECTRIC. (DIELECTRIC CONSTANTS AND
	35-3504	CRYSTAL STRUCTURE OF FERROELECTRIC LEAD(5) GERMANIUM(3) OXIDE(11).
	35-3515	X-RAY TOPOGRAPHIC OBSERVATION OF 180 DEG DOMAINS IN FERROELECTRIC LEAD GERMANATE SINGL
	35-3516	X-RAY TOPOGRAPHIC OBSERVATION OF 180 DEG DOMAINS IN FERROELECTRIC LEAD(5) GERMANIUM
	36-3570	SWITCHING OF OPTICAL ROTATORY POWER IN FERROELECTRIC LEAD(5) GERMANIUM(3) OXIDE(11) SI
NIKIFOROV IYA	15-2179	BAND STRUCTURE OF THE FERROELECTRIC SEMICONDUCTOR ANTIMONY SULFOIODIDE.
	15-2204	ELECTRON STATE DENSITY AND OPTICAL PROPERTIES OF ANTIMONY SULFOIODIDE. (IN PARAELECTRI
	36-3575	BAND STRUCTURE OF THE FERROELECTRIC SEMICONDUCTOR ANTIMONY TRISULFIDE.
NIKIFOROV LG	36-3599	POSSIBLE METHOD FOR ESTIMATING THE PARAMETER X IN COMPOUNDS OF THE PYROCHLORE TYPE WIT
NIMURA N	31-3331	FERROELECTRICITY AND PHASE TRANSITIONS IN SOLID HYDROGEN HALIDES (SINGLE CRYSTAL X-RAY
NINOMIYA Y	11-1781	GROWTH OF LITHIUM NIOBATE SINGLE DOMAIN CRYSTALS.
	11-1782	LITHIUM NIOBATE LIGHT MODULATOR.
NISENOFF M	17-2579	EFFECTS OF DISPERSION AND FOCUSING ON THE PRODUCTION OF OPTICAL HARMONICS. (KDP)
NISHI M	04-0871	CHANGE IN THE DECAY CONSTANT OF TECHNETIUM-99 IN BARIUM TITANATE BY THE FERROELECTRIC
NITSCHE R	15-2205	NEW FERROELECTRIC A(V)B(VII)C(VII) COMPOUNDS OF ANTIMONY SULFOIODIDE TYPE. (PARTIAL OR
NITTA I	19-2777	ON THE PHASE TRANSITION OF AMMONIUM HYDROGEN SULFATE AND THE STRUCTURE OF THE FERROELE
NITTA T	04-0872	CHEMICAL CHANGES IN BARIUM TITANATE FIRED UNDER NITROGEN IN THE PRESENCE OF CARBON.
	13-1999	X-RAY AND THERMAL EXPANSION STUDY OF A SODIUM(0.88) LITHIUM(0.12) NIOBATE CERAMIC.
NIX WD	04-0836	FERROELECTRIC TRANSFORMATION IN BARIUM TITANATE. (MODEL GIVES NEARLY COMPLETE DESCRIPT
NIZEKI N	35-3520	ELASTIC AND PIEZOELECTRIC PROPERTIES OF FERROELECTRIC LEAD(5) GERMANIUM(3) OXIDE(11) C
NOACK F	34-3488	PROTEIN SPIN RELAXATION IN HEXAGONAL ICE. PART-2: THE T(1 RHO) MINIMUM.
NOAKE H	09-1540	PRECIPITATION OF ZIRCONIA PHASE IN NIOBIUM MODIFIED CERAMICS OF LEAD ZIRCONATE TITANAT
NODA J	11-1742	PRECISION MACHINING OF LITHIUM NIOBATE CRYSTALS FOR ELECTROOPTIC ELEMENTS.
	11-1783	LAPPING CHARACTERISTICS OF LITHIUM NIOBATE SINGLE CRYSTAL.
	11-1784	LAPPING CHARACTERISTICS OF LITHIUM NIOBATE SINGLE CRYSTALS.
NOLTA JP	16-2285	SWITCHING BEHAVIOR IN FERROELECTRIC POTASSIUM NITRATE.
	16-2356	DIELECTRIC BEHAVIOR OF (FERROELECTRIC) FILMS OF VACUUM DEPOSITED POTASSIUM NITRATE.
	37-3668	FERROELECTRIC HYSTERESIS TRACER FEATURING COMPENSATION AND SAMPLE GROUNDING.
NOMURA S	05-1070	DIELECTRIC AND PIEZOELECTRIC PROPERTIES IN THE TERNARY SYSTEM LEAD (ZINCATE NIOBATE)-
	08-1331	HIGH PRESSURE SYNTHESIS OF LEAD METAL- OXYGEN(3) TYPE PEROVSKITE.
	08-1473	THE EFFECT OF PRESSURE ON THE TRANSITION OF LEAD ZINCATE(0.33) NIOBATE(0.67).
	13-1900	ELECTROOPTIC EFFECTS OF HOT PRESSED CERAMICS IN THE SYSTEM LEAD (ZINCATE(0.333) NIOBAT
	13-2000	PTC EFFECT (POSITIVE TEMPERATURE COEFFICIENT OF RESISTIVITY) IN (DOPED AND UNDOPED CER
	13-2001	PREPARATION AND DIELECTRIC PROPERTIES OF POTASSIUM BISMUTH ZINC NIOBATE.
	13-2065	DIELECTRIC AND OPTICAL PROPERTIES OF LEAD ZINCATE(0.33) NIOBATE(0.67) SINGLE CRYSTAL.
	13-2066	FERROELECTRIC PROPERTIES OF LEAD ZINCATE(0.33) NIOBATE(0.67) (SINGLE CRYSTALS GROWN FR
	13-2067	CRYSTAL STRUCTURE AND PIEZOELECTRICITY OF THE SYSTEM LEAD (ZINCATE NIOBATE)- LEAD TITA
	17-2559	CUTTING AND POLISHING IN UNIAXIAL CRYSTALS OF THE TYPE X DIHYDROGEN PHOSPHATE FOR LIGH
	24-2902	ANOMALOUS SPECIFIC HEAT OF FERROELECTRIC THIOUREA AND POTASSIUM FERROCYANIDE TRIHYDRAT
	37-3648	ELECTROOPTIC EFFECT IN FERROELECTRICS.
NORBERT A	17-2598	(PREPARATION AND STRUCTURE OF) RUBIDIUM AND CESIUM SEMIMETALLIC ORTHOPHOSPHATES.
	32-3355	CRYSTALLOGRAPHIC STUDY OF SEVERAL HYDRAZINIUM ORTHO FLUOBERYLLATES.
NORDAL PE	17-2599	NUCLEAR DOUBLE RESONANCE STUDY OF POTASSIUM-39 IN FERROELECTRIC POTASSIUM DIHYDROGEN P
NORDLAND WA	02-0095	ABSOLUTE SIGNS OF NONLINEAR OPTICAL (SECOND HARMONIC GENERATION) COEFFICIENTS OF POLAR
	02-0096	ABSOLUTE SIGNS OF SECOND HARMONIC GENERATION COEFFICIENTS OF PIEZOELECTRIC CRYSTALS.
	02-0098	RELATIVE SIGNS OF NONLINEAR OPTICAL COEFFICIENTS OF POLAR CRYSTALS (LITHIUM NIOBATE OR
	05-1066	FURTHER MEASUREMENTS OF ABSOLUTE SIGNS OF SECOND HARMONIC GENERATION COEFFICIENTS OF P

NORDLAND WA	11-1769	DEPENDENCE OF SECOND HARMONIC GENERATION COEFFICIENTS OF LITHIUM NIOBATE ON MELT COMPO
	17-2597	ABSOLUTE SIGNS OF THE NONLINEAR OPTICAL COEFFICIENTS OF POTASSIUM DIHYDROGEN PHOSPHATE
	30-3305	NONLINEAR OPTICAL PROPERTIES OF GADOLINIUM MOLYBDATE AND TERBIUM MOLYBDATE.
	35-3506	NONLINEAR OPTICAL PROPERTIES OF FERROELECTRIC LEAD(5) GERMANIUM(3) OXIDE(11).
	36-3600	ANOMALIES IN THE (1 KHZ) DIELECTRIC CONSTANT (5-300K) OF THE PRASEODYMIUM(1-X) NEODYM
NORMAN SL	11-1785	WAVELENGTH DEPENDENCE OF HOLOGRAPHIC PHENOMENA IN DOPED LITHIUM NIOBATE.
NORTHROP DA	09-1578	VAPORIZATION CF LEAD ZIRCONATE- LEAD TITANATE MATERIALS. PART-2: HOT PRESSED COMPOSITI
NOVAK J	02-0089	EFFECT OF CRYSTALLIZATION CONDITIONS ON THE SHAPE AND COMPOSITION OF FERROELECTRIC CRY
	04-0540	CHEMICAL PROCESSES DURING HYDROGEN REDUCTION OF BARIUM TITANATE CRYSTALS.
	04-0618	SOME CRYSTAL CHEMICAL AND PHYSICAL PROPERTIES OF SINGLE CRYSTALS OF BARIUM TITANATE WI
	04-0633	CRYSTAL CHEMICAL PROBLEMS OF (BARIUM, STRONTIUM) TITANATE SOLID SOLUTIONS.
	04-0635	DIELECTRIC PROPERTIES OF BARIUM(1-X) STRONTIUM(X) TITANATE SINGLE CRYSTALS GROWN FROM
	04-0636	HYDROXYL AS DEFECT IN BARIUM TITANATE.
	04-0736	DETERMINATION OF THE DEVIATION FROM OXYGEN STOICHIOMETRY OF BARIUM TITANATE SINGLE CRY
	04-0873	DEFECT STRUCTURE OF (BARIUM(1-X) STRONTIUM(X)) TITANATE- MIXED CRYSTALS FROM A POTASSI
	04-0874	HETEROVALENT SUBSTITUTION AND OXYGEN NONSTOICHIOMETRY OF BARIUM TITANATE AND HEMATITE
	04-0875	OPTICAL AND DIELECTRIC PROPERTIES OF SINGLE CRYSTAL BARIUM TITANATE WITH NIOBIUM ADDIT
NOVAKOVIC L	03-0401	FREQUENCY SPECTRUM OF HYDROGEN BONDED FERROELECTRICS. (INFRARED ABSORPTION, NEUTRON IN
	03-0402	QUANTUM THEORY OF FERROELECTRICITY, PART-1: SPECTRAL ANALYSIS (PROTON DYNAMICS IN HYDR
	03-0403	QUANTUM THEORY OF FERROELECTRICITY, PART-2: THERMODYNAMIC BEHAVIOR OF FERROELECTRIC SP
	03-0475	DYNAMICAL THEORY OF HYDROGEN BONDED FERROELECTRICS. PART-1: DAMPINGS AND (FREQUENCY) S
	17-2600	DYNAMICAL BEHAVIOR OF FERROELECTRIC SPINS. (KDP TYPE CRYSTALS)
	17-2601	POSSIBLE THEORETICAL EXPLANATION OF SHIFT OF CURIE TEMPERATURE IN POTASSIUM DIDEUTERIU
NOVIK VK	02-0083	PYROELECTRIC INVESTIGATIONS OF THE DOMAIN STRUCTURE STABILITY IN FERROELECTRICS. (ROCH
	04-0686	STUDY OF THE PYROELECTRIC EFFECT IN THIN FILMS OF TRIGLYCINE SULFATE AND BARIUM TITANA
	27-3115	VARIATION OF THE UNIPOLARITY OF TRIGLYCINE SULFATE CRYSTALS WITH TIME.
	36-3559	FERROELECTRIC BEHAVIOR OF THE MINERAL STIBOTANTALITE.
	36-3601	DOMAIN STRUCTURE AND TEMPERATURE DEPENDENCE OF THE PIEZOELECTRIC CONSTANTS OF STIBOTAN
	37-3673	GENERATOR OF BIPOLAR PULSES FOR INVESTIGATION OF FERROELECTRIC CRYSTALS.
NOVOSELOV AS	04-0656	SIZE EFFECTS IN FERROELECTRICS. (BARIUM TITANATE)
NOVOSILTSEV VN		
NOVOTNY J	02-0088	INFLUENCE OF THE QUALITY OF TRIGLYCINE SULFATE SINGLE CRYSTALS ON THEIR DIELECTRIC PRO
	27-3063	INFLUENCE OF DEFECTS ON SOME DIELECTRIC PROPERTIES OF TGS SINGLE CRYSTALS.
	27-3081	AGING PROCESSES IN TRIGLYCINE SULFATE. (TANDEL OR CAPACITANCE PROPERTIES DOMAIN STRUCT
	27-3083	GROWTH OF TRIGLYCINE SULFATE SINGLE CRYSTALS.
	27-3084	INFLUENCE OF IMPURITIES ON THE GROWTH AND SOME PHYSICAL PROPERTIES OF TGS SINGLE CRYST
	27-3096	GROWTH OF TRIGLYCINE SULFATE FROM SLIGHTLY SUPERSATURATED SOLUTIONS.
NOZAKI K	27-3127	DIP ETCH PATTERNS ON THE FACES OF TRIGLYCINE SULFATE CRYSTALS.
NOZIK YUZ	19-2776	STRUCTURE OF AMMONIUM FLUOBERYLLATE AT AMBIENT TEMPERATURE.
NUNES AC	08-1407	NEUTRON STUDY OF DIFFUSE SCATTERING IN CUBIC POTASSIUM NIOBATE.
NYIKOS P	04-0865	CHANGE IN THE DECAY CONSTANT OF ZIRCONIUM-89 IN BARIUM TITANATE (BARIUM CHLORIDE FLUX
	04-0876	VARIATION OF THE DECAY CONSTANT OF STRONTIUM-85 IN BARIUM TITANATE.
OBUKHOVSKII VV	11-1787	SPONTANEOUS RAMAN SCATTERING BY POLARITONS. PART-2: THE CASE OF CADMIUM SULFIDE AND LI
ODAKA T	26-2950	PROTON RESONANCE STUDY OF SUBLATTICE ROTATIONS AND SPIN DEVIATION IN COPPER FORMATE TE
ODEHNAL M	04-0877	FORBIDDEN LINES IN THE HYPERFINE SPECTRUM OF THE ELECTRON PARAMAGNETIC RESONANCE OF MA
OETTEL RE	17-2437	EFFECT CF DEUTERON SUBSTITUTION ON DOMAIN DYNAMICS IN POTASSIUM DIHYDROGEN PHOSPHATE.
OFFENBACHER EL	34-3460	HARDNESS ANISOTROPY OF SINGLE CRYSTALS OF ICE I(H).
OGADZHANOVA VV	16-2266	FERROELECTRIC PHASE TRANSITION IN SILVER SODIUM DINITRITE
	16-2370	PROPERTIES OF SILVER SODIUM DINITRITE AND SILVER SODIUM NITRITE. (DIFFERENTIAL THERMAL
OGAWA K	19-2777	ON THE PHASE TRANSITION OF AMMONIUM HYDROGEN SULFATE AND THE STRUCTURE OF THE FERROELE
OGG NR	11-1698	SURFACE WAVE DIFFRACTION IN LITHIUM NIOBATE.
OGUCHI T	22-2823	ANOMALOUS SPECIFIC HEAT OF SODIUM TRIHYDROGEN SELENITE.
	22-2835	RESIDUAL ENTROPY OF SODIUM TRIHYDROGEN SELENATE.
OGURTSOV SI	04-0818	UHF DIELECTRIC PROPERTIES OF SOME POLYCRYSTALLINE FERROELECTRICS (BARIUM TITANATE AND
OHAMA N	04-0878	POLARITON DISPERSION RELATION IN CUBIC BARIUM TITANATE.
OHANESSIAN H	17-2689	DIELECTRIC PERMITTIVITY OF FINELY DIVIDED POTASSIUM AND AMMONIUM DIHYDROGEN PHOSPHATE.
OHANESSIAN SP	03-0240	THERMODYNAMICS OF AN ELASTIC FERROELECTRIC MODEL.
OHARA G	13-2013	DIELECTRIC AND CERAMIC PROPERTIES OF (BISMUTH(0.5) (SODIUM OR POTASSIUM)´(0.5)(X) LEAD(
OHASHI Y	09-1624	STUDIES ON THE SINTERING BETWEEN LEAD MONOXIDE AND PZT. PART-1. PART-2: EFFECTS OF AVE
OHI K	08-1408	PHOTOCONDUCTION OF POTASSIUM TANTALATE.
	15-2173	POTENTIAL DISTRIBUTION IN ANTIMONY SULFOIODIDE CRYSTAL (IN THE PARAELECTRIC PHASE, UND
	15-2206	REFRACTIVE INDICES OF ANTIMONY SULFOIODIDE. (DEPENDENCE ON TEMPERATURE, WAVELENGTH, AP
	28-3215	DOMAIN WALLS CAUGHT IN (FINE LINE) "SUDARES" IN ROCHELLE SALT CRYSTAL.
OHKOHCHI K	03-0308	DISTRIBUTION OF ZEROES OF THE PARTITION FUNCTION FOR THE SLATER MODEL OF FERROELECTRIC
OHMACHI Y	35-3509	ACCUSTOOPTIC PROPERTY OF SINGLE CRYSTAL LEAD(5) GERMANIUM(3) OXIDE(11).
OHMORI Y	11-1861	NATURE OF INTERNAL ELECTRIC FIELD DURING OPTICAL DAMAGE PROCESS IN LITHIUM NIOBATE.
OHNEDA Y	16-2254	NUCLEAR QUADRUPOLE RELAXATION OF NITROGEN-14 IN SODIUM NITRITE.
OHNO T	16-2321	ACOUSTIC NUCLEAR MAGNETIC RESONANCE IN FERROELECTRIC SODIUM NITRITE.
OHSUMI K	30-3308	FERROELECTRIC AND FERROELASTIC CRYSTAL, GADOLINIUM MOLYBDATE.
OHTA T	13-2002	SUPERLATTICE STRUCTURE OF STRONTIUM SODIUM LITHIUM NIOBATE (TUNGSTEN BRONZE) TYPE FERR
	13-2062	ELECTROOPTIC PROPERTIES OF STRONTIUM LITHIUM NIOBATE TYPE FERROELECTRICS (SINGLE CRYST
OHYA S	03-0286	A THEORY OF THE PHASE TRANSITION IN AMMONIUM DIHYDROGEN PHOSPHATE.
OJA T	32-3345	NITROGEN-14 NUCLEAR QUADRUPOLE RESONANCE STUDY OF LITHIUM HYDRAZINIUM SULFATE.
OKA K	08-1409	CRYSTAL GROWTH OF SODIUM NIOBATE.
OKADA K	03-0404	PHENOMENOLOGICAL THEORY OF ANTIFERROELECTRIC TRANSITION. PART-2: FIRST ORDER TRANSITIO
	17-2602	DIELECTRIC EVIDENCE OF A FIRST ORDER TRANSITION IN POTASSIUM DIHYDROGEN PHOSPHATE.
	17-2603	EXPERIMENTS ON THE ORDER OF THE TRANSITION IN KDP.
	17-2604	INTRACAVITY SECOND HARMONIC GENERATION WITH RUBY LASER BY KDP.
	17-2682	THERMAL HYSTERESIS OF THE FERROELECTRIC TRANSITION IN POTASSIUM DIHYDROGEN PHOSPHATE.
	27-3098	FERROELECTRIC BEHAVIOR OF RADIATION DAMAGED TRIGLYCINE SULFATE AND ROCHELLE SALT UP TO
OKADA M	17-2605	LIGHT MODULATORS USING 45 DEGREE X CUT AND 45 DEGREE Y CUT ADP CRYSTALS.
OKAI B	04-0853	PRESSURE DEPENDENCE OF TRANSITION TEMPERATURES AND ELECTROSTRICTIONS IN PEROVSKITE BAR
	08-1473	THE EFFECT CF PRESSURE ON THE TRANSITION OF LEAD ZINCATE(0.33) NIOBATE(0.67).
OKAMOTO Y	04-0878	POLARITON DISPERSION RELATION IN CUBIC BARIUM TITANATE.
OKANO T	09-1539	PIEZOELECTRIC CERAMICS OF LEAD ZIRCONATE TITANATE MODIFIED WITH SODIUM OR POTASSIUM AN
	13-1961	ANTIFERROELECTRIC PHASE IN THE SYSTEM LEAD TITANATE- LEAD ZIRCONATE- LANTHANUM FERRATE
OKAYAMA H	03-0324	90 DEGREE TYPE DOMAIN WALL OF BARIUM TITANATE CRYSTAL. (THEORY, CALCULATIONS)
OKAZAKI K	03-0487	EFFECTS OF THERMALLY DIFFUSED IMPURITIES ON PROPAGATION OF ELASTIC SURFACE WAVES ON FE
	04-0879	ELECTRICAL PROPERTIES OF HOT PRESSED FERROELECTRIC (BARIUM TITANATE TYPE) CERAMICS. (I
	04-0880	SPACE CHARGE POLARIZATION AND AGING OF BARIUM TITANATE CERAMICS.
	13-2004	EFFECTS OF GRAIN SIZE AND POROSITY ON THE ELECTRICAL AND OPTICAL PROPERTIES OF TRANSPA

OKUMA H	36-3538	FERROELECTRICITY AND PHASE TRANSITION IN AMMONIUM DICHLORO ACETATE AND DEUTERATED AMMO
OLECH J	07-1267	THERMOELECTRIC EFFECT IN SEMICONDUCTIVE LEAD ZIRCONATE.
OLIKH YAM	15-2114	ABSORPTION OF ULTRASOUND IN A(V)B(VI)C(VII) FERROELECTRIC SEMICONDUCTORS AT TEMPERATUR
OLIVER DS	37-3661	USE OF CUBIC PHOTOSENSITIVE CRYSTALS AS POCKELS EFFECT LIGHT VALVES IN DISPLAY AND MOD
OLSON BL	13-1948	(TETRAGONAL) TUNGSTEN BRONZE TYPE CRYSTALS GROWN IN MOLTEN SOLUTIONS OF THE PSEUDOSYST
	13-1949	(EXTENT OF) TUNGSTEN BRONZE (LIQUIDUS AND SOLIDUS) FIELD AND MELT GROWTH OF CRYSTALS I
	13-2017	TUNGSTEN BRONZE FIELD IN THE (OXIDE) SYSTEM POTASSIA LITHIA NIOBIA (POTASSIUM LITHIUM
OLSON DH	06-1242	LIGHT SCATTERING STUDIES OF THE SOFT MODES NEAR THE CUBIC TO TETRAGONAL PHASE TRANSITI
	08-1365	ELECTRON TUNNELING INTO POTASSIUM TANTALATE SCHOTTKY BARRIER JUNCTIONS.
OMACHI Y	12-1885	BEHAVIOR OF LITHIUM TANTALATE SINGLE CRYSTAL NEAR ITS CURIE POINT, PART-2, DIELECTRIC
OMOTOSO EA	35-3510	ULTRASONIC ATTENUATION IN SINGLE CRYSTALS OF BISMUTH(12) GERMANIUM OXIDE(20) AT LOW TE
ONODERA Y	03-0405	DYNAMIC SUSCEPTIBILITY OF CLASSICAL ANHARMONIC OSCILLATOR. A UNIFIED OSCILLATOR MODEL
	03-0406	FREE ENERGY AND SPECIFIC HEAT IN FERROELECTRIC PHASE TRANSITION IN TERMS OF A SINGLE M
	13-1964	SOME COMPOUNDS OF TUNGSTEN BRONZE TYPE A(6)B(10)O(30). (WHERE B IS (NIOBIUM, TITANIUM)
ONOPRIYENKO LG	03-0509	RESONANCE EFFECTS IN MAGNETICALLY EQUIAXIAL FERRODIELECTRIC SINGLE CRYSTALS HAVING DOM
ONSAGER L	34-3461	PROTONIC SEMICONDUCTORS. (ICE)
ONUKI K	35-3518	ELECTROOPTIC PROPERTIES OF FERROELECTRIC LEAD GERMANATE SINGLE CRYSTAL.
OONO K	04-0717	ATTEMPT TO PREPARE HIGH DOPED BARIUM STRONTIUM TITANATE SEMICONDUCTOR WITH PZT.
OPPELT A	14-2080	(OPTICAL) FREQUENCY DOUBLING IN (MAGNESIUM CHLORINE, ZINC CHLORINE, CADMIUM CHLORINE,
ORIKASA K	04-0881	PIEZOELECTRIC CHARACTERISTIC OF BARIUM TITANATE CERAMIC EXHIBITED WHEN IT IS GIVEN AN
ORLYANSKII YUN	22-2863	PREPARATION AND ANALYSIS OF SODIUM AMMONIUM SELENATE DIHYDRATE CRYSTALS.
ORLYUKAS AS	15-2117	DIELECTRIC DISPERSION IN ANTIMONY SULFOIODIDE CRYSTALS.
ORMANCEY G	04-0699	"VIBRATING WIRE" STUDY OF POTENTIAL DISTRIBUTION (STABILITY) IN BARIUM(X) STRONTIUM(1-
	04-0882	INFLUENCE OF WATER VAPOR ON BARIUM TITANATE CERAMICS.
	04-0883	INFLUENCE OF HUMIDITY ON THE POTENTIAL DISTRIBUTION IN BARIUM TITANATE CERAMICS AND CR
O'BRIEN EJ	27-3097	ULTRASONIC RELAXATION NEAR THE CURIE TEMPERATURE OF FERROELECTRIC TRIGLYCINE SULFATE.
O'BRIEN RJ	11-1786	BRILLOUIN SCATTERING IN LITHIUM NIOBATE.
O'BRYAN HM	02-0103	PREPARATION OF TRANSPARENT FERROELECTRIC CERAMICS. (HOT PRESSING)
	09-1518	PROPERTIES OF NEW TRANSPARENT FERROELECTRIC CERAMICS. (LEAD ZIRCONATE- LEAD TITANATE T
	09-1545	ACOUSTIC DETECTION OF FERROELECTRIC PHASE TRANSITIONS IN PLZT CERAMICS.
	09-1570	ENHANCED ORDERING IN PLZT CERAMICS.
	09-1571	FERROELASTIC BEHAVIOR OF LEAD LANTHANUM ZIRCONATE TITANATE CERAMICS WHEN SUBJECTED TO
	09-1573	LANTHANUM DEPENDENCE OF ELASTIC AND PIEZOELECTRIC PROPERTIES OF PLZT CERAMICS WITH A Z
	09-1579	ENHANCED ORDERING OF FERROELECTRIC DOMAINS IN PLZT CERAMICS.
	09-1580	ENHANCED ORDERING OF FERROELECTRIC DOMAINS IN PLZT CERAMICS.
	13-1901	GROWTH OF UNCRACKED BARIUM SODIUM NIOBATE CRYSTALS.
	13-1908	EFFECTS OF CHANGES IN MELT COMPOSITION ON CRYSTAL GROWTH OF BARIUM SODIUM NIOBATE.
O'KANE DF	02-0053	PROPERTIES OF TUNGSTEN BRONZE FERROELECTRICS. (REVIEW, 33 REFS)
	13-1948	(TETRAGONAL) TUNGSTEN BRONZE TYPE CRYSTALS GROWN IN MOLTEN SOLUTIONS OF THE PSEUDOSYST
	13-1949	(EXTENT OF) TUNGSTEN BRONZE (LIQUIDUS AND SOLIDUS) FIELD AND MELT GROWTH OF CRYSTALS I
	13-2003	CRYSTAL GROWTH AND ELECTROOPTIC PROPERTIES OF TUNGSTEN BRONZE CRYSTALS FROM THE POTASS
	13-2016	PHASE EQUILIBRIA IN THE POTASSIUM NIOBATE- STRONTIUM DINIOBATE AND POTASSIUM NIOBATE-
	13-2017	TUNGSTEN BRONZE FIELD IN THE (OXIDE) SYSTEM POTASSIA LITHIA NIOBIA (POTASSIUM LITHIUM
	13-2028	OPTICAL AND FERROELECTRIC PROPERTIES OF (THE TUNGSTEN BRONZE SYSTEM) POTASSIUM SODIUM
O'REILLY DE	16-2320	NITROGEN-14 NUCLEAR QUADRUPOLE RESONANCE IN FERROELECTRIC SODIUM NITRITE.
	17-2439	RUBIDIUM-87 AND ARSENIC-75 QUADRUPOLAR COUPLING IN FERROELECTRIC RUBIDIUM DIHYDROGEN P
	25-2918	DEUTERON MAGNETIC RESONANCE OF THE HIGH TEMPERATURE PHASE OF FERROELECTRIC THIOUREA.
	25-2919	STUDIES OF FERROELECTRIC SOLIDS BY MAGNETIC RESONANCE PART-18: PROTON AND DEUTERON RES
	29-3240	NUCLEAR MAGNETIC RESONANCE OF FERROELECTRIC LITHIUM AMMONIUM TARTRATE.
	29-3241	NUCLEAR MAGNETIC RESONANCE OF FERROELECTRIC LITHIUM AMMONIUM TARTRATE.
	31-3332	BROMINE NUCLEAR QUADRUPOLE RESONANCE IN FERROELECTRIC HYDROGEN BROMIDE.
	31-3335	NUCLEAR QUADRUPOLE RESONANCE OF IODINE-127 IN SOLID HYDROGEN IODIDE.
	31-3336	NUCLEAR QUADRUPOLE FREQUENCIES AND RELAXATION TIMES IN FERROELECTRIC HYDROGEN CHLORIDE
O'SHEA CC	06-1177	ELASTIC CONSTANT ANOMALIES IN STRONTIUM TITANATE.
OSAKA K	19-2777	ON THE PHASE TRANSITION OF AMMONIUM HYDROGEN SULFATE AND THE STRUCTURE OF THE FERROELE
OSAKA T	19-2778	DOMAIN STRUCTURE OF FERROELECTRIC SODIUM AMMONIUM SULFATE DIHYDRATE.
	19-2779	THERMAL AND SOME OTHER STUDIES OF SODIUM AMMONIUM SULFATE DIHYDRATE, SODIUM TRIHYDROGE
OSIKO VV	11-1733	ELECTRICAL CONDUCTIVITY OF LITHIUM NIOBATE SINGLE CRYSTALS.
OSIPOV VN	17-2419	MACROSCOPIC STIMULATED POLARIZATION ECHO IN FERROELECTRICS. (POTASSIUM DIHYDROGEN PHOS
OSKEDKAR R	19-2749	NITROGEN-14 QUADRUPOLE COUPLING IN PARAELECTRIC AMMONIUM SULFATE.
	27-2969	PULSED DOUBLE RESONANCE STUDY OF THE NUCLEAR QUADRUPOLE INTERACTIONS OF NITROGEN-14 IN
OSSWALD R	31-3325	PARAELECTRIC RESONANCE OF LITHIUM ION IN POTASSIUM CHLORIDE.
OSTERINK LM	08-1328	OPTICALLY INDUCED PHYSICAL DAMAGE TO LITHIUM NIOBATE, PROUSTITE, AND LITHIUM IODATE.
OSTROVSKII BI	30-3265	ABNORMAL ABSORPTION OF ELASTIC WAVES NEAR THE PHASE TRANSITION OF GADOLINIUM MOLYBDATE
OSTROWSKY D	08-1410	EFFECT OF INTENSE LIGHT ON THE REFRACTIVE INDEX OF A CRYSTALLINE POTASSIUM NIOBATE.
OTA K	16-2357	ELASTIC CONSTANTS AND ULTRASONIC ABSORPTION OF SODIUM NITRITE SINGLE CRYSTALS.
OTNES K	06-1178	TEMPERATURE DEPENDENCE OF THE SOFT MODE IN STRONTIUM TITANATE ABOVE THE 105 DEG K TRAN
	06-1194	CRITICAL BEHAVIOR OF STRONTIUM TITANATE NEAR THE 105 K PHASE TRANSITION.
	06-1195	CRITICAL NEUTRON SCATTERING FROM STRONTIUM TITANATE.
	17-2608	LOW FREQUENCY HYDROGEN VIBRATIONS IN POTASSIUM DIHYDROGEN PHOSPHATE.
OTTO OW	11-1788	LITHIUM NIOBATE SILICON SURFACE WAVE CONVOLUTER.
OUSEPH PJ	06-1153	MOSSBAUER STUDY OF COLOR CENTERS IN STRONTIUM TITANATE.
OVANDER LN	03-0407	RAMAN SCATTERING IN PIEZOELECTRIC CRYSTALS WITH TETRAGONAL SYMMETRY.
OWENS BB	16-2265	MELTING PROPERTIES OF POTASSIUM NITRATE UNDER HIGH PRESSURE.
OZAWA K	16-2294	HYDROSTATIC PRESSURE EFFECT ON THE DIELECTRIC RELAXATION TIME IN FERROELECTRIC SILVER
OZDEMIR FS	13-2005	FORMATION OF THIN LAYERS OF MIXED TITANATES BY SOLID SOLID REACTIONS.
PACHEROVA O	17-2527	INFLUENCE OF TUNNELING ON DIELECTRIC BEHAVIOR OF POTASSIUM DIHYDROGEN PHOSPHATE.
PADMANABHAN VM	27-3099	NEUTRON DIFFRACTION STUDY OF TRIGLYCINE SULFATE.
PAKHOMOV VI	17-2606	EFFECT OF IMPURITIES ON THE STRUCTURE OF POTASSIUM DIHYDROGEN PHOSPHATE CRYSTALS.
	17-2607	STRUCTURE OF FERROELECTRIC CRYSTALS OF THE POTASSIUM DIHYDROGEN PHOSPHATE GROUP.
	17-2613	DIELECTRIC ANOMALIES IN CRYSTALS OF POTASSIUM DIHYDROGEN PHOSPHATE POTASSIUM DIDEUTERI
PALAU JM	32-3349	DETERMINATION OF FERROELECTRIC PROPERTIES OF LITHIUM HYDRAZINIUM FLUOBERYLLATE.
PALETTO J	04-0778	CHARACTERISTICS OF THIN, MULTILAYERED FERROELECTRIC CERAMICS OBTAINED BY PROJECTION WI
	04-0779	LOW LOSS (FERROELECTRIC CERAMICS) WITH GREAT DIELECTRIC PERMITTIVITY. (BARIUM TITANATE
	09-1581	PHASE DIAGRAM OF LEAD ZIRCONATE(1-X) TITANATE(X) PLUS 0.8 PERCENT TUNGSTEN TRIOXIDE.
	17-2689	DIELECTRIC PERMITTIVITY OF FINELY DIVIDED POTASSIUM AND AMMONIUM DIHYDROGEN PHOSPHATE.
PALEVSKY H	17-2608	LOW FREQUENCY HYDROGEN VIBRATIONS IN POTASSIUM DIHYDROGEN PHOSPHATE.
PALMER CH	04-0653	BINARY LIGHT BEAM DEFLECTION IN SINGLE CRYSTAL BARIUM TITANATE.
	04-0654	BINARY LIGHT BEAM DETECTION IN SINGLE CRYSTAL BARIUM TITANATE.
PANCHENKO TV	04-0792	DYNAMIC FATIGUE OF SINGLE CRYSTALS OF BARIUM TITANATE.

PANCHENKO TV	04-0795	STABILIZATION OF SPONTANEOUS POLARIZATION OF BARIUM TITANATE SINGLE CRYSTALS.
PANCHENKOV GM	04-0884	CURRENT EXCITATION IN GAMMA-IRRADIATED BARIUM TITANATE.
PANDEY SD	16-2358	ELECTRON PARAMAGNETIC RESONANCE OF MANGANESE(2+) IN THE FERROELECTRIC PHASE OF SODIUM
	16-2359	FERROELECTRIC PHASE TRANSITION IN SODIUM NITRITE AS STUDIED BY EPR (PHASE TRANSITION A
PANKRATOV VM	11-1863	LONGITUDINAL ELECTROOPTICAL EFFECT IN OBLIQUE SECTIONS OF LITHIUM NIOBATE.
PANLENER RJ	04-0885	HIGH TEMPERATURE THERMODYNAMIC AND X-RAY DIFFRACTION STUDY OF NONSTOICHIOMETRIC BARIUM
	04-0886	TITANIUM RICH NONSTOICHIOMETRIC BARIUM TITANATE. PART-3: HIGH TEMPERATURE THERMODYNAMI
PANNETIER J	36-3529	NEW FERROELECTRIC (PYROCHLORE TYPE) COMPOUND (AMMONO THIO CADMIUM NIOBATE, WITH 4 ELEC
	36-3602	NEW DESCRIPTION OF THE PYROCHLORE STRUCTURE (CADMIUM(2) NIOBIUM(2) OXIDE(6) SULFIDE).
PANTELL RH	11-1858	EFFECT TUNABLE OPTICAL EMISSION FROM LITHIUM NIOBATE WITHOUT A RESONATOR.
	12-1873	TEMPERATURE DEPENDENCE OF POLARITON DISPERSION IN LITHIUM TANTALATE. (OPTICAL MODULATI
	13-1899	RAMAN SCATTERING BY BARIUM STRONTIUM NIOBATE.
PARASHCHUKOV MP	05-1065	SEIGNETTO MAGNETIC SOLID SOLUTIONS OF THE LEAD FERRATE(0.67) TUNGSTATE(0.33) - LEAD TIT
PARFENOVA NN	08-1378	DIELECTRIC PROPERTIES AND THERMAL EXPANSION OF SOME FERROELECTRICS AND ANTIFERROELECTR
PARISI AJ	37-3662	COLOR IMAGING WITH FERROELECTRIC CERAMICS.
PARISOT G	33-3395	IMPURITY CONCENTRATION DEPENDENCE OF RAMAN SCATTERING BY MAGNONS IN NICKEL DOPED RUBID
PARKER RD	32-3350	DIPOLAR RELAXATION OF LITHIUM-7 BY HINDERED ROTATORS IN LITHIUM HYDRAZINIUM SULFATE.
PARKER RS	22-2849	NMR STUDY OF DEUTERIUM DIFFUSION AND DEUTERIUM AND SODIUM-23 RELAXATION IN SODIUM TRID
	32-3351	ANALYSIS OF HYSTERESIS LOOPS IN LITHIUM HYDRAZINIUM SULFATE.
PARKER TJ	04-0606	INTERPRETATION OF THE DYNAMICS OF THE BARIUM TITANATE TRANSITION IN TERMS OF NUCLEATIO
	04-0608	NUCLEATION MODELS FOR A FIRST ORDER TRANSITION FRONT, AND OBSERVATIONS IN BARIUM TITAN
	04-0887	CUBIC TETRAGONAL (MARTENSITIC TYPE) PHASE TRANSITIONS IN HIGH RESISTIVITY BARIUM TITAN
	04-0888	DIFFUSIONLESS PHASE TRANSITIONS AND THE STABLE DOMAIN STRUCTURE OF (HIGH SENSITIVITY)
	17-2609	DETERMINATION OF THE COMPLEX PERMITTIVITY OF ADP AND KDP IN THE FAR INFRARED BY DISPER
PARSHTENDIKER VL	11-1668	HEAT TREATMENT INDUCED POINT DEFECTS IN LITHIUM METANIOBATE SINGLE CRYSTALS.
PARUZINA IYA	19-2762	FERROELECTRIC PROPERTIES OF SOLID SOLUTIONS IN THE SYSTEM AMMONIUM SULFATE- AMMONIUM F
PASHKOV VA	11-1664	INHOMOGENEITY OF THE OPTICALLY INDUCED INDEX OF REFRACTION IN LITHIUM NIOBATE AND LITH
	11-1865	AUTOFOCUSING OF LASER RADIATION IN ACTIVE MATERIALS AND NONLINEAR CRYSTALS. (KDP, ADP,
	11-1866	DAMAGE TO THE SURFACE OF LITHIUM NIOBATE BY LIGHT.
PASTERNAK NA	17-2630	PREPARATION (MORPHOLOGY) AND MECHANICAL PROPERTIES OF FILAMENTARY CRYSTALS OF ADP.
PASTO AE	08-1411	INFRARED AND RAMAN SPECTRA OF SEVERAL PEROVSKITE ZIRCONATES.
PASTORI-PARRAVICINI G	34-3462	SYMMETRY ANALYSIS AND ELECTRONIC STATES IN CUBIC ICE.
PASTRNAK J	14-2093	OPTICAL ANISOTROPY OF CUBIC NICKEL IODINE BORACITE DUE TO QUADRUPOLE TRANSITIONS.
	14-2094	OPTICAL ANISOTROPY OF THE CUBIC COBALT IODINE BORACITE.
PASYNKOV RE	01-0029	FERROELECTRICS AND ANTIFERROELECTRICS.
	03-0408	EFFECT OF FREE CARRIERS ON THE INFRARED SPECTRUM OF A FERROELECTRIC. (THEORY, ABSTRACT
	03-0409	THERMODYNAMIC THEORY OF FERROELECTRIC SEMICONDUCTORS.
	03-0410	THERMODYNAMIC THEORY OF FERROELECTRIC SEMICONDUCTORS.
	15-2247	NONLINEAR EFFECTS AND THE LOW TEMPERATURE TRANSITION IN ANTIMONY SULFOIODIDE CRYSTALS
PATEL AR	21-2816	OPTICAL AND INTERFEROMETRIC STUDIES ON FERROELECTRIC GUANIDINIUM ALUMINUM SULFATE HEXA
PATIL PV	04-0889	SOLUBILITY LIMIT OF LANTHANUM(0.5) LITHIUM(0.5) TITANATE IN BARIUM TITANATE.
PAUDERT R	04-0890	INFLUENCE OF VIBRATIONS ON THE DIELECTRIC PROPERTIES OF BARIUM TITANATE CERAMICS.
PAUL GL	17-2586	SYMMETRY OF ATOMIC VIBRATIONS IN POTASSIUM DIHYDROGEN PHOSPHATE.
	17-2610	FERROELECTRIC CRITICAL SCATTERING FROM DEUTERATED KDP. (NEUTRON INELASTIC SCATTERING)
	17-2611	FERROELECTRIC TRANSITION IN POTASSIUM DIDEUTERIUM PHOSPHATE.
	33-3366	ANTIFERROELECTRICITY IN (PURE AND) COBALT DOPED POTASSIUM MANGANESE TRIFLUORIDE. (20 P
PAUL R	08-1412	SPACE CHARGE LIMITED CURRENTS IN SEMICONDUCTING POTASSIUM NIOBATE CRYSTALS.
PAVLOV VS	09-1582	INFLUENCE OF THE CONCENTRATION OF POINT DEFECTS ON THE INTERNAL FRICTION IN POLYCRYSTA
	09-1584	INTERNAL FRICTION (AND CURIE TEMPERATURES) IN (LEAD ZIRCONATE TITANATE) FERROELECTRICS
	11-1798	INTERNAL FRICTION OF LITHIUM NIOBATE SINGLE CRYSTALS AT 1-40 HZ.
PAVLUSHKIN NM	13-2006	PROPERTIES OF FERROGLASS CERAMICS BASED ON LEAD BARIUM NIOBATE COMPOUNDS.
PAWLACZYK C	21-2813	DEFECTS AND THE FERROELECTRIC PROPERTIES OF GUANIDINIUM ALUMINUM SULFATE HEXAHYDRATE.
PAWLEY GS	36-3553	LATTICE DYNAMICS OF IONIC CRYSTALS WITH HIGH DIELECTRIC CONSTANTS. (LEAD AND TIN TELL
PAYNE DA	04-0891	EFFECT OF INTERNAL BOUNDARIES ON THE DIELECTRIC PROPERTIES OF BARIUM TITANATE CERAMICS
PEARSON JE	08-1413	MEASUREMENT OF THE RELATIVE NONLINEAR COEFFICIENTS OF POTASSIUM DIHYDROGEN PHOSPHATE,
PECKHAM G	11-1704	GROUP THEORETICAL SELECTION RULES FOR INELASTIC NEUTRON SCATTERING WITH APPLICATIONS T
PEDERSEN T	04-0713	X-RAY AND NEUTRON DIFFRACTION STUDY OF TETRAGONAL BARIUM TITANATE.
	04-0740	X-RAY AND NEUTRON DIFFRACTION STUDY OF TETRAGONAL BARIUM TITANATE.
	04-0892	TEMPERATURE DEPENDENCE OF THE DEBYE- WALLER FACTORS OF BARIUM TITANATE (BRAGG REFLECTI
PEDINOFF ME	11-1789	OPTICAL STUDIES OF REFRACTION AND REFLECTION OF ULTRASONIC SURFACE WAVES BY THIN GOLD
	11-1790	REFRACTION AND REFLECTION OF ULTRASONIC SURFACE WAVES BY THIN GOLD FILM LAYERS ON LITH
	11-1846	FUNDAMENTAL AND HIGHER ORDER MODE SURFACE WAVE PROPAGATION ON LAYERED ANISOTROPIC SUBS
	11-1849	DIFFRACTION SPREADING OF SURFACE WAVES ON LITHIUM NIOBATE.
	11-1850	(FAR FIELD) DIFFRACTION SPREADING OF SURFACE (ACOUSTIC) WAVES ON (PIEZOELECTRIC) LITHI
PEERCY PS	04-0893	PRESSURE DEPENDENCE OF ACOUSTIC MODE- SOFT OPTIC MODE INTERACTIONS IN FERROELECTRIC BA
	22-2864	RAMAN SPECTRUM OF SODIUM TRIHYDROGEN DISELENITE (PARAELECTRIC PHASE, COMPARISON WITH F
PEINADO F	37-3641	ON THE SWITCHING PROPERTIES OF THE FERROELECTRIC SURFACES. (NEW MODEL REPRODUCES EXPER
PEKAR SI	15-2248	EFFECTS OF TEMPERATURE, ELECTRIC FIELD, AND ILLUMINATION ON THE ABSORPTION OF ULTRASOU
PELAH I	17-2465	MOSSBAUER EFFECT OBSERVATION OF A VERY LOW FREQUENCY MODE AT THE TRANSITION TEMPERATUR
	17-2489	HIGH TEMPERATURE PHASE TRANSITION IN RUBIDIUM DIHYDROGEN PHOSPHATE BY SCATTERING OF CO
	17-2490	HYDROGEN DENSITY DISTRIBUTION IN PARAELECTRIC POTASSIUM DIHYDROGEN PHOSPHATE (NEUTRON
	17-2513	HIGH TEMPERATURE PHASE TRANSITIONS AND METASTABILITY IN POTASSIUM DIHYDROGEN PHOSPHATE
	17-2612	DETECTION OF ATOM TUNNELING IN SOLIDS. (KDP)
	17-2706	ANTIFERROELECTRIC TRANSITIONS IN ADP AND AMMONIUM DIHYDROGEN ARSENATE STUDIED BY INFRA
	17-2708	PROTON DYNAMICS IN KDP TYPE FERROELECTRICS STUDIED BY INFRARED ABSORPTION.
PELIKH IN	30-3309	TEMPERATURE DEPENDENCE OF DIELECTRIC CONSTANT IN THE SYSTEM POTASSIUM YTTRIUM MOLYBDAT
PELL RF	27-3100	EFFECT OF EVAPORATED METALLIC ELECTRODES ON THE TRANSITION TEMPERATURE OF TRIGLYCINE S
PELLACH E	25-2915	VIBRATIONAL LINE WIDTHS OF THIOUREA.
PENIN AN	11-1746	PARAMETRIC LUMINESCENCE AND LIGHT SCATTERING BY POLARITONS. (LITHIUM NIOBATE)
	17-2553	MEASUREMENT OF THE REFRACTIVE INDEX OF ADP AND KDP CRYSTALS IN THE INFRARED RANGE BY P
PENNEBAKER WB	06-1154	FAR INFRARED REFLECTIVITY AND TRANSMISSION MEASUREMENTS ON STRONTIUM TITANATE FILMS.
PEPINSKY R	03-0411	SOME STRUCTURAL MECHANISMS IN FERROELECTRICITY.
PERDUIJN DJ	13-2007	CONTINUOUS HOT PRESSING OF POTASSIUM SODIUM NIOBATE.
PEREL J	03-0414	EXTENDED FINE STRUCTURE IN X-RAY ABSORPTION SPECTRA OF CERTAIN PEROVSKITES (STRONTIUM
	06-1179	DETERMINATION OF X-RAY ABSORPTION COEFFICIENTS AND THE THICKNESS OF NONUNIFORM THIN SA
PEREVERZEVA LP	03-0412	DIELECTRIC PERMITTIVITY DISPERSION PECULIARITIES IN FERROELECTRICS WITH ORDER DISORDER
	17-2613	DIELECTRIC ANOMALIES IN CRYSTALS OF POTASSIUM DIHYDROGEN PHOSPHATE POTASSIUM DIDEUTERI
PERISON J	36-3523	GEL GROWTH OF SILVER PERIODATE CRYSTALS.
PERLIN YUE	02-0127	FUNDAMENTAL PROPERTIES OF METAL- FERROELECTRIC- METAL JUNCTION SYSTEMS. (INFLUENCE OF
PERLMAN SS	09-1566	AN ADAPTIVE FERROELECTRIC TRANSFORMER - A SOLID STATE ANALOG MEMORY DEVICE. (USING, FO

PERLMAN SS   09-1583   ADAPTIVE FERROELECTRIC TRANSFORMERS WITH IMPROVED TEMPERATURE CHARACTERISTICS. (PZT-5H
                          37-3663   ANALOG MEMORY DEVICES EMPLOYING PIEZOELECTRIC- FERROELECTRIC INTERACTIONS FOR ADAPTIVE
PERON A   13-1921   EFFECT OF THE SUBSTITUTION OF TANTALUM FOR NIOBIUM ON THE CRYSTALLOGRAPHIC AND DIELECT
PERRIGOT J   03-0177   VOLTAGES AND CURRENTS DURING DEPOLARIZATION OF A FERROELECTRIC MATERIAL. (THEORY)
                          09-1615   METHOD OF MEASURING POLARIZATION VIBRATIONS OF A PREPOLARIZED CERAMIC FERROELECTRIC. (
                          27-2962   SPECIFIC HEAT ANOMALY IN THE NEIGHBORHOOD OF A FERRO- PARAELECTRIC TRANSITION. (TGS)
PERRIN J   27-3008   BLOCKING OF SPONTANEOUS POLARIZATION IN TRIGLYCINE SULFATE.
PERRINO CT   18-2737   PROTONIC CONDUCTION IN POTASSIUM DIHYDROGEN ARSENATE.
PERRO IT   04-0894   SPECIFIC FEATURES OF THE TEMPERATURE DEPENDENCE OF THE VELOCITY OF ULTRASOUND IN POLYC
PERRON A   13-2010 · CRYSTAL CHEMISTRY AND ELECTRICAL PROPERTIES OF TWO NEW PHASES OF QUADRATIC TUNGSTEN BR
PERRY CH   02-0104   STUDY OF OPTICAL PROPERTIES AND COLLECTIVE OSCILLATIONS IN NEW SOLID STATE MATERIALS (
                          03-0413   DIELECTRIC PROPERTIES AND OPTICAL PHONONS IN PARA- AND FERROELECTRIC PEROVSKITES. (THE
                          05-1083   TEMPERATURE DEPENDENT OPTICAL PHONONS IN LEAD TITANATE.
                          15-2106   LONG WAVELENGTH OPTICAL PHONONS AND PHASE TRANSITIONS IN ANTIMONY SULFOIODIDE. (POLARI
                          15-2207   (POLARIZED) RAMAN SPECTRUM OF (PARAELECTRIC AND) FERROELECTRIC ANTIMONY SULFOIODIDE. (
                          15-2208   OPTICAL PHONONS AND PHASE TRANSITIONS IN SOME ORDER DISORDER AND DISPLACIVE FERROELECT
                          17-2409   TEMPERATURE DEPENDENT RAMAN SPECTRA OF POTASSIUM DIHYDROGEN PHOSPHATE, POTASSIUM DIDEU
PERRY GS   08-1414   DIELECTRIC PROPERTIES OF TITANATE- ALUMINA CERAMICS.
PERSON JS   06-1180   ELASTIC PROPERTIES OF YTTRIUM DOPED STRONTIUM TITANATE SINGLE CRYSTALS.
PESHIKOV EV   17-2614   EFFECT OF STRUCTURAL DEFECTS PRODUCED BY GAMMA AND REACTOR IRRADIATION ON PHASE TRANSI
                          17-2615   STRUCTURAL SENSITIVITY OF THE FERROELECTRIC PHASE TRANSITION AND DIELECTRIC PROPERTIES
                          22-2865   RADIATION EFFECTS IN SODIUM TRIHYDROGEN SELENATE.
                          27-3101   RESULTS OF EXPERIMENTAL STUDY OF GAMMA-RADIATION EFFECTS IN ROCHELLE SALT AND TRIGLYCI
                          27-3102   UNUSUAL EFFECT OF ISOTOPIC AND ISOMORPHOUS SUBSTITUTION IN FERROELECTRIC CRYSTALS OF T
PESKIN VF   36-3589   NEW FERROELECTRIC: ANTIMONY ORTHONIOBATE. (CURIE TEMPERATURE)
PETCH HE   19-2746   CRYSTAL STRUCTURE OF FERROELECTRIC RUBIDIUM HYDROGEN SULFATE. (ABSTRACT)
                          19-2761   DEUTERON QUADRUPOLE COUPLING CONSTANT IN DEUTERATED AMMONIUM SULFATE.
                          19-2800   PROTON MAGNETIC RESONANCE STUDY OF THE SPIN SYMMETRY STATES OF AMMONIUM IONS IN SOLIDS
                          23-2887   PROTON DYNAMICS IN FERROELECTRIC COLEMANITE.
                          32-3343   NUCLEAR MAGNETIC RESONANCE STUDY OF FERROELECTRIC LITHIUM HYDRAZINIUM SULFATE.
                          32-3347   PROTON ROTATING FRAME RELAXATION IN LITHIUM HYDRAZINIUM SULFATE.
PETERS CW   27-2958   OPTICAL MIXING. (OF PULSED RUBY LASER EMISSIONS BY TRIGLYCINE SULFATE CRYSTAL)
PETERS RD   06-1180   ELASTIC PROPERTIES OF YTTRIUM DOPED STRONTIUM TITANATE SINGLE CRYSTALS.
                          06-1181   ULTRASONIC THIRD HARMONIC GENERATION IN STRONTIUM TITANATE SINGLE CRYSTALS (NONLINEAR
PETERSEN PE   08-1415   PHOTOEFFECTS IN POTASSIUM TANTALATE.
PETERSON DG   17-2616   PARAMETRIC FREQUENCY CONVERSION OF COHERENT LIGHT BY THE ELECTROOPTIC EFFECT IN POTASS
PETERSON EM   17-2439   RUBIDIUM-87 AND ARSENIC-75 QUADRUPOLAR COUPLING IN FERROELECTRIC RUBIDIUM DIHYDROGEN P
                          25-2918   DEUTERON MAGNETIC RESONANCE OF THE HIGH TEMPERATURE PHASE OF FERROELECTRIC THIOUREA.
                          25-2919   STUDIES OF FERROELECTRIC SOLIDS BY MAGNETIC RESONANCE PART-18: PROTON AND DEUTERON RES
                          29-3240   NUCLEAR MAGNETIC RESONANCE OF FERROELECTRIC LITHIUM AMMONIUM TARTRATE.
                          29-3241   NUCLEAR MAGNETIC RESONANCE OF FERROELECTRIC LITHIUM AMMONIUM TARTRATE.
                          31-3335   NUCLEAR QUADRUPOLE RESONANCE OF IODINE-127 IN SOLID HYDROGEN IODIDE.
PETERSON GE   11-1691   NONSTOICHIOMETRY AND CRYSTAL GROWTH OF LITHIUM NIOBATE.
                          11-1716   CONTROL OF LASER DAMAGE IN LITHIUM NIOBATE.
                          11-1791   CONTROL OF THE SUSCEPTIBILITY OF LITHIUM NIOBATE TO LASER INDUCED REFRACTIVE INDEX CHA
                          11-1793   LOW FREQUENCY (510 MHZ) PARAMAGNETIC RESONANCE INVESTIGATION OF LITHIUM NIOBATE.
                          11-1794   NIOBIUM-93 NMR STUDY OF THE LITHIUM NIOBATE TANTALATE SOLID SOLUTION SYSTEM.
                          11-1795   NIOBIUM-93 NMR LINE WIDTHS IN NONSTOICHIOMETRIC LITHIUM NIOBATE.
                          11-1811   NIOBIUM-93 NUCLEAR QUADRUPOLE RESONANCE INVESTIGATION OF (STOICHIOMETRIC) LITHIUM NIOB
PETERSSON J   01-0023   EUROPEAN MEETING ON FERROELECTRICITY, PROC.
                          03-0415   DIELECTRIC DISPERSION AND THERMODYNAMICS OF IRREVERSIBLE PROCESSES. (APPLIED TO FERROE
                          03-0420   THERMODYNAMIC THEORY OF DISPERSION PHENOMENA. (FERROELECTRICITY)
                          16-2264   NUCLEAR SPIN LATTICE RELAXATION OF SODIUM-23 IN THE PARAELECTRIC AND FERROELECTRIC PHA
                          16-2350   INFLUENCE OF THE ELECTRICAL BOUNDARY CONDITIONS ON THE NMR RELAXATION RATE OF SODIUM-2
                          24-2889   STUDY OF HYDROGEN-1 IN PARAELECTRIC AND FERROELECTRIC POTASSIUM FERROCYANIDE TRIHYDRAT
                          27-3117   LIGHT EMISSION FROM TGS. (CRYSTALS DURING HEATING)
PETROFF JF   11-1671   X-RAY TOPOGRAPHICAL STUDY OF THE CONTRAST AT FERROELECTRIC DOMAIN BOUNDARIES (DOPED AN
                          19-2747   X-RAY TOPOGRAPHIC STUDY OF DISLOCATIONS AND FERROELECTRIC DOMAINS IN GLYCOL SULFATE.
                          27-3025   X-RAY TOPOGRAPHIC STUDY OF GROWTH DEFECTS IN TRIGLYCINE SULFATE CRYSTALS IN RELATION T
                          27-3026   X-RAY TOPOGRAPHIC STUDY OF DOMAIN WALL MOVEMENT IN TGS.
PETROV EG   03-0317   ADMIXTURE OF THE 4T1G STATE TO THE 6A1G(6S) STATE OF THE MANGANESE(2+) ION IN A CUBIC
                          03-0416   EXCITON MODEL OF SPIN AND OPTICAL EXCITATIONS OF AN ANTIFERRODIELECTRIC.
                          03-0418   ELECTRONIC AND SPIN EXCITATIONS IN ANTIFERRODIELECTRIC ELECTRIC AND MAGNETIC MOMENTS O
                          03-0419   ON THE THEORY OF LIGHT ABSORPTION BY ANTIFERRODIELECTRICS IN THE FREQUENCY RANGE OF DO
PETROV MP   11-1792   ELECTRON PARAMAGNETIC RESONANCE SPECTRUM IN (MANGANESE DOPED) LITHIUM NIOBATE AND (COP
PETROV SV   33-3396   ANTIFERROMAGNETIC RESONANCE SPECTRUM IN BARIUM MANGANESE TRIFLUORIDE.
PETROV TG   08-1367   CRYSTALLIZATION OF POTASSIUM IODATE.
PETROV VM   03-0417   EFFECT OF COLLECTIVE INTERACTION OF DIPOLES ON THE NATURE OF DIELECTRIC DISPERSION IN
                          04-0707   RELAXATION AND FERROELECTRIC PROPERTIES OF SOLID SOLUTIONS IN THE SYSTEM BARIUM TITANA
                          04-0895   DEPENDENCE OF DIELECTRIC DISPERSION IN BARIUM TITANATE SINGLE CRYSTALS ON THEIR DOMAIN
                          04-0896   EFFECT OF LINEAR COMPRESSION ON THE UHF DIELECTRIC PROPERTIES OF BARIUM TITANATE SINGL
                          08-1460   STRUCTURE AND PROPERTIES OF SOME NEW FERROELECTRIC MATERIALS WITH PEROVSKITE STRUCTURE
                          27-3103   RELAXATION OF DOMAIN WALLS IN TRIGLYCINE SULFATE. (TEMPERATURE DEPENDENCES OF THE COMP
                          37-3664   POSSIBLE REALIZATION OF ULTRA LOW NOISE LEVEL PARAELECTRIC MICROWAVE AMPLIFIER (BASED
PETROVSKII AN   11-1663   FRACTURE OF NONLINEAR CRYSTALS (KDP AND LITHIUM NIOBATE) BY RADIATION FROM A RUBY LASE
PETZELT J   03-0230   SOFT MODE BEHAVIOR IN IMPROPER FERROELECTRICS. (THEORY, KDP)
                          15-2209   FAR INFRARED REFLECTIVITY OF ANTIMONY SULFOIODIDE.
                          30-3310   NEW TYPE OF FAR INFRARED SOFT MODE IN FERROELECTRIC GADOLINIUM MOLYBDATE.
                          30-3311   NEW TYPE OF FERROELECTRIC SOFT MODE IN GADOLINIUM MOLYBDATE.
PETZSCH K   02-0105   PIEZOELECTRIC CERAMIC SUBSTANCES. (REVIEW, 23 REFS)
PEUZIN JC   03-0421   PIEZOELECTRIC SURFACE WAVES ON A C-DOMAIN WALL OF A FERROELECTRIC (CLASS 6MM).
                          03-0422   THEORY OF POLARIZATION REVERSAL IN FERROELECTRICS. (CALCULATIONS OF SWITCHING CURRENT)
                          36-3542   FERROELECTRIC PROPERTIES OF HEXAGONAL ORTHOMANGANITES OF YTTRIUM AND RARE EARTHS. (YTT
                          36-3588   SURFACE EFFECT IN FERROELECTRIC YTTRIUM MANGANATE.
PFANNSCHMIDT P   27-3118   PIEZOELECTRICITY AND ELECTROSTRICTION OF TRIGLYCINE SULFATE.
PFEIFFER ER   06-1182   EFFECT OF STRESS ON THE SUPERCONDUCTING TRANSITION TEMPERATURE OF STRONTIUM TITANATE.
PFEUTY P   03-0423   CRITICAL BEHAVIOR OF THE PSEUDOSPIN MODEL FOR ORDER DISORDER TYPE FERROELECTRICS WITH
PHATAK SC   08-1416   ELASTIC CONSTANT OF ORTHORHOMBIC POTASSIUM NIOBATE BY X-RAY DIFFUSE SCATTERING.
PHELAN RJ   09-1560   HIGH D*, FAST, LEAD ZIRCONATE TITANATE PYROELECTRIC DETECTORS.
PHILIPPE G   04-0897   STUDY OF CRYSTALS OF BARIUM TITANATE HAVING PARALLEL INTERFERENCE FRINGES.

PHILLIPS RA	16-2272	NONLINEAR OPTICAL PROPERTIES OF FERROELECTRIC SODIUM NITRITE.
	16-2273	OPTICAL ACTIVITY IN NON-ENANTIOMORPHOUS BIAXIAL CRYSTALS: METHYL MESITYL OXIDE OXALATE
	16-2274	TEMPERATURE DEPENDENCE OF NONLINEAR OPTICAL COEFFICIENTS IN FERROELECTRIC SODIUM NITRI
PHILLIPS W	11-1796	OPTICAL AND HOLOGRAPHIC STORAGE PROPERTIES OF TRANSITION METAL DOPED LITHIUM NIOBATE.
	11-1827	MULTIPLE STORAGE OF THICK PHASE HOLOGRAMS IN LITHIUM NIOBATE.
PIA G	04-0622	SOME PROPERTIES OF THE THERMOCURRENTS IN BARIUM TITANATE AND TITANIUM DIOXIDE SINGLE C
	04-0697	(PHASE TRANSITIONS AND) THERMOCURRENTS IN TITANIUM DIOXIDE AND BARIUM TITANATE SINGLE
PICK MA	34-3463	SPECIFIC HEAT OF PURE AND DOPED ICE NEAR 120 DEGREES K.
	34-3464	SPECIFIC HEAT OF ICE I(H).
PIECH T	04-0838	ELECTRET EFFECT IN A MIXTURE OF POLY METHACRYLATE AND BARIUM TITANATE
PIECH TJ	05-1057	THERMOELECTRIC EFFECTS IN THE SYSTEM METAL- LEAD TITANATE- METAL. (METHOD OF STUDYING
PIERRE MC	17-2617	ELECTRONIC EFFECT AND ELASTIC PROPERTIES OF RUBIDIUM DIHYDROGEN PHOSPHATE CRYSTALS.
PIETRASS B	03-0424	SELF CONSISTENT LATTICE DYNAMICAL THEORY OF STRUCTURAL PHASE TRANSITIONS IN PEROVSKITE
	04-0898	INFLUENCE OF PRESSURE ON PHASE TRANSITIONS AT LOW TEMPERATURES: STRONTIUM TITANATE AND
	06-1183	ANALYSIS OF THE ELASTIC ANOMALIES AT THE STRUCTURAL PHASE TRANSITION OF STRONTIUM TITA
	06-1184	FIELD INDUCED SWITCHING OF TETRAGONAL DOMAINS IN STRONTIUM TITANATE AT LOW TEMPERATURE
	17-2495	INFLUENCE OF HYDROSTATIC PRESSURE ON THE PHASE TRANSITION TEMPERATURE OF FERROELECTRIC
PIKKA T	15-2210	REFLECTION SPECTRA OF SOME FERROELECTRIC CRYSTALS OF GROUPS V, VI AND VII. (ANTIMONY A
PILET JC	04-0899	DIELECTRIC ABSORPTION IN THE ORTHORHOMBIC AND QUADRATIC PHASES OF BARIUM TITANATE.
	04-0900	DIELECTRIC ABSORPTION IN THE ORTHORHOMBIC AND QUADRATIC PHASES OF BARIUM TITANATE.
PINES MY	13-2030	CHARACTERISTICS OF KTN (POTASSIUM TANTALATE NIOBATE) PYROELECTRIC DETECTORS.
PINK DA	03-0426	IMPURITY MODES IN AN INDUCED MOMENT SYSTEM AND HYDROGEN BONDED FERROELECTRICS (KDP TYP
	17-2618	RESOLUTION OF A DISCREPANCY CONCERNING THE AVERAGE DIPOLE MOMENT IN KDP TYPE FERROELEC
PINNOW DA	02-0106	ELECTROOPTICAL MATERIALS.
	03-0425	GUIDE LINES FOR THE SELECTION OF ACOUSTOOPTIC MATERIALS.
	08-1464	PIEZOELECTRIC AND PHOTOELASTIC PROPERTIES OF LITHIUM IODATE.
	08-1465	PIEZOELECTRIC AND PHOTOELASTIC PROPERTIES OF LITHIUM IODATE.
PINSKER ZG	19-2797	ELECTRON DIFFRACTION STUDY OF THE STRUCTURE OF AMMONIUM SULFATE.
PINTAR MM	19-2761	DEUTERON QUADRUPOLE COUPLING CONSTANT IN DEUTERATED AMMONIUM SULFATE.
	19-2800	PROTON MAGNETIC RESONANCE STUDY OF THE SPIN SYMMETRY STATES OF AMMONIUM IONS IN SOLIDS
	23-2887	PROTON DYNAMICS IN FERROELECTRIC COLEMANITE.
	27-2967	NUCLEAR SPIN LATTICE RELAXATION IN FERROELECTRIC TRIGLYCINE SULFATE.
	34-3480	STRONG COLLISION LIMIT OF SPIN LATTICE RELAXATION IN HEXAGONAL ICE.
PIRC R	17-2570	PROTON LATTICE INTERACTIONS AND THE DYNAMICAL SUSCEPTIBILITY OF POTASSIUM DIHYDROGEN P
PIRKMAJER E	19-2752	ROTATIONAL MOTIONS IN SOLIDS: THE FERROELECTRIC TRANSITION IN AMMONIUM SULFATE STUDIED
PIROZHKOV VA	11-1797	INVESTIGATION OF CERTAIN PARAMETERS OF LITHIUM NIOBATE SINGLE CRYSTALS USING SCATTERIN
	11-1843	FREQUENCY DEPENDENCES OF SOME PARAMETERS OF LITHIUM NIOBATE.
PIRS J	17-2441	ULTRASLOW HYDROGEN MOTION IN POTASSIUM DIHYDROGEN PHOSPHATE AND SODIUM TRIHYDROGEN SEL
PISAREV RV	11-1743	MANDELSTAM BRILLOUIN SCATTERING OF LIGHT IN LITHIUM NIOBATE (FROM TRANSVERSE (MORE INT
	11-1744	SCATTERING OF LIGHT ON ACOUSTIC PHONONS IN LITHIUM NIOBATE AND LITHIUM TANTALATE.
	12-1878	SCATTERING OF LIGHT BY ACOUSTIC PHONONS AND POLARITONS IN LITHIUM TANTALATE.
	14-2095	CRYSTAL FIELD THEORY AND OPTICAL ABSORPTION (CALCULATIONS) OF FERROELECTRIC BORACITES.
	14-2096	CRYSTAL FIELD THEORY AND OPTICAL ABSORPTION OF COBALT AND NICKEL BORACITES.
	14-2097	OPTICAL ABSORPTION (625-25 MICRONS, 77-650 DEG K) OF FERROELECTRIC IRON(2+) BORACITES
PISKARSKAS A	17-2468	PARAMETRIC LIGHT AMPLIFICATION AND OSCILLATION IN KDP WITH MODE LOCKED PUMP.
PISLEWSKI N	27-3104	INVESTIGATIONS OF NUCLEAR RELAXATION OF TRIGLYCINE FLUOBERYLLATE.
PIVOVAR TL	08-1417	ELECTRICAL CONDUCTIVITY OF STRONTIUM ZIRCONATE AND HAFNATE AT HIGH TEMPERATURES.
PLANT JG	13-1904	BARIUM SODIUM NIOBATE: CONSTITUTIONAL STUDIES AND (CZOCHRALSKI) CRYSTAL GROWTH.
	13-1928	CRUCIBLE BASE COOLING: AN AID TO CZOCHRALSKI CRYSTAL GROWTH. (BARIUM SODIUM NIOBATE AN
	13-2047	PROGRESS TOWARDS THE GROWTH OF GOOD OPTICAL QUALITY BARIUM SODIUM NIOBATE (BY USING CO
PLATONOV GL	04-1038	STRUCTURAL ELECTRON MICRODIFFRACTION STUDIES OF FERROELECTRICS AND FERROELECTRIC MAGNE
PLATONOVA LM	03-0503	PHENOMENOLOGICAL THEORY OF MICROWAVE LOSSES IN FERROELECTRICS. (AT TEMPERATURES ABOVE
	04-1017	EFFECT OF CHARGED LATTICE IMPERFECTIONS ON THE DIELECTRIC PROPERTIES OF MATERIALS. (BA
PLATZ W	36-3565	SPECTRUM AND ENERGY LEVELS OF CHROMIUM(3+) IONS AND EXCHANGE COUPLED CHROMIUM(3+) PAIR
PLESHAKOV IA	27-3105	SECOND HARMONIC GENERATION IN THE TRIGLYCINE CRYSTAL IN THE PARAELECTRIC PHASE.
PLESSER TH	17-2510	EXPERIMENTAL PROOF FOR PROTON TUNNELING IN POTASSIUM DIHYDROGEN PHOSPHATE.
	17-2511	NEUTRON MEASUREMENTS ON THE HYDROGEN BOND POTENTIAL IN PARAELECTRIC POTASSIUM DIHYDROG
PLETS YUM	04-0534	STUDYING SMALL DEFORMATIONS OF A CRYSTAL LATTICE BASED ON THE SHADOW EFFECT. (NEUTRON
PLICQUE F	08-1399	DETERMINATION OF THE ELASTIC CONSTANTS OF POTASSIUM NITRATE.
PLOKARZ H	27-3047	THERMAL EXPANSION OF TRIGLYCINE SULFATE CRYSTALS STUDIED WITH TELLURIUM THIN FILMS.
PLOTNIKOVA MV	08-1401	MOSSBAUER STUDY AND THEORETICAL ESTIMATION OF INTERNAL ELECTRIC FIELD GRADIENTS IN FER
PLOVNICK RH	33-3392	ANOMALOUS ELASTIC BEHAVIOR OF POTASSIUM MANGANESE TRIFLUORIDE NEAR A STRUCTURAL PHASE
	33-3397	PREPARATION AND DISLOCATION DENSITY DETERMINATION OF LARGE, PURE RUBIDIUM MANGANESE TR
PLUMLEE RH	08-1418	HIGH FIELD POLARIZATION REVERSAL IN POLYCRYSTALLINE FERROELECTRICS. (8 SUBSTITUTED LEA
	28-3216	ZERO SHIFT IN PIEZOELECTRIC ACCELEROMETERS. (POLARIZATION SWITCHING IN POLYCRYSTALLINE
PLUZHNIKOV VM	04-0884	CURRENT EXCITATION IN GAMMA-IRRADIATED BARIUM TITANATE.
PLYUSHKO NYA	03-0207	FERROMAGNETIC RESONANCE LINE WIDTH IN FERROELECTRIC FERROMAGNETS.
POCHAEA AM	28-3190	EFFECT OF ELECTRIC FIELD ON DOMAIN STRUCTURE FORMATION IN SEIGNETTES SALT (ROCHELLE SA
PODLESSKAYA AF	28-3229	PIEZOELECTRIC EFFECT AND INTERNAL FRICTION IN GAMMA-IRRADIATED ROCHELLE SALT CRYSTALS.
POGOSSKAYA NZ	17-2613	DIELECTRIC ANOMALIES IN CRYSTALS OF POTASSIUM DIHYDROGEN PHOSPHATE POTASSIUM DIDEUTERI
POHANKA RC	08-1419	THERMODYNAMICS OF FERROELECTRICITY IN BISMUTH TITANATE.
	13-1932	FERROELECTRICITY IN BISMUTH OXIDE TYPE LAYER STRUCTURE COMPOUNDS.
POKATILOV EP	03-0427	EFFECTS OF ELECTRON PHONON INTERACTION IN FERROELECTRIC CRYSTALS (CYCLOTRON RESONANCE,
POLANDOV IN	05-1071	DIELECTRIC PROPERTIES OF LEAD TITANATE- LEAD (FERRATE NIOBATE) SOLID SOLUTIONS AT HIGH
	15-2237	CRITICAL POINT IN FERROELECTRIC ANTIMONY SULFOIODIDE.
	16-2360	EFFECT OF HYDROSTATIC PRESSURE ON THE FERROELECTRIC PHASE TRANSITION OF AMMONIUM SULFA
	24-2896	FERROELECTRIC TRANSITION IN A POTASSIUM FERROCYANIDE TRIHYDRATE CRYSTAL UNDER HIGH HYD
	24-2897	FERROELECTRIC BEHAVIOR OF POTASSIUM FERROCYANIDE TRIHYDRATE CRYSTALS AT HIGH PRESSURE.
POLCAROVA M	27-3067	THE EFFECT OF GROWTH RATE ON THE DEFECT STRUCTURE AND DIELECTRIC PROPERTIES OF TGS SIN
	27-3106	X-RAY TOPOGRAPHIC OBSERVATIONS OF RADIATION DAMAGE IN TRIGLYCINE SULFATE SINGLE CRYSTA
POLINKOVA MV	08-1400	CHANGE OF RESONANCE ABSORPTION SPECTRA OF 23.8 KEV GAMMA-RAYS OF TIN-119 DURING PHASE
POLKOVNIKOV BF	11-1746	PARAMETRIC LUMINESCENCE AND LIGHT SCATTERING BY POLARITONS. (LITHIUM NIOBATE)
	17-2553	MEASUREMENT OF THE REFRACTIVE INDEX OF ADP AND KDP CRYSTALS IN THE INFRARED RANGE BY P
POLLACK MA	02-0107	DETERMINATION OF ABSOLUTE SIGNS OF MICROWAVE NONLINEAR SUSCEPTIBILITIES. (BARIUM TITAN
POLLARD G	16-2280	ATR INFRARED SPECTRA OF UNIAXIAL NITRATE CRYSTALS.
POLOGNYAGIN VA	11-1719	TEMPERATURE DEPENDENCES OF THE ABSORPTION OF 9.4 GHZ HYPERSOUND IN ALUMINUM OXIDE AND
POLOTSKII IG	15-2248	EFFECTS OF TEMPERATURE, ELECTRIC FIELD, AND ILLUMINATION ON THE ABSORPTION OF ULTRASOU
POLYAKOV SM	03-0505	DOMAIN STRUCTURE OF POLYCRYSTALLINE FERROELECTRICS WITH DIFFUSE PHASE TRANSITIONS.
POMEROY WRM	13-2008	TEMPERATURE DEPENDENCE OF THE REFRACTIVE INDICES OF BARIUM SODIUM NIOBATE.
POMPE G	17-2619	MEASUREMENT OF THE SPECIFIC HEAT OF KDP BY A DYNAMIC METHOD.

PAGE 675

PONATH H   11-1787   SPONTANEOUS RAMAN SCATTERING BY POLARITONS. PART-2: THE CASE OF CADMIUM SULFIDE AND LI
PONOMAREV VI   22-2838   CRYSTAL STRUCTURES OF POTASSIUM TRIHYDROGEN SELENITE AND SODIUM TRIHYDROGEN SELENITE.
                         30-3271   PHASE TRANSITICNS OF GADOLINIUM MOLYBDATE AND ISOSTRUCTURAL COMPOUNDS (EUROPIUM TERBIU
POOLE CP   22-2833   ESR STUDY OF GAMMA-IRRACIATED CESIUM TRIHYDROGEN SELENITE AT ROOM TEMPERATURE.
POOLE RR   27-3119   PERFCRMANCE CHARACTERISTICS OF A SMALL TRIGLYCINE SULFATE DETECTOR OPERATED IN THE PYR
POPEKO GS   17-2620   FERROELECTRIC TRANSITION IN RUBIDIUM DIHYDROGEN PHOSPHIDE USING RAMAN SCATTERING SPECT
POPKOV YUA   33-3372   RAMAN SCATTERING AND PHASE TRANSITIONS IN POTASSIUM MANGANESE TRIFLUORIDE CRYSTALS.
                         33-3398   MANDELSTAM BRILLOUIN SCATTERING OF LIGHT IN MANGANESE FLUORIDE CRYSTALS (POTASSIUM TRI
                         33-3399   MANDELSTAM BRILLOUIN SCATTERING OF LIGHT IN MANGANESE FLUORIDE, COBALT FLUORIDE, POTAS
                         33-3400   RAMAN SCATTERING AND PHASE TRANSITIONS IN A POTASSIUM MANGANESE TRIFLUORIDE CRYSTAL.
                         33-3401   TWO MAGNON SCATTERING OF LIGHT IN ANTIFERROMAGNETIC POTASSIUM MANGANESE TRIFLUORIDE.
POPLAVKO YUM   02-0108   STUDY OF MICROWAVE DIELECTRIC DISPERSION IN FERROELECTRICS OF VARIOUS TYPES. (BARIUM T
                         02-0124   MILLIMETER WAVELENGTH DISPERSION IN THE DIELECTRIC CONSTANT ABOVE THE CURIE POINT IN V
                         03-0428   CIELECTRIC DISPERSION IN FERROELECTRICS.
                         03-0429   PROBLEM OF THE ACCURATE MEASUREMENT OF THE DIELECTRIC PROPERTIES OF FERROELECTRICS AT
                         04-0767   ULTRAHIGH FREQUENCY DIELECTRIC ANISOTROPY OF A POLARIZED FERROELECTRIC CERAMIC. (BARIU
                         04-0855   MEASUREMENT OF DIELECTRIC PROPERTIES OF FERROELECTRIC MATERIALS AT CENTIMETER AND DECI
                         04-0901   FERROELECTRIC PROPERTIES OF BARIUM TITANATE IN THE CENTIMETER RANGE OF SUPERHIGH FREQU
                         04-0902   MICROWAVE DISPERSION MECHANISM IN BARIUM TITANATE TYPE FERROELECTRICS.
                         04-0903   FERMITTIVITY DISPERSION OF BARIUM TITANATE TYPE FERROELECTRICS. (SINGLE CRYSTALS AND C
                         04-0904   THERMAL CHARACTERISTICS OF A FERROELECTRIC (BARIUM TITANATE TYPE) CERAMIC WITH CONTROL
                         04-0905   UHF (WAVEGUIDE RESONATOR) METHOD OF STUDYING TEMPERATURE CHARACTERISTICS OF ROCHELLE T
                         06-1250   INFRARED SPECTRA OF SINGLE AND POLYCRYSTALLINE STRONTIUM TITANATE.
                         17-2613   DIELECTRIC ANOMALIES IN CRYSTALS OF POTASSIUM DIHYDROGEN PHOSPHATE POTASSIUM DIDEUTERI
                         17-2627   THERMCELASTIC PROPERTIES OF POTASSIUM DIHYDROGEN PHOSPHATE NEAR THE HIGH TEMPERATURE P
POPOLITOV VI   15-2180   MOSSBAUER STUDIES OF ANTIMONY SULFOIODIDE TYPE CRYSTALS.
                         15-2194   HYDROTHERMAL METHOD FOR PREPARING A(V)B(VI)C(VII) COMPOUNDS. (USING ANTIMONY OR BISMUT
                         15-2212   CRYSTALLIZATION OF BISMUTH SULFOIODIDE. (HYDROTHERMAL SYNTHESIS)
                         15-2213   CRYSTALLIZATION OF SEMICONDUCTORS OF THE COMPOSITION A(V)B(VI)C(VII) UNDER HYDROTHERMA
                         15-2214   HYDROTHERMAL CRYSTALLIZATION OF SEMICONDUCTING COMPOUNDS OF GROUP A(V)B(VI)C(VII). (HI
                         15-2215   KINETICS OF CRYSTALLIZATION OF ANTIMONY SULFOIODIDE (CRYSTAL GROWTH RATE, ANISOTROPY)
                         36-3589   NEW FERROELECTRIC: ANTIMONY ORTHONIOBATE. (CURIE TEMPERATURE)
POPOV ES   28-3228   LCW FREQUENCY DIELECTRIC DISPERSION IN ROCHELLE SALT CRYSTALS.
POPOV MA   33-3396   ANTIFERROMAGNETIC RESONANCE SPECTRUM IN BARIUM MANGANESE TRIFLUORIDE.
POPOV SN   04-0986   NONLINEARITY OF ELECTROMECHANICAL AND MECHANICAL PROPERTIES CF FERROELECTRICS. (BARIUM
                         15-2189   NUCLEAR QUADRUPOLE AND ELECTROACOUSTIC ECHOES IN FERROELECTRIC ANTIMONY SULFOIODIDE.
                         15-2211   ANOMALOUS ECHO IN FERROELECTRIC ANTIMONY SULFOIODIDE. (PIEZOELECTRIC DOMAIN OSCILLATIO
                         15-2216   NUCLEAR CUADRUPOLE RESONANCE IN FERROELECTRICS OF THE ANTIMONY SULFOIODIDE (AND BROMID
POPOVA EA   17-2620   FERROELECTRIC TRANSITION IN RUBIDIUM DIHYDROGEN PHOSPHIDE USING RAMAN SCATTERING SPECT
                         17-2621   ISOTOPE EFFECT IN THE RAMAN SPECTRA OF A DEUTERATED KDP CRYSTAL.
                         17-2622   RAMAN SPECTRA OF THE FERROELECTRIC MODIFICATION OF A POTASSIUM DIHYDROGEN PHOSPHATE CR
                         17-2623   SYMMETRY CF NCRMAL OSCILLATIONS IN A FERROELECTRIC CRYSTAL OF KDP.
                         17-2624   STRUCTURE OF DIHYDROGEN PHOSPHATE ION IN THE FERROELECTRIC POTASSIUM SALT.
                         17-2625   SPECTRUM OF RAMAN SCATTERED LASER LIGHT IN FERROELECTRICS OF THE KDP GROUP.
                         17-2626   TEMPERATURE DEFENDENCE OF RAMAN SCATTERING SPECTRUM OF FERROELECTRIC DEUTERATED KDP DU
POPOVIC S   36-3610   SILVER LITHORIUM PHOSPHATE- A NEW CASE OF THE NON-HYDROGEN BONDED PHOSPHATE FERROELECT
POPOVKIN BA   15-2230   CRYSTAL GROWTH IN THE ANTIMONY SULFOIODIDE- ANTIMONY TRISULFIDE SYSTEM. (BRIDGMAN- STO
POPVUC S   36-3611   TEMPERATURE DEPENDENCE OF SOME PROPERTIES OF SODIUM DITHORIUM PHOSPHATE FERROELECTRIC
PORTER SG   13-1896   FERROELECTRICS IN THE POTASSIUM OXIDE- STRONTIUM OXIDE- NIOBIUM PENTOXIDE.
PORTNOV VN   17-2628   CONDITIONS FOR THE GROWTH OF POTASSIUM DIHYDROGEN PHOSPHATE WHISKERS.
                         28-3217   GROWTH RATE OF THE (010) PLANE OF ROCHELLE SALT IN THE PRESENCE OF PURE DIRECT SKY BLU
PORTO SPS   04-0548   TEMPERATURE DEPENDENCE OF THE RAMAN CROSS SECTIONS IN BARIUM TITANATE AND STRONTIUM TI
                         04-0619   COUPLED POLARITIONS OF A1 SYMMETRY IN BARIUM TITANATE.
                         04-0923   AUGER-LIKE RESONANT INTERFERENCE IN RAMAN SCATTERING FROM ONE AND TWO PHONON STATES OF
                         08-1344   RAMAN SCATTERING FROM FERROELECTRIC SODIUM NITRITE.
                         16-2299   ANALYSIS OF THE TEMPERATURE DEPENDENT PHONON STRUCTURE IN SODIUM NITRITE BY RAMAN SPEC
POSHIN VG   04-0686   STUDY OF THE PYROELECTRIC EFFECT IN THIN FILMS OF TRIGLYCINE SULFATE AND BARIUM TITANA
POSTNIKOV VS   09-1582   INFLUENCE OF THE CONCENTRATION OF POINT DEFECTS ON THE INTERNAL FRICTION IN POLYCRYSTA
                         09-1584   INTERNAL FRICTION (AND CURIE TEMPERATURES) IN (LEAD ZIRCONATE TITANATE) FERROELECTRICS
                         11-1798   INTERNAL FRICTION OF LITHIUM NIOBATE SINGLE CRYSTALS AT 1-40 HZ.
FOTAPOV EV   13-1952   NONLINEAR PROPERTIES AND PHASE TRANSITIONS OF STRONTIUM BISMUTH TITANATES.
POTOPOWICZ JR   05-1079   NONLINEAR OPTICAL PROPERTIES OF FERROELECTRIC LEAD TITANATE.
                         05-1080   NONLINEAR OPTICAL PROPERTIES OF FERROELECTRIC LEAD TITANATE.
                         13-2023   NONLINEAR OPTICAL FROPERTIES OF FERROELECTRIC LEAD NIOBIUM(4) OXIDE(12).
                         26-2945   NONLINEAR OPTICAL SUSCEPTIBILITY OF LITHIUM FORMATE MONOHYDRATE.
POTTHARST J   27-3088   MICROWAVE BEHAVIOR OF ROCHELLE SALT AND TGS (18 REFS).
                         28-3218   FERROELECTRICITY IN ROCHELLE SALT.
                         28-3219   INFLUENCE OF CRYSTAL DEFECTS ON THE DIELECTRIC BEHAVIOR OF ROCHELLE SALT IN MICROWAVE
                         28-3220   TEMPERATURE AND FREQUENCY DEPENDENCE OF THE LOSS FACTOR IN FERROELECTRIC MATERIALS (RO
POULET H   17-2480   VIBRATION SPECTRA OF POTASSIUM DIHYDROGEN PHOSPHATE IN PARA- AND FERROELECTRIC PHASES.
                         26-2942   THE ANOMALOUS BEHAVIOR OF PHONON NEAR THE 60 DEGREES C PHASE TRANSITION IN LEAD DICALC
POULIQUEN J   04-0938   BARIUM TITANATE PIEZOELECTRIC MICROPROBE FOR STUDYING ULTRASONIC FIELDS.
POWELL CG   11-1752   OPTICAL PROBING OF ACOUSTIC SURFACE WAVE HARMONIC GENERATION (IN LITHIUM NIOBATE)
POZDNYAKOVA LA   33-3402   SUPERSTRUCTURE ARISING DURING THE PHASE TRANSITION IN POTASSIUM MANGANESE TRIFLUORIDE.
PRAKASH A   34-3438   DEUTERON ARRANGEMENTS IN THE HIGH PRESSURE FORMS OF ICE.
PRASAD R   04-0798   INFLUENCE OF MICROSTRUCTURES (PREPARATION CONDITIONS) ON DIELECTRIC PROPERTIES OF BARI
PRASAD RAO AD   16-2257   RELATIONSHIP OF THE TEMPERATURE DEPENDENCE OF B1 PHONON AND DIELECTRIC RELAXATION IN S
PRASAD VCS   08-1420   DEFORMATION AND RELAXATION BEHAVIOR IN FERROELECTRIC POTASSIUM NIOBATE SINGLE CRYSTALS
                         08-1421   MECHANICAL SWITCHING OF 60 DEGREE DOMAIN WALLS IN FERROELECTRIC POTASSIUM NIOBATE SING
PRASK HJ   34-3465   ICE I: LATTICE DYNAMICS AND INCOHERENT NEUTRON SCATTERING.
PRATT IH   04-0906   CHARACTERISTICS OF RF SPUTTERED BARIUM TITANATE THIN FILMS.
                         04-0907   FABRICATION OF RF SPUTTERED BARIUM TITANATE THIN FILMS.
PRATT TE   17-2629   PARAMAGNETIC RELAXATION AND DIVALENT COPPER IONS IN FERROELECTRIC KDP.
PREDVODITELEV AA   17-2630   PREPARATION (MORPHOLOGY) AND MECHANICAL PROPERTIES OF FILAMENTARY CRYSTALS OF ADP.
PRELESNIK A   19-2749   NITROGEN-14 QUADRUPOLE COUPLING IN PARAELECTRIC AMMONIUM SULFATE.
                         27-2969   PULSED DOUBLE RESONANCE STUDY OF THE NUCLEAR QUADRUPOLE INTERACTIONS OF NITROGEN-14 IN
PRELESNIK B   22-2866   A NEUTRON DIFFRACTION STUDY OF POTASSIUM TRIHYDROGEN SELENITE.
PRESSLEY RJ   06-1102   EFFECT OF UNIAXIAL STRESS ON RAMAN SCATTERING IN STRONTIUM TITANATE.
                         06-1103   RAMAN SCATTERING AND PHASE TRANSITIONS IN STRESSED STRONTIUM TITANATE.
                         06-1104   STRESS INDUCED FERROELECTRICITY IN STRONTIUM TITANATE.

PREUDHOMME JM	30-3296	RAMAN SPECTRUM AND STRUCTURE OF TERBIUM MOLYBDATE.
PREWITT CT	30-3312	X-RAY DIFFRACTION STUDY OF GADOLINIUM MOLYBDATE (CRYSTALLOGRAPHIC PHASE CHANGE AT CURI
PRIMAK W	11-1799	DECORATION OF ION BOMBARDED LITHIUM NIOBATE SURFACES.
	11-1800	EXPANSION, CRAZING AND EXFOLIATION OF LITHIUM NIOBATE ON ION BOMBARDMENT AND COMPARISO
	11-1801	SECONDARY (ELECTRO) PHOTOELASTIC EFFECT IN LITHIUM NIOBATE (ON ELECTRON IRRADIATION).
PRISOVA NA	11-1844	ELECTROOPTIC EFFECT IN LITHIUM NIOBATE IN THE (SUB) MILLIMETER RANGE.
PRIVOROTSKII IA	03-0430	CONTRIBUTION TO THE THEORY OF DOMAIN STRUCTURES IN MAGNETS AND FERROELECTRICS.
PROCHOROVA SD	14-2095	CRYSTAL FIELD THEORY AND OPTICAL ABSORPTION (CALCULATIONS) OF FERROELECTRIC BORACITES.
	14-2096	CRYSTAL FIELD THEORY AND OPTICAL ABSORPTION OF COBALT AND NICKEL BORACITES.
	14-2097	OPTICAL ABSORPTION (625-25 MICRONS, 77-650 DEG K) OF FERROELECTRIC IRON(2+) BORACITES
PROKHOROV VM	11-1706	DISPERSION OF ELASTIC WAVES IN FERROELECTRICS. (LITHIUM NIOBATE)
PROKHORTSEVA TM	30-3255	CERTAIN GROWTH, LASING AND SPECTROSCOPIC PROPERTIES OF NEODYMIUM(3+) DOPED GADOLINIUM
	30-3321	EFFECT OF HYDROSTATIC PRESSURE ON PHASE TRANSITION IN GADOLINIUM MOLYBDATE.
PROKOPALO OI	04-0568	ELECTRICAL PROPERTIES OF CRYSTALLITES AND INTERCRYSTALLITE LAYERS IN THE SEMICONDUCTOR
	04-0570	THERMOSTABILIZING PROPERTIES OF SEMICONDUCTING BARIUM TITANATE.
	04-0908	MECHANISM OF ELECTROCONDUCTIVITY IN BARIUM TITANATE.
	05-1051	POSITIVE TEMPERATURE COEFFICIENT OF RESISTANCE IN DOPED LEAD TITANATE.
PROZOROVA LA	33-3396	ANTIFERROMAGNETIC RESONANCE SPECTRUM IN BARIUM MANGANESE TRIFLUORIDE.
PSHEDMOISKII YT	04-0909	EFFECT OF SOME ADMIXED NICKEL ON SOME PHYSICAL PROPERTIES OF BARIUM TITANATE SINGLE CR
PUCHKOVSKAYA GA	08-1278	POTASSIUM IODATE CRYSTAL STRUCTURE STUDIED BY INFRARED SPECTROSCOPY AND NUCLEAR QUADRU
PUGA GD	15-2217	ABSORPTION SPECTRA OF BISMUTH TELLUROBROMIDE AND BISMUTH TELLUROIODIDE CRYSTALS.
PUGA PP	15-2142	PRODUCTION AND SOME OPTICAL PROPERTIES OF ANTIMONY SULFOBROMIDE IN GLASSY AND CRYSTALL
PUGOVKIN AV	11-1802	NONLINEAR INTERACTION OF HYPERSONIC WAVES IN A LITHIUM NIOBATE CRYSTAL.
PULVARI CF	08-1422	COEXISTENCE OF THE ANTIPOLARIZED AND POLARIZED STATE IN BISMUTH TITANATE.
	13-2009	FERROELECTRIC DIELECTRIC ANOMALY IN BISMUTH(4) TITANIUM(3) OXIDE(12) SINGLE CRYSTALS.
	37-3665	FERROELECTRICS AND THEIR APPLICATION IN SOLID STATE DEVICES AS ADAPTIVE CONTROL.
PUPKEVICH PA	27-3157	APPLICABILITY OF THE PIPPARD RELATIONS TO THE PHASE TRANSITION IN TRIGLYCINE SULFATE.
	27-3158	NONLINEAR PROPERTIES OF TRIGLYCINE SELENATE. (DIELECTRIC CONSTANT VERSUS ALTERNATING F
	27-3167	EFFECT OF CONSTANT ELECTRICAL FIELD ON PHASE TRANSITIONS IN TRIGLYCINE FLUOBERYLLATE.
	27-3169	THERMAL EXPANSION OF TRIGLYCINE FLUOBERYLLATE CRYSTALS IN THE FERROELECTRIC TRANSITION
PURDES AJ	03-0352	DIFFUSION POTENTIALS IN BARIUM TITANATE AND THE THEORY OF PTC MATERIALS.
PUSHKINA NI	03-0431	STIMULATED COMBINATION SCATTERING OF SOUND IN PIEZOELECTRICS. (THEORY, LITHIUM NIOBATE
PUTHOFF HE	11-1736	POWER AND LINE WIDTH OF TUNABLE STIMULATED FAR INFRARED EMISSION IN LITHIUM NIOBATE.
	11-1858	EFFECT TUNABLE OPTICAL EMISSION FROM LITHIUM NIOBATE WITHOUT A RESONATOR.
PUTLEY EH	17-2703	(REVERSIBLE) OPTICAL DAMAGE IN KDP (WITH RELAXATION TIME OF SECONDS WHEN ILLUMINATED U
	17-2517	NEW METHOD OF MEASURING PYROELECTRIC COEFFICIENTS. (TGS EXAMPLE)
	27-3107	PYROELECTRIC THERMAL IMAGING DEVICES. (TGS)
PYLYPYSHIN BP	17-2697	ELECTROOPTICAL AND PIEZOOPTICAL PROPERTIES OF POTASSIUM DIHYDROGEN PHOSPHATE AND AMMON
PYTTE E	03-0241	THEORY OF A STRUCTURAL PHASE TRANSITION IN PEROVSKITE- TYPE CRYSTALS, PART-2, INTERACT
	03-0432	THEORY OF PEROVSKITE FERROELECTRICS (HAMILTONIAN IN TERMS OF LOCALIZED D-STRAIN AND SO
	08-1450	STRUCTURAL PHASE TRANSITIONS IN PEROVSKITE TYPE CRYSTALS. (EXTENDED ABSTRACT)
	30-3313	MODEL FOR FERROELECTRIC GADOLINIUM MOLYBDATE.
QUADFLIEG P	10-1650	GROWTH, CRYSTALLOGRAPHY AND DIELECTRIC PROPERTIES OF BISMUTH TUNGSTATE.
QUATE CF	11-1837	(BACKWARD WAVE) ACOUSTIC PARAMETRIC OSCILLATIONS IN LITHIUM NIOBATE (60 DB GAIN).
	11-1838	NONLINEAR INTERACTION OF MICROWAVE ELECTRIC FIELDS AND SOUND IN LITHIUM NIOBATE.
QUERROU M	17-2631	TRANSIENT PHENOMENA IN THE DISCHARGE OF CONDENSERS WITH POLAR DIELECTRICS. (POTASSIUM
QUILICHINI M	26-2942	THE ANOMALOUS BEHAVIOR OF PHONON NEAR THE 60 DEGREES C PHASE TRANSITION IN LEAD DICALC
QUITTET AM	04-0537	DIELECTRIC PROPERTIES AND DOMAIN TEXTURE OF BARIUM TITANATE CRYSTALS.
	04-0676	STRUCTURE DISORDER AND RAMAN SCATTERING IN BARIUM TITANATE AND POTASSIUM NIOBATE.
	04-0910	DIELECTRIC MEASUREMENTS ON PILE-IRRADIATED BARIUM TITANATE SINGLE CRYSTALS.
RABENAU A	15-2218	CRYSTAL GROWTH AND CHEMICAL SYNTHESIS UNDER HYDROTHERMAL CONDITIONS. (BISMUTH SULFOBRO
RABIDEAU SW	34-3466	STRUCTURAL STUDIES OF ICE POLYMORPHS BY NEUTRON DIFFRACTION, PROTON AND DEUTERON NUCLE
RABINOVICH AZ	02-0064	FERROELECTRIC RARE EARTH MOLYBDATES. (REVIEW, FORMATION, CRYSTAL GROWTH, PHASE TRANSIT
	12-1884	PRIMARY AND SECONDARY PYROELECTRIC EFFECTS AND SPONTANEOUS POLARIZATION IN LITHIUM TAN
	30-3314	PYROELECTRIC EFFECT AND SPONTANEOUS POLARIZATION OF GADOLINIUM MOLYBDATE (TEMPERATURE
	30-3315	PYROELECTRIC EFFECT AND DOMAIN STRUCTURE OF GADOLINIUM MOLYBDATE.
RABSON TA	37-3638	OPTICALLY ERASABLE AND REWRITABLE SOLID STATE HOLOGRAMS.
RAEVSKAYA EB	03-0505	DOMAIN STRUCTURE OF POLYCRYSTALLINE FERROELECTRICS WITH DIFFUSE PHASE TRANSITIONS.
RAI R	20-2806	POSSIBLE EXPLANATION OF THE UNUSUAL TEMPERATURE VARIATION OF THE MEAN MAGNETIC MOMENT
RAJAGOPAL AK	06-1185	LATTICE DYNAMICS OF CUBIC PEROVSKITE STRUCTURES, IN PARTICULAR STRONTIUM TITANATE.
RAMACHANDRA RS	09-1585	ULTRASONIC INTERFEROMETER WITH LEAD ZIRCONATE TITANATE AS TRANSDUCER.
RAMANUJAM PS	26-2936	RAMAN AND INFRARED SPECTRA OF COPPER FORMATE TETRAHYDRATE.
	26-2937	RAMAN AND INFRARED SPECTRA OF INORGANIC FORMATES.
RAMASWAMY V	17-2632	AN APPARATUS FOR POLISHING WATER SOLUBLE CRYSTALS (APPLICATION TO ADP AND KDP).
	17-2633	EPITAXIAL ELECTROOPTIC MIXED CRYSTAL AMMONIUM POTASSIUM DIHYDROGEN PHOSPHATE FILM WAVE
RAMDAS AK	35-3519	RAMAN SPECTRUM OF BISMUTH(12) GERMANIUM OXIDE(20).
RAO CNR	02-0109	ELECTRICAL CONDUCTION IN METAL OXIDES. (REVIEW, 319 REFS)
	36-3608	ELECTRICAL TRANSPORT IN (SEMICONDUCTING AND MAGNETIC) RARE EARTH ORTHOCHROMITES, MANGA
RAO KV	08-1272	DIELECTRIC PROPERTIES AND OPTICAL ABSORPTION OF POTASSIUM TANTALATE SINGLE CRYSTALS AN
RAPOPORT E	16-2361	PHASE DIAGRAMS OF SODIUM NITRITE AND POTASSIUM NITRITE TO 40 KBAR.
	17-2634	PHASE TRANSFORMATIONS AND MELTING IN KDP TO 40 KBAR (BY DTA).
RAPOPORT SL	04-0911	AGING PROCESSES IN BARIUM TITANATE (DECREASE IN NUMBER OF 180 DEG DOMAINS).
	28-3228	LOW FREQUENCY DIELECTRIC DISPERSION IN ROCHELLE SALT CRYSTALS.
RASHIDI K	17-2635	PULSE RESPONSE OF ZERO DEGREE CUT ADP MODULATORS.
RASHKOVICH LN	11-1708	DOMAIN STRUCTURE OF LITHIUM METANIOBATE CRYSTALS.
	17-2583	CRYSTALLIZATION OF RUBIDIUM DIDEUTERIUM PHOSPHATE FROM SOLUTIONS IN HEAVY WATER.
	17-2698	ELECTROOPTICAL AND OPTICAL PROPERTIES OF PARTIALLY DEUTERATED RUBIDIUM DIHYDROGEN PHOS
RASTORGUEV LN	13-2046	EFFECTS OF ROASTING CONDITIONS ON THE STRUCTURE AND PROPERTIES OF LEAD BARIUM NIOBATE
RATAJCZAK H	18-2738	STRUCTURAL STUDIES OF HYDROGEN BONDED FERROELECTRICS USING POLARIZED IR RADIATION. PAR
	19-2768	INFRARED SPECTRA OF RUBIDIUM HYDROGEN SULFATE AND RUBIDIUM DEUTERIUM SULFATE IN THE PA
RATNER MA	03-0244	THEORY OF THE MOBILITY OF STRUCTURAL DEFECTS IN ICE.
RAU H	15-2218	CRYSTAL GROWTH AND CHEMICAL SYNTHESIS UNDER HYDROTHERMAL CONDITIONS. (BISMUTH SULFOBRO
RAVEZ J	04-0912	COMPARATIVE STUDY OF THE CRYSTALLOGRAPHIC, DIELECTRIC, AND NONLINEAR OPTICAL PROPERTIE
	13-1921	EFFECT OF THE SUBSTITUTION OF TANTALUM FOR NIOBIUM ON THE CRYSTALLOGRAPHIC AND DIELECT
	13-2010	CRYSTAL CHEMISTRY AND ELECTRICAL PROPERTIES OF TWO NEW PHASES OF QUADRATIC TUNGSTEN BR
	13-2011	FIRST FERROELECTRIC OXYFLUORIDE PHASES OF THE OXYGENATED QUADRATIC TUNGSTEN BRONZE TYP
REBIC M	19-2795	DEUTERON SPIN LATTICE RELAXATION IN FERROELECTRIC DEUTERATED AMMONIUM HYDROGEN SULFATE
RECHESTER AB	03-0433	THEORY OF SECOND ORDER PHASE TRANSITIONS AT LOW TEMPERATURES. (TIN TELLURIDE, POTASSIU
RECHOWICZ M	20-2807	THERMAL CONDUCTIVITY OF METHYL AMMONIUM ALUMINUM ALUM.
RECHZIEGEL E	27-3005	TEMPERATURE AUTOSTABILIZATION EFFECT OF TRIGLYCINE SULFATE SINGLE CRYSTALS IN AN AC EL

REDDIG H	02-0110	ADVANCES IN PRODUCTION OF (COMMERCIAL) PIEZOELECTRIC CERAMICS AND (TABLES OF) THEIR PR
REDFIELD D	06-1186	FUNDAMENTAL ABSORPTION EDGE OF STRONTIUM TITANATE.
	06-1187	OPTICAL ABSORPTION EDGE OF STRONTIUM TITANATE.
	06-1188	REINTERPRETATION OF WAVELENGTH MODULATED ABSORPTION IN STRONTIUM TITANATE WITHOUT COEX
REDIN RD	03-0434	SYMMETRY LIMITATIONS TO POLARIZATION OF POLYCRYSTALLINE FERROELECTRICS.
REES LVC	04-0827	INVESTIGATION OF THE FERROELECTRIC PHASE CHANGE IN BARIUM TITANATE USING ELECTRON SPIN
REESE RL	03-0435	CALORIMETRIC INVESTIGATIONS OF SECOND ORDER FERROELECTRIC TRANSITIONS.
	17-2497	LIGHT SCATTERING STUDIES OF THE SOFT OPTIC AND ACOUSTIC MODES OF POTASSIUM DIHYDROGEN
	17-2636	SIMULTANEOUS OBSERVATION OF THE SOFT FERROELECTRIC AND ACOUSTIC MODES OF POTASSIUM DID
REESE W	17-2429	ELECTRONIC STUDIES OF POTASSIUM DIHYDROGEN PHOSPHATE. (ELECTROCALORIC STUDIES, SPONTAN
	18-2730	THERMODYNAMIC PROPERTIES OF POTASSIUM DIHYDROGEN ARSENATE AND POTASSIUM DIDEUTERIUM AR
REHME H	04-0913	IMAGING ELECTRIC MICRO FIELDS (OF SOLID SURFACES) IN THE EMISSION ELECTRON MICROSCOPE
REHWALD W	06-1189	ANOMALOUS ULTRASONIC ATTENUATION AT THE 105K TRANSITION IN STRONTIUM TITANATE.
	06-1190	LOW TEMPERATURE ELASTIC MODULI OF STRONTIUM TITANATE (10 TO 130 DEG K).
	06-1191	ULTRASONIC PROPERTIES OF STRONTIUM TITANATE AT THE 105 DEGREE K TRANSITION.
REIBER LM	04-0914	PREPARATION AND PROPERTIES OF THIN FILM MATERIALS WITH PEROVSKITE TYPE STRUCTURE. (MIX
REIBER M	04-0915	REMARKS ABOUT SWITCHING IN BARIUM TITANATE AND TRIGLYCINE SULFATE SINGLE CRYSTALS.
REMEIKA JP	05-1072	(FLUX) GROWTH AND FERROELECTRIC PROPERTIES OF HIGH RESISTIVITY SINGLE CRYSTALS OF LEAD
	05-1079	NONLINEAR OPTICAL PROPERTIES OF FERROELECTRIC LEAD TITANATE.
	05-1080	NONLINEAR OPTICAL PROPERTIES OF FERROELECTRIC LEAD TITANATE.
REMOISSENET M	17-2617	ELECTRONIC EFFECT AND ELASTIC PROPERTIES OF RUBIDIUM DIHYDROGEN PHOSPHATE CRYSTALS.
	17-2637	CONTRIBUTION TO THE PREPARATION ON NON-WEDGE SHAPED RUBIDIUM DIHYDROGEN OR DIDEUTERIUM
	17-2638	FAR INFRARED SPECTRA OF RUBIDIUM DIHYDROGEN PHOSPHATE CRYSTALS.
	19-2753	PREPARATION OF AMMONIUM SULFATE SINGLE CRYSTALS.
RENARD JP	26-2929	ZERO POINT SPIN DEVIATION AND SPONTANEOUS SUBLATTICE MAGNETIZATION IN THE TWO DIMENSIO
RENKER KB	34-3467	LATTICE DYNAMICS OF ICE.
	34-3468	LATTICE DYNAMICS OF ICE I(H).
REPELIN Y	08-1303	INFRARED AND RAMAN SPECTROSCOPY OF POTASSIUM METANIOBATE.
RES IS	17-2578	X-RAY DIFFRACTION TOPOGRAPHY STUDY OF DEFECTS (DISLOCATIONS) IN KDP AND ADP SINGLE CRY
RESCA L	34-3462	SYMMETRY ANALYSIS AND ELECTRONIC STATES IN CUBIC ICE.
RESHCHIKOVA LM	08-1274	DISPLACEMENT PHASE TRANSITIONS IN PEROVSKITES.
	17-2584	EFFECT OF DEUTERATION ON THE ELASTIC PROPERTIES OF RUBIDIUM DIHYDROGEN PHOSPHATE (RDP)
REWAJ T	28-3221	EFFECT OF THE CUPRIC ION ON DELAYED PHENOMENA IN ROCHELLE SALT.
	28-3222	ON THE EFFECTS OF X-RAY IRRADIATION ON TRANSIENT PHENOMENA IN ROCHELLE SALT.
REX IS	17-2627	THERMOELASTIC PROPERTIES OF POTASSIUM DIHYDROGEN PHOSPHATE NEAR THE HIGH TEMPERATURE P
REXFORD DG	11-1803	ELECTRON SPIN RESONANCE STUDIES OF CRYSTAL FIELD PARAMETERS IN MANGANESE(2+) DOPED LIT
	11-1805	FORBIDDEN HYPERFINE LINES IN ESR SPECTRA OF MANGANESE(2+) DOPED LITHIUM NIOBATE.
REXFORD OB	11-1804	ELECTRON SPIN RESONANCE STUDIES OF CHROMIUM(3+) IN LITHIUM NIOBATE.
REYNOLDS CE	04-0916	TWO WAVE SHOCK STRUCTURES IN THE FERROELECTRIC CERAMICS BARIUM TITANATE AND LEAD ZIRCO
REZ IS	02-0111	SOME ASPECTS OF THE CURRENT STUDY AND USE OF SEVERAL TECHNICALLY IMPORTANT PROPERTIES
	03-0379	NONLINEAR OPTICAL PROPERTIES OF CRYSTALS.
	08-1278	POTASSIUM IODATE CRYSTAL STRUCTURE STUDIED BY INFRARED SPECTROSCOPY AND NUCLEAR QUADRU
	08-1423	ELECTROOPTICAL PROPERTIES OF LITHIUM IODATE SINGLE CRYSTALS.
	11-1663	FRACTURE OF NONLINEAR CRYSTALS (KDP AND LITHIUM NIOBATE) BY RADIATION FROM A RUBY LASE
	17-2613	DIELECTRIC ANOMALIES IN CRYSTALS OF POTASSIUM DIHYDROGEN PHOSPHATE POTASSIUM DIDEUTERI
	17-2648	STUDY OF THE (MECHANICAL) STRENGTH PROPERTIES OF POTASSIUM DIHYDROGEN PHOSPHATE CRYSTA
	17-2649	TEMPERATURE DEPENDENCE OF THE MECHANICAL CHARACTERISTICS OF POTASSIUM DIHYDROGEN PHOSP
RHODES E	16-2278	STUDIES OF PHASE TRANSFORMATIONS IN NITRATES AND NITRITES. PART-1: CHANGE IN ULTRAVIOL
RIBARIC M	03-0181	PROTON LATTICE INTERACTIONS IN HYDROGEN BONDED FERROELECTRIC CRYSTALS. (THEORY, KDP TY
RICE RW	04-1026	INFLUENCE OF ADDITIVES ON DENSIFICATION AND PROPERTIES OF SINTERED AN PRESSURE SINTERE
RICHARD M	03-0237	ELASTIC ENERGY AND THE FERROELECTRIC PARAELECTRIC TRANSITION.
	04-0778	CHARACTERISTICS OF THIN, MULTILAYERED FERROELECTRIC CERAMICS OBTAINED BY PROJECTION WI
	04-0917	EFFECT OF GRAIN SIZE ON THE FERROELECTRIC PARAELECTRIC TRANSITION OF BARIUM TITANATE.
	27-2962	SPECIFIC HEAT ANOMALY IN THE NEIGHBORHOOD OF A FERRO- PARAELECTRIC TRANSITION. (TGS)
RICHARDS JL	02-0112	FLASH EVAPORATION. (REVIEW OF TECHNIQUE AND APPLICATION TO THIN FILMS OF PEROVSKITES A
RICHARDS PL	11-1857	GENERATION OF FAR INFRARED RADIATION BY PICOSECOND LIGHT PULSES IN LITHIUM NIOBATE.
RICHMOND P	03-0436	ELECTRICAL SUSCEPTIBILITY OF ADP (THEORY).
RICHTER K	17-2540	DEPOLARIZATION OF LINEARLY POLARIZED LIGHT DUE TO HEATING OF PARALLEL PLANES OF POTASS
RIDPATH DL	04-0918	ELECTRICAL CONDUCTIVITY (SEEBECK COEFFICIENT, CURIE POINT) OF REDUCED BARIUM TITANATE
	04-0919	SOME SEMICONDUCTING PROPERTIES OF BARIUM TITANATE. (RESISTIVITY, SEEBECK COEFFICIENT,
RIEDBERG AL	08-1275	CRYSTALLIZATION OF LEAD METANIOBATE FROM GLASS.
RIEDE V	15-2219	OPTICAL PROPERTIES OF ANTIMONY SELENOIODIDE IN THE INFRARED.
RIEHL N	01-0026	PHYSICS OF ICE. (PROC OF INT SYMP, 9-14 SEP 1968 MUNICH)
	34-3414	PROTONIC CONDUCTION OF ICE. PART-1: HIGH TEMPERATURE REGION. PART-2: LOW TEMPERATURE R
RIGAMONTI A	03-0187	NUCLEAR QUADRUPOLE SPIN LATTICE RELAXATION AND DYNAMICAL PROPERTIES OF FERROELECTRIC F
	08-1283	NMR STUDY OF PHASE TRANSITIONS IN PEROVSKITES CRYSTALS. (NIOBIUM-93 AND SODIUM-23 RESO
	13-1910	TRANSIENT NMR STUDY OF PHASE TRANSITIONS IN METALLIC SODIUM TUNGSTEN BRONZES.
	16-2268	NUCLEAR QUADRUPOLE SPIN LATTICE RELAXATION AND CRITICAL DYNAMICS OF FERROELECTRIC CRYS
	16-2362	QUADRUPOLE SPIN PHONON RELAXATION AND FERROELECTRIC TRANSITION. (IN SODIUM NITRITE)
RIGERMAN LG	08-1438	PREPARATION AND INVESTIGATION OF FERROELECTRICS OF ILMENITE STRUCTURE. (ABO(3) NIOBATE
RIMAI L	04-0920	ELECTRON PARAMAGNETIC RESONANCE OF TRIVALENT GADOLINIUM IONS IN STRONTIUM AND BARIUM T
	06-1192	ELECTRON PARAMAGNETIC RESONANCE OF SOME RARE EARTH IMPURITIES IN STRONTIUM TITANATE.
	06-1193	EFFECT OF TEMPERATURE AND PRESSURE ON THE ELECTRON PARAMAGNETIC RESONANCE SPECTRA OF S
RISTE T	06-1178	TEMPERATURE DEPENDENCE OF THE SOFT MODE IN STRONTIUM TITANATE ABOVE THE 105 DEG K TRAN
	06-1194	CRITICAL BEHAVIOR OF STRONTIUM TITANATE NEAR THE 105 K PHASE TRANSITION.
	06-1195	CRITICAL NEUTRON SCATTERING FROM STRONTIUM TITANATE.
RISTIC VM	11-1806	INTERACTION OF VOLUME ACOUSTIC MODES IN LITHIUM NIOBATE.
	11-1864	VISUALIZATION OF ACOUSTIC RADIATION IN LITHIUM NIOBATE.
RITTER G	19-2751	RELAXATION PHENOMENA AND MOSSBAUER SPECTRA OF MAGNETIZED PARAMAGNETIC SUBSTANCES. PART
RIVERA JM	03-0262	FERROELECTRIC FREE ENERGY EXPANSION COEFFICIENTS FROM DOUBLE HYSTERESIS LOOPS.
	04-0921	THERMAL HYSTERESIS IN THE TETRAGONAL CUBIC TRANSITION OF BARIUM TITANATE.
	27-3098	FERROELECTRIC BEHAVIOR OF RADIATION DAMAGED TRIGLYCINE SULFATE AND ROCHELLE SALT UP TO
RIVIERE R	04-0917	EFFECT OF GRAIN SIZE ON THE FERROELECTRIC PARAELECTRIC TRANSITION OF BARIUM TITANATE.
ROACH WR	06-1091	PHOTOCONDUCTIVITY AND TRAPPING PHENOMENA IN STRONTIUM TITANATE.
ROBBRECHT GG	08-1356	PREPARATION AND PHYSICAL PROPERTIES OF PEROVSKITE TYPE COMPOUNDS, LANTHANUM(1-X) CALCI
ROBERTSON BK	08-1287	PRECISION DETERMINATION OF THE LATTICE PARAMETERS AND THE COEFFICIENTS OF THERMAL EXPA
ROBERTSON DS	35-3494	ELECTROOPTIC MEASUREMENTS ON LITHIUM GERMANATE AND LEAD GERMANATE. (LITHIUM GERMANATE
ROCARIES JC	36-3603	NEW FERROELECTRIC: SEMICARBAZIDE HYDROCHLORIDE.
ROCCHICCIOLI-DELTCHEFF C	02-0113	INFRARED ABSORPTION SPECTROPHOTOMETRIC STUDY OF MONOVALENT METAL NIOBATES.
RODICHEVA EN	08-1424	ABOUT THE CURIE POINT AND NATURE OF DIELECTRIC PROPERTIES IN PEROVSKITE MODIFICATION O

RODICHEVA EN	19-2788	POLARIZATION REVERSAL CHARACTERISTICS OF AMMONIUM HYDROGEN SULFATE.
RODIN AI	04-0552	PHOTOFERROELECTRIC EFFECTS IN A(V) B(VI) C(VII) AND BARIUM TITANATE TYPE FERROELECTRICS.
	15-2145	INFLUENCE OF NONEQUILIBRIUM CARRIERS ON THE PHASE TRANSITION IN FERROELECTRIC SEMICOND
	15-2147	TEMPERATURE DEPENDENCE OF THE CROSS SECTION OF CAPTURE CENTERS IN FERROELECTRIC SEMICO
	15-2158	INFLUENCE OF NONEQUILIBRIUM CARRIER SCREENING ON POLARIZATION PROCESSES IN FERROELECTR
	15-2159	INDUCED IMPURITY PHOTOCONDUCTIVITY IN THE FERROELECTRIC SEMICONDUCTOR ANTIMONY SULFOIO
	15-2160	METASTABLE OPTICAL CHARGE EXCHANGE AND "FROZEN" SHIFT OF THE CURIE TEMPERATURE OF A FE
RODIONOVA NA	17-2658	SOLUBILITY ISOBAR IN THE AMMONIUM DIHYDROGEN PHOSPHATE- POTASSIUM DIHYDROGEN PHOSPHATE
ROETSCHI H	08-1322	DIELECTRIC PROPERTIES OF NIOBATES AND TUNGSTATES.
	15-2205	NEW FERROELECTRIC A(V) B(VII) C(VII) COMPOUNDS OF ANTIMONY SULFOIODIDE TYPE. (PARTIAL OR
	37-3666	NEW TYPE CF LOCP TRACER FOR FERROELECTRICS.
ROGACH ED	15-2155	DILATOMETRIC INVESTIGATION OF ANTIMONY SULFOIODIDE SINGLE CRYSTALS.
	15-2156	DILATCMETRIC INVESTIGATION OF PHOTOFERROELECTRIC EFFECTS IN SINGLE CRYSTALS OF ANTIMON
ROGACH TV	09-1507	TERNARY SYSTEM LEAD TITANATE- LEAD ZIRCONATE- LEAD NIOBATE(0.67) ZINCATE(0.33).
ROGERS JD	08-1276	NUCLEAR QUADRUPOLE INTERACTIONS IN PEROVSKITE TYPE COMPOUNDS OF HAFNIUM-181 STUDIED BY
ROGINSKAYA YUE	08-1425	COEXISTENCE OF ANTIFERROMAGNETIC AND SPECIAL DIELECTRIC PROPERTIES IN THE BISMUTH- LAN
	36-3604	STRUCTURE AND MAGNETIC PROPERTIES OF FERROELECTRIC FERROMAGNETIC SOLID SOLUTIONS IN TH
ROHRER GA	16-2269	PROGRESS TOWARD A FAST, NONVOLATILE, NONDESTRUCTIVE READOUT MEMORY ELEMENT UTILIZING P
ROITBERG MB	12-1884	PRIMARY AND SECONDARY PYROELECTRIC EFFECTS AND SPONTANEOUS POLARIZATION IN LITHIUM TAN
	30-3315	PYROELECTRIC EFFECT AND DOMAIN STRUCTURE OF GADOLINIUM MOLYBDATE.
ROKNI M	06-1196	SOFT MODE STUDY OF THE STRESS INDUCED PHASE TRANSITIONS NEAR THE TRANSITION TEMPERATUR
	06-1235	CUBIC TO TRIGONAL STRESS INDUCED PHASE TRANSITION IN STRONTIUM TITANATE.
	06-1236	RAMAN STUDIES OF STRESS INDUCED PHASE TRANSITION NEAR CURIE TEMPERATURE IN STRONTIUM T
	11-1807	TEMPERATURE STUDY OF THE POLARITON ASSOCIATED WITH THE 248 CM(-1) SOFT MODE IN LITHIUM
	12-1873	TEMPERATURE DEPENDENCE OF POLARITON DISPERSION IN LITHIUM TANTALATE. (OPTICAL MODULATI
ROLOV BN	01-0011	PHASE TRANSITIONS IN FERROELECTRICS.
	03-0437	AC POLARIZATION AND DIELECTRIC HYSTERESIS NEAR THE TRANSITION POINT OF DIFFUSE FERROEL
	03-0438	ACCOUNTING FOR INTERACTION IN THE KANZIG REGIONS MODEL FOR DIFFUSE FERROELECTRIC PHASE
	03-0439	A MODEL OF KANZIG REGIONS TAKING INTO ACCOUNT DIFFERENT ORIENTATIONS OF THE SPONTANEOU
	03-0440	DIELECTRIC CONSTANT OF FERROELECTRIC SOLID SOLUTIONS WITH ALLOWANCE FOR CONCENTRATION
	03-0441	EFFECT CF AN EXTERNAL ELECTRIC FIELD ON A DIFFUSE FERROELECTRIC PHASE TRANSITION, TALK
	03-0442	EFFECT OF HYDROSTATIC PRESSURE ON THE NATURE OF DIFFUSE FERROELECTRIC PHASE TRANSITION
	03-0443	HEAT CAPACITY IN THE FERROELECTRIC TRANSITION REGION (THEORY, RELATION TO KAENZIG DOMA
	03-0444	SOME POSSIBILITIES OF DETERMINING THE VOLUME OF KANZIG REGIONS IN FERROELECTRICS.
	03-0445	THERMODYNAMICS AND STATISTICAL MECHANICS OF DIFFUSE FERROELECTRIC PHASE TRANSITIONS.
	03-0446	ESTIMATION OF THE DIMENSIONS OF KANZIG REGIONS ON THE BASIS OF THE DYNAMIC THEORY OF C
	03-0522	DIELECTRIC CONSTANT BEHAVIOR IN THE NEIGHBORHOOD OF THE PHASE TRANSITION IN CHARACTERI
	03-0523	THERMODYNAMICS OF FERROELECTRIC SOLID SOLUTIONS. (THEORY)
	03-0524	THERMODYNAMICS OF FERROELECTRIC SOLID SOLUTIONS. PART-1. PART-2.
ROMANOV VP	04-0573	CHANGE OF THE MOSSBAUER EFFECT PROBABILITY ON TIN-119 IN BARIUM TITANATE STANNATE SOLI
ROMANOVSKII TB	03-0443	HEAT CAPACITY IN THE FERROELECTRIC TRANSITION REGION (THEORY, RELATION TO KAENZIG DOMA
	03-0445	THERMODYNAMICS AND STATISTICAL MECHANICS OF DIFFUSE FERROELECTRIC PHASE TRANSITIONS.
	03-0446	ESTIMATION OF THE DIMENSIONS OF KANZIG REGIONS ON THE BASIS OF THE DYNAMIC THEORY OF C
ROMANOVSKIS T	03-0437	AC POLARIZATION AND DIELECTRIC HYSTERESIS NEAR THE TRANSITICN POINT OF DIFFUSE FERROEL
ROMANYUK NA	27-3108	IONIZATION OF THE (X-RADIATION AND COPPER AND IRON IMPURITY) ABSORPTION BANDS OF DEFEC
	27-3159	NATURE OF RADIATION INDUCED DEFECTS IN CRYSTALS OF ROCHELLE SALT AND TRIGLYCINE SULFAT
	27-3160	POSITION OF THE FUNDAMENTAL ABSORPTION EDGE AS A FUNCTION OF TEMPERATURE FOR ROCHELLE
ROOS J	08-1426	NUCLEAR QUADRUPOLE RESONANCE OF IODINE-127 IN SILVER PERIODATE.
ROPER JG	13-2012	SODIUM ORDERING IN SODIUM TUNGSTEN BRONZES.
ROSASCO GJ	11-1786	BRILLOUIN SCATTERING IN LITHIUM NIOBATE.
ROSELMAN I	34-3460	HARDNESS ANISOTROPY OF SINGLE CRYSTALS OF ICE I(H).
ROSENBERG LA	03-0501	MICROSCOPIC THEORY OF SURFACE PHENOMENA IN FERROELECTRIC CRYSTALS. (CALCULATION OF FIE
ROSENGREEN A	13-2021	DEVELOPMENT OF SINGLE CRYSTAL BARIUM SODIUM NIOBATE AS A NONLINEAR OPTICAL MATERIAL. (
ROSENSTEIN G	05-1059	X-RAY STRUCTURE INVESTIGATION OF LANTHANUM MODIFIED LEAD TITANATE WITH A-SITE AND B-SI
ROSENTHAL A	13-2018	PYROELECTRIC VOLTAGE RESPONSE TO SHORT INFRARED LASER PULSES IN TRIGLYCINE SULFATE AND
ROSHCHINA ZV	22-2867	HEAT OF FORMATION OF LITHIUM TRIHYDROGEN DISELENITE.
	22-2868	THERMAL STABILITY OF LITHIUM HYDROGEN SELENITES
ROSS SD	11-1808	VIBRATIONAL SPECTRA OF LITHIUM NIOBATE, BARIUM SODIUM NIOBATE AND BARIUM SODIUM TANTAL
ROSSNER R	15-2201	CRYSTALLIZATION BEHAVIOR OF ANTIMONY SULFOIODIDE. (SINGLE CRYSTAL GROWTH, CHEMICAL TRA
ROST A	06-1197	INERTIAL ELECTRICAL POLARIZATION OF CRYSTALS (ALKALI HALIDES, STRONTIUM TITANATE).
ROSTUNTSEVA AI	22-2863	PREPARATION AND ANALYSIS OF SODIUM AMMONIUM SELENATE DIHYDRATE CRYSTALS.
ROTHWELL WS	04-0922	ETCH PITS IN BARIUM TITANATE SINGLE CRYSTALS.
ROUNDY CB	13-1919	PYROELECTRIC COEFFICIENT DIRECT MEASUREMENT TECHNIQUE AND APPLICATION TO A NANOSECOND
ROUSE KD	13-1957	SOFT MODES AND STRUCTURE OF FERROELECTRIC TETRAGONAL POTASSIUM TANTALATE NIOBATE.
	13-2068	POTASSIUM TANTALATE NIOBATE, STRUCTURAL AND DYNAMICAL STUDIES.
ROUSSEAU DL	04-0923	AUGER-LIKE RESONANT INTERFERENCE IN RAMAN SCATTERING FROM ONE AND TWO PHONON STATES OF
ROUSSEAU M	17-2524	STUDY OF THE LIMITATION OF THE ANGULAR FIELD BY THE COMMUTATOR OF POLARIZATION IN KDP,
ROUVAEN JM	11-1680	COLLINEAR INTERACTION BETWEEN A LONGITUDINAL ACOUSTIC WAVE AND TWO LUMINOUS WAVES ALON
	11-1681	DETERMINATION OF A NONLINEAR PARAMETER FOR ACOUSTIC SURFACE WAVE CONVOLUTION IN LITHIU
ROY R	08-1403	CRYSTAL CHEMISTRY CF THE ABX(3) FAMILY WITH PARTICULAR EMPHASIS ON PEROVSKITES.
ROYAL G	19-2780	OPTICAL ABSORPTION, ELECTRIC CONDUCTIVITY AND PHOTOEMISSION OF GLYCOL SULFATE.
ROYER O	08-1410	EFFECT OF INTENSE LIGHT ON THE REFRACTIVE INDEX OF A CRYSTALLINE POTASSIUM NIOBATE.
ROZENBAUM LB	03-0447	NATURE OF DISSIPATIVE EFFECTS WHICH ACCOMPANY DISACCOMODATION PHASE PHENOMENA IN FERRIT
ROZENSHTEIN LD	36-3539	FIRST ORDER NEGATIVE PHOTODIELECTRIC EFFECT IN PHTHALOCYANINE IN THE FERROELECTRIC STA
	36-3540	PYROELECTRIC EFFECT IN METAL-FREE PHTHALOCYANINE. (IN THE POLAR, OR FERROELECTRIC STAT
	36-3621	DEBYE TYPE ABSORPTION IN POLAR PHTHALOCYANINE.
	36-3622	ELECTRICAL PROPERTIES OF PHTHALOCYANINE AS A SEMICONDUCTOR IN THE FERROELECTRIC STATE.
	36-3623	LOW TEMPERATURE PHASE TRANSITION IN THE POLAR STATE OF THE ORGANIC SEMICONDUCTOR PHTHA
ROZHDESTVENSKAYA MV	06-1198	ELECTRICAL AND OPTICAL PROPERTIES (CONDUCTIVITY AND MOBILITY VERSUS TEMPERATURE, ABSOR
ROZORENOVA LA	03-0505	DOMAIN STRUCTURE OF POLYCRYSTALLINE FERROELECTRICS WITH DIFFUSE PHASE TRANSITIONS.
RUBIN B	17-2629	PARAMAGNETIC RELAXATION AND DIVALENT COPPER IONS IN FERROELECTRIC KDP.
RUBIN JJ	13-2045	GROWTH OF BARIUM SODIUM NIOBATE SINGLE CRYSTALS FOR OPTICAL APPLICATIONS. (BY PULLING
RUBINS RS	06-1144	STRONG AXIAL ELECTRON PARAMAGNETIC RESONANCE SPECTRUM OF IRON(3+) IN STRONTIUM TITANAT
RUBISH ID	15-2186	EFFECT OF PHOTOELECTRET CHARGE ON VALUE OF FERROELECTRIC POLARIZATION IN ANTIMONY SULF
	15-2187	EFFECT OF OXYGEN ADSORPTION ON THE FERROELECTRIC PHASE TRANSITION IN ANTIMONY SULFOIOD
RUCHNOVA SA	17-2658	SOLUBILITY ISOBAR IN THE AMMONIUM DIHYDROGEN PHOSPHATE- POTASSIUM DIHYDROGEN PHOSPHATE
RUDIGER U	19-2798	FERROELECTRIC TRANSITION OF AMMONIUM SULFATE.
RUDYAK VM	03-0448	MECHANISMS FOR BARKHAUSEN JUMPS AND BEHAVIOR OF THE BARKHAUSEN EFFECT IN FERROELECTRIC
	04-0924	EFFECT OF ULTRASOUND ON THE POLARIZATION AND REPOLARIZATION PROCESSES IN TRIGLYCINE SU
	15-2220	EFFECT OF ILLUMINATION ON REPOLARIZATION PROCESS IN ANTIMONY SULFOIODIDE SINGLE CRYSTA

RUDYAK VM	27-3109	BARKHAUSEN JUMPS AND SWITCHING CURRENT IN TGS CRYSTALS.
	27-3110	DISTINCTIVE FEATURES OF BARKHAUSEN EFFECT IN ROCHELLE SALT AND TRIGLYCINE SULFATE CRYS
	27-3111	INVESTIGATION OF BARKHAUSEN EFFECT IN TRIGLYCINE SULFATE CRYSTALS.
	27-3123	INFLUENCE OF GAMMA-IRRADIATION ON BARKHAUSEN EFFECT IN FERROELECTRIC MATERIALS (TGS AN
RUEPP R	34-3469	DIELECTRIC RELAXATION, BULK AND SURFACE CONDUCTIVITY OF ICE SINGLE CRYSTALS.
RUKIN EI	06-1105	ANOMALOUS BEHAVIOR OF THE DIELECTRIC NONLINEARITY COEFFICIENT IN STRONTIUM TITANATE NE
RUNCK AH	34-3483	DIELECTRIC AND MECHANICAL RESPONSE OF ICE I(H) SINGLE CRYSTALS AND ITS INTERPRETATION.
RUNK RB	09-1553	ISOTHERMAL GRAIN GROWTH AND ELECTRICAL BEHAVIOR OF FULLY DENSE PLZT CERAMICS.
RUNNELS IK	34-3470	DIFFUSION AND RELAXATION PHENOMENA IN ICE.
RUPPEL W	03-0519	PHOTOEFFECTS AT THE PHOTOCONDUCTOR- FERROELECTRIC BOUNDARY.
	27-3018	TEMPERATURE DEPENDENCE OF THE FERROELECTRIC FIELD EFFECT IN TELLURIUM FILMS ON TRIGLYC
RUPPRECHT G	04-0925	PHOTON PHONON INTERACTION IN THE NEAR INFRARED. (BARIUM TITANATE)
	06-1199	MICROWAVE LOSSES IN STRONTIUM TITANATE ABOVE THE PHASE TRANSITION.
RUSSELL R	05-1052	TIME EFFECTS IN THE HYSTERESIS LOOP OF (LEAD, STRONTIUM) TITANATE.
RUSTAMOV SR	11-1665	PHASE MATCHING ANGLES AND TEMPERATURE OF LITHIUM METANIOBATE CRYSTALS OF DIFFERENT STO
RUTT TC	04-0926	DIFFUSION COEFFICIENT OF NIOBIUM IN BARIUM TITANATE BY A DIELECTRIC MEASUREMENT TECHNI
RUVALDS J	04-0806	ACOUSTIC PHONON SOFT OPTIC PHONON INTERACTION IN BARIUM TITANATE.
RYABINKIN LN	17-2410	ELASTIC PROPERTIES OF AMMONIUM DIHYDROGEN PHOSPHATE AND THE LAVAL RAMAN ELASTICITY THE
RYABOV VA	04-0533	(FAST ION SCATTERING) SHADOW EFFECT IN (DISPLACIVE) PHASE TRANSITIONS. (BARIUM TITANAT
RYAN JF	17-2639	RAMAN SCATTERING FROM FERROELECTRIC MODES IN THE KDP ISOMORPHOUS PHOSPHATES AND ARSENA
	18-2723	DIELECTRIC RESPONSE IN PIEZOELECTRIC CRYSTALS. (RAMAN SCATTERING IN POTASSIUM DIHYDROG
	18-2733	PROTON PHONON COUPLING IN CESIUM DIHYDROGEN ARSENATE AND POTASSIUM DIHYDROGEN ARSENATE
	35-3497	QUASI ELASTIC SCATTERING IN LEAD GERMANATE.
	35-3498	RAMAN SCATTERING FROM THE FERROELECTRIC MODE IN LEAD(5) GERMANIUM(3) OXIDE(11).
RYAZANOV GV	03-0449	EFFECT OF VIOLATION OF ELECTRIC NEUTRALITY ON THE PHASE TRANSITION IN A FERROELECTRIC
	03-0450	SECOND ORDER PHASE TRANSITION IN THE THREE DIMENSIONAL MODEL OF A FERROELECTRIC.
	03-0451	TYPE-II PHASE CHANGE IN THREE DIMENSIONAL FERROELECTRIC MODEL. (THEORY)
RYCHGORSKII VV	04-0601	DISPERSION AT MILLIMETER WAVELENGTHS IN BARIUM TITANATE ABOVE THE TRANSITION POINT.
RYKHLYUK AV	28-3217	GROWTH RATE OF THE (010) PLANE OF ROCHELLE SALT IN THE PRESENCE OF PURE DIRECT SKY BLU
RYSAVA N	04-0875	OPTICAL AND DIELECTRIC PROPERTIES OF SINGLE CRYSTAL BARIUM TITANATE WITH NIOBIUM ADDIT
RYZHAKOV AG	15-2199	PREPARATION AND PHYSICAL PROPERTIES OF NONSTOICHIOMETRIC ANTIMONY SULFOIODIDE SINGLE C
RYZHKOVA TM	17-2628	CONDITIONS FOR THE GROWTH OF POTASSIUM DIHYDROGEN PHOSPHATE WHISKERS.
SABUROVA RV	03-0452	ON THE QUANTUM THEORY OF THE FERROELECTRIC STATE IN ELECTRIC DIPOLE SYSTEMS (SPIN WAVE
SACCENTI JC	09-1567	MEASUREMENT OF THE PYROELECTRIC COEFFICIENT OF 65-35 PZT. (LEAD ZIRCONATE TITANATE)
SADOC A	16-2363	CALCULATION OF THE VARIATION OF THE RAMAN FREQUENCIES IN SODIUM NITRITE AT ORDINARY TE
SAEKI H	13-2004	EFFECTS OF GRAIN SIZE AND POROSITY ON THE ELECTRICAL AND OPTICAL PROPERTIES OF TRANSPA
SAFIN IA	03-0310	MACROSCOPIC ANALOG OF THE SPIN- ECHO EFFECT IN POLYCRYSTALLINE FERROELECTRICS. (THEORY
SAFONOV AI	30-3314	PYROELECTRIC EFFECT AND SPONTANEOUS POLARIZATION OF GADOLINIUM MOLYBDATE (TEMPERATURE
SAFRANKOVA M	03-0246	(FACTORS INFLUENCING) TYPE AND ORIENTATION OF THE EQUILIBRIUM DOMAIN STRUCTURES IN FER
	27-3112	EQUILIBRIUM DOMAIN STRUCTURE OF TRIGLYCINE SULFATE IN THE VICINITY OF THE CURIE POINT.
SAIFI MA	06-1200	DIELECTRIC PROPERTIES OF STRONTIUM TITANATE AT LOW TEMPERATURES.
	06-1201	DIELECTRIC PROPERTIES OF STRONTIUM TITANATE (ANNEALED SINGLE CRYSTALS) AT LOW TEMPERAT
SAIFULLIN LI	03-0153	AN APPARATUS FOR INVESTIGATING THE ELECTRICAL PROPERTIES OF FERROELECTRICS.
SAILER E	03-0495	INFLUENCE OF POINT DEFECTS ON DIELECTRIC PROPERTIES, DOMAIN DYNAMICS, AND PHASE TRANSI
	36-3605	DIELECTRIC PROPERTIES OF LITHIUM IODATE. (FOUND NOT FERROELECTRIC)
SAITO S	17-2640	MICROWAVE MODULATOR OF RUBY LASER LIGHT USING POTASSIUM DIHYDROGEN PHOSPHATE CRYSTAL.
SAITO T	13-1964	SOME COMPOUNDS OF TUNGSTEN BRONZE TYPE A(6)B(10)O(30). (WHERE B IS (NIOBIUM, TITANIUM)
SAKAGUCHI M	11-1809	DEFLECTED BEAM DIAMETER AND DEFLECTION CAPACITIES OF A DIGITAL LIGHT DEFLECTOR UTILIZI
SAKALAS AP	15-2193	TEMPERATURE DEPENDENCES OF THE HALL AND DRIFT MOBILITIES OF HOLES IN ANTIMONY TRISULFI
SAKAMOTO A	17-2550	TEMPERATURE DEPENDENCE OF RAMAN SPECTRUM AND SECOND HARMONIC GENERATION IN KDP.
SAKAMOTO S	13-2063	LONGITUDINAL ELECTROOPTIC EFFECT IN OBLIQUE CUT STRONTIUM BARIUM NIOBATE PLATES.
SAKATA K	04-0880	SPACE CHARGE POLARIZATION AND AGING OF BARIUM TITANATE CERAMICS.
	13-2013	DIELECTRIC AND CERAMIC PROPERTIES OF (BISMUTH(0.5)(SODIUM OR POTASSIUM)(0.5))(X)LEAD(
	26-2947	ANOMALY OF THE PROTON SPIN LATTICE RELAXATION TIME NEAR THE CRITICAL TEMPERATURE OF DI
SAKHNENKO VP	03-0269	THEORY OF PHASE TRANSITIONS IN BORACITES.
	08-1377	RELATIONSHIP BETWEEN THE ATOMIC DISPLACEMENTS AND SPONTANEOUS DEFORMATION IN THE FERRO
SAKOWSKI-COWLEY AC	02-0094	ROLE OF THE REGULAR OCTAHEDRON AND LARGER POLYHEDRA IN FERROELECTRIC MATERIALS. (HETER
SAKU T	35-3518	ELECTROOPTIC PROPERTIES OF FERROELECTRIC LEAD GERMANATE SINGLE CRYSTAL.
SAKUDO T	04-0927	ELECTRON PARAMAGNETIC RESONANCE OF IRON(3+) IN BARIUM TITANATE AT LOW TEMPERATURES.
	04-0928	ELECTRON PARAMAGNETIC RESONANCE IN BARIUM TITANATE NEAR THE CURIE TEMPERATURE.
	04-0929	ELECTRON PARAMAGNETIC RESONANCE OF IRON(3+) IN BARIUM TITANATE IN THE RHOMBOHEDRAL AND
	04-1016	PHASE TRANSITIONS IN THE MIXED CRYSTAL SYSTEM STRONTIUM(1-X) BARIUM(X) TITANATE.
	06-1130	INTERFEROMETRIC DETERMINATION OF THE QUADRATIC ELECTROOPTIC COEFFICIENTS IN STRONTIUM
	06-1131	SECOND HARMONIC GENERATION IN STRESS INDUCED FERROELECTRIC STRONTIUM TITANATE.
	06-1202	DIELECTRIC PROPERTIES OF STRONTIUM TITANATE AT LOW TEMPERATURES. (TEMPERATURE DEPENDEN
	06-1203	ELECTRIC FIELD INDUCED OPTICAL HARMONIC GENERATION IN STRONTIUM TITANATE AND POTASSIUM
	06-1229	ELECTRIC FIELD EFFECT OF ELECTRON SPIN RESONANCE IN STRONTIUM TITANATE. (DOPED WITH GA
	06-1230	NEW ESR EVIDENCE FOR THE 65 DEGREES K ANOMALY IN STRONTIUM TITANATE.
	08-1409	CRYSTAL GROWTH OF SODIUM NIOBATE.
	30-3316	PHASE TRANSITION IN GADOLINIUM MOLYBDATE.
SAKURAI J	16-2284	CRYSTAL DYNAMICS OF SODIUM NITRITE. (NEUTRON SCATTERING)
SAKURAI M	16-2364	CRYSTAL DYNAMICS AND THE FERROELECTRIC PHASE TRANSITION OF SODIUM NITRITE (STUDIED BY
SAKURAI S	09-1612	ELASTIC SURFACE WAVE ON THE PZT CERAMIC PLATE.
SAKURAI T	36-3557	PIEZOELECTRICITY IN POLARIZED POLYVINYLIDENE FLUORIDE FILMS.
SALAMA CAT	04-0930	CHARACTERISTICS OF RF SPUTTERED BARIUM TITANATE FILMS ON SILICON.
SALANECK WR	08-1432	SOME FATIGUING EFFECTS IN 8/65/35 PLZT FINE GRAINED FERROELECTRIC CERAMIC.
	09-1586	TEMPERATURE DEPENDENT FERROELECTRIC DOMAIN ALIGNMENT IN LEAD ZIRCONATE- LEAD TITANATE-
SALIM AJ	27-3062	FREQUENCY DEPENDENCE OF THE COERCIVE FIELD OF TRIGLYCINE SULFATE CRYSTALS.
SALJE E	08-1427	ELECTRICAL AND OPTICAL CONSTANTS OF FERROELECTRIC POTASSIUM IODATE.
	08-1430	POLAR AND LATTICE DYNAMIC CHARACTERISTICS OF FERROELECTRIC POTASSIUM IODATE CRYSTALS.
	08-1431	PIEZOELECTRIC AND ELECTROOPTIC MODULI OF FERROELECTRIC POTASSIUM IODATE.
SALLEE CF	11-1720	BISMUTH(12) GERMANIUM OXIDE(20) VERSUS LITHIUM NIOBATE IN THE SURFACE WAVE AMPLIFIER.
SALNIKOV AN	33-3387	SPECIFIC HEAT OF POTASSIUM MANGANESE TRIFLUORIDE.
SALNIKOV VD	02-0128	NEW FERROELECTRICS AND SEIGNETTO MAGNETS OF THE PEROVSKITE AND PYROCHLORE TYPE. SYNTHE
	08-1428	GROWTH OF SINGLE CRYSTALS OF VARIOUS PEROVSKITES WITH COMPLEX COMPOSITIONS, AND THE ST
	08-1429	GROWTH AND STUDY OF NEW SINGLE CRYSTAL CADMIUM COMPOUNDS WITH PEROVSKITE STRUCTURE.
	08-1458	CRYSTAL STRUCTURE AND PROPERTIES OF NEW CADMIUM AND THALLIUM CONTAINING PEROVSKITES. (
	13-2014	DIELECTRIC AND STRUCTURAL PROPERTIES OF FERROELECTRICS WITH DIFFUSE PHASE TRANSITIONS.
SALVETTI G	34-3417	THERMODIELECTRIC EFFECT AND FREEZING POTENTIAL IN GROWING ICE.
SAMARA GA	02-0114	THE GRUNEISEN PARAMETER OF THE SOFT FERROELECTRIC MODE IN THE CUBIC PEROVSKITES. (STRO

SAMARA GA
03-0453 RELATIONSHIP BETWEEN IONIC DISPLACEMENT AND THE CURIE TEMPERATURE AND SPONTANEOUS POLA
04-0893 PRESSURE DEPENDENCE OF ACOUSTIC MODE- SOFT OPTIC MODE INTERACTIONS IN FERROELECTRIC BA
06-1096 PRESSURE DEPENDENCE OF THE ELASTIC CONSTANTS OF STRONTIUM TITANATE.
06-1097 PRESSURE DEPENDENCE OF THE ELASTIC CONSTANT OF STRONTIUM TITANATE.
07-1263 EFFECTS OF HYDROSTATIC PRESSURE ON FERROELECTRIC PROPERTIES. (SUMMARY WITH NEW RESULTS
08-1433 PRESSURE AND TEMPERATURE DEPENDENCE OF THE DIELECTRIC PROPERTIES AND PHASE TRANSITIONS
08-1434 PRESSURE AND TEMPERATURE DEPENDENCE OF THE DIELECTRIC PROPERTIES AND PHASE TRANSITIONS
08-1435 PRESSURE STUDIES OF FERROELECTRIC PROPERTIES. (PEROVSKITES, KDP TYPE)
17-2588 PRESSURE EFFECTS ON THE LATTICE PARAMETERS AND STRUCTURE OF POTASSIUM DIHYDROGEN PHOSP
17-2641 PRESSURE DEPENDENCE OF THE FERROELECTRIC PROPERTIES OF POTASSIUM DIDEUTERIUM PHOSPHATE
17-2642 VANISHING OF THE FERROELECTRIC AND ANTIFERROELECTRIC STATES IN POTASSIUM DIHYDROGEN PH
SAMARTSEV RG
11-1797 INVESTIGATION OF CERTAIN PARAMETERS OF LITHIUM NIOBATE SINGLE CRYSTALS USING SCATTERIN
SAMCHENKO YUI
04-0672 LIGHT EMISSION FROM BARIUM TITANATE CRYSTALS SUBJECTED TO UNIAXIAL STRESS PULSES.
SAMOILOV VB
02-0127 FUNDAMENTAL PROPERTIES OF METAL- FERROELECTRIC- METAL JUNCTION SYSTEMS. (INFLUENCE OF
SAMUELSEN EJ
06-1194 CRITICAL BEHAVIOR OF STRONTIUM TITANATE NEAR THE 105 K PHASE TRANSITION.
06-1195 CRITICAL NEUTRON SCATTERING FROM STRONTIUM TITANATE.
SAMULIONIS VI
03-0353 ACOUSTOELECTRIC INTERACTION IN FERROELECTRIC SUBSTANCES.
15-2139 FREQUENCY DEPENDENCE OF ANOMALOUS ULTRASOUND DAMPING IN ANTIMONY SULFOIODIDE SINGLE CR
15-2221 ANOMALOUS ABSORPTION OF SOUND NEAR THE FERROELECTRIC PHASE TRANSITION (ANTIMONY SULFOI
15-2222 ANISOTROPY OF ANOMALOUS ABSORPTION OF ULTRASOUND IN SINGLE CRYSTALS OF ANTIMONY SULFOI
15-2223 GENERATION OF THE SECOND ACOUSTIC HARMONIC IN ANTIMONY SULFOIODIDE SINGLE CRYSTALS.
15-2224 GENERATION OF THE SECOND ACOUSTIC HARMONIC IN ANTIMONY SULFOIODIDE SINGLE CRYSTALS.
SANCHEZ AS
05-1073 TESTING A POINT DIPOLE ELECTROOPTIC MODEL BY AN APPLICATION TO LEAD TITANATE.
SANDERCOCK JR
15-2225 (ELASTIC CONSTANTS FROM) BRILLOUIN SCATTERING STUDY OF ANTIMONY SULFOIODIDE USING A DO
SANDSTROM DB
09-1589 A SEMICONDUCTOR- FERROELECTRIC MEMORY DEVICE. (PLZT)
SANNIKOV DG
03-0312 PIEZOELECTRIC EFFECT ON THE REFLECTION OF A TRANSVERSE ACOUSTIC WAVE FROM DOMAIN BOUND
03-0313 REFLECTION AND REFRACTION OF TRANSVERSE SOUND WAVES AT DOMAIN BOUNDARIES IN FERROELECT
03-0314 REFLECTION AND REFRACTION OF QUASI LONGITUDINAL AND QUASI TRANSVERSE SOUND WAVES BY 18
03-0315 SURFACE ELASTOPOLARIZATION WAVES AT DOMAIN BOUNDARIES IN A FERROELECTRIC.
03-0363 PHENOMENOLOGICAL THEORY OF DIELECTRIC ANOMALIES IN FERROELECTRIC MATERIAL WITH SEVERAL
03-0454 DISPERSION IN FERROELECTRICS. (THEORY, BARIUM TITANATE)
03-0455 ELECTROMAGNETIC AND ACOUSTIC WAVES IN FERROELECTRICS. (THEORY, BARIUM TITANATE, ROCHEL
19-2765 (EXAMINATION OF) ANOMALIES OF DIELECTRIC PROPERTIES AT THE PHASE TRANSITIONS. (BASED O
30-3303 PHENOMENOLOGICAL THEORY OF THE FERROELECTRIC PHASE TRANSITION IN GADOLINIUM MOLYBDATE.
SANVORDENKER V
04-0603 ELECTRICAL AND OPTICAL PROPERTIES OF (BARIUM, STRONTIUM) MIXED TITANATE THIN SECTIONS.
SARGENT M
11-1773 EFFECT OF OPTICAL INHOMOGENEITIES ON PHASE MATCHING IN NONLINEAR CRYSTALS (LITHIUM NIO
SARNATSKII VM
11-1666 ACOUSTIC DIPOLE AND QUADRUPOLE SATURATION OF THE LITHIUM-7 NMR SIGNALS IN LITHIUM NIOB
SARTAIN CC
28-3207 RADIATION DAMAGE AND THE FERROELECTRIC EFFECT IN ROCHELLE SALT.
SARVACA A
17-2643 MICROSCOPIC OBSERVATION OF ANTIFERROELECTRIC DOMAINS OF AMMONIUM DIHYDROGEN PHOSPHATE.
SASAKI H
04-0735 EFFECT OF GRAIN SIZE ON THE RESISTIVITY ANOMALY IN SEMICONDUCTIVE BARIUM TITANATE CERA
04-0835 EXAGGERATED GRAIN GROWTH IN LIQUID PHASE SINTERING OF BARIUM TITANATE.
04-0837 PTCR (POSITIVE TEMPERATURE COEFFICIENT OF RESISTANCE) BEHAVIOR OF BARIUM TITANATE WITH
04-0872 CHEMICAL CHANGES IN BARIUM TITANATE FIRED UNDER NITROGEN IN THE PRESENCE OF CARBON.
04-0931 A NEW PTC THERMISTOR MATERIAL. (BARIUM TITANATE)
08-1393 HIGH PRESSURE SYNTHESIS (SOLID STATE REACTION, SINTERING) AND PROPERTIES OF PEROVSKITE
09-1587 OXIDATION REDUCTION PHENOMENA IN THE LEAD TITANATE- LANTHANUM TITANATE SYSTEM.
17-2693 SECOND HARMONIC GENERATION OF LIGHT IN THE KDP CRYSTAL BY THE INDEX MATCHING METHOD.
SASAKI K
33-3403 RESONANCE BEHAVIOR IN CANTED ANTIFERROMAGNETIC POTASSIUM MANGANESE TRIFLUORIDE.
SASAKI R
05-1069 LEAD TITANATE CERAMIC RESONATORS OPERATING IN VHF BAND.
SASAKI Y
15-2226 PHOTOCONDUCTIVITY OF FERROELECTRIC PHOTOCONDUCTOR ANTIMONY SULFOIODIDE.
SASTRY MD
17-2644 MOSSBAUER STUDY OF ANTIFERROELECTRIC TRANSITION IN IRON-57(3X) AMMONIUM DIHYDROGEN PHO
SATO H
04-0714 POLARIZATION REFLECTION TYPE LIGHT MODULATOR USING FERROELECTRICS. (BARIUM TITANATE)
04-1015 DIELECTRIC BREAKDOWN OF POLYCRYSTALLINE BARIUM TITANATE.
SATO R
37-3674 AN APPROXIMATIVE ANALYSIS OF A FERROELECTRIC RECORDING SYSTEM.
SATO S
16-2312 CRYSTAL STRUCTURE AND PHASE TRANSITION OF DIGLYCINE NITRATE.
22-2869 NEUTRON SCATTERING ANALYSIS OF CESIUM TRIHYDROGEN SELENITE AT ROOM TEMPERATURE.
SATO T
28-3234 SURFACE STRUCTURES AND CORRESPONDING LIGHT FIGURES REVEALED BY WATER ETCHED ROCHELLE S
SATO Y
13-2051 ELECTROOPTICAL PROPERTIES OF STRONTIUM POTASSIUM LITHIUM NIOBATE. (TUNGSTEN BRONZE TYP
14-2084 PHENOMENOLOGICAL THEORY OF DIELECTRIC AND MECHANICAL PROPERTIES OF IMPROPER FERROELECT
14-2086 X-RAY STUDY ON PHASE TRANSITIONS OF FERROELECTRIC IRON IODINE BORACITE AT LOW TEMPERAT
17-2645 DEUTERON MOTIONS IN POTASSIUM DIDEUTERIUM PHOSPHATE (FROM POLARIZED REFLECTION SPECTRA
17-2646 OPTICAL STUDY ON THE PHASE TRANSITIONS IN POTASSIUM DIHYDROGEN OR DEUTERIUM PHOSPHATE
30-3294 X-RAY STUDY OF THERMAL EXPANSION OF FERROELECTRIC GADOLINIUM MOLYBDATE.
SATOH M
16-2321 ACOUSTIC NUCLEAR MAGNETIC RESONANCE IN FERROELECTRIC SODIUM NITRITE.
SATYANARAYANA BS
36-3606 A NOTE ON THE APPARENT CONTACT FERROELECTRIC EFFECT OF MANGANESE DIOXIDE.
SAUER HA
36-3592 SYSTEM LANTHANA TITANIA PHASE EQUILIBRIA AND ELECTRICAL PROPERTIES (LANTHANUM TITANATE
SAUKOV AI
17-2460 CHARACTERISTICS OF AN ELECTROOPTICAL SHUTTER BASED ON Z CUT POTASSIUM DIDEUTERIUM PHOS
SAVAGE A
04-0852 QUANTITATIVE STUDIES OF OPTICAL HARMONIC GENERATION IN CADMIUM SULFIDE, BARIUM TITANAT
11-1770 TEMPERATURE DEPENDENCE OF THE OPTICAL PROPERTIES OF FERROELECTRIC LITHIUM NIOBATE AND
SAVAGE CM
17-2579 EFFECTS OF DISPERSION AND FOCUSING ON THE PRODUCTION OF OPTICAL HARMONICS. (KDP)
SAVATINOVA IT
15-2143 PHASE TRANSITION IN FERROELECTRIC ANTIMONY SULFOIODIDE CRYSTALS USING THE RAMAN SCATTE
17-2470 EFFECT OF TEMPERATURE AND PHASE TRANSITION ON THE HYDROXIDE VIBRATION SPECTRUM OF THE
17-2471 POLARIZATION OF THE RAMAN BANDS OF THE HYDROXYL VIBRATIONS AND STRUCTURE OF THE HYDROG
17-2472 VIBRATION SPECTRUM OF THE PHOSPHATE ANION IN A RUBIDIUM DIHYDROGEN PHOSPHATE CRYSTAL I
17-2620 FERROELECTRIC TRANSITION IN RUBIDIUM DIHYDROGEN PHOSPHIDE USING RAMAN SCATTERING SPECT
17-2621 ISOTOPE EFFECT IN THE RAMAN SPECTRA OF A DEUTERATED KDP CRYSTAL.
17-2626 TEMPERATURE DEPENDENCE OF RAMAN SCATTERING SPECTRUM OF FERROELECTRIC DEUTERATED KDP DU
17-2647 RAMAN SPECTRUM OF POTASSIUM DIDEUTERIUM PHOSPHATE CRYSTAL IN PARA- AND FERROELECTRIC P
27-3113 LOW FREQUENCY RAMAN SCATTERING BY FERROELECTRIC TRIGLYCINE FLUOBERYLLATE CRYSTALS.
27-3114 LOW FREQUENCY RAMAN SPECTRA OF FERROELECTRIC CRYSTALS OF THE TRIGLYCINE SULFATE TYPE.
SAVCHENKO EA
15-2202 RELAXATION PROCESSES IN ANTIMONY SULFOIODIDE SINGLE CRYSTALS EXCITED WITH MICROSECOND
15-2229 LUMINESCENCE OF SINGLE CRYSTALS OF THE FERROELECTRIC SEMICONDUCTOR ANTIMONY SULFOIODID
SAVCHENKO VG
04-0955 ELECTROOPTICAL PROPERTIES OF BARIUM TITANATE- TANTALUM PENTOXIDE SOLID SOLUTION SINGLE
04-0958 ELECTROOPTICAL EFFECT IN SINGLE CRYSTALS OF BARIUM TITANATE- ZINC TITANATE SOLID SOLUT
04-0959 EFFECT OF NONISOVALENT SUBSTITUTION OF TITANIUM(4+) IONS ON THE ELECTROOPTICAL PROPERT
SAVENKOVA GE
09-1544 NEW PIEZOELECTRIC CERAMICS. (LEAD TITANATE- LEAD FERRATE(0.5) NIOBATE(0.5), LEAD TITAN
09-1588 EFFECT OF BISMUTH AND MANGANESE CONTAINING ADDITIVES ON THE PROPERTIES OF SOLID SOLUTI
36-3587 PREPARATION OF THE TERNARY TITANATE BARIUM(0.8) LEAD(0.12) CALCIUM(0.08) TITANATE.
SAVIN W
10-1651 STRUCTURAL AND DIELECTRIC MODIFICATIONS OF TUNGSTEN TRIOXIDE DUE TO NEUTRON IRRADIATIO
SAVINKOV AI
17-2648 STUDY OF THE (MECHANICAL) STRENGTH PROPERTIES OF POTASSIUM DIHYDROGEN PHOSPHATE CRYSTA

SAVINKOV AI	17-2649	TEMPERATURE DEPENDENCE OF THE MECHANICAL CHARACTERISTICS OF POTASSIUM DIHYDROGEN PHOSP
SAVOSHCHENKO VS	04-0869	FLUORINE CONTAINING BARIUM TITANATE. (ELECTRICAL PROPERTIES)
	04-0932	FORMATION OF FLUORINE CONTAINING SOLID SOLUTIONS BASED ON BARIUM TITANATE.
	04-0983	INFRARED SPECTRA OF FLUORINE CONTAINING BARIUM TITANATE.
SAVTCHENKO EA	04-0681	EXTRINSIC PHOTOCONDUCTIVITY IN FERROELECTRICS DUE TO SURFACE LAYERS. (BARIUM TITANATE,
SAVVINOV AM	27-3115	VARIATION OF THE UNIPOLARITY OF TRIGLYCINE SULFATE CRYSTALS WITH TIME.
SAWADA A	17-2534	DIELECTRIC DISPERSION OF TGS AND DEUTERATED KDP (THEORY, CALCULATIONS).
	27-3070	NONLINEARITY OF THE DIELECTRIC CONSTANT OF FERROELECTRIC TRIGLYCINE SULFATE.
	27-3116	DIELECTRIC STUDY OF CRITICAL BEHAVIOR OF FERROELECTRIC TRIGLYCINE SULFATE BY A DIGITAL
	28-3223	MECHANISM OF FERROELECTRIC PHASE TRANSITION IN AMMONIUM ROCHELLE SALT. (THEORY)
SAWADA S	10-1658	PHASE TRANSITION IN TUNGSTEN TRIOXIDE UNDER HIGH PRESSURE.
	16-2258	ELECTRICAL RESISTIVITIES OF SODIUM NITRITE AND POTASSIUM NITRATE CRYSTALS.
	16-2261	QUASI STATIC SWITCHING CURRENT IN SODIUM NITRITE.
	16-2301	FERROELECTRIC POLARIZATION SWITCHING IN SODIUM NITRITE.
	16-2351	THERMAL EXPANSION IN SODIUM NITRITE CRYSTALS.
	16-2373	TEMPERATURE DEPENDENCE OF FAR INFRARED REFLECTIVITY SPECTRA OF SODIUM NITRITE CRYSTALS
	16-2401	FERROELECTRICITY IN AMMONIUM(X) POTASSIUM(1-X) NITRATE MIXED CRYSTAL.
	16-2403	STUDY OF PHASE TRANSITION OF SODIUM NITRITE BY SECOND HARMONIC GENERATION. (TEMPERATUR
	24-2902	ANOMALOUS SPECIFIC HEAT OF FERROELECTRIC THIOUREA AND POTASSIUM FERROCYANIDE TRIHYDRAT
SAWAGUCHI E	03-0456	DOUBLE HYSTERESIS LOOPS AND FIRST ORDER PHASE TRANSITIONS IN FERROELECTRIC CRYSTALS.
	08-1436	DIELECTRIC BEHAVIOR (PERMITTIVITY, SPONTANEOUS POLARIZATION, COERCIVE FIELD) OF BISMUT
	15-2197	EFFECT OF PHOTOCONDUCTION ON THE FERROELECTRIC PROPERTIES OF ANTIMONY SULFOIODIDE. (PH
	29-3247	CRYSTAL DATA FOR LITHIUM THALLIUM TARTRATE MONOHYDRATE.
	29-3249	CONTROL OF THE ELASTIC COMPLIANCE OF FERROELECTRIC LITHIUM THALLIUM TARTRATE HYDRATE (
	29-3250	ELECTRIC FIELD CONTROL OF THE ELASTIC COMPLIANCE IN LITHIUM THALLIUM TARTRATE (TUNING
	29-3251	ELECTROMECHANICAL COUPLING EFFECTS ON THE DIELECTRIC PROPERTIES AND FERROELECTRIC PHAS
	36-3550	FERROELECTRIC OPTICAL ROTATION DOMAINS IN SINGLE CRYSTAL LEAD(5) GERMANIUM(3) OXIDE (11
SAWAMOTO K	11-1810	ZERO TEMPERATURE COEFFICIENT OF RESONANT FREQUENCY IN LITHIUM TANTALATE LENGTH EXPANDE
	12-1868	TEMPERATURE DEPENDENCE OF AN X CUT LITHIUM TANTALATE CRYSTAL RESONATOR VIBRATING IN TH
	12-1885	BEHAVIOR OF LITHIUM TANTALATE SINGLE CRYSTAL NEAR ITS CURIE POINT, PART-2, DIELECTRIC
SAWASA A	28-3224	SUPERSTRUCTURE IN THE FERROELECTRIC PHASE OF AMMONIUM ROCHELLE SALT.
SAWYER DE	09-1589	A SEMICONDUCTOR- FERROELECTRIC MEMORY DEVICE. (PLZT)
SAYAR M	36-3563	MAGNETOELECTRIC EFFECT IN THE ANTIFERROMAGNET IRON ANTIMONY(2) OXIDE(4) (MAY BE FERROE
SCALAPINO DJ	33-3376	EXCHANGE NARROWING: MAGNETIC RESONANCE LINE SHAPES AND SPIN CORRELATION IN PARAMAGNETI
SCHACHER GE	03-0220	INTERNAL FIELD IN GENERAL DIPOLE LATTICES. (THEORY APPLIED TO POTASSIUM FERROCYANIDE T
	27-3120	SHAPE DEPENDENCE OF PROPERTIES (CURIE TEMPERATURE, CURIE CONSTANT) OF FERROELECTRIC CR
SCHACHNER H	14-2085	X-RAY STUDY ON THE LATTICE STRAINS OF (PARAELECTRIC AND) (IRON- IODINE- BORACITE).
SCHAFER G	04-0937	STATIC ELECTRIC QUADRUPOLE INTERACTION OF TANTALUM AND HAFNIUM IONS IN BARIUM AND LEAD
SCHAFER H	28-3226	THERMAL CONDUCTIVITY OF ROCHELLE SALT.
SCHAGINA NM	17-2666	PHASE DIAGRAM OF SODIUM TRIDEUTERIUM(X) TRIHYDROGEN(1-X) PHOSPHATE SYSTEM VERSUS CONCE
	22-2848	(FERROELECTRIC AND OPTICAL PROPERTIES OF) CRYSTALS OF THE ALKALINE TRIHYDROGEN DISELEN
SCHARA M	22-2870	DIELECTRIC PROPERTIES AND PRESSURE EFFECTS ON SELENITE (ION). RADIATION DAMAGE ELECTRO
	22-2871	ESR STUDY OF THE FERROELECTRIC PHASE TRANSITIONS IN SODIUM TRIHYDROGEN DISELENITE.
SCHAWLOW AL	06-1112	DIRECT OBSERVATION OF SINGLE DOMAIN STRONTIUM TITANATE (OPTICAL STUDY UNDER UNIAXIAL S
	06-1115	POLARIZED FLUORESCENCE STUDY OF CHROMIUM(3+) THROUGH A STRESS INDUCED PHASE TRANSITION
	06-1235	CUBIC TO TRIGONAL STRESS INDUCED PHASE TRANSITION IN STRONTIUM TITANATE.
	06-1236	RAMAN STUDIES OF STRESS INDUCED PHASE TRANSITION NEAR CURIE TEMPERATURE IN STRONTIUM T
SCHEIDING C	15-2227	PIEZOELECTRICITY AND ELECTROSTRICTION OF ANTIMONY SULFOIODIDE SINGLE CRYSTALS.
	30-3317	PIEZOELECTRIC COEFFICIENT OF GADOLINIUM MOLYBDATE.
SCHEIN BJB	21-2818	FERROELECTRIC MOMENT IN GUANIDINIUM ALUMINUM SULFATE HEXAHYDRATE (AND MECHANISM OF REV
	21-2819	REDETERMINATION OF STRUCTURE OF FERROELECTRIC CRYSTAL GUANIDINIUM ALUMINUM SULFATE HEX
SCHELKENS R	04-0933	FERROELECTRIC AND ANTIFERROELECTRIC SYMMETRY GROUPS. (AND THEIR COMPARISON, APPLICATIO
SCHEMPP E	11-1811	NIOBIUM-93 NUCLEAR QUADRUPOLE RESONANCE INVESTIGATION OF (STOICHIOMETRIC) LITHIUM NIOB
SCHIEBER M	28-3200	CRYSTALLIZATION OF ROCHELLE SALT IN ELECTRIC FIELDS (KINETICS OF CRYSTAL GROWTH)
SCHINKE DP	30-3280	FERROELECTRIC DOMAIN SHIFTING DEVICES. (GADOLINIUM MOLYBDATE)
SCHLEGEL LH	29-3247	CRYSTAL DATA FOR LITHIUM THALLIUM TARTRATE MONOHYDRATE.
SCHLOSSBERG H	11-1729	LUMINESCENCE FROM LITHIUM NIOBATE.
SCHLOSSER PA	37-3667	A SELF SCANNED FERROELECTRIC IMAGE SENSOR.
	37-3669	THEORY AND APPLICATION OF FERROELECTRIC RADIATION DETECTORS. (PYROELECTRICS)
SCHMELZ H	04-0934	INCORPORATION OF ANTIMONY INTO THE BARIUM TITANATE LATTICE. (STRUCTURE, AND PCT RESIST
SCHMID H	14-2082	ELECTROOPTICAL PROPERTIES OF FERROELECTRIC (IRON IODINE) BORACITE. (SPONTANEOUS LATTIC
	14-2085	X-RAY STUDY ON THE LATTICE STRAINS OF (PARAELECTRIC AND) (IRON- IODINE- BORACITE).
	14-2086	X-RAY STUDY ON PHASE TRANSITIONS OF FERROELECTRIC IRON IODINE BORACITE AT LOW TEMPERAT
	14-2098	TRIGONAL BORACITES: A NEW TYPE OF FERROELECTRIC AND FERROMAGNETOELECTRIC MATERIAL THAT
	14-2103	OBSERVATIONS OF FERROELECTRIC DOMAINS IN BORACITES.
SCHMIDT G	03-0457	THERMODYNAMICS AND THE PHENOMENOLOGY OF FERROELECTRICITY.
	04-0550	TIME DEPENDENCE OF THE DIELECTRIC PERMITTIVITY OF HYDROXYL DOPED BARIUM TITANATE.
	06-1197	INERTIAL ELECTRICAL POLARIZATION OF CRYSTALS (ALKALI HALIDES, STRONTIUM TITANATE).
	06-1206	ELECTROMECHANICAL PROPERTIES OF STRONTIUM TITANATE SINGLE CRYSTALS.
	06-1218	MECHANICAL RELAXATION AND NONLINEARITY IN STRONTIUM TITANATE SINGLE CRYSTALS.
	06-1218	PRESSURE DEPENDENCE OF THE ELASTIC COMPLIANCE OF STRONTIUM TITANATE NEAR THE TRANSITIO
	15-2227	PIEZOELECTRICITY AND ELECTROSTRICTION OF ANTIMONY SULFOIODIDE SINGLE CRYSTALS.
	27-3117	LIGHT EMISSION FROM TGS. (CRYSTALS DURING HEATING)
	27-3118	PIEZOELECTRICITY AND ELECTROSTRICTION OF TRIGLYCINE SULFATE.
	28-3225	EFFECT OF MECHANICAL STRESS ON DIELECTRIC PROPERTIES OF PARAELECTRIC ROCHELLE SALT.
	30-3317	PIEZOELECTRIC COEFFICIENT OF GADOLINIUM MOLYBDATE.
SCHMIDT H	14-2071	DIELECTRIC PROPERTIES OF BORACITES AND EVIDENCE FOR FERROELECTRICITY.
SCHMIDT P	20-2803	INFRARED AND RAMAN SPECTRA OF FERROELECTRIC ALUMS: METHYL AMMONIUM ALUMINUM (SULFATE O
SCHMIDT VH	17-2651	RANDOM MOTION OF DEUTERONS IN DEUTERATED POTASSIUM DIHYDROGEN PHOSPHATE.
	17-2652	THEORY OF DOMAIN WALL MOTION IN POTASSIUM DIHYDROGEN PHOSPHATE TYPE CRYSTALS.
	19-2781	ELECTRICAL CONDUCTIVITY OF AMMONIUM SULFATE SINGLE CRYSTALS.
	22-2849	NMR STUDY OF DEUTERIUM DIFFUSION AND DEUTERIUM AND SODIUM-23 RELAXATION IN SODIUM TRID
	32-3350	DIPOLAR RELAXATION OF LITHIUM-7 BY HINDERED ROTATORS IN LITHIUM HYDRAZINIUM SULFATE.
	32-3351	ANALYSIS OF HYSTERESIS LOOPS IN LITHIUM HYDRAZINIUM SULFATE.
	32-3352	DIELECTRIC PROPERTIES OF LITHIUM HYDRAZINIUM SULFATE.
	32-3353	DIELECTRIC PROPERTIES OF LITHIUM HYDRAZINIUM SULFATE.
	32-3354	DEUTERON NMR STUDIES OF LITHIUM HYDRAZINIUM SULFATE.
	32-3357	MICROWAVE DIELECTRIC STUDY OF LITHIUM HYDRAZINIUM SULFATE.
	34-3471	NMR INVESTIGATIONS OF ICE CRYSTALS. PART-1: PROTON MAGNETIC RESONANCE IN WATER ICE. PA
SCHMIDT W	15-2140	DETERMINATION OF THE TEMPERATURE DEPENDENCE OF LATTICE CONSTANTS IN FERROELECTRIC ANTI

SCHMIT J   17-2650   INFRARED VIBRATION FREQUENCIES OF THE HYDROGEN BOND IN KDP FERROELECTRIC (CALCULATIONS
SCHMITT H   27-2995   NEW MEASUREMENTS OF DIVERSE THERMODYNAMICAL COEFFICIENTS OF ROCHELLE SALT AND TRIGLYCI
                        28-3213   MEASUREMENTS OF COMPLEX PIEZOELECTRIC CONSTANTS IN FERROELECTRIC CRYSTALS. (ROCHELLE S
SCHMITZ W   15-2201   CRYSTALLIZATION BEHAVIOR OF ANTIMONY SULFOIODIDE. (SINGLE CRYSTAL GROWTH, CHEMICAL TRA
SCHNEIDER J   11-1725   AN EPR INVESTIGATION CF IRON(3+) AND MANGANESE(2+) IN LITHIUM NIOBATE.
SCHNEIDER VE   08-1475   LATTICE DYNAMICS OF PHASE TRANSITIONS IN PEROVSKITE TYPE ANTIFERROELECTRICS.
SCHOIJET M   04-0817   REJUVENATION OF A BARIUM(0.72) STRONTIUM(0.28) TITANATE CERAMIC.
                        04-0935   ON THE ANOMALOUS CRYSTALLOGRAPHIC PROPERTIES OF SMALL BARIUM TITANATE PARTICLES.
                        04-0936   SPACE CHARGE LIMITED CURRENTS IN HEXAGONAL BARIUM TITANATE(0.75) PLATINATE(0.25) CRYST
SCHOOLEY JF   06-1182   EFFECT OF STRESS ON THE SUPERCONDUCTING TRANSITION TEMPERATURE OF STRONTIUM TITANATE.
SCHOUMACHER P   27-3008   ELOCKING OF SPONTANEOUS POLARIZATION IN TRIGLYCINE SULFATE.
SCHRADER B   25-2907   MOLECULAR AND LATTICE VIBRATIONS CF ORTHORHOMBIC AND FERROELECTRIC THIOUREA. PART-2: C
                        25-2920   MOLECULAR AND LATTICE VIBRATIONS OF ORTHORHOMBIC AND FERROELECTRIC THIOUREA. PART-1: T
SCHRODER FA   10-1652   A NEW FCRM OF TUNGSTEN TRIOXIDE.
                        10-1653   SOME ASPECTS OF THE STRUCTURE CF TUNGSTEN TRIOXIDE AND A CONTRIBUTION TO THE UNDERSTAN
SCHROTTER HW   11-1696   DIRECTIONAL DISPERSION OF PHONON FREQUENCIES AND POLARITON SCATTERING IN LITHIUM NIOBA
SCHUBRING NW   16-2285   SWITCHING BEHAVIOR IN FERROELECTRIC POTASSIUM NITRATE.
                        16-2356   DIELECTRIC BEHAVIOR OF (FERROELECTRIC) FILMS OF VACUUM DEPOSITED POTASSIUM NITRATE.
                        37-3668   FERROELECTRIC HYSTERESIS TRACER FEATURING COMPENSATION AND SAMPLE GROUNDING.
SCHUELE D   33-3361   LOW FREQUENCY CIELECTRIC CONSTANTS OF THE ALKALINE EARTH FLUORIDES BY THE METHOD OF SU
SCHULZ BS   11-1812   TEMPERATURE DEPENDENCE OF RESONANT FREQUENCIES OF LITHIUM NIOBATE PLATE RESONATORS. (C
SCHULZ MP   12-1886   TEMPERATURE DEPENDENCE OF SURFACE ACOUSTIC WAVE VELOCITY IN LITHIUM TANTALATE.
SCHULZE WA   04-0566   DC BIAS EFFECTS IN BARIUM TITANATE CERAMICS.
                        09-1490   DC BIAS EFFECTS IN LEAD ZIRCONATE TITANATE CERAMICS.
                        09-1491   TEMPERATURE AND FIELD DEPENDENCE OF THE ANTIFERROELECTRIC FERROELECTRIC PHASE BOUNDARY
                        09-1590   PHASE BOUNDARY AND DEFECT STRUCTURE STUDIES IN PLZT CERAMICS.
SCHUTTE CJH   19-2782   LOW TEMPERATURE INFRARED STUDIES. PART-8: FERROELECTRIC AMMONIUM HYDROGEN SULFATE AND
                        19-2783   LOW TEMPERATURE STUDIES. PART-4: PHASE TRANSITIONS OF AMMONIUM SULFATE: THE NATURE OF
SCHWABL F   06-1204   CRITICAL DYNAMICS OF STRONTIUM TITANATE ABOVE THE CRITICAL TEMPERATURE.
                        06-1205   DYNAMICS AT STRUCTURAL TRANSITIONS ABOVE FERROELECTRIC CRITICAL TEMPERATURE (ROTATIONS
SCHWARTZ RJ   13-2005   FORMATICN OF THIN LAYERS OF MIXED TITANATES BY SOLID SOLID REACTIONS.
SCHWARTZ RW   21-2815   MAGNETIC SUSCEPTIBILITY OF GUANIDINIUM VANADIUM SULFATE HEXAHYDRATE AT LOW TEMPERATURE
                        21-2817   ELECTRON PARAMAGNETIC RESONANCE STUDY OF SOME METAL IONS IN GUANIDINIUM ALUMINUM SULFA
SCHWARZ F   27-3119   PERFORMANCE CHARACTERISTICS OF A SMALL TRIGLYCINE SULFATE DETECTOR OPERATED IN THE PYR
SCOTT BA   02-0053   PROPERTIES OF TUNGSTEN BRONZE FERROELECTRICS. (REVIEW, 33 REFS)
                        04-0610   POWDER RAMAN STUDY FOR BARIUM TITANATE.
                        04-0611   RAMAN SCATTERING IN THE FERROELECTRIC SYSTEM LEAD BARIUM TITANATE.
                        05-1041   RAMAN STUDIES (TEMPERATURE DEPENDENCE) OF UNDER DAMPED SOFT MODES IN LEAD TITANATE.
                        05-1074   CRYSTAL GROWTH AND TEMPERATURE DEPENDENT RAMAN SPECTRA OF LEAD TITANATE.
                        07-1264   CRYSTAL GROWTH AND OBSERVATION OF THE FERROELECTRIC PHASE OF LEAD ZIRCONATE.
                        07-1265   GROWTH AND DIELECTRIC PROPERTIES OF SINGLE CRYSTAL LEAD ZIRCONATE
                        09-1497   RAMAN SPECTRA OF POLYCRYSTALLINE SOLIDS: APPLICATION TO THE LEAD TITANATE ZIRCONATE SY
                        11-1765   ELECTRON PARAMAGNETIC RESONANCE OF IRON(3+) DOPED LITHIUM NIOBATE.
                        11-1813   STOICHICMETRY VARIATIONS IN LITHIUM NIOBATE AND LITHIUM TANTALATE BY RAMAN POWDER SPEC
                        13-1948   (TETRAGONAL) TUNGSTEN BRONZE TYPE CRYSTALS GROWN IN MOLTEN SOLUTIONS OF THE PSEUDOSYST
                        13-1949   (EXTENT OF) TUNGSTEN BRONZE (LIQUIDUS AND SOLIDUS) FIELD AND MELT GROWTH OF CRYSTALS I
                        13-2015   CRYSTAL GROWTH AND NONLINEAR OPTICAL PROPERTIES OF POTASSIUM LITHIUM NIOBATES.
                        13-2016   PHASE EQUILIBRIA IN THE POTASSIUM NIOBATE- STRONTIUM DINIOBATE AND POTASSIUM NIOBATE-
                        13-2017   TUNGSTEN BRONZE FIELD IN THE (OXIDE) SYSTEM POTASSIA LITHIA NIOBIA (POTASSIUM LITHIUM
                        13-2026   NONLINEAR OPTICAL PROPERTIES OF POTASSIUM LITHIUM NIOBATES (TRANSPARENT, TETRAGONAL TU
                        13-2027   NONLINEAR OPTICAL PROPERTIES OF POTASSIUM LITHIUM NIOBATE.
                        35-3495   SOFT OPTIC PHONON MODE IN FERROELECTRIC LEAD(5) GERMANIUM(3) OXIDE(11).
SCOTT JF   06-1207   SOFT MODE RAMAN SPECTROSCOPY: COUPLED MODES (FERROELECTRICS, CESIUM DIHYDROGEN ARSENAT
                        10-1654   RAMAN STUDY OF METALLIC TUNGSTEN BRONZES. (SODIUM AND RUBIDIUM TUNGSTATES)
                        17-2653   PROTON PHONON INTERACTIONS IN POTASSIUM DIHYDROGEN PHOSPHATE.
                        18-2723   DIELECTRIC RESPONSE IN PIEZOELECTRIC CRYSTALS. (RAMAN SCATTERING IN POTASSIUM DIHYDROG
                        18-2733   PROTON PHONON COUPLING IN CESIUM DIHYDROGEN ARSENATE AND POTASSIUM DIHYDROGEN ARSENATE
                        18-2739   PHONON PHONON AND PROTON INTERACTIONS IN ALUMINUM PHOSPHATE AND CESIUM DIHYDROG
SCZANIECKI B   21-2820   EPR STUDY OF MAGNETIC COPPER(2+) COMPLEXES IN CRYSTALS OF GUANIDINIUM ALUMINUM SULFATE
SEAGER CH   33-3404   ELECTRONIC HALL MOBILITY IN THE ALKALINE EARTH FLUORIDES.
SEARLE CW   33-3389   MAGNETOELASTIC EFFECTS IN POTASSIUM MANGANESE TRIFLUORIDE.
SEAY GE   04-0916   TWO WAVE SHOCK STRUCTURES IN THE FERROELECTRIC CERAMICS BARIUM TITANATE AND LEAD ZIRCO
SEBER GA   11-1814   SEPARATE EXCITATION OF LONGITUDINAL AND TRANSVERSE HYPERSONIC PULSES IN PIEZOELECTRIC
SEDLAKOVA LN   17-2567   GROWTH AND ETCHING OF POTASSIUM DIHYDROGEN PHOSPHATE CRYSTALS IN SOLUTIONS CONTAINING
SEEHRA MS   26-2943   STUDY OF THE ORDERED MAGNETIC STATE OF COPPER FORMATE TETRAHYDRATE BY ANTIFERROMAGNETI
                        33-3377   SHIFT OF NEEL TEMPERATURE AND EPR LINE WIDTH OF POTASSIUM MANGANESE TRIFLUORIDE WITH M
SEELY K   29-3242   SOME PROPERTIES OF THE FERROELECTRIC LITHIUM THALLIUM TARTRATE.
SEGALLA AG   09-1514   TIME AND TEMPERATURE DEPENDENCES OF THE PARAMETERS OF A FERROELECTRIC CERAMIC.
                        09-1591   EFFECT OF FERROELECTRIC PHASE TRANSITION CONDITIONS ON DIELECTRIC AND PIEZOELECTRIC PR
SEGARD N   04-0938   BARIUM TITANATE PIEZOELECTRIC MICROPROBE FOR STUDYING ULTRASONIC FIELDS.
SEGEL SL   11-1829   TONE BURSTING OF QUADRUPOLAR NUCLEI. DETERMINATION OF THE SPIN LATTICE RELAXATION TIM
SEIDENSTICKER RG   34-3472   IMPURITY STATISTICS IN ICE.
SEKI S   34-3436   RELAXATIONAL PROTON ORDERING AND GLASSY CRYSTALLINE STATE IN HEXAGONAL ICE.
                        34-3476   CALORIMETRIC STUDY OF GLASS TRANSITION OF THE AMORPHOUS ICE AND OF THE PHASE TRANSFORM
SEKIDO T   02-0099   DIELECTRIC AND X-RAY STUDIES OF THE PHASE TRANSITIONS IN ROCHELLE SALT, BARIUM TITANAT
SEKINE E   04-0939   EXPERIMENTAL STUDY OF EVAPORATED BARIUM TITANATE FILMS.
SELIGER J   19-2749   NITROGEN-14 QUADRUPOLE COUPLING IN PARAELECTRIC AMMONIUM SULFATE.
SELIVANOVA NM   22-2867   HEAT OF FORMATION OF LITHIUM TRIHYDROGEN DISELENITE.
                        22-2868   THERMAL STABILITY OF LITHIUM HYDROGEN SELENITES.
SELLA C   10-1646   STUDY CF DOMAINS OF FERROELECTRIC CRYSTALS WITH SCANNING ELECTRON MICROSCOPE. (TUNGSTE
SELYUK BV   03-0458   INFLUENCE OF COMPENSATING CHARGES ON THE C-DOMAIN STRUCTURE OF FERROELECTRICS.
SEMAK DG   15-2127   DOMAIN STRUCTURE AND LOCAL STATES IN ANTIMONY SULFOIODIDE TYPE SEMICONDUCTOR FERROELEC
                        15-2181   POLARIZATION NATURE IN SEMICONDUCTOR FERROELECTRICS OF THE ANTIMONY SULFOIODIDE TYPE.
                        15-2182   SOME FEATURES OF TRAPPING LEVELS IN ANTIMONY SULFOIODIDE TYPE SEMICONDUCTING FERROELEC
                        15-2186   EFFECT OF PHOTOELECTRET CHARGE ON VALUE OF FERROELECTRIC POLARIZATION IN ANTIMONY SULF
                        15-2199   PREPARATION AND PHYSICAL PROPERTIES OF NONSTOICHIOMETRIC ANTIMONY SULFOIODIDE SINGLE C
SEMILETOVA DV   04-1002   INFLUENCE OF FIRING OF BARIUM TITANATE SINTERS ON ELECTRICAL AND TECHNOLOGICAL PROPERT
                        04-1003   REACTION SEQUENCE IN FORMING BARIUM METATITANATE IN AN INDUSTRIAL ROTARY FURNACE.
SEMIN GK   18-2722   STARK EFFECT IN THE NUCLEAR QUADRUPOLE RESONANCE SPECTRA OF THE FERROELECTRIC CRYSTAL
                        18-2740   ARSENIC-75 NUCLEAR QUADRUPOLE RESONANCE SPECTRA OF PARTIALLY DEUTERATED FERROELECTRIC

SENHOUSE LS	08-1437	CALCIUM CONCENTRATION VS NET IONIZED DONOR CONCENTRATION IN SINGLE CRYSTAL POTASSIUM T
SENSIK M	04-0897	STUDY OF CRYSTALS OF BARIUM TITANATE HAVING PARALLEL INTERFERENCE FRINGES.
SENTJURC M	28-3185	EFFECT OF APPLIED ELECTRIC FIELD ON EPR SPECTRA OF (COPPER(2+)) DOPED FERROELECTRIC RO
SEO I	02-0087	CRITICAL SLOWING DOWN PROCESS OF (ANOMALOUS) DIELECTRIC RELAXATION IN SEVERAL CRYSTALS
SEPLYUK PV	03-0459	WEIGHT OF SHOTTKY BARRIERS IN FERROELECTRICS.
SERAFIN RJ	11-1738	ELECTRON BEAM SENSING OF SURFACE ELASTIC WAVES FOR SIGNAL PROCESSING PURPOSES. (TECHNI
SERDYUKOV AN	03-0186	NONLINEAR SADOVSKII EFFECT (THEORY, PROPAGATION OF RADIATION AND FIRST HARMONIC, CALCU
SERGEEV NA	03-0460	NATURE OF THE TEMPERATURE DEPENDENCE OF NMR SPECTRA SECOND MOMENT IN ORDER DISORDER TY
SESSELMANN I	34-3473	INFLUENCE OF IMPURITIES ON THE ORDER DISORDER REACTION IN HEXAGONAL ICE I(H).
SETHARES JC	35-3512	ELASTIC, PIEZOELECTRIC, AND DIELECTRIC CONSTANTS OF BISMUTH(12) GERMANIUM OXIDE(20).
SEUFERT SW	09-1592	THE ROLE OF IRON OXIDE ADULTERANT IN A LEAD ZIRCONATE LEAD TITANATE CERAMICS.
SEVER GA	11-1815	INDEPENDENT EXCITATION OF LONGITUDINAL AND TRANSVERSE HYPERSONIC WAVE PULSES IN LITHIU
SEWARD WD	06-1209	EFFECTS OF TEMPERATURE AND PHASE TRANSITIONS ON THE HYDROXYL STRETCHING VIBRATION IN S
	06-1226	SPECIFIC HEAT ANOMALY ASSOCIATED WITH THE 110 DEG K PHASE TRANSITION IN STRONTIUM TITA
SFORZA PM	27-3031	A NEW FERROELECTRIC TRANSDUCER FOR USE IN HEAT TRANSFER AND FLOW STUDIES. (TRIGLYCINE
SHABALIN KN	16-2316	RATE OF ROTATING CRYSTALS AND CHARACTERISTICS OF THEIR GROWTH. (AMMONIUM DIHYDROGEN PH
SHABANOV VF	22-2827	MECHANISMS OF FERROELECTRIC PHASE TRANSITION IN CRYSTALS WITH REORIENTATING STRUCTURAL
SHABANOVA LA	22-2884	ELECTROMECHANICAL PROPERTIES OF THE FERROELECTRIC SODIUM AMMONIUM SELENATE DIHYDRATE.
SHACK R	25-2912	DIELECTRIC DISPERSION OF UREA AND THIOUREA.
SHAH A	03-0462	THEORETICAL AND PRACTICAL STUDIES OF DIELECTRIC APPLICATIONS OF FERROELECTRICS.
SHAKIN OV	11-1754	LIGHT SCATTERING BY ULTRASONIC WAVES IN LITHIUM NIOBATE CRYSTALS. (PHOTOELASTIC CONSTA
SHAKMANOV VV	04-0940	MOTION OF DOMAIN BOUNDARIES DURING SWITCHING IN SINGLE CRYSTAL FILMS OF BARIUM TITANAT
	04-0942	OBSERVATION OF THE CHANGE IN THE POLARIZATION OF SINGLE CRYSTAL FILMS OF BARIUM TITANA
	04-1031	MICROSTRUCTURE OF DOMAINS AND DOMAIN WALLS IN SINGLE CRYSTAL FILMS OF BARIUM TITANATE.
SHALDIN YUV	04-0941	NONLINEAR OPTICAL PROPERTIES OF LEAD TITANATE AND BISMUTH TITANATE CRYSTALS.
	08-1279	HYDROTHERMAL CRYSTALLIZATION AND SOME PROPERTIES OF BISMUTH TITANATES
	17-2654	MEASUREMENT OF THE PRIMARY LINEAR ELECTROOPTIC EFFECT IN AMMONIUM DIHYDROGEN PHOSPHATE
SHAMBUROV VA	03-0461	ELECTROOPTIC EFFECT IN OPTICALLY HETEROGENEOUS CRYSTALS (THEORY AND METHOD OF DETERMIN
SHANDARCV SM	11-1802	NONLINEAR INTERACTION OF HYPERSONIC WAVES IN A LITHIUM NIOBATE CRYSTAL.
SHANER JW	06-1209	EFFECTS OF TEMPERATURE AND PHASE TRANSITIONS ON THE HYDROXYL STRETCHING VIBRATION IN S
SHANKS HR	13-2057	THERMAL CONDUCTIVITY OF CUBIC SODIUM TUNGSTEN BRONZES. (SODIUM TUNGSTATES)
SHAPIRO SM	06-1208	CRITICAL NEUTRON SCATTERING IN STRONTIUM TITANATE AND POTASSIUM MANGANESE TRIFLUORIDE.
	06-1210	NEW OBSERVATIONS FROM NEUTRON SCATTERING STUDIES OF THE STRUCTURAL PHASE TRANSITION IN
	15-2110	LATTICE MODES AND PHASE TRANSITION IN ANTIMONY SULFOIODIDE.
	15-2111	RAMAN SCATTERING IN FERROELECTRIC SEMICONDUCTOR ANTIMONY SULFOIODIDE.
	15-2177	NOTES ON ABSORPTION BAND EDGE OF (PARAELECTRIC) ANTIMONY SULFOIODIDE.
SHAPIRO ZI	08-1424	ABOUT THE CURIE POINT AND NATURE OF DIELECTRIC PROPERTIES IN PEROVSKITE MODIFICATION O
	08-1438	PREPARATION AND INVESTIGATION OF FERROELECTRICS OF ILMENITE STRUCTURE. (ABO(3) NIOBATE
	12-1887	CURIE TEMPERATURE OF FERROELECTRIC LITHIUM TANTALATE.
SHAPKIN VV	04-0607	MAGNITUDE OF SPONTANEOUS POLARIZATION IN THIN CRYSTALS AND FINE PARTICLES OF BARIUM TI
SHARIPOV RZ	17-2419	MACROSCOPIC STIMULATED POLARIZATION ECHO IN FERROELECTRICS. (POTASSIUM DIHYDROGEN PHOS
	17-2564	POLARIZATION ECHO IN THE FERROELECTRIC SINGLE CRYSTAL POTASSIUM DIHYDROGEN PHOSPHATE.
SHARMA BS	09-1593	RETENTION IN THE THIN FERROELECTRIC FILMS. (PLZT)
	37-3670	EFFECT OF GAUSSIAN IRRADIANCE PROFILE ON THE SWITCHING CHARACTERISTICS AND THE BIT SIZ
	37-3671	PHOTOCONDUCTOR AND ELECTRODE REQUIREMENTS FOR THIN FILM FERROELECTRIC PHOTOCONDUCTOR M
SHARMA ML	07-1260	TEMPERATURE DEPENDENT OPTICAL MODE IN ANTIFERROELECTRIC LEAD ZIRCONATE BY THE MOSSBAUE
SHARMA PK	03-0297	INELASTIC SCATTERING CROSS SECTION OF NEUTRONS IN FERROELECTRICS.
	06-1142	ATOMIC DISPLACEMENTS IN PEROVSKITE STRONTIUM TITANATE (AT OPTICAL FREQUENCIES, CALCULA
SHARP AR	19-2800	PROTON MAGNETIC RESONANCE STUDY OF THE SPIN SYMMETRY STATES OF AMMONIUM IONS IN SOLIDS
SHASKOLSKAYA MP	17-2648	STUDY OF THE (MECHANICAL) STRENGTH PROPERTIES OF POTASSIUM DIHYDROGEN PHOSPHATE CRYSTA
	17-2649	TEMPERATURE DEPENDENCE OF THE MECHANICAL CHARACTERISTICS OF POTASSIUM DIHYDROGEN PHOSP
SHAULOV A	03-0469	MEASUREMENT OF THE PYROELECTRIC COEFFICIENT AND PERMITTIVITY FROM THE PYROELECTRIC RES
	13-2018	PYROELECTRIC VOLTAGE RESPONSE TO SHORT INFRARED LASER PULSES IN TRIGLYCINE SULFATE AND
	13-2019	PYROELECTRIC VOLTAGE RESPONSE TO RECTANGULAR INFRARED SIGNALS IN TRIGLYCINE SULFATE AN
	13-2022	PYROELECTRIC VOLTAGE RESPONSE TO STEP SIGNALS OF INFRARED RADIATION IN TRIGLYCINE SULF
SHAW EL	31-3339	INTERMOLECULAR POTENTIAL AND FERROELECTRIC TRANSITIONS IN HYDROGEN HALIDES.
SHAWYER RE	34-3474	ATOMIC VIBRATIONS IN ORIENTATIONALLY DISORDERED SYSTEMS. PART-1: A TWO DIMENSIONAL MOD
SHCHAGINA NM	28-3204	ATTENUATION OF SOUND WAVES IN DEUTERATED ROCHELLE SALT AT THE PHASE TRANSITION POINT.
	28-3205	ATTENUATION OF SOUND WAVES IN DEUTERATED ROCHELLE SALT AT THE PHASE TRANSITION POINT.
SHCHEDRINA NV	03-0361	FREQUENCY AND TEMPERATURE DEPENDENCES OF THE COMPLEX DIELECTRIC CONSTANT OF FERROELECT
SHCHELKANOGOV VV	15-2119	INFLUENCE OF ILLUMINATION ON THE FORM OF THE VOLTAGE AMPERE CHARACTERISTICS OF ANTIMON
SHCHERBAKOV VN	19-2744	NUCLEAR MAGNETIC RESONANCE OF DEUTERIUM IN THE PERDEUTERO AMMONIUM FLUOBERYLLATE CRYST
	22-2828	NUCLEAR MAGNETIC RESONANCE OF DEUTERIUM IN THE FERROELECTRIC SODIUM DEUTERO AMMONIUM S
	22-2883	NUCLEAR MAGNETIC RESONANCE OF SODIUM-23 IN THE SODIUM AMMONIUM SELENATE DIHYDRATE CRYS
SHE CY	17-2461	LONG WAVELENGTH POLARIZATION FLUCTUATIONS IN ANTIFERROELECTRIC AMMONIUM DIHYDROGEN PHO
	17-2655	EFFECT OF PROTON- PHONON COUPLING ON THE FERROELECTRIC MODE IN POTASSIUM DIHYDROGEN PH
	17-2657	RAMAN SPECTRA OF TETRAGONAL POTASSIUM DIHYDROGEN PHOSPHATE.
	17-2701	MODE INTERACTIONS IN POTASSIUM DIHYDROGEN PHOSPHATE.
SHECHTER H	19-2770	MOSSBAUER STUDY ON FERROELECTRIC PROPERTIES OF FERRIC AMMONIUM SULFATE DODECAHYDRATE S
	24-2901	MOSSBAUER STUDY OF THE FERROELECTRIC PHASE TRANSITION IN POTASSIUM FERROCYANIDE TRIHYD
SHEFELBINE HC	09-1594	FERROELECTRIC MEMORY DEVICE PARAMETER STUDY. (LEAD ZIRCONATE TITANATE)
SHEFER NI	04-0944	RELATIONSHIP BETWEEN THE BREAKDOWN MECHANISM IN BARIUM TITANATE AND COLORATION. (INJEC
SHEFTEL IT	04-0943	ELECTROCONDUCTIVITY AND HALL EFFECT IN SEMICONDUCTING SOLID SOLUTIONS OF ISO- AND HETE
	04-0995	SEMICONDUCTING BARIUM AND STRONTIUM TITANATES WITH POSITIVE TEMPERATURE COEFFICIENTS O
	06-1198	ELECTRICAL AND OPTICAL PROPERTIES (CONDUCTIVITY AND MOBILITY VERSUS TEMPERATURE, ABSOR
SHEINKMAN MK	15-2104	INFLUENCE OF NONEQUILIBRIUM CARRIERS ON THE DOMAIN STRUCTURE OF THE FERROELECTRIC SEMI
	15-2129	CHARACTERISTICS OF THE PHOTOELECTRIC EMF OF ANTIMONY SULFOIODIDE.
	15-2132	PHOTOEMF'S IN THE FERROELECTRIC SEMICONDUCTOR ANTIMONY SULFOIODIDE.
	15-2133	PHOTODOMAIN EFFECT IN NATURALLY POLARIZED ANTIMONY SULFOIODIDE CRYSTALS.
	15-2134	PHOTOELECTRIC PROPERTIES (PHOTOCONDUCTIVITY) OF ANTIMONY SULFOIODIDE SINGLE CRYSTALS.
	15-2135	PHOTODOMAIN EFFECT IN NATURALLY POLARIZED ANTIMONY SULFOIODIDE CRYSTALS.
	15-2153	CHARACTERISTICS OF THE PHOTOELECTRIC EMF OF ANTIMONY SULFOIODIDE.
	15-2228	INFLUENCE OF LIGHT ON SELF POLARIZATION OF SINGLE CRYSTALS OF ANTIMONY SULFOIODIDE.
	15-2229	LUMINESCENCE OF SINGLE CRYSTALS OF THE FERROELECTRIC SEMICONDUCTOR ANTIMONY SULFOIODID
SHEN YR	11-1857	GENERATION OF FAR INFRARED RADIATION BY PICOSECOND LIGHT PULSES IN LITHIUM NIOBATE.
SHENKIN YAS	17-2658	SOLUBILITY ISOBAR IN THE AMMONIUM DIHYDROGEN PHOSPHATE- POTASSIUM DIHYDROGEN PHOSPHATE
SHENOY GK	04-0563	IONIC CHARACTERS FROM MOSSBAUER ISOMER SHIFTS. (DEGREE OF COVALENCY IN IRON DOPED BARI
SHEPHERD IW	30-3318	DOMAIN WALL STRUCTURE IN GADOLINIUM MOLYBDATE BY RAMAN SCATTERING.
	30-3319	RAMAN SCATTERING TECHNIQUES APPLIED TO PROBLEMS IN SOLID STATE PHYSICS (DOMAIN WALL TH
	30-3320	WAVEVECTOR DEPENDENT RELAXATION OF AN OPTICAL PHONON IN GADOLINIUM MOLYBDATE.

ShER ES	08-1441	ANTIFERROMAGNETIC PROPERTIES OF SOME PEROVSKITES. (BISMUTH FERRATE)
SHERMAN AB	11-1753	GENERATION AND PROPAGATION OF HYPERSONIC WAVES IN LITHIUM NIOBATE CRYSTALS. (80-950 DE
	11-1816	ABSORPTION OF ELASTIC WAVES IN REDUCED LITHIUM NIOBATE.
	17-2656	LANDAU-KHALATNIKOV ATTENUATION IN POTASSIUM DIDEUTERIUM PHOSPHATE CRYSTALS.
	17-2659	ULTRASONIC STRESS WAVE INTERACTION WITH DOMAIN WALLS IN POTASSIUM DIDEUTERIUM PHOSPHAT
SHEVCHIK VN	11-1719	TEMPERATURE DEPENDENCES OF THE ABSORPTION OF 9.4 GHZ HYPERSOUND IN ALUMINUM OXIDE AND
ShIBATA H	04-0949	THICKNESS DEPENDENCE OF POLARIZATION REVERSAL IN BARIUM TITANATE SINGLE CRYSTALS.
	04-1006	BARIUM TITANATE SINGLE CRYSTALS SUBJECTED TO MECHANICAL IMPULSES.
SHIBAYAMA K	11-1832	PIEZOELECTRIC LEAKY SURFACE WAVE IN LITHIUM NIOBATE (ELECTROMAGNETIC COUPLING).
	11-1856	PROPAGATION AND AMPLIFICATION OF RAYLEIGH WAVES AND PIEZOELECTRIC LEAKY SURFACE WAVES
SHIBUKAWA K	02-0099	DIELECTRIC AND X-RAY STUDIES OF THE PHASE TRANSITIONS IN ROCHELLE SALT, BARIUM TITANAT
SHIBUYA I	16-2365	NEUTRON DIFFRACTION STUDIES OF FERROELECTRIC POLARIZATION REVERSALS IN SODIUM NITRITE.
	16-2397	PHASE TRANSITION IN SODIUM NITRITE.
	17-2661	NEUTRON DIFFRACTION STUDIES OF FERROELECTRIC PHASE TRANSITIONS IN KDP AND POTASSIUM DI
	27-3121	REEXAMINATION OF THE THERMAL EXPANSION OF THE FERROELECTRIC TRIGLYCINE SULFATE.
SHIELDS M	10-1651	STRUCTURAL AND DIELECTRIC MODIFICATIONS OF TUNGSTEN TRIOXIDE DUE TO NEUTRON IRRADIATIO
	36-3586	TRANSITION IN TIN TELLURIDE- GERMANIUM TELLURIDE ALLOYS. (NEUTRON SCATTERING)
SHIEVER JW	10-1648	STRUCTURAL AND PHASE RELATIONSHIPS AMONG TRIVALENT TUNGSTATES AND MOLYBDATES.
	10-1649	STRUCTURAL AND PHASE RELATIONSHIPS AMONG TRIVALENT TUNGSTATES AND MOLYBDATES.
	14-2092	MODIFIED TECHNIQUE FOR THE GROWTH OF BORACITE CRYSTALS.
	15-2200	GROWTH OF LARGE ANTIMONY SULFOIODIDE CRYSTALS (TO 1 CM DIAMETER): CONTROL OF NEEDLE MO
SHIGENARI T	17-2662	OSCILLATORY BEHAVIOR OF RAMAN SCATTERING INTENSITY WITH TEMPERATURE IN THE FERROELECTR
	17-2663	RAMAN SPECTRUM OF FERROELECTRIC MODE IN A POTASSIUM DIHYDROGEN PHOSPHATE CRYSTAL.
SHILNIKOV AV	28-3228	LOW FREQUENCY DIELECTRIC DISPERSION IN ROCHELLE SALT CRYSTALS.
SHIMADA K	04-0945	CATALYTIC OXIDATION OF CARBON MONOXIDE ON SEMICONDUCTIVE BARIUM TITANATE.
SHIMAOKA K	22-2872	THE STRUCTURE AND PHASE TRANSITION IN POTASSIUM SELENATE.
	31-3334	CRYSTAL STRUCTURE AND PHASE TRANSITION OF HYDROGEN CHLORIDE.
	31-3337	STRUCTURE AND PHASE TRANSITION IN HYDROGEN CHLORIDE.
SHIMAVKA K	31-3331	FERROELECTRICITY AND PHASE TRANSITIONS IN SOLID HYDROGEN HALIDES (SINGLE CRYSTAL X-RAY
SHIMIZU K	09-1624	STUDIES ON THE SINTERING BETWEEN LEAD MONOXIDE AND PZT. PART-1. PART-2: EFFECTS OF AVE
SHIMIZU S	04-0871	CHANGE IN THE DECAY CONSTANT OF TECHNETIUM-99 IN BARIUM TITANATE BY THE FERROELECTRIC
SHIMIZU T	04-0945	CATALYTIC OXIDATION OF CARBON MONOXIDE ON SEMICONDUCTIVE BARIUM TITANATE.
	04-0947	OXIDATION OF CARBON MONOXIDE ON LANTHANUM OXIDE DOPED BARIUM TITANATE.
SHIMOMURA O	17-2499	ELECTROOPTIC CHARACTERISTICS OF DEUTERATED POTASSIUM DIHYDROGEN PHOSPHATE CRYSTAL AND
SHIMOMURA S	09-1624	STUDIES ON THE SINTERING BETWEEN LEAD MONOXIDE AND PZT. PART-1. PART-2: EFFECTS OF AVE
SHIMONY U	08-1282	MOSSBAUER MEASUREMENTS OF BISMUTH FERRATE- BISMUTH FERRATE- LEAD ZIRCONATE SYSTEMS.
	24-2901	MOSSBAUER STUDY OF THE FERROELECTRIC PHASE TRANSITION IN POTASSIUM FERROCYANIDE TRIHYD
SHIMSHONI M	17-2660	ANOMALOUS PROPAGATION OF LONGITUDINAL ULTRASONIC WAVES IN POTASSIUM DIDEUTERIUM PHOSPH
	17-2664	ULTRASONIC MEASUREMENT OF THE ELECTROCALORIC EFFECT IN KDP (NEAR THE CURIE TEMPERATURE
SHINDO Y	15-2239	STRAIN ALONG C-AXIS OF ANTIMONY SULFOIODIDE CAUSED BY ILLUMINATION IN DC ELECTRIC FIEL
SHINMI T	15-2165	ELECTROSTRICTION, PIEZOELECTRICITY AND ELASTICITY IN FERROELECTRIC ANTIMONY SULFOIODID
SHINNAKA Y	16-2366	X-RAY STUDY ON THE DISORDERED STRUCTURE ABOVE THE FERROELECTRIC CURIE POINT IN POTASSI
SHINTANI Y	04-0948	PREPARATION OF THIN BARIUM TITANATE FILMS BY DC DIODE SPUTTERING (ONTO PLATINUM OR FUS
	04-0988	ANOMALY OF RESISTANCE IN REDUCED BARIUM TITANATE CRYSTAL.
SHIONOYA S	33-3381	LUMINESCENCE OF EUROPIUM(3+) ION IN ANTIFERROMAGNETIC POTASSIUM MANGANESE TRIFLUORIDE.
SHIOZAKI Y	02-0077	ACCURATE STRUCTURE ANALYSIS OF FERROELECTRICS IN CONNECTION WITH THEIR PHASE TRANSITIO
	02-0099	DIELECTRIC AND X-RAY STUDIES OF THE PHASE TRANSITIONS IN ROCHELLE SALT, BARIUM TITANAT
	25-2921	SATELLITE X-RAY SCATTERING AND STRUCTURAL MODULATION OF THIOUREA.
	25-2922	STUDIES ON THE MECHANISM OF THE PHASE TRANSITIONS IN THIOUREA. (SATTELITE SCATTERING)
SHIPKO MN	15-2180	MOSSBAUER STUDIES OF ANTIMONY SULFOIODIDE TYPE CRYSTALS.
SHIRAI Y	16-2261	QUASI STATIC SWITCHING CURRENT IN SODIUM NITRITE.
SHIRAISHI T	32-3344	DIELECTRIC PROPERTIES OF LITHIUM HYDRAZINIUM SULFATE.
SHIRAIWA T	04-0683	CERAMIC DIELECTRICS. (BARIUM TITANATE, STRONTIUM TITANATE)
SHIRAKI H	26-2944	PHASE TRANSITION IN DIVALENT METAL DICALCIUM PROPIONATES SUBSTITUTED PARTIALLY BY ACET
SHIRAND G	04-0712	NEUTRON SCATTERING STUDY OF SOFT MODES IN CUBIC BARIUM TITANATE.
SHIRANE G	02-0115	NEUTRON INELASTIC SCATTERING STUDY ON SOFT MODES. (REVIEW OF BROOKHAVEN WORK ON POTASS
	04-0946	INELASTIC NEUTRON SCATTERING FROM SINGLE DOMAIN BARIUM TITANATE. (LATTICE DYNAMICS)
	04-1032	CRITICAL (POLARIZATION) FLUCTUATIONS IN BARIUM TITANATE (ABOVE THE FERROELECTRIC CURIE
	05-1077	SOFT FERROELECTRIC MODES IN LEAD TITANATE (NEUTRON INELASTIC SCATTERING).
	06-1139	DETERMINATION OF THE NORMAL VIBRATIONAL DISPLACEMENTS OF SEVERAL PEROVSKITES (POTASSIU
	06-1208	CRITICAL NEUTRON SCATTERING IN STRONTIUM TITANATE AND POTASSIUM MANGANESE TRIFLUORIDE.
	06-1210	NEW OBSERVATIONS FROM NEUTRON SCATTERING STUDIES OF THE STRUCTURAL PHASE TRANSITION IN
	08-1407	NEUTRON STUDY OF DIFFUSE SCATTERING IN CUBIC POTASSIUM NIOBATE.
	13-2064	NEUTRON SCATTERING STUDY OF THE SOFT MODES IN CUBIC POTASSIUM TANTALATE NIOBATE.
	17-2668	FERROELECTRIC (SOFT) MODE MOTION IN POTASSIUM DIDEUTERIUM PHOSPHATE . (BY NEUTRON TRIP
	30-3254	MECHANISM OF THE FERROELECTRIC PHASE TRANSFORMATION IN RARE EARTH MOLYBDATES. (TERBIUM
	30-3268	NEUTRON SCATTERING STUDY OF THE FERROELECTRIC PHASE TRANSFORMATION IN TERBIUM MOLYBDAT
	30-3269	ZONE BOUNDARY SOFT MODE IN TERBIUM MOLYBDATE.
	33-3374	DISPERSION AND DAMPING OF SOFT ZONE BONDARY PHONONS IN POTASSIUM MANGANESE TRIFLUORIDE
	33-3405	NEUTRON SCATTERING STUDY OF THE LATTICE DYNAMICAL PHASE TRANSITIONS IN POTASSIUM MANGA
	36-3546	NEUTRON SCATTERING ANALYSIS OF THE LINEAR DISPLACEMENT CORRELATIONS IN POTASSIUM TANTA
SHIRASAKI S	05-1075	CHARACTERISTICS OF DEFECT LEAD TITANATE.
	05-1076	DEFECT LEAD TITANATES WITH DIVERSE CURIE TEMPERATURES.
SHIRN GA	04-0814	CONDUCTIVITY INJECTION AND EXTRACTION IN POLYCRYSTALLINE BARIUM TITANATE.
SHIROKOV AM	17-2666	PHASE DIAGRAM OF SODIUM TRIDEUTERIUM(X) TRIHYDROGEN(1-X) PHOSPHATE SYSTEM VERSUS CONCE
	22-2848	(FERROELECTRIC AND OPTICAL PROPERTIES OF) CRYSTALS OF THE ALKALINE TRIHYDROGEN DISELEN
	22-2875	UNUSUAL EFFECT OF AN ELECTRIC FIELD ON THE PERMITTIVITY OF A NEW DELTA PHASE OF SODIUM
	27-3122	AMPLITUDE DEPENDENCE OF INTERNAL FRICTION IN SINGLE CRYSTALS OF FERROELECTRICS. (ROCHE
	28-3204	ATTENUATION OF SOUND WAVES IN DEUTERATED ROCHELLE SALT AT THE PHASE TRANSITION POINT.
	28-3205	ATTENUATION OF SOUND WAVES IN DEUTERATED ROCHELLE SALT AT THE PHASE TRANSITION POINT.
	28-3227	INTERNAL FRICTION AS A FUNCTION OF TEMPERATURE FOR ROCHELLE SALT.
	30-3321	EFFECT OF HYDROSTATIC PRESSURE ON PHASE TRANSITION IN GADOLINIUM MOLYBDATE.
SHKLOVSKII BI	03-0159	ATTENUATION OF CRITICAL VIBRATIONS AND DIELECTRIC LOSSES IN DISPLACIVE TYPE FERROELECT
SHNEERSON VL	03-0320	LOW TEMPERATURE DISPLACEMENT TYPE PHASE TRANSITION IN CRYSTALS. (THEORY, FERROELECTRIC
SHOIJET M	04-0627	VARIATION IN CERTAIN ELECTRICAL PROPERTIES OF SINTERED BARIUM TITANATE AS A FUNCTION O
SHOLOKHOVICH ML	08-1373	FERROELECTRIC SINGLE CRYSTALS WITH LARGE NONLINEARITY. (BARIUM TITANATE HAFNATE GROWN
ShPAK VG	04-0596	SURFACE BREAKDOWN IN VACUUM ON BARIUM TITANATE.
SHPINEL VS	03-0276	ABRUPT (STEPWISE) CHANGE IN PROBABILITY OF THE MOSSBAUER EFFECT AT THE PHASE TRANSITIO
	08-1400	CHANGE OF RESONANCE ABSORPTION SPECTRA OF 23.8 KEV GAMMA-RAYS OF TIN-119 DURING PHASE
	08-1401	MOSSBAUER STUDY AND THEORETICAL ESTIMATION OF INTERNAL ELECTRIC FIELD GRADIENTS IN FER

PAGE 685

SHPINEL VS	36-3566	RESONANCE ABSORPTION OF GAMMA QUANTA IN BARIUM, STRONTIUM, AND CALCIUM STANNATE.
SHRING1 SN	07-1260	TEMPERATURE DEPENDENT OPTICAL MODE IN ANTIFERROELECTRIC LEAD ZIRCONATE BY THE MOSSBAUE
	07-1262	ANTIFERROELECTRIC MODE IN LEAD ZIRCONATE. (LATTICE VIBRATIONS)
SHRIVASTAVA KN	19-2784	(NMR STUDY OF) FERROELECTRIC TRANSITION IN (MANGANESE DOPED) AMMONIUM SULFATE.
	19-2785	MANGANESE(2+) ABSORPTION NEAR THE CRITICAL REGION IN AMMONIUM SULFATE.
SHTRIKMAN S	36-3563	MAGNETOELECTRIC EFFECT IN THE ANTIFERROMAGNET IRON ANTIMONY(2) OXIDE(4) (MAY BE FERROE
SHTUREINA NA	06-1227	THERMALLY STIMULATED CONDUCTIVITY IN STRONTIUM TITANATE SINGLE CRYSTALS.
SHTYRKOV EI	11-1797	INVESTIGATION OF CERTAIN PARAMETERS OF LITHIUM NIOBATE SINGLE CRYSTALS USING SCATTERIN
SHUGUROV VK	15-2117	DIELECTRIC DISPERSION IN ANTIMONY SULFOIODIDE CRYSTALS.
SHUKLA GC	28-3236	DYNAMICS OF FERROELECTRIC ROCHELLE SALT.
SHULENIN BM	04-0760	X-RAY INVESTIGATION OF THE EFFECT OF HYDROSTATIC PRESSURE ON THE STRUCTURE OF BARIUM T
SHULKA GC	28-3237	DYNAMICS OF FERROELECTRIC ROCHELLE SALT.
SHULTIN AA	16-2323	PECULIARITIES IN THE VIBRATIONAL SPECTRA OF CESIUM AND RUBIDIUM NITRATE CRYSTALS IN IT
SHUNCZEWSKI JC	06-1170	ORDER PARAMETER AND PHASE TRANSITIONS OF STRESSED STRONTIUM TITANATE. (EPR OF IRON- VA
SHUR MS	01-0029	FERROELECTRICS AND ANTIFERROELECTRICS.
	03-0268	ANOMALIES IN THE ELECTRICAL CONDUCTIVITY OF A FERROELECTRIC SEMICONDUCTOR NEAR THE CUB
	03-0466	THEORY OF ELECTRIC INDUCTION WAVES IN FERROELECTRIC SEMICONDUCTOR CRYSTALS.
	16-2275	INVESTIGATION OF FERROELECTRIC TRANSITION IN SODIUM NITRITE CRYSTAL BY OBSERVATION OF
SHUSTIN OA	28-3183	FREQUENCY DEPENDENCE OF THE ABSORPTION OF SOUND IN ROCHELLE SALT NEAR ITS UPPER CURIE
SHUTILOV VA	11-1666	ACOUSTIC DIPOLE AND QUADRUPOLE SATURATION OF THE LITHIUM-7 NMR SIGNALS IN LITHIUM NIOB
SHUTZ J	34-3488	PROTEIN SPIN RELAXATION IN HEXAGONAL ICE. PART-2: THE T(1 RHO) MINIMUM.
SHUVAEV AT	08-1439	INTERATOMIC DISTANCES IN OXIDES WITH THE PEROVSKITE STRUCTURE.
SHUVAEVA ET	08-1439	INTERATOMIC DISTANCES IN OXIDES WITH THE PEROVSKITE STRUCTURE.
	13-1940	PHASE TRANSITION IN BARIUM BISMUTH(5+) BISMUTH(3+) OXIDE.
	36-3607	SYNTHESIS AND X-RAY INVESTIGATION OF THE PEROVSKITES BARIUM OR STRONTIUM METAPLUMBATES
SHUVALOV LA	03-0312	PIEZOELECTRIC EFFECT ON THE REFLECTION OF A TRANSVERSE ACOUSTIC WAVE FROM DOMAIN BOUND
	03-0313	REFLECTION AND REFRACTION OF TRANSVERSE SOUND WAVES AT DOMAIN BOUNDARIES IN FERROELECT
	03-0314	REFLECTION AND REFRACTION OF QUASI LONGITUDINAL AND QUASI TRANSVERSE SOUND WAVES BY 18
	03-0330	(EFFECT OF TEMPERATURE ON) QUASI SYMMETIRICAL HARD IONS (PHOSPHATE, TITANATE, SULFATE,
	03-0463	CRYSTALLOPHYSICAL CLASSIFICATION OF FERROELECTRICS. (BASED ON DOMAIN STRUCTURE AND ITS
	03-0464	(MACROSCOPIC) SYMMETRY ASPECTS OF FERROELECTRICITY. (DESCRIPTIVE AND PREDICTIVE POSSIB
	03-0465	SYMMETRY OF CRYSTALS IN WHICH FERROMAGNETIC AND FERROELECTRIC PROPERTIES APPEAR SIMULT
	13-2020	ANOMALIES OF ELASTICITY AND INTERNAL FRICTION IN THE REGION OF THE ANTIFERROELECTRIC C
	17-2665	ANOMALIES IN THE INTERNAL STRESS FOR FERROELECTRIC PHOSPHATES NEAR THEIR CURIE POINTS.
	17-2666	PHASE DIAGRAM OF SODIUM TRIDEUTERIUM(X) TRIHYDROGEN(1-X) PHOSPHATE SYSTEM VERSUS CONCE
	22-2830	DOMAIN STRUCTURE OF THE BETA PHASE OF SODIUM TRIHYDROGEN SELENATE IN ELECTRIC FIELDS.
	22-2836	INFRARED SPECTRA OF ISOTOPIC NONISOMORPHISM OF SODIUM TRIHYDROGEN SELENITE (SHS) OR SO
	22-2840	PHASE TRANSFORMATION AND THERMOOPTICAL AND ELECTROOPTICAL EFFECT IN RUBIDIUM TRIHYDROG
	22-2841	THERMOOPTIC AND ELECTROOPTIC PROPERTIES OF TRIHYDROGEN SELENITES.
	22-2873	ISOTOPIC NONISOMORPHISM OF RUBIDIUM TRIHYDROGEN(1-X) TRIDEUTERIUM(X) DISELENATE CRYSTA
	22-2874	THE PECULIAR DOMAIN STRUCTURE AND ITS UNUSUAL BEHAVIOR IN THE BETA PHASE OF THE FERROE
	22-2875	UNUSUAL EFFECT OF AN ELECTRIC FIELD ON THE PERMITTIVITY OF A NEW DELTA PHASE OF SODIUM
	22-2882	EFFECT OF HIGH PRESSURE ON PHASE TRANSITIONS OF ALKALI TRIHYDROGEN SELENITE CRYSTALS.
	27-3110	DISTINCTIVE FEATURES OF BARKHAUSEN EFFECT IN ROCHELLE SALT AND TRIGLYCINE SULFATE CRYS
	27-3122	AMPLITUDE DEPENDENCE OF INTERNAL FRICTION IN SINGLE CRYSTALS OF FERROELECTRICS. (ROCHE
	27-3123	INFLUENCE OF GAMMA-IRRADIATION ON BARKHAUSEN EFFECT IN FERROELECTRIC MATERIALS (TGS AN
	28-3204	ATTENUATION OF SOUND WAVES IN DEUTERATED ROCHELLE SALT AT THE PHASE TRANSITION POINT.
	28-3205	ATTENUATION OF SOUND WAVES IN DEUTERATED ROCHELLE SALT AT THE PHASE TRANSITION POINT.
	28-3227	INTERNAL FRICTION AS A FUNCTION OF TEMPERATURE FOR ROCHELLE SALT.
	28-3229	PIEZOELECTRIC EFFECT AND INTERNAL FRICTION IN GAMMA-IRRADIATED ROCHELLE SALT CRYSTALS.
	30-3255	CERTAIN GROWTH, LASING AND SPECTROSCOPIC PROPERTIES OF NEODYMIUM(3+) DOPED GADOLINIUM
SHUVAYEV AT	03-0243	PEROVSKITE FAMILY AND CHEMICAL BOND. (COVALENCY, DIRECTIVITY CONDITIONS FOR ELECTRIC O
SHVORNEVA LI	02-0128	NEW FERROELECTRICS AND SEIGNETTO MAGNETS OF THE PEROVSKITE AND PYROCHLORE TYPE. SYNTHE
	08-1460	STRUCTURE AND PROPERTIES OF SOME NEW FERROELECTRIC MATERIALS WITH PEROVSKITE STRUCTURE
SIAPKAS D	25-2916	INFRARED AND RAMAN STUDIES IN THE FERROELECTRIC TRANSITION OF THIOUREA.
SICIUNAS E	04-0930	CHARACTERISTICS OF RF SPUTTERED BARIUM TITANATE FILMS ON SILICON.
SIDNENKO EV	17-2503	DOUBLE DIELECTRIC HYSTERESIS LOOP OF POTASSIUM DIDEUTERIUM PHOSPHATE CRYSTALS.
	17-2504	POLARIZATION OF A RUBIDIUM DIHYDROGEN PHOSPHATE CRYSTAL IN THE PHASE TRANSITION REGION
	17-2505	POLARIZATION (DOMAIN STRUCTURE) AND PHASE TRANSITION IN KDP SINGLE CRYSTAL. (ABSTRACT
	17-2506	POLARIZATION OF A RUBIDIUM DIHYDROGEN PHOSPHATE CRYSTAL IN THE PHASE TRANSITION REGION
SIDORENKO EN	04-1012	INFLUENCE OF A CONSTANT ELECTRIC FIELD ON THE DIELECTRIC PROPERTIES OF POLYCRYSTALLINE
	04-1014	UHF REVERSAL PROPERTIES OF BARIUM TITANATE TYPE FERROELECTRIC CERAMICS.
SIEGEL RW	13-1955	SCANNING ELECTRON MICROSCOPIC EXAMINATION OF SINTERED BARIUM SODIUM NIOBATES.
SIEGLE G	34-3475	INTERPRETATION OF THE PROTON SPIN LATTICE RELAXATION IN HEXAGONAL ICE.
SIENKO MJ	10-1627	EXAMINATION OF SUBSTOICHIOMETRIC TUNGSTEN OXIDE(3-X) CRYSTALS BY ELECTRON MICROSCOPY.
	10-1629	EFFECT OF OXYGEN DEFICIENCY ON ELECTRICAL TRANSPORT PROPERTIES OF TUNGSTEN TRIOXIDE CR
SILBERGLITT R	06-1211	EFFECT OF THE QUARTIC ANHARMONICITY ON THE SOFT PHONON INSTABILITY OF STRONTIUM TITANA
	06-1213	MICROSCOPIC CALCULATION OF THE RESPONSE FUNCTION OF THE SOFT ZONE BOUNDARY PHONON IN S
SILBERNAGEL BG	33-3376	EXCHANGE NARROWING: MAGNETIC RESONANCE LINE SHAPES AND SPIN CORRELATION IN PARAMAGNETI
SILIN AT	33-3387	SPECIFIC HEAT OF POTASSIUM MANGANESE TRIFLUORIDE.
SILNITSKAYA GB	17-2606	EFFECT OF IMPURITIES ON THE STRUCTURE OF POTASSIUM DIHYDROGEN PHOSPHATE CRYSTALS.
	17-2607	STRUCTURE OF FERROELECTRIC CRYSTALS OF THE POTASSIUM DIHYDROGEN PHOSPHATE GROUP.
	17-2613	DIELECTRIC ANOMALIES IN CRYSTALS OF POTASSIUM DIHYDROGEN PHOSPHATE POTASSIUM DIDEUTERI
SILSBEE HB	17-2436	ABSENCE OF A LINE WIDTH TRANSITION FOR THE PROTONS IN POTASSIUM DIHYDROGEN PHOSPHATE T
SILVA WJ	13-2021	DEVELOPMENT OF SINGLE CRYSTAL BARIUM SODIUM NIOBATE AS A NONLINEAR OPTICAL MATERIAL. (
SILVERMAN BD	02-0116	PHONONS, ELECTRONS, AND PROTONS IN FERROELECTRIC MATERIALS.
	03-0167	POLARIZATION DISTRIBUTION IN THE VICINITY OF A FERROELECTRIC SEMICONDUCTOR INTERFACE.
	03-0216	DOMAIN STATISTICS AND FERROELECTRIC TRANSIENTS.
	03-0296	ASYMMETRIC MINOR LOOP AND TRANSIENT OPEN LOOP COMPUTATIONS FOR FERROELECTRICS.
	03-0304	OPTICAL MODES OF THE PEROVSKITE STRUCTURE. (THEORY, STRONTIUM TITANATE DATA)
	03-0467	DIELECTRIC RESONANCE AND RELAXATION IN ELECTRICALLY BIASED KDP AND DEUTERATED ISOMORPH
	03-0468	TEMPERATURE DEPENDENT SPECTRUM OF AN ANTIFERROELECTRIC LINEAR CHAIN MODEL. (PEROVSKITE
	04-0950	FERROELECTRICITY (IN BARIUM TITANATE TYPE STRUCTURES) AND LATTICE DYNAMICS.
	05-1078	APPARENT SOFT MODE LINE WIDTH DIVERGENCE IN LEAD TITANATE.
	06-1193	EFFECT OF TEMPERATURE AND PRESSURE ON THE ELECTRON PARAMAGNETIC RESONANCE SPECTRA OF S
	06-1212	MICROWAVE ABSORPTION IN CUBIC STRONTIUM TITANATE.
	08-1366	INFRARED ACTIVE MODES IN POTASSIUM TANTALATE.
	17-2667	COUPLED ORDER DISORDER PHONON SYSTEM WITH DAMPING: KDP (FREQUENCY DEPENDENT SUSCEPTIBI
	27-2959	THERMODYNAMIC STABILITY OF THIN FERROELECTRIC FILMS. (TRIGLYCINE SULFATE)
	37-3644	EQUIVALENT CIRCUIT ANALYSIS OF A FERROELECTRIC PHOTOCONDUCTOR MEMORY DEVICE.

SILVERMAN BD   37-3659 DEPOLARIZATION FIELD IN THIN FERROELECTRIC FILMS.
SILVESTROVA IM   15-2118 EFFECT OF COMPOSITION ON GROWTH, DIELECTRIC, AND PHOTOELECTRIC PROPERTIES OF ANTIMONY
                    15-2138 PREPARATION OF ANTIMONY SULFOIODIDE TEXTURES FROM THE MELT UNDER PRESSURE AND INVESTIG
                    15-2248 EFFECTS OF TEMPERATURE, ELECTRIC FIELD, AND ILLUMINATION ON THE ABSORPTION OF ULTRASOU
                    28-3229 PIEZOELECTRIC EFFECT AND INTERNAL FRICTION IN GAMMA-IRRADIATED ROCHELLE SALT CRYSTALS.
SILVIDI AA   22-2876 PROTON MAGNETIC RESONANCE STUDY OF SODIUM TRIHYDROGEN SELENATE AND POTASSIUM TRIHYDROG
SIMANEK E   04-0977 PARAMAGNETIC RESONANCE OF PLATINUM IONS IN NIOBIUM DOPED BARIUM TITANATE SINGLE CRYSTA
                    06-1214 FREE CARRIER OPTICAL ABSORPTION IN NIOBIUM DOPED STRONTIUM TITANATE.
SIMEONOV K   04-0966 EFFECT OF EXCESS TITANIA ON THE ELECTRIC RESISTANCE OF SEMICONDUCTIVE BARIUM TITANATE
SIMHONY M   03-0469 MEASUREMENT OF THE PYROELECTRIC COEFFICIENT AND PERMITTIVITY FROM THE PYROELECTRIC RES
                    13-2018 PYROELECTRIC VOLTAGE RESPONSE TO SHORT INFRARED LASER PULSES IN TRIGLYCINE SULFATE AND
                    13-2019 PYROELECTRIC VOLTAGE RESPONSE TO RECTANGULAR INFRARED SIGNALS IN TRIGLYCINE SULFATE AN
                    13-2022 PYROELECTRIC VOLTAGE RESPONSE TO STEP SIGNALS OF INFRARED RADIATION IN TRIGLYCINE SULF
SIMONOV VI   22-2881 CRYSTAL STRUCTURE OF RUBIDIUM TRIHYDROGEN SELENITE.
SIMOVA P   27-3114 LOW FREQUENCY RAMAN SPECTRA OF FERROELECTRIC CRYSTALS OF THE TRIGLYCINE SULFATE TYPE.
SIMOVA PD   27-3113 LOW FREQUENCY RAMAN SCATTERING BY FERROELECTRIC TRIGLYCINE FLUOBERYLLATE CRYSTALS.
SINGAL SP   37-3672 ACOUSTIC TRANSIENTS GENERATED BY ANTIFERROELECTRIC- FERROELECTRIC TRANSDUCERS.
SINGH K   11-1703 DOMAIN STRUCTURE IN LITHIUM NIOBATE SINGLE CRYSTALS.
SINGH S   03-0223 DIELECTRIC SUSCEPTIBILITY AND THE ORDER OF FERROELECTRIC PHASE TRANSITIONS.
                    05-1079 NONLINEAR OPTICAL PROPERTIES OF FERROELECTRIC LEAD TITANATE.
                    05-1080 NONLINEAR OPTICAL PROPERTIES OF FERROELECTRIC LEAD TITANATE.
                    13-2023 NONLINEAR OPTICAL PROPERTIES OF FERROELECTRIC LEAD NIOBIUM(4) OXIDE(12).
                    13-2024 OPTICAL AND FERROELECTRIC PROPERTIES OF BARIUM SODIUM NIOBATE. (OPTICAL ABSORPTION, RE
                    13-2025 ROLE OF HYDROGEN IN POLARIZATION REVERSAL OF FERROELECTRIC BARIUM SODIUM NIOBATE.
                    26-2945 NONLINEAR OPTICAL SUSCEPTIBILITY OF LITHIUM FORMATE MONOHYDRATE.
SINHA AP   04-0964 ROLE OF JAHN TELLER EFFECTS IN ORIGIN OF FERROELECTRICITY, OCCURRENCE OF ORDERED PHASE
SINHA KP   04-0584 RESISTIVITY ANOMALY IN SEMICONDUCTING BARIUM TITANATE. (POSITIVE TEMPERATURE COEFFICIE
                    04-0964 ROLE OF JAHN TELLER EFFECTS IN ORIGIN OF FERROELECTRICITY, OCCURRENCE OF ORDERED PHASE
SINYAKOV EV   01-0027 FERROELECTRICS AND OXIDE SEMICONDUCTORS.
                    04-0655 ANTIPARALLEL SWITCHING IN ALPHA DOMAIN BARIUM TITANATE CRYSTALS.
                    04-0951 CRITICAL FIELD IN CERAMIC BARIUM TITANATE.
                    04-0952 DIRECT OBSERVATION OF 180 DEG DOMAINS IN BARIUM TITANATE BY MEANS OF A POLARIZING MICR
                    04-0953 ELECTROOPTICAL PROPERTIES OF SINGLE CRYSTALS OF SOLID SOLUTIONS OF BARIUM TITANATE AND
                    04-0954 EFFECT OF EXTERNAL ACTIONS ON THE DOMAIN STRUCTURE OF SINGLE CRYSTALS FORMED BY BARIUM
                    04-0955 ELECTROOPTICAL PROPERTIES OF BARIUM TITANATE- TANTALUM PENTOXIDE SOLID SOLUTION SINGLE
                    04-0956 EFFECT OF TRANSVERSE COMPRESSION OF DIELECTRIC PROPERTIES OF BARIUM TITANATE SINGLE CR
                    04-0957 EFFECT OF EXTERNAL ACTIONS ON THE REPOLARIZATION OF BARIUM TITANATE SINGLE CRYSTALS.
                    04-0958 ELECTROOPTICAL EFFECT IN SINGLE CRYSTALS OF BARIUM TITANATE- ZINC TITANATE SOLID SOLUT
                    04-0959 EFFECT OF NONISOVALENT SUBSTITUTION OF TITANIUM(4+) IONS ON THE ELECTROOPTICAL PROPERT
                    04-0960 GROWING SINGLE CRYSTALS OF SOLID SOLUTION BASED ON BARIUM TITANATE, WHERE BARIUM(2+) A
                    04-0961 INFLUENCE OF UNIAXIAL TENSILE STRESSES ON THE ELECTRICAL PROPERTIES AND DOMAIN STRUCTU
                    04-0962 PIEZOELECTRIC PROPERTIES OF POLARIZED BARIUM TITANATE CERAMICS SUBJECTED TO UNIAXIAL T
                    04-0965 REPOLARIZATION PROCESSES OF BARIUM TITANATE SINGLE CRYSTALS IN ORTHORHOMBIC AND RHOMBO
                             SEMICONDUCTOR PROPERTIES OF BARIUM (TITANATE, STANNATE) SOLID SOLUTIONS DOPED BY TUNGS
SIOUFFI JC   16-2367 ELECTRICAL PROPERTIES OF RHOMBOHEDRAL POTASSIUM NITRATE.
                    16-2368 TWO MODES OF ELECTRICAL CONDUCTIVITY IN POTASSIUM NITRATE.
SIROTA NN   09-1595 CURIE TEMPERATURES OF SOLID SOLUTIONS IN A LEAD ZIRCONATE- LEAD TITANATE- LEAD METANIO
                    09-1596 CRYSTAL STRUCTURE AND PHASE RELATIONS IN THE LEAD ZIRCONATE LEAD TITANATE- LEAD METANI
                    09-1597 THERMAL VARIATION OF SPONTANEOUS POLARIZATION AND COERCIVE FORCE OF THE PEROVSKITES IN
                    27-3124 BAND GAP OF TRIGLYCINE SELENATE, TRIGLYCINE FLUOBERYLLATE, TRIGLYCINE SULFATE AND ITS
                    27-3125 DIELECTRIC PERMEABILITY AND DIELECTRIC LOSS ANGLE TANGENT OF TRIGLYCINE SELENATE DEPEN
                    27-3126 THERMAL HYSTERESIS OF THE DIELECTRIC CONSTANT OF TRIGLYCINE SULFATE AS A FUNCTION OF T
SIVOVA R   04-0966 EFFECT OF EXCESS TITANIA ON THE ELECTRIC RESISTANCE OF SEMICONDUCTIVE BARIUM TITANATE
SIXOU P   34-3447 DIPOLAR RELAXATION AT LOW TEMPERATURE OF ICE SINGLE CRYSTAL.
                    34-3448 ELECTRICAL BEHAVIOR OF ICE AT LOW TEMPERATURES, FERROELECTRICITY AND SPACE CHARGE.
                    34-3458 STUDY OF CONDUCTIVITY AND DIPOLAR RELAXATION IN DOPED ICE SINGLE CRYSTALS.
SIZOVA LH   17-2542 DIELECTRIC PROPERTIES OF GAMMA-IRRADIATED POTASSIUM DIHYDROGEN PHOSPHATE CRYSTALS.
SKAAR DL   04-0831 ENERGY LOSS NONPOLAR STATE POLYCRYSTALLINE BARIUM TITANATE. (150-500 DEG C, 20-50 KILO
SKALYO J   17-2668 FERROELECTRIC (SOFT) MODE MOTION IN POTASSIUM DIDEUTERIUM PHOSPHATE . (BY NEUTRON TRIP
SKAPSKI AC   13-1978 BISMUTH(3) NIOBIUM(17) OXIDE(47): POTENTIALLY FERROELECTRIC CRYSTAL STRUCTURE OF THE
                    27-2996 CRYSTALLOGRAPHIC STUDIES OF IRRADIATION FIELD TREATED TRIGLYCINE SULFATE: NEW STRUCTU
SKHNENKO VF   08-1439 INTERATOMIC DISTANCES IN OXIDES WITH THE PEROVSKITE STRUCTURE.
SKINNER JG   13-1924 USE OF PEROVSKITE PARAELECTRICS IN BEAM DEFLECTORS AND LIGHT MODULATORS. (POTASSIUM TA
SKINNER SM   02-0117 MAGNETICALLY ORDERED FERROELECTRIC MATERIALS (REVIEW, 251 REFS).
SKUBITSKII VN   13-1967 ELASTIC AND PIEZOELECTRIC PROPERTIES OF CADMIUM PYRONIOBATE IN STRONG ELECTRIC FIELDS.
SLACK JR   04-0604 GROWTH AND STRUCTURE OF BARIUM TITANATE EVAPORATED FILMS.
                    04-0605 GROWTH STRUCTURE (MICROSTRUCTURE) AND ELECTRICAL PROPERTIES OF FLASH EVAPORATED BARIUM
                    04-0967 ELECTRICAL PROPERTIES OF FLASH EVAPORATED FERROELECTRIC BARIUM TITANATE THIN FILMS. (T
                    04-0968 FLASH EVAPORATION OF FERROELECTRIC THIN FILMS OF POLYCRYSTALLINE BARIUM TITANATE.
SLAK J   18-2720 CESIUM-133 SPIN LATTICE RELAXATION AND RESONANCE IN FERROELECTRIC CESIUM DIDEUTERIUM A
                    27-2965 BERYLLIUM-9 QUADRUPOLE PERTURBED NMR STUDY OF THE FERROELECTRIC TRANSITION IN DEUTERAT
SLATER JC   03-0470 INTRODUCTION TO THE THEORY OF FERROELECTRICITY.
SLEIGHT AW   30-3262 PRECISION PARAMETERS OF SOME LANTHANON MOLYBDATE TYPE RARE EARTH MOLYBDATES. (CRYSTAL
                    30-3264 PRECISION PARAMETERS OF THE FERROELECTRIC RARE EARTH MOLYBDATES: LANTHANUM MOLYBDATE.
SLEPTSOV AI   15-2188 INJECTION CURRENTS AND ACCUMULATION OF POLARIZATION CHARGE IN ANTIMONY SULFOIODIDE.
SLIVKA VYU   15-2114 ABSORPTION OF ULTRASOUND IN A(V)B(VI)C(VII) FERROELECTRIC SEMICONDUCTORS AT TEMPERATUR
                    15-2115 ABSORPTION OF ULTRASOUND IN AN ANTIMONY SULFOIODIDE FERROELECTRIC SEMICONDUCTOR NEAR I
                    15-2116 DETERMINATION OF THE LATTICE PARAMETERS AND CURIE POINTS OF SOLID SOLUTIONS OF A(V)B(V
                    15-2121 PIEZOELECTRIC VIBRATIONS IN ANTIMONY SULFOIODIDE CRYSTALS NEAR THE CURIE POINT (EFFECT
                    15-2125 ANISOTROPY OF THE REFLECTION SPECTRA OF ANTIMONY SULFOIODIDE CRYSTALS.
                    15-2128 PIEZOELECTRIC RESISTANCE EFFECT IN BISMUTH SELENOIODIDE CRYSTALS.
                    15-2142 PRODUCTION AND SOME OPTICAL PROPERTIES OF ANTIMONY SULFOBROMIDE IN GLASSY AND CRYSTALL
SLOBODNIK AJ   01-0028 MICROWAVE ACOUSTICS HANDBOOK. VOL 1: SURFACE WAVE VELOCITIES. (BARIUM SODIUM NIOBATE,
                    11-1817 MICROWAVE FREQUENCY ACOUSTIC SURFACE WAVE LOSS MECHANISMS ON LITHIUM NIOBATE (TEMPERAT
                    11-1818 NONLINEAR EFFECTS IN MICROWAVE ACOUSTIC LITHIUM NIOBATE SURFACE DELAY LINES.
                    11-1819 NEW HIGH FREQUENCY HIGH COUPLING LOW BEAM STEERING CUT FOR ACOUSTIC SURFACE WAVES ON L
                    11-1820 NEW HIGH COUPLING LOW DIFFRACTION CUT FOR ACOUSTIC SURFACE WAVES ON LITHIUM NIOBATE.
                    11-1821 TEMPERATURE COEFFICIENTS OF ACOUSTIC SURFACE WAVE VELOCITY AND DELAY ON LITHIUM NIOBAT
                    35-3511 ACOUSTIC SURFACE WAVE LOSS MECHANISMS ON BISMUTH(12) GERMANIUM OXIDE(20) AT MICROWAVE
                    35-3512 ELASTIC, PIEZOELECTRIC, AND DIELECTRIC CONSTANTS OF BISMUTH(12) GERMANIUM OXIDE(20).

SLONCZEWSKI JC	06-1103	RAMAN SCATTERING AND PHASE TRANSITIONS IN STRESSED STRONTIUM TITANATE.
	06-1215	ANALYSIS OF STRESS AND TEMPERATURE DEPENDENCE OF FLUORESCENCE IN CHROMIUM(3+) DOPED ST
	06-1216	INTERACTION OF ELASTIC STRAIN WITH THE STRUCTURAL TRANSITION OF STRONTIUM TITANATE.
SLY FAW	04-0587	EFFECT OF OXIDE IMPURITIES ON THE ELECTRICAL RESISTIVITY OF LANTHANUM DOPED BARIUM TIT
SLYUDKIN OP	17-2425	RAMAN SPECTRA OF POLYMORPHIC MODIFICATIONS OF POTASSIUM DIHYDROGEN PHOSPHATE IN AN ELE
SMAZHEVSKAYA EG	09-1514	TIME AND TEMPERATURE DEPENDENCES OF THE PARAMETERS OF A FERROELECTRIC CERAMIC.
	09-1591	EFFECT OF FERROELECTRIC PHASE TRANSITION CONDITIONS ON DIELECTRIC AND PIEZOELECTRIC PR
SMIRNOV VP	10-1655	NEUTRON DIFFRACTION STUDY OF (TETRAGONAL- CUBIC PEROVSKITE) STRONTIUM IRON TUNGSTATE .
SMIRNOVA NP	04-0607	MAGNITUDE OF SPONTANEOUS POLARIZATION IN THIN CRYSTALS AND FINE PARTICLES OF BARIUM TI
	04-0612	SINGLE CRYSTAL BARIUM TITANATE FILMS GROWN FROM THE MELT IN AN OXYGEN ATMOSPHERE.
SMITH AB	12-1892	ANOMALOUS (LONGITUDINAL WAVE) ATTENUATION IN LITHIUM TANTALATE (EFFECT OF VACANCIES) .
SMITH AM	16-2340	DIELECTRIC AND ELECTRICAL CONDUCTIVITY STUDIES IN POTASSIUM NITRITE AND POTASSIUM NITR
	16-2341	DIELECTRIC DISPERSION IN THE PARAELECTRIC PHASE OF POTASSIUM NITRATE.
	16-2369	PHASE TRANSFORMATIONS IN POTASSIUM NITRATE AND POTASSIUM NITRITE. (DOES NOT SEEM FERRO
SMITH AW	02-0053	PROPERTIES OF TUNGSTEN BRONZE FERROELECTRICS. (REVIEW, 33 REFS)
	13-1948	(TETRAGONAL) TUNGSTEN BRONZE TYPE CRYSTALS GROWN IN MOLTEN SOLUTIONS OF THE PSEUDOSYST
	13-2003	CRYSTAL GROWTH AND ELECTROOPTIC PROPERTIES OF TUNGSTEN BRONZE CRYSTALS FROM THE POTASS
	13-2015	CRYSTAL GROWTH AND NONLINEAR OPTICAL PROPERTIES OF POTASSIUM LITHIUM NIOBATES.
	13-2017	TUNGSTEN BRONZE FIELD IN THE (OXIDE) SYSTEM POTASSIA LITHIA NIOBIA (POTASSIUM LITHIUM
	13-2026	NONLINEAR OPTICAL PROPERTIES OF POTASSIUM LITHIUM NIOBATES (TRANSPARENT, TETRAGONAL TU
	13-2027	NONLINEAR OPTICAL PROPERTIES OF POTASSIUM LITHIUM NIOBATE.
	13-2028	OPTICAL AND FERROELECTRIC PROPERTIES OF (THE TUNGSTEN BRONZE SYSTEM) POTASSIUM SODIUM
	17-2669	OPTICAL MIXING OF COHERENT AND INCOHERENT LIGHT. (KDP CRYSTAL FILTER)
SMITH DK	13-1935	CRYSTAL STRUCTURE OF BISMUTH TITANIUM OXIDE.
	13-2058	CRYSTAL STRUCTURE OF BISMUTH TITANIUM NIOBIUM OXIDE.
SMITH GV	13-1897	THE SEARCH FOR NEW FERROELECTRICS WITH THE TUNGSTEN BRONZE STRUCTURE.
SMITH GW	37-3649	DEPENDENCE OF THE SMALL SIGNAL PARAMETERS OF FERROELECTRIC CERAMIC RESONATORS UPON STA
SMITH HI	11-1853	REFLECTION OF ELASTIC SURFACE WAVE FROM PERIODIC ARRAYS ON LITHIUM NIOBATE.
SMITH HL	04-0841	FRACTURE TOUGHNESS OF COMMERCIAL BARIUM TITANATE TRANSDUCER MATERIAL.
SMITH JS	06-1217	PREPARATION AND CHARACTERIZATION OF ALKOXY-DERIVED STRONTIUM ZIRCONATE AND STRONTIUM T
SMITH JW	08-1440	A MICROSCOPIC THEORY FOR THE DIELECTRIC PROPERTIES OF LEAD MAGNESIUM NIOBATE.
SMITH PL	02-0062	AGING OF BARIUM TITANATE AND LEAD ZIRCONATE TITANATE FERROELECTRIC CERAMICS. (REVIEW,
SMITH RC	11-1764	OPTIMUM SECOND HARMONIC GENERATION IN LITHIUM NIOBATE.
SMITH RT	11-1724	LITHIUM TANTALATE AND LITHIUM NIOBATE PIEZOELECTRIC RESONATORS IN THE MEDIUM FREQUENCY
	11-1823	TEMPERATURE DEPENDENCE OF THE ELASTIC, PIEZOELECTRIC, AND DIELECTRIC CONSTANTS OF LITH
	11-1851	MEDIUM FREQUENCY RESONATORS OF LITHIUM NIOBATE AND LITHIUM TANTALATE.
SMITH RW	11-1822	GAMMA-RADIATION EFFECTS IN LITHIUM NIOBATE.
SMITH SH	36-3558	FLUX GROWTH OF MAGNESIUM OXIDE AND LANTHANUM ALUMINATE CRYSTALS DOPED WITH ISOTOPE 170
SMITH WD	09-1598	ELECTROELASTIC CONSTANT MEASUREMENTS FOR PLZT CERAMICS.
	09-1599	SCATTERING MODE FERROELECTRIC PHOTOCONDUCTOR IMAGE STORAGE AND DISPLAY DEVICES. (PLZT
	09-1618	ELECTROOPTIC CERAMICS AS WAVELENGTH SELECTION DEVICES IN DYE LASERS (PLZT).
SMOKE EJ	04-0551	METHOD FOR ACHIEVING HIGH DIELECTRIC CONSTANT TEMPERATURE STABLE FERROELECTRICS. (BARI
SMOLENSKII GA	01-0029	FERROELECTRICS AND ANTIFERROELECTRICS.
	01-0030	FERROELECTRICS AND ANTIFERROELECTRICS.
	02-0118	NEW CLASSES OF FERROELECTRICS OF DISPLACEMENT TYPE. (BORACITES, MOLYBDATES, TUNGSTEN B
	02-0119	PHYSICAL PHENOMENA IN FERROELECTRICS WITH DIFFUSED PHASE TRANSITION. (POLARIZATION MEC
	03-0471	LIGHT SCATTERING BY ACOUSTIC PHONONS IN FERROELECTRIC AND PIEZOELECTRIC CRYSTALS.
	03-0472	THERMODYNAMIC THEORY OF CRYSTALS POSSESSING FERROELECTRIC AND FERROMAGNETIC PROPERTIES
	08-1441	ANTIFERROMAGNETIC PROPERTIES OF SOME PEROVSKITES. (BISMUTH FERRATE)
	11-1753	GENERATION AND PROPAGATION OF HYPERSONIC WAVES IN LITHIUM NIOBATE CRYSTALS. (80-950 DE
	11-1754	LIGHT SCATTERING BY ULTRASONIC WAVES IN LITHIUM NIOBATE CRYSTALS. (PHOTOELASTIC CONSTA
	11-1792	ELECTRON PARAMAGNETIC RESONANCE SPECTRUM IN (MANGANESE DOPED) LITHIUM NIOBATE AND (COP
	36-3569	X-RAY AND ELECTRIC INVESTIGATIONS OF THE SYSTEMS YTTRIUM MANGANESE(1-X) (B(X)) OXYGEN(
SMOLYAKOV BP	17-2419	MACROSCOPIC STIMULATED POLARIZATION ECHO IN FERROELECTRICS. (POTASSIUM DIHYDROGEN PHOS
	17-2564	POLARIZATION ECHO IN THE FERROELECTRIC SINGLE CRYSTAL POTASSIUM DIHYDROGEN PHOSPHATE.
SMOLYANINOV NP	13-2029	THERMOGRAPHIC STUDY OF THE REACTION OF BISMUTH(4) TITANIUM(3) OXIDE(12) WITH VANADIUM
SMUTNY F	14-2078	FERROELECTRIC TRANSITION IN COBALT- IODINE- BORACITE UNDER PRESSURE.
	14-2099	FERROELECTRIC TRANSITION IN COBALT IODINE BORACITE.
	14-2100	PIEZOELECTRIC PROPERTIES OF COBALT IODINE BORACITE.
	14-2100	PIEZOELECTRIC PROPERTIES OF COBALT IODINE BORACITE.
	14-2101	SPECIFIC HEAT ANOMALY OF FERROELECTRIC COBALT IODINE BORACITE.
	17-2458	REGULAR BEHAVIOR OF SOLID SOLUTIONS IN POTASSIUM DIHYDROGEN(1-X) DIDEUTERIUM(X) PHOSPH
	29-3239	SOME PROPERTIES OF NONDEUTERATED AND DEUTERATED LITHIUM THALLIUM TARTRATE MONOHYDRATE.
SMYTH DM	04-0602	ENERGY STORAGE IN CERAMIC DIELECTRICS. (STRONTIUM TITANATE, BARIUM TITANATE)
	04-0663	ELECTRICAL CONDUCTIVITY OF SINGLE CRYSTALLINE BARIUM TITANATE.
	04-0664	OXYGEN STOICHIOMETRY OF DONOR DOPED BARIUM TITANATE AND TITANIUM DIOXIDE. (STOICHIOMET
	04-0969	CATION NONSTOICHIOMETRY IN TERNARY OXIDES. (BARIUM TITANATE)
SNOW GS	09-1600	FABRICATION OF ELECTROOPTIC PLZT CERAMICS BY ATMOSPHERE SINTERING.
SOBOLEEV CHS	17-2413	PRODUCTION OF LARGE SEED CRYSTALS OF KDP.
SOBOLEVA LV	16-2266	FERROELECTRIC PHASE TRANSITION IN SILVER SODIUM DINITRITE
	16-2370	PROPERTIES OF SILVER SODIUM DINITRITE AND SILVER SODIUM NITRITE. (DIFFERENTIAL THERMAL
SODA G	36-3538	FERROELECTRICITY AND PHASE TRANSITION IN AMMONIUM DICHLORO ACETATE AND DEUTERATED AMMO
	17-2594	ABSENCE OF MOTIONAL NARROWING IN THE PROTON NMR IN POTASSIUM DIHYDROGEN PHOSPHATE AT H
SOEST JF	18-2736	ARSENIC-75 NMR RESONANCE IN AMMONIUM DIHYDROGEN ARSENATE AND DEUTERO AMMONIUM DIDEUTER
SOEYA T	27-3127	LIP ETCH PATTERNS ON THE FACES OF TRIGLYCINE SULFATE CRYSTALS.
SOKOLOVA LS	08-1461	ELECTRICAL PROPERTIES OF NONLINEAR FERROELECTRIC CERAMICS WITH THE PEROVSKITE STRUCTUR
SOLIMAN MA	17-2679	SPECIFIC HEAT OF SOME KDP TYPE CRYSTALS. (POTASSIUM DIHYDROGEN PHOSPHATE, RUBIDIUM AND
SOLODUKHIN AV	13-1952	NONLINEAR PROPERTIES AND PHASE TRANSITIONS OF STRONTIUM BISMUTH TITANATES.
SOLOMON AL	04-0669	PREPARATION AND PROPERTIES OF THIN BARIUM TITANATE FILMS.
SOLOVEV AA	36-3528	X-RAY PHASE STUDY OF THE CADMIUM BORATE- ABO(3) SYSTEMS. (WHERE A IS STRONTIUM OR BARI
SOLOVEV LA	02-0047	SYNTHESIS AND CERTAIN PROPERTIES OF NEW COMPOUNDS WITH A PYROCHLORE STRUCTURE. (NIOBAT
SOLOVEV SP	04-0645	DIELECTRIC PERMEABILITY OF BARIUM TITANATE SINGLE CRYSTALS IN THE PHASE TRANSITION REG
	04-0646	DIELECTRIC DISPERSION IN BARIUM TITANATE.
	04-0647	DIELECTRIC SPECTRUM OF BARIUM TITANATE. (AND MICROSCOPIC MODEL OF LOCALLY ORDERED "NUC
	04-0648	DIELECTRIC DISPERSION OF IRRADIATED BARIUM TITANATE NEAR THE PHASE TRANSITION.
	04-0652	DIELECTRIC DISPERSION OF IRRADIATED BARIUM TITANATE IN THE PHASE TRANSITION REGION.
	04-0970	EFFECTS OF (MONOENERGETIC) ELECTRON AND (COBALT-60) GAMMA-IRRADIATION ON BARIUM TITANA
	04-0971	(IN PILE, FAST NEUTRON) RADIATION INDUCED PHASE TRANSFORMATIONS IN (CERAMIC) BARIUM AN
	04-0972	RADIATION PHYSICS OF FERROELECTRICS OF BARIUM TITANATE TYPE.
	05-1081	X-RAY INVESTIGATION OF IRRADIATED LEAD TITANATE.

SOLOVEV SP	07-1261	DIPOLE STRUCTURE AND INTERNAL ELECTRIC FIELDS IN LEAD ZIRCONATE.
	08-1442	CALCULATIONS OF THE INTERNAL ELECTRIC FIELDS AND ELECTRIC FIELD GRADIENTS IN THE PEROV
	08-1443	X-RAY AND DIELECTRIC STUDIES OF IRRADIATED PEROVSKITE TYPE COMPOUNDS.
	08-1455	CALCULATION OF INTERNAL ELECTRIC FIELDS AND THEIR GRADIENTS IN ABO(3) PEROVSKITE COMPO
	08-1474	EFFECT OF IRRADIATION OF DIELECTRIC PROPERTIES OF SOME FERRO-, ANTIFERRO- AND PARAELE
	10-1655	NEUTRON DIFFRACTION STUDY OF (TETRAGONAL- CUBIC PEROVSKITE) STRONTIUM IRON TUNGSTATE .
SOLOVEVA ES	09-1514	TIME AND TEMPERATURE DEPENDENCES OF THE PARAMETERS OF A FERROELECTRIC CERAMIC.
	09-1591	EFFECT OF FERROELECTRIC PHASE TRANSITION CONDITIONS ON DIELECTRIC AND PIEZOELECTRIC PR
SOLOVEVA NM	11-1664	INHOMOGENEITY OF THE OPTICALLY INDUCED INDEX OF REFRACTION IN LITHIUM NIOBATE AND LITH
SOMEYA T	03-0473	ELECTRON MIRROR MICROSCOPIC OBSERVATION OF FERROELECTRIC DOMAINS.
	04-0973	ELECTRON MIRROR MICRCSCOPIC ASPECTS OF FERROELECTRIC DOMAINS OF BARIUM TITANATE AND DI
	04-0974	ELECTRON MIRROR MICROSCCPIC OBSERVATION OF FERROELECTRIC DOMAINS. (GOLD COATED BARIUM
	26-2946	ELECTRON MIRROR MACROSCOPIC OBSERVATION OF FERROELECTRIC DOMAINS OF DICALCIUM STRONTIU
	30-3293	APPLICATION OF ELECTRON MIRROR MICROSCOPY TO DIRECT OBSERVATION OF MOVING FERROELECTRI
SOMOV VG	11-1863	LONGITUDINAL ELECTRCOPTICAL EFFECT IN OBLIQUE SECTIONS OF LITHIUM NIOBATE.
SONIN AS	17-2696	RELATIONSHIP BETWEEN THE DIELECTRIC AND ELECTROOPTICAL PROPERTIES OF FERROELECTRIC CRY
	26-2925	ANTIPOLARIZATION IN CCPPER FORMATE CRYSTALS BY THE ELECTROOPTICAL METHOD. (ANTIFERROEL
	27-3138	SECOND HARMONIC GENERATION IN TRIGLYCINE SULFATE CRYSTALS.
	28-3209	SPONTANEOUS ELECTROOPTICAL EFFECT IN ROCHELLE SALT CRYSTALS.
SOOD RK	25-2924	SEMICONDUCTIVITY OF THIOUREA.
SOOHOO J	11-1736	POWER AND LINE WIDTH OF TUNABLE STIMULATED FAR INFRARED EMISSION IN LITHIUM NIOBATE.
SORGE G	06-1218	MECHANICAL RELAXATION AND NONLINEARITY IN STRONTIUM TITANATE SINGLE CRYSTALS.
	06-1219	PRESSURE DEPENDENCE OF THE ELASTIC COMPLIANCE OF STRONTIUM TITANATE NEAR THE TRANSITIO
	06-1220	TEMPERATURE DEPENDENCE OF THE ELASTIC COMPLIANCE OF STRONTIUM TITANATE SINGLE CRYSTALS
SOROKIN NG	30-3265	ABNORMAL ABSORPTION OF ELASTIC WAVES NEAR THE PHASE TRANSITION OF GADOLINIUM MOLYBDATE
SOROKINA IA	04-1000	PREPARATION ANC STUDY OF THIN FILMS OF (BARIUM, STRONTIUM) TITANATE.
SORRELL CA	10-1626	THERMAL EXPANSION AND THE ORTHORHOMBIC- TETRAGONAL TRANSITION OF THE TUNGSTEN TRIOXIDE
SOSNOWSKA I	17-2416	INCOHERENT NEUTRON SCATTERING FROM HYDROGEN BOND IN KDP AND ADP.
SOULES TF	06-1221	ENERGY BAND STRUCTURE OF STRONTIUM TITANATE FROM A SELF CONSISTENT FIELD TIGHT BINDING
SOZONTOVA GN	17-2628	CONDITIONS FOR THE GROWTH OF POTASSIUM DIHYDROGEN PHOSPHATE WHISKERS.
SPAIGHT RN	11-1824	PIEZCELECTRIC SURFACE WAVES ON LITHIUM NIOBATE.
SPANN JR	04-1026	INFLUENCE OF ADDITIVES ON DENSIFICATION AND PROPERTIES OF SINTERED AN PRESSURE SINTERE
SPEARS DL	11-1825	LOW POWER DENSITY GALLIUM ARSENIDE- LITHIUM NIOBATE SURFACE WAVE AMPLIFIER.
SPENCER EG	33-3406	BARIUM MANGANESE FLUORIDE, A NEW CRYSTAL FOR MICROWAVE ULTRASONICS.
SPERANSKAYA EI	10-1656	THE BISMUTH OXIDE- TUNGSTEN TRIOXIDE SYSTEM.
SPINKO RI	05-1047	GROWTH OF LEAD TITANATE CRYSTALS AND EXAMINATION OF THEIR DOMAIN STRUCTURE.
	05-1050	SPONTANEOUS POLARIZATION AND COERCIVE FIELD OF LEAD TITANATE (FLUX GROWN SINGLE CRYSTA
SPITSYNA VD	15-2161	PHOTOFERROELECTRIC EFFECTS IN FERROELECTRIC SEMICONDUCTORS OF THE A(V)B(VI)C(VII) TYPE
	15-2230	CRYSTAL GROWTH IN THE ANTIMONY SULFOIODIDE- ANTIMONY TRISULFIDE SYSTEM. (BRIDGMAN- STO
SPITZ E	08-1410	EFFECT OF INTENSE LIGHT ON THE REFRACTIVE INDEX OF A CRYSTALLINE POTASSIUM NIOBATE.
SPITZER WG	04-0975	FAR INFRARED DIELECTRIC DISPERSION IN BARIUM AND STRONTIUM TITANATES AND TITANIUM DIOX
SPIVAK GV	03-0474	CBSERVATION OF FERROELECTRIC DOMAIN STRUCTURES, MODULATED BY A PHASE TRANSITION, IN A
	04-0940	MOTION OF DOMAIN BOUNDARIES DURING SWITCHING IN SINGLE CRYSTAL FILMS OF BARIUM TITANAT
	04-0942	CBSERVATION OF THE CHANGE IN THE POLARIZATION OF SINGLE CRYSTAL FILMS OF BARIUM TITANA
	04-1031	MICRCSTRUCTURE CF DOMAINS AND DOMAIN WALLS IN SINGLE CRYSTAL FILMS OF BARIUM TITANATE.
SQUIRE PT	17-2517	NEW METHOD OF MEASURING PYROELECTRIC COEFFICIENTS. (TGS EXAMPLE)
SRINIVASAN R	06-1185	LATTICE DYNAMICS OF CUBIC PEROVSKITE STRUCTURES, IN PARTICULAR STRONTIUM TITANATE.
	18-2726	ELECTRON PARAMAGNETIC RESONANCE AND ELECTRON NUCLEAR DOUBLE RESONANCE OBSERVATIONS OF
	18-2727	ELECTRON PARAMAGNETIC RESONANCE AND ELECTRON NUCLEAR DOUBLE RESONANCE OF THE ARSENATE
	18-2728	HYDROGEN BONDED FERROELECTRICS: EVIDENCE FOR SLATER CONFIGURATIONS FROM EPR STUDIES OF
SRIVASTAVA RC	08-1416	ELASTIC CONSTANT OF ORTHORHOMBIC POTASSIUM NIOBATE BY X-RAY DIFFUSE SCATTERING.
SROUBEK Z	04-0976	PARAMAGNETIC RESONANCE OF AN F CENTER IN NONSTOICHIOMETRIC BARIUM TITANATE.
	04-0977	PARAMAGNETIC RESONANCE OF PLATINUM IONS IN NIOBIUM DOPED BARIUM TITANATE SINGLE CRYSTA
	06-1222	ELECTRON TUNNELING AND BAND STRUCTURE OF (NIOBIUM DOPED) STRONTIUM TITANATE AND (CALCI
STABINIS A	17-2468	PARAMETRIC LIGHT AMPLIFICATION AND OSCILLATION IN KDP WITH MODE LOCKED PUMP.
STADELMAIER HH	04-0980	RELATION BETWEEN ELECTRICAL PROPERTIES AND MICROSTRUCTURE OF BARIUM TITANATE.
STADLER HL	04-0979	NUCLEATION AND GROWTH OF FERROELECTRIC DOMAINS IN BARIUM TITANATE AT FIELDS FROM 2 TO
	04-0981	SUSCEPTIBILITY INCREASE DURING POLARIZATION REVERSAL. (IN BARIUM TITANATE AND TRIGLYCI
STADNIK B	17-2675	TEMPERATURE INSTABILITY OF ELECTROOPTIC LASER MODULATORS WITH AMMONIUM DIHYDROGEN PHOS
STAEBLER DL	06-1098	ELECTROCOLORATION IN STRONTIUM TITANATE: VACANCY DRIFT AND OXIDATION REDUCTION OF TRAN
	06-1140	PHOTOCHROMIC MATERIALS RESEARCH. (STRONTIUM TITANATE PHOTOCONDUCTIVITY, CALCIUM FLUORI
	11-1661	HCLCGRAPHIC PATTERN FIXING IN ELECTROOPTIC CRYSTALS. (LITHIUM NIOBATE, BARIUM SODIUM N
	11-1662	HOLOGRAPHIC RECORDING IN LITHIUM NIOBATE.
	11-1796	OPTICAL AND HOLOGRAPHIC STORAGE PROPERTIES OF TRANSITION METAL DOPED LITHIUM NIOBATE.
	11-1826	COUPLED WAVE ANALYSIS OF HOLOGRAPHIC STORAGE IN LITHIUM NIOBATE.
	11-1827	MULTIPLE STORAGE OF THICK PHASE HOLOGRAMS IN LITHIUM NIOBATE.
	11-1828	THERMALLY FIXED HOLOGRAMS IN LITHIUM NIOBATE.
	13-1898	HCLOGRAPHIC STORAGE IN DOPED BARIUM SODIUM NIOBATE.
STAFSUDD OM	13-2030	CHARACTERISTICS OF KTN (POTASSIUM TANTALATE NIOBATE) PYROELECTRIC DETECTORS.
STAHR S	13-1907	EPITAXY OF PLATINUM ON BARIUM STRONTIUM NIOBATE.
STAMENKOVIC S	03-0403	QUANTUM THEORY OF FERROELECTRICITY. PART-2: THERMODYNAMIC BEHAVIOR OF FERROELECTRIC SP
	03-0475	DYNAMICAL THEORY OF HYDROGEN BONDED FERROELECTRICS. PART-1: DAMPINGS AND (FREQUENCY) S
	03-0478	THEORY OF COHERENT NEUTRON SCATTERING BY HYDROGEN BONDED FERROELECTRICS AT LOW TEMPERA
	17-2673	PROPERTIES OF THE KDP TYPE FERROELECTRICS WITH IMPURITIES.
STANDLEY RD	12-1888	PERFORMANCE OF AN 11 GHZ OPTICAL MODULATOR USING LITHIUM TANTALATE.
STANFORD AL	04-0544	EFFECT OF FERRIC OXIDE ON THE 120 DEGREE PHASE TRANSITION IN BARIUM TITANATE SINGLE CR
	04-0978	DETECTION OF ELECTROMAGNETIC RADIATION USING PYROELECTRIC EFFECT. (BARIUM TITANATE, TR
STANKOVSKAYA Y	27-3043	DOMAIN STRUCTURE OF TRIGLYCINE SELENATE CRYSTAL (DURING COOLING THROUGH CURIE POINT)
STANKOWSKA J	17-2561	INVESTIGATION OF THE DOMAIN STRUCTURE OF A TRIGLYCINE SULFATE CRYSTAL DURING AGING.
	27-3128	AGING PROCESS IN TRIGLYCINE SULFATE. (HYSTERESIS LAPS, DOMAIN STRUCTURE)
	27-3129	DEPENDENCE OF THE DOMAIN STRUCTURE OF TRIGLYCINE SULFATE CRYSTALS ON TEMPERATURE AND A
	27-3131	EFFECT OF A DC FIELD ON THE DIELECTRIC PROPERTIES OF TRIGLYCINE SELENATE. (AGING AND P
	27-3132	ELECTRON PARAMAGNETIC RESONANCE INVESTIGATIONS OF (COPPER) DOPED FERROELECTRIC CRYSTAL
	28-3230	EFFECT OF A DC FIELD CN DOMAIN STRUCTURE IN SEIGNETTE SALT. (EFFECT OF AGING AND COPPE
STANKOWSKI J	03-0477	EPR STUCY OF PARAMAGNETIC ION COMPLEXES IN FERROELECTRIC CRYSTALS.
	27-3130	DOMAIN STRUCTURE OF PURE AND DOPED TRIGLYCINE SULFATE CRYSTALS GROWN BELOW AND ABOVE T
	27-3133	EPR CF THE CHRCMIUM COMPLEX IN TRIGLYCINE FLUOBERYLLATE.
	27-3134	EPR OF THE FOUR COPPER ION COMPLEX WITH SPIN S=2 IN TRIGLYCINE FLUOBERYLLATE MONOCRYST
	27-3135	EPR STUDIES OF CHROMIUM(3+), COPPER(2+), AND MANGANESE(2+) PARAMAGNETIC ION COMPLEXES
	27-3136	INFLUENCE OF SPONTANECUS POLARIZATION ON THE EPR SPECTRA OF CHROMIUM(3+) IONS IN TRIGL

STANLEY HE	06-1223	STATIC SCALING HYPOTHESIS AND ITS IMPLICATIONS FOR THE PLANAR SPIN MODEL OF STRONTIUM
STARK AH	13-1956	SOLID SOLUTIONS IN ANTIFERROELECTRIC REGION OF THE SYSTEM LEAD HAFNATE TITANATE, LEAD
STARODUBTSEV SV	27-3101	RESULTS OF EXPERIMENTAL STUDY OF GAMMA-RADIATION EFFECTS IN ROCHELLE SALT AND TRIGLYCI
STAROV EN	04-0609	POLARON MECHANISM OF CONDUCTIVITY IN BARIUM TITANATE.
	04-0613	SMALL POLARONS IN CONDUCTING BARIUM TITANATE CRYSTALS.
	04-0614	TEMPERATURE DEPENDENCE OF CARRIER MOBILITY IN BARIUM TITANATE.
STASYUK IV	03-0476	DYNAMICAL THEORY OF AMMONIUM DIHYDROGEN PHOSPHATE TYPE ANTIFERROELECTRICS WITH HYDROGE
	17-2670	COUPLED VIBRATIONS OF THE PROTON ION SYSTEM IN HYDROGEN BONDED FERROELECTRICS OF THE K
	17-2671	INFLUENCE OF POLAR STATES OF PROTONS ON THE PHASE TRANSITION IN FERROELECTRICS WITH HY
	17-2672	PROTON PHONON INTERACTION IN FERROELECTRICS WITH HYDROGEN BONDS. (KDP TYPE)
	17-2674	ROLE OF PROTON PHONON INTERACTION IN THE PHASE TRANSITION OF FERROELECTRICS WITH HYDRO
STEADMAN R	13-2031	STRUCTURE OF TETRAGONAL TIN TUNGSTEN BRONZES.
STEEPLE H	20-2801	HEAT CAPACITY OF METHYL AMMONIUM CHROMIUM ALUM.
	20-2807	THERMAL CONDUCTIVITY OF METHYL AMMONIUM ALUMINUM ALUM.
STEFFAN L	11-1694	ASSIGNMENTS OF OPTICAL PHONON MODES IN LITHIUM NIOBATE.
	11-1695	DIRECTIONAL DISPERSION AND ASSIGNMENT OF OPTICAL PHONONS IN LITHIUM NIOBATE.
STEGUN IA	03-0239	FERROELECTRIC SWITCHING AND THE SIEVERT INTEGRAL.
STEIER WH	11-1689	ROTATING WAVEPLATE OPTICAL FREQUENCY SHIFTING IN LITHIUM NIOBATE.
STEIGMEIER EF	02-0120	RAMAN SCATTERING IN FERROELECTRICS. (REVIEW)
	15-2167	SOFT PHONON MODE AND MODE COUPLING IN ANTIMONY SULFOIODIDE (BY RAMAN SCATTERING).
	15-2231	CRITICAL PHENOMENA IN ANTIMONY SULFOIODIDE.
	15-2232	SOFT MODE COUPLING AND CRITICAL RAYLEIGH SCATTERING IN FERROELECTRIC ANTIMONY SULFOIOD
	15-2233	SOFT LATTICE VIBRATIONS AND VIBRATION COUPLING IN ANTIMONY SULFOIODIDE.
STEINFINK H	30-3261	PHASE TRANSITIONS IN COMPLEX PEROVSKITES OF THE TYPE BARIUM LANTHANON MOLYBDATE.
STEKHANOV AI	17-2623	SYMMETRY OF NORMAL OSCILLATIONS IN A FERROELECTRIC CRYSTAL OF KDP.
	17-2624	STRUCTURE OF DIHYDROGEN PHOSPHATE ION IN THE FERROELECTRIC POTASSIUM SALT.
	19-2786	RAMAN SCATTERING IN A FERROELECTRIC AMMONIUM SULFATE CRYSTAL.
	28-3231	RAMAN LIGHT SCATTERING IN ROCHELLE SALT.
STELLA A	10-1634	OPTICAL TRANSITION TO THE FERMI LEVEL IN SODIUM TUNGSTATE.
STELLENBERGER K	04-0890	INFLUENCE OF VIBRATIONS ON THE DIELECTRIC PROPERTIES OF BARIUM TITANATE CERAMICS.
STEPANOV VG	28-3176	ELECTRON PARAMAGNETIC RESONANCE OF COPPER IONS IN ROCHELLE SALT.
STEPANOVA NS	17-2427	EFFECT OF ANNEALING ON OPTICAL ANOMALIES OF POTASSIUM DIHYDROGEN PHOSPHATE CRYSTALS.
	17-2496	GROWTH LAYERS AND OTHER MACRODEFECTS IN POTASSIUM DIHYDROGEN PHOSPHATE CRYSTALS GROWN
STEPHENS AW	13-1898	HOLOGRAPHIC STORAGE IN DOPED BARIUM SODIUM NIOBATE.
STEPHENS JS	22-2845	CRYSTAL STRUCTURE OF POTASSIUM SELENATE.
STEPHENSON NC	13-1931	STRUCTURE OF THE BRONZE SODIUM(13) NIOBIUM(35) OXIDE(94) AND THE GEOMETRY OF FERROELEC
STEPISNIK J	17-2438	DEUTERON MAGNETIC RESONANCE AND RELAXATION IN FERROELECTRIC POTASSIUM DIDEUTERIUM PHOS
	18-2720	CESIUM-183 SPIN LATTICE RELAXATION AND RESONANCE IN FERROELECTRIC CESIUM DIDEUTERIUM A
	18-2721	DEUTERON QUADRUPOLE COUPLING IN FERROELECTRIC CESIUM DIDEUTERIUM ARSENATE.
	22-2832	NUCLEAR MAGNETIC RESONANCE STUDY OF THE FERROELECTRIC TRANSITIONS IN SODIUM TRIHYDROGE
	27-2965	BERYLLIUM-9 QUADRUPOLE PERTURBED NMR STUDY OF THE FERROELECTRIC TRANSITION IN DEUTERAT
STERIN KE	13-1992	POLARITONS IN A BIAXIAL CRYSTAL OF BARIUM SODIUM NIOBATE.
STERIN KHE	11-1761	FERMI RESONANCE OF POLARITON WITH BIPHONON IN AN LITHIUM NIOBATE CRYSTAL.
	11-1762	TRANSVERSE POLARITONS IN A LITHIUM NIOBATE CRYSTAL. (RAMAN SPECTRUM, PERMITTIVITY)
STEVENSON RWH	33-3407	MAGNETIC EXCITATIONS IN NICKEL DOPED POTASSIUM MANGANESE TRIFLUORIDE.
STEWART JM	21-2819	REDETERMINATION OF STRUCTURE OF FERROELECTRIC CRYSTAL GUANIDINIUM ALUMINUM SULFATE HEX
STEWART WC	09-1601	SOME PHYSICAL AND ELECTRICAL SWITCHING CHARACTERISTICS OF A LEAD ZIRCONATE TITANATE FE
	37-3675	ACTIVE COMPENSATORS FOR FERROELECTRIC OPTICAL CIRCUITS.
STIGLITZ ME	09-1602	PREPARATION AND ELECTRICAL PROPERTIES OF SOME CERAMIC MATERIALS AT MICROWAVE FREQUENCI
STILLER H	17-2418	NEUTRON MEASUREMENTS ON THE FERROELECTRIC PHASE TRANSFORMATION IN POTASSIUM DIHYDROGEN
	17-2510	EXPERIMENTAL PROOF FOR PROTON TUNNELING IN POTASSIUM DIHYDROGEN PHOSPHATE.
	17-2511	NEUTRON MEASUREMENTS ON THE HYDROGEN BOND POTENTIAL IN PARAELECTRIC POTASSIUM DIHYDROG
STIRLING WG	06-1224	LATTICE DYNAMICS OF STRONTIUM TITANATE.
	06-1225	NEUTRON INELASTIC SCATTERING STUDY OF THE LATTICE DYNAMICS OF STRONTIUM TITANATE: HARM
STOCKLER HA	04-1027	CORRELATION BETWEEN RHEOLOGICAL AND ELECTRICAL BEHAVIOR IN SINTERED BARIUM TITANATE.
STOGOVA VA	04-0943	ELECTROCONDUCTIVITY AND HALL EFFECT IN SEMICONDUCTING SOLID SOLUTIONS OF ISO- AND HETE
	06-1198	ELECTRICAL AND OPTICAL PROPERTIES (CONDUCTIVITY AND MOBILITY VERSUS TEMPERATURE, ABSOR
STOKOWSKI SE	06-1123	SHARED HOLES TRAPPED BY CHARGE DEFECTS IN STRONTIUM TITANATE.
STOLSHTEIN DI	36-3587	PREPARATION OF THE TERNARY TITANATE BARIUM(0.8) LEAD(0.12) CALCIUM(0.08) TITANATE.
STOLYPIN YUE	05-1063	PHYSICAL PHENOMENA IN PIEZOCERAMICS BASED ON LEAD TITANATE AND ZIRCONATE. (SOLID SOLUT
	08-1441	ANTIFERROMAGNETIC PROPERTIES OF SOME PEROVSKITES. (BISMUTH FERRATE)
	09-1603	DIELECTRIC SYSTEM POLARIZATION AND THERMAL EXPANSION OF POLARIZED CERAMICS IN THE LEAD
STONE CD	17-2580	RELAXATION OF FERROELECTRIC PHOSPHATES (AMMONIUM DIHYDROGEN PHOSPHATE AND POTASSIUM DI
STORY HS	08-1469	NIOBIUM-93 AND SODIUM-23 IN POLYCRYSTALLINE SODIUM NIOBATE.
	11-1804	ELECTRON SPIN RESONANCE STUDIES OF CHROMIUM(3+) IN LITHIUM NIOBATE.
	11-1839	EPR STUDIES OF CRYSTAL FIELD PARAMETERS IN IRON(3+) DOPED LITHIUM NIOBATE.
STRAJBLOVA J	02-0088	INFLUENCE OF THE QUALITY OF TRIGLYCINE SULFATE SINGLE CRYSTALS ON THEIR DIELECTRIC PRO
	03-0479	ON THE FREQUENCY DEPENDENCE OF THE TEMPERATURE AUTOSTABILIZATION EFFECT IN FERROELECTR
	27-3063	INFLUENCE OF DEFECTS ON SOME DIELECTRIC PROPERTIES OF TGS SINGLE CRYSTALS.
	27-3067	THE EFFECT OF GROWTH RATE ON THE DEFECT STRUCTURE AND DIELECTRIC PROPERTIES OF TGS SIN
	27-3068	TEMPERATURE AND FIELD STRENGTH DEPENDENCE OF LARGE SIGNAL PERMITTIVITY OF TGS. (10 KHZ
STRIMER P	27-3007	OPTICAL CONSTANTS OF TRIGLYCINE SELENATE IN THE NEAR INFRARED TO FAR INFRARED AND IN T
	28-3210	FAR INFRARED ABSORPTION SPECTRA OF ROCHELLE SALT.
STRINGFELLOW M	17-2700	DETERMINATION OF THE EIGENVECTORS OF SOFT MODES IN POTASSIUM DIDEUTERIUM PHOSPHATE AND
STRIZHEVSKII VL	11-1787	SPONTANEOUS RAMAN SCATTERING BY POLARITONS. PART-2: THE CASE OF CADMIUM SULFIDE AND LI
STROMME KO	16-2371	EVIDENCE OF ADDITIONAL DISORDER IN THE RHOMBOHEDRRAL FORM OF POTASSIUM NITRITE.
STROUD JD	11-1829	TONE BURSTING OF QUADRUPOLAR NUCLEI. DETERMINATION OF THE SPIN LATTICE RELAXATION TIM
STRUKOV BA	17-2414	EFFECTS OF DEUTERATION ON THE SPECIFIC HEAT OF ADP CRYSTALS.
	17-2415	HEAT CAPACITY OF RUBIDIUM DIHYDROGEN PHOSPHATE SINGLE CRYSTALS.
	17-2676	CRITICAL BEHAVIOR OF KDP AND RUBIDIUM DIHYDROGEN PHOSPHATE CRYSTALS.
	17-2677	FIRST ORDER PHASE TRANSITION IN POTASSIUM DEUTERIUM PHOSPHATE CRYSTALS.
	17-2678	SPONTANEOUS POLARIZATION OF A POTASSIUM DIHYDROGEN PHOSPHATE CRYSTAL NEAR THE CURIE PO
	17-2679	SPECIFIC HEAT OF SOME KDP TYPE CRYSTALS. (POTASSIUM DIHYDROGEN PHOSPHATE, RUBIDIUM AND
	17-2680	SOME THERMODYNAMICAL PROPERTIES OF POTASSIUM DIHYDROGEN PHOSPHATE- POTASSIUM DIHYDROGE
	19-2787	EXPERIMENTAL INVESTIGATION OF THE FERROELECTRIC PROPERTIES OF AMMONIUM BISULFATE IN TH
	19-2788	POLARIZATION REVERSAL CHARACTERISTICS OF AMMONIUM HYDROGEN SULFATE.
	19-2789	SPECIFIC HEAT OF AMMONIUM HYDROGEN SULFATE IN THE TEMPERATURE RANGE FROM -70 TO +14 DE
	27-3055	ANISOTROPY OF ULTRASONIC ATTENUATION IN UNIAXIAL FERROELECTRICS.(TGS, ABSTRACT ONLY)
	27-3078	ANISOTROPY OF SOUND ABSORPTION IN TRIGLYCINE SULFATE SINGLE CRYSTALS. (THEORY OF ANIS
	27-3079	FREQUENCY DEPENDENCE OF THE FLUCTUATION ABSORPTION OF ULTRASOUND IN TRIGLYCINE SULFATE

PAGE 690

STRUKOV BA   27-3137   ANOMALY OF THERMAL PROPERTIES OF FERROELECTRICS (SINGLE CRYSTAL TRIGLYCINE SULFATE AND
                   27-3141   THERMAL AND ELECTRICAL PROPERTIES OF TRIGLYCINE SULFATE SINGLE CRYSTALS (HEAT CAPACITY
                   28-3211   ABSORPTION OF LONGITUDINAL ULTRASONIC WAVES IN ROCHELLE SALT CRYSTALS NEAR THE CURIE P
SUBBARAO EC   08-1416   ELASTIC CONSTANT OF ORTHORHOMBIC POTASSIUM NIOBATE BY X-RAY DIFFUSE SCATTERING.
                   08-1420   DEFORMATION AND RELAXATION BEHAVIOR IN FERROELECTRIC POTASSIUM NIOBATE SINGLE CRYSTALS
                   08-1421   MECHANICAL SWITCHING OF 60 DEGREE DOMAIN WALLS IN FERROELECTRIC POTASSIUM NIOBATE SING
                   09-1604   MOSSBAUER STUDIES OF LEAD ZIRCONATE TITANATE STANNATE. (CERAMIC DOPED WITH NIOBIUM AND
                   13-2032   CRYSTAL CHEMISTRY OF MIXED BISMUTH OXIDES WITH LAYER TYPE STRUCTURE.
SUBBARAO GV   02-0109   ELECTRICAL CONDUCTION IN METAL OXIDES. (REVIEW, 319 REFS)
                   36-3608   ELECTRICAL TRANSPORT IN (SEMICONDUCTING AND MAGNETIC) RARE EARTH ORTHOCHROMITES, MANGA
SUBEVA EV   17-2647   RAMAN SPECTRUM OF POTASSIUM DIDEUTERIUM PHOSPHATE CRYSTAL IN PARA- AND FERROELECTRIC P
SUGA H   34-3436   RELAXATIONAL PROTON ORDERING AND GLASSY CRYSTALLINE STATE IN HEXAGONAL ICE.
                   34-3476   CALORIMETRIC STUDY OF GLASS TRANSITION OF THE AMORPHOUS ICE AND OF THE PHASE TRANSFORM
SUGAHARA M   17-2693   SECOND HARMONIC GENERATION OF LIGHT IN THE KDP CRYSTAL BY THE INDEX MATCHING METHOD.
SUGAI T   13-2033   CRYSTAL GROWTH AND DIELECTRIC PROPERTIES OF BISMUTH NIOBATE.
                   13-2034   SINGLE CRYSTAL GROWTH OF STRONTIUM POTASSIUM TANTALATE FROM MOLTEN SALTS.
SUGAWARA F   15-2234   FAR INFRARED REFLECTIVITY SPECTRA OF ANTIMONY SULFOIODIDE.
                   15-2235   FAR INFRARED REFLECTIVITY SPECTRA OF ANTIMONY SULFOIODIDE.
                   16-2373   TEMPERATURE DEPENDENCE OF FAR INFRARED REFLECTIVITY SPECTRA OF SODIUM NITRITE CRYSTALS
                   17-2681   FAR INFRARED REFLECTIVITY SPECTRA OF KDP CRYSTAL (16-400 PER CM, 83-300K).
                   22-2877   FAR INFRARED REFLECTIVITY SPECTRA OF SODIUM TRIHYDROGEN DISELENITE CRYSTAL.
SUGAYA S   09-1549   DIELECTRIC AND PIEZOELECTRIC PROPERTIES OF LEAD COBALTATE(0.33) NIOBATE(0.67)- TITANAT
SUGH K   35-3515   X-RAY TOPOGRAPHIC OBSERVATION OF 180 DEG DOMAINS IN FERROELECTRIC LEAD GERMANATE SINGL
SUGIBUCHI K   12-1874   ELECTROOPTIC AND FERROELECTRIC PROPERTIES OF LITHIUM TANTALATE SINGLE CRYSTALS AS A FU
                   13-2043   DEPENDENCE OF SECOND HARMONIC GENERATION ON CRYSTAL INHOMOGENEITY (THEORY AND TEST ON
SUGIE H   17-2602   DIELECTRIC EVIDENCE OF A FIRST ORDER TRANSITION IN POTASSIUM DIHYDROGEN PHOSPHATE.
                   17-2682   THERMAL HYSTERESIS OF THE FERROELECTRIC TRANSITION IN POTASSIUM DIHYDROGEN PHOSPHATE.
SUGII K   11-1830   X-RAY TOPOGRAPHIC STUDIES ON LITHIUM NIOBATE AND LITHIUM TANTALATE SINGLE CRYSTALS.
                   35-3501   FERROELECTRIC AND OPTICAL PROPERTIES OF LEAD(5) GERMANIUM(3) OXIDE(11) AND ITS ISOMORP
                   35-3502   LEAD(5) GERMANIUM(3) OXIDE(11) CRYSTAL, A NEW FERROELECTRIC. (DIELECTRIC CONSTANTS AND
                   35-3503   OPTICAL ACTIVITY OF FERROELECTRIC LEAD(5) GERMANIUM(3) OXIDE(11) SINGLE CRYSTALS.
                   35-3513   CRYSTAL GROWTH AND PROPERTIES OF LEAD(5) GERMANIUM(3) OXIDE(11) SINGLE CRYSTALS.
                   35-3514   GROWTH OF SINGLE CRYSTALS IN LEAD MONOXIDE- GERMANIUM DIOXIDE BINARY SYSTEM.
                   35-3516   X-RAY TOPOGRAPHIC OBSERVATION OF 180 DEGREE DOMAINS IN FERROELECTRIC LEAD(5) GERMANIUM
                   36-3570   SWITCHING OF OPTICAL ROTATORY POWER IN FERROELECTRIC LEAD(5) GERMANIUM(3) OXIDE(11) SI
SUGISAKI M   34-3476   CALORIMETRIC STUDY OF GLASS TRANSITION OF THE AMORPHOUS ICE AND OF THE PHASE TRANSFORM
SUGIURA A   15-2236   PHOTODIELECTRIC EFFECT IN ANTIMONY SULFOIODIDE SINGLE CRYSTALS.
SUGIYAMA H   35-3508   FERROELECTRICITY IN LEAD(5) GERMANIUM(3) OXIDE(11).
SUJAK B   04-0982   EMISSION OF LIGHT BY SURFACE MICROPLASMA OF CERAMIC (BARIUM, LEAD) TITANATE NEAR THE C
                   28-3208   CORRELATION BETWEEN BARKHAUSEN NOISE AND SURFACE MICRODAMAGES ON SINGLE CRYSTALS OF SE
SUKHINSKII AN   15-2190   RELAXATION PROCESSES OF ANTIMONY SULFOIODIDE SINGLE CRYSTALS.
                   15-2191   TEMPERATURE DEPENDENCE OF THE SWITCHING PROCESS IN ANTIMONY SULFOIODIDE SINGLE CRYSTAL
                   15-2245   INFLUENCE OF ONE DIMENSIONAL MECHANICAL STRESS ON THE PHASE TRANSITION OF ANTIMONY SUL
                   15-2246   POLARIZATION REVERSAL OF ANTIMONY SULFOIODIDE SINGLE CRYSTALS IN THE REGION OF PHASE C
SUKIYAZOV AG   15-2156   DILATOMETRIC INVESTIGATION OF PHOTOFERROELECTRIC EFFECTS IN SINGLE CRYSTALS OF ANTIMON
SULCEK Z   27-3085   PREPARATION OF PURE GLYCINE USED FOR GROWING OF TRIGLYCINE SULFATE SINGLE CRYSTALS.
                   27-3086   PREPARATION OF PURE TGS SINGLE CRYSTALS AND SOME INVESTIGATIONS OF THEIR QUALITY. (VOL
                   27-3087   PREPARATION OF PURE TRIGLYCINE SULFATE. (USING ION EXCHANGE)
SULTANOV GD   13-2035   MOSSBAUER STUDIES OF IRON-57 IN SOME PEROVSKITE LIKE LAYER TYPE FERROELECTRICS. (IN BI
SUMI M   08-1359   RH DOPED LITHIUM NIOBATE AS AN IMPROVED NEW MATERIAL FOR REVERSIBLE HOLOGRAPHIC STORAG
                   11-1731   ACOUSTOOPTICAL MEASUREMENTS OF ACOUSTIC WAVE GENERATION IN PIEZOELECTRICS COUPLED TO A
SUMITA M   02-0087   CRITICAL SLOWING DOWN PROCESS OF (ANOMALOUS) DIELECTRIC RELAXATION IN SEVERAL CRYSTALS
                   20-2804   DIELECTRIC RELAXATION IN FERROELECTRIC MASD, METHYL AMMONIUM ALUMINUM SULFATE DODECAHY
SUMMITT R   04-0823   EFFECT OF PARTICLE SIZE AND SHAPE ON THE INFRARED ABSORPTION SPECTRA OF BARIUM TITANAT
SUMUTNY F   27-2973   PROPERTIES OF DEUTERATED TRIGLYCINE FLUOBERYLLATE SINGLE CRYSTALS.
SUN TS   31-3338   CRYSTAL SIZE EFFECTS IN THE RAMAN SPECTRA OF ORTHORHOMBIC HYDROGEN CHLORIDE CRYSTALS.
SUNDARAM VS   09-1604   MOSSBAUER STUDIES OF LEAD ZIRCONATE TITANATE STANNATE. (CERAMIC DOPED WITH NIOBIUM AND
SURAKKA J   11-1758   ACOUSTIC CONVOLUTION AND CORRELATION AND THE ASSOCIATED NONLINEARITY PARAMETERS IN LIT
SUSSMAN SS   11-1736   POWER AND LINE WIDTH OF TUNABLE STIMULATED FAR INFRARED EMISSION IN LITHIUM NIOBATE.
                   11-1858   EFFECT TUNABLE OPTICAL EMISSION FROM LITHIUM NIOBATE WITHOUT A RESONATOR.
SUSSMANN JA   34-3477   ELECTRIC RESONANCE: APPLICATION TO THE HYDROGEN BOND. (ICE)
SUVAOV LA   03-0480   CRYSTALLOGRAPHICAL CLASSIFICATION OF FERROELECTRICS, FERROELECTRIC PHASE TRANSITIONS A
SUVOROV VS   27-3105   SECOND HARMONIC GENERATION IN THE TRIGLYCINE CRYSTAL IN THE PARAELECTRIC PHASE.
                   27-3138   SECOND HARMONIC GENERATION IN TRIGLYCINE SULFATE CRYSTALS.
SUZUKI H   09-1624   STUDIES ON THE SINTERING BETWEEN LEAD MONOXIDE AND PZT. PART-1. PART-2: EFFECTS OF AVE
SUZUKI I   17-2683   ESR STUDY OF ADP- KDA MIXED CRYSTALS.
                   17-2684   ESR STUDY OF AMMONIUM IONS IN IRRADIATED AMMONIUM DIDEUTERIUM PHOSPHATE- 5 PERCENT POT
                   19-2790   ESR OF GAMMA-IRRADIATED AMMONIUM SULFATE.
                   27-3035   ESR IN GAMMA-IRRADIATED TGS.
                   27-3139   ELECTRON SPIN RESONANCE OF GAMMA-IRRADIATED TRIGLYCINE SELENATE.
                   28-3180   ELECTRON PARAMAGNETIC RESONANCE OF GAMMA-IRRADIATED ROCHELLE SALT.
                   29-3252   ESR OF COPPER(2+) DOPED SODIUM AMMONIUM TARTRATE.
SUZUKI K   03-0323   ON SURFACE STATE AND POLARIZATION REVERSAL OF FERROELECTRIC CRYSTALS.
                   16-2373   TEMPERATURE DEPENDENCE OF FAR INFRARED REFLECTIVITY SPECTRA OF SODIUM NITRITE CRYSTALS
                   30-3322   OBSERVATION OF DIFFRACTED LIGHT FROM DOMAIN WALLS IN GADOLINIUM MOLYBDATE.
SUZUKI M   03-0481   ZEROES OF THE PARTITION FUNCTION FOR THE HEISENBERG, FERROELECTRIC, AND GENERAL ISING
SUZUKI S   02-0121   X-RAY DIFFRACTION TOPOGRAPHY OF FERROELECTRIC CRYSTALS.
                   16-2372   TOPOGRAPHIC STUDY ON FERROELECTRIC SODIUM NITRITE CRYSTALS. PART-1: STRUCTURE OF 180
                   16-2374   TOPOGRAPHIC STUDY OF FERROELECTRIC SODIUM NITRITE CRYSTALS. PART-2: MECHANISM OF POLAR
                   16-2375   APPLICATION OF THE X-RAY DIFFRACTION TOPOGRAPHIC METHOD TO FERROELECTRIC SODIUM NITRIT
                   26-2938   X-RAY DIFFRACTION STUDY OF PHASE TRANSITION IN COPPER FORMATE.
SUZUKI T   04-0862   FORMATION PROCESS OF BARIUM TITANATE THIN FILM EVAPORATED IN VACUO.
SVENSSON EC   33-3407   MAGNETIC EXCITATIONS IN NICKEL DOPED POTASSIUM MANGANESE TRIFLUORIDE.
SVETINA S   17-2442   VANISHING OF FERROELECTRICITY IN POTASSIUM DIHYDROGEN PHOSPHATE AT HIGH PRESSURES.
                   17-2685   ORDER DISORDER THEORY OF POTASSIUM DIHYDROGEN PHOSPHATE TYPE FERROELECTRICS.
SVIRIDOV DT   11-1831   CALCULATION OF THE (ABSORPTION) SPECTRA OF LITHIUM NIOBATE CRYSTALS CONTAINING CHROMIU
SVIRIDOVA RK   11-1831   CALCULATION OF THE (ABSORPTION) SPECTRA OF LITHIUM NIOBATE CRYSTALS CONTAINING CHROMIU
SWEEDLER AR   10-1654   RAMAN STUDY OF METALLIC TUNGSTEN BRONZES. (SODIUM AND RUBIDIUM TUNGSTATES)
SYCH AM   04-0869   FLUORINE CONTAINING BARIUM TITANATE. (ELECTRICAL PROPERTIES)
                   04-0983   INFRARED SPECTRA OF FLUORINE CONTAINING BARIUM TITANATE.

Author	Ref	Title
SYCH AM	04-0984	PREPARATION OF FLUORINE CONTAINING BARIUM TITANATE.
SYRBU NN	15-2125	ANISOTRCPY OF THE REFLECTION SPECTRA OF ANTIMONY SULFOIODIDE CRYSTALS.
SYRKIN LN	04-0985	INFLUENCE OF DOMAIN STRUCTURE OF CERAMIC FERROELECTRICS ON THEIR MECHANICAL PROPERTIES
	04-0986	NONLINEARITY OF ELECTROMECHANICAL AND MECHANICAL PROPERTIES OF FERROELECTRICS. (BARIUM
	09-1605	ELECTROMECHANICAL FRCPERTIES OF CERAMIC FERROELECTRICS (BARIUM TITANATE ZIRCONATES) IN
	15-2237	CRITICAL POINT IN FERROELECTRIC ANTIMONY SULFOIODIDE.
	15-2249	DIELECTRIC AND PIEZOELECTRIC PROPERTIES OF ANTIMONY SULFOIODIDE IN A CONSTANT ELECTRIC
	15-2250	INVESTIGATION OF THE PIEZORESISTANCE OF (POLYCRYSTALLINE) ANTIMONY SULFOIODIDE (DARK C
	36-3589	NEW FERROELECTRIC: ANTIMONY ORTHONIOBATE. (CURIE TEMPERATURE)
SZABO TL	11-1820	NEW HIGH COUPLING LOW DIFFRACTION CUT FOR ACOUSTIC SURFACE WAVES ON LITHIUM NIOBATE.
SZCZEPANIAK B	27-3045	EFFECT OF A POIARIZING FIELD AND OF THE ADDITION OF A MIXTURE OF COPPER AND CHROMIUM I
SZWEYCER H	03-0327	CURIE PCINT SHIFT WITH APPLIED FIELD IN FERROELECTRICS POSSESSING FIRST ORDER TRANSITI
	03-0328	ELASTIC GIBBS FUNCTION FOR A FERROELECTFIC CRYSTAL.
	03-0332	TWO POLARIZATION MODEL THEORY OF FERROELECTRICS.
SZYMANSKA J	04-0987	ON THE INFLUENCE OF DEFECTS ON THE TEMPERATURE HYSTERESIS OF BARIUM TITANATE.
SZYMCZAK H	11-1759	CRYSTAL FIELD IN CHROMIUM(3+) DOPED LITHIUM NIOBATE.
SZYWYCER H	03-0329	INDUCED POLARIZATION AND A ONE POLARIZATION MODEL OF FERROELECTRICITY.
TABATA Y	22-2852	PROTON SPIN LATTICE RELAXATION IN FERROELECTRIC SODIUM TRIHYDROGEN SELENATE.
TADA K	13-2036	ELECTROOPTIC LIGHT BEAM DEFLECTION WITH BARIUM STRONTIUM NIOBATE PRISM.
TADA O	04-0948	PREPARATION OF THIN BARIUM TITANATE FILMS BY DC DIODE SPUTTERING (ONTO PLATINUM OR FUS
	04-0988	ANOMALY OF RESISTANCE IN REDUCED BARIUM TITANATE CRYSTAL.
TAGUCHI I	19-2777	CN THE PHASE TRANSITION OF AMMONIUM HYDROGEN SULFATE AND THE STRUCTURE OF THE FERROELE
TAGUCHI N	03-0324	90 DEGREE TYPE DOMAIN WALL OF BARIUM TITANATE CRYSTAL. (THEORY, CALCULATIONS)
TAISEVA VA	04-1012	INFLUENCE OF A CONSTANT ELECTRIC FIELD ON THE DIELECTRIC PROPERTIES OF POLYCRYSTALLINE
TAKADA T	04-0990	TWC SETS OF ELECTRON SPIN RESONANCE OF GADOLINIUM(3+) IN BARIUM TITANATE CERAMIC SEMIC
TAKAGI M	02-0121	X-RAY DIFFRACTION TOPOGRAPHY OF FERROELECTRIC CRYSTALS.
	16-2372	TOFOGRAPHIC STUDY ON FERROELECTRIC SODIUM NITRITE CRYSTALS. PART-1: STRUCTURE OF 180
	16-2374	TOFCGRAPHIC STUDY OF FERROELECTRIC SODIUM NITRITE CRYSTALS. PART-2: MECHANISM OF POLAR
	16-2375	APPLICATION OF THE X-RAY DIFFRACTION TOPOGRAPHIC METHOD TO FERROELECTRIC SODIUM NITRIT
TAKAGI Y	01-0031	INTERNATIONAL MEETING ON FERROELECTRICITY, SECOND PROC.
	03-0286	A THEORY OF THE PHASE TRANSITION IN AMMONIUM DIHYDROGEN PHOSPHATE.
	03-0287	COMPUTER EXPERIMENT ON LINEAR CHAIN OF ATOMS LYING IN DOUBLE MINIMUM POTENTIAL. (BARIU
	03-0288	DISTRIBUTION OF RELAXATION TIMES IN SOME FERROELECTRICS.
	03-0289	THE ROLE CF LORENTZ FACTORS IN ROCHELLE SALT. (THEORY)
	03-0482	MICRCSCOPIC THEORY OF THE PHASE TRANSITIONS IN ROCHELLE SALT.
	04-0748	FORCE CONSTANTS IN BARIUM TITANATE AND STRONTIUM TITANATE.
	16-2317	LATTICE VIBRATION CF SODIUM NITRITE. (CALCULATIONS TO RELATE TO SPECIFIC ANTIFERROELEC
	16-2344	EFFECT OF HYDROSTATIC PRESSURE ON FERROELECTRICITY OF POTASSIUM NITRATE.
	16-2345	PHASE TRANSITIONS IN POTASSIUM NITRATE NITRITE MIXED CRYSTALS (GROWTH BY DOUBLE BRIDGM
	16-2376	DIELECTRIC ANOMALIES CF SODIUM NITRATE ABOVE THE FERROELECTRIC CURIE TEMPERATURE.
	17-2534	DIELECTRIC DISPERSION OF TGS AND DEUTERATED KDP (THEORY, CALCULATIONS).
	17-2643	MICROSCOPIC OBSERVATION OF ANTIFERROELECTRIC DOMAINS OF AMMONIUM DIHYDROGEN PHOSPHATE.
	17-2662	OSCILIATORY BEHAVIOR OF RAMAN SCATTERING INTENSITY WITH TEMPERATURE IN THE FERROELECTR
	17-2663	RAMAN SPECTRUM OF FERROELECTRIC MODE IN A POTASSIUM DIHYDROGEN PHOSPHATE CRYSTAL.
	19-2796	EFFECT OF THE HYDROSTATIC PRESSURE ON THE TRANSITION TEMPERATURE IN AMMONIUM SULFATE.
	26-2933	DIELECTRIC RESPONSE FUNCTION OF COPPER FORMATE TETRAHYDRATES.
	26-2934	TEMPERATURE DEPENDENCE CF SCUND VELOCITY IN DICALCIUM STRONTIUM PROPIONATE.
	27-3116	DIELECTRIC STUDY OF CRITICAL BEHAVIOR OF FERROELECTRIC TRIGLYCINE SULFATE BY A DIGITAL
	28-3223	MECHANISM OF FERROELECTRIC PHASE TRANSITION IN AMMONIUM ROCHELLE SALT. (THEORY)
	28-3224	SUPERSTRUCTURE IN THE FERROELECTRIC PHASE OF AMMONIUM ROCHELLE SALT.
	33-3383	MICRCSCCPIC OBSERVATIONS OF PHASE TRANSITIONS IN CESIUM LEAD TRICHLORIDE.
	33-3393	FRESSURE DEPENDENCE CF CUBIC TETRAGONAL TRANSITION TEMPERATURE OF CESIUM LEAD TRICHLOR
	36-3612	DIELECTRIC PROPERTIES OF DIAMMCNIUM HEXABROMO STANNATE (DIELECTRIC TRANSITIONS NOT FER
	37-3642	FERROELECTRIC DOMAIN SWITCHING.
TAKAHASHI H	03-0483	THECRY OF BARIUM TITANATE.
	25-2907	MOLECULAR ANC IATTICE VIBRATIONS CF ORTHORHOMBIC AND FERROELECTRIC THIOUREA. PART-2: C
TAKAHASHI K	03-0487	EFFECTS OF THERMALLY DIFFUSED IMPURITIES ON PROPAGATION OF ELASTIC SURFACE WAVES ON FE
	04-0879	ELECTRICAL FROPERTIES OF HOT PRESSED FERROELECTRIC (BARIUM TITANATE TYPE) CERAMICS. (I
	05-1075	CHARACTERISTICS OF DEFECT LEAD TITANATE.
TAKAHASHI M	09-1606	DETERIORATION OF THE ELECTROMECHANICAL PROPERTIES OF LEAD ZIRCONATE(0.52) TITANATE(0.4
	09-1607	ELECTRICAL RESISTIVITY OF LEAD ZIRCONATE TITANATE CERAMICS CONTAINING IMPURITIES.
	09-1608	EFFECTS CF IMPURITIES ON THE MECHANICAL QUALITY FACTOR OF LEAD ZIRCONATE TITANATE CERA
	09-1609	SPACE CHARGE EFFECT IN LEAD ZIRCONATE TITANATE CERAMICS CAUSED BY THE ADDITION OF IMPU
	09-1623	INTERNAL FRICTION OF MODIFIED LEAD ZIRCONATE- LEAD TITANATE CERAMICS. (PROPRIETARY COM
	13-1997	A NEW FERROELECTRIC SINGLE CRYSTAL STRONTIUM(2) NIOBIUM(2) OXIDE(7).
	13-2037	GROWTH OF FERROELECTRIC SINGLE CRYSTAL STRONTIUM(2) NIOBIUM(2) OXIDE(7) BY MEANS OF TH
	13-2067	CRYSTAL STRUCTURE AND PIEZOELECTRICITY OF THE SYSTEM LEAD (ZINCATE NIOBATE)- LEAD TITA
	36-3576	ELECTROOPTIC AND PIEZOELECTRIC PROPERTIES OF LANTHANUM TITANATE SINGLE CRYSTAL. (FOUND
TAKAHASHI S	08-1408	FHOTOCONDUCTION OF POTASSIUM TANTALATE.
	09-1606	DETERIORATION OF THE ELECTROMECHANICAL PROPERTIES OF LEAD ZIRCONATE(0.52) TITANATE(0.4
	09-1608	EFFECTS OF IMPURITIES ON THE MECHANICAL QUALITY FACTOR OF LEAD ZIRCONATE TITANATE CERA
TAKAHASHI T	02-0076	NEW FIEZOELECTRIC MATERIALS.
	03-0283	THE BEHAVIORS OF THE MICRODOMAINS ON THE DIFFUSE PHASE TRANSITION FERROELECTRICS.
	06-1141	IONIC CONDUCTION IN SINTERED OXIDES BASED ON CALCIUM TITANATE OR STRONTIUM TITANATE.
	08-1444	IONIC CONDUCTION IN PEROVSKITE TYPE OXIDE SOLID SOLUTION AND ITS APPLICATION TO THE SO
	10-1638	ELECTRCMECHANICAL FRCFERTIES IN THE BINARY SYSTEM LEAD CADMATE(0.5) TUN
	13-1959	CRYSTAL GROWTH OF LEAD (CADMATE(0.33) NIOBATE(0.67)) AND ITS DIELECTRIC PROPERTIES. (F
	13-2066	FERROELECTRIC PROPERTIES OF LEAD ZINCATE(0.33) NIOBATE(0.67) (SINGLE CRYSTALS GROWN FR
TAKAMA T	02-0099	DIELECTRIC AND X-RAY STUDIES OF THE PHASE TRANSITIONS IN ROCHELLE SALT, BARIUM TITANAT
TAKASU M	04-0881	PIEZOELECTRIC CHARACTERISTIC OF BARIUM TITANATE CERAMIC EXHIBITED WHEN IT IS GIVEN AN
TAKAYANAGI A	11-1832	PIEZOELECTRIC LEAKY SURFACE WAVE IN LITHIUM NIOBATE (ELECTROMAGNETIC COUPLING).
TAKEI H	13-1958	NEW (UNIAXIAL) NONLINEAR OPTIC CRYSTAL BARIUM LITHIUM NIOBATE HAVING NO MICROTWINNING
TAKEI NJ	13-2038	PREPARATION AND EPITAXY OF SPUTTERED FILMS OF FERROELECTRIC BISMUTH TITANIUM OXIDE.
TAKEI WJ	08-1329	GROWTH AND PROPERTIES OF EPITAXIAL FILMS OF FERROELECTRIC BISMUTH TITANATE. (ELECTROOP
	08-1470	DOMAIN STRUCTURE AND POLARIZATION REVERSAL IN FILMS OF FERROELECTRIC BISMUTH TITANATE.
	09-1515	FERROELECTRIC DISPLAYS. (BISMUTH TITANATE, LEAD ZIRCONATE TITANATE)
TAKEMURA M	04-0989	THERMAL CONDUCTIVITY OF STRONTIUM TITANATE, POTASSIUM TANTALATE AND BARIUM TITANATE NE
	04-0992	DYNAMICAI BEHAVIOR OF DISPLACIVE TYPE FERROELECTRICS NEAR THE TRANSITION POINT. (BARIU
TAKEOKA K	09-1624	STUDIES ON THE SINTERING BETWEEN LEAD MONOXIDE AND PZT. PART-1. PART-2: EFFECTS OF AVE
TAKETA Y	10-1647	EVAPORATED FILMS OF TUNGSTEN OXIDE.

TAKI T	16-2321	ACOUSTIC NUCLEAR MAGNETIC RESONANCE IN FERROELECTRIC SODIUM NITRITE.
TAKIUCHI M	04-1015	DIELECTRIC BREAKDOWN OF POLYCRYSTALLINE BARIUM TITANATE.
TALLAN NM	04-0704	ELECTRICAL PROPERTIES OF HIGH PURITY POLYCRYSTALLINE BARIUM TITANATE. (DIELECTRIC CONS
TALYANSKII VI	17-2686	CURIE-WEISS LAW FOR THE NONLINEAR SUSCEPTIBILITY OF A POTASSIUM DIHYDROGEN PHOSPHATE C
TAM CP	11-1763	EPR OF GADOLINIUM(3+) IN LITHIUM NIOBATE.
	11-1833	ELECTRON PARAMAGNETIC RESONANCE OF GADOLINIUM(3+) DOPED LITHIUM NIOBATE.
TAMATANI M	21-2821	ZEEMAN EFFECT AND DILUTION EFFECT OF ABSORPTION LINES IN GUANIDINIUM CHROMIUM SULFATE
TAMBOVTSEV DA	37-3673	GENERATOR OF BIPOLAR PULSES FOR INVESTIGATION OF FERROELECTRIC CRYSTALS.
TAMURA H	15-2197	EFFECT OF PHOTOCONDUCTION ON THE FERROELECTRIC PROPERTIES OF ANTIMONY SULFOIODIDE. (PH
TANAKA H	37-3674	AN APPROXIMATIVE ANALYSIS OF A FERROELECTRIC RECORDING SYSTEM.
TANAKA M	04-0853	PRESSURE DEPENDENCE OF TRANSITION TEMPERATURES AND ELECTROSTRICTIONS IN PEROVSKITE BAR
	04-0991	BARIUM TITANATE BY CONVERGENT BEAM ELECTRON DIFFRACTION.
	04-0993	ELECTRON MICROSCOPIC STUDIES ON FERROELECTRIC DOMAINS OF PEROVSKITE TYPE OXIDES. (LEAD
	10-1657	DIELECTRIC (AND TEMPERATURE) CHARACTERISTICS OF SOLID SOLUTIONS OF TUNGSTEN OXIDE.
	10-1658	PHASE TRANSITION IN TUNGSTEN TRIOXIDE UNDER HIGH PRESSURE.
TANAKA R	15-2175	RELATION BETWEEN THE ELECTRIC POLARIZATION AND THE ABSORPTION EDGE IN ANTIMONY SULFOIO
TANAKA T	09-1612	ELASTIC SURFACE WAVE ON THE PZT CERAMIC PLATE.
	09-1614	SURFACE WAVE DELAY LINES WITH INTERDIGITAL TRANSDUCERS ON UNPOLARIZED PZT CERAMIC PLAT
	27-3149	DIELECTRIC BEHAVIOR OF GAMMA-RAY IRRADIATED TRIGLYCINE SULFATE.
TANAKA Y	09-1540	PRECIPITATION OF ZIRCONIA PHASE IN NIOBIUM MODIFIED CERAMICS OF LEAD ZIRCONATE TITANAT
TANG CL	08-1291	(CONTINUOUS WAVE) SPONTANEOUS PARAMETRIC SCATTERING OF LIGHT IN LITHIUM IODATE (INDEX
TANI K	03-0484	THEORY OF NEUTRON SCATTERING FROM LATTICE VIBRATIONS (REDEVELOPED FROM FIRST PRINCIPLE
	04-0989	THERMAL CONDUCTIVITY OF STRONTIUM TITANATE, POTASSIUM TANTALATE AND BARIUM TITANATE NE
	04-0992	DYNAMICAL BEHAVIOR OF DISPLACIVE TYPE FERROELECTRICS NEAR THE TRANSITION POINT. (BARIU
TANIGUCHI I	09-1533	FLUX GROWN KTN SINGLE CRYSTALS AND THEIR ELECTRICAL PROPERTIES.
TANISAKI S	25-2923	CRYSTAL STRUCTURE OF THE IV PHASE OF THIOUREA. (SATELLITE SCATTERING)
TAR D	14-2071	DIELECTRIC PROPERTIES OF BORACITES AND EVIDENCE FOR FERROELECTRICITY.
TARAN JP	17-2690	PULSE DISTORTIONS IN MISMATCHED SECOND HARMONIC GENERATION. (POTASSIUM DIHYDROGEN PHOS
TARASENKO VV	03-0485	ENERGY SPECTRUM OF AN ORDER DISORDER FERROELECTRIC CONTAINING DISLOCATIONS.
	03-0486	EFFECT OF POINT DEFECTS ON THE ENERGY SPECTRUM OF HYDROGEN BONDED FERROELECTRICS.
TARASKIN SA	15-2238	HEAT CAPACITY OF POLYCRYSTALLINE ANTIMONY SULFOIODIDE.
	27-3140	EFFECTS ON ELECTRICAL BOUNDARY CONDITIONS ON THE THERMAL CAPACITY OF SINGLE CRYSTALS O
	27-3141	THERMAL AND ELECTRICAL PROPERTIES OF TRIGLYCINE SULFATE SINGLE CRYSTALS (HEAT CAPACITY
TARNA T	16-2322	X-RAY MEASUREMENTS ON THERMAL EXPANSION OF FERROELECTRIC POTASSIUM NITRATE.
TASHIRO M	04-0829	BARIUM TITANATE CERAMICS WITH HIGH DIELECTRIC CONSTANT. PART-1: EFFECTS OF ADDITION OF
TATARENKO LN	15-2202	RELAXATION PROCESSES IN ANTIMONY SULFOIODIDE SINGLE CRYSTALS EXCITED WITH MICROSECOND
	15-2229	LUMINESCENCE OF SINGLE CRYSTALS OF THE FERROELECTRIC SEMICONDUCTOR ANTIMONY SULFOIODID
	15-2239	STRAIN ALONG C-AXIS OF ANTIMONY SULFOIODIDE CAUSED BY ILLUMINATION IN DC ELECTRIC FIEL
TATSUZAKI I	16-2307	SPLITTING OF THE SODIUM-23 NMR CENTRAL LINE IN FERROELECTRIC SILVER SODIUM DINITRITE.
	16-2324	NMR STUDY OF SODIUM-23 IN SILVER SODIUM DINITRITE.
	16-2395	ELECTRIC FIELD GRADIENT AND LONG RANGE ORDER PARAMETER OF FERROELECTRIC SODIUM NITRITE
	16-2396	NUCLEAR MAGNETIC RESONANCE STUDY ON SODIUM-23 IN SODIUM NITRITE IN THE VICINITY OF THE
	19-2760	EFFECT OF DEUTERATION IN AMMONIUM BISULFATE (CHANGE IN TRANSITION TEMPERATURES).
	19-2791	ELECTRON PARAMAGNETIC RESONANCE OF MANGANESE(2+) DOPED AMMONIUM CADMIUM SULFATE.
	19-2792	TEMPERATURE DEPENDENCE OF THE ELECTRON PARAMAGNETIC RESONANCE OF AMMONIUM CADMIUM SULF
	26-2944	PHASE TRANSITION IN DIVALENT METAL DICALCIUM PROPIONATES SUBSTITUTED PARTIALLY BY ACET
	26-2947	ANOMALY OF THE PROTON SPIN LATTICE RELAXATION TIME NEAR THE CRITICAL TEMPERATURE OF DI
	26-2948	ULTRASONIC ATTENUATION NEAR THE CURIE TEMPERATURE IN DICALCIUM STRONTIUM PROPIONATE.
	26-2949	ULTRASONIC ATTENUATION NEAR THE PHASE TRANSITION IN DICALCIUM LEAD PROPIONATE.
TAUBENBERGER R	34-3478	DIELECTRIC CONSTANT OF PURE ICE I(H) SINGLE CRYSTALS.
	34-3479	PRESSURE DEPENDENCE OF THE COMPLEX DISPERSION COEFFICIENT OF ICE I(H) SINGLE CRYSTALS.
TAUREL L	16-2305	BRILLOUIN SCATTERING IN SODIUM NITRITE AT ROOM TEMPERATURE.
	17-2520	BRILLOUIN DIFFUSION IN A SINGLE CRYSTAL OF RUBIDIUM DIHYDROGEN PHOSPHATE.
	17-2521	INTENSITY OF SOME BRILLOUIN AND RAYLEIGH LINES IN RUBIDIUM DIHYDROGEN PHOSPHATE.
	17-2522	STUDY OF THE BRILLOUIN EFFECT IN ADP (ALL ELASTIC CONSTANTS).
	27-3002	DIELECTRIC RELAXATION IN MULTIDOMAIN CRYSTALS OF TRIGLYCINE SULFATE.
	27-3004	INFLUENCE OF DISLOCATIONS, SURFACE LAYER AND X-RAY IRRADIATION ON DOMAIN STRUCTURE IN
	27-3142	CORROSION FIGURES DUE TO THE FERROELECTRIC DOMAIN BOUNDARIES OF TRIGLYCINE SULFATE.
	27-3143	ETCHING AND ULTRAMICROSCOPIC STUDIES OF FERROELECTRIC DOMAINS IN TRIGLYCINE SULFATE.
TAWFEIK A	04-0536	PIEZOELECTRIC EFFECT IN ALUMINUM DOPED BARIUM TITANATE CERAMIC.
TAYLOR BE	13-2053	IMPURITY EFFECTS IN SODIUM TUNGSTEN BRONZES: NIOBIUM AND TANTALUM DOPING.
TAYLOR DJ	06-1226	SPECIFIC HEAT ANOMALY ASSOCIATED WITH THE 110 DEG K PHASE TRANSITION IN STRONTIUM TITA
TAYLOR GW	04-0769	FERROELECTRIC PHOTOCONDUCTOR OPTICAL STORAGE MEDIUM UTILIZING BISMUTH TITANATE.
	04-0994	HIGH FIELD (1.5-6 KV/CM) POLARIZATION REVERSALS IN LIQUID ELECTRODED BARIUM TITANATE C
	08-1371	BISMUTH TITANATE FERROELECTRIC- PHOTOCONDUCTOR OPTICAL STORAGE MEDIUM FURTHER DEVELOPM
	08-1445	BISMUTH TITANATE PHOTOCONDUCTOR OPTICAL STORAGE MEDIUM. (HIGH RESOLUTION READ WRITE)
	08-1446	DEPOLING OF SINGLE DOMAIN BISMUTH TITANATE.
	08-1447	FEASIBILITY OF ELECTROOPTIC DEVICES UTILIZING FERROELECTRIC BISMUTH TITANATE.
	08-1448	SELF REVERSAL AND WAITING TIME EFFECTS IN SINGLE CRYSTAL BISMUTH TITANATE.
	11-1705	RADIATION DAMAGE IN LITHIUM NIOBATE. (CAPACITANCE, OPTICAL SPECTRA, ESR, CELL PARAMETE
	13-1977	PHASE HOLOGRAMS IN A FERROELECTRIC PHOTOCONDUCTOR DEVICE. (BISMUTH(4) TITANIUM(3) OXID
	27-3144	UTILIZATION OF THE (CRITICAL PULSE WIDTH) PARTIAL SWITCHING PROPERTIES OF FERROELECTRI
	35-3517	ELECTRICAL AND OPTICAL SWITCHING PROPERTIES OF SINGLE CRYSTAL BISMUTH(4) TITANIUM(3) O
	37-3646	STORAGE OF HOLOGRAMS IN A FERROELECTRIC PHOTOCONDUCTOR DEVICE (ERASABLE, NONDESTRUCTIV
	37-3675	ACTIVE COMPENSATORS FOR FERROELECTRIC OPTICAL CIRCUITS.
	37-3676	FERROELECTRIC LIGHT VALVE ARRAYS FOR OPTICAL MEMORIES.
TAYLOR JC	24-2903	NEUTRON DIFFRACTION STUDY OF FERROELECTRIC POTASSIUM FERROCYANIDE TRIDEUTERO HYDRATE A
TAYLOR W	17-2639	RAMAN SCATTERING FROM FERROELECTRIC MODES IN THE KDP ISOMORPHOUS PHOSPHATES AND ARSENA
	17-2705	RAMAN SCATTERING BY THE TEMPERATURE DEPENDENT FERROELECTRIC MODE IN POTASSIUM DIDEUTER
TEAGUE JR	08-1449	DIELECTRIC HYSTERESIS IN (FLUX GROWN) SINGLE CRYSTAL BISMUTH FERRATE (CONFIRMS FERROEL
TEBEEK JB	16-2389	DIPOLE CORRELATIONS IN MOLTEN ALKALI METAL NITRATES AND SODIUM NITRITE.
TEDENAC JC	32-3355	CRYSTALLOGRAPHIC STUDY OF SEVERAL HYDRAZINIUM ORTHO FLUOBERYLLATES.
TEKSTER-PROSKURYAKOVA GN	04-0943	ELECTROCONDUCTIVITY AND HALL EFFECT IN SEMICONDUCTING SOLID SOLUTIONS OF ISO- AND HETE
	04-0995	SEMICONDUCTING BARIUM AND STRONTIUM TITANATES WITH POSITIVE TEMPERATURE COEFFICIENTS O
TELLE F	27-3145	ELECTROMECHANICAL PROPERTIES OF TRIGLYCINE SULFATE. (MEASUREMENTS VERIFY THEORY APPROX
TELLGREN R	22-2878	HYDROGEN BOND STUDIES. PART-54: A NEUTRON DIFFRACTION STUDY OF THE FERROELECTRIC LITHI
	22-2879	HYDROGEN BOND STUDIES. PART-58: CRYSTAL STRUCTURE OF AMMONIUM TRIHYDROGEN SELENITE.
TELLO MJ	27-3146	FERROELECTRIC SPECIFIC HEAT OF TRIGLYCINE SULFATE.
TEMME FP	17-2687	PROTON MOTIONS IN HYDROGEN BONDED FERROELECTRIC AND ANTIFERROELECTRIC SOLIDS. PART-1:
TENG MK	03-0163	RAMAN SPECTROSCOPY OF PHONON INSTABILITY IN FIRST ORDER FERROELECTRIC PHASE TRANSITION

TENG MK	15-2110	LATTICE MODES AND PHASE TRANSITION IN ANTIMONY SULFOIODIDE.
	15-2111	RAMAN SCATTERING IN FERROELECTRIC SEMICONDUCTOR ANTIMONY SULFOIODIDE.
	15-2240	SOFT PHONON COUPLINGS IN THE PRESSURE INDUCED PHASE TRANSITION IN ANTIMONY SULFOIODIDE
	16-2377	LATTICE DYNAMICS MODEL OF POTASSIUM NITRATE IN ITS FERROELECTRIC PHASE.
	16-2378	PRESSURE INDUCED FERROELECTRIC PHASE TRANSITION IN POTASSIUM NITRATE.
	16-2379	STRUCTURE OF THE DISORDERED PHASE OF POTASSIUM NITRATE.
TENNERY VJ	07-1266	HIGH TEMPERATURE (TO 236 DEG C) PHASE TRANSITIONS IN (HIGH PURITY) LEAD ZIRCONATE.
	08-1466	ISOTHERMAL PHASE TRANSITIONS IN CERAMIC LEAD ZIRCONATE.
	36-3609	STRUCTURAL AND DIELECTRIC PROPERTIES OF BARIUM ZINCATE(0.33) NIOBATE(0.67) (NOT FERROE
TERAO T	17-2688	MEASUREMENT OF CHEMICAL SHIFT IN SOLID POTASSIUM DIHYDROGEN PHOSPHATE.
TERAUCHI H	16-2380	X-RAY STUDY OF PHASE TRANSITION IN SODIUM NITRATE.
	16-2400	X-RAY CRITICAL SCATTERING IN TRIGLYCINE SULFATE AND SODIUM NITRATE. (ORDER DISORDER TR
TERENTEV YUI	04-0596	SURFACE BREAKDOWN IN VACUUM ON BARIUM TITANATE.
TERHUNE RW	17-2579	EFFECTS OF DISPERSION AND FOCUSING ON THE PRODUCTION OF OPTICAL HARMONICS. (KDP)
TERMARTIROSYAN LT	37-3677	SOME PROPERTIES AND APPLICATION OF FERROELECTRICS AT MICROWAVES.
TERPILOWSKI J	04-0752	ANOMALOUS DECAY EFFECT IN POLYCRYSTALLINE BARIUM TITANATE.
THACHER PD	09-1550	ELECTROOPTIC CERAMICS- NEW MATERIALS FOR INFORMATION STORAGE AND DISPLAY (HOT PRESSED
	09-1610	LINEAR ELECTROOPTIC EFFECT IN FERROELECTRIC CERAMICS: PLZT 12-40-60. (USED FOR SENSING
	13-2039	ELECTROOPTIC G COEFFICIENTS (RELATING INDUCED BIREFRINGENCE AND POLARIZATION) OF LEAD
THERY P	11-1682	GENERATION OF SECOND AND THIRD HARMONICS BY A TRANSVERSE ACOUSTIC WAVE PROPAGATING ITS
	11-1834	EXPERIMENTAL STUDY OF LIGHT DIFFRACTION BY HIGH FREQUENCY ACOUSTIC WAVES. APPLICATION
	11-1835	GENERATION OF THE THIRD HARMONIC OF A LONGITUDINAL HIGH FREQUENCY (950 MHZ) ACOUSTIC W
	11-1836	INTERACTIONS BETWEEN HIGH FREQUENCY ACOUSTICAL WAVES IN LITHIUM NIOBATE.
THIAGARAJAN K	04-0797	DEVELOPMENT OF BARIUM TITANATE WITH CONSISTENT DIELECTRIC PROPERTIES.
THOMANN H	09-1611	STABILIZATION EFFECTS AT THE PHASE TRANSITION TETRAGONAL- RHOMBOHEDRAL IN LEAD ZIRCONA
THOMAS H	06-1216	INTERACTION OF ELASTIC STRAIN WITH THE STRUCTURAL TRANSITION OF STRONTIUM TITANATE.
	06-1233	FLUCTUATIONS AND CORRELATIONS IN STRONTIUM TITANATE ABOVE THE CRITICAL TEMPERATURE.
	08-1450	STRUCTURAL PHASE TRANSITIONS IN PEROVSKITE TYPE CRYSTALS. (EXTENDED ABSTRACT)
THOMAS JM	17-2690	PULSE DISTORTIONS IN MISMATCHED SECOND HARMONIC GENERATION. (POTASSIUM DIHYDROGEN PHOS
THOMAS LA	02-0122	APPLICATIONS OF FERROELECTRICS AND RELATED MATERIALS: REVIEW OF DEVELOPMENTS IN EUROPE
THOMAS R	27-3007	OPTICAL CONSTANTS OF TRIGLYCINE SELENATE IN THE NEAR INFRARED TO FAR INFRARED AND IN T
	27-3008	BLOCKING OF SPONTANEOUS POLARIZATION IN TRIGLYCINE SULFATE.
	27-3010	IRREVERSIBLE AND SPONTANEOUS POLARIZATION REVERSAL IN TRIGLYCINE SULFATE BY MEANS OF A
	27-3011	LASER STUDY OF REVERSIBLE NUCLEATION SITES IN TRIGLYCINE SULFATE AND APPLICATIONS TO P
	27-3012	REVERSIBLE DOMAIN REVERSAL IN TRIGLYCINE SULFATE CAUSED BY A LASER BEAM.
THOMAS RT	04-0997	TIME DEPENDENCE OF THE ELECTRICAL CONDUCTIVITY OF BARIUM TITANATE SINGLE CRYSTALS HEAT
THOME JR	04-0996	EFFECT OF OXIDE ADDITIONS ON THE SINTERING OF BARIUM TITANATE COMPACT
THOMPSON RB	11-1837	(BACKWARD WAVE) ACOUSTIC PARAMETRIC OSCILLATIONS IN LITHIUM NIOBATE (60 DB GAIN).
	11-1838	NONLINEAR INTERACTION OF MICROWAVE ELECTRIC FIELDS AND SOUND IN LITHIUM NIOBATE.
THOMSON J	02-0103	PREPARATION OF TRANSPARENT FERROELECTRIC CERAMICS. (HOT PRESSING)
THOUY G	17-2689	DIELECTRIC PERMITTIVITY OF FINELY DIVIDED POTASSIUM AND AMMONIUM DIHYDROGEN PHOSPHATE.
THRANE N	19-2771	MOSSBAUER STUDY OF ELECTRONIC SPIN FLIP PROCESSES IN AMMONIUM IRON SULFATE DODECAHYDRA
THU HC	27-3079	FREQUENCY DEPENDENCE OF THE FLUCTUATION ABSORPTION OF ULTRASOUND IN TRIGLYCINE SULFATE
THURBER WR	06-1129	ELECTRONIC TRANSPORT (CONDUCTIVITY AND HALL AND SEEBECK COEFFICIENTS) IN STRONTIUM TIT
TICHY J	27-2984	DYNAMIC DETERMINATION OF (ELASTIC AND PIEZOELECTRIC) MATERIAL CONSTANTS OF TRIGLYCINE
TIEN TY	04-0998	A FLOATING ZONE SINGLE CRYSTAL GROWING APPARATUS. (BARIUM TITANATE)
	04-0999	INFLUENCE OF OXYGEN PARTIAL PRESSURE ON PROPERTIES OF SEMICONDUCTING BARIUM TITANATE.
TIKHOMIROVA NA	15-2138	PREPARATION OF ANTIMONY SULFOIODIDE TEXTURES FROM THE MELT UNDER PRESSURE AND INVESTIG
	16-2266	FERROELECTRIC PHASE TRANSITION IN SILVER SODIUM DINITRITE.
TILLEY RJD	10-1627	EXAMINATION OF SUBSTOICHIOMETRIC TUNGSTEN OXIDE(3-X) CRYSTALS BY ELECTRON MICROSCOPY.
	10-1659	FORMATION OF SHEAR STRUCTURES IN SUBSTOICHIOMETRIC TUNGSTEN TRIOXIDE.
TINDERMANS-VAN EYNDHOVEN	33-3388	A PHASE TRANSITION IN A COMPOUND WITH HELICAL ELECTRIC DIPOLE STRUCTURE: CESIUM COPPER
TINKHAM M	06-1094	FAR INFRARED FERROELECTRIC VIBRATION MODE IN STRONTIUM TITANATE.
	17-2423	FAR INFRARED DIELECTRIC MEASUREMENTS ON POTASSIUM DIHYDROGEN PHOSPHATE, TRIGLYCINE SUL
TITOVA SA	08-1428	GROWTH OF SINGLE CRYSTALS OF VARIOUS PEROVSKITES WITH COMPLEX COMPOSITIONS, AND THE ST
TITTEL FK	11-1701	MEASUREMENTS OF PHOTON CORRELATIONS OF SECOND HARMONIC GENERATED LIGHT (FROM LITHIUM N
	13-1984	VISIBLE CW PARAMETRIC OSCILLATOR USING BARIUM SODIUM NIOBATE.
	37-3638	OPTICALLY ERASABLE AND REWRITABLE SOLID STATE HOLOGRAMS.
TITTMANN BR	37-3647	APPLICATION OF NONLINEAR INTERACTIONS IN FERROELECTRIC CERAMICS TO MICROWAVE SIGNAL PR
TIVARI HW	22-2880	ANOMALOUS THERMAL EXPANSION OF POTASSIUM TRIHYDROGEN SELENATE CRYSTALS.
TIWARY HV	29-3245	IS SODIUM RUBIDIUM TARTRATE TETRAHYDRATE A FERROELECTRIC. (ANSWER-NO)
TODA K	03-0487	EFFECTS OF THERMALLY DIFFUSED IMPURITIES ON PROPAGATION OF ELASTIC SURFACE WAVES ON FE
	09-1612	ELASTIC SURFACE WAVE ON THE PZT CERAMIC PLATE.
	09-1613	PROPAGATION CHARACTERISTICS OF SURFACE ELASTIC WAVES ON BARIUM TITANATE(0.95) ZIRCONAT
	09-1614	SURFACE WAVE DELAY LINES WITH INTERDIGITAL TRANSDUCERS ON UNPOLARIZED PZT CERAMIC PLAT
TODD LT	13-2040	SOFT PHONONS IN POTASSIUM TANTALATE NIOBATE.
TODO I	16-2396	NUCLEAR MAGNETIC RESONANCE STUDY ON SODIUM-23 IN SODIUM NITRITE IN THE VICINITY OF THE
	26-2947	ANOMALY OF THE PROTON SPIN LATTICE RELAXATION TIME NEAR THE CRITICAL TEMPERATURE OF DI
	26-2948	ULTRASONIC ATTENUATION NEAR THE CURIE TEMPERATURE IN DICALCIUM STRONTIUM PROPIONATE.
	26-2949	ULTRASONIC ATTENUATION NEAR THE PHASE TRANSITION IN DICALCIUM LEAD PROPIONATE
TOKAGI Y	16-2357	ELASTIC CONSTANTS AND ULTRASONIC ABSORPTION OF SODIUM NITRITE SINGLE CRYSTALS.
TOKAR LF	19-2776	STRUCTURE OF AMMONIUM FLUOBERYLLATE AT AMBIENT TEMPERATURE.
TOKITO Y	09-1624	STUDIES ON THE SINTERING BETWEEN LEAD MONOXIDE AND PZT. PART-1. PART-2: EFFECTS OF AVE
TOKMAKOVA EI	27-2987	ELECTRICAL PROPERTIES OF THE SURFACES OF IONIC CRYSTALS (TGS AND OTHERS, BY SELECTIVE
TOKUGAWA F	04-0862	FORMATION PROCESS OF BARIUM TITANATE THIN FILM EVAPORATED IN VACUO.
TOKUGAWA Y	16-2381	POLARIZATION REVERSAL IN A WIDE RANGE OF IMPRESSING RATE OF FIELD IN SODIUM NITRITE.
TOKUMARU Y	02-0035	MEASUREMENT OF DIELECTRIC CONSTANTS FOR A FEW FERROELECTRIC MATERIALS AT LOW TEMPERATU
	17-2691	EFFECT OF GAMMA-IRRADIATION ON THE DIELECTRIC CONSTANTS OF KDP.
	28-3179	DIELECTRIC CONSTANT AND DIELECTRIC RELAXATION TIME OF ROCHELLE SALT.
TOKUNAGA M	03-0488	ANOMALY OF THE SPIN LATTICE RELAXATION TIME NEAR THE TRANSITION TEMPERATURE IN THE ORD
	03-0489	MODELS OF THE PHASE TRANSITION IN POTASSIUM DIHYDROGEN PHOSPHATE TYPE FERROELECTRICS.
	16-2365	NEUTRON DIFFRACTION STUDIES OF FERROELECTRIC POLARIZATION REVERSALS IN SODIUM NITRITE.
	26-2947	ANOMALY OF THE PROTON SPIN LATTICE RELAXATION TIME NEAR THE CRITICAL TEMPERATURE OF DI
TOLSTAYA-BELIK VYA	08-1417	ELECTRICAL CONDUCTIVITY OF STRONTIUM ZIRCONATE AND HAFNATE AT HIGH TEMPERATURES.
TOMASHPOLSKII YUYA	02-0128	NEW FERROELECTRICS AND SEIGNETTO MAGNETS OF THE PEROVSKITE AND PYROCHLORE TYPE. SYNTHE
	03-0490	(METHODS OF) STRUCTURAL STUDIES OF FERROELECTRICS AND FERROELECTRIC MAGNETS. (ELECTRON
	04-1000	PREPARATION AND STUDY OF THIN FILMS OF (BARIUM, STRONTIUM) TITANATE.
	04-1001	VACUUM EVAPORATION OF BARIUM TITANATE.
	04-1038	STRUCTURAL ELECTRON MICRODIFFRACTION STUDIES OF FERROELECTRICS AND FERROELECTRIC MAGNE
	05-1082	STRUCTURAL MICROELECTRON DIFFRACTION STUDY OF FERROELECTRIC AND FERROMAGNETIC SUBSTANC

PAGE 694

TOMASHPOLSKII YUYA	08-1401	MOSSBAUER STUDY AND THEORETICAL ESTIMATION OF INTERNAL ELECTRIC FIELD GRADIENTS IN FER
	08-1428	GROWTH OF SINGLE CRYSTALS OF VARIOUS PEROVSKITES WITH COMPLEX COMPOSITIONS, AND THE ST
	08-1451	SUPERSTRUCTURE OF BISMUTH METAL OXYGEN(3) PEROVSKITES. (WHERE METAL IS SCANDIUM, CHROM
	08-1452	STRUCTURE AND DIELECTRIC LOSSES OF FERROELECTRIC PEROVSKITES SYNTHESIZED AT HIGH PRESS
	30-3271	PHASE TRANSITIONS OF GADOLINIUM MOLYBDATE AND ISOSTRUCTURAL COMPOUNDS (EUROPIUM TERBIU
TOMINAGA Y	16-2382	STUDY OF THE CRITICAL SLOWING DOWN OF THE DIELECTRIC RELAXATION IN SILVER SODIUM DINIT
	27-3147	STUDY OF THE CRITICAL BEHAVIOR OF TRIGLYCINE SULFATE BY THERMAL NOISE MEASUREMENTS.
TOMISHIMA K	17-2604	INTRACAVITY SECOND HARMONIC GENERATION WITH RUBY LASER BY KDP.
TOMORA EL	04-0789	DEPENDENCE OF DIELECTRIC PROPERTIES OF BARIUM TITANATE CERAMICS ON THE PHYSICO- CHEMIC
TOPF KJ	02-0105	PIEZOELECTRIC CERAMIC SUBSTANCES. (REVIEW, 23 REFS)
	27-2995	NEW MEASUREMENTS OF DIVERSE THERMODYNAMICAL COEFFICIENTS OF ROCHELLE SALT AND TRIGLYCI
TOPIC M	36-3610	SILVER DITHORIUM PHOSPHATE- A NEW CASE OF THE NON-HYDROGEN BONDED PHOSPHATE FERROELECT
	36-3611	TEMPERATURE DEPENDENCE OF SOME PROPERTIES OF SODIUM DITHORIUM PHOSPHATE FERROELECTRIC
TOPP W	15-2201	CRYSTALLIZATION BEHAVIOR OF ANTIMONY SULFOIODIDE. (SINGLE CRYSTAL GROWTH, CHEMICAL TRA
TORNBERG NE	05-1083	TEMPERATURE DEPENDENT OPTICAL PHONONS IN LEAD TITANATE.
TOROPOV AN	04-1002	INFLUENCE OF FIRING OF BARIUM TITANATE SINTERS ON ELECTRICAL AND TECHNOLOGICAL PROPERT
	04-1003	REACTION SEQUENCE IN FORMING BARIUM METATITANATE IN AN INDUSTRIAL ROTARY FURNACE.
TORRE LP	14-2102	FERROELECTRIC AND FERROELASTIC PROPERTIES OF MAGNESIUM CHLORINE BORACITE.
TORRIE BH	19-2793	RAMAN AND INFRARED STUDIES OF THE FERROELECTRIC TRANSITION IN AMMONIUM SULFATE.
TOSHEV SD	03-0341	OBSERVATION OF DOMAIN STRUCTURE IN LOW TEMPERATURE FERROELECTRICS BY SOLID DEW TECHNIQ
	24-2895	DOMAIN STRUCTURE IN THE FERROELECTRIC POTASSIUM FERROCYANIDE.
	24-2904	APPLICATION OF THE DEW METHOD FOR REVEALING THE DOMAIN STRUCTURE IN POTASSIUM FERROCYA
	27-3148	EFFECT OF SAMPLE TREATMENT AND ELECTRODE NATURE ON SOME TRIGLYCINE SULFATE PARAMETERS.
TOSI M	33-3363	TWO PHONON PROCESSES AND PHASE TRANSITIONS IN POTASSIUM MANGANESE TRIFLUORIDE: TEMPERA
TOURNEUR D	13-2011	FIRST FERROELECTRIC OXYFLUORIDE PHASES OF THE OXYGENATED QUADRATIC TUNGSTEN BRONZE TYP
TOVBIS AE	22-2881	CRYSTAL STRUCTURE OF RUBIDIUM TRIHYDROGEN SELENITE.
TOWNER HH	11-1839	EPR STUDIES OF CRYSTAL FIELD PARAMETERS IN IRON(3+) DOPED LITHIUM NIOBATE.
TOWNSEND RL	04-1004	OPTICALLY INDUCED REFRACTIVE INDEX CHANGES IN BARIUM TITANATE (EXPERIMENTS TO CHARACTE
	04-1005	OPTICALLY INDUCED REFRACTIVE INDEX CHANGES IN BARIUM TITANATE.
TOYODA H	04-0939	EXPERIMENTAL STUDY OF EVAPORATED BARIUM TITANATE FILMS.
	04-0949	THICKNESS DEPENDENCE OF POLARIZATION REVERSAL IN BARIUM TITANATE SINGLE CRYSTALS.
	04-1006	BARIUM TITANATE SINGLE CRYSTALS SUBJECTED TO MECHANICAL IMPULSES.
	04-1007	DISLOCATIONS IN BARIUM TITANATE SINGLE CRYSTALS.
	12-1877	BEHAVIOR OF LITHIUM TANTALATE SINGLE CRYSTAL NEAR ITS CURIE POINT. PART-1: PIEZOELECTR
	36-3570	SWITCHING OF OPTICAL ROTATORY POWER IN FERROELECTRIC LEAD(5) GERMANIUM(3) OXIDE(11) SI
TOYODA K	01-0032	BIBLIOGRAPHY OF FERROELECTRICS.
	01-0033	BIBLIOGRAPHY ON FERROELECTRICS. (1949-1964, 3048 REFERENCES, SUBJECT AND AUTHOR INDEXE
	15-2175	RELATION BETWEEN THE ELECTRIC POLARIZATION AND THE ABSORPTION EDGE IN ANTIMONY SULFOIO
	15-2241	TRANSPORT PHENOMENA IN ANTIMONY SULFOIODIDE. (CONDUCTION MECHANISM FOR THERMOELECTRIC
	16-2312	CRYSTAL STRUCTURE AND PHASE TRANSITION OF DIGLYCINE NITRATE.
	27-3149	DIELECTRIC BEHAVIOR OF GAMMA-RAY IRRADIATED TRIGLYCINE SULFATE.
TRAPPE KI	06-1159	ANOMALOUS SPECIFIC HEAT OF STRONTIUM TITANATE.
TREDGOLD RH	04-0580	DIELECTRIC BREAKDOWN IN BARIUM TITANATE.
	04-0638	DOUBLE SPACE CHARGE INJECTION IN SOLIDS. (USING BARIUM TITANATE ABOVE CURIE POINT)
TREFLER M	19-2794	DYNAMICS OF AMMONIUM IONS IN AMMONIUM SULFATE.
TREILLEUX M	04-0843	PREPARATION AND PROPERTIES OF THIN FILMS OF BARIUM TITANATE.
TREIVUS EB	08-1367	CRYSTALLIZATION OF POTASSIUM IODATE.
TREVINO SF	34-3465	ICE I: LATTICE DYNAMICS AND INCOHERENT NEUTRON SCATTERING.
TROCCAZ M	03-0491	SHOCK WAVE INDUCED ELECTROMECHANICAL CONVERSION IN FERROELECTRIC MATERIALS.
	09-1508	FERROELECTRIC CERAMICS AND ENERGY CONVERSION. (LEAD ZIRCONATE TITANATE)
	09-1581	PHASE DIAGRAM OF LEAD ZIRCONATE(1-X) TITANATE(X) PLUS 0.8 PERCENT TUNGSTEN TRIOXIDE. (
	09-1615	METHOD OF MEASURING POLARIZATION VIBRATIONS OF A PREPOLARIZED CERAMIC FERROELECTRIC. (
TRONTELJ Z	19-2795	DEUTERON SPIN LATTICE RELAXATION IN FERROELECTRIC DEUTERATED AMMONIUM HYDROGEN SULFATE
	28-3232	STUDY OF SLOW MOTION OF WATER MOLECULES IN ROCHELLE SALT AND AMMONIUM ROCHELLE SALT BY
TRUBICYN AM	16-2319	EFFECT OF ADDING TITANIUM(+) AND STRONTIUM(2+) IONS ON THE ELECTRICAL CONDUCTIVITY OF
TSAI J	06-1243	EFFECTS OF LONGITUDINAL MAGNETIC FIELD ON ULTRASOUND IN STRONTIUM TITANATE.
TSALLIS C	04-0578	ELECTROSTATIC FIELD IN A SLIGHTLY ORTHORHOMBIC IONIC CRYSTAL. APPLICATION TO THE CALCU
	04-1008	ELECTROSTATIC FIELD IN A PSEUDOCUBIC IONIC CRYSTAL. APPLICATION TO ALL BARIUM TITANA
	17-2692	DYNAMICS OF POTASSIUM DIHYDROGEN PHOSPHATE TYPE FERROELECTRIC PHASE TRANSITIONS.
TSANG T	31-3339	INTERMOLECULAR POTENTIAL AND FERROELECTRIC TRANSITIONS IN HYDROGEN HALIDES.
TSEDRIK MS	27-2978	PERMITTIVITY OF SINGLE CRYSTALS OF TRIGLYCINE SULFATE IN A STRONG ELECTRIC FIELD.
	27-3069	INFLUENCE OF GROWTH CONDITIONS OF TRIGLYCINE SULFATE AND TRIGLYCINE SELENATE CRYSTALS
	27-3125	DIELECTRIC PERMEABILITY AND DIELECTRIC LOSS ANGLE TANGENT OF TRIGLYCINE SELENATE DEPEN
	27-3126	THERMAL HYSTERESIS OF THE DIELECTRIC CONSTANT OF TRIGLYCINE SULFATE AS A FUNCTION OF T
	27-3150	DIELECTRIC CONSTANT OF TRIGLYCINE SULFATE SINGLE CRYSTALS IN A STRONG ELECTRIC FIELD.
	27-3151	EFFECT OF TRIGLYCINE FLUOBERYLLATE SINGLE CRYSTAL GROWTH CONDITIONS ON A CHANGE IN THE
	27-3152	EFFECT OF GROWTH CONDITIONS ON DIELECTRIC LOSSES IN TRIGLYCINE SULFATE SINGLE CRYSTALS
	27-3153	EFFECT OF GROWTH CONDITIONS OF TRIGLYCINE SULFATE SINGLE CRYSTALS ON A CHANGE IN THE T
	27-3154	MAXIMUM REPOLARIZATION CURRENT FOR TRIGLYCINE SULFATE CRYSTALS DEPENDENT ON GROWTH CON
TSEKVAVA BE	03-0492	RELAXATION ABSORPTION OF SOUND NEAR THE CURIE POINT OF FERROELECTRIC SEMICONDUCTORS.
TSENG CC	11-1752	OPTICAL PROBING OF ACOUSTIC SURFACE WAVE HARMONIC GENERATION (IN LITHIUM NIOBATE)
	13-2041	PIEZOELECTRIC SURFACE WAVES IN CUBIC AND ORTHORHOMBIC CRYSTALS. (BISMUTH(12) GERMANIUM
TSIKIN AN	06-1227	THERMALLY STIMULATED CONDUCTIVITY IN STRONTIUM TITANATE SINGLE CRYSTALS.
TSIVILEV RP	11-1840	MECHANISM AND KINETICS OF LITHIUM METANIOBATE FORMATION BY A SOLID PHASE REACTION.
	13-2042	MECHANISM AND KINETICS OF SOLID PHASE FORMATION OF LEAD METANIOBATE.
TSUBOUCHI N	08-1453	SYNTHESIS OF PEROVSKITE COMPLEX COMPOSITIONS: LEAD MANGANATE(0.5) NIOBATE(0.5) AND LEA
	13-2067	CRYSTAL STRUCTURE AND PIEZOELECTRICITY OF THE SYSTEM LEAD (ZINCATE NIOBATE)- LEAD TITA
TSUCHIYA A	17-2693	SECOND HARMONIC GENERATION OF LIGHT IN THE KDP CRYSTAL BY THE INDEX MATCHING METHOD.
TSUDA N	22-2872	THE STRUCTURE AND PHASE TRANSITION IN POTASSIUM SELENATE.
TSUJIKAWA I	21-2821	ZEEMAN EFFECT AND DILUTION EFFECT OF ABSORPTION LINES IN GUANIDINIUM CHROMIUM SULFATE
TSUNEKAWA S	19-2796	EFFECT OF THE HYDROSTATIC PRESSURE ON THE TRANSITION TEMPERATURE IN AMMONIUM SULFATE.
	36-3612	DIELECTRIC PROPERTIES OF DIAMMONIUM HEXABROMO STANNATE (DIELECTRIC TRANSITIONS NOT FER
TSUNODA H	04-0781	EFFECT OF THERMAL NEUTRON IRRADIATION ON THE RELATIVE PERMITTIVITY OF BARIUM TITANATE
TSUYA H	11-1841	DETERMINATION OF THE DISTRIBUTION COEFFICIENTS OF LITHIUM OXIDE INTO LITHIUM NIOBATE B
	12-1874	ELECTROOPTIC AND FERROELECTRIC PROPERTIES OF LITHIUM TANTALATE SINGLE CRYSTALS AS A FU
	12-1889	SELF INDUCED THERMAL EFFECTS ON LIGHT EXTINCTION OF LITHIUM TANTALATE.
TSUYA T	13-2043	DEPENDENCE OF SECOND HARMONIC GENERATION ON CRYSTAL INHOMOGENEITY (THEORY AND TEST ON
TSYAHCHENKO YUP	17-2512	USE OF KDP TYPE CRYSTALS AS ELECTROOPTICAL REFLECTORS.
TSYKALOV VG	02-0124	MILLIMETER WAVELENGTH DISPERSION IN THE DIELECTRIC CONSTANT ABOVE THE CURIE POINT IN V
TUFTE ON	09-1558	PYROELECTRIC PROPERTIES OF THE LANTHANUM DOPED FERROELECTRIC PLZT CERAMICS.

PAGE 695

TUKHTASUNOV IT	22-2840	PHASE TRANSFORMATION AND THERMOOPTICAL AND ELECTROOPTICAL EFFECT IN RUBIDIUM TRIHYDROG
TULINOV AF	04-0534	STUDYING SMALL DEFORMATIONS OF A CRYSTAL LATTICE BASED ON THE SHADOW EFFECT. (NEUTRON
TUREK RF	30-3274	ELASTIC CONSTANTS OF GADOLINIUM MOLYBDATE (NEAR THE CURIE POINT).
TURIK AV	03-0493	STATISTICAL METHOD FOR INVESTIGATING (DYNAMICS OF) POLARIZATION REVERSAL IN FERROELECT
	03-0494	THEORY OF POLARIZATION AND HYSTERESIS OF FERROELECTRICS.
	04-1009	DIELECTRIC DISPERSION IN BARIUM TITANATE TYPE FERROELECTRICS IN THE METER BAND.
	04-1010	DIELECTRIC, ELASTIC AND PIEZOELECTRIC PROPERTIES OF BARIUM TITANATE SINGLE CRYSTALS WI
	04-1011	EFFECT OF GRAIN SIZE ON THE PIEZOELECTRIC AND ELASTIC PROPERTIES OF BARIUM TITANATE.
	04-1012	INFLUENCE OF A CONSTANT ELECTRIC FIELD ON THE DIELECTRIC PROPERTIES OF POLYCRYSTALLINE
	04-1013	PROBLEM OF THE SURFACE LAYER IN BARIUM TITANATE SINGLE CRYSTALS.
	04-1014	UHF REVERSAL PROPERTIES OF BARIUM TITANATE TYPE FERROELECTRIC CERAMICS.
TURILE AV	04-0785	EFFECT OF CRYSTALLINE GRAIN SIZE ON THE PHYSICAL PROPERTIES OF CERAMIC BARIUM TITANATE
TURKOV SK	09-1582	INFLUENCE OF THE CONCENTRATION OF POINT DEFECTS ON THE INTERNAL FRICTION IN POLYCRYSTA
	09-1584	INTERNAL FRICTION (AND CURIE TEMPERATURES) IN (LEAD ZIRCONATE TITANATE) FERROELECTRICS
TURNER CW	11-1692	ACOUSTIC SURFACE WAVE CONVOLUTION ON CRYSTALS OF CADMIUM SULFIDE, LITHIUM NIOBATE AND
TURNER EH	01-0016	LINEAR ELECTROOPTICAL MATERIALS. (REVIEW, 138 REFS)
	02-0107	DETERMINATION OF ABSOLUTE SIGNS OF MICROWAVE NONLINEAR SUSCEPTIBILITIES. (BARIUM TITAN
	11-1842	DEPENDENCE OF LINEAR ELECTROOPTIC EFFECT AND DIELECTRIC CONSTANT ON MELT COMPOSITION (
	13-1998	MEASUREMENTS OF SECOND HARMONIC GENERATION AND THE VARIATIONS IN THE FREE AND CLAMPED
TURNER ES	01-0004	FERROELECTRICITY AND FERROELECTRIC MATERIALS: A COMPREHENSIVE BIBLIOGRAPHY.
TURYANITSA ID	15-2114	ABSORPTION OF ULTRASOUND IN A(V)B(VI)C(VII) FERROELECTRIC SEMICONDUCTORS AT TEMPERATUR
	15-2115	ABSORPTION OF ULTRASOUND IN AN ANTIMONY SULFOIODIDE FERROELECTRIC SEMICONDUCTOR NEAR I
	15-2116	DETERMINATION OF THE LATTICE PARAMETERS AND CURIE POINTS OF SOLID SOLUTIONS OF A(V)B(V
	15-2119	INFLUENCE OF ILLUMINATION ON THE FORM OF THE VOLTAGE AMPERE CHARACTERISTICS OF ANTIMON
	15-2121	PIEZOELECTRIC VIBRATIONS IN ANTIMONY SULFOIODIDE CRYSTALS NEAR THE CURIE POINT (EFFECT
	15-2125	ANISOTROPY OF THE REFLECTION SPECTRA OF ANTIMONY SULFOIODIDE CRYSTALS.
	15-2128	PIEZOELECTRIC RESISTANCE EFFECT IN BISMUTH SELENOIODIDE CRYSTALS.
	15-2141	PRODUCTION METHODS AND SOME OPTICAL PROPERTIES OF BISMUTH TELLUROIODIDE.
	15-2142	PRODUCTION AND SOME OPTICAL PROPERTIES OF ANTIMONY SULFOBROMIDE IN GLASSY AND CRYSTALL
	15-2154	EFFECT OF THE ADSORPTION OF OXYGEN ON THE PROPERTIES OF ANTIMONY SULFOIODIDE SINGLE CR
	15-2187	EFFECT OF OXYGEN ADSORPTION ON THE FERROELECTRIC PHASE TRANSITION IN ANTIMONY SULFOIOD
	15-2199	PREPARATION AND PHYSICAL PROPERTIES OF NONSTOICHIOMETRIC ANTIMONY SULFOIODIDE SINGLE C
	15-2217	ABSORPTION SPECTRA OF BISMUTH TELLUROBROMIDE AND BISMUTH TELLUROIODIDE CRYSTALS.
	15-2242	DIELECTRIC PROPERTIES OF ANTIMONY SULFOIODIDE- SULFOBROMIDE SOLID SOLUTIONS.
TUTOV AG	08-1292	STRUCTURE AND DIELECTRIC PROPERTIES OF LITHIUM CONTAINING PEROVSKITES. (STRONTIUM LITH
	36-3613	SYNTHESIS AND X-RAY DIFFRACTION STUDY OF SINGLE CRYSTALS OF A NEW PYROCHLORE CONTAININ
TYULYAEVA NG	09-1514	TIME AND TEMPERATURE DEPENDENCES OF THE PARAMETERS OF A FERROELECTRIC CERAMIC.
UBBELOHDE AR	16-2278	STUDIES OF PHASE TRANSFORMATIONS IN NITRATES AND NITRITES. PART-1: CHANGE IN ULTRAVIOL
UCHIDA N	11-1712	GROWTH AND PROPERTIES OF LITHIUM NIOBATE AND LITHIUM TANTALATE FILMS.
	12-1876	SINGLE CRYSTAL GROWTH AND PHYSICAL PROPERTIES OF LITHIUM TANTALATE. (DIELECTRIC CONSTA
	35-3509	ACCUSTOOPTIC PROPERTY OF SINGLE CRYSTAL LEAD(5) GERMANIUM(3) OXIDE(11).
	35-3518	ELECTROOPTIC PROPERTIES OF FERROELECTRIC LEAD GERMANATE SINGLE CRYSTAL.
UCHIDATE M	04-1025	STUDIES ON THE BARIUM STRONTIUM TITANATE BOUNDARY LAYER CERAMIC DIELECTRICS.
UDALOVA VV	19-2797	ELECTRON DIFFRACTION STUDY OF THE STRUCTURE OF AMMONIUM SULFATE.
UEDA I	04-0744	FAR INFRARED REFLECTIVITY OF BARIUM TITANATE.
	04-1015	DIELECTRIC BREAKDOWN OF POLYCRYSTALLINE BARIUM TITANATE.
	05-1060	ELECTROMECHANICAL PROPERTIES OF LEAD TITANATE CERAMICS CONTAINING LANTANUM AND MANGANE
	05-1062	PIEZOELECTRIC PROPERTIES OF LEAD TITANATE CERAMICS.
	05-1084	EFFECTS ON ADDITIVES ON PIEZOELECTRIC AND RELATED PROPERTIES OF LEAD TITANATE CERAMICS
	05-1085	ELECTROMECHANICAL PROPERTIES AND APPLICATION OF LEAD TITANATE CERAMICS.
	09-1616	LEAD TITANATE PIEZOELECTRIC CERAMICS. (MATSUSHITA LEAD TITANATE)
UEDA J	05-1061	POLARIZATION AND PIEZOELECTRIC PROPERTIES OF LEAD TITANATE CERAMICS. (CONTAINING 1 MOL
UEDA R	04-0862	FORMATION PROCESS OF BARIUM TITANATE THIN FILM EVAPORATED IN VACUO.
	16-2325	DISORDERED STRUCTURE OF SODIUM NITRITE AT 185 DEG C.
	16-2328	X-RAY AND NEUTRON STUDY ON THE PHASE TRANSFORMATION OF SODIUM NITRITE.
UEDA S	15-2239	STRAIN ALONG C-AXIS OF ANTIMONY SULFOIODIDE CAUSED BY ILLUMINATION IN DC ELECTRIC FIEL
UEHLING EA	17-2436	ABSENCE OF A LINE WIDTH TRANSITION FOR THE PROTONS IN POTASSIUM DIHYDROGEN PHOSPHATE T
	17-2575	POLARIZATION RELAXATION AND SUSCEPTIBILITY IN THE FERROELECTRIC TRANSITION REGION OF P
	17-2651	RANDOM MOTION OF DEUTERONS IN DEUTERATED POTASSIUM DIHYDROGEN PHOSPHATE.
UEMATSU Y	08-1332	PREPARATION OF POTASSIUM NIOBATE SINGLE CRYSTAL FOR OPTICAL APPLICATION.
	08-1454	NONLINEAR OPTICAL PROPERTIES OF POTASSIUM NIOBATE SINGLE CRYSTALS. (REFRACTIVE INDICES
	12-1890	PIEZOELECTRIC PROPERTIES OF FERROELECTRIC POTASSIUM LITHIUM TANTALATE(0-0.5) NIOBATE(1
UESU Y	17-2555	ORDER OF THE FERROELECTRIC PHASE TRANSFORMATION OF POTASSIUM DIHYDROGEN PHOSPHATE.
	17-2556	X-RAY STUDY ON THERMAL EXPANSION OF FERROELECTRIC POTASSIUM DIHYDROGEN PHOSPHATE. (NEA
	17-2557	X-RAY DILATOMETRIC STUDY OF THE FERROELECTRIC PHASE TRANSITION OF POTASSIUM DIHYDROGEN
UGODENKO AA	17-2460	CHARACTERISTICS OF AN ELECTROOPTICAL SHUTTER BASED ON Z CUT POTASSIUM DIDEUTERIUM PHOS
UGRYUMOVA MA	09-1617	FERROELECTRIC AND PIEZOELECTRIC PROPERTIES IN A LEAD TITANATE- LEAD ZIRCONATE- LEAD (Z
	13-1951	ELECTROPHYSICAL PROPERTIES OF THE SOLID SOLUTION SYSTEM LEAD TITANATE- LEAD ZIRCONATE-
UJMA Z	07-1267	THERMOELECTRIC EFFECT IN SEMICONDUCTIVE LEAD ZIRCONATE.
ULLMAN FG	30-3279	PYROELECTRIC BEHAVIOR OF FERROELECTRIC GADOLINIUM MOLYBDATE.
	30-3282	PRODUCTION OF SPONTANEOUS POLARIZATION BY ELASTIC INSTABILITIES IN PIEZOELECTRIC MATER
	30-3285	TEMPERATURE DEPENDENCE OF THE RAMAN SPECTRUM OF GADOLINIUM MOLYBDATE.
	30-3323	PYROELECTRIC EFFECT IN GADOLINIUM MOLYBDATE.
	30-3324	PYROELECTRIC DETECTION PROPERTIES OF GADOLINIUM MOLYBDATE.
UNO T	12-1885	BEHAVIOR OF LITHIUM TANTALATE SINGLE CRYSTAL NEAR ITS CURIE POINT, PART-2, DIELECTRIC
UNOKI H	04-0928	ELECTRON PARAMAGNETIC RESONANCE IN BARIUM TITANATE NEAR THE CURIE TEMPERATURE.
	04-0929	ELECTRON PARAMAGNETIC RESONANCE OF IRON(3+) IN BARIUM TITANATE IN THE RHOMBOHEDRAL AND
	04-1016	PHASE TRANSITIONS IN THE MIXED CRYSTAL SYSTEM STRONTIUM(1-X) BARIUM(X) TITANATE.
	06-1131	SECOND HARMONIC GENERATION IN STRESS INDUCED FERROELECTRIC STRONTIUM TITANATE.
	06-1202	DIELECTRIC PROPERTIES OF STRONTIUM TITANATE AT LOW TEMPERATURES. (TEMPERATURE DEPENDEN
	06-1228	ELECTRON SPIN RESONANCE STUDY IN STRONTIUM TITANATE.
	06-1229	ELECTRIC FIELD EFFECT OF ELECTRON SPIN RESONANCE IN STRONTIUM TITANATE. (DOPED WITH GA
	06-1230	NEW ESR EVIDENCE FOR THE 65 DEGREES K ANOMALY IN STRONTIUM TITANATE.
	08-1409	CRYSTAL GROWTH OF SODIUM NIOBATE.
UNRUH HG	02-0125	BEHAVIOR OF THE COMPLEX PERMITTIVITY (AUDIO FREQUENCIES) OF SOME FERROELECTRIC CRYSTAL
	03-0495	INFLUENCE OF POINT DEFECTS ON DIELECTRIC PROPERTIES, DOMAIN DYNAMICS, AND PHASE TRANSI
	03-0496	ON THE INFLUENCE OF CRYSTAL DEFECTS UPON FERROELECTRIC PROPERTIES. (DISPERSION, RELAXA
	19-2798	FERROELECTRIC TRANSITION OF AMMONIUM SULFATE.
	19-2799	(TEMPERATURE DEPENDENCE OF) THE SPONTANEOUS POLARIZATION OF AMMONIUM SULFATE.
	27-3017	OPTICAL ACTIVITY OF SOME FERROELECTRIC CRYSTALS (PYROELECTRICITY, TGS, ROCHELLE SALT).

UNRUH HG	27-3155	CRITICAL SLOWING DOWN AT FERROELECTRIC TRANSITIONS (TRIGLYCINE SULFATE).
	27-3156	SHORT LIVED FERROELECTRIC "AFTER EFFECT" PHENOMENA. (HYSTERESIS LOOP CONSTRICTIONS, RO
UPADHYAYA DD	04-0797	DEVELOPMENT OF BARIUM TITANATE WITH CONSISTENT DIELECTRIC PROPERTIES.
UPRETI GC	16-2358	ELECTRON PARAMAGNETIC RESONANCE OF MANGANESE(2+) IN THE FERROELECTRIC PHASE OF SODIUM
	16-2359	FERROELECTRIC PHASE TRANSITION IN SODIUM NITRITE AS STUDIED BY EPR (PHASE TRANSITION A
	16-2383	ALLOWED AND FORBIDDEN HYPERFINE TRANSITIONS IN THE ELECTRON PARAMAGNETIC RESONANCE OF
USACHEV EP	03-0153	AN APPARATUS FOR INVESTIGATING THE ELECTRICAL PROPERTIES OF FERROELECTRICS.
USHATKIN EF	06-1089	MEASUREMENT OF DIELECTRIC PROPERTIES OF FERROELECTRICS AT SUBMILLIMETER WAVELENGTHS. (
USMANOV RG	11-1797	INVESTIGATION OF CERTAIN PARAMETERS OF LITHIUM NIOBATE SINGLE CRYSTALS USING SCATTERIN
	11-1843	FREQUENCY DEPENDENCES OF SOME PARAMETERS OF LITHIUM NIOBATE.
UWE H	04-1016	PHASE TRANSITIONS IN THE MIXED CRYSTAL SYSTEM STRONTIUM(1-X) BARIUM(X) TITANATE.
	06-1131	SECOND HARMONIC GENERATION IN STRESS INDUCED FERROELECTRIC STRONTIUM TITANATE.
VACHER R	17-2454	MEASUREMENT OF THE ELASTIC CONSTANTS OF POTASSIUM DIHYDROGEN PHOSPHATE BY BRILLOUIN DI
VACHERAND C	10-1643	(PREPARATION AND ELECTRICAL AND OPTICAL) STUDY OF TUNGSTEN TRIOXIDE AT LOW TEMPERATURE
	10-1644	PREPARATION OF TUNGSTIC OXIDE SINGLE CRYSTALS (SUBLIMATION IN AIR OF TUNGSTEN TRIOXIDE
VACACCHINO M	17-2416	INCOHERENT NEUTRON SCATTERING FROM HYDROGEN BOND IN KDP AND ADP.
VAIDA C	17-2656	LANDAU-KHALATNIKOV ATTENUATION IN POTASSIUM DIDEUTERIUM PHOSPHATE CRYSTALS.
VAJDA D	17-2659	ULTRASONIC STRESS WAVE INTERACTION WITH DOMAIN WALLS IN POTASSIUM DIDEUTERIUM PHOSPHAT
VAKS VG	03-0159	ATTENUATION OF CRITICAL VIBRATIONS AND DIELECTRIC LOSSES IN DISPLACIVE TYPE FERROELECT
	03-0497	CORRELATION EFFECTS IN DISPLACEMENT TYPE PHASE TRANSITIONS IN FERROELECTRICS. (THEORY
VALASEK J	02-0126	EARLY HISTORY OF FERROELECTRICITY (1920-1943).
VALATHUR M	03-0222	ENERGY GENERATION FROM IMPACT ON PIEZOELECTRIC (OR FERROELECTRIC) MATERIALS.
VALENTA MW	09-1548	FERROELECTRIC PROPERTIES OF LEAD (STANNATE(0.5) NIOBATE(0.5)) (1-X) ZIRCONATE(X) PEROVS
	13-2044	DIFFUSE TO SHARP FERROELECTRIC PHASE TRANSITION SYSTEMS LEAD (B(0.5) NIOBIUM(0.5)) (1-X
VALIC MI	34-3480	STRONG COLLISION LIMIT OF SPIN LATTICE RELAXATION IN HEXAGONAL ICE.
VALITOVA NR	03-0498	NONLINEAR EFFECTS IN THE PROPAGATION OF FINITE AMPLITUDE HYPERSONIC WAVES IN FERROELEC
	12-1891	INTERACTION OF TWO HYPERSOUND WAVES IN LITHIUM TANTALATE.
VALLADE M	02-0063	GENERATION OF THE SECOND HARMONIC AND FERROELECTRICITY. (THEORY)
	27-2991	SECOND HARMONIC LIGHT SCATTERING BY DOMAINS IN FERROELECTRIC TRIGLYCINE SULFATE.
	27-2390	STUDY OF THE PHASE TRANSITION OF TGS BY SECOND HARMONIC SCATTERING.
VAN DEN BERG CB	36-3614	FERROELECTRICITY IN THE IRON DEFICIENT FERROUS SULFIDE SYSTEM. (EXISTS TO DEFICIENCIES
	36-3615	MODEL FOR THE SEMICONDUCTING UNIAXIAL FERROELECTRIC IRON(1-X) SULFIDE. PART-1. PART-2.
	36-3617	PHASE DIAGRAM OF THE SEMICONDUCTING UNIAXIAL FERROELECTRIC IRON(1-X) SULFIDE.
VAN DEN HENDE JH	32-3356	CRYSTAL STRUCTURE OF FERROELECTRIC LITHIUM HYDRAZINIUM SULFATE.
VAN DEN HEUVEL AP	11-1738	ELECTRON BEAM SENSING OF SURFACE ELASTIC WAVES FOR SIGNAL PROCESSING PURPOSES. (TECHNI
VAN DER ELSKEN J	06-1146	(TEMPERATURE AND) ELECTRIC FIELD DEPENDENCE OF A HYDROGEN IMPURITY MODE IN STRONTIUM T
	16-2388	ANHARMONICITY IN SODIUM NITRATE AND SODIUM NITRITE.
	16-2389	DIPOLE CORRELATIONS IN MOLTEN ALKALI METAL NITRATES AND SODIUM NITRITE.
	16-2390	GRUNEISEN PARAMETERS AROUND THE PHASE TRANSITIONS IN SODIUM NITRITE AND SODIUM NITRATE
VAN DER LINDEN H	16-2389	DIPOLE CORRELATIONS IN MOLTEN ALKALI METAL NITRATES AND SODIUM NITRITE.
VAN DER ZIEL A	15-2243	MODE FOR THE ANTIMONY SULFOIODIDE- TIN DIOXIDE HETEROSTRUCTURE SWITCH.
VAN DER ZIEL JP	17-2694	ELECTROOPTIC AMPLITUDE MODULATION OF LASER GENERATED SECOND HARMONICS IN POTASSIUM DIH
	17-2695	TEMPERATURE DEPENDENCE OF OPTICAL (SECOND) HARMONIC GENERATION IN POTASSIUM DIHYDROGEN
	36-3616	OPTICAL SPECTRA OF CHROMIUM(3+) PAIRS IN LANTHANUM ALUMINATE.
VAN LANDUYT J	36-3522	USE OF ELECTRON MICROSCOPY IN THE STUDY OF EXTENDED DEFECTS RELATED TO NONSTOICHIOMETR
VAN UITERT LG	04-0690	ELECTROOPTIC PROPERTIES OF SOME ABO(3) PEROVSKITES IN THE PARAELECTRIC PHASE. (POTASSI
	13-2023	NONLINEAR OPTICAL PROPERTIES OF FERROELECTRIC LEAD NIOBIUM(4) OXIDE(12).
	13-2025	ROLE OF HYDROGEN IN POLARIZATION REVERSAL OF FERROELECTRIC BARIUM SODIUM NIOBATE.
	13-2045	GROWTH OF BARIUM SODIUM NIOBATE SINGLE CRYSTALS FOR OPTICAL APPLICATIONS. (BY PULLING
	26-2945	NONLINEAR OPTICAL SUSCEPTIBILITY OF LITHIUM FORMATE MONOHYDRATE.
	36-3600	ANOMALIES IN THE (1 KHZ) DIELECTRIC CONSTANT (5-300K) OF THE PRASEODYMIUM(1-X) NEODYM
VAN VOORST JDW	06-1147	ESR OF CHROMIUM(3+) IN CHROMIUM DOPED STRONTIUM TITANATE SINGLE CRYSTALS.
VAREKAMP RPP	13-2007	CONTINUOUS HOT PRESSING OF POTASSIUM SODIUM NIOBATE.
VARGA AJ	17-2538	EFFICIENT SECOND HARMONIC GENERATION IN ADP WITH TWO NEW FLUORESCEIN DYE LASERS.
VARIKASH VM	27-2986	DEUTERATION EFFECT ON THERMAL AND ELASTIC PROPERTIES OF FERROELECTRIC CRYSTALS THE TRI
	27-3157	APPLICABILITY OF THE PIPPARD RELATIONS TO THE PHASE TRANSITION IN TRIGLYCINE SULFATE.
	27-3158	NONLINEAR PROPERTIES OF TRIGLYCINE SELENATE. (DIELECTRIC CONSTANT VERSUS ALTERNATING F
	27-3166	DIELECTRIC PERMEABILITY OF TRIGLYCINE SELENATE NEAR THE PHASE TRANSITION.
	27-3167	EFFECT OF CONSTANT ELECTRICAL FIELD ON PHASE TRANSITIONS IN TRIGLYCINE FLUOBERYLLATE.
	27-3168	RELATION BETWEEN ELECTRICAL AND THERMAL PROPERTIES OF TRIGLYCINE SELENATE NEAR THE CUR
	27-3169	THERMAL EXPANSION OF TRIGLYCINE FLUOBERYLLATE CRYSTALS IN THE FERROELECTRIC TRANSITION
VARNADO SG	09-1618	ELECTROOPTIC CERAMICS AS WAVELENGTH SELECTION DEVICES IN DYE LASERS (PLZT).
VARNSTORFF K	27-3078	ANISOTROPY OF SOUND ABSORPTION IN TRIGLYCINE SULFATE SINGLE CRYSTALS. (THEORY OF ANIS
VASILCHENKO VV	11-1749	DETERMINATION OF ELASTIC AND PIEZOELECTRIC CONSTANTS OF LITHIUM NIOBATE SINGLE CRYSTAL
VASILEVA NV	04-0942	OBSERVATION OF THE CHANGE IN THE POLARIZATION OF SINGLE CRYSTAL FILMS OF BARIUM TITANA
	04-1031	MICROSTRUCTURE OF DOMAINS AND DOMAIN WALLS IN SINGLE CRYSTAL FILMS OF BARIUM TITANATE.
VASILEVSKAYA AS	17-2696	RELATIONSHIP BETWEEN THE DIELECTRIC AND ELECTROOPTICAL PROPERTIES OF FERROELECTRIC CRY
VAUGHT DM	06-1221	ENERGY BAND STRUCTURE OF STRONTIUM TITANATE FROM A SELF CONSISTENT FIELD TIGHT BINDING
VAZQUEZ F	15-2244	REFLECTIVITY ON ANTIMONY SULFOIODIDE.
VEHSE WE	33-3377	SHIFT OF NEEL TEMPERATURE AND EPR LINE WIDTH OF POTASSIUM MANGANESE TRIFLUORIDE WITH M
VEKHTER BG	02-0127	FUNDAMENTAL PROPERTIES OF METAL- FERROELECTRIC- METAL JUNCTION SYSTEMS. (INFLUENCE OF
	03-0175	MICROTHEORY OF SPONTANEOUS POLARIZATION AND PHASE TRANSITIONS IN CRYSTALS OF PEROVSKIT
	17-2433	ORIGIN OF THE SPONTANEOUS POLARIZATION AND THE ISOTOPE EFFECTS IN FERROELECTRIC POTASS
VELICHKINA TS	28-3183	FREQUENCY DEPENDENCE OF THE ABSORPTION OF SOUND IN ROCHELLE SALT NEAR ITS UPPER CURIE
VELICHKO IA	17-2620	FERROELECTRIC TRANSITION IN RUBIDIUM DIHYDROGEN PHOSPHIDE USING RAMAN SCATTERING SPECT
	17-2621	ISOTOPE EFFECT IN THE RAMAN SPECTRA OF A DEUTERATED KDP CRYSTAL.
	17-2656	LANDAU-KHALATNIKOV ATTENUATION IN POTASSIUM DIDEUTERIUM PHOSPHATE CRYSTALS.
	17-2677	FIRST ORDER PHASE TRANSITION IN POTASSIUM DEUTERIUM PHOSPHATE CRYSTALS.
VELLA GJ	17-2518	POCKELS EFFECT IN AMMONIUM DIHYDROGEN PHOSPHATE ABOVE ROOM TEMPERATURE.
VELVARSKY J	27-3029	LARGE SIGNAL PERMITTIVITY OF TRIGLYCINE SULFATE AT 100 MHZ. (NEAR THE CURIE TEMPERATUR
	37-3640	VARIOUS HYSTERESIS LOOP TRACERS FOR A WIDE FREQUENCY RANGE. (CIRCUITRY DESCRIBED)
VENDIK OG	03-0499	DEBYE SCREENING OF A NONUNIFORM ELECTRIC FIELD IN A FERROELECTRIC.
	03-0500	ELECTRICAL NONLINEARITY OF A FERROELECTRIC SEMICONDUCTOR IN A NONUNIFORM ELECTRIC FIEL
	03-0501	MICROSCOPIC THEORY OF SURFACE PHENOMENA IN FERROELECTRIC CRYSTALS. (CALCULATION OF FIE
	03-0502	PARAELECTRIC SEMICONDUCTOR IN A NONUNIFORM ELECTRIC FIELD.
	03-0503	PHENOMENOLOGICAL THEORY OF MICROWAVE LOSSES IN FERROELECTRICS. (AT TEMPERATURES ABOVE
	04-1017	EFFECT OF CHARGED LATTICE IMPERFECTIONS ON THE DIELECTRIC PROPERTIES OF MATERIALS. (BA
	37-3677	SOME PROPERTIES AND APPLICATION OF FERROELECTRICS AT MICROWAVES.
VENEVTSEV YUN	02-0064	FERROELECTRIC RARE EARTH MOLYBDATES. (REVIEW, FORMATION, CRYSTAL GROWTH, PHASE TRANSIT
	02-0128	NEW FERROELECTRICS AND SEIGNETTO MAGNETS OF THE PEROVSKITE AND PYROCHLORE TYPE. SYNTHE

VENEVTSEV YUN	03-0276	ABRUPT (STEPWISE) CHANGE IN PROBABILITY OF THE MOSSBAUER EFFECT AT THE PHASE TRANSITIO
	03-0372	FERROELECTRICITY AND ANTIFERROELECTRICITY IN POLAR CRYSTALS. (THEORY, LEAD ZIRCONATE)
	03-0490	(METHODS OF) STRUCTURAL STUDIES OF FERROELECTRICS AND FERROELECTRIC MAGNETS. (ELECTRON
	03-0504	SOME ASPECTS OF THE PROBLEM OF FERROELECTRICITY AND ANTIFERROELECTRICITY IN PEROVSKITE
	04-0824	CALCULATION OF ELECTRIC FIELDS IN IONIC CRYSTALS. (BARIUM TITANATE TYPE)
	04-1000	PREPARATION AND STUDY OF THIN FILMS OF (BARIUM, STRONTIUM) TITANATE.
	04-1018	FERROELECTRIC FAMILY OF BARIUM TITANATE.
	04-1038	STRUCTURAL ELECTRON MICRODIFFRACTION STUDIES OF FERROELECTRICS AND FERROELECTRIC MAGNE
	05-1043	EFFECT OF GERMANIUM ON THE STRUCTURE AND PROPERTIES OF FERROELECTRIC LEAD TITANATE (GE
	05-1044	THE SYSTEM LEAD TITANATE- STRONTIUM COPPER NIOBATE. (CERAMIC SYNTHESIS)
	05-1045	THE SYSTEM LEAD TITANATE- STRONTIUM CUPRATE(0.33) NIOBATE(0.67).
	05-1082	STRUCTURAL MICROELECTRON DIFFRACTION STUDY OF FERROELECTRIC AND FERROMAGNETIC SUBSTANC
	07-1261	DIPOLE STRUCTURE AND INTERNAL ELECTRIC FIELDS IN LEAD ZIRCONATE.
	08-1307	CHARACTER OF PHASE TRANSITION OF FERROELECTRIC POTASSIUM TANTALATE. (SECOND ORDER)
	08-1308	POTASSIUM TANTALATE LEAD TITANATE SYSTEM IN THE TEMPERATURE RANGE 4.2-300 DEG K (LATTI
	08-1309	SYNTHESIS, X-RAY ANALYSIS, AND STUDY OF THE DIELECTRIC PROPERTIES OF NEW CADMIUM-CONTA
	08-1310	X-RAY ANALYSIS AND DIELECTRIC PROPERTIES OF SOLID SOLUTIONS BASED ON THE FERROELECTRIC
	08-1362	X-RAY DETERMINATION OF THE SYMMETRY OF THE FERROELECTRIC COMPOUNDS (POTASSIUM OR SODIU
	08-1368	ELECTRON SPIN RESONANCE (SPECTRA, TEMPERATURE DEPENDENCE) IN SOME FERROELECTRICS AND F
	08-1387	HIGH TEMPERATURE X-RAY INVESTIGATION OF THE PEROVSKITE MODIFICATION OF CADMIUM TITANAT
	08-1389	INTERNAL ELECTRIC FIELDS IN CRYSTALS OF SODIUM TANTALATE AND CADMIUM TITANATE.
	08-1390	ORIGIN OF DIPOLE CONFIGURATIONS IN SOME (PEROVSKITE) STRUCTURES WITH SPECIAL DIELECTRI
	08-1400	CHANGE OF RESONANCE ABSORPTION SPECTRA OF 23.8 KEV GAMMA-RAYS OF TIN-119 DURING PHASE
	08-1401	MOSSBAUER STUDY AND THEORETICAL ESTIMATION OF INTERNAL ELECTRIC FIELD GRADIENTS IN FER
	08-1424	ABOUT THE CURIE POINT AND NATURE OF DIELECTRIC PROPERTIES IN PEROVSKITE MODIFICATION O
	08-1425	COEXISTENCE OF ANTIFERROMAGNETIC AND SPECIAL DIELECTRIC PROPERTIES IN THE BISMUTH- LAN
	08-1428	GROWTH OF SINGLE CRYSTALS OF VARIOUS PEROVSKITES WITH COMPLEX COMPOSITIONS, AND THE ST
	08-1429	GROWTH AND STUDY OF NEW SINGLE CRYSTAL CADMIUM COMPOUNDS WITH PEROVSKITE STRUCTURE.
	08-1438	PREPARATION AND INVESTIGATION OF FERROELECTRICS OF ILMENITE STRUCTURE. (ABO(3) NIOBATE
	08-1442	CALCULATIONS OF THE INTERNAL ELECTRIC FIELDS AND ELECTRIC FIELD GRADIENTS IN THE PEROV
	08-1451	SUPERSTRUCTURE OF BISMUTH METAL OXYGEN(3) PEROVSKITES. (WHERE METAL IS SCANDIUM, CHROM
	08-1452	STRUCTURE AND DIELECTRIC LOSSES OF FERROELECTRIC PEROVSKITES SYNTHESIZED AT HIGH PRESS
	08-1455	CALCULATION OF INTERNAL ELECTRIC FIELDS AND THEIR GRADIENTS IN ABO(3) PEROVSKITE COMPO
	08-1456	CERTAIN ASPECTS OF THE PROBLEM OF FERROELECTRICITY AND ANTIFERROELECTRICITY IN THE PER
	08-1457	CONDITIONS FAVORING THE ORIGINATION OF SPONTANEOUS POLARIZATION IN PEROVSKITES. (ABSTR
	08-1458	CRYSTAL STRUCTURE AND PROPERTIES OF NEW CADMIUM AND THALLIUM CONTAINING PEROVSKITES. (
	08-1459	PEROVSKITE TYPE SEIGNETTO MAGNETS.
	08-1460	STRUCTURE AND PROPERTIES OF SOME NEW FERROELECTRIC MATERIALS WITH PEROVSKITE STRUCTURE
	08-1462	NEW FERROELECTRICS WITH PEROVSKITE AND PYROCHLORE STRUCTURES. (BARIUM AND BISMUTH(0.5)
	08-1463	SYNTHESIS OF THALLIUM CONTAINING PEROVSKITS AT HIGH PRESSURES AND THEIR X-RAY ANALYSI
	09-1510	PIEZOCERAMICS BASED ON LEAD ZIRCONATE TITANATE WITH COMPLEX ADDITIVES CONTAINING GERMA
	09-1544	NEW PIEZOELECTRIC CERAMICS. (LEAD TITANATE- LEAD FERRATE(0.5) NIOBATE(0.5), LEAD TITAN
	10-1640	THALLIUM PEROVSKITES. THALLIUM A(1-X) TUNGSTATE OR MOLYBDATE. (WHERE A IS ZIRCONIUM, H
	12-1887	CURIE TEMPERATURE OF FERROELECTRIC LITHIUM TANTALATE.
	13-2014	DIELECTRIC AND STRUCTURAL PROPERTIES OF FERROELECTRICS WITH DIFFUSE PHASE TRANSITIONS.
	30-3271	PHASE TRANSITIONS OF GADOLINIUM MOLYBDATE AND ISOSTRUCTURAL COMPOUNDS (EUROPIUM TERBIU
	36-3566	RESONANCE ABSORPTION OF GAMMA QUANTA IN BARIUM, STRONTIUM, AND CALCIUM STANNATE.
	36-3604	STRUCTURE AND MAGNETIC PROPERTIES OF FERROELECTRIC FERROMAGNETIC SOLID SOLUTIONS IN TH
	36-3618	A NEW MAGNETOFERROELECTRIC: CADMIUM IRON NIOBATE.
	36-3619	NEW FERROELECTRICS. (BARIUM(2) BISMUTH (NIOBIUM OR TANTALUM OR VANADIUM) OXIDE(6), AND
VENKATESH GM	20-2803	INFRARED AND RAMAN SPECTRA OF FERROELECTRIC ALUMS: METHYL AMMONIUM ALUMINUM (SULFATE O
	20-2808	SPECTROSCOPIC STUDIES OF FERROELECTRIC METHYL AMMONIUM ALUMS.
VENUGOPALAN S	35-3519	RAMAN SPECTRUM OF BISMUTH(12) GERMANIUM OXIDE(20).
VERBITSKAYA TN	03-0505	DOMAIN STRUCTURE OF POLYCRYSTALLINE FERROELECTRICS WITH DIFFUSE PHASE TRANSITIONS.
	04-0542	DETERMINATION OF CHROMIUM VALENCY FROM STUDIES OF EPR SPECTRA AND STATIC MAGNETIC SUSC
	04-0543	STATE OF CHROMIUM ADDED TO BARIUM METATITANATE.
	04-1019	DIELECTRIC AND NONLINEAR MICROWAVE PROPERTIES OF POLYCRYSTALLINE SOLID SOLUTIONS BASED
	08-1461	ELECTRICAL PROPERTIES OF NONLINEAR FERROELECTRIC CERAMICS WITH THE PEROVSKITE STRUCTUR
VERE AW	13-1894	DILATOMETRIC STUDY OF THE ORTHORHOMBIC- TETRAGONAL PHASE TRANSITION IN BARIUM SODIUM N
	13-1904	BARIUM SODIUM NIOBATE: CONSTITUTIONAL STUDIES AND (CZOCHRALSKI) CRYSTAL GROWTH.
	13-1928	CRUCIBLE BASE COOLING: AN AID TO CZOCHRALSKI CRYSTAL GROWTH. (BARIUM SODIUM NIOBATE AN
	13-2047	PROGRESS TOWARDS THE GROWTH OF GOOD OPTICAL QUALITY BARIUM SODIUM NIOBATE (BY USING CO
VERESHCHAGIN LF	04-0760	X-RAY INVESTIGATION OF THE EFFECT OF HYDROSTATIC PRESSURE ON THE STRUCTURE OF BARIUM T
	05-1064	AN X-RAY STUDY OF THE EFFECT OF HYDROSTATIC PRESSURE ON THE STRUCTURE OF LEAD TITANATE
VERIGINA ZS	13-2046	EFFECTS OF ROASTING CONDITIONS ON THE STRUCTURE AND PROPERTIES OF LEAD BARIUM NIOBATE
	36-3554	MICROSTRUCTURE OF TSTS-19 TYPE FERROELECTRIC CERAMICS.
VERJANS HC	13-2007	CONTINUOUS HOT PRESSING OF POTASSIUM SODIUM NIOBATE.
VERKHOVSKAYA KA	16-2266	FERROELECTRIC PHASE TRANSITION IN SILVER SODIUM DINITRITE
	22-2839	BEHAVIOR OF THE FUNDAMENTAL ABSORPTION EDGE IN TRIHYDROGEN SELENITE CRYSTALS AT THEIR
VERKIN BI	04-0573	CHANGE OF THE MOSSBAUER EFFECT PROBABILITY ON TIN-119 IN BARIUM TITANATE STANNATE SOLI
VERLEUR HW	36-3526	INFRARED OPTICAL PROPERTIES OF VANADIUM DIOXIDE ABOVE AND BELOW TRANSITION TEMPERATURE
VERREAULT R	36-3620	CRYSTALLOGRAPHIC, OPTICAL AND MAGNETIC PROPERTIES OF (MONOCLINIC) EUROPIUM SILICATE. (
VEST RW	08-1314	ELECTRICAL PROPERTIES AND DEFECT STRUCTURE OF POTASSIUM TANTALATE.
VETSSILE ZA	04-0586	CONCENTRATION DEPENDENCE OF THE CURIE-WEISS CONSTANT IN POLYCRYSTALLINE SOLID SOLUTION
VEZZETTI CF	36-3544	PYROELECTRICITY AND PIEZOELECTRICITY IN ORIENTED FILMS OF POLYVINYL FLUORIDE AND POLYV
VIBLYI IF	27-3108	IONIZATION OF THE (X-RADIATION AND COPPER AND IRON IMPURITY) ABSORPTION BANDS OF DEFEC
	27-3159	NATURE OF RADIATION INDUCED DEFECTS IN CRYSTALS OF ROCHELLE SALT AND TRIGLYCINE SULFAT
	27-3160	POSITION OF THE FUNDAMENTAL ABSORPTION EDGE AS A FUNCTION OF TEMPERATURE FOR ROCHELLE
VIDADI YUA	36-3539	FIRST ORDER NEGATIVE PHOTODIELECTRIC EFFECT IN PHTHALOCYANINE IN THE FERROELECTRIC STA
	36-3540	PYROELECTRIC EFFECT IN METAL-FREE PHTHALOCYANINE. (IN THE POLAR, OR FERROELECTRIC STAT
	36-3621	DEBYE TYPE ABSORPTION IN POLAR PHTHALOCYANINE.
	36-3622	ELECTRICAL PROPERTIES OF PHTHALOCYANINE AS A SEMICONDUCTOR IN THE FERROELECTRIC STATE.
	36-3623	LOW TEMPERATURE PHASE TRANSITION IN THE POLAR STATE OF THE ORGANIC SEMICONDUCTOR PHTHA
VIJAYARAGHAVAN R	18-2732	PURE QUADRUPOLE RESONANCE OF ARSENIC-75 IN AMMONIUM DIHYDROGEN ARSENATE.
VILLAIN J	03-0506	ENERGY PARAMETERS AND PROTON TUNNELING IN POTASSIUM DIHYDROGEN PHOSPHATE TYPE FERROELE
	03-0507	SELF CONSISTENCY OF LANDAU'S MODEL IN THE TRANSITION FROM PIEZO TO FERROELECTRICITY (T
	34-3481	THE THREE DIMENSIONAL EIGHT VERTEX MODEL AND THE PROTON PROTON CORRELATION FUNCTIONS I
VILLELA G	04-0657	PREPARATION OF SODIUM AND BARIUM NIOBATES AND LEAD MOLYBDATE.
VILLERMAIN-LECOLLIER G	17-2587	FAR INFRARED ABSORPTION SPECTRA FOR AMMONIUM DIHYDROGEN PHOSPHATE.
VILMINOT S	32-3355	CRYSTALLOGRAPHIC STUDY OF SEVERAL HYDRAZINIUM ORTHO FLUOBERYLLATES.

VINCENT SM	09-1619	X-RAY ANALYSIS OF MODIFIED LEAD ZIRCONATE- LEAD TITANATE CERAMICS.
VINETSKII VL	03-0508	NATURE CF THE ANOMALOUS ELECTRIC CONDUCTIVITY OF FERROELECTRICS IN THE PHASE TRANSITIO
VINOGRADOV EA	11-1844	ELECTROOPTIC EFFECT IN LITHIUM NIOBATE IN THE (SUB) MILLIMETER RANGE.
VINOGRADOVA IS	20-2809	INVESTIGATION OF PHASE TRANSITION IN FERROELECTRIC ALUMS (METHYL AMMONIUM ALUMINUM SUL
	20-2810	PHYSICAL PROPERITES OF FERROELECTRIC ALUMS. (ESPECIALLY METHYL AMMONIUM ALUMINUM SULFA
VINTRUFF V	04-0543	STATE OF CHROMIUM ADDED TO BARIUM METATITANATE.
VISKOV AS	02-0128	NEW FERROELECTRICS AND SEIGNETTO MAGNETS OF THE PEROVSKITE AND PYROCHLORE TYPE. SYNTHE
	03-0276	ABRUPT (STEPWISE) CHANGE IN PROBABILITY OF THE MOSSBAUER EFFECT AT THE PHASE TRANSITIO
	04-0707	RELAXATION AND FERROELECTRIC PROPERTIES OF SOLID SOLUTIONS IN THE SYSTEM BARIUM TITANA
	08-1400	CHANGE OF RESONANCE ABSORPTION SPECTRA OF 23.8 KEV GAMMA-RAYS OF TIN-119 DURING PHASE
	08-1401	MOSSBAUER STUDY AND THEORETICAL ESTIMATION OF INTERNAL ELECTRIC FIELD GRADIENTS IN FER
	08-1460	STRUCTURE ANC PROPERTIES OF SOME NEW FERROELECTRIC MATERIALS WITH PEROVSKITE STRUCTURE
	08-1462	NEW FERROELECTRICS WITH PEROVSKITE AND PYROCHLORE STRUCTURES. (BARIUM AND BISMUTH(0.5)
	08-1463	SYNTHESIS OF THALLIUM CONTAINING PEROVSKITES AT HIGH PRESSURES AND THEIR X-RAY ANALYSI
	36-3566	RESONANCE ABSORPTION OF GAMMA QUANTA IN BARIUM, STRONTIUM, AND CALCIUM STANNATE.
	36-3619	NEW FERROELECTRICS. (BARIUM(2) BISMUTH (NIOBIUM OR TANTALUM OR VANADIUM) OXIDE(6), AND
VISNOVSKA M	17-2474	THERMAL VARIATION OF REFRACTIVE INDEX IN AMMONIUM DIHYDROGEN PHOSPHATE AND TRIGLYCINE
VISWAMITRA MA	16-2333	CRYSTAL STRUCTURE OF FERROELECTRIC GLYCINE SILVER NITRATE.
	16-2348	CRYSTAL STRUCTURE OF GLYCINE SILVER NITRATE.
	22-2859	REFINEMENT OF THE CRYSTAL STRUCTURE OF FERROELECTRIC ACID LITHIUM SELENITE, POSITION O
	22-2860	REFINEMENT OF THE CRYSTAL STRUCTURE OF FERROELECTRIC LITHIUM HYDROGEN SELENITE: POSITI
	26-2940	CRYSTAL STRUCTURE OF PIEZOELECTRIC LITHIUM FORMATE MONOHYDRATE.
VIVAS E	27-2993	DIGLYCINE SULFATE AN INTERESTING NEW DIELECTRIC CRYSTAL SPECIES.
VLAHOV A	03-0403	QUANTUM THEORY OF FERROELECTRICITY. PART-2: THERMODYNAMIC BEHAVIOR OF FERROELECTRIC SP
VLASOV KB	03-0509	RESONANCE EFFECTS IN MAGNETICALLY EQUIAXIAL FERRODIELECTRIC SINGLE CRYSTALS HAVING DOM
VLOKH OG	17-2697	ELECTROOPTICAL AND PIEZOOPTICAL PROPERTIES OF POTASSIUM DIHYDROGEN PHOSPHATE AND AMMON
VOGT H	16-2384	OPTICAL SECOND HARMONIC GENERATION DURING FERROELECTRIC POLARIZATION REVERSAL. (SODIUM
	16-2385	SECOND HARMONIC GENERATION IN FERROELECTRIC SODIUM NITRITE.
	16-2386	TEMPERATURE DEPENDENCE OF THE OPTICAL NONLINEAR SUSCEPTIBILITY OF SODIUM NITRITE.
	16-2387	TEMPERATURE DEPENDENCE OF ABSORPTION OF SODIUM NITRITE IN THE FAR INFRARED. (350-1100
	16-2392	MODULATION OF THE MILLIMETER WAVE ABSORPTION IN SODIUM NITRITE BY POLARIZATION REVERSA
VOITKO II	04-0932	FORMATICN OF FLUORINE CONTAINING SOLID SOLUTIONS BASED ON BARIUM TITANATE.
VOLGER J	04-0830	PHONON TRANSPORT IN BARIUM TITANATE.
VOLIOTIS SD	04-0590	BINARY AND PSEUDOBINARY PHASES BETWEEN NIOBIUM PENTOXIDE AND LEAD, BARIUM, AND STRONTI
VOLK TR	04-0681	EXTRINSIC PHOTOCONDUCTIVITY IN FERROELECTRICS DUE TO SURFACE LAYERS. (BARIUM TITANATE,
	16-2266	FERROELECTRIC PHASE TRANSITION IN SILVER SODIUM DINITRITE
VOLKEL G	28-3233	EPR INVESTIGATION OF THE INTERNAL ELECTRIC FIELD IN COPPER DOPED ROCHELLE SALT.
VOLKONSKII VB	17-2407	DIFFRACTION OF POLARIZED LIGHT BY ULTRASONIC WAVES IN POTASSIUM DIHYDROGEN PHOSPHATE C
VOLKOVA EN	17-2698	ELECTROOPTICAL AND OPTICAL PROPERTIES OF PARTIALLY DEUTERATED RUBIDIUM DIHYDROGEN PHOS
VOLKOVA LS	13-2050	PHASE TRANSITIONS IN TANTALATES WITH THE STRUCTURE OF POTASSIUM TUNGSTEN BRONZE. (POTA
VOLNYANSKII MD	15-2245	INFLUENCE OF ONE DIMENSIONAL MECHANICAL STRESS ON THE PHASE TRANSITION OF ANTIMONY SUL
	15-2246	POLARIZATION REVERSAL OF ANTIMONY SULFOIODIDE SINGLE CRYSTALS IN THE REGION OF PHASE C
VON HIPPEL AR	03-0510	DO WE REALLY UNDERSTAND (THE CAUSE OF) FERROELECTRICITY. (ITS NATURE AND THE PREDICTIO
	34-3455	TRANSFER OF PROTONS THROUGH "PURE" ICE I(H), SINGLE CRYSTALS. PART-2: MOLECULAR MODELS
	34-3482	DIELECTRIC RELAXATION SPECTRA OF WATER, ICE, AND AQUEOUS SOLUTIONS AND THEIR INTERPRET
	34-3483	DIELECTRIC AND MECHANICAL RESPONSE OF ICE I(H) SINGLE CRYSTALS AND ITS INTERPRETATION.
	34-3484	MOLECULAR UNDERSTANDING OF ELECTROCHEMICAL PROCESSES BY ICE RESEARCH.
	34-3485	MOLECULAR PHENOMENA IN WATER SYSTEMS. PART-1: MOLECULAR INTERPRETATION OF THE PHASE DI
	34-3486	TRANSFER OF PROTONS THROUGH PURE ICE I(H) SINGLE CRYSTALS. PART-1: POLARIZATION SPECTR
	37-3678	HIGH CIELECTRIC CONSTANT MATERIALS AND FERROELECTRICITY AS CAPACITOR DIELECTRICS, A ST
VON WALDKIRCH TH	06-1171	PHOTOCHROMIC IRON IN STRONTIUM TITANATE. EVIDENCE FROM PARAMAGNETIC RESONANCE.
	06-1231	ANALYSIS OF THE IRON OXYGEN VACANCY CENTER IN THE TETRAGONAL PHASE OF STRONTIUM TITANA
	06-1232	EPR STUDIES ANC DYNAMICS IN STRONTIUM TITANATE IN THE NEIGHBORHOOD OF THE PHASE TRANSF
	06-1233	FLUCTUATIONS AND CORRELATIONS IN STRONTIUM TITANATE ABOVE THE CRITICAL TEMPERATURE.
	06-1234	THE IRON(3+)- VACANCY CENTER IN THE TETRAGONAL PHASE OF STRONTIUM TITANATE.
VOS A	08-1370	NCTE ON THE SPACE GROUP OF POTASSIUM HYDROGEN IODATE.
VOTINOV MP	04-1020	PECULIARITIES OF THE ELECTRON PARAMAGNETIC RESONANCE SPECTRA IN A FERROELECTRIC CERAMI
VROBEV AF	17-2585	ENTHALPIES OF SOLUTION OF RUBIDIUM DIHYDROGEN PHOSPHATE IN WATER AND RUBIDIUM DIDEUTER
VUL BM	03-0511	FIELD EFFECT AT THE CCNTACT BETWEEN A SEMICONDUCTOR AND A C-DOMAIN FERROELECTRIC (THEO
	04-1021	CEPENDENCE OF THE PIEZOEFFECT IN BARIUM TITANATE ON THE MEANS OF SCREENING THE SPONTAN
VULKAN U	17-2489	HIGH TEMPERATURE PHASE TRANSITION IN RUBIDIUM DIHYDROGEN PHOSPHATE BY SCATTERING OF CO
VYAS BR	04-0798	INFLUENCE OF MICRCSTRUCTURES (PREPARATION CONDITIONS) ON DIELECTRIC PROPERTIES OF BARI
WACHERNIG H	15-2164	OPTICAL SECOND HARMONIC GENERATION IN ANTIMONY SULFOIODIDE.
WADA M	13-2033	CRYSTAL GROWTH AND DIELECTRIC PROPERTIES OF BISMUTH NIOBATE.
	13-2034	SINGLE CRYSTAL GROWTH OF STRONTIUM POTASSIUM TANTALATE FROM MOLTEN SALTS.
WADA S	16-2382	STUDY OF THE CRITICAL SLOWING DOWN OF THE DIELECTRIC RELAXATION IN SILVER SODIUM DINIT
WADA T	24-2894	NUCLEAR MAGNETIC RESONANCE AND X-RAY STUDIES OF POTASSIUM FERROCYANIDE TRIHYDRATE CRYS
WADA Y	36-3598	PIEZOELECTRICITY, PYROELECTRICITY, AND THE ELECTROSTRICTION CONSTANT OF POLYVINYLIDENE
WADDINGTON TC	17-2687	PROTON MOTIONS IN HYDROGEN BONDED FERROELECTRIC AND ANTIFERROELECTRIC SOLIDS. PART-1:
WAHL HJ	27-3155	CRITICAL SLOWING DOWN AT FERROELECTRIC TRANSITIONS (TRIGLYCINE SULFATE).
WAKAKI N	02-0099	DIELECTRIC AND X-RAY STUDIES OF THE PHASE TRANSITIONS IN ROCHELLE SALT, BARIUM TITANAT
WAKU S	04-0833	DIELECTRIC CERAMICS WITH BOUNDARY LAYER STRUCTURE FOR HIGH FREQUENCY APPLICATION. (BAR
	04-1007	DISLOCATIONS IN BARIUM TITANATE SINGLE CRYSTALS.
	04-1022	DISCUSSION ON THE CRYSTAL GROWTH OF BARIUM TITANATE SINGLE CRYSTALS IN POTASSIUM FLUOR
	04-1023	ETCH PITS CORRESPONDING TO DISLOCATIONS IN BARIUM TITANATE SINGLE CRYSTALS.
	04-1024	FERROELECTRIC PROPERTIES OF (INSULATING, BOUNDARY LAYER) BARIUM TITANATE CERAMICS AT G
	04-1025	STUDIES ON THE BARIUM STRONTIUM TITANATE BOUNDARY LAYER CERAMIC DIELECTRICS.
	06-1248	STUDIES ON STRONTIUM TITANATE BOUNDARY LAYER DIELECTRICS.
WAKUTA Y	17-2608	LOW FREQUENCY HYDROGEN VIBRATICNS IN POTASSIUM DIHYDROGEN PHOSPHATE.
	17-2699	LOW FREQUENCY HYDROGEN VIBRATIONS IN POTASSIUM DIHYDROGEN PHOSPHATE.
WALDNER M	11-1789	CPTICAL STUDIES OF REFRACTION AND REFLECTION OF ULTRASONIC SURFACE WAVES BY THIN GOLD
	11-1790	REFRACTION ANC REFLECTION OF ULTRASONIC SURFACE WAVES BY THIN GOLD FILM LAYERS ON LITH
	11-1846	FUNDAMENTAL AND HIGHER ORDER MODE SURFACE WAVE PROPAGATION ON LAYERED ANISOTROPIC SUBS
WALKER BE	04-1026	INFLUENCE OF ADDITIVES ON DENSIFICATION AND PROPERTIES OF SINTERED AN PRESSURE SINTERE
WALKER JC	24-2892	MOSSBAUER EFFECT STUDY OF THE FERROELECTRIC TRANSITIONS IN FERRIC AMMONIUM SULFATE DOD
WALL LS	06-1196	SOFT MODE STUDY OF THE STRESS INDUCED PHASE TRANSITIONS NEAR THE TRANSITION TEMPERATUR
	06-1235	CUBIC TO TRIGONAL STRESS INDUCED PHASE TRANSITION IN STRONTIUM TITANATE.
	06-1236	RAMAN STUDIES OF STRESS INDUCED PHASE TRANSITION NEAR CURIE TEMPERATURE IN STRONTIUM T
	11-1807	TEMPERATURE STUDY OF THE POLARITON ASSOCIATED WITH THE 248 CM(-1) SOFT MODE IN LITHIUM
	12-1873	TEMPERATURE DEPENDENCE CF POLARITON DISPERSION IN LITHIUM TANTALATE. (OPTICAL MODULATI

WALL LS	17-2461	LONG WAVELENGTH POLARIZATION FLUCTUATIONS IN ANTIFERROELECTRIC AMMONIUM DIHYDROGEN PHO
	17-2655	EFFECT OF PROTON- PHONON COUPLING ON THE FERROELECTRIC MODE IN POTASSIUM DIHYDROGEN PH
	17-2701	MODE INTERACTIONS IN POTASSIUM DIHYDROGEN PHOSPHATE.
WALLACE CA	11-1845	DISPLAY OF TWINNING IN LITHIUM NIOBATE BY X-RAY DIFFRACTION TOPOGRAPHY.
WALLACE DC	04-0694	THERMAL STUDIES OF CALCIUM(X) BARIUM(1-X) TITANATE.
WALLACE EA	17-2700	DETERMINATION OF THE EIGENVECTORS OF SOFT MODES IN POTASSIUM DIDEUTERIUM PHOSPHATE AND
WAN P	32-3357	MICROWAVE DIELECTRIC STUDY OF LITHIUM HYDRAZINIUM SULFATE.
WANG FFY	11-1848	DIELECTRIC STUDIES OF THE CATION SUBSTITUTED MIXED CRYSTALS OF NIOBATES.
	13-1953	MICROCHEMICAL ANALYSIS OF GADOLINIUM DOPED BARIUM SODIUM NIOBATES.
	13-1954	SOLIDIFICATION STUDY OF BARIUM STRONTIUM NIOBATE.
	13-1955	SCANNING ELECTRON MICROSCOPIC EXAMINATION OF SINTERED BARIUM SODIUM NIOBATES.
	13-1996	LINEAR THERMAL EXPANSIONS OF GADOLINIUM SUBSTITUTED BARIUM SODIUM NIOBATES.
	13-2048	CAUSES OF THERMAL EXPANSION OF GADOLINIUM SUBSTITUTED BARIUM SODIUM NIOBATES.
	13-2049	EMPIRICAL FACTORS FOR CALCULATION OF THE FERROELECTRIC TRANSITION TEMPERATURES OF TUNG
WANG J	06-1180	ELASTIC PROPERTIES OF YTTRIUM DOPED STRONTIUM TITANATE SINGLE CRYSTALS.
WANG SY	03-0200	CIRCLE THEOREM FOR ICE RULE FERROELECTRIC MODELS.
WANG T	13-1912	ELECTRICAL PROPERTIES OF TERNARY BARIUM OXIDE- LANTHANUM OXIDE- TITANIUM DIOXIDE COMPO
WANG WC	11-1847	CONVOLUTION OF SURFACE WAVES IN A STRUCTURE OF SEMICONDUCTOR ON LITHIUM NIOBATE.
WANG-TYAU C	13-2050	PHASE TRANSITIONS IN TANTALATES WITH THE STRUCTURE OF POTASSIUM TUNGSTEN BRONZE. (POTA
WANKLYN BM	02-0129	FLUX GROWTH OF SOME COMPLEX OXIDE MATERIALS. (NIOBATES, TANTALATES, STANNATES, TITANAT
	36-3558	FLUX GROWTH OF MAGNESIUM OXIDE AND LANTHANUM ALUMINATE CRYSTALS DOPED WITH ISOTOPE 170
	36-3608	ELECTRICAL TRANSPORT IN (SEMICONDUCTING AND MAGNETIC) RARE EARTH ORTHOCHROMITES, MANGA
WAPLAK S	27-3133	EPR OF THE CHROMIUM COMPLEX IN TRIGLYCINE FLUOBERYLLATE.
	27-3135	EPR STUDIES OF CHROMIUM(3+), COPPER(2+), AND MANGANESE(2+) PARAMAGNETIC ION COMPLEXES
	27-3136	INFLUENCE OF SPONTANEOUS POLARIZATION ON THE EPR SPECTRA OF CHROMIUM(3+) IONS IN TRIGL
WARD JF	02-0067	OPTICAL HARMONICS AND NONLINEAR PHENOMENA. (THEORY, REVIEW)
WARING RK	30-3256	CONTROL AND APPLICATION OF DOMAIN WALL MOTION IN GADOLINIUM MOLYBDATE.
WARKUSZ F	27-3175	CRYSTAL STRUCTURE OF TRIGLYCINE FLUOBERYLLATE IN THE FERRO- AND PARAELECTRIC PHASE.
WARNER AW	08-1464	PIEZOELECTRIC AND PHOTOELASTIC PROPERTIES OF LITHIUM IODATE.
	08-1465	PIEZOELECTRIC AND PHOTOELASTIC PROPERTIES OF LITHIUM IODATE.
WARNER J	17-2702	SIMULATION OF A DOUBLE 45 DEGREES, Z CUT POTASSIUM DIDEUTERIUM PHOSPHATE. ELECTROOPTIC
WARNKE DF	09-1520	EFFECTS OF RADIATION INDUCED DAMAGE CENTERS IN LEAD ZIRCONATE TITANATE CERAMICS.
WARREN D	04-0660	EFFECT OF SILICA ADDITIONS ON THE SEMICONDUCTING PROPERTIES OF DOPED BARIUM TITANATE.
WARTEWIG S	27-3161	EPR STUDIES OF CHROMIUM- AND VANADYL- DOPED TRIGLYCINE SULFATE SINGLE CRYSTALS.
WATANABE A	04-0990	TWO SETS OF ELECTRON SPIN RESONANCE OF GADOLINIUM(3+) IN BARIUM TITANATE CERAMIC SEMIC
	11-1812	TEMPERATURE DEPENDENCE OF RESONANT FREQUENCIES OF LITHIUM NIOBATE PLATE RESONATORS. (C
	13-1946	ELASTIC AND PIEZOELECTRIC CONSTANTS OF STRONTIUM(4) POTASSIUM LITHIUM NIOBATE TYPE FER
	13-2002	SUPERLATTICE STRUCTURE OF STRONTIUM SODIUM LITHIUM NIOBATE (TUNGSTEN BRONZE) TYPE FERR
	13-2051	ELECTROOPTICAL PROPERTIES OF STRONTIUM POTASSIUM LITHIUM NIOBATE. (TUNGSTEN BRONZE TYP
	13-2062	ELECTROOPTIC PROPERTIES OF STRONTIUM LITHIUM NIOBATE TYPE FERROELECTRICS (SINGLE CRYST
WATANABE H	16-2375	APPLICATION OF THE X-RAY DIFFRACTION TOPOGRAPHIC METHOD TO FERROELECTRIC SODIUM NITRIT
WATANABE J	28-3234	SURFACE STRUCTURES AND CORRESPONDING LIGHT FIGURES REVEALED BY WATER ETCHED ROCHELLE S
WATANABE T	19-2777	ON THE PHASE TRANSITION OF AMMONIUM HYDROGEN SULFATE AND THE STRUCTURE OF THE FERROELE
WATERWORTH P	33-3359	PREPARATION AND PROPERTIES OF CESIUM LEAD HALIDES. (FERROELECTRICITY INDETERMINITE)
WATTON A	19-2800	PROTON MAGNETIC RESONANCE STUDY OF THE SPIN SYMMETRY STATES OF AMMONIUM IONS IN SOLIDS
	23-2887	PROTON DYNAMICS IN FERROELECTRIC COLEMANITE.
WATTON R	27-3107	PYROELECTRIC THERMAL IMAGING DEVICES. (TGS)
WAUGH JS	19-2769	NUCLEAR SPIN LATTICE RELAXATION IN SOME FERROELECTRIC AMMONIUM SALTS. (AMMONIUM SULFAT
	23-2885	PROTON SPIN LATTICE RELAXATION IN FERROELECTRIC COLEMANITE.
WAX SI	17-2703	(REVERSIBLE) OPTICAL DAMAGE IN KDP (WITH RELAXATION TIME OF SECONDS WHEN ILLUMINATED U
WAYE BE	02-0130	CERAMIC FERROELECTRICS.
WEAVER H	06-1100	BAND STRUCTURE OF STRONTIUM TITANATE AS MEASURED BY X-RAY PHOTOELECTRON SPECTROSCOPY (
WEBB R	13-2052	CATASTROPHIC SURFACE DAMAGE PRODUCED IN BARIUM SODIUM NIOBATE CRYSTALS DURING INTRACAV
WEBB SL	27-2974	CRITICAL FLUCTUATIONS IN TRIGLYCINE SULFATE.
WEBER A	11-1786	BRILLOUIN SCATTERING IN LITHIUM NIOBATE.
WEBER MJ	06-1237	NUCLEAR MAGNETIC RESONANCE STUDY OF THE PHASE TRANSITION IN STRONTIUM TITANATE.
WEBSTER JC	17-2715	AMPLITUDE MODULATORS BASED ON THE MICHELSON INTERFEROMETER. (AMMONIUM DIHYDROGEN PHOSP
WEGDAM GH	16-2388	ANHARMONICITY IN SODIUM NITRATE AND SODIUM NITRITE.
	16-2389	DIPOLE CORRELATIONS IN MOLTEN ALKALI METAL NITRATES AND SODIUM NITRITE.
	16-2390	GRUNEISEN PARAMETERS AROUND THE PHASE TRANSITIONS IN SODIUM NITRITE AND SODIUM NITRATE
WEGENER H	19-2751	RELAXATION PHENOMENA AND MOSSBAUER SPECTRA OF MAGNETIZED PARAMAGNETIC SUBSTANCES. PART
WEGLEIN RD	11-1849	DIFFRACTION SPREADING OF SURFACE WAVES ON LITHIUM NIOBATE.
	11-1850	(FAR FIELD) DIFFRACTION SPREADING OF SURFACE (ACOUSTIC) WAVES ON (PIEZOELECTRIC) LITHI
WEGMANN L	37-3679	OBSERVATION OF THE SURFACES OF FERROELECTRIC CRYSTALS IN A PHOTOEMISSION ELECTRON MICR
WEHNER RK	15-2167	SOFT PHONON MODE AND MODE COUPLING IN ANTIMONY SULFOIODIDE (BY RAMAN SCATTERING).
	15-2232	SOFT MODE COUPLING AND CRITICAL RAYLEIGH SCATTERING IN FERROELECTRIC ANTIMONY SULFOIOD
	15-2233	SOFT LATTICE VIBRATIONS AND VIBRATION COUPLING IN ANTIMONY SULFOIODIDE.
WEIDENTHALER P	16-2391	CHANGES IN ELECTRICAL CONDUCTIVITY ACCOMPANYING THE POLYMORPHIC TRANSITIONS IN POTASSI
WEIDMANN EJ	16-2394	THIN FILM NUCLEATION ON FERROELECTRIC SUBSTRATES. (TELLURIUM ON TRIGLYCINE SULFATE AND
	27-3162	INDUCED GROWTH ANISOTROPY IN TGS CRYSTALS.
WEIK H	06-1238	LATTICE DISTORTIONS IN A STRONTIUM TITANATE SINGLE CRYSTAL ELECTRET.
	08-1467	SPLIT INTERSTITIAL CONFIGURATION INDUCED IN PEROVSKITE TYPE SINGLE CRYSTALS BY NEUTRON
WEINMANN D	16-2384	OPTICAL SECOND HARMONIC GENERATION DURING FERROELECTRIC POLARIZATION REVERSAL. (SODIUM
	16-2392	MODULATION OF THE MILLIMETER WAVE ABSORPTION IN SODIUM NITRITE BY POLARIZATION REVERSA
WEINREICH G	27-2958	OPTICAL MIXING. (OF PULSED RUBY LASER EMISSIONS BY TRIGLYCINE SULFATE CRYSTAL)
WEINTRITT DJ	04-1027	CORRELATION BETWEEN RHEOLOGICAL AND ELECTRICAL BEHAVIOR IN SINTERED BARIUM TITANATE.
WEIRAUCH DF	08-1466	ISOTHERMAL PHASE TRANSITIONS IN CERAMIC LEAD ZIRCONATE.
	09-1620	STUDY OF PHASE TRANSITIONS IN THE LEAD ZIRCONATE TITANATE SYSTEM.
	09-1621	STUDY OF PHASE TRANSITIONS IN THE LEAD ZIRCONATE- LEAD TITANATE SYSTEM.
WEIS J	17-2704	GROWTH BEHAVIOR OF SMALL SEEDS OF AMMONIUM DIHYDROGEN PHOSPHATE IN A MODIFIED CIRCULAT
WEISS A	16-2393	TEMPERATURE DEPENDENCE OF THE SODIUM-23 NUCLEAR QUADRUPOLE COUPLING CONSTANT IN SODIUM
WEISS JJ	34-3467	FORMATION OF COLOR CENTERS IN IRRADIATED ICE.
WEISSBACH G	28-3235	GRAVIMETRIC MEASUREMENTS OF THE CHANGES OF WATER OF CRYSTALLIZATION OF ROCHELLE SALT.
WEITHASE M	34-3475	INTERPRETATION OF THE PROTON SPIN LATTICE RELAXATION IN HEXAGONAL ICE.
	34-3488	PROTEIN SPIN RELAXATION IN HEXAGONAL ICE. PART-2: THE T(1 RHO) MINIMUM.
WELLER EF	01-0034	SYMPOSIUM ON FERROELECTRICITY, 1966, GENERAL MOTORS RES LABS, WARREN, MICHIGAN, PROC,
WELLER PF	13-2053	IMPURITY EFFECTS IN SODIUM TUNGSTEN BRONZES: NIOBIUM AND TANTALUM DOPING.
	13-2054	KYROPOULOS SINGLE CRYSTAL GROWTH OF SODIUM TUNGSTEN BRONZES. (SODIUM TUNGSTATE)
WELLES SJ	01-0021	LINEAR ELECTROOPTIC MODULATOR MATERIALS. (POTASSIUM OR AMMONIUM DIHYDROGEN PHOSPHATE O
WELSH FS	11-1823	TEMPERATURE DEPENDENCE OF THE ELASTIC, PIEZOELECTRIC, AND DIELECTRIC CONSTANTS OF LITH

WELSH FS	11-1851	MEDIUM FREQUENCY RESONATORS OF LITHIUM NIOBATE AND LITHIUM TANTALATE.
	11-1852	SURFACE WAVE TEMPERATURE COEFFICIENTS ON LITHIUM TANTALATE.
WELZ F	04-0560	PREPARATION AND PROPERTIES OF (BARIUM, STRONTIUM) TITANATE SINGLE CRYSTALS.
WEMPLE SH	03-0512	ELECTRICAL CONTACT TO N- AND P-TYPE FERROELECTRIC OXIDES.
	03-0513	THEORY OF THE ELASTOPTIC EFFECT IN NONMETALLIC CRYSTALS. (STRAIN DEPENDENT CURIE TEMP
	04-0616	DIELECTRIC PROPERTIES OF SINGLE DOMAIN MELT GROWN BARIUM TITANATE.
	04-0628	INFLUENCE OF CRITICAL POLARIZATION FLUCTUATIONS ON THE PHOTOELASTIC BEHAVIOR OF BARIUM
	04-0690	ELECTROOPTIC PROPERTIES OF SOME ABO(3) PEROVSKITES IN THE PARAELECTRIC PHASE. (POTASSI
	04-0737	OPTICAL EVIDENCE FOR POLARIZATION FLUCTUATIONS IN PARAELECTRIC BARIUM TITANATE.
	04-1028	POLARIZATION FLUCTUATIONS AND THE OPTICAL ABSORPTION EDGE IN BARIUM TITANATE.
	13-1924	USE OF PEROVSKITE PARAELECTRICS IN BEAM DEFLECTORS AND LIGHT MODULATORS. (POTASSIUM TA
WENZL H	34-3463	SPECIFIC HEAT OF PURE AND DOPED ICE NEAR 120 DEGREES K.
WERNICKE R	04-0711	LOWERING THE CURIE TEMPERATURE IN REDUCED BARIUM TITANATE.
WESSEL GK	05-1086	ELECTRON SPIN RESOURCE STUDIES AT DEFECTS IN FLUORITES. (LEAD TITANATE)
WESSON RA	06-1239	ULTRASONIC DISPERSION IN STRONTIUM TITANATE.
WESTGATE CR	37-3649	DEPENDENCE OF THE SMALL SIGNAL PARAMETERS OF FERROELECTRIC CERAMIC RESONATORS UPON STA
WESTPHAL WB	34-3455	TRANSFER OF PROTONS THROUGH "PURE" ICE I(H), SINGLE CRYSTALS. PART-2: MOLECULAR MODELS
	34-3482	DIELECTRIC RELAXATION SPECTRA OF WATER, ICE, AND AQUEOUS SOLUTIONS AND THEIR INTERPRET
	34-3483	DIELECTRIC AND MECHANICAL RESPONSE OF ICE I(H) SINGLE CRYSTALS AND ITS INTERPRETATION.
	34-3486	TRANSFER OF PROTONS THROUGH PURE ICE I(H) SINGLE CRYSTALS. PART-1: POLARIZATION SPECTR
	37-3678	HIGH DIELECTRIC CONSTANT MATERIALS AND FERROELECTRICITY AS CAPACITOR DIELECTRICS, A ST
WHALLEY E	34-3489	DIPOLE MOMENT DERIVATIVE OF THE HYDROGEN BOND IN ICE.
	34-3490	INFRARED SPECTRUM OF ICE I(H) IN THE RANGE 4000 TO 15 CM(-1).
WHIFFIN PAC	13-1913	CZOCHRALSKI GROWTH OF BARIUM STRONTIUM NIOBATE CRYSTALS.
	13-1914	MODIFICATION TO THE CZOCHRALSKI METHOD OF CRYSTAL PULLING. (CONTROL OF CRYSTAL DIAMETE
WHIPPS PW	13-1942	LONGITUDINAL ELECTROOPTIC EFFECT IN KTN.
	13-2055	GROWTH OF HIGH QUALITY CRYSTALS OF POTASSIUM TANTALATE NIOBATE.
	13-2056	STABILITY REGIONS FOR THE GROWTH OF BARIUM STRONTIUM NIOBATE CRYSTALS.
	27-2977	HIGH INTERNAL BIAS FIELDS IN L-ALANINE SUBSTITUTED TGS.
	27-3037	EFFECTS OF ADDITIVES ON THE PYROELECTRIC PROPERTIES OF TGS.
	27-3038	STRUCTURAL INHIBITION OF FERROELECTRIC SWITCHING IN TRIGLYCINE SULFATE. PART-1: ADDITI
	36-3624	ORTHORHOMBIC DIGLYCINE SULFATE. (NOT PIEZOELECTRIC, NOT FERROELECTRIC)
WHIROKOV AM	22-2882	EFFECT OF HIGH PRESSURE ON PHASE TRANSITIONS OF ALKALI TRIHYDROGEN SELENITE CRYSTALS.
WHITE EAD	13-1916	LOW FREQUENCY LINEAR ELECTROOPTIC EFFECT IN LEAD DITANATLATE.
	27-3162	INDUCED GROWTH ANISOTROPY IN TGS CRYSTALS.
WHITE KI	17-2705	RAMAN SCATTERING BY THE TEMPERATURE DEPENDENT FERROELECTRIC MODE IN POTASSIUM DIDEUTER
WHITEMAN AE	13-2057	THERMAL CONDUCTIVITY OF CUBIC SODIUM TUNGSTEN BRONZES. (SODIUM TUNGSTATES)
WHITFIELD HJ	25-2917	FAR INFRARED TRANSMISSION SPECTRUM OF THIOUREA.
WICHANSKY H	04-1029	SINGLE FIRE PROCESS FOR BARIUM TITANATE BARRIER LAYER DIELECTRICS.
WIEDER HH	23-2888	SWITCHING PROPERTIES OF BIASED FERROELECTRIC COLEMANITE.
WIEDERKEHR H	17-2707	CRYSTALLIZATION POTENTIALS IN AQUEOUS SOLUTIONS. (POTASSIUM BROMIDE, POTASSIUM CHLORID
WIENAR EA	08-1344	RAMAN SCATTERING FROM FERROELECTRIC SODIUM NITRITE.
WIENER E	16-2299	ANALYSIS OF THE TEMPERATURE DEPENDENT PHONON STRUCTURE IN SODIUM NITRITE BY RAMAN SPEC
	17-2465	MOSSBAUER EFFECT OBSERVATION OF A VERY LOW FREQUENCY MODE AT THE TRANSITION TEMPERATUR
	17-2706	ANTIFERROELECTRIC TRANSITIONS IN ADP AND AMMONIUM DIHYDROGEN ARSENATE STUDIED BY INFRA
	17-2708	PROTON DYNAMICS IN KDP TYPE FERROELECTRICS STUDIED BY INFRARED ABSORPTION.
WIESENDANGER E	08-1468	OPTICAL PROPERTIES (BIREFRINGENCE AND REFRACTIVE INDICES) OF POTASSIUM NIOBATE (LINEAR
	11-1694	ASSIGNMENTS OF OPTICAL PHONON MODES IN LITHIUM NIOBATE.
	11-1695	DIRECTIONAL DISPERSION AND ASSIGNMENT OF OPTICAL PHONONS IN LITHIUM NIOBATE.
WILCOX PD	09-1494	EFFECTS OF CALCINING ON SINTERING OF LEAD ZIRCONATE TITANATE CERAMICS.
	09-1622	EFFECTS OF CALCINING ON THE FIRING OF LEAD ZIRCONATE TITANATE CERAMIC
WILD RL	06-1214	FREE CARRIER OPTICAL ABSORPTION IN NIOBIUM DOPED STRONTIUM TITANATE.
WILDEY CG	17-2477	PULSE MEASUREMENT OF R(63) IN KDP.
WILKE KT	13-1907	EPITAXY OF PLATINUM ON BARIUM STRONTIUM NIOBATE.
WILKINSON GR	25-2916	INFRARED AND RAMAN STUDIES IN THE FERROELECTRIC TRANSITION OF THIOUREA.
WILKINSON JA	13-1913	CZOCHRALSKI GROWTH OF BARIUM STRONTIUM NIOBATE CRYSTALS.
WILL G	14-2073	NEUTRON AND ELECTRON DIFFRACTION STUDY IN NICKEL IODINE BORACITE.
WILLIAMS HCWL	03-0384	THEORY OF HYDROGEN BONDED FERROELECTRICS. PART-1: (FORMALISM OF LOCAL MODE MODEL.) PAR
WILLIAMS R	06-1240	PHOTOCONDUCTIVITY (TRANSIENTS) IN PHOTOCHROMIC STRONTIUM TITANATE (ON COLORING AND BLE
WILLIAMSON RC	11-1853	REFLECTION OF ELASTIC SURFACE WAVE FROM PERIODIC ARRAYS ON LITHIUM NIOBATE.
WILSON CM	17-2497	LIGHT SCATTERING STUDIES OF THE SOFT OPTIC AND ACOUSTIC MODES OF POTASSIUM DIHYDROGEN
	17-2709	RAMAN SCATTERING IN POTASSIUM DIHYDROGEN PHOSPHATE.
WINDSCH W	27-3013	CHANGES OF THE DOMAIN STRUCTURE IN VANADIUM DOPED TRIGLYCINE SULFATE OBSERVED BY EPR.
	27-3161	EPR STUDIES OF CHROMIUM- AND VANADYI- DOPED TRIGLYCINE SULFATE SINGLE CRYSTALS.
	28-3233	EPR INVESTIGATION OF THE INTERNAL ELECTRIC FIELD IN COPPER DOPED ROCHELLE SALT.
	36-3525	EPR INVESTIGATIONS OF MANGANESE(2+) COMPLEXES IN FERROELECTRIC TRIS-SARCOSINE CALCIUM
WINSTON H	11-1849	DIFFRACTION SPREADING OF SURFACE WAVES ON LITHIUM NIOBATE.
	11-1850	(FAR FIELD) DIFFRACTION SPREADING OF SURFACE (ACOUSTIC) WAVES ON (PIEZOELECTRIC) LITHI
WINTER FX	11-1854	OBSERVATION OF ORDINARY POLARITONS IN LITHIUM NIOBATE.
WINTRUFF W	04-0542	DETERMINATION OF CHROMIUM VALENCY FROM STUDIES OF EPR SPECTRA AND STATIC MAGNETIC SUSC
WISSEL C	04-1030	FERROELECTRIC PHASE TRANSITION OF FIRST ORDER IN BARIUM TITANATE.
WOJCIK TJ	09-1509	REALIZATION OF A 40 MHZ COLOR TELEVISION INFRARED RESPONSE USING SURFACE WAVE TRANSDUC
WOLBECK B	04-0937	STATIC ELECTRIC QUADRUPOLE INTERACTION OF TANTALUM AND HAFNIUM IONS IN BARIUM AND LEAD
WOLD A	02-0042	SYNTHESIS OF OXIDE BRONZES. (REVIEW, 88 REFS)
WOLD P	15-2205	NEW FERROELECTRIC A(V)B(VII)C(VII) COMPOUNDS OF ANTIMONY SULFOIODIDE TYPE. (PARTIAL OR
WOLF FJ	08-1469	NIOBIUM-93 AND SODIUM-23 IN POLYCRYSTALLINE SODIUM NIOBATE.
	13-2059	NUCLEAR MAGNETIC RESONANCE STUDIES OF OXYGEN OCTAHEDRA FERROELECTRICS.
WOLF HC	31-3325	PARAELECTRIC RESONANCE OF LITHIUM ION IN POTASSIUM CHLORIDE.
WOLFE RW	08-1406	STRUCTURAL BASIS OF FERROELECTRICITY IN THE BISMUTH TITANATE FAMILY.
	13-2058	CRYSTAL STRUCTURE OF BISMUTH TITANIUM NIOBIUM OXIDE.
WOLFF GA	05-1042	SOLVENT ZONE GROWTH OF FERROELECTRIC CRYSTALS. (STRONTIUM TITANATE, LEAD(X) STRONTIUM(
WOLFF H	34-3491	VAPOR PRESSURE ISOTOPE EFFECT OF ICE AND ITS ISOMERS.
WOOD VE	02-0131	PREPARATION AND PROPERTIES OF RARE EARTH COMPOUNDS. (WITH MAGNETIC AND FERROELECTRIC O
	36-3545	MAGNETIC PROPERTIES OF HEAVY RARE EARTH ORTHOMANGANATES.
WOOD VM	27-3162	INDUCED GROWTH ANISOTROPY IN TGS CRYSTALS.
WOODBURY EJ	11-1779	OBSERVATION OF THE SIMULTANEOUS GENERATION OF THE SECOND, THIRD, AND FOURTH HARMONICS
WOOLLETT RS	03-0514	MEASUREMENT OF THE DISSIPATION ASSOCIATED WITH THE ELECTROMECHANICAL COUPLING IN PIEZO
WORKMAN DT	22-2876	PROTON MAGNETIC RESONANCE STUDY OF SODIUM TRIHYDROGEN SELENATE AND POTASSIUM TRIHYDROG
WORLEY JC	12-1892	ANOMALOUS (LONGITUDINAL WAVE) ATTENUATION IN LITHIUM TANTALATE (EFFECT OF VACANCIES).
WORLOCK JM	06-1241	ELECTRIC FIELD DEPENDENCE OF OPTICAL PHONON FREQUENCIES (RAMAN SCATTERING IN CUBIC PER

WORLOCK JM   06-1242   LIGHT SCATTERING STUDIES OF THE SOFT MODES NEAR THE CUBIC TO TETRAGONAL PHASE TRANSITI
                   08-1326   ELECTRIC FIELD INDUCED RAMAN EFFECT IN PARAELECTRIC CRYSTALS. (POTASSIUM TANTALATE)
WORZ O   34-3419   DIELECTRIC PROPERTIES OF ICE I.
WRIGHT DA   04-0918   ELECTRICAL CONDUCTIVITY (SEEBECK COEFFICIENT, CURIE POINT) OF REDUCED BARIUM TITANATE
WRIGHT HC   27-2964   THERMAL ANALYSIS OF PYROELECTRIC DETECTORS. (TRIGLYCINE SULFATE EXAMPLE)
WROBEL Z   07-1267   THERMOELECTRIC EFFECT IN SEMICONDUCTIVE LEAD ZIRCONATE.
WU CC   06-1243   EFFECTS OF LONGITUDINAL MAGNETIC FIELD ON ULTRASOUND IN STRONTIUM TITANATE.
WU FY   03-0365   CIRCLE THEOREM FOR ICE RULE FERROELECTRIC MODELS.
                   03-0515   TWO DIMENSIONAL FERROELECTRIC MODELS.
                   03-0515   APPARENT VIOLATION OF THE UNIVERSALITY HYPOTHESIS IN A LATTICE MODEL OF A PHASE TRANSI
                   03-0516   CRITICAL BEHAVIOR OF HYDROGEN BONDED FERROELECTRICS.
                   03-0517   PHASE TRANSITION IN A SIXTEEN VERTEX LATTICE MODEL (FERROELECTRIC SQUARE LATTICE).
                   17-2710   MODIFIED POTASSIUM DIHYDROGEN PHOSPHATE MODEL IN A STAGGERED FIELD.
WU SY   08-1329   GROWTH AND PROPERTIES OF EPITAXIAL FILMS OF FERROELECTRIC BISMUTH TITANATE. (ELECTROOP
                   08-1470   DOMAIN STRUCTURE AND POLARIZATION REVERSAL IN FILMS OF FERROELECTRIC BISMUTH TITANATE.
                   09-1515   FERROELECTRIC DISPLAYS. (BISMUTH TITANATE, LEAD ZIRCONATE TITANATE)
WUNSCH DC   03-0518   PHENOMENOLOGICAL THEORY OF REFRACTIVE INDEX AND ELECTROOPTICAL PHENOMENA IN CRYSTALS O
WURFEL P   03-0519   PHOTOEFFECTS AT THE PHOTOCONDUCTOR- FERROELECTRIC BOUNDARY.
                   27-3018   TEMPERATURE DEPENDENCE OF THE FERROELECTRIC FIELD EFFECT IN TELLURIUM FILMS ON TRIGLYC
                   27-3163   THERMODYNAMIC STABILITY OF POLARIZATION IN THIN FERROELECTRIC FILMS ON SEMICONDUCTING
WYNNE JJ   11-1855   NONLINEAR OPTICAL SPECTROSCOPY OF THIRD ORDER SUSCEPTIBILITY IN LITHIUM NIOBATE.
XENNIKOVA TI   09-1625   CHROMATOGRAPHIC METHOD FOR ANALYZING LEAD ZIRCONATE TITANATES CONTAINING BISMUTH AND S
YADAV VS   27-3099   NEUTRON DIFFRACTION STUDY OF TRIGLYCINE SULFATE.
YAGI T   16-2395   ELECTRIC FIELD GRADIENT AND LONG RANGE ORDER PARAMETER OF FERROELECTRIC SODIUM NITRITE
                   16-2396   NUCLEAR MAGNETIC RESONANCE STUDY ON SODIUM-23 IN SODIUM NITRITE IN THE VICINITY OF THE
                   26-2944   PHASE TRANSITION IN DIVALENT METAL DICALCIUM PROPIONATES SUBSTITUTED PARTIALLY BY ACET
YAGNIK CM   07-1253   MOSSBAUER EFFECT STUDIES OF THE PHASE TRANSITIONS IN ANTIFERROELECTRIC LEAD ZIRCONATE
                   08-1471   MOSSBAUER STUDIES OF BISMUTH FERRATE LEAD TITANATE PEROVSKITE TYPE SOLID SOLUTIONS (RO
YAHATA Y   27-3127   DIP ETCH PATTERNS ON THE FACES OF TRIGLYCINE SULFATE CRYSTALS.
YAHIA J   06-1151   ELECTRONIC CONDUCTION IN SLIGHTLY REDUCED STRONTIUM TITANATE AT LOW TEMPERATURES.
                   06-1152   TRANSPORT PROPERTIES IN REDUCED STRONTIUM TITANATE. (ABSTRACT)
YAKABE M   22-2850   THE ELECTRON PARAMAGNETIC RESONANCE OF POTASSIUM SELENATE.
YAKOVLEV IA   28-3183   FREQUENCY DEPENDENCE OF THE ABSORPTION OF SOUND IN ROCHELLE SALT NEAR ITS UPPER CURIE
YAKUNIN SI   04-0940   MOTION OF DOMAIN BOUNDARIES DURING SWITCHING IN SINGLE CRYSTAL FILMS OF BARIUM TITANAT
                   04-0942   OBSERVATION OF THE CHANGE IN THE POLARIZATION OF SINGLE CRYSTAL FILMS OF BARIUM TITANA
                   04-0952   DIRECT OBSERVATION OF 180 DEG DOMAINS IN BARIUM TITANATE BY MEANS OF A POLARIZING MICR
                   04-1031   MICROSTRUCTURE OF DOMAINS AND DOMAIN WALLS IN SINGLE CRYSTAL FILMS OF BARIUM TITANATE.
YAKUPOV RG   08-1360   MAGNETOELECTRIC EFFECT IN FERROELECTRIC- ANTIFERROMAGNETIC BISMUTH TITANIUM FERRATE.
                   36-3568   MAGNETOELECTRIC EFFECT IN FERROELECTRIC- ANTIFERROMAGNETIC LITHIUM IRON(0.5) TANTALUM(
YAMADA H   06-1244   DEFECT STUDIES IN STRONTIUM TITANATE.
                   06-1245   OPTICAL ABSORPTION SPECTRA OF REDUCED STRONTIUM TITANATE.
YAMADA N   14-2082   ELECTROOPTICAL PROPERTIES OF FERROELECTRIC (IRON IODINE) BORACITE. (SPONTANEOUS LATTIC
YAMADA T   12-1876   SINGLE CRYSTAL GROWTH AND PHYSICAL PROPERTIES OF LITHIUM TANTALATE. (DIELECTRIC CONSTA
                   12-1877   BEHAVIOR OF LITHIUM TANTALATE SINGLE CRYSTAL NEAR ITS CURIE POINT. PART-1: PIEZOELECTR
                   12-1893   PIEZOELECTRIC AND ELASTIC PROPERTIES OF LITHIUM TANTALATE CHARACTERISTICS.
                   13-1969   DIELECTRIC PROPERTY OF BARIUM TITANATE NIOBATE SINGLE CRYSTAL (TETRAGONAL TUNGSTEN BRO
                   13-2060   ELASTIC ANOMALY OF BARIUM SODIUM NIOBATE (ELASTIC, PIEZOELECTRIC, DIELECTRIC, AND OPTI
                   13-2061   ELECTROMECHANICAL PROPERTIES OF OXYGEN OCTAHEDRAL FERROELECTRIC CRYSTALS.
                   17-2536   ELECTROCHEMISTRY ON THE POLYCRYSTAL OF POTASSIUM DIHYDROGEN PHOSPHATE.
                   35-3502   LEAD(5) GERMANIUM(3) OXIDE(11) CRYSTAL, A NEW FERROELECTRIC. (DIELECTRIC CONSTANTS AND
                   35-3520   ELASTIC AND PIEZOELECTRIC PROPERTIES OF FERROELECTRIC LEAD(5) GERMANIUM(3) OXIDE(11) C
                   36-3626   NEW FERROELECTRIC COMPOUND STRONTIUM TELLURATE.
                   36-3627   PHASE TRANSITION OF A NEW FERROELECTRIC OXIDE, STRONTIUM OXIDE- TELLURIUM DIOXIDE.
YAMADA Y   02-0132   FERRO- AND ANTIFERROELECTRIC PHASE TRANSITIONS AND RELATED SCATTERING EXPERIMENTS. (RE
                   04-1032   CRITICAL (POLARIZATION) FLUCTUATIONS IN BARIUM TITANATE (ABOVE THE FERROELECTRIC CURIE
                   16-2380   X-RAY STUDY OF PHASE TRANSITION IN SODIUM NITRATE.
                   16-2397   PHASE TRANSITION IN SODIUM NITRITE.
                   16-2400   X-RAY CRITICAL SCATTERING IN TRIGLYCINE SULFATE AND SODIUM NITRATE. (ORDER DISORDER TR
                   19-2777   ON THE PHASE TRANSITION OF AMMONIUM HYDROGEN SULFATE AND THE STRUCTURE OF THE FERROELE
                   27-2998   X-RAY CRITICAL SCATTERING IN FERROELECTRIC TRIGLYCINE SULFATE.
                   33-3394   X-RAY SCATTERING AND THE PHASE TRANSITION OF POTASSIUM MANGANESE TRIFLUORIDE AT 184 DE
YAMAGATA K   26-2950   PROTON RESONANCE STUDY OF SUBLATTICE ROTATIONS AND SPIN DEVIATION IN COPPER FORMATE TE
YAMAGUCHI O   09-1624   STUDIES ON THE SINTERING BETWEEN LEAD MONOXIDE AND PZT. PART-1. PART-2: EFFECTS OF AVE
YAMAJI A   04-1024   FERROELECTRIC PROPERTIES OF (INSULATING, BOUNDARY LAYER) BARIUM TITANATE CERAMICS AT G
                   04-1033   MEASUREMENT OF OXYGEN DIFFUSION CONSTANT IN SOME TITANATES. (STRONTIUM TITANATE, BARIU
                   06-1246   STUDIES ON THE (STRONTIUM, CALCIUM) TITANATE SOLID SOLUTION CERAMIC WITH BOUNDARY LAYE
                   06-1247   STUDIES ON THE (STRONTIUM, CALCIUM) TITANATE BOUNDARY LAYER CERAMIC DIELECTRICS.
                   06-1248   STUDIES ON STRONTIUM TITANATE BOUNDARY LAYER DIELECTRICS.
YAMAKA E   05-1087   LEAD TITANATE PYROELECTRIC INFRARED DETECTOR.
YAMAMOTO M   28-3234   SURFACE STRUCTURES AND CORRESPONDING LIGHT FIGURES REVEALED BY WATER ETCHED ROCHELLE S
YAMAMOTO O   04-0945   CATALYTIC OXIDATION OF CARBON MONOXIDE ON SEMICONDUCTIVE BARIUM TITANATE.
YAMAMOTO T   22-2823   ANOMALOUS SPECIFIC HEAT OF SODIUM TRIHYDROGEN SELENITE.
                   36-3625   NUCLEAR QUADRUPOLE RESONANCE STUDY OF PHASE TRANSITION IN AMMONIUM MONOCHLORO ACETATE,
YAMANAKA M   17-2711   AN OBSERVATION OF THE CHANGE IN COOLED KDP TEMPERATURE DUE TO 337 MICRON INCIDENT RADI
YAMANAKA S   37-3681   MANUFACTURING METHOD AND PROPERTIES OF FERROELECTRIC THIN FILM.
YAMANISHI M   37-3680   ACOUSTIC SURFACE WAVE GUIDE, AVAILING FERROELECTRIC POLARIZATIONS.
YAMANOUCHI K   11-1832   PIEZOELECTRIC LEAKY SURFACE WAVE IN LITHIUM NIOBATE (ELECTROMAGNETIC COUPLING).
                   11-1856   PROPAGATION AND AMPLIFICATION OF RAYLEIGH WAVES AND PIEZOELECTRIC LEAKY SURFACE WAVES
YAMAOTO T   36-3538   FERROELECTRICITY AND PHASE TRANSITION IN AMMONIUM DICHLORO ACETATE AND DEUTERATED AMMO
YAMASHITA H   16-2398   VACUUM ULTRAVIOLET ABSORPTION OF THE ALKALI NITRITES AND NITRATES.
                   16-2399   VACUUM ULTRAVIOLET ABSORPTION (5-23 EV AT ROOM AND LIQUID NITROGEN TEMPERATURES IN ALK
YAMAUCHI F   09-1623   INTERNAL FRICTION OF MODIFIED LEAD ZIRCONATE- LEAD TITANATE CERAMICS. (PROPRIETARY COM
YAMZIN II   02-0133   NEUTRON DIFFRACTION INVESTIGATIONS OF THE ATOMIC STRUCTURES OF INORGANIC MATERIALS. (R
YANAGI T   16-2301   FERROELECTRIC POLARIZATION SWITCHING IN SODIUM NITRITE.
                   16-2401   FERROELECTRICITY IN AMMONIUM(X) POTASSIUM(1-X) NITRATE MIXED CRYSTAL.
                   16-2402   SOME (DIELECTRIC MEASUREMENTS AND INFRARED ABSORPTION) STUDIES ON FERROELECTRICITY IN
                   16-2403   STUDY OF PHASE TRANSITION OF SODIUM NITRITE BY SECOND HARMONIC GENERATION. (TEMPERATUR
YANAGIDA H   04-0799   LANTHANUM DOPED BARIUM TITANATE CERAMICS WITH BISMUTH OXIDE AS A GRAIN BOUNDARY ELEMEN
                   04-0800   TWO-STEP ANOMALOUS INCREASE OF RESISTIVITY ON LANTHANUM DOPED BARIUM TITANATE- BISMUTH
                   04-0801   TWO STEP ANOMALOUS INCREASE OF RESISTANCE ON LANTHANUM(0.002) BARIUM(0.998) TITANATE-

YANAI H   04-0945 CATALYTIC OXIDATION OF CARBON MONOXIDE ON SEMICONDUCTIVE BARIUM TITANATE.
YANASE M   04-0717 ATTEMPT TO PREPARE HIGH DOPED BARIUM STRONTIUM TITANATE SEMICONDUCTOR WITH PZT.
YANCHEVSKAYA IS   08-1355 NATURE OF THE HIGH PERMITTIVITY OF CERIUM ALUMINATES WITH PEROVSKITE TYPE STRUCTURE.
YANG CP   03-0520 EXACT SOLUTION OF MODEL OF TWO DIMENSIONAL FERROELECTRICS IN ARBITRARY EXTERNAL FLECTR
YANG GC   13-1930 FHOTCCHROMIC EFFECT AND ELECTRON SPIN RESONANCE OF IRON AND MOLYBDENUM IN BARIUM SODIU
                     17-2629 PARAMAGNETIC RELAXATION AND DIVALENT COPPER IONS IN FERROELECTRIC KDP.
YANG KH   11-1857 GENERATION OF FAR INFRARED RADIATION BY PICOSECOND LIGHT PULSES IN LITHIUM NIOBATE.
YANKELEVICH FP   03-0157 INTERACTION OF LIGHT WITH MAGNETOACOUSTIC WAVES IN A FERROELECTRIC (THEORY).
                     03-0158 INTERACTION OF LIGHT WITH SPIN WAVES IN FERROELECTRICS. (THEORY)
YANKOVSKAYA LI   04-1003 REACTION SEQUENCE IN FORMING BARIUM METATITANATE IN AN INDUSTRIAL ROTARY FURNACE.
YANO T   13-2051 ELECTROOPTICAL PROPERTIES CF STRONTIUM POTASSIUM LITHIUM NIOBATE. (TUNGSTEN BRONZE TYP
                     13-2062 ELECTROOPTIC PROPERTIES OF STRONTIUM LITHIUM NIOBATE TYPE FERROELECTRICS (SINGLE CRYST
YARBOROUGH JM   11-1858 EFFECT TUNABLE OPTICAL EMISSION FROM LITHIUM NIOBATE WITHOUT A RESONATOR.
                     11-1859 SIMULTANEOUS OPTICAL FARAMETRIC OSCILLATION, SECOND HARMONIC GENERATION (IN LITHIUM NI
                     17-2712 EFFICIENT HIGH GRAIN PARAMETRIC GENERATION IN ADP CONTINUOUSLY TUNABLE ACROSS THE VISI
YARIV A   08-1413 MEASUREMENT OF THE RELATIVE NONLINEAR COEFFICIENTS OF POTASSIUM DIHYDROGEN PHOSPHATE,
                     17-2616 FARAMETRIC FREQUENCY CONVERSION OF COHERENT LIGHT BY THE ELECTROOPTIC EFFECT IN POTASS
YASAKI T   13-2063 LONGITUCINAL ELECTROOPTIC EFFECT IN OBLIQUE CUT STRONTIUM BARIUM NIOBATE PLATES.
YASHCHIN EG   17-2686 CURIE-WEISS LAW FOR THE NONLINEAR SUSCEPTIBILITY OF A POTASSIUM DIHYDROGEN FIIOSPHATE C
YASOJIMA Y   11-1861 NATURE OF INTERNAL ELECTRIC FIELD DURING OPTICAL DAMAGE PROCESS IN LITHIUM NIOBATE.
YASUDA H   11-1860 GROWTH OF CRATERS ON ION ETCHED SURFACE OF LITHIUM NIOBATE AND ION ETCHING WITHOUT CRA
                     11-1862 THINNING OF LITHIUM NIOBATE FOR ACOUSTOOPTIC DEFLECTORS BY ION ETCHING.
YASUMOTO M   09-1624 STUDIES ON THE SINTERING BETWEEN LEAD MONOXIDE AND PZT. PART-1. PART-2: EFFECTS OF AVE
YASUNAGA H   06-1249 PHOTOCONDUCTIVITY ANOMALY AS AN EVIDENCE OF THE LOCAL FERROELECTRIC PHASE TRANSITION A
YATABE K   04-0782 X-RAY IRRADIATION EFFECT ON DIELECTRIC PROPERTIES OF SOLID SOLUTIONS OF BARIUM TITANAT
YATOM H   03-0234 ANHARMONIC MODE MIXING IN IONIC FERROELECTRICS. (LATTICE DYNAMICS)
YATSENKO AF   16-2335 LUMINESCENCE OF POTASSIUM NITRATE IRRADIATED WITH GAMMA-RAYS.
                     19-2772 CHANGES IN THE INFRARED SPECTRA OF AMMONIUM OR RUBIDIUM BISULFATE, AND AMMONIUM SULFAT
YATSKOVICH II   15-2141 PRODUCTION METHODS AND SOME OPTICAL PROPERTIES OF BISMUTH TELLUROIODIDE.
YATSUHASHI T.   04-0993 ELECTRON MICROSCOPIC STUDIES ON FERROELECTRIC DOMAINS OF PEROVSKITE TYPE OXIDES. (LEAD
                     04-1034 ELECTRON MICROSCOPIC STUDY OF FERROELECTRIC DOMAINS OF BARIUM TITANATE IN LOW TEMPERAT
YAZAKI T   08-1472 LONGITUDINAL ELECTROOPTIC EFFECT IN OBLIQUE CUT STRONTIUM BARIUM NIOBATE PLATES.
                     09-1549 CIELECTRIC AND PIEZOELECTRIC PROPERTIES OF LEAD COBALTATE(0.33) NIOBATE(0.67)- TITANAT
YAZYTSKII BYA   04-0905 UHF (WAVEGUIDE RESCNATOR) METHOD OF STUDYING TEMPERATURE CHARACTERISTICS OF ROCHELLE T
YELON WB   13-2064 NEUTRON SCATTERING STUDY OF THE SOFT MODES IN CUBIC POTASSIUM TANTALATE NIOBATE.
YI PN   17-2713 FERROELECTRIC MODE AND NQR (NUCLEAR QUADRUPOLE RESONANCE) FREQUENCY. (POTASSIUM DIHYDR
                     31-3340 MOLECULAR MOTION IN THE FERROELECTRIC PHASE OF HYDROGEN CHLORIDE. (CALCULATED IN THE H
YOCKEY HP   21-2812 THERMOLUMINESCENCE STUDIES OF THE GAMMA-RAY IRRADIATED FERROELECTRICS ROCHELLE SALT AN
                     28-3207 RADIATION DAMAGE AND THE FERROELECTRIC EFFECT IN ROCHELLE SALT.
YOGANARASIMHAN SR   25-2924 SEMICONDUCTIVITY OF THIOUREA.
YOKOMIZO Y   03-0283 THE BEHAVIORS OF THE MICRODOMAINS ON THE DIFFUSE PHASE TRANSITION FERROELECTRICS.
                     13-1959 CRYSTAL GROWTH OF LEAD (CADMATE(0.33) NIOBATE(0.67)) AND ITS DIELECTRIC PROPERTIES. (F
                     13-2065 DIELECTRIC AND OPTICAL PROPERTIES OF LEAD ZINCATE(0.33) NIOBATE(0.67) SINGLE CRYSTAL.
                     13-2066 FERROELECTRIC PROPERTIES OF LEAD ZINCATE(0.33) NIOBATE(0.67) (SINGLE CRYSTALS GROWN FR
YOKOYAMA S   03-0326 THEORY OF POLARIZATION REVERSAL OF BARIUM TITANATE CRYSTAL.
YONEZAWA M   08-1453 SYNTHESIS OF PEROVSKITE COMPLEX COMPOSITIONS: LEAD MANGANATE(0.5) NIOBATE(0.5) AND LEA
                     13-2067 CRYSTAL STRUCTURE AND PIEZOELECTRICITY OF THE SYSTEM LEAD (ZINCATE NIOBATE)- LEAD TITA
YOSHIDA I   16-2258 ELECTRICAL RESISTIVITIES OF SODIUM NITRITE AND POTASSIUM NITRATE CRYSTALS.
YOSHIDA K   37-3680 ACOUSTIC SURFACE WAVE GUIDE, AVAILING FERROELECTRIC POLARIZATIONS.
YOSHIDA M   15-2166 SWITCHING AND MEMORY EFFECT IN ANTIMONY SULFOIODIDE- TIN DIOXIDE HETEROSTRUCTURE THIN
YOSHIHARA T   04-0988 ANOMALY OF RESISTANCE IN REDUCED BARIUM TITANATE CRYSTAL.
YOSHIKAWA A   03-0324 90 DEGREE TYPE DOMAIN WALL OF BARIUM TITANATE CRYSTAL. (THEORY, CALCULATIONS)
YOSHIMITSU K   03-0488 ANOMALY OF THE SPIN LATTICE RELAXATION TIME NEAR THE TRANSITION TEMPERATURE IN THE ORD
                     03-0521 DISTRIBUTION OF RELAXATION TIME IN FERROELECTRICS NEAR THE CURIE TEMPERATURE (POLY VER
YOSHIMOTO J   08-1473 THE EFFECT OF PRESSURE ON THE TRANSITION OF LEAD ZINCATE(0.33) NIOBATE(0.67).
YOSHIMURA Y   22-2872 THE STRUCTURE AND PHASE TRANSITION IN POTASSIUM SELENATE.
YOSHINAGA H   04-0744 FAR INFRARED REFLECTIVITY OF BARIUM TITANATE.
                     17-2549 FAR INFRARED SPECTRA OF KDP AND ADP. (20-550 PER CM, 200 AND 300 DEG K)
YOUNG JF   11-1688 GROWTH OF HIGH QUALITY LITHIUM NIOBATE CRYSTALS (FOR NONLINEAR OPTICS WITH LOW BIREFRI
YOUNG KF   33-3368 PHASE TRANSITIONS IN CESIUM LEAD TRICHLORIDE. (LOSS OF A CENTER OF SYMMETRY AT 194 DEG
                     33-3408 DETERMINATION OF TRANSITION TEMPERATURES IN CESIUM LEAD TRICHLORIDE USING EPR.
YOUNG RA   36-3591 MONOCLINIC STRUCTURE OF SYNTHETIC CALCIUM CHLORAPATITE. (DEVELOPMENT OF FERROELECTRIC
YOUSSEF H   28-3197 ULTRASONIC ANOMALIES IN ROCHELLE SALT.
YUDIN VM   08-1441 ANTIFERROMAGNETIC PROPERTIES OF SOME PEROVSKITES. (BISMUTH FERRATE)
YUMOTO H   30-3297 FERROELECTRIC FERROELASTIC CRYSTAL GADOLINIUM MOLYBDATE.
YURIKE YAYA   04-0596 SURFACE BREAKDOWN IN VACUUM ON BARIUM TITANATE.
YURIN VA   15-2180 MOSSBAUER STUDIES OF ANTIMONY SULFOIODIDE TYPE CRYSTALS.
                     27-3164 DOMAIN STRUCTURE OF GAMMA-IRRADIATED TGS CRYSTALS.
                     27-3165 PHASE STABILIZATION (HINDERED TRANSITION) IN GAMMA-IRRADIATED FERROELECTRIC MATERIALS.
                     28-3229 FIEZOELECTRIC EFFECT AND INTERNAL FRICTION IN GAMMA-IRRADIATED ROCHELLE SALT CRYSTALS.
YURKEVICH VE   03-0438 ACCOUNTING FOR INTERACTION IN THE KANZIG REGIONS MODEL FOR DIFFUSE FERROELECTRIC PHASE
                     03-0441 EFFECT OF AN EXTERNAL ELECTRIC FIELD ON A DIFFUSE FERROELECTRIC PHASE TRANSITION, TALK
                     03-0522 CIELECTRIC CONSTANT BEHAVIOR IN THE NEIGHBORHOOD OF THE PHASE TRANSITION IN CHARACTERI
                     03-0523 THERMODYNAMICS OF FERROELECTRIC SOLID SOLUTIONS. (THEORY)
                     03-0524 THERMODYNAMICS OF FERROELECTRIC SOLID SOLUTIONS. PART-1. PART-2.
YUSHIN NK   11-1678 COLLINEAR INTERACTION OF LONGITUDINAL ELASTIC WAVES IN LITHIUM NIOBATE CRYSTALS.
YUSHKA GB   15-2193 TEMPERATURE CEPENCENCES OF THE HALL AND DRIFT MOBILITIES OF HOLES IN ANTIMONY TRISULFI
YUZVAK VI   22-2827 MECHANISMS OF FERROELECTRIC PHASE TRANSITION IN CRYSTALS WITH REORIENTATING STRUCTURAL
                     22-2828 NUCLEAR MAGNETIC RESONANCE OF DEUTERIUM IN THE FERROELECTRIC SODIUM DEUTERO AMMONIUM S
                     22-2883 NUCLEAR MAGNETIC RESONANCE OF SODIUM-23 IN THE SODIUM AMMONIUM SELENATE DIHYDRATE CRYS
ZABARA YUV   04-0708 INVESTIGATION OF THE FINE STRUCTURE OF BARIUM TITANATE CRYSTALS FLUX GROWN FROM POTASS
                     04-1035 ANISCTRCPY OF THE ELECTRICAL CONDUCTIVITY OF BARIUM TITANATE SINGLE CRYSTALS GROWN FRO
                     04-1036 POSITIVE TEMPERATURE COEFFICIENT OF RESISTANCE OF BARIUM TITANATE SINGLE CRYSTALS IN T
ZACCAI G   13-1957 SOFT MODES AND STRUCTURE OF FERROELECTRIC TETRAGONAL POTASSIUM TANTALATE NIOBATE.
                     13-2068 FOTASSIUM TANTALATE NIOBATE, STRUCTURAL AND DYNAMICAL STUDIES.
ZACHMANIDIS PJ   04-0979 NUCLEATION AND GROWTH OF FERROELECTRIC DOMAINS IN BARIUM TITANATE AT FIELDS FROM 2 TO
ZAFRIR H   17-2489 HIGH TEMPERATURE PHASE TRANSITION IN RUBIDIUM DIHYDROGEN PHOSPHATE BY SCATTERING OF CO
ZAIKOVSKII OI   03-0195 POLARIZATION OF FERROELECTRIC FILM BY BENDING AND BENDING DEFORMATION IN THE ELECTRIC
                     04-0600 DIELECTRIC CONSTANT AND LATTICE DYNAMICS OF A THIN FILM OF A FERROELECTRIC (BARIUM TIT
                     04-0607 MAGNITUDE OF SPONTANEOUS POLARIZATION IN THIN CRYSTALS AND FINE PARTICLES OF BARIUM TI

PAGE 703

ZAIONTS LR	13-1951	ELECTROPHYSICAL PROPERTIES OF THE SOLID SOLUTION SYSTEM LEAD TITANATE- LEAD ZIRCONATE-
ZAITOV MM	28-3176	ELECTRON PARAMAGNETIC RESONANCE OF COPPER IONS IN ROCHELLE SALT.
ZAITSEVA MP	20-2809	INVESTIGATION OF PHASE TRANSITION IN FERROELECTRIC ALUMS (METHYL AMMONIUM ALUMINUM SUL
	20-2810	PHYSICAL PROPERITES OF FERROELECTRIC ALUMS. (ESPECIALLY METHYL AMMONIUM ALUMINUM SULFA
	22-2884	ELECTROMECHANICAL PROPERTIES OF THE FERROELECTRIC SODIUM AMMONIUM SELENATE DIHYDRATE.
ZAKHAROV VP	06-1250	INFRARED SPECTRA OF SINGLE AND POLYCRYSTALLINE STRONTIUM TITANATE.
ZAKHAROVA MV	17-2630	PREPARATION (MORPHOLOGY) AND MECHANICAL PROPERTIES OF FILAMENTARY CRYSTALS OF ADP.
ZAKS PL	15-2247	NONLINEAR EFFECTS AND THE LOW TEMPERATURE TRANSITION IN ANTIMONY SULFOIODIDE CRYSTALS
	15-2249	DIELECTRIC AND PIEZOELECTRIC PROPERTIES OF ANTIMONY SULFOIODIDE IN A CONSTANT ELECTRIC
	15-2250	INVESTIGATION OF THE PIEZORESISTANCE OF (POLYCRYSTALLINE) ANTIMONY SULFOIODIDE (DARK C
ZAKURKIN VV	04-0971	(IN PILE, FAST NEUTRON) RADIATION INDUCED PHASE TRANSFORMATIONS IN (CERAMIC) BARIUM AN
	05-1081	X-RAY INVESTIGATION OF IRRADIATED LEAD TITANATE.
	08-1443	X-RAY AND DIELECTRIC STUDIES OF IRRADIATED PEROVSKITE TYPE COMPOUNDS.
	08-1474	EFFECT OF IRRADIATION OF DIELECTRIC PROPERTIES OF SOME FERRO-, ANTIFERRO- AND PARAELE
ZANNE M	17-2457	LOW FREQUENCY INFRARED ABSORPTION SPECTRA OF MONOCRYSTALS OF POTASSIUM DIHYDROGEN PHOS
ZAPOROZHETS OI	15-2248	EFFECTS OF TEMPERATURE, ELECTRIC FIELD, AND ILLUMINATION ON THE ABSORPTION OF ULTRASOU
ZAREMBO LK	11-1707	GENERATION OF SUPERHIGH FREQUENCY ACOUSTIC HARMONICS IN A (SINGLE DOMAIN) LITHIUM NIOB
ZAREMBOVSKAYA TA	27-3157	APPLICABILITY OF THE PIPPARD RELATIONS TO THE PHASE TRANSITION IN TRIGLYCINE SULFATE.
	27-3166	DIELECTRIC PERMEABILITY OF TRIGLYCINE SELENATE NEAR THE PHASE TRANSITION.
	27-3167	EFFECT OF CONSTANT ELECTRICAL FIELD ON PHASE TRANSITIONS IN TRIGLYCINE FLUOBERYLLATE.
	27-3168	RELATION BETWEEN ELECTRICAL AND THERMAL PROPERTIES OF TRIGLYCINE SELENATE NEAR THE CUR
	27-3169	THERMAL EXPANSION OF TRIGLYCINE FLUOBERYLLATE CRYSTALS IN THE FERROELECTRIC TRANSITION
ZAROCHENTSEV EV	05-1046	FERROELECTRIC PROPERTIES OF LEAD TITANATE SINGLE CRYSTALS.
ZASOVIN EA	11-1768	INVESTIGATION OF A SYSTEM FOR BEAM DEFLECTION BY MEANS OF LITHIUM NIOBATE CRYSTALS.
ZAVYANLOVA AM	15-2249	DIELECTRIC AND PIEZOELECTRIC PROPERTIES OF ANTIMONY SULFOIODIDE IN A CONSTANT ELECTRIC
	15-2250	INVESTIGATION OF THE PIEZORESISTANCE OF (POLYCRYSTALLINE) ANTIMONY SULFOIODIDE (DARK C
ZAYACHKOVSKII MP	15-2126	CONDUCTIVITY, THERMOELECTRIC PROPERTIES, AND BAND STRUCTURE OF CRYSTALLINE BISMUTH SEL
	15-2128	PIEZOELECTRIC RESISTANCE EFFECT IN BISMUTH SELENOIODIDE CRYSTALS.
ZDANOV GS	08-1442	CALCULATIONS OF THE INTERNAL ELECTRIC FIELDS AND ELECTRIC FIELD GRADIENTS IN THE PEROV
ZDANSKY K	04-0977	PARAMAGNETIC RESONANCE OF PLATINUM IONS IN NIOBIUM DOPED BARIUM TITANATE SINGLE CRYSTA
ZEER EP	24-2893	ORDERING OF DIPOLE MOMENTS OF WATER MOLECULES AND FERROELECTRIC PHASE TRANSITION IN PO
	24-2898	POLYTYPE AND TWINNING OF THE FERROELECTRIC CRYSTALS POTASSIUM FERROCYANIDE TRIHYDRATE
	24-2899	ANOMALOUS SPIN LATTICE RELAXATION BY QUASI SPIN WAVES IN POTASSIUM FERROCYANIDE TRIHYD
	24-2900	RADIO SPECTROSCOPIC INVESTIGATION OF POTASSIUM FERROCYANIDE TRIHYDRATE AND ISOMORPHOUS
ZEHENTNER J	17-2714	RELATIVE PERMITTIVITY OF POTASSIUM DIHYDROGEN PHOSPHATE AND AMMONIUM DIHYDROGEN PHOSPH
ZEIDLER JR	30-3324	PYROELECTRIC DETECTION PROPERTIES OF GADOLINIUM MOLYBDATE.
ZEIN NE	08-1475	LATTICE DYNAMICS OF PHASE TRANSITIONS IN PEROVSKITE TYPE ANTIFERROELECTRICS.
ZEINALLY AKH	15-2104	INFLUENCE OF NONEQUILIBRIUM CARRIERS ON THE DOMAIN STRUCTURE OF THE FERROELECTRIC SEMI
	15-2105	SHORT CIRCUIT PHOTOEMF IN FERROELECTRIC SEMICONDUCTORS. (ANTIMONY SULFOIODIDE CRYSTALS
	15-2129	CHARACTERISTICS OF THE PHOTOELECTRIC EMF OF ANTIMONY SULFOIODIDE.
	15-2130	LIGHT INDUCED PYROCURRENTS IN FERROSEMICONDUCTOR ANTIMONY SULFOIODIDE.
	15-2131	PYROCURRENTS HIGHER THAN THE CURIE POINT IN ANTIMONY SULFOIODIDE.
	15-2132	PHOTOEMF'S IN THE FERROELECTRIC SEMICONDUCTOR ANTIMONY SULFOIODIDE.
	15-2133	PHOTODOMAIN EFFECT IN NATURALLY POLARIZED ANTIMONY SULFOIODIDE CRYSTALS.
	15-2134	PHOTOELECTRIC PROPERTIES (PHOTOCONDUCTIVITY) OF ANTIMONY SULFOIODIDE SINGLE CRYSTALS.
	15-2135	PHOTODOMAIN EFFECT IN NATURALLY POLARIZED ANTIMONY SULFOIODIDE CRYSTALS.
	15-2136	SOME FEATURES OF THE PHOTOEMF SPECTRUM OF ANTIMONY SULFOIODIDE.
	15-2153	CHARACTERISTICS OF THE PHOTOELECTRIC EMF OF ANTIMONY SULFOIODIDE.
	15-2228	INFLUENCE OF LIGHT ON SELF POLARIZATION OF SINGLE CRYSTALS OF ANTIMONY SULFOIODIDE.
ZEKOVIC S	17-2673	PROPERTIES OF THE KDP TYPE FERROELECTRICS WITH IMPURITIES.
ZEKS B	03-0179	DYNAMICS OF ORDER DISORDER TYPE FERROELECTRICS AND ANTIFERROELECTRICS
	17-2442	VANISHING OF FERROELECTRICITY IN POTASSIUM DIHYDROGEN PHOSPHATE AT HIGH PRESSURES.
	17-2569	LASER RAMAN STUDY OF QUASI SPIN WAVE HYDROGEN TUNNELING MODES IN KDP AND DEUTERATED KD
	17-2570	PROTON LATTICE INTERACTIONS AND THE DYNAMICAL SUSCEPTIBILITY OF POTASSIUM DIHYDROGEN P
	28-3187	SPECIFIC HEAT ANOMALY IN FERROELECTRIC ROCHELLE SALT (CALCULATION OF TEMPERATURE DEPEN
	28-3236	DYNAMICS OF FERROELECTRIC ROCHELLE SALT
	28-3237	DYNAMICS OF FERROELECTRIC ROCHELLE SALT.
ZELDES H	16-2404	PARAMAGNETIC RESONANCE STUDY OF PRODUCTION OF NITROGEN DIOXIDE AND NITRITE ION IN IRRA
	16-2405	PARAMAGNETIC SPECIES IN IRRADIATED NITRATE.
ZELENKA J	27-2984	DYNAMIC DETERMINATION OF (ELASTIC AND PIEZOELECTRIC) MATERIAL CONSTANTS OF TRIGLYCINE
ZELJIC Z	36-3611	TEMPERATURE DEPENDENCE OF SOME PROPERTIES OF SODIUM DITHORIUM PHOSPHATE FERROELECTRIC
ZENTKOVA A	06-1128	NORMAL VIBRATIONS OF STRONTIUM TITANATE LATTICE.
ZEREM JZ	17-2716	LIGHT AND CURRENT PULSES FROM X-RAYED POTASSIUM DIHYDROGEN PHOSPHATE CRYSTALS.
ZERNIKE F	17-2715	AMPLITUDE MODULATORS BASED ON THE MICHELSON INTERFEROMETER. (AMMONIUM DIHYDROGEN PHOSP
ZHABITENKO NK	04-1037	ABSORPTION AND AMPLIFICATION OF ULTRASOUND IN A TWO-LAYER SYSTEM CONSISTING OF A PIEZO
ZHABOTINSKII VA	13-2006	PROPERTIES OF FERROGLASS CERAMICS BASED ON LEAD BARIUM NIOBATE COMPOUNDS.
ZHASHKOV AA	11-1863	LONGITUDINAL ELECTROOPTICAL EFFECT IN OBLIQUE SECTIONS OF LITHIUM NIOBATE.
ZHDAN AG	03-0525	MEASUREMENT OF THE TEMPERATURE DEPENDENCE OF THE ELECTRICAL CONDUCTIVITY OF FERROELECT
	15-2251	ELECTRICAL AND PHOTOELECTRIC PROPERTIES OF THE CONTACTS BETWEEN ANTIMONY SULFOIODIDE S
	15-2252	OPPOSING FERROELECTRIC DOMAINS IN SINGLE CRYSTALS OF ANTIMONY SULFOIODIDE.
ZHDANOV GS	02-0128	NEW FERROELECTRICS AND SEIGNETTO MAGNETS OF THE PEROVSKITE AND PYROCHLORE TYPE. SYNTHE
	03-0372	FERROELECTRICITY AND ANTIFERROELECTRICITY IN POLAR CRYSTALS. (THEORY, LEAD ZIRCONATE)
	03-0490	(METHODS OF) STRUCTURAL STUDIES OF FERROELECTRICS AND FERROELECTRIC MAGNETS. (ELECTRON
	04-1038	STRUCTURAL ELECTRON MICRODIFFRACTION STUDIES OF FERROELECTRICS AND FERROELECTRIC MAGNE
	05-1044	THE SYSTEM LEAD TITANATE- STRONTIUM COPPER NIOBATE. (CERAMIC SYNTHESIS)
	05-1045	THE SYSTEM LEAD TITANATE- STRONTIUM CUPRATE(0.33) NIOBATE(0.67).
	05-1082	STRUCTURAL MICROELECTRON DIFFRACTION STUDY OF FERROELECTRIC AND FERROMAGNETIC SUBSTANC
	07-1261	DIPOLE STRUCTURE AND INTERNAL ELECTRIC FIELDS IN LEAD ZIRCONATE.
	08-1308	POTASSIUM TANTALATE LEAD TITANATE SYSTEM IN THE TEMPERATURE RANGE 4.2-300 DEG K (LATTI
	08-1310	X-RAY ANALYSIS AND DIELECTRIC PROPERTIES OF SOLID SOLUTIONS BASED ON THE FERROELECTRIC
	08-1362	X-RAY DETERMINATION OF THE SYMMETRY OF THE FERROELECTRIC COMPOUNDS (POTASSIUM OR SODIU
	08-1387	HIGH TEMPERATURE X-RAY INVESTIGATION OF THE PEROVSKITE MODIFICATION OF CADMIUM TITANAT
	08-1389	INTERNAL ELECTRIC FIELDS IN CRYSTALS OF SODIUM TANTALATE AND CADMIUM TITANATE.
	08-1425	COEXISTENCE OF ANTIFERROMAGNETIC AND SPECIAL DIELECTRIC PROPERTIES IN THE BISMUTH- LAN
	08-1455	CALCULATION OF INTERNAL ELECTRIC FIELDS AND THEIR GRADIENTS IN ABO(3) PEROVSKITE COMPO
	08-1458	CRYSTAL STRUCTURE AND PROPERTIES OF NEW CADMIUM AND THALLIUM CONTAINING PEROVSKITES. (
	08-1459	PEROVSKITE TYPE SEIGNETTO MAGNETS.
	08-1460	STRUCTURE AND PROPERTIES OF SOME NEW FERROELECTRIC MATERIALS WITH PEROVSKITE STRUCTURE
	08-1462	NEW FERROELECTRICS WITH PEROVSKITE AND PYROCHLORE STRUCTURES. (BARIUM AND BISMUTH(0.5)
	30-3271	PHASE TRANSITIONS OF GADOLINIUM MOLYBDATE AND ISOSTRUCTURAL COMPOUNDS (EUROPIUM TERBIU

PAGE 704

ZHDANOV GS	36-3604	STRUCTURE AND MAGNETIC PROPERTIES OF FERROELECTRIC FERROMAGNETIC SOLID SOLUTIONS IN TH
	36-3619	NEW FERROELECTRICS. (BARIUM(2) BISMUTH (NIOBIUM OR TANTALUM OR VANADIUM) OXIDE(6), AND
ZHDANOVA VV	08-1378	DIELECTRIC PROPERTIES AND THERMAL EXPANSION OF SOME FERROELECTRICS AND ANTIFERROELECTR
ZHECHKOVA LA	09-1625	CHROMATOGRAPHIC METHOD FOR ANALYZING LEAD ZIRCONATE TITANATES CONTAINING BISMUTH AND S
ZHELUDEV IS	02-0134	DIPOLE PATTERNS IN THE STRUCTURES OF SOME FERROELECTRICS AND ANTIFERROELECTRICS. (BARI
	03-0526	CHARACTERISTICS OF THE STRUCTURAL CHANGES INVOLVED IN PHASE TRANSFORMATIONS IN ANTIFER
	03-0527	CHANGE OF COMPLETE SYMMETRY OF THE CRYSTALS DURING THE PHASE TRANSITIONS IN FERROELECT
	03-0528	FERROELECTRICITY AND SYMMETRY. (41 REFS)
	17-2505	POLARIZATION (DOMAIN STRUCTURE) AND PHASE TRANSITION IN KDP SINGLE CRYSTAL. (ABSTRACT
	27-3164	DOMAIN STRUCTURE OF GAMMA-IRRADIATED TGS CRYSTALS.
ZHEREPTSOVA LI	19-2742	INVESTIGATION OF FERROELECTRICITY IN SODIUM AMMONIUM SELENATE AND SODIUM AMMONIUM SULF
	20-2809	INVESTIGATION OF PHASE TRANSITION IN FERROELECTRIC ALUMS (METHYL AMMONIUM ALUMINUM SUL
	20-2810	PHYSICAL PROPERITES OF FERROELECTRIC ALUMS. (ESPECIALLY METHYL AMMONIUM ALUMINUM SULFA
	22-2884	ELECTROMECHANICAL PROPERTIES OF THE FERROELECTRIC SODIUM AMMONIUM SELENATE DIHYDRATE.
ZHESTKOV VE	04-1012	INFLUENCE OF A CONSTANT ELECTRIC FIELD ON THE DIELECTRIC PROPERTIES OF POLYCRYSTALLINE
ZHESTKOV VG	04-1014	UHF REVERSAL PROPERTIES OF BARIUM TITANATE TYPE FERROELECTRIC CERAMICS.
ZHUKOV AP	18-2740	ARSENIC-75 NUCLEAR QUADRUPOLE RESONANCE SPECTRA OF PARTIALLY DEUTERATED FERROELECTRIC
ZHUKOV OK	04-0818	UHF DIELECTRIC PROPERTIES OF SOME POLYCRYSTALLINE FERROELECTRICS (BARIUM TITANATE AND
	15-2180	MOSSBAUER STUDIES OF ANTIMONY SULFOIODIDE TYPE CRYSTALS.
	17-2466	DIELECTRIC PROPERTIES OF A (SMALL DOSE) GAMMA-IRRADIATED KDP CRYSTAL (TEMPERATURE DEPE
	17-2467	DIELECTRIC PROPERTIES OF A GAMMA-IRRADIATED POTASSIUM DIHYDROGEN PHOSPHATE CRYSTAL.
	17-2542	DIELECTRIC PROPERTIES OF GAMMA-IRRADIATED POTASSIUM DIHYDROGEN PHOSPHATE CRYSTALS.
	17-2543	NONLINEAR PROPERTIES OF POTASSIUM DIHYDROGEN PHOSPHATE. (DETERMINATION OF BETA IN THE
ZHUKOV OV	27-2976	DIELECTRIC LOSSES OF TRIGLYCINE SULFATE CRYSTALS EXPOSED TO SMALL DOSES OF X- AND GAMM
ZHUKOVSKII VI	04-0572	INVERSION OF ELECTRICAL PROPERTIES OF SEMICONDUCTING BARIUM TITANATE.
ZHURAVLEV AK	13-2006	PROPERTIES OF FERROGLASS CERAMICS BASED ON LEAD BARIUM NIOBATE COMPOUNDS.
ZIEBERT V	27-3170	AFTER EFFECTS IN SEMICONDUCTOR ELECTRODES EVAPORATED ON INSULATORS ESPECIALLY ON TRIGL
	27-3171	AFTER EFFECTS IN THE FIELD EFFECT AT VERY LARGE GATE CHARGES APPLIED TO TELLURIUM FILM
	27-3172	FIELD EFFECT AND HALL MOBILITY OF TELLURIUM AND LEAD TELLURIDE ON UHV CLEAVED TRIGLYCI
	27-3173	THIN FILMS ON TGS CLEAVED IN ULTRAHIGH VACUUM.
ZIMMERMANN A	14-2103	OBSERVATIONS OF FERROELECTRIC DOMAINS IN BORACITES.
ZINENKO VI	08-1274	DISPLACEMENT PHASE TRANSITIONS IN PEROVSKITES.
	08-1475	LATTICE DYNAMICS OF PHASE TRANSITIONS IN PEROVSKITE TYPE ANTIFERROELECTRICS.
	33-3360	ELASTIC PROPERTIES OF CESIUM LEAD TRICHLORIDE.
ZIOLKIEWICZ MK	15-2110	LATTICE MODES AND PHASE TRANSITION IN ANTIMONY SULFOIODIDE.
ZIOLKIEWICZ S	15-2144	BOSON ECHOES. (ANTIMONY SULFOIODIDE)
ZKDANSKY K	04-0976	PARAMAGNETIC RESONANCE OF AN F CENTER IN NONSTOICHIOMETRIC BARIUM TITANATE.
ZOLOTARYOV VM	34-3492	OPTICAL CONSTANTS OF ICE I OVER THE ENTIRE INFRARED REGION.
ZOLOTOTRUBOV YUS	27-2976	DIELECTRIC LOSSES OF TRIGLYCINE SULFATE CRYSTALS EXPOSED TO SMALL DOSES OF X- AND GAMM
ZORENKO VP	11-1665	PHASE MATCHING ANGLES AND TEMPERATURE OF LITHIUM METANIOBATE CRYSTALS OF DIFFERENT STO
	11-1687	INFLUENCE OF A TRANSVERSE INHOMOGENEITY OF THE REFRACTIVE INDEX OF A NONLINEAR CRYSTAL
ZORIN RV	13-2069	SUPPRESSION OF THE SPONTANEOUS MAGNETOELECTRIC MAGNETIZATION OF LEAD FERRATE(0.5) NIOB
	33-3409	LOW FREQUENCY MAGNETOELECTRIC RESONANCES IN BARIUM COBALT TETRAFLUORIDE.
ZPANCIC I	28-3186	SECOND ORDER QUADRUPOLE SHIFTS OF THE SODIUM-23 MAGNETIC RESONANCE IN ROCHELLE SALT.
ZUBOVA YEV	02-0128	NEW FERROELECTRICS AND SEIGNETTO MAGNETS OF THE PEROVSKITE AND PYROCHLORE TYPE. SYNTHE
	08-1452	STRUCTURE AND DIELECTRIC LOSSES OF FERROELECTRIC PEROVSKITES SYNTHESIZED AT HIGH PRESS
	08-1463	SYNTHESIS OF THALLIUM CONTAINING PEROVSKITES AT HIGH PRESSURES AND THEIR X-RAY ANALYSI
ZUCKER J	27-3174	ELECTROMECHANICAL COUPLING NEAR THE FERROELECTRIC PHASE TRANSITION OF TRIGLYCINE SULFA
ZUKASZEWICZ K	27-3175	CRYSTAL STRUCTURE OF TRIGLYCINE FLUOBERYLLATE IN THE FERRO- AND PARAELECTRIC PHASE.
ZULIANI M	11-1864	VISUALIZATION OF ACOUSTIC RADIATION IN LITHIUM NIOBATE.
ZUMER S	17-2438	DEUTERON MAGNETIC RESONANCE AND RELAXATION IN FERROELECTRIC POTASSIUM DIDEUTERIUM PHOS
	17-2440	SPIN LATTICE RELAXATION BY THE FERROELECTRIC MODE IN POTASSIUM DIHYDROGEN PHOSPHATE.
	18-2720	CESIUM-183 SPIN LATTICE RELAXATION AND RESONANCE IN FERROELECTRIC CESIUM DIDEUTERIUM A
	26-2928	SPIN LATTICE RELAXATION BY QUASI SPIN WAVES IN ORDER DISORDER TYPE FERROELECTRICS: POL
ZUPANCIC I	19-2749	NITROGEN-14 QUADRUPOLE COUPLING IN PARAELECTRIC AMMONIUM SULFATE.
	22-2832	NUCLEAR MAGNETIC RESONANCE STUDY OF THE FERROELECTRIC TRANSITIONS IN SODIUM TRIHYDROGE
	27-2967	NUCLEAR SPIN LATTICE RELAXATION IN FERROELECTRIC TRIGLYCINE SULFATE.
	27-2969	PULSED DOUBLE RESONANCE STUDY OF THE NUCLEAR QUADRUPOLE INTERACTIONS OF NITROGEN-14 IN
ZUREK R	06-1149	BRILLOUIN SCATTERING IN STRONTIUM TITANATE SINGLE CRYSTALS IN THE TEMPERATURE RANGE 5
ZUSMAN A	36-3583	ELECTRON SPIN RESONANCE SPECTRA OF CHROMIUM(3+) AND GADOLINIUM(3+) IN LANTHANUM ALUMIN
ZVEREV GM	11-1865	AUTOFOCUSING OF LASER RADIATION IN ACTIVE MATERIALS AND NONLINEAR CRYSTALS. (KDP, ADP,
	11-1866	DAMAGE TO THE SURFACE OF LITHIUM NIOBATE BY LIGHT.
	17-2717	THERMAL SELF FOCUSING IN POTASSIUM DIHYDROGEN PHOSPHATE AND AMMONIUM DIHYDROGEN PHOSPH
ZVIRGZD YUA	04-0686	STUDY OF THE PYROELECTRIC EFFECT IN THIN FILMS OF TRIGLYCINE SULFATE AND BARIUM TITANA
	04-1039	INVESTIGATION OF THE EFFECT OF SAMPLE THICKNESS ON THE DOMAIN STRUCTURE AND THE PYROEL
ZVONIK VA	04-0780	EFFECT OF LITHIUM OXIDE ON THE PROPERTIES OF A SERIES OF FERROELECTRIC MATERIALS. (BAR
ZVYAGIN AI	30-3309	TEMPERATURE DEPENDENCE OF DIELECTRIC CONSTANT IN THE SYSTEM POTASSIUM YTTRIUM MOLYBDAT
ZYURYUKIN YUA	11-1718	TEMPERATURE DEPENDENCE OF HYPERSONIC WAVE DAMPING IN RUBY AND LITHIUM NIOBATE IN THE T
	11-1719	TEMPERATURE DEPENDENCES OF THE ABSORPTION OF 9.4 GHZ HYPERSOUND IN ALUMINUM OXIDE AND